3-26-76 To "Uncle Gene" a real car enthusiast— Hope you're out & on your way soon Love & Cal

MOTOR
AUTO REPAIR MANUAL

39th Edition

First Printing

Editor
Louis C. Forier, S.A.E.

Managing Editor
Larry Solnik, S.A.E.

Associate Editors
Michael Kromida · Dan Irizarry · Warren Schildknecht

Editorial Assistant
Michele Berkowitz

Published by

MOTOR

1790 Broadway, New York, N. Y. 10019

The Automotive Business Magazine

ISBN 0-910992-46-0

Books Published by MOTOR

For the Automotive Service Trade

Auto Repair Manual

Mechanical specifications and service procedures for 1969-76 American built cars. Special sections on Engine Service, Wiper Motor Service, Emission System Service, Oscilloscopes, Power Steering, Power Brakes, Rear Axles.

Auto Repair Manual—*Early Model Edition*

This volume covers mechanical specifications and service procedures for 1962-68 American built cars and Volkswagen. Also includes chapters on Ford Bronco, International Scout and Jeep. Special sections on Generators, Starting Motors, Speed Controls, Power Steering, Power Brakes, Tune Up & Wheel Alignment Data for 1941-61 models and Foreign Car Tune Up Data.

Imported Car Repair Manual

Includes specifications and service procedures for 25 popular imports plus a valuable parts and tool availability list. An additional feature is the Operational Time Guide for these popular models.

Truck & Diesel Repair Manual

Mechanical specifications and service procedures on 1962-76 models of popular trucks plus special Diesel Engine section. Also includes Farm Tractor and Stock Engine specifications.

Emission Control Manual—Vols. I & II

Covers servicing and specifications of 1968-75 car and truck emission systems. Special sections on Retrofit Emission Control Equipment, Alternative Fuel Systems, Emission Control Law Summary, Imported Cars. Also includes a valuable Parts and Time Guide.

Parts & Time Guide

Part numbers and manufacturers suggested retail prices for American built cars from 1969-76. Interchangeable Parts information is provided for G.M., Ford and Chrysler cars. Operational Times for all cars listed plus popular truck models.

Automatic Transmission Manual—6th Edition

Covers hydraulic oil circuits (most in full color), describes operation of unit and details complete servicing and Trouble Shooting. Also covers popular truck units.

Vacuum & Wiring Diagrams

Lists fuse and lamp bulb data and flasher locations in addition to schematic diagrams of vacuum and electrical circuits. A large 10″ x 13″ book containing 800 pages.

Air Conditioner Service Manual

Covers basic theory, explains Automatic Temperature Control systems, Maintenance, Diagnosis & Testing, Unit Replacement, Compressor Service, vacuum and wiring diagrams. Special sections on aftermarket hang-on units, and truck and trailer refrigeration systems. Complete with parts listings and operational times.

Automobile Trouble Shooter

Lists more than 2000 causes of car and truck troubles on engines, fuel and electrical systems, clutches, transmissions, brakes, rear axles, suspension and more. Pocket size edition.

INDEX

This Edition Covers Mechanical Specifications and Service Procedures on 1969-76 Models

***For your convenience in locating this section, a black bar has been positioned beneath all odd page numbers.**

CAR INFORMATION SECTION—1

GENERAL SERVICE INFORMATION SECTION—2*

AMERICAN MOTORS

OLD CAR SPECIFICATIONS: For 1946-68 Tune Up and Wheel Alignment Specifications see main index.

INDEX OF SERVICE OPERATIONS

SERIAL NUMBER LOCATION: 1969-76: Plate is attached to top of instrument panel, on driver's side.

ENGINE IDENTIFICATION

6-199 & 6-232 (1969): The engine code number is located on a machined pad adjacent to the distributor. The letter contained in the code number identifies the cubic inch displacement of the engine. The letter "J" denotes the 199 engine with 8.5 compression ratio. The letter "L" denotes the 232 engine with 8.5 compression ratio.

6-199 & 6-232 (1970): The engine code number is located on a machined pad adjacent to the distributor. The letter "A" denotes the 199 engine. The letter "E" denotes the 232 engine with 1 barrel carburetor while the letter "G" denotes the 232 engine with 2 barrel carburetor.

6-232 & 6-258 (1971-75): The engine code is located on a pad between number two and three cylinders. The letter "A" denotes the 258 engine. The letter "E" denotes the 232 engine.

V8-290, 343 & 390 (1969): The engine code is located on a plate attached to the front of the right-hand rocker arm cover. The letter "H" four-barrel carburetor. The letter "Z" is for the 343 engine with four-barrel carburetor; "S" for two-barrel carburetor. The letter "W" is for the 390 engine with four-barrel in 1969.

V8-304, 360, 390, & 401 (1970-76): The engine code is located on a tag attached to the right bank rocker cover. The letter "H" denotes the 304 engine. The letter "N" denotes the 360 engine with 2 barrel carburetor while the letter "P" denotes the 360 engine with 4 barrel carburetor. The letter "X" denotes the 390 325 H.P. and letter Y the 390 340 H.P. engine. The letter "Z" denotes the 401 engine.

GRILLE IDENTIFICATION

1969 Javelin

1969 AMX

1969 Rambler

1969 Rebel

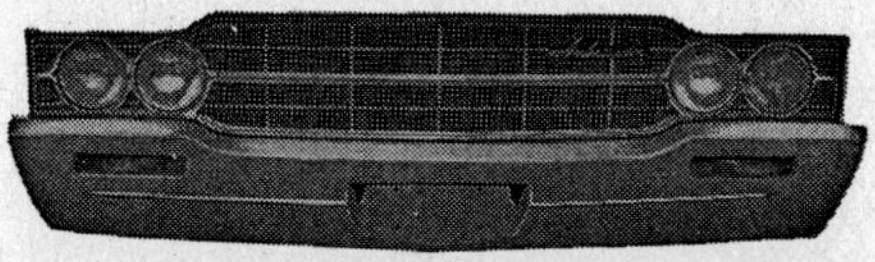
1969 Ambassador

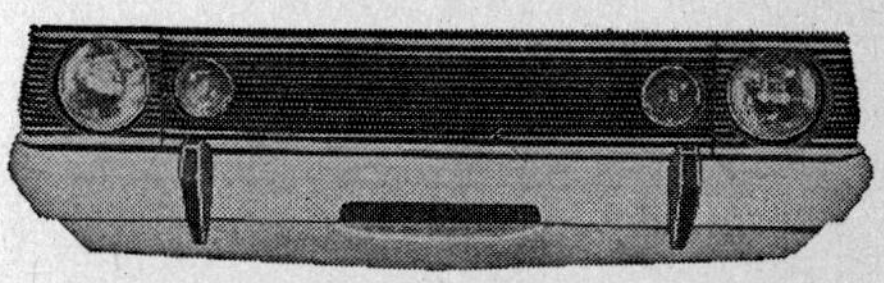
1970-72 Hornet

1970 Rebel

1970 Ambassador

1970 AMX

1970 Javelin

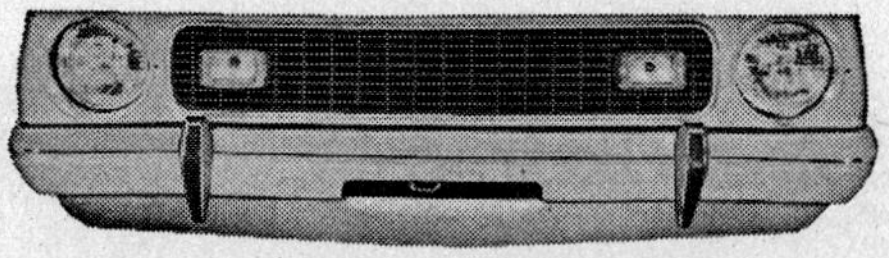
1970-72 Gremlin

1971 Javelin

1971-72 AMX

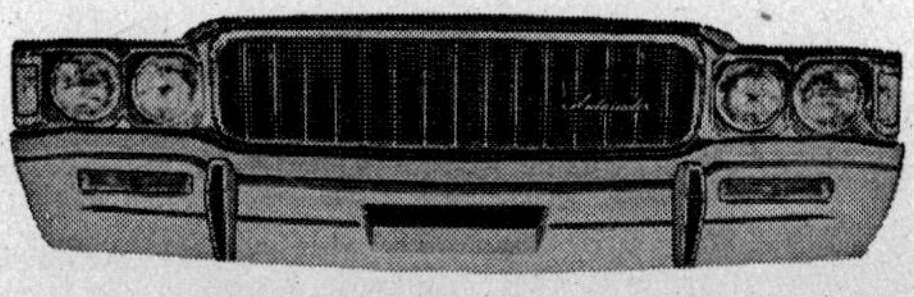
1971-72 Ambassador

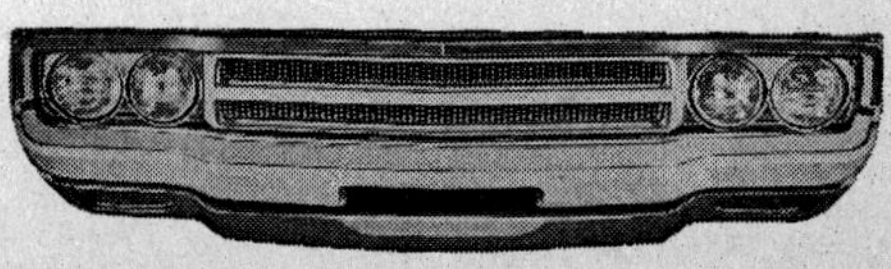
1971 Matador

GRILLE IDENTIFICATION—Continued

1972 Javelin

1972 Matador

1973 Gremlin

1973 Hornet

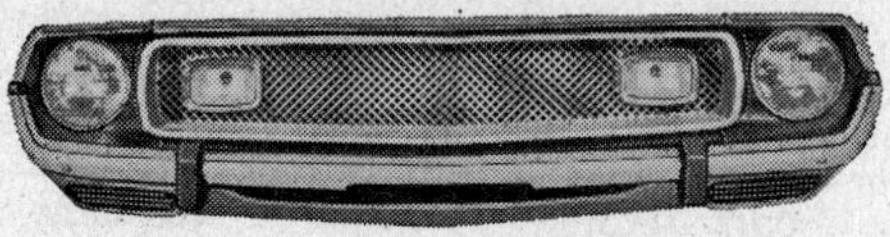
1973 Javelin

1973 AMX

1973 Matador

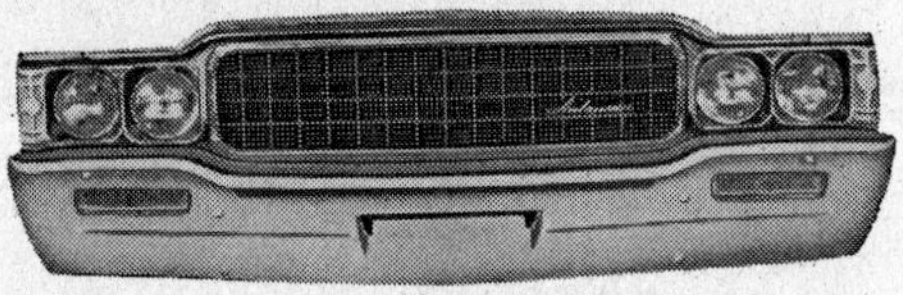
1973 Ambassador

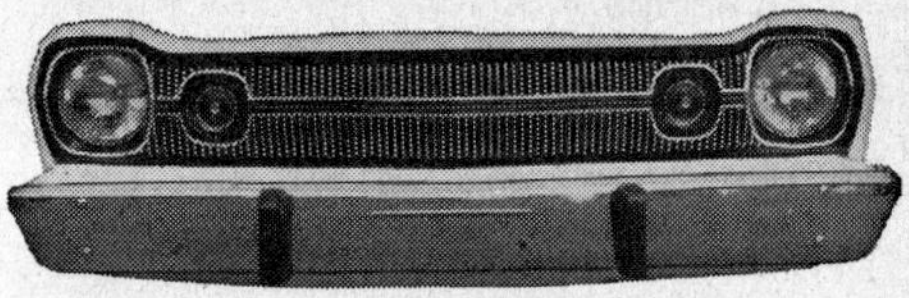
1974 Hornet

1974 Javelin

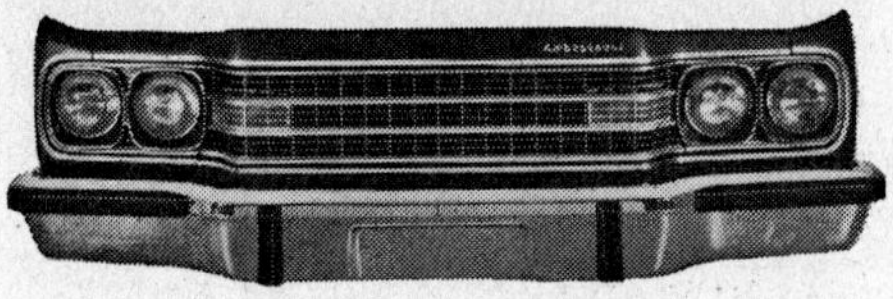
1974 Ambassador

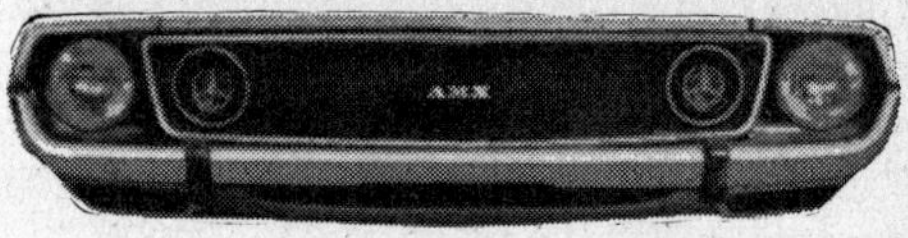

1974 AMX

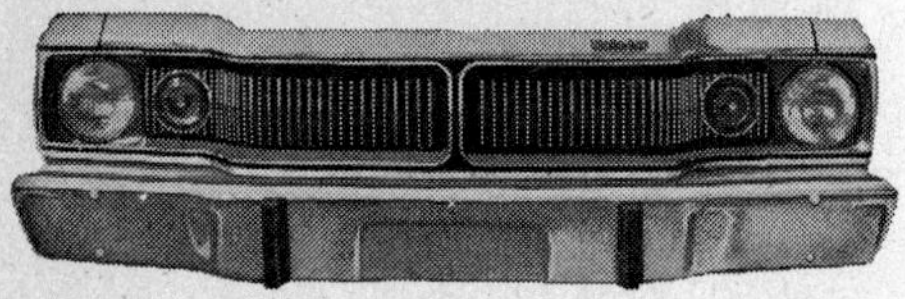
1974 Matador 4 Door

1974-75 Matador 2 Door

1975 Matador 4 Door

1974-75 Gremlin

1974-75 Gremlin X

1975-76 Hornet

1975-76 Pacer

1976 Gremlin

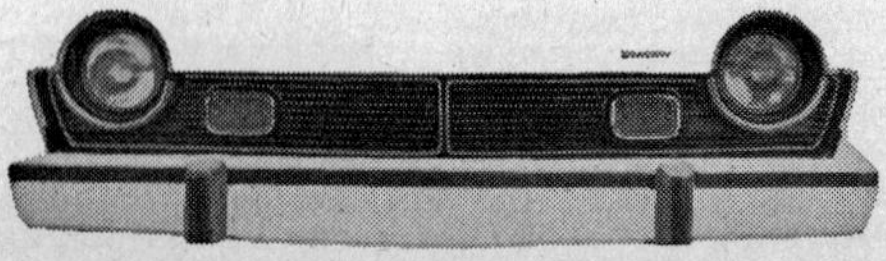
1976 Matador 2 Door

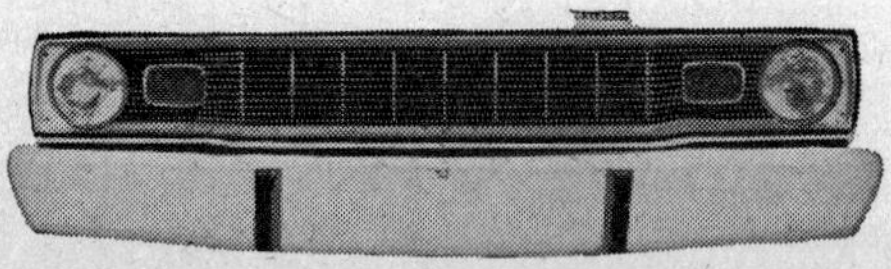
1976 Matador 4 Door

GENERAL ENGINE SPECIFICATIONS

Year	Engine	Carburetor	Bore and Stroke	Piston Displacement, Cubic Inches	Compression Ratio	Maximum Brake H.P. @ R.P.M.	Maximum Torque Lbs. Ft. @ R.P.M.	Normal Oil Pressure Pounds
1969	128 Horsepower................6-199	1 Barrel	3.75 x 3.00	199	8.5	128 @ 4400	182 @ 1600	75
	145 Horsepower................6-232	1 Barrel	3.75 x 3.50	232	8.5	145 @ 4300	215 @ 1600	75
	155 Horsepower................6-232	2 Barrel	3.75 x 3.50	232	8.5	155 @ 4400	222 @ 1600	75
	200 Horsepower..............V8-290	2 Barrel	3.75 x 3.28	290	9.0	200 @ 4600	285 @ 2800	75
	225 Horsepower..............V8-290	4 Barrel	3.75 x 3.28	290	10.0	225 @ 4700	300 @ 3200	75
	235 Horsepower..............V8-343	2 Barrel	4.08 x 3.28	343	9.0	235 @ 4400	345 @ 2600	75
	280 Horsepower..............V8-343	4 Barrel	4.08 x 3.28	343	10.2	280 @ 4800	365 @ 3000	75
	315 Horsepower..............V8-390	4 Barrel	4.165 x 3.574	390	10.2	315 @ 4600	425 @ 3200	75
1970	128 Horsepower................6-199	1 Barrel	3.75 x 3.00	199	8.5	128 @ 4400	182 @ 1600	75
	145 Horsepower................6-232	1 Barrel	3.75 x 3.50	232	8.5	145 @ 4300	215 @ 1600	75
	155 Horsepower................6-232	2 Barrel	3.75 x 3.50	232	8.5	155 @ 4400	222 @ 1600	75
	210 Horsepower..............V8-304	2 Barrel	3.75 x 3.44	304	9.0	210 @ 4400	305 @ 2800	75
	245 Horsepower..............V8-360	2 Barrel	4.08 x 3.44	360	9.0	245 @ 4400	365 @ 2400	75
	290 Horsepower..............V8-360	4 Barrel	4.08 x 3.44	360	10.0	290 @ 4800	395 @ 3200	75
	325 Horsepower..............V8-390	4 Barrel	4.156 x 3.574	390	10.0	325 @ 5000	420 @ 3200	75
	340 Horsepower①.............V8-390	4 Barrel	4.165 x 3.574	390	10.0	340 @ 5100	430 @ 3600	75
1971	135 Horsepower................6-232	1 Barrel	3.75 x 3.50	232	8.0	135 @ 4000	210 @ 1600	75
	150 Horsepower................6-258	1 Barrel	3.75 x 3.90	258	8.0	150 @ 3800	240 @ 1800	75
	210 Horsepower..............V8-304	2 Barrel	3.75 x 3.44	304	8.4	210 @ 4400	300 @ 2600	75
	245 Horsepower..............V8-360	2 Barrel	4.08 x 3.44	360	8.5	245 @ 4400	365 @ 2600	75
	285 Horsepower..............V8-360	4 Barrel	4.08 x 3.44	360	8.5	285 @ 4800	390 @ 3200	75
	330 Horsepower..............V8-401	4 Barrel	4.17 x 3.68	401	9.5	330 @ 5000	430 @ 3400	75
1972	100 Horsepower②..............6-232	1 Barrel	3.75 x 3.50	232	8.0	100 @ 3600	185 @ 1800	75
	110 Horsepower②..............6-258	1 Barrel	3.75 x 3.895	258	8.0	110 @ 3500	195 @ 2000	75
	150 Horsepower②............V8-304	2 Barrel	3.75 x 3.44	304	8.3	150 @ 4200	245 @ 2500	75
	175 Horsepower②............V8-360	2 Barrel	4.08 x 3.44	360	8.3	175 @ 4000	285 @ 2400	75
	195 Horsepower②............V8-360	4 Barrel	4.08 x 3.44	360	8.3	195 @ 4400	295 @ 2900	75
	220 Horsepower②③..........V8-360	4 Barrel	4.08 x 3.44	360	8.3	220 @ 4400	315 @ 3100	75
	255 Horsepower②............V8-401	4 Barrel	4.165 x 3.68	401	8.5	255 @ 4600	345 @ 3300	75
1973	100 Horsepower②..............6-232	1 Barrel	3.75 x 3.50	232	8.0	100 @ 3600	185 @ 1800	75
	110 Horsepower②..............6-258	1 Barrel	3.75 x 3.895	258	8.0	110 @ 3500	195 @ 2000	75
	150 Horsepower②............V8-304	2 Barrel	3.75 x 3.44	304	8.4	150 @ 4200	245 @ 2500	75
	175 Horsepower②............V8-360	2 Barrel	4.08 x 3.44	360	8.5	175 @ 4000	285 @ 2400	75
	195 Horsepower②............V8-360	4 Barrel	4.08 x 3.44	360	8.5	195 @ 4400	295 @ 2900	75
	220 Horsepower②③..........V8-360	4 Barrel	4.08 x 3.44	360	8.5	220 @ 4400	315 @ 3100	75
	255 Horsepower②............V8-401	4 Barrel	4.165 x 3.68	401	8.5	255 @ 4600	345 @ 3300	75
1974	100 Horsepower②..............6-232	1 Barrel	3.75 x 3.50	232	8.0	100 @ 3600	185 @ 1800	75
	110 Horsepower②..............6-258	1 Barrel	3.75 x 3.90	258	8.0	110 @ 3500	195 @ 2000	75
	150 Horsepower②............V8-304	2 Barrel	3.75 x 3.44	304	8.4	150 @ 4200	245 @ 2500	75
	175 Horsepower②............V8-360	2 Barrel	4.08 x 3.44	360	8.5	175 @ 4000	285 @ 2400	75
	195 Horsepower②............V8-360	4 Barrel	4.08 x 3.44	360	8.5	195 @ 4400	295 @ 2900	75
	220 Horsepower②③..........V8-360	4 Barrel	4.08 x 3.44	360	8.5	220 @ 4400	315 @ 3100	75
	255 Horsepower②............V8-401	4 Barrel	4.165 x 3.68	401	8.5	235 @ 4600	335 @ 3200	75
1975	100 Horsepower②..............6-232	1 Barrel	3.75 x 3.50	232	8.0	90 @ 3050	163 @ 2200	37–75
	110 Horsepower②..............6-258	1 Barrel	3.75 x 3.90	258	8.0	95 @ 3050	179 @ 2100	37–75
	150 Horsepower②............V8-304	2 Barrel	3.75 x 3.44	304	8.4	150 @ 4200	245 @ 2500	37–75
	175 Horsepower②............V8-360	2 Barrel	4.08 x 3.44	360	8.25	175 @ 4000	285 @ 2400	37–75
	195 Horsepower②............V8-360	4 Barrel	4.08 x 3.44	360	8.25	195 @ 4400	295 @ 2900	37–75
	220 Horsepower②③..........V8-360	4 Barrel	4.08 x 3.44	360	8.25	220 @ 4400	315 @ 3100	37–75
1976	90 Horsepower②..............6-232	1 Barrel	3.75 x 3.50	232	8.0	90 @ 3050	163 @ 2200	37–75
	95 Horsepower②..............6-258	1 Barrel	3.75 x 3.90	258	8.0	95 @ 3050	179 @ 2100	37–75
	120 Horsepower②..............6-258	2 Barrel	3.75 x 3.90	258	8.0	120 @ 3600	200 @ 2000	37–75
	120 Horsepower②............V8-304	2 Barrel	3.75 x 3.44	304	8.4	120 @ 3200	220 @ 2000	37–75
	140 Horsepower②............V8-360	2 Barrel	4.08 x 3.44	360	8.25	140 @ 3300	251 @ 1600	37–75
	180 Horsepower②............V8-360	4 Barrel	4.08 x 3.44	360	8.25	180 @ 3600	280 @ 2800	37–75

①—Rebel Machine. ②—Ratings are net (as installed in the vehicle). ③—With dual exhausts.

TUNE UP SPECIFICATIONS

OLD CAR SPECIFICATIONS: For 1946-68 Tune Up Specifications see main index.

The following specifications are published from the latest information available. This data should be used only in the absence of a decal affixed in the engine compartment.

★When using a timing light, disconnect vacuum hose or tube at distributor and plug opening in hose or tube so idle speed will not be affected.

●When checking compression, lowest cylinder must be within 80 percent of highest.

▲Before removing wires from distributor cap, determine location of the No. 1 wire in cap, as distributor position may have been altered from that shown at the end of this chart.

Year	Spark Plug		Distributor		Ignition Timing★			Carb. Adjustments					
	Type	Gap Inch	Point Gap Inch	Dwell Angle Deg.	Firing Order Fig. ▲	Timing BTDC ①	Mark Fig.	Hot Idle Speed③ Std. Trans.	Hot Idle Speed③ Auto. Trans.②	Air Fuel Ratio Std. Trans.	Air Fuel Ratio Auto. Trans.	Idle CO% Std. Trans.	Idle CO% Auto. Trans.
1969													
6-199, 232 Std. Tr.	N14Y	.035	.016	31–34	A	TDC	D	600⑤	—	—	—	—	—
6-199, 232 Auto. Tr.	N14Y	.035	.016	31–34	A	5°	D	—	525D⑤	—	—	—	—
V8-290, 343, 390	N12Y	.035	.016	29–31	B	TDC	C	650⑤	550D⑤	—	—	—	—
1970													
6-199, 232	N14Y	.035	.016	31–34	A	3°	D	600	550D	14.0 to 1	14.0 to 1	—	—
V8-304, 360, 390	N12Y	.035	.016	29–31	B	5°④	C	650	600D	14.0 to 1	14.0 to 1	—	—
1971													
6-232 Std. Tr.	N12Y	.035	.016	31–34	A	3°	D	700	—	14.0 to 1	14.0 to 1	—	—
6-232 Auto. Tr.	N12Y	.035	.016	31–34	A	5°	D	—	600D	14.0 to 1	14.0 to 1	—	—
6-258	N12Y	.035	.016	31–34	A	5°	D	700	600D	14.0 to 1	14.0 to 1	—	—
V8-304, 360, 401	N12Y	.035	.016	29–31	B	2½°	C	750	650D	14.0 to 1	14.0 to 1	—	—
1972													
6-232 Exc. Calif.	N12Y	.035	.016	31–34	A	5°	D	600	550D	14.0 to 1	14.0 to 1	—	—
6-232 Calif.	N12Y	.035	.016	31–34	A	3°	D	700	600D	14.0 to 1	14.0 to 1	—	—
6-258 Exc. Calif.	N12Y	.035	.016	31–34	A	5°	D	600	550D	14.0 to 1	14.0 to 1	—	—
6-258 Calif.	N12Y	.035	.016	31–34	A	3°	D	700	600D	14.0 to 1	14.0 to 1	—	—
V8-304 Exc. Calif.	N12Y	.035	.016	29–31	B	5°	C	750	650D	13.5 to 1	14.0 to 1	—	—
V8-304 Calif.	N12Y	.035	.016	29–31	B	5°	C	750	700D	13.5 to 1	14.0 to 1	—	—
V8-360	N12Y	.035	.016	29–31	B	5°	C	750	700D	14.0 to 1	14.0 to 1	—	—
V8-401 Exc. Calif.	N12Y	.035	.016	29–31	B	5°	C	700	650D	13.5 to 1	13.5 to 1	—	—
V8-401 Calif.	N12Y	.035	.016	29–31	B	5°	C	700	700D	14.0 to 1	14.0 to 1	—	—
1973													
6-232	N12Y	.035	.016	31–34	A	5°	D	700⑦	600D⑧	—	—	⑨	⑨
6-258	N12Y	.035	.016	31–34	A	3°	D	700⑦	600D⑧	—	—	⑨	⑨
V8-304	N12Y	.035	.016	29–31	B	5°	C	750	700D	—	—	0.5–1.0	0.5–1.0
V8-360	N12Y	.035	.016	29–31	B	5°	C	750	700D	—	—	0.5–1.0	0.5–1.0
V8-401	N12Y	.035	.016	29–31	B	5°	C	750	700D	—	—	0.5–1.0	0.5–1.0
1974													
6-232 L/EGR	N12Y	.035	.016	31–34	A	5°	D	700	600D	—	—	⑨	⑨
6-232 W/EGR	N12Y	.035	.016	31–34	A	5°	D	600	550D	—	—	⑨	⑨
6-232 Calif.	N12Y	.035	.016	31–34	A	5°	D	600	700D	—	—	⑨	⑨
6-258 L/EGR	N12Y	.035	.016	31–34	A	3°	D	550	700D	—	—	⑨	⑨
6-258 W/EGR	N12Y	.035	.016	31–34	A	3°	D	550	600D	—	—	⑨	⑨
6-258 Calif.	N12Y	.035	.016	31–34	A	3°	D	600	700D	—	—	⑨	⑨
V8-304	N12Y	.035	.016	29–31	B	⑩	C	750	700D	—	—	0.5–1.0	0.5–1.0
V8-360	N12Y	.035	.016	29–31	B	5°	C	750	700D	—	—	0.5–1.0	0.5–1.0
V8-401	N12Y	.035	.016	29–31	B	5°	C	750	700D	—	—	0.5–1.0	0.5–1.0

Continued

TUNE UP SPECIFICATIONS—Continued

OLD CAR SPECIFICATIONS: For 1946-68 Tune Up Specifications see main index.

The following specifications are published from the latest information available. This data should be used only in the absence of a decal affixed in the engine compartment.

★When using a timing light, disconnect vacuum hose or tube at distributor and plug opening in hose or tube so idle speed will not be affected.

●When checking compression, lowest cylinder must be within 80 percent of highest.

▲Before removing wires from distributor cap, determine location of the No. 1 wire in cap, as distributor position may have been altered from that shown at the end of this chart.

Year	Spark Plug		Distributor		Ignition Timing★			Carb. Adjustments					
								Hot Idle Speed(3)		Air Fuel Ratio		Idle CO%	
	Type	Gap Inch	Point Gap Inch	Dwell Angle Deg.	Firing Order Fig. ▲	Timing BTDC (1)	Mark Fig.	Std. Trans.	Auto. Trans.(2)	Std. Trans.	Auto. Trans.	Std. Trans.	Auto. Trans.
1975													
6-232	N12Y	.035	—	—	A	5°	E	600	(6)	—	—	—	(11)
6-258 Calif.	N12Y	.035	—	—	A	3°	E	600	(6)	—	—	—	(11)
V8-304	N12Y	.035	—	—	B	5°	C	750	700D	—	—	—	—
V8-360	N12Y	.035	—	—	B	5°	C	—	700D	—	—	—	—
1976													
6-232	N12Y	.035	—	—	A	8°	E	850	(6)	—	—	(12)	(13)
6-258 Std. Tr.	N12Y	.035	—	—	A	6°	E	(14)	—	—	—	(12)	(13)
6-258 Auto. Tr.	N12Y	.035	—	—	A	8°	E	—	(15)	—	—	(12)	(13)
V8-304	N12Y	.035	—	—	B	(16)	C	750	700D	—	—	—	—
V8-360	N12Y	.035	—	—	B	(16)	C	—	700D	—	—	—	—
V8-401(17)	N12Y	.035	—	—	B	(16)	C	—	700D	—	—	—	—

(1)—BTDC: Before top dead center.
(2)—D: Drive. N: Neutral.
(3)—Where two speeds are listed, lower speed indicates idle solenoid disconnected.
(4)—V8-390 with distributor 1111948 set at TDC.
(5)—If air conditioned turn A/C switch full on.
(6)—Exc. Calif., 550D; Calif., 700D.
(7)—Set Matador wagon at 600 R.P.M.
(8)—Set Matador at 650D.
(9)—W/Air Guard 0.5–1.0%. W/O Air Guadr 1.0–1.5%.
(10)—Exc. Calif. auto. trans., 5° BTDC; Calif. auto. trans., 2½° BTDC.
(11)—W/Air Guard, 0.8% maximum; Less Air Guard, 1.0% maximum.
(12)—Exc. Gremlin, 0.5% maximum; Gremlin, 1.0% maximum.
(13)—Exc. Matador, 1.0% maximum; Matador, 0.8% maximum.
(14)—W/Idle solenoid & all Calif., 850 RPM; Less idle solenoid, 600 RPM.
(15)—Exc. Calif. & 2 bbl. carb., 550D; Calif. & 2 bbl. carb., 700D.
(16)—Exc. Calif. auto. tr. & all std. tr., 10° BTDC; Calif. & all std. tr., 5° BTDC.
(17)—Police.

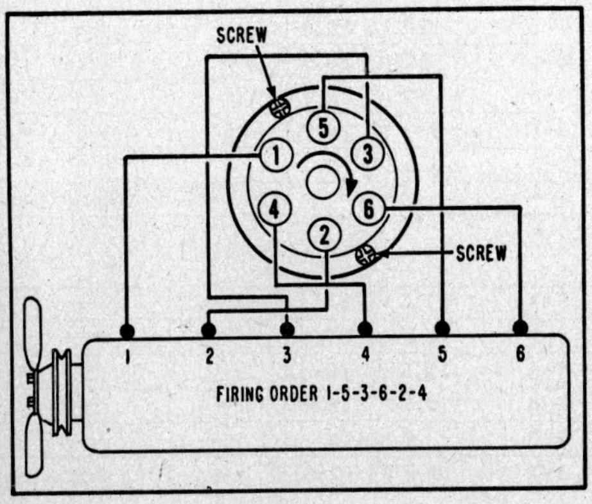

Fig. A

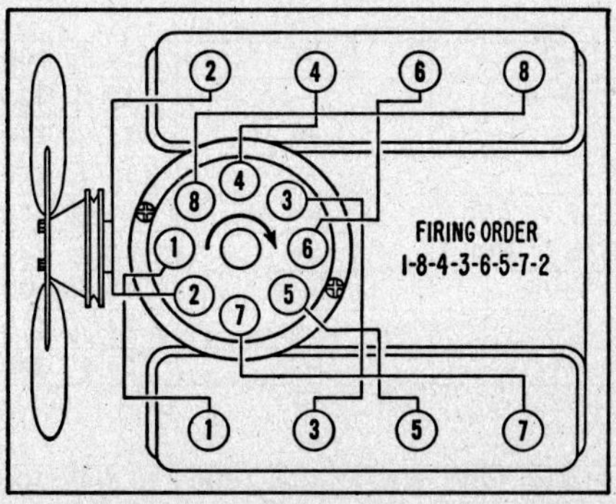

Fig. B

Fig. C

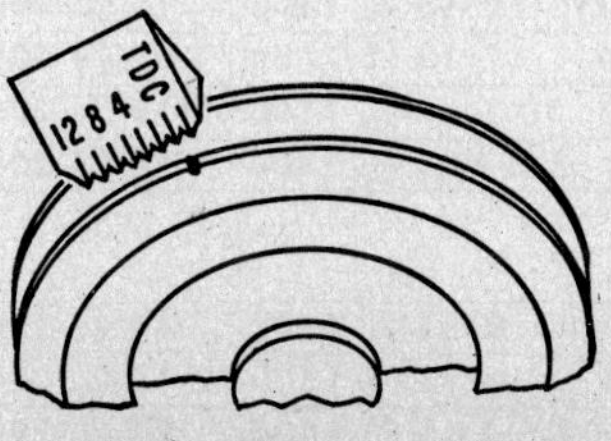

Fig. D

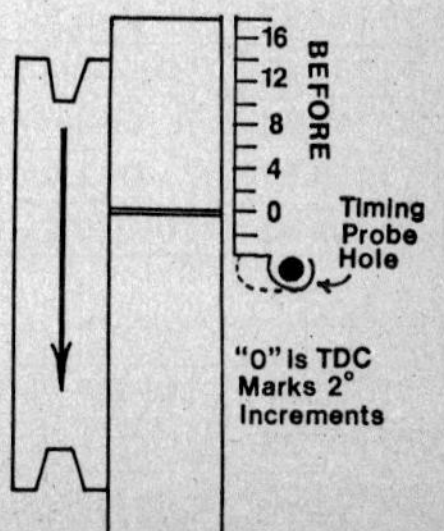

Fig. E

VALVE SPECIFICATIONS

Year	Model	Valve Lash Int.	Valve Lash Exh.	Valve Angles Seat	Valve Angles Face	Valve Spring Installed Height	Valve Spring Pressure Lbs. @ In.	Stem Clearance Intake	Stem Clearance Exhaust	Stem Diameter Intake	Stem Diameter Exhaust
1969	6-199, 232	Hydraulic③		④	②	1 13/16	195 @ 1 7/16	.001–.003	.001–.003	.3715–.3725	.3715–.3725
	8-290, 343, 390	Hydraulic③		①	⑤	1 13/16	200 @ 1 25/64	.001–.003	.001–.003	.3715–.3725	.3715–.3725
1970	6-199, 232	Hydraulic③		④	②	1 13/16	195 @ 1 7/16	.001–.003	.001–.003	.3715–.3725	.3715–.3725
	V8-304, 360, 390	Hydraulic③		①	⑦	1 13/16	200 @ 1 25/64	.001–.003	.001–.003	.3715–.3725	.3715–.3725
	8-390⑥	Hydraulic③		①	⑦	1 13/16	189 @ 1 23/64	.001–.003	.001–.003	.3715–.3725	.3715–.3725
1971	6-232, 258	Hydraulic③		⑦	②	1 13/16	195 @ 1 7/16	.001–.003	.001–.003	.3715–.3725	.3715–.3725
	V8-304, 360, 401	Hydraulic③		①	⑤	1 13/16	189 @ 1 23/64	.001–.003	.001–.003	.3715–.3725	.3715–.3725
1972	6-232, 258	Hydraulic③		⑦	②	1 13/16	195 @ 1 7/16	.001–.003	.001–.003	.3715–.3725	.3715–.3725
	V8-304, 360, 401	Hydraulic③		①	⑤	1 13/16	218 @ 1 23/64	.001–.003	.001–.003	.3715–.3725	.3715–.3725
1973	6-232, 258	Hydraulic③		⑦	②	1 13/16	195 @ 1 7/16	.001–.003	.001–.003	.3715–.3725	.3715–.3725
	V8-304, 360, 401	Hydraulic③		⑦	②	1 13/16	218 @ 1 23/64	.001–.003	.001–.003	.3715–.3725	.3715–.3725
1974	6-232, 258	Hydraulic③		⑦	②	⑧	⑨	.001–.003	.001–.003	.3715–.3725	.3715–.3725
	V8-304, 360, 401	Hydraulic③		⑦	②	⑧	⑩	.001–.003	.001–.003	.3715–.3725	.3715–.3725
1975–76	6-232, 258	Hydraulic③		⑦	②	1 13/16	195 @ 1 7/16	.001–.003	.001–.003	.3715–.3725	.3715–.3725
	V8-304, 360	Hydraulic③		⑦	②	1 13/16	213 @ 1 23/64	.001–.003	.001–.003	.3715–.3725	.3715–.3725

①—Intake 30°, exhaust 45°.
②—Intake 29°, exhaust 44°.
③—No adjustment.
④—Intake 30°, exhaust 44°.
⑤—Intake 29°, exhaust 44½°.
⑥—Rebel Machine.
⑦—Intake 30°, exhaust 44½°.
⑧—With valve rotator, 1 5/8"; less valve rotator, 1 13/16"
⑨—With valve rotator, 218 lbs. @ 1 3/16"; less valve rotator, 195 lbs. @ 1 7/16".
⑩—With valve rotator, 213 lbs. @ 1 3/16"; less valve rotator, 213 lbs. @ 1 23/64.

DISTRIBUTOR SPECIFICATIONS

★If unit is checked on vehicle, double the RPM and degrees to get crankshaft figures.

Breaker arm spring tension 17–21.

Distributor Part No.①	Centrifugal Advance Degrees @ RPM of Distributor: Advance Starts	Intermediate Advance			Full Advance	Vacuum Advance: Inches of Vacuum to Start Plunger	Max. Adv. Dist. Deg. @ Vacuum	Distributor Retard: In. of Mercury Start Retard	Max. Retard Dist. Deg. @ Vacuum
1969-70									
1110444	2–3 @ 450	8–10 @ 1000	—	—	14 @ 2000	5–7	11 @ 17	—	—
1110481	0–2 @ 450	7–9 @ 1050	—	—	13 @ 2250	5–7	11 @ 17	5	4½ @ 14
1111191	0–1 @ 450	7.5–9.5 @ 1000	—	—	15 @ 2200	8–10	12 @ 20.5	—	—
1111472	0–1 @ 450	7.5–9.5 @ 1000	—	—	15 @ 2200	4–6	12 @ 19½	—	—
1111473	0–1 @ 400	0–2.5 @ 450	8.5–10.5 @ 800	—	16 @ 2200	8–10	12 @ 20.5	—	—
1111948	0–1 @ 400	0–2.5 @ 450	8.5–10.5 @ 800	—	16 @ 2200	4–6	12 @ 19.5	—	—
1111987	0–2 @ 450	6–8 @ 750	—	—	13 @ 2000	5–7	9½ @ 14.5	5	5½ @ 14
1111988	0–2 @ 500	5–7 @ 825	—	—	14 @ 2100	5–7	9½ @ 14.5	5	5½ @ 14
1112018	0–2 @ 500	4.5–6.5 @ 800	—	—	14 @ 2200	—	—	5	5½ @ 14
1971-72									
1110340	0–2 @ 450	7–9 @ 950	8–10 @ 1200	8.5–11.5 @ 1500	13.5 @ 2000	5–7	11.5 @ 17	—	—
1110497	0–5 @ 500	5–7 @ 950	7.5–9.5 @ 1200	9–11 @ 1500	14 @ 2000	5–7	9 @ 11	—	—
1111948	0–2.5 @ 450	8.9–10.5 @ 800	9–11.5 @ 1100	11–13 @ 1500	15 @ 2000	5–6	12.5 @ 19	—	—
1112028	0–2 @ 450	7–9 @ 1100	9.5–11.5 @ 1500	11–13 @ 1750	14.5 @ 2000	5–6	12.5 @ 19	—	—
1112111	0–2 @ 550	7.5–9.5 @ 1100	9.5–11.5 @ 1300	10.5–12 @ 1500	15 @ 2000	5–7	9 @ 16	—	—
1112112	0–1.5 @ 500	3.5–5.5 @ 750	7.5–9.5 @ 1000	9.5–11.5 @ 1500	14 @ 2000	5–7	8¼ @ 12½	—	—
1973									
1110522	0–2.5 @ 500	4.5–6.5 @ 800	7–9 @ 1000	9–11 @ 1500	14 @ 2200	5–7	9 @ 13¼	—	—
1110523	0–1 @ 550	2–4.5 @ 750	6.5–8.5 @ 1000	9–11 @ 1500	15 @ 2200	5–7	9 @ 13¼	—	—
1112112	0–1.5 @ 500	3.5–5.5 @ 750	7.5–9.5 @ 1000	9.5–11.5 @ 1500	14 @ 2000	5–7	8¼ @ 12½	—	—

Continued

DISTRIBUTOR SPECIFICATIONS—Continued

★If unit is checked on vehicle, double the RPM and degrees to get crankshaft figures.

Breaker arm spring tension 17–21.

Distributor Part No.①	Centrifugal Advance Degrees @ RPM of Distributor					Vacuum Advance		Distributor Retard	
	Advance Starts	Intermediate Advance			Full Advance	Inches of Vacuum to Start Plunger	Max. Adv. Dist. Deg. @ Vacuum	In. of Mercury Start Retard	Max. Retard Dist. Deg. @ Vacuum
1973 Continued									
1112179	0–1 @ 400	4.5–6.5 @ 750	8–9.75 @ 1000	10¾–12¼ @ 1500	17 @ 2200	5–7	8¼ @ 12½	—	—
1112214	0–2¼ @ 450	6¾–8¾ @ 750	8–10 @ 1000	11–13 @ 1500	17 @ 2200	4–6	8¼ @ 13	—	—
1112215	0–3¼ @ 400	5–7 @ 750	6¼–8½ @ 1000	9½–11½ @ 1500	16 @ 2300	4–6	8¼ @ 13	—	—
1974									
1110528	0–2 @ 600	3–5.5 @ 800	6–8.5 @ 1000	6.5–9 @ 1500	15 @ 2200	5–7	9 @ 13	—	—
1110529	0–2 @ 500	4.5–7 @ 800	7–9 @ 1000	7–9.5 @ 1500	14 @ 2300	5–7	9 @ 13	—	—
1112112	0–2 @ 500	4.5–6.5 @ 800	7.5–9.5 @ 1000	9¾–11¾ @ 1500	14 @ 2200	5–7	8¼ @ 12¾	—	—
1112179	0–2 @ 500	4.5–6.5 @ 750	8–10 @ 1000	12.5–13.5 @ 1600	17 @ 2200	5–7	8¼ @ 12¾	—	—
1112214	0–2.5 @ 500	6.5–8.5 @ 700	—	11–13 @ 1500	17 @ 2200	4–6	8¼ @ 13	—	—
1112215	0–1.5 @ 400	4–6¾ @ 600	6.5–8.5 @ 1000	10–12 @ 1600	16 @ 2200	4–6	8¼ @ 13	—	—
1975									
3224746	0–1½ @ 400	5–7 @ 750	6½–8½ @ 1000	—	9¼–14 @ 1500	—	16¾ @ 1000	—	—
3224965	0–1 @ 400	4½–6 @ 750	8–10 @ 1000	—	11–13 @ 1500	—	18 @ 1000	—	—
3224966	0–2½ @ 450	6¼–8¾ @ 750	8¼–10¼ @ 1000	—	11–13 @ 1500	—	18¼ @ 1000	—	—
3224968	0–1 @ 500	2–4½ @ 750	6½–8½ @ 1500	—	6–9 @ 1500	5–7	9 @ 13.1	—	—
3224969	0–2 @ 500	3½–6 @ 750	7–9 @ 1000	—	7–9½ @ 1500	5–7	9 @ 13.1	—	—
1976									
3227331	0–4 @ 500	2–4 @ 750	4–6 @ 1000	4.65–6.65 @ 1500	5¼–7¼ @ 2000	5–7	9 @ 12.6	—	—
3228263	0–2.8 @ 500	2.8–4.85 @ 750	4.2–6.25 @ 1000	7.05–9.05 @ 1500	11–13 @ 2000	4–6	8½ @ 12.7	—	—
3228264	0–2.4 @ 500	3.4–6.4 @ 750	6.3–8.3 @ 1000	9.3–11.3 @ 1500	13½–15½ @ 2000	5–7	8½ @ 12.5	—	—
3228265	0–2.7 @ 500	2.4–4.4 @ 750	3.3–5.3 @ 1000	5.1–7.1 @ 1500	7½–9½ @ 2000	4–6	8½ @ 12.5	—	—
3228266	0–3¾ @ 500	4.4–6.4 @ 750	5.4–7.4 @ 1000	7.3–9.3 @ 1500	10–12 @ 2000	4–6	8½ @ 12.7	—	—

①—Stamped on distributor housing plate.

PISTONS, PINS, RINGS, CRANKSHAFT & BEARINGS

Year	Model	Piston Clearance Top of Skirt	Ring End Gap①		Wrist-pin Diameter	Rod Bearings		Main Bearings			
			Comp.	Oil		Shaft Diameter	Bearing Clearance	Shaft Diameter	Bearing Clearance	Thrust on Bear. No.	Shaft End Play
1969	6-199, 232	.0005–.0013	.010	.015	.9306	2.0934–2.0955	.001–.002	2.4981–2.5001	.001–.002	3	.0015–.007
	V8-290	.001–.0018	.010	.015	.9306	2.0934–2.0955	.001–.002	②	.001–.002	3	.003–.008
	V8-343	.0012–.002	.010	.015	.9306	2.0934–2.0955	.001–.002	②	.001–.002	3	.003–.008
	V8-390	.001–.0018	.010	.015	1.000	2.2471–2.2492	.001–.002	②	.001–.002	3	.003–.008

Continued

PISTONS, PINS, RINGS, CRANKSHAFT & BEARINGS—Continued

Year	Model	Piston Clearance Top of Skirt	Ring End Gap①		Wrist-pin Diameter	Rod Bearings		Main Bearings			
			Comp.	Oil		Shaft Diameter	Bearing Clearance	Shaft Diameter	Bearing Clearance	Thrust on Bear. No.	Shaft End Play
1970	6-199, 232	.0005–.0013	.010	.015	.9306	2.0934–2.0955	.001–.002	2.4981–2.5001	⑤	3	.0015–.007
	V8-304	.001–.0018	.010	.015	.9306	2.0934–2.0955	.001–.002	④	⑤	3	.003–.008
	V8-360	.0012–.002	.010	.015	.9306	2.0934–2.0955	.001–.002	④	⑤	3	.003–.008
	V8-390	.0012–.002	.010	.015	1.000	2.2402–2.2471	.001–.002	④	⑤	3	.003–.008
	V8-390③	.001–.0018	.010	.015	1.000	2.2402–2.2471	.001–.002	④	⑤	3	.003–.008
1971	6-232, 258	.0005–.0013	.010	.015	.9306	2.0934–2.0955	.001–.002	2.4986–2.5001	.001–.002	3	.0015–.007
	V8-304	.001–.0018	.010	.015	.9306	2.0934–2.0955	.001–.002	④	⑤	3	.003–.008
	V8-360	.0012–.002	.010	.015	.9306	2.0934–2.0955	.001–.002	④	⑤	3	.003–.008
	V8-401	.001–.0018	.010	.015	1.000	2.2471–2.2485	.001–.002	④	⑤	3	.003–.008
1972-73	6-232, 258	.0009–.0017	.010	.010	.9306	2.0934–2.0955	.001–.002	2.4986–2.5001	.001–.002	3	.0015–.007
	V8-304	.001–.0018	.010	.010	.9306	2.0934–2.0955	.001–.002	④	⑤	3	.003–.008
	V8-360	⑥	.010	.015	.9306	2.0934–2.0955	.001–.002	④	⑤	3	.003–.008
	V8-401	⑦	.010	.015	1.000	2.2464–2.2485	.001–.002	④	⑤	3	.003–.008
1974	6-232, 250	.0009–.0017	.010	.010	.9304	2.0934–2.0955	.001–.003	2.4986–2.5001	.001–.003	3	.0015–.0065
	V8-304	.0010–.0018	.010	.010	.9308	2.0934–2.0955	.001–.003	④	.001–.003	3	.003–.008
	V8-360	.0012–.0020	.010	.015	.9308	2.0934–2.0955	.001–.003	④	.001–.003	3	.003–.008
	V8-401	.0010–.0018	.010	.015	1.0009	2.2464–2.2485	.001–.003	④	.001–.003	3	.003–.008
1975-76	6-232, 258	.0009–.0017	.010	.010	.9306	2.0934–2.0955	.001–.003	2.4986–2.5001	.001–.003	3	.0015–.0065
	V8-304	.0010–.0018	.010	.010	.9310	2.0934–2.0955	.001–.003	④	.001–.003	3	.003–.008
	V8-360	.0012–.0020	.010	.015	.9310	2.0934–2.0955	.001–.003	④	.001–.003	3	.003–.008

①—Fit rings in tapered bores for clearance listed in tightest portion of ring travel.
②—Rear main 2.7464–2.7479″, others 2.7469–2.7489″.
③—Rebel machine.
④—Rear main 2.7479–2.7464″, others 2.7489–2.7474″.
⑤—Rear main .002–.003″, others .001–.002″.
⑥—Except police, .0012–.002″. Police, .0016–0024″.
⑦—Except police, .001–.0018″. Police, .0014–.0022″.

ENGINE TIGHTENING SPECIFICATIONS★

★Torque specifications are for clean and lightly lubricated threads only. Dry or dirty threads produce increased friction which prevents accurate measurement of tightness.

Year	Engine Model	Spark Plugs Ft. Lbs.	Cylinder Head Bolts Ft. Lbs.	Intake Manifold Ft. Lbs.	Exhaust Manifold Ft. Lbs.	Rocker Arm Shaft Bracket Ft. Lbs.	Rocker Arm Cover Ft. Lbs.	Connecting Rod Cap Bolts Ft. Lbs.	Main Bearing Cap Bolts Ft. Lbs.	Flywheel to Crankshaft Ft. Lbs.	Vibration Damper or Pulley Ft. Lbs.
1969	6-199, 232	25–30	80–85	20–25	20–25	—	45–55①	26–30	75–85	100–110	50–60
1970	6-199, 232, 258	25–30	75–85	20–25	20–25	—	45–55①	26–30	75–85	100–110	50–60
1971–72	6-232, 258	25–30	80–85	20–25	20–25	—	45–55①	26–30	75–85	100–110	50–60
1973–74	6-232, 258	25–30	105	23	23	21③	50①	28	80	105	55
1975–76	6-232, 258	28	105	23	23	—	50①	28	80	105	55
1969	V8-290, 343	25–30	90–100	40–45	30–35	65–70②	20–30①	26–30	95–105	100–110	50–60
1970–71	V8-304, 360	25–30	105–115	40–45	30–35	65–70②	20–30①	26–30	95–105	100–110	50–60
1972	V8-304, 360	25–30	105–115	40–45	30–35	65–70②	45–55①	26–30	95–105	100–110	50–60
1973–74	V8-304, 360	28	110	43	25	67②	50①	28	100	105	55
1969	V8-390	25–30	90–100	40–45	30–35	65–70②	20–30①	35–40	95–105	100–110	50–60
1970–71	V8-390, 401	25–30	105–115	40–45	30–35	65–70②	20–30①	35–40	95–105	100–110	50–60
1972	V8-401	25–30	105–115	40–45	30–35	65–70②	45–55①	35–40	95–105	100–110	50–60
1973–74	V8-401	28	110	43	25	67②	50①	38	100	105	55
1975–76	V8-304, 360	28	110	43	25	19③	50①	28	100	105	55

①—Inch pounds. ②—Rocker arm stud. ③—Rocker arm cap screw.

REAR AXLE SPECIFICATIONS

Year	Model	Carrier Type ②	Ring Gear & Pinion Backlash		Pinion Bearing Preload			Differential Bearing Preload		
			Method	Adjustment	Method	New Bearings Inch-Lbs.	Used Bearings Inch-Lbs.	Method	New Bearings Inch-Lbs.	Used Bearings Inch-Lbs.
1969	6 Cyl.	Integral	Shims	.005–.009	Shims	15–25①	15–25①	Shims	.008	.008
	V8s	Integral	Shims	.005–.009	Spacer	17–28①	17–28①	Shims	.008	.008
1970	7 7/16" Dr. Gr.	Integral	Shims	.005–.009	Sleeve	15–25①	15–25①	Shims	.008	.008
	8 7/8" Dr. Gr.	Integral	Shims	.005–.009	Sleeve	17–28①	17–28①	Shims	.008	.008
1971–76	7 9/16" Dr. Gr.	Integral	Shims	.005–.009	Sleeve	15–25①	15–25①	Shims	.008	.008
	8 7/8" Dr. Gr.	Integral	Shims	.005–.009	Sleeve	17–28①	17–28①	Shims	.008	.008

①—Adjust at drive pinion flange nut with inch-pound torque wrench.

②—Axle shaft end play .006".

WHEEL ALIGNMENT SPECIFICATIONS

OLD CAR SPECIFICATIONS: For 1946–68 Wheel Alignment Specifications see main index.

Year	Model	Caster Angle, Degrees		Camber Angle, Degrees				Toe-In. Inch	Toe-Out on Turns, Deg.①	
		Limits	Desired	Limits Left	Limits Right	Desired Left	Desired Right		Outer Wheel	Inner Wheel
1969	Rebel, Amb.	0 to −1	−1/2	−3/8 to +3/8	−3/8 to +3/8	Zero	Zero	1/8	22	25
	Others②	−1/2 to +1/2	Zero	−3/8 to +3/8	−3/8 to +3/8	Zero	Zero	1/8	22	25
	Others③	+1/2 to +1 1/2	+3/4	−3/8 to +3/8	−3/8 to +3/8	Zero	Zero	1/8	22	25
1970	Man. Steer.④	−1/2 to +1/2	Zero	−3/8 to +3/8	−3/8 to +3/8	Zero	Zero	1/16–3/16	22	25
	Power Steer.④	+1/2 to +1 1/2	+1	−3/8 to +3/8	−3/8 to +3/8	Zero	Zero	1/16–3/16	22	25
	Javelin	+1/2 to +1 1/2	+1	−3/8 to +3/8	−3/8 to +3/8	Zero	Zero	1/16–3/16	22	25
1971	All Models	+1/2 to +1 1/2	+1	−3/8 to +3/8	−3/8 to +3/8	Zero	Zero	1/16–3/16	22	25
1972	All Models	—	+1	—	—	+3/8	+1/8	1/8	22	25
1973–74	Hornet, Gremlin	−1/2 to +1/2	Zero	+1/8 to +5/8	0 to +1/2	+3/8	+1/8	1/8	22	25
	Others	+1/2 to +1 1/2	+1	+1/8 to +5/8	0 to +1/2	+3/8	+1/8	1/8	22	25
1975–76	Hornet, Gremlin	−1/2 to +1/2	Zero	+1/8 to +5/8	0 to +1/2	+3/8	+1/8	1/8	⑤	38
	Others	+1/2 to +1 1/2	+1	+1/8 to +5/8	0 to +1/2	+3/8	+1/8	1/8	⑤	⑥

①—Incorrect toe-out when other adjustments are correct, indicates bent steering arms.

②—Manual steering.

③—Power steering.

④—Except Javelin.

⑤—Wheels at full turn.

⑥—Matador 38°; Pacer 35°.

ALTERNATOR SPECIFICATIONS

Year	Alternator							Regulator			
	Make	Model	Ground Polarity	Rated Output		Field Current		Model	Regulator Test @ 120°F.		
				Amperes	Volts	Amperes ①	Volts		Ampere Load	Altern. R.P.M.	Volts
1969–71	Amer. Mtrs.	3195534	Negative	35	15	2.4–2.5	10	3195003②	10	2000	13.7–14.5
	Motorola	A-12NAM 456	Negative	35	14.2	2.0–2.6	—	R2AM 4②	10	2000	13.7–14.5
	Motorola	A-12NAM 606	Negative	55	15	1.8–2.4	—	R2AM 4②	10	2000	13.7–14.5
1972	Motorola	—	Negative	37	—	1.8–2.5	—	8RB2005②	10	2000	13.4–13.9
	Motorola	—	Negative	55	—	1.8–2.5	—	8RB2005②	10	2000	13.4–13.9
1973	Motorola	—	Negative	37	—	1.8–2.5	—	8RD2001②	10	2000	13.4–13.9
	Motorola	—	Negative	51	—	1.8–2.5	—	8RD2001②	10	2000	13.4–13.9
	Motorola	—	Negative	62	—	1.8–2.5	—	8RD2001②	10	2000	13.4–13.9
1974–75	Motorola	—	Negative	37	—	1.8–2.5	—	8RH2003②	10	2000	13.1–14.3
	Motorola	—	Negative	51	—	1.8–2.5	—	8RH2003②	10	2000	13.1–14.3
	Motorola	—	Negative	62	—	1.8–2.5	—	8RH2003②	10	2000	13.1–14.3
1975-76	Delco	—	Negative	37	—	—	—	1116389②	10③	3000③	14.0-14.3③
	Delco	—	Negative	55	—	—	—	1116389②	10③	3000③	14.0-14.3③
	Delco	—	Negative	63	—	—	—	1116389②	10③	3000③	14.0-14.3③
1976	Motorcraft	—	Negative	40	—	2.5–3	12	—	—	—	—
	Motorcraft	—	Negative	60	—	2.5–3	12	—	—	—	—

①—Excessive current drawn indicates shorted field winding. No current draw indicates an open winding.
②—Regulator is a sealed assembly, requiring no adjustments.

STARTING MOTOR SPECIFICATIONS

Year	Part No. ②	Rotation ①	Brush Spring Tension, Ounces	No Load Test			Torque Test		
				Amperes	Volts	R.P.M.	Amperes	Volts	Torque, Lbs. Ft.
1969–70	C7FF-11001-B	C	40	65	12	9250	500	4.5	—
	C9FF-11001-A	C	40	65	12	9250	500	4.5	—
1971	—	C	40	65	12	9250	600	3.4	13
1972	—	C	40	65	12	9250	600	4.5	13
1973–76	—	C	40	65	12	9250	600	3.4	13

①—As viewed from drive end. C—Clockwise. ②—Stamped on plate riveted to side of housing.

BRAKE SPECIFICATIONS

Year	Model	Brake Drum Inside Diameter	Wheel Cylinder Bore Diameter: Disc Brake	Wheel Cylinder Bore Diameter: Front Drum Brake	Wheel Cylinder Bore Diameter: Rear Drum Brake	Master Cylinder Bore Diameter: Disc Brakes	Master Cylinder Bore Diameter: Drum Brakes	Master Cylinder Bore Diameter: Power Brakes
1969	American, Javelin 6-Cyl.	9	—	1 1/8	15/16	—	1	1
	Rebel (Except Wagon) 6-Cyl.	9	—	1 1/8	15/16①	—	1	1
	All V8s & Rebel 6 Wagon	10	—	1 3/16③	15/16④	—	1	1
	All V8s with Disc Brakes	10	2	—	1	1	—	1
1970	Gremlin	9	—	1 1/8	15/16	—	1	1
	Hornet, Javelin 6-Cyl.	9	—	1 1/8	15/16	—	1	1
	Ambassador, Rebel 6-Cyl.	10	—	1 1/8	15/16	—	1	1
	AMX, Hornet, Javelin V8	10	—	—	—	—	1	1
	Rebel V8	10	—	—	—	—	1	1
	All with Disc Brakes	10	2	—	1	1	—	1
1971	Gremlin	9	—	1 1/8	7/8	—	1	1
	Hornet, Javelin 6-Cyl.	9	—	1 1/8	7/8	—	1	1
	Matador, Amb. 6-Cyl.	10	—	1 3/32	15/16	—	1	1
	Hornet, Javelin V8	10	2 3/4	1 3/16	7/8	⑤	1	1
	Matador, Amb. V8	10	2 3/4	1 3/16	15/16⑥	1 1/8	1	1
1972	Gremlin 6-Cyl.	9	2 3/4	1 1/8	13/16	1①	1	1
	Gremlin V8	9	2 3/4	1 3/16	7/8	1 1/16	1	1
	Hornet, Javelin 6-Cyl.	9	2 3/4	1 1/8	7/8	1①	1	1
	Hornet, Javelin V8	10	2 3/4	1 3/16	7/8	1①	1	1
	Matador, Amb. 6-Cyl.	10	2 3/4	1 3/32	15/16	1 1/8	1	1
	Matador, Amb. V8	10	2 3/4	1 3/16	15/16	1 1/8	1	1
1973	Gremlin 6-Cyl.	9	2 3/4	1 1/8	13/16	1 1/16	1	1
	Gremlin V8	10	2 3/4	1 3/16	7/8	1 1/16	1	1
	Hornet, Javelin 6-Cyl.	9	2 3/4	1 1/8	7/8	1 1/16	1	1
	Hornet, Javelin V8	10	2 3/4	1 3/16	7/8	1 1/16	1	1
	Matador 6-Cyl.	10	2 3/4	1 3/32	15/16	1 1/16	1	②
	Matador, Amb. V8	10	2 3/4	1 3/32	15/16	1 1/8	1	②
1974	Gremlin 6-Cyl.	9	2 3/4	1 1/8	13/16	1 1/8①	1	1
	Gremlin V8	10	2 3/4	1 3/16	7/8	1 1/8①	1	1
	Hornet, Javelin 6-Cyl.	9	2 3/4	1 1/8	7/8	1	1	1
	Hornet, Javelin V8	10	2 3/4	1 3/16	7/8	1	1	1
	Matador, Ambassador	10	2 3/4	1 3/32	15/16	1 1/8①	1	1
	Station Wagons	10	2 3/4	1 3/32	15/16	1 1/8①	1	1
1975–76	Pacer	⑧	3.1	1 3/32	7/8	1①	1	1
	Gremlin⑨	9	3.1	1 1/8	7/8	1①	1	1
	Hornet⑨	9	3.1	1 1/8	13/16	1①	1	1
	Gremlin, Hornet⑩	10	3.1	1 3/16	7/8	1①	1	1
	Matador	10	3.1	—	⑦	1 1/16	—	1 1/8

①—Non power disc brakes 1 1/16".
②—Power drum 1", power disc 1 1/8".
③—3/32" on V8-343, 390 wagons.
④—7/8" on Rambler, Javelin and AMX.
⑤—Manual 1 1/16", power 1".
⑥—1" on wagons.
⑦—Exc. sta. wag., 7/8"; sta. wag. 15/16".
⑧—Front drums, 10"; rear drums, 9".
⑨—1975 & Early 1976 6 cyl. models.
⑩—1975–76 V8 models & 1976 mid-year 6 cyl. models.

AMERICAN MOTORS

COOLING SYSTEM & CAPACITY DATA

Year	Model or Engine	Cooling Capacity, Qts.			Radiator Cap Relief Pressure, Lbs.		Thermo. Opening Temp. ①	Fuel Tank Gals.	Engine Oil Refill Qts. ②	Transmission Oil			Rear Axle Oils Pint
		No Heater	With Heater	With A/C	With A/C	No. A/C				3 Speed Pints	4 Speed Pints	Auto. Trans. Qts. ⑩	
1969	6-199, 232	9½	10½	10½	14	14	195⑪	⑫	4	1½⑬	—	9	3
	V8-290	13	14	14	14	14	195	⑫	4	3	2½	⑭	4
	V8-343, 390	12	13	13	14	14	195	⑫	4	—	2½	11	4
1970	6-199, 232	9½	10½	10½	14	14	195	⑮	4	1½⑬	—	9½	3
	V8-304	13	14	14	14	14	195	⑮	4	3	2½	9½	4
	V8-360, 390	12	13	13	14	14	195	⑮	4	—	2½	10	4
1971	6-232	9½	10½	10½	14	14	205	③	4	1½⑬	—	9½	④
	6-258	9½	10½	10½	14	14	205	③	4	2½	—	9½	④
	V8-304	13	14	14	14	14	195	③	4	2½	2½	9½	4
	V8-360, 401	12	13	13	14	14	195	③	4	3	2½	10	4
1972	6-232	9½	10½	10½	14	14	205	⑤	4	1½⑬	—	8½	④
	6-258	9½	10½	10½	14	14	205	⑤	4	1½⑬	—	8½	④
	V8-304	13	14	14	14	14	195	⑤	4	3	2½	9½	4
	V8-360, 401	12	13	13	14	14	195	⑤	4	3	2½	9½	4
1973	6-332, 258	9½	10½	10½	14	14	205	⑥	4	2½	2½	8½	⑦
	V8-304	13	14	14	14	14	195	⑥	4	2½	2½	9½	⑦
	V8-360, 401	12	13	13	14	14	195	⑥	4	2½	2½	9½	⑦
1974	6-232, 258	10	11	11½	14	14	205	⑯	4	2½	—	8½	⑦
	6-232, 258 Mat. 2 Dr. Coupe	10	11	13½⑰	14	14	205	⑯	4	2½	—	8½	⑦
	V8-304 Hornet, Gremlin	15	16	16	14	14	195	⑯	4	2½	—	8½	⑦
	V8-304 Mat., Amb.	15½	16½	16½	14	14	195	⑯	4	2½	—	8½	⑦
	V8-304, Mat. 2 Dr. Coupe	17½	18½⑰	18½⑰	14	14	195	⑯	4	2½	—	8½	⑦
	V8-304 Javelin	—	15	15½	14	14	195	⑯	4	1½	—	8½	⑦
	V8-360 Hornet	14	15	15	14	14	195	⑯	4	1½	—	9½	⑦
	V8-360, 401 Matador	14½	15½	15½⑰	14	14	195	⑯	4	—	—	9½	⑦
	V8-360, 401 Mat. 2 Dr. Coupe	16½	17½⑰	17½⑰	14	14	195	⑯	4	1½	—	9½	⑦
	V8-360, 401 Javelin	14	15	15	14	14	195	⑯	4	—	2½	9½	⑦
	V8-360, 401 Ambassador	14½	15½	15½	14	14	195	⑯	4	1½	—	9½	⑦
1975	6-232-258	—	⑱	⑥	14	14	195	⑨	4	3½⑲	—	8½	⑦
	Gremlin, Hornet V8-304	—	16	16	14	14	195	⑨	4	3½	—	8½	⑦
	Matador⑳ V8-304	—	16½	16½	14	14	195	⑨	4	3½	—	8½	⑦
	Matador㉑ V8-304	—	18½⑰	18½⑰	14	14	195	24½	4	3½	—	8½	⑦
	Matador⑳ V8-360	—	15½	15½	14	14	195	⑨	4	3½	—	9½	⑦
	Matador㉑ V8-360	—	17½⑰	17½⑰	14	14	195	24½	4	3½	—	9½	⑦
1976	6-232, 258	—	㉒	㉓	14	14	195	㉔	4	㉕	—	8½	⑦
	V8-304 Gremlin, Hornet	—	⑯	⑯	14	14	195	㉔	4	3½	—	9½	4
	V8-304 Matador⑳	—	16½⑰	16½⑰	14	14	195	㉔	4	3½	—	9½	4
	V8-304 Matador㉑	—	18½	18½	14	14	195	24½	4	3½	—	9½	4
	V8-360 Matador⑳	—	15½⑰	15½⑰	14	14	195	㉔	4	—	—	9½	4
	V8-360 Matador㉑	—	17½	17½	14	14	195	24½	4	—	—	9½	4

①—With permanent type anti-freeze.

②—Add one quart with filter change.

③—Gremlin, 21; Hornet & Javelin, 16; Ambassador wagons & Matador 3 seat wagon, 17; all others, 19½.

④—Matador & Ambassador, 4; all others, 3.

⑤—Hornet, 16; Gremlin, 21; Javelin; 16; Matador & Ambassador sedans, 19½; Matador 2 seat wagons, 19½; Matador 3 seat wagons, 21½; Ambassdor wagons, 21½.

⑥—Hornet, 16; Gremlin, 21; Javelin, 16; Matador & Amb. sedans, 19½; Matador 2 seat wagon, 19½; Matador 3 seat wagon, 20; Ambassador wagon, 20.

⑦—7⁹⁄₁₆″ axle, 3 pints. 8⅞″ axle, 4 pints.

⑧—Matador 2 dr. coupe 13½ qts., all others 11½ qts.

⑨—Gremlin, 21 gals.; Hornet, early 17 gals.; late 22 gals.; Matador exc. wagon, 24½ gals.; wagon, 21 gals.; Pacer, 22 gals.

⑩—Approximate. Make final check with dipstick.

⑪—Rambler Rogue with 6-232 uses 205°.

⑫—Rambler 16.
AMX and Javelin 19.
Rebel 3-seat wagon and Amb. wagon 19.
All Others 21½.

⑬—Fully synchronized 2½.

⑭—With 2 barrel carb. 9; with 4 barrel carb. 11.

⑮—Rebel and Ambassador sedans, 21½, Rebel 2 seat wagons, 21½. All others 19. California vehicles about 2 gal. less.

Continued

COOLING SYSTEM & CAPACITY DATA NOTES—Continued

⑯—Hornet & Javelin, 16; Gremlin, 21; Matador Sedan & 2 Dr. Coupe, 24¾; Ambassador Sedan, 24¾; Mat. & Amb. Wagons, 21.
⑰—Add two quarts with coolant recovery system.
⑱—Exc. Pacer, 11 qts.; Pacer, 10½ qts.
⑲—With overdrive exc. Pacer, add 1 pt.; Pacer with overdrive, add ½ pt.
⑳—Sedan & wagon.
㉑—2 dr. coupe.
㉒—Exc. Pacer, 11 qts.; Pacer, 14 qts.
㉓—Exc. Pacer & Matador 2 dr. coupe, 11½ qts.; Pacer, 14 qts.; Matador 2 dr. coupe, 13½ qts.
㉔—Gremlin, 21 gals.; Hornet, 22 gals.; Matador exc. wagon, 24½ gals.; Wagon, 21 gals.; Pacer, 22 gals.
㉕—Exc. Pacer, Matador & overdrive, 2½ pts.; Pacer & Matador, 3½ pts.; all models with overdrive, 4 pts.

Electrical Section

DISTRIBUTOR, REPLACE

1. Disconnect distributor primary wire from coil terminal.
2. Remove distributor cap and rotor. *Mark position of rotor arm on distributor housing so distributor can be installed in same position.*
3. Remove vacuum line from distributor.
4. Remove distributor hold-down clamp.
5. Note relative position of distributor in block, then work it out of the engine.

Installation

1. Turn rotor about ⅛ of a turn counterclockwise past the mark previously placed on the distributor housing.
2. Push the distributor down into the block with the housing in the normal "installed" position. *On gear-driven distributors, it may be necessary to move the rotor slightly to start gear into mesh with camshaft gear, but rotor should line up with mark when distribuor is down in place.*
3. Tighten distributor clamp screw snugly and connect vacuum line, primary wire to coil, and install cap.

V8 Note

If the engine was disturbed while the distributor was removed from the engine, first crank the engine to bring No. 1 piston up on its compression stroke and continue cranking until the timing mark is adjacent to the timing indicator. Then rotate the distributor cam until the rotor is in position to fire No. 1 cylinder. Install the distributor as outlined above and set the ignition timing as directed elsewhere in this manual.

STARTER, REPLACE

To remove starter, disconnect cable from battery. Disconnect cable and solenoid lead wire from solenoid switch. Remove starter attaching bolts and take off starter.

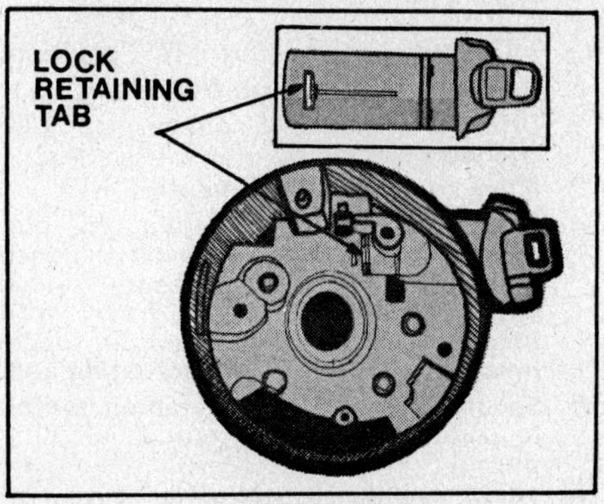

Fig. 1 Lock cylinder removal

IGNITION LOCK

1970-76

1. Remove turn signal switch as described further on.
2. Place key lock in "Lock" position and using a small flat blade screwdriver to depress the lock cylinder retaining tab, remove the lock cylinder, Fig. 1.

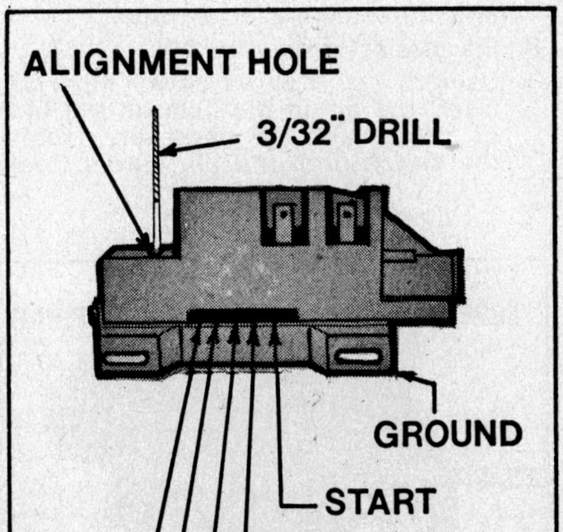

Fig. 2 Ignition switch alignment. 1970-76 (typical)

IGNITION SWITCH, REPLACE

1970-76

The ignition switch on all models is mounted on the lower section of the steering column and is connected to the lock by a remote control rod. To remove switch, place key in Off-Lock position and remove mounting screws. Disconnect switch from remote control rod, remove wire harness and remove switch.

To install switch on 1970-71 models, place switch and ignition lock in Off-Lock position and insert a $^{3}/_{32}$ inch drill into the switch aligning hole, Fig. 2. With drill in place, install switch on remote control rod. Remove all slack by sliding switch toward steering wheel. Install mounting screws, remove drill and connect wiring harness.

To install switch on 1972-76 models equipped with standard steering column, move slider to extreme left position (left side of switch pointing inward toward steering column). Place actuator rod in slider hole and install switch on column. Tighten retaining screws.

To install switch on 1972-76 models equipped with tilt steering column, move slider to extreme right position (right side of switch pointing downward from steering wheel). Place actuator rod in slider hole and install switch on column without tightening screws. Remove lash from actuator rod by pushing downward lightly on switch and tighten retaining screws.

1969 Rebel & Ambassador

1. Remove escutcheon nut from switch.
2. Disconnect wires and remove switch.

1969 AMX, Javelin & Rambler

Remove ignition switch from rear of instrument panel by holding switch escutcheon while pressing switch assembly toward instrument panel and turning counterclockwise.

The multi-wire connector on the switch should be removed after the switch is removed from the instrument panel. The multi-wire connector is locked to the switch by plastic fingers which are part of the connector and must be released before removing the connector.

NOTE: The AMX and Javelin have a separate ground wire added to the ignition switch plate.

LIGHT SWITCH, REPLACE

1971-74 Javelin

1. Remove knob from toggle switch by depressing a small spring steel retaining tab up toward the handle.
2. Remove seven screws from steering column lower cover and remove cover.
3. Remove wire connections, two retaining screws and remove switch.

1969-70 & 1971-76 Except Javelin

1. Disconnect a battery cable.
2. With switch in full "On" position, press button on side or top of switch to release shaft and knob assembly.
3. Remove switch mounting sleeve nut.
4. Disconnect wire harness connector.
5. Reverse procedure to install. Position switch so that shaft is lined up properly before tightening.

STOP LIGHT SWITCH, REPLACE

1972-76

1. Disconnect wire connector from switch.
2. Remove brake pedal pivot bolt, nylon retaining rings, sleeve and remove switch.
3. When installing switch, be sure dimple on switch is opposite the bushing collar, Fig. 3.

1969-71

The switch is a mechanical, pedal operated switch. When replacing, be sure switch is adjusted so it is in "off" position when pedal is pulled back to stop. If switch is "on", check for binding linkage or improperly adjusted "Cruise Command" switch.

NEUTRAL SAFETY SWITCH

1972-76

A non-adjustable combination neutral safety and back-up light switch is located at the lower left hand side of the automatic transmission case.

1969-71 Console Type

The switch is located on the right side of the selector shaft under the console. It also functions as a back-up light switch, therefore, the following adjustments will automatically adjust the back-up light switch for proper contact when

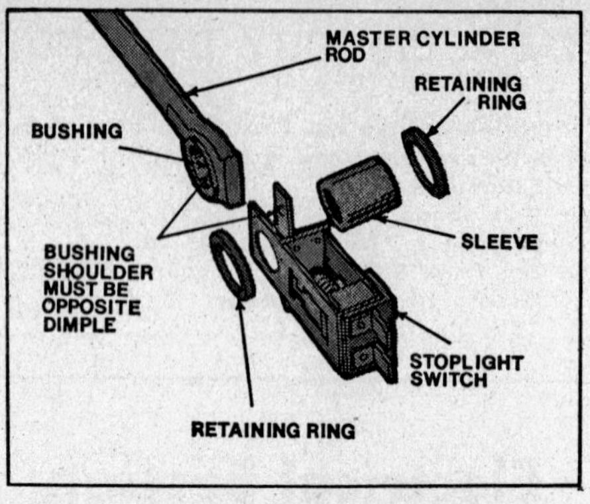

Fig. 3 Stoplight switch. 1972-76 (typical)

the shift lever is in the "R" position.

1. Remove selector knob from shaft.
2. Unfasten and lift console up and over selector shaft.
3. Place selector lever in neutral.
4. Use a 3/32" drill as an aligning pin. Insert the pin in the hole on the face of the switch. If necessary, move switch until pin enters freely into hole in switch toggle. *Switch toggle must enter into slot of actuating tab.*
5. Secure two screws that fasten switch to pivot bracket and remove aligning pin.
6. When switch is adjusted properly, engine will start only in Park or Neutral.

1969-71 Steering Column Type

The neutral safety switch also functions as a back-up light switch (when the car is so equipped). Therefore, the following adjustments will automatically adjust the back-up light switch for proper contact when the gear selector lever is in the "R" position.

1. Loosen two screws that secure switch to steering jacket tube.
2. Place selector lever in "N".
3. Using a 3/32" drill as an aligning pin, insert the pin in the hole on the face of the switch. If necessary, rotate the switch until the pin enters freely in the hole in the switch toggle.
4. Tighten the switch screws and remove aligning pin.

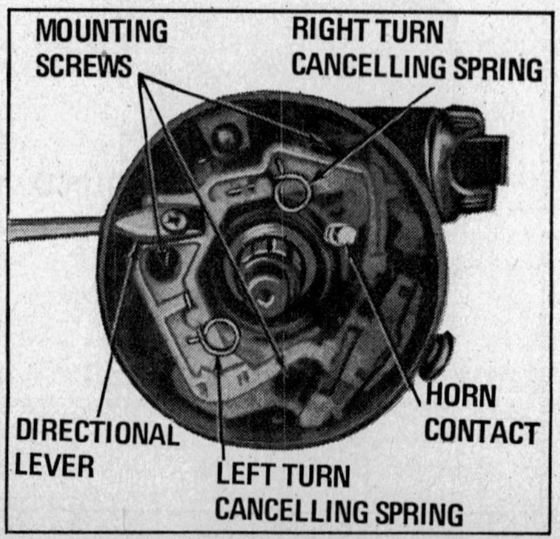

Fig. 4 Turn signal switch. 1970-76

TURN SIGNAL SWITCH, REPLACE

1970-76

1. Disconnect negative battery cable.
2. Remove steering wheel.
3. Loosen anti-theft cover retaining screws and lift cover from column. *It is not necessary to remove these screws completely as they are held on the cover by plastic retainers.*
4. Using a suitable tool, depress lock plate and pry out snap ring from steering shaft groove.
5. Remove tool, lock plate, cancelling cam, upper bearing preload spring and thrust washer.
6. Place directional signal lever in right turn position and remove lever.
7. Depress hazard warning light switch and remove the button by turning in a counterclockwise direction.
8. Remove directional signal wire harness from mounting bracket on lower column. On Shift Command, column shift, use a stiff wire to depress lock tab which retains the shift quadrant light wire in connector block.
9. Remove directional signal switch screws and remove switch, Fig. 4.

1969 All

1. Disconnect battery ground cable then remove steering wheel using a suitable puller.
2. Remove cancelling cam, turn signal switch lever, hazard warning knob and light shield (if equipped).
3. Remove plastic trim covering column support plate. Remove signal switch wiring harness protector from lower side of column. Do not remove column support plate.
4. Disconnect signal switch connector from main harness and remove terminals from switch harness. Do not cut switch harness wires.
5. Remove turn signal cover, switch retaining screws and switch.
6. Reverse procedure to install.

HORN SOUNDER & STEERING WHEEL

Disconnect battery. Remove horn button or ring. This is retained by screws from under the steering wheel on some models and is rubber mounted and can be pulled up on others. Remove steering wheel with a puller. Note line up of dash marks on steering shaft and wheel and align these marks when wheel is installed.

INSTRUMENT CLUSTER

1975-76 Pacer

1. Disconnect battery ground cable.
2. Remove cluster bezel.
3. Remove radio knobs and nuts, then remove radio overlay.
4. Remove headlamp switch overlay retaining screws, then pull overlay back so that speedometer cable can be removed.
5. Remove cluster retaining screws, disconnect wiring harness and gear selector cable if equipped, then remove cluster assembly.

1974-76 Matador & Ambassador

1. Disconnect battery ground cable.
2. Remove radio knobs, attaching nuts and bezel retaining screws.
3. Tilt bezel forward and disconnect wiring.
4. Remove bezel, then remove clock or economy fuel gauge (if used) attaching screws, pull assembly away from cluster and disconnect bulbs and electrical leads. If equipped with fuel economy gauge, disconnect vacuum line and remove assembly.
5. Using clock opening, disconnect speedometer cable from instrument cluster, and disconnect gear selector dial cable from steering column.
6. Remove cluster mounting screws, disconnect any remaining electrical connections and remove cluster.

1972-76 Hornet & Gremlin

1. Disconnect battery ground cable.
2. Remove package tray, if so equipped, to gain access to wiper control and speedo cable.
3. Remove wiper control knob and spanner nut. To remove control knob, rotate knob until slot in neck of knob is visible. Insert a small diameter tool in slot and apply pressure toward knob to release pressure on spring metal clip. Remove knob.
4. Remove speedometer cable.
5. Remove four top and two side screws from instrument panel and tilt panel forward slightly to gain access to light switch and connectors.
6. Remove headlight switch knob, retaining nut and switch.
7. Cover steering column with a cloth to prevent scratches and disconnect harness connectors and remove cluster.

1972-73 Matador & Ambassador

1. Disconnect battery ground cable.
2. Remove screws from cluster overlay. Remove radio knobs and remove cluster overlay.
3. Remove screws securing cluster to panel.
4. Disconnect speedo cable and harness connector plug and remove cluster.

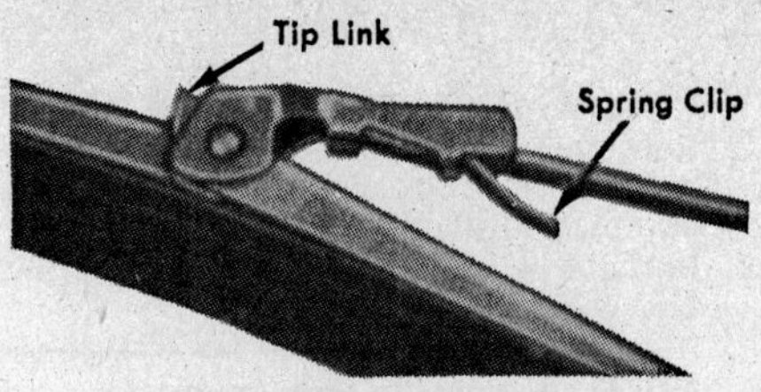

Fig. 5 Removing wiper blade. 1969-76 Exc. Matador X

1971 Hornet & Gremlin

1. If so equipped, the package tray must be removed to gain access to speedometer cable, light switch and wiper control.
2. Disconnect battery.
3. Remove wiper control knob and spanner nut.
4. Disconnect speedometer cable at rear of cluster.
5. Remove top and side screws from cluster overlay.
6. Partially remove cluster from instrument panel to gain access to harness connectors. **NOTE:** If equipped with low fuel warning system, remove relay mounting screw at lower left side of panel and remove complete system with the cluster.

1971 Matador & Ambassador

1. Disconnect battery and remove screws from cluster overlay.
2. Remove A/C thermostat control knob if so equipped.
3. Remove overlay.
4. Remove screws securing cluster to panel.
5. Disconnect speedometer cable from behind panel.
6. Remove cluster disconnect plug and flasher unit.
7. Remove low fuel warning relay mounting screw at lower side of panel, if so equipped.
8. Remove cluster assembly.

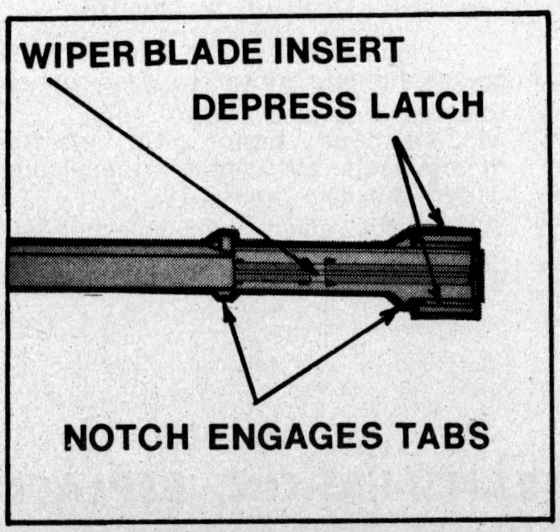

Fig. 6 Removing blade element. 1969-76 Exc. Matador X

1971-74 Javelin

1. Disconnect battery.
2. Cover painted surface of column with a cloth.
3. Remove the top, side and lower screws around bezel.
4. If equipped with radio, remove control knobs and retaining nuts.
5. Remove knobs from four instrument panel switches.
6. If equipped with A/C, release the speedometer cable hold-down clip on wheelhouse panel to allow movement of cable.
7. Move bezel and cluster out of opening far enough to reach in and disconnect speedometer cable, wire harness plug and wire connections.
8. The cluster can be removed from bezel by removing eight screws and the clock set knob.

1970 Hornet & Gremlin

1. Disconnect negative battery cable.
2. Cover steering column to prevent scratching and remove package tray if so equipped.
3. Disconnect speedometer cable.
4. Remove control knobs and retaining nuts from wiper and headlight switch.
5. Remove bezel retaining screws and remove bezel and cluster toward the center of the car as an assembly.
6. Disconnect all wires and lamps from rear of cluster.

1969 Rambler

1. Disconnect battery.
2. Remove cigar lighter and ignition switch.
3. Remove W/S wiper switch knob.
4. Remove light switch knob and shaft.
5. Remove flasher unit.
6. Disconnect speedometer cable housing at rear of cluster.
7. Remove two Phillips head screws at top of cluster overlay.
8. Partially remove cluster from instrument panel to gain access for removal of cluster pin plug.
9. Remove cluster assembly.
10. Reverse procedure to install.

1969 Except Rambler
1970 Except Hornet & Gremlin

1. Disconnect battery.
2. Remove screws from cluster overlay.
3. Remove overlay.
4. Remove screws securing cluster to instrument panel.
5. Remove pin plug and cluster.
6. Disconnect speedometer cable and parking brake light.

IMPORTANT: Spring metal grounding clip is required between cluster and Weather-eye controls. Also between clock or tachometer of AMX, Javelin. Be sure to install clip upon assembly of cluster if it was removed.

W/S WIPER BLADES

1974-76 Matador X & Pacer

Insert an appropriate tool into spring release opening of blade saddle, depress spring clip and pull blade from arm. To install, push blade saddle onto pin so spring engages pin.

1969-76 Exc. Matador X & Pacer

The wiper blade assembly can be removed from the wiper arm by lifting the blade off the windshield and tipping the blade to arm connecting link toward the glass. This will disengage the embossing tab on the top of the arm out of the hole in the link.

At the same time, slide the blade away from the end of the arm so the embossing does not index with the hole. Then push the spring tab, Fig. 5, downward away from the depression in the bottom side of the arm and slide the blade off the arm.

The blade element can be removed by compressing the blade latch and sliding it from the bridge, Fig. 6 When installing, the metal backing must engage the tabs on the bridge.

W/S WIPER ARMS

The wiper arms are set on the serrated pivot shafts and held securely by spring tension on the arm. To remove the arm, lift the arm against the spring tension and with a screwdriver, slide the cap away from the serrated pivot shaft.

NOTE: Arms are marked "R" or "L" on the underside of the arm to designate right or left arm.

W/S WIPER MOTOR, REPLACE

1975-76 Pacer

1. Remove vacuum canister and mounting bracket.
2. Disconnect linkage from wiper motor output arm.
3. Remove three nuts and one screw from heater housing. On models with A/C, pull housing forward.
4. Remove screws from wiper motor mounting plate and disconnect wiring harness.
5. Remove attaching screws and wiper motor.

NOTE: Ensure output arm is in parked position before installing wiper motor.

1975-76 Pacer Liftgate Wiper

1. Remove wiper arm and blade.
2. Remove liftgate trim pad.
3. Disconnect wiring harness and ground wire.
4. Remove nut and pad securing wiper motor shaft to liftgate.
5. Remove screws securing wiper motor bracket, then separate bracket from motor.

1974 Matador X

1. Remove wiper arms and cowl screen.
2. Remove retaining clip from linkage drive arm and disconnect electrical connectors from motor.
3. Remove wiper motor retaining screws and wiper motor.

NOTE: If output arm contacts dash panel, preventing wiper motor removal, hand turn output arm so arm clears dash opening.

4. Before installing motor, be sure output arm is in park position.

1973-74 Exc. Matador X & All 1975-76

1. Remove wiper arms and blades.
2. Remove four screws holding motor to dash.
3. Separate harness connector at the motor.
4. Pull motor and linkage out of opening to expose the drive link to crank stud retaining clip. Raise up the lock tab of the clip with a flat bladed screwdriver and slide clip off stud.

1969-72 Rebel, Ambassador & Matador

1. Remove wiper arms and blades and cowl air intake cover.
2. Slide link to motor retainer clip off motor arm stud which is accessible through cowl top opening. Remove link from motor.
3. On vacuum type motor, disconnect control cable and hose from motor, then remove motor and mounting plate from dash panel.
4. On models with electric wiper, disconnect switch-to-motor female connectors at switch and remove through main wire harness grommet. Remove wiper motor and mounting plate.

1969-72 Rambler, AMX, Javelin, Gremlin & Hornet

1. To remove wiper motor, remove four screws holding motor to dash panel.
2. Remove hose and control cable from vacuum type motor. On electric motors, separate wiper harness plug under instrument panel.
3. Tilt motor and slide link-to-motor retaining clip off of stud and remove motor.

W/S WIPER TRANSMISSION, REPLACE

1975-76 Pacer

1. Remove wiper arms and blades.
2. Remove pivot shaft bodies to cowl screws using tool J-25359.
3. Disconnect linkage from motor output arm, then remove pivot shaft body assembly.

1974-76 Matador X

1. Remove wiper arms and cowl screen.
2. Remove clip securing left linkage arm to drive arm.
3. Remove left pivot shaft body to cowl retaining screws and remove left arm and pivot shaft body through cowl opening.
4. Remove clip securing drive arm from motor arm crankpin.
5. Remove right pivot shaft body to cowl retaining screws and remove right arm and pivot shaft body through cowl opening.

1969-76 Rebel, Ambassador & Matador Exc. Matador X

1. Remove wiper arms and blades.
2. Remove cowl air intake cover.
3. Disconnect link-to-motor retainer and link from wiper arm through cowl top opening.
4. Close hood and remove two capscrews holding each pivot shaft body to cowl top.
5. Remove both pivot body and link assemblies as a unit through cowl top opening.
6. Pivot shaft bodies may than be removed by sliding retainer off stud.

NOTE

When installing pivot shaft bodies to cowl top, the assist spring must be in a position to engage rubber sleeved stud on vacuum wiper. There are no assist springs on electric wipers.

AMX & Javelin; 1969-76 Rambler, AMX, Javelin, Hornet & Gremlin

1. Remove wiper arms and blades.
2. Remove pivot shaft-to-cowl top nuts.
3. Remove wiper motor.
4. Slide pivot shaft body and link assembly to the left to clear right pivot shaft opening and move assembly to the right side of car to remove as a unit.

NOTE

When installing pivot shafts to cowl top, flat side of pivot shaft indexes flat side of hole in cowl top when pivot shaft is in up position.

W/S WIPER CONTROL, REPLACE

1975-76 Pacer

1. Remove instrument cluster bezel, then the four screws from headlamp over-

lay.
2. Remove control knob, using a small screw driver to release spring tension on clip.
3. Remove wire connector, then the two screws attaching switch to instrument panel.

1974-76 All Except Javlin & Pacer

1. Locate small notch at base of knob and insert a small screwdriver and apply pressure to release spring and pull knob from shaft.
2. Remove slotted trim nut from front of switch.
3. Push switch through instrument panel then disconnect wiring harness and remove switch.

1973 Except Javelin

1. Remove knob and slotted trim nut from front of switch.
2. Push switch through instrument panel, disconnect harness and remove switch.

1973-74 Javelin

1. Remove control knob.
2. Access can be gained to switch screws by removing steering column cover.

1972 Ambassador & Matador

The control is mounted behind the instrument panel with two sheet metal screws. These are accessible after removing the cardboard shield. Then disconnect washer and wiper motor wires from switch.

1970-71 Ambassador, Rebel & Matador

1. Disconnect negative battery cable.
2. Cover steering column to prevent scratching.
3. If so equipped, remove package tray and disconnect speedometer cable.
4. Remove control knob and retaining nut from the wiper control.
5. On vacuum wiper systems, disconnect control at the wiper motor.
6. Remove instrument cluster retaining screws and pull out cluster as a unit.
7. Remove vacuum wiper control.
8. On electrical systems, disconnect the electrical connections to the switch and remove switch.

1969 All, 1970-72 AMX, Javelin, Hornet & Gremlin

1. Remove control knob. The knob is retained either by a set screw or by a spring retainer. The spring retainer is released as follows: turn knob fully to right and insert Allen wrench in notch at small end of knob. Push toward shaft to raise spring out of groove and at same time pull knob off shaft.
2. On all models, after the knob has been removed, disconnect control cable from motor on vacuum wipers. On electric wipers, disconnect three

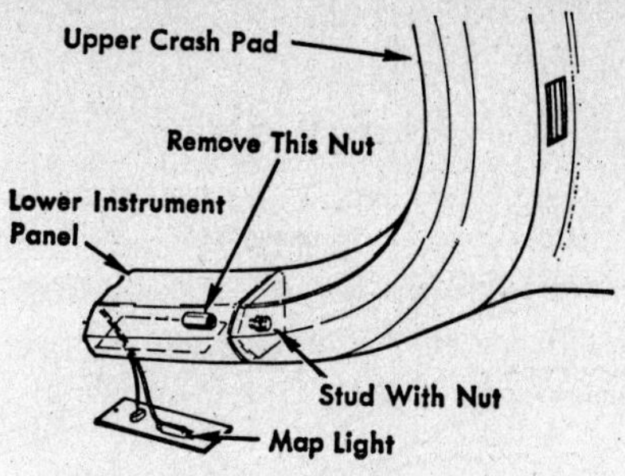

Fig. 7 Location of hidden stud nut. 1972-74 Javelin

female connectors from switch. Remove French nut to disengage switch from instrument panel.

NOTE: On vacuum wipers the control cable must be installed with the washer at the end of the control wire inserted in slot in slide valve. The conduit must butt against the shoulders of the anchoring slot and be fastened securely to assure positive opening and closing of the valve for efficient operation.

RADIO

NOTE: When installing radio, be sure to adjust antenna trimmer for peak performance.

1975-76 Pacer

1. Remove radio knobs, nuts and bezels, then the radio overlay.
2. Loosen radio attaching screws and lift radio from mounting bracket.
3. Disconnect antenna lead and wire connectors and remove radio.

1974 Ambassador & 1974-76 Matador

1. Remove radio knobs, retaining nuts and bezel.
2. Loosen upper radio retaining screw and lift radio disengaging bracket from screw, and pull radio forward.
3. Disconnect antenna lead, electrical wiring and remove radio.

1971-74 Javelin

To remove the radio, the entire crash pad will have to be removed.
1. Disconnect battery ground cable.
2. Remove the six attaching screws that lie next to lower edge of windshield.
3. Remove radio knobs and retaining nuts.
4. Open right door and remove two panel attaching screws at door pillar area.
5. Remove the five attaching screws in the upper flange of the cluster bezel. Remove moulding attaching screws and assist handle at lower right finish panel.

NOTE: On 1972-74 remove map light to gain access to hidden stud nut which retains crash pad to lower instrument panel, Fig. 7.

6. Remove entire panel top cover to expose radio and speaker.
7. Disconnect and remove speaker except on 1971 models with FM radio. Slide radio back and lift up to disconnect speaker and bulb wires.
8. Remove radio input wire from fuse panel. Tie a heavy string to wire before removing radio to assist in dressing wire back through wiring and duct assemblies.
9. Remove radio.

1971-73 Matador & Ambassador

1. The radio is retained to the cluster by three screws adjacent to the face of the radio.
2. Disconnect battery.
3. Disconnect antenna and speaker leads from radio.
4. Remove instrument cluster overlay.
5. Remove radio lead from fuse panel.
6. Remove radio mounting screws and remove radio.

1970-76 Hornet & Gremlin

1. Disconnect battery.
2. Remove package tray if so equipped.
3. Remove ash tray and bracket.
4. Remove radio knobs and shaft nuts.
5. Remove bezel retaining screws and bezel.
6. Disconnect antenna, speaker and power lead.
7. Remove radio.

1970 AMX & Javelin

1. Disconnect battery.
2. Remove four bezel retaining screws.
3. Remove knobs and shaft nuts.
4. Remove bezel.
5. Removing retaining screw from rear or radio.
6. Disconnect leads and remove radio.

1969 Javelin & AMX

1. Remove the ash tray.
2. Remove the retaining bolt in the ash tray which is threaded into the radio. With AM-Tapeplayer combination this bolt is not used. A retaining nut is used on the rear of the radio and must be removed.
3. Remove the shaft retaining nuts.
4. Disconnect all leads from radio and tip the rear of the radio up and to the toe board and remove from the rear edge of the panel.

NOTE: If equipped with A/C the air discharge duct must be removed to gain clearance.

1969 Rambler

1. Disconnect battery and remove antenna lead from radio.
2. Disconnect power lead from radio and unplug radio-to-speaker cable.
3. Remove control knobs and control

shaft bushing retainer nuts.
4. Remove ash tray.
5. Remove Phillips head screw which is threaded into cage nut on radio.
6. Radio can now be removed toward back and down from panel.

NOTE: If equipped with A/C it will be necessary to remove the glove box.

1969-70 Rebel & Ambassador

1. The radio is retained to the instrument cluster by four screws adjacent to the face of the radio and a brace rod from the rear of the radio down to the instrument panel flange.
2. Disconnect the battery.
3. Disconnect antenna, power, ground and speaker leads from radio.
4. Remove instrument panel overlay.
5. Remove radio mounting screws and remove radio.

NOTE: Leads attached to radio must be drawn carefully through the opening while removing the radio to prevent damage to the wiring.

HEATER CORE REMOVAL

1975-76 Pacer, Fig. 8

1. Drain approximately 2 qts. from cooling system.
2. Remove heater hoses and install plugs in hoses and core openings.

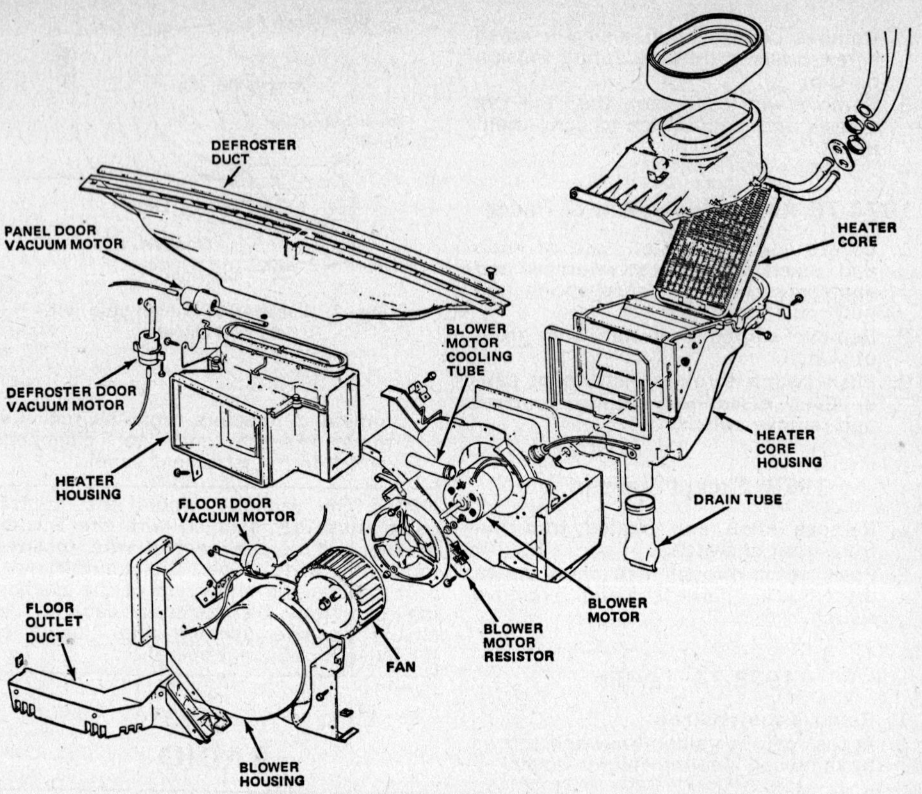

Fig. 8 Heater and blower housing assembly. 1975-76 Pacer

3. Position vacuum hoses away from core housing.
4. Remove core housing cover screws, disconnect overcenter spring and remove cover.
5. Remove screws attaching core to housing and remove core.

1974 Ambassador & 1974-76 Matador, Fig. 9

1. Disconnect battery ground cable.
2. Drain about 2 quarts of coolant then disconnect heater hoses and plug ends of hoses and core openings.
3. Remove lower instrument finish panel and glove box.
4. Disconnect control cables and vacuum motor hoses.
5. Remove retaining screws and remove heater core housing assembly.

1971-74 Javelin

1. Drain 2 qts. of water from cooling system and disconnect hoses from core tubes in engine compartment. Install corks in hoses and core tubes.
2. Disconnect blower motor wires.
3. Remove ground wire from dash panel ground stud and remove ground stud nut and washer.
4. Remove instrument panel top cover, right side mouldings, assist handle, lower right finish panel and right hand panel support brace.
5. Disconnect air vent cable at heater housing.
6. Remove heater housing attaching screws in front compartment.
7. After removal of housing assembly the heater core can be removed.

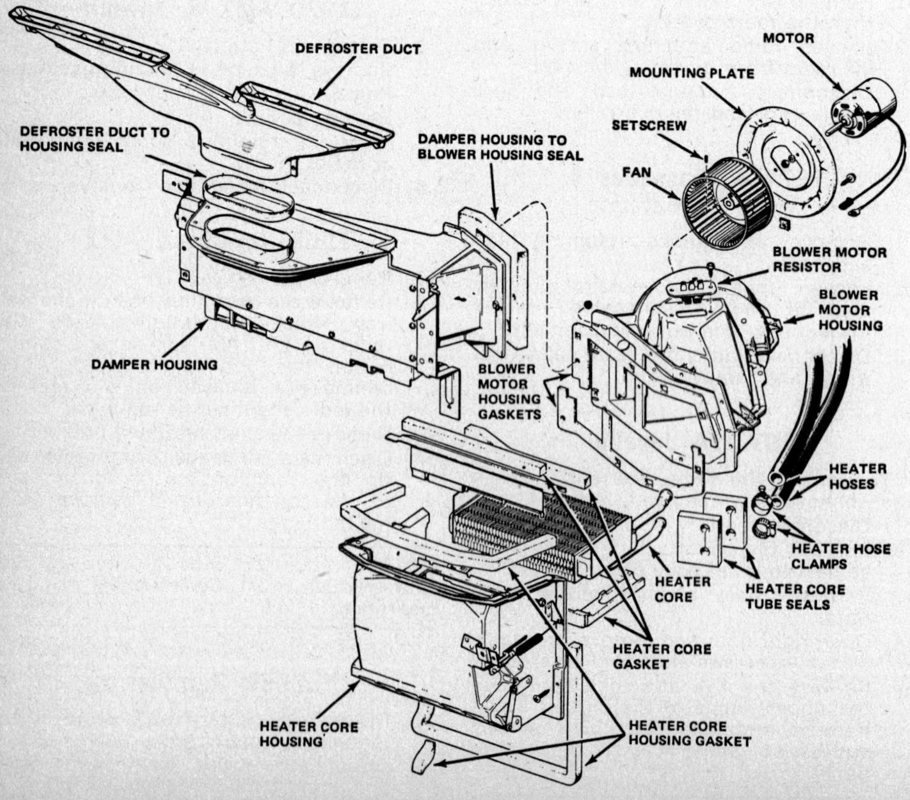

Fig. 9 Heater and blower housing assembly. 1974 Ambassador & 1974-76 Matador

1970-76 Hornet & Gremlin

1. Open heater valve and drain about 2 qts. from cooling system.
2. Disconnect heater hoses and plug heater core tubes.
3. Remove blower motor and fan.
4. Remove package tray if so equipped.
5. Disconnect wire connector at resistor.
6. Remove instrument panel bezel, outlet and duct.
7. Disconnect control cables from damper levers.
8. Remove right side windshield pillar moulding and the instrument panel upper attaching screws and right side cap screw at the door hinge post.
9. Remove right side kick panel and heater housing attaching screws.
10. Pull the right side of the instrument panel slightly rearward and remove the housing.

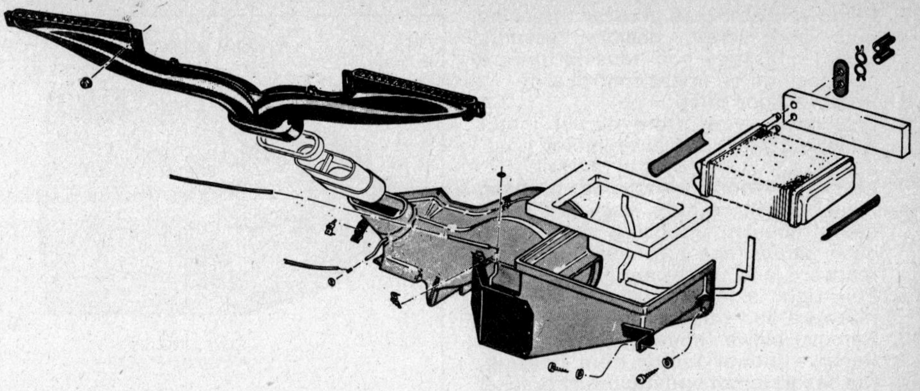

Fig. 10 Heater core and housing assembly. 1969 Rambler, 1969-70 AMX and Javelin

1969 Rambler, 1969-70 AMX & Javelin, Fig. 10

1. Open heater water valve and drain about 2 qts from cooling system.
2. After disconnecting hoses from heater core, install cork plugs in hoses and core tubes.
3. Disconnect blower motor wires and ground wire to dash panel.
4. Remove glove box and its door. On AMX and Javelin, it is necessary to remove glove box hinge bracket.
5. Disconnect outside air control cable at damper lever.
6. Remove three heater housing attaching screws in front compartment and lift out heater core and blower housing assembly.
7. Heater core is now accessible for removal.

1969-73 Rebel, Ambassador & Matador, Fig. 11

1. Disconnect outside air damper cable at heater housing under dash.
2. Drain system and disconnect hoses.
3. Remove two lower attaching nuts for blower housing in engine compartment.
4. Remove glove box and door in order to remove remaining heater housing screws and lower heater and core housing as an assembly.
5. Slide heater core out of housing.

BLOWER MOTOR REMOVE

1975-76 Pacer

Less Air Conditioning

1. Disconnect battery ground cable.
2. Remove right side windshield molding, then the instrument panel crash pad.
3. Remove right scuff plate and cowl panel.
4. Remove lower instrument panel to pillar attaching screws, then pull instrument panel rearward and reinstall attaching screws, allowing panel to rest on screws.
5. Remove heater core housing attaching nuts and screw, position vacuum lines aside, disconnect air door, then pull housing forward and position on upper control arm.
6. Remove blower motor attaching screws, disconnect relay ground wire and resistor wiring.
7. Remove blower motor brace.
8. Loosen heater housing to instrument panel attaching nuts.
9. Pull blower motor housing rearward and downward, disconnect vacuum hoses from motors and remove blower housing.
10. Remove housing cover, disconnect wire and remove blower to housing attaching screws.
11. Remove blower from housing, then separate fan and mounting bracket from motor.

With Air Conditioning

1. Disconnect battery ground cable.
2. Remove right scuff plate and cowl panel.
3. Remove radio overlay, then the instrument panel crash pad.
4. Remove instrument panel to right pillar attaching screws, then the upper to lower instrument panel attaching screws, located above glove box.
5. Disconnect air door cable from heater core housing and remove housing to floor pan brace screw.
6. Disconnect wires from blower motor resistor and vacuum lines from vacuum motors.

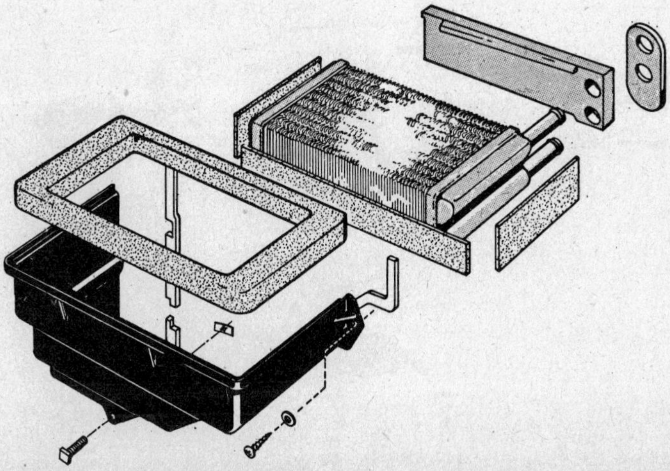

Fig. 11 Heater core assembly. 1969-73 Ambassador and Rebel

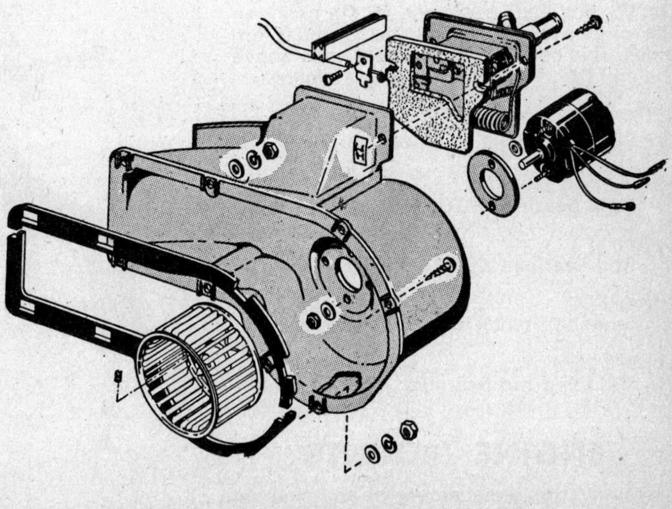

Fig. 12 Blower and housing assembly. 1969-73 Ambassador and Rebel

7. Remove heater core housing attaching nuts and screw, position vacuum lines aside, then pull housing forward and position on upper control arm.
8. Remove floor duct.
9. Disconnect wires from blower motor relay and remove blower motor housing attaching screw from firewall.
10. Loosen evaporator housing to dash panel attaching nuts, remove blower motor housing attaching screw from dash panel, then pull blower housing rearward and downward.
11. Pull right side of instrument panel rearward and remove blower motor.
12. Remove blower housing cover.
13. Remove blower motor from housing, the separate fan and mounting bracket from motor.

1974 Ambassador & 1974-76 Matador, Fig. 9

1. Disconnect blower motor wiring.
2. Remove blower motor attaching screws and remove blower motor.

1969-73 Rebel, Matador & Ambassador, Fig. 12

1. Remove heater valve from blower housing. It is not necessary to disconnect hoses and cable.
2. In engine compartment, unfasten blower housing from dash panel.
3. After blower housing is removed, the fan and motor are accessible.

1969-76 Except Rebel, Matador & Ambassador

1. Working in engine compartment, disconnect blower motor wire.

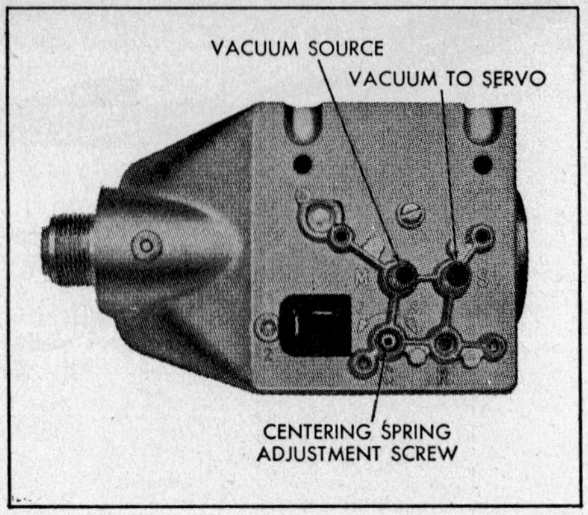

Fig. 13 Centering spring adjustment. 1969-76

2. Remove three retaining nuts for blower scroll cover to which blower assembly is attached and remove blower motor and fan.

SPEED CONTROL

1969-76

Brake Release Switch, Adjust

1. Disconnect multiple connector at regulator.
2. Turn ignition switch to accessory position.
3. Using a test lamp, ground one test lamp lead and touch the other to terminal No. 2 in harness connector.
4. Adjust switch so that lamp will light when brake pedal is fully released and will go out when brake pedal is depressed 3/8 inch.
5. If switch cannot be adjusted, it is defective and should be replaced. Install new switch and repeat Step 4.
6. Remove test lamp, turn off ignition key and plug connector to regulator.

Chain Linkage, Adjust

Chain linkage should never be taut. To adjust, start engine and set carburetor at hot idle with anti-stall plunger backed off so as not to affect idle speed and on 1973-76 models, disconnect idle stop solenoid. Hook chain to accelerator linkage, pull taut, then loosen one ball at a time until a slight chain deflection is obtained without moving carburetor throttle or servo. After chain has been adjusted, bend servo hook tabs together and chain must be free in hook.

NOTE: Whenever adjusting chain linkage, be sure chain does not hold carburetor throttle open.

Centering Springs, Adjust

If speed control system holds speed three or more mph higher than selected speed, turn centering spring adjusting screw (C) toward (S) 1/32″ or less, Fig. 13.

If speed control system holds speed three or more mph below selected speed, turn centering spring adjusting screw (C) toward (F) 1/32″ or less. *Do not move adjustment screw (R).*

Engine Section

ENGINE MARKINGS

A letter code is used to denote size of the bore, main bearings and rod bearings. On V8 engines, this code is stamped on the engine code tag. On six cylinder engines, this code is located on a boss above the oil filter. This letter code is as follows:

Letter "B"
Cyl. bore .010″ oversize

Letter "M"
Main bearings .010″ undersize

Letter "P"
Rod bearings .010″ undersize

Letter "C"
Camshaft block bore .010″ oversize

Letters "PM"
Main and rod bearings .010″ undersize

ENGINE MOUNTS

Removal or replacement of any cushion can be accomplished by supporting the weight of the engine or transmission at the area of the cushion to be replaced, Fig. 1.

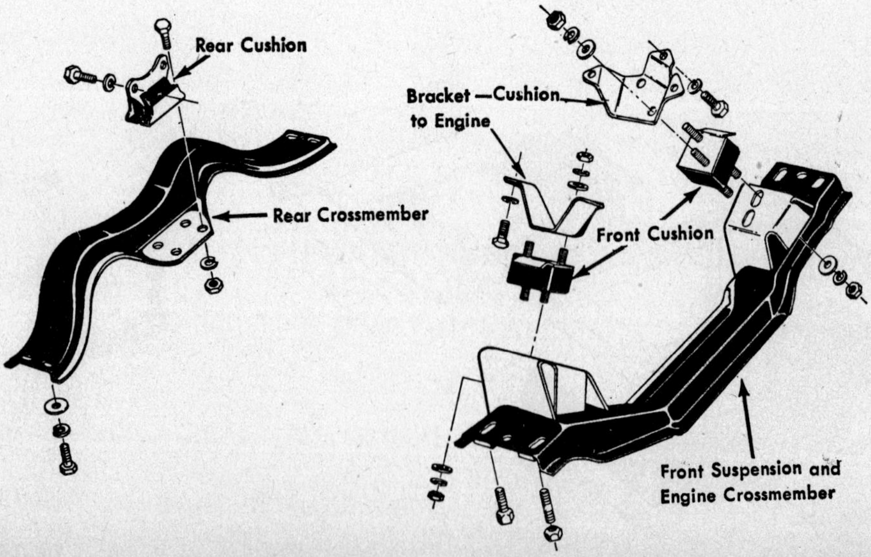

Fig. 1 Typical V8 engine mounts

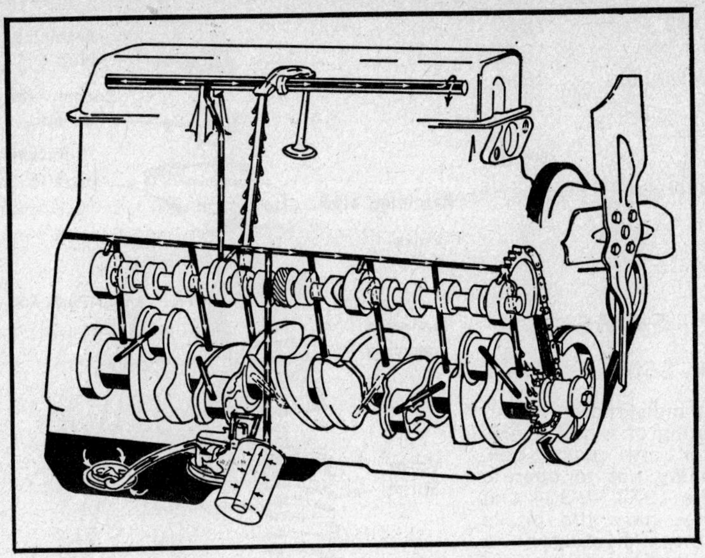

Engine oiling system. 1969-72 & early 1974 6-199, 232, 258 engines

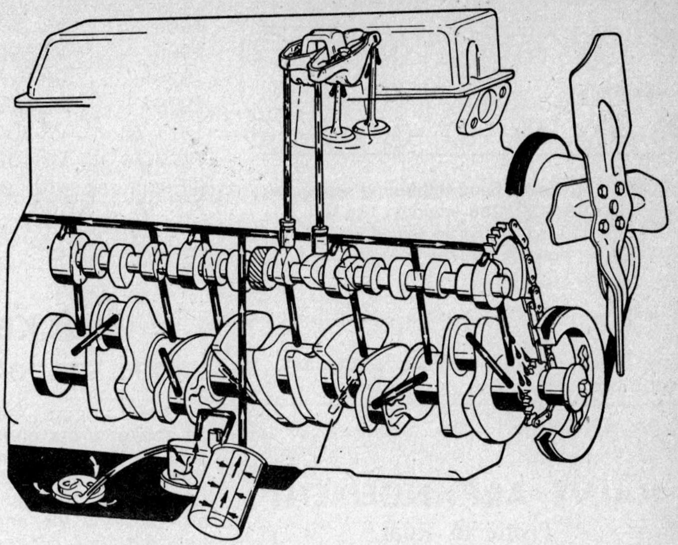

Engine oiling system. 1973, late 1974 & 1975 6-232, 258

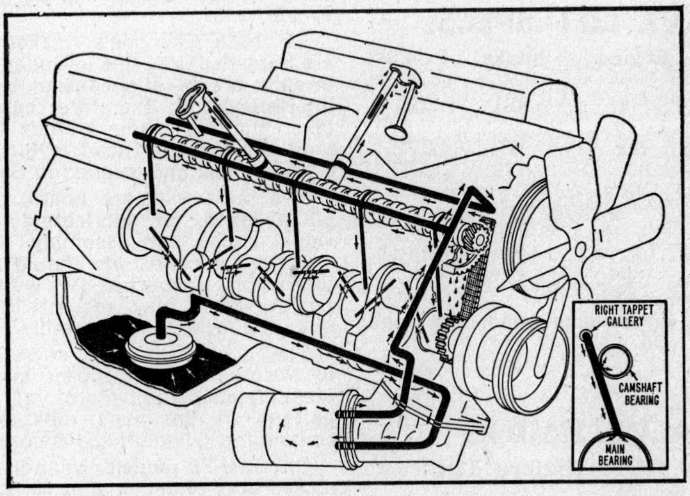

Engine oiling system. V8-290, 304, 343, 360, 390, 401 engines

ENGINE REPLACE

All Models

The engine and transmission is removed as an assembly as follows:

1. Disconnect battery negative and positive cables.
2. Remove hood. Mark hinges for alignment at installation.
3. Remove fender braces, if equipped.
4. Remove air cleaner.
5. Drain cooling system, crankcase and transmission.
6. Remove radiator upper air baffle, if equipped.
7. Remove power steering pump and position aside, do not disconnect hoses.
8. If equipped with A/C, discharge refrigerant system. Disconnect refrigerant lines from compressor and the receiver outlet at disconnect coupling. Then, remove condenser and receiver assembly.
9. On all models, all electrical connections and lines and hoses from engine.
10. Disconnect accelerator and transmission linkage.
11. Disconnect exhaust pipes from manifolds and the speedometer cable from transmission.
12. Support engine with suitable lifting equipment, then disconnect engine front support cushions and ground strap from engine brackets.
13. Remove rear crossmember.
14. Remove engine and transmission by pulling forward and upward. Support propeller shaft when disengaging transmission output shaft from slip joint.

CYLINDER HEAD

Tighten cylinder head bolts a little at a time in three steps in the sequence shown in the illustrations. Final tightening should be to the torque specifications listed in the *Engine Tightening* table.

Fig. 2 Cylinder head tightening sequence on V8-290, 304, 343, 360, 390 & 401. The No. 7 bolt indicated (second from front on left bank only) must be sealed to prevent coolant leakage

V8-290, 304, 343, 360, 390, 401

The cylinder block has two locating dowels on each bank to assist in lining up and holding the cylinder head and gasket in position during installation.

IMPORTANT: The No. 7 bolt shown in Fig. 2, second from front on the left bank, must have the threads sealed to prevent coolant leakage. Permatex No. 2 or equivalent is recommended.

6-199, 6-232, 258 Engines

IMPORTANT: The cylinder head bolt located at the left front corner of the head (No. 11, Fig. 3) must have the threads sealed to prevent coolant leak-

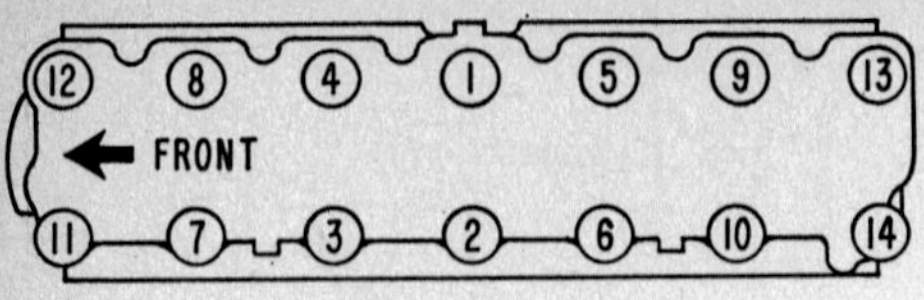

Fig. 3 Cylinder head tightening sequence 6-199, 232, 258 engines. The No. 11 bolt must be sealed to prevent coolant leakage

age. Permatex No. 2 or equivalent is recommended.

VALVE ARRANGEMENT

Front to Rear

All V8s . . . . E-I-I-E-E-I-I-E
6-199, 232, 258 . . . . . E-I-I-E-I-E-E-I-E-I-I-E

VALVE LIFT SPECS.

Year	Engine	Intake	Exhaust
1969-71 & Early 1974	Six	.381	.381
Late 1974[2] & 1975-76	Six	.372	.372
1972-73	Six	.372	.372
1969-76	V8's exc. 401 & 1970 390	.425[1]	.425[1]
1970	V8-390	.457[1]	.457[1]
1971-74	V8-401	.457[1]	.457[1]

[1]Hi-Perf. cam, .477.
[2]Beginning with engine build date code 704 (A or E) 04.

VALVE TIMING

Intake Opens Before TDC

Year	Engine	Degrees
1969-73 & Early 1974	Six	12½
Late 1974[1] & 1975-76	Six	12.12

Fig. 4A Rocker arms, push rod and pivot assembly. 1973, late 1974 & 1975-76 6-232, 258, 1973 V8-304 and all 1974-76 V8's

1969	V8-290, 343	18½
1969	V8-390	18½
1970	V8 Exc. Machine	18½
1970	Machine	18
1971-76	V8-304, 360	14¾
1971-74	V8-401	25.57

[1]Beginning with engine build date code 704 (A or E) 04.

ROCKER ARMS

V8-290, 304, 343, 360, 390, 401

These engines have individually mounted rocker arms consisting of a rocker arm retaining stud, rocker arm pivot ball, rocker arm and retaining nut to operate each valve, Fig. 4. The 1973 V8-304 and all 1974-76 V8 engines have the intake and exhaust rocker arms pivoting on a bridged pivot assembly which is secured to the cylinder head by two cap screws, Fig. 4A. When installing cap screws, turn each screw one turn at a time to avoid breaking the bridge. Torque cap screws to 19 ft. lbs.

On 1969-73 models, rocker arm studs are threaded into the cylinder head. The threads are of such design to cause an interference fit; therefore, care must be taken that replacement studs be installed until the hexagon head is flush with the cylinder head and torqued to 65-70 ft-lbs.

The push rods are hollow, serving as oil galleries for lubricating each individual rocker arm assembly. Prior to installing, the push rods should be cleaned thoroughly, inspected for wear and deposits which may restrict the flow of oil to the rocker arm assembly.

The push rods also serve as guides to maintain correct rocker arm to valve stem relationship; therefore, a contact pattern on the push rods where they contact the cylinder head is normal.

On 1969-73 models, when installing the rocker arm retaining nut, it is important that they be tightened until bottomed, using a torque of 20-25 ft-lbs.

Lubrication to each rocker arm is supplied by the corresponding hydraulic valve lifter. A metering system located in each valve lifter consists of a stepped lower surface on the push rod cap that contacts a flat plate, causing a restriction, Fig. 5. The restriction meters the amount of oil flow through the push rod cap, hollow push rod, and upper valve train components. A loss of lubrication to the rocker arm could be caused by a restricted or plugged push rod or a defective hydraulic valve lifter.

CAUTION: Correct installation of push rods in these engines is critical and more than normal care must be taken upon installation. When placing the push rods through the guide hole in the cylinder head, it is important that the push rod end is inserted in the plunger cap socket. It is possible that the push rod may seat itself on the edge of the plunger cap which will restrict valve lifter rotation and lubrication to rocker arms.

It is recommended that, just prior to installation of the cylinder head covers, the engine be operated and the supply of lubrication to each rocker arm be visually inspected. If inspection reveals that an individual rocker arm is not being supplied with lubrication, the push rod and/or valve lifter must be inspected to determine the cause.

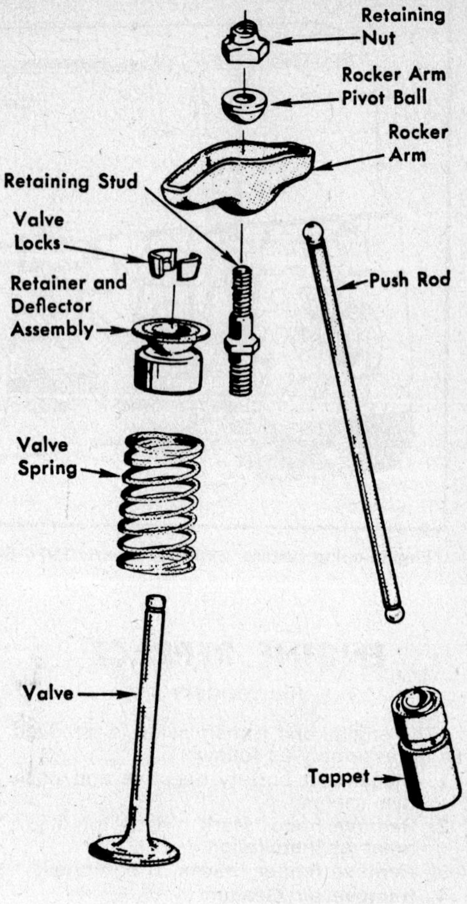

Fig. 4 Valve train on 1969-73 V8-290, 343, 360, 390, 401 and 1969-72 V8-304 engines

Fig. 4B 1973 (7300) and 1974 (7400) rocker arm identification

6-199, 6-232, 258 Engines

On 1969-72 and early 1974 engines, the pressure supply for each rocker arm is obtained from No. 3 camshaft bearing location where the camshaft meters the flow of oil from the main lubricating gallery through a groove in the camshaft bearing surface to a gallery extending upwards to the cylinder head gasket surface, Fig. 6. On 1974, late 1974 and 1975-76 232, 258, oil is supplied through the valve lifters and hollow push rods. On 1969-72 and early 1974 199, 232 and 258, install the rocker arm shaft with oil holes facing down toward the cylinder head, Fig. 7. The 1973, late 1974 and 1975-76 232, 258 engines have the intake and exhaust rocker arms pivoting on a bridged pivot assembly which is secured to the cylinder head by two cap screws, Fig. 4A.

NOTE: The new rocker arms used on late 1974 engines (beginning with build date code 704 (A or E) 04, have been redesigned with the oil hole on the push rod end relocated, pointing away from the valve assemblies allowing for improved oil control, Fig. 4B. Also, these rocker arms are not interchangeable with the 1973 type, since the camshaft and rocker arm ratio has been modified.

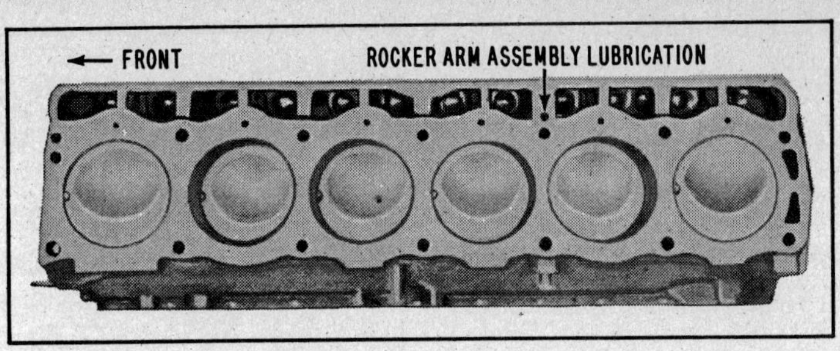

Fig. 6 Rocker arm lubrication gallery. 1969-72 & early 1974 6-199, 232, 258 engines

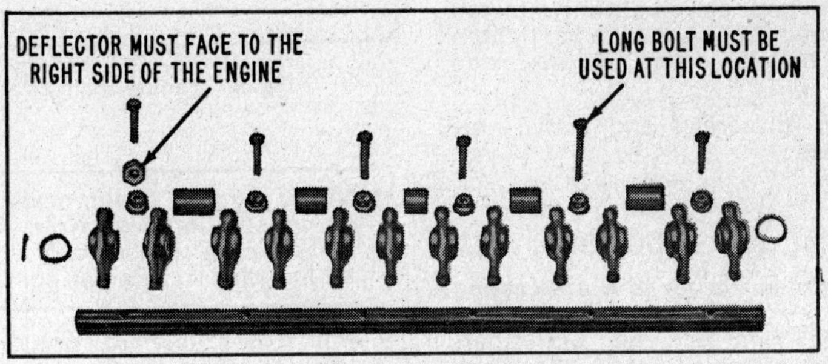

Fig. 7 Rocker arm assembly. 1969-72 & early 1974 6-199, 232, 258 engines

6-232 & 258 Valve Train

The exhaust valve train on certain engines, has been modified to incorporate valve rotators to improve valve durability. The engines involved are those equipped with EGR systems less Air Guard.

NOTE: When replacing rocker arms on these engines, use only the cast type rocker arms, as they are designed to accommodate the valve rotators.

VALVE GUIDES

Excessive valve stem-to-guide clearance will cause lack of power, rough idling and noisy valves, and may cause valve breakage. Insufficient clearance will result in noisy and sticky functioning of valves and disturb engine smoothness of operation.

Valve stem-to-guide clearances are listed in the *Engine Valve Specifications* table. By using a micrometer and a suitable telescope hole gauge, check the diameter of the valve stem in three places (top, center and bottom). Insert telescope hole gauge in valve guide bore, measuring at the center. Subtract the highest reading of valve stem diameter from valve guide bore center diameter to obtain valve-to-guide clearance. If clearance is not within specified limits, use the next oversize valve and ream bore to fit. Valves with oversize stems are available in .003", .015" and .030".

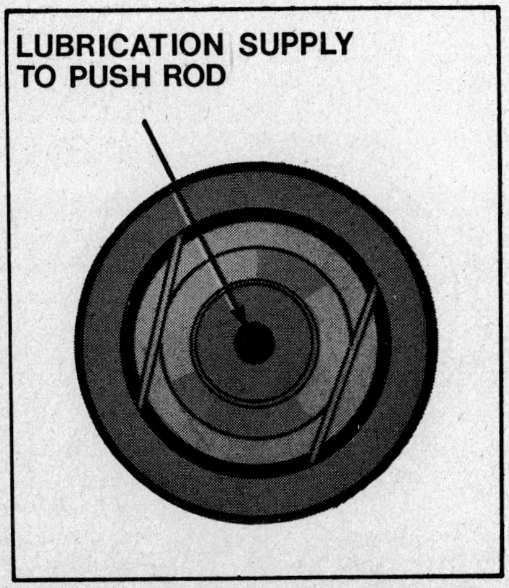

Fig. 5 Hydraulic lifter identification. V8-290, 304, 343, 360, 390, 401 engines

Hydraulic Lifters

Valve lifters may be removed from their bores after removing the rocker arms and push rods. Adjustable pliers with taped jaws may be used to remove lifters that are stuck due to varnish, carbon, etc. Fig. 8 illustrates the type of lifter used.

TIMING CHAIN COVER

6-199, 232, 258

1. Remove drive belts, fan and pulley.

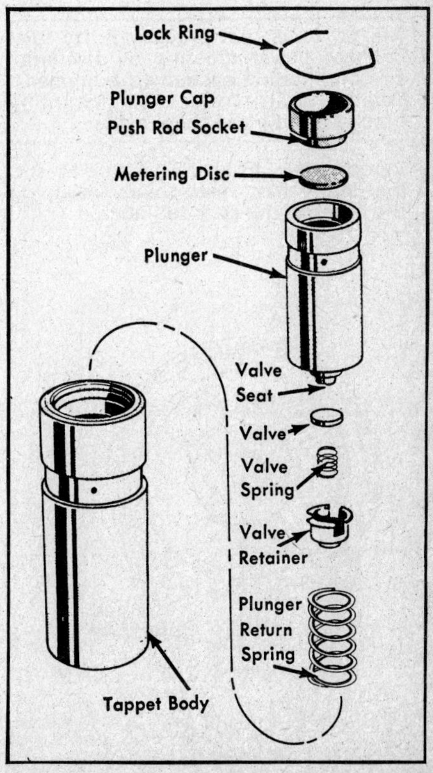

Fig. 8 Hydraulic valve lifter

2. Remove vibration damper.
3. Remove oil pan-to-timing chain cover screws and cover-to-block screws.
4. Raise the cover and pull the oil pan front seal up enough to pull the retaining nibs from the holes in the cover.
5. Remove timing chain cover gasket from block. Cut off seal tab flush with front face of cylinder block. Clean gasket surfaces.
6. Remove oil seal.
7. Place gasket in position on cylinder block. Install new oil pan front seal, cut off protruding tab of seal to match portion of the original seal.
8. Insert suitable aligning tool in cover seal bore and on crankshaft. Install cover-to-oil pan screws and tighten lightly. Install cover screws and tighten.
9. Retighten all screws and install new cover seal.

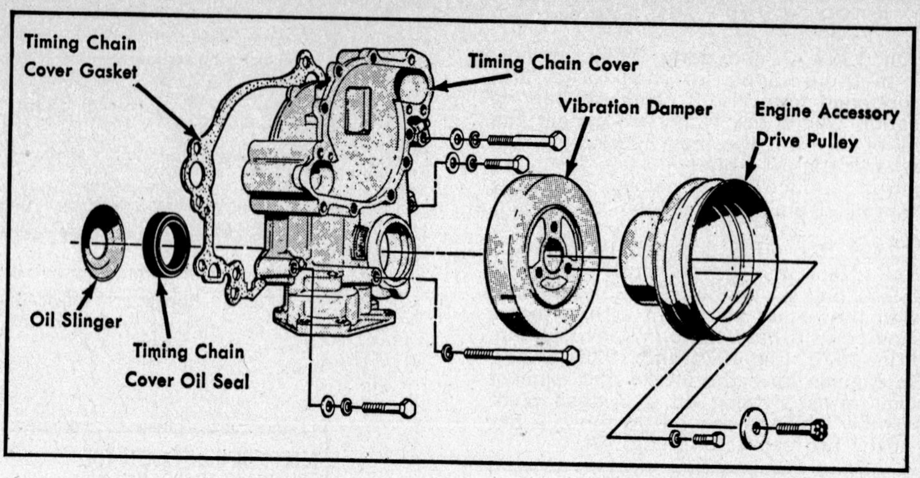

Fig. 9 Timing chain cover assembly. V8-290, 304, 343, 360, 390, 401 engines

V8-290, 304, 343, 360, 390, 401

The timing chain cover is a die casting incorporating an oil seal at the vibration damper hub, Fig. 9. The crankshaft front seal is installed from the back side of the cover, therefore, it is necessary to remove the cover when replacement of the seal is required. To remove cover, proceed as follows:

1. Drain cooling system completely.
2. Remove lower radiator hose and bypass hose from cover.
3. Remove distributor, fuel pump, drive belts, fan and hub assembly and vibration damper, using a suitable puller.

NOTE: It is not necessary to disconnect power steering or discharge air conditioning system (if equipped). Remove units from their mounting brackets and place them aside.

4. Remove two front oil pan bolts and the eight hex head bolts retaining the cover to the cylinder block.

NOTE: Timing chain cover bolts are of various lengths. Note location of bolts during disassembly so they can be installed in original position.

5. Pull cover forward until free from locating dowel pins.
6. Remove used seal and clean seal bore and gasket surface of cover.
7. Apply sealing compound to outer surface of seal and a film of Lubriplate or equivalent to seal lips. Drive seal into cover bore until seal contacts outer flange of cover.

Installation of Cover

1. Prior to installation of cover, remove lower dowel pin from cylinder block.
2. Using a sharp knife or razor blade, cut oil pan gasket flush with cylinder block on both sides of oil pan.
3. Cut corresponding pieces of gasket from the replacement oil pan gasket set. Cement gasket to cover. Install replacement Neoprene oil pan seal into cover and align cork gasket tabs to the pan seal.
4. Apply a strip of sealing compound to both the cut-off oil pan gaskets at the oil pan to cylinder block location.
5. Place cover in position, install oil pan bolts in cover, tighten evenly and slowly until cover aligns with upper dowel. Then install lower dowel through cover. Drive dowel in corresponding hole in cylinder block.
6. Install cover attaching bolts and torque to 20-30 ft-lbs.

Six Cylinder

When installing the cover it is important that the cover be properly aligned when installing the vibration damper to prevent damage to the oil seal. This is accomplished by leaving the cover-to-block cap screws loose until the vibration damper has been partially installed. Then tighten the cover screws.

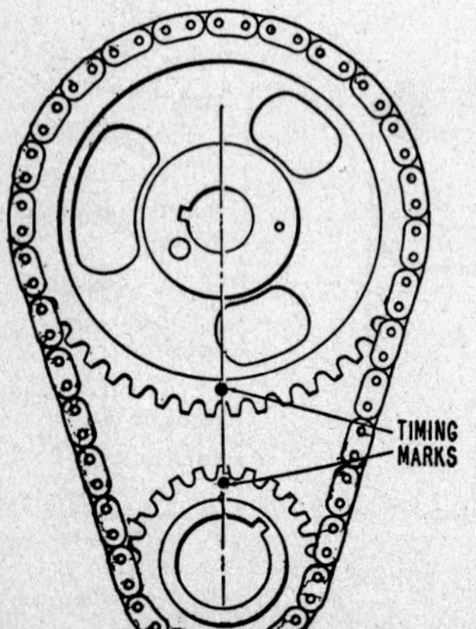

Fig. 10 Valve timing. V8 engines

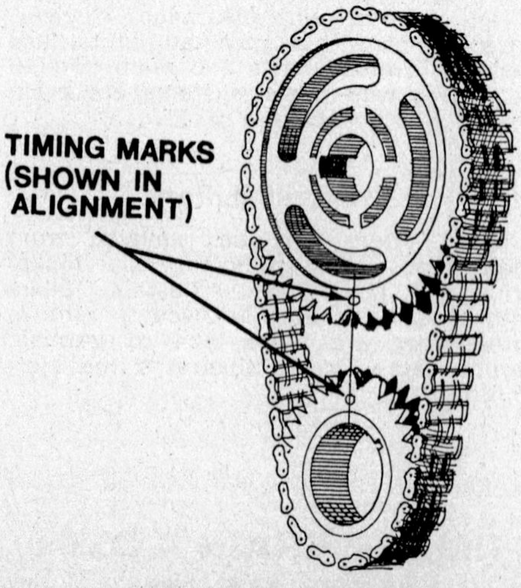

Fig. 10A Valve timing. V8-360 & 401 with double row roller timing chain

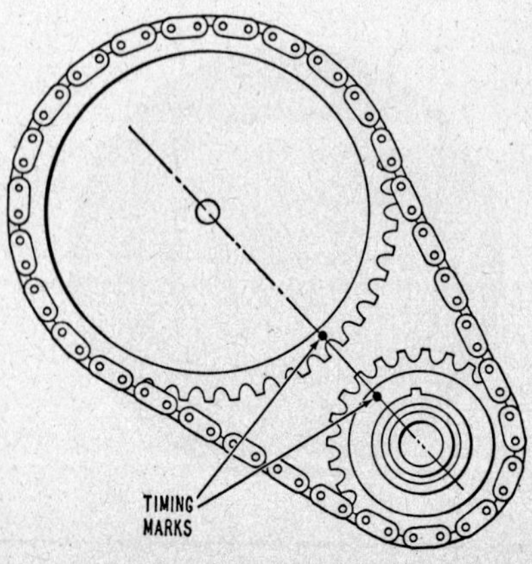

Fig. 11 Valve timing. 6-199, 232, 258 engines

TIMING CHAIN

When installing a timing chain, see that the timing marks on the sprockets are in line as shown in Figs. 10, 10A and 11.

NOTE: All V8-360 and 401 heavy duty engines (Fleet) built after April 22, 1974 use a new double row roller type timing chain to improve timing chain durability. This new timing chain is available as a service replacement item and should be installed when timing chain failures occur to eliminate recurrence of the problem. Installation of the new timing chain requires the use of a new heavy duty camshaft sprocket and crankshaft sprocket.

CAMSHAFT

6-199, 232, 258

1. Remove cylinder head.
2. Remove value lifters.
3. Remove radiator and, if so equipped, air conditioning condenser.
4. Remove timing chain cover.
5. Rotate crankshaft until timing marks on sprockets are aligned, Fig. 11.
6. Remove sprockets and chain.
7. Lower front bumper by removing forward back bar-to-side sill bolts.
8. Remove crankshaft.

V8-290, 304, 343, 360, 390, 401

1. Disconnect battery ground cable.
2. Disconnect transmission cooler lines at radiator if so equipped.
3. Remove radiator.
4. Remove distributor, wires and coil.
5. Remove intake manifold and carburetor as an assembly.
6. Remove cylinder head covers, loosen rocker arms and remove push rods and lifters.
7. Dismount power steering pump.
8. Remove fan and hub, fuel pump and heater hose at water pump.
9. Remove alternator.
10. Remove vibration damper and pulley and lower radiator hose at water pump.
11. Remove timing chain cover, distributor-oil pump drive gear, fuel pump eccentric sprockets and chain.
12. Remove hood latch support bracket upper retaining screws and move bracket as required to allow removal of camshaft.

PISTONS & RODS, ASSEMBLE

V8 Engines

Assemble piston to connecting rod as shown in Figs. 12 to 19.

Overhead Valve 6-Cyl.

Pistons are marked with a depression notch on the top perimeter, Fig. 20. When installed in the engine this notch must be toward the front of the engine. Always assemble rods and caps with the cylinder numbers facing the camshaft side of engine.

PISTONS, PINS & RINGS

Pistons are furnished in standard sizes and oversizes of .002, .005. .010 and .020".

Piston pins are furnished in oversizes of .003 and .005".

Piston rings are available in .020" oversizes.

MAIN & ROD BEARINGS

Both main and rod bearings are supplied in undersizes of .001, .002, .010 and .012".

CRANKSHAFT REAR OIL SEAL

199, 232, 290, 304, 343, 360, 390, 401 Engines

1. To replace the seal, Fig. 21, remove oil pan and scrape oil pan surfaces clean.
2. Remove rear main bearing cap.
3. Remove and discard old seals.
4. Clean cap throughly.
5. Loosen all remaining main bearing cap screws.
6. With a brass drift and hammer, tap upper seal until sufficient seal is protruding to permit pulling seal out completely with pliers.
7. Wipe seal surface of crankshaft clean, then oil lightly.
8. Coat back surface of upper seal with soap, and lip of seal with engine oil.
9. Install upper seal into cylinder block. *Lip of seal must face to front of engine.*
10. Coat cap and cylinder block mating surface portion of seal with Permatex No. 2 or equivalent, being careful not to apply sealer on lip of seal.
11. Coat back surface of lower seal with soap, and lip of seal with No. 40 engine oil. Place into cap, seating seal firmly into seal recess in cap.
12. Place Permatex No. 2 or equivalent on both chamfered edges of rear main bearing cap.
13. Install main bearings and install cap. Tighten all caps to correct torque as listed in the *Engine Tightening Specifications* table.
14. Cement oil pan gasket to cylinder block with tongue of gasket at each end coated with Permatex or equivalent before installing into rear main bearing cap at joint of tongue and oil pan front neoprene seal.
15. Coat oil pan rear seal with soap. Place into recess of rear main bearing cap, making certain seal is firmly and evenly seated.
16. Install oil pan and tighten drain plug securely.

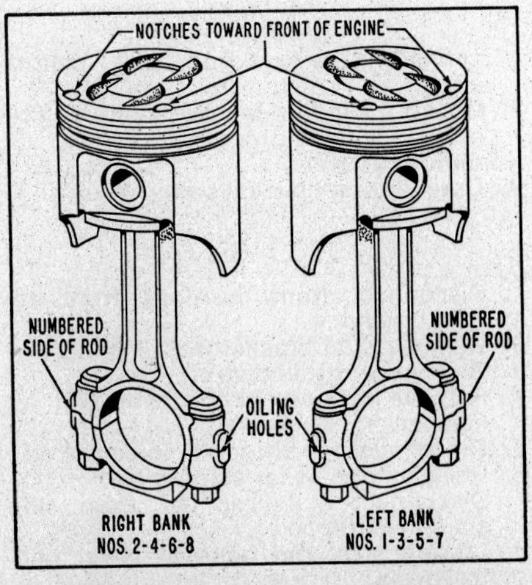

Fig. 12 Piston and rod assembly. 1969 Low Compression V8-290, 343

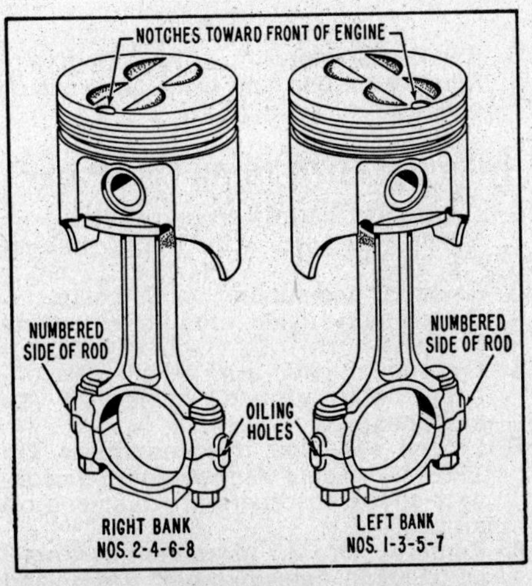

Fig. 13 Piston and rod assembly. 1969 High Comp. V8-343

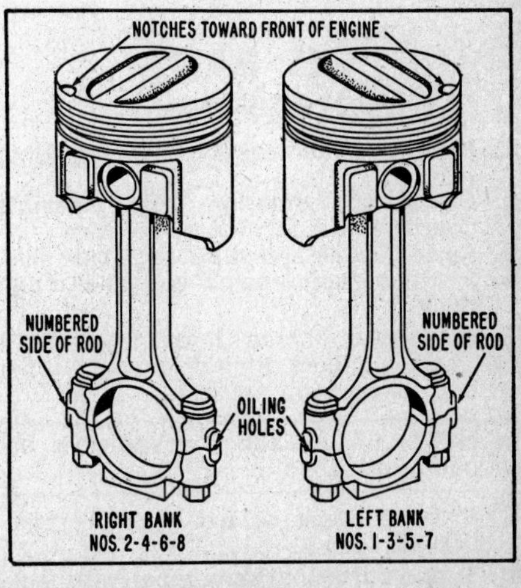

Fig. 14 Piston and rod assembly. 1969 V8-390

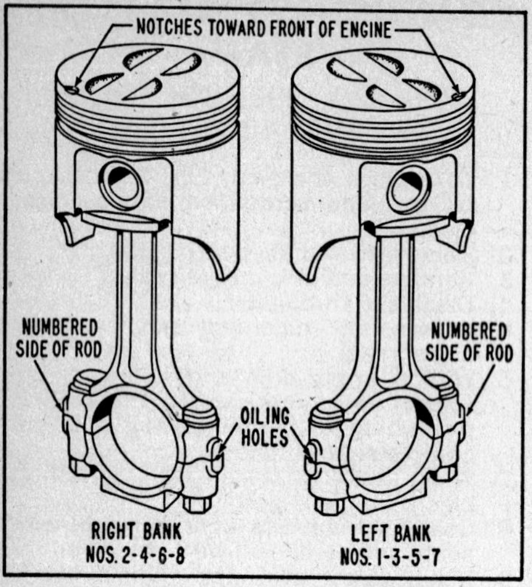

Fig. 15 Piston and rod assembly. 1969 V8-290 High Comp.

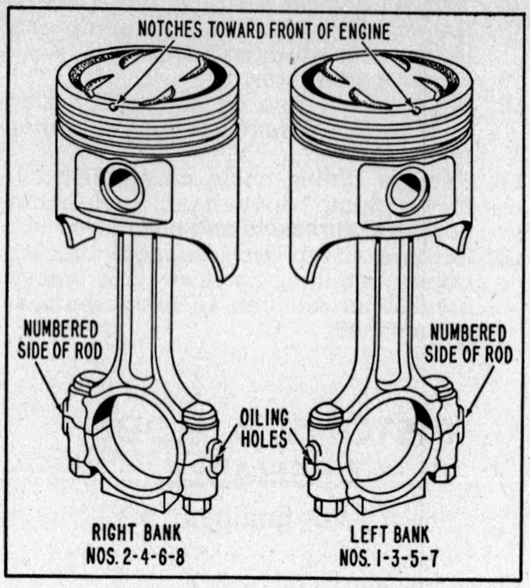

Fig. 16 Piston and rod assembly. 1970 V8-304

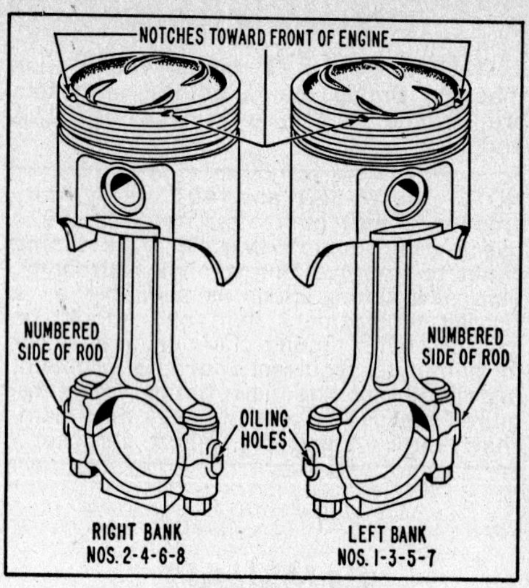

Fig. 17 Piston and rod assembly. 1970-76 V8-360 Std. Compression and 1971-76 V8-304

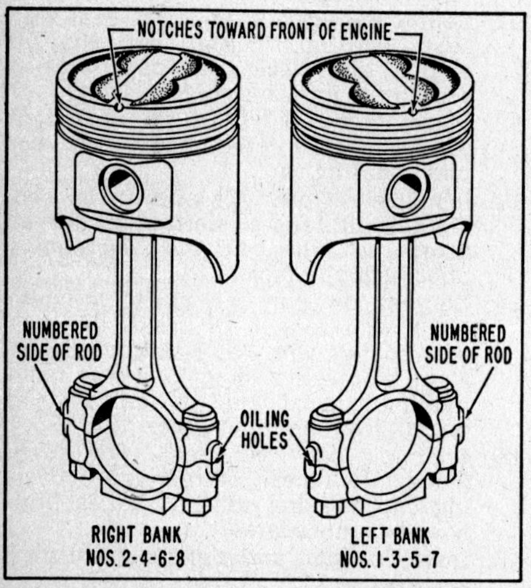

Fig. 18 Piston and rod assembly. 1970 V8-360 High Compression

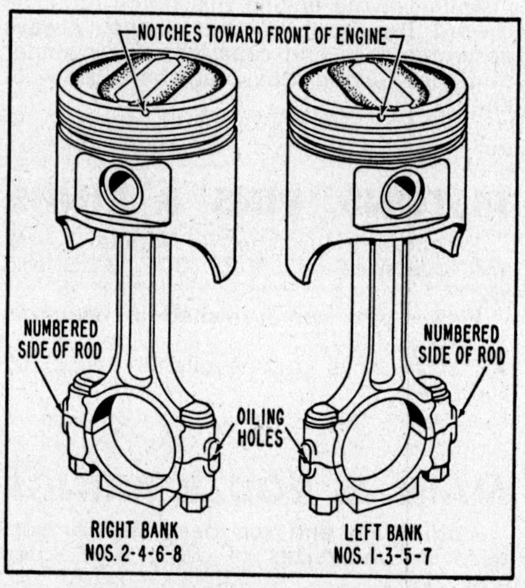

Fig. 19 Piston and rod assembly. 1970-74 V8-390, 401

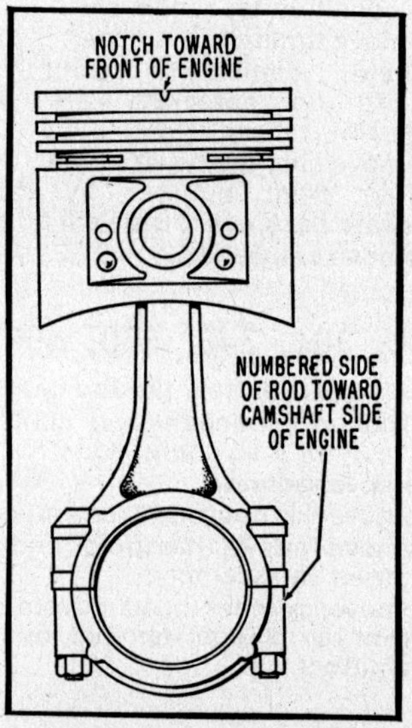

Fig. 20 Piston and rod assembly. 6-199, 232, 258 engines

OIL PAN

1975-76 Pacer

1. Drain oil, then install engine holding fixture.
2. Disconnect steering shaft flexible coupling and position out of way.
3. Raise vehicle and support on side sills.
4. Remove front engine mount through bolts.
5. Disconnect brake lines from front wheel cylinders, then disconnect upper ball joints from spindles.

NOTE: Ensure shock absorber is securely attached.

6. Remove upper control arm and position aside.
7. Support front crossmember with jack, then remove nuts from rear mounts and swing crossmember forward.
8. Remove starter.
9. Remove oil pan and front and rear oil pan seals.

1969 V8; 1970-76 All Exc. Pacer

1. Disconnect battery ground cable.
2. Support engine with lifting fixture, Fig. 22.
3. Raise car and support on side sills.
4. Disconnect engine mounts at engine brackets.
5. Disconnect idler arm at sill. Disconnect body ground cables if so equipped.
6. Loosen sway bar if so equipped. On 1969 V8 except Ambassador, disconnect shock absorbers at lower control arms.
7. Remove front crossmember-to-sill bolts, pull crossmember down and place 2"x4"x6" blocks between crossmember and sills. On 1971-76 V8, disconnect strut rods from lower control arms.
8. On six cylinder, remove right engine mount bracket from engine. On V8, remove starter.
9. Drain and remove oil pan.

1969 Six

1. Disconnect front cushions from engine bracket.
2. Remove right bracket from engine.
3. Disconnect ground strap.
4. Remove cylinder head cover and air cleaner.
5. Disconnect fan shroud if so equipped.
6. Raise engine as far as possible.
7. Disconnect stabilizer bar from side sill if so equipped.
8. Loosen strut rod bolts at lower control arms.
9. Remove crossmember-to-side sill bolts.

10. With weight of car on wheels pry down crossmember, use wooden blocks to hold crossmember down.
11. Drain engine oil, remove oil pan.

OIL PUMP

6-199, 6-232, 258 & V8s

NOTE: When servicing oil pump on 6 cylinder engines if inlet tube is moved out of position a new inlet tube and screen assembly must be installed.

The oil pump on six cylinder models is located in the oil pan thus necessitating removal of the pan to gain access to the pump. The pump on V8 engines is an integral part of the timing case cover and it can be serviced after removal of the oil filter adapter body.

Oil pump removal or replacement will not affect distributor timing as the distributor drive gear remains in mesh with the camshaft gear.

Upon disassembly of the oil pump, place a straightedge across gears and pump body and check clearance between straightedge and pump body which should be .003 inch for all 6 cylinder engines and .004 inch for all V8 engines. Clearance between gears and pump housing should be .003 inch for 1969-71 V8 engines and .0005-.0025 for 1972-76 V8 engines and all 6 cylinder engines.

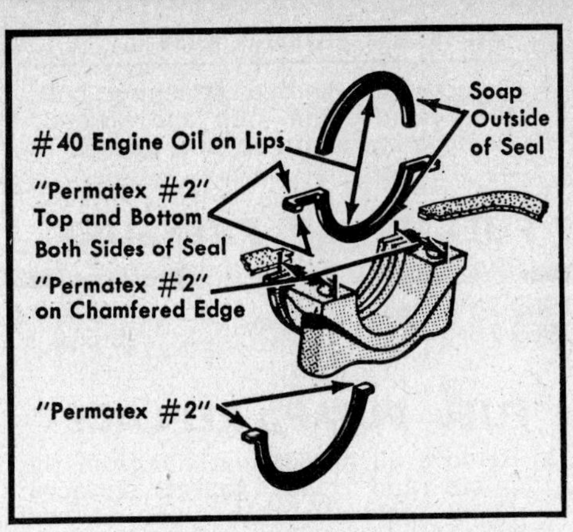

Fig. 21 Rear main bearing sealing. 199, 232, 290, 304, 343, 360, 390, 401 engines

The pump cover should be installed with the pump out of the engine and pump checked for freedom of operation before installation.

The oil pressure relief valve, which is built into the pump, is not adjustable, the correct pressure being built into the relief valve spring.

BELT TENSION DATA

	New Lbs.	Used Lbs.
1969-70—		
Air Condition	125-145	90-110
Fan and Alternator	125-145	90-110
Power Steering	125-145	90-110
1971-72—		
Air Condition	125-145	105-125
Fan and Alternator—		
Exc. V8 With A/C	125-145	90-110
V8 With A/C	125-145	105-125
Power Steering—		
Exc. V8 With A/C	125-145	90-110
V8 With A/C	125-145	105-120
1973-74—		
Air Condition	125-155	105-130
Air Pump—		
Exc. 6 cyl. With A/C	125-155	90-115
6 cyl. With A/C (1/4 inch belt)	40-50	35-40
Fan and Alternator—		
Exc. V8 With A/C	125-155	90-115
V8 With A/C	125-155	105-130
Power Steering—		
Exc. V8 With A/C	125-145	90-110
V8 with A/C	125-145	105-120
1975—		
Air Condition	125-155	105-130

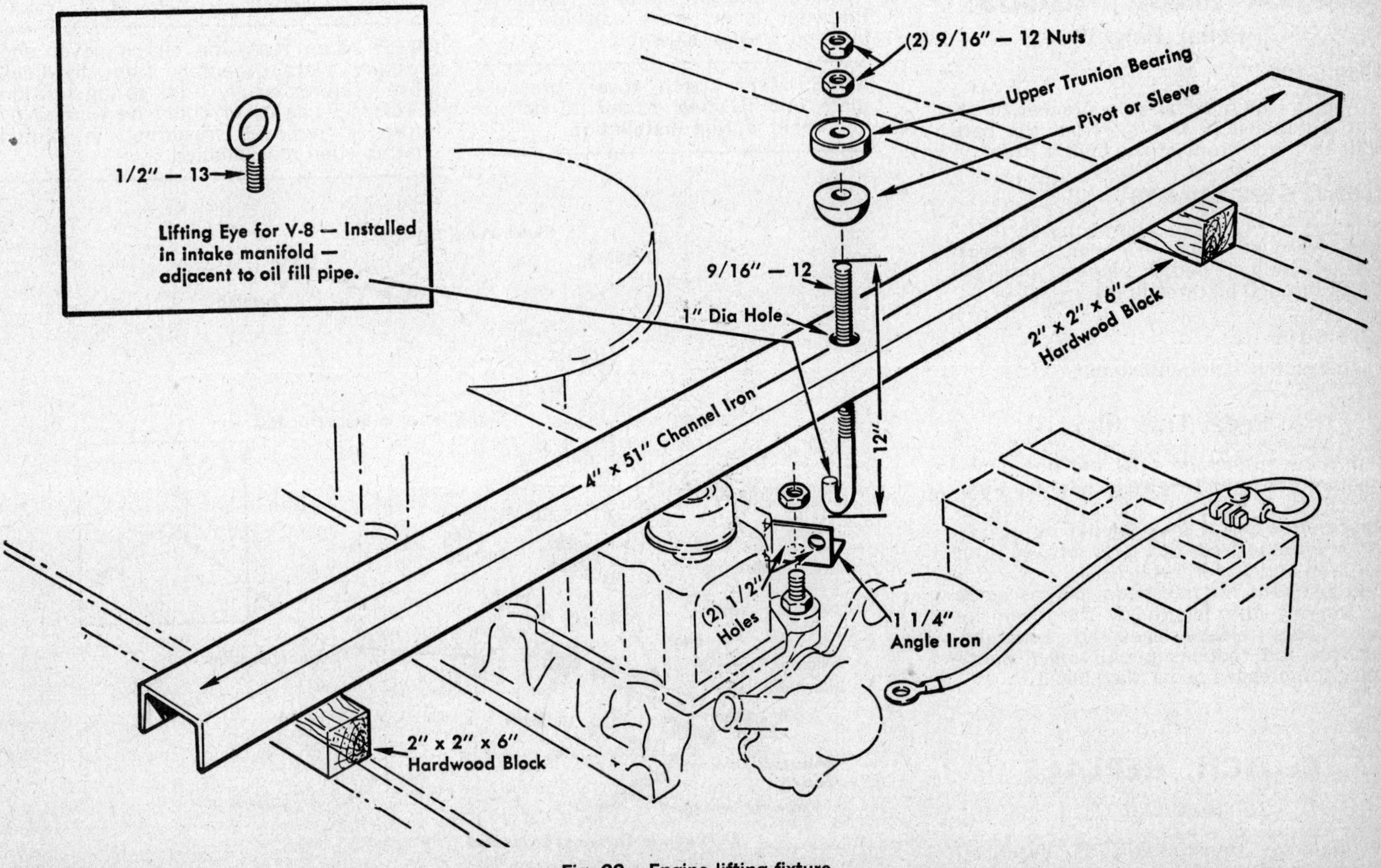

Fig. 22 Engine lifting fixture

Air Pump—		
Exc. 6 cyl.		
With A/C	125-155	90-115
6 cyl. With A/C		
(¼ inch belt)	40-50	35-40
Fan and Power		
Steering	125-155	90-115

WATER PUMP

1. Disconnect battery ground cable.
2. Drain cooling system and disconnect radiator and heater hoses from pump.
3. Remove drive belts.

NOTE: On some models it will be necessary to remove alternator front bracket and place alternator aside, without disconnecting wires.

4. Remove fan shroud attaching bolts, then remove fan, hub and shroud.
5. Remove water pump and gasket.

FUEL PUMP PRESSURE

Year	Engine	Pressure lbs.
1969-76	Six cyl.	4-5½
1969-76	V8	5-6½

FUEL PUMP, REPLACE

1. Remove all gasket material from the pump and block gasket surfaces. Apply sealer to both sides of new gasket.
2. Position gasket on pump flange and hold pump in position against its mounting surface. Make sure rocker arm is riding on camshaft eccentric.
3. Press pump tight against its mounting. Install retaining screws and tighten them alternately.
4. Connect fuel lines. Then operate engine and check for leaks.

SERVICE NOTE: Before installing the pump, it is good practice to crank the engine so that the nose of the camshaft eccentric is out of the way of the fuel pump rocker arm when the pump is installed. In this way there will be the least amount of tension on the rocker arm, thereby easing the installation of the pump.

Clutch, Transmission & Overdrive Section

NOTE: 1976 Linkage adjustment information is in this section. Repair procedures on both automatic and manual shift transmissions are covered elsewhere in this manual. Procedures for removing automatic transmissions as well as linkage adjustments on 1969-75 models are included in the automatic transmission chapters. See Chapter Index.

CLUTCH PEDAL, ADJUST

Pedal Height

1969 6-199

Adjust clutch pedal stop bracket to obtain a dimension of 6½" from the bare floor to the bottom of the clutch pedal.

1969-71 6-232, 258 & V8

Insert a 5/16" pin, approximately 4½" long, through holes in pedal support bracket. Adjust pedal support until pin slides through all three holes.

1970 6-199

This clutch is not adjustable.

Pedal Free Play

In order to provide sufficient free movement of the clutch release bearing when the clutch is engaged and pedal fully released, free pedal play should be ⅞" to 1⅛" with desired free play of 1⅛" for 1973-76 and 1" for 1969-72.

Adjustment for free pedal play is made by varying the length of the beam or link to the release lever rod. Lengthening this rod reduces pedal travel; shortening it increases pedal play, Fig. 1.

CLUTCH, REPLACE

Removal

1. Remove transmission as described further on.
2. Remove starter, clutch housing, throwout lever, wave washer, bearing and sleeve assembly.

NOTE: Mark clutch cover, pressure plate and flywheel to insure correct alignment during installation.

3. Remove clutch cover and pressure plate assembly.

NOTE: When removing clutch cover and pressure plate assembly from flywheel, loosen screws evenly until spring tension is released, as cover could be warped by improper removal, resulting in clutch chatter when reassembled.

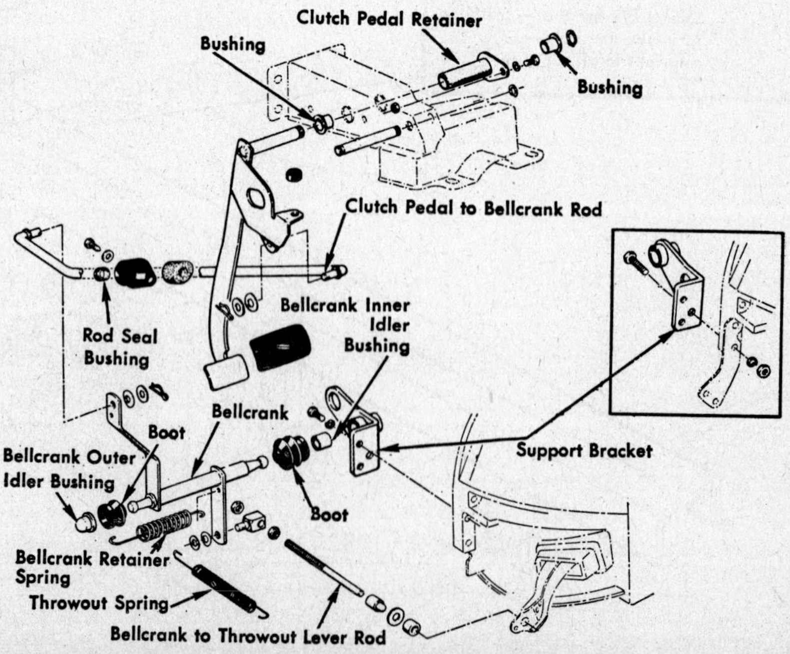

Fig. 1 Typical clutch linkage

NOTE: Unless special clutch rebuilding equipment is available, it is recommended that the clutch assembly be exchanged for a rebuilt unit should the clutch require rebuilding. The driven disc, however, may be replaced without special equipment. If clutch rebuilding equipment is available, follow the equipment manufacturer's instructions.

SERVICE BULLETIN

Lubricates Clutch Cover: Three felt pads in the clutch cover are lubricated in production before they are assembled over the eyebolts. A squeak that may occur during clutch operation after mileage has accumulated can be eliminated by relubricating the felt pads.

To do this, remove the cover pan from the clutch housing. Next, with a squirt-type oil can, carefully apply a mixture of heavy oil and mineral spirits to the three felt pads. Use only a small amount of lubricant, and make certain the lubricant does not get on the clutch driven member. Then apply a thin coating of a suitable light grease to the sides of the pressure plate lugs, where the lugs extend through the clutch cover. Install the pan to complete the job.

Clutch, Install

1. Very sparingly apply front wheel bearing lubricant to the clutch shaft pilot bearing in the crankshaft. If too much lubricant is used, it will run out of face of flywheel when hot and ruin driven plate facings. Make certain that flywheel surface is clean and dry.
2. Make sure that splines in driven plate hub are clean, and apply a light coating of lubriplate. Driven plate facings must be clean and dry.
3. Place driven plate on pressure plate, then place clutch assembly in position on flywheel, being sure to align marks made on flywheel and cover before removal.
4. Install cover bolts with washers but do not tighten.
5. Insert a spare clutch shaft through hub of driven plate and into pilot bearing.
6. Tighten each clutch cover bolt several turns at a time to draw cover evenly to flywheel and avoid distortion of cover.
7. While tightening cover bolts, move clutch shaft from side to side to center driven plate with pilot bearing. If driven plate is not centered, it will be difficult to slide the transmission into place. Make sure all cover bolts are uniformly tightened.
8. Remove aligning clutch shaft and install transmission, clutch linkage and adjust clutch pedal free play.

MANUAL TRANS.

Three Speed, Replace

1971-76

NOTE: On overdrive equipped vehicles, depress clutch then engage and disengage overdrive to relieve overrunning clutch and pinion carrier torque loading.

IMPORTANT: It is necessary to open the hood to avoid damage to the hood and air cleaner whenever the rear crossmember is removed.

NOTE: On cars equipped with floor shift, remove bezel, boot and shift selector lever.

1. Raise the vehicle.
2. Mark rear universal joint yoke and bearing prior to removal to insure proper alignment at time of installation.
3. Disconnect propeller shaft from rear axle.
4. Slide front universal joint yoke from transmission.
5. Unfasten and lower exhaust pipes on V8 models.
6. Support the engine and detach rear support cushion from transmission.

NOTE: On overdrive equipped vehicles, disconnect front cable from parking brake equalizer, remove adjuster and hooks from floor pan and lower parking brake equalizer to provide working clearance.

7. Remove bolts securing rear crossmember to side sills and remove rear crossmember.
8. Disconnect speedo cable, back up light switch wires and any other wiring.
9. Disconnect column shift rods and reverse lock up rod.
10. Lower engine until clearance is obtained to permit transmission removal.

NOTE: Care must be taken not to damage clutch shaft, pilot bushing or clutch disc.

1969-70

1. Raise car and support engine.
2. Mark rear U-joint yoke to aid in assembly, and disconnect propeller shaft from rear axle.
3. Slide front U-joint from transmission mainshaft.
4. Remove screws from rear engine support cushion at extension housing.
5. Remove bolts that secure rear corssmember to sills and remove crossmember.

NOTE: Before removing rear member on Javelin and AMX with power steering, open hood to avoid damage from pump reservoir wing nut.

6. Disconnect speedometer cable and shift rods from transmission.
7. On models with floor shift, remove shift lever bezel, boot, retainer and shift lever.
8. When removing transmission, use guide pins in place of two lower cap screws so as not to damage clutch shaft.

Four Speed, Replace

1969 All Models

1. Remove chrome trim ring that secures rubber boot to floor pan and slide boot up on shift lever.
2. Unfasten and remove shift lever from cross shaft (2 capscrews).
3. Raise vehicle.
4. Unfasten and remove exhaust pipes from exhaust manifolds.
5. Jack up or support engine.
6. Unfasten rear engine support cushion from crossmember.
7. Unfasten and remove crossmember.

NOTE: Before removing crossmember on Javelin and AMX models with power steering, open hood to avoid damage from pump reservoir wing nut.

8. Remove propeller shaft.
9. Disconnect speedometer cable from transmission.
10. Lower engine until adequate clearance is obtained to permit transmission removal.
11. Reverse procedure to install.

1970-74 All Models

The procedure to remove the transmission is the same as that given previously for the three speed transmission.

OVERDRIVE REPLACE

NOTE: In order to ease removal, operate vehicle, then engage and disengage overdrive with clutch pedal depressed. This procedure will relieve torque loading on overrunning clutch and pinion carrier.

1. On floorshift vehicles, remove gearshift lever bezel, boot, insulator, retaining bolts, crossover spring, and gearshift lever.
2. Raise and support vehicle on hoist.
3. On catalytic converter equipped vehicles, disconnect converter from exhaust pipes, and lower converter, muffler and tailpipes to obtain clearance.
4. Disconnect solenoid valve wires and disconnect speedometer cable from governor speed switch, then remove speedometer clamp, governor speed switch, speedometer support, adaptor and driven gear from speedometer bore.
5. Mark rear universal joint and pinion yoke for correct alignment at installation and remove propeller shaft.
6. On floorshift vehicles, disconnect shift rods at transmission shifter levers and slide rods forward out of gearshift lever retainer bushings, then remove retainer to transmission adapter bolts and remove retainer.
7. Place a support under clutch housing, then remove rear support cushion to transmission adaptor bolts and remove support cushion and crossmember.
8. Remove the eight locknuts and washers from overdrive main case-to-transmission adaptor studs and remove overdrive. Discard old adaptor gasket.
9. When installing overdrive, use a long screwdriver to align overrunning clutch splines with transmission output shaft splines, then align pump strap with drive cam and install over-

drive. Torque locknuts to 18 ft. lbs.

NOTE: Make certain that transmission output shaft splines are aligned with splines in overrunning clutch hub. Do not force engagement when installing.

MANUAL TRANS. SHIFT LINKAGE

1969-76 Three Spd. Floor Shift

NOTE: For 1969 All & 1973-75 exc. tilt column models omit Steps 1-4.

1. Loosen reverse trunnion nuts.
2. Shift transmission into reverse and lock steering column.

NOTE: It may be necessary to move the lower column shift lever upward until it is in the locked position.

3. Tighten lower trunnion lock nut until it contacts the trunnion then tighten upper lock nut while holding trunnion centered in lever.
4. Unlock steering column.
5. Shift transmission into neutral.
6. Loosen 2-3 lever attaching nut and adjusting bolt.
7. With 1st-Reverse shift rod in neutral, align the 2-3 shift rod so the shift notch is exactly aligned with the 1st-Reverse shift notch. Tighten adjusting bolt and attaching nut.
8. On 1970-72 units, shift transmission into reverse and lock steering column. Column must lock without any binding.

1969-76 Three Spd. Column Shift

NOTE: For 1969 models, omit Steps 1-3.

1. With shift rods disconnected from levers, place column lever in reverse position and lock column.
2. Place transmission levers in reverse position and adjust shift rod trunnion for a free pin fit in the lever. Tighten trunnion. Unlock column.
3. Shift transmission into neutral.
4. Insert a 3/16" drill bit through the aligning holes of the two shift levers, shift gate and the bracket on the jacket tube.
5. With the shift levers in the neutral position, carefully adjust the trunnions to a free fit in the shift levers and secure the trunnion in this position with the lock nuts. Use care when tightening the lock nuts so a binding condition does not exist.

1969-74 Four Speed

NOTE: On some units, it is necessary to lower the rear of the transmission before adjusting the linkage.

1. Loosen the transmission shift lever nuts (two on each lever) and loosen the lock nuts on the reverse shift rod at the trunnion.
2. Install a 1/4" drill bit through the selector lever retainer, through the levers, spacer plate, and through the aligning hole in the mounting bracket.
3. This is the neutral position. Place all three transmission levers in the neutral position.
4. Adjust the trunnion on the reverse shift rod to enter freely into the reverse lever. Lock the check nuts on the reverse rod and secure with washer and cotter pin.
5. Tighten the lower nuts of the shift levers first, being careful not to move the outer levers out of position.
6. Tighten the upper "hug nuts" to 10 ft. lbs. Do not overtighten or the shift shafts could be broken. Remove the aligning pin.

1976 AUTO. TRANS. LINKAGE ADJUST

Linkage adjustments for the 1976 models are the same as for 1976 models and can be found elsewhere in this manual. 1969-71 models use the Shift Command Transmission, 1972-76 models use the Torque Command Transmission.

Rear Axle, Propeller Shaft & Brakes

REAR AXLES

Fig. 1 illustrates the rear axle assembly used on these cars. When necessary to overhaul the unit, refer to the *Rear Axle Specifications* table in this chapter.

DESCRIPTION

Except Twin Grip

In these rear axles, Fig. 1, the drive pinion is mounted in two tapered roller bearings. These bearings are preloaded by a washer behind the front bearing. The pinion is positioned by shims located in front of the rear bearing. The differential is supported in the carrier by two tapered roller side bearings. These bearings are preloaded by shims located between the bearings and carrier housing. The differential assembly is positioned for proper ring gear and pinion backlash by varying the position of these shims. The differential case houses two side gears in mesh with two pinions mounted on a pinion shaft which is held in place by a lock pin. The side gears and pinions are backed by thrust washers.

It is not necessary to remove the rear axle assembly. However, the underbody should be washed to prevent particles of road dirt from contaminating the parts.

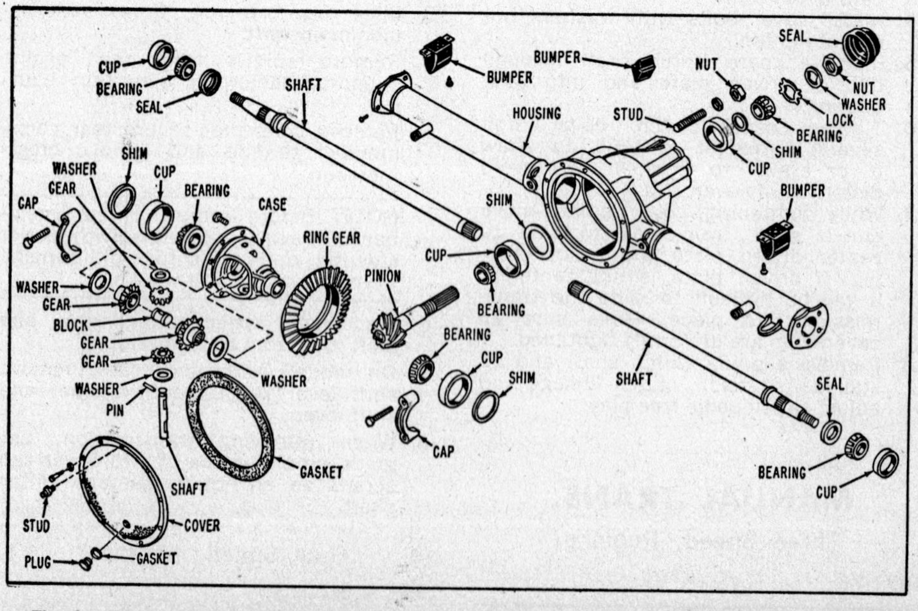

Fig. 1 Rear axle assembly, except Twin Grip. For 1969-76 V8s the pinion bearing preload is adjusted by means of a collapsible spacer instead of the shim shown

Twin Grip

Conventional type differentials divide torque equally between both wheels, causing the wheel with the least traction to slip first. Twin Grip locking type differential, Fig. 2, provides the torque of the slipping wheel to the driving wheel. Unlike the positive lock type, this unit will release before excessive driving force can be directed to one rear wheel.

Locking action on the 7⁹⁄₁₆ inch axle, is accomplished through cone brakes which are spring loaded to allow adequate driving force at the high traction wheel and at the same time not interfere with steering characteristics or differential action. Under extremely unbalanced traction conditions, wheel spin may occur if over-acceleration is attempted. This spinning produces a whirring sound caused by clutch cone overrun. This spinning condition or sound does not indicate a unit failure.

NOTE: Twin Grip differentials used on 7⁹⁄₁₆ inch axles, are serviced as an assembly only.

Locking action on 8⅞ inch axles, is accomplished through multiple disc clutch packs between the differential gears and case. The concentrically grooved clutch discs are splined to the differential gears and are engaged primarily by spring preload pressure on the clutch discs. The locking action is increased by reaction between the differential gears and pinions when torque and traction increase. When turning corners, the wheel travel differential torque will overcome clutch spring load and permit normal differential action.

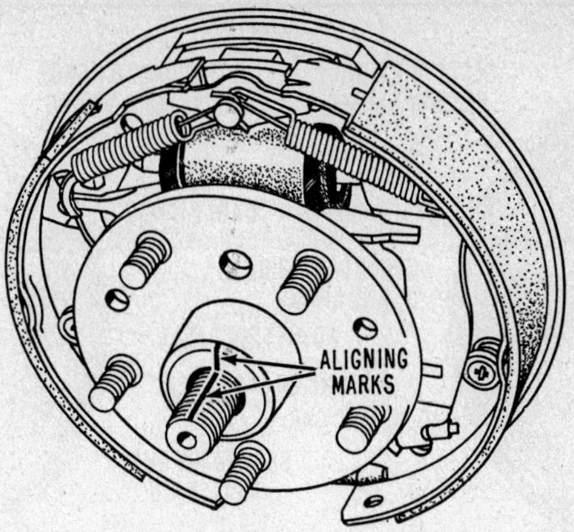

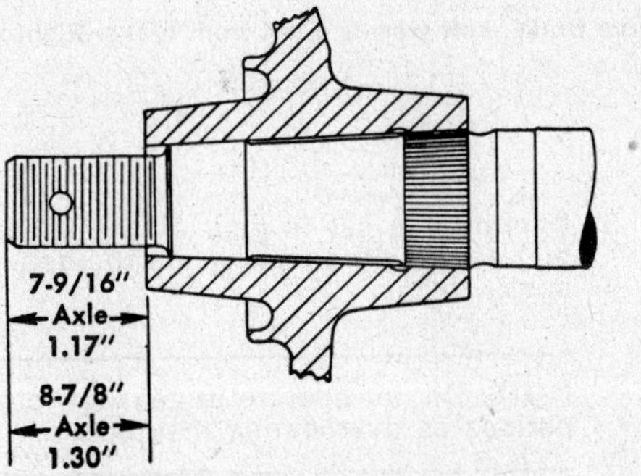

Fig. 3 Installing hub on axle shaft

REAR AXLE & PROP. SHAFT, REPLACE

1969-76 Hornet, Rambler, AMX, Javelin, Gremlin, & 1975-76 Pacer

1. Remove axle shaft nuts prior to raising the car weight from the wheels.
2. Remove the axle housing cover to drain the lubricant.
3. Raise and support the rear of the body.
4. Remove the rear wheels and rear wheel hubs and drums.
5. Disconnect rear parking brake cables at equalizer.
6. Remove the brake support plates.

NOTE: Retain the shims located between the left support plate and axle tube for use on reassembly.

7. Remove the axle shafts from the axle.
8. Mark the universal joint yoke and bearing before separating to insure same alignment at time of assembly. Disconnect the propeller shaft at the rear universal joint.
9. Disconnect the rear shocks at the axle.
10. Disconnect the brake line at the body floor pan bracket.
11. Remove the rear spring U bolts and the axle may now be removed.

1969-76 Rebel, Ambassador & Matador

1. Remove axle shaft nuts prior to raising the car weight from the wheels.
2. Raise and support the rear of the car.
3. Remove the axle housing cover and drain the lubricant.
4. Mark the rear universal joint and bearing and disconnect the propeller shaft at the rear universal joint.
5. Disconnect the parking brake cable at the equalizer.

NOTE: The left cable is routed to upper long end of equalizer. Disconnect the brake lines at the support plates.

6. Remove the wheels, hubs, drums, support plates, seals, axle shafts and bearings.
7. Support the axle assembly and disconnect the shocks at the axle tubes.
8. Lower axle assembly until it is supported by the control arms.
9. Pull one axle tube down and remove the spring. Pull the other axle tube and remove the other spring.
10. Support the axle assembly and disconnect the upper control arms at the axle housing.
11. Disconnect the lower control arms at the axle tubes and the axle may now be removed.

AXLE SHAFTS

The hub and drum are separate units, and the hub and axle shaft are serrated to mate and fit together on the taper. Both are punched marked to insure correct assembly, Fig. 3. The axle shaft and bearing may be removed as follows:

1. Remove rear wheel, drum and hub, then disconnect parking brake cable at equalizer.
2. Disconnect brake tube from wheel cylinder and remove brake support plate assembly, oil seal and axle shims from axle shaft.

NOTE: Axle shaft end play shims are located on the left side only.

3. Using suitable puller, pull axle shaft and bearing from axle tube, then remove and discard inner oil seal.

NOTE: The bearing cone must be pressed off the shaft, using an arbor press.

When installing hub onto axle, install two well lubricated thrust washers and axle shaft nut. Tighten axle shaft nut until hub is installed to the dimensions shown in Fig. 3. Remove axle shaft nut and one thrust washer. Reinstall axle shaft nut and tighten to 250 ft. lbs. If cotter pin hole is not aligned, tighten the nut to the next castellation and install cotter pin.

NOTE

Do not use an original hub on a replacement axle shaft; use a new hub. A new hub may be installed on an original axle shaft providing the serrations on the shaft are not worn or damaged. Be certain that the hub and axle shaft are punch marked to insure proper alignment on installation. A replacement hub, which is not serrated, can be installed and serrations will be cut in the hub when installed on the shaft due to the difference in hardness of the shaft and the hub.

Assembly

Replace the parts in the reverse order of their removal. If the old parts are replaced and the shims have not been disturbed, the axle shaft end play should be correct when the parts are assembled. However, if a new shaft, bearing, differential carrier or housing has been in-

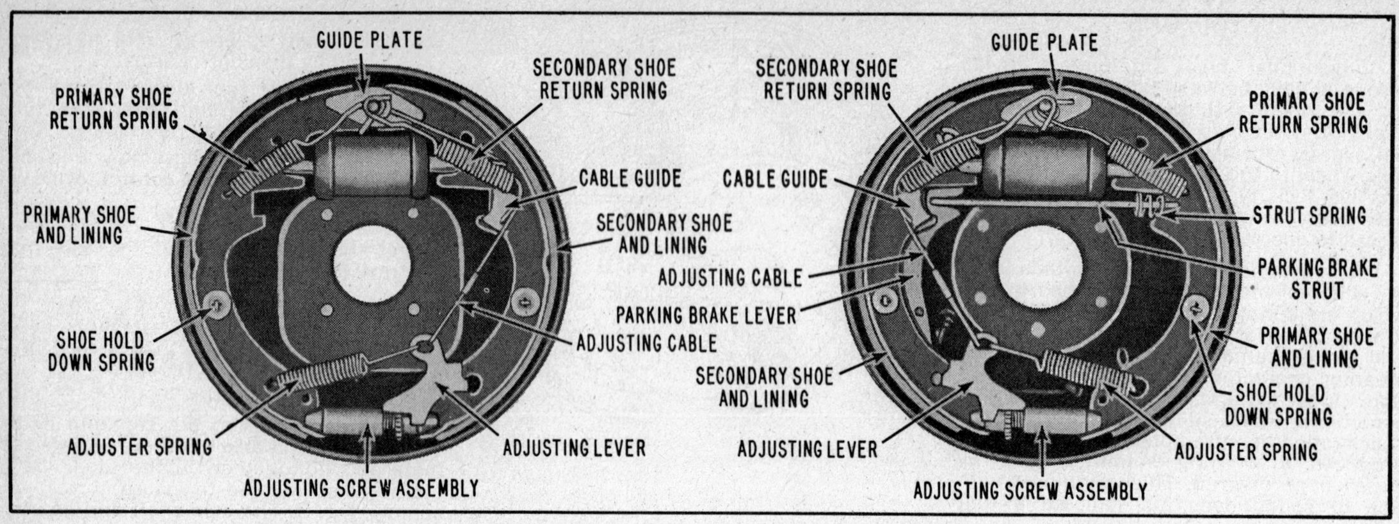

Fig. 4 Bendix Duo-Servo Brake. Left view is a left front brake. Right view is a right rear brake

stalled, it will be necessary to check the end play.

The end play can be checked when all parts have been replaced except the wheel and hub. To make this check, rap each axle shaft after the nuts are tight to be sure the bearing cups are seated. Then place a dial indicator so that its stem contacts the end of the shaft and work the shaft in and out to determine the amount of existing end play. If an adjustment is necessary, remove the outer oil seal and brake support and add or remove shims as required. When making this adjustment, an equal thickness of shims should be removed or added on each side of the axle housing to maintain a central position of the differential thrust block.

NOTE

The application of a bead of sealing material such as "Pliobond" or "Permatex" to the outer diameter of axle tube flange and the brake support contact area is recommended. The sealing material will be used in addition to the gasket for improved sealing.

PROPELLER SHAFT VIBRATION

With the use of an electronic wheel balancer, the propeller shaft can be balanced as follows:

1. Raise and support the rear of the car at the axle and remove rear wheels.
2. Place electronic pick up unit under axle housing as close as possible to pinion yoke. Use crayon or chalk to mark four equally spaced horizontal lines on the propeller shaft. To aid in identifying lines, it is suggested they be of unequal length.
3. Operate the car in gear at the speed of greatest vibration. Locate the heavy spot.

NOTE: Do not operate in gear for long periods as overheating may occur.

4. Place two worm type hose clamps on propeller shaft with heads of clamps located 180° from heavy spot noted previously. Slide clamps as far to rear as possible.
5. Again operate car in gear and if vibration still exists move both clamp heads an equal distance in opposite directions toward the heavy spot until balancer indicator remains within acceptable range of scale.
6. Replace rear wheels and road test car.

BRAKE ADJUSTMENTS

1969-76 Self-Adjusting Brakes

These brakes, Figs. 4 and 5 have self-adjusting mechanisms that assure correct lining-to-drum clearances at all times. The automatic adjusters operate only when the brakes are applied as the car is moving rearward.

Although the brakes are self-adjusting, an initial adjustment is necessary after the brake shoes have been relined or replaced, or when the length of the star wheel adjusting screw has been changed during some other service operation.

Frequent usage of an automatic transmission forward range to halt reverse vehicle motion may prevent the automatic adjusters from functioning, thereby inducing low pedal heights. Should low pedal heights be encountered on these models, it is recommended that numerous forward and reverse stops be made until satisfactory pedal height is obtained.

NOTE

If a low pedal condition cannot be corrected by making numerous stops (provided the hydraulic system is free of air) it indicates that the self-adjusting mechanism is not functioning. Therefore, it will be necessary to remove the brake drum, clean, free up and lubricate the adjusting mechanisms. Then adjust the brakes as follows, being sure the parking brake is fully released.

Adjustment

1. When the brake parts have been installed in their correct position, initially adjust the star wheel assembiles to a point where 3/16" to 1/4" of threads are exposed between star wheel and star wheel nut on Series 10, 3/8" of threads exposed on American, Ambassador and Rebel models.
2. Following the initial adjustment and final assembly, check the brake pedal height to insure brake operation. Then drive the car forward and reverse, making 10 to 15 brake applications prior to road testing. This action balances the adjustment of the four brake units and raises the brake pedal.

PARKING BRAKE, ADJUST

1969 Rambler

1. With service brakes properly adjusted, pull parking brake handle to the third notch from the released position.
2. Tighten the parking brake cable at the equalizer to a point where the rear wheels are locked from forward rotation.
3. Release brake handle and check for rear brake drag—wheels should rotate freely.

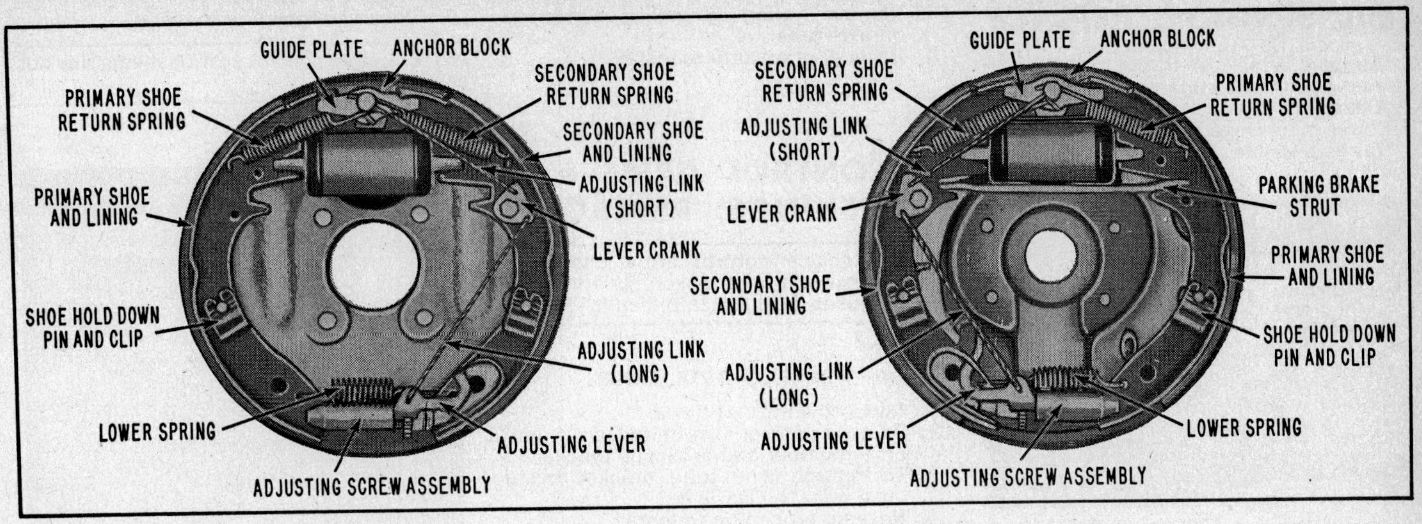

Fig. 5 Wagner Compound Shoe-Type Brake. Left view is a left front brake. Right view is a right rear brake

1969 Except Rambler; 1970-76 All

1. With service brakes properly adjusted, set parking brake pedal on the first notch from fully released position.
2. Tighten parking brake cable at equalizer to a point where the rear wheels are locked in forward rotation.
3. Release pedal and check for rear wheel drag—wheels should rotate freely.

BRAKE MASTER CYLINDER, REPLACE

To remove the master cylinder, disconnect the brake lines from the connections on the master cylinder. Unfasten the cylinder from its mounting and remove from the car.

Install in the reverse order of removal and bleed the brake system.

Rear Suspension

SHOCK ABSORBER, REPLACE

1. With the rear axle supported properly, disconnect lower end of shock absorber from stud on mounting bracket.
2. On models with leaf springs, upper attaching nut is accessible from trunk. On models with coil springs, remove upper mounting bracket from underbody.
3. Reverse procedure to install.

LEAF SPRINGS, REPLACE

1. Support rear axle, removing tension from springs.
2. Disconnect lower end of shock absorber from stud on mounting bracket.
3. Remove "U" bolts securing spring plate and spring to axle tube.
4. Disassemble rear shackle and remove eye bolt from spring forward mounting bracket.
5. Reverse procedure to install. Replace bushings as necessary.

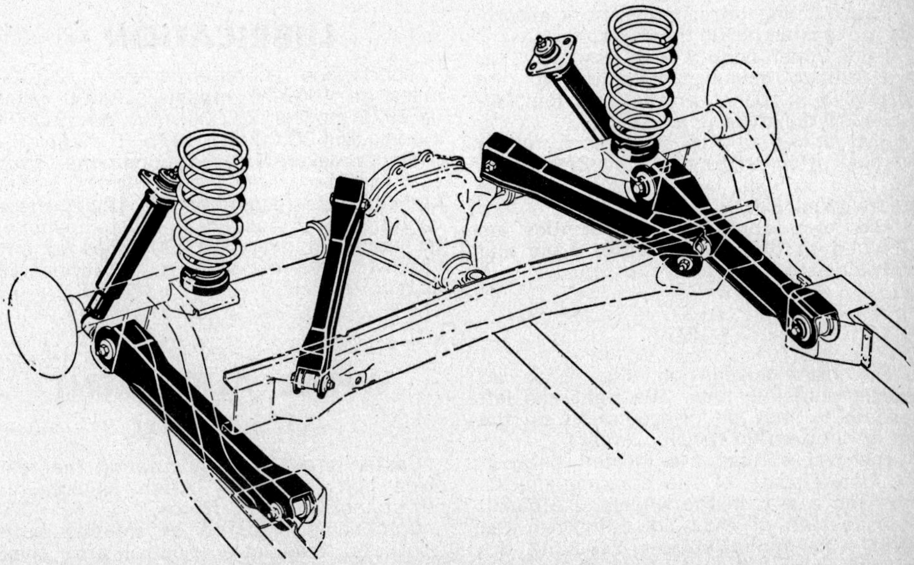

Fig. 1 Coil spring suspension (typical)

COIL SPRINGS, REPLACE

1. Support vehicle at frame and support rear axle with a suitable jack.
2. Disconnect shock absorbers from lower mountings and lower the axle as far as possible, Fig. 1.
3. Remove spring by pulling downward on axle tube.
4. Reverse procedure to install.

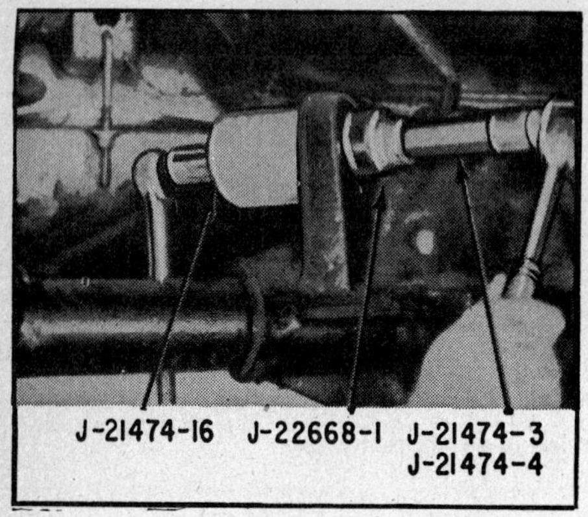

Fig. 2 Upper control arm rear bushing removal

CONTROL ARMS & BUSHINGS, REPLACE

NOTE: Replace control arms one at a time to prevent axle assembly misalignment, making installation difficult.

Upper Control Arms

1. Support vehicle at frame.
2. Remove control arm bolts from frame crossmember and axle tube bracket.
3. To replace axle tube bracket bushings, refer to Figs. 2 & 3.
4. Reverse procedure to install.

Lower Control Arms

1. Remove control arm mount bolts from frame and axle tube brackets.
2. Reverse procedure to install.

NOTE: Lower control arm bushings are not serviceable.

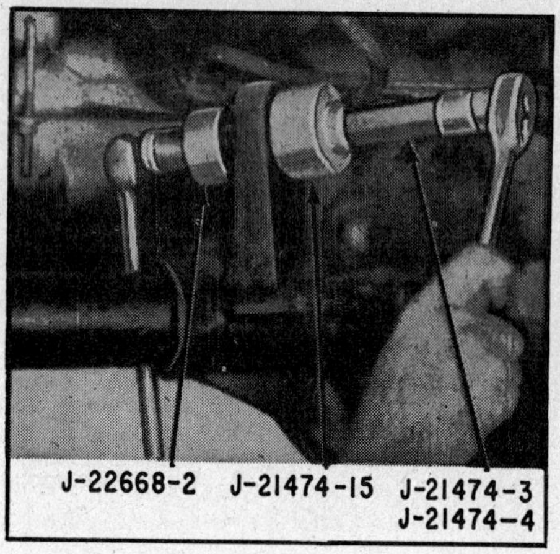

Fig. 3 Upper control arm rear bushing installation

Front End and Steering Section

1975-76 Pacer

These models have an independent front coil spring front suspension system using unequal length upper and lower control arms. The coil springs are mounted between the lower control arms and the front crossmember with the lower arm functioning as the loaded member.

1970-76 Exc. Pacer

The front suspension, Fig. 1, is an independent linked type, with the coil springs located between seats in the wheel house panels and seats attached to the upper control arms.

Direct acting telescoping shock absorbers are located inside the coil springs.

Each upper control arm assembly has two rubber bushings attached to the wheel house panel and a ball joint attached to the steering knuckle.

Each lower control arm has a rubber bushing attached to the front crossmember and a ball joint attached to the steering knuckle.

The lower control arm strut rods are attached to the lower control arms and body side sill brackets.

1969

The front suspension, Fig. 2, is an independent link type. The right and left assemblies may be disassembled on the car or removed for bench overhaul.

The coil springs are located between the upper seats of the steering knuckle pins and a seat in the wheelhouse panel. Transmission of road noise through the springs is minimized by insulating the springs from the body with rubber cushions.

The shock absorbers are the direct-acting type with built-in rebound bumper control. The end mountings of the shocks are retained in rubber grommets.

The upper control arms contain rubber insulated bushings installed in the inner end of the arms. The control arms are attached to the mounting bracket on the wheelhouse panel and the trunnion at the outer end.

The lower control arms are attached to a removable crossmember at the pivot ends. The outer ends are attached to the steering knuckle pin with a ball joint stud.

LUBRICATION

Lubrication is required every 24,000 miles on 1969-71 models, 18,000 miles on 1972 models, 25,000 miles on 1973-74 models and 30,000 on 1975-76 models.

Under severe driving conditions, such as dusty or extreme wet conditions, earlier lubrication is recommended. Under these conditions the suspension system should be inspected every 12,000 miles or one year whichever occurs first, and lubricated as required.

WHEEL ALIGNMENT

1975-76 Pacer

Caster is adjusted by rotating the rear pivot bolt eccentric. After adjustment, torque locknut to 95 ft. lbs.

Camber is adjusted by rotating both front and rear lower control arm pivot bolt eccentrics. After adjustment, torque locknuts to 95 ft. lbs.

1969-76 Exc. Pacer

Caster is obtained by moving the two adjusting nuts on the threaded strut rod, Fig. 2. One nut is on each side of the mounting bracket. Therefore, moving the nuts on the rod will move the lower control arm to front or rear for desired caster angle. After adjustment, torque nuts to 85 ft. lbs. on all 1969-73 models. On 1974-76 models, torque adjusting nuts to 65 ft. lbs. and locknut to 55 ft. lbs.

Camber is obtained by turning on the eccentric lower control arm bolt, Fig. 3. After adjustment, torque lock nut to 95 ft-lbs.

TOE-IN, ADJUST

To adjust toe-in, loosen the clamps at both ends of the adjustable tubes on each tie rod. Turn the tubes an equal amount until the toe-in is correct. Turning the right tube in the direction the wheels revolve when the car is going forward increases the toe-in and turning the left tube in the opposite direction increases toe-in. To decrease toe-in turn the right tube backward and the left tube forward. It is important that both tubes be turned an equal amount in order to maintain the correct position of the steering wheel. When adjustment is complete, tighten all clamp bolts.

NOTE: *In performing service operations on the steering linkage or when adjusting toe-in, be sure to square the tie rod ball sockets on the studs and align the tie rod stud in the center, or slightly above center, of the cross tube opening, before tightening the steering linkage adjusting* *tube. This will prevent the stud from contacting the side of the cross tube opening, which would otherwise result in noise problems or damage.*

WHEEL BEARINGS, ADJUST

1969-76

1. To adjust bearings, tighten spindle nut to 22 ft-lbs torque while rotating the wheel to seat bearings.
2. Then loosen spindle nut 1/3 turn and with wheel rotating, retorque spindle nut to 12 inch pound on 1969-72, 2-10 inch pound on 1973-74, and 6 inch pounds on 1975-76.
3. Place the nut retainer on spindle nut with the slots of the retainer aligned with the cotter pin hole on the spindle.
4. Install cotter pin and dust cap.

WHEEL BEARINGS, REPLACE

(Disc Brakes)

1. Remove two thirds of the total fluid capacity of the master cylinder reservoir to prevent fluid overflow when the caliper pistons are pushed back on their bores.
2. Raise car and remove front wheels.
3. Disconnect hydraulic tube from mounting bracket. Do not disconnect any hydraulic fitting.
4. Holding the lower edge of the caliper, remove the lower bolt. Any shims that fall out at this point should be labeled to insure that they be replaced in their original position.
5. Holding the upper edge of the caliper, remove the upper bolt, tag these shims.
6. Hang caliper from upper suspension to prevent strain being placed on brake hose.
7. Remove spindle nut and hub and disc assembly. Grease retainer and inner bearing can now be removed.

CHECKING BALL JOINTS FOR WEAR

1969-76

Before checking ball joints for wear, make sure the front wheel bearings are properly adjusted and that the control arms are tight.

Referring to Fig. 4, raise wheel with a jack placed under the frame as shown. Then test by moving the wheel up and down to check axial play, and rocking it at the top and bottom to measure radial play.

The upper ball joint should be replaced if total travel when rocking wheel and tire exceeds .160".

The lower ball joint is spring loaded and should be replaced if there is any noticeable lateral shake.

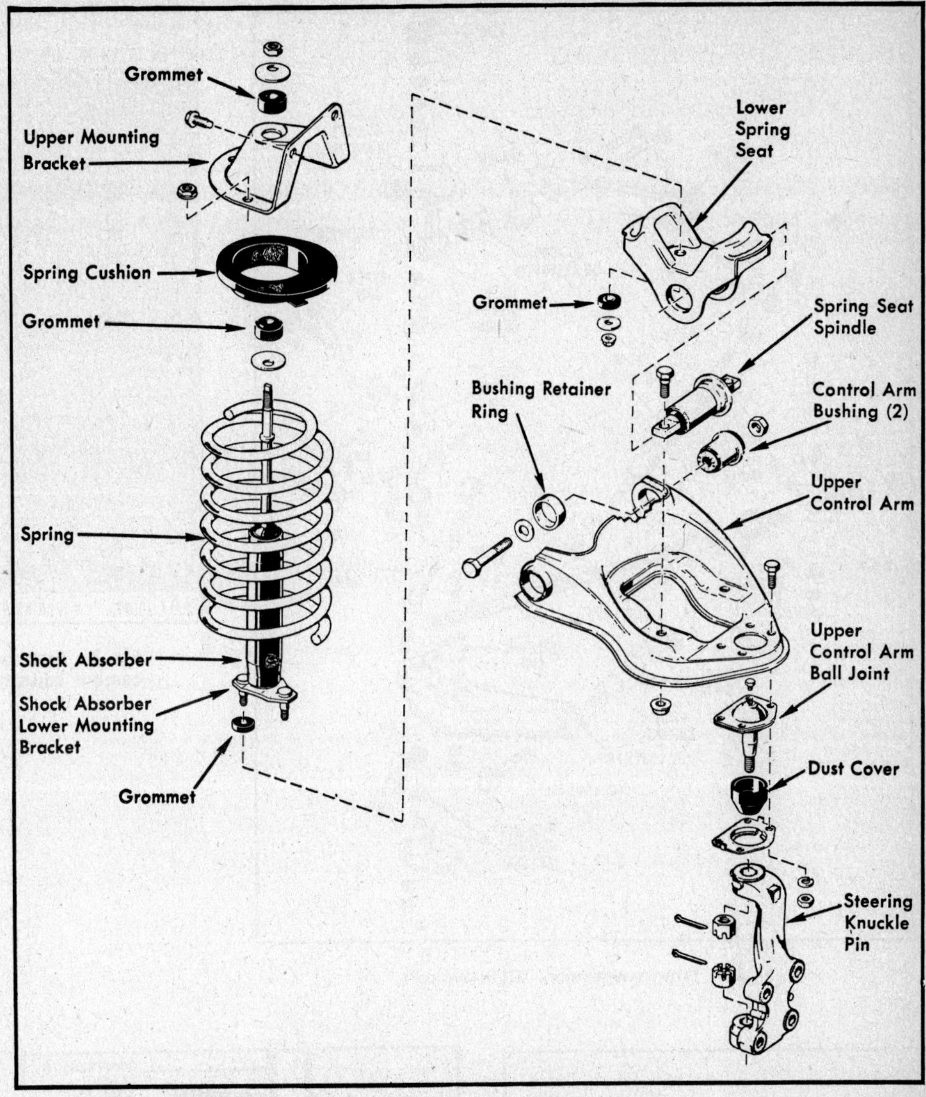

Fig. 1 1970-76 Exc. Pacer front suspension. Lower control arm (not shown) is the same as used on previous models

BALL JOINTS, REPLACE

NOTE: Upper and lower ball joint can be replaced using the following procedure.

1. Drill out the rivets that attach ball joint to control arm.
2. Remove two strut rod mounting bolts.
3. Remove steering knuckle arm and stud nut and remove ball joint from knuckle pin with a suitable ball joint remover.
4. When installing new ball joints, note that two bolts, lockwashers and nuts are furnished to replace rivets that were removed.

SHOCK ABSORBER, REPLACE

1970-76

After disconnecting shock absorber from wheelhouse panel at top and lower spring seat at the bottom, withdraw shock absorber out of top of wheelhouse.

1969

After disconnecting the shock from the wheelhouse panel at the top and from the lower control arm at the bottom, collapse the unit, work to one side and remove from out of the lower control arm.

Reverse the foregoing procedure to install.

SPRING, REPLACE

1975-76 Pacer

1. Remove shock absorber, then the stabilizer bar link bolt at lower control arm, if equipped.

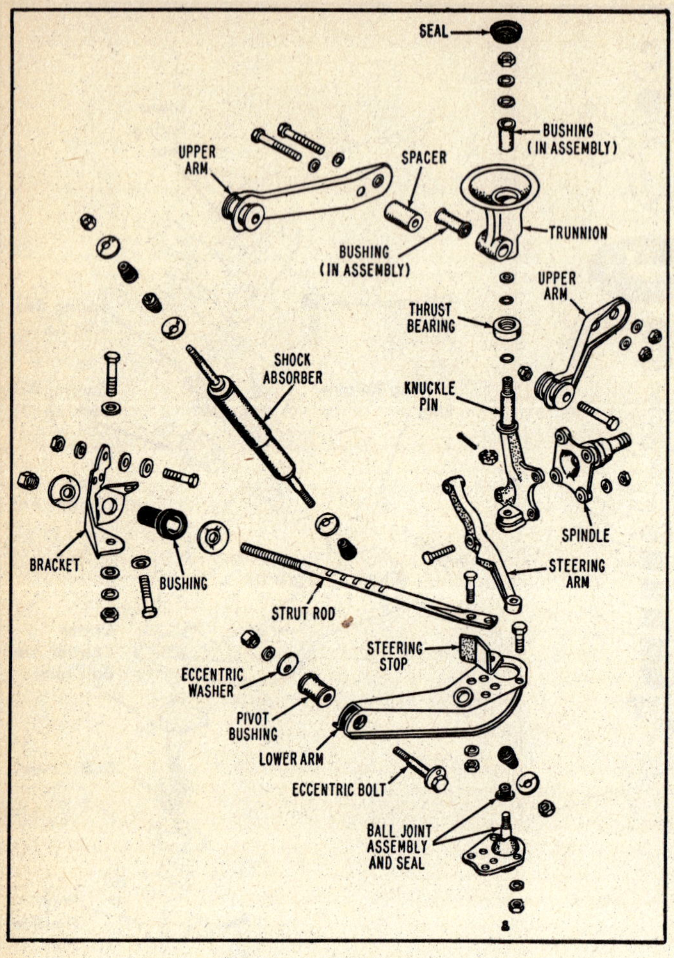

Fig. 2 Front suspension. 1969 (typical)

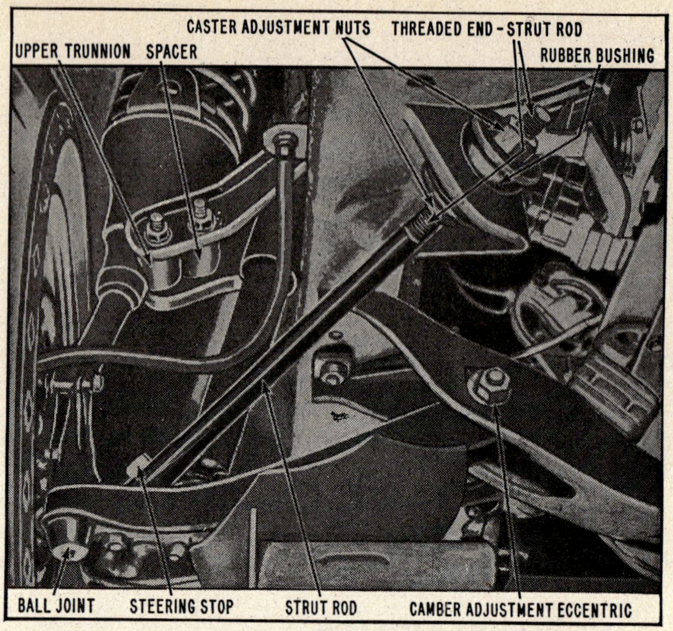

Fig. 3 Front suspension (rear view) showing caster and camber adjustments. 1969-76 All exc. Pacer (typical)

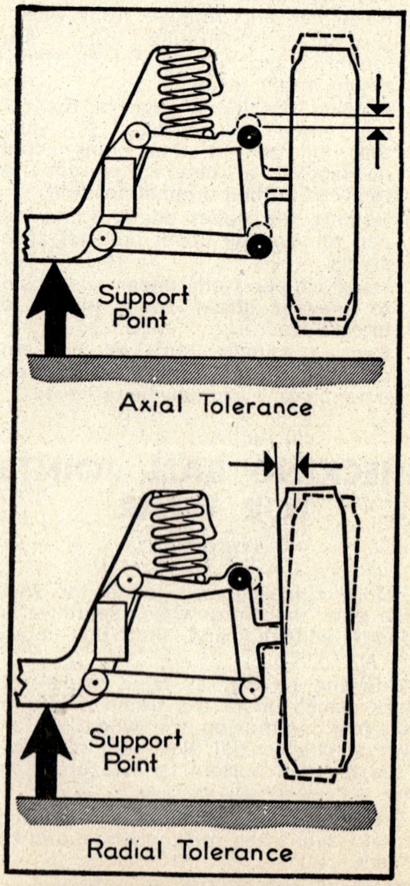

Fig. 4 Checking ball joints for wear. 1969-76

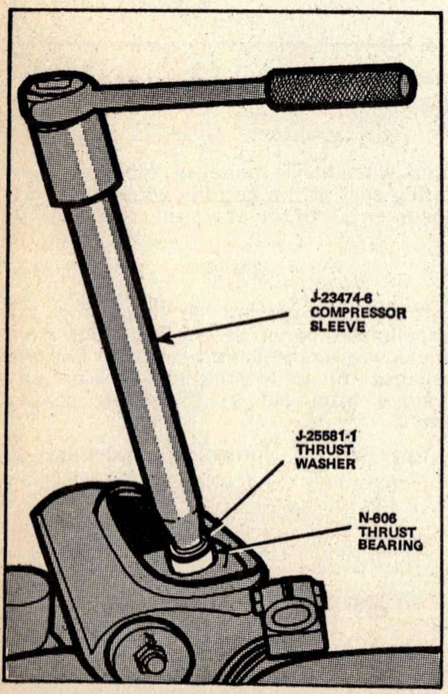

Fig. 5 Installing spring compressor. 1975-76 Pacer

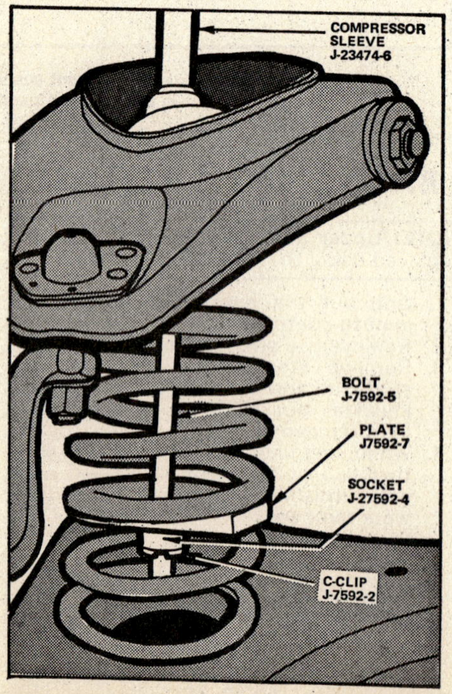

Fig. 6 Compressing spring. 1975-76 Pacer

2. On models with drum brakes, remove wheel, tire and drum. On disc brake models, remove wheel, tire, caliper and rotor.
3. Remove steering arm to knuckle attaching bolts and position arm aside.
4. Compress spring, Figs. 5 and 6.
5. Remove cotter pin and nut from lower ball joint, then using tool J-9656 and a hammer, disengage stud from steering knuckle and position knuckle aside.

NOTE: Use wire to support steering knuckle otherwise brake hoses may be damaged.

6. Position lower control arm aside and relieve spring tension. When all tension is relieved, remove tool and spring.

1970-76 Exc. Pacer

1. Remove shock absorber.
2. Install spring compressor (J-23474) through upper spring seat opening, Fig. 7. Place tool lower attaching screws through shock absorber mounting holes in the lower spring seat. Install tool lower retainer.
3. Remove lower spring seat pivot retaining nuts.
4. Tighten compressor until spring is compressed approximately 1".
5. Raise and support front of car under frame allowing control arms to fall free of lower spring seat. Remove wheel.
6. Pull lower spring seat away from car. Loosen compressor and allow lower spring seat to come out.
7. When all spring tension is released, remove tool lower retainer spring seat and spring.

1969

1. To remove a spring, raise rear end of car opposite from side from which spring is to be removed. Additional pressure may be gained by leaning on fender over the spring.

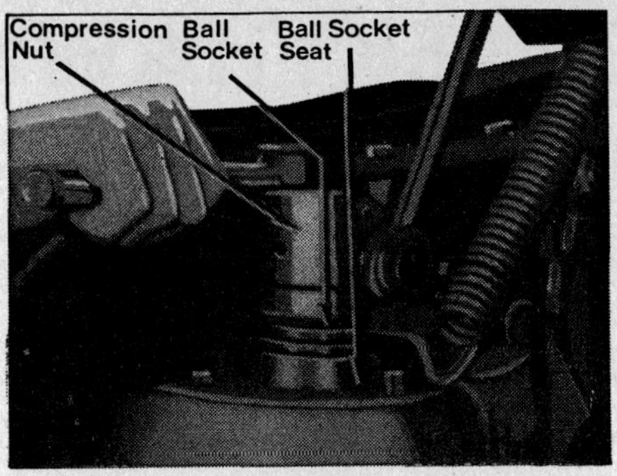

Fig. 7 Installation of spring compressor. 1970-76 exc. Pacer

2. Install hooks in holes on ears of spring seats. Hooks will hold spring in compressed position to allow removal from vehicle. A service spring may be installed as follows:
3. Install upper and lower cushions and upper and lower spring seats on spring. Align holes in ears of upper and lower spring seats.
4. Compress spring by suitable means (arbor press or hydraulic jack) and install hook on spring seats. Spring can then be installed on spring support.

CAUTION: Lip of lower spring seat must engage seat support to prevent spring from shifting during operation.

5. Hooks are released from spring seat by raising opposite near end of vehicle.

STEERING GEAR, REPLACE

1975-76 Pacer

1. Unlock steering column and raise vehicle.
2. Remove flexible coupling to intermediate shaft flange attaching nuts.
3. Turn wheels to stop in direction of tie rod being disconnected, then using a floor jack, raise lower control arm at least 2 inches. Remove cotter pin and retaining nut from tie rod end and using tool J-24319-01 on models with drum brakes or tool J-3295 on models with disc brakes, disconnect tie rod end from steering arm.
4. On models with power steering, disconnect and cap hoses.
5. Remove steering gear attaching bolts and the gear.

1969-76 Exc. Pacer

1. Remove flexible coupling bolts.
2. Remove pitman arm, using a suitable puller.
3. Remove mounting screws and lower steering gear from vehicle.

POWER STEERING

1975-76 Pacer

Refer to the "Steering Gear Replace", 1975-76 Pacer procedure for replacement procedure.

1969-76 Exc. Pacer
Steering Gear, Replace

1. Disconnect pressure and return hoses from gear. Raise hoses above pump level to keep oil from draining out of pump.
2. Remove flexible coupling bolt nuts, noting the different nut sizes to insure correct assembly.
3. Remove pitman arm with a suitable puller.
4. Remove gear attaching bolts.
5. Slide lower shaft free of coupling flange, then remove gear.
6. Reverse procedure to install.

BUICK
EXC. SKYHAWK

OLD CAR SPECIFICATIONS: For 1946-68 Tune Up and Wheel Alignment Specifications see main index.

INDEX OF SERVICE OPERATIONS

GRILLE IDENTIFICATION

1969 Sportwagon, Special

1969 Skylark

1969 Gran Sport

1969 LeSabre

1969 Wildcat

1969 Electra

1969 Riviera

1970 Skylark, Sportwagon

1970 Gran Sport

1970 LeSabre

1970 Estate Wagon

1970 Wildcat

1970 Electra

1970 Riviera

1971 Skylark

1971 Gran Sport 400

1971 Gran Sport Stage 1

1971 LeSabre

1971 Centurion

1971 Electra

1971 Riviera

1972 Skylark

1972 Skylark Custom

1972 Gran Sport

GRILLE IDENTIFICATION—Continued

1972 LeSabre

1972 Centurion

1972 Electra

1972 Riviera

1973 Apollo

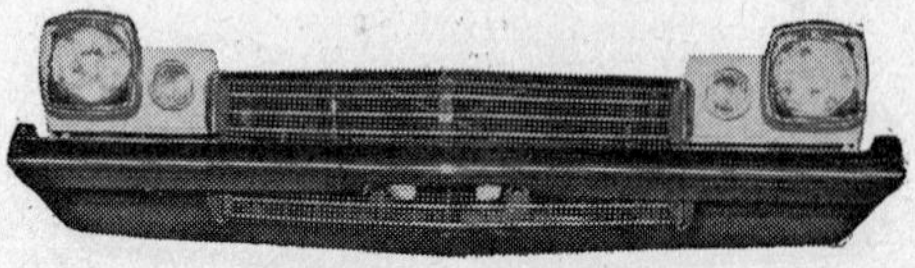
1973 Century

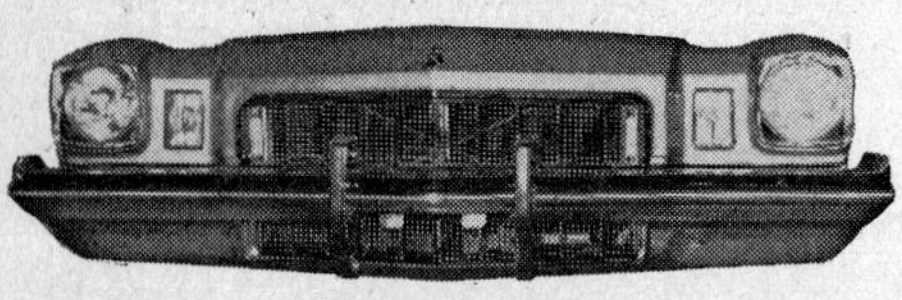
1973 Regal

1973 LeSabre & Centurion

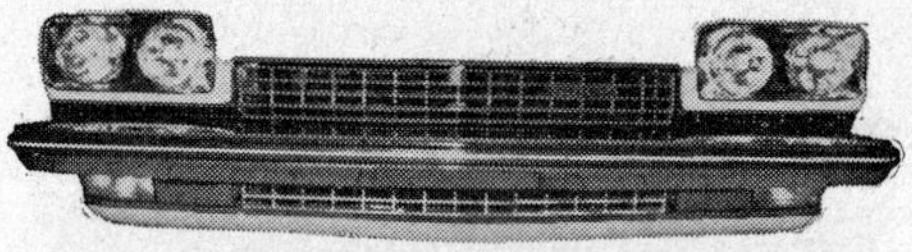
1973 Electra

1973 Riviera

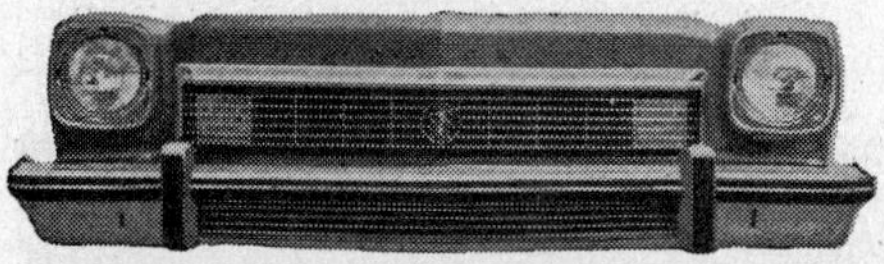
1974 Apollo

1974 Century

1974 Regal

1974 LeSabre

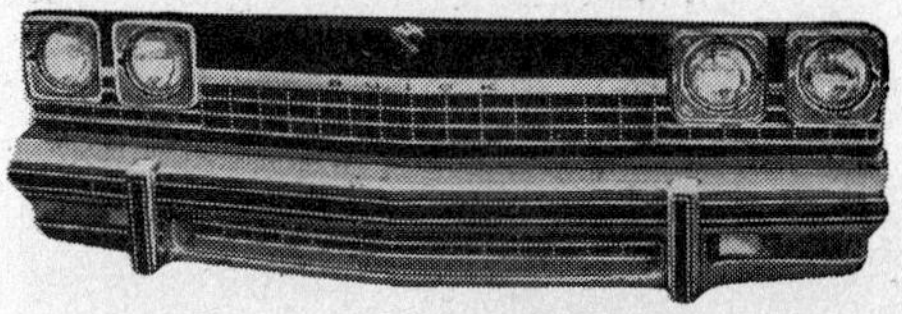
1974 Electra & Estate Wagon

1974 Riviera

1975 Apollo

1975 Skylark

1975 Century

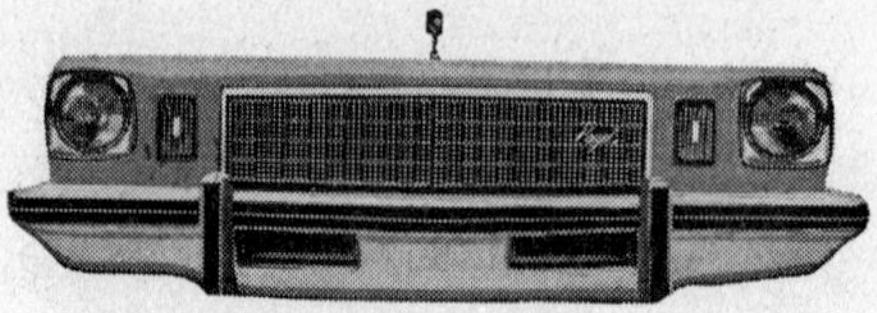
1975 Regal

1975 LeSabre

1975 Electra

1975 Riviera

1976 Skylark

GRILLE IDENTIFICATION—Continued

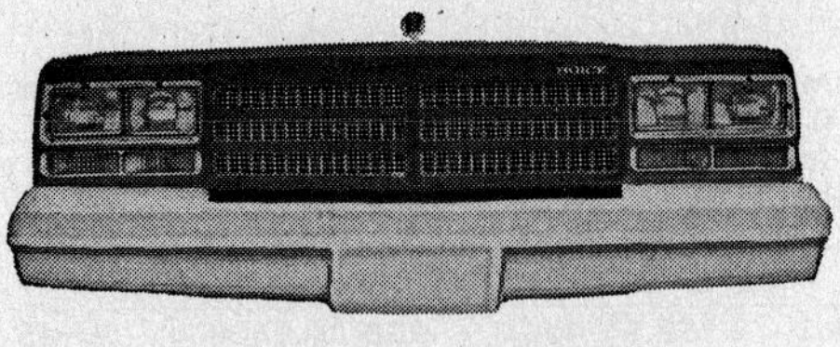

1976 Special & Century 2 Door

1976 Century 4 Door

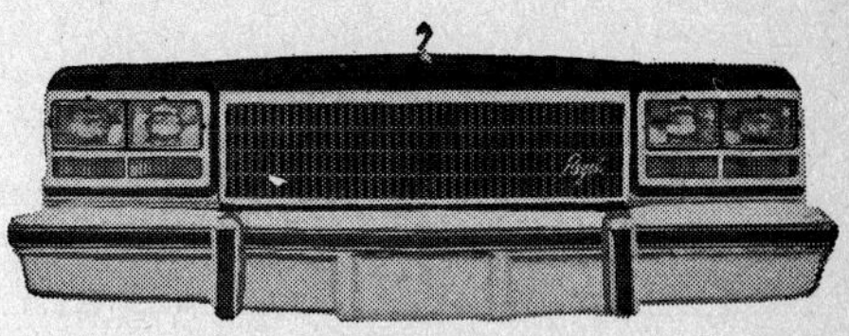

1976 Regal 2 Door

1976 Regal 4 Door

1976 Electra

1976 Riviera

ENGINE IDENTIFICATION

Buick engines are stamped with two different sets of numbers. One is the engine production code which identifies the engine and its approximate production date. The other is the engine serial number which is the same number that is found on the vehicle identification plate. To identify an engine, look for the production code prefix letters, then refer to the following table for its identification.

On Six 250 code is stamped on cylinder block next to distributor.

On 1969-74 V8 350, 400, 430, 455 on left bank cylinder head.

On 1975-76 models the fifth digit in the VIN denotes the engine used.

Year	Engine	Code Prefix
1969	6-250	
	V8-350 2 Bar. Carb.	RO
	V8-350 4 Bar. Carb.	RP
	V8-400	RR
	V8-430	RD
1970	6-250	
	V8-350 2 Bar. Carb.	SO
	V8-350 4 Bar. Carb.	SB, P
	V8-455	SF, R, S
1971	6-250	
	V8-350 2 Bar. Carb.	TC, TO
	V8-350 4 Bar. Carb.	TB, TP
	V8-455	TA, R, S
1972	V8-350 2 Bar. Carb.	WC
	V8-350 4 Bar. Carb.	WB
	V8-455	WF
	V8-455 (Stage I)	WS
	V8-455 (Riviera G.S.)	WA
1973	6-250	
	V8-350 2 Bar. Carb.	XC
	V8-350 4 Bar. Carb.	XB
	V8-455	XF
	V8-455 (Stage 1)	XS
	V8-455 (Riviera G.S.)	XA
1974	6-250 Auto. Trans.	CCW, CCX
	6-250 Manual Trans.	CCR
	V8-350 2 Bar. Carb.	XC
	V8-350 4 Bar. Carb.	XB
	V8-455 4 Bar. Carb.	XF
	V8-455 (Stage 1)	XS
	V8-455 (Riviera G.S.)	XA
1975	V6-231	C
	6-250	D
	V8-260	F
	V8-350 2 Bar. Carb.	H
	V8-350 4 Bar. Carb.	J
	V8-400	S
	V8-455	T
1976	V6-231	C
	V8-260	F
	V8-350 2 Bar. Carb.	H
	V8-350 4 Bar. Carb.	J
	V8-455	T

GENERAL ENGINE SPECIFICATIONS

Year	Engine	Carburetor	Bore and Stroke	Piston Displacement, Cubic Inches	Compression Ratio	Maximum Brake H.P. @ R.P.M.	Maximum Torque Lbs. Ft. @ R.P.M.	Normal Oil Pressure Pounds
1969	155 Horsepower ① 6-250	1 Barrel	3.875 x 3.53	250	8.5	155 @ 4200	235 @ 1600	30–45
	230 Horsepower V8-350	2 Barrel	3.800 x 3.85	350	9.0	230 @ 4400	350 @ 2400	37
	280 Horsepower V8-350	4 Barrel	3.800 x 3.85	350	10.25	280 @ 4600	375 @ 3200	37
	340 Horsepower V8-400	4 Barrel	4.040 x 3.90	400	10.25	340 @ 5000	440 @ 3200	40
	360 Horsepower V8-430	4 Barrel	4.1875 x 3.90	430	10.25	360 @ 5000	475 @ 3200	40

Continued

GENERAL ENGINE SPECIFICATIONS—Continued

Year	Engine	Carburetor	Bore and Stroke	Piston Displacement, Cubic Inches	Compression Ratio	Maximum Brake H.P. @ R.P.M.	Maximum Torque Lbs. Ft. @ R.P.M.	Normal Oil Pressure Pounds
1970	155 Horsepower..............①6-250	1 Barrel	3.875 x 3.53	250	8.5	155 @ 4200	235 @ 1600	30–45
	260 Horsepower..............V8-350	2 Barrel	3.800 x 3.85	350	9.0	260 @ 4600	360 @ 2600	37
	285 Horsepower..............V8-350	4 Barrel	3.800 x 3.85	350	9.0	285 @ 4600	375 @ 3000	37
	315 Horsepower..............V8-350	4 Barrel	3.800 x 3.85	350	10.25	315 @ 4800	410 @ 3200	37
	350 Horsepower..............V8-455	4 Barrel	4.3125 x 3.90	455	10.00	350 @ 4600	510 @ 2800	40
	360 Horsepower..............V8-455	4 Barrel	4.3125 x 3.90	455	10.00	360 @ 4600	510 @ 2800	40
	370 Horsepower..............V8-455	4 Barrel	4.3125 x 3.90	455	10.00	370 @ 4600	510 @ 2800	40
1971	145 Horsepower..............①6-250	1 Barrel	3.875 x 3.53	250	8.5	145 @ 4000	235 @ 2400	30–45
	230 Horsepower..............V8-350	2 Barrel	3.800 x 3.85	350	8.5	230 @ 4400	350 @ 2400	37
	260 Horsepower..............V8-350	4 Barrel	3.800 x 3.85	350	8.5	260 @ 4600	360 @ 3000	37
	315 Horsepower..............V8-455	4 Barrel	4.3125 x 3.90	455	8.5	315 @ 4400	450 @ 2800	40
	330 Horsepower..............V8-455	4 Barrel	4.3125 x 3.90	455	8.5	330 @ 4600	455 @ 2800	40
	345 Horsepower..............V8-455	4 Barrel	4.3125 x 3.90	455	8.5	345 @ 5000	460 @ 3000	40
1972	150 Horsepower②............V8-350	2 Barrel	3.800 x 3.85	350	8.5	150 @ 3800	265 @ 2400	37
	155 Horsepower②............V8-350	2 Barrel	3.800 x 3.85	350	8.5	155 @ 3800	270 @ 2400	37
	175 Horsepower②............V8-350	4 Barrel	3.800 x 3.85	350	8.5	175 @ 3800	270 @ 2400	37
	180 Horsepower②............V8-350	4 Barrel	3.800 x 3.85	350	8.5	180 @ 3800	275 @ 2400	37
	190 Horsepower②............V8-350	4 Barrel	3.800 x 3.85	350	8.5	190 @ 4000	285 @ 2800	37
	195 Horsepower②............V8-350	4 Barrel	3.800 x 3.85	350	8.5	195 @ 4000	290 @ 2800	37
1972	225 Horsepower②............V8-455	4 Barrel	4.3125 x 3.90	455	8.5	225 @ 4000	360 @ 2600	40
	250 Horsepower②............V8-455	4 Barrel	4.3125 x 3.90	455	8.5	250 @ 4000	375 @ 2800	40
	260 Horsepower②............V8-455	4 Barrel	4.3125 x 3.90	455	8.5	260 @ 4400	380 @ 2800	40
	270 Horsepower②............V8-455	4 Barrel	4.3125 x 3.90	455	8.5	270 @ 4400	390 @ 3000	40
1973	100 Horsepower②............6-250①	1 Barrel	3.875 x 3.53	250	8.25	100 @ 3600	175 @ 1600	40
	150 Horsepower②............V8-350	2 Barrel	3.800 x 3.85	350	8.5	150 @ 3800	265 @ 2400	37
	175 Horsepower②............V8-350	4 Barrel	3.800 x 3.85	350	8.5	175 @ 3800	270 @ 2400	37
	190 Horsepower②............V8-350	4 Barrel	3.800 x 3.85	350	8.5	190 @ 4000	285 @ 2800	37
	225 Horsepower②............V8-455	4 Barrel	4.3125 x 3.90	455	8.5	225 @ 4000	360 @ 2600	40
	250 Horsepower②............V8-455	4 Barrel	4.3125 x 3.90	455	8.5	250 @ 4000	375 @ 2800	40
	260 Horsepower②............V8-455	4 Barrel	4.3125 x 3.90	455	8.5	260 @ 4400	380 @ 2800	40
	270 Horsepower②............V8-455	4 Barrel	4.3125 x 3.90	455	8.5	270 @ 4400	390 @ 3000	40
1974	100 Horsepower②............6-250①	1 Barrel	3.875 x 3.53	250	8.25	100 @ 3600	175 @ 1600	40
	150 Horsepower②............V8-350	2 Barrel	3.800 x 3.85	350	8.5	150 @ 3600	270 @ 2000	37
	165 Horsepower②..........V8-350③	2 Barrel	3.800 x 3.85	350	8.5	165 @ 3800	285 @ 2000	37
	175 Horsepower②............V8-350	4 Barrel	3.800 x 3.85	350	8.5	175 @ 3800	260 @ 2000	37
	195 Horsepower②..........V8-350③	4 Barrel	3.800 x 3.85	350	8.5	195 @ 4000	280 @ 2000	37
	175 Horsepower②............V8-455	2 Barrel	4.3125 x 3.90	455	8.5	175 @ 3400	355 @ 2000	40
	190 Horsepower②..........V8-455③	2 Barrel	4.3125 x 3.90	455	8.5	190 @ 3600	370 @ 2000	40
	210 Horsepower②............V8-455	4 Barrel	4.3125 x 3.90	455	8.5	210 @ 3600	335 @ 2200	40
	230 Horsepower②..........V8-455③	4 Barrel	4.3125 x 3.90	455	8.5	230 @ 3800	355 @ 2200	40
	245 Horsepower②..........V8-455④	4 Barrel	4.3125 x 3.90	455	8.5	245 @ 4000	360 @ 2400	40
	255 Horsepower②..........V8-455④	4 Barrel	4.3125 x 3.90	455	8.5	255 @ 4400	370 @ 2800	40
1975	105 Horsepower②............6-250①	1 Barrel	3.875 x 3.53	250	8.25	105 @ 3800	185 @ 1200	40
	110 Horsepower②............V6-231	2 Barrel	3.80 x 3.40	231	8.0	110 @ 4000	175 @ 2000	37
	110 Horsepower②..........V8-260⑤	2 Barrel	3.50 x 3.385	260	8.0	110 @ 3400	205 @ 1600	35
	145 Horsepower②............V8-350	2 Barrel	3.80 x 3.85	350	8.0	145 @ 3200	270 @ 2000	37
	165 Horsepower②............V8-350	4 Barrel	3.80 x 3.85	350	8.0	165 @ 3800	260 @ 2200	37
	190 Horsepower②..........V8-400⑥	4 Barrel	4.12 x 3.75	400	7.6	190 @ 3400	350 @ 2400	55–60
	205 Horsepower②............V8-455	4 Barrel	4.3125 x 3.90	455	7.9	205 @ 3800	345 @ 2000	40

Continued

GENERAL ENGINE SPECIFICATIONS—Continued

Year	Engine	Carburetor	Bore and Stroke	Piston Displacement, Cubic Inches	Compression Ratio	Maximum Brake H.P. @ R.P.M.	Maximum Torque Lbs. Ft. @ R.P.M.	Normal Oil Pressure Pounds
1976	110 Horsepower②.............V6-231	2 Barrel	3.80 x 3.40	231	8.0	110 @ 4000	175 @ 2000	37
	110 Horsepower②...........V8-260⑤	2 Barrel	3.50 x 3.385	260	8.5	110 @ 3400	210 @ 1600	—
	145 Horsepower②...........V8-350⑦	2 Barrel	3.80 x 3.85	350	8.0	145 @ 3200	270 @ 2000	37
	160 Horsepower②...........V8-350⑧	4 Barrel	3.80 x 3.85	350	8.0	160 @ 3800	260 @ 2200	37
	165 Horsepower②.............V8-350	4 Barrel	3.80 x 3.85	350	8.0	165 @ 3800	260 @ 2200	37
	Horsepower②.............V8-455	4 Barrel	4.3125 x 3.90	455	7.9	—	—	40
	Horsepower②.............V8-455	4 Barrel	4.3125 x 3.90	455	7.9	—	—	40
	Horsepower②.............V8-455	4 Barrel	4.3125 x 3.90	455	7.9	—	—	40

①—See Chevrolet chapter for service procedures on this engine.
②—Net Rating—As installed in vehicle.
③—Dual exhaust.
④—Stage 1.
⑤—See Oldsmobile chapter for service procedures on this engine.
⑥—See Pontiac chapter for service procedure on this engine.
⑦—Except California.
⑧—California.

TUNE UP SPECIFICATIONS

The following specifications are published from the latest information available. This data should be used only in the absence of a decal affixed in the engine compartment.

OLD CAR SPECIFICATIONS: For 1946–68 Tune Up Specifications see main index.

★When using a timing light, disconnect vacuum hose or tube at distributor and plug opening in tube or hose so idle speed will not be affected.

●When checking compression, lowest cylinder must be within 80 percent of highest.

▲Before removing wires from distributor cap, determine location of the No. 1 wire in cap, as distributor position may have been altered from that shown at the end of this chart.

Year	Spark Plug		Distributor		Ignition Timing★			Carb. Adjustments					
								Hot Idle Speed		Air Fuel Ratio		Idle CO %	
	Type	Gap Inch	Point Gap Inch	Dwell Angle Deg.	Firing Order Fig. ▲	Timing BTDC ①	Mark Fig.	Std. Trans.	Auto. Trans. ②	Std. Trans.	Auto. Trans.	Std. Trans.	Auto. Trans.
1969													
6-250⑧ Std. Tr.	R46N	.035	.019	32	A	TDC	C	700	—	—	—	—	—
6-250⑧ Auto. Tr.	R46N	.035	.019	32	A	4°	C	—	500D④	—	—	—	—
V8-350	R45TS	.030	.016	30	B	TDC⑨	E	700	600D	—	—	—	—
V8-400 Std. Tr.	R44TS	.030	.016	30	B	2½° ATDC	E	700	—	—	—	—	—
V8-400 Auto. Tr.	R44TS	.030	.016	30	B	TDC	E	—	600D	—	—	—	—
V8-400 G.S. Stage 1	R44TS	.030	.016	30	B	10°	E	700	600D	—	—	—	—
V8-430	R44TS	.030	.016	30	B	TDC	E	—	550D	—	—	—	—
1970													
6-250⑧	R46N	.035	.019	32	A	⑥	C	700	550D	—	—	—	—
V8-350	R45TS	.030	.016	30	B	6°	G	700	600D	—	—	—	—
V8-455	R44TS	.030	.016	30	B	6°⑩	G	700	600D	—	—	—	—
1971													
6-250⑧	R46TS	.035	.019	32	A	4°	D	550	500D	—	—	—	—
V8-350 Std. Tr.	R45TS	.030	.016	30	B	6°	F	800	—	—	—	—	—
V8-350 Auto. Tr.	R45TS	.030	.016	30	B	⑪	F	—	600D	—	—	—	—
V8-455 Std. Tr.	R44TS	.030	.016	30	B	6°	F	700	—	—	—	—	—
V8-455 Auto. Tr.	R44TS	.030	.016	30	B	4°	F	—	600D	—	—	—	—
V8-455 Stage 1	R44TS	.030	.016	30	B	10°	F	700	600D	—	—	—	—

Continued

TUNE UP SPECIFICATIONS—Continued

The following specifications are published from the latest information available. This data should be used only in the absence of a decal affixed in the engine compartment.

OLD CAR SPECIFICATIONS: For 1946–68 Tune Up Specifications see main index.

★When using a timing light, disconnect vacuum hose or tube at distributor and plug opening in tube or hose so idle speed will not be affected.

●When checking compression, lowest cylinder must be within 80 percent of highest.

▲Before removing wires from distributor cap, determine location of the No. 1 wire in cap, as distributor position may have been altered from that shown at the end of this chart.

Year	Spark Plug		Distributor		Ignition Timing★			Carb. Adjustments					
								Hot Idle Speed		Air Fuel Ratio		Idle CO %	
	Type	Gap Inch	Point Gap Inch	Dwell Angle Deg.	Firing Order Fig. ▲	Timing BTDC ①	Mark Fig.	Std. Trans.	Auto. Trans.②	Std. Trans.	Auto. Trans.	Std. Trans.	Auto. Trans.
1972													
V8-350	R45TS	.040	.016	30	B	4°	F	800	650D	—	—	—	—
V8-455	R45TS	.040	.016	30	B	4°	F	900	650D	—	—	—	—
V8-455 Stage 1	R45TS	.040	.016	30	B	⑦	F	900	650D	—	—	—	—
1973													
6-250⑧	R46T	.035	.019	31–34	A	6°	D	700	600D	—	—	—	—
V8-350	R45TS	.040	.016	30	B	4°	F	800	650D	—	—	—	—
V8-455	R45TS	.040	.016	30	B	4°	F	900	650D	—	—	—	—
V8-455 Stage I	R45TS	.040	.016	30	B	10°	F	900	650D	—	—	—	—
1974													
6-250⑧	R46T	.035	.019	31–34	A	③	D	950	600D	—	—	—	—
V8-350	R45TS	.040	.016	29–31	⑭	4°	F	—	650D	—	—	—	—
V8-455	R45TS	.040	.016	29–31	⑭	⑤	F	—	650D	—	—	—	—
1975													
6-250⑧	R46TX	.060	—	—	I	10°	D	800	⑮	—	—	—	—
V6-231	R44SX	.060	—	—	J	12°	K	800	700D	—	—	—	—
V8-260⑫	R46SX	.080	—	—	L	16°	M	—	650D	—	—	—	—
V8-350	R45TSX	.060	—	—	H	12°	F	—	600D	—	—	—	—
V8-400 4 B. Carb.⑬	R45TSX	.060	—	—	N	16°	O	—	650D	—	—	—	—
V8-455	R45TSX	.060	—	—	H	12°	F	—	600D	—	—	—	—
1976													
V6-231	R44SX	.060	—	—	J	12°	K	800	600D	—	—	—	—
V8-260⑫	R46SX	.080	—	—	L	⑯	M	—	650D⑰	—	—	—	—
V8-350	R45TSX	.060	—	—	H	12°	F	—	600D	—	—	—	—
V8-455	R45TSX	.060	—	—	H	12°	F	—	600D	—	—	—	—

①—BTDC: Before top dead center.

②—D: Drive. N: Neutral.

③—Manual trans., 8° BTDC; auto. trans., 6° BTDC.

④—If air conditioned, turn A/C switch to "Full On" position.

⑤—Exc. intermediate model Stage 1 eng., 4° BTDC; intermediate model Stage 1 eng., 10° BTDC.

⑥—Std. trans. TDC: Auto. Trans. 4°.

⑦—Manual trans., 8°; Automatic trans., 10°.

⑧—See Chevrolet chapter for service procedures on this engine.

⑨—5° BTDC for LeSabre.

⑩—Early Gran Sport models set at TDC.

⑪—4 Barrel carburetors and all LeSabres 4°; All others 10°.

⑫—See Oldsmobile chapter for service procedures on this engine.

⑬—See Pontiac chapter for service procedures on this engine.

⑭—Exc. H.E.I., Fig. B; H.E.I., Fig. H.

⑮—Exc. Calif., 550D with A/C On & compressor clutch wires disconnected; Calif., 600D.

⑯—Exc. Calif., 18° BTDC; Calif., 14° BTDC. At 1100 RPM.

⑰—Calif. models with A/C ON & compressor clutch wires disconnected.

Continued

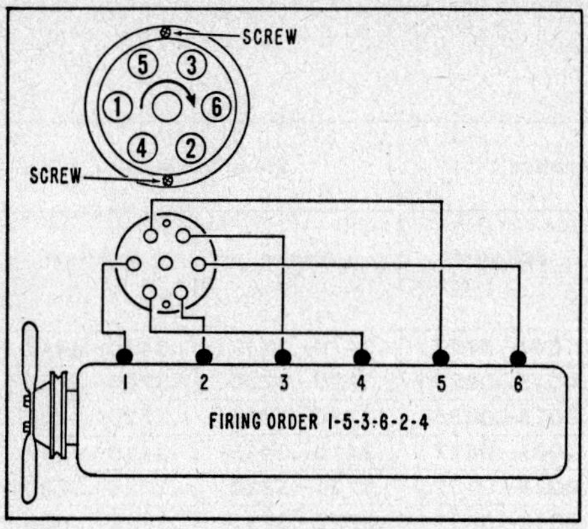

Fig. A

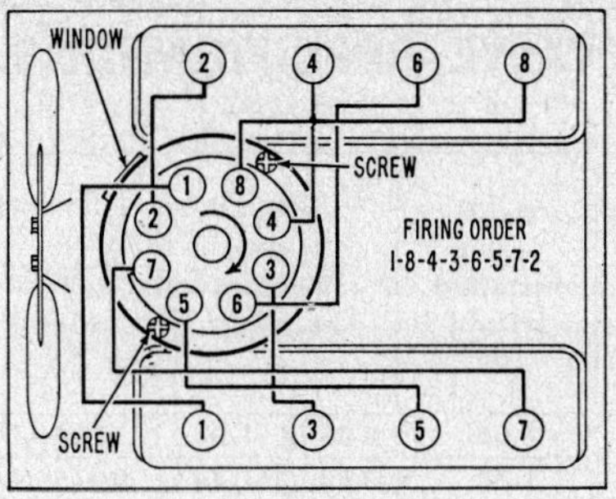

Fig. B

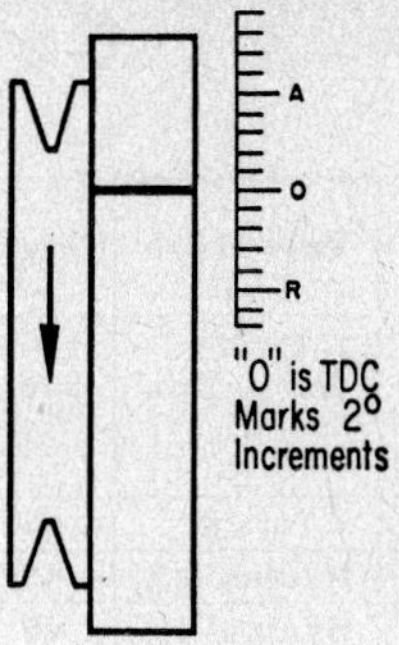

Fig. C

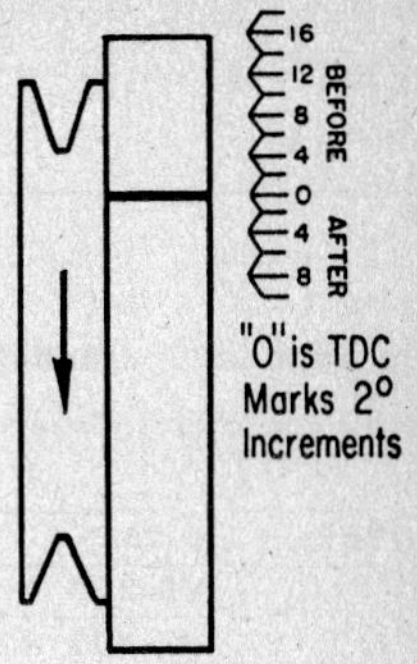

Fig. D

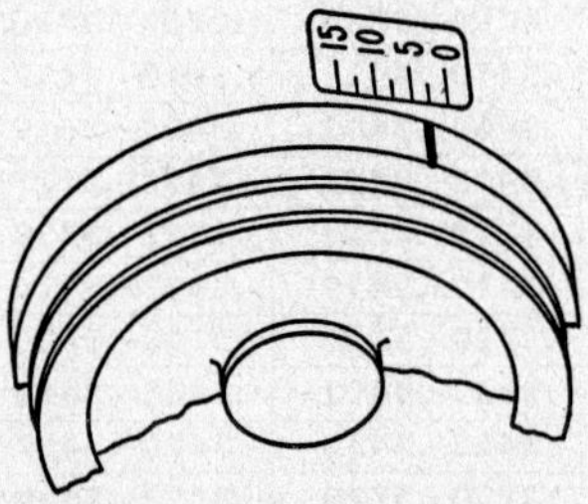
Fig. E

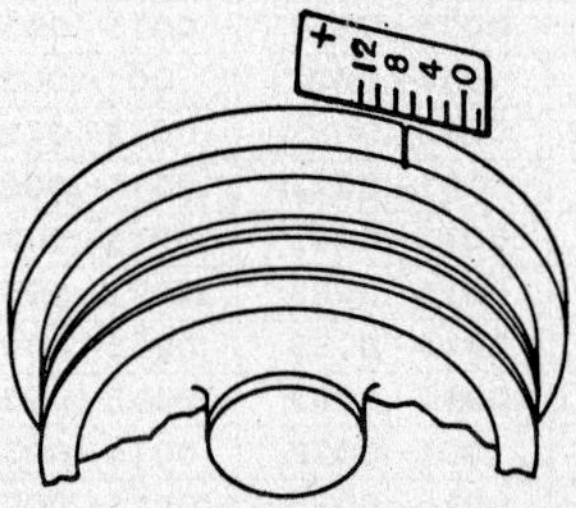
Fig. F

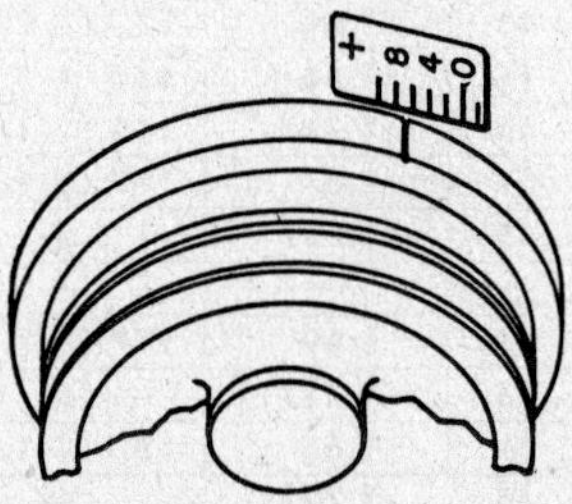
Fig. G

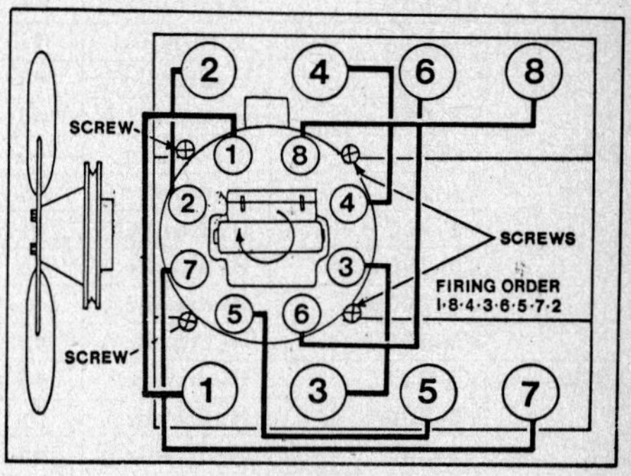

Fig. H

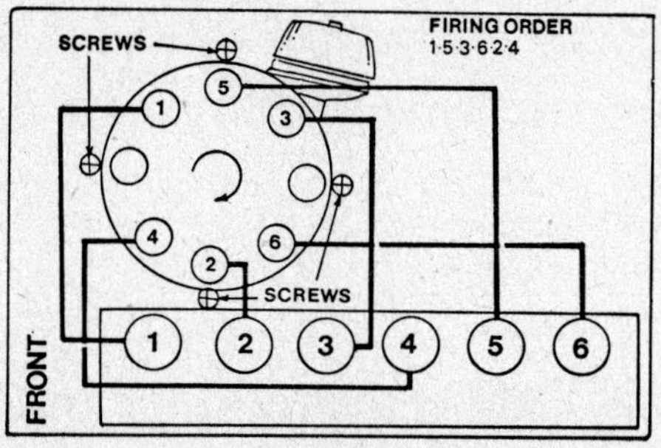

Fig. I

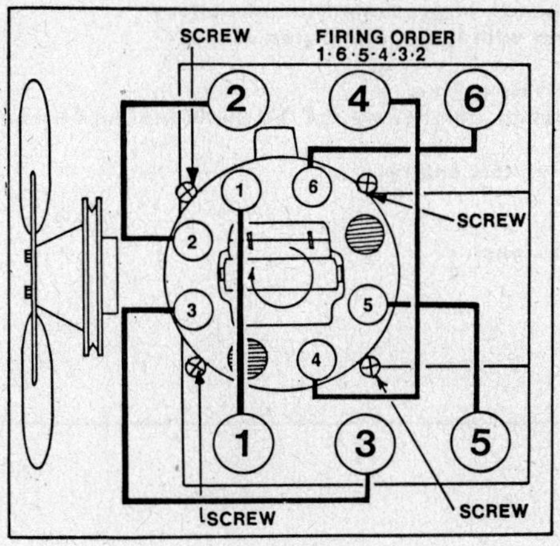

Fig. J

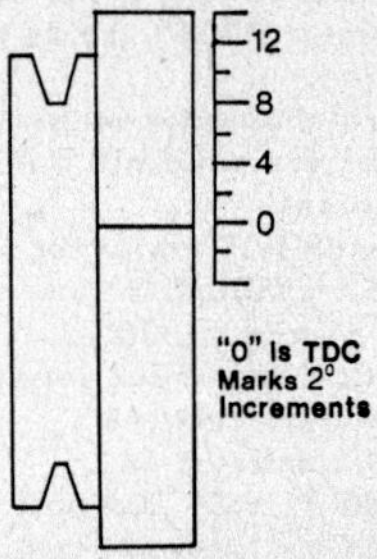

Fig. K

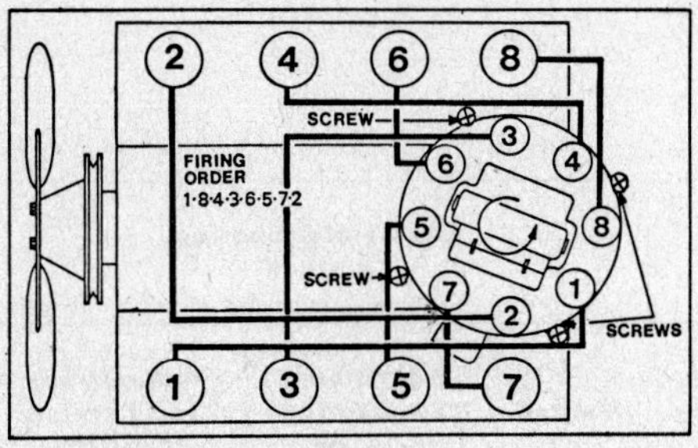

Fig. L

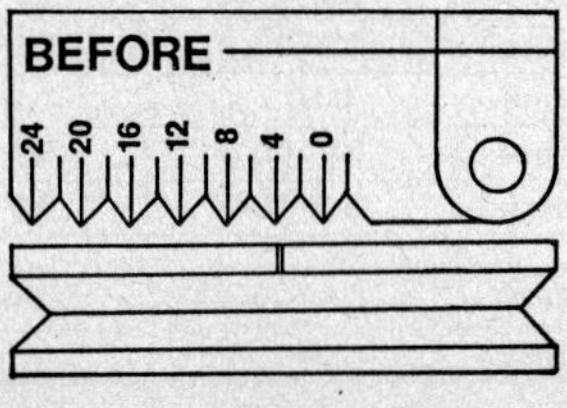

Fig. M

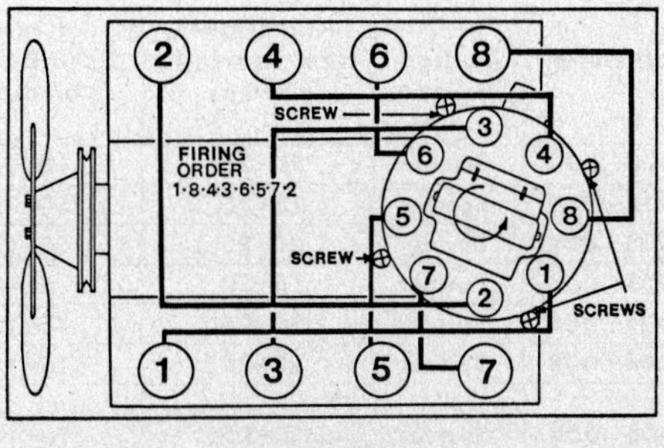

Fig. N

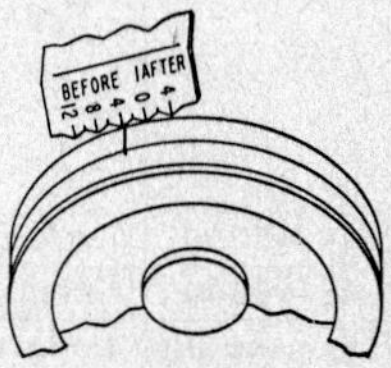

Fig. O

VALVE SPECIFICATIONS

Year	Model	Valve Lash		Valve Angles		Valve Spring Installed Height ③	Valve Spring Pressure Lbs. @ In.	Stem Clearance		Stem Diameter	
		Int.	Exh.	Seat	Face			Intake	Exhaust	Intake	Exhaust
1969	6-250④	1 Turn⑤		46	45	1.66	185 @ 1.27	.001–.0027	.001–.0027	.3410–.3417	.3410–.3417
	V8-350	Hydraulic⑥		45	45	1.72	180 @ 1.34	.0015–.0035	.0015–.0035①	.3720–.3730	.3720–.3730②
	V8-400, 430	Hydraulic⑥		45	45	1.89	177 @ 1.45	.0015–.0035	.0015–.0035①	.3720–.3730	.3720–.3730②
1970–71	6-250④	1 Turn⑤		46	45	1.66	185 @ 1.27	.001–.0027	.001–.0027	.3410–.3417	.3410–.3417
	V8-350	Hydraulic⑥		45	45	1.72	180 @ 1.34	.0015–.0035	.0015–.0035①	.3720–.3730	.3720–.3730②
	V8-455	Hydraulic⑥		45	45	1.89	177 @ 1.45	.0015–.0035	.0015–.0035①	.3720–.3730	.3720–.3730②
1972–74	6-250④	1 Turn⑤		46	45	1.66	186 @ 1.27	.001–.0027	.001–.0027	.3410–.3417	.3410–.3417
	V8-350	Hydraulic⑥		45	45	1.72	180 @ 1.34	.0015–.0035	.0015–.0032①	.3720–.3730	.3723–.3730②
	V8-455	Hydraulic⑥		45	45	1.89	177 @ 1.45	.0015–.0035	.0015–.0032①	.3720–.3730	.3723–.3730②
1975	6-250	1 Turn⑤		46	45	1.66	186 @ 1.27	.001–.0027	.001–.0027	.3410–.3417	.3410–.3417
	V6-231	Hydraulic⑥		45	45	1.727	168 @ 1.327	.0015–.0032	.0015–.0032	.3405–.3412	.3405–.3412
	V8-260⑦	Hydraulic⑥		⑧	⑨	1.67	187 @ 1.27	.001–.0027	.0015–.0032	.3425–.3432	.3420–.3427
	V8-350	Hydraulic⑥		45	45	1.727	⑬	.0015–.0035	.0015–.0032	.3720–.3730	.3723–.3730
	V8-400⑩	Hydraulic⑥		⑪	⑫	—	—	.0016–.0033	.0021–.0038	.3412–.3419	.3407–.3414
	V8-455	Hydraulic⑥		45	45	1.89	177 @ 1.45	.0015–.0035	.0015–.0032	.3720–.3730	.3723–.3730
1976	V6-231	Hydraulic⑥		45	45	1.727	⑭	.0015–.0032	.0015–.0032	.3402–.3412	.3405–.3412
	V8-260⑦	Hydraulic⑥		⑧	⑨	1.67	187 @ 1.27	.001–.0027	.0015–.0032	.3425–.3432	.3420–.3427
	V8-350	Hydraulic⑥		45	45	1.727	⑮	.0015–.0035	.0015–.0032	.3720–.3730	.3723–.3030
	V8-455	Hydraulic⑥		45	45	1.89	177 @ 1.45	.0015–.0035	.0015–.0032	.3720–.3730	.3723–.3730

①—Plus or minus .001″. Guide tapers top to bottom with larger dimension at bottom.
②—Plus or minus .0005″. Guide tapers top to bottom with larger dimension at top.
③—Outer spring.
④—See Chevrolet chapter for service procedures on this engine.
⑤—Turn rocker arm stud nut until all lash is eliminated, then tighten nut the additional turn listed.
⑥—No adjustment.
⑦—See Oldsmobile Chapter for service procedures on this engine.
⑧—Intake, 45°; exhaust, 31°.
⑨—Intake 44°; exhaust, 30°.
⑩—See Pontiac Chapter for service procedure on this engine.
⑪—Intake, 30°; exhaust, 45°.
⑫—Intake, 29°; exhaust 44°.
⑬—Intake, 180 @ 1.34; exhaust, 177 @ 1.45.
⑭—Intake, 164 @ 1.34; exhaust, 182 @ 1.34.
⑮—Intake, 180 @ 1.34; exhaust, 175 @ 1.34.

REAR AXLE SPECIFICATIONS

Year	Model	Carrier Type	Ring Gear & Pinion Backlash		Pinion Bearing Preload			Differential Bearing Preload		
			Method	Adjustment	Method	Adjustment New Bearings Inch-Lbs.	Adjustment Used Bearings Inch-Lbs.	Method	Adjustment New Bearings Inch-Lbs.	Adjustment Used Bearings Inch-Lbs.
1969–70	43-44-45000	Integral	Shims	.006–.008	Spacer	20–25①	10–15①	Shims	35–40②	20–25②
	46-48-49000	Integral	Shims	.007–.009	Spacer	25–30①	10–15①	Shims	35–40②	25–30②
1971–74	All Exc. Apollo	Integral	Shims	.006–.008	Spacer	20–25①	10–15①	Shims	35–40②	20–25②
1973	Apollo	Integral	Shims	.005–.008	Spacer	15–30	5–10	Shims	.010	.010
1974	Apollo	Integral	Shims	.006–.008	Spacer	20–25①	10–15①	Shims	30–40②	20–25②
1975	All	Integral	Shims	.006–.008	Spacer	20–25①	10–15①	Shims	30–40②	20–25②
1976	All Exc. Skylark	Integral	Shims	.006–.008	Spacer	20–25①	10–15①	Shims	—	—
	Skylark	Integral	Shims	.005–.008	Spacer	10–25①	8–12①	Shims	—	—

①—Measured with torque wrench at pinion flange nut.

②—Total preload measured with torque wrench at pinion flange nut with new seal installed.

STARTING MOTOR SPECIFICATIONS

Year	Model	Starter Number	Brush Spring Tension Oz①	Free Speed Test			Resistance Test③	
				Amps.	Volts	R.P.M.	Amps. ①	Volts
1969–70	6-250	1108365	35	49–87②	10.6	6200–10700	290–425②	4.2
	V8-350	1108391	35	55–85②	9	3100–4900	—	—
	V8-400, 430, 455	1108392	35	48–74②	9	4100–6300	—	—
1971	V8-455	1108392	35	45–80②	9	4000–6500	—	—
1971–73	V8-350	1108391	35	80②	9	3500–6000	—	—
1971–74	6-250	1108365	35	50–80②	9	5500–10500	—	—
1972–73	V8-455	1108392	35	45–80②	9	4000–6500	—	—
1973	V8-350	1108503	35	55–80②	9	3500–6000	—	—
1974	V8-350	1108506	35	55–80②	9	3500–6000	—	—
	V8-455	1108507	35	45–80②	9	4000–6500	—	—
1975	6-250	1108366	35	50–80	9	5500–10500	—	—
	V6-231	1108763	35	55–80	9	3500–6000	—	—
	V8-260	230232	35	55–80	9	3500–6000	—	—
	V8-350	1108763	35	55–80	9	3500–6000	—	—
	V8-400	—	35	65–95	9	7500–10500	—	—
	V8-455	1108763	35	45–80	9	4000–6500	—	—
1976	V6-231	1108770	35	50–80	9	5500–10500	—	—
	V8-260	—	35	55–80	9	3500–6000	—	—
	V8-350	1108762	35	55–80	9	3500–6000	—	—
	V8-455	1108763	35	50–80	9	4400–6500	—	—

①—Minimum. ②—Includes solenoid.
③—Check capacity of motor by using a 500 ampere meter and a carbon pile rheostat to control voltage. Apply volts listed across motor with armature locked. Current should be as listed.

DISTRIBUTOR SPECIFICATIONS

★Note: If unit is checked on the vehicle, double the RPM and degrees to get crankshaft figures.

Breaker arm spring tension—19–23.

Distributor Part No.①	Centrifugal Advance Degrees @ RPM of Distributor					Vacuum Advance	
	Advance Starts	Intermediate Advance			Full Advance	Inches of Vacuum to Start Plunger	Max. Adv. Dist. Deg. @ Vacuum
1969							
1110463	0–1 @ 500	9½–11½ @ 975	—	—	17 @ 2100	6–8	12 @ 17
1110464	0–1 @ 500	7½–9½ @ 950	—	—	15 @ 2100	6–8	12 @ 17
1111334	0–1 @ 600	8½–10½ @ 900	—	—	15 @ 2300	6–8	9 @ 16
1111335	0–1 @ 600	9½–11½ @ 900	—	—	17 @ 2300	6–8	9 @ 16
1111938	0–1 @ 475	9½–11½ @ 975	—	—	17 @ 2300	6–8	9 @ 16
1970							
1111962	0–1 @ 600	4½–6½ @ 900	—	—	12 @ 2300	6–8	9 @ 16
1111984	0–1 @ 600	9½–11½ @ 900	—	—	17 @ 2300	6–8	9 @ 16
1111986	0–1 @ 475	9½–11½ @ 975	—	—	17 @ 2300	6–8	9 @ 16
1112006	0–1 @ 600	8½–10½ @ 900	—	—	15 @ 2300	6–8	9 @ 16

Continued

DISTRIBUTOR SPECIFICATIONS—Continued

★Note: If unit is checked on the vehicle, double the RPM and degrees to get crankshaft figures.

Breaker arm spring tension—19–23.

Distributor Part No.①	Centrifugal Advance Degrees @ RPM of Distributor					Vacuum Advance	
	Advance Starts	Intermediate Advance			Full Advance	Inches of Vacuum to Start Plunger	Max. Adv. Dist. Deg. @ Vacuum
1971							
1110489	0–1 @ 500	4–6 @ 1150	—	—	11 @ 2050	6–8	12 @ 16
1112006	0–1 @ 550	3½–5½ @ 1000	—	—	10 @ 2300	6–8	9 @ 16
1112037	0–1 @ 800	2½–3½ @ 925	—	—	7 @ 2300	6–8	9 @ 16
1112077	0–1 @ 800	7–9 @ 1000	—	—	10 @ 1500	6–8	10 @ 16
1112080	0–1 @ 800	6–8 @ 1050	—	—	9 @ 1450	6–8	9 @ 16
1972							
1112016	0 @ 350	4½–6½ @ 900	—	—	12 @ 2300	6–8	9 @ 16
1112109	0 @ 400	4½–6½ @ 900	—	—	8 @ 1500	6–8	9 @ 16
1112110	0 @ 500	6–8 @ 1050	—	—	9 @ 1450	6–8	9 @ 13
1973							
1112016	0 @ 350	4½–6½ @ 900	—	—	12 @ 2300	6–8	9 @ 16
1112109	0 @ 375	5½–7½ @ 900	—	—	9 @ 1450	6–8	9 @ 16
1112110	0 @ 375	6–8 @ 1050	—	—	10 @ 1500	6.5–8.5	9.5 @ 13
1110499	0 @ 550	7 @ 1150	—	—	11½ @ 2100	6	11 @ 14
1974							
1110499	0–1 @ 550	7 @ 1150	—	—	10½ @ 2050	6	11 @ 14
1112520②	0–1 @ 450	7–9 @ 1200	—	—	10 @ 1500	6.5–8.5	9.5 @ 13
1112521②	0–1 @ 400	4½–6½ @ 900	—	—	12 @ 2300	6–8	9 @ 16
1112541	0–1 @ 450	5½–7½ @ 900	—	—	12 @ 1800	6–8	9 @ 16
1112542	0–1 @ 450	7–9 @ 1050	—	—	14 @ 2050	6.5–8.5	9.5 @ 13
1112802②	0–1 @ 450	5½–7½ @ 900	—	—	12 @ 1800	6–8	9 @ 16
1112803②	0–1 @ 450	7–9 @ 1050	—	—	14 @ 2050	6.5–8.5	9.5 @ 13
1975							
V6-231②	0 @ 500	5 @ 1000	—	—	8 @ 2050	5–7	9 @ 10
6-250②	0 @ 550	3½ @ 1150	—	—	8 @ 2100	4	③
V8-260②	0 @ 325	9½ @ 1200	—	—	14 @ 2200	4	12 @ 15
V8-350②	—	2–4 @ 1050	—	—	7 @ 2250	6½–8½	8 @ 11½
V8-400②	0 @ 600	2 @ 700	—	—	8 @ 2200	6–8	12½ @ 13½
V8-455②	—	4½–6 @ 1500	—	—	9 @ 2200	4–6	10 @ 11
1976							
V6-231②	0 @ 500	5 @ 1800	—	—	8 @ 2050	5–7	9 @ 10
V8-260②	0 @ 325	9½ @ 1200	—	—	14 @ 2200	4	12 @ 15
V8-350②	0 @ 375	4 @ 600	—	—	7 @ 2250	6.5–8.5	8 @ 11½
V8-455②	0 @ 375	6 @ 1800	—	—	9 @ 2200	4–6	10 @ 11

①—Stamped on distributor housing plate.
②—High Energy Ignition.
③—Exc. Calif., 9 @ 12; Calif., 9 @ 15.

ALTERNATOR & REGULATOR SPECIFICATIONS

Year	Alternator					Regulator						
				Cold Output @ 14 Volts			Field Relay			Voltage Regulator		
	Model	Rated Hot Output Amps.	Field Current 12 Volts @ 80 F.	2000 R.P.M. Amps.	5000 R.P.M. Amps.	Model	Air Gap In.	Point Gap In.	Closing Voltage	Air Gap In.	Point Gap In.	Voltage @ 125° F.
1969	1100762	37	2.2–2.6	25	35	1119515	.015	.030	1.5–3.2	.067	.014	13.6–14.4
	1100761	37	2.2–2.6	25	35	1119515	.015	.030	1.5–3.2	.067	.014	13.6–14.4
	1100802	55	2.2–2.6	32	50	1119515	.015	.030	1.5–3.2	.067	.014	13.6–14.4
	1100691	42	2.2–2.6	28	40	1119515	.015	.030	1.5–3.2	.067	.014	13.6–14.4
	1100774	55	2.2–2.6	32	50	1119515	.015	.030	1.5–3.2	.067	.014	13.6–14.4
1970	1100761	37	2.2–2.6	7①	29②	1119515	.015	.030	1.5–3.2	.067	.014	13.6–14.4
	1100888	37	2.2–2.6	7①	29②	1119515	.015	.030	1.5–3.2	.067	.014	13.6–14.4
	1100691	42	2.2–2.6	9①	32②	1119515	.015	.030	1.5–3.2	.067	.014	13.6–14.4
	1100774	55	2.2–2.6	9①	44②	1119515	.015	.030	1.5–3.2	.067	.014	13.6–14.4
	1100892	55	2.2–2.6	9①	44②	1119515	.015	.030	1.5–3.2	.067	.014	13.6–14.4
	1100860	61	2.2–2.6	12①	47②	1119515	.015	.030	1.5–3.2	.067	.014	13.6–14.4
1971	1100891	—	2.2–2.6	—	—	1119515	.015	.030	1.5–3.2	.067	.014	13.6–14.4
	1100905	37	2.2–2.6	7①	29②	1119515	.015	.030	1.5–3.2	.067	.014	13.6–14.4
	1100924	55	4.0–4.5	9①	44②	1116384	—	—	—	—	—	13.5–14.5
	1100926	42	4.0–4.5	9①	32②	1116384	—	—	—	—	—	13.5–14.5
	1100931	42	2.2–2.6	9①	44②	1119515	.015	.030	1.5–3.2	.067	.014	13.6–14.4
	1100932	—	2.2–2.6	12①	47②	1119515	.015	.030	1.5–3.2	.067	.014	13.6–14.4
	1100933	63	2.8–3.2	13①	51②	1119519	.015	.030	1.5–3.2	.067	.014	—
	1100943	42	2.2–2.6	9①	32②	1119515	.015	.030	1.5–3.2	.067	.014	13.6–14.4
1972	1102449	37	2.2–2.6	7①	29②	1119515	.015	.030	1.5–3.2	.067	.014	13.5–14.5
	1102443	42	2.2–2.6	9①	32②	1119515	.015	.030	1.5–3.2	.067	.014	13.5–14.5
	1102448	55	2.2–2.6	9①	44②	1119515	.015	.030	1.5–3.2	.067	.014	13.5–14.5
	1102442	55	2.2–2.6	9①	44②	1119515	.015	.030	1.5–3.2	.067	.014	13.5–14.5
	1102450	61	2.2–2.6	12①	47②	1119515	.015	.030	1.5–3.2	.067	.014	13.5–14.5
	1102447	63	2.8–3.2	13①	51②	1119519	.015	.030	1.5–3.2	.067	.014	13.5–14.5
1973	1100497	37	4.0–4.5	13③	37	Integral	—	—	—	—	—	13.5–14.5
	1100925	63	4.0–4.5	36③	58	Integral	—	—	—	—	—	13.5–14.5
	1100926	42	4.0–4.5	15③	37	Integral	—	—	—	—	—	13.5–14.5
	1100946	55	4.0–4.5	19③	50	Integral	—	—	—	—	—	13.5–14.5
	1100947	37	4.0–4.5	12③	32	Integral	—	—	—	—	—	13.5–14.5
	1100948	61	4.0–4.5	24③	55	Integral	—	—	—	—	—	13.5–14.5
	1101018	80	4.0–4.5	40③	74	Integral	—	—	—	—	—	13.5–14.5
1974	1100497	37	4.0–4.5	13③	37	Integral	—	—	—	—	—	13.5–14.5
	1100925	63	4.0–4.5	36③	58	Integral	—	—	—	—	—	13.5–14.5
	1100926	42	4.0–4.5	15③	37	Integral	—	—	—	—	—	13.5–14.5
	1100946	55	4.0–4.5	—	50	Integral	—	—	—	—	—	13.5–14.5
	1100947	37	4.0–4.5	12③	32	Integral	—	—	—	—	—	13.5–14.5
	1100948	61	4.0–4.5	24③	55	Integral	—	—	—	—	—	13.5–14.5
	1101018	80	4.0–4.5	—	74	Integral	—	—	—	—	—	13.5–14.5
1975	1101016	80	4.4–4.9	—	80④	Integral	—	—	—	—	—	—
	1101024	80	4.4–4.9	—	80④	Integral	—	—	—	—	—	—
	1101031	80	4.4–4.9	—	80④	Integral	—	—	—	—	—	—
	1102388	37	4.0–4.5	—	37④	Integral	—	—	—	—	—	13.6–14.2
	1102389	42	4.0–4.5	—	42④	Integral	—	—	—	—	—	13.6–14.2
	1102390	55	4.0–4.5	—	55④	Integral	—	—	—	—	—	13.6–14.2
	1102391	61	4.0–4.5	—	61④	Integral	—	—	—	—	—	13.6–14.2
	1102392	63	4.0–4.5	—	63④	Integral	—	—	—	—	—	—
	1102394	37	4.0–4.5	—	37④	Integral	—	—	—	—	—	13.6–14.2
	1102399	37	4.0–4.5	—	37④	Integral	—	—	—	—	—	—
	1102400	42	4.0–4.5	—	42④	Integral	—	—	—	—	—	—

Continued

ALTERNATOR & REGULATOR SPECIFICATIONS—Continued

Year	Alternator					Regulator						
				Cold Output @ 14 Volts			Field Relay			Voltage Regulator		
	Model	Rated Hot Output Amps.	Field Current 12 Volts @ 80 F.	2000 R.P.M. Amps.	5000 R.P.M. Amps.	Model	Air Gap In.	Point Gap In.	Closing Voltage	Air Gap In.	Point Gap In.	Voltage @ 125° F.
1975	1102457	55	4.0–4.5	—	55④	Integral	—	—	—	—	—	—
	1102460	61	4.0–4.5	—	61④	Integral	—	—	—	—	—	—
	1102461	63	4.0–4.5	—	63④	Integral	—	—	—	—	—	—
	1102467	63	4.0–4.5	—	63④	Integral	—	—	—	—	—	—
	1102478	55	4.0–4.5	—	55④	Integral	—	—	—	—	—	—
	1102483	37	4.0–4.5	—	37④	Integral	—	—	—	—	—	—
	1102488	55	4.0–4.5	—	55④	Integral	—	—	—	—	—	—
	1102491	37	4.0–4.5	—	37④	Integral	—	—	—	—	—	—
	1102494	55	4.0–4.5	—	55④	Integral	—	—	—	—	—	—
	1102495	55	4.0–4.5	—	55④	Integral	—	—	—	—	—	13.6–14.2
	1102497	37	4.0–4.5	—	37④	Integral	—	—	—	—	—	—
	1102861	42	4.0–4.5	—	42④	Integral	—	—	—	—	—	—
1976	1101024	80	4.4–4.9	—	80④	Integral	—	—	—	—	—	—
	1102388	37	4.0–4.5	—	37④	Integral	—	—	—	—	—	—
	1102389	42	4.0–4.5	—	42④	Integral	—	—	—	—	—	—
	1102390	55	4.0–4.5	—	55④	Integral	—	—	—	—	—	—
	1102391	61	4.0–4.5	—	61④	Integral	—	—	—	—	—	—
	1102392	63	4.0–4.5	—	63④	Integral	—	—	—	—	—	—
	1102394	37	4.0–4.5	—	37④	Integral	—	—	—	—	—	—
	1102495	55	4.0–4.5	—	55④	Integral	—	—	—	—	—	—
	1102840	55	4.0–4.5	—	55④	Integral	—	—	—	—	—	—

①—At 500 engine R.P.M. ②—At 1500 engine R.P.M. ③—At 600 engine R.P.M. ④—At 5500 RPM & 80° F.

PISTONS, PINS, RINGS, CRANKSHAFT & BEARINGS

Year	Engine	Piston Clearance	Ring End Gap①		Wrist-pin Diameter	Rod Bearings		Main Bearings			
			Comp.	Oil		Shaft Diameter	Bearing Clearance	Shaft Diameter	Bearing Clearance	Thrust on Bear. No.	Shaft End Play
1969	6-250②	.0005–.0011	.010	.015	.9271	1.999–2.000	.0007–.0027	2.3004	.0003–.0029	7	.002–.006
	V8-350	.0008–.0014	.013	.015	.9393	2.0000	.0002–.0023	2.9995	.0004–.0015	3	.003–.009
	V8-400, 430	.0007–.0013	.013	.015	.9993	2.249–2.250	.0002–.0023	3.2500	.0007–.0018	3	.003–.009
1970	6-250②	.0005–.0011	.010	.015	.9271	1.999–2.000	.0007–.0027	2.3004	.0003–.0029	7	.002–.006
	V8-350	.0008–.0014	.013	.015	.9393	2.0000	.0002–.0023	2.9995	.0004–.0015	3	.003–.009
	V8-455	.0010–.0022	.013	.015	.9993	2.249–2.250	.0002–.0023	3.2500	.0007–.0018	3	.003–.009
1971–74	6-250②	.0005–.0015	.010	.015	.927	1.999–2.000	.0007–.0027	2.3004	.0003–.0029	7	.002–.006
	V8-350	.0008–.0020	.010	.015	.9393	2.000	.0002–.0023	③	.0004–.0015	3	.003–.009
	V8-455	.0010–.0016	.013	.015	.9993	2.249–2.250	.0002–.0023	3.250	.0007–.0018	3	.003–.009
1975	6-250②	.0005–.0015	.010	.015	.927	1.999–2.000	.0007–.0027	2.3004	.0003–.0029	7	.002–.006
	V6-231	.0008–.0020	.010	.015	.9393	1.991–2.000	.0005–.0026	2.4995	.0004–.0015	2	.004–.008
	V8-260④	.001–.002	.010	.015	.9805	2.1238–2.1248	.0004–.0033	⑤	.0005–.0021⑥	3	.004–.008
	V8-350	.0008–.0014	.010	.015	.9393	1.991–2.000	.0005–.0026	2.9995	.0004–.0015	3	.003–.009
	V8-400⑦	.003–.005	⑧	.035	.9802	2.25	.0005–.0025	3.00	.0002–.0017	4	.003–.009
	V8-455	.0007–.0013	.013	.015	.9991	2.487–2.495	.0005–.0026	3.250	.0007–.0018	3	.003–.009

Continued

PISTONS, PINS, RINGS, CRANKSHAFT & BEARINGS—Continued

Year	Engine	Piston Clearance	Ring End Gap(1)		Wrist-pin Diameter	Rod Bearings		Main Bearings			
			Comp.	Oil		Shaft Diameter	Bearing Clearance	Shaft Diameter	Bearing Clearance	Thrust on Bear. No.	Shaft End Play
1976	V6-231	.0008–.0014	.013	.015	.9393	2.000	.0002–.0023	2.4995	.0004–.0015	2	.004–.008
	V8-260(4)	.001–.002	.010	.015	.9805	2.1238–2.1248	.0004–.0033	(5)	.0005–.0021(6)	3	.004–.008
	V8-350	.0008–.0014	.013	.015	.9393	1.9991–2.000	.0005–.0026	2.9995	.0004–.0015	3	.003–.009
	V8-455	.0010–.0016	.013	.015	.9993	2.2487–2.2495	.0005–.0026	3.250	.0007–.0018	3	.003–.009

(1)—Fit rings in tapered bores for clearance given in tightest portion of ring travel. Clearances specified are minimum gaps.
(2)—See Chevrolet Chapter for service procedures on this engine.
(3)—1971–73, 2.9995 inch; 1974, 3.00 inch.
(4)—See Oldsmobile Chapter for service procedures on this engine.
(5)—No. 1: 2.4988–2.4998 inch; Nos. 2, 3, 4, 5: 2.4985–2.4995 inch.
(6)—Rear, .0015–.0031 inch.
(7)—See Pontiac Chapter for service procedures on this engine.
(8)—No. 1, .019; No. 2, .015.

ENGINE TIGHTENING SPECIFICATIONS★

★Torque specifications are for clean and lightly lubricated threads only. Dry or dirty threads produce increased friction which prevents accurate measurement of tightness.

Year	Engine	Spark Plugs Ft. Lbs. (3)	Cylinder Head Bolts Ft. Lbs.	Intake Manifold Ft. Lbs.	Exhaust Manifold Ft. Lbs.	Rocker Arm Shaft Bracket Ft. Lbs.	Rocker Arm Cover Ft. Lbs.	Connecting Rod Cap Bolts Ft. Lbs.	Main Bearing Cap Bolts Ft. Lbs.	Flywheel to Crankshaft Ft. Lbs.	Vibration Damper or Pulley Ft. Lbs.
1969	6-250	25	95	(1)	(1)	—	5	35	65	60	(2)
	V8-350	15	75	55	18	30	4	35	95	60	140 min.
	V8-400, 430	15	100	55	18	30	4	45	110	60	200 min.
1970	6-250	25	95	(1)	(1)	—	5	35	65	60	(2)
	V8-350	15	75	55	18	25	4	35	95	60	120 min.
	V8-455	15	100	55	18	25	4	45	110	58	200 min.
1971–72	V8-350	15	75	55	18	25	4	35	95	60	120 min.
	V8-455	15	100	65	18	25	4	45	110	58	200 min.
1973	6-250	15	95	35	30	—	4	35	65	60	(2)
	V8-350	15	80	55	18	30	4	35	115	60	140 min.
	V8-455	15	100	65	18	30	4	45	115	60	200 min.
1974	6-250	15	95	35	30	—	4	35	65	60	(2)
	V8-350	15	80	45	28	30	4	40	115	60	140 min.
	V8-455	15	100	45	28	30	4	45	115	60	200 min.
1975	6-250	15	95	35	30	—	4	35	65	60	60
	V6-231	15	75	45	25	30	5	40	115	55	150 min.
	V8-260	15	85	40	25	—	7	42	(4)	60	255
	V8-350	15	95	45	28	30	4	40	115	60	150
	V8-400	15	95	40	30	—	8	45	(5)	95	160
	V8-455	15	100	45	28	30	4	45	115	60	200
1976	V6-231	20	80	45	25	30	4	40	115	60	175
	V8-260	20	85	40	25	—	7	42	(4)	60	255
	V8-350	20	80	45	25	30	4	40	115	60	175
	V8-455	20	100	45	25	30	4	45	115	60	225

(1)—End clamps 20, center bolts 30.
(2)—Pressed on.
(3)—Dry threads.
(4)—Nos. 1, 2, 3, 4–80 ft. lbs., No. 5–120 ft. lbs.
(5)—Rear Main, 120 ft. lbs.; all others, 100 ft. lbs.

WHEEL ALIGNMENT SPECIFICATIONS

OLD CAR SPECIFICATIONS: For 1946–68 Wheel Alignment Specifications see main index.

Year	Model	Caster Angle, Degrees: Limits	Caster Angle, Degrees: Desired	Camber Angle, Degrees: Limits: Left	Camber Angle, Degrees: Limits: Right	Camber Angle, Degrees: Desired: Left	Camber Angle, Degrees: Desired: Right	Toe-In, Inch	Toe-Out on Turns, Deg.: Outer Wheel	Toe-Out on Turns, Deg.: Inner Wheel
1969–70	Intermediates	−1 to 0	−½	0 to +1	0 to +1	+½	+½	⅛–¼	18½	20
	Le Sabre	+¼ to +1¼	+¾	−½ to +½	−½ to +½	0	0	7/32–5/16	19½	20
	Estate Wagon	+¼ to +1¼	+¾	−½ to +½	−½ to +½	0	0	7/32–5/16	19½	20
	Wildcat	+¼ to +1¼	+¾	−½ to +½	−½ to +½	0	0	7/32–5/16	19½	20
	Electra 225	+¼ to +1¼	+¾	−½ to +½	−½ to +½	0	0	7/32–5/16	19½	20
	Riviera	+½ to +1½	+1	−¼ to +¾	−¼ to +¾	+¼	+¼	5/32–¼	16¾	20
1971	Intermediates	−1 to 0	−½	0 to +1	0 to 1	+½	+½	⅛–¼	18½	20
	Others	+½ to +1½	+1	−¼ to +¾	−¼ to +¾	+¼	+¼	⅛–¼	18½	20
1972	Intermediates	−1 to 0	−½	0 to +1	0 to +1	+½	+½	⅛–¼	18½	20
	Others	+½ to +1½	+1	0 to +1	0 to +1	+½	+½	⅛–¼	18½	20
1973	Apollo	0 to +1	+½	−¼ to +¾	−¼ to +¾	+¼	+¼	⅛–¼	—	—
	Intermediates①	−½ to −1½	−1	+½ to +1½	0 to +1	+1	+½	0–⅛	③	20
	Intermediates②	−½ to +½	0	+½ to +1½	0 to +1	+1	+½	0–⅛	③	20
	Others	+½ to +1½	+1	+½ to +1½	0 to +1	+1	+½	0–⅛	18½	20
1974	Apollo	0 to +1	+½	−¼ to +¾	−¼ to +¾	+¼	+¼	⅛–¼	—	20
	Intermediates	−½ to +½	0	+½ to +1½	0 to +1	+1	+½	0–⅛	③	20
	Others	+½ to +1½	+1	+½ to +1½	0 to +1	+1	+½	0–⅛	18½	20
1975	Apollo, Skylark①	−½ to −1½	−1	+¼ to +1¼	+¼ to +1¼	+¾	+¾	0–⅛	18½	—
	Apollo, Skylark②	+½ to +1½	+1	+¼ to +1¼	+¼ to +1¼	+¾	+¾	0–⅛	—	—
	Century	+1½ to +2½	+2	+½ to +1½	0 to +1	+1	+½	0–⅛	③	20
	Others	+½ to +2½	+1½	+½ to +1½	0 to +1	+1	+½	0–⅛	18½	20
1976	Skylark①	−½ to −1½	−1	+¼ to +1¼	+¼ to +1¼	+¾	+¾	0–⅛	—	—
	Skylark	+½ to +1½	+1	+¼ to +1¼	+¼ to +1¼	+¾	+¾	0–⅛	—	—
	Century, Regal	+1½ to +2½	+2	+½ to +1½	0 to +1	+1	+½	0–⅛	③	20
	Others	+1 to +2	+1½	+½ to +1½	0 to +1	+1	+½	0–⅛	18½	20

①—Manual steering. ②—Power steering.
③—Manual steering & all Sta. Wag.—right turn, $19\frac{3}{16}°$; left turn, $18\frac{13}{16}°$: power steering Exc. Sta. Wag.—right turn, 19°; left turn, $18\frac{11}{16}°$.

BRAKE SPECIFICATIONS

Year	Model	Brake Drum Inside Diameter	Wheel Cylinder Bore Diameter: Front Disc Brake	Wheel Cylinder Bore Diameter: Front Drum Brake	Wheel Cylinder Bore Diameter: Rear Drum Brake	Master Cylinder Bore Diameter: With Disc Brakes	Master Cylinder Bore Diameter: With Drum Brakes	Master Cylinder Bore Diameter: With Power Brakes
1969	Special, Skylark	9.495–9.505	—	1⅛	⅞	1⅛	1.00	①
	Sportwagon	9.495–9.505	—	1⅛	1.00	1⅛	1.00	①
	GS-350, GS-400	9.495–9.505	—	1⅛	⅞	1⅛	1.00	①
	LeSabre, Wildcat, Electra	11.997–12.022	2 15/16	1 3/16	1.00	1⅛	1.00	①
	Riviera	11.997–12.022	2 15/16	1 3/16	15/16	1⅛	1.00	①
1970	Skylark, GS, GS-455	9.495–9.505	—	1⅛	⅞	1⅛	1.00	①
	Sportwagon	9.495–9.505	—	1⅛	⅞	1⅛	1.00	①
	Le Sabre, Wildcat, Electra, Wagon	11.997–12.022	2 15/16	1 3/16	1.00	1⅛	1.00	①
	Riviera	11.997–12.022	2 15/16	1 3/16	15/16	1⅛	1.00	①
1971–72	Skylark, GS, GS-455	9.495–9.505	—	1⅛	⅞	1⅛	1.00	①
	Sportwagon	9.495–9.505	—	1⅛	⅞	1⅛	1.00	①
	Le Sabre, Centurion, Electra	10.997–11.007	2 15/16	—	15/16	1⅛	—	①
	Estate Wagon	11.997–12.007	2 15/16	—	15/16	1⅛	—	①
	Riviera	10.997–11.007	2 15/16	—	15/16	1⅛	—	①

Continued

BRAKE SPECIFICATIONS—Continued

Year	Model	Brake Drum Inside Diameter	Wheel Cylinder Bore Diameter			Master Cylinder Bore Diameter		
			Front Disc Brake	Front Drum Brake	Rear Drum Brake	With Disc Brakes	With Drum Brakes	With Power Brakes
1973–74	Apollo	9.50	2 15/16	1⅛	⅞	1⅛	1.00	①
	Century, Luxus, Regal	9.495–9.505	2 15/16	—	⅞	1⅛	—	1⅛
	Century Wagon	10.997–11.007	2 15/16	—	15/16	1⅛	—	1⅛
	LeSabre, Centurion, Electra, Riviera	10.997–11.007	2 15/16	—	15/16	1⅛	—	1⅛
	Estate Wagon	11.997–12.007	2 15/16	—	1	1⅛	—	1⅛
1975	Apollo, Skylark	9.495–9.505	2 15/16	—	⅞	1⅛	—	1⅛
	Century, Regal, Custom	9.495–9.505	2 15/16	—	⅞	1⅛	—	1⅛
	Century Wagon	10.997–11.007	2 15/16	—	15/16	1⅛	—	1⅛
	LeSabre	10.997–11.007	2 15/16	—	15/16	1⅛	—	1⅛
	Electra, Riviera	10.997–11.007	2 15/16	—	15/16	1⅛	—	1⅛
	Estate Wagon	11.997–12.007	2 15/16	—	1	1⅛	—	1⅛
1976	Skylark	9.495–9.505	2 15/16	—	⅞	1⅛	—	1⅛
	Century, Regal	10.997–11.007	2 15/16	—	③	15/16	—	1⅛
	Century Wagon	10.997–11.007	2 15/16	—	15/16	15/16	—	1⅛
	LeSabre	10.997–11.007	2 15/16	—	1	15/16	—	1⅛
	Electra, Riviera	10.997–11.007	2 15/16	—	1	15/16	—	1⅛
	Estate Wagon	11.997–12.007	2 15/16	—	1	15/16	—	1⅛

①—Drum brakes 1", disc brakes 1⅛".

②—Manual brakes, 1.00 inch; power brakes, 1⅛ inch.

③—Manual brakes, 1.00 inch; power brakes, 15/16 inch.

COOLING SYSTEM & CAPACITY DATA

Year	Model or Engine	Cooling Capacity, Qts.			Radiator Cap Relief Pressure, Lbs.		Thermo. Opening Temp.	Fuel Tank Gals.	Engine Oil Refill Qts. ①	Transmission Oil			Rear Axle Oil Pints
		No Heater	With Heater	With A/C	With A/C	No A/C				3 Speed Pints	4 Speed Pints	Auto. Trans. Qts. ⑬	
1969	6-250	10.0	11.3	13.0	15	15	195	20	4	3⅜	—	⑫	2.9
	V8-350⑪	12.6	13.5	13.5	15	15	190	20	4	3⅜	—	⑭	2.9
	GS-350	12.6	13.5	13.5	15	15	190	20	4	3.4	3	⑮	2.9
	GS-400	15.3	16.2	16.7	15	15	190	20	4	3.5	3	⑯	2.9
	V8-350⑩	12.3	13.2	13.6	15	15	190	25	4	3.5	—	⑰	2.9
	V8-430	16.0	16.7	17.0	15	15	190	25	4	—	—	⑯	4¼
	Riviera	16.0	16.7	17.0	15	15	190	21	4	—	—	⑯	4¼
1970	6-250	—	16.04	16.04	15	15	190	20	4	3½	3	⑮	3
	V8-350⑪	—	16.45	16.52	15	15	190	20⑱	4	3½	3	⑮	3
	GS-455	—	19.17	19.67	15	15	190	20	4	4	3	⑯	3
	V8-350⑩	—	16.20	16.55	15	15	190	25⑲	4	3½	—	⑮	3
	V8-455	—	19.70	20.0	15	15	190	25⑲	4	3½	—	⑯	4¼
	Riviera	—	19.70	20.0	15	15	190	21	4	—	—	⑯	4¼
1971	6-250	—	16.04	16.04	15	15	195	20	4	3½	—	⑮	4¼
	V8-350⑪	—	16.5	16.5	15	15	190	20⑱	4	3½	3	⑮	4¼
	GS-455	—	19.0	19.5	15	15	190	20	4	3½	3	⑯	4¼
	V8-350⑩	—	16.2	16.55	15	15	190	25⑲	4	3½	—	⑮	4¼
	V8-455	—	19.7	20.0	15	15	190	25⑲	4	3½	—	⑯	5½
	Riviera	—	19.7	20.0	15	15	190	21	4	—	—	⑯	5½
1972	V8-350⑪	—	16.45	16.85	15	15	190	20⑱	4	3½	3	⑮	②
	V8-350⑩	—	18.9	19.3	15	15	190	25	4	—	—	⑮	②
	GS-455	—	16.2	16.3	15	15	190	20	4	3½	3	⑯	②
	V8-455	—	18.7	19.0	15	15	190	25⑲	4	—	—	⑯	②

Continued

COOLING SYSTEM & CAPACITY DATA—Continued

Year	Model or Engine	Cooling Capacity, Qts.			Radiator Cap Relief Pressure, Lbs.		Thermo. Opening Temp.	Fuel Tank Gals.	Engine Oil Refill Qts. ①	Transmission Oil			Rear Axle Oil Pints
		No Heater	With Heater	With A/C	With A/C	No A/C				3 Speed Pints	4 Speed Pints	Auto. Trans. Qts. ⑬	
1973	6-250	—	14.0	14.0	15	15	195	21	4	3½	—	⑮	②
	V8-350 Apollo	—	19.5	20.0	15	15	190	21	4	3½	—	⑮	②
	V8-350⑪	—	16.45	16.85	15	15	190	22	4	3½	—	⑮	②
	V8-350⑩	—	18.9	19.3	15	15	190	26	4	—	—	⑮	②
	GS-455	—	16.2	16.6	15	15	190	22	4	3½	—	⑯	②
	V8-455	—	18.7	19.0	15	15	190	26③	4	—	—	⑯	②
1974	6-250	—	14.0	14.0	15	15	195	21	4	3½	—	⑮	②
	V8-350 Apollo	—	16.5	17.0	15	15	190	21	4	3½	—	⑮	②
	V8-350⑪	—	17.3	④	15	15	190	22	4	—	—	⑮	②
	V8-350⑩	—	17.3	17.2	15	15	190	26	4	—	—	⑮	②
	V8-455⑪	—	19.4	19.9	15	15	190	22	4	—	—	⑯	②
	V8-455⑥	—	19.6	19.8⑤	15	15	190	26③	4	—	—	⑯	②
1975	6-250	—	16.3	16.4	15	15	190	21	4	3½	—	⑮	4.25
	V6-231⑦	—	16.5	16.6	15	15	190	21	4	3½	—	⑮	4.25
	V6-231⑪	—	15.3	15.3	15	15	190	22	4	3½	—	⑮	4.25
	V8-260⑦	—	22.4	22.9	—	—	195	21	4	3½	—	⑮	4.25
	V8-350⑦	—	17.9	18.6	15	15	190	21	4	—	—	⑮	4.25
	V8-350⑪	—	16.9	17.2	15	15	190	22	4	3½	—	⑮	⑳
	V8-350⑩	—	16.9	17.2	15	15	190	26	4	—	—	⑮	4.25
	V8-400	—	23.6⑧	⑨	15	15	190	26③	4	—	—	—	⑳
	V8-455	—	19.6	20⑤	15	15	190	26③	4	—	—	⑯	⑳
1976	V6-231⑦	—	16.6	16.7	15	15	195	21	4	3½	—	⑮	3.5
	V6-231⑪	—	15.5	15.4	15	15	195	22	4	3½	—	⑮	4.25
	V8-260⑦	—	22.4	22.9	15	15	195	21	4	3½	—	⑮	3.5
	V8-350⑦	—	17.9	18.5	15	15	195	21	4	—	—	⑮	3.5
	V8-350⑪	—	16.9	17.2	15	15	195	22	4	—	—	⑮	4.25
	V8-350⑩	—	16.9	17.2	15	15	195	26③	4	—	—	⑮	4.25
	V8-455	—	19.7	20①	15	15	195	26	4	—	—	⑥	5.4

①—Add one quart with filter change.
②—8½ inch axle, 4¼ pts.; 8⅞ inch axle, 5¼ pts.; 9⅜ inch axle, 5½ pts.
③—Estate Wagon 22 gallons.
④—With 20 inch fan shroud, 17.6 qts.; with 22 inch fan shroud, 17.2 qts.
⑤—With heavy duty cooling system 21.6 qts.
⑥—Estate Wagon, Electra & Riviera.
⑦—Apollo & Skylark.
⑧—With heavy duty cooling system, 25.6 qts.
⑨—Exc. wagon & heavy duty, 24.4 qts;. wagon & heavy duty, 26.1 qts.
⑩—LeSabre.
⑪—Intermediates
⑫—Total 9½ qts. Oil pan only 2½ qts.
⑬—Approximate. Make final check with dipstick.
⑭—Two speed unit 9½ qts. total. Oil pan only 2½ qts.
Three speed unit 10 qts. total. Oil pan only 3 qts.
⑮—Total 10 qts. pan only 3 qts.
⑯—Total 11½ qts. Oil pan only 3½ qts.
⑰—Two speed unit 9½ qts. total. Oil pan only 2½ qts.
Three speed unit 11½ qts. total. Oil pan only 3½ qts.
⑱—Sportwagon 23 gallons.
⑲—Estate Wagon 24 gallons.
⑳—Exc. wagon, 4¼ pts.; wagon, 5.4 pts.
㉑—With heavy duty cooling system, 21.45 qts.

Electrical Section

DISTRIBUTOR, REPLACE

1. Disconnect primary wire from distributor and disconnect hose from vacuum control unit.

NOTE: On H.E.I. systems, disconnect feed and module connectors from distributor cap.

2. Remove distributor cap.
3. Crank engine until distributor rotor is in position to fire No. 1 cylinder and the timing mark (see *Tune Up Chart*) is aligned with the timing indicator.
4. Remove distributor clamp and lift out distributor.

NOTE: Before installation of either a new or repaired distributor apply a few drops of engine oil to the drain hole near the lower end of the housing and apply oil to the oiler on the housing. Rotate the distributor shaft several times by hand to distribute the oil and to make sure that the shaft turns freely.

1. Check to make sure that the timing mark is aligned with the timing indicator with No. 1 piston on the compression stroke in position to fire.
2. Place a new seal on distributor housing.
3. Rotate distributor cam in direction of arrow on cam until rotor is in position to fire No. 1 cylinder.
4. Rotate oil pump shaft with screwdriver to align slot in shaft with tongue on lower end of distributor shaft.
5. Install distributor in crankcase with vacuum control pointing to right side of engine, in position to connect to vacuum hose.
6. Install distributor clamp and bolt with lockwasher, leaving bolt just loose enough to permit movement of distributor.
7. Rotate distributor housing until breaker points just start to open and tighten clamp bolt. This will permit starting engine for setting timing.
8. Connect pipe to vacuum control and primary wire to terminal stud.
9. Install distributor cap. If spark plug wires are disconnected from cap make certain that wires are connected in accordance with firing order.
10. Check and set ignition timing.

NOTE: When using a timing light to adjust ignition timing, the connection should be made at the No. 1 spark plug. Forcing foreign objects through the boot at the No. 1 terminal of the distributor cap will damage the boot and could cause engine misfiring.

STARTER, REPLACE

To remove the starter, disconnect bat-

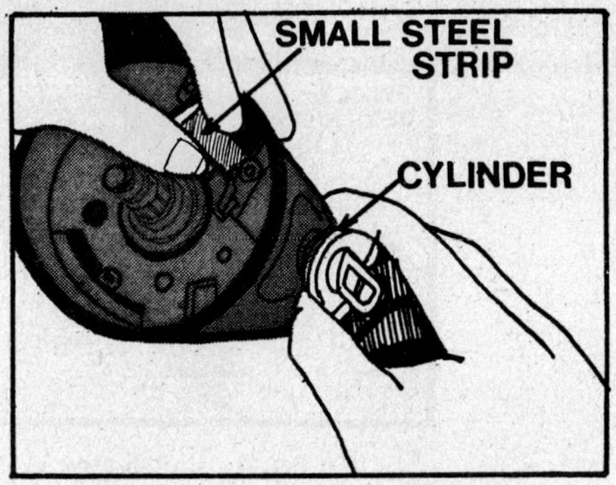

Fig. 1 Ignition lock. 1969-76

tery cable from battery. Disconnect cable and solenoid lead wire from solenoid switch. Remove flywheel inspection cover. Remove starter attaching bolts and remove starter.

LIGHT SWITCH, REPLACE

1969-76

CAUTION: On vehicles equipped with an Air Cushion Restraint System, turn ignition switch to "Lock," disconnect battery ground cable and tape end, thereby deactivating system.

1. Disconnect battery ground cable.
2. On 1969-72 intermediate models equipped with A/C, remove left hand duct.
3. On 1971-76 full size models, remove left hand trim panel.
4. On all models, pull switch knob to full "On" position, depress latch button on rear of switch and pull knob and rod from switch. Remove switch escutcheon or retaining nut, if used.
5. Pull switch down, disconnect electrical connector and remove switch.

NOTE: On 1975-76 Full Size Models, remove switch from mounting bracket.

6. Reverse procedure to install.

Fig. 2 Ignition lock disassembly

IGNITION LOCK

1969-76

1. Follow procedure to remove turn signal switch as described further on.
2. The lock cylinder may be removed in any position from "accessory" to "run." However, the "accessory" position is desired because of its positive position.
3. Place a thin tool (small screw-driver or knife blade) into the slot next to the switch mounting screw boss (right hand slot) and depress spring latch at bottom of slot which releases lock Fig. 1, Remove lock.
4. To separate the lock from the cylinder, Fig. 2, use a paper clip or similar quality wire, place a 90 deg. bend about $\frac{1}{8}$" from one end of the wire.
5. Using the wire as described in step 4, depress the plunger pin (above the stake mark) approx. $\frac{1}{8}$" vertically in sleeve housing, at the same time exert light upward pressure on the lock assembly, rotate the cylinder counter-clockwise until it pops up from the sleeve assembly.

IGNITION SWITCH, REPLACE

1969-76

CAUTION: On vehicles equipped with an Air Cushion Restraint System, turn ignition switch to "Lock" position, disconnect battery ground cable and tape end, thereby deactivating system.

The ignition switch is located on the top of the steering column under the instrument panel. To replace it the steering column must be lowered as follows:

1. Disconnect shift indicator link.
2. Remove nuts securing bracket to dash panel and carefully lower column.
3. Disconnect electrical connector from switch. Ensure switch is in "Accessory" position on 1969-73 models, Fig. 3, or "Off-Unlock" position on 1974-76 models.
4. Remove two screws securing switch and remove switch.
5. On 1969-73 models, position switch and lock in the "Lock" position for installation, Fig. 8. On 1974-76 models, the switch and lock are placed in the "Off-Unlock" position.
6. Fit actuator rod into switch and assemble to column.
7. Complete assembly in reverse of removal procedure.

Fig. 3 Ignition switch in "Lock" position. 1969-76

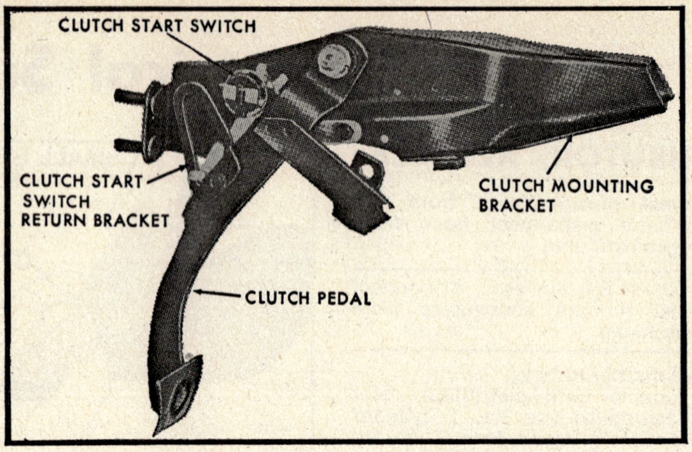

Fig. 4 Clutch start switch in start position. 1969-76

STOP LIGHT SWITCH, REPLACE

1969-76

The stop lights are controlled by a mechanical switch mounted on the brake pedal bracket. This spring loaded switch makes contact whenever the brake pedal is applied. When the brake pedal is released it depresses the switch to open the contacts and turn brake lights off.

CLUTCH START SWITCH

1969-76

A clutch start switch is installed on all manual transmission cars. The switch is mounted on the clutch pedal bracket and it prevents the car from being started until the clutch pedal is depressed, Fig. 4.

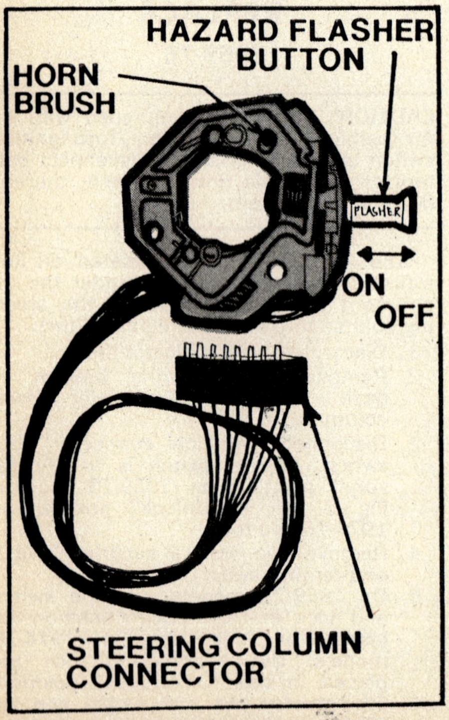

Fig. 5 Turn signal and hazard warning flasher switch assembly. 1969-76

TURN SIGNAL SWITCH, REPLACE

1969-76

CAUTION: On vehicles equipped with an Air Cushion Restraint System, turn ignition switch to "Lock," disconnect battery ground cable and tape end, thereby deactivating system.

As shown in Fig. 5, the assembly is a turn signal switch and hazard warning switch. It is mounted in a housing at the upper end of the steering column mast jacket, just below the steering wheel. Therefore to get at the switch the steering wheel will have to be removed.

Also, on models equipped with tilt steering columns, it is necessary to lower the column assembly from instrument panel.

1. Disconnect battery ground cable.
2. Remove steering wheel and lock plate cover. On tilt columns, remove tilt lever.
3. With a suitable compressor, compress lock plate and spring, then remove snap ring from shaft.

NOTE: On 1976 models, remove lock plate cover with a suitable screwdriver.

4. Remove lock plate, cancelling cam, preload spring and thrust washer.
5. Remove turn signal lever and hazard warning switch knob.
6. Disconnect switch wiring connector and wrap a piece of tape around connector upper end and wiring harness, preventing snagging when removing switch.
7. Remove three switch retaining screws and switch.
8. Reverse procedure to install.

NEUTRAL START & BACK-UP LIGHT SWITCH

To check operation of switch after adjustments are made as outlined below, proceed as follows:

1. With shift lever in Park starter should operate.
2. With shift lever in Reverse back-up lights should light but starter should not operate.
3. With shift lever in Neutral starter should operate but back-up lights should be out.
4. With shift lever in Drive, starter should not operate and back-up lights should be out.

1969-76

1. Place shift lever in Drive (Neutral for 1972-76).
2. Attempt to insert a 3/32" or a No. 42 drill through gauging hole in switch body into inner hole in sliding part of switch, Fig. 6.

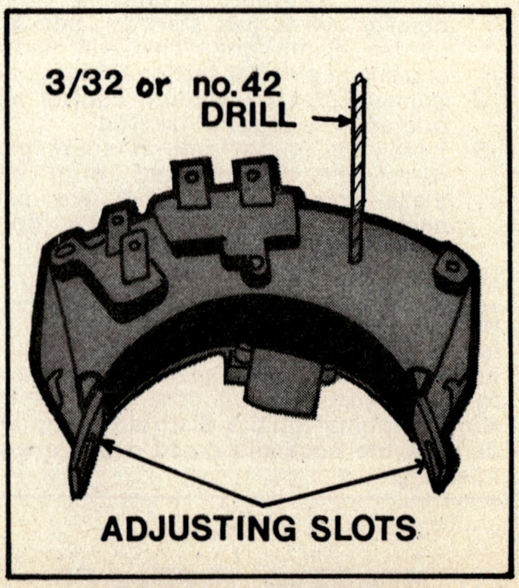

Fig. 6 Adjusting neutral safety switch. 1969-76

3. If drill does not enter inner hole, loosen two switch mounting screws and slide switch body as required to allow drill to enter inner hole. Tighten screws and remove drill.

HORN SOUNDER & STEERING WHEEL

1969-76

CAUTION: On vehicles equipped with an Air Cushion Restraint System, turn ignition switch to "Lock," disconnect battery ground cable and tape end, thereby deactivating system. Also, on these vehicles, it is necessary to remove the drivers cushion module before removing steering wheel. With tool J-24628-2, remove the module to steering wheel screws, lift module and disconnect horn wire. Then, with tool J-24628-3, disconnect module wire connector from slip ring.

1. Remove horn cap or actuator bar.
2. On 1975-76 models, remove steering wheel nut retainer.
3. On all models, back off nut until flush with top of steering shaft.
4. Use a suitable puller to remove wheel.

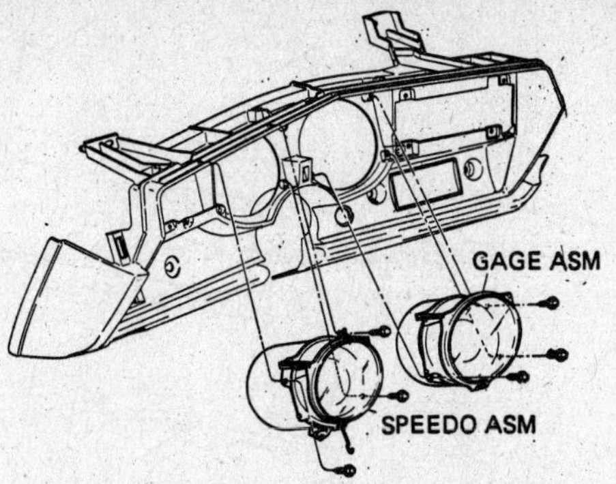

Fig. 7 Speedometer & telltale assembly. 1973-76 Century & Regal

INSTRUMENT CLUSTER

CAUTION: On vehicles equipped with an Air Cushion Restraint System, turn ignition switch to "Lock," disconnect battery ground cable and tape end, thereby deactivating system.

1975-76 Full Size

1. Disconnect battery ground cable.
2. Remove instrument cluster bezel and lens.
3. Remove screws securing speedometer head assembly and pull assembly from housing.
4. Remove screws securing fuel gauge assembly and pull assembly from housing.
5. Reverse procedure to install.

1973-76 Apollo & Skylark

1. Disconnect battery ground cable.
2. Disconnect heater or A/C control panel from the instrument panel carrier.
3. Remove radio control knobs, bezels and nuts, leaving the radio attached to the instrument panel reinforcement.
4. Disconnect instrument panel pad from the carrier and disconnect the shift quadrant indicator cable at the shift bowl. On automatic transmission equipped vehicles remove the two nuts securing the steering column to instrument panel.
5. Remove toe plate cover and discon-

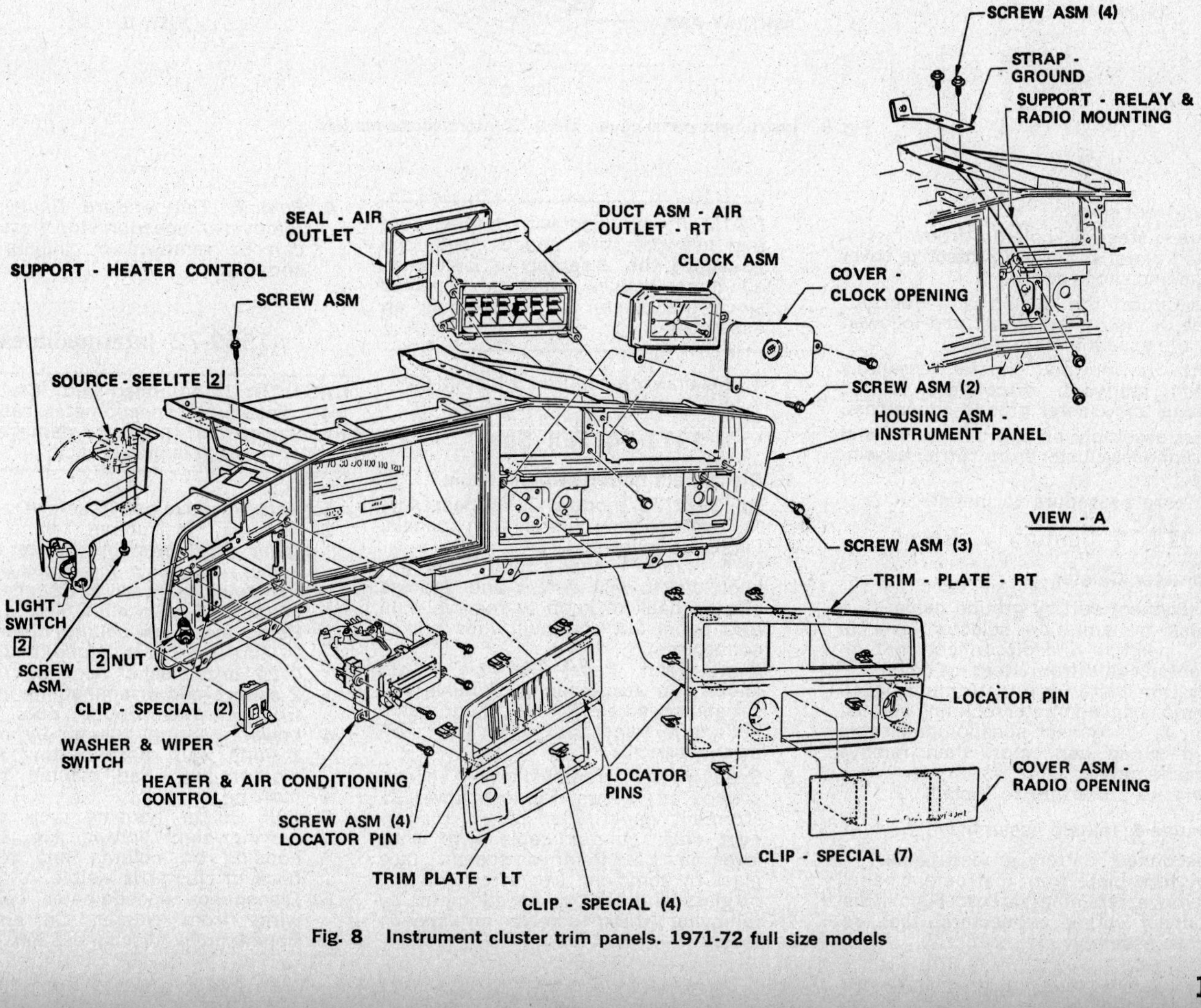

Fig. 8 Instrument cluster trim panels. 1971-72 full size models

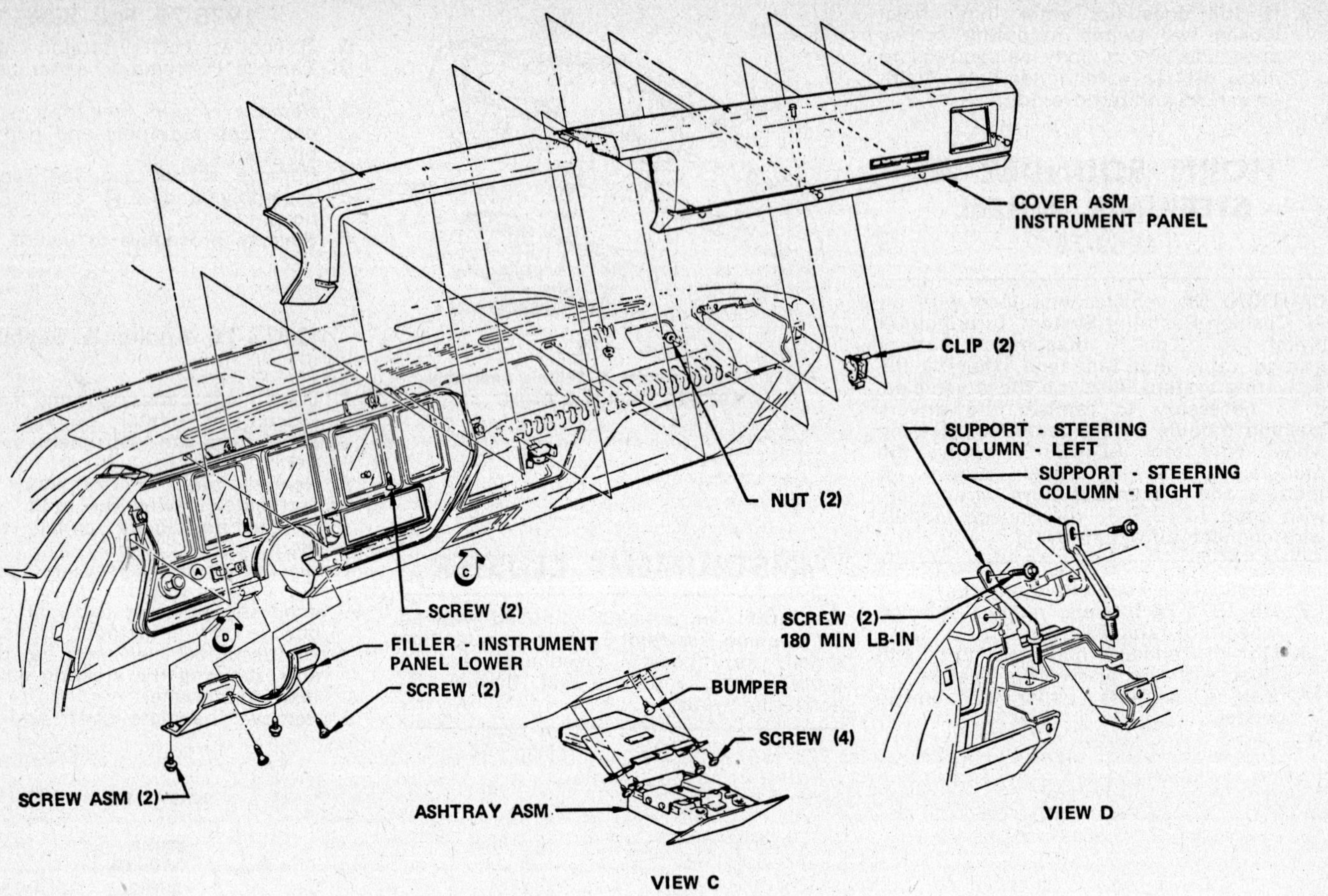

Fig. 9 Instrument panel cover. 1969-72 intermediate models

nect from cowl.

6. Lower steering column from instrument panel and use a protective cover (such as shop towel).
7. Disconnect the ground wire from left side of instrument panel pad followed by the speedometer cable.
8. With carrier and cluster assembly tilted rearward, disconnect printed circuit and cluster ground connectors.
9. Rest assembly on top of column and disconnect cluster from carrier assembly.
10. Reverse procedure to install.

1973-76 Century & Regal

Speedometer Cluster

1. Disconnect battery ground cable.
2. Place transmission selector lever in "L" position and disconnect shift indicator cable from steering column.
3. Pry trim plate from instrument panel.
4. Remove speedometer retaining screws, Fig. 7, disconnect speedometer cable and wiring connector, then remove speedometer.
5. Reverse procedure to install.

Fuel Gauge & Telltale Assembly

1. Disconnect battery ground cable.
2. Pry trim plate from instrument panel.
3. Remove retaining screws, Fig. 7, disconnect wiring connectors, then remove assembly.

NOTE: If equipped with temperature and oil pressure gauges and one gauge of the assembly is defective, all three gauges must be replaced since the cluster is serviced as an assembly.

---

4. Reverse procedure to install.

1971-74 Full Size

1. Disconnect battery ground cable.
2. On 1971-72 models, it is necessary to remove the cluster trim panels, Fig. 8, as follows:
   Left Trim Panel—pull headlight knob out to last detent and depress clip on back of knob to remove. Pull trim panel out and pull wires out of connectors.
   Right Trim Panel—Remove radio knobs and escutcheons. Remove trip set and speed alert knobs, if equipped. Pull trim panel out and pull wire from connector.
3. *Speedometer* removal is accomplished as follows: Remove lower instrument panel filler panel. Disconnect shift pointer cable with shift lever in L1. Remove speedo face glass by removing two screws on top of glass and remove speedometer by removing three screws on speedo face.
4. *Fuel & Temperature Gauges:* After removal of speedometer these gauges can be removed by pulling out of connector.

1969-72 Intermediates

CAUTION: If equipped with Cruise Control, disconnect speedometer cable from transducer to prevent damage when cluster housing is pulled back.

---

1. Remove glove box (9 screws).
2. Remove instrument panel upper cover (2 screws thru cluster housing and 2 nuts above glove box opening). Pull cover rearward to disengage three guide pins from clips, Fig. 9.
3. Remove steering column opening filler (4 screws). On air conditioned units, drop left plastic duct by removing 2 screws and disconnecting inner end from center distribution duct.
4. Lower steering column by removing 2 nuts and disconnecting shift indicator link. Pad column to avoid marring paint.
5. Pull cluster housing back after removing eight screws, Fig. 10. Rest housing on column and rotate so back of cluster is visible.
6. Disconnect speedometer cable and wires from cluster. On air conditioned units, disconnect left air hose

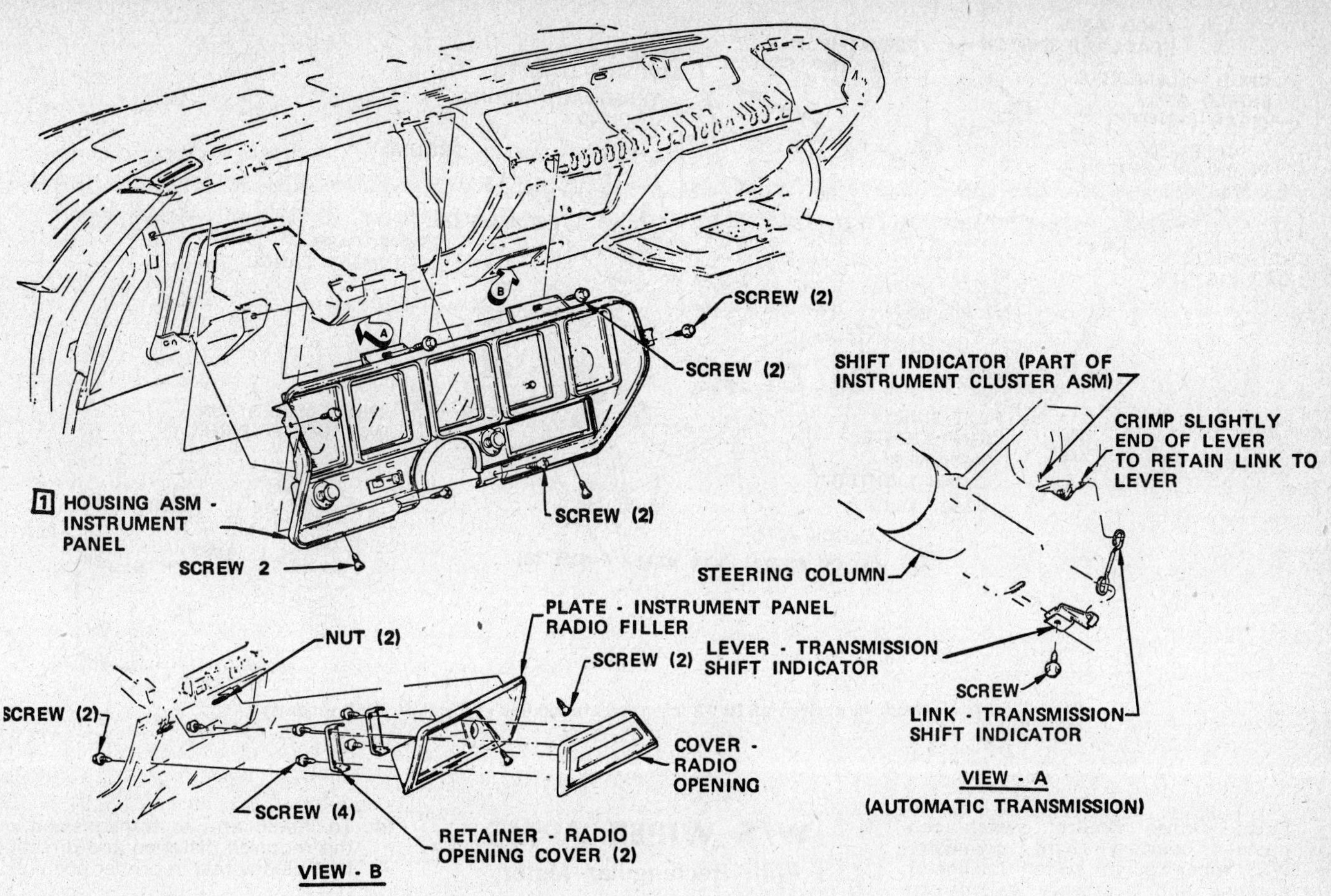

Fig. 10 Instrument panel housing. 1969-72 intermediate models

from left outlet by twisting "quick-connect" coupling counterclockwise. Disconnect heater control panel.

7. Remove cluster by removing cluster to housing screws, Fig. 11.
8. Reverse above procedure to install.

1969-70 Full Size Exc. Riviera

CAUTION: If equipped with Cruise Control, disconnect speedometer cable from transducer to prevent damage when cluster housing is pulled back.

1. Remove lower instrument panel filler (4 screws), then slide filler forward and down.
2. Remove glove box (5 screws).

NOTE: Do not remove glove box door as it comes off with cover along with ash tray assembly.

3. Remove instrument panel cover (2 nuts above glove box opening and 3 screws thru cluster housing) and remove all screws along bottom edge of cover.
4. Lower steering column by removing two nuts and disconnecting shift indicator link. Pad column to avoid marring paint.
5. Pull cluster housing back after removing eight screws. Rest housing on column and rotate so back of housing is visible.
6. Disconnect speedometer cable and wires.
7. Disconnect heater-air conditioner control panel.

NOTE: Do not disturb cables and vacuum hoses or adjustment will be required after reassembly.

8. Remove instrument cluster (6 cluster to housing screws).
9. Reverse procedure to install.

1969-70 Riviera

CAUTION: If equipped with Cruise Control the upper speedometer cable must be disconnected from the transducer before cluster housing is pulled back.

1. Remove instrument panel compartment body (8 screws).
2. Remove four nuts at right underside of dash and four screws at housing. Pull instrument panel upper cover rearward to remove.
3. Remove steering column filler (2 screws). If column shift, disconnect shift quadrant link wire at steering column. Remove two nuts from column mounting bracket and one nut from column wedge. Lower steering column.
4. Remove two nuts from lower edge of instrument panel housing at steering column. Remove four screws across upper edge of instrument panel housing.
5. Remove one nut at lower left side of instrument housing.
6. Remove ash receiver (4 screws).
7. Remove one nut at lower right side of instrument housing.
8. Remove two screws at heater control installation and separate from instrument panel housing.
9. Protect steering column so panel housing will not mar column when housing is tilted back.
10. Disconnect from cluster the speedometer cable, two wiring harness clips, printed circuit connector (all from above).

NOTE: If equipped with Cruise Control, disconnect speedometer cable at Cruise Control transducer located at rear of engine compartment. This will allow instrument housing to be pulled rearward.

11. Disconnect from instrument housing the clock connector and two clock

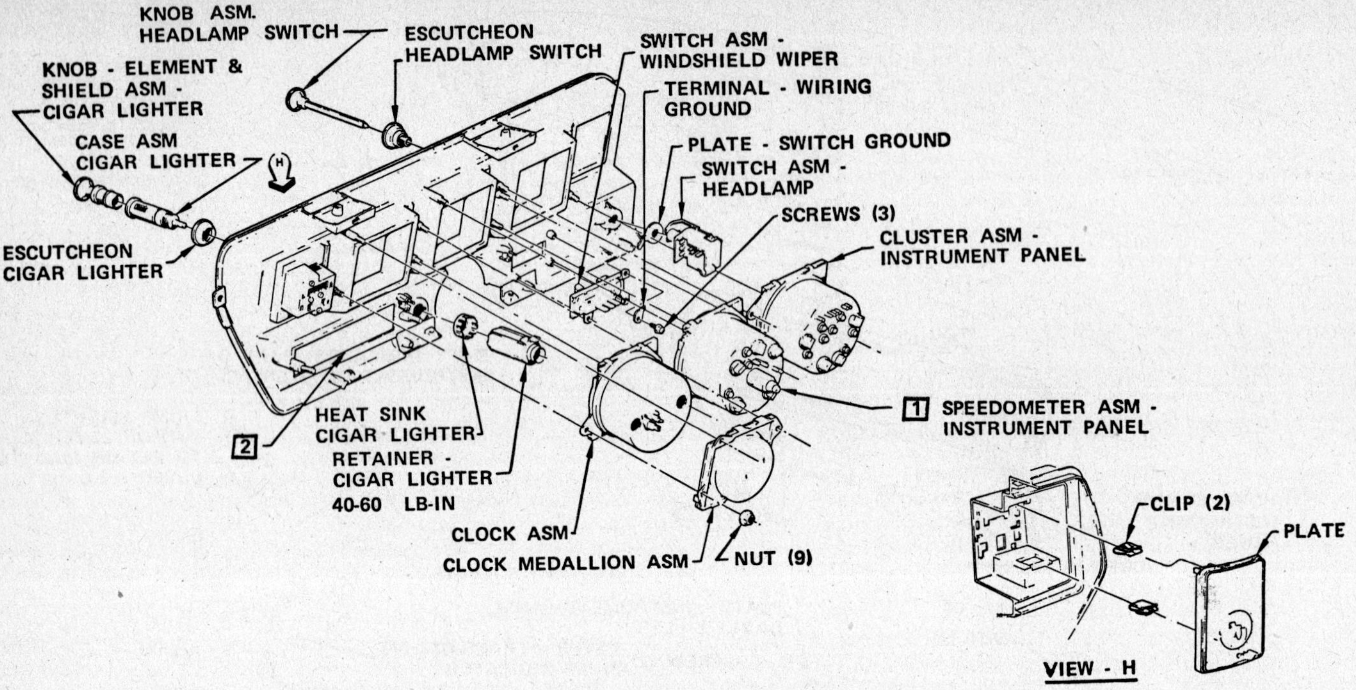

Fig. 11 Instrument cluster. 1970-72 intermediate models (Typical of 1969 models)

bulbs, Cruise Control switch connector, courtesy light connector, W/S wiper-washer switch connector, antenna and accessory switch connectors, cluster ground wire, A/C hose and headlight connector (all from above).

12. Remove instrument panel housing.
13. Separate panel housing from cluster (6 screws), Fig. 12.
14. Reverse procedure to install.

W/S WIPER BLADES

Two methods are used to retain wiper blades to wiper arms, Fig. 13. One method uses a press type release tab. When the release tab is depressed the blade assembly can be slid off the wiper arm pin. The other method uses a coil spring retainer. A screw driver must be inserted on top of the spring and the spring pushed downward. The blade assembly can then be slid off the wiper arm pin. Two methods are also used to retain the blade element in the blade assembly, Fig. 13. One method uses a press type release button. When the button is depressed, the two piece blade assembly can be slid off the blade element. The other method uses a spring type retainer clip in the end of the blade element. When the retainer clip is squeezed together, the blade element can be slid out of the blade assembly.

NOTE: To be sure of correct installation, the element release button, or the spring element retaining clip should be at the end of the wiper blade assembly nearest the wiper transmission.

W/S WIPER ARMS

With Rectangular Motor

1. Wiper motor must be in park position.
2. Use suitable tool to minimize the possibility of windshield or paint finish damage during arm removal.
3. Remove arm by prying up with tool to disengage arm from serrated transmission shaft.
4. To install arm to transmission rotate the required distance and direction so that blades rest in proper position.

With Round Motor

1. Wiper motor must be in park position.
2. Raise hood to gain access to wiper arm.
3. *On Intermediate Models:* lift arm off transmission shaft. On left arm, slide

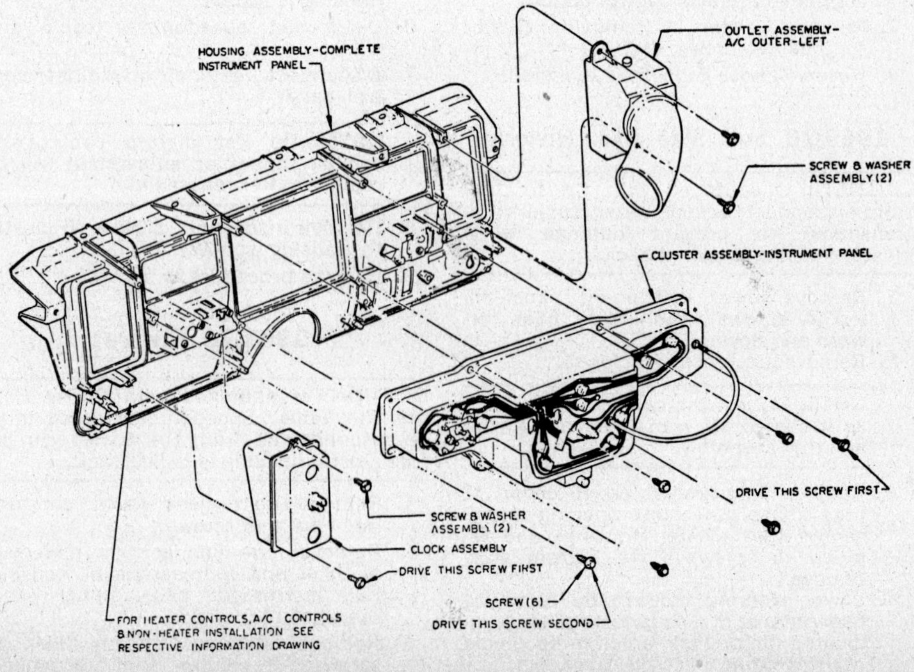

Fig. 12 Instrument cluster. 1969-70 Riviera

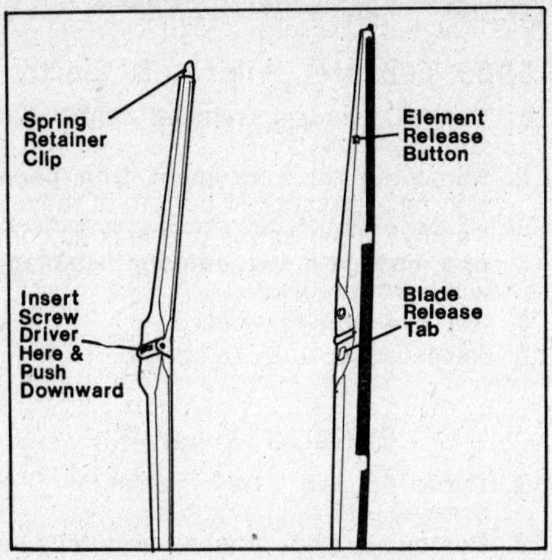

Fig. 13 Wiper blade assy

articulating arm lock clip, Fig. 14, away from transmission pivot pin and lift arm off pin. *On Full Size Models:* lift wiper arm and slide latch clip, Fig. 15, out from under wiper arm.

4. Release wiper arm and lift arm assembly off transmission shaft.

W/S WIPER MOTOR, REPLACE

1972-76

1. Disconnect battery and remove cowl screen.
2. Loosen nuts on wiper drivelink to motor cranking arm and slip drivelink off cranking arm.
3. Disconnect washer hoses and electrical connections.
4. Unfasten motor and remove.

NOTE: On models with round motor, the motor must be in "Park" position when assembling crank arm to transmission drive link.

1969-71

1. With wiper motor in Park position, remove washer hoses and electrical connections from motor.
2. On Special and Riviera, remove the air intake grille. On LeSabre, Wildcat and Electra, remove plastic access cover.
3. Loosen nuts which retain drive link to crank arm ball stud or remove nut retaining crank arm to motor, depending on model.
4. Remove motor retaining bolts and remove motor.

CAUTION: Wiper motor must be in Park position prior to installation on the cowl.

W/S WIPER TRANSMISSION

1971-76

1. Disconnect battery ground cable and remove cowl vent screen or grille.
2. Disconnect wiper motor electrical connector.
3. Remove wiper arm and blade assemblies.

NOTE: On full size models equipped with round motor, remove wiper arm and blade from transmission being removed.

4. Loosen transmission drive link to motor crankarm attaching nuts, then disconnect drive link from crankarm.

NOTE: On full size models equipped with round motor, if left hand transmission is being removed, it is not necessary to disconnect the right hand transmission drive link.

5. Remove right and left transmission to body retaining screws and guide transmission and linkage through cowl opening.

NOTE: On full size models equipped with round motor, remove transmission retaining screws from transmission being removed.

6. Reverse procedure to install.

1969-70

1. Make sure motor is in Park position.
2. Remove arm and blade assemblies.
3. Remove air intake grille or screen (if equipped).
4. Disconnect drive rod from motor crank.
5. Remove transmission retaining nuts or bolts and lower drive rods into plenum chamber.
6. Remove transmission assemblies through cowl opening.

W/S WIPER SWITCH, REPLACE

1973-76 Apollo & Skylark

1. Remove electrical connector, three attaching screws and switch.

1973-76 Century & Regal

1. Insert blade of a small screwdriver into slots above knobs, bend retaining clips down and pull top of switch outward.
2. Remove electrical connector from switch.

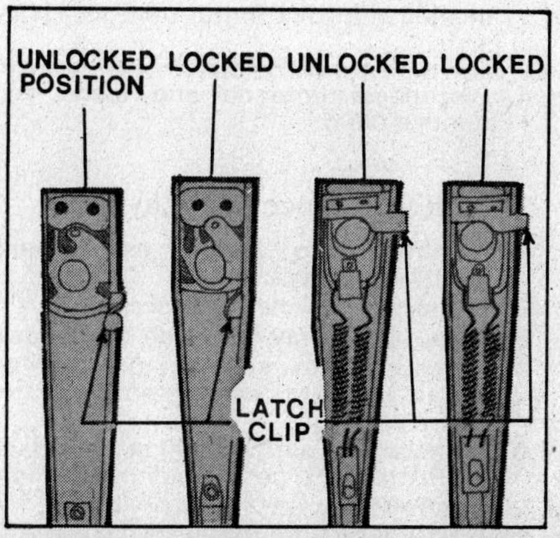

Fig. 15 Wiper arm latch clips

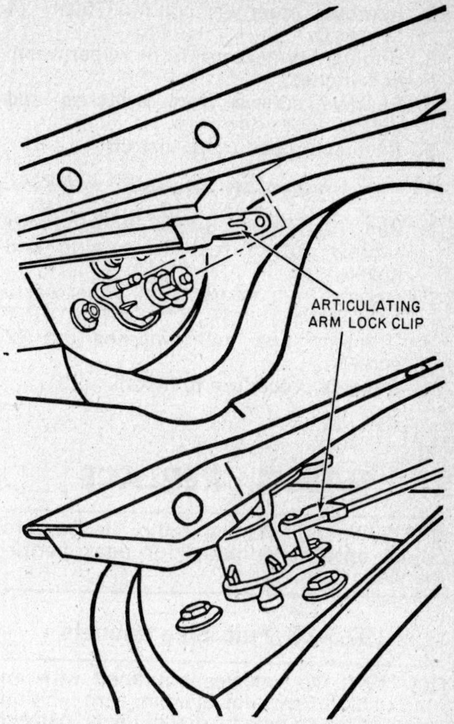

Fig. 14 Articulating arm lock clips

1971-76 Senior Models

CAUTION: On vehicles equipped with an Air Cushion Restraint System, turn ignition switch to "Lock," disconnect battery ground cable and tape end, thereby deactivating system.

1. Remove left trim panel as outlined under the "Instrument Cluster, 1971-74 Full Size" procedure.
2. Using a screwdriver, depress switch retaining clips and remove switch.
3. Disconnect seelite cable and electrical connector.
4. Reverse procedure to install.

1970 LeSabre, Estate Wagon, Wildcat & Electra

1. Remove lower instrument panel filler (4 screws) then slide filler forward and down.
2. Lower steering column by removing two nuts and disconnecting shift indicator link.
3. Unplug connector from switch.
4. Remove wiper knob.
5. Remove switch (two screws).

1969 LeSabre, Wildcat & Electra

1. Pry wiper-washer switch from cluster housing.
2. Remove switch by unplugging switch connector.
3. Reverse procedure to install.

1969-72 Special & Skylark

1. If air conditioned, remove two screws at left A/C distribution duct and remove duct.

2. Remove steering column filler (4 screws).
3. Unplug connectors from wiper-washer switches.
4. Remove screws from switches and pull switches down.
5. Reverse procedure to install.

1969-70 Riviera

1. Pull instrument panel housing assembly out to rest on column and knees.
2. Unplug connectors from wiper-washer switches.
3. Unfasten and pull switches out (2 screws).
4. Reverse procedure to install.

RADIO, REPLACE

NOTE: When installing radio, be sure to adjust antenna trimmer for peak performance.

1975-76 Full Size Models

CAUTION: On vehicles equipped with an Air Cushion Restraint System, turn ignition switch to "Lock," disconnect battery ground cable and tape end, thereby deactivating system.

1. Disconnect battery ground cable.
2. Pry out lower right instrument panel trim.
3. Remove radio knobs, escutcheons and retaining nuts from control shafts.
4. Remove plastic trim plate.
5. Disconnect leads from radio and remove rear radio attaching nut, then pull radio from panel.
6. Reverse procedure to install.

1974 Full Size Models with Air Cushion Restraint System

1. Turn ignition switch to "Lock" and disconnect battery ground cable and tape end.
2. Remove both lower instrument panel cover trim plates.
3. Disconnect parking brake release cable, then remove lower left instrument panel cover retaining screws and cover.
4. Remove two horizontal screws below instrument panel, four screws from upper horizontal surface of instrument panel, two outer screws from glove box door hinge and screw from instrument panel cover right side.
5. Disconnect leads from radio.
6. Release four clips behind instrument panel by grasping tongue of right hand side clip, squeezing and pulling forward.
7. Remove radio knobs, escutcheons and instrument panel trim plate.
8. Remove retaining nuts from control shafts and power antenna relay retaining screws.
9. Loosen left radio support nut, remove right radio support nut and lower radio from instrument panel.
10. Reverse procedure to install.

1973-76 Apollo & Skylark

1. Remove radio knobs, bezels, nuts and side brace screw.
2. Disconnect antenna and leads.
3. Remove radio from under dash.
4. Reverse procedure to install.

1973-76 Century & Regal

1. Remove radio knobs and escutcheons. If equipped with center air duct assembly control, remove screws.
2. Disconnect antenna and leads.
3. Remove radio support nut.
4. Remove attaching nuts and slide radio to front of car and downward.
5. Reverse procedure to install.

1972-74 Senior Models less Air Cushion Restraint System

1. Remove knobs and escutcheons. If equipped with trip-set and speed alert, unscrew the cone shaped knobs.
2. Remove right trim panel.
3. Remove retaining nuts from shafts.
4. Remove ash tray and separate connectors and antenna lead in from radio.

1971 Senior Models

1. Remove right trim panel as described previously.
2. Remove ash tray assembly.
3. Disconnect antenna lead-in and radio connector.
4. Loosen radio side brace.
5. Remove radio retaining nuts on front of panel.
6. Disconnect center A/C duct behind ash tray.
7. Remove radio.

1970 LeSabre, Estate Wagon, Wildcat & Electra

1. Remove instrument panel lower filler (4 screws) then slide filler forward and down.
2. Remove radio ground strap screws.
3. Remove radio knobs, escutcheons, hex nuts and lower radio downward.

NOTE: On air condition cars, remove center distributor duct for clearance.

4. Disconnect antenna and leads and remove radio.

1969-72 Special & Skylark

1. Remove radio knobs, escutcheons and two hex nuts.
2. Remove filler plate (2 screws).
3. Remove ash tray and slide (4 screws).
4. Remove radio support (2 screws). If air conditioned, remove center distributor duct.
5. Remove two nuts attaching radio face to instrument panel and move radio downward.
6. Disconnect antenna lead and wiring and remove radio.
7. Reverse procedure to install.

1969 LeSabre, Wildcat & Electra

1. On A/C models, remove center distribution duct.
2. Remove right instrument trim panel and remove screw in bottom of radio.
3. Remove radio knobs and escutcheons and nuts. Unplug antenna lead and wires from radio.
4. Remove radio downward.
5. Reverse procedure to install.

1969-70 Riviera

1. Remove ash tray assembly (4 screws).
2. Remove radio knobs, escutcheons and hex nuts.
3. Unplug antenna and leads from radio.
4. Remove radio downward through ash tray opening.
5. Reverse procedure to install.

HEATER CORE REMOVAL

After draining radiator and disconnecting heater hoses, proceed as follows:

1973-76 Apollo & Skylark

Without Air Conditioning

1. Disconnect battery ground cable and drain radiator.
2. Disconnect heater hoses from core and plug hoses and core openings to prevent coolant spillage.
3. Remove core case retaining nuts from engine side of dash, Fig. 16.
4. Remove glove compartment and door.
5. Using a 1/4 inch twist drill, drill out lower right hand heater case stud from inside vehicle.
6. Pull heater core and case assembly from dash.
7. Disconnect heater cables and blower motor resistor connector from case, then remove heater core and case assembly from vehicle.
8. Remove core tube seal and retaining strips, then heater core from case.
9. Reverse procedure to install.

With Air Conditioning

1. Disconnect battery ground cable and drain radiator.
2. Disconnect upper heater hose and plug hose and core openings. Remove accessible heater core and case attaching nuts.
3. Remove right front fender skirt bolts, lower skirt and disconnect lower heater hose from core tube. Plug hose and core openings.
4. Remove lower right hand heater core and case attaching nuts, Fig. 16.
5. Remove glove compartment and door.
6. Remove right hand kick panel recirculation vacuum diaphragm.
7. Remove heater outlet from bottom of heater case, then cold air distributor duct.

8. Remove heater case extension screws, then separate extension from heater case.
9. Disconnect heater cables and electrical connectors from heater case, then remove heater core and case assembly. Separate core from case.
10. Reverse procedure to install.

1973-76 Century & Regal

Without Air Conditioning

1. Disconnect control cables from door levers.
2. Remove four nuts securing heater to dash, Fig. 17.
3. Remove screw securing defroster outlet to heater.
4. Work heater assembly rearward until studs clear dash.

With Air Conditioning

1. Disconnect control cables from temperature door guides and vacuum hoses from actuator diaphragms.
2. Remove resistor assembly and reach through opening and remove one attaching nut to dash. Remove one attaching nut to dash directly over transmission and two attaching nuts to upper and lower inboard evaporator case half.
3. From inside car remove one screw in lower righthand corner on passenger side.
4. Remove lower attaching outlets and work assembly rearward until studs clear dash.

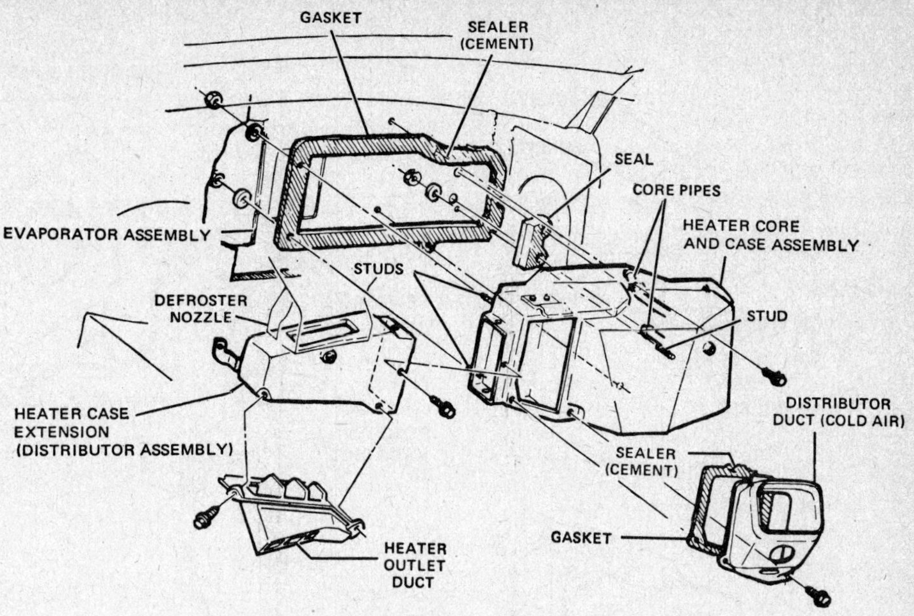

Fig. 16 Heater core. 1973-76 Apollo & Skylark (Typical)

1969-72 Special & Skylark

Without Air Conditioning

1. Remove right front fender skirt.
2. Disconnect control cables from lever of defroster door and outside air inlet door on heater.
3. Disconnect temperature control cable

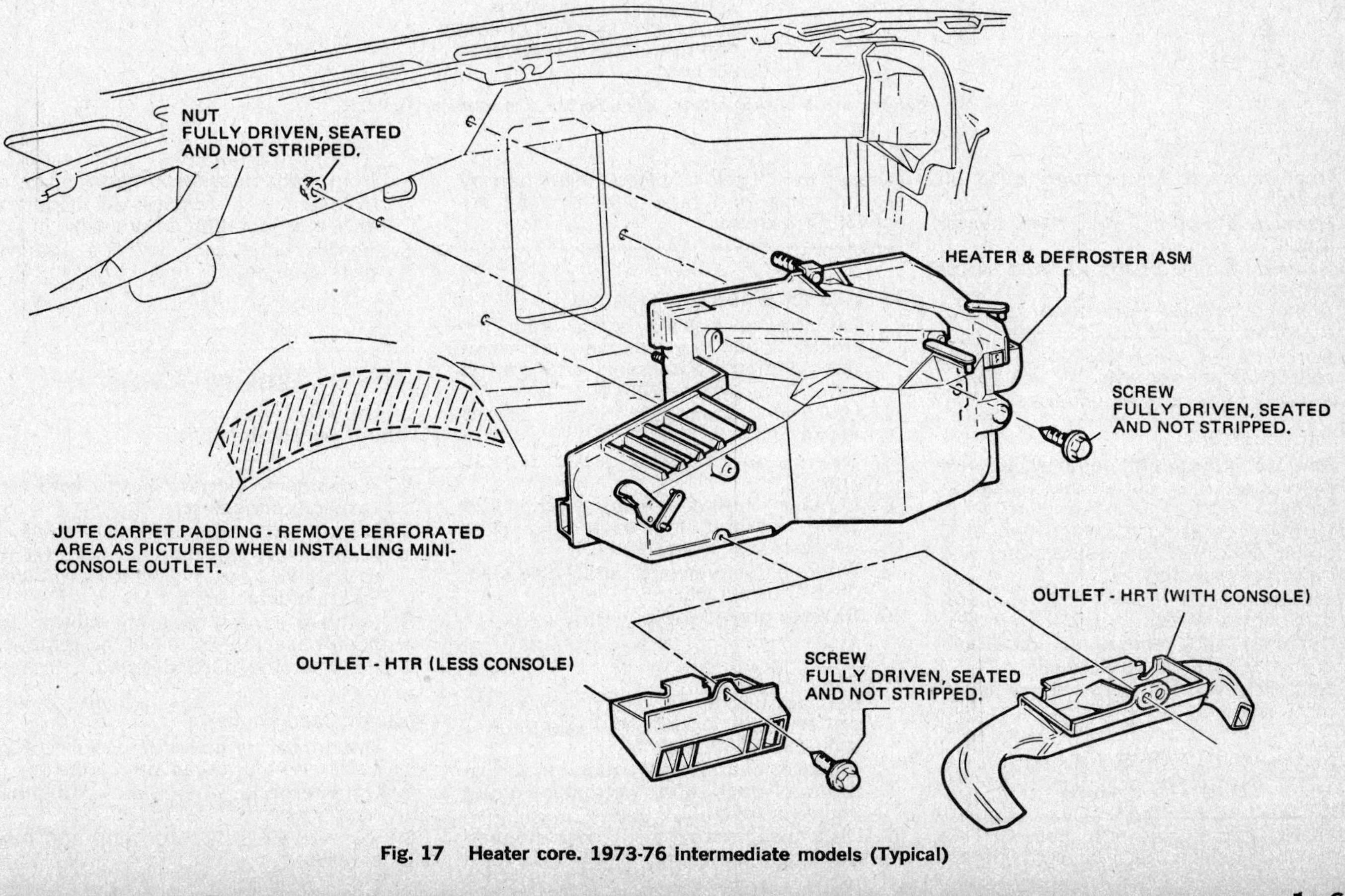

Fig. 17 Heater core. 1973-76 intermediate models (Typical)

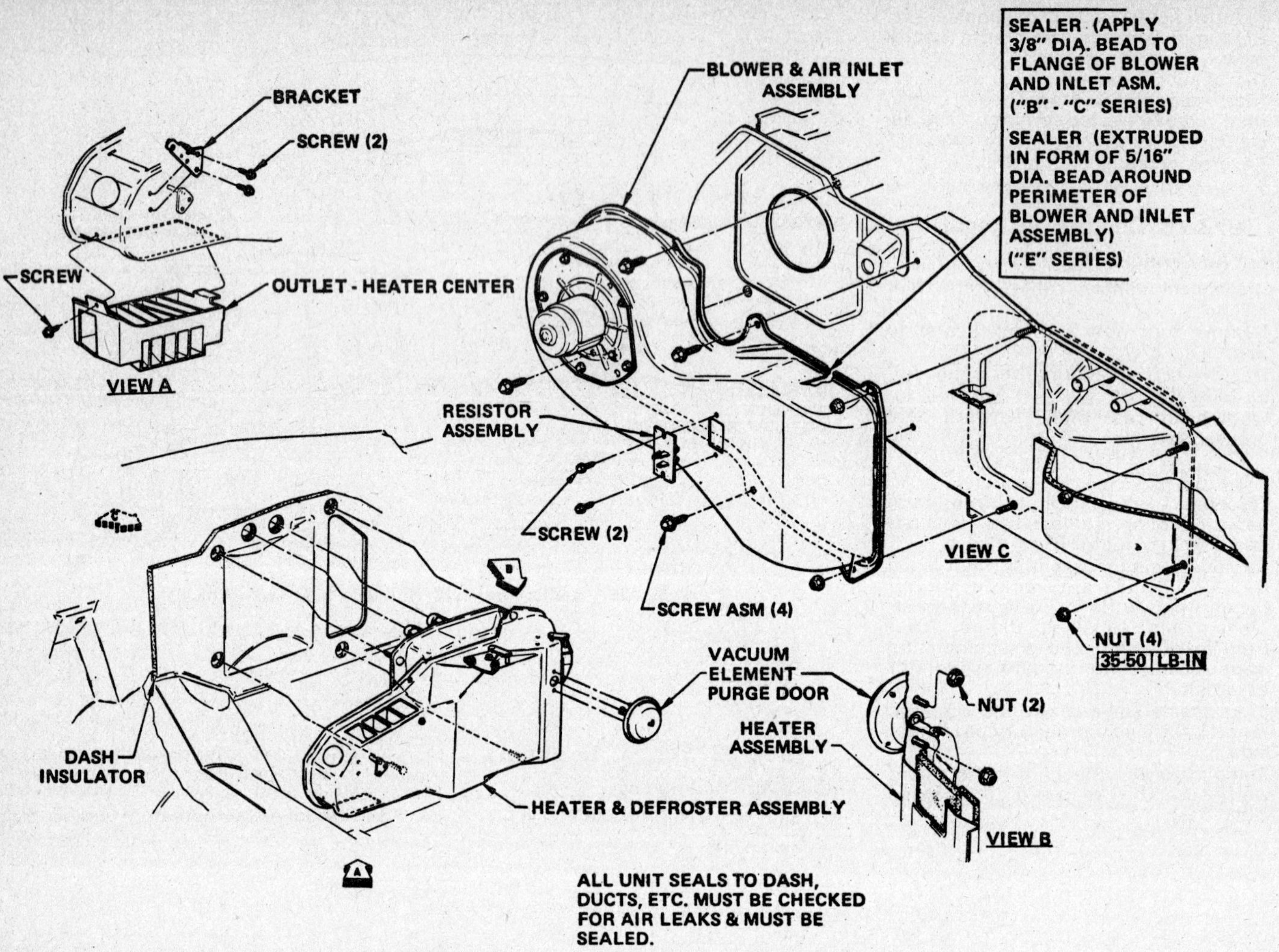

Fig. 18 Heater core & blower motor. 1971-76 full size models (Typical)

from lever of temperature door on heater.
4. Remove attaching nuts from heater studs.
5. Remove connector from blower motor resistor.
6. Remove screws from lower part of defroster outlet to top of heater.
7. Work heater rearward until studs clear dash and remove.
8. Reverse procedure to install.

With Air Conditioning

1. Remove instrument panel cover with right side A/C outlet and hose attached.
2. Remove center A/C duct, left A/C outlet duct, A/C distributor duct and defroster assembly.
3. Disconnect defroster and temperature control wires.
4. Unfasten and remove air conditioner-heater assembly from dash.
5. Heater core can now be removed from assembly.

1969-76 Full Size Exc. 1969-70 Riviera

CAUTION: On vehicles equipped with an Air Cushion Restraint System, turn ignition switch to "Lock," disconnect battery ground cable and tape end, thereby deactivating system.

1971-76 All & 1969-70 Less A/C

1. Disconnect vacuum hoses from defroster door and outside air inlet door actuator diaphragms and control cable from temperature door lever.
2. Unfasten connector from blower motor resistor.
3. Remove nuts securing heater to dash, Fig. 18.
4. Remove screws securing defroster outlet adapter to heater and raise adapter away from heater.
5. Work heater rearward until studs clear dash and remove.
6. Reverse procedure to install.

1969-70 With A/C

1. Remove instrument panel cover with center A/C outlet and right A/C outlet and hose attached.
2. Remove center A/C duct, A/C distributor duct and defroster outlet manifold assembly.
3. Disconnect defroster and temperature control wires and pink hose from mode door diaphragm.
4. Unfasten and remove air conditioner-heater assembly from dash.
5. Heater core can now be removed from assembly.

1969-70 Riviera

Without Air Conditioning

1. Remove right front fender.
2. Disconnect blower motor wire and resistor connectors.
3. Disconnect temperature door cable.
4. Disconnect vacuum hoses attached to outside door and vent heater door vacuum diaphrams.
5. Remove screws securing blower and heater assembly to dash and remove.
6. Reverse procedure to install.

With Air Conditioning

1. Disconnect temperature door control cable and blower resistor connector.
2. Unfasten and remove air conditioner-heater assembly from dash.
3. Remove air distributor duct and heater core.

BLOWER MOTOR REMOVE

1969-70

1. On Intermediates and Riviera, remove right front fender skirt.
2. On all models, disconnect wires and unfasten and remove motor.

1971-72 All & 1973-76 Full Size Models

CAUTION: On vehicles equipped with an Air Cushion Restraint System, turn ignition switch to "Lock", disconnect battery ground cable and tape end, thereby deactivating system.

1. On Intermediate models, remove right front fender skirt.
2. On 1971-73 full size models and 1974-76 Riviera, support hood and loosen hood hinge from extension. Remove extension and plate assembly.
3. Disconnect all wires, unfasten and remove motor, Fig. 18.

1973-76 Century & Regal

1. Disconnect blower motor wire.
2. Remove screws securing motor to air inlet and remove blower motor, Fig. 19.

1973-76 Apollo & Skylark

1. Disconnect battery ground cable and raise vehicle on hoist.
2. Remove fender skirt attaching bolts except those attaching skirt to radiator support.
3. Pull skirt out and down, then place a block of wood between skirt and fender to provide clearance for motor removal.
4. Disconnect blower motor electrical connections and remove motor attaching screws and blower motor.
5. Remove blower motor retaining nut and separate impeller from motor.
6. Reverse procedure to install.

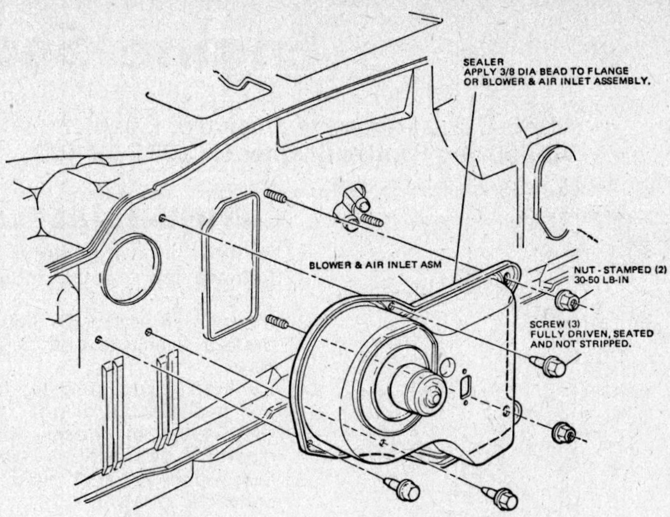

Fig. 19 Blower motor. 1973-76 intermediate (Typical)

SPEED CONTROL

1969-76 Cruise Master

Bead Chain Adjustment

1. Adjust engine hot idle speed and mixture, then shut off engine.
2. Check slack in chain by unsnapping swivel from ball stud and holding chain taut at ball stud; center of swivel should extend 1/8" beyond center of ball stud.
3. Adjust bead chain slack by sliding sleeve back on chain and removing loose rivet. Move swivel on ball chain until slack is correct. Then reinstall rivet and slide sleeve over rivet. Beginning with 1973 models the chain is secured with a retainer which does not require removing and replacing a rivet to adjust slack.

Cruise Speed Adjustment

The cruise speed adjustment can be set as follows:

1. If car cruises below engagement speed, screw orifice tube on transducer outward.
2. If car cruises above engagement speed, screw orifice tube inward.

NOTE: Each 1/4 turn of the orifice tube will change cruise speed about one mile per hour. Snug up lock nut after each adjustment.

1971-76 Brake Release Switch Adjustment

Fully depress brake pedal, then push switch and valve forward to contact bracket or arm. Pull pedal rearward with approximately 15 to 20 pounds of force to adjust switch and valve properly.

Engine Section

NOTE: For service see Chevrolet Chapter for 6-250; Oldsmobile Chapter for V8-260 and Pontiac Chapter for 1975 V8-400.

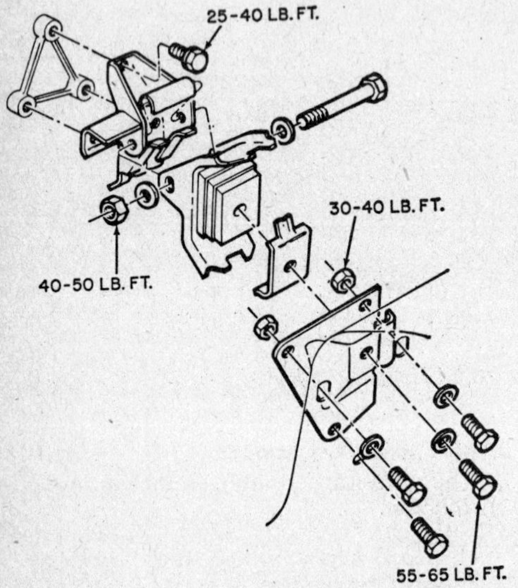

Fig. 1 Engine mounts. 1969-70 Six

ENGINE, REPLACE

1. Remove hood and drain radiator.
2. Remove fan shroud, radiator and air cleaner.
3. If equipped with A/C, disconnect compressor brackets and position out of way.
4. Remove power steering from bracket and position out of way.
5. Disconnect all hoses, linkages and electrical connections from engine.
6. Disconnect exhaust pipes and remove converter cover.
7. Remove flywheel to converter or pressure plate bolts, engine to transmission bolts and motor mount bolts.
8. Support transmission and remove engine.

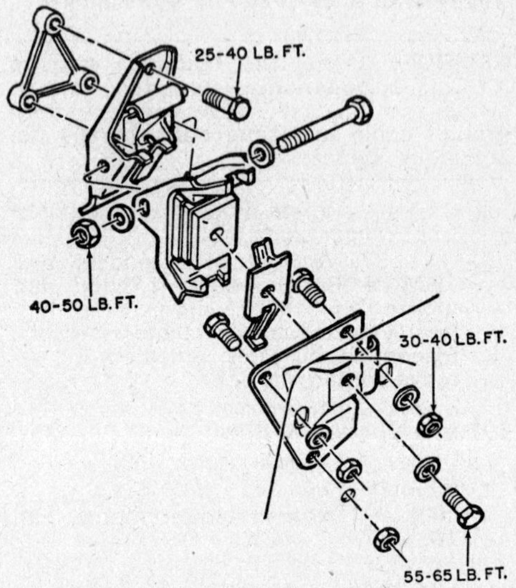

Fig. 2 Engine mounts. 1971 Six

ENGINE MOUNTS REPLACE

1. Raise car and provide frame support at front of car.
2. Support weight of engine at forward edge of pan.
3. Remove mount to engine block bolts. Raise engine slightly and remove mount to mount bracket bolt and nut. Remove mount.
4. Reverse above procedure to install.

NOTE: On 1973-74 V8-350, 455 torque mount to engine bolt to 50-75 ft. lbs., torque through bolt to 40-55 ft. lbs. On 1975-76 V6-231 & V8-350, 455 torque mount to engine bolt to 50-65 ft. lbs., torque through bolt to 40-55 ft. lbs. On 1969-72 models torque to specifications as shown in Figs. 1 thru 8.

CYLINDER HEADS

Some cylinder head gaskets are coated with a special lacquer to provide a good seal once the parts have warmed up. Do not use any additional sealer on such gaskets. If the gasket does not have this lacquer coating, apply suitable sealer to both sides.

1969-76

1. Drain coolant and disconnect battery.
2. Remove intake manifold.
3. When removing right cylinder head, remove Delcotron and/or A/C compressor with mounting bracket and move out of the way. *Do not disconnect hoses from air compressor.*
4. When removing left cylinder head, remove oil dipstick, power steering pump and move out of the way with hoses attached.
5. Disconnect exhaust manifold from head to be removed.
6. Remove rocker arm shaft and lift out push rods. On 1973-76 Apollo & Skylark models equipped with V8-350 engines. To replace left cylinder head, disconnect power brake unit hose at rear of cylinder head. Remove left front engine mount bolt and loosen right front engine mount bolt. Raise engine until exhaust manifold clears steering gear.
7. Remove cylinder head.
8. Reverse procedure to install and tighten bolts gradually and evenly in the sequence shown in Figs. 9, 10, and 11.

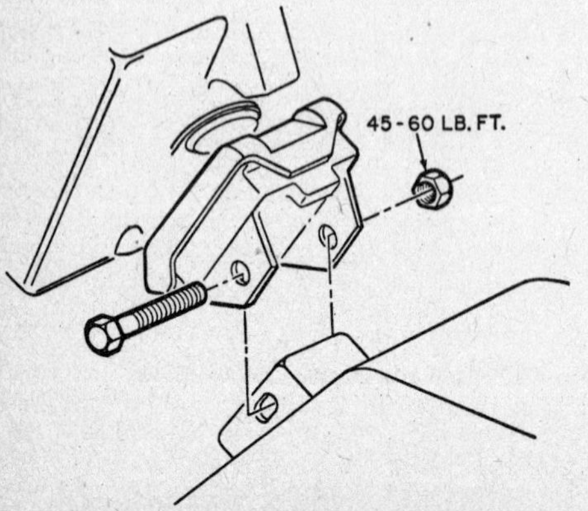

Fig. 3 Engine mounts. 1969-70 LeSabre V8-350

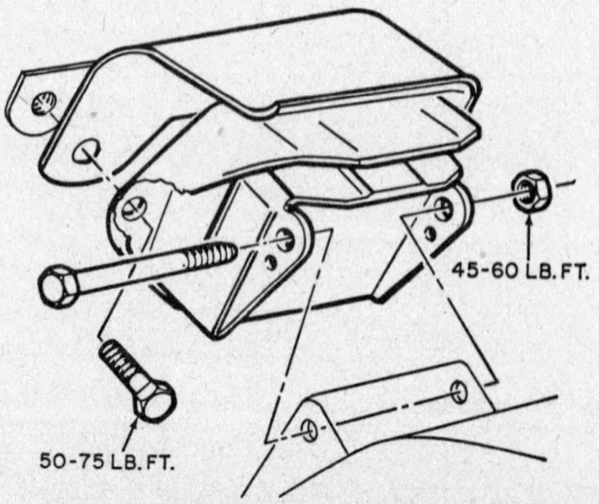

Fig. 4 Engine mounts. 1969 Wildcat, Electra, Riviera V8-400, 430; 1970 V8-455 LeSabre & Riviera

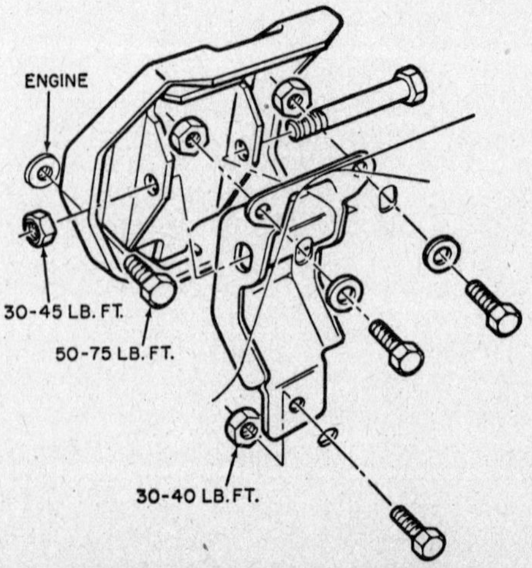

Fig. 5 Engine mounts. 1969-72 Skylark and all 1973-76 with V8-350. Typical

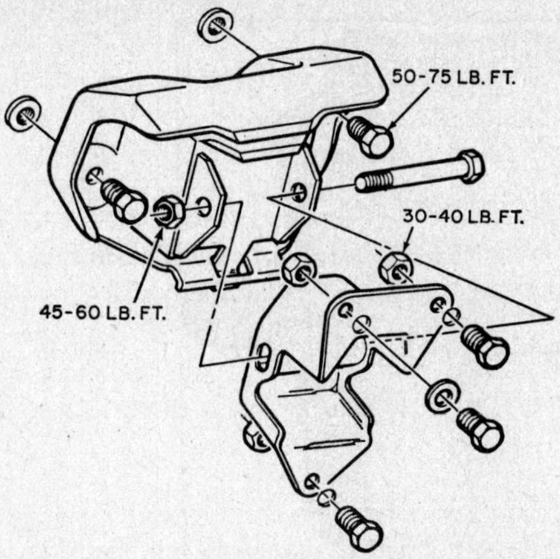

Fig. 6 Engine mounts. 1971-72 V8-350 (Exc. Skylark)

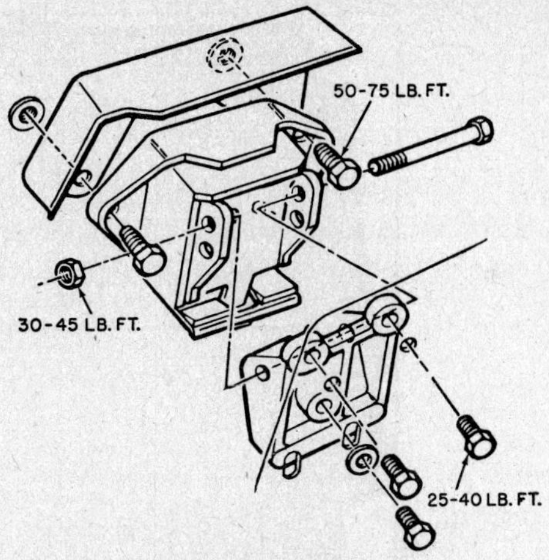

Fig. 7 Engine mounts. 1969-70 Special & Skylark V8-400; 1971-72 Skylark and all 1973-76 with V8-455. Typical

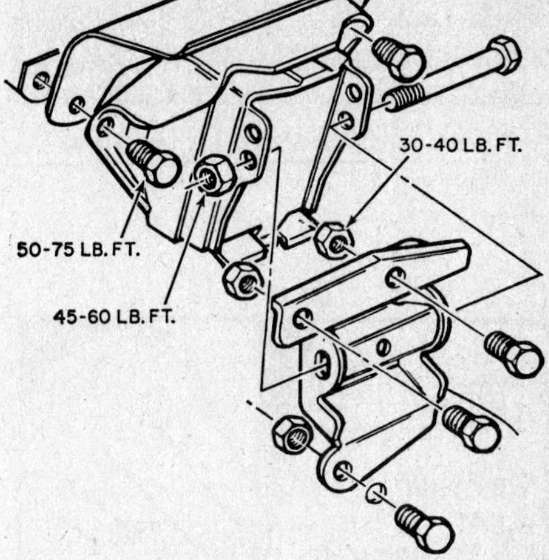

Fig. 8 Engine mounts. 1971 V8-455 Riviera; 1972 V8-455 (Exc. Skylark)

ROCKER ARMS

1970-76, V8-350, 455 & 1975-76 V6-231

A nylon retainer is used to retain the rocker arm. Break them below their heads with a chisel, Fig. 13, or pry out with channel locks. On 1970-72 engines, install rocker arms so the external rib on each pair of arms point away from the rocker arm shaft bolt, Fig. 14. On 1973-76 V8 engines, production rocker arms can be installed in any sequence since the arms are identical. V6 engine production rocker arms are left hand and right hand and must never be interchanged.

Replacement rocker arms for all engines are identified with a stamping, right (R) and left (L), Fig. 14A and must be installed as shown in Fig. 14B.

1969 V8-350, 400, 430

When installing rocker arm shaft, be sure that drill mark is facing up and toward rear of left cylinder head and toward front on right cylinder head, Fig. 15.

VALVE ARRANGEMENT

Front to Rear

V6-231 .... E-I-I-E-I-E
V8-350, 400, 430, 455 ..... E-I-I-E-E-I-I-E

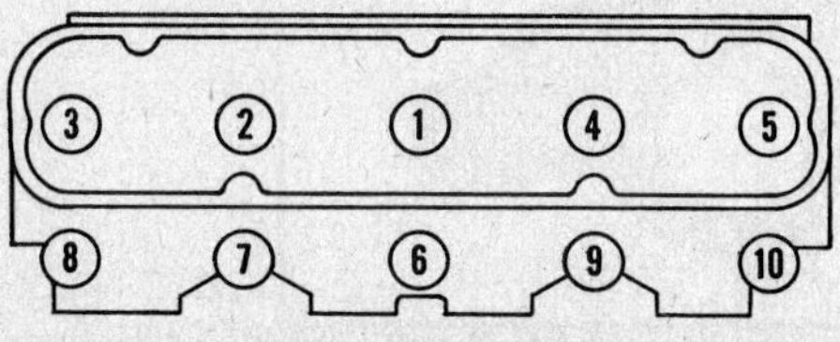

Fig. 9 Cylinder head tightening sequence. 1969 V8-350, 400, 430

VALVE LIFT SPECS.

Engine	Year	Intake	Exhaust
6-250	1969-74	.3880	.3880
6-250	1975	.3880	.4051
V6-231	1975-76	.4011	.3768
V8-350	1969	.3766	.3840
V8-400, 430	1969	.4187	.4482
V8-350	1970-76	.3818	.3984
V8-350	1976	.3873	.4930
V8-455	1970	.3873	.4584
V8-455	1971-72	.3873	.456
V8-455	1973-74	.3873	.455
V8-455[1]	1975	.3873	.445
V8-455[2]	1975	.3873	.403

[1]—LeSabre.
[2]—Exc. LeSabre.

VALVE TIMING

Intake Opens Before TDC

Engine	Year	Degrees
6-250	1969-70	62
6-250	1971-75	16
V6-231	1975-76	17
V8-350	1969-74	24
V8-350	1975	19
V8-350	1976	13.5
V8-400, 430	1969	14
V8-455	1970	18
V8-455	1971-72	12
V8-455	1973-74	14
V8-455	1975-76	10

VALVE GUIDES

The valve guides are an integral part of the cylinder head and cannot be replaced.

If valve stem clearance is excessive, the valve guide must be reamed and an oversize valve installed. Valves are available in the oversize of .010 inch.

HYDRAULIC VALVE LIFTERS

Failure of an hydraulic valve lifter, Fig. 16, is generally caused by an inadequate oil supply or dirt. An air leak at the intake side of the oil pump or too much oil in the engine will cause air bubbles in the oil supply to the lifters, causing them to collapse. This is a probable cause of trouble if several lifters fail to function, but air in the oil is an unlikely cause of failure of a single unit.

The valve lifters may be lifted out of their bores after removing the rocker arms, push rods and intake manifold. Adjustable pliers with taped jaws may be used to remove lifters that are stuck due to varnish, carbon, etc. Fig. 16 illustrates the type of lifter used.

NOTE: 1974-75 in production .010" oversize lifters are being used for oversize lifter bores. The lifter bore will be marked with an "O" and the lifter will have two grooves in the lifter body. When replacing lifters, check bores for oversize markings.

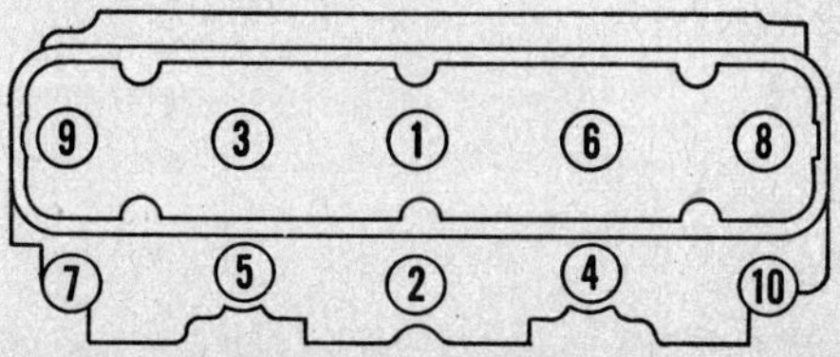

Fig. 10 Cylinder head tightening sequence. 1970-76 V8-350, 455

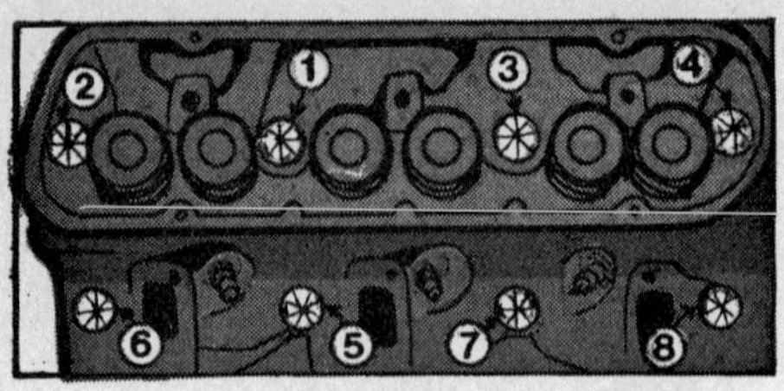

Fig. 11 Cylinder tightening sequence. 1975-76 V6-231

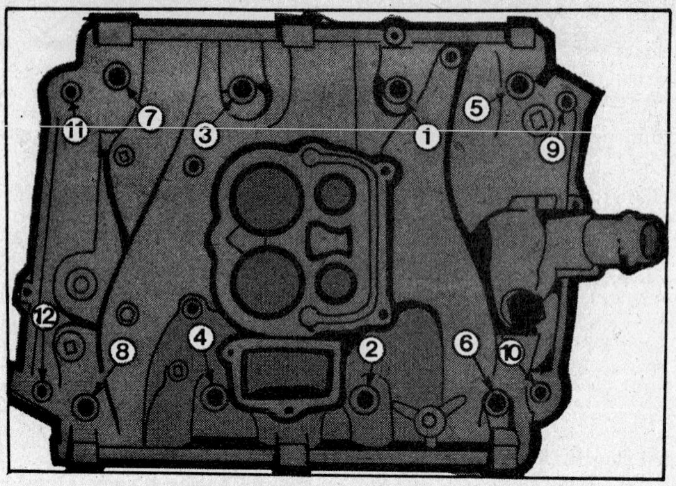

Fig. 12 Intake manifold tightening sequence. V8-350, 400, 430, 455

Fig. 13 Removing nylon retainer. 1970-76 V8-350, 455 & 1975-76 V6-231

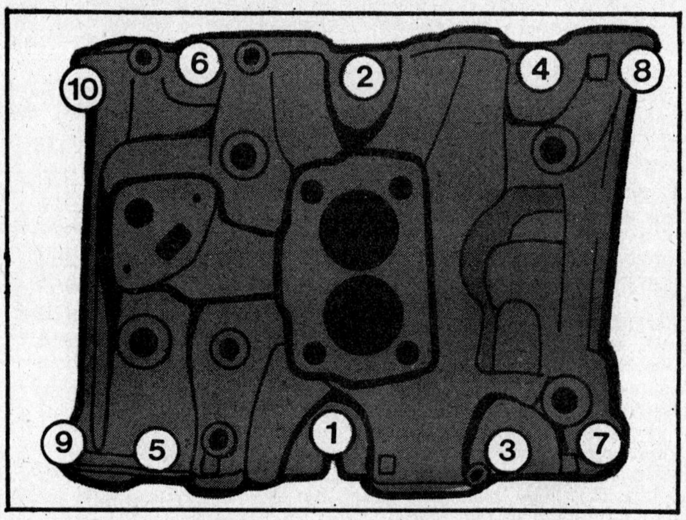

Fig. 12A Intake manifold tightening sequence. V6-231

Fig. 14 Rocker arm positioned on shaft. 1970-72 V8-350, 455

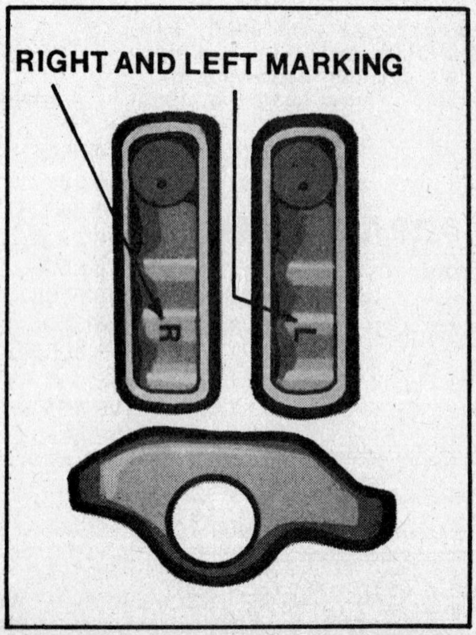

Fig. 14A Service rocker arms identification. V6-231 & 1973-75 V8-350, 455

Fig. 15 Rocker arms positioned on shaft. 1969 V8-350, 400, 430

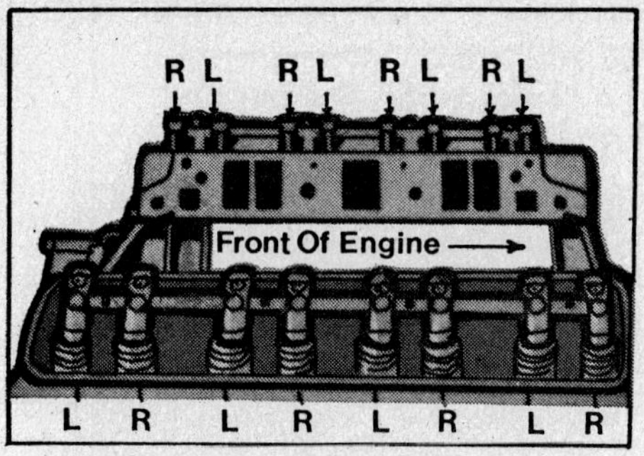

Fig. 14B Service rocker arms installation. V6-231 & 1973-76 V8-350, 455

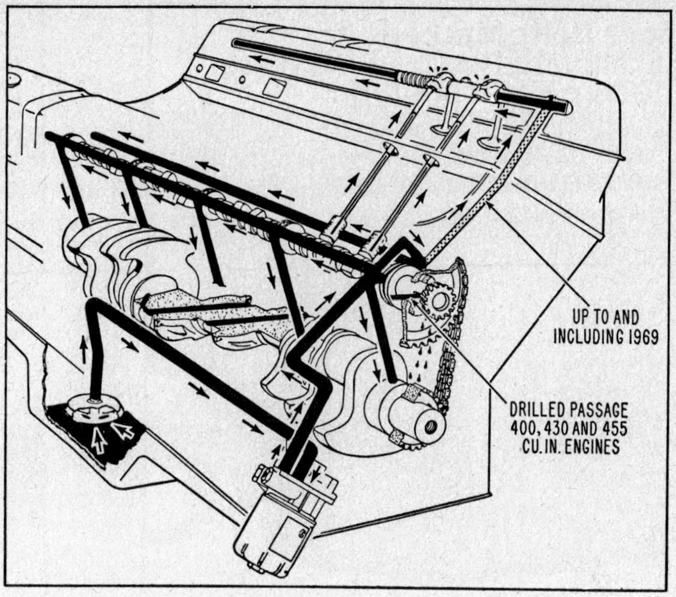

Engine lubrication system. V8-350, 400, 430, 455

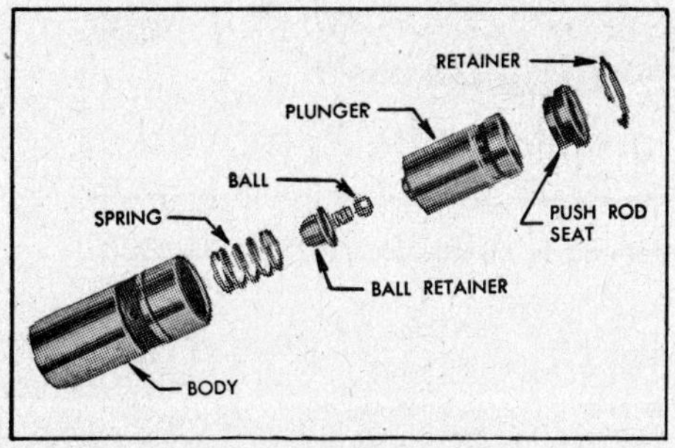

Fig. 16 Hydraulic valve lifter parts

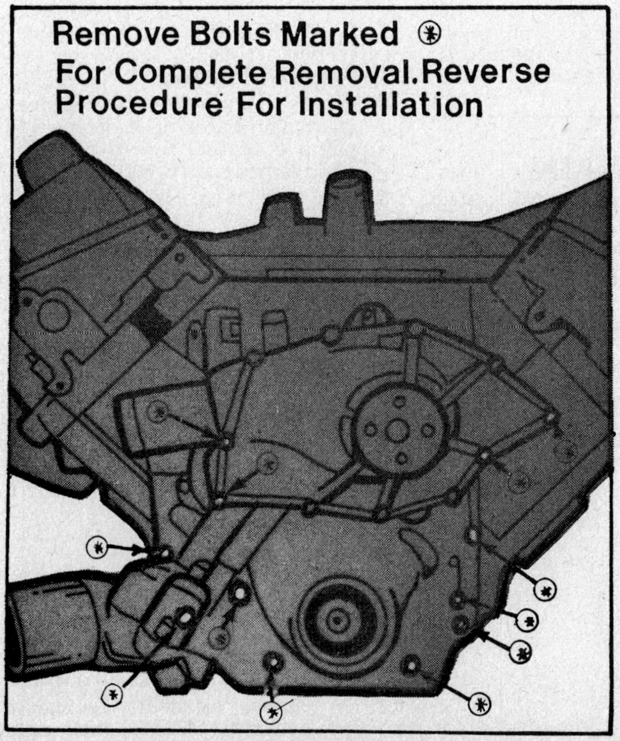

Fig. 17 Timing chain cover installation. 1969-71 V8-400, 430, 455

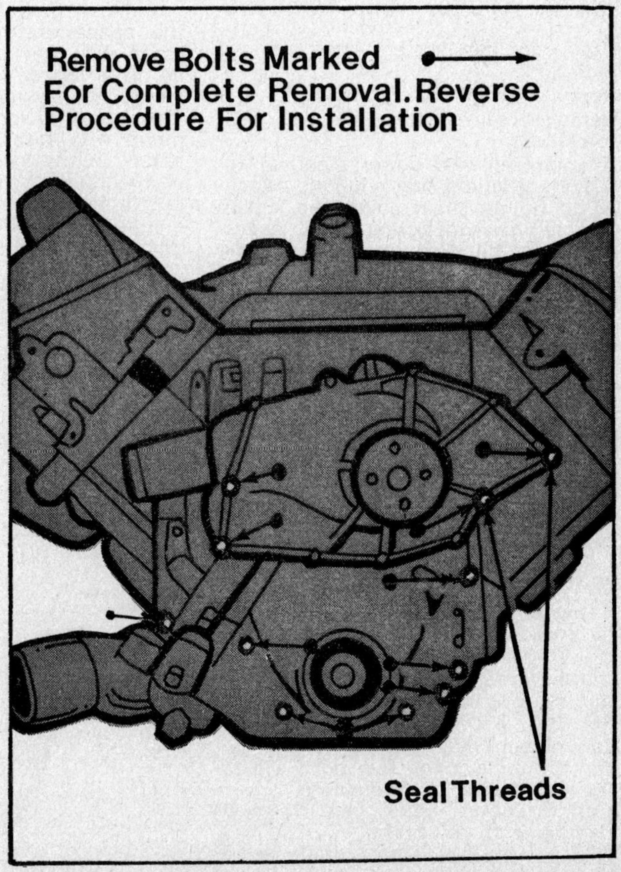

Fig. 17A Timing chain cover installation, 1972-76 V8-455

Fig. 18 Timing chain cover installation. 1969-72 V8-350

Fig. 18A Timing chain cover installation. 1973-76 V8-350

TIMING CHAIN COVER

V6-231 & V8-350, 400, 430, 455

1. Drain cooling system and remove radiator.
2. Remove fan, pulleys and belts.
3. Remove crankshaft pulley and reinforcement.
4. If equipped with power steering, remove any pump bracket bolts attached to timing chain cover and loosen and remove any other bolts necessary that will allow pump and brackets to be moved out of the way.
5. Remove fuel pump.
6. Remove Delcotron and brackets.
7. Remove distributor cap and pull spark plug wire retainers off brackets on rocker arm cover. Swing distributor cap with wires attached out of the way. Disconnect distributor primary lead. On 400, 430, remove coil.
8. Remove distributor. *If chain and sprockets are not to be disturbed, note position of distributor rotor for installation in the same position.*
9. Loosen and slide clamp on thermostat by-pass hose rearward.
10. Remove bolts attaching chain cover to block.
11. On V6-231 and V8-350 engines, remove two oil pan-to-chain cover bolts and remove cover.
12. On 400, 430, 455 engines, remove four oil pan-to-chain cover bolts. *Do not remove the five bolts attaching water pump to chain cover.* Remove cover, using care to avoid damaging oil pan gasket.
13. Reverse procedure to install, noting data shown in Figs. 17, 17A, 18 and 18A.

IMPORTANT

Remove the oil pump cover and pack the space around the oil pump gears completely full of vaseline. There must be no air space left inside the pump. Reinstall the cover using a new gasket. This step is very important as the oil pump may lose its prime whenever the pump, pump cover or timing chain cover is disturbed. If the pump is not packed it may not begin to pump oil as soon as the engine is started.

TIMING CHAIN

V6-231 & V8-350, 400, 430, 455

1. With the timing case cover removed as outlined above, temporarily install the vibration damper bolt and washer in end of crankshaft.
2. Turn crankshaft so sprockets are positioned as shown in Fig. 19. Use a sharp rap on a wrench handle to start the vibration damper bolt out without disturbing the position of the sprockets.
3. On V8-400, 430 and 455, remove the oil pan.
4. Remove oil slinger.
5. On V6-231 and V8-350 remove camshaft distributor drive gear and fuel pump eccentric. On V8-400, 430 and 455, remove sprocket bolts.
6. Use two large screwdrivers to alternately pry the camshaft sprocket then the crankshaft sprocket forward until the camshaft sprocket is free. Then remove camshaft sprocket and chain, and crankshaft sprocket off crankshaft.
7. To install, assemble chain on sprockets and slide sprockets on their respective shafts with the "O" marks on the sprockets lined up as shown.
8. Complete the installation in the reverse order of removal.

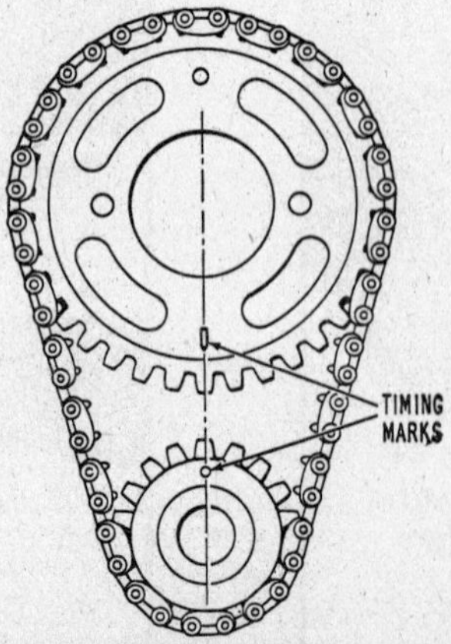

Fig. 19 Valve timing marks. V6-231 & V8-350, 400, 430, 455

CAMSHAFT

NOTE: If engine is in the car, the radiator, grille and A/C components will have to be removed. If engine is out of car, proceed as follows:

1. To remove camshaft, remove rocker arm shaft assemblies, push rods and valve lifters.
2. Remove timing chain and sprockets.

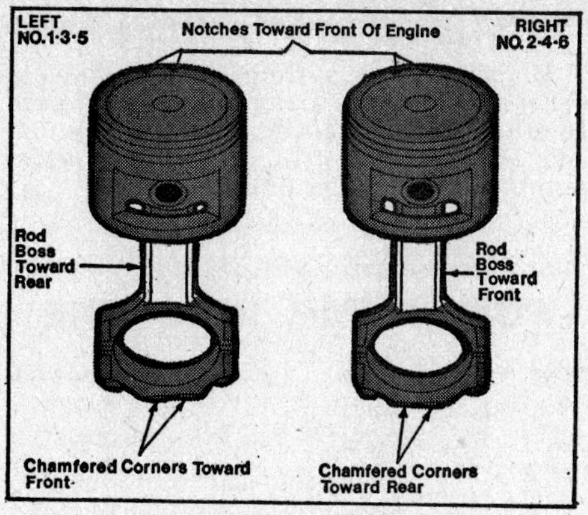

Fig. 20 Piston and rod assembly. 1975-76 V6-231

3. Slide camshaft out of engine, using care not to mar the bearing surfaces.

PISTONS & RODS, ASSEMBLE

Rods and pistons should be assembled and installed as shown in Figs. 20 through 23.

PISTONS, PINS & RINGS

Pistons are available in standard sizes and oversizes of .001, .005, .010, .020

Fig. 24 Location of undersize mark on main and rod bearing shell

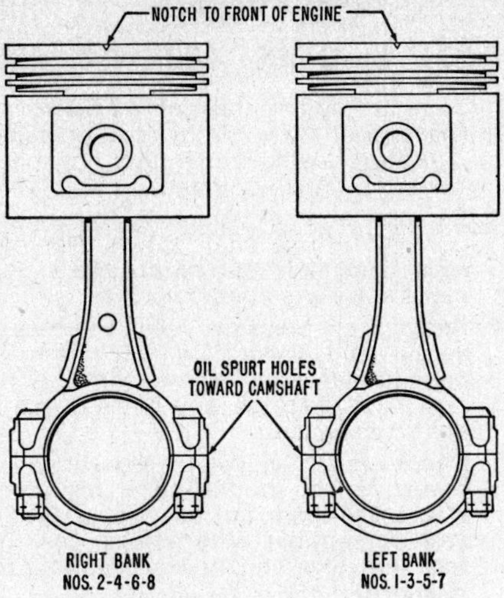

Fig. 21 Pistons and rod assembly. 1969-72 350, 400, 430, 455

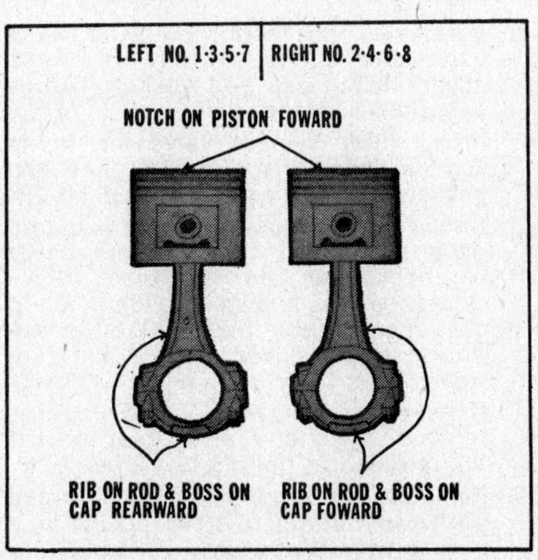

Fig. 23 Piston and rod assembly. 1973-75 V8-455

Rings are furnished in standard sizes and oversizes of .010, .020 and .030 inch.

Piston pins are supplied in standard sizes and oversizes of .003 and .005 inch.

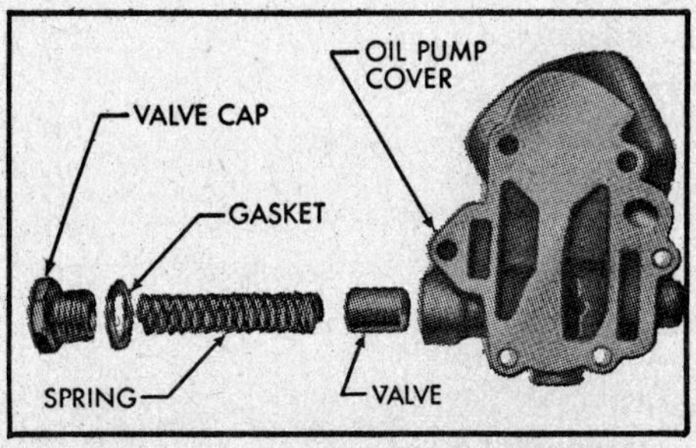

Fig. 25 Oil pump cover and by-pass valve. V6-231 & V8-350, 400, 430, 455

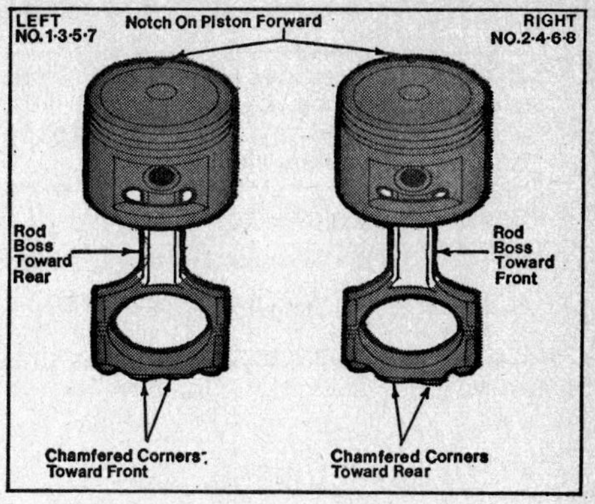

Fig. 22 Piston and rod assembly. 1973-76 V8-350

MAIN & ROD BEARINGS

Main bearings are available in standard sizes and undersizes of .001, .002, .003 and .010 inch.

Rod bearings are furnished in standard sizes and undersizes of .001, .002 and .010", Fig. 24.

CRANKSHAFT OIL SEAL

A braided oil seal is pressed into the upper and lower grooves behind the rear main bearing.

OIL PAN

6-250 Exc. 1973-75 Apollo

To remove oil pan it is necessary to remove the engine from the chassis.

1975-76 V6-231

1. Support vehicle on hoist and drain oil.
2. Remove transmission dust cover and exhaust crossover pipe.
3. Remove oil pan bolts and allow pan

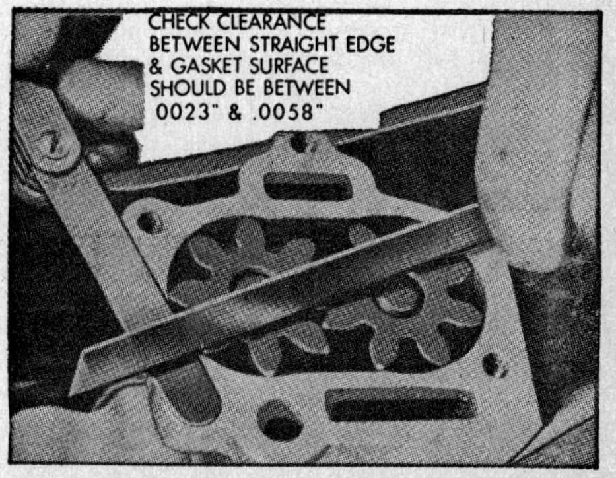

Fig. 26 Checking oil pump gear and clearance. V6-231 & V8-350, 400, 430, 455

to drop.

NOTE: To remove oil pan, front wheels must be turned to the left, also the crankshaft may have to be turned to provide clearance.

4. Reverse procedure to install.

1973-75 Apollo 6-250

Refer to the 1973-75 Six cylinder procedure in the Chevrolet Chapter.

1970-76 V8-350, 455

NOTE: On 1973-76 Intermediate models, it is no longer necessary to disconnect the idler arm. Otherwise, proceed as follows:

1. Disconnect battery and drain oil.
2. Remove fan shroud to radiator tie bar screws.
3. Remove air cleaner and disconnect linkage to throttle.
4. Raise and support car on stands.
5. With manual transmission, loosen clutch equalizer bracket to frame bolts. Disconnect crossover pipe at engine.
6. With automatic transmission, remove lower flywheel housing. Remove shift linkage attaching bolt and swing out of way. Disconnect crossover pipe at engine. Disconnect idler arm at frame and push steering linkage forward to crossmember. Remove front engine mount bolts and raise engine by placing jack under crankshaft pulley mounting.

NOTE: If car is air conditioned, at this point it will be necessary to place a support under right side of transmission prior to raising engine to prevent transmission from cocking to the right when raised.

7. Remove oil pan. It may be necessary to position crankshaft so 1 and 2 crankpin and counterweight will not interfere with front of pan.

1969 V8-350, 400, 430

1. Disconnect battery and drain oil.
2. Raise and support car on stands.
3. With manual transmission, loosen clutch equalizer bracket-to-frame bolts. Remove exhaust crossover pipe, and front engine mounting bolts. Remove fan shroud-to-radiator tie bar screws.
4. With automatic transmission, remove lower flywheel housing. Remove shift linkage attaching bolts and swing out of the way (LeSabre only). Remove front engine mounting bolts. Remove fan shroud-to-radiator tie bar screws.
5. Raise engine by placing jack under crankshaft pulley mounting.
6. Unfasten and remove pan.
7. Reverse procedure to install. loosen clutch equalizer bracket-to-frame bolts. Remove lower flywheel housing, exhaust crossover pipe, front engine mount bolts and fan shroud-to-radiator tie bar screws.

OIL PUMP

V6-231 & V8-350, 400, 430, 455

1. To remove pump, take off oil filter.
2. Disconnect wire from oil pressure indicator switch in filter by-pass valve cap (if so equipped).
3. Remove screws attaching oil pump cover to timing chain cover. Remove cover and slide out pump gears. Replace any parts not serviceable.
4. Remove oil pressure relief valve cap, spring and valve, Fig. 25. Remove oil filter by-pass valve cap, spring and valve. Replace any parts of valve not serviceable.
5. Check relief valve in its bore in cover. Valve should have no more clearance than an easy slip fit. If any perceptible side shake can be felt, the valve and/or cover should be replaced.
6. The filter by-pass valve should be flat and free of nicks and scratches.

Assembly & Installation

1. Lubricate and install pressure relief valve and spring in bore of pump cover. Install cap and gasket. Torque cap to 30-35 ft-lbs.
2. Install filter by-pass valve flat in its seat in cover. Install spring, cap and gasket. Torque cap to 30-35 ft-lbs.
3. Install pump gears and shaft in pump body section of timing chain cover to check gear end clearance. Check clearance as shown in Fig. 26. If clearance is less than .0018″ check timing chain cover for evidence of wear.
4. If gear end clearance is satisfactory, remove gears and pack gear pocket *full* of vaseline, not chassis lube.
5. Reinstall gears so vaseline is forced into every cavity of gear pocket and between teeth of gears. *Unless pump is packed with vaseline, it may not prime itself when engine is started.*
6. Install cover and tighten screws alternately and evenly. Final tightening is 10-15 ft-lbs. torque. Install filter on nipple.

BELT TENSION DATA

	New lbs.
1976—	
Generator	125
1975—	
Air Condition	150
Air Pump	70
Generator	125
Power Steering	150
1969-74—	
Air Condition	100
Generator	80
Power Steering—	
Exc. V8-Models	80
V8-Models	90

WATER PUMP, REPLACE

Drain cooling system, being sure to drain into a clean container if antifreeze solution is to be saved. Remove the fan belt and disconnect all hoses from water pump. Remove water pump.

FUEL PUMP PRESSURE

Year	Engine	Pressure, Lbs.
1969-71	6-250	4-5
1973	6-250	4-5
1974	6-250	3½-4½
1975	6-250	3½-4½
1975	V6-231	3 Min.
1969-70	V8-350	4¼-5¾
1969	V8-400 (Exc. Stage I)	5½-7
1969	V8-400 Stage I	6-8
1969	V8-430	5½-7
1970	V8-455	4¼-5¾
1971	V8 All	4¼-5¾
1972-74	V8-350	3
1972-74	V8-455	4½
1975	V8-260	5½-6½
1975-76	V8-350	3 Min.
1975	V8-400	5½-6½
1975-76	V8-455	4½ Min.

FUEL PUMP, REPLACE

1969-70 Riviera Electric Pump

These models have a turbine type electric pump located at the lower end of the fuel pick-up pipe in the bottom of the tank. This pump runs continuously whenever the engine is running thus maintaining a steady pressure whether fuel is needed or not. To replace, proceed as follows:

1. Raise car, disconnect two terminal connector at tank and remove ground wire screw.
2. Lower car and pull back trunk floor mat.
3. Remove five screws from access hole cover and remove cover.
4. Disconnect fuel hose from tank unit.
5. Unscrew retaining cam ring and remove fuel pump-tank unit assembly.
6. To remove pump from tank unit, remove flat wire conductor from plastic clip on fuel tube.
7. Squeeze clamp and pull pump straight back about ½″.
8. Remove two nuts and washers from pump terminals.
9. Squeeze clamp and pull pump straight back, take care to prevent bending of circular support bracket.
10. Reverse procedure to install.

All Models-Mechanical Pump

NOTE: Before installing the pump, it is good practice to crank the engine so that the nose of the camshaft eccentric is out of the way of the fuel pump rocker arm when the pump is installed. In this way there will be the least amount of tension on the rocker arm, thereby easing the installation of the pump.

1. Remove all gasket material from the pump and block gasket surfaces. Apply sealer to both sides of new gasket.
2. Position gasket on pump flange and hold pump in position against its mounting surface. Make sure rocker arm is riding on camshaft eccentric.
3. Press pump tight against its mounting. Install retaining screws and tighten them alternately.
4. Connect fuel lines. Then operate engine and check for leaks.

Clutch and Transmission Section

NOTE: 1976 linkage adjustment information is in this section. Repair procedures on both automatic and manual shift transmissions are covered elsewhere in this manual. Procedures for removing automatic transmissions as well as linkage adjustments on 1969-75 models are included in the automatic transmission chapters. See Chapter Index.

CLUTCH PEDAL, ADJUST

1975-76 Apollo, Skylark & Century

1. Disconnect return spring from clutch fork.
2. Rotate clutch lever and shaft assembly until pedal is against rubber bumper on dash brace.
3. Push outer end of clutch fork rearward until throwout bearing lightly contacts pressure plate fingers.
4. Install lower push rod in gauge hole and increase length until all lash is removed.
5. Install swivel or rod in hole furthest from centerline of lever and shaft assembly, then install retainer.
6. Tighten lock nut and spacer against swivel and connect clutch fork return spring.
7. Check clutch pedal free travel which should be 55/64 inch to 1 29/64 inch.

1969-74

Adjust linkage to provide 5/8" to 7/8" pedal free play. On LeSabre, Wildcat and G.S. 400 units, remove swivel retainer, disconnect swivel from equalizer arm and turn swivel to obtain proper adjustment. On other units, loosen lock nut and turn rod. When proper clearance is reached, tighten lock nut.

CLUTCH, REPLACE

1969-76

1. Remove transmission.
2. Remove pedal return spring from clutch fork. *On LeSabre, Wildcat and G.S. 400, disconnect rod assembly from clutch fork.*
3. Remove flywheel housing.
4. Remove clutch throw-out bearing from clutch fork.
5. Disconnect clutch fork from ball stud by moving it toward center of flywheel housing.
6. Mark clutch cover and flywheel so it can be installed in the same position.
7. Loosen clutch cover to flywheel bolts one turn at a time to avoid bending of clutch cover flange until spring pressure is released.
8. Support pressure plate and cover assembly while removing last bolts, then remove pressure plate and driven plate.
9. Reverse procedure to install being sure to line up marks made in removal.

3 SPEED TRANS. REPLACE

All 1969-76

1. Disconnect speedometer cable from driven gear fitting.
2. Disconnect shift control rods from shifter levers at transmission.
3. Remove propeller shaft.
4. Support rear of engine and remove transmission crossmember.
5. Remove two top transmission attaching bolts and insert guide pins in these holes.
6. Remove two lower bolts and slide transmission straight back and out of vehicle.
7. Reverse procedure to install.

3 SPEED TRANS. SHIFT LINKAGE

1969-76

Column Shift

1. Place column selector lever in Reverse detent, making sure the steering column selector plate engages lower most column lever (1st-reverse).
2. Loosen 1st-reverse adjusting clamp.
3. Shift transmission lever into reverse and tighten the 1st-reverse clamp to 17-23 ft-lbs.
4. Shift transmission levers into neutral and loosen 2nd-3rd clamp.
5. Install 3/16" diameter rod through 2nd-3rd lever selector plate and the 1st-reverse lever and alignment plate.
6. Tighten 2nd-3rd shift rod clamp to 17-23 ft-lbs.

Floor Shift

1. Place transmission levers in Neutral.
2. Loosen shift rod adjusting clamp bolts on shifter assembly.
3. Insert 1/4" drill rod through shift assembly and shift levers.
4. Tighten clamp bolts to 17-23 ft-lbs.
5. To adjust back drive linkage, shift transmission into Reverse. *This must be done using selector lever inside the car.* Loosen clamp bolt.
6. Push back drive rod up against stop in steering column. Tighten clamp bolt.

4 SPEED TRANS. REPLACE

1969-73 G.S. 350, 400, 455

1. Disconnect speedometer cable and remove driven gear.
2. Disconnect shift control rods from transmission.
3. Remove propeller shaft.
4. Support rear of engine and remove transmission support.
5. Remove two top transmission-to-flywheel housing bolts and insert guide pins.
6. Remove two lower bolts.
7. Slide transmission back and out
8. Reverse procedure to install.

4 SPEED TRANS. SHIFT LINKAGE

1969-73 G.S. 350, 400, 455

Floor Shift

1. Place transmission in neutral.
2. Adjust all three shift rods so a 1/4" drill rod can be installed through

shifter assembly and shift levers.
3. Tighten swivel nuts to 17-23 ft-lbs.
4. On 1969-73 models, using shift handle inside car only, shift transmission into Reverse, loosen clamp bolt and push back drive rod up against stop in steering column. Tighten clamp bolt.

1976 AUTO. TRANS. LINKAGE ADJUST

This adjustment is the same as that described for previous models and is found elsewhere in this manual.

Rear Axle, Propeller Shaft & Brakes

REAR AXLES

Figure 1 illustrates the type rear axle assemblies used on Buicks. When necessary to overhaul any of these units, refer to the *Rear Axle Specifications* table in this chapter.

NOTE: Canadian built Specials and Skylarks may use a "C" type axle. This design axle employs "C" locks to retain the axle shafts. Service on this type axle is covered in the Chevrolet section of this manual.

1969-76 All

In this rear axle, Fig. 1, the drive pinion is mounted in two tapered roller bearings which are preloaded by two selected spacers at assembly. The pinion is positioned by shims located between a shoulder on the drive pinion and the rear bearing. The front bearing is held in place by a large nut.

The differential is supported in the carrier by two tapered roller side bearings. These are preloaded by inserting shims between the bearings and the pedestals. The differential assembly is positioned for proper ring gear and pinion backlash by varying these shims. The ring gear is bolted to the case. The case houses two side gears in mesh with two pinions mounted on a pinion axle which is anchored in the case by a spring pin. The pinions and side gears are backed by thrust washers.

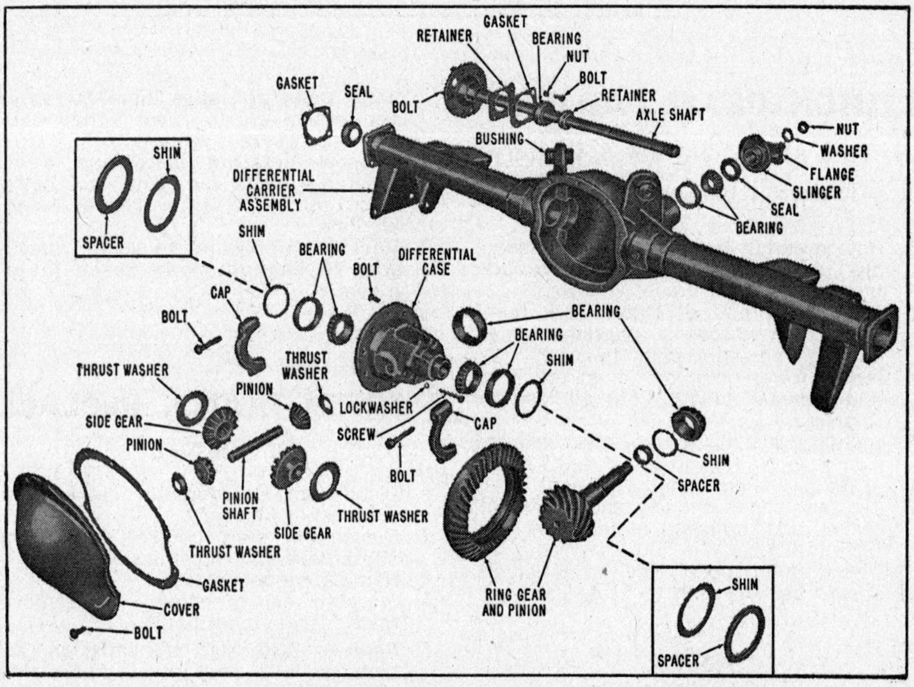

Fig. 1 Rear axle assembly. Roller bearings are used in the axle shafts of some 1969-76 models

REAR AXLE ASSEMBLY, REPLACE

1970-76

It is not necessary to remove the rear axle assembly for any normal repairs but if the housing must be replaced the assembly may be removed as follows:

1. Raise car high enough and support using jack stands under both frame side rails.
2. Mark rear universal joint and flange for proper reassembly and disconnect rear joint.
3. Push propeller shaft as far forward as possible and wire up out of way.
4. Disconnect parking brake cables and rear brake hose. Cover brake hose opening to prevent entrance of dirt.
5. Support axle with jack and disconnect shock absorbers at lower ends.

   On 1971-76 Estate Wagons, disconnect exhaust system right side by removing exhaust hanger screw at rear frame crossmember. On Apollo, Skylark and 1971-76 Estate Wagons, remove leaf spring as outlined under "Leaf Spring, Replace" procedure in the Rear Suspension section.
6. On 1969-70 Riviera, disconnect track bar at axle housing.
7. Disconnect upper control arms at axle housing.
8. Disconnect lower control arms and remove axle assembly.

1969 All

It is not necessary to remove the rear axle assembly for any normal repairs. The axle shafts and carrier assembly can easily be removed from the vehicle, leaving the axle housing in place.

1. Raise rear end of car with rear axle hanging on shock absorbers.
2. Mark rear universal joint and pinion flange for proper reassembly. *These parts are carefully balanced in production and assembled with heavy sides opposite. For this reason they should be reassembled the same way.*
3. Disconnect rear universal joint from pinion flange by removing two U-bolts. Wire propeller shaft to exhaust pipe to support it out of the way.
4. Remove rear wheels and brake drums.

5. Remove axle shafts as outlined below.
6. Remove all cover bolts and break cover loose at the bottom to allow lubricant to drain.
7. Remove carrier assembly by prying or by use of a slide hammer.
8. Reverse foregoing procedure to install the carrier assembly, being sure to connect the rear universal joint to the pinion flange according to the alignment marks made previously.

AXLE SHAFT, REPLACE

1969-76

IMPORTANT

Design allows for axle shaft end play of .018″ on all 1969-72 and 1973-76 "B" and "O" axles. On 1973-76 "C,," "G" and "K" axles, allowable end play is .022″ and .032″ on "P" axles. These axles may be identified by the third letter located on the right rear tube on the forward side. This end play can be checked with the wheel and brake drum removed by measuring the difference between the end of the housing and the axle shaft flange while moving the axle shaft in and out by hand.

End play over this is excessive. Compensating for all the end play by inserting a shim inboard of the bearing in the housing is not recommended since it ignores the end play of the bearing itself, and may result in improper seating of the gasket or backing plate against the housing. If end play is excessive, the axle shaft and bearing assembly should be removed and the cause of the excessive end play determined and corrected.

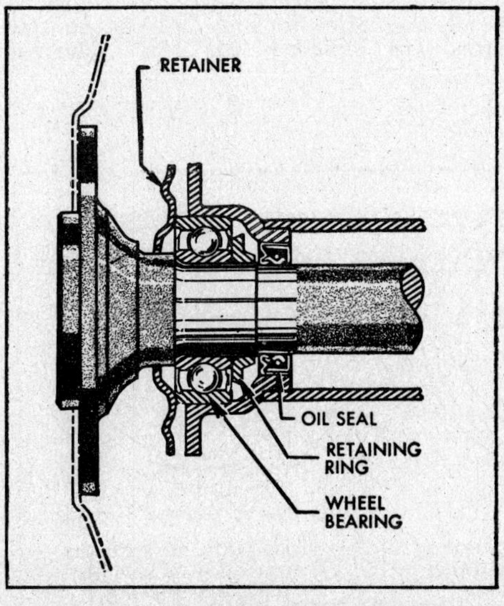

Fig. 3 Axle shaft bearing and oil seal. 1969-76 Senior Series except Estate Wagon

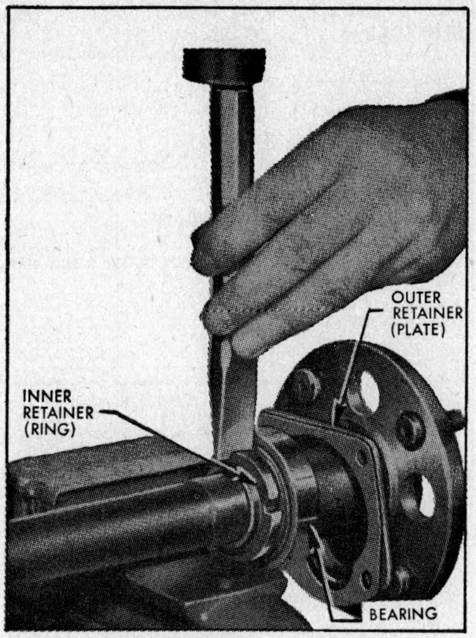

Fig. 2 Removing axle shaft bearing retainer

Removing Axle Shaft

1971-76 C, G, K & P Axles

1. Remove wheels and brake drums.
2. Remove bolts and differential cover and allow lubricant to drain.
3. Remove pinion shaft lock bolt and pinion shaft then push axle shafts inward, remove C-lock and axle shaft.

1969-70 All & 1971-76 B & O Axles

1. Remove wheels and brake drums.
2. Remove retainer plate nuts. Pull retainers clear of bolts and reinstall two opposite nuts to hold brake backing plate in place.
3. Pull axle shaft assembly using a puller.

Replacing Axle Shaft Bearings & Seals

1971-76 C, G, K & P Axles

To remove the axle shaft bearing and seal, use Remover Tool J-23689 and to install, use Installer Tool J-23690.

1971-76 C, G, K & P Axles

1. Place axle shaft in a vise so that the retainer ring rests on vise jaws. Use a chisel and a hammer to crack the ring, Fig. 2.
2. Press bearing off shaft and remove seal. Inspect seal running surface for bad spots and replace if necessary, Figs. 3 and 4.

NOTE: Before installing seal, apply grease between seal lips to avoid damaging the seal.

3. Press bearing against shoulder on shaft and retainer ring against bearing using Installer tool J-21022 for intermediate models and J-8609 for full size models.

Axle Shaft, Install

1. Apply a coat of wheel bearing grease in wheel bearing and seal recess.
2. For 1969-70 all and B & O axles, apply a coat of wheel bearing grease on seal surface of shaft and install shaft.

NOTE: If Posi-Traction, use only positive traction lubricant.

3. For C, G, K and P axles:
   a. Install axle shaft through seal and bearing and through side gear as far in as possible.

   NOTE: Do not let shaft drag across seal lip and apply grease between seal lips.

   b. Install C-lock and move axle shaft outward to bottom C-lock in recess of side gear.
   c. Install pinion shaft and torque lock bolt to 15 ft. lbs.
   d. Install gasket and cover and torque bolts to 30 ft. lbs. After 20 minutes, retorque bolts to 30 ft. lbs.
   e. Install correct type and amount of lubricant.
4. Install retainer nuts and torque to 35 ft. lbs. on Century models and 55 ft. lbs. on all other models.
5. Install drum and wheel and torque nuts to 70-75 ft. lbs.

PROPELLER SHAFT

NOTE

When service is required, the propeller

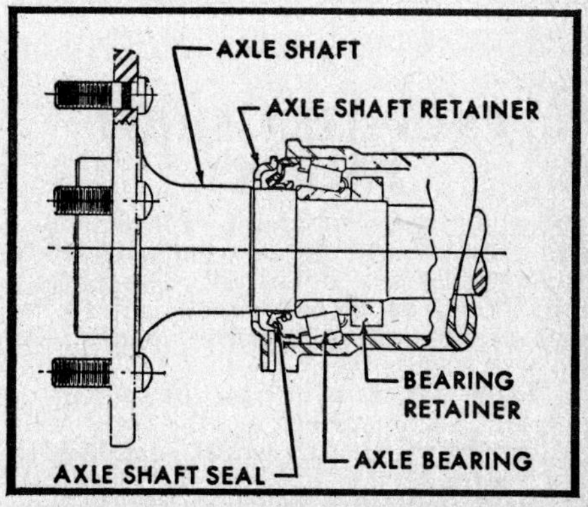

Fig. 4 Axle shaft, bearing and oil seal. 1969-76 Intermediate Series and 1970-76 Estate Wagon

shaft must be removed from the car as a complete assembly. While handling it out of the car, the assembly must be supported on a straight line as nearly as possible to avoid jamming or bending any of the parts.

The propeller shaft used on 1975 vehicles equipped with V6-231 engines are constructed of concentric steel tubes with rubber elements between them, Fig. 5.

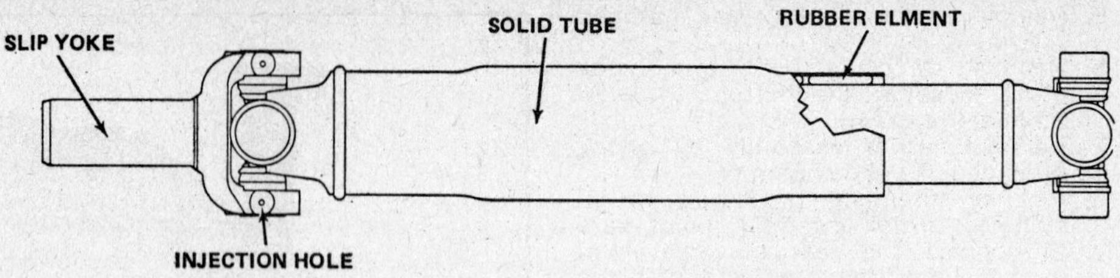

Fig. 5 Propeller shaft used with 1975-76 V6-231 engines

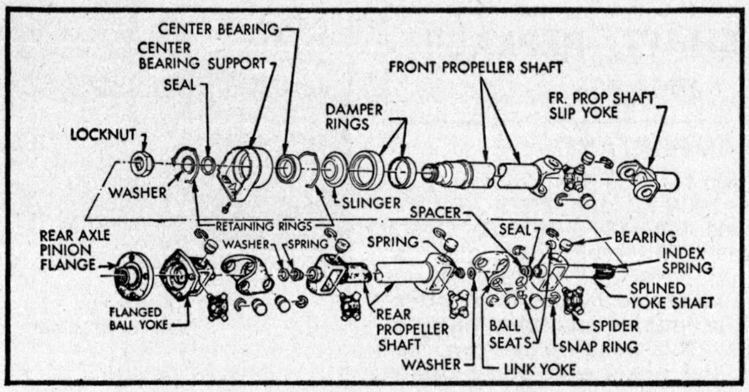

Fig. 6 Propeller shaft assembly. 1969-70 Riviera

1969-76 Exc. 1969-70 Riviera

NOTE: Two attachment methods are used to secure the propeller shaft to the pinion flange or end yoke, a pair of bolted straps or a set of bolted flanges.

1. Scribe alignment marks between propeller shaft and pinion flange or end yoke to aid reassembly.
2. Remove strap or flange bolts at rear of propeller shaft. Tape bearing cups to prevent loss of needle bearings.
3. Slide shaft assembly rearward, disengaging front yoke from transmission output shaft splines and lower from vehicle.
4. Reverse procedure to install.

NOTE: Apply grease (Grade EP 1) to internal splines of front yokes having a .060 inch vent hole (automatic transmission models). Apply engine oil to same area of front yokes without vent holes (manual transmission models).

1969-70 Riviera

1. Scribe alignment marks between rear flange and pinion flange to aid reassembly.
2. Remove center bearing support attaching bolts and the four CV joint to pinion flange bolts, Fig. 6.
3. Support rear of propeller shaft and slide assembly rearward, disengaging front yoke from transmission output shaft splines, and out from frame tunnel.
4. Reverse procedure to install.

NOTE: Lubricate front yoke internal splines with automatic transmission fluid. Fill space between lips of transmission seal with wheel bearing grease and coat seal surface of front U-joint with same grease.

PROPELLER SHAFT BALANCE

A wheel balancer of the type equipped with a strobe light can be used to facilitate balancing of the driveshaft. The pick-up unit should be placed directly under the nose of the rear axle carrier and as far forward as possible.

1. Place car on twin post lift so rear of car is supported on the rear axle housing and rear wheels are free to rotate.
2. Remove both rear wheels and tire assemblies and reinstall wheel lug nuts with flat side next to drum.
3. Mark and number driveshaft at four points 90° apart at rear of shaft just forward of balance weights.
4. Place strobe light pick-up under nose of differential.
5. With car running in gear at car speed where unbalance is at its peak, allow driveline to stabilize by holding at constant speed. Point strobe light at spinning shaft and note position of one of the reference marks.
6. Shut off engine and position shaft so reference mark will be in position noted when car was running.

CAUTION: Do not run car on hoist for extended periods due to danger of overheating transmission or engine.

7. When strobe light flashed, the heaviest point of the shaft was down. To balance shaft it will be necessary to apply weight 180° away. Screw type hose clamps can be used as weights as shown in Fig. 7.

REAR U-JOINT ANGLE

If drive line shudder, roughness, vibration, or rumble is experienced, it may be due to incorrect rear universal joint angle and this angle should be checked. Also, if there is a severe rear end collision, or if the axle housing or any control arms are replaced, the rear universal

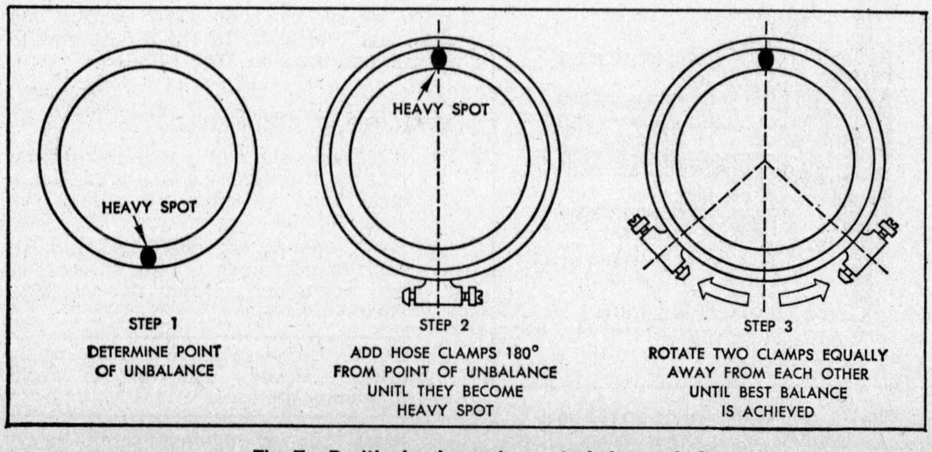

Fig. 7 Positioning hose clamps to balance shaft

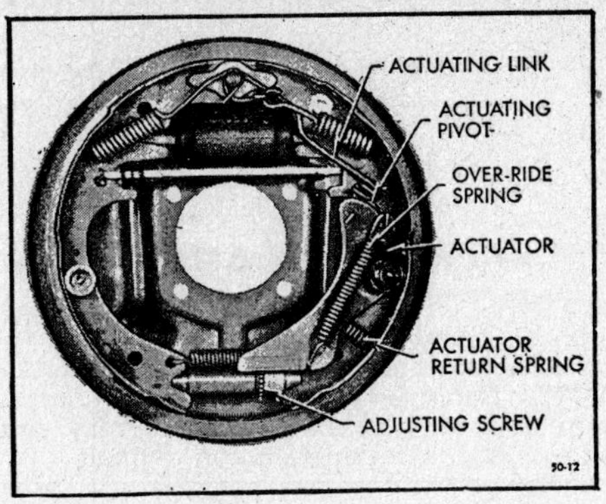

Fig. 8 Left rear wheel brake. 1969-76

joint angle should be checked and corrected if necessary. To make the check, however, special Alignment Set No. J-8973 must be used. Inasmuch as this equipment is not likely to be found in general repair shops, it is recommended that a Buick dealer having this equipment do the work.

BRAKE ADJUSTMENTS

1969-76 Self-Adjusting Brakes

These brakes, Fig. 8, have self-adjusting shoe mechanisms that assure correct lining-to-drum clearances at all times. The automatic adjusters operate only when the brakes are applied as the car is moving rearward.

Although the brakes are self-adjusting, an initial adjustment is necessary after the brake shoes have been relined or replaced, or when the length of the adjusting screw has been changed during some other service operation.

Frequent usage of an automatic transmission forward range to halt reverse vehicle motion may prevent the automatic adjusters from functioning, thereby inducing low pedal heights. Should low pedal heights be encountered, it is recommended that numerous forward and reverse stops be made until satisfactory pedal height is obtained.

NOTE

If a low pedal condition cannot be corrected by making numerous reverse stops (provided the hydraulic system is free of air) it indicates that the self-adjusting mechanism is not functioning. Therefore, it will be necessary to remove the brake drum, clean, free up and lubricate the adjusting mechanism. Then adjust the brakes as follows, being sure the parking brake is fully released.

Adjustment

1. Remove adjusting hole cover from backing plate. Turn brake adjusting screw to expand shoes until wheel can just be turned by hand.
2. Using suitable tool to hold actuator away from adjuster, Fig. 9, back off adjuster 30 notches. If shoes still drag, back off one or two additional notches.

NOTE: Brakes should be free of drag when adjuster has been backed off approximately 12 notches. Heavy drag at this point indicates tight parking brake cables.

3. Install adjusting hole cover and check parking brake adjustment.

CAUTION

If finger movement will not turn the screw, free it up. If this is not done, the actuator will not turn the screw during subsequent vehicle operation. Lubricate the screw with oil and coat with wheel bearing grease. Any other adjustment procedure may cause damage to the adjusting screw with consequent self-adjuster problems.

4. Install wheel and drum, and adjusting hole cover. Adjust brakes on remaining wheels in the same manner.
5. If pedal height is not satisfactory, drive the vehicle and make sufficient reverse stops until proper pedal height is obtained.

PARKING BRAKE, ADJUST

1969-76 All

Need for parking brake adjustment is indicated if the service brake operates with good pedal reserve but the parking brake pedal can be depressed more than eight ratchet clicks for intermediate models and sixteen for full size models, under heavy foot pressure. After making sure that the service brakes are properly adjusted, adjust the parking brake as follows:

1. Depress parking brake exactly three ratchet clicks on all except Apollo and 1974-76 Estate Wagon, two ratchet clicks on Apollo and six ratchet clicks on the 1974-76 Estate Wagon.
2. Loosen jam nut, and tighten adjusting nut until rear wheels can just be turned rearward using both hands but are locked when forward motion is attempted.
3. Tighten jam nut and release parking brake. Rear wheels should turn freely in either direction with no brake drag.

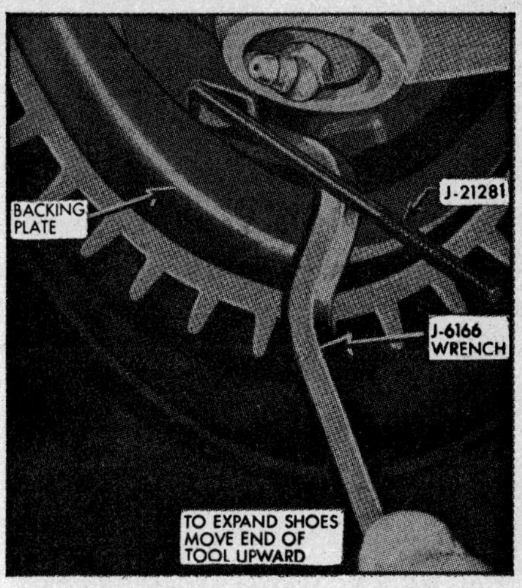

Fig. 9 Adjusting drum brakes. 1969-76

MASTER CYLINDER, REPLACE

1969-76

1. Disconnect brake pipes from master cylinder and tape end of pipes to prevent entrance of dirt.
2. On manual brakes, disconnect brake pedal from master cylinder push rod.
3. Remove two nuts holding master cylinder to dash or power cylinder and remove master cylinder from car.

POWER BRAKE UNIT

1969-76

1. Disconnect brake pipes from master cylinder and tape ends of pipes to prevent entrance of dirt. Disconnect vacuum hose from cylinder.
2. Remove four nuts holding power unit to dash.
3. Remove retainer and washer from brake pedal pin and disengage push rod eye or clevis.
4. Remove power unit from car.

Rear Suspension

SHOCK ABSORBER, REPLACE

1. With the rear axle supported properly, disconnect shock absorber from lower mounting bracket.
2. Disconnect shock absorber upper end from underbody attachment.
3. Reverse procedure to install.

LEAF SPRINGS & BUSHINGS, REPLACE

1973-76 Apollo & Skylark

1. Support vehicle at frame and rear axle, relieving tension from spring.
2. Disconnect shock absorber from lower mounting.
3. Loosen spring front mount bolt.
4. Remove spring front mounting bracket attaching screws, lower axle and remove bracket.
5. Disconnect parking brake cable from spring plate bracket.
6. Remove "U" bolts and spring plate.
7. Support spring, remove front mount bolt and disassemble rear shackle.
8. Replace rear shackle and spring eye bushings as necessary, Figs. 1 and 2.
9. Reverse procedure to install.

1971-76 Estate Wagon

1. Support vehicle at frame and support rear axle, removing tension from springs.
2. Disconnect shock absorbers from lower mounting bracket and the exhaust system right side from hanger and support system, Fig. 3.
3. Remove "U" bolts and lower spring plates.
4. Disconnect spring at front and rear attachments.
5. Reverse procedure to install.

LEAF SPRING SERVICE

1973-76 Apollo & Skylark

NOTE: The spring leaves are not serviced separately, however, the spring leaf inserts may be replaced.

1. Clamp spring in a vise and remove spring clips.
2. File peened end of center bolt to permit nut removal, remove nut and open vise slowly, allowing spring to expand.
3. Replace spring leaf inserts as necessary.
4. Use a drift to align center bolt holes, compress spring in vise and install new center bolt and nut. Peen end of bolt to retain nut.
5. Align springs and bend spring clips into position.

NOTE: Overtightening of spring clips will cause spring binding.

COIL SPRINGS, REPLACE

1. Support vehicle at frame and support rear axle with a suitable jack.
2. On 1969-72 models, place an alignment mark between rear universal joint and pinion flange for use during installation. Disconnect and slide drive shaft just far enough forward to clear rear companion flange and support drive shaft.
3. Disconnect shock absorbers from lower mounting bracket, Fig. 4, and the brake lines from wheel cylinders.
4. Lower jack fully extending springs and remove springs.
5. Reverse procedure to install, Fig. 5.

CONTROL ARMS, REPLACE

NOTE: Remove and replace one control arm at a time as axle assembly may slip sideways, making installation difficult.

Upper Control Arms

1. Support vehicle at frame and rear axle.
2. Remove control arm mount bolts from frame and axle housing attachments, Fig. 4.
3. Reverse procedure to install.

NOTE: Control arm bolts must be tightened with vehicle at curb height.

Lower Control Arm

Lower control arms may be removed and replaced using the "Upper Control Arms" procedure. However, it may be necessary to reposition the jack farther forward under carrier to aid in removing rear mount bolt. Also, a brass drift may be needed to remove mount bolts.

TRACK BAR BUSHING, REPLACE

1969-70 Riviera

1. Support rear axle so vehicle weight will be on springs.
2. Remove track bar attaching bolts from brackets on frame and axle housing, Fig. 6.
3. Reverse procedure to install.
4. If bushing is damaged follow replacement procedure below:

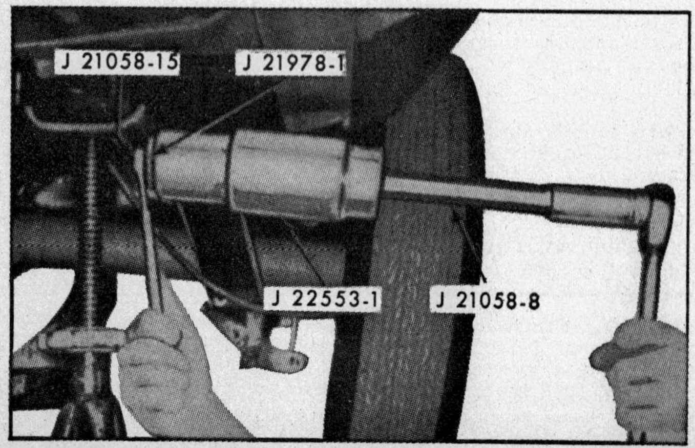

Fig. 1 Leaf spring bushings removal. 1973-76 Apollo & Skylark

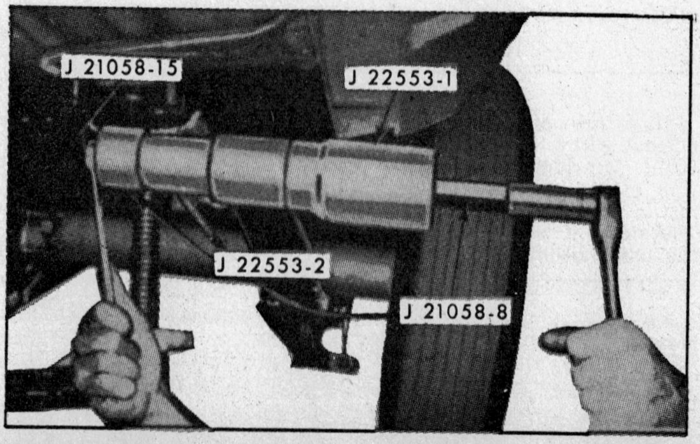

Fig. 2 Leaf spring bushings installation. 1973-76 Apollo & Skylark

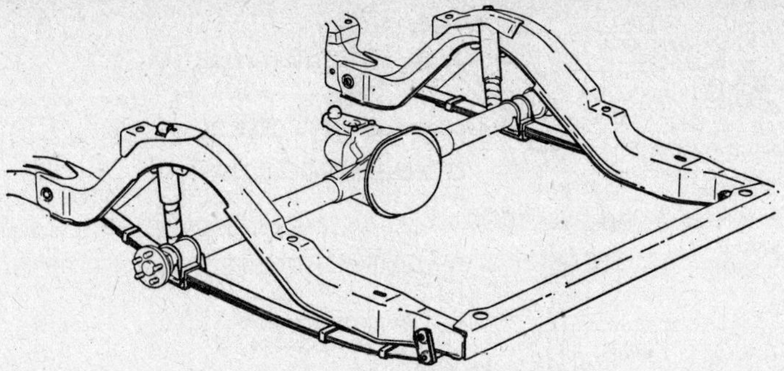

Fig. 3 Leaf spring suspension (typical)

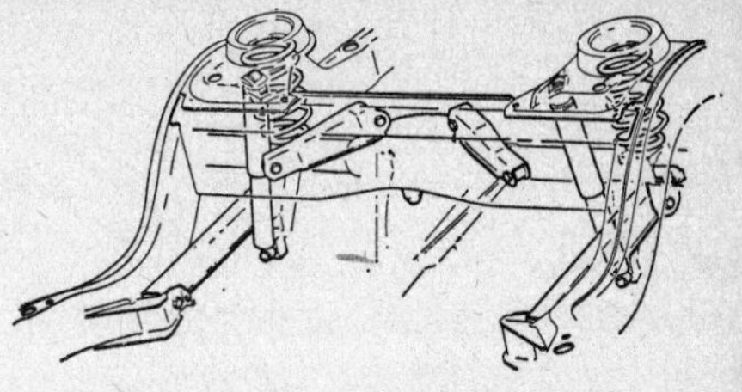

Fig. 4 Coil spring suspension. Except 1969-70 Riviera

a. Remove enough material from bushing flange by grinding or some other suitable means so track bar eye is supported by press plate.

b. Press out bushing from side with no flange using a ram with an outside diameter of 1⅞ inch.

c. Press new bushing into eye by pressing on flanged side until flange contacts eye of track bar. Bushing must be installed so slots in rubber bushing are in line with centerline of track bar, Fig. 7.

STABILIZER BAR, REPLACE

1973-76 Apollo & Skylark

1. Support vehicle at rear axle.
2. Disconnect stabilizer bar from spring plate brackets, Fig. 8.
3. Disconnect stabilizer bar from body brackets.
4. Reverse procedure to install. Tighten attaching bolts with vehicle at curb height.

1969-76 Exc. Apollo

1. Support vehicle at rear axle.
2. Remove stabilizer bar attaching bolts from brackets on lower control arms, Fig. 9.
3. Reverse procedure to install.

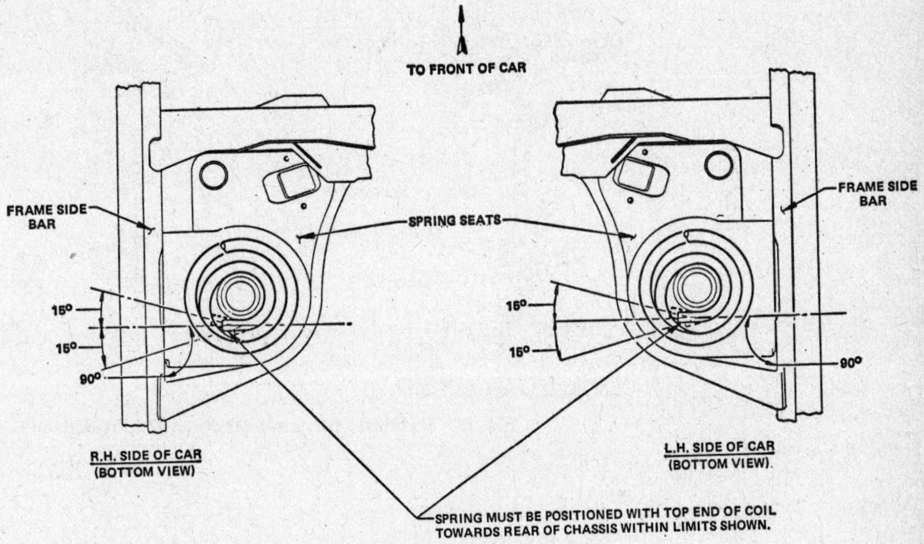

Fig. 5 Coil spring installation

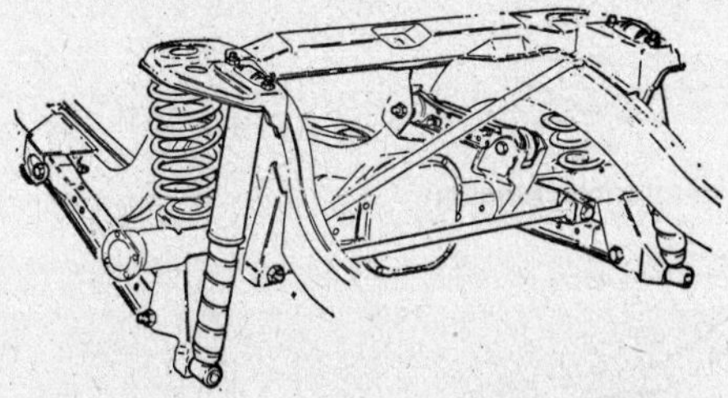

Fig. 6 Rear suspension. 1969-70 Riviera

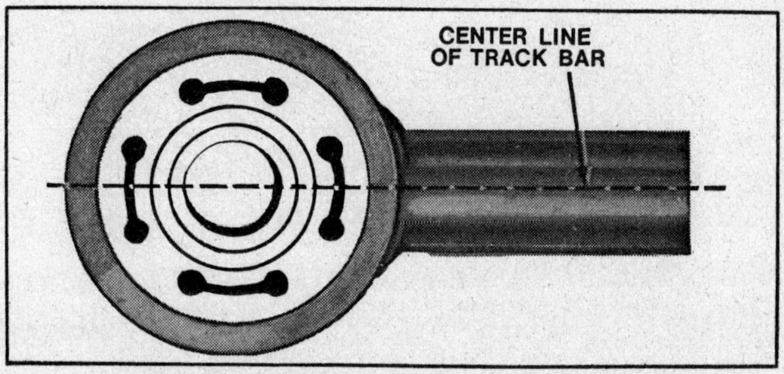

Fig. 7 Track bar bushing installation. 1969-70 Riviera

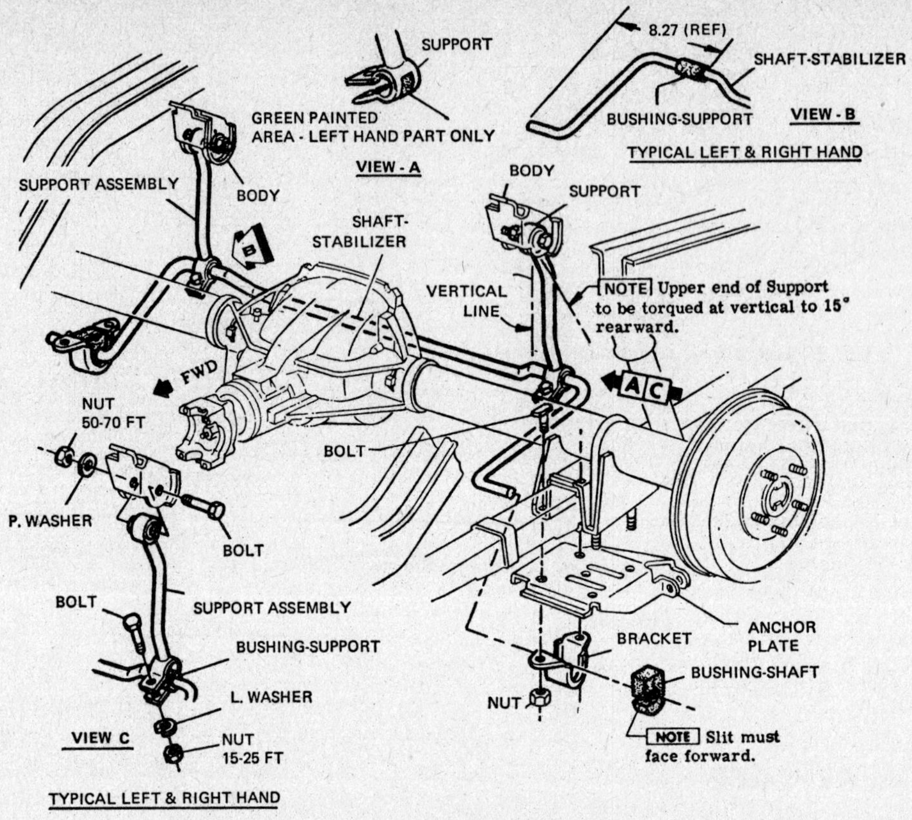

Fig. 8 Stabilizer bar installation. 1973-76 Apollo & Skylark

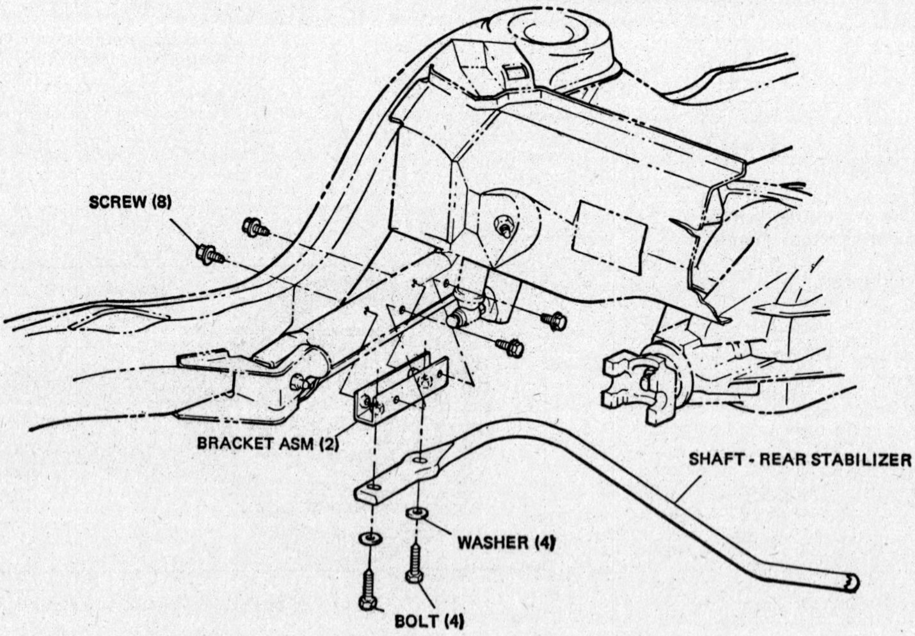

Fig. 9 Stabilizer bar installation exc. 1973-76 Apollo & Skylark

Front End and Steering Section

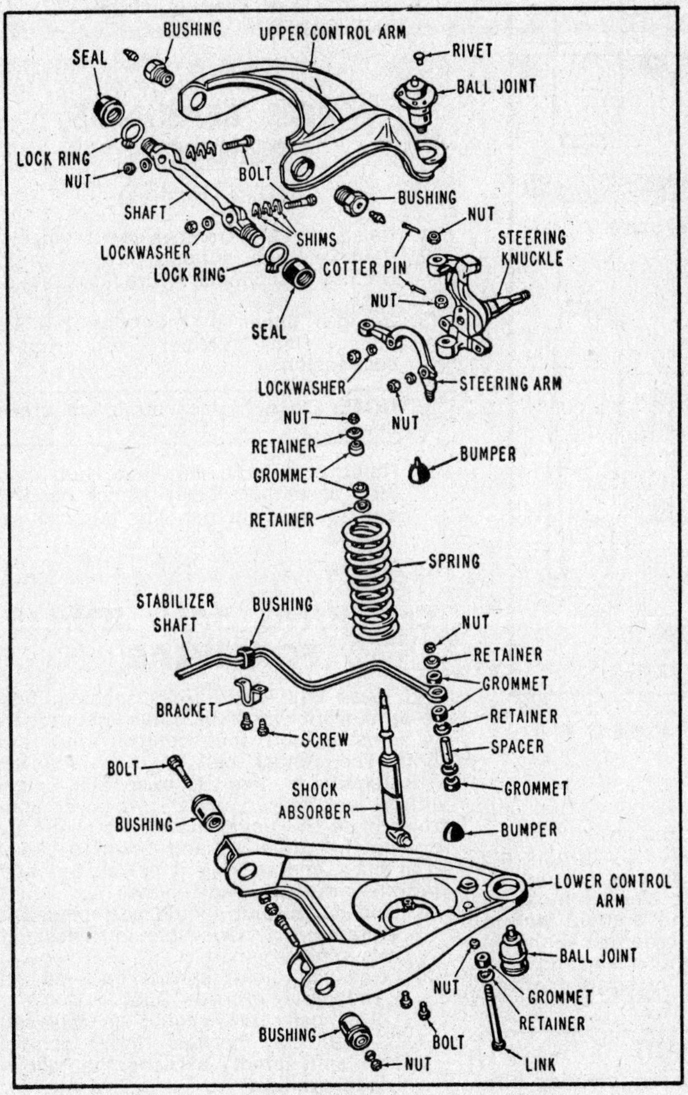

Fig. 1 Front suspension. 1969-76 intermediate models (Typical)

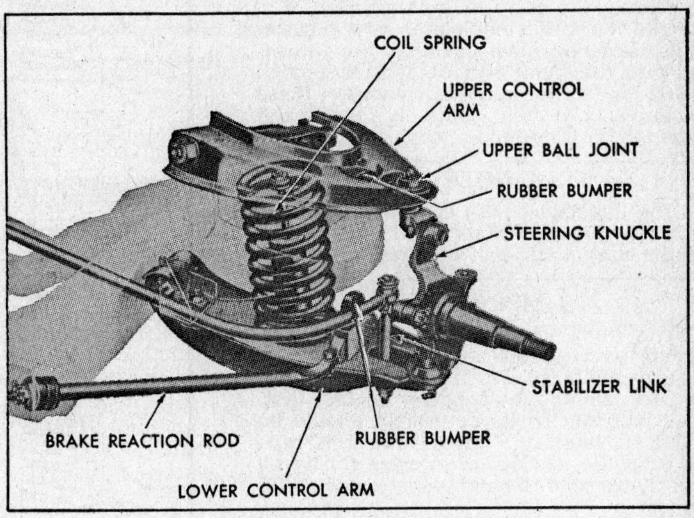

Fig. 2 Front suspension. 1969-70 Full Size. Rubber bushings are used at upper shaft ends

FRONT SUSPENSION

1971-76 Full Size

The strut rod and lower control arm used previously are replaced by a wide span lower control arm. The brake reaction rod is no longer used.

1969-76 Intermediate Models

NOTE: Rubber bushings are used at the upper shaft ends in place of the threaded steel bushings shown in Fig. 1.

A ball joint is riveted to the outer end of the upper arm and is spring loaded to insure proper alignment of the ball in the socket.

The inner end of the lower control arm has pressed-in bushings. Two bolts, passing through the bushings, attach the arm to the frame. The lower ball joint is a press fit in the arm and attaches to the steering knuckle with a castellated nut that is retained with a cotter pin.

Rubber seals are provided on upper and lower shafts and at ball socket assemblies to exclude dirt and moisture from bearing surfaces. Grease fittings are provided at all bearing locations.

SERVICE BULLETIN

Wheel Bolt Replaced: Wheel bolts should not be pressed out of a front hub. A shoulder is formed on each bolt by a swaging operation when the bolts are pressed into the hub and drum during manufacture. Pressing out a swaged bolt enlarges the bolt hole in the hub and drum, making it impossible to install the new bolt tightly.

The method recommended to remove the bolt is to secure the hub and drum in a vise, and mark the center of the bolt head with a center punch. Drill a 1/8 in. pilot hole in the head of the bolt, and then redrill with a 9/16 in. bit. Use a chisel to cut off a portion of the bolt head, and then drive out the bolt with a drift.Press the new wheel bolt into place to complete the job.

1969-70 Full Size

Referring to Fig. 2, the lower control arm assembly consists of two stamped steel plates welded together. The inner ends of the lower control arms are bolted to the frame front crossmember through rubber bushings. The outer end of each arm is connectd to the steering knuckle with a ball joint assembly pressed into the lower control arm and bolted to the steering knuckle. The lower ball joint can be removed for service replacement. Position of the lower control arms is maintained by a brake reaction rod mounted between the lower control arm and frame.

To resist fore and aft movement of the lower control arm in relation to the frame, two solid steel brake reaction rods are positioned between the lower control arms and front of frame side rails. The forward ends of the rods are rubber mounted to hold securely to the frame bracket with nuts and cotter pins. The rearward end of the brake reaction rod attaches to the lower control arm with two bolts.

Special hardened flat washers are used under the bolts and nuts to aid in maintaining required torque. The brake reaction rod must be properly installed and secured prior to checking caster and camber.

The upper control arms consist of a single stamped steel plate. Two replaceable hardened steel bushings are threaded into the inner end of each assembly. 1969 models use rubber bushings at these locations. A ball joint is positioned through the outer end of each arm.

NOTE

The upper ball joint is pressed into the control arm and is serviced only as part of the control arm-ball joint assembly.

IMPORTANT

The front suspension is initially lubricated with a special lubricant (Buick Specification No. 742). Every 6000 miles or six months, whichever occurs first, this lubricant or its equivalent should be used. If lubricants other than this type is used the lubrication interval should be shortened and should not exceed 2000 miles.

WHEEL ALIGNMENT

1969-76

Caster and camber are adjusted by shimming at the upper control arm shaft attaching points.

Adding shims at the front locations will change caster toward negative with practically no change in camber. Adding shims at the rear locations will change caster toward positive and camber toward negative. Adding equal shims at both front and rear locations will not change caster but will change camber toward negative.

To adjust, loosen both front and rear bolts to free shims for removal or addition. The maximum dimension for one shim pack is .600 inch for 1969-72 full models and .750 inch for 1969-72 intermediate models and all 1973-76 models. After installing or removing shims, torque shaft nuts to 50 ft. lbs on 1969-72 intermediate models; 75 ft. lbs. on 1969-72 full size models and all 1973-76 models except Apollo and 45-55 ft. lbs. on 1973-76 Apollo & Skylark.

TOE-IN, ADJUST

1969-76

IMPORTANT

Car must be at curb weight and running height; bounce front end and allow it to settle at running height. Steering gear and front wheel bearings must be properly adjusted with no looseness at tie rod ends. The car should be moved forward one complete revolution of the wheels before the toe-in check and adjustment is started and the car should never be moved backward while making the check and adjustment.

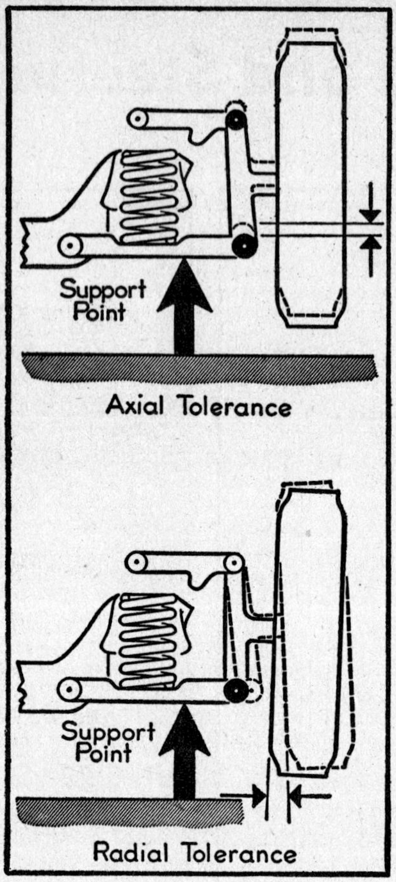

Fig. 3 Checking ball joints for wear

With front wheels in the straight ahead position, toe-in is adjusted by turning the tie rod adjusting sleeves as required. Left and right adjusting sleeves must be turned exactly the same amount but in opposite directions in order to maintain front wheels in straight ahead position when steering wheel is in straight ahead position.

IMPORTANT

The steering knuckle and steering arm "rock" or tilt as front wheel rises and falls. Therefore, it is vitally important to position the bottom face of the tie rod end parallel with the machined surface at the outer end of the steering arm when tie rod length is adjusted. Severe damage and possible failure can result unless this precaution is taken. The tie rod sleeve clamps must be straight down to 45° forward to provide clearance.

WHEEL BEARINGS, ADJUST

1969-76

1. Hand spin wheel in a forward direction and while wheel is spinning, snug up spindle nut to 19 ft. lbs. to fully seat bearings.
2. Back off nut 1/4 to 1/2 turn.
3. Snug up spindle nut by hand. Do not install cotter pin if hole in spindle lines up with a slot in spindle nut.
4. Loosen spindle nut 1/12 to 1/6 turn then insert cotter pin.
5. With bearings properly adjusted, on 1969-73 models the end play must be .002-.006 inch. On 1974-76 models the end play must be .001 to .005 inch.

WHEEL BEARINGS, REPLACE

(Disc Brakes)

1. Raise car and remove front wheels.
2. Remove tube support bracket bolt. Do not disconnect hydraulic tube or hose.
3. Remove cliper to mounting bracket bolts. Hang caliper from upper suspension.

NOTE: Do not place strain on brake line.

4. Remove spindle nut and hub and disc assembly. Inner wheel bearing and grease retainer can now be removed.

CHECKING BALL JOINTS FOR WEAR

If loose ball joints are suspected, first be sure front wheel bearings are properly adjusted and that control arms are tight. Then check ball joints as follows:

Referring to Fig. 3, raise the wheel with a jack placed under the lower control arm at the point shown. Then test by moving the wheel up and down to check axial play, and rocking it at the top and bottom to measure radial play.

1. Upper ball joint should be replaced if there is any noticeable looseness at this joint.
2. Lower ball joint should be replaced if radial play exceeds .250".
3. Lower ball joint should be replaced if axial play between lower control arm and spindle exceeds the following tolerances:

1969-70	Senior Series	.100"
1969-72	Intermediate Models	.070"
1971-72	Senior Series	1
1973	Intermediate Models	2
1973-76	Senior Series	3
1974-76	Intermediate Models	3

1—With ball joint dislodged from steering knuckle, install stud nut. Rotating torque should be 2-10 ft. lbs.
2—No looseness.
3—Visually inspect wear indicator. See Fig. 4.

BALL JOINTS, REPLACE

NOTE: On all models the upper ball joint is spring-loaded in its socket. If the ball stud has any perceptible shake or if it can be twisted with the fingers, the ball joint should be replaced.

On all models except 1973-76 Century and Regal, the lower ball joint is not spring-loaded and depends upon car weight to load the ball. The lower ball joint should never be replaced merely because it "feels" loose when in an unloaded condition.

Upper ball joints on 1969-70 full size models are pressed into the control arm and are not serviced separately. Upper ball joints on 1969-70 intermediate models and all 1971-76 models are riveted to the control arm and can be replaced.

Lower ball joints on all 1969-76 models are pressed into the control arm and can be replaced with a suitable ball joint tool.

CAUTION: When servicing lower ball joints, be sure to support lower control arm with a suitable jack. If lower control arm is not supported and steering knuckle is disconnected from control arm, the heavily compressed front spring will be completely released.

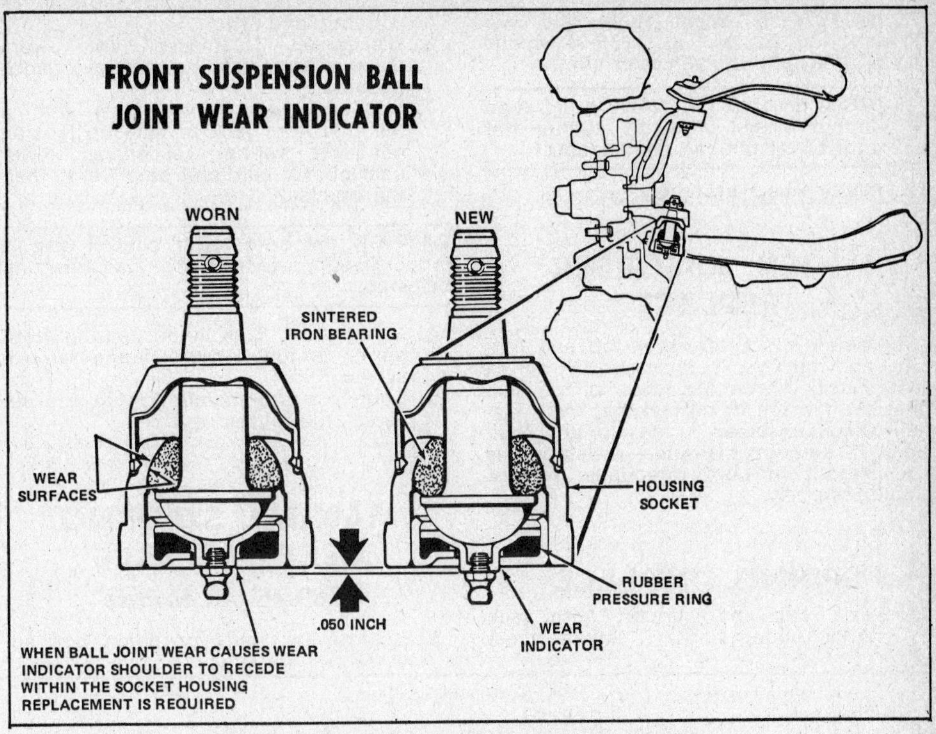

Fig. 4 Lower ball joint check. 1973 Full Size Models, 1974 Models exc. Apollo & all 1975-76 Models

Upper Ball Joint

1969-76

1. Support vehicle at frame and remove wheel assembly.
2. Remove cotter pin from upper ball joint stud, loosen nut approximately 2 turns, do not remove nut.
3. On 1969-70 models, strike steering knuckle in area of ball joint stud to disengage stud from knuckle.
4. On 1971-76 models, position tool as shown in Fig. 5, turn threaded end of tool until stud disengages knuckle.
5. Position jack under lower control arm at spring seat, raise jack until compression is relieved from upper control arm bumper.
6. Remove stud nut, life control arm from steering knuckle and place a block of wood between control arm and frame.
7. On 1969-70 full size models, remove upper control arm to frame bracket nuts, note location and thickness of adjusting shims between shaft and frame bracket, remove control arm. The control arm and ball joint are serviced as an assembly.
8. On 1969-70 except full size and all 1971-76 models,
   a. Center punch rivet head as close to center as possible.
   b. Using a 1/8 inch twist drill, drill through center of rivet approximately 1/2 rivet length deep.
   c. Enlarge hole using a 7/32 inch drill.
   d. Using a chisel, remove rivet heads, then drive rivets out using a suitable punch.
9. On 1969-70 full size models,
   a. Assemble upper control arm and shaft assembly to frame bracket.
   b. Ensure adjusting shims are properly positioned.
   c. Install bumper on upper control arm.
   d. Install and torque frame bracket bolts to 75 ft. lbs.
10. On 1969-70 except full size and all 1971-76 models,
    a. Position ball joint on upper control arm. Install bolts through bottom of control arm with nuts on top, torque to 8 ft. lbs.
11. On all models, position ball joint stud so cotter pin hole is facing forward and remove wooden block from between control arm and frame.
12. Place wheels in straight ahead position, raise jack under lower control until ball joint stud can be installed in steering knuckle.
13. Install castellated nut on ball joint stud, torque to 35 ft. lbs. on 1969-70 full size models, 50 ft. lbs. on 1969-70 except full size, 1971-73 all, and 1974 Apollo, 60 ft. lbs. on 1974 except Apollo and 1975-76 all. Install cotter pin.

NOTE: If cotter pin holes do not align, do not loosen nut, however, tighten until cotter pin can be installed.

14. Install wheel assembly and lower vehicle.
15. On 1969-70 full size models, loosen upper control arm bushing to shaft bolts, bounce front end of car, then retorque bolts to 35 ft. lbs.

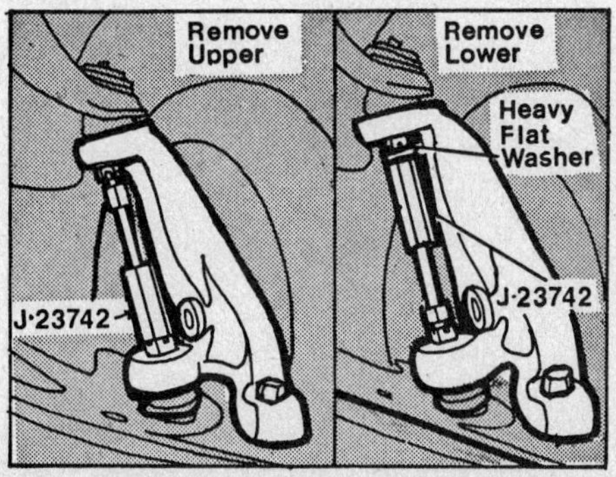

Fig. 5 Removing ball joint stud from steering knuckle.

Lower Ball Joint

1969-76

1. Support vehicle under frame side rails.

NOTE: Position jack under lower control at outboard end, raise jack until it is 1/2 inch below control arm.

2. Remove cotter pin, loosen ball joint stud nut approximately 2 turns, do not remove nut.
3. On 1969-70 models, strike steering knuckle in area of ball joint stud to disengage stud from knuckle.
4. On 1971-76 models position tool as shown in Fig. 5, turn threaded end of tool until ball joint stud is disengaged from steering knuckle.
5. Position jack under lower control arm at spring seat and raise jack until compression is removed from upper control arm bumper.
6. Remove lower ball joint stud nut and position steering knuckle aside.
7. Using tool No. J-9519-10, remove ball joint from lower control arm.
8. Position ball joint on control arm with bleed vent on rubber boot facing inward.
9. Install tools J-9519-17 and J-9519-10 and turn threaded end of tool until ball joint is fully seated in lower control arm.
10. Position new ball joint stud so cotter pin hole is facing forward.
11. Place wheels in the straight ahead position and install ball joint stud on steering knuckle.
12. Install stud nut, torque to 85 ft. lbs. on 1969-73 except Apollo, 90 ft. lbs.

on 1974-76 except Apollo and Skylark, 80 ft. lbs. on 1973-76 Apollo and Skylark. Install cotter pin.

NOTE: If cotter pin holes do not align do not loosen stud nut, tighten nut until cotter pin can be installed.

13. Install wheel assembly.

SHOCK ABSORBER, REPLACE

Unfasten shock absorber top and bottom and remove it through the spring seat. Check shock absorber for obvious physical damage or oil leakage. Push and pull shock absorber in an upright position. If smooth hydraulic resistance is not present in both directions, replace shock absorber.

SPRING, REPLACE

1. Raise car and support with jack stands under frame. Remove wheel with hub and drum.
2. Disconnect stabilizer link from lower control arm and remove shock absorber.
3. Support lower control arm with a suitable floor jack to take up tension of front spring. Disconnect lower control arm ball joint stud from steering knuckle.

CAUTION: Be sure lower control arm is properly supported before disconnecting ball stud.

4. Lower floor jack under spring until spring is fully extended and remove spring.
5. Complete the installation in the reverse order of removal.

MANUAL STEERING GEAR, REPLACE

1969-76 All Models

1. Remove two nuts or pinch bolt securing lower coupling to steering shaft flange.
2. Use a suitable puller to remove pitman arm.
3. Unfasten gear (3 bolts) from frame and remove from car.

POWER STEERING, REPLACE

1969-76

1. Disconnect pressure and return line hoses at steering gear and elevate ends of hoses higher than pump to prevent oil from draining out of pump.
2. Remove pinch bolt securing coupling to steering gear.
3. Jack up car and remove pitman shaft nut, then use a suitable puller to remove pitman arm.
4. On Senior models, remove sheet metal baffle that covers frame-to-gear attaching bolts.
5. Loosen the three frame-to-steering gear bolts and remove steering gear.

CADILLAC

OLD CAR SPECIFICATIONS: For 1946-68 Tune Up and Wheel Alignment Specifications see main index.

INDEX OF SERVICE OPERATIONS

VEHICLE IDENTIFICATION NUMBER LOCATION: On 1969-76 models it is located on rear upper portion of cylinder block, behind intake manifold and on left side of transmission.

ENGINE UNIT NUMBER LOCATION: On 1969-76 at rear of cylinder block.

GRILLE IDENTIFICATION

1969 Eldorado

1969 Except Eldorado

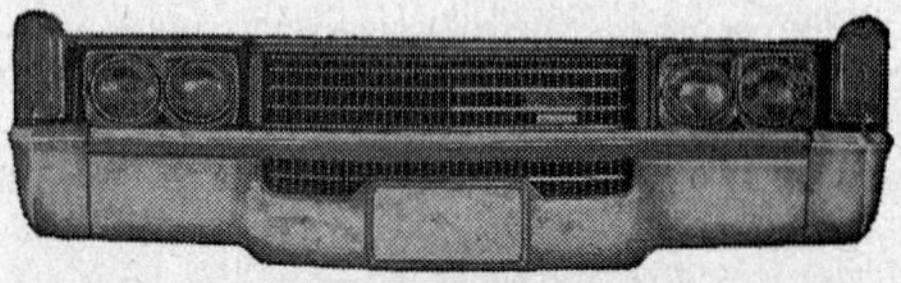
1970 Eldorado

1970 Except Eldorado

1971 Except Eldorado

1971 Eldorado

1972 Eldorado

1972 Except Eldorado

1973 Except Eldorado

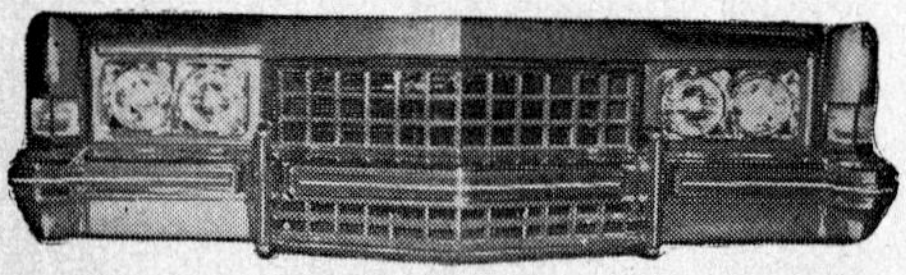
1973 Eldorado

1974 Eldorado

1974 Except Eldorado

1975 Eldorado

1975 Except Eldorado

1976 Seville

1976 Eldorado

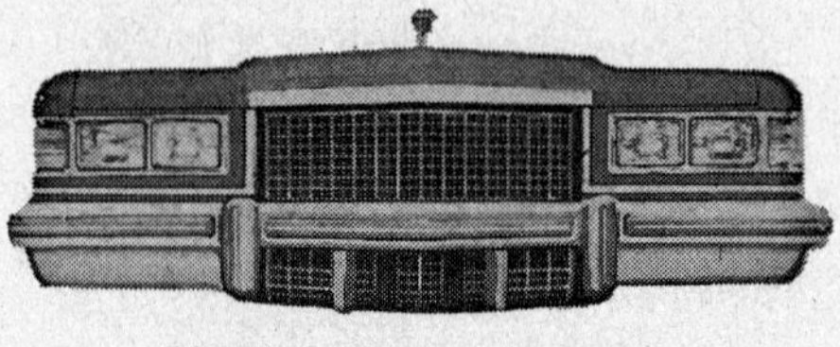
1976 Exc. Eldorado & Seville

GENERAL ENGINE SPECIFICATIONS

Year	Engine	Carburetor	Bore and Stroke	Piston Displacement, Cubic Inches	Compression Ratio	Maximum Brake H.P. @ R.P.M.	Maximum Torque Lbs. Ft. @ R.P.M.	Normal Oil Pressure Pounds
1969	375 Horsepower..............V8-472	4 Barrel	4.3000 x 4.060	472	10.50	375 @ 4400	525 @ 3000	30–35
1970	375 Horsepower..............V8-472	4 Barrel	4.3000 x 4.060	472	10.00	375 @ 4400	525 @ 3000	35–40
	400 Horsepower..............V8-500	4 Barrel	4.3000 x 4.304	500	10.00	400 @ 4400	550 @ 3000	35–40
1971	345 Horsepower..............V8-472	4 Barrel	4.3000 x 4.060	472	8.50	345 @ 4400	500 @ 2800	35–40
	365 Horsepower..............V8-500	4 Barrel	4.3000 x 4.304	500	8.50	365 @ 4400	535 @ 2800	35–40
1972–73	220 Horsepower①............V8-472	4 Barrel	4.300 x 4.060	472	8.50	220 @ 4000	365 @ 2400	35–40
	235 Horsepower①............V8-500	4 Barrel	4.300 x 4.304	500	8.50	235 @ 3800	385 @ 2400	35–40
1974	205 Horsepower①............V8-472	4 Barrel	4.300 x 4.060	472	8.25	205 @ 3600	365 @ 2000	35–40
	210 Horsepower①............V8-500	4 Barrel	4.300 x 4.304	500	8.25	210 @ 3600	380 @ 2000	35–40
1975-76	190 Horsepower①............V8-500	4 Barrel	4.300 x 4.304	500	8.5	190 @ 3600	360 @ 2000	35–40
1976	180 Horsepower①............V8-350	Fuel Inj.	4.057 x 3.385	350	8.0	180 @ 4400	275 @ 2000	30–45
	215 Horsepower①............V8-500	Fuel Inj.	4.300 x 4.304	500	8.5	215 @ 3600	400 @ 2000	35

①—Net rating—as installed in the vehicle.

DISTRIBUTOR SPECIFICATIONS

★Note: If unit is checked on the vehicle, double the RPM and degrees to get crankshaft figures.

Breaker arm spring tension—19-23.

Distributor Part No.①	Centrifugal Advance Degrees @ RPM of Distributor: Advance Starts	Intermediate Advance			Full Advance	Vacuum Advance: Inches of Vacuum to Start Plunger	Max. Adv. Dist. Deg. @ Vacuum
1103202②	0 @ 300	0–4 @ 500	2–6 @ 650	—	10 @ 1400	5½	9½–10½ @ 14
1111262	0–2¼ @ 400	3¼–7¼ @ 800	4¾–8½ @ 1000	6¼–10 @ 1200	16 @ 2000	10–12	13¼–15¼ @ 11¼
1111939	0–2¼ @ 400	5.15–7.15 @ 800	7.40–9.40 @ 1000	8.40–10.30 @ 1200	14 @ 2000	8–10	11.30–12.60 @ 13
1112065	0–1 @ 400	3–5 @ 800	4¾–6¾ @ 1000	6–8 @ 1200	13 @ 2000	8–10	11.30–12.60 @ 13
1112108	0–1 @ 400	3–5 @ 800	4¾–6¾ @ 1000	6–8 @ 1200	13 @ 2000	8–10	11.30–12.60 @ 13
1112219	0 @ 400	—	4½–6½ @ 950	—	13 @ 2000	5–7	12 @ 13
1112835	0 @ 400	5 @ 600	—	—	10 @ 2500	5	16 @ 11.5
1112836	0 @ 400	5 @ 600	—	—	10 @ 2500	7	16 @ 13.5
1112837	0 @ 400	4 @ 600	—	—	9 @ 2500	5	16 @ 11.5
1112838	0 @ 400	4 @ 600	—	—	9 @ 2500	7	16 @ 13.5
1112839	0 @ 400	5 @ 600	—	—	10 @ 2500	5	16 @ 11.5
1112840	0 @ 400	5 @ 600	—	—	10 @ 2500	7	16 @ 13.5
1112841	0 @ 400	4 @ 600	—	—	9 @ 2500	5	16 @ 11.5
1112842	0 @ 400	4 @ 600	—	—	9 @ 2500	7	16 @ 13.5
1112845	1 @ 500	—	—	—	12 @ 2000	5	16 @ 11.5
1112855	1 @ 500	—	—	—	12 @ 2000	5	16 @ 11.5
1112892②③	−¼–0 @ 330	−½ to 2½ @ 450	2½–5 @ 600	5–7 @ 1400	10 @ 3000	5.5	9.5–10.5 @ 14
1112954②④	−¼–0 @ 330	−½ to 2½ @ 450	2½–5 @ 600	5–7 @ 1400	10 @ 3000	4.5	13.5–14.5 @ 16
1112897②	0 @ 200	−½ to +2½ @ 450	2½–5 @ 600	5–7 @ 1400	10 @ 2500	5½	9½–10½ @ 13½
1112924②	0 @ 200	−1 to +1¾ @ 450	1½–3 @ 800	4–6 @ 1700	8 @ 2500	5	13½–14½ @ 16
1112931②	0 @ 200	½ to 1¾ @ 450	6¼–8½ @ 800	13½–15¾ @ 1700	22 @ 2500	5	13½–14½ @ 18
1112932②	0 @ 200	½ to 1¾ @ 450	6¼–8½ @ 800	13½–15¾ @ 1700	22 @ 2500	5	8½–9½ @ 18
1112954②	−¼–0 @ 330	−½ to +2½ @ 450	2½–5 @ 600	5–7 @ 1400	10 @ 3000	4½	13½–14½ @ 16

①—Stamped on distributor housing plate.
②—High energy ignition.
③—California.
④—Exc. California.

TUNE UP SPECIFICATIONS

The following specifications are published from the latest information available. This data should be used only in the absence of a decal affixed in the engine compartment.

OLD CAR SPECIFICATIONS: For 1946-68 Tune Up Specifications see main index.

★When using a timing light, disconnect vacuum hose or tube at distributor and plug opening in hose or tube so idle speed will not be affected.

●When checking compression, lowest cylinder must be within 80 percent of highest.

▲Before removing wires from distributor cap, determine location of No. 1 wire in cap, as distributor position may have been altered from that shown at the end of this chart.

Year	Spark Plug		Distributor		Ignition Timing★			Carb. Adjustments					
								Hot Idle Speed		Air Fuel Ratio		Idle "CO" %	
	Type	Gap Inch	Point Gap Inch	Dwell Angle Deg.	Firing Order Fig. ▲	Timing BTDC ①	Mark Fig.	Std. Trans.	Auto. Trans. ②	Std. Trans.	Auto. Trans.	Std. Trans.	Auto. Trans.
1969	R44N	.035	③	30	C	5°	B	—	550D④	—	—	—	—
1970	R46N	.035	③	30	C	7½°	A	—	600D④	—	—	—	—
1971–72	R46N	.035	③	30	C	8°	D	—	600D④	—	—	—	—
1973	R46N	.035	③	30	C	8°	D	—	600D④	—	—	—	.5
1974	R45NS	.035	③	30	⑤	10°	D	—	600D④	—	—	—	.4
1975–76 V8-500⑥	R45NSX	.060	—	—	E	6°	F	—	600D	—	—	—	—
V8-500⑦	R45NSX	.060	—	—	E	12°	F	—	600D	—	—	—	—
1976 V8-350	R46SX	.080	—	—	G	10°	H	—	—	—	—	—	—

①—BTDC: Before top dead center.
②—D: Drive. N: Neutral.
③—Turn adjusting screw in (clockwise) until engine begins to misfire; then back screw out ½ turn.
④—When making adjustments, air conditioner must be turned off (if equipped). Also, hose must be disconnected at vacuum release cylinder. The hot idle compensator must be closed; this can be done by pressing finger or eraser end of pencil on compensator.
⑤—Exc. H.E.I., Fig. C; H.E.I., Fig. E.
⑥—Exc. Electronic Fuel Injection models.
⑦—Electronic Fuel Injection models.

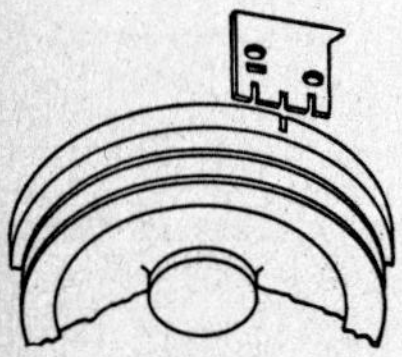

Fig. A

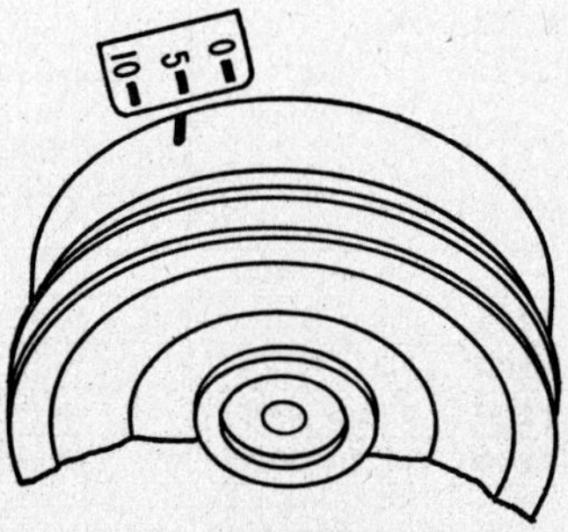

Fig. B

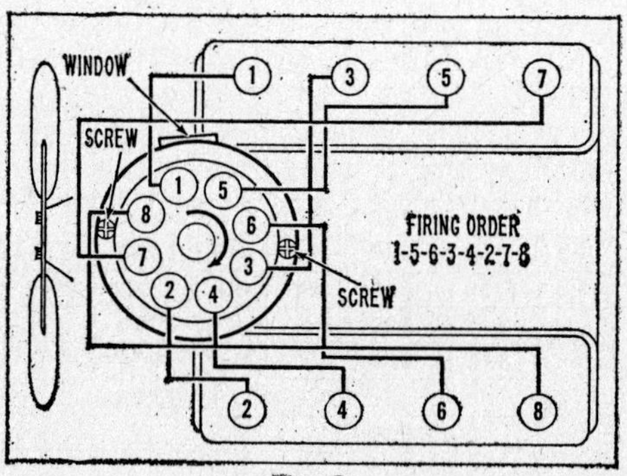

Fig. C

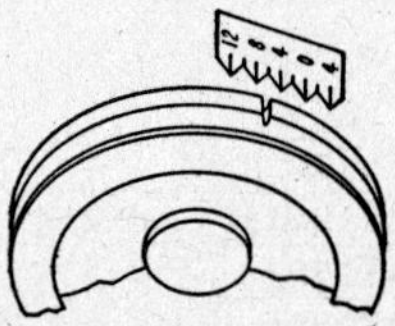
Fig. D

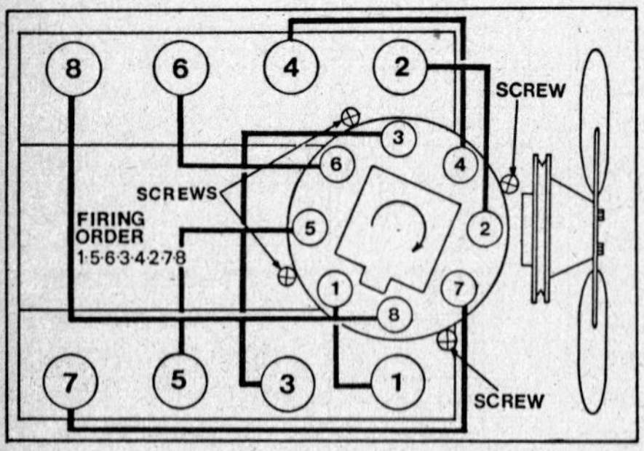

Fig. E

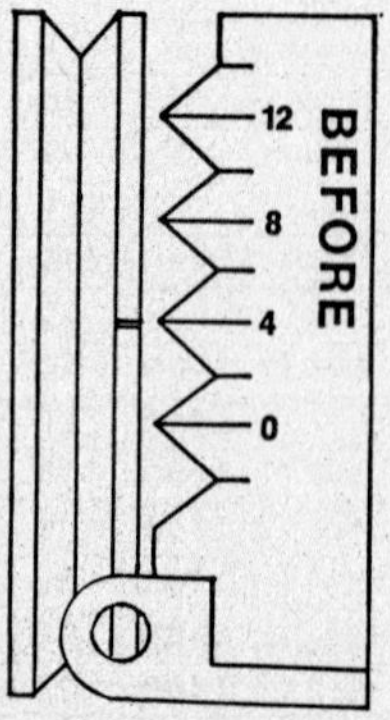

Fig. F

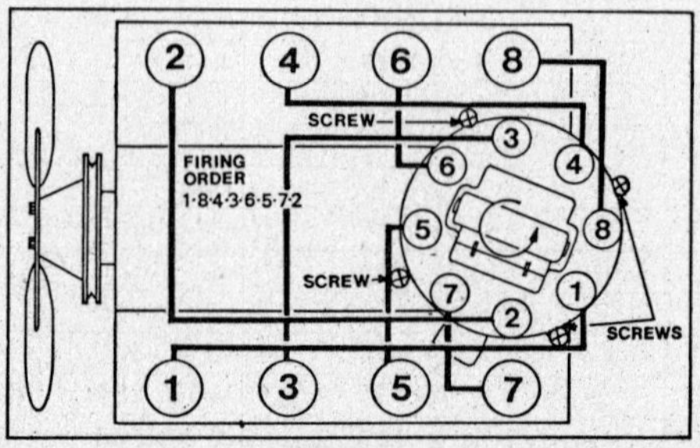

Fig. G

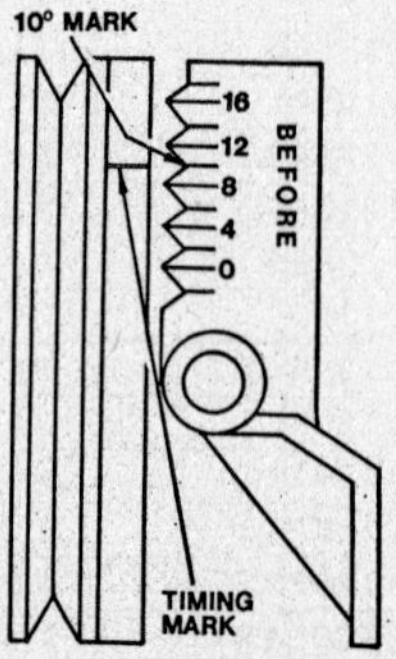

Fig. H

STARTING MOTOR SPECIFICATIONS

Year	Model	Starter Number	Brush Spring Tension Oz①	Free Speed Test			Resistance Test③	
				Amps.	Volts	R.P.M.	Amps.	Volts
1969–70;	Std. Cars	1108381	35	70–99②	10.6	7800–12000	435–455②	3.0
1972	Eldorado	1108352	35	70–99②	10.6	7800–12000	435–455②	3.0
1971	Std. Cars	1108371	35	65–95②	—	7000–10500	—	—
	Eldorado	1107389	35	70–99②	10.6	7800–12000	435–535②	3.0
1973	Std. Cars	1108371	35	65–95②	—	7000–10500	—	—
	Eldorado	1108352	35	70–99②	10.6	7800–12000	435–535②	3.0
1974	Std. Cars	1108521	35	65–95②	—	7000–10500	—	—
	Eldorado	1108522	35	70–99②	10.6	7800–12000	435–535②	3.0
1975–76	All	—	35	65–95②	9	7000–10500	—	—

①—Minimum.
②—Includes solenoid.
③—Check capacity of motor by using a 500-ampere meter and a carbon pile rheostat to control voltage. Apply volts listed across motor with armature locked. Current should be as listed.

VALVE SPECIFICATIONS

Year	Model	Valve Lash	Valve Angles		Valve Spring Installed Height	Valve Spring Pressure Lbs. @ In.	Stem Clearance		Stem Diameter	
			Seat	Face			Intake	Exhaust	Intake	Exhaust
1969–70	All	Hydraulic①	45	44	1 15/16	160 @ 1 1/2	.0005–.0025	.001–.0025	.3415–.3425	.3415–.3420
1971	All	Hydraulic①	45	44	1 15/16	160 @ 1 1/2	.0005–.0025	.001–.0025	.3413–.3420	.3415–.3420
1972–73	All	Hydraulic①	45	44	1 61/64	168 @ 1 1/2	.0010–.0027	.0012–.0027	.3413–.3420	.3413–.3418
1974–76	V8-472, 500	Hydraulic①	45	44	1 61/64	160 @ 1 1/2	.0010–.0027	.0010–.0027	.3413–.3420	.3413–.3420
1976	V8-350	Hydraulic①	②	③	1 43/64	187 @ 1 9/32	.0010–.0027	.0015–.0032	.3425–.3432	.3420–.3427

①—No adjustment.
②—Intake 45°; exhaust 31°.
③—Intake 44°; exhaust 30°.

BRAKE SPECIFICATIONS

Year	Model	Brake Drum Inside Diameter	Wheel Cylinder Bore Diameter			Master Cylinder Bore Diameter		
			Disc Brake	Front Drum Brake	Rear Drum Brake	Disc Brakes	Drum Brakes	Power Brakes
1969–70	Series 75	12	2 3/4	—	7/8	1	—	1
	Eldorado	11	2 15/16	—	7/8	1	—	1
	All Others	12	2 3/4	—	13/16	1	—	1
1971–75	Eldorado	11	2 15/16	—	15/16	1 1/8	—	1 1/8
	Others	12	2 15/16	—	15/16	1 1/8	—	1 1/8
1976	Seville	11	2 15/16	—	15/16	1 1/8	—	1 1/8
	Eldorado	—	2 15/16	—	2 1/2①	1 1/8	—	1 1/8
	Others	12	2 15/16	—	15/16②	1 1/8	—	1 1/8

①—Rear disc brake.
②—Commercial Chassis 1″.

WHEEL ALIGNMENT SPECIFICATIONS

OLD CAR SPECIFICATIONS: For 1946-68 Wheel Alignment Specifications see main index.

Year	Model	Caster Angle, Degrees		Camber Angle, Degrees				Toe-In. Inch	Toe-Out on Turns, Deg.	
		Limits	Desired	Limits		Desired			Outer Wheel	Inner Wheel
				Left	Right	Left	Right			
1969–70	Eldorado	−1½ to −2½	−2	+⅜ to −⅜	+⅜ to −⅜	Zero	Zero	0 to ⅛	18⅙	20
	All Others	−½ to −1½	−1	+⅜ to −⅜	+⅜ to −⅜	Zero	Zero	⅛ to ¼	18⅙	20
1971–72	Eldorado	−½ to −1½	−1	+⅜ to −⅜	+⅜ to −⅜	Zero	Zero	0 to 1/16	—	—
	All Others	−½ to −1½	−1	+⅜ to −⅜	−⅝ to +⅛	Zero	−¼	⅛ to ¼	—	—
1973	Eldorado	−½ to +½	Zero	−⅜ to +⅜	−⅜ to +⅜	Zero	Zero	0 to 1/16	—	—
	Others	−½ to −1½	−1	−⅜ to +⅜	−⅝ to +⅛	Zero	−¼	⅛ to ¼	—	—
1974	Eldorado	−½ to +½	Zero	−⅜ to +⅜	−⅝ to +⅛	Zero	−¼	0 to 1/16	—	—
	Series 75	−½ to +½	Zero	−⅜ to +⅜	−⅝ to +⅛	Zero	−¼	1/16 to 3/16	—	—
	Others	−½ to −1½	−1	−⅜ to +⅜	−⅝ to +⅛	Zero	−¼	1/16 to 3/16	—	—
1975–76	Eldorado	−½ to +½	Zero	−⅜ to +⅜	−⅝ to +⅛	Zero	−¼	0 to 1/16	—	—
	Series 75	−½ to −1½	−1	−⅜ to +⅜	−⅝ to +⅛	Zero	−¼	1/16 to 3/16	—	—
	Others Exc. Seville	−½ to −1½	Zero	−⅜ to +⅜	−⅝ to +⅛	Zero	−¼	1/16 to 3/16	—	—
1976	Seville	+1½ to +2½	+2	+⅛ to ⅞	−⅛ to +⅝	+½	+¼	0 to ⅛	—	—

COOLING SYSTEM & CAPACITY DATA

Year	Model or Engine	Cooling Capacity, Qts.			Radiator Cap Relief Pressure, Lbs.		Thermo. Opening Temp. ①	Fuel Tank Gals.	Engine Oil Refill Qts. ③	Transmission Oil			Rear Axle Oil Pints
		No Heater	With Heater	With A/C	With A/C	No. A/C				3 Speed Pints	4 Speed Pints	Auto. Trans. Qts. ⑦	
1969	Eldorado	—	21.3	21.8	15	15	195	24	5	—	—	⑥	4½⑧
	Series 75	—	24.8	24.8	15	15	195	20	4	—	—	⑤	5
	Others	—	21.3	21.8	15	15	195	26	4	—	—	⑤	5
1970	Eldorado	—	21.3	21.3	15	15	180	24⑨	5	—	—	⑩	4½⑧
	Series 75	—	21.8	21.8	15	15	180	26⑨	4	—	—	⑪	5
	Others	—	21.3	21.3	15	15	180	26⑨	4	—	—	⑪	5
1971–72	Eldorado	—	21.3	21.8	15	15	180	27	5	—	—	⑩	4⑧
	Series 75	—	—	24.8	15	15	180	27	4	—	—	⑪	5
	Others	—	21.3	21.8	15	15	180	27	4	—	—	⑪	5
1973–74	Eldorado	—	21.3	23.8	15	15	180	27	5	—	—	⑩	4⑧
	Series 75	—	—	26.8	15	—	180	27	4	—	—	⑪	5
	Others	—	21.3	23.8	15	15	180	27	4	—	—	⑪	5
1975	Eldorado	—	23	23	15	15	180	27½	5	—	—	⑥	4⑧
	Series 75	—	25.8	25.8	15	15	180	27½	4	—	—	⑪	5
	Others	—	23	23	15	15	180	27½	4	—	—	⑪	5
1976	Eldorado	—	23	23	15	15	180	27½	5	—	—	②	4④⑧
	Series 75	—	25.8	25.8	15	15	180	27½	4	—	—	⑪	4④
	Seville	—	18.9	18.9	15	15	180	21	4	—	—	⑪	4¼
	Others	—	23	23	15	15	180	27½	4	—	—	⑪	4④

①—For permanent anti-freeze.
②—Oil pan 5¾ qts. Total capacity 13½ qts.
③—Add one quart with filter change.
④—Exc. 3.15:1 ratio axle; 3.15:1 ratio axle, 5 pts.
⑤—Oil pan 2 qts. Total capacity 12½ qts.
⑥—Oil pan 5 qts. Total capacity 13 qts.
⑦—Approximate. Make final check with dipstick.
⑧—Front drive axle.
⑨—California vehicles approx. 2 gallons less.
⑩—Oil pan 6 qts. Total capacity 13½ qts.
⑪—Oil pan 4 qts. Total capacity 12½ qts.

ALTERNATOR & REGULATOR SPECIFICATIONS

Year	Alternator					Regulator							
	Model	Rated Hot Output Amps.	Field Current 12 Volts @ 80° F.	Output @ 14 Volts: 2000 R.P.M. Amps.	Output @ 14 Volts: 5000 R.P.M. Amps.	Model	Field Relay: Air Gap In.	Field Relay: Point Gap In.	Field Relay: Closing Voltage	Voltage Regulator: Air Gap In.	Voltage Regulator: Point Gap In.	Voltage Regulator: Voltage @ 125 F.
1969–70	1100742	63	2.8–3.2	35	59	1119519	.015	.030	1.5–2.7	.067	.014	13.5–14.3
	1100696	42	2.2–2.6	28	40	1119515	.015	.030	2.3–3.7	.060	.014	13.5–14.4
	1100694	55	2.2–2.6	32	50	1119515	.015	.030	2.3–3.7	.060	.014	13.5–14.4
1971–74	1100558	42	2.2–2.6	—	—	1119515	.015	.030	2.3–3.7	.060	.014	13.8–14.8
	1100940	42	4.0–4.5	—	—	—	—	—	—	—	—	—
	1100557	63	2.8–3.2	—	—	1119519	—	—	—	—	—	13.8–14.8
	1100937	63	4.0–4.5	—	—	—	—	—	—	—	—	—
	1101015	80	4.0–4.5	—	—	—	—	—	—	—	—	—
1975–76	—	63	4.0–4.5	—	—	Integral	—	—	—	—	—	—
	—	80	4.0–4.5	—	—	Integral	—	—	—	—	—	—
	—	145	4.0–4.5	—	—	Integral	—	—	—	—	—	—

PISTONS, PINS, RINGS, CRANKSHAFT & BEARINGS

Year	Model	Fitting Pistons: Shim To Use	Fitting Pistons: Pounds Pull On Scale	Ring End Gap①: Comp.	Ring End Gap①: Oil	Wrist-pin Diameter	Rod Bearings: Shaft Diameter	Rod Bearings: Bearing Clearance	Main Bearings: Shaft Diameter	Main Bearings: Bearing Clearance	Main Bearings: Thrust on Bear. No.	Main Bearings: Shaft End Play
1969–70	All	②	②	.013	.015	.9995	2.500	.0005–.0028	3.250	.0003–.0026	3	.002–.012
1971–75	All	②	②	.013	.015	.9995	2.500	.0005–.0028	3.250	.0001–.0026	3	.002–.012
1976	V8-350	③	—	.010	.015	.9805	2.1238–2.1248	.0004–.0033	④	⑤	3	.004–.008
	V8-500	②	②	.013	.015	.9997	2.500	.0005–.0028	3.250	.0001–.0026	3	.002–.012

①—Fit rings in tapered bores for clearance given in tightest portion of ring travel.
②—See text under "Pistons".
③—Piston clearance .001–.002.
④—Exc. No. 1, 2.4985–2.4995; No. 1, 2.4988–2.4998.
⑤—Exc. No. 5, .0005–.0021; No. 5 .0015–.0031.

ENGINE TIGHTENING SPECIFICATIONS★

★Torque specifications are for clean and lightly lubricated threads only. Dry or dirty threads produce increased friction which prevents accurate measurement of tightness.

Year	Spark Plugs Ft. Lbs.	Cylinder Head Bolts Ft. Lbs.	Intake Manifold Ft. Lbs.	Exhaust Manifold Ft. Lbs.	Rocker Arm Shaft Bracket Ft. Lbs.	Rocker Arm Cover Ft. Lbs.	Connecting Rod Cap Bolts Ft. Lbs.	Main Bearing Cap Bolts Ft. Lbs.	Flex Plate to Crankshaft Ft. Lbs.	Vibration Damper or Pulley Ft. Lbs.
1969–75	25	115	30	35	—	24①	40	90	75	17
1976 V8-350	25	85	40	25	—	7	42	②	60	310
1976 V8-500	25	115	30	35	—	30①	40	90	75	17

①—Inch pounds. Retorque after engine has been run.
②—Exc. No. 5, 80 ft. lbs.; No. 5, 120 ft. lbs.

Electrical Section

DISTRIBUTOR, REPLACE

1. Remove distributor cap.
2. On vehicles without H.E.I. system, disconnect primary wire from coil. On vehicles equipped with H.E.I. system, disconnect harness connector from side of distributor cap.
3. Disconnect vacuum advance pipe or hose from distributor.
4. Crank engine until rotor is pointing to No. 1 spark plug wire position on cap.
5. Remove distributor hold-down nut and clamp.
6. Lift distributor from engine.
7. Note that the rotor will turn slightly as the drive gear becomes disengaged from the camshaft gear. Therefore, when installing the distributor, the rotor should be turned slightly counter-clockwise from No. 1 spark plug position to insure proper engagement of gears. When properly installed, rotor should point directly to No. 1 spark plug position.

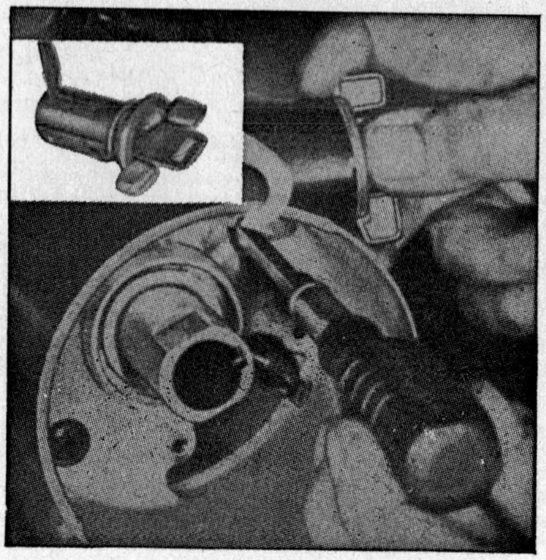

Fig. 1 Lock cylinder removal

STARTER, REPLACE

1972-76

1. Disconnect battery ground cable at battery.
2. On Eldorado, disconnect starter harness at connector at right rear of engine.
3. Raise front end of car.
4. On all except 1972-73 Eldorado models, disconnect battery lead at starter solenoid and disconnect neutral switch wire and coil feed wire at starter solenoid terminals.
5. On all except 1974 Eldorado and H.E.I. equipped vehicles, disconnect interlock relay wire and solenoid wire at starter solenoid terminals.
6. On Eldorado, remove spring clip securing wire to solenoid housing.
7. On Seville, remove crossover pipe.
8. Remove screw and nut securing support bracket to starter and crankcase.
9. Unfasten starter motor from crankcase and remove starter by pulling it forward, then toward RH front wheel and up over steering linkage toward rear of car.

1969-71

1. Disconnect ground strap at battery.
2. Raise and support front end of car.
3. Disconnect battery lead at starter solenoid terminal.
4. Disconnect neutral safety switch wire and coil feed wire at starter solenoid.
5. Remove spring clip securing wires to solenoid housing.
6. Unfasten starter (2 screws) and remove starter by pulling it forward and then lowering it straight down.
7. Reverse procedure to install.

IGNITION LOCK

CAUTION: On vehicles equipped with an Air Cushion Restraint system, turn ignition switch to "Lock," disconnect battery ground cable and tape end, thereby deactivating system.

1. Follow procedure to remove the turn signal switch as described further on.
2. Turn ignition switch to the "On" or "Run" position.
3. Insert a small screw-driver into slot next to the switch mounting screw boss, Fig. 1. Gently tap on screw-driver until screw-driver breaks through thin wall casting. Depress lock cylinder retaining tab with screw-driver and remove cylinder.
4. To separate the lock from the cylinder Fig. 2, use a paper clip or similar quality wire, place a 90 deg. bend about ⅛" from one end of the wire.
5. Using the wire as described in step 4, depress the plunger pin (above the stake mark) approx. ⅛" vertically in sleeve housing, at the same time exert light pressure on the lock assembly, rotate the cylinder counter-clockwise until it pops up from the sleeve assembly.

IGNITION SWITCH, REPLACE

CAUTION: On vehicles equipped with an Air Cushion Restraint system, turn ignition switch to "Lock," disconnect battery ground cable and tape end, thereby deactivating system.

1969-76

1. Disconnect battery cable and position ignition key in "Lock".
2. Remove steering column lower cover.
3. Loosen two upper column support nuts and allow column to drop as far

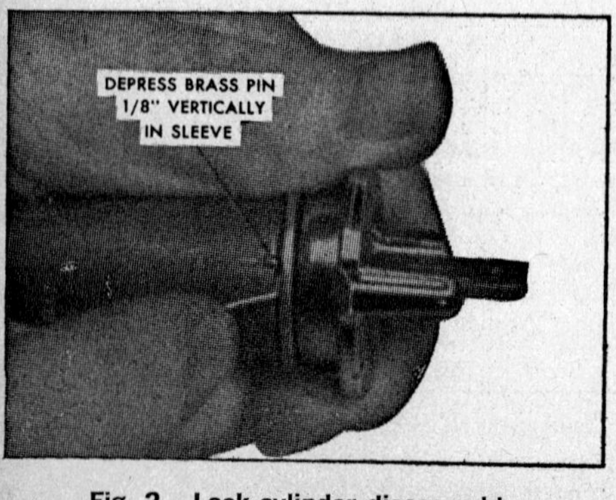

Fig. 2 Lock cylinder disassembly

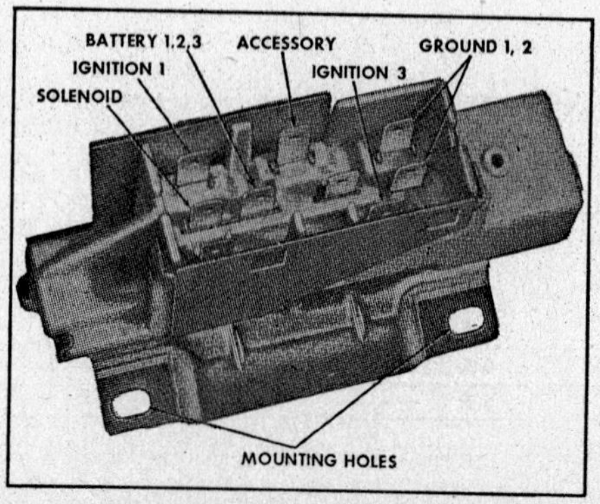

Fig. 3 Ignition switch. 1969-72

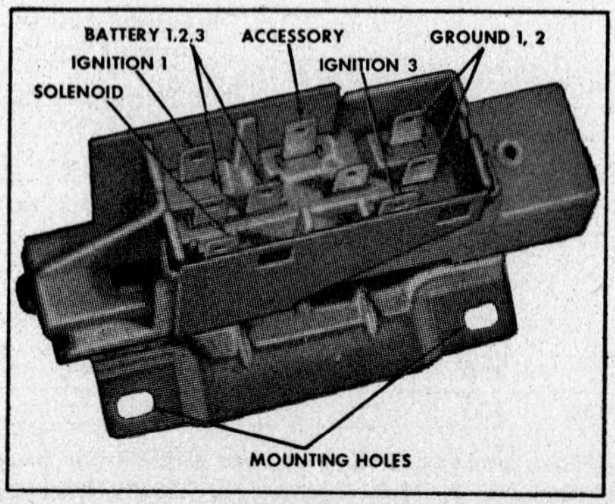

Fig. 4 Ignition switch. 1973-76

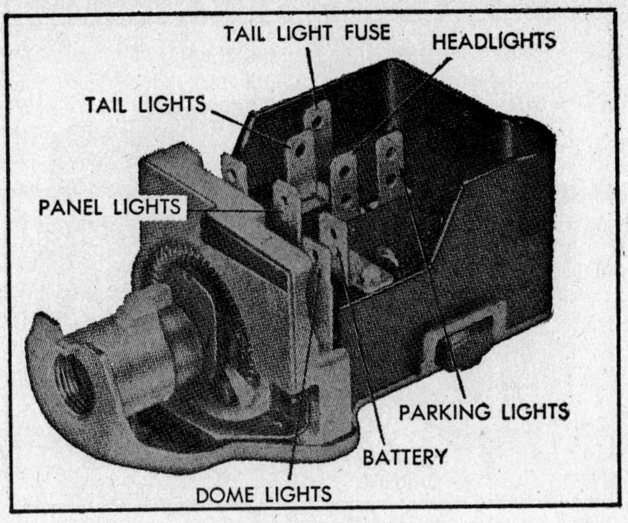

Fig. 5 Headlight switch. 1969

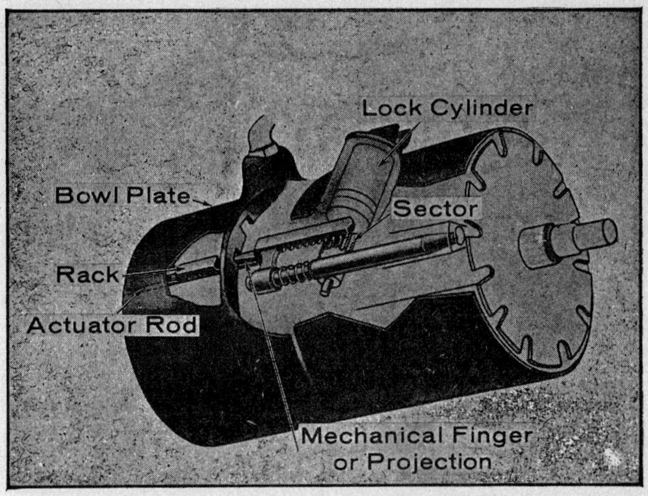

Fig. 6 Mechanical neutral start system late 1973 with Standard column and all 1974-76 models

as possible without removing the nuts.

NOTE: Do not remove nuts as column may bend under its own weight.

4. Disconnect switch connector and remove switch, Figs. 3 and 4.
5. When reassembling, make sure the ignition key is in the "Lock" position. Assemble switch on actuator rod. Hold rod stationary and move switch towards bottom of column then back off one detent on standard steering column models. On models with tilt column, move switch toward upper end of column and back off one detent.

LIGHT SWITCH, REPLACE

CAUTION: On vehicles equipped with an Air Cushion Restraint system, turn ignition switch to "Lock," disconnect battery ground cable and tape end, thereby deactivating system.

1976 Seville

1. Disconnect battery ground cable.
2. Remove steering column lower cover and cluster bezel.
3. Remove lower left instrument panel assembly.
4. Pull knob to "ON" position, depress button on bottom of switch and remove knob and shaft assembly.
5. Remove headlamp case to lower instrument panel attaching screws.
6. Remove sleeve attaching headlamp switch to case.

1974-76 Exc. Seville

1. Disconnect battery ground cable.
2. Remove steering column lower cover.
3. Disconnect headlamp electrical connector and lower bulb.
4. Pull knob to "ON" position, depress spring loaded button and remove knob and shaft, escutcheon and washer.
5. Remove case to instrument support screw.
6. Lower switch assembly, disconnect upper bulb and remove switch.

1970-73

1. Remove steering column lower cover.
2. Unfasten wiring harness retainer running below switch.
3. Depress button on top of switch and remove knob and rod.
4. Remove screw with ground wire at bottom of switch housing.
5. Pull switch down and rearward and disconnect wire and bulbs.
6. Remove hex head sleeve securing switch to housing case and remove switch.

1969

1. Remove instrument panel top cover and steering column lower cover.
2. Remove left A/C duct and outlet and disconnect wiring harness below headlight switch.
3. Depress button on top of switch and remove knob and rod assembly.
4. Remove attaching screws, pull switch rearward and disconnect bulbs and wires and remove switch, Fig. 5.

STOP LIGHT SWITCH

1970-76

The stoplight is retained to the brake pedal bracket. To adjust, pull the brake pedal fully up to its stop. This action automatically adjusts the switch.

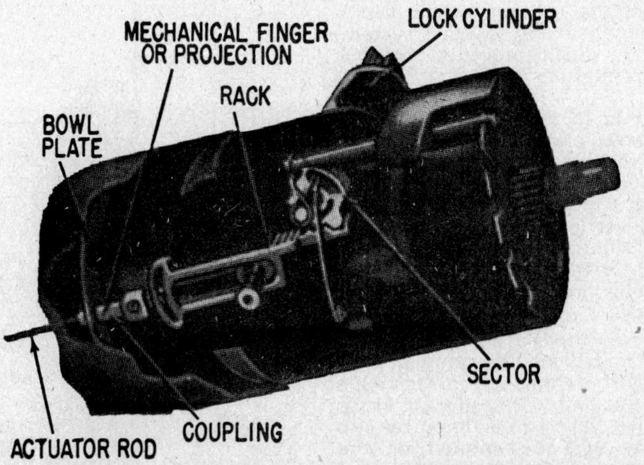

Fig. 7 Mechanical neutral switch system with Tilt-Telescope column. 1974-76

Fig. 8 Mechanical neutral start system in Park position

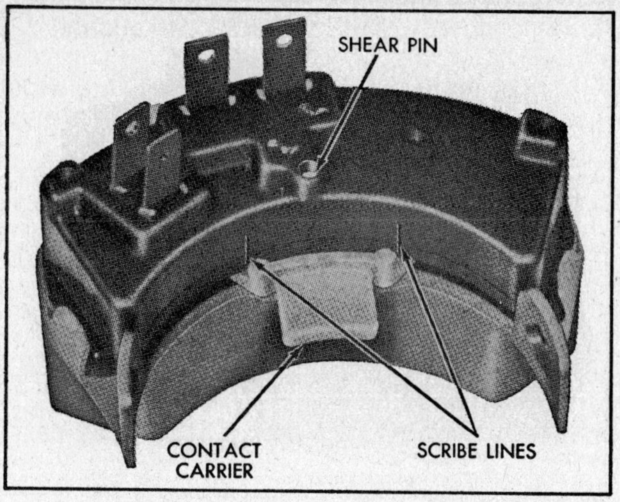

Fig. 9 Switch in neutral position. 1969-72

1969

1. Disconnect two leads wires from switch on brake pedal flange.
2. Remove locking nut from switch.
3. Remove stop light switch.
4. Reverse procedure to install.

NOTE: Adjust switch action so that stop light is on when brake pedal is depressed ½". Loosen front and rear nuts that hold switch and move switch up or down until this action is obtained. Tighten switch lock nuts securely to prevent loss of adjustment.

NEUTRAL START SWITCH

1973 Late Std. Column & All 1974-76 Models

Actuation of the ignition switch is prevented by a mechanical lockout system, Figs. 6 and 7, which prevents the lock cylinder from rotating when the selector lever is out of Park or Neutral. When the selector lever is in Park or Neutral, the slots in the bowl plate and the finger on the actuator rod align allowing the finger to pass through the bowl plate in turn actuating the ignition switch, Fig. 8. If the selector lever is in any position other than Park or Neutral, the finger contacts the bowl plate when the lock cylinder is rotated, thereby preventing full travel of the lock cylinder.

NOTE: On all models incorporating an electric neutral start switch, this switch plus the back-up light switch and parking brake vacuum release valve are combined into one unit. This unit is mounted on the steering column under the instrument panel.

1969-73 Exc. Late 1973 W/Std. Column

1. Place transmission shift lever in neutral.
2. Remove switch from column, being careful not to disturb neutral position of contact carrier, Fig. 9.
3. Mark neutral position of contact carrier.
4. Mark top vacuum hose for identification and remove hoses.
5. Disconnect wiring connectors from switch.

Installation

1. Place transmission shift lever in neutral detent.
2. Connect wires to switch.
3. Move contact carrier on switch to neutral position as marked during removal.

NOTE: If necessary to install a new switch, the switch will be secured in neutral by a shear pin. Do not break pin.

4. Install switch on steering column, aligning contact carrier blade with slot in shift tube.
5. Secure switch to steering column with two mounting nuts, making sure that shift lever is in neutral detent while this operation is performed.
6. Connect vacuum hoses to switch as marked during removal.
7. Switch should now be properly adjusted. If new switch was installed, a slightly greater effort to position

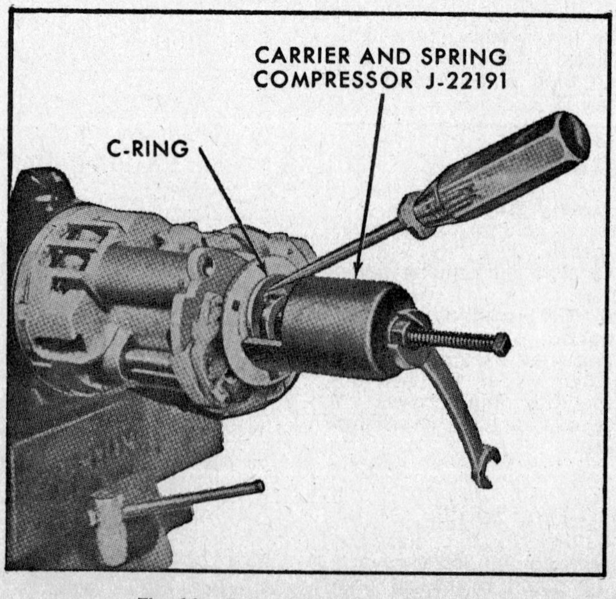

Fig. 10 Removing C-ring. 1969-73

the shift lever in any position besides neutral will be necessary to break the shear pin.

TURN SIGNAL SWITCH, REPLACE

CAUTION: On vehicles equipped with an Air Cushion Restraint system, turn ignition switch to "Lock", disconnect battery ground cable and tape end, thereby deactivating system.

1974-76

Standard column

1. Disconnect battery cable and remove steering wheel.
2. Remove lock plate cover screws and cover.
3. With a suitable compressor, compress lock plate and spring and remove snap ring from shaft.
4. Remove lock plate, cancelling cam, preload spring and thrust washer.
5. Remove steering lower cover and the signal lever.
6. On car equipped with cruise control proceed as follows:
   a. Disconnect cruise control wire from harness.
   b. Remove harness protector from cruise control.
   c. Wrap wire around turn signal lever until lever is disconnected. Do not remove wire from column.
7. On all models remove bolts at upper support.
8. Remove four screws securing upper mounting bracket to column and remove bracket.
9. Disconnect turn signal harness and remove wires from plastic protector.
10. Remove screw securing turn signal switch to column and pull switch out.

Tilt & Telescope Wheel

1. Disconnect battery cable.
2. Remove steering wheel and slide rubber sleeve from steering shaft.
3. Remove plastic retainer from C-ring.
4. With a suitable compressor, thread bolt into steering shaft lock hole.
5. Compress preload spring and remove C-ring.
6. Remove compressor and remove lock plate, horn contact carrier and preload spring.
7. Remove steering lower cover and the signal lever.
8. On cars equipped with cruise control proceed as follows:
   a. Disconnect cruise control wire from harness.
   b. Remove harness protector from cruise control.
   c. Wrap wire around turn signal lever until lever is disconnected. Do not remove wire from column.
9. On all models remove bolts at upper support.
10. Remove four screws securing upper mounting bracket to column and remove bracket.
11. Disconnect turn signal harness and remove wires from plastic protector.

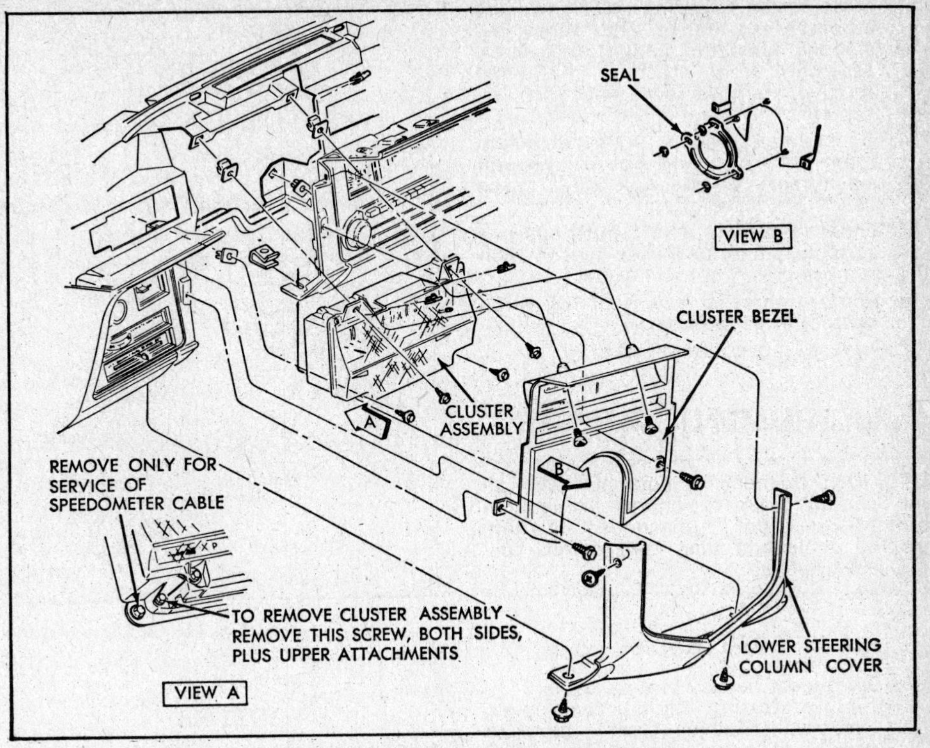

Fig. 10A Instrument cluster. 1976 Seville

12. Remove screw securing turn signal switch to column and pull switch out.

1969-73

Standard Wheel

1. Remove battery cable.
2. Remove steering wheel.
3. Remove lock plate cover screws and cover.
4. With a suitable compressor, compress lock plate and spring and remove snap ring from shaft.
5. Remove lock plate, cancelling cam, preload spring and thrust washer.
6. Remove turn signal lever.
7. Disconnect switch wiring connector and wrap a piece of tape around connector and harness to facilitate removal.
8. Remove upper mounting bracket from column.
9. Remove three retaining screws and remove switch. Reverse procedure to install.

Tilt & Telescope Wheel

1. Disconnect battery cable and remove steering wheel and rubber bump stop.
2. With a suitable tool, compress upper steering shaft pre-load spring and remove C-ring, Fig. 10.
3. Remove compressor, lock plate, horn contact, and upper shaft pre-load spring.
4. Disconnect switch wiring connector and wrap a piece of tape around connector and harness to facilitate removal.
5. Remove upper mounting bracket and switch wiring protector.
6. Position shift bowl in "Park" and remove turn signal lever.
7. On Eldorado, disconnect switch and Cruise Control connectors on column and attach a piece of piano wire to connector on Cruise Control harness, unscrew signal lever and gently pull Cruise Control harness up through and out of column. Secure piano wire in column to aid in reassembly.
8. Push hazard warning flasher button in then unscrew and remove button.
9. Remove retaining screws and remove switch.
10. Reverse procedure to install. On Eldorado, use the piece of piano wire previously installed to draw the Cruise Control harness through the column.

HORN SOUNDER & STEERING WHEEL

1969-76

CAUTION: On vehicles equipped with an Air Cushion Restraint system, turn ignition switch to "Lock," disconnect battery ground cable and tape end, thereby deactivating system.

1. Disconnect battery ground cable.
2. On vehicles equipped with Air Cushion Restraint System, remove the driver module attaching screws using tool J-24628-2. On vehicles not equipped with Air Cushion Restraint System, remove three screws from back of spokes and lift pad assembly from wheel.
3. On tilt and telescope wheels remove

three screws securing lever and knob assembly to flange and screw assembly. Unscrew flange and screw assembly from steering shaft and remove. Remove lever and knob assembly.

4. On standard wheels, scribe an alignment mark on wheel hub in line with slash mark on steering shaft to be used upon installation.
5. Loosen steering shaft nut, apply a suitable puller to loosen wheel; then remove puller, nut and wheel.
6. Remove three screws securing three contact wires to wheel.
7. Reverse procedure to install.

INSTRUMENT CLUSTER

CAUTION: On vehicles equipped with an Air Cushion Restraint system, turn ignition switch to "Lock," disconnect battery ground cable and tape end, thereby deactivating system.

1976 Seville

1. Disconnect battery ground cable.
2. Remove steering column lower cover, 4 screws attaching cluster bezel to instrument panel, then press bezel downward, rotate top outward and carefully remove bezel.
3. Position shift lever in "Park" and remove screw attaching indicator cable to steering column.
4. Remove two screws securing top of cluster to instrument panel support and two screws securing speedometer cable retainer to cluster, Fig. 10A.

NOTE: Do not remove speedometer cable retainer.

5. Pull cluster outward to disengage speedometer cable, then rotate cluster downward and disconnect wire connector. Remove cluster.

1974-76 Exc. Seville

1. Disconnect battery ground cable.
2. Remove cluster bezel and with shift lever in "Park" remove screw securing shift indicator cable to steering column.
3. Remove four screws securing cluster to instrument horizontal support, Fig. 11.

NOTE: Do not remove the 2 outboard screws which retain speedometer cable mounting.

4. Disengage speedometer cable at neck by pulling cluster straight out and depressing retaining spring.

NOTE: To remove cluster, place shift lever in low range and on cars with tilt wheel, place wheel in lowest position.

5. Rotate cluster downward, disconnect printed circuit connector and remove cluster.

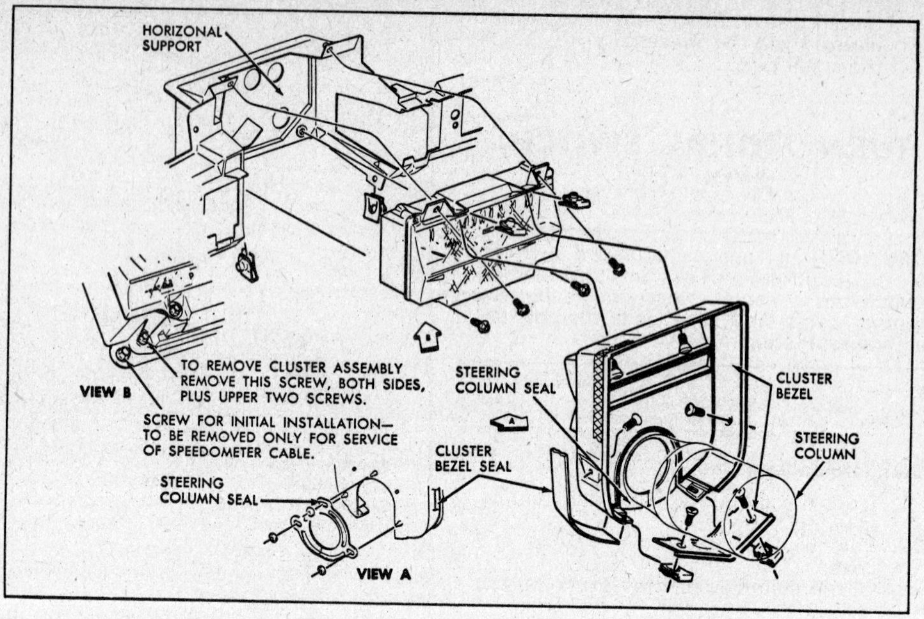

Fig. 11 Instrument panel cluster. 1974-76. Exc. Seville

1971-73

1. Disconnect battery ground cable.
2. Remove instrument panel top cover.
3. Remove steering column lower cover.
4. Disconnect speedo cable from speedometer.
5. Remove left hand air conditioner outlet hose from behind cluster.
6. Disconnect cluster connector.
7. Disengage wiring harness from clip at

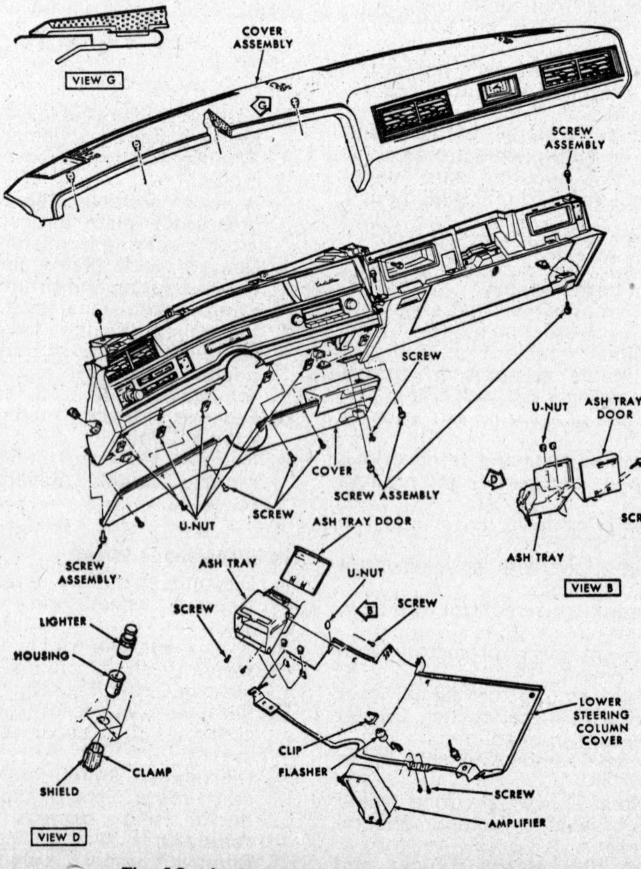

Fig. 12 Instrument panel cover. 1971-73

back of cluster and position out of way.
8. Remove screw securing shift pointer to column and remove pointer, Fig. 12.
9. Remove four screws (2 upper and 2 lower) securing cluster to bezel.
10. Move cluster back and to right. Tip right hand corner of cluster upward and remove cluster out through top of panel.

1969-70

1. Disconnect battery cable.
2. Remove instrument panel top cover, clock, radio and steering column lower cover.
3. Remove shift indicator pointer.
4. Remove odometer reset knob by pulling if off shaft.
5. Remove four cluster-to-bezel screws, Fig. 13.
6. Move cluster forward and to the right. Tip right hand corner of cluster downward and remove cluster from under panel.

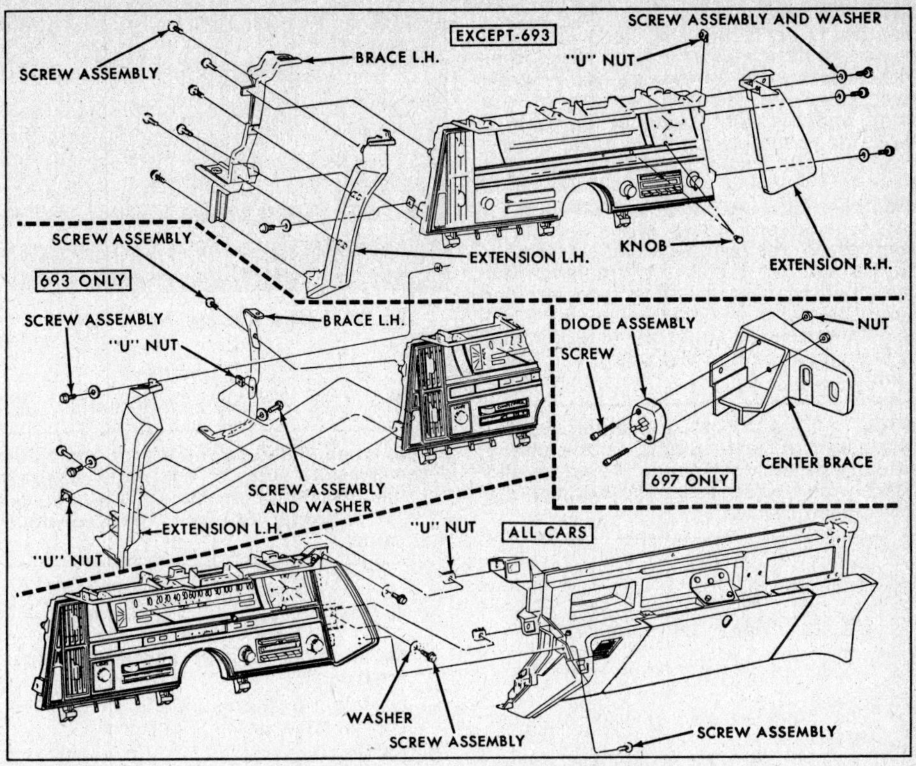

Fig. 13 Instrument panel. 1969-70 (Typical)

RADIO, REPLACE

NOTE: When installing radio, be sure to adjust antenna trimmer for peak performance.

1976 Seville

1. Remove E.F.I. Electronic Control Unit as described in "Electronic Fuel Injection Section".
2. Remove screw attaching A/C outlet to heater case.
3. Disconnect antenna lead, then remove nut securing radio to rear mounting bracket.
4. Remove control knobs, anti-rattle springs, washers and nuts.

NOTE: On models with tape player, control knobs are retained by a 5/64" allen head screw.

5. Disconnect wiring harness and remove radio.

1974-76 Exc. Seville

CAUTION: On vehicles equipped with an Air Cushion Restraint system, turn ignition switch to "Lock," disconnect battery ground cable and tape end, thereby deactivating system.

1. Disconnect battery ground cable.
2. Remove ash tray, radio knobs, washers, control rings and nuts.
3. Remove radio to lower support brace nut from rear and rotate brace to the right.
4. Disconnect wires, antenna lead and the dial bulb from radio.
5. Lower left hand side of radio and remove radio through the ash tray opening.

1970-73

1. Remove steering column lower cover.
2. If necessary, remove defroster hose from behind radio.
3. Remove radio knobs, washers and control rings.
4. Remove spanner nuts that hold shafts to panel.
5. Disconnect wires and antenna lead from radio.
6. Remove two hex head screws that hold lower support bracket to radio and instrument panel center support and remove bracket.
7. Pull radio back from panel to disengage shafts and lower to gain access to dial bulb.
8. Disconnect dial bulb and remove radio.

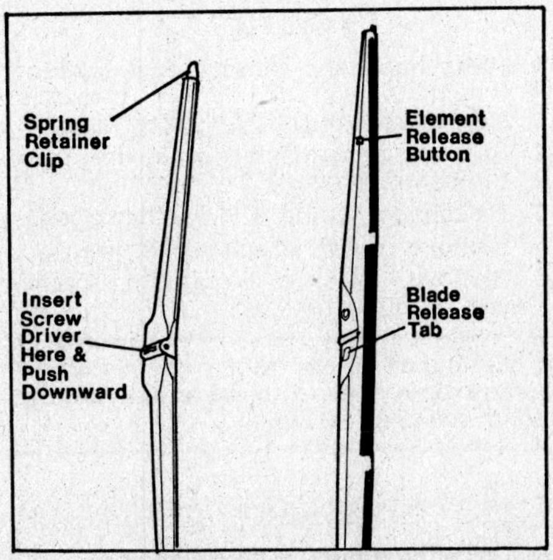

Fig. 14 Wiper Blades

1969

1. Remove steering column lower cover. If necessary, remove defroster hose from behind radio.
2. Remove radio knobs, springs and rings.
3. Remove spanner nuts that hold radio control shafts to instrument panel.
4. Remove screw on right side that secures radio to bracket and loosen screw that secures bracket to brace.
5. Pull radio rearward to disengage control shafts and drop radio slightly to gain access to wiring connectors.
6. On AM/FM Stereo radio, disconnect audio-amplifier unit connector.
7. Disconnect antenna lead-in cable and connector at radio.
8. Disconnect light bulb socket from radio.
9. Disconnect foot control cable plug from radio if equipped.
10. On stereo radio, remove tape securing speaker leads: Two pieces on instrument panel cluster and two pieces at instrument panel frame above glove box door.
11. Remove radio.

W/S WIPER BLADES

Two methods are used to retain wiper blades to wiper arms, Fig. 14. One method uses a press type release tab. When the release tab is depressed the blade assembly can be slid off the wiper arm pin. The other method uses a coil spring retainer. A screw driver must be inserted on top of the spring and the spring pushed downward. The blade assembly can then slide off the wiper arm pin. Two methods are also used to retain the blade element in the blade assembly, Fig. 14. One method uses a press type release button. When the button is depressed, the two piece blade assembly can be slid off the blade element. The other method uses a spring type retainer clip in the end of the blade element. When the retainer clip is squeezed together, the blade element can be slid out of the blade assembly.

NOTE: To be sure of correct installation, the element release button, or the spring element retaining clip should be at the end of the wiper blade assembly nearest the wiper transmission.

W/S WIPER ARMS

1969-70

1. Open hood to gain access to wiper arms.
2. Lift wiper arms off windshield and insert a 3/32" drill (#43) in hole behind wiper arm, Fig. 15.
3. Release wiper arm, leaving drill in hole, and lift wiper arm assembly off transmission shaft.
4. On left side, slide drag link clip toward end of arm enough to disengage drag link from pivot pin and remove wiper arm and drag link from vehicle.

CAUTION: Do not remove drill installed in step 2.

5. To install left wiper arm, position drag link on pivot pin and secure by sliding retainer down to pivot pin until it locks.
6. Install wiper arm on transmission shaft.
7. Pull up on wiper arm and remove drill from hole in wiper arm assembly.

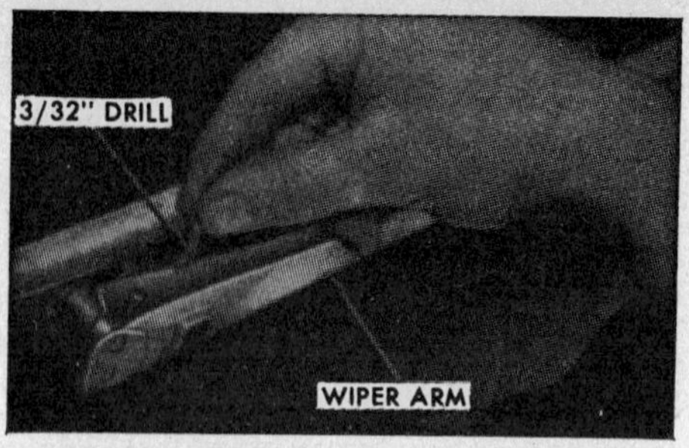

Fig. 15 Wiper arm removal. 1969-70

NOTE: When installing wiper arm and blade assemblies, be sure to "overpark" the arm and blade assemblies below the windshield so that a proper return to park position will result.

1971-76

1. Raise hood to gain access to wiper arms.
2. Lift wiper arm and slide latch clip, Fig. 16, out from under wiper arm.
3. Release wiper arm and lift wiper arm assembly off transmission shaft.
4. On left arm, slide articulating arm lock clip away from transmission pivot pin, Fig. 17, and lift arm off pin.
5. To install left wiper arm assembly, position the articulating arm over the transmission pivot pin and slide the lock clip toward the pivot pin until it locks in place on the pin. Install the left wiper arm assembly to the transmission shaft aligning the keyway to the shaft.
6. Align keyway in right wiper arm assembly to transmission shaft and install arm assembly to shaft.
7. Lift the wiper arm assemblies and slide latch clips under the arms. Release wiper arms and check wipe pattern and park position.

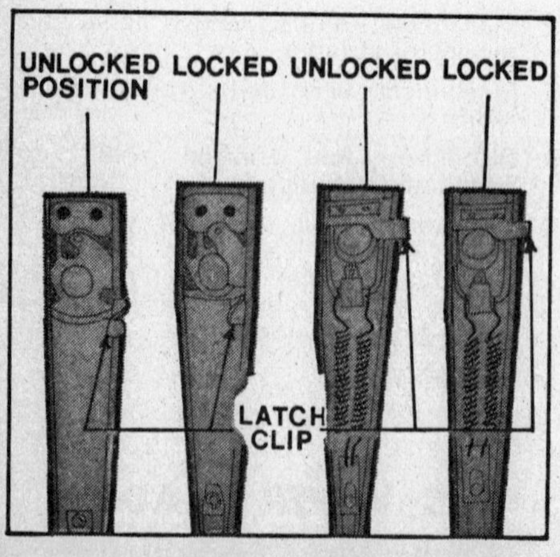

Fig. 16 Wiper arm removal. 1971-76

W/S WIPER MOTOR

1972-76

1. Raise hood and remove cowl screen.
2. Reach opening and loosen transmission drive link to crank arm nuts.
3. Remove transmission drive link from motor crankarm.
4. Disconnect wiring and washer hoses.
5. Remove motor attaching screws.
6. Remove motor while guiding crankarm through hole.

NOTE: Ensure wiper motor is in "Park" position before assembling crank arm to transmission drive link.

1969-71

1. Disconnect battery ground cable.
2. Disconnect three washer hoses from washer control valve. *Mark small outlet hoses and corresponding control valve nozzles for identification.*
3. Disconnect two-way wire connector at washer unit and three-way wire connector at wiper unit.
4. Remove cover on 1969-70 from opening in left side of cowl to gain access to wiper crank arm. *Cover is located above wiper-washer assembly.*
5. Loosen two lock nuts securing wiper unit crank arm to ball socket on end of transmission drive linkage, then disengage crank arm from ball socket. *Do not remove lock nuts from ball sockets.*

NOTE: On 1969-70 Eldorado, access to the crank arm is gained by removing the air inlet screen.

6. Remove three screws that hold wiper-washer assembly to cowl and remove assembly.
7. Reverse procedure to install.

W/S WIPER TRANSMISSION

1972-76

1. Raise hood and remove cowl vent screen.
2. Remove wiper arm and blade from transmission to be removed.
3. Loosen attaching nuts securing transmission drive link to motor crankarm.

NOTE: If only left side is to be removed, it will not be necessary to loosen nuts securing right drive link to motor crankarm.

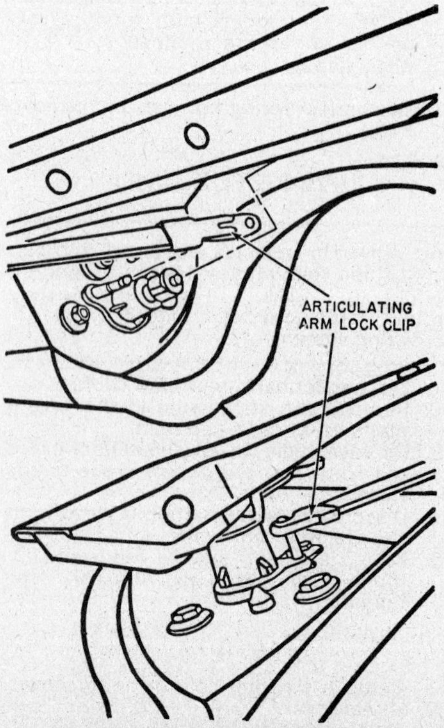

Fig. 17 Articulating arm lock clip

4. Disconnect transmission drive link from crankarm.
5. Remove attaching screws securing transmission to body.
6. Remove transmission and linkage assembly through plenum chamber opening.

1969-71 Except Eldorado

1. Remove both wiper arms.
2. Remove six clips securing rubber hood seal to cowl and position seal out of the way.
3. Remove cowl ventilator screen (12 screws).
4. Unfasten left wiper transmission from cowl (3 screws).
5. Repeat Step 4 for right side.
6. Allow transmissions, linkage and bellcrank to lie in cowl plenum.
7. Remove ball socket cover (2 screws). Cover is located on top of cowl directly behind wiper-washer assembly.
8. Loosen two lock nuts securing crank arm to ball socket on end of drive linkage, then disengage crank arm from ball socket. Do not remove locknuts from ball socket studs.
9. Slide linkage to one side to allow one end of linkage to be drawn out through opening in cowl and remove linkage.
10. Reverse procedure to install.

1969-71 Eldorado

1. Remove both wiper arms.
2. Remove cowl air inlet screen.
3. Remove access hole cover from opening in center of cowl to gain access to wiper unit crank arm.
4. Remove lock nut securing crank arm to ball socket stud.
5. Remove three transmission mounting screws on right and left transmissions.
6. Disengage ball socket stud at wiper unit and remove transmissions and linkages as a complete assembly.
7. Reverse procedure to install.

W/S WIPER SWITCH

1974-76

1. Remove left hand climate control outlet grille.
2. Remove screw securing switch to instrument panel.
3. Pull control switch and electrical connector out and disconnect from panel.

1969-73

1. Open left front door to gain access to screw securing control switch to instrument panel extension on door.
2. Loosen screw securing control switch to extension. *Screw is trapped and cannot be removed.*
3. Pull control switch out and disconnect electrical connector.

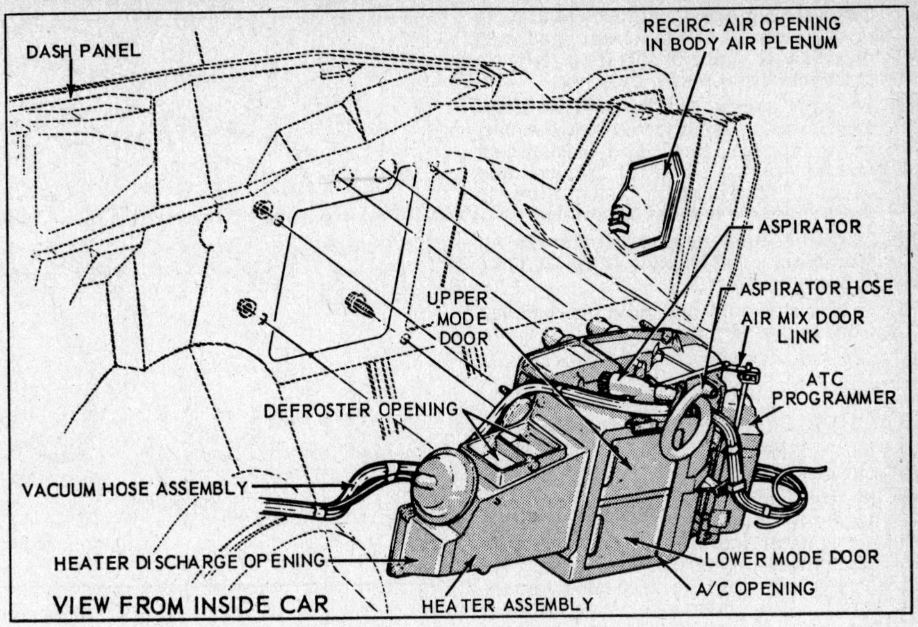

Fig. 18 Heater core. 1971-76 (Typical Exc. Seville)

HEATER CORE REMOVAL

CAUTION: On vehicles equipped with an Air Cushion Restraint system, turn ignition switch to "Lock," disconnect battery ground cable and tape end, thereby deactivating system.

After draining radiator and disconnecting heater hoses proceed as follows:

1971-76 Exc. Seville

With Air Conditioning

1. Remove instrument panel top cover.
2. Remove right and left outlet A/C hoses and center outlet connector. On 1974 models, remove right and left outlet A/C hoses from A/C distributor.

NOTE: On 1974-76 models, remove center outlet support and connector from position between cowl and horizontal support.

3. Remove two screws securing A/C distributor to heater case and remove distributor.
4. Remove one screw securing defroster nozzle to heater case and remove nozzle from under clips and out of car.
5. Remove glove compartment liner.
6. Disconnect vacuum hoses at recirc door, water valve, control head and supply hose.
7. Disconnect aspirator hose from in-car sensor.
8. Remove instrument panel center vertical brace.
9. Remove instrument panel horizontal brace.
10. Remove four nuts securing heater case to cowl on engine side of dash, Fig. 18.
11. Remove heater case from position under panel being careful to hold case as upright as possible to avoid spilling coolant.
12. Remove rubber seal from around core water nipples.
13. Remove screw and clip from under seal.
14. Remove two screws and clip from opposite end of case and remove core.

Without Air Conditioning

1. Remove instrument panel top cover.
2. Remove two screws and position center ventilation duct and sleeve out of way.
3. Remove vacuum hoses from diverter door and defroster door vacuum actuators.
4. Remove Bowden cable from temperature door arm and remove screw securing cable to heater case. Reposition cable out of way.
5. Remove two screws and nuts securing heater case to cowl and pull case away from cowl.
6. Remove heater case from position under panel being careful to hold in upright position to avoid spilling coolant.
7. Remove two screws and retainer at each side of core securing core to case and remove core.

1969-70 All Models

With Air Conditioning

1. Working in engine compartment, disconnect vacuum hoses and wiring from servo units being sure to mark them for proper installation.

2. On 1969 models remove fender to cowl tie struts. If car is equipped with Auto Level Control position left tie strut with compressor attached.
3. To gain access to the Heater & Air Modulator assembly, remove the vacuum storage tank and disconnect master switch, control cables and vacuum hoses again being sure to mark them for ease of assembly.
4. Unfasten and remove Heater & Air Modulator assembly from engine compartment.
5. The heater core can now be removed from this assembly.

1969-70 Except Eldorado

Without Air Conditioning

1. Blower motor and heater-blower assembly must be removed to get at the heater core. To do this, first disconnect electrical connector to blower on cowl in engine compartment. Then remove five screws securing blower motor to its case and remove motor.
2. Disconnect cable at temperature valve at pivot point on heater and remove cable clamp.
3. Remove screw securing vacuum manifold to heater and move manifold and hoses out of the way.
4. Remove seven screws securing bottom of heater to cowl and six screws securing top of heater to cowl, and take out heater assembly.
5. To remove heater core, remove four screws (two each side) that secure wire retaining clamp to heater blower case and remove clamps.
6. Heater core can now be pulled out of case and rubber grommets removed from inlet and outlet fittings.

1969-70 Eldorado

Without Air Conditioning

1. Before removing heater blower and motor, disconnect rubber cooling hose from nipple on blower motor.
2. The left cowl-to-fender shield strut rod must also be removed before taking out heater-blower assembly.

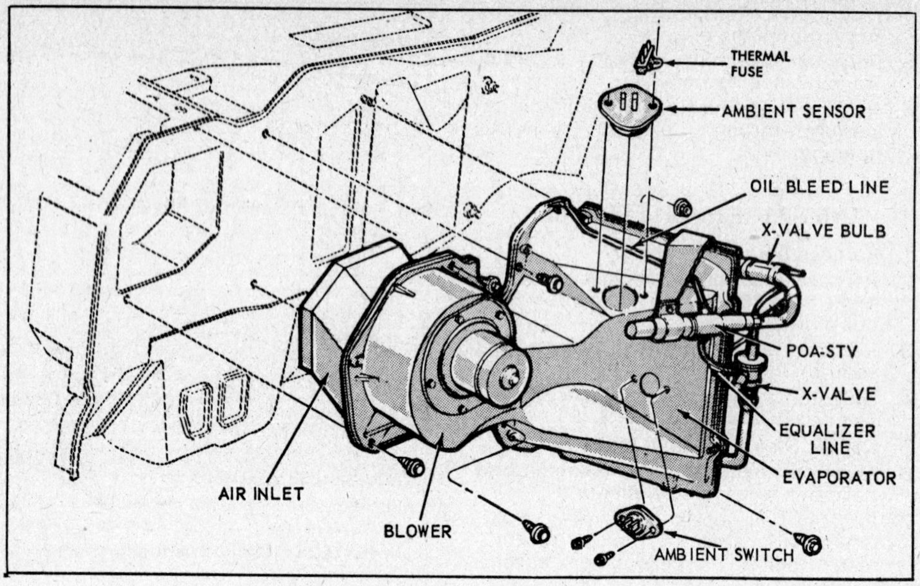

Fig. 19 Blower motor. 1971-76 (Typical)

BLOWER MOTOR REPLACE

1969-70

1. Disconnect battery ground cable.
2. Remove screw securing antenna bracket to wheel housing.
3. Remove cooling hose from nipple and blower motor.
4. Disconnect feed wire at motor.
5. Unfasten motor, rotate 180° and remove.

1971-76 All

CAUTION: On vehicles equipped with an Air Cushion Restraint system, turn ignition switch to "Lock," disconnect battery ground cable and tape end, thereby deactivating system.

1. Disconnect battery ground cable.
2. Remove cooling hose from nipple and blower.
3. Disconnect electrical connector at lead to motor.
4. Unfasten and remove motor, Fig. 19.

SPEED CONTROL

1970-76

Bead Chain Adjustment

1. Adjust engine hot idle speed and mixture, then shut off engine.
2. Check slack in chain by unsnapping swivel from ball stud and holding chain taut at ball stud; center of swivel should extent ⅛" beyond center of ball stud.
3. Adjust bead chain slack by sliding sleeve back on chain and removing

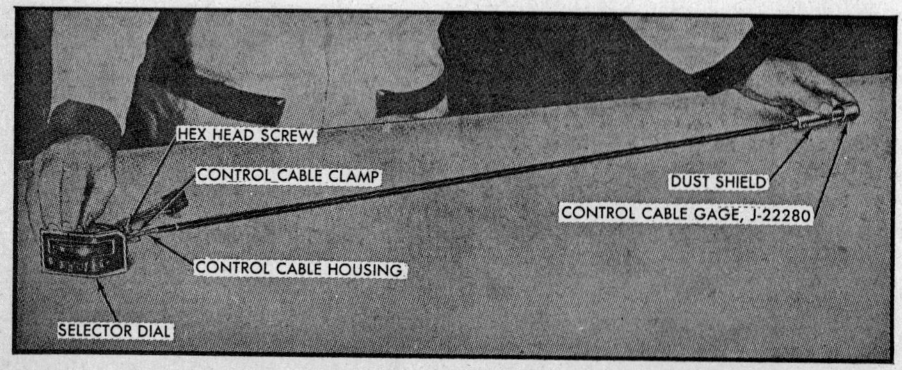

Fig. 20 Control cable adjustment for Slide Switch Speedostat. 1969

Fig. 21 Accelerator linkage adjustment for Slide Switch type Speedostat. 1969

loose rivet. Move swivel on ball chain until slack is correct. Then reinstall rivet and slide sleeve over rivet.

On late models, install bead chain with second ball on the inboard slot of the throttle plate clip.

Cruise Speed Adjustment

The cruise speed adjustment can be set as follows:

1. If car cruises below engagement speed, screw orifice tube on transducer outward.
2. If car cruises above engagement speed, screw orifice tube inward.

NOTE: Each ¼ turn of the orifice tube will change cruise speed about one mile per hour. Snug up lock nut after each adjustment.

Brake Release Switch Adjustment

1. Turn on ignition switch and connect a test light between one terminal of brake release switch and ground; select terminal where light goes out when pedal is depressed.
2. Loosen screw that retains switch to pedal support bracket. Position switch so circuit opens (light goes out) when pedal is depressed ¼". Tighten screw and recheck.

Control Cable Adjustment, 1969

The cable is preset at the factory and should not require adjustment unless a new cable is installed. This adjustment must be performed off the car as follows:

1. Remove the selector assembly.
2. Rotate selector dial to low speed position until it is positioned against its stop but do not force beyond its stop.
3. Position assembly flat on workbench and make certain there are no kinks in cable.
4. Loosen hex head set screw at cable clamp on selector control.
5. Pull cable housing until it is approximately half-way out of cable clamp. Position control cable gauge shown in Fig. 20 in end of dust shield. Hold dust shield and gauge and push toward selector control assembly until gauge bottoms. While holding in this position, tighten set screw at cable clamp.

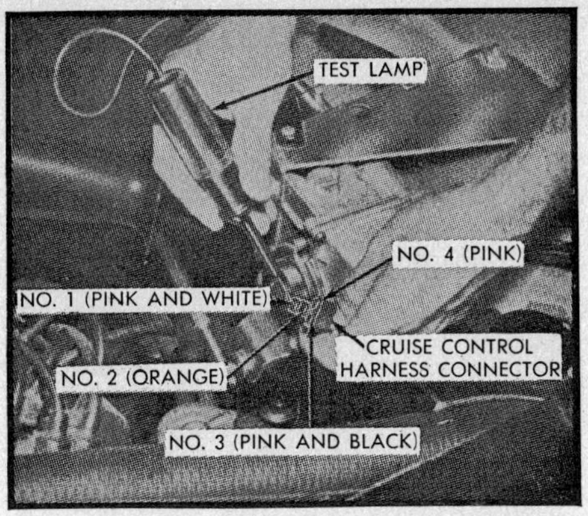

Fig. 22 Electrical connections on Slide Switch type Speedostat. 1969

Accelerator Linkage

1. Adjust throttle rod.
2. Start engine and operate at slow idle with transmission lever in "Park".
3. Separate linkage from exterior arm.
4. Adjust trunnion so that when it is installed through exterior arm, the stop stud will be aligned with locating notch and throttle valves will be closed.
5. Install washer on trunnion and secure with cotter pin.

NOTE: Due to the angle at which the trunnion enters hole in exterior arm, it is necessary to rotate the exterior arm slightly forward when inserting the trunnion. Repeat this operation until proper alignment is obtained. Be careful not to turn trunnion too far back or throttle valves will unseat and cause an incorrect adjustment. Insert the gauge shown in Fig. 21 (or small diameter pipe) over stop stud to check alignment.

Brake Release Switch

1. Turn on ignition but do not start engine.
2. Momentarily move slide switch to AUTO position until red indicator light glows.
3. Using a test lamp, ground one lead and touch the other lead to terminal No. 4, Fig. 22.
4. Loosen mounting screw securing release switch to brake pedal mounting bracket.
5. Adjust release switch so that lamp will light when brake pedal is fully released, and will go out when brake pedal is depressed about ¼ inch. Tighten switch mounting screw. If switch cannot be adjusted, it is defective and should be replaced.

Engine Section

ENGINE MOUNTS

1976 Seville

1. Remove fan shroud, then the wheel housing struts from both sides.
2. Raise vehicle, then remove crossover pipe and engine mount through bolts.
3. Support vehicle, raise engine until support can be removed from engine block.
4. Working through opening between lower control arm and frame, remove bolts attaching engine mount to frame and the mount, Fig. 1.

1971-76 Except Eldorado & Seville

1. On 1973-75 vehicles, open hood and remove radiator cover.
2. Raise and support vehicle on hoist.
3. Remove two through bolts from mount to be replaced, Fig. 2.
4. If left hand (driver side) mount is being replaced, remove nut and washer from both mounts.
5. If right hand (passenger side) mount

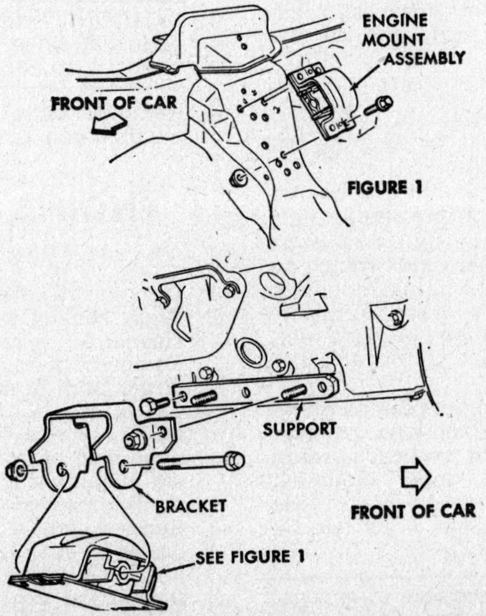

Fig. 1 Front engine mounts. 1976 Seville

is being replaced, remove nut and washer from left hand mount and loosen right hand nut until no threads are visible on stud.

6. Support car with stands at each front frame horn.
7. Apply pressure (hoist, jack, etc.) to oil pan. Apply pressure over entire bottom with no concentration of pressure due to bumps or knobs on lift.
8. Raise engine until mount may be removed.

CAUTION: Amount engine may be raised is limited due to clearance between fan and shroud. Use caution when raising engine to avoid damage to these components.

1973-76 Eldorado

Left Front Mount

1. Remove radiator cover and disconnect battery ground cable.
2. Raise car and remove nut and washer from mount stud.
3. Remove capscrew securing transmission cooler lines to final drive bracket.
4. Remove screw securing top of steering gear flex coupling shroud to frame.
5. Remove nut from large through bolt securing top of engine mount and top of final drive bracket to engine.
6. Remove nut from lower engine mount through bolt.

NOTE: Attach box end wrench to nut, then use pry bar between wrench and engine.

7. Remove remaining screw securing flex coupling shroud to frame and remove shroud.
8. Remove cross bar bolt, nut and washers securing bottom of final drive bracket to final drive. Remove bracket by tipping back and to left of car.
9. Using block of wood and jack stand, lift engine from frame at crankshaft pulley to relieve load on engine mount bolts and remove bolts from mount.
10. Loosen, but do not remove, nut and washer from right engine mount stud.
11. Raise engine enough to remove mount. Work mount down and forward between fuel pump and frame.

Right Front Mount

1. Remove radiator cover.
2. Remove nut and washer securing left engine mount to frame.
3. Raise car and remove nut and washer securing right mount to frame.
4. Reach between right mount and oil pump to attach open end wrench to nut on lower bolt securing mount to engine. Holding this wrench in place, remove lower mounting bolt from rear, using ratchet, U-joint socket and 18″ extension.
5. Work end wrench up between oil pump and crankshaft pulley and position wrench in a straight vertical position to engage nut on upper mounting bolt. Holding wrench in place, use same tools as in Step 4 to remove upper mounting bolt.

NOTE: If nut is impossible to engage, an alternative is to first perform the following Step, then use a 5″ extension and a 9/16″ crowfoot adapter and engage tool on nut by slipping between oil pump and A.I.R. pump.

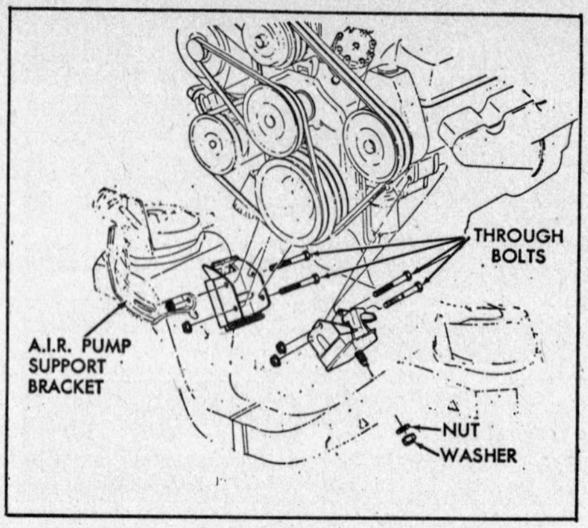

Fig. 2 Front engine mounts. Exc. Seville

6. Place jackstand under crankshaft pulley and, using wood block to avoid damage to pulley, either raise engine or lower chassis to remove mount.

CAUTION: When separating engine and chassis, observe relative position of frame rail and right drive axle tri-pot housing. Interference between these parts requires removal of right drive axle before mount can be removed. Furthermore, if tri-pot housing does clear the frame, vertical movement is limited by presence of fuel lines. Do not allow tri-pot housing to contact fuel lines.

7. Work mount free and pull rearward over tri-pot housing to remove.

1971-72 Eldorado

1. Remove fan and then raise car.
2. Remove nuts securing engine mount to crossmember.
3. Position a stand under crankshaft pulley and lower car until mount studs are disengaged from frame.

CAUTION: Transmission cooler lines and fuel lines must receive special handling to avoid damage.

4. Remove bolts securing mount to adapter and remove mount.

ENGINE, REPLACE

1976 Seville

1. Disconnect battery ground cable.
2. Drain cooling system and remove air cleaner.
3. Remove hood and wheel housing struts from both sides.

NOTE: Scribe outline of hinge on underside of hood before removing to aid alignment at installation.

4. Remove fan shroud and disconnect power brake vacuum hose at steel line.
5. Disconnect wiring harness at fuel injectors, throttle switch, EGR solenoid, fast idle valve, air temperature sensor, A/C compressor, MAP sensor, economy lite hose and two compressor ground wires.
6. Disconnect throttle linkage and position aside, remove vapor canister hose from throttle body.
7. Remove fuel inlet line from fuel rail.

NOTE: Use backup wrench on fuel rail to prevent damage to fitting. Ensure fuel pressure is relieved before loosening fittings.

8. Disconnect power steering hoses at steering gear and cap hoses and connections.
9. Disconnect fuel return line from pressure regulator outlet.
10. Remove compressor from engine and position aside.

NOTE: Do not disconnect refrigerant hoses from compressor.

11. Raise vehicle and remove exhaust pipe from right bank manifold.
12. Remove torque converter cover, then remove starter.
13. Remove clip securing transmission oil cooler lines to oil pan.
14. Remove bolts securing flex plate to converter.
15. Remove engine mount through bolts, then the crossover pipe.
16. Remove engine to transmission attaching bolts.
17. Remove screws attaching heater water valve to evaporator and position valve aside.
18. Support transmission, then raise engine off mounts and reposition transmission mount. Raise engine, then pull forward to disengage transmission and remove engine.

1970-76 Exc. Seville

1. Disconnect battery ground cable.
2. Remove hood and carburetor air cleaner with hoses.
3. Disconnect carburetor linkage at pedal lever.
4. Disconnect Cruise Control vacuum hoses if so equipped.
5. Remove clamp securing upper radiator hose to cradle.
6. Disconnect and/or remove all necessary wiring, pipes, hoses, linkage, etc.
7. Remove fan and A/C compressor.
8. Partially remove power steering pump, laying it aside so hoses are above fluid level in reservoir.
9. Drain radiator and disconnect hoses.
10. Proceed with steps 13 through 32 as described for 1969 models.

1969

NOTE: Disregard optional equipment items mentioned in the following if the car being serviced is not so equipped.

1. Disconnect battery ground cable.
2. Remove hood, air cleaner and automatic level control hoses.
3. Disconnect carburetor linkage at pedal lever.
4. Disconnect Cruise Control linkage.
5. Remove radiator shroud and clamp securing radiator upper hoses to cradle.
6. Disconnect and/or remove all neces-

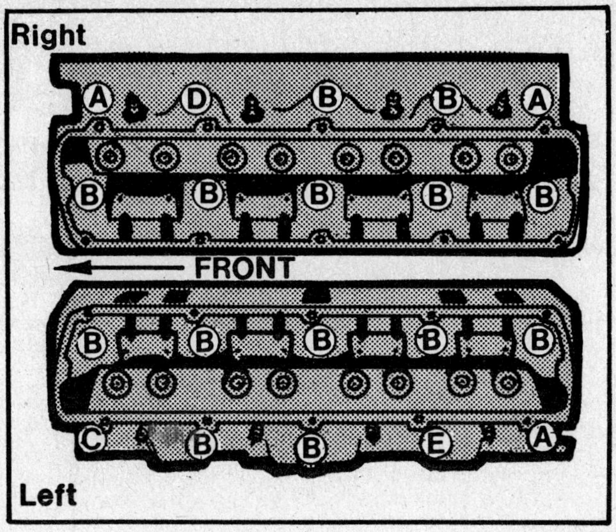

Bolt Location	Length
A (Bolt)	2.96 (Short)
B (Bolt)	4.16 (Long)
C (Bolt/Stud Head)	2.96 (Short)
D (Bolt/Stud Head)	4.16 (Long)
E (Bolt-Special)*	4.16 (Long)*

* This special bolt has an extra thick hex head which is drilled and tapped to accept the temperature switch.

Fig. 3 Location and length of cylinder head bolts. V8-350

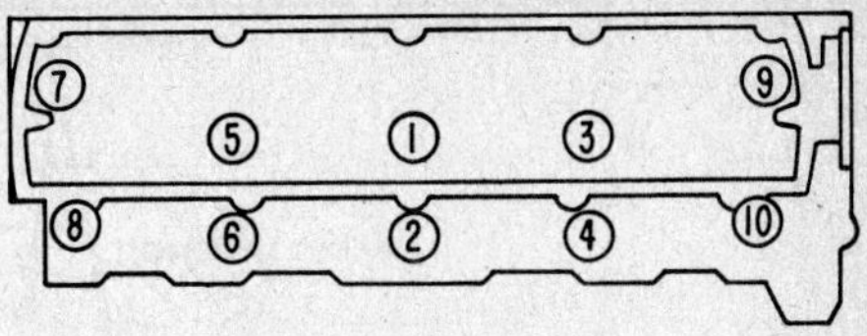

Fig. 4 Cylinder head tightening sequence. V8-350

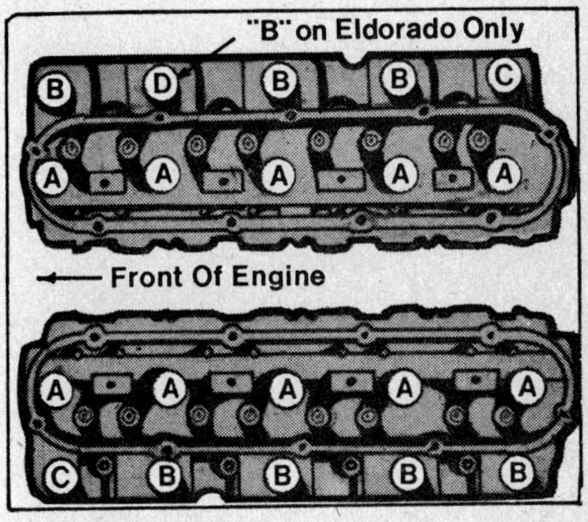

Bolt Location	Length
A (Bolt)	4.36" (Medium)
B (Bolt)	4.77" (Long)
C (Bolt)	3.02" (Short)
D (Bolt/stud)	4.77" (Long)

Fig. 5 Location and length of cylinder head screws. 1969-76 V8-472, 500

sary wiring, pipes, hoses, linkage, etc.

7. Remove fan. On Eldorado, remove studs from water pump shaft hub.
8. Remove Cruise Control power head.
9. Remove A/C compressor.
10. Partially remove power steering pump, laying it aside so that pressure and return hoses are above the fluid level in pump reservoir.
11. Position power steering cooler out of the way.
12. Loosen A.I.R. pump and remove belt.
13. Remove water pump pulley and alternator.
14. Remove two upper transmission-to-engine screws. Right screw secures transmission dipstick and modulator line to engine.
15. Remove A/C power servo from heater air selector. On cars other than Eldorado, remove blower relay and master switch from heater air selector.
16. Remove dust shield-to-cowl tie struts.
17. Position a lifting bracket to rear intake manifold attaching screws on Cruise Control cars. Bracket is in place on all others.
18. Raise car and place on jack stands.
19. Remove front engine mount nuts.
20. Remove oil filter.
21. On Eldorado, remove right output shaft.
22. Unfasten starter and allow it to hang by its cables.
23. Remove converter inspection pan.
24. Disconnect exhaust pipes from manifolds.
25. Place a jack under exhaust pipes to support system while engine is removed.
26. Support transmission with a jack.
27. Remove four lower transmission-to-engine screws.
28. Remove three screws securing converter to flex plate.
29. With lifting rig attached, pry engine forward while raising it as far as possible.
30. Lower transmission and again raise engine.
31. Remove flex plate and lift engine free of car.
32. Reverse procedure to install.

INTAKE MANIFOLD, REPLACE

With E.F.I.

1. Disconnect battery ground cable and remove air cleaner and crankcase filter.
2. Disconnect throttle cable and cruise control linkage, if equipped, from throttle body. Remove cable from bracket and position aside.
3. Working on engine left side, disconnect coolant temperature tell-tale switch wire, HEI lead wire, speed sensor connector, downshift switch and injector wires. Disconnect harness from fuel rail brackets and position aside.
4. Disconnect vacuum hoses and power brake pipe from throttle body.

NOTE: Use a back-up wrench when removing power brake pipe to prevent damage to fitting.

5. Disconnect fuel line from fuel rail.

NOTE: Use a back-up wrench on fuel rail to prevent damage to fitting.

6. Working on engine right side, disconnect injector wires, EGR solenoid leads, air temperature sensor connector and MAP sensor hose. Disconnect harness from fuel rail brackets and position aside.
7. Remove PCV valve from rocker cover and position aside.
8. Remove spark plug leads and distributor cap.
9. Remove front fuel rail as outlined under "Fuel Rail, Replace" procedure.
10. Position A/C compressor aside, do not disconnect refrigerant lines.
11. Remove fuel return hose from pressure regulator.
12. Remove manifold bolts and lift manifold from engine. Do not pry or lift manifold by the fuel rails or the mounting brackets.
13. Reverse procedure to install. On V8-350, torque manifold bolts to 40 ft.

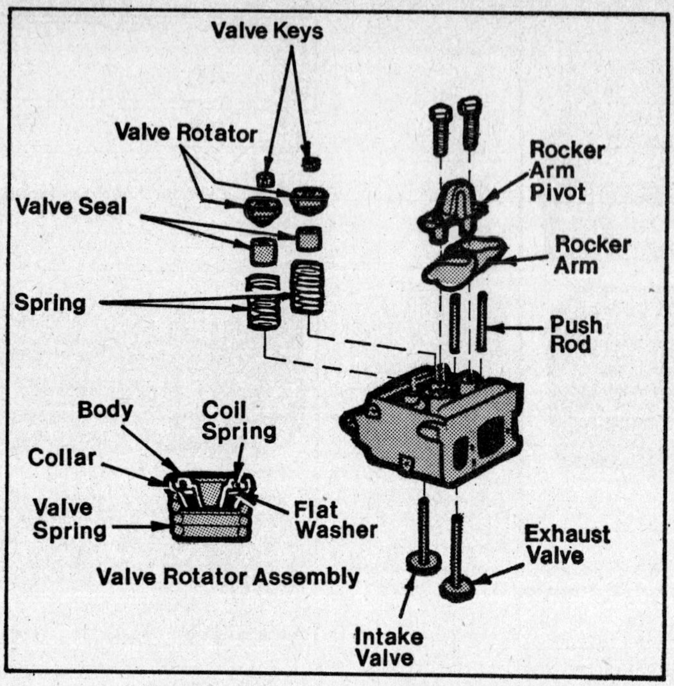

Fig. 6 Rocker arm components disassembled. V8-350

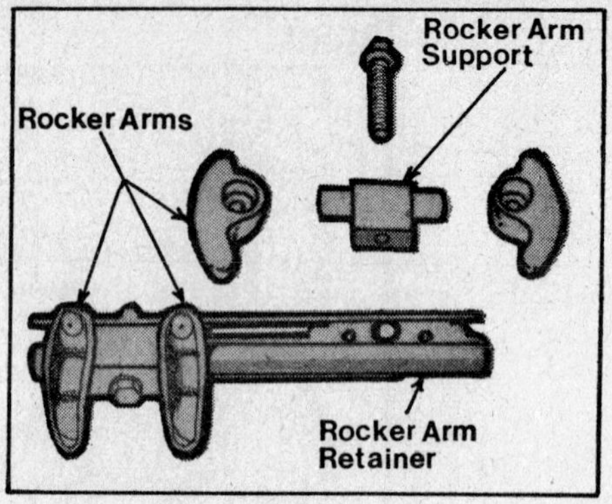

Fig. 7 Rocker arm components disassembled. 1969-76. V8-472, 500

lbs. and on V8-500, torque bolts to 30 ft. lbs.

Less E.F.I.

1. Disconnect battery ground cable and remove air cleaner.
2. Disconnect throttle linkage and Cruise Control linkage, if equipped.
3. Disconnect coil wires at H.E.I. connector, then remove distributor cap.
4. Disconnect wire connectors from temperature sender, down shift switch and anti-dieseling solenoid.
5. Remove coil, anti-dieseling solenoid and solenoid bracket.
6. Disconnect vacuum lines from carburetor, remove Cruise Control servo top mounting bolt and position brake hose aside, if equipped.
7. Disconnect wiring connector from A/C compressor, if equipped.
8. Disconnect fuel line at carburetor and remove vacuum lines from intake manifold.
9. Loosen power steering mounting bolts and pivot pump toward engine. Remove A/C compressor drive belt, then the screws attaching compressor rear mounting to engine and the screws attaching compressor mounting flange to front mounting bracket. Position compressor aside.
10. Remove PCV valve from rocker cover.
11. Remove manifold to cylinder head bolts, intake manifold, manifold shield and gaskets.

CYLINDER HEAD, REPLACE

With E.F.I.

1. Drain cooling system and intake manifold as described under "Intake Manifold Replace".
2. Remove exhaust manifolds.
3. Remove rocker cover, rocker arm bolts, pivots, rocker arms and push rods.

NOTE: Note location of rocker arms and pivots to ensure installation in original position.

4. Remove cylinder head bolts and the cylinder head.
5. When installing cylinder head it is important that cylinder head bolts are installed in the correct position, refer to Fig. 3. Torque bolts to 60 ft. lbs. in sequence, then retorque to 85 ft. lbs., Fig. 4.

Less E.F.I.

1. Remove intake manifold.
2. Drain coolant from radiator.
3. Disconnect ground strap at rear of cylinder heads from cowl on 1969-73 models.
4. Disconnect wiring connector for high engine temperature warning system from sending unit at rear of left cylinder head.
5. Remove alternator if working on right cylinder head, or partially remove power steering pump if working on left cylinder head.
6. Remove A.I.R. manifold from both cylinder heads if equipped.
7. Disconnect wiring harness from cylinder head and position out of the way.
8. Remove exhaust manifold from cylinder head.
9. Remove rocker arm cover.
10. Remove rocker arm assemblies and lift out push rods.
11. Remove cylinder head bolts, remove head.
12. After carefully removing all gasket material from mating surfaces of head and block, position new gasket over dowels and install cylinder head in reverse order of removal, being sure to install the bolts as indicated in Fig. 5.

NOTE: When installing cylinder head bolts no torque sequence is required, start at center and work from side to side outwards and towards the ends.

VALVE ARRANGEMENT

Front to Rear

V8-472, 500 .... E-I-I-E-E-I-I-E
V8-350 .... I-E-I-E-E-I-E-I

VALVE LIFT SPECS.

Engine	Year	Intake	Exhaust
V8-472	1969-71	.440	.454
V8-500	1970-71	.440	.454
V8-472	1972-73	.490	.490
V8-500	1972-73	.490	.490
V8-472	1974	.457	.473
V8-500	1974-76	.475	.473
V8-350	1976	.400	.400

VALVE TIMING

Intake Opens Before TDC

Engine	Year	Degrees
V8-472	1969-70	18
V8-472	1971	38
V8-500	1970	18
V8-500	1971-73	34
V8-472	1972-73	34
V8-472	1974	21
V8-500	1974-76	21
V8-350	1976	16

ROCKER ARMS

V8-350

1. Remove rocker cover, rocker arm bolts, pivot and rocker arms.

NOTE: Remove each set as a unit.

2. Position rocker arms on cylinder head, apply lubricant 1050169 or equivalent to wear points on pivot, then install pivot.
3. Install rocker arm bolts, tighten alternately, torque to 25 ft. lbs., Fig. 6.

V8-472, 500

When disassembling the rocker arm assembly, be sure to keep the supports and rocker arms in order so they can be installed in the exact same position.

Install rocker arms on supports and place supports in retainers as shown in Fig. 7. *Be sure that the "EX" on support is positioned toward the exhaust valve and "IN" toward the intake valve.*

Place capscrews through the reinforcements, supports and retainers and position assemblies on cylinder head. Make sure that push rods are properly seated in the lifter seats and in the rocker arms. Lubricate rocker arm bearing surfaces before assembling in order to prevent wear.

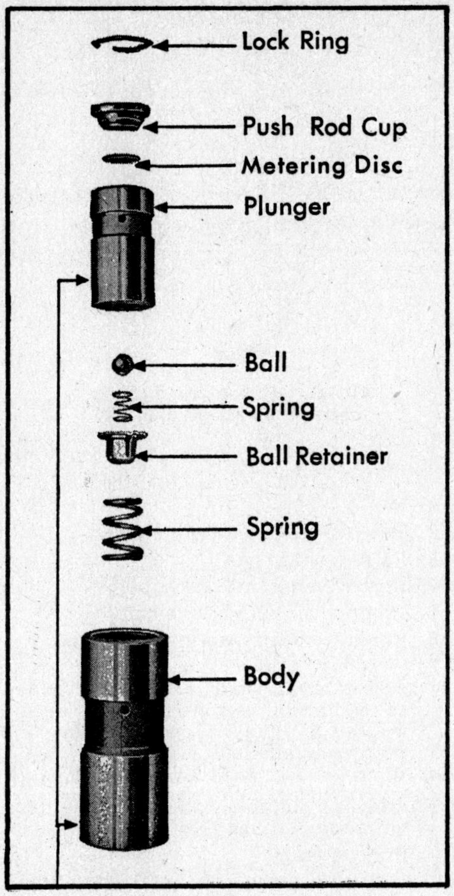

Fig. 8 Hydraulic valve lifter

VALVE GUIDES

V8-350

Valve stem to guide clearance should be .001-.004 inch. Service valves are available in standard (.343 inch) and .003, .005, .010, .013 inch oversizes. If stem to guide clearance is excessive, ream valve guide bore to the next oversize using appropriate reamer. Install proper oversize valve and valve seal. Valve seals can be identified as follows,

Intake
Std.-.005 O.S. . . . . Gray Colored
.010-.013 O.S. . Orange Colored

Exhaust
Std.-.005 O.S. . . . Ivory Colored
.010-.013 O.S. . . . Blue Colored

On occasion an engine will be manufactured with an oversize valve and valve guide bore. A number stamped on the inboard side of the head will indicate which valve and valve guide bore are oversize. The number 10 would indicate that a .010 inch oversize valve is installed in the adjacent valve guide bore.

V8-472, 500

Check valve stem to valve guide clearance, clearance should be no more than .005 inch. Service valves are available in standard (.343 inch) and .003, .006 and .013 inch oversizes. If clearance is found to be excessive, valve guide should be reamed out to next oversize using appropriate reamer, and a corresponding oversize valve installed. On some engines, valves with a .003 inch oversize diameter and .003 inch oversize valve guides are installed at the factory. Engines so fitted will be identified by a "3" stamped on the cylinder head gasket surface inline with the oversize valve.

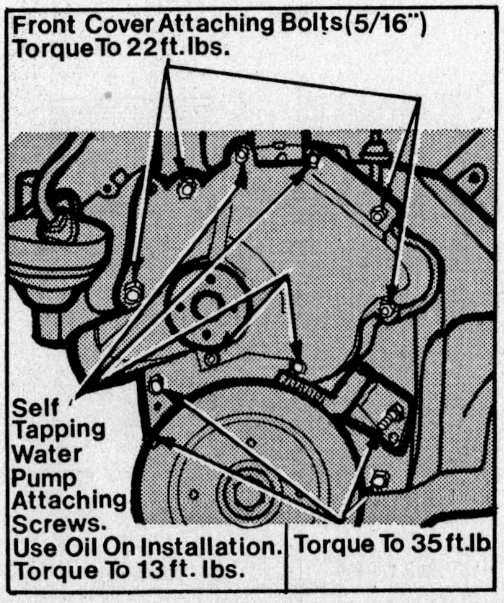

Fig. 9 Engine front cover attaching screws. V8-350

VALVE LIFTERS

The valve lifters may be lifted out of their bores after removing the rocker arms, push rods and intake manifold. Adjustable pliers with taped jaws may be used to remove lifters that are stuck due to varnish, carbon, etc. Fig. 8 illustrates the type of lifter used.

ENGINE FRONT COVER

V8-350

1. Drain cooling system, disconnect radiator hoses, heater hose and bypass hose.
2. Remove belts, fan, fan pully, crankshaft pully and harmonic balancer.
3. Remove oil pan, front cover attaching bolts, front cover and water pump.
4. Reverse procedure to install, refer to Fig. 9.

V8-472, 500

1. On Eldorado models, remove engine from vehicle.
2. Remove harmonic balancer.
3. Loosen starter sufficiently to gain access to oil pan screws and lower front of oil pan until it clears front studs.

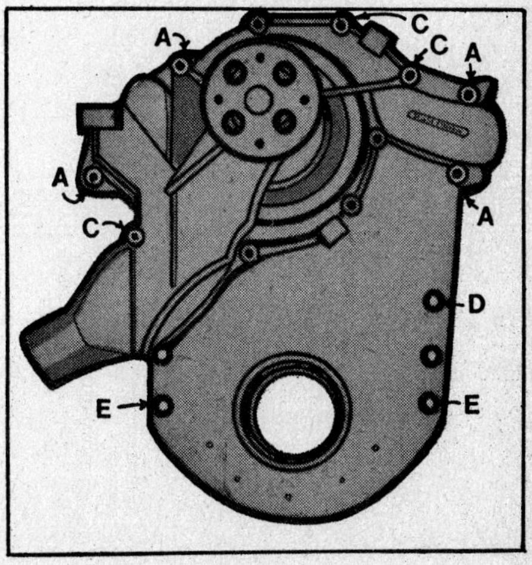

Key	(No.)	Size	Torque
A	(4)	3/8-16 x 1-3/8	25 Foot-Pounds
C	(3)	5/16-18 x 1-1/4	15 Foot-Pounds
D	(1)	5/16-18 x 5/8	15 Foot-Pounds
E	(2)	3/8-16 x 5/8	25 Foot-Pounds

Fig. 10 Engine front cover attaching screws. 1969-76 V8-472, 500

4. Drain radiator and remove lower hose from water pump.
5. Remove the 10 screws securing front cover to the cylinder block and remove cover with water pump as an assembly.
6. Reverse removal procedure, to install, referring to Fig. 10.

FRONT COVER OIL SEAL

V8-350

1. Remove drive belts, crankshaft pulley and harmonic balancer.
2. Using tool J-1859-03 or a suitable puller, remove front cover oil seal.
3. Apply sealer to outside diameter of seal.
4. Position seal on front cover, then using tools No. J-25264 and J-21150, drive seal into place.
5. Install harmonic balancer and crankshaft pulley, install and adjust drive belts.

1970-76 V8-472, 500

1. Remove vibration damper.
2. With a thin blade screw driver or similar tool, pry out front cover oil seal.
3. Lubricate new seal and fill cavity with wheel bearing grease. Position seal on end of crankshaft with garter spring side toward engine.
4. Using a suitable installer drive seal into cover until it bottoms against cover.

1969 V8-472

1. Disconnect battery ground cable.

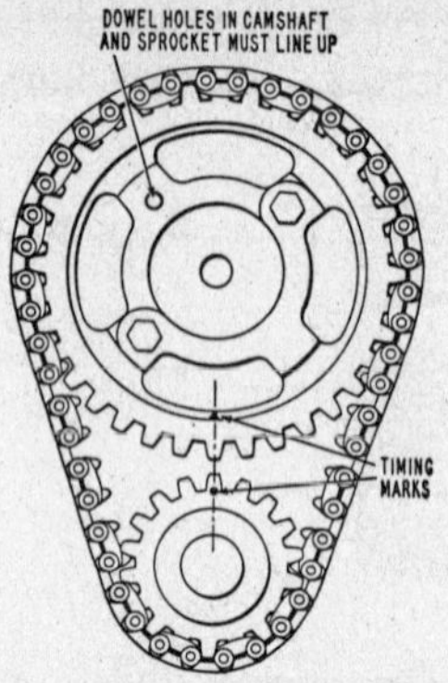

Fig. 11 Timing gear locating marks. 1969-75. V8-472, 500

2. Remove carburetor air cleaner.
3. Raise front of car.
4. Remove alternator drive belt.
5. Remove air pump drive belt.
6. Remove power steering pump belt.
7. Working under car, remove four screws that hold crankshaft pulley to vibration damper. Place scribe marks on pulley and damper for proper installation.
8. Remove plug from end of crankshaft.
9. Use a suitable puller to remove vibration damper.
10. Pry out seal.
11. Lubricate new oil seal by filling cavity between lips with wheel bearing grease. Position seal on end of crankshaft with garter spring side toward engine.
12. Using a suitable seal installer, drive seal into front cover until it bottoms against cover.

TIMING CHAIN

1. Remove engine front cover as outlined above.
2. Remove two capscrews and washers that hold sprocket to camshaft.
3. Remove camshaft sprocket with chain.
4. Remove crankshaft sprocket.
5. To install, reverse removal procedure, being sure to line up the timing marks as shown in Fig. 11.

CAMSHAFT

1976 V8-350

1. Disconnect battery, remove air cleaner.
2. Drain cooling system and remove radiator, four bolts attaching A/C condenser to radiator and position condenser aside.
3. Remove crankshaft pulley and balancer, then the front cover.
4. Remove A/C compressor from engine and position aside.

NOTE: Do not disconnect refrigerant hoses from compressor.

5. Remove intake manifold as described under "Intake Manifold Replace".
6. Remove rocker covers, rocker arms, push rods and lifters.

NOTE: Note position of rocker arms, pivots, push rods and lifters to ensure installation in original position.

7. Remove camshaft.

1970-76 V8-472, 500

1. Remove engine front cover.
2. Remove distributor and oil pump.
3. Remove oil slinger from crankshaft.
4. Remove fuel pump and fuel pump eccentric.
5. Unfasten and remove camshaft sprocket with chain attached.
6. Remove valve lifters as previously outlined.
7. Remove radiator.
8. Carefully slide camshaft forward until it is out of engine.

1969 V8-472

In order to replace the camshaft, it is necessary to remove the engine on 1969 Eldorado. Then follow the procedure outlined below.

1. To remove camshaft, take off engine front cover, timing chain and sprockets.
2. If equipped with air conditioner, remove condenser.
3. Remove hood lock plate support.
4. Remove valve lifters as previously outlined.
5. Slide camshaft out of engine carefully. *Use extreme care to keep cam lobes from scratching camshaft bearings.*

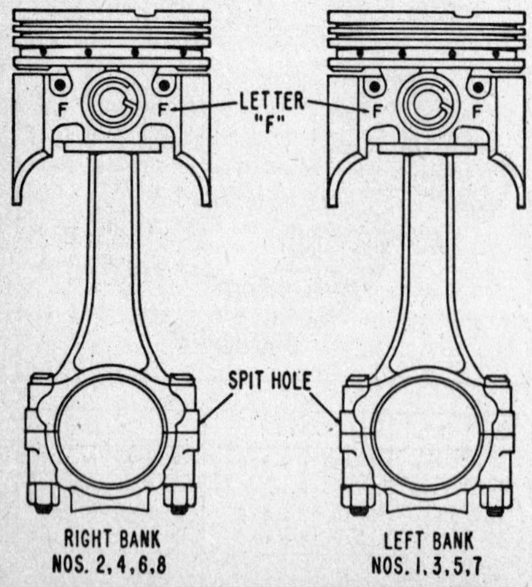

Fig. 12 Piston and rod assembly. V8-350

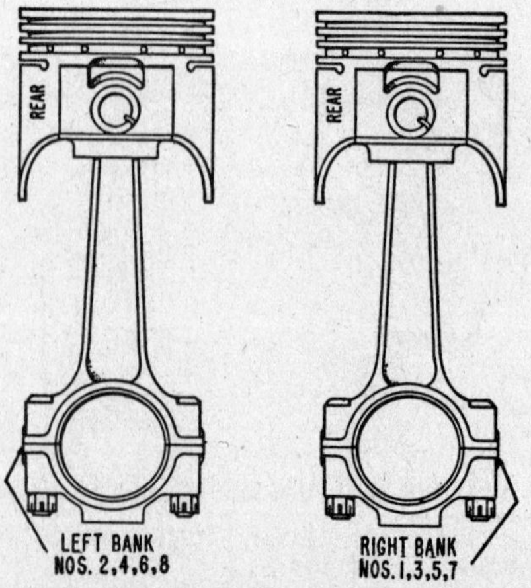

Fig. 13 1969-76 Piston and rod assembly. V8-472, 500

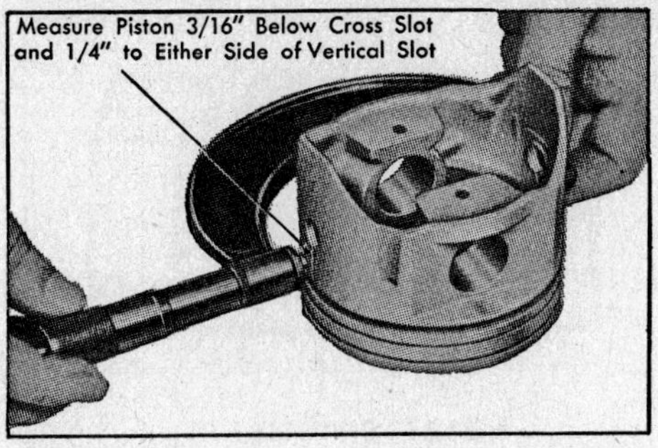

Fig. 14 Measuring piston diameter. 1969-76

6. When installing camshaft, apply a coating of rear axle lubricant to camshaft bearing journals and camshaft lobes.

PISTONS & RODS ASSEMBLE

On all engines, assemble and install the piston and rod assemblies as shown in Figs. 12 & 13.

PISTONS

V8-350

Measure pistons for size, Fig. 14. Measure cylinder bore, piston to cylinder bore clearance should be .00075-.00175 inch. Measure piston for taper, the largest reading must be at the bottom of the skirt. Allowable taper is .0005-.0015 inch.

Cylinder and piston sizes are indicated by a letter stamped on the cylinder gasket surface of the block. The letters that denote cylinder and piston sizes are listed in Fig. 15.

V8-472, 500

The V8-472 and 500 engines both have identical bore sizes (4.300 inch), therefore it is important that the correct piston be installed for the engine being worked on. Refer to Figs. 16 and 17, for correct piston identification.

Pistons should be measured for size as shown in Fig. 14. Cylinders should be measured 1⅛" from the top, crosswise to the cylinder block. The clearance should be .0006-.0010" in this position at room temperature (70°F). Subtract .0001" from measurement for every 6° above 70°.

An identification letter is stamped on the valve lifter compartment cover next to lower inside edge of cylinder head. The letters are in groups of two for adjacent cylinders (such as "A" "B") midway between the two cylinders. This letter denotes the cylinder size as shown in Fig. 18. The table indicates ten piston sizes to match ten bore sizes. This makes it possible to maintain the proper clearance between block and piston.

If double letters (such as "AA" "BB") appear, it indicates that the cylinder has been bored .010" over the diameter indicated by the single letter in the chart.

Cylinder bores must not be reconditioned to more than .0100 inch oversize, as pistons are not available over this range.

Letter	Cylinder Size (Diameter in inches)	Piston Size (Diameter in inches)
A	4.0560-4.0565	4.05475-4.05525
B	4.0565-4.0570	4.05525-4.05575
C	4.0570-4.0575	4.05575-4.05625
D	4.0575-4.0580	4.05625-4.05675
J	4.0660-4.0665	4.06475-4.06525
K	4.0665-4.0670	4.06525-4.06575
L	4.0670-4.0675	4.06575-4.06625
M	4.0675-4.0680	4.06625-4.06675

Fig. 15 Cylinder and piston sizes. V8-350

PISTON RINGS

Replacement rings are available from Cadillac in standard size and .010"oversize.

PISTON PINS

V8-350

Piston pin to piston clearance should be .0003-.0005 inch. If clearance is more than .0005 inch, a new piston pin should be installed. Piston pin to connecting rod is a press fit and clearance should be .0008-.0018 inch tight fit.

V8-472, 500

Piston pins are a matched fit with the piston and are not available separately. Piston pins are pressed in the connecting rods and will not become loose enough to cause a knock or tapping until after very high mileages. In such cases a new piston and pin assembly should be installed.

MAIN & ROD BEARINGS

V8-350

Main bearing clearance should not exceed .0035 inch. If clearance is more than .0035 inch replace both bearing shells. Main bearings are available in oversizes, Fig. 19.

Rod bearing clearance should not exceed .0035 inch. If clearance is more than .0035 inch replace both bearing shells and recheck clearance.

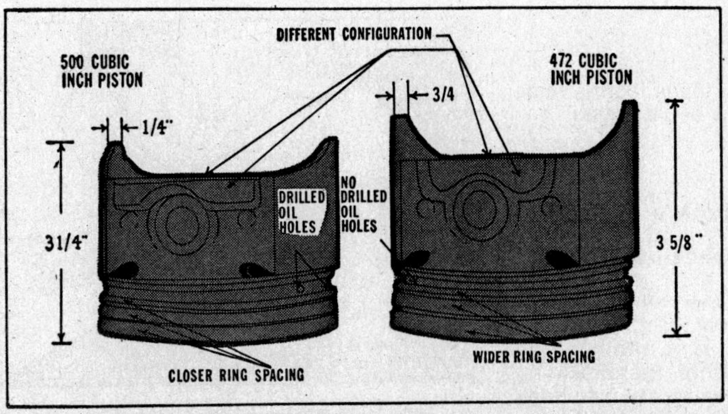

Fig. 16 Piston identification. 1970-73 V8-472 and 500

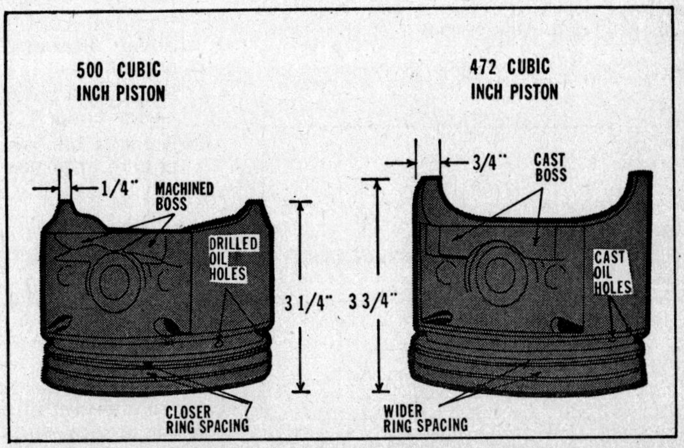

Fig. 17 Piston identification. 1974 V8-472 and 1974-76 V8-500

Cylinder and piston sizes (as indicated by letters stamped on the cylinder head gasket surface). The letters are in groups of two for adjacent cylinders (such as "H" and "B") midway between the two cylinders. The letters denote the cylinder piston sizes as shown below).

Letter	Cylinder Size (Diameter in Inches)	Piston Size (Diameter in Inches)
A	4.3000 - 4.3002	4.2992 - 4.2994
B	4.3002 - 4.3004	4.2994 - 4.2996
C	4.3004 - 4.3006	4.2996 - 4.2998
D	4.3006 - 4.3008	4.2998 - 4.3000
E	4.3008 - 4.3010	4.3000 - 4.3002
H	4.3010 - 4.3012	4.3002 - 4.3004
J	4.3012 - 4.3014	4.3004 - 4.3006
K	4.3014 - 4.3016	4.3006 - 4.3008
L	4.3016 - 4.3018	4.3008 - 4.3010
M	4.3018 - 4.3020	4.3010 - 4.3012
AA	4.3100 - 4.3102	4.3092 - 4.3094
BB	4.3102 - 4.3104	4.3094 - 4.3096
CC	4.3104 - 4.3106	4.3096 - 4.3098
DD	4.3106 - 4.3108	4.3098 - 4.3100
EE	4.3108 - 4.3110	4.3100 - 4.3102
HH	4.3110 - 4.3112	4.3102 - 4.3104
JJ	4.3112 - 4.3114	4.3104 - 4.3106
KK	4.3114 - 4.3116	4.3106 - 4.3108
LL	4.3116 - 4.3118	4.3108 - 4.3110
MM	4.3118 - 4.3120	4.3110 - 4.3112

Fig. 18 Cylinder and piston sizes. 1969-76 V8-472, 500

V8-472, 500

Main and rod bearings are supplied by Cadillac in standard sizes only.

CRANKSHAFT OIL SEAL

V8-350

To replace upper rear main bearing seal, the crankshaft must be removed. After removing crankshaft use tool No. J-25285 to install seal, Fig. 20.

Lower rear main bearing seal can be installed with crankshaft in place and engine in vehicle. Remove oil pan, then the rear main bearing cap. Remove bearing and seal from cap. Position new seal in groove in bearing cap. Using tool J-25285, hammer seal into groove. Rotate tool slightly, then cut seal ends flush with bearing cap surface. Pack seal end fibers away from edges, using a screw driver. Install bearing in cap, then install cap, lubricate bolt threads and torque to 120 ft. lbs. Install oil pan and lower flywheel cover.

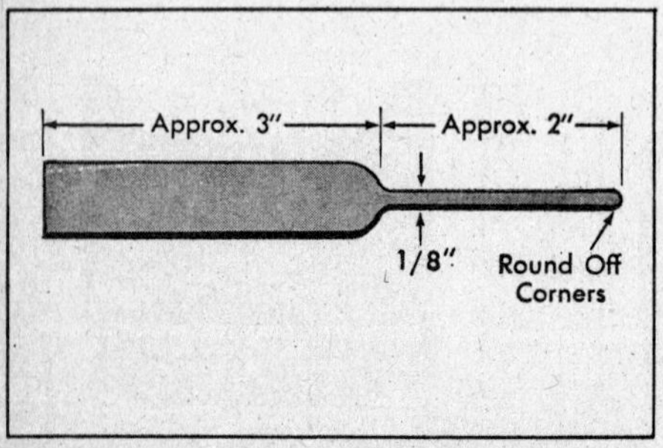

Fig. 21 Main bearing oil seal tool. V8-472, 500 Engine oiling system. V8-350

V8-472, 500

Rear main bearing installation tool can be made from shim stock or a metal banding strap using dimensions in Fig. 21.

The two seal halves are identical and can be used in either the lower or upper location. However, both seal halves are pre-lubricated with a film of wax for break-in. Do not remove or damage this film.

To install the lower half of the seal into the bearing cap, slide either end of seal into position at one end of bearing cap and place tool on seal land at other end of bearing. Make sure seal is positioned over bearing ridge and lip of seal is facing forward (car position).

Hold thumb over end of seal that is flush with split line to prevent it from slipping upward, and push seal into seated position by applying pressure to the other end. Make sure seal is pressed down firmly and is flush on each side to avoid possibility of a leak at seal split line. Avoid pressing on lip as damage to sealing edge could result.

To install upper half of seal in cylinder block (with crankshaft in car), position "shoehorn" tool on land of block, Fig. 22. Start seal into groove in block with lip facing forward and rotate seal into position. Do not press on lip or sealing edge may be damaged. Both ends of seal must be flush at seal split line to avoid leaks. If necessary, Lubriplate or its equivalent may be used to facilitate installation of both upper and lower seal halves. Do not use silicone or a leak may result.

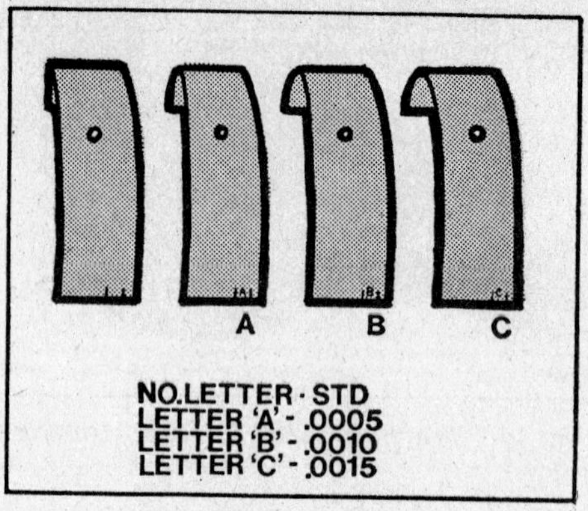

Fig. 19 Main bearing identification. V8-350

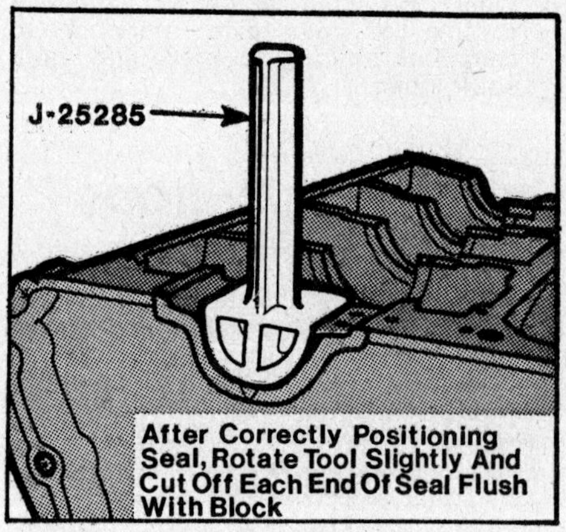

Fig. 20 Rear main bearing oil seal installation. V8-350

OIL PAN, REPLACE

1976 V8-350 Seville

1. Remove wheel housing struts from both fenders.
2. Remove fan shroud.
3. Remove front motor mount through bolts, disconnect crossover pipe at manifold.
4. Remove starter, then remove converter cover.
5. Remove oil pan attaching bolts, raise engine until pan can be removed.

1969-76 Except Eldorado & Seville

NOTE: For easier removal of the oil pan past the stabilizer bar, first remove the two *front* dowel studs from the block by running a jam nut on each stud to lock the pan nut on the stud. Working over the front frame crossmember, use a socket to remove the stud with the two nuts attached.

1. Disconnect battery ground cable.
2. Drain engine oil.
3. Remove "Y" exhaust pipe at exhaust manifold.
4. Remove starter.

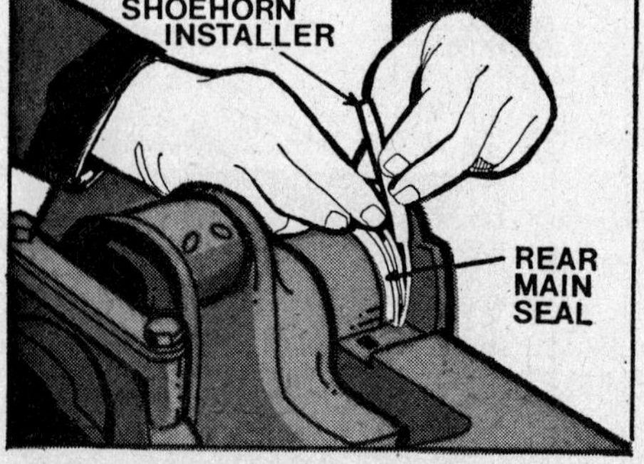

Fig. 22 Installing rear main bearing oil seal. 1969-76 V8-472, 500

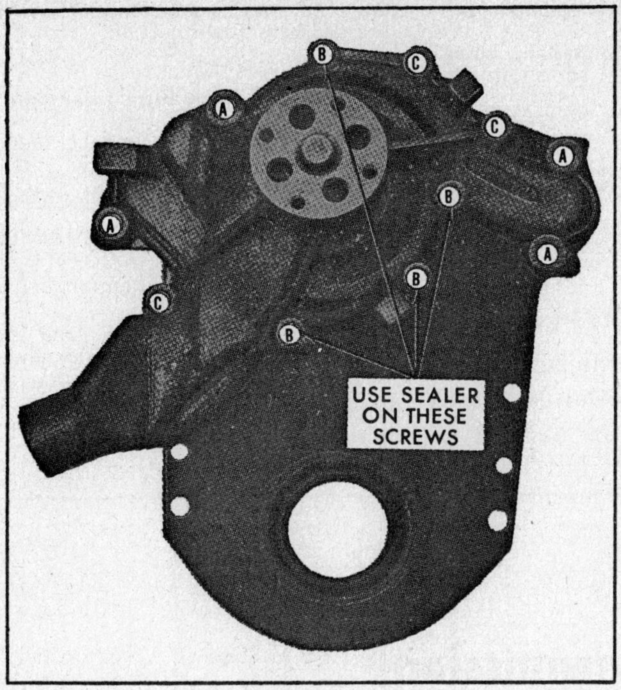

Fig. 23 Water pump attaching screws. 1969-76 V8-472, 500

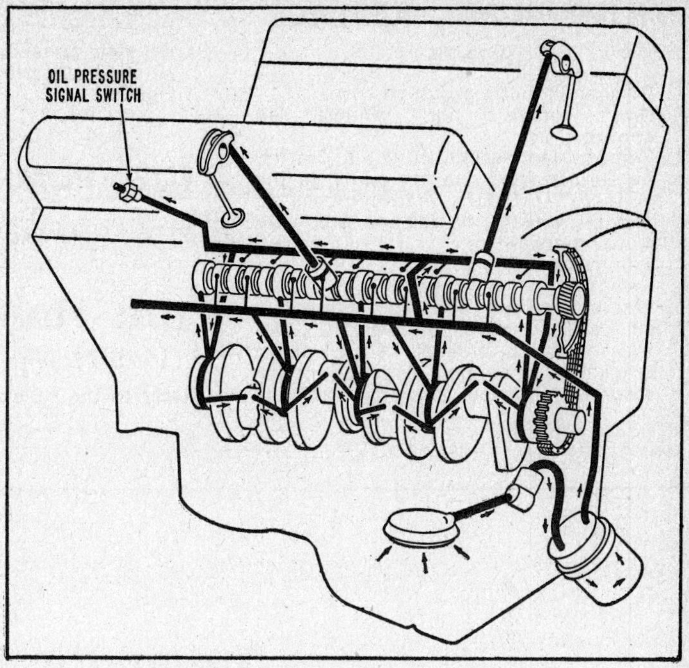

Engine oiling system. 1969-76 V8-472, 500

5. Unfasten and lower idler arm support.
6. Disconnect pitman arm at center link and lower steering linkage.
7. Remove transmission lower cover.
8. Unfasten and lower oil pan.

NOTE: To align the oil pan at the rear and prevent damage from the flywheel teeth during installation, first locate the oil pan on the crankcase with two screws at mid-point. Then install two screws at the rear while checking gasket alignment.

1969-76 Eldorado

To remove the oil pan it is necessary to remove the engine as described previously.

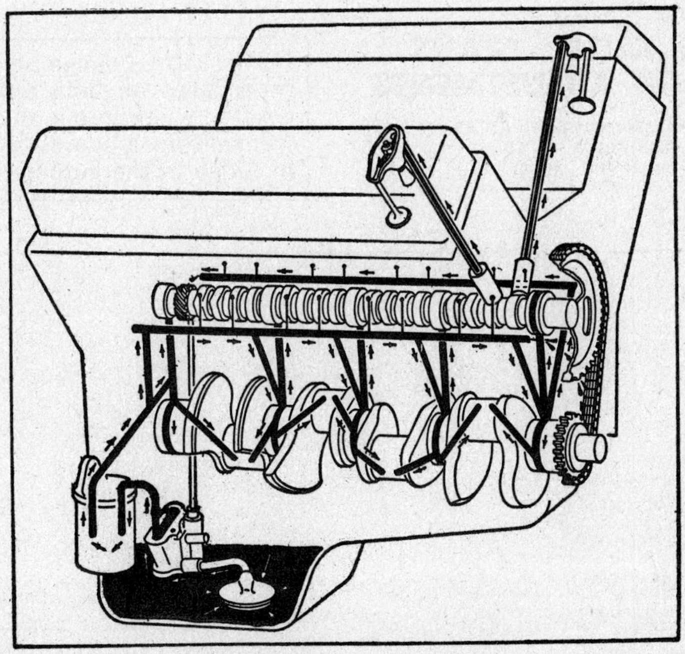

Engine oiling system. 1976 V8-350

OIL PUMP

1976 V8-350

1. Remove oil pan, then remove pump to rear main bearing attaching screws, remove pump.

1969-76 V8-472, 500

1. Raise car and remove oil filter.
2. Remove five screws securing pump to engine. *The screw nearest the pressure regulator should be removed last, allowing the pump to come down with screw.*
3. Remove pump drive shaft.
4. Reverse procedure to install, being sure to pack the pump with petrolatum.

BELT TENSION DATA

	New Lbs.	Used Lbs.
1969-73—		
Air Condition	100	55-70
A.I.R. Pump	100	55-70
Generator	100	55-70
Power Steering	100	55-70
1974-76 Exc. Seville—		
Air Condition	170	120
A.I.R. Pump	170	120
Generator—		
Exc. 145 Amp.	100	70
145 Amp.	170	120
Power Steering	170	120
1976 Seville—		
Air Condition	125	90
A.I.R. Pump	125	90
Generator	125	90
Power Steering	125	90

WATER PUMP, REPLACE

1969-76

1. Disconnect battery ground cable.
2. Drain radiator and remove fan shroud.
3. Remove fan assembly. *On A/C cars, be sure to keep the fan clutch in the "on car" position when removed to prevent leakage of silicone fluid into clutch mechanism.*
4. Remove all drive belts.
5. Pull pump pulley off shaft.
6. Disconnect water inlet from pump.
7. On Seville models, remove A/C Compressor front bracket, A.I.R. pump bracket and support rod and power steering pump adjusting bracket.
8. Unfasten and remove pump from front cover.
9. Reverse procedure to install, being sure to install bolts as shown in Fig. 23.

FUEL PUMP PRESSURE

Year	Engine	Pressure, Lbs.
1969-76	All Exc. E.F.I.	5¼-6½

FUEL PUMP, REPLACE

1975-76 With Fuel Injection

Refer to the "Electronic Fuel Injection" chapter for service procedures for vehicles equipped with fuel injection.

1969-76 Less Fuel Injection

1. Raise car and disconnect fuel line at pump and plug line.
2. Disconnect fuel pipe to fuel filter at pump.
3. Remove mounting screw on upper pump flange.
4. Remove nut from mounting stud at lower pump flange.
5. Tipping pump upward, pull pump straight out from engine and remove.
6. Reverse procedure to install.

Automatic Transmission

1976 AUTO. TRANS. LINKAGE ADJUSTMENTS

Linkage adjustment procedures for 1976 models are the same as those for 1975 models as outlined elsewhere in this manual.

NOTE: 1976 linkage adjustment information is in this section. Repair procedures on both automatic and manual shift transmissions are covered elsewhere in this manual. Procedures for removing automatic transmissions as well as linkage adjustments on 1969-75 models are included in the automatic transmission chapters. See Chapter Index.

Eldorado Drive Link Belt

LINK BELT OR SPROCKETS

Removal

After removal of transmission as described elsewhere in this manual, proceed as follows:

1. Remove sprocket housing cover attaching bolts and cover.
2. Remove sprocket bearing retaining snap rings from retaining grooves in support housing located under drive and driven sprockets, Fig. 1.

NOTE: Do not remove the snap rings from beneath the sprockets, leave them in a loose position between the sprockets and the bearing assemblies.

3. Remove drive and driven sprockets, link belt, bearings and shafts simultaneously by alternately pulling upwards and driven support housing, Fig. 2.

NOTE: If sprockets and link belt are difficult to remove, place a small piece of masonite, or similar material between the sprocket and a short pry bar. Alternately pry upward under each sprocket. Do not pry on links or aluminum case, Fig. 3.

4. Remove link belt from drive and driven sprockets.

Installation

1. Place link belt around the drive and driven sprockets so that the links engage the teeth of the sprockets, colored guide link which has etched numerals facing link cover.
2. Simultaneously place link belt, drive and driven sprockets into support housing, Fig. 1.
3. Using a plastic mallet, gently seat the sprocket bearing assemblies into the support housings.
4. Install sprocket assembly to support housing snap rings, Fig. 2.
5. Install new case to cover and plate assembly sprocket housing gasket.

NOTE: Important: One sprocket cover housing attaching bolt is ¼ inch longer. This bolt must be installed in the tapped hole located directly over the cooler fittings on the transmission case.

6. Install sprocket housing cover and plate assembly and eighteen attaching bolts. Torque bolts to 8 ft. lbs.

Fig. 1 Removing or installing retaining rings

Fig. 2 Removing or installing sprockets and link assembly

Fig. 3 Removing tight sprockets

Rear Axle, Propeller Shaft & Brakes

REAR AXLE

1969-76 Eldorado

1969-70 models use the beam type rear axle, while 1971-76 models use the hollow tube type axle. On both axles the rear wheel spindles are a press fit and bolted to the rear axle, Fig. 1. As shown, tapered roller bearings are used in the rear wheels. These bearings do not require regularly scheduled repacking. When major brake service work is to be performed, however, it is recommended that the bearings be cleaned and repacked.

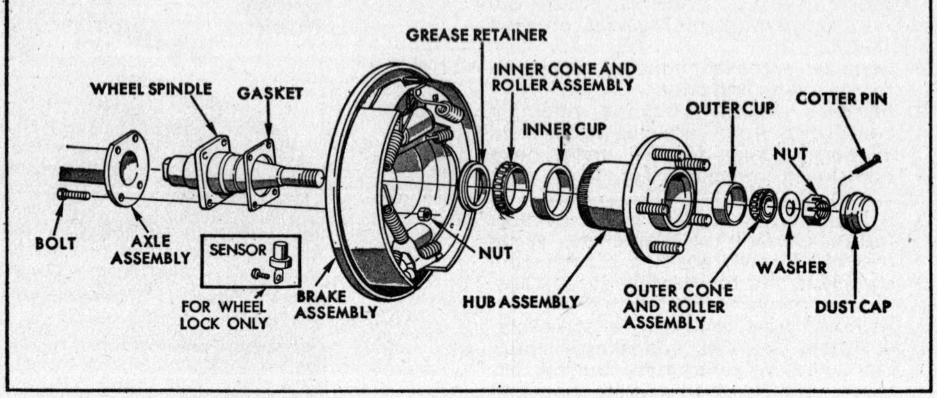

Fig. 1 Rear wheel spindle disassembled. 1969-76 Eldorado (typical)

Wheel Bearing Adjustment

Adjustment of the rear wheel bearings should be made while revolving the wheel at least three times the speed of the nut rotation when taking torque readings.

1. Check to make sure that hub is completely seated on wheel spindle.
2. While rotating wheel, tighten spindle nut to 30 ft-lbs. Make certain all parts are properly seated and that threads are free.
3. Back nut off ½ turn, then retighten nut to 2 ft. lbs. and install cotter pin.
4. If cotter pin cannot be installed in either of the two holes in the spindle, with the nut at 2 ft. lbs., back nut off until cotter pin can be installed.
5. The rear hub must be rotated at least three revolutions during tightening of spindle nut. The final adjustment to be 2 ft. lbs. to provide .004" bearing end play.
6. Peen end of cotter pin snug against side of nut. If it can be moved with a finger, vibration may cause it to wear and break.

Wheel Spindle, Replace

1. Raise and support rear of car and remove hub.
2. Disconnect brake line at wheel cylinder.
3. Unfasten and remove brake backing plate and position out of the way.
4. Remove bolts securing spindle to axle.
5. On 1971-75 models, use a suitable puller to remove spindle from axle tube.
6. On 1969-70 models:
   a. Place jack under rear axle.
   b. Remove four nuts from center spring clamp and lower rear axle until spindle is accessible.
   c. Remove lower spring insulators from rear axle.
   d. Drive spindle from rear axle.

Installation

1. On 1971-75 models, use a suitable slide hammer to drive spindle into axle tube until fully seated.
2. On 1969-70 models;
   a. Position spindle on axle with keyway facing up, then install four backing plate to spindle nuts.
   b. Alternately tighten nuts until spindle is fully seated, then remove nuts and bolts.
   c. Position lower spring insulator on rear axle.
   d. Position axle to spring clamp, ensure aligning pin locates into rear axle, lower insulator is properly positioned and center spring clamp bolts engage axle mounting holes.
   e. Install four nuts securing center spring to rear axle, torque to 30 ft. lbs.
3. Install new gasket on wheel spindle.
4. Install brake backing plate and tighten nuts to 40 ft-lbs.
5. Connect brake line to wheel cylinder, tightening fitting to 14 ft-lbs.
6. Install rear hub.

Rear Axle, Replace

1. Raise and support rear of car with jack stands at rear frame pads ahead of rear wheel opening.
2. Remove rear wheels and hubs.
3. Disconnect brake lines at wheel cylinders.
4. Disconnect parking brake cable at equalizer.
5. Disconnect rubber brake hose at underbody connector.
6. Disconnect overtravel lever link from bracket on rear axle.
7. Remove spring guides retaining parking brake cable to center spring clamp.
8. If rear axle is being replaced, remove brake backing plates.
9. Supporting rear axle at center with a jack, remove eight nuts (4 each side) from center spring clamp assemblies.
10. Lower rear axle with jack and remove from car.
11. Remove lower spring insulators from rear axle.
12. If rear axle is being replaced, remove bolt securing brake line junction fitting to axle. Remove brake line, overtravel lever link bracket, and drive spindles from axle.
13. Reverse procedure to install.

1976 Seville

In this axle, Fig. 2, the rear axle housing and differential carrier are cast into an integral assembly. The drive pinion assembly is mounted in two opposed tapered roller bearings. The pinion bearings are preloaded by a spacer behind the front bearing. The pinion is positioned by a washer between the head of the pinion and the rear bearing.

The differential is supported in the carrier by two tapered roller side bearings. These bearings are preloaded by spacers located between the bearings and carrier housing. The differential assembly is positioned for proper ring gear and pinion backlash by varying these spacers. The differential case houses two side gears in mesh with two pinions mounted on a pinion shaft which is held in place by a lock pin. The side gears and pinions are backed by thrust washers.

Differential Removal

1. Raise vehicle and support so axle can be raised and lowered.
2. Disconnect leveling control lever from link and hold lever down until shock absorbers are deflated.
3. Remove shock absorber lower mounting bolts and position shock aside.
4. Mark propeller shaft flange and pinion flange to ensure installation in original position. Remove attaching screws and position propeller shaft aside.

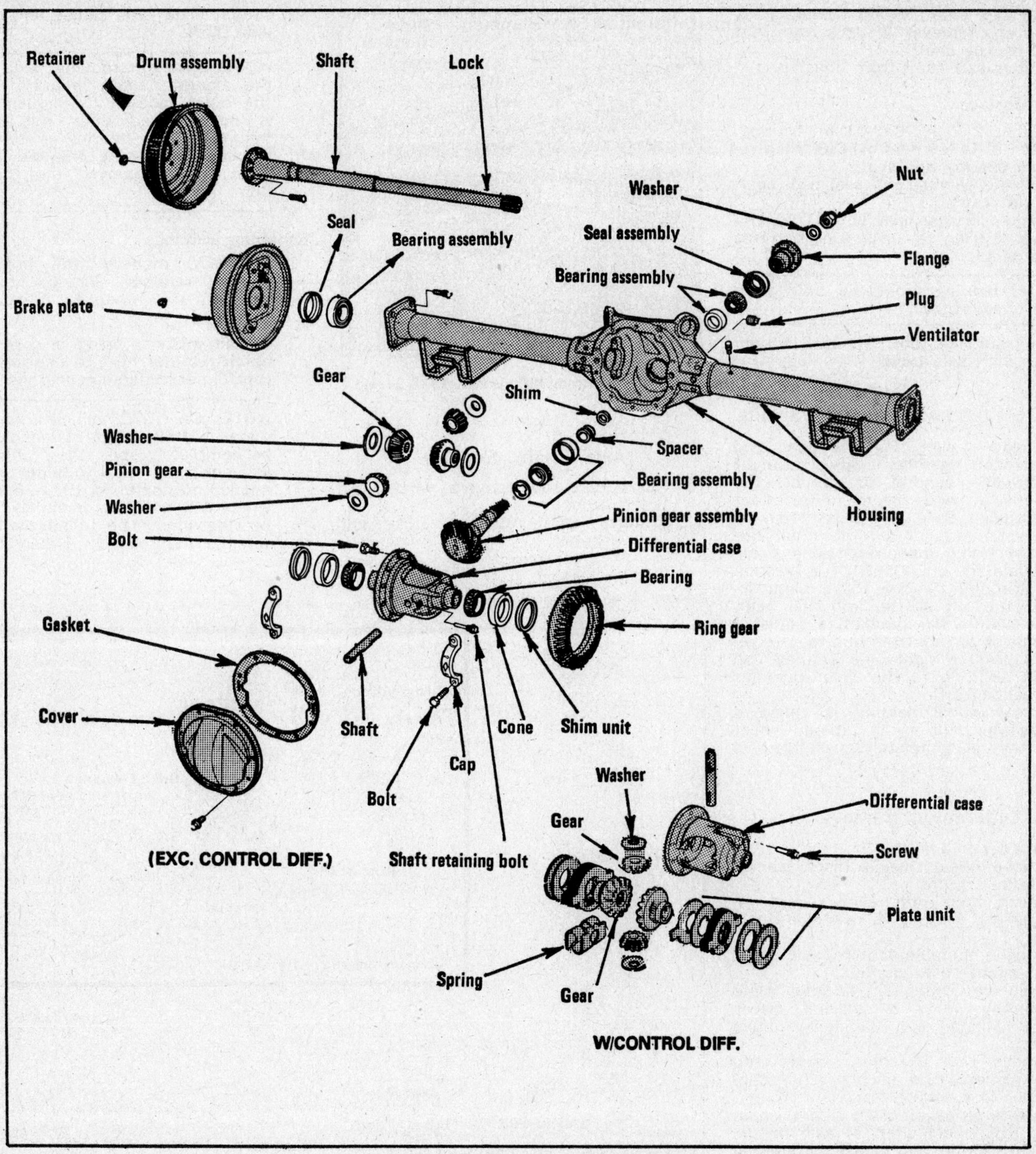

Fig. 2 Rear axle assembly. 1976 Seville

5. Remove stabilizer bar link nuts, retainer and bushings.
6. Remove nut from parking brake equalizer, then disconnect cables at connectors.
7. Remove clips securing brake tubing to axle, then lower axle slightly.
8. Position drain pan under axle, loosen rear cover bolts and allow axle to drain, then remove cover.
9. Remove wheel and brake drum.
10. Remove differential cross shaft through bolt, then the shaft. Push in on axle shafts, then remove "C" locks and pull axle shafts out about 1 inch.
11. Install cross shaft and bolt, then remove side bearing cap and bearing.

NOTE: Mark bearing caps to ensure installation in original position.

12. Remove one ring gear attaching bolt, position tools, Fig. 3, and pull differential case from housing until side shims can be removed, then remove differential.

NOTE: Mark shims to ensure installation in original position.

Axle Shaft Replace

1. Raise vehicle and remove wheel and brake drum.
2. Place drain pan under differential and remove cover.
3. Remove differential cross shaft.

4. Push axle shaft toward center of vehicle and remove "C" lock from butt end of axle shaft.
5. Remove axle shaft from housing.

Bearing Replace

1. Position tool J-22813-01 on bearing, attach slide hammer J-2619, then remove bearing and seal.
2. Lubricate bearing and seal with bearing lubricant.
3. Position bearing on tool J-23765 and drive bearing in until tool bottoms against tube.

NOTE: Bearing should be .550" from end of axle tube.

4. Using suitable tool, tap seal in until flush with axle tube.

1970-76 Standard Cars Exc. Seville

This axle design, Fig. 4, uses two tapered roller bearings and a straight roller straddle bearing to support the pinion and provide rigidity. Adjustment of the pinion is done by the use of shims at the pinion retainer to differential carrier connection. Pinion-ring gear backlash adjustment of .005-.010" is accomplished through the use of a shim located to the left of the left side bearing. The differential preload is adjusted by means of an adjuster nut that is retained under the right side bearing cap which in addition to the shim provides the correct backlash.

The rear wheel bearing is called a "unit bearing". It is a tapered roller bearing that is completely assembled as a unit.

Differential Removal

1. Raise car on a hoist.
2. Remove wheel shields, discs, wheels and brake drums.
3. Remove four nuts that secure retainer and backing to rear axle housing.
4. Attach a suitable tool to axle shaft and remove axle shafts.
5. Install two nuts on backing plate mounting studs to prevent plates from falling and damaging brake lines.
6. Remove two attaching screws and lockwashers that secure differential carrier nose bumper arm.
7. Support propeller shaft with a chain and disconnect shaft at axle flange.
8. Place a drain pan under differential and remove axle housing cover.
9. Remove five pinion retainer to carrier screws and loosen remaining screw.

CAUTION: Do not remove remaining screw.

10. Install guide pins, Fig. 5.
11. Position drain pan under pinion retainer and attach puller.
12. Using slide hammer, unseat pinion from carrier, removing remaining screw and slide pinion assembly over guide pins and out of carrier.
13. Remove adjuster lock tab, bearing caps, adjuster nut and remove carrier case.

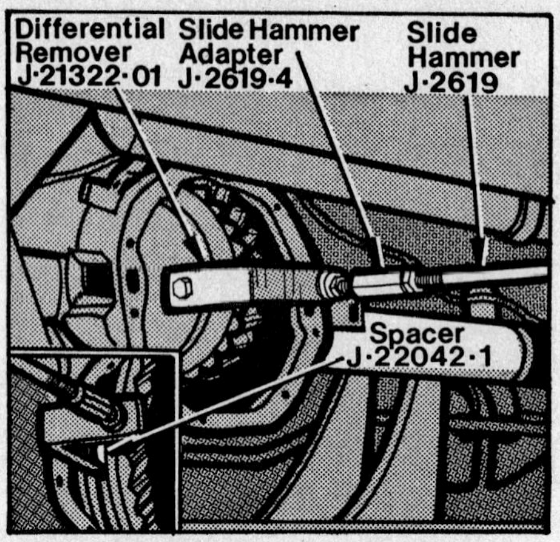

Fig. 3 Removing differential. 1976 Seville

Axle Shaft, Replace

1. Raise car and remove wheel and brake drum.
2. Remove axle retaining nuts and lockwashers.
3. Attach slide type puller and remove axle shaft.

NOTE: When removing axle shaft, the bearing may separate leaving the outer race in the housing. This is normal and does not indicate failure. If the bearing is to be replaced, make sure the old bearing outer race is removed from the housing.

Removing Bearings

1. On 1970 models, drill holes into bearing retainer, Fig. 6, not more than ½ inch deep to prevent axle shaft damage.
2. On all models, using a cold chisel, notch retainer next to bearing, being careful not to damage bearing.

NOTE: On 1970 vehicles where retainer is drilled, the retainer need not be completely split. Drive chisel into retainer only until retainer can be slipped off shaft. On 1971-76 vehicles where it is not recommended that the bearing retainer be drilled, the bearing retainer must be completely split by

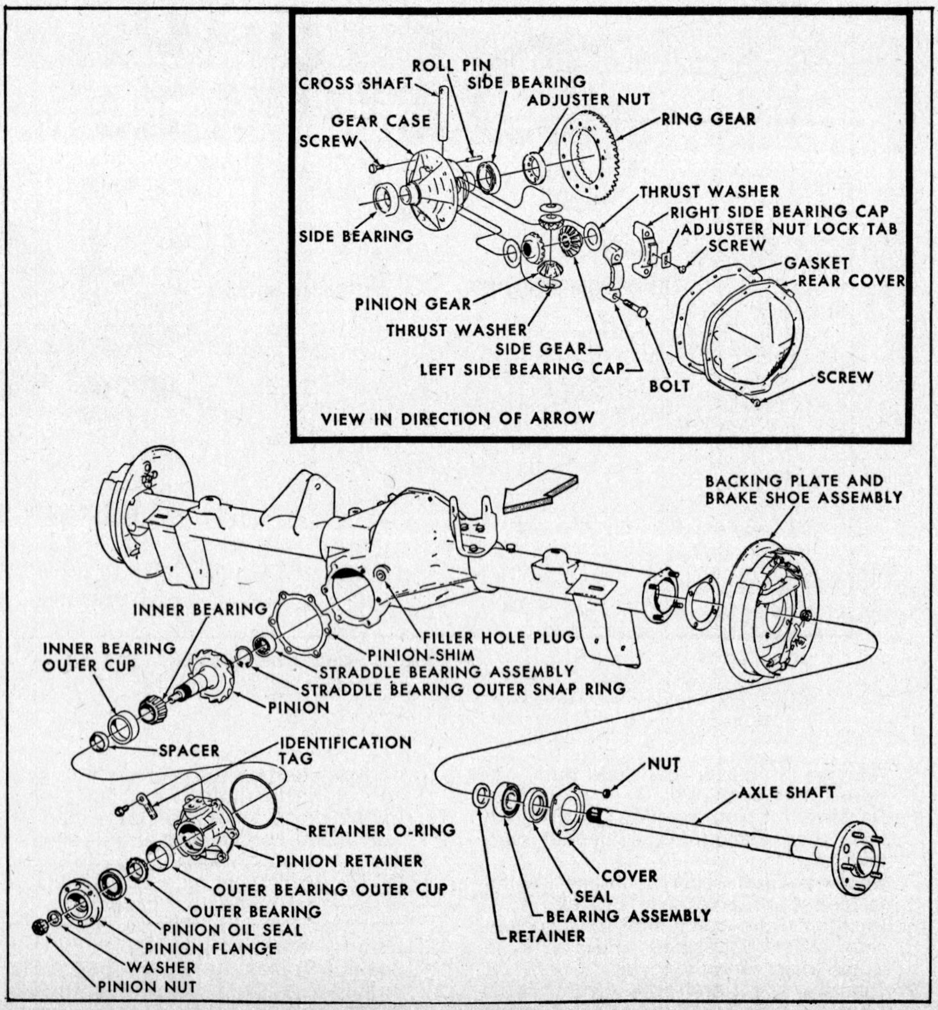

Fig. 4 Rear axle assembly. 1970-76 standard cars exc. Seville

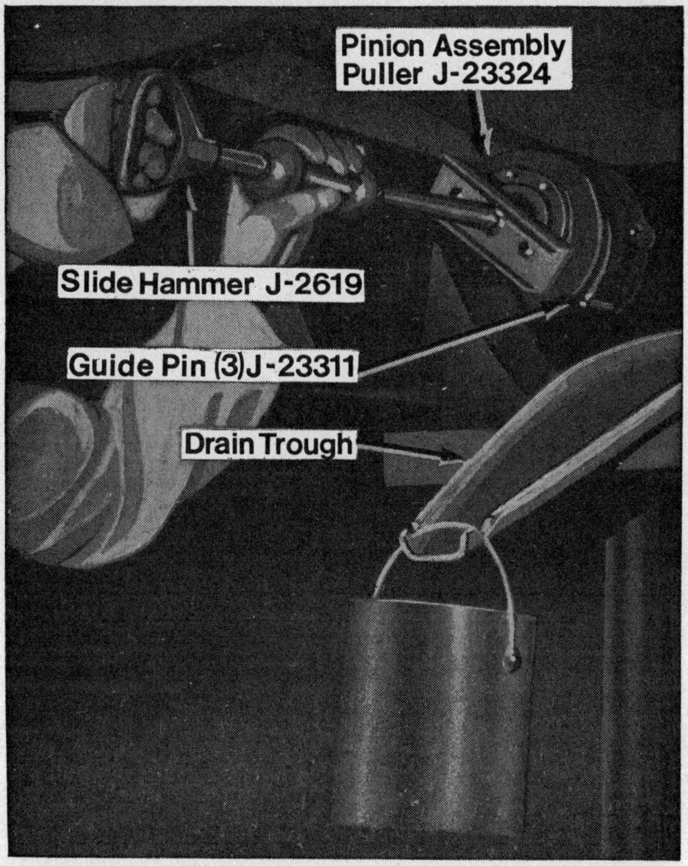

Fig. 5 Removing pinion assembly. 1970-76 standard cars exc. Seville

driving chisel into retainer until it separates.

---

3. Stand axle upright on flanged end and using two screwdrivers pry seal away from bearing.
4. Position shaft in a press and press shaft out of bearing.

Fig. 7 Installing wheel bearing. 1970-76 standard cars exc. Seville

Installing Bearing

1. Install bearing cover on shaft with raised side of cover against shaft flange, Fig. 7.
2. Install seal protector on shaft with small end of protector toward splined end of shaft, Fig. 8.
3. Lubricate lip of seal and install by pressing down over protector. Seal is properly installed when the lip of the seal clears the large end of the protector.
4. Position bearing on shaft with narrow ring of bearing facing flanged end of shaft.
5. Press bearing on shaft until bearing bottoms against shoulder on shaft.
6. Position retainer on shaft with chamfer on retainer next to bearing.
7. Press retainer on shaft until retainer bottoms against bearing.

1969 Standard Cars

The design of the rear axle carrier assembly is shown in Fig. 9. The axle shafts are supported at the outer ends by ball bearings which are lubricated from the differential carrier. The axle shaft oil seals are an integral part of the bearings and they are not servicable separately from the bearing. An "O" ring seal is located in the grooved outer surface of the bearing.

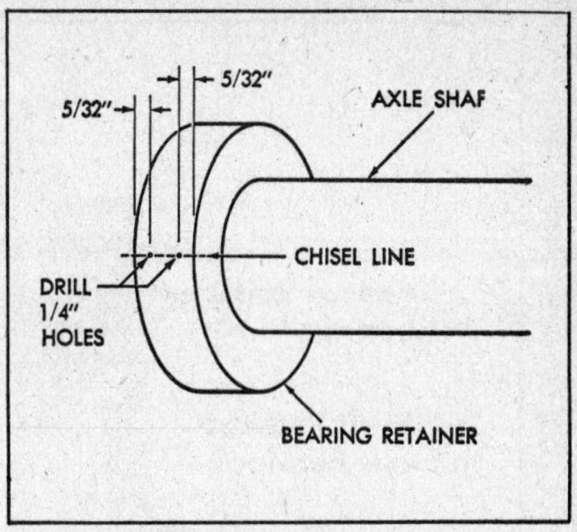

Fig. 6 Removing bearing retainer. 1970 standard cars

Any service on the differential carrier assembly, except drive pinion oil seal or yoke replacement, should be handled by replacement of the complete assembly. No disassembly or adjustment of this unit should be attempted because special equipment is used at the factory for selection of mating parts and setting side bearing preload.

Whenever a carrier is removed because of scored gears, worn bearings, or any failure which causes dirt or metal chips, it will be necessary to remove the housing from the car for thorough cleaning before the new carrier is installed. Also check axle shaft bearings for metal chips and clean if necessary.

In case of lubricant leakage between differential carrier and axle housing, check first to make sure that the nuts are tightened to the recommended torque of 37 ft. lbs. If tightening the nuts does not stop the leak, an extra gasket should be installed, using a non-hardening sealer. The additional sealing effect of the extra gasket should prevent further leakage. If a replacement differential is installed, the special lubricant shipped with the new differential must be used.

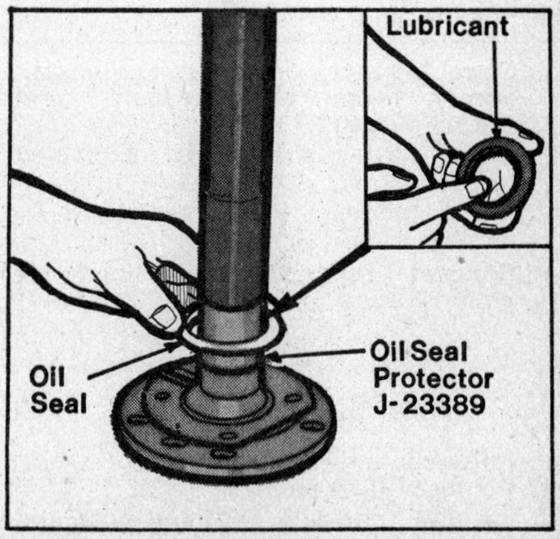

Fig. 8 Installing rear wheel oil seal. 1970-76 standard cars exc. Seville

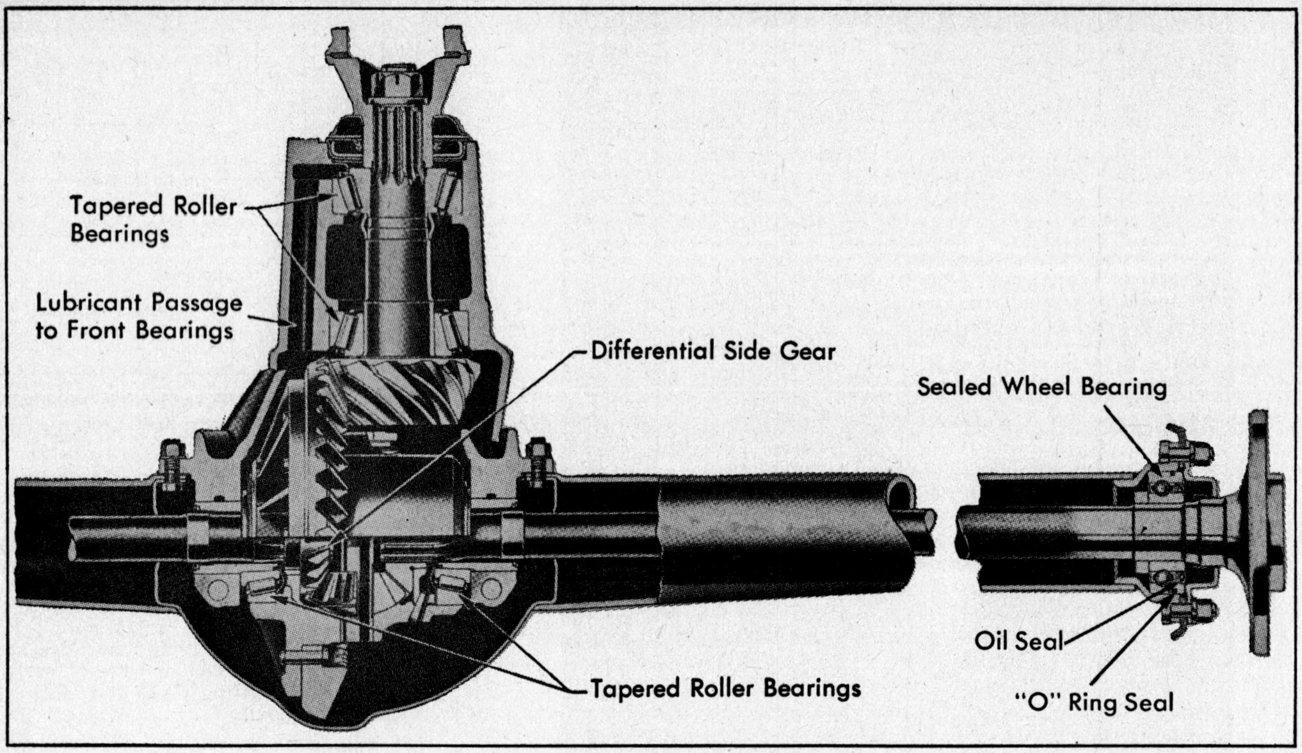

Fig. 9 Rear axle assembly. 1969 standard cars

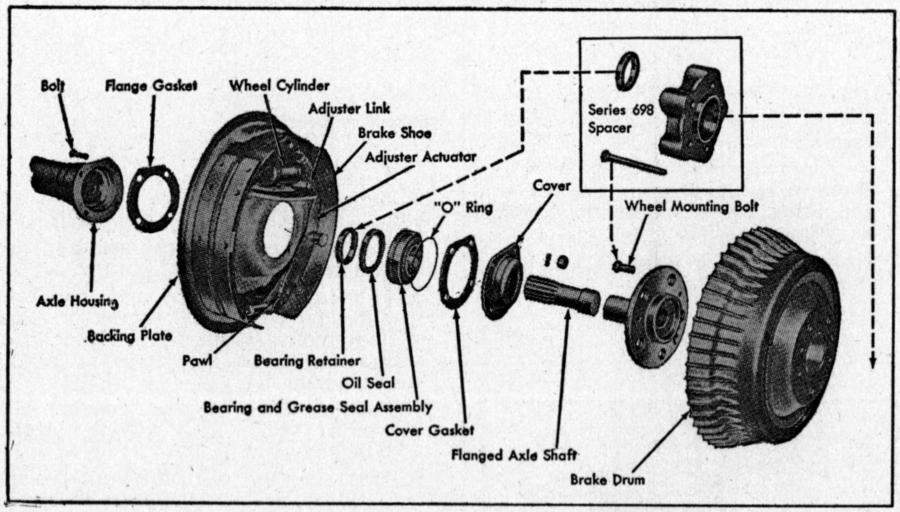

Fig. 10 Rear wheel disassembled. 1969

Removal

1. Disconnect rear universal joint at pinion yoke. Use suction gun to remove grease from differential carrier.
2. Remove axle shafts as outlined below.
3. Remove nuts and washers that hold carrier to axle housing and remove entire assembly with gasket.

Installation

NOTE: Use a new carrier-to-housing gasket and new "O" ring seals on axle shaft bearings.

1. Place gasket on housing, install carrier, nuts and washers, and torque nuts to 37 ft. lbs.
2. Install axle shafts as explained below.
3. Install lubricant in differential to filler plug level.
4. Connect rear universal joint at pinion yoke.

Axle Shaft, Replace

1. Raise car and remove wheel.
2. Remove brake drum.
3. Remove four axle retaining nuts and lockwashers.
4. Using slide hammer type puller, remove axle shaft.

Removing Bearings

1. Using a chisel and hammer, groove spacer next to bearing. The spacer need not be split. Drive chisel into spacer only enough to allow spacer to be slipped from shaft.

NOTE: On commercial chassis models, before notching retainer, drill two 1/4" holes along chisel line. Be sure not to drill any deeper than 1/2" as axle shaft may be damaged.

2. Press shaft through bearing. *If bearing has been removed because of failure, inspect axle housing and differential carrier for metal chips and, if necessary, clean thoroughly.*

Installing Bearing

1. Install new bearing on shaft so that bearing seal is toward flange end of shaft.
2. Press bearing on shaft so that it is

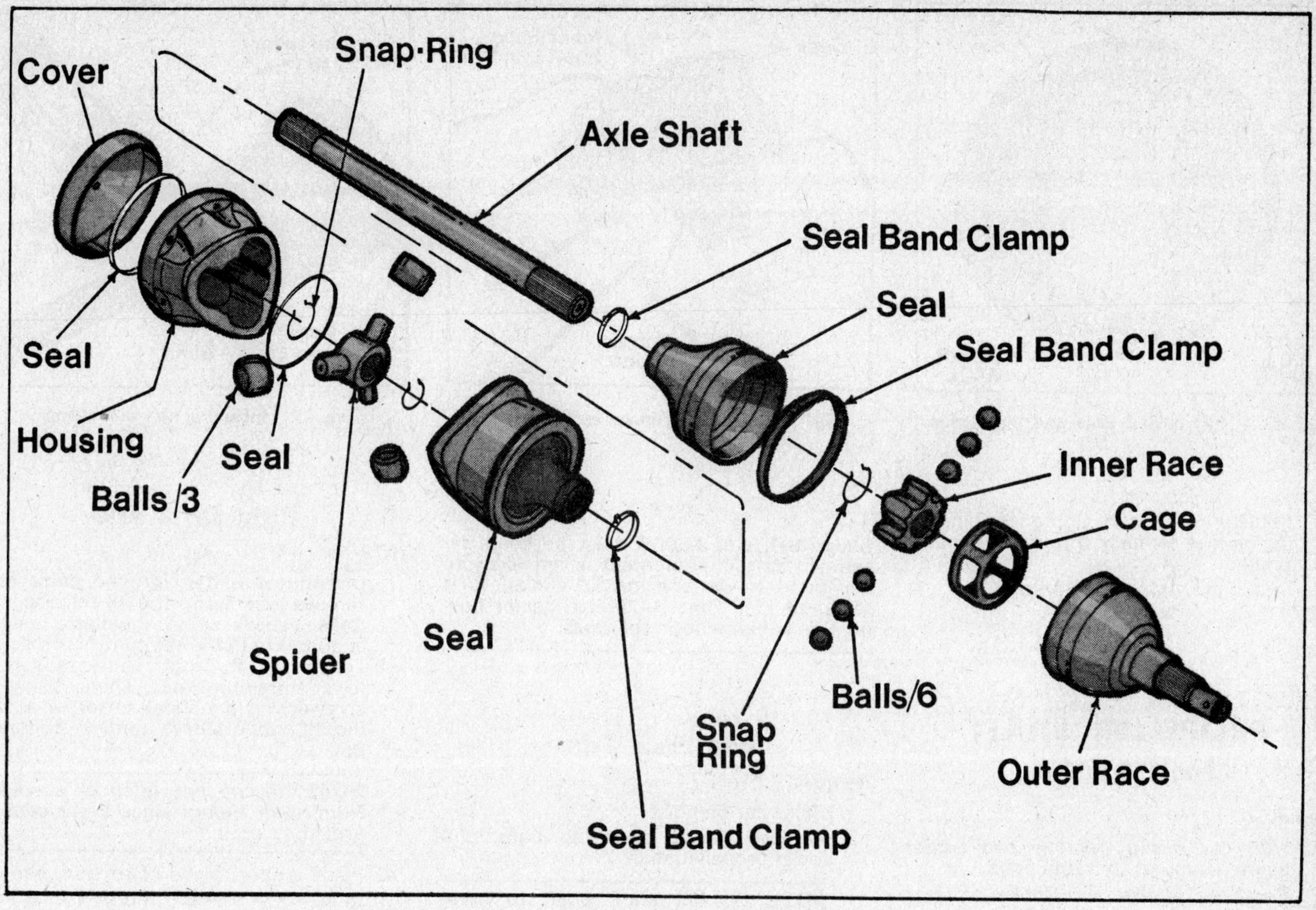

Fig. 11 Eldorado drive axle, exploded

located 3.180″ from outer surface of axle shaft flange to inner end of bearing inner race.

3. Install oil ring seal on bearing.

Install Axle Shaft

1. Apply film of differential lube to wheel bearing bore in axle housing after checking for burrs and nicks.
2. Use a new "O" ring seal on wheel bearing and install axle shaft (shorter one on left side), being careful not to damage "O" ring seal.
3. Install brake backing plate and bearing retainer.
4. Install nuts and lockwashers on

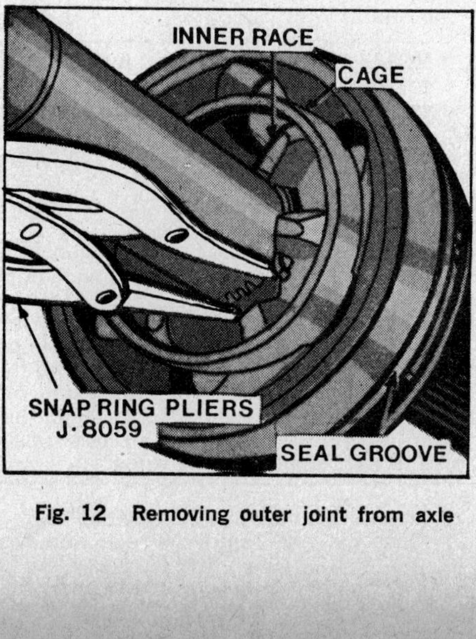

Fig. 12 Removing outer joint from axle

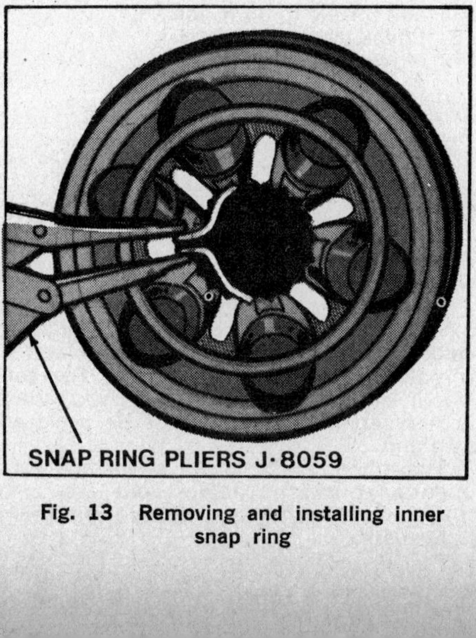

Fig. 13 Removing and installing inner snap ring

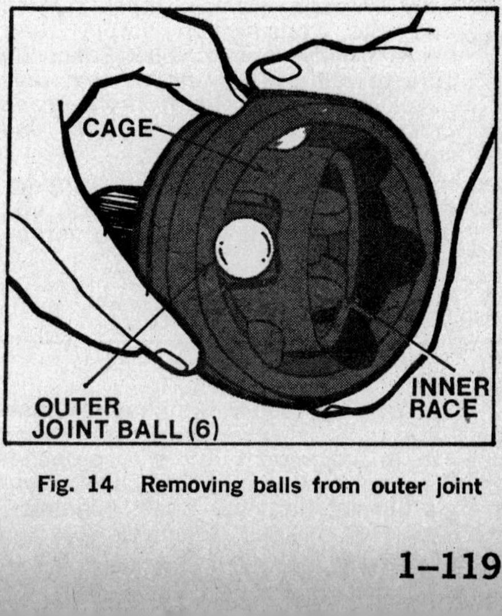

Fig. 14 Removing balls from outer joint

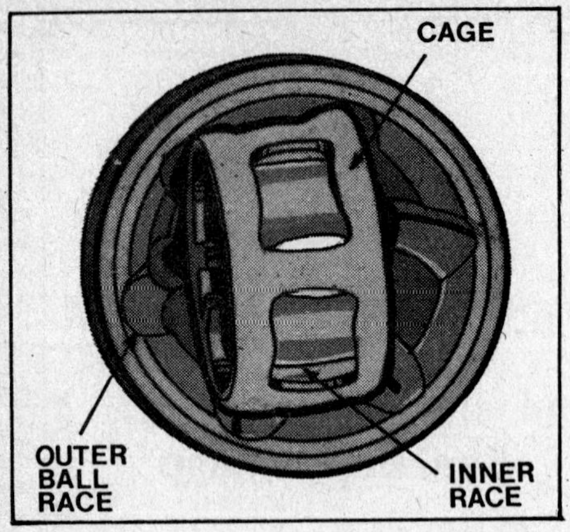

Fig. 15 Removing cage and inner race

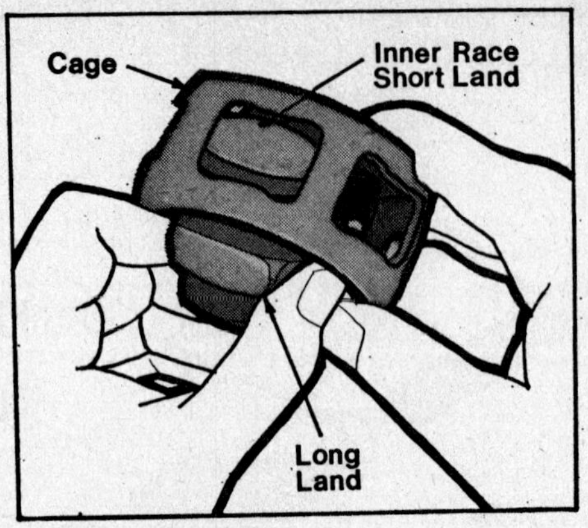

Fig. 16 Removing inner race from cage

Fig. 17 Installing new seal clamp

housing bolts and tighten by inserting socket wrench through rear axle flange.

5. Install brake drum and wheel.

PROPELLER SHAFT

Two Piece Type

Removal

1. With car raised, remove two center bearing support-to-frame bolts.
2. Remove U-bolts and locks at rear axle pinion.
3. Slide propeller shaft and front yoke off transmission output shaft and through frame tunnel section, removing it at the rear. *Slide spare yoke into transmission extension housing to prevent oil from leaking out.*

Installation

1. Lubricate front propeller shaft yoke with automatic transmission oil and remove spare yoke from transmission extension housing that had been installed to prevent oil leaking out.
2. Slide propeller shaft and front yoke through frame tunnel section and into transmission output shaft being careful not to nick yoke. *Be sure to reinstall any shims that may have been removed from between propeller shaft center support and frame tunnel.*
3. Install two center bearing support-to-frame bolts. Install U-bolts and locks at rear axle pinion and torque U-bolt nuts to 15 ft. lbs.

IMPORTANT

If drive line shudder, roughness, vibration or rumble is experienced, it may be due to misalignment of the propeller shaft assembly. To make this check, however, a special Propeller Shaft Alignment Gauge Set. No. J-8905 must be used. Inasmuch as this equipment is not likely to be found in general repair shops, it is recommended that a Cadillac dealer having this equipment do the work.

One Piece Type

Removal

1. Raise car on hoist.
2. Remove flange attaching bolts and lower propeller shaft.

NOTE: Do not allow shaft to hang on front constant velocity joint.

---

3. Push shaft forward so rear universal joint flange clears pinion shaft, then remove shaft by pulling rearward to disengage slip yoke. Install spare yoke into transmission extension housing to prevent loss of oil.

Installation

1. Remove spare yoke and insert front yoke of propeller shaft into extension housing, engaging transmission shaft splines.
2. Install rear flange attaching bolts and torque to 65 ft. lbs.
3. Check transmission oil.

ELDORADO DRIVE AXLES

General Description

Each drive axle, Fig. 11, consists of an axle shaft, with a ball type constant velocity joint at the outboard end and a tri-pot type at the inboard end. The torsional damper on the right hand shaft is not serviceable and must be replaced as a unit.

The inboard joint is not only flexible to operate at various angles, but can also move in and out as required by suspension movement.

Right Drive Axle

1972-76

1. Disconnect battery ground cable and remove wheel and the wheel disc.
2. Raise vehicle so front vehicle weight is supported at lower control arms.
3. Loosen right front shock absorber lower mounting nut, then, using a screwdriver, pry shock absorber at the mount inner sleeve until it contacts nut.

NOTE: Do not remove shock absorber from lower mount since lower control arm may drop.

---

4. Place a short piece of rubber hose on both lower control arm torsion bar connectors.
5. Remove screws and lockwashers securing right drive axle to output shaft.
6. Position inboard end of axle rearward toward starter to provide access to output shaft.
7. Remove output shaft support strut to final drive housing screw, then the two screws securing right output shaft support to the engine.
8. Slide output shaft outboard to disengage splines, then move inboard end of the assembly forward and downward until clear of vehicle.
9. Using a block of wood and a hammer, strike end of drive axle to unseat axle at hub.

NOTE: Loosen spindle nut just enough to allow axle to unseat.

---

10. Rotate axle inboard and forward, guiding axle over front cross-member and from under vehicle.

1969-71

NOTE: When attaching right hand output shaft support to the engine, do not allow the shaft and support assembly to hang in the final drive unit. Install support bolts and washers loosely and by moving the flange end of the shaft up and down, and back and forth, find the center location. Hold shaft in this position and tighten bolts to 50 ft-lbs.

---

1. Disconnect negative battery cable.
2. Raise car and remove wheel. Remove

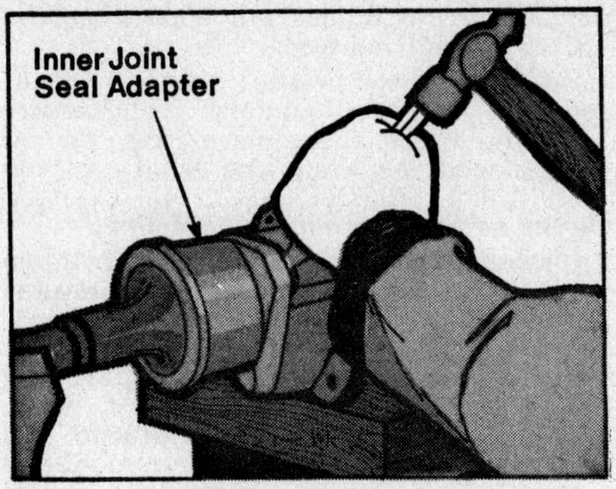

Fig. 18 Removing seal adapter

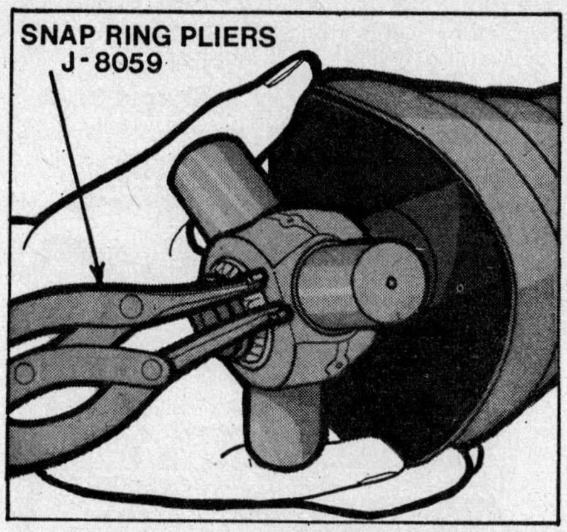

Fig. 19 Removing spider snap ring

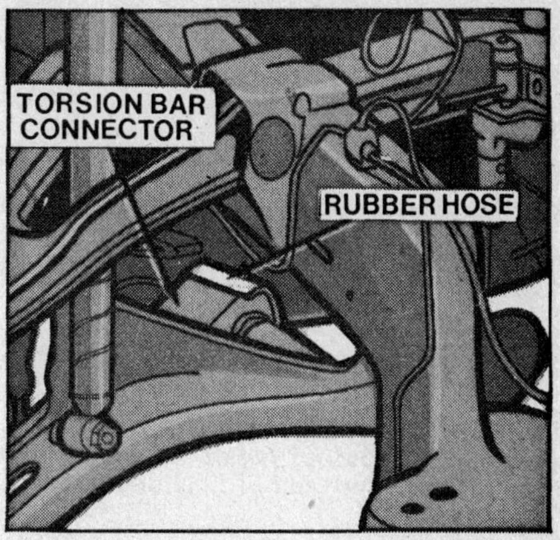

Fig. 21 Rubber hose location

drive axle spindle nut.
3. Using a block of wood and a hammer, tap on end of drive axle to unseat axle at hub assembly.

NOTE: Install a short piece of rubber hose on the lower control arm torsion bar connector to prevent damage to the drive axle seals.

4. Remove six drive axle-to-output shaft screws and lock washers.
5. Remove two output shaft support-to-engine bolts and one support-to-brace self tapping screw.
6. Rotate inboard end of drive axle rearward toward starter.
7. Slide shaft straight out toward side of car and remove output shaft from underside of car.
8. Remove drive axle by rotating axle inboard and toward front of car.
9. Reverse procedure to install.

Left Drive Axle

Perform steps 1-4 of removal of right axle then proceed as follows:

1. Loosen shock absorber upper mounting bolt.
2. Remove stabilizer bar link bolt cotter pin and nut.
3. Remove upper ball joint cotter pin and nut. Using a hammer, strike knuckle to disengage upper ball joint then remove ball joint stud from knuckle and brake line clip from stud.
4. Remove brake hose bracket from frame.
5. Carefully tip disc and knuckle assembly out at upper end to extent of brake hose.

NOTE: Wire assembly to upper control arm so brake does not support weight of knuckle assembly.

6. Rotate inner end of drive axle toward front of car.
7. Guide drive axle out of knuckle and remove axle from car.
8. Reverse procedure to install, being careful not to damage brake line when installing upper ball joint stud.

Fig. 20 Installing seal adapter on joint housing

Outer Constant Velocity Joint

1. Remove inner and outer seal clamps by cutting with a chisel and slide seal down axle shaft to gain access to joint.
2. Wipe excess grease from joint and spread snap ring and slide joint off spline, Fig. 12.
3. Remove inner race snap ring, Fig. 13.
4. Hold constant velocity joint in one hand then tilt cage and inner race so that one ball can be removed, Fig. 14. Continue until all balls have been removed.

NOTE: It may be necessary to tap the outer cage to rotate it.

5. Turn cage 90° with slot in cage aligned with short land on outer race and lift cage out of race, Fig. 15.
6. Turn short land of inner race 90° in line with hole in cage. Lift land on inner race up through hole in cage, then turn up and out to separate, Fig. 16.
7. Reverse procedure to install, using new seal clamps, Fig. 17. Cut off excess strap.

Inner Constant Velocity Joint

1. Remove small seal clamp bank by cutting it with a chisel.
2. Remove large end of seal from joint housing by prying up crimped edge on seal adapter. Drive seal adapter and seal off of joint housing with hammer and chisel, Fig. 18.

NOTE: Use care when removing seal adapter not to damage it or the seal.

3. Slide seal and adapter down axle shaft until joint is exposed.

NOTE: Do not allow spider leg balls to fall off by accident as adapter is moved.

4. Remove spider leg balls.
5. Remove spider outer snap ring, Fig. 19.
6. Tap spider assembly from shaft.
7. Reverse procedure to install, being sure to stake seal adapter to joint housing, Fig. 20.

Right Hand Output Shaft

1. Disconnect negative battery cable.
2. Raise car and install a short length of rubber hose on lower control arm torsion bar connector, Fig. 21.
3. Remove six drive axle-to-output shaft bolts and lock washers.
4. Remove output shaft support mounting bolts.
5. Rotate inboard end of drive axle rearward.
6. Pull output shaft out until it clears final drive then lower splined end and remove from car.

Right Hand Output Shaft Bearing, Replace

1. Remove output shaft as described above.
2. Remove three output shaft bearing retainer-to-support bolts.
3. Using two fabricated steel plates and four 3/8 x 24 bolts five inches long illustrated in Fig. 22, tighten bolts alternately to press out bearing.

Left Hand Output Shaft

1. Remove left drive axle.
2. Remove output shaft retaining bolt and pull shaft out of final drive.

Final Drive, Replace

The final drive unit is not serviced but is replaced as a unit.

1. Disconnect negative battery cable.
2. Remove approximately one gallon of fluid from transmission then remove transmission filler tube and plug tube hole.
3. Remove bolts "A" and "B" and nut "H", Fig. 23.
4. Remove bolt securing transmission oil cooler lines to final drive.
5. Remove nut from large through bolt, final drive support bracket to final drive.
6. Disconnect left front engine mount support bracket from engine. And final drive support bracket from left front engine mount support.
7. Remove right output shaft as outlined above.
8. Remove six left output shaft-to-drive axle bolts and lock washers.
9. Loosen final drive cover screws and drain fluid, then remove cover.
10. Compress left hand inner constant velocity joint and secure drive axle to frame with a piece of wire to provide clearance.

NOTE: On 1972-76 models, disconnect left tie strut from crossmember and loosen strut to side rail bolt. Then rotate strut outboard until clear of final drive.

11. Remove final drive support bracket.
12. Remove remaining final drive-to-transmission bolts and nut "G".
13. Disengage final drive splines from transmission.

NOTE: To avoid damage to seals, final drive unit should be supported by a suitable jack and proper alignment must be maintained throughout removal.

14. Remove final drive unit from underside of car by sliding unit toward front of car, permitting ring gear to rotate up over steering linkage and work unit free from car.
15. Reverse procedure to install, being very careful to maintain proper final drive-to-transmission alignment to prevent seal damage.

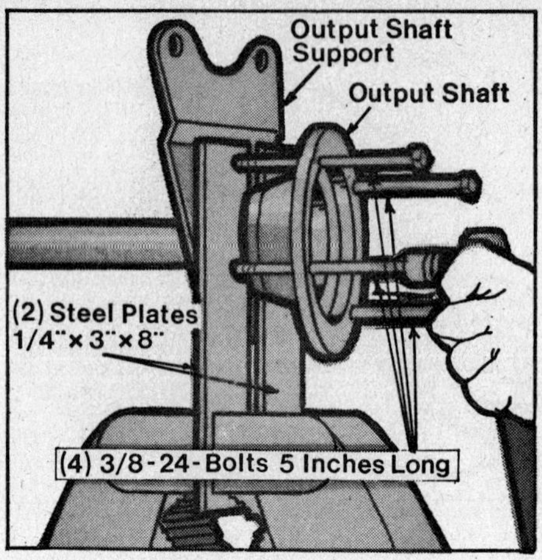

Fig. 22 Removing right hand output shaft support and bearing

BRAKE ADJUSTMENTS

1969-76 Self-Adjusting Brakes

These brakes, Figs. 24, 25, have self-adjusting shoe mechanisms that assure correct lining-to-drum clearances at all times. The automatic adjusters operate only when the brakes are applied as the car is moving rearward.

Although the brakes are self-adjusting, an initial adjustment is necessary after the brake shoes have been relined or replaced, or when the length of the star wheel adjuster has been changed during some other service operation.

Frequent usage of an automatic transmission forward range to halt reverse vehicle motion may prevent the automatic adjusters from functioning, thereby inducing low pedal heights. Should low pedal heights be encountered, it is recommended that numerous forward and reverse stops be made with a moderate pedal effort until satisfactory pedal height is obtained.

NOTE

If a low pedal height condition cannot be corrected by making numerous reverse stops (provided the hydraulic system is free of air) it indicates that the self-adjusting mechanism is not functioning. Therefore, it will be necessary to remove the brake drum, clean, free up and lubricate the adjusting mechanism. Then adjust the brakes, being sure the parking brake is fully released.

Manual Shoe Adjustment

1. Check fluid level in master cylinder and add fluid as necessary to a level

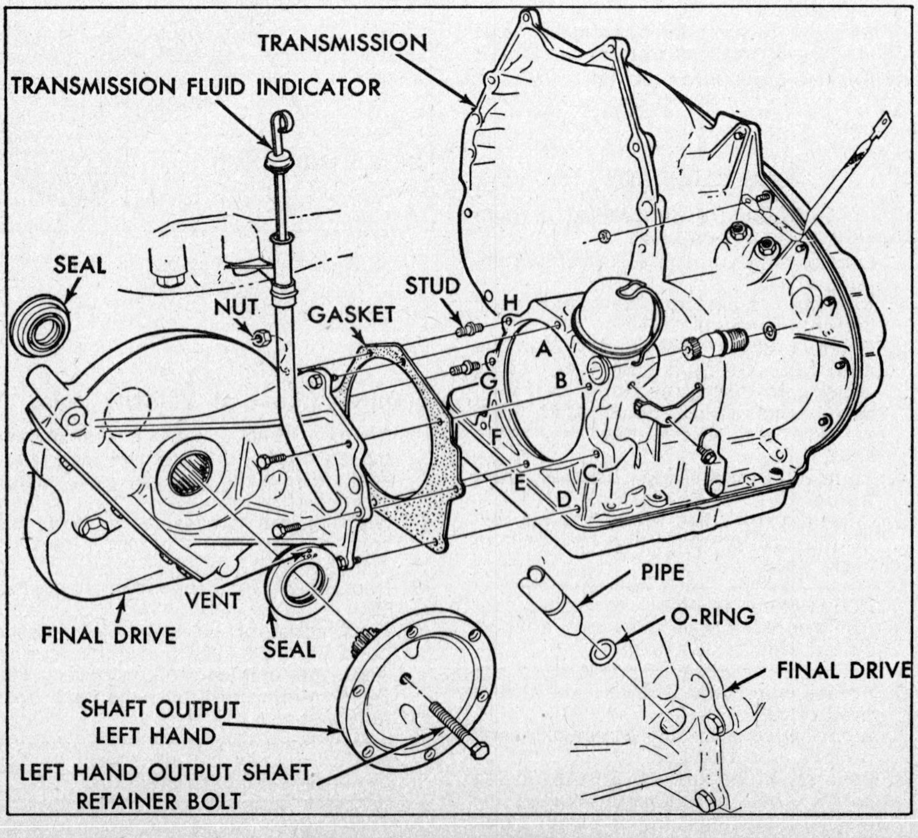

Fig. 23 Final drive attachment

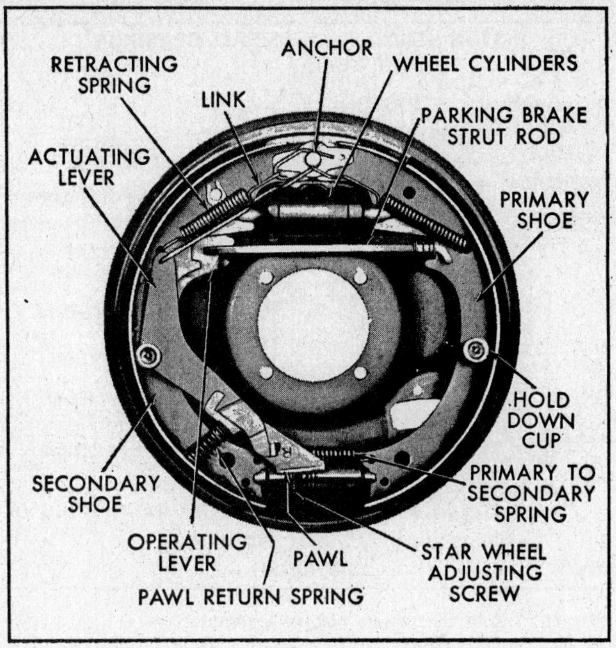

Fig. 24 Rear drum brake. 1974-76 all and 1969-73 Exc. Eldorado

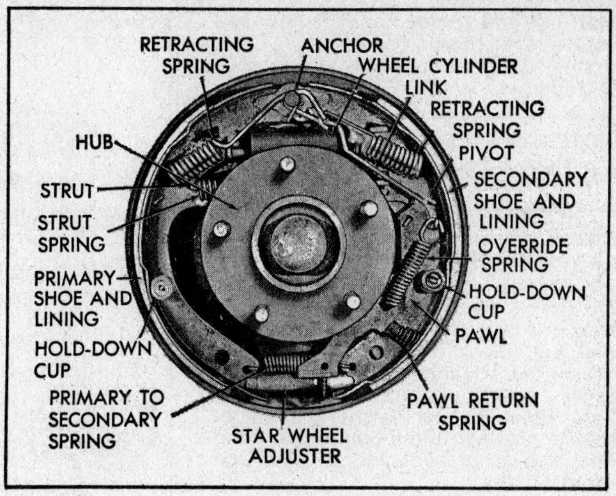

Fig. 25 Rear drum brake. 1969-73 Eldorado

¼" below top of reservoir.

2. Check to make certain that parking brake cable and linkage, including levers on rear secondary shoes, are free.
3. Tighten star wheel until brake drums can just be rotated forward with a two-foot bar placed between wheel studs.
4. Disengage adjusting pawl from star wheel with a hooked tool and back off star wheel 40 notches, Fig. 26.
5. Install wheels and drive car alternately forward and backward, applying brakes moderately in each direction until pedal travel is normal and brakes are adjusted satisfactorily.

REAR WHEEL DISC BRAKE & PARKING BRAKE SERVICE

Rear wheel disc brakes are used on 1976 Eldorado models. This single piston type caliper incorporates a cable actuated parking brake mechanism with an automatic adjusting provision.

A hydraulically powered brake booster (Hydroboost) provides power assist for this four wheel disc brake system.

Operation, Fig. 27

Upon application of brake, the cone and piston move out as one part. The nut remains stationary on the high lead screw and a gap develops between the cone and nut. When lining wear occurs, the cone and piston do not return to their original position, thereby leaving a small gap equal to the lining wear between the nut and cone. The adjusting spring causes the nut to rotate on the high lead screw to close the gap and adjust the caliper.

Upon application of parking brake, the lever rotation causes the high lead screw to turn and the nut to move down the screw, thereby loading through the cone and the cone-clutch interface of the piston, resulting in a clamp load on the linings. When the parking brake is released, the cone rotates on the clutch interface to adjust the caliper. The clutch interface prevents the cone from turning when the parking brake is applied.

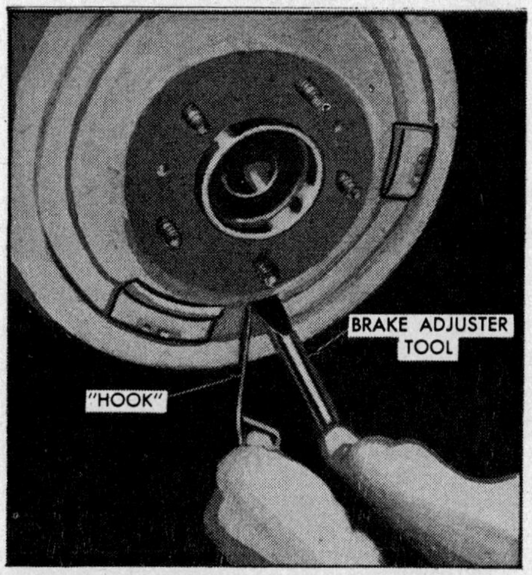

Fig. 26 Adjusting rear brakes through holes in brake drums. Hooked tool shown is used to hold pawl free of star wheel while adjusting rear brakes on 1969-71 models

Caliper Removal

CAUTION: Do not mix power steering fluid with brake fluid. If brake seals contact steering fluid or steering seals contact brake fluid, damage will result.

1. Remove two thirds of the total brake fluid capacity from the master cylinder front reservoir, to prevent overflow of brake fluid.
2. Support vehicle on a hoist and remove tire and wheel assembly.
3. Install one nut with flat side facing rotor to prevent rotor from falling out when caliper is removed.
4. Loosen parking brake cable tension at equalizer, then remove cable from parking brake lever and remove return spring, lock nut, lever, lever seal and anti-friction washer.

NOTE: Lever must be held in place while removing nut.

5. Clean surface in area of lever seal, then using a 7 inch (or larger) C-clamp, with the solid end on lever stop and screw end on back of outboard lining, turn clamp until piston is bottomed in caliper.

NOTE: Do not position C-clamp on actuator screw.

6. Before removing clamp, lubricate housing surface under lever seal with silicone lubricant.
7. Install a new anti-friction washer, a new lever seal and lever.

NOTE: Install lever on hex with arm pointing downward.

8. Rotate lever toward front of vehicle and while holding in this position, in-

stall nut and torque to 25 ft. lbs. and rotate lever back to stop.
9. Install lever return spring and remove C-clamp.

NOTE: Springs are color coded, red for the right hand caliper and black for the left hand caliper.

10. Disconnect brake line from caliper and plug openings to prevent loss of fluid and entry of dirt.

NOTE: If brake line nut is seized, brass bolt and block can be removed with brake line attached by removing bolt. Block copper washers after removing bolt. Block copper washers after removing caliper mounting bolts. Plug openings to prevent loss of fluid and entry of dirt.

11. Remove caliper mounting bolts and remove caliper and linings as an assembly.
12. Reverse procedure to install and torque caliper mounting bolts to 30 ft. lbs.

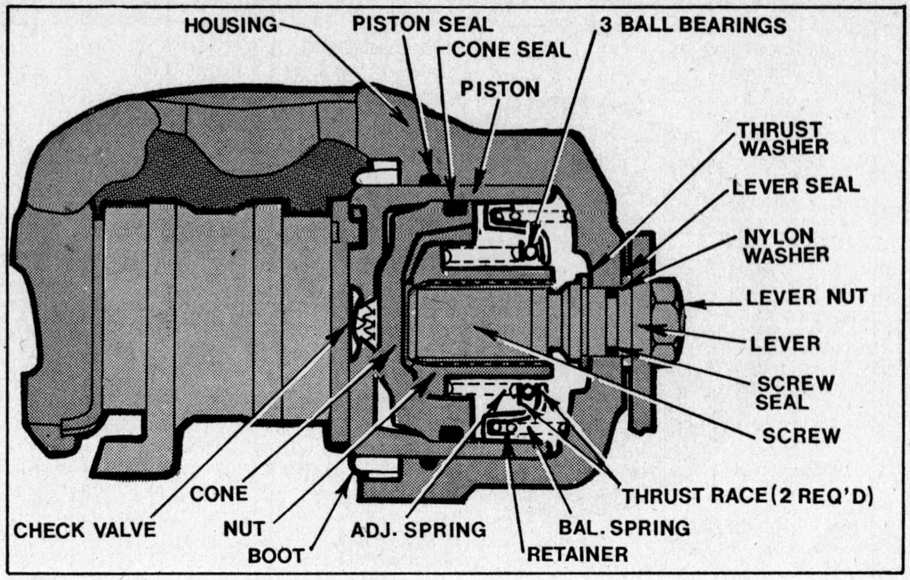

Fig. 27 1976 Rear disc brake caliper assembly

Inspection

1. Clean corrosion and dirt from face of piston. Inspect piston and check valve area for fluid leakage, indicated by excessive moisture around boot area.
2. Inspect dust boot for cuts, cracks or other damage which may affect its sealing ability. If leaks are present, replace dust boot.

NOTE: Do not use compressed air to clean caliper as it may unseat the dust boot.

3. Inspect piston boot seal. Replace boot seal if leakage is indicated.
4. Inspect for leaks at threaded end of actuator screw. Replace seal if leakage is indicated. If bore is nicked or scratched, replace caliper.

Shoe & Lining Replacement

1. Remove caliper as described previously and remove shoe and lining.
2. Remove and discard the two caliper mounting sleeves and the four bushings. Using silicone lubricant, install new bushings and seals.

NOTE: Sleeves are installed in inner bushings.

3. Remove and discard piston check valve and install a new one.

CAUTION: Front brake shoes must not be installed on rear calipers.

4. Position new inboard shoe assembly on piston. The D-shaped tab must fit into indentation in piston. If piston requires rotation, use too J-7642 to rotate it.
5. Install new outboard shoe assembly onto caliper. Install caliper and torque mounting bolts to 30 ft. lbs.

PARKING BRAKE, ADJUST

1976 W/Rear Disc Brakes

1. Lubricate parking brake cables at equalizer and underbody rub points. Check all cables for freedom of operation.
2. Fully release parking brake and raise vehicle.
3. Hold cable stud from turning and tighten equalizer nut until cable slack is removed and levers are against stops on caliper housing. If levers are off stops, loosen cable until levers return to stop.
4. Operate parking brake several times to check adjustment. When properly adjusted, the parking brake pedal should move 4-5 inches when a force of about 125 pounds is applied.

1969-76 W/Rear Drum Brakes

1. With service brakes properly adjusted, lubricate parking brake linkage at equalizer and cable stud with heat-resistant lubricant, and check for free movement of cables.
2. Depress parking brake pedal about 1¾" on 1969-72 models and 1" on 1973-76 models from full released position.
3. Raise rear wheels off floor.
4. Hold brake cable and stud from turning and tighten equalizer nut until a slight drag is felt on either wheel (going forward). After each turn of equalizer nut, check to see if either wheel begins to drag.
5. Release parking brake. No brake drag should be felt at either rear wheel. Operate several times to check adjustment. After adjustment is completed, parking brake pedal should travel 1¾" to 2¾".

VACUUM RELEASE PARKING BRAKE

1969-76

The foot-operated parking brake is mounted on the cowl to the left of the steering column. It incorporates a vacuum release, Fig. 28, operated by a vacuum diaphragm that is connected to the parking brake mechanism. When the transmission selector is moved into any Drive position, a vacuum valve in the neutral safety switch opens, allowing diaphragm to be actuated by engine vacuum.

The diaphragm is connected by a link to a release mechanism on the parking brake. Vacuum acting on the diaphragm unlocks the parking brake pedal, permitting it to return to the release position by spring action. Any abnormal leaks in the vacuum release system will prevent proper brake release. A manual release is provided and may be used if the automatic release is inoperative or if manual release is desired at any time.

Testing Vacuum Release

1. If the mechanism is inoperative, first check for damaged or kinked vacuum hoses and for loose hose connections at the diaphragm, vacuum release valve at neutral safety switch, and at engine manifold connection.
2. Check adjustment of neutral safety switch and operation of vacuum release valve.
3. Check diaphragm piston travel by running engine and moving transmission selector lever from drive to

neutral. The manual release lever should move up and down as vacuum is applied and released. If no movement is observed, or if movement is slow (more than 1 or 2 seconds to complete the full stroke) diaphragm is leaking and should be replaced.

4. Check brake release with vacuum applied. If diaphragm piston completes full stroke but does not release brake, a malfunction of the pedal assembly is indicated, and the complete parking brake assembly should be replaced.
5. Check operation of parking brake with engine off. Parking brake should remain engaged regardless of transmission selector lever position. If not, replace parking brake assembly.

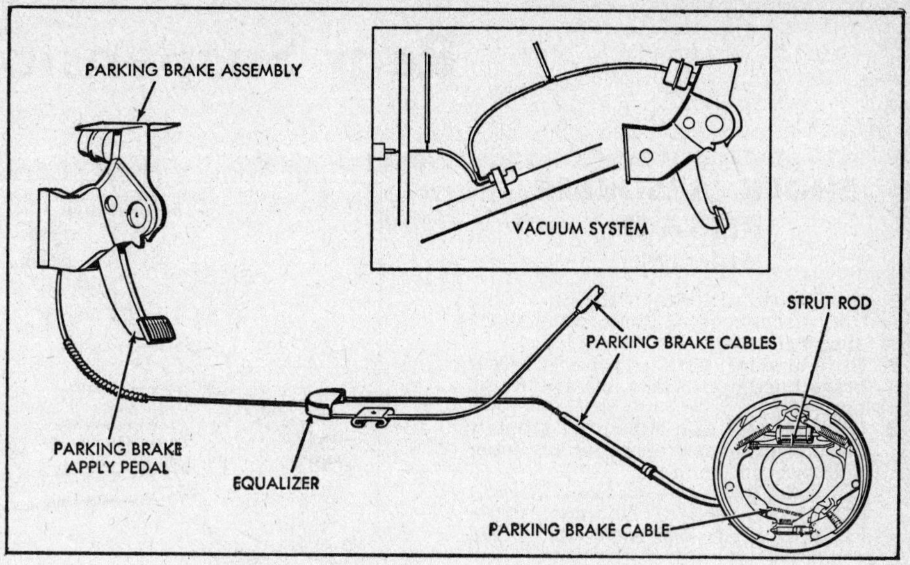

Fig. 28 Parking brake linkage. 1969-76 (typical)

POWER BRAKE, REPLACE

Service Bulletin

A sign of brake fluid dampness below the master cylinder at the power brake unit or on wheel cylinders at the bottom of the boot, does not necessarily indicate that these cylinders are leaking.

A small amount of fluid leakage at these areas can occur due to the creeping action of a very light film of fluid on the cylinder bores around the seals. This action provides proper seal lubrication. In addition, normal brake heat will produce a slight escape of lubricant from the impregnated, porous-metal wheel cylinder pistons.

Normal dampness at the master cylinder or wheel cylinders is not easily distinguishable from a definite leak. Therefore, this condition must be checked carefully.

If there is sufficient dampness to form a "teardrop" of fluid at the bottom of the master cylinder or on the bottom of the wheel cylinders at the boot area, the rate of fluid seepage is too high and the cause should be determined and corrected.

1976 Seville

1. Disconnect and cap brake lines, remove wiring harness from combination valve.
2. Remove master cylinder to brake booster attaching nuts, then the master cylinder and combination valve.
3. Disconnect vacuum line from check valve.
4. Remove clip and washer from push rod pin at brake pedal, however, do not remove push rod.
5. Remove twilight sentinel amplifier attaching screws and wiring harness and lower unit, if equipped.
6. Remove brake booster to cowl attaching nuts, slide booster out and disconnect push rod from brake pedal pin.

1969-76 Exc. Seville

1. Disconnect hydraulic lines from master cylinder on power unit. Cap line fittings to prevent dirt entering system.
2. Disconnect vacuum hose from vacuum check valve on power unit. Remove steering column lower cover.
3. Remove cotter pin and spring spacer that attach power unit push rod to brake pedal relay lever.
4. Unfasten and remove power unit from cowl.

Rear Suspension

SHOCK ABSORBER, REPLACE

1969-76

1. If equipped with Automatic Level Control, disconnect air lines from shock absorber fittings.
2. On standard cars, disconnect rear brake hose from brake line and frame bracket.
3. With the rear axle supported properly disconnect shock absorber at upper and lower mountings.

NOTE: On 1969-70 Eldorado, upper retaining nut is accessible from trunk.

4. Reverse procedure to install.

NOTE: 1969-70 Eldorado models incorporate horizontal shock absorbers in addition to the conventional type, Fig. 1. To replace, remove front and rear attaching bolts and lift unit from vehicle.

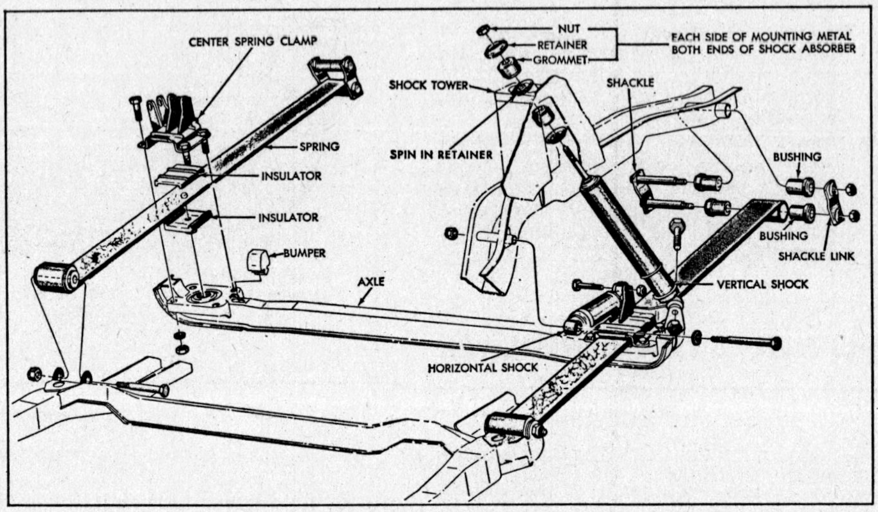

Fig. 1 1969-70 Eldorado rear suspension

COIL SPRINGS, REPLACE

1971-76

1. Support vehicle at frame and support rear axle with a suitable jack.
2. If equipped with Automatic Level Control, disconnect overtravel lever link and place lever in center position.
3. Disconnect shock absorbers from lower mountings and rear brake hose from brake line.
4. On standard cars, disconnect propeller shaft from differential pinion flange.
5. Remove upper control arm rear mount bolt and lift arm from axle bracket, Figs. 2 and 3.
6. Lower axle assembly slowly until springs are free.

NOTE: When lowering axle assembly, use care to prevent axle from rotating, as springs may snap from the seats. If necessary, compress springs by hand to remove.

7. Reverse procedure to install. Note spring positions, Figs. 4 and 5. Tighten control arm bolt with vehicle at curb height.

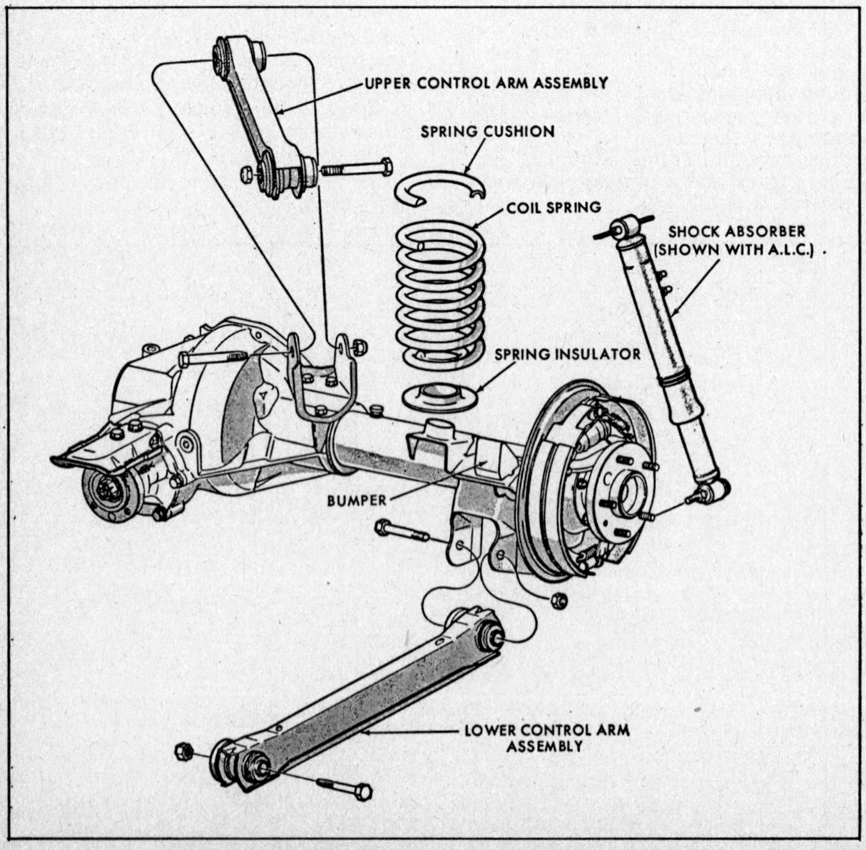

Fig. 2 1969-76 rear suspension (typical). Except Eldorado

1969-70 Except Eldorado

1. Perform steps 1, 2, 3 and 5 as outlined in above procedure and remove rear wheels.

NOTE: Install a second jack under differential carrier nose to aid removal of control arm bolt.

2. Remove screw securing parking brake cable strap.
3. Remove jack from carrier nose and lower axle assembly.
4. Install a jack under lower control arm rear mounting on side opposite spring being removed and raise jack until spring can be removed.
5. Reverse procedure to install. Tighten control arm bolt with vehicle at curb height.

LEAF SPRING, REPLACE

1976 Seville

1. Raise vehicle on hoist.

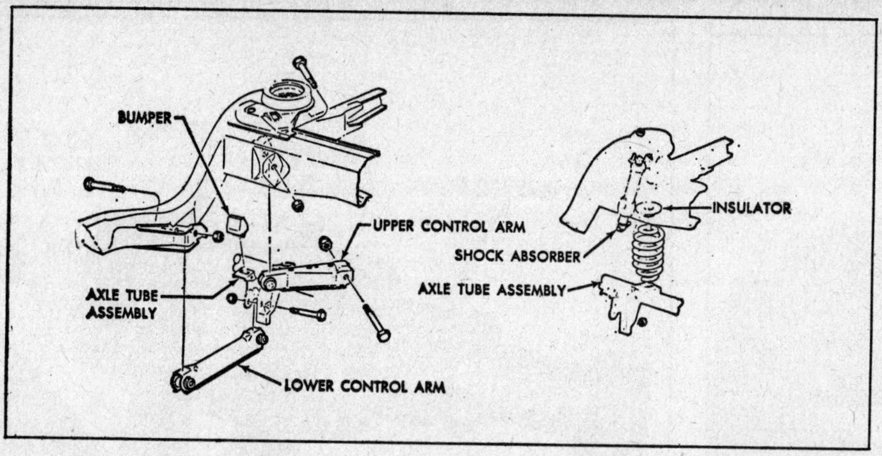

Fig. 3 1971-76 Eldorado rear suspension

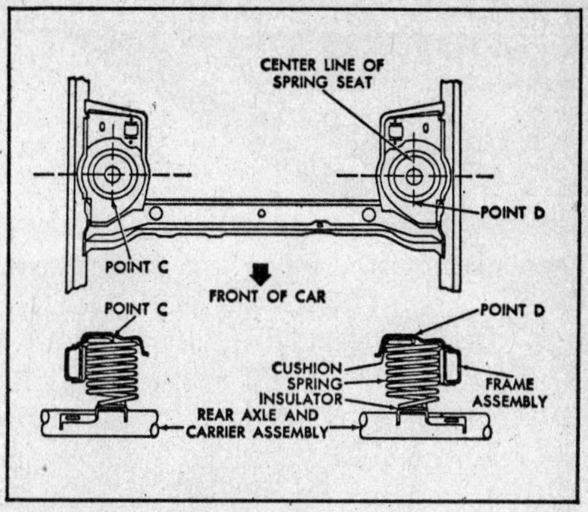

Fig. 4 Coil spring installation. 1969-76 Except Eldorado

2. Using a suitable jack, raise axle until spring tension is relieved, Fig. 5A.
3. Disconnect Automatic Level Control overtravel lever from link.
4. Push overtravel lever down to deflate shock absorbers, then remove shock lower mounting nut and bolt and position shock absorber aside.
5. Back off parking brake adjustment at equalizer and remove cable clip from spring front retaining bracket and cable clamps from bottom of spring.
6. Loosen spring front eye to retaining bracket bolt.
7. Remove spring front bracket to underbody attaching screws.
8. Lower axle until bolt can be removed from spring front eye and remove bracket from spring.
9. Remove nuts securing lower spring plate to axle and stabilizer bar brackets.

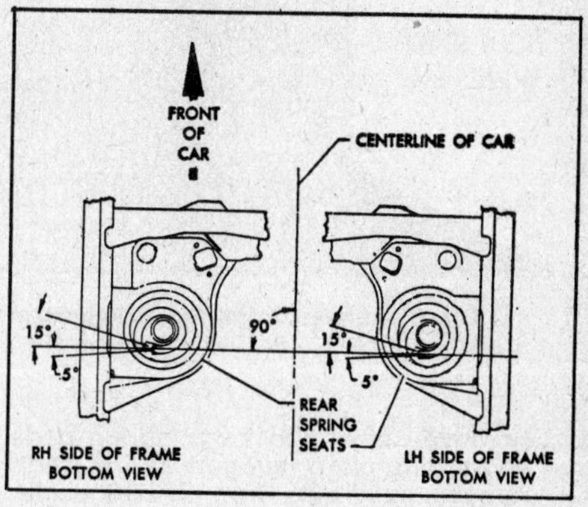

Fig. 5 Coil spring installation. 1971-76 Eldorado

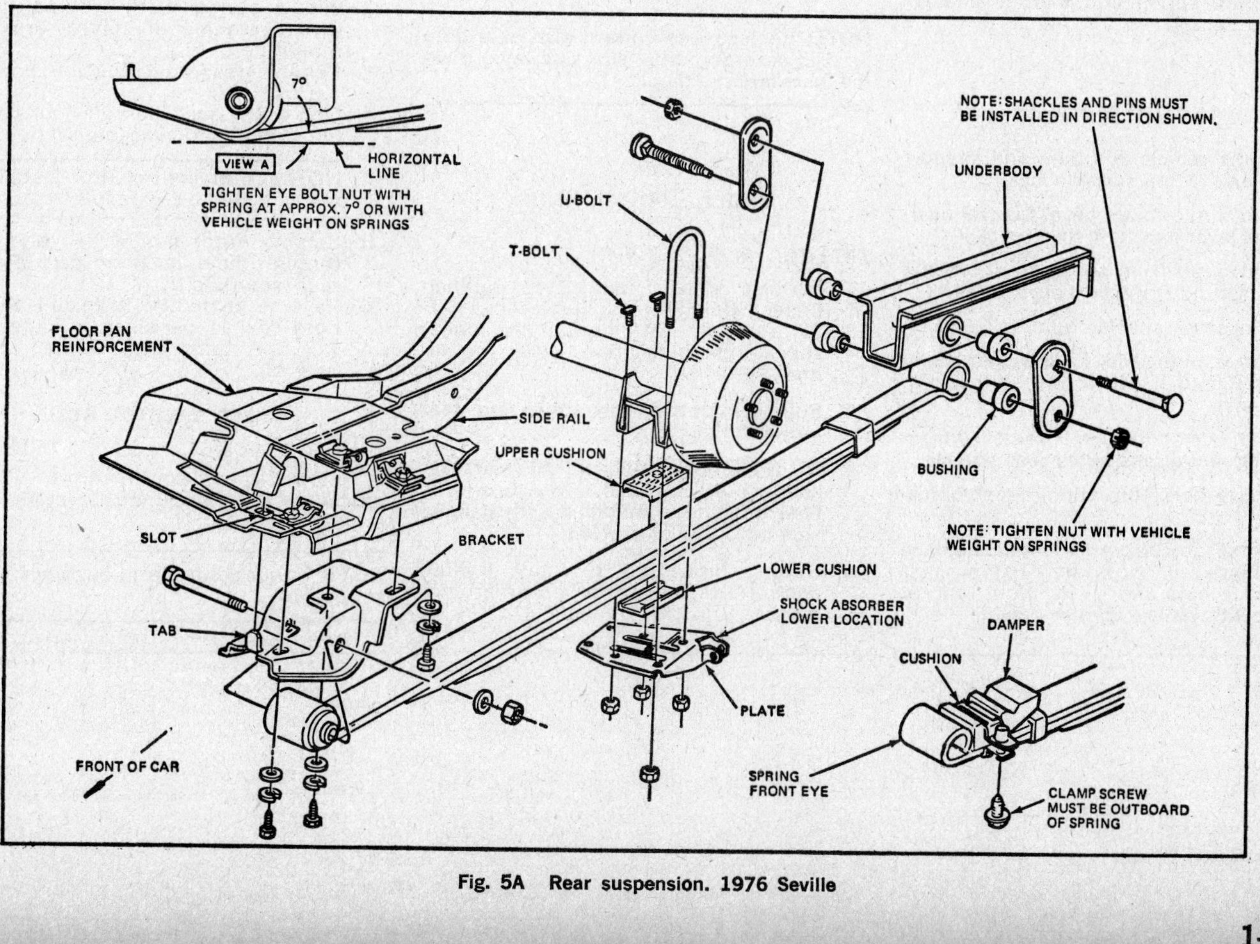

Fig. 5A Rear suspension. 1976 Seville

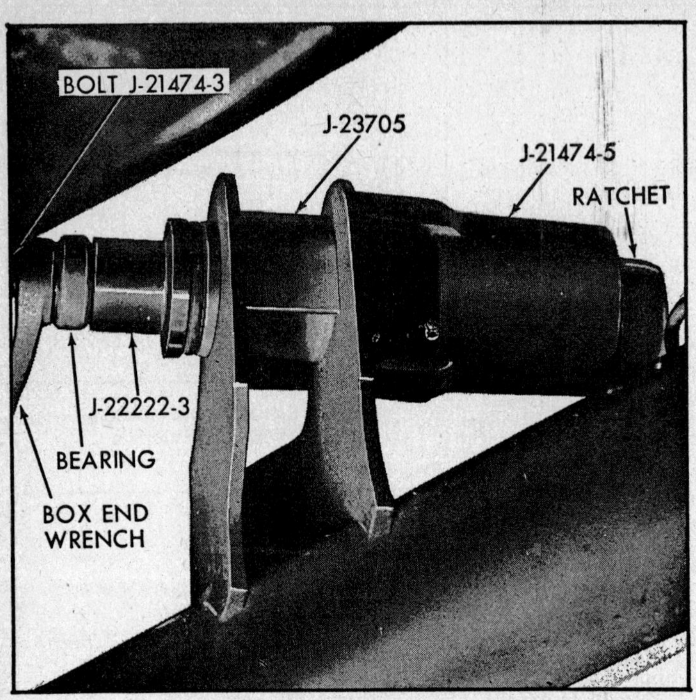

Fig. 6 Upper control arm axle bracket bushing removal, 1971-76 Eldorado

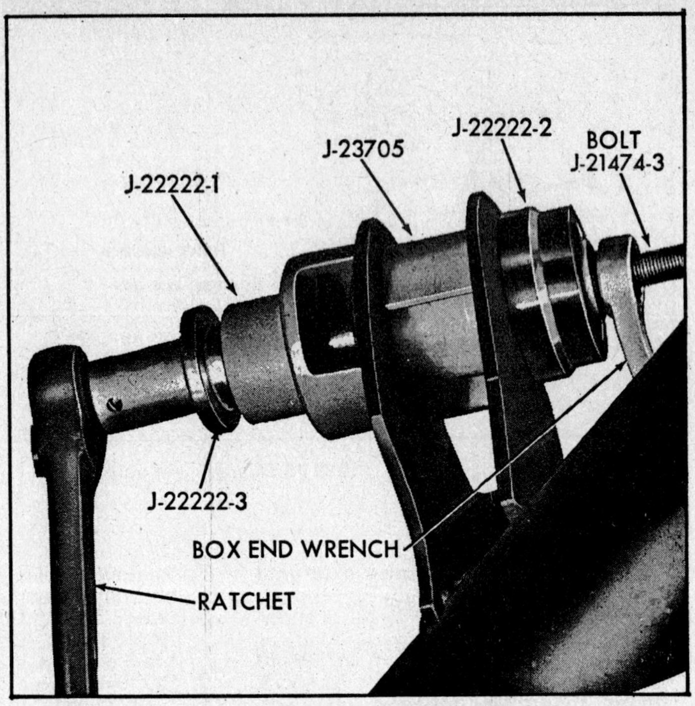

Fig. 7 Upper control arm axle bracket bushing installation, 1971-76 Eldorado

10. Remove upper and lower spring pads and spring plate, support spring, then remove both nuts from rear shackle.
11. Separate spring and shackle and remove spring.

1969-70 Eldorado

1. Support vehicle at frame and support rear axle with a suitable jack.
2. Remove Automatic Level Control overtravel lever link from axle bracket.
3. Remove nut from spring front bracket bolt. Do not remove bolt.
4. Remove rear shackle outerlink, Fig. 1.
5. Remove nuts and lockwashers from center spring clamp and shift clamp to a side.
6. Lower axle so tension is removed from spring and disassemble rear shackle.
7. Remove bolt from spring front bracket.
8. Reverse procedure to install. Replace bushings as necessary. Tighten rear shackle nuts and spring front bracket bolt with vehicle at curb height.

CONTROL ARMS & BUSHINGS, REPLACE

NOTE: Replace one control arm at a time as axle assembly may slip sideways, making installation difficult.

Upper Control Arms

1971-76

1. Perform steps 1 and 2 as outlined under "Coil Springs, Replace" 1971-75 procedure. On Eldorado disconnect shock absorbers from lower mountings.
2. Remove control arm front and rear mount bolts.
3. On Eldorado, replace axle bracket bushing as required, Figs. 6 and 7. New bushing is installed with flanged side outboard of bracket.
4. Reverse procedure to install. Tighten control arm bolts with vehicle at curb height.

1969-70 Standard Cars

1. Perform steps 1 and 2 as outlined under "Coil Springs, Replace" 1971-76 procedure.
2. With a punch, mark floor pan in line with centerline of control arm front mount bolt.
3. Remove rear seat cushion and seat back.
4. Drill a $1\frac{1}{4}$ inch hole in floor pan using punch mark as center of hole.

NOTE: After control arm installation, install a suitable plug in hole.

5. Remove front mount bolt and guide through hole in floor pan. Remove rear mount bolt.
6. Reverse procedure to install. Replace bushings as necessary. Tighten control arm bolts with vehicle at curb height.

Lower Control Arms

1969-76 Standard Cars & 1971-76 Eldorado

Follow "Upper Control Arms" 1971-76 procedure for replacement of lower control arms.

NOTE: Lower control arm bushings are not serviceable.

Front End and Steering Section

FRONT SUSPENSION EXC. ELDORADO

1976 Seville

The front suspension consist of two upper and lower control arm assemblies, steel coil springs, shock absorbers, stabilizer bar, two integral steering arms, and knuckles, Fig. 1.

Ball joints are used at outer ends of upper and lower control arms. The upper ball joints is riveted to the upper control arm. The lower ball joint is pressed into the lower control arm.

A one inch diameter stabilizer bar is mounted in rubber bushings to the front frame side rails and is attached to the lower control arms by means of steel links.

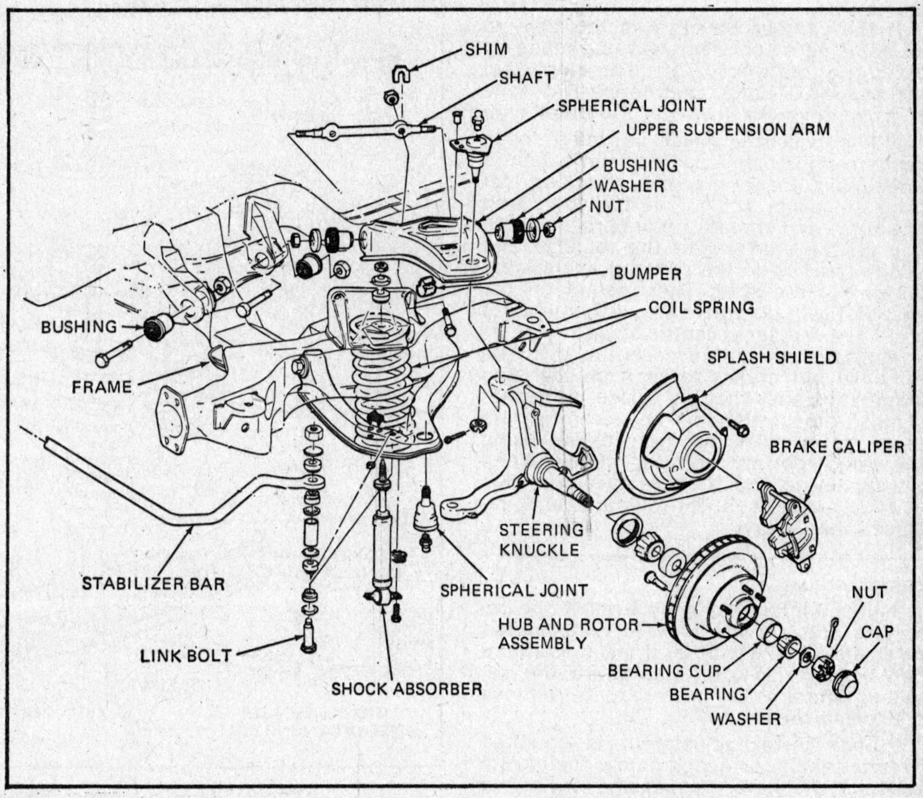

Fig. 1 Front suspension disassembled. 1976 Seville

1969-76 Exc. Seville

The front suspension system consists of two upper and lower control arm assemblies, steel coil springs, shock absorbers, front diagonal tie struts, and a stabilizer bar. Rubber bushings are used at all frame attaching points. Fig. 2.

Ball joints are used at the outer ends of the upper and lower control arms. The upper ball joint is pressed into the upper control arm and tack-welded to the arm at two points. It connects the upper control arm to the steering knuckle through a camber adjustment eccentric. The lower ball joint, a tension type joint, is pressed into the lower control arm. It connects the lower control arm to the steering knuckle.

Beginning with 1970 models the steering knuckle is a combination steering knuckle, brake caliper support and steering arm.

The upper control arms pivot at their inner ends on two flanged rubber bushings, one at each end of the one-piece control arm shaft which is bolted to the top surface of the spring tower on the front frame crossmember. The lower control arms pivot on a single rubber bushing that is bolted to the front suspension frame crossmember.

Diagonal tie struts are used to control the fore and aft movement of the wheels. The struts are bolted to the outer ends of the lower control arms and extend through the frame crossmember. Rubber bushings and a steel spacer are used at the frame mount.

Lubrication

The ball joints are packed with lubricant and sealed at assembly and should not require further lubrication throughout their service life under normal driving conditions. The only maintenance they normally require is an inspection of the seals for physical damage each time the engine oil is changed.

Service plugs are provided in the ball joint covers so that the joints may be packed in the event a seal should become damaged and require replacement. Both the seals and plugs are serviceable.

Wheel Alignment 1976 Seville

Caster and camber are adjusted by adding or removing shims from between the upper control arm and frame bracket, Fig. 3. Caster is adjusted by transfering shims from front to rear and rear to front. Transfering shims from front bolt to rear bolt will increase positive caster, from rear bolt to front bolt will increase negative caster.

Camber is adjusted by adding or removing an equal number of shims from the front and rear bolts. To increase positive camber remove an equal amount of shims from front and rear bolts, to increase negative camber add an equal amount of shims to front and real bolts.

After adjusting caster and camber torque nut to 70-80 ft-lbs. Tighten bolt with the least amount of shims first. After adjusting caster and camber toe-in must be adjusted.

NOTE: The differance in thickness between front and rear shim packs should not excede .040 inch. If differance is greater than .040 inch check arms, frame and related parts for damage.

Exc. Seville

Camber, Adjust

Adjustment is made at the camber eccentric located in the steering knuckle upper support, Fig. 4. The upper ball joint stud fits through the camber eccentric and knuckle support. Turning the eccentric repositions the upper ball joint stud.

To adjust camber, loosen locknut on ball joint stud one turn and strike steering knuckle in area of ball joint stud to free eccentric from steering knuckle, Fig. 5.

CAUTION: Use extreme care to prevent striking and damaging brake hose or ball joint seal.

Using adjustable wrench shown in Fig. 6, turn eccentric as required to obtain the camber specifications listed in the Wheel Alignment chart. The final position of the stud should be in the rear portion of the camber eccentric in order to keep steering angle correct. After proper adjustment has been established, torque stud nut to 60 ft. lbs.

NOTE

If the camber eccentric is too tight to be adjusted a tool can easily be made by cutting a piece of 7/16" diameter steel rod about 20" long, chamfering it on one end and rounding it off on the other end.

If the eccentric must be freed for a camber adjustment, position the car on a wheel alignment machine, backing off the self-locking nut on the ball joint stud one turn, and thread a standard nut halfway on the stud. Insert the rounded end of the tool inside the nut and against the bottom of the stud. Then pound on the end of the tool with a heavy hammer to break the camber eccentric loose.

In cases where the eccentric is to be removed but comes loose from the stud instead of the knuckle, place a washer against the bottom of the eccentric and drive the eccentric out from below, using the same tool as described above. When the eccentric must be removed in this manner, inspect it for damage and replace if necessary.

Caster, Adjust

Adjustment is made by turning the retaining nuts on the forward ends of the tie-struts at the frame front crossmember, Fig. 7. To gain access to the retaining nuts, it is necessary to remove the splash shield.

Proper caster adjustment is obtained by shortening or lengthening the struts between the lower suspension arms and the frame front crossmember. To provide more negative caster, lengthen the struts by loosening the front bushing retaining nuts and tightening the rear bushing retaining nuts. One turn of the nuts results in approximately ½° change in caster.

To provide more positive caster, shorten the struts by loosening the rear bushing retaining nuts and tightening the front bushing retaining nuts.

After proper adjustment has been made, tighten front retaining nuts to 60 ft. lbs., being sure to hold the rear nut with a wrench so as not to disturb the adjustment.

Toe-In, Adjust

Toe-in is adjusted by turning the tie rod adjusters at the outer ends of each tie rod after loosening the clamp bolts. (Both right and left pivot ends have right-hand threads.) Be sure to turn both adjusters an equal amount so that the relation of the steering gear high spot to the straight ahead position of the front wheels will not be changed.

When adjustment has been completed according to the specification listed in the *Wheel Alignment* chart, tighten nuts on clamp bolts to 20 ft. lbs. torque.

NOTE

Make certain that both inner and outer tie rod ball pivots are centered in their respective housings prior to tightening the tie rod adjusting clamps.

If the tie rods are not properly positioned, a binding condition may occur, resulting in poor return of wheels to the straight ahead position.

Each tie rod should be checked after adjustment by grasping the center of the tie rod and rotating it fore and aft. The movement should be equal in both directions. If not, it indicates that the pivot studs are not properly positioned.

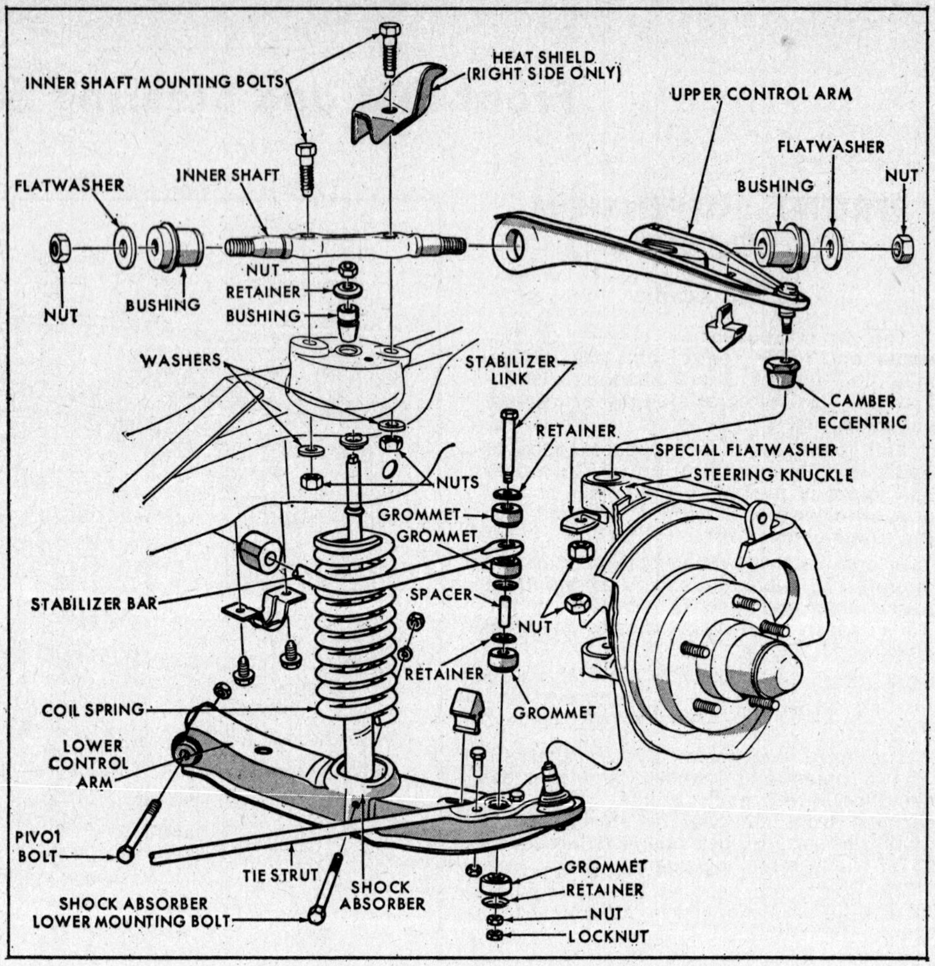

Fig. 2 Front suspension disassembled. Except Eldorado & Seville

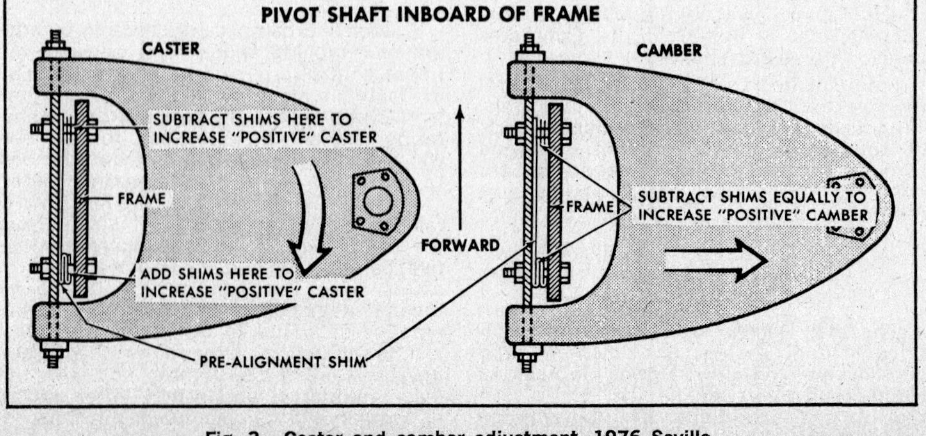

Fig. 3 Caster and camber adjustment. 1976 Seville

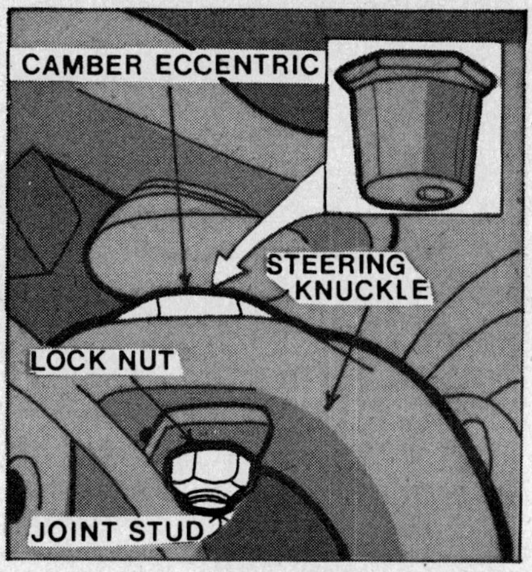

Fig. 4 Camber adjustment eccentric. 1969-76

Wheel Bearings, Adjust

Service Bulletin

Looseness at a front wheel does not necessarily indicate worn bearings or a loose spindle nut, since the tapered roller bearings used on front wheels of all standard Cadillacs should not be preloaded and normally can have up to .004″ end play.

1. While rotating wheel and tire assembly, tighten spindle nut to 30 ft. lbs. for 1969-71 vehicles and 15 ft. lbs. for 1972-76 vehicles, making certain that hub is fully seated on spindle.
2. Back off spindle nut until free, and on 1969-71 vehicles, retorque nut to 6 ft. lbs. On 1972-76 vehicles tighten nut finger tight.
3. Install new cotter pin. If pin cannot be installed, back off nut to next hole and install pin.

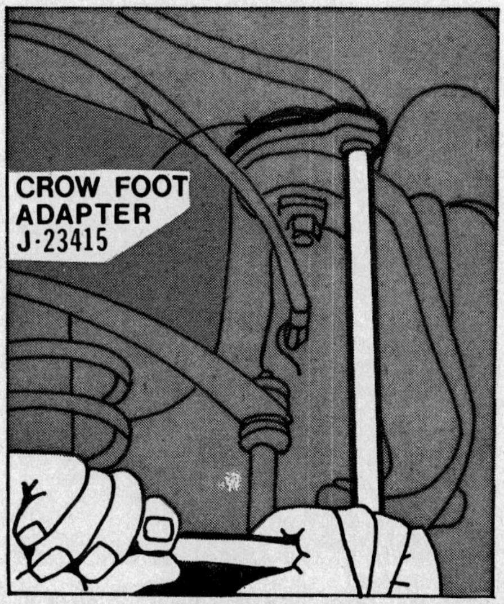

Fig. 6 Adjusting camber. 1969-76

NOTE: Cotter pin must be tight after installation, as vibration can break pin.

Wheel Bearings, Replace

1972-76

1. Remove caliper retaining bolts, then slide caliper off disc and using a length of wire, attach caliper to upper control arm.

NOTE: Never allow caliper to hang from brake hose.

2. Remove dust cap, cotter pin, spindle nut, washer and outer bearing assembly.
3. Remove hub and disc assembly, being careful to avoid damage to spindle threads or grease seal.
4. Remove inner bearing grease seal and bearing assembly.

NOTE: Inner and outer bearing cups are press fit in hub and can be driven out from the opposite side using a brass drift. Tap alternately on opposite sides to prevent cocking cup and damaging hub.

1969-71

1. Remove two thirds of the total fluid capacity in the front master cylinder reservoir to prevent fluid overflow when the piston is pushed back in its bore.
2. Raise car and remove front wheels.
3. Position 7 inch "C" clamp on the caliper so that solid side rests against the back of the caliper. The screw end rests against the back of the outboard shoe.
4. Tighten "C" clamp until caliper moves out far enough to bottom the piston in its bore. This will release the pressure on the shoe and lining assemblies.
5. Remove "C" clamp.
6. Remove two bolts which hold caliper to support plate.

NOTE: It is not necessary to remove brake hose from caliper during this operation.

7. Slide caliper off of disc. Do not allow caliper to hang from brake hose.
8. Remove spindle nut and hub and disc assembly. Grease retainer and inner bearing can now be removed.

Checking Ball Joints For Wear 1969-76 Std. Cars

Upper Ball Joint

1976 Seville

If ball joint has any noticeable lateral movement or can be twisted within its socket using finger pressure, the joint must be replaced.

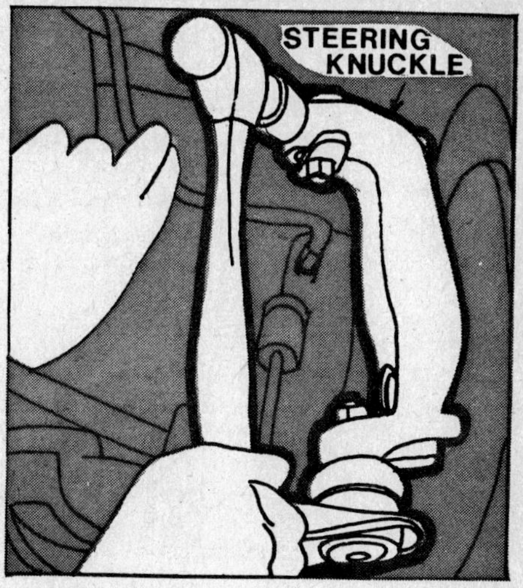

Fig. 5 Loosening camber eccentric

1969-76 Exc. Seville

Using the regular ball joint stud nut and a second nut as a lock nut, turn joint in its socket with a torque wrench. If the torque is not within the limits of 2 to 4 ft-lbs, the joint should be replaced.

Lower Ball Joint

1969-73

The lower ball joint is designed to turn freely in its socket and cannot be checked with a torque wrench. It should be checked by noting the amount of free play as the joint is worked vertically in its socket. Free play should not exceed 1/16″. Replace joint if it exceeds this limit.

1974-76

A wear indicator is built in the ball joint. Remove dirt deposits around service plug and observe position of nipple. Refer to Fig. 8, for wear tolerance.

Fig. 7 Adjusting caster. 1969-76

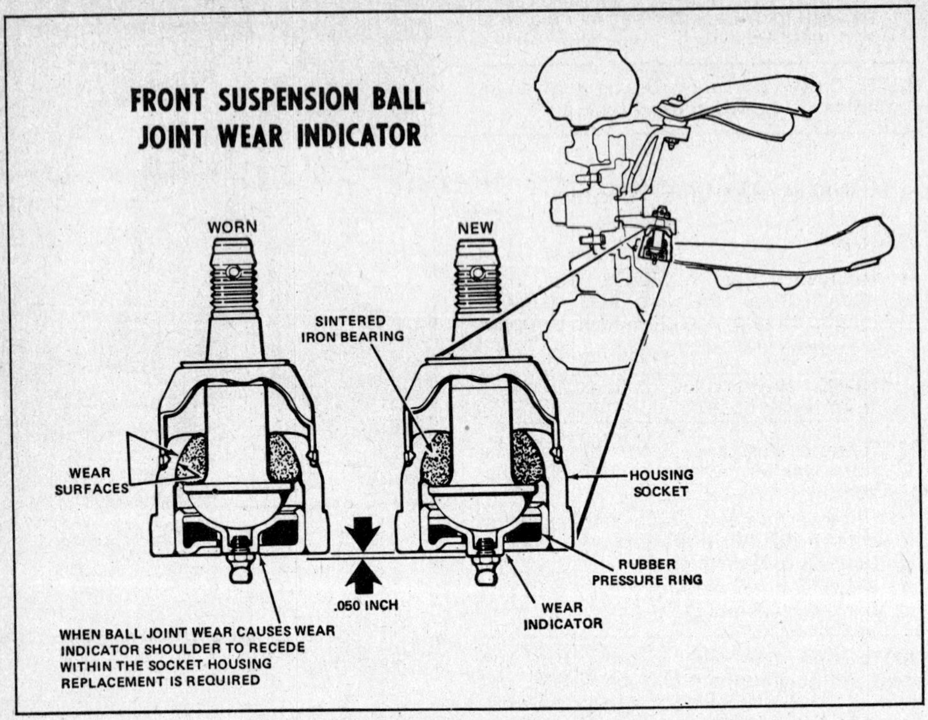

Fig. 8 Ball joint wear indicator. 1974-76 except Eldorado

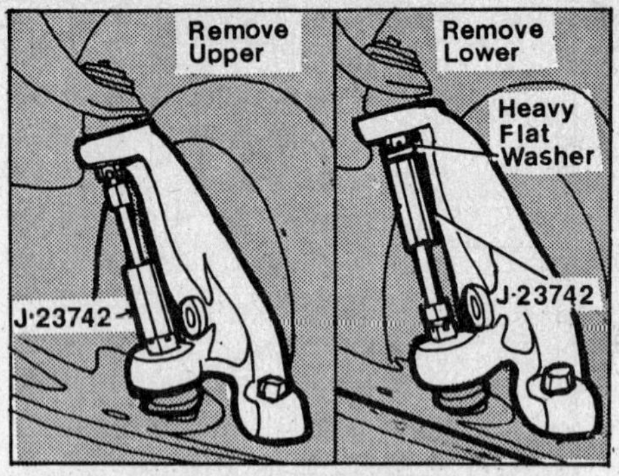

Fig. 9 Removing ball joint studs from steering knuckle. 1976 Seville

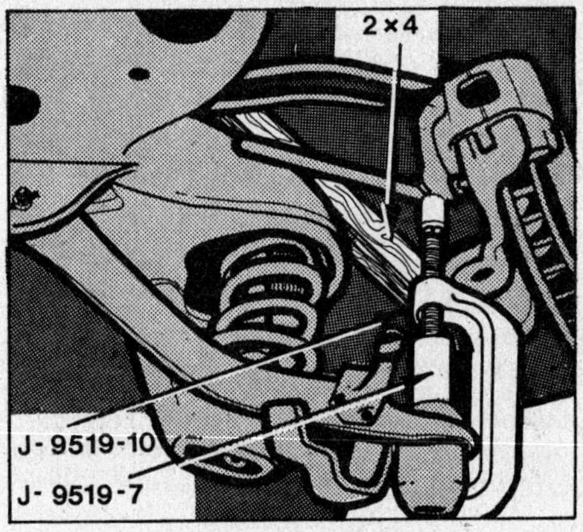

Fig. 10 Removing ball joint from lower control arm. 1976 Seville

Upper Ball Joint, Replace

1976 Seville

1. Raise vehicle and remove wheel.
2. Remove cotter pin from upper ball joint stud.
3. Remove caliper and position aside.

NOTE: When removing caliper use care not to damage brake tubing or hose. Secure caliper to frame with wire.

4. Loosen stud nut, however, not more than one turn.
5. Using tool No. J-23742, free ball joint stud from steering knuckle, Fig. 9.

NOTE: The lower control arm must be supported so spring can not force arm down.

6. Remove upper ball joint stud nut, allow steering knuckle to swing out of way and place block of wood between frame and upper control arm.
7. To remove rivets securing ball joint to upper control arm, grind rivet heads off, then, using a punch, drive rivets out.

NOTE: Use care not to damage ball joint seat or upper control arm when removing rivets.

8. Position ball joint on upper control arm, install bolts through bottom of control arm, install and torque nuts to 25 ft. lbs.
9. Remove block of wood from between frame and upper control arm, position upper ball stud to steering knuckle, install and torque nut to 60 ft. lbs. Install cotter pin.

NOTE: If cotter pin hole is not aligned do not back nut off. Nut may be torqued to a maximum of 100 ft. lbs. (1/6 additional turn) to align cotter pin hole.

10. Install caliper, lubricate ball joint and install wheel.
11. Check wheel alignment.

1969-76 Exc. Seville

The upper ball joints are pressed into the upper control arm and tack-welded to the arm. Therefore, the upper ball joints are supplied only with the upper control arm.

Lower Ball Joint, Replace

1976 Seville

1. Raise vehicle and remove wheel.
2. Remove cotter pin from lower ball joint stud and loosen nut one turn.
3. Using tool No. J-23742, free ball joint stud from steering knuckle, Fig. 9.

NOTE: Lower control arm must be supported so spring can not force arm down.

4. Remove lower stud nut and pull upward and outward on bottom of brake disc to free ball joint stud from steering knuckle.
5. Lift upper control arm up with steering knuckle and hub attached and

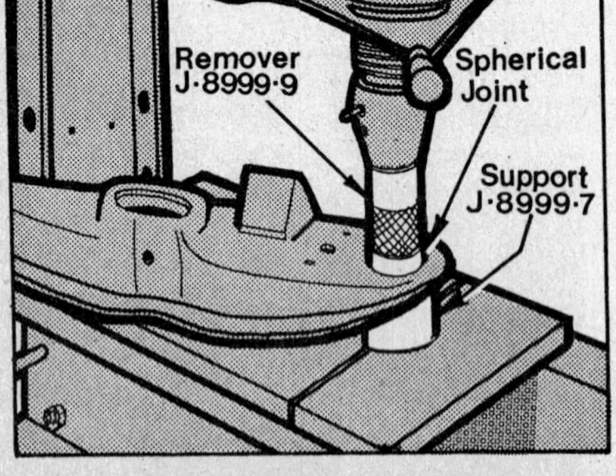

Fig. 11 Removing ball joint from lower control arm. 1969-76 Exc. Seville

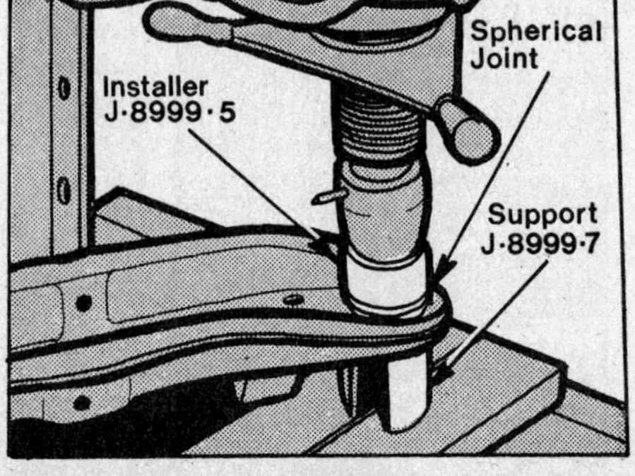

Fig. 12 Installing ball joint in lower control arm. 1969-76 Exc. Seville

Fig. 13 Removing and installing front spring. 1976 Seville

place a block of wood between frame and upper control arm.

NOTE: It maybe necessary to remove tie rod from steering knuckle.

6. Using tools J-9519-10 and J-9519-7, remove ball joint from lower control arm, Fig. 10.
7. Install ball joint in lower control arm using tools J-9519-9 and J-9519-10.

NOTE: Position bleed vent in ball joint rubber boot to face inward and the stud cotter pin hole to face forward.

8. Remove block of wood from between frame and upper control arm, then install lower ball joint stud in steering knuckle. Install and torque nut to 80 ft. lbs. and install cotter pin.

NOTE: If cotter pin hole is not aligned do not back nut off. Nut maybe torqued to a maximum of 125 ft. lbs. (1/6 additional turn) to align cotter pin holes.

9. Lubricate ball joint and install tie rod, if removed. Torque tie rod nut to 35 ft. lbs. and install cotter pin.
10. Install wheel and check wheel alignment.

1969-76 Exc. Seville

1. Remove coil spring as outlined under "Coil Spring, Replace" procedure.
2. Remove pivot bolt and washer from lower control arm at frame mount, then remove lower control arm.
3. Remove band and seal from lower ball joint.
4. Using a suitable press and tools J-8999-9 and J-8999-7, Fig. 11, press lower ball joint from lower control arm.
5. Install new ball joint with tools J-8999-7 and J-8999-5, Fig. 12. Press ball joint into lower control arm until the ball joint flange bottoms on mounting hole flange.
6. Reverse procedure to install lower control arm.

Shock Absorber, Replace

The shock absorbers are removed through the bottom of the lower control arm after unfastening it at the top and bottom.

To install, place the retainer and rubber grommet on the upper stem and fully extend the shock absorber rod. Insert the shock absorber up into the coil spring and guide the stem through the tower in the crossmember. Then place the lower end in position on the lower control arm. Install bolt, washer and nut and torque nut to install bolt, washer and nut. Torque nut to 60 ft. lbs. on 1969-71 models, 55 ft. lbs. on 1972-76 models except Seville and 19 ft. lbs. on 1976 Seville. Install shock absorber upper retaining nuts. Fasten shock absorber at top.

Tie-Strut & Bushings 1970-76 Exc. Seville

Raise car and disconnect stabilizer link from lower arm on side from which tie-strut is to be removed. Remove strut and bushings and replace as follows:

1. Replace rear locknut on threaded end of tie-strut and run nut about ¾" from end of thread.
2. Install rear bushing retainer on tie-strut with concave side against nut.
3. Insert metal spacer part way through conical shaped bushing from small end and install on tie-strut with small end toward front of car.
4. With strut held in horizontal position install threaded end through frame front cross member.
5. Position opposite end of strut on lower arm with pointed end inward, and install attaching bolts and nuts loosely.
6. Install front bushing on end of strut, cupped side toward frame, and slide bushing against cross member.
7. Install front bushing retainer on strut with concave side against bushing.
8. Start new locknut on threaded end of strut and connect stabilizer link to lower arm.
9. Lower car and with car weight on all four wheels, position front bushing on metal spacer and torque locknut on front end of strut, on 1969-73 models to 35 ft. lbs. and on 1974-76 models to 50 ft. lbs.
10. Tighten tie-strut to lower arm nuts and bolts to 55 ft. lbs.

Tie-Strut & Bushings 1969 Std. Cars

Raise car and jack up under frame side rails. Remove splash shield and disconnect stabilizer link from lower control arm on side tie-strut is to be removed. Remove strut and bushings. Then replace as follows:

1. Insert spacer in rear bushing and install bushing and spacer in frame through rear side of crossmember.
2. Install nut on tie-strut and run nut to bottom of thread.
3. Install rear bushing retainer on strut with concave side toward nut.
4. With strut held in horizontal position, install threaded end of strut through rear bushing in frame crossmember.
5. Install front bushing on end of strut and slide bushing into position in frame crossmember. Front bushing should lock in rear bushing.
6. Install front bushing retainer on threaded end of strut with flat side against bushing. Start front nut on end of strut but do not tighten.
7. Secure opposite end of strut to lower control arm and torque bolts to 55 ft. lbs.
8. Connect stabilizer link.
9. Lower car and with car weight on all four wheels, tighten front nut to 55-70 ft. lbs.
10. Tighten rear nut, compressing bushing until retainer bottoms on metal spacer in bushing.
11. Adjust caster as outlined previously.
12. Install splash shield.

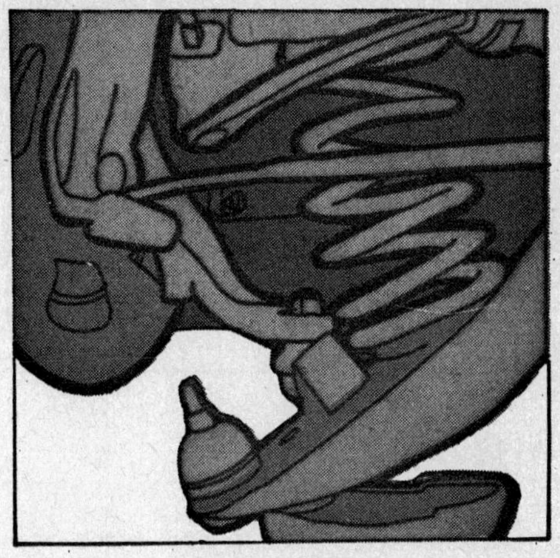

Fig. 14 Coil spring installation. 1969-76

Coil Spring, Replace

1976 Seville

1. Raise vehicle on hoist, remove shock absorber lower mounting bolts and push shock through control arm up into spring.
2. Support vehicle so control arms hang free and disconnect stabilizer bar from lower control arm.
3. Secure tool No. J-23028-01 to a suitable jack, position tool so lower control arm is supported by inner bushings, Fig. 13.
4. Raise jack to relieve spring tension from lower control arm pivot and install a safety chain around spring and through lower control arm.

5. Remove bolt from rear of lower control arm, then remove other bolt and slowly lower control arm until all spring tension is relieved.
6. Remove safety chain, spring and spring insulator.
7. Reverse procedure to install.

NOTE: When installing spring, bottom coil must cover all or part of one inspection hole on the lower control arm, the other inspection hole must be fully or partialy uncovered.

1969-76 Exc. Seville

1. Disconnect shock absorber at upper end.
2. Raise front of car and place jack stands under frame side rails.
3. Disconnect stabilizer link from side from which spring is to be removed.
4. Disconnect tie strut at lower arm.
5. Remove shock absorber.
6. Remove wheel and brake drum.
7. Remove nut from pivot bolt in lower control arm at frame mount.
8. Position jack under outer end of lower control arm and install a safety chain around spring and through lower control arm.
9. Remove stud nut from lower ball joint, install a standard nut on stud and tighten nut to within two threads of steering knuckle.
10. Using a suitable hammer, strike steering knuckle to free from ball joint stud.

NOTE: It maybe necessary to raise opposite rear wheel of vehicle to further compress spring when freeing ball joint stud from steering knuckle.

11. Raise jack to relieve spring tension from ball joint stud nut, remove nut and slowly lower jack until all spring tension is relieved, then remove safety chain and spring, Fig. 14.
12. Reverse procedure to install.

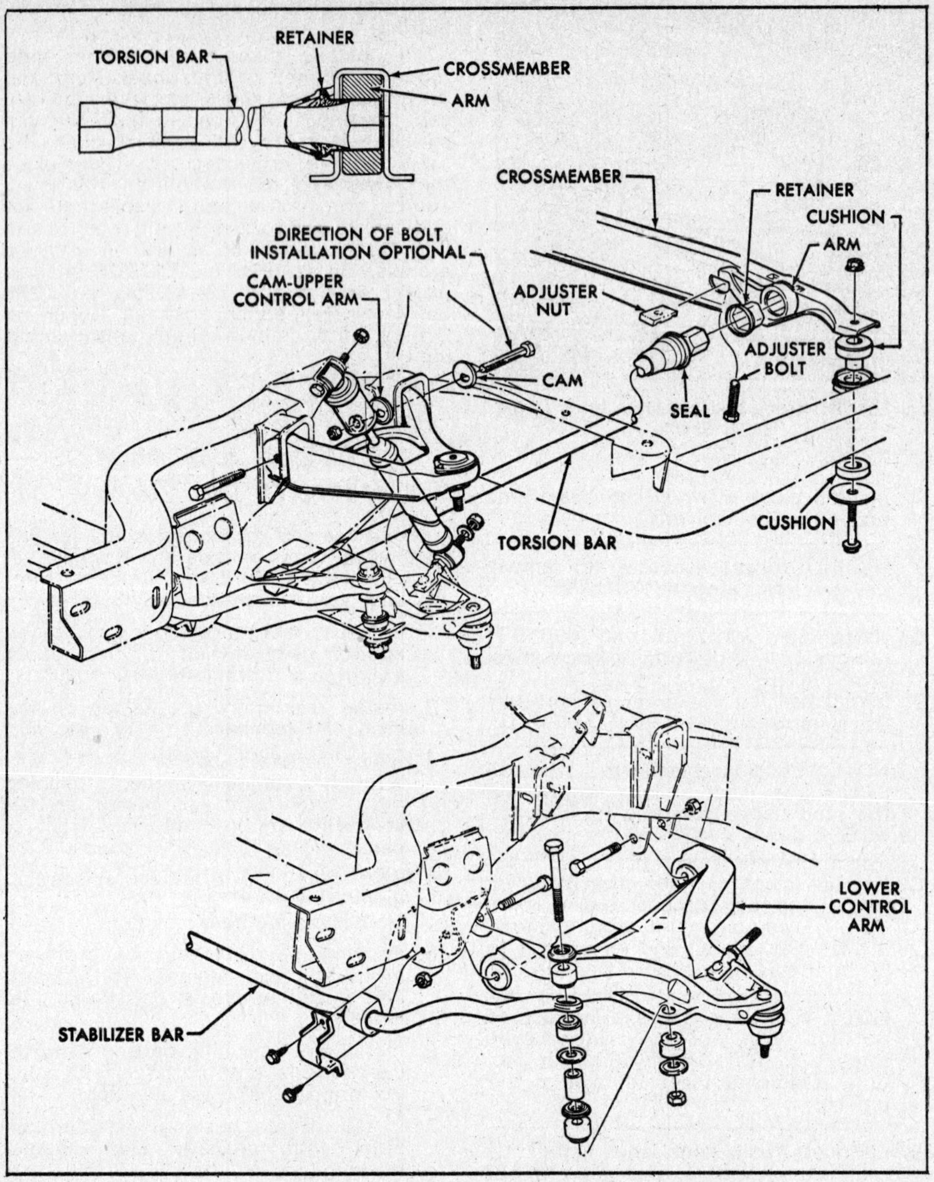

Fig. 15 Front suspension disassembled. 1969-76 Eldorado

ELDORADO FRONT SUSPENSION

The front suspension consists of two upper and two lower control arms, a stabilizer bar, shock absorbers and a right and left torsion bar, Fig. 15. Torsion bars are used instead of the conventional coil springs. The front end of the torsion bar is attached to the lower control arm. The rear of the torsion bar is mounted into an adjustable arm in the torsion bar crossmember. The standing height of the car is controlled by this adjustment.

Standing Height, Adjust

The standing height must be checked and adjusted if necessary before checking and adjusting front wheel alignment. The standing height is controlled by the adjustment setting of the torsion bar adjusting bolt, Fig. 15. Clockwise rotation of the bolt increases standing height: counterclockwise rotation decrease standing height.

To check height on 1969-70 vehicles measure from center of the upper vehicles absorber mounting bolt to the center of the lower shock mount bolt. This dimension should be 14⅝-14⅞". If dimensions are not correct, adjust as required.

To check height on 1971-76 vehicles, measure from lower edge of front shock absorber dust tube (A) to centerline of lower attachment (B), Fig. 16, this dimension between (A) and (B) should be 8-8¼ inch on 1971-73 vehicles, and 8³⁄₁₆-8⁷⁄₁₆ inch on 1974-76 vehicles.

Wheel Alignment, Adjust

Caster and camber can be adjusted from under hood or under car. If under hood method is used, adjustments must be rechecked due to the change in weight distribution. After checking vehicle standing height adjust caster and camber as follows:

1. Loosen front and read adjusting cam nuts, Fig. 17.
2. Rotate front cam to correct for ½ of incorrect camber reading.
3. Rotate rear cam to bring camber reading to 0°.
4. Tighten front and rear cam nuts, then check caster.

NOTE: If caster is within specifications, procede to step 7.

5. If caster is to be adjusted, loosen front and rear cam nuts.
6. Rotate front cam bolt so that camber will change ¼ of desired caster change.

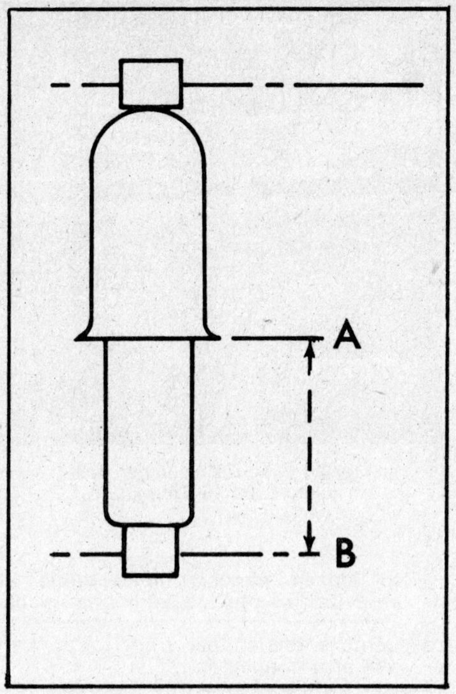

Fig. 16 Checking standing height. 1971-75 Eldorado

NOTE: A change of 1° in camber reading will change caster reading about 2°. To correct for excessive negative caster rotate front cam bolt to increase positive camber. To correct for excessive positive caster rotate front cam bolt to increase negative camber.

7. Set camber to specifications by rotating rear cam bolt.
8. Torque front and rear cam nuts to 95 ft. lbs.

NOTE: When tightening cam nuts ensure cam bolt does not move, any movement of cam bolt will effect wheel alignment.

Toe-In, Adjust

Toe-in is adjusted by turning the tie rod adjusting tubes at outer ends of each tie rod after loosening clamp bolts. Readings should be taken only when front wheels are straight ahead and steering gear is on its high spot.

1. Center steering wheel, raise car and check wheel run-out.
2. Loosen tie rod adjuster nuts and adjust tie rods to obtain the specified toe-in.
3. Tighten tie rod adjuster nuts to 20-22 ft-lbs.
4. Position adjuster clamps so that opening of clamps are facing up. Interference with front suspension components could occur while turning if clamps are facing down.

Wheel Bearing & Steering Knuckle, Replace

1. Raise and support vehicle under lower control arms.
2. Remove drive axle nut and washer and remove wheel and tire assembly.
3. Remove brake hose clip from ball joint and replace nut, then remove brake caliper off disc, and using a

Fig. 17 Caster-camber cam locations. 1969-76 Eldorado

length of wire support caliper on suspension.

NOTE: Do not allow caliper to hang from brake hose as this could cause damage and premature failure of hose.

4. Mark hub and disc assembly for alignment during assembly and remove disc, then strike steering knuckle in area of upper ball joint until upper ball joint is loose.

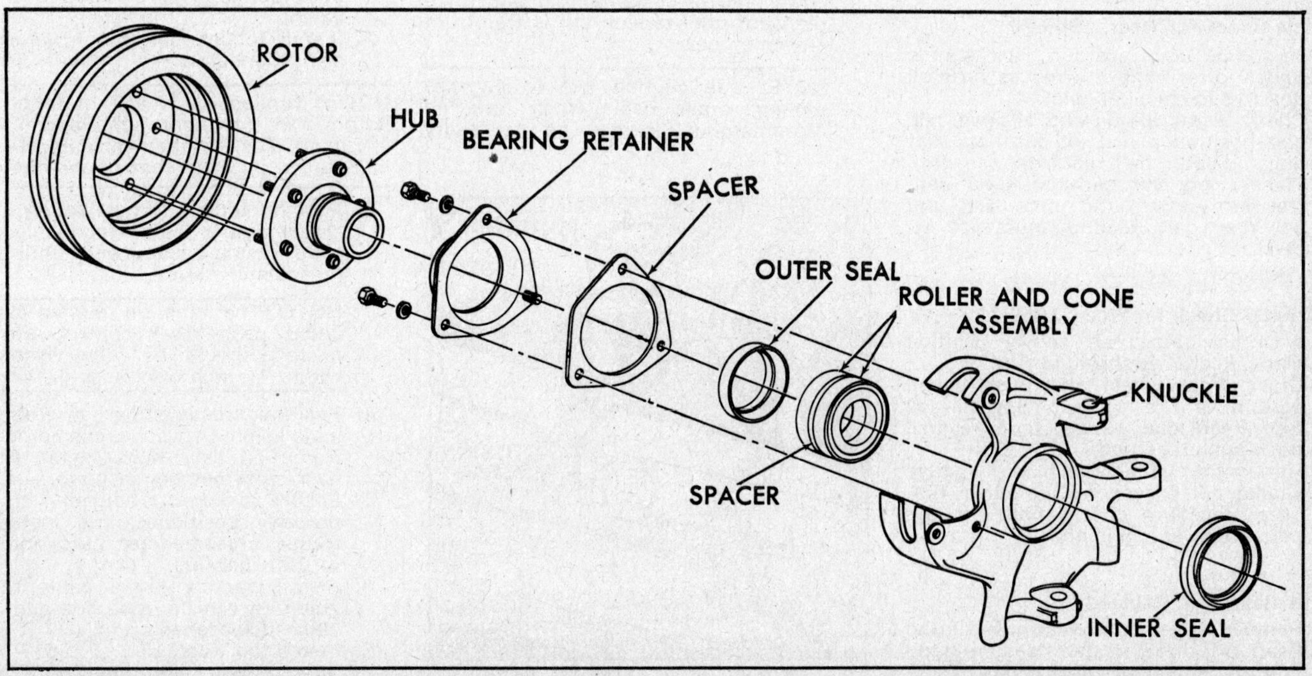

Fig. 18 Wheel bearing and steering knuckle assembly. Eldorado

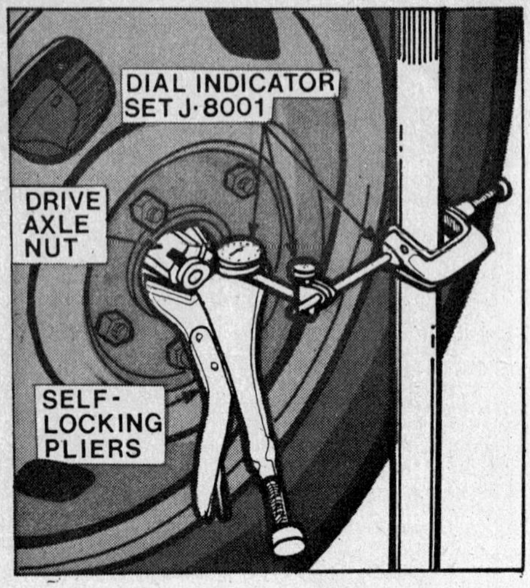

Fig. 19 Horizontal check for ball joint wear. 1969-76 Eldorado

CAUTION: Use extreme care to prevent striking and damaging brake hose or ball joint seal.

5. Place a short length of rubber hose over lower control arm torsion bar connector to avoid damage to inboard tri-pot joint seal when hub and knuckle are removed.
6. Using appropriate puller, disconnect tie rod end, upper and lower ball joints and remove steering knuckle and hub assembly, Fig. 18.

Ball Joints

Verticle Check for Wear, 1969-76

1. Raise car and position jack stands under lower control arms as near as possible to each ball joint.
2. Clamp vise grips on end of drive axle and position a dial indicator so that dial indicator ball rests on vise grip.
3. Place a pry bar between lower control arm and outer race and pry down on bar. Reading must not exceed 1/8".

Horizontal Check for Wear, 1969-71

1. With car raised as above, position dial indicator as shown in Fig. 19.
2. Grasp front wheel and push in on bottom of tire while pulling out at top. Read dial gauge, then reverse push-pull procedure.
3. Horizontal deflection on gauge should not exceed 1/8" at wheel rim. This procedure checks both the upper and lower ball joints.

Upper Ball Joint, Replace

1. Remove upper control arm and grind head off three rivets. Using a hammer and punch, drive out rivets.
2. Install new ball joint, securing it in place with three bolts and nuts contained in the kit.
3. Install upper control arm and lubricate ball joint fitting until grease escapes between seal and steering knuckle, Fig. 20.

Lower Ball Joint, Replace

1. Remove lower control arm and cut off two rivet heads from sides of control arm. Grind off head of rivet at bottom of control arm, then drive rivet out of arm.
2. Install service ball joint, securing it to control arm with bolts and nuts contained in kit.

Torsion Bar, Replace

Removal

1. Support vehicle so front suspension hangs at full rebound position.
2. Remove adjusting bolts from torsion bar adjuster nuts.
3. Install tool J-22517 on 1969-73 models or tool J-22517-01 on 1974-76 models, on torsion bar crossmember, Fig. 21. It may be necessary to pry the crossmember downward to install the tool "U"-bolt.
4. Tighten tool center bolt until torsion bar adjusting arm is raised enough to permit removal of adjuster nut, then remove nut.
5. Loosen tool center bolt until clear of adjusting arm.
6. Repeat steps 3 through 5 on other end of crossmember.
7. Remove parking brake cable from guide at right side of vehicle.
8. Remove torsion bar crossmember bolts, nuts, shim and retainer from both ends of crossmember. Move crossmember as far as possible toward the side opposite torsion bar being removed. One end of crossmember should be clear of frame.
9. Lower end of crossmember and drive rearward until torsion bar is free from crossmember.

NOTE: Use caution not to damage parking brake cable when lowering crossmember. It may be necessary to loosen parking brake cable at equalizer to provide additional slack.

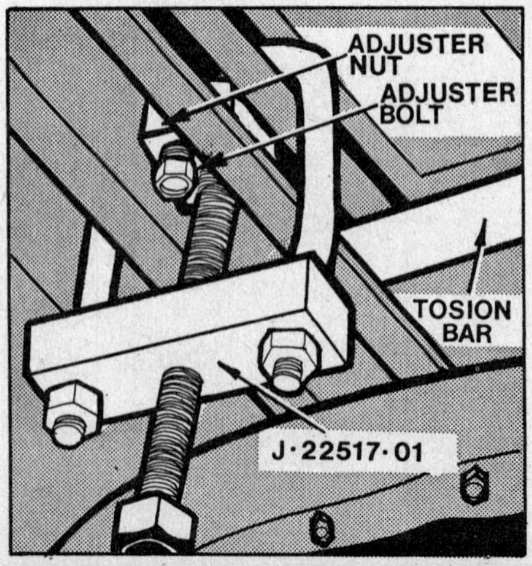

Fig. 21 Torsion bar removal tool installation

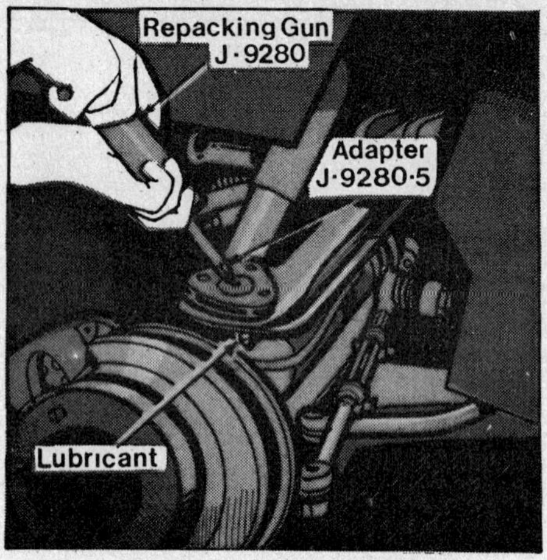

Fig. 20 Repacking upper ball joint. 1969-76 Eldorado

10. Remove torsion bar from lower control arm connector.

NOTE: It is possible to remove both torsion bars at one time, however, installation is easier with one torsion mounted at all times.

Inspection

1. Check torsion bar for nicks, scratches or dents. If any of these conditions exist the torsion bar must be replaced.

Installation

1. Lubricate approximately three inches of each end of torsion bar with suitable lubricant, P/N 9985092 or equivalent.
2. Install torsion bar into lower control arm connector.

NOTE: Torsion bars are stamped with letters "R" for right hand and "L" for left hand. The stamped end of torsion bar is installed in lower control chamber.

3. Position torsion bar adjusting arm in crossmember, hold arm in place and slide torsion bar rearward until seated in adjusting arm.

NOTE: The tape on torsion bar, located near lower control arm connector, should be fully visable but within 1/4 inch of connector.

4. Position crossmember on frame and install shim between crossmember and frame on each side. Install retainer over crossmember insulator.
5. Ensure torsion bar adjusting arms are properly positioned and install and torque crossmember nuts and bolts to 120 inch lbs.
6. Install parking brake cable through guide on vehicle right side and adjust cable if loosened.
7. Install tool J-22517 or J-22517-01 and tighten tool center bolt until torsion bar adjusting arm is raised enough

to install adjustor nut and install nut. Remove tool.

8. Lubricate threads on new torsion bar adjusting bolts and install bolts into adjuster nuts.
9. Check vehicle standing height and wheel alignment. Adjust if necessary.

POWER STEERING, REPLACE

1969-76 Standard Cars

1. Disconnect pressure and return line hoses at steering gear. Have container ready to catch dripping oil. Secure hoses in raised position to prevent loss of fluid.
2. Raise car and use a puller to disconnect pitman arm from steering linkage.
3. Remove screws or nuts securing flexible coupling to steering shaft.
4. Unfasten gear from frame side rail, lower gear down and out of car with pitman arm attached.
5. Reverse procedure to install.

1969-76 Eldorado

1. Disconnect hydraulic hoses at rear of pump reservoir. Cap pump fittings to prevent drainage of fluid from pump. Also, cap or tape hose fittings. If equipped with steering pump cooler, disconnect return hose at cooler.
2. Pull pitman arm from drag link.
3. Remove two bolts holding flexible coupling together.
4. Unfasten gear from frame and move gear forward and down out of car.
5. Reverse procedure to install. Check fluid level and bleed hydraulic system as outlined for standard cars.

CHECKER MOTORS

OLD CAR SPECIFICATIONS: For 1946-68 Tune Up and Wheel Alignment Specifications see main index.

INDEX OF SERVICE OPERATIONS

SERIAL NUMBER LOCATION: Plate on left front door pillar, cowl left side above master cylinder or top of left side instrument panel

ENGINE NUMBER LOCATION

SIX CYL.: Pad at front righthand side of cylinder block at rear of distributor

V8 ENGINES: Pad at front righthand side of cylinder block

ENGINE IDENTIFICATION CODE

Engines are identified in the following table by the code letter or letters immediately following the engine serial number.

CODE	TRANS.	ENGINE
1969		
6BG	①	6-250
6BH	②	6-250
6BS	①	V8-327
6BT	②	V8-327
6BV	②	V8-350③
6BV	①	V8-350④
6BX	②	V8-350
1970		
1CE	②	6-250
2CA	①	V8-350④
2CB	②	V8-350④
2CC	②	V8-350③
2CD	②	V8-350
1971		
CAD	②	6-250
2DI	②	V8-350
2DJ	①	V8-350④
2DK	②	V8-350④
1972		
CBK	②	6-250
EAJ	②	6-250⑤
EBS	②	V8-350
EBT	②	V8-350④
EBW	②	V8-350④⑤
1973		
EBD	②	6-250
ECH	②	6-250
ECJ	②	V8-350
ECK	②	V8-350④
EDA	②	V8-350
EDB	②	V8-350④
1974		
4SB	②	6-250
4SC	②	6-250⑤
4WD	②	V8-350
4WF	②	V8-350④
1975		
5SB	②	6-250
5SC	②	6-250⑤
5WD	②	V8-350
1976		
6SB		6-250
6SC		6-250⑤
5WD		V8-350

①—Std. trans.
②—Auto. trans.
③—High performance.
④—Aerobus.
⑤—California.

GENERAL ENGINE SPECIFICATIONS

Note: See Chevrolet chapter for engine service procedures.

Year	Engine	Carburetor	Bore and Stroke	Piston Displacement, Cubic Inches	Compression Ratio	Maximum Brake H.P. @ R.P.M.	Maximum Torque Lbs. Ft. @ R.P.M.	Normal Oil Pressure Pounds
1969	155 Horsepower....6-250	1 Barrel	3.875 x 3.53	250	8.5	155 @ 4200	235 @ 1600	30–45
	235 Horsepower....V8-327	2 Barrel	4.001 x 3.25	327	9.0	235 @ 4800	325 @ 2800	30–45
	215 Horsepower....V8-350	2 Barrel	4.001 x 3.48	350	8.1	215 @ 4400	320 @ 2400	30–45
	300 Horsepower....V8-350	4 Barrel	4.001 x 3.48	350	10.25	300 @ 4800	380 @ 3200	30–45
1970	155 Horsepower....6-250	1 Barrel	3.875 x 3.53	250	8.5	155 @ 4200	235 @ 1600	30–45
	215 Horsepower....V8-350	2 Barrel	4.001 x 3.48	350	8.1	215 @ 4400	320 @ 2400	30–45
	250 Horsepower....V8-350	2 Barrel	4.001 x 3.48	350	9.0	250 @ 4800	345 @ 2800	30–45
1971	145 Horsepower....6-250	1 Barrel	3.875 x 3.53	250	8.5	145 @ 4200	230 @ 1600	30–45
	215 Horsepower....V8-350	2 Barrel	4.001 x 3.48	350	8.1	215 @ 4000	335 @ 2800	30
	245 Horsepower....V8-350	2 Barrel	4.001 x 3.48	350	8.5	245 @ 4800	350 @ 2800	35–45
1972	110 Horsepower①....6-250	1 Barrel	3.875 x 3.53	250	8.5	110 @ 3800	185 @ 1600	30–45
	165 Horsepower①....V8-350	2 Barrel	4.001 x 3.48	350	8.5	165 @ 4000	280 @ 2400	35–45
	175 Horsepower①....V8-350	4 Barrel	4.001 x 3.48	350	8.5	175 @ 4000	280 @ 2400	35–45
1973	100 Horsepower①....6-250	1 Barrel	3.875 x 3.53	250	8.25	100 @ 3800	175 @ 1600	40
	145 Horsepower①....V8-350	2 Barrel	4.001 x 3.48	350	8.5	145 @ 4000	255 @ 2400	40
	155 Horsepower①....V8-350	4 Barrel	4.001 x 3.48	350	8.5	155 @ 4000	255 @ 2400	40
1974	100 Horsepower①....6-250	1 Barrel	3.875 x 3.53	250	8.25	100 @ 3600	175 @ 1800	40
	145 Horsepower①....V8-350	2 Barrel	4.000 x 3.48	350	8.5	145 @ 3800	250 @ 2200	40
	160 Horsepower①....V8-350	4 Barrel	4.000 x 3.48	350	8.5	160 @ 3800	250 @ 2400	40

Continued

GENERAL ENGINE SPECIFICATIONS—Continued

Note: See Chevrolet chapter for engine service procedures.

Year	Engine	Carburetor	Bore and Stroke	Piston Displacement, Cubic Inches	Compression Ratio	Maximum Brake H.P. @ R.P.M.	Maximum Torque Lbs. Ft. @ R.P.M.	Normal Oil Pressure Pounds
1975	105 Horsepower①..............6-250	1 Barrel	3.875 x 3.53	250	8.25	105 @ 3800	185 @ 1200	40
	145 Horsepower①............V8-350	2 Barrel	4.000 x 3.48	350	8.5	145 @ 3800	250 @ 2200	40
1976	105 Horsepower①..............6-250	1 Barrel	3.875 x 3.53	250	8.25	105 @ 3800	185 @ 1200	40
	145 Horsepower①............V8-350	2 Barrel	4.000 x 3.48	350	8.5	145 @ 3800	250 @ 2200	40

①—Ratings are net—As installed in vehicle.

TUNE UP SPECIFICATIONS

★When using a timing light, disconnect vacuum hose or tube at distributor and plug opening in hose or tube so idle speed will not be affected.

●When checking compression, lowest cylinder must be within 80 percent of highest.

▲Before removing wires from distributor cap, determine location of the No. 1 wire in cap, as distributor position may have been altered from that shown at the end of this chart.

Year	Spark Plug		Distributor		Ignition Timing★			Carb. Adjustments					
								Hot Idle Speed		Air Fuel Ratio		Idle "CO" %	
	Type	Gap Inch	Point Gap Inch	Dwell Angle Deg.	Firing Order Fig. ▲	Timing BTDC ①	Mark Fig.	Std. Trans.	Auto. Trans.②	Std. Trans.	Auto. Trans.	Std. Trans.	Auto. Trans.
1969													
6-250③	R46N	.035	⑤	31–34	A	TDC	C	700	—	—	—	—	—
6-250④	R46N	.035	⑤	31–34	A	4	C	—	500D	—	—	—	—
V8-327③	R44	.035	⑤	28–32	B	2 ATC	C	700	—	—	—	—	—
V8-327④	R44	.035	⑤	28–32	B	2	C	—	600D	—	—	—	—
V8-350⑥	CR43	.035	⑤	28–32	B	4	C	—	600D	—	—	—	—
V8-350⑦	R44	.035	⑤	28–32	B	4	C	—	600D	—	—	—	—
1970													
6-250	R46T	.035	⑤	31–34	A	4	C	700	600D	—	—	—	—
V8-350	R44	.035	⑤	28–32	B	4	C	—	600D	—	—	—	—
1971													
6-250	R46TS	.035	⑤	31–34	A	4	D	700	600D	—	—	—	—
V8-350⑧	R44TS	.035	⑤	29–31	B	6	D	—	550D	—	—	—	—
V8-350⑨	R44T	.035	⑤	29–31	B	4	D	—	550D	—	—	—	—
1972													
6-250	R46T	.035	⑤	31–34	A	4	D	—	600D	—	—	—	—
V8-350⑧	R44T	.035	⑤	29–31	B	6	D	—	600D	—	—	—	—
V8-350⑨	R44T	.035	⑤	29–31	B	8	D	—	600D	—	—	—	—
1973													
6-250	R46T	.035	⑤	31–34	A	6	D	—	600D	—	—	—	—
V8-350⑧	R44T	.035	⑤	29–31	B	8	D	—	600D	—	—	—	—
V8-350⑨	R44T	.035	⑤	29–31	B	12	D	—	600D	—	—	—	—
1974													
6-250	R46T	.035	⑤	31–34	A	6	D	—	600D	—	—	—	—
V8-350	R44T	.035	⑤	29–31	B	8	D	—	600D	—	—	—	—

Continued

TUNE UP SPECIFICATIONS—Continued

★When using a timing light, disconnect vacuum hose or tube at distributor and plug opening in hose or tube so idle speed will not be affected.
●When checking compression, lowest cylinder must be within 80 percent of highest.
▲Before removing wires from distributor cap, determine location of the No. 1 wire in cap, as distributor position may have been altered from that shown at the end of this chart.

Year	Spark Plug		Distributor		Ignition Timing★			Carb. Adjustments					
								Hot Idle Speed		Air Fuel Ratio		Idle "CO" %	
	Type	Gap Inch	Point Gap Inch	Dwell Angle Deg.	Firing Order Fig. ▲	Timing BTDC ①	Mark Fig.	Std. Trans.	Auto. Trans.②	Std. Trans.	Auto. Trans.	Std. Trans.	Auto. Trans.
1975													
6-250	R46TX	.060	—	—	E	10	D	—	550D⑩	—	—	—	—
V8-350	R44TX	.060	—	—	F	6	D	—	600D	—	—	—	—
1976													
6-250	R46TS	.035	—	—	E	10	D	—	550D⑩	—	—	—	—
V8-350	R45TS	.045	—	—	F	6	D	—	600D	—	—	—	—

①—BTDC: Before top dead center.
②—D: Drive. N: Neutral.
③—Std. Trans.
④—Auto. Trans.
⑤—New points, .019″, used, .016″. On V8's, turn adjusting screw in (clockwise) until engine misfires, then back off ½ turn.
⑥—215 H.P.
⑦—300 H.P.
⑧—Marathon & Taxicab.
⑨—Aerobus.
⑩—Calif. 600D RPM.

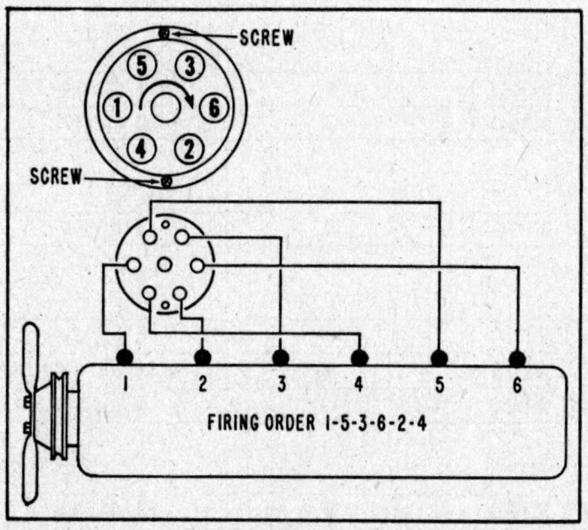

Fig. A

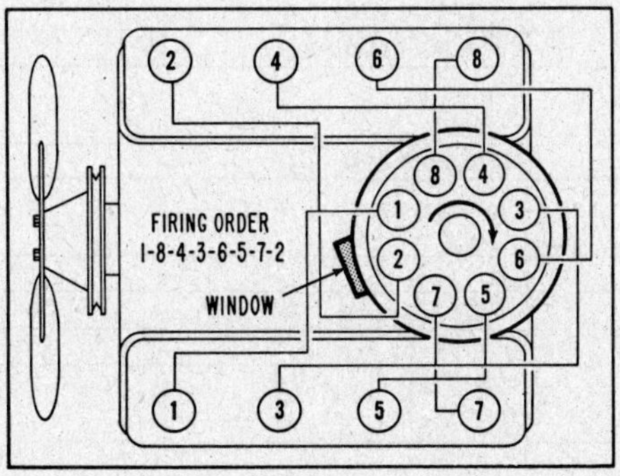

Fig. B

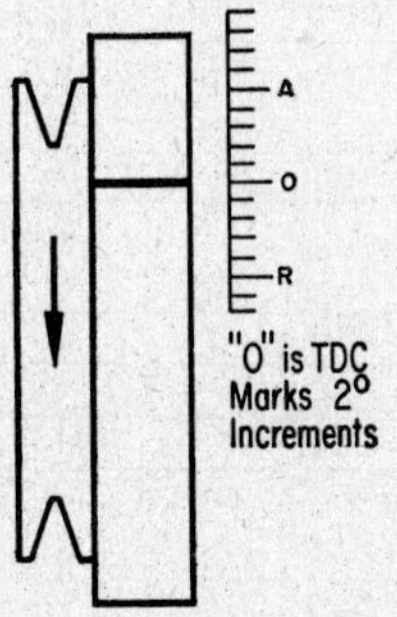

Fig. C

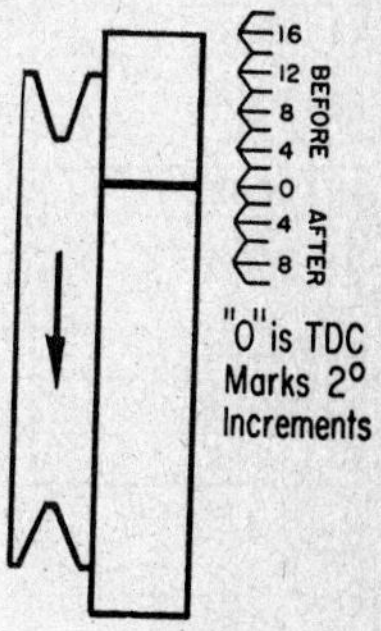

Fig. D

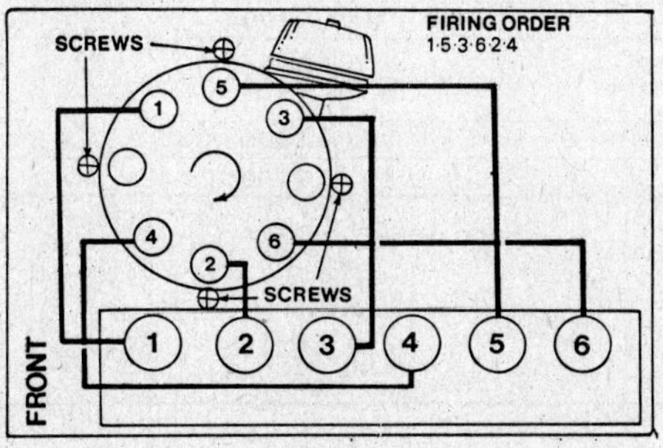

Fig. E

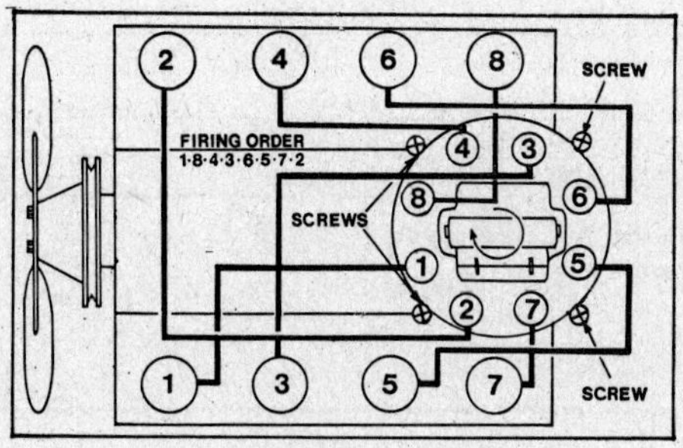

Fig. F

DISTRIBUTOR SPECIFICATIONS

★Note: If unit is checked on vehicle, double the RPM and degrees to get crankshaft figures.

Distributor Part No.①	Centrifugal Advance Degrees @ RPM of Distributor					Vacuum Advance	
	Advance Starts	Intermediate Advance			Full Advance	Inches of Vacuum to Start Plunger	Max. Adv. Dist. Deg. @ Vacuum
1969							
1110469	0 @ 500	3 @ 600	8½ @ 975	—	14 @ 2100	6	7½ @ 12
1110470	0 @ 500	5 @ 580	10½ @ 975	—	16 @ 2100	6	7½ @ 12
1111338	0 @ 450	4½ @ 625	8½ @ 1000	—	16 @ 2500	—	—
1111482	0 @ 500	5 @ 700	—	—	16 @ 2150	6	7½ @ 12
1111483	0 @ 450	6 @ 1000	—	—	14 @ 2150	6	7½ @ 12
1111489	0 @ 450	4½ @ 650	7½ @ 850	—	13 @ 2350	10	5 @ 17
1970							
1110482	0 @ 500	5 @ 700	—	—	16 @ 2150	6	7½ @ 12
1111495	0 @ 375	1 @ 525	7 @ 1000	—	14 @ 2050	8	7½ @ 15½
1112002	0 @ 450	1 @ 550	4 @ 700	—	16 @ 2200	7	12 @ 17
1971							
1112048	0 @ 600	—	5 @ 850	—	10 @ 1900	8	8 @ 17
1971-72							
1110489	0 @ 475	1 @ 635	7 @ 1150	—	12 @ 2050	8	11½ @ 16
1112005	0 @ 400	1 @ 600	6 @ 1100	—	12 @ 2150	8	10 @ 17
1972							
1112047	0 @ 650	—	6 @ 1200	—	9 @ 2100	8	8 @ 17
1973							
1110499	0 @ 550	—	—	—	11½ @ 2100	6	11 @ 14
1112094	0 @ 600	—	—	—	7 @ 2100	6	7½ @ 14
1112168	0 @ 500	6 @ 1500	—	—	9 @ 2100	4	8 @ 7
1974							
1110499	0 @ 550	—	—	—	11½ @ 2100	6	11 @ 14
1112093	0 @ 550	5½ @ 1200	—	—	9 @ 2100	6	7½ @ 13½
1112844	0 @ 500	—	—	—	9 @ 2100	4	7 @ 7
1975							
1110650②	0 @ 550	3½ @ 1150	—	—	8 @ 2100	4	7½ @ 12
1975-76							
1112863②	0 @ 800	3 @ 1100	—	—	7 @ 1900	4	9 @ 12
1112880②	0 @ 600	6 @ 1000	—	—	11 @ 2100	4	9 @ 12

①—Stamped on distributor housing cover.
②—High Energy Ignition.

VALVE SPECIFICATIONS

Year	Engine Model	Valve Lash Int.	Valve Lash Exh.	Valve Angles Seat	Valve Angles Face	Valve Spring Installed Height	Valve Spring Pressure Lbs. @ In.	Stem Clearance Intake	Stem Clearance Exhaust	Stem Diameter Intake	Stem Diameter Exhaust
1969	8-327, 350	Hydraulic①		46	45	1.70	200 @ 1.25	.001–.0027	.001–.0027	.3410–.3417	.3410–.3417
1969–76	6-250	Hydraulic①		46	45	1.66	186 @ 1.27	.001–.0027	.0015–.0032	.3410–.3417	.3410–.3417
1970	8-350	Hydraulic①		46	45	1.70	200 @ 1.25	.001–.0027	.0012–.0029	.3410–.3417	.3410–.3417
1971–72	8-350	Hydraulic①		46	45	1.70	200 @ 1.25	.001–.0027	.001–.0027	.3410–.3417	.3410–.3417
1973–76	8-350	Hydraulic①		46	45	1.66	189 @ 1.20	.001–.0027	.0012–.0029	.3410–.3417	.3410–.3417

①—1 turn down from zero lash.

PISTONS, PINS, RINGS, CRANKSHAFT & BEARINGS

Year	Engine Model	Piston Clearance	Ring End Gap① Comp.	Ring End Gap① Oil	Wrist-pin Diameter	Rod Bearings Shaft Diameter	Rod Bearings Bearing Clearance	Main Bearings Shaft Diameter	Main Bearings Bearing Clearance	Main Bearings Thrust on Bear. No.	Main Bearings Shaft End Play
1969–72	6-250	.0015–.0025	.010	.015	.9272	1.999–2.000	.0007–.0027	2.2983–2.2993	.0003–.0029	7	.002–.006
1973–76	6-250	.0015–.0025	.010	.015	.9272	1.9928–2.000	.0007–.0027	2.2983–2.2993	.0003–.0029	7	.002–.006
1969	8-327	.0005–.0011	.013	.015	.9272	2.099–2.100	.0007–.0028	2.4479–2.4488	③	5	.003–.011
1969	8-350	.0005–.0011	.013	.015	.9272	2.099–2.100	.0007–.0028	2.4479–2.4488	③	5	.003–.011
1970	8-350	.0007–.0013	④	.015	.9272	2.099–2.100	.0007–.0028	②	⑤	5	.002–.006
1971	8-350	.0007–.0017	.010	.015	.9272	2.099–2.100	.0013–.0035	②	⑥	5	.002–.006
1972	8-350	.0007–.0017	.010	.015	.9272	2.099–2.100	.0013–.0035	⑦	⑥	5	.002–.006
1973–76	8-350	.0007–.0017	④	.015	.9272	2.099–2.100	.0013–.0035	⑦	⑥	5	.002–.006

①—Fit rings in tapered bores to the clearance listed in tightest portion of ring travel. Clearance specified are mimimum gaps.
②—No. 1, 2, 3, 4: 2.4484–2.4493; No. 5: 2.4479–2.4488.
③—No. 1, 2, 3, 4: .0008–.0020; No. 5: .0018–.0034.
④—Top ring, .010; lower ring, .013.
⑤—No. 1: .0003–.0015; No. 2, 3, 4: .0006–.0018; No. 5: .0008–.0028.
⑥—No. 1: .0008–.0020; No. 2, 3, 4: .0011–.0023; No. 5: .0017–.0033.
⑦—No. 1: 2.4484–2.4493; No. 2, 3, 4: 2.4481–2.4490; No. 5: 2.4475–2.4488.

ALTERNATOR SPECIFICATIONS

Year	Alternator Make	Alternator Model	Ground Polarity	Rated Output Amperes	Rated Output Volts	Field Current Amperes ①	Field Current Volts	Regulator② Model	Regulator Test @ 120°F Ampere Load	Regulator Test @ 120°F Altern. R.P.M.	Regulator Test @ 120°F Volts
1969–70	Motorola	8AL-2001	Negative	35	15	1.8–2.4	12.6	R2CC1	10	1500	13.7–14.5
1969–74	Motorola	8AL-2002	Negative	55	15	1.8–2.4	12.6	8RD2009	10	2000	13.4–13.9
1975	Delco-Remy	1100542	Negative	63	—	4.0–4.5	12	Integral	—	—	—
	Delco-Remy	1102854	Negative	63	—	4.0–4.5	12	Integral	—	—	—
1976	Delco-Remy	1102854	Negative	63	—	4.0–4.5	12	Integral	—	—	—
	Delco-Remy	1102860	Negative	63	—	4.0–4.5	12	Integral	—	—	—

①—Excessive current drain indicates shorted field windings.
②—Regulator is a sealed assembly, requiring no adjustments.

ENGINE TIGHTENING SPECIFICATIONS★

★Torque specifications are for clean and lightly lubricated threads only. Dry or dirty threads produce increased friction which prevents accurate measurement of tightness.

Year	Engine Model	Spark Plugs Ft. Lbs.	Cylinder Head Bolts Ft. Lbs.	Intake Manifold Ft. Lbs.	Exhaust Manifold Ft. Lbs.	Rocker Arm Stud Ft. Lbs.	Rocker Arm Cover Ft. Lbs.	Connecting Rod Cap Bolts Ft. Lbs.	Main Bearing Cap Bolts Ft. Lbs.	Flywheel to Crankshaft Ft. Lbs.	Vibration Damper or Pulley Ft. Lbs.
1969-70	6-230, 250	25①	95	②	②	—	55③	35	65	60	④
1971-76	6-250	15	95	②	②	—	45③	35	65	60	60
1969	8-327	25	65	30	20	—	55③	⑥	80	60	60
1969-70	8-350	25	65	30	20⑤	—	55③	45	75	60	60
1971-76	8-350	15	65	30	20⑤	50	45③	45	75⑦	60	60

①—1970 6-250, 15 ft. lbs.
②—End clamp bolts 20 ft. lbs.; center bolts, 30 ft. lbs.
③—Inch lbs.
④—Pressed on.
⑤—Inside bolts, 30 ft. lbs.
⑥—3/8 inch bolts, 45 ft. lbs.; 11/32 inch bolts, 35 ft. lbs.
⑦—Outer bolts on engines with 4 bolt caps, 65 ft. lbs.

STARTING MOTOR SPECIFICATIONS

Year	Engine	Starter Number	Brush Spring Tension Oz.①	Free Speed Test			Resistance Test③	
				Amps.②	Volts	R.P.M.①	Amps.	Volts
1969	6-230, 250	1108360	35	55–80	9	3500–6000	—	—
	V8-327, 350	1108361	35	65–100	10.6	3600–5100	300–360	3.5
	V8-307, 327	1108367	35	49–87	10.6	6200–10700	290–425	4.2
	V8-307, 327	1108368	35	50–80	9	5500–10500	—	—
1969–73	6-230, 250	1108365	35	49–87	10.6	6200–10700	290–425	4.2
1970	V8-350	1108420	35	65–95	9	7500–10500	—	—
1971–74	V8-350	1108361	35	65–100	10.6	3600–5100	300–360	3.5
1973–74	6-250	1108480	35	50–80	9	5500–10500	—	—
1973–74	V8-350	1108430	35	65–95	9	7500–10500	—	—
1975–76	6-250	1108778	—	50–80②	9	5500–10500	—	—
	V8-350	1108776	—	65–95②	9	7500–10500	—	—

①—Minimum.
②—Includes solenoid.
③—Check capacity of motor by using a 500 ampere meter and a carbon pile rheostat to control voltage. Apply the volts listed across motor with armature locked. Current should be as listed.

WHEEL ALIGNMENT SPECIFICATIONS

Year	Model	Caster Angle, Degrees		Camber Angle, Degrees				Toe-In Inch	Toe-Out on Turns, Deg.①	
		Limits	Desired	Limits		Desired			Outer Wheel	Inner Wheel
				Left	Right	Left	Right			
1969–74	Aerobus	+2	+2	+1 to +2	+1 to +2	+1½	+1½	1/8 to 3/16	—	—
1969–76	Exc. Aerobus	+2	+2	0 to +1	0 to +1	+½	+½	1/16 to 1/8	—	—

①—Incorrect toe-out when other adjustments are correct, indicates bent steering arms.

REAR AXLE SPECIFICATIONS

Year	Model	Carrier Type	Ring Gear & Pinion Backlash		Pinion Bearing Preload			Differential Bearing Preload		
			Method	Adjustment	Method	New Bearings Inch-Lbs.	Used Bearings Inch-Lbs.	Method	New Bearings Inch-Lbs.	Used Bearings Inch-Lbs.
1969–76	44	Integral	Shims	.005–.010	Shims	20–40①	10–20①	Shims	.015	.015
1969–76	62	Integral	Shims	.005–.009	Shims	20–40①	10–20①	Shims	.015	.015

①—Use inch-pound torque wrench on pinion shaft nut.

COOLING SYSTEM & CAPACITY DATA

Year	Model or Engine	Cooling Capacity, Qts.			Radiator Cap Relief Pressure, Lbs.		Thermo. Opening Temp. ①	Fuel Tank Gals.	Engine Oil Refill Qts. ②	Transmission Oil			Rear Axle Oil Pints
		No Heater	With Heater	With A/C	With A/C	No A/C				3 Speed Pints	4 Speed Pints	Auto. Trans. Qts.③	
1969	6-250	11	12	12	13	13	195	23	4	2.6	—	8½	3
1969	V8-350④	16	17	17	13	13	195	23	4	2.6	—	8½	3
1969	V8-350⑤	—	17⑥	17⑥	13	13	180	23	4	2.6	—	8½	6
1970	6-250	11	12	12	13	13	195	23	4	—	—	8½	3
1970	V8-350④	16	17	17	13	13	195	23	4	—	—	8½	3
1970	V8-350⑤	—	17	17	13	13	180	23	4	—	—	8½	6
1971	6-250	—	12	12	13	13	195	21½	4	—	—	8½	3
1971	V8-350④	—	17	17	13	13	195	21½	4	—	—	8½	3
1971	V8-350⑤	—	21	21	13	13	195	23	4	—	—	8½	6
1972–76	6-250	—	12	12	15	15	195	21½	4	—	—	⑦	3
1972–76	V8-350④	—	17	17	15	15	195	21½	4	—	—	⑦	3
1972–76	V8-350⑤	—	21	21	15	15	195	21½	4	—	—	⑦	6

①—For permanent type anti-freeze.
②—Add one quart with filter change.
③—Approximate. Make final check with dipstick.
④—Marathon & Taxicab.
⑤—Aerobus.
⑥—Add 9 qts. with underseat heaters.
⑦—Exc. late 1973, 8½ qts. Late 1973, Turbo Hydra-Matic, 9 qts.

BRAKE SPECIFICATIONS

Year	Model	Brake Drum Inside Diameter	Wheel Cylinder Bore Diameter			Master Cylinder Bore Diameter		
			Disc Brake	Front Drum Brake	Rear Drum Brake	Disc Brakes	Drum Brakes	Power Brakes
1969–70	All Exc. Aerobus	11	2 15/16	1 1/8	15/16	1 1/8	1	1
	Aerobus	11	—	1 1/8	1	—	—	1
1971–76	All Exc. Aerobus	11	2 15/16	—	15/16	1 1/8	—	1
	Aerobus	11	—	1 1/8	1	—	—	1

Electrical Section

DISTRIBUTOR

Refer to the Chevrolet chapter for removal and installation procedures.

STARTER, REPLACE

Refer to the Chevrolet chapter for replacement procedure.

IGNITION LOCK

Refer to the Chevrolet chapter for replacement procedure.

IGNITION SWITCH, REPLACE

Refer to the Chevrolet chapter for replacement procedure.

LIGHT SWITCH, REPLACE

1969-76 All

1. Disconnect battery ground cable.
2. Pull switch knob to "ON" position, reach under instrument panel and depress shaft retainer, then pull knob and shaft from switch.
3. Remove ferrule nut and switch from instrument panel and disconnect wiring.
4. Reverse procedure to install.

STOP LIGHT SWITCH, REPLACE

1969-76 All

1. Disconnect electrical connector from switch on brake pedal support.
2. Remove retaining nut, if so equipped, then remove switch from bracket.
3. Install and adjust new switch so stop lights are actuated when brake pedal is depressed 3/8 inch from fully released position.

NEUTRAL SAFETY SWITCH, REPLACE

Refer to the Chevrolet chapter, "1971-76 Column Shift" and "1969-70 Column Shift" for replacement procedures.

TURN SIGNAL SWITCH, REPLACE

Refer to the Chevrolet chapter, "1969-75" replacement procedures.

HORN SOUNDER & STEERING WHEEL

Refer to the Chevrolet chapter, "1969 Deluxe & 1970-76 Standard Wheel" procedures.

INSTRUMENT CLUSTER

1969-76 All

1. Disconnect battery ground cable.
2. Remove instrument cluster retaining screws.
3. Pull cluster forward, disconnect wiring and remove instrument cluster.
4. Reverse procedure to install.

W/S WIPER BLADES

Two methods are used to retain wiper blades to wiper arms, Fig. 1. One method uses a press type release tab. When the release tab is depressed the blade assembly can be slid off the wiper arm pin. The other method uses a coil spring retainer. A screwdriver must be inserted on top of the spring and the spring pushed downward. The blade assembly can then be slid off the wiper arm pin. Two methods are also used to retain the blade element in the blade assembly, Fig. 1. One method uses a press type release button. When the button is depressed the two piece blade assembly can be slid off the blade element. The other method uses a spring type retainer clip in the end of the blade element. When the retainer clip is squeezed together, the blade element can be slid out of the blade assembly.

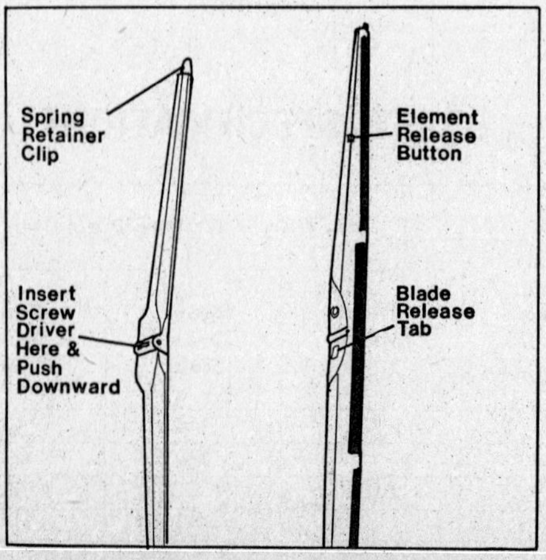

Fig. 1 Wiper blades

NOTE: To be sure of correct installation, the element release button, or the spring element retaining clip should be at the end of the wiper blade assembly nearest the wiper transmission.

W/S WIPER ARMS

1. Wiper motor must be in park position.
2. Use suitable tool to minimize the possibility of windshield or paint finish damage during arm removal.
3. Remove arm by prying up with tool to disengage arm from serrated transmission shaft.
4. To install arm to transmission shaft rotate the required distance and direction so that the blades rest in proper position.

W/S WIPER MOTOR

1. Remove nut and lock washer through center hole of pivot arm drive disc and separate drive disc from motor.
2. Remove bolts securing motor to cowl, lower transmission and disconnect motor electrical connector.
3. Remove motor from vehicle.

W/S WIPER TRANSMISSION

1. Perform step 1 as outlined under "W/S Wiper Motor".
2. Remove retaining clip from stud and disconnect right hand arm from transmission.
3. Remove wiper arm assemblies, outside linkage nut and transmission and linkage from under dash.

RADIO, REPLACE

1. Remove inner and outer knobs and right hand shaft insert.
2. Remove mounting nuts, escutcheon inserts and escutcheon.
3. Disconnect antenna, speaker and power feed leads.
4. Remove mounting nut from bracket and remove radio.

BLOWER MOTOR, REMOVAL

1. Disconnect wiring from motor.
2. Remove blower to heater housing screws and remove blower motor.

HEATER CORE, REMOVAL

1. Disconnect resistor wires and remove heater core cover.
2. Disconnect heater hoses from core and plug hoses and core openings.
3. Slide heater core from housing.

Engine Section

Refer to Chevrolet Chapter for Engine Service Procedures.

Clutch and Transmission Section

CLUTCH PEDAL, ADJUST

Proper clutch pedal adjustment is obtained by adjusting the length of the throwout lever rod. Clutch pedal free travel for all models is 1 to 1⅜ inch.

CLUTCH, REPLACE

Refer to Chevrolet chapter for replacement procedure.

THREE SPEED TRANSMISSION, REPLACE

1. Support vehicle on jack stands and remove propeller shaft.
2. Disconnect shift linkage and speedometer cable from transmission.
3. Support transmission, remove transmission to clutch housing bolts, slide transmission rearward and lower transmission from vehicle.
4. Reverse procedure to install.

THREE SPEED SHIFT LINKAGE, ADJUST

1. Disconnect shift rods from shift arms on steering column.
2. Position shift arms parallel one above another, check shift lever movement through the steering column.
3. Hold shift arms in straight position and adjust one shift rod between shift arm and transmission lever.
4. Insert a wedge between the 2nd and 3rd selector lever and selector gate and check for proper cross-over. If necessary, adjust lower steering column bearing by loosening screws, forcing bearing upward and adding shims if necessary.
5. Check adjustment for proper cross-over and gear engagement.

TURBO-HYDRAMATIC 400, REPLACE

Late 1973-76

1. Disconnect battery negative cable and raise vehicle.
2. Remove filler tube and drain transmission.
3. Support engine and remove converter underpan.
4. Remove converter to flywheel bolts.
5. Disconnect speedometer cable, kickdown switch wiring and shift linkage.
6. Disconnect vacuum modulator line and remove vacuum modulator.
7. Disconnect oil cooler lines.
8. Reinstall vacuum modulator with retaining bolt finger tight.
9. Disconnect parking brake cable from frame lever and on all models except Aerobus, remove parking brake lever bracket.
10. Remove propeller shaft.
11. Place a jack under transmission and remove rear engine support.
12. Lower engine and transmission slightly and remove engine to transmission bolts. Move transmission rearward and install a suitable converter holding tool. Lower transmission and remove from vehicle.
13. Reverse procedure to install.

IN-CAR ADJUSTMENTS

NOTE: Refer to the Automatic Transmission section, Turbo-Hydramatic 400 chapter, for other adjustments and minor service.

Manual Linkage, Adjust

Refer to Turbo-Hydramatic 400 chapter, "Chevrolet Column Shift" 1973-76 manual linkage adjustment.

Kickdown Switch, Adjust

Adjust switch, located under accelerator pedal, so plunger is depressed when carburetor throttle is in the wide open position.

WARNER AUTOMATIC, REPLACE

1969-Early 1973

1. Support vehicle on jack stands.
2. Remove filler tube from pan, then drain transmission.
3. Disconnect speedometer cable from transmission and remove vacuum line from vacuum control unit.
4. Disconnect electrical wiring and oil cooler lines from transmission.
5. Remove propeller shaft.
6. Support transmission, remove transmission to engine bolts and slide transmission rearward, disengaging transmission from converter. Remove transmission from vehicle.
7. Remove converter.
8. Reverse procedure to install.

IN-CAR ADJUSTMENTS

NOTE: Refer to the Automatic Transmission section, "American Motors Shift-Command" chapter, for other adjustments and minor service.

Manual Linkage, Adjust

Refer to Turbo-Hydramatic 400 chapter, "Chevrolet Column Shift" 1969-72 manual linkage adjustment.

Kickdown Switch, Adjust

Adjust switch by loosening the two nuts on the switch and turn the upper nut clockwise or counter-clockwise so plunger is depressed by the accelerator pedal when carburetor throttle is the wide open position.

Rear Axle, Propeller Shaft & Brakes

REAR AXLE, REPLACE

1. Support vehicle at frame and rear axle, relieving tension from springs and remove rear wheels.
2. Disconnect parking brake cable at equalizer and brake line at axle tube connector.
3. Disconnect shock absorbers from lower mountings.
4. Place an alignment mark between rear propeller shaft yoke, universal joint and pinion flange yoke for alignment during assembly. Disconnect propeller shaft from pinion flange.
5. Remove axle shafts and brake support plates.
6. Remove leaf spring "U" bolts and spring plates.
7. Remove axle assembly from vehicle.
8. Reverse procedure to install.

AXLE SHAFT

1969-76 All

1. Raise vehicle and remove wheel and brake drum.
2. Remove axle shaft flange dust cap, then brake backing plate attaching nuts or screws.
3. Using a suitable puller, remove axle shaft.
4. Remove bearing cup and inner axle shaft seal with a suitable puller.
5. Reverse procedure to install.

PROPELLER SHAFT

Marathon and Taxicab models use a one piece propeller shaft while the Aerobus has a two piece shaft. Universal joints must be lubricated every 2500 miles with a suitable lithium soap base grease. When lubricating these joints, lubricant must flow from all four bearing seals.

BRAKE ADJUSTMENTS

Refer to the Chevrolet chapter for brake adjustments on the Marathon and Taxicab models.

Refer to the Chrysler chapter for brake adjustments on Aerobus models.

PARKING BRAKE, ADJUST

Depress parking brake pedal approximately ½ inch, then remove slack from cable at clevis. Check cable adjustment to ensure against brake drag.

MASTER CYLINDER, REPLACE

Disconnect hydraulic lines from master cylinder. Remove bolts securing master cylinder to mounting.

POWER BRAKE UNIT

Refer to the Chevrolet chapter for replacement procedures.

Rear Suspension Section

SHOCK ABSORBER, REPLACE

1. With rear axle properly supported, remove nuts securing shock absorber to spring plate bracket, Fig. 1.
2. Remove nuts securing shock absorber to upper mounting bracket.
3. Reverse procedure to install.

LEAF SPRING, REPLACE

1. Support vehicle at frame and rear axle, removing tension from spring.
2. Disconnect shock absorber from spring plate mounting bracket.
3. Remove "U" bolt nuts, spring plate and "U" bolts.
4. Disassemble rear shackle.
5. Remove nuts securing spring front mounting bracket to the frame, pull spring assembly rearward and remove from vehicle. Remove spring front mounting bracket from spring.
6. Reverse procedure to install.

HELPER SPRING, REPLACE

1. Perform steps 1 to 3 as outlined under "Leaf Spring, Replace" procedure.
2. Remove rubber bumper "U" bolt nuts, plate, "U" bolt and rubber bumper, Fig. 2.
3. Reverse procedure to install.

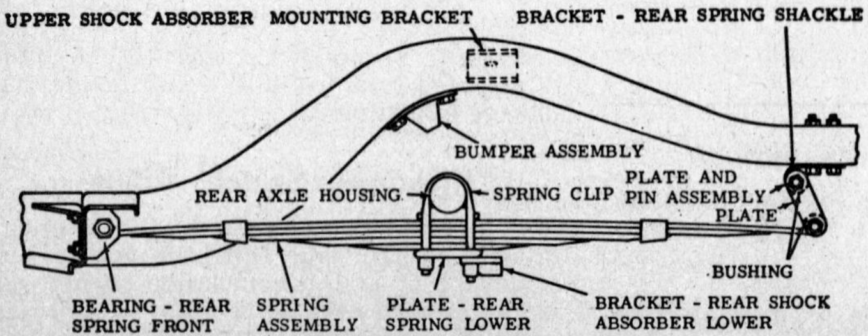

Fig. 1 Leaf spring suspension. 1969-75

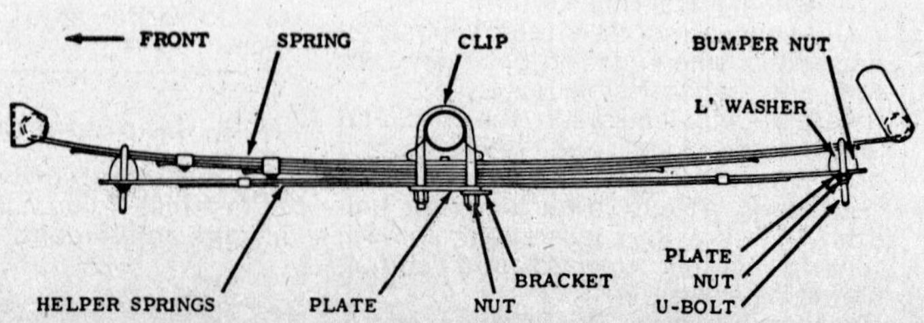

Fig. 2 Helper spring installation. 1969-76

Front End and Steering Section

FRONT SUSPENSION

All front suspension systems incorporate independent coil springs riding on the lower control arms. Ball joints connect the upper and lower control arms to the steering knuckles, Figs. 1 and 2.

LUBRICATION

Lubrication is recommended at 2,500 mile intervals with a suitable lithium base grease. When lubricating the ball joints, the joint must be unloaded.

WHEEL ALIGNMENT

Caster and camber adjustments are made by means of shims between the upper control arm shaft and the frame. Shims may be added, subtracted or transferred to change the readings as follows:

Caster, Adjust

Transfer shims from front to rear or rear to front. The transfer of one shim (.06 inch) from rear to front will increase positive caster by ½°. In any case, the total thickness of the shim pack should not exceed .562 inch.

Camber, Adjust

Add or subtract shims at both bolts (front and rear). Adding one shim (.06 inch) at front and rear bolts will increase positive camber by ¼°.

TOE-IN, ADJUST

Toe-in can be adjusted by loosening the clamp bolts at each end of the tie rods and turning each tie rod to increase or decrease its length as necessary until proper toe-in is obtained with the steering gear on the high point for straight ahead driving and the steering wheel centered.

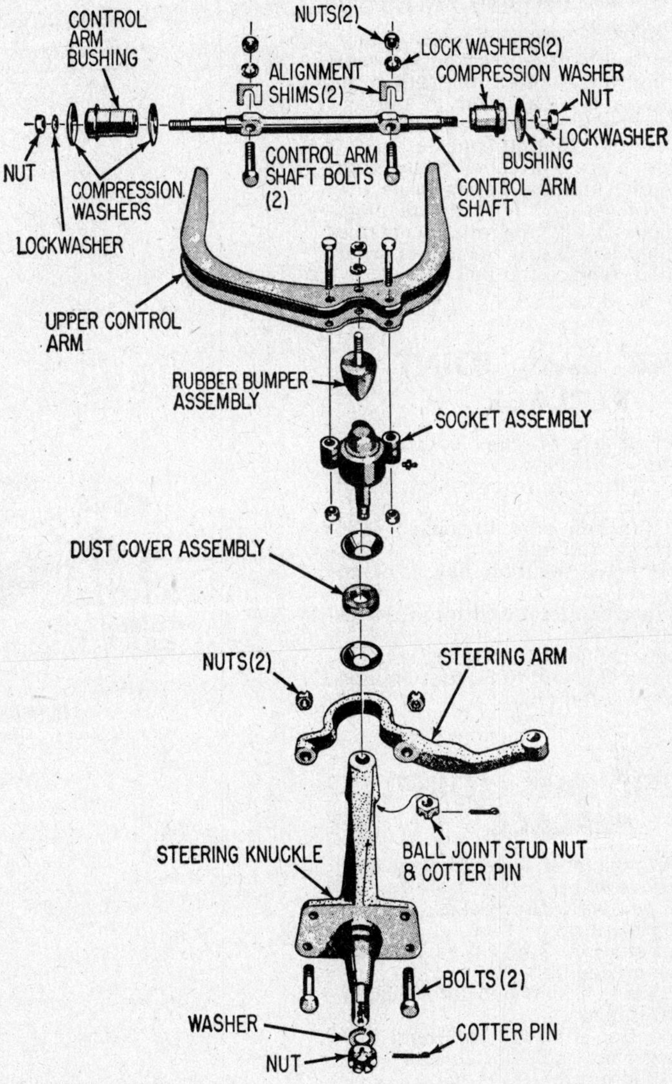

Fig. 1 Upper control arm & steering knuckle installation. 1969-76 All

WHEEL BEARINGS, ADJUST

Drum Brakes

1. While rotating wheel, tighten spindle nut until a slight bind is felt.
2. Back nut off to nearest cotter pin hole or enough so wheel rotates freely and insert cotter pin.
3. When bearings are properly adjusted, there should be .001-.010 inch end play present.

Disc Brakes

1. While rotating wheel, snug-up spindle nut.
2. Back nut off, then snug-up nut by hand.
3. Loosen nut to nearest cotter pin hole and insert cotter pin.
4. When bearings are properly adjusted, there should be .001-.005 inch end play present.

WHEEL BEARINGS, REPLACE

Disc Brakes

1. Raise front of vehicle and remove front wheels.
2. Remove brake caliper mounting bolts and caliper. Once caliper is removed, support caliper away from disc to avoid damage to hydraulic brake line.
3. Remove grease cup, spindle nut, thrust washer and hub and disc assembly. The grease retainer and inner wheel bearing can now be removed.

CHECKING BALL JOINTS FOR WEAR

Upper Ball Joint

The upper ball joint is checked for wear by measuring the amount of torque required to rotate the ball stud in the assembly. To make this type of check, it is necessary to remove the stud from the steering knuckle.

Install a nut on the ball stud and using a torque wrench, turn the stud. Acceptable torque for an upper ball joint is 28-40 inch lbs. If torque readings are excessively low or high, replace the ball joint.

Lower Ball Joint

The lower ball joint is checked for wear by measuring the amount of "pull" required to move the ball stud in the assembly. This check is made with the stud removed from the steering knuckle.

With a spring scale attached to the stud cotter pin hole, pull the spring scale until the stud moves and note the reading. The specified "pull" for an acceptable lower ball joint is 8 lbs. If reading is lower than specified, replace the ball joint.

UPPER BALL JOINT, REPLACE

1. Support vehicle at lower control arm and remove wheel.
2. Remove cotter pin from ball stud and loosen stud nut.
3. Lightly tap ball stud to loosen stud and remove stud nut.
4. Remove bolts securing ball joint to control arm.
5. Install new ball joint and torque bolts to 20-30 ft. lbs.
6. Assemble ball joint to steering knuckle and torque stud nut to 80 foot pounds and install cotter pin.

LOWER BALL JOINT, REPLACE

1. Support vehicle at lower control arm and remove wheel.
2. If equipped with disc brakes, remove caliper assembly.
3. Perform steps 2, 3 and 4 as outlined in above procedure.
4. Install new ball joint and torque bolts to 20-30 ft. lbs.
5. Raise control arm with jack until ball stud contacts spindle, install and torque stud nut to 100 ft. lbs.

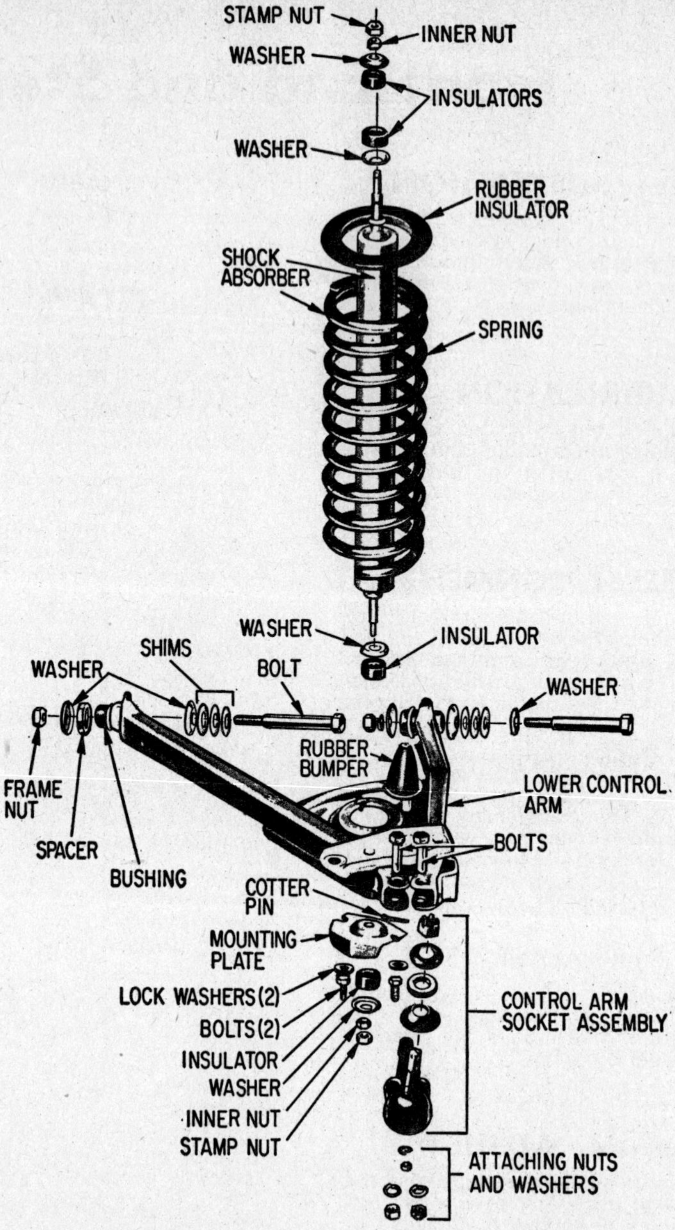

Fig. 2 Lower control arm, coil spring & shock absorber installation. 1969-76 All

SHOCK ABSORBER, REPLACE

1. Support vehicle at lower control arm.
2. Remove shock absorber upper mounting nuts.
3. Remove shock absorber mounting plate to lower control arm bolts. After shock absorber is removed, disconnect mounting plate from shock absorber.
4. Reverse procedure to install.

COIL SPRING, REPLACE

1. Support vehicle at frame and lower control arm with a suitable jack.
2. Disconnect stabilizer bar from lower control arm and loosen lower ball joint stud nut two turns. Lightly tap steering knuckle to loosen ball stud, then remove stud nut.
3. Lower jack slowly until tension is relieved from spring and remove spring and insulator.

NOTE: Install a safety chain through coil spring and lower control arm to prevent injury or damage if spring slips from mounting.

STEERING GEAR, REPLACE

1. Remove pitman arm nut and washer from sector shaft and mark position of arm on shaft.
2. With a suitable puller, remove pitman arm from sector shaft.
3. Remove steering shaft coupling bolts and steering gear to frame bolts. Remove steering gear.

INTEGRAL POWER STEERING

To remove gear assembly, disconnect pressure and return hoses, cap hoses and steering gear outlet ports, then follow procedure as outlined under "Steering Gear, Replace".

CAMARO • CHEVELLE • CHEVROLET CORVETTE • MONTE CARLO • NOVA

OLD CAR SPECIFICATIONS: For 1946-68 Tune Up and Wheel Alignment Specifications see main index.

INDEX OF SERVICE OPERATIONS

CHEVROLET—Exc. Chevette, Monza & Vega

SERIAL NUMBER LOCATION: Plate on left front door pillar or top of left side instrument panel

ENGINE NUMBER LOCATION

4 & 6 CYL.: Pad at front righthand side of cylinder block at rear of distributor

V8 ENGINES: Pad at front righthand side of cylinder block

ENGINE IDENTIFICATION CODE

Engines are identified in the following table by the code letter or letters immediately following the engine serial number.

CAMARO

CODE		
AM	6-230 with M/T	1969
AN	6-230 with P/G, T/D	1969
AO	6-230 with T/H	1969
AP	6-230 with A/C	1969
AQ	6-230 with P/G, T/D, A/C	1969
AR	6-230 with A/C, T/H	1969
BB	6-250 with P/G, T/D	1969
BC	6-250 with P/G, T/D, A/C	1969
BD	6-250 with T/H	1969
BE	6-250 with M/T	1969
BF	6-250 with A/C	1969
BH	6-250 with T/H, A/C	1969
CAA	6-250 with M/T	1971
CAB	6-250 with P/G	1971
CAY	8-307 with M/T, E.E.C.	1972
CAZ	8-307 with P/G, E.E.C.	1972
CBA	6-250 with M/T, E.E.C.	1972
CBG	6-250 with M/T	1972
CBJ	6-250 with P/G	1972
CCA	6-250 with T/H	1973
CCB	6-250 with E.E.C.	1973
CCC	6-250 with M/T	1973
CCD	6-250 with M/T & E.E.C.	1973
CCR	6-250 with M/T	1974
CCW	6-250 with T/H, E.E.C.	1974
CCX	6-250 with T/H	1974
CCA	8-307	1971
CCG	6-250 with M/T	1970
CCK	6-250 with T/H, taxi & police	1970
CCL	6-250 with M/T, taxi & police	1970
CCM	6-250 with P/G, taxi & police	1970
CCZ	6-250 with M/T	1970
CDM	6-250 with M/T, AIR	1972
CDL	6-250 with P/G, AIR	1972
CGB	8-350 with P/G, 250 H.P.	1971
CGB	8-350 with P/G	1972
CGK	8-350 with M/T, 300 H.P.	1971
CGL	8-350 with T/H, 300 H.P.	1971
CGP	8-350 with M/T, 360 H.P.	1971
CGR	8-350 with T/H, 360 H.P.	1971
CJD	8-350 with T/H, 270 H.P.	1971
CJF	8-396 with M/T, 350 H.P.	1970
CJG	8-350 with M/T, 270 H.P.	1971
CJH	8-396 with M/T, 375 H.P.	1970
CJI	8-396 with T/H, 350 H.P.	1970
CJL	8-396 with T/H, 375 H.P.	1970
CJL	6-250	1975
CJM	6-250	1975
CJR	6-250	1975
CJT	6-250	1975
CJU	6-250	1975
CJF	6-250	1975
CJY	6-250	1975
CJZ	6-250	1975
CKG	8-307 with M/T	1972
CKH	8-307 with P/G	1972

CODE		
CHB	8-307 with M/T	1973
CHH	8-307 with T/H	1973
CHJ	8-307 with M/T & E.E.C.	1973
CHK	8-307 with T/H & E.E.C.	1973
CKO	8-400 with M/T, 375 H.P.	1970
CKT	8-350 with T/H, 255 H.P.	1972
CKB	8-350 165 H.P.	1972
CKK	8-350 with M/T, 175, 200 H.P.	1972
CKD	8-350 with T/H, 175, 200 H.P.	1972
CKS	8-350 with M/T, 255 H.P.	1972
CLJ	8-350 with 4 sp. tr. & 245 H.P.	1973
CLK	8-350 with T.H. 400	1973
CKA	8-350 4 sp. tr. & 145 H.P.	1973
CKU	8-350 with T/H	1973
CKB	8-350 with M/T & 175 H.P.	1973
CKW	8-350 with T/H & 145 H.P.	1973
CLL	8-350 with T/H 400 & E.E.C.	1973
CLM	8-350 with M/T & E.E.C. & 245 H.P.	1973
CKX	8-350 with T/H, E.E.C. & 145 H.P.	1973
CKY	8-350 with M/T, E.E.C. & 145 H.P.	1973
CKD	8-350 with T/H, E.E.C. & 175 H.P.	1973
CKH	8-350 with M/T, E.E.C. & 175 H.P.	1973
CMA	8-307 with T/H, E.E.C.	1972
CMA	8-350 with T/H	1974
CMC	8-350 with M/T	1974
CKB	8-350 with M/T, 185 H.P.	1974
CKH	8-350 with 4 BC, M/T	1974
CKH	8-350 with M/T, E.E.C., 185 H.P.	1974
CLJ	8-350 with M/T, 245 H.P.	1974
CKU	8-350 with T/H, 185 H.P.	1974
CKD	8-350 with 4BC, T/H	1974
CKD	8-350 with T/H, E.E.C., 185 H.P.	1974
CLK	8-350 with 4BC, T/H	1974
CLD	8-400 with T/H, 350 H.P.	1971
CLC	8-400 with M/T, 350 H.P.	1971
CMU	8-350	1975
CRC	8-350	1975
CMF	8-350	1975
CMH	8-350	1975
CRX	8-350	1975
CHW	8-350	1975
CHS	8-350	1975
CHT	8-350	1975
CNC	8-307 with M/T	1970
CND	8-307 with 4 sp. tr.	1970
CNE	8-307 with P/G	1970
CNF	8-307 with T/H	1970
CNI	8-350 with M/T, 250 H.P.	1970
CNJ	8-350 with M/T, 300 H.P.	1970
CNK	8-350 with P/G, 300 H.P.	1970

CODE		
CNM	8-350 with P/G, 250 H.P.	1970
CNN	8-350 with T/H, 250 H.P.	1970
CRE	8-350 with T/H, 300 H.P.	1970
CRF	6-250 with M/T	1970
CRG	6-250 with M/T	1970
CSD	6-250 with P/G, E.E.C.	1972
CTB	8-350 with M/T, 320 H.P.	1970
CTC	8-350 with T/H, 320 H.P.	1970
CTK	8-307 with T/H	1972
CTW	8-400 with T/H, 350 H.P.	1970
CTX	8-400 with M/T, 350 H.P.	1970
CTY	8-400 with T/H, 375 H.P.	1970
DA	8-307 with M/T	1969
DC	8-307 with P/G	1969
DD	8-307 with T/H	1969
DE	8-307 with 4 sp. tr.	1969
DZ	8-302 with 4 BC	1969
FA	8-327 with M/T	1969
FB	8-327 with P/G	1969
FC	8-327 with T/H	1969
FH	8-327 with T/H	1969
HA	8-350 with M/T	1969
HB	8-350 with T/H	1969
HC	8-350 with 2 BC	1969
HD	8-350 with 2 BC, T/H	1969
HE	8-350 with P/G	1969
HF	8-350 with 2 BC, P/G	1969
HQ	8-350 with M/T	1969
HR	8-350 with P/G	1969
HS	8-350 with T/H	1969
JB	8-396 with P/G	1969
JF	8-396 with HPE	1969
JG	8-396 with T/H	1969
JH	8-396 with SHPE	1969
JI	8-396 with HPE, T/H	1969
JJ	8-396 with M/T, A/H	1969
JL	8-396 with SHPE, T/H	1969
JM	8-396 with A/H, T/H	1969
JU	8-396 with M/T	1969
KA	8-396 with SHPE, M/T	1969
KC	8-396 with SHPE, M/T	1969
KE	8-396 with A/H, M/T	1969
CLA	8-400 with M/T	1972
CLB	8-400 with T/H	1972
CCC	6-250	1976
CCD	6-250	1976
CCF	6-250	1976
CHT	8-350	1976
CHU	8-350	1976
CML	8-350	1976
CPA	8-305	1976
CPB	8-305	1976
CPJ	8-305	1976
9M	8-350	1976
9W	8-305	1976

Continued

ENGINE IDENTIFICATION CODE—Continued

CHEVROLET

CODE		
BA	6-250 with M/T	1969
BG	6-250 with M/T, A/C	1969
CAA	6-250 with M/T	1971
CAB	6-250 with P/G, T/D	1971
CAC	6-250 with M/T, taxi & police	1971
CAD	6-250 with P/G, taxi & police	1971
CAT	8-400 with T/H	1972
CBH	6-250 with M/T, taxi & police	1972
CBJ	6-250 with P/G	1972
CBK	6-250 with P/G, taxi & police	1972
CNJ	6-250 with M/T	1972
CCL	6-250 with M/T	1973
CCM	6-250 with M/T, E.E.C.	1973
CCM	6 cyl. with P/G, taxi & police	1970
CCG	6 cyl. with M/T	1970
CCH	6 cyl. with M/T	1970
CCK	6 cyl. T/H, taxi & police	1970
CCZ	6 cyl. with M/T	1970
CCL	6 cyl. with M/T, taxi & police	1970
CAR	8-350 with T/H, taxi & police	1972
CDB	8-350 with T/H, E.E.C.	1972
CDL	6-250 with P/G, AIR	1972
CDL	8-400 with T/H, Police	1972
CDM	8-400 with T/H, Police, E.E.C.	1972
CKB	8-350 with T/H	1972
CSH	8-350 with T/H, taxi & police	1972
CSJ	8-350 with M/T	1972
CJH	8-350 with M/T, Police, 270 H.P.	1971
CKL	8-350 with T/H & 145 H.P.	1973
CKS	8-350 with T/H, 145 H.P., taxi & police	1973
CKK	8-350 with T/H, 145 H.P., E.E.C.	1973
CLT	8-350 with T/H, 145 H.P., E.E.C.	1973
CLU	8-350 with T/H & 145 H.P.	1973
CLW	8-350 with T/H, 145 H.P., E.E.C., police	1973
CLX	8-350 with T/H, 145 H.P., police	1973
CKR	8-350 with T/H, E.E.C., 175 H.P., police	1973
CKD	8-350 with T/H, E.E.C. & 175 H.P.	1973
CKJ	8-350 with T/H & 175 H.P.	1973
CLR	8-400 with M/T, Police, 300 H.P.	1971
CMK	8-350 with 4BC, E.E.C., taxi & police	1974
CMA	8-350 with E.E.C., taxi & police	1974
CMD	8-350 with taxi & police	1974
CMH	8-350 with 4BC, E.E.C., police	1974
CMJ	8-350 with 4BC, police	1974
CMJ	8-350	1975
CKD	8-350 with 4BC, E.E.C.	1974
CGA	8-350, with M/T, 245 H.P.	1971
CGB	8-350, with P/G, 245 H.P.	1971
CGC	8-350 with M/T, 245 H.P.	1971
CGJ	8-350 with T/H, taxi & police, 245 H.P.	1971

CODE		
CJB	8-350 with M/T, police, 275 H.P.	1971
CKP	8-400 with T/H	1972
CLB	8-400 with M/T	1972
CLR	8-400 with M/T, police	1972
CRU	8-350	1975
CRW	8-454 T/H, AIR	1972
CRW	8-350	1975
CRY	8-454 with T/H, AIR, Police	1972
CRY	8-350	1975
CSA	8-400	1975
CSA	8-400 with T/H	1973
CSB	8-400 with T/H, police	1973
CSC	8-400 with T/H, E.E.C., police	1973
CSD	8-400 with T/H & E.E.C.	1973
CSK	8-400 with T/H	1973
CSL	8-400 with T/H, E.E.C., police	1973
CSM	8-400 with T/H & police	1973
CTC	8-400 with 4BC, E.E.C., 180 H.P.	1974
CTD	8-400 with 4BC, 180 H.P.	1974
CTK	8-400 with 4BC, E.E.C., 180 H.P., police	1974
CTJ	8-400 with 4BC, 180 H.P., police	1974
CTA	8-400 with 150 H.P.	1974
CTB	8-400 with 150 H.P., police	1974
CLK	8-400 with T/H, 255 H.P.	1971
CLP	8-400 with T/H, 300 H.P.	1971
CPD	8-454 with M/T, 365 H.P.	1971
CPG	8-454 with M/T, police, 365 H.P.	1971
CPD	8-454 with T/H	1972
CPG	8-454 with T/H, police	1972
CTL	8-400	1975
CTM	8-400	1975
CTU	8-400	1975
CTW	8-400	1975
CTY	8-400	1975
CTZ	8-400	1975
CWD	8-454 with T/H, E.E.C.	1973
CWJ	8-454 with T/H, E.E.C., police	1973
CWK	8-454 with T/H, police	1973
CWL	8-454 with T/H	1973
CWU	8-454 with police	1974
CWW	8-454 with E.E.C., police	1974
CWY	8-454 with E.E.C.	1974
CXA	8-454	1974
CRF	6 cyl. with M/T	1970
CRG	6 cyl. with M/T	1970
CND	8-350 with M/T, 250 H.P.	1970
CNP	8-350 with M/T, taxi & police, 250 H.P.	1970
CNQ	8-350 with M/T, 300 H.P.	1970
CNR	8-350 with T/H, 300 H.P.	1970
CNS	8-350 with P/G, police, 300 H.P.	1970
CNT	8-350 with T/H, police, 300 H.P.	1970
CNU	8-350 with P/G, 250 H.P.	1970
CNV	8-350 with T/H, 250 H.P.	1970

CODE		
CNW	8-350 P/G, taxi & police, 250 H.P.	1970
CNX	8-350 T/H, taxi & police, 250 H.P.	1970
CGR	8-400 with M/T, 265 H.P.	1970
CGV	8-454 with M/T, 345 H.P.	1970
CGS	8-454 M/T, police, 345 H.P.	1970
CGT	8-454 with M/T, police, 390 H.P.	1970
CGU	8-454 with M/T, 390 H.P.	1970
CXX	8-454	1975
CXY	8-454	1975
FA	8-327 with M/T	1969
FB	8-327 with P/G	1969
FC	8-327 with T/H	1969
FH	8-327 with T/H	1969
FJ	8-327 with M/T	1969
FK	8-327 with P/G	1969
FL	8-327 with T/H	1969
GE	8-327 with P/G	1969
HD	8-350 with 2 bar. carb., M/T, T/M	1969
HF	8-350 with 2 bar. carb., P/G	1969
HG	8-350 with M/T	1969
HH	8-350 with T/H	1969
HI	8-350 with 2 BC	1969
HJ	8-350 with 2 BC, T/H	1969
HK	8-350 with P/G	1969
HL	8-350 with 2 BC, P/G	1969
HM	8-350 with 2 BC, T/H	1969
HN	8-350 with T/H	1969
HP	8-350 with M/T	1969
HT	8-350 with M/T	1969
HU	8-350 with P/G	1969
HY	8-350 with T/H	1969
IA	8-350 with T/H	1969
JN	8-396 with 2 BC, M/T	1969
JQ	8-396 with 2 BC, T/H	1969
JT	8-396 with M/T	1969
LA	8-427 with HPE	1969
LB	8-427 with 4 BC	1969
LC	8-427 with HPE, T/H	1969
LD	8-427 with SHPE, M/T	1969
LE	8-427 with 4 BC, T/H	1969
LH	8-427 with HPE	1969
LI	8-427 with T/H	1969
LS	8-427 with SHPE, T/H	1969
MA	8-427 with M/T	1969
MC	8-427 with HPE, M/T	1969
MD	8-427 with SHPE, M/T	1969
CHS	8-350	1976
CMJ	8-350	1976
CMM	8-350	1976
CSF	8-400	1976
CSX	8-400	1976
CTL	8-400	1976
CXX	8-454	1976

CHEVELLE & MONTE CARLO

CODE		
AD	6-230 with T/H	1969
AM	6-230 with M/T	1969
AN	6-230 with P/G, T/D	1969
AP	6-230 with A/C	1969
AQ	6-230 with P/G, T/D, A/C	1969
AR	6-230 with T/H, A/C	1969
BB	6-250 with P/G, T/D	1969

CODE		
BC	6-250 with P/G, T/D, A/C	1969
BD	6-250 with T/H	1969
BE	6-250 with M/T	1969
BF	6-250 with A/C	1969
BH	6-250 with T/H, A/C	1969
CAA	6-250 with M/T	1971
CBA	6-250 with M/T, E.E.C.	1972

CODE		
CBD	6-250 with M/T, Taxi, E.E.C.	1972
CBG	6-250 with M/T	1972
CBJ	6-250 with P/G	1972
CBK	6-250 with P/G, Taxi	1972
CSD	6-250 with P/G, E.E.C.	1972
CCA	6-250 with T/H	1973
CCB	6-250 with T/H, E.E.C.	1973

Continued

ENGINE IDENTIFICATION CODE—Continued

CHEVELLE & MONTE CARLO—Continued

CODE		
CCC	6-250 with M/T	1973
CCD	6-250 with M/T, E.E.C.	1973
CCR	6-250 with M/T	1974
CCW	6-250 with T/H, E.E.C.	1974
CCX	6-250 with T/H	1974
CCK	6-250	1974
CCA	8-307 with M/T	1971
CCM	6-250 P/G, police & taxi	1970
CCK	6-250 T/H, police & taxi	1970
CCL	6-250 M/T, police & taxi	1970
CAR	8-350 with T/H, Police, E.E.C.	1972
CAY	8-307 with M/T, E.E.C.	1972
CAZ	8-307 with P/G, E.E.C.	1972
CDA	8-350 with M/T	1972
CDB	8-350 with P/G	1972
CDD	8-350 with T/H	1972
CDG	8-350 with M/T	1972
CJD	8-350 with T/H, 270 H.P.	1971
CJJ	8-350 with M/T, 270 H.P.	1971
CJL	6-250	1975
CLM	6-250	1975
CJR	6-250	1975
CJT	6-250	1975
CJU	6-250	1975
CJF	6-250	1975
CJZ	6-250	1975
CKA	8-350 with M/T	1972
CKA	8-350 with M/T, 145 H.P.	1973
CKL	8-350 with T/H, 145 H.P.	1973
CKB	8-350 with M/T, 145 H.P.	1973
CKB	8-350 with P/G	1972
CKC	8-350 with M/T, E.E.C., 145 H.P.	1973
CKD	8-350 with T/H	1972
CKK	8-350 with T/H, E.E.C., 145 H.P.	1973
CKH	8-350 with M/T, E.E.C., 175 H.P.	1973
CKD	8-350 with T/H, E.E.C., 175 H.P.	1973
CKJ	8-350 with T/H, 175 H.P.	1973
CKK	8-350 with M/T	1972
CLR	8-400 with M/T, Police	1971
CLS	8-400 with M/T	1971
CMC	8-350 with M/T	1974
CMA	8-350 with T/H	1974
CKH	8-350 with 4BC, M/T	1974
CKD	8-350 with 4BC, T/H	1974
CGA	8-350 with M/T, 245 H.P.	1971
CGB	8-350 with P/G, 245 H.P.	1971
CGK	8-350 with M/T, 270 H.P.	1971
CGL	8-350 with T/H, 270 H.P.	1971
CKG	8-307 with M/T	1972
CKH	8-307 with P/G	1972
CHA	8-307 with T/H	1973
CHB	8-307 with M/T	1973
CHC	8-307 with T/H, E.E.C.	1973
CLA	8-400 with M/T, 330 H.P.	1971
CLB	8-400 with T/H, 300 H.P.	1971
CLJ	8-400 with M/T, 255 H.P.	1971
CLK	8-400 with T/H, 255 H.P.	1971
CLL	8-400 with M/T, 300 H.P.	1971
CLP	8-400 with T/H, 300 H.P.	1971

CODE		
CLA	8-400 with M/T	1972
CLB	8-400 with T/H	1972
CLJ	8-400 with M/T	1972
CLK	8-400 with T/H	1972
CLL	8-400 with 4 sp. tr.	1972
CLR	8-400 with M/T, police	1972
CLS	8-400 with M/T	1972
CMA	8-307 with T/H, E.E.C.	1972
CMD	8-350 with T/H, E.E.C.	1972
CMF	8-350	1975
CMH	8-350	1975
CMJ	8-350	1975
CMU	8-350	1975
CTC	8-400 with T/H, E.E.C., 180 H.P.	1974
CTA	8-400 with T/H, 150 H.P.	1974
CNC	8-307 with M/T	1970
CND	8-307 with 4 sp. tr.	1970
CNE	8-307 with P/G	1970
CNF	8-307 with T/H	1970
CNI	8-350 with M/T, 250 H.P.	1970
CNJ	8-350 with M/T, 300 H.P.	1970
CNK	8-350 with P/G, 300 H.P.	1970
CPA	8-454 with M/T, 365 H.P.	1971
CPG	8-454 with M/T, 365 H.P.	1971
CPP	8-454 with M/T, 460 H.P.	1971
CPR	8-454 with T/H, 460 H.P.	1971
CPA	8-454 with M/T	1972
CPD	8-454 with T/H	1972
CWC	8-454 with M/T, E.E.C.	1973
CWD	8-454 with T/H, E.E.C.	1973
CWA	8-454 with M/T	1973
CWB	8-454 with T/H	1973
CWA	8-454 with M/T	1974
CWX	8-454 with T/H	1974
CWD	8-454 with T/H, E.E.C.	1974
CRE	8-350 with T/H, 300 H.P.	1970
CNM	8-350 with P/G, 250 H.P.	1970
CTW	8-396 with T/H, 350 H.P.	1970
CTX	8-396 with M/T, 350 H.P.	1970
CTY	8-396 with T/H, 375 H.P.	1970
CTZ	8-396 with HDC, 350 H.P.	1970
CKN	8-400 with T/H, 325 H.P.	1970
CKD	8-396 with M/T, 375 H.P.	1970
CKP	8-396, with T/H, A/H, 375 H.P.	1970
CKQ	8-396 with HDC, 375 H.P.	1970
CKR	8-400 with M/T, 4BC, 330 H.P.	1970
CKS	8-400 with HDC, 4BC, 330 H.P.	1970
CKT	8-396 with M/T, 375 H.P.	1970
CKU	8-396 with HDC, 375 H.P.	1970
CZX	8-400 with M/T, 265 H.P.	1970
CRH	8-400 with T/H, 265 H.P.	1970
CRN	8-454 with M/T, 390 H.P.	1970
CRQ	8-454 with T/H, 390 H.P.	1970
CRR	8-454 with T/H, 450 H.P.	1970
CRS	8-454 T/H alum. heads, 450 H.P.	1970
CRT	8-454 with M/T, 390 H.P.	1970
CRT	8-350	1975
CRU	8-350	1975

CODE		
CRU	8-454 with HDC, 390 H.P.	1970
CRV	8-454 with M/T, 450 H.P.	1970
CRW	8-454 M/T, alum. heads, 450 H.P.	1970
CRW	8-454 with T/H, AIR	1972
CRX	8-350	1975
CRX	8-454 with T/H, AIR	1972
CRX	8-454 with HDC, 450 H.P.	1970
CRY	8-454 HDC, alum. heads, 450 H.P.	1970
CSM	8-400	1975
CSH	8-350 with T/H, Police	1972
CTA	8-400 M/T, AIR, Taxi & Police	1972
CTB	8-400 T/H, AIR	1972
CTH	8-400 HDC, AIR	1972
CTJ	8-400 T/H, AIR, Taxi & Police	1972
CTK	8-307 with T/H	1972
CTL	8-350 with T/H	1972
CTL	8-400	1975
CTU	8-400	1975
CTX	8-400	1975
CXW	8-454	1975
DA	8-307 with M/T	1969
DC	8-307 with P/G	1969
DD	8-307 with T/H	1969
DE	8-307 with 4 sp. tr.	1969
HA	8-350 with M/T	1969
HB	8-350 with T/H	1969
HC	8-350 with 2 BC	1969
HD	8-350 with 2 BC, T/H	1969
HE	8-350 with P/G	1969
HF	8-350 with 2 BC, P/G	1969
HP	8-350 with M/T	1969
HR	8-350 with M/T, P/G	1969
HS	8-350 with T/H	1969
JA	8-396 with M/T	1969
JC	8-396 with HPE	1969
JD	8-396 with SHPE	1969
JE	8-396 with HPE, T/H	1969
JK	8-396 with T/H	1969
JV	8-396 with M/T	1969
KB	8-396 with HPE, M/T	1969
KD	8-396 with SHPE, M/T	1969
KF	8-396 with SHPE, T/H	1969
KG	8-396 with M/T	1969
KH	8-396 with T/H	1969
KI	8-396 with M/T	1969
CCC	6-250	1976
CCD	6-250	1976
CCF	6-250	1976
CPB	8-305	1976
CMH	8-350	1976
CMJ	8-350	1976
CMM	8-350	1976
CSB	8-400	1976
CSF	8-400	1976
CSX	8-400	1976
9R	8-350	1976
9W	8-305	1976

CHEVY NOVA

CODE		
AA	4-153 with M/T	1969
AB	4-153 with T/D, P/G	1969
AM	6-230 with M/T	1969
AN	6-230 with P/G, T/D	1969
AO	6-230 with T/H	1969
AP	6-230 with A/C	1969

CODE		
AR	6-230 with T/H, A/C	1969
AQ	6-230 with T/D, P/G, A/C	1969
BB	6-250 with P/G, T/D	1969
BC	6-250 with P/G, T/D, A/C	1969
BD	6-250 with T/H	1969
BE	6-250 with M/T	1969

CODE		
BF	6-250 with A/C	1969
BH	6-250 with T/H, A/C	1969
CAA	6-250 with M/T	1971
CAB	6-250 with P/G, T/D	1971
CAL	6-250 with M/T, Taxi	1972

Continued

ENGINE IDENTIFICATION CODE—Continued

CHEVY NOVA—Continued

CODE		
CAY	8-307 with T/H	1972
CAZ	8-307 with M/T, E.E.C.	1972
CBA	6-250 with M/T, E.E.C.	1972
CBG	6-250 with M/T	1972
CBJ	6-250 with P/G, T/D	1972
CBK	6-250 with P/G, Taxi	1972
CBL	6-250 with P/G, AIR	1972
CBM	6-250 with M/T, AIR	1972
CCA	6-250 with P/G	1973
CCB	6-250 with P/G, E.E.C.	1973
CCC	6-250 with M/T	1973
CCD	6-250 with M/T, E.E.C.	1973
CCR	6-250 with M/T	1974
CCW	6-250 with T/H, E.E.C.	1974
CCX	6-250 with T/H	1974
CCK	6-250	1974
CCA	8-307 with M/T	1971
CCA	4-153 with M/T	1970
CCB	4-153 with T/D	1970
CCC	8-307 with P/G	1971
CCD	6-230 with T/D	1970
CCM	6-250 with P/G, police & taxi	1970
CCG	6-250 with M/T	1970
CCK	6-250 with T/H, police & taxi	1970
CCL	6-250 with M/T, police & taxi	1970
CCS	6-250	1975
CCT	6-250	1975
CCU	6-250	1975
CCW	6-250	1975
CCZ	6-250 with M/T	1970
CDA	8-350 with M/T, E.E.C.	1972
CDD	8-350 with T/H, E.E.C.	1972
CDG	8-350 with M/T, E.E.C.	1972
CGB	8-350 with P/G	1972
CGC	8-262	1975
CGD	8-262	1975
CGF	8-262	1975
CGH	8-262	1975
CHW	8-350	1975
CJF	6-250	1975
CJL	6-250	1975
CJM	6-250	1975
CJR	6-250	1975
CJS	6-250	1975
CJT	6-250	1975
CJU	6-250	1975
CJW	6-250	1975
CJX	6-250	1975
CJZ	6-250	1975
CKA	8-350 with M/T, 165 H.P.	1972
CKB	8-350 with T/H, 165 H.P.	1972
CKD	8-350 with T/H	1972
CKK	8-350	1972

CODE		
CKA	8-350 with M/T, 145 H.P.	1973
CKU	8-350 with T/H, 175 H.P.	1973
CKB	8-350 with M/T, 175 H.P.	1973
CKW	8-350 with T/H, 145 H.P.	1973
CKC	8-350 with M/T, E.E.C., 145 H.P.	1973
CKK	8-350 with T/H, E.E.C., 145 H.P.	1973
CKD	8-350 with T/H, E.E.C., 175 H.P.	1973
CKH	8-350 with M/T, E.E.C., 175 H.P.	1973
CMC	8-350 with M/T	1974
CMA	8-350 with T/H	1974
CKB	8-350 with M/T, 185 H.P.	1974
CKH	8-350 with 4BC, M/T	1974
CKH	8-350 with M/T, E.E.C., 185 H.P.	1974
CKU	8-350 with T/H, 185 H.P.	1974
CKD	8-350 with 4 BC, T/H, E.E.C.	1974
CGB	8-350 with P/G, 250 H.P.	1971
CGK	8-350 with M/T, 300 H.P.	1971
CGL	8-350 with T/H, 300 H.P.	1971
CKG	8-307 with M/T	1972
CKH	8-307 with P/G	1972
CHB	8-307 with M/T	1973
CHH	8-307 with T/H	1973
CHC	8-307 with T/H, E.E.C.	1973
CHD	8-307 with M/T, E.E.C.	1973
CKO	8-400 with M/T, 375 H.P.	1970
CKP	8-400 T/H, alum. heads, 375 H.P.	1970
CKQ	8-400 with HDC, 375 H.P.	1970
CKR	8-400 with M/T, 330 H.P.	1970
CKS	8-400 with HDC, 330 H.P.	1970
CKT	8-400 M/T, alum. heads, 375 H.P.	1970
CKU	8-400 HDC, alum. heads, 375 H.P.	1970
CMA	8-307 with P/G, E.E.C.	1972
CMD	8-350 with T/H, AIR	1972
CMF	8-350	1975
CMH	8-350	1975
CMU	8-350	1975
CNC	8-307 with M/T	1970
CND	8-307 with 4 sp. tr.	1970
CNE	8-307 with P/G	1970
CNF	8-307 with T/H	1970
CNI	8-350 with M/T, 250 H.P.	1970
CNJ	8-350 with M/T, 300 H.P.	1970
CNK	8-350 with P/G, 300 H.P.	1970
CNM	8-350 with P/G, 250 H.P.	1970
CNN	8-350 with T/H, 250 H.P.	1970
CPT	8-454 with T/H	1971
CPS	8-454 with M/T	1971
CRC	8-350	1975
CRE	8-350 with T/H, 300 H.P.	1970
CRF	6-250 with M/T	1970
CRK	8-350 with T/H, AIR	1972
CRL	8-350 with M/T, AIR	1972

CODE		
CRX	8-350	1975
CTB	8-350 with M/T, 320 H.P.	1970
CTC	8-350 with T/H, 320 H.P.	1970
CTK	8-307 with T/H	1972
CTL	8-350 with T/H	1972
CTW	8-400 with T/H, 350 H.P.	1970
CTX	8-400 with M/T, 350 H.P.	1970
CTY	8-400 with T/H, 375 H.P.	1970
CTZ	8-400 with HDC, 350 H.P.	1970
CZF	8-262	1975
CZH	8-262	1975
CZJ	8-262	1975
CZK	8-262	1975
CZL	8-262	1975
CZM	8-262	1975
CZY	8-262	1975
CZZ	8-262	1975
DA	8-307 with M/T	1969
DC	8-307 with P/G	1969
DD	8-307 with T/H	1969
DE	8-307 with 4 sp. tr.	1969
HA	8-350 with M/T	1969
HB	8-350 with T/H	1969
HC	8-350 with 2 BC	1969
HD	8-350 with 2 BC, T/H	1969
HE	8-350 with P/G	1969
HF	8-350 with 2 BC, P/G	1969
HQ	8-350 with M/T	1969
HR	8-350 with P/G	1969
HS	8-350 with T/H	1969
JF	8-396 with HPE	1969
JH	8-396 with SHPE	1969
JI	8-396 with HPE, T/H	1969
JL	8-396 with SHPE, T/H	1969
JM	8-396 with T/H	1969
JU	8-396 with P/G	1969
KA	8-396 with SHPE	1969
KC	8-396 with SHPE	1969
KE	8-396 with M/T	1969
CCB	6-250	1976
CCC	6-250	1976
CCD	6-250	1976
CCF	6-250	1976
CHT	8-350	1976
CHU	8-350	1976
CML	8-350	1976
CPA	8-305	1976
CPB	8-305	1976
CPJ	8-305	1976
9M	8-350	1976
9W	8-305	1976

CORVETTE

CODE		
CHA	8-350	1975
CHB	8-350	1975
CHC	8-350	1975
CHR	8-350	1975
CHU	8-350	1975
CHZ	8-350	1975
CKZ	8-350 with M/T	1974
CLB	8-350 with M/T, E.E.C.	1974
CLR	8-350 with M/T, 245 H.P.	1974
CLA	8-350 with T/H	1974
CLC	8-350 with T/H, E.E.C.	1974
CLD	8-350 with T/H, 245 H.P.	1974
CKZ	8-350 with M/T, 175 H.P.	1973

CODE		
CLA	8-350 with T/H	1973
CLR	8-350 with M/T, 250 H.P.	1973
CLB	8-350 with M/T, E.E.C., 175 H.P.	1973
CLC	8-350 with T/H, E.E.C., 175 H.P.	1973
CLS	8-350 with M/T, E.E.C., 250 H.P.	1973
CLD	8-350 with T/H, 250 H.P.	1973
CLH	8-350 with T/H, E.E.C., 250 H.P.	1973
CWM	8-454 with M/T	1974
CWR	8-454 with T/H	1974
CWS	8-454 with T/H, E.E.C.	1974
CWS	8-454 with T/H, E.E.C.	1973
CWT	8-454 with M/T, E.E.C.	1973
CWM	8-454 with M/T	1973

CODE		
CWR	8-454 with T/H	1973
CKW	8-350 with 4 sp. trs., 200 H.P.	1972
CDH	8-350 with 4 sp. trs., E.E.C., 200 H.P.	1972
CKX	8-350 with T/H 400, 200 H.P.	1972
CDJ	8-350 with T/H 400, E.E.C., 200 H.P.	1972
CKY	8-350 with 4 sp. trs., 255 H.P.	1972
CKZ	8-350 with H.D. 4 sp. trs., 255 H.P.	1972
CRT	8-350 with 4 sp. trs., A.I.R., 255 H.P.	1972
CRS	8-350 with T/H 400, A.I.R., 255 H.P.	1972

Continued

ENGINE IDENTIFICATION CODE—Continued

CORVETTE—Continued

CODE		
CPH	8-454 with 4 sp. trs.	1972
CPJ	8-454 with T/H-400	1972
CSR	8-454 with A.I.R.	1972
CSS	8-454 with A.I.R.	1972
CGS	8-350 with M/T, 300 H.P.	1971
CGT	8-350 with T/H, 270 H.P.	1971
CGW	8-454 with 4 BC HPE, T/H	1970
CGW	8-350 with M/T, 350 H.P.	1971
CGX	8-350 with M/T, 350 H.P.	1971
CGY	8-350 with M/T, 330 H.P.	1971
CGZ	8-350 with M/T, 330 H.P.	1971
CPH	8-454 with M/T, 365 H.P.	1971
CPJ	8-454 with T/H, 365 H.P.	1971
CPW	8-454 with M/T, 425 H.P.	1971
CPX	8-454 with T/H, 425 H.P.	1971
CPK	8-454 with M/T, 460 H.P.	1970
CPL	8-454 with T/H, 460 H.P.	1971
CRI	8-454 with HPE, T/I	1970

CODE		
CTL	8-350 with M/T	1970
CTM	8-350 with T/H	1970
CTN	8-350 with HPE	1970
CTO	8-350 with HPE, A/C	1970
CTP	8-350 with HPE, T/I	1970
CTQ	8-350 with HPE, T/I, A/C	1970
CTR	8-350 with SHPE	1970
CTU	8-350 with SHPE, T/I	1970
CTV	8-350 with SHPE, T/I, 4 sp. tr.	1970
CZL	8-454 with H/D	1970
CZN	8-454 with H/D, T/H	1970
CZU	8-454 with HPE	1970
HW	8-350 with HPE	1969
HX	8-350 with HPE, A/C	1969
HY	8-350 with M/T	1969
HZ	8-350 with T/H	1969
LL	8-427 with HPE, T/H	1969

CODE		
LM	8-427 with HPE	1969
LN	8-427 with HPE, T/H, 3 Carbs.	1969
LO	8-427 with HD	1969
LP	8-427 with A/H	1969
LQ	8-427 with HPE, 3 Carbs.	1969
LR	8-427 with SHPE, 3 Carbs.	1969
LT	8-427 with SHPE, 3 Carbs., HDC	1969
LU	8-427 with A/H, HDC	1969
LV	8-427 with T/H	1969
LW	8-427 with A/H, T/H	1969
LX	8-427 with SHPE, 3 Carbs, T/H	1969
CLS	8-350	1976
CHC	8-350	1976
CKC	8-350	1976
CKW	8-350	1976
CKX	8-350	1976

NOTES

A/C: Air Conditioned
A/H: Aluminum Heads
A/S: Air suspension
AIR: Air injection reactor
A/T: Automatic transmission
4BC: Four barrel carburetor
E.E.C.: Exhaust emission control.
F/I: Fuel injection
H/D: Heavy Duty
HDC: Heavy duty clutch
HPE: High performance engine
H/L: Hydraulic lifters
M/T: Manual transmission
O/D: Overdrive
P/G: Powerglide
P/S: Power steering
P/V: Positive crankcase ventilation
SHPE: Special high perf. engine
S.S.: Super Sport
T/D: Torque Drive
T/H: Turbo Hydramatic
T/I: Transistor ignition

GRILLE IDENTIFICATION

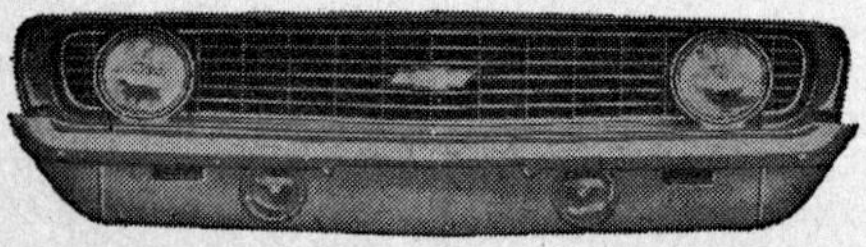

1969 Camaro

1969 Camaro S. S.

1969 Chevy Nova

1969 Chevelle

1969 Chevelle S. S.

1969 Chevrolet

1969 Corvette

1970 Chevrolet

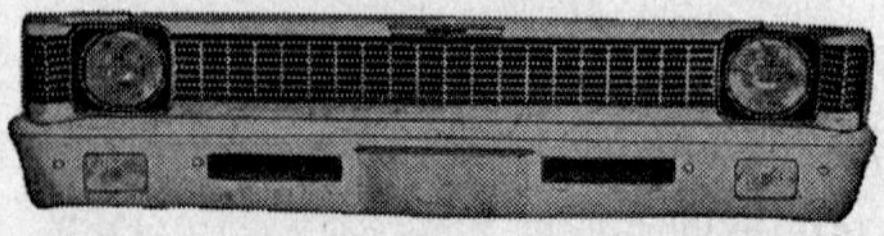

1970-72 Chevy Nova

1970-71 Camaro

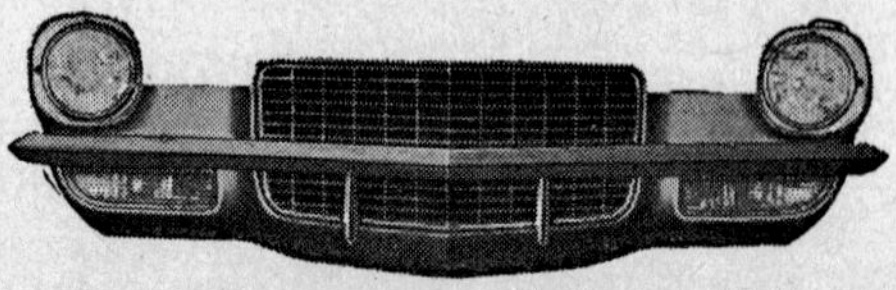

1970-71 Camaro S. S.

1970 Monte Carlo

GRILLE IDENTIFICATION—Continued

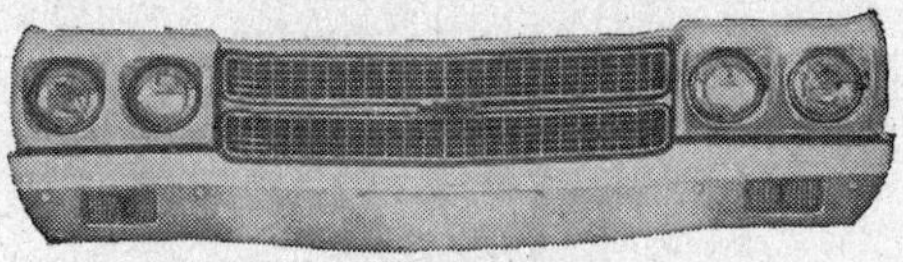
1970 Chevelle

1970 Chevelle S. S.

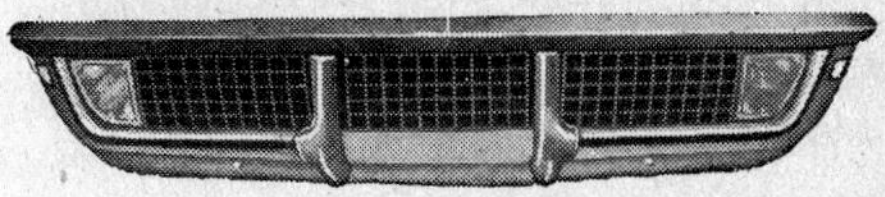
1970-72 Corvette

1971 Chevelle

1971 Chevelle S. S.

1971 Monte Carlo

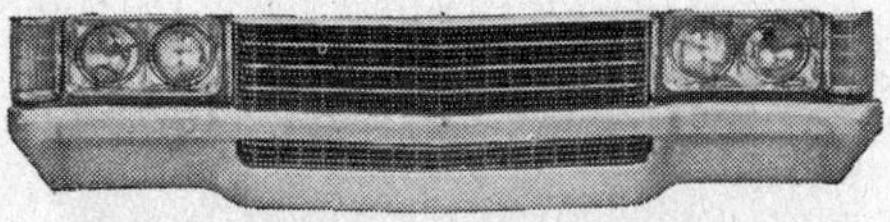
1971 Bel Air & Impala

1971 Caprice & Estate Wagon

1972 Camaro

1972 Chevelle

1972 Monte Carlo

1972 Bel Air & Impala

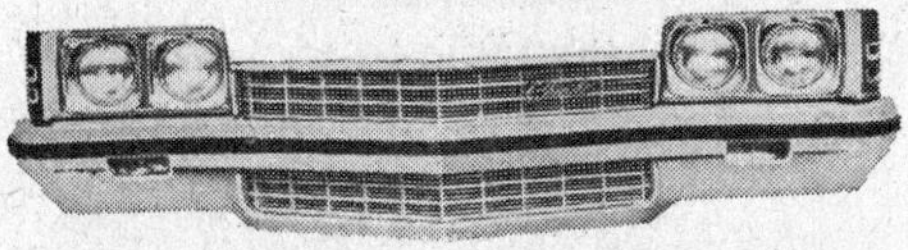
1972 Caprice & Estate Wagon

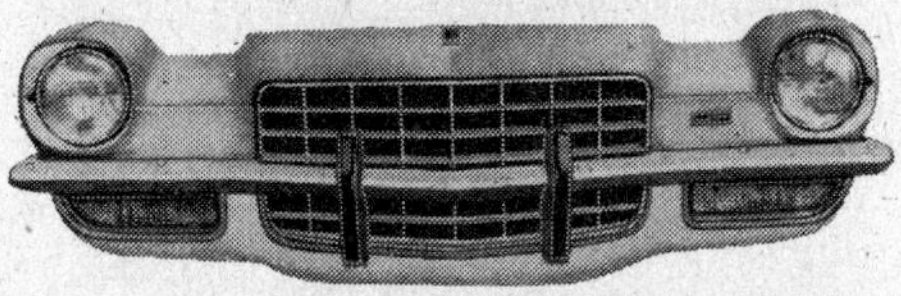
1973 Camaro

1973 Chevy Nova

1973 Chevelle Malibu

1973 Chevelle Laguna

1973 Monte Carlo

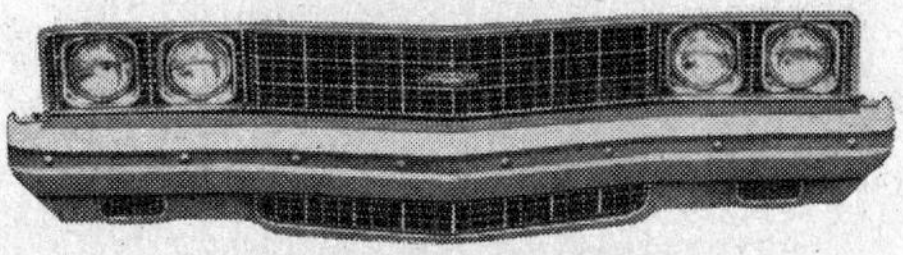
1973 Bel Air & Impala

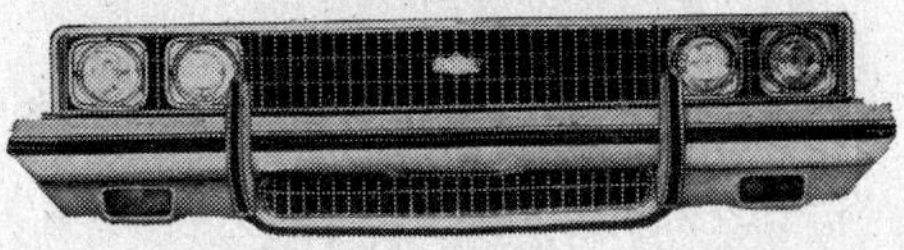
1973 Caprice & Estate Wagon

1973-74 Corvette

1974 Camaro

1974 Chevelle Malibu

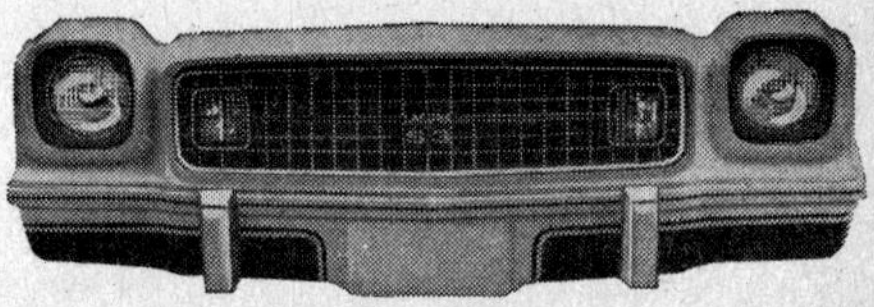
1974 Chevelle Laguna

GRILLE IDENTIFICATION—Continued

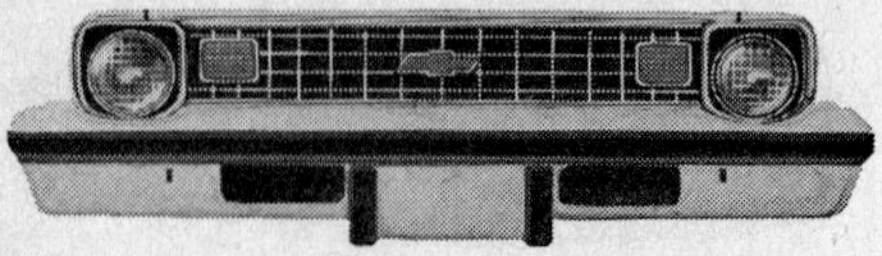
1974 Chevy Nova

1974 Monte Carlo

1974 Bel Air & Impala

1974 Caprice & Estate Wagon

1975 Camaro

1975 Chevelle

1975 Nova

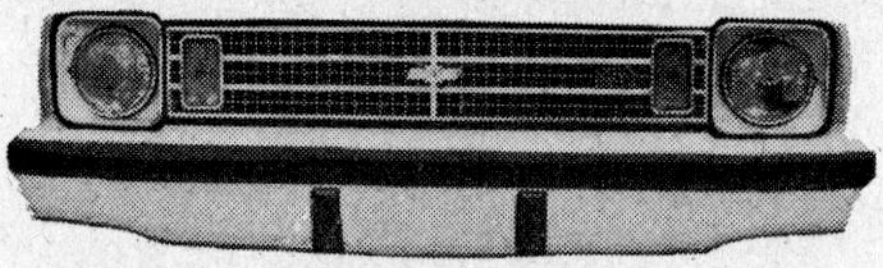
1975 Nova Custom

1975 Monte Carlo

1975 Bel Air & Impala

1975 Caprice & Estate Wagon

1975-76 Corvette

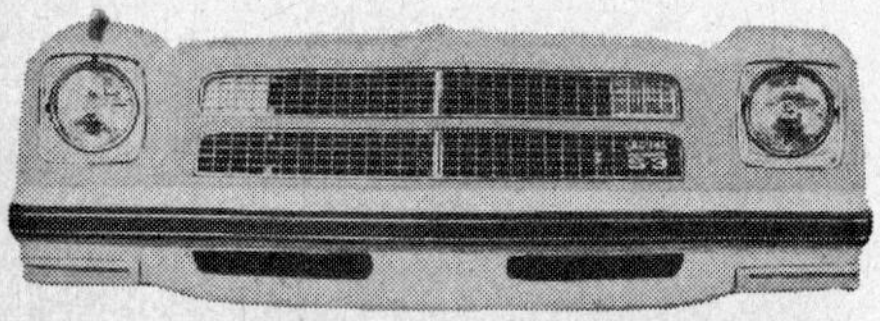
1975 Laguna S-3

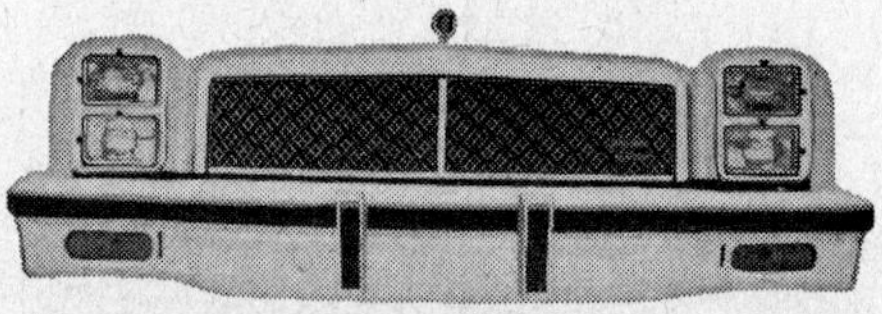
1976 Malibu Classic

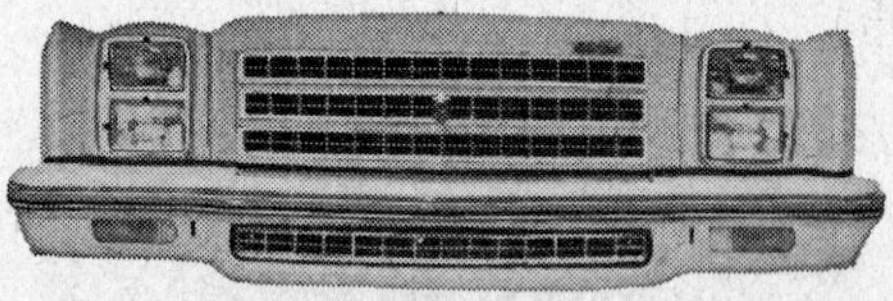
1976 Monte Carlo

1976 Nova

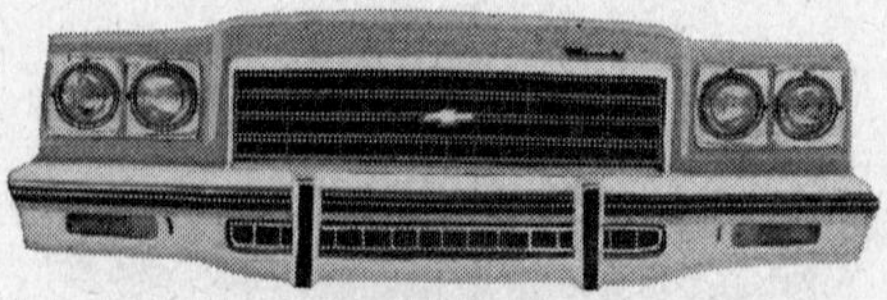
1976 Impala

1976 Caprice & Estate Wagon

GENERAL ENGINE SPECIFICATIONS

Year	Engine	Carburetor	Bore and Stroke	Piston Displacement, Cubic Inches	Compression Ratio	Maximum Brake H.P. @ R.P.M.	Maximum Torque Lbs. Ft. @ R.P.M.	Normal Oil Pressure Pounds
1969	90 Horsepower................4-153	1 Barrel	3.875 x 3.25	153	8.50	90 @ 4000	152 @ 2400	30–45
	140 Horsepower................6-230	1 Barrel	3.875 x 3.25	230	8.50	140 @ 4400	220 @ 1600	30–45
	155 Horsepower................6-250	1 Barrel	3.875 x 3.53	250	8.50	155 @ 4200	235 @ 1600	30–45
	290 Horsepower.............V8-302	4 Barrel	4.000 x 3.00	302	11.00	290 @ 5800	290 @ 4200	30–45
	200 Horsepower.............V8-307	2 Barrel	3.875 x 3.25	307	9.00	200 @ 4600	300 @ 2400	30–45
	210 Horsepower.............V8-327	2 Barrel	4.001 x 3.25	327	9.00	210 @ 4600	320 @ 2400	30–45
	235 Horsepower.............V8-327	2 Barrel	4.001 x 3.25	327	9.00	235 @ 4800	325 @ 2800	30–45
	255 Horsepower.............V8-350	4 Barrel	4.001 x 3.48	350	9.00	255 @ 4800	365 @ 3200	30–45
	300 Horsepower.............V8-350	4 Barrel	4.001 x 3.48	350	10.25	300 @ 4800	380 @ 3200	30–45
	350 Horsepower.............V8-350	4 Barrel	4.001 x 3.48	350	11.00	350 @ 5600	380 @ 3600	30–45
	370 Horsepower.............V8-350	4 Barrel	4.001 x 3.48	350	11.00	370 @ 5800	380 @ 4000	30–45
	265 Horsepower.............V8-396	2 Barrel	4.094 x 3.76	396	9.00	265 @ 4800	400 @ 2800	30–35
	325 Horsepower.............V8-396	4 Barrel	4.094 x 3.76	396	10.25	325 @ 4800	410 @ 3200	30–35
	350 Horsepower.............V8-396	4 Barrel	4.094 x 3.76	396	10.25	350 @ 5200	415 @ 3400	30–35
	375 Horsepower.............V8-396	4 Barrel	4.094 x 3.76	396	11.00	375 @ 5600	415 @ 3600	30–35
	335 Horsepower.............V8-427	4 Barrel	4.251 x 3.76	427	10.25	335 @ 4800	470 @ 3200	30–35
	390 Horsepower.............V8-427	4 Barrel	4.251 x 3.76	427	10.25	390 @ 5400	460 @ 3600	30–35
	400 Horsepower.............V8-427	4 Barrel	4.251 x 3.76	427	10.25	400 @ 5400	460 @ 3600	30–35
	425 Horsepower.............V8-427	4 Barrel	4.251 x 3.76	427	11.00	425 @ 5600	460 @ 4000	30–35
	430 Horsepower.............V8-427	4 Barrel	4.251 x 3.76	427	12.00	430 @ 5200	450 @ 4400	30–35
	435 Horsepower.............V8-427	3 Carbs.	4.251 x 3.76	427	11.00	435 @ 5800	460 @ 4000	30–35
1970	90 Horsepower................4-153	1 Barrel	3.875 x 3.25	153	8.50	90 @ 4000	152 @ 2400	30–45
	140 Horsepower................6-230	1 Barrel	3.875 x 3.25	230	8.50	140 @ 4400	220 @ 1600	30–45
	155 Horsepower................6-250	1 Barrel	3.875 x 3.53	250	8.50	155 @ 4200	235 @ 1600	30–45
	200 Horsepower.............V8-307	2 Barrel	3.875 x 3.25	307	9.00	200 @ 4600	300 @ 2400	30–45
	250 Horsepower.............V8-350	2 Barrel	4.001 x 3.48	350	9.00	250 @ 4800	345 @ 2800	30–45
	300 Horsepower.............V8-350	4 Barrel	4.001 x 3.48	350	10.25	300 @ 4800	380 @ 3200	30–45
	350 Horsepower.............V8-350	4 Barrel	4.001 x 3.48	350	11.00	350 @ 5600	380 @ 3600	30–45
	360 Horsepower.............V8-350	4 Barrel	4.001 x 3.48	350	11.00	360 @ 6000	380 @ 4000	35–45
	370 Horsepower.............V8-350	4 Barrel	4.001 x 3.48	350	11.00	370 @ 6000	380 @ 4000	30–45
	350 Horsepower.............V8-396	4 Barrel	4.125 x 3.76	①	10.25	350 @ 5200	415 @ 3400	30–35
	375 Horsepower.............V8-396	4 Barrel	4.125 x 3.76	①	11.00	375 @ 5600	415 @ 3600	30–35
	265 Horsepower.............V8-400	2 Barrel	4.125 x 3.75	400	9.00	265 @ 4400	400 @ 2400	30–35
	330 Horsepower.............V8-400	4 Barrel	4.125 x 3.75	400	10.25	330 @ 4800	410 @ 3200	30–35
	345 Horsepower.............V8-454	4 Barrel	4.251 x 4.00	454	10.25	345 @ 4400	500 @ 3000	30–35
	360 Horsepower.............V8-454	4 Barrel	4.251 x 4.00	454	10.25	360 @ 4400	500 @ 3200	30–35
	390 Horsepower.............V8-454	4 Barrel	4.251 x 4.00	454	10.25	390 @ 4800	500 @ 3400	30–35
	450 Horsepower.............V8-454	4 Barrel	4.251 x 4.00	454	11.25	450 @ 5600	500 @ 3600	30–35
	460 Horsepower.............V8-454	4 Barrel	4.251 x 4.00	454	11.25	460 @ 5600	490 @ 3000	30–35
1971	145 Horsepower................6-250	1 Barrel	3.875 x 3.53	250	8.50	145 @ 4200	230 @ 1600	30–45
	200 Horsepower.............V8-307	2 Barrel	3.875 x 3.25	307	8.50	200 @ 4600	300 @ 2400	30–45
	245 Horsepower.............V8-350	2 Barrel	4.00 x 3.48	350	8.50	245 @ 4800	350 @ 2800	35–45
	270 Horsepower.............V8-350	4 Barrel	4.00 x 3.48	350	8.50	270 @ 4800	360 @ 3200	35–45
	330 Horsepower.............V8-350	4 Barrel	4.00 x 3.48	350	9.0	330 @ 5600	360 @ 4000	35–45
	255 Horsepower.............V8-400	2 Barrel	4.125 x 3.76	400	8.50	255 @ 4400	390 @ 2400	35–45
	300 Horsepower.............V8-396	4 Barrel	4.126 x 3.76	①	8.50	300 @ 4800	400 @ 3200	35–45
	365 Horsepower.............V8-454	4 Barrel	4.251 x 4.00	454	8.50	365 @ 4800	465 @ 3200	35–45
	425 Horsepower.............V8-454	4 Barrel	4.251 x 4.00	454	9.0	425 @ 5600	475 @ 4000	35–45
1972	110 Horsepower②.............6-250	1 Barrel	3.875 x 3.53	250	8.50	110 @ 3800	185 @ 1600	30–45
	130 Horsepower②............V8-307	2 Barrel	3.875 x 3.25	307	8.50	130 @ 4000	230 @ 2400	30–45
	165 Horsepower②............V8-350	2 Barrel	4.00 x 3.48	350	8.50	165 @ 4000	280 @ 2400	35–45
	175 Horsepower②............V8-350	4 Barrel	4.00 x 3.48	350	8.50	175 @ 4000	280 @ 2400	35–45
	200 Horsepower②............V8-350	4 Barrel	4.00 x 3.48	350	8.50	200 @ 4400	300 @ 2800	35–45
	255 Horsepower②............V8-350	4 Barrel	4.00 x 3.48	350	9.0	255 @ 5600	280 @ 4000	35–45
	170 Horsepower②............V8-400	2 Barrel	4.126 x 3.75	400	8.50	170 @ 3400	325 @ 2000	35–45

Continued

GENERAL ENGINE SPECIFICATIONS—Continued

Year	Engine	Carburetor	Bore and Stroke	Piston Displacement, Cubic Inches	Compression Ratio	Maximum Brake H.P. @ R.P.M.	Maximum Torque Lbs. Ft. @ R.P.M.	Normal Oil Pressure Pounds
1972	210 Horsepower②............V8-402	4 Barrel	4.126 x 3.76	402	8.50	210 @ 4400	320 @ 2400	35–45
	240 Horsepower②............V8-402	4 Barrel	4.126 x 3.76	402	8.50	240 @ 4400	345 @ 3200	35–45
	230 Horsepower②............V8-454	4 Barrel	4.251 x 4.00	454	8.50	230 @ 4000	360 @ 3200	35–45
	270 Horsepower②............V8-454	4 Barrel	4.251 x 4.00	454	8.50	270 @ 4000	390 @ 3200	35–45
1973	100 Horsepower②..............6-250	1 Barrel	3.875 x 3.53	250	8.25	100 @ 3800	175 @ 1600	40
	115 Horsepower②............V8-307	2 Barrel	3.875 x 3.25	307	8.50	115 @ 4000	205 @ 2000	40
	145 Horsepower②............V8-350	2 Barrel	4.00 x 3.48	350	8.50	145 @ 4000	255 @ 2400	40
	175 Horsepower②............V8-350	4 Barrel	4.00 x 3.48	350	8.50	175 @ 4000	270 @ 2400	40
	190 Horsepower②............V8-350	4 Barrel	4.00 x 3.48	350	8.50	190 @ 4400	270 @ 2800	40
	245 Horsepower②............V8-350	4 Barrel	4.00 x 3.48	350	9.0	245 @ 5200	280 @ 4000	40
	250 Horsepower②............V8-350	4 Barrel	4.00 x 3.48	350	9.0	250 @ 5200	285 @ 4000	40
	150 Horsepower②............V8-400	2 Barrel	4.126 x 3.75	400	8.50	150 @ 3200	295 @ 2000	40
	215 Horsepower②............V8-454	4 Barrel	4.251 x 4.00	454	8.50	215 @ 4000	345 @ 2400	40
	245 Horsepower②............V8-454	4 Barrel	4.251 x 4.00	454	8.50	245 @ 4000	375 @ 2800	40
	275 Horsepower②............V8-454	4 Barrel	4.251 x 4.00	454	8.50	275 @ 4000	395 @ 2800	40
1974	100 Horsepower②..............6-250	1 Barrel	3.875 x 3.53	250	8.25	100 @ 3600	175 @ 1800	40
	145 Horsepower②............V8-350	2 Barrel	4.00 x 3.48	350	8.50	145 @ 3600	250 @ 2200	40
	160 Horsepower②............V8-350	4 Barrel	4.00 x 3.48	350	8.50	160 @ 3800	250 @ 2400	40
	185 Horsepower②............V8-350	4 Barrel	4.00 x 3.48	350	8.50	185 @ 4000	270 @ 2600	40
	195 Horsepower②............V8-350	4 Barrel	4.00 x 3.48	350	8.50	195 @ 4400	275 @ 2800	40
	245 Horsepower②............V8-350	4 Barrel	4.00 x 3.48	350	9.0	245 @ 5200	280 @ 4000	40
	250 Horsepower②............V8-350	4 Barrel	4.00 x 3.48	350	9.0	250 @ 5200	285 @ 4000	40
	150 Horsepower②............V8-400	2 Barrel	4.125 x 3.75	400	8.50	150 @ 3200	295 @ 2000	40
	180 Horsepower②............V8-400	4 Barrel	4.125 x 3.75	400	8.50	180 @ 3800	290 @ 2400	40
	235 Horsepower②............V8-454	4 Barrel	4.250 x 4.00	454	8.50	235 @ 4000	360 @ 2800	45
	270 Horsepower②............V8-454	4 Barrel	4.250 x 4.00	454	8.50	270 @ 4400	380 @ 2800	45
1975	105 Horsepower②..............6-250	1 Barrel	3.875 x 3.53	250	8.25	105 @ 3800	185 @ 1200	36–41
	110 Horsepower②............V8-262	2 Barrel	3.671 x 3.10	262	8.5	110 @ 3600	200 @ 2000	32–40
	145 Horsepower②............V8-350	2 Barrel	4.00 x 3.48	350	8.5	145 @ 3800	250 @ 2200	32–40
	155 Horsepower②............V8-350	4 Barrel	4.00 x 3.48	350	8.5	155 @ 3800	250 @ 2400	32–40
	165 Horsepower②............V8-350	4 Barrel	4.00 x 3.48	350	8.5	165 @ 3800	255 @ 2400	32–40
	205 Horsepower②............V8-350	4 Barrel	4.00 x 3.48	350	9.0	205 @ 4800	255 @ 3600	32–40
	175 Horsepower②............V8-400	4 Barrel	4.125 x 3.75	400	8.5	175 @ 3600	305 @ 2000	42–46
	215 Horsepower②............V8-454	4 Barrel	4.251 x 4.00	454	8.15	215 @ 4000	350 @ 2400	42–46
1976	105 Horsepower②..............6-250	1 Barrel	3.875 x 3.53	250	8.25	105 @ 3800	185 @ 1200	36–41
	140 Horsepower②............V8-305	2 Barrel	3.736 x 3.48	305	8.5	140 @ 3800	245 @ 2000	32–40
	145 Horsepower②............V8-350	2 Barrel	4.00 x 3.48	350	8.5	145 @ 3800	250 @ 2200	32–40
	165 Horsepower②............V8-350	4 Barrel	4.00 x 3.48	350	8.5	165 @ 3800	260 @ 2400	32–40
	185 Horsepower②............V8-350	4 Barrel	4.00 x 3.48	350	8.5	185 @ 4000	275 @ 2400	32–40
	195 Horsepower②............V8-350	4 Barrel	4.00 x 3.48	350	8.5	195 @ 4400	275 @ 2800	32–40
	210 Horsepower②............V8-350	4 Barrel	4.00 x 3.48	350	9.0	210 @ 5200	255 @ 3600	32–40
	270 Horsepower②............V8-350	4 Barrel	4.00 x 3.48	350	8.5	270 @ 4400	380 @ 2800	32–40
	175 Horsepower②............V8-400	4 Barrel	4.125 x 3.75	400	8.5	175 @ 3600	305 @ 2000	32–40
	235 Horsepower②............V8-454	4 Barrel	4.251 x 4.00	454	8.5	235 @ 4000	360 @ 2800	42–46

①—Marketed as 396 cu. in. but actually 402 cu. in.

②—Ratings are net—As installed in the vehicle.

TUNE UP SPECIFICATIONS

The following specifications are published from the latest information available. This data should be used only in the absence of a decal affixed in the engine compartment.

OLD CAR SPECIFICATIONS: For 1946-68 Tune Up Specifications see main index.

★When using a timing light, disconnect vacuum hose or tube at distributor and plug opening in hose or tube so idle speed will not be affected.

●When checking compression, lowest cylinder must be within 80 percent of highest.

▲Before removing wires from distributor cap, determine location of the No. 1 wire in cap, as distributor position may have been altered from that shown at the end of this chart.

Year	Spark Plug		Distributor		Ignition Timing★			Carb. Adjustments					
	Type	Gap Inch	Point Gap Inch	Dwell Angle Deg.	Firing Order Fig. ▲	Timing BTDC ①	Mark Fig.	Hot Idle Speed③ Std. Trans.	Hot Idle Speed③ Auto. Trans. ②	Air Fuel Ratio Std. Trans.	Air Fuel Ratio Auto. Trans.	Idle "CO" % Std. Trans.	Idle "CO" % Auto. Trans.
CAMARO													
1969													
6-230, 250⑱	R46N	.035	④	31–34	D	TDC	A	700⑯	—	—	—	—	—
6-230, 250⑲	R46N	.035	④	31–34	D	4°	A	—	550D⑯	—	—	—	—
8-302	R43	.035	④	28–32	E	4°	A	900⑯	—	—	—	—	—
8-307	R45S	.035	④	28–32	E	2°	A	700⑯	600D⑯	—	—	—	—
8-327, 210 H.P.⑱	R45S	.035	④	28–32	E	2° ATC	A	700⑯	—	—	—	—	—
8-327, 210 H.P.⑲	R45S	.035	④	28–32	E	2°	A	—	600D⑯	—	—	—	—
8-350, 255 H.P.⑱	R44	.035	④	28–32	E	TDC	A	700⑯	—	—	—	—	—
8-350, 255 H.P.⑲	R44	.035	④	28–32	E	4°	A	—	600D⑯	—	—	—	—
8-350, 300 H.P.⑱	R44	.035	④	28–32	E	TDC	A	700⑯	—	—	—	—	—
8-350, 300 H.P.⑲	R44	.035	④	28–32	E	4°	A	—	600D⑯	—	—	—	—
8-396, 325 H.P.	R44N	.035	④	28–32	E	4°	A	800⑯	600D⑯	—	—	—	—
8-396, 350 H.P.⑱	R43N	.035	④	28–32	E	TDC	A	800⑯	—	—	—	—	—
8-396, 350 H.P.⑲	R43N	.035	④	28–32	E	4°	A	—	600D⑯	—	—	—	—
8-396, 375 H.P.	R43N	.035	④	28–32	E	4°	A	750⑯	700D⑯	—	—	—	—
1970													
6-250 Std. Tr.	R46T	.035	④	31–34	D	TDC	A	750	—	—	—	—	—
6-250 Auto. Tr.	R46T	.035	④	31–34	D	4°	A	—	600D/400	—	—	—	—
8-307 Std. Tr.	R43	.035	④	29–31	E	2°	A	700	—	—	—	—	—
8-307 Auto. Tr.	R43	.035	④	29–31	E	8°	A	—	600D/450	—	—	—	—
8-350, 250 H.P.⑱	R44	.035	④	29–31	E	TDC	A	750	—	—	—	—	—
8-350, 250 H.P.⑲	R44	.035	④	29–31	E	4°	A	—	600D/450	—	—	—	—
8-350, 300 H.P.⑱	R44	.035	④	29–31	E	TDC	A	700	—	—	—	—	—
8-350, 300 H.P.⑲	R44	.035	④	29–31	E	4°	A	—	600D	—	—	—	—
8-350, 360 H.P.	R43	.035	④	29–31	E	8°	A	800	750D/500	—	—	—	—
8-396, 350 H.P.⑱㉑	R44T	.035	④	29–31	E	TDC	A	700	—	—	—	—	—
8-396, 350 H.P.⑲㉑	R44T	.035	④	29–31	E	4°	A	—	600D	—	—	—	—
8-396, 375 H.P.㉑	R43T	.035	④	29–31	E	4°	A	750	700D	—	—	—	—
8-454, 450 H.P.	R43T	.035	④	29–31	E	4°	A	750	700D	—	—	—	—
1971													
8-250	R46TS	.035	④	31–34	D	4°	B	550	550D	—	—	1.0	1.0
6-307 Std. Tr.	R45TS	.035	④	29–31	E	4°	B	550	—	—	—	0.5	—
8-307 Auto. Tr.	R45TS	.035	④	29–31	E	8°	B	—	550D	—	—	—	0.5
8-350, 245 H.P.⑱	R45TS	.035	④	29–31	E	2°	B	600	—	—	—	0.5	—
8-350, 245 H.P.⑲	R45TS	.035	④	29–31	E	6°	B	—	550D	—	—	—	0.5
8-350, 270 H.P.⑱	R44TS	.035	④	29–31	E	4°	B	600	—	—	—	1.0	—
8-350, 270 H.P.⑲	R44TS	.035	④	29–31	E	8°	B	—	550D	—	—	—	0.5
8-350, 330 H.P.⑱	R43TS	.035	④	29–31	E	8°	B	700	—	—	—	—	—
8-350, 330 H.P.⑲	R43TS	.035	④	29–31	E	12°	B	—	700D	—	—	—	—
8-396, 300 H.P.	R44TS	.035	④	29–31	E	8°	B	600	600D	—	—	1.0	1.0

Continued

TUNE UP SPECIFICATIONS—Continued

The following specifications are published from the latest information available. This data should be used only in the absence of a decal affixed in the engine compartment.

OLD CAR SPECIFICATIONS: For 1946-68 Tune Up Specifications see main index.

★When using a timing light, disconnect vacuum hose or tube at distributor and plug opening in hose or tube so idle speed will not be affected.

●When checking compression, lowest cylinder must be within 80 percent of highest.

▲Before removing wires from distributor cap, determine location of the No. 1 wire in cap, as distributor position may have been altered from that shown at the end of this chart.

Year	Spark Plug		Distributor		Ignition Timing★			Carb. Adjustments					
								Hot Idle Speed③		Air Fuel Ratio		Idle "CO" %	
	Type	Gap Inch	Point Gap Inch	Dwell Angle Deg.	Firing Order Fig. ▲	Timing BTDC ①	Mark Fig.	Std. Trans.	Auto. Trans. ②	Std. Trans.	Auto. Trans.	Std. Trans.	Auto. Trans.
CAMARO—Continued													
1972													
6-250	R46T	.035	④	31–34	D	4°	B	700	600D	—	—	—	—
8-307 Std. Tr.	R44T	.035	④	29–31	E	4°	B	900	—	—	—	—	—
8-307 Auto. Tr.	R44T	.035	④	29–31	E	8°	B	—	600D	—	—	—	—
8-350, 165 H.P.	R44T	.035	④	29–31	E	6°	B	900	600D	—	—	—	—
8-350, 200 H.P.⑱	R44T	.035	④	29–31	E	4°	B	800	—	—	—	—	—
8-350, 200 H.P.⑲	R44T	.035	④	29–31	E	8°	B	—	600D	—	—	—	—
8-350, 255 H.P.⑱	R44T	.035	④	29–31	E	4°	B	900	—	—	—	—	—
8-350, 255 H.P.⑲	R44T	.035	④	29–31	E	8°	B	—	700D	—	—	—	—
8-402	R44T	.035	④	29–31	E	8°	B	750	600D	—	—	—	—
1973													
6-250	R46T	.035	④	31–34	D	6°	B	700	600D	—	—	—	—
8-307 Std. Tr.	R44T	.035	④	29–31	E	4°	B	900	—	—	—	—	—
8-307 Auto. Tr.	R44T	.035	④	29–31	E	8°	B	—	600D	—	—	—	—
8-350, 145 H.P.	R44T	.035	④	29–31	E	8°	B	900	600D	—	—	—	—
8-350, 175 H.P.⑱	R44T	.035	④	29–31	E	8°	B	900	—	—	—	—	—
8-350, 175 H.P.⑲	R44T	.035	④	29–31	E	12°	B	—	600D	—	—	—	—
8-350, 245 H.P.	R44T	.035	④	29–31	E	8°	B	900	700D	—	—	—	—
1974													
6-250⑱	R46T	.035	④	31–34	D	8°	B	850	—	—	—	.3	.3
6-250⑲	R46T	.035	④	31–34	D	6°	B	—	600D	—	—	.3	.3
8-350, 145 H.P.⑱	R44T	.035	④	29–31	E	TDC	B	900	—	—	—	.5	.5
8-350, 145 H.P.⑲	R44T	.035	④	29–31	E	8°	B	—	600D	—	—	.5	.5
8-350, 160 H.P.⑱	R44T	.035	④	29–31	E	4°	B	900	—	—	—	.5	.5
8-350, 160 H.P.⑲	R44T	.035	④	29–31	E	8°	B	—	600D	—	—	.5	.5
8-350, 185 H.P.⑱	R44T	.035	④	29–31	E	8°⑥	B	900	—	—	—	.5	.5
8-350, 185 H.P.⑲	R44T	.035	④	29–31	E	8°	B	—	600D	—	—	.5	.5
8-350, 245 H.P.	R44T	.035	④	29–31	⑫	8°	B	900	700D	—	—	.5	.5
1975													
6-250	R46TX	.060	—	—	H	8°	B	850	600D	—	—	—	—
6-250⑦	R46TX	.060	—	—	H	10°	B	850	550D⑨	—	—	—	—
8-350 2 Bbl. Carb.	R44TX	.060	—	—	I	6°	B	800	600D	—	—	—	—
8-350⑱	R44TX	.060	—	—	I	6°⑥	B	800	—	—	—	—	—
8-350⑲	R44TX	.060	—	—	I	8°⑧	B	—	600D	—	—	—	—
1976													
6-250	R46TS	.035	—	—	H	6°	B	850	550D⑨	—	—	—	—
8-305⑱	R45TS	.045	—	—	I	6°	B	800	—	—	—	—	—
8-305⑲	R45TS	.045	—	—	I	8°⑬	B	—	600D	—	—	—	—
8-350 2 Bbl. Carb.	R45TS	.045	—	—	I	6°	B	—	600D	—	—	—	—
8-350	R45TS	.045	—	—	I	8°⑧	B	800	600D	—	—	—	—

Continued

TUNE UP SPECIFICATIONS—Continued

The following specifications are published from the latest information available. This data should be used only in the absence of a decal affixed in the engine compartment.

OLD CAR SPECIFICATIONS: For 1946-68 Tune Up Specifications see main index.

★When using a timing light, disconnect vacuum hose or tube at distributor and plug opening in hose or tube so idle speed will not be affected.

●When checking compression, lowest cylinder must be within 80 percent of highest.

▲Before removing wires from distributor cap, determine location of the No. 1 wire in cap, as distributor position may have been altered from that shown at the end of this chart.

Year	Spark Plug		Distributor		Ignition Timing★			Carb. Adjustments					
								Hot Idle Speed③		Air Fuel Ratio		Idle "CO" %	
	Type	Gap Inch	Point Gap Inch	Dwell Angle Deg.	Firing Order Fig. ▲	Timing BTDC ①	Mark Fig.	Std. Trans.	Auto. Trans. ②	Std. Trans.	Auto. Trans.	Std. Trans.	Auto. Trans.
CHEVELLE & MONTE CARLO													
1969													
6-230, 250⑱	R46N	.035	④	31–34	D	TDC	A	700⑯	—	—	—	—	—
6-230, 250⑲	R46N	.035	④	31–34	D	4°	A	—	550D⑯	—	—	—	—
8-307	R45S	.035	④	28–32	E	2°	A	700⑯	600D⑯	—	—	—	—
8-350, 255 H.P.⑱	R44	.035	④	28–32	E	TDC	A	700⑯	—	—	—	—	—
8-350, 255 H.P.⑲	R44	.035	④	28–32	E	4°	A	—	600D⑯	—	—	—	—
8-350, 300 H.P.⑱	R44	.035	④	28–32	E	TDC	A	700⑯	—	—	—	—	—
8-350, 300 H.P.⑲	R44	.035	④	28–32	E	4°	A	—	600D⑯	—	—	—	—
8-396, 325 H.P.	R44N	.035	④	28–32	E	4°	A	800⑯	600D⑯	—	—	—	—
8-396, 350 H.P.⑱	R43N	.035	④	28–32	E	TDC	A	800⑯	—	—	—	—	—
8-396, 350 H.P.⑲	R43N	.035	④	28–32	E	4°	A	—	600D⑯	—	—	—	—
8-396, 375 H.P.	R43N	.035	④	28–32	E	4°	A	750⑯	750D⑯	—	—	—	—
1970													
6-250⑱	R46T	.035	④	31–34	D	TDC	A	750	—	—	—	—	—
6-250⑲	R46T	.035	④	31–34	D	4°	A	—	600D/400	—	—	—	—
8-307⑱	R43	.035	④	28–32	E	2°	A	700	—	—	—	—	—
8-307⑲	R43	.035	④	28–32	E	8°	A	—	600D/450	—	—	—	—
8-350⑱	R44	.035	④	28–32	E	TDC	A	700	—	—	—	—	—
8-350⑲	R44	.035	④	28–32	E	4°	A	—	600D	—	—	—	—
8-400, 265 H.P.⑱	R44	.035	④	28–32	E	4°	A	700	—	—	—	—	—
8-400, 265 H.P.⑲	R44	.035	④	28–32	E	8°	A	—	600D	—	—	—	—
8-400, 330 H.P.	R44T	.035	④	28–32	E	4°	A	700	600D	—	—	—	—
8-396, 350 H.P.⑱	R44T	.035	④	28–32	E	TDC	A	700	—	—	—	—	—
8-396, 350 H.P.⑲	R44T	.035	④	28–32	E	4°	A	—	600D	—	—	—	—
8-396, 375 H.P.	R43T	.035	④	28–32	E	4°	A	750	700D	—	—	—	—
8-454, 360 H.P.	R43T	.035	④	28–32	E	6°	A	700	600	—	—	—	—
8-454, 390 H.P.	R43T	.035	④	28–32	E	6°	A	700	600D	—	—	—	—
8-454, 450 H.P.	R43T	.035	④	28–32	E	4°	A	700	700D	—	—	—	—
1971													
6-250	R46TS	.035	④	31–34	D	4°	B	550	500D	—	—	1.0	1.0
8-307⑱	R45TS	.035	④	29–31	E	4°	B	550	—	—	—	0.5	—
8-307⑲	R45TS	.035	④	29–31	E	8°	B	—	500D	—	—	—	0.5
8-350, 245 H.P.⑱	R45TS	.035	④	29–31	E	2°	B	600	—	—	—	0.5	—
8-350, 245 H.P.⑲	R45TS	.035	④	29–31	E	6°	B	—	600D	—	—	—	0.5
8-350, 270 H.P.⑱	R44TS	.035	④	29–31	E	4°	B	600	—	—	—	1.0	—
8-350, 270 H.P.⑲	R44TS	.035	④	29–31	E	8°	B	—	600D	—	—	—	0.5
8-400	R44TS	.035	④	29–31	E	8°	B	600	600D	—	—	0.5	0.5
8-454, 365 H.P.	R43TS	.035	④	29–31	E	8°	B	600	600D	—	—	1.0	1.0
8-454, 425 H.P.⑱	R44TS	.035	④	29–31	E	8°	B	700	—	—	—	—	—
8-454, 425 H.P.⑲	R44TS	.035	④	29–31	E	12°	B	—	700D	—	—	—	—

Continued

TUNE UP SPECIFICATIONS—Continued

The following specifications are published from the latest information available. This data should be used only in the absence of a decal affixed in the engine compartment.

OLD CAR SPECIFICATIONS: For 1946-68 Tune Up Specifications see main index.

★When using a timing light, disconnect vacuum hose or tube at distributor and plug opening in hose or tube so idle speed will not be affected.

●When checking compression, lowest cylinder must be within 80 percent of highest.

▲Before removing wires from distributor cap, determine location of the No. 1 wire in cap, as distributor position may have been altered from that shown at the end of this chart.

Year	Spark Plug		Distributor		Ignition Timing★			Carb. Adjustments					
								Hot Idle Speed③		Air Fuel Ratio		Idle "CO" %	
	Type	Gap Inch	Point Gap Inch	Dwell Angle Deg.	Firing Order Fig. ▲	Timing BTDC ①	Mark Fig.	Std. Trans.	Auto. Trans. ②	Std. Trans.	Auto. Trans.	Std. Trans.	Auto. Trans.
CHEVELLE & MONTE CARLO—Continued													
1972													
6-250	R46T	.035	④	31–34	D	4°	B	700	600D	—	—	—	—
8-307 Std. Tr.	R44T	.035	④	29–31	E	4°	B	900	—	—	—	—	—
8-307 Auto. Tr.	R44T	.035	④	29–31	E	8°	B	—	600D	—	—	—	—
8-350, 165 H.P.	R44T	.035	④	29–31	E	6°	B	900	600D	—	—	—	—
8-350, 175 H.P.⑱	R44T	.035	④	29–31	E	4°	B	800	—	—	—	—	—
8-350, 175 H.P.⑲	R44T	.035	④	29–31	E	8°	B	—	600D	—	—	—	—
8-402	R44T	.035	④	29–31	E	8°	B	750	600D	—	—	—	—
8-454	R44T	.035	④	29–31	E	8°	B	750	600D	—	—	—	—
1973													
6-250	R46T	.035	④	31–34	D	6°	B	700	600D	—	—	—	—
8-307⑱	R44T	.035	④	29–31	E	4°	B	900	—	—	—	—	—
8-307⑲	R44T	.035	④	29–31	E	8°	B	—	600D	—	—	—	—
8-350, 145 H.P.	R44T	.035	④	29–31	E	8°	B	900	600D	—	—	—	—
8-350, 175 H.P.⑱	R44T	.035	④	29–31	E	8°	B	900	—	—	—	—	—
8-350, 175 H.P.⑲	R44T	.035	④	29–31	E	12°	B	—	600D	—	—	—	—
8-454, 245 H.P.	R44T	.035	④	29–31	E	10°	B	900	600D	—	—	—	—
1974													
6-250⑱	R46T	.035	④	31–34	D	8°	B	850	—	—	—	.3	.3
6-250⑲	R46T	.035	④	31–34	D	6°	B	—	600D	—	—	.3	.3
8-350, 145 H.P.⑱	R44T	.035	④	29–31	E	TDC	B	900	—	—	—	.5	.5
8-350, 145 H.P.⑲	R44T	.035	④	29–31	E	8°	B	—	600D	—	—	.5	.5
8-350, 160 H.P.⑱	R44T	.035	④	29–31	E	4°	B	900	—	—	—	.5	.5
8-350, 160 H.P.⑲	R44T	.035	④	29–31	E	8°	B	—	600D	—	—	.5	.5
8-350, 185 H.P.⑱	R44T	.035	④	29–31	E	8°⑥	B	900	—	—	—	.5	.5
8-350, 185 H.P.⑲	R44T	.035	④	29–31	E	8°	B	—	600D	—	—	.5	.5
8-400	R44T	.035	④	29–31	E	8°	B	—	600D	—	—	.5	.5
8-454	R44T	.035	④	29–31	E	10°	B	800	600D	—	—	.5	.5
1975													
6-250	R46TX	.060	—	—	H	8°	B	850	600D	—	—	—	—
6-250⑦	R46TX	.060	—	—	H	10°	B	850	550D⑨	—	—	—	—
8-350 2 BBl. Carb.	R44TX	.060	—	—	I	6°	B	800	600D	—	—	—	—
8-350⑱	R44TX	.060	—	—	I	6°⑥	B	800	—	—	—	—	—
8-350⑲	R44TX	.060	—	—	I	8°⑧	B	—	600D	—	—	—	—
8-400	R44TX	.060	—	—	I	8°	B	—	600D	—	—	—	—
8-454	R44TX	.060	—	—	I	16°	B	—	600D	—	—	—	—
1976													
6-250	R46TS	.035	—	—	H	6°	B	850	550D9	—	—	—	—
8-305⑱	R45TS	.045	—	—	I	6°	B	800	—	—	—	—	—
8-305⑲	R45TS	.045	—	—	I	8°⑬	B	—	600D	—	—	—	—
8-350	R45TS	.045	—	—	I	6°	B	800	600D	—	—	—	—
8-400	R45TS	.045	—	—	I	8°	B	—	600D	—	—	—	—

Continued

TUNE UP SPECIFICATIONS—Continued

The following specifications are published from the latest information available. This data should be used only in the absence of a decal affixed in the engine compartment.

OLD CAR SPECIFICATIONS: For 1946-68 Tune Up Specifications see main index.

★When using a timing light, disconnect vacuum hose or tube at distributor and plug opening in hose or tube so idle speed will not be affected.

●When checking compression, lowest cylinder must be within 80 percent of highest.

▲Before removing wires from distributor cap, determine location of the No. 1 wire in cap, as distributor position may have been altered from that shown at the end of this chart.

Year	Spark Plug		Distributor		Ignition Timing★			Carb. Adjustments					
								Hot Idle Speed③		Air Fuel Ratio		Idle "CO" %	
	Type	Gap Inch	Point Gap Inch	Dwell Angle Deg.	Firing Order Fig. ▲	Timing BTDC ①	Mark Fig.	Std. Trans.	Auto. Trans. ②	Std. Trans.	Auto. Trans.	Std. Trans.	Auto. Trans.
CHEVY NOVA													
1969													
4-153⑱	R46N	.035	④	31-34	F	TDC	A	750⑯	—	—	—	—	—
4-153⑲	R46N	.035	④	31-34	F	4°	A	—	600D⑯	—	—	—	—
6-230, 250⑱	R46N	.035	④	31-34	D	TDC	A	700⑯	—	—	—	—	—
6-230, 250⑲	R46N	.035	④	31-34	D	4°	A	—	550D⑯	—	—	—	—
8-307	R45S	.035	④	28-32	E	2°	A	700⑯	600D⑯	—	—	—	—
8-327, 210 H.P.⑱	R45S	.035	④	28-32	E	2° ATC	A	700⑯	—	—	—	—	—
8-350, 255 H.P.⑱	R44	.035	④	28-32	E	TDC	A	700⑯	—	—	—	—	—
8-350, 255 H.P.⑲	R44	.035	④	28-32	E	4°	A	—	600D⑯	—	—	—	—
8-350, 300 H.P.⑱	R44	.035	④	28-32	E	TDC	A	700⑯	—	—	—	—	—
8-350, 300 H.P.⑲	R44	.035	④	28-32	E	4°	A	—	600D⑯	—	—	—	—
8-396, 350 H.P.⑱	R43N	.035	④	28-32	E	TDC	A	800⑯	—	—	—	—	—
8-396, 350 H.P.⑲	R43N	.035	④	28-32	E	4°	A	—	600D⑯	—	—	—	—
8-396, 375 H.P.	R43N	.035	④	28-32	E	4°	A	750⑯	750D⑯	—	—	—	—
1970													
4-153⑱	R46N	.035	④	31-34	F	TDC	A	750	—	—	—	—	—
4-153⑲	R46N	.035	④	31-34	F	4°	A	—	650D	—	—	—	—
6-230, 250⑱	R46T	.035	④	31-34	D	TDC	A	750	—	—	—	—	—
6-230, 250⑲	R46T	.035	④	31-34	D	4°	A	—	600D/400	—	—	—	—
8-307⑱	R43	.035	④	28-32	E	2°	A	700	—	—	—	—	—
8-307⑲	R43	.035	④	28-32	E	8°	A	—	600D/450	—	—	—	—
8-350, 250 H.P.⑱	R44	.035	④	28-32	E	TDC	A	750	—	—	—	—	—
8-350, 250 H.P.⑲	R44	.035	④	28-32	E	4°	A	—	600D/450	—	—	—	—
8-350, 300 H.P.⑱	R44	.035	④	28-32	E	TDC	A	700	—	—	—	—	—
8-350, 300 H.P.⑲	R44	.035	④	28-32	E	4°	A	—	600D	—	—	—	—
8-396, 350 H.P.	R44T	.035	④	28-32	E	TDC	A	700	—	—	—	—	—
8-396, 375 H.P.⑱	R43T	.035	④	28-32	E	4°	A	750	—	—	—	—	—
8-396, 375 H.P.⑲	R43T	.035	④	28-32	E	4°	A	—	700D	—	—	—	—
1971													
6-250	R46TS	.035	④	31-34	D	4°	B	550	500D	—	—	1.0	1.0
8-307, 200 H.P.⑱	R45TS	.035	④	29-31	E	4°	B	550	—	—	—	0.5	0.5
8-307, 200 H.P.⑲	R45TS	.035	④	29-31	E	8°	B	—	550D	—	—	0.5	0.5
8-350, 245 H.P.⑱	R44TS	.035	④	29-31	E	2°	B	600	—	—	—	0.5	0.5
8-350, 245 H.P.⑲	R44TS	.035	④	29-31	E	6°	B	—	550D	—	—	0.5	0.5
8-350, 270 H.P.⑱	R44TS	.035	④	29-31	E	4°	B	600	—	—	—	1.0	—
8-350, 270 H.P.⑲	R44TS	.035	④	29-31	E	8°	B	—	550D	—	—	—	0.5
1972													
6-250	R46T	.035	④	31-34	D	4°	B	700	600D	—	—	—	—
8-307, Std. Tr.	R44T	.035	④	29-31	E	4°	B	900	—	—	—	—	—
8-307, Auto. Tr.	R44T	.035	④	29-31	E	8°	B	—	600D	—	—	—	—

Continued

TUNE UP SPECIFICATIONS—Continued

The following specifications are published from the latest information available. This data should be used only in the absence of a decal affixed in the engine compartment.

OLD CAR SPECIFICATIONS: For 1946-68 Tune Up Specifications see main index.

★When using a timing light, disconnect vacuum hose or tube at distributor and plug opening in hose or tube so idle speed will not be affected.

●When checking compression, lowest cylinder must be within 80 percent of highest.

▲Before removing wires from distributor cap, determine location of the No. 1 wire in cap, as distributor position may have been altered from that shown at the end of this chart.

Year	Spark Plug		Distributor		Ignition Timing★			Carb. Adjustments					
	Type	Gap Inch	Point Gap Inch	Dwell Angle Deg.	Firing Order Fig. ▲	Timing BTDC (1)	Mark Fig.	Hot Idle Speed (3) Std. Trans.	Hot Idle Speed (3) Auto. Trans. (2)	Air Fuel Ratio Std. Trans.	Air Fuel Ratio Auto. Trans.	Idle "CO" % Std. Trans.	Idle "CO" % Auto. Trans.
CHEVY NOVA—Continued													
1972													
8-350, 165 H.P.	R44T	.035	(4)	29–31	E	6°	B	900	600D	—	—	—	—
8-350, 200 H.P.(18)	R44T	.035	(4)	29–31	E	4°	B	800	—	—	—	—	—
8-350, 200 H.P.(19)	R44T	.035	(4)	29–31	E	8°	B	—	600D	—	—	—	—
1973													
6-250	R46T	.035	(4)	31–34	D	6°	B	700	600D	—	—	—	—
8-307 Std. Tr.	R44T	.035	(4)	29–31	E	4°	B	900	—	—	—	—	—
8-307 Auto. Tr.	R44T	.035	(4)	29–31	E	8°	B	—	600D	—	—	—	—
8-350, 145 H.P.	R44T	.035	(4)	29–31	E	8°	B	900	600D	—	—	—	—
8-350, 175 H.P.(18)	R44T	.035	(4)	29–31	E	8°	B	900	—	—	—	—	—
8-350, 175 H.P.(19)	R44T	.035	(4)	29–31	E	12°	B	—	600D	—	—	—	—
1974													
6-250(18)	R46T	.035	(4)	31–34	D	8°	B	850	—	—	—	.3	.3
6-250(19)	R46T	.035	(4)	31–34	D	6°	B	—	600D	—	—	.3	.3
8-350, 145 H.P.(18)	R44T	.035	(4)	29–31	E	TDC	B	900	—	—	—	.5	.5
8-350, 145 H.P.(19)	R44T	.035	(4)	29–31	E	8°	B	—	600D	—	—	.5	.5
8-350, 160 H.P.(18)	R44T	.035	(4)	29–31	E	4°	B	900	—	—	—	.5	.5
8-350, 160 H.P.(19)	R44T	.035	(4)	29–31	E	8°	B	—	600D	—	—	.5	.5
8-350, 185 H.P.(18)	R44T	.035	(4)	29–31	E	8°(6)	B	900	—	—	—	.5	.5
8-350, 185 H.P.(19)	R44T	.035	(4)	29–31	E	8°	B	—	600D	—	—	.5	.5
1975													
6-250	R46TX	.060	—	—	H	8°	B	850	600D	—	—	—	—
6-250(7)	R46TX	.060	—	—	H	10°	B	850	550D(9)	—	—	—	—
8-262	R44TX	.060	—	—	I	8°	G	800	600D	—	—	—	—
8-350 2 BBl. Carb.	R44TX	.060	—	—	I	6°	B	800	600D	—	—	—	—
8-350(18)	R44TX	.060	—	—	I	6°(6)	B	800	—	—	—	—	—
8-350(19)	R44TX	.060	—	—	I	8°(8)	B	—	600D	—	—	—	—
1976													
6-250	R46TS	.035	—	—	H	6°(10)	B	850	550D(9)	—	—	—	—
8-305(18)	R45TS	.045	—	—	I	6°	B	800	—	—	—	—	—
8-305(19)	R45TS	.045	—	—	I	8°(13)	B	—	600D	—	—	—	—
8-350 2 Bbl. Carb.	R45TS	.045	—	—	I	6°	B	—	600D	—	—	—	—
8-350	R45TS	.045	—	—	I	8°(8)	B	800	600D	—	—	—	—
CHEVROLET													
1969													
6-250(18)	R46N	.035	(4)	31–34	D	TDC	A	700(16)	—	—	—	—	—
6-250(19)	R46N	.035	(4)	31–34	D	4°	A	—	550D(10)	—	—	—	—
8-327, 235 H.P.(18)	R45S	.035	(4)	28–32	E	2° ATC	A	700(16)	—	—	—	—	—
8-327, 235 H.P.(19)	R45S	.035	(4)	28–32	E	2°	A	—	600D(16)	—	—	—	—

Continued

TUNE UP SPECIFICATIONS—Continued

The following specifications are published from the latest information available. This data should be used only in the absence of a decal affixed in the engine compartment.

OLD CAR SPECIFICATIONS: For 1946-68 Tune Up Specifications see main index.

★When using a timing light, disconnect vacuum hose or tube at distributor and plug opening in hose or tube so idle speed will not be affected.

●When checking compression, lowest cylinder must be within 80 percent of highest.

▲Before removing wires from distributor cap, determine location of the No. 1 wire in cap, as distributor position may have been altered from that shown at the end of this chart.

Year	Spark Plug		Distributor		Ignition Timing★			Carb. Adjustments					
								Hot Idle Speed③		Air Fuel Ratio		Idle "CO" %	
	Type	Gap Inch	Point Gap Inch	Dwell Angle Deg.	Firing Order Fig. ▲	Timing BTDC ①	Mark Fig.	Std. Trans.	Auto. Trans. ②	Std. Trans.	Auto. Trans.	Std. Trans.	Auto. Trans.
CHEVROLET—Continued													
1969													
8-350, 255 H.P.⑱	R44	.035	④	28–32	E	TDC	A	700⑯	—	—	—	—	—
8-350, 255 H.P.⑲	R44	.035	④	28–32	E	4°	A	—	600D⑯	—	—	—	—
8-350, 300 H.P.⑱	R44	.035	④	28–32	E	TDC	A	700⑯	—	—	—	—	—
8-350, 300 H.P.⑲	R44	.035	④	28–32	E	4°	A	—	600D⑯	—	—	—	—
8-396, 265 H.P.⑱	R44N	.035	④	28–32	E	TDC	A	700⑯	—	—	—	—	—
8-396, 265 H.P.⑲	R44N	.035	④	28–32	E	4°	A	—	600D⑯	—	—	—	—
8-427, 335 H.P.	R44N	.035	④	28–32	E	4°	A	800⑯	600D⑯	—	—	—	—
8-427, 390 H.P.	R43N	.035	④	28–32	E	4°	A	800⑪	600D⑪	—	—	—	—
8-427, 425 H.P.	R43N	.035	④	28–32	E	4°	A	750⑯	750D⑯	—	—	—	—
1970													
6-250⑱	R46T	.035	④	31–34	D	TDC	A	750	—	—	—	—	—
6-250⑲	R46T	.035	④	31–34	D	4°	A	—	600D/400	—	—	—	—
8-350, 250 H.P.⑱	R44	.035	④	28–32	E	TDC	A	750	—	—	—	—	—
8-350, 250 H.P.⑲	R44	.035	④	28–32	E	4°	A	—	600D/450	—	—	—	—
8-350, 300 H.P.⑱	R44	.035	④	28–32	E	TDC	A	700	—	—	—	—	—
8-350, 300 H.P.⑲	R44	.035	④	28–32	E	4°	A	—	600D	—	—	—	—
8-400, 265 H.P.⑱	R44	.035	④	28–32	E	4°	A	700	—	—	—	—	—
8-400, 265 H.P.⑲	R44	.035	④	28–32	E	8°	A	—	600D/450	—	—	—	—
8-454, 345 H.P.	R44T	.035	④	28–32	E	6°	A	—	600D	—	—	—	—
8-454, 360 H.P.	R43T	.035	④	28–32	E	6°	A	700	600	—	—	—	—
8-454, 390 H.P.	R43T	.035	④	28–32	E	6°	A	700	600D	—	—	—	—
1971													
6-250	R46TS	.035	④	31–34	D	4°	B	550	550D	—	—	1.0	1.0
8-350, 245 H.P.⑱	R44TS	.035	④	29–31	E	2°	B	550	—	—	—	0.5	0.5
8-350, 245 H.P.⑲	R44TS	.035	④	29–31	E	6°	B	—	550D	—	—	0.5	0.5
8-350, 270 H.P.	R44TS	.035	④	29–31	E	8°	B	—	550D	—	—	1.0	0.5
8-400, 255 H.P.⑱	R44TS	.035	④	29–31	E	4°	B	550	—	—	—	0.5	0.5
8-400, 255 H.P.⑲	R44TS	.035	④	29–31	E	8°	B	—	550D	—	—	0.5	0.5
8-400, 300 H.P.	R44TS	.035	④	29–31	E	8°	B	—	600D	—	—	1.0	1.0
8-454, 365 H.P.	R43TS	.035	④	29–31	E	8°	B	—	600D	—	—	1.0	1.0
1972													
6-250	R46T	.035	④	31–34	D	4°	B	700	600D	—	—	—	—
8-350	R44T	.035	④	29–31	E	6°	B	—	600D	—	—	—	—
8-400	R44T	.035	④	29–31	E	6°	B	—	600D	—	—	—	—
8-402	R44T	.035	④	29–31	E	8°	B	—	600D	—	—	—	—
8-454	R44T	.035	④	29–31	E	8°	B	—	600D	—	—	—	—
1973													
6-250	R46T	.035	④	31–34	D	6°	B	700	600D	—	—	—	—
8-350, 145 H.P.	R44T	.035	④	29–31	E	8°	B	—	600D	—	—	—	—

Continued

TUNE UP SPECIFICATIONS—Continued

The following specifications are published from the latest information available. This data should be used only in the absence of a decal affixed in the engine compartment.

OLD CAR SPECIFICATIONS: For 1946-68 Tune Up Specifications see main index.

★When using a timing light, disconnect vacuum hose or tube at distributor and plug opening in hose or tube so idle speed will not be affected.

●When checking compression, lowest cylinder must be within 80 percent of highest.

▲Before removing wires from distributor cap, determine location of the No. 1 wire in cap, as distributor position may have been altered from that shown at the end of this chart.

Year	Spark Plug		Distributor		Ignition Timing★			Carb. Adjustments					
								Hot Idle Speed(3)		Air Fuel Ratio		Idle "CO" %	
	Type	Gap Inch	Point Gap Inch	Dwell Angle Deg.	Firing Order Fig. ▲	Timing BTDC (1)	Mark Fig.	Std. Trans.	Auto. Trans. (2)	Std. Trans.	Auto. Trans.	Std. Trans.	Auto. Trans.
CHEVROLET—Continued													
1973													
8-350, 175 H.P.	R44T	.035	(4)	29–31	E	12°	B	—	600D	—	—	—	—
8-400	R44T	.035	(4)	29–31	E	8°	B	—	600D	—	—	—	—
8-454	R44T	.035	(4)	29–31	E	10°	B	—	600D	—	—	—	—
1974													
8-350, 145 H.P.	R44T	.035	(4)	29–31	E	8°	B	—	600D	—	—	.5	.5
8-350, 160 H.P.	R44T	.035	(4)	29–31	E	8°	B	—	600D	—	—	.5	.5
8-400	R44T	.035	(4)	29–31	E	8°	B	—	600D	—	—	.5	.5
8-454	R44T	.035	(4)	29–31	(12)	10°	B	—	600D	—	—	.5	.5
1975													
8-350 2 BBl. Carb.	R44TX	.060	—	—	I	6°	B	—	600D	—	—	—	—
8-350	R44TX	.060	—	—	I	8°(8)	B	—	600D	—	—	—	—
8-400	R44TX	.060	—	—	I	8°	B	—	600D	—	—	—	—
8-454	R44TX	.060	—	—	I	16°	B	—	600D	—	—	—	—
1976													
8-350 2 Bbl. Carb.	R45TS	.045	—	—	I	6°	B	—	600D	—	—	—	—
8-350	R45TS	.045	—	—	I	8°(8)	B	—	600D	—	—	—	—
8-400	R45TS	.045	—	—	I	8°	B	—	600D	—	—	—	—
8-454	R45TSX	.060	—	—	I	12°	B	—	550D	—	—	—	—
CORVETTE													
1969													
8-350, 300 H.P.(18)	R44	.035	(4)	28–32	E	4°	A	700(16)	—	—	—	—	—
8-350, 300 H.P.(19)	R44	.035	(4)	28–32	E	4°	A	—	600D(16)	—	—	—	—
8-350, 350 H.P.	R44	.035	(4)	28–32	E	8°	A	750(16)	—	—	—	—	—
8-350, 370 H.P.	R43	.035	(4)	28–32	E	(20)	A	750(16)	—	—	—	—	—
8-427, 390 H.P.	R43N	.035	(4)	28–32	E	4°	A	800(11)	600D(11)	—	—	—	—
8-427, 400 H.P.	R43N	.035	(4)	28–32	E	4°	A	800(11)	600D(11)	—	—	—	—
8-427, 425 H.P.	R43N	.035	(4)	28–32	E	4°	A	750(16)	750D(16)	—	—	—	—
8-427, 430 H.P.	R43XL	.035	—	—	E	12°(5)	A	1000(16)	—	—	—	—	—
8-427, 435 H.P.	R43N	.035	—	—	E	4°	A	750(16)	750D(16)	—	—	—	—
1970													
8-350, 300 H.P.	R44	.035	(4)	29–31	E	4°	A	700	600	—	—	—	—
8-350, 350 H.P.	R44	.035	(4)	29–31	E	8°	A	750	—	—	—	—	—
8-350, 370 H.P.	R43	.035	—	—	E	8°	A	900	—	—	—	—	—
8-454, 360 H.P.	R43T	.035	(4)	28–32	E	6°	A	700	600	—	—	—	—
8-454, 390 H.P.	R43T	.035	(4)	29–31	E	6°	A	700	600	—	—	—	—
8-454, 460 H.P.	R43XL	.035	—	—	E	8°	A	700	600	—	—	—	—

Continued

TUNE UP SPECIFICATIONS—Continued

The following specifications are published from the latest information available. This data should be used only in the absence of a decal affixed in the engine compartment.

OLD CAR SPECIFICATIONS: For 1946-68 Tune Up Specifications see main index.

★When using a timing light, disconnect vacuum hose or tube at distributor and plug opening in hose or tube so idle speed will not be affected.

●When checking compression, lowest cylinder must be within 80 percent of highest.

▲Before removing wires from distributor cap, determine location of the No. 1 wire in cap, as distributor position may have been altered from that shown at the end of this chart.

Year	Spark Plug		Distributor		Ignition Timing★			Carb. Adjustments					
	Type	Gap Inch	Point Gap Inch	Dwell Angle Deg.	Firing Order Fig. ▲	Timing BTDC ①	Mark Fig.	Hot Idle Speed③ Std. Trans.	Hot Idle Speed③ Auto. Trans. ②	Air Fuel Ratio Std. Trans.	Air Fuel Ratio Auto. Trans.	Idle "CO" % Std. Trans.	Idle "CO" % Auto. Trans.
CORVETTE—Continued													
1971													
8-350, 270 H.P.	R44TS	.035	④	29–31	E	8°	B	600	550D	—	—	1.0	0.5
8-350, 330 H.P.	R43TS	.035	—	—	E	8°	B	700	—	—	—	—	—
8-454, 365 H.P.	R43TS	.035	④	29–31	E	8°	B	600	600D	—	—	1.0	1.0
8-454, 425 H.P.⑱	R44TS	.035	—	—	E	8°	B	700	—	—	—	—	—
8-454, 425 H.P.⑲	R44TS	.035	—	—	E	12°	B	—	700D	—	—	—	—
1972													
8-350, 200 H.P.	R44T	.035	④	29–31	E	8°	B	800	600D	—	—	—	—
8-350, 255 H.P.	R44T	.035	④	29–31	E	4°	B	900	—	—	—	—	—
8-454	R44T	.035	④	29–31	E	8°	B	750	600D	—	—	—	—
1973													
8-350, 190 H.P.	R44T	.035	④	29–31	C	12°	B	900	600D	—	—	—	—
8-350, 250 H.P.	R44T	.035	④	29–31	C	8°	B	900	700D	—	—	—	—
8-454, 275 H.P.	R44T	.035	④	29–31	E	10°	B	900	600D	—	—	—	—
1974													
8-350, 195 H.P.⑱	R44T	.035	④	29–31	C	8°⑥	B	900	—	—	—	.5	.5
8-350, 195 H.P.⑲	R44T	.035	④	29–31	C	8°	B	—	600D	—	—	.5	.5
8-350, 250 H.P.	R44T	.035	④	29–31	C	8°	B	900	700D	—	—	.5	.5
8-454	R44T	.035	④	29–31	E	10°	B	800	600D	—	—	.5	.5
1975													
8-350⑱	R44TX	.060	—	—	I	6°⑥	B	800	—	—	—	—	—
8-350⑲	R44TX	.060	—	—	I	6°	B	—	600D	—	—	—	—
1976													
8-350	R45TS	.045	—	—	I	8°⑧	B	1000	700D	—	—	—	—
8-350⑭	R45TS	.045	—	—	I	12°	B	1000	700D	—	—	—	—

①—BTDC: Before top dead center.
②—D: Drive. N: Neutral.
③—Where two speeds are listed, lower speed indicates idle solenoid disconnected.
④—New points, .019", used .016". On V8s, turn adjusting screw in (clockwise) until engine misfires; then back off ½ turn.
⑤—Adjust timing at 800 R.P.M.
⑥—For California set at 4° BTDC.
⑦—With integral intake manifold and all California models.
⑧—For California set at 6° BTDC.
⑨—For California 600 R.P.M.
⑩—With Distributor 1110662 set at 8° BTDC.
⑪—With A/C "ON".
⑫—Exc. H.E.I., Fig. E; H.E.I., Fig. I.
⑬—For California set at 0°.
⑭—With dual exhaust.
⑯—With A/C "OFF".
⑱—With standard transmission.
⑲—With automatic transmission.
⑳—With distributor 1111496 set at 14°. All others set at 4°.
㉑—Marketed as 396 but actually 402 cu. in.

Continued

TUNE-UP NOTES—Continued

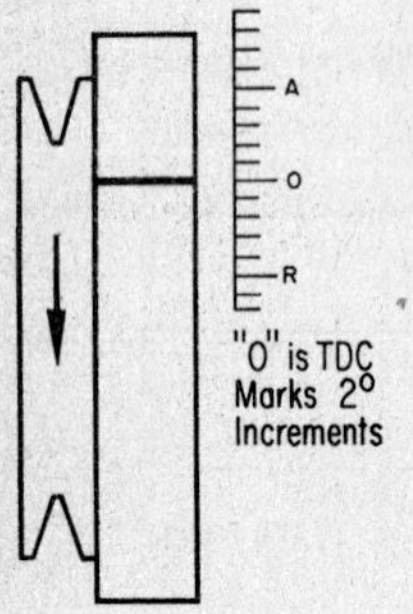

Fig. A

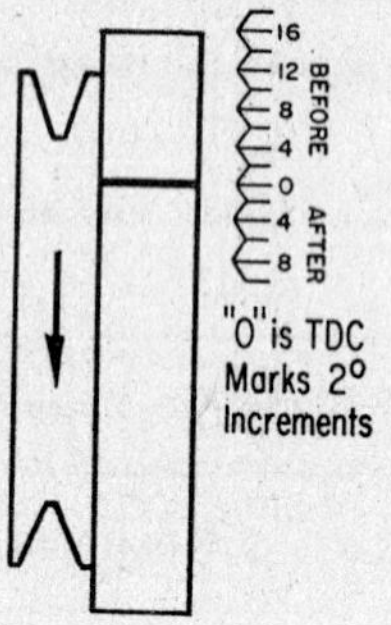

Fig. B

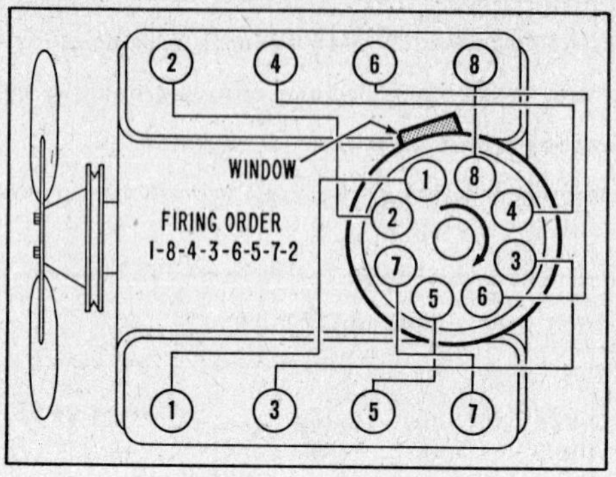

Fig. C

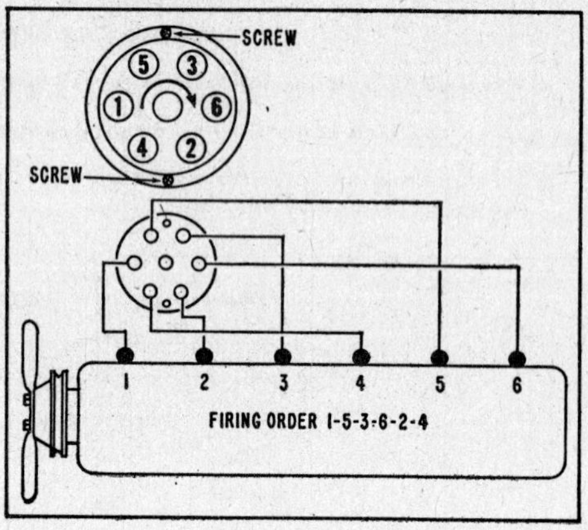

Fig. D

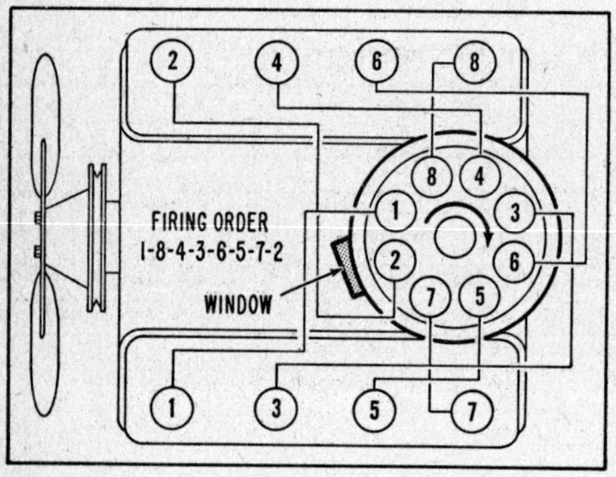

Fig. E

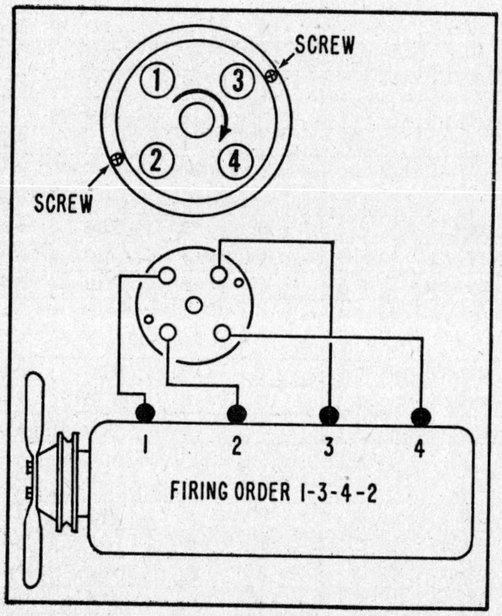

Fig. F

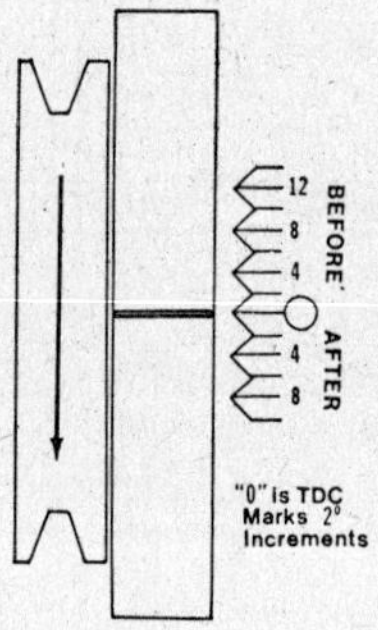

Fig. G

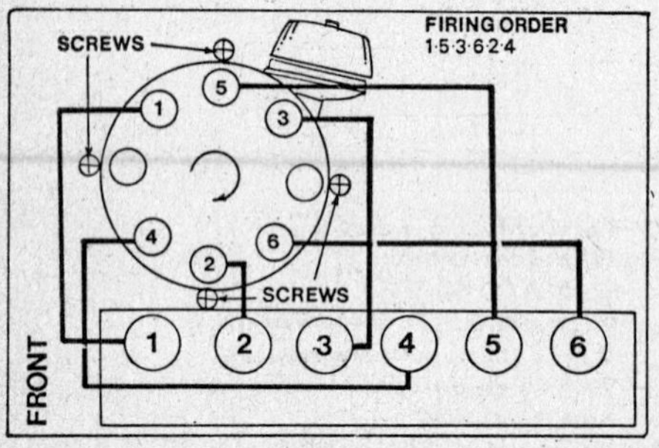

Fig. H

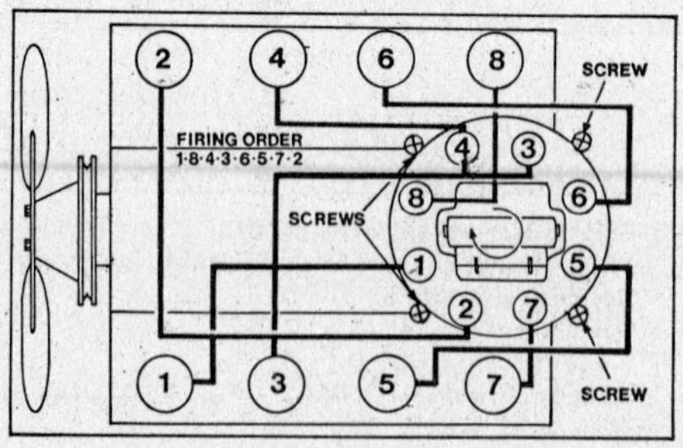

Fig. I

DISTRIBUTOR SPECIFICATIONS

★Note: If unit is checked on vehicle, double the RPM and degrees to get crankshaft figures.

Distributor Part No.①	Centrifugal Advance Degrees @ RPM of Distributor					Vacuum Advance	
	Advance Starts	Intermediate Advance			Full Advance	Inches of Vacuum to Start Plunger	Max. Adv. Dist. Deg. @ Vacuum
1969-70							
1110457②	0 @ 450	8½ @ 850	—	—	14 @ 1850	7	12 @ 15
1110458②	0 @ 450	7 @ 850	—	—	12 @ 1800	7	12 @ 15
1110459②	0 @ 500	3½ @ 600	10½ @ 1050	—	18 @ 2300	7	11½ @ 16
1110460②	0 @ 500	8½ @ 1050	—	—	16 @ 2300	7	11½ @ 16
1110463②	0 @ 450	4½ @ 610	10½ @ 975	—	16 @ 2100	7	11½ @ 16
1110464②	0 @ 450	8½ @ 975	—	—	14 @ 2100	7	11½ @ 16
1111436②	0 @ 400	1 @ 540	8½ @ 1050	—	13 @ 2000	8	7½ @ 15
1111437②	0 @ 400	4 @ 625	9½ @ 1000	—	13 @ 1900	—	—
1111480②	0 @ 625	11 @ 1100	—	—	16 @ 2200	8	7½ @ 15
1111481②	0 @ 500	5 @ 800	—	—	14 @ 2150	6	7½ @ 12
1111482②	0 @ 500	5 @ 700	—	—	16 @ 2150	6	7½ @ 12
1111483②	0 @ 450	6 @ 1000	—	—	14 @ 2150	6	7½ @ 12
1111486②	0 @ 400	1½ @ 500	7½ @ 900	—	18 @ 2050	7	6½ @ 19
1111487②	0 @ 450	1 @ 550	4 @ 700	—	16 @ 2200	7	6½ @ 17
1111488②	0 @ 475	7 @ 700	10 @ 900	—	15 @ 2350	10	5 @ 17
1111489②	0 @ 450	4½ @ 650	7½ @ 850	—	13 @ 2350	10	5 @ 17
1111490②	0 @ 450	4½ @ 600	7½ @ 750	—	15 @ 2550	8	5 @ 17
1111491②	0 @ 500	5 @ 650	—	—	13 @ 2500	7	7½ @ 12
1111492②	0 @ 400	1 @ 500	10 @ 1225	—	16 @ 2200	6	7½ @ 12
1111493②	0 @ 500	5 @ 850	—	—	13 @ 2500	7	7½ @ 12
1111494②	0 @ 350	1 @ 550	6 @ 1000	—	14 @ 2200	6	7½ @ 12
1111496②	0 @ 600	6 @ 1000	—	—	10 @ 2300	7	6 @ 12
1111497②	0 @ 450	4½ @ 625	8½ @ 1000	—	16 @ 2500	8	7½ @ 15
1111498②	0 @ 450	6½ @ 635	10½ @ 1000	—	18 @ 2500	8	7½ @ 15
1111499②	0 @ 450	4½ @ 625	8½ @ 1000	—	16 @ 2500	6	7½ @ 12
1111925②	0 @ 400	4 @ 625	9½ @ 1000	—	13 @ 1900	8	7½ @ 15
1111926②	0 @ 400	4 @ 625	8½ @ 1000	—	13 @ 1900	7	6 @ 12
1111927	0 @ 600	8 @ 950	—	—	14½ @ 2500	—	—
1111928	0 @ 450	1 @ 550	—	—	15 @ 1900	8	7½ @ 15
1111949	0 @ 450	6½ @ 650	11 @ 1000	—	19 @ 2100	8	7½ @ 15
1111950	0 @ 450	1½ @ 550	8½ @ 950	—	17 @ 2150	8	7½ @ 15
1111954	0 @ 450	4 @ 600	8½ @ 1000	—	13 @ 1850	7	6 @ 12
1111955	0 @ 450	8½ @ 950	11 @ 1000	—	16 @ 2200	7	6½ @ 17
1111956	0 @ 450	5 @ 600	10½ @ 1000	—	16 @ 2200	7	12 @ 13
1111963	0 @ 400	1 @ 540	8½ @ 1050	—	13 @ 2000	8	7½ @ 15
1111971	0 @ 475	1 @ 600	6 @ 1000	—	10 @ 2300	7	6 @ 12
1111995	0 @ 500	1½ @ 600	5 @ 800	—	14 @ 2150	6	7½ @ 12
1111996	0 @ 475	7 @ 700	10 @ 900	—	15 @ 2350	10	7½ @ 17
1111997	0 @ 450	4½ @ 650	7½ @ 850	—	13 @ 2350	10	10 @ 17
1111998	0 @ 450	4½ @ 625	8½ @ 1000	—	16 @ 2500	8	7½ @ 15
1111999	0 @ 450	6½ @ 635	10½ @ 1000	—	18 @ 2500	8	7½ @ 15
1112000	0 @ 450	4½ @ 625	8½ @ 1000	—	16 @ 2500	6	7½ @ 12
1112001	0 @ 400	1½ @ 500	7½ @ 900	—	18 @ 2050	7	12 @ 17
1112002	0 @ 450	1 @ 550	4 @ 700	—	16 @ 2200	7	12 @ 17
1112005	0 @ 400	1 @ 600	6 @ 1100	—	12 @ 2150	8	10 @ 17
1971-72							
1110489②	0 @ 475	1 @ 635	7 @ 1150	—	12 @ 2050	8	11½ @ 16
1112005②	0 @ 400	1 @ 600	6 @ 1100	—	12 @ 2150	8	10 @ 17
1112038②	0 @ 530	1 @ 670	8½ @ 1200	—	12 @ 2400	8	7½ @ 15

Continued

DISTRIBUTOR SPECIFICATIONS—Continued

★Note: If unit is checked on vehicle, double the RPM and degrees to get crankshaft figures.

Distributor Part No.①	Centrifugal Advance Degrees @ RPM of Distributor: Advance Starts	Intermediate Advance			Full Advance	Vacuum Advance: Inches of Vacuum to Start Plunger	Max. Adv. Dist. Deg. @ Vacuum
1971-72—Continued							
1112039②	0 @ 340	1 @ 670	—	—	10 @ 2100	8	10 @ 17
1112042②	0 @ 440	1 @ 560	5 @ 800	7½ @ 1100	14 @ 2150	8	10 @ 17
1112044②	0 @ 420	1 @ 580	5 @ 900	7½ @ 1200	11 @ 2100	8	7½ @ 15
1112045②	0 @ 435	1 @ 670	5½ @ 1200	—	9 @ 2100	8	7½ @ 15
1112049②	0 @ 535	1 @ 665	8 @ 1125	—	12 @ 2500	8	7½ @ 15
1112050②	0 @ 435	1 @ 670	5½ @ 1200	—	9 @ 2100	8	6 @ 15
1112051②	0 @ 430	1 @ 570	7 @ 1000	—	11 @ 1950	8	10 @ 17
1112052③	0 @ 430	1 @ 670	7 @ 1000	—	11 @ 1950	8	10 @ 17
1112053②	0 @ 545	1 @ 655	11 @ 1200	—	14 @ 2750	7	6 @ 12
1112054③	0 @ 545	1 @ 660	10½ @ 1175	—	14 @ 2500	7	6 @ 12
1112055②	0 @ 500	1 @ 600	4½ @ 760	8½ @ 1150	14 @ 2250	10	9 @ 17
1112056②	0 @ 475	1 @ 635	7 @ 1150	—	12 @ 2250	10	9 @ 17
1112057③	0 @ 475	1 @ 630	8 @ 1200	—	15 @ 2200	8	10 @ 17
1112074②	0 @ 520	1 @ 680	6 @ 1100	—	10 @ 2500	8	7½ @ 15
1112075③	0 @ 650	2½ @ 750	6 @ 1200	—	8 @ 2500	7	6 @ 12
1112076②	0 @ 550	1½ @ 650	5.4 @ 795	12½ @ 1175	16 @ 2500	7	6 @ 12
1112095②	0 @ 545	1 @ 655	10½ @ 1175	—	14 @ 2500	8	7½ @ 15
1112099②	0 @ 475	1 @ 635	7 @ 1150	7 @ 1150	12 @ 2250	8	10 @ 17
1973							
1110499②	0 @ 465	1 @ 635	7 @ 1150	—	12 @ 2050	7	12 @ 15
1112093②	0 @ 550	1 @ 660	3 @ 900	5½ @ 1200	9 @ 2100	6	7½ @ 14
1112094②	0 @ 550	1 @ 775	3 @ 1205	6 @ 1650	7 @ 2100	6	7½ @ 14
1112098②	0 @ 550	1 @ 775	3 @ 1205	6 @ 1650	7 @ 2100	6	7½ @ 14
1112102②	0 @ 500	1 @ 660	5 @ 1050	—	10 @ 2100	6	7½ @ 12
1112113③	0 @ 550	1 @ 660	5½ @ 1200	—	9 @ 2100	6	10 @ 15
1112114③	0 @ 550	1 @ 670	5½ @ 1200	—	9 @ 2100	6	10 @ 15
1112148②	0 @ 600	1 @ 730	6 @ 1100	—	10 @ 2500	6	7½ @ 12
1112150②	0 @ 600	1 @ 730	6 @ 1100	—	10 @ 2500	6	7½ @ 12
1112166②	0 @ 500	1 @ 650	5 @ 1300	—	10 @ 2100	8	7½ @ 15.5
1112168②	0 @ 500	1 @ 650	5 @ 1300	—	10 @ 2100	4	7 @ 7
1112227②	0 @ 500	1 @ 600	6 @ 1100	—	12 @ 2150	6	7½ @ 12
1112230②	0 @ 500	1 @ 650	5 @ 1300	—	10 @ 2100	4	5 @ 5.8
1974							
1110499②	0 @ 550	7 @ 1150	—	—	12 @ 2050	7	12 @ 15.5
1112093②	0 @ 550	5½ @ 1200	—	—	9 @ 2100	6	7½ @ 14
1112113③	0 @ 550	5½ @ 1200	—	—	9 @ 2100	6	10 @ 15.7
1112114③	0 @ 550	5½ @ 1200	—	—	9 @ 2100	6	10 @ 15.7
1112168	0 @ 500	—	—	—	10 @ 1100	4	7 @ 8.5
1112247②	0 @ 550	5½ @ 1200	—	—	9 @ 2100	6	7½ @ 14
1112250②	0 @ 550	5½ @ 1200	—	—	9 @ 2100	10	5 @ 15.5
1112504③	0 @ 550	5½ @ 1200	—	—	9 @ 2100	8	8 @ 16
1112846	0 @ 500	—	—	—	10 @ 2100	4	15 @ 10.5
1112847	0 @ 550	5½ @ 1200	—	—	9 @ 2100	6	14 @ 8.5
1112849	0 @ 500	5 @ 900	7½ @ 1200	—	11 @ 2100	3	14 @ 8.5
1112850	0 @ 500	5 @ 900	7½ @ 1200	—	11 @ 2100	3	14 @ 8.5
1112851	0 @ 550	5½ @ 1200	—	—	9 @ 2100	3	14 @ 8.5
1112852	0 @ 600	1 @ 730	6 @ 1100	—	10 @ 2500	3	14 @ 8.5

Continued

DISTRIBUTOR SPECIFICATIONS—Continued

★Note: If unit is checked on vehicle, double the RPM and degrees to get crankshaft figures.

Distributor Part No.①	Centrifugal Advance Degrees @ RPM of Distributor: Advance Starts	Centrifugal Advance: Intermediate Advance			Centrifugal Advance: Full Advance	Vacuum Advance: Inches of Vacuum to Start Plunger	Vacuum Advance: Max. Adv. Dist. Deg. @ Vacuum
1974—Continued							
1112853	0 @ 500	6 @ 1100	—	—	10 @ 2500	3	14 @ 8.5
1112854	0 @ 500	—	—	—	10 @ 2100	4	15 @ 10.5
1975							
1110650	0 @ 550	3½ @ 1150	—	—	8 @ 2100	4	9 @ 12
1110652	0 @ 550	7 @ 1150	—	—	12 @ 2050	7	12 @ 15
1112863	0 @ 550	3½ @ 1150	—	—	8 @ 2100	4	9 @ 12
1112880	0 @ 600	6 @ 1000	—	—	11 @ 2100	4	9 @ 12
1112882	0 @ 500	4 @ 800	—	—	7½ @ 1400	8	7½ @ 15.5
1112883	0 @ 550	6 @ 800	8 @ 1200	—	11 @ 2300	4	7½ @ 10
1112886	0 @ 650	—	—	—	6 @ 2100	4	9 @ 7
1112888	0 @ 550	6 @ 800	—	—	8 @ 2100	4	9 @ 12
1112933	0 @ 600	4½ @ 1000	—	—	11 @ 2000	3	8 @ 8
1976							
1103200	0 @ 600	6½ @ 800	—	—	8 @ 1000	4	5 @ 8
1110662	0 @ 550	7 @ 1150	—	—	12 @ 2050	4	9 @ 12
1110666	0 @ 500	3½ @ 800	—	—	10 @ 2100	4	12 @ 15
1112863	0 @ 550	3½ @ 1150	—	—	8 @ 2100	4	9 @ 12
1112880	0 @ 600	6 @ 1000	—	—	11 @ 2100	4	9 @ 12
1112882	0 @ 500	4 @ 800	—	—	7½ @ 1400	8	7½ @ 15.5
1112886	0 @ 650	—	—	—	6 @ 2100	4	9 @ 7
1112888	0 @ 550	6 @ 800	8 @ 1200	—	11 @ 2300	4	9 @ 12
1112905	0 @ 600	6 @ 1000	—	—	11 @ 2100	6	7½ @ 12
1112977	0 @ 500	5 @ 850	—	—	10 @ 1900	4	9 @ 12
1112999	0 @ 500	5 @ 850	—	—	10 @ 1900	4	5 @ 8

①—Stamped on distributor housing cover. ②—Breaker arm spring tension—19–23. ③—Breaker arm spring tension—28–32.

REAR AXLE SPECIFICATIONS

Year	Model	Carrier Type	Ring Gear & Pinion Backlash: Method	Ring Gear & Pinion Backlash: Adjustment	Pinion Bearing Preload: Method	Pinion Bearing Preload: New Bearings Inch-Lbs.	Pinion Bearing Preload: Used Bearings Inch-Lbs.	Differential Bearing Preload: Method	Differential Bearing Preload: New Bearings Inch-Lbs.	Differential Bearing Preload: Used Bearings Inch-Lbs.
1969	Exc. Corvette	Integral	Shims	.005–.008	Spacer	20–30①	5–15①	Shims	.010	.010
1969	Corvette	Integral	Shims	.005–.008	Spacer	20–25①	5–15①	Shims	.010	.010
1970	All	Integral	Shims	.005–.008	Spacer	20–25①	5–10①	Shims	.010	.010
1971–76	Corvette	Integral	Shims	.005–.008	Spacer	20–25①	5–10①	Shims	.010	.010
1971–72	Exc. Corvette	Integral	Shims	.005–.008	Spacer	25①	10①	Shims	.010	.010
1973–76	Exc. Corvette	Integral	Shims	.005–.008	Spacer	15–30①	5–10①	Shims	.010	.010

①—Use inch-pound torque wrench on pinion shaft nut.

VALVE SPECIFICATIONS

★Adjust hydraulic lifters by tightening rocker arm stud nut just to the point where all lash is eliminated. Then turn nut the additional turns listed. See Valves Adjust text for details.

Year	Engine Model	Valve Lash★		Valve Angles		Valve Spring Installed Height	Valve Spring Pressure Lbs. @ In.	Stem Clearance		Stem Diameter	
		Int.	Exh.	Seat	Face			Intake	Exhaust	Intake	Exhaust
1969	4-153	1 Turn(5)		46	45	1.66	175 @ 1.26	.001-.0027	.0015-.0032	.3410-.3417	.3410-.3417
	6-230	1 Turn(5)		46	45	1.66	175 @ 1.33	.001-.0027	.0015-.0032	.3410-.3417	.3410-.3417
	6-250	1 Turn(5)		46	45	1.66	186 @ 1.27	.001-.0027	.0015-.0032	.3410-.3417	.3410-.3417
	8-302	.030H	.030H	46	45	1.70	200 @ 1.25	.001-.0027	.001-.0027	.3410-.3417	.3410-.341/
	8-307, 327, 350(1)	1 Turn(5)		46	45	1.70	200 @ 1.25	.001-.0027	.001-.0027	.3410-.3417	.3410-.3417
	8-350, 370 H.P.	.030H	.030H	46	45	1.70	200 @ 1.25	.001-.0027	.001-.0027	.3410-.3417	.3410-.3417
	8-396, 265, 325 H.P.	1 Turn(5)		46	45	1.88	220 @ 1.46	.001-.0025	.0012-.0027	.3715-.3722	.3713-.3722
	8-396, 350	1 Turn(5)		46	45	1.88	312 @ 1.38	.001-.0025	.0012-.0027	.3715-.3722	.3713-.3722
	8-396, 375	.024H	.028H	46	45	1.88	312 @ 1.38	.001-.0025	.0012-.0027	.3715-.3722	.3713-.3722
	8-427(6)	1 Turn		46	45	1.88	312 @ 1.38	.001-.0025	.0012-.0027	.3715-.3722	.3713-.3722
	8-427, 425 H.P.	.024H	.028H	46	45	1.88	312 @ 1.38	.001-.0025	.0012-.0027	.3715-.3722	.3713-.3722
	8-427, 430 H.P.	.022H	.024H	46(4)	45	1.88	198 @ 1.32	.001-.0025	.0012-.0027	.3715-.3722	.3713-.3722
	8-427, 435 H.P.	.024H	.028H	46(4)	45	1.88	312 @ 1.38	.001-.0025	.0012-.0027	.3715-.3722	.3713-.3722
1970	4-153	1 Turn(5)		46	45	1.66	175 @ 1.26	.001-.0027	.0015-.0032	.3410-.3417	.3410-.3417
	6-230	1 Turn(5)		46	45	1.66	177 @ 1.33	.001-.0027	.0015-.0032	.3410-.3417	.3410-.3417
	6-250	1 Turn(5)		46	45	1.66	186 @ 1.27	.001-.0027	.0015-.0032	.3410-.3417	.3410-.3417
	8-307	1 Turn(5)		46	45	1.70	200 @ 1.25	.001-.0027	.0012-.0029	.3410-.3417	.3410-.3417
	8-350, 250, 300 H.P.	1 Turn(5)		46	45	1.70	200 @ 1.25	.001-.0027	.0012-.0029	.3410-.3417	.3410-.3417
	8-350, 360 H.P.	.024H	.030H	46	45	1.70	200 @ 1.25	.001-.0027	.0012-.0029	.3410-.3417	.3410-.3417
	8-350, 370 H.P.	.020H	.025H	46	45	1.70	200 @ 1.25	.001-.0027	.0012-.0029	.3410-.3417	.3410-.3417
	8-396, 350 H.P.(7)	1 Turn(5)		46	45	1.88	240 @ 1.38	.001-.0027	.0012-.0027	.3715-.3722	.3715-.3722
	8-396, 375 H.P.(7)	.024H	.028H	46	45	1.88	240 @ 1.38	.001-.0027	.0012-.0027	.3715-.3722	.3715-.3722
	8-400, 265 H.P.	1 Turn(5)		46	45	1.70	200 @ 1.25	.001-.0027	.0012-.0027	.3410-.3417	.3410-.3417
	8-400, 330 H.P.	1 Turn(5)		46	45	1.88	240 @ 1.38	.001-.0027	.0012-.0027	.3715-.3722	.3715-.3722
	8-454, 345 H.P.	1 Turn(5)		46	45	1.88	240 @ 1.38	.001-.0027	.0012-.0027	.3715-.3722	.3715-.3722
	8-454, 360, 390 H.P.	1 Turn(5)		46	45	1.88	240 @ 1.38	.001-.0027	.0012-.0027	.3715-.3722	.3715-.3722
	8-454, 450 H.P.	.024H	.028H	46	45	1.88	240 @ .138	.001-.0027	.0012-.0027	.3715-.3722	.3715-.3722
1971	6-250	1 Turn(5)		46	45	1.66	186 @ 1.27	.001-.0027	.0015-.0032	.3410-.3417	.3410-.3417
	8-307	1 Turn(5)		46	45	1.70	200 @ 1.25	.001-.0027	.001-.0027	.3410-.3417	.3410-.3417
	8-350, 245, 270 H.P.	1 Turn(5)		46	45	1.70	200 @ 1.25	.001-.0027	.001-.0027	.3410-.3417	.3410-.3417
	8-350, 330 H.P.	.024H	.030H	46	45	1.70	200 @ 1.25	.001-.0027	.001-.0027	.3410-.3417	.3410-.3417
	8-396(7)	1 Turn(5)		46	45	1.88	240 @ 1.38	.001-.0027	.001-.0027	.3715-.3722	.3715-.3722
	8-400	1 Turn(5)		46	45	1.70	200 @ 1.25	.001-.0027	.001-.0027	.3410-.3417	.3410-.3417
	8-454, 365 H.P.	1 Turn(5)		46	45	1.88	240 @ 1.38	.001-.0027	.001-.0027	.3715-.3722	.3715-.3722
	8-454, 425 H.P.	.024H	.028H	46	45	1.88	240 @ 1.38	.001-.0027	.001-.0027	.3713-.3720	.3713-.3720
1972	6-250	1 Turn(5)		46	45	1.66	186 @ 1.27	.001-.0027	.001-.0027	.3410-.3417	.3410-.3417
	8-307	1 Turn(5)		46	45	1.68	200 @ 1.17	.001-.0027	.001-.0027	.3410-.3417	.3410-.3417
	8-350, 165 H.P.	1 Turn(5)		46	45	1.70	200 @ 1.25	.001-.0027	.001-.0027	.3410-.3417	.3410-.3417
	8-350, 175, 200 H.P.	1 Turn(5)		46	45	1.70	200 @ 1.25	.001-.0027	.001-.0027	.3410-.3417	.3410-.3417
	8-350, 255 H.P.	.024H	.030H	46	45	1.70	200 @ 1.25	.001-.0027	.001-.0027	.3410-.3417	.3410-.3417
	8-400	1 Turn(5)		46	45	1.70	200 @ 1.25	.001-.0027	.001-.0027	.3410-.3417	.3410-.3417
	8-402	1 Turn(5)		46	45	1.88	215 @ 1.48	.001-.0027	.001-.0027	.3715-.3722	.3713-.3720
	8-454	1 Turn(5)		46	45	1.88	240 @ 1.38	.001-.0027	.001-.0027	.3715-.3722	.3713-.3720
1973	6-250	1 Turn(5)		46	45	1.66	186 @ 1.27	.001-.0027	.0015-.0032	.3410-.3417	.3410-.3417
	8-307	1 Turn(5)		46	45	1.61	189 @ 1.20	.001-.0027	.0012-.0029	.3410-.3417	.3410-.3417
	8-350, 145, 175 H.P.	1 Turn(5)		46	45	1.61	189 @ 1.20	.001-.0027	.0012-.0029	.3410-.3417	.3410-.3417
	8-350, 190 H.P.	1 Turn(5)		46	45	1.61	189 @ 1.20	.001-.0027	.0012-.0029	.3410-.3417	.3410-.3417
	8-350, 245, 250 H.P.	1 Turn(5)		46	45	1.70	200 @ 1.25	.001-.0027	.0012-.0029	.3410-.3417	.3410-.3417
	8-400	1 Turn(5)		46	45	1.70	200 @ 1.25	.001-.0027	.0012-.0027	.3410-.3417	.3410-.3417
	8-454	1 Turn(5)		46	45	1.88	300 @ 1.38	.001-.0027	.0012-.0027	.3715-.3722	.3715-.3722
1974	6-250	1 Turn(5)		46	45	1.66	186 @ 1.27	.001-.0027	.0015-.0032	.3410-.3417	.3410-.3417
	8-350, 145, 160 H.P.	1 Turn(5)		46	45	(2)	(3)	.001-.0027	.0012-.0029	.3410-.3417	.3410-.3417
	8-350, 185, 245 H.P.	1 Turn(5)		46	45	(2)	(3)	.001-.0027	.0012-.0029	.3410-.3417	.3410-.3417
	8-350, 250 H.P.	1 Turn(5)		46	45	1.70	200 @ 1.25	.001-.0027	.0012-.0029	.3410-.3417	.3410-.3417

Continued

VALVE SPECIFICATIONS—Continued

★Adjust hydraulic lifters by tightening rocker arm stud nut just to the point where all lash is eliminated. Then turn nut the additional turns listed. See Valves Adjust text for details.

Year	Engine Model	Valve Lash★ Int.	Valve Lash★ Exh.	Valve Angles Seat	Valve Angles Face	Valve Spring Installed Height	Valve Spring Pressure Lbs. @ In.	Stem Clearance Intake	Stem Clearance Exhaust	Stem Diameter Intake	Stem Diameter Exhaust
1974	8-400	1 Turn⑤		46	45	1.70	200 @ 1.25	.001–.0027	.0012–.0027	.3410–.3417	.3410–.3417
	8-454	1 Turn⑤		46	45	1.88	300 @ 1.38	.001–.0027	.0012–.0027	.3715–.3722	.3715–.3722
1975	6-250	1 Turn⑤		46	45	1.66	186 @ 1.27	.001–.0027	.0015–.0032	.3410–.3417	.3410–.3417
	V8-262	1 Turn⑤		46	45	②	③	.001–.0027	.001–.0027	.3410–.3417	.3410–.3417
	V8-350	1 Turn⑤		46	45	1.70	200 @ 1.25	.001–.0027	.0012–.0029	.3410–.3417	.3410–.3417
	V8-400	1 Turn⑤		46	45	1.70	200 @ 1.25	.001–.0027	.0012–.0027	.3410–.3417	.3410–.3417
	V8-454	1 Turn⑤		46	45	1.88	300 @ 1.38	.001–.0027	.0012–.0027	.3715–.3722	.3713–.3720
1976	6-250	¾ Turn⑤		46	45	1.66	175 @ 1.26	.0010–.0027	.0015–.0032	.3410–.3417	.3410–.3417
	8-305	¾ Turn⑤		46	45	1.70	200 @ 1.25	.0010–.0027	.0010–.0027	.3410–.3417	.3410–.3417
	8-350	¾ Turn⑤		46	45	1.70	200 @ 1.25	.0010–.0027	.0010–.0027	.3410–.3417	.3410–.3417
	8-400	¾ Turn⑤		46	45	1.70	200 @ 1.25	.0010–.0027	.0010–.0027	.3410–.3417	.3410–.3417
	8-454	¾ Turn⑤		46	45	1.88	300 @ 1.38	.0010–.0027	.0010–.0027	.3715–.3722	.3713–.3720

①—255, 300, 350 H.P.
②—Intake 1.70; exhaust 1.61.
③—Intake 200 @ 1.25; exhaust 189 @ 1.20.
④—Aluminum heads 45°.
⑤—Turn rocker arm stud nut until all lash is eliminated, then tighten nut the additional turn listed.
⑥—335, 385, 390, 400 H.P.
⑦—Marketed as 396 but actually 402 cu. in.
⑧—Corvette Hi Perf.

PISTONS, PINS, RINGS, CRANKSHAFT & BEARINGS

Year	Engine Model	Piston Clearance	Ring End Gap① Comp.	Ring End Gap① Oil	Wrist-pin Diameter	Rod Bearings Shaft Diameter	Rod Bearings Bearing Clearance	Main Bearings Shaft Diameter	Main Bearings Bearing Clearance	Thrust on Bear. No.	Shaft End Play
1969–70	4-153	.0005–.0015	.010	.015	.9272	1.999–2.000	.0007–.0027	2.2983–2.2993	.0003–.0029	5	.002–.006
1969–70	6-230, 250	.0005–.0015⑪	.010	.015	.9272	1.999–2.000	.0007–.0027	2.2983–2.2993	.0003–.0029	7	.002–.006
1971–72	6-250	.0015–.0025	.010	.015	.9272	1.999–2.000	.0007–.0027	2.2983–2.2993	.0003–.0029	7	.002–.006
1973–74	6-250	.0005–.0015	.010	.015	.9272	1.9928–2.000	.0007–.0027	2.2983–2.2993	.0003–.0029	7	.002–.006
1975	6-250	.0005–.0015	.010	.015	.9272	1.9928–2.000	.0007–.0027	2.2983–2.2993	.0003–.0029	7	.002–.006
1976	6-250	.0010–.0020	.010	.015	.9272	1.9928–2.000	.0007–.0027	2.2983–2.2993	.0003–.0029	7	.002–.006
1975	V8-262	.0007–.0013	⑭	.015	.9272	2.098–2.099	.0013–.0035	㉒	㊴	5	.002–.007
1976	8-305	.0007–.0017	.010	.015	.9272	2.099–2.100	.0013–.0035	㉒	⑮	5	.002–.006
1969	8-302	.0024–.0030	.013	.015	.9272	1.999–2.000	.0007–.0028	2.4479–2.4488	.0008–.0030	5	.003–.011
1969	8-307	.0005–.0011	.010	.015	.9272	2.099–2.100	.0007–.0027	2.4479–2.4488	.0008–.0020	5	.003–.011
1970	8-307	.0005–.0011	.010	.015	.9272	2.099–2.100	.0007–.0028	②	㉘	5	.002–.006
1971	8-307	.0005–.0011	.010	.015	.9272	2.099–2.100	.0013–.0035	②	㊴	5	.002–.006
1972–73	8-307	.0005–.0011	.010	.015	.9272	2.099–2.100	.0013–.0035	㉒	㊴	5	.002–.006
1969	8-327	.0005–.0011	.013	.015	.9272	2.099–2.100	.0007–.0028	2.4479–2.4488	㉗	5	.003–.011
1969	8-350㉕	.0005–.0011	.013	.015	.9272	2.099–2.100	.0007–.0028	2.4479–2.4488	㉗	5	.003–.011
	8-350㉖	.0024–.0030	⑭	.015	.9272	2.099–2.100	.0007–.0028	2.4479–2.4488	㉗	5	.003–.011
1970	8-350㉕	.0007–.0013	⑭	.015	.9272	2.099–2.100	.0007–.0028	②	㉘	5	.002–.006
	8-350㉙	.0020–.0026	⑭	.015	.9272	2.009–2.100	.0007–.0028	②	㉘	5	.002–.006
	8-350㉚	.0036–.0042	⑭	.015	.9272	2.099–2.100	.0007–.0028	②	㉘	5	.002–.006

Continued

PISTONS, PINS, RINGS, CRANKSHAFT & BEARINGS—Continued

Year	Engine Model	Piston Clearance	Ring End Gap①		Wrist-pin Diam-eter	Rod Bearings		Main Bearings			
			Comp.	Oil		Shaft Diameter	Bearing Clearance	Shaft Diameter	Bearing Clearance	Thrust on Bear. No.	Shaft End Play
1971	8-350㊵	.0007–.0017	.010	.015	.9272	2.099–2.100	.0013–.0035	②	㊴	5	.002–.006
	8-350㊶	.0036–.0042	.010	.015	.9272	2.099–2.100	.0013–.0035	⑫	⑮	5	.002–.006
1972	8-350㊸	.0007–.0013	⑭	.015	.9272	2.099–2.100	.0013–.0035	㉒	㊴	5	.002–.006
	8-350⑬	.0036–.0042	⑭	.015	.9272	2.099–2.100	.0013–.0035	㉒	⑮	5	.002–.006
1973–75	8-350㊶	.0007–.0013	⑭	.015	.9272	2.099–2.100	.0013–.0035	㉒	㊴	5	.002–.006
	8-350⑯	.0036–.0042	⑭	.015	.9272	2.099–2.100	.0013–.0035	㉒	⑮	5	.002–.006
1976	8-350	.0007–.0017	㊺	.015	.9272	2.099–2.100	.0013–.0035	㉒	⑮	5	.002–.006
	8-350㊻	.0046–.0056	.010	.015	.9272	2.099–2.100	.0013–.0035	㉒	⑮	5	.002–.006
1969	8-396⑰	.0010–.0018	.010	.010	.9897	2.199–2.200	.0009–.0025	⑲	⑤	5	.006–.010
	8-396⑱	.0036–.0044	.010	.010	.9897	2.1985–2.1995	.0014–.0030	④	⑥	5	.006–.010
1970	8-396㉙㊳	.0018–.0026	.010	.010	.9897	2.199–2.200	.0009–.0025	③	㉝	5	.006–.010
	8-396⑱㊳	.0036–.0046	.010	.010	.9897	2.1985–2.1995	.0014–.0030	⑨	㉞	5	.006–.010
1971	8-396㊳	.0018–.0026	.010	.015	.9897	2.199–2.200	.0009–.0025	③	㊷	5	.006–.010
1970	8-400㉛	.0014–.0020	.010	.015	.9897	2.099–2.100	.0009–.0030	⑩	㉘	5	.002–.006
	8-400㉜	.0018–.0026	.010	.015	.9897	2.199–2.200	.0009–.0025	⑩	㉝	5	㊹
1971	8-400	.0014–.0020	.010	.015	.9272	2.099–2.100	.0013–.0035	⑩	㊴	5	.006–.010
1972–76	8-400	.0014–.0020	.010	.015	.9272	2.099–2.100	.0013–.0035	⑩	㊴	5	.002–.006
1972	8-402	.0018–.0026	.010	.010	.9897	2.199–2.200	.0009–.0025	2.7504	㊷	5	.006–.010
1969	8-427⑳	.0012–.0020	.010	.010	.9897	2.199–2.200	.0009–.0025	㉑	⑥	5	.006–.010
	8-427㉒	.0037–.0043	.010	.010	.9897	2.1985–2.1995	.0014–.0030	㉑	⑥	5	.006–.010
	8-427㉓	.0058–.0066	.010	.010	.9897	2.1985–2.1995	.0014–.0030	㉑	⑥	5	.006–.010
	8-427㉔	.0040–.0048	.010	.010	.9897	2.1985–2.1995	.0014–.0030	㉑	⑥	5	.006–.010
1970	8-454㉟	.0024–.0034	.010	.010	.9897	2.199–2.200	.0009–.0025	④	㊲	5	.006–.010
	8-454㊱	.0040–.0050	.010	.010	.9897	2.1985–2.1995	.0014–.0030	㉑	㉞	5	.006–.010
1971	8-454⑦	.0024–.0034	.010	.015	.9897	2.199–2.200	.0009–.0025	④	㊲	5	.006–.010
	8-454⑧	.0040–.0050	.010	.015	.9897	2.199–2.200	.0009–.0025	㉑	㉞	5	.006–.010
1972	8-454	.0024–.0034	.010	.010	.9897	2.199–2.200	.0009–.0025	④	㊲	5	.006–.010
1973–76	8-454	.0018–.0028	.010	.015	.9897	2.199–2.200	.0009–.0025	④	㊲	5	.006–.010

①—Fit rings in tapered bores to the clearance listed in tightest portion of ring travel. Clearances specified are minimum gaps.
②—No. 1, 2, 3, 4: 2.4484–2.4493; No. 5: 2.4479–2.4488.
③—No. 1 & 2: 2.7487–2.7496; No. 3 & 4: 2.7481–2.7490; No. 5: 2.7478–2.7488.
④—1: 2.7485–2.7494; 2, 3 & 4: 2.7481–2.7490; 5: 2.7478–2.7488.
⑤—No. 1 & 2: .0010–.0022; No. 3 & 4: .0013–.0025; No. 5: .0015–.0031.
⑥—No. 1, 2, 3 & 4: .0013–.0025; No. 5: .0015–.0031.
⑦—Except 425 H.P.
⑧—425 H.P.
⑨—No. 1, 2, 3 & 4: 2.7481–2.7490; No. 5: 2.7473–2.7483.
⑩—No. 1, 2, 3 & 4: 2.6484–2.6493; No. 5: 2.6479–2.6488.
⑪—6-250 Piston Clearance .0015–.0025.
⑫—Manual/Trans.; No. 1, 2, 3 & 4: 2.7481–2.7490; No. 5: 2.7473–2.7483. Auto/Trans.; No. 1: 2.745–2.7484; No. 2, 3 & 4: 2.7481–2.7490; No. 5: 2.7473–2.7483.
⑬—255 H.P.
⑭—Top ring, .010, lower ring, .013″.
⑮—Manual/Trans.; No. 1, 2, 3 & 4: .0013–.0025; No. 5: .0023–.0033. Auto/Trans.; No. 1: .0019–.0031; No. 2, 3 & 4: .0013–.0025; No. 5: .0023–.0033.
⑯—245, 250 H.P.
⑰—Except 375 H.P.
⑱—375 H.P.
⑲—1 & 2: 2.7484–2.7493; 3 & 4: 2.7481–2.7490; 5: 2.7478–2.7488.
⑳—335, 390, 400 H.P.
㉑—1, 2, 3 & 4: 2.7481–2.7490; 5: 2.7478–2.7488.
㉒—No. 1: 2.4484–2.4493; No. 2, 3 & 4: 2.4481–2.4490; No. 5: 2.4479–2.4488.
㉓—430 H.P.
㉔—435 H.P.
㉕—250, 255 & 300 H.P.
㉖—350 & 370 H.P.
㉗—No. 1, 2, 3, 4: .0008–.002; No. 5: .0018–.0034.
㉘—No. 1: .0003–.0015; No. 2, 3, 4: .0006–.0018; No. 5. .0008–.0023.
㉙—350 H.P.
㉚—360 & 370 H.P.
㉛—265 H.P.
㉜—330 H.P.
㉝—No. 1: .0007–.0019; No. 2, 3, 4: .0013–.0025; No. 5: .0024–.0040.
㉞—No. 1, 2, 3, 4: .0013–.0025; No. 5: .0029–.0045.
㉟—Except 450 H.P.
㊱—450 H.P.
㊲—No. 1, 2, 3, 4: .0013–.0025; No. 5: .0024–.0040.
㊳—Marketed as 396 but actually 402 cu. in.
㊴—No. 1: .0008–.0020; No. 2, 3, 4: .0011–.0023; No. 5: .0017–.0033.
㊵—245 & 270 H.P.
㊶—Except 245, 250 H.P.
㊷—No. 1: .0007–.0019; No. 2, 3, 4: .0013–.0025; No. 5: .0019–.0035.
㊸—165, 175, 200 H.P.
㊹—8-400 2 bbl. carb., .002–.006; 8-400 4 bbl. carb., .006–.010.
㊺—Exc. 145 H.P., top ring .010, bottom ring .013; 145 H.P. top & bottom ring .010.
㊻—210 H.P.

ENGINE TIGHTENING SPECIFICATIONS★

★Torque specifications are for clean and lightly lubricated threads only. Dry or dirty threads produce increased friction which prevents accurate measurement of tightness.

Year	Engine Model	Spark Plugs Ft. Lbs.	Cylinder Head Bolts Ft. Lbs.	Intake Manifold Ft. Lbs.	Exhaust Manifold Ft. Lbs.	Rocker Arm Stud Ft. Lbs.	Rocker Arm Cover Ft. Lbs.	Connecting Rod Cap Bolts Ft. Lbs.	Main Bearing Cap Bolts Ft. Lbs.	Flywheel to Crankshaft Ft. Lbs.	Vibration Damper or Pulley Ft. Lbs.
1969–70	4-153	25	95	③	③	—	55④	35	65	60	②
1969	6-230	25	95	③	③	—	55④	35	65	60	②
1970	6-230	15	95	③	③	—	55④	35	65	60	②
1969	6-250	25	95	③	③	—	55④	35	65	60	②
1970	6-250	15	95	③	③	—	55④	35	65	60	②
1971–74	6-250	15	95	③	③	—	45④	35	65	60	60
1975	6-250	15	95	⑬	30⑭	—	45④	35	65	60	60
1976	6-250	15	95⑮	⑬	30	—	80④	35	65	60	60
1975	8-262	15	65	30	20	—	45④	45	70	60	60
1969	8-302	25	65	30	20	—	55④	45	80	60	60
1976	8-305	15	65	30	20	50	45④	45	70	60	60
1969–70	8-307	25	65	30	20①	—	55④	45	75	60	②
1971–73	8-307	15	65	30	20①	—	45④	45	75⑫	60	60
1969	8-327	25	65	30	20	—	55④	⑩	80	60	60
1969–70	8-350	25	65	30	20①	—	55④	45	75	60	60
1971–74	8-350	15	65	30	20①	50	45④	45	75⑫	60	60
1975–76	8-350	15	65	30	20①	—	45④	45	70⑫	60	60
1969	8-396	25	80	30	20	50	50④	45	⑤	60	85
1970	8-396⑪	15	80	30	20	50	50④	50	105	65	85
1970–74	8-400⑥	25	65	30	20	—	55④	45	75	60	60
1970	8-400⑦	15	80	30	20	50	50④	50	105	65	85
1975–76	8-400	15	65	30	20	50	45④	45	70	60	60
1971–72	8-402	15	80	30	20	50	50④	50	105	65	85
1969	8-427	25	80⑧	30	20	50⑨	50④	45	⑤	60	85
1970–73	8-454	15	80	30	20	50	50④	50	105	65	85
1974–76	8-454	15	80	30	20	50	50④	50	110	65	85

①—Inside bolts 30 ft. lbs.
②—Pressed on.
③—End clamp bolts 20, center bolts 30.
④—Inch lbs.
⑤—2 bolt caps 95 ft.-lbs., 4 bolt caps 105 ft.-lbs.
⑥—Exc. 330 H.P.
⑦—330 H.P.
⑧—Aluminum Head–Short bolts 65 ft. lbs Long bolts 75 ft. lbs.
⑨—Aluminum Head–60 ft. lbs.
⑩—For 3/8" bolts 45 ft.-lbs.; for 11/32" bolts 35 ft.-lbs.
⑪—Marketed as 396 but actually 402 cu. in.
⑫—Outer bolts on engines with 4 bolt caps 65 ft. lbs.
⑬—Integral Intake Manifold.
⑭—Outer bolts 20 ft. lb.
⑮—Left hand front head bolt 85 ft. lbs.

BRAKE SPECIFICATIONS

Year	Model	Brake Drum Inside Diameter	Wheel Cylinder Bore Diameter			Master Cylinder Bore Diameter		
			Disc Brake	Front Drum Brake	Rear Drum Brake	Disc Brakes	Drum Brakes	Power Brakes
1969	Camaro	9½	2 1/16④	1⅛	⅞	—	1	1
	Chevelle	9½	2 1/16	1⅛	⅞	1⅛	1	1
	Nova	9½	2 1/16	1⅛	⅞	—	1	1
	Chevrolet	11	2 1/16	1 3/16	1	1	1	1
	Corvette	—	③	—	—	1	—	1
1970–72	Camaro	9½	2 15/16	—	⅞	1⅛	—	1⅛
	Chevelle & Monte Carlo	9½	2 15/16	1⅛	⅞	1⅛	1	1
	Nova	9½	2 15/16	1⅛	⅞	1⅛	1	1
	Chevrolet	11⑦	2 15/16	1 3/16	15/16⑤	1⅛	—	1
	Corvette	—	③	—	—	1	—	1②

Continued

BRAKE SPECIFICATIONS—Continued

Year	Model	Brake Drum Inside Diameter	Wheel Cylinder Bore Diameter			Master Cylinder Bore Diameter		
			Disc Brake	Front Drum Brake	Rear Drum Brake	Disc Brakes	Drum Brakes	Power Brakes
1973–74	Camaro	9½	2 15/16	—	⅞	1	—	1⅛
	Chevelle & Monte Carlo	9½⑥	2 15/16	—	⅞①	1	1	1⅛
	Nova	9½	2 15/16	1⅛	⅞	1	1	1⅛
	Chevrolet	11⑦	2 15/16	—	15/16⑤	1⅛	—	1⅛
	Corvette	—	③	—	—	1	—	1⅛
1975	Camaro	9½	2 15/16	—	⅞	1	—	1⅛
	Chevelle & Monte Carlo	9½⑥	2 15/16	—	⅞①	1	—	1⅛
	Nova	9½	2 15/16	—	⅞	1	—	1⅛
	Chevrolet	11⑦	2 15/16	—	15/16⑤	1⅛	—	1⅛
	Corvette	—	③	—	—	1	—	1⅛
1976	Camaro	9½	2 15/16	—	15/16	1.00	—	1⅛
	Chevelle & Monte Carlo	11	2 15/16	—	1⑧	15/16⑨	—	1⅛
	Nova	9½	2 15/16	—	15/16	1	—	1⅛
	Chevrolet	11⑦	—	—	1	1⅛	—	1⅛
	Corvette	—	③	—	—	1	—	1⅛

①—Sta. Wagon 15/16″.
②—1972 power brakes 1⅛″.
③—Front 1⅞; Rear 1⅜.
④—Front discs only. For 4 wheel disc option, see Corvette.
⑤—1971–76 Sta. Wag., 1″.
⑥—Sta. Wagon 11″.
⑦—1971–76 Wagon, 12.
⑧—Power brakes 15/16″.
⑨—Malibu Classic and all V-8; 1⅛″.

STARTING MOTOR SPECIFICATIONS

Year	Model	Starter Number	Brush Spring Tension Oz.①	Free Speed Test			Resistance Test③	
				Amps.	Volts	R.P.M.①	Amps.	Volts
1969	8-327, 427	1108351	35	70–99②	10.6	6200–10700	300–360	3.5
	8-327, 350	1108361	35	65–100②	10.6	6200–10700	300–360②	3.5
	4-153, 6-230, 250	1108365	35	49–87②	10.6	6200–10700	290–425②	4.2
	4-153	1108366	35	49–87②	10.6	6200–10700	290–425②	4.2
	8-307, 302, 327	1108367	35	49–87②	10.6	6200–10700	290–425②	4.2
	8-350	1108338	35	55–85②	9	3100–4900	—	—
	8-396, 427	1108418	35	65–95②	9	7500–10500	—	—
	8-327	1108382	35	53–69②	9	6400–8600	—	—
	6-250	1107372	35	55–95②	9	3000–4800	—	—
	8-427	1108400	35	65–95②	9	7500–10500	—	—
1970	4 & 6 Cyl.	1108365	35	50–80②	9	5500–10500	—	—
	6-250	1107372	35	55–95②	9	3000–4800	—	—
	8-307	1108367	35	50–80②	9	5500–10500	—	—
	8-350	1108338	35	55–80②	9	3500–6000	—	—
	8-350	1108427	35	55–80②	9	3500–6000	—	—
	8-454	1108430	35	65–95②	9	7500–10500	—	—
	8-400	1108418	35	65–95②	9	7500–10500	—	—
	8-454	1108400	35	65–95②	9	7500–10500	—	—
1971–74	6-250	1108365	35	50–80②	9	5500–10500	—	—
	8-307	1108367	35	50–80②	9	5500–10500	—	—
	8-307⑤	1108512	35	50–80②	9	5500–10500	—	—
	④	1108418	35	65–95②	9	7500–10500	—	—
	④	1108430	35	65–95②	9	7500–10500	—	—
	8-454	1108429	35	65–95②	9	7500–10500	—	—
	8-454	1108400	35	65–95②	9	7500–10500	—	—

Continued

STARTING MOTOR SPECIFICATIONS—Continued

Year	Model	Starter Number	Brush Spring Tension Oz.①	Free Speed Test			Resistance Test③	
				Amps.	Volts	R.P.M.①	Amps.	Volts
1975	6-250	1108365	—	50–80②	9	5500–10500	—	—
	6-250	1108774	—	50–80②	9	5500–10500	—	—
	V8-262	1108512	—	55–80②	9	3500–6000	—	—
	V8-262	1108790	—	55–80②	9	3500–6000	—	—
	8-350, 400, 454	1108430	—	65–95②	9	7500–10500	—	—
	8-350, 400, 454	1108776	—	65–95②	9	7500–10500	—	—
	8-350, 454	1108418	—	65–95②	9	7500–10500	—	—
	8-350, 454	1108775	—	65–95②	9	7500–10500	—	—
1976	6-250	1108774	—	50–80②	9	5500–10500	—	—
	8-305	1108798	—	50–80②	9	5500–10500	—	—
	8-305	1108799	—	50–80②	9	5500–10500	—	—
	8-350, 400, 454	1108776	—	65–95②	9	7500–10500	—	—
	8-350, 454	1108775	—	65–95②	9	7500–10500	—	—

①—Minimum.

②—Includes solenoid.

③—Check capacity of motor by using a 500 ampere meter and a carbon pile rheostat to control voltage. Apply the volts listed across motor with armature locked. Current should be as listed.

④—Used on V8-350, 400, 402, and 454 engines.

⑤—1973 only.

ALTERNATOR & REGULATOR SPECIFICATIONS

Year	Alternator					Regulator						
	Model	Rated Hot Output Amps.	Field Current 12 Volts @ 80° F.	Output @ 14 Volts		Model	Field Relay			Voltage Regulator		
				2000 R.P.M. Amps.	5000 R.P.M. Amps.		Air Gap In.	Point Gap In.	Closing Voltage	Air Gap In.	Point Gap In.	Voltage @ 125° F.
1969–70	1100825①	61	4.0–4.5	—	—	—	—	—	—	—	—	—
	1100833①	42	4.0–4.5	—	—	—	—	—	—	—	—	—
	1100834	37	2.2–2.6	—	—	1119515	.015	.030	1.5–3.2	.067	.014	13.4–14.3
	1100836	37	2.2–2.6	25	35	1119515	.015	.030	1.5–3.2	.067	.014	13.4–14.3
	1100837	37	2.2–2.6	25	35	1119515	.015	.030	1.5–3.2	.067	.014	13.4–14.3
	1100839	42	2.2–2.6	—	—	1119515	.015	.030	1.5–3.2	.067	.014	13.4–14.3
	1100841	42	2.2–2.6	—	—	1119515	.015	.030	1.5–3.2	.067	.014	13.4–14.3
	1100843	61	2.2–2.6	33	58	1119515	.015	.030	1.5–3.2	.067	.014	13.4–14.3
	1100845	61	2.2–2.6	—	—	1119515	.015	.030	1.5–3.2	.067	.014	13.4–14.3
	1100846	63	2.2–2.6	—	—	1119515	.015	.030	1.5–3.2	.067	.014	13.4–14.3
	1100847	61	2.2–2.6	—	—	1119515	.015	.030	1.5–3.2	.067	.014	13.4–14.3
	1100859①	42	4.0–4.5	—	—	—	—	—	—	—	—	—
	1100896	61	2.2–2.6	—	—	1119515	.015	.030	1.5–3.2	.067	.014	13.4–14.3
	1100897	61	2.2–2.6	—	—	1119515	.015	.030	1.5–3.2	.067	.014	13.4–14.3
	1100900①	42	—	—	—	—	—	—	—	—	—	—
	1100901①	42	—	—	—	—	—	—	—	—	—	—
	1100950①	42	—	—	—	—	—	—	—	—	—	—
1971	1100543①	42	4–4.5	—	37	—	—	—	—	—	—	—
	1100544①	61	4–4.5	—	55	—	—	—	—	—	—	—
	1100566	37	2.2–2.6	25	35	1119515	.015	.030	1.5–3.2	.067	.014	13.8–14.8
	1100567	42	2.2–2.6	28	40	1119515	.015	.030	1.5–3.2	.067	.014	13.8–14.8
	1100836	37	2.2–2.6	25	35	1119515	.015	.030	1.5–3.2	.067	.014	13.8–14.8
	1100837	37	2.2–2.6	25	35	1119515	.015	.030	1.5–3.2	.067	.014	13.8–14.8
	1100843	61	2.2–2.6	33	58	1119515	.015	.030	1.5–3.2	.067	.014	13.8–14.8
	1100917	63	2.8–3.2	35	59	1119519	.030	.030	1.5–3.2	.067	.014	13.8–14.8
	1100950①	42	4–4.5	—	37	—	—	—	—	—	—	—

Continued

ALTERNATOR & REGULATOR SPECIFICATIONS—Continued

Year	Alternator					Regulator						
	Model	Rated Hot Output Amps.	Field Current 12 Volts @ 80° F.	Output @ 14 Volts		Model	Field Relay			Voltage Regulator		
				2000 R.P.M. Amps.	5000 R.P.M. Amps.		Air Gap In.	Point Gap In.	Closing Voltage	Air Gap In.	Point Gap In.	Voltage @ 125° F.
1972	1100543	42	4–4.5	—	37	—	—	—	—	—	—	—
	1100544	61	4–4.5	—	55	—	—	—	—	—	—	—
	1100950	42	4–4.5	—	37	—	—	—	—	—	—	—
	1102440	37	2.2–2.6	25	35	1119515	.015	.030	1.5–3.2	.067	.014	13.5–14.4
	1102452	37	2.2–2.6	25	35	1119515	.015	.030	1.5–3.2	.067	.014	13.5–14.4
	1102453	37	2.2–2.6	25	35	1119515	.015	.030	1.5–3.2	.067	.014	13.5–14.4
	1102454	37	2.2–2.6	25	35	1119515	.015	.030	1.5–3.2	.067	.014	13.5–14.4
	1102456	37	2.2–2.6	25	35	1119515	.015	.030	1.5–3.2	.067	.014	13.5–14.4
	1102458	42	2.2–2.6	28	40	1119515	.015	.030	1.5–3.2	.067	.014	13.5–14.4
	1102459	42	2.2–2.6	28	40	1119515	.015	.030	1.5–3.2	.067	.014	13.5–14.4
	1102463	61	2.2–2.6	33	58	1119515	.015	.030	1.5–3.2	.067	.014	13.5–14.4
	1102464	63	2.8–3.2	35	59	1119519	.030	.030	1.5–3.2	.067	.014	13.5–14.5
1973	1100542	63	4–4.5	—	58	Integral	—	—	—	—	—	—
	1102346	42	4–4.5	—	37	Integral	—	—	—	—	—	—
	1102354	63	4–4.5	—	58	Integral	—	—	—	—	—	—
1973–75	1100497	37	③	—	④	Integral	—	—	—	—	—	13.8–14.8
	1100934	37	4–4.5	—	32	Integral	—	—	—	—	—	13.8–14.8
	1100950	42	4–4.5	—	38	Integral	—	—	—	—	—	13.8–14.8
	1100544	61	4–4.5	—	55	Integral	—	—	—	—	—	—
	1100573	42	4–4.5	—	37	Integral	—	—	—	—	—	—
	1100597	61	4–4.5	—	55	Integral	—	—	—	—	—	—
	1102353	42	4–4.5	—	37	Integral	—	—	—	—	—	—
1974-75	1100560	55	4–4.5	—	50	Integral	—	—	—	—	—	—
	1100575	55	4–4.5	—	50	Integral	—	—	—	—	—	—
	1102347	61	4–4.5	—	55	Integral	—	—	—	—	—	—
1975	1102397	37	4–4.5	—	33	Integral	—	—	—	—	—	—
	1102483	37	4–4.5	—	33	Integral	—	—	—	—	—	—
	1102493	42	4–4.5	—	38	Integral	—	—	—	—	—	—
1976	1102394	37	4–4.5	—	33	Integral	—	—	—	—	—	—
	1102474	61	4–4.5	—	57	Integral	—	—	—	—	—	—
	1102478	55	4–4.5	—	51	Integral	—	—	—	—	—	—
	1102479	55	4–4.5	—	51	Integral	—	—	—	—	—	—
	1102480	61	4–4.5	—	57	Integral	—	—	—	—	—	—
	1102484	42	4–4.5	—	38	Integral	—	—	—	—	—	—
	1102486	61	4–4.5	—	57	Integral	—	—	—	—	—	—
	1102491	37	4–4.5	—	33	Integral	—	—	—	—	—	—

①—Integral System.
②—At 7000 RPM.
③—1973; 4.4–4.9; 1974–75: 4–4.5.
④—1973–74 36 @ 7000 RPM; 1975 33 @ 5000 RPM.

Continued

WHEEL ALIGNMENT SPECIFICATIONS

OLD CAR SPECIFICATIONS: For 1946-68 Wheel Alignment Specifications see main index.

Year	Model	Caster Angle, Degrees		Camber Angle, Degrees				Toe-In, Inch	Toe-Out on Turns, Deg.①	
		Limits	Desired	Limits		Desired			Outer Wheel	Inner Wheel
				Left	Right	Left	Right			
CAMARO										
1969	All	0 to +1	+½	−¼ to +¾	−¼ to +¾	+½	+½	⅛ to ¼	—	20
1970	Z-28	−¾ to +¼	−¼	+¼ to +1¼	+¼ to +1¼	+¾	+¾	⅛ to ¼	—	—
	Others	0 to +1	+½	+½ to +1½	+½ to +1½	+1	+1	⅛ to ¼	—	—
1971-72	Z-28	−½ to +½	Zero	−1¼ to −¼	−1¼ to −¼	−¾	−¾	⅛ to ¼	—	—
	Others	−1 to +1	Zero	+¼ to +1¾	+¼ to +1¾	+1	+1	1/16 to 5/16	—	—
1973-74	Z-28	−½ to −1½	−1	+¼ to +1¼	+¼ to +1¼	+¾	+¾	⅛ to ¼	—	—
	Others	−½ to +½	Zero	+½ to +1½	+½ to +1½	+1	+1	⅛ to ¼	—	—
1975	All	−½ to ½	Zero	+½ to +1½	+½ to +1½	+1	+1	0 to ⅛	—	—
1976	All	+½ to +1½	+1	+½ to +1½	+½ to +1½	+1	+1	0 to ⅛	—	—
CHEVELLE & MONTE CARLO										
1969-70	S.S., M. Carlo	−1 to 0	−½	0 to +1	0 to +1	+½	+½	⅛ to ¼	18.4	20
	Others	−1½ to −½	−1	0 to +1	0 to +1	+½	+½	⅛ to ¼	18.4	20
1971	All	−½ to −1½	−1	+¼ to +1¼	+¼ to +1¼	+¾	+¾	⅛ to ¼	—	—
1972	All	−2 to 0	−1	0 to +1½	0 to +1½	+¾	+¾	1/16–5/16	—	—
1973-74	Chevelle③	−½ to −1½	−1	+½ to +1½	0 to +1	+1	+½	0 to ⅛	—	—
	Chevelle④	−½ to +½	Zero	+½ to +1½	0 to +1	+1	+½	0 to ⅛	—	—
	Monte Carlo	+4½ to +5½	+5	+½ to +1½	0 to +1	+1	+½	0 to ⅛	—	—
1975-76	Chevelle③	+½ to +1½	+1	+½ to +1½	0 to +1	+1	+½	0 to ⅛	—	—
	Chevelle④	+1½ to +2½	+2	+½ to +1½	0 to +1	+1	+½	0 to ⅛	—	—
	Monte Carlo	+4½ to +5½	+5	+½ to +1½	0 to +1	+1	+½	0 to ⅛	—	—
CHEVY NOVA										
1969-71	All	0 to +1	+½	−¼ to +¾	−¼ to +¾	+½	+½	⅛ to ¼	—	20
1972	All	−1 to +1½	+½	−½ to +1	−½ to +1	+¼	+¼	1/16–5/16	—	—
1973-74	All	0 to +1	+½	−¼ to +¾	−¼ to +¾	+¼	+¼	⅛ to ¼	—	—
1975-76	All③	−½ to −1½	−1	+¼ to +1¼	+¼ to +1¼	+¾	+¾	0 to ⅛	—	—
	All④	+½ to +1½	+1	+¼ to +1¼	+¼ to +1¼	+¾	+¾	0 to ⅛	—	—
CHEVROLET										
1969-70	All	+¼ to +1¼	+¾	−¼ to +¾	−¼ to +¾	+¼	+¼	⅛ to ¼	20	22½
1971	All	−1½ to −½	−1	0 to +1	0 to +1	+½	+½	⅛ to ¼	—	—
1972	All	+½ to +1½	+1	0 to +1	0 to +1	+½	+½	⅛ to ¼	—	—
1973-74	All	+½ to +1½	+1	+½ to +1½	0 to +1	+1	+½	0 to ⅛	—	—
1975	All	+1 to +2	+1½	+½ to +1½	0 to +1	+1	+½	0 to ⅛	—	—
1976	All	⑤	⑥	+½ to +1½	0 to +1	+1	+½	0 to ⅛	—	—

Continued

WHEEL ALIGNMENT SPECIFICATIONS—Continued

OLD CAR SPECIFICATIONS: For 1946-68 Wheel Alignment Specifications see main index.

Year	Model	Caster Angle, Degrees		Camber Angle, Degrees				Toe-In. Inch	Toe-Out on Turns, Deg.①	
		Limits	Desired	Limits		Desired			Outer Wheel	Inner Wheel
				Left	Right	Left	Right			
CORVETTE										
1969–74	Manual Steer.	+½ to +1½	+1	+¼ to +1¼	+¼ to +1¼	+¾	+¾	3/16 to 5/16②	—	20
	Power Steer.	+1¾ to +2¾	+2¼	+¼ to +1¼	+¼ to +1¼	+¾	+¾	3/16 to 5/16②	—	20
	Rear Wheel Align.	—	—	−⅝ to −1⅛	−⅝ to −1⅛	−⅞	−⅞	1/32 to 3/32	—	—
1975	Manual Steer.	+½ to +1½	+1	+¼ to +1¼	+¼ to +1¼	+¾	+¾	3/16 to 5/16	—	—
	Power Steer.	+1¾ to +2¾	+2¼	+¼ to +1¼	+¼ to +1¼	+¾	+¾	3/16 to 5/16	—	—
	Rear Whl. Align.	—	—	−1 to 0	−1 to 0	−½	−½	0 to 3/16	—	—
1976	Manual Steer.	+½ to +1½	+1	+¼ to +1¼	+¼ to +1¼	+¾	+¾	3/16 to 5/16	—	—
	Power Steer.	+1¾ to +2¾	+2¼	+¼ to +1¼	+¼ to +1¾	+¾	+¾	3/16 to 5/16	—	—
	Rear Whl. Align.	—	—	−1⅛ to −⅝	−1⅛ to −⅝	−⅞	−⅞	1/32 to 3/32	—	—

①—Incorrect toe-out, when other adjustments are correct, indicates bent steering arms.

②—1971 ⅛" to ⅜".

③—Manual steering.

④—Power steering.

⑤—Equipped with radial tires, +1° to +2°; equipped with bias belted tires, +½° to +1½°.

⑥—Equipped with radial tires, +1½°; equipped with bias belted tires, +1°.

COOLING SYSTEM & CAPACITY DATA

Year	Model or Engine	Cooling Capacity, Qts.			Radiator Cap Relief Pressure, Lbs.		Thermo. Opening Temp. ①	Fuel Tank Gals.	Engine Oil Refill Qts. ②	Transmission Oil			Rear Axle Oil Pints
		No Heater	With Heater	With A/C	With A/C	No A/C				3 Speed Pints	4 Speed Pints	Auto. Trans. Qts. ⑮	
CAMARO													
1969	6-230, 250	11	13	13	15	15	195	18	4	3⑧	3	⑲	3½
	8-327	16	17	17	15	15	195	18	4	3⑧	3	⑲	3½
	8-350	15	16	17	15	15	195	18	4	3⑧	3	⑲	3½
	8-396	22	23	23	15	15	195	18	4	3⑧	3	⑲	4
1970	6-250	11	12	13	15	15	195	19	4	3	—	⑲	3½
	8-307	14	15	16	15	15	195	19	4	3	—	⑲	3½
	8-350 Exc. 360 H.P.	15	16	16	15	15	195	19	4	—	3	⑲	3½
	8-350, 360 H.P.	15	16	16	15	15	180	19	4	—	3	⑲	3½
	8-396㉕	22	23	23	15	15	195	19	4	—	3	⑲	3½
1971	6-250	11	12	12	15	15	195	18	4	3	—	⑲	3½
	8-307	14	15	15	15	15	195	18	4	3	—	⑲	3½
	8-350 Exc. 330 H.P.	14	15	15	15	15	195	18	4	—	3	⑲	㉖
	8-350, 330 H.P.	14	15	15	15	15	180	18	4	—	3	⑲	㉖
	8-396㉕	23	24	24	15	15	195	18	4	—	3	⑲	㉖
1972	6-250	11	12	13	15	15	195	18	4	3	—	⑲	4¼
	8-307	14	15	17	15	15	195	18	4	3	—	⑲	4¼
	8-350 Exc. 255 H.P.	15	16	17	15	15	195	18	4	—	3	⑲	4¼
	8-350, 255 H.P.	15	16	17	15	15	180	18	4	—	3	⑲	4¼
	8-402	23	24	25	15	15	195	18	4	—	3	⑲	4¼
1973	6-250	11	12	13	15	15	195	18	4	3	—	⑲	4¼
	V8-307	15	16	17	15	15	195	18	4	3	—	⑲	4¼
	V8-350, 145 H.P.	15	16	17	15	15	195	18	4	3	3	⑲	4¼
	V8-350, 175 H.P.	15	16	17	15	15	195	18	4	3	3	⑲	4¼
	V8-350, 245 H.P.	15	16	17	15	15	180	18	4	3	3	⑲	4¼

Continued

COOLING SYSTEM & CAPACITY DATA—Continued

Year	Model or Engine	Cooling Capacity, Qts.			Radiator Cap Relief Pressure, Lbs.		Thermo. Opening Temp. ①	Fuel Tank Gals.	Engine Oil Refill Qts. ②	Transmission Oil			Rear Axle Oil Pints
		No Heater	With Heater	With A/C	With A/C	No A/C				3 Speed Pints	4 Speed Pints	Auto. Trans. Qts. ⑮	
CAMARO—Continued													
1974	6-250	14	14	—	15	15	195	21	4	3	—	⑦	4¼
	8-350, 145 H.P.	18	18	—	15	15	195	21	4	3	3	⑦	4¼
	8-350, 160 H.P.	18	18	—	15	15	195	21	4	3	3	⑦	4¼
	8-350, 185 H.P.	18	18	—	15	15	195	21	4	3	3	⑦	4¼
	8-350, 245 H.P.	18	18	—	15	15	180	21	4	3	3	⑦	4¼
1975	6-250	—	12½	12½	15	15	195	21	4	3	—	4⑥	4¼
	8-350	—	15½	16½	15	15	195	21	4	3	3	4⑥	4¼
	8-350, Z28	—	15½	16½	15	15	180	21	4	—	3	4⑥	4¼
1976	6-250	—	15	16	15	15	195	21	4	3	—	⑥	4¼
	8-305	—	17½	18½	15	15	195	21	4	3	—	⑥	4¼
	8-350	—	17½	18½	15	15	195	21	4	—	3	⑥	4¼
CHEVELLE & MONTE CARLO													
1969	6-230, 250	11	13	13	15	15	195	20⑳	4	3⑧	—	⑲	3½
	8-307	16	17	18	15	15	195	20⑳	4	3⑧	3	⑲	3½
	8-350	15	16	17	15	15	195	20⑳	4	3⑧	3	⑲	4
	8-396	22	23	23	15	15	195	20⑳	4	3⑧	3	⑲	4
1970	6-250	11	12	13	15	15	195	20㉒	4	3⑧	—	⑲	⑨
	8-307	14	15	16	15	15	195	20㉒	4	3⑧	3	⑲	⑨
	8-350	15	16	16	15	15	195	20㉒	4	3⑧	3	⑲	⑨
	8-400	22	23	24	15	15	195	20㉒	4	3⑧	3	⑲	⑨
	8-454	21	22	23	15	15	195	20㉒	4	3⑧	3	⑲	⑨
1971	6-250	11	12	12	15	15	195	18	4	3	—	⑲	㉖
	8-307	15	16	16	15	15	195	18	4	3	—	⑲	㉖
	8-350	15	16	16	15	15	195	18	4	3	3	⑲	㉖
	8-396㉕	22	23	23	15	15	195	18	4	3	3	⑲	㉖
	8-454, 365 H.P.	21	22	22	15	15	195	18	4	—	3	⑲	㉖
	8-454, 425 H.P.	21	22	22	15	15	180	18	4	—	3	⑲	㉖
1972	6-250	11	12	12	15	15	195	㉘	4	3	—	⑲	㉙
	8-307	14	15	16	15	15	195	㉘	4	3	—	⑲	㉙
	8-350	15	16	17	15	15	195	㉘	4	3	3	⑲	㉙
	8-402	22	23	24	15	15	195	㉘	4	3	3	⑲	㉙
	8-454	22	23	24	15	15	195	㉘	4	—	3	⑲	㉙
1973	6-250	11	12	12	15	15	195	22	4	3	—	⑲	㉙
	8-307	15	16	17	15	15	195	22	4	3	3	⑲	㉙
	8-350, 145 H.P.	15	16	17	15	15	195	22	4	3	—	⑲	㉙
	8-350, 175 H.P.	15	16	17	15	15	195	22	4	—	4	⑲	㉙
	8-454	22	23	24	15	15	195	22	4	—	4	⑲	㉙
1974	6-250	—	14	14	15	15	195	22⑩	4	3	—	⑦	㉙
	8-350, 145 H.P.	—	18	18③	15	15	195	22⑩	4	3	—	⑦	㉙
	8-350, 160 H.P.	—	18	18③	15	15	195	22⑩	4	3	—	⑦	㉙
	8-400, 150 H.P.	—	18	18③	15	15	195	22⑩	4	—	—	⑦	㉙
	8-400, 180 H.P.	—	18	18③	15	15	195	22⑩	4	—	—	⑦	㉙
	8-454, 235 H.P.	—	24	18⑤	15	15	195	22⑩	4	—	3	⑦	㉙
1975	6-250	—	12½	12½	15	15	195	22⑩	4	3	—	4⑥	㉙
	8-350	—	16	17	15	15	195	22⑩	4	3	—	4⑥	㉙
	8-400	—	16	17	15	15	195	22⑩	4	—	—	4⑥	㉙
	8-454	—	23	24	15	15	195	22⑩	4	—	—	4½⑥	㉙
1976	6-250	—	15	17	15	15	195	22	4	3	—	⑥	4¼
	8-305	—	17½	18½	15	15	195	22	4	—	—	⑥	4¼
	8-350	—	17½	18½	15	15	195	22	4	—	—	⑥	4¼
	8-400	—	17½	18½	15	15	195	22	4	—	—	⑥	4¼

Continued

COOLING SYSTEM & CAPACITY DATA—Continued

Year	Model or Engine	Cooling Capacity, Qts.			Radiator Cap Relief Pressure, Lbs.		Thermo. Opening Temp. ①	Fuel Tank Gals.	Engine Oil Refill Qts. ②	Transmission Oil			Rear Axle Oil Pints
		No Heater	With Heater	With A/C	With A/C	No A/C				3 Speed Pints	4 Speed Pints	Auto. Trans. Qts. ⑮	
CHEVY NOVA													
1969	4-153	8	9	9	15	15	195	18	4	3⑧	—	⑲	3½
	6-230, 250	11	13	13	15	15	195	18	4	3⑧	—	⑲	3½
	8-307	16	17	17	15	15	195	18	4	3⑧	3	⑲	3½
	8-350	15	16	16	15	15	195	18	4	3½	3	⑲	4
	8-396	22	23	23	15	15	195	18	4	3½	3	⑲	4
1970	4-153	8	9	9	15	15	195	18㉓	3½⑪	3	—	⑲	⑨
	6-230, 250	11	12	13	15	15	195	18㉓	4	3	—	⑲	⑨
	8-307	14	15	16	15	15	195	18㉓	4	3⑧	3	⑲	⑨
	8-350	15	16	16	15	15	195	18㉓	4	3⑧	3	⑲	⑨
1971	6-250	11	12	12	15	15	195	18	4	3	—	⑲	㉖
	8-307	15	16	16	15	15	195	18	4	3	—	⑲	㉖
	8-350	15	16	16	15	15	195	18	4	3	3	⑲	㉖
1972	6-250	11	12	12	15	15	195	16	4	3	—	⑲	4¼
	8-307	14	15	16	15	15	195	16	4	3	—	⑲	4¼
	8-350	15	16	17	15	15	195	16	4	3	3	⑲	4¼
1973	6-250	11	12	12	15	15	195	21	4	3	—	⑲	4¼
	8-307	16	15	16	15	15	195	21	4	3	—	⑲	4¼
	8-350, 145 H.P.	17	16	17	15	15	195	21	4	3	—	⑲	4¼
	8-350, 175 H.P.	17	16	17	15	15	195	21	4	—	4	⑲	4¼
1974	6-250	14	14	—	15	15	195	21	4	3	—	⑦	4¼
	8-350, 145 H.P.	18	18	—	15	15	195	21	4	3	—	⑦	4¼
	8-350, 160 H.P.	18	18	—	15	15	195	21	4	3	—	⑦	4¼
	8-350, 185 H.P.	18	18	—	15	15	195	21	4	—	3	⑦	4¼
1975	6-250	—	12½	12½	15	15	195	21	4	3	—	4⑥	4¼
	8-262	—	—	—	15	15	195	21	4	3	—	4⑥	4¼
	8-350	—	15½	16½	15	15	195	21	4	3	—	4⑥	4¼
1976	6-250	—	14	15	15	15	195	21	4	3	—	⑥	4¼
	8-305	—	17	18	15	15	195	21	4	3	—	⑥	4¼
	8-350	—	17	18	15	15	195	21	4	3	3	⑥	4¼
CHEVROLET													
1969	6-250	11	12	13	15	15	195	24	4	3⑧	—	⑲	3½
	8-327	16	17	18	15	15	195	24	4	3⑧	3	⑲	3½
	8-350	14	15	16	15	15	195	24	4	3½	3	⑲	4
	8-396	22	23	23	15	15	195	24	4	3½	3	⑲	4
	8-427	21	22	22	15	15	195	24	4	3½	3	⑲	4
1970	6-250	11	12	12	15	15	195	25㉔	4	3⑧	—	⑲	⑨
	8-350, 250 H.P.	15	16	16	15	15	195	25㉔	4	3⑧	—	⑲	⑨
	8-350, 300 H.P.	15	16	17	15	15	195	25㉔	4	3⑧	—	⑲	⑨
	8-400	15	16	17	15	15	195	25㉔	4	3⑧	—	⑲	⑨
	8-454	21	22	22	15	15	195	25㉔	4	3⑧	—	⑲	⑨
1971	6-250	11	12	12	15	15	195	23㉗	4	3	—	⑲	㉖
	8-350	15	16	16	15	15	195	23㉗	4	3	—	⑲	㉖
	8-396㉕	22	23	23	15	15	195	23㉗	4	—	—	⑲	㉖
	8-400	15	16	16	15	15	195	23㉗	4	3	—	⑲	㉖
	8-454	21	22	22	15	15	195	23㉗	4	—	—	⑲	㉖
1972	6-250	11	12	12	15	15	195	23⑳	4	3	—	⑲	㉙
	8-350	15	16	17	15	15	195	23⑳	4	—	—	⑲	㉙
	8-400	15	16	17	15	15	195	23⑳	4	—	—	⑲	㉙
	8-402	22	23	24	15	15	195	23⑳	4	—	—	⑲	㉙
	8-454	21	22	23	15	15	195	23⑳	4	—	—	⑲	㉙

Continued

COOLING SYSTEM & CAPACITY DATA—Continued

Year	Model or Engine	Cooling Capacity, Qts. No Heater	Cooling Capacity, Qts. With Heater	Cooling Capacity, Qts. With A/C	Radiator Cap Relief Pressure, Lbs. With A/C	Radiator Cap Relief Pressure, Lbs. No A/C	Thermo. Opening Temp. (1)	Fuel Tank Gals.	Engine Oil Refill Qts. (2)	Transmission Oil 3 Speed Pints	Transmission Oil 4 Speed Pints	Transmission Oil Auto. Trans. Qts. (15)	Rear Axle Oil Pints
CHEVROLET—Continued													
1973	6-250	—	12½	—	15	15	195	(30)	4	3	—	—	(29)
	8-350, 145 H.P.	—	16	17	15	15	195	(30)	4	—	—	4	(29)
	8-350, 175 H.P.	—	16	17	15	15	195	(30)	4	—	—	4	(29)
	8-400	—	16½	17	15	15	195	(30)	4	—	—	4	(29)
	8-454	—	23	24	15	15	195	(30)	4	—	—	4½	(29)
1974	8-350, 145 H.P.	—	16	17	15	15	195	26(20)	4	—	—	(7)	(29)
	8-350, 160 H.P.	—	16	17	15	15	195	26(20)	4	—	—	(7)	(29)
	8-400, 150 H.P.	—	16½	17½	15	15	195	26(20)	4	—	—	(7)	(29)
	8-400, 180 H.P.	—	16½	17½	15	15	195	26(20)	4	—	—	(7)	(29)
	8-454, 235 H.P.	—	23	24	15	15	195	26(20)	4	—	—	(7)	(29)
1975	8-350	—	16	17	15	15	195	26(20)	4	—	—	4(6)	(29)
	8-400	—	16½	17½	15	15	195	26(20)	4	—	—	4½(6)	(29)
	8-454	—	23	24	15	15	195	26(20)	4	—	—	4½(6)	(29)
1976	8-350	—	18	20	15	15	195	26(20)	4	—	—	(6)	(29)
	8-400	—	18	20	15	15	195	26(20)	4	—	—	(6)	(29)
	8-454	—	23	25	15	15	195	26(20)	4	—	—	(6)	(29)
CORVETTE													
1969	8-350	14	15	15	15	15	195	20	4	3	3	(21)	4
	8-427	21	22	22	15	15	195	20	5	3	3	(21)	4
1970	8-350 Exc. 370 H.P.	14	15	21	15	15	195	20	4	—	3	(4)	4
	8-350, 370 H.P.	17	18	22	15	15	180	20	4	—	3	(4)	4
	8-454	21	22	—	15	15	195	20	5	—	3	(4)	4
1971	8-350 Exc. 330 H.P.(13)	14	15	15	15	15	195	18	4	—	3	—	4
	8-350 Exc. 330 H.P.(14)	17	18	18	15	15	195	18	4	—	—	(19)	4
	8-350, 330 H.P.	17	18	18	15	15	180	18	4	—	3	(19)	4
	8-454, 365 H.P.	21	22	22	15	15	195	18	5	—	3	(19)	4
	8-454, 425 H.P.(13)	19	20	20	15	15	180	18	5	—	3	—	4
	8-454, 425 H.P.(14)	21	22	22	15	15	180	18	5	—	—	(19)	4
1972	8-350, 200 H.P.	16	17	18	15	15	195	18	4	—	3	4	4
	8-350, 255 H.P.	16	17	18	15	15	180	18	4	—	3	—	4
	8-454	22	23	24	15	15	195	18	5	—	3	4	4
1973	8-350, 190 H.P.	17	18	18	15	15	195	18	4	—	3	4	4
	8-350, 250 H.P.	17	18	18	15	15	180	18	4	—	3	4	4
	8-454, 275 H.P.	23	24	24	15	15	195	18	5	—	3	4	4
1974	8-350, 195 H.P.	17	19	—	15	15	195	18	4	—	3	(7)	4
	8-350, 250 H.P.	17	19	—	15	15	180	18	4	—	3	(7)	4
	8-454, 270 H.P.	24	24	—	15	15	195	18	4	—	3	(7)	4
1975	8-350	—	18	18	15	15	180	18	4	—	3	4(6)	4
1976	8-350	—	21	21	15	15	195(12)	17	4	—	3	11	4

(1)—For permanent type anti-freeze.
(2)—Add one quart with filter change.
(3)—Monte Carlo 20 qts.
(4)—Refill 1½ qts. Total capacity 7½ qts.
(5)—Monte Carlo 26 qts.
(6)—Turbo Hydramatic 250 & 350 total capacity 10 qts., T.H. 400 total capacity 11 qts.
(7)—Turbo Hydramatic 250 & 350 refill 2½ qts., T.H. 400 refill 3¾ qts.
(8)—Heavy duty unit 3½.
(9)—3¾ for 8⅛″ ring gear and 4¼ for 8⅞″ ring gear.
(10)—El Camino 26 gal.
(11)—Add one pint with filter change.
(12)—Optional 350 engine: 180°.
(13)—Standard trans.
(14)—Auto. trans.
(15)—Approximate. Make final check with dipstick.
(19)—Powerglide & Torque Drive: Refill 3 qts. Total capacity 8½ qts. Turbo-Hydramatic 350: Refill 2½ qts. Total capacity 10 qts. Turbo-Hydramatic 400: Refill 4 qts. Total capacity 11 qts.
(20)—Wagons 22 gallons.
(21)—Refill 4 qts. Total capacity 11 qts.
(22)—California vehicles about 2 gallons less.
(23)—California vehicles about 1 gallon less.
(24)—Wagons 22 gallons. California vehicles about 2 gallons less.
(25)—Marketed as 396 but actually 402 cu. in.
(26)—3½ for 8⅛″ ring gear and 4 for 8⅞″ ring gear.
(27)—Wagons 20 gallons.
(28)—Wagons 19 gallons; others 18 gallons.
(29)—4¼ for 8⅛″ and 8½″ ring gears and 4.9 for 8⅞″ ring gear.
(30)—Station wagons 22 gals.; others 26 gals.

Electrical Section

DISTRIBUTOR

Removal

1. Disconnect distributor primary wire from coil terminal.

NOTE: On H.E.I. systems, disconnect feed and module connectors from distributor cap.

2. Remove distributor cap and rotor. *Mark position of rotor arm on distributor housing so distributor can be installed in same position.*
3. Remove vacuum line and distributor hold down clamp. On Corvette, units, remove tachometer drive cable.
4. Note relative position of distributor in block, then work it out of the engine.

Installation

1. Turn rotor about 1/8 of a turn counterclockwise past the mark previously placed on the distributor housing.
2. Push the distributor down into the block with the housing in the normal "installed" position. *It may be necessary to move the rotor slightly to start gear into mesh with camshaft gear, but rotor should line up with mark when distributor is down in place.*

NOTE: Because the lower end of the distributor shaft drives the oil pump, use extra care when installing the distributor to be sure it is completely seated and engaged in oil pump.

NOTE

If the engine was disturbed while the distributor was removed from the engine, first crank the engine to bring No. 1 piston up on its compression stroke and continue cranking until the timing mark is adjacent to the timing indicator. Then rotate the distributor cam until the rotor is in position to fire No. 1 cylinder. Install the distributor.

3. Tighten distributor clamp screw snugly and connect vacuum line, primary wire to coil, and install cap.
4. Set ignition timing.

NOTE: When using a timing light to adjust ignition timing, the connection should be made at the No. 1 spark plug. Forcing foreign objects through the boot at the No. 1 terminal of the distributor cap will damage the boot and could cause engine misfiring.

STARTER, REPLACE

1969-76 All Models

1. Disconnect ground cable at battery.

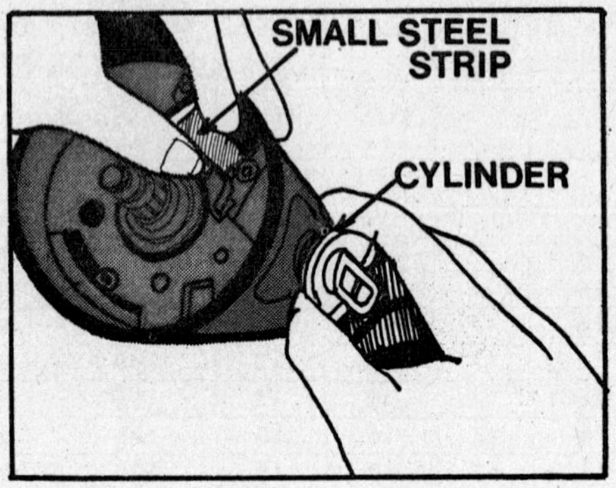

Fig. 1 Ignition lock removal

2. Raise vehicle to working height.
3. Disconnect all wires at solenoid.

NOTE: Reinstall terminal nuts as each wire is disconnected as thread size is different but may be mixed and stripped.

4. Loosen starter front bracket (nut on V8 and bolt on Sixes) then remove two mounting bolts.

NOTE: On V8 engines using solenoid heat shield, remove front bracket upper bolt and detach bracket from starter.

5. Remove front bracket bolt or nut and rotate bracket clear of work area. Then lower starter from vehicle by lowering front end first (hold starter against bell housing and sort of roll end-over end).
6. Reverse removal procedure to install and torque mount bolts to 25-35 ft-lbs. On 1973-76 Corvette, apply a suitable sealing compound around starter where it enters engine splash shield.

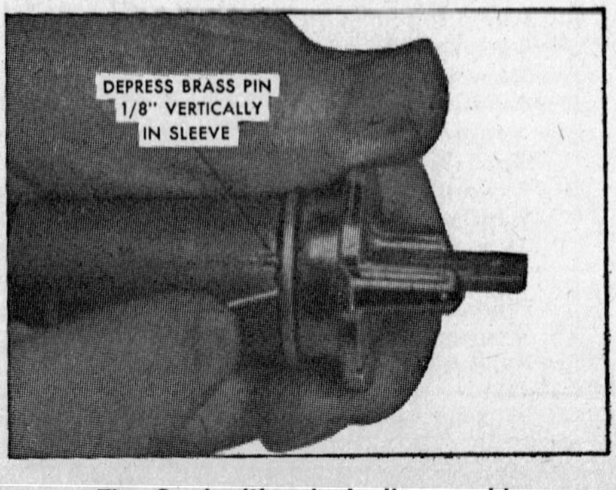

Fig. 2 Ignition lock disassembly

IGNITION LOCK

1969-76

1. Follow the procedure to remove the turn signal switch as described further on.
2. On *1969-70* models the lock cylinder may be removed in any position from "Accessory" to "ON". However, the "LOCK" position is desired because of its positive location. On 1971-76 models the lock cylinder should be removed in the "RUN" position only.
3. Place a thin tool (small screwdriver or knife blade) into the slot, Fig. 1, next to the switch mounting screw boss (right hand slot) and depress spring latch at bottom of slot which releases lock. Remove lock.
4. To separate the lock from the cylinder, Fig. 2, use a paper clip or similar wire, place a 90° bend about 1/8" from one end of the wire.
5. Place the cylinder in "LOCK" position. Using the wire as described in step 4, depress the plunger pin (above the stake mark) about 1/8" vertically in sleeve housing, at the same time exert light upward pressure on the lock assembly, rotate the cylinder counter-clockwise until it pops up from the sleeve assembly.

IGNITION SWITCH, REPLACE

1969-76

The ignition switch is mounted on top of the mast jacket inside the brake pedal support and is actuated by a rod and rack assembly.

1. Disconnect battery cable.
2. Disconnect and lower steering column.

NOTE: It may be necessary, on some models, to remove the upper column mounting bracket if it hinders servicing of switch.

3. The switch should be in the "Lock" position before removal. If lock cylinder has been removed, the actuating rod is pulled up to the stop then back one detent, to place it in the "Lock" position. Remove retaining screws and switch.
4. Reverse procedure to install being sure that the switch is in the "Lock" position before installing.

LIGHT SWITCH, REPLACE

1973-76 Chevelle & Monte Carlo

1. Disconnect battery ground cable.
2. Remove instrument panel pad.

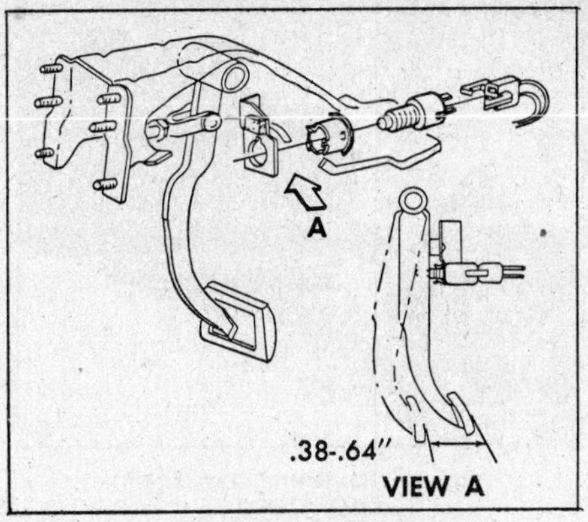

Fig. 3 Stop light switch. 1969-72 Chevelle

3. Remove left radio speaker to one side.
4. Pull headlamp control knob to "ON" position and while standing outside vehicle on left side, reach in and behind instrument panel and depress switch shaft retainer and pull knob and shaft assembly out.
5. Remove ferrule nut and switch from panel.

1969-76 Corvette

1. Disconnect battery cable.
2. Remove screws securing mast jacket trim covers and remove covers.
3. Remove left side console forward trim panel.
4. Lower steering column.
5. Remove screws securing left instrument panel to door opening, top of dash and left side of center instrument panel.
6. Pull cluster down and tip forward for access.
7. Depress switch shaft retainer and remove the knob and shaft assembly. Remove switch retaining bezel.
8. Disconnect vacuum hoses from switch, tagging them for assembly. Pry the connector from the switch and remove the switch.

1970-76 Camaro

1. Disconnect battery ground cable.
2. Remove steering column lower cover.
3. Reach up under cluster and depress lighting switch shaft retainer while pulling gently on shaft.
4. Remove nut securing switch to carrier.
5. Remove cluster carrier screws and tilt right side of cluster out.

NOTE: On 1970-71 models, grounding ring at cigar lighter may be disconnected to permit further carrier movement.

6. Unplug connector and remove switch.

1973-76 Chevrolet & Nova; 1969-72 Except Corvette & 1970-72 Camaro

1. Disconnect battery ground cable.

NOTE: On Chevy Nova remove the parking brake bracket and lower assembly to the floor. Heater on air conditioning control head must also be removed.

2. Pull switch knob to "ON" position.
3. Reach up under instrument panel and depress switch shaft retainer, then remove knob and shaft assembly.
4. Remove ferrule nut and switch from panel.

NOTE: Remove vacuum hoses from Camaro and Chevrolet optional headlight switches. Tag location of hoses for assembly.

5. Disconnect multi-contact connector from light switch.
6. Reverse procedure to install.

STOP LIGHT SWITCH, REPLACE

1973-76 Except Corvette

1. Disconnect wiring connector at switch.
2. Remove retaining nut, if so equipped, and unscrew switch from bracket.
3. To install: On Nova, install inboard adjusting nut, install switch through bracket and install retaining nut. On all other models, depress brake pedal and push new switch into clip until shoulder bottoms out.
4. Plug connector onto switch and check operation. Electrical contact should be made when pedal is depressed 3/8" to 5/8" from fully released position.

1973-76 Corvette

1. Disconnect wiring connector at switch.
2. Remove retaining nut and unscrew switch from bracket.
3. Upon installation, check for proper operation. Electrical contact should be made when pedal is depressed 1/4" to 5/8". Switch bracket has a slotted screw hole for adjustment.

1969-72

1. Disconnect wiring harness connector from switch, Figs. 3 and 4.
2. Remove retaining nut (if equipped) and unscrew switch from bracket.

NOTE: On Corvettes, remove screw holding switch bracket to brake pedal housing.

3. On Chevelles, depress brake pedal and push new switch into clip until shoulder bottoms out.
4. On Corvettes, align switch bracket on brake pedal housing and install screw.
5. On all other models, install inboard adjusting nut, install switch through bracket and install retaining nut.
6. Plug connector onto switch.

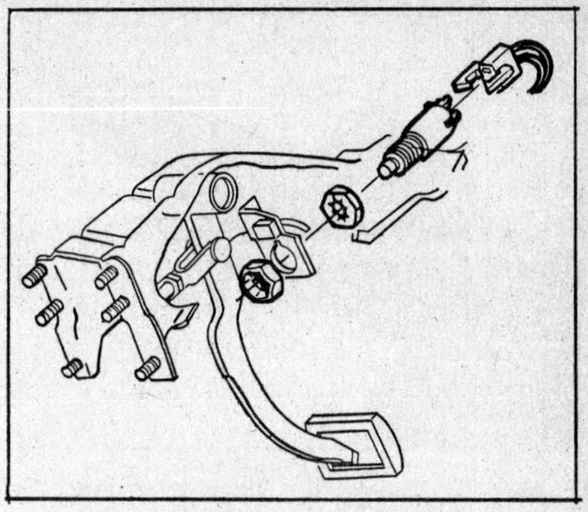

Fig. 4 Stop light switch. 1969-72 Camaro, Chevrolet and Chevy Nova

CLUTCH START SWITCH

1969 All; 1973-76 Corvette

1. Unplug connector from switch.
2. Remove retainer from pins or link on clutch pedal arm.
3. Remove retaining screw and switch.

1970-72 All; 1973-76 Except Corvette

1. Unplug connector from switch.
2. Compress switch actuating shaft retainer and remove shaft with switch attached from switch bracket.

NEUTRAL SAFETY SWITCH, REPLACE

1971-76 Column Shift & 1974 Floor Shift Exc. Corvette

NOTE: On 1974-75 floor shift vehicles, position shift lever in "Park." Also, the new switch assembly is pinned in the "Park" position.

1. Disconnect wiring connector at switch terminals.
2. Unfasten and remove switch from mast jacket.
3. To install: position shift lever against NEUTRAL gate by rotating lower lever counterclockwise as viewed from drivers seat.
4. Assemble switch to column by inserting actuating tang in shifter tube slot.
5. Tighten screws, connect wiring connector and move selector lever out of NEUTRAL to shear pin which is part of new switch.

1969-70 Column Shift

1. Disconnect wiring connectors at switch terminals.
2. Unfasten and remove switch from mast jacket.
3. To install, locate shift lever in

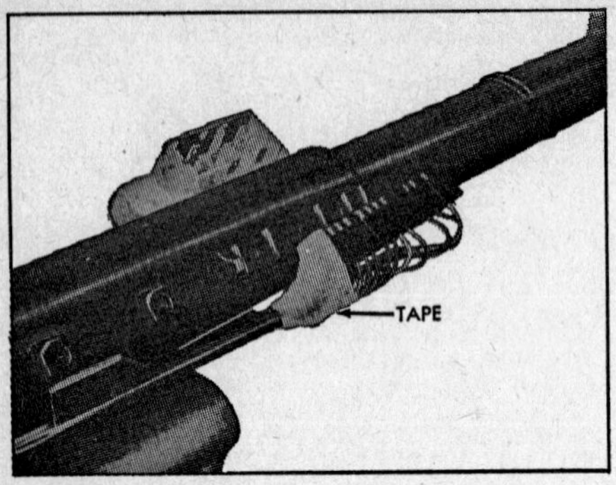

Fig. 5 Taping turn signal connector and wires. 1969-76

"Drive" and locate lever tang against transmission selector plate.
4. If provided for, align slot in contact support with hole in switch and insert a 3/32" rod to hold support in place. Switch is now in drive position.
5. Place contact support drive slot over shifter tube drive tang and tighten screws. Remove clamp and 3/32" rod.
6. Connect wiring harness and check operation of switch.

1974 Corvette & 1969-73 All Floor Shift

1. Disconnect shift control lever arm from transmission control rod.
2. Remove shift control knob.
3. Remove trim plate.
4. Remove control assembly from seal and disconnect switch wiring.
5. Remove switch from control assembly.
6. To install, position gearshift in Drive position, align hole in contact support with hole in switch and insert a pin (3/32") to hold support in place.
7. Place contact support drive slot over drive tang and tighten switch mounting screws.
8. Connect wiring harness to switch wiring.
9. Install trim plate control knob and connect shift lever arm to transmission control rod.

TURN SIGNAL SWITCH, REPLACE

1969-76

1. Disconnect battery cable and remove steering wheel.
2. Remove cover from shaft. *The cover retaining screws need not be completely removed from the cover.*
3. Using a suitable tool, compress lock plate (horn contact carrier on tilt models) and remove snap ring ("C" ring on tilt models).

NOTE: On 1976 models, remove lock plate cover with a suitable screwdriver.

4. Remove lock plate, cancelling cam, spring, thrust washer and signal lever.
5. Push hazard warning knob in and unscrew knob.
6. Pull connector from bracket and wrap upper part of connector with tape to prevent snagging the wires during removal. On Tilt models, position shifter housing in "Low" position. Remove harness cover.
7. Remove retaining screws and remove switch, Figs. 5 and 6.

HORN SOUNDER & STEERING WHEEL

1970-76 Cushioned Rim Wheel

1. Disconnect battery ground cable.
2. Remove two steering wheel shroud screws at underside of steering wheel and remove shroud.
3. Remove three spacer screws, spacer, plate and belleville spring.
4. Remove steering wheel nut and washer and, using a suitable puller, remove steering wheel.

NOTE: On 1975-76 models, remove snap ring before steering wheel nut.

1969 Deluxe Wheel 1970-76 Standard Wheel

1. Disconnect battery ground cable.
2. Remove attaching screws on underside of steering wheel, Fig. 7.
3. Lift steering wheel shroud and pull horn wires from cancelling cam tower.
4. Remove steering wheel nut and washer, and use a suitable puller to remove steering wheel.

NOTE: On 1975-76 models, remove snap ring before steering wheel nut.

1969 Standard Wheel Exc. Corvette

1. Disconnect battery ground cable.

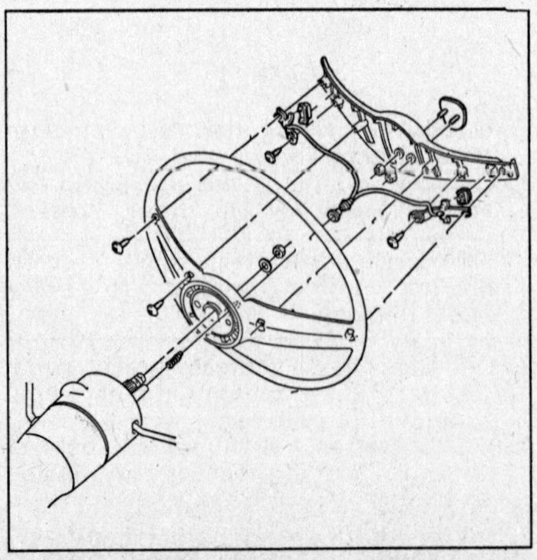

Fig. 7 Steering wheel and horn attachments. 1969-76 (typical)

Fig. 6 Removing turn signal switch. 1969-76

2. Pull out horn button cap or center ornament and retainer.
3. Remove three screws from receiving cup.
4. Remove receiving cup, belleville spring, bushing and pivot ring.
5. Remove wheel nut and washer and use a suitable puller to remove wheel.

CAUTION: Turn signal control assembly must be in neutral position when assembling steering wheel to prevent damage to cancelling cam and control assembly.

1969 Simulated Wood Wheel

1. Disconnect steering column harness from chassis wiring harness at connector.
2. Disconnect battery ground cable.
3. Remove horn cap by pulling up.
4. Remove screws and contact assembly.

NOTE: If steering wheel only is to be replaced, perform Step 4. If turn signal cancelling cam is to be replaced, omit Step 4 and proceed with Steps 5 and 6.

5. Remove retaining screws and remove wheel from hub assembly.
6. Remove wheel nut and washer.
7. Use a suitable puller to remove wheel.

NOTE: Turn signal control assembly must be in neutral position when assembling the hub to prevent damage to cancelling cam and control assembly.

8. Reverse procedure to install.

1969-76 Corvette (Exc. Simulated Wood Wheel)

1. Disconnect steering column harness at wiring connector. Disconnect battery ground cable.
2. Pry off horn button cap.
3. Remove three screws securing horn contact to spacer and hub.
4. On telescoping wheels, remove two screws securing lock screw to lock

knob and remove screw, knob and spacer.

NOTE: If wheel only is to be replaced, perform Step 5. If turn signal cancelling cam is to be replaced, omit Step 5 and proceed with Steps 6 and 7.

5. Remove wheel from hub (6 screws).
6. Remove nut and washer from shaft and use a suitable puller to remove wheel and hub.

NOTE: On 1975-76 models, remove snap ring before steering wheel nut.

7. Slide cancelling cam and spring off shaft.

INSTRUMENT CLUSTER

1973-76 Chevelle & Monte Carlo

1. Disconnect battery ground cable.
2. Remove radio knobs and clock stem set knob, if equipped.
3. Remove instrument bezel retaining screws, Figs. 8 and 9.
4. Pull bezel out to disconnect tail gate release or rear defogger switch, if equipped.
5. Remove instrument bezel.

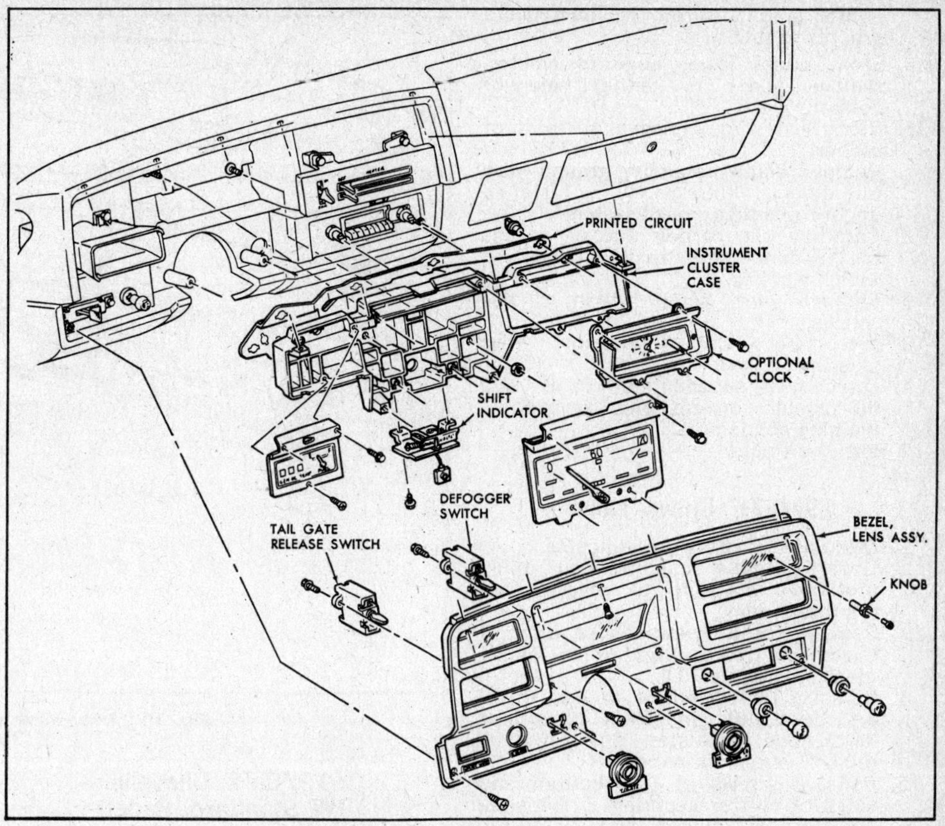

Fig. 8 Instrument cluster. 1973-76 Chevelle with standard cluster

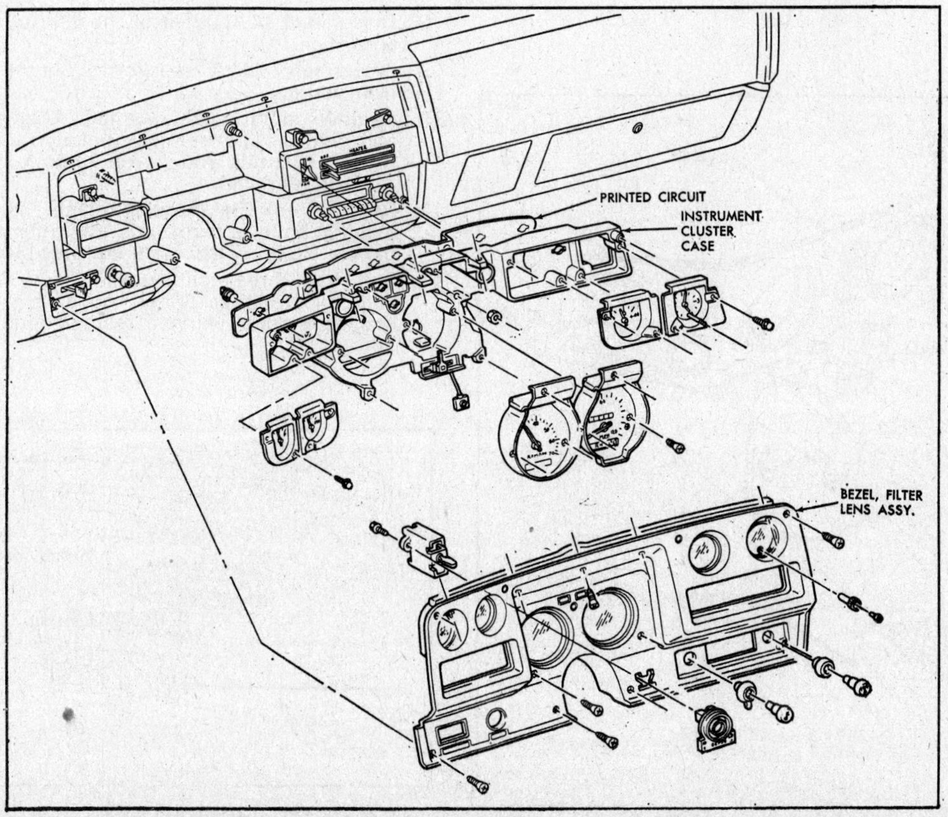

Fig. 9 Instrument cluster. 1973-76 Chevelle with optional cluster & Monte Carlo

1971-76 Chevrolet

1. Disconnect battery ground cable.
2. Remove cigar lighter knob and hidden screw in shroud where knob was.
3. Pull on headlamp switch shaft then remove hidden screw above middle of shaft.
4. Remove two screws at bottom corners of shroud and lift off shroud, Fig. 10.
5. To service instruments, remove clock stem set knob.
6. Remove lens retaining strip secured by three screws at top of lens. Be careful not to mar lens.
7. Lift off lens carefully—guide pins are on bottom of lens.
8. Gently lift up on bottom of filter housing and rotate housing up and rearward, toward technician on front seat. Top of housing should clear top of instrument carrier. Use care around PRNDL housing and lift filter assembly off. Speedometer, fuel gauge and clock can now be removed.

1970-76 Camaro

1. Disconnect battery ground cable.
2. Remove 6 screws securing trim cover beneath steering column. Two of these screws are located above the ash tray.
3. Remove headlamp switch retaining nut.
4. From behind panel, disconnect cigar

lighter and unscrew retainer. Note grounding ring.

5. From under lower edge of cluster, remove screw on either side of column, Fig. 11.
6. Remove 4 screws visible on front of carrier.
7. Remove screw retaining ground wire for wiper switch. Screw is fastened under top left corner of switch.
8. Carefully tilt carrier out of access to the connectors on headlamp and wiper switches.
9. Remove lens screws then cluster screws.
10. Disconnect shift indicator from steering column.
11. Disconnect speedometer cable and tilt cluster forward and remove remaining connectors.
12. Lift cluster out.

1970-76 Chevy Nova

1. Disconnect battery ground cable.
2. Lower steering column and apply protective covering to mast jacket to protect paint.
3. Remove three screws above front of heater control securing it to instrument cluster, Fig. 12.
4. Remove radio control knobs, washers, bezel nuts and front support at lower edge of cluster. This will allow radio to remain in panel.
5. Remove screws at top, bottom and sides of cluster securing it to panel.
6. Tilt console forward and reach behind to disconnect speedometer cable and all other connections and lift instrument panel out of carrier after removing screws.

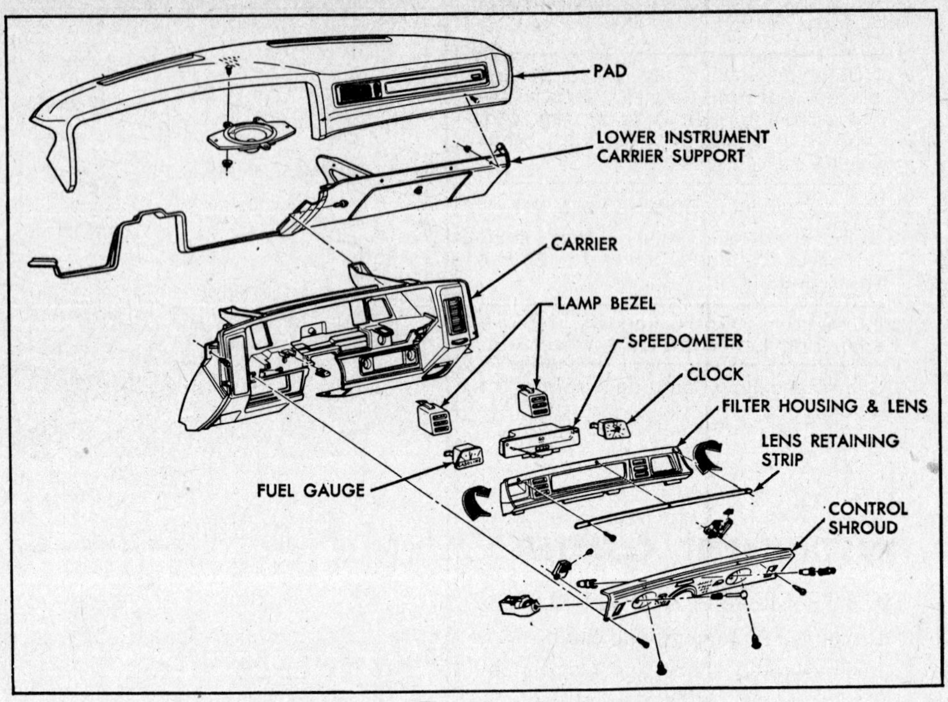

Fig. 10 Instrument cluster. 1971-76 Chevrolet

1970-72 Chevelle (W/Standard Panel)

1. Disconnect battery ground cable.
2. Lower steering column.
3. Disconnect parking brake hand release rod attachment.
4. Disconnect speedometer cable from speedometer.
5. Remove instrument panel pad (six screws), Fig. 13.
6. Disconnect radio speaker bracket from instrument panel.
7. Disconnect A/C center outlet (three screws).
8. Disconnect A/C control head (four screws).
9. Disconnect radio speaker connector from radio.
10. Remove radio knobs, washers, bezels and wiring and remove bolts from braces securing radio; roll radio out from under panel.
11. Remove six instrument panel attaching bolts and roll out instrument panel. **NOTE: Two men are needed.**
12. Pop out three telltale snap covers and remove three telltale housing attaching screws, remove telltale housing.

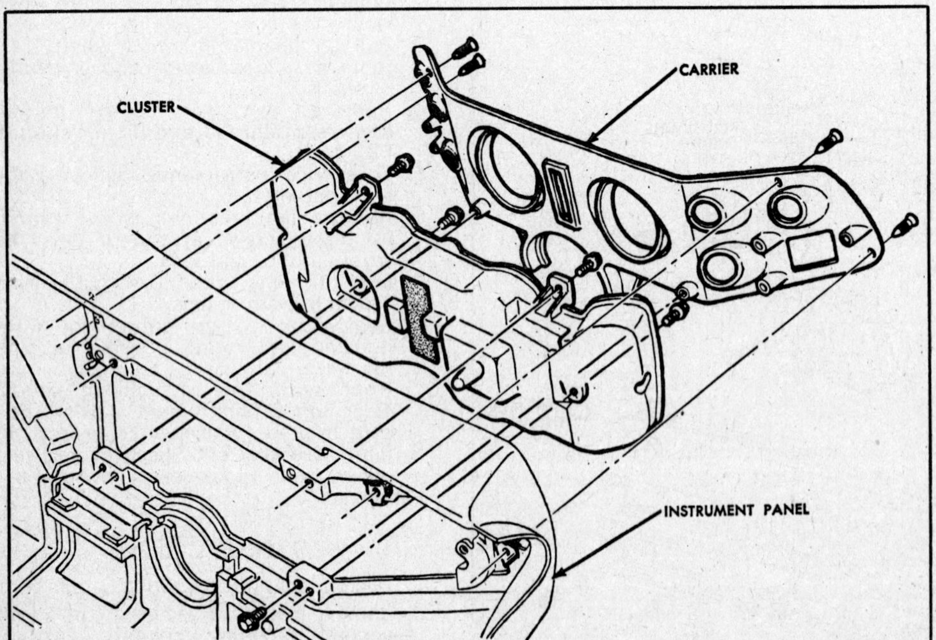

Fig. 11 Instrument cluster. 1970-76 Camaro

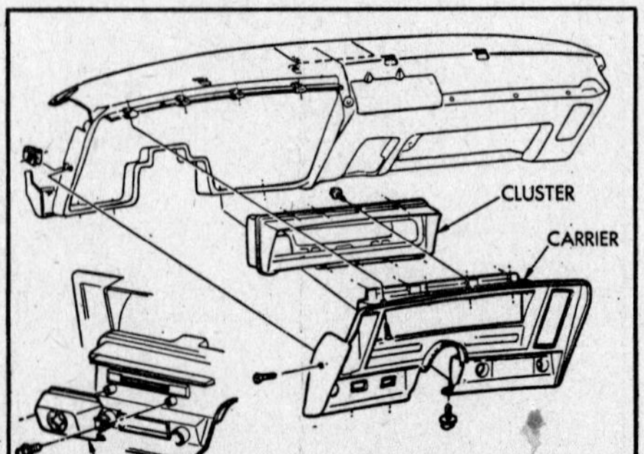

Fig. 12 Instrument cluster. 1970-76 Chevy Nova

13. Remove clock stem set knob and separate cluster from carrier (8 screws).

1970-72 Chevelle S.S. & Monte Carlo & Optional Panel

1. Disconnect battery ground cable.
2. Remove instrument panel pad (6 screws).
3. Disconnect A/C center outlet (3 screws).
4. Disconnect A/C control head (4 screws).
5. Disconnect radio speaker and brackets from cluster.
6. Disconnect speaker leads from radio and remove radio knobs, washers, bezels and wiring including antenna lead.
7. Unfasten radio and roll radio out from under instrument panel.
8. Remove steering column cover.
9. Remove two steering column attaching bolts. Note shims.
10. Disconnect PRNDL cable from column housing and lower column.
11. Disconnect parking brake hand release rod attachment.
12. Remove six instrument panel attaching bolts and roll out panel, Fig. 14.

NOTE: Two men are needed.

13. Disconnect speedometer cable.
14. Remove cluster lamp sockets, clock, fuel gauge, and all connectors from rear of panel.
15. Remove screws retaining printed circuit and lift off circuit panel.

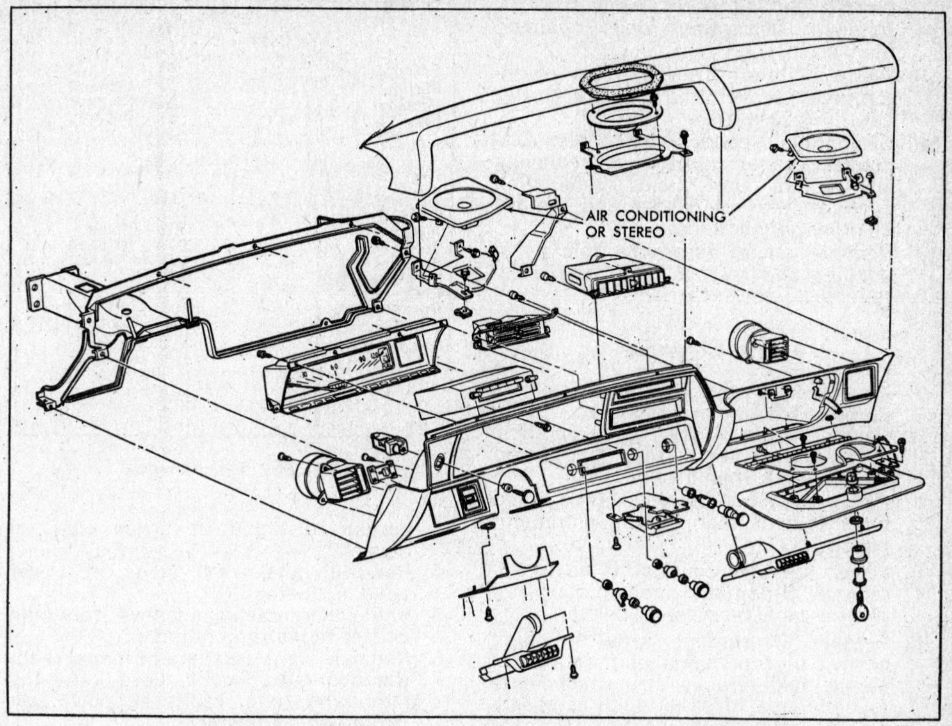

Fig. 13 Instrument cluster. 1970-72 Chevelle with standard cluster

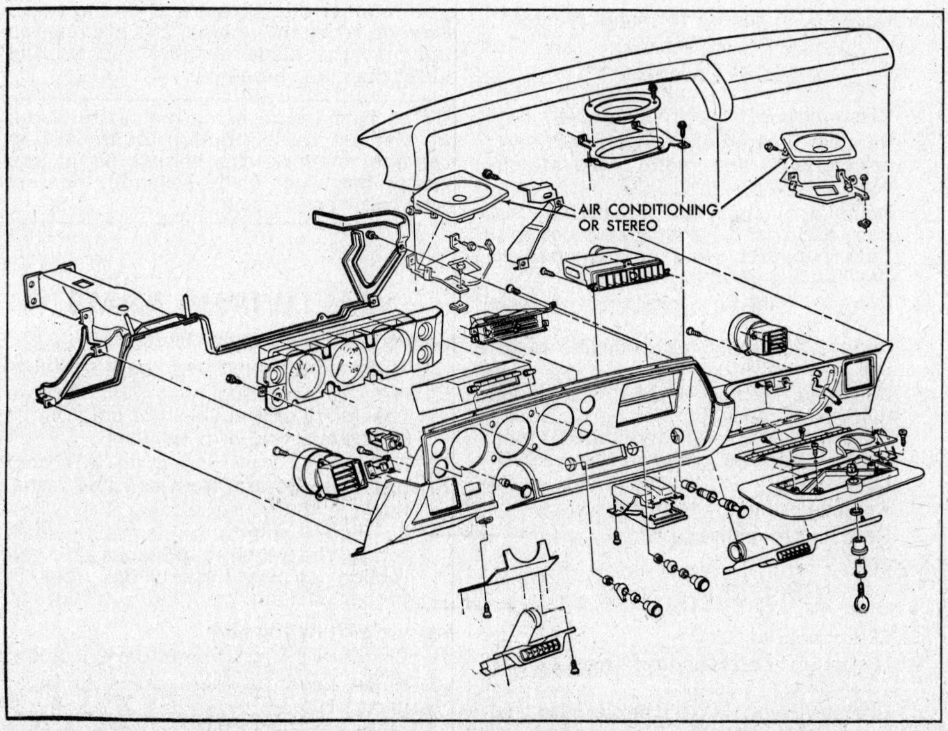

Fig. 14 Instrument cluster. 1970-72 Chevelle with optional cluster & Monte Carlo

1969-70 Chevrolet

1. Disconnect battery ground cable.
2. If air conditioned, remove lap cooler from under steering column.
3. Remove column mast jacket trim cover.
4. Lower steering column.

CAUTION: *Do not allow column to hang from dash or distortion to the column will result.*

5. If equipped with Comfortron, remove the in-car sensor.
6. Remove instrument panel pad and if air conditioned, disconnect center outlet hose when removing pad.
7. Disconnect transmission indicator cable on column.
8. Remove indicator bulb bezel by pushing in at right side of the four bezel screw covers to access to screws.
9. Remove transmission indicator lamp housing.
10. Remove radio knobs and nuts. Remove rear brace, antenna lead and move radio back away from instrument carrier assembly.
11. Remove instrument panel trim plate by inserting a hooked tool such as a cotter pin remover, behind the trim plate and pulling the plate forward enough to release it from its retainers.
12. Remove air conditioning and/or heater control retaining screws and move control back away from instrument carrier assembly.
13. Remove screws retaining instrument carrier to panel lower reinforcement

and to parking brake pedal bracket.
14. Remove ash tray and retaining bracket.
15. Remove three screws at top of instrument carrier and tilt the carrier assembly forward.
16. Disconnect speedometer cable. Disconnect panel harness by removing six screws that hold illumination cover to rear of carrier and remove all other connections.
17. Remove carrier assembly from panel opening and remove eight screws retaining cluster to carrier.

1969 Chevelle

1. Disconnect battery ground cable.
2. Remove ash tray and retainer.
3. Remove radio knobs, nuts, electrical connections, aerial plug and radio rear support, then lift out radio.
4. Remove heater control screws, then push control head out of instrument panel.
5. Lower steering column. If equipped, remove automatic transmission indicator cable on steering column.
6. Remove instrument panel retaining screws at top, sides and bottom of panel. Also remove any attachments to underside of panel such as speedometer stem, defogger or tail gate control switch.
7. Lift loosened instrument panel up and back slightly, reach behind cluster and remove speedometer cable housing retaining nut, then support instrument panel on protected steering column.
8. Remove clips on top of instrument cluster rear cover and remove all connectors at back of cover and oil pressure pipe fitting from rear of oil pressure gauge (if so equipped).
9. Remove five screws securing twin window cluster to back of instrument panel and remove cluster.
10. Reverse procedure to install.

1969-76 Corvette

Left Hand Side

1. Disconnect battery ground cable.
2. Lower steering column.
3. Remove screws and washers securing left instrument panel to door opening, top of dash and left side of center instrument panel.
4. Unclip and remove floor console trim panel.
5. Pull cluster slightly forward to obtain clearance for removal of speedometer cable housing nut, tachometer cable housing nut, headlamp and ignition switch connectors and panel illuminating lamps.
6. Reverse procedure to install.

Center Cluster

1. Disconnect battery ground cable at battery.
2. Remove wiper switch trim plate screws and tip plate forward for access to switch connector. Lift trim plate out from cluster.
3. Unclip and remove right and left console forward trim pads to gain access to studs at lower edge of cluster.
4. Remove nuts from studs at lower edge of cluster.
5. Remove remaining screws retaining cluster to instrument panel.
6. Remove right instrument panel pad.
7. Remove radio knobs, bezel retaining nuts and one radio support bolt (from behind cluster).
8. Slide radio back towards firewall and pull cluster forward. Reach behind cluster, disconnect oil pressure line, wiring harness and bulbs.

CAUTION: The center cluster trim plate is designed to collapse under impact, Consequently, do not try to deflect the cluster plate forward to gain access to back of gauges.

9. Lift cluster assembly up and forward to remove.
10. Reverse procedure to install cluster.

1969 Camaro

1. Disconnect battery ground cable.
2. Remove instrument panel pad, A/C attachments and radio brace attachments.
3. Remove mast jacket supports at toe pan and dash, and lower column. *Both supports must be detached to prevent distortion of mast jacket.*
4. Remove cluster retaining screws from face of panel and partially remove assembly from console opening.
5. Reaching behind cluster assembly, disconnect speedometer cable, speed warning device (if equipped) and chassis harness connector at rear of panel.
6. Remove assembly from console.
7. Reverse procedure to install.

1969 Chevy Nova

1. Disconnect battery.
2. Unfasten heater control from cluster (2 screws).
3. Remove radio knobs, bezel nuts and front support at lower edge of cluster. This will allow radio to remain in instrument panel during cluster removal.
4. Disconnect and lower column.
5. Remove toe pan trim cover.
6. Remove bolts securing steering column retainer to toe pan. Loosen bolts securing retainer halves.
7. Remove bolts at top, bottom and sides of cluster securing it to instrument panel.
8. Remove steering column-to-dash panel bracket nuts and carefully lower column. Be careful not to jar or apply any load on column during this operation.
9. Remove ignition switch tumbler and bezel through cluster and leave hanging.
10. Tilt cluster forward and disconnect speedometer cable and necessary electrical connectors. Then remove cluster.
11. Reverse procedure to install and align steering column.

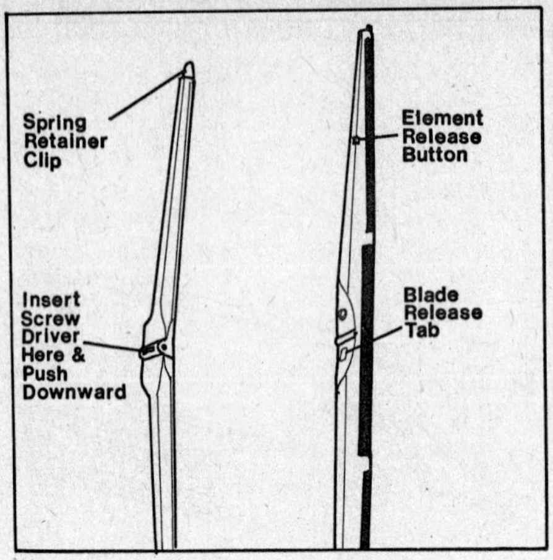

Fig. 15 Wiper blades

W/S WIPER BLADES

Two methods are used to retain wiper blades to wiper arms, Fig. 15. One method uses a press type release tab. When the release tab is depressed the blade assembly can be slid off the wiper arm pin. The other method uses a coil spring retainer. A screwdriver must be inserted on top of the spring and the spring pushed downward. The blade assembly can then be slid off the wiper arm pin. Two methods are also used to retain the blade element in the blade assembly, Fig. 15. One method uses a press type release button. When the button is depressed the two piece blade assembly can be slid off the blade element. The other method uses a spring type retainer clip in the end of the blade element. When the retainer clip is squeezed together, the blade element can be slid out of the blade assembly.

NOTE: To be sure of correct installation, the element release button, or the spring element retaining clip should be at the end of the wiper blade assembly nearest the wiper transmission.

W/S WIPER ARMS

Models with Rectangular Motor

1. Wiper motor must be in park position.
2. Use suitable tool to minimize the possibility of windshield or paint finish damage during arm removal.
3. Remove arm by prying up with tool to disengage arm from serrated transmission shaft.
4. To install arm to transmission shaft rotate the required distance and direction so that the blades rest in proper position.

Models with Round Motor

1. Wiper motor must be in park position.
2. Raise hood to gain access to wiper arms.
3. *Corvette* only: Remove the rubber plug from the front of the wiper door

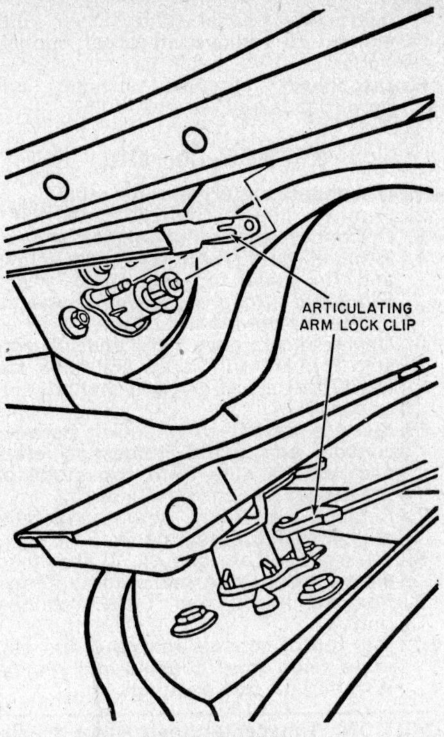

Fig. 16 Wiper articulating arm lock clip

actuator then insert a screwdriver, pushing the internal piston rearward to actuate the wiper door open.
4. Lift the wiper arm off the transmission shaft. On left arm slide the articulating arm lock clip, Fig. 16, away from the transmission pivot pin and lift arm off pin.
5. Slide latch clip, Fig. 17, out from under the wiper arm.
6. Release the wiper arm and lift assembly off transmission shaft.

W/S WIPER MOTOR

1971-76 Except Corvette

1. Raise hood and remove cowl screen or grille.
2. Disconnect wiring and washer hoses.
3. Reaching through cowl opening, loosen transmission drive link attaching nuts to motor crankarm.
4. Disconnect drive link from motor crankarm.
5. Remove motor attaching screws.
6. Remove motor while guiding crankarm through hole.

1970 Camaro

1. Remove two cowl screen attaching screws from center of cowl then pry up screen at eight clips and remove.
2. Reaching through cowl opening, loosen the two crank arm attaching nuts.
3. Remove transmission crank arm.
4. Disconnect wiring and washer hoses.
5. Remove motor attaching screws and remove motor while guiding crank arm through hole.

1970 Except Camaro & Corvette

The procedure is the same as for previous models except that the plastic access cover is now used on Chevrolet, Chevelle and Monte Carlo models.

1969 All; 1970-76 Corvette

1. Make sure wiper motor is in park position.
2. Disconnect washer hoses and electrical connectors from assembly.
3. Remove the air intake grille on Nova, Camaro and Corvette models. On Chevrolet models, remove plastic access cover.
4. Loosen nuts which retain the drive link to the crank arm ball stud on Chevrolet models. On all other models, remove the nut which retains the crank arm to the motor.
5. On Corvette, it is necessary to remove the ignition shield and distributor cap to gain access to the motor retaining screws or nuts.

NOTE: *Remove left bank spark plug wires from the cap and mark both cap and wires for aid in reinstallation.*

6. Remove three motor retaining screws or nuts and remove motor.

CAUTION: Wiper motor must be in the park position prior to installation on the cowl. Do not install a motor that was dropped or hung by the drive link.

W/S WIPER TRANSMISSION

1971-76 W/Rectangular Motor

1. Remove wiper arms and blades.
2. Raise hood and remove cowl vent screen or grille.
3. Disconnect wiring from motor.
4. Loosen, but do not remove, transmission drive link to motor crankarm attaching nuts and disconnect drive link from crankarm.
5. Remove right and left transmission to body attaching screws and guide transmission and linkage through cowl opening.

NOTE: When installing, motor must be in Park position.

1971-76 W/Round Motor

1. Raise hood and remove cowl vent screen.
2. On Chevelle, Monte Carlo and Camaro models, remove right and left wiper arm and blades.
3. On Chevrolet models, remove arm and blade only from transmission to be removed.
4. Loosen, do not remove, attaching nuts securing drive link to motor crankarm.
5. Disconnect transmission drive link from motor crankarm.
6. On Chevelle, Monte Carlo and Camaro models, remove right and left transmission to body screws.
7. On Chevrolet models, remove attaching screws securing only the transmission to be removed.
8. Remove transmission and linkage through cowl opening.

NOTE: When installing, motor must be in Park position.

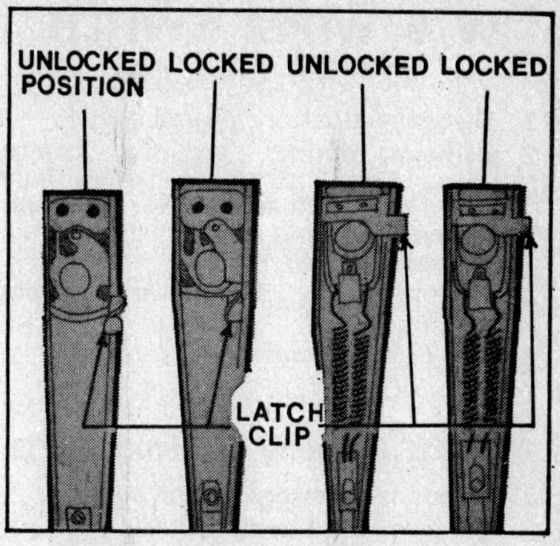

Fig. 17 Wiper arm latch clips

1973-76 Corvette

1. Make sure motor is in PARK position.
2. Disconnect battery ground cable.
3. Open hood and remove chamber screen.
4. Loosen nuts retaining ball sockets to crankarm and detach drive rod from crankarm.
5. Remove transmission nuts, then lift rod assemblies from chamber.
6. Remove transmission linkage from chamber.

1969-70 All & 1971-72 Corvette

1. Make sure wiper motor is in park position.
2. On Corvette only, remove rubber plug from front of wiper motor actuator, then insert a screwdriver, pushing internal piston rearward to actuate wiper door open.
3. On all models, remove wiper arm and blade assembly from one transmission. On articulated left-hand arm assemblies, remove clip retaining pinned arm to blade arm.
4. Remove air intake grille or screen (if equipped).
5. Loosen nuts retaining drive rod ball stud-to-crank arm and detach rod from arm.
6. Remove transmission retaining screws, or nuts, then lower drive rod assemblies into plenum chamber.
7. Remove transmission and linkage through cowl opening.
8. Reverse procedure to install. Make sure wiper blades are installed in the park position (plus or minus ⅜″ from top of reveal molding on recessed wiper arms).

W/S WIPER SWITCH

1971-76 Chevrolet

1. Disconnect battery ground cable.
2. Remove screws securing control shroud on instrument panel. (One screw is hidden above headlight switch shaft and one is hidden above cigarette lighter knob.)
3. Lift off shroud and remove remaining screws.
4. Unplug wiper switch and remove.

1973-76 Chevelle & Monte Carlo

1. Disconnect battery ground cable.
2. Remove instrument cluster.
3. Pull electric connector off rear of wiper switch.
4. Remove screws from front of switch and lift switch out front of panel.

1970-76 Camaro

1. Disconnect battery ground cable.
2. Remove trim plate and A/C outlet from below steering column if so equipped.
3. Remove light switch.
4. Remove 6 screws securing instrument carrier. Two of these are behind the cluster on either side of steering column.

NOTE: Cigar lighter grounding ring may have to be removed with lighter housing to gain access to left side of carrier.

5. Disconnect wiper switch wiring.
6. Tilting carrier forward, reach behind and remove 3 switch retaining screws and lift out switch.

1969-70 Chevrolet
1970-72 Chevelle & Monte Carlo

1. Disconnect battery ground cable.
2. On 1969-72, remove trim plate and lap cooler outlet.
3. Disconnect and remove switch from rear of panel.
4. Reverse procedure to install.

1969 Camaro, Chevelle, Nova
1970-76 Chevy Nova

1. Disconnect battery ground cable.
2. On Chevelle, remove left air outlet and headlight switch.
3. Disconnect and remove switch from rear of panel.
4. Reverse procedure to install.

1969-76 Corvette

1. Disconnect battery ground cable.
2. Remove screws from upper part of center console marked "Corvette".
3. Disconnect and remove switch and plate.
4. Carefully pry knob from switch then remove switch from plate.

Fig. 18 Blower Motor. 1969-76 All (Typical)

5. To install, insert a small rod in the switch arm before pushing the knob on the arm outside of the trim plate then reverse the above procedure.

RADIO, REPLACE

NOTE: When installing radio, be sure to adjust antenna trimmer for peak performance.

NOTE: When installing, be sure to connect speaker before applying power to radio.

1973-76 Chevrolet

1. Disconnect battery ground cable.
2. On A/C equipped vehicles, it may be necessary to remove the lap cooler duct.
3. Turn radio knobs until slot in base in knob is visible. Depress the metal retainer and remove knobs and bezels.
4. Remove control shaft nuts and washers.
5. Remove right hand bracket to instrument panel bolt and the stud nut on left side of radio.
6. Push radio forward and then lower it far enough to gain access to electrical connections.
7. Disconnect antenna, speaker and power feed and remove radio.

1973-76 Chevelle & Monte Carlo

1. Disconnect battery ground cable.
2. On A/C vehicles, remove left lap cooler duct.
3. Pull off radio knobs and bezels.
4. Remove control shaft nuts and washers.
5. Disconnect antenna, speaker and power wires.
6. Remove radio to support bracket stud nut.
7. Push radio forward until shafts clear instrument panel and then lower from car.

1973-76 Camaro & Nova

1. Disconnect battery ground cable.
2. Pull off radio control knobs and bezels.
3. Remove control shaft nuts and washers.
4. Remove screws or nuts from radio brackets.
5. Push radio forward until shafts clear instrument panel and lower unit enough to remove electrical connections.
6. Disconnect antenna, speaker and power leads and remove radio.

1973-76 Corvette

1. Disconnect battery ground cable.
2. Remove right instrument panel pad.
3. Disconnect speaker connectors.
4. Remove wiper switch trim plate screws and tip plate for access to switch connector. Remove switch connector and trim plate from cluster.
5. Unclip and remove right and left console forward trim pads. Remove forward most screw on right and left side of console.
6. Insert a flexible drive socket between console and metal horseshoe brace and remove nuts from two studs on lower edge of console cluster.
7. Remove remaining screws retaining cluster to instrument panel.
8. From rear of console, disconnect electric connector, brace and antenna.
9. Remove knobs and bezel retaining nuts.
10. Pull top of console rearward and separate radio from console and remove it from right side opening.

CAUTION: The center instrument cluster trim plate is designed to collapse under impact. DO NOT try to deflect the cluster plate forward to gain more access to remove radio. Also use care so as not to damage the plastic oil pressure line when pulling console forward.

1970-72 Chevelle, Chevrolet & Nova

1. Disconnect battery ground cable.
2. Remove ash tray and retainer.
3. Remove radio knobs, washers, controls, trim plate and nuts from radio bushings.
4. Remove hoses from center A/C duct if equipped.
5. Disconnect all leads to radio.
6. Remove screw from radio rear mounting bracket and lower radio.

1969 Except Corvette

1. Disconnect battery ground cable.
2. On Chevelle and Chevrolet, remove ash tray and retainer.
3. Remove radio knobs, nuts, electrical connections, rear support and lift radio out.

1969-72 Corvette

Coupe

1. Disconnect battery ground cable.
2. Remove left and right door sill plates and kick pads. Disconnect radio to speaker connectors (left and right side).
3. Remove right side dash pad.
4. Remove right and left console forward trim pads.
5. Remove one bolt securing heater floor outlet duct to center distributor assembly.
6. Remove floor outlet duct by pulling it through left hand opening.
7. From front of console, tape radio

push buttons in depressed position. From rear of console, disconnect electrical connector, brace and antenna.
8. Remove radio knobs and bezel retaining nuts. Push radio forward (towards front of car). From rear of console, tip back of radio up and remove from right side opening.
9. Reverse procedure to install.

Convertible

1. Disconnect battery cable.
2. Remove right instrument panel pad.
3. Disconnect speaker connectors.

NOTE: On some vehicles it may be necessary to remove kick panels to gain access to speaker connectors.

4. Remove wiper switch trim plate screws and tip plate for access to switch connector. Remove switch connector and trim plate from cluster assembly.
5. Unclip and remove right and left console forward trim pads. Remove forwardmost screw on each side of console.
6. Insert a flexible drive socket between the console and metal horseshoe brace, remove the nuts from the two studs on the lower edge of the console cluster.
7. Remove remaining screws retaining cluster to instrument panel.
8. From rear of console, disconnect radio connector, brace and antenna.
9. Remove radio knobs and bezel retaining nuts.
10. Pull top of console forward. Separate radio from console and remove it from the right side opening.

NOTE: The center instrument cluster trim plate is designed to collapse under impact. Consequently, do not try to deflect the cluster plate forward to gain more access to remove the radio. Also use care so as not to damage the plastic oil pressure line when pulling console forward.

BLOWER MOTOR REMOVAL

1969 Chevrolet

1. Remove battery and tray.
2. Remove all wires, hoses, etc. attached to skirt.
3. Raise front of vehicle on hoist.
4. Remove inner skirt to cowl brace.
5. Remove skirt attaching bolts.
6. Remove lower fender attaching bolt.
7. Pull lower section of fender outward slightly, pushing skirt upward and inboard. When skirt is free, lower and rest on wheel.
8. Disconnect the blower motor lead wire.
9. Remove motor attaching screws and remove motor, Fig. 18.

1969 Chevelle

1. Disconnect the battery ground cable.
2. Remove battery. Remove all heater hoses, wires, etc. attached to the right fender skirt.
3. Raise front of vehicle on hoist.
4. Remove all skirt-to-fender attaching bolts and lower skirt on tire.
5. Disconnect blower motor wire and remove screws retaining blower motor to case. Remove blower motor, Fig. 18.

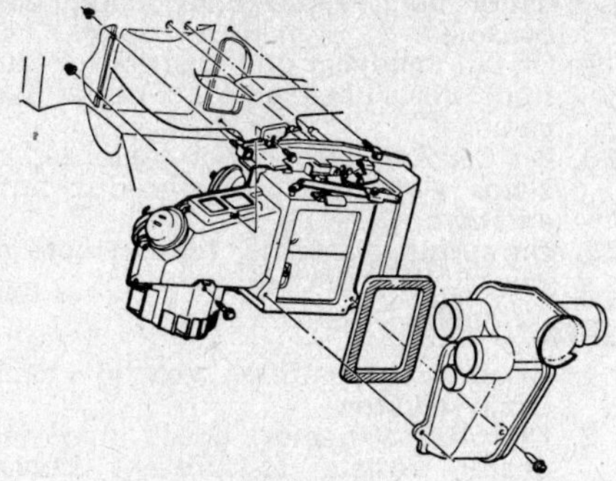

Fig. 19 Heater core. 1971-76 Chevrolet (Typical)

1969 Chevy Nova

1. Disconnect the battery ground cable.
2. Remove all wires, hoses, etc. attached to the right fender skirt.
3. Remove all skirt-fender screws and rest skirt on wheel.
4. Remove blower motor wire.
5. Remove blower to case attaching screws and remove blower assembly, Fig. 18.

1969 Camaro

1. Remove battery and battery tray.
2. Disconnect hoses, wiring, etc., at fender skirt.
3. Remove wheel opening trim, if so equipped.
4. Remove rocker panel moulding to fender and rear quarter panel screws and remove moulding.
5. Loosen rear lower fender to body bolt.
6. Remove all fender skirt attaching screws.
7. Pull out on lower edge of fender at rear of wheelhouse, push in on skirt until flange clears fender and then lower fender on wheel.
8. Remove the blower to case attaching screws and remove the blower assembly, Fig. 18.

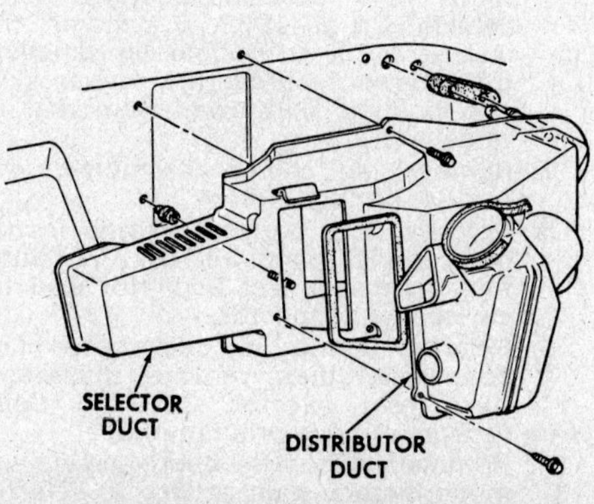

Fig. 20 Heater core. 1970-72 Chevelle & Monte Carlo (Typical)

1969-72 Corvette

1. Disconnect the battery ground cable, blower motor wire and the rubber air cooling tube.
2. Remove the three front sill plate moulding screws and pry out to gain access to the right splash shield retaining bolts.
3. Remove the splash shield retaining bolts and remove splash shield.
4. Remove motor to case mounting screws and lower motor assembly through splash shield opening, Fig. 18.

1970-72 Except Corvette

1. Disconnect the battery ground cable.
2. Remove all heater hoses and wires attached to the fender skirt.
3. Raise the vehicle on hoist.
4. Remove all fender skirt attaching bolts except those attaching the skirt to the radiator support.
5. Pull out then down on the fender skirt and place a block of wood between the skirt and the fender to allow room to remove the blower and motor.
6. Disconnect the blower motor cooling tube at the motor. Remove the blower-case attaching screws and remove the blower assembly.
7. Remove the blower wheel retaining nut and separate the motor and wheel, Fig. 18.

1973-76 Corvette

1. Disconnect battery ground cable.
2. Remove radiator supply tank, if so equipped.
3. Disconnect blower motor wires.
4. Remove blower motor to case mounting screws and remove motor, Fig. 18.

1973 Chevrolet & Camaro

1. Disconnect battery ground cable.
2. Remove all heater hoses and wires attached to fender skirt.
3. Raise vehicle on a hoist.
4. Remove all fender skirt attaching bolts except those attaching skirt to radiator support.
5. Pull out then down on fender skirt and place a 2″ × 4″ block of wood between skirt and fender to allow room to remove blower.
6. Remove blower to case screws and remove blower, Fig. 18.

1973-76 Chevelle & Monte Carlo

1. Disconnect battery ground cable.
2. Disconnect blower lead wire at motor.
3. Remove blower to case screws and remove blower, Fig. 18.

1974-76 Chevrolet, Camaro & 1973-76 Nova

1. Disconnect battery ground cable.
2. Disconnect blower lead and remove all hoses and wires connected to fender skirt then raise vehicle.
3. On all except Nova, remove all fender skirt retaining bolts except those retaining skirt to radiator support. On Nova, remove the eight rearmost fender skirt to fender retaining screws.
4. Pull out and down on fender skirt and

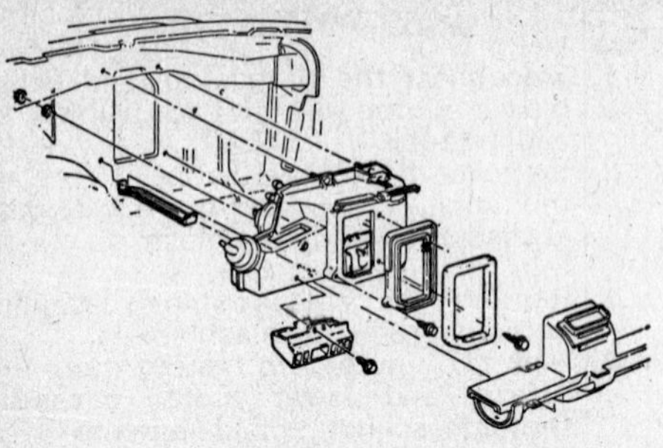

Fig. 21 Heater core. 1973-76 Chevelle & Monte Carlo (Typical)

wedge a 2x4 inch block of wood between skirt and fender to allow room for blower motor removal.

5. Remove blower to case retaining screws and remove blower assembly. Gently pry flange, since sealer will act as an adhesive, Fig. 18.

NOTE: On Nova, remove blower retaining nut and separate wheel and motor before removing through fender and skirt.

---

6. Reverse procedure to install.

HEATER CORE REMOVAL

L/Air Cond.

1971-76

1. Disconnect battery ground cable.
2. Drain radiator, disconnect heater hoses at core and plug openings to prevent spillage of water.
3. Remove nuts from air distributor duct studs on engine side of firewall, Figs. 19 through 24.
4. On Chevrolet and 1973-76 Chevelle and Monte Carlo, remove distributor duct retaining screw.
5. On 1971-72 Chevelle and Monte Carlo, from under dash, drill out lower right hand distributor stud with a 1/4" drill.
6. On Nova, remove glove box and door and drill out lower right hand distributor stud with a 1/4" drill.
7. On Camaro, remove glove box, radio, and defroster duct to distributor duct screw.
8. On Corvette, remove right instrument panel pad, right hand dash braces, center dash console duct and floor outlet duct, radio and center dash console.
9. On all models, pull distributor duct from firewall being careful not to bend cable.
10. On Camaro, disconnect cable and resistor wires and remove distributor and core.
11. On all other models, remove core assembly from duct.

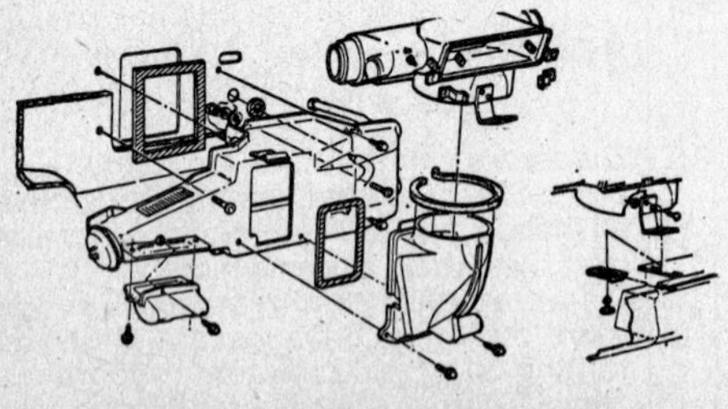

Fig. 22 Heater core. 1969-76 Chevy Nova (Typical)

1969-70 All

1. Disconnect battery ground cable. Drain radiator.
2. Disconnect heater hoses and plug heater outlets to prevent coolant spillage.
3. Remove nuts from air distributor duct studs on engine side of firewall.
4. On Chevrolet and Nova, remove glove box and door assembly. On Corvette, remove right instrument panel pad, right hand dash braces, center dash console duct and floor outlet duct.
5. Except Corvette: from under dash, drill out lower right distributor stud. Corvette: remove radio and center dash console.
6. Pull distributor duct from firewall mounting, remove resistor wires and lay distributor duct on floor.

NOTE: Use care to avoid bending bowden cables.

---

7. Remove core from distributor duct.

W/Air Cond.

Chevrolet 1971-76 Comfortron & 1972-76 4 Seasons

1. Disconnect battery ground cable and compressor clutch connector.
2. Drain radiator and disconnect heater hoses from core. Plug hoses and core openings.
3. On 1971 Comfortron, disconnect control head vacuum hoses from in-line connector. On all 1972-76 systems, disconnect vacuum hose from vacuum check valve and push hose grommet through firewall.
4. Remove accessible screws and stud nuts from heater and air selector duct from dash side of firewall, Fig. 19.
5. On 1971 Comfortron, remove three stud nuts and on all 1972-76 systems, remove two screws and three stud nuts, accessible from engine side of dash.
6. On 1971 Comfortron, drill out lower right hand case stud from inside vehicle. On all 1972-76 systems, the last screw is accessible by removing the screws securing the fender skirt and holding skirt away from rear of wheel-well.
7. Remove lap cooler assembly, glove box door and glove box.
8. Remove floor outlet duct and instrument panel pad. On 1972-76 Comfortron, disconnect aspirator and in-car sensor connector.
9. Remove distributor duct hoses and connector, then remove distributor duct from selector. On 1971 Comfortron, remove programmer.
10. Remove defroster duct screw and move duct to gain access to selector and heater core. On 1972-76 Comfortron, disconnect in-line vacuum connector and line from outside air diaphragm, then disconnect electrical and vacuum connections from programmer.
11. Disconnect cables and vacuum lines from selector, then remove heater core and selector. Separate heater core from selector by removing clamp screws.

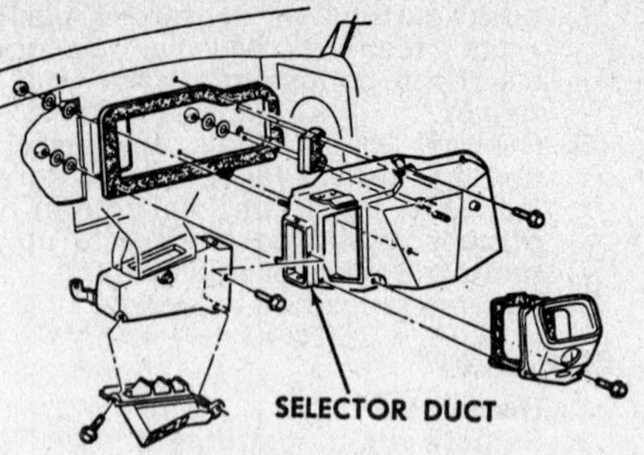

Fig. 23 Heater core. 1970-76 Camaro (Typical)

Chevrolet, 1969-71 Four Seasons

1. Disconnect battery ground cable.
2. Drain radiator and disconnect heater hoses from core. Plug openings to prevent spillage of coolant.
3. Remove nuts from selector duct studs projecting through firewall.
4. Remove glove box and door.
5. Remove center distributor duct hoses, duct cable, center distributor duct to selector duct screws and remove center duct.
6. Inside car, drill out lower right hand case stud using a 1/4" drill.
7. Remove floor distributor duct.
8. Unfasten and remove selector duct from firewall.
9. Disconnect wiring, vacuum lines and cables and remove from vehicle.
10. Scribe location of temperature door camming plate on selector duct and remove camming plate. Remove core and core housing from duct.

Chevrolet, 1969-70 Comfortron

1. Disconnect battery ground cable and drain radiator.
2. Disconnect heater hoses and plug hoses and core openings.

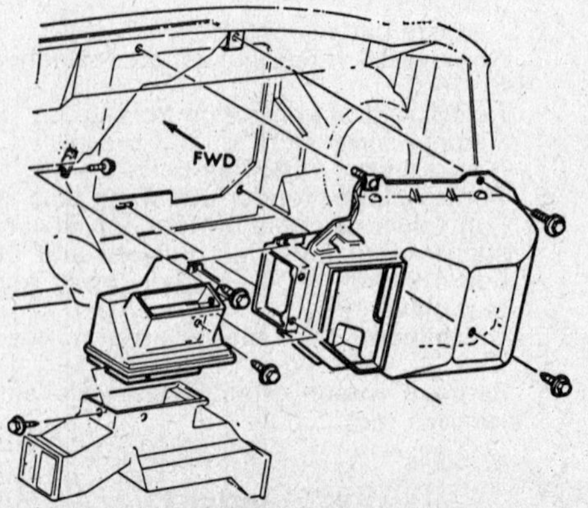

Fig. 24 Heater core. 1969-76 Corvette (Typical)

3. Disconnect vacuum hoses from under dash and push large firewall grommet into passenger compartment.
4. Remove accessible stud nuts from heater and air selector case from dash side of firewall and then the three case stud nuts from engine compartment.
5. Drill out lower right hand case stud from inside vehicle.
6. On all models remove right hand courtesy light retaining screw.
7. Remove glove box door and glove box.
8. On all models remove floor distributor duct.
9. Remove center distributor duct hoses, shut-off cables and center duct.
10. On all models disconnect power servo electrical leads, then remove power retaining screws. Tilt servo forward and remove cover screws. Disconnect vacuum line from servo diaphragm and remove vacuum head and servo cover. Disconnect link from servo and remove servo assembly.
11. Remove transducer retaining screws and without removing vacuum hoses, pull transducer to front of dash.

NOTE: One transducer retaining screw secures a ground wire.

12. Remove heater, air distributor and control head as an assembly. Remove screws securing die cast plate to air inlet side of case and disconnect temperature door linkage from door shaft. Remove die cast plate with core from case.

Chevelle & Monte Carlo, 1969-76

1. Disconnect battery ground cable and drain radiator.
2. Disconnect heater hoses from core and plug hoses and core openings.
3. Remove distributor case stud nuts projecting through firewall, Figs. 20 and 21. On 1973-76, remove resistor assembly and the last stud nut through its opening.
4. On all except 1970-72, remove glove box door and glove box.
5. On all models, remove right hand lap cooler and kick pad cover. On 1969, remove recirculating air valve.
6. On all models, disconnect center duct hoses and remove center duct to selector duct screws, then remove center duct.
7. Remove floor distributor duct.
8. On 1969-72, drill out lower right hand selector duct stud from inside vehicle. On 1973-76, remove lower right hand distributor duct to dash panel screw.
9. On all models, remove air selector retaining screws, then lower selector assembly. Disconnect all wiring, vacuum lines and cables and on 1969-72, scribe location of temperature door camming plate on selector duct, then remove camming plate.
10. On all models, remove core and housing from selector assembly, then core from housing.

Nova, 1969-76 & Camaro 1969-70

1. Disconnect battery ground cable and drain radiator.
2. Disconnect heater hoses and plug hoses and core openings to prevent coolant spillage.

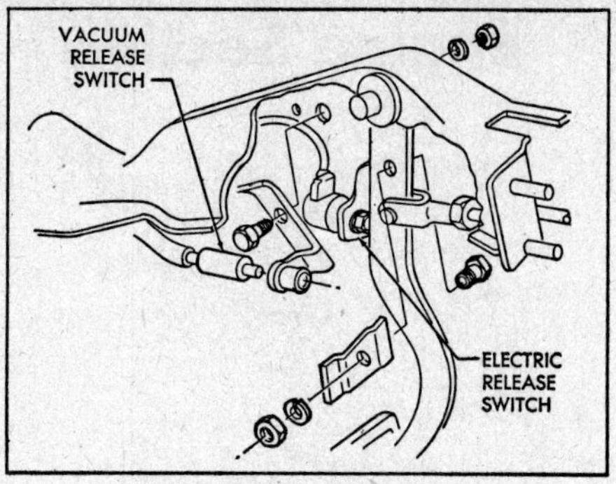

Fig. 25 Release switches and brackets. 1969-70

3. Remove accessible stud nut from air selector duct, Fig. 22.
4. On Camaro, remove right side lower rocker molding and the lower fender attaching bolts.
5. On all models, remove right hand fender skirt to fender and skirt reinforcing screws. Lower skirt to wheel and remove remaining stud nut.
6. Remove glove box door, glove box and right hand kick pad recirculating air valve.
7. Remove center and floor ducts. Remove screws securing selector duct halves together, then separate.
8. On Nova, remove selector duct right half to firewall screws and remove duct. On Camaro, remove screws securing selector duct to firewall and remove duct.
9. On all models, disconnect all wiring and cables. Scribe location of temperature camming plate on selector duct and remove plate and core.

Camaro, 1971-76

1. Disconnect battery ground cable.
2. Drain radiator and disconnect heater hoses from core. Plug openings to prevent spillage of coolant.
3. Remove nuts from distributor studs on engine side of firewall, Fig. 23.
4. Remove glove box and radio.
5. Remove defroster duct to distributor duct screw. With radio removed, the defroster duct can be pulled rearward to gain clearance for distributor duct removal.
6. Carefully pull distributor from firewall and disconnect wiring and cables. Remove duct and core from vehicle.

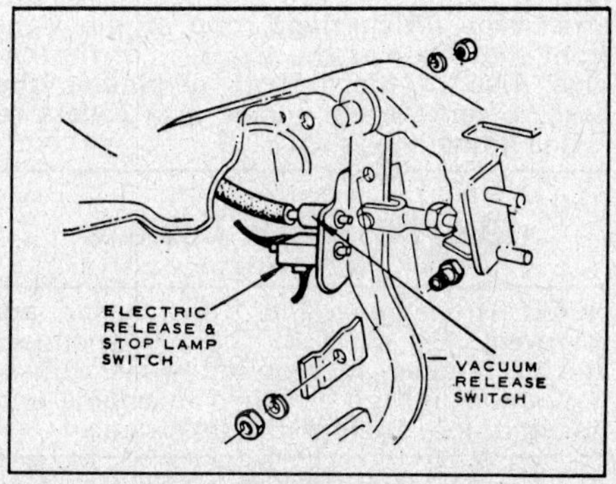

Fig. 26 Release switches and brackets. 1971-76

Corvette, 1969-76

1. Disconnect battery ground cable. Drain radiator and disconnect heater hoses from core. Plug openings to prevent spillage of coolant.
2. Remove nuts from engine side distributor duct, Fig. 24.
3. Remove right hand dash pad and center instrument cluster. Remove dash braces.
4. Disconnect right dash outlet duct from center duct. Remove screws attaching center duct to selector duct and remove center duct.
5. Remove screws attaching selector duct to firewall and pull selector rearward and to the right. Disconnect cables and wiring.
6. Remove selector duct from car. Remove temperature door camming plate from duct and remove core and housing.

SPEED CONTROLS

1969-76 Cruise Master

Servo Unit Adjustment

Adjust the bead chain or cable so that it is as tight as possible without holding the throttle open when the carburetor is set as its lowest idle throttle position. The cable is adjusted by turning the hex portion of servo. On 1973-75 units, the bead chain or cable is adjusted so there is 1/16 inch of lost motion in servo cable.

When connecting the bead chain or cable (engine stopped) manually set the fast idle cam at its lowest step and connect the chain so that it does not hold the idle screw off the cam. If the chain needs to be cut, cut it three beads beyond the bead that pulls the linkage.

Regulator Unit Adjustment

To remove any difference between engagement and cruising speed, one adjustment is possible. However, no adjustment should be made until the following items have been checked or serviced.

1. Bead chain or cable properly adjusted.
2. All hoses in good condition, properly attached, not leaking, pinched or cracked.
3. Regulator air filter cleaned and properly oiled.
4. Electric and vacuum switches properly adjusted.

Engagement - Cruising Speed Zeroing

If the cruising speed is lower than the engagement speed, loosen the orifice tube locknut and turn the tube outward; if higher turn the tube inward. Each 1/4 turn will alter the engagement-cruising speed difference one mph. Tighten locknut after adjustment and check the system operation at 50 mph.

Brake Release Switch

The electric brake switch is actuated when the brake pedal is depressed .25 inch on 1969-70 units, Fig. 25, and .38-.64 inch on 1971-76 units, Fig. 26. The vacuum release switch is actuated when brake pedal is moved 5/16 inch on all units.

Engine Section

NOTE

Throughout the engine section there will be references to Small V8s and Mark IV V8s. These can be distinguished as follows:

Small V8s: V8-262, 302, 305, 307, 327, 350 and 170 H.P. and 265 H.P. V8-400.

Mark IV V8s: V8-396, other V8-400s, V8-402, 427, 454.

ENGINE MOUNTS, REPLACE

1973-76 All 6 & 1969-72 Chevy Nova 4 & 6

1. Remove nut, washer and engine mount through-bolt.
2. Raise engine to release weight from mount.
3. Remove bracket-to-mount bolt, then remove mount.
4. Install new mount on bracket.
5. Lower engine, install through-bolt and tighten all mount bolts, Figs. 1, 2 and 6.

1969-72 Camaro, Chevelle & Chevrolet 6

1. Remove nut, washer, spacer then engine mount through-bolt.
2. Raise engine to release weight from mount.
3. Remove mount, stop bracket and frame bracket assembly from crossmember, then remove stop bracket and mount from frame bracket.
4. Install stop bracket and new mount on frame bracket, then install assembly on crossmember.
5. Lower engine, install through-bolt and tighten all mount bolts, Fig. 2.

Fig. 2 Engine mounts. 1969-72 6 cyl. (typical)

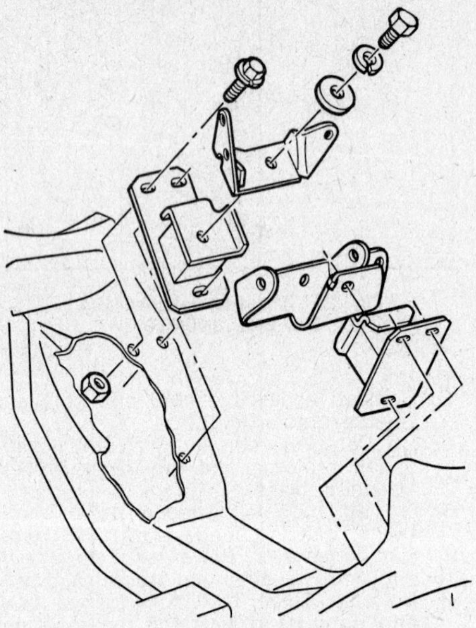

Fig. 1 Engine mounts. 1969-70 4 cyl.

1969-76 All V-8

NOTE: On some V8 engines, the mount on the right side of the engine cannot be removed without first removing the fuel pump.

1. Remove nut, washer and engine mount through-bolt.
2. Raise engine to release weight from mount.
3. Remove mount from engine.
4. Install new mount on engine.
5. Lower engine, install through-bolt and tighten all mount bolts, Figs. 3 thru 10.

ENGINE, REPLACE

NOTE: V8 engines are equipped with two strap type lifting rings, one at the right front the other at the left rear of the engine. Use of these rings eliminates the need to remove the rocker arm covers to install lifting adapters.

1969-76 Except Corvette

NOTE: The engine and transmission are removed as a unit. First disconnect and/or remove as required wires, tubes, hoses and linkage attached to engine and transmission. Then do the following:

1. Remove hood, radiator and fan. Also power steering pump and A/C compressor (if equipped).
2. Remove drive shaft.
3. Unfasten and lower exhaust pipes.
4. Remove rocker arm cover (s) and attach lifting device.
5. Remove front mount through bolts.
6. Raise engine to take weight off mounts, then remove rear mount bolts and crossmember. On Chevrolet models it will be necessary to remove mount from transmission before crossmember can be removed. On Chevrolet it will be necessary to remove mount from transmission and loosen rear frame cushion bolts before crossmember can be removed. On 1969-76 Camaro, it is not necessary to remove the crossmember completely; after removing the bolts, slide it back.
7. Remove engine and transmission assembly.

1969-76 Corvette

When necessary to remove the engine, it is recommended that the engine and transmission assembly be lifted from the vehicle as a unit. It takes considerably less time to remove the engine and transmission as a unit than it would to "split" the engine from the transmission in the vehicle and remove the engine alone.

CYLINDER HEAD

1969-76 Four & Six Cylinder

1. Drain cooling system and remove air cleaner.
2. Disconnect choke rod or choke cable, accelerator pedal rod at bellcrank on manifold, and fuel and vacuum lines at carburetor.
3. Disconnect exhaust pipe at manifold flange, then unfasten and remove manifolds and carburetor as an assembly.
4. Remove fuel and vacuum line retaining clip from water outlet and disconnect wire harness from temperature sending unit and coil, leaving

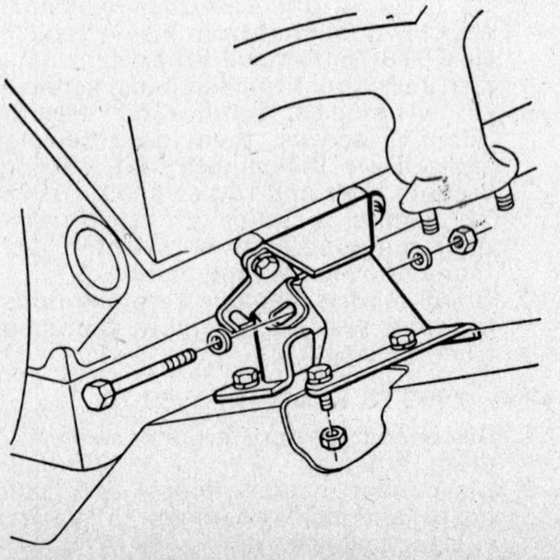

Fig. 3 Engine mounts. 1969 Chevrolet V8

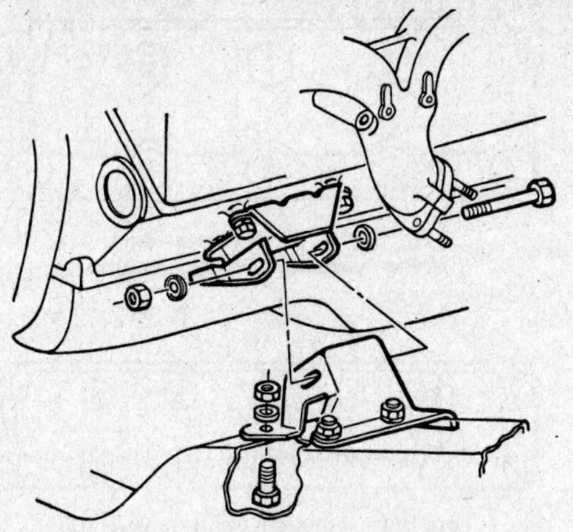

Fig. 4 Engine mounts. 1970 Chevrolet V8

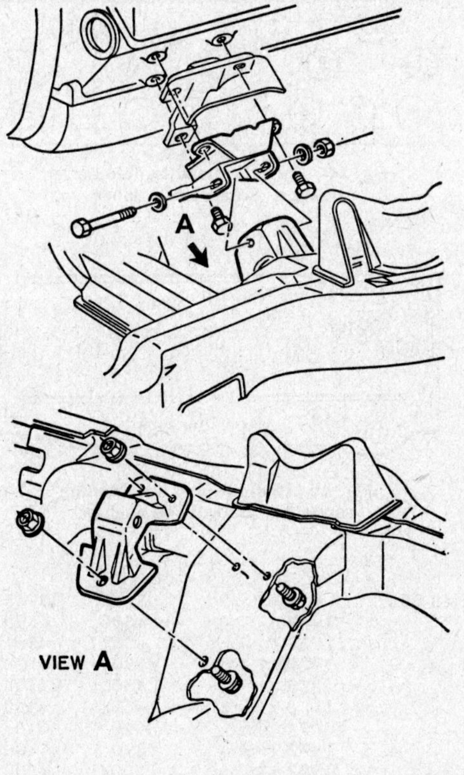

Fig. 5 Engine mounts. 1971-72 Chevrolet V8

harness clear of clips on rocker arm cover.

5. Disconnect radiator hose at water outlet housing and battery ground strap at cylinder head.
6. Remove spark plugs and coil.
7. Remove rocker arm cover. Back off rocker arm nuts, pivot rocker arms to clear push rods and lift out push rods.
8. Unfasten and remove cylinder head.
9. Reverse procedure to install and tighten head bolts in the sequence shown in Figs. 12 and 13.

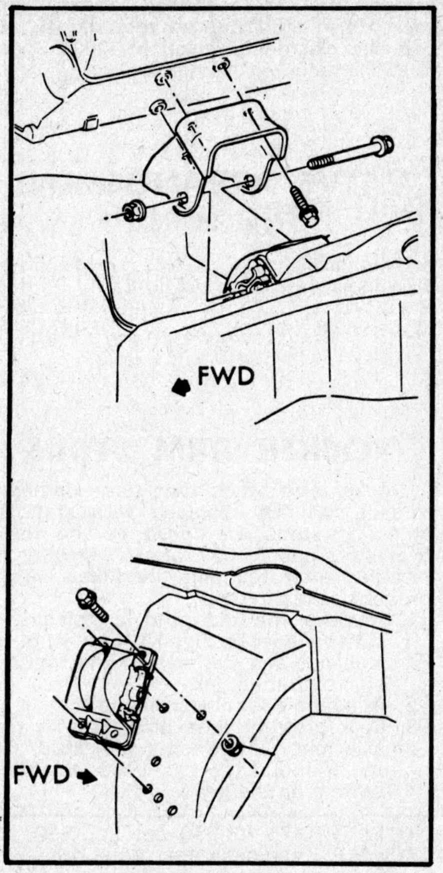

Fig. 6 Engine mounts. 1973-76 Except Corvette (typical)

1969-76 V8s

NOTE: On Camaro with V8-396 and air conditioning it is necessary to remove the battery, A/C compressor, radiator shroud, air injector pump and starter. As the engine must be raised 2½" to remove the exhaust manifold it will be necessary to remove the engine mount through bolts and loosen transmission mount bolts.

When removing left head, power steering pump or air suspension compressor must be removed if so equipped.

On all other models, proceed as follows:

1. Remove intake and exhaust manifolds.
2. Remove rocker arm covers.
3. Back off rocker arm nuts, pivot rocker arms to clear push rods and remove push rods.
4. Unfasten and remove cylinder heads.
5. Reverse procedure to install and tighten head bolts in the sequence shown in Figs. 14 to 16.

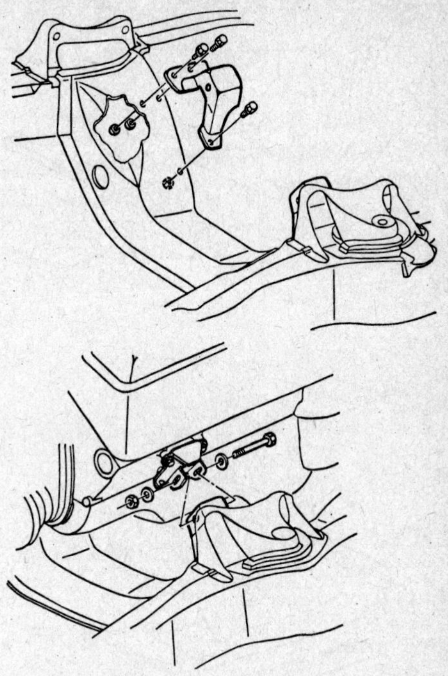

Fig. 8 Engine mounts. 1969 Camaro V8

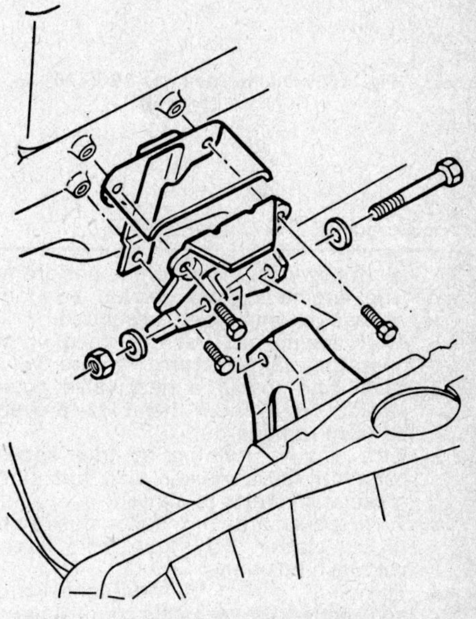

Fig. 9 Engine mounts. 1971-72 Camaro V8

VALVES, ADJUST

Hydraulic Lifters

NOTE: *After the engine has been thoroughly warmed up the valves may be adjusted with the engine shut off as follows: With engine in position to fire No. 1 cylinder the following valves may be adjusted: Exhaust 1-3-4-8, intake 1-2-5-7. Then crank the engine one more complete revolution which will bring No. 6 cylinder to the firing position at which time the following valves may be adjust-*

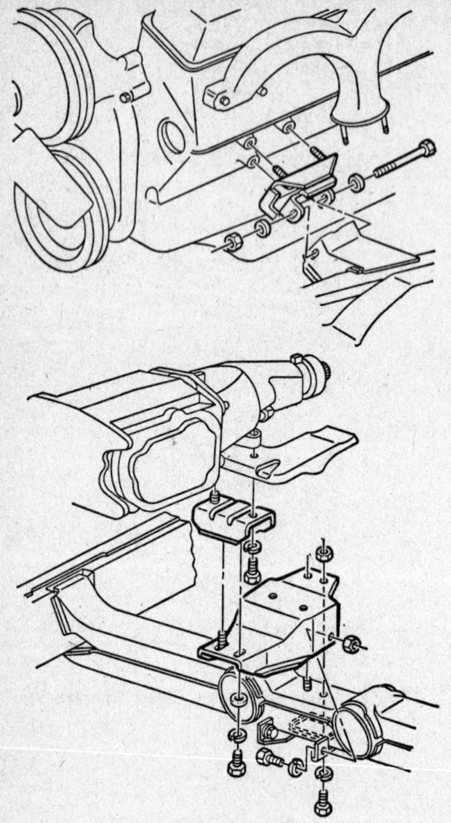

Fig. 10 Engine mounts. 1969-76 Corvette (typical)

ed: Exhaust 2-5-6-7, intake 3-4-6-8.

The following procedure, performed with the engine running should be done only in case readjustment is required.

1. After engine has been warmed up to operating temperature, remove valve cover and install a new valve cover gasket on cylinder head to prevent oil from running out.
2. With engine running at idle speed, back off valve rocker arm nut until rocker arm starts to clatter.
3. Turn rocker arm nut down slowly until the clatter just stops. This is the zero lash position.
4. Turn nut down 1/4 additional turn and pause 10 seconds until engine runs smoothly. Repeat additional 1/4 turns, pausing 10 seconds each time, until nut has been turned down the number of turns listed in the *Valve Specifications Chart* from the zero lash position.

NOTE

This preload adjustment must be done slowly to allow the lifter to adjust itself to prevent the possibility of interference between the intake valve head and top of piston, which might result in internal damage and/or bent push rods. Noisy lifters should be replaced.

Mechanical Lifters

With the engine warmed to operating temperature as described above, turn rocker arm stud nut as required to obtain the clearances given in *Valve Specifications* table.

VALVE ARRANGEMENT

Front to Rear

4 Cylinder . . . . E-I-I-E-E-I-I-E
6 Cylinder . . . . E-I-I-E-E-I-I-E-E-I-I-E
Small V8 . . . . E-I-I-E-E-I-I-E
Mark IV V8 . . . . I-E-I-E-I-E-I-E

ROCKER ARM STUDS

Rocker arm studs that have damaged threads may be replaced with standard studs. If studs are loose in the head, oversize studs (.003" or .013") may be installed after reaming the holes with a proper size reamer.

1. Remove the old stud by placing a suitable spacer, Fig. 17, over stud. Install nut and flat washer and remove stud by turning nut.
2. Ream hole for oversize stud.
3. Coat press-fit area of stud with rear axle lube. Then install new stud, Fig. 18. If tool shown is used, it should bottom on the head.

NOTE: 1970-71 V8-350 cu. in., 360 and 370 H.P. engines are equipped with threaded rocker arm studs.

VALVE LIFT SPECS

Engine	Year	Intake	Exhaust
4-153	1969-70	.3973	.3973
6-230	1969-70	.3317	.3317
6-250	1969-74	.388	.388
	1972㉗	.388	.4051
	1974㉗	.388	.4051
	1975-76	.388	.4051
8-262	1975	.3727	.3900
8-302	1969	.4851	.4851
8-305	1976	.3727	.4100
8-307	1969③⑯	.3945	.3945
	1970-73	.390	.410
	1972㉗	.4006	.4100
8-327	1969	.3945	.3945
8-350	1969③⑯	.3945	.3945
	1969④	.4500	.4600
	1969⑰	.4851	.4851
	1970②③	.390	.410
	1970④	.4500	.4600
	1970-71⑰㉒㉔	.4586	.4850
	1971㉕㉖	.390	.410
	1972-73㉘	.390	.4100
	1972㉗㉘	.4006	.4100
	1972㉙	.4586	.4850
	1974㉜	.390	.410
	1974㉕	.4500	.4600
	1975-76⑥	.3900	.4100
	1975-76⑦	.4500	.4600
8-396	1969⑨⑱	.398	.398
	1969④	.4614	.4800
	1969⑲	.5197	.5197
8-400	1970⑱	.390	.410
	1970㉒	.3983	.3983
	1970④	.4614	.480
	1971③	.3983	.430
	1972-76	.390	.410
	1972㉗	.4006	.410
8-402	1972	.3983	.430
8-427	1969⑭	.4614	.4800
	1969⑫	.4614	.4800
	1969⑬	.5917	.5917
	1969⑮	.5197	.5197
	1969⑳	.398	.398
	1969㉑	.5586	.580

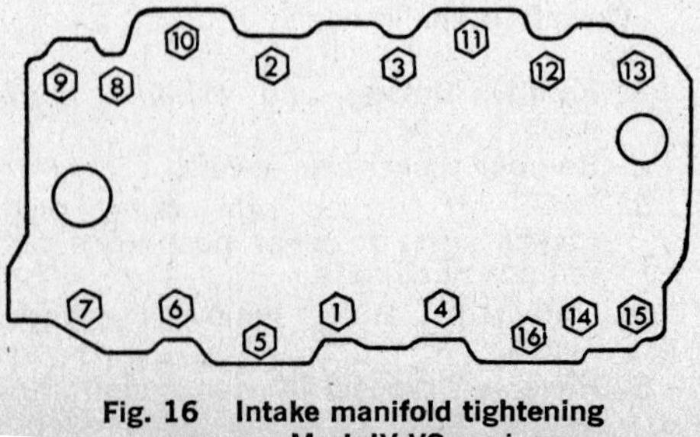

Fig. 16 Intake manifold tightening sequence. Mark IV V8 engines

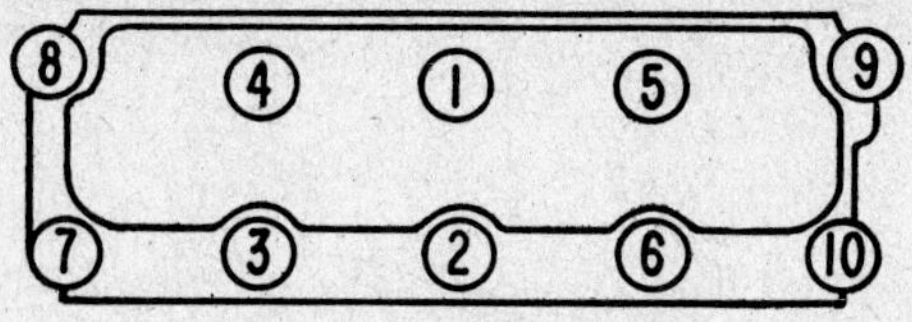

Fig. 12 Cylinder head tightening sequence. 4-153 engine

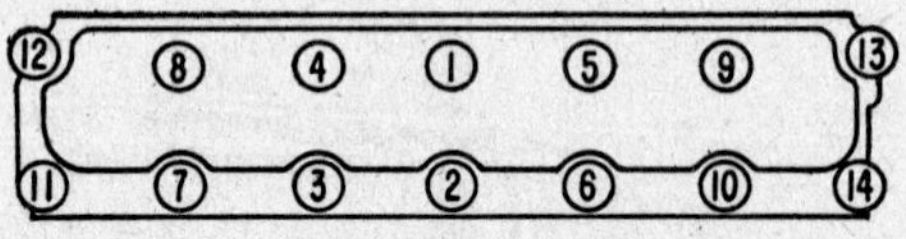

Fig. 13 Cylinder head tightening sequence. 6-230, 250

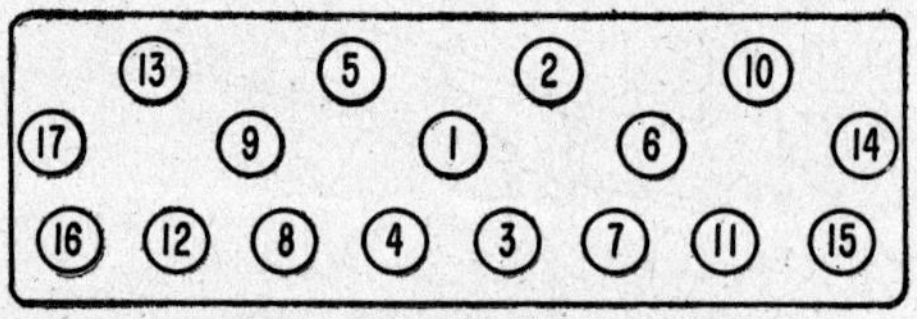

Fig. 14 Cylinder head tightening sequence. Small V8 engines

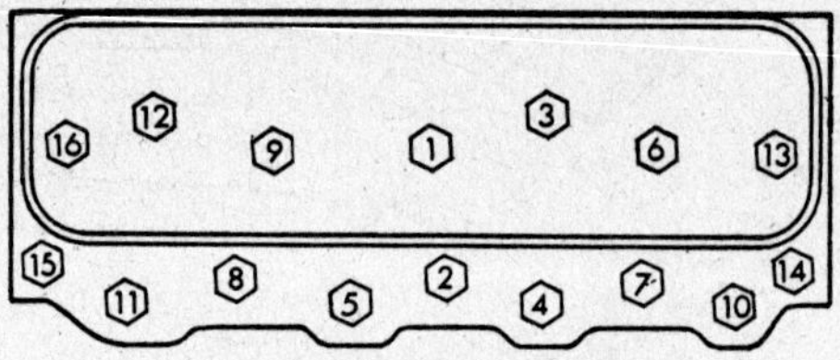

Fig. 15 Cylinder head tightening sequence. Mark IV V8 engines

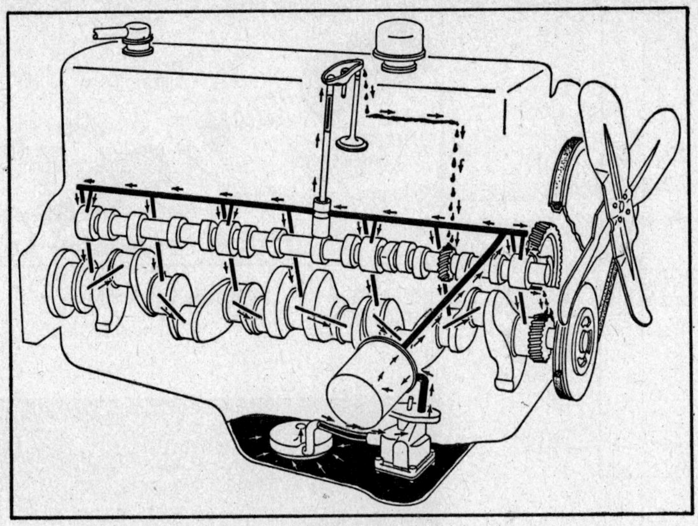

Engine oiling system. 6-230, 250. Four cylinder is similar

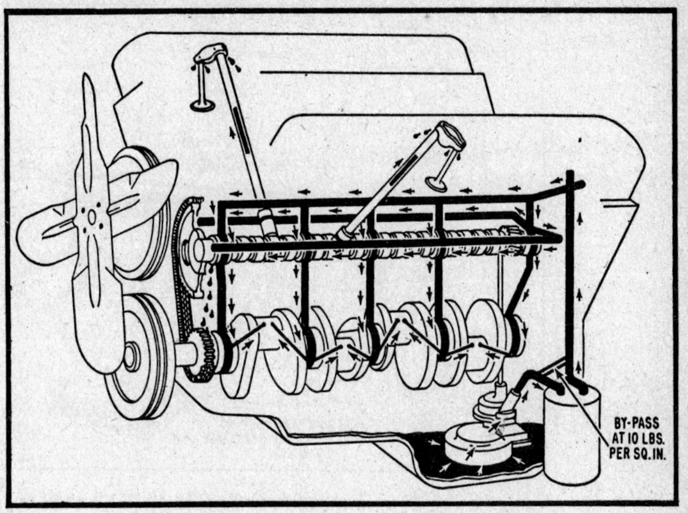

Engine oiling system. Small V8 engines

8-454	1970 (23)	.3983	.4300
	1970 (14)(5)	.4614	.480
	1971 (10)	.5197	.5197
	1972 (30)	.4614	.480
	1972 (31)	.3983	.480
	1973-75	.440	.440
	1976	.3983	.4300

(2)—250 H.P.
(3)—300 H.P.
(4)—350 H.P.
(5)—365 H.P.
(6)—Exc. L82 & Z28
(7)—L82 & Z28
(9)—325 H.P.
(10)—425 H.P.
(11)—340 H.P.
(12)—400 H.P.
(13)—385 H.P.
(14)—390 H.P.
(15)—435 H.P.
(16)—255 H.P.
(17)—370 H.P.
(18)—265 H.P.
(19)—375 H.P.
(20)—335 H.P.
(21)—430 H.P.
(22)—330 H.P.
(23)—345 H.P.
(24)—360 H.P.
(25)—245 H.P.
(26)—270 H.P.
(27)—California vehicles
(28)—165, 200 H.P.
(29)—255 H.P.
(30)—230 H.P.
(31)—270 H.P.
(32)—145, 160, 185, 195 H.P.

Fig. 17 Removing valve rocker arm stud

VALVE TIMING

Intake Opens Before TDC
All Except Corvette

Engine	Year	Degrees
4-153	1969-70	17½
6-230	1969-70	16
6-250	1969-76	16
8-262	1975	26
8-305	1976	28
8-307	1968-72; 1973	28
	1972 Calif.	44
8-327	1969	28
8-350	1969-70 255, 300 H.P.	28
	1969 350 H.P.	52
	1970 250, 300 H.P.	28
	1970 360 H.P.	42½
	1971 245, 270 H.P.	28
	1971 330 H.P.	42⅔
	1972-73 165, 200 H.P.	28
	1972 165, 200 H.P. Calif.	44
	1972 255 H.P.	42⅔
	1974 145, 160, 185, 195 H.P.	28
	1974 145, 160, 185, 195 H.P. Calif.	44
	1974 245 H.P.	52
	1975-76 Exc. Z28	28
	1975-76 Z28	52
8-396	1969 265, 325 H.P.	28
	1969 350 H.P.	56
8-400	1970 265, 330 H.P.	28
	1970 350 H.P.	56
	1971 300 H.P.	28
	1972-76	28
	1972 Calif	44
	1974 Calif.	44
8-402	1972	30
8-427	1969 390, 400 H.P.	56
	1969 435 H.P.	44
8-454	1970 360 H.P.	56
	1971 365 H.P.	56
	1971 425 H.P.	44
	1972	56
	1973-75	55
	1976	42

Corvette

Engine	Year	Degrees
8-350	1969-70 300 H.P.	28
	1969-70 350 H.P.	52
	1970 370 H.P.	42½
	1971 270 H.P.	28
	1971 330 H.P.	42⅔
	1972 200 H.P.	28
	1972 200 H.P. Calif.	44
	1972 255 H.P.	42¾
	1973-74 195 H.P.	28
	1973-74 250 H.P.	52
	1975-76 Exc. L82	28
	1975-76 L82	52
8-427	1969 390, 400 H.P.	56
	1969 435 H.P.	44
8-454	1970 390 H.P.	56
	1970 460 H.P.	62
	1971 365 H.P.	56
	1971 425 H.P.	44
	1972	56
	1973-74	55

PUSH RODS

On engines that use push rods with a hardened insert at one end, the hardened end is identified by a color stripe and should always be installed toward the rocker arm during assembly.

Service Bulletin

On 6-cylinder engines with air conditioning, it is not necessary to remove the distributor wires, etc. to replace the valve

Fig. 18 Installing valve rocker arm stud

Engine lubrication. Mark IV V8 engines

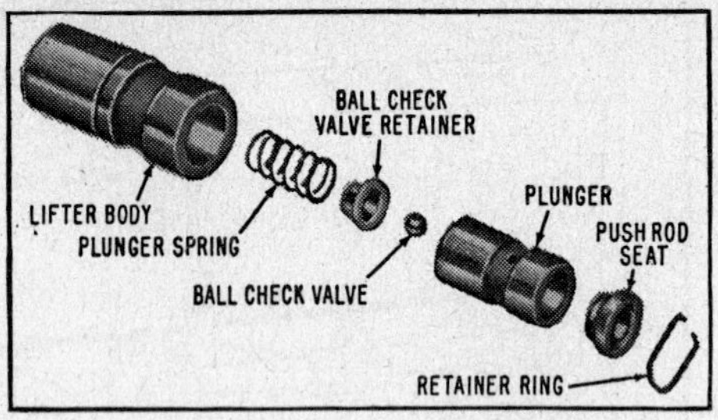

Fig. 19 Hydraulic valve lifter

push rod cover and/or gasket.

1. Remove coil and bracket from block.
2. Remove distributor hold-down clamp.
3. Lift distributor up for clearance (do not disengage from cam gear), then remove push rod cover.
4. Use new gasket and reverse procedure to install.

VALVE GUIDES

On all engines valves operate in guide holes bored in the head. If clearance becomes excessive, use the next oversize valve and ream the bore to fit. Valves with oversize stems are available in .003, .015 and .030".

HYDRAULIC LIFTERS

Valve lifters may be lifted from their bores after removing rocker arms and push rods. Adjustable pliers with taped jaws may be used to remove lifters that are stuck due to varnish, carbon, etc. Fig. 19 illustrates the type of lifter used.

TIMING CASE COVER

NOTE: On all engines the cover oil seal may be replaced without taking off the timing gear cover. After removing the vibration damper, pry out the old seal with a screwdriver. Install the new seal with the lip or open end toward inside of cover and drive it into position.

1969-76 Four & Six Cylinder

1. To remove cover, remove radiator.
2. Remove vibration damper (6-cyl.) or pulley (4-cyl.).

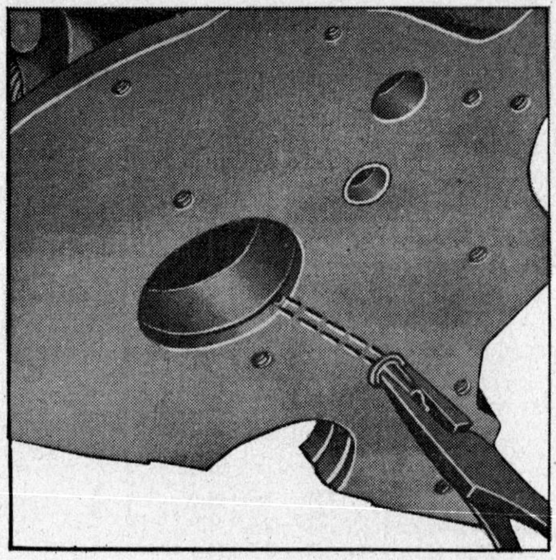

Fig. 20 Timing gear oil nozzle removal. 4-153, 6-230, 250

3. Remove oil pan.
4. Unfasten and remove cover.
5. Pry oil seal out of cover with a large screwdriver. Install new seal with open side of seal inside of cover and drive or press seal into place.
6. If oil nozzle is to be replaced, remove it with pliers as shown in Fig. 20. Drive new nozzle in place, using a suitable light plastic or rubber hammer.
7. Clean gasket surfaces.
8. Install a suitable centering tool over

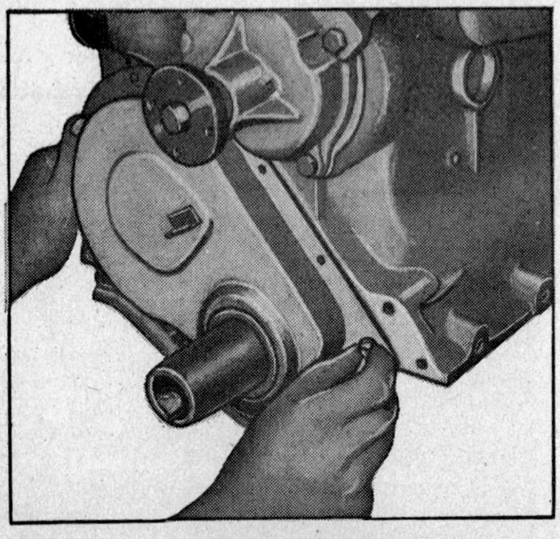

Fig. 21 Installing timing case cover. 4-153, 6-230, 250

Fig. 22 Removing camshaft gear from camshaft. 4-153 and all 6 cylinder engines

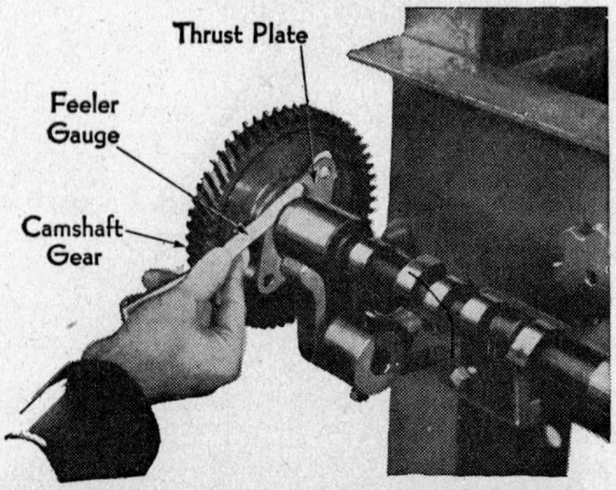

Fig. 23 Checking camshaft end play which should be .001 to .005". 4-153 and all 6 cylinder engines

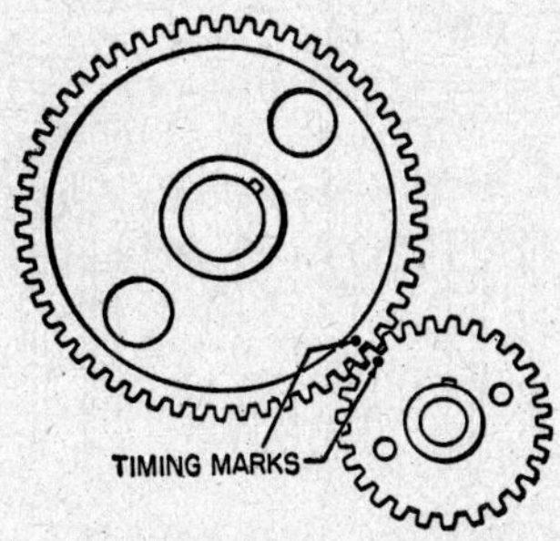

Fig. 24 Timing gear locating marks. 4-153 and all 6 cylinder engines

end of crankshaft.

9. Coat gasket with light grease and stick it in position on block.
10. Install cover over centering tool, Fig. 21, and install cover screws, tightening them to 6 to 8 ft. lbs. *It is important that the centering tool be used to align the cover so the vibration damper installation will not damage the seal and position seal evenly around damper hub surface.*

1969-76 V8s

NOTE: The timing chain cover has been redesigned on 1974-76 small V8 engines to facilitate removal. The inner flange seal retainer has been shortened enough to clear the oil pan lip, making oil pan removal unnecessary.

Remove vibration damper, oil pan, heater hose from water pump, and water pump from cylinder block. Unfasten and remove cover and gaskets.

Fig. 25 Checking timing gear backlash with feeler gauge. Lash should be .004 to .006". 4-153 and all 6 cylinder engines

Pry old seal out of cover from the front with a large screwdriver. Install the new seal so the open end of the seal is toward the inside of the cover and drive it into position with a suitable driver, being sure to support cover at sealing area.

1. Make certain that the mating faces of cover and block are clean and flat.
2. Make certain oil slinger is in place against crankshaft sprocket.
3. Coat oil seal with light grease and, using a new cover gasket, install cover and gasket over dowel pins in cylinder block.
4. Install and tighten cover screws to 6-8 lb. ft. torque.
5. Install oil pan, harmonic balancer and water pump.
6. Start engine and check for leaks.

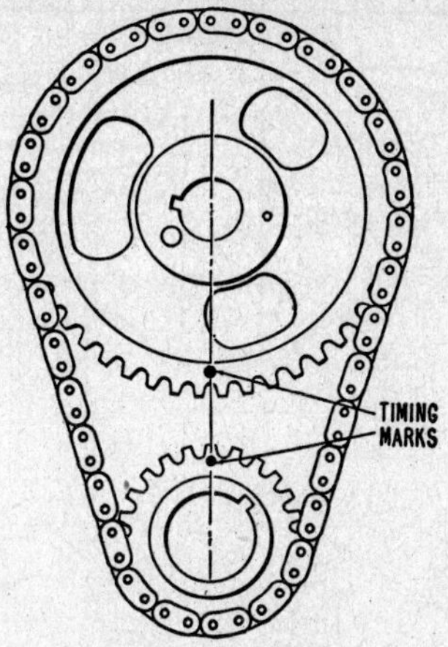

Fig. 26 Timing gear locating marks. V8 engines

TIMING GEARS

4-153 & All 6 Cyl. Engines

When necessary to install a new camshaft gear, the camshaft will have to be removed as the gear is a pressed fit on the shaft. The camshaft is held in position by a thrust plate which is fastened to the crankcase by two capscrews which are accessible through two holes in the gear web.

Use an arbor press to remove the gear and when doing so, a suitable sleeve, Fig. 22, should be employed to support the gear properly on its steel hub.

Before installing a new gear, assemble a new thrust plate on the shaft and press the gear on just far enough so that the thrust plate has practically no clearance, yet is free to turn. The correct clearance

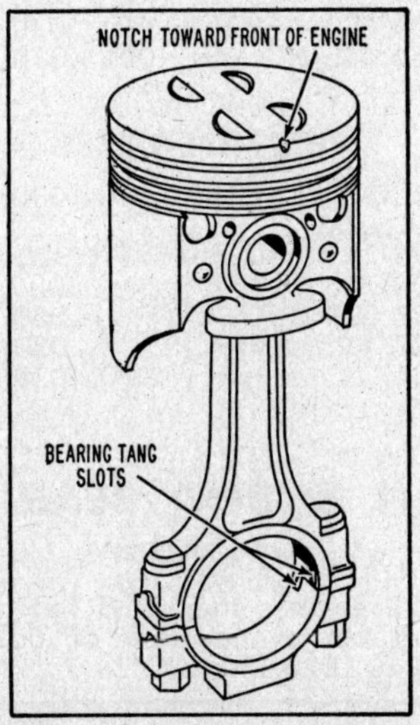

Fig. 28 Piston and rod assembly. 4-153 and 6-230 engines

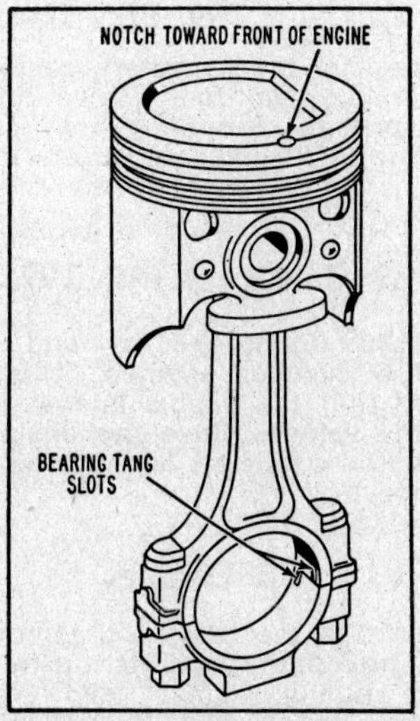

Fig. 29 Piston and rod assembly. 6-250 engine

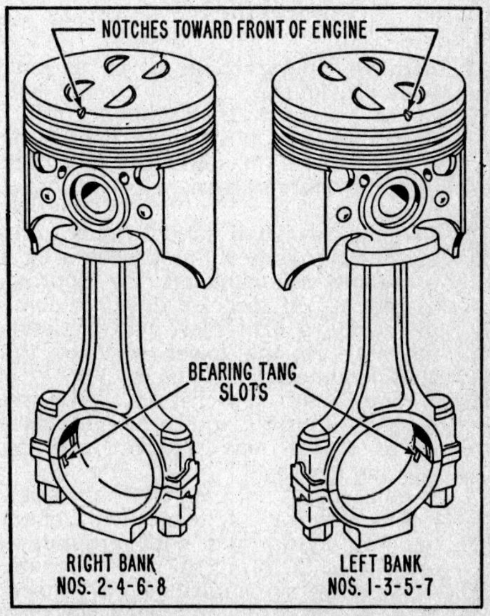

Fig. 30 V8-307, 350 and 327 (except 325 and 350 H.P.)

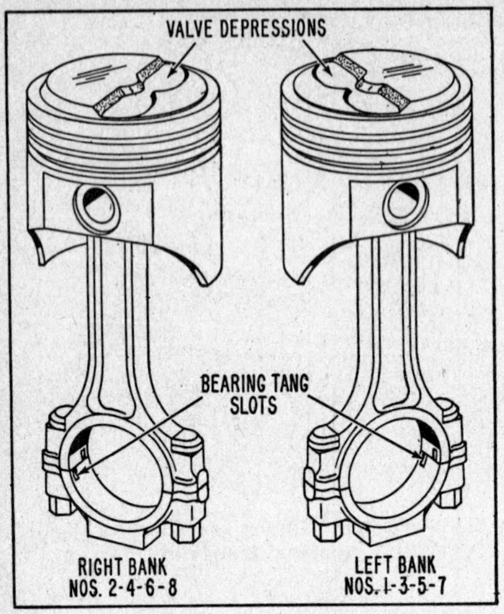

Fig. 31 V8-302, 327 (325 and 350 H.P.)

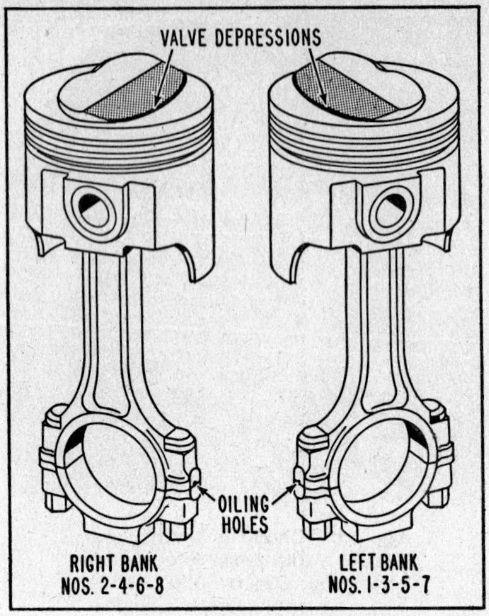

Fig. 32 V8-302, 350 Z-28 High performance

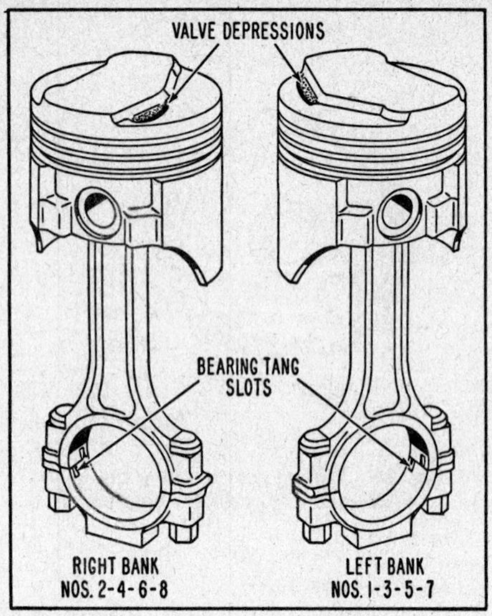

Fig. 33 V8-396, 454 engines

is from .001″ to .005″, Fig. 23.

The crankshaft gear can be removed by utilizing the two tapped holes in conjunction with a gear puller.

When the timing gears are installed, be sure the punch-marks on both gears are in mesh, Fig. 24. Backlash between the gears should be from .004″ to .006″, Fig. 25. Check the run-out of the gears, and if the camshaft gear run-out exceeds .004″ or the crank gear run-out is in excess of .003″, remove the gear (or gears) and examine for burrs, dirt or some other fault which may cause the run-out. If these conditions are not the cause, replace the gear (or gears).

TIMING CHAIN

V8 Engines

1. Remove timing chain cover as outlined previously.
2. Remove crankshaft oil slinger.
3. Crank engine until "O" marks on sprockets are in alignment, Fig. 26.
4. Remove three camshaft-to-sprocket bolts.
5. Remove camshaft sprocket and timing chain together. Sprocket is a light press fit on camshaft for approximately ⅛″. If sprocket does not come off easily, a light blow with a plastic hammer on the lower edge of the sprocket should dislodge it.
6. If crankshaft sprocket is to be replaced, remove it with a suitable gear puller. Install new sprocket, aligning key and keyway.
7. Install chain on camshaft sprocket. Hold sprocket vertical with chain hanging below and shift around to align the "O" marks on sprockets.
8. Align dowel in camshaft with dowel hole in sprocket and install sprocket on camshaft. *Do not attempt to drive sprocket on camshaft as welch plug at rear of engine can be dislodged.*
9. Draw sprocket onto camshaft, using the three mounting bolts. Tighten to 15-20 lb. ft. torque.
10. Lubricate timing chain and install cover.

CAMSHAFT

If necessary to replace bearings, the engine will have to be removed from the chassis. Then remove camshaft, flywheel and crankshaft. Drive out the expansion plug at the rear of the rear camshaft bearing and use suitable equipment to remove and replace the bearings.

6 Cyl. & 1969 Exc. Camaro

To remove the camshaft, remove radiator, grille, push rods, valve lifters, oil pan, vibration damper (6 cyl) or pulley (4 cyl), and timing case cover. Then pull out camshaft and gear assembly.

1969 Camaro & All 1970-76

On 1969 Camaro 6 cyl. and 1970-76 4 and 6 cylinder engines, it is recommended that the engine is first removed from the vehicle. Once the engine is removed, the camshaft is removed as outlined above.

V8 Engines

To remove the camshaft, remove valve lifters, fuel pump and its push rod, radiator, timing chain and camshaft sprocket. Install two $5/16$″-18 x 4″ bolts in two camshaft bolt holes. Using these bolts as a puller, remove camshaft.

PISTONS & RODS, ASSEMBLE

1969-76

Assemble pistons to connecting rods as shown in Fig. 28 to 36.

PISTON OVERSIZES

4-153	.001, .020, .030, .040″
6 Cyl.	.001, .020, .030, .040″
V8-302, 307, 327, 350	.001, .020, .030″
V8-396, 310, 350 H.P.	.001, .020, .030″
V8-396 375 H.P.	.001, .030″
V8-400	.001, .020, .030″
V8-427 390 H.P.	.001, .020, .030″
V8-427, 425, 435 H.P.	.001, .030, .060″
V8-454, 230, 245 H.P.	.001, .030″
V8-454, 270, 275 H.P.	.001, .030″
V8-454 345, 390 H.P.	.001, .020, .030″
V8-454, 365, 425 H.P.	.001, .020, .030″
V8-454 450, 460 H.P.	.001, .030, .060″
V8-454 465 H.P.	.001, .030, .060″

RING OVERSIZES

4-153, 6-230, 6-250	.020, .030, .040″
V8-327, 350	.020, .030″
V8-396	.020, .030″
V8-400	.020, .030″
V8-427 390 H.P.	.020, .030″
V8-427, 425, 435 H.P.	.030, .060″
V8-454	.020, .030, .060″

MAIN & ROD BEARINGS

Undersizes

On all engines main and rod bearings are available in undersizes of .001, .002, .009, .010, .020 and .030″.

SERVICE BULLETIN

Connecting rod type noise correction on

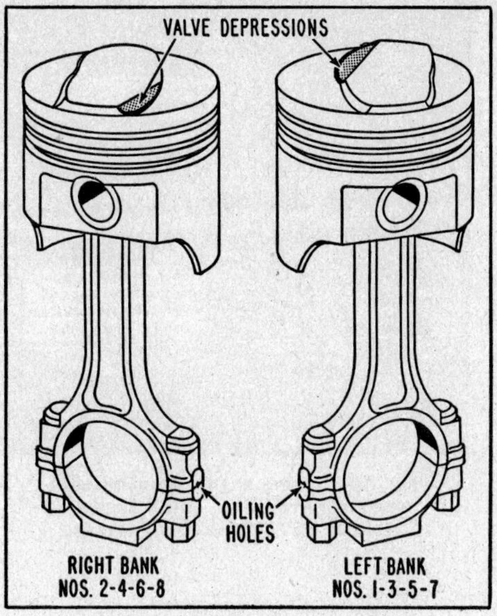

Fig. 34 V8-400 High performance

6-230 engines may be made as follows:

Remove connecting rod cap and Plastigage the bearings. If clearance is within .001 to .002", reinstall cap and retaining nuts. Then tighten nuts to the proper torque as listed in the *Engine Tightening Specifications* table.

NOTE 4-153 & All 6 Cyl.

The rear main bearing journal has no oil hole drilling. To remove the upper bearing half (bearing half with oil hole) proceed as follows after cap is removed:

1. Use a small drift punch and hammer to start the bearing rotating out of the block.
2. Use a pair of pliers (tape jaws) to hold the bearing thrust surface to the oil slinger and rotate the crankshaft to pull the bearing out.
3. To install, start the bearing (side not notched) into side of block by hand, then use pliers as before to turn bearing half into place.
4. The last 1/4" movement may be done by holding just the slinger with the pliers or tap in place with a drift punch.

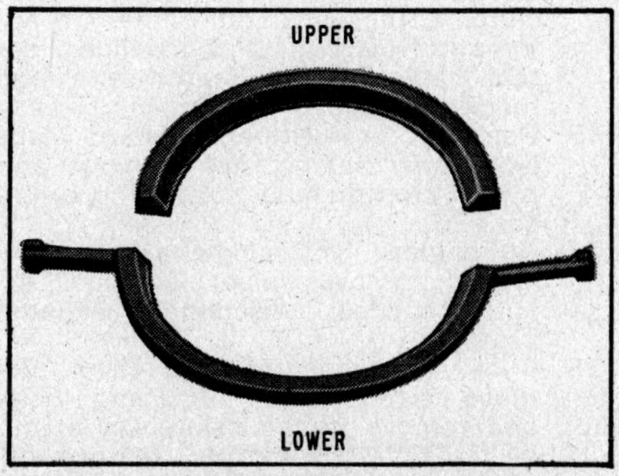

Fig. 37 Crankshaft and rear oil seal (typical)

CRANKSHAFT REAR OIL SEAL

NOTE: 1971-76 engines are equipped with helix type rear seal. A seal starting tool, Fig. 38, must be used to prevent the upper seal half from coming into contact with the sharp edge of the block. Place the tip of the tool into the seal channel and "shoehorn" the seal into the upper seal channel.

1969-76

When necessary to correct an oil leak due to a defective seal, always replace the upper and lower seal halves as a unit, Fig. 37. *When installing either half, lubricate the lip portion only with engine oil, keeping oil off the parting line surface as this is treated with glue.* Always clean crankshaft surface berore installing a new seal. Be careful of seal retainer tang while inserting a new seal so that it doesn't cut the seal.

1. To replace the lower seal, remove seal from groove in bearing cap, using a small screwdriver to pry it out.
2. Insert new seal and roll it in place with finger and thumb.
3. To replace the upper seal (with engine in car) use a small hammer and tap a brass pin punch on one end of the seal until it protrudes far enough to be removed with pliers.
4. Insert the new seal, gradually pushing with a hammer handle until seal is rolled into place.
5. Install bearing cap with new seal and tighten bearing cap bolts.

OIL PAN

1973-76 Six

1. Disconnect battery positive cable.
2. Remove radiator upper mounting panel or side mount bolts.
3. Place a piece of heavy cardboard between fan and radiator.
4. Disconnect fuel suction line at fuel pump.
5. Raise vehicle on a hoist and drain engine oil.
6. Disconnect and remove starter and flywheel underpan or converter housing underpan and splash shield.
7. On Nova, disconnect steering rod at idler lever and position linkage to one side for pan clearance.
8. Rotate crankshaft until timing mark on damper is at 6 o'clock.
9. Remove bolts attaching brake line to front crossmember and move brake line away from crossmember.
10. Remove through bolts from engine front mounts.
11. Remove oil pan bolts.

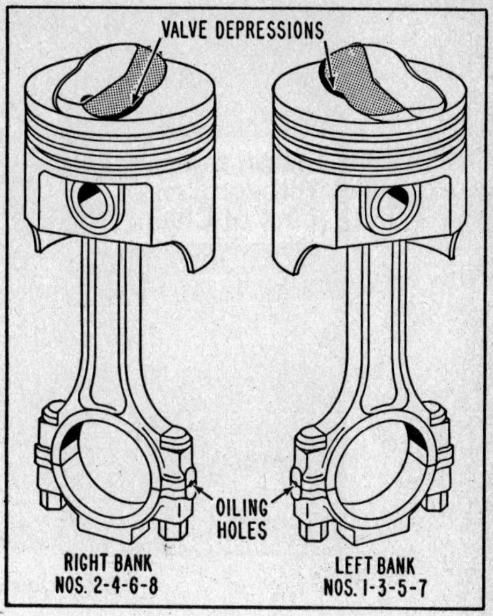

Fig. 36 V8-427 engine, 1969-70

12. On Chevrolet models, raise engine slowly until motor mount through bolts can be removed, remove bolts, then continue to raise engine about 3".
13. On Nova, remove left engine mount and frame bracket. Remove oil pan by lowering slightly and then rolling it into opening created by removal of left engine mount. Tilt front of pan upward and remove by pulling pan down and to rear.
14. On Camaro, raise engine enough to insert 2" x 4" block of wood under engine mounts. Then lower engine onto blocks and lower and remove pan.

1973-76 V8 Except Corvette

NOTE: On Chevelle equipped with V8-454 and manual transmission, it is necessary to remove the starter and the flywheel shield. Also, remove rear transmission mount to crossmember nut and raise rear of transmission.

1. Disconnect battery ground cable.
2. Remove distributor cap and fan shroud retaining bolts.

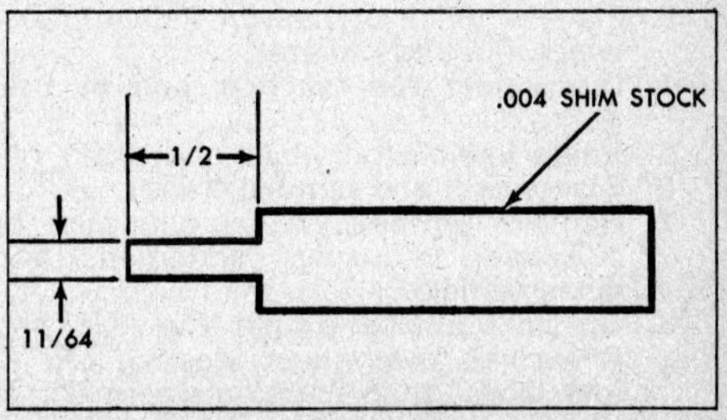

Fig. 38 Fabricated seal starting tool for helix type seal

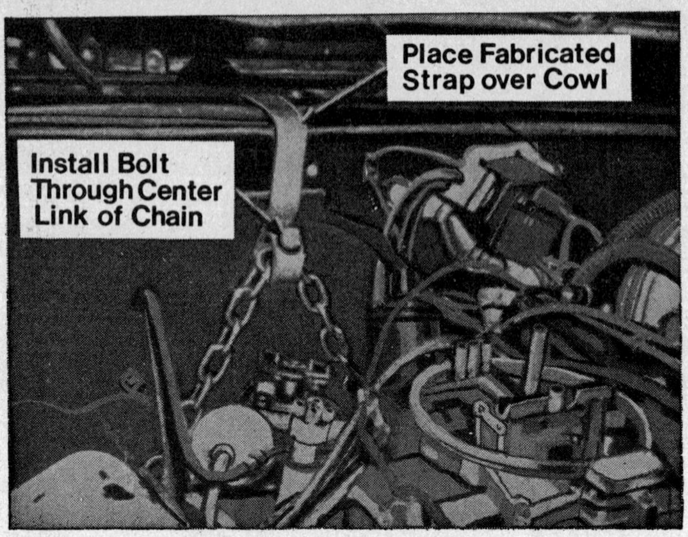

Fig. 39 Install strap over cowl

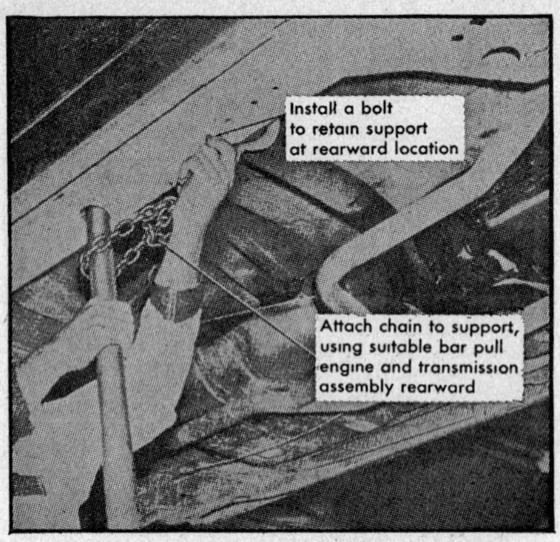

Fig. 40 Prying engine/transmission assembly rearward. 1969 Chevelle V8-307, 327, 350 with Std. Trans.

3. On Mark IV engines, place a piece of heavy cardboard between fan and radiator.
4. Raise vehicle on hoist and drain oil.
5. Disconnect exhaust or crossover pipes.
6. If equipped with automatic transmission, remove converter underpan and splash shield.
7. On 1973-76 Nova, disconnect steering idler lever at frame and swing linkage down for pan clearance.
8. Rotate crankshaft until timing mark on damper is at 6 o'clock.
9. On all except V8-350 and Mark IV engines with automatic transmission, disconnect starter brace at starter. Remove inboard starter bolt and loosen outboard starter bolt. Swing starter outward to gain clearance.
10. Remove both through bolts from engine front mounts.
11. Using suitable equipment, raise engine until wood blocks of 2″ on Chevrolet and Nova and 3″ on Chevelle and Monte Carlo can be inserted under engine mounts. Lower engine onto blocks.
12. Remove oil pan bolts and lower oil pan.

1971-72 Six

1. Disconnect battery positive cable.
2. Remove radiator upper mounting panel or side mount bolts.
3. Place a piece of heavy cardboard between fan and radiator.
4. Disconnect fuel suction line at fuel pump.
5. Raise vehicle on hoist and drain oil.
6. Disconnect and remove starter.
7. Remove either flywheel underpan or converter housing underpan and splash shield.
8. On all vehicles except Chevelle and Chevrolet, disconnect steering rod at idler lever, than position steering linkage to one side for clearance.
9. Rotate crankshaft until timing mark on damper is at 6 o'clock position.
10. Remove bolts attaching brake line to front crossmember and move brake line away from member.
11. Remove through bolts from engine front mounts.
12. Remove oil pan bolts.
13. On Chevelle and Chevrolet, raise engine slowly until motor mounts can be removed from frame brackets, remove mount and continue to raise engine until it is about three inches up.
14. On all models except Chevelle and Chevrolet, raise engine enough to insert 2″ x 4″ wood blocks under engine mounts, then lower engine onto blocks.
15. Lower oil pan and remove.

1971-72 V8 Except Chevelle with Mark IV Engines

1. Disconnect battery ground cable.
2. Remove distributor cap.
3. Remove fan shroud retaining bolts.
4. On Mark IV engines, place a heavy piece of cardboard between fan and radiator.
5. Raise vehicle on hoist and drain oil.
6. Disconnect exhaust or crossover pipes.
7. On vehicles with automatic transmission, remove converter housing underpan and splash shield.
8. On all models except Chevelle and Chevrolet, disconnect steering idler lever at frame and swing linkage down for clearance.
9. Rotate crankshaft until timing mark on damper is at 6 o'clock position.
10. On small V8 engines, disconnect starter brace at starter. Remove inboard starter bolt and loosen outboard starter bolt. Starter can be swung out for clearance.
11. Remove through bolts from engine front mounts.
12. Use a suitable jack and a block of wood and raise engine until blocks of wood 2″ on Chevrolet and Nova and 3″ on Chevelle can be inserted under mounts. Lower engine until supported by blocks.
13. Remove oil pan.

1971-72 Chevelle with Mark IV Engines

1. Disconnect battery positive cable.
2. Remove air cleaner, dipstick and disconnect distributor cap from distributor.
3. Disconnect radiator shroud and upper mounting panel. This allows radiator to move up and down freely in later steps.
4. Place a heavy piece of cardboard between fan and radiator.
5. Disconnect engine ground straps at engine.
6. Disconnect accelerator control cable from engine.
7. Place a hook and chain over cowl, Fig. 39.
8. Raise vehicle on hoist and drain oil.
9. Remove propeller shaft and plug rear of transmission.
10. On floor shift models, remove two bolts securing shift lever to linkage. On all other models, disconnect linkage at transmission.
11. Disconnect speedometer cable and back-up lamp switch connector.
12. On manual transmissions, disconnect clutch cordon shaft at frame. On automatics, disconnect cooler lines, detent cable, rod or switch wire and modulator pipe.
13. Remove crossmember bolts and place jack under engine. Raise engine and move crossmember toward rear of car.
14. On engines with single exhaust, remove crossover pipe. On dual exhaust models, disconnect exhaust pipes.
15. Remove flywheel housing cover. Remove transmission attaching bolts and remove transmission. On manual transmission models, remove flywheel housing, throwout bearing and flywheel.
16. Remove engine mount through bolts.

17. Raise rear of engine about 4 inches. Attach each end of chain to engine using bell housing bolts or transmission mounting bolts. Lower engine jack and move to front of engine.
18. Raise front of engine and insert 2" blocks of wood under front mounts.
19. Rotate crankshaft until timing mark on damper is at 6 o'clock position.
20. Remove oil pan.

1970 Camaro

1. Disconnect battery ground cable.
2. Remove distributor cap on V8 units to prevent damage when engine is raised.
3. On Mark IV engines, remove dipstick and tube.
4. Raise car and drain crankcase.
5. Disconnect exhaust crossover pipe at manifolds if necessary.
6. Remove automatic transmission converter under pan if so equipped.
7. Rotate crankshaft so timing mark is at 6 o'clock position.
8. Remove front engine mount through bolts.
9. Raise engine and insert 3" wood blocks under engine mounts. Lower engine until it is supported by blocks.
10. Remove oil pan bolts and oil pan.

1970 Chevrolet Six

1. Disconnect battery ground cable, starter wires and starter brace attaching bolt.
2. Loosen fan belt.
3. Raise vehicle and drain crankcase.
4. Remove starter, flywheel underpan and engine mount through bolts. Rotate crankshaft so timing mark is positioned at 6 o'clock position.
5. Position fan so the widest openings between the blades are at the top and bottom.
6. Raise the front of the engine and remove the engine mounts at the frame brackets.
7. Insert 5" wood blocks between the engine mount brackets and the frame brackets.
8. Remove oil pan bolts and oil pan.

1970 Chevrolet V8, 1969 Chevrolet All, 1969 Camaro 1969-70 Chevelle V8 with Auto. Trans. Exc. V8-396 1969-70 Chevy Nova

1. Disconnect battery ground cable.
2. On V8s, lift cap off distributor to prevent damaging cap when engine is raised.
3. Remove through bolts from engine front mounts.
4. Remove fan blade and starter.
5. Disconnect cooler lines (if equipped) at transmission and remove converter housing underpan.
6. Disconnect steering linkage at idler lever and swing linkage for pan clearance.
7. Rotate crankshaft until timing mark on vibration damper is at 6 o'clock position.
8. Use a suitable jack and a block of wood to prevent damage to oil pan and raise engine enough to insert 2" x 4" wood blocks under engine mounts; then lower engine to blocks.

On 1969 Chevy Nova, attach a cantilever hoist or chain-fall to engine. If chainfall is used, disconnect and move hood out of the way.

NOTE: If 2" x 4" wood blocks are cut 5½" long they can be used on all Chevrolet engines. The 5½" length up for 4 and 6 cylinder engines and the 4" side for V8 engines.

9. On Camaro six-cylinder models, after oil pan bolts are removed and oil pan lowered, remove the oil pump, then remove the pan. On all other models, remove the pan without removing the pump.

NOTE: On 396 and 427 engines the oil pan has three ¼"-20 attaching bolts at the crankcase front cover—one at each corner and one at lower center.

1969-70 Chevelle V8-307, 327, 350 with Std. Trans.

1. Disconnect battery positive cable and remove air cleaner. Remove distributor cap.
2. Remove radiator upper mounting panel.
3. Remove fuel pump and left rear engine mount bolt. Remove fan belt.
4. On floor shift, remove shift lever and floor plate.
5. Raise vehicle on hoist and drain oil.
6. Remove crankshaft pulley and starter motor.
7. Remove flywheel underpan, disconnect transmission linkage and remove shift bracket assembly.
8. Remove drive shaft, disconnect exhaust pipes from manifolds and remove front engine mounts to block bolts.
9. Rotate crankshaft until timing mark on harmonic balancer is at 6 o'clock position.
10. Remove bolts attaching transmission crossmember to frame. Remove two bolts attaching brake line to front crossmember and move brake line from crossmember.
11. Raise front of engine at harmonic balancer until clearance between engine mounts and block is approximately ½ inch.
12. Using a chain and bar as illustrated in Fig. 40, pull the engine/transmission assembly rearward and insert a bolt through the forward mounting hole in the crossmember and the rearward hole in the frame on each side.

NOTE: Use extreme caution during this step as the engine is "free" in the chassis.

13. Raise front of engine approximately 2 additional inches. Remove oil pan bolts and lower oil pan.
14. Remove front main bearing cap and remove oil pan.

1969-70 Chevelle V8-396

The engine must be completely disconnected from its mounts, front and rear, and the crossmember disconnected and moved rearward to allow the engine to be raised high enough to provide clearance for the oil pan to be removed. The engine therefore, must be supported by a fabricated hook and a suitable chain, hung over the cowl, Fig. 39, or by removing the hood and attaching a chain hoist to raise the engine. *Because the engine is free and the crossmember removed, use caution.*

Once the engine has been raised and secured, proceed as outlined for Chevelle V8 with auto. trans.

1969 Chevelle Six

1. Remove radiator upper mounting panel. Disconnect battery positive cable.
2. Disconnect fuel line. Raise vehicle and remove splash shield, underpan and starter motor.
3. Remove through bolts from front engine mounts.
4. Rotate crankshaft until timing mark on harmonic balancer is at 6 o'clock position.
5. Remove brake line retainer bolts at crossmember and move line out of the way. Remove oil pan bolts.
6. Raise engine slowly until engine mounts can be removed from frame brackets. Continue to raise engine until it has been raised approximately 3 inches.
7. Lower oil pan and remove.

1969-76 Corvette

1. Disconnect battery and remove oil dipstick and tube.
2. Disconnect steering linkage idler arm at frame and lower linkage.
3. Remove oil pan.
4. On high performance engines, the oil baffle must be removed before additional operations can be performed.

NOTE: On 427 engines the oil pan has three ¼"-20 bolts at the crankcase front cover—one at each corner and one at lower center.

OIL PUMP

4-153, 6-230, 6-250

The pumps used in these engines are of the positive gear type. After disassembling the pump, examine the shaft and gears for excessive wear and replace where necessary, or better still, install a new pump. When assembling the pump, be sure the ground side of the idler gear is toward the cover.

The gasket used between the pump cover and the body is special in that it controls the clearance in the pump. If the relief valve parts show wear, install new parts. Be sure that the tapered set screw which holds the pump in place is fully seated and locked with its lock nut.

V8 Engines

After removing the oil pan, unfasten pump from rear main bearing cap. Disconnect pump shaft from extension by removing clip from collar. Remove pump cover and take out idler gear, drive gear and shaft.

Should any of the following conditions be found it is advisable to replace the pump assembly.

1. Inspect pump body for cracks or wear.
2. Inspect gears for wear or damage.
3. Check shaft for looseness in housing.
4. Check inside of cover for wear that would permit oil to leak past the ends of gear.
5. Check oil pick-up screen for damage to screen, by-pass valve or body.
6. Check for oil in air chamber.

BELT TENSION DATA

	New Lbs.	Used Lbs.
1969-76		
Air Condition	135-145	95
Air Pump—		
1969	70-80	55
1970-76	120-130	75
Fan and Power Steering	120-130	75

WATER PUMP, REPLACE

All Models

1. Drain radiator and break loose fan pulley bolts.
2. Disconnect heater hose at water pump.
3. Loosen Delcotrón and remove fan belt, then unfasten and remove pump. On 6-cylinder engines, pull pump straight out of block first to prevent damage to impeller.
4. Reverse procedure to install.

FUEL PUMP PRESSURE

Year	Engine	Pressure Lbs.
Camaro		
1969	Six	3-4½
1969	V8 (Exc. V8-396)	5-6½
1969-70	V8-396	5-8½
1970-76	Six	3½-4½
1976	V8-305	7-8½
1970-73	V8-307	5-6½
1970-76	V8-350	7-8½
1971	V8-396	7-8½
1972	V8-402	7-8½
Chevelle & Monte Carlo		
1969-70	Six	3-4½
1969-70	V8-307, 350	5-6½
1969-70	V8-396, 400, 454	5-8½
1971-76	Six	3½-4½
1971-76	V-8-850, 400, 454	7-8½
1971-73	V8-307	5-6½
1976	V8-305	7-8½
Chevy Nova		
1969-70	Four & Six	3-4½
1969-70	V8-307, 327, 350	5-6½
1969-70	V8-396	5-8½
1971-76	Six	3½-4½
1975	V8-262	7-8½
1976	V8-305	7-8½
1971-73	V8-307	5-6½
1971-76	V8-350	7-8½
Chevrolet		
1969-70	Six	3-4½
1969	V8-327, 350, 396	5-6½
1969	V8-427	5-8½
1970	V8-350	5-6½
1970	V8-400, 454	5-8½
1971-73	Six	3½-4½
1971-76	V8	7-8½
Corvette		
1969	V8-350	5-6½
1969	V8-427	5-8½
1970	Exc. V8-454, 360 H.P.	7½-9
1970	V8-454, 360 H.P.	5-8½
1971-76	All	7-8½

FUEL PUMP, REPLACE

1969-76

1. Disconnect fuel lines at pump.
2. Unfasten and remove pump.
3. If push rod is to be removed on 396, 400, 427 and 454 engines, remove pipe plug, then remove push rod. On 262, 302, 305, 307, 327, 350 and 265 H.P. V8-400, remove fuel pump adapter and gasket then remove push rod.
4. Reverse procedure to install. *On V8 engines, a pair of mechanical fingers may be used to hold fuel pump push rod up while installing pump.*

NOTE

Installation of the pump on the V8-327 engine can be performed faster by using the following means of retaining the fuel pump push rod in the engine during pump replacement.

1. Before removing the pump, remove upper bolt from engine front mount boss located to the right of the timing chain cover on the front of the engine. This bolt enters the push rod bore acting as a plug.
2. Remove the original bolt and install one that is 1¾" long. Turn it down by hand until it is stopped by the push rod. *Do not use a wrench to turn in this bolt as it may damage the push rod, necessitating its replacement.*
3. The fuel pump can then be removed without the push rod slipping from its installed position. After installing the pump, the original shorter bolt must be reinstalled to avoid oil leakage.

Service Bulletin

On Mark IV engines the access hole for removal of the fuel pump push rod is located below the fuel pump mounting bolts. This access hole, which permits removal of the push rod after the pump has been removed, has a threaded plug with a square recessed hole. The plug may be removed by using a hand-made tool. A 5/16" square end for inserting and turning the plug for removal is required. A piece of bar stock, approximately 1½" long is suggested.

Clutch and Transmission Section

NOTE: 1976 linkage adjustment information is in this section. Repair procedures on both automatic and manual shift transmissions are covered elsewhere in this manual. Procedures for removing automatic transmissions as well as linkage adjustments on 1969-75 models are included in the automatic transmission chapters. See Chapter Index.

CLUTCH PEDAL, ADJUST

1972-76 Except Corvette

1. Disconnect return spring at clutch fork.
2. Rotate clutch lever and shaft assembly until pedal is against rubber bumper on dash brace.
3. Push outer end of clutch fork rearward until throwout bearing lightly contacts pressure plate fingers.
4. Install push rod in gauge hole and increase length until all lash is removed, Fig. 1.

NOTE: On 1973-76 Chevelle, it may be necessary to move rubber bumper rearward to obtain additional clutch pedal travel.

5. Remove swivel or rod from gauge hole and insert into lower hole on lever. Install retainer and tighten lock nut being careful not to change rod length.
6. Reinstall return spring and check pedal free travel:
   1975-76 Camaro & Nova. 55/64″ to 1 29/64″.
   Chevelle & Monte Carlo, 45/64″ to 1 5/16″.
   1972–74 Camaro, 1″ to 1 3/8″.
   Chevelle, 3/4″ to 1 5/16″.
   Nova, 3/4″ to 1 3/8″.
   1972–73 Chevrolet, 1 1/4″ to 1 3/4″.

1975-76 Corvette

1. Disconnect return spring between toe pan brace and cross shaft lever, Fig. 2.
2. Rotate clutch lever and shaft assembly until pedal is against dash brace rubber bumper.
3. Install swivel (C) into clutch lever hole and install retainer.
4. Loosen nuts (A) and (B), then apply a 5 pound load in direction of arrow (F) until bearing lightly contacts plate fingers.
5. Rotate nut (B) toward swivel until dimension "X" is approximately 3/8 to 7/16 inch, then tighten nut (A) to lock swivel against nut (B).
6. Reinstall return spring and adjust clutch pedal free travel which should be 1 to 1 1/2 inch.

1972-74 Corvette

1. Referring to Fig. 2, disconnect spring (E) between toe pan brace and cross shaft lever.
2. With pedal against stop, loosen jam nuts sufficiently to allow adjusting rod to move against clutch fork until release bearing contacts pressure plate fingers lightly.
3. Rotate upper nut (B) against swivel and back off 4 1/2 turns. Tighten lower nut (A) to lock swivel against nut (B).
4. Install return spring (E) and check pedal free travel. It should be 1″ to 1 1/2″.

1969-70 Chevrolet & 1969-71 Corvette

1. Disconnect spring between clutch push rod and cross shaft lever.
2. With clutch pedal against stop, loosen jam nuts just enough to allow adjusting rod to move against the clutch fork until the release bearing contacts the pressure plate fingers lightly.
3. Rotate upper nut against swivel and back off 4 1/2 turns. Tighten lower nut to lock swivel against nut.
4. Install return spring and check clutch pedal free travel:
   Chevrolet 1 to 1 1/2″.
   Corvette standard clutch 1 1/4 to 2″.
   Corvette heavy duty clutch 2 to 2 1/2″.

1971 Chevrolet & 1969-71 Chevelle, Camaro & Chevy Nova

1. Disconnect return spring at clutch fork.
2. With clutch pedal against stop, loosen lock nut just enough to allow the adjusting rod to be turned out off swivel and against clutch fork until release bearing contacts pressure plate fingers lightly.
3. Rotate push rod into swivel three turns and tighten lock nut.

NOTE: Nova and Camaro V8 models use a two-piece push rod. Turn adjusting rod portion of push rod three turns into rod end and tighten lock nut.

4. Reinstall return spring and check clutch pedal free travel:
   Exc. Nova, 1 1/8″ to 1 3/4″.
   Nova, 1″ to 1 1/8″.

CLUTCH, REPLACE

Clutch Disc Installation

On all 4 and 6-cylinder engines the clutch disc is installed with the damper springs to the flywheel side. On V8's install clutch disc with damper springs and grease slinger to transmission side.

1969-76

1. Support engine and remove transmission as outlined further on.
2. Disconnect clutch fork push rod and spring.
3. Remove flywheel housing.
4. Slide clutch fork from ball stud and remove fork from dust boot.

NOTE: Look for "X" mark on flywheel and on clutch cover. If "X" mark is not evident, prick punch marks on flywheel and clutch cover for indexing purposes during installation.

5. Loosen clutch-to-flywheel attaching bolts evenly one turn at a time until spring pressure is released. Then remove bolts and clutch assembly.
6. Reverse procedure to install.

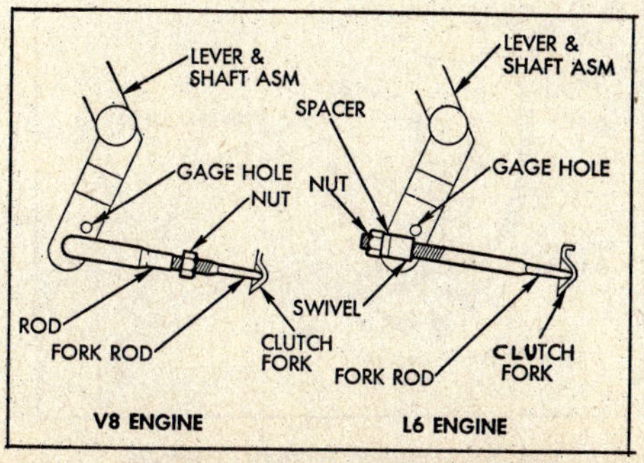

Fig. 1 Clutch pedal adjust. Except Corvette

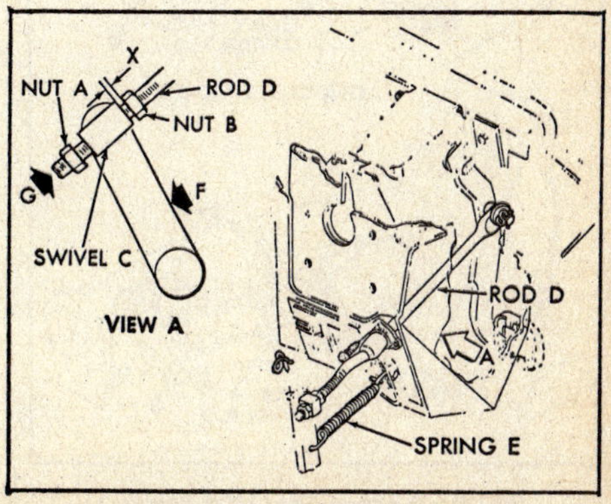

Fig. 2 Clutch pedal adjust. Corvette

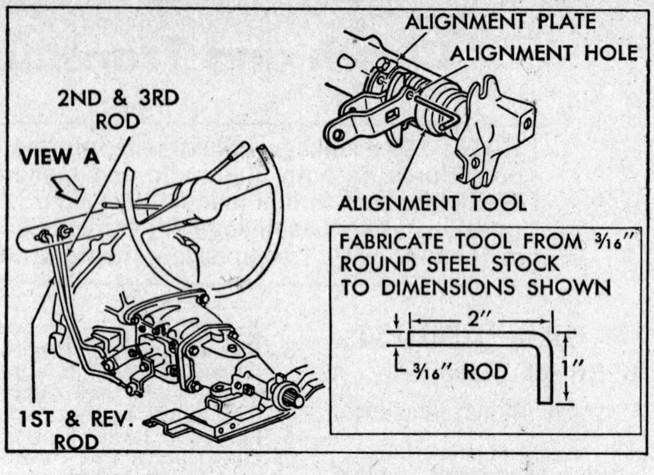

Fig. 3 Three-speed column shift linkage adjustment. 1969-74 Chevrolet, Chevy Nova, Chevelle and Camaro

THREE SPEED TRANSMISSION, REPLACE

1969 Corvette

1. Disconnect battery ground cable.
2. Remove shift ball.
3. Remove console trim plate.
4. Raise vehicle on a hoist.
5. Remove right and left exhaust pipes. *In order to remove the exhaust pipes on 8-427 engines, it will be necessary to remove the forward stud on each manifold.*
6. Disconnect and lower front of propeller shaft. Remove slip yoke from transmission.
7. Remove bolts retaining rear mounts to bracket. Raise engine, lifting transmission off mount bracket.
8. Unfasten transmission linkage mounting bracket from frame.
9. Detach gearshift from mounting bracket and remove bracket. Remove shifter mechanism with rods attached.
10. Disconnect shifter levers at transmission. Disconnect speedometer cable.
11. Unfasten and remove transmission mount bracket. Unfasten rear mount cushion and exhaust pipe yoke.
12. Unfasten transmission from clutch housing.
13. Pull transmission rearward and rotate clockwise until it is clear of clutch housing. *To allow room for transmission removal, slowly lower rear of engine until tachometer drive cable at the distributor just clears horizontal ledge across front of dash.*

CAUTION: The tachometer cable can be easily damaged by heavy contact with the dash. Slide transmission rearward out of the clutch, then tip front end of transmission downward and lower assembly from vehicle.

14. Reverse procedure to install.

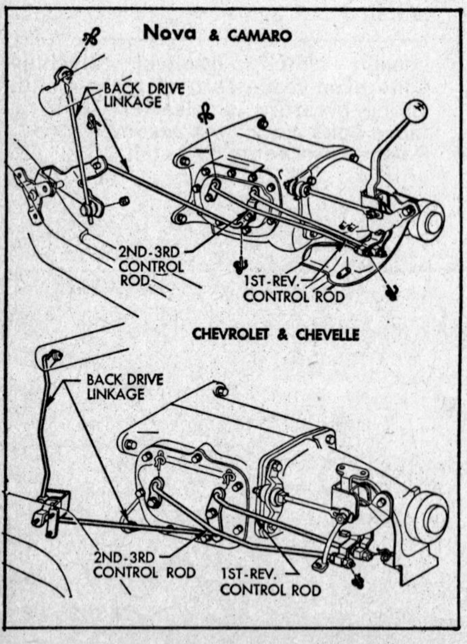

Fig. 4 3-Speed floor shift linkage. 1969-76 except Corvette (typical)

1969-76 Except Corvette

NOTE: On 1975-76 models, it may be necessary to remove the catalytic converter and its support bracket to facilitate transmission removal.

1. On floor shift models, remove shift knob and console trim plate.
2. Disconnect speedometer cable at transmission and on floor shift models, disconnect back-up lamp switch. Disconnect T.C.S. switch wire if so equipped.
3. Remove drive shaft.
4. On 1969 Camaro, disconnect exhaust pipe at manifold.
5. Remove crossmember-to-frame attaching bolts. On floor shift models, remove crossmember-to-control lever support attaching bolts. On Chevelle, also remove control lever brace-to-crossmember attaching bolt.
6. Remove bolts retaining transmission mount to crossmember.
7. Support engine and raise slightly until crossmember may be slid rearward or removed.

NOTE: On some Camaros, it may be necessary to remove the right rear body mount bolt then, using a suitable tool, pry the right rear portion of the frame downward and insert a block of wood to maintain enough of a gap so the crossmember can be removed.

8. Remove shift levers at transmission side cover. On floor shift models, remove stabilizer-to-control lever retaining nut. Push bolt toward transmission until stabilizer rod may be disconnected.
9. Remove transmission-to-clutch housing upper retaining bolts and install guide pins in holes, then remove lower bolts.
10. Slide transmission rearward and remove from vehicle.
11. Reverse procedure to install.

FOUR SPEED TRANSMISSION, REPLACE

1969-76 All Models

The procedure for removing this transmission is similar to that described for the "1969-76 Exc. Corvette" 3-speed unit.

THREE SPEED SHIFT LINKAGE, ADJUST

1969-75 Except Corvette

Column Shift

1. Place shift lever in "Reverse" position and ignition switch in "Lock".
2. Raise vehicle on a hoist.

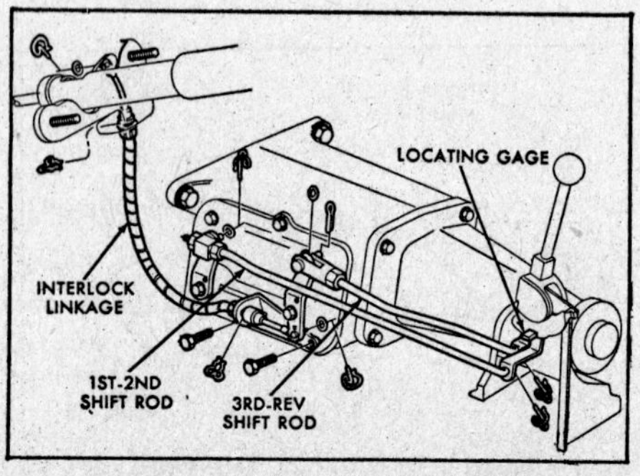

Fig. 5 3 Speed linkage. 1969-70 Corvette (typical)

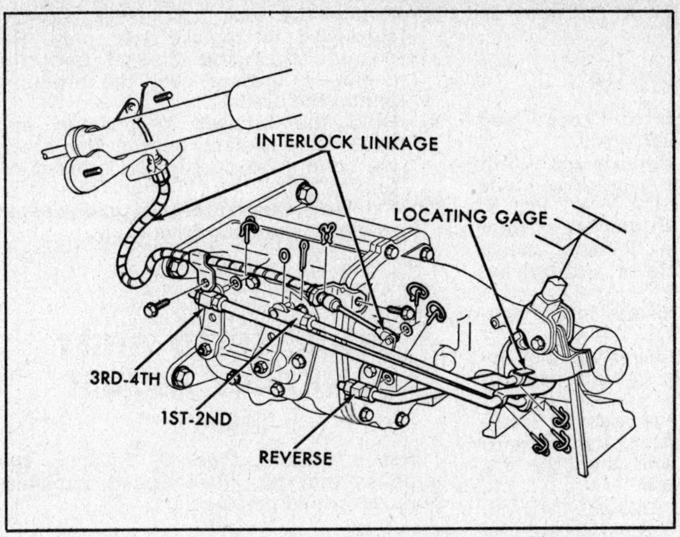

Fig. 6 4 Speed Linkage. 1969-76 Corvette (typical)

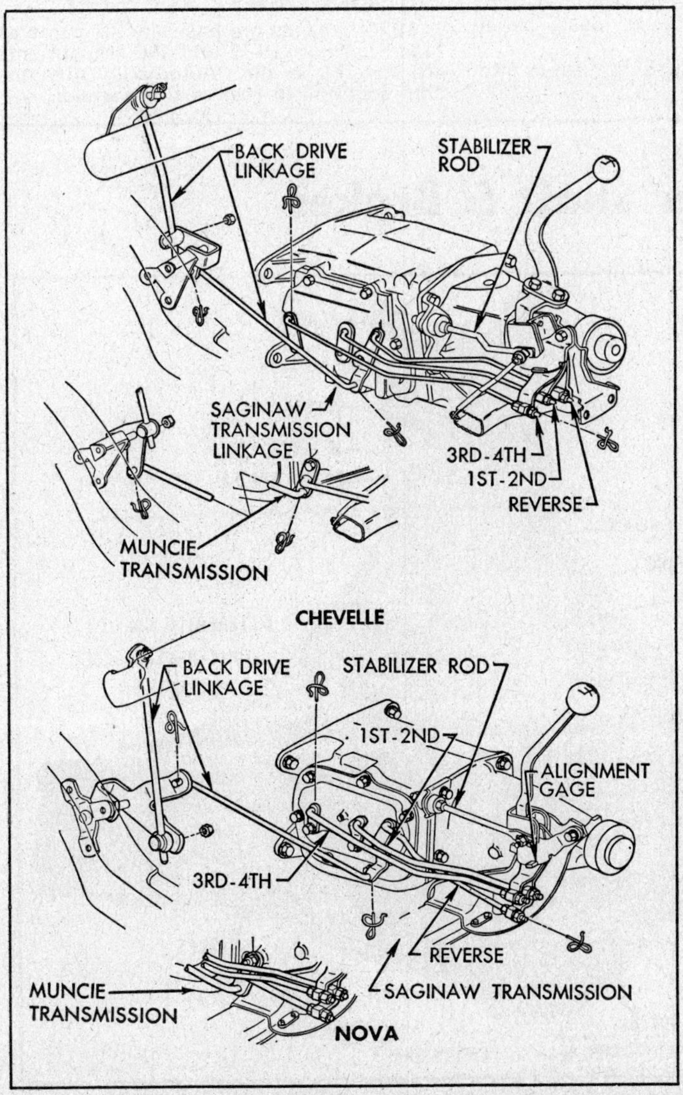

Fig. 7 4 Speed linkage. 1969-72 except Corvette

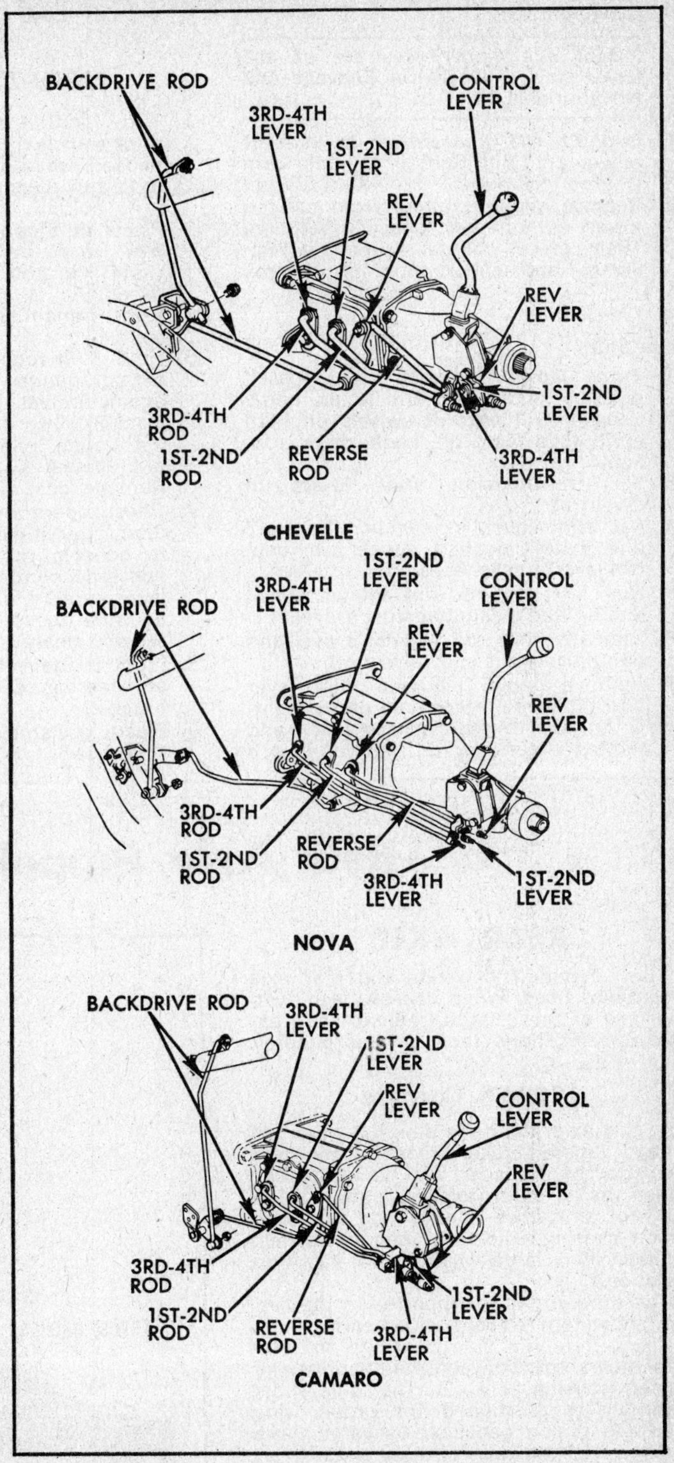

Fig. 8 4 Speed linkage. 1973-76 except Corvette

3. Loosen shift control rod swivel lock nuts. Pull down slightly on 1/R rod attached to column lever to remove any slack and then tighten clevis lock nut at transmission lever.
4. Unlock ignition switch and shift column lever to "Neutral". Position column lower levers in "Neutral", align gauge holes in levers and insert

gauge pin.

NOTE: Alignment holes are on the lower side of levers on Chevelle and Nova models.

5. Support rod and swivel to prevent movement and tighten 2/3 shift control rod lock nut.
6. Remove alignment tool from column lower levers and check operation. Then place column lever in "Reverse" and check interlock control.

Floor Shift

1. Turn ignition switch to "Lock" position and raise vehicle on hoist.
2. Loosen lock nuts at swivels on both shift rods and on back drive rod, Fig. 4.
3. Set transmission shift levers in "Neutral".
4. Set shift lever in car in "Neutral" and install locating gauge into control lever bracket assembly.
5. Run 1/R shift rod nut against swivel and tighten locknut against swivel.
6. Run 2/3 shift rod against swivel and tighten locknut against swivel.
7. Remove gauge pin and shift lever into "Reverse", then pull down slightly on rod to remove any slack and tighten clevis jam nut. Ignition switch should move freely in and out of "Lock".

1969-70 Corvette

1. Turn ignition switch to "Lock" and raise vehicle on a hoist.
2. Loosen swivel lock nuts on both shift rods and disconnect back drive cable from column lock tube lever, Fig. 5.
3. Place shift control in car into "Neutral" and insert a locking gauge (.644") in notch of lever and bracket assembly.
4. Place transmission levers into "Neutral".
5. Hold 1/R rod and lever against locating gauge and tighten locknut against swivel.
6. Hold 2/3 rod and lever against locating gauge and tighten the forward nut against swivel and then the aft. Remove locating gauge.
7. Check operation of linkage. *Caution: Shift lever must be properly centered in console to prevent fore and aft and side contact with console. Shim lever support bracket as required but be sure to maintain support to extension clearance.*
8. From inside vehicle, loosen two nuts at steering column to dash panel bracket.
9. Place transmission shift lever into "Reverse".
10. Rotate the lock tube lever counterclockwise to remove free play. Reposition the cable bracket until the cable eye passes over the retaining pin on the bracket.
11. Hold the bracket in position and have an assistant tighten the steering column to dash panel bracket retaining nuts.
12. Reinstall the cotter pin and washer retaining the back drive cable.

FOUR SPEED SHIFT LINKAGE, ADJUST

1969-76

The procedure, Figs. 6, 7, 8, is the same as that for Three Speed transmissions described previously.

1976 AUTO. TRANS. LINKAGE, ADJUST

1976 linkages are basically the same as those used in 1975 and the adjustments are covered in the "Automatic Transmission section" in rear of this manual.

Rear Axle, Propeller Shaft & Brakes

REAR AXLE

Figs. 1 and 2 illustrate the rear axle assemblies used. When necessary to overhaul any of these units, refer to the *Rear Axle Specifications* table in this chapter.

1969-76 Corvette

In this axle, the drive pinion is mounted in two tapered roller bearings that are preloaded by a spacer. The pinion is positioned by a shim located between the head of the drive pinion and the rear pinion bearing. The front bearing is held in place by a large washer and a locking pinion nut.

The differential is supported in the carrier by two tapered roller side bearings.

The differential side bearings are preloaded by shims between the bearings and carrier housing, Fig. 2. The differential assembly is positioned for proper ring gear and pinion backlash by varying the position and thickness of these shims.

The ring gear is bolted to the case. The case houses two side gears in mesh with two pinions mounted on a pinion shaft which is held in place by a lock screw. The side gears are backed by thrust washers.

The differential side gears drive two splined yokes which are retained by snap rings located on the yoke splined end. The yokes are supported on caged needle bearings pressed into the carrier, adjacent to the differential bearings. A lip seal, pressed into the carrier outboard of the bearing, prevents oil leakage and dirt entry.

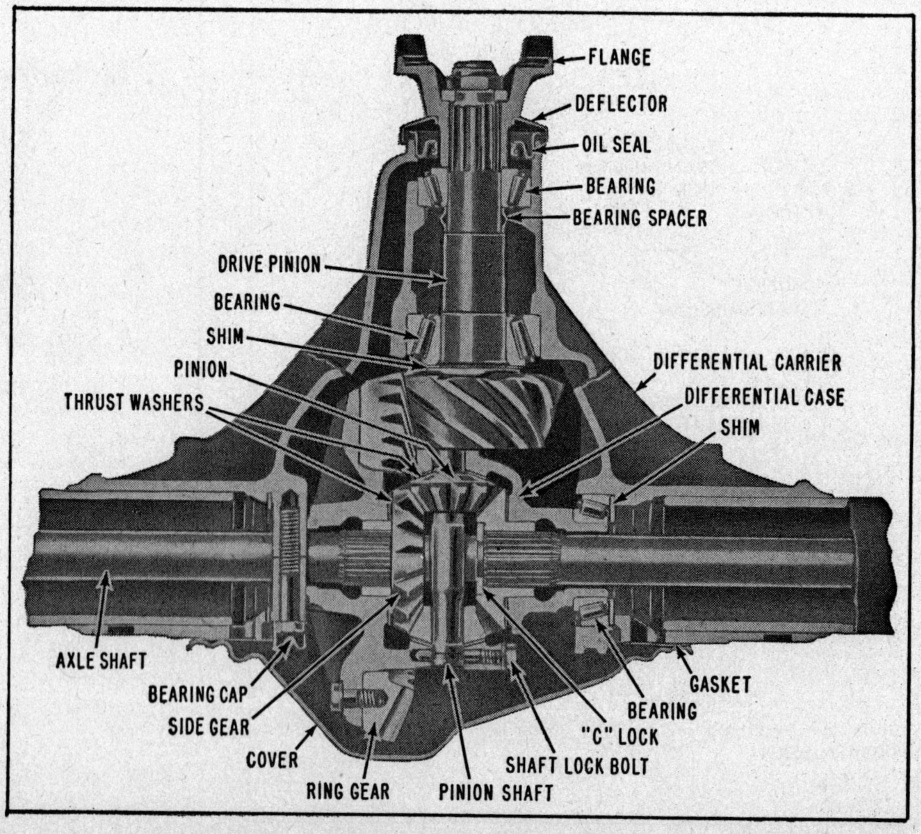

Fig. 1 Rear axle. 1969-76 Camaro, Chevrolet, Chevelle, Monte Carlo and Nova. "C" Type

On 1969 models, Positraction type differential is optional. Positraction is standard on 1970-76 units.

Remove & Replace

It is not necessary to remove the rear axle assembly for any normal repairs. The axle shafts and carrier assembly can easily be removed from the vehicle, leaving the rear axle housing in place.

Corvette 1969-76

1. Disconnect spring and spring links.
2. Disconnect axle drive shafts at carrier by removing U-bolts securing trunnion to side gear yoke, Fig. 2.
3. Disconnect carrier front support bracket at frame crossmember. Remove bolts and bracket.
4. Disconnect propeller shaft at transmission and at companion flange. Slide transmission yoke forward into transmission. Drop propeller shaft down and out toward the rear.
5. Mark camber cam and bolt relative location on strut rod bracket and loosen cam bolts, Fig. 3.
6. Remove four bolts securing bracket to carrier lower surface and drop bracket. Remove camber cam bolts and swing strut rods up and out of the way.
7. Loosen carrier-to-cover bolts gradually to allow grease to drain out.
8. With mounting bolts removed, pull carrier partially out of cover, drop nose to clear crossmember and gradually work carrier down and out.

CAUTION: *When removing carrier, use care so as not to scrape or gouge gasket mounting surface on cover with ring gear. A deep scratch or gouge at this point may cause a lubricant leak after assembly.*

9. Reverse removal procedure to install carrier, *being sure to move the camber cams to marked location before tightening nuts.*

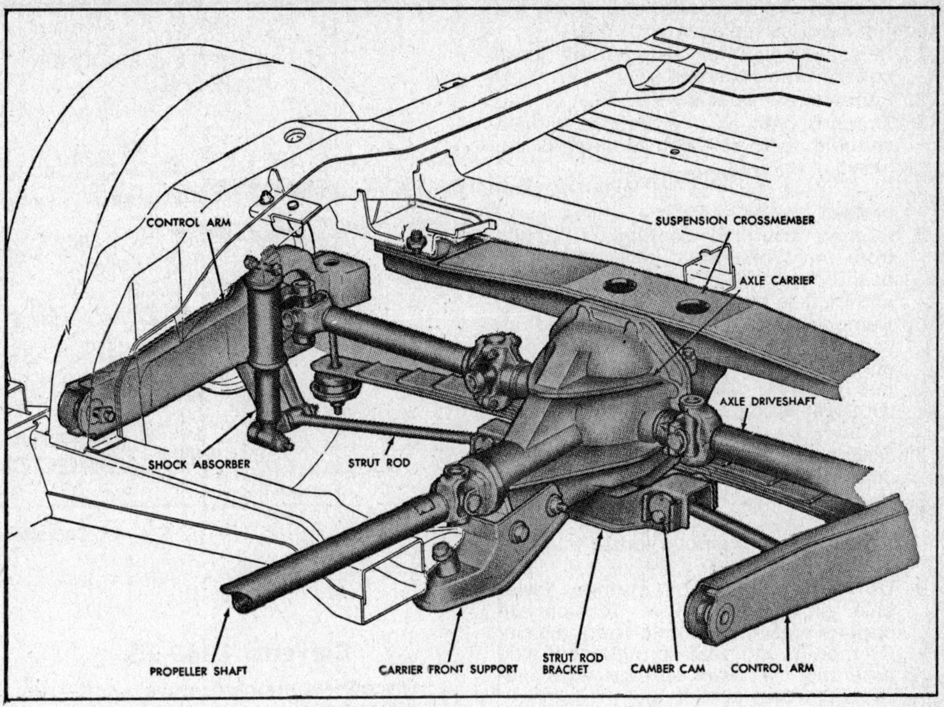

Fig. 3 Rear suspension. 1969-76 Corvette

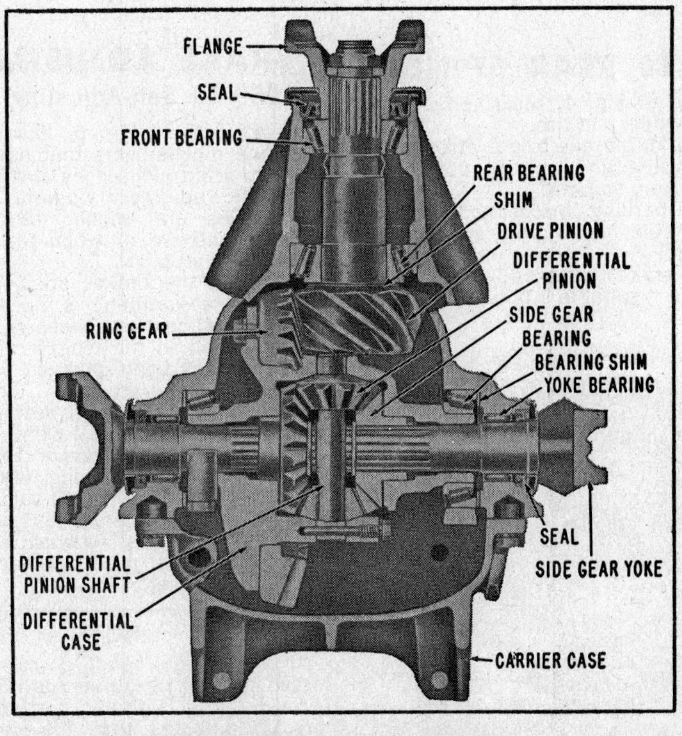

Fig. 2 Rear axle. 1969-76 Corvette

REAR AXLE

1969-76 Camaro, Chevrolet, Chevelle, Nova & Monte Carlo

In these rear axles, Fig. 1, the rear axle housing and differential carrier are cast into an integral assembly. The drive pinion assembly is mounted in two opposed tapered roller bearings. The pinion bearings are preloaded by a spacer behind the front bearing. The pinion is positioned by a washer between the head of the pinion and the rear bearing.

The differential is supported in the carrier by two tapered roller side bearings. These bearings are preloaded by spacers located between the bearings and carrier housing. The differential assembly is positioned for proper ring gear and pinion backlash by varying these spacers. The differential case houses two side gears in mesh with two pinions mounted on a pinion shaft which is held in place by a lock pin. The side gears and pinions are backed by thrust washers.

Remove & Replace

Construction of the axle assembly is such that service operations may be performed with the housing installed in the vehicle or with the housing removed and installed in a holding fixture. The following procedure is necessary only when the housing requires replacement.

1. Raise vehicle and place stand jacks under frame side rails.
2. Remove rear wheels.
3. Support rear axle assembly with a suitable jack so that tension is relieved in springs and tie rod.
4. Disconnect tie rod at axle housing bracket.
5. Remove trunnion bearing "U" bolts from rear yoke, split universals joint, position propeller shaft to one side and tie it to frame side rail.
6. Remove axle "U" bolt nuts and allow control arm and shock absorbers to hang freely so that they do not interfere with axle. On models with leaf springs, remove spring plate nuts.
7. Disconnect hydraulic brake hose at connector on axle housing.
8. Remove brake drum and disconnect parking brake cable at lever and at flange plate. Remove brake support plate.
9. On models with coil springs, lower axle slowly to relieve tension on springs and remove axle from vehicle. On models with leaf springs, shift axle assembly to clear springs when removing from vehicle.
10. Reverse foregoing procedure to install axle assembly.

AXLE SHAFT

1969-76 Camaro, Chevrolet, Chevelle & Chevy Nova "C" Type

1. Raise vehicle and remove wheel and brake drum.
2. Drain lube from carrier and remove cover.
3. Remove differential pinion shaft lock screw and remove differential pinion shaft.
4. Pull flanged end of axle shaft toward center of vehicle and remove "C" lock from button end of shaft.
5. Remove axle shaft from housing, being careful not to damage seal.
6. Reverse foregoing procedure to install the axle shaft.

1971-76 Chevrolet "B" & "O" Type

1. Raise vehicle and remove wheel and brake drum.
2. Remove bolts attaching axle shaft retainer plate to backing plate.
3. Using a slide hammer/puller, the axle shaft can now be removed.

NOTE: You may find the wheel bearing will come out in pieces as you remove the shaft. The inner race with bearing and one retainer plate will come out with the shaft. The outer race and inner retainer plate will remain in the axle housing. These pieces can easily be removed. Even though the bearing is not in one piece it is no indication that the bearing has failed.

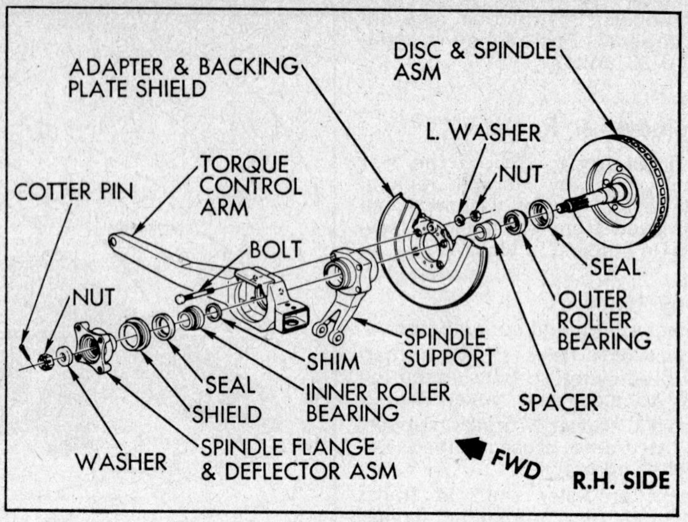

Fig. 4 Corvette rear wheel spindle

Corvette 1969-76

1. Disconnect inboard drive shaft trunnion from side gear yoke, Fig. 3.
2. Remove four bolts securing shaft flange to spindle drive flange.
3. Pry drive shaft out of outboard drive flange pilot and remove by withdrawing outboard end first.
4. If necessary, overhaul universal joints as described in the *Universal Joint* chapter.
5. Install axle shaft in reverse order of removal, *being certain to rotate yokes so that trunnion seats are phased 90 degrees apart.* Torque drive flange bolts to 70-90 ft-lbs and U-bolt nuts to 14-18 ft-lbs.

CORVETTE REAR SPINDLE

1. Referring to Fig. 4, raise vehicle and remove wheel and tire.
2. Apply parking brake to prevent spindle from turning and remove cotter pin and nut from spindle.
3. Release parking brake and remove drive spindle flange from splined end of spindle.
4. Remove brake caliper and disc.
5. Using a suitable puller, remove spindle.
6. When spindle is removed, the outer bearing will remain on the spindle. The inner bearing, tubular spacer, end play shim and both outer races will remain in the spindle support.

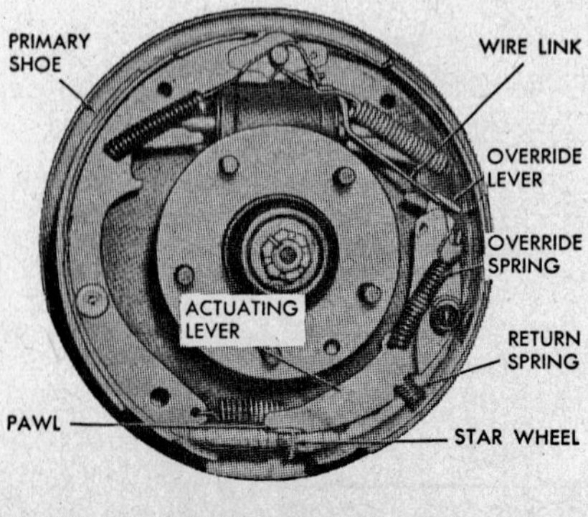

Fig. 5 Left front brake. 1969-76

PROPELLER SHAFT

1969-76

These models use a one-piece propeller shaft. Powerglide models use a propeller shaft tube which incorporates rubber insulators for quieter operation. These insulators, which are bonded to the smaller tube, press fit to the larger tube.

BRAKE ADJUSTMENTS

1969-76 Self-Adjusting Brakes

These brakes, Fig. 5, have self-adjusting shoe mechanisms that assure correct lining-to-drum clearances at all times. The automatic adjusters operate only when the brakes are applied as the car is moving rearward or when the car comes to an uphill stop.

Although the brakes are self-adjusting, an initial adjustment is necessary after the brake shoes have been relined or replaced, or when the length of the adjusting screw has been changed during some other service operation.

Frequent usage of an automatic transmission forward range to halt reverse vehicle motion may prevent to automatic adjusters from functioning, thereby inducing low pedal heights. Should low pedal heights be encountered, it is recommended that numerous forward and reverse stops be made until satisfactory pedal height is obtained.

NOTE

If a low pedal condition cannot be corrected by making numerous reverse stops (provided the hydraulic system is free of air) it indicates that the self-adjusting mechanism is not functioning. Therefore

Fig. 6 Brake drum access hole

Fig. 7 Aligning drum tang with wheel hub. 1969-76

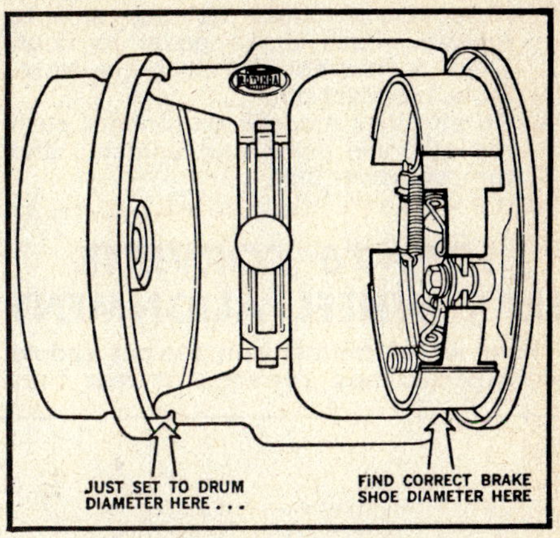

Fig. 8 Use of Drum-to-Brake Shoe Clearance Gauge (J-21177)

it will be necessary to remove the brake drum, clean, free up and lubricate the adjusting mechanism. Then adjust the brakes, being sure the parking brake is fully released.

Adjustment

A lanced "knock out" area, Fig. 6, is provided in the web of the brake drum for servicing purposes in the event retracting of the brake shoes is required in order to remove the drum.

1. With brake drum off, disengage the actuator from the star wheel and rotate the star wheel by spinning or turning with a screwdriver.
2. Using the brake drum as an adjustment fixture, turn the star wheel until the drum slides over the brake shoes with a slide drag.
3. Turn the star wheel 1¼ turns to retract the brake shoes. This will allow sufficient lining-to-drum clearance so final adjustment may be made.
4. Install drum and wheel.

NOTE: *If lanced area in brake drum was knocked out, be sure all metal has been removed from brake compartment. Install new hole cover in drum to prevent contamination of brakes. Make certain that drums are installed in the same position as when removed with the drum locating tang in line with the locating hole in the wheel hub, Fig. 7.*

5. Make final adjustment by driving and stopping in forward and reverse until satisfactory pedal height is obtained.

NOTE: The recommended method of adjusting the brakes is by using the Drum-to-Brake Shoe Clearance Guage shown in Fig. 8 to check the diameter of the brake drum inner surface. Turn the tool to the opposite side and fit over the brake shoes by turning the star wheel until the gauge just slides over the linings. Rotate the gauge around the brake shoe lining surface to assure proper clearance.

PARKING BRAKE, ADJUST

1969-76 Except Corvette

1. Jack up both rear wheels.
2. Apply parking brake two notches from fully released position.
3. Loosen equalizer forward check nut and tighten rear nut until a light to moderate drag is felt when rear wheels are rotated.
4. Tighten check nuts securely.
5. Fully release parking brake and rotate rear wheels; no drag should be present.

1969-76 Corvette

1. With car on lift or jack stands, remove wheel. (On optional knock-off wheels, the adapter bracket must be removed to gain access to the hole in the hat section of the disc.)
2. Turn disc until the adjusting screw can be seen through hole in disc.
3. Insert a screwdriver through hole in disc and tighen adjusting screw by moving your hand away from the floor on both the left and right sides, Fig. 9.
4. Tighten until disc will not move, then back off 6 to 8 notches.
5. Apply emergency brake two notches from inside of car.
6. Tighten brake cables at equalizer to produce a light drag with wheels mounted.
7. Fully released parking brake handle and rotate rear wheels. No drag should be evident with handle released.

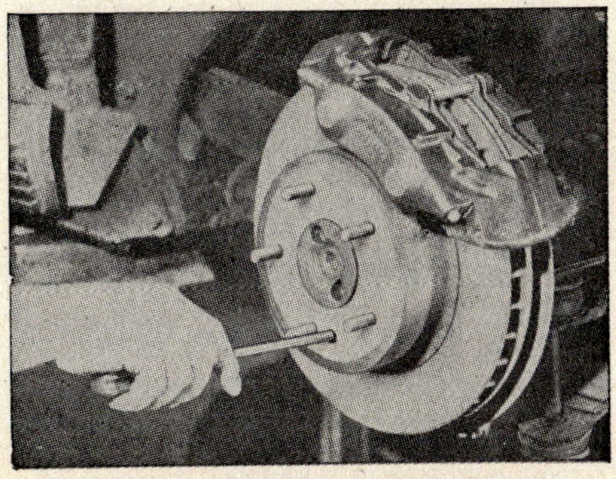

Fig. 9 Adjusting parking brake shoes. 1969-76 Corvette

MASTER CYLINDER, REPLACE

1969-76

Disconnect hydraulic lines from master cylinder. Unfasten and remove cylinder from its mounting.

POWER BRAKE UNIT

1973-76

1. Remove vacuum hose from check valve and master cylinder retaining nuts.
2. Pull master cylinder forward so it clears mounting studs and move to one side. Support cylinder to avoid stress on hydraulic lines.
3. Remove power unit to dash nuts. On Chevelle and Monte Carlo, remove brake line clip from power unit.
4. Remove brake pedal push rod retainer and disconnect push rod from pin.

NOTE: On Chevelle and Monte Carlo, push brake pedal to floor. This pushes power unit away from dash, providing clearance to remove push rod.

5. Remove power unit from vehicle.

1969-72

1. Remove vacuum hose from vacuum check valve.
2. Disconnect hydraulic line at main cylinder.
3. On 1969-72 Chevelle with standard transmission, remove main cylinder

from power unit and from inside vehicle remove nuts holding unit to firewall. Push brake pedal to floor. There is now enough clearance to remove the pivot pin.
4. On all other models, disconnect push rod at brake pedal and unfasten and remove power unit.

1969-76 CORVETTE REAR WHEEL ALIGNMENT

Rear wheel camber and toe-out should be inspected and corrected if rear tires show unusual wear.

Camber, Adjust

Wheel camber is obtained by adjusting the eccentric cam and bolt assembly located at the inboard mounting of the strut rod (see Fig. 2). Place rear wheels on alignment machine and determine camber angle.

To adjust, loosen camber bolt nut and rotate cam and bolt assembly until the camber angle is within specifications listed in the front of this chapter.

Toe-Out, Adjust

Rear wheel toe-out is adjusted by inserting slotted shims of varying thickness inside the frame side member on both sides of the torque control arm pivot bushing. Shims are available in thicknesses of 1/32″, 1/8″ and 1/4″.

To adjust, loosen torque arm pivot bolts until shims are free enough to remove. Position torque arm assembly to obtain toe-out of 1/32″ to 3/32″ per wheel. Shim gap toward vehicle centerline between end of control arm bushing and frame side inner wall.

Rear Suspension

SHOCK ABSORBER, REPLACE

1969-76 All

1. If equipped with Superlift shock absorbers, disconnect air lines from shock absorber fittings.
2. With rear axle properly supported, disconnect shock absorber from upper and lower mountings, Figs. 1, 2 and 3.
3. Reverse procedure to install.

COIL SPRINGS, REPLACE

1971-76 Chevrolet, 1969-76 Chevelle & Monte Carlo

1. Support vehicle at frame and rear axle.
2. Disconnect shock absorbers at lower mountings.
3. Lower axle until it reaches end of its travel and using a suitable tool, pry lower pigtail over retainer on axle bracket. Remove spring and insulator.
4. Reverse procedure to install. Springs must be installed with an insulator between upper seat and spring and positioned properly, Fig. 4.

1969-70 Chevrolet

1. Perform steps 1 and 2 as outlined in above procedure. Remove rear wheels.
2. Loosen rear mounting bolts on upper and lower control arms. Do not remove nuts.
3. Disconnect suspension tie rod from stud on axle tube.
4. Slightly loosen bolt retaining lower seat and spring to control arm.

CAUTION: Do not remove nut at this time. Nut should still have all threads engaged on bolt.

5. Lower axle to carry springs out of upper seats and remove nut and bolt retaining spring to lower seat.
6. Reverse procedure and install spring with an insulator between upper seat and spring.

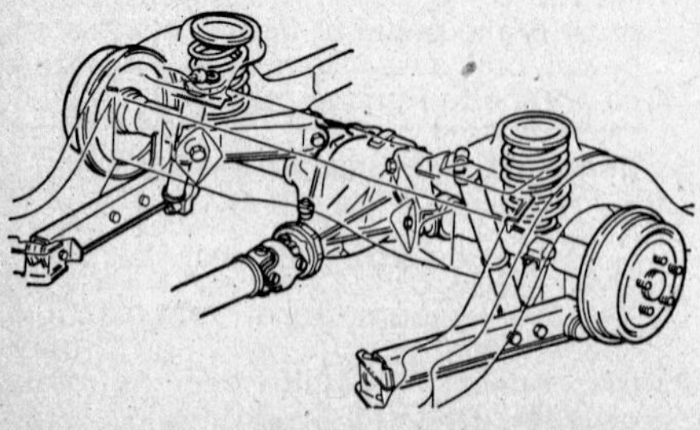

Fig. 1 Rear suspension. 1969-75 Chevrolet Chevelle & Monte Carlo (typical)

LEAF SPRINGS & BUSHINGS, REPLACE

1971-76 125″ Wheel Base Wagon

1. Support vehicle at frame and rear axle, removing tension from spring.
2. Disconnect shock absorber from lower mounting.
3. Remove rear shackle upper bolt and spring front mount bolt.
4. Support spring, remove "U" bolts, spring plate and spring. Remove rear shackle from spring.
5. Replace shackle and spring eye bushings as necessary. To replace spring eye bushing, use a press and a suitably sized piece of pipe, tubing or similar tool as a ram, press out bushing. Press in new bushing with a suitable tool pressing on outer steel shell of bushing. Bushing is correctly installed when it protrudes equally on both sides of eye.
6. Reverse procedure to install.

1969-76 Camaro & Nova

1. Perform steps 1 and 2 as outlined in above procedure.
2. Loosen spring front mount bolt.
3. Remove spring front mounting bracket attaching screws, lower axle and remove bracket.
4. Disconnect parking brake cable from spring plate bracket.
5. Remove spring plate by removing axle bracket nuts on single leaf models and the "U" bolts on multi-leaf models.
6. Support spring, remove front mount bolt and disassemble rear shackle.
7. Replace rear shackle and spring eye bushings as necessary, Figs. 5 and 6.
8. Reverse procedure to install.

1969-76 Corvette

1. Support vehicle at frame and remove rear wheels.
2. Install a "C" clamp approximately 9 inches from end of spring.
3. Place a suitable jack under spring, Fig. 7, and place a wooden block between "C" clamp and jack pad.
4. Raise jack until load is off spring link, then remove cotter pin, link nut and spring cushion. Lower jack, removing tension from spring.
5. Repeat steps 2, 3 and 4 on opposite side of spring.
6. Remove bolts from spring center clamp plate, then remove clamp plate.
7. Remove spring from vehicle, Fig. 8.
8. Reverse procedure to install.

LEAF SPRING SERVICE

NOTE: On 1971-76 models, the spring leaves are not serviced separately, how-

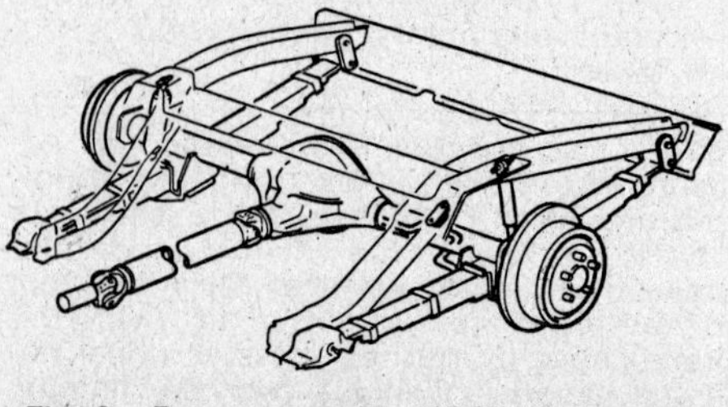

Fig. 2 Rear suspension. 1969-76 Camaro, Nova & 125″ Wheelbase station wagon (typical)

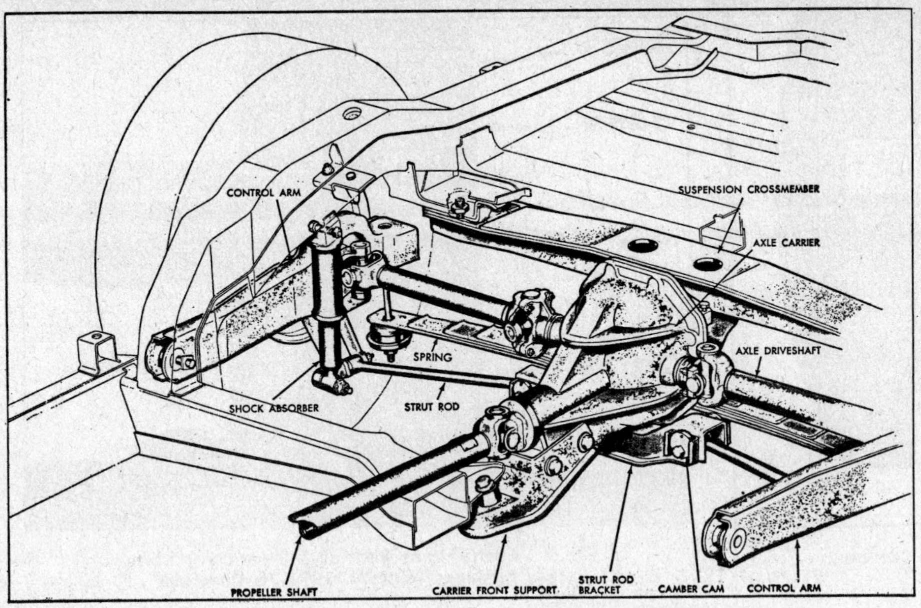

Fig. 3 Rear suspension. 1969-76 Corvette

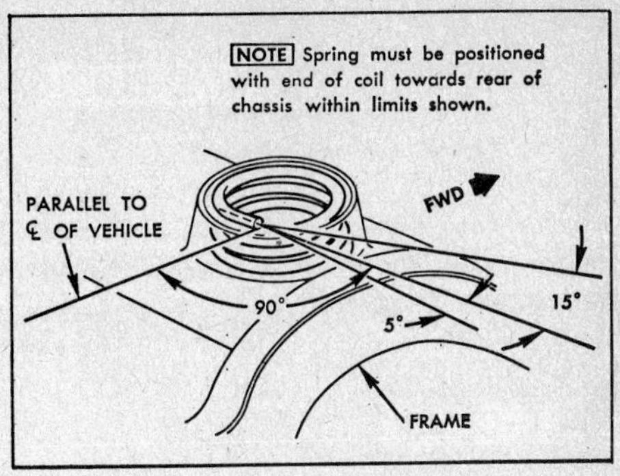

Fig. 4 Coil spring installation. 1969-76 all models

ever, the spring leaf inserts may be replaced.

1. Clamp spring in a vise and remove spring clips.
2. File peened end of center bolt to permit nut removal, remove nut and open vise slowly, allowing spring to expand.
3. Replace spring leaves or leaf inserts.
4. On Corvette, to replace main leaf cushion retainers, chisel flared portion until retainer can be removed from leaf. Install new retainers and flare over with a hammer.
5. On all models use a drift to align center bolt holes, compress spring in vise and install new center bolt and nut. Peen end of bolt to retain nut.
6. Align springs and bend spring clips into position.

NOTE: Overtightening of spring clips will cause spring binding.

CONTROL ARMS & BUSHINGS, REPLACE

NOTE: On Chevelle, Monte Carlo & 119" wheelbase station wagons, if both upper or lower control arms are to be removed at the same time, remove coil springs as outlined previously.

Upper Control Arms

1971-76 Chevrolet, 1969-76 Chevelle, Monte Carlo

1. Support vehicle at frame and rear axle with a suitable jack.
2. Remove control arm front and rear mounting bolts.
3. Replace bushings as necessary, Figs., 9 thru 15.

NOTE: To remove bushings, Fig. 9, use tool J-21830-2. To install bushings, Fig. 10, reverse tool J-21474-2.

4. Reverse procedure to install. Control arm bolts must be tightened with vehicle at curb height.

1969-70 Chevrolet

1. Support vehicle at rear axle.
2. Remove control arm rear mount bolt.
3. Remove bolts securing control arm forward bracket to crossmember, then remove control arm mounting bolt.
4. Replace bushings as necessary, Fig. 16.

NOTE: To remove bushings, Fig. 16, reverse tool J-21830-2 and on 1970 V8 models, omit tool J-21830-4 and use a similar tool of appropriate size.

5. Reverse procedure to install. Control arm bolts must be tightened with vehicle at curb height.

Lower Control Arms

1971-76 Chevrolet, 1969-76 Chevelle, Monte Carlo

Follow "Upper Control Arms" 1971-75 procedure for replacement of lower control arms.

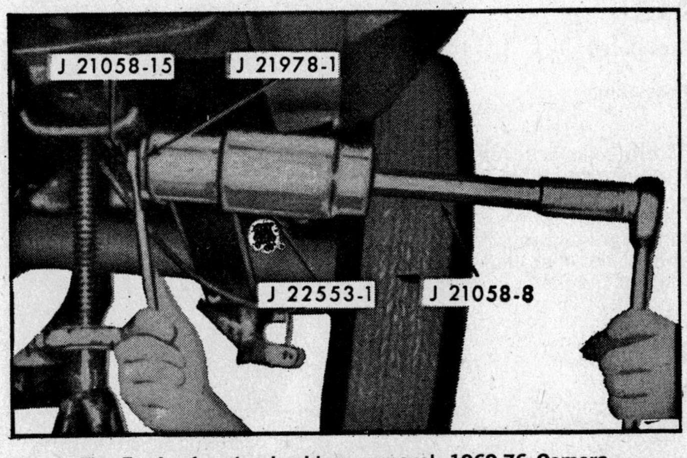

Fig. 5 Leaf spring bushings removal. 1969-76 Camaro. Nova & 125" Wheelbase station wagon

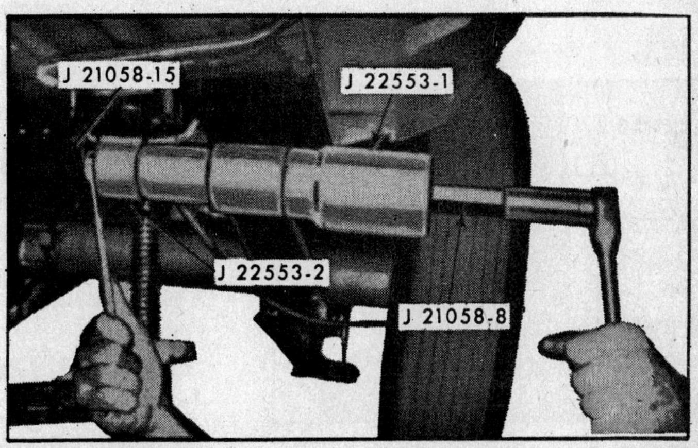

Fig. 6 Leaf springs bushing installation. 1969-76 Camaro, Nova & 125" Wheelbase station wagon

Fig. 7 Supporting leaf spring. 1969-76 Corvette

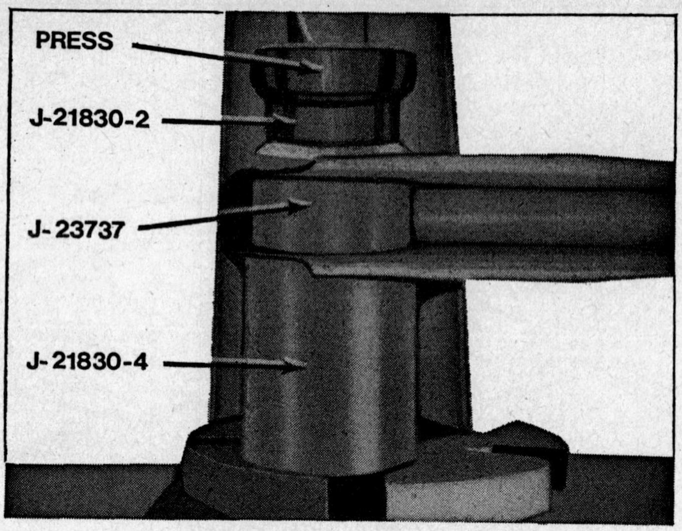

Fig. 9 Control arms front all & lower control arm rear bushings removal. 1971-76 Chevrolet

NOTE: To remove bushings, Fig. 9, use tool J-21830-2. To install bushings, Fig. 10, reverse tool J-21474-2.

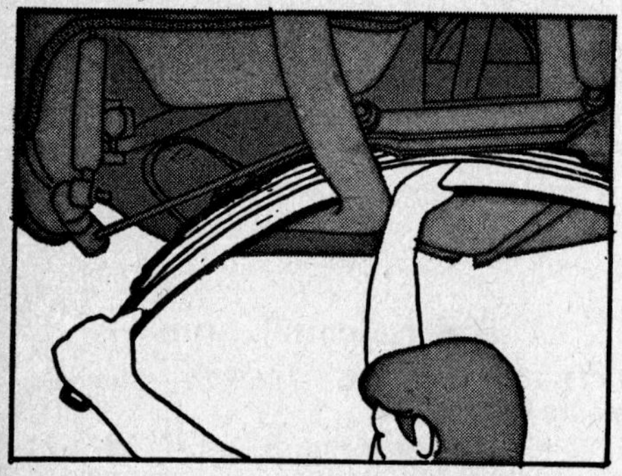

Fig. 8 Removing leaf spring. 1969-76 Corvette

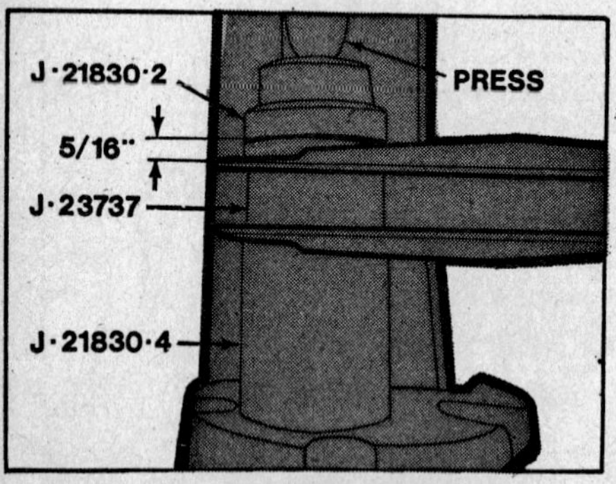

Fig. 10 Control arms front all & lower control arm rear bushings installation. 1971-76 Chevrolet

1969-70 Chevrolet

1. Remove shock absorbers and coil springs as outlined previously.
2. Note control arm position in forward attachment bracket to facilitate reassembly, then disconnect control arm from front and rear mountings.
3. Replace bushings as necessary, Fig. 19.

NOTE: To remove control arm front bushings on 1969 models and 1970 6 cylinder models, substitute tool J-7574-1 for tool J-21830-2, Fig. 19. To install control arm front bushings on 1969 models and 1970 6 cylinder models, substitute tool J-7574-2 for tool J-21830-2, Fig. 19. To install control arm rear bushings and control arm front bushings on 1970 V8 models, reverse tool J-21830-2, Fig. 19.

4. Reverse procedure to install control arm.

TIE ROD & BUSHING, REPLACE

1969-71 Chevrolet

1. Support vehicle at rear axle.
2. Disconnect tie rod from left side frame bracket.
3. Disconnect tie rod from axle bracket mounting stud and replace bushing if necessary, Fig. 20.
4. Reverse procedure to install. On all models except station wagons, flanged side of bushing must be installed against frame attachment bracket.

STABILIZER BAR, REPLACE

1971-76 Chevrolet & Chevelle

1. Support vehicle at rear axle.
2. Remove bolts securing stabilizer bar

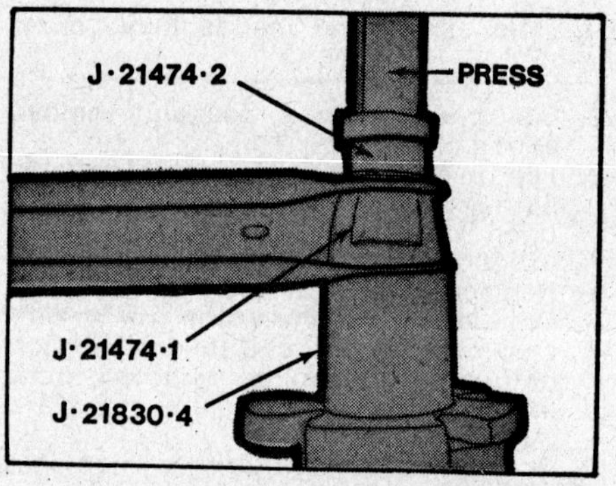

Fig. 11 Upper control arm front & all lower control arm bushing. 1969-76 Chevelle & Monte Carlo removal. Also upper control arm rear bushing removal. 1971-76 Chevrolet

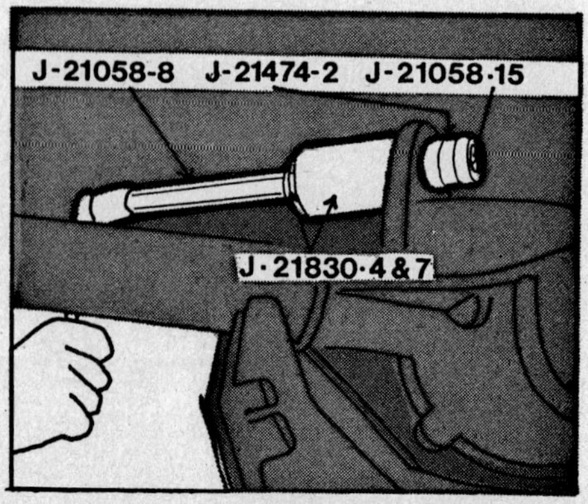

Fig. 12 Upper control arm rear bushing removal. 1969-72 Chevelle & Monte Carlo with 10 bolt carrier

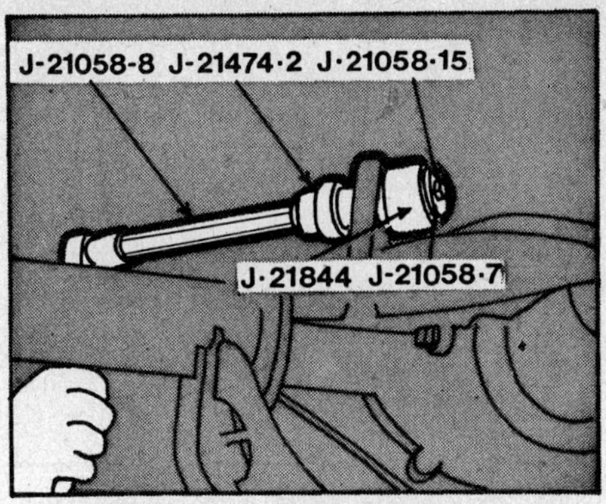

Fig. 13 Upper control arm rear bushing installation. 1969-72 Chevelle & Monte Carlo with 10 bolt carrier

Fig. 15 Upper control arm rear bushing installation. 1969-76 Chevelle & Monte Carlo with 12 bolt carrier

Fig. 14 Upper control arm rear bushing removal. 1969-76 Chevelle & Monte Carlo with 12 bolt carrier

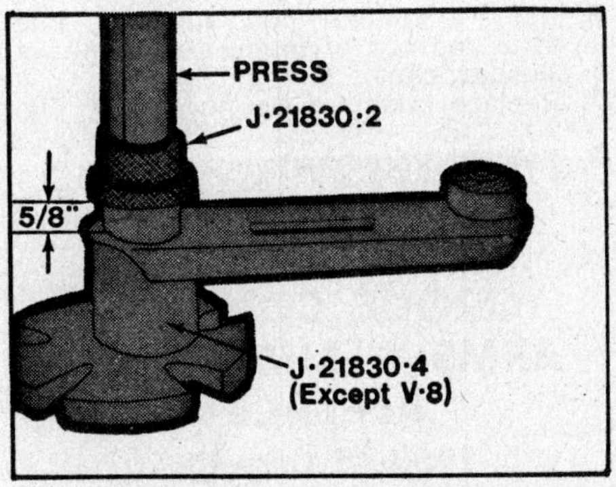

Fig. 16 Upper control arm bushings installation. 1969-70 Chevrolet

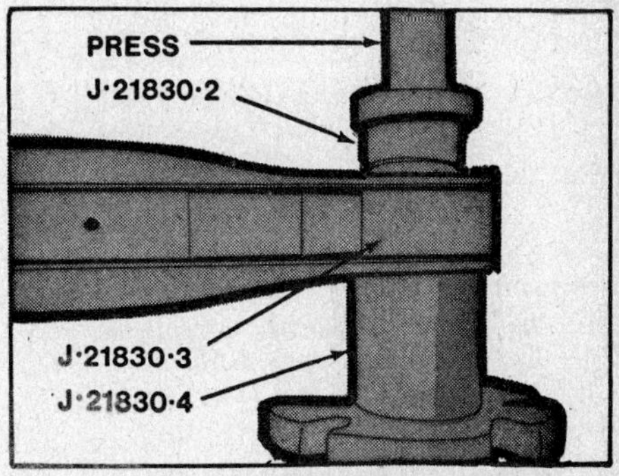

Fig. 19 Lower control arm bushings removal. 1969-70 Chevrolet

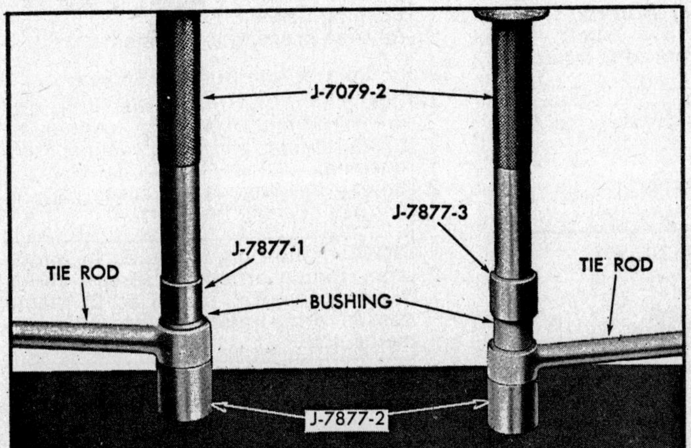

Fig. 20 Suspension tie rod bushing. 1969-71 Chevrolet, also strut rod bushing. 1969-76 Corvette replacement

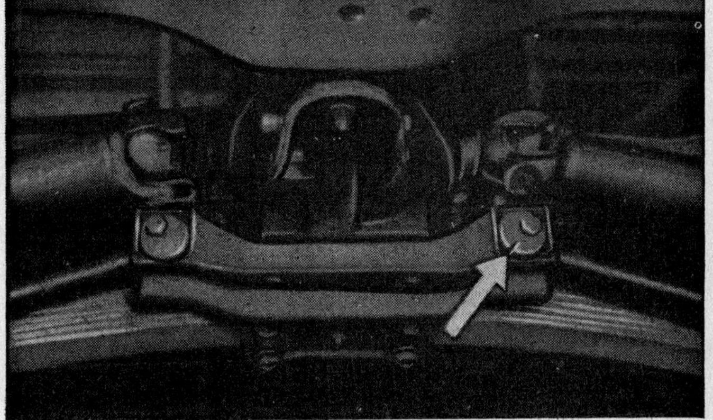

Fig. 24 Indexing camber cam & bracket. 1969-76 Corvette

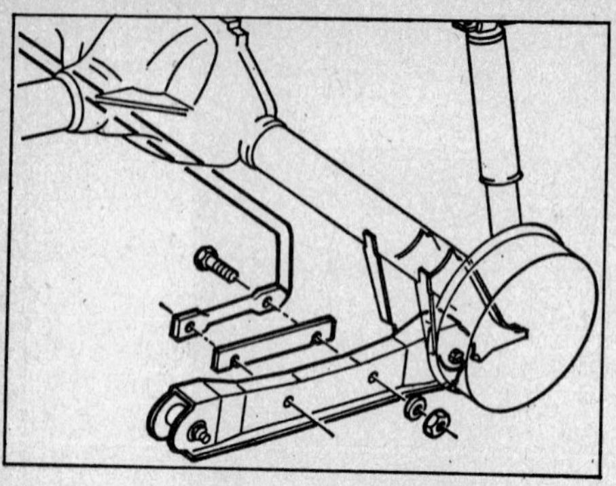

Fig. 21 Stabilizer bar installation. 1969-76 Chevrolet & Chevelle

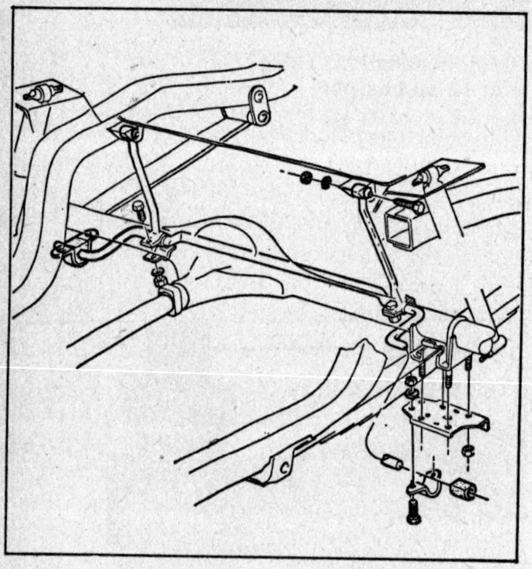

Fig. 22 Stabilizer bar installation. 1969-76 Camaro & Nova

to lower control arms, Fig. 21.
3. Reverse procedure to install. Use spacer shims, if needed, placed equally on each side of stabilizer bar. Tighten attaching bolts with vehicle at curb height.

1971-73 Camaro & Nova

1. Support vehicle at rear axle.
2. Disconnect stabilizer bar from spring plate brackets, Fig. 22.
3. Disconnect stabilizer bar from body brackets.
4. Reverse procedure to install. Tighten attaching bolts with vehicle at curb height.

1969-76 Corvette

1. Disconnect stabilizer bar from torque control arms and remove stabilizer bar frame brackets, Fig. 23.
2. Replace bushings as necessary, Fig. 23.
3. Reverse procedure to install.

STRUT ROD, REPLACE

1969-76 Corvette

1. Support vehicle at frame.

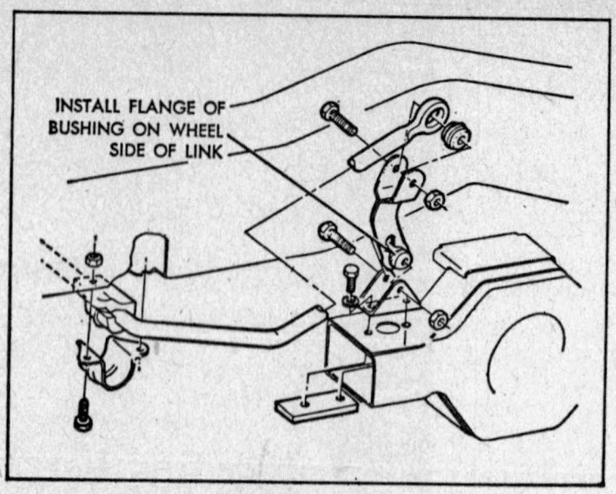

Fig. 23 Stabilizer bar installation. 1969-76 Corvette

2. Disconnect shock frame from lower mounting.
3. Remove cotter pin and nut from strut rod shaft. Pull shaft toward front of vehicle and remove from bracket.
4. Mark position of camber adjusting cam to ensure proper installation, Fig. 24, and loosen camber bolt nut.
5. Remove bolts securing strut rod bracket to carrier.
6. Remove camber bolt and nut, pull strut rod out of bracket and remove bushing caps.
7. Replace bushings as necessary, Fig. 20.
8. Reverse procedure to install.

TORQUE CONTROL ARMS & BUSHINGS, REPLACE

1969-76 Corvette

1. Perform steps 1 thru 4 as outlined under "Leaf Spring, Replace" 1969-75 Corvette procedure.
2. If equipped with a stabilizer shaft, disconnect shaft at torque arms.
3. Disconnect shock absorber at lower mounting.
4. Disconnect and lower strut rod shaft.
5. Disconnect axle drive shaft from spindle flange by removing attaching bolts.

NOTE: It may be necessary to force torque arm outboard providing clearance to lower axle drive shaft.

Fig. 27 Torque control arm bushing installation. 1969-76 Corvette

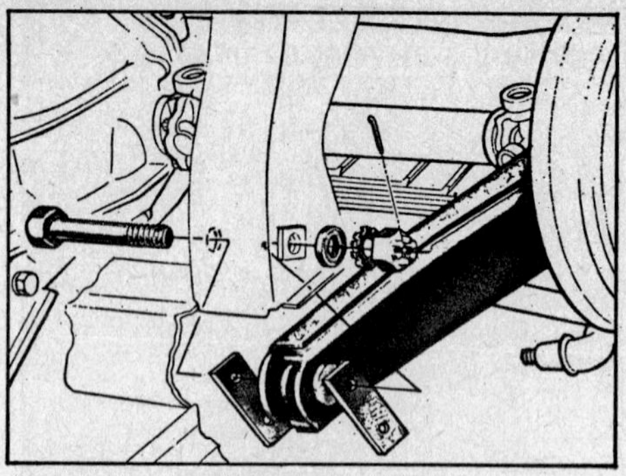

Fig. 25 Torque control arm installation. 1969-76 Corvette

Fig. 26 Torque control arm bushing removal. 1969-76 Corvette

6. Disconnect brake line from caliper and from torque arm. Disconnect parking brake cable.
7. Remove torque arm forward mounting bolt and toe-in shims, Fig. 25, and pull torque arm out of frame attachment.
8. Replace bushings if necessary as described under "Torque Control Arm Bushing Service."
9. Reverse procedure to install.

Torque Control Arm Bushing Service

1. Using an 11/16 inch twist drill, drill out flared end of bushing retainer, remove retainer plate and retainer from bushing.
2. Spread bushing with a chisel, Fig. 26, and tap bushing from arm.

NOTE: If bushing is rusted in torque arm, torque arm may spread during bushing removal. Install a "C" clamp torque arm, preventing torque arm spreading.

3. Oil steel portion of new bushing and press bushing into arm, Fig. 27.
4. Place retainer plate over flared portion of bushing retainer and insert retainer into bushing.
5. Make a flaring tool back-up plate, Fig. 28, with ½ inch bolt holes.
6. Place back-up plate on flared end of bushing retainer and assemble tool to plate, Fig. 29, with ½ x 5 inch bolts.

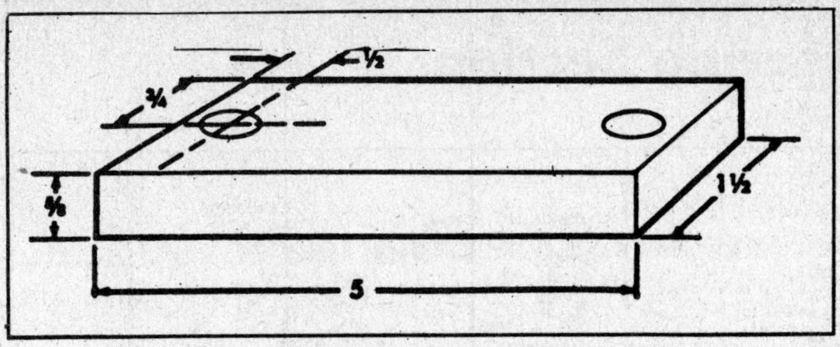

Fig. 28 Flaring tool back-up plate dimension

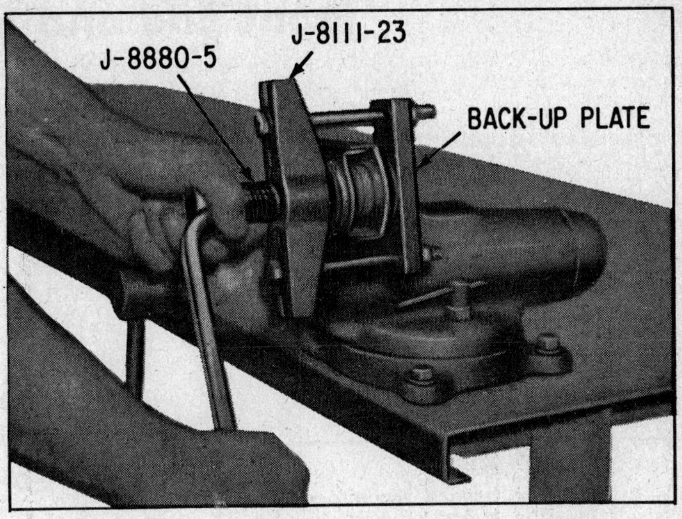

Fig. 29 Flaring torque control arm bushing retainer

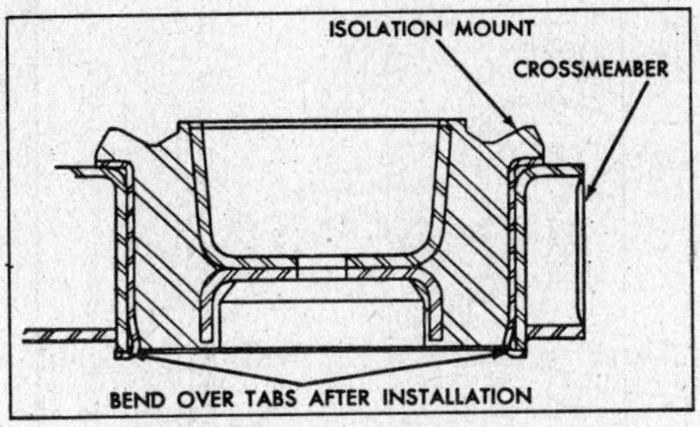

Fig. 30 Crossmember isolation mount replacement. 1969-76 Corvette

Center threaded hole in tool # J-8111-23 over unflared end of bushing retainer. Also center chamfered retainer plate over retainer tube.

7. Lubricate end of tool # J-8880-5 and thread screw into tool, flaring retainer.

CROSSMEMBER & ISOLATION MOUNT, REPLACE

1969-76 Corvette

1. Remove leaf spring as outlined under "Leaf Spring, Replace" 1969-76 Corvette procedure.
2. Remove differential carrier and cover as outlined in "Rear Axle, Propeller Shaft & Brakes" section 1969-76 Corvette procedure.
3. Support crossmember and remove bolts securing isolation mounts to frame and lower the crossmember.
4. To replace isolation mount, straighten isolation mount tabs and using a suitable ram, press on outer steel shell or inner steel insert, removing mount from crossmember. Install new mount into position, compress outer sleeve, press mount into crossmember and bend over locking tabs, Fig. 30.
5. Reverse procedure to install crossmember.

Front End and Steering Section

FRONT SUSPENSION

All front suspension systems are basically similar, being of the S.L.A. (short-long arm) type with independent coil springs riding on the lower control arms. Ball joints connect the upper and lower control arms to the steering knuckles. See Figs. 1 and 2.

LUBRICATION

IMPORTANT

On models that have a recommended chassis lubrication period of 6000 miles, the car should be warmed up before lubricating the front suspension ball joints. If the car has been outdoors in extreme cold weather it should be allowed to warm up to at least 10°F. above zero before this job is started. Inadequate ball joint lubrication or seal damage can result should the job be done while the parts are at lower temperatures.

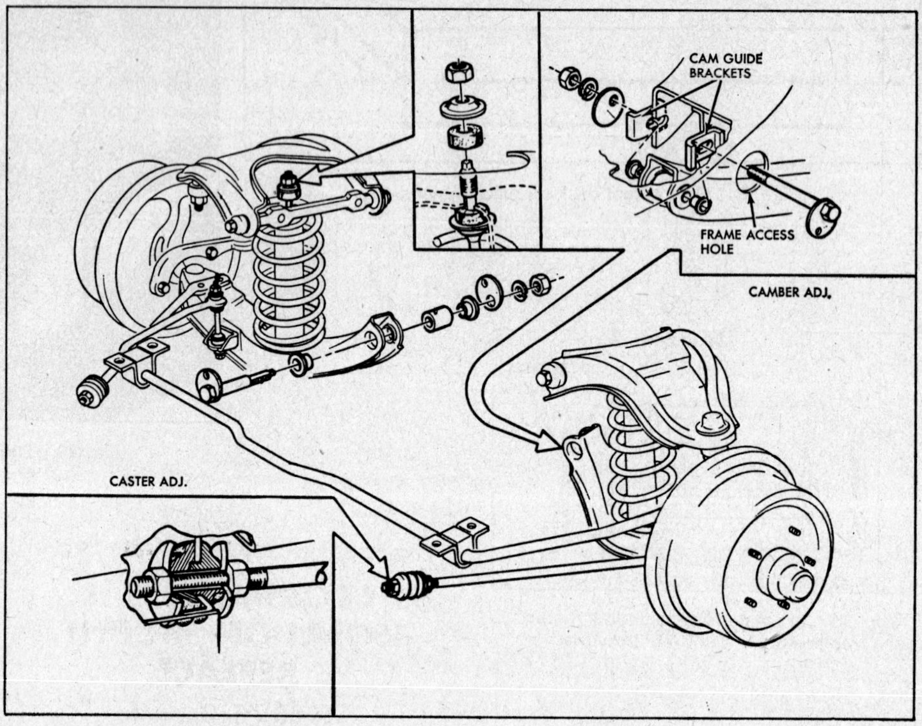

Fig. 1 Front suspension. 1969 Chevrolet

WHEEL ALIGNMENT

NOTE: Before adjusting caster and camber angles after complaint of excessive tire wear or poor handling, the front bumper should be raised and quickly released to allow car to return to its normal height.

1969 Chevrolet

The caster angle is adjusted by turning the two nuts at the front of the lower control arm strut rod, Fig. 1. Shortening this rod will increase caster, lengthening it will decrease caster.

Camber angle is adjusted by loosening the lower control arm pivot bolt and rotating the cam located on the pivot, Fig. 1. This eccentric cam action will move the lower control arm in or out, thereby varying camber.

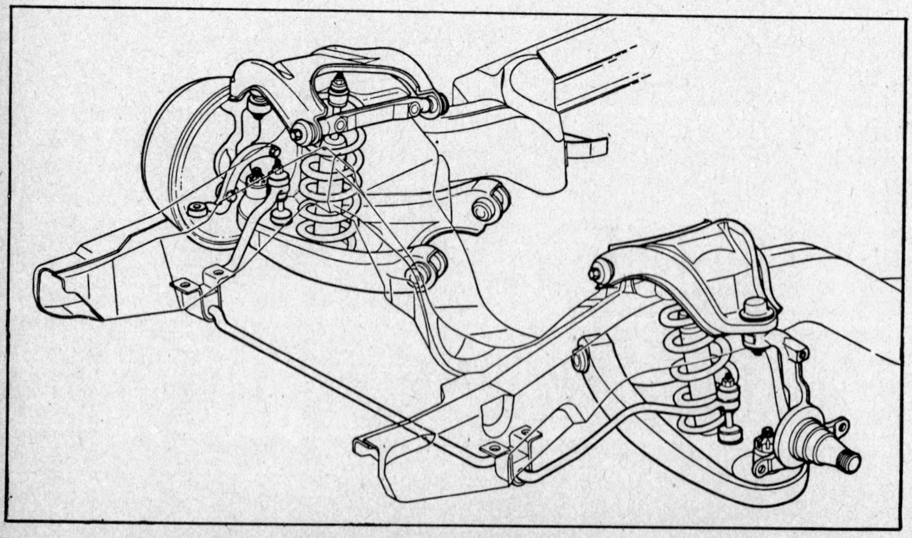

Fig. 2 Front suspension. 1969-76 Nova, 1969-76 Camaro, 1969-76 Chevelle. Typical of 1969-76 Corvette and 1970-76 Chevrolet

All Exc. 1969 Chevrolet

Caster and camber adjustments are made by means of shims between the upper control arm inner support shaft and the support bracket attached to the frame. Shims may be added, subtracted or transferred to change the readings as follows:

Caster, Adjust

Transfer shims from front to rear or rear to front. The transfer of one shim to the front bolt from the rear bolt will decrease positive caster. On shim ($1/32''$) transferred from the rear bolt to the front bolt will change caster about ½ degree.

Camber, Adjust

Change shims at both the front and rear of the shaft. Adding an equal number of shims at both front and rear of the support shaft will decrease positive camber. One shim ($1/32''$) at each location will move camber approximately 1/5 degree on Camaro, Chevelle and Nova, on Chevrolet and Corvette the change will be about 1/6 degree.

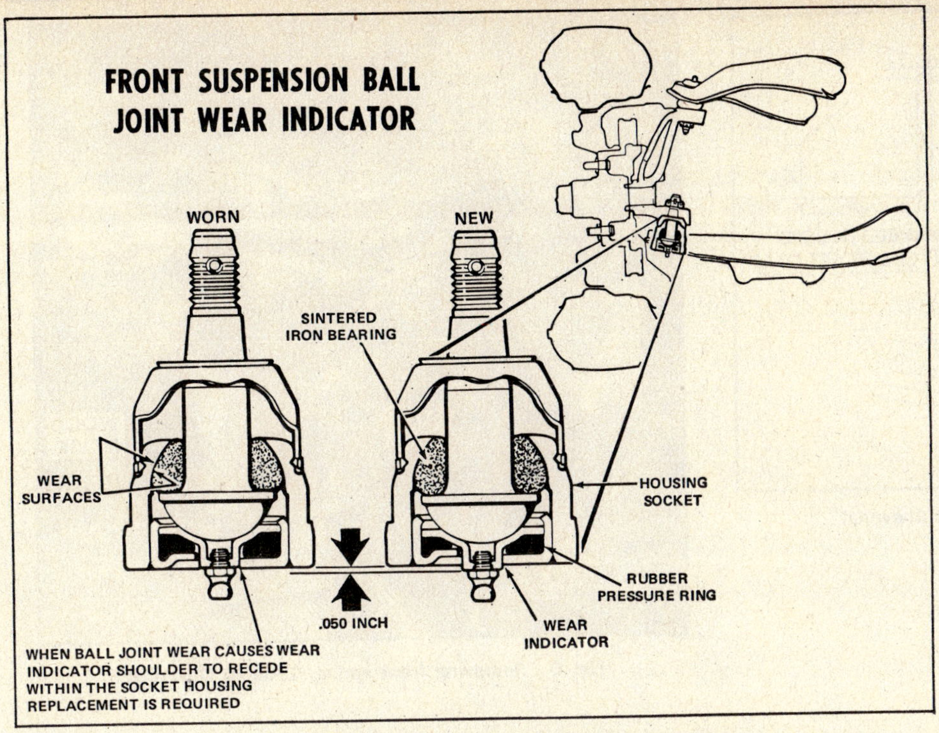

Fig. 3 Lower ball joint wear indicator. 1973 Chevrolet, 1974 Exc. Nova & 1975-76 All

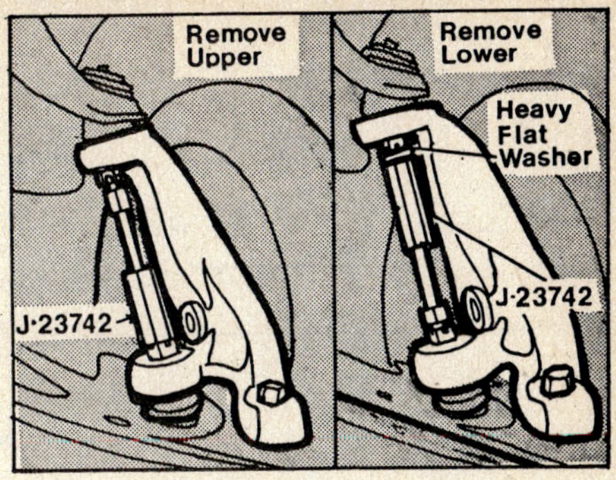

Fig. 3A Removing upper & lower ball joint stud from knuckle.

TOE-IN, ADJUST

Toe-in can be adjusted by loosening the clamp bolts at each end of each tie rod and turning each tie rod to increase or decrease its length as necessary until proper toe-in is secured and the steering gear is on the high point for straight-ahead driving.

WHEEL BEARINGS, ADJUST

1974-76

1. While rotating wheel forward, torque spindle nut to 12 ft. lbs.
2. Back off nut until "just loose" then hand tighten nut and back it off again until either hole in spindle lines up with hole in nut.

NOTE: Do not back off nut more than ½ flat.

3. Install new cotter pin. With wheel bearing properly adjusted, there will be .001-.005 inch end play.

1972-73

1. While rotating wheel, snug-up spindle nut, then back off until just loose (¼-½ turn).
2. Tighten nut by hand and then loosen nut so cotter pin can be installed.

NOTE: Bearing must not be even finger tight.

3. With bearing properly adjusted, there will be .001″-.008″ end play.

1969-71

1. While rotating wheel, tighten spindle nut to 12 ft-lbs torque.
2. Back off adjusting nut one flat and insert cotter pin. If slot and pin hole do not line up, back off adjusting nut an additional ½ flat or less as required to insert cotter pin.
3. Spin wheel to see that it turns freely, then spread cotter pin.
4. Bearings should have zero preload and .001 to .008″ end movement when properly adjusted.

WHEEL BEARINGS, REPLACE

Disc Brakes

1. Raise car and remove front wheels.
2. Remove bolts holding brake caliper to its mounting and insert a fabricated block (1 1/16 x 1 1/16 x 2 inches in length) between the brake pads as the caliper is being removed. Once removed, the caliper can be wired or secured in some manner away from the disc.
3. Remove spindle nut and hub and disc

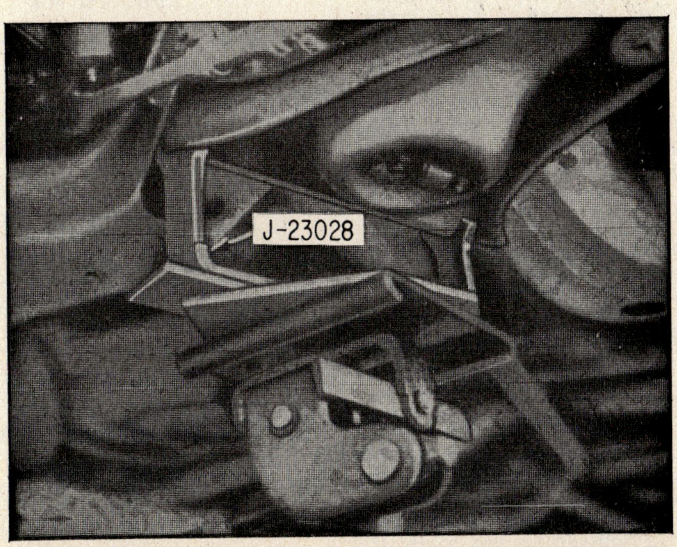

Fig. 4 Supporting coil spring. 1969-76 All except 1969-70 Chevrolet

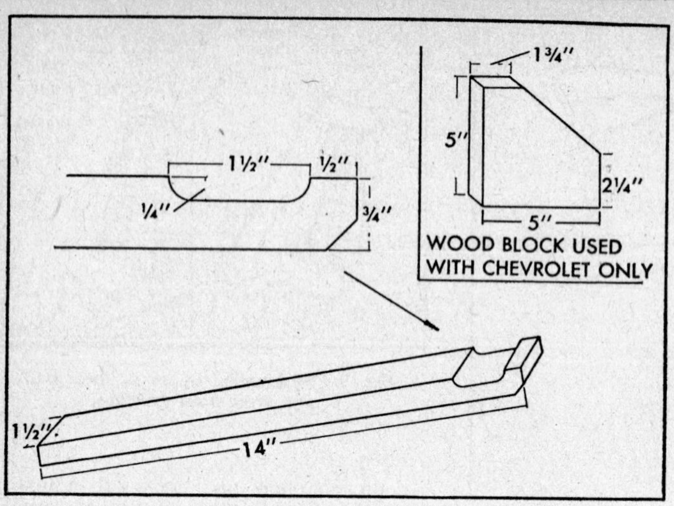

Fig. 5 Spring removal tools. 1969-70 Chevrolet

Fig. 6 Removing front spring. 1969-70 Chevrolet

assembly. Grease retainer and inner wheel bearing can now be removed.

CHECKING BALL JOINTS FOR WEAR

1969-76 All

Upper Ball Joint, 1976

1. Raise front of vehicle with jacks placed between the coil spring pocket and ball joint of lower control arm.
2. Shake top of wheel in and out. Observe steering knuckle for any movement relative to the control arm.
3. Replace upper ball joint if looseness is indicated.

Upper Ball Joint, 1969-75

The upper ball joint is checked for wear by checking the torque required to rotate the ball stud in the assembly. To make this type of check it will be necessary to remove the stud from the steering knuckle.

Install a nut on the ball stud and measure the torque required to turn the stud in the assembly with a torque wrench. Specified torque for a new ball joint is 3 to 10 ft-lbs. If torque readings are excessively high or low, replace the ball joint. If excessive wear is indicated in the upper ball joint, both upper and lower ball joints should be replaced.

Lower Ball Joint

Raise car and support lower control arm so spring is compressed in the same manner as if the wheels were on the ground and check axial (up and down) play at ball joint. If play exceeds $\frac{1}{16}$″, replace the joint.

Another indication of lower ball joint excessive wear is when difficulty is experienced when lubricating the joint. If the liner has worn to the point where the lubrication grooves in the liner have been worn away, then abnormal pressure is required to force lubricant through the joint. Should this condition be evident, replace both lower ball joints.

NOTE: 1973 Chevrolet, 1974 except Nova and all 1975-76 models have a wear indicator built into the lower ball joint, Fig. 3.

UPPER BALL JOINT, REPLACE

1969-76 All

1. Support weight of vehicle at outer end of lower control arm.
2. Remove wheel assembly.
3. Remove nut from upper ball joint stud.
4. Remove stud from knuckle, Fig. 3A.
5. Cut off ball rivets with a chisel.
6. Enlarge ball stud attaching holes in control arm (if necessary) to accept the bolts included in the ball joint replacement kit.
7. Install and tighten new ball joint.
8. Reassemble ball joint to steering knuckle.

LOWER BALL JOINT, REPLACE

1971-76 Exc. Corvette

1. Support vehicle at frame and lower control arm with suitable jacks.
2. Remove wheel and, if equipped with disc brakes, caliper assembly.
3. Loosen ball joint stud nut not more than one turn and, using a suitable tool, Fig. 3A, press ball stud free of steering knuckle, then remove stud nut.
4. Lower the jack from under the control arm.
5. With a suitable tool, replace lower ball joint.
6. Reverse procedure to assemble.

1969-70 Chevrolet & 1969-76 Corvette

1. Support lower control arm at outer end on floor jack with hoist or jack pad clear of lower ball stud nut and seal.
2. If equipped with disc brakes, remove caliper assembly.
3. Remove upper and lower ball stud nuts, free ball studs from steering knuckle and wire knuckle and brake drum or disc assembly out of the way.
4. Being careful not to enlarge holes in control arm, cut off rivets.
5. Install new joint against underside of control arm and retain in place with special bolts supplied with replacement ball joint kit. Use only the alloy bolts supplied for this operation. The special thick headed bolt must be installed in the forward side of the control arm.
6. Tighten bolts and nut on ball stud and lubricate joint.

1969-70 Chevelle, Camaro & Nova

1. Support lower control arm at outer end of floor jack with hoist or jack

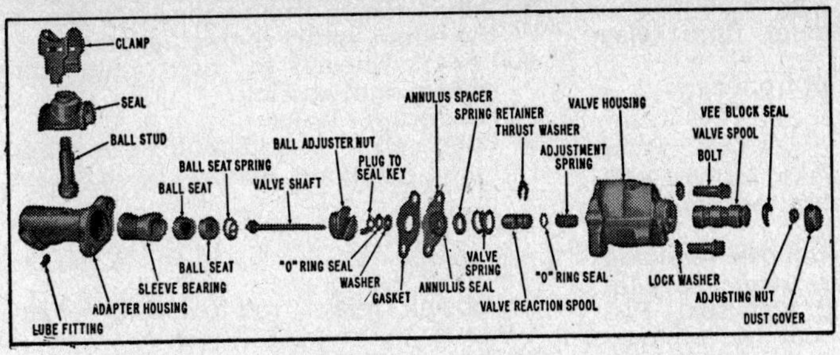

Fig. 7 Power steering control valve and adapter assembly. 1969-76 Corvette

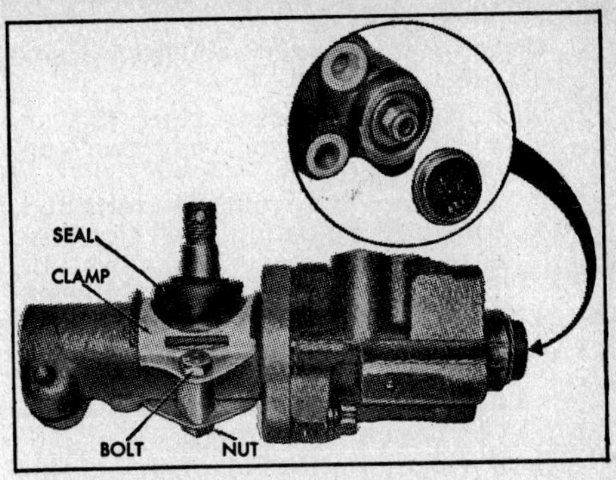

Fig. 8 Power steering control valve ball stud seal replacement. 1969-76 Corvette

pad clear of lower ball stud and remove wheel. If equipped with disc brakes, remove caliper assembly.
2. Remove upper and lower ball stud nuts, free ball studs from steering knuckle and wire knuckle and brake drum or disc assembly out of the way.
3. Use a screwdriver to pry out seal and retainer. Use a suitable extractor to push out ball joint.
4. Start replacement ball joint into control arm and use a suitable puller tool to pull the ball joint into position. Be sure to position air vent in rubber boot inboard.
5. Install stud into steering knuckle and secure in place.

SHOCK ABSORBER, REPLACE

1969-76 All

1. Hold shock upper stem from turning with a suitable wrench and remove nut and grommet.
2. Unfasten lower shock pivot from lower control arm and pull shock and mounting out through bottom of spring housing.
3. Reverse removal procedure to install. Tighten upper retaining nut until it bottoms on shoulder of stem.

COIL SPRING, REPLACE

1969-76 All Exc. 1969-70 Chevrolet

1. Remove shock absorber on 1969-70 models as outlined under "Shock Absorber, Replace" procedure. On 1971-76 models, disconnect shock absorber from lower mounting, push shock absorber through hole in lower control arm and compress into spring.
2. Support vehicle by frame so control arms hang free.
3. On 1969-70 models, remove stabilizer bar-to-lower control arm link.
4. Install a safety chain through spring and lower control arm.
5. Install tool J-23028 onto a suitable jack and position jack so control arm is supported by bushings seated in grooves of tool, Fig. 4.
6. Raise jack to relieve tension on control arm bolts and remove bolts.
7. Lower jack until tension is removed from spring, remove chain and spring from vehicle.

1969-70 Chevrolet

1. With suitable wrench, hold shock absorber upper stem from turning, then remove retaining nut, retainer and upper grommet.
2. With car supported by the frame so that control arms hang free, remove wheel assembly. Replace one wheel nut to hold brake drum.
3. Remove shock absorber, stabilizer bar-to-lower control arm link, strut rod-to-lower control arm attaching nuts and tie rod end.
4. Install a steel bar (made to the dimensions shown in Fig. 5) through shock absorber mounting hole in lower control arm so that notch seats over bottom spring coil and bar extends inboard and under inner bushing.
5. Fit a 5″ wood block, Fig. 6, between bar and bushing.
6. With suitable jack or hoist, lift up slightly on end of bar to remove tension from inner pivot cam bolt, which can then be removed.
7. Lower inner end of lower control arm. Tension on spring will be removed before spring can be removed from vehicle.
8. Reverse procedure to install and check wheel alignment.

CAUTION

The spring force under compression is very great. Exercise every safety precaution when performing this operation to see that individuals and materials subject to damage are removed from the path of the spring when the control arm is being lowered. Also, the compressed spring should be relaxed immediately after lowering the control arm to reduce the time of exposure to the great compressive force.

STEERING GEAR, REPLACE

1969-76 All Models

NOTE: On models where shield is installed, remove shield from coupling.

1. Remove nuts, washers and bolts at steering coupling.
2. Remove pitman arm nut and washer from sector shaft and mark relation of arm position to shaft.
3. Use a suitable puller to remove pitman arm. On Camaro models, use care not to bend or facture brake pipe when removing steering gear.
4. Unfasten gear from frame and remove assembly.
5. Reverse procedure to install.

INTEGRAL POWER STEERING

1969-76 Exc. Corvette

To remove gear assembly, disconnect pressure and return hoses from gear housing and cap both hoses and steering gear outlets to prevent foreign material from entering system, then follow procedure as outlined under *Steering Gear, Replace*.

LINKAGE TYPE POWER STEERING

1969-76 Corvette

Power steering equipment consists of a recirculating ball type steering gear and linkage to which a hydraulic power mechanism has been added as part of the steering linkage. The hydraulic mechanism furnishes additional power to *assist* the manual operation so that the turning effort at the steering wheel is greatly reduced. The hydraulic mechanism consists of three basic units: a hydraulic pump and reservoir, a control valve, and a power cylinder.

Control Valve, Adjust

1. Disconnect cylinder rod from frame bracket.
2. With car on a hoist, start the engine. One of the following two conditions will exist:
   a. If piston rod remains retracted, turn the adjusting nut clockwise until the rod begins to move out. Then turn the nut counterclockwise until the rod just begins to move in. Now turn the nut clockwise to exactly one half the rotation needed to change the direction of shaft movement.
   b. If the rod extends upon starting the pump, move the nut counterclockwise until the rod begins to retract, then clockwise until the rod begins to move out again. Now turn the rod to exactly one half the rotation needed to change the direction of shaft movement.

Do not turn the nut back and forth more than is absolutely necessary to balance the valve.

3. Restart engine. Front wheels should not turn from center if valve has been properly balanced.

Power Cylinder Repairs

Removal

1. Disconnect two hydraulic lines at power cylinder.
2. Unfasten power cylinder rod from brace at frame.
3. Unfasten power cylinder from relay rod bracket.
4. Remove power cylinder from car.

Inspection

1. Inspect seals for leaks around cylinder rod and if leaks are present, replace seals as follows:
2. Use a hook tool to remove retaining ring. Remove wiper ring, back-up washer, back-up ring and seal. *Piston rod seal should not be removed unless there are signs of leakage along the piston shaft at shaft seal.*
3. Examine brass fitting hose connection seats for cracks or damage and replace if necessary.
4. For service other than seat or seal replacement, replace the power cylinder.

Installation

1. Install power cylinder on car by reversing removal procedure.
2. Reconnect two hoses, fill system with fluid and bleed system as outlined below.

Filling & Bleeding System

1. Fill reservoir to proper level with Automatic Transmission Fluid and let fluid remain undisturbed for about two minutes.
2. Raise front wheels off floor.
3. Run engine at idle for two minutes.
4. Increase engine speed to about 1500 rpm.
5. Turn wheels from one extreme to the other, lightly contacting stops.
6. Lower wheels to floor and turn wheels right and left.
7. Recheck for leaks.
8. Check oil level and refill as required. Pump pressure should be 870 lbs.

Control Valve Repairs

Removal, Fig. 7

1. Loosen relay rod-to-control valve clamp.
2. Disconnect hose connections at control valve.
3. Disconnect control valve from pitman arm.
4. Unscrew control valve from relay rod.
5. Remove control valve from car.

Ball Stud Seal, Replace

In servicing the control valve, refer to Fig. 7. To replace the ball stud seal, refer to Fig. 8 and proceed as follows:

1. Remove pitman arm with a suitable puller.
2. Remove clamp by removing nut, bolt and spacer. If crimped type clamp is used, straighten clamp end and pull clamp and seal off end of stud.
3. Install new seal and clamp over stud so lips on seal mate with clamp. (A nut and bolt attachment type clamp replaces the crimped type for service, Fig. 8).
4. Center the ball stud, seal and clamp in opening in adapter housing, then install spacer, bolt and nut.

CHEVROLET CHEVETTE

OLD CAR SPECIFICATIONS: For 1946-68 Tune Up and Wheel Alignment Specifications see main index.

CAUTION

This vehicle contains many parts dimensioned in the metric system. Most fasteners are metric and many are very close in dimension to familiar customary measurements in the inch system. However, it is important to note that, during any vehicle maintenance procedures, replacement fasteners must have the same measurements as those removed, whether metric or customary. Mismatched or incorrect fasteners can result in vehicle damage or malfunction or possible personal injury. Therefore, fasteners removed from the vehicle should be saved for re-use or care should be taken to select a replacement that matches the original.

INDEX OF SERVICE OPERATIONS

SERIAL NUMBER LOCATION

On top of instrument panel, left front.

ENGINE NUMBER LOCATION

On pad at right side of cylinder block, forward of exhaust manifold.

ENGINE IDENTIFICATION CODE

Engines are identified in the following table by the code letter or letters immediately following the engine serial number.

Year	Engine	Code
1976	4-85	CDA, CDB, CDC
	4-85	CDD, CDF, CDH
	4-85	CDJ, CDK, CDL
	4-85	CDM, CDN, CDR
	4-85	CDS, CDT, CDU
	4-85	CDW, CDX, CDY
	4-85	CVA, CVB
	4-97	CNA, CNB
	4-97	CYA, CYB, CYC
	4-97	CYD, CYE, CYH
	4-97	CYJ, CYK, CYL
	4-97	CYM, CYR, CYS
	4-97	CYT, CYU, CYW
	4-97	CYX, CYY, CYZ

GRILLE IDENTIFICATION

1976

GENERAL ENGINE SPECIFICATIONS

Year	Engine	Carburetor	Bore and Stroke	Piston Displacement, Cubic Inches	Compression Ratio	Maximum Brake H.P. @ R.P.M.	Maximum Torque Lbs. Ft. @ R.P.M.	Normal Oil Pressure Pounds
1976	52 Horsepower① .... 4-85 1400 c.c.	1 Barrel —	3.228 x 2.606 82.0 x 66.2 mm.	85 1.4 ltr.	8.5 —	52 @ 5300 —	67 @ 3400 97 Joules @ 3400	55 —
	60 Horsepower① .... 4-97 1600 c.c.	1 Barrel —	3.228 x 2.90 82.0 x 75.7 mm.	97.6 1.6 ltr.	8.5 —	60 @ 5300 —	77 @ 3400 104 Joules @ 3200	55 —

①—Ratings are net—as installed in vehicle.

TUNE UP SPECIFICATIONS

The following specifications are published from the latest information available. This data should be used only in the absence of a decal affixed in the engine compartment.

OLD CAR SPECIFICATIONS: For 1946-68 Tune Up Specifications see main index.

★When using a timing light, disconnect vacuum hose or tube at distributor and plug opening in hose or tube so idle speed will not be affected.

●When checking compression, lowest cylinder must be within 80 percent of highest.

▲Before removing wires from distributor cap, determine location of the No. 1 wire in cap, as distributor position may have been altered from that shown at the end of this chart.

Year	Spark Plug Type	Spark Plug Gap Inch	Distributor Point Gap Inch	Distributor Dwell Angle Deg.	Ignition Timing★ Firing Order Fig. ▲	Ignition Timing★ Timing BTDC ①	Ignition Timing★ Mark Fig.	Hot Idle Speed Std. Trans.	Hot Idle Speed Auto. Trans.	Air Fuel Ratio Std. Trans.	Air Fuel Ratio Auto. Trans.	Idle "CO" % Std. Trans.	Idle "CO" % Auto. Trans.
1976													
4-85	R43TS	.035 (.889 mm.)	—	—	A	10	B	③	④	—	—	—	—
4-97.6	R43TS	.035 (.889 mm.)	—	—	A	②	B	③	④	—	—	—	—

①—BTDC—Before top dead center.
②—Man. trans., 8° BTDC; auto. trans., 10° BTDC.
③—All with solenoid energized—exc. Calif., 800 RPM; Calif., 1000 RPM. With solenoid de-energized—less A/C, 600 RPM; with A/C, 700 RPM.
④—All with solenoid energized—exc. Calif., 800 RPM; Calif., 850 RPM. With solenoid de-energized—all less A/C, Exc. Calif., 700 RPM; Calif., 600 RPM.

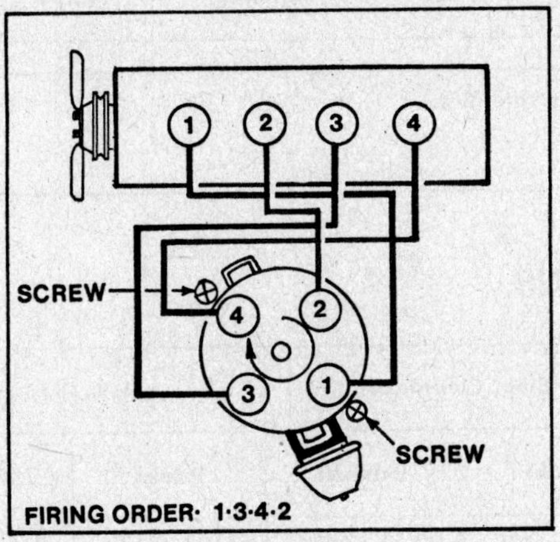

Fig. A

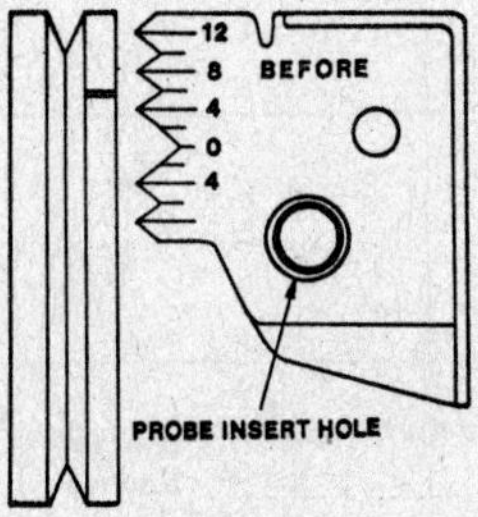

Fig. B

ALTERNATOR & REGULATOR SPECIFICATIONS

Year	Alternator Model	Rated Hot Output Amps.	Field Current 12 Volts @ 80° F.	Output @ 14 Volts 2000 R.P.M. Amps.	Output @ 14 Volts 2000 R.P.M. Amps.	Regulator Model	Field Relay Air Gap In.	Field Relay Point Gap In.	Field Relay Closing Voltage	Voltage Regulator Air Gap In.	Voltage Regulator Point Gap In.	Voltage Regulator Voltage @ 125° F.
1976	1102845	32	4.0–4.5	—	31	Integral	—	—	—	—	—	13.8–14.8①
	1102846	55	4.0–4.5	—	51	Integral	—	—	—	—	—	—

①—At 85° F.

DISTRIBUTOR SPECIFICATIONS

★Note: If unit is checked on vehicle, double the RPM and degrees to get crankshaft figures.

Distributor Part No.①	Centrifugal Advance Degrees @ RPM of Distributor					Vacuum Advance	
	Advance Starts	Intermediate Advance			Full Advance	Inches of Vacuum to Start Plunger	Max. Adv. Dist. Deg. @ Vacuum
1110654	0 @ 600	4 @ 1000	—	—	10 @ 2400	4 (14 kPa②)	7 @ 8 (27 kPa②)
1110655	0 @ 600	4 @ 1000	—	—	10 @ 2400	5 (17 kPa②)	6 @ 12 (41 kPa②)
1110657	0 @ 600	4 @ 1000	—	—	10 @ 2400	5 (17 kPa②)	12 @ 12 (41 kPa②)
1110658	0 @ 600	4 @ 1000	—	—	10 @ 2400	5 (17 kPa②)	13 @ 14.5 (49 kPa②)
1110659	0 @ 750	1 @ 800	—	—	8 @ 2400	4 (14 kPa②)	13 @ 12 (41 kPa②)

①—Stamped on distributor housing.

②—kPa—kilopascals.

STARTING MOTOR SPECIFICATIONS

Year	Model	Starter Number	Brush Spring Tension Oz①	Free Speed Test			Resistance Test③	
				Amps.	Volts	R.P.M.①	Amps.	Volts
1976	All	1110941	—	50–75②	9	6500	—	—

①—Minimum.

②—Includes solenoid.

VALVE SPECIFICATIONS

Year	Model	Valve Lash		Valve Angles		Valve Spring Installed Height	Valve Spring Pressure Lbs. @ In.	Stem Clearance		Stem Diameter	
		Int.	Exh.	Seat	Face			Intake	Exhaust	Intake	Exhaust
1976	All	Hydraulic		46	45	1.25 32 mm.	173 @ .886 770N @ 22 mm.	.0006–.0017 .015–.045 mm.	.0014–.0025 .035–.065 mm.	.3138–.3144 7.972 mm.	.3130–.3136 7.952–7.965 mm.

ENGINE TIGHTENING SPECIFICATIONS★

★Torque specifications are for clean and lightly lubricated threads only. Dry or dirty threads produce increased friction which prevents accurate measurement of tightness.

Year	Engine Model	Spark Plugs Ft. Lbs.	Camshaft Carrier Bolts Ft. Lbs.	Intake Manifold Ft. Lbs.	Exhaust Manifold Ft. Lbs.	Camshaft Sprocket Bolt Ft. Lbs.	Cam Cover In. Lbs.	Connecting Rod Cap Bolts Ft. Lbs.	Main Bearing Cap Bolts Ft. Lbs.	Flywheel to Crankshaft Ft. Lbs.	Vibration Damper or Pulley Ft. Lbs.
1976	All	15 20.3 N-m	75 101.7 N-m	16 21 N-m	①	75 101.7 N-m	14 1.6 N-m	37 50 N-m	46 65 N-m	46 65 N-m	75 101.7 N-m

①—Center bolts, 16 ft. lbs. (21 Newton-meters); end legs, 22 ft. lbs. (30 Newton-meters).

PISTONS, PINS, RINGS, CRANKSHAFT & BEARINGS

Year	Model	Piston Clearance Top of Skirt	Ring End Gap		Wrist-pin Diameter	Rod Bearings		Main Bearings			
			Comp.	Oil		Shaft Diameter	Bearing Clearance	Shaft Diameter	Bearing Clearance	Thrust on Bear. No.	Shaft End Play
1976	All	.0008–.0016	.009	.015	.9052	1.8093–1.8103	.0014–.0030	2.0075–2.0085	.0009–.0026	4	.004–.008
		.020–.040 mm	.229 mm	.381 mm	22.994 mm	45.958–45.984 mm	.036–.078 mm	50.987–51.013 mm	.024–.066 mm		.100–.202 mm.

BRAKE SPECIFICATIONS

Year	Model	Brake Drum Inside Diameter	Wheel Cylinder Bore Diameter			Master Cylinder Bore Diameter		
			Disc Brake	Front Drum Brake	Rear Drum Brake	Disc Brakes	Drum Brakes	Power Brakes
1976	All	7.88 200.15 mm	1⅞ 47.625 mm	— —	¾ 19.05 mm	¾ 19.05 mm	— —	¾ 19.05 mm

REAR AXLE SPECIFICATIONS

Year	Model	Carrier Type ②	Ring Gear & Pinion Backlash		Pinion Bearing Preload			Differential Bearing Preload		
			Method	Adjustment	Method	New Bearings Inch-Lbs.	Used Bearings Inch-Lbs.	Method	New Bearings Inch-Lbs.	Used Bearings Inch-Lbs.
1976	All	Integral	Shims	—	Spacer	5–11 .5–1.2 N-m	3–6 .3–.6 N-m	Shims	—	—

WHEEL ALIGNMENT SPECIFICATIONS

Year	Model	Caster Angle, Degrees		Camber Angle, Degrees				Toe-In. Inch	Toe-Out on Turns, Deg.①	
		Limits	Desired	Limits		Desired			Outer Wheel	Inner Wheel
				Left	Right	Left	Right			
1976	All	+4 to +5	+4½	−¼ to +¾	−¼ to +¾	+¼	+¼	1/16 1.5 mm	—	—

COOLING SYSTEM & CAPACITY DATA

Year	Model or Engine	Cooling Capacity, Qts.			Radiator Cap Relief Pressure, Lbs.		Thermo. Opening Temp. ①	Fuel Tank Gals.	Engine Oil Refill Qts. ②	Transmission Oil			Rear Axle Oils Pints
		No Heater	With Heater	With A/C	With A/C	No. A/C				3 Speed Pints	4 Speed Pints	Auto. Trans. Qts. ③	
1976	4-All	—	8½	9	15	15	190	13	4	—	3	10	2
		—	8 ltr.	8.5 ltr.	—	—	—	49.2 ltr.	3.8 ltr.	—	1.5 ltr.	9.5 ltr.	.9 ltr.

①—Use with permanent type anti-freeze.
②—Add 1 qt. with filter change.
③—Approximate. Make final check with dipstick.

Electrical Section

ELECTRICAL DIAGNOSIS

This vehicle is equipped with either one or two "Master Electrical Connectors," Fig. 1. One connector is the "Engine Electrical Master Diagnostic Connector" and the other is the "Air Conditioning Electrical Master Diagnostic Connector." These connectors are used to diagnose electrical Malfunctions in the air conditioning, charging, cranking and ignition systems.

CAUTION: When performing any of the following tests, ensure transmission is in Neutral on manual transmission models or Park on automatic transmission models. Also, fully apply the parking brake.

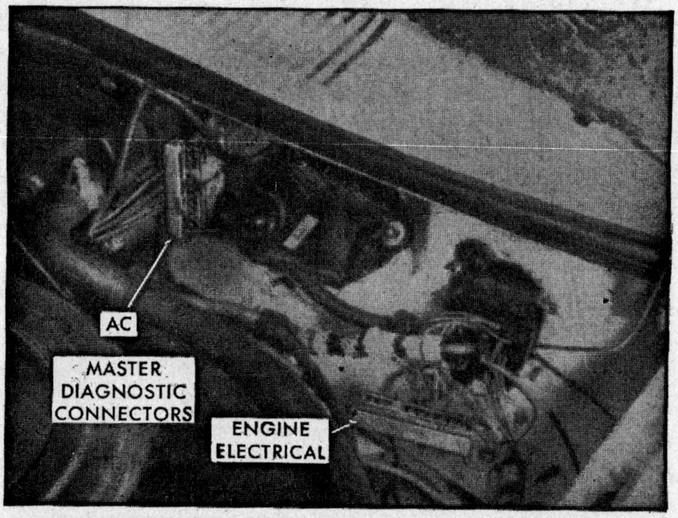

Fig. 1 Master electrical diagnostic connector locations

ENGINE ELECTRICAL DIAGNOSTIC TESTS

NOTE: The following tests are performed with a 14 gauge jumper wire and the ignition switch in the "Off" position.

Starter Does Not Crank Or Cranks Slowly

Test T-1

1. Connect jumper wire between terminal Nos. 1 and 8.

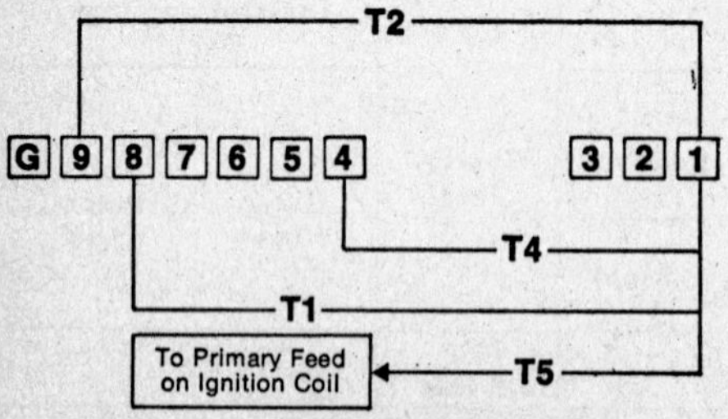

Fig. 2 Engine electrical diagnostic test connections

2. If malfunction remains there may be an open wire between terminal No. 8 and the starter solenoid "S" terminal, a faulty connection at the starter solenoid "S" terminal or a poor connection at battery or starter. If the wiring and connections are satisfactory, the starter solenoid may be defective.
3. If starter cranks in step 1, proceed to Test T-2.

Test T-2

1. Connect jumper wire between terminals Nos. 1 and 9.
2. If malfunction remains, the neutral starter switch may be defective or faulty neutral starter switch wiring.
3. If starter cranks in step 1, the ignition switch may be defective or improperly connected.

Engine Cranks But Will Not Start

Test T-3

1. Check for ignition output by removing a spark plug wire, insert an extension into the boot and crank en-

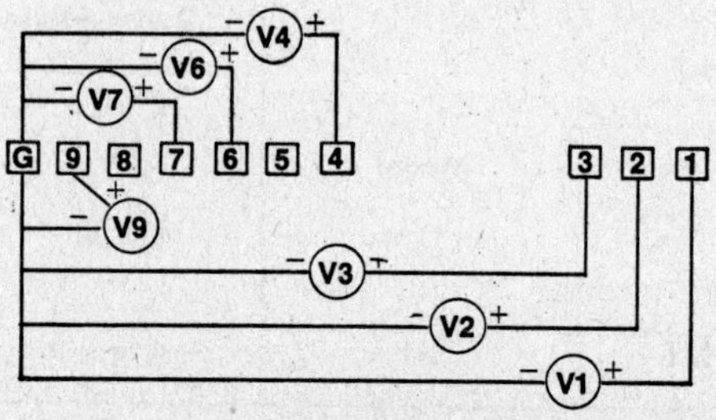

Fig. 3 A/C electrical diagnostic test connections

gine while holding the spark plug wire with insulated pliers approximately ¼ inch (6mm) from engine block.

2. If spark is noted in step 1, the malfunction is not in ignition system (excluding the spark plugs).
3. If no spark occurs in step 1, proceed to Test T-4.

Test T-4

1. Connect jumper wire between terminals Nos. 1 and 4 and crank engine.
2. If engine starts, the ignition switch may be defective or not connected properly.
3. If engine does not start in step 1, proceed to Test T-5.

Test T-5

1. Connect jumper wire between terminal No. 1 and the ignition coil primary feed terminal and crank engine.
2. If engine starts, there is an open wire between test terminal No. 4 and the coil or a faulty coil connection.
3. If engine does not start, the H.E.I. system is malfunctioning.

A/C ELECTRICAL DIAGNOSTIC TESTS FIG. 3

NOTE: The following tests are performed with a voltmeter.

Blower Inoperative/Compressor OK

Place A/C control in "Off", blower switch in "Low" and the ignition switch "On".

Test V-1

1. Connect voltmeter between terminals Nos. 1 and G.
2. If voltage obtained is over 3 volts, the blower motor may be improperly connected. If not, replace blower motor.
3. If voltage obtained in step 1 is under 3 volts, proceed to Test V-2.

Test V-2

1. Connect voltmeter between terminals Nos. 2 and G.
2. If voltage obtained is over 3 volts, the high blower relay is disconnected or defective.
3. If voltage obtained in step 1 is under 3 volts, proceed to Test V-3.

Test V-3

1. Connect voltmeter between terminal Nos. 3 and G.
2. If voltage is over 10 volts, the blower resistor may be open or an open circuit between the blower resistor and high blower relay is present.
3. If voltage obtained in step 1 is under 10 volts, there may be an open circuit between the time-temperature relay and the blower resistor.

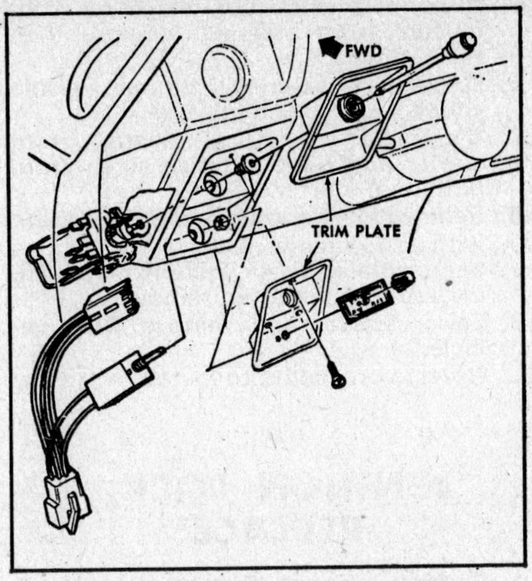

Fig. 4 Light switch replacement

Blower & Compressor Inoperative

Place A/C control in "Off", blower switch in "Low" and ignition switch "On".

Test V-4

1. Connect voltmeter between terminal Nos. 4 and G.
2. If voltage obtained is over 10 volts, there may be a grounded circuit in the temperature switch or the wiring.
3. If voltage obtained in step 1 is under 10 volts, there may be a blown fuse or an open circuit between the fuse and time-temperature relay.

No Hi Blower In Hi Blower Or Max A/C Switch Positions

Test V-9

1. Place A/C control in "Vent", blower switch in "Hi" and ignition switch in "On".
2. Connect a voltmeter between terminal Nos. 9 and G.
3. If voltage obtained is over 10 volts, the high blower fuse may be blown or the feed wire disconnected. Also, the high blower relay may be defective or the feed wire disconnected.
4. If the voltage obtained is under 10 volts, there may be an open circuit between the A/C control and the high blower relay coil.

No Low Blower Speed Or High Speed Only

1. Place A/C control in "Vent", blower switch in "Low" and ignition switch "On".
2. Connect voltmeter between terminal Nos. 3 and G (Test V-3).
3. If over 10 volts is obtained in Step 2, the blower resistor may be open or an open circuit between the blower resistor and high blower relay is present.

NOTE: If the blower runs at all speeds except "Low", the resistor is faulty. If the blower runs on high speed only, the resistor may be faulty or an open circuit is present between the resistor and high blower relay.

4. If voltage obtained in step 2 is under 10 volts, there may be an open circuit between the time-temperature relay and the blower resistor.

No Medium Blower Speeds

Check for a defective blower switch or an open circuit between the blower switch and blower resistor.

Blower Runs In Hi Speed In All Switch Positions

1. Place A/C control in "Vent", blower switch in "Low" and ignition switch "On".
2. Connect voltmeter between terminal Nos. 9 and G (Test V-9).
3. If any voltage reading is obtained in Step 2, there may be a short in the high blower relay feed wire, blower switch, A/C control or high blower relay.
4. If no voltage reading is obtained in step 2, a defective high blower relay or a short in the relay to blower motor wiring.

NOTE: If zero voltage is obtained in A/C control positions "Norm", "Heat" or "Defrost", indicates a malfunction in the A/C control. Zero voltage obtained in the "Off" indicates a malfunction in the A/C control or faulty wiring between the control and blower switch. If the voltage drops to zero when the blower switch is moved indicates a faulty blower switch.

Compressor Inoperative/Blower OK

Place A/C control in "Norm", blower switch in "Low" and ignition switch "On".

Test V-6

1. Connect voltmeter between terminal Nos. 6 and G.
2. If voltage obtained is over 10 volts, check compressor wiring and connections. If wiring and connections are satisfactory, check for defective compressor clutch.
3. If voltage obtained in step 1 is under 10 volts, proceed to Test V-7.

Test V-7

1. Connect voltmeter between terminals Nos. 7 and G.
2. If voltage obtained is over 10 volts, there may be an open in the compressor feed circuit. Note that this circuit includes the pressure switch and the thermostatic switch.
3. If voltage obtained in step 1 is under 10 volts, an open circuit in the A/C control is indicated.

DISTRIBUTOR

1. Disconnect wiring connectors from distributor cap.

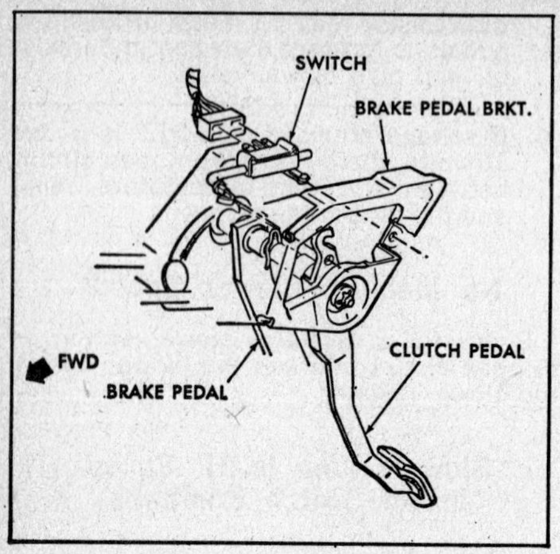

Fig. 5 Clutch start switch replacement

2. Remove distributor cap and position out of way, then disconnect vacuum advance hose.
3. Scribe a mark on engine in line with rotor, noting approximate position of distributor housing in relation to engine.
4. Remove hold-down nut and clamp and lift distributor from engine. Do not crank engine after distributor is removed, otherwise the distributor will have to be initially timed to the engine.
5. Install distributor in reverse order of removal, making certain to align scribe marks, then start engine and adjust timing.

STARTER, REPLACE

Less A/C

1. Disconnect battery ground cable and remove air cleaner.
2. Remove distributor cap and position aside.
3. Remove the fuel line from fuel pump and carburetor.
4. Disconnect electrical connectors from ignition coil, remove coil bracket retaining screws and the coil and bracket assembly.
5. Disconnect vacuum hose from vacuum advance unit and the electrical connector from the oil pressure sender, then remove the sender.
6. Disconnect electrical leads from starter and remove the brace screw from bottom of starter housing.
7. Remove starter retaining screws and the starter from vehicle.
8. Reverse procedure to install.

With A/C

1. Disconnect battery ground cable and remove air cleaner.
2. Remove starter upper retaining screw.
3. Remove steering column lever cover attaching screws.
4. Remove mast jacket lower bracket screw and the steering column upper mounting bracket.
5. Disconnect the four electrical connectors from steering column, then raise vehicle.
6. Disconnect steering column flexible coupling and position aside.
7. Disconnect electrical leads from starter and remove brace screw from bottom of starter.
8. Remove starter lower mounting screw.
9. With a suitable jack, raise engine approximately ½ inch to provide clearance for starter removal.
10. Lower starter and remove from vehicle.
11. Reverse procedure to install.

IGNITION LOCK, REPLACE

1. Perform steps 1 through 8 outlined under "Turn Signal, W/S Wiper, Dimmer Or Ignition Switches" procedure. However, do not completely remove turn signal switch.
2. Insert a small screwdriver or similar tool into turn signal housing slot. Keeping tool to right side of slot, break housing flash loose and at same time depress spring latch at lower end of lock cylinder.
3. Remove lock cylinder from housing.

LIGHT SWITCH, REPLACE

1. Disconnect ground cable at battery.
2. Pull headlamp switch knob to "ON" position, Fig. 4.
3. Reach under instrument panel and depress switch shaft retainer button while pulling on the switch control shaft knob.
4. With a large bladed screwdriver, remove the light switch ferrule nut from front of instrument panel.
5. Disconnect the multi-contact connector from side of switch and remove switch.
6. Reverse procedure to install.

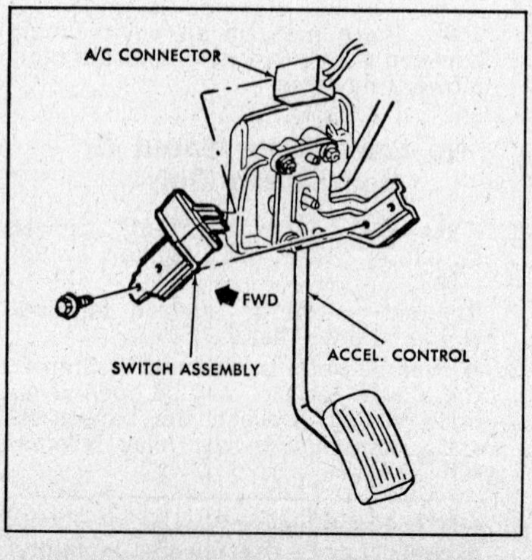

Fig. 7 A/C compressor cut-out switch

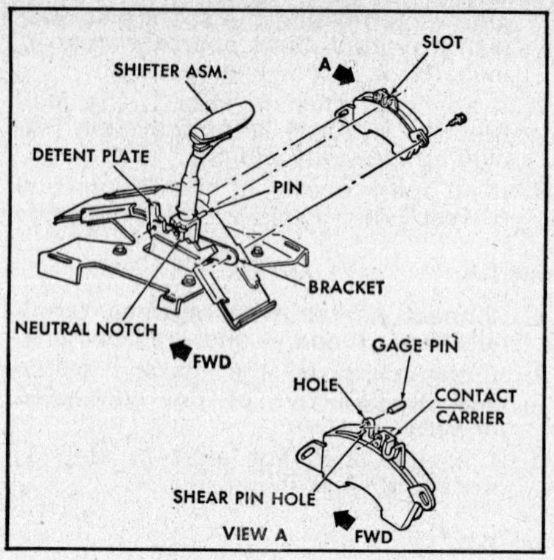

Fig. 6 Neutral start switch replacement

STOP LIGHT SWITCH

1. Reach under right side of instrument panel at brake pedal support and release wiring harness connector at switch.
2. Pull switch from mounting bracket.
3. When installing switch, adjust by bringing brake pedal to normal position. Electrical contact should be made when pedal is depressed .53 inch (13.5 mm). To adjust, the switch may be rotated or pulled in the clip.

CLUTCH START SWITCH

NOTE: The clutch pedal must be fully depressed and the ignition switch in START position for the vehicle to start.

The clutch switch assembly mounts with two tangs to the clutch pedal brace switch pivot bracket and the clutch pedal arm, Fig. 5.

1. Under the instrument panel on the clutch pedal support remove the multi-contact connector from switch.
2. Compress switch assembly actuating shaft barb retainer and push out of clutch pedal.
3. Compress switch assembly pivot bracket barb and lift off switch.
4. When installing new switch, no adjustments are necessary as the switch is self aligning.

NEUTRAL SAFETY SWITCH, REPLACE

1. Remove floor console cover.
2. Disconnect electrical plugs on back-up contacts and neutral start contacts of switch assembly, Fig. 6.
3. Place shift lever in Neutral.
4. Remove two screws securing switch to lever assembly.
5. When installing switch, make sure it is in Neutral position. When switch is installed, shifting out of Neutral will shear the switch plastic locating pin.

Key No.	Part Name
1	NUT, HEXAGON JAM
2	COVER ASSEMBLY, SHAFT LOCK
3	RING, RETAINING
4	LOCK, STEERING SHAFT
5	CAM ASSEMBLY, TURN SIGNAL CANCELLING
6	SPRING, UPPER BEARING
7	SCREW, PAN HEAD CROSS RECESS
8	SCREW, FLAT HEAD CROSS RECESS
9	SCREW, ROUND HEAD CROSS RECESS
10	ARM, PIVOT
11	SPACER, TURN SIGNAL SCREW
12	SWITCH ASSEMBLY, TURN SIGNAL
13	SWITCH ASEMBLY, PIVOT &
14	SCREW HEX WASHER HEAD TAPPING
15	WASHER, THRUST
16	SWITCH ASSEMBLY, BUZZER
17	CLIP, BUZZER SWITCH RETAINING
18	HOUSING, STEERING COLUMN
19	BEARING ASSEMBLY
20	RETAINER, BEARING
21	SECTOR, SWITCH ACTUATOR
22	SPRING, RACK PRELOAD
23	RACK, SWITCH ACTUATOR
24	BOLT ASSEMBLY, SPRING &
25	WASHER, SPRING THRUST
26	WASHER, WAVE
27	LEVER, KEY RELEASE
28	SPRING, LOCK INHIBITER
29	HOUSING ASSEMBLY, SHROUD &
30	SCREW, PAN HEAD CROSS RECESS
31	ROD, SWITCH ACTUATOR
32	ROD, DIMMER SWITCH ACTUATOR
33	SWITCH ASSEMBLY, DIMMER
34	SWITCH ASSEMBLY, IGNITION
35	SCREW, WASHER HEAD
36	JACKET ASSEMBLY, STEERING COLUMN
37	SEAL, STEERING SHAFT
38	BUSHING, STEERING COLUMN JACKET
39	SHAFT ASSEMBLY, STEERING
40	RETAINER

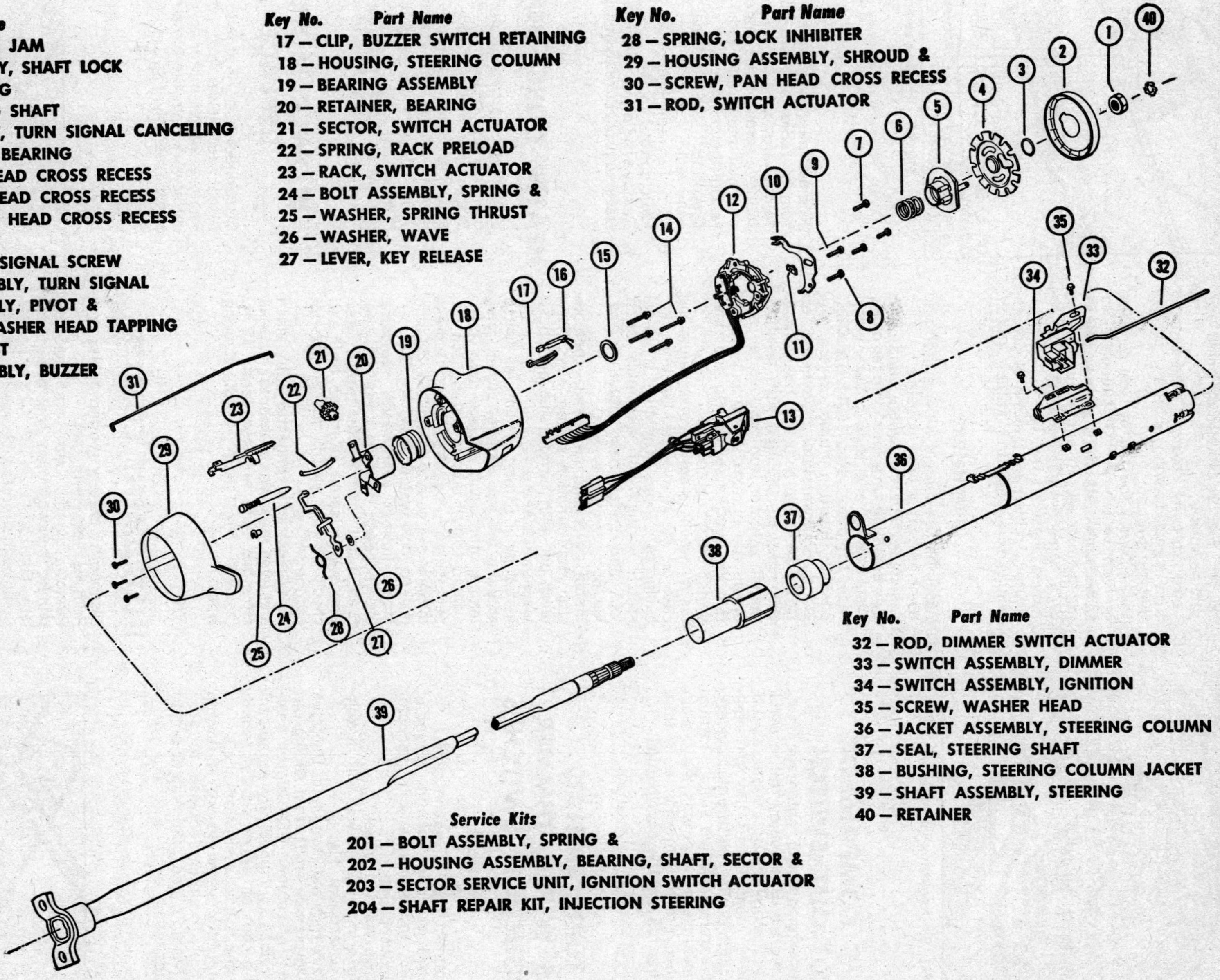

Service Kits

201 — BOLT ASSEMBLY, SPRING &
202 — HOUSING ASSEMBLY, BEARING, SHAFT, SECTOR &
203 — SECTOR SERVICE UNIT, IGNITION SWITCH ACTUATOR
204 — SHAFT REPAIR KIT, INJECTION STEERING

Fig. 8 Chevette steering column, disassembled

Fig. 9 Removing pivot arm

A/C COMPRESSOR CUT-OUT SWITCH

Vehicles equipped with automatic transmission and air conditioning utilize a full throttle A/C compressor cut-out switch, Fig. 7, which de-energizes the A/C compressor clutch during full throttle acceleration. A pressure sensitive switch, located in the transmission, overrides the cut-out switch when the transmission is in third gear during full throttle acceleration.

TURN SIGNAL, W/S WIPER, DIMMER OR IGNITION SWITCHES

Removal

1. Disconnect battery ground cable.
2. Remove steering column mounting bracket and unsnap switch connector from jacket.
3. Remove steering wheel.
4. Remove lock plate cover with a suitable screwdriver, Fig. 8.
5. Remove ring and lock plate. Use caution to prevent shaft from sliding out bottom of column. Slide upper bearing preload spring and turn signal cancelling cam off upper steering shaft, then the thrust washer off shaft.
6. Rotate turn signal lever-W/S switch assembly counter-clockwise to stop (Off position) and pull straight out to disengage.
7. Remove two screws, pivot arm and spacer, Fig. 9. Note that pivot arm retains spacer.
8. Lower steering column, then remove turn signal switch mounting screws and the switch.
9. Pull actuator rod to the stop to place ignition switch in "Off-Unlock". Remove upper attaching screw, releasing dimmer switch and actuator. The W/S wiper switch may now be removed, Figs. 10 and 11.
10. Remove remaining ignition switch retaining screw and the ignition switch.

Installation

1. Assemble windshield wiper switch and pivot assembly onto housing.
2. Assemble buzzer switch and lock cylinder, then turn lock cylinder clockwise to stop and then counter-clockwise to other stop (OFF-UNLOCK) position. Position ignition switch, Fig. 10, then move slider to extreme left (ACC) and slide back two positions to the right to OFF-UNLOCK position. Install actuator rod into slider and install the bottom screw only, to retain the ignition switch. Do not move switch out of detent.
3. Install washer, spring and cancelling cam on steering shaft. Position cancelling cam lobes in relation to signal switch springs and assemble shaft lock and install new retaining ring, Fig. 8.
4. Install cover and snap ring, then install multi-function switch lever. Align lever pin with switch slot and push lever until it seats.
5. Install pinched end of dimmer switch actuator rod into dimmer switch, then install other end of rod into pivot switch. Install but do not tighten upper ignition switch screw. Depress dimmer switch and insert a 3/32 inch drill to lock switch body, Fig. 12.
6. Move dimmer switch up, removing lash between both switches and rod, then install and tighten upper ignition switch screw. Remove drill and check dimmer switch for proper operation.
7. Snap electrical connector into place and raise steering column, then install mounting nuts and torque to 22 ft. lbs (30 Nm) and install steering wheel.

Fig. 11 Dimmer & ignition switch replacement

INSTRUMENT CLUSTER

1. Disconnect battery ground cable.
2. Remove clock stem knob.
3. Remove cluster bezel and lens retaining screws, then the bezel and lens, Fig. 13.
4. Remove instrument cluster to instrument panel retaining nuts and pull cluster toward vehicle rear.
5. Disconnect all electrical connectors and speedometer cable from cluster, then remove cluster.
6. Reverse procedure to install.

RADIO, REPLACE

1. Disconnect battery ground cable.

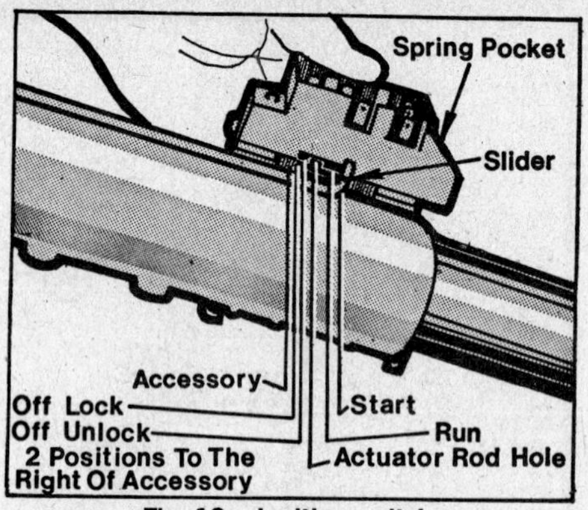

Fig. 10 Ignition switch

2. Remove mounting stud nut from bottom of radio, and the control knobs from shafts.
3. Remove screws from center trim panel and pull panel and radio toward the rear of the vehicle.
4. Disconnect all electrical connectors from radio.
5. Remove radio retaining nuts from radio control shafts.
6. Remove radio from vehicle.
7. Reverse procedure to install.

HEATER CORE, REMOVAL

1. Disconnect battery ground cable and drain cooling system.
2. Disconnect heater hoses from core and plug openings in core.
3. Remove heater core housing to dash panel attaching screws, then the housing.
4. Remove core from housing.

BLOWER MOTOR, REPLACE

1. Disconnect battery ground cable.
2. Disconnect blower lead wire.
3. Remove blower motor to case attaching screws then the blower and wheel as an assembly, Fig. 1.

Fig. 12 Dimmer switch alignment

NOTE: Scribe mark on blower motor and case so motor is installed in original position.

4. Remove nut and separate motor from wheel.

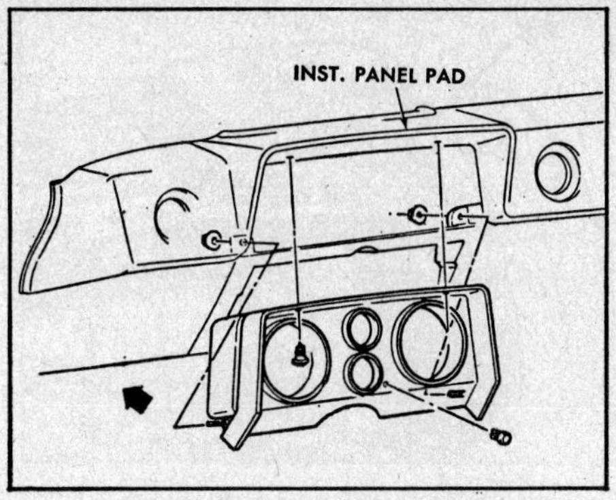

Fig. 13 Instrument cluster

5. Reverse procedure to install.

NOTE: At assembly position open end of blower wheel away from blower motor. Replace sealer at blower motor flange if necessary.

Engine Section

ENGINE MOUNT, REPLACE

Front

1. Remove heater assembly and position on top of engine.
2. Remove radiator upper support.
3. Remove engine mount nuts and the restraint cable.
4. Raise vehicle, install engine lifting device and raise engine to relieve weight from mounts.
5. Remove mount to engine bracket, then using tool J-25510, remove mount.

Rear

1. Raise vehicle and remove crossmember to mount bolts.
2. Raise transmission at extension housing to relieve weight from mount.
3. Remove mount to transmission bolt and the mount.

ENGINE, REPLACE

1. Remove hood.
2. Disconnect battery cables and remove clips securing battery cable to right side frame rail.
3. Drain cooling system and disconnect radiator and heater hoses.
4. Disconnect engine wiring harness.
5. Remove radiator upper support, radiator and fan.
6. Remove air cleaner.
7. Disconnect fuel line at rubber hose located along left side frame rail.
8. Disconnect accelerator and automatic transmission throttle valve linkage, if equipped.
9. Remove A/C compressor from mounting bracket and position aside, if equipped.
10. Raise vehicle and disconnect exhaust pipe at manifold.
11. Remove flywheel dust cover.
12. On models with automatic transmission, remove converter to flywheel bolts.
13. Remove converter housing to engine bolts on automatic transmission models or flywheel housing to engine bolt on manual transmission models, then lower vehicle.
14. Support transmission using a suitable jack.
15. Remove safety straps from engine mounts, then remove engine mount bolts.
16. Install engine lifting device, raise engine slowly, pull engine forward to clear transmission and remove engine from vehicle.

CYLINDER HEAD

1. Remove timing belt.
2. Drain cooling system, remove upper radiator hose and heater hose at intake manifold.
3. Remove air cleaner, then remove accelerator support bracket.
4. Disconnect spark plug wires.
5. Disconnect wiring harnesses at idle solenoid, choke, temperature sending switch and alternator.
6. Raise vehicle and disconnect exhaust pipe at manifold.
7. Lower vehicle and remove bolt retaining dipstick bracket to manifold.
8. Disconnect fuel line at carburetor.
9. Remove coil bracket bolts and position coil aside.
10. Remove camshaft cover, then remove camshaft cover to housing attaching studs.
11. Remove rocker arms, guides and lash adjusters.

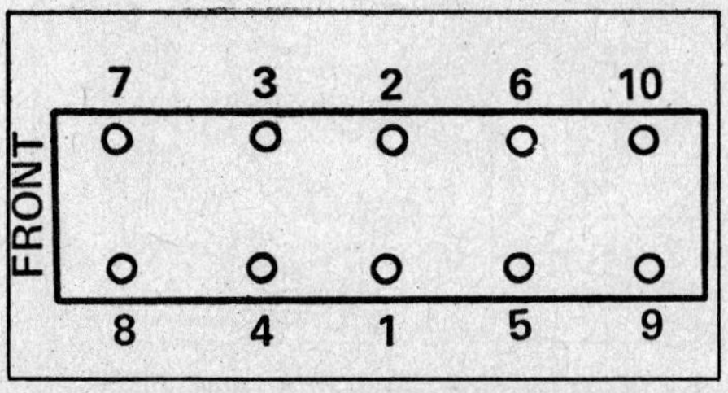

Fig. 1 Cylinder head tightening sequence

NOTE: Rocker arms, guides and lash adjusters must be installed in original location during assembly.

12. Remove camshaft carrier from cylinder head.

NOTE: It may be necessary to use a wedge to separate camshaft carrier from cylinder head.

13. Remove cylinder head and manifold as an assembly.
14. Reverse procedure to install, tighten cylinder bolts in the sequence as shown in Fig. 1.

HYDRAULIC VALVE LIFTERS

Failure of an hydraulic valve lifter is generally caused by an inadequate oil supply or dirt. An air leak at the intake side of the oil pump or too much oil in the engine will cause air bubbles in the oil supply to the lifters, causing them to collapse. This is a probable cause of trouble if several lifters fail to function, but air in the oil is an unlikely cause of failure of a single unit.

ROCKER ARMS

1. Remove camshaft cover.
2. Using tool J-25477, depress valve spring and remove rocker arm, guide and lash adjuster, Fig. 2.

NOTE: Rocker arms, guides and lash adjusters must be installed in the same location during assembly.

VALVE GUIDES

Valve guides are an integral part of the cylinder head. If stem to guide clearance is excessive, the guide should be reamed to the next oversize and the appropriate oversize valve installed. Valves are available in standard size and oversizes of .003 in. (.075mm), .006 in. (.150mm), and .012 in. (.300mm).

Fig. 2 Depressing valve spring

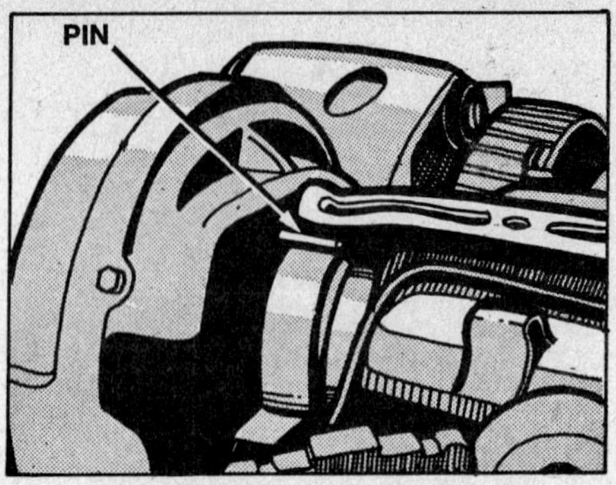

Fig. 3 Aligning camshaft sprocket

Fig. 4 Timing belt idler arm and pulley

CAMSHAFT COVER

1. Raise hood to fully open position.
2. Disconnect battery ground cable.
3. Remove air cleaner, PCV valve, air cleaner snorkle and heat tube assembly.
4. Remove spark plug wires from retainer on camshaft cover.
5. Remove accelerator cable support and position aside.
6. Remove nut and gasket from stud and the camshaft cover.

CAMSHAFT SPROCKET

1. Remove drive belts, then the fan and pulley.
2. Remove engine front cover.
3. Loosen idler pulley and remove timing belt from camshaft sprocket.
4. Remove camshaft sprocket bolt and washer, remove camshaft sprocket.

CAMSHAFT, REMOVAL

1. Remove camshaft sprocket as described under "Camshaft Sprocket" procedure.
2. Remove rocker arms.
3. Remove heater assembly and position aside.
4. Remove camshaft carrier rear cover.
5. Remove camshaft thrust plate bolts, slide camshaft rearward and remove thrust plate.
6. Raise engine, then carefully slide camshaft from carrier.

CAM LOBE LIFT SPECS.

Year	Intake	Exhaust
All 1976	.232 (5.893mm)	.232 (5.893mm)

VALVE TIMING

Intake Open Before TDC

Engine	Year	Degrees
All	1976	43⅓

ENGINE FRONT COVER

Upper Cover

1. Raise hood and disconnect battery ground cable.
2. Remove fan.
3. Remove upper cover retaining screws and the upper cover.

Lower Cover

1. Remove crankshaft pulley.
2. Remove upper front cover.
3. Remove lower cover attaching nut and the lower cover.

UPPER REAR TIMING BELT COVER, REMOVE

1. Remove engine front cover, timing belt and camshaft sprocket.
2. Remove three upper rear timing belt cover to camshaft carrier attaching screws.
3. Inspect camshaft seal and replace, if necessary.

TIMING BELT

1. Remove engine front upper and lower covers and crankshaft pulley.
2. Loosen idler pulley bolt and remove timing belt from camshaft and crankshaft sprockets.

Fig. 5 Checking timing belt tension

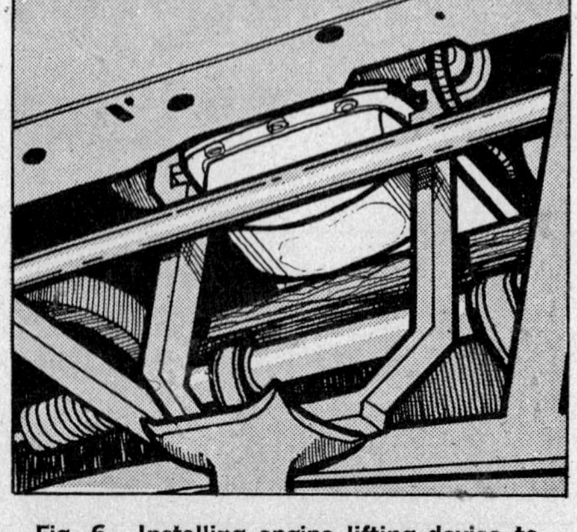

Fig. 6 Installing engine lifting device to facilitate oil pan removal

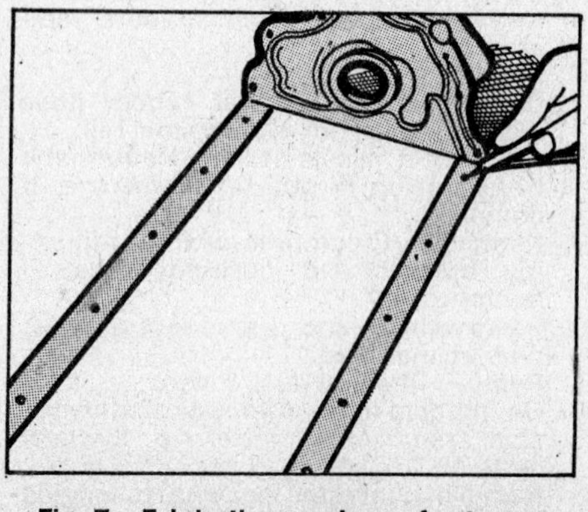

Fig. 7 Fabricating crankcase front cover to oil pan seal

3. Position timing belt over crankshaft sprocket, then install crankshaft pulley.
4. Position crankshaft at TDC number 1 cylinder.
5. Align timing mark on camshaft sprocket with hole in upper rear cover, Fig. 3.
6. Install timing belt on crankshaft and camshaft sprockets, then adjust belt tension.

TIMING BELT TENSION, ADJUST

1. Disconnect battery cables.
2. Remove radiator upper support brackets.
3. Remove A/C compressor drive belt and upper part of fan shroud, if equipped.
4. Remove fan belt, fan and pulley.
5. Remove engine front cover.
6. Rotate crankshaft clockwise at least 1 turn and position at TDC number 1 cylinder.
7. Loosen idler attaching bolt, then rotate idler pulley counterclockwise on its attaching bolt until slack is removed from belt, Fig. 4.
8. Apply sealer to idler attaching bolt and torque bolt to 15 ft. lbs., (21 N-m.).
9. Check belt tension using strand tension gauge. Belt tension should be 55 lbs. Fig. 5.

CRANKSHAFT SPROCKET

1. Remove engine front cover, crankshaft pulley and timing belt.
2. Remove crankshaft sprocket.

PISTONS, & RINGS

Pistons and rings are available in standard size and .030 in. (.750mm) oversize.

MAIN & ROD BEARINGS

Main bearings are available standard size and undersizes of .001 in. (.026mm), .002 in. (.050mm), .010 in. (.250mm) and .020 in. (.500mm).

Rod bearings are available in standard size and undersizes of .001 in. (.026mm), .010 in. (.250mm) and .020 in. (.500mm).

Fig. 8 Installing crankcase front cover

OIL PAN

1. Remove heater housing from dash and position on top of engine.
2. Remove motor mount nuts, then pull back motor mount restraint cables.
3. Remove radiator upper support, on models with A/C remove upper fan shroud.
4. Remove oil pan bolts.
5. Install engine lifting device and raise engine until oil pan can be removed, Fig. 6.

OIL PUMP

1. Remove coil bracket attaching bolts and position coil aside.
2. Remove fuel pump and push rod.
3. Remove distributor.
4. Remove oil pan, then remove oil pump screen and pipe assembly and oil pump.

CRANKSHAFT REAR OIL SEAL

1. Remove engine and place on stand.
2. Remove oil pan and rear main bearing cap.
3. Clean bearing cap and case. Inspect crankshaft seal for damage.
4. Install seal in case, ensure seal is fully seated.
5. Coat bearing cap split line with head bolt sealer.
6. Install rear main bearing cap.
7. Install sealer in bearing cap vertical grooves, remove any excess sealer.
8. Install oil pan, then install engine in vehicle.

CRANKCASE FRONT COVER

1. Remove engine upper and lower front covers, crankshaft pulley, timing belt and crankshaft sprocket.
2. Remove three crankcase front cover to oil pan attaching bolts, then the crankcase cover to engine attaching bolts.
3. Remove crankcase front cover, cover gasket and front portion of oil pan gasket.

NOTE: To fabricate a replacement crankcase front cover to oil pan gasket, position crankcase cover over a new oil pan gasket as shown in Fig. 7. When installing gasket apply sealer to cut off portion of gasket.

4. Inspect crankshaft oil seal, replace if necessary.
5. Install crankcase front cover using tool J-26434, Fig. 8. to 92 in. lbs. (10.4 N-m.).

BELT TENSION DATA

	New Lbs.	Used Lbs.
1976—		
Air Condition	120-150	—
A.I.R. Pump	70-100	—
Generator	70-100	—
Timing Belt	55	—

FUEL PUMP PRESSURE

Year	Engine	Pressure Lbs.
1976	All	5-6½

FUEL PUMP, REMOVAL

1. Raise vehicle and remove ignition coil.
2. Remove A/C compressor rear bracket, if equipped.
3. Remove fuel pump attaching bolts and fuel pump.

WATER PUMP, REPLACE

1. Disconnect battery ground cable.
2. Remove drive belts, fan and pulley.
3. Remove engine front cover.
4. Drain cooling system and disconnect radiator hose and heater hose from water pump.
5. Loosen idler pulley bolt and remove belt from camshaft pulley.
6. Remove water pump attaching bolts and the water pump.

CARBURETOR SECTION

MONOJET 1ME CARBURETOR ADJUSTMENT SPECIFICATIONS

See Tune Up Chart for hot idle speeds.

Year	Carb. Part No. ①	Float Level	Metering Rod	Choke Rod	Vacuum Break	Unloader	Choke Setting	Fast Idle R.P.M.
1976	17056030	5/32	.072	.065	.070	.165	3 Rich	2200
	17056031	5/32	.072	.065	.070	.165	3 Rich	2000
	17056032	5/32	.073	.045	.070	.200	3 Rich	2200
	17056033	5/32	.073	.045	.070	.200	3 Rich	2000
	17056034	5/32	.073	.045	.070	.200	3 Rich	2200
	17056035	5/32	.073	.045	.070	.200	3 Rich	2000
	17056036	5/32	.072	.065	.070	.165	3 Rich	2200
	17056037	5/32	.072	.065	.070	.165	3 Rich	2000
	17056330	5/32	.072	.065	.070	.165	3 Rich	2200
	17056331	5/32	.072	.065	.070	.165	3 Rich	2000
	17056332	5/32	.073	.045	.070	.200	3 Rich	2200
	17056333	5/32	.073	.045	.070	.200	3 Rich	2000
	17056334	5/32	.073	.045	.070	.200	3 Rich	2200
	17056335	5/32	.073	.045	.120	.200	3 Rich	2000

①—On tag attached to carburetor.

MONOJET 1ME CARBURETOR ADJUSTMENTS

Float Level Adjustment

Fig. 1

1. Hold float retaining pin firmly in place and float arm against top of float needle by pushing downward on float arm at point between needle seat and hinge pin as shown.
2. With adjustable T-scale, measure distance from top of float at toe to float bowl gasket surface (gasket removed). Measurement should be made at a point 1/16" in from end of flat surface at float toe (not on radius).
3. Bend float pontoon up or down at float arm junction to adjust.

Metering Rod Adjustment

Fig. 2

1. Remove metering rod by holding throttle valve wide open. Push downward on metering rod against spring tension, then slide metering rod out of slot in holder and remove from main metering jet.
2. To check adjustment, back out slow idle screw and rotate fast idle cam so that fast idle cam follower is not contacting steps on cam.
3. With throttle valve completely closed, apply pressure to top of power piston and hold piston down against its stop.
4. While holding downward pressure on power piston, swing metering rod holder over flat surface of bowl casting next to carburetor bore.
5. Use specified size drill and insert between bowl casting sealing bead and lower surface of metering rod holder. Drill should have a slide fit between both surfaces as shown.
6. To adjust, carefully bend metering rod holder up or down at point shown.
7. After adjustment, install metering rod.

Choke Coil Lever Adjustment

Fig. 3

1. With fast idle adjusting screw on fast idle cam high step, close choke valve.
2. Insert a .120 inch gauge pin through hole in lever and into casting. If holes do not align, bend link to adjust.

Fast Idle Adjustment

Fig. 4

1. Set normal engine idle speed.
2. Place fast idle cam follower tang on highest step of cam.
3. To adjust, insert screwdriver in slot provided in fast idle cam follower tang and bend inwards (towards cam) or outward to obtain specified dimension.

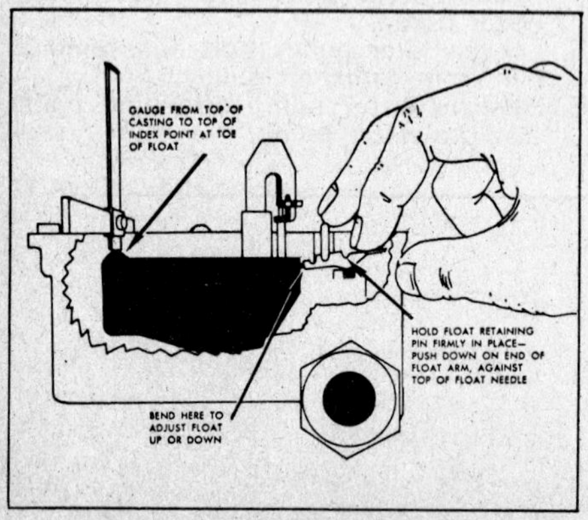

Fig. 1 Float level adjustment

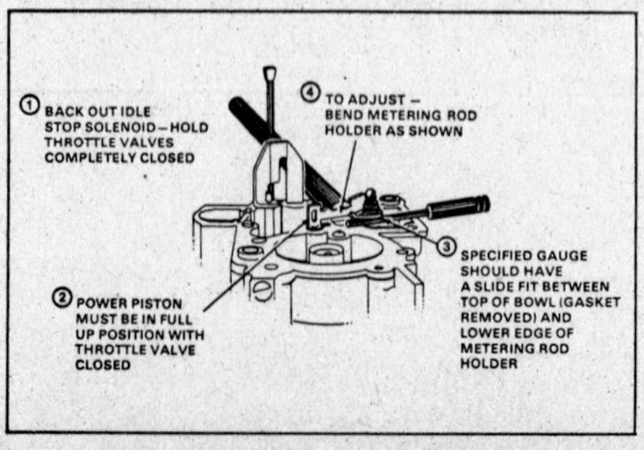

Fig. 2 Metering rod adjustment

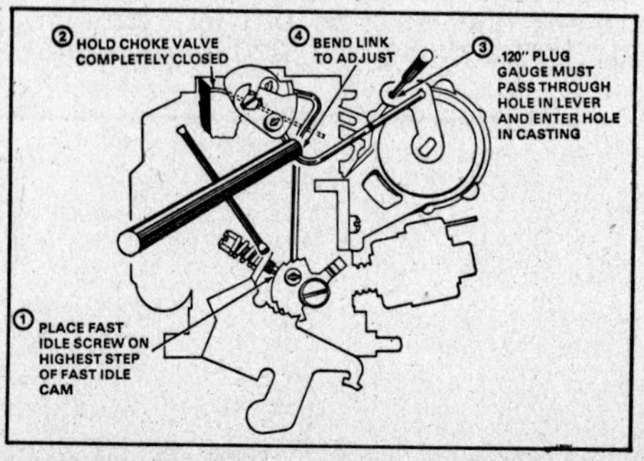

Fig. 3 Choke coil lever adjustment

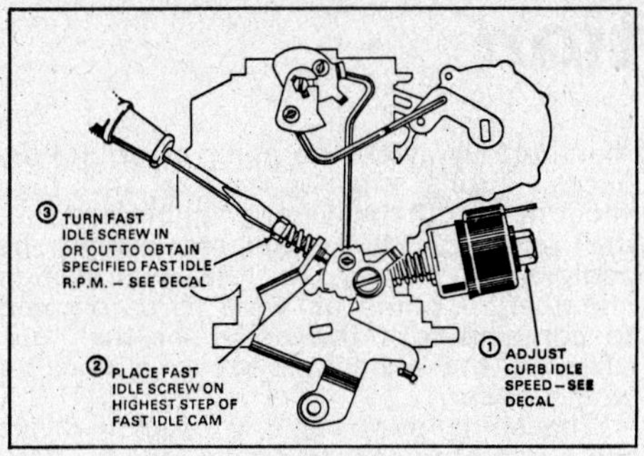

Fig. 4 Fast idle adjustment

Choke Rod Adjustment

Fig. 5

1. With fast idle adjustment made, place fast idle cam follower on second step of fast idle cam and hold firmly against the rise to the high step.
2. Rotate choke towards direction of closed choke by applying force to choke coil lever.
3. Bend choke rod at point shown to give specified opening between upper edge of choke valve (at center of valve) and inside air horn wall.

Vacuum Break Adjustment

Fig. 6

1. With fast idle screw on fast idle cam high step, seat diaphragm using an outside vacuum source. Cover purge hole with a piece of tape.
2. Close choke valve to compress plunger bucking spring and seat plunger stem.
3. Place specified gauge between upper edge of choke valve and air horn wall. Bend link to adjust.

Unloader Adjustment

Fig. 7

1. With throttle valve wide open, hold choke valve closed with specified gauge between upper edge of choke valve and air horn wall.
2. If distance is not within specifications, bend unloader tang to adjust.

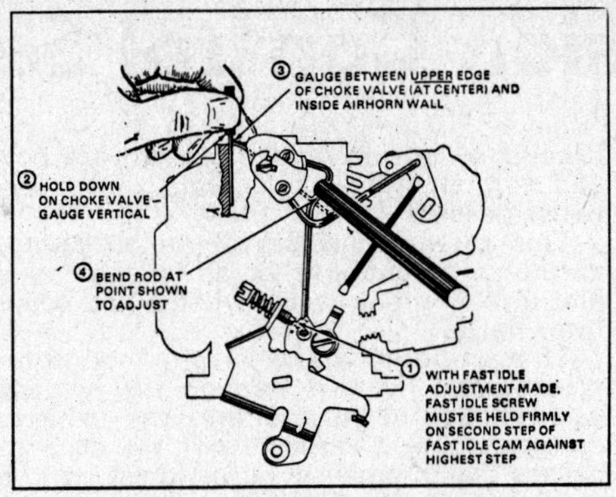

Fig. 5 Choke rod adjustment

Choke Coil Adjustment

Fig. 8

1. Install electric choke assembly into housing, ensuring coil tang contacts lower side of coil lever pick-up arm.
2. Place fast cam follower on high step of cam.
3. Rotate cover and coil assembly counter-clockwise until choke valve just closes, then align cover index point with the specified mark on housing.
4. Install and tighten cover retainers and screws.

NOTE: Do not install a choke cover gasket between the choke assembly and housing since the ground contact for the choke is provided by a metal plate at the rear of the choke assembly.

Low Idle & Curb Idle Speeds Adjustment

NOTE: On all models except vehicles equipped with automatic transmission and air conditioning, the idle solenoid wire will be color-coded tan. On vehicles equipped with automatic transmission and air conditioning, the solenoid wire is color-coded green with a double white stripe. The tan wire will be left disconnected.

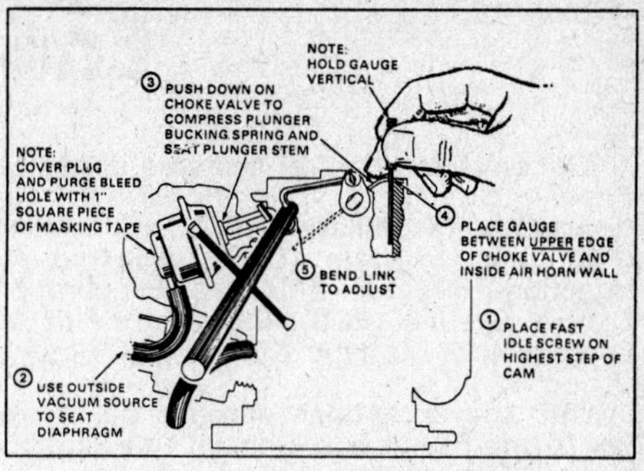

Fig. 6 Vacuum break adjustment

Fig. 9

1. The engine should be at normal operating temperature with the air cleaner on, choke open and air conditioning off. Apply parking brake and chock wheels.
2. Disconnect and plug PCV hose at vapor canister and the vacuum advance hose from distributor.
3. On all models except equipped with automatic transmission and air conditioning, rotate solenoid body to adjust curb idle speed. Then, disconnect solenoid electrical connector. With automatic transmission in Drive or manual transmission in Neutral, turn 1/8 inch hex screw in end of solenoid to adjust low idle speed. Connect solenoid electrical connector.
4. On models equipped with automatic transmission and air conditioning, turn solenoid hex screw inward until fully bottomed. Rotate solenoid body to obtain 950 RPM. Then, rotate hex screw to obtain curb idle RPM.
5. On all models, unplug and connect PCV vapor canister hose and vacuum advance hose.

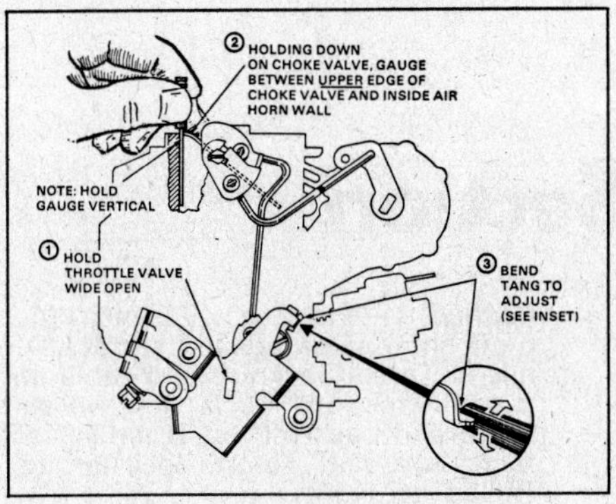

Fig. 7 Unloader adjustment

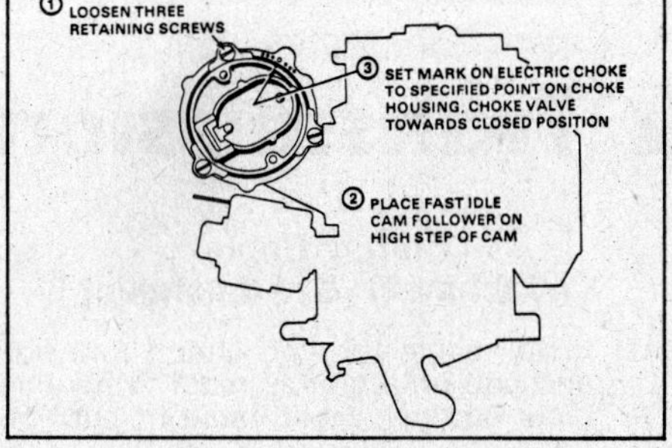

Fig. 8 Choke coil adjustment

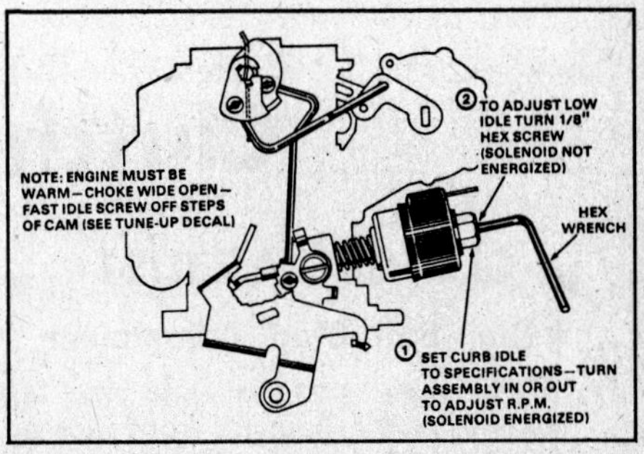

Fig. 9 Low & curb idle speeds adjustment

Emission Control Section

The emission control systems used are, Positive Crankcase Ventilation (PCV), Exhaust Gas Recirculation (EGR), Controlled Combustion System (CCS), Manifold Air Injection Reactor (MAIR), Evaporation Control System (ECS), Carburetor Hot Air System and Catalytic Converters, Figs. 1, 2 and 3.

With the exceptions outlined below, refer to the "Emission Control" section for service information.

MANIFOLD AIR INJECTION REACTOR (MAIR)

This system is basically the same as the Air Injection Reactor (AIR). It compresses, regulates and distributes fresh air to the exhaust system to reduce the hydrocarbon and carbon monoxide content of the exhaust gasses.

This system consists of an air pump, vacuum differential valve, air by-pass valve and a differential vacuum delay and separator valve.

The air pump draws in air, then compresses and feeds it through the by-pass valve, check valve and into the exhaust system, where it is mixed with the exhaust gasses. This process burns most of the unburned hydrocarbon and carbon monoxide, thereby reducing exhaust emissions.

The vacuum differential valve is used to momentarily divert the pump air to atmosphere, during engine overrun and high vacuum conditions. During engine overrun, the richer air/fuel mixtures cannot be completely burned in the combustion chamber. If pump air were to be allowed to combine with the richer air/fuel mixtures in the exhaust system, a backfire would occur.

The air by-pass valve is used to divert the air to atmosphere during low vacuum. conditions. This valve also diverts the pump air to atmosphere when there is a restriction in the MAIR system and pump air becomes excessive.

The differential vacuum delay and separator valve is used to control the amount of pump air during hard acceleration and low manifold vacuum. During these conditions, the air is normally vented to atmosphere. The differential vacuum delay and separator valve provides a one-way restriction which results in delaying the venting of pump air to atmosphere.

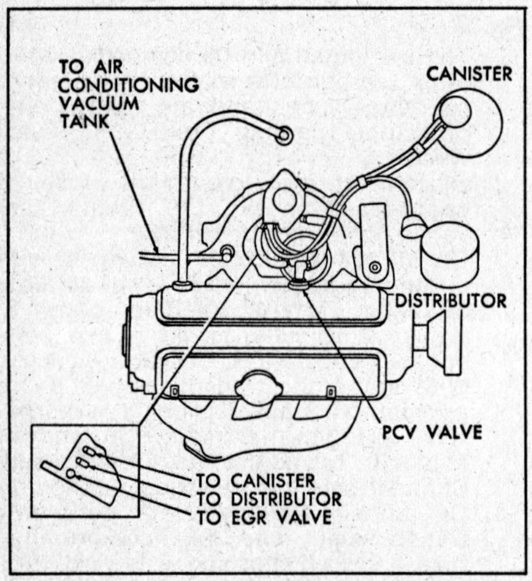

Fig. 1 Vacuum hose routing without MAIR System

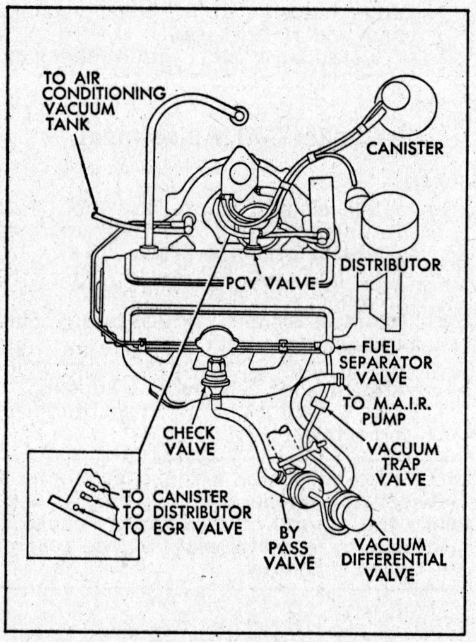

Fig. 2 Vacuum hose routing with MAIR system

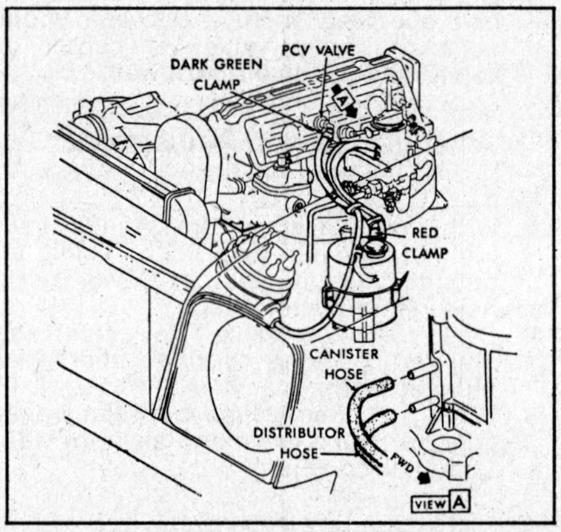

Fig. 3 Evaporation Emission Control system

Clutch & Transmission Section

CLUTCH, ADJUST

Initial Ball Stud Adjustment

1. Place gage J-23644 with flat end against clutch housing front face and the hooked end is positioned in the bottom depression of the clutch fork, Fig. 1.
2. Turn ball inward until clutch release bearing contacts clutch spring.
3. Install and torque lock nut to 25 ft. lbs. (33 N-m).
4. Remove gage.

Clutch Cable Attachment & Adjustment

1. Install cable through clutch fork hole and pull until clutch pedal firmly contacts bumper. Hold cable in position.
2. Push clutch fork forward until throwout bearing contacts clutch spring fingers. Then, thread nut on cable until bottomed against spherical surface of clutch fork.
3. To adjust clutch pedal free play, follow either of the two methods outlined below Fig. 2:

Method 1—Place a .171 inch (4.35 mm) shim stock against surface D of nut B. Thread locknut onto cable until it contacts shim stock. Remove shim stock and back off nut B until it contacts lock nut. Torque lock nut to 4 ft. lbs. (6 N-m).

Method 2—Rotate nut B 4.35 turns counter-clockwise. Thread lock nut on cable until it contacts nut B. Torque lock nut to 4 ft. lbs. (6 N-m).

4. Attach return spring.
5. Clutch pedal free travel should be .812 inch (20.6mm).

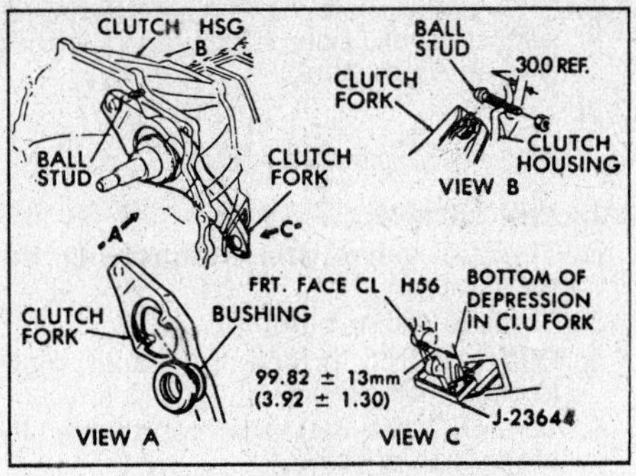

Fig. 1 Ball stud adjustment

Clutch Pedal Adjustment

1. Loosen cable ball stud lock nut, located on the clutch housing right side.
2. Adjust ball stud to obtain .812 inch (20.6mm) clutch pedal free travel.
3. Torque lock nut to 25 ft. lbs. (33 N-m).
4. Check adjustment.

CLUTCH, REPLACE

1. Raise vehicle and remove transmission as outlined under "Manual Transmission, Replace" procedure.
2. Remove release bearing from the clutch fork and sleeve by sliding lever off ball stud and against spring force. If ball stud is to be replaced, remove cap, lock nut and stud from housing.
3. Make sure alignment marks on clutch assembly and flywheel are distinguishable.
4. Loosen clutch cover to flywheel bolts one turn at a time until spring pressure is released, to avoid bending the clutch cover flange.
5. Support the pressure plate and cover assembly while removing bolts and clutch assembly.

NOTE: Do not disassemble the clutch cover, spring and pressure plate for repairs. If defective replace complete assembly.

6. Reverse procedure to install making sure to index alignment marks.

CAUTION: Check position of engine in front mounts and realign as necessary.

MANUAL TRANSMISSION, REPLACE

1. Remove shifter assembly.
2. Raise vehicle and drain transmission lubricant.
3. Remove propeller shaft.
4. Disconnect speedometer cable and the back-up lamp switch.
5. Disconnect return spring and clutch cable from clutch fork.
6. Remove transmission to crossmember bolts.
7. Remove exhaust manifold nuts and the catalytic converter to transmission bracket bolts.
8. Remove crossmember to frame bolts, then the crossmember from vehicle.
9. Remove dust cover.
10. Remove clutch housing to engine retaining bolts, slide transmission rearward and remove from vehicle.
11. Reverse procedure to install.

MANUAL TRANSMISSION, SERVICE

Refer to the "Chevrolet Monza, Buick Skyhawk, Olds. Starfire & Pontiac Sunbird" chapter under "1976 70mm (Brazil) 4 Speed Transmission" for service procedures.

TURBO HYDRA-MATIC 200 AUTOMATIC TRANSMISSION

Identification

This transmission may be identified by the following codes located on a blue tag attached to the right side of the transmission: BH, BR, BZ, CH, CN, CS, CQ, CU, CX, OZ.

Description

The Turbo Hydra-Matic 200 transmission is fully automatic and consists of a three element torque converter and a compound planetary gear set. Three multiple disc clutches, a roller clutch and a band provide the required friction elements to obtain the desired function of the planetary gear set.

TROUBLE SHOOTING GUIDE

No Drive in Drive Range

1. Low oil level.
2. Manual linkage maladjusted.
3. Low oil pressure due to:
   a. Restricted or plugged oil screen.
   b. Oil screen gasket improperly installed.
   c. Oil pump pressure regulator.
   d. Pump drive gear tangs damaged by converter.
   e. Case porosity in intake bore.
4. Forward clutch malfunctioning due to:
   a. Forward clutch not applying due to cracked piston, damaged or missing seals, burned clutch plates, snap ring not in groove.
   b. Forward clutch seal rings damaged or missing on turbine shaft, leaking feed circuits due to damaged or mispositioned gasket.
   c. Clutch housing check ball stuck or missing.
   d. Cup plug leaking or missing from rear of turbine shaft in clutch apply passage.
   e. Incorrect forward clutch piston assembly or incorrect number of clutch plates.

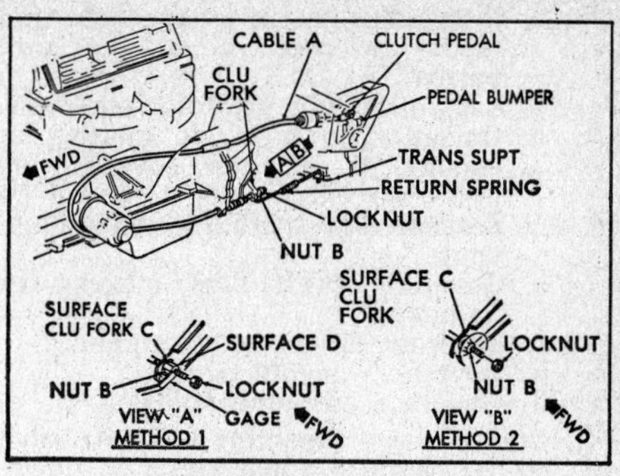

Fig. 2 Clutch cable adjustment

5. Roller clutch malfunctioning due to missing rollers or springs or possibly galled rollers.

Oil Pressure High Or Low

1. Throttle valve cable maladjusted, binding, disconnected or broken.
2. Throttle lever and bracket improperly installed, disconnected or binding.
3. Throttle valve shift valve, throttle valve or plunger binding.
4. Pressure regulator valve and spring malfunctioning due to:
   a. Binding valve.
   b. Incorrect spring.
   c. Oil pressure control orifice in pump cover plugged, causing high oil pressure.
   d. Pressure regulator bore plug leaking.
5. Manual valve disconnected.
6. Intermediate boost valve binding, causing oil pressures to be incorrect in 2nd and low ranges.
7. Orifice in spacer plate at end of intermediate boost valve plugged.
8. Reverse boost valve binding, causing pressure to be incorrect in reverse only.
9. Orifice in spacer plate at end of reverse boost valve plugged.

1-2 Shift At Full Throttle Only

1. Throttle valve cable maladjusted, binding, disconnected or broken.
2. Throttle lever and bracket assembly binding or disconnected.
3. Throttle valve exhaust ball lifter or number 5 check ball binding, mispositioned or disconnected.

NOTE: If number 5 ball is fully seated, it will cause full throttle valve pressure regardless of throttle valve position.

4. Throttle valve and plunger binding.
5. Valve body gaskets leaking, damaged or incorrectly installed.
6. Porous control valve assembly.

First Speed Only, No 1-2 Shift

1. Due to governor and governor feed passages:
   a. Plugged governor oil feed orifice in spacer plate.

b. Plugged orifice in spacer plate that feeds governor oil to the shift valves.
c. Balls missing in governor assembly.
d. Governor cover O-ring missing or leaking. If governor cover O-ring leaks, an external oil leak will be present and there will be no upshift.
e. Governor shaft seal missing or damaged.
f. Governor driven gear stripped.
g. Governor weights binding.
h. Governor assembly missing.

2. Control valve assembly 1-2 shift valve or 1-2 throttle valve stuck in downshift position.
3. Porosity in case channels or undrilled 2nd speed feed holes.
4. Excessive leakage between case bore and intermediate band apply ring.
5. Intermediate band anchor pin missing or disconnected from band.
6. Missing or broken intermediate band.
7. Due to intermediate servo assembly:
   a. Servo to cover oil seal ring damaged or missing.
   b. Porous servo cover or piston.
   c. Incorrect intermediate band apply pin.
   d. Incorrect cover and piston.

1st & 2nd Only, No 2-3, Shift

1. 2-3 shift valve or 2-3 throttle valve stuck in downshift position.
2. Direct clutch feed orifice in spacer plate plugged.
3. Valve body gaskets leaking, damaged or incorrectly installed.
4. Porosity between case passages.
5. Pump passages plugged or leaking.
6. Pump gasket incorrectly installed.
7. Rear seal on pump cover leaking or missing.
8. Direct clutch oil seals missing or damaged.
9. Direct clutch piston or housing cracked.
10. Direct clutch plates damaged or missing.
11. Direct clutch backing plate snap ring out of groove.
12. Intermediate servo to case oil seal broken or missing on intermediate servo piston.
13. Intermediate servo exhaust hole in case between servo piston seals plugged or undrilled.

Moves Forward In Neutral

1. Manual linkage maladjusted.
2. Forward clutch does not release.
3. Cross leakage between pump passages.
4. Cross leakage to forward clutch through clutch passages.

No Drive in Reverse or Slips in Reverse

1. Throttle valve cable binding or maladjusted.
2. Manual linkage maladjusted.
3. Throttle valve binding.
4. Reverse boost valve binding in bore.
5. Low overrun clutch valve binding in bore.
6. Reverse clutch piston cracked, broken or has missing seals.
7. Reverse clutch plates burned.
8. Reverse clutch has incorrect selective spacer ring.
9. Porosity in passages to direct clutch.
10. Pump to case gasket improperly installed or missing.
11. Pump passages cross leaking or restricted.
12. Pump cover seals damaged or missing.
13. Direct clutch piston or housing cracked.
14. Direct clutch piston seals cut or missing.
15. Direct clutch housing ball check, stuck, leaking or missing.
16. Direct clutch plates burned.
17. Incorrect direct clutch piston.
18. Direct clutch orifices plugged in spacer plate.
19. Intermediate servo to case seal cut or missing.

Slips 1-2 Shift

1. Aerated oil due to low level.
2. 2nd speed feed orifice in spacer plate partially blocked.
3. Improperly installed or missing spacer plate gasket.
4. 1-2 accumulator valve stuck, causing low 1-2 accumulator pressure.
5. Weak or missing 1-2 accumulator valve spring.
6. 1-2 accumulator piston seal leaking or spring missing or broken.
7. Leakage between 1-2 accumulator piston and pin.
8. Incorrect intermediate band apply pin.
9. Excessive leakage between intermediate band apply pin and case.
10. Porous intermediate servo piston.
11. Servo cover to servo seal damaged or missing.
12. Incorrect servo and cover.
13. Throttle valve cable improperly adjusted.
14. Shift throttle valve or throttle valve binding.
15. Intermediate band worn or burned.
16. Case porosity in 2nd clutch passages.

Rough 1-2 Shift

1. Throttle valve cable improperly adjusted or binding.
2. Throttle valve or plunger binding.
3. Shift throttle or 1-2 accumulator valve binding.
4. Incorrect intermediate servo pin.
5. Intermediate servo piston to case seal damaged or missing.
6. 1-2 accumulator oil ring damaged, piston stuck, bore damaged or spring broken or missing.

Slips 2-3 Shift

1. Low oil level.
2. Throttle valve cable improperly adjusted.
3. Throttle valve binding.
4. Direct clutch orifice in spacer plate partially blocked.
5. Spacer plate gaskets improperly installed or missing.
6. Intermediate servo to case seal damaged.
7. Porous direct clutch feed passages in case.
8. Pump to case gasket improperly installed or missing.
9. Pump passages cross feeding, leaking or restricted.
10. Pump cover oil seal rings damaged or missing.
11. Direct clutch piston or housing cracked.
12. Direct clutch piston seals cut or missing.
13. Direct clutch plates burned.

Rough 2-3 Shift

1. Throttle valve cable improperly installed or missing.
2. Throttle valve or throttle valve plunger binding.
3. Shift throttle valve binding.
4. Intermediate servo exhaust hole undrilled or plugged between intermediate servo piston seals.
5. Direct clutch exhaust valve number 4 check ball missing or improperly installed.

No Engine Braking In 2nd Speed

1. Intermediate boost valve binding in valve body.
2. Intermediate-Reverse number 3 check ball improperly installed or missing.
3. Shift throttle valve number 3 check ball improperly installed or missing.
4. Intermediate servo to cover seal missing or damaged.
5. Intermediate band off anchor pin, broken or burned.

No Engine Braking In 1st Speed

1. Low overrun clutch valve binding in valve body.

NOTE: The following conditions will also cause no reverse.

2. Low-reverse clutch piston seals broken or missing.
3. Porosity in low-reverse piston or housing.
4. Low-reverse clutch housing snap ring out of case.
5. Cup plug or rubber seal missing or damaged between case and low-reverse clutch housing.

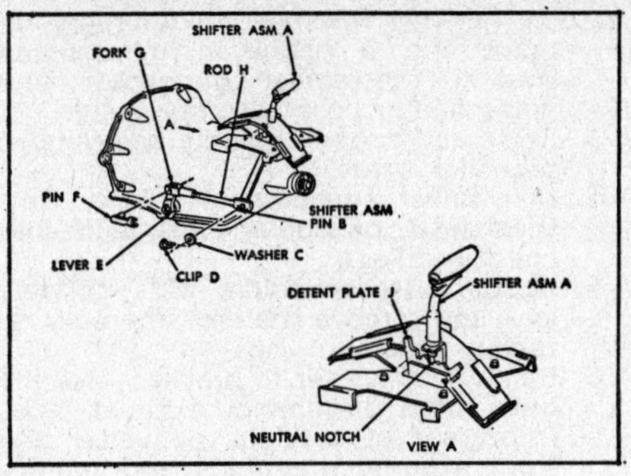

Fig. 3 Manual linkage adjust

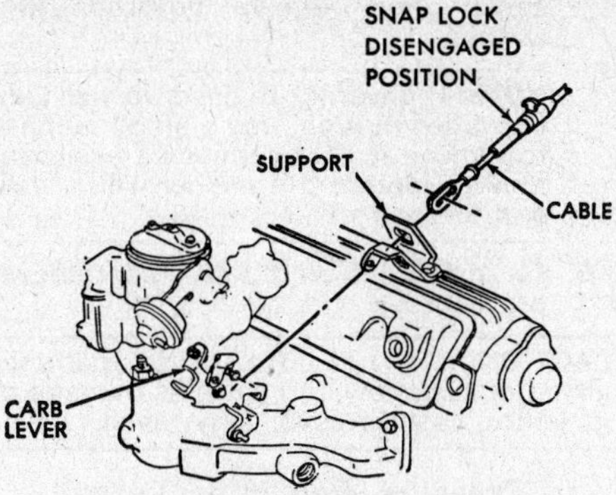

Fig. 4 Detent downshift cable adjust

Fig. 5 Removing or installing pressure regulator

No Part Throttle Downshift

1. Throttle plunger bushing passages obstructed.
2. 2-3 throttle valve bushing passages obstructed.
3. Valve body gaskets improperly installed or damaged.
4. Spacer plate hole obstructed or undrilled.
5. Throttle valve cable maladjusted.
6. Throttle valve or shift throttle valve binding.

Low or High Shift Points

1. Throttle valve cable binding or disconnected.
2. Throttle valve or shift throttle valve binding.
3. Number 1 throttle shift check ball improperly installed or missing.
4. Throttle valve plunger, 1-2 or 2-3 throttle valves binding.
5. Valve body gaskets improperly installed or missing.
6. Pressure regulator valve binding.
7. Throttle valve exhaust number 5 check ball and lifter, improperly installed, disconnected or missing.
8. Throttle lever binding, disconnected or loose at valve body mounting bolt or not positioned at the throttle valve plunger bushing pin locator.
9. Governor shaft to cover seal broken or missing.
10. Governor cover O-rings broken or missing.

NOTE: Outer ring will leak externally and the inner ring will leak internally.

11. Case porosity.

Will Not Hold In Park

1. Manual linkage maladjusted.
2. Parking pawl binding in case.
3. Actuator rod or plunger damaged.
4. Parking pawl damaged.
5. Parking bracket loose or damaged.
6. Detent lever nut loose.
7. Detent lever hole worn or damaged.
8. Detent roller to valve body bolt loose.
9. Detent roller or pin damaged, incorrectly installed or missing.

MAINTENANCE

To check fluid, drive vehicle for at least 15 minutes to bring fluid to operating temperature (200° F). With vehicle on a level surface and engine idling in Park and parking brake applied, the level on the dipstick should be at the "F" mark. To bring the fluid level from the ADD mark to the FULL mark requires 1 pint of fluid. If vehicle cannot be driven sufficiently to bring fluid to operating temperature, the level on the dipstick should be between the two dimples on the dipstick with fluid temperature at 70° F.

If additional fluid is required, use only Dexron or Dexron II automatic transmission fluid.

NOTE: An early change to a darker color from the usual red color and or a strong odor that is usually associated with overheated fluid is normal and should not be considered as a positive sign of required maintenance or unit failure.

CAUTION: When adding fluid, do not overfill, as foaming and loss of fluid through the vent may occur as the fluid heats up. Also, if fluid level is too low, complete loss of drive may occur especially when cold, which can cause transmission failure.

Every 60,000 miles, the oil should be drained, the oil pan removed, the screen cleaned and fresh fluid added. For vehicles subjected to more severe use such as heavy city traffic especially in hot weather, prolonged periods of idling or towing, this maintenance should be performed every 15,000 miles.

Draining Bottom Pan

1. Remove front and side oil pan attaching bolts, then loosen the rear oil pan attaching bolts.
2. Carefully pry oil pan loose and allow fluid to drain into a suitable container.
3. Remove the oil pan and gasket, then remove the screen attaching bolts and remove screen.
4. Thoroughly clean oil screen and oil pan with solvent.
5. Install oil screen using a new gasket and torque attaching bolts to 6-10 ft. lbs., then install oil pan using a new gasket and torque attaching bolts to 10-13 bolts.
6. Add 3 quarts of fluid, then with engine idling and parking brake applied, move selector lever through each range and return selector lever to PARK.
7. Check fluid level and add fluid as required to bring level between the two dimples on the dipstick.

Adding Fluid To Dry Transmission and Converter

1. Add 4½ quarts of fluid.
2. With transmission in PARK and parking brake applied, start engine and place carburetor on fast idle cam.
3. Move shifter lever through each range, then with transmission in PARK, add additional fluid as required to bring the level between the two dimples on the dipstick.

Manual Linkage, Adjust

1. Place shift lever (A) in neutral position, Fig. 3.
2. Move lever (E) clockwise to maximum detent to PARK, then move lever counterclockwise two detent positions to NEUTRAL.
3. With lever (E) in NEUTRAL, insert pin (F) on fork (G). Adjust rod (H) until hole in rod aligns with shifter assembly pin (B) and install rod on pin.

Detent Downshift Cable Adjustment

1. Disconnect cable from support, Fig. 4, and check that cable is free to slide through snap lock.
2. With cable installed in support and connected to transmission and carburetor lever, move throttle valve to wide open position.
3. Push snap lock until flush with support and throttle valve.

IN VEHICLE REPAIRS

Valve Body Assembly

1. Drain transmission fluid, then remove oil pan and screen.
2. Remove detent cable retaining bolt and disconnect cable.
3. Remove throttle lever and bracket assembly. Use care to avoid bending throttle lever link.
4. Remove detent roller and spring assembly.

5. Support valve body and remove retaining bolts, then while holding manual valve, remove valve assembly, spacer plate and gaskets as an assembly to prevent dropping the five check balls.

NOTE: After removing valve body assembly, the intermediate band anchor band pin, and reverse cup plug may be removed.

6. To install control valve, reverse removal procedure and torque all valve body bolts to 8 ft. lbs.

CAUTION: Assure that intermediate band anchor pin is located on intermediate band prior to installation of valve body, as damage will result.

Governor

1. Disconnect battery ground cable and remove air cleaner.
2. On vehicles with air conditioning, remove the five heater core cover screws, then disconnect the electrical connectors and position heater core aside.
3. Disconnect exhaust pipe and allow to hang down.
4. Support transmission, then remove transmission rear support bolts and propeller shaft and lower transmission until enough clearance is obtained to remove governor.
5. Remove governor retainer ring and cover, then remove governor and washer.

NOTE: If governor to case washer falls into transmission, use a small magnet to remove it. If it cannot be easily removed, replace the washer with a new one.

6. To install governor, reverse removal procedure.

CAUTION: Do not attempt to hammer governor assembly into case, as damage to governor, case or cover may result.

Pressure Regulator Valve

1. Drain transmission fluid, then remove oil pan and screen.
2. Using a small screwdriver or tool J-24684, Fig. 5, compress regulator spring.
3. Remove retaining ring and slowly release spring tension.
4. Remove pressure regulator bore plug, valve, spring and guide.
5. To assemble, install pressure regulator spring, guide and valve with stem end first and bore plug with hole side out.
6. Using a small screwdriver or tool J-24684, Fig. 5, compress regulator spring and install retaining ring.

TRANSMISSION REPLACE

1. Disconnect battery ground cable then disconnect detent downshift cable from bracket.
2. Remove air cleaner and dipstick, then on vehicles with air conditioning, remove the 5 heater core retaining screws, disconnect connector and place heater core assembly aside.
3. Raise and support vehicle and remove propeller shaft.
4. Disconnect speedometer cable, electrical lead, oil cooler lines, and shift control linkage.
5. Support transmission with suitable jack and remove the crossmember retaining bolts.
6. Remove converter to bracket retaining nuts, then disconnect exhaust pipes from rear of catalytic converter and from exhaust manifolds and remove catalytic converter and converter bracket as an assembly.
7. Remove converter dust shield and remove converter to flywheel bolts.
8. Lower transmission until it is barely supported by jack and remove transmission to engine bolts.
9. Raise transmission to its normal position, then support engine with jack and lower and remove transmission from vehicle.

NOTE: Use converter holding tool J-5384, or keep rear of transmission lower than the front to prevent the converter from sliding out.

10. Before installing transmission, place two inch blocks between rack and pinion housing assembly and oil pan to permit correct alignment of engine and transmission.
11. Reverse removal procedure to install and torque converter to flywheel bolts to 30-40 ft. lbs. (40-54 N-m).

Rear Axle, Propeller Shaft & Brakes

REAR AXLE

Description

The rear axle, Fig. 1, is a semi-floating type consisting of a cast carrier and large bosses on each end into which two welded steel tubes are fitted. The carrier contains an overhung hypoid pinion and ring gear. The differential is a two pinion arrangement.

The overhung hypoid drive pinion, is supported by two preloaded tapered roller bearings. The pinion shaft is sealed by means of a molded, spring loaded, rubber seal. The seal is mounted on the pinion shaft flange which is splined and bolts to the hypoid pinion shaft.

The ring gear is bolted to a one piece differential case and is supported by two preloaded tapered roller bearings.

A rear axle extension housing is bolted to the axle housing and is attached to the underbody by a center bearing support. An extension shaft inside the housing is splined to the drive pinion at the rear end and to the companion flange at the forward end.

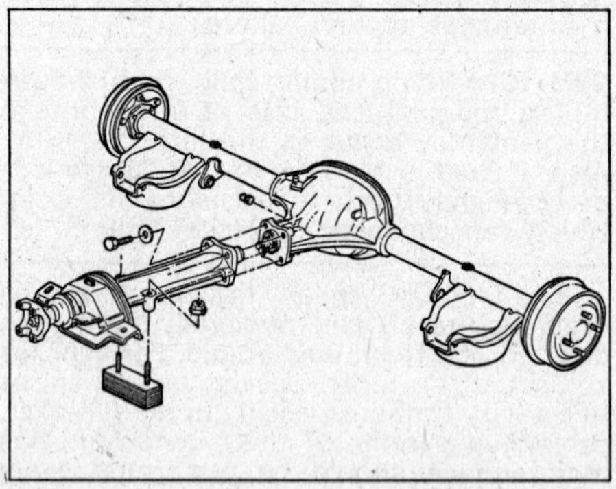

Fig. 1 Rear axle assembly

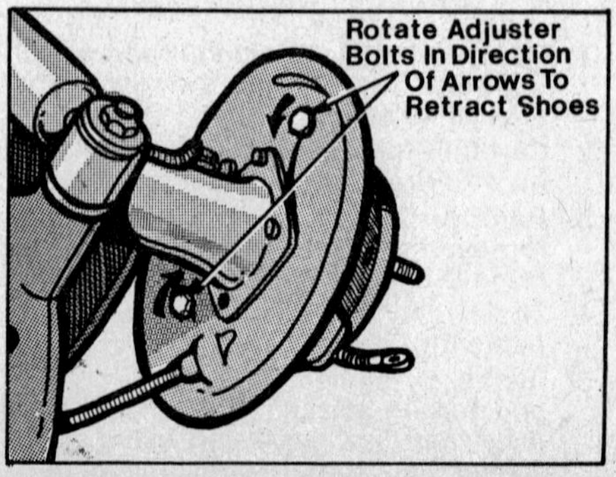

Fig. 2 Retracting brake shoes

Removal

1. Raise vehicle on a hoist.
2. Place adjustable lifting device under axle.
3. Disconnect rear shock absorbers from axle and remove propeller shaft and extension housing.
4. Remove both rear wheels.
5. Retract shoes and remove right and left brake drums, Fig. 2.
6. Disconnect brake lines from clips on axle tubes.
7. Disconnect track and stabilizer bars from axle tube.
8. Remove differential cover and drain lubricant.
9. Unscrew differential lock screw, remove pinion shaft and axle shaft "C" locks. Reinstall pinion shaft and tighten lock screw to retain differential gears.
10. Remove both axle shafts.
11. Remove brake backing plate retaining nuts and remove backing plates, with shoes and brake lines attached, and wire to frame.
12. Remove right and left lower control arm pivot bolts at axle.
13. Lower axle assembly slowly until coil spring tension is released, then remove axle.

AXLE SHAFT

1. Raise vehicle on a hoist and remove wheel and tire assembly and brake drum.
2. Drain lubricant from axle by removing carrier cover.
3. Unscrew pinion shaft lock screw and remove pinion shaft.
4. Push flanged end of exle shaft toward center of car and remove "C" lock from button end of shaft.

5. Remove axle shaft from housing being careful not to damage seal.

Oil Seal &/or Bearing Replacement

1. If replacing seal only, remove the seal by using the button end of axle shaft. Insert the button end of shaft behind the steel case of the seal and pry seal out of bore being careful not to damage housing.
2. If replacing bearings, insert tool J-22813 into bore so tool head grasps behind bearing. Slide washer against seal, or bearing, and turn nut against washer. Attach slide hammer J-2619 and remove bearing.
3. Pack cavity between seal lips with a high melting point wheel bearing lubricant. Position seal on tool J-21491 and position seal in axle housing bore, tap seal in bore just below end of housing.

REAR AXLE EXTENSION

Replace

1. Raise vehicle and support rear axle.
2. Disconnect propeller shaft from the rear yoke and remove from transmission, Fig. 1.
3. Support front of rear axle carrier housing. Ensure the rear axle extension is also supported.
4. Disconnect center support bracket from underbody and the extension housing flange from axle housing.
5. Remove axle extension housing from vehicle. Pry extension housing from axle housing with a suitable screwdriver, if necessary.
6. Reverse procedure to install.

Service

1. Remove bolt securing extension housing to center support bracket.
2. Mount companion flange in a vise and loosen lock nut.
3. Drive splined companion flange off shaft by tapping on lock nut end of shaft.
4. Remove lock nut, companion flange and thrust washer, then pull shaft from housing. If centering bearing, located in the rubber cushion, remains on extension shaft, drive off shaft with a suitable drift.

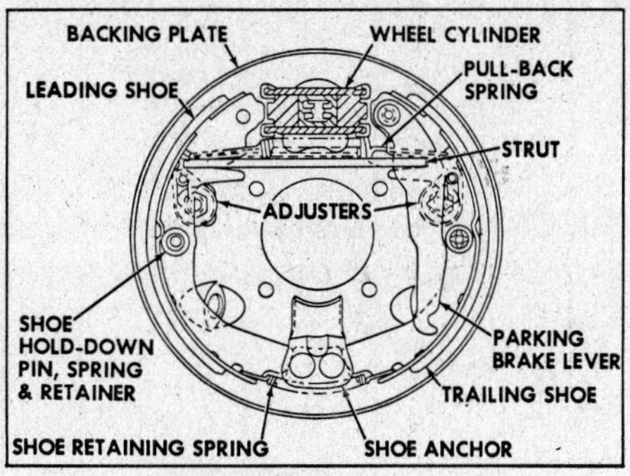

Fig. 3 Rear drum brake

5. Note position of bearing in rubber cushion and the cushion in center support. Then, using a suitable screwdriver, separate rubber cushion from center support bracket and remove bearing from cushion.
6. Clean, inspect and replace components, if necessary.
7. Install rubber cushion into center support bracket and place center support assembly over extension housing.
8. Press bearing onto extension shaft and insert shaft into housing through rubber cushion. Install thrust washer with circular cavity facing toward bearing, then the companion flange, using the lock nut to press flange onto shaft splines.
9. Install bolts retaining bracket to housing.

PROPELLER SHAFT

1. Raise vehicle on a hoist. Mark relationship of shaft to companion flange and disconnect the rear universal joint by removing trunnion bearing "U" bolts. Tape bearing cups to trunnion to prevent dropping and loss of bearing rollers.
2. Withdraw propeller shaft front yoke from transmission.
3. When installing, be sure to align marks made in removal to prevent driveline vibration.

BRAKE ADJUSTMENTS

Disc brakes are used on front wheels and drum brakes are used on rear wheels. Rear brake adjustment is automatic. Adjustment takes place whenever brakes are applied. An adjuster is attached to each brake shoe by means of a pin, Fig. 3. This pin is smaller than the slot in the brake shoe. When brakes are applied and brake shoes move outward the automatic adjuster follows. When brakes are released the brake shoe moves inward until it contacts the automatic adjuster pin. The space between adjuster pin and slot in brake shoe provides shoe to drum clearance.

NOTE: If manual adjustment or shoe retraction is necessary refer to Fig. 2.

PARKING BRAKE, ADJUST

1. Raise vehicle.
2. Apply parking brake one notch from fully released position.
3. Tighten equalizer adjusting nut until a light drag is felt when rear wheels are rotated in a forward direction.
4. Release parking brake and rotate rear wheels. No drag should be felt.

MASTER CYLINDER, REPLACE

1. Disconnect push rod from brake pedal and remove push rod boot.
2. Remove air cleaner.
3. Disconnect brake lines from master cylinder. Cap ends of lines to prevent entry of dirt.
4. Remove master cylinder to dash attaching nuts and the master cylinder.

FRONT DISC BRAKE SERVICE

This single piston sliding caliper assembly, Figs. 1 and 2, incorporates a one piece housing with the inboard side of the housing bored for the piston. A seal within the housing bore provides a hydraulic seal between the piston and housing wall.

A spring steel scraper (wear sensor) is incorporated on each inboard shoe. When shoe lining has worn to within .030 inch of the shoe, the sensor scrapes the rotor and emits an audible high frequency sound indicating that the linings should be replaced.

The caliper assembly slides on a mounting sleeve which is secured by a mounting bolt. Upon brake application fluid pressure against the piston forces the inboard shoe against the inboard side of the rotor. This action causes the caliper to slide until the outboard shoe comes in contact with the rotor.

Caliper Removal

1. Siphon brake fluid from master cylinder to bring level to ⅓ full, discard brake fluid removed.

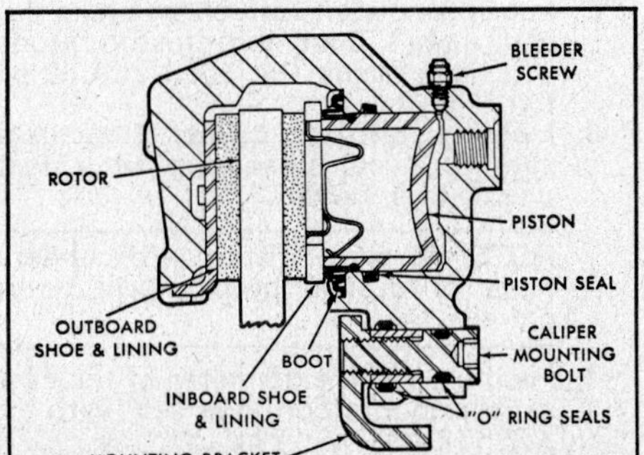

Fig. 1 Single piston disc brake assembly

2. Raise vehicle and remove wheel and tire assembly.
3. Install a 7 inch C-clamp on caliper with solid end of clamp on caliper housing and screw end on metal portion of outboard brake shoe. Tighten clamp until piston bottoms in caliper bore, then remove clamp.
4. Disconnect brake hose from caliper and remove copper gaskets, cap end of brake hose.

NOTE: If only brake shoes are to be replaced do not disconnect brake hose.

5. Remove the two mounting bracket to steering knuckle bolts, Fig. 3.

NOTE: Do not remove socket head retaining bolt. Support caliper when removing second bolt to prevent caliper from falling.

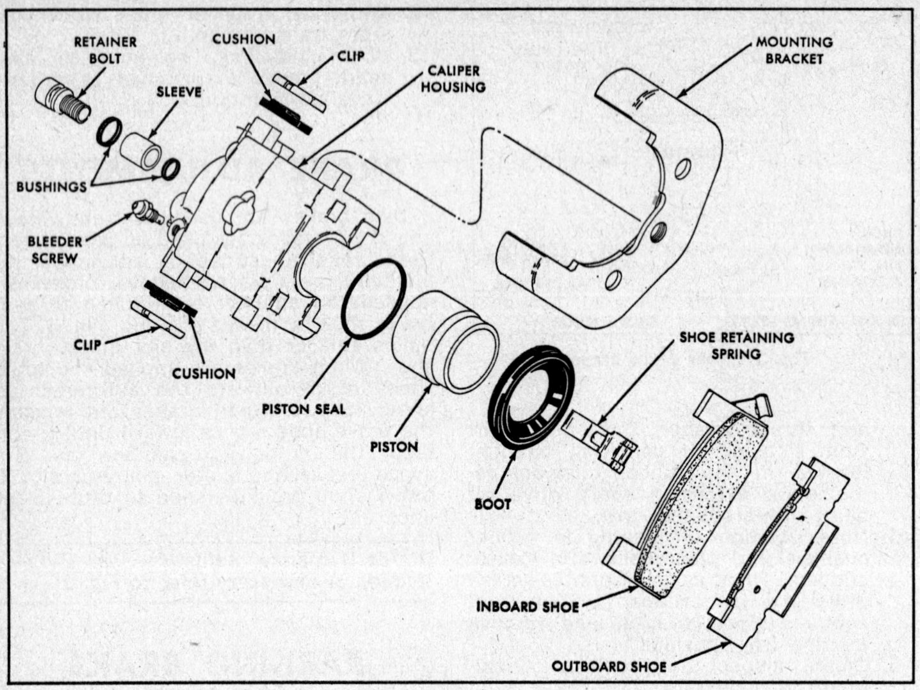

Fig. 2 Single piston caliper disassembled

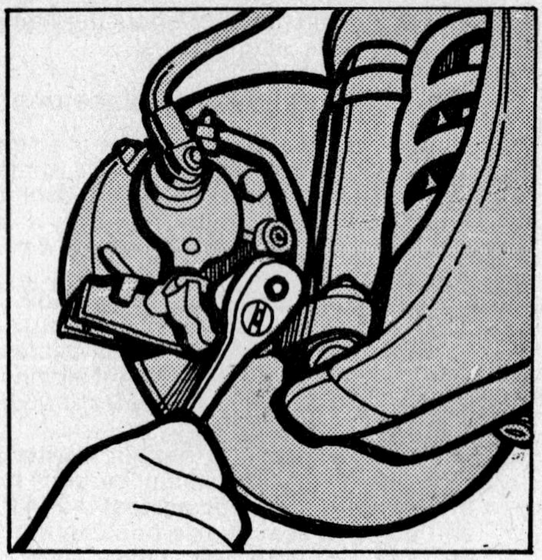

Fig. 3 Removing mounting bracket from steering knuckle

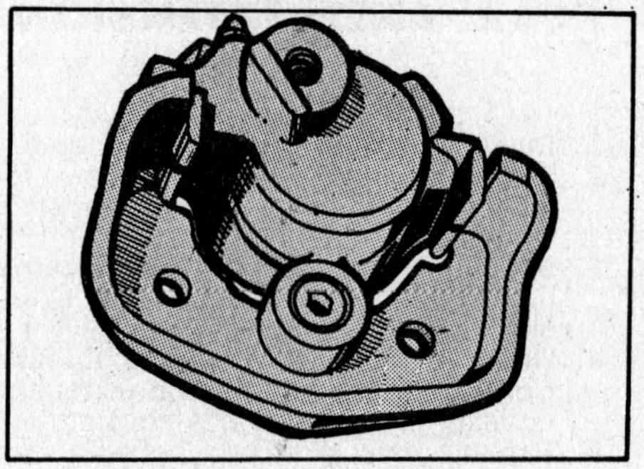

Fig. 4 Bracket assembly installed on caliper

6. Slide caliper from rotor.

NOTE: If only brake shoes are to be replaced support caliper from suspension using wire. Do not stretch or kink brake hose.

Brake Shoe Removal

1. Remove caliper as described under "Caliper Removal".
2. Remove brake shoes, if retaining spring does not come off with inboard shoe remove it from piston.

Caliper Disassembly

1. Using clean brake fluid clean exterior of caliper.
2. Drain brake fluid from caliper.
3. Remove caliper mounting bracket bolt and slide bracket from caliper, Fig. 4, then remove sleeve and bushing from bolt and bushing from caliper mounting hole, Fig. 2.
4. Remove clips if still in place, then remove cushions, Fig. 5.

Fig. 5 Removing and installing cushions

5. Pad interior of caliper with clean shop towels, then direct compressed air through caliper inlet hole to remove piston.

NOTE: Use only enough air pressure to ease piston out of bore.

CAUTION: Do not place fingers in front of piston for any reason when applying compressed air. This could result in serious personal injury.

6. Using a screwdriver carefully pry boot from caliper.
7. Remove piston seal from caliper bore using a piece of wood or plastic.

NOTE: Do not use metal tool to remove piston seal as it may damage caliper bore.

8. Remove bleeder valve.

Caliper Assembly

1. Lubricate caliper bore and piston seal with clean brake fluid, then position seal in caliper bore groove.
2. Lubricate piston with clean brake fluid and install boot into piston groove with fold facing the open end of piston, Fig. 2.
3. Insert piston into caliper, then using care to avoid unseating seal, force piston into caliper.

NOTE: To force piston into caliper a force of 50-100 pounds will be required.

4. Position outside diameter of boot into caliper counterbore and seat with tool J-23572.

CAUTION: Assure that retaining ring molded into boot is not bent and that boot is installed fully and evenly below and around the caliper, as dirt and moisture may enter caliper and cause damage and corrosion.

5. Install bleeder screw.
6. Position and stretch cushions over caliper lugs, fitting the heavy section in the lug recess and saw-tooth edges of cushions facing out, Fig. 5.
7. Using silicone lubricant, liberally lubricate sleeve and bushings and the unthreaded portion of the retainer bolt. Install the larger bushing in the caliper hole groove and install the sleeve. Install the smaller bushing in the retainer bolt groove.
8. With caliper clamped in a vise, position clips over cushions and squeeze mounting bracket over clips, aligning the bolt hole. Move bracket against

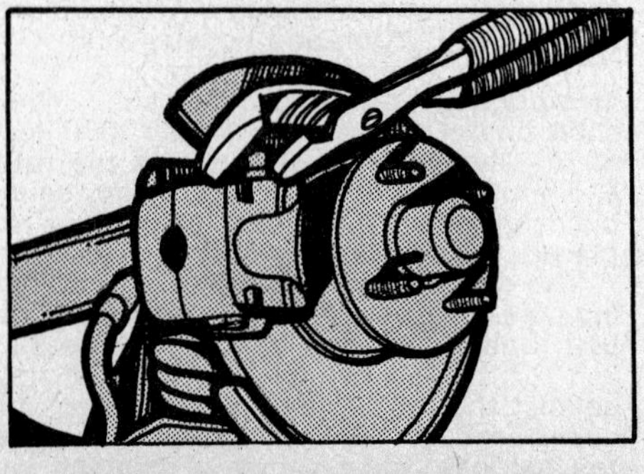

Fig. 6 Fitting brake pad to caliper

retainer boss on caliper and install retainer bolt. Torque bolt to 28 ft. lbs., (38 N-m).

NOTE: Considerable force may be required to squeeze bracket over cushions and clips on caliper. Start open end of bracket over ends of clips near the boot and move the bracket toward the closed end of caliper.

Brake Shoes Installation

1. Position retaining spring on inboard shoe, place single leg in brake shoe hole, then snap two other legs over notch in shoe.
2. Install shoe in caliper.
3. Position caliper over rotor, align mounting holes, install and torque bolts to 70 ft. lbs., (95 N-m).
4. Using suitable pliers clinch outboard shoe to caliper, place lower jaw of pliers on bottom edge of shoe, place upper jaw of pliers on shoe tab, squeeze pliers and bend tab, Fig. 6. Clinch other end of shoe in same manner. Outboard end play should be zero to .005 inch., (zero to 0.127mm).
5. Install wheel and tire assembly and lower vehicle.
6. Add brake fluid to within 1/4 inch from top of master cylinder.

NOTE: Pump brake pedal several times to ensure it is firm before moving vehicle.

DISC BRAKE ROTOR SPECIFICATIONS

Nominal Thickness	Thickness Variation Parallelism	Runout (T.I.R.)	Finish (Micro-In.)
.50 (12.7 mm)	.003 (.08 mm)	.005 (.13 mm)	20–60 (.5–1.6 micro/m)

Caliper Installation

1. Install caliper as described under "Brake Shoe Installation", then install brake hose with new copper gaskets, torque fitting to 21 ft. lbs., (29 N-m), if removed, bleed brake system.

NOTE: Brake hose fitting must be against machined surface on caliper to ensure proper hose positioning.

Rear Suspension Section

DESCRIPTION

This suspension system, Fig. 1, incorporates two tubular lower control arms, a straight track rod, two shock absorbers, two coil springs and a stabilizer bar.

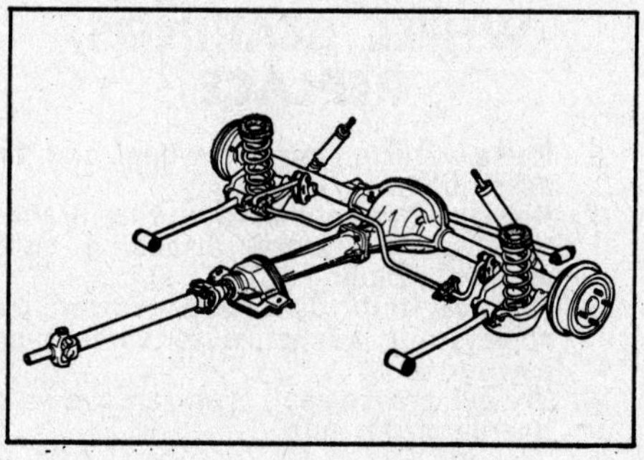

Fig. 1 Rear suspension

SHOCK ABSORBER, REPLACE

1. Remove upper attaching nut from underneath cargo compartment.
2. Remove stud nut attaching shock absorber to axle.
3. Remove shock absorber, noting position of washers and grommets.
4. Reverse procedure to install and tighten attaching nuts securely.

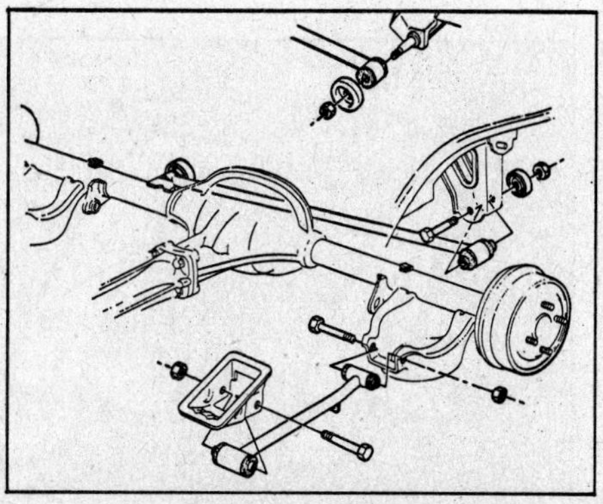

Fig. 2 Track rod & lower control arm installation

COIL SPRING, REPLACE

1. Support vehicle at frame and rear axle with a suitable jack.
2. Disconnect shock absorbers from lower mountings.
3. Lower rear axle until springs can be removed.
4. Reverse procedure to install.

TRACK ROD, REPLACE

1. Raise vehicle and support axle assembly.
2. Remove bolts at underbody end of rod, Fig. 2.
3. Remove bolts at axle bracket and remove track rod.
4. Reverse removal procedure to install and torque bolts to specifications.

LOWER CONTROL ARMS, REPLACE

1. Raise vehicle and support rear axle.
2. Disconnect the front and rear control arm attaching bolts and remove control arm from vehicle, Fig. 2.
3. Reverse procedure to install.

STABILIZER BAR, REPLACE

1. Raise vehicle and support rear axle.
2. Disconnect stablizer bar from underbody and axle tube connections, Fig. 3, then remove stabilizer bar from vehicle.
3. Reverse procedure to install.

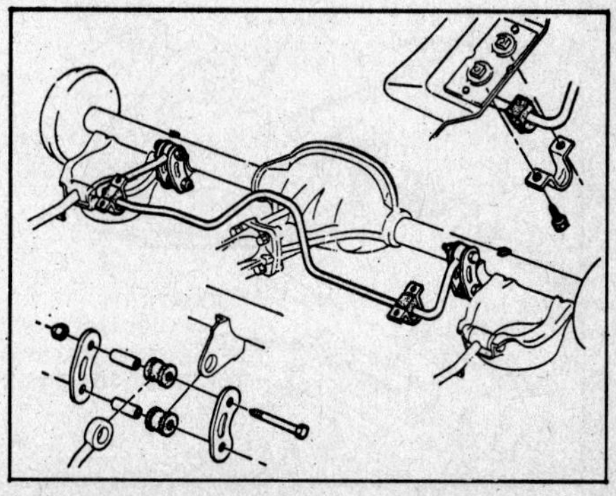

Fig. 3 Stabilizer bar installation

Front End & Steering Section

DESCRIPTION

This front suspension system, Fig. 1, uses long and short control arms with coil springs mounted between the lower control arms and front suspension crossmember.

WHEEL ALIGNMENT

Caster

Caster angle is adjusted by rearranging washers located at both ends of the upper control arm, Fig. 2. A kit consisting of two washers, one of 3 mm thickness and one of 9mm thickness, must be used when adjusting caster angle.

Camber

Camber is adjusted by removing the upper ball joint, rotating it ½ turn and reinstalling it with flat of upper flange on inboard side of control arm, Fig. 3. This will increase positive camber by approximately 1°.

TOE-IN ADJUST

To adjust toe-in, loosen tie rod jam nuts at tie rod ends and loosen clamp at rubber bellows. Turn each tie rod to increase or decrease its length until proper toe-in is obtained.

WHEEL BEARINGS, ADJUST

1. While rotating wheel in the forward direction, torque spindle nut to 12 ft. lbs, Fig. 2.
2. Back spindle nut off to the just loose position.

Fig. 1 Front suspension

3. Hand tighten spindle nut, then loosen until cotter holes are aligned, install cotter pin.

NOTE: Do not loosen nut more than ½ flat to align cotter pin holes.

4. With bearings properly adjusted, end play must be .001-.005 inch.

WHEEL BEARINGS, REPLACE

1. Raise vehicle, remove wheel and tire assembly.
2. Remove two caliper mounting bracket to steering knuckle attaching bolts, remove caliper.
3. Remove hub dust cap, cotter pin, spindle nut, washer, outer wheel bearing and hub.
4. Pry out grease seal, then remove inner bearing from hub.

SHOCK ABSORBER, REPLACE

1. Hold upper stem of shock absorber, and remove nut, retainer and grommet.
2. Raise vehicle and remove lower shock absorber retaining bolts.
3. Lower shock absorber from vehicle.
4. Reverse procedure to install. Torque shock absorber lower retaining bolts to 35-50 ft. lbs. and the upper retaining nut to 60-120 in. lbs.

BALL JOINTS, REPLACE

Upper Ball Joint

1. Raise vehicle and remove wheel and tire assembly.
2. Support lower control arm using a suitable jack.
3. Loosen upper ball joint stud nut, however, do not remove nut.
4. Position tool J-26407 with cupped end over lower ball joint stud and turn threaded end of tool until upper ball joint stud is free of steering knuckle, Fig. 4.
5. Remove tool, then the upper stud nut.
6. Remove two nuts securing ball joint to upper control arm and the ball joint.

NOTE: Inspect tapered hole in steering knuckle. If out of round or damaged, the knuckle must be replaced.

7. Install ball joint on upper control arm and torque bolts to 29 ft. lbs.
8. Position upper ball joint stud on steering knuckle, then install and torque stud nut to 29-36 ft. lbs.
9. Install wheel and tire assembly and lower vehicle.

Lower Ball Joint

1. Raise vehicle and remove wheel and tire assembly.
2. Support lower control arm using a suitable jack.
3. Loosen lower ball joint stud nut, however, do not remove nut.
4. Position tool J-26407 with cupped end of tool over upper ball joint stud and turn threaded end of tool until lower ball joint stud is free of steering knuckle, Fig. 4.
5. Remove tool, then the lower ball joint stud nut.

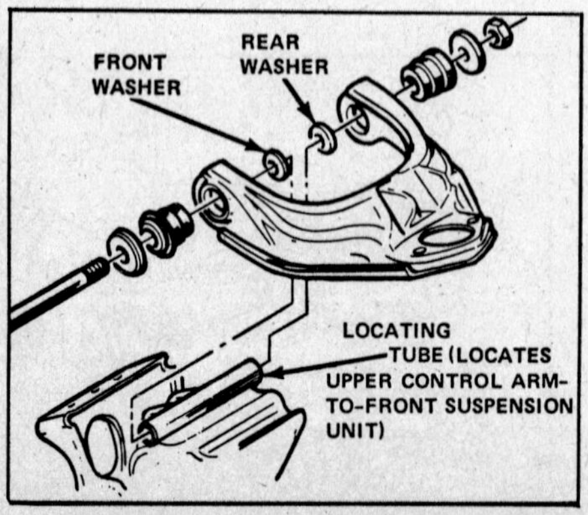

Fig. 2 Caster adjustment

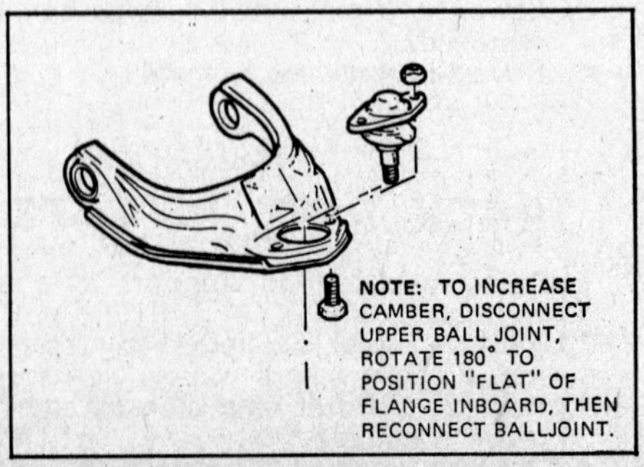

Fig. 3 Camber adjustment

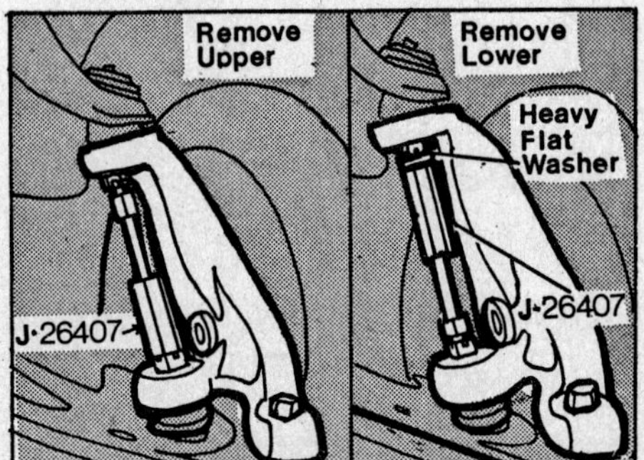

Fig. 4 Separating ball joint studs from steering knuckle

6. Remove ball joint from lower control arm.

NOTE: Inspect tapered hole in steering knuckle, if out of round or damaged, the knuckle must be replaced.

7. Insert ball joint through lower control arm and into steering knuckle.
8. Install ball joint stud nut and torque to 41-54 ft. lbs.
9. Install wheel and tire assembly and lower vehicle.

COIL SPRING, REPLACE

1. Support vehicle by frame, remove wheel and tire assembly.
2. Disconnect stabilizer bar from lower control arm and tie rod from steering knuckle.
3. Support lower control arm using a suitable jack.
4. Loosen lower ball joint stud nut, then use tool J-26407 to free stud from steering knuckle, position knuckle and hub out of way. Remove stud nut.
5. Loosen lower control arm pivot bolts.
6. Install a safety chain around spring and through lower control arm.
7. Slowly lower control arm until spring is extended as far as possible, then use pry bar to carefully lift spring over lower control arm spring seat.

STEERING GEAR, REPLACE

1. Raise vehicle and remove bolts and shield.
2. Remove tie rod cotter pins and nuts from tie rod ends and separate tie rods from steering knuckles.
3. Remove flexible coupling pinch bolt.
4. Remove steering gear clamp bolts, then the steering gear assembly from vehicle.
5. Reverse procedure to install.

RACK & PINION STEERING GEAR SERVICE

Disassembly

1. Clamp assembly in a soft jawed vise.
2. Loosen jam nuts and remove outer tie rods, Fig. 1.

NOTE: Hold housing when loosening nut to prevent damage to internal gear components.

3. Cut off inner boot clamps, remove outer boot clamps and pull boots from gear.
4. Clamp rack to prevent rotation and turn inner tie rod housings counterclockwise to separate inner tie rod assemblies from rack, Fig. 2.
5. Remove adjuster plug lock nut, Fig. 3, adjuster plug and spring, Fig. 4, and rack bearing, Fig. 5, from housing.
6. Remove pinion seal, Fig. 6, and retaining snap-ring, Fig. 7.
7. Clamp end of pinion shaft in a soft jawed vise and tap housing with a mallet to separate pinion assembly from housing, Fig. 8. Use caution since rack may slide from housing.
8. Remove rack from housing.

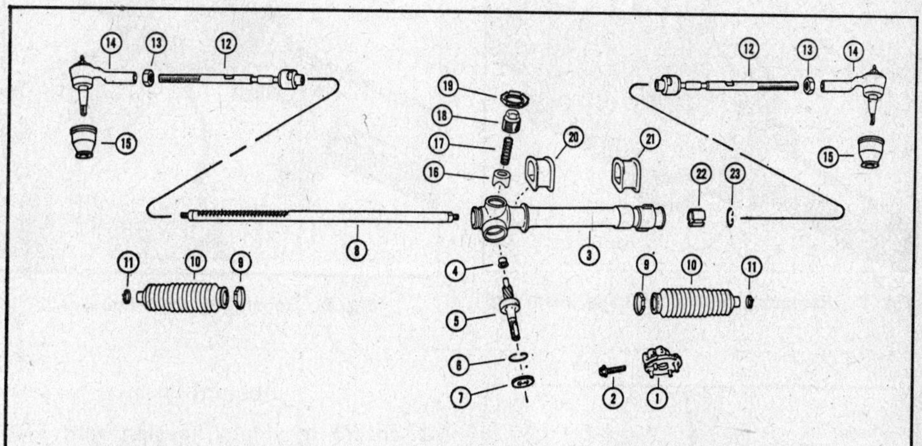

Fig. 1 Rack & pinion steering gear, disassembled

Component Service

Mounting Grommets

1. Cut grommet from housing.
2. Lubricate inside diameter of grommets with chassis lubricant.
3. Slide left hand grommet over right side boss, then the right hand grommet.

Guide Bushings

The guide bushings are not serviceable and if damaged, the housing must be replaced.

Rack Bushing

1. Remove retaining ring, Fig. 9, and using a suitable size socket, drive bushing from housing. If a puller is used, position fingers behind bushing.
2. Using a socket, press new bushing into housing until bottomed and install retaining ring.

Roller Bearing Assembly

1. The assembly may be removed with a drift and mallet or a press, Fig. 10.
2. With a suitable size socket, press or drive new bearing into housing until bottomed.

Bearing & Pinion Assembly

Inspect roller bearing pilot, pinion teeth and roller bearing. If pilot is scored, teeth are chipped or loose on pinion shaft, the bearing and pinion assembly should be replaced.

Fig. 2 Removing inner tie rod assemblies

Fig. 3 Removing adjuster plug lock nut

Fig. 4 Removing adjuster plug & spring

Fig. 5 Removing rack bearing

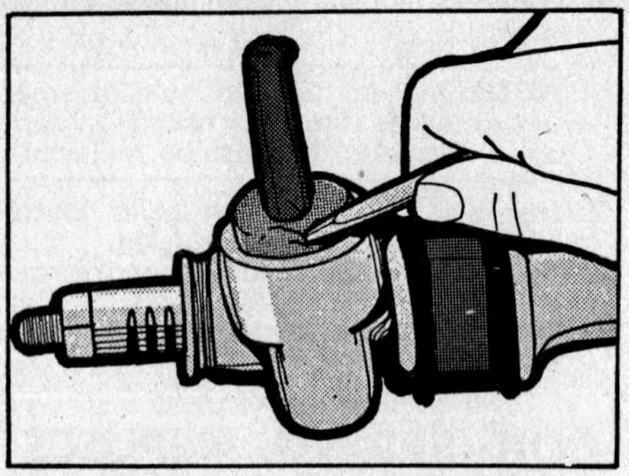

Fig. 6 Removing pinion seal

Fig. 7 Removing pinion retaining snap ring

Fig. 8 Separating from housing

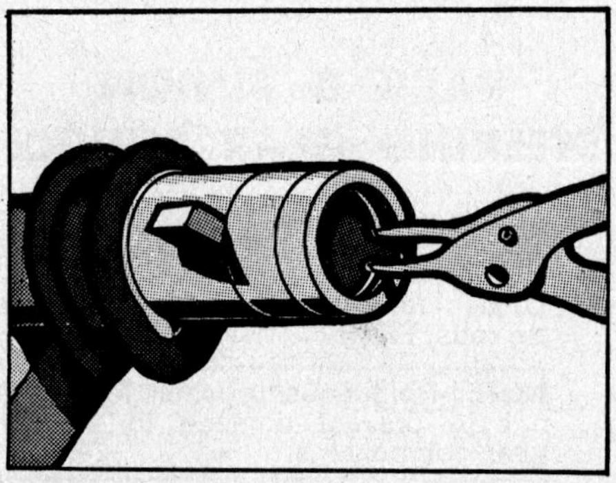

Fig. 9 Removing rack bushing retaining ring

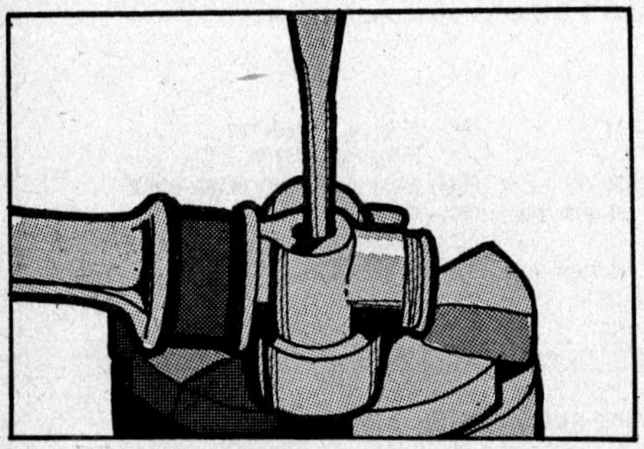

Fig. 10 Removing roller bearing assembly

Inner Tie Rods

The inner tie rod assemblies are not serviceable. If pivot is loose or the joint rocking or rotating torque exceeds 150 inch lbs. (17 N-m), replace inner tie rod assembly.

Assembly

1. Install rack into housing with teeth facing toward pinion. The flat on the teeth should be parallel with pinion shaft. Set distance from lip of housing to end of rack to 2.70 inch (68.5mm), Fig. 11.
2. Insert pinion with flat approximately 75° from the vertical. Tap on pinion shaft until pinion seats. Reset rack position outlined in step 1. The flat should now be vertical. If flat is at plus or minus 30° from the vertical repeat procedure.

NOTE: The rack must be properly centered to prevent unequal turning radii.

3. Install pinion retaining ring with beveled edge facing upward.
4. Lubricate top of pinion bearing with anhydrous calcium grease, then seat piinon seal flush with housing.
5. Lubricate and install rack bearing.
6. Lubricate ends of preload spring and the adjuster plug threads. Install spring and adjuster plug into housing. Rotate adjuster plug until bottomed, then counter-clockwise 45 to 60 degrees. The pinion preload should be 8 to 12 inch lbs. (0.9 to 1.1 N-m), Fig. 12. Turn adjuster plug to set proper torque. Torque adjuster plug lock nut to 50 ft. lbs. (68 N-m).

Fig. 11 Installing rack

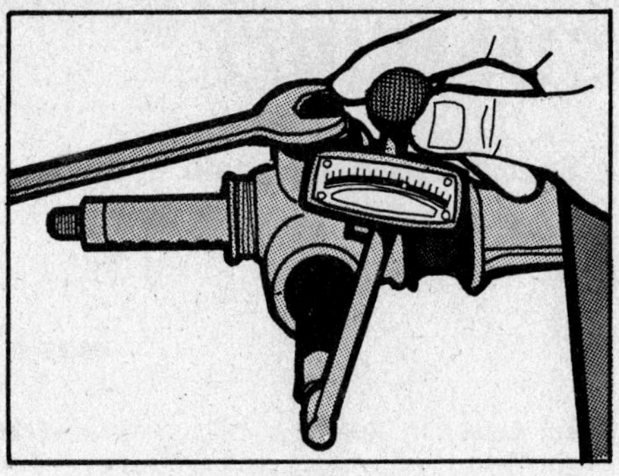

Fig. 12 Adjusting pinion preload

7. Lubricate rack ends and teeth, then turn pinion shaft in both directions several times, adding grease to rack teeth.
8. Install inner tie rod assemblies onto rack until bottomed, then stake inner tie rod housings to rack flat, Fig. 13.
9. Install clamps and boots onto assembly. Do not secure outer end of boot until toe-in is adjusted.
10. Thread jam nuts and outer tie rods onto inner tie rods. Do not tighten jam nuts until toe-in is adjusted.
11. Install coupling assembly. The flat on coupling inside diameter mates with flat on pinion shaft.

Fig. 13 Staking inner tie rods

CHEV. MONZA • BUICK SKYHAWK
OLDS STARFIRE • PONT. SUNBIRD

OLD CAR SPECIFICATIONS: For 1946-68 Tune Up and Wheel Alignment Specifications see main index.

INDEX OF SERVICE OPERATIONS

SERIAL NUMBER LOCATION

On top of instrument panel, left front.

ENGINE NUMBER LOCATION

4 Cyl: On pad at right side of cylinder block, above starter.
V6 & V8: Pad at front righthand side of cylinder block.

ENGINE INDENTIFICATION CODE

4 cyl. & V8 engines are identified in the following table by the code letters immediately following the engine serial numbers.
V6 engines used in Buick vehicles are identified in the following table by the fifth digit in the Vehicle Identification Number.
V6 engines used in Oldsmobile and Pontiac vehicles are identified in the following table by the code letters preceeding the engine serial number or on a code letter tape located on the left bank valve cover.

Monza

1975	4-140	CAM, CAR, CAS
	4-140	CAT, CBB, CBC
	V8-262 Std. Tr.	CZA, CZB, CZC
	V8-262 Std. Tr.	CZD, CZT, CZU
	V8-262 Auto. Tr.	CZE, CZG
	V8-262 Auto. Tr.	CGA, CGJ, CGK
	V8-350	CHY
1976	4-140	CAY, CAZ, CBK, CBL
	4-140	CBS, CBT, CBU, CBW
	4-140	CBX, CBY, CBZ
	V8-262	CGL, CZU, CZT

Skyhawk

1975-76	V6-231	C

Starfire

1975	V6-231 Std. Tr.①	FP
	V6-231 Auto. Tr.①	FR
	V6-231 Auto. Tr.②	FS
1976	V6-231 Std. Tr.①	FH
	V6-231 Std. Tr.②	FO
	V6-231 Auto. Tr.①	FI
	V6-231 Auto. Tr.②	FJ

Sunbird

1976	4-140	CBH, CBJ
	4-140	CBK, CBL, CBS, CBT
	V6-231 Std. Tr.①	FH
	V6-231 Auto. Tr.①	FI
	V6-231 Std. Tr.②	FO
	V6-231 Auto. Tr.②	FJ

GRILLE IDENTIFICATION

1975-76 Chevrolet Monza Hatchback

1975-76 Chevrolet Monza Towne Coupe

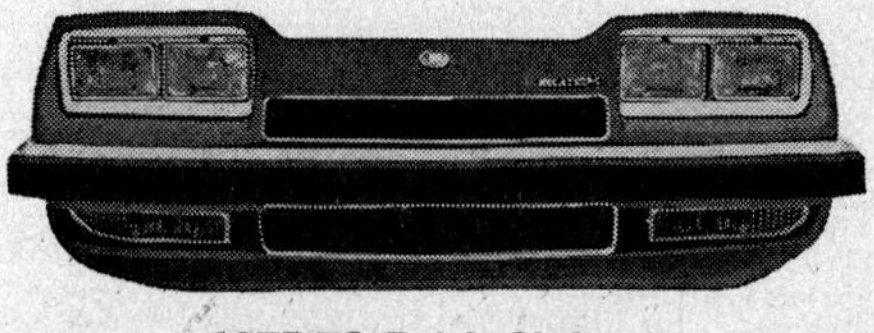

1975-76 Buick Skyhawk

1975-76 Oldsmobile Starfire

1976 Pontiac Sunbird

GENERAL ENGINE SPECIFICATIONS

Year	Engine	Carburetor	Bore and Stroke	Piston Displacement, Cubic Inches	Compression Ratio	Maximum Brake H.P. @ R.P.M.	Maximum Torque Lbs. Ft. @ R.P.M.	Normal Oil Pressure Pounds
1975	87 Horsepower①........4-140	2 Barrel	3.5 x 3.625	140	8.0	87 @ 4400	122 @ 2800	40
	110 Horsepower①.......V6-231	2 Barrel	3.8 x 3.4	231	8.0	110 @ 4000	175 @ 2000	37
	110 Horsepower①.......V8-262	2 Barrel	3.67 x 3.10	262	8.5	110 @ 3600	200 @ 2000	32–40
	125 Horsepower①.......V8-350	2 Barrel	4.00 x 3.48	350	8.5	125 @ 3600	235 @ 2000	32–40
1976	69 Horsepower①........4-140	1 Barrel	3.501 x 3.625	140	8.0	69 @ 4000	113 @ 2400	27–45
	80 Horsepower①②......4-140	2 Barrel	3.501 x 3.625	140	8.0	80 @ 4000	116 @ 2800	27–45
	87 Horsepower①........4-140	2 Barrel	3.501 x 3.625	140	8.0	87 @ 4400	122 @ 2800	27–45
	110 Horsepower①.......V6-231	2 Barrel	3.8 x 3.4	231	8.0	110 @ 4000	175 @ 2000	37
	110 Horsepower①.......V8-262	2 Barrel	3.671 x 3.100	262	8.5	110 @ 3600	195 @ 2000	32–40

①—Ratings are net—As installed in vehicle.
②—California.

ALTERNATOR & REGULATOR SPECIFICATIONS

Year	Alternator					Regulator						
	Model	Rated Hot Output Amps.	Field Current 12 Volts @ 80 F.	Cold Output @ 14 Volts		Model	Field Relay			Voltage Regulator		
				2000 R.P.M. Amps.	5000 R.P.M. Amps.		Air Gap In.	Point Gap In.	Closing Voltage	Air Gap In.	Point Gap In.	Voltage @ 125°F.
1975	1100545	32	4–4.5	—	31	Integral	—	—	—	—	—	13.5–14.5
	1100546	55	4–4.5	—	50	Integral	—	—	—	—	—	13.5–14.5
	1100559	32	4–4.5	—	31	Integral	—	—	—	—	—	13.5–14.5
	1100560	55	4–4.5	—	50	Integral	—	—	—	—	—	13.5–14.5
	1102394	37	4–4.5	—	37①	Integral	—	—	—	—	—	13.5–14.5
	1102483	37	4–4.5	—	37①	Integral	—	—	—	—	—	13.5–14.5
	1102494	55	4–4.5	—	55①	Integral	—	—	—	—	—	13.5–14.5
	1102495	55	4–4.5	—	55①	Integral	—	—	—	—	—	13.5–14.5
	1102500	55	4–4.5	—	50	Integral	—	—	—	—	—	13.5–14.5
	1102854	63	4–4.5	—	60	Integral	—	—	—	—	—	13.5–14.5
	1102856	37	4–4.5	—	33	Integral	—	—	—	—	—	13.5–14.5
	1102857	63	4–4.5	—	60	Integral	—	—	—	—	—	13.5–14.5
1976	1102394	37	4–4.5	—	33	Integral	—	—	—	—	—	—
	1102479	55	4–4.5	—	51	Integral	—	—	—	—	—	—
	1102495	55	4–4.5	—	55①	Integral	—	—	—	—	—	—
	1102840	55	4–4.5	—	51	Integral	—	—	—	—	—	—
	1102845	37	4–4.5	—	33	Integral	—	—	—	—	—	—
	1102846	55	4–4.5	—	51	Integral	—	—	—	—	—	—
	1102851	55	4–4.5	—	51	Integral	—	—	—	—	—	—
	1102854	63	4–4.5	—	60	Integral	—	—	—	—	—	—
	1102858	37	4–4.5	—	33	Integral	—	—	—	—	—	—
	1102891	55	4–4.5	—	51	Integral	—	—	—	—	—	—
	1102893	63	4–4.5	—	60	Integral	—	—	—	—	—	—
	1102894	37	4–4.5	—	33	Integral	—	—	—	—	—	—

①—At 5500 RPM.

TUNE UP SPECIFICATIONS

The following specifications are published from the latest information available. This data should be used only in the absence of a decal affixed in the engine compartment.

OLD CAR SPECIFICATIONS: For 1946–68 Tune Up Specifications see main index.

★When using a timing light, disconnect vacuum hose or tube at distributor and plug opening in hose or tube so idle speed will not be affected.

●When checking compression, lowest cylinder must be within 80 percent of highest.

▲Before removing wires from distributor cap, determine location of No. 1 wire in cap, as distributor position may have been altered from that shown at the end of this chart.

Year	Spark Plug		Distributor		Ignition Timing★			Carb. Adjustments					
								Hot Idle Speed		Air Fuel Ratio		Idle "CO" %	
	Type	Gap Inch	Point Gap Inch	Dwell Angle Deg.	Firing Order Fig. ▲	Timing BTDC (1)	Mark Fig.	Std. Trans.	Auto. Trans.(2)	Std. Trans.	Auto. Trans.	Std. Trans.	Auto. Trans.
1975													
4-140	R43TSX(4)	.060	—	—	A	(3)	B	700	750D	—	—	.5	.5
V6-231	R44SX	.060	—	—	C	12°	D	800	650D	—	—	—	—
V8-262	R44TX	.060	—	—	E	8°	F	800	600D	—	—	—	—
V8-350	R44TX	.060	—	—	E	6°	G	800	600	—	—	—	—
1976													
4-140 1 Bar. Carb.	R43TS	.035	—	—	A	(7)	B	(8)	750D	—	—	—	—
4-140 2 Bar. Carb.(5)	R43TS	.035	—	—	A	(9)	B	700	750D	—	—	—	—
4-140 2 Bar. Carb.(6)	R43TS	.035	—	—	A	(9)	B	(10)	750D	—	—	—	—
V6-231	R44TSX	.060	—	—	C	12°	D	800	600D	—	—	—	—
V8-262	R45TS	.035	—	—	E	8°	F	800	600D	—	—	—	—

(1)—BTDC—Before Top Dead Center.
(2)—D—Drive.
(3)—Standard trans. 10°, Automatic trans. 12°.
(4)—If cold weather starting problems are encountered, use R43TS spark plug, gapped at .035 inch.
(5)—Exc. Calif.
(6)—California.
(7)—Man. trans., 8° BTDC; auto. trans., 10° BTDC.
(8)—Monza, 1200 RPM; Sunbird, 700 RPM.
(9)—Monza man. trans., 10° BTDC; auto. trans. 12° BTDC. Sunbird man. trans., 8° BTDC; auto. trans., 10° BTDC.
(10)—Monza; 1000 RPM; Sunbird, 700 RPM.

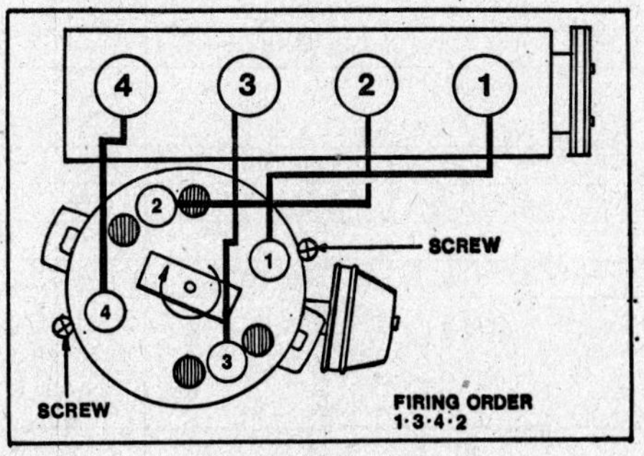

Fig. A

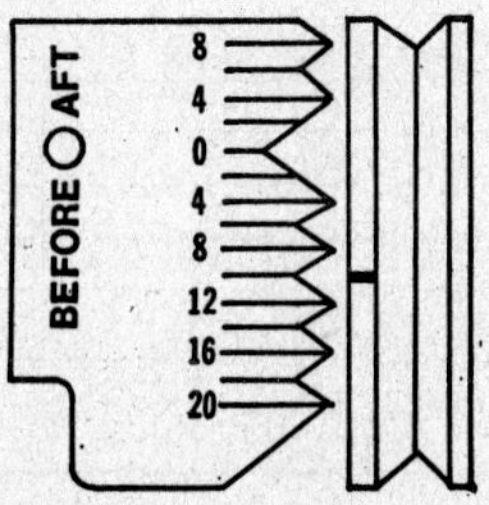

Fig. B

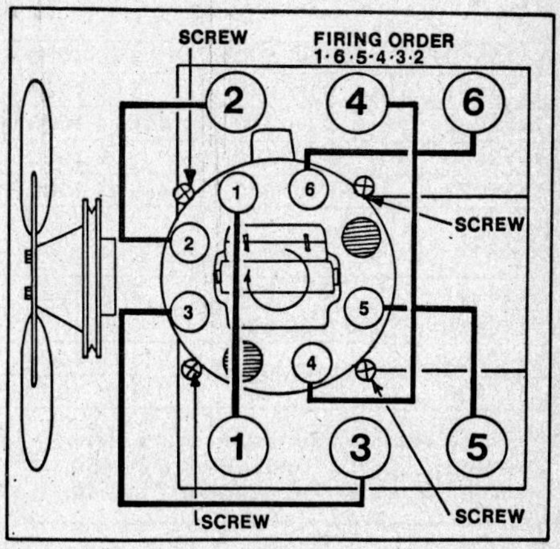

Fig. C

Continued

TUNE UP NOTES—Continued

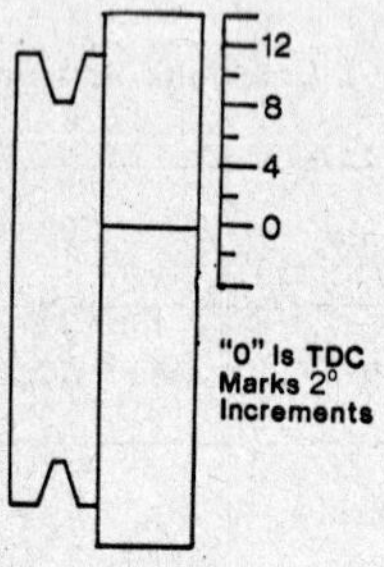

Fig. D

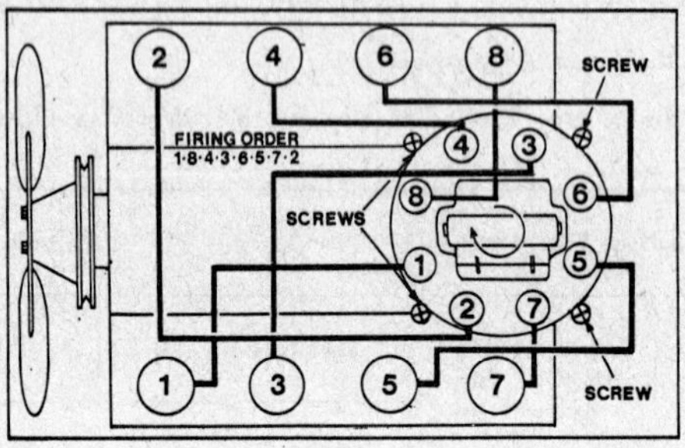

Fig. E

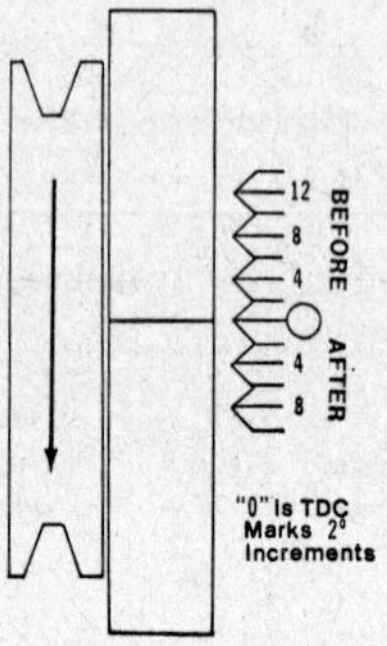

Fig. F

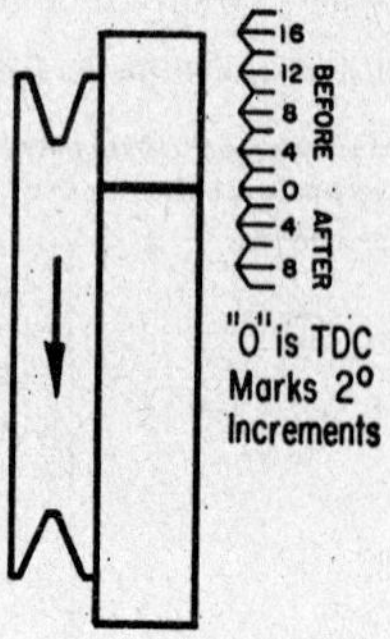

Fig. G

DISTRIBUTOR SPECIFICATIONS

★If unit is checked on vehicle double the RPM and degrees to get crankshaft figures.

Distributor Part No.①	Centrifugal Advance Degrees @ RPM of Distributor: Advance Starts	Intermediate Advance			Full Advance	Vacuum Advance: Inches of Vacuum To Start Plunger	Max. Adv. Dist. Deg. @ Vacuum	Distributor Retard: Max. Ret. Dist. Deg. @ Vacuum
1975								
1110650	0 @ 500	5 @ 1000	—	—	8 @ 2050	5–7	9 @ 10	—
1110651	0 @ 540	3 @ 1000	4½ @ 1750	—	8 @ 2050	5–7	9 @ 10	—
1112862	0 @ 810	2½ @ 1000	—	—	11 @ 2400	5	12 @ 12	—
1112880	0 @ 600	6 @ 1000	—	—	11 @ 2100	4	9 @ 12	—
1112933	0 @ 600	4½ @ 1000	—	—	11 @ 2000	3	8 @ 8	—
1976								
1110661	0 @ 525	3 @ 1000	—	—	8 @ 2050	6	9 @ 10	—
1110668	0 @ 638	—	—	—	8 @ 1600	6	12 @ 12	—
1112862	②	2½ @ 1000	—	—	11 @ 2400	5	③	—
1112983	0 @ 600	4½ @ 1000	—	—	11 @ 2000	4	7½ @ 10	—

①—Located on distributor housing plate.
②—Monza, 0 @ 810; Sunbird, 0 @ 450.
③—Monza, 12 @ 12; Sunbird, 10 @ 14.

VALVE SPECIFICATIONS

Year	Engine Model	Valve Lash		Valve Angles		Valve Spring Installed Height	Valve Spring Pressure Lbs. @ In.	Stem Clearance		Stem Diameter	
		Int.	Exh.	Seat	Face			Intake	Exhaust	Intake	Exhaust
1975	4-140	.015C	.030C	46	45	1.75	①	.0010–.0027	.0010–.0027	.3410–.3417	.3410–.3417
	V6-231	Hydraulic④		45	—	1.727	168 @ 1.327	.0015–.0032	.0015–.0032	.3405–.3412	.3405–.3412
	V8-262	1 Turn		46	45	②	③	.0010–.0027	.0010–.0027	.3410–.3417	.3410–.3417
	V8-350	1 Turn		46	45	1.70	200 @ 1.25	.0010–.0027	.0012–.0029	.3410–.3417	.4310–.3417
1976	4-140	Hydraulic④		46	45	1.746	190 @ 1.31	.0010–.0027	.0010–.0027	.3410–.3417	.3410–.3417
	V6-231	Hydraulic④		45	45	1.727	168 @ 1.327	.0015–.0035	.0015–.0032	.3405–.3412	.3405–.3412
	V8-262	1 Turn		46	45	②	⑤	.0010–.0027	.0010–.0027	.3410–.3417	.3410–.3417

①—Intake 190 @ 1.31, Exhaust 186 @ 1.29.
②—Intake 1.70, Exhaust 1.61.
③—Intake 200 @ 1.25, Exhaust 189 @ 1.20.
④—No adjustment.
⑤—Intake 200 @ 1.25, Exhaust 200 @ 1.16.

STARTING MOTOR SPECIFICATIONS

Year	Model	Starter Number	Brush Spring Tension Oz.①	Free Speed Test			Resistance Test	
				Amps.	Volts	R.P.M.①	Amps.	Volts
1975	4-140 Std. Trans.	1108771	35	50–75②	9	6500–10000	—	—
	4-140 Auto. Trans.	1108772	35	50–75②	9	6500–10000	—	—
	V6-231	1108770	35	50–80②	9	3500–6000	—	—
	V8-262	1108790	35	50–80②	9	3500–6000	—	—
	V8-350	1108775	35	65–95	9	7500–10500	—	—
	V8-350	1108776	35	65–95	9	7500–10500	—	—
1976	4-140 Std. Trans.	1108771	35	50–75	9	6500–10000	—	—
	4-140 Auto. Trans.	1108772	35	50–75	9	6500–10000	—	—
	V6-231	1108797	35	50–80	9	5500–10500	—	—
	V8-262	1108790	35	50–80	9	3500–6000	—	—

①—Minimum.
②—Includes solenoid.

REAR AXLE SPECIFICATIONS

Year	Model	Carrier Type	Ring Gear & Pinion Backlash		Pinion Bearing Preload			Differential Bearing Preload		
			Method	Adjustment	Method	New Bearings Inch-Lbs.	Used Bearings Inch-Lbs.	Method	New Bearings Inch-Lbs.	Used Bearings Inch-Lbs.
1975–76	Monza	Integral	Shim	.005–.008	Spacer	10–25	8–12	Shim	—	—
	Skyhawk	Integral	Shim	.006–.008	Spacer	10–25	8–12	Shim	35–40	20–25
	Starfire	Integral	Shim	.005–.009	Spacer	24–32	8–12	Shim	—	—

PISTONS, PINS, RINGS, CRANKSHAFT & BEARINGS

Year	Engine Model	Piston Clearance	Ring End Gap①		Wrist-pin-Diameter	Rod Bearings		Main Bearings			
			Comp.	Oil		Shaft Diameter	Bearing Clearance	Shaft Diameter	Bearing Clearance	Thrust on Bear. No.	Shaft End Play
1975	4-140	.0018–.0028	②	.010	.9272	1.999–2.000	.0007–.0038	2.2983–2.2993	③	4	.002–.007
	V6-231	.0008–.0020	.010	.015	.9393	1.991–2.000	.0005–.0026	2.4995	.0004–.0015	2	.004–.008
	V8-262	.0007–.0013	④	.015	.9272	2.098–2.099	.0013–.0035	⑤	⑥	5	.002–.007
	V8-350	.0007–.0013	④	.015	.9272	2.098–2.099	.0013–.0035	⑤	⑦	5	.002–.007
1976	4-140	.0018–.0028	②	.010	.9272	1.999–2.000	.0007–.0027	2.3004	.0003–.0029	4	.002–.008
	V6-231	.0008–.0020	.010	.013	.9393	1.991–2.000	.0005–.0026	2.4995	.0004–.0015	2	.003–.008
	V8-262	.0007–.0013	④	.015	.9272	2.098–2.099	.0013–.0035	⑤	⑥	5	.002–.007

①—Fit rings in tapered bore for clearance listed in tightest portion of ring travel.
②—#1—.015″, #2—.009″.
③—#1—.0003–.002, #2, 3, 4 & 5—.0003–.0027.
④—#1—.010″, #2—.013″.
⑤—#1—2.4484–2.4493, #2, 3 & 4—2.4481–2.4490, #5—2.4479–2.4488.
⑥—#1—.0008–.0020, #2, 3 & 4—.0011–.0023, #5—.0017–.0033.
⑦—#1,2,3,4—.0013–.0025, #5—.0023–.0033.

ENGINE TIGHTENING SPECIFICATIONS★

★Torque specifications are for clean and lightly lubricated threads only. Dry or dirty threads produce increased friction which prevents accurate measurement of tightness.

Year	Engine Model	Spark Plugs Ft. Lbs.	Cylinder Head Bolts Ft. Lbs.	Intake Manifold Ft. Lbs.	Exhaust Manifold Ft. Lbs.	Rocker Arm Stud Ft. Lbs.	Cam Cover Ft. Lbs. ①	Connecting Rod Cap Bolts Ft. Lbs.	Main Bearing Cap Bolts Ft. Lbs.	Flywheel to Crankshaft Ft. Lbs.	Vibration Damper or Pulley Ft. Lbs.
1975	4-140	15	60	30	30	—	35②	35	65	60	80
	V6-231	20	75	45	25	30③	5	40	115	55	150 min.
	V8-262,350	15	65	30	20④	—	45②	45	70	60	60
1976	4-140	15	60	30	30	—	35②	35	65	60	80
	V6-231	20	80	45	25	30③	4	40	115	60	175
	V8-262	15	65	30	20④	—	45②	45	70	60	60

①—Rocker arm cover ft. lbs.
②—In. lbs.
③—Rocker arm shaft to cylinder head.
④—Inside bolts 30 ft. lbs.

BRAKE SPECIFICATIONS

Year	Model	Brake Drum Inside Diameter	Wheel Cylinder Bore Diameter			Master Cylinder Bore Diameter		
			Disc Brake	Front Drum Brake	Rear Drum Brake	Disc Brakes	Drum Brakes	Power Brakes
1975	All	9	1⅞	—	¾	¾	—	¾
1976	All	9½	2½	—	①	②	—	②

①—Exc. Sunbird, ¾″; Sunbird, 11/16″.
②—Exc. Sunbird, ¾″; Sunbird, ⅞″.

WHEEL ALIGNMENT SPECIFICATIONS

OLD CAR SPECIFICATIONS: For 1946–69 Wheel Alignment Specifications see main index.

Year	Model	Caster Angle, Degrees		Camber Angle, Degrees				Toe-In. Inch	Toe-Out on Turns, Deg.①	
		Limits	Desired	Limits		Desired			Outer Wheel	Inner Wheel
				Left	Right	Left	Right			
1975–76	All	−¼ to −1¼	−¾	−¼ to +¾	−¼ to +¾	+¼	+¼	0 to ⅛	—	—

①—Incorrect toe out when other adjustments are correct indicates bent steering arms.

COOLING SYSTEM & CAPACITY DATA

Year	Model or Engine	Cooling Capacity, Qts.		Radiator Cap Releif Pressure, Lbs.		Thermo. Opening Temp. ①	Fuel Tank Gals.	Engine Oil Refill Qts. ②	Transmission Oil				Rear Axle Oil Pints
		With Heater	With A/C	With A/C	No A/C				3 Speed Pints	4 Speed Pints	5 Speed Pints	Auto. Trans. Qts.	
1975	4-140, 350	8	8	15	15	195	18½	3	—	3	—	④	2¾
	V6-231	13¼	13¾	15	15	190	18½	4	—	⑤	—	④	2¾
	V8-262, 350	18	18	15	15	195	18½	4	—	3	3½	④	2¾
1976	4-140	8½	8½	15	15	195	18½	3½	3	3	3	⑥	2¾
	V6-231	13½	14	15	15	195	18½	4	—	⑦	⑧	⑥	⑨
	V8-262	18½	18½	15	15	195	18½	4	—	3	3	④	2¾

①—For permanent type anti freeze.
②—Add 1 qt. with filter change.
③—Approximate. Make final check with dip stick.
④—Refill 3 qts., total capacity 10 qts.
⑤—Exc. Skyhawk, 2½ pts.; Skyhawk, 3½ pts.
⑥—Exc. Sunbird, refill 3 qts., total capacity 10 qts.; Sunbird, refill 2½ qts.; total capacity 10½ qts.
⑦—Skyhawk, 3½ pts.; Starfire, 2½ pts.; Sunbird, 3 pts.
⑧—Exc. Sunbird, 3½ pts.; Sunbird, 3 pts.
⑨—Exc. Sunbird, 3½ pts.; Sunbird, 2¾ pts.

Electrical Section

DISTRIBUTOR

Removal

1. Disconnect wiring connectors from distributor cap.
2. Remove distributor cap and position out of way, then disconnect vacuum advance hose.
3. Scribe a mark on engine in line with rotor, noting approximate position of distributor housing in relation to engine.
4. Remove hold-down nut and clamp and lift distributor from engine. Do not crank engine after distributor is removed, otherwise the distributor will have to be initially timed to the engine.
5. Install distributor in reverse order of removal, making certain to align scribe marks, then start engine and adjust timing.

STARTER, REPLACE

4-140 & V8-262, 350

1. Disconnect battery ground cable and disconnect all wires at solenoid terminals.

NOTE: Reinstall the nuts on the terminals as each wire is removed, as thread size is different and if mixed, stripping of threads may occur.

2. Loosen starter front bracket and remove the two mounting bolts.
3. Remove from bracket bolt or nut and rotate brace clear of work area, then remove starter from vehicle by lowering front end of starter first.
4. Reverse procedure to install.

V6-231

1. Disconnect battery ground cable.
2. Remove exhaust crossover pipe and transmission flywheel cover.
3. Remove transmission mount bolts, then position adjustable jack under transmission extension housing.
4. Remove right transmission support to underbody bolt, then loosen left bolt and pivot support downward.
5. Lower transmission and disconnect oil cooler lines at transmission.
6. Remove the two starter attaching bolts, then lower starter and disconnect wires and remove starter.
7. Reverse procedure to install.

IGNITION LOCK, REPLACE

1. Remove steering wheel as outlined under "Horn Sounder & Steering Wheel" procedure, then *with ignition lock in "Run" position,* pull directional switch rearward far enough to slip it over end of shaft. Do not pull harness out of column.

NOTE: Do not remove buzzer switch as damage to lock cylinder will result.

2. Insert a small screwdriver or similar tool into turn signal housing slot, Fig. 1. Keeping tool to right side of slot, break housing flash loose and at same time depress spring latch at lower end of lock cylinder.
3. Remove lock cylinder from housing.

IGNITION SWITCH, REPLACE

1. Disconnect battery ground cable.
2. Remove left A/C outlet duct.
3. Remove steering column to support retaining nuts and allow column to lower.
4. Disconnect connector from switch, then remove switch retaining screws and remove switch.
5. Reverse procedure to install with ignition switch in "lock" position, Fig. 2.

NEUTRAL SAFETY SWITCH, REPLACE

1. Disconnect battery ground cable.
2. Remove console cover.
3. Disconnect connector from switch, then remove switch retaining screws and remove switch.
4. Reverse procedure to install and make certain that switch is correctly adjusted.

Fig. 1 Ignition lock removal

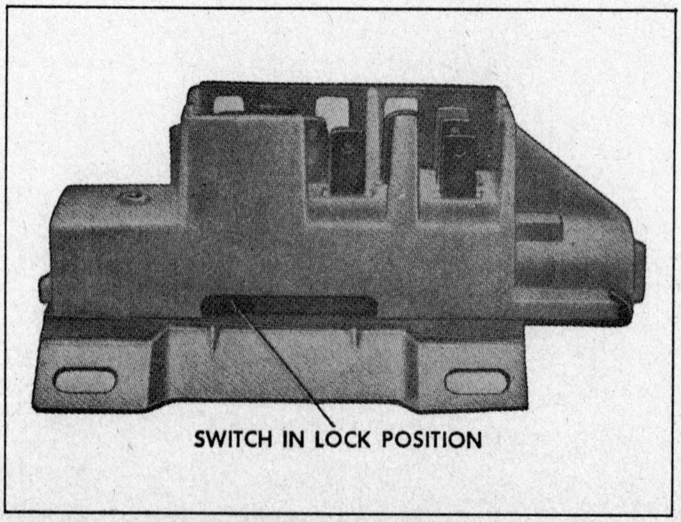

Fig. 2 Ignition switch assembly

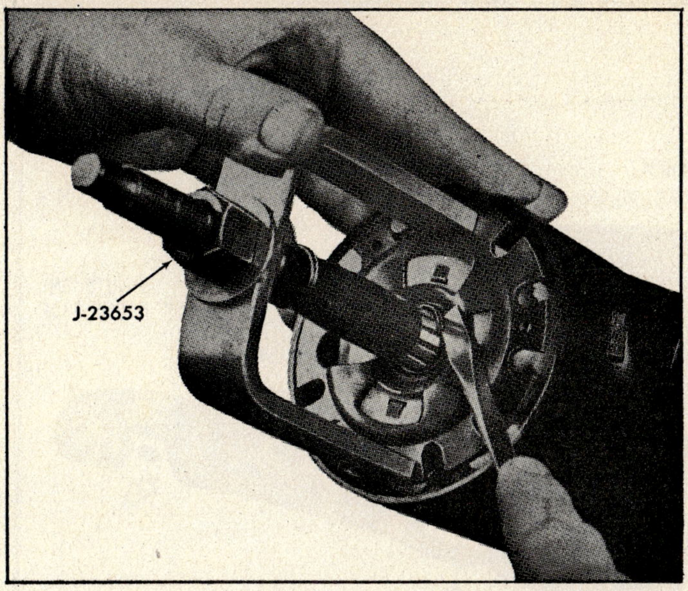

Fig. 3 Compressing lock plate and removing retaining ring

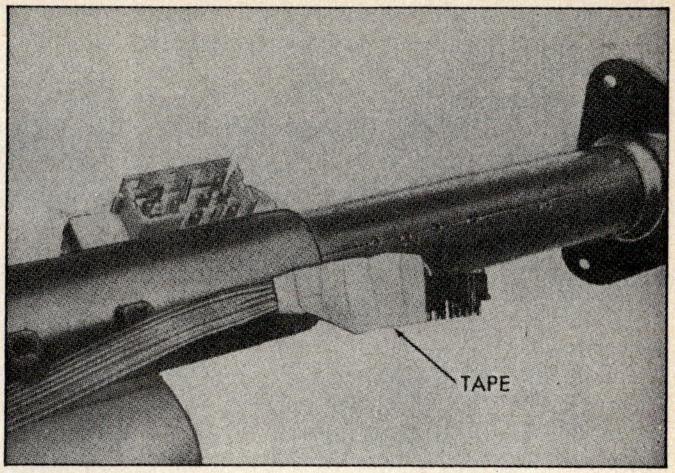

Fig. 4 Taping turn signal connector and wires

LIGHT SWITCH, REPLACE

1. Disconnect battery ground cable.
2. Remove left A/C duct if equipped.
3. Reaching under instrument panel, release and pull switch knob and shaft assembly out of switch.
4. Remove switch bezel retaining nut, then disconnect switch connector and remove switch.
5. Reverse procedure to install.

STOP LIGHT SWITCH

1. Reach under right side of instrument panel at brake pedal support and lease wiring harness connector at switch.
2. Pull switch from mounting bracket.
3. When installing switch, adjust by bringing brake pedal to normal position. Electrical contact should be made when pedal is depressed ⅜ to ⅝ inch. To adjust, the switch may be rotated or pulled in the clip.

CLUTCH START SWITCH

NOTE: The clutch pedal must be fully depressed and the ignition switch in START position for the vehicle to start.

The clutch switch assembly mounts with two tangs to the clutch pedal brace switch pivot bracket and the clutch pedal arm.

1. Under the instrument panel on the clutch pedal support remove the multi-contact connector from switch.
2. Compress switch assembly actuating shaft barb retainer and push out of clutch pedal.
3. Compress switch assembly pivot bracket barb and lift off switch.
4. When installing new switch, no adjustments are necessary as the switch is self aligning.

TURN SIGNAL SWITCH, REPLACE

1. Remove steering wheel with suitable puller.
2. Remove three cover screws and lift cover off the shaft.

NOTE: These screws have plastic retainers on the back of the cover so it is not necessary to completely remove these screws.

3. Place Lock Plate Compressing Tool J-23653, Fig. 3, on end of steering shaft and compress the lock plate as far as possible using the shaft nut as shown. Pry the round wire snap ring out of the shaft groove and discard the ring. Remove tool and lift lock plate off end of shaft.

NOTE: On 1976 models, remove lock plate cover.

4. Slide the signal cancelling cam, upper bearing preload spring and thrust washer off the end of shaft.
5. Remove turn signal lever screw and remove the lever.
6. Push hazard warning knob in and unscrew the knob.
7. Wrap upper part of connector with tape, Fig. 4 to prevent snagging of wires during switch removal.
8. Remove three screws on switch and pull switch straight up through housing, Fig. 5.

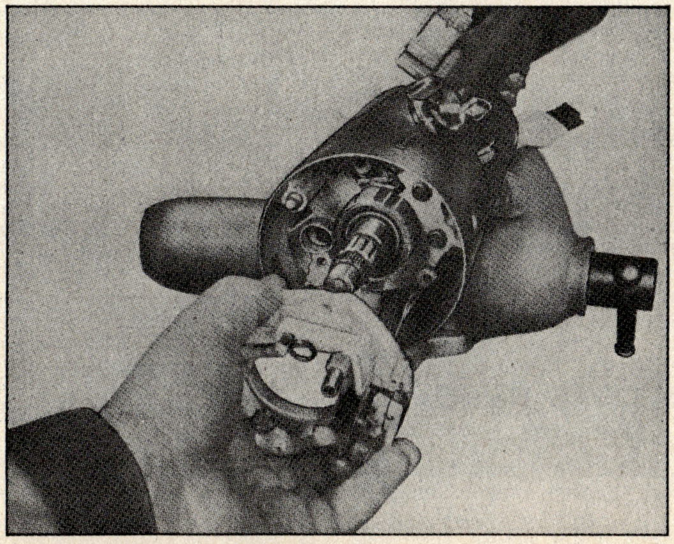

Fig. 5 Removing turn signal switch

HORN SOUNDER & STEERING WHEEL

1. Disconnect battery ground cable.
2. On regular production steering wheel models, remove the two screws securing the steering wheel shroud from beneath the wheel and remove the shroud. On optional wheels models, pry off horn button cap.
3. Remove steering wheel nut and use a suitable puller to remove the steering wheel.

RADIO, REPLACE

1975 Monza Towne Copue & 1976 All

1. Disconnect battery ground cable.
2. Remove radio control knobs and bezels, then the shaft nuts and washers.
3. Disconnect antenna lead from radio rear and remove screws securing radio to instrument panel reinforcement.
4. Lower radio from instrument panel and disconnect electrical connectors from radio.
5. Remove mounts from radio.
6. Reverse procedure to install.

1975 Exc. Monza Towne Coupe

1. Disconnect battery ground cable.
2. Remove clock stem knob, instrument panel bezel and glove box.
3. Remove instrument panel pad and radio knobs.
4. Remove the two lower screws from radio mounting bracket and remove the left lap cooler and duct, if equipped with A/C.
5. Remove two nuts from steering column mounting bracket, three screws from top of instrument panel and three bolts from instrument panel carrier reinforcement.

CAUTION: The two steering column mounting nuts are important attaching parts as they could affect the operation of the collapsible steering column. If replacement becomes necessary, they must be replaced with one of the same or equivalent part number. Also, the 15-25 ft. lb. torque must be adhered to.

6. Disconnect speedometer cable from instrument cluster, then pull complete instrument panel slightly forward and disconnect all electrical leads and antenna.
7. Remove radio from instrument panel.
8. Reverse procedure to install.

HEATER CORE REMOVAL

Less A/C

1. Disconnect battery ground cable and disconnect blower motor electrical lead.
2. Place a pan under vehicle, then disconnect heater hoses from core and secure hoses in a raised position.
3. Remove blower inlet to dash panel screws and nuts, then remove blower inlet and blower motor and wheel assembly.
4. Remove core retaining strap screws and remove core.
5. Reverse procedure to install making certain that blower inlet sealer is intact, replace as necessary.

With A/C

1. Disconnect battery ground cable.
2. Remove glove box and right air duct.
3. Remove instrument panel bezel and pad.
4. Remove left air duct and feed duct.
5. Lower steering column and remove instrument panel.
6. Remove control assembly, radio, defroster duct and large distribution duct.
7. Purge system of refrigerent.
8. Disconnect heater hoses at core.
9. Clean VIR exterior surface and line connections, then disconnect compressor inlet line, oil bleed line and condenser outlet line from VIR assembly.

NOTE: Cap all line and VIR ports.

10. Loosen evaporator inlet and outlet lines and remove VIR mounting clamp, then slide VIR assembly from evaporator lines.
11. Remove heater—distributer case to cowl attaching nuts and remove case, then disconnect wiring harness and vacuum lines.
12. Separate heater case from distributor, then remove heater core from case.

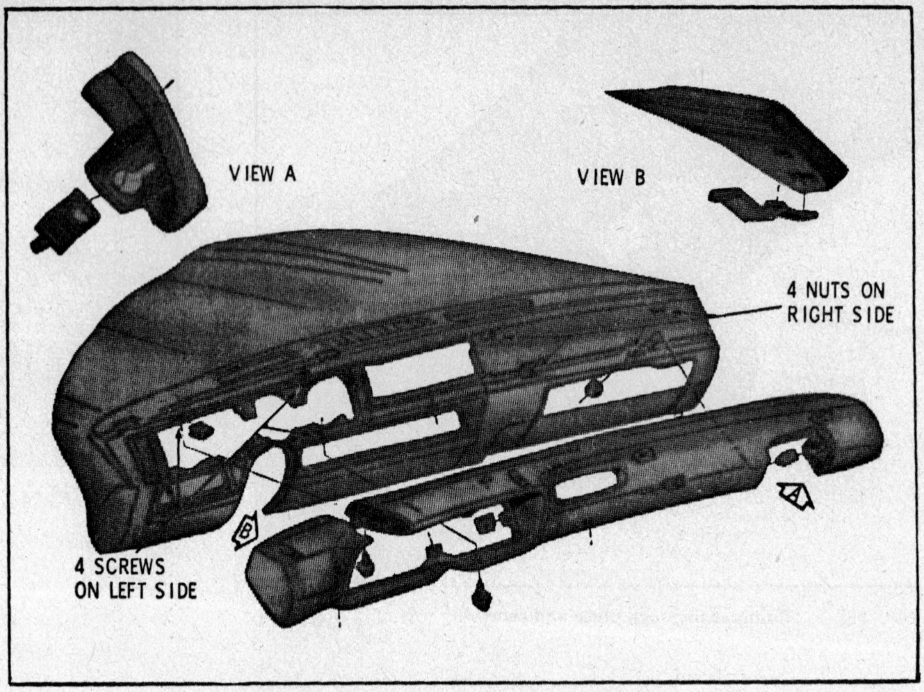

Fig. 5A Instrument panel pad (Typical)

BLOWER MOTOR, REPLACE

1. Disconnect battery ground cable and disconnect blower motor electrical leads.
2. Scribe blower motor flange to case position.
3. Remove blower motor retaining screws and remove blower motor and wheel assembly. Pry flange gently, if sealer acts as an adhesive.
4. Reverse removal procedure to install aligning scribe marks made during removal.

INSTRUMENT PANEL PAD

The instrument panel pad must be removed to service all gauges and instruments.

1. Remove glove box, then reaching through glove box opening, remove the nuts retaining the right side of pad, Fig. 5A.
2. Remove clock set knob and instrument panel bezel.
3. Remove two screws at top of cluster, then remove screws along lower edge of cover and remove instrument panel pad.
4. Reverse procedure to install.

W/S WIPER SWITCH

1975 Monza Towne Coupe & 1976 All

1. Disconnect battery ground cable.

2. Remove screws securing light shield to wiper switch and position light shield aside.

NOTE: The light shield retaining screws also retains the upper portions of the wiper switch to the mounting bosses.

3. Remove lower switch mounting screw, lower switch from panel and disconnect electrical connector, then remove switch.
4. Reverse procedure to install.

1975 Starfire

1. Disconnect battery ground cable and remove instrument panel pad as outlined under "Instrument Panel Pad" procedure.
2. Remove left A/C duct.
3. Remove two nuts securing steering column to upper bracket and lower steering column.
4. Remove screw securing upper column bracket to instrument cluster and pull cluster toward rear of vehicle.
5. Disconnect electrical connector and remove switch retaining screws, then the switch from vehicle.
6. Reverse procedure to install.

1975 Monza 2+2 & Skyhawk

1. Disconnect battery ground cable.
2. Remove left A/C duct if equipped, then disconnect headlight switch from instrument panel.
3. Disconnect electrical connectors from wiper switch and seat belt buzzer and remove wiper switch.
4. Reverse procedure to remove.

W/S WIPER BLADES

1. Remove the wiper blade from the arm by depressing the spring type blade clip, Fig. 6, away from the under side of the arm and slide the arm out of the blade clip.
2. To install wiper blade to wiper arm, slide tip end of arm into blade clip, until pin on tip end of arm engages hole in clip.
3. The blade element is retained in the blade assembly by a spring type retainer clip in the end of the blade element. When the retainer clip is squeezed together, the blade element can be slid out of the blade assembly.
4. When installing a blade element into a blade assembly, be certain to engage the metal insert in the element into all retaining tabs of the blade assembly.

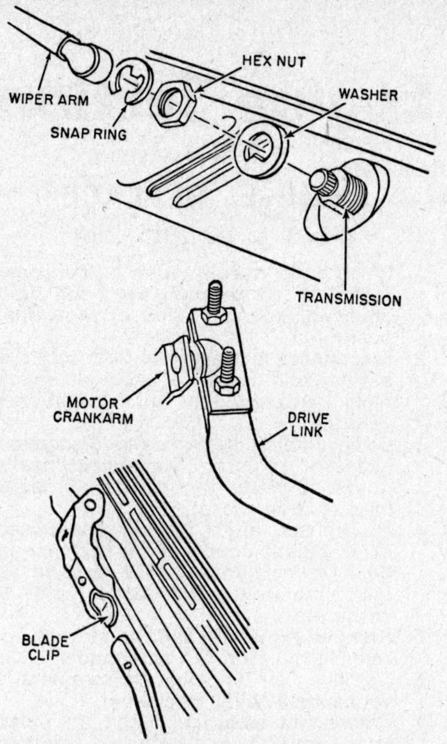

Fig. 6 Wiper blade clip location

NOTE: When properly installed, the spring type element retaining clip should be at the end of the wiper blade assembly nearest the wiper transmission.

W/S WIPER ARMS

1. Wiper motor must be in park position.
2. Use suitable tool to minimize the possibility of windshield or paint finish damage during arm removal.
3. Remove arm by prying up with tool to disengage arm from serrated transmission shaft.
4. To install arm to transmission shaft rotate the required distance and direction so that blades rest in proper parked position.

CAUTION: The parked position for the left blade tip is approx. 2" above lower windshield reveal mounting and the right blade tip within 2" of the lower windshield reveal moulding.

W/S WIPER MOTOR

1. Raise hood, remove cowl screen.
2. Reaching through cowl opening, loosen the two transmission drive link attaching nuts to motor crankarm.
3. Remove transmission drive link from motor crankarm.
4. Disconnect wiring and unfasten motor and remove.

Engine Section

ENGINE MOUNTS, REPLACE

4-140

1. Raise vehicle on hoist and support front of engine to take weight off front mounts.
2. If only one mount is being replaced, remove the mount-to-engine bracket nut on the mount not being replaced.
3. Remove the stud nut and two bolts securing mount to housing support.
4. Remove the three stud nuts securing the bracket to the engine. On the right side remove the starter brace at starter and on air conditioned equipped vehicles, remove the compressor rear lower brace at the compressor.
5. Raise front of engine to provide maximum clearance without imposing stress on other engine components.

V6-231

1. Raise car and provide frame support at front of car.
2. Support weight of engine at forward edge of pan.
3. Remove mount to engine block bolts. Raise engine slightly and remove mount to mount bracket bolt and nut. Remove mount.
4. Reverse above procedure to install and torque to specifications as shown on Fig. 1.

V8-262, 350

1. Remove lower mount bolts from frame bracket, Fig. 1A, and raise engine to relieve weight from mount.
2. Remove mount from engine.
3. Install new mount on engine, lower engine and install lower mount bolts to frame bracket.

ENGINE, REPLACE

4-140 & V8-262, 350

1. Raise hood to fully open position and install a bolt through the hood hold-open link to hold hood in wide open position.
2. Disconnect battery positive cable at battery and negative cable at engine block (except on air conditioned vehicles).
3. Drain cooling system and disconnect hoses at radiator. Disconnect heater hoses at water pump and at heater inlet (bottom hose).
4. Disconnect emission system hoses: PCV at cam cover; cannister vacuum hose at carburetor; PCV vacuum at inlet manifold and bowl vent at carburetor.
5. Remove radiator panel or shroud and remove radiator, fan and spacer.
6. Remove air cleaner, disconnecting vent tube at base of cleaner.
7. Disconnect electrical leads at: Delcotron, ignition coil, starter solenoid, oil pressure switch, engine temperature switch, transmission controlled spark switch at transmission, transmission controlled spark solenoid and engine ground strap at cowl.
8. Disconnect: Fuel line at rubber hose to rear of carburetor, transmission vacuum modulator and air conditioning vacuum line at inlet manifold, accelerator cable at manifold bellcrank.
9. On air conditioned cars, disconnect compressor at front support, rear support, rear lower bracket and remove drive belt from compressor.
10. Move compressor slightly forward and allow front of compressor to rest on frame forward brace, then secure rear of compressor to engine compartment so it is out of way.
11. Disconnect power steering pump, if equipped, and position it out of way.
12. Raise car on a hoist and disconnect exhaust pipe at manifold.
13. Remove engine flywheel dust cover or converter underpan.
14. On automatic transmission cars, remove converter-to-flywheel retaining bolts and nuts and install coverter safety strap.
15. Remove converter housing or flywheel housing-to-engine retaining bolts.
16. Loosen engine front mount retaining bolts at frame attachment and lower vehicle.
17. Install floor jack under transmission.
18. Install suitable engine lifting equipment and raise engine slightly to take weight from engine mounts and remove engine front mount retaining bolts.
19. Remove engine and pull forward to clear transmission while slowly lifting engine from car.

V6-231

1. Remove hood and drain radiator.
2. Remove fan shroud, radiator and air cleaner.
3. If equipped with A/C, disconnect compressor brackets and position out of way.
4. Remove power steering from bracket and position out of way.
5. Disconnect all hoses, linkages and electrical connections from engine.
6. Disconnect exhaust pipes and remove converter cover.
7. Remove flywheel to converter or pressure plate bolts, engine to transmission bolts and motor mount bolts.
8. Support transmission and remove engine.

CYLINDER HEADS

Some cylinder head gaskets are coated with a special lacquer to provide a good

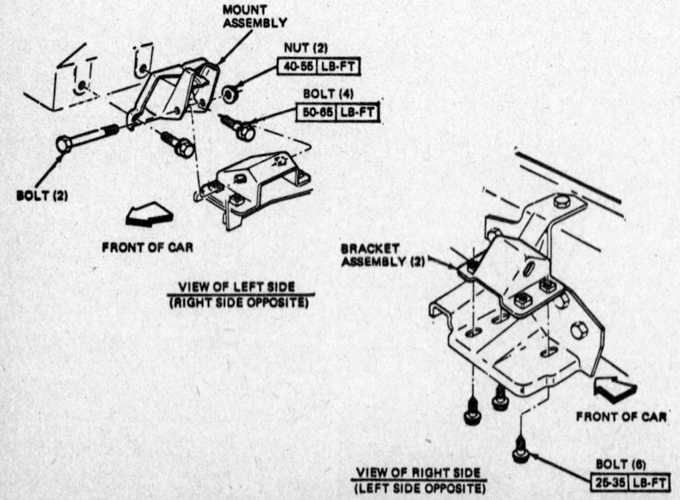

Fig. 1 Engine mounts. V6-231

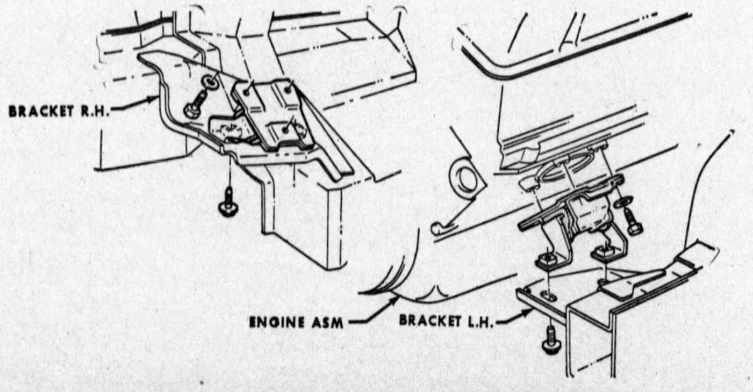

Fig. 1A Engine mounts, V8-262, 350

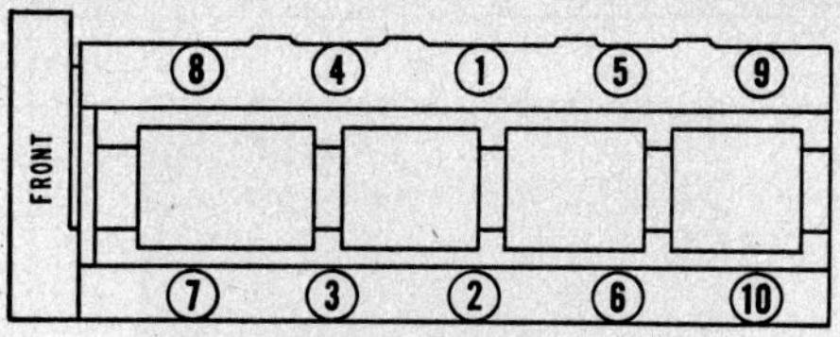

Fig. 2 Cylinder head tightening sequence. 4-140

Fig. 3 Cylinder head tightening sequence. V6-231

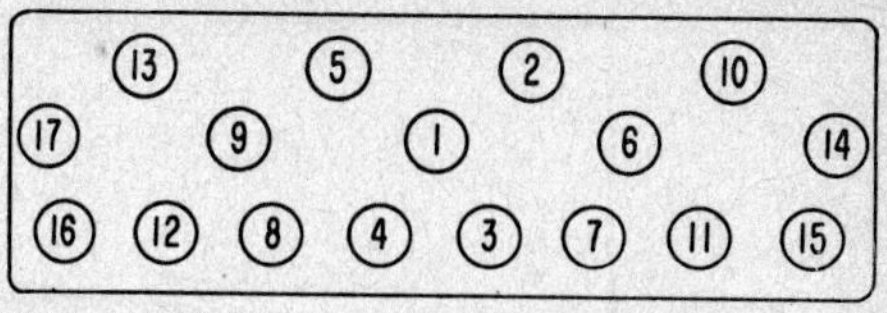

Fig. 4 Cylinder head tightening sequence. V8-262, 350

seal once the parts have warmed up. Do not use any additional sealer on such gaskets. If the gasket does not have this lacquer coating, apply suitable sealer to both sides.

4-140

1. Remove engine front cover and camshaft cover as outlined further on.
2. Remove timing belt and camshaft sprocket.
3. Remove intake and exhaust manifolds.
4. Disconnect hose at thermostat housing.
5. Remove cylinder head bolts and with the aid of an assistant lift head and gasket from engine. Place head on two blocks of wood to prevent damage to valves.
6. Reverse procedure to install and tighten head bolts in sequence shown in Fig. 2.

V6-231

1. Drain coolant and disconnect battery.
2. Remove intake manifold.
3. When removing right cylinder head, remove Delcotron and/or A/C compressor with mounting bracket and move out of the way. *Do not disconnect hoses from air compressor.*
4. When removing left cylinder head, remove oil dipstick, power steering pump and move out of the way with hoses attached.
5. Disconnect exhaust manifold from head to be removed.
6. Remove rocker arm shaft and lift out push rods.
7. Remove cylinder head.
8. Reverse procedure to install and tighten bolts gradually and evenly to specifications and in order shown in Fig. 3.

V8-262, 350

1. Remove intake and exhaust manifolds.
2. Remove rocker arm covers.
3. Remove rocker arm nuts, rocker arm balls and rocker arms.

NOTE: Rocker arms, rocker arm balls and push rods must be installed in the original position.

4. Drain coolant from cylinder block, then remove cylinder head bolts and cylinder head.
5. Reverse procedure to install and tighten head bolts in sequence shown in Fig. 4.

ROCKER ARMS

V6-231

A nylon retainer is used to retain the rocker arms. Break them below their heads with a chisel, Fig. 5, or pry out with channel locks. Production rocker arms can be installed in any sequence since the arms are identical, however, replacement arms are identified with a stamping right (R) and left (L).

ROCKER ARM STUDS

V8-262, 350

Rocker arm studs that have damaged threads may be replaced with standard studs. If studs are loose in the head, oversize studs (.003" or .013") may be installed after reaming the holes with a proper size reamer.

1. Remove old stud by placing a suitable spacer, Fig. 6 over stud. Install nut and flat washer and remove stud by turning nut.
2. Ream hole for oversize stud.
3. Coat press-fit area of stud with rear axle lube. Then install new stud, Fig. 7. If tool J-6880 shown is used, it should bottom on the head.

VALVES, ADJUST

4-140

NOTE: On 1976 engines, new type hydraulic valve lifters are used and no valve lash adjustment is required. These new type lifters are not interchangeable with the type previously used since the cylinder head has been re-designed to facilitate installation of the new type lifter.

Fig. 5 Removing nylon rocker arm retainer. V6-231

Fig. 6 Removing rocker arm stud. V8-262, 350

Fig. 7 Installing rocker arm stud. V8-262, 350

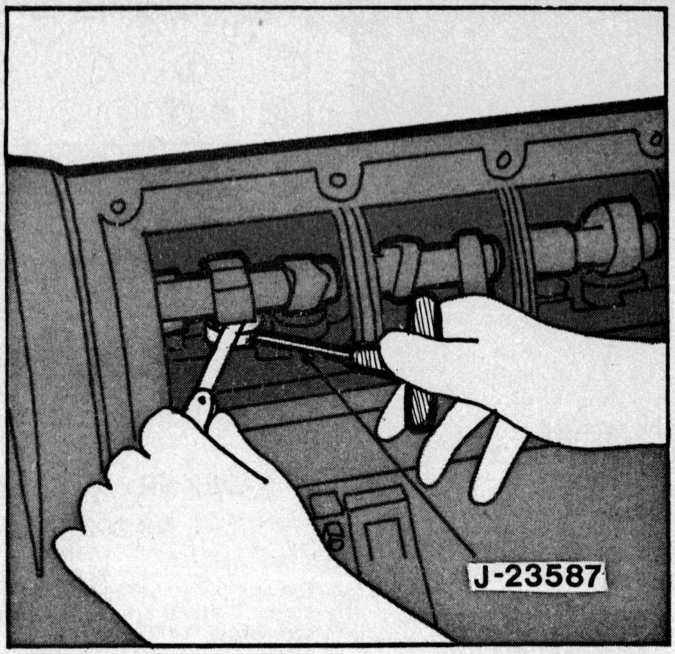

Fig. 8 Adjusting valve lash. 1975 4-140

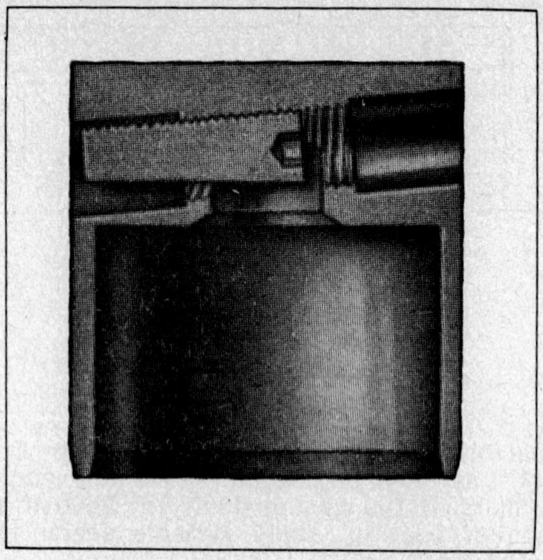

Fig. 9 Tappet and adjusting screw assembly. Screw is threaded in all areas except in valve stem contact surface. 1975 4-140

1. To adjust valves, the tappet must be on the base circle of the cam lobe. Do this as follows:
   a. Rotate camshaft timing sprocket to align timing mark on sprocket with inverted "V" notch on timing belt upper cover. The following valves can be adjusted with cam in this position (number one firing).
      Number one cylinder—Intake and Exhaust
      Number two cylinder—Intake
      Number three cylinder—Exhaust
   b. Use a feeler gauge and measure clearance between tappet and cam lobe. Adjust clearance by turning adjusting screw in tappet, Fig. 8.

   NOTE: It is mandatory that the adjusting screw, Fig. 9, be turned one complete revolution to maintain proper stem-to-screw relationship. Each revolution of screw alters clearance by .003".

   c. Rotate camshaft timing sprocket 180 degrees so timing mark is at 12 o'clock position and in line with notch on timing belt upper cover. The following valves can be adjusted with camshaft in this position (number four firing).
      Number two cylinder—Exhaust
      Number three cylinder—Intake
      Number four cylinder—Intake and Exhaust

V8-262, 350

NOTE: *After the engine has been thoroughly warmed up the valves may be adjusted with the engine shut off as follows: With engine in position to fire No. 1 cylinder the following valves may be adjusted: Exhaust 1-3-4-8, intake 1-2-5-7. Then crank the engine one more complete revolution which will bring No. 6 cylinder to the firing position at which time the following valves may be adjusted: Exhaust 2-5-6-7, intake 3-4-6-8.*

The following procedure, performed with the engine running should be done only in case readjustment is required.

1. After engine has been warmed up to operating temperature, remove valve cover and install a new valve cover gasket on cylinder head to prevent oil from running out.
2. With engine running at idle speed, back off valve rocker arm nut until rocker arm starts to clatter.
3. Turn rocker arm nut down slowly until the clatter just stops. This is the zero lash position.
4. Turn nut down $1/4$ additional turn and pause 10 seconds until engine runs smoothly. Repeat additional $1/4$ turns, pausing 10 seconds each time, until nut has been turned down the number of turns listed in the *Valve Specifications Chart* from the zero-lash position.

NOTE

This preload adjustment must be done slowly to allow the lifter to adjust itself to prevent the possibility of interference between the intake valve head and top of piston, which might result in internal damage and/or bent push rods. Noisy lifters should be replaced.

VALVE ARRANGEMENT

Front to Rear

4-140 . . . . I-E-I-E-I-E-I-E
V6-231 . . . . E-I-E-I-I-E
V8-262, 350 . . . . E-I-I-E-E-I-I-E

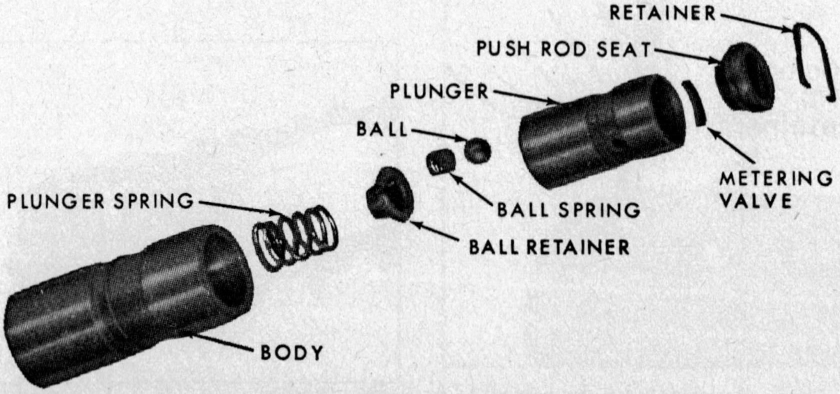

Fig. 10 Hydraulic valve lifter. V6-231 and V8-262, 350

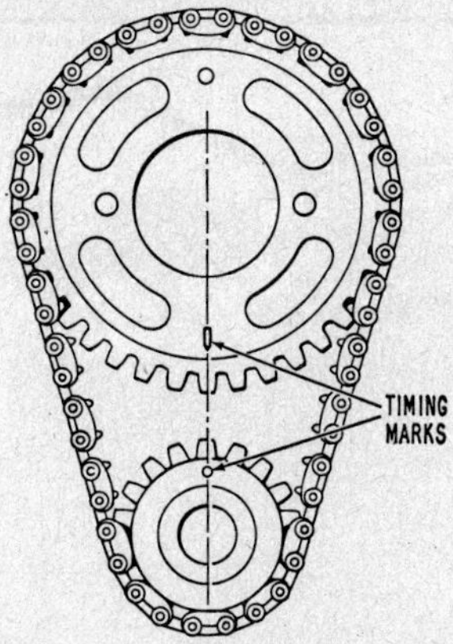

Fig. 11 Valve timing marks aligned for correct valve timing. V6-231

HYDRAULIC VALVE LIFTERS

V6-231 & V8-262, 350

Failure of an hydraulic valve lifter, is generally caused by an inadequate oil supply or dirt. An air leak at the intake side of the oil pump or too much oil in the engine will cause air bubbles in the oil supply to the lifters, causing them to collapse. This is a probable cause of trouble if several lifters fail to function, but air in the oil is an unlikely cause of failure of a single unit.

The valve lifters may be lifted out of their bores after removing the rocker arms, push rods and intake manifold. Adjustable pliers with taped jaws may be used to remove lifters that are stuck due to varnish, carbon, etc. Fig. 10 illustrates the type of lifter used.

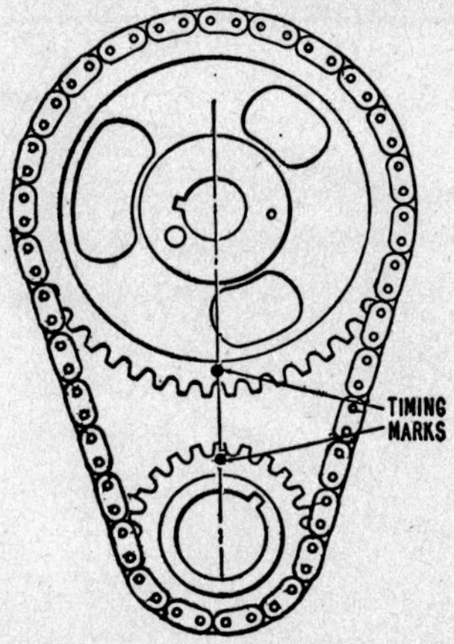

Fig. 12 Valve timing marks aligned for correct valve timing. V8-262, 350

CAM LOBE LIFT SPECS.

	Year	Intake	Exhaust
4-140 1 bar. carb.	1975	.4199	.4302
4-140 2 bar. carb.	1975	.4369	.4379
4-140	1976	.4000	.4150

VALVE LIFT SPECS.

Engine	Year	Intake	Exhaust
V6-231	1975	.4011	.3768
V6-231	1976	.3768	.3768
V8-262	1975-76	.3727	.3900
V8-350	1975-76	.3900	.4100

VALVE TIMING

Intake Opens Before TDC

Engine	Year	Degrees
4-140 1 bar. carb.	1975	22
4-140 2 bar. carb.	1975	28
4-140	1976	34
V6-231	1975-76	17
V8-262	1975-76	26
V8-350	1976	28

VALVE GUIDES

Valve guides are an integral part of the cylinder head and are not removable. If valve stem clearance is excessive, the valve stem should be reamed to the next oversize and the appropriate oversize valves installed. Valves are available in .003, .015 and .030 inch oversize for the 4-140 and V8-262, 350 and .010 inch for the V6-231.

CAMSHAFT COVER

4-140

1. Raise hood to fully open position and install bolt through the hood holdopen link.
2. Disconnect battery negative cable at battery.
3. Remove air cleaner wing nut. Disconnect ventilation tube at camshaft cover or at air cleaner; then remove air cleaner.
4. Remove PCV valve from grommet at front of cover.
5. Remove cover-to-cylinder head screws and withdraw cover from head.

ENGINE FRONT COVER

4-140

1. Raise hood to fully open position and install hood hold-open bolt.
2. Disconnect battery ground cable at battery.
3. Remove fan and spacer.
4. Loosen, but do not remove, the two cover lower screws. Cover is slotted to permit easy removal.
5. Remove the two cover upper retaining screws and remove cover.

V6-231

1. Drain cooling system and remove radiator.
2. Remove fan, pulleys and belts.
3. Remove crankshaft pulley and reinforcement.
4. If equipped with power steering, remove any pump bracket bolts attached to timing chain cover and loosen and remove any other bolts necessary that will allow pump and brackets to be moved out of the way.
5. Remove fuel pump.
6. Remove Delcotron and brackets.
7. Remove distributor cap and pull spark plug wire retainers off brackets on rocker arm cover. Swing distributor cap with wires attached out of the way.
8. Remove distributor. *If chain and sprockets are not to be disturbed, note position of distributor rotor for installation in the same position.*
9. Loosen and slide clamp on thermostat by-pass hose rearward.
10. Remove bolts attaching chain cover to block.
11. Remove two oil pan-to-chain cover bolts and remove cover.
12. If seal replacement is required, drive seal from cover using a suitable punch.
13. Position packing around opening with ends of packing facing upward. Drive shedder into position using a suitable punch and stake in three locations.
14. Rotate a hammer handle around packing until balancer hub can be inserted through opening.

V8-262, 350

1. Drain cooling system.
2. Remove alternator and A.I.R. pump, if necessary.
3. Remove fan, fan shroud and radiator.
4. Remove crankshaft damper.
5. Remove front cover attaching bolts and front cover.
6. If seal replacement is required, pry seal from cover using a screw driver.
7. Position seal so open end faces toward inside of cover.
8. Drive seal into position using tool No. J-23042.

TIMING CHAIN

V6-231 & V8-262, 350

1. Remove timing chain cover.
2. Turn crankshaft so that sprockets are aligned as shown in Figs. 11 and 12.
3. Remove oil slinger.
4. On V6-231, remove camshaft distributor drive gear and fuel pump eccentric, then using two large screwdrivers, alternately pry camshaft sprocket and crankshaft sprocket until camshaft is free. Remove camshaft

Fig. 13 Camshaft removal tool installed. 4-140

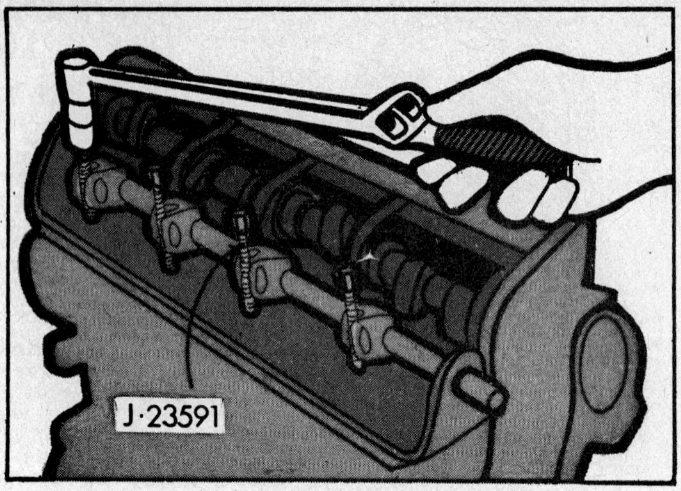

Fig. 14 Depressing valve tappets. 4-140

sprocket and chain, then slide crankshaft sprocket off crankshaft.

5. On V8-262, 350 remove three camshaft to sprocket bolts, then remove camshaft sprocket and timing chain together. Sprocket is a light press fit on camshaft, if sprocket does not come off easily, a light blow with a plastic hammer should dislodge it. If crankshaft sprocket is to be replaced, remove it with a suitable puller.
6. To install, assemble chain on sprockets with timing marks aligned as shown in Figs. 11 & 12.
7. Complete installation in reverse order of removal.

CAMSHAFT, REMOVAL

4-140

1. Remove hood.
2. Remove camshaft timing sprocket.
3. Remove timing belt upper cover and cam retainer and seal assembly.
4. Remove camshaft cover.
5. Disconnect fuel line at carburetor and remove idle solenoid from bracket.
6. Remove carburetor choke coil, cover and rod assembly.
7. Remove distributor.
8. Raise vehicle on a hoist, disconnect engine front mounts at body attachment, raise front of engine and install wood blocks, about 1½" thick, between engine mounts and body. Lower vehicle.
9. Install camshaft removal tool as shown in Fig. 13, to cylinder head as follows:
   a. Position tool to cylinder head so attaching holes align with cam cover lower attaching holes.
   b. Align tappet depressing levers on tool so each lever will depress both intake and exhaust valve for their respective cylinder. Lever should fit squarely in notches adjacent to valve tappets.
   c. With tool aligned, make sure screws in bottom of tool are backed off so they do not make contact with bosses beneath tool.
   d. Install hardened screws supplied with tool, to attach tool to head. Torque screws securely.
   e. Turn screws in bottom of tool downward until they just seat against corresponding bosses on head.
   f. Apply a heavy body lubricant to ball end of lever depressing screws and proceed to tighten screws to depress tappets, Fig. 14.

NOTE: Use a torque wrench to tighten screws the final few turns. About 10 ft. lbs. is required to depress tappets.

10. At this point the camshaft can be removed by sliding it forward from the head.

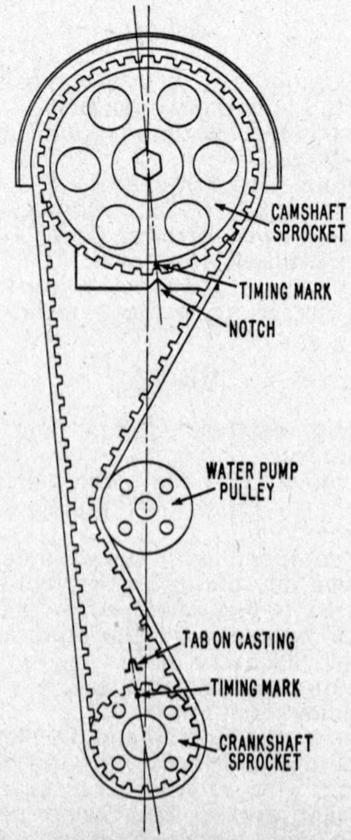

Fig. 15 Sprocket alignment marks. 4-140

V6-231 & V8-262, 350

NOTE: If engine is in the car, the radiator, grille and A/C components will have to be removed. If engine is out of car, proceed as follows:

1. To remove camshaft, remove rocker arm assemblies, push rods and valve lifters.
2. Remove timing chain and sprockets.
3. On V8-262, 350 engines install two 5/16" (18x4") bolts in camshaft bolt holes.
4. On all engines, slide camshaft out of engine using care not to damage the bearing surfaces.

TIMING BELT

4-140

1. Remove engine front cover as previously described.
2. Remove accessory drive pulley or damper.
3. Drain coolant and loosen water pump bolts to relieve tension on belt.
4. Remove timing belt lower cover, then remove belt from sprockets.

CAMSHAFT SPROCKET

4-140

After removal of timing belt, the cam-

Fig. 16 Checking sprocket alignment. 4-140

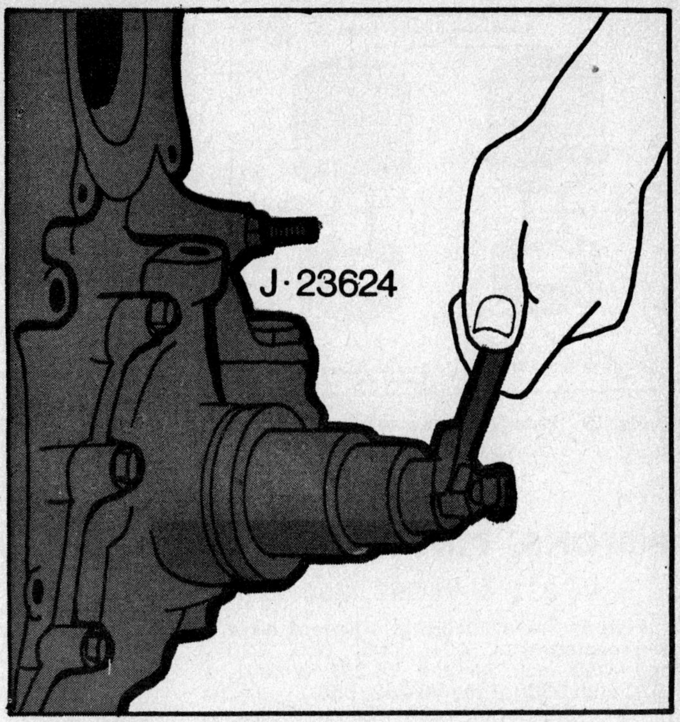

Fig. 17 Installing crankcase front oil seal. 4-140

shaft sprocket can be removed as follows:

1. Align one hole in sprocket with bolt head behind sprocket, then using a socket on bolt head to prevent cam from turning, remove sprocket retaining bolt and withdraw sprocket from camshaft.
2. When installing, be sure timing marks are aligned as in Fig. 15.

NOTE: A simplified method is provided for checking camshaft and crankshaft alignment. Proper alignment is assured by making sure hole in left rear of the timing belt upper cover is in line with the corresponding hole in the camshaft sprocket. Check alignment by inserting a pencil or other similar tool through hole in cover. If alignment is correct, the tool will enter small hole in cam gear, Fig. 16.

CRANKSHAFT SPROCKET

4-140

1. Remove engine front cover, accessory drive pulley, timing belt and timing belt lower cover.
2. Install suitable puller to crankshaft sprocket and remove sprocket.

OIL PUMP (CRANKCASE FRONT COVER) SEAL

4-140

1. Remove engine front cover, accessory drive pulley, timing belt and timing belt lower cover and crankshaft sprocket.
2. Pry old seal from front cover being careful not to damage seal housing or seal lip contact surfaces.
3. Coat seal with light engine oil and apply an approved sealing compound to outside diameter of seal.
4. Position seal, closed end outward, onto crankshaft. Then install seal into bore using tool J-23624, Fig. 17.

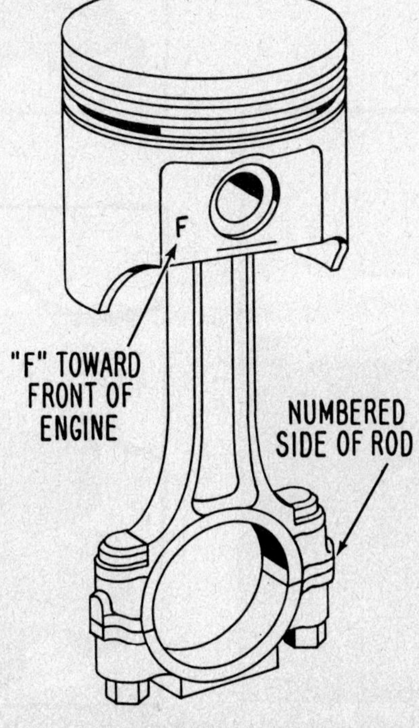

Fig. 18 Piston and rod assembly. 4-140

PISTONS & RODS, ASSEMBLE

4-140

The "F" on the front of the piston must face the front of the engine when the piston and rod assembly is installed in its proper cylinder, Fig. 18.

Oversize pistons are not available since bore reconditioning is not recommended. Production pistons are available in four size ranges so pistons can be select fitted to the bores. Cylinder bore repairs, such as glaze busting, honing or reboring will destroy the electro-chemically treated finish necessary for proper piston-to-bore operation. Therefore, if cylinder bores are unserviceable, a new cylinder case and piston assembly should be installed.

V6-231 & V8-262, 350

Rods and pistons should be assembled and installed as shown in Figs. 19 & 19A.

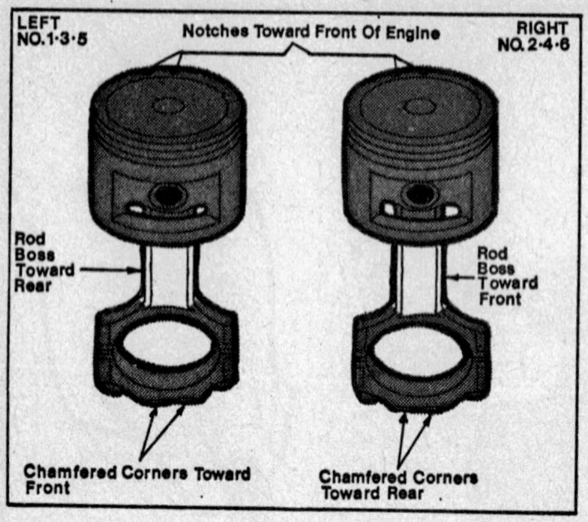

Fig. 19 Piston and rod assembly. V6-231

PISTONS, PINS & RINGS

V6-231 & V8-262, 350

Pistons are available in standard sizes and oversizes of .001, .005, .010, .020 and .030 inch for the V6-231 & .001, .020 and .030 for the V8-262, 350.

Rings are furnished in standard sizes and oversizes of .010, .020 and .030 inch for the V6-231 and .020 and .030 for the V8-262, 350.

Pistons pins are furnished in standard sizes and oversizes of .003 and .005 inch for the V6-231.

MAIN & ROD BEARINGS

Main and rod bearings are available in standard sizes and undersizes of .001, .002, .020 and .030 inch for the 4-140, .001, .002, .003 and .010 inch for the V6-231 and .001, .002, .009, .010 and .020 inch for the V8-262, 350.

NOTE: If for any reason the bearing caps are replaced on the 4-140, shimming may be necessary. Laminated shims for each cap are available for service. Shim requirements will be determined by bearing clearance.

OIL PAN

4-140

1. Raise vehicle on a hoist and drain engine oil.
2. Support front of engine so weight is off front mounts, and remove frame crossmember and both front crossmember braces.
3. Disconnect idler arm at frame side rail.

NOTE: On air conditioned vehicles, disconnect idler arm at relay rod.

4. Mark relationship of steering linkage pitman arm to steering gear pitman shaft and remove pitman arm.

NOTE: Do not rotate steering gear pitman shaft while arm is disconnected as this will change steering wheel alignment.

5. Remove flywheel cover or converter underpan.
6. Remove oil pan to cylinder case screws, tap pan lightly to break sealing bond and remove oil pan.
7. Remove pick up screen to baffle support bolts, then remove support from baffle.
8. Remove bolt securing oil pan drain

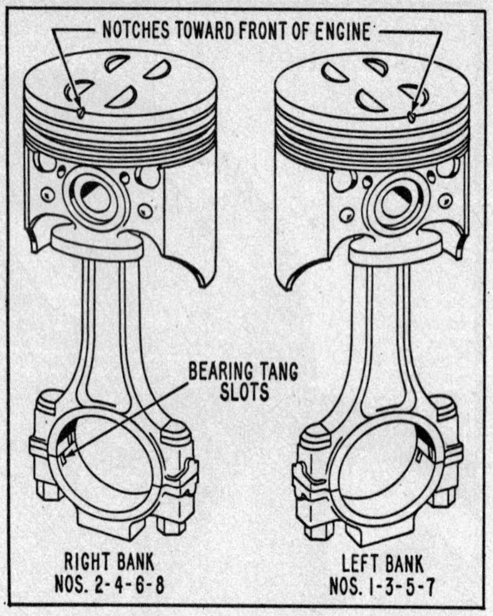

Fig. 19A Piston and rod assembly. V8-262, 350

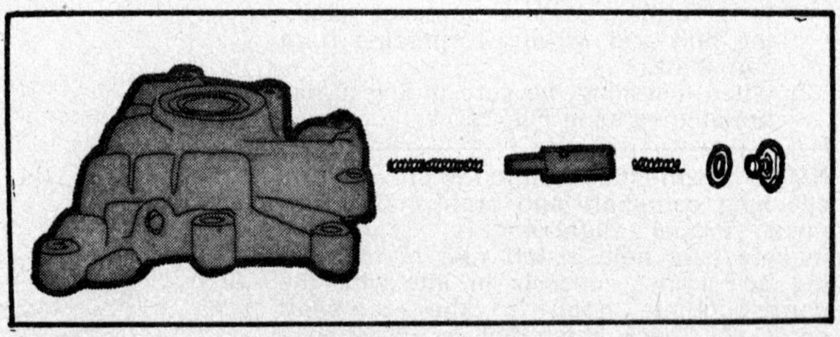

Fig. 21 Oil pump pressure regulator. 4-140

Fig. 20 Removing oil pan baffle. 4-140

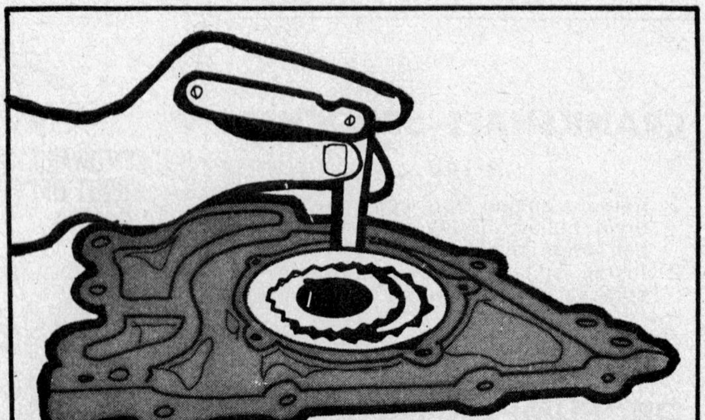

Fig. 22 Checking driven gear to housing clearance. 4-140

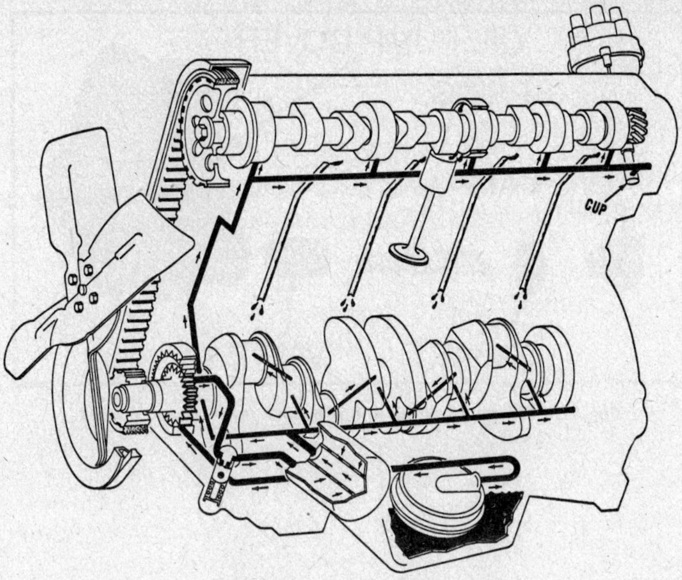

Engine oiling system. 4-140

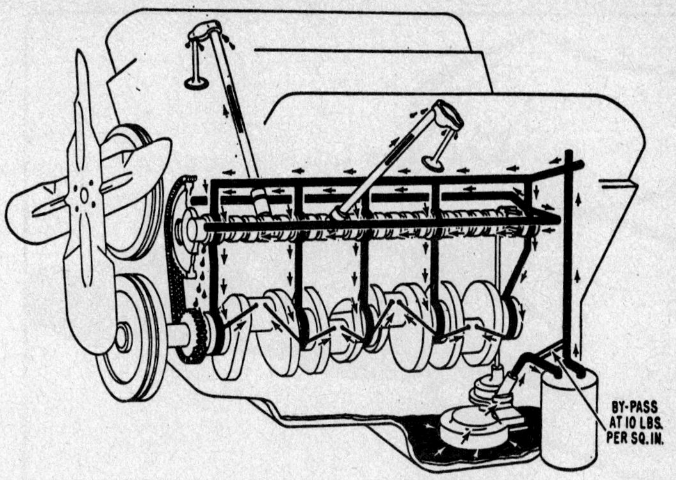

Engine oil system. V8-262, 350

back tube to baffle. Then rotate baffle 90 degrees towards left side of car and remove baffle from pick up screen, Fig. 20.

V6-231

1. Support vehicle on hoist and drain oil.
2. Remove transmission dust cover and exhaust crossover pipe.
3. Remove oil pan bolts and allow pan to drop.
4. Reverse removal procedure to install.

V8-262, 350

1. Disconnect battery ground cable.
2. Raise vehicle, drain oil pan and disconnect crossover pipe.
3. Remove underpan and splash shield from converter housing.
4. Support engine, then remove crossmember and braces.

NOTE: Scribe reference marks on each side of brace to ensure crossmember installation in original position.

5. Disconnect idler arm at frame side rail.
6. Remove starter, then the oil pan.

OIL PUMP

4-140

1. Remove engine front cover, accessory drive pulley, timing belt, timing belt lower cover and crankshaft sprocket.
2. Raise vehicle on a hoist and remove oil pan and baffle.
3. Remove bolts and stud securing oil pump to cylinder case.

Inspection

1. Clean gasket surfaces, then wash parts in approved solvent and blow out all passages.
2. Check pressure regulator for free operation, Fig. 21.
3. Inspect pump gears for nicks, broken parts and other damage.
4. Check clearance between outside diameter, of driven gear and pump. Clearance should be .0038-.0068", Fig. 22.
5. Check clearance between outside diameter of drive gear and crescent. Clearance should be .0023-.0093", Fig. 23.

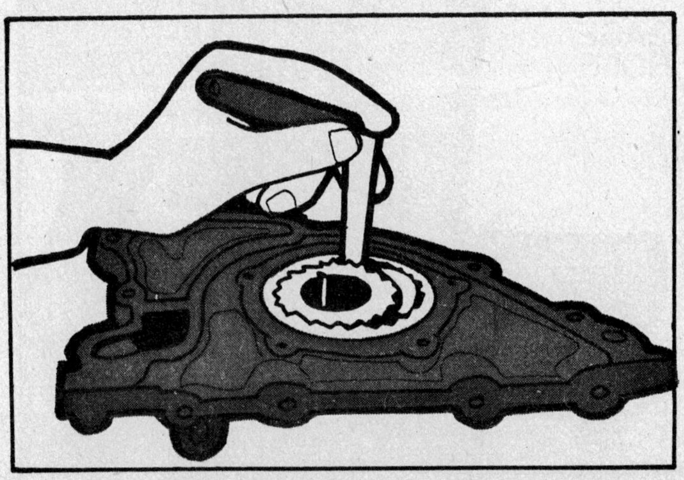

Fig. 23 Checking drive gear to crescent clearance. 4-140

Fig. 24 Checking driven gear to crescent clearance. 4-140

Fig. 25 Checking gear end clearance. 4-140

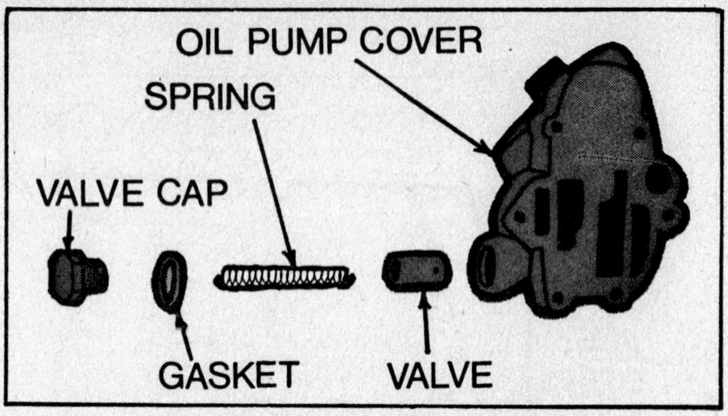

Fig. 26 Oil pump cover and pressure relief valve. V6-231

6. Check clearance between inside diameter of driven gear and pump crescent. Clearance should be .0068-.0148″, Fig. 24.
7. Check gear end clearance. It should be .0009-.0023″, Fig. 25.

NOTE: The pump gears and body are not serviced separately. If pump gears or body are worn, replacement of the entire oil pump is necessary.

V6-231

1. To remove pump, take off oil filter.
2. Disconnect wire from oil pressure indicator switch in filter by-pass valve cap (if so equipped).
3. Remove screws attaching oil pump cover to timing chain cover. Remove cover and slide out pump gears. Replace any parts not serviceable.
4. Remove oil pressure relief valve cap, spring and valve, Fig. 26. Remove oil filter by-pass valve cap, spring and valve. Replace any parts of valve not serviceable.
5. Check relief valve in its bore in cover. Valve should have no more clearance than an easy slip fit. If any perceptible side shake can be felt, the valve and/or cover should be replaced.
6. The filter by-pass valve should be flat and free of nicks and scratches.

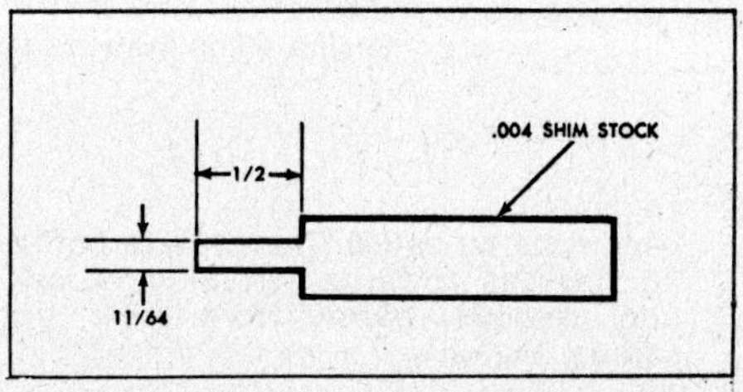

Fig. 28 Fabricated seal starting tool for helix type seal. V6-231, V8-262, 350

Fig. 27 Checking oil pump gear end clearance. V6-231

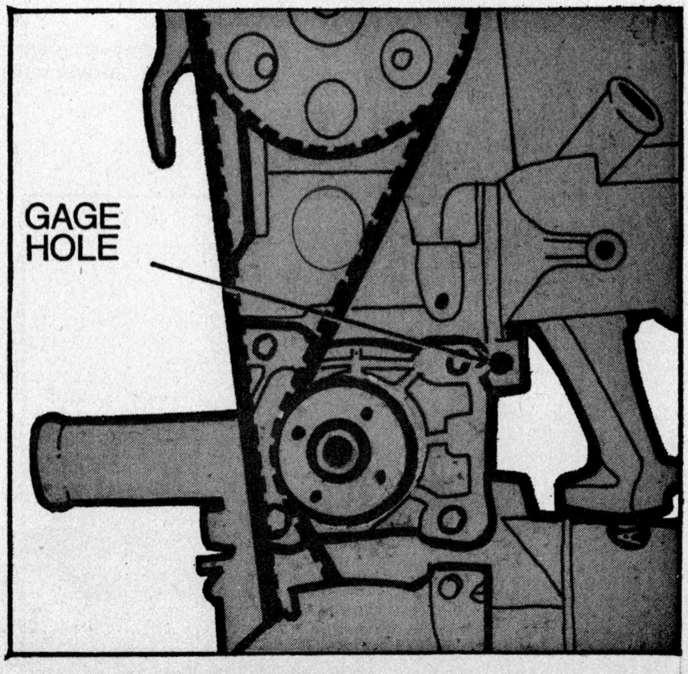

Fig. 29 Tensioning adaptor locating hole

Fig. 30 Adjusting timing belt

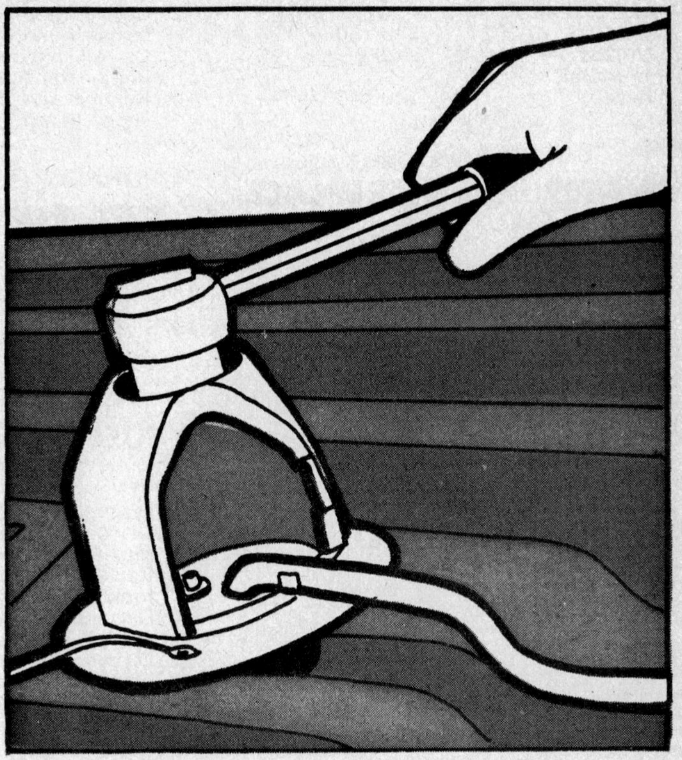

Fig. 31 Removing fuel pump and gauge unit

Assembly & Installation

1. Lubricate and install pressure relief valve and spring in bore of pump cover. Install cap and gasket. Torque cap to 30-35 lbs.
2. Install filter by-pass valve flat in its seat in cover. Install spring, cap and gasket. Torque cap to 30-35 ft-lbs.
3. Install pump gears and shaft in pump body section of timing chain cover to check gear end clearance. Check clearance as shown in Fig. 27. If clearance is less than .0018" check timing chain cover for evidence of wear.
4. If gear end clearance is satisfactory, remove gears and pack gear pocket *full* of vaseline, not chassis lube.
5. Reinstall gears so vaseline is forced into every cavity of gear pocket and between teeth of gears. *Unless pump is packed with vaseline, it may not prime itself when engine is started.*
6. Install cover and tighten screws alternately and evenly. Final tightening is 10-15 ft-lbs. torque. Install filter on nipple.

V8-262, 350

After removing the oil pan, remove pump from rear main bearing cap. Disconnect pump shaft from extension by removing clip from collar. Remove pump cover and take out idler gear, drive gear and shaft.

Should any of the following conditions be found it is advisable to replace the pump assembly.

1. Inspect pump body for cracks or wear.
2. Inspect gears for wear or damage.
3. Check shaft for looseness in housing.
4. Check inside of cover for wear that would permit oil to leak past the ends of gear.
5. Check oil pick-up screen for damage to screen, by-pass valve or body.
6. Check for oil in air chamber.

CRANKSHAFT REAR OIL SEAL

V6-231 & V8-262, 350

NOTE: V8-262, 350 engines are equipped with helix type rear seal. A seal starting tool, Fig. 28, must be used to prevent the upper seal half from coming into contact with the sharp edge of the block. Place the tip of the tool into the seal channel and "shoehorn" the seal into the upper seal channel.

When necessary to correct an oil leak due to a defective seal, always replace the upper and lower halves as a unit. *When installing either half, lubricate the lip portion only with engine oil, keeping oil off the parting line surface as this is treated with glue.* Always clean crankshaft surface before installing a new seal. Be careful of seal retainer tang while inserting a new seal so that it doesn't cut the seal.

1. To replace the lower seal, remove seal from groove in bearing cap, using a small screwdriver to pry it out.
2. Insert new seal and roll it in place with finger and thumb.
3. To replace the upper seal (with engine in car) use a small hammer and tap a brass pin punch on one end of the seal until it protrudes far enough to be removed with pliers.
4. Insert the new seal, gradually pushing with a hammer handle until seal is rolled into place.
5. Install bearing cap with new seal and tighten bearing cap bolts.

TIMING BELT TENSION, ADJUST

4-140

1. Drain coolant at engine block and remove fan and extension.
2. Remove engine front cover.
3. Remove water pump retaining bolts, clean gasket surfaces on block and pump. Install new gasket and loosely install water pump bolts.

NOTE: Apply an approved anti-seize compound to the water pump bolts before installation.

4. Position tool J-23564 in gauge hole adjacent to left side of pump, Fig. 29.
5. Apply 15 ft. lbs. of torque to water pump as shown in Fig. 30. Tighten water pump bolts while maintaining torque on side of pump.
6. Reinstall front cover, fan, extension and fill cooling system.

BELT TENSION DATA

	New lbs.	Used lbs.
Air Condition		
4-140 & V8-262	135-145	95
V6-231	150	100
Air Pump		
4-140 & V8-262	120-130	75
V6-231	70	65
Delcotron		
4-140 & V8-262	120-130	75
V6-231	125	80

Power Steering		
4-140 & V8-262	120-130	75
V6-231	150	90
Timing Belt		
4-140	100-140	—

WATER PUMP, REPLACE

4-140

1. Raise hood to fully open position and install hold-open bolt.
2. Disconnect battery ground cable at battery.
3. Remove engine fan and spacer.
4. Loosen, but do not remove, the two cover lower screws. Cover is slotted to permit easy removal.
5. Remove the two cover upper retaining screws and remove cover.
6. Drain coolant and loosen water pump bolts to relieve tension in timing belt.
7. Remove radiator lower hose and heater hose at water pump.
8. Remove water pump bolts and water pump.

V6-231 & V8-262, 350

1. Drain cooling system and loosen fan pulley bolts.
2. On V8-262, 350 engines, remove alternator and A.I.R. pump, if necessary.
3. On all engines, disconnect radiator and heater hoses from water pump.
4. Remove fan and pulley, then remove water pump.

FUEL PUMP PRESSURE

Year	Engine	Pressure lbs.
1975	4-140, V6-231	3-4½
	V8-262, 350	7-8½
1976	All	3-4½

FUEL PUMP, REPLACE

Removal

1. Disconnect battery ground cable.
2. Disconnect meter and pump wires at rear wiring harness connector.
3. Raise vehicle on hoist and drain fuel tank.
4. Disconnect fuel line hose at gauge unit pick up line.
5. Disconnect tank vent lines to vapor separator.
6. Remove gauge ground wire screw at underbody floorpan.
7. Remove tank straps bolts and lower tank carefully.
8. Unscrew retaining ring using spanner wrench J-22554, Fig. 31, and remove pump-tank unit assembly.

Replacement

1. Remove flat wire conductor from plastic clip on fuel tube.
2. Squeeze clamp and pull pump straight back about ½ inch.
3. Remove two nuts and lockwashers and conductor wires from pump terminals.
4. Squeeze clamp and pull pump straight back to remove it from tank unit. Take care to prevent bending of circular support bracket.
5. Slide replacement pump through circular support bracket until it rests against rubber coupling. Make sure pump has rubber isolator and saran strainer attached.
6. Attach two conductor wires to pump terminals using lockwashers and nuts being certain flat conductor is attached to terminal located on side away from float arm.
7. Squeeze clamp and push pump into rubber coupling.
8. Replace flat wire conductor in plastic clip on fuel pick up tube.
9. Install unit into tank and replace fuel tank.

Clutch & Transmission Section

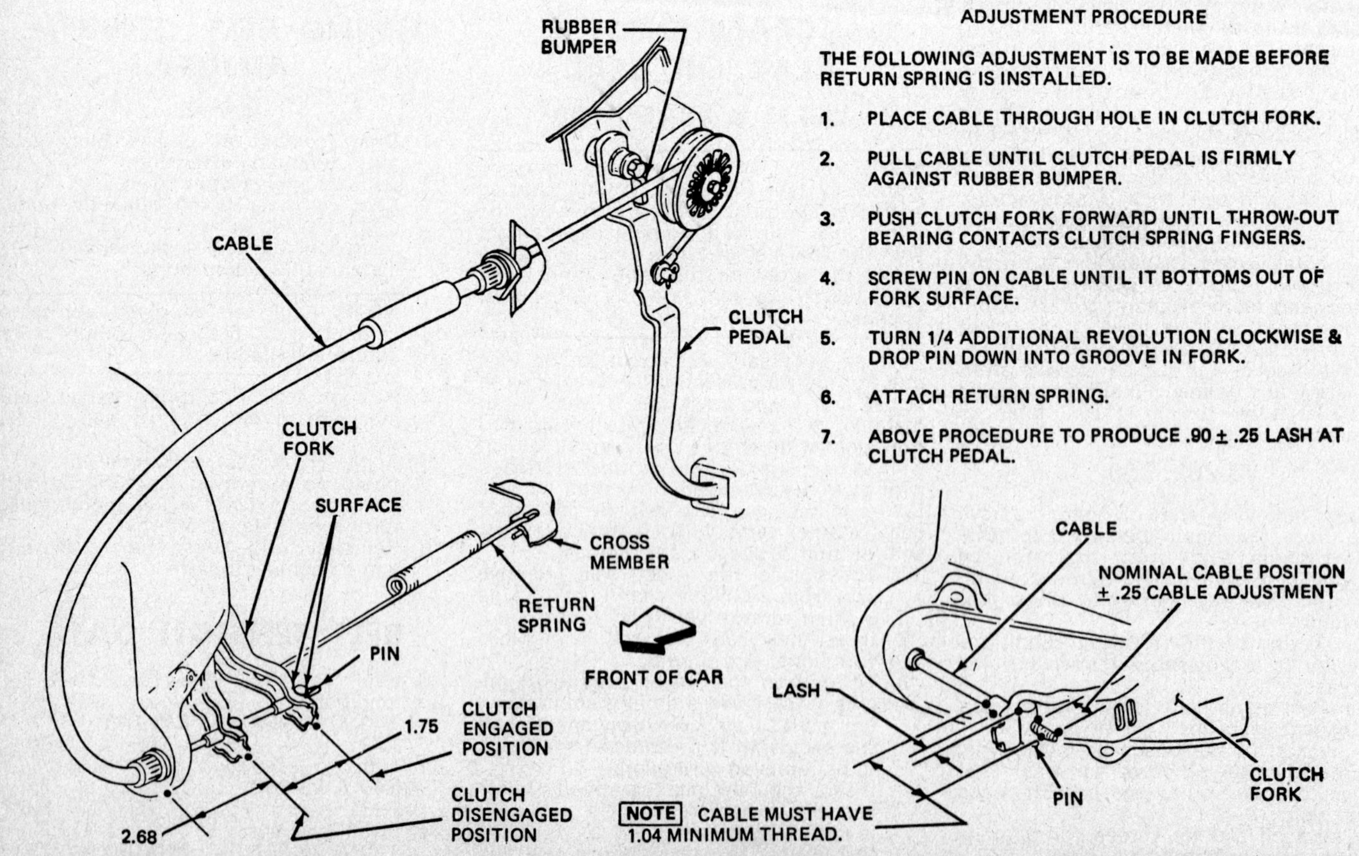

Fig. 1 Cable actuated clutch adjustment

CLUTCH PEDAL, ADJUST

The clutch used on these vehicles is cable actuated. For proper adjustment, refer to Fig. 1.

CLUTCH REPLACE

1. Remove transmission.
2. Remove clutch fork cover then disconnect clutch return spring and control cable from clutch fork.
3. Remove flywheel housing lower cover and flywheel housing.
4. Remove release bearing from the clutch fork and sleeve by sliding lever off ball stud and against spring pressure. If ball stud is to be replaced, remove cap, lock nut and stud from housing.
5. Make sure alignment marks on clutch assembly and flywheel are distinguishable.
6. Loosen clutch cover to flywheel bolts one turn at a time until spring pressure is released to avoid bending clutch cover flange, Fig. 2.
7. Support the pressure plate and cover assembly while removing bolts and clutch assembly.

NOTE: Do not disassemble the clutch cover, spring and pressure plate for repairs. If defective, replace complete assembly.

8. Reverse procedure to install making sure to index alignment marks.
9. After installing crossmember, loosely install retaining bolts, then the crossmember to transmission mount bolts. Tighten all bolts to specifications and remove the engine support.

CAUTION: Check position of engine in front mounts and realign as necessary.

FOUR SPEED TRANSMISSION, REPLACE

1976 70MM (Brazil)

1. Push shift lever boot down and loosen lock nut, remove upper portion of shift lever.
2. Remove console, retainer and boot, then remove shift control from extension housing.
3. Raise vehicle and drain lubricant.
4. Remove torque arm, referring to the "Rear Suspension" section, then remove propeller shaft, and disconnect speedometer cable and back-up light from transmission.
5. Disconnect clutch return spring and cable from fork.
6. Remove transmission to crossmember bolts.
7. Remove converter bracket and damper, if equipped.

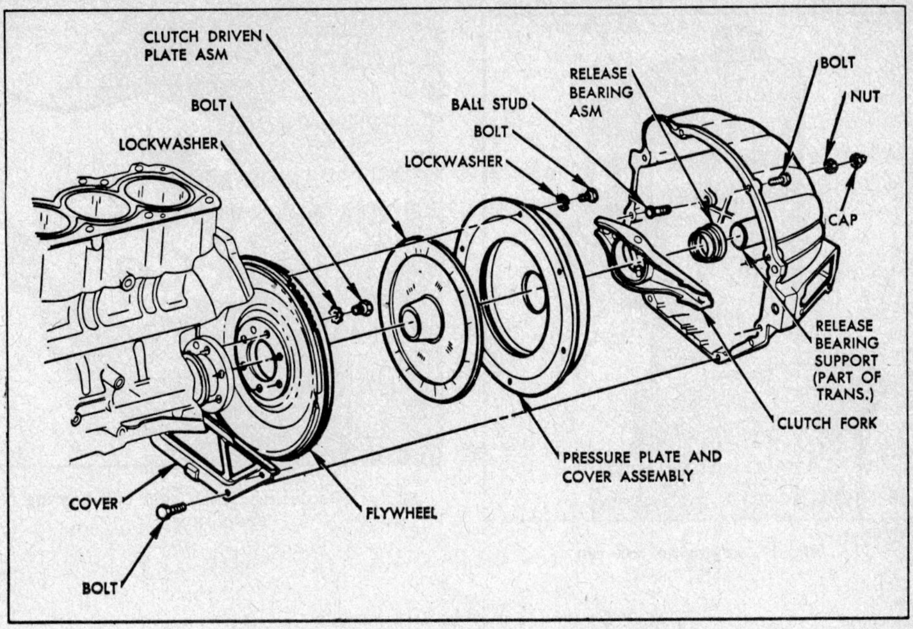

Fig. 2 Four cylinder engine clutch & components

8. Remove crossmember to frame bolts, support transmission, then remove crossmember.
9. Remove clutch housing to engine attaching bolts, slide transmission and clutch housing rearward and remove from vehicle.

1975-76 Saginaw

1. Place shift lever in neutral and remove shift lever.
2. Raise vehicle on a hoist and drain lubricant.
3. Remove propeller shaft and torque arm.
4. Remove catalytic converter bracket from transmission and disconnect exhaust pipe and converter.
5. Disconnect speedo cable, TCS switch and back-up lamp switch.
6. Remove crossmember to transmission mount bolts.
7. Support engine with an appropriate jack stand and remove crossmember to frame bolts and remove crossmember.
8. Remove transmission to clutch housing upper retaining bolts and install guide pins in holes.
9. Remove lower bolts, then slide transmission rearward and remove from vehicle.

NOTE: Inspect throwout bearing support gasket located beneath lip of support. If necessary, replace gasket before installing transmission.

FIVE SPEED TRANS. REPLACE

1. Remove shifter assembly and raise vehicle.
2. Remove propeller shaft and disconnect speedometer cable from transmission.
3. Remove crossmember to transmission bolts and the catalytic converter support bracket.
4. Support engine and remove crossmember.
5. Remove transmission to clutch housing upper retaining bolts and install guide pins.
6. Remove lower transmission to clutch housing bolts, slide transmission rearward and lower from vehicle.
7. Remove back-up lamp switch and fill plug, then drain transmission.

1976 AUTO. TRANS. LINKAGE, ADJUST

Linkage adjustments for these models are the same as the 1975 linkage adjustments found in the "Automatic Transmission" chapter, Turbo Hydra-matic 250, 350 section.

Fig. 1 Replacing roll pin

Fig. 4 Replacing clutch gear ball bearing snap ring

Fig. 5 Replacing 3-4 synchronizer hub snap ring

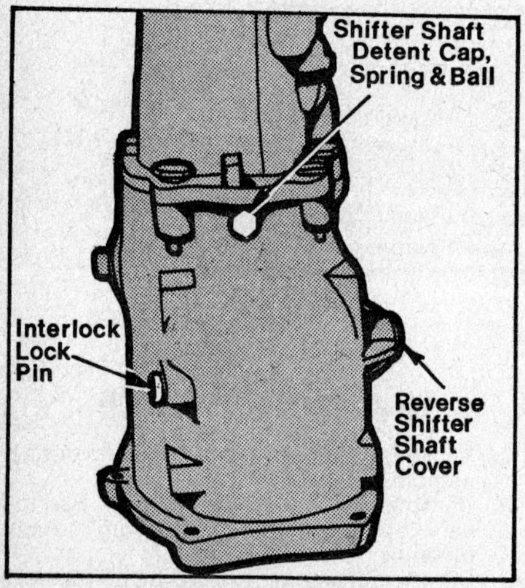

Fig. 2 Cover, cap, spring, ball and interlock pin removal

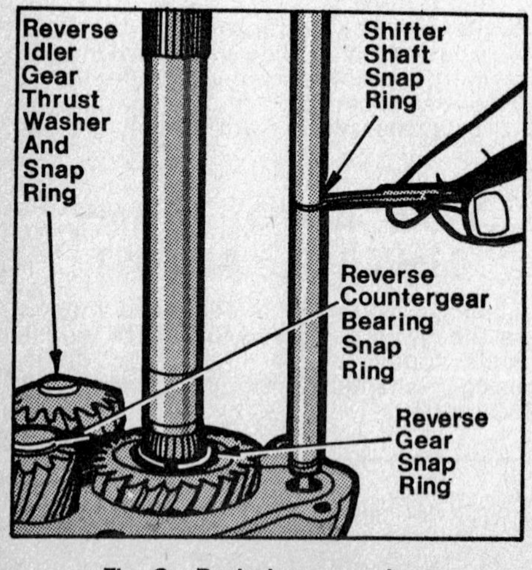

Fig. 3 Replacing snap rings

FOUR SPEED 70MM (BRAZIL) TRANSMISSION

Transmission Disassembly

1. With transmission in a vertical position, drive roll pin out and remove shifter shaft lever assembly, Fig. 1.
2. Using tools J-25295 and J-2619-01, remove reverse shifter shaft cover, shifter shaft detent cap, spring and ball and interlock lock pin, Fig. 2.
3. Remove extension housing retaining bolts and remove extension housing, then press down on speedometer gear retainer and remove gear and retainer from mainshaft.
4. Remove shifter shaft snap rings, Fig. 3, then using a pair of pliers, pull exposed end of reverse shifter shaft outward to disengage reverse lever from idler shaft. Remove idler shaft with gear attached.
5. Remove reverse gear and reverse countergear snap rings, Fig. 3, and remove gears.
6. With transmission in a horizontal position, remove clutch gear bearing retainer bolts, retainer and gasket, then remove snap ring retaining clutch gear ball bearing to housing, Fig. 4, and remove the bellhousing retaining bolts.
7. With transmission in a vertical position, expand snap ring in mainshaft bearing opening and remove case by lifting it off mainshaft. Assure that mainshaft, countergear and shifter shaft assemblies remain with bellhousing.

NOTE: It may be necessary to tap case with a plastic hammer to remove it.

8. As an assembly, remove from the bellhousing, the mainshaft with the shifter forks attached and the countergear meshed with gear teeth in the mainshaft.

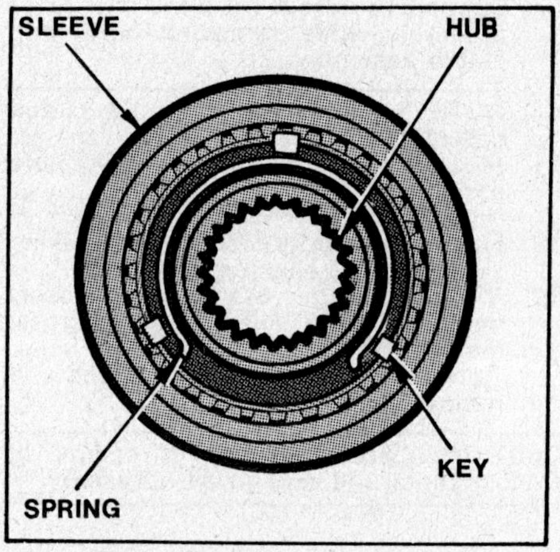

Fig. 6 Synchronizer assembly

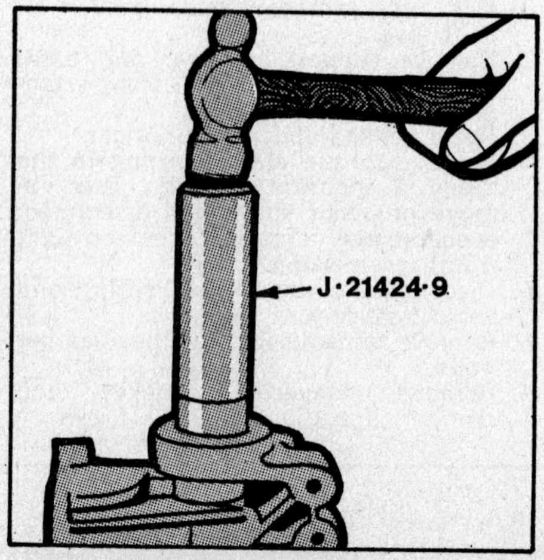

Fig. 7 Replacing extension housing bushing

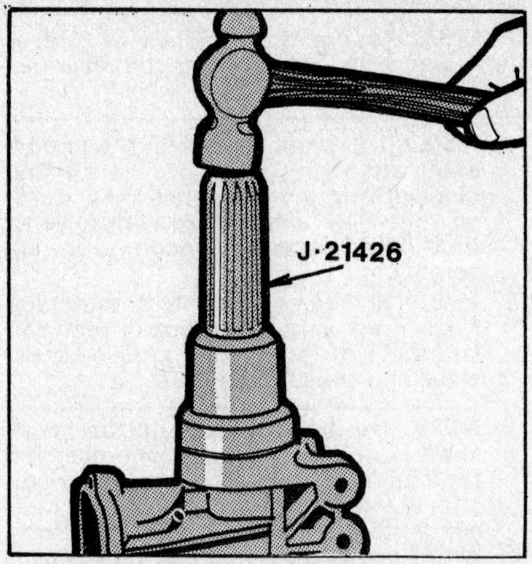

Fig. 8 Installing extension housing oil seal

Fig. 11 Installing clutch gear ball bearing snap ring

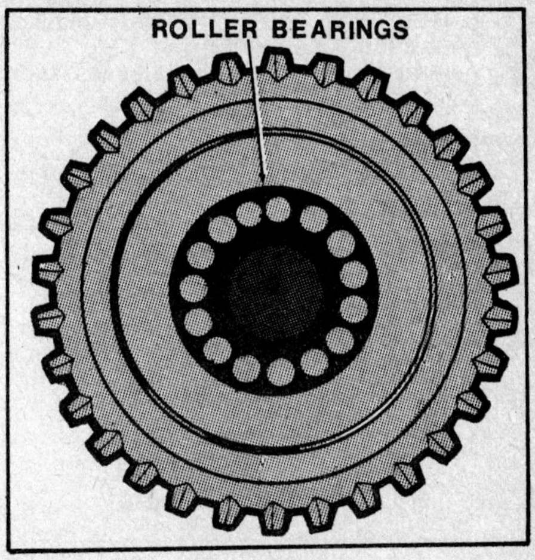

Fig. 12 Roller bearings installed in clutch gear

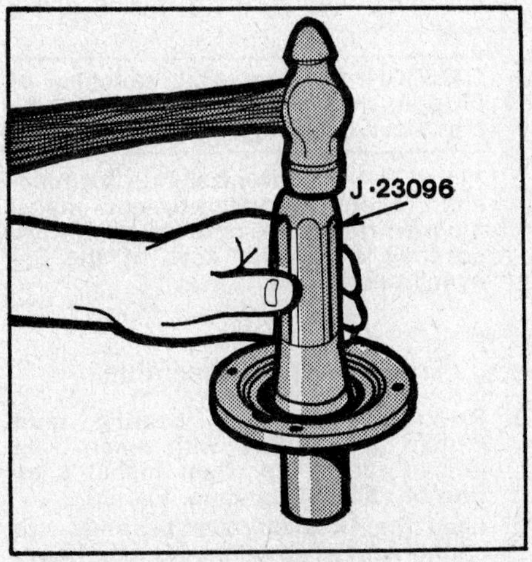

Fig. 9 Installing drive gear bearing retainer oil seal

Mainshaft Disassembly

1. Separate shifter shaft and countergear assemblies from mainshaft, then remove clutch gear and blocker ring from mainshaft.

NOTE: The clutch gear contains 15 roller bearings.

2. Remove 3-4 synchronizer hub snap ring and remove synchronizer assembly, Fig. 5. Use press if required.
3. Remove blocker ring and third speed gear, then using a press, remove ball bearing from mainshaft.
4. The remaining components can be removed as required.

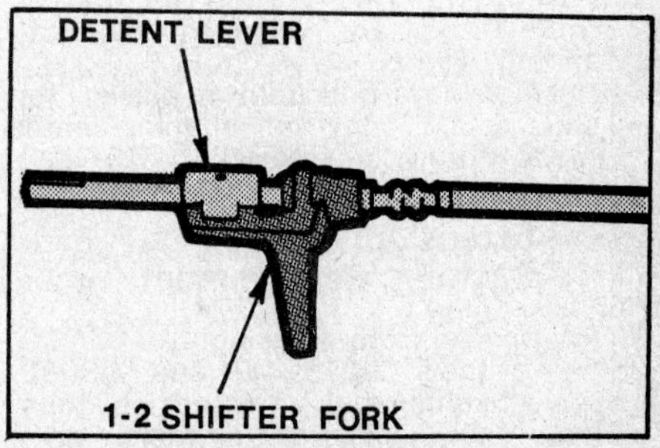

Fig. 13 Engaging detent lever to 1-2 shifter fork

Inspection

Transmission Case

1. Thoroughly clean transmission case and inspect for cracks.
2. Inspect front and rear mating faces for burrs. Smooth surfaces using a fine file, if needed.
3. Clean bearings and lubricate with oil, then inspect for roughness by turning with hand.

Roller Bearing & Shafts

Inspect countergear bearings carefully and replace if signs of wear are evident. Inspect countershaft and reverse idler shaft and replace if necessary.

Gears

1. Inspect gears for excessive wear, chips or cracks and replace as necessary.
2. Inspect reverse idler gear bushing. If the bushing is worn or damaged, the

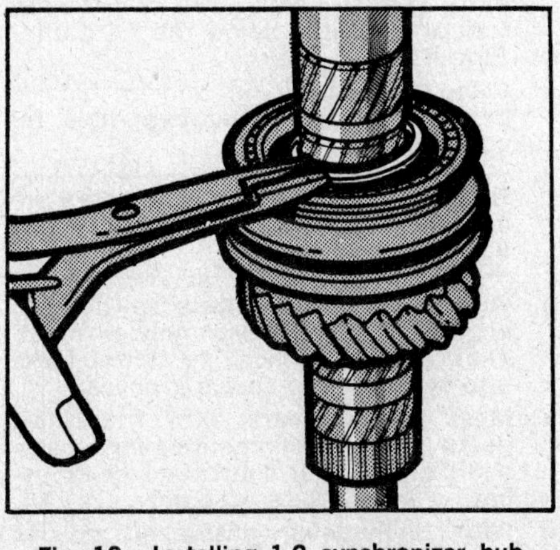

Fig. 10 Installing 1-2 synchronizer hub snap ring

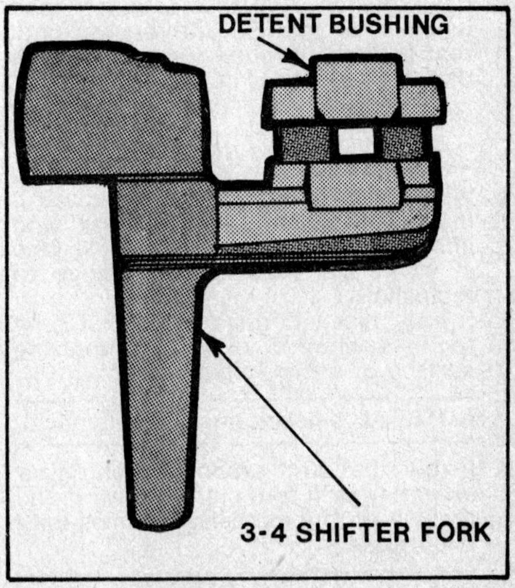

Fig. 14 Assembling detent bushing to 3-4 shifter fork

reverse idler gear must be replaced, as it cannot be serviced.
3. Assure that both clutch sleeves slide freely on their hubs.

Front & Rear Bearings

1. Thoroughly clean bearings in clean solvent and dry with compressed air.

CAUTION: Do not allow bearings to spin as this will damage race and balls.

2. Lubricate bearings with oil and check for roughness by turning by hand. Replace as necessary.

Synchronizer Keys & Springs Replacement

NOTE: Synchronizer hubs and sliding sleeves should be kept together as an assembly, however, the keys and springs may be replaced, if needed.

1. Mark hub and sleeve for alignment during assembly.
2. Push hub from sliding sleeve. The keys will become free and the springs can be removed.
3. Position the two springs so that all three keys are engaged by both springs, Fig. 6.
4. Position keys and hold in place, then slide sleeve onto hub aligning marks made during disassembly.

Extension Oil Seal & Bushing Replacement

1. Pry oil seal from extension.
2. Using tools J-21524-9 and J-8092, drive bushing out of extension housing, then using same tools, drive bushing into housing, Fig. 7.
3. Coat inside diameter of bushing and seal with transmission lubricant, then using tool J-21426, install new seal, Fig. 8.

Drive Gear Bearing Retainer Oil Seal Replacement

1. Pry out seal from retainer.
2. Position new seal into retainer, then using tool J-23096, drive seal until tool bottoms in bore. Lubricate inside diameter of seal, Fig. 9.

Mainshaft Assemble

1. With rear of mainshaft facing upward, install the second speed gear with clutching teeth upward. The rear face of gear will butt against flange of mainshaft.
2. Install blocker rings with clutching teeth downward over synchronizing surface of second speed gear.

NOTE: All blocker rings are identical.

3. Install first and second synchronizer assembly with fork slot downward and press it on the mainshaft splines until it bottoms.

NOTE: Both synchronizers are identical.

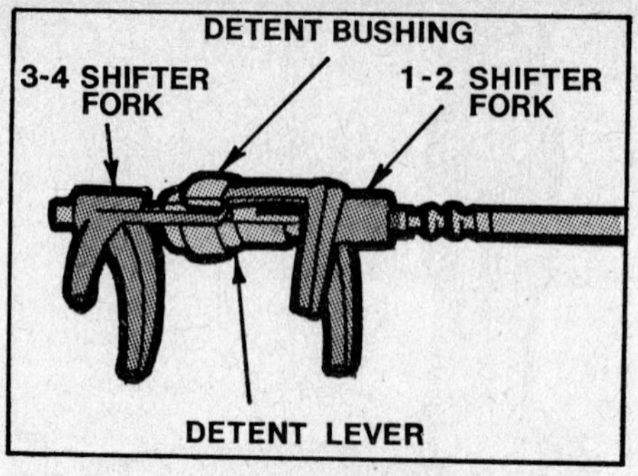

Fig. 15 Assembling 1-2 and 3-4 shifter forks on shifter shaft

Fig. 16 Shifter shaft, mainshaft gears and countergear meshed and installed on bellhousing

Fig. 17 Reverse lever installation

CAUTION: Assure that notches of blocker ring align with keys of synchronizer assembly.

4. Install 1-2 synchronizer hub to mainshaft snap ring, Fig. 10, then install blocker ring with notches downward, so that they are aligned with keys of first and second synchronizer assembly.
5. Install first speed gear with clutching teeth downward, then install rear ball bearing with snap ring groove downward and press onto shaft.

NOTE: The bearing used on the mainshaft is shielded and the one used on the clutch gear is not. These bearings should not be intermixed.

6. With front of mainshaft facing upward, install third speed gear with clutching teeth upward. The front face of gear will butt against mainshaft flange.
7. Install blocker ring with clutching teeth downward over synchronizer surface of third speed gear, then install third and fourth synchronizer assembly with fork slot facing downward.

CAUTION: Assure that notches of blocker ring align with keys of synchronizer assembly.

8. Install 3-4 synchronizer hub to mainshaft snap ring, Fig. 5, and install blocker ring with notches downward so they align with keys of the 3-4 synchronizer.

Transmission Assemble

1. Press shielded ball bearing onto clutch gear shaft with snap ring groove facing up, then install snap ring on clutch gear hub, Fig. 11.
2. Load the 15 pilot roller bearings into clutch gear cavity, Fig. 12. Use heavy grease to retain them in place.
3. Install clutch gear to mainshaft, then install detent lever to shifter shaft with roll pin an dslide 1-2 shifter fork on shaft so its arm engages detent lever, Fig. 13.
4. Assemble 3-4 shifter fork onto detent bushing, Fig. 14, then slide assembly onto shifter shaft, with 3-4 shifter fork arm located below the 1-2 shifter fork arm, Fig. 15.
5. Place bellhousing on wooden blocks, place a thrust washer over hole for countergear shaft.

NOTE: Assure that thrust washers are located in holes provided in bellhousing.

6. With both synchronizers in neutral, align shifter shaft assembly to mainshaft assembly, indexing shifter forks into synchronizer sleeve grooves.
7. Mesh countergears with mainshaft gears and install countergear, mainshaft and shifter shaft onto bellhousing as a complete assembly, Fig. 16.
8. With bellhousing positioned on its side, install snap ring onto ball bearing on clutch gear, Fig. 4.

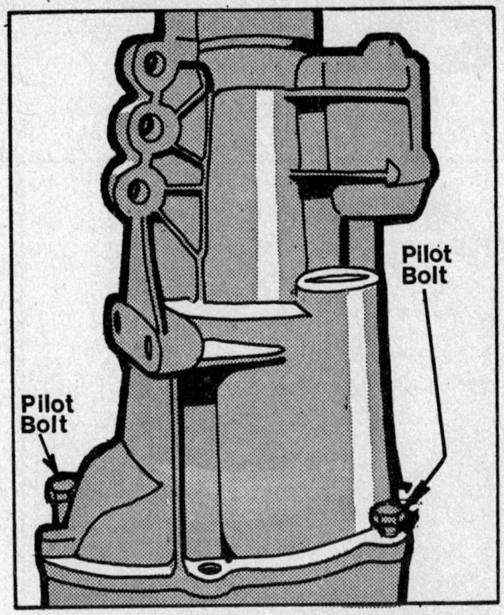

Fig. 18 Extension housing to case pilot bolt locations

9. Install gasket and bearing retainer onto bellhousing. Use sealer on the 4 retaining bolts. Reposition bellhousing on wooden blocks.

CAUTION: Assure that drain hole in retainer is aligned with hole at end of countershaft.

10. If removed, install reverse lever into case and install snap ring, Fig. 17.

NOTE: When reverse lever is installed, screwdriver slot should be parallel with front of case.

11. Install roller bearing into countergear opening with snap ring groove inside of case.

NOTE: Snap ring will be assembled to roller bearing.

12. Place gasket on bellhousing and use sealer to retain it in place.

CAUTION: Before installing case, assure that synchronizers are in neutral position, detent bushing slot is facing outward and reverse lever is flush with inside wall of case.

13. While installing case over mainshaft, countergear and shifter shaft, expand snap ring in mainshaft bearing and allow snap ring to pilot over mainshaft.

NOTE: It may be necessary to tap case using a plastic hammer to aid assembly.

14. Install interlock pin to hold detent bushing in place on shifter shaft, Fig. 2, then install idler shaft so that it engages reverse lever within case.
15. Install cover over reverse shifter shaft to retain reverse lever in place, then install detent ball, spring and cap, Fig. 2.
16. Install large reverse gear with chamfer on gear teeth upward, then push reverse gear onto mainshaft splines and retain with snap ring, Fig. 3.
17. Install small reverse gear with chamfer on gear teeth upward onto countershaft, with shoulder resting against countergear bearing and retain with snap ring, Fig. 3.
18. If removed, install snap ring, thrust washer and reverse idler gear with chamfer on gear teeth facing downward onto idler shaft. Retain thrust washer with snap ring, Fig. 3.
19. Install snap rings onto shifter shaft, Fig. 3.
20. Position speedometer gear retainer in hole in mainshaft with retainer loop forward and slide speedometer gear over mainshaft and into place.
21. Position extension housing and gasket on case, then install the 2 pilot bolts followed by the 3 regular bolts, Fig. 18.

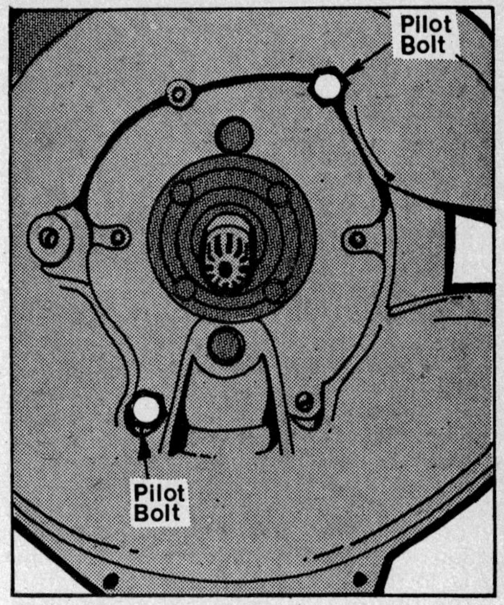

Fig. 19 Bellhousing to case pilot bolt locations

NOTE: Pilot bolts are partially threaded and are installed in the top right hand corner and bottom left hand corner of the case.

22. Install shifter shaft lever onto shifter shaft, aligning drilled hole near end of shaft. Drive roll pin into shifter shaft lever and shifter shaft, Fig. 1.
23. Install the 2 bellhousing retaining pilot bolts followed by the 4 regular bolts, Fig. 19.

NOTE: Pilot bolts are partially threaded and are installed in the right hand top corner and left hand bottom corner of the bellhousing.

Rear Axle, Propeller Shaft & Brakes

REAR AXLE

Description

The rear axle, Fig. 1, is a semi-floating type consisting of a cast carrier and large bosses on each end into which two welded steel tubes are fitted. The carrier contains an overhung hypoid pinion and ring gear. The differential is a two pinion arrangement.

The overhung hypoid drive pinion is supported by two preloaded tapered roller bearings. The pinion shaft is sealed by means of a molded, spring loaded, rubber seal. The seal is mounted on the pinion shaft flange which is splined and bolts to the hypoid pinion shaft.

The ring gear is bolted to a one piece differential case and is supported by two preloaded tapered roller bearings.

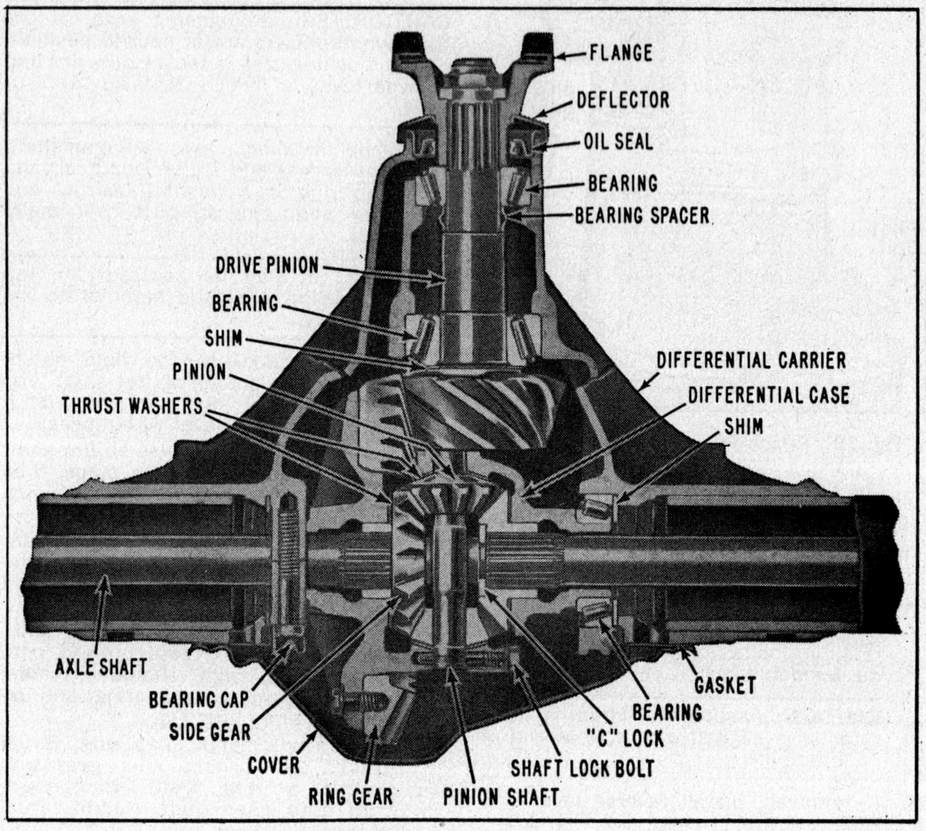

Fig. 1 Rear axle cross section

Removal

1. Support vehicle at frame and using a suitable jack support axle housing.
2. Remove wheels, brake drums and axle shafts.
3. Disconnect brake lines from axle tube clips and remove bolt securing brake line junction block to rear axle housing.

NOTE: Do not disconnect brake lines from wheel cylinders or junction block.

4. Remove brake backing plates and secure backing plates to frame.
5. Remove parking brake cable adjusting nuts at equalizer, pull center cable rearward and disconnect two rear cables from body connectors.
6. Disconnect rear brake hose at floor pan and cap ends of hose and line.
7. Disconnect shock absorber from rear axle.
8. Disconnect track rod, then slowly lower axle until all spring tension is relieved and pry springs from axle housing pads.
9. Disconnect propeller shaft and torque arm and position aside.

NOTE: Support axle at companion flange to prevent housing from rotating.

10. Remove lower control arm attaching bolts from axle housing.
11. Remove support at companion flange, then the axle assembly.

AXLE SHAFT

1. Raise vehicle on a hoist and remove wheel and tire assembly and brake drum.
2. Drain lubricant from axle by removing carrier cover.
3. Unscrew pinion shaft lock screw and remove pinion shaft, Fig. 2.
4. Push flanged end of axle shaft toward center of car and remove "C" lock from button end of shaft.
5. Remove axle shaft from housing being careful not to damage seal.

Oil Seal &/or Bearing Replacement

1. If replacing seal only, remove the

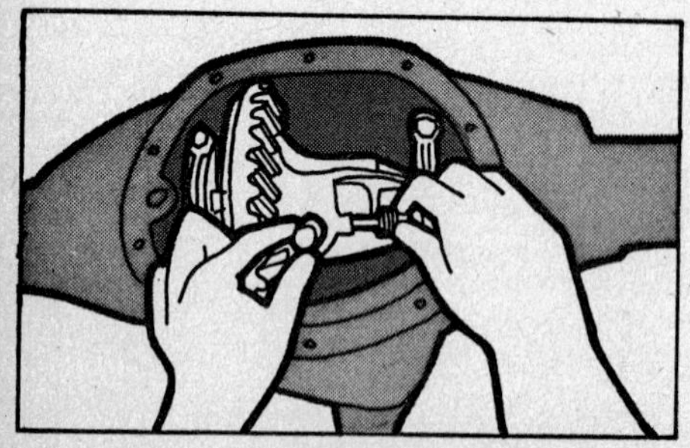

Fig. 2 Differential pinion shaft removal

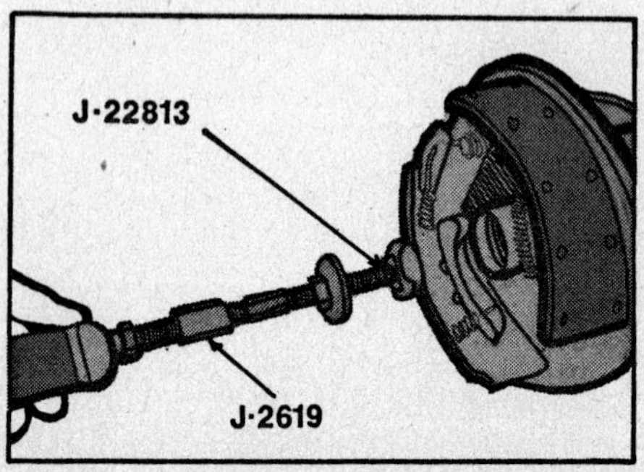

Fig. 3 Removing wheel bearing and seal

Fig. 4 Installing seal and wheel bearing

seal by using the button end of axle shaft. Insert the button end of shaft behind the steel case of the seal and pry seal out of bore being careful not to damage housing.

2. If replacing bearings, insert tool J-22813 into bore so tool head grasps behind bearing, Fig. 3. Slide washer against seal, or bearing, and turn nut against washer. Attach slide hammer J-2619 and remove bearing.
3. Pack cavity between seal lips with a high melting point wheel bearing lubricant. Position seal on tool J-21491 and position seal in axle housing bore, tap seal in bore just below end of housing, Fig. 4.

PINION FLANGE

Removal

1. Support vehicle and remove both rear wheels and drums.
2. Disconnect propeller shaft and support it by wiring shaft to exhaust pipe.

NOTE: Use a heavy rubber band or tape to hold bearings onto journal to prevent losing bearing rollers if tie wire has been removed.

3. Using an inch pound torque wrench, check pinion bearing pre-load.
4. Remove pinion flange.

Installation

1. Coat pinion flange splines with silactic and install pinion flange on pinion by tapping with a plastic hammer until a few threads project through.
2. Install pinion flange washer and nut.
3. While holding pinion flange with holder tighten the nut slowly while intermittantly rotating pinion to seat bearing. Check preload until it is 3-5 inch pounds more than reading obtained in step 3 under "Removal."
4. Connect propeller shaft and install drums and wheels.
5. Check and add lubricant as necessary.

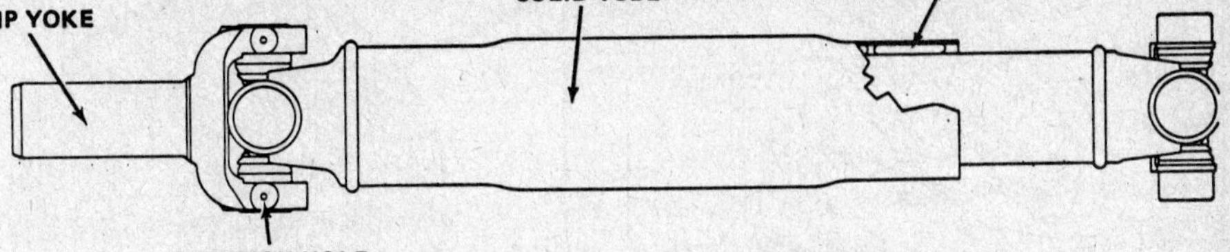

Fig. 5 Propeller shaft cross section

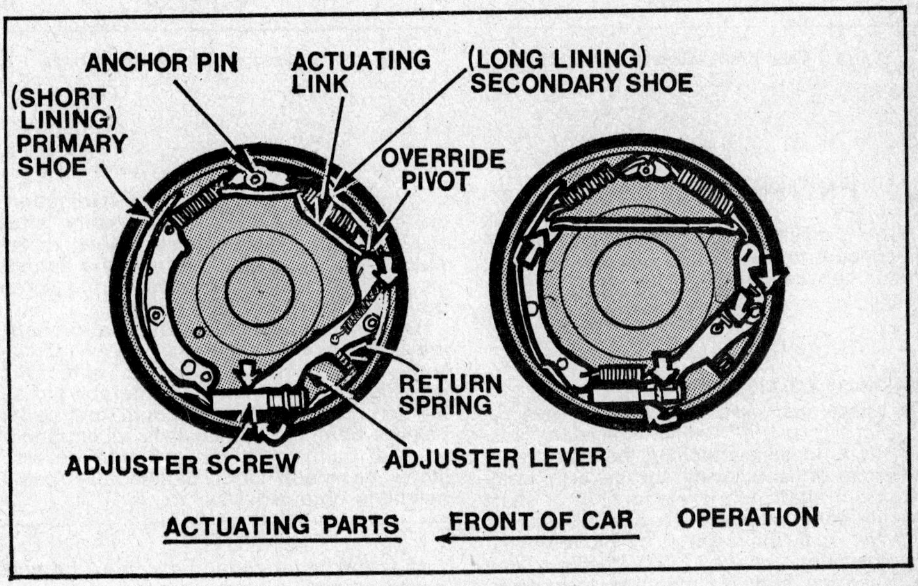

Fig. 5A Self-adjusting brakes. 1976

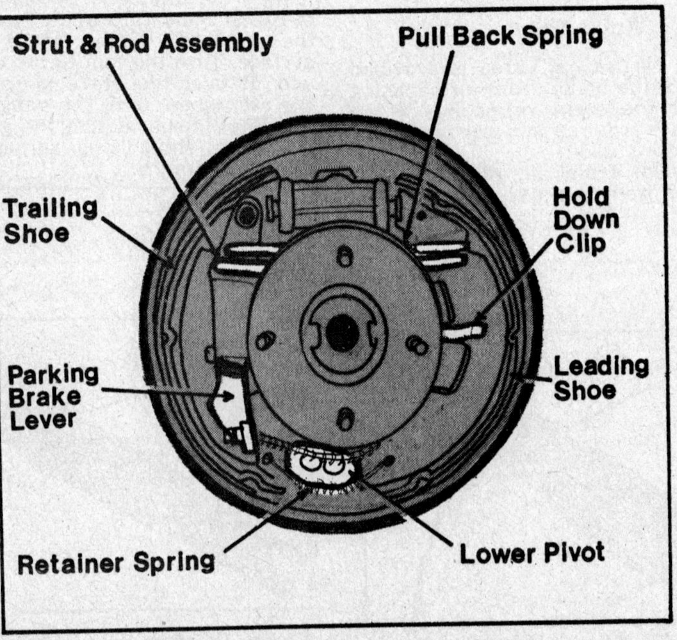

Fig. 6 Rear brake assembly. 1975

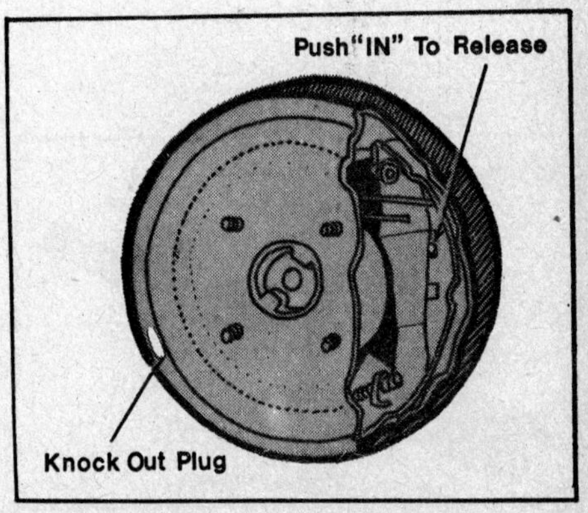

Fig. 7 Drum knock out provision

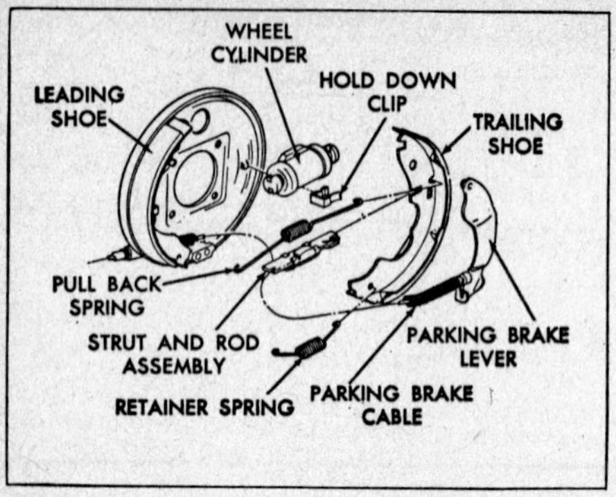

Fig. 8 Rear brake disassembled. 1975

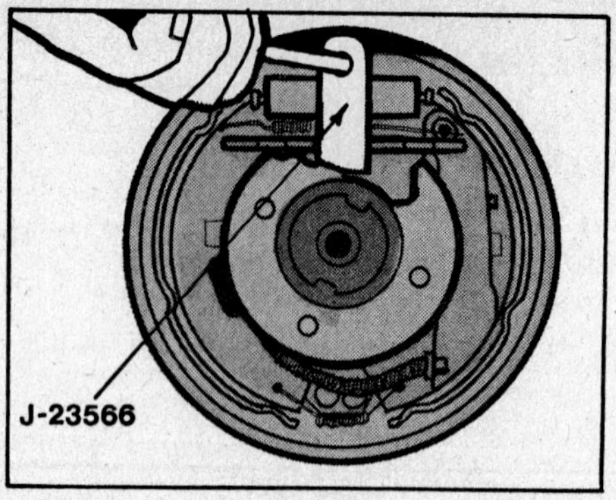

Fig. 9 Releasing adjuster on vehicle

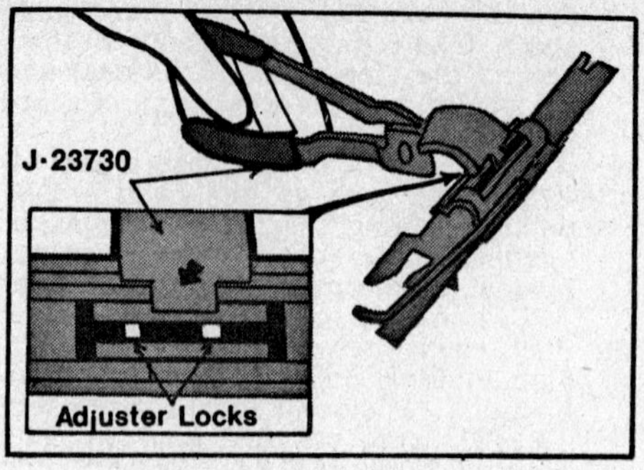

Fig. 10 Releasing adjuster locks

PROPELLER SHAFT

The propeller shaft used is made up of concentric steel tubes with rubber elements between, Fig. 5.

Propeller Shaft, Replace

1. Raise vehicle on a hoist.
2. Disconnect arm at rear axle, then referring to the "Rear Suspension" section, loosen attaching bolt at transmission and swing torque arm away from shaft. Mark relationship of shaft to companion flange and disconnect the rear universal joint by removing trunnion bearing "U" bolts. Tape bearing cups to trunnion to prevent dropping and loss of bearing rollers.
3. Withdraw propeller shaft front yoke from transmission.
4. When installing, be sure to align marks made in removal to prevent driveline vibration.

BRAKE ADJUSTMENTS

1976

These rear wheel brakes, Fig. 5A, have self-adjusting shoe mechanisms that assure correct lining-to-drum clearances at all times. The automatic adjusters operate only when the brakes are applied as the car is moving rearward or when the car comes to an uphill stop.

Although the brakes are self-adjusting, an initial adjustment is necessary after the brake shoes have been relined or replaced, or when the length of the adjusting screw has been changed during some other service operation.

Frequent usage of an automatic transmission forward range to halt reverse vehicle motion may prevent to automatic adjusters from functioning, thereby inducing low pedal heights. Should low pedal heights be encountered, it is recommended that numerous forward and reverse stops be made until satisfactory pedal height is obtained.

NOTE

If a low pedal condition cannot be corrected by making numerous reverse stops (provided the hydraulic system is free of air) it indicates that the self-adjusting mechanism is not functioning. Therefore it will be necessary to remove the brake drum, clean, free up and lubricate the adjusting mechanism. Then adjust the brakes, being sure the parking brake is fully released.

Adjustment

A lanced "knock out" area is provided in the web of the brake drum for servicing purposes in the event retracting of the brake shoes is required in order to remove the drum.

1. With brake drum off, disengage the actuator from the star wheel and rotate the star wheel by spinning or turning with a screwdriver.
2. Using the brake drum as an adjustment fixture, turn the star wheel until the drum slides over the brake shoes with a slide drag.
3. Turn the star wheel $1\frac{1}{4}$ turns to retract the brake shoes. This will allow sufficient lining-to-drum clearance so final adjustment may be made.
4. Install drum and wheel.

NOTE: *If lanced area in brake drum was knocked out, be sure all metal has been removed from brake compartment. Install new hole cover in drum to prevent contamination of brakes. Make certain that drums are installed in the same position as when removed with the drum locating tang in line with the locating hole in the wheel hub.*

---

5. Make final adjustment by driving and stopping in forward and reverse until satisfactory pedal height is obtained.

NOTE: The recommended method of adjusting the brakes is by using the Drum-to-Brake Shoe Clearance Gauge to check the diameter of the brake drum inner surface. Turn the tool to the opposite side and fit over the brake shoes by turning the star wheel until the gauge just slides over the linings. Rotate the gauge around the brake shoe lining surface to assure proper clearance.

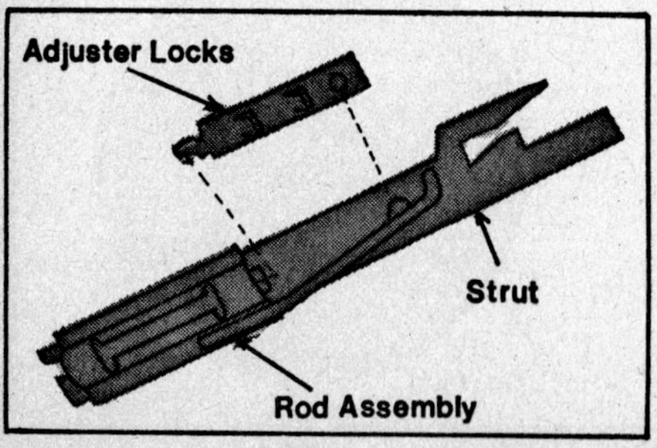

Fig. 11 Adjuster lock positioning

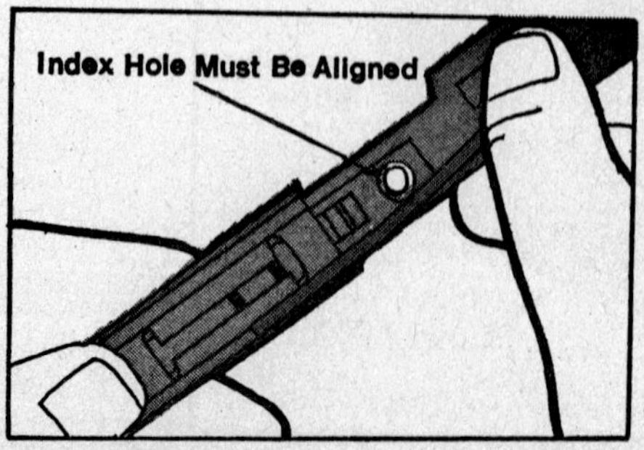

Fig. 12 Installing rod assembly

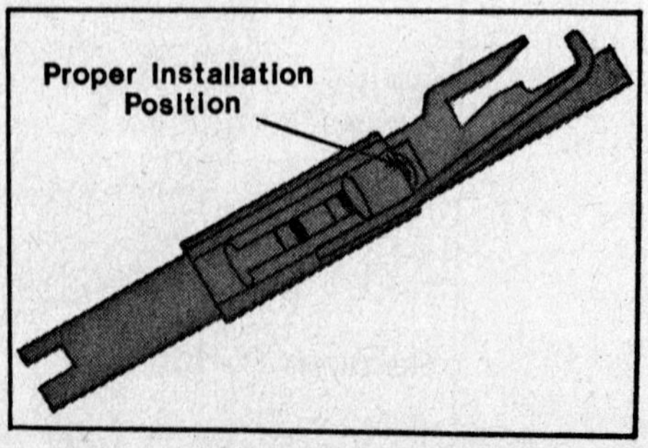

Fig. 13 Properly positioned adjuster rod assembly

1975

Power disc brakes are used on the front wheels and drum brakes, Fig. 6, are used on the rear wheels. Rear brake adjustment is not automatic. Adjustment takes place, if needed, only when the parking brake is applied. When the parking brake is applied, the strut is pushed against the front shoe and the rod is pulled against the rear shoe. As the shoes spread, a spring lock mounted within the strut and rod assembly, allows the strut and rod assembly to lengthen. When the parking brake is released, the rod connected to the rear shoe is relaxed and brake shoe pressure on the drum is released providing running clearance. Clearance is obtained by the difference in the diameter of the rod and the diameter of the hole in the rear shoe.

NOTE: If the brake drum cannot be removed, it will be necessary to release the brake adjuster. To gain access to the adjuster, knock out the lanced area in web of brake drum, Fig. 7, using a chisel or similar tool. Release the rod from the trailing shoe by pushing in on the rod until it is clear of the shoe. The pull back spring will then pull the shoes toward each other and the drum may be removed.

CAUTION: After knocking out the lanced area, be sure to remove the piece of metal from inside the drum.

When installing new brake shoes, Fig. 8, the adjuster assembly must be released and repositioned. To release adjuster on vehicle, position tool J-23566 as shown in Fig. 9. The tool releases the adjuster assembly by unseating the clips from the rod assembly. Both clips must be unseated at the same time.

To release adjuster off vehicle, position tool J-23730 so that tang of tool rests on flat portion of locks (between lock tangs) Fig. 10, then press down on adjuster locks. Work rod assembly free of adjuster locks, then when both adjuster tangs are clear of rod assembly, slide rod off lever. Subassemble rod assembly to strut making certain that index hole is lined up and seated Fig. 11, then slide rod assembly over adjuster locks, Fig. 12, until both locks are positioned as shown in Fig. 13. Note that adjuster lock index is about ½ covered by rod assembly.

PARKING BRAKE, ADJUST

1. Raise vehicle and remove propeller shaft.
2. Apply parking brake one notch from fully released position and raise hoist.
3. Loosen equalizer check nut and tighten the adjusting nut until a slight drag is felt when rear wheels are rotated.
4. Tighten check nut securely.
5. Release parking brake and rotate rear wheels. No drag should be present.

MASTER CYLINDER, REPLACE

1. Disconnect cylinder push rod from brake pedal.
2. Disconnect brake lines from two outlets on cylinder and cover ends of lines to prevent entry of dirt.
3. Unfasten and remove cylinder from brake booster.

POWER BRAKE UNIT

1975-76

1. Remove the vacuum hose from check valve and then remove the master cylinder retaining nuts.
2. Remove the bolt securing the brake pipe distributor and switch assembly to fender skirt.
3. Pull master cylinder forward until it just clears mounting studs and move aside. Support cylinder to avoid stress on hydraulic lines.
4. Remove power unit to dash nuts.
5. Remove brake pedal pushrod retainer and disconnect push rod from pin.
6. Remove power unit from vehicle.

Rear Suspension

DESCRIPTION

This rear suspension system does not use upper control arms, instead, it uses a torque arm mounted rigidly to the differential housing at the rear and to the transmission through a rubber bushing at the front. This torque arm prevents axle housing rotation caused by starting and stopping. Along with the torque arm a track rod is used to connect the axle housing to the body to control side sway and a rear stabilizer shaft is used for improved handling, Fig. 1.

SHOCK ABSORBER, REPLACE

1. With the rear axle supported properly disconnect shock absorber from upper and lower mounting, Fig. 1.

NOTE: Use a wrench to hold lower mounting stud from turning.

2. Reverse procedure to install.

COIL SPRING, REPLACE

1. Support vehicle at frame and rear axle with a suitable jack.
2. Disconnect shock absorbers from lower mountings.
3. Lower rear axle until springs can be removed.
4. Reverse procedure to install.

STABILIZER BAR, REPLACE

1. Remove bolts securing stabilizer bar to lower control arms, Fig. 2.
2. Reverse procedure to install and torque bolts to specifications.

TRACK ROD, REPLACE

1. Raise vehicle and support axle assembly.
2. Remove bolts at underbody end of rod.
3. Remove bolts at axle bracket and remove track rod, Fig. 2.
4. Reverse removal procedure to install and torque bolts to specifications.

TORQUE ARM, REPLACE

1. Raise vehicle by axle assembly and support underbody with jackstands.
2. Lower axle assembly slightly and remove torque arm to differential bolts.
3. Disconnect mounting bracket from transmission and then remove through bolt from bracket and remove torque arm, Figs. 3 and 4.
4. To replace bushing, use an arbor press and tool J-25317-2 as a receiver and press bushing out of arm, then position new bushing in torque arm with bushing sleeve aligned with the length of the torque arm. Press bushing into place using tool J-25317-1 over bushing to properly locate bushing in arm.
5. Reverse removal procedure to install and torque nuts and bolts to specifications.

LOWER CONTROL ARM, REPLACE

1. Raise and support vehicle at rear axle.
2. Disconnect stabilizer bar from lower control arms.
3. Remove control arm front and rear mount bolts and remove control arm, Fig. 5.
4. Reverse removal procedure to install and torque bolts to specifications.

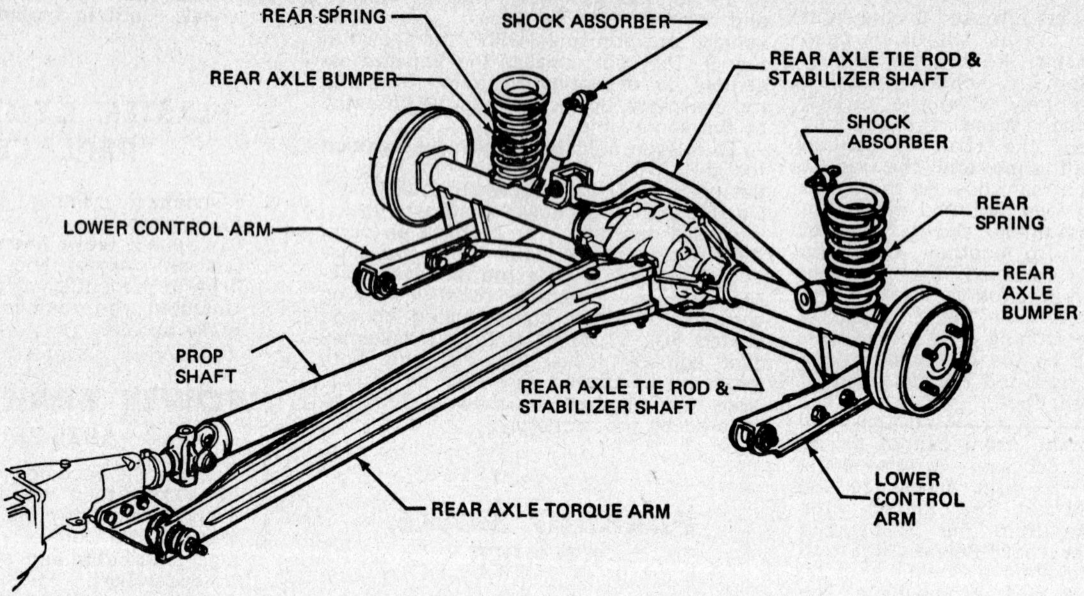

Fig. 1 Rear suspension

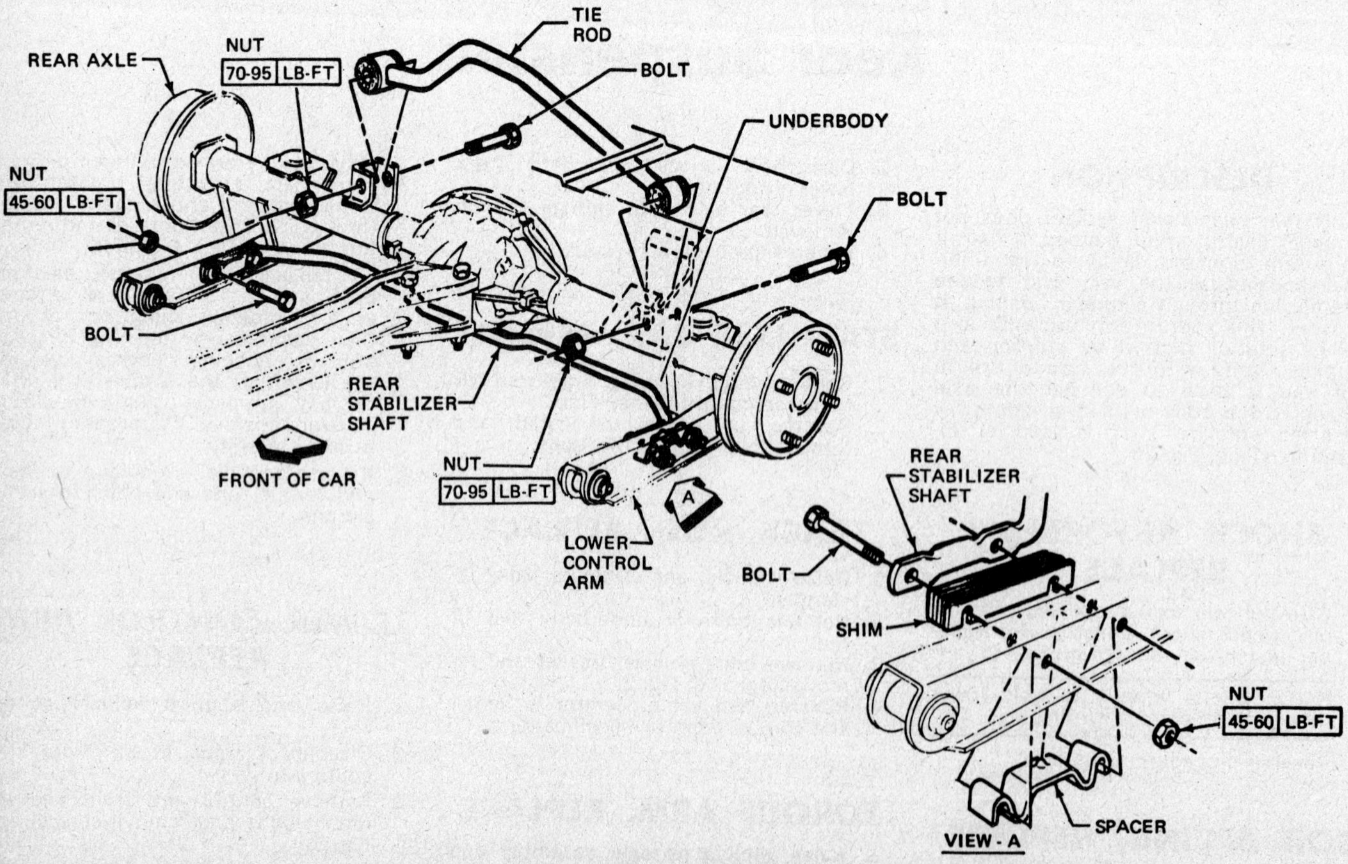

Fig. 2 Track rod and stabilizer bar

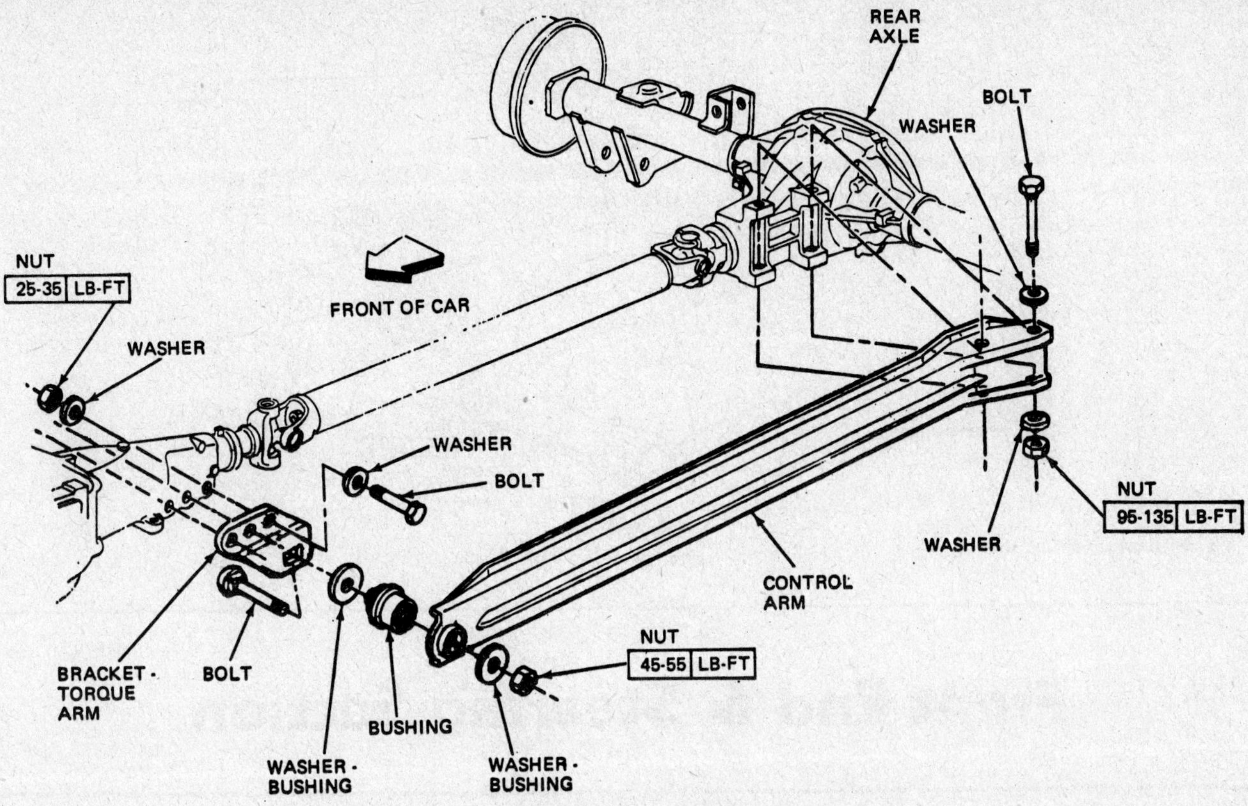

Fig. 3 Torque arm removal. Manual transmission

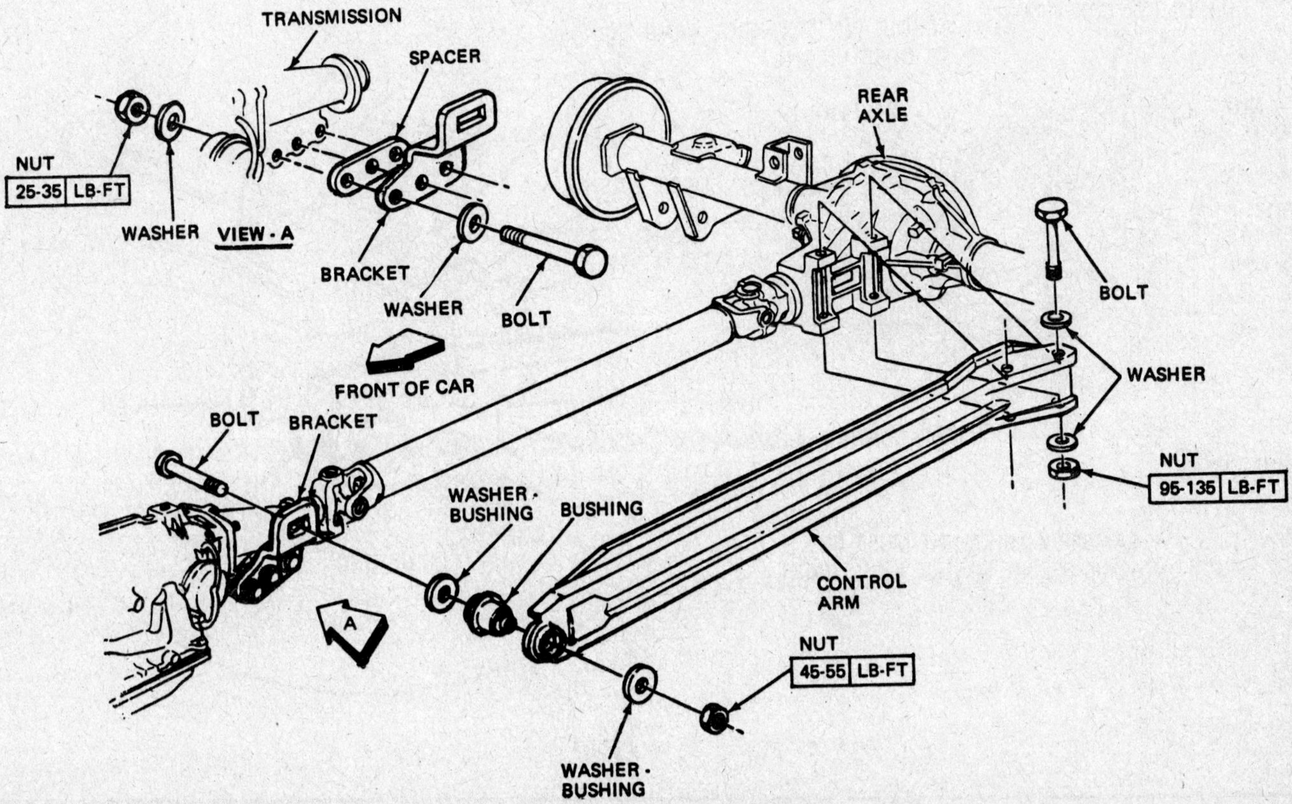

Fig. 4 Torque arm removal. Automatic transmission

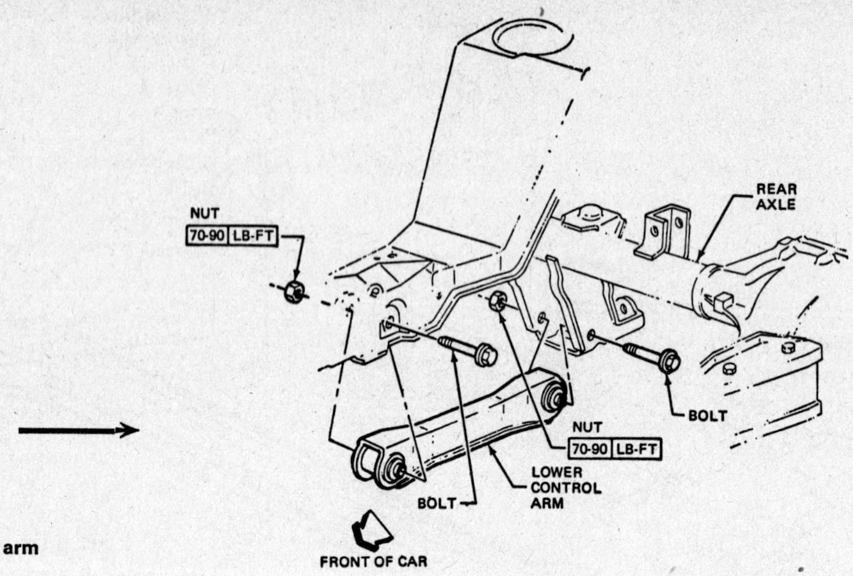

Fig. 5 Lower control arm

Front End & Steering Section

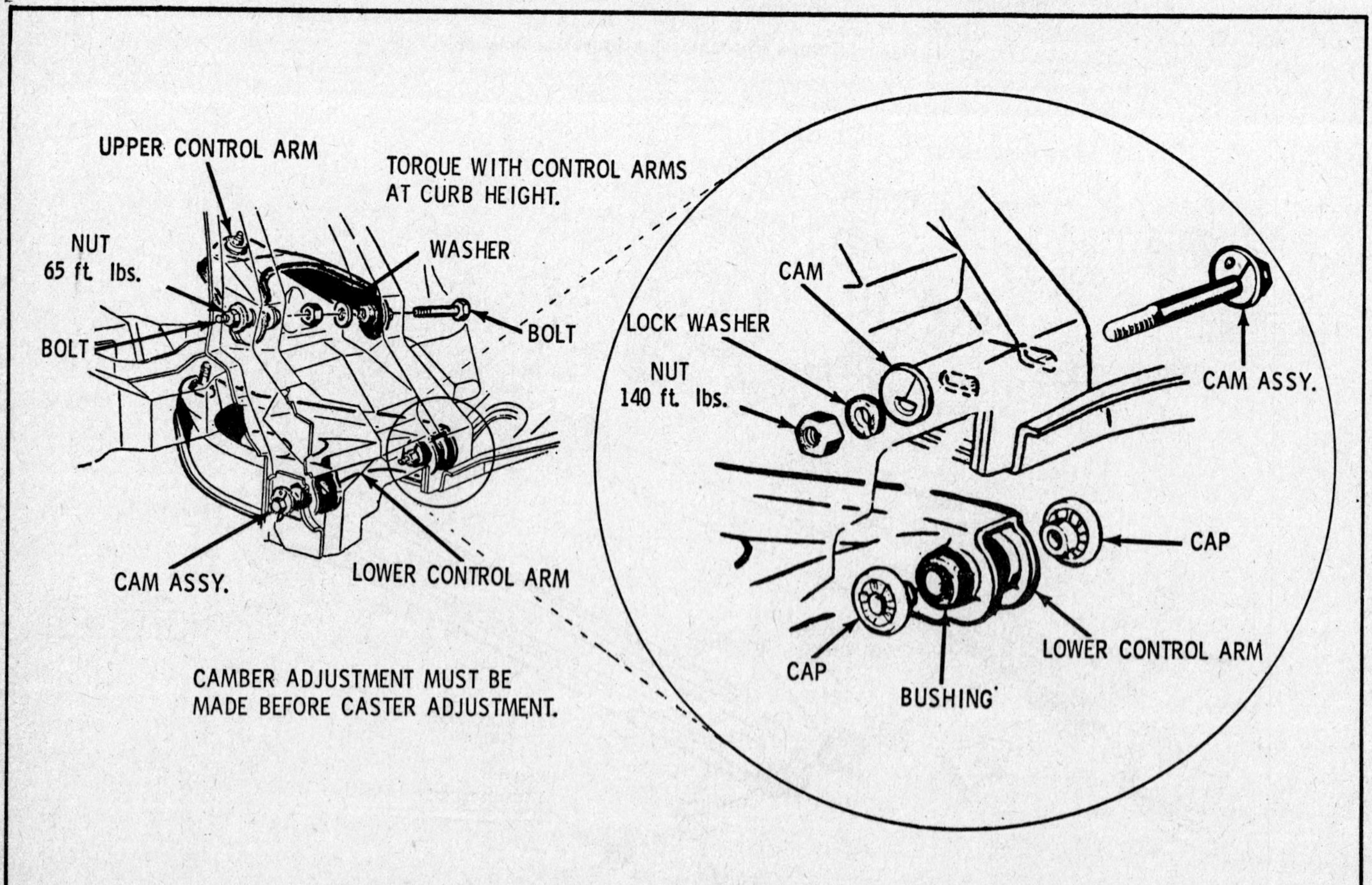

Fig. 1 Front suspension

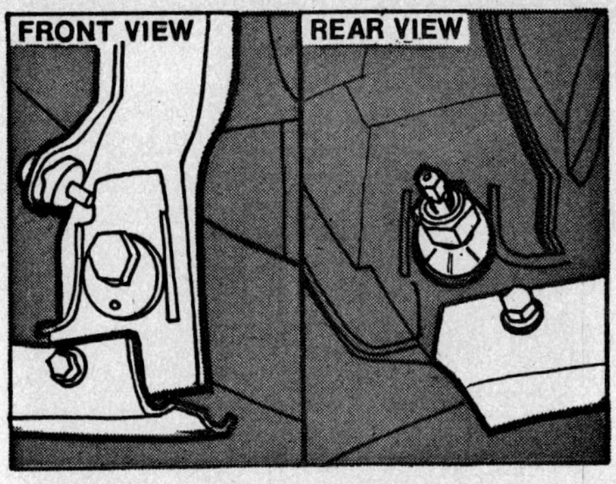

Fig. 2 Caster and camber adjustment

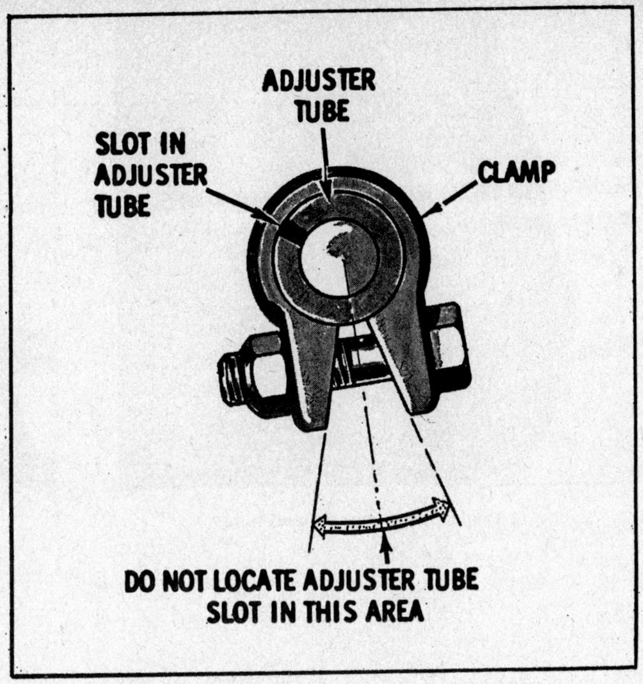

Fig. 3 Adjuster sleeve and clamp location

FRONT SUSPENSION

The front suspension, Fig. 1, is of the A frame type with short and long control arms. The upper control arm is bolted to the front end sheet metal at each inner pivot point. Rubber bushings are used for mounting.

The lower control arms attach to the front end sheet metal with cam type bolts through rubber bushings. The cam bolts adjust caster and camber.

The upper ball joint is riveted in the upper arm and the lower ball joint is pressed into the lower arm.

Coil springs are mounted between the lower control arms and the shock absorber tower.

WHEEL ALIGNMENT

Caster

Caster angle is adjusted by loosening the rear lower control arm pivot nut and rotating the cam until proper setting is reached, Fig. 2.

NOTE: This eccentric cam action will tend to move the lower control arm fore or aft thereby varying the caster. Hold the cam bolt while tightening the nut.

Camber

Camber angle is adjusted by loosening the front lower control arm pivot nut and rotating the cam until setting is reached, Fig. 2.

NOTE: This eccentric cam action will move the lower arm in or out thereby varying the setting. Hold the cam bolt head while tightening the nut.

TOE-IN, ADJUST

1. Loosen clamp bolt nut at each end of each tie rod and rotate the sleeve until proper toe-in is reached.
2. Position tie rod ball stud assembly straight on a center line through their attaching points.
3. Position clamp as shown in Fig. 3 and tighten clamp nuts.

WHEEL BEARINGS, ADJUST

Monza

1. While rotating wheel, torque spindle nut to fully seat bearing.
2. Back off adjusting nut one flat and insert cotter pin. If slot and pin hole do not line up, back off the adjusting nut an additional ½ flat or less as required to insert cotter pin.
3. Spin wheel to check that it rolls freely and then lock the cotter pin.

NOTE: Bearing should have zero preload and .001-.005″ end movement when properly adjusted.

Skyhawk

1. While rotating wheel, snug up spindle nut to fully seat bearing.
2. Back off nut ¼-½ turn to the just loose position.
3. Tighten spindle nut finger tight, then back off nut a minimum of 1/12 turn or a maximum of 1/6 turn to align cotter pin hole. Insert cotter pin.
4. With bearings properly adjusted, end play should be .001-.005 inch.

Starfire

1. While rotating wheel, torque spindle nut to 30 ft. lbs.
2. Back off nut ½ turn.
3. Tighten spindle nut finger tight and insert cotter pin. If cotter pin cannot be installed, back off nut not more than 1/24 turn to align slot and pin hole.
4. With bearings properly adjusted, end play should be .001-.008 inch.

WHEEL BEARINGS, REPLACE

1. Raise vehicle on a hoist and remove the wheel and tire assembly.
2. Remove the brake caliper from the disc by removing the mounting pins and stamped nuts, Figs. 4 & 5.
3. Remove hub grease cap, cotter pin, spindle nut and washer and remove hub and bearings.
4. Remove inner bearing by prying out the grease seal.

CHECKING BALL JOINTS FOR WEAR

Upper Ball Joint

The upper ball joint is checked for wear by checking the torque required to rotate the ball joint stud in the assembly. This is done after first dislodging the ball joint from the steering knuckle.

1. Install the stud nut to the ball stud in the seat.
2. Check the torque required to turn the ball stud.

NOTE: Specified torque for a new joint is 2 to 4 ft. lbs. rotating torque. If readings are excessively high or low, replace the joint.

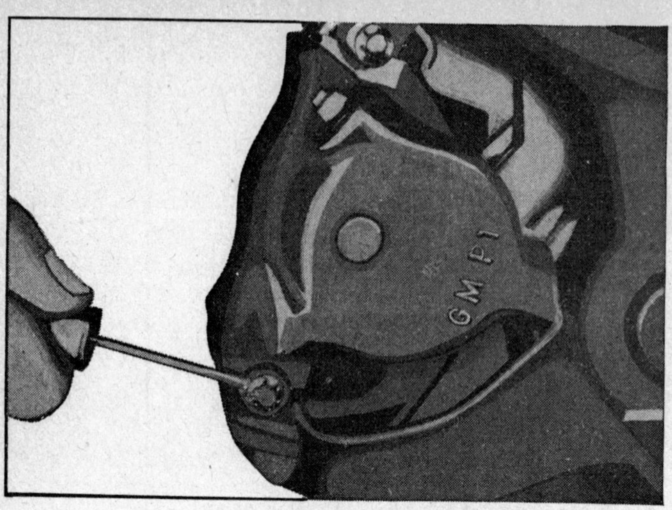

Fig. 4 Removing caliper retaining stamped nuts

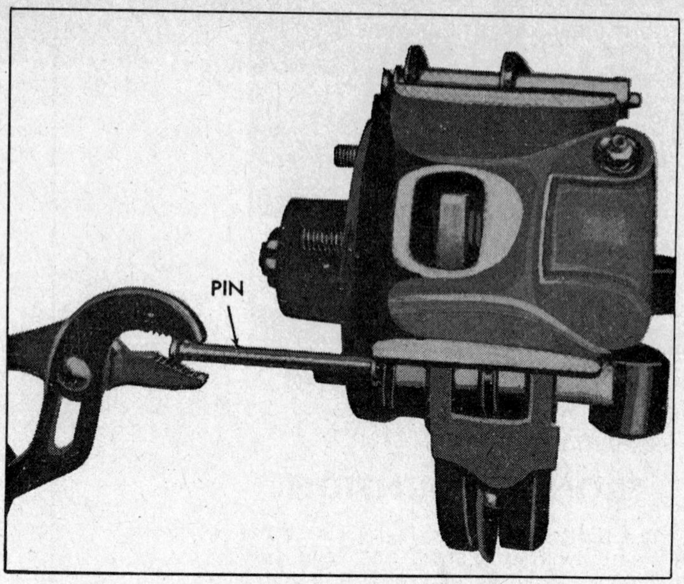

Fig. 5 Removing caliper mounting pins

Lower Ball Joint

The lower ball joints incorporate wear indicators for visual inspection, Fig. 5.

UPPER BALL JOINT, REPLACE

1. Raise vehicle on a hoist and remove wheel and tire assembly.
2. Support lower control arm with a floor jack.
3. Remove upper ball stud nut and remove ball stud from knuckle, Fig. 7.
4. Remove control arm pivot bolts and remove control arm.
5. Grind off rivets securing ball joint to arm.
6. Install new ball joint using bolts and nuts supplied in kit.

LOWER BALL JOINT, REPLACE

1. Place vehicle on hoist and support lower control arm at outer end on a jack.
2. Free lower ball stud from knuckle, Fig. 7.
3. Using tool shown in Fig. 8, press ball joint out of control arm.
4. Position new ball joint so that grease bleed vent in rubber boot is facing inboard.
5. Install tool shown in Fig. 9 and press new joint into arm.
6. Install lube fitting in joint and install stud into steering knuckle.

SHOCK ABSORBER, REPLACE

1. Hold shock absorber stem and remove the nut, upper retainer and rubber grommet, Fig. 10.
2. Raise vehicle on a hoist.
3. Remove bolts from lower end of shock absorber and lower the shock from the vehicle.

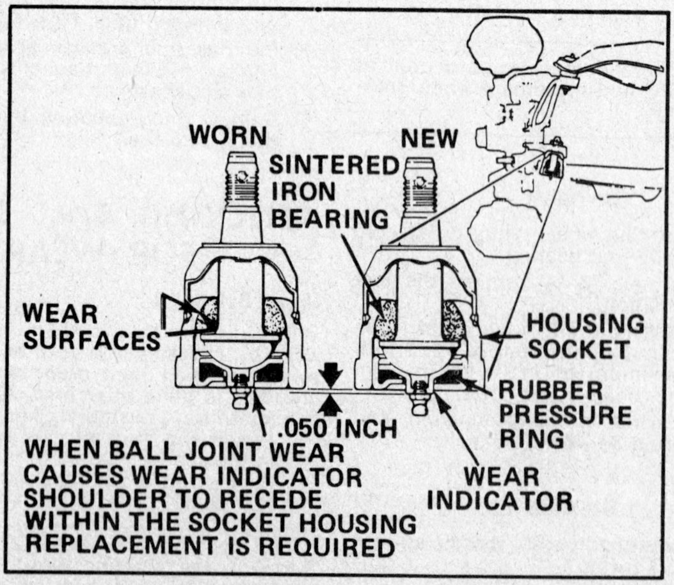

Fig. 6 Lower ball joint wear indicator

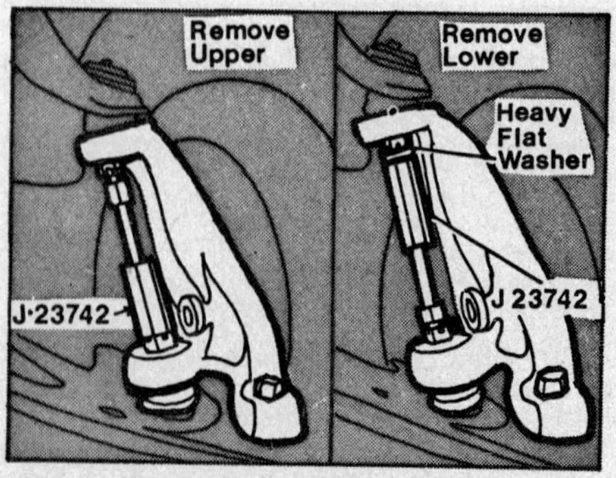

Fig. 7 Removing upper and lower ball joint from steering knuckle

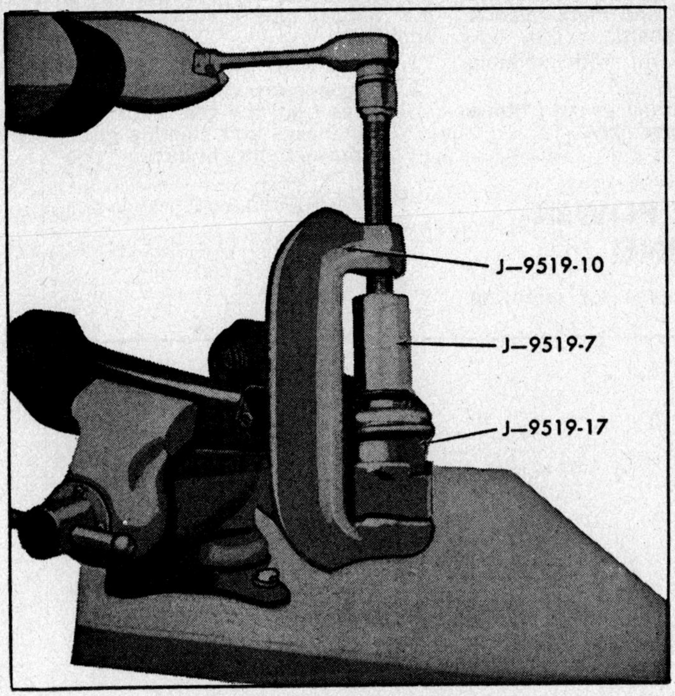

Fig. 8 Removing lower ball joint

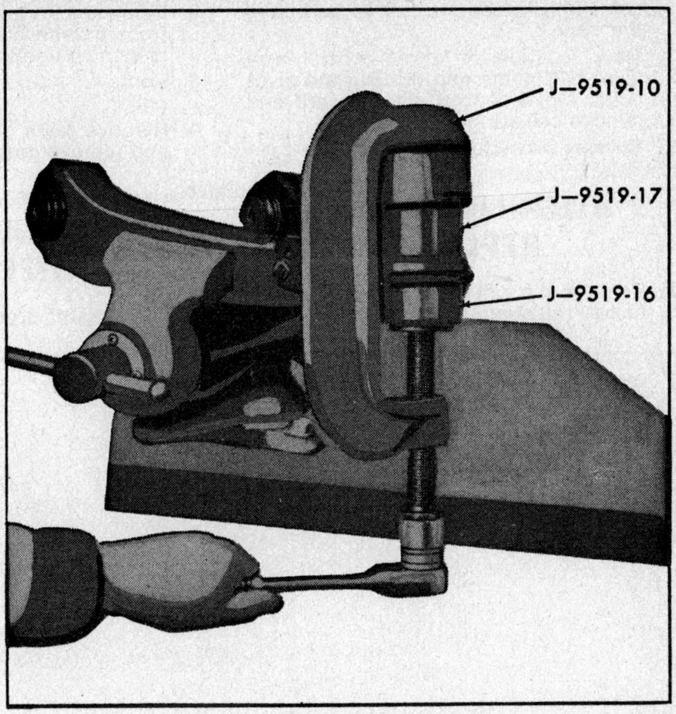

Fig. 9 Installing lower ball joint

COIL SPRING, REPLACE

Exc. Starfire

1. With shock absorber removed and stabilizer bar removed, raise the vehicle and place jackstands under front braces.
2. Remove the wheel and tire assembly.
3. Place a floor jack under the lower arm and support the arm. Use a block of wood between the control arm and the jack, Fig. 11.
4. Remove the lower ball stud from the knuckle.
5. Remove the tie rod end from the knuckle.
6. Lower the control arm by slowly lowering the jack until the spring can be removed.

Starfire

1. Place transmission in Neutral.
2. Disconnect shock absorber from upper mounting.
3. Raise vehicle, support at frame and remove wheel.
4. Remove stabilizer link nut, grommets, washers and bolt.
5. Remove shock absorber.
6. Install lower plate BT-7522, Fig. 12, with pivot ball seat facing downward into spring coils. Rotate plate to fully seat it in lower control arm spring seat.
7. Install upper plate BT-7522, Fig. 12, with pivot ball seat facing upward into spring coils. Insert ball nut BT-7408-4 through spring coils and onto upper plate.
8. Install rod BT-7408-5 through shock absorber opening in lower control arm and through the lower and upper plates. Depress lock pin on shaft and thread shaft into upper ball nut BT-7408-4. Ensure lock pin is fully extended above ball nut upper surface.
9. With ball nut tang engaged in upper plate slot, rotate upper plate until it contacts upper spring seat.
10. Install lower pivot ball, thrust bearing and nut on rod, then rotate nut until

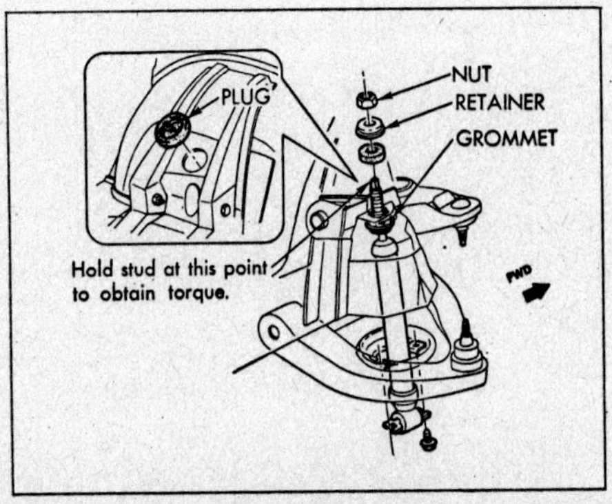

Fig. 10 Front shock absorber mountings

Fig. 11 Removing front coil spring. Exc. Starfire

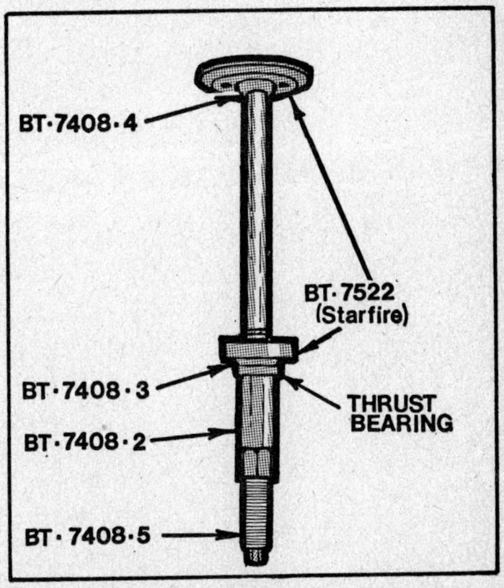

Fig. 12 Coil spring replacement tool. Starfire

coil spring is compressed to be free in the seat.

11. Mark location of lower control arm pivot bolt cams and remove the pivot bolts. Move control arm forward and remove coil spring.
12. Reverse procedure to install.

STEERING GEAR, REPLACE

1. Remove the pot joint coupling clamp bolt at the steering gear wormshaft.
2. Remove pitman arm nut and washer from pitman shaft and mark relation of arm position to shaft.
3. Remove pitman arm with suitable puller.
4. Remove bolts securing gear to frame and remove gear assembly.

INTEGRAL POWER STEERING

Replacement procedures for removing the gear assembly are the same as for the manual type gear with the following additions:

1. Remove left front crossmember brace.
2. Disconnect both pressure and return hoses from the gear housing and cap both hoses and steering gear outlets to prevent entry of dirt.

CHEVROLET VEGA • PONTIAC ASTRE

OLD CAR SPECIFICATIONS: For 1946-68 Tune Up and Wheel Alignment Specifications see main index.

INDEX OF SERVICE OPERATIONS

CHEVROLET VEGA • PONTIAC ASTRE

SERIAL NUMBER LOCATION

On top of instrument panel, left front.

ENGINE NUMBER LOCATION

On pad at right side of cylinder block, above starter.

ENGINE IDENTIFICATION CODE

Engines are identified in the following table by the code letter or letters immediately following the engine serial number.

Code	Year
CHA with P/G, T/D	1971
CHC with M/T	1971
CHB with P/G, T/D	1971
CHD with M/T	1971
CND with P/G, T/D, T.H. 350	1972
CNA with M/T	1972
CG13 with M/T	1972
CGD with P/G, T/D	1972
CSK with P/G, T.H. 350	1972
CNB with M/T	1972
CBM with P/G, T.H. 350	1972
CAA with M/T, L/E.E.C.	1973
CAB with P/G, T. H. 350 L/E.E.C.	1973
CAH with P/G, T.H. 350 W/E.E.C.	1973
CAJ with M/T W/E.E.C.	1973
CAC with P/G., T.H. 350 L/E.E.C.	1973
CAD with M/T L/E.E.C.	1973
CAK W/P.G., T.H. 350 W/E.E.C.	1973
CAL W/M.T. W/E.E.C.	1973
CAB with T.H. 350, W/E.E.C.	1974
CAA with M/T, W/E.E.C.	1974
CAH with T.H. 350, W/E.E.C.	1974
CAJ with M/T, W/E.E.C.	1974
CAC with T.H. 350, W/E.E.C.	1974
CAD with M/T, W/E.E.C.	1974
CAK with T.H. 350, W/E.E.C.	1974
CAL with M/T, W/E.E.C.	1974
CAM 4-140	1975
CAR 4-140	1975
CAS 4-140	1975
CAT 4-140	1975
CBB 4-140	1975
CBC 4-140	1975
ZCA 4-122	1975
CAY 4-140	1976
CAZ 4-140	1976
CBH 4-140	1976
CBJ 4-140	1976
CBK 4-140	1976
CBL 4-140	1976
CBS 4-140	1976
CBT 4-140	1976
CBU 4-140	1976
CBW 4-140	1976
CBX 4-140	1976
CBY 4-140	1976
CBZ 4-140	1976
ZCB 4-122	1976

M/T: Manual transmission
P/G: Powerglide
T/D: Torque drive
T/H: Turbo Hydramatic
L/E.E.C.: Less Exhaust Emission Control
W/E.E.C.: With Exhaust Emission Control

GRILLE IDENTIFICATION

1971-72

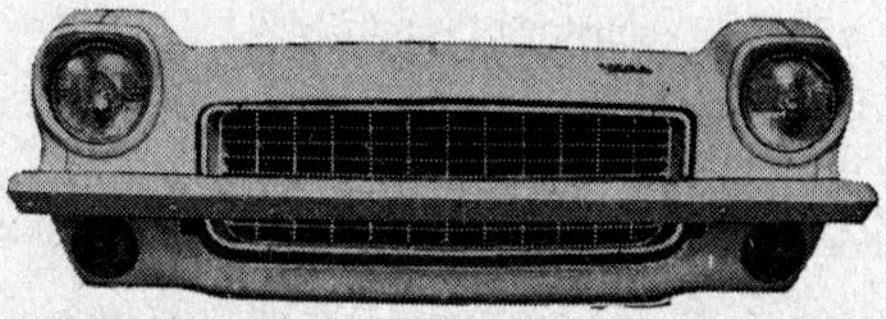

1973

1974

1975 Vega

1975 Astre

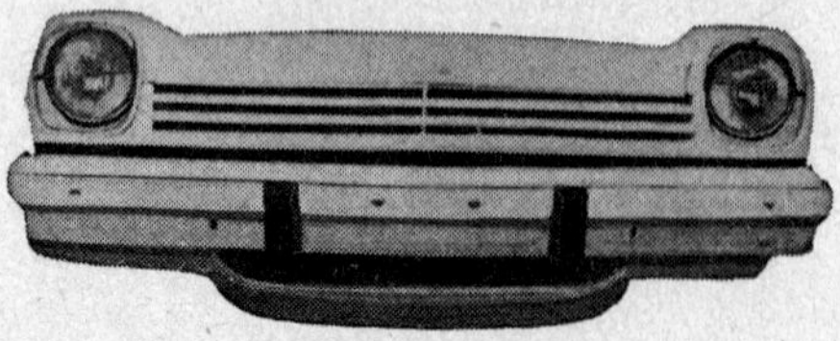

1976 Vega

1976 Astre

GENERAL ENGINE SPECIFICATIONS

Year	Engine	Carburetor	Bore and Stroke	Piston Displacement, Cubic Inches	Compression Ratio	Maximum Brake H.P. @ R.P.M.	Maximum Torque Lbs. Ft. @ R.P.M.	Normal Oil Pressure Pounds
1971–72	80 Horsepower①..............4-140	1 Barrel	3.500 x 3.625	140	8.00	80 @ 4400	121 @ 2400	40
	90 Horsepower①..............4-140	2 Barrel	3.500 x 3.625	140	8.00	90 @ 4800	121 @ 2800	40
1973	72 Horsepower①..............4-140	1 Barrel	3.500 x 3.625	140	8.00	72 @ 4400	100 @ 2000	40
	85 Horsepower①..............4-140	2 Barrel	3.500 x 3.625	140	8.00	85 @ 4800	115 @ 2400	40
1974	75 Horsepower①..............4-140	1 Barrel	3.500 x 3.625	140	8.00	75 @ 4400	115 @ 2400	40
	85 Horsepower①..............4-140	2 Barrel	3.500 x 3.625	140	8.00	85 @ 4400	122 @ 2400	40
1975	78 Horsepower①..............4-140	1 Barrel	3.500 x 3.625	140	8.00	78 @ 4200	120 @ 2000	40
	87 Horsepower①..............4-140	2 Barrel	3.500 x 3.625	140	8.00	87 @ 4400	122 @ 2800	40
1976	110 Horsepower①..............4-122	②	3.500 x 3.160	122	8.50	110 @ 5600	107 @ 4800	40
	69 Horsepower①..............4-140	1 Barrel	3.500 x 3.625	140	8.00	69 @ 4000	113 @ 2400	40
	80 Horsepower①③............4-140	2 Barrel	3.500 x 3.625	140	8.00	80 @ 4000	116 @ 2800	40
	87 Horsepower①..............4-140	2 Barrel	3.500 x 3.625	140	8.00	87 @ 4400	122 @ 2800	40
	110 Horsepower①..............4-122	②	3.500 x 3.160	122	8.50	110 @ 5600	107 @ 4800	40

①—Ratings Net—as installed in the vehicle. ②—Fuel Injection. ③—California.

DISTRIBUTOR SPECIFICATIONS

★If unit is checked on vehicle double the RPM and degrees to get crankshaft figures.

Distributor Part No.①	Centrifugal Advance Degrees @ RPM of Distributor					Vacuum Advance		Distributor Retard
	Advance Starts	Intermediate Advance			Full Advance	Inches of Vacuum To Start Plunger	Max. Adv. Dist. Deg. @ Vacuum	Max. Ret. Dist. Deg. @ Vacuum
1971-72								
1110492	0 @ 590	1 @ 800	—	—	12 @ 2000	7	12 @ 15	—
1110435	0 @ 470	1 @ 725	—	—	11 @ 2000	7	12 @ 15	—
1973-74								
1110496	0 @ 800	1 @ 875	6 @ 1600	—	11 @ 2400	7½	12 @ 15	—
1975-76								
1110649②	0 @ 1000	—	—	—	8½ @ 1825	—	—	—
1112862②	0 @ 810③	2½ @ 1000	—	—	11 @ 2400	5	12 @ 12④	—

①—Located on distributor housing plate.
②—High Energy Ignition.
③—Exc. 1976 Astre; 1976 Astre, 0 @ 450.
④—Exc. 1976 Astre; 1976 Astre, 10 @ 14.

TUNE UP SPECIFICATIONS

The following specifications are published from the latest information available. This data should be used only in the absence of a decal affixed in the engine compartment.

OLD CAR SPECIFICATIONS: For 1946-68 Tune Up Specifications see main index.

★When using a timing light, disconnect vacuum hose or tube at distributor and plug opening in hose or tube so idle speed will not be affected.

●When checking compression, lowest cylinder must be within 80 percent of highest.

▲Before removing wires from distributor cap, determine location of the No. 1 wire in cap, as distributor position may have been altered from that shown at the end of this chart.

Year	Spark Plug		Distributor		Ignition Timing★			Carb. Adjustments					
	Type	Gap Inch	Point Gap Inch	Dwell Angle Deg.	Firing Order Fig. ▲	Timing BTDC ①	Mark Fig.	Hot Idle Speed Std. Trans.	Hot Idle Speed Auto. Trans. ②	Air Fuel Ratio Std. Trans.	Air Fuel Ratio Auto. Trans.	Idle "CO" % Std. Trans.	Idle "CO" % Auto. Trans.
1971													
80 Horsepower	R42TS	.035	③	31–34	A	6°	B	850⑤	650D⑤	—	—	2.0	2.0
90 Horsepower	R42TS	.035	③	31–34	A	④	B	1200⑤	700D⑤	—	—	2.0	2.0
1972													
80 Horsepower	R42TS	.035	③	31–34	A	6°⑦	B	850⑥	700D⑤	—	—	—	—
90 Horsepower	R42TS	.035	③	31–34	A	8°	B	1200	700D⑤	—	—	—	—
1973													
72 Horsepower	R42TS	.035	③	31–34	A	8°	B	1000	750D	—	—	—	—
85 Horsepower	R42TS	.035	③	31–34	A	10°⑨	B	1200	750D⑧	—	—	—	—
1974													
All	R42TS	.035	③	31–34	A	⑩	B	700	750D	—	—	0.5	0.5
1975													
78 Horsepower	R43TSX	.060	—	—	C	⑪	D	700	550D	—	—	—	—
87 Horsepower	R43TSX⑫	.060	—	—	C	⑩	D	700	600D	—	—	—	—
Cosworth Vega	R43LTSX	.060	—	—	E	12°⑬	F	1600	—	—	—	—	—
1976													
1 Bar. Carb.	R43TS	.035	—	—	C	⑪	D	⑯	750D	—	—	—	—
2 Bar. Carb.⑭	R43TSX	.035	—	—	C	⑰	D	700	750D	—	—	—	—
2 Bar. Carb.⑮	R43TS	.035	—	—	C	⑰	D	⑱	750D	—	—	—	—
Cosworth Vega	R43LTS	.035	—	—	E	12°⑬	F	1600	—	—	—	—	—

①—BTDC—Before top dead center.
②—D—Drive.
③—New points .019″, used .016″.
④—With synchromesh trans. 6°; with automatic trans. 10°.
⑤—Solenoid disconnected.
⑥—1200 R.P.M. in California.
⑦—4° for California vehicles w/manual trans.
⑧—With Air Cond. 800 RPM.
⑨—With Auto. Trans. 12°.
⑩—Synchromesh trans. 10° BTDC; automatic trans. 12° BTDC.
⑪—Synchromesh trans. 8° BTDC, automatic trans. 10° BTDC.
⑫—If cold weather starting problems are encountered, use R43TS spark plug, gapped at .035 inch.
⑬—At 1600 RPM.
⑭—Exc. California.
⑮—California.
⑯—Astre 700 RPM; Vega, 1200 RPM.
⑰—Astre man. trans., 8° BTDC; auto. trans., 10° BTDC. Vega man. trans., 10° BTDC; auto. trans., 12° BTDC.
⑱—Astre, 700 RPM; Vega, 1000 RPM.

Continued

TUNE UP NOTES—Continued

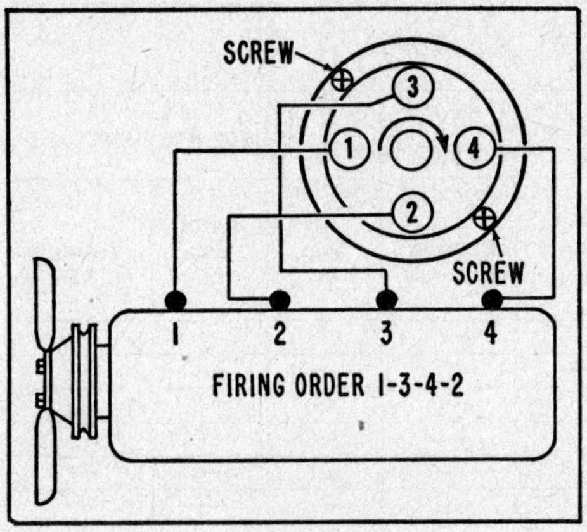

Fig. A

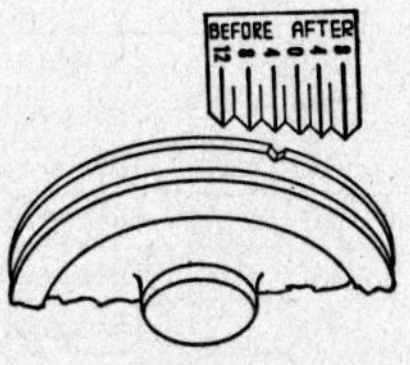

Fig. B

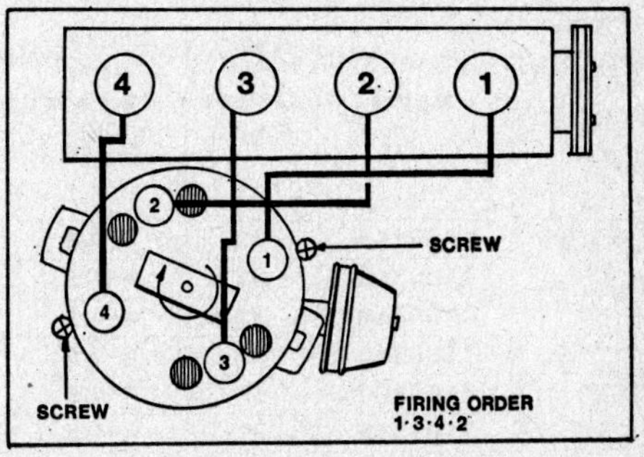

Fig. C

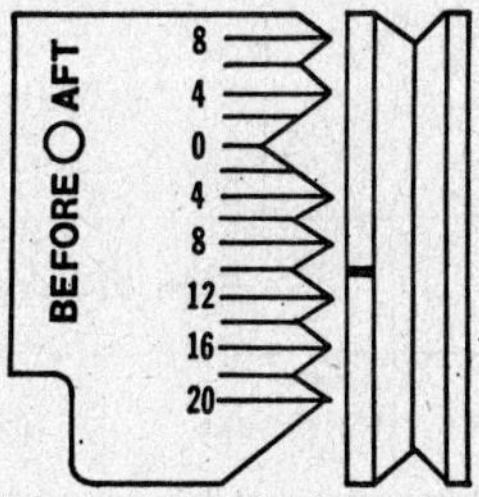

Fig. D

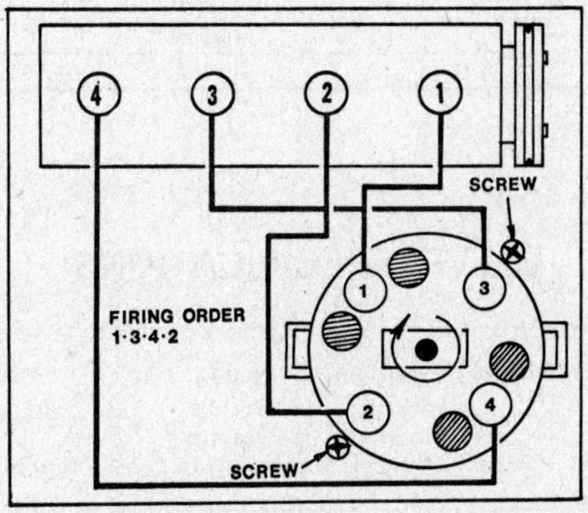

Fig. E

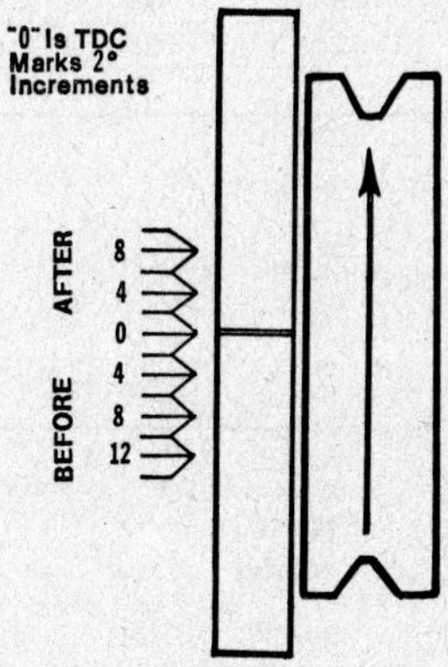

Fig. F

STARTING MOTOR SPECIFICATIONS

Year	Model	Starter Number	Brush Spring Tension Oz.①	Free Speed Test			Resistance Test	
				Amps.	Volts	R.P.M.①	Amps.	Volts
1971–74	Std. Trans.	1108195	—	50–75	9	6500–10000	—	—
	Auto. Trans.	1108196	—	50–75	9	6500–10000	—	—
1975–76	Std. Trans.	1108771	—	50–75	9	6500–10000	—	—
	Auto. Trans.	1108772	—	50–75	9	6500–10000	—	—
	Cosworth Vega	1108773	—	50–75	9	6500–10000	—	—

①—Minimum.

ALTERNATOR & REGULATOR SPECIFICATIONS

Year	Alternator					Regulator						
	Model	Rated Hot Output Amps.	Field Current 12 Volts @ 80 F.	Cold Output @ 14 Volts		Model	Field Relay			Voltage Regulator		
				2000 R.P.M. Amps.	5000 R.P.M. Amps.		Air Gap In.	Point Gap In.	Closing Voltage	Air Gap In.	Point Gap In.	Voltage @ 125° F.
1971–75	1100545	32	4–4.5	—	31	Integral	—	—	—	—	—	—
	1100546	55	4–4.5	—	50	Integral	—	—	—	—	—	—
	1100559	32	4–4.5	—	31	Integral	—	—	—	—	—	—
	1100560	55	4–4.5	—	50	Integral	—	—	—	—	—	—
1975	1102500	55	4–4.5	—	50	Integral	—	—	—	—	—	—
	1102856	37	4–4.5	—	33	Integral	—	—	—	—	—	—
1976	1102845	37	4–4.5	—	33	Integral	—	—	—	—	—	—
	1102846	55	4–4.5	—	51	Integral	—	—	—	—	—	—
	1102851	55	4–4.5	—	51	Integral	—	—	—	—	—	—
	1102858	37	4–4.5	—	33	Integral	—	—	—	—	—	—
	1102891	55	4–4.5	—	51	Integral	—	—	—	—	—	—
	1102893	55	4–4.5	—	51	Integral	—	—	—	—	—	—

VALVE SPECIFICATIONS

Year	Engine Model	Valve Lash		Valve Angles		Valve Spring Installed Height	Valve Spring Pressure Lbs. @ In.	Stem Clearance		Stem Diameter	
		Int.	Exh.	Seat	Face			Intake	Exhaust	Intake	Exhaust
1971–72	80 Horsepower	.015C	.030C	46	45	1.746	186 @ 1.29	.001–.0027	.001–.0027	.3410–.3417	.3410–.3417
	90 Horsepower	.015C	.030C	46	45	1.746	190 @ 1.31	.001–.0027	.001–.0027	.3410–.3417	.3410–.3417
1973	72 Horsepower	.015C	.030C	46	45	1.750	186 @ 1.29	.001–.0027	.001–.0027	.3410–.3417	.3410–.3417
	85 Horsepower	.015C	.030C	46	45	1.750	190 @ 1.31	.001–.0027	.001–.0027	.3410–.3417	.3410–.3417
1974	All	.015C	.030C	46	45	1.746	190 @ 1.31	.001–.0027	.001–.0027	.3410–.3417	.3410–.3417
1975	4-140 1 BBl.	.015C	.030C	46	45	1.75	186 @ 1.29	.001–.0027	.001–.0027	.3410–.3417	.3410–.3417
	4-140 2 BBl.	.015C	.030C	46	45	1.75	190 @ 1.31	.001–.0027	.001–.0027	.3410–.3417	.3410–.3417
1975–76	4-122	.014	.014	46	45	1.30	110 @ .92	.001–.0027	.001–.0027	.2788–.2795	.2788–.2795
1976	4-140	Hydraulic①		46	45	1.56	185 @ 1.15	.001–.0027	.001–.0027	.3410–.3417	.3410–.3417

①No adjustment.

ENGINE TIGHTENING SPECIFICATIONS★

★Torque specifications are for clean and lightly lubricated threads only. Dry or dirty threads produce increased friction which prevents accurate measurement of tightness.

Year	Engine Model	Spark Plugs Ft. Lbs.	Cylinder Head Bolts Ft. Lbs.	Intake Manifold Ft. Lbs.	Exhaust Manifold Ft. Lbs.	Rocker Arm Stud Ft. Lbs.	Cam Cover Ft. Lbs.	Connecting Rod Cap Bolts Ft. Lbs.	Main Bearing Cap Bolts Ft. Lbs.	Flywheel to Crankshaft Ft. Lbs.	Vibration Damper or Pulley Ft. Lbs.
1971–76	4-140	15	60	30	30	—	35①	35	65	60	80
1975–76	4-122	15	60	20	15	—	65①	40–45	65	60	15

①—Inch pounds.

PISTONS, PINS, RINGS, CRANKSHAFT & BEARINGS

Year	Engine Model	Piston Clearance	Ring End Gap①		Wrist-pin-Diameter	Rod Bearings		Main Bearings			
			Comp.	Oil		Shaft Diameter	Bearing Clearance	Shaft Diameter	Bearing Clearance	Thrust on Bear. No.	Shaft End Play
1971–72	4-140	.0018–.0028	④	.010	.9272	1.999–2.000	.0007–.0027	2.2983–2.2992	②	4	.002–.007
1973	4-140	.0018–.0028	④	.010	.9272	1.999–2.000	.0007–.0038	2.2983–2.2993	③	4	.002–.007
1974–75	4-140	.0018–.0028	④	.010	.9272	1.999–2.000	.0007–.0038	2.2983–2.2993	③	4	.002–.007
1975	4-122	.002–.003	④	.010	.9272	1.999–2.000	.0007–.0027	2.2983–2.2993	.0003–.0019	4	.002–.007
1976	4-140	.0018–.0028	④	.010	.9272	1.999–2.000	.0007–.0027	2.2983–2.2993	.0003–.0029	4	.002–.008
	4-122	.002–.003	④	.010	.9272	1.999–2.000	.0007–.0027	⑤	⑥	4	.002–.008

①—Fit rings in tapered bores for clearance listed in tightest portion of ring travel.
②—#1 .0003–.002; others .0003–.003.
③—#1 .0003–.002; others .0003–.0027.
④—Top ring .015; lower ring .009.
⑤—No. 1,2,3,5: 2.3011; No. 4: 2.3006.
⑥—No. 1,2,3,5: .0008–.0034; No. 4: .0002–.0029.

WHEEL ALIGNMENT SPECIFICATIONS

OLD CAR SPECIFICATIONS: For 1946-68 Wheel Alignment Specifications see main index.

Year	Model	Caster Angle, Degrees		Camber Angle, Degrees				Toe-In. Inch	Toe-Out on Turns, Deg.①	
		Limits	Desired	Limits		Desired			Outer Wheel	Inner Wheel
				Left	Right	Left	Right			
1971–74	All	−1/4 to −1 1/4	−3/4	−1/4 to +3/4	−1/4 to +3/4	+1/4	+1/4	3/16 to 5/16	—	—
1975–76	All	−1/4 to −1 1/4	−3/4	−1/4 to +3/4	−1/4 to +3/4	+1/4	+1/4	0 to 1/8	—	—

①—Incorrect toe-out when other adjustments are correct, indicates bent steering arms.

BRAKE SPECIFICATIONS

Year	Model	Brake Drum Inside Diameter	Wheel Cylinder Bore Diameter			Master Cylinder Bore Diameter		
			Disc Brake	Front Drum Brake	Rear Drum Brake	Disc Brakes	Drum Brakes	Power Brakes
1971–75	All	9	1 7/8	—	3/4	3/4	—	3/4
1976	All	9 1/2	1 7/8	—	11/16	3/4	—	3/4

COOLING SYSTEM & CAPACITY DATA

Year	Model or Engine	Cooling Capacity, Qts.		Radiator Cap Relief Pressure, Lbs.		Thermo. Opening Temp. ①	Fuel Tank Gals.	Engine Oil Refill Qts. ②	Transmission Oil				Rear Axle Oil Pints
		With Heater	With A/C	With A/C	No A/C				3 Speed Pints	4 Speed Pints	5 Speed Pints	Auto. Trans. Qts. ⑥	
1971–72	4-140	6.5	6.5	15	15	195	11	3	2.4	3	—	③	2.5
1973	4-140	8.6	9.0	15	15	195	11	3	3	3	—	④	2.8
1974	4-140	7.6	8.0	15	15	195	16	3	3	3	—	4	2.8
1975	4-140	7	7½	15	15	195	16	3	2.4	2.4	—	⑤	2¼
1975–76	4-122	6.8	6.8	15	15	195	16	3½	—	3	—	—	2¼
1976	4-140	8	8	15	15	195	16	3½	3	3	3	⑤	2¼

①—For permanent type anti-freeze.
②—Add 1 quart with filter change.
③—Refill 2 qts. Total capacity 8½ qts.
④—Powerglide 2½ qts.; turbo hydra-matic 4 qts.
⑤—Refill 4 qts. Total capacity 10 qts.
⑥—Approximate; make final check with dipstick.

REAR AXLE SPECIFICATIONS

Year	Model	Carrier Type	Ring Gear & Pinion Backlash		Pinion Bearing Preload			Differential Bearing Preload		
			Method	Adjustment	Method	New Bearings Inch-Lbs.	Used Bearings Inch-Lbs.	Method	New Bearings Inch-Lbs.	Used Bearings Inch-Lbs.
1971–72	All	Integral	Shims	.005–.008	Spacer	25	10	Shims	—	—
1973–76	All	Integral	Shims	.005–.008	Spacer	10–25	8–12	Shims	—	—

Electrical Section

DISTRIBUTOR

Removal

1. Release distributor cap hold-down screws and remove cap.
2. Disconnect distributor primary lead from coil terminal.

NOTE: On H.E.I. systems, disconnect feed and module connectors from distributor cap.

3. Scribe an alignment mark on the distributor and engine in line with the rotor.
4. Remove distributor hold-down bolt and clamp and remove distributor.

CAUTION: Avoid rotating engine while distributor is removed.

Installation

1. Turn rotor about ⅛ turn in a clockwise direction past the mark previously made to locate rotor.
2. Push distributor down into position in cylinder head with distributor housing in a normal installed position.

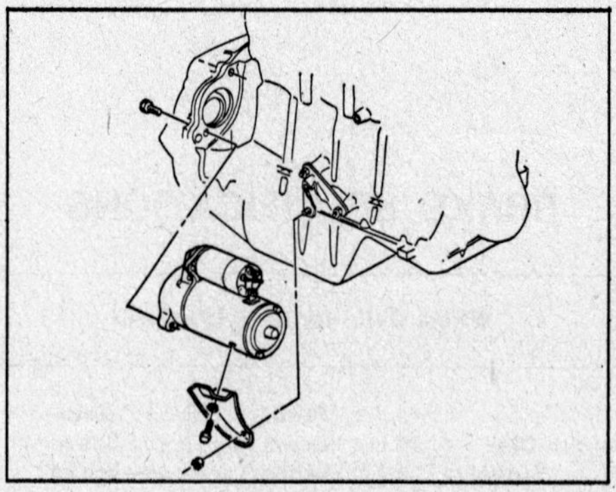

Fig. 1 Starter motor installation

NOTE: It may be necessary to move rotor slightly to start gear into mesh with camshaft gear, but rotor should line up with the mark when distributor is down in place.

NOTE

If the engine was disturbed while the distributor was removed, it will be necessary to crank the engine to bring No. 1 cylinder piston up on the compression stroke. Continue cranking until timing mark is adjacent to pointer. Then rotate the distributor cam until rotor is in position to fire No. 1. Install distributor in this position.

3. Tighten distributor clamp bolt and connect vacuum line, primary wire and install the cap.
4. Set ignition timing.

NOTE: When using a timing light to ad-

Fig. 2 Ignition lock removal

just ignition timing, the connection should be made at the No. 1 spark plug. Forcing foreign objects through the boot at the No. 1 terminal of the distributor cap will damage the boot and could cause engine misfiring.

STARTER, REPLACE

NOTE: The following procedure may vary slightly depending on model and series of vehicle.

1. Disconnect battery ground cable at battery.
2. Disconnect all wires at solenoid terminals.

NOTE: Reinstall the nuts on the terminals as each wire is removed as thread size is different and if mixed, stripping of threads may occur.

3. Loosen starter front bracket then remove two mount bolts, Fig. 1.
4. Remove front bracket bolt and rotate bracket clear of work area then lower starter from vehicle by lowering front end first.
5. Reverse procedure to install.

IGNITION LOCK, REPLACE

1. Remove the turn signal switch as described further on.
2. On *1971 Models* the lock cylinder may be removed in any position from "Accessory" to "On" however, the "Lock" position is recommended because of its positive position. *On 1972-76 Models* the lock cylinder should be removed in the "Run" position only. Removal in any other position will damage the buzzer switch.
3. Insert a small screw driver or similar tool into the turn signal housing slot, Fig. 2. Keeping the tool to the right side of the slot, break the housing flash loose, and at the same time depress the spring latch at the lower end of the lock cylinder. With the latch depressed, the lock cylinder can be removed from the housing.
4. To separate the lock from the cylinder, Fig. 3, use a paper clip or similar quality wire, place a 90° bend about ⅛" from one end of the wire.
5. Place the cylinder in the "LOCK" position. Using the wire as described in step 4, depress the plunger pin (above the stake mark) approx. ⅛" vertically in sleeve housing, at the same time exert light upward pressure on the lock assembly, rotate the cylinder counter-clockwise until it pops up from the sleeve assembly.

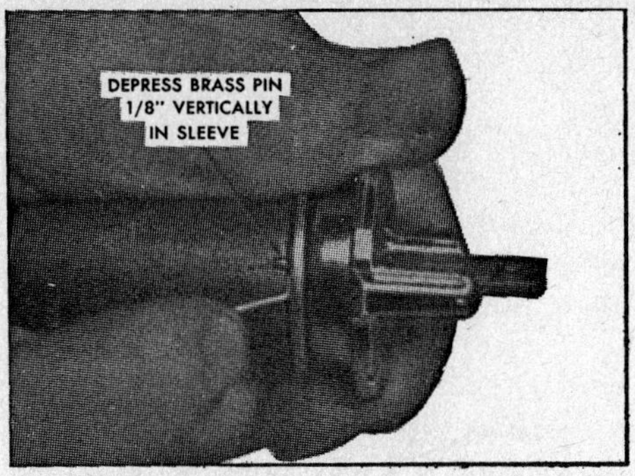

Fig. 3 Disassembling ignition lock

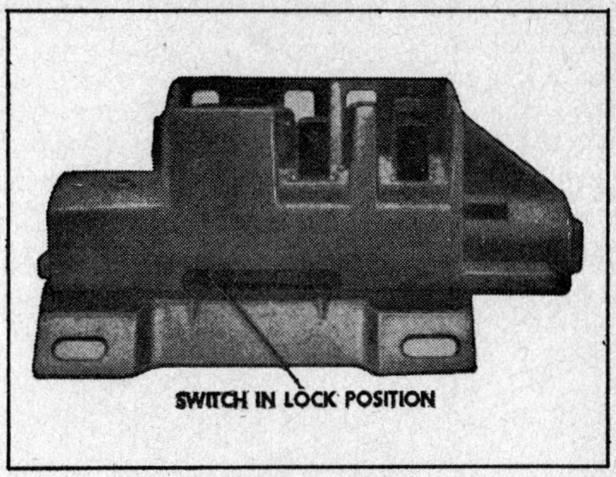

Fig. 4 Ignition switch assembly

IGNITION SWITCH, REPLACE

The ignition switch is mounted on the top of the steering column jacket near the front of the dash. It is located inside the channel section of the brake pedal support and is completely inaccessible without first lowering the steering column.

1. Lower the steering column and be sure it is properly supported before proceeding.
2. The switch should be positioned in Lock position before removing, Fig. 4.
3. Unfasten and remove the switch, detaching it from the actuating rod.
4. When installing, make sure the lock and the switch are in the Lock position. Then install the activating rod into the switch and fasten the switch.

LIGHT SWITCH, REPLACE

1. Disconnect ground cable at battery.
2. Pull headlamp switch knob to "ON" position.
3. Reach under instrument panel and depress switch shaft retainer button while pulling on the switch control shaft knob.
4. With a large bladed screwdriver, remove the light switch ferrule nut

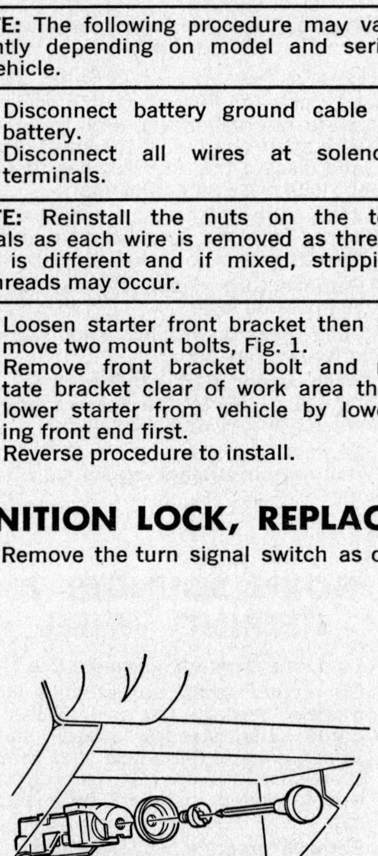

Fig. 5 Light switch replacement

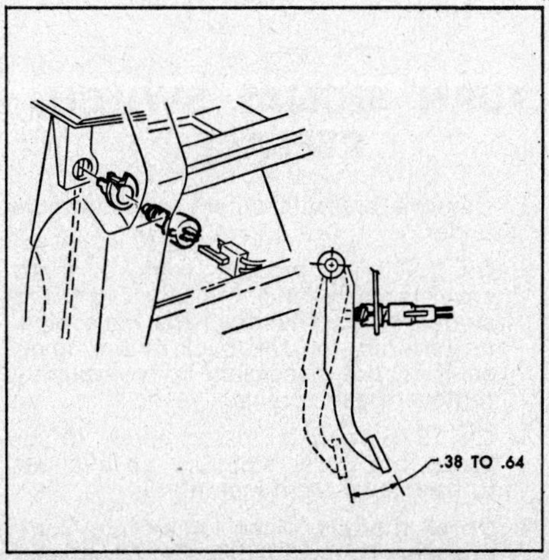

Fig. 6 Stop light switch replacement

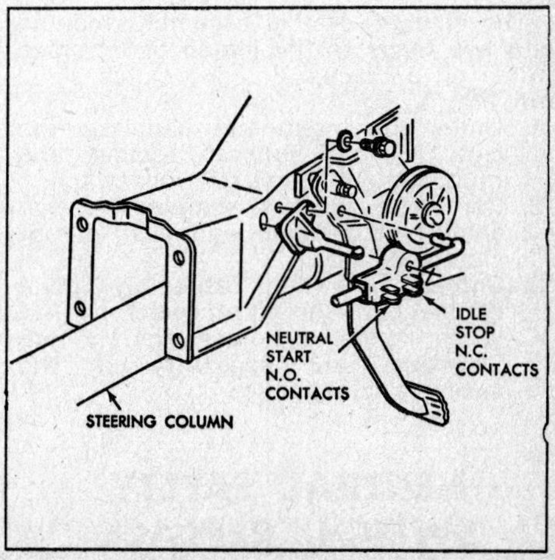

Fig. 7 Clutch operated neutral start switch

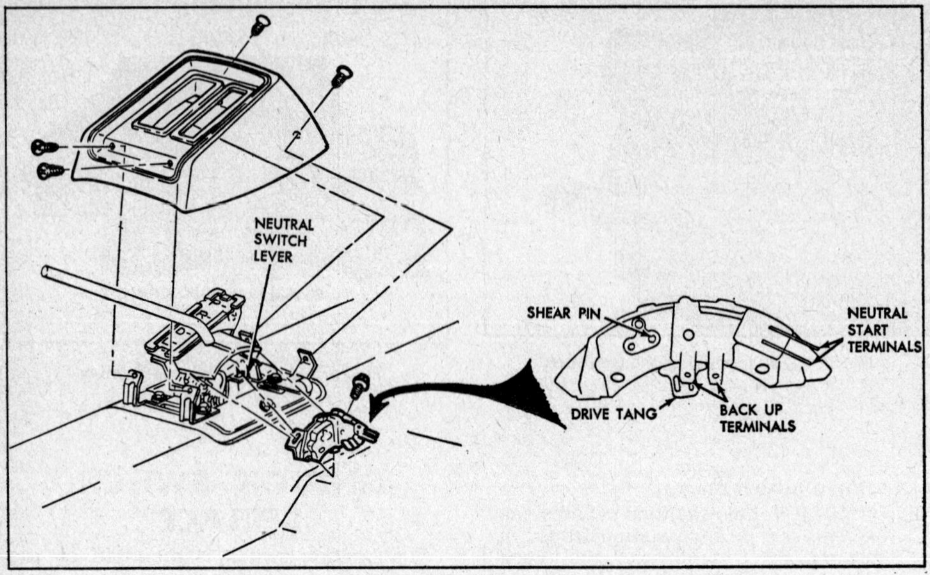

Fig. 8 Neutral safety switch installation (Typical)

from front of instrument panel.

5. Disconnect the multi-contact connector from side of switch and remove switch.

STOP LIGHT SWITCH

1. Reach under right side of instrument panel at brake pedal support and release wiring harness connector at switch.
2. Pull switch from mounting bracket.
3. When installing switch, adjust by bringing brake pedal to normal position. Electrical contact should be made when pedal is depressed as shown in Fig. 6. To adjust, the switch may be rotated or pulled in the clip.

CLUTCH START SWITCH

NOTE: The clutch pedal must be fully depressed and the ignition switch in START position for the vehicle to start.

The clutch switch assembly mounts with two tangs to the clutch pedal brace switch pivot bracket and the clutch pedal arm, Fig. 7.

1. Under the instrument panel on the clutch pedal support remove the multi-contact connector from switch.
2. Compress switch assembly actuating shaft barb retainer and push out of clutch pedal.
3. Compress switch assembly pivot bracket barb and lift off switch.
4. When installing new switch, no adjustments are necessary as the switch is self aligning.

NEUTRAL SAFETY SWITCH, REPLACE

1. Remove screws securing floor console, Fig. 8.
2. Disconnect electrical plugs on back-up contacts and neutral start contacts of switch assembly.
3. Place shift lever in Neutral.
4. Remove two screws securing shift indicator plate.
5. Remove two screws securing shift lever curved cover.
6. Remove two screws securing switch to lever assembly.

NOTE: Screws are hidden beneath lever cover.

7. Tilt switch to right as you lift switch out of lever hole.
8. When installing switch, make sure it is in Neutral position. When switch is installed, shifting out of Neutral will shear the switch plastic locating pin.

TURN SIGNAL SWITCH, REPLACE

1. Remove steering wheel with suitable puller.
2. On 1971-75 models remove three cover screws and lift cover off the shaft. **NOTE:** These screws have plastic retainers on the back of the cover so it is not necessary to completely remove these screws.
3. On 1976 models, place screw driver blade into cover slot, pry up and out to free cover from lock plate.
4. On all models, place Lock Plate Compressing Tool J-23653, Fig. 9, on end of steering shaft and compress the

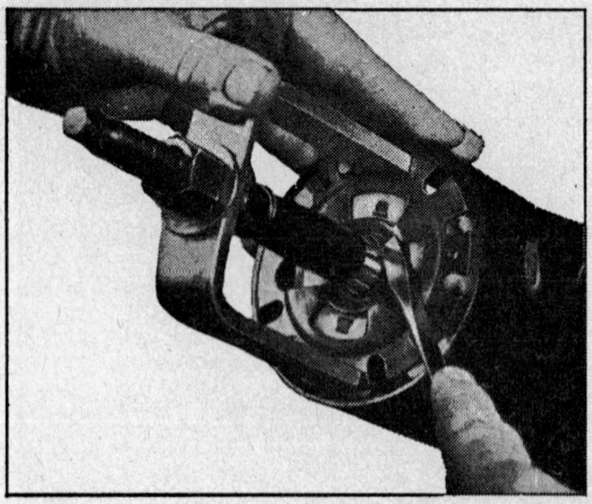

Fig. 9 Removing lock plate retaining ring

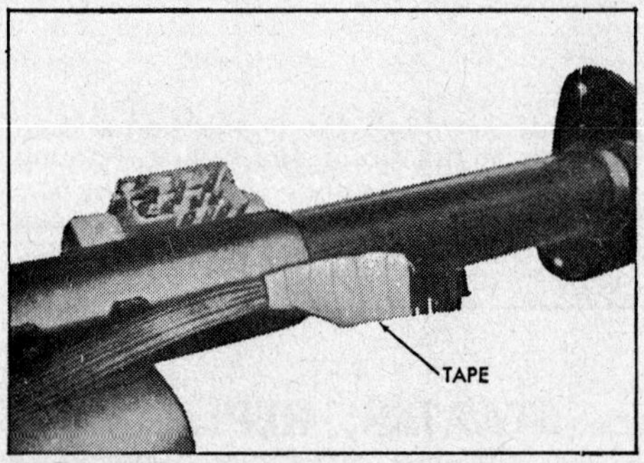

Fig. 10 Taping turn signal connector and wires

lock plate as far as possible using the shaft nut as shown. Pry the round wire snap ring out of the shaft groove and discard the ring. Remove tool and lift lock plate off end of shaft.

5. Slide the signal cancelling cam, upper bearing preload spring and thrust washer off the end of shaft.
6. Remove turn signal lever screw and remove the lever.
7. Push hazard warning knob in and unscrew the knob.
8. Wrap upper part of connector with tape, Fig. 10 to prevent snagging of wires during switch removal.
9. Remove three screws on switch and pull switch straight up through housing.

HORN SOUNDER & STEERING WHEEL

1. Disconnect battery ground cable.
2. On regular production steering wheel models, remove the two screws securing the steering wheel shroud from beneath the wheel and remove the shroud. On GT or optional wheel models, pry off horn button cap.
3. Remove steering wheel nut and use a suitable puller to remove the steering wheel.

NOTE: On 1975-76 models, remove snap ring before steering wheel nut.

INSTRUMENT CLUSTER

Standard Cluster

The instrument cluster bezel is retained by nine screws, Fig. 11. After removal of bezel, remove cluster lens-light shield combination (2 screws at top of lens and 2 screws at bottom of light shield). The lens tips out at the top and then lifts off. Instruments are then easily removed.

GT Cluster

The cluster bezel is retained by six screws, Fig. 12. After removal of bezel, remove the lens light shield (6 screws). Then lift lens and light shield straight out. Instruments are then accessible for replacement.

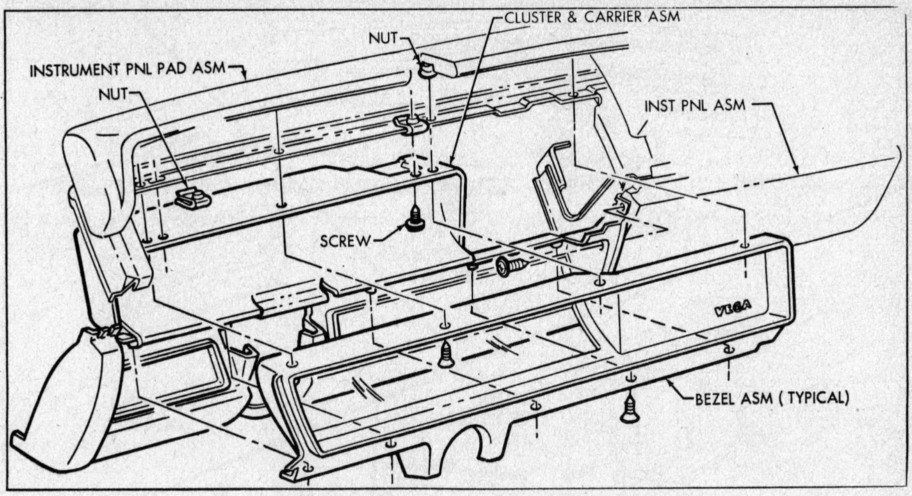

Fig. 11 Instrument panel. Standard cluster

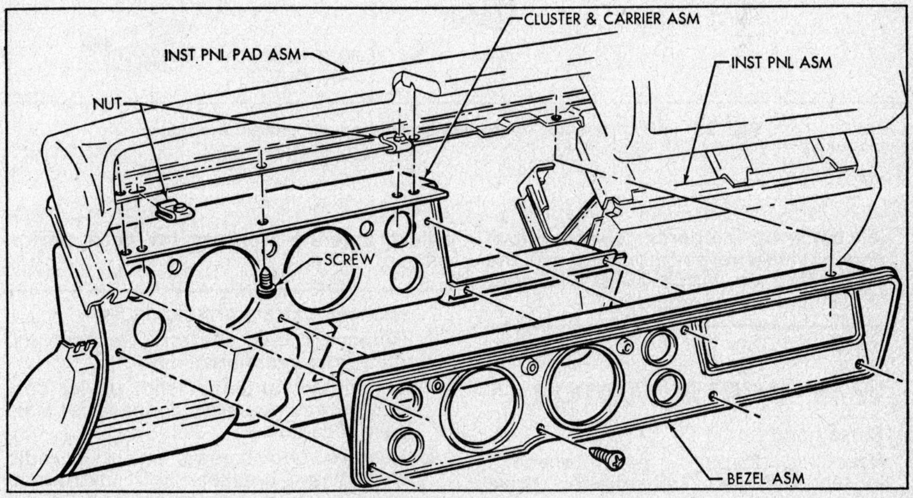

Fig. 12 Instrument panel. GT cluster

W/S WIPER BLADES

1. Remove the wiper blade from the arm by depressing the spring type blade clip, Fig. 13, away from the under side of the arm and slide the arm out of the blade clip.
2. To install wiper blade to wiper arm, slide tip end of arm into blade clip, until pin on tip end of arm engages hole in clip.
3. The blade element is retained in the blade assembly by a spring type retainer clip in the end of the blade element. When the retainer clip is squeezed together, the blade element can be slid out of the blade assembly.
4. When installing a blade element into a blade assembly, be certain to engage the metal insert of the element into all retaining tabs of the blade assembly.

NOTE: When properly installed, the spring type element retaining clip should be at the end of the wiper blade assembly nearest the wiper transmission.

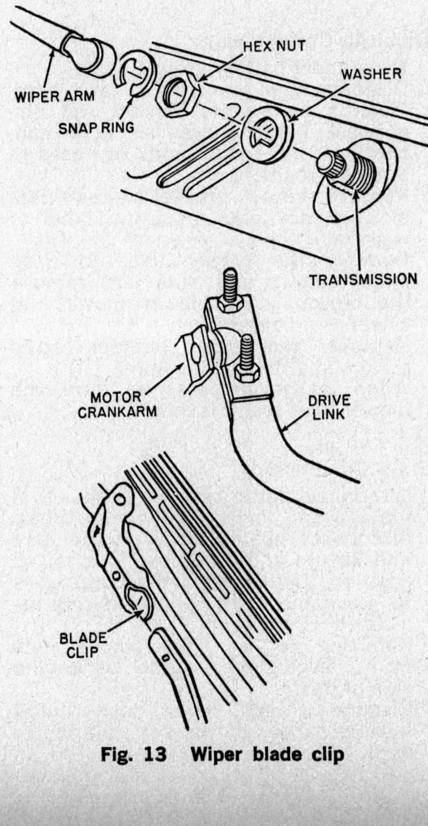

Fig. 13 Wiper blade clip

W/S WIPER ARMS

1. Wiper motor must be in park position.
2. Use suitable tool to minimize the possibility of windshield or paint finish damage during arm removal.
3. Remove arm by prying up with tool to disengage arm from serrated transmission shaft.
4. To install arm to transmission shaft rotate the required distance and direction so that blades rest in proper parked position.

CAUTION: The parked position for the

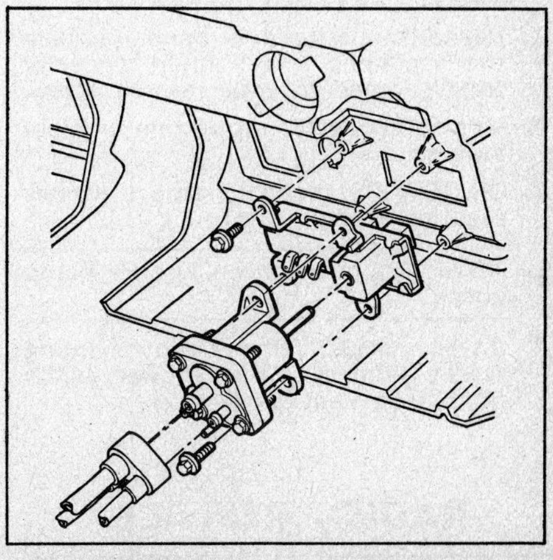

Fig. 14 Wiper-washer control switch 1971-73

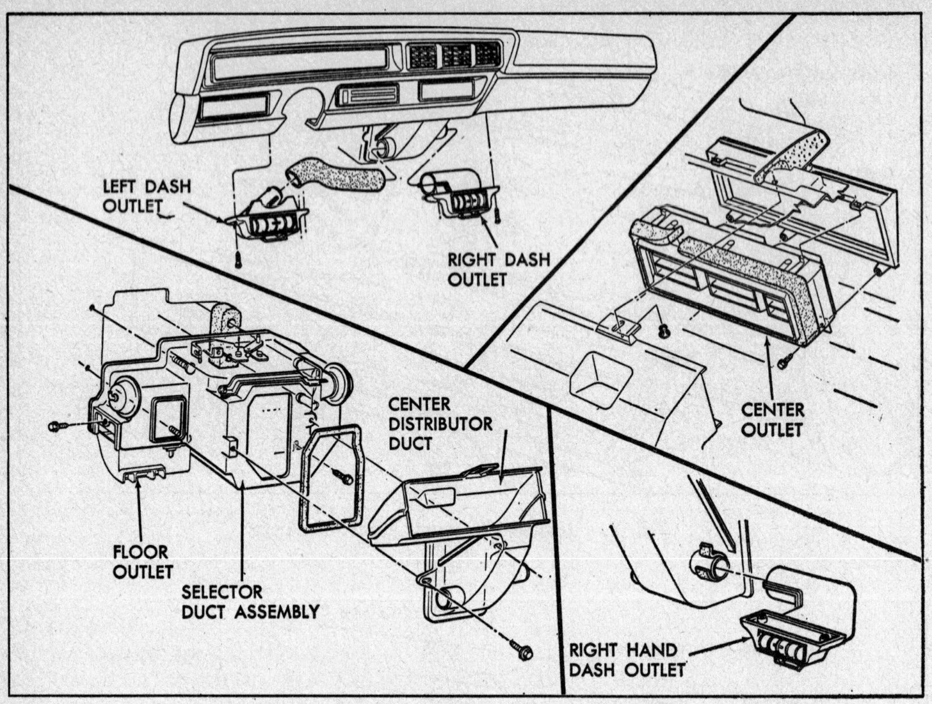

Fig. 15 Air distributor ducts and outlets on air conditioned cars

Fig. 16 Removing temperature door bell crank on air conditioned vehicles

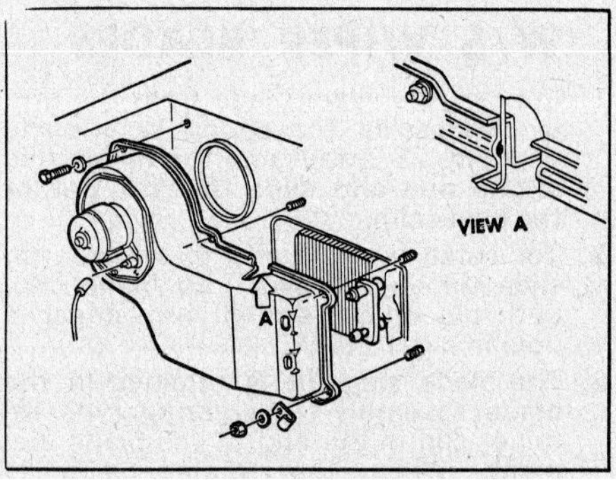

Fig. 17 Blower motor (Typical)

left blade tip is approx. 2″ above lower windshield reveal moulding and the right blade tip within 2″ of the lower windshield reveal moulding.

W/S WIPER MOTOR

1. Raise hood.
2. Reaching through cowl opening, loosen the two transmission drive link attaching nuts to motor crankarm.
3. Remove transmission drive link from motor crankarm.
4. Disconnect wiring and unfasten motor and remove.

W/S WIPER SWITCH

1. Beneath instrument panel, unplug the headlamp switch multi-connector for clearance to wiper switch screw.
2. Unplug connector on bottom of wiper switch.
3. On 1971-73 models, remove screws securing washer pump, Fig. 14.

NOTE: Do not remove washer pump hoses.

4. On all models, remove two mounting screws from switch and lower switch from instrument panel.

RADIO, REPLACE

NOTE: When installing radio, be sure to adjust antenna trimmer for peak performance.

1. Remove battery ground cable.
2. Remove knobs, controls, washers and nuts from radio bushings.
3. Disconnect antenna lead, power connector and speaker connectors from rear of radio.
4. Remove two screws securing radio mounting bracket to instrument panel lower reinforcement and lift out radio.

HEATER CORE REMOVAL

Cosworth Vega

1. Disconnect battery ground cable.
2. Disconnect heater hoses from heater core. Plug hoses and core tubes.
3. Remove nuts and washers from heater air distributor studs on engine side of dash.
4. Remove floor outlet to air distributor duct screw and the outlet.
5. Remove air distributor to dash screws, then pull duct assembly rearward until core tubes clear dash openings. Tilt assembly downward and disconnect bowden cables from distributor assembly.
6. Disconnect resistor wiring harness and remove air distributor assembly from vehicle.
7. Remove core retaining strap screw and the heater core.

All Exc. Cosworth Vega

Without Air Conditioning

1. Disconnect battery ground cable.
2. Disconnect blower motor lead wire.
3. Place a pan under vehicle and disconnect heater hoses at core connections and secure ends of hoses in a raised position.
4. Remove the coil bracket to dash panel stud nut and move coil out of way.
5. Remove the blower inlet to dash panel screws and nuts and remove the blower inlet, blower motor and wheel as an assembly.
6. Remove the core retainer strap screws and remove the core.
7. When replacing core, be sure the blower inlet sealer is intact.

With Air Conditioning

1. Disconnect battery ground cable.
2. Place a pan under heater core tubes, disconnect heater hoses at the core and secure ends of hoses in a raised position. Cap or tape the core tubes to prevent coolant spillage during selector duct removal.
3. Referring to Fig. 15, remove nuts from selector duct studs on engine side of dash.
4. Disconnect left hand dash outlet flexible hose at center distributor duct.

5. Remove the right hand dash outlet assembly.
6. Remove instrument cluster bezel screws and remove the bezel and center outlet as an assembly.
7. Remove ash tray and retainer.
8. Remove the radio as outlined previously.
9. Remove control assembly to instrument panel screws and lower the control assembly.
10. Remove cigarette lighter assembly.
11. Remove the screw securing the right end of the instrument panel carrier reinforcement.
12. Pry out the clip retaining the center distributor duct to the instrument panel. Remove the center duct to selector duct screws and remove the center distributor duct. Rotate the duct clockwise and then pull down and to the left to remove. It may be necessary to pull rearward on the instrument panel to remove the duct.
13. Remove the defroster duct to selector duct screws.
14. Remove the remaining selector duct to dash panel screws and pull the duct rearward far enough to disconnect all electrical and vacuum lines. Mark these lines for proper installation.
15. Disconnect temperature door cable, all vacuum and electrical connections and remove selector duct assembly. Mark vacuum hoses for proper installation.
16. If the core must be replaced, pry off or punch out the temperature door bell crank, Fig. 16.

CAUTION: Use care to prevent bending the arm or damaging the selector case. Then remove the screws securing the backing plate and the temperature door cable retainer and remove the core and backing plate as an assembly.

BLOWER MOTOR, REPLACE

1. Disconnect the battery ground cable.
2. On Cosworth Vega, loosen air cleaner seal to intake manifold clamp, then remove air cleaner bracket screws and rotate filter assembly upward.
3. On all models, disconnect blower motor lead wire.
4. Scribe the blower motor flange to case position.
5. Remove the blower to case attaching screws and remove the blower wheel and motor assembly, Fig. 17. Pry the flange gently if the sealer acts as an adhesive.
6. Remove the blower motor wheel retaining nut and separate the motor and wheel.
7. To install, reverse steps 1 thru 5, lining up the scribe marks on the motor flange which were made at removal.

NOTE: Assemble the blower wheel to the motor with the open end of the blower away from the motor. Replace sealer at the motor flange if necessary.

4-140 Engine Section

CHECKING ENGINE MOUNTS

Raise engine to remove weight from mounts leaving a slight tension in the rubber. Observe both mounts while raising engine. Check for the following and replace if:

a. Hard rubber surface is covered with heat check cracks.
b. Rubber separated from metal plate of mount.
c. Rubber split through center.

If there is relative movement between a metal plate of the mount and its attaching points lower the engine on the mounts and tighten the screws or nuts attaching the mounts to the engine, frame, or bracket.

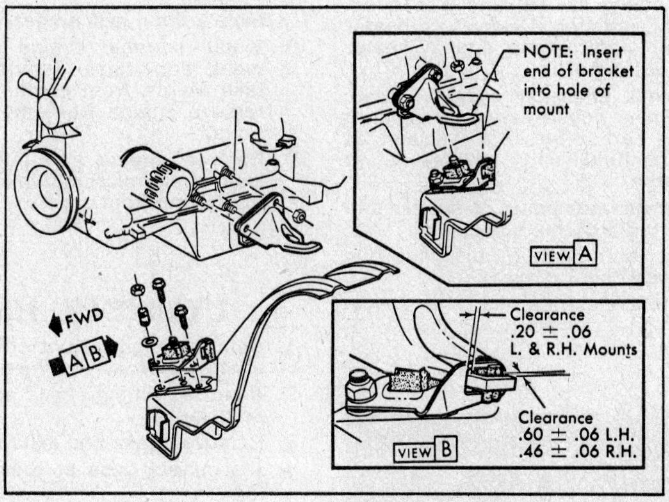

Fig. 1 Front mounts. 1971-76 (Typical)

ENGINE MOUNTS, REPLACE

1971-76 Front

1. Raise vehicle on hoist and support front of engine to take weight off front mounts.
2. If only one mount is being replaced, remove the mount-to-engine bracket nut on the mount not being replaced, Fig. 1.
3. Remove the stud nut and two bolts securing mount to housing support.
4. Remove the three stud nuts securing the bracket to the engine. On the right side remove the starter brace at starter and on air conditioned equipped vehicles, remove the compressor rear lower brace at the compressor.
5. Raise front of engine to provide maximum clearance without imposing stress on other engine components.
6. Remove mount and bracket as a unit and separate by removing stud nut.

1971-76 Rear

CAUTION: The rear mount serves to locate the power train fore and aft and side to side; therefore mark relationship of rear mount to transmission support cross member to ensure proper alignment when new mount is installed.

1. Remove crossmember-to-mount bolts, Figs. 2 and 3.
2. Raise transmission at extension housing to release weight from mount.
3. Remove mount-to-transmission bolts, then remove mount.

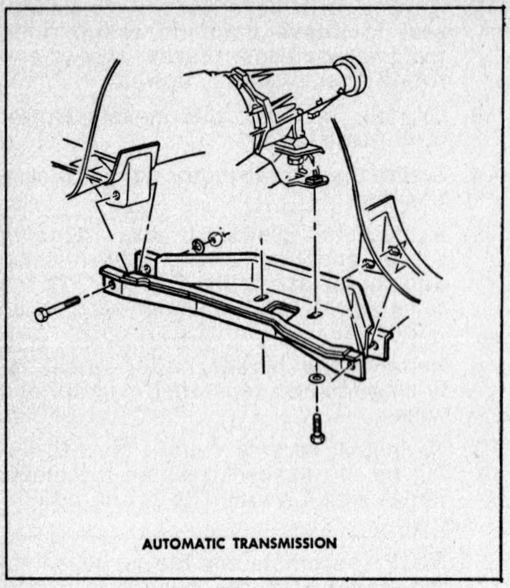

Fig. 2 Rear mount. Automatic trans

ENGINE, REPLACE

1. On 1973-76 models, remove hood. On 1971-72 models, raise hood to fully open position and install a bolt through the hood-open link, Fig. 4.
2. Disconnect battery positive cable at battery and negative cable at engine block (except on air conditioned vehicles).
3. Drain cooling system and disconnect hoses at radiator. Disconnect heater hoses at water pump and at heater inlet (bottom hose).
4. Disconnect emission system hoses: PCV at cam cover; cannister vacuum hose at carburetor; PCV vacuum at inlet manifold and bowl vent at carburetor.
5. Remove radiator panel or shroud and remove radiator, fan and spacer.
6. Remove air cleaner, disconnecting vent tube at base of cleaner.
7. Disconnect electrical leads at: Delcotron, ignition coil, starter solenoid, oil pressure switch, engine temperature switch, transmission controlled spark switch at transmission, transmission controlled spark solenoid and engine ground strap at cowl.
8. Disconnect: Automatic transmission throttle valve linkage at manifold mounted bellcrank, fuel line at rubber hose to rear of carburetor, transmission vacuum modulator and air conditioning vacuum line at inlet manifold, accelerator cable at manifold bellcrank.
9. On air conditioned cars, disconnect compressor at front support, rear support, rear lower bracket and remove drive belt from compressor.
10. Move compressor slightly forward and allow front of compressor to rest on frame forward brace, then secure rear of compressor to engine compartment so it is out of way.
11. Disconnect power steering pump, if equipped, and position it out of way.
12. Raise car on a hoist and disconnect exhaust pipe at manifold.
13. Remove engine flywheel dust cover or converter underpan.
14. On automatic transmission cars, remove converter-to-flywheel retaining bolts and nuts and install coverter safety strap.
15. Remove converter housing or flywheel housing-to-engine retaining bolts.
16. Loosen engine front mount retaining bolts at frame attachment and lower vehicle.
17. Install floor jack under transmission.
18. Install suitable engine lifting equipment and raise engine slightly to take weight from engine mounts and remove engine front mount retaining bolts.
19. Remove engine and pull forward to clear transmission while slowly lifting engine from car.

Fig. 4 Hood hold-open bolt installed

CYLINDER HEAD

1. Remove engine front cover and camshaft cover as outlined further on.
2. Remove timing belt and camshaft sprocket.
3. Remove intake and exhaust manifolds.
4. Disconnect hose at thermostat housing.
5. Remove cylinder head bolts and with the aid of an assistant lift head and gasket from engine. Place head on two blocks of wood to prevent damage to valves.
6. Reverse procedure to install and tighten head bolts in sequence shown in Fig. 5.

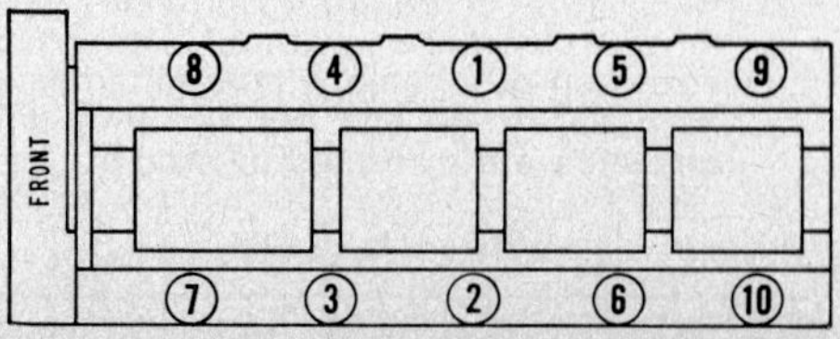

Fig. 5 Cylinder head tightening sequence

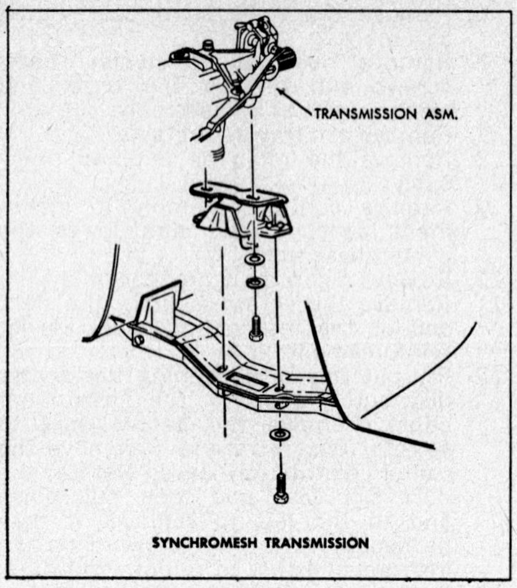

Fig. 3 Rear mount. Synchromesh trans

NOTE: Cylinder head bolts are of different lengths. Install the longer bolts adjacent to intake manifold.

VALVES, ADJUST

1976

These engines use new hydraulic valve lifters which require no lash adjustment. Also, these lifters are not interchangeable with the previous type valve lifters since the cylinder head has been re-designed to facilitate the use of the new type lifters.

1971-75

1. To adjust valves, the tappet must be on the base circle of the cam lobe.

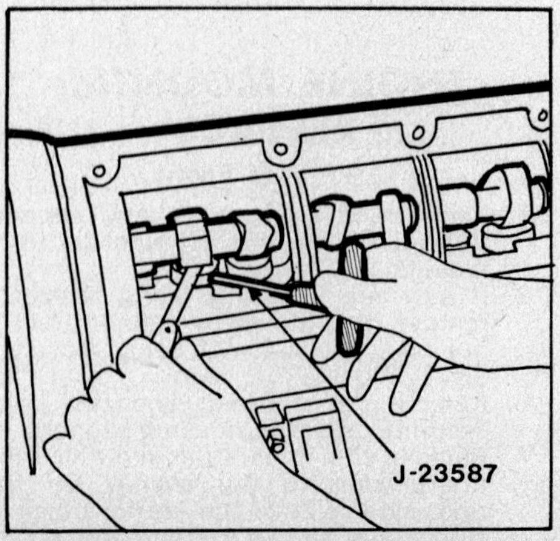

Fig. 6 Adjusting valve lash. 1971-75

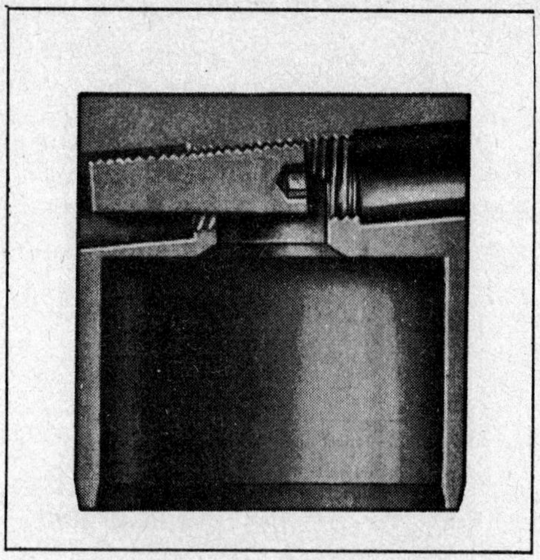

Fig. 7 Tappet and adjusting screw assembly. Screw is threaded in all areas except in valve stem contact surface. 1971-75

Fig. 8 Camshaft removal tool installed

Do this as follows:

a. Rotate camshaft timing sprocket to align timing mark on sprocket with inverted "V" notch on timing belt upper cover. The following valves can be adjusted with cam

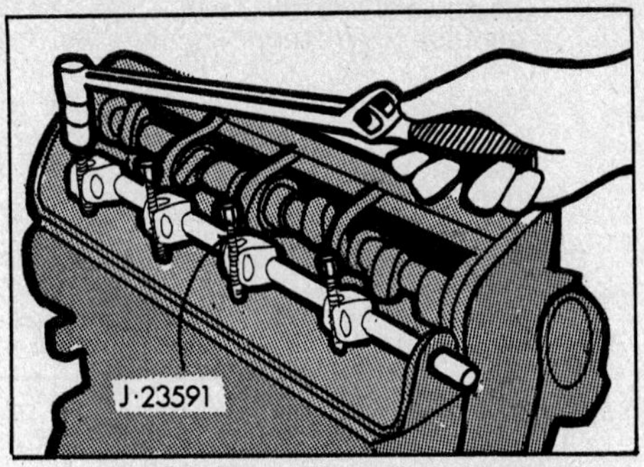

Fig. 9 Depressing valve tappets

in this position (number one firing).
Number one cylinder—Intake and Exhaust
Number two cylinder—Intake
Number three cylinder—Exhaust

b. Use a feeler gauge and measure clearance between tappet and cam lobe. Adjust clearance by turning adjusting screw in tappet, Fig. 6.

NOTE: It is mandatory that the adjusting screw, Fig. 7, be turned one complete revolution to maintain proper stem-to-screw relationship. Each revolution of screw alters clearance by .003".

c. Rotate camshaft timing sprocket 180 degrees so timing mark is at 12 o'clock position and in line with notch on timing belt upper cover. The following valves can be adjusted with camshaft in this position (number four firing).
Number two cylinder—Exhaust
Number three cylinder—Intake
Number four cylinder—Intake and Exhaust

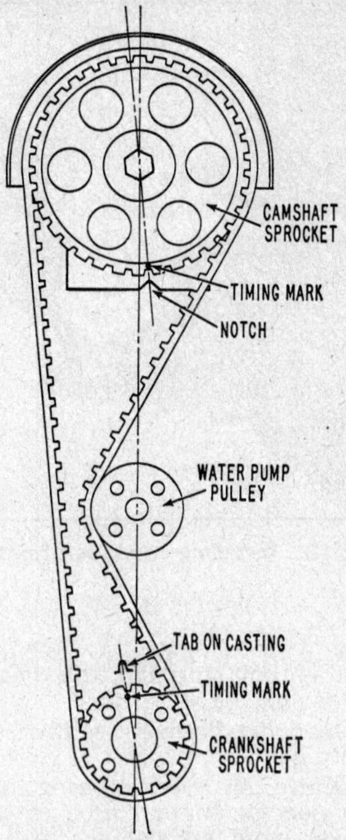

Fig. 10 Sprocket alignment marks

VALVE ARRANGEMENT

Front to Rear

4-cylinder . . . . I-E-I-E-I-E-I-E

CAMSHAFT COVER

1. Raise hood to fully open position and

Fig. 11 Checking sprocket alignment. 1972-76

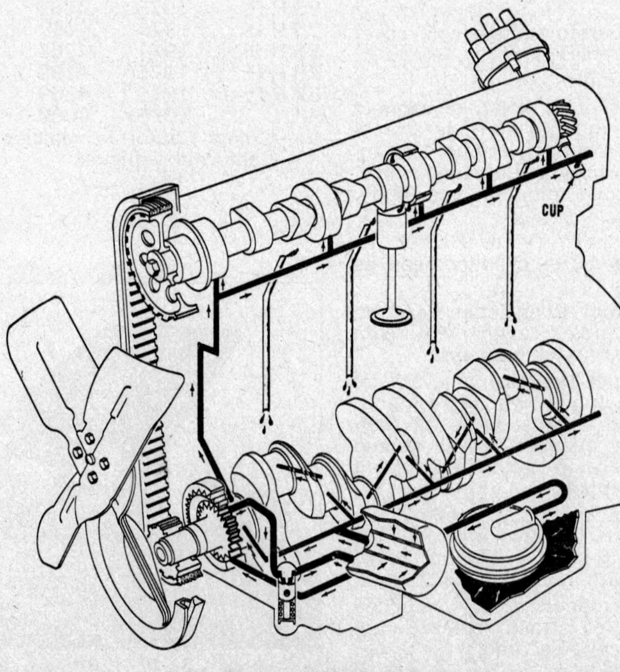

Engine lubrication system

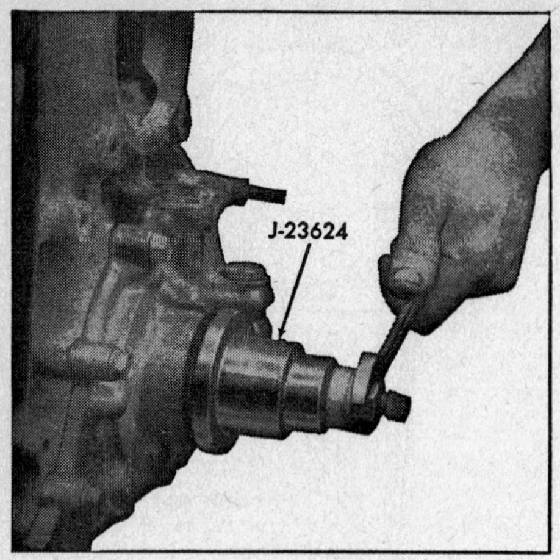

Fig. 12 Installing crankshaft front seal

install bolt through the hood hold-open link, Fig. 4.
2. Disconnect battery negative cable at battery.
3. Remove air cleaner wing nut. Disconnect ventilation tube at camshaft cover or at air cleaner; then remove air cleaner.
4. Remove PCV valve from grommet at front of cover.
5. Remove cover-to-cylinder head screws and withdraw cover from head.

CAMSHAFT, REMOVAL

1. Remove hood.
2. Remove camshaft timing sprocket.
3. Remove timing belt upper cover and cam retainer and seal assembly.
4. Remove camshaft cover.
5. Disconnect fuel line at carburetor and remove idle solenoid from bracket.
6. Remove carburetor choke coil, cover and rod assembly.
7. Remove distributor.
8. Raise vehicle on a hoist, disconnect engine front mounts at body attachment, raise front of engine and install wood blocks, about 1½" thick, between engine mounts and body. Lower vehicle.
9. Install camshaft removal tool as shown in Fig. 8, to cylinder head as follows:
   a. Position tool to cylinder head so attaching holes align with cam cover lower attaching holes.
   b. Align tappet depressing levers on tool so each lever will depress both intake and exhaust valve for their respective cylinder. Lever should fit squarely in notches adjacent to valve tappets.
   c. With tool aligned, make sure screws in bottom of tool are backed off so they do not make contact with bosses beneath tool.
   d. Install hardened screws supplied with tool, to attach tool to head. Torque screws securely.
   e. Turn screws in bottom of tool downward until they just seat against corresponding bosses on head.
   f. Apply a heavy body lubricant to ball end of lever depressing screws and proceed to tighten screws to depress tappets, Fig. 9.

NOTE: Use a torque wrench to tighten screws the final few turns. About 10 ft. lbs. is required to depress tappets.

10. At this point the camshaft can be removed by sliding it forward from the head.

CAMSHAFT BEARINGS

After removal of the camshaft as described previously, the bearings can be removed without removing the camshaft end plug.

NOTE: Install bearings from rear to front. Align oil holes in the rear three bearings with oil holes in bearing bore. Position oil holes in the first two bearings at 11 O'clock with oil groove in front bearing facing forward.

CAM LOBE LIFT SPECS.

		Intake	Exhaust
80 H.P.	1971	.4199	.4301
90 H.P.	1971	.4365	.4365
80 H.P.①	1972	.4199	.4302
80 H.P.②	1972	.4367	.4379
90 H.P.	1972	.4367	.4379
72 H.P.	1973	.4199	.4302
85 H.P.	1973	.4369	.4379
75 H.P.	1974	.4199	.4302
85 H.P.	1974	.4367	.4379
78 H.P.	1975	.4199	.4302
87 H.P.	1975	.4369	.4379
All	1976	.4000	.4150

①—Except California vehicles.
②—California vehicles.

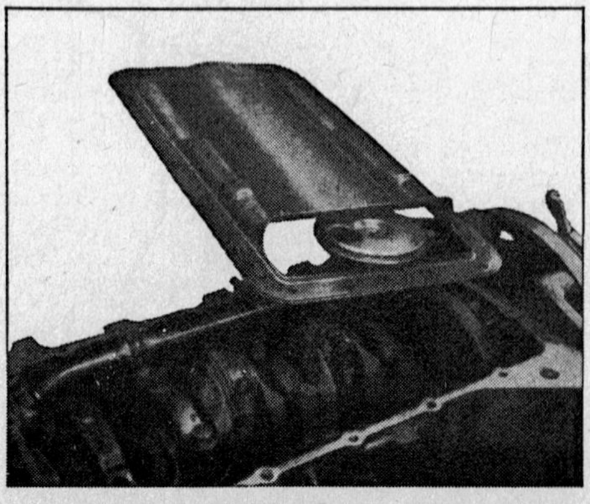

Fig. 14 Removing oil pan baffle

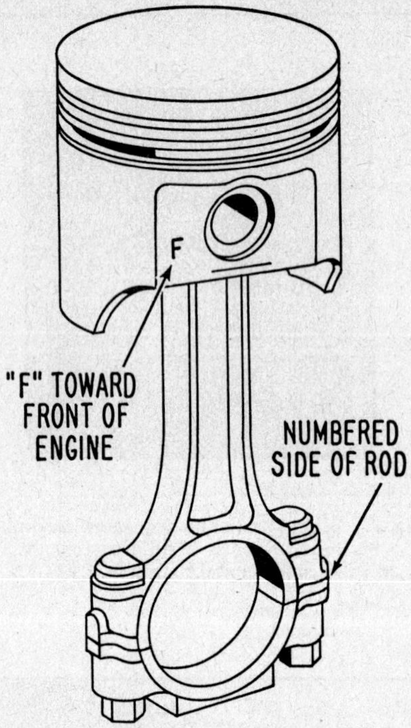

Fig. 13 Piston and rod assembly

VALVE TIMING

Intake Open Before TDC

80 H.P.	1971	22
90 H.P.	1971	25
80 H.P.①	1972	22
80 H.P.②	1972	28
90 H.P.	1972	28
72 H.P.	1973	22
85 H.P.	1973-74	28
75 H.P.	1974	22
78 H.P.	1975	22
87 H.P.	1975	28
All	1976	34

①—Except California vehicles.
②—California vehicles.

VALVE GUIDES

On all engines, valves operate in guide holes bored in the cylinder head. If clearance becomes excessive, use the next oversize valve and ream toe bore to fit. Valves with oversize stems are available in .003, .015 and 030".

ENGINE FRONT COVER

1. Raise hood to fully open position and install hood hold-open bolt.
2. Disconnect battery ground cable at battery.
3. Remove fan and spacer.
4. Loosen, but do not remove, the two cover lower screws. Cover is slotted to permit easy removal.
5. Remove the two cover upper retaining screws and remove cover.

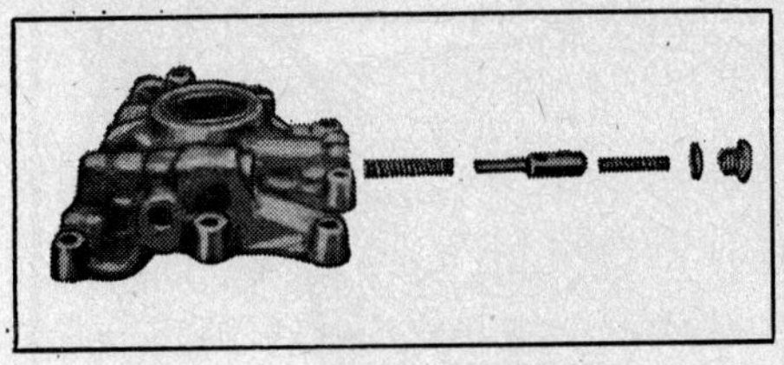

Fig. 15 Oil pump pressure regulator

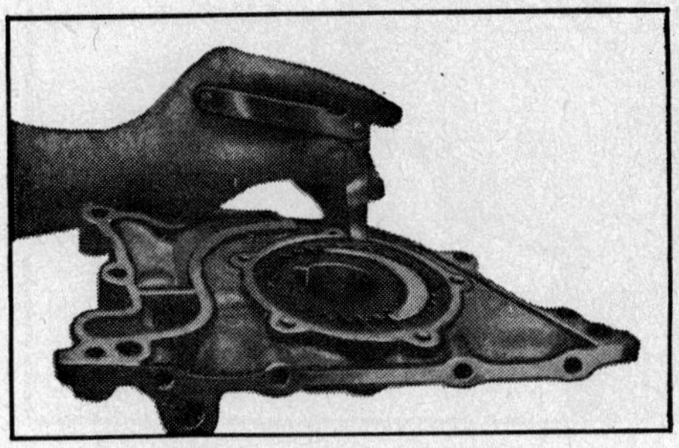

Fig. 16 Checking driven gear-to-housing clearance

TIMING BELT & WATER PUMP

1. Remove engine front cover as previously described.
2. Remove accessory drive pulley or damper.
3. Drain coolant and loosen water pump bolts to relieve tension on belt.
4. Remove timing belt lower cover, then remove belt from sprockets.
5. Remove water pump bolts and lift off pump after removing hoses.

CAMSHAFT SPROCKET

After removal of timing belt, the camshaft sprocket can be removed as follows:

1. Align one hole in sprocket with bolt head behind sprocket, then using a socket on bolt head to prevent cam from turning, remove sprocket retaining bolt and withdraw sprocket from camshaft.
2. When installing, be sure timing marks are aligned as in Fig. 10.

NOTE: For 1972-76 a simplified method is provided for checking camshaft and crankshaft alignment. Proper alignment is assured by making sure hole in left rear of the timing belt upper cover is in line with the corresponding hole in the camshaft sprocket. Check alignment by inserting a pencil or other similar tool through the hole in cover. If alignment is correct, tool will also enter small hole in cam gear, Fig. 11.

CRANKSHAFT SPROCKET

1. Remove engine front cover, accessory drive pulley, timing belt and timing belt lower cover.
2. Install suitable puller to crankshaft sprocket and remove sprocket.

OIL PUMP (CRANKCASE FRONT COVER) SEAL

1. Remove engine front cover, accessory dirve pulley, timing belt and timing belt lower cover and crankshaft sprocket.
2. Pry old seal from front cover being careful not to damage seal housing or seal lip contact surfaces.
3. Coat seal with light engine oil and apply an approved sealing compound to outside diameter of seal.
4. Position seal, closed end outward, onto crankshaft. Then install seal into bore using tool J-23624, Fig. 12.

PISTONS & RODS, ASSEMBLE

The "F" on the front of the piston must face the front of the engine when the piston and road assembly is installed in its proper cylinder, Fig. 13.

Piston Oversizes

Oversize pistons are available in oversizes of .010 and .020 inch.

MAIN & ROD BEARINGS

Undersizes

Bearings are available in .001, .002, .010 and .020" undersizes.

NOTE: If for any reason main bearing caps are replaced, shimming may be necessary. Laminated shims for each cap are available for service. Shim requirements will be determined by bearing clearance.

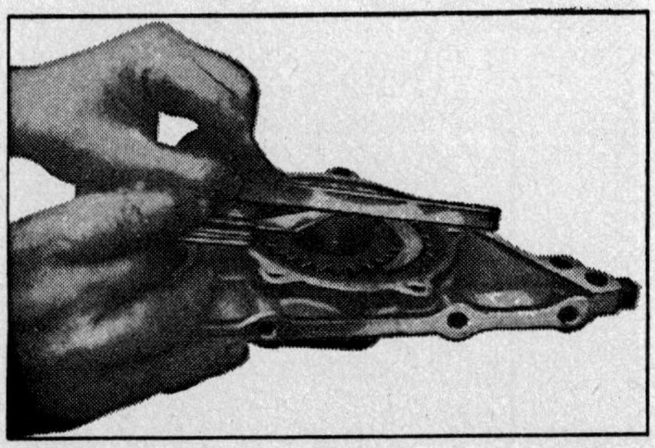

Fig. 19 Checking gear end clearance

Fig. 17 Checking drive gear-to-crescent clearance

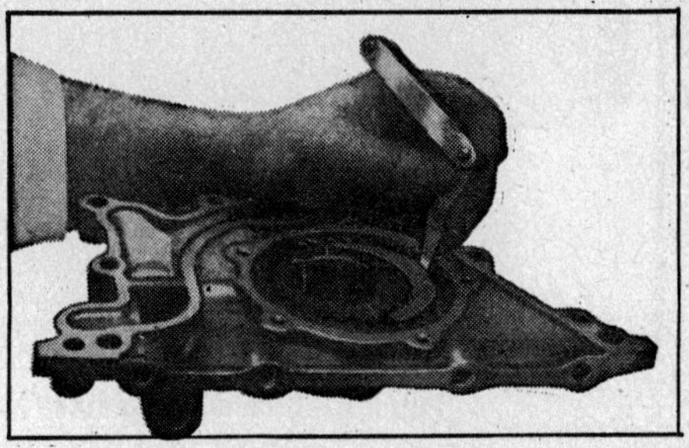

Fig. 18 Checking driven gear-to-crescent clearance

OIL PAN

1. Raise vehicle on a hoist and drain engine oil.
2. Support front of engine so weight is off front mounts, and remove frame crossmember and both front crossmember braces.
3. Disconnect idler arm at frame side rail.

NOTE: On air conditioned vehicles, disconnect idler arm at relay rod.

---

4. Mark relationship of steering linkage pitman arm to steering gear pitman shaft and remove pitman arm.

NOTE: Do not rotate steering gear pitman shaft while arm is disconnected as this will change steering wheel alignment.

---

5. Remove flywheel cover or converter underpan.
6. Remove oil pan to cylinder case screws, tap pan lightly to break sealing bond and remove oil pan.
7. Remove pick up screen to baffle support bolts, then remove support from baffle.
8. Remove bolt securing oil pan drain back tube to baffle. Then rotate baffle 90 degrees towards left side of car and remove baffle from pick up screen, Fig. 14.

OIL PUMP (CRANKCASE FRONT COVER)

1. Remove engine front cover, acces-

Fig. 20 Tensioning adapter locating hole

Fig. 21 Adjusting timing belt

sory drive pulley, timing belt, timing belt lower cover and crankshaft sprocket.
2. Raise vehicle on a hoist and remove oil pan and baffle.
3. Remove bolts and stud securing oil pump to cylinder case.

Inspection

1. Clean gasket surfaces, then wash parts in approved solvent and blow out all passages.
2. Check pressure regulator for free operation, Fig. 15.
3. Inspect pump gears for nicks, broken parts and other damage.
4. Check clearance between outside diameter of driven gear and pump. Clearance should be .0038-.0068", Fig. 16.
5. Check clearance between outside diameter of drive gear and crescent. Clearance should be .0023-.0093", Fig. 17.
6. Check clearance between inside diameter of driven gear and pump crescent. Clearance should be .0068-.0148", Fig. 18.
7. Check gear end clearance. It should be .0009-.0023", Fig. 19.

NOTE: The pump gears and body are not serviced separately. If pump gears or body are worn, replacement of the entire is necessary.

BELT TENSION DATA

	New Lbs.	Used Lbs.
1971-76—		
Air Condition	135-145	95
A.I.R. Pump	120-130	75
Generator	120-130	75
Power Steering	120 130	75
Timing Belt	100-140	—

WATER PUMP, REPLACE

1. Raise hood to fully open position and install hold-open bolt.
2. Disconnect battery ground cable at battery.
3. Remove engine fan and spacer.
4. Loosen, but do not remove, the two cover lower screws. Cover is slotted to permit easy removal.
5. Remove the two cover upper retaining screws and remove cover.
6. Drain coolant and loosen water pump bolts to relieve tension in timing belt.
7. Remove radiator lower hose and heater hose at water pump.
8. Unfasten and remove water pump.

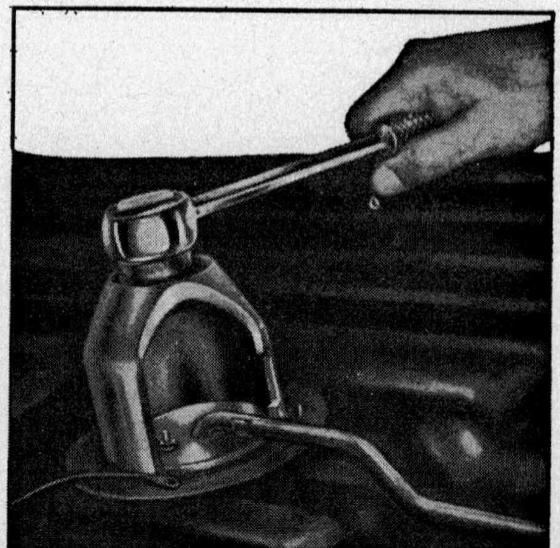

Fig. 22 Removing fuel pump and gauge unit

TIMING BELT TENSION, ADJUST

1. Drain coolant at engine block and remove fan and extension.
2. Remove engine front cover.
3. Remove water pump retaining bolts, clean gasket surfaces on block and pump. Install new gasket and loosely install water pump bolts.

NOTE: Apply an approved anti-seize compound to the water pump bolts before installation.

4. Position tool J-23564 in gauge hole adjacent to left side of pump, Fig. 20.
5. Apply 15 ft. lbs. of torque to water pump as shown in Fig. 21. Tighten water pump bolts while maintaining torque on side of pump.
6. Reinstall front cover, fan, extension and fill cooling system.

FUEL PUMP PRESSURE

Year	Engine	Pressure, Lbs.
1971-76	All	3-4½

FUEL PUMP, REPLACE

Removal

1. Disconnect battery ground cable.
2. Disconnect meter and pump wires at rear wiring harness connector.
3. Raise vehicle on hoist and drain fuel tank.
4. Disconnect fuel line hose at gauge unit pick up line.

5. Disconnect tank vent lines to vapor separator.
6. Remove gauge ground wire screw at underbody floorpan.
7. Remove tank straps bolts and lower tank carefully.
8. Unscrew retaining ring using spanner wrench J-22554, Fig. 22, and remove pump-tank unit assembly.

Replacement

1. Remove flat wire conductor from plastic clip on fuel tube.
2. Squeeze clamp and pull pump straight back about ½ inch.
3. Remove two nuts and lockwashers and conductor wires from pump terminals.
4. Squeeze clamp and pull pump straight back to remove it from tank unit. Take care to prevent bending of circular support bracket.
5. Slide replacement pump through circular support bracket until it rests against rubber coupling. Make sure pump has rubber isolator and saran strainer attached.
6. Attach two conductor wires to pump terminals using lockwashers and nuts being certain flat conductor is attached to terminal located on side away from float arm.
7. Squeeze clamp and push pump into rubber coupling.
8. Replace flat wire conductor in plastic clip on fuel pick up tube.
9. Install unit into tank and replace fuel tank.

Cosworth Engine Section

ENGINE, REPLACE

1. Remove hood.
2. Disconnect battery positive cable from battery and the ground cable from engine block.
3. Drain cooling system and disconnect hoses from radiator and water pump.
4. Remove fan and spacer, then the radiator shroud and radiator.
5. Disconnect electrical leads from oil pressure switch and water temperature switch, then disconnect wiring harness from block and position aside.
6. Disconnect EFI coolant temperature sensor lead and position aside.
7. Disconnect air cleaner bellows from throttle body and remove air cleaner to fender skirt attaching bolts, then the air cleaner from vehicle.
8. Disconnect accelerator cable from throttle body and position aside.
9. Remove intake manifold to support bolts from underside of manifold, then the manifold attaching bolts and nuts. Position intake manifold with electrical leads and hoses attached, on fender skirt.
10. Disconnect electrical leads from starter solenoid, starter and alternator.
11. Disconnect fuel lines, PCV hose and the electrical leads from distributor.
12. Raise vehicle and disconnect exhaust pipe from manifold.
13. Remove flywheel housing to engine attaching bolts and loosen engine front mount retaining bolts at frame, then lower vehicle.
14. Support transmission with a suitable jack.
15. Install suitable engine lifting equipment and raise engine slightly to relieve weight from engine mounts. Remove engine front mount retaining bolts.
16. Pull engine forward to clear transmission and raise engine slowly. Also, turn engine sideways (engine front toward vehicle left side) to obtain maximum clearance. Remove engine from vehicle.
17. Reverse procedure to install.

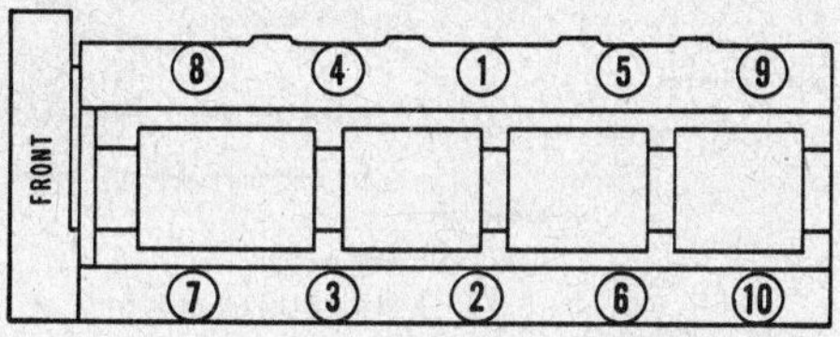

Fig. 1 Cylinder head tightening sequence

CYLINDER HEAD

1. Remove engine front cover and camshaft carrier cover.
2. Remove timing belt and sprockets, then timing belt upper cover and camshaft carrier.
3. Remove intake manifold and position manifold with electrical leads and hoses attached on fender skirt.
4. Disconnect exhaust pipe from header.
5. Remove exhaust header and cylinder head.
6. Reverse procedure to install. Torque cylinder head bolts in sequence, Fig. 1, to 60 ft. lbs.

Fig. 2 Checking valve lash

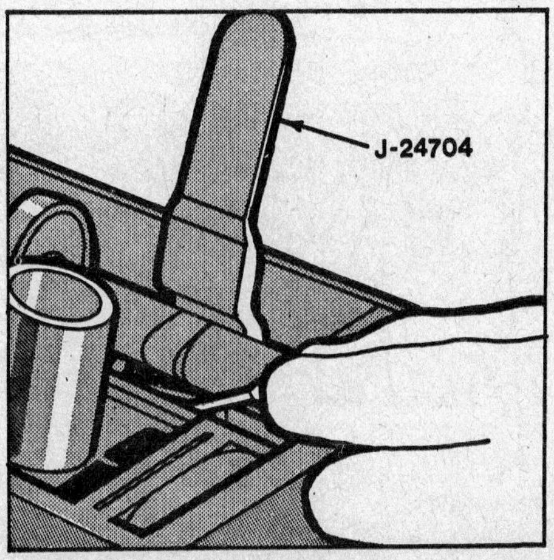

Fig. 3 Adjusting valve lash

VALVES, ADJUST

1. Remove camshaft carrier cover and gasket.
2. Mark distributor housing, with chalk, at plug wire locations Nos. 1 and 4. Then, disconnect plug wires at spark plugs and all other wiring from distributor cap. Remove distributor cap and plug wire assembly.
3. Crank engine to align distributor rotor with the No. 1 cylinder position mark on distributor housing. Check valve lash, Fig. 2, on the following cylinders with the engine at the No. 1 firing position:
   No. 1 cylinder—Intake and Exhaust
   No. 2 cylinder—Intake
   No. 3 cylinder—Exhaust
4. To adjust valve lash, crank engine until camshaft lobe fully depresses tappet of valve being adjusted. Install tool J-24704, Fig. 3, to hold tappet in depressed position and rotate camshaft to provide clearance to remove tappet adjusting shim. Install proper selective thickness tappet adjusting shim to bring valve lash within specifications as listed in the Valve Specifications Chart found in the front of this chapter. Rotate camshaft to fully depress tappet and remove tool J-24704.
5. Crank engine to align distributor rotor with the No. 4 cylinder position mark on distributor housing. Check valve lash on the following cylinders with the engine at the No. 4 firing position:
   No. 2 cylinder—Exhaust
   No. 3 cylinder—Intake
   No. 4 cylinder—Intake and Exhaust
   Adjust valve lash as outlined previously.
6. Install distributor cap and spark plug wire assembly, then the camshaft carrier cover gasket and cover.

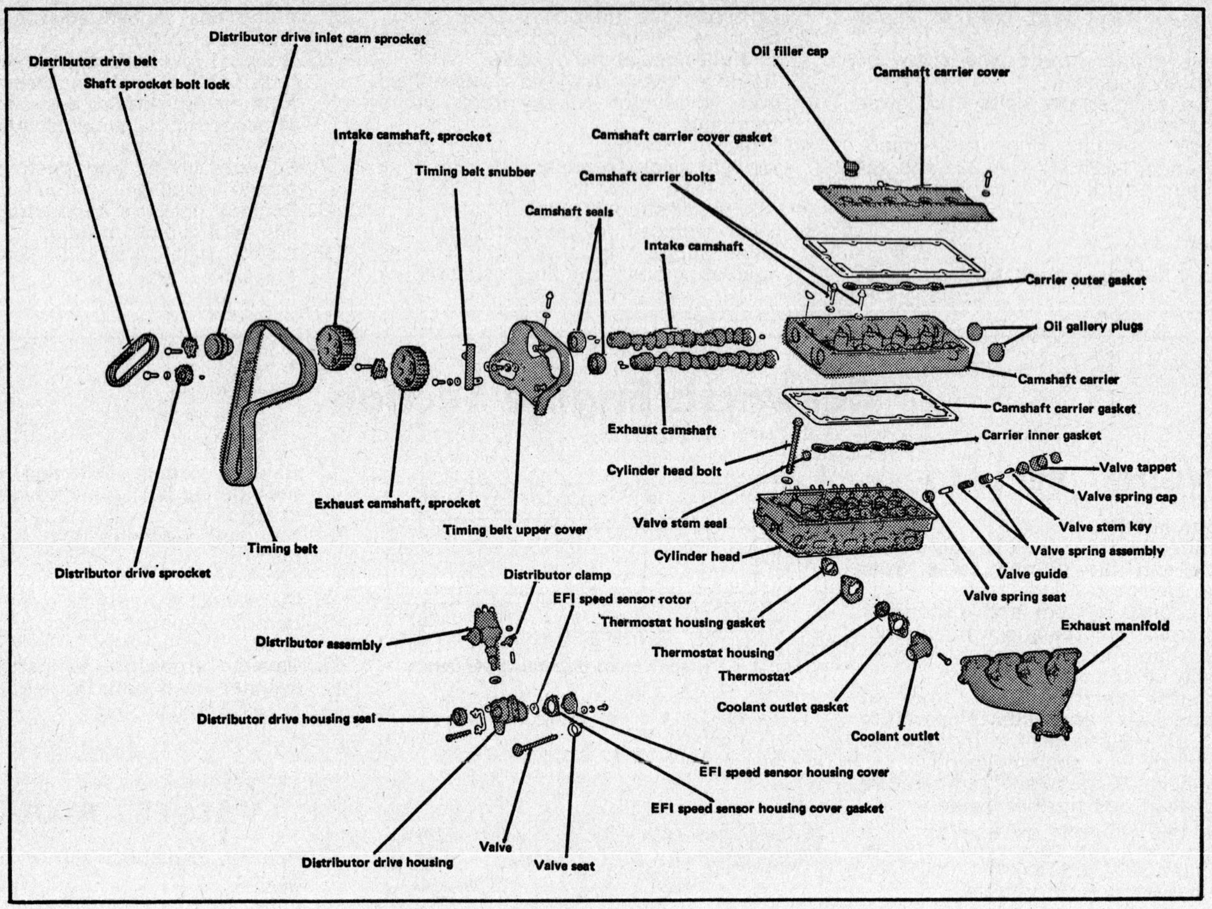

Exploded view of Cosworth engine cylinder head assembly

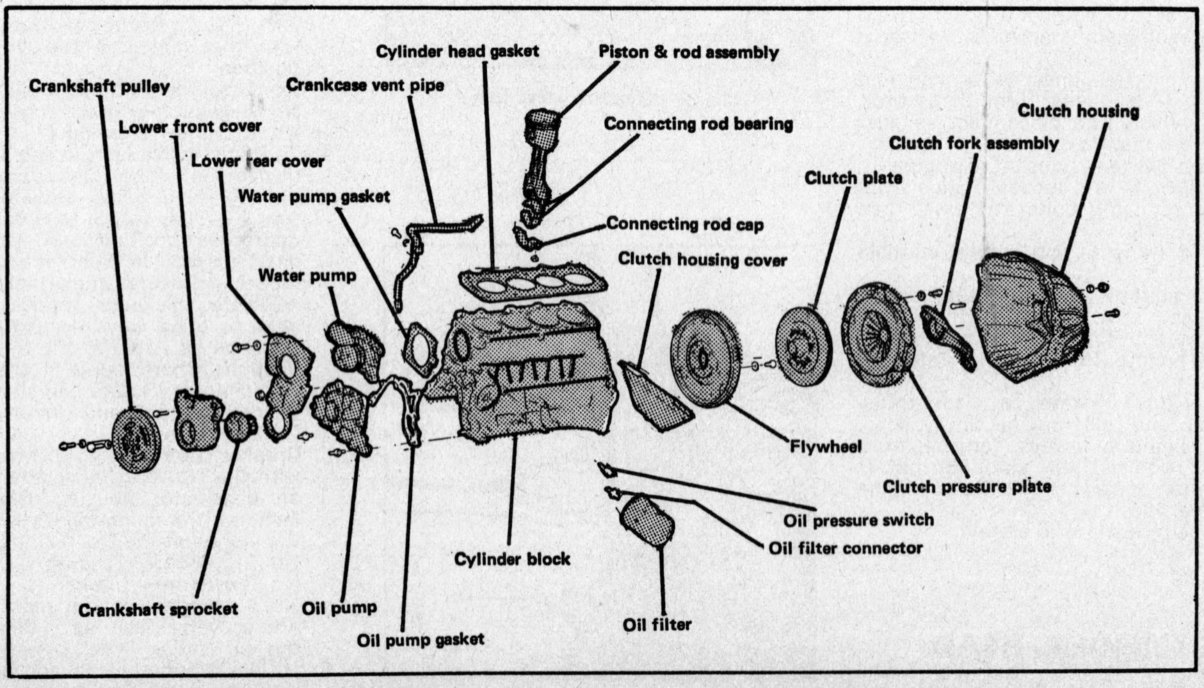

Exploded view of Cosworth engine cylinder block assembly

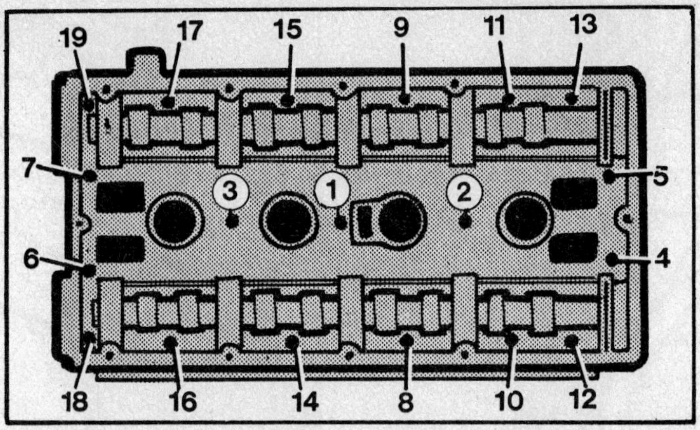

Fig. 4 Camshaft carrier tightening sequence

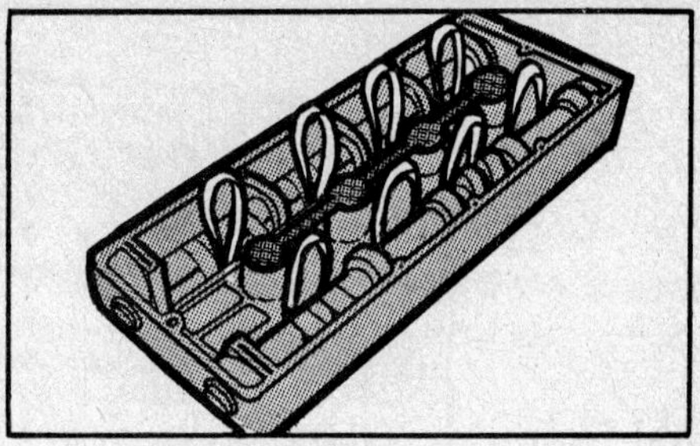

Fig. 5 Valve tappet retaining tool installation

Fig. 6 Installing camshaft oil seals

VALVE ARRANGEMENT

Right side: all intake
Left side: all exhaust

CAMSHAFT CARRIER COVER

1. Raise hood and install hold-open link.
2. Disconnect battery ground cable.
3. Disconnect PCV hose from cover right side.
4. Remove spark plug leads from spark plugs and position aside.
5. Remove screws securing pulse air pipe brackets to cover, then the four hoses from pulse air manifold to pulse air pipe check valves.
6. Remove cover attaching screws, then the cover and gasket.
7. Reverse procedure to install.

CAMSHAFT CARRIER

1. Remove engine front cover, timing belts and camshaft sprockets.
2. Remove timing belt upper cover.
3. Remove camshaft carrier cover as outlined previously.
4. Remove cam carrier to cylinder head bolts.

NOTE: When removing bolts, loosen bolts in small increments in sequence, Fig. 4, thereby preventing carrier distortion due to valve spring tension.

5. Rotate camshafts to expose the holes in tapped body wall.
6. Position tappets so body holes are approximately 15 degrees from carrier center, then install tool J-24705 into tappet body hole and over camshaft,

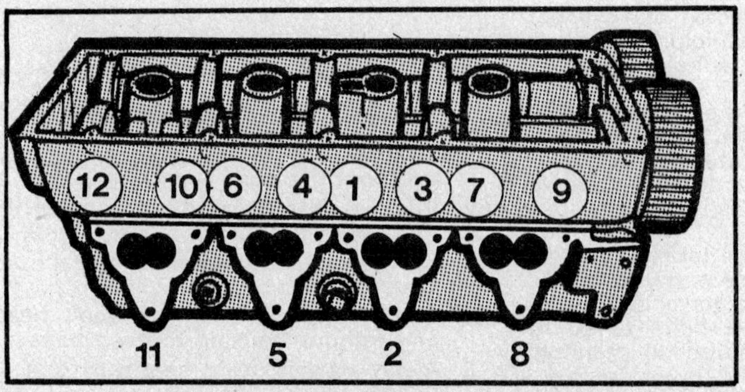

Fig. 7 Intake manifold tightening sequence

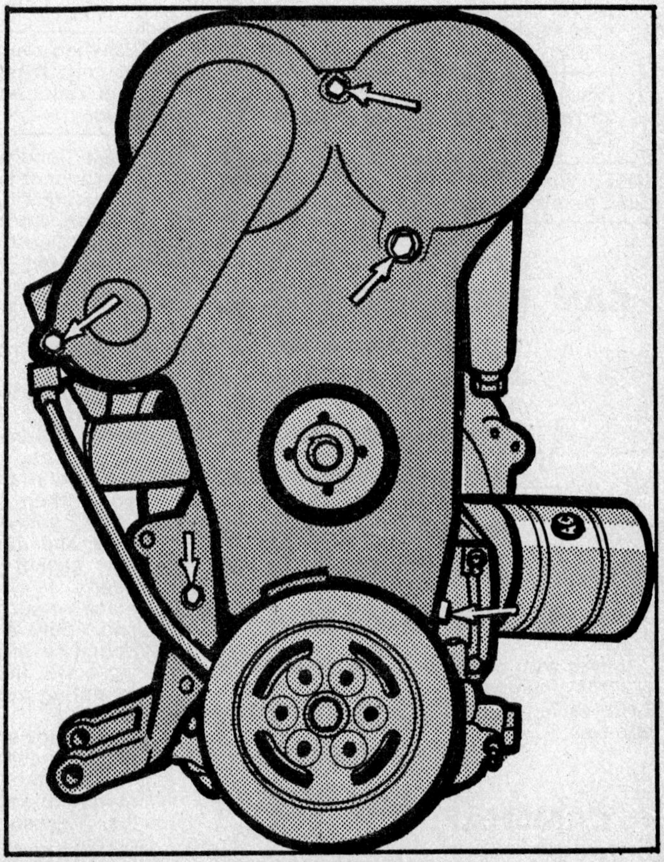

Fig. 8 Engine front cover

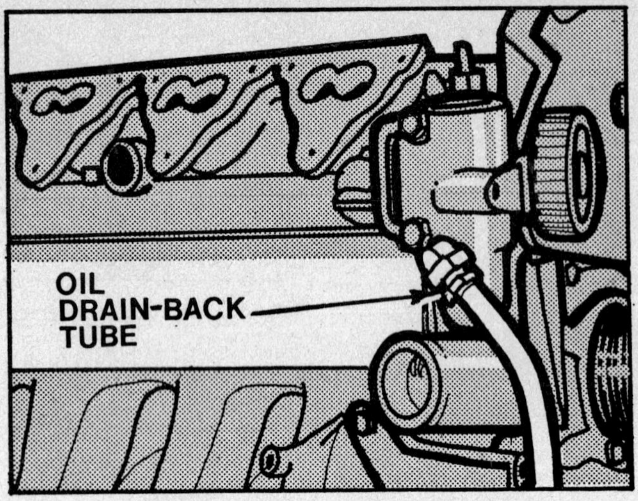

Fig. 9 Distributor drive housing oil drain-back tube

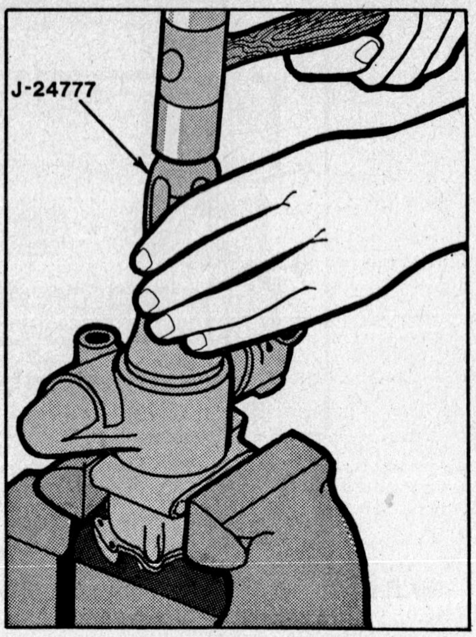

Fig. 10 Installing distributor drive housing seal

Fig. 5. One tool J-24705 is used for each pair of tappets.

7. Lift cam carrier from cylinder head.
8. Remove tools J-24705 from tappets and identify each tappet and adjusting shim to facilitate re-installation in original position.
9. Pry camshaft seal from carrier and remove camshaft retainer (thrust) key from keyway at front of carrier.
10. Remove camshaft from carrier.

NOTE: The camshafts are identified by cast raised letters near front of camshaft.

11. Reverse procedure to install. Torque carrier mounting bolts in sequence, Fig. 4, to 65 inch lbs.

NOTE: When installing carrier, use two guide pins fabricated from 1/4-20 bolts.

CAM LOBE LIFT SPECS.

		Intake	Exhaust
4-122	1975-76	.355	.355

VALVE TIMING

Intake Open Before TDC

4-122 1975-76 38

VALVE GUIDES

Valves with oversize stems are available in .003 inch oversize for replacement. Ream valve guide bores for oversize valves with tool J-24703.

CAMSHAFT OIL SEAL

1. Remove engine front cover and timing belts as outlined previously.
2. Remove sprocket from cam.
3. Pry seal from cam carrier, using caution not to damage carrier surfaces.
4. Coat outside diameter of new seal with a suitable sealant, then using tool J-24777, Fig. 6, install seal into cam carrier.

INTAKE MANIFOLD, REPLACE

NOTE: When replacing the intake manifold gasket only, it is not necessary to remove the fuel rail and throttle body from the manifold.

1. Raise hood and install hold-open link.
2. Disconnect battery ground cable.
3. Disconnect harness at air temperature sensor, throttle body air enrichment valve, distributor and fuel injectors.
4. Disconnect air cleaner bellows and MAP sensor lines from throttle body.
5. Disconnect fuel lines from fuel rail.
6. Remove bolts securing intake manifold to engine brace.
7. Disconnect accelerator cable from throttle body and position aside.
8. Disconnect PCV hose and position aside, then remove fuel rail and throttle body from intake manifold.
9. Remove manifold retaining nuts and bolts, then the intake manifold.
10. Clean gasket surfaces on cylinder head and intake manifold, then place new gasket over manifold studs on head.
11. Place intake manifold on cylinder head and apply a suitable anti-seize compound to manifold studs and mounting bolts. Install and torque manifold mounting nuts and bolts to 20 ft. lbs., Fig. 7.
12. Loosen lower nuts on intake manifold brace, ensuring brace is free to move. Then, attach brace to underside of manifold and torque bolts to 20 ft. lbs. Tighten lower nuts on manifold brace.
13. Install throttle body and connect accelerator cable.
14. Install fuel rail and fuel injectors, then connect harness to fuel injectors, air temperature sensor, throttle body air enrichment valve and the distributor.
15. Connect fuel inlet and return lines to fuel rail, then install air cleaner bellows on throttle body.
16. Connect MAP sensor lines to throttle body and install PCV hose in camshaft cover.
17. Connect battery ground cable, start engine and check for proper operation.

ENGINE FRONT COVER

1. Raise hood and install hold-open link. Disconnect battery ground cable.
2. Remove fan and spacer.
3. Remove the two lower cover screws and the three, cover attaching bolts, Fig. 8. Lift upward on cover right side and pull forward to clear accessory drive belt and water pump hub. Push downward on cover left side, disengaging cover from support and lift cover from engine.
4. Reverse procedure to install.

DISTRIBUTOR DRIVE HOUSING

1. Remove engine front cover and distributor drive belt.
2. Remove distributor drive housing oil drain-back tube from carrier, Fig. 9.
3. Disconnect speed sensor terminal from housing rear.
4. Bend bolt lock tabs from bolt head and remove housing to head bolts.
5. Install new housing seal with tool J-24777, Fig. 10.
6. Reverse procedure to install.

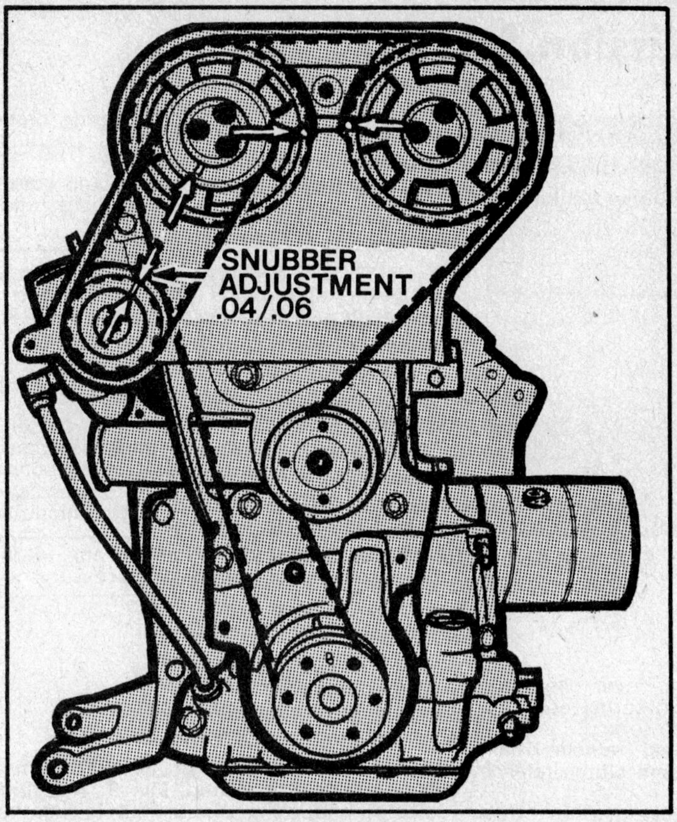

Fig. 11 Camshaft sprocket alignment

Fig. 12 Timing belt lower rear cover

TIMING BELT

1. Remove engine front cover.
2. Align timing marks, Fig. 11.
3. Drain cooling system and loosen water pump mounting bolts, thereby relieving timing belt tension.
4. Pry distributor drive belt from drive sprocket with a suitable tool.
5. Remove timing belt lower cover, then the timing belt from camshaft and crankshaft sprockets.
6. If engine was rotated with timing belt off, position No. 1 piston at TDC compression stroke and align all timing marks, Fig. 11.

NOTE: The camshaft sprockets may be aligned by using a straight edge or by aligning the small hole in the sprocket tooth space with the "V" notch of the upper sheet metal cover.

7. Slide water pump to extreme left hand position and install timing belt over crankshaft sprocket and water pump pulley. Stretch belt tight and mesh belt with camshaft sprockets, starting with the inlet sprocket.
8. Using tool J-23654, slide water pump to the right and torque two pump attaching bolts to 15 ft. lbs.
9. Rotate engine clockwise two or three revolutions, then rotate engine counter-clockwise approximately 10 degrees to place the long slack side of the belt in tension.
10. Loosen the water pump attaching bolts and apply 15 ft. lbs. of torque to tool J-23654. Then, while holding tool J-23654 at specified torque, torque water pump attaching bolts to 15 ft. lbs.
11. Rotate engine clockwise to position No. 1 piston on TDC, compression stroke, and check timing mark alignment.
12. Install engine front cover.

TIMING BELT LOWER REAR COVER

1. Remove engine front cover and timing belt.
2. Remove alternator bracket to block bolt that passes through the rear cover.
3. Remove nut from oil pump mounting stud.
4. Remove water pump bolts, Fig. 12, and hold water pump in position. Then, remove rear cover and re-install two water pump bolts.
5. Reverse procedure to install.

PISTONS

Oversize pistons are available in .020 inch oversize.

MAIN BEARINGS

Bearings are available in undersizes of .001, .002, .010 and .020 inch.

FUEL PUMPS, REPLACE

Refer to the "Electronic Fuel Injection" chapter, EFI Service section for service procedures.

FUEL PUMP PRESSURE

1975-76 In-tank Unit .... 7-8½ PSI
Chassis-mounted Unit 39 PSI

Refer to the "Electronic Fuel Injection" chapter for trouble-shooting and service procedures.

Clutch & Transmission Section

CLUTCH PEDAL, ADJUST

Initial adjustment after clutch and/or cable replacement is made at two points, at the ball stud and the lower end of the cable.

1. Ball stud adjustment is made before attaching the cable as follows: using gauge J-23644, place it so flat end is against the front face of the clutch housing and the hooked end is located at the point of cable attachment on the fork, Fig. 1.
2. Turn ball stud inward until clutch release bearing lightly contacts the clutch spring levers, then install the lock nut and tighten being careful not to change the adjustment. Remove the gauge by pulling outward at the housing end.
3. To adjust the cable, place it through the hole in the fork and pull it until the clutch pedal is firmly against the rubber bumper.
4. Push the clutch fork forward until the throwout bearing lightly contacts the clutch spring levers, then screw the lock pin on the cable until it bottoms out on the fork. Turn it 1/4 additional clockwise revolution, set pin into groove in the fork and attach return spring. This procedure will produce .90" lash at the pedal.

Adjustment for normal clutch wear is accomplished by loosening the lock nut and by turning the clutch fork ball stud counterclockwise until .90" of free play is obtained at the pedal.

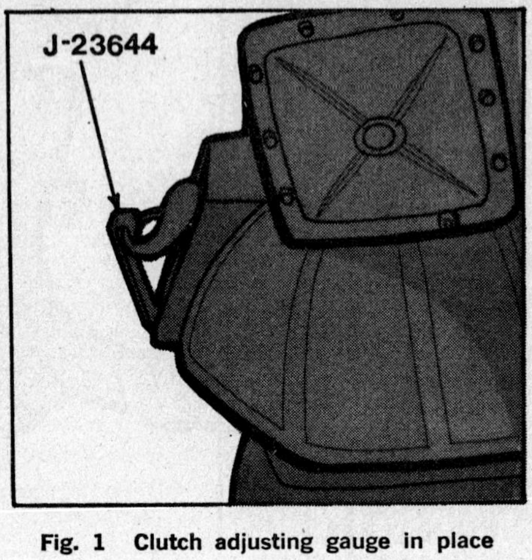

Fig. 1 Clutch adjusting gauge in place

CLUTCH REPLACE

1. Remove transmission.
2. Remove clutch fork cover then disconnect clutch return spring and control cable from clutch fork.
3. On 1971-72 vehicles, remove main drive gear oil seal from clutch release bearing sleeve.
4. On all vehicles, remove flywheel housing lower cover and flywheel housing, Fig. 2.
5. Remove release bearing from the clutch fork and sleeve by sliding lever off ball stud and against spring force. If ball stud is to be replaced, remove cap, lock nut and stud from housing.
6. Make sure alignment marks on clutch assembly and flywheel are distinguishable.
7. Loosen clutch cover to flywheel bolts one turn at a time until spring pressure is released, to avoid bending the clutch cover flange.
8. Support the pressure plate and cover assembly while removing bolts and clutch assembly.

NOTE: Do not disassemble the clutch cover, spring and pressure plate for repairs. If defective replace complete assembly.

---

9. Reverse procedure to install making sure to index alignment marks.
10. After installing crossmember, loosely install retaining bolts, then the crossmember to transmission mount bolts. Tighten all retaining bolts to specifications and remove the engine support.

CAUTION: Check position of engine in front mounts and realign as necessary.

GEARSHIFT

1971-72

The shift controls for both transmissions are floor mounted, Fig. 3. The shift selector shaft, Fig. 4, located inside the case with the ends extending through the case perpendicular to the mainshaft, has both side to side and angular movement.

Two intermediate shift levers, which select the gear required, are attached to the selector shaft by means of spiral pins. The side movement of the selector shaft allows the intermediate shift levers to pick up the proper shift shaft. The angular movement positions the shift shaft such that the shift forks (attached to shafts by spiral pins) move the synchronizer sleeves to obtain the required gear.

The 4-speed shift control operation is similar to the 3-speed except for the gear shift lever which has a reverse lockout feature, Fig. 5.

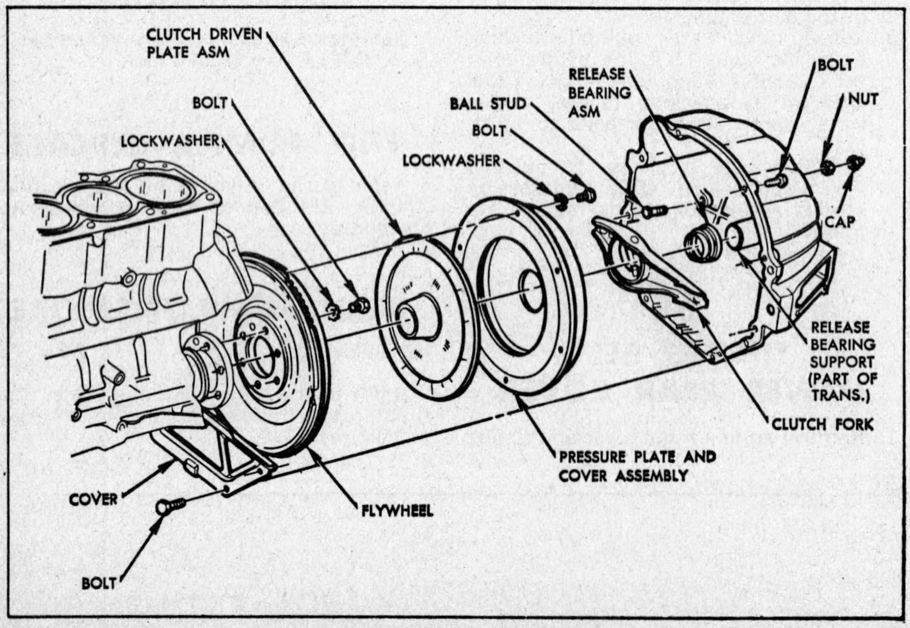

Fig. 2 1971-76 clutch & housing

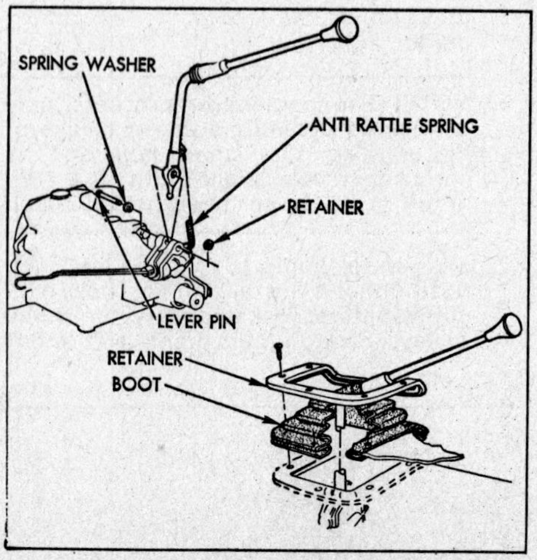

Fig. 3 Gearshift lever installation. 1971-72

LINKAGE ADJUST

1973-76 3 & 4 Speed Saginaw

1. Place ignition switch in "Off" position.
2. Raise vehicle and loosen shift rod nuts at swivels.
3. Position shift levers in neutral.
4. Place shift control in neutral position, align control levers and install gauge pin into levers and brackets.
5. On 3 speed models, tighten 1st-Rev. shift rod nut against swivel, then tighten 2-3 shift rod nut against swivel, torque nuts to 120 in. lbs.
6. On 4 speed models, tighten 1-2 shift rod nut against swivel, then tighten 3-4 shift rod nut against swivel, torque nuts to 120 in. lbs.
7. On 4 speed models, torque reverse shift rod nut to 120 in. lbs.
8. On all models, remove gauge pin and check linkage for operation.

TRANSMISSION, REPLACE

3 & 4 Speed Exc. 70mm (Brazil)

1. Place shift lever in neutral and remove shift lever.
2. Raise vehicle on a hoist and drain lubricant.
3. On 1976 models, remove torque arm, referring to the "Rear Suspension" section. On all models, remove the propeller shaft.
4. Disconnect speedo cable, TCS switch and back-up lamp switch.
5. Remove crossmember to transmission mount bolts.
6. Support engine with an appropriate jack stand and remove crossmember to frame bolts and remove crossmember.
7. Remove transmission to clutch housing upper retaining bolts and install guide pins in holes.
8. Remove lower bolts, then slide transmission rearward and remove from vehicle.

NOTE: Inspect throwout bearing support gasket located beneath lip of support. If necessary, replace gasket before installing transmission.

4 Speed 70mm (Brazil)

1. Push shift lever boot down and loosen lock nut, remove upper portion of shift lever.
2. Remove console, retainer and boot, then remove shift control from extension housing.
3. Raise vehicle and drain lubricant.
4. Remove torque arm, referring to the "Rear Suspension" section.
5. Remove propeller shaft, then disconnect speedometer cable and back-up light switch.
6. Disconnect clutch return spring and cable at fork.
7. Remove transmission to crossmember bolts.
8. Remove converter bracket and damper, if equipped.
9. Remove crossmember to frame bolts, support transmission, then remove crossmember.
10. Remove clutch housing to engine attaching bolts, slide transmission and clutch housing rearward and remove from vehicle.

Fig. 4 Shift selector shaft

5 Speed Unit

1. Remove shifter assembly and raise vehicle.
2. Remove torque arm, referring to the "Rear Suspension" section, then remove propeller shaft and disconnect speedometer cable from transmission.
3. Remove crossmember to transmission bolts and the catalytic converter support bracket.
4. Support engine and remove crossmember.
5. Remove transmission to clutch housing upper retaining bolts and install guide pins.
6. Remove lower transmission to clutch housing bolts, slide transmission rearward and lower from vehicle.
7. Remove back-up lamp switch and fill plug, then drain transmission.

1976 70mm (BRAZIL) 4 SPEED TRANS.

Refer to the "Chevrolet Monza, Buick Skyhawk, Olds Starfire & Pontiac Sunbird" chapter for service procedures.

1976 AUTO. TRANS. LINKAGE, ADJUST

Linkage adjustments for 1976 models are the same as 1975 Turbo Hydra-matic 250 adjustments found in the Automatic Transmission section.

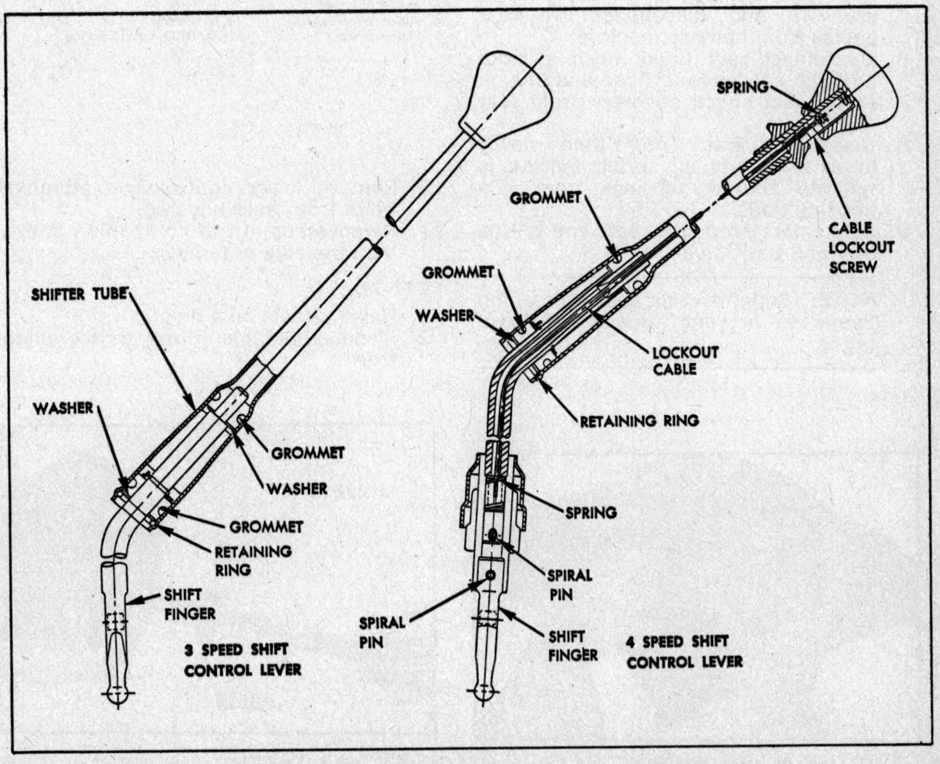

Fig. 5 Shift control lever cross section. 1971-72

Rear Axle, Propeller Shaft & Brakes

REAR AXLE

Description

The rear axle, Fig. 1, is a semi-floating type consisting of a cast carrier and large bosses on each end into which two welded steel tubes are fitted. The carrier contains an overhung hypoid pinion and ring gear. The differential is a two pinion arrangement.

The overhung hypoid drive pinion is supported by two preloaded tapered roller bearings. The pinion shaft is sealed by means of a molded, spring loaded, rubber seal. The seal is mounted on the pinion shaft flange which is splined and bolts to the hypoid pinion shaft.

The ring gear is bolted to a one piece differential case and is supported by two preloaded tapered roller bearings.

Removal

1976

1. Support vehicle at frame and using a suitable jack support axle housing.
2. Remove wheels, brake drums and axle shafts.
3. Disconnect brake lines from axle tube clips and remove bolt securing brake line junction block to rear axle housing.

NOTE: Do not disconnect brake lines from wheel cylinders or junction block.

4. Remove brake backing plates and secure backing plates to frame.
5. Remove parking brake cable adjusting nuts at equalizer, pull center cable rearward and disconnect two rear cables from body connectors.
6. Disconnect rear brake hose at floor pan and cap ends of hose and line.
7. Disconnect shock absorber from rear axle.
8. Disconnect track rod, then slowly lower axle until all spring tension is relieved and pry springs from axle housing pads.
9. Disconnect propeller shaft and torque arm and position aside.

NOTE: Support axle at companion flange to prevent housing from rotating.

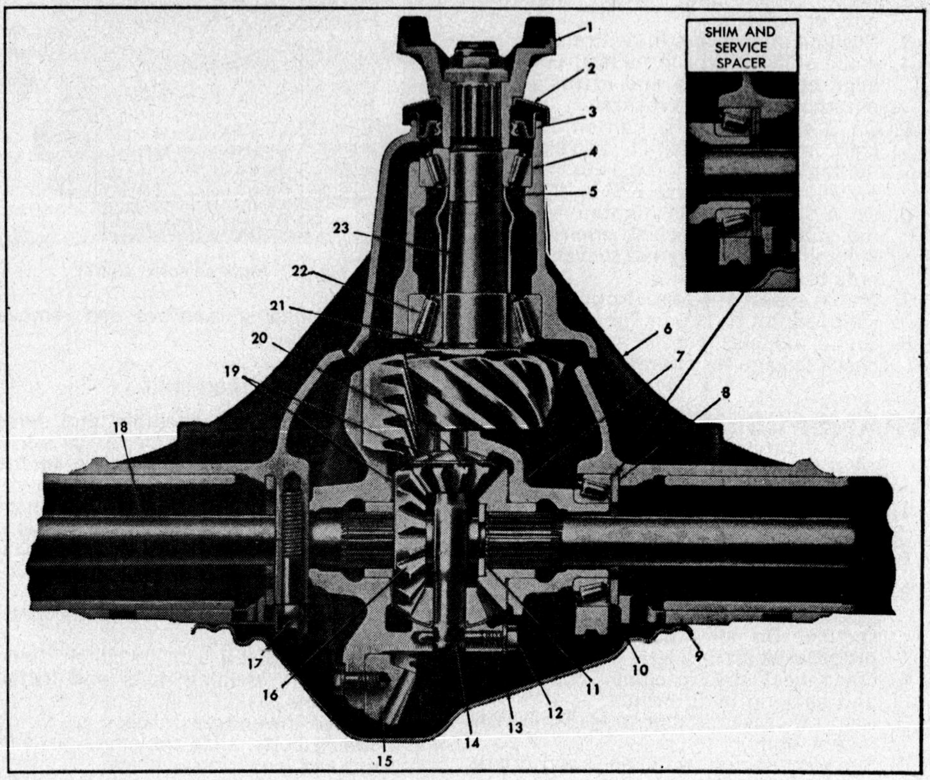

Fig. 1 Rear axle cross section

1. Companion Flange
2. Deflector
3. Pinion Oil Seal
4. Pinion Front Bearing
5. Pinion Bearing Spacer
6. Differential Carrier
7. Differential Case
8. Shim
9. Gasket
10. Differential Bearing
11. "C" Lock
12. Pinion Shaft Lock Bolt
13. Cover
14. Pinion Shaft
15. Ring Gear
16. Side Gear
17. Bearing Cap
18. Axle Shaft
19. Thrust Washer
20. Differential Pinion
21. Shim
22. Pinion Rear Bearing
23. Drive Pinion

10. Remove lower control arm attaching bolts from axle housing.
11. Remove support at companion flange, then the axle assembly.

1971-75

1. Raise vehicle on a hoist.
2. Place adjustable lifting device under axle.
3. Disconnect rear shock absorbers from axle and remove propeller shaft.
4. Disconnect upper control arm from axle and remove both rear wheels.
5. Remove right and left brake drums.
6. Disconnect brake lines from clips on axle tubes.

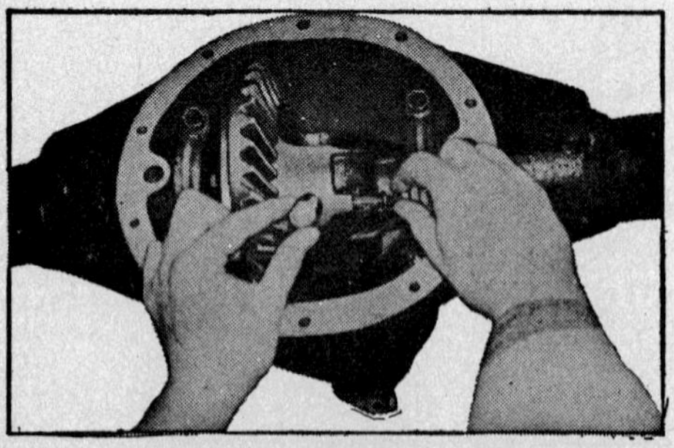

Fig. 2 Differential pinion shaft removal

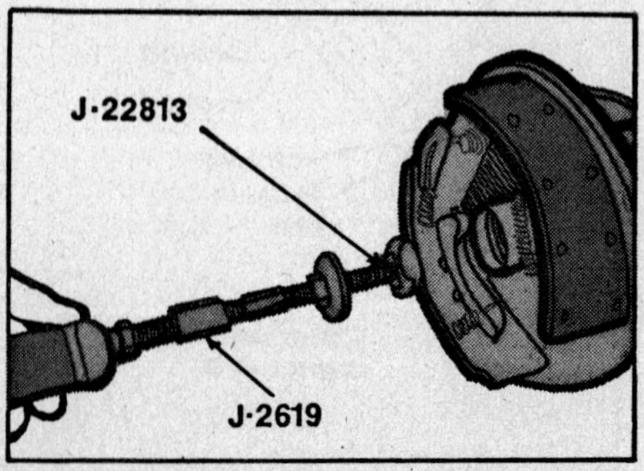

Fig. 3 Removing wheel bearing and seal

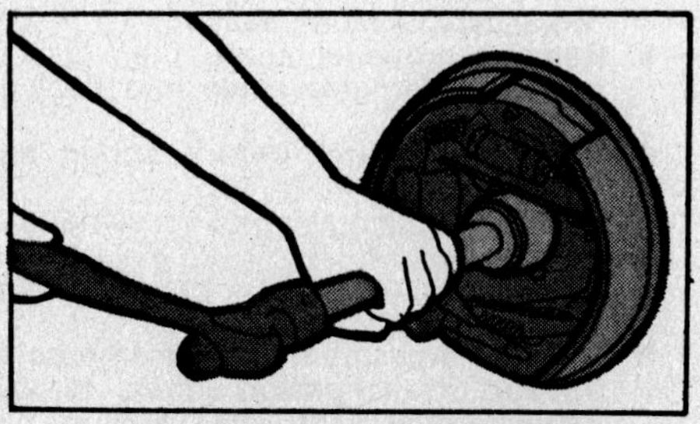

Fig. 4 Installation seal and wheel bearing

7. Remove differential cover and drain lubricant.
8. Unscrew differential lock screw, remove pinion shaft and axle shaft "C" locks. Reinstall pinion shaft and tighten lock screw to retain differential gears, Fig. 2.
9. Remove both axle shafts.
10. Remove brake backing plate retaining nuts and remove backing plates, with shoes and brake lines attached, and wire to frame.
11. Remove right and left lower control arm pivot bolts at axle.
12. Lower axle assembly slowly until coil spring tension is released, then remove axle.

AXLE SHAFT

1. Raise vehicle on a hoist and remove wheel and tire assembly and brake drum.
2. Drain lubricant from axle by removing carrier cover.
3. Unscrew pinion shaft lock screw and remove pinion shaft, Fig. 2.
4. Push flanged end of axle shaft toward center of car and remove "C" lock from button end of shaft.
5. Remove axle shaft from housing being careful not to damage seal.

Oil Seal &/or Bearing Replacement

1. If replacing seal only, remove the seal by using the button end of axle shaft. Insert the button end of shaft behind the steel case of the seal and pry seal out of bore being careful not to damage housing.
2. If replacing bearings, insert tool J-22813 into bore so tool head grasps behind bearing, Fig. 3. Slide washer against seal, or bearing, and turn nut against washer. Attach slide hammer J-2619 and remove bearing.
3. Pack cavity between seal lips with a high melting point wheel bearing lubricant. Position seal on tool J-21491 and position seal in axle housing bore, tap seal in bore just below end of housing, Fig. 4.

PROPELLER SHAFT

1. Raise vehicle on a hoist.
2. On 1976 models, remove torque arm bolts at rear axle, then referring to the "Rear Suspension" section, loosen attaching bolt at transmission and swing arm away.
3. On all models mark relationship of shaft to companion flange and disconnect the rear universal joint by removing trunnion bearing "U" bolts. Tape bearing cups to trunnion to prevent dropping and loss of bearing rollers.
4. Withdraw propeller shaft front yoke from transmission.
5. When installing, be sure to align marks made in removal to prevent driveline vibration.

BRAKE ADJUSTMENTS

1976

These brakes, Fig. 6A, have self-adjusting shoe mechanisms that assure correct lining-to-drum clearances at all times. The automatic adjusters operate only when the brakes are applied as the car is moving rearward.

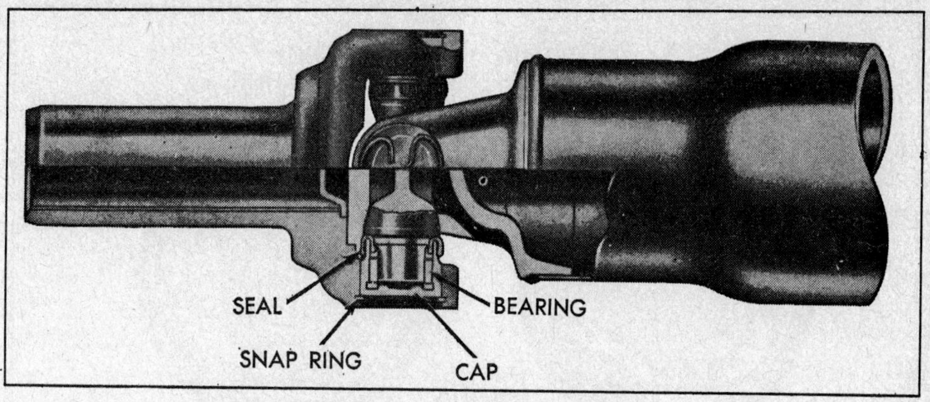

Fig. 5 Propeller shaft cross section

Although the brakes are self-adjusting, an initial adjustment is necessary after the brake shoes have been relined or replaced, or when the length of the adjusting screw has been changed during some other service operation.

Frequent usage of an automatic transmission forward range to halt reverse vehicle motion may prevent to automatic adjusters from functioning, thereby inducing low pedal heights. Should low pedal heights be encountered, it is recommended that numerous forward and reverse stops be made until satisfactory pedal height is obtained.

NOTE

If a low pedal condition cannot be corrected by making numerous reverse stops (provided the hydraulic system is free of air) it indicates that the self-adjusting mechanism is not functioning. Therefore it will be necessary to remove the brake drum, clean, free up and lubricate the adjusting mechanism. Then adjust the brakes, being sure the parking brake is fully released.

Adjustment

A lanced "knock out" area is provided in the web of the brake drum for servicing purposes in the event retracting of the brake shoes is required in order to remove the drum.

1. With brake drum off, disengage the actuator from the star wheel and rotate the star wheel by spinning or turning with a screwdriver.
2. Using the brake drum as an adjustment fixture, turn the star wheel until the drum slides over the brake shoes with a slide drag.
3. Turn the star wheel 1¼ turns to retract the brake shoes. This will allow sufficient lining-to-drum clearance so final adjustment may be made.
4. Install drum and wheel.

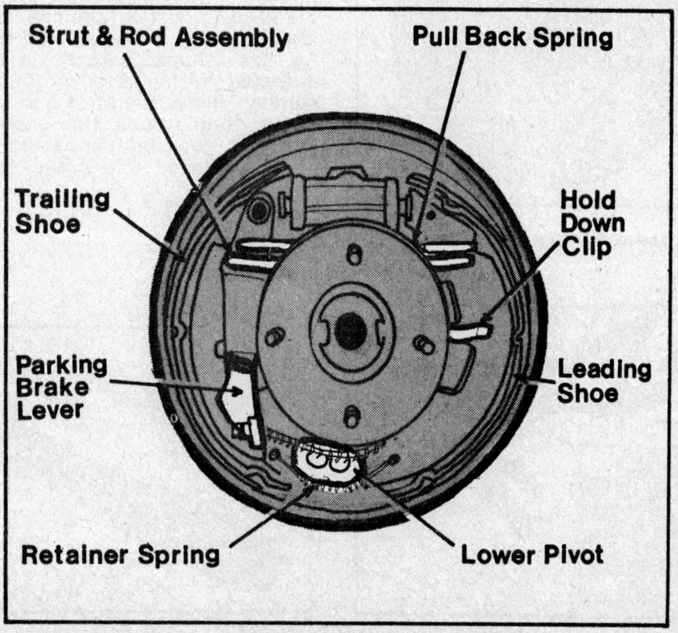

Fig. 6 Rear brake assembly. 1971-75

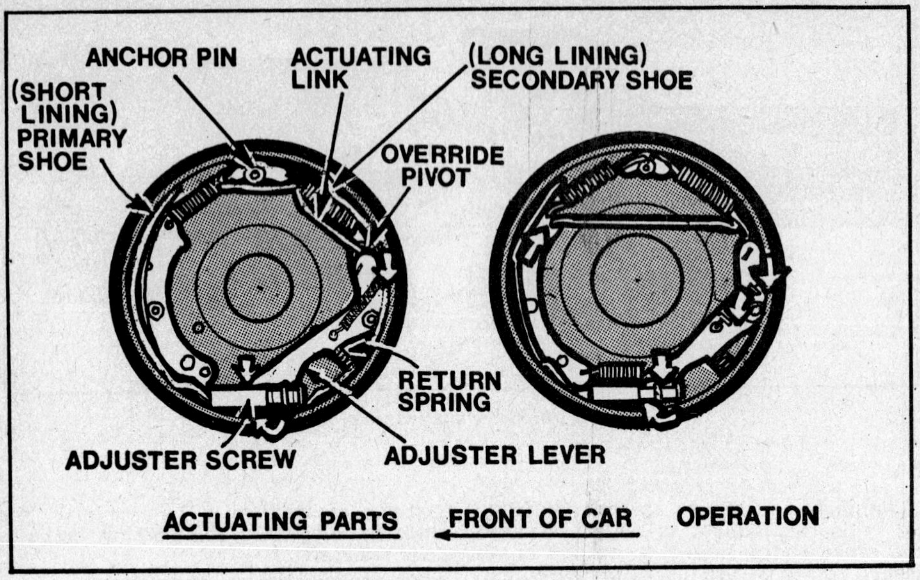

Fig. 6A Self-adjusting brakes. 1976

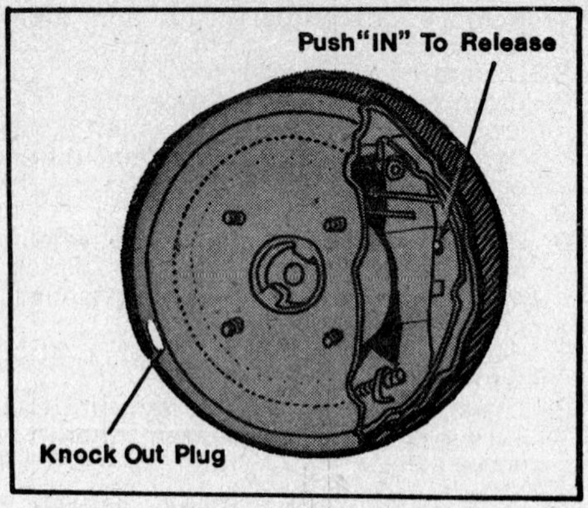

Fig. 7 Drum knock out provision. 1971-75

NOTE: *If lanced area in brake drum was knocked out, be sure all metal has been removed from brake compartment. Install new hole cover in drum to prevent contamination of brakes. Make certain that drums are installed in the same position as when removed with the drum locating tang in line with the locating hole in the wheel hub.*

5. Make final adjustment by driving and stopping in forward and reverse until satisfactory pedal height is obtained.

NOTE: The recommended method of adjusting the brakes is by using a Drum-to-Brake Shoe Clearance Guage to check the diameter of the brake drum inner surface. Turn the tool to the opposite side and fit over the brake shoes by turning the star wheel until the gauge just slides over the linings. Rotate the gauge around the brake shoe lining surface to assure proper clearance.

1971-75

Drum brakes, Fig. 6, are used on the rear wheels. Rear brake adjustment is not automatic. Adjustment takes place, if needed, only when the parking brake is applied. When the parking brake is applied, the strut is pushed against the front shoe and the rod is pulled against the rear shoe. As the shoes spread, a spring lock mounted within the strut and rod assembly, allows the strut and rod assembly to lengthen. When the parking brake is released, the rod connected to the rear shoe is relaxed and brake shoe pressure on the drum is released providing running clearance. Clearance is obtained by the difference in the diameter of the rod and the diameter of the hole in the rear shoe.

NOTE: If the brake drum cannot be removed, it will be necessary to release the brake adjuster. To gain access to the adjuster, knock out the lanced area in web of brake drum, Fig. 7, using a chisel or similar tool. Release the rod from the trailing shoe by pushing in on the rod until it is clear of the shoe. The pull back spring will then pull the shoes toward each other and the drum may be removed.

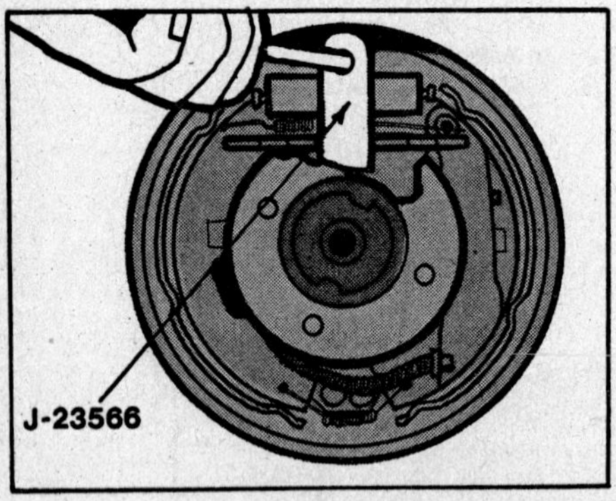

Fig. 8 Releasing adjuster on car. 1971-75

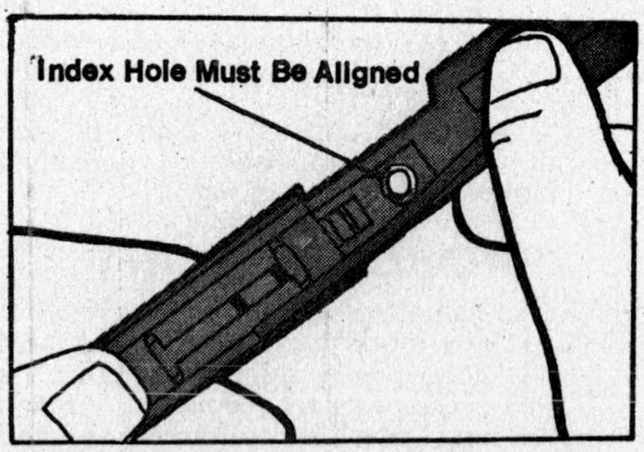

Fig. 11 Installing rod assembly. 1971-75

Fig. 9 Releasing adjuster off car. 1971-75

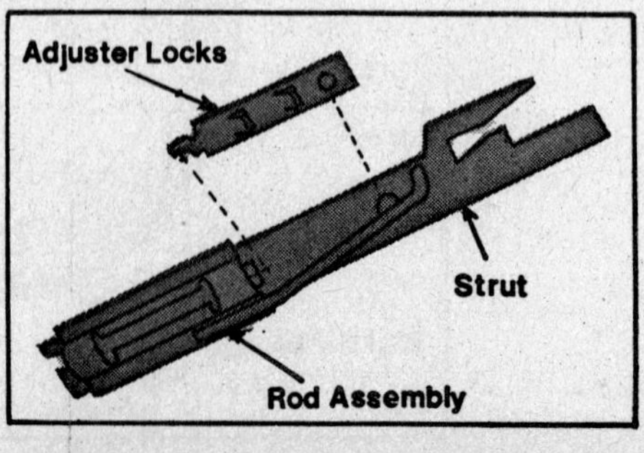

Fig. 10 Adjuster lock positioning. 1971-75

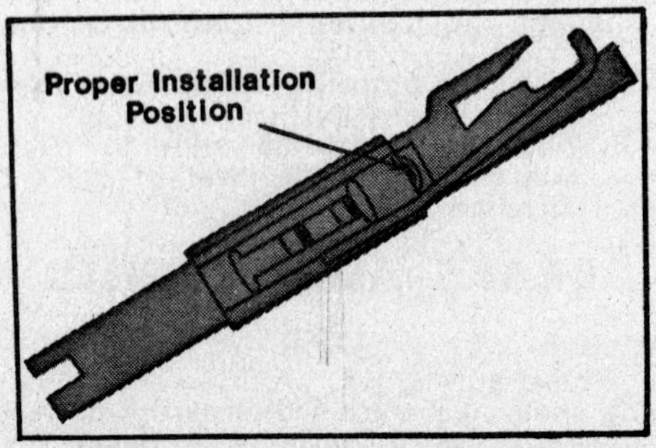

Fig. 12 Adjuster position for new shoes. 1971-75

CAUTION: After knocking out the lanced area, be sure to remove the piece of metal from inside the drum.

Installation

When installing new brake shoes the adjuster assembly must be released and repositioned. To release adjuster on vehicle, position tool J-23566, Fig. 8. The tool releases the adjuster assembly by unseating the clips from the rod assembly. Both clips must be unseated at the same time.

To release adjuster off vehicle, position tool J-23730 so that tang of tool rests on flat portion of locks (between lock tangs) Fig. 9, then press down on adjuster locks. Work rod assembly free of adjuster locks, then when both adjuster tangs are clear of rod assembly, slide rod off lever. Subassemble rod assembly to strut making certain that index hole is lined up and seated, Fig. 10, then slide rod assembly over adjuster locks, Fig. 11, until both locks are positioned as shown in Fig. 12. Note that adjuster lock index is about ½ covered by rod assembly.

PARKING BRAKE, ADJUST

1. Raise vehicle and on 1975 Cosworth Vega and all 1976 models, remove propeller shaft.
2. Apply parking brake two notches from fully released position and raise hoist.
3. Loosen equalizer check nut and tighten the adjusting nut until a slight drag is felt when rear wheels are rotated.
4. Tighten check nut securely.
5. Release parking brake and rotate rear wheels. No drag should be present.

MASTER CYLINDER, REPLACE

1. Disconnect cylinder push rod from brake pedal.
2. Disconnect brake lines from two outlets on cylinder and cover ends of lines to prevent entry of dirt.
3. Unfasten and remove cylinder from dash.

POWER BRAKE UNIT

1975-76

1. Remove the vacuum hose from check valve and then remove the master cylinder retaining nuts.
2. Remove the bolt securing the brake pipe distributor and switch assembly to fender skirt.
3. Pull master cylinder forward until it just clears mounting studs and move aside. Support cylinder to avoid stress on hydraulic lines.
4. Remove power unit to dash nuts.
5. Remove brake pedal pushrod retainer and disconnect push rod from pin.
6. Remove power unit from vehicle.

Rear Suspension

DESCRIPTION

1975 Cosworth Vega & All 1976 Models

This rear suspension system does not use upper control arms, instead, it uses a torque arm mounted rigidly to the differential housing at the rear and to the transmission through a rubber bushing at the front. This torque arm prevents axle housing rotation caused by starting and stopping. Along with the torque arm a track rod is used to connect the axle housing to the body to control side sway and a rear stabilizer shaft is used for improved handling, Fig. 1.

1971-75 Exc. Cosworth Vega

The rear suspension system on these models use two upper and two lower control arms mounted between the axle tube

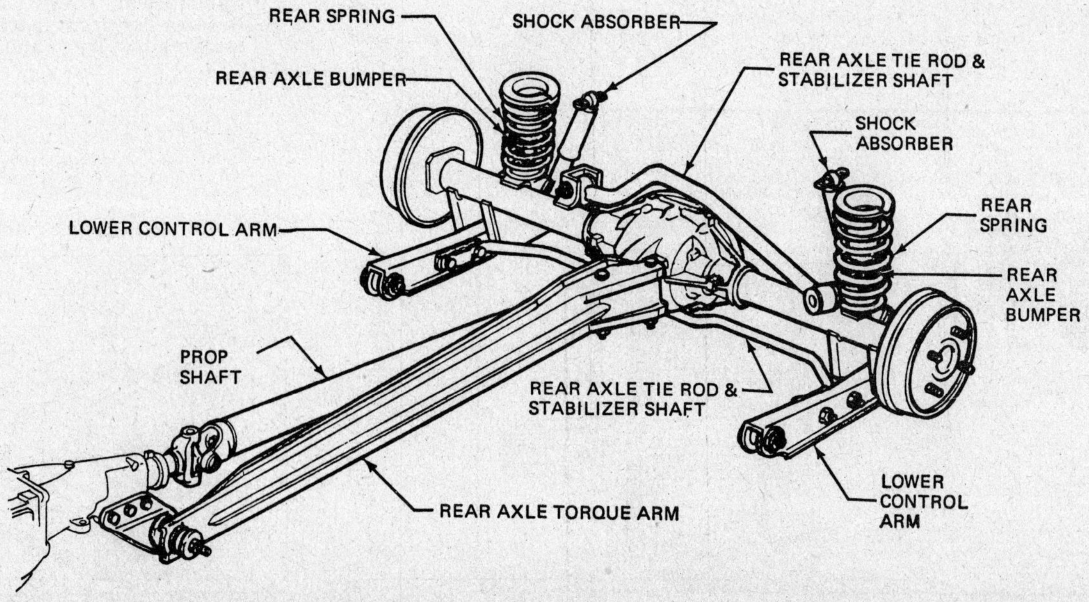

Fig. 1 Rear suspension. 1975 Cosworth Vega & all 1976 models

and frame, Fig. 2. The upper control arms are shorter than the lower arm, thereby allowing the axle tube assembly to "Rock" or tilt forward on compression. The upper control arms control drive forces, side sway and pinion nose angle. The lower control arms maintain fore and aft relationship of the axle assembly to the chassis.

SHOCK ABSORBER, REPLACE

1. With the rear axle supported properly disconnect shock absorber from upper and lower mounting, Figs. 1 and 2.

NOTE: Use a wrench to hold lower mounting stud from turning.

2. Reverse procedure to install.

COIL SPRING, REPLACE

1. Support vehicle at frame and rear axle with a suitable jack.
2. Disconnect shock absorbers from lower mountings.
3. Lower rear axle until springs can be removed.

NOTE: Do not stretch or damage brake hoses.

4. Reverse procedure to install.

CONTROL ARMS & BUSHINGS, REPLACE

NOTE: Replace control arms one at a time to prevent axle misalignment, making installation difficult.

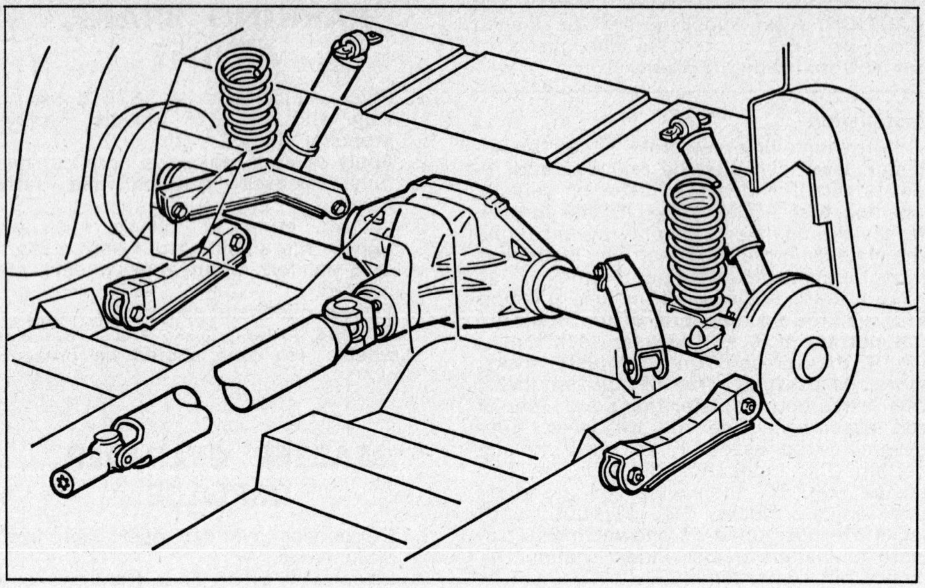

Fig. 2 Rear suspension. 1971-75

Upper Control Arms, 1971-75

1. Support vehicle at rear axle.
2. Remove control arm front and rear mount bolts.
3. Replace bushings as necessary, Fig. 3.
4. Reverse procedure to install. Control arm bolts must be tightened with vehicle at curb height.

Lower Control Arms, 1971-76

1. Support vehicle at rear axle.
2. Disconnect stabilizer bar, if used, from lower arms.
3. Remove control arm front and rear mount bolts.
4. Replace bushings as necessary, Fig. 3.
5. Reverse procedure to install. Control arm bolts must be tightened with vehicle at curb height.

TRACK ROD, REMOVAL

1. Raise vehicle and support axle assembly.
2. Remove bolts at underbody end of rod.
3. Remove bolts at axle bracket and remove track rod, Fig. 4.
4. Reverse removal procedure to install and torque bolts to specifications.

TORQUE ARM, REPLACE

1. Raise vehicle by axle assembly and support underbody with jackstands.
2. Lower axle assembly slightly and remove torque arm to differential bolts.
3. Disconnect mounting bracket from transmission and then remove through bolt from bracket and remove torque arm, Figs. 5 and 6.

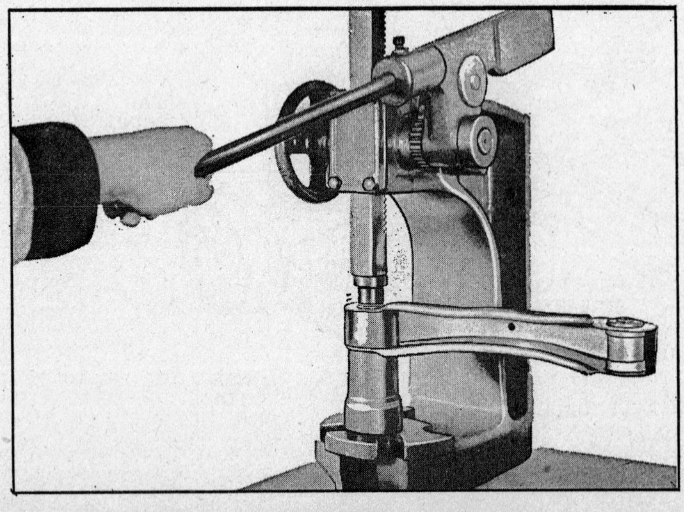

Fig. 3 Control arm bushing replacement

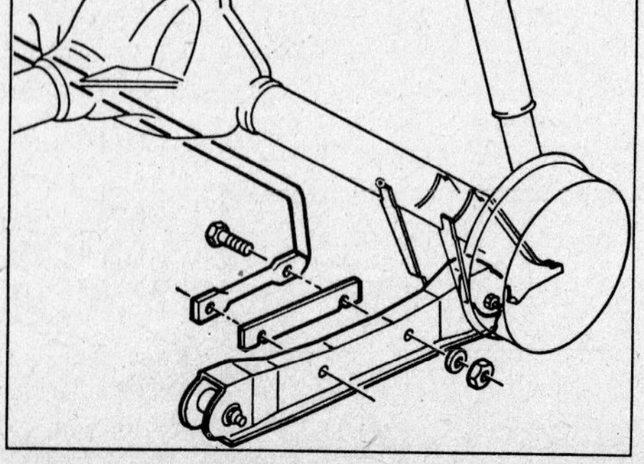

Fig. 7 Stabilizer bar installation. 1971-75

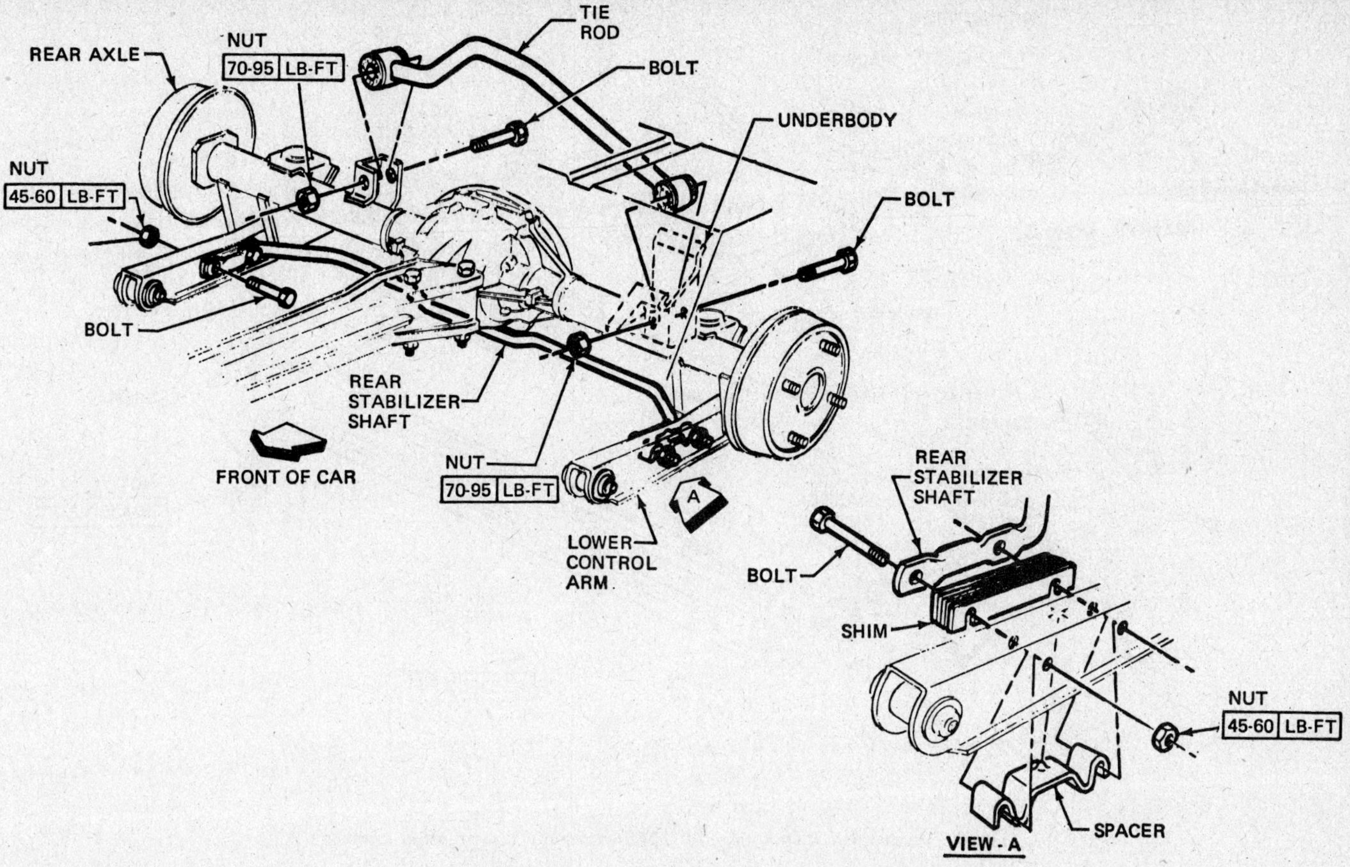

Fig. 4 Track rod & stabilizer bar installation. 1975 Cosworth Vega & all 1976 models

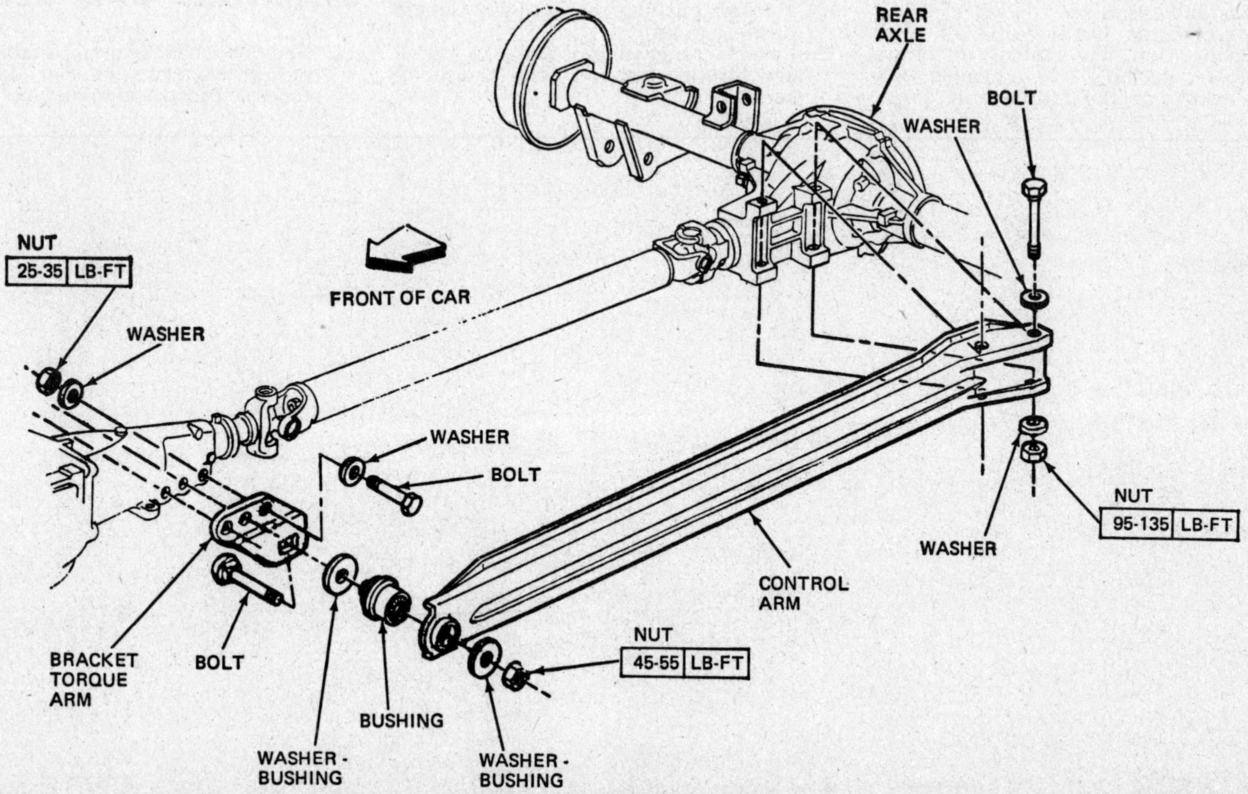

Fig. 5 Torque arm installation. 1975 Cosworth Vega & all 1976 manual transmission models

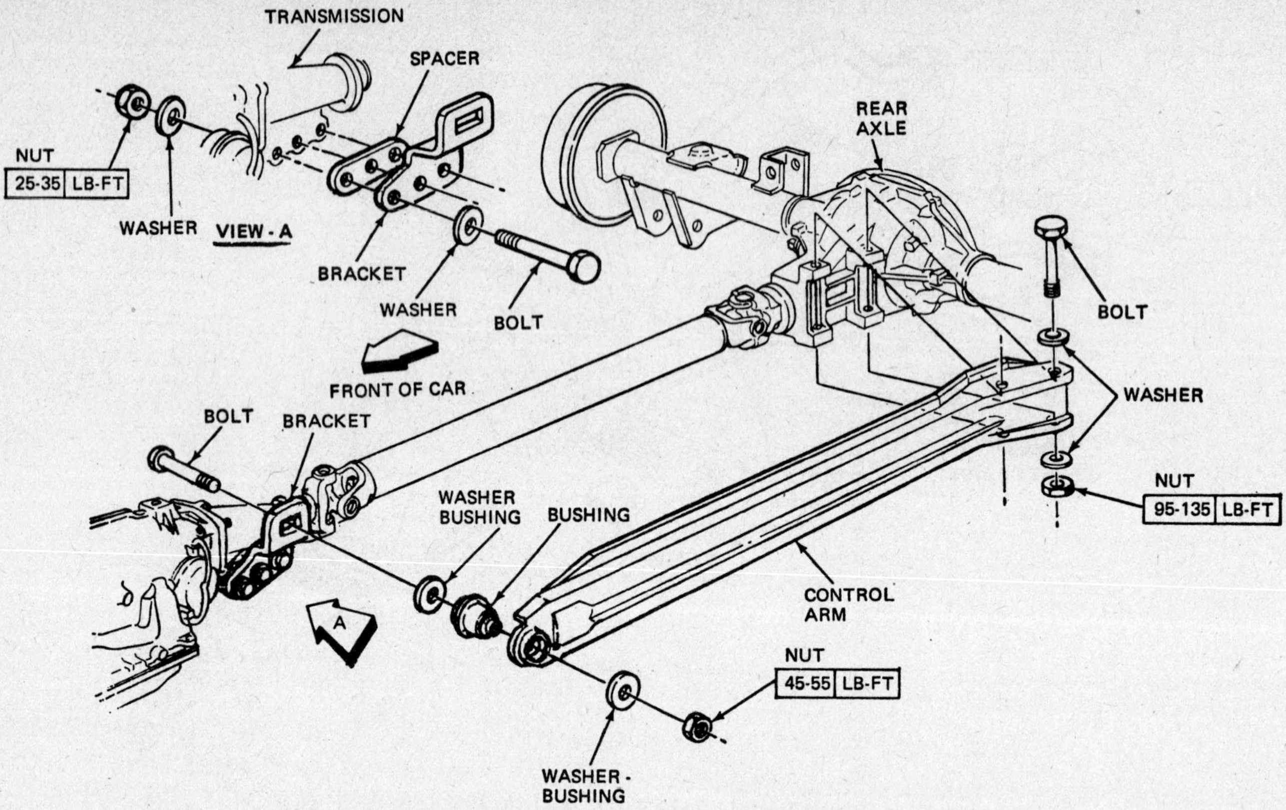

Fig. 6 Torque arm installation. All 1976 automatic transmission models

4. To replace bushing, use an arbor press and tool J-25317-2 as a receiver and press bushing out of arm, then position new bushing in torque arm with bushing sleeve aligned with the length of the torque arm. Press bushing into place using tool J-25317-1 over bushing to properly locate bushing in arm.
5. Reverse removal procedure to install and torque nuts and bolts to specifications.

STABILIZER BAR, REPLACE

1. Remove bolts securing stabilizer bar to lower control arms, Figs. 4 and 7.
2. Reverse procedure to install.

Front End & Steering Section

FRONT SUSPENSION

The front suspension, Fig. 1, is of the A frame type with short and long control arms. The upper control arm is bolted to the front end sheet metal at each inner pivot point. Rubber bushings are used for mounting.

The lower control arms attach to the front end sheet metal with cam type bolts through rubber bushings. The cam bolts adjust caster and camber.

The upper ball joint is riveted in the upper arm and the lower ball joint is pressed into the lower arm.

Coil springs are mounted between the lower control arms and the shock absorber tower.

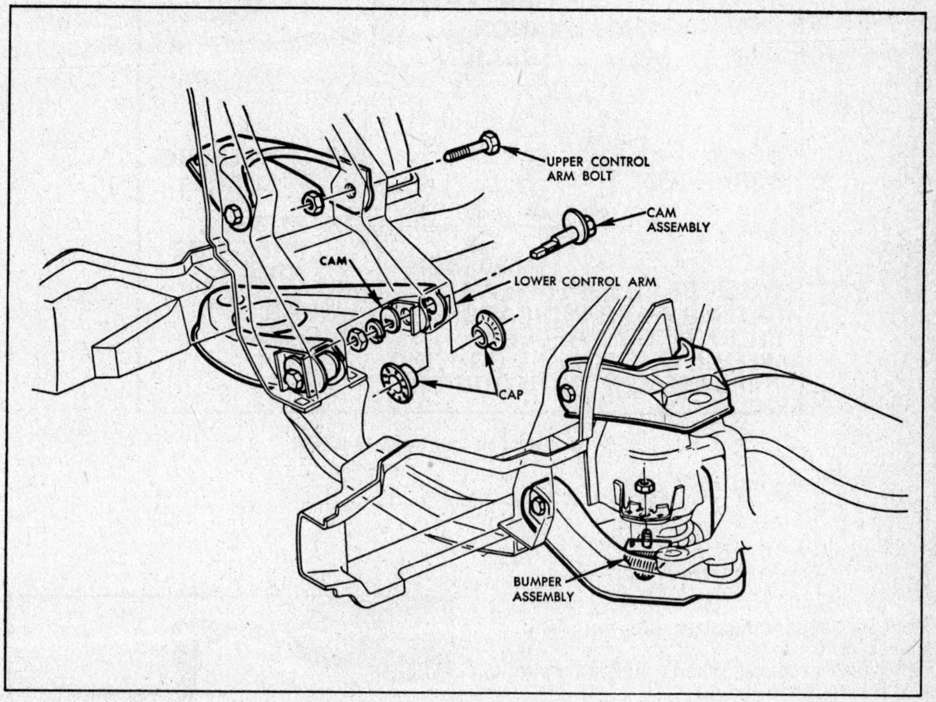

Fig. 1 Front suspension

WHEEL ALIGNMENT

Caster

Caster angle is adjusted by loosening the rear lower control arm pivot nut and rotating the cam until proper setting is reached.

NOTE: This eccentric cam action will tend to move the lower control arm fore or aft thereby varying the caster. Hold the cam bolt while tightening the nut.

Camber

Camber angle is adjusted by loosening the front lower control arm pivot nut and rotating the cam until setting is reached.

NOTE: This eccentric cam action will move the lower arm in or out thereby varying the setting. Hold the cam bolt head while tightening the nut.

TOE-IN, ADJUST

1. Loosen clamp bolt nut at each end of each tie rod and rotate the sleeve until proper toe-in is reached.
2. Position tie rod ball stud assembly straight on a center line through their attaching points.
3. Tighten clamp nuts.

WHEEL BEARINGS, ADJUST

1. With wheel raised, remove hub cap,

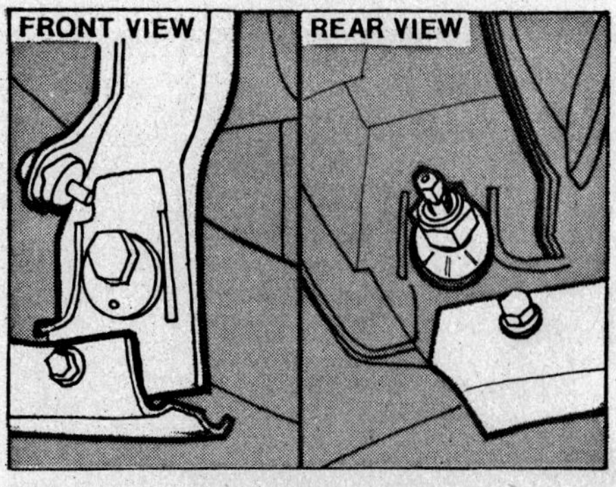

Fig. 2 Caster and camber adjustment

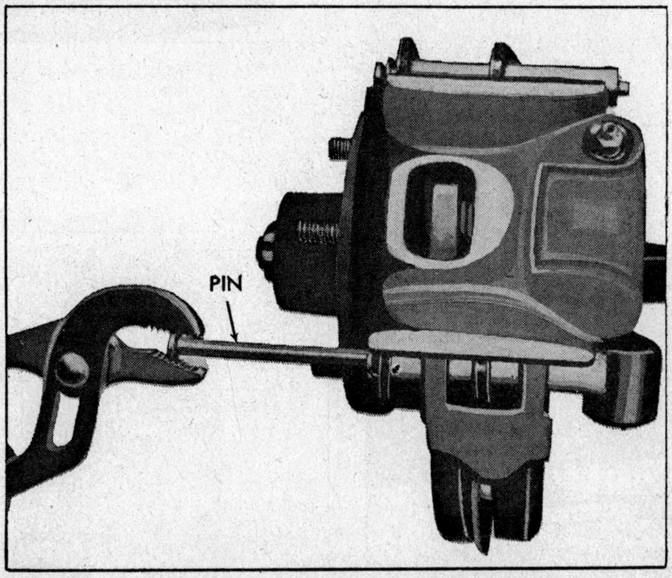

Fig. 3 Removing caliper mounting pins

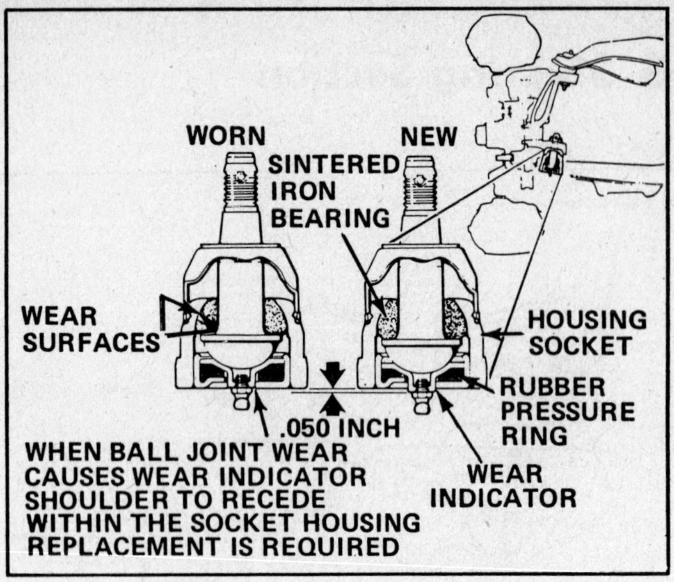

Fig. 3A Lower ball joint wear indicator

Fig. 4 Removing upper ball joint stud from knuckle. 1971

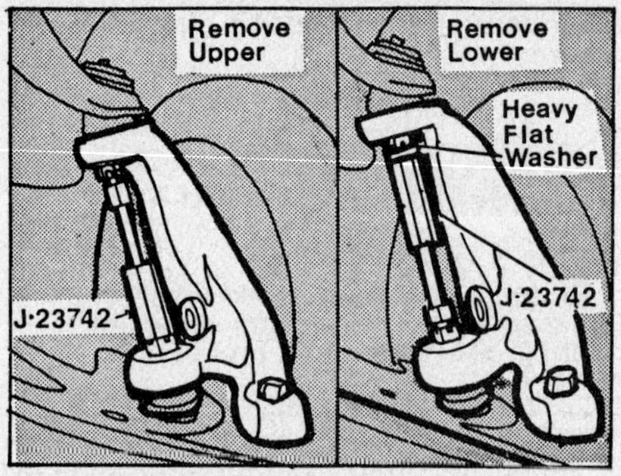

Fig. 4A Removing upper & lower ball joint stud from knuckle. 1972-76

dust cap and cotter pin from end of spindle.
2. While rotating wheel, tighten spindle nut to 12 ft. lbs.
3. Back off adjusting nut one flat and insert cotter pin. If slot and pin hole do not line up, back off the adjusting nut an additional ½ flat or less as required to insert cotter pin.
4. Spin wheel to check that it rolls freely and then lock the cotter pin.

NOTE: Wheel bearings should have zero preload and the allowable end play is .001-.008 inch for 1971-73 models or .001-.005 inch for 1974-76 models.

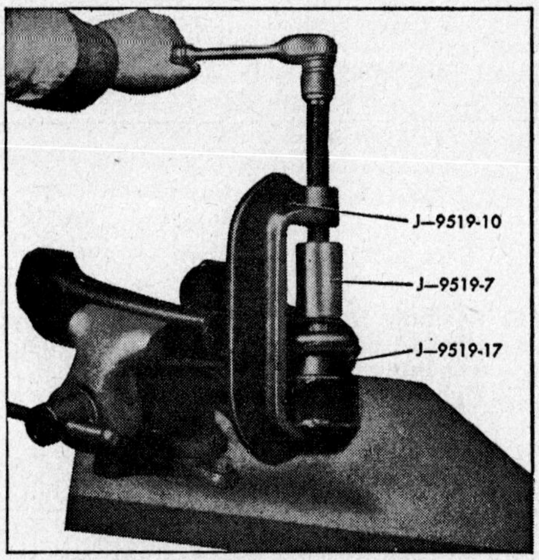

Fig. 5 Removing lower ball joint

WHEEL BEARINGS, REPLACE

1. Raise vehicle on a hoist and remove the wheel and tire assembly.
2. Remove the brake caliper from the disc by removing the mounting pins and stamped nuts, Fig. 3.
3. Remove hub grease cap, cotter pin, spindle nut and washer and remove hub and bearings.
4. Remove inner bearing by prying out the grease seal.

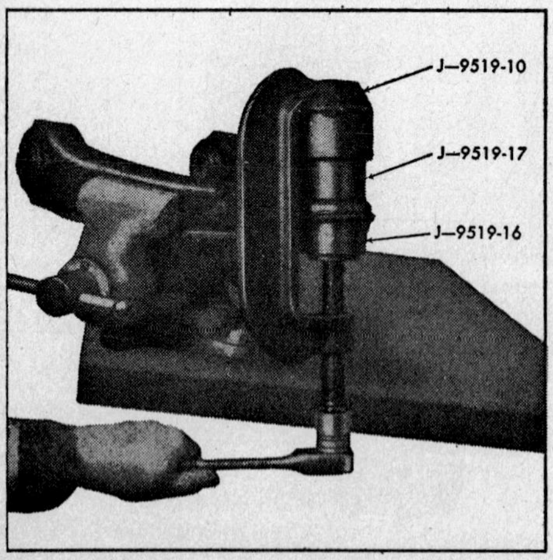

Fig. 6 Installing lower ball joint

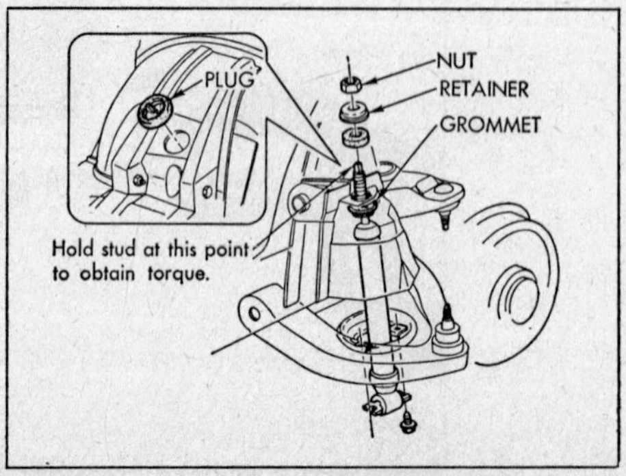

Fig. 7 Front shock absorber mountings

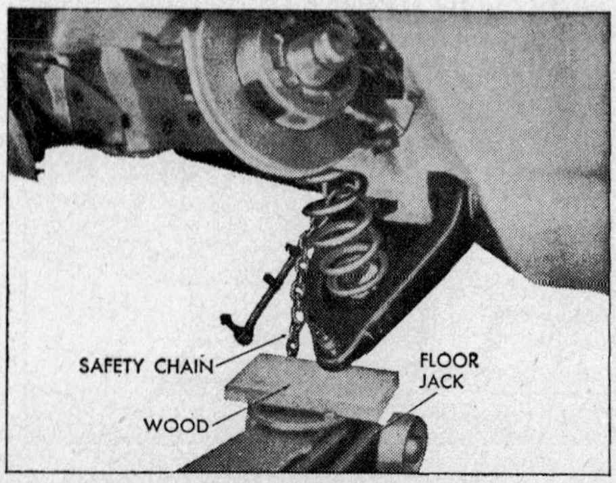

Fig. 8 Removing front coil spring

CHECKING BALL JOINTS FOR WEAR

Upper Ball Joint

The upper ball joint is checked for wear by checking the torque required to rotate the ball joint stud in the assembly. This is done after first dislodging the ball joint from the steering knuckle.

1. Install the stud nut to the ball stud. in the seat.
2. Check the torque required to turn the ball stud.

NOTE: Specified rotating torque for a new joint for 1971 models is 2 to 4 ft. lbs. and for 1972-76 models, 1 to 4 ft. lbs. If readings are excessively high or low replace joint.

Lower Ball Joint

NOTE: All 1975-76 vehicles use lower ball joints incorporating wear indicators, Fig. 3A.

The lower ball joint is checked for wear by checking to be sure torque is present. After dislodging ball joint from steering knuckle perform the following:

1. Install the stud nut to the ball stud.
2. Check the torque required to turn the ball stud in the seat.

NOTE: Some torque is required, if zero torque is observed, sufficient wear has taken place for replacement of the ball joint.

UPPER BALL JOINT, REPLACE

1. Raise vehicle on a hoist and remove wheel and tire assembly.
2. Support lower control arm with a floor jack.
3. Remove upper ball stud nut and remove ball stud from knuckle, Figs. 4 and 4A.
4. Remove control arm pivot bolts and remove control arm.
5. Grind off rivets securing ball joint to arm.
6. Install new ball joint using bolts and nuts supplied in kit.

LOWER BALL JOINT, REPLACE

1. Place vehicle on hoist and support lower control arm at outer end on a jack.
2. Free lower ball stud from knuckle, Fig. 4A.
3. Using tool shown in Fig. 5, press ball joint out of control arm.
4. Position new ball joint so that grease bleed vent in rubber boot is facing inboard.
5. Install tool shown in Fig. 6 and press new joint into arm.
6. Install lube fitting in joint and install stud into steering knuckle.

SHOCK ABSORBER, REPLACE

1. Hold shock absorber stem and remove the nut, upper retainer and rubber grommet, Fig. 7.
2. Raise vehicle on a hoist.
3. Remove bolts from lower end of shock absorber and lower the shock from the vehicle.

COIL SPRING, REPLACE

1. With shock absorber removed and stabilizer bar removed, raise the vehicle and place jackstands under front braces.
2. Remove the wheel and tire assembly.
3. Place a floor jack under the lower arm and support the arm. Use a block of wood between the control arm and the jack, Fig. 8.
4. Remove the lower ball stud from the knuckle.
5. Remove the tie rod end from the knuckle.
6. Lower the control arm by slowly lowering the jack until the spring can be removed.

STEERING GEAR, REPLACE

1. Remove the pot joint coupling clamp bolt at the steering gear wormshaft.
2. On 1973-76 models remove "K" brace.
3. Remove pitman arm nut and washer from pitman shaft and mark relation of arm position to shaft.
4. Remove pitman arm with suitable puller.
5. Remove bolts securing gear to frame and remove gear assembly.

INTEGRAL POWER STEERING

Replacement procedures for removing the gear assembly are the same as for the manual type gear with the following additions:

1. Disconnect both pressure and return hoses from the gear housing and cap both hoses and steering gear outlets to prevent entry of dirt.

CHRYSLER • DODGE
IMPERIAL • PLYMOUTH

OLD CAR SPECIFICATIONS: For 1946-68 Tune Up and Wheel Alignment Specifications see main index.

INDEX OF SERVICE OPERATIONS

VEHICLE NUMBER LOCATION

1969-76: ON PLATE ATTACHED TO DASH PAD AND VISIBLE THROUGH WINDSHIELD.

ENGINE NUMBER LOCATION

1969-76 Six: Right front of block below cylinder head.
1969-76 V8-273, 318, 340, 360: Left front of block below cylinder head.
1969-71 V8-383 426, 440: Left side rear of block near oil pan flange.
1971-72 V8-400: Right front of block below cylinder head.
1972 V8-440: Upper left front of cylinder block.
1973-76 V8-400, 440: Upper right front of cylinder block.

ENGINE IDENTIFICATION CODE

1969-76 engines are identified by the cubic inch displacement found within the engine number stamped on the pad.

CHRYSLER GRILLE IDENTIFICATION

1969 Newport

1969 "300"

1969 New Yorker

1970 Newport

1970 "300"

1970 New Yorker

1971 Newport

1971 "300" & New Yorker

1971 New Yorker & Town Country

1972 Newport

1972 New Yorker & Town Country

1973 Newport

1973 New Yorker & Town Country

1974 Newport

1974 New Yorker & Town Country

1975 Cordoba

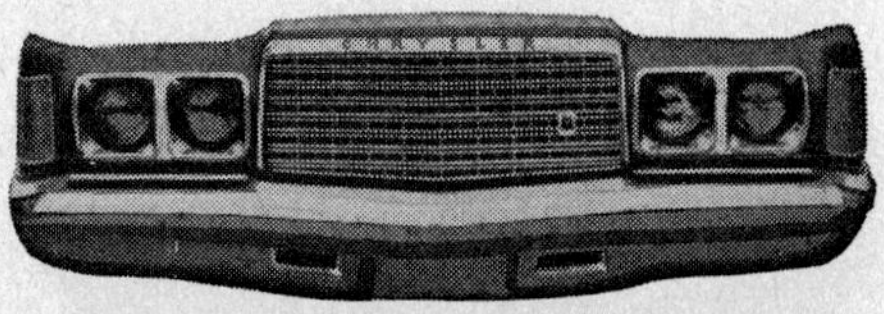
1975-76 Newport

1975 New Yorker & Town Country
1976 Newport Custom & Town Country

CHRYSLER GRILLE IDENTIFICATION—Continued

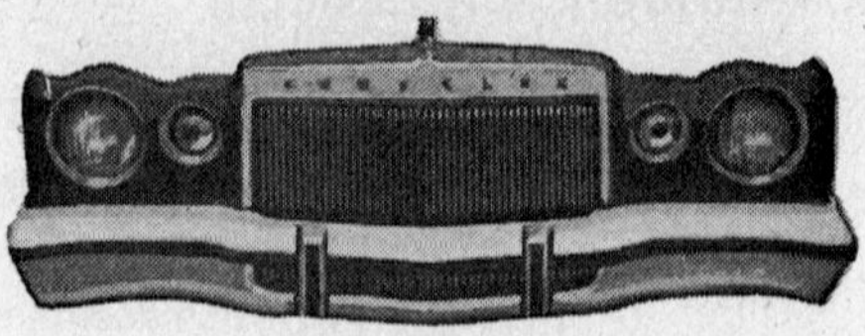
1976 Cordoba

1976 New Yorker

DODGE GRILLE IDENTIFICATION

1969 Dart

1969 Coronet

1969 Coronet "500"

1969 Charger

1969 Polara

1969 Monaco

1970 Dart

1970 Coronet, R/T, Super Bee

1970 Challenger

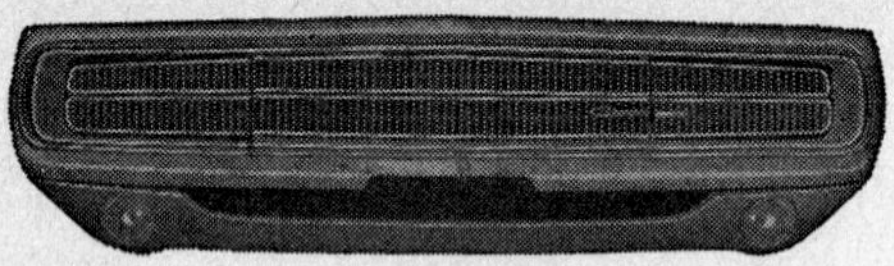
1970 Charger

1970 Polara

1970 Monaco

1971 Dart & Demon

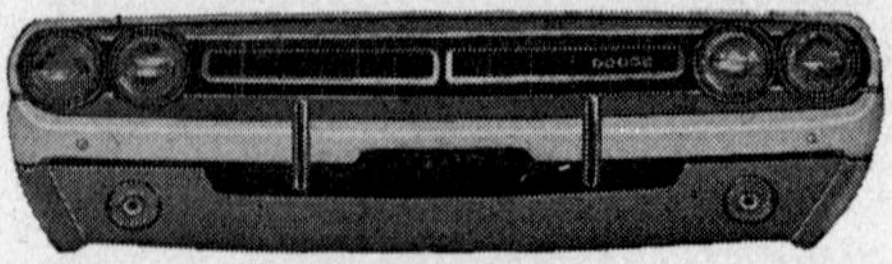
1971 Challenger

1971 Charger

1971 Coronet

1971 Polara

1971 Monaco

DODGE GRILLE IDENTIFICATION—Continued

1972 Dart

1972 Challenger

1972 Charger

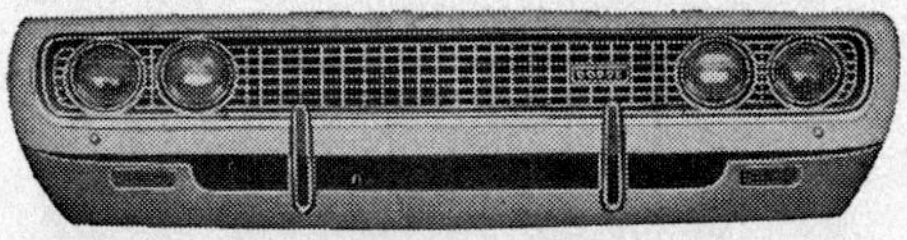
1972 Coronet

1972 Polara

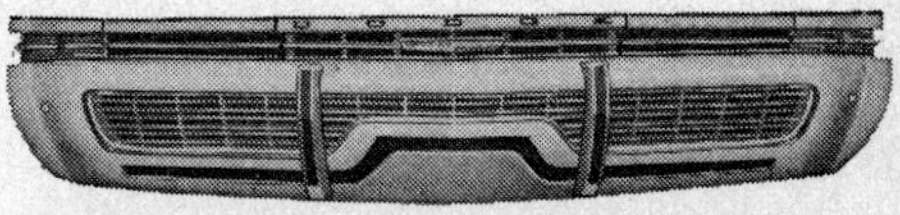
1972 Monaco

1973 Dart

1973 Coronet

1973 Charger

1973 Challenger

1973 Polara

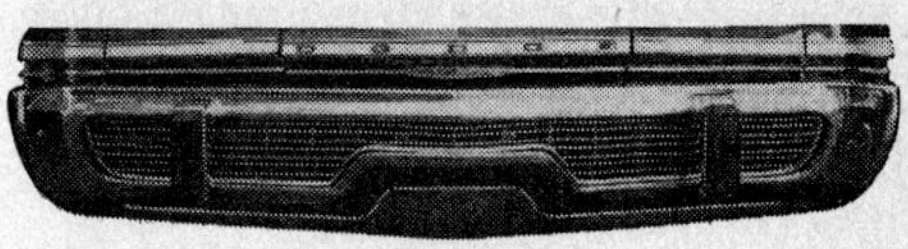
1973 Monaco

1974 Dart

1974 Challenger

1974 Charger

1974 Charger SE

1974 Coronet

1974 Monaco

1975-76 Dart

1975 Coronet
1976 Charger

1975 Coronet Brougham & Crestwood
1976 Coronet

DODGE GRILLE IDENTIFICATION—Continued

1975 Charger S.E.

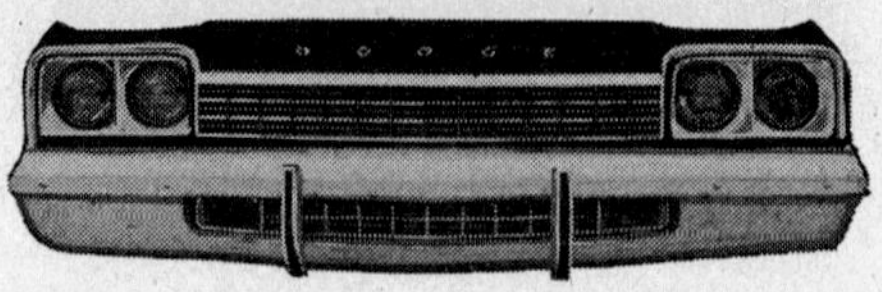
1975 Monaco

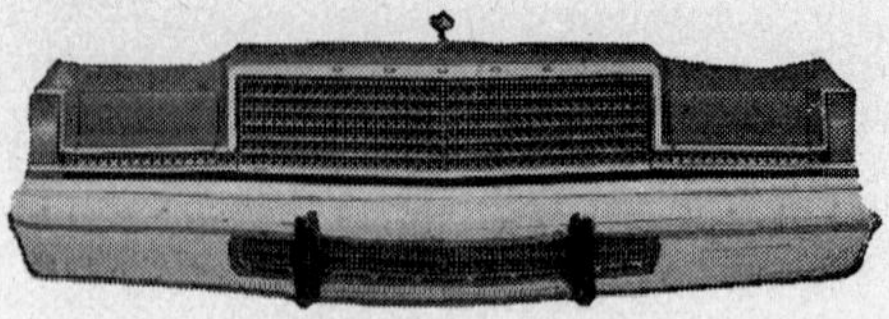
1975 Royal Monaco
1976 Monaco

1976 Aspen

1976 Charger SE

1976 Charger Daytona

IMPERIAL GRILLE IDENTIFICATION

1969

1970

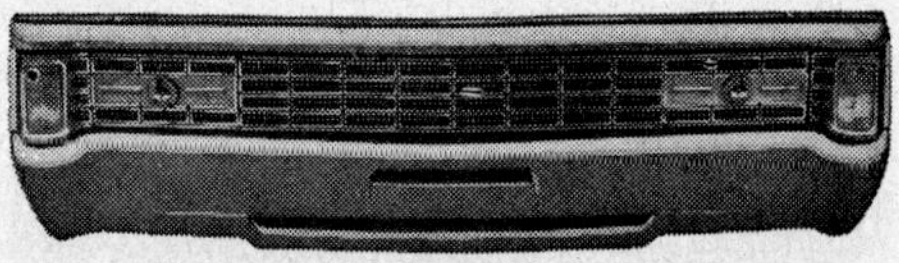
1971

1972

1973

1974

1975

PLYMOUTH GRILLE IDENTIFICATION

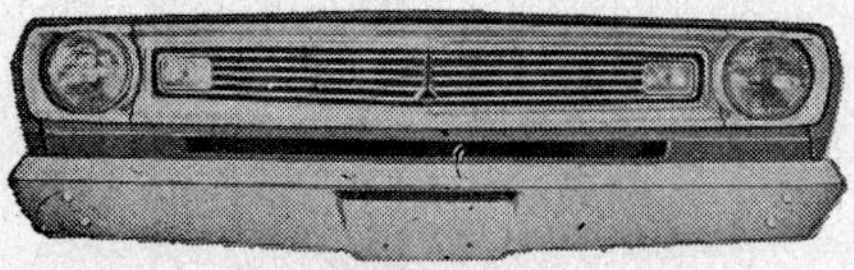
1969 Valiant

1969 Barracuda

1969 Belvedere, Road Runner & Satellite

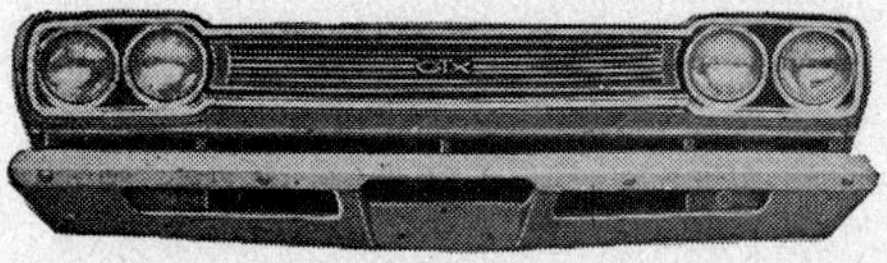
1969 "GTX"

1969 Sport Satellite

1969 Fury

1969 Sport Fury & VIP

1970 Valiant & Duster

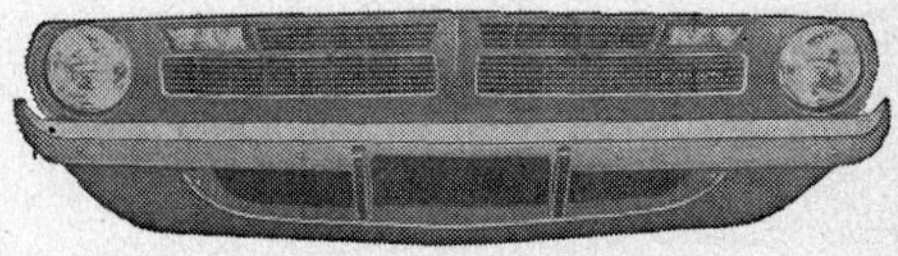
1970 Barracuda

1970 Hemi 'Cuda

1970 Belvedere, Road Runner & Satellite

1970 Sport Satellite

1970 "GTX"

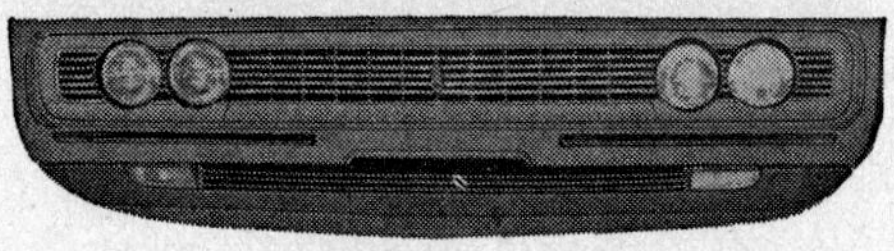
1970 Fury

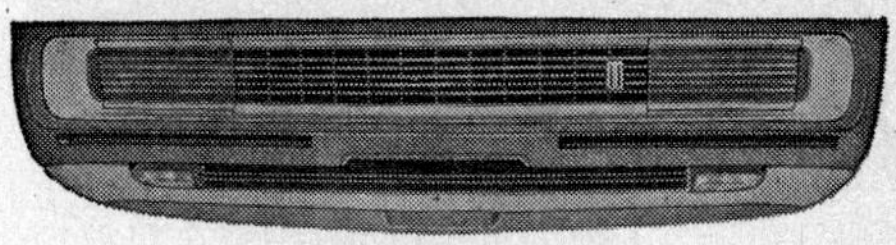
1970 Sport Fury

1971 Valiant & Duster

1971 Duster 340

1971 Barracuda

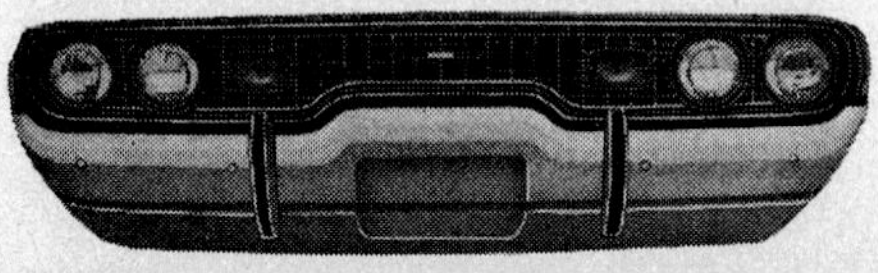
1971 Satellite

1971 Road Runner

1971 Fury

PLYMOUTH GRILLE IDENTIFICATION—Continued

1971 Sport Fury

1972 Valiant & Duster

1972 Barracuda

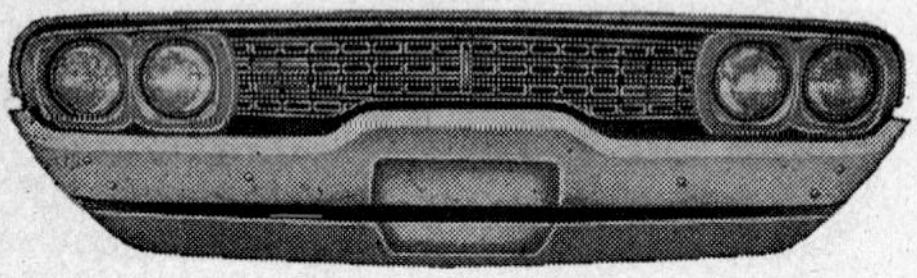
1972 Satellite

1972 Road Runner

1972 Sebring

1972 Fury

1973 Valiant & Duster

1973 Barracuda

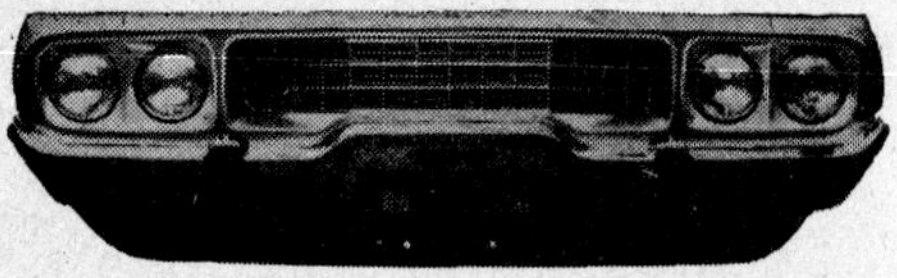
1973 Satellite

1973 Sebring

1973 Fury

1974 Valiant & Duster

1974 Barracuda

1974 Satellite 2-Door,
Sebring & Road Runner

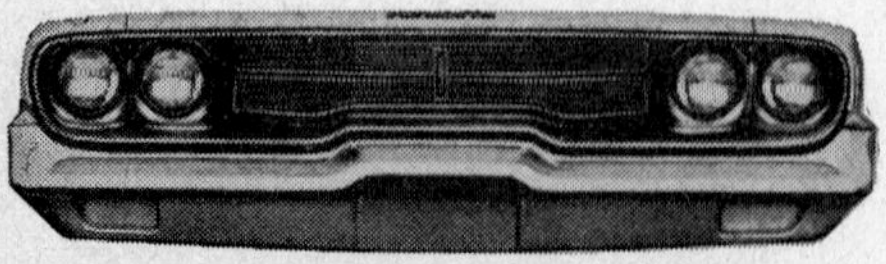
1974 Satellite 4-Door

1974 Fury

1975-76 Valiant & Duster

1975 Road Runner

1975-76 Fury & Custom

1975-76 Fury Sport

PLYMOUTH GRILLE IDENTIFICATION—Continued

1975 Gran Fury

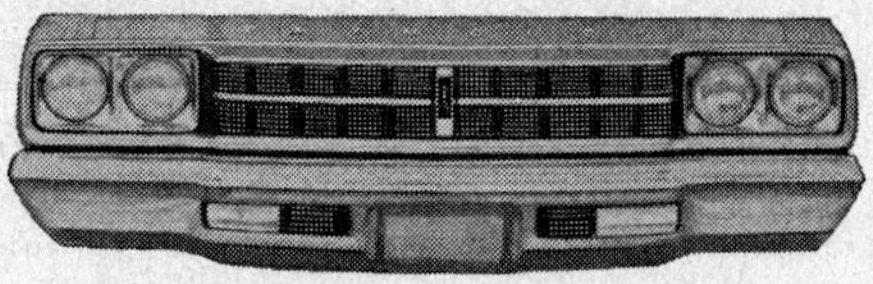

1975 Gran Fury Custom

1975 Gran Fury Sport & Brougham
1976 Gran Fury

1976 Volare

GENERAL ENGINE SPECIFICATIONS

Year	Engine	Carburetor	Bore and Stroke	Piston Displacement, Cubic Inches	Compression Ratio	Maximum Brake H.P. @ R.P.M.	Maximum Torque Lbs. Ft. @ R.P.M.	Normal Oil Pressure Pounds
CHRYSLER AND IMPERIAL								
1969	290 Horsepower..............V8-383	2 Barrel	4.25 x 3.375	383	9.2	290 @ 4400	380 @ 2400	45–65
	330 Horsepower..............V8-383	4 Barrel	4.25 x 3.375	383	10.0	330 @ 5000	425 @ 3200	45–65
	350 Horsepower..............V8-440	4 Barrel	4.32 x 3.75	440	10.1	350 @ 4400	480 @ 2800	45–65
	375 Horsepower..............V8-440	4 Barrel	4.32 x 3.75	440	10.1	375 @ 4600	480 @ 3200	45–65
1970	290 Horsepower..............V8-383	2 Barrel	4.25 x 3.375	383	8.7	290 @ 4400	380 @ 2800	45–65
	330 Horsepower..............V8-383	4 Barrel	4.25 x 3.375	383	9.5	330 @ 5000	425 @ 3200	45–65
	350 Horsepower..............V8-440	4 Barrel	4.32 x 3.75	440	9.7	350 @ 4400	480 @ 2800	45–65
	375 Horsepower..............V8-440	4 Barrel	4.32 x 3.75	440	9.7	375 @ 4600	480 @ 3200	45–65
1971	275 Horsepower..............V8-383	2 Barrel	4.25 x 3.375	383	8.7	275 @ 4400	375 @ 2800	45–65
	300 Horsepower..............V8-383	4 Barrel	4.25 x 3.375	383	8.7	300 @ 4800	410 @ 3400	45–65
	190 Horsepower①............V8-400	2 Barrel	4.342 x 3.375	400	8.2	190 @ 4400	310 @ 2400	45–65
	335 Horsepower..............V8-440	4 Barrel	4.32 x 3.75	440	9.0	335 @ 4400	460 @ 3200	45–65
	370 Horsepower..............V8-440	4 Barrel	4.32 x 3.75	440	9.7	370 @ 4600	480 @ 3200	45–65
1972	175 Horsepower①............V8-360	2 Barrel	4.00 x 3.58	360	8.8	175 @ 4000	285 @ 2400	45–65
	190 Horsepower①............V8-400	2 Barrel	4.342 x 3.375	400	8.2	190 @ 4400	310 @ 2400	45–65
	225 Horsepower①............V8-440	4 Barrel	4.32 x 3.75	440	8.2	225 @ 4400	345 @ 3200	45–65
	245 Horsepower①②..........V8-440	4 Barrel	4.32 x 3.75	440	8.2	245 @ 4400	360 @ 3200	45–65
1973	185 Horsepower①............V8-400	2 Barrel	4.34 x 3.38	400	8.2	185 @ 3600	310 @ 2400	45–65
	215 Horsepower①............V8-440	4 Barrel	4.32 x 3.75	440	8.2	215 @ 3600	345 @ 2000	45–65
1974	185 Horsepower①............V8-400	2 Barrel	4.34 x 3.38	400	8.2	185 @ 4000	315 @ 2400	45–65
	200 Horsepower①..........V8-400④	4 Barrel	4.34 x 3.38	400	8.2	200 @ 4400	310 @ 2400	45–65
	205 Horsepower①..........V8-400③	4 Barrel	4.34 x 3.38	400	8.2	205 @ 4400	310 @ 2400	45–65
	220 Horsepower①..........V8-440④	4 Barrel	4.32 x 3.75	440	8.2	220 @ 4000	345 @ 3200	45–65
	230 Horsepower①..........V8-440③	4 Barrel	4.32 x 3.75	440	8.2	230 @ 4000	350 @ 3200	45–65

Continued

GENERAL ENGINE SPECIFICATIONS—Continued

Year	Engine	Carburetor	Bore and Stroke	Piston Displacement, Cubic Inches	Compression Ratio	Maximum Brake H.P. @ R.P.M.	Maximum Torque Lbs. Ft. @ R.P.M.	Normal Oil Pressure Pounds
CHRYSLER AND IMPERIAL—Continued								
1975	135 Horsepower①..........V8-318④	2 Barrel	3.91 x 3.31	318	8.5	135 @ 3600	245 @ 1600	30–80
	150 Horsepower①..........V8-318③	2 Barrel	3.91 x 3.31	318	8.5	150 @ 4000	255 @ 1600	30–80
	180 Horsepower①..........V8-360③	2 Barrel	4.00 x 3.58	360	8.4	180 @ 4000	290 @ 2400	30–80
	190 Horsepower①..........V8-360④	4 Barrel	4.00 x 3.58	360	8.4	190 @ 4000	270 @ 3200	30–80
	165 Horsepower①..........V8-400③	2 Barrel	4.34 x 3.38	400	8.2	165 @ 4000	295 @ 3200	30–80
	175 Horsepower①..........V8-400③	2 Barrel	4.34 x 3.38	400	8.2	175 @ 4000	300 @ 2400	30–80
	185 Horsepower①..........V8-400④	4 Barrel	4.34 x 3.38	400	8.2	185 @ 4000	285 @ 3200	30–80
	190 Horsepower①..........V8-400③	4 Barrel	4.34 x 3.38	400	8.2	190 @ 4000	290 @ 3200	30–80
	195 Horsepower①..........V8-400④	4 Barrel	4.34 x 3.38	400	8.2	195 @ 4000	285 @ 3200	30–80
	235 Horsepower①②.........V8-400③	4 Barrel	4.34 x 3.38	400	8.2	235 @ 4000	320 @ 3200	30–80
	215 Horsepower①............V8-440	4 Barrel	4.32 x 3.75	440	8.2	215 @ 4000	330 @ 3200	30–80
1976	140 Horsepower①..........V8-318④	2 Barrel	3.91 x 3.31	318	8.5	140 @ 3600	250 @ 2000	40–65
	150 Horsepower①..........V8-318③	2 Barrel	3.91 x 3.31	318	8.5	150 @ 4000	255 @ 1600	40–65
	170 Horsepower①..........V8-360③	2 Barrel	4.00 x 3.58	360	8.4	170 @ 4000	280 @ 2400	40–65
	175 Horsepower①..........V8-360④	4 Barrel	4.00 x 3.58	360	8.4	175 @ 4000	270 @ 1600	40–65
	175 Horsepower①..........V8-400③	2 Barrel	4.34 x 3.38	400	8.2	175 @ 4000	300 @ 2400	50–75
	185 Horsepower①..........V8-400④	4 Barrel	4.34 x 3.38	400	8.2	185 @ 3600	285 @ 3200	50–75
	210 Horsepower①..........V8-400③	4 Barrel	4.34 x 3.38	400	8.2	210 @ 4000	305 @ 3200	50–75
	240 Horsepower①②.........V8-400③	4 Barrel	4.34 x 3.38	400	8.2	240 @ 4400	325 @ 3200	50–75
	200 Horsepower①..........V8-440④	4 Barrel	4.32 x 3.75	440	8.2	200 @ 3600	310 @ 2400	50–75
	205 Horsepower①..........V8-440③	4 Barrel	4.32 x 3.75	440	8.2	205 @ 3600	320 @ 2000	50–75
DODGE								
1969	115 Horsepower...............6-170	1 Barrel	3.40 x 3.125	170	8.5	115 @ 4400	115 @ 2400	45–65
	145 Horsepower...............6-225	1 Barrel	3.40 x 4.125	225	8.4	145 @ 4000	215 @ 2400	45–65
	190 Horsepower.............V8-273	2 Barrel	3.63 x 3.31	273	9.0	190 @ 4400	260 @ 2000	45–65
	230 Horsepower.............V8-318	2 Barrel	3.91 x 3.31	318	9.2	230 @ 4400	340 @ 2400	45–65
	275 Horsepower.............V8-340	4 Barrel	4.04 x 3.31	340	10.5	275 @ 5000	340 @ 3200	45–65
	290 Horsepower.............V8-383	2 Barrel	4.25 x 3.38	383	9.2	290 @ 4400	390 @ 2800	45–65
	330 Horsepower.............V8-383	4 Barrel	4.25 x 3.38	383	10.0	330 @ 5000	425 @ 3200	45–65
	335 Horsepower.............V8-383	4 Barrel	4.25 x 3.38	383	10.0	335 @ 5000	425 @ 3400	45–65
	350 Horsepower.............V8-440	4 Barrel	4.32 x 3.75	440	10.1	350 @ 4400	480 @ 2800	45–65
	375 Horsepower.............V8-440	4 Barrel	4.32 x 3.75	440	10.1	375 @ 4600	480 @ 3200	45–65
	390 Horsepower.............V8-440	Three 2 Bar.	4.32 x 3.75	440	10.5	390 @ 4700	490 @ 3200	45–65
	425 Horsepower.............V8-426	Two 4 Bar.	4.25 x 3.75	426	10.25	425 @ 5000	490 @ 4000	45–65
1970	125 Horsepower...............6-198	1 Barrel	3.40 x 3.64	198	8.4	125 @ 4400	180 @ 2000	45–65
	145 Horsepower...............6-225	1 Barrel	3.40 x 4.12	225	8.4	145 @ 4000	215 @ 2400	45–65
	230 Horsepower.............V8-318	2 Barrel	3.91 x 3.31	318	8.8	230 @ 4400	320 @ 2000	45–65
	275 Horsepower.............V8-340	4 Barrel	4.04 x 3.31	340	8.8	275 @ 5000	340 @ 3200	45–65
	290 Horsepower.............V8-383	2 Barrel	4.25 x 3.38	383	8.7	290 @ 4400	390 @ 2800	45–65
	330 Horsepower.............V8-383	4 Barrel	4.25 x 3.38	383	9.5	330 @ 5000	425 @ 3200	45–65
	335 Horsepower.............V8-383	4 Barrel	4.25 x 3.38	383	9.5	335 @ 5200	425 @ 3400	45–65
	425 Horsepower.............V8-426	Two 4 Bar.	4.25 x 3.75	426	10.2	425 @ 5000	490 @ 4000	45–65
	350 Horsepower.............V8-440	4 Barrel	4.32 x 3.75	440	9.7	350 @ 4400	480 @ 2800	45–65
	375 Horsepower.............V8-440	4 Barrel	4.32 x 3.75	440	9.7	375 @ 4600	480 @ 3200	45–65
	390 Horsepower.............V8-440	Three 2 Bar.	4.32 x 3.75	440	10.5	390 @ 4700	490 @ 3200	45–65
1971	125 Horsepower...............6-198	1 Barrel	3.40 x 3.64	198	8.4	125 @ 4400	180 @ 2000	45–65
	145 Horsepower...............6-225	1 Barrel	3.40 x 4.12	225	8.4	145 @ 4000	215 @ 2400	45–65
	230 Horsepower.............V8-318	2 Barrel	3.91 x 3.31	318	8.6	230 @ 4400	320 @ 2000	45–65
	275 Horsepower.............V8-340	4 Barrel	4.04 x 3.31	340	10.2	340 @ 5000	340 @ 3200	45–65
	290 Horsepower.............V8-340	Three 2 Bar.	4.04 x 3.31	340	10.2	290 @ 5000	340 @ 3200	45–65
	255 Horsepower.............V8-360	2 Barrel	4.00 x 3.58	360	8.7	255 @ 4400	360 @ 2400	45–65

Continued

GENERAL ENGINE SPECIFICATIONS—Continued

Year	Engine	Carburetor	Bore and Stroke	Piston Displacement, Cubic Inches	Compression Ratio	Maximum Brake H.P. @ R.P.M.	Maximum Torque Lbs. Ft. @ R.P.M.	Normal Oil Pressure Pounds
DODGE—Continued								
1971	275 Horsepower.............V8-383	2 Barrel	4.25 x 3.38	383	8.5	275 @ 4400	375 @ 2800	45–65
	300 Horsepower.............V8-383	4 Barrel	4.25 x 3.38	383	8.5	300 @ 4800	410 @ 3400	45–65
	425 Horsepower.............V8-426	Two 4 Bar.	4.25 x 3.75	426	10.2	425 @ 5000	490 @ 4000	45–65
	335 Horsepower.............V8-440	4 Barrel	4.32 x 3.75	440	9.0	335 @ 4400	460 @ 3200	45–65
	370 Horsepower.............V8-440	4 Barrel	4.32 x 3.75	440	9.7	370 @ 4600	280 @ 3200	45–65
	385 Horsepower.............V8-440	Three 2 Bar.	4.32 x 3.75	440	10.5	385 @ 4700	490 @ 3200	45–65
1972	100 Horsepower[1].............6-198	1 Barrel	3.40 x 3.64	198	8.4	100 @ 4400	160 @ 2400	45–65
	110 Horsepower[1].............6-225	1 Barrel	3.40 x 4.12	225	8.4	110 @ 4000	185 @ 2000	45–65
	150 Horsepower[1]............V8-318	2 Barrel	3.91 x 3.31	318	8.6	150 @ 4000	260 @ 1600	45–65
	240 Horsepower[1]............V8-340	4 Barrel	4.04 x 3.31	340	8.5	240 @ 4800	290 @ 3600	45–65
	175 Horsepower[1]............V8-360	2 Barrel	4.00 x 3.58	360	8.8	175 @ 4000	285 @ 2400	45–65
	190 Horsepower[1]............V8-400	2 Barrel	4.34 x 3.38	400	8.2	190 @ 4400	310 @ 2400	45–65
	255 Horsepower[1][2]..........V8-400	4 Barrel	4.34 x 3.38	400	8.2	255 @ 4800	340 @ 3200	45–65
	225 Horsepower[1]............V8-440	4 Barrel	4.32 x 3.75	440	8.2	225 @ 4400	345 @ 3200	45–65
	245 Horsepower[1][2]..........V8-440	4 Barrel	4.32 x 3.75	440	8.2	245 @ 4400	360 @ 3200	45–65
	330 Horsepower[1]............V8-440	Three 2 Bar.	4.32 x 3.75	440	10.3	330 @ 4800	410 @ 3600	45–65
1973	95 Horsepower[1].............6-198	1 Barrel	3.40 x 3.64	198	8.4	95 @ 4000	150 @ 1600	45–65
	105 Horsepower[1].............6-225	1 Barrel	3.40 x 4.12	225	8.4	105 @ 4000	185 @ 1600	45–65
	150 Horsepower[1]............V8-318	2 Barrel	3.91 x 3.31	318	8.6	150 @ 3600	265 @ 2000	45–65
	240 Horsepower[1]............V8-340	4 Barrel	4.04 x 3.31	340	8.5	240 @ 4800	295 @ 3600	45–65
	170 Horsepower[1]............V8-360	2 Barrel	4.00 x 3.58	360	8.4	170 @ 4000	285 @ 2400	45–65
	175 Horsepower[1]............V8-400	2 Barrel	4.34 x 3.38	400	8.2	175 @ 3600	305 @ 2400	45–65
	185 Horsepower[1]............V8-400	2 Barrel	4.34 x 3.38	400	8.2	185 @ 3600	310 @ 2400	45–65
	260 Horsepower[1]............V8-400	4 Barrel	4.34 x 3.38	400	8.2	260 @ 4800	335 @ 3600	45–65
	220 Horsepower[1]............V8-440	4 Barrel	4.32 x 3.75	440	8.2	220 @ 3600	350 @ 2400	45–65
	280 Horsepower[1]............V8-440	4 Barrel	4.32 x 3.75	440	8.2	280 @ 4800	380 @ 3200	45–65
1974	95 Horsepower[1].............6-198	1 Barrel	3.40 x 3.64	198	8.4	95 @ 4000	145 @ 2000	45–65
	105 Horsepower[1].............6-225	1 Barrel	3.40 x 4.12	225	8.4	105 @ 3600	180 @ 1600	45–65
	150 Horsepower[1]............V8-318	2 Barrel	3.91 x 3.31	318	8.6	150 @ 4000	255 @ 2200	45–65
	180 Horsepower[1]..........V8-360[3]	2 Barrel	4.00 x 3.58	360	8.4	180 @ 4000	290 @ 2400	45–65
	200 Horsepower[1]..........V8-360[4]	4 Barrel	4.00 x 3.58	360	8.4	200 @ 4000	290 @ 3200	45–65
	245 Horsepower[1][2].........V8-360[3]	4 Barrel	4.00 x 3.58	360	8.4	245 @ 4800	320 @ 3600	45–65
	185 Horsepower[1]..........V8-400[3]	2 Barrel	4.34 x 3.38	400	8.2	185 @ 4000	315 @ 2400	45–65
	200 Horsepower[1]..........V8-400[4]	4 Barrel	4.34 x 3.38	400	8.2	200 @ 4400	310 @ 2400	45–65
	205 Horsepower[1]..........V8-400[3]	4 Barrel	4.34 x 3.38	400	8.2	205 @ 4400	310 @ 2400	45–65
	250 Horsepower[1][2].........V8-400[3]	4 Barrel	4.34 x 3.38	400	8.2	250 @ 4800	330 @ 3400	45–65
	230 Horsepower[1]..........V8-440[3]	4 Barrel	4.32 x 3.75	440	8.2	230 @ 4000	350 @ 3200	45–65
	220 Horsepower[1]..........V8-440[4]	4 Barrel	4.32 x 3.75	440	8.2	220 @ 4000	345 @ 3200	45–65
	275 Horsepower[1][2]..........V8-440	4 Barrel	4.32 x 3.75	440	8.2	275 @ 4400	375 @ 3200	45–65
1975	90 Horsepower[1]............6-225[4]	1 Barrel	3.40 x 3.64	225	8.4	90 @ 3600	165 @ 1600	30–70
	95 Horsepower[1]............6-225[3]	1 Barrel	3.40 x 3.64	225	8.4	95 @ 3600	170 @ 1600	30–70
	135 Horsepower[1]..........V8-318[4]	2 Barrel	3.91 x 3.31	318	8.5	135 @ 3600	245 @ 1600	30–80
	140 Horsepower[1]..........V8-318[4]	2 Barrel	3.91 x 3.31	318	8.5	140 @ 3600	255 @ 1600	30–80
	145 Horsepower[1]..........V8-318[3]	2 Barrel	3.91 x 3.31	318	8.5	145 @ 4000	255 @ 1600	30–80
	150 Horsepower[1]..........V8-318[3]	2 Barrel	3.91 x 3.31	318	8.5	150 @ 4000	255 @ 1600	30–80
	180 Horsepower[1]..........V8-360[3]	2 Barrel	4.00 x 3.58	360	8.4	180 @ 4000	290;@ 2400	30–80
	190 Horsepower[1]..........V8-360[4]	4 Barrel	4.00 x 3.58	360	8.4	190 @ 4000	270 @ 3200	30–80
	230 Horsepower[1][2].........V8-360[3]	4 Barrel	4.00 x 3.58	360	8.4	230 @ 4200	320 @ 3200	30–80
	165 Horsepower[1]..........V8-400[3]	2 Barrel	4.34 x 3.38	400	8.2	165 @ 4000	295 @ 3200	30–80
	175 Horsepower[1]..........V8-400[3]	2 Barrel	4.34 x 3.38	400	8.2	175 @ 4000	300 @ 2400	30–80
	185 Horsepower[1]..........V8-400[4]	4 Barrel	4.34 x 3.38	400	8.2	185 @ 4000	285 @ 3200	30–80
	190 Horsepower[1]..........V8-400[3]	4 Barrel	4.34 x 3.38	400	8.2	190 @ 4000	290 @ 3200	30–80
	195 Horsepower[1]..........V8-400[4]	4 Barrel	4.34 x 3.38	400	8.2	195 @ 4000	285 @ 3200	30–80
	235 Horsepower[1][2].........V8-400[3]	4 Barrel	4.34 x 3.38	400	8.2	235 @ 4200	320 @ 3200	30–80
	240 Horsepower[1][2].........V8-400[3]	4 Barrel	4.34 x 3.38	400	8.2	240 @ 4400	325 @ 3200	30–80

Continued

GENERAL ENGINE SPECIFICATIONS—Continued

DODGE—Continued

Year	Engine	Carburetor	Bore and Stroke	Piston Displacement, Cubic Inches	Compression Ratio	Maximum Brake H.P. @ R.P.M.	Maximum Torque H.P. @ R.P.M.	Normal Oil Pressure Pounds
1975	215 Horsepower[1] .... V8-440[3]	4 Barrel	4.32 x 3.75	440	8.2	215 @ 4000	330 @ 3200	30–80
	250 Horsepower[1][2] .... V8-440[4]	4 Barrel	4.32 x 3.75	440	8.2	250 @ 4000	350 @ 3200	30–80
	260 Horsepower[1][2] .... V8-440[3]	4 Barrel	4.32 x 3.75	440	8.2	260 @ 4400	355 @ 3200	30–80
1976	90 Horsepower[1] .... 6-225[4]	1 Barrel	3.40 x 4.12	225	8.4	90 @ 3600	165 @ 1600	40–65
	100 Horsepower[1] .... 6-225[3]	1 Barrel	3.40 x 4.12	225	8.4	100 @ 3600	170 @ 1600	40–65
	140 Horsepower[1] .... V8-318[4]	2 Barrel	3.91 x 3.31	318	8.5	140 @ 3600	250 @ 2000	40–65
	150 Horsepower[1] .... V8-318[3]	2 Barrel	3.91 x 3.31	318	8.5	150 @ 4000	255 @ 1600	40–65
	170 Horsepower[1] .... V8-360[3]	2 Barrel	4.00 x 3.58	360	8.4	170 @ 4000	280 @ 2400	40–65
	175 Horsepower[1] .... V8-360[4]	4 Barrel	4.00 x 3.58	360	8.4	175 @ 4000	270 @ 1600	40–65
	220 Horsepower[1][2] .... V8-360[3]	4 Barrel	4.00 x 3.58	360	8.4	220 @ 4400	280 @ 3200	40–65
	175 Horsepower[1] .... V8-400[3]	2 Barrel	4.34 x 3.38	400	8.2	175 @ 4000	300 @ 2400	50–75
	185 Horsepower[1] .... V8-400[4]	4 Barrel	4.34 x 3.38	400	8.2	185 @ 3600	285 @ 3200	50–75
	210 Horsepower[1] .... V8-400[3]	4 Barrel	4.34 x 3.38	400	8.2	210 @ 4000	305 @ 3200	50–75
	240 Horsepower[1][2] .... V8-400[3]	4 Barrel	4.34 x 3.38	400	8.2	240 @ 4400	325 @ 3200	50–75
	200 Horsepower[1] .... V8-440[4]	4 Barrel	4.32 x 3.75	440	8.2	200 @ 3600	310 @ 2400	50–75
	205 Horsepower[1] .... V8-440[3]	4 Barrel	4.32 x 3.75	440	8.2	205 @ 3600	320 @ 2000	50–75
	250 Horsepower[1][2] .... V8-440[4]	4 Barrel	4.32 x 3.75	440	8.2	250 @ 4000	350 @ 3200	50–75
	255 Horsepower[1][2] .... V8-440[3]	4 Barrel	4.32 x 3.75	440	8.2	255 @ 4400	355 @ 3200	50–75

PLYMOUTH

Year	Engine	Carburetor	Bore and Stroke	Piston Displacement, Cubic Inches	Compression Ratio	Maximum Brake H.P. @ R.P.M.	Maximum Torque H.P. @ R.P.M.	Normal Oil Pressure Pounds
1969	115 Horsepower 6-170	1 Barrel	3.40 x 3.125	170	8.5	115 @ 4400	155 @ 2400	45–65
	145 Horsepower 6-225	1 Barrel	3.40 x 4.125	225	8.4	145 @ 4000	215 @ 2400	45–65
	190 Horsepower V8-273	2 Barrel	3.63 x 3.31	273	9.0	190 @ 4400	260 @ 2000	45–65
	230 Horsepower V8-318	2 Barrel	3.91 x 3.31	318	9.2	230 @ 4400	340 @ 2400	45–65
	275 Horsepower V8-340	4 Barrel	4.04 x 3.31	340	10.5	275 @ 5000	340 @ 3200	45–65
	290 Horsepower V8-383	2 Barrel	4.25 x 3.38	383	9.2	290 @ 4400	390 @ 2800	45–65
	330 Horsepower V8-383	4 Barrel	4.25 x 3.38	383	10.0	330 @ 5000	425 @ 3200	45–65
	335 Horsepower V8-383	4 Barrel	4.25 x 3.38	383	10.0	335 @ 5000	425 @ 3400	45–65
	350 Horsepower V8-440	4 Barrel	4.32 x 3.75	440	10.1	350 @ 4400	480 @ 2800	45–65
	375 Horsepower V8-440	4 Barrel	4.32 x 3.75	440	10.1	375 @ 4600	480 @ 3200	45–65
	390 Horsepower V8-440	Three 2 Bar.	4.32 x 3.75	440	10.5	390 @ 4700	490 @ 3200	45–65
	425 Horsepower V8-426	Two 4 Bar.	4.25 x 3.75	426	10.25	425 @ 5000	490 @ 4000	45–65
1970	125 Horsepower 6-198	1 Barrel	3.40 x 3.64	198	8.4	125 @ 4400	180 @ 2000	45–65
	145 Horsepower 6-225	1 Barrel	3.40 x 4.12	225	8.4	145 @ 4000	215 @ 2400	45–65
	230 Horsepower V8-318	2 Barrel	3.91 x 3.31	318	8.8	230 @ 4400	320 @ 2000	45–65
	275 Horsepower V8-340	4 Barrel	4.04 x 3.31	340	10.5	275 @ 5000	340 @ 3200	45–65
	290 Horsepower V8-383	2 Barrel	4.25 x 3.38	383	8.7	290 @ 4400	390 @ 2800	45–65
	330 Horsepower V8-383	4 Barrel	4.25 x 3.38	383	9.5	330 @ 5000	425 @ 3200	45–65
	335 Horsepower V8-383	4 Barrel	4.25 x 3.38	383	9.5	335 @ 5200	425 @ 3400	45–65
	425 Horsepower V8-426	Two 4 Bar.	4.25 x 3.75	426	10.2	425 @ 5000	490 @ 4000	45–65
	350 Horsepower V8-440	4 Barrel	4.32 x 3.75	440	9.7	350 @ 4400	480 @ 2800	45–65
	375 Horsepower V8-440	4 Barrel	4.32 x 3.75	440	9.7	375 @ 4600	480 @ 3200	45–65
	390 Horsepower V8-440	Three 2 Bar.	4.32 x 3.75	440	10.5	390 @ 4700	490 @ 3200	45–65
1971	125 Horsepower 6-198	1 Barrel	3.40 x 3.64	198	8.4	125 @ 4400	180 @ 2000	45–65
	145 Horsepower 6-225	1 Barrel	3.40 x 4.12	225	8.4	145 @ 4000	215 @ 2400	45–65
	230 Horsepower V8-318	2 Barrel	3.91 x 3.31	318	8.6	230 @ 4400	320 @ 2000	45–65
	275 Horsepower V8-340	4 Barrel	4.04 x 3.31	340	10.3	340 @ 5000	340 @ 3200	45–65
	290 Horsepower V8-340	Three 2 Bar.	4.04 x 3.31	340	10.2	290 @ 5000	340 @ 3200	45–65
	255 Horsepower V8-360	2 Barrel	4.00 x 3.58	360	8.7	255 @ 4400	360 @ 2400	45–65
	275 Horsepower V8-383	2 Barrel	4.25 x 3.38	383	8.5	275 @ 4400	375 @ 2800	45–65
	300 Horsepower V8-383	4 Barrel	4.25 x 3.38	383	8.5	300 @ 4800	410 @ 3400	45–65
	425 Horsepower V8-426	Two 4 Bar.	4.25 x 3.75	426	10.2	425 @ 5000	490 @ 4000	45–65
	335 Horsepower V8-440	4 Barrel	4.32 x 3.75	440	9.0	335 @ 4400	460 @ 3200	45–65
	370 Horsepower V8-440	4 Barrel	4.32 x 3.75	440	9.7	370 @ 4600	280 @ 3200	45–65
	385 Horsepower V8-440	Three 2 Bar.	4.32 x 3.75	440	10.5	385 @ 4700	490 @ 3200	45–65

Continued

GENERAL ENGINE SPECIFICATIONS—Continued

Year	Engine	Carburetor	Bore and Stroke	Piston Displacement, Cubic Inches	Compression Ratio	Maximum Brake H.P. @ R.P.M.	Maximum Torque H.P. @ R.P.M.	Normal Oil Pressure Pounds
PLYMOUTH—Continued								
1972	100 Horsepower①.............6-198	1 Barrel	3.40 x 3.64	198	8.4	100 @ 4400	160 @ 2400	45–65
	110 Horsepower①.............6-225	1 Barrel	3.40 x 4.12	225	8.4	110 @ 4000	185 @ 2000	45–65
	150 Horsepower①............V8-318	2 Barrel	3.91 x 3.31	318	8.6	150 @ 4000	260 @ 1600	45–65
	240 Horsepower①............V8-340	4 Barrel	4.04 x 3.31	340	8.5	240 @ 4800	290 @ 3600	45–65
	175 Horsepower①............V8-360	2 Barrel	4.00 x 3.58	360	8.8	175 @ 4000	285 @ 2400	45–65
	190 Horsepower①............V8-400	2 Barrel	4.34 x 3.38	400	8.2	190 @ 4400	310 @ 2400	45–65
	255 Horsepower①②..........V8-400	4 Barrel	4.34 x 3.38	400	8.2	255 @ 4800	340 @ 3200	45–65
	255 Horsepower①............V8-440	4 Barrel	4.32 x 3.75	440	8.2	225 @ 4400	345 @ 3200	45–65
	245 Horsepower①②..........V8-440	4 Barrel	4.32 x 3.75	440	8.2	245 @ 4400	360 @ 3200	45–65
	330 Horsepower①............V8-440	Three 2 Bar.	4.32 x 3.75	440	10.3	330 @ 4800	410 @ 3600	45–65
1973	95 Horsepower①.............6-198	1 Barrel	3.40 x 3.64	198	8.4	95 @ 4000	150 @ 1600	45–65
	105 Horsepower①.............6-225	1 Barrel	3.40 x 4.12	225	8.4	105 @ 4000	185 @ 1600	45–65
	150 Horsepower①............V8-318	2 Barrel	3.91 x 3.31	318	8.6	150 @ 3600	265 @ 2000	45–65
	240 Horsepower①............V8-340	4 Barrel	4.04 x 3.31	340	8.5	240 @ 4800	295 @ 3600	45–65
	170 Horsepower①............V8-360	2 Barrel	4.00 x 3.58	360	8.4	170 @ 4000	285 @ 2400	45–65
	175 Horsepower①............V8-400	2 Barrel	4.34 x 3.38	400	8.2	175 @ 3600	305 @ 2400	45–65
	185 Horsepower①............V8-400	2 Barrel	4.34 x 3.38	400	8.2	185 @ 3600	310 @ 2400	45–65
	260 Horsepower①............V8-400	4 Barrel	4.34 x 3.38	400	8.2	260 @ 4800	335 @ 3600	45–65
	220 Horsepower①............V8-440	4 Barrel	4.32 x 3.75	440	8.2	220 @ 3600	350 @ 2400	45–65
	280 Horsepower①............V8-440	4 Barrel	4.32 x 3.75	440	8.2	280 @ 4800	380 @ 3200	45–65
1974	95 Horsepower①.............6-198	1 Barrel	3.40 x 3.64	198	8.4	95 @ 4000	145 @ 2000	45–65
	105 Horsepower①.............6-225	1 Barrel	3.40 x 4.12	225	8.4	105 @ 3600	180 @ 1600	45–65
	150 Horsepower①............V8-318	2 Barrel	3.91 x 3.31	318	8.6	150 @ 4000	255 @ 2200	45–65
	170 Horsepower①②..........V8-318	2 Barrel	3.91 x 3.31	318	8.6	170 @ 4000	265 @ 2600	45–65
	180 Horsepower①............V8-360	2 Barrel	4.00 x 3.58	360	8.4	180 @ 4000	290 @ 2400	45–65
	200 Horsepower①..........V8-360④	4 Barrel	4.00 x 3.58	360	8.4	200 @ 4000	290 @ 3200	45–65
	245 Horsepower①②........V8-360③	4 Barrel	4.00 x 3.58	360	8.4	245 @ 4800	320 @ 3600	45–65
	185 Horsepower①............V8-400	2 Barrel	4.34 x 3.38	400	8.2	185 @ 4000	315 @ 2400	45–65
	205 Horsepower①............V8-400	4 Barrel	4.34 x 3.38	400	8.2	205 @ 4400	310 @ 2400	45–65
	250 Horsepower①②..........V8-400	4 Barrel	4.34 x 3.38	400	8.2	250 @ 4800	330 @ 3400	45–65
	230 Horsepower①..........V8-440③	4 Barrel	4.32 x 3.75	440	8.2	230 @ 4000	350 @ 3200	45–65
	220 Horsepower①..........V8-440④	4 Barrel	4.32 x 3.75	440	8.2	220 @ 4000	345 @ 3200	45–65
	275 Horsepower①②..........V8-440	4 Barrel	4.32 x 3.75	440	8.2	275 @ 4400	375 @ 3200	45–65
1975	90 Horsepower①...........6-225④	1 Barrel	3.40 x 4.12	225	8.4	90 @ 3600	165 @ 1600	30–70
	95 Horsepower①...........6-225③	1 Barrel	3.40 x 4.12	225	8.4	95 @ 3600	170 @ 1600	30–70
	135 Horsepower①..........V8-318④	2 Barrel	3.91 x 3.31	318	8.5	135 @ 3600	245 @ 1600	30–80
	140 Horsepower①..........V8-318④	2 Barrel	3.91 x 3.31	318	8.5	140 @ 3600	255 @ 1600	30–80
	145 Horsepower①..........V8-318③	2 Barrel	3.91 x 3.31	318	8.5	145 @ 4000	255 @ 1600	30–80
	150 Horsepower①..........V8-318③	2 Barrel	3.91 x 3.31	318	8.5	150 @ 4000	255 @ 1600	30–80
	180 Horsepower①..........V8-360③	2 Barrel	4.00 x 3.58	360	8.4	180 @ 4000	290 @ 2400	30–80
	190 Horsepower①..........V8-360④	4 Barrel	4.00 x 3.58	360	8.4	190 @ 4000	270 @ 3200	30–80
	230 Horsepower①②........V8-360③	4 Barrel	4.00 x 3.58	360	8.4	230 @ 4400	300 @ 3600	30–80
	165 Horsepower①..........V8-400③	2 Barrel	4.34 x 3.38	400	8.2	165 @ 4000	295 @ 3200	30–80
	175 Horsepower①..........V8-400③	2 Barrel	4.34 x 3.38	400	8.2	175 @ 4000	300 @ 2400	30–80
	185 Horsepower①..........V8-400④	4 Barrel	4.34 x 3.38	400	8.2	185 @ 4000	285 @ 3200	30–80
	190 Horsepower①..........V8-400③	4 Barrel	4.34 x 3.38	400	8.2	190 @ 4000	290 @ 3200	30–80
	195 Horsepower①..........V8-400④	4 Barrel	4.34 x 3.38	400	8.2	195 @ 4000	285 @ 3200	30–80
	235 Horsepower①②........V8-400③	4 Barrel	4.34 x 3.38	400	8.2	235 @ 4200	320 @ 3200	30–80
	240 Horsepower①②........V8-400③	4 Barrel	4.34 x 3.38	400	8.2	250 @ 4400	325 @ 3200	30–80
	215 Horsepower①............V8-440	4 Barrel	4.32 x 3.75	440	8.2	215 @ 4000	330 @ 3200	30–90
	250 Horsepower①②........V8-440④	4 Barrel	4.32 x 3.75	440	8.2	250 @ 4000	350 @ 3200	30–80
	260 Horsepower①②........V8-440③	4 Barrel	4.32 x 3.75	440	8.2	260 @ 4400	355 @ 3200	30–80

Continued

GENERAL ENGINE SPECIFICATIONS—Continued

Year	Engine	Carburetor	Bore and Stroke	Piston Displacement, Cubic Inches	Compression Ratio	Maximum Brake H.P. @ R.P.M.	Maximum Torque H.P. @ R.P.M.	Normal Oil Pressure Pounds
PLYMOUTH—Continued								
1976	90 Horsepower①............6-225④	1 Barrel	3.40 x 4.12	225	8.4	90 @ 3600	165 @ 1600	40–65
	100 Horsepower①............6-225③	1 Barrel	3.40 x 4.12	225	8.4	100 @ 3600	170 @ 1600	40–65
	140 Horsepower①..........V8-318④	2 Barrel	3.91 x 3.31	318	8.5	140 @ 3600	250 @ 2000	40–65
	150 Horsepower①..........V8-318③	2 Barrel	3.91 x 3.31	318	8.5	150 @ 4000	255 @ 1600	40–65
	170 Horsepower①..........V8-360③	2 Barrel	4.00 x 3.58	360	8.4	170 @ 4000	280 @ 2400	40–65
	175 Horsepower①..........V8-360④	4 Barrel	4.00 x 3.58	360	8.4	175 @ 4000	270 @ 1600	40–65
	200 Horsepower①②.........V8-360③	4 Barrel	4.00 x 3.58	360	8.4	220 @ 4400	280 @ 3200	40–65
	175 Horsepower①..........V8-400③	2 Barrel	4.34 x 3.38	400	8.2	175 @ 4000	300 @ 2400	50–75
	185 Horsepower①..........V8-400④	4 Barrel	4.34 x 3.38	400	8.2	185 @ 3600	285 @ 3200	50–75
	210 Horsepower①..........V8-400③	4 Barrel	4.34 x 3.38	400	8.2	210 @ 4000	305 @ 3200	50–75
	240 Horsepower①②.........V8-400③	4 Barrel	4.34 x 3.38	400	8.2	240 @ 4400	325 @ 3200	50–75
	200 Horsepower①..........V8-440④	4 Barrel	4.32 x 3.75	440	8.2	200 @ 3600	310 @ 2400	50–75
	205 Horsepower①..........V8-440③	4 Barrel	4.32 x 3.75	440	8.2	205 @ 3600	320 @ 2000	50–75
	250 Horsepower①②.........V8-440④	4 Barrel	4.32 x 3.75	440	8.2	250 @ 4000	350 @ 3200	50–75
	255 Horsepower①②.........V8-440③	4 Barrel	4.32 x 3.75	440	8.2	255 @ 4400	355 @ 3200	50–75

)—Ratings are NET—as installed in the vehicle.
)—With dual exhausts.
③—Exc. California.
④—California.

TUNE UP SPECIFICATIONS

The following specifications are published from the latest information available. This data should be used only in the absence of a decal affixed in the engine compartment.

OLD CAR SPECIFICATIONS: For 1946-68 Tune Up Specifications see main index.

★When using a timing light, disconnect vacuum hose or tube at distributor and plug opening in hose or tube so idle speed will not be affected.

●When checking compression, lowest cylinder must be within 80 percent of highest.

▲Before removing wires from distributor cap, determine location of the No. 1 wire in cap, as distributor position may have been altered from that shown at the end of this chart.

Year	Spark Plug		Distributor		Ignition Timing★			Carb. Adjustments					
								Hot Idle Speed		Air Fuel Ratio		Idle CO % ③	
	Type ⑦	Gap Inch	Point Gap Inch	Dwell Angle Deg.	Firing Order Fig. ▲	Timing BTDC ①	Mark Fig.	Std. Trans.	Auto. Trans. ②	Std. Trans.	Auto. Trans.	Std. Trans.	Auto. Trans.
CHRYSLER & IMPERIAL													
1969													
V8-383 Std. Tr.㉑	J14Y	.035	.017	30–35	J	TDC	F	700⑥	—	—	—	—	—
V8-383 Auto. Tr.㉑	J14Y	.035	.017	30–35	J	7½°	F	—	600⑥	—	—	—	—
V8-383 Auto. Tr.④	J11Y	.035	.017	30–35	J	5°	F	700⑥	650⑥	—	—	—	—
V8-440 Auto. Tr.	J13Y	.035	.017	30–35	J	7½°	F	—	650⑥	—	—	—	—
V8-440 Auto. Tr.㉒	J11Y	.035	.017	30–35	J	5°	F	—	650⑥	—	—	—	—
V8-440 Std. Tr.㉒	J11Y	.035	.017	⑰	J	TDC	F	700⑥	—	—	—	—	—
1970-71													
V8-360	N13Y	.035	.017	30–34	H	2½	E	750	700	14.2 to 1	14.2 to 1	—	—
V8-383 Std. Tr.㉑	J14Y	.035	.019	28½–32½	J	TDC	F	750	—	14.2 to 1	—	—	—
V8-383 Auto. Tr.㉑	J14Y	.035	.019	28½–32½	J	2½°	F	—	650N	—	14.2 to 1	—	—
V8-383 Std. Tr.④	J11Y	.035	.019	28½–32½	J	TDC	F	900	—	14.2 to 1	—	—	—
V8-383 Auto. Tr.④	J11Y	.035	.019	28½–32½	J	2½°	F	—	700N	—	14.2 to 1	—	—
V8-400 Auto. Tr.	J11Y	.035	.019	28½–32½	J	2½°	F	—	700N	—	14.2 to 1	—	—
V8-440 Auto. Tr.	J13Y	.035	.019	28½–32½	J	5°	F	—	650N	—	14.2 to 1	—	—
V8-440 Std. Tr.㉒	J11Y	'035	.019	28½–32½	J	TDC	F	900	—	14.2 to 1	—	—	—
V8-440 Auto. Tr.㉒	J11Y	.035	.019	28½–32½	J	2½°	F	—	800N	—	14.2 to 1	—	—

Continued

TUNE UP SPECIFICATIONS—Continued

The following specifications are published from the latest information available. This data should be used only in the absence of a decal affixed in the engine compartment.

OLD CAR SPECIFICATIONS: For 1946-68 Tune Up Specifications see main index.

★When using a timing light, disconnect vacuum hose or tube at distributor and plug opening in hose or tube so idle speed will not be affected.

●When checking compression, lowest cylinder must be within 80 percent of highest.

▲Before removing wires from distributor cap, determine location of the No. 1 wire in cap, as distributor position may have been altered from that shown at the end of this chart.

Year	Spark Plug		Distributor		Ignition Timing★			Carb. Adjustments					
								Hot Idle Speed		Air Fuel Ratio		Idle "CO" % ③	
	Type ⑦	Gap Inch	Point Gap Inch	Dwell Angle Deg.	Firing Order Fig. ▲	Timing BTDC ①	Mark Fig.	Std. Trans.	Auto. Trans. ②	Std. Trans.	Auto. Trans.	Std. Trans.	Auto. Trans.
CHRYSLER & IMPERIAL—Continued													
1972													
V8-360	N13Y	.035	.017	30–34	H	TDC	E	—	750N	14.2 to 1	14.2 to 1	—	—
V8-400	J13Y	.035	.019	28½–32½	J	5°	F	—	700N	14.2 to 1	14.2 to 1	—	—
V8-440 Chrysler	J11Y	.035	.019	28½–32½	J	10°㉗	F	—	750N㉘	14.2 to 1	14.2 to 1	—	—
V8-440 Imperial	J11Y	.035	—	—	J	10°	F	—	750N㉘	14.2 to 1	14.2 to 1	—	—
1973													
V8-400㉑	J13Y	.035	—	—	J	10°	F	—	700	—	14.2 to 1	—	—
V8-400④	J11Y	.035	—	—	J	7½°	F	—	750	—	14.2 to 1	—	—
V8-440	J11Y	.035	—	—	J	10°	F	—	700	—	14.2 to 1	—	—
1974													
V8-400㉑	J13Y	.035	—	—	J	⑨	F	—	750	—	14.3	—	—
V8-400④㉔	J13Y	.035	—	—	J	5°	F	—	750	—	14.3	—	—
V8-400④㉕	J13Y	.035	—	—	J	5°	F	—	750	—	14.1	—	—
V8-440㉔	J11Y	.035	—	—	J	10°	F	—	750	—	14.3	—	—
V8-440㉕	J11Y	.035	—	—	J	5°	F	—	750	—	14.1	—	—
1975													
V8-318⑤	N13Y	.035	—	—	H	㊱	B	—	750N	—	—	.3⑲	㉟
V8-318⑧	N13Y	.035	—	—	H	2°⑮	B	—	900N	—	—	—	.5⑳
V8-360	N12Y	.035	—	—	H	6°	B	—	750N	—	—	—	㉝
V8-400㉑	J13Y	.035	—	—	J	10°	F	—	750N	—	—	—	.3⑲
V8-400④	J13Y	.035	—	—	J	8°	F	—	750N	—	—	—	㉜
V8-400㉒	RJ87P	.035	—	—	J	6°	F	—	850N	—	—	—	.5⑳
V8-440	RJ87P	.040	—	—	J	6°	D	—	750N	—	—	—	㉞
1976													
V8-318⑤	RN12Y	.035	—	—	H	㊱	B	—	750	—	—	—	㊲
V8-318⑧	RN12Y	.035	—	—	H	2°⑮	B	—	900	—	—	—	.5⑳
V8-360㉑	RN12Y	.035	—	—	H	6°	B	—	700	—	—	—	.3⑲
V8-360④	RN12Y	.035	—	—	H	6°	B	—	750	—	—	—	2.0⑲
V8-400㉑	RJ13Y	.035	—	—	J	10°	F	—	700	—	—	—	.3⑲
V8-400④㉔	㊳	.035	—	—	J	6°	F	—	850	—	—	—	.5⑳
V8-400④㉕	RJ13Y	.035	—	—	J	8°	F	—	750	—	—	—	.5⑲
V8-440	RJ13Y	.035	—	—	J	8°	D	—	750	—	—	—	.3⑲

TUNE UP SPECIFICATIONS—Continued

The following specifications are published from the latest information available. This data should be used only in the absence of a decal affixed in the engine compartment.

OLD CAR SPECIFICATIONS: For 1946-69 Tune Up Specifications see main index.

★When using a timing light, disconnet vacuum hose or tube at distribtor and plug opening in hose or tube so idle speed will not be affected.

●When checking compression, lowest cylinder must be within 80 percent of highest.

▲Before removing wires from distributor cap, determine location of the No. 1 wire in cap, as distributor position may have been altered from that shown at the end of this chart.

Year	Spark Plug		Distributor		Ignition Timing★			Carb. Adjustments					
	Type ⑦	Gap Inch	Point Gap Inch	Dwell Angle Deg.	Firing Order Fig. ▲	Timing BTDC ①	Mark Fig.	Hot Idle Speed Std. Trans.	Hot Idle Speed Auto. Trans. ②	Air Fuel Ratio Std. Trans.	Air Fuel Ratio Auto. Trans.	Idle "CO" % ③ Std. Trans.	Idle "CO" % ③ Auto. Trans.
DODGE													
1969													
6-170 Std. Tr.	N14Y	.035	.020	42–47	G	5° ATC	E	750⑭	—	14.2 to 1	—	—	—
6-170 Auto. Tr.	N14Y	.035	.020	42–47	G	TDC	E	—	750N⑭	—	14.2 to 1	—	—
6-225	N14Y	.035	.020	42–47	G	TDC	E	650⑭	650N⑭	14.2 to 1	14.2 to 1	—	—
V8-273	N14Y	.035	.017	30–35	H	2½° ATC	F	700⑭	650N⑭	14.2 to 1	14.2 to 1	—	—
V8-318	N14Y	.035	.017	30–35	H	TDC	F	700⑭	650N⑭	14.2 to 1	14.2 to 1	—	—
V8-340 Std. Tr.	N9Y	.035	.017	⑰	H	TDC	F	750⑭	—	14.2 to 1	—	—	—
V8-340 Auto. Tr.	N9Y	.035	.017	30–35	H	5°	F	—	700N⑭	—	14.2 to 1	—	—
V8-383 Std. Tr.㉑	J14Y	.035	.017	30–35	J	TDC	F	700⑭	—	14.2 to 1	—	—	—
V8-383 Auto. Tr.㉑	J14Y	.035	.017	30–35	J	7½°	F	—	600N⑭	—	14.2 to 1	—	—
V8-383 Std. Tr.④	J11Y	.035	.017	30–35	J	TDC	F	700	—	14.2 to 1	—	—	—
V8-383 Auto. Tr.④	J11Y	.035	.017	30–35	J	5°	F	—	650N⑭	—	14.2 to 1	—	—
V8-383 Std. Tr.㉓	J11Y	.035	.017	⑰	J	TDC	F	700	—	14.2 to 1	—	—	—
V8-383 Auto. Tr.㉓	J11Y	.035	.017	⑰	J	5°	F	—	650N⑭	—	14.2 to 1	—	—
V8-440, 350 H.P.	J13Y	.035	.017	30–35	J	7½°	F	—	600N⑭	14.2 to 1	14.2 to 1	—	—
V8-440 Std. Tr.㉒	J11Y	.035	.017	⑰	J	TDC	F	700	—	14.2 to 1	—	—	—
V8-440 Auto. Tr.㉒	J11Y	.035	.017	30–35	J	5°	F	—	650N⑭	—	14.2 to 1	—	—
V8-426 Hemi.	N10Y	.035	.017	⑰	J	TDC	F	750	750N⑭	14.2 to 1	—	—	—
V8-440, 390H.P.	J11Y	.035	.017	⑰	J	5°	F	900	900	14.2 to 1	14.2 to 1	—	—
1970													
6-198 Std. Tr.	N14Y	.035	.020	41–46	G	2½°	C	750⑥	—	14.2 to 1	—	—	—
6-198 Auto. Tr.	N14Y	.035	.020	41–46	G	TDC	C	—	750⑥	—	14.2 to 1	—	—
6-225	N14Y	.035	.020	41–46	G	TDC	C	700⑥	650⑥	14.2 to 1	14.2 to 1	—	—
V8-318	N14Y	.035	.017	30–34	H	TDC	E	750⑥	700⑥	14.2 to 1	14.2 to 1	—	—
V8-340 Std. Tr.	N9Y	.035	.017	⑰	H	5°	E	950	—	14.2 to 1	—	—	—
V8-340 Auto. Tr.	N9Y	.035	.017	30–34	H	5°	E	—	900	—	14.2 to 1	—	—
V8-383 Std. Tr.㉑	J14Y	.035	.017	28–32	J	TDC	F	750	—	14.2 to 1	—	—	—
V8-383 Auto. Tr.㉑	J14Y	.035	.017	28–32	J	2½°	F	—	650	—	14.2 to 1	—	—
V8-383 Std. Tr.④	J11Y	.035	.017	28–32	J	TDC	F	750	—	14.2 to 1	—	—	—
V8-383 Auto. Tr.④	J11Y	.035	.017	28–32	J	2½°	F	—	750	—	14.2 to 1	—	—
V8-440, 350 H.P.	J13Y	.035	.017	28–32	J	5°	F	—	600	14.2 to 1	14.2 to 1	—	—
V8-440 Std. Tr.㉒	J11Y	.035	.017	28–32	J	TDC	F	900	—	14.2 to 1	—	—	—
V8-440 Auto. Tr.㉒	J11Y	.035	.017	28–32	J	2½°	F	—	800	—	14.2 to 1	—	—
V8-440, 390 H.P.	J11Y	.035	.017	⑰	J	5°	F	900	900	14.2 to 1	14.2 to 1	—	—
V8-426 Std. Tr.	N10Y	.035	.017	⑰	J	TDC	F	900	—	14.2 to 1	—	—	—
V8-426 Auto. Tr.	N10Y	.035	.017	⑰	J	5°	F	—	900	—	14.2 to 1	—	—
1971													
6-198	N14Y	.035	.020	41–46	G	2½°	C	800	800	14.2 to 1	14.2 to 1	—	—
6-225㉔	N14Y	.035	.020	41–46	G	TDC	C	750	750	14.2 to 1	14.2 to 1	—	—
6-225㉕	N14Y	.035	.020	41–46	G	2½°	C	750	750	14.2 to 1	14.2 to 1	—	—
V8-318	N14Y	.035	.017	30–34	H	TDC	E	750	700	14.2 to 1	14.2 to 1	—	—

Continued

TUNE UP SPECIFICATIONS—Continued

The following specifications are published from the latest information available. This data should be used only in the absence of a decal affixed in the engine compartment.

OLD CAR SPECIFICATIONS: For 1946-68 Tune Up Specifications see main index.

★When using a timing light, disconnect vacuum hose or tube at distributor and plug opening in hose or tube so idle speed will not be affected.

●When checking compression, lowest cylinder must be within 80 percent of highest.

▲Before removing wires from distributor cap, determine location of the No. 1 wire in cap, as distributor position may have been altered from that shown at the end of this chart.

Year	Spark Plug		Distributor		Ignition Timing★			Carb. Adjustments					
	Type ⑦	Gap Inch	Point Gap Inch	Dwell Angle Deg.	Firing Order Fig. ▲	Timing BTDC ①	Mark Fig.	Hot Idle Speed Std. Trans.	Hot Idle Speed Auto. Trans. ②	Air Fuel Ratio Std. Trans.	Air Fuel Ratio Auto. Trans.	Idle "CO" % ③ Std. Trans.	Idle "CO" % ③ Auto. Trans.
DODGE—Continued													
1971													
V8-340 Std. Tr.④	N9Y	.035	.017	⑰	H	5°	E	900	—	14.2 to 1	—	—	—
V8-340 Auto. Tr.④	N9Y	.035	.017	30–34	H	5°	E	—	900	—	14.2 to 1	—	—
V8-340 Std. Tr.㉖	N9Y	.035	.017	⑰	H	2½°	E	1000	—	14.2 to 1	—	—	—
V8-340 Auto. Tr.㉖	N9Y	.035	.017	30–34	H	2½°	E	—	950	—	14.2 to 1	—	—
V8-340③	N9Y	.035	—	—	H	5°	E	900	—	14.2 to 1	14.2 to 1	—	—
V8-360	N13Y	.035	.017	30–34	H	2½°	E	750	700	14.2 to 1	14.2 to 1	—	—
V8-383 Std. Tr.㉑	J14Y	.035	.017	30–34	J	TDC	F	750	—	14.2 to 1	—	—	—
V8-383 Auto. Tr.㉑	J14Y	.035	.017	30–34	J	2½°	F	—	700	—	14.2 to 1	—	—
V8-383 Std. Tr.④	J11Y	.035	.017	30–34	J	TDC	F	750	—	14.2 to 1	—	—	—
V8-383 Auto. Tr.④	J11Y	.035	.017	30–34	J	2½°	F	—	700	—	14.2 to 1	—	—
V8-426 Std. Tr.	N10Y	.035	.017	⑰	J	TDC	F	900	—	14.2 to 1	—	—	—
V8-426 Auto. Tr.	N10Y	.035	.017	⑰	J	2½°	F	—	900	—	14.2 to 1	—	—
V8-440	J13Y	.035	.017	28½–32½	J	5°	F	—	900	14.2 to 1	14.2 to 1	—	—
V8-440 Std. Tr.㉒	J11Y	.035	.017	28½–32½	J	TDC	F	900	—	14.2 to 1	—	—	—
V8-440 Auto. Tr.㉒	J11Y	.035	.017	28½–32½	J	2½°	F	—	900	—	14.2 to 1	—	—
V8-440㉖	J11Y	.035	.017	⑰	J	5°	F	900	900	14.2 to 1	—	—	—
1972													
6-198	N14Y	.035	.020	41–46	G	2½°	C	800㉘	800N㉘	14.2 to 1	14.2 to 1	—	—
6-225	N14Y	.035	.020	41–46	G	TDC	C	750㉘	750N㉘	14.2 to 1	14.2 to 1	—	—
V8-318	N13Y	.035	.017	30–34	H	TDC	E	750	750N㉘	14.2 to 1	14.2 to 1	—	—
V8-340	N9Y	.035	—	—	H	2½°	E	900㉙	750N	14.2 to 1	14.2 to 1	—	—
V8-360	N13Y	.035	.017	30–34	H	TDC	E	—	750N	14.2 to 1	14.2 to 1	—	—
V8-400㉑	J13Y	.035	.019	28½–32½	J	5°	F	—	700N	14.2 to 1	—	—	—
V8-400 Std. Tr.④	J11Y	.035	—	—	J	2½°	F	900㉚	—	14.2 to 1	—	—	—
V8-400 Auto. Tr.④	J11Y	.035	—	—	J	10°㉗	F	—	750N	—	14.2 to 1	—	—
V8-440	J11Y	.035	.019	28½–32½	J	10°㉗	F	—	750N㉘	14.2 to 1	14.2 to 1	—	—
V8-440 Std. Tr.㉒	J11Y	.035	—	—	J	2½°	F	900㉚	—	14.2 to 1	—	—	—
V8-440 Auto. Tr.㉒	J11Y	.035	—	—	J	10°㉗	F	—	900N	—	14.2 to 1	—	—
V8-440㉖	J11Y	.035	—	—	J	2½°	F	—	900N	14.2 to 1	—	—	—
1973													
6-198	N14Y	.035	—	—	G	TDC	C	800	750N	14.2 to 1	14.2 to 1	—	—
6-225 Std. Trans.	N14Y	.035	—	—	G	2½°	C	750	—	14.2 to 1	—	—	—
6-225 Auto. Trans.	N14Y	.035	—	—	G	TDC	C	—	750N	—	14.2 to 1	—	—
8-318	N13Y	.035	—	—	H	TDC	E	750	750N	14.2 to 1	14.2 to 1	—	—
8-340 Std. Trans.	N12Y	.035	—	—	H	5°	E	900	—	14.2 to 1	—	—	—
8-340 Auto. Trans.	N12Y	.035	—	—	H	2½°⑱	E	—	750N	—	14.2 to 1	—	—
8-360	N13Y	.035	—	—	H	⑩	E	750	750N	14.2 to 1	14.2 to 1	—	—
8-400 2 Bar. Carb.	J13Y	.035	—	—	J	10°	F	700	700N	14.2 to 1	14.2 to 1	—	—
8-400 4 Bar. Carb.㉛	J11Y	.035	—	—	J	10°	F	900	—	14.2 to 1	—	—	—
8-400 4 Bar. Carb.⑫	J11Y	.035	—	—	J	7½°	F	—	750N	—	14.2 to 1	—	—
8-440 Std. Trans.	J11Y	.035	—	—	J	10°	F	800	—	14.2 to 1	—	—	—
8-440 Auto. Trans.	J11Y	.035	—	—	J	10°	F	—	700N	—	14.2 to 1	—	—

Continued

TUNE UP SPECIFICATIONS—Continued

The following specifications are published from the latest information available. This data should be used only in the absence of a decal affixed in the engine compartment.

OLD CAR SPECIFICATIONS: For 1946-68 Tune Up Specifications see main index.

★When using a timing light, disconnect vacuum hose or tube at distributor and plug opening in hose or tube so idle speed will not be affected.

●When checking compression, lowest cylinder must be within 80 percent of highest.

▲Before removing wires from distributor cap, determine location of the No. 1 wire in cap, as distributor position may have been altered from that shown at the end of this chart.

Year	Spark Plug		Distributor		Ignition Timing★			Carb. Adjustments					
	Type (7)	Gap Inch	Point Gap Inch	Dwell Angle Deg.	Firing Order Fig. ▲	Timing BTDC (1)	Mark Fig.	Hot Idle Speed		Air Fuel Ratio		Idle "CO" % (3)	
								Std. Trans.	Auto. Trans. (2)	Std. Trans.	Auto. Trans.	Std. Trans.	Auto. Trans.
DODGE—Continued													
1974													
6-198	N14Y	.035	—	—	G	2½°	C	800	750	14.3 to 1	14.3 to 1	—	—
6-225	N14Y	.035	—	—	G	TDC	C	800	750	14.3 to 1	14.3 to 1	—	—
V8-318	N13Y	.035	—	—	H	TDC	E	750	750	14.3 to 1	14.3 to 1	—	—
V8-360	N12Y	.035	—	—	H	5°(11)	E	850	850	14.3 to 1	14.3 to 1	—	—
V8-400 Auto. Trans.(21)	J13Y	.035	—	—	J	7½°(9)	F	—	750	—	14.3 to 1	—	—
V8-400 Auto. Trans.(4)	J13Y	.035	—	—	J	5°	F	—	750	—	14.3 to 1	—	—
V8-400 Std. Trans.(22)	J11Y	.035	—	—	J	5°	F	900	—	14.3 to 1	—	—	—
V8-400 Auto. Trans.(16)	J11Y	.035	—	—	J	5°(13)	F	—	850	—	14.3 to 1	—	—
V8-440	J11Y	.035	—	—	J	(35)	F	—	850	—	14.3 to 1	—	—
1975													
6-225	BL13Y	.035	—	—	G	TDC	A	800	750N	—	—	.3(19)	(34)
V8-318(5)	N13Y	.035	—	—	H	(36)	B	750	750N	—	—	.3(19)	(33)
V8-318	N13Y	.035	—	—	H	2°(15)	B	—	900N	—	—	—	.5(20)
V8-360	N12Y	.035	—	—	H	6°	B	—	750N	—	—	—	(33)
V8-360(22)	N12Y	.035	—	—	H	2°	B	—	850N	—	—	—	.5(20)
V8-400(21)	J13Y	.035	—	—	J	10°	F	—	750N	—	—	—	.3(19)
V8-400(4)	J13Y	.035	—	—	J	8°	F	—	750N	—	—	—	(32)
V8-400(22)	RJ87P	.035	—	—	J	6°	F	—	850N	—	—	—	.5(20)
V8-440	RJ87P	.040	—	—	J	6°	D	—	750N	—	—	—	(38)
V8-440(22)	J11Y	.035	—	—	J	10°	D	—	750N	—	—	—	(33)
1976													
6-225 Std. Tr.	RBL13Y	.035	—	—	G	(39)	A	750(30)	—	—	—	(37)	—
6-225 Auto. Tr.	RBL13Y	.035	—	—	G	2°	A	—	750	—	—	—	(37)
V8-318(5)	RN12Y	.035	—	—	H	(36)	B	750	750	—	—	.3(19)	(37)
V8-318(8)	RN12Y	.035	—	—	H	2°(15)	B	—	900	—	—	—	.5(20)
V8-360(21)	RN12Y	.035	—	—	H	6°	B	—	700	—	—	—	.3(19)
V8-360(4)	RN12Y	.035	—	—	H	6°	B	—	750	—	—	—	2.0(19)
V8-360(22)	RN12Y	.035	—	—	H	2°	B	—	850	—	—	—	.5(20)
V8-400(21)	RJ13Y	.035	—	—	J	10°	F	—	700	—	—	—	.3(19)
V8-400(4)(24)	(38)	.035	—	—	J	6°	F	—	850	—	—	—	.5(20)
V8-400(4)(25)	RJ13Y	.035	—	—	J	8°	F	—	750	—	—	—	.5(19)
V8-440	RJ13Y	.035	—	—	J	8°	D	—	750	—	—	—	.3(19)
V8-440(22)	RJ11Y	.035	—	—	J	(40)	D	—	750	—	—	—	(32)

Continued

TUNE UP SPECIFICATIONS—Continued

The following specifications are published from the latest information available. This data should be used only in the absence of a decal affixed in the engine compartment.

OLD CAR SPECIFICATIONS: For 1946-68 Tune Up Specifications see main index.

★When using a timing light, disconnect vacuum hose or tube at distributor and plug opening in hose or tube so idle speed will not be affected.

●When checking compression, lowest cylinder must be within 80 percent of highest.

▲Before removing wires from distributor cap, determine location of the No. 1 wire in cap, as distributor position may have been altered from that shown at the end of this chart.

Year	Spark Plug		Distributor		Ignition Timing★			Carb. Adjustments					
	Type (7)	Gap Inch	Point Gap Inch	Dwell Angle Deg.	Firing Order Fig. ▲	Timing BTDC (1)	Mark Fig.	Hot Idle Speed		Air Fuel Ratio		Idle "CO" % (8)	
								Std. Trans.	Auto. Trans.(2)	Std. Trans.	Auto. Trans.	Std. Trans.	Auto. Trans.
PLYMOUTH													
1969													
6-170 Std. Tr.	N14Y	.035	.020	42–47	G	5° ATC	E	750(14)	—	14.2 to 1	—	—	—
6-170 Auto. Tr.	N14Y	.035	.020	42–47	G	2½° ATC	E	—	750N(14)	—	14.2 to 1	—	—
6-225	N14Y	.035	.020	42–47	G	TDC	E	650(14)	650N(14)	14.2 to 1	14.2 to 1	—	—
V8-273	N14Y	.035	.017	30–35	H	2½° ATC	F	700(14)	650N(14)	14.2 to 1	14.2 to 1	—	—
V8-318	N14Y	.035	.017	30–35	H	TDC	F	700(14)	650N(14)	14.2 to 1	14.2 to 1	—	—
V8-340 Std. Tr.	N9Y	.035	.017	(17)	H	TDC	F	750(14)	—	14.2 to 1	—	—	—
V8-340 Auto. Tr.	N9Y	.035	.017	30–35	H	5°	F	—	700N(14)	—	14.2 to 1	—	—
V8-383 Std. Tr.(21)	J14Y	.035	.017	30–35	J	TDC	F	700(14)	—	14.2 to 1	—	—	—
V8-383 Auto. Tr.(21)	J14Y	.035	.017	30–35	J	7½°	F	—	600N(14)	—	14.2 to 1	—	—
V8-383 Std. Tr.(4)	J11Y	.035	.017	30–35	J	TDC	F	700	—	14.2 to 1	—	—	—
V8-383 Auto. Tr.(4)	J11Y	.035	.017	30–35	J	5°	F	—	650N(14)	—	14.2 to 1	—	—
V8-383 Std. Tr.(23)	J11Y	.035	.017	(17)	J	TDC	F	700	—	14.2 to 1	—	—	—
V8-383 Auto. Tr.(23)	J11Y	.035	.017	(17)	J	5°	F	—	650N(14)	—	14.2 to 1	—	—
V8-440, 350 H.P.	J13Y	.035	.017	30–35	J	7½°	F	—	600N(14)	14.2 to 1	14.2 to 1	—	—
V8-440 Std. Tr.(22)	J11Y	.035	.017	(17)	J	TDC	F	700	—	14.2 to 1	—	—	—
V8-440 Auto. Tr.(22)	J11Y	.035	.017	30–35	J	5°	F	—	650N(14)	—	14.2 to 1	—	—
V8-440, 390 H.P.	J11Y	.035	.017	(17)	J	5°	F	900	900(14)	14.2 to 1	14.2 to 1	—	—
V8-426 Hemi	N10Y	.035	.017	(17)	J	TDC	F	750	750(14)	14.2 to 1	—	—	—
1970													
6-198 Std. Tr.	N14Y	.035	.020	41–46	G	2½°	C	750(6)	—	14.2 to 1	—	—	—
6-198 Auto. Tr.	N14Y	.035	.020	41–46	G	TDC	C	—	750(6)	—	14.2 to 1	—	—
6-225	N14Y	.035	.020	41–46	G	TDC	C	700(6)	650(6)	14.2 to 1	14.2 to 1	—	—
V8-318	N14Y	.035	.017	30–34	H	TDC	E	750(6)	700(6)	14.2 to 1	14.2 to 1	—	—
V8-340 Std. Tr.	N9Y	.035	.017	(17)	H	5°	E	950	—	14.2 to 1	—	—	—
V8-340 Auto. Tr.	N9Y	.035	.017	30–34	H	5°	E	—	900	—	14.2 to 1	—	—
V8-383 Std. Tr.(21)	J14Y	.035	.017	28–32	J	TDC	F	750	—	14.2 to 1	—	—	—
V8-383 Auto. Tr.(21)	J14Y	.035	.017	28–32	J	2½°	F	—	650	—	14.2 to 1	—	—
V8-383 Std. Tr.(4)	J11Y	.035	.017	28–32	J	TDC	F	750	—	14.2 to 1	—	—	—
V8-383 Auto. Tr.(4)	J11Y	.035	.017	28–32	J	2½°	F	—	750	—	14.2 to 1	—	—
V8-440, 350 H.P.	J13Y	.035	.017	28–32	J	5°	F	—	600	14.2 to 1	14.2 to 1	—	—
V8-440 Std. Tr.(22)	J11Y	.035	.017	28–32	J	TDC	F	900	—	14.2 to 1	—	—	—
V8-440 Auto. Tr.(22)	J11Y	.035	.017	28–32	J	2½°	F	—	800	—	14.2 to 1	—	—
V8-440, 390 H.P.	J11Y	.035	.017	(17)	J	5°	F	900	900	14.2 to 1	14.2 to 1	—	—
V8-426 Std. Tr.	N10Y	.035	.017	(17)	J	TDC	F	900	—	14.2 to 1	—	—	—
V8-426 Auto. Tr.	N10Y	.035	.017	(17)	J	5°	F	—	900	—	14.2 to 1	—	—
1971													
6-198	N14Y	.035	.020	41–46	G	2½°	C	800	800	14.2 to 1	14.2 to 1	—	—
6-225(24)	N14Y	.035	.020	41–46	G	TDC	C	750	750	14.2 to 1	14.2 to 1	—	—
6-225(25)	N14Y	.035	.020	41–46	G	2½°	C	750	750	14.2 to 1	14.2 to 1	—	—
V8-318	N14Y	.035	.017	30–34	H	TDC	E	750	700	14.2 to 1	14.2 to 1	—	—

Continued

TUNE UP SPECIFICATIONS—Continued

The following specifications are published from the latest information available. This data should be used only in the absence of a decal affixed in the engine compartment.

OLD CAR SPECIFICATIONS: For 1946–68 Tune Up Specifications see main index.

★When using a timing light, disconnect vacuum hose or tube at distributor and plug opening in tube or hose so idle speed will not be affected.

●When checking compression, lowest cylinder must be within 80 percent of highest.

▲Before removing wires from distributor cap, determide location of the No. 1 wire in cap, as distributor position may have been altered from that shown at the end of this chart.

Year	Spark Plug		Distributor		Ignition Timing★			Carb. Adjustments					
	Type (7)	Gap Inch	Point Gap Inch	Dwell Angle Deg.	Firing Order Fig. ▲	Timing BTDC (1)	Mark Fib.	Hot Idle Speed Std. Trans.	Hot Idle Speed Auto. Trans. (2)	Air Fuel Ratio Std. Trans.	Air Fuel Ratio Auto. Trans.	Idle "CO" ★ (3) Std. Trans.	Idle "CO" ★ (3) Auto. Trans.
PLYMOUTH—Continued													
1971													
V8-340 Std. Tr.(4)	N9Y	.035	.017	(17)	H	5°	E	900	—	14.2 to 1	—	—	—
V8-340 Auto. Tr.(4)	N9Y	.035	.017	30–34	H	5°	E	—	900	—	14.2 to 1	—	—
V8-340 Std. Tr.(26)	N9Y	.035	.017	(17)	H	2½°	E	1000	—	14.2 to 1	—	—	—
V8-340 Auto. Tr.(26)	N9Y	.035	.017	30–34	H	2½°	E	—	950	—	14.2 to 1	—	—
V8-340(3)	N9Y	.035	—	—	H	5°	E	900	—	14.2 to 1	14.2 to 1	—	—
V8-360	N13Y	.035	.017	30–34	H	2½°	E	750	700	14.2 to 1	14.2 to 1	—	—
V8-383 Std. Tr.(21)	J14Y	.035	.017	30–34	J	TDC	F	750	—	14.2 to 1	—	—	—
V8-383 Auto. Tr.(21)	J14Y	.035	.017	30–34	J	2½°	F	—	700	—	14.2 to 1	—	—
V8-383 Std. Tr.(4)	J11Y	.035	.017	30–34	J	TDC	F	750	—	14.2 to 1	—	—	—
V8-383 Auto. Tr.(4)	J11Y	.035	.017	30–34	J	2½°	F	—	700	—	14.2 to 1	—	—
V8-426 Std. Tr.	N10Y	.035	.017	(17)	J	TDC	F	900	—	14.2 to 1	—	—	—
V8-426 Auto. Tr.	N10Y	.035	.017	(17)	J	2½°	F	—	900	—	14.2 to 1	—	—
V8-440	J13Y	.035	.017	28½–32½	J	5°	F	—	900	14.2 to 1	14.2 to 1	—	—
V8-440 Std. Tr.(22)	J11Y	.035	.017	28½–32½	J	TDC	F	900	—	14.2 to 1	—	—	—
V8-440 Auto. Tr.(22)	J11Y	.035	.017	28½–32½	J	2½°	F	—	900	—	14.2 to 1	—	—
V8-440(26)	J11Y	.035	.017	(17)	J	5°	F	900	900	14.2 to 1	—	—	—
1972													
6-198	N14Y	.035	.020	41–46	G	2½°	C	800(28)	800N(28)	14.2 to 1	14.2 to 1	—	—
6-225	N14Y	.035	.020	41–46	G	TDC	C	750(28)	750N(28)	14.2 to 1	14.2 to 1	—	—
V8-318	N13Y	.035	.017	30–34	H	TDC	E	750	750N(28)	14.2 to 1	14.2 to 1	—	—
V8-340	N9Y	.035	—	—	H	2½°	E	900(29)	750N	14.2 to 1	14.2 to 1	—	—
V8-360	N13Y	.035	.017	30–34	H	TDC	E	—	750N	14.2 to 1	14.2 to 1	—	—
V8-400(21)	J13Y	.035	.019	28½–32½	J	5°	F	—	700N	14.2 to 1	14.2 to 1	—	—
V8-400 Std. Tr.(4)	J11Y	.035	—	—	J	2½°	F	900(30)	—	14.2 to 1	—	—	—
V8-400 Auto. Tr.(4)	J11Y	.035	—	—	J	10°(27)	F	—	750N	—	14.2 to 1	—	—
V8-440	J11Y	.035	.019	28½–32½	J	10°(27)	F	—	750N(28)	14.2 to 1	14.2 to 1	—	—
V8-440 Std. Tr.(22)	J11Y	.035	—	—	J	2½°	F	900(30)	—	14.2 to 1	—	—	—
V8-440 Auto. Tr.(22)	J11Y	.035	—	—	J	10°(27)	F	—	900N	—	14.2 to 1	—	—
V8-440(26)	J11Y	.035	—	—	J	2½°	F	—	900N	14.2 to 1	—	—	—
1973													
6-198	N14Y	.035	—	—	G	TDC	C	800	750N	14.2 to 1	14.2 to 1	—	—
6-225 Std. Trans.	N14Y	.035	—	—	G	2½°	C	750	—	14.2 to 1	—	—	—
6-225 Auto. Trans.	N14Y	.035	—	—	G	TDC	C	—	750N	—	14.2 to 1	—	—
8-318	N13Y	.035	—	—	H	TDC	E	750	750N	14.2 to 1	14.2 to 1	—	—
8-340 Std. Trans.	N12Y	.035	—	—	H	5°	E	900	—	14.2 to 1	—	—	—
8-340 Auto. Trans.	N12Y	.035	—	—	H	2½°(18)	E	—	750N	—	14.2 to 1	—	—
8-360	N13Y	.035	—	—	H	(10)	E	750	750N	14.2 to 1	14.2 to 1	—	—
8-400 2 Bar. Carb.	J13Y	.035	—	—	J	10°	F	700	700N	14.2 to 1	14.2 to 1	—	—
8-400 4 Bar. Carb.(31)	J11Y	.035	—	—	J	10°	F	900	—	14.2 to 1	—	—	—
8-400 4 Bar. Carb.(12)	J11Y	.035	—	—	J	7½°	F	—	750N	—	14.2 to 1	—	—
8-440 Std. Trans.	J11Y	.035	—	—	J	10°	F	800	—	14.2 to 1	—	—	—
8-440 Auto. Trans.	J11Y	.035	—	—	J	10°	F	—	700N	—	14.2 to 1	—	—

Continued

TUNE UP SPECIFICATIONS—Continued

The following specifications are published from the latest information available. This data should be used only in the absence of a decal affixed in the engine compartment.

OLD CAR SPECIFICATIONS: For 1946–68 Tune Up Specifications see main index.

★When using a timing light, disconnect vacuum hose or tube at distributor and plug opening in tube or hose so idle speed will not be affected.

●When checking compression, lowest cylinder must be within 80 percent of highest.

▲Before removing wires from distributor cap, determide location of the No. 1 wire in cap, as distributor position may have been altered from that shown at the end of this chart.

Year	Spark Plug		Distributor		Ignition Timing★			Carb. Adjustments					
								Hot Idle Speed		Air Fuel Ratio		Idle "CO" ★ ③	
	Type ⑦	Gap Inch	Point Gap Inch	Dwell Angle Deg.	Firing Order Fig. ▲	Timing BTDC ①	Mark Fib.	Std. Trans.	Auto. Trans.②	Std. Trans.	Auto. Trans.	Std. Trans.	Auto. Trans.
PLYMOUTH—Continued													
1974													
6-198	N14Y	.035	—	—	G	2½°	C	800	750	14.3 to 1	14.3 to 1	—	—
6-225	N14Y	.035	—	—	G	TDC	C	800	750	14.3 to 1	14.3 to 1	—	—
V8-318	N13Y	.035	—	—	H	TDC	E	750	750	14.3 to 1	14.3 to 1	—	—
V8-360	N12Y	.035	—	—	H	5°⑪	E	850	850	14.3 to 1	14.3 to 1	—	—
V8-400 Auto. Trans.㉑	J13Y	.035	—	—	J	⑨	F	—	750	—	14.3 to 1	—	—
V8-400 Auto. Trans.④	J13Y	.035	—	—	J	5°	F	—	750	—	14.3 to 1	—	—
V8-400 Std. Trans.㉒	J11Y	.035	—	—	J	5°	F	900	—	14.3 to 1	—	—	—
V8-400 Auto. Trans.⑯	J11Y	.035	—	—	J	5°⑬	F	—	850	—	14.3 to 1	—	—
V8-440	J11Y	.035	—	—	J	㉟	F	—	850	—	14.3 to 1	—	—
1975													
6-225	BL13Y	.035	—	—	G	TDC	A	800	750N	—	—	.3⑲	㉞
V8-318⑤	N13Y	.035	—	—	H	㊱	B	750	750N	—	—	.3⑲	㉝
V8-318	N13Y	.035	—	—	H	2 ⑮	B	—	900N	—	—	—	.5⑳
V8-360	N12Y	.035	—	—	H	6°	B	—	750N	—	—	—	㉝
V8-360㉒	N12Y	.035	—	—	H	2°	B	—	850N	—	—	—	.5⑳
V8-400㉑	J13Y	.035	—	—	J	10°	F	—	750N	—	—	—	.3⑲
V8-400④	J13Y	.035	—	—	J	8°	F	—	750N	—	—	—	㉜
V8-400㉒	RJ87P	.035	—	—	J	6°	F	—	850N	—	—	—	.5⑳
V8-440	RJ87P	.040	—	—	J	6°	D	—	750N	—	—	—	㉞
V8-440㉒	J11Y	.035	—	—	J	10°	D	—	750N	—	—	—	㉝
1976													
6-225 Std. Tr.	RBL13Y	.035	—	—	G	㊴	A	750㉚	—	—	—	㊲	—
6-225 Auto. Tr.	RBL13Y	.035	—	—	G	2°	A	—	750	—	—	—	㊲
V8-318⑤	RN12Y	.035	—	—	H	㊱	B	750	750	—	—	.3⑲	㊲
V8-318⑧	RN12Y	.035	—	—	H	2°⑮	B	—	900	—	—	—	.5⑳
V8-360㉑	RN12Y	.035	—	—	H	6°	B	—	700	—	—	—	.3⑲
V8-360④	RN12Y	.035	—	—	H	6°	B	—	750	—	—	—	2.0⑲
V8-360㉒	RN12Y	.035	—	—	H	2°	B	—	850	—	—	—	.5⑳
V8-400㉑	RJ13Y	.035	—	—	J	10°	F	—	700	—	—	—	.3⑲
V8-400④㉔	㊳	.035	—	—	J	6°	F	—	850	—	—	—	.5⑳
V8-400④㉕	RJ13Y	.035	—	—	J	8°	F	—	750	—	—	—	.5⑲
V8-440	RJ13Y	.035	—	—	J	8°	D	—	750	—	—	—	.3⑲
V8-440㉒	RJ11Y	.035	—	—	J	㊵	D	—	750	—	—	—	㉜

TUNE UP NOTES

(1)—BTDC: Before top dead center.
(2)—D: Drive. N: Neutral.
(3)—On 1974–76 models before adjusting idle "CO", disconnect A.I.R. pump outlet hose, if equipped.
(4)—Four barrel carburetor.
(5)—With catalytic converter.
(6)—Set idle speed with air conditioning compressor operating.
(7)—Champion.
(8)—With air pump.
(9)—Early production—Exc. sta. wag., 7½° BTDC; sta. wag., 5° BTDC. Late production—Exc. sta. wag., 10° BTDC; sta. wag., 7½° BTDC.
(10)—Dist. No. 3656780, TDC.
Dist. No. 3755336, 7½°.
Dist. No. 3755337 & 3755365, 5°.
(11)—Calif. V8-360 Hi Perf. Manual Trans. 2½° BTDC.
(12)—Auto. trans.
(13)—Exc. Calif. auto. trans. & Police; Calif. auto. trans. & Police, 2½°.
(14)—Adjust idle speed with headlights on. If air conditioned turn A/C switch to "Full On" position.
(15)—ATDC: after top dead center.
(16)—High performance & Police.
(17)—Each set of points 27–32°; total dwell both sets 37–42°.
(18)—Exc. Calif. late production; Calif. late production, TDC.
(19)—Measured ahead of catalytic converter.
(20)—Measured in tailpipe.
(21)—Two barrel carburetor.
(22)—High performance engine.
(23)—Formula "S" and Super Bee only.
(24)—Exc. California.
(25)—California only.
(26)—Three Carbs.
(27)—California vehicles with electronic ignition 5°.
(28)—California vehicles 700N.
(29)—California vehicles 850.
(30)—California vehicles 800.
(31)—Std. trans.
(32)—Exc. Calif., .3 (see note 19); Calif., .5 (see note 19).
(33)—Except Calif. .3 (see note 19), Calif. .5 (see note 19).
(34)—Except Calif. .3 (see note 19), Calif. 1.5 (see note 20).
(35)—Exc. Calif., 10° BTDC; Calif., 5° BTDC.
(36)—Exc. Calif., 2° BTDC; Calif., TDC.
(37)—Exc. Calif., .3 (see note 19); Calif., 1.0 (see note19).
(38)—Exc. high performance engine, RJ13Y; high performance engine, RJ87P.
(39)—Exc. Calif., 6° BTDC; Calif., 4° BTDC.
(40)—Exc. Calif., 10° BTDC; Calif., 8° BTDC.

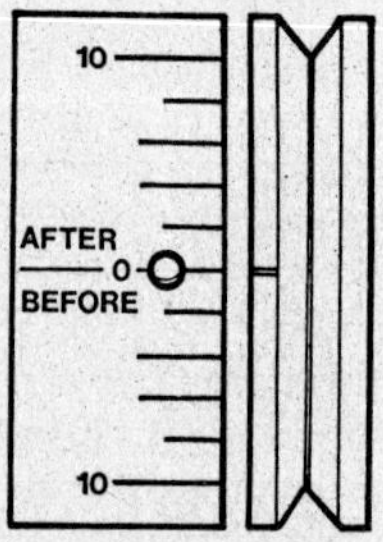

Fig. A

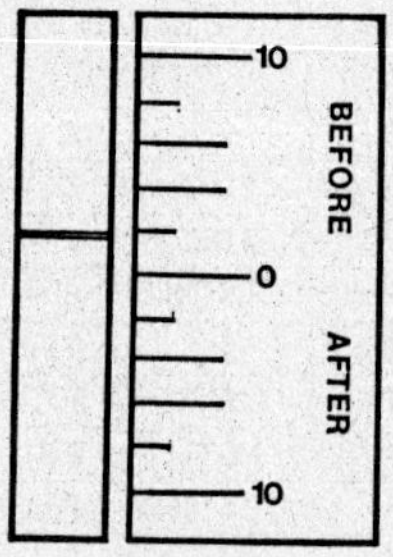

Fig. B

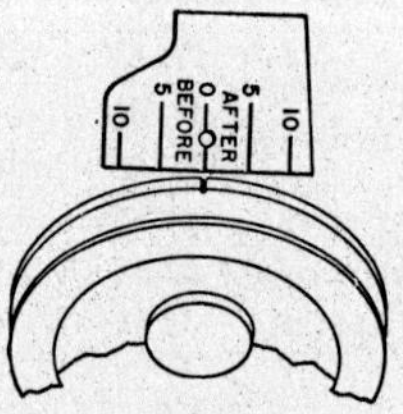

Fig. C

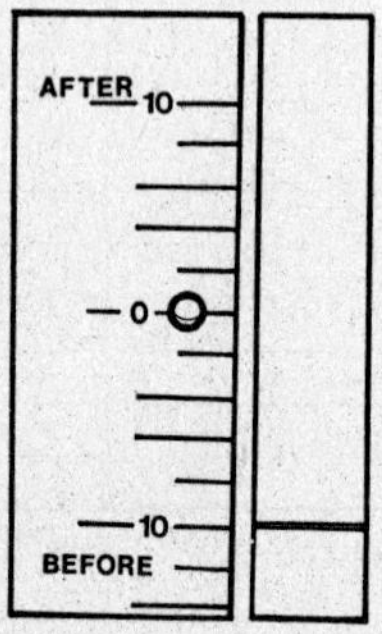

Fig. D

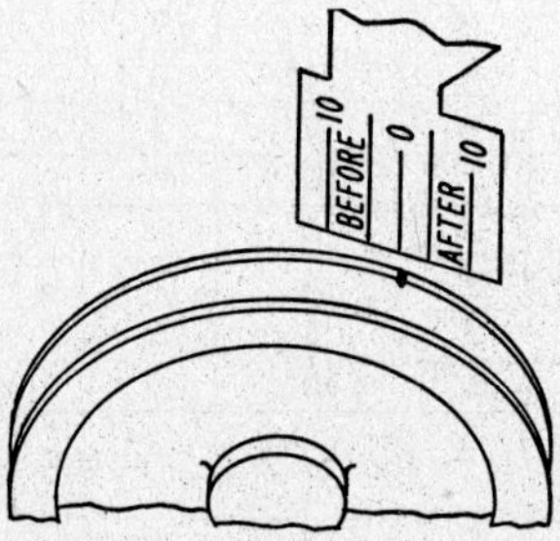

Fig. E

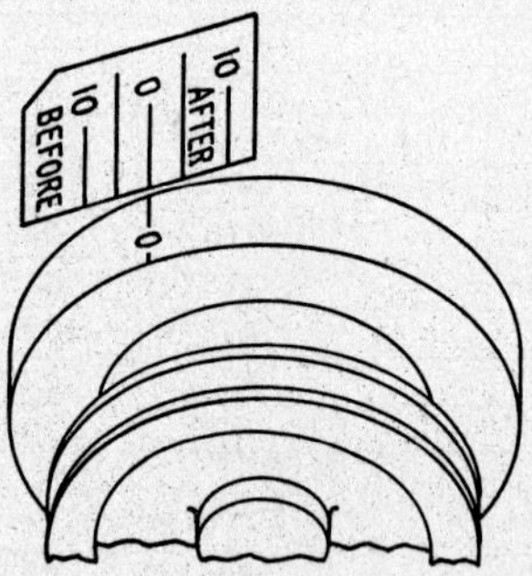

Fig. F

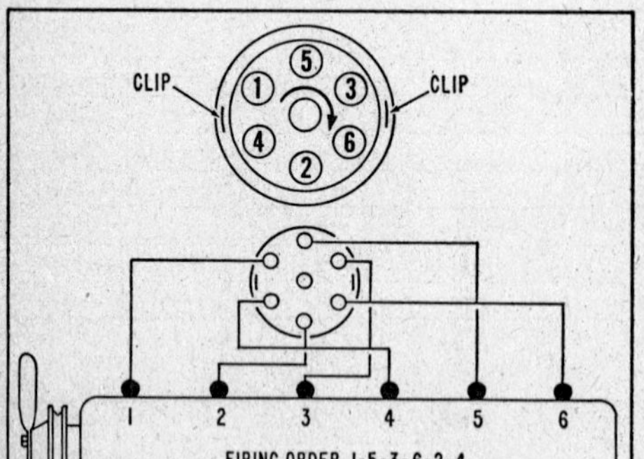

Fig. G

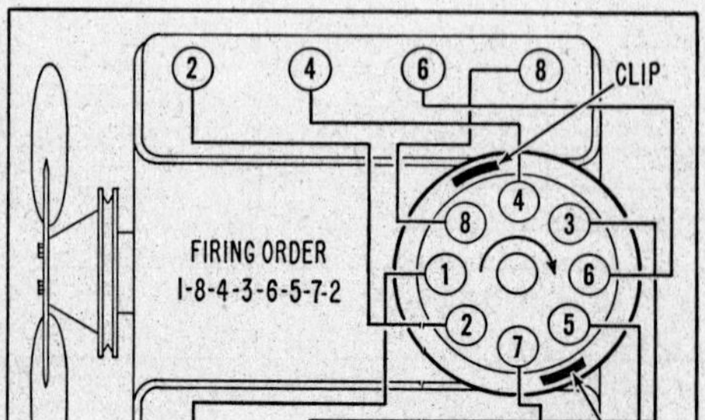

Fig. H

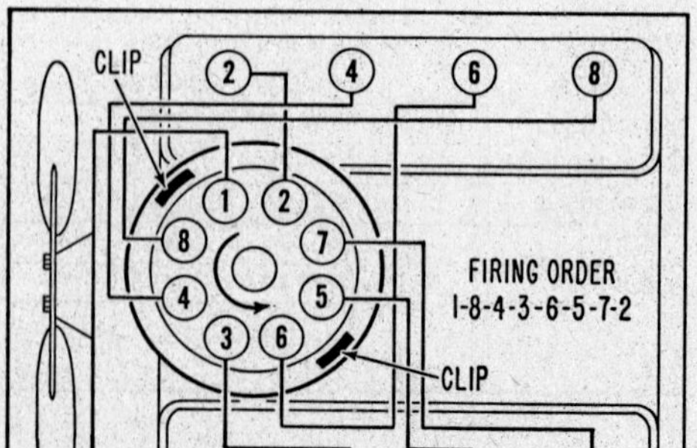

Fig. J

DISTRIBUTOR SPECIFICATIONS

★Note: If unit is checked on vehicle, double the RPM and degrees to get crankshaft figures.

Breaker arm spring tension—17–20.

Distributor Part No.①	Centrifugal Advance Degrees @ RPM of Distributor					Vacuum Advance		Distributor Retard
	Advance Starts	Intermediate Advance			Full Advance	Inches of Vacuum to Start Plunger	Max. Adv. Dist. Deg. @ Vacuum	Max. Retard Dist. Deg. @ Vacuum
CHRYSLER & IMPERIAL								
1969								
2875731	1–7.5 @ 550	10.5–12.5 @ 850	—	—	17 @ 2500	8	12 @ 15	—
2875742	1–9.5 @ 550	11–13 @ 800	—	—	23 @ 2350	8.5	13.5 @ 13.5	—
2875747	1–6 @ 500	8–10 @ 750	—	—	19 @ 2300	8.5	13.5 @ 13.5	—
2875758	2–7 @ 550	9–11 @ 850	—	—	16 @ 2400	10.5	12 @ 16	—
2875764	0–3.5 @ 500	6.5–8.5 @ 850	—	—	14 @ 2250	11.5	13.5 @ 16	—
2875772	1–8.5 @ 550	11.5–13.5 @ 850	—	—	19 @ 2500	10.5	12 @ 16	—
1970								
3438219	0.5–3.7 @ 650	5.7–7.7 @ 900	—	—	14 @ 2300	10.5	12 @ 15.5	2¾③
3438222	0–4.6 @ 600	5.6–7.6 @ 800	—	—	12 @ 2300	10.5	12 @ 15.5	2¾③
3438231	0–3.8 @ 550	7.5–9.5 @ 850	—	—	16 @ 2200	7.5	11.8 @ 12	2¾③
3438233	0–4.6 @ 600	5.6–7.6 @ 800	—	—	12 @ 2300	10.5	12 @ 15.5	2¾③
1971								
3438534	1–4.5 @ 600	5–7 @ 800	—	—	14 @ 2000	9	10 @ 15	—
3438544	0.5–3.5 @ 700	5–7.5 @ 900	—	—	14 @ 2000	9	10 @ 15	—
3438559	0.5–3.5 @ 700	3.5–5.5 @ 900	—	—	12 @ 2400	12	10 @ 16	—
3438572	0–3.5 @ 600	4.5–6.5 @ 800	—	—	10 @ 2200	10.5	12 @ 15.5	—
3438690	0.5–3.5 @ 650	6–8 @ 900	—	—	14 @ 2400	10.5	10 @ 15	—
3438694	0.5–4 @ 700	5–7 @ 900	—	—	10 @ 2200	10.5	12 @ 15.5	—
1972								
3656272	1–6 @ 550	8.5–10.5 @ 800	—	—	16 @ 2100	9.5	10.5 @ 15	—
3656329	1–5 @ 650	8–10 @ 900	—	—	14 @ 2000	10.5	10.5 @ 15.5	—
3656332	0.5–4 @ 650	6–8 @ 950	—	—	12 @ 2000	10.5	10.5 @ 15.5	—
3656335	1–5 @ 650	8–10 @ 900	—	—	15 @ 2250	10.5	10.5 @ 15.5	—
3656338	0.5–4 @ 650	6–8 @ 950	—	—	12.5 @ 2500	10.5	10.5 @ 15.5	—
3656341	0.5–4 @ 650	6–8 @ 950	—	—	12.5 @ 2500	10.5	10.5 @ 15.5	—
3656344	0.5–4 @ 650	6–8 @ 950	—	—	12 @ 2000	10.5	10.5 @ 15.5	—
3656347	1–5 @ 650	8–10 @ 900	—	—	15 @ 2250	10.5	10.5 @ 15.5	—
3656350	1–5 @ 650	8–10 @ 900	—	—	14 @ 2000	10.5	10.5 @ 15.5	—
3656429	1–6 @ 550	8.5–10.5 @ 800	—	—	15 @ 1900	9	10 @ 14	—
3656435	0–4 @ 600	9–11 @ 1100	—	—	15 @ 1900	9	10 @ 14	—
3656593	1–4.5 @ 550	7.5–9.5 @ 800	—	—	16 @ 1900	10.5	10.5 @ 15.5	—
3656596	1–4.5 @ 550	7.5–9.5 @ 800	—	—	15 @ 1800	10.5	10.5 @ 15.5	—
1973								
3656791	0.5–3.5 @ 550	8–10.5 @ 850	—	—	15 @ 1950	10.5	10.5 @ 15.5	—
3656802	0.5–4.0 @ 650	6–8 @ 950	—	—	12.5 @ 2500	10.5	10.5 @ 15.5	—
3755157	0.5–4.0 @ 650	6–8 @ 950	—	—	12.5 @ 2500	10.5	10.5 @ 15.5	—
1974								
3755518	0.5–3.5 @ 650	5.5–8 @ 900	—	—	12 @ 2000	8	10 @ 14	—
3755522	0.5–3 @ 650	4–6.5 @ 900	—	—	10 @ 2000	8	10 @ 14	—
3755681	1–4.5 @ 650	8.5–11 @ 950	—	—	16 @ 2150	8	10 @ 14	—

Continued

DISTRIBUTOR SPECIFICATIONS—Continued

★Note: If unit is checked on vehicle, double the RPM and degrees to get crankshaft figures.

Breaker arm spring tension—17-20.

Distributor Part No.①	Centrifugal Advance Degrees @ RPM of Distributor					Vacuum Advance		Distributor Retard
	Advance Starts	Intermediate Advance			Full Advance	Inches of Vacuum to Start Plunger	Max. Adv. Dist. Deg. @ Vacuum	Max. Retard Dist. Deg. @ Vacuum
CHRYSLER & IMPERIAL—Continued								
1975-76								
3874090	1.5–5.5 @ 550	5.5–8 @ 700	—	—	14 @ 2200	7	12 @ 12.5	—
3874101	.5–3 @ 600	6–8.5 @ 950	—	—	12 @ 2000	8	11 @ 14	—
3874110	1–3.5 @ 600	6–8.5 @ 900	—	—	12 @ 2000	8	11 @ 14	—
3874115	1–3.5 @ 600	6–8.5 @ 900	—	—	12 @ 2000	7	12 @ 12.5	—
3874119	1–3.5 @ 600	8–10 @ 900	—	—	11.5 @ 2400	7	12 @ 12.5	—
3874173	1–3.5 @ 600	3.5–6 @ 750	—	—	10 @ 2000	8	11 @ 14	—
3874298	.5–3.5 @ 700	7–10 @ 1100	—	—	16 @ 2100	9	12.5 @ 15.5	—
3874596	.5–4 @ 600	6–9 @ 850	—	—	10.9 @ 2000	8	11 @ 14	—
3874754	0–4 @ 500	4.5–7 @ 650	—	—	13.5 @ 2400	8	12 @ 13.5	—

①—Stamped on distributor housing. ②—Solenoid controlled.

Distributor Part No.①	Advance Starts	Intermediate Advance			Full Advance	Inches of Vacuum to Start Plunger	Max. Adv. Dist. Deg. @ Vacuum	Max. Retard Dist. Deg. @ Vacuum
DODGE & PLYMOUTH								
1969								
2875140③	2.5–9.5 @ 650	10–12 @ 850	—	—	16 @ 1550	10	11 @ 15	—
2875715③	1–7 @ 550	12–14 @ 900	—	—	19 @ 2500	8	12 @ 15	—
2875731②	1–7.5 @ 550	10.5–12.5 @ 850	—	—	17 @ 2500	8	12 @ 15	—
2875742②	1–9.5 @ 550	11–13 @ 800	—	—	23 @ 2350	8.5	13.5 @ 13.5	—
2875747②	1–6 @ 500	8–10 @ 750	—	—	19 @ 2300	8.5	13.5 @ 13.5	—
2875750②	1–7 @ 550	12–14 @ 900	—	—	19 @ 2500	8	12 @ 15	—
2875758②	2–7 @ 550	9–11 @ 850	—	—	16 @ 2400	10.5	12 @ 16	—
2875764②	0.5–3.5 @ 500	6.5–8.5 @ 850	—	—	14 @ 2250	11.5	13.5 @ 16	—
2875772②	1–8.5 @ 550	11.5–13.5 @ 850	—	—	19 @ 2500	10.5	12 @ 16	—
2875779②	2–6 @ 600	7–9 @ 900	—	—	12 @ 2000	8	10 @ 10.5	—
2875782②	1–6.5 @ 550	9.5–11.5 @ 900	—	—	14 @ 1800	8	10 @ 10.5	—
2875790②	2–8.5 @ 550	11.5–13.5 @ 900	—	—	16 @ 1900	8.5	13.5 @ 13.5	—
2875796②	1.5–6.5 @ 550	7.5–9.5 @ 850	—	—	19 @ 2400	10.5	10¾ @ 15	—
2875813②	2–6 @ 600	12.5–14.5 @ 1100	—	—	19 @ 2200	7.5	8.5 @ 10	—
2875822②	1–5 @ 550	9.5–11.5 @ 1000	—	—	14 @ 2000	10	7¾ @ 16	—
2875826②	1–5 @ 550	9.5–11.5 @ 1000	—	—	14 @ 2000	7	7¾ @ 10	—
2875846②	1–7.5 @ 550	10.5–12.5 @ 850	—	—	17 @ 2500	8	12 @ 15	—
2875855②	1–8 @ 550	9–11 @ 900	—	—	16 @ 2200	7.5	8 @ 10	—
1970								
2875822②	1–5 @ 550	9.2–11.2 @ 900	—	—	14 @ 2000	10	7¾ @ 15	—
2875826②	1–5 @ 550	9.2–11.2 @ 900	—	—	14 @ 2000	7	7¾ @ 10	—
2875982④	0–5.3 @ 600	9–11 @ 850	—	—	14 @ 2400	11	12.5 @ 15.5	—
2875987④	0–4.5 @ 650	12.2–14.2 @ 1050	—	—	16 @ 1600	9	9.2 @ 13.5	—
2875989④	0–4.2 @ 600	9.7–11.7 @ 950	—	—	13.5 @ 1600	9	9.2 @ 13.5	—
3438219②	0.5–3.7 @ 650	5.7–7.7 @ 900	—	—	14 @ 2300	10.5	12 @ 15.5	—

Continued

DISTRIBUTOR SPECIFICATIONS—Continued

★**Note: If unit is checked on vehicle, double the RPM and degrees to get crankshaft figures.**

Breaker arm spring tension—17-20.

Distributor Part No.①	Centrifugal Advance Degrees @ RPM of Distributor					Vacuum Advance		Distributor Retard
	Advance Starts	Intermediate Advance			Full Advance	Inches of Vacuum to Start Plunger	Max. Adv. Dist. Deg. @ Vacuum	Max. Retard Dist. Deg. @ Vacuum
DODGE & PLYMOUTH—Continued								
1970								
3438222②	0–4.6 @ 600	5.6–7.6 @ 800	—	—	12 @ 2300	10.5	12 @ 15.5	—
3438225②	1–6 @ 550	8.5–10.5 @ 800	—	—	14 @ 2100	12	10¾ @ 15	—
3438231②	0–3.8 @ 550	7.5–9.5 @ 850	—	—	16 @ 2200	7.5	11.8 @ 12	—
3438233②	0–4.6 @ 600	5.6–7.6 @ 800	—	—	12 @ 2300	10.5	12 @ 15.5	—
3438237②	0–7.5 @ 525	9.3–11.3 @ 750	—	—	14 @ 1600	7	7¾ @ 10	—
3438255②	1–6 @ 550	8.5–10.5 @ 800	—	—	16 @ 2100	10.5	10¾ @ 15	—
3438314④	0–4.5 @ 650	9–11 @ 950	—	—	14 @ 2400	11	12.5 @ 15.5	—
3438317④	1.5–6.5 @ 700	—	—	—	10 @ 900	7.7	10 @ 10.5	—
3438325②	1.5–5.5 @ 650	—	—	—	10 @ 850	7.7	10 @ 10.5	—
1971								
2875822②	1–5 @ 550	9.2–11.2 @ 900	—	—	14 @ 2000	10	7¾ @ 15	—
2875826②	1–5 @ 550	9.2–11.2 @ 900	—	—	14 @ 2000	7	7¾ @ 10	—
2875987④	0–4.5 @ 650	12.2–14.2 @ 1050	—	—	16 @ 1600	9	9.2 @ 13.5	—
3438225②	1–6 @ 550	8.5–10.5 @ 800	—	—	16 @ 2100	12	10¾ @ 15	—
3438255②	1–6 @ 550	8.5–10.5 @ 800	—	—	16 @ 2100	10.5	10¾ @ 15	—
3438422②	1–6 @ 550	8.5–10.5 @ 800	—	—	16 @ 2100	9.5	10.5 @ 15	—
3438440②	0.5–3.5 @ 800	8–10.5 @ 1200	—	—	14 @ 2200	10	7¾ @ 15	—
3438442②	0.5–3.5 @ 800	8–10.5 @ 1200	—	—	14 @ 2200	7	7¾ @ 10	—
3438453②	0–4 @ 600	8–10.5 @ 1000	—	—	16 @ 2100	9.5	10.5 @ 15	—
3438509②	1–6.5 @ 550	8–10 @ 750	—	—	14 @ 2000	7	8.5 @ 11	—
3438517②	0.5–4 @ 600	9.5–12 @ 1000	—	—	12 @ 2100	9	10 @ 12.5	—
3438522④	0.5–4 @ 650	5.5–8.5 @ 850	—	—	12 @ 2000	7	10 @ 11	—
3438524②	2–6 @ 700	8.5–10.5 @ 900	—	—	14 @ 2000	7	7¾ @ 10	—
3438534②	1–4.5 @ 600	5–7 @ 800	—	—	14 @ 2000	9	10 @ 15	—
3438544②	0.5–3.5 @ 700	5–7.5 @ 900	—	—	14 @ 2000	9	10 @ 15	—
3438559②	0.5–3.5 @ 700	3.5–5.5 @ 900	—	—	12 @ 2400	12	10 @ 16	—
3438572②	0–3.5 @ 600	4.5–6.5 @ 800	—	—	10 @ 2200	10.5	12 @ 15.5	—
3438577④	1–3 @ 700	4.5–6.5 @ 950	—	—	6.5 @ 2200	10.5	10 @ 15	—
3438579④	0–4 @ 600	9.5–12.5 @ 1250	—	—	15 @ 1600	9	9.2 @ 13.5	—
3438615④	0.5–4 @ 700	9–11 @ 1100	—	—	15 @ 2200	7	10 @ 11	—
3438617②	0.5–4 @ 700	9–11 @ 1100	—	—	15 @ 2200	9	10 @ 12.5	—
3438690②	0.5–3.5 @ 650	6–8 @ 900	—	—	14 @ 2400	10.5	10 @ 15	—
3438694②	0.5–4 @ 700	5–7 @ 900	—	—	10 @ 2200	10.5	12 @ 15.5	—
3656151⑤	0.5–4 @ 650	5.5–8.5 @ 850	—	—	12 @ 2000	7	10 @ 11	—
1972								
3656237②	0–5 @ 550	8–10 @ 800	—	—	14 @ 2000	7	8.5 @ 11	—
3656243②	1–4.5 @ 650	8.5–10.5 @ 950	—	—	14 @ 2000	7	8.5 @ 11	—
3656252②	0.5–4.5 @ 550	9–11 @ 900	—	—	14 @ 2000	10	8 @ 15	—
3656257②	0.5–4.5 @ 550	9–11 @ 900	—	—	14 @ 2000	7	7.5 @ 9.5	—
3656260②	0.5–3.5 @ 700	9.5–11.5 @ 1150	—	—	14 @ 2000	10	8 @ 15	—
3656266②	0.5–3.5 @ 700	9.5–11.5 @ 1150	—	—	14 @ 2000	7	7.5 @ 9.5	—
3656272②	1–6 @ 550	8.5–10.5 @ 800	—	—	16 @ 2100	9.5	10.5 @ 15	—
3656275②	0–4 @ 600	9–11 @ 1000	—	—	16 @ 2100	9.5	10.5 @ 15	—
3656278②	1–5 @ 650	12–14 @ 1050	—	—	13.5 @ 2400	9	10 @ 12.5	—

Continued

DISTRIBUTOR SPECIFICATIONS—Continued

★**Note: If unit is checked on vehicle, double the RPM and degrees to get crankshaft figures.**

Distributor Part No.[1]	Centrifugal Advance Degrees @ RPM of Distributor					Vacuum Advance		Distributor Retard
	Advance Starts	Intermediate Advance			Full Advance	Inches of Vacuum to Start Plunger	Max. Adv. Dist. Deg. @ Vacuum	Max. Retard Dist. Deg. @ Vacuum
DODGE AND PLYMOUTH—Continued								
1972								
3656329[2]	1–5 @ 650	8–10 @ 900	—	—	14 @ 2000	10.5	10.5 @ 15.5	—
3656332[2]	0.5–4 @ 650	6–8 @ 950	—	—	12 @ 2000	10.5	10.5 @ 15.5	—
3656335	1–5 @ 650	8–10 @ 900	—	—	15 @ 2250	10.5	10.5 @ 15.5	—
3656338	0.5–4 @ 650	6–8 @ 950	—	—	12.5 @ 2500	10.5	10.5 @ 15.5	—
3656341	0.5–4 @ 650	6–8 @ 950	—	—	12.5 @ 2500	10.5	10.5 @ 15.5	—
3656344[3]	0.5–4 @ 650	6–8 @ 950	—	—	12 @ 2000	10.5	10.5 @ 15.5	—
3656347	1–5 @ 650	8–10 @ 900	—	—	15 @ 2250	10.5	10.5 @ 15.5	—
3656350	1–5 @ 650	8–10 @ 900	—	—	14 @ 2000	10.5	10.5 @ 15.5	—
3656390	1–6 @ 550	8.5–10.5 @ 800	—	—	16 @ 2100	12	10.5 @ 15	—
3656429	1–6 @ 550	8.5–10.5 @ 800	—	—	15 @ 1900	9	10 @ 14	—
3656435	0–4 @ 600	9–11 @ 1100	—	—	15 @ 1900	9	10 @ 14	—
3656587	1–6 @ 550	8.5–10.5 @ 800	—	—	16 @ 2100	12	10 @ 14.5	—
3656593	1–4.5 @ 550	7.5–9.5 @ 800	—	—	16 @ 1900	10.5	10.5 @ 15.5	—
3656596	1–4.5 @ 550	7.5–9.5 @ 800	—	—	15 @ 1800	10.5	10.5 @ 15.5	—
1973								
3755037	0.5–4.5 @ 550	9–11 @ 900	—	—	14 @ 2000	9.5	10 @ 15	—
3755042	0.5–4.5 @ 550	9–11 @ 900	—	—	14 @ 2000	7	10 @ 11	—
3755157	0.5–4.0 @ 650	6–8 @ 950	—	—	12.5 @ 2500	10.5	10.5 @ 15.5	—
3755308	0–2.5 @ 600	8–10.5 @ 1000	—	—	15.5 @ 2150	10.5	10.5 @ 15.5	—
3755336	0–2.5 @ 600	6.5–8.5 @ 900	—	—	12 @ 2000	9	12.5 @ 15	—
3755337	0–2.5 @ 600	6.5–8.5 @ 900	—	—	12 @ 2000	9	12.5 @ 15	—
3755365	0.5–4.0 @ 550	8–10 @ 800	—	—	14 @ 2000	9	12.5 @ 15	—
3656859	0–5 @ 550	8–10 @ 800	—	—	14 @ 2000	7	10 @ 11	—
3656763	1–4.5 @ 525	6.5–8.5 @ 750	—	—	15 @ 1900	9	12.5 @ 15	—
3656771	0.5–3.5 @ 600	8.5–11 @ 1000	—	—	13 @ 2000	9	10 @ 12.5	—
3656780	0.5–3.5 @ 500	10–12.5 @ 800	—	—	15 @ 1750	9	12.5 @ 15	—
3656791	0.5–3.5 @ 550	8–10.5 @ 850	—	—	15 @ 1950	10.5	10.5 @ 15.5	—
3656802	0.5–4.0 @ 650	6–8 @ 950	—	—	12.5 @ 2500	10.5	10.5 @ 15.5	—
1974								
3755037	1–4 @ 550	9–11.5 @ 900	—	—	14 @ 2000	9	8.5 @ 15.5	—
3755042	1–4 @ 550	9–11.5 @ 900	—	—	14 @ 2000	7	8.5 @ 11.5	—
3755467	0.5–3.5 @ 550	7–9.5 @ 950	—	—	14 @ 2000	9	8.5 @ 15.5	—
3755470	0.5–3.5 @ 550	7–9.5 @ 950	—	—	14 @ 2000	7	8.5 @ 11.5	—
3755475	0.5–4 @ 550	8–11.5 @ 800	—	—	14 @ 2000	7	11 @ 12.5	—
3755486	1–5 @ 600	8–10 @ 850	—	—	14 @ 2000	8	11 @ 13.5	—
3755503	0.5–4 @ 550	8–10.5 @ 850	—	—	16 @ 2200	8	10 @ 14	—
3755508	1–4.5 @ 650	8–10.5 @ 950	—	—	14 @ 2000	8	10¼ @ 16	—
3755512	0.5–3.5 @ 650	5.5–8 @ 900	—	—	12 @ 2000	8	10 @ 14	—
3755518	0.5–3.5 @ 650	5.5–8 @ 900	—	—	12 @ 2000	8	10 @ 14	—
3755522	0.5–3 @ 650	4–6.5 @ 900	—	—	10 @ 2000	8	10 @ 14	—
3755681	1–4.5 @ 650	8.5–11 @ 950	—	—	16 @ 2150	8	10 @ 14	—
3755686	0.5–4 @ 550	8–10.5 @ 850	—	—	16 @ 2200	8	10¼ @ 16	—
3656763	1.5–5.5 @ 550	6–8.5 @ 700	—	—	16 @ 2100	9	11 @ 15.5	—
3656859	0.5–4 @ 550	8–10.5 @ 800	—	—	14 @ 2000	7	8.5 @ 11.5	—

Continued

DISTRIBUTOR SPECIFICATIONS—Continued

★Note: If unit is checked on vehicle, double the RPM and degrees to get crankshaft figures.

Distributor Part No.①	Centrifugal Advance Degrees @ RPM of Distributor					Vacuum Advance		Distributor Retard
	Advance Starts	Intermediate Advance			Full Advance	Inches of Vacuum to Start Plunger	Max. Adv. Dist. Deg. @ Vacuum	Max. Retard Dist. Deg. @ Vacuum
DODGE & PLYMOUTH—Continued								
1975-76								
3874082	1-4.5 @ 600	8-10.5 @ 900	—	—	14 @ 2200	7	10 @ 11.5	—
3874090	1.5-5.5 @ 550	5.5-8 @ 700	—	—	14 @ 2200	7	12 @ 12.5	—
3874097	1-4.5 @ 650	8.5-11 @ 950	—	—	16 @ 2150	7	12 @ 12.5	—
3874101	.5-3 @ 600	6-8.5 @ 950	—	—	12 @ 2000	8	11 @ 14	—
3874110	1-3.5 @ 600	6-8.5 @ 900	—	—	12 @ 2000	8	11 @ 14	—
3874115	1-3.5 @ 600	6-8.5 @ 900	—	—	12 @ 2000	7	12 @ 12.5	—
3874173	1-3.5 @ 600	3.5-6 @ 750	—	—	10 @ 2000	8	11 @ 14	—
3874298	.5-3.5 @ 700	7-10 @ 1100	—	—	16 @ 2100	9	12.5 @ 15.5	—
1976								
3874596	.5-4.0 @ 600	6-9 @ 850	—	—	10.9 @ 2000	8	11 @ 14	—
3874598	1-4.5 @ 600	6.5-8.5 @ 800	—	—	11.5 @ 2300	7	10 @ 11.5	—
3874714	1-4.5 @ 450	4-6.5 @ 650	—	—	9 @ 2500	7	10 @ 11.5	—
3874754	0-4 @ 500	4.5-7 @ 650	—	—	13.5 @ 2400	8	12 @ 13.5	—

①—Stamped on distributor housing.
②—Breaker arm spring tension—17-20.
③—California vehicles.
④—Breaker arm spring tension—17-21½.
⑤—Electronic ignition.

VALVE SPECIFICATIONS

Year	Model	Valve Lash		Valve Angles		Valve Spring Installed Height	Valve Spring Pressure Lbs. @ In.	Stem Clearance		Stem Diameter	
		Int.	Exh.	Seat	Face			Intake	Exhaust	Intake	Exhaust
CHRYSLER & IMPERIAL											
1969–70	8-383, 290 H.P.	Hydraulic①		45	45	1.86	200 @ 1.43	.001–.003	.002–.004	.372–.373	.371–.372
	8-383, 330 H.P.	Hydraulic①		45	45	1.86	225 @ 1.43	.001–.003	.002–.004	.372–.373	.371–.372
	8-440, 350 H.P.	Hydraulic①		45	45	1.86	200 @ 1.43	.001–.003	.002–.004	.372–.373	.371–.372
	8-440, 375 H.P.	Hydraulic①		45	45	1.86	230 @ 1.41	.001–.003	.002–.004	.372–.373	.371–.372
1971	8-383, 270 H.P.	Hydraulic①		45	45	1 55/64	200 @ 1 7/16	.001–.003	.002–.004	.372–.373	.371–.372
	8-383, 300 H.P.	Hydraulic①		45	45	1 55/64	246 @ 1 23/64	.001–.003	.002–.004	.372–.373	.371–.372
	8-400	Hydraulic①		45	45	1 55/64	200 @ 1 7/16	.001–.003	.002–.004	.372–.373	.371–.372
	8-440, 335 H.P.	Hydraulic①		45	45	1 55/64	200 @ 1 7/16	.001–.003	.002–.004	.372–.373	.371–.372
	8-440, 370 H.P.	Hydraulic①		45	45	1 55/64	246 @ 1 23/64	.001–.003	.002–.004	.372–.373	.371–.372
1972	8-360	Hydraulic①		45	45	1.65	195 @ 1.24	.001–.003	.002–.004	.372–.373	.371–.372
1972–74	8-400, 440	Hydraulic①		45	45	1.86	200 @ 1.42	.001–.003	.002–.004	.372–.373	.371–.372
1975–76	V8-318	Hydraulic①		45	②	1 11/16	177 @ 1 5/16	.001–.003	.002–.004	.372–.373	.371–.372
	V8-360	Hydraulic①		45	②	1 11/16	⑧	.001–.003	.002–.004	.372–.373	.371–.372
	V8-400, 440	Hydraulic①		45	45	1 55/64	200 @ 1 7/16	.0011–.0028	④	.3723–.373	⑥
	V8-400, 440③	Hydraulic①		45	45	1 55/64	246 @ 1 23/64	.0016–.0033	⑤	.3718–.3725	⑦
DODGE & PLYMOUTH											
1969–70	6-170, 198, 225	.010H	.020H	45	②	1 11/16	145 @ 1 5/16	.001–.003	.002–.004	.372–.373	.371–.372
	8-273, 318	Hydraulic①		45	②	1 11/16	177 @ 1 5/16	.001–.003	.002–.004	.372–.373	.371–.372
	8-340	Hydraulic①		45	②	1 11/16	242 @ 1 7/32	.001–.003	.002–.004	.372–.373	.371–.372
	8-383, 290 H.P.	Hydraulic①		45	45	1 57/64	200 @ 1 7/16	.001–.003	.002–.004	.372–.373	.371–.372
	8-383, 330 H.P.	Hydraulic①		45	45	1 57/64	246 @ 1 23/64	.001–.003	.002–.004	.372–.373	.371–.372
	8-383, 335 H.P.	Hydraulic①		45	45	1 57/64	246 @ 1 23/64	.001–.003	.002–.004	.372–.373	.371–.372
	8-440, 350 H.P.	Hydraulic①		45	45	1 57/64	200 @ 1 7/16	.001–.003	.002–.004	.372–.373	.371–.372
	8-440, 375 H.P.	Hydraulic①		45	45	1 57/64	246 @ 1 23/64	.001–.003	.002–.004	.372–.373	.371–.372
	8-440, 390 H.P.	Hydraulic①		45	45	1 57/64	310 @ 1 3/8	.001–.003	.002–.004	.372–.373	.371–.372
	8-426	.028C	.032C	45	45	1 57/64	280 @ 1 3/8	.001–.003	.002–.004	.372–.373	.371–.372
1971	6-198, 225	.010H	.020H	45	②	1 11/16	145 @ 1 5/16	.001–.003	.002–.004	.372–.373	.371–.372
	8-318	Hydraulic①		45	②	1 11/16	177 @ 1 5/16	.001–.003	.002–.004	.372–.373	.371–.372
	8-340	Hydraulic①		45	②	1 11/16	238 @ 1 5/16	.001–.003	.002–.004	.372–.373	.371–.372
	8-360	Hydraulic①		45	②	1 11/16	177 @ 1 5/16	.001–.003	.002–.004	.372–.373	.371–.372
	8-383, 2 B. Carb.	Hydraulic①		45	45	1 55/64	200 @ 1 7/16	.001–.003	.002–.004	.372–.373	.371–.372
	8-383, 4 B. Carb.	Hydraulic①		45	45	1 55/64	246 @ 1 23/64	.001–.003	.002–.004	.372–.373	.371–.372
	8-440	Hydraulic①		45	45	1 55/64	200 @ 1 7/16	.001–.003	.002–.004	.372–.373	.371–.372
	8-440, Hi Perf.	Hydraulic①		45	45	1 55/64	246 @ 1 23/64	.001–.003	.002–.004	.372–.373	.371–.372
	8-440, 3 Carbs.	Hydraulic①		45	45	1 55/64	246 @ 1 23/64	.001–.003	.002–.004	.372–.373	.371–.372
	8-426	Hydraulic①		45	45	1 55/64	310 @ 1 3/8	.002–.004	.003–.005	.3085–.3095	.3075–.3085
1972	6-198, 225	.010H	.020H	45	②	1.65	160 @ 1.24	.001–.003	.002–.004	.372–.373	.371–.372
	8-318	Hydraulic①		45	②	1.65	189 @ 1.28	.001–.003	.002–.004	.372–.373	.371–.372
	8-340	Hydraulic①		45	②	1.65	238 @ 1.22	.001–.003	.002–.004	.372–.373	.371–.372
	8-360	Hydraulic①		45	②	1.65	195 @ 1.24	.001–.003	.002–.004	.372–.373	.371–.372
	8-400, 2 B. Carb.	Hydraulic①		45	45	1.86	200 @ 1.42	.001–.003	.002–.004	.372–.373	.371–.372
	8-400, 4 B. Carb.	Hydraulic①		45	45	1.86	246 @ 1.36	.001–.003	.002–.004	.372–.373	.371–.372
	8-440	Hydraulic①		45	45	1.86	200 @ 1.42	.001–.003	.002–.004	.372–.373	.371–.372
	8-440, Hi Perf.	Hydraulic①		45	45	1.86	246 @ 1.36	.001–.003	.002–.004	.372–.373	.371–.372
	8-440, 3 Carbs.	Hydraulic①		45	45	1.86	246 @ 1.36	.001–.003	.002–.004	.372–.373	.371–.372

Continued

VALVE SPECIFICATIONS—Continued

Year	Model	Valve Lash Int.	Valve Lash Exh.	Valve Angles Seat	Valve Angles Face	Valve Spring Installed Height	Valve Spring Pressure Lbs. @ In.	Stem Clearance Intake	Stem Clearance Exhaust	Stem Diameter Intake	Stem Diameter Exhaust
DODGE & PLYMOUTH—Continued											
1973	6-198, 225	.010H	.020H	45	②	1.65	160 @ 1.25	.001–.003	.002–.004	.372–.373	.371–.372
	8-318	Hydraulic①		45	②	1.65	193 @ 1.25	.001–.003	.002–.004	.372–.373	.371–.372
	8-340	Hydraulic①		45	②	1.65	241 @ 1.21	.001–.003	.002–.004	.371–.372	.370–.371
	8-360	Hydraulic①		45	②	1.65	195 @ 1.24	.001–.003	.002–.004	.372–.373	.371–.372
	8-400, 2 Bar. Carb.	Hydraulic①		45	45	1.86	200 @ 1.42	.001–.003	.002–.004	.372–.373	.371–.372
	8-400, 4 Bar. Carb.	Hydraulic①		45	45	1.86	246 @ 1.36	.001–.003	.002–.004	.372–.373	.371–.372
	8-440	Hydraulic①		45	45	1.86	200 @ 1.42	.001–.003	.002–.004	.372–.373	.371–.372
1974	6-198, 225	.010H	.020H	45	②	1.65	147 @ 1 5/16	.001–.003	.002–.004	.372–.373	.371–.372
	8-318	Hydraulic①		45	②	1.65	177 @ 1 5/16	.001–.003	.002–.004	.372–.373	.371–.372
	8-360, 2 Bar. Carb.	Hydraulic①		45	②	1.65	177 @ 1 5/16	.001–.003	.002–.004	.372–.373	.371–.372
	8-360, 4 Bar. Carb.	Hydraulic①		45	②	1.65	238 @ 1 5/16	.001–.003	.002–.004	.372–.373	.371–.372
	8-400, 440	Hydraulic①		45	45	1.86	200 @ 1 7/16	.001–.003	.002–.004	.372–.373	.371–.372
	8-400, 440 Hi Perf.	Hydraulic①		45	45	1.86	246 @ 1 23/64	.001–.003	.002–.004	.372–.373	.371–.372
1975–76	6-225	.010H	.020H	45	45	1 11/16	144 @ 1 5/16	.001–.003	.002–.004	.372–.373	.371–.372
	V8-318	Hydraulic①		45	②	1 11/16	177 @ 1 5/16	.001–.003	.002–.004	.372–.373	.371–.372
	V8-360	Hydraulic①		45	②	1 11/16	⑧	.001–.003	.002–.004	.372–.373	.371–.372
	V8-360③	Hydraulic①		45	②	1 11/16	1 5/16	.001–.003	.002–.004	.372–.373	.371–.372
	V8-400, 440	Hydraulic①		45	45	1 55/64	200 @ 1 7/16	.0011–.0028	④	.3723–.373	⑥
	V8-400, 440③	Hydraulic①		45	45	1 55/64	246 @ 1 23/64	.0016–.0033	⑤	.3718–.3725	⑦

①—No adjustment.
②—Intake 45°, exhaust 43°.
③—High Performance.
④—Hot end .0021-.0038, Cold end .0011-.0028.
⑤—Hot end .0026-.0043, Cold end .0016-.0033.
⑥—Hot end .3713-.372, Cold end .3723-.373.
⑦—Hot end .3708-.3715, Cold end .3718-.3725.
⑧—1975, 208 @ 1 5/16; 1976, 183 @ 1 5/16.

PISTONS, PINS, RINGS, CRANKSHAFT & BEARINGS

Year	Model	Piston Clearance Top of Skirt	Ring End Gap①		Wrist-pin Diameter	Rod Bearings		Main Bearings			
			Comp.	Oil		Shaft Diameter	Bearing Clearance	Shaft Diameter	Bearing Clearance	Thrust on Bear. No.	Shaft End Play
CHRYSLER & IMPERIAL											
1969	8-383	.0003–.0013	.013	.015	1.0936	2.374–2.375	.001–.002	2.6245–2.6255	.0005–.0015	3	.002–.007
	8-440	.0003–.0013	.013	.015	1.0936	2.374–2.375	.001–.002	2.7495–2.7505	.0005–.0015	3	.002–.007
1970	8-383, 2 B. C.	.0003–.0012	.013	.015	1.0936	2.374–2.375	.0005–.0015	2.6245–2.6255	.0005–.0015	3	.002–.007
	8-383, 4 B. C.	.0003–.0012	.013	.015	1.0936	2.374–2.375	.001–.002	2.6245–2.6255	.0005–.0015	3	.002–.007
	8-440	.0003–.0012	.013	.015	1.0936	2.374–2.375	.001–.002	2.7495–2.7505	.0005–.0015	3	.002–.007
1971	8-383, 2 B. C.	.0003–.0012	.013	.015	1.0936	2.374–2.375	.0005–.0025	2.6245–2.6255	.0005–.0015	3	.002–.007
	8-383, 4 B. C.	.0003–.0012	.013	.015	1.0936	2.374–2.375	.0007–.0032	2.6245–2.6255	.0005–.0015	3	.002–.007
	8-400	.0003–.0013	.013	.015	1.0936	2.374–2.375	.0005–.002	2.6245–2.6255	.0005–.002	3	.002–.007
	8-440	.0003–.0012	.013	.015	1.0936	2.374–2.375	.0001–.0032	2.7495–2.7505	.0005–.0015	3	.002–.007
1972	8-360	.0005–.0015	.010	.015	.9842	2.124–2.125	.0005–.0025	2.8095–2.8105	.0005–.0025	3	.002–.010
	8-400	.0003–.0013	.013	.015	1.0936	2.374–2.375	②	2.6245–2.6255	.0005–.0025	3	.002–.010
	8-440	.0003–.0013	.013	.015	1.0936	2.374–2.375	.001–.0025	2.7495–2.7505	.0005–.0025	3	.002–.010
1973–74	8-400	.0003–.0013	.013	.015	1.0935	2.374–2.375	②	2.6245–2.6255	.0005–.0025	3	.002–.007
	8-440	.0003–.0013	.013	.015	1.0935	2.374–2.375	.001–.0025	2.7495–2.7505	.0005–.0025	3	.002–.007
1975–76	V8-318	.0005–.0015	.010	.015	.9482	2.124–2.125	.0005–.0025	2.4995–2.5005	.0005–.0020	3	.002–.010
	V8-360	.0005–.0015	.010	.015	.9482	2.124–2.125	.0005–.0025	2.8095–2.8105	.0005–.0020	3	.002–.010
	V8-400	.0003–.0013	.013	.015	1.0936	2.375–2.376	③	2.6245–2.6255	.0005–.0020	3	.002–.007
	V8-440	.0003–.0013	.013	.015	1.0936	2.375–2.376	.0004–.0029	2.7495–2.7505	.0005–.0020	3	.002–.007
DODGE & PLYMOUTH											
1969–70	6-170	.0005–.0015	.010	.010	.9008	2.1865–2.1875	.0005–.001	2.7495–2.7505	.0005–.0015	3	.002–.007
	6-198, 225	.0005–.0015	.010	.015	.9008	2.1865–2.1875	.0005–.0015	2.7495–2.7505	.0005–.0015	3	.002–.007
	8-273, 318	.0005–.0015	.010	.015	.9842	2.124–2.125	.0005–.0015	2.4995–2.5005	.0005–.0015	3	.002–.007
	8-340	.0005–.0015	.010	.015	.9842	2.124–2.125	.0005–.0020	2.4995–2.5005	.0005–.0015	3	.002–.007
	8-383, 290 H.P.	.0003–.0013	.013	.015	1.0936	2.374–2.375	.0005–.0015	2.6245–2.6255	.0005–.0015	3	.002–.007
	8-383, 330 H.P.	.0003–.0013	.013	.015	1.0936	2.374–2.375	.001–.002	2.6245–2.6255	.0005–.0015	3	.002–.007
	8-383, 335 H.P.	.0003–.0013	.013	.015	1.0936	2.374–2.375	.001–.002	2.6245–2.6255	.0005–.0015	3	.002–.007
	8-440	.0003–.0013	.013	.015	1.0936	2.374–2.375	.001–.002	2.7495–2.7505	.0005–.0015	3	.002–.007
	8-426	.0025–.0035	.013	.015	1.0311	2.374–2.375	.0015–.0025	2.7495–2.7505	.0015–.0025	3	.002–.007
1971	6-198, 225	.0005–.0015	.010	.015	.9008	2.1865–2.1875	.0005–.002	2.7495–2.7505	.0005–.002	3	.002–.007
	8-318, 340, 360	.0005–.0015	.010	.015	.9842	2.124–2.125	.0005–.002	2.4995–2.5005	.0005–.002	3	.002–.006
	8-383	.0003–.0013	.013	.015	1.0936	2.374–2.375	.0005–.002	2.6245–2.6255	.0005–.002	3	.002–.007
	8-440	.0003–.0013	.013	.015	1.0936	2.374–2.375	.0005–.002	2.7495–2.7505	.0005–.002	3	.002–.007
	8-426	.0025–.0035	.013	.015	1.0311	2.374–2.375	.0015–.0025	2.7490–2.75005	.0015–.003	3	.002–.007
1972–74	6-198, 225	.0005–.0015	.010	.015	.9008	2.1865–2.1875	.0005–.0025	2.7495–2.7505	.0005–.0025	3	.002–.010
	8-318	.0005–.0015	.010	.015	.9842	2.124–2.125	.0005–.0025	2.4995–2.5005	.0005–.0025	3	.002–.010
	8-340	.0005–.0015	.010	.015	.9842	2.124–2.125	.0005–.0030	2.4995–2.5005	.0005–.0025	3	.002–.010
	8-360	.0005–.0015	.010	.015	.9842	2.124–2.125	.0005–.0025	2.8095–2.8105	.0005–.0025	3	.002–.010
	8-400	.0003–.0013	.013	.015	1.0936	2.374–2.375	②	2.6245–2.6255	.0005–.0025	3	.002–.010
	8-440	.0003–.0013	.013	.015	1.0936	2.374–2.375	.001–.0025	2.7495–2.7505	.0005–.0025	3	.002–.010
1975–76	6-225	.0005–.0015	.010	.015	.9008	2.1865–2.1875	.0005–.002	2.7495–2.7505	.0005–.002	3	.002–.007
	V8-318	.0005–.0015	.010	.015	.9482	2.124–2.125	.0005–.0025	2.4995–2.5005	.0005–.002	3	.002–.010
	V8-360	.0005–.0015	.010	.015	.9482	2.124–2.125	.0005–.0025	2.8095–2.8105	.0005–.002	3	.002–.010
	V8-400	.0003–.0013	.013	.015	1.0936	2.375–2.376	③	2.6245–2.6255	.0005–.002	3	.002–.007
	V8-440	.0003–.0013	.013	.015	1.0936	2.375–2.376	.0004–.0029	2.7495–2.7505	.0005–.002	3	.002–.007

①—Fit rings in tapered bores for clearance listed in tightest portion of ring travel.
②—W/2 bbl. carb., .0005–.0020″; W/4 bbl. carb., .001–.0025″.
③—1975: 2 bbl. carb., .0002–.0022, 4 bbl. carb., .004–.0029; 1976: 2 bbl. carb., .0005–.0025, 4 bbl. carb., .0005–.0030.

ALTERNATOR & REGULATOR SPECIFICATIONS

Year	Unit Number	Ground Polarity	Field Coil Draw Amperes	Current Output			Operating Voltage			Voltage Regulator Point Gap	Regulator Armature Air Gap
				Engine R.P.M.	Amperes	Volts	Engine R.P.M.	Volts	Voltage @ 120° (1)		
1969	(6)	Negative	2.38–2.75(2)	1250	26(3)	15	1250	15	13.3–14.3	.012–.016	.048–.052
	(7)	Negative	2.38–2.75(2)	1250	34.5(3)	15	1250	15	13.3–14.3	.012–.016	.048–.052
	(8)	Negative	2.38–2.75(2)	1250	41(3)	15	1250	15	13.3–14.3	.012–.016	.048–.052
	(9)	Negative	2.38–2.75(2)	1250	51(3)	15	1250	15	13.3–14.3	.012–.016	.048–.052
1970–71	(6)	Negative	2.38–2.75(2)	1250	26(3)	15	1250	15	13.3–14.4	—	—
	(7)	Negative	2.38–2.75(2)	1250	34.5(3)	15	1250	15	13.3–14.4	—	—
	(8)	Negative	2.38–2.75(2)	1250	44.5(3)	15	1250	15	13.3–14.4	—	—
	(9)	Negative	2.38–2.75(2)	1250	51(3)	15	1250	15	13.3–14.4	—	—
1972–73	(6)	Negative	2.5–3.1(2)	1250	39(3)	15	1250	15	13.8–14.4(10)	—	—
	(7)	Negative	2.5–3.1(2)	1250	41(3)	15	1250	15	13.8–14.4(10)	—	—
	(8)	Negative	2.5–3.1(2)	1250	50(3)	15	1250	15	13.8–14.4(10)	—	—
	(9)	Negative	2.5–3.1(2)	1250	60(3)	15	1250	15	13.8–14.4(10)	—	—
1973-74	Black Tag	Negative	2.5–3.7	1250	65(3)	15	1250	15	13.8–14.4(10)	—	—
1974	Yellow Tag	Negative	2.5–3.7	1250	36(3)	15	1250	15	13.8–14.4(10)	—	—
	Red Tag	Negative	2.5–3.7	1250	41(3)	15	1250	15	13.8–14.4(10)	—	—
	Green Tag	Negative	2.5–3.7	1250	50(3)	15	1250	15	13.8–14.4(10)	—	—
	Blue or Natural Tag	Negative	2.5–3.7	1250	60(3)	15	1250	15	13.8–14.4(10)	—	—
1975	Yellow Tag(4)	Negative	2.5–3.7(2)	1250	36(3)	15	1250	15	13.9–14.6(10)	—	—
1975–76	Red Tag	Negative	2.5–3.7(2)	1250	40(3)	15	1250	15	13.9–14.6(10)	—	—
	Green Tag	Negative	2.5–3.7(2)	1250	47(3)	15	1250	15	13.9–14.6(10)	—	—
	Blue Tag	Negative	2.5–3.7(2)	1250	57(3)	15	1250	15	13.9–14.6(10)	—	—
	Natural Tag	Negative	2.5–3.7(2)	1250	57(3)	15	1250	15	13.9–14.6(10)	—	—
	Black Tag	Negative	2.5–3.7(2)	1250	62(3)	15	1250	15	13.9–14.6(10)	—	—
	Yellow Tag(5)	Negative	4.75–6(2)	900	84(11)	13	900	13	13.9–14.6(10)	—	—

(1)—For each 10 degree rise in temperature subtract .04 volt. Temperature is checked with thermometer two inches from installed voltage regulator cover.
(2)—Current draw at 12 volts while turning rotor shaft by hand.
(3)—Plus or minus three amperes. If output is low, stator or rectifier is shorted.
(4)—34 amp rating.
(5)—100 amp rating.
(6)—Standard with 6-cyl. engines.
(7)—Standard with V8 engines.
(8)—Heavy duty and/or air conditioning.
(9)—Special equipment.
(10)—At 80 degrees F.
(11)—Minimum output, rating 100 amps.

STARTING MOTOR SPECIFICATIONS

Year	Part No.	Rotation (1)	Brush Spring Tension, Ounces	No Load Test			Torque Test		
				Amperes	Volts	R.P.M.	Amperes	Volts	Torque, Lbs. Ft.
1969–72	2095150	Clockwise	32–36	90	11	1925–2600	400–450	4	—
	2098500	Clockwise	32–36	90	11	2950	340–420	4	—
	1889100	Clockwise	32–36	78	11	3800	310–445	4	—
	2642930	Clockwise	32–36	78	11	3800	310–445	4	—
	2875560	Clockwise	32–36	90	11	1925–2600	400–450	4	—
1973	2875560	Clockwise	32–36	90	11	1925–2600	400–450	4	—
1973-74	3656575	Clockwise	32–36	90	11	(2)	475–550	4	—
1975–76	3755250	Clockwise	32–36	90	11	5700	475–550	4	—
	3755900	Clockwise	32–36	90	11	3700	475–550	4	—

(1)—Viewed from drive end.
(2)—1973 4300 R.P.M., 1974 3700–4200 R.P.M.

ENGINE TIGHTENING SPECIFICATIONS★

★Torque specifications are for clean and lightly lubricated threads only. Dry or dirty threads produce increased friction which prevents accurate measurement of tightness.

Year	Engine	Spark Plugs Ft. Lbs.	Cylinder Head Bolts Ft. Lbs.	Intake Manifold Ft. Lbs.	Exhaust Manifold Ft. Lbs.	Rocker Arm Shaft Bracket Ft. Lbs.	Rocker Arm Cover Ft. Lbs.	Connecting Rod Cap Bolts Ft. Lbs.	Main Bearing Cap Bolts Ft. Lbs.	Flywheel to Crank-shaft Ft. Lbs.	Vibration Damper or Pulley Ft. Lbs.
1969–72	6-170, 198	30	65	200①	10	25	40①	45	85	55	②
1969–72	6-225	30	65	200①	10	25	40①	45	85	55	②
1969	8-273	30	85	35	30	210①	36①	45	85	55	135
1969–72	8-318	30	④	35	30	200①	36①	45	85	55	135⑤
1969–72	8-340, 360	30	95	35	30	200①	36①	45	85	55	135⑤
1969–72	8-383, 400	30	70	50	30	25	40①	45	85	55	135
1969–71	8-426 Hemi	30	75	③	35	30	40①	75	100	70	135
1969–72	8-440	30	70	50	30	25	40①	45	85	55	135
1973–74	6-198, 225	30	70	240①	10	25	40①	45	85	55	②
1973–74	8-318, 360	30	95	35	20	200①	40①	45	85	55	100
1973	8-340	30	95	35	20	200①	40①	45	85	55	100
1973–74	8-400, 440	30	70	40	30	25	40①	45	85	55	135
1975–76	6-225	10	70	240①	120①	25	40①	45	85	55	②
	V8-318, 360	30	95	40	20	200①	40①	45	85	55	100
	V8-400, 440	30	70	45	30	25	40①	45	85	55	135

①—Inch pounds.
②—Press fit.
③—Tighten 8 (4 each side) center screws to 72 inch-lbs; all others 48 inch-lbs.
④—Composition gasket 95 ft.-lbs. Steel gasket 85 ft.-lbs.
⑤—1971 100 ft. lbs.

WHEEL ALIGNMENT SPECIFICATIONS

NOTE: See that riding height is correct before checking wheel alignment.

OLD CAR SPECIFICATIONS: For 1946-68 Wheel Alignment Specifications see main index.

Year	Model	Caster Angle, Degrees		Camber Angle, Degrees				Toe-In. Inch	Toe-Out on Turns, Deg.	
		Limits	Desired	Limits		Desired			Outer Wheel	Inner Wheel
				Left	Right	Left	Right			
CHRYSLER & IMPERIAL										
1969	Manual Steer.	+1/16 to −1 1/16	−1/2	+1/4 to +3/4	0 to +1/2	+1/2	+1/4	1/8	18.8①	20
	Power Steer.	+1/16 to −1 1/16②	−1/2	+1/4 to +3/4	0 to +1/2	+1/2	+1/4	1/8	18.8①	20
1970	Manual Steer.	0 to −1	−1/2	+1/4 to +3/4	0 to +1/2	+1/2	+1/4	1/8	18.8①	20
	Power Steer.	0 to −1③	−1/2	+1/4 to +3/4	0 to +1/2	+1/2	+1/4	1/8	18.8①	20
1971	Manual Steer.	0 to −1	−1/2	+1/4 to +3/4	0 to +1/2	+1/2	+1/4	1/8	—	—
	Power Steer.	+1/4 to +1 1/4③	+3/4	+1/4 to +3/4	0 to +1/2	+1/2	+1/4	1/8	—	—
1972	All	−1/16 to +1 5/16	+5/8	+1/8 to +7/8	−1/8 to +5/8	+1/2	+1/4	1/8	—	—
1973-74	All	+1/4 to +1 1/4	+3/4	+1/4 to +3/4	0 to +1/2	+1/2	+1/4	1/8	18.3	20
1975	All	−1/2 to +1 3/4	+3/4	0 to +1	−1/4 to +3/4	+1/2	+1/4	1/8	④	20
1976	All	+1/4 to +1 1/4	+3/4	+1/4 to +3/4	0 to +1/2	+1/2	+1/4	1/8–3/16	④	20

①—1969–70 Imperial 17.9°.
②—Imperial +3/16 to +1 5/16 with +3/4 preferred.
③—Imperial +1/4 to +1 1/4 with +3/4 preferred.
④—Exc. Cordoba, 18.3°; Cordoba, 18°.

Continued

WHEEL ALIGNMENT SPECIFICATIONS—Continued

NOTE: See that riding height is correct before checking wheel alignment.

OLD CAR SPECIFICATIONS: For 1946-68 Wheel Alignment Specifications see main index.

DODGE

Year	Model	Caster Angle, Degrees: Limits	Caster Angle, Degrees: Desired	Camber Angle, Degrees: Limits Left	Camber Angle, Degrees: Limits Right	Camber Angle, Degrees: Desired Left	Camber Angle, Degrees: Desired Right	Toe-In, Inch	Toe-Out on Turns, Deg.: Outer Wheel	Toe-Out on Turns, Deg.: Inner Wheel
1969–70	Man. Steer.①	0 to −1	−½	+¼ to +¾	0 to +½	+½	+¼	⅛	17.6	20
	Power Steer.①	+¼ to +1¼	+¾	+¼ to +¾	0 to +½	+½	+¼	⅛	17.6	20
	Man. Steer.②	0 to −1	−½	+¼ to +¾	0 to +½	+½	+¼	⅛	17.8	20
	Power Steer.②	+¼ to +1¼	+¾	+¼ to +¾	0 to +½	+½	+¼	⅛	17.8	20
	Man. Steer.③	0 to −1	−½	+¼ to +¾	0 to +½	+½	+¼	⅛	18.8	20
	Power Steer.③	0 to −1	−½	+¼ to +¾	0 to +½	+½	+¼	⅛	18.8	20
1971–73	Man. Steer.	0 to −1	−½	+¼ to +¾	0 to +½	+½	+¼	⅛	④	20
	Power Steer.	+¼ to +1¼	+¾	+¼ to +¾	0 to +½	+½	+¼	⅛	④	20
1974	Man. Steer.	−1¾ to +½	−½	0 to +1	−¼ to +¾	+½	+¼	⅛ to 3/16	⑤	20
	Power Steer.	−½ to +1¾	+¾	0 to +1	−¼ to +¾	+½	+¼	⅛ to 3/16	⑤	20
1975	Man. Steer.	−1¾ to +½	−½	0 to +1	−¼ to +¾	+½	+¼	⅛	⑥	20
	Power Steer.	−½ to +1¾	+¾	0 to +1	−¼ to +¾	+½	+¼	⅛	⑥	20
1976	Man. Steer.	0 to −1	−½	+¼ to +¾	0 to +½	+½	+¼	⅛–3/16	⑦	20
	Power Steer.	+¼ to +1¼	+¾	+¼ to +¾	0 to +½	+½	+¼	⅛–3/16	⑦	20

①—Dart.
②—Coronet, Charger and Challenger.
③—Monaco, Polara.
④—Dart & Challenger 17.5°, 1971–72 Charger & Coronet 17.8°, 1973 Charger & Coronet 18.5°, Polara & Monaco 18.8°.
⑤—Dart 18.5°, Challenger 18.4°, Charger & Coronet 18°, Monaco 18.3°.
⑥—Dart, 18.5°; Charger SE & Coronet, 18° Monaco, 18.3°.
⑦—Dart, 18.5°; Aspen, 16.1°; Charger SE & Coronet, 18°; Monaco, 18.3°.

PLYMOUTH

Year	Model	Caster Angle, Degrees: Limits	Caster Angle, Degrees: Desired	Camber Angle, Degrees: Limits Left	Camber Angle, Degrees: Limits Right	Camber Angle, Degrees: Desired Left	Camber Angle, Degrees: Desired Right	Toe-In, Inch	Toe-Out on Turns, Deg.: Outer Wheel	Toe-Out on Turns, Deg.: Inner Wheel
1969–70	Man. Steer.	+1/16 to −11/16	−½	+¼ to +¾	0 to +½	+½	+¼	⅛	①	20
	Power Steer.	+3/16 to +15/16	+¾	+0¼ to +¾	0 to +½	+½	+¼	⅛	①	20
1971–73	Man. Steer.	0 to −1	−½	+¼ to +¾	0 to +½	+½	+¼	⅛	②	20
	Power Steer.	+¼ to +1¼	+¾	+¼ to +¾	0 to +½	+½	+¼	⅛	②	20
1974	Man. Steer.	−1¾ to +½	−½	0 to +1	−¼ to +¾	+½	+¼	5/32	③	20
	Power Steer.	−½ to +1¾	+¾	0 to +1	−¼ to +¾	+½	+¼	5/32	③	20
1975	Man. Steer.	−1¾ to +½	−½	0 to +1	−¼ to +¾	+½	+¼	⅛	④	20
	Power Steer.	−½ to +1¾	+¾	0 to +1	−¼ to +¾	+½	+¼	⅛	④	20
1976	Man. Steer.	0 to −1	−½	+¼ to +¾	0 to +1½	+½	+1	⅛–3/16	⑤	20
	Power Steer.	+¼ to +1¼	+¾	+¼ to +¾	0 to +1½	+½	+1	⅛–3/16	⑤	20

①—Fury 17.7°, others 18°.
④—Valiant, 18.5°; Fury, 18°; Gran Fury, 18.3°.
②—Barracuda & Valiant, 17.5°; 1971–72 Satellite, 17.8°; 1973 Satellite, 18.5°; Fury, 18.8°.
③—Valiant 18.5°, Barracuda 18.4°, Satellite 18°, Fury 18.3°
⑤—Valiant, 18.5°; Volaré, 16.1°; Fury, 18.5°; Gran Fury, 18.3°.

BRAKE SPECIFICATIONS

Year	Model	Brake Drum Inside Diameter	Wheel Cylinder Bore Diameter			Master Cylinder Bore Diameter		
			Disc Brake	Front Drum Brake	Rear Drum Brake	Disc Brakes	Drum Brakes	Power Brakes
CHRYSLER & IMPERIAL								
1969	Chrysler	11	2.750	1 1/8	15/16	1 1/8	1	1
	Imperial	11	2.375	—	15/16	1 1/8	—	—
1970	All	11	2.750	1 1/8	15/16	1 1/8	1	1
1971	All	11	2.750	1 3/16	15/16	1	1	1
1972–73	All	11	2.750	—	15/16	1 1/32	—	1 1/32
1974	Chrysler	11	3.10	—	15/16	1 1/32	—	1 1/32
	Imperial	—	①	—	—	1 1/16	—	1 1/16
1975	Cordoba	10	2 3/4	—	15/16	1.03	—	1.03
	Chrysler	11	3.1	—	15/16	1.03	—	1.03
	Imperial	—	①	—	—	1.06	—	1.06
1976	Cordoba	10	2 3/4	—	15/16	1.03	—	1.03
	Chrysler	11	3.1	—	15/16	1.03	—	1.03

①—Front 3.10; Rear 2.60.

Year	Model	Brake Drum Inside Diameter	Disc Brake	Front Drum Brake	Rear Drum Brake	Disc Brakes	Drum Brakes	Power Brakes
DODGE								
1969	Dart 6	9	1.638	1	13/16	1	1	1
	Dart V8	10	1.638	1 1/8	15/16	1	1	1
	Coronet Deluxe, 440	10	2.00	1 1/8	15/16	1 1/8	1	1
	Coronet R/T, 500	11	2.00	1 1/8	15/16	1 1/8	1	1
	Charger	10	2.00	1 1/8	15/16	1 1/8	1	1
	Polara, Monaco	11	②	1 1/8	15/16	1 1/8	1	1
1970	Dart 6	③	1.638	1	13/16	1	1	1
	Dart V8	10	1.638	1 1/8	15/16	1	1	1
	Challenger	10④	2.750	1 3/16	15/16	1 1/8	1	1
	Coronet, Charger	10①	2.750	1 1/8	15/16	1 1/8	1	1
	Monaco, Polara	11	2.750	1 1/8	15/16	1 1/8	1	1
1971	Dart, Demon 6	9	—	1	13/16	—	1	1
	Dart, Demon V8	10	1 5/8	1 3/16	15/16	1	1	1
	Challenger	10④	2.750	1 3/16	15/16	1 1/8	1	1
	Coronet, Charger	10④	2.750	1 3/16	15/16	1 1/8	1	1
	Monaco, Polara	11	2.750	1 3/16	15/16	1	1	1
1972	Dart, Demon 6	9	—	1	13/16	—	15/16	15/16
	Dart, Demon V8	10	1 5/8	1 3/16	15/16	1 1/32	1 1/32	1 1/32
	Challenger	10	2.750	1 3/16	15/16	1 1/32	1 1/32	1 1/32
	Coronet, Charger	10④	2.750	1 3/16	15/16	1 1/32	1 1/32	1 1/32
	Monaco, Polara	11	2.750	1 3/16	15/16	1 1/32	1 1/32	1 1/32
1973	Dart, 6 cyl. (Early)	9	—	1 1/16	13/16	—	15/16	15/16
	Dart, 6 cyl. (Late)	10	—	1 13/16	13/16	—	1 1/32	1 1/32
	Dart, V8	10	2.599	1 1/16	15/16	1 1/32	1 1/32	1 1/32
	Challenger	10	2.750	—	15/16	1 1/32	1 1/32	1 1/32
	Coronet, Charger	10④	2.750	—	15/16	1 1/32	1 1/32	1 1/32
	Monaco, Polara	11	2.750	—	15/16	1 1/32	1 1/32	1 1/32
1974	Dart, 6 cyl.	③	—	1 1/8	13/16	—	1 1/32	1 1/32
	Dart, V8	10	2.60	—	15/16	1 1/32	—	15/16
	Challenger	10	2.75	—	15/16	1	—	1 1/32
	Charger, Coronet Exc. Sta. Wag.	10	2.75	—	15/16	1	—	1 1/32
	Coronet Sta. Wag.	11	2.75	—	15/16	1	—	1 1/32
	Monaco, Polara	11	3.10	—	15/16	1 1/32	—	1 1/32

Continued

BRAKE SPECIFICATIONS—Continued

Year	Model	Brake Drum Inside Diameter	Wheel Cylinder Bore Diameter			Master Cylinder Bore Diameter		
			Disc Brake	Front Drum Brake	Rear Drum Brake	Disc Brakes	Drum Brakes	Power Brakes
DODGE—Continued								
1975	Dart 6 cyl.	③	—	1 1/8	13/16	—	1.03	—
	Dart V8	10	2.6	—	15/16	1.03	—	15/16
	Coronet, Charger SE	10⑤	2.75	—	15/16	1	—	1.03
	Monaco	11	3.1	—	15/16	1.03	—	1.03
1976	Aspen	10⑤	2 3/4	—	15/16	1.00	—	1.03
	Dart 6 cyl.	③	—	1 1/8	13/16	—	1.03	—
	Dart V8	10	2 3/4	—	15/16	1.03	—	.938
	Coronet, Charger	10⑤	2 3/4	—	15/16	1.03	—	1.03
	Monaco	11	3.1	—	15/16	1.03	—	1.03
PLYMOUTH								
1969	Valiant, Barracuda 6	9	1.638	1	13/16	1	1	1
	Valiant, Barracuda V8	10	1.638	1 1/8	15/16	1	1	1
	Belvedere, Satellite	10①	2.00	1 1/8	15/16	1 1/8	1	1
	Fury, VIP	11	②	1 1/8	15/16	1 1/8	1	1
1970	Valiant, Duster 6	③	1.638	1	13/16	1	1	1
	Valiant, Duster V8	10	1.638	1 3/16	15/16	1	1	1
	Barracuda	10④	2.750	1 3/16	15/16	1 1/8	1	1
	Belvedere, Satellite	10①	2.750	1 1/8	15/16	1 1/8	1	1
	Fury	11	2.750	1 1/8	15/16	1 1/8	1	1
1971	Valiant, Duster 6	9	—	1	13/16	—	1	1
	Valiant, Duster V8	10	1 5/8	1 3/16	15/16	1	1	1
	Barracuda	10④	2.750	1 3/16	15/16	1 1/8	1	1
	Satellite	10④	2.750	1 3/16	15/16	1 1/8	1	1
	Fury	11	2.750	1 3/16	15/16	1	1	1
1972	Valiant, Duster 6	9	—	1	13/16	—	15/16	15/16
	Valiant, Duster V8	10	1 5/8	1 3/16	15/16	1 1/32	1 1/32	1 1/32
	Barracuda	10	2.750	1 3/16	15/16	1 1/32	1 1/32	1 1/32
	Satellite	10④	2.750	1 3/16	15/16	1 1/32	1 1/32	1 1/32
	Fury	11	2.750	1 3/16	15/16	1 1/32	1 1/32	1 1/32
1973	Valiant, 6 cyl. (Early)	9	—	1 1/16	13/16	—	15/16	15/16
	Valiant, 6 cyl. (Late)	10	—	1 3/16	13/16	—	1 1/32	1 1/32
	Valiant, V8	10	2.599	1 1/16	15/16	1 1/32	1 1/32	1 1/32
	Barracuda	10	2.750	—	15/16	1 1/32	1 1/32	1 1/32
	Satellite	10④	2.750	—	15/16	1 1/32	1 1/32	1 1/32
	Fury	11	2.750	—	15/16	1 1/32	1 1/32	1 1/32
1974	Valiant, 6 cyl.	③	—	1 1/8	13/16	—	1 1/32	1 1/32
	Valiant, V8	10	2.60	—	15/16	1 1/32	—	15/16
	Barracuda	10	2.75	—	15/16	1	—	1 1/32
	Satellite Exc. Sta. Wag.	10	2.75	—	15/16	1	—	1 1/32
	Satellite Sta. Wag.	11	2.75	—	15/16	1	—	1 1/32
	Fury	11	3.10	—	15/16	1 1/32	—	1 1/32
1975	Valiant 6 cyl.	③	—	1 1/8	13/16	—	1.03	—
	Valiant V8	10	2.6	—	15/16	1.03	—	15/16
	Fury	10⑤	2.75	—	15/16	1	—	1.03
	Gran Fury	11	3.1	—	15/16	1.03	—	1.03
1976	Volare	10⑤	2 3/4	—	15/16	1	—	1.03
	Valiant 6 cyl.	③	—	1 1/8	13/16	—	1.03	1.03
	Valiant V8	10	2 3/4	—	15/16	1.03	—	.938
	Fury	10⑤	2 3/4	—	15/16	1.03	—	1.03
	Gran Fury	11	3.1	—	15/16	1.03	—	1.03

①—With 383 Hi Perf., 426 or 440 engine, 11″ drums. ②—For 1969, 2.757. ③—Front 10″, rear 9″. ④—Heavy duty 11″.
⑤—Wagon 11″.

COOLING SYSTEM & CAPACITY DATA

Year	Model or Engine	Cooling Capacity, Qts. No Heater	Cooling Capacity, Qts. With Heater	Cooling Capacity, Qts. With A/C	Radiator Cap Relief Pressure, Lbs. With A/C	Radiator Cap Relief Pressure, Lbs. No A/C	Thermo. Opening Temp. (1)	Fuel Tank Gals.	Engine Oil Refill Qts. (2)	Transmission Oil 3 Speed Pints	Transmission Oil 4 Speed Pints	Transmission Oil Auto. Trans. Qts. (12)	Rear Axle Oil Pints
CHRYSLER													
1969	8-383	—	16(9)	17	16	16	190	24(3)	4	6	—	9¼	4
	8-440	—	17(9)	18	16	16	190	24(3)	4	6	—	9¼	4
1970	8-383, 2 B. Carb.	—	14½(9)	15	16	16	195	24(17)	4	5	—	9½	4.4
	8-383, 4 B. Carb.	—	14½(9)	16	16	16	195	24(17)	4	5	—	8	4.4
	8-440	—	15½(9)	17	16	16	195(4)	24(17)	4	5	—	9½	4.4
1971	8-383, 2 B. Carb.	13½	14½	15	16	16	185	23	4	4¾	—	9½	4.5
	8-383, 4 B. Carb.	13½	14½	15	16	16	185	23	4	—	—	8	4.5
	8-440	14½	15½	17	16	16	185	23	4	—	—	9½	4.5
1972	8-360	14½	15½	15	16	16	185	23	4	—	—	8	4.5
	8-400	13½	14½	14	16	16	185	23	4	—	—	9½	4.5
	8-440	14½	15½	16	16	16	185	23	4	—	—	9½	4.5
1973	8-400, 440	—	16	17	16	16	185	23	4	—	—	9½	4.4
1974	8-400, 440	—	16½	16½	16	16	195	25(27)	4	—	—	9½	4.5
1975–76	V8-318	—	16	18	16	16	195	25½	4	—	—	8½(6)	4½
	V8-360	—	16	16	16	16	195	(5)	4	—	—	9½	4½
	V8-400	—	16½	16½	16	16	195	(21)	4	—	—	9½	4½
	V8-440	—	16	16	16	16	195	26½(27)	4(30)	—	—	9½	4½
IMPERIAL													
1969	All	—	19(9)	19	16	16	190	24	4	—	—	9¼	4
1970	All	—	17½(9)	17½	16	16	195	24	4	—	—	9½	4.4
1971–72	All	16½	17½	17½	16	16	185	23	4	—	—	9½	4.5
1973	All	—	18	18	16	16	185	23	4	—	—	9½	4.4
1974	All	—	17	17	16	16	195	25	4	—	—	9½	4.5
1975	All	—	16	16	16	16	195	26½	4	—	—	9½	4½
DODGE													
1969	Dart 6-170	—	12(15)	—	—	16	200	18	4	6½	—	7¾	2
	Dart 6-225	—	13	15	16	16	190	18	4	6½	—	7¾	2
	Dart 8-273	—	17	19	16	16	190	18	4	6	7	7¾	2
	Dart 318	—	16	18	16	16	190	18	4	6	7	7¾	2(10)
	Dart 8-340	—	16	16	16	16	190	18	4	—	7	7¾	4
	Dart 8-383	—	16	16	16	16	190	18	4	—	7	7¾(16)	4
	Coronet 6-225	—	13	15	16	16	190	19	4	6½	—	7¾	2(7)
	Coro. Charger 8-318	—	16	19	16	16	190	19	4	6	—	7¾	4
	Coro. Charger 8-383	—	16	17	16	16	190	19	4	—	7½	7¾(16)	4
	Coro. Charger 8-440	—	17	18	16	16	190	19	4	—	7½	9¼	4(13)
	Coro. Charger 8-426	—	18	—	16	16	190	19	6	—	7½	8	4(13)
	Polara, Monaco 8-318	—	16	19	16	16	190	24(3)	4	6	—	7¾	4
	Polara, Monaco 8-383	—	16	17	16	16	190	24(3)	4	6	—	7¾(16)	4
	Polara, Monaco 8-440	—	17	18	16	16	190	24(3)	4	—	—	9¼	4(14)
1970	Dart 6 Cyl.	12	13	13	16	16	190	18	4	6½	—	8½	2
	Dart 8-318	15	16	16	16	16	195	18	4	4¾	—	8	4
	Dart 8-340	14	15	15	16	16	190	18	4	4¾	7	8	4
	Challenger 6 Cyl.	—	13	14	16	16	190	18	4	6½	—	8½	2
	Challenger 8-318	—	16	17½	16	16	195	18	4	4¾	7	8	4
	Challenger 8-340	—	15½	15½	16	16	190	18	4	4¾	7	8	4

Continued

COOLING SYSTEM & CAPACITY DATA—Continued

Year	Model or Engine	Cooling Capacity, Qts.			Radiator Cap Relief Pressure, Lbs.		Thermo. Opening Temp. ①	Fuel Tank Gals.	Engine Oil Refill Qts. ②	Transmission Oil			Rear Axle Oil Pints
		No Heater	With Heater	With A/C	With A/C	No A/C				3 Speed Pints	4 Speed Pints	Auto. Trans. Qts. ⑫	
DODGE—Continued													
1970	Challenger 8-383	—	14½	15	16	16	190	18	4	4¾	7	8⑱	4
	Challenger 8-440	—	17	17	16	16	190	18	6	—	7	9	5½
	Challenger 8-426	—	17	17	16	16	190	18	6	—	7	8½	5½
	Coro., Charger 6-225	12	13	13	16	16	190	19	4	4¾	—	8½	2
	Coro., Charger 8-318	15	16	16	16	16	195	19	4	4¾	—	8	4
	Coro., Charger 8-383	13½	14½	15	16	16	190	19	4	4¾	7	8⑱	4
	Coro., Charger 8-440	16	17	17	16	16	190	19	6	—	7	8½	5½
	Coro., Charger 8-426	16	17	17	16	16	190	19	6	—	7	8½	5½
	Polara, Monaco 8-318	16	17	17	16	16	195	24⑰	4	4¾	—	8	4
	Polara, Monaco 8-383	13½	14½	14½	16	16	195	24⑰	4	4¾	—	9½	4
	Polara, Monaco 8-440	14½	15½	15½	16	16	195	24⑰	4	—	—	9½	4
1971	Dart 6 Cyl.	12	13	13	16	16	185	17	4	6½	—	8½	2
	Dart 8-318	15	16	16½	16	16	185	17	4	4¾	—	8½	4½
	Dart 8-340	14	15	15	16	16	185	17	4	4¾	7	8	4½
	Challenger 6 Cyl.	12	13	13	16	16	185	18	4	4¾	—	8½	2
	Challenger 8-318	15	16	16½	16	16	185	18	4	4¾	—	8½	2
	Challenger 8-340	14	15	15	16	16	185	18	4	4¾	7½	8	4½
	Challenger 8-383	13½	14½	15	16	16	185	18	4	4¾	7½	8⑱	4½
	Challenger 8-440	16⑲	17⑲	17⑳	16	16	185	18	4	—	7½	9½	5½
	Challenger 8-426	14½	15½	—	16	16	185	18	6	—	7½	8	5½
	Coro., Charger 6-225	12	13	13	16	16	185	21	4	6½	—	8½	4
	Coro., Charger 8-318	15	16	16½	16	16	185	21	4	4¾	—	8½	4
	Coro., Charger 8-383	13½	14½	15	16	16	185	21	4	4¾	7½	8⑱	4
	Coro., Charger 8-440	14½	15½	17	16	16	185	21	4	—	7½	9½	5½
	Coro., Charger 8-426	14½	15½	—	16	16	185	21	6	—	7½	8	5½
	Polara, Monaco 6-225	12	13	13	16	16	185	23	4	4¾	—	8½	4½
	Polara, Monaco 8-318	15	16	16½	16	16	185	23	4	4¾	—	8½	4½
	Polara, Monaco 8-360	14	15	15	16	16	185	23	4	4¾	—	8	4½
	Polara, Monaco 8-383	13½	14½	15	16	16	185	23	4	4¾	—	8⑱	4½
	Polara, Monaco 8-440	14½	15½	17	16	16	185	23	4	—	—	9½	4½
1972	Dart 6 Cyl.	12	13	13	16	16	185	16	4	6½	—	8½	2
	Dart 8-318	15	16	16	16	16	185	16	4	4¾	—	8½	4½
	Dart 8-340	14	15	15	16	16	185	16	4	4¾	7	8	4¼
	Challenger 6 Cyl.	12	13	13	16	16	185	18	4	4¾	—	8½	2
	Challenger 8-318	15	16	16½	16	16	185	18	4	4¾	—	8½	4½
	Challenger 8-340	14	15	14½	16	16	185	18	4	4¾	7½	8	4½
	Coro., Charger 6-225	12	13	13	16	16	185	21	4	6½	—	8½	4½
	Coro., Charger 8-318	15	16	16½	16	16	185	21	4	4¾	—	8½	4½
	Coro., Charger 8-340	14	15	14½	16	16	185	21	4	—	7½	8	4½
	Coro., Charger 8-400	13½	14½	15	16	16	185	21	4	—	7½	8⑱	4½⑭
	Coro., Charger 8-440	14	15	15	16	16	185	21	4㉖	—	7½	9½	5½
	Polara, Monaco 8-318	15	16	16½	16	16	185	23	4	—	—	8½	4½
	Polara, Monaco 8-360	14½	15½	15	16	16	185	23	4	—	—	8	4½
	Polara, Monaco 8-400	13½	14½	14	16	16	185	23	4	—	—	9½	4½
	Polara, Monaco 8-440	14½	15½	16	16	16	185	23	4	—	—	9½	4½

Continued

COOLING SYSTEM & CAPACITY DATA—Continued

Year	Model or Engine	Cooling Capacity, Qts.			Radiator Cap Relief Pressure, Lbs.		Thermo. Opening Temp. ①	Fuel Tank Gals.	Engine Oil Refill Qts. ②	Transmission Oil			Rear Axle Oil Pints
		No Heater	With Heater	With A/C	With A/C	No A/C				3 Speed Pints	4 Speed Pints	Auto. Trans. Qts. ⑫	
DODGE—Continued													
1973	Dart 6 Cyl.	12	13	13	16	16	185	16	4	6½	—	8½	2
	Dart 8-318	15	16	18	16	16	185	16	4	4¾	—	8½	2
	Dart 8-340	15	16	16	16	16	185	16	4	4¾	7	8	4½
	Challenger 6 Cyl.	12	13	13	16	16	185	18	4	4¾	—	8½	4½
	Challenger 8-318	15	16	18	16	16	185	18	4	4¾	—	8½	4½
	Challenger 8-340	14	15	16	16	16	185	18	4	4¾	7	8	4½
	Coro, Charger 6-225	12	13	13	16	16	185	21	4	6½	—	8½	2
	Coro, Charger 8-318	15	16	18	16	16	185	21	4	4¾	7½	8½	4½
	Coro, Charger 8-340	14	15	16	16	16	185	21	4	4¾	7½	8	4½
	Coro, Charger 8-400	15	16	17	16	16	185	21	4	4¾	7½	8⑱	4½
	Coro, Charger 8-440	16	17	17	16	16	185	21	4	—	—	9½	4½
	Polara, Monaco 8-318	15	16	19	16	16	185	23	4	—	—	8½	4½
	Polara, Monaco 8-360	15	16	16	16	16	185	23	4	—	—	8½	4½
	Polara, Monaco 8-400	15㉒	16㉓	17㉓	16	16	185	23	4	—	—	9½㉔	4½
	Polara, Monaco 8-440	15㉓	16㉓	17㉓	16	16	185	23	4	—	—	9½	4½
1974	Dart 6 cyl.	—	13	14	16	16	195	16	4	6½	—	8¼	2
	Dart 8-318	—	16	17½	16	16	195	16	4	4¾	7	8¼	4½
	Dart 8-360	—	16	16	16	16	195	16	4	4¾	7	8	4½
	Challenger 8-318	—	16	17½	16	16	195	18	4	4¾	—	8¼	4½
	Challenger 8-360	—	16	16	16	16	195	18	4	4¾	7½	8	4½
	Coro., Charger 6-225	—	13	15	16	16	195	19½㉘	4	4¾	—	8¼	4½
	Coro., Charger 8-318	—	16	18	16	16	195	19½㉘	4	4¾	7½	8¼	4½
	Coro., Charger 8-360	—	16½	16½	16	16	195	19½㉘	4	—	7½	8	4½
	Coro., Charger 8-400	—	16½	16½	16	16	195	19½㉘	4	—	7½	9½㉙	4½
	Coro., Charger 8-440	—	16	16	16	16	195	19½㉘	4	—	—	8	4½
	Monaco 8-360	—	16	16	16	16	195	25㉗	4	—	—	8	4½
	Monaco 8-400	—	16½	16½	16	16	195	25㉗	4	—	—	9½	4½
	Monaco 8-440	—	16	16	16	16	195	25㉗	4	—	—	9½	4½
1975	Dart 6-225	—	13	14	16	16	195	16	4	3½	—	8½	2.1
	Dart 8-318	—	16	17½	16	16	195	16	4	4¾	7	8½⑥	4½
	Dart 8-360	—	16	16	16	16	195	16	4	4¾	—	8¼	4½
	Coronet 6-225	—	13	—	16	16	195	25½	4	4¾	—	8½	4½
	Coro., Charger 8-318	—	16½	18	16	16	195	25½⑪	4	4¾	—	8½⑥	4½
	Coro., Charger 8-360	—	16	16	16	16	195	25½⑪	4	—	—	9½	4½
	Coro., Charger 8-400	—	16	16	16	16	195	25½⑪	4	—	—	9½	4½
	Coronet 8-440	—	16	16	16	16	195	25½⑪	4㉚	—	—	8¼	4½
	Monaco 8-318	—	17½	17½	16	16	195	26½㉗	4	—	—	8½⑥	4½
	Monaco 8-360	—	16	16	16	16	195	26½㉗	4	—	—	9½	4½
	Monaco 8-400	—	16½	16½	16	16	195	26½㉗	4	—	—	9½	4½
	Monaco 8-440	—	16	16	16	16	195	26½㉗	4㉚	—	—	9½	4½
1976	Aspen, Dart 6-225	—	13	14	16	16	195	㉛	4	㉜	7	8½	2
	Aspen, Dart 8-318	—	16	17	16	16	195	㉛	4	4¾	7	8½	4½
	Aspen, Dart 8-360	—	16	16	16	16	195	㉛	4	4¾	7	8½	4½
	Charger, Coronet 6-225	—	13	14½	16	16	195	25½⑪	4	4¾	—	8½	4½
	Charger, Coronet 8-318	—	16½	18	16	16	195	25½⑪	4	4¾	—	8½	4½
	Charger, Coronet 8-360	—	16	16	16	16	195	25½⑪	4	4¾	—	8½	4½
	Charger, Coronet 8-400	—	16½	16½	16	16	195	25½⑪	4	4¾	—	㉝	4½
	Coronet 8-440	—	16	16	16	16	195	25½⑪	4	4¾	—	㉝	4½
	Monaco 8-400	—	16½㉞	16½㉞	16	16	195	26½㉗	4㉚	—	—	㉝	4½
	Monaco 8-440	—	16㉞	16㉞	16	16	195	26½㉗	4㉚	—	—	㉝	4½

Continued

COOLING SYSTEM & CAPACITY DATA—Continued

Year	Model or Engine	Cooling Capacity, Qts.			Radiator Cap Relief Pressure, Lbs.		Thermo. Opening Temp. (1)	Fuel Tank Gals.	Engine Oil Refill Qts. (2)	Transmission Oil			Rear Axle Oil Pints
		No Heater	With Heater	With A/C	With A/C	No A/C				3 Speed Pints	4 Speed Pints	Auto. Trans. Qts. (12)	
PLYMOUTH													
1969	Fury, VIP 6-225	—	13	—	16	—	190	24(3)	4	6½	—	7¾	4
	Others 6-225	—	13	15	16	16	190	19	4	6½	—	7¾	2(7)
	Fury, VIP 8-318	—	16	19	16	16	190	24(3)	4	6½	—	7¾	4
	Others 8-318	—	16	19	16	16	190	19	4	6½	—	7¾	4
	Fury, VIP 8-383 2 B.C.	—	16	17	16	16	190	24(3)	4	—	7¾	9¼	4
	Others 8-383 2 B.C.	—	16	17	16	16	190	19	4	—	7½	9¼	4
	Fury, VIP 8-383 4 B.C.	—	16	17	16	16	190	24(3)	4	—	7¾	7¾	4
	Others 8-383 4 B.C.	—	16	17	16	16	190	19	4	—	7½	7¾	4
	8-426 Hemi.	—	18	—	16	—	190	19	6	—	7½	8	4(13)
	Fury, VIP 8-440	—	17	18	16	16	190	24(3)	4	—	7¾	9¼	4
	Others 8-440	—	17	18	16	16	190	19	4	—	7½	9¼	4(13)
1970	Fury 6-225	—	12	14	16	16	190	24(3)	4	4¾	—	8½	4
	Others 6-225	—	12	14	16	16	190	19	4	4¾	—	8½	2
	Fury 8-318	—	16	17½	16	16	195	24(2)	4	4¾	—	8	4
	Others 8-318	—	16	17½	16	16	195	19	4	4¾	—	8	4
	Fury 8-383 2 B.C.	—	14½	16	16	16	190	24(3)	4	4¾	—	9½	4
	Others 8-383 2 B.C.	—	14½	15	16	16	190	19	4	4¾	—	9½	4
	Fury 8-383 4 B.C.	—	14½	16	16	16	190	24(3)	4	—	—	8	4
	Others 8-383 4 B.C.	—	14½	16	16	16	190	19	4	4¾	7	8	4
	Fury 8-440	—	15½	17	16	16	190	24(3)	4	—	—	9½	4
	Others 8-440	—	15½	17	16	16	190	19	4	—	7	9½	5½
	8-426 Hemi	—	17	—	16	16	190	19	6	—	7	8½	5½
1971	Fury 6-225	12	13	13	16	16	185	23	4	6½	—	8½	4½
	Others 6-225	12	13	13	16	16	185	21	4	4¾	—	8½	4
	Fury 8-318	15	16	16½	16	16	185	23	4	4¾	—	8½	4½
	Others 8-318	15	16	16½	16	16	185	21	4	4¾	—	8½	4
	Fury 8-360	14	15	15	16	16	185	23	4	4¾	—	8	4½
	Fury 8-383	13½	14½	15	16	16	185	23	4	4¾	—	8(18)	4½
	Others 8-383	13½	14½	15	16	16	185	21	4	4¾	7½	8(18)	4
	Others 8-426	14½	15½	—	16	16	185	21	6	—	7½	8½	5½
	Fury 8-440	14½	15½	17	16	16	185	23	4	—	—	9½	4½
	Others 8-440	14½	15½	17	16	16	185	21	4	—	7½	9½	5½
1972	Satellite 6-225	12	13	13	16	16	185	21	4	6½	—	8½	4½
	Satellite 8-318	15	16	16½	16	16	185	21	4	4¾	—	8½	4½
	Satellite 8-340	14	15	14½	16	16	185	21	4	4¾	7½	8	4½
	Satellite 8-400	13½	14½	15	16	16	185	21	4	4¾	7½	8(18)	4½
	Satellite 8-440	14	15	15	16	16	185	21	4(26)	—	7½	9½	5½
	Fury 8-318	15	16	16½	16	16	185	23	4	—	—	8½	4½
	Fury 8-360	14½	15½	15	16	16	185	23	4	—	—	8	4½
	Fury 8-400	13½	14½	14	16	16	185	23	4	—	—	9½	4½
	Fury 8-440	14½	15½	16	16	16	185	23	4	—	—	9½	4½
1973	Satellite 6-225	12	13	13	16	16	185	21	4	6½	—	8½	4½
	Satellite 8-318	15	16	18	16	16	185	21	4	4¾	7.5	8½	4½
	Satellite 8-340	14	15	16	16	16	185	21	4	4¾	7.5	8	4½
	Satellite 8-400	15	16	17	16	16	185	21	4	4¾	7.5	8	4½
	Satellite 8-440	16	17	17	16	16	185	21	4	—	—	9½	4½
	Fury 8-318	15	16	19	16	16	185	23	4	—	—	8½	4½
	Fury 8-360	15	16	16	16	16	185	23	4	—	—	8	4½
	Fury 8-400	15(22)	16(23)	17(25)	16	16	185	23	4	—	—	8(18)	4½
	Fury 8-440	15(22)	16(23)	17(25)	16	16	185	23	4	—	—	9½	4½

Continued

COOLING SYSTEM & CAPACITY DATA—Continued

Year	Model or Engine	Cooling Capacity, Qts.			Radiator Cap Relief Pressure, Lbs.		Thermo. Opening Temp. [1]	Fuel Tank Gals.	Engine Oil Refill Qts. [2]	Transmission Oil			Rear Axle Oil Pints
		No Heater	With Heater	With A/C	With A/C	No A/C				3 Speed Pints	4 Speed Pints	Auto. Trans. Qts. [12]	
PLYMOUTH—Continued													
1974	Satellite 6-225	—	13	15	16	16	195	19½[28]	4	4¾	—	8½	4½
	Satellite 8-318	—	16	18	16	16	195	19½[28]	4	4¾	7½	8½	4½
	Satellite 8-360	—	16½	16½	16	16	195	19½[28]	4	—	7½	8	4½
	Satellite 8-400	—	16½	16½	16	16	195	19½[28]	4	—	7½	9[29]	4½
	Satellite 8-440	—	16	16	16	16	195	19½[28]	4	—	—	8	4½
	Fury 8-360	—	16	16	16	16	195	25[27]	4	—	—	8	4½
	Fury 8-400	—	16½	16½	16	16	195	25[27]	4	—	—	9[18]	4½
	Fury 8-440	—	16	16	16	16	195	25[27]	4	—	—	9½	4½
1975–76	Fury 6-225	—	13	—	16	16	195	25½	4	4¾	—	8½	4½
	Fury 8-318	—	16½	18	16	16	195	25½[11]	4	4¾	—	[8]	4½
	Fury 8-360	—	16	16	16	16	195	25½[11]	4	—	—	9½	4½
	Fury 8-400	—	16½	16½	16	16	195	25½[11]	4	—	—	9½	4½
	Fury 8-440	—	16	16	16	16	195	25½[11]	4[30]	—	—	8¼	4½
	Gran Fury 8-318	—	17½	17½	16	16	195	26½[27]	4	—	—	[6]	4½
	Gran Fury 8-360	—	16	16	16	16	195	26½[27]	4	—	—	9½	4½
	Gran Fury 8-400	—	16½	16½	16	16	195	26½[27]	4	—	—	9½	4½
	Gran Fury 8-440	—	16	16	16	16	195	26½[27]	4[30]	—	—	9½	4½
1976	Fury 6-225	—	13	14½	16	16	195	25½[11]	4	4¾	—	8½	4½
	Fury 8-318	—	16½	18	16	16	195	25½[11]	4	4¾	—	8½	4½
	Fury 8-360	—	16	16	16	16	195	25½[11]	4	4¾	—	8½	4½
	Fury 8-400	—	16½	16½	16	16	195	25½[11]	4	4¾	—	[33]	4½
	Fury 8-440	—	16	16	16	16	195	25½[11]	4	4¾	—	[33]	4½
	Gran Fury 8-318	—	17½	17½	16	16	195	19½	4	—	—	[33]	4½
	Gran Fury 8-360	—	16	16	16	16	195	26½[27]	4	—	—	[33]	4½
	Gran Fury 8-400	—	16½[34]	16½[34]	16	16	195	26½[27]	4	—	—	[33]	4½
	Gran Fury 8-440	—	16[34]	16[34]	16	16	195	26½[27]	4	—	—	[33]	4½
VALIANT, BARRACUDA & VOLARÉ													
1969	6-170	—	12	14	16	16	200	18	4	6½	—	7¾	2
	6-225	—	13	15	16	16	190	18	4	6½	—	7¾	2
	8-273	—	17	19	16	16	190	18	4	6½	7	7¾	4
	8-318	—	16	18	16	16	190	18	4	6½	7	9¼	4
	8-340	—	16	—	16	—	190	18	4	6½	7	9¼	4
	8-383 2 Bar. Carb.	—	16	—	16	—	190	18	4	6½	7	9¼	4
	8-383 4 Bar. Carb.	—	16	—	16	—	190	18	4	6½	7	7¾	4
1970	Valiant 6-198, 225	—	13	14	16	16	190	18	4	6½	—	8½	2
	Barracuda 6-225	—	13	14	16	16	190	18	4	4¾	—	8½	2
	8-318	—	16	17	16	16	195	18	4	4¾	7	8½	4
	Valiant 8-340	—	15	17	16	16	190	18	4	4¾	7	8	4
	Barracuda 8-340	—	15½	15½	16	16	190	18	4	4¾	7	8	4
	8-383 2 Bar. Carb.	—	14½	15	16	16	190	18	4	—	—	9½	4
	8-383 4 Bar. Carb.	—	14½	15	16	16	190	18	4	4¾	7	8	4
	8-440	—	17	17	16	16	190	18	6	—	7	9½	5½
	8-426	—	17	—	16	16	190	18	6	—	7	8	5½
1971	Valiant 6 Cyl.	12	13	13	16	16	185	17	4	6½	—	8½	2
	Barracuda 6 Cyl.	12	13	13	16	16	185	18	4	4¾	—	8½	2
	Valiant 8-318	15	16	16½	16	16	185	17	4	4¾	—	8½	4½
	Barracuda 8-318	15	16	16½	16	16	185	18	4	4¾	—	8½	2
	Valiant 8-340	14	15	15	16	16	185	17	4	4¾	7	8	4½
	Barracuda 8-340	14	15	15	16	16	185	18	4	4¾	7½	8	4½

Continued

COOLING SYSTEM & CAPACITY DATA—Continued

Year	Model or Engine	Cooling Capacity, Qts.			Radiator Cap Relief Pressure, Lbs.		Thermo. Opening Temp. (1)	Fuel Tank Gals.	Engine Oil Refill Qts. (2)	Transmission Oil			Rear Axle Oil Pints
		No Heater	With Heater	With A/C	With A/C	No A/C				3 Speed Pints	4 Speed Pints	Auto. Trans. Qts. (12)	
VALIANT BARRACUDA & VOLARÉ—Continued													
1971	8-383	13½	14½	15	16	16	185	18	4	4¾	7½	8(18)	4½
	8-440	16(19)	17(19)	17(20)	16	16	185	18	4	—	7½	9½	5½
	8-426	14½	15½	—	16	16	185	18	6	—	7½	8	5½
1972	Valiant 6 Cyl.	12	13	13	16	16	185	16	4	6½	—	8½	2
	Barracuda 6 Cyl.	12	13	13	16	16	185	16½	4	4¾	—	8½	2
	Valiant 8-318	15	16	16	16	16	185	16	4	4¾	—	8½	4½
	Barracuda 8-318	15	16	16½	16	16	185	16½	4	4¾	—	8½	4½
	Valiant 8-340	14	15	15	16	16	185	16	4	4¾	7	8	4½
	Barracuda 8-340	14	15	14½	16	16	185	16½	4	—	7½	8	4½
1973	Valiant 6 Cyl.	12	13	13	16	16	185	16	4	6½	—	8½	2
	Valiant 8-318	15	16	18	16	16	185	16	4	4¾	—	8½	2
	Valiant 8-340	15	16	16	16	16	185	16	4	4¾	7	8	4½
	Barracuda 6 Cyl.	12	13	13	16	16	185	16½	4	4¾	7½	8	4½
	Barracuda 8-318	15	16	18	16	16	185	16½	4	4¾	—	8½	4½
	Barracuda 8-340	14	15	16	16	16	185	16½	4	4¾	7½	8	4½
1974	Valiant 6-198	—	13	14	16	16	195	16	4	6½	—	8¼	2
	Valiant 6-225	—	13	14	16	16	195	16	4	6½	—	8¼	2
	Valiant 8-318	—	16	17½	16	16	195	16	4	4¾	7	8¼	4½
	Valiant 8-360	—	16	16	16	16	195	16	4	4¾	7	8	4½
	Barracuda 8-318	—	16	17½	16	16	195	16½	4	4¾	—	8¼	4½
	Barracuda 8-360	—	16	16	16	16	195	16½	4	4¾	7½	8	4½
1975	Valiant 6-225	—	13	14	16	16	195	16	4	3½	—	8½	2.1
	Valiant V8-318	—	16	17½	16	16	195	16	4	4¾	7	8½(6)	4½
	Valiant V8-360	—	16	16	16	16	195	16	4	—	—	8¼	4½
1976	Valiant, Volaré 6-225	—	13	14	16	16	195	(35)	4	(32)	7	8½	2
	Valiant, Volaré V8-318	—	16	17	16	16	195	(35)	4	4¾	7	8½	4½
	Valiant, Volaré V8-360	—	16	16	16	16	195	(35)	4	4¾	7	8½	4½

(1)—With permanent type anti-freeze.
(2)—Add one qt. with filter change.
(3)—Wagons 22 gals.
(4)—440 Hi Perf. uses 190°.
(5)—Cordoba 25½ gals., Chrysler 26½, Wagon 24 gals.
(6)—With 727 transmission (heavy duty), 9½ qts.
(7)—Station Wagon 4 pints.
(8)—With 727 transmission (heavy duty), 9½ qts.; Roadrunner models 8¼ qts.
(9)—Add 1½ qts. for rear seat heater.
(10)—With manual transmission 4 pints.
(11)—Wagon 20 gals., Sedan Models with dual exhaust 20½ gals.
(12)—Approximate. Make final check with dipstick.
(13)—With manual trans. 5½ pints.
(14)—5½ pints for High Perf. engine.
(15)—Add 2 qts. for 22″ radiator.
(16)—9¼ qts. for High Perf.
(17)—Wagons 23 gals.
(18)—With 2 bar. carb. 9½ qts.
(19)—Auto. Trans., 1½ qts. less.
(20)—Auto. Trans. 18 qts.
(21)—Chrysler exc. wagon, 26½ gals.; wagon, 24 gals.; Cordoba exc. dual exhaust, 25½ gals.; dual exhaust 20.5 gals.
(22)—16 quarts with 4 Bar. Carb. or Hi-Perf.
(23)—17 quarts with 4 Bar. Carb. or Hi-Perf.
(24)—8 quarts with 4 Bar. Carb. or Hi-Perf.
(25)—18 quarts with 4 Bar. Carb. or Hi-Perf.
(26)—Hi Perf. uses 6 qts. plus filter.
(27)—Wagons 24 gals.
(28)—Wagons 21 gals.
(29)—Hi Perf. 8 qts.
(30)—High performance engine 5 qts.
(31)—Aspen, 18 gals.; Dart, 16 gals.
(32)—Exc. floorshift, 3.6 pts.; floor shift, 4¾ pts.
(33)—Exc. High Perf., 9½ qts.; High Perf., 8¼ qts.
(34)—Add 1 qt. with maximum cooling or trailer towing package.
(35)—Volare, 18 gal.; Valiant, 16 gals.

REAR AXLE SPECIFICATIONS

Year	Model	Carrier Type	Ring Gear & Pinion Backlash		Pinion Bearing Preload			Differential Bearing Preload		
			Method	Adjustment	Method	New Bearings Inch-Lbs.	Used Bearings Inch-Lbs.	Method	New Bearings Inch-Lbs.	Used Bearings Inch-Lbs.
1969–73	All	Removable①	②	.006–.008	Shims	20–30④	10–15④	②	③	③
1969–73	Exc. Below	Integral	Shims	.004–.007	Shims	15–25④	8–12④	Shims	③	③
1969–73	⑤	Integral	Shims	.004–.009	Shims	10–20④	10–20④	Shims	③	③
1974	8¾″	Integral	②	.006–.008	⑦	20–30	—	②	③	③
1969–76	8¼″⑥	Integral	②	.006–.008	⑦	20–35④	10–25④	②	③	③
1973–76	9¼⑥	Integral	②	.006–.008	⑦	20–35④	10–25④	②	③	③
1974–76	7¼″	Integral	②	.004–.006	⑦	15–25	—	②	③	③

①—Adjust axle shaft end play to .008–.018″.
②—Threaded adjusters.
③—Preload is correct when ring gear and pinion backlash is properly adjusted.
④—Adjust by turning pinion shaft nut with an inch-pound torque wrench and seal removed.
⑤—9¼″ and 9¾″ axles.
⑥—"C" lock type.
⑦—Collapsible spacer.

Electrical Section

DISTRIBUTOR, REPLACE

6-170, 198, 225

The distributor rotates clockwise. To remove, take off cap, disconnect primary wire and vacuum line. Remove hold-down bolt and lift out distributor. Install in the following manner.

1. Rotate crankshaft to bring No. 1 piston up on its compression stroke, and position mark on inner edge of crankshaft pulley in line with the "O" (TDC) mark on timing chain cover.
2. With distributor gasket in position, hold distributor over mounting pad.
3. Turn rotor to a position just ahead of the No. 1 distributor cap terminal.
4. Install distributor, engaging distributor gear with drive gear on camshaft. With distributor fully seated on engine, rotor should be under No. 1 cap terminal and on conventional systems, breaker points just opening.
5. Install hold-down bolt.
6. Adjust timing with timing light.

1969-76 V8s

To remove the distributor, disconnect the vacuum control line and low tension wire and remove the cap and lock plate hold-down screw.

When installing the distributor, make sure that No. 1 piston is on top dead center on compression stroke and that the distributor rotor is in No. 1 firing position.

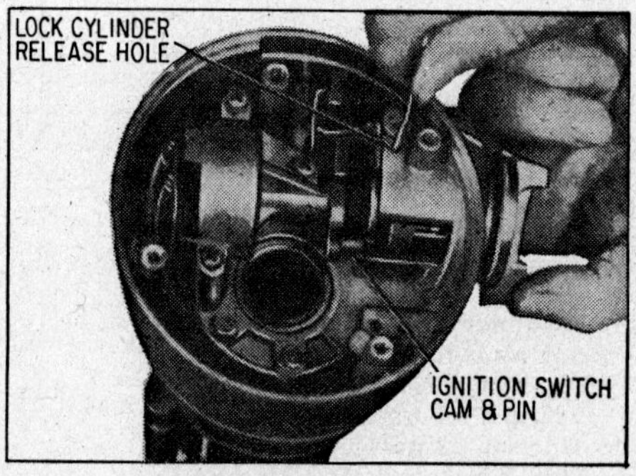

Fig. 1 Lock cylinder removal. 1970-76

SERVICE BULLETIN

FUSIBLE LINK REPAIR: Alternator equipped 1969-76 cars have charging circuits protected by a fuse-type wire. This fusible link is installed on the starter relay battery terminal.

If the charging circuit becomes overloaded, the inner fuse wire of this link burns out and the insulation heats up and breaks apart. This cuts off the battery from the charging system.

In the event one of these cars has none of its electrical parts functioning, check for a burned out fusible link. After locating and correcting the short, a new fusible link should be installed. Do not allow the insulation to contact any other wiring.

In situations like this, never use an uninsulated wire as a jumper if a fusible link replacement is not available. This can cause a fire in the electrical system.

STARTER, REPLACE

Reduction Gear Type

1. To remove starter, disconnect ground cable at battery.
2. Remove cable at starter.
3. Disconnect wires at solenoid.
4. Remove one stud nut and one bolt attaching starter motor to flywheel housing.
5. Slide transmission oil cooler bracket off stud (if so equipped).
6. Remove starter motor and removeable seal.
7. Reverse above procedure to install.

NOTE: When tightening attaching bolt and nut be sure to hold starter away from engine to insure proper alignment.

Direct Drive Type

1. Disconnect battery ground cable.
2. Remove cable at starter.
3. Disconnect lead wire from solenoid.
4. Unfasten and remove starter and removable seal.

NOTE

Noisy or erratic starter operation may

be caused by lack of lubrication or deposits of foreign material on the Bendix driveshaft.

To correct this condition, remove the inspection plate at the bottom of the torque converter or clutch housing. Then apply a suitable upper-cylinder rust inhibitor or SAE 5W or SAE 10W oil to the shaft by means of a 7" piece of tubing attached to the spout of a pressure oil can. In extreme cases, it may be necessary to remove, disassemble and clean the starter.

CLUTCH SWITCH

1970-76

A clutch switch is used which necessitates depressing the clutch pedal before the engine can be started.

IGNITION SWITCH & LOCK, REPLACE

1971-76

1. Disconnect battery ground cable.
2. Remove steering column cover and remove two screws attaching wiring cover from column.
3. Disconnect wiring connectors at column.
4. Remove horn ring ornament, horn ring or rim blow switch pad and ornament if so equipped.
5. Disconnect horn wires at steering wheel hub.
6. Remove horn ring.
7. Remove steering wheel with suitable puller.
8. Remove screw attaching turn signal lever and remove lever.

NOTE: On Tilt & Tel columns, lever screws out.

9. Attach a string or fine wire to signal switch wiring before removing switch from column. When switch is removed leave wire in column jacket tube as an aid in replacement.
10. Remove screws attaching signal switch and upper bearing retainer screws and remove retainer and signal switch and flasher switch.
11. Remove screw and lift out ignition key lamp assembly.
12. Remove snap ring from upper end of steering shaft.
13. Remove three bearing housing attaching screws.
14. With tool C-3044 attached to threaded holes for signal switch retaining screws, pull bearing and housing from steering shaft.
15. Remove lower snap ring from steering shaft.
16. Remove lock plate pin retaining ring from lock plate hub.
17. Use tool C-4113 and press steering shaft lock plate pin out of shaft and plate and remove lock plate.

NOTE: Do not use hammer as damage to column may result.

18. Remove lock lever guide plate screws and plate.
19. Depress key cylinder retainer toward the cylinder to disengage it from slot in housing bore, then withdraw key cylinder from lock housing, Fig. 1.

1970 With Standard Steering Column

1. Disconnect battery ground cable.
2. Remove steering column cover and remove two screws attaching wiring cover to steering column. Disconnect wiring connectors at column.
3. Remove steering wheel.
4. Remove turn signal lever.
5. Attach a piece of fine wire to signal switch wiring before removing switch. When switch is removed, leave wire in column to aid in replacement of wire.
6. Remove turn signal switch and upper bearing retainer screws and remove switch.
7. Remove ignition key lamp retaining screws and lift out lamp.
8. Remove snap ring from upper end of column.
9. Remove three bearing housing attaching screws and with a suitable puller, pull bearing and housing from steering shaft.
10. Remove lower snap ring from steering shaft.
11. Remove lock plate retaining pin from lock plate hub. *This pin must be pressed out. Do not attempt to hammer on pin or shaft as damage to the collapsible column may result.*
12. Remove lock plate and lock lever guide.
13. Insert a small wire in access hole, depress lock cylinder retainer and withdraw cylinder, Fig. 1.
14. Remove switch retaining screws and remove switch.

1970 With Tilt-O-Scope

The ignition switch is mounted on the steering column lower cover and is actuated by a rod connected to the rack. Before removal or installation, be sure lock cylinder is in the Lock position.

1969 Polara & Monaco

1. Remove center air conditioning duct and left cooler duct.
2. Remove ignition switch bezel nut.
3. Remove ignition switch from under side of panel and disconnect main harness.

1969 Chrysler & Imperial

1. Remove switch bezel nut from front of panel.
2. Push switch into and down below panel.
3. Disconnect wiring harness multiple connector and remove switch.

1969 Charger

1. Remove lower center A/C duct and left A/C duct (if equipped).
2. Remove steering column cover for visual assist and remove ignition switch spanner.
3. Disconnect wiring connector and remove switch.

1969 Coronet & Dart

1. Remove A/C duct (if equipped).
2. Remove switch bezel.
3. On Coronet, loosen harness from clip for access.
4. Disconnect wiring connector and pull switch out from under panel.

1969 Plymouth & Valiant

1. Where necessary, remove air conditioning elbow. Remove bezel nut and lower switch behind panel far enough to remove multiple connector.
2. Before installing switch, connect multiple connector and position switch in panel.
3. Install and tighten bezel nut.

LIGHT SWITCH, REPLACE

1976 Aspen & Volaré

1. Remove cluster bezel switch.
2. Remove switch mounting plate attaching screws and pull switch and plate assembly outward.
3. Depress headlight switch stem, then depress release button and pull knob and stem from switch.
4. Remove switch mounting nut, then disconnect electrical connector and remove switch.

1975-76 Cordoba, Coronet, Charger SE & Fury

1. Disconnect battery ground cable and fusible link.
2. Remove instrument cluster upper bezel and escutcheon mounting screw.
3. Remove switch mounting plate to cluster housing screws, pull switch from housing and disconnect electrical connector.
4. Depress release button on rear of switch and pull knob and stem from switch.
5. Remove switch escutcheon and mounting nut.
6. Reverse procedure to install.

1974-76 Chrysler Newport, New Yorker & Imperial

1. Remove headlight switch lens.
2. Remove headlight switch to mounting plate retaining nut.
3. Remove switch.

1974-76 Fury & Monaco

1. Remove instrument cluster bezel.

2. Remove windshield wiper switch mounting screws and headlight switch mounting screws.
3. Pull switch outwards and disconnect electrical leads, then pull switch to the "ON" position and depress release button on side of switch. Remove knob and stem from switch.
4. Remove escutcheon and mounting plate retaining nut and remove switch.

1973 Polara & Monaco

1. Remove column cover. Remove A/C duct.
2. Reach up under instrument panel and depress stem release button. At the same time remove switch stem from front panel.
3. Disconnect wiring from switch. Remove switch mounting nut and remove switch.

1973-76 Dart & Valiant

1. On 1973 models, remove fuse block attaching screw and position fuse block aside.
2. On all models, depress release button on rear of switch and remove knob and stem.
3. Remove spanner nut and drop switch below panel and disconnect wiring from switch.

1971-73 Chrysler & Imperial

1. Remove left spotcooler duct.
2. Remove headlamp switch knob and shaft by pulling the switch to on, depressing the release button on bottom of switch case, and pulling on the knob.
3. Remove headlamp sentinel and automatic dimmer control knobs if so equipped.
4. Remove headlamp switch mounting nut.
5. Remove headlamp switch from under instrument panel and disconnect wiring.

1971-74 Coronet, Charger, & 1971-72 Polara & Monaco

1. Reaching under instrument panel, disconnect wiring to switch.
2. Remove switch mounting screws and remove switch.

1971-73 Fury & 1971-74 Satellite

1. Reaching under instrument panel, disconnect switch wiring.
2. Unfasten and remove switch.

1970 Coronet & Belvedere

1. Disconnect battery ground cable.
2. Remove mounting screws and move safety relay and vent controls to one side.
3. Remove left air conditioner duct if necessary.
4. Remove two mounting screws, disconnect wire and remove switch. Place wires on switch before installing switch.

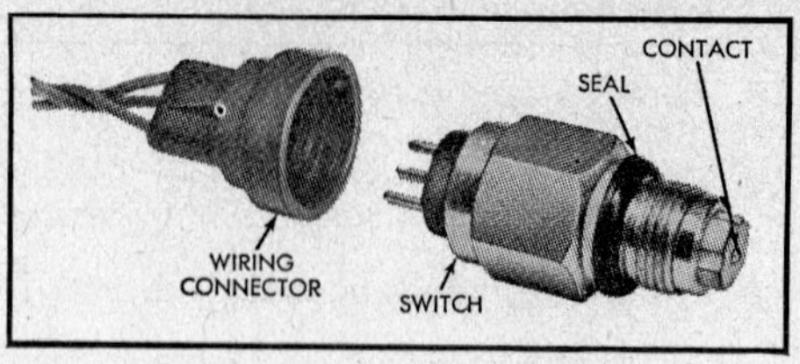

Fig. 2 Neutral safety switch

1969 Coronet, Belvedere & Satellite

1. Disconnect battery ground cable at battery.
2. Disconnect all wiring connectors from back of switches.
3. Remove switch bezel mounting screws.
4. Remove switches and bezel assembly and remove headlight switch.

NOTE: *Carefully pull trim bezel straight to avoid damaging trim pad.*

---

5. Reverse procedure to install.

1969-70 Monaco & Polara

1. Remove switch bezel.
2. Remove two mounting screws and remove switch.

1969-70 Chrysler & Imperial

1. Remove lamp panel.
2. Remove instrument cluster to gain access to switch.
3. Remove switch mounting screws and switch.

1969-70 Charger

1. Remove instrument cluster as described further on.
2. Disconnect wiring connector.
3. Disconnect heater vacuum hoses for accessibility.
4. Remove two switch mounting screws.
5. Reverse procedure to install.

1970-74 Barracuda, Challenger

1. Disconnect battery ground cable.

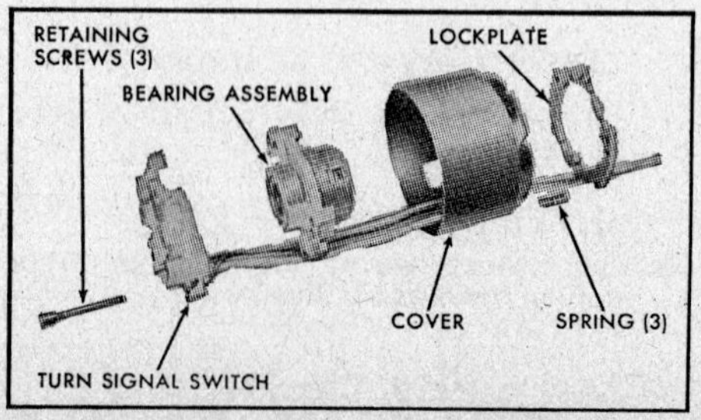

Fig. 3 Turn signal switch. 1969-76

2. Remove lamp panel retaining screws and carefully slide panel out and lay it on top of instrument panel.
3. Remove bezel retaining screws and slide assembly out and disconnect wiring harness.
4. Remove switch mounting screws and remove switch.

1969-72 Valiant, 1969-72 Dart 1969 Barracuda

1. From under instrument panel, remove screw retaining fuse block to panel and move fuse block out of the way.
2. Reaching under panel, depress release button on right side of switch and pull knob out of switch.
3. Remove bezel nut and lower switch under panel to disconnect multiple connector and remove switch.
4. Reverse procedure to install. Insert switch knob into switch until a "click" is heard.

1969-70 Fury & V.I.P.

1. On models with air conditioning, remove left spot cooler hose from duct and move out of the way.
2. If necessary, remove fuse block retaining screw and move fuse block away to gain access to switch.
3. From under panel, use a magnetic screwdriver to remove two screws retaining switch to rear of panel sheet metal.
4. Move switch down from panel reinforcement far enough to disconnect multiple connector and remove switch.
5. Reverse procedure to install.

STOP LIGHT SWITCH REPLACE

1969-76 All Cars

To remove the switch, disconnect wires from switch and remove switch from its mounting. Install the new switch and connect the wires.

NEUTRAL SAFETY & BACK-UP SWITCH

1969-76 All Cars

1. Unscrew switch from transmission case, allowing fluid to drain into a container, Fig. 2.
2. Move shift lever to "Park" and then to "Neutral" positions and inspect to see that switch operating lever is centered in switch opening in case.
3. Screw switch into transmission case and torque to 24 ft-lbs.
4. Add fluid to proper level.
5. Check to see that switch operates only in "Park" and "Neutral".

HORN SOUNDER & STEERING WHEEL

1969-76 All Cars

1. Disconnect ground cable at battery.
2. Remove horn ring ornament by turning counterclockwise or remove rim blow switch pad and ornament if so equipped.
3. Disconnect wires at horn switch.
4. Remove three screws attaching horn ring and switch to steering wheel, then remove horn ring and switch.
5. Remove wheel nut and use a suitable puller to remove steering wheel.

CAUTION: Do not bump or hammer on steering shaft to remove wheel as damage to shaft may result. See *Steering Gear, Replace* for other precautions.

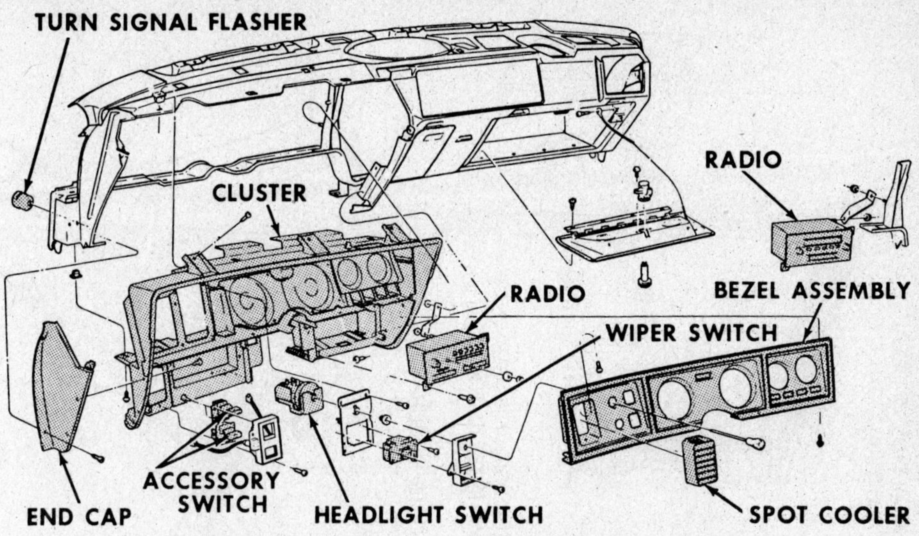

Fig. 4 Instrument cluster: 1975-76 Cordoba, Coronet, Charger SE & Fury

TURN SIGNAL SWITCH

1974-76

1. Disconnect battery ground cable.
2. Remove steering wheel as previously outlined.
3. Remove turn signal lever mounting nut and lever.

NOTE: On vehicles equipped with cruise control, do not remove lever, but let lever hang.

4. Remove the switch and upper bearing retainer attaching screws, then remove retainer.
5. Remove column cover, if equipped, and disconnect horn wire from mounting nut.
6. Remove column bracket to lower panel reinforcement attaching nuts, then bracket to column attaching nuts.
7. Remove wiring harness through from steering column and the tape securing wiring harness to column, then disconnect wiring harness connector.
8. Pull switch outward from column, guiding wires and connector through column opening, Fig. 3.

1971-73

1. Disconnect battery ground cable.
2. Remove steering column cover and remove screws attaching wiring through cover from column, Fig. 3.
3. Disconnect wiring connectors at column.
4. Remove horn ring ornament, horn ring or rim blow switch pad and ornament, if so equipped.
5. Disconnect horn wires at steering wheel hub.
6. Remove horn ring.
7. Remove steering wheel with a suitable puller.
8. Remove screw attaching turn signal lever and remove lever. On Tilt and Tel columns, the lever screws out.
9. Attach a piece of string or fine wire, to turn signal wiring before removing switch from column.
10. Remove screws attaching signal switch and upper bearing retainer screws and remove retainer and signal and flasher switch.

1969-70 All Cars

1. Disconnect battery ground cable.
2. Remove steering wheel as outlined above and remove signal operating lever.
3. Disconnect switch wiring multiple connector at steering column jacket. Remove each terminal from connector, or, tying them together with a piece of string or fine wire, Fig. 3.
4. Remove screws attaching turn signal switch to steering column and remove switch and wires from column. Leave string or wire in column to aid installation.
5. Reverse procedure to install, using wire or string to pull wires through column. Tighten switch lever to 30 inch pounds, steering wheel nut to 24 foot pounds.

INSTRUMENT CLUSTER REMOVAL

1976 Aspen & Volaré

1. Remove steering column cover and instrument panel end cap.
2. Remove lower left reinforcement and remove gearshift indicator.
3. Remove steering column toe plate bolts, then remove the two upper mounting nuts and lower steering column.
4. Disconnect speedometer cable, then remove 1 screw attaching radio to mid-reinforcement, 4 screws at bottom of cluster and 3 screws at top of cluster.
5. Pull cluster from panel, then disconnect all electrical connectors, control cables and vacuum hoses and remove cluster.

1975-76 Cordoba, Coronet, Charger SE & Fury

1. Disconnect battery ground cable.
2. Remove trim pad, radio and heater or A/C controls, Fig. 4.
3. Remove cluster housing reinforcement bracket.
4. Disconnect speedo cable, all electrical connectors and three wiring through clips from cluster.
5. Remove upper cluster bezel and instrument panel end cap.
6. Remove steering column to support bracket nuts.
7. Remove cluster housing to instrument panel retaining screws, then remove cluster.
8. Reverse procedure to install.

1976 Gran Fury & Monaco

Refer to the "1974-75 Fury & Monaco" procedure for instrument cluster removal.

1974-76 Chrysler Newport, New Yorker & Imperial

1. Disconnect battery ground cable.
2. Remove instrument panel upper cover, then working through top of panel, disconnect speedometer cable and printed circuit multiple connector.
3. Remove instrument cluster bezel, Fig. 5.
4. On Chrysler, remove gear selector and warning lamp bezel.
5. On all models disconnect instrument cluster lens and cluster housing from carrier.
6. Pull cluster out and disconnect two illumination and warning light modules.
7. Disconnect remaining electrical leads and remove cluster assembly.

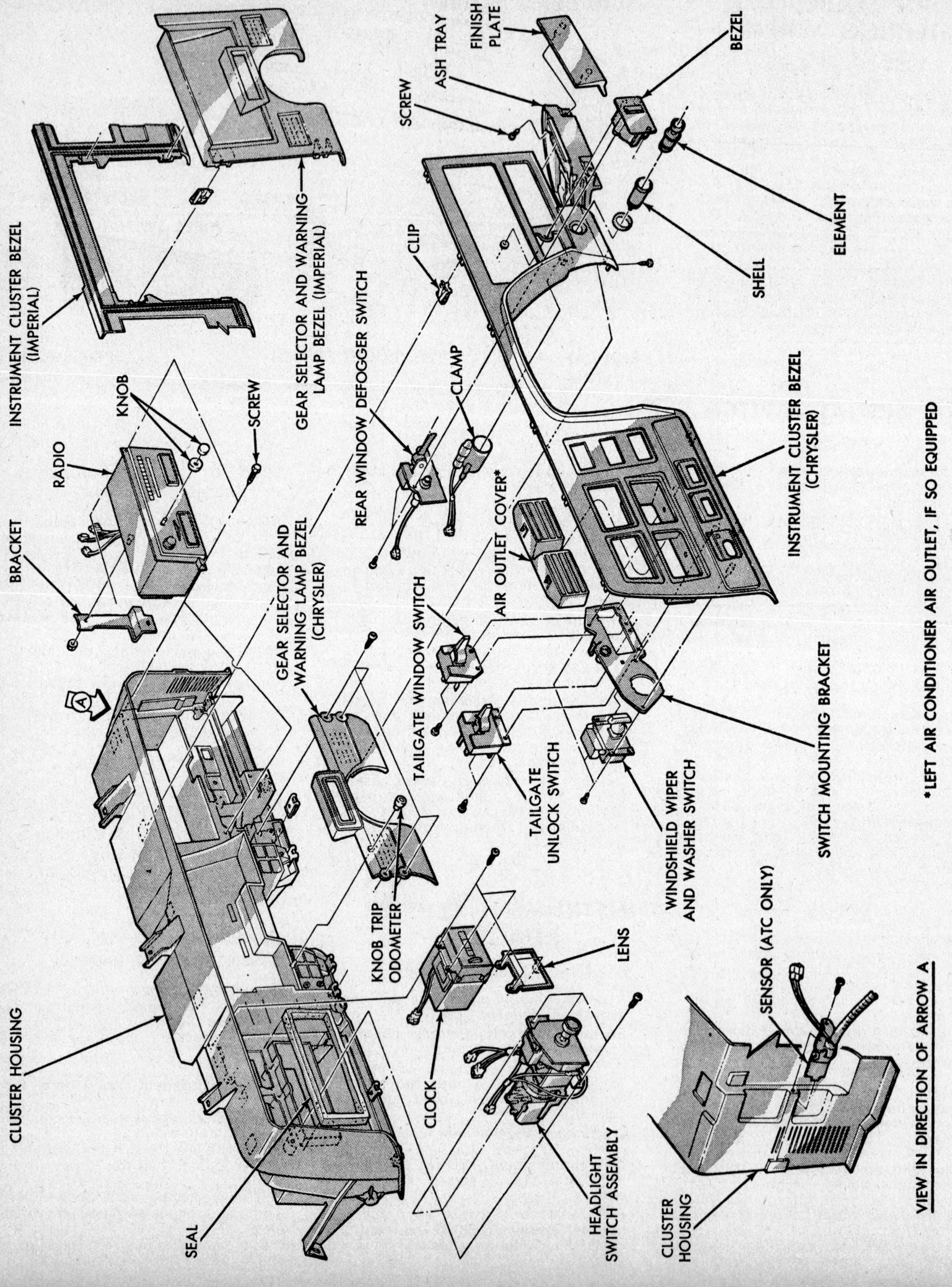

Fig. 5 Instrument cluster. 1974-76 Chrysler Newport, New Yorker & Imperial

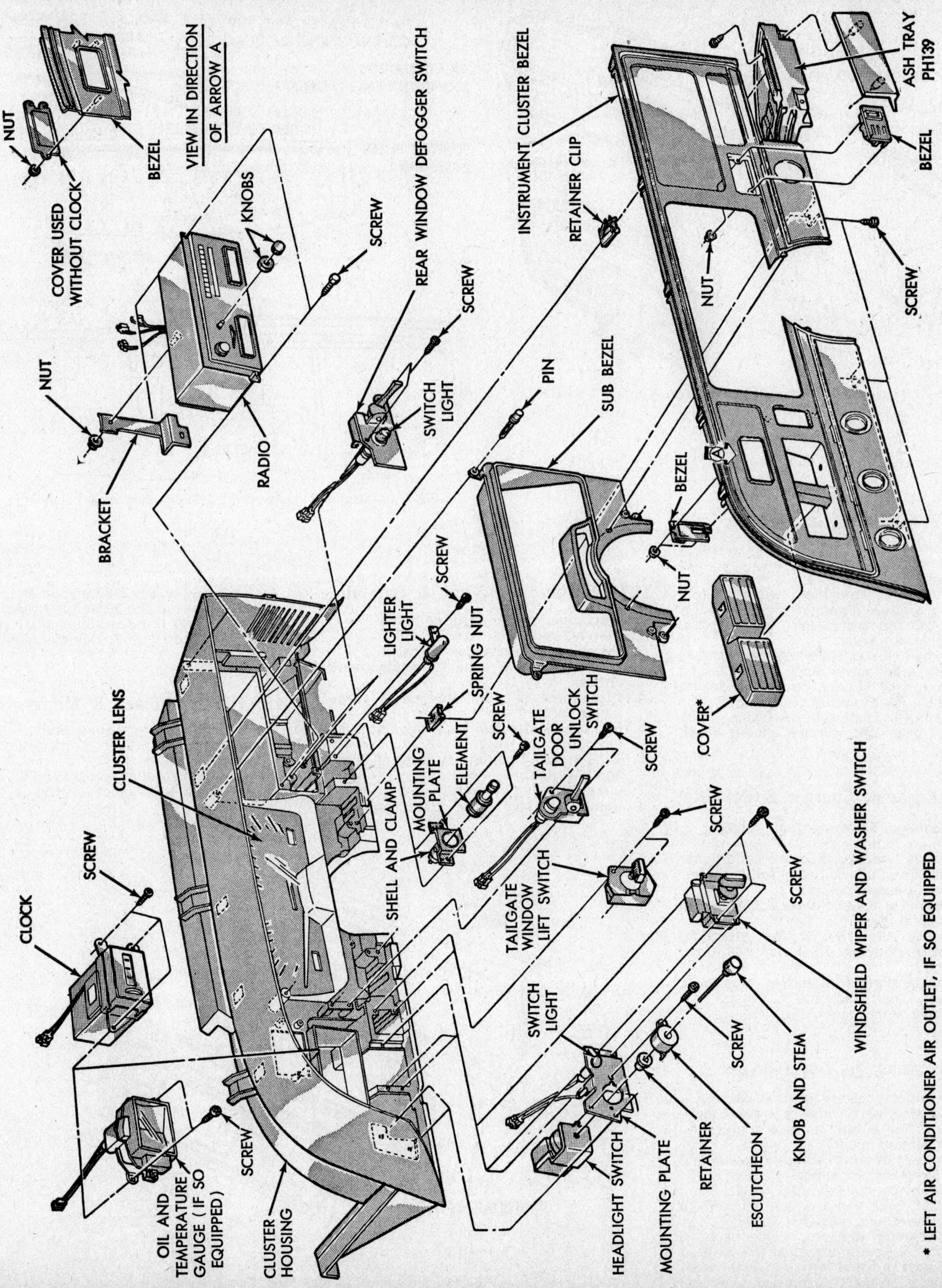

* LEFT AIR CONDITIONER AIR OUTLET, IF SO EQUIPPED

Fig. 6 Instrument cluster: 1974 Fury, 1974-76 Monaco

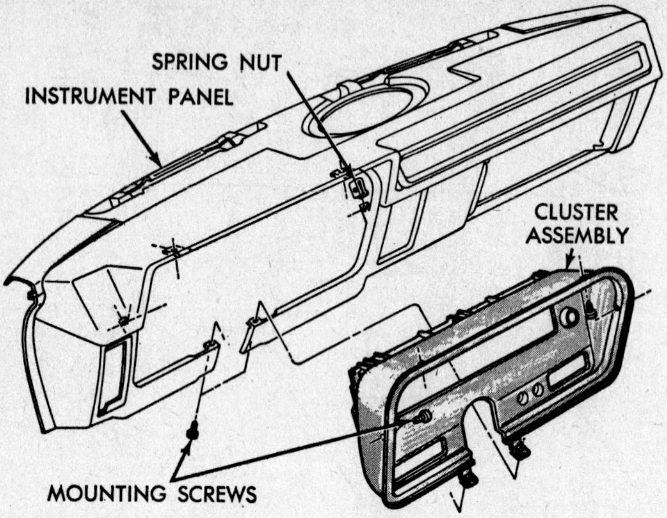

Fig. 7 Instrument cluster. 1971-74 Charger, Coronet & Satellite

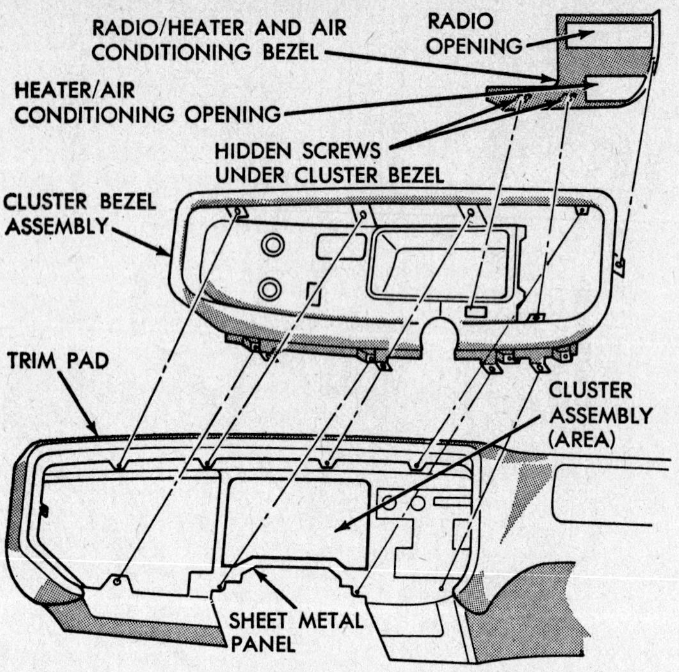

Fig. 8 Instrument cluster. 1972-73 Fury, Monaco & Polara

1974-75 Fury & Monaco

1. Place selector lever in low position and remove ash tray and cigarette lighter.
2. Center the windshield wiper switch, rear window defogger switch, heater or A/C fan switch and the temperature control lever.
3. Remove cluster bezel retaining screws from under lower edge of bezel and pull top of cluster bezel outward to release upper spring clips, Fig. 6.
4. Disengage cluster bezel locking tabs from bezel and remove cluster bezel.

1974 Coronet, Charger & Satellite

1. Disconnect battery ground cable.
2. Remove radio.
3. Remove heater and/or air conditioner control and allow control to hang.
4. Lower steering column.
5. Disconnect speedometer cable and all electrical leads.
6. Remove the two screws attaching cluster housing to lower reinforcement.
7. Remove screws mounting cluster to panel, then disconnect cluster assembly from panel, Fig. 7.

1973-76 Dart & Valiant

1. Disconnect battery ground cable.
2. Tape steering column to prevent damage to finish and remove center A/C duct (if equipped).
3. Remove steering column cover and lower steering column.
4. Disconnect speedometer cable from cluster and remove six cluster to panel retaining screws.
5. Pull cluster outward, disconnect all electrical leads attached to printed circuit board and remove cluster from panel.

1973 Coronet, Charger & Satellite

1. Disconnect battery ground cable.
2. Remove radio.
3. Remove all electrical leads and all lamps on right side of instrument cluster.
4. Disconnect speedometer cable from under dash.
5. Remove heater control screws and allow control to hang.
6. Tape steering column to prevent damage to finish.
7. Lower steering column.
8. Remove cluster scews attaching cluster to panel, lower cluster out at top, and disconnect the remaining electrical leads and remove lamps from left rear of cluster, Fig. 7. Remove cluster from panel.

1973 Fury, Polara & Monaco

1. Disconnect battery ground cable.
2. Tape column to prevent damage upon removal of cluster.
3. Remove lamp panel screws and lower panel down; then place on top of instrument panel.

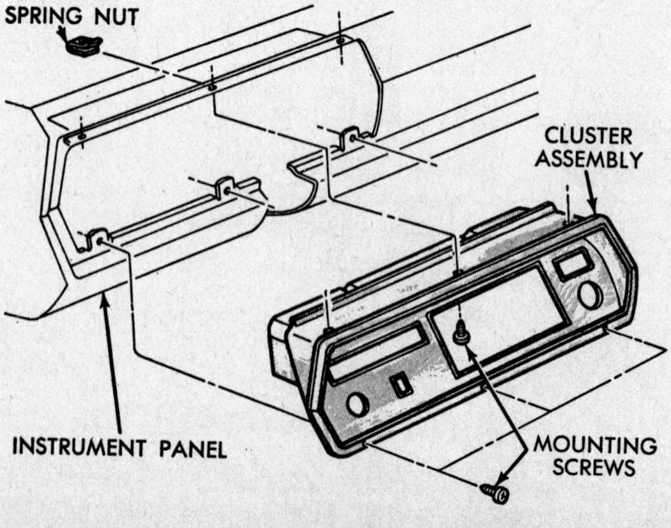

Fig. 9 Instrument cluster. 1972 Dart & Valiant

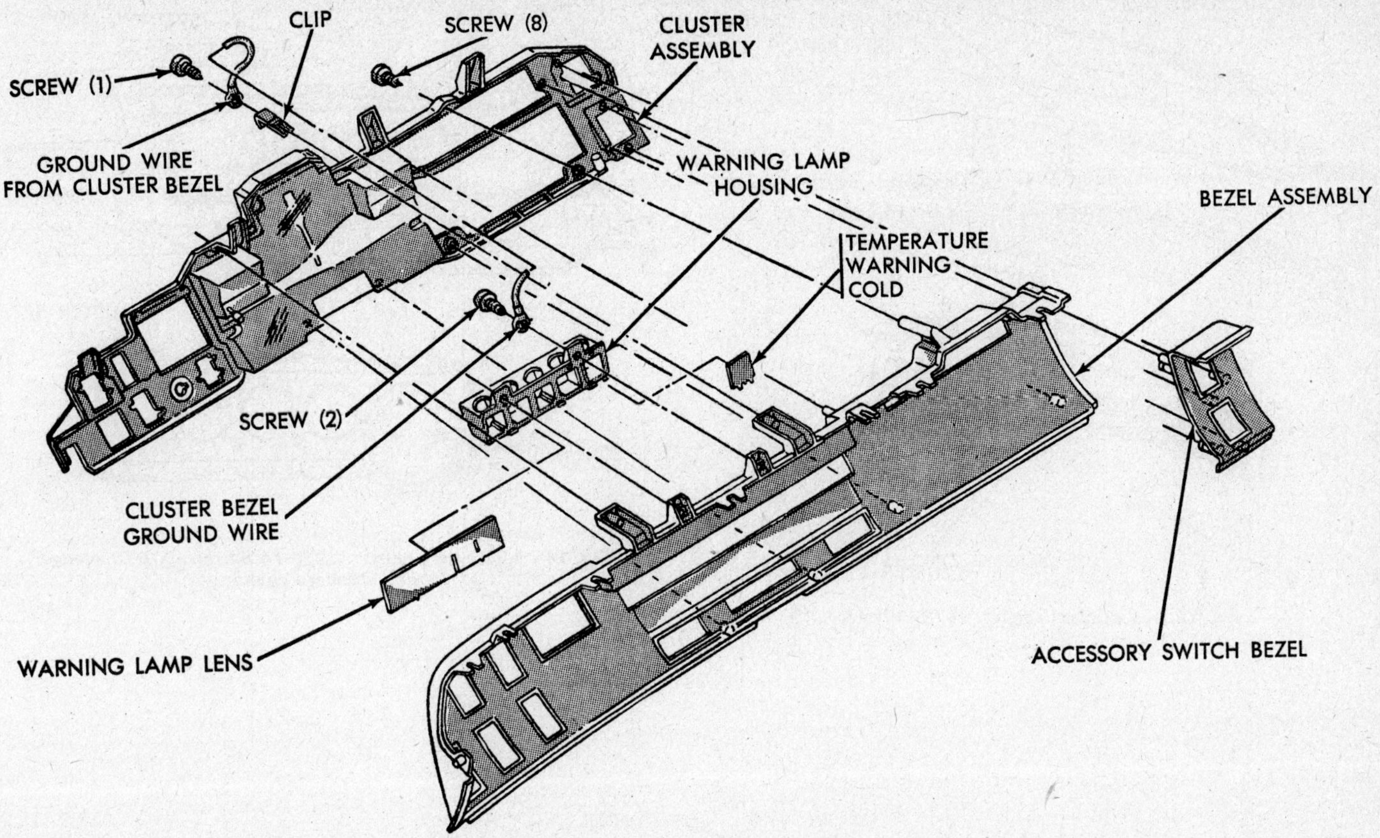

Fig. 10 Instrument cluster. 1969-71 Chrysler (Typical)

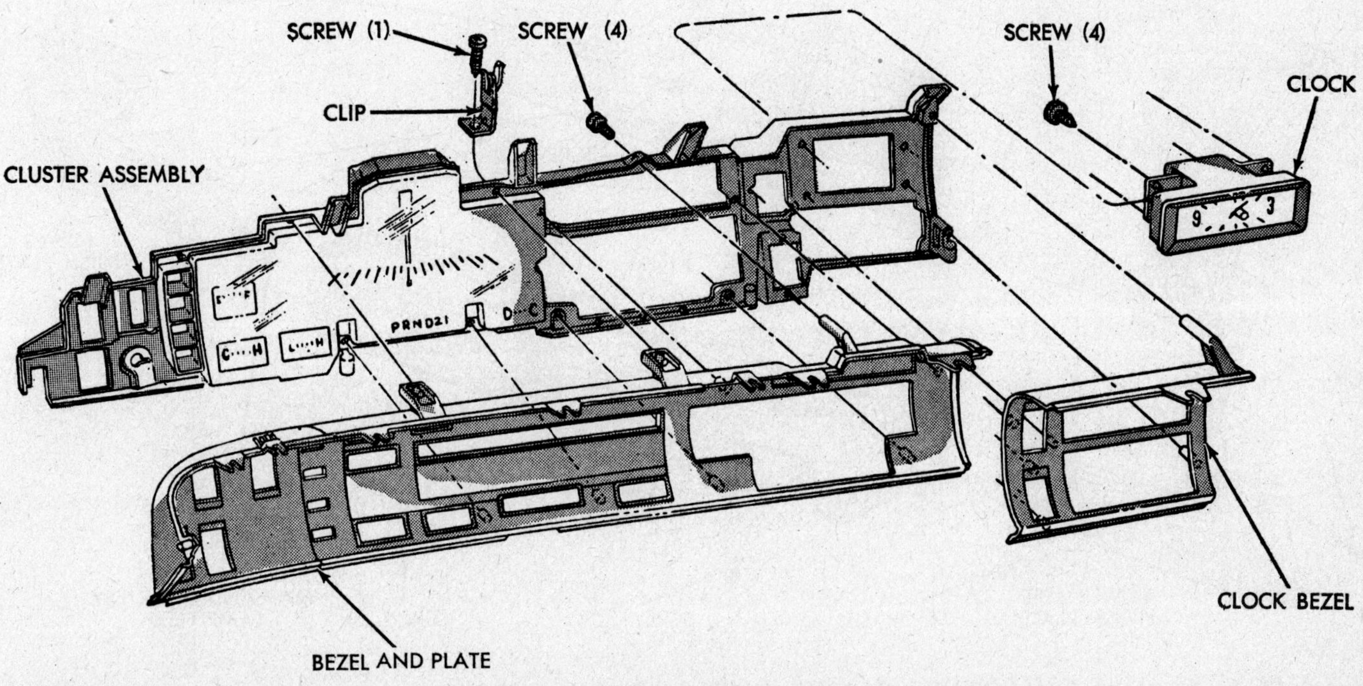

Fig. 11 Instrument cluster. 1969-71 Imperial (Typical)

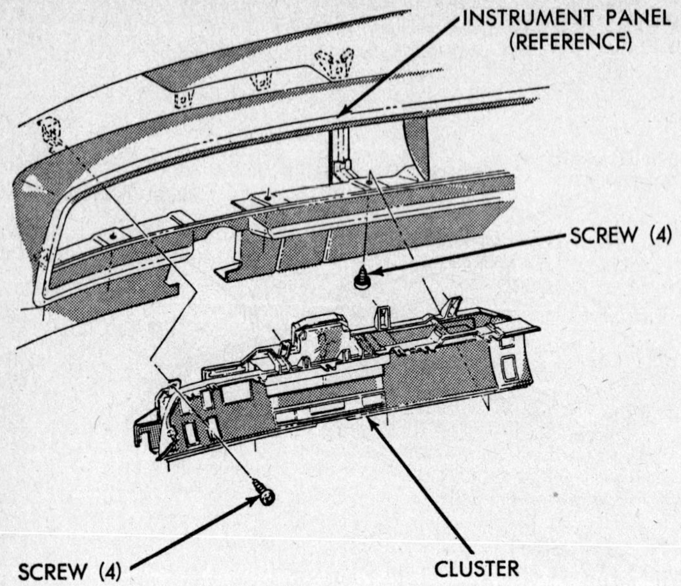

Fig. 12 Instrument cluster. 1972 Chrysler

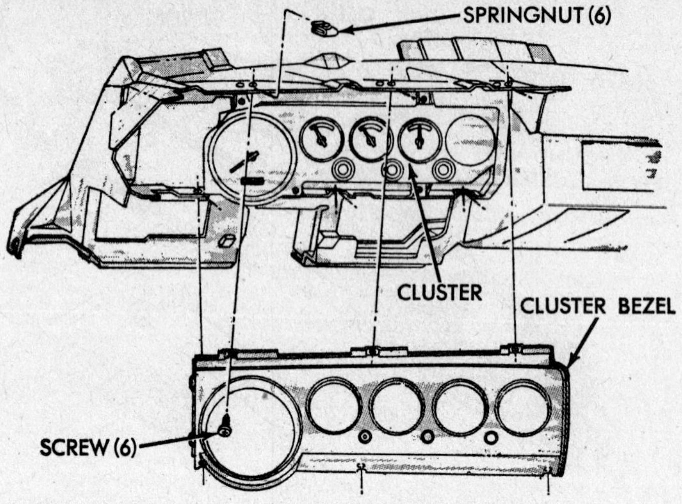

Fig. 14 Instrument cluster. 1970-74 Barracuda & Challenger with standard cluster

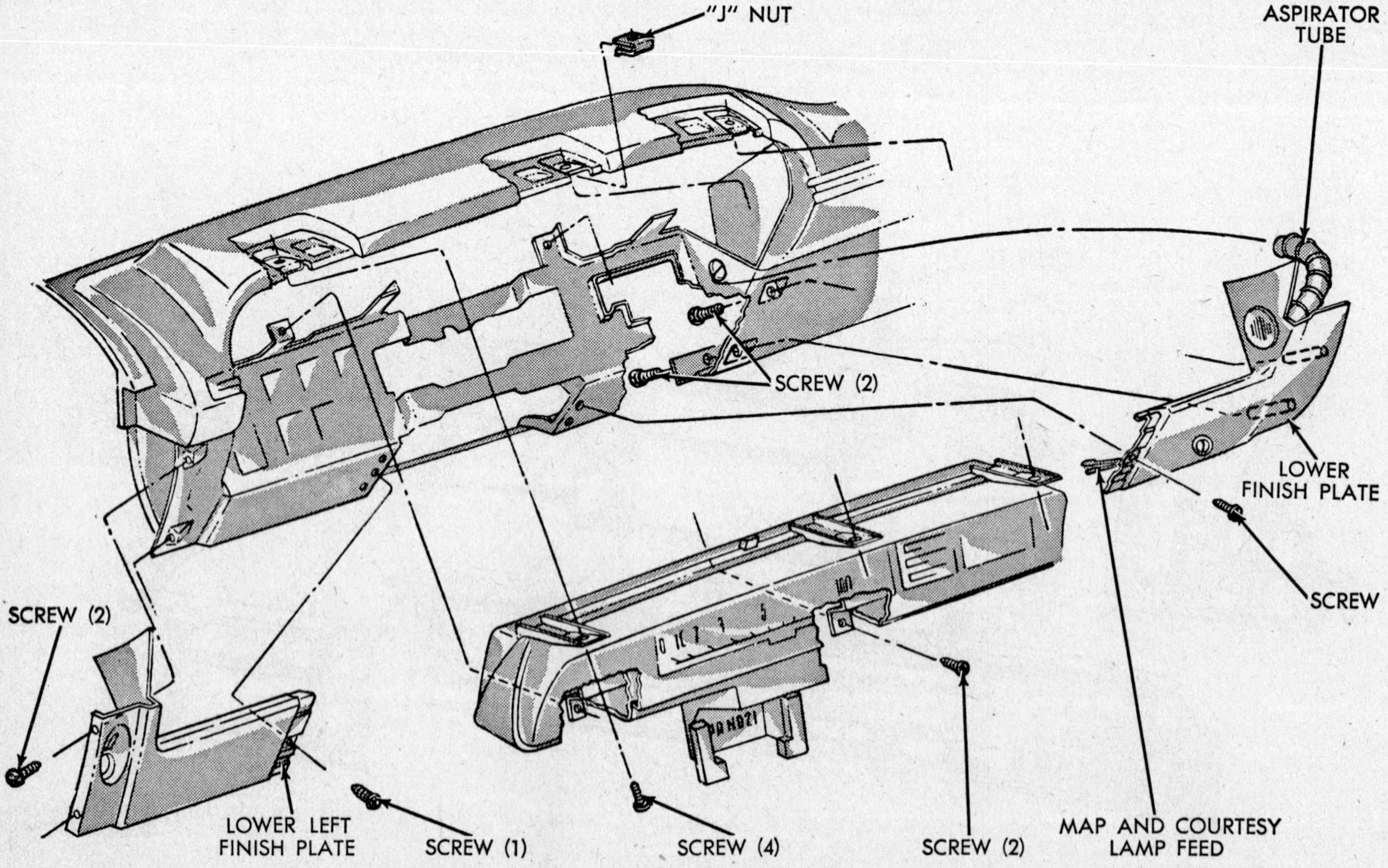

Fig. 13 Instrument cluster. 1969-71 Monaco & Polara

4. Remove screws holding hood release handle and allow handle to lay on floor.
5. Remove column cover screws and drop cover down.
6. Disconnect rear speaker fader control wiring, if equipped.
7. Snip the plastic rib that retains the vent control cable in column cover.
8. Remove the cable retaining clip. Remove the vent control cable through the opening in column cover.
9. Remove the center air conditioner duct by removing the screw from the duct support bracket.
10. Lower the steering column, and disconnect the speedometer cable from under the dash panel.
11. Remove two hidden screws from lower edge of cluster bezel, one screw from side and one screw from top of bezel, Fig. 8.
12. Remove bezel mounting screws and roll cluster down and towards you far enough to disconnect all electrical leads and lamps.
13. Remove cluster bezel from panel.
14. Remove four cluster mounting screws. Remove cluster assembly.

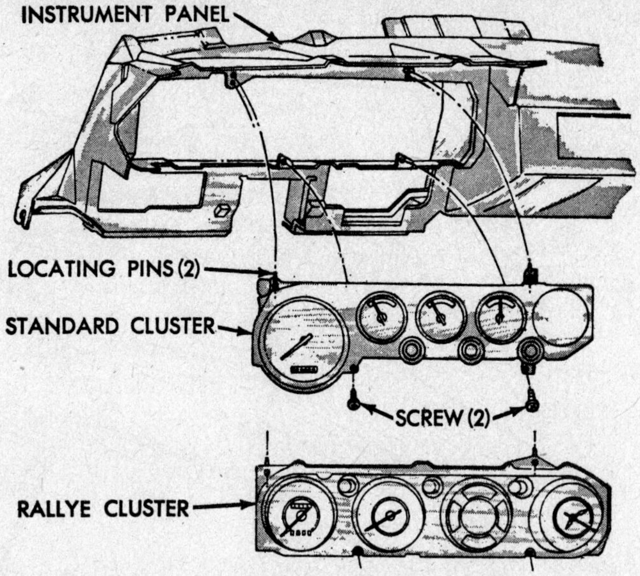

Fig. 15 Instrument cluster. 1970-74 Barracuda & Challenger with Rallye cluster

1972 Fury, Polara & Monaco

1. Disconnect battery ground cable.
2. Remove lamp panel and remove column cover.
3. Lower steering column.
4. Disconnect speedometer cable.
5. Remove radio bezel.
6. Remove bezel mounting screws.
7. Roll cluster bezel down and toward you and disconnect wiring. Remove bezel, Fig. 8.

1972 Dart & Valiant

1. Disconnect battery ground cable.
2. Disconnect speedometer cable and printed circuit board connector from under dash panel.
3. Remove column cover and loosen steering column floor plate attaching screws, remove steering column upper clamp and allow column to rest in lowered position.
4. Tape top of column to prevent damage to painted surface.
5. Remove six mounting screws from cluster, Fig. 9.
6. Before rolling cluster out of instrument panel, reach up behind and above cluster and bend three wire harness clips out of the way. Roll upper edge of cluster out far enough to disconnect wire to ammeter. Disconnect windshield wiper and headlight connectors.
7. Remove cluster assembly.

1971-73 Chrysler & Imperial

1. Remove map lamp, lamp panel end and panel.
2. Remove headlamp switch, radio and A.T.C. controls.
3. Lower steering column cover by removing four screws. Do not disconnect vent control cables.
4. Remove gearshift indicator pointer.
5. Lower steering column and allow steering wheel to rest on seat cushion.
6. Disconnect speedometer cable.
7. Remove five lower cluster mounting screws working through access holes in lower instrument panel.
8. Remove screw holding A.T.C. sensor to instrument cluster bezel and take sensor loose from bezel. On Imperial, remove clock.
9. Working over top of cluster, remove five upper cluster mounting screws, Figs. 10 through 12.
10. Carefully remove cluster by pulling it rearward in car and rolling the bottom slightly up to clear upper legs on cluster. Disconnect wiring as access is gained.

1971-72 Coronet, Charger & Satellite

1. Disconnect battery ground cable.
2. Remove ash tray, radio and heater control panel.
3. Remove upper steering column clamp and lower steering column.
4. Remove instrument cluster mounting screws.
5. Disconnect speedometer cable and wiring and remove cluster.

1971 Polara & Monaco

1. Disconnect battery ground cable.
2. Remove lamp panel and steering column cover.
3. Remove radio trim bezel.
4. Remove left trim bezel and/or spot cooler if so equipped.
5. Remove left A/C duct and center A/C connector if so equipped.
6. From under panel disconnect wiring and speedometer cable.
7. Remove gearshift indicator pointer.
8. Tape column to protect finish and remove three column upper clamp nuts and three bolts at lower support at floor.
9. Carefully lower column and allow steering wheel to rest on seat cushion.
10. Remove eight screws mounting cluster to panel and roll cluster out, disconnecting remaining wiring to high beam indicator, fuel gauge, ammeter and temperature gauge and remove cluster to padded work bench for service, Fig. 13.

1969-71 Dart

1. Disconnect battery ground cable.
2. Remove column opening cover (4 screws).
3. Remove lower column plate (3 bolts) and upper mounting clamp.
4. Remove six cluster mounting screws.
5. Disconnect speedometer cable.
6. Remove fuse block (1 screw).
7. Rock cluster out and release wiring harness from spring clip at back of cluster, then continue to rock cluster out while using a screwdriver to hold harness clear of speedometer.
8. Disconnect left printed circuit board connector and brake system warning light.
9. From front of panel, disconnect right printed circuit board connector and ammeter leads, then complete cluster roll-out.

1970-74 Barracuda, Challenger

1. Disconnect battery ground cable.
2. Remove lamp panel mounting screws and carefully slide panel out and lay it on top of instrument panel. It is not necessary to disconnect wiring.
3. Remove switch bezel mounting screws and allow bezel to hang loose.
4. Remove steering column plate and disconnect column clamps and allow column to rest on seat.
5. Disconnect speedometer cable.
6. Remove six cluster bezel mounting screws, angle bezel out to clear clock button. Reach behind bezel and dis-

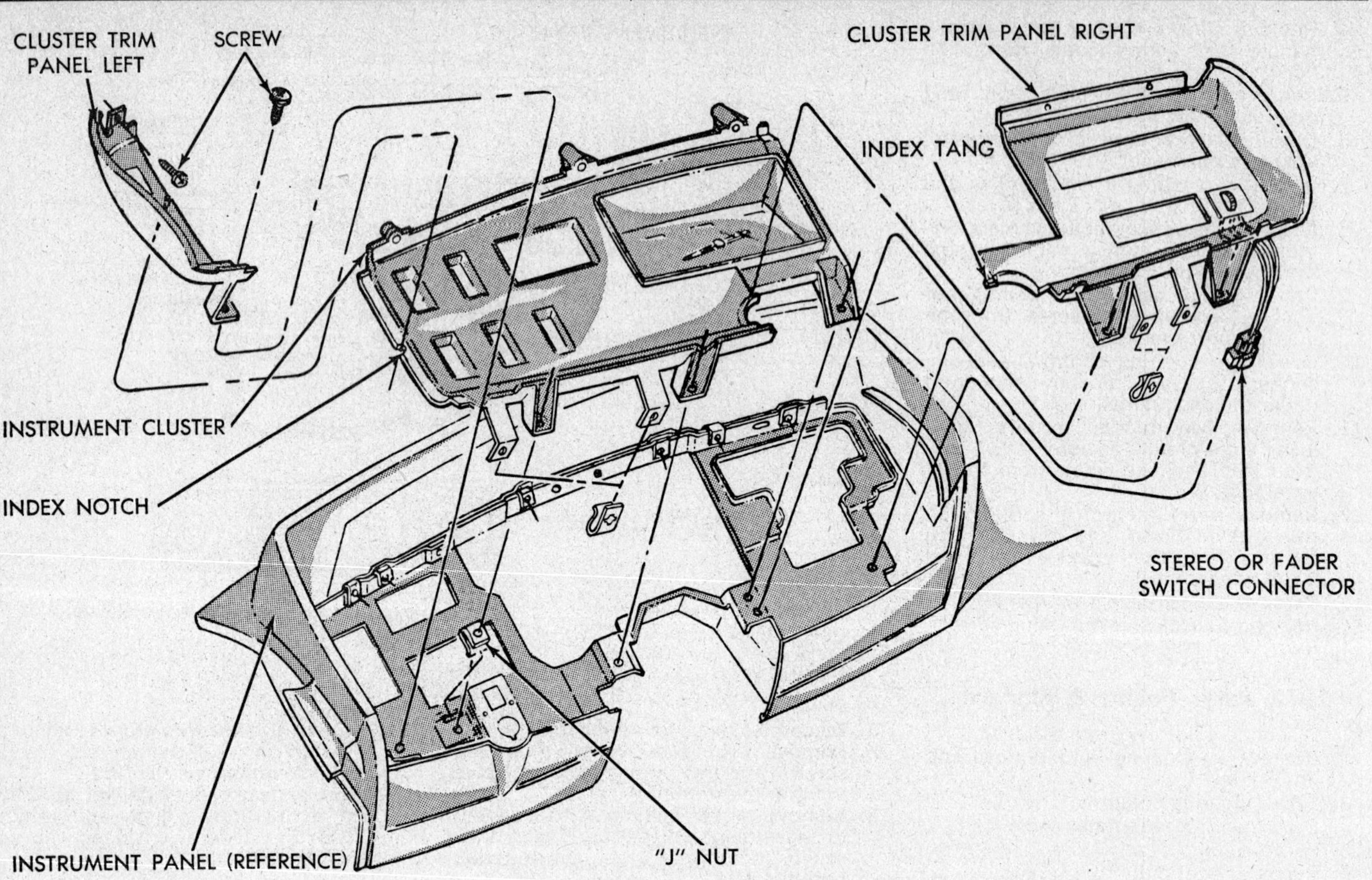

Fig. 16 Instrument cluster. 1971 Fury

connect stereo control wiring, if so equipped.

7. On Rallye cluster, remove clock and odometer reset knobs.
8. Remove four cluster-to-panel mounting screws, disconnect wiring harnesses and remove cluster.

1969-70 Belvedere, Coronet & Satellite

1. Disconnect battery ground cable.
2. Remove steering column cover (4 screws).
3. Roll carpeting down and remove steering column cover plate (4 bolts).
4. Remove column clamp at instrument panel (2 nuts).
5. Remove upper trim molding (6 screws if equipped).
6. Remove left side trim molding (1 screw if equipped).
7. Remove left side trim plate (1 screw and 1 T-bolt, if equipped).
8. Remove radio trim plate (2 screws).
9. Remove switch bezel (4 screws).

NOTE: Steps 10-12 do not apply to 1970 models.

10. Remove ignition switch.
11. Remove A/C center opening cover (if equipped).
12. Remove lower left trim pad (6 screws from under panel and 4 screws from front of instrument panel).
13. Disconnect speedometer cable.
14. Remove six screws attaching cluster to panel, rock cluster out far enough to reach and disconnect wiring harness and connectors and remove cluster.
15. Reverse procedure to install.

1969-70 Charger

1. Disconnect battery ground cable.
2. Remove steering column opening cover (4 screws).
3. Remove steering column lower support plate (3 bolts).
4. Remove upper mounting bracket support bolts.
5. Disconnect speedometer cable.
6. Remove five screws mounting cluster to panel. A small screwdriver may be used in removing the medallion from the cluster bezel.
7. Release wire harness from retainer clips and roll cluster out of panel far enough to disconnect wiring from ammeter, switches, tachometer, clock, light bulbs and printed circuit board connectors, then complete cluster roll-out.

1969-70 Monaco & Polara

1. Disconnect battery ground cable.
2. Remove steering column cover.
3. Remove gear shift indicator from the column.
4. Remove lower column floor plate.
5. Remove upper column mounting nuts and lower steering column down to seat.
6. Remove switch and radio bezels, cluster trim pad and trim bezel.
7. Disconnect clock reset cable at the instrument panel lower reinforcement.
8. Disconnect clock lead and remove five cluster mounting screws.
9. Roll cluster out slightly and disconnect the speedometer cable.
10. Disconnect gear shift indicator lamp, main harness and alternator gauge leads and remove cluster.

1970 Chrysler

1. Disconnect battery ground cable.
2. Remove lower steering column cover.
3. Remove three outside floor plate mounting screws.
4. Remove steering column ground strap, disconnect clamp and lower steering column.
5. Remove left ash tray, radio and heater controls.
6. Remove two vent control screws and allow assembly to hang free.
7. Remove map light.

4. Remove motor mounting nuts, pull motor away from bulkhead and remove motor crank arm retaining nut. After crank arm is removed from motor shaft, remove drive link assembly from under left side of panel.

NOTE: In heater equipped models, remove motor drive crank arm retaining nut and pry crank arm off motor shaft. Gently pry drive link and bushing from left pivot crank arm pin and withdraw assembly from under panel. Remove motor drive crank arm from drive link after removal of assembly from vehicle.

5. To remove connecting link from pivots, remove glove box, reach through opening and gently pry bushing and link from right pivot pin. Lift the link from the pivot crank arm pin and repeat operation at left pivot. Withdraw from under left side of panel. The wiper pivots can be removed if required by the removal of the ½" mounting nuts at this time.

1971-72 With Concealed Wipers

1. Remove wiper arms and blade assemblies and the cowl screen to provide access.
2. Disconnect battery ground cable.
3. Remove crank arm nut and crank arm from motor shaft.
4. Remove bolts mounting left and right pivots to body.
5. Remove links and pivots through cowl top opening. The linkage and pivots can be serviced on bench after removal from vehicle.

1969-70 Chrysler, Imperial, Polara & Monaco

To service either the drive link or the connecting link, it is necessary to remove the wiper arms and blades, windshield lower moulding and cowl grille panel to gain access to the wiper system. Before starting the installation procedure, make certain the wiper system is in the "Park" position and the battery ground cable disconnected.

1. With connecting and drive links assembled as a unit, insert links through cowl top opening and bolt pivots in position.
2. Position motor crank arm on motor shaft and tighten to 140 inch-pounds. Reconnect battery ground cable.
3. Test wiper system operation and then replace cowl screen and windshield lower moulding.
4. Using a pin or drill, install and adjust wiper arm and blade assemblies.

1970 Barracuda, Challenger, Fury

1. Disconnect battery ground cable.
2. Remove motor crank arm.
3. Remove pivot bolts and remove links and pivots through top cowl opening.

NOTE: If servicing of the mechanism on the 2-speed motor crank is required, be sure that during reassembly, the link is positioned between the ears of the cover retainer.

1969-70 Dart, Coronet, Charger, Valiant, Belvedere, Satellite & 1969 Barracuda

NOTE: On air conditioned models, after disconnecting the battery ground cable, remove left spot cooler duct and carefully pry the link and bushing from the pivot pin. Remove motor mounting nuts, pull motor away from bulkhead and remove motor crank arm retaining nut and crank arm. Remove drive link assembly from under panel.

On models without air conditioning, remove crank arm from motor, remove drive link from left pivot pin and withdraw assembly from under panel. Remove crank arm from drive link after assembly is removed from vehicle. Connecting link can be removed from pivots by removing glove box and reaching through the opening to pry bushing and link from pivot pin. Withdraw from under left side of panel.

1. Install bushing on motor crank arm pin, position drive link on bushing so large side of pivot bushing faces away from drive crank arm. Large side of bushing will be on same side of link as crank arm retainer.
2. Install spring washer with convex side towards link and install retainer. *If retainer was distorted in removal, it should be replaced.*
3. In heater equipped vehicles, insert drive link assembly under left side of panel, position crank arm on motor shaft, indexing flats on shaft with flats on arm and install retaining nut.

NOTE: On air conditioned models, install drive link from under instrument panel. Install crank arm on motor shaft from engine side of bulkhead. Secure motor with three nuts. Press plastic bushing over pin on left pivot.

4. Insert connecting link into place with "R" (right side) and "L" (left side) facing instrument panel side. Press link bushings onto pivot crank pins.
5. Reconnect battery cable and test operation. Install glove box if necessary.

W/S WIPER SWITCH

1976 Aspen & Volar'e

1. Remove cluster bezel.
2. Remove switch mounting plate attaching screws and pull switch and plate assembly outward.
3. Pull knob off switch and remove switch retaining nut
4. Disconnect electrical connector and remove switch.

1975-76 Cordoba, Coronet, Charger SE & Fury

1. Disconnect battery ground cable.
2. Remove instrument cluster upper bezel and the switch escutcheon mounting screw.
3. Remove switch to cluster housing retaining screws and pull headlamp switch to the "On" position.
4. Slide escutcheon on shaft toward rear of vehicle and rotate upward, thereby gaining clearance for switch removal.
5. Pull switch from cluster housing, disconnect electrical connector and remove switch.
6. Reverse procedure to install.

1975-76 Chrysler Newport, New Yorker, Gran Fury, Imperial & Monaco

1. Disconnect battery ground cable.
2. Remove instrument cluster bezel.
3. Remove switch mounting screws.
4. Disconnect electrical connector and remove switch from bezel.

1974 Chrysler, Fury, Imperial, Monaco

1. Remove instrument cluster bezel.
2. Remove switch mounting screws.
3. Disconnect electric leads and remove switch from bezel.

1973-74 Charger, Coronet, & Satellite

1. Remove wiper switch knob.
2. Remove wiper switch bezel nut.
3. Remove wiper switch from instrument cluster and disconnect electrical leads.

1973 Fury, Polara & Monaco

1. Remove column cover and A/C duct.
2. Remove left spot cooler duct.
3. Reach up under dash panel and disconnect wiring from switch.
4. Remove two switch mounting screws and remove switch.

1971-73 Chrysler & Imperial

1. Remove instrument cluster.
2. Remove screws securing switch and plate to cluster housing.
3. Remove the assembly.
4. Remove screws holding switch to mounting plate and remove switch.

1971-72 Fury, Polara & Monaco

1. Roll cluster down and towards you.

NOTE: It is not necessary to remove bezel assembly, only roll it far enough to gain access to switch mounting screws and wiring.

2. Disconnect leads to switch and unfasten switch.

1970-74 Barracuda, Challenger

1. Disconnect battery ground cable.
2. Remove lamp panel retaining screws and carefully slide panel out and lay it on top of instrument panel.
3. Remove bezel retaining screws and slide assembly out and disconnect wiring harness.
4. Pull control knob from shaft. Remove mounting nut and remove switch from bezel.

1970-72 Coronet & Satellite

1. Remove steering column cover.
2. Lower steering column and allow it to rest on seat.
3. Remove lower trim bezel.
4. Remove switch mounting screws and disconnect wiring. *Place wires on switch before installing switch.*

1969 Coronet, Belvedere & Satellite

The wiper switch is serviced in the same manner as the headlight switch.

1969-76 Dart, 1969 Barracuda & 1969-76 Valiant

1. Remove air conditioning duct (if so equipped).
2. On 1969-71 models, loosen knob set screw and on all models, pull knob from switch.
3. Remove spanner nut.
4. Remove wiring harness from clip. Disconnect wiring and remove switch.

1969-70 Polara & Monaco

The wiper switch is serviced in the same manner as the headlight switch.

1969-70 Fury & V.I.P.

1. Remove all A/C ducts from underside of panel if so equipped.
2. From under panel, disconnect electrical leads.
3. Remove two mounting screws and remove switch.

1969-70 Chrysler & Imperial

1. Remove lamp panel.
2. Remove instrument panel to gain access to switch.
3. Remove two switch mounting screws and remove switch.

1969-72 Charger

The wiper switch is serviced in the same manner as the headlight switch.

RADIO, REPLACE

NOTE: When installing radio, be sure to adjust antenna trimmer for peak performance.

1975-76 Cordoba, Coronet, Charger SE & Fury

1. Disconnect battery ground cable.
2. Remove lower cluster bezel.
3. Remove radio knobs and mounting screws.
4. Remove radio rear support bracket mounting nut on radio right side.
5. Disconnect electrical leads from radio and pull radio from cluster housing.
6. Reverse procedure to install.

1975-76 Chrysler Newport, New Yorker, Gran Fury, Imperial & Monaco

Refer to the "1974 Fury, Monaco, Chrysler & Imperial" procedure for radio replacement.

1974 Barracuda & Challenger

1. Disconnect battery ground cable.
2. Remove radio retaining screws and disconnect antenna lead.
3. Remove radio support bracket nut and move radio forward and down.
4. Disconnect electrical and speaker leads and remove radio.

1974 Charger, Coronet, & Satellite

1. Disconnect battery ground cable.
2. Remove ash tray and right radio retaining screw which is accessible through lower left corner of ash tray housing.
3. Loosen radio support bracket nut and remove radio knobs and mounting nuts.
4. Disconnect antenna and electrical leads and remove radio.

1974-76 Valiant & Dart

1. Disconnect battery ground cable.
2. If equipped with A/C, remove outlet duct assembly, ash tray and ash tray housing mounting screws.
3. Remove radio knobs, mounting nuts and radio support bracket screw.
4. Lift radio, relocate ash tray housing for additional clearance and disconnect antenna lead.
5. Move radio forward and down, disconnect electrical and speaker leads and remove radio.

1974 Fury, Monaco, Chrysler & Imperial

1. Disconnect battery ground cable.
2. Remove bezel. On Fury and Monaco, also remove sub bezel.
3. Remove lamp assembly from radio (monaural only).
4. Remove radio to panel retaining screws and instrument panel upper cover.
5. Working through top of panel, disconnect antenna lead and remove radio mounting bracket nut.
6. On monaural radio, disconnect speaker lead from speaker. On stereo radio, disconnect speaker lead from radio.
7. Pull radio out from panel and disconnect electrical lead.

1971-73 Coronet, Charger & Satellite

1. Disconnect battery ground cable.
2. Remove ash tray.
3. Remove radio knobs and mounting nuts.
4. Remove radio.

1971-73 Polara & Monaco

1. Disconnect battery ground cable.
2. Remove nine lamp panel mounting screws, lower lamp panel slightly, disconnect lamp harness from main harness and remove lamp panel.
3. Remove steering column cover and radio trim bezel.
4. Remove center lower A/C duct, if so equipped.
5. Disconnect wiring and antenna from radio.
6. Remove radio support bracket.
7. Remove radio mounting bolts.
8. Remove radio down through bottom of instrument panel carefully to avoid damage to vacuum hoses and wiring.

1969-70 Polara & Monaco

1. Disconnect battery ground cable.
2. Remove Auto-Temp. control (if so equipped).
3. Remove radio bezel and two radio mounting bolts at front of instrument panel.
4. Remove air conditioning duct (if so equipped).
5. Disconnect electrical leads, loosen radio mounting bracket stud nut and slide radio and stud towards front of car from mounting bracket. Carefully remove radio from under panel.

1969-73 Dart, Challenger, Barracuda, Valiant

1. Disconnect battery ground cable.
2. Disconnect electrical leads and remove knob.
3. From under panel, remove two radio mounting nuts and remove radio mounting bracket. Remove radio down and out from under panel.

1969-73 Chrysler & Imperial

1. Disconnect battery ground cable.
2. Remove left ash receiver and steering column cover.
3. Unscrew stereo tape reset knob if so equipped.
4. Disconnect all electrical leads.
5. Loosen defroster vacuum actuator and move it to facilitate radio removal.

6. Remove two radio mounting screws through access openings in lower panel. On search-tune radios, remove knobs bezels and nuts.
7. Remove support bracket screw from lower reinforcement. *Support radio.*
8. Working through ashtray opening, remove support bracket from radio.
9. Remove radio from under panel.

1969-70 Charger

1. Disconnect battery ground cable.
2. Remove radio finish plate.
3. On A/C models, remove lower center air duct, left air duct and upper center duct.
4. Remove radio mounting bracket.
5. Unfasten radio from instrument panel (2 screws).
6. Disconnect antenna and speaker leads and remove radio.

1969-72 Fury & V.I.P.

1. Disconnect battery ground cable.
2. Remove lamp panel and steering column cover.
3. Remove radio trim bezel.
4. Remove center lower A/C duct if so equipped.
5. Disconnect electrical leads.
6. Remove radio support bracket and mounting bolts.
7. Remove radio down through bottom of panel carefully to avoid damage to vacuum and electrical connections.

1969-70 Belvedere, Coronet & Satellite

1. Disconnect battery ground cable.
2. Remove radio upper trim panel.
3. Remove radio finish plate.
4. Remove radio rear mounting nut from bracket.
5. Disconnect electrical wiring and antenna lead.
6. Remove two screws from front of instrument panel and remove radio.

HEATER CORE REMOVAL

Before attempting to remove a heater core, disconnect the battery ground cable, drain the radiator and remove inlet and outlet hoses from heater assembly in engine compartment. Refer to Figs. 21 and 22.

1975-76 Charger SE, Cordoba, Coronet & Fury

Less Air Conditioning

1. Disconnect fusible link and drain coolant.
2. Disconnect heater hoses from core tubes and plug tubes to prevent coolant from spilling into interior of vehicle.
3. Remove nuts from around blower motor and one from center of dash panel, then remove lower instrument panel bezel, glove box and door.
4. Disconnect antenna lead from radio, then remove screw from housing to plenum support rod on right side of housing above outside air opening.
5. Disconnect air door control cables and wire from blower motor resistor.
6. Tip heater assembly down and out from under instrument panel.
7. Remove screws retaining front cover to heater assembly, then cut rubber seal in two places where front cover separates cover from housing.
8. Remove core tube retaining screw located behind housing and between core tubes.
9. Remove sponge rubber gaskets from core tlbes and remove heater core from housing.

With Air Conditioning

1. Disconnect fusible link and drain coolant.
2. Disconnect heater hoses from core tubes and plug tubes to prevent coolant from spilling into interior of vehicle.
3. Discharge A/C refrigerant and disconnect line from H-valve and cover tubing sealing plate. Remove expansion valve from evaporator and cover evaporator sealing plate and both sealing surfaces of expansion valve.
4. Disconnect blower motor wires and remove blower motor cooling tube.
5. Remove glove box, ash tray and housing bezel, right lap cooler, right side cowl panel and air distribution duct.
6. Remove A/C mode door vacuum actuator from bracket and shift actuator forward on top of unit.
7. Disconnect connector from seat belt interlock control module and disconnect wires from blower motor resistor.
8. Disconnect antenna lead from radio and remove radio.
9. Remove housing retaining nuts through engine compartment, then remove rubber drain tube.
10. Remove support bracket from rear of unit to plenum.
11. Pull unit back so that tubes clear dash panel, then rotate so that right end of unit comes out first from under the instrument panel first.
12. Carefully remove plenum air seal, then disconnect vacuum hose from inlet air door actuator.
13. Remove the 14 clamps and 4 screws and separate housings.
14. Carefully lift left half of housing seal from rear cover. Do not remove entire seal as the lower portion is a water seal.
15. Remove two retaining screws from mounting plate and one from between heater core tubes and lift heater core from housing.

1974-76 Chrysler exc. Cordoba, 1974-76 Imperial, Monaco, 1974 Fury, 1975-76 Gran Fury

1. Disconnect battery and drain radiator.
2. Disconnect heater hoses and plug hose fittings on heater core.
3. Move front seat rearward and remove instrument panel lower cover. Unplug antenna lead from radio and disconnect upper level vent vacuum line.
4. Disconnect the upper level vent from heater housing and the mounting bracket from dash, then swing duct back.
5. Disconnect electrical connectors from blower motor resistor and control cable from clip and crank.
6. Disconnect support bracket and swing bracket out of way.
7. From engine compartment, remove five retaining nuts from studs.
8. Roll or tip housing out from under instrument panel and disconnect control cable from clip and the blend air door crank.
9. Remove core tube locating screw from between core tubes, remove housing retaining nuts, separate housing and remove core.

1971-76 Dart & Valiant

Less Air Conditioning

1. Remove heater hoses to dash panel seal and retainer plate.
2. Remove heater motor seal retainer plate and seal from dash panel.
3. Disconnect control cables from heater.
4. Remove heater motor resistor wire from resistor.
5. Remove defroster tubes from heater.
6. Disconnect heater housing support rod from outside air duct.
7. Remove heater assembly.

With Air Conditioning: 1973-76

1. Remove battery, drain radiator and disconnect heater from unit.
2. Remove core tube seal nut, bracket and seal.
3. Remove A/C duct, ash tray and housing and radio.
4. Remove heat-defrost vacuum actuator pot; let hang by rod.
5. To remove heat distribution duct, remove three screws on front cover, two on each end and work housing out of lip and remove to left side.
6. Remove left defroster duct. Remove right defroster duct and let it hang from top.
7. To remove the rear distribution housing, reach through radio opening and remove three screws top and bottom and one at left end.
8. With housing off, core will be settling loose. Separate seal and lift out.

With Air Conditioning: 1971-72

1. Disconnect battery and remove air cleaner.
2. Remove air outlet assembly, glove box and right defroster tube.
3. Disconnect wiring from resistor block, vacuum hoses from outside recirculating air door actuator, temperature control cable, evaporator temperature control switch control cable and heater core ground wire.
4. Remove screw securing heater to evaporator assembly.
5. Disconnect heater housing support rod from outside air duct.
6. Remove heater assembly.
7. Remove outside recirculating air door

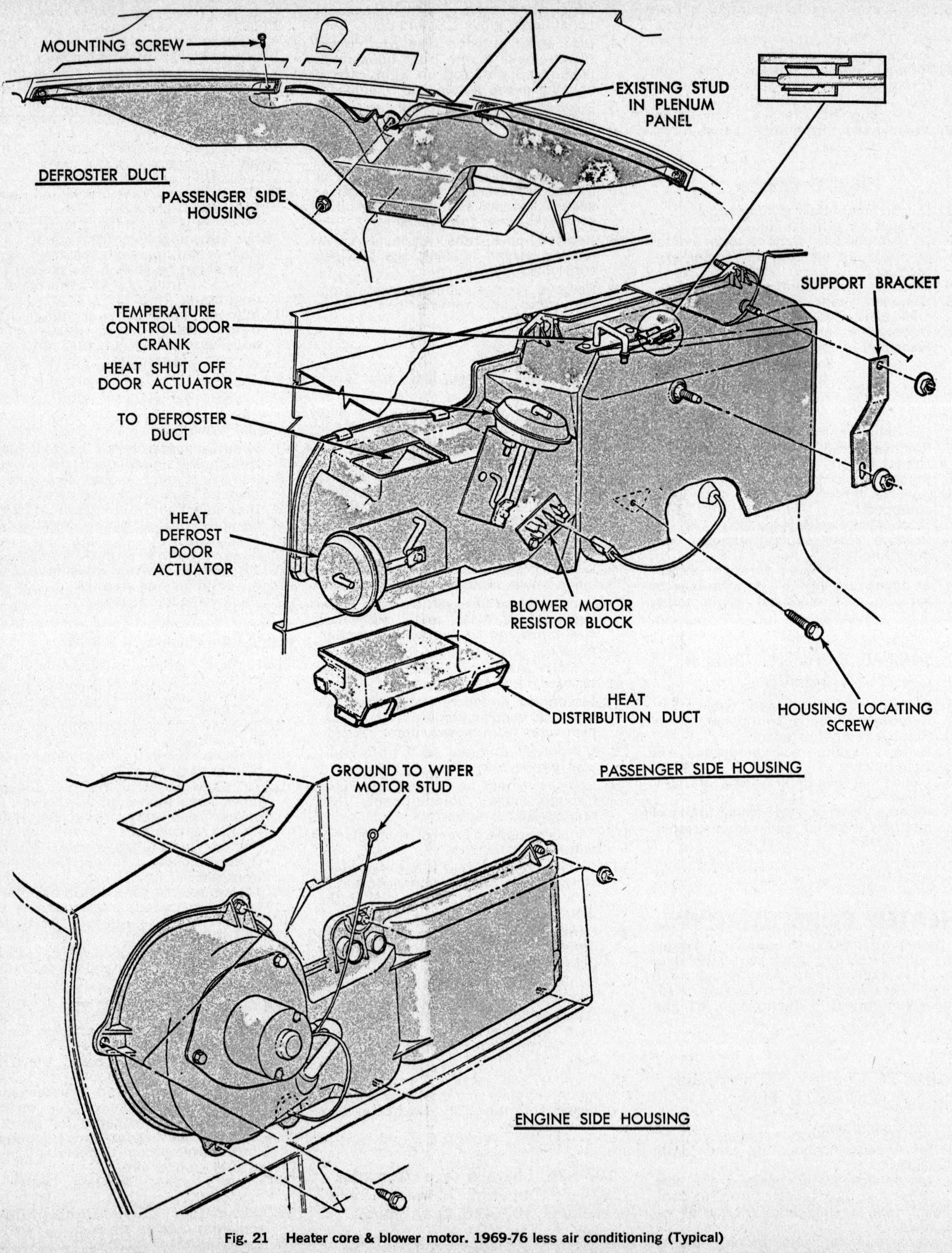

Fig. 21 Heater core & blower motor. 1969-76 less air conditioning (Typical)

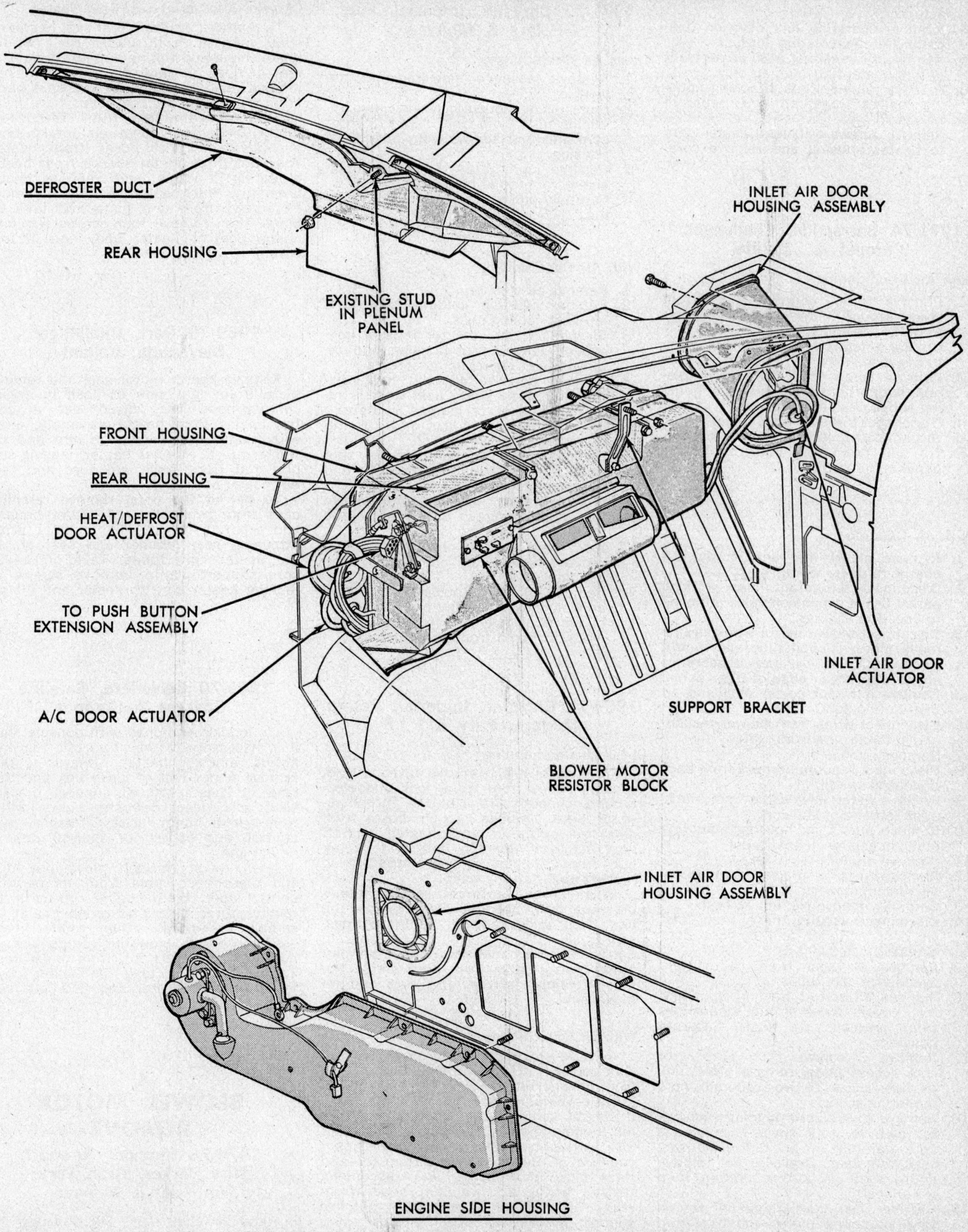

Fig. 22 Heater core & blower motor. 1969-76 with air conditioning (Typical)

actuator.

8. Remove operating link between bell-crank and recirculating door.
9. Remove air inlet seal from either front or rear heater housing half only.
10. Remove retainer clips attaching housing halves together and separate halves.
11. Remove screws attaching heater core to heater housing and remove core.

1971-74 Barracuda, Challenger, Coronet & Satellite

Less Air Conditioning

1. Remove three mounting nuts from studs around blower motor and remove flange and air seal.
2. Unplug antenna from radio and place to one side.
3. Remove screw from housing to plenum support rod on right side of housing above fresh air opening.
4. Disconnect three air door cables.
5. Disconnect wires from blower resistor.
6. Tip unit down and out from under panel.

With Air Conditioning

1. Remove air cleaner and slowly discharge refrigerant from system.
2. Disconnect refrigerant lines at dash panel. Leave expansion valve attached to line. Cap all lines.
3. Disconnect blower motor wires and remove motor cooling tube and motor.
4. Remove glove box and appearance shield from lower edge of dash panel.
5. Remove left spot cooler duct and air distribution duct.
6. Disconnect wires from blower resistor and antenna wire from radio.
7. Remove radio.
8. Disconnect vacuum harness from back of control switch.
9. Remove water valve cable from bracket on left end of housing.
10. Remove nuts from housing mounting studs in engine compartment.
11. Remove rubber drain tube.
12. Remove support bracket from housing plenum panel.
13. Carefully remove plenum air seal.
14. Disconnect vacuum hose from inlet air door actuator and by-pass door actuator.
15. Remove air seal from heater and evaporator core tubes.
16. Remove 18 screws holding front and rear covers together and one screw from between core tubes. Separate housings.
17. Remove 3 screws from evaporator core access plate, remove plate, this provides access to two evaporator core mounting screws.
18. Remove four screws holding evaporator core to front cover and remove core. Carefully lift left half of housing seal from rear cover. Do not remove entire seal as bottom portion is a water seal.
19. Remove two core retaining screws from mounting plates and one from between core tubes in back of rear cover. Lift core out of housing.

1971-73 Chrysler, Imperial, Fury, Polara & Monaco

Less Air Conditioning

1. Remove passenger side housing from vehicle.
2. From inside housing, remove two retaining nuts from right side of heater core and four screws from outside of housing.
3. Remove core tube locating metal screw from top of housing.
4. Carefully pull heater core out of housing.

With Air Conditioning

1. Remove air cleaner.
2. Remove steering column cover and left spot cooling duct.
3. Disconnect two actuator rods at linkage on left side of housing and remove two cover retaining screws.
4. Remove screws retaining heat distribution duct. When heat duct is removed, screws or clamps in bottom lip of front cover can now be removed.
5. Remove glove box and center spot cooler, air distribution housing and right spot cooler duct.
6. From glove box opening, remove two top retaining screws and three screws from right side of housing.
7. Disconnect wires at resistor block and vacuum hoses from air inlet housing actuator.
8. Remove nut from housing end of support bracket and swing bracket up out of way. Carefully roll front cover and heater core out from under panel.

1969-70 Chrysler, Imperial, Polara, Monaco, Fury & V.I.P

Less Air Conditioning

From under dash, remove antenna lead, vacuum hoses from trunk lock and electrical connectors from blower motor resistor block. Remove vacuum hoses from defroster actuator and heater shutoff door actuator. Remove bottom retaining nut from support bracket and swing bracket up out of the way.

In engine compartment, remove retaining nuts from housing. Remove locating bolt from under bottom center of passenger side housing. Roll or tip housing out from under panel and remove temperature control cable from door crank. Remove retaining screws and nuts and remove core.

With Air Conditioning

Remove steering column cover and left spot cooler duct. Disconnect actuator rods from linkage on left side of housing. Remove cover and distribution duct screws, duct and screws from front cover lower lip. Remove glove box, center spot cooler, air distribution housing and right spot cooler duct. Remove screws from top and right side of housing and disconnect resistor wiring and vacuum hoses from recirculating housing actuator. Remove support bracket to housing nut, swing bracket up and roll front cover and heater core from under instrument panel.

Chrysler & Imperial Rear Seat Heater

To remove core from rear seat heater on 1969 models, from under car, remove heater hose and tubing support clamps at rear of floor pan and drain system. Then remove rear seat cushion and seat back. From inside car remove hoses from heater, and spare tire from trunk. Disconnect motor feed wire, fresh air intake hose and floor air duct hoses from heater. Take out three metal screws from heater mounting brackets and remove heater assembly. With heater on bench, remove 11 screws from end plate and take off plate. Then remove four screws retaining heater core to heater body and lift out core.

1969-70 Dart, Challenger, Barracuda, Valiant

Remove heater motor seal and retainer plate from right side of dash in engine compartment. Then disconnect all control cables from heater assembly, along with heater motor resistor wire and defroster tubes. Remove heater housing support rod from fresh air duct and take out heater assembly.

To get at the core, remove retaining clips from two halves of heater housing and separate halves. Then remove screw attaching seal retainer and seal around the heater core tubes. Take off heater core support clamp, remove screws attaching heater core to heater and lift out core.

1969-70 Belvedere, Satellite, Charger & Coronet

On models equipped with console shift, it is necessary to move console rearward before removing heater. After doing this, remove upper half of glove box and then take off heater-to-cowl support bracket. Next, disconnect defroster hoses, wiring from heater motor resistor, fresh air vent control, and cables for shut-off door at heater end.

Now, reach through glove box door and disconnect cable from temperature control door. From engine compartment, remove three nuts that retain heater assembly to firewall. Then rotate heater assembly until mounting studs are up and lift heater assembly from under dash. To remove core, take off heater cover from front of heater and take out core mounting screws.

BLOWER MOTOR REMOVE

1975-76 Chrysler Newport, New Yorker, Gran Fury, Imperial & Monaco

Refer to the "1969-74 Chrysler, Imperial, Fury, Polara & Monaco" procedure for blower motor removal.

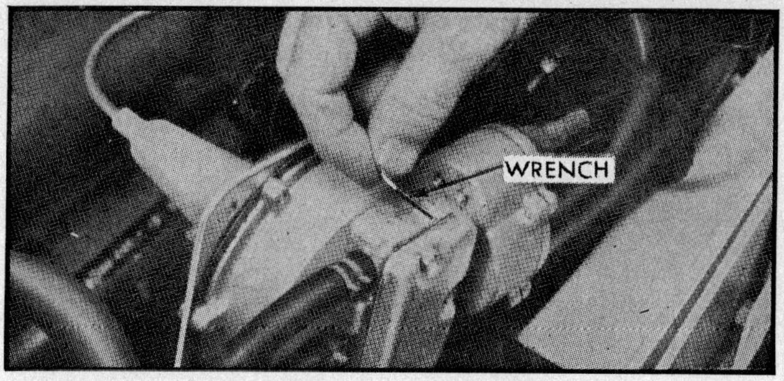

Fig. 23 Speed Control lock-in screw adjustment. 1969-76

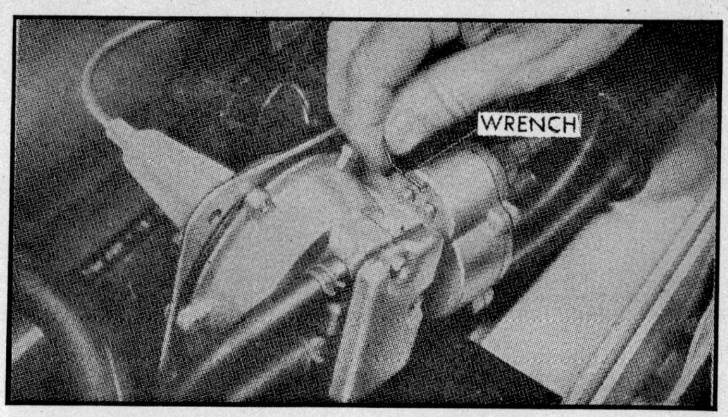

Fig. 24 Speed Control cut-in screw adjustment. 1969-70

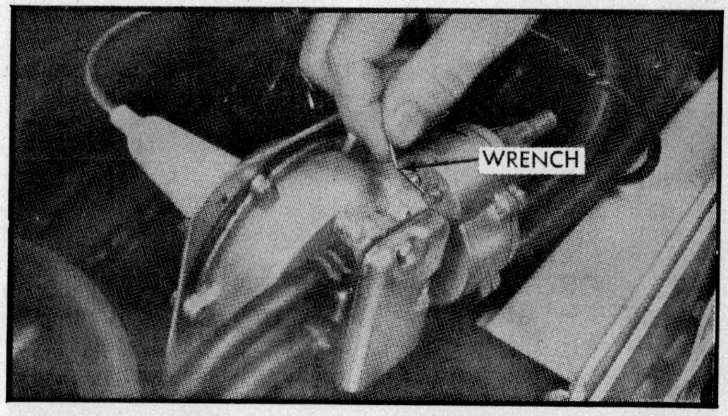

Fig. 25 Speed Control cut-out screw adjustment. 1969-70

1970-74 Challenger, 1971-76 Coronet, Charger, 1975-76 Cordoba & Gran Fury, 1971-74 Barracuda & Satellite

Less Air Conditioning

The heater assembly must be removed to service the blower motor. See procedure for *Heater Core Removal* to remove assembly, then proceed to *Remove Blower Motor* as follows:

1. Disconnect blower motor lead from resistor block. Disconnect the ground wire from the mounting plate, Fig. 21.
2. Remove six screws and retaining clips holding blower motor mounting plate to housing. Separate the blower motor and mounting plate from the housing, Fig. 21.
3. Remove the blower wheel from the motor shaft.
4. Remove the two retaining nuts and separate the motor from the mounting plate. Remove the motor.

With Air Conditioning

NOTE: All service to the blower motor is made from the engine compartment side.

1. Disconnect the feed wire at the connector and the ground wire. Remove the air tube.
2. Remove three screws located on the outer surface of the mounting plate.
3. Remove the mounting plate and blower motor as an assembly, Fig. 22.

1969-74 Chrysler, Imperial, Fury, Polara & Monaco

The blower motor is mounted to the engine side housing under the right front fender between the inner fender shield and the fender, Figs. 21 and 22. The inner fender shield must be removed to service the blower motor.

1969-70 Coronet, Charger & Satellite; 1969-74 Dart & Valiant

Less Air Conditioning

The heater assembly must be removed to service the blower motor. See procedure for *Heater Core Removal* to remove assembly, then proceed to *Remove Blower Motor* as follows:

1. Disconnect lead wire from the blower motor to the heater assembly.
2. Remove motor cooler tube.
3. Remove heater backplate assembly from heater.
4. Remove fan from motor shaft. Remove blower motor from backplate, Fig. 21.

With Air Conditioning

NOTE: All service to the blower motor is made from the engine compartment side.

1. Disconnect the feed wire at the connector and the ground wire. Remove the air tube.
2. Remove three screws located on the outer surface of the mounting plate.
3. Remove the mounting plate and blower motor as an assembly, Fig. 22.

SPEED CONTROLS

1969-76 Speed Control

Servo Adjustments

There are three adjustment set screws in the servo housing, Figs. 23, 24 and 25. The adjustment of these set screws have been factory set and under normal conditions there should be no need for altering the factory setting during the life of the vehicle.

Need for adjustment can be determined only after accurate diagnosis of the system operation. If adjustment is found to be necessary, perform the appropriate adjustment outlined below; if screw is loose stake side of servo housing adjacent to screw to insure a snug fit.

Lock-in Screw Adjustment, Fig. 23

Lock-in accuracy will be affected by poor engine performance (need for tune-up), loaded gross weight of car (trailering), improper slack in control cable. After the foregoing items have been considered and the speed sags or drops more than 2 to 3 mph when the speed control is activated, the lock-in adjusting screw should be turned counter-clockwise approximately $1/4$ turn per one mph correction required.

If a speed increase of more than 2 to 3 mph occurs, the lock-in adjusting screw should be turned clockwise $1/4$ turn per one mph correction required.

CAUTION: This adjustment must not exceed two turns in either direction or damage to the unit may occur.

Cut-in Speed Adjustment, Fig. 24

This adjustment regulates the minimum road speed at which the low speed inhibit switch allows the speed control to be

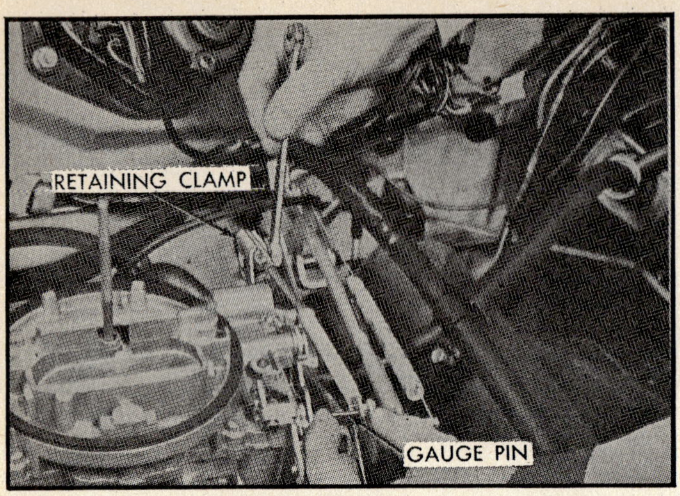

Fig. 26 Speed Control servo cable throttle adjustment. 1969-76

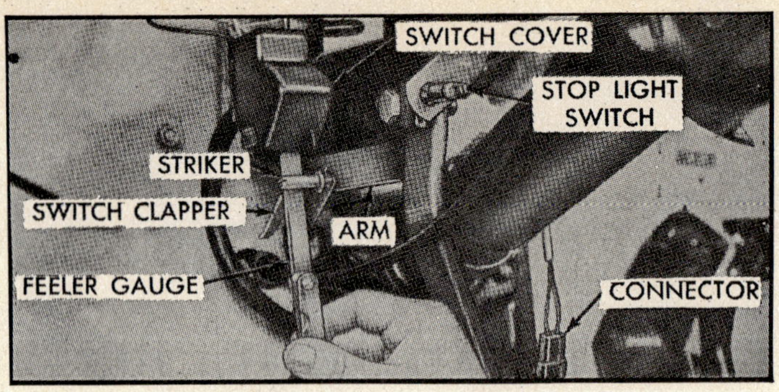

Fig. 28 Checking clapper and striker pin clearance on intermediate size cars. 1969 Speed Control

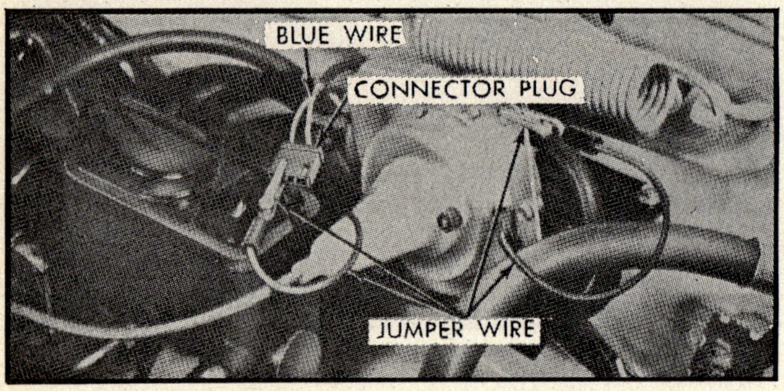

Fig. 27 Speed Control brake switch adjustment. 1969-70

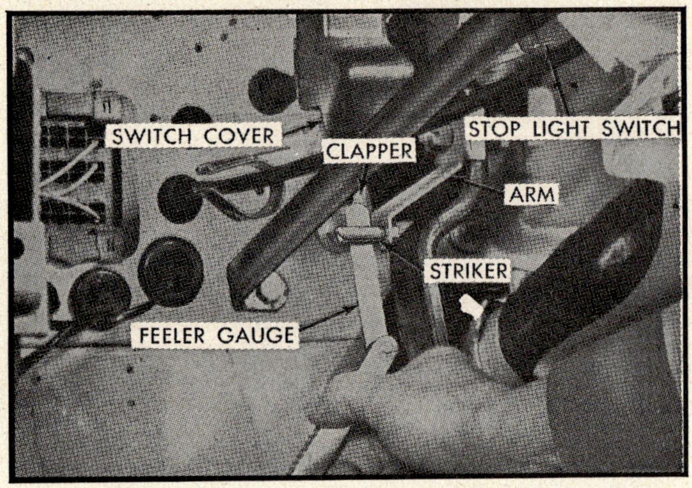

Fig. 29 Checking clapper and striker pin clearance on full size cars. 1969 Speed Control

activated. This should range from 25 to 33 mph. If cut-in speed is too low, turn set screw counter-clockwise. If too high, turn set screw clockwise. Make adjustments in ⅛ turn increments; total adjustment must not exceed two turns.

Cut-out Adjustment, Fig. 25

This adjustment affects the road speed at which the system is deactivated during deceleration. Turning the screw clockwise increases road speed at which the system deactivates (cuts out). A counterclockwise adjustment decreases the cut-out speed. The desired cut-out speed should occur approximately 5 mph below the cut-in setting. Make adjustments in ⅛ turn increments; total adjustment must not exceed two turns.

Throttle Cable Adjustment, Fig. 26

Optimum servo performance is obtained with a given amount of free play in the throttle control cable. To obtain proper free play, insert a $\frac{1}{16}$" diameter pin between forward end of slot in cable end of carburetor linkage pin (hair pin clip removed from linkage pin). With choke in full open position and carburetor at curb idle, pull cable back toward dash panel without moving carburetor linkage until all free play is removed. Tighten cable clamp bolt to 45 inch-pounds, remove $\frac{1}{16}$" pin and install hair pin clip.

Brake Switch Adjustment, 1969

1. Disconnect harness connector at speed control servo and run a jumper wire from the blue wire terminal of connector to a good ground, Fig. 27.
2. Turn ignition key to accessory position, depress and release turn signal lever push button and check clearance between engaged actuator arm and striker pin on brake pedal, Figs. 28 and 29. Clearance should be .070" to .100".
3. If clearance is not correct, loosen striker pin attaching nut and move pin to obtain this clearance. Then tighten nut securely, insuring that the adjusted clearance is maintained.

NOTE: Before making the low speed inhibit switch adjustment, check speedometer cables to assure proper core length so that both cable drive ends are properly engaged in servo shaft keyways without binding. Also be sure that cable ferrule nuts are properly positioned on servo pilot diameters and nuts properly tightened.

Brake Switch Adjustment, 1970

1. Loosen switch bracket.
2. Insert .140"-.150" spacer gauge between brake push rod and switch with pedal in free position.
3. Push switch bracket assembly toward brake push rod until plunger is fully depressed and switch contacts spacer.
4. Tighten bracket bolt to 100 in-lbs. and remove spacer.

Brake Switch Adjustment, 1971-76

This procedure is the same as that for 1970 models, but the spacer must be .110 inch for 1971-73 models, .120 inch for 1974-76 Charger, Coronet and Satellite models and .130 inch for 1974-76 Fury, Dart, Chrysler and Imperial models.

Engine Section

ENGINE MOUNTS REPLACE

1969-76 Exc. V8-426

1. Disconnect throttle linkage at transmission and at carburetor.
2. Raise hood and position fan to clear radiator hose and radiator top tank.
3. Remove torque nuts from insulator studs.
4. Raise engine just enough to remove front engine mount.
5. Reverse above to install.

1969-71 V8-426

1. Disconnect throttle linkage from bellcrank at rear of engine.
2. Position fan to clear radiator top tank and hoses.
3. Remove engine mount to frame bolts; on the right side disconnect flexible line and fitting from fuel pump and move line to side; on left side remove splash shield.
4. Raise engine and remove engine mounts wth brackets. Remove insulators from brackets.
5. Reverse above to install.

ENGINE, REPLACE

1969-76 Chrysler & Imperial

In addition to the usual items such as fuel lines, linkage, propeller shaft, etc., perform the following:

1. Scribe a line on hinge brackets on hood to assure proper adjustments when installing. Then remove hood.
2. Remove battery, drain cooling system, remove all hoses, fan shroud, disconnect oil cooler lines and remove radiator.
3. Attach lifting fixture.
4. Disconnect torque converter drive plate from engine.
5. Support transmission in its normal position in relation to the vehicle, to insure ease of installation.
6. Remove engine front mounting bolts Then raise and work engine out of chassis.

Dodge & Plymouth 1969-76 V8s

1. Scribe hood hinge outlines on hood and remove hood.
2. Drain cooling system and remove battery.
3. Remove fan shroud (if equipped) and radiator.
4. Disconnect fuel lines and wiring to engine.
5. Remove carburetor. Attach engine lifting fixture to carburetor flange studs on intake manifold.
6. Remove engine front mounting nuts.
7. Disconnect exhaust pipes at manifold.

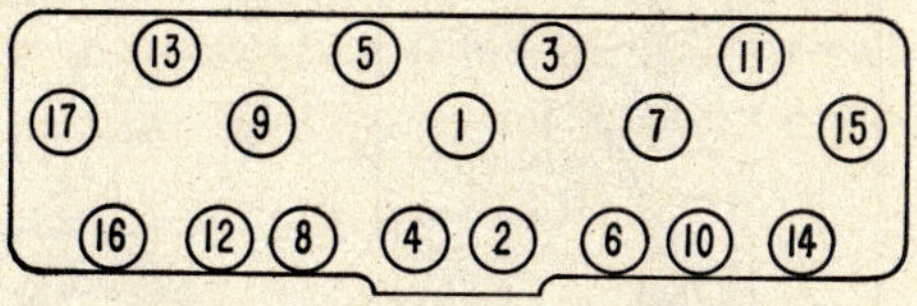

Fig. 1 Cylinder head tightening sequence. V8-383, 400, 440

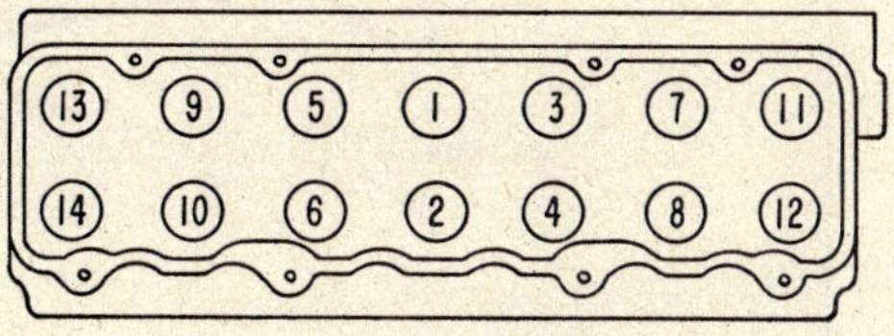

Fig. 2 Cylinder head tightening sequence. 6-170, 198, 225

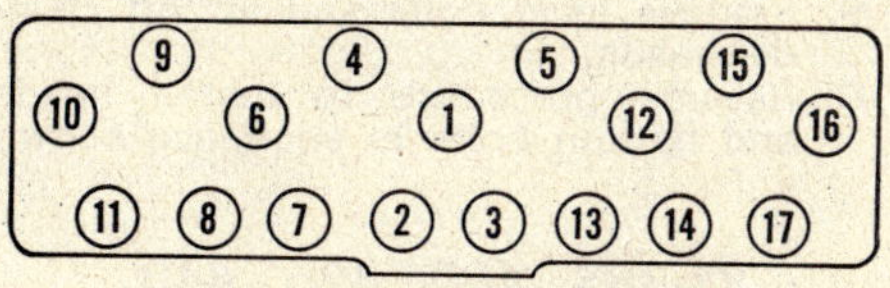

Fig. 3 Cylinder head tightening sequence. V8-426 engine

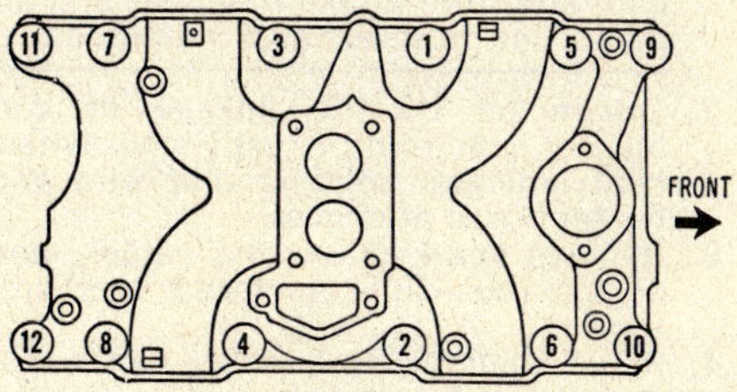

Fig. 4 Intake manifold tightening sequence. V8-273, 318, 340, 360

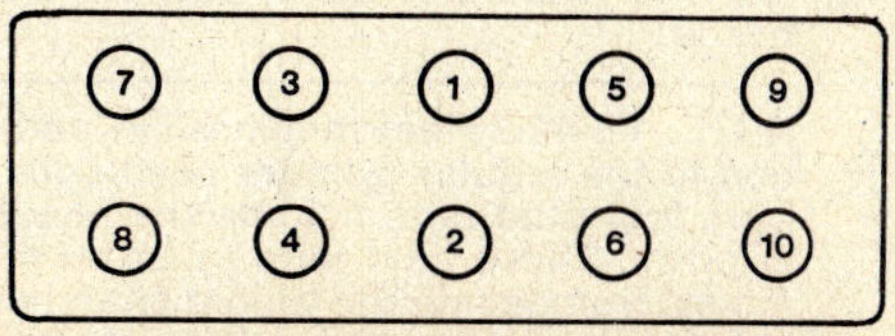

Fig. 5 Cylinder head tightening sequence. V8-273, 318, 340, 360

8. On manual transmission equipped vehicles disconnect propeller shaft, tie out of the way and disconnect wires and linkage at transmission.
9. On manual transmission equipped vehicles attach engine support fixture, remove engine rear crossmember and remove transmission.
10. On automatic transmission equipped vehicles disconnect torque converter drive plate from engine. Support transmission in its normal position in relation to the vehicle, to insure ease of installation.
11. Lift engine out of chassis.

1969-76 6-170, 198, 225

1. Scribe hood hinge outlines on hood and remove hood.
2. Drain cooling system and remove battery and carburetor air cleaner.
3. Disconnect transmission cooler lines at radiator (if equipped).
4. Remove radiator and hoses.
5. Remove outlet vent pipe or closed vent system and rocker arm cover.
6. Disconnect fuel lines, carburetor linkage and wiring to engine.
7. Disconnect exhaust pipe at manifold.
8. Remove converter cover plate.
9. On manual transmission equipped vehicles disconnect propeller shaft, tie out of the way and disconnect wires and linkage at transmission.
10. On manual transmission equipped vehicles attach engine support fixture, remove engine rear crossmember and remove transmission.
11. On automatic transmission equipped vehicles disconnect torque converter drive plate from engine. Support transmission in its normal position in relation to the vehicle, to insure ease of installation.
12. Attach lifting fixture to cylinder head and attach chain hoist.
13. Remove engine support and front engine mounting bolts and lift engine from chassis.

CYLINDER HEAD

Some cylinder head gaskets are coated with a special lacquer to provide a good seal once the parts have warmed up. Do not use any additional sealer on such gaskets. If the gasket does not have this lacquer coating, apply suitable sealer to both sides.

6-170, 198, 225

1. Drain cooling system.

2. Remove carburetor air cleaner and fuel line.
3. Disconnect accelerator linkage.
4. Remove vacuum control tube at carburetor and distributor.
5. Disconnect spark plug wires, heater hose and clamp holding by-pass hose.
6. Disconnect heat indicator sending unit wire.

NOTE: On models equipped with air pump, disconnect diverter valve vacuum line from intake manifold and remove air tubes from cylinder head.

7. Disconnect exhaust pipe at manifold.
8. Remove intake and exhaust manifold and carburetor as a unit.
9. Remove closed vent system and rocker arm cover.
10. Remove rocker shaft assembly and push rods.

NOTE: During disassembly note location of push rods so they can be installed in the same position.

11. Remove head bolts and lift off head.
12. Install the head in the reverse order of removal, and tighten the bolts in the sequence shown in Fig. 2.
13. *When installing the manifolds, loosen the three bolts holding the intake and exhaust manifolds together. This is required to maintain proper alignment.* Install intake and exhaust manifolds with cup side of the conical washers against the manifolds.

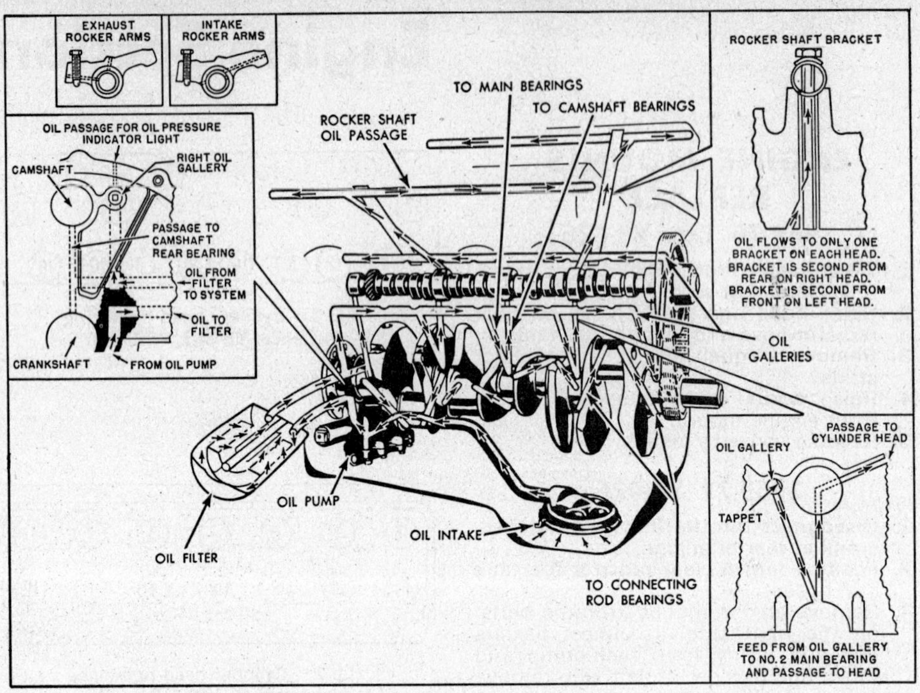

Engine oiling system, V8-273, 318, 340, 360

V8-273, 318, 340, 360

NOTE: The intake manifold attaching bolts on some engines are tilted upward about 30 degrees at an angle to the manifold-to-cylinder head gasket face. The purpose of this design is to provide more effective sealing at the cylinder block end gaskets. If the intake manifold is removed the installation should be such that the bolt tightening is done evenly and in the sequence shown in Fig. 4.

With gaskets in place start all bolts, leaving them loose. Run bolts 1 through 4 down so the heads just touch manifold. Then tighten these four bolts to 25 foot-pounds torque. After checking to see that gaskets are properly seated at all surfaces, tighten remaining bolts to 25 foot-pounds. Finally tighten all bolts in the sequence shown to 35 foot-pounds.

1. Drain cooling system and disconnect battery ground cable.
2. Remove alternator, carburetor air cleaner and fuel line. Disconnect accelerator linkage.
3. Remove vacuum advance hose and distributor cap and wires.
4. Disconnect coil wires, heat indicator wire, heater and by-pass hoses.
5. Remove closed ventilation system and rocker arm covers.
6. Remove intake manifold, coil and carburetor as an assembly.
7. Remove exhaust manifolds.
8. Remove rocker arm and shaft assemblies. Remove push rods.

NOTE: During disassembly note location of push rods so they can be installed in the same position.

9. Remove head bolts and lift off cylinder heads.
10. Reverse procedure to install heads and tighten bolts in sequence shown in Fig. 5.

V8-383, 400, 426, 440

1. Drain cooling system, remove air cleaner, fuel line from pump and carburetor, distributor vacuum tube and alternator.

NOTE: On models equipped with air pump, disconnect diverter valve vacuum line from intake manifold and air pump line from exhaust manifolds.

2. Disconnect throttle linkage at carburetor, distributor cap, coil wires, heat indicator sending unit wire and heater hoses at engine.
3. Remove spark plugs and cables, and engine vent pipe or closed vent system.
4. Remove intake manifold, carburetor and coil as an assembly.
5. Remove exhaust manifolds.
6. Remove cylinder head covers and spark plug cable support brackets.
7. Remove rocker shaft assemblies. *Do not remove bolts from end brackets.*
8. Remove push rods and valve lifter chamber cover.
9. Remove attaching bolts and lift off heads.

NOTE: V8-426 Hemi engines, in addition to the regular cylinder head bolts, have four stud nuts holding the heads in place. These stud nuts must be removed from inside the tappet chamber before any attempt is made to remove the head.

10. Reverse the foregoing procedure to install the heads and tighten bolts in the sequence shown in Figs. 1 and 3.

V8-426 Hemi Engine

Cylinder head bolt torque is critical on this engine and in order to obtain proper cylinder head gasket compression, Lubriplate should be applied to the bolt threads and between the bolt head and hardened washer.

NOTE: Care must be taken so as not to get any of the Lubriplate between the hardened washer and bolt boss as this can cause excessive bolt tension and may result in head bolt breakage.

VALVES, ADJUST

6-170, 198, 225

Before the final valve lash adjustment is made, operate the engine for 30 minutes at a fast idle to stabilize engine temperatures.

Before starting the adjustment procedure, make two chalk marks on the vibration damper. Space the marks approximately 120° apart (1/3 of circumference) so that with the timing mark the damper is divided into three equal parts. Adjust the valves for No. 1 cylinder. Repeat the procedure for the remaining valves, turning the crankshaft 1/3 turn in the direction of normal rotation while adjusting the valves in the firing order sequence of 153624.

V8s With Mechanical Lifters

Engines with mechanical lifters can be identified by the rocker arm adjusting screws. These screws are self-locking and when turning them during the process of adjustment they should indicate some resistance to turning (a minimum of 3 lb. ft. tension). If any screw turns too easily it should be replaced and, if necessary, the rocker arm as well.

Valve clearances should be set up after the engine is warmed up to operating temperature and to the clearances listed in the *Valve Specifications* table.

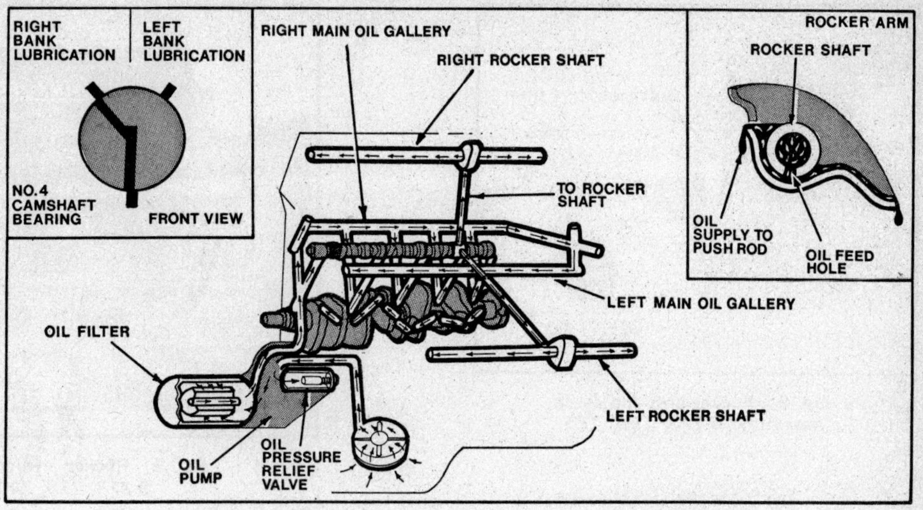

Engine oiling system. V8-383, 400, 426, 440 engines

VALVE ARRANGEMENT

Front to Rear

8-318:
Right Bank ....I-E-I-E-I-E-I-E
Left Bank ....E-I-E-I-E-I-E-I
8-273, 340, 360 ....E-I-I-E-E-I-I-E
8-383, 426, 400, 440 ......E-I-I-E-E-I-I-E
6-170, 198, 225 .....E-I-E-I-E-I-I-E-I-E-I-E

VALVE LIFT SPECS

Engine	Year	Intake	Exhaust
6-170	1969	.395	.395
6-198	1970	.395	.395
	1971-74	.406	.414
6-225	1969-70	.395	.395
	1971-76	.406	.414
8-273	1969	.373	.399
8-318	1969-76	.373	.399
8-340	1969-73	.429	.444
8-360	1971-74	.410	.412
	1974[2]	.429	.444
	1975-76 Exc. Hi Perf.	.410	.410
	1975-76 Hi Perf.	.429	.444
8-383	1969-71[1][6]	.425	.435
	1969-71[2][9]	.450	.458
8-400	1972-74[1]	.434	.430
	1972-73[2]	.449	.464
	1974[2]	.450	.464
	1975-76 All	.434	.430
8-426	1969	.467	.473
	1970-71	.490	.481
8-440	1969-71[9]	.425	.435
	1969-71[13]	.450	.458
	1972	.434	.430
	1972-73[13]	.449	.464
	1972-73[14]	.449	.464
	1974-76	.434	.430
	1974[13]	.450	.464

[1]—2 bar. carb. [2]—4 bar. carb.
[3]—340 H.P.
[6]—330 H.P. [7]—360 H.P.
[8]—Exc. 330 H.P. [9]—Exc. 330 H.P.
[10]—Exc. Hi. Perf.
[13]—Hi Perf.
[14]—3 Carbs.

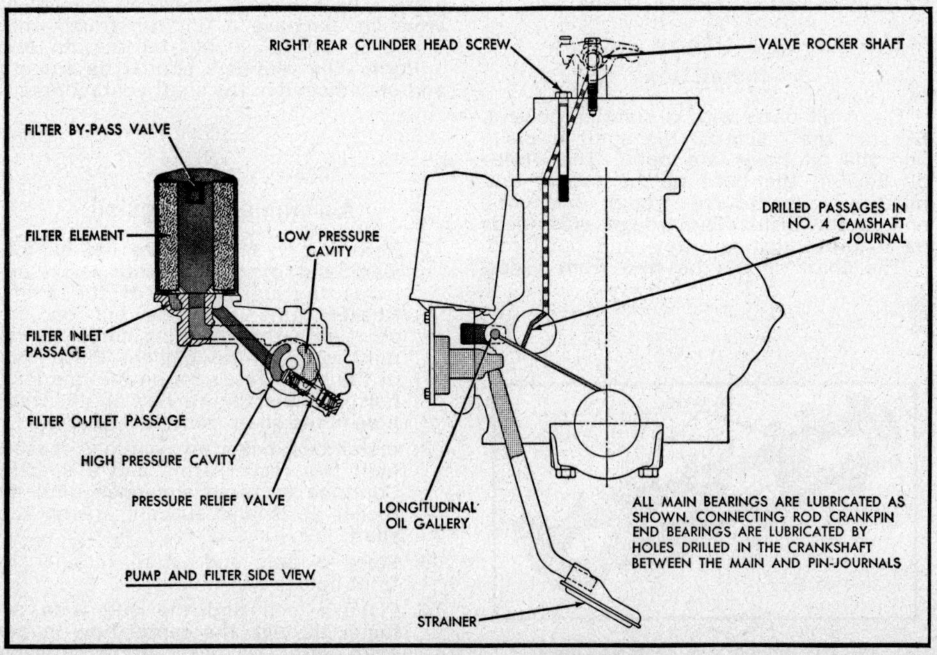

Engine oiling system. 6-170, 198, 225

VALVE TIMING SPECS.

Intake Opens Before TDC

Engine	Year	Degrees
6-170	1969	10
6-198	1970	10
	1971-74	16
6-225	1969-70	10
	1971-76	16
8-273	1969	10
8-318	1969-76	10
8-340	1969-73	22

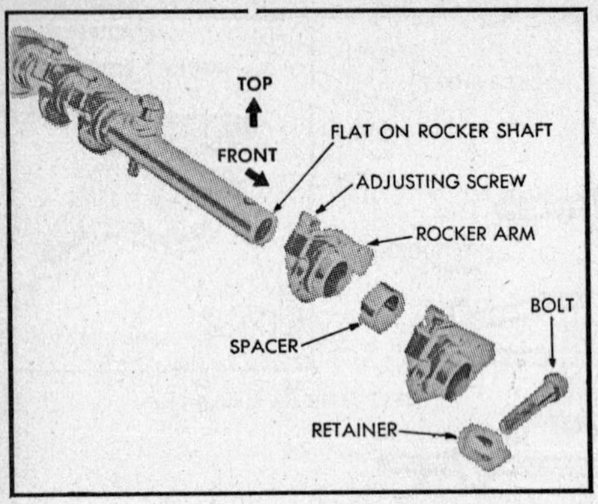

Fig. 6 Rocker arm and shaft assembly. 6-170, 198, 225

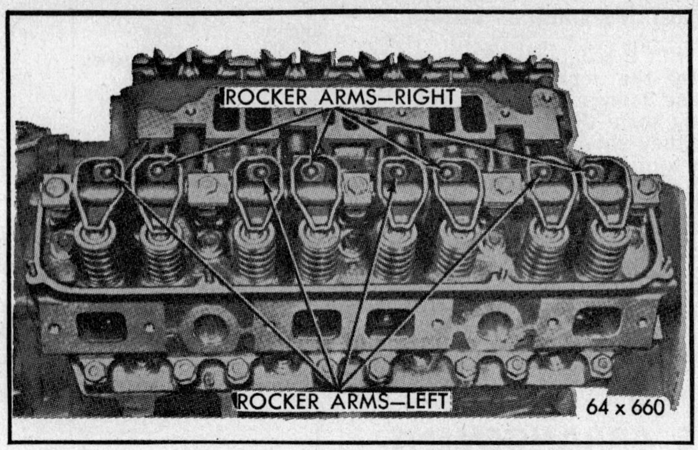

Fig. 8 Rocker arm and shaft assembly installed. 8-273, 318, 340, 360, 383, 400, 440

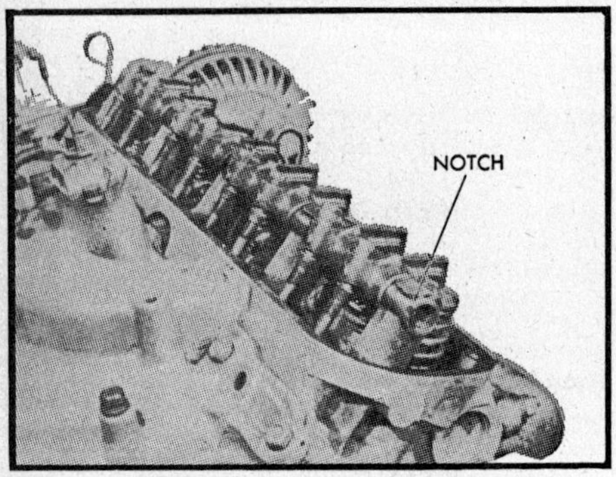

Fig. 7 Notches at end of both rocker arm shafts must face toward center of engine. V8-273, 318, 340, 360

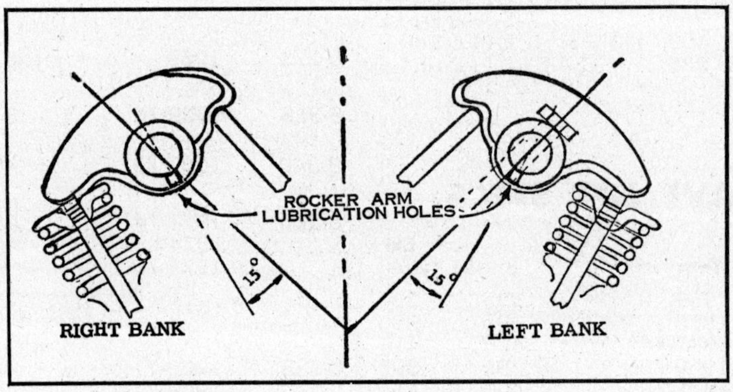

Fig. 9 Rocker arm shaft installation. V8-383, 400, 440

8-360	1971-74	16
	1974 4 B. C.	20
	1975 Exc. Hi Perf.	18
	1975-76 Hi Perf.	22
8-383	1969-71①	18
	1969-71②	21
8-400	1972-74 2 bar. carb.	18
	1972-74 4 bar. carb.	21
	1975-76 All	18
8-426	1969-71	36
8-440	1969-76 L/Hi. Perf.	18
	1969-74 W/Hi. Perf.	21

①—2 bar. carb. & 330 H.P.
②—4 bar. carb. exc. 330 H.P.

ROCKER ARMS

6-170, 198, 225

1. To remove rocker arms, take off head cover outlet tube.
2. Remove rocker arm cover.
3. Remove rocker shaft bolts and retainers.
4. Lift off rocker arms and shaft.

Inspection

Clean all parts with a suitable solvent. Be sure the inside of the shaft is clean and the oil holes are open. The drilled oil hole in the bore of the rocker arm must be open to the trough and valve end of the arms. The trough also feeds oil to the adjusting screw and push rod.

The shaft should be free from excessive wear in arm contact areas. The shaft should be smooth in retainer contact areas. The adjusting screws in the rocker arms should have a uniform round end. The drag torque should be smooth and uniform. The retainers should be smooth and undamaged in the shaft contact area.

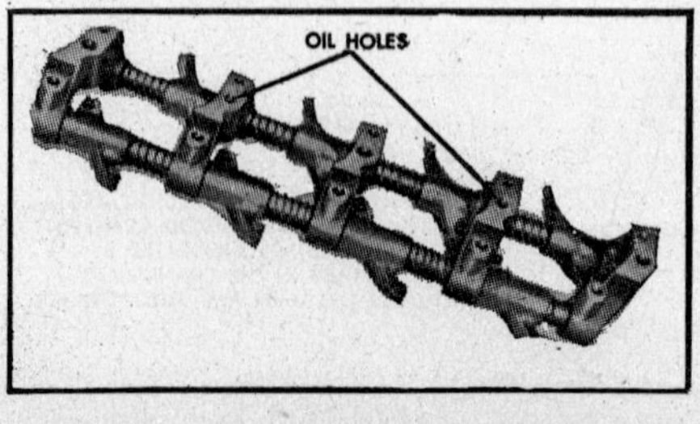

Fig. 10 Rocker arm shaft assembly. V8-426 engine

Assemble and Install

1. Referring to Fig. 6, note flat on forward end of rocker shaft which denotes the upper side of the shaft. Rocker arms must be put on the shaft with the adjusting screw to the right side of the engine. Place one of the small retainers on the one long bolt and install the bolt in the rear hole in the shaft from the top side.
2. Install one rocker arm and one spacer; then two rocker arms and a spacer. Continue in same sequence until all rocker arms and spacers are on the shaft.
3. Place a bolt and small retainer in front hole in shaft.
4. Place a bolt and the one *wide* retainer through the center hole in the shaft with six rocker arms on each side of center.

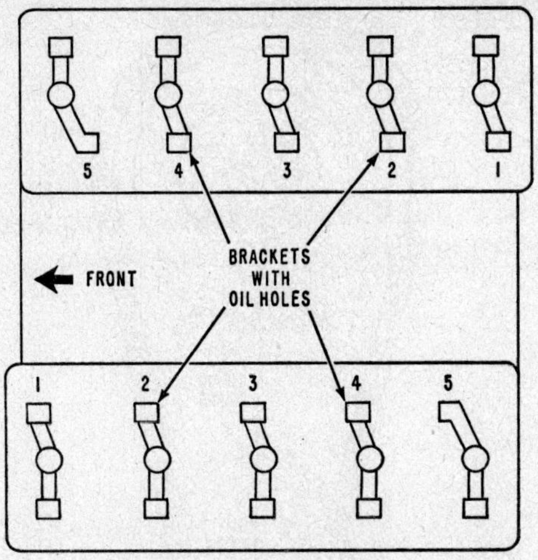

Fig. 11 Valve rocker shaft bracket locations. V8-426 Hemi engine

5. Install remaining bolts and retainers, separating the four pairs of rocker arms.
6. Locate the assembly on the cylinder head and position rocker arm adjusting screws in push rods.
7. Tighten bolts finger tight, bringing retainers in contact with the shaft *between rocker arms.*
8. Tighten bolts to specified torque.
9. After running engine to normal operating temperature, adjust valve lash to specifications.
10. Complete the job by installing the remaining parts removed.

V8-273, 318, 340, 360

To provide correct lubrication for the rocker arms on these engines, the rocker shafts have a small notch machined at one end, Fig. 7, and these notches must always face inward toward the center of the engine when installed. In other words, the notched end must be toward the rear of the engine on the right bank, and to the front of the engine on the left bank.

Rocker arms must be correctly positioned on the shaft prior to installation on cylinder head, Fig. 8.

It is also important when installing the rocker shaft assembly on the cylinder head to position the short retainers at each end and in the center, and to place long retainers in the two remaining positions.

V8-383, 400, 440

1. Install rocker shafts so that the $^3/_{16}$″ diameter rocker arm lubrication holes point downward into rocker arm, and so that the 15 degree angle of these holes point outward toward valve end of rocker arm, Fig. 9. The 15 degree angle is determined from the center line of the bolt holes through the shaft which are used to attach the shaft assembly to the cylinder head.
2. On all engines, install rocker arms and shaft assembly, making sure to install long stamped steel retainers in No. 2 and 4 positions.

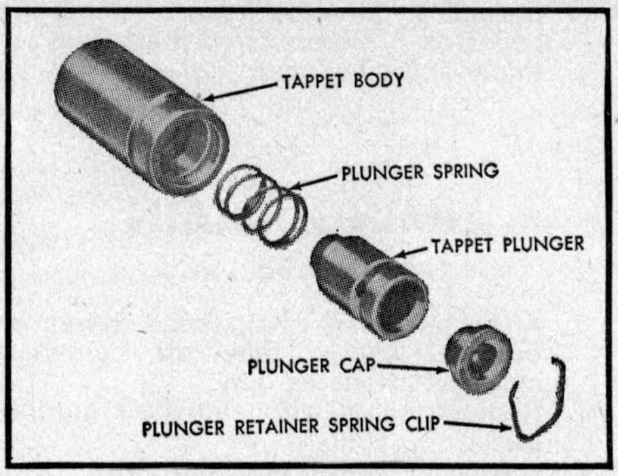

Fig. 12 Hydraulic valve lifter

NOTE: Use extreme care in tightening the bolts so that valve lifters have time to bleed down to their operating length. Bulged lifter bodies, bent push rods and permanent noisy operation may result if lifters are forced down too rapidly.

3. Installation should be as shown in Fig. 8.

V8-426

1. Remove rocker arm covers, then remove rocker shaft bracket attaching bolts.
2. Remove rocker arms and brackets from shaft, then remove push rods from bores.

NOTE: Tag push rods so they can be installed in the same position.

4. Install short push rods in upper holes and long push rods in lower holes.
5. Position rocker arms and brackets on shafts, Fig. 10.

NOTE: Position shaft brackets with oil holes in No. 2 & 4 positions, Fig. 10, see Service Note.

6. Position rocker shaft assembly on head, Fig. 11, ensure push rods engage rocker arms.
7. Install rocker shaft bracket to cylinder head attaching bolts, torque to 50 ft. lbs.
8. Adjust valve clearance, then install rocker covers.

426 Hemi Service Note

This engine has three different rocker shaft brackets and it is very important that they are positioned as shown in Figs. 10 & 11. Because the oil feed holes in the cylinder block are the number 2 and 4 positions, and the number 1 and 3 positions do not have cylinder head gasket beads, mis-location of the brackets can cause either of the following conditions:

1. No. 1 and 3 brackets installed in No. 2 and 4 positions will cause a dry cylinder head.

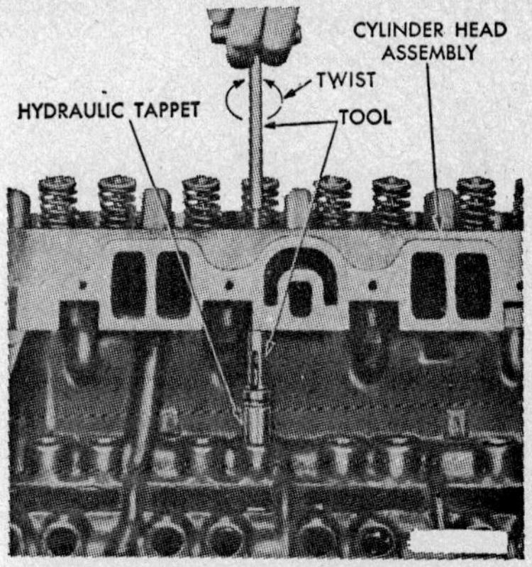

Fig. 13 Removing valve lifter

2. No. 2 and 4 brackets installed in No. 1 and 3 positions will cause a cylinder head gasket leak.

VALVE GUIDES

Non-Removable Type

Valves operate in guide holes bored directly in the cylinder head. When valve stem-to-guide clearance becomes excessive, valves with oversize stems of .005″ .015″ and .030″ are available for service replacement. When necessary to install valves with oversize stems the valve bores should be reamed to provide the proper operating clearance.

VALVE LIFTERS

6-170, 198, 225

After taking off rocker arm and shaft assembly, lift out push rods. The valve lifters may then be removed with a suitably long magnet rod. If the lifters cannot be removed with the magnet rod, a special tool (C-3661) may be used, Fig. 13. Insert the tool through the push rod opening in the cylinder head and into lifter. Turn the handle to expand the tool in the lifter, then with a twisting motion remove the lifter from its bore.

HYDRAULIC LIFTERS

Lifter, Replace

Chrysler Tool is available for this operation. To remove the lifter, insert the tool in the lifter body. (This portion of the tool can be used to remove lifters without a varnish build-up around the bottom of the body.) Lift the lifter out of the bore, Fig. 12. If they are struck proceed as follows:

Slide the puller portion of the tool through the cylinder head push rod holes

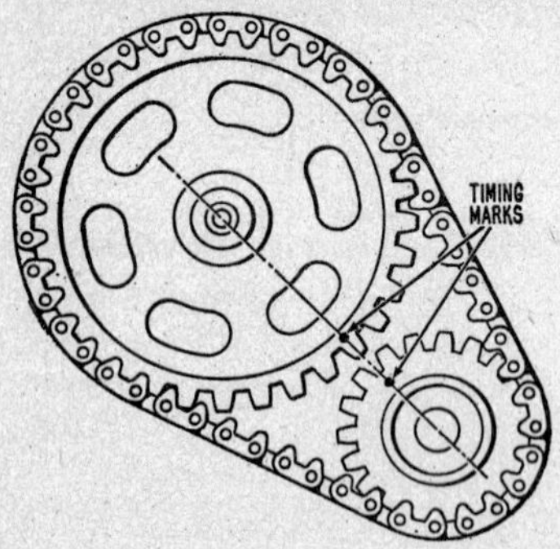

Fig. 15 Valve timing marks aligned for correct valve timing. All Sixes

and seat it firmly in the top of the lifter. Insert the puller pin through the body and tool shaft in the holes provided, Fig. 13. Grasp the tool handle and pull the lifter out of the bore as shown.

TIMING CHAIN COVER

NOTE: In order to replace the cover oil seal the cover must be removed from the engine.

6-170, 198, 225

1. To remove cover, drain cooling system and remove radiator and fan.
2. Remove vibration damper with a puller.
3. Loosen oil pan bolts to allow clearance and remove chain case cover.
4. Reverse above procedure to install cover.

V8-383, 400, 426, 440

1. Drain cooling system.
2. Remove radiator, fan and belt.
3. Remove water pump and housing as an assembly.
4. Remove crankshaft bolt and pulley from vibration damper and remove damper with a puller.
5. Remove key from crankshaft. On V8-426 Hemi, remove two front pan bolts.
6. Remove chain case cover and gasket. *Use extreme caution to avoid damaging the oil pan gasket; if damaged it will be necessary to remove the oil pan in order to install a new pan gasket.*

V8-273, 318, 340, 360

1. Remove radiator, fan and belt.
2. Remove water pump and housing as a unit.
3. Remove crankshaft pulley.
4. Remove key from crankshaft.
5. Remove fuel pump.
6. Remove chain case cover and gasket, *using extreme caution to avoid damaging oil pan gasket otherwise oil pan will have to be removed. It is normal to find particles of neoprene collected between crankshaft seal retainer and oil slinger.*

TIMING CHAIN

6-170, 198, 225

1. After removing chain case cover as outlined above, take off camshaft sprocket attaching bolt.
2. Remove chain with camshaft sprocket.
3. Clean all parts and dry with compressed air.
4. Inspect timing chain for broken or damaged links. Inspect sprockets for cracks and chipped, worn or damaged teeth.

Installation

1. Turn crankshaft so sprocket timing mark is toward and directly in line with centerline of camshaft.
2. Temporarily install camshaft sprocket. Rotate camshaft to position sprocket timing mark toward and directly in line with centerline of crankshaft; then remove camshaft sprocket.
3. Place chain on crankshaft sprocket and position camshaft sprocket in chain so sprocket can be installed with timing marks aligned without moving camshaft, Fig. 15.
4. Install parts removed in reverse order of removal.

V8 Engines

To install chain and sprockets, lay both the camshaft and crankshaft sprockets on the bench. Position the sprockets so that the timing marks are next to each other. Place the chain on both sprockets, then push the gears apart as far as the chain will permit. Use a straightedge to form a line through the exact centers of both gears. The timing marks must be on this line, Fig. 16.

This is the same procedure as in previous models, except that now the alignment is done on the bench rather than on the engine.

Slide the chain with both sprockets on the camshaft and crankshaft at the same time; then recheck the alignment.

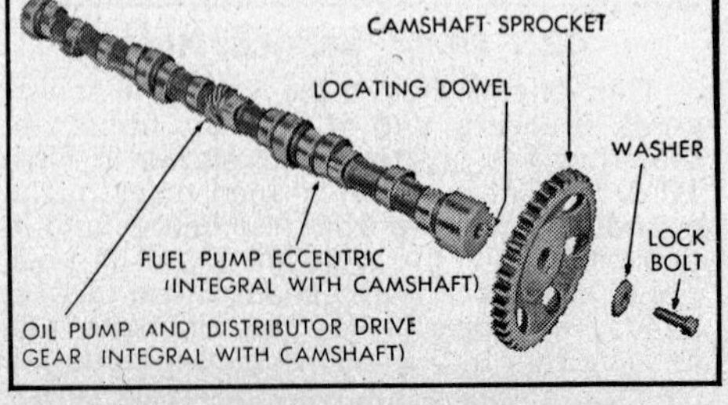

Fig. 17 Camshaft and related parts. 6-170, 198, 225

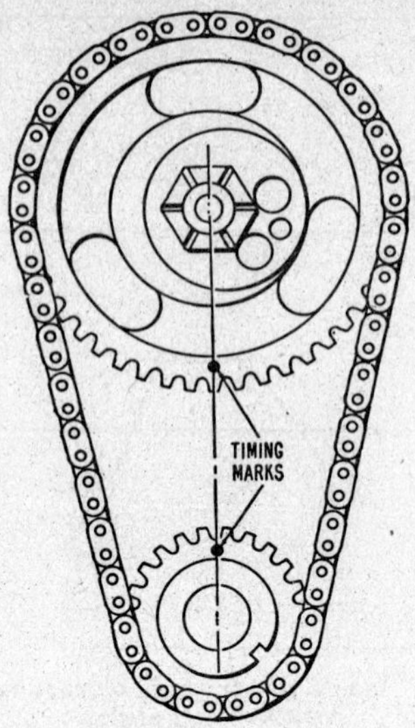

Fig. 16 Valve timing marks aligned for correct valve timing. All V8s

CAMSHAFT & BEARINGS

6-170, 198, 225

The camshaft is supported by four precision type, steel backed, babbitt-lined bearings. Rearward thrust is taken by the rear face of the sprocket hub contacting the front of the engine block.

The camshaft, Fig. 17, can be removed after removing the grille, radiator and timing chain. To remove the camshaft bearings, the torque converter (or flywheel) must also be removed.

1. Remove valve lifters, oil pump and distributor.
2. Slide camshaft out of engine.
3. Remove welch plug back of rear camshaft bearing.
4. Remove bearings with suitable puller equipment.
5. Install new bearings, being sure the oil holes in bearings line up with the corresponding oil holes in the crankcase.

V8 Engines

To remove the camshaft, remove all valve lifters, timing chain and sprockets. Remove distributor and oil pump-distributor drive gear. Remove fuel pump and see that push rod has moved away from eccentric drive cam. On V8-273, 318, 340 & 360 engine, remove thrust plate, 1972-75 engines incorporate a timing chain oil tab, before removing, note position of tab so it can be installed in the same manner. On all engines withdraw camshaft from engine, using care to see that camshaft lobes do not damage the camshaft bearings.

If camshaft bearings are to be replaced, it is recommended that the engine be removed from the chassis and the crankshaft taken out in order that any chips or foreign material may be removed from the oil passages.

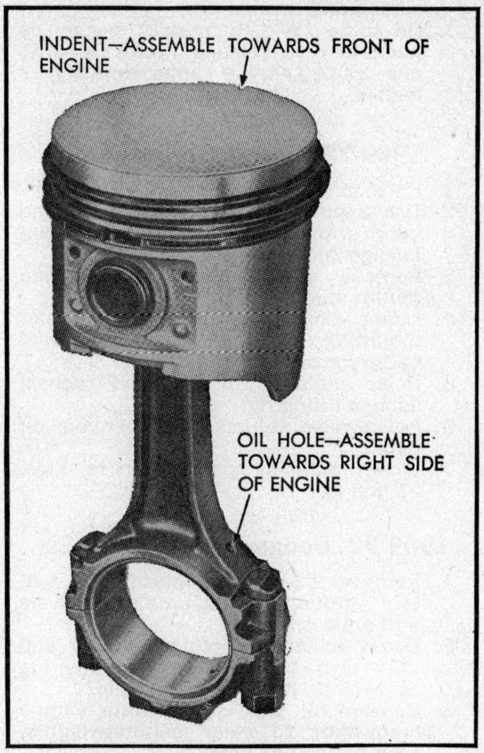

Fig. 19 Piston and rod assembly. 6-170, 198, 225 engines with cast iron block.

PISTON & ROD, ASSEMBLE

6-170, 198, 225

Piston and rod assemblies must be installed as shown in Fig. 19.

V8 Engines

When installing piston and rod assemblies in the cylinders, the compression ring gaps should be diametrically opposite one another and not in line with the oil ring gap. The oil ring expander gap should be toward the outside of the "V" of the engine. The oil ring gap should be turned toward the inside of the engine "V".

Immerse the piston head and rings in clean engine oil and, with a suitable piston ring compressor, insert the piston and rod assembly into the bore. Tap the piston down into the bore, using the handle of a hammer.

Assembly the pistons to the rods as shown in Fig. 20.

PISTONS, PINS & RINGS

Pistons are available in standard sizes and the following oversizes: V8-440: 005, .020". All others: .005, .020, .040".

Pins are available in the following oversizes: V8-273, 318, 426, .003, .008". Not furnished on all other engines.

Rings are available in the following oversizes: 1969 Six and all V8s, std. to .009, .020-.029, .040-.049".

MAIN & ROD BEARINGS

Main bearings are furnished in standard sizes and the following undersizes: 1969-75 Six, .001, .002, .010". 1969-75 V8s, 001, .002, .003, .010, .012".

Rod bearings are furnished in standard sizes and the following undersizes: Six, .001, .002, .003, .010". V8s, .001, .002, .003, .010, .012".

NOTE

Some 1974 V8-400 and V8-440 engines were built with low thrust wall cylinder blocks. These engines require a small thrust wall or flange (No. 3) crankshaft main bearing. The dimension "A" in Fig. 20A, must be noted if the thrust bearing (No. 3) is to be replaced.

Engines using the small thrust wall (flange) bearing, Fig. 20A, can be identified by the yellow "X" painted on the engine identification pad and also engine will be date coded 8-12 on the pad.

All other 1974 V8-400 and V8-440 engines require the large thrust wall bearing.

1972-76 V8-400 and 1972-75 built prior to January 1975 use either forged or cast crankshafts. V8-440 engines built starting January 1975, use cast crankshafts. These engines require that a matching torque converter, short block, damper and connecting rods be used.

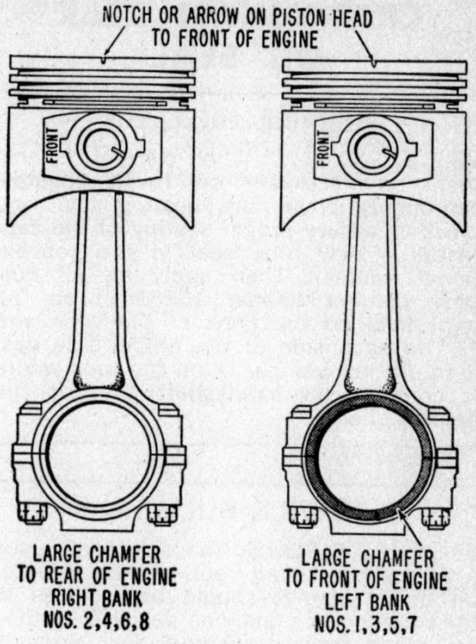

Fig. 20 Piston and rod assembly, V8. On V8-426 Hemi, bearing tangs must face outboard

If replacement of the crankshaft, torque converter, crankshaft damper or short engine is required, it is important that matching parts are used otherwise severe engine vibration will result, (Consult Chrysler Parts Dept.).

The cast crankshaft engine can be easily identified since it has the letter "E" stamped on the engine numbering pad following the built date.

Also, V8-440 engines built starting January 1975, will have external balance weights at the front damper and torque converter and a clock face will appear on the number 1 counterweight of the cast crankshaft.

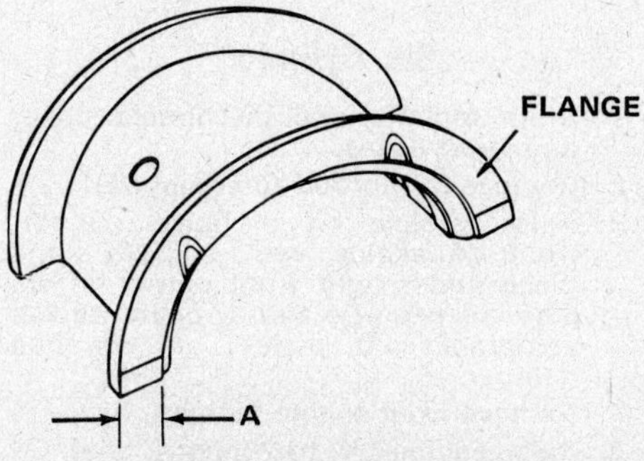

Fig. 20A Main bearing with small thrust wall flange (A)

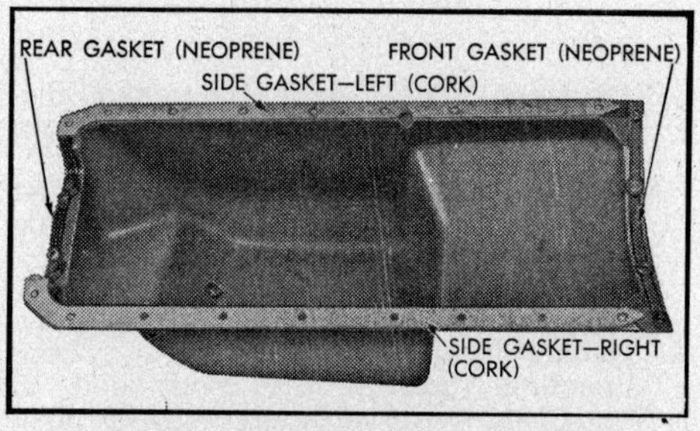

Fig. 21 Location of oil pan gaskets. 6-170, 198, 225

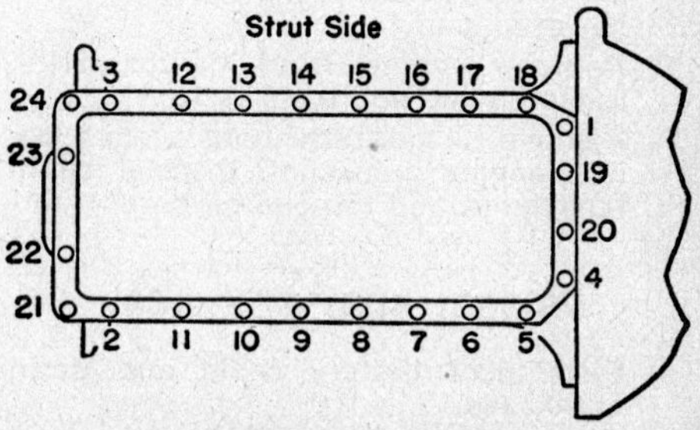

Fig. 22 Tighten oil pan screws to 200 inch pounds in the sequence shown. 6-170, 198, 225

CRANKSHAFT REAR OIL SEAL

SERVICE BULLETIN

V8-273, 318 SEAL: When oil seal replacement is necessary on these engines, thoroughly clean the bearing cap and block to assure proper seating of the cap. Install a new rope seal in the conventional manner. Then apply an All Purpose cement (Mopar 1316241) on the joint face on the ends of the rope and 1/4" to each side of the rubber side gaskets. Do not use sealer on the rope where it contacts the crankshaft or near the bearing shell.

SERVICE BULLETIN

V8-383 SIDE SEALS: The side seals used with the crankshaft rear bearing retainer on these engines should be installed in the retainer as rapidly as possible as they are made from a material that expands rapidly when oiled. Apply mineral spirits or kerosene to the seals and install them in the grooves immediately. Install seal retainer and torque to 30 ft-lbs. Failure to pre-oil seals will result in an oil leak.

OIL PAN

CAUTION: *Engine oil pan bolts on all V8-383, 426, 440 engines are 13/16" long with the exception of two bolts at the rear center of the oil pan. The two rear center bolts are 9/16" long and thread into the aluminum seal retainer. Do not use longer bolts than 9/16" at this location as they will bottom in the aluminum seal retainer and, if forced in may strip the threads and damage the seal retainer, causing an oil leak.*

1973-76 Six

1. Drain radiator, disconnect battery and radiator hoses and remove oil dipstick.
2. Remove shroud attaching screws, separate shroud from radiator and position rearward on engine.
3. Raise vehicle and drain oil pan.
4. Remove center link from steering arm and idler arm ball joints.
5. Support front of engine with a jack stand placed under the right front corner of oil pan.
6. Remove engine front mount bolts. Raise engine approx. 1½ to 2".
7. Remove oil pan attaching screws, rotate engine crankshaft to clear counterweights and remove oil pan.

V8-383, 400, 426, 440

1. Disconnect battery cable and drain crankcase.
2. Raise car on hoist and disconnect steering linkage from idler arm and pitman arm.

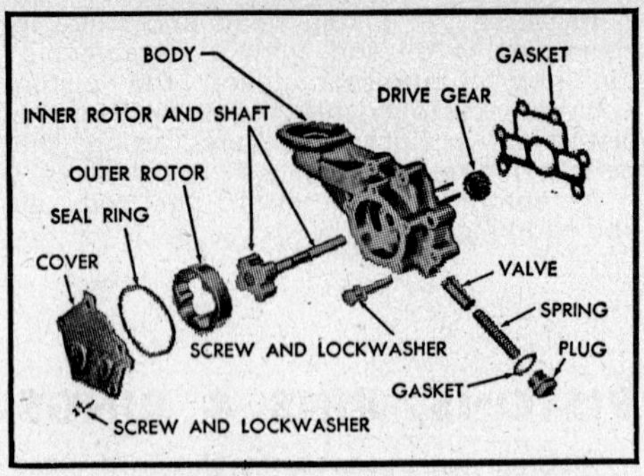

Fig. 23 Oil pump. 6-170, 198, 6-225

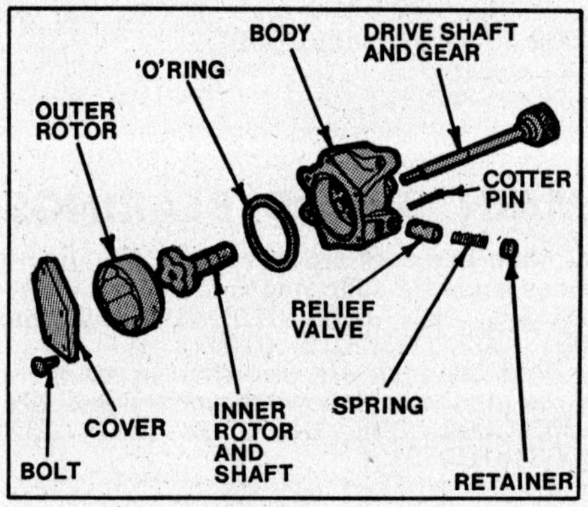

Fig. 24 Oil pump. V8-273, 318, 340 & 360

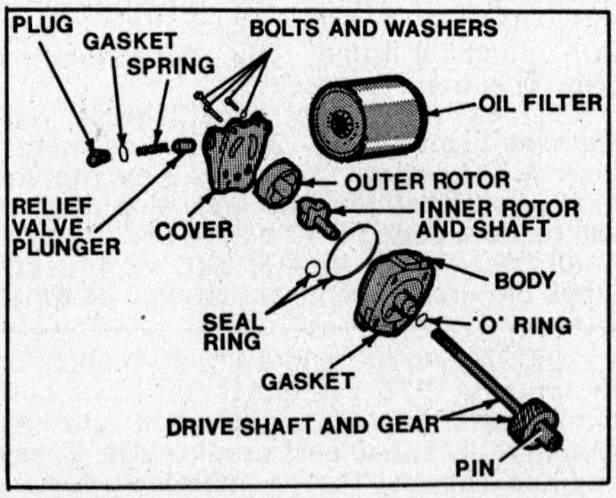

Fig. 24A Oil pump. V8-383, 400, 426 & 440

3. Remove outlet vent pipe and disconnect exhaust pipe branches from manifolds.
4. Remove clamp attaching exhaust pipe to extension and remove exhaust pipe.
5. Remove converter dust shield.
6. Remove oil pan bolts and turn flywheel until counterweight and connecting rods at the front end of crankshaft are at their highest position to provide clearance, and lower the pan. Turn the pan to clear oil screen and suction pipe.
7. On some 1974-76 models, it may be necessary to release motor mounts and raise engine approximately 3 inches.

1969-72 Dart & Valiant Six

1. Raise car and drain oil pan.
2. Use a puller to remove steering and idler arm ball joints from steering linkage center link.
3. Remove dust shield and engine mount stud nuts.
4. Lower vehicle and remove horns and mounting brackets, then disconnect battery ground cable.
5. Raise engine from 1½ to 2 inches, using a lifting rig.
6. Again raise vehicle, then remove oil pan.
7. When installing pan, refer to Figs. 21 and 22.

1969-72 Dodge & Plymouth Six

1. Remove oil dipstick, disconnect battery ground cable. Raise vehicle and drain oil.
2. Use a puller to remove steering and idler arm ball joints from steering center link. Remove dust shield.
3. Remove oil pan bolts, rotate engine crankshaft to clear counterweights, then remove oil pan.
4. When installing pan, refer to Figs. 21 and 22.

V8-273, 318, 340, 360

1. Disconnect battery ground cable.
2. Remove oil level dipstick.
3. Raise vehicle and drain oil.
4. Remove engine-to-torque converter left housing brace.
5. Remove steering and idler arm ball joints from steering center link.
6. Disconnect exhaust crossover pipe from exhaust manifolds and leave it hang without disconnecting it from muffler.
7. On some models it will be necessary to remove crossover pipe.
8. On some 1974-76 models, it may be necessary to release motor mounts and raise engine approximately 2 inches.
9. Unfasten and remove oil pan.

OIL PUMP, REPLACE

Six Cylinder

1. Drain radiator and disconnect upper and lower hoses.
2. Remove fan shroud (if equipped).
3. Raise vehicle on a hoist, support front of engine with a jack stand place under right front corner of engine oil pan. *Do not support engine at crankshaft pulley or vibration damper.*
4. Remove front engine mounts.
5. Raise engine 1½ to 2 inches.
6. Remove oil filter, pump attaching bolts and remove pump assembly.

V8 Engines

On 273, 318, 340, 360 engines, remove oil pump from rear main bearing cap.

On 383, 400, 426, 440 engines, unfasten oil pump from engine and remove pump and filter assembly from bottom of engine.

OIL PUMP REPAIRS

6-170, 198, 225

To disassemble, remove the pump cover seal ring, Fig. 23. Press off the drive gear, supporting the gear to keep load off aluminum body. Remove rotor and shaft and lift out outer pump rotor. Remove oil pressure relief valve plug and lift out spring and plunger. Remove oil pressure sending unit.

Inspection

1. The rotor contact area and the bores for the shaft and valve in the pump body should be smooth, free from scratches, scoring or excessive wear.
2. The pump cover should be smooth, flat and free from scoring or ridges. Lay a straightedge across the cover. If a .0015″ feeler gauge can be inserted under the straightedge, the cover should be replaced.
3. All surfaces of the outer rotor should be smooth and uniform, free from ridges, scratches or uneven wear. Discard a rotor less than .649″ thick and/or less than 2.469″ in diameter.
4. The inner rotor and shaft assembly should be smooth, free from scoring and uneven wear. Discard rotors less than .649″ thick.
5. Place outer rotor in pump body and measure clearance between rotor and body. Discard pump body if clearance is more than .014″.
6. Install inner rotor and shaft in pump body. Shaft should turn freely but without side play. If clearance between rotor teeth is more than .010″, replace both rotors.
7. Measure rotor end clearance. If feeler gauge of more than .004″ can be inserted between straightedge and rotors, install a new pump body.
8. The oil pressure relief valve should be smooth, free from scratches or scoring, and should be a free fit in its bore.
9. Relief valve springs are painted either gray, red or brown to denote free lengths of 2.19, 2.29 and 2.39 inches. Rather than change the length, replace a spring with one of the same color.

Assemble and Install

1. With pump rotors in body, press drive gear on shaft, flush with end of shaft.
2. Install seal ring in groove in body and install cover. Tighten bolts to 10 ft. lbs. Test pump for free turning.
3. Install oil pressure relief valve spring. Use new washer (gasket) and tighten plug securely.
4. If pump shaft turns freely, remove pump cover and outer rotor before installation of pump on engine.
5. Install oil pressure sending unit and tighten to 60 inch lbs. (5 ft. lbs.)
6. Using a new gasket, install pump on engine and tighten bolts to 200 inch lbs. (16 ft. lbs.)
7. Install oil filter reservoir on pump. Install filter element and tighten cover nuts to 25 ft. lbs.
8. Connect oil pressure sending unit wire.
9. Complete the installation by reversing steps as given under Oil Pump, Replace.

V8 Engines

After removing the pump from the engine it should be disassembled, cleaned and inspected for wear, Figs. 24 & 24A.

1. Remove the cotter pin holding the oil strainer to the oil suction pipe. Then remove the pipe from the pump body.
2. Remove the pump cover and discard the oil seal ring.
3. Remove pump rotor and shaft and lift out rotor body.
4. Remove oil pressure relief valve plug and lift out the spring and plunger.
5. Wash all parts in cleaning solvent and inspect carefully for damage or wear.
6. The mating face of the oil pump cover should be smooth. If it is scratched or grooved, the cover should be replaced with a new one.
7. Check for excessive cover-to-rotor wear by laying a straight edge across the cover surface. If a .0015″ feeler gauge can be inserted between cover and straight edge, the cover should be discarded and a new one installed.
8. Slide rotor body and rotor into pump body and then place a straight edge across the face of the pump body between the bolt holes. If a feeler gauge of less than .003″ or more than .006″ can be inserted between the rotors and straight edge, install a new pump body.
9. Remove the pump rotor and shaft, leaving rotor body in pump cavity. Press rotor body to one side with the fingers and measure the clearance between rotor and pump bodies. If it is more than .014″, install a new pump body.
10. Check the clearance between the pump rotor and rotor body. If the measurement is more than .014″, install a new pump rotor and rotor body.
11. Check the oil pump relief valve plunger for scoring and free operation in its bore. If the plunger is scored, install a new one.

BELT TENSION DATA

	New[1]	Used[1]
1969-70—		
Air Condition—		
Exc. Six Cylinder	1/8	3/16
Six Cylinder	3/32	1/8
Alternator—		
With A/C—		
Exc. V8-273, 318 & 340	3/16	9/32
V8-273, 318 & 340	1/8	3/16
Without A/C—		
Six Cylinder	3/16	9/32
V8-273, 318 & 340	1/8	3/16
V8-383, 426 Hemi & 440	3/32	3/16
Fan	1/16	3/32
Power Steering	3/32	5/32
1971-76 All	120[2]	70[2]

[1]—Deflection measured in inches.
[2]—Belt tension in ft. lbs.

WATER PUMP, REPLACE

CAUTION: When it becomes necessary to remove a fan clutch of the silicone type, the assembly must be supported in the vertical position to prevent leaks of silicone fluid from the clutch mechanism. This loss of fluid will render the fan clutch inoperative.

Drain cooling system, and on air conditioned cars only remove upper half of fan shroud. Loosen power steering pump or idler pulley, and generator. Remove all belts, fan, space and pulley.

On air conditioned cars, remove pulley from water pump fan hub, loosen all nuts from fan and remove the fan drive.

On some air conditioned cars, it may be necessary to remove the compressor bracket and/or air pump bracket and tie the compressor and/or air pump out of the way.

On all models, remove bolts holding water pump body to housing and remove water pump.

Service Bulletin

CORE HOLE PLUG SIZES: When replacing a cup-type core hole plug in an engine, the size of the hole in the cylinder head, water jacket or rear bearing bore for the camshaft should be checked. At these locations a 1/16″ oversize hole is sometimes bored in production and an oversize core plug installed.

Core plugs 1/16″ oversize are available for replacement should they be required at these locations.

FUEL PUMP PRESSURE

Year	Engine	Pressure Lbs.
CHRYSLER & IMPERIAL		
1969-71	All	3½-5
1972	V8-360	5-7
1972-76	V8-400	3½-5
1972-76	V8-440 Chrysler	3½-5
1972-73	V8-440 Imperial	5½-7
1974-76	V8-440 Imperial	3½-5

DODGE & PLYMOUTH

1969-76	Six	3½-5
1969-76	V8-273, 318, 340, 360	5-7
1969-71	V8-383	3½-5
1969-71	V8-440 (Exc. Hi-Perf.)	3½-5
1969-71	V8-440 Hi-Perf.	6-7½
1969-71	V8-426	7-8½
1972-76	V8-400, 440[1]	3½-5

[1] V8-440 Hi-Perf. 6-7½ lbs.

FUEL PUMP, REPLACE

Electric Fuel Pump

The 1972 Imperial equipped with an Air Injection pump employs a Bendix electric fuel pump. It is mounted on the left side of the frame in the kick up area about even with the front of the fuel tank. When replacing the pump it must be mounted so that the outlet is higher than the inlet. This is critical to pump operation.

Mechanical Fuel Pump

SERVICE NOTE: Before installing the pump, it is good practice to crank the engine so that the nose of the camshaft eccentric is out of the way of the fuel pump rocker arm when the pump is installed. In this way there will be the least amount of tension on the rocker arm, thereby easing the installation of the pump.

1. Remove all gasket material from the pump and block gasket surfaces. Apply sealer to both sides of new gasket.
2. Position gasket on pump flange and hold pump in position against its mounting surface. Make sure rocker arm is riding on camshaft eccentric.
3. Press pump tight against its mounting. Install retaining screws and tighten them alternately.
4. Connect fuel lines. Then operate engine and check for leaks.

Clutch and Transmission Section

NOTE: 1976 linkage adjustment information is in this section. Repair procedures on both automatic and manual shift transmissions are covered elsewhere in this manual. Procedures for removing automatic transmissions as well as linkage adjustments on 1969-75 models are included in the automatic transmission chapters. See Chapter Index.

CLUTCH PEDAL, ADJUST

1969-76 All Cars

1. Inspect condition of clutch pedal rubber stop, if stop is damaged install a new one.
2. Where necessary, disconnect interlock clutch rod at transmission end.
3. Adjust linkage by turning self-locking adjusting nut to provide 5/32" free movement at outer end of fork. This movement will provide the prescribed one-inch free play at pedal.
4. Assemble interlock clutch rod (if used) to transmission pawl.

CLUTCH, REPLACE

Unless special clutch rebuilding equipment is available, it is recommended that the clutch assembly be exchanged for a rebuilt unit should the clutch require rebuilding. The driven disc, however, may be replaced without special equipment. If clutch rebuilding equipment is available, follow the equipment manufacturer's instructions.

Removal

1. Remove transmission and clutch pan.
2. Pull out release bearing and sleeve.
3. Mark clutch cover and flywheel so they may be assembled in the same relative position and thus maintain original balance.
4. Remove cap screws which retain clutch cover to flywheel. Loosen each screw a few turns in succession until cover is free.
5. Clutch assembly and driven disc may now be removed from the housing.

Installation

1. Coat the pilot bearing in crankshaft with wheel bearing grease.
2. Clean surfaces of flywheel and pressure plate, making certain no oil or grease remains on these parts.
3. Hold cover plate and disc in place and insert a special clutch aligning tool or a spare clutch shaft through the hub of the disc and into the crankshaft pilot bearing.
4. Bolt clutch cover loosely to flywheel, being sure marks previously made are lined up.
5. To avoid distortion of clutch cover, tighten cover bolts a few turns each in progression until all are tight. The final tightening should be 15-20 ft-lbs. for 5/16" bolts and 30 ft-lbs. for 3/8" bolts.
6. Install transmission by guiding it into place with guide studs inserted in the two top holes of the housing.
7. Adjust clutch pedal free travel.

THREE SPEED TRANSMISSION, REPLACE

1969-71 Chrysler, 1969-76 Dodge & Plymouth

1. Drain lubricant from transmission.
2. Disconnect propeller shaft, speedometer cable and housing and gearshift control rods.
3. Remove speedometer cable with hand so that housing is not crushed.
4. Disconnect back-up lamp switch leads (if so equipped) and if necessary, loosen exhaust system for clearance.
5. Support engine with a jack or suitable fixture against underside of oil pan flange.
6. Raise engine slightly and disconnect extension housing from removable center crossmember.
7. Support transmission with a suitable jack. Then tap out four long bolts and remove center crossmember. Remove bolts that attach transmission to clutch housing.
8. Slide transmission rearward until clutch shaft clears clutch disc before lowering transmission. Lower and remove transmission.
9. Reverse removal procedure to install.

FOUR SPEED TRANSMISSION, REPLACE

1969-76 Dodge & Plymouth

1. Remove console and shift components.

NOTE: On 1969 models, the shift lever is removed by using a .010" feeler gauge as shown in Fig. 1 to release internal spring clip. When reinstalling lever, push it down into shift unit far enough for spring to click and lock lever in place.

2. Drain fluid from transmission.

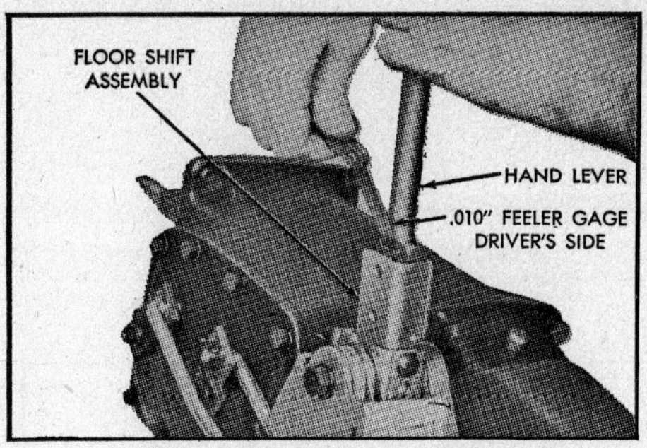

Fig. 1 Removing floor shift handle. 1969-76

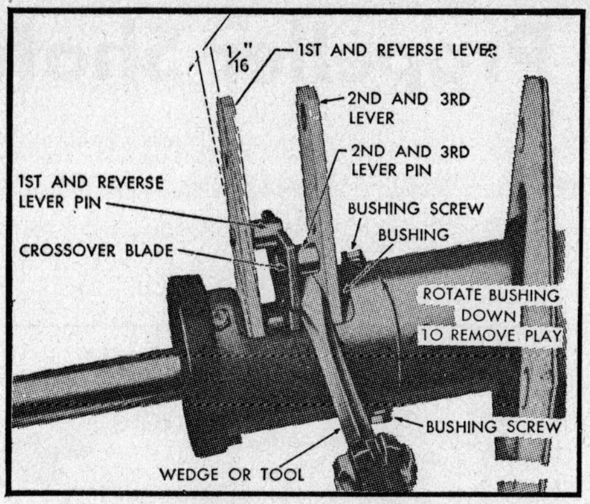

Fig. 2 Gearshift lever adjustment. 1969 Three Speed Transmission

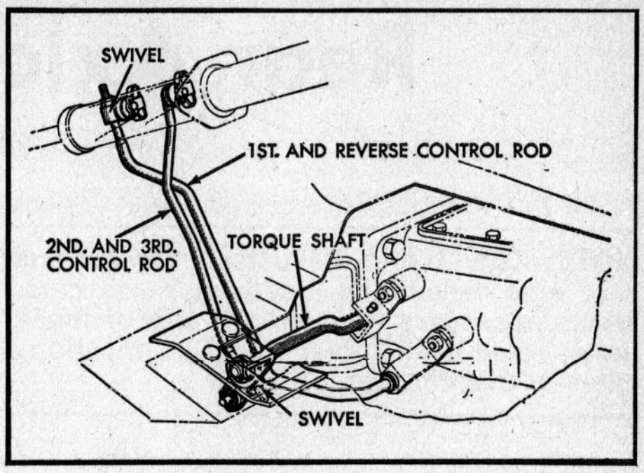

Fig. 3 Gearshift controls. 1969-76 Three Speed Transmission

3. Disconnect propeller shaft at rear universal joint and carefully pull yoke out of extension housing. *Be careful not to scratch or nick ground surface on sliding spline yoke during removal and installation of shaft.*
4. Disconnect speedometer cable and stop light switch leads.
5. Disconnect left-hand exhaust pipe (dual exhaust) from manifold.
6. Disconnect parking cable where necessary.
7. Support rear of engine with a jack.
8. Raise engine slightly and disconnect extension housing from removable center crossmember.
9. Support transmission with a suitable jack and remove center crossmember.
10. Remove transmission-to-clutch housing bolts.
11. Slide transmission rearward and out of vehicle.
12. Reverse procedure to install.

SHIFT LINKAGE, ADJUST THREE SPEED TRANS.

1969-76

1. With 2-3 control rod disconnected from lever on steering column and 1st-reverse rod disconnected from transmission lever, position both transmission levers in neutral.

NOTE: The neutral detent balls must be engaged to make this adjustment. To check this, start engine (clutch disengaged) then release clutch slowly.

2. Inspect fore and aft movement of shift levers in steering column. If movement at outer end of levers exceeds 1/16", loosen two upper bushing screws, Fig. 2, and rotate bushing downward until all free play of levers has been removed. Then retighten bushing screws.
3. Wedge a screwdriver between crossover blade and the 2-3 lever so that crossover blade is engaged with both lever crossover pins.
4. Adjust length of 2-3 control rod until stud shaft of control rod and swivel enters hole in column lever, Fig. 3. Install washer and clip and tighten swivel lock nut. During the above setting the 2-3 control rod should be adjusted also to position selector lever on column 5°-10° above horizontal.
5. Slide clamp and swivel (on end of 1st-reverse control rod) either in or out until swivel stub shaft enters hole in transmission lever, Fig. 3. Install washers and clip. Determine middle backlash position in linkage, then tighten control rod lock nut.
6. Remove screwdriver from crossover blade and lever. Then move selector lever through all positions to check adjustments and to insure crossover smoothness.

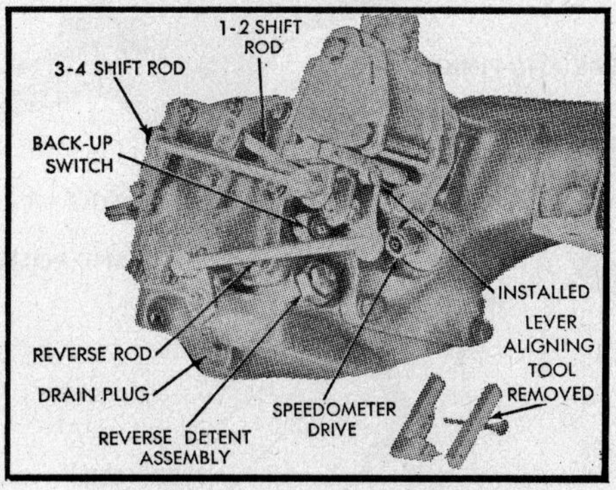

Fig. 4 Gearshift linkage adjustment. 1969 four speed transmission

SHIFT LINKAGE, ADJUST FOUR SPEED TRANS.

1969-76 Four Speed Trans.

NOTE: 1970-76 floor shift linkages incorporate a transmission lock rod. This rod is adjusted after the shift linkage as follows:

1. Loosen rod swivel clamp bolt.
2. Place transmission in reverse.
3. Align steering column locating slots and install a suitable tool to hold position.
4. Tighten clamp bolt. Column should now lock in reverse but not in any other gear.

1. Install lever alignment tool to hold levers in neutral, Fig. 4.
2. Shift transmission into neutral, then disconnect all control rods from transmission levers.
3. Insert lever alignment tool through slots in levers, making sure it is through all levers and against the back plate.
4. Adjust length of control rods so they enter transmission levers freely without any rearward or forward movement.
5. Secure adjustment and remove tool.
6. Check linkage for ease of shifting into all gears and for crossover smoothness.

1976 AUTO. TRANS. LINKAGE, ADJUST

Column Shift

Linkage adjustment procedures are the same as 1975 models described in the Automatic Transmission section of this manual.

Rear Axle, Propeller Shaft & Brakes

NOTE: Figs. 1 and 2 illustrate the various rear axle assemblies used on these cars. When necessary to overhaul any of these units, refer to the *Rear Axle Specifications* table in this chapter.

INTEGRAL TYPE REAR AXLE

Two types of integral carrier axles are used. In both types, the drive pinion is mounted in two opposing tapered roller bearings which are preloaded by a spacer positioned between them.

In the unit shown in Fig. 1, used from 1969 through 1976, the differential is supported by two tapered roller side bearings. These bearings are preloaded by spacers located between the bearings and carrier housing. The differential assembly is positioned for ring and pinion backlash by varying these spacers.

Axle shafts in this unit are held in place by retainers at the outer ends of the shafts. These retainers are bolted

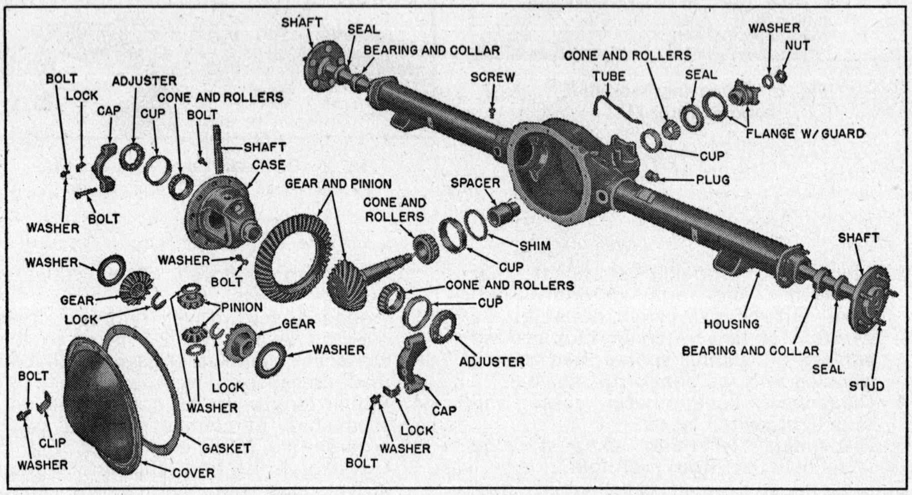

Fig. 2 Integral "C" washer type rear axle. 1969-76

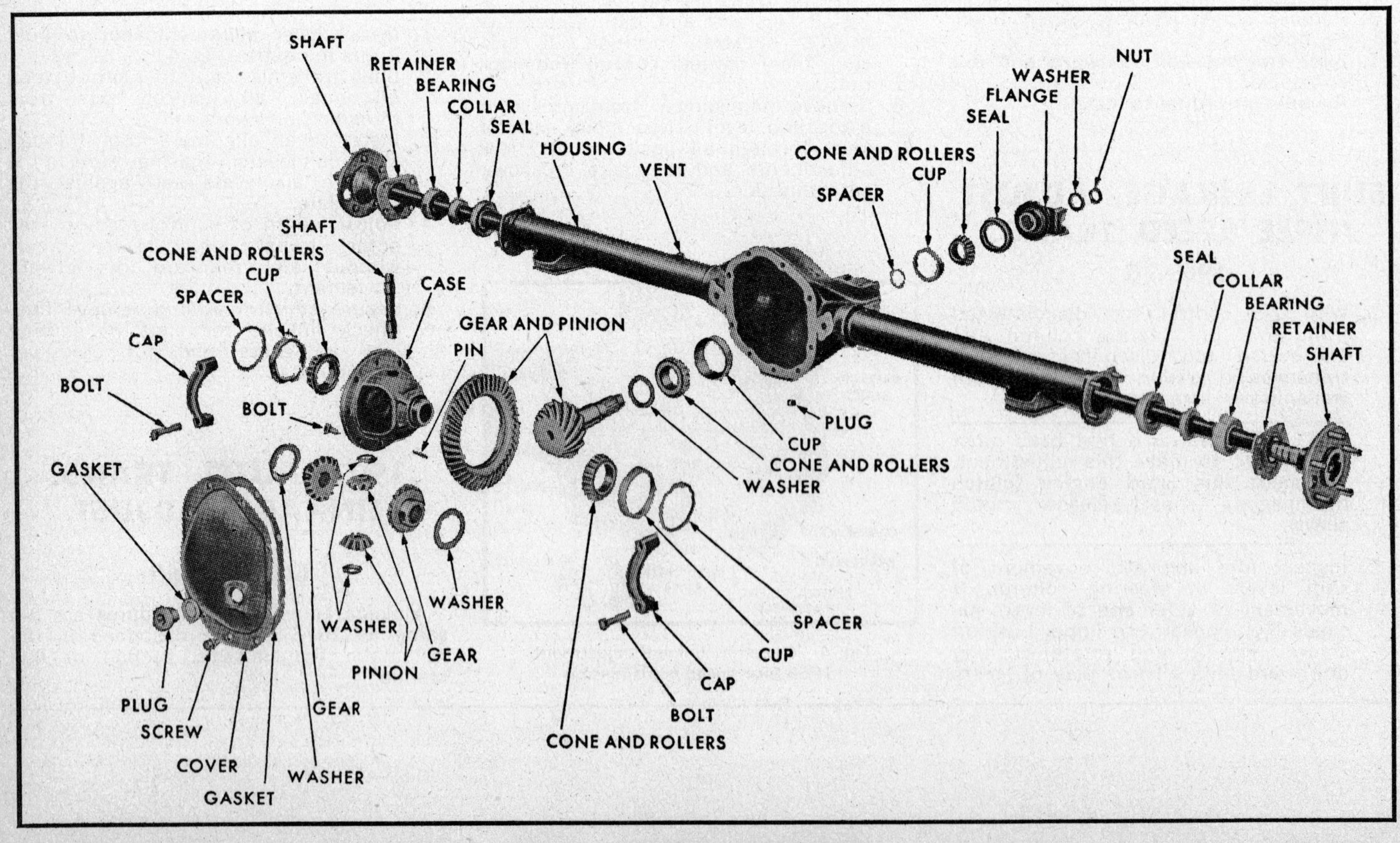

Fig. 1 Integral rear axle. 1969-76

through the brake backing plates to the rear axle tubes.

In the unit shown in Fig. 2, introduced in 1969, the differential is also supported by two tapered roller bearings. A threaded differential bearing adjuster is located in each bearing pedestal cap to eliminate differential side play, adjust and maintain ring and pinion backlash and provide a means of obtaining differential bearing preload.

Axles are retained by means of a "C" washer which is installed into a groove in the inner end of the axle shaft inside the differential unit.

On both these units, a removable stamped steel cover, bolted to the rear of the carrier, permits inspection and service of the differential without removal of the complete axle assembly from the vehicle.

Axle Shaft, Renew (Fig. 1 Type)

1. With wheel removed, remove clips holding brake drum on wheel studs and remove drum.
2. Disconnect brake lines at wheel cylinders.
3. Using access hole in axle flange, remove retainer nuts from end of housing.
4. Remove axle shaft and brake assembly, using a slide hammer-type puller.
5. Remove brake assembly from axle shaft with care to avoid damaging shaft in seal contact area.
6. Remove oil seal from axle housing.
7. *Remove axle shaft bearings only when necessary. Removal of bearings makes them unfit for further use.*
8. *Axle shaft end play is pre-set and not adjustable. End play is accomplished by the amount of end play built into the bearings. The two axle housing brake support plate gaskets on each side are used for sealing purposes only. Always replace the gaskets once they have been removed.*

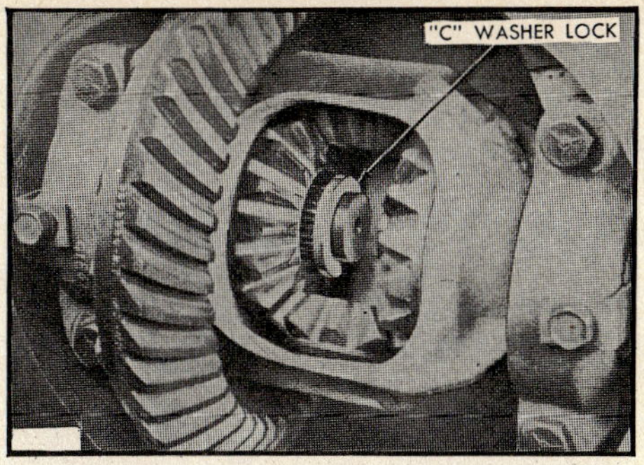

Fig. 3 Location of "C" washer locks

9. Press bearing and collar on shaft firmly against shoulders on shaft.
10. Install new oil seal in housing.
11. Install brake assembly on axle housing and carefully slide axle shaft through oil seal and into side gear splines.
12. Tap end of axle shaft lightly to position axle shaft bearing into bearing bore and attach retainer plate to housing.
13. Install brake drums and wheels.

Axle Shaft, Renew (Fig. 2 Type)

1. With wheel and brake drum removed, or caliper and rotor assembly removed, loosen differential housing cover and drain lubricant. Remove cover.
2. Turn differential case to make pinion shaft lock screw accessible and remove lock screw and shaft.
3. Push axle shaft inward toward center of car and remove "C" washer from groove in axle shaft, Fig. 3.
4. Remove axle shaft from housing, being careful not to damage the axle bearing, which will remain in the housing.
5. The axle bearing and/or seal can now be removed if necessary.
6. Reverse procedure to install.

REMOVABLE CARRIER TYPE

In these rear axles, Figs. 4 and 5, the drive pinion is mounted in two tapered roller bearings. The bearings are preloaded by a spacer and shims behind the front bearing. The drive pinion is positioned by an adjusting washer between the head of the drive pinion and the rear pinion bearing. The front bearing is held in place by a large washer and nut.

The differential is supported in the carrier by two tapered roller side bearings. These bearings are preloaded by two threaded ring nuts between the bearings and the pedestals. The differential assembly is positioned for proper ring gear and pinion backlash by varying the adjustment of these ring nuts. The differential case houses two side gears in mesh with two pinions mounted on a pinion shaft which is held in place by a lock pin. The side gears and pinions are backed up by thrust washers. Side thrust of the wheels is transferred from one axle shaft to the other by means of a thrust block in the center of the differential case.

Carrier, Remove & Replace

It is not necessary to remove the rear

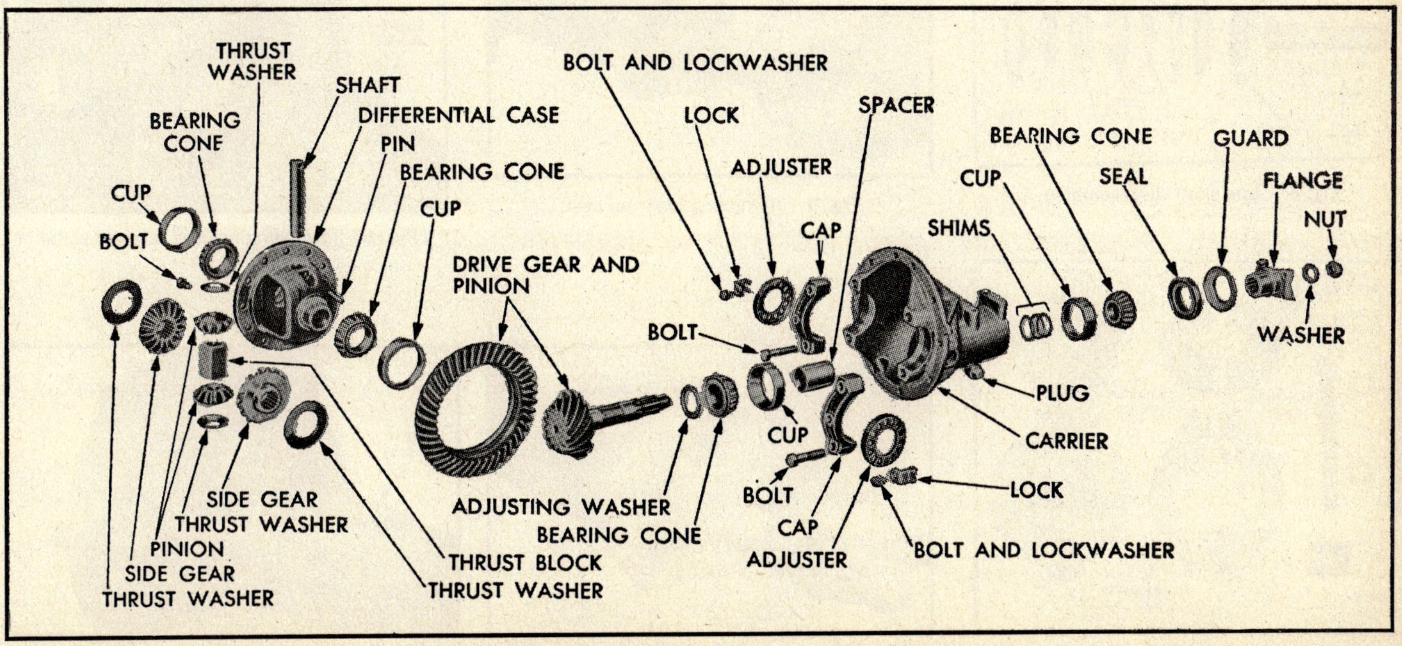

Fig. 4 Removable type rear axle assembly with large pinion

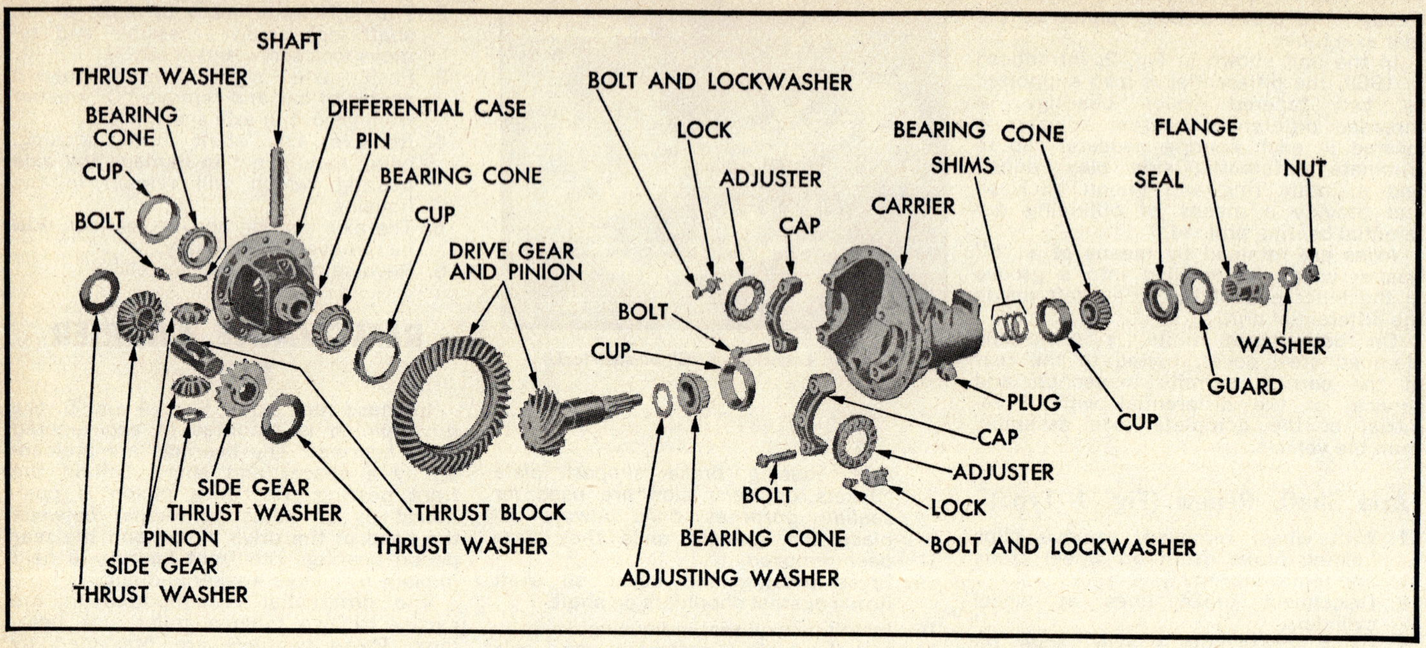

Fig. 5 Removable type rear axle assembly with small pinion

axle assembly for any normal repairs. The axle shafts and carrier assembly can easily be moved from the vehicle, leaving the rear axle housing in place.

1. Remove axle shafts as outlined below.
2. Disconnect rear universal joint and move propeller shaft out of the way. Support shaft to relieve strain on the front universal joint.
3. Remove lubricant from axle housing with a suction gun.
4. Remove attaching nuts and lift carrier assembly out of axle housing.
5. Reverse removal procedure to install.

Axle Shaft, Renew (1969-70)

1. With wheels removed, remove clips holding brake drum on axle shaft studs and remove brake drum.
2. Using access hole in axle shaft flange, remove retainer nuts. The right shaft with threaded adjuster in retainer plate will have a lock

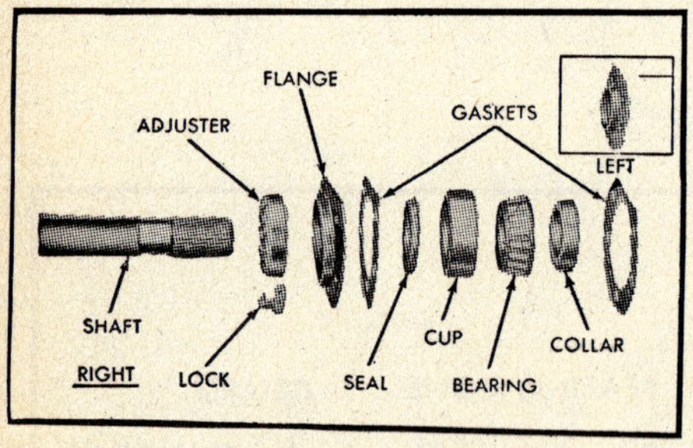

Fig. 6 Axle shaft disassembled

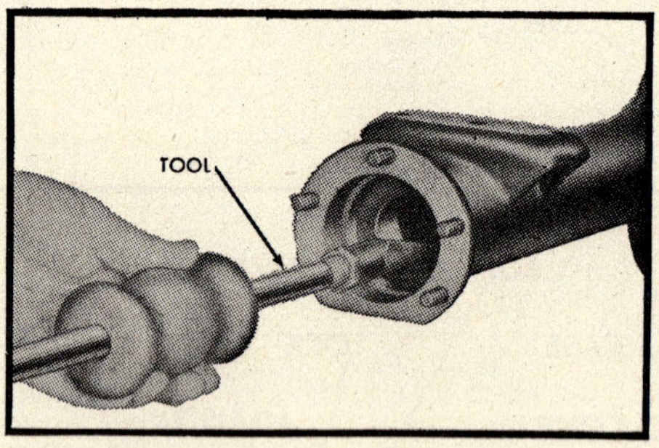

Fig. 8 Removing inner oil seal

Fig. 10 Notching bearing retainer collar

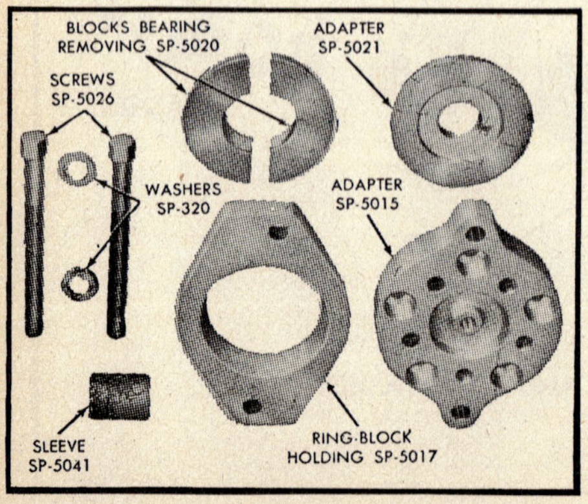

Fig. 7 Tool set for removing axle shaft

Fig. 9 Installing inner oil seal

Fig. 11 Removing roller retainer

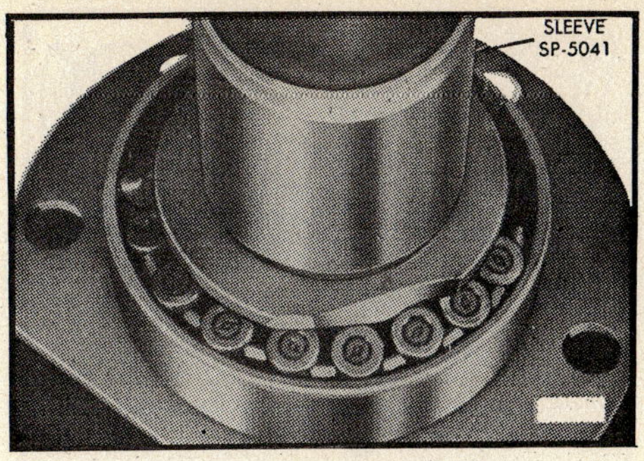

Fig. 12 Flange ground off inner bearing cone

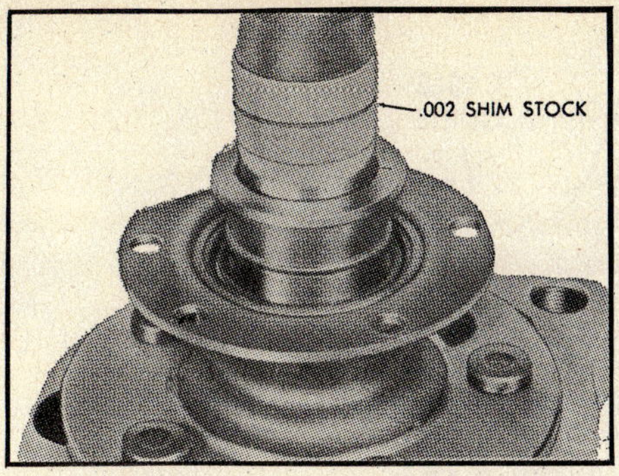

Fig. 15 Seal journal protection

Fig. 16 Removing bearing cone

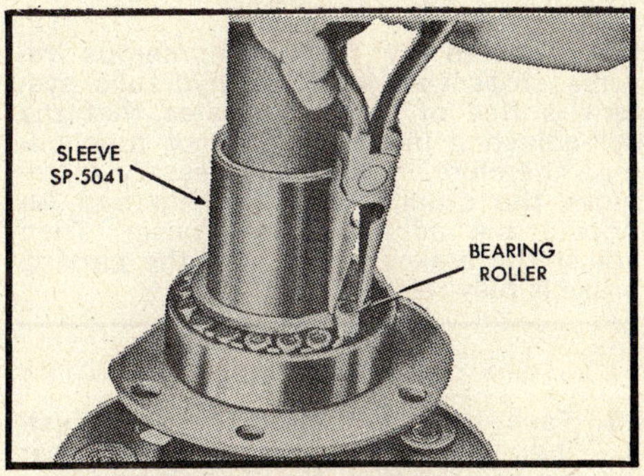

Fig. 13 Removing bearing rollers

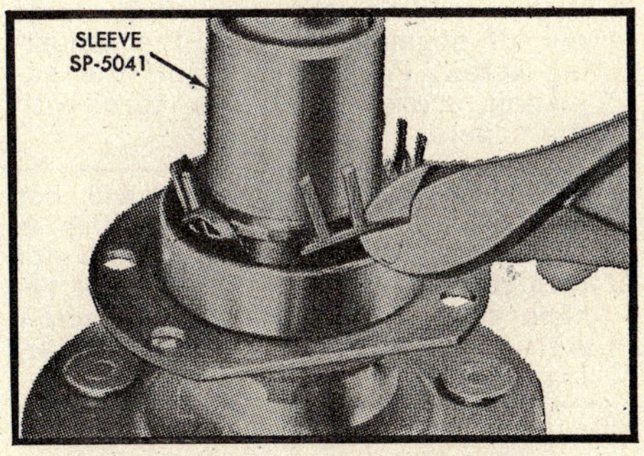

Fig. 14 Cutting out bearing retainer

Fig. 17 Installing bearing and collar

Fig. 18 Measuring axle shaft end play

under one of the studs that should be removed at this time, Fig. 6.

3. Remove parking brake strut.
4. Attach axle shaft remover tool, Fig. 7, to axle shaft flange and remove axle shaft. Remove brake assembly and foam gaskets.
5. Remove oil seal, Fig. 8.
6. Wipe axle shaft housing seal bore clean and install a new seal, Fig. 9.

Disassembly

To prevent the possibility of damaging axle shaft seal surface, slide a protective sleeve over seal surface next to bearing collar, Fig. 10.

1. Position axle shaft bearing collar on a heavy vise and, using a chisel, cut deep grooves into retaining collar at 90-degree intervals, as shown in Fig. 10. This will enlarge bore of collar and permit it to be driven off axle shaft.
2. Remove bearing roller retainer flange by cutting off lower edge with a chisel, Fig. 11.
3. Grind a section off flange of inner bearing cone, Fig. 12, and remove bearing rollers, Fig. 13.
4. Pull bearing roller retainer down as far as possible and cut with side cutters and remove, Fig. 14.
5. Remove roller bearing cup and protective sleeve from axle shaft.

CAUTION: Sleeve should not be used as a protector for the seal journal when pressing off bearing cone as it was not designed for this purpose.

6. To avoid scuffing seal journal when bearing cone is being removed, it should be protected by a single wrap of .002" shim stock held in place by a rubber band, Fig. 15.
7. Remove bearing cone with tool set shown in Fig. 7. Tighten bolts of tool alternaterly until cone is removed, Fig. 16.
8. Remove seal in bearing retainer plate and replace with a new seal.

Assembly

1. Install retainer plate and seal on axle shaft.
2. Install new axle shaft bearing cup, cone and collar on shaft, using tool shown in Fig. 17. Tighten bolts of tool alternately until bearing and collar are seated properly.
3. Inspect axle shaft seal journal for scratches and polish with #600 crocus cloth if necessary.
4. Lubricate wheel bearings with approved grease.

Installation

1. Clean axle shaft flange face and install new gasket followed by brake support plate on left side of axle housing.
2. Install foam gasket on studs of axle housing and slide shaft through oil seal and engage splines in differential side gear.
3. Tap end of axle shaft lightly with a plastic mallet to position axle shaft bearing in housing bearing bore. Position retainer plate over axle housing studs. Install retainer nuts and torque to 30-35 ft-lbs. Start by tightening bottom nut.
4. Repeat Step 1 on right side of axle housing.
5. Back off threaded adjuster on right axle shaft until inner face of adjuster is flush with inner face of retainer plate. Carefully slide axle shaft through oil seal and engage splines in differential side gears.
6. Repeat Step 3.

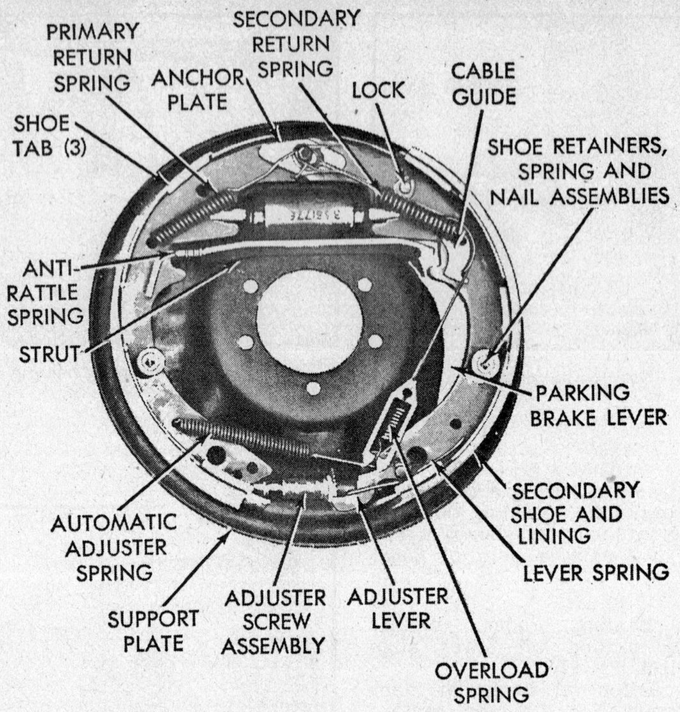

Fig. 19 Left rear brake (10-11 inch). 1969-76

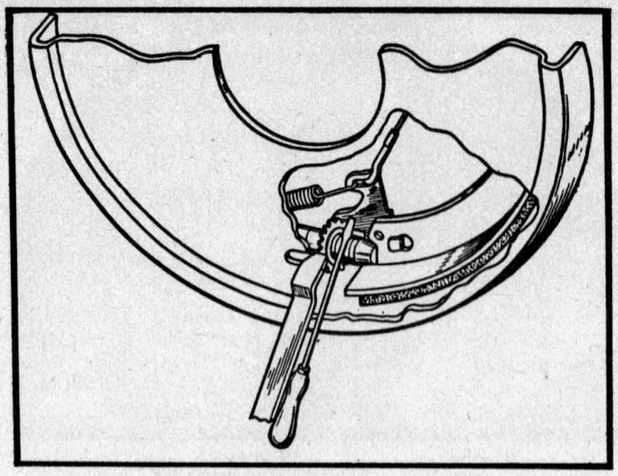

Fig. 20 Releasing brake lever with screwdriver while adjusting star wheel

Axle Shaft End Play

When setting end play both rear wheels must be off the ground, otherwise a false end play setting will occur.

1. Using a dial indicator mounted as shown in Fig. 18, turn the adjuster clockwise until both wheel bearings are seated and there is zero end play in axle shafts. Back off adjuster counterclockwise four notches to establish an axle shaft end play of .013-.023".
2. Tap end of axle shaft lightly with a plastic mallet to seat right wheel bearing cup against adjuster, and rotate axle shaft several revolutions so that a true end play reading is indicated.
3. Remove one retainer plate nut and install adjuster lock. If tab on lock does not mate with notch in adjuster, turn adjuster slightly until it does. Install nut and torque to 30-35 ft-lbs.
4. Recheck axle shaft end play. If not within prescribed limits, repeat adjustment procedure.
5. Remove dial indicator and install brake drum, drum retaining clips and wheel.

PROPELLER SHAFT

1. Remove both rear universal joint roller and bushing assembly clamps from pinion yoke. Do not disturb retaining strap holding roller assemblies on cross.
2. Lower front of vehicle slightly to prevent loss of transmission oil and pull drive shaft out as an assembly.
3. To install, carefully slide yoke into splines on transmission output shaft.
4. Align rear of propeller shaft with pinion yoke and position roller and bushing assemblies into seats of pinion yoke.
5. Install bushing clamps and tighten clamp bolts to 170 inch lbs.

BRAKE ADJUSTMENTS

1969-76 Self Adjusting Brakes

These brakes, Fig. 19, have self-adjusting shoe mechanisms that assure correct lining-to-drum clearances at all times. The automatic adjusters operate only when the brakes are applied as the car is moving rearward.

Although the brakes are self-adjusting, an initial adjustment is necessary when the brake shoes have been relined or replaced, or when the length of the star wheel adjuster has been changed during some other service operation.

Frequent usage of an automatic transmission forward range to halt reverse vehicle motion may prevent the automatic adjusters from functioning, thereby inducing low pedal heights. Should low pedal heights be encountered, it is recommended that numerous forward and reverse stops be made until satisfactory pedal height is obtained.

Service Note

If a low pedal height condition cannot be corrected by making numerous reverse stops (provided the hydraulic system is free of air) it indicates that the self-adjusting mechanism is not functioning. Therefore, it will be necessary to remove the drum, clean, free up and lubricate the adjusting mechanism. Then adjust the brakes, being sure the parking brake is fully released.

Adjustment

1. Each backing plate has two adjusting hole covers; remove the rear cover and turn the adjusting screw upward with a screwdriver or other suitable tool to expand the shoes until a slight drag is felt when the drum is rotated.
2. Remove the drum.
3. While holding the adjusting lever out of engagement with the adjusting screw, Fig. 20; back off the adjusting screw about one turn with the fingers.

NOTE: *If finger movement will not turn the screw, free it up. If this is not done, the adjusting lever will not turn the screw during subsequent vehicle operation. Lubricate the screw with oil and coat with wheel bearing grease.*

---

4. Install wheel and drum, and adjusting hole cover. Adjust brakes on remaining wheels in the same manner.
5. If pedal height is not satisfactory, drive the vehicle and make sufficient reverse stops until proper pedal height is obtained.

PARKING BRAKE, ADJUST

Exc. 1974-75 Imperial

1. Release parking brake lever and loosen cable adjusting nut to be sure cable is slack.
2. With rear wheel brakes properly adjusted, tighten cable adjusting nut until a slight drag is felt when the rear wheels are rotated. Then loosen the cable adjusting nut until both rear wheels can be rotated freely.

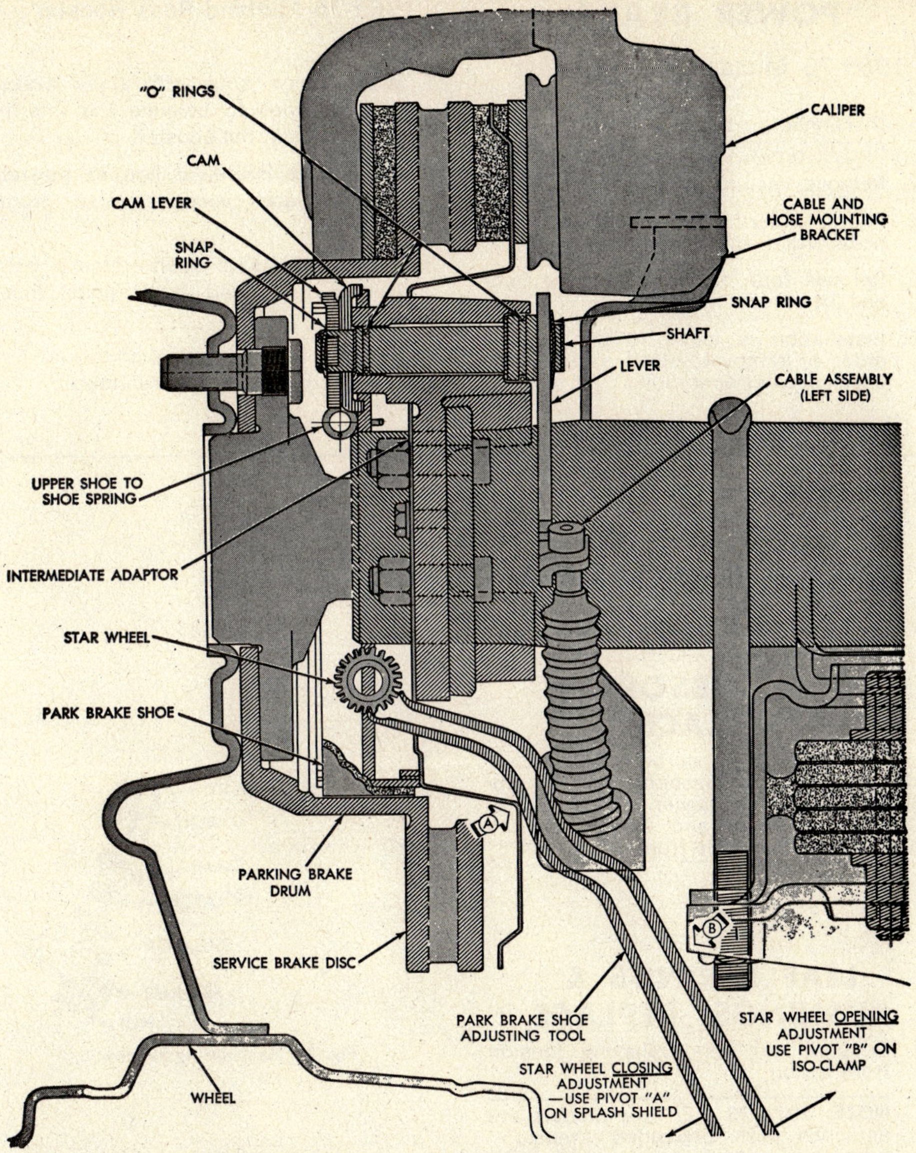

Fig. 21 Rear wheel disc brake drum type parking brake shoe adjustment. 1974-75 Imperial

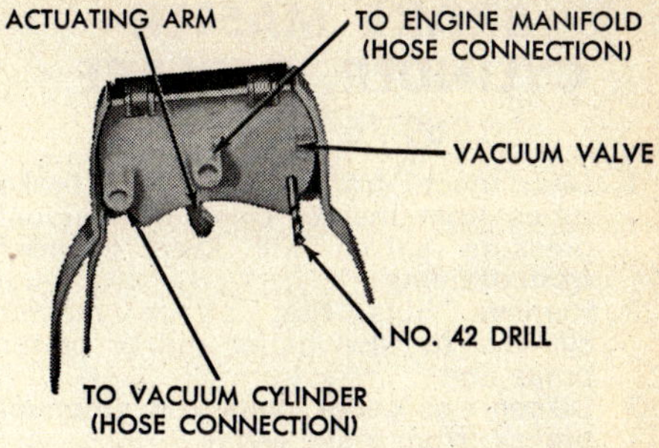

Fig. 22 Parking brake vacuum release valve. 1969-76 (typical)

3. To complete the operation, back off an additional two turns of the cable adjusting nut.
4. Apply and release parking brake several times to be sure rear wheels are not dragging when cable is in released position.

1974-75 Imperial

Adjust parking brake by rotating adjuster wheel with tool C-4223, Fig. 21, until brakes seat against drums, then back off adjuster wheel 12 clicks to prevent brake drag.

VACUUM RELEASE PARKING BRAKE

The parking brake is pedal applied and released by a vacuum chamber. When the engine is started and vacuum is developed, energy is then available to release the parking brake. This is controlled by the transmission push buttons or shift linkage. When the transmission is in "Neutral", vacuum is cut off from the release chamber and there is no action of the parking brake pedal.

When the transmission is shifted into a drive gear (forward or reverse), the vacuum control valve is opened, actuating the vacuum release chamber mounted on the parking brake assembly.

NOTE: *In the event of engine failure and no vacuum, the brake may be released by a manual release lever mounted on the left side of the parking brake pedal assembly. This assembly prevents the vehicle from being driven with the parking brake in the applied position.*

Testing Vacuum Release

1. If the mechanism is inoperative, first check for damaged or kinked vacuum hoses and for loose hose connections at the vacuum chamber, vacuum release valve at neutral safety switch, and at engine manifold connection.
2. Check adjustment of neutral safety switch and operation of vacuum release valve.
3. Check vacuum chamber piston travel by running engine and shifting transmission selector from drive to neutral. The manual release lever should move up and down as vacuum is applied and released. If no movement is observed or if movement is slow (more than 1 or 2 seconds to complete full stroke), the vacuum chamber is leaking and should be replaced.
4. Check brake release with vacuum applied. If vacuum chamber piston completes full stroke but does not release brake, a malfunction of the pedal assembly is indicated.
5. Check operation of parking brake with engine off. Parking brake should remain engaged regardless of transmission selector position.

Parking Brake Vacuum Valve, Replace

1. With engine off, place shift lever in drive, remove vacuum hoses and retaining screws and remove valve from steering column.
2. To install, move actuating arm on valve against spring to extreme position or until locating holes line up and install a #42 drill in hole, Fig. 22.
3. Place shift lever in park and install valve but do not tighten screws. Rotate valve clockwise until actuating arm contacts tab in steering column then tighten screws.
4. Remove drill and install vacuum hoses making sure that hose from engine manifold is attached to center fitting on valve, Fig. 22.
5. Start engine and check that parking brake can be set in neutral and park and will release in reverse and drive positions.

BRAKE MASTER CYLINDER, REPLACE

1969-76 All Cars

1. Disconnect front and rear brake tubes from master cylinder (residual pressure valves will keep cylinder from draining).
2. Remove nuts that attach master cylinder to cowl panel and/or power brake unit.
3. Disconnect pedal push rod (manual brakes) from brake pedal.
4. Slide master cylinder straight out from cowl panel and/or power brake unit.
5. Reverse procedure to install.

POWER BRAKE

1969-76 Midland-Ross Booster

1. Disconnect brake line(s) from master cylinder.
2. Remove vacuum hose from booster.
3. From under dash, remove brake pedal and push rod attaching bolt.
4. Remove four booster attaching nuts and lift booster from vehicle.
5. Installation is made in the reverse order of removal. Bleed system and check booster operation.

1969-76 Midland-Ross Booster

1. With engine shut off, apply brake several times to balance the internal pressure of the booster.
2. Disconnect hydraulic line at master cylinder and vacuum hose from booster.
3. From underneath dash, remove bolt from plunger and brake pedal linkage.
4. Remove four attaching bolts and lift off booster and master cylinder.

Rear Suspension

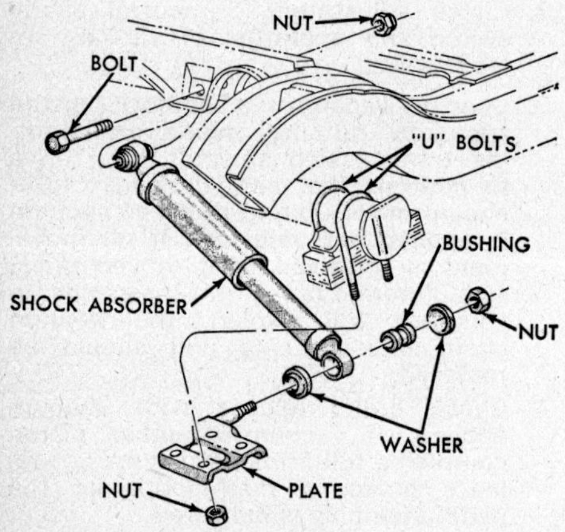

Fig. 1 Spring plate (typical)

SHOCK ABSORBER, REPLACE

To replace shock absorber, support rear axle properly and disconnect shock absorber at upper and lower mountings. On 1970-74 Barracuda and Challenger, the upper nut is accessible from trunk.

LEAF SPRINGS & BUSHINGS, REPLACE

1. Support rear axle, relieving tension from spring.

NOTE: 1974-76 Full Size Models are equipped with preloaded "tension" type springs and a spring stretcher (C-4211) must be used during spring removal.

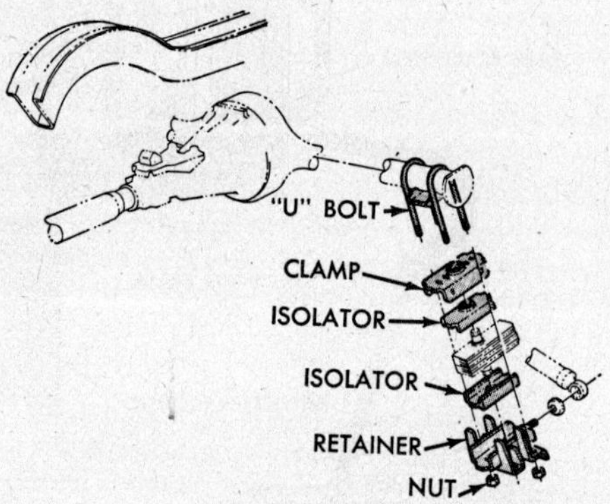

Fig. 2 Rear spring isolator

2. Disconnect shock absorber from lower mounting.

NOTE: On 1968 Imperial, disconnect track bar from axle stud.

3. Remove "U" bolts and spring plate, Fig. 1, or lower spring seat isolator retainer and isolator, Fig. 2.

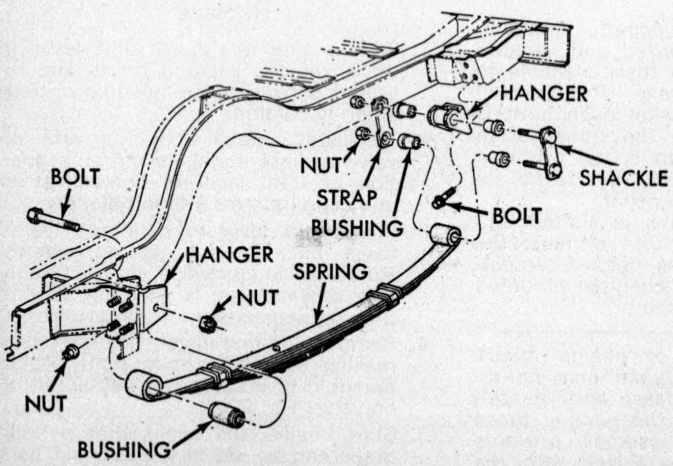

Fig. 3 Rear spring (typical)

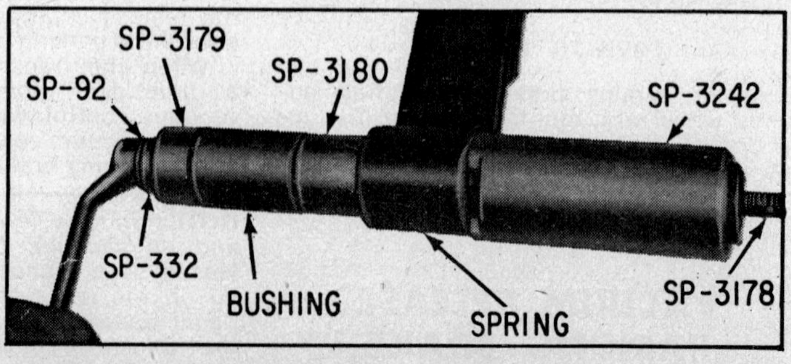

Fig. 4 Spring front bushing replacement

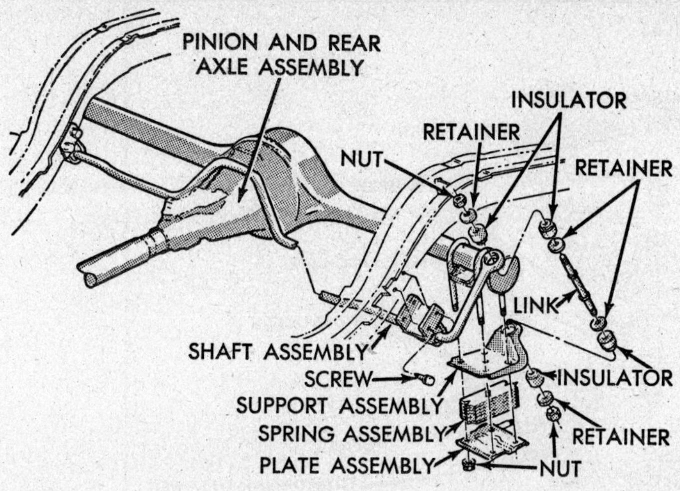

Fig. 5 Sway bar installation. 1973-74 Barracuda & Challenger

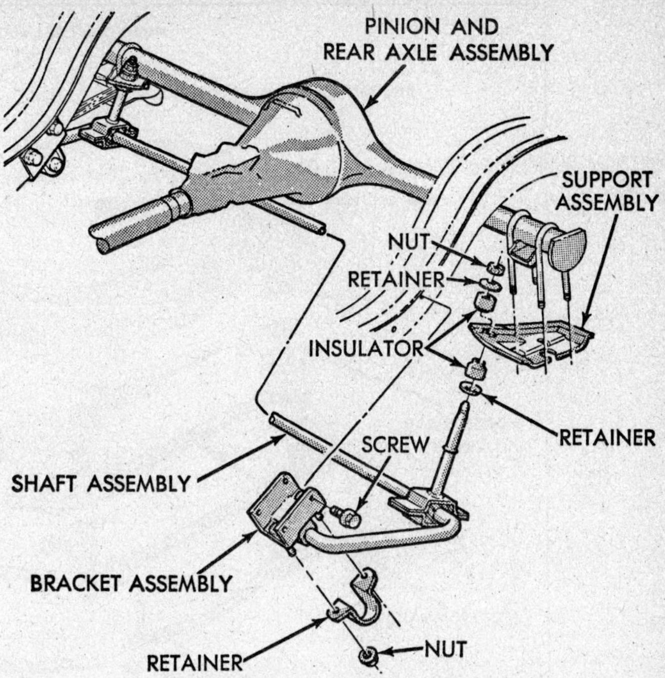

Fig. 6 Sway bar installation. 1973-76 Charger, Coronet & Satellite

4. Remove spring front hanger to body mount bracket nuts, Fig. 3.
5. Remove rear shackle bolts, lower spring, thus pulling spring front hanger bolts out of holes.
6. Remove front hanger and rear shackle from spring. To replace eye bushings, refer to Fig. 4. Bushing replacement is accomplished in one operation.
7. Reverse procedure to install.

Leaf Spring Service

To replace interliners, remove spring alignment clips and on all models except Imperial, discard alignment clips. Separate spring leaves with a screwdriver or other suitable tool and remove interliners. Thoroughly clean spring surfaces before installation of new interliners.

To replace zinc interleaves, clamp spring in a vise and remove center bolt. Open vise carefully, allowing spring to expand. Interleaves can now be serviced. Install a drift through spring center bolt holes and clamp spring in a vise. Remove drift and install center bolt.

SWAY BAR, REPLACE

1. Remove nuts, retainers and rubber insulators from sway bar upper links, Figs. 5 and 6.
2. Disconnect sway bar brackets from frame.
3. Remove link from support assembly and replace insulators.
4. Reverse procedure to install.

Front End and Steering Section

FRONT SUSPENSION

1976 Aspen & Volaré

This front suspension, Fig. 1, incorporates two transverse torsion bars which react on the outboard end of the lower control arms. The torsion bars are anchored in the front crossmember opposite the affected wheel. The torsion bars are mounted parallel to the front crossmember through a "Pivot Cushion Bushing", attached to the crossmember, and turns and extends rearward to the lower control arm. The torsion bar ends are provided with an isolated bushing, bolted to the lower control arm and sway bar, which acts as the lower control arm strut.

Riding height is controlled by the torsion bar adjusting bolts on the anchor end of the torsion bar. The right torsion bar is adjusted from the left side and the left torsion bar is adjusted from the right side.

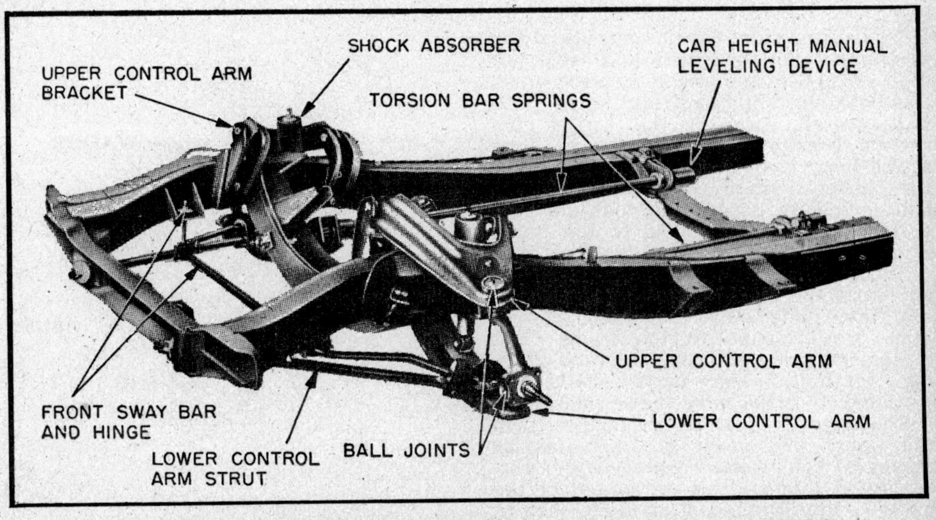

Fig. 2 Longitudinal torsion bar front suspension. Typical for all exc. 1969-73 Imperial & 1976 Aspen & Volaré

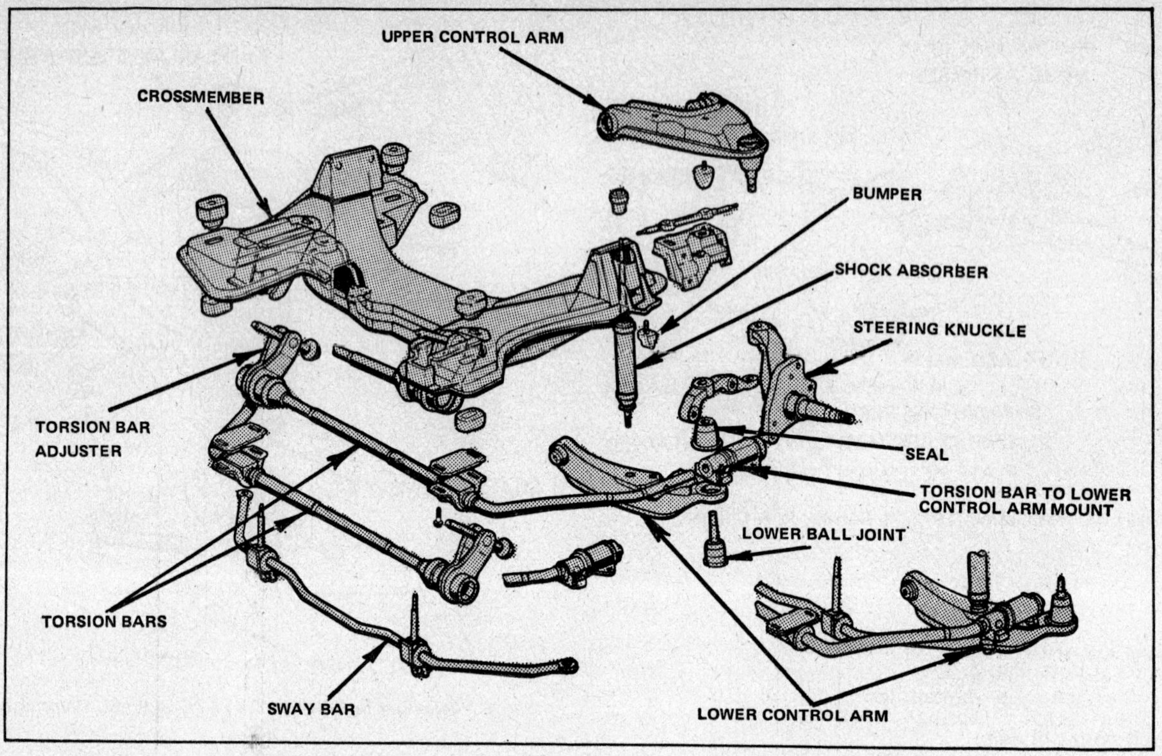

Fig. 1 Transverse torsion bar front suspension. 1976 Aspen & Volaré

The torsion bar assembly incorporates the "Pivot Cushion Bushing" and "Bushing to Lower Control Arm". The lower control arm inner ends are bolted to the crossmember and pivots through bushings.

All Exc. 1969-73 Imperial & 1976 Aspen & Volaré

This suspension, Fig. 2, consists of two torsion bar springs (right and left), two sets of upper and lower control arms, four ball joints and two struts.

The front ends of the torsion bar springs engage the lower control arms at the inner pivot points. The rear end of the torsion bars engage adjustable anchor and cam assemblies that are supported by brackets welded to the frame side rails and a removable crossmember.

The upper control arms are mounted on removable brackets that are bolted to the frame side rails. The lower control arms are attached to the frame front crossmember by a pivot shaft and bushing assembly. The pivot shafts are mounted in replaceable rubber bushings.

The steering knuckles are connected to the upper and lower control arms by means of ball joints. To prevent the possibility of fore and aft movement of the lower control arms, a strut is attached to the front crossmember and to the lower control arm.

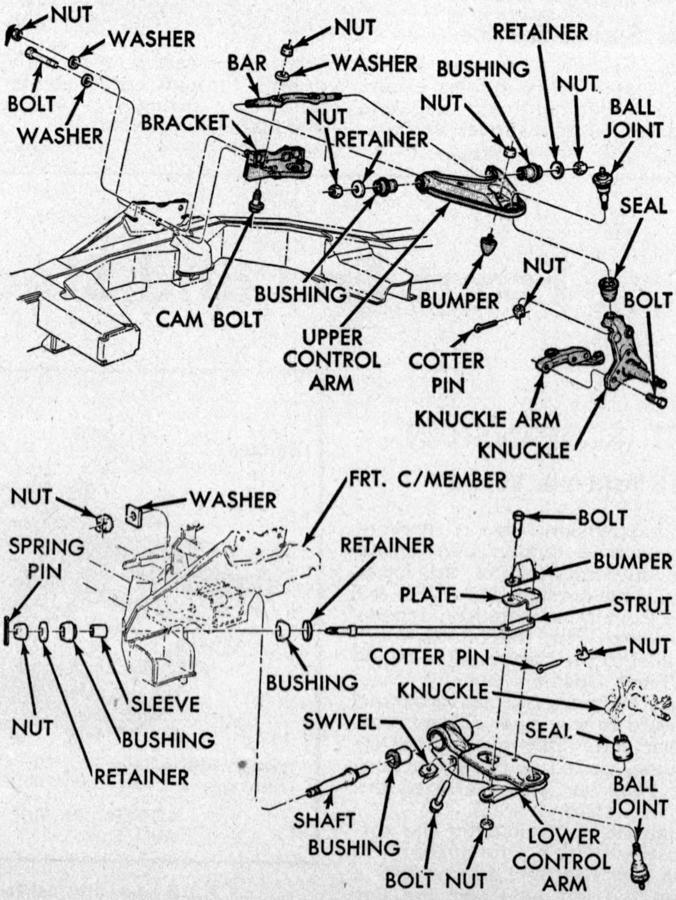

Fig. 3 1969-73 Imperial front suspension.

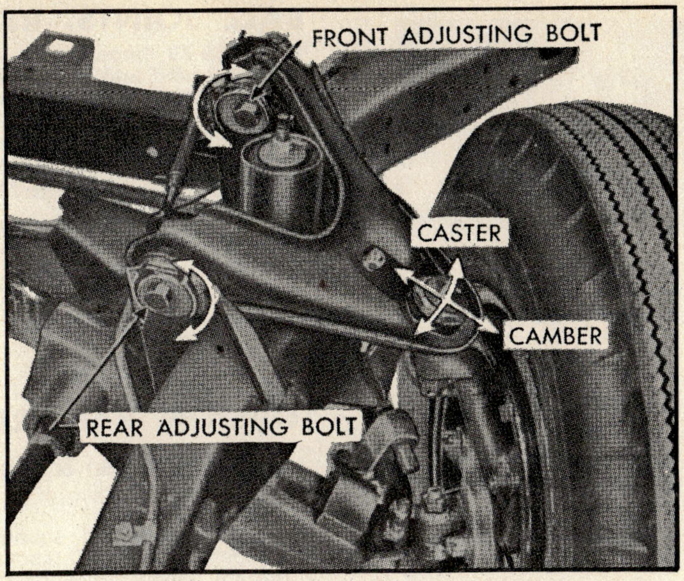

Fig. 4 Camber and caster adjusting bolts.

1974 Fury; 1974-76 Imperial, Monaco & Chrysler Exc. Cordoba; 1975-76 Gran Fury

This suspension has new lower control arms with pressed in ball joints and more serviceable struts. Caster and camber settings are made by loosening the upper control arm pivot bar bolt nuts and adjusting as necessary.

1973-74 Satellite; 1973-76 Charger & Coronet; 1975-76 Cordoba & Fury

This suspension has a rubber mounted crossmember, a torsion bar crossmember, more serviceable struts and a lower control arm with screw in ball joints for 1973 and pressed in for 1974-76 models. Caster and camber settings are made by loosening the upper control arm pivot bar bolt nuts and adjusting as necessary.

1969-73 Imperial

The front suspension has a front "K" crossmember that is isolated from the stub frame by four rubber bushing type isolators. The torsion bar rear anchor crossmember is isolated from the stub frame crossmember by two sandwich type rubber insulators. The front anchors are part of the lower control arms and provide the means of adjusting the vehicle front height. The upper control arm is mounted on a pivot bar and the front *wheel alignment is set by the adjustment of two vertically mounted cam bolts,* Fig. 3.

LUBRICATION

1969-76

All ball joints and torsion bars are effectively sealed against road splash by tightly fitted balloon type flexible type seals. The ball joints are semi-permanently lubricated with special lubricant, and should not under normal conditions require lubrication before 36,000 miles.

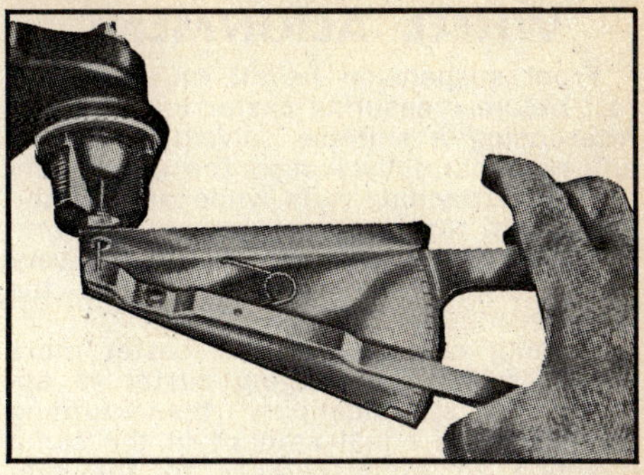

Fig. 5 Scale used to check lower ball joint for wear (tool C-3911)

All ball joints, tie rod end seals and protectors should be inspected at all oil change periods. Damaged seals must be replaced to prevent lubricant leakage or contamination and subsequent component failure.

WARNING

Do not use pressure type lubrication equipment as the pressure may damage the balloon type seals. Use a hand type lubrication gun filled only with the special lubricant specified for the job. Fill each unit slowly to avoid rupturing the seal.

Every 36,000 miles remove the plug from the ball joint and install a grease fitting. Using a hand gun, pump the grease into the unit until the seal balloons—indicating fullness. Remove the grease fitting and reinstall the plug.

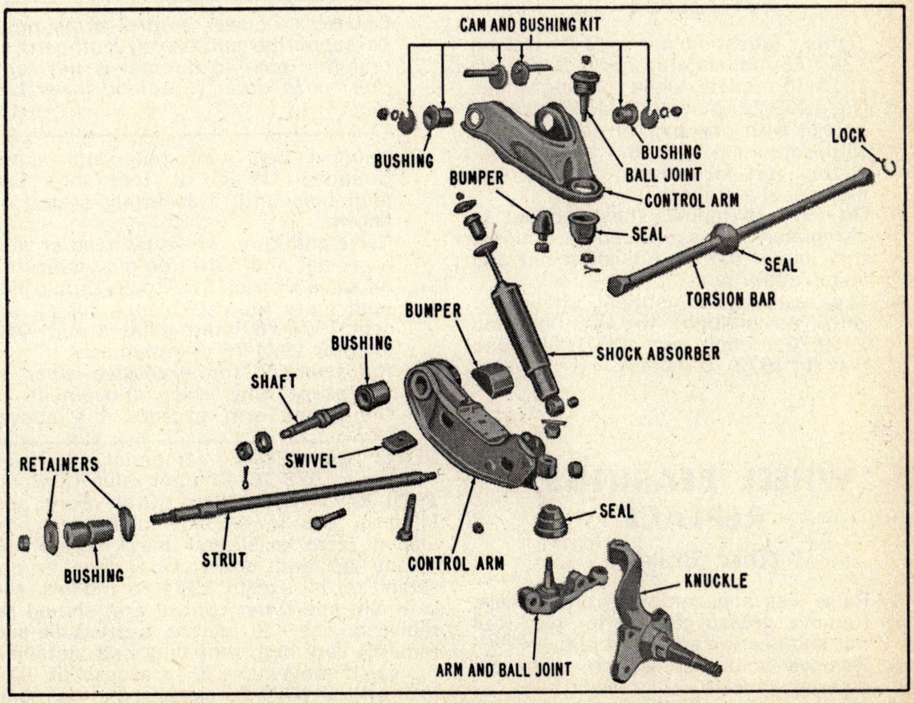

Fig. 6 Front suspension. All except 1969-73 Imperial & 1976 Aspen & Volaré

WHEEL ALIGNMENT

Front suspension height must be correct before measuring caster and camber. After using a suitable solvent to loosen any rust, carefully loosen the upper control arm attaching nuts while holding the bolts from turning. Once caster and camber have been adjusted, Fig. 3, a very small turn of the bolts will affect the gauge readings.

Turning one bolt affects caster more than camber. By bringing caster to approximate specifications, then turning both bolts an equal amount in the same direction to bring camber to the preferred specification, will usually bring caster to the preferred setting.

NOTE: Turning both cams in the same direction an equal amount will change camber with little or no caster change. Turning both cams an equal amount in opposite directions will change caster with little or no change of camber.

TOE-IN, ADJUST

1969-76

With the front wheels in straight ahead position, loosen the clamps at each end of both adjusting tubes. Adjust toe-in by turning the tie rod sleeve which will "center" the steering wheel spokes. If the steering wheel was centered, make the toe-in adjustment by turning both sleeves an equal amount. Position the clamps so they are on the bottom and tighten bolts to 15 ft. lbs.

WHEEL BEARINGS, ADJUST

1. Tighten adjusting nut to 8 ft. lbs. on 1969-72 models and 20-25 ft. lbs. on 1973-76 models, while rotating wheel.
2. On 1969-72 models, place nut lock on nut with one pair of slots aligned with cotter pin hole and back off adjusting nut lock assembly one slot and install cotter pin.
3. On 1973-76 models, bask off nut to completely release bearing preload, then finger tighten adjusting nut and install cotter pin.
4. The resulting adjustment should be zero (no preload) to .003 inch on 1969-72 models and .0001-.003 end play for 1973-76 models.

WHEEL BEARINGS, REPLACE

(Disc Brakes)

1. Raise car and remove front wheels.
2. Remove grease cap, cotter pin, lock nut and bearing adjusting nut.
3. Remove bolts that attach caliper to steering knuckle.
4. Slowly slide caliper up and away from disc and support caliper on steering knuckle arm.

NOTE: Do not allow caliper to hang by brake hose.

5. Remove thrust washer and outer bearing cone. Remove hub and disc assembly. Grease retainer and inner bearing can now be removed.

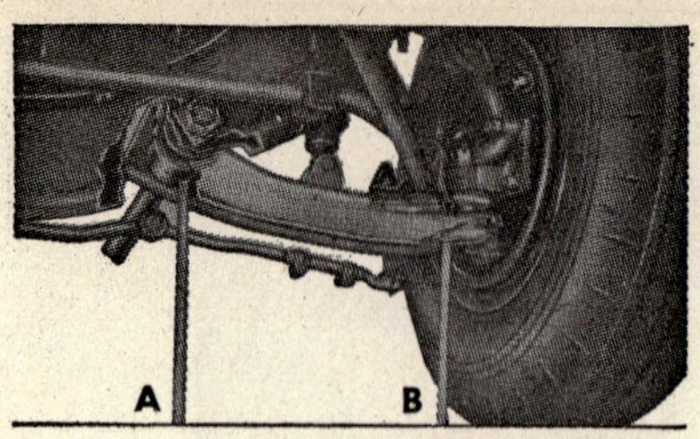

Fig. 7 Measuring front suspension height (typical)

CHECKING BALL JOINTS FOR WEAR

If loose ball joints are suspected, first make sure the front wheel bearings are properly adjusted and that the control arms are tight.

Fig. 5 illustrates tool No. C-3911 which has been developed to measure accurately lower ball joint wear and to eliminate needless replacement of ball joints not excessively worn. Checking procedure is as follows:

1. Raise front of vehicle at lower control arms to bring wheels clear of floor.

CAUTION—*Lower control arms must be supported sufficiently outboard so that the rebound bumper is not compressed in order to unload lower ball joint.*

2. Remove ball joint plug and screw threaded fitting of tool into ball joint hole until it is firmly seated as shown.
3. Raise and lower wheel by hand or with a pry bar and note free play indicated on scale of tool. The scale should not read more than .070 inch for all except 1974-76 intermediates and .040 inch for 1974-76 intermediates.
4. Replacement for excessive wear is necessary only when movement of the gauge arm exceeds the above.

NOTE: On 1969-76 Imperial, 1973-74 Satellite, 1973-76 Charger and Coronet, 1975-76 Chrysler, Fury, Gran Fury and Monaco, the lower ball joints are preloaded (zero axial end play). Therefore, if any up and down movement is observed on all except 1974-76 models, the ball joint and lower control arm should be replaced. The ball joint is a press fit and requires very high removing and installing forces. If movement is in excess of .020 inch on the 1974-76 models, the ball joint assembly should be replaced.

BALL JOINTS, REPLACE

On 1969-73 Imperial, the upper ball joint is threaded into the control arm and the lower ball joint is serviced as an assembly with the lower control arm. On all other models, the upper ball joint is threaded into the control arm whereas the lower ball joint is furnished as an assembly with the steering arm.

Use a suitable tool to press the ball joints from the steering knuckles, and when installing a ball joint, be sure to start it squarely into the control arm threads.

TORSION BAR, REPLACE

1976 Aspen & Volaré

Removal

1. Raise vehicle and support so front suspension is in full rebound position.
2. Rotate anchor adjusting bolt, located in frame crossmember, counter-clockwise to release load on torsion bar. Then, remove anchor adjusting bolt from torsion bar.
3. Raise lower control arms until 2⅞ inch clearance is obtained between crossmember ledge at jounce bumper and the torsion bar end bushing and support lower arms at this height.

NOTE: This procedure will align the sway bar and the lower control arm attaching points for disassembly and component realignment and attachment during assembly.

4. Remove sway bar to control arm attaching bolt and retainers, then the two bolts securing torsion bar end bushing to lower control arm.
5. Remove two bolts securing torsion bar pivot cushion bushing to crossmember, then the torsion bar and anchor assembly from crossmember.
6. Separate anchor from torsion bar.

Inspection

1. Inspect seal for damage and replace, if necessary.
2. Inspect bushing to lower control arm and pivot cushion bushing. Inspect seals on cushion bushing for cuts, tears or severe deterioration that may allow moisture to enter under cushion. If corrosion is evident, replace torsion bar assembly.
3. Inspect torsion bars for paint damage and touch up, if necessary.
4. Clean anchor hex openings and torsion bar hex ends.
5. Inspect torsion bar adjusting bolt and swivel for damage or corrosion and replace, if necessary.

Installation

1. Slide balloon seal over torsion bar end with cupped end facing toward hex.
2. Lubricate torsion bar hex end with lubricant, P/N 2525035, and install hex end into anchor bracket. With the torsion bar in horizontal position, the anchor bracket ears should be positioned nearly straight upward. Position swivel into anchor bracket ears.

3. Install torsion bar anchor bracket assembly into crossmember anchor retainer, then the anchor adjusting bolt and bearing.
4. Install the two bolt and washer assemblies securing pivot cushion bushing to crossmember and tighten bolts finger tight.
5. With lower control arms supported as outlined in step 3 under "Removal", install the two bolt and nut assemblies securing torsion bar bushing to lower control arm and torque nuts to 50 ft. lbs.
6. Ensure anchor bracket is fully seated in crossmember, then torque pivot cushion retainer to crossmember bolts to 75 ft. lbs. Place balloon seal over anchor bracket.
7. Install new bolt through sway bar, retainer cushions and sleeve and attach to lower control arm end bushing, then torque bolt to 50 ft. lbs.
8. Rotate anchor adjusting bolt clockwise to load torsion bar.
9. Lower vehicle and adjust riding height.

1969-76 All Exc. 1976 Aspen & Volaré

The torsion bars are not interchangeable side for side. The bars are marked either right or left by an "R" or an "L" stamped on one end of the bar. The general procedure for replacing a torsion bar is as follows:

Removal

1. Remove upper control arm rebound bumper.
2. If vehicle is to be raised on a hoist, make sure it is lifted on the body only so suspension is in full rebound position (no load).
3. Release all load from torsion bar by turning anchor adjusting bolt counterclockwise.

NOTE: On 1969-73 Imperial models load on *both* torsion bars will have to be released by turning the anchor adjusting bolts counterclockwise. This is necessary because the rubber isolator rear crossmember would be under load and could possibly cause severe damage or personal injury.

---

4. Slide rear anchor balloon seal off of rear anchor and remove lock ring from anchor.
5. Remove torsion bar, by sliding bar out through rear of rear anchor. Use care not to damage balloon seal when it is removed from torsion bar.

Inspection

1. Inspect balloon seal for damage and replace if necessary.
2. Inspect torsion bar for scores or nicks. Dress down all scratches and nicks to remove sharp edges, then paint repaired areas with a rust preventive.
3. Remove all foreign material from hex openings in anchors and from hex ends of torsion bars.
4. Inspect adjusting bolt and swivel and replace if there is any sign of corrosion or other damage. Lubricate for easy operation.

Installation

1. Insert torsion bar through rear anchor.
2. Slide balloon seal over torsion bar with cupped end toward rear of bar.
3. Coat both ends of torsion bar with a long mileage lubricant.
4. Slide torsion bar in hex opening of lower control arm.
5. Install lock ring, making sure it is seated in groove.
6. Pack annular opening in rear anchor completely full of a long mileage lubricant.
7. Position lip of balloon seal in groove of anchor. *On 1969-73 Imperial models, install balloon seal clamp.*
8. On all models except 1969-73 Imperial, turn adjusting bolt clockwise to place a load on torsion bar. *On 1969-73 Imperial, turn both adjusting bolts clockwise to place a load on both torsion bars.*
9. Lower vehicle to floor and adjust front suspension height.
10. Install upper control arm rebound bumper.

RIDING HEIGHT, ADJUST

Before taking measurements, grasp the bumpers at the center (rear bumper first) and jounce the car up and down several times. Jounce the car at the front bumper the same number of times and release the bumper at the same point in the cycle each time.

1. On 1969-74 models, measure from control arm torsion bar anchor to floor (Measurement A) and from ball joint to floor (Measurement B), Fig. 7. Subtract Measurement B from Measurement A. The difference should be as listed below (plus or minus ⅛ inch).
2. On 1975-76 models except 1976 Aspen and Volaré, measure distance from a point 1 inch forward of rear face of torsion bar anchor to the floor (Measurement A), Fig. 7. The distance should be as listed below.
3. On 1976 Aspen and Volaré, measure distance from lowest point of lower control arm inner pivot bushing to the floor. The distance should be as listed below.
4. Measure the other side in the same manner.
5. Adjust by turning the torsion bar anchor adjusting nut *clockwise* to *increase* the height and *counterclockwise* to *decrease* the height. The difference from side-to-side should not exceed ⅛".
6. After adjusting, jounce the car and recheck the measurements on both sides, even if only one side may have been adjusted.

Dodge Exc. Aspen, Challenger & Dart

1969-70	Coronet, Charger	1⅞"
	Polara, Monaco	1⅜"
1971-72	Coronet & Charger	1⅝"
	Polara & Monaco	1⅜"
1973	Coronet & Charger	1⅞"
	Polara & Monaco	1½"
1974	Charger, Coronet	1⅞"
	Monaco	1"
1975-76	Charger	10¾" [1]
	Coronet Exc. Wag.	10¾" [1]
	Coronet Sta. Wag.	11¼" [1]
	Monaco	10⅛" [1]

Plymouth Exc. Barracuda, Valiant & Volaré

1969-70	Belvedere, Satellite, Roadrunner	1⅞"
	Fury, V.I.P.	1⅜"
1971-72	Satellite	1⅝"
	Fury	1⅜"
1973	Satellite	1⅞"
	Fury	1½"
1974	Satellite	1⅞"
	Fury	1"
1975-76	Fury	10¾" [1]
	Gran Fury	10⅛" [1]
	Suburban	11¼" [1]

Aspen & Volaré

1976	All	10¼" [1]

Dart & Valiant

1969-70	All	2⅛"
1971-72	2 Dr	2⅛"
	4 Dr	1⅝"
1973	2 Dr	1⅞"
	4 Dr	2⅛"
1974	All	1⅞"
1975-76	All	10$^{15}/_{16}$" [1]

Barracuda & Challenger

1969	Dart	1⅜"
1970	All	1$^{3}/_{16}$"
1971-72	All	1"
1973-74	All	1⅛"

Chrysler & Imperial

1969-72	Chrysler	1⅛"
	Imperial	1¾"
1973	Chrysler	1¼"
	Imperial	1¾"
1974	Chrysler	1"
	Imperial	1"
1975	Cordoba	10¾" [1]
	Newport, New Yorker	10⅛" [1]
	Imperial	10⅛" [1]
1976	Cordoba	10¾" [1]
	Newport, New Yorker	10⅛" [1]

[1] Car riding heights checked at "Dimension A" only.

MANUAL STEERING GEAR, REPLACE

1970-76

CAUTION: To avoid damage to the energy absorbing steering column, it is recommended that the steering column be completely detached from floor and instrument panel before steering gear is removed.

1. Use a suitable puller to remove steering arm from under vehicle.
2. Remove gear to frame retaining bolts and remove gear.

1969 Chrysler, Monaco, Polara

1. Use a suitable puller to remove steering arm.
2. Disconnect transmission gear selector linkage (if column mounted).
3. Remove pin from coupling clamp at upper end of steering gear worm shaft.
4. To provide sufficient clearance at coupling, loosen column jacket-to-instrument panel clamp bolts enough to disengage tab on clamp from slot in column jacket. Slide column up far enough to disengage coupling from worm shaft.
5. Raise carpet and remove column lower support plate-to-floor pan bolts.
6. Remove three gear housing mounting bolts and remove gear from under vehicle.

1969-70 Dodge & Plymouth (Except Monaco & Polara)

1. Perform Steps 1 through 4 as described above for other 1969-70 models.
2. Then on Dart and Valiant with 273 engines, from under vehicle remove left front engine mount stud nut and washer. Using a suitable jack, raise engine about 1½ inches. Remove starter. After removing the three steering gear mounting bolts, lower gear through opening.
3. On models with 426 engine, remove battery and battery tray. Remove left front engine mount stud nut and washer. Using a suitable jack, raise engine about 1½ inches. After removing the three gear mounting bolts, rotate gear forward between cylinder head and shock absorber tower, and up through opening left by battery tray removal.
4. On models with 273, 318, 340, 383 and 440 engines (except Dart and Valiant), remove the three steering gear mounting bolts and lower gear from under vehicle.
5. On all 6-cylinder models, remove steering gear mounting bolts and remove gear through engine compartment.

POWER STEERING, REPLACE

1969-75 Imperial

1. Disconnect battery ground cable.
2. Disconnect pressure and return hoses at steering gear and fasten ends of hoses above oil level in pump reservoir.
3. Remove rubber coupling heat shield.
4. Remove two capscrews attaching rubber coupling to upper flange of steering column shaft.
5. Remove roll pin from pot coupling.
6. Move upper end of intermediate shaft until rubber coupling clears upper flange and carefully tap lower coupling up and off steering gear worm shaft.
7. Using a suitable puller, remove steering gear arm.
8. Disconnect exhaust pipe at ball coupling and exhaust manifold flanges and remove pipe.
9. Disconnect transmission cooler lines from transmission and clamp at starter flange bolt.
10. Remove two steering gear mounting bolts and stud nut. Then remove gear from under vehicle.

1969-76 Chrysler, Plymouth & Dodge

1. Disconnect battery ground cable.
2. Remove column coupling-to-worm shaft roll pin.
3. Loosen column jacket clamp nuts enough to allow column to be pulled up two inches. On some models it will be necessary to remove column finish plate to get at the clamp.
4. Remove three bolts in lower column support plate at floor pan.
5. Tap coupling and column upward and lift off end of worm shaft.
6. Disconnect pressure and return hoses at gear. Fasten hose ends above fluid level in pump.
7. Use a suitable puller to remove steering arm.
8. Remove gear-to-frame bolts and remove gear from vehicle as follows:
9. On six-cylinder models, remove gear from top of engine compartment.
10. Dart and Valiant with 8-273 engine, remove left front engine mount stud nut. Using suitable jack, raise engine slightly to provide clearance between left exhaust manifold and body sheet metal. Remove starter and then take steering gear out from under vehicle.
11. Dodge and Plymouth with 8-426 engine, remove battery and battery tray. From top side of engine compartment, remove left engine mount insulator stud nut, through bolt and bracket upper bolt. Jack up left side of engine about 1½". Separate engine from engine mount and allow insulator to rest on frame. Rotate steering gear (worm shaft end) up between cylinder head and shock absorber tower and out through battery tray opening.
12. On all other models remove gear from under vehicle.

FORD & MERCURY

Full Size Models

OLD CAR SPECIFICATIONS: For 1946-68 Tune Up and Wheel Alignment Specifications see main index.

INDEX OF SERVICE OPERATIONS

ENGINE & SERIAL NUMBER LOCATION

Plate On Left Front Door Pillar

ENGINE IDENTIFICATION

★Serial number on Vehicle Warranty Plate

Engine code for 1969-76 is the last letter in the serial number.

Year	Engine	Engine Code★	Year	Engine	Engine Code★	Year	Engine	Engine Code★	Year	Engine	Engine Code★
1969–70	6-240	V	1971	V8-302①	F	1972	V8-351①	H	1974–76	V8-351	H
	V8-302①	F		V8-351①	H		V8-351④	Q		V8-400	S
	V8-351 (W)①	H		V8-390①	Y		V8-400	S		V8-460	A
	V8-390①	Y		V8-400①	S		V8-429	N			
	V8-390⑤	X		V8-429①	K	1973	V8-351	H			
	V8-429①	K		V8-429②	N		V8-400	S			
	V8-429②	N	1972	6-240	V		V8-429	N			
1971	6-240	V		V8-302①	F		V8-460	A			

①—Two barrel carburetor.
②—Four barrel carburetor.
③—Two 4 barrel carburetors.
④—Four barrel special.
⑤—Premium fuel.
⑥—Two barrel special.

GRILLE IDENTIFICATION

1969 Ford Custom, Galaxie

1969 Ford "LTD", XL, Squire

1969 Mercury Monterey

1969 Mercury Marauder, Marquis

1970 Ford Custom, Galaxie

1970 Ford "LTD", XL, Squire

1970 Mercury Marauder, Marquis

1970 Mercury Monterey

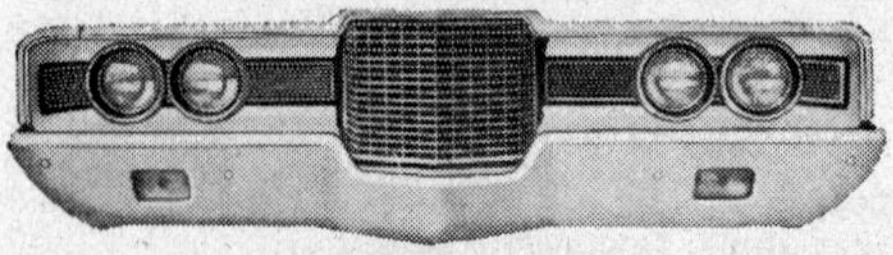

1971 Ford Custom, Galaxie

1971 Ford "LTD", Squire

1971 Mercury Brougham, Marquis

1971 Mercury Monterey

1972 Ford Custom, Galaxie

1972 Ford "LTD", Squire

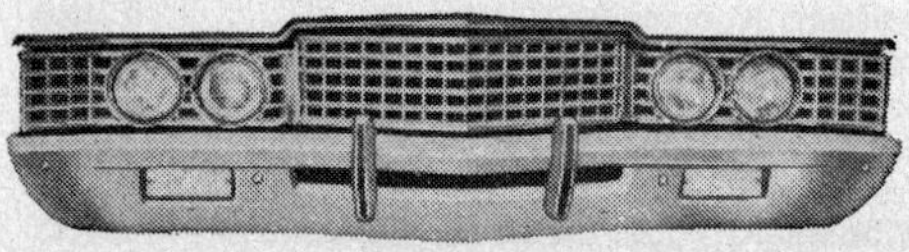

1972 Mercury Monterey

GRILLE IDENTIFICATION—Continued

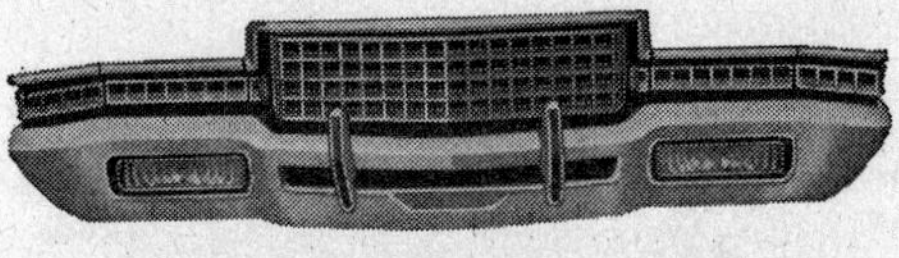

1972 Mercury Brougham, Marquis

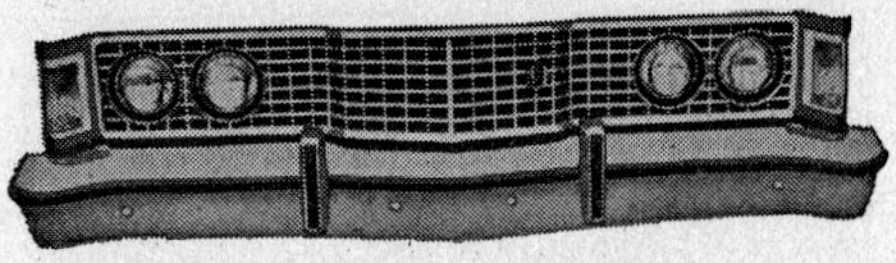

1973 Ford Custom, Galaxie

1973 Ford "LTD"

1973 Mercury Monterey

1973 Mercury Marquis

1974 Ford Custom, Galaxie

1974 Ford "LTD"

1974 Mercury Monterey

1974 Mercury Marquis

1975-76 Ford "LTD" & Custom

1975-76 Ford "LTD" Landau

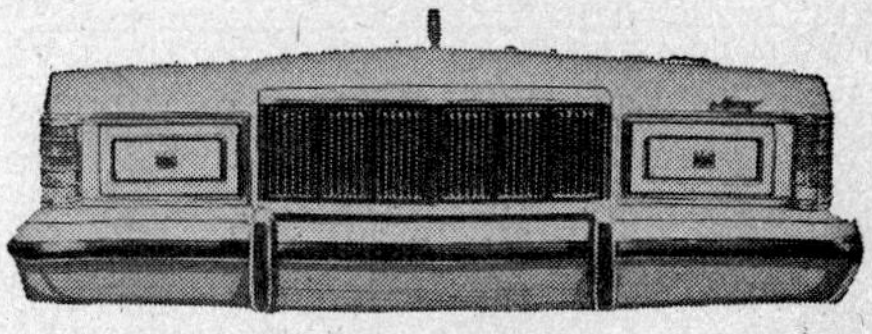

1975-76 Mercury

GENERAL ENGINE SPECIFICATIONS

Year	Engine	Carburetor	Bore and Stroke	Piston Displacement, Cubic Inches	Compression Ratio	Maximum Brake H.P. @ R.P.M.	Maximum Torque Lbs. Ft. @ R.P.M.	Normal Oil Pressure Pounds
FORD								
1969–70	150 Horsepower..............6-240	1 Barrel	4.00 x 3.18	240	9.2	150 @ 4000	234 @ 2200	35–60
	210 Horsepower............V8-302	2 Barrel	4.00 x 3.00	302	9.5	210 @ 4400	295 @ 2400	35–60
	250 Horsepower............V8-351	2 Barrel	4.00 x 3.50	351	9.5	250 @ 4600	355 @ 2600	35–60
	270 Horsepower............V8-390	2 Barrel	4.05 x 3.78	390	9.5	270 @ 4400	390 @ 2600	35–60
	320 Horsepower............V8-429	2 Barrel	4.36 x 3.59	429	10.5	320 @ 4400	460 @ 2200	35–60
	360 Horsepower............V8-429	4 Barrel	4.36 x 3.59	429	11.0	360 @ 4600	476 @ 2800	35–60
1971	140 Horsepower..............6-240	1 Barrel	4.00 x 3.18	240	8.9	140 @ 4000	230 @ 2200	35–60
	210 Horsepower............V8-302	2 Barrel	4.00 x 3.00	302	9.0	210 @ 4600	296 @ 2600	35–60
	240 Horsepower............V8-351	2 Barrel	4.00 x 3.50	351	9.0	240 @ 4600	350 @ 2600	35–60
	255 Horsepower............V8-390	2 Barrel	4.05 x 3.78	390	8.6	255 @ 4400	376 @ 2600	35–60
	260 Horsepower............V8-400	2 Barrel	4.00 x 4.00	400	9.0	260 @ 4400	400 @ 2200	35–60
	320 Horsepower............V8-429	2 Barrel	4.36 x 3.59	429	10.5	320 @ 4400	460 @ 2200	35–60
	360 Horsepower............V8-429	4 Barrel	4.36 x 3.59	429	10.5	360 @ 4600	480 @ 2800	35–60

Continued

GENERAL ENGINE SPECIFICATIONS—Continued

Year	Engine	Carburetor	Bore and Stroke	Piston Displacement, Cubic Inches	Compression Ratio	Maximum Brake H.P. @ R.P.M.	Maximum Torque Lbs. Ft. @ R.P.M.	Normal Oil Pressure Pounds
FORD—Continued								
1972	103 Horsepower①............6-240	1 Barrel	4.00 x 3.18	240	8.5	103 @ 3800	170 @ 2200	35–60
	140 Horsepower①...........V8-302	2 Barrel	4.00 x 3.00	302	8.5	140 @ 4000	239 @ 2000	35–60
	153 Horsepower①...........V8-351	2 Barrel	4.00 x 3.50	351	8.3	153 @ 3800	265 @ 2000	35–60
	163 Horsepower①...........V8-351	2 Barrel	4.00 x 3.50	351	8.6	163 @ 3800	277 @ 2000	35–60
	172 Horsepower①...........V8-400	2 Barrel	4.00 x 4.00	400	8.4	172 @ 4000	298 @ 2200	35–60
	208 Horsepower①...........V8-429	4 Barrel	4.36 x 3.59	429	8.5	208 @ 4400	322 @ 2800	35–60
1973	154 Horsepower①...........V8-351	2 Barrel	4.00 x 3.50	351	8.0	154 @ 3800	256 @ 2400	45–65
	157 Horsepower①...........V8-351	2 Barrel	4.00 x 3.50	351	8.0	157 @ 4000	246 @ 2400	45–75
	158 Horsepower①...........V8-351	2 Barrel	4.00 x 3.50	351	8.0	158 @ 3800	264 @ 2400	45–65
	161 Horsepower①...........V8-351	2 Barrel	4.00 x 3.50	351	8.0	161 @ 4000	254 @ 2400	45–75
	167 Horsepower①...........V8-400	2 Barrel	4.00 x 4.00	400	8.0	167 @ 3600	312 @ 2200	45–75
	171 Horsepower①...........V8-400	2 Barrel	4.00 x 4.00	400	8.0	171 @ 3600	314 @ 2000	45–75
	198 Horsepower①...........V8-429	4 Barrel	4.36 x 3.59	429	8.0	198 @ 4400	320 @ 2800	35–65
	202 Horsepower①...........V8-429	4 Barrel	4.36 x 3.59	429	8.0	202 @ 4400	320 @ 2800	35–65
	198 Horsepower①...........V8-460	4 Barrel	4.36 x 3.85	460	8.0	198 @ 4400	328 @ 2800	35–65
	202 Horsepower①...........V8-460	4 Barrel	4.36 x 3.85	460	8.0	202 @ 4400	330 @ 2800	35–65
1974	162 Horsepower①...........V8-351	2 Barrel	4.00 x 3.50	351	8.0	162 @ 4000	275 @ 2200	45–75
	163 Horsepower①...........V8-351	2 Barrel	4.00 x 3.50	351	8.0	163 @ 4200	278 @ 2000	45–65
	170 Horsepower①...........V8-400	2 Barrel	4.00 x 4.00	400	8.0	170 @ 3400	330 @ 2000	45–75
	195 Horsepower①...........V8-460	4 Barrel	4.36 x 3.85	460	8.0	195 @ 3800	335 @ 2600	35–65
	275 Horsepower①...........V8-460	4 Barrel	4.36 x 3.85	460	8.8	275 @ 4400	395 @ 2800	35–65
1975	148 Horsepower①...........V8-351	2 Barrel	4.00 x 3.50	351	8.0	148 @ 3800	243 @ 2400	50–75
	150 Horsepower①...........V8-351	2 Barrel	4.00 x 3.50	351	8.0	150 @ 3800	244 @ 2800	50–75
	158 Horsepower①...........V8-400	2 Barrel	4.00 x 4.00	400	8.0	158 @ 3800	276 @ 2000	50–75
	144 Horsepower①...........V8-400	2 Barrel	4.00 x 4.00	400	8.0	144 @ 3600	255 @ 2200	50–75
	218 Horsepower①...........V8-460	4 Barrel	4.36 x 3.85	460	8.0	218 @ 4000	369 @ 2600	40–65
1976	Horsepower①...........V8-351	2 Barrel	4.00 x 3.50	351	8.0	—	—	45–65
	Horsepower①...........V8-400	2 Barrel	4.00 x 4.00	400	8.0	—	—	45–65
	Horsepower①...........V8-460	4 Barrel	4.36 x 3.85	460	8.0	—	—	45–65
MERCURY								
1969–70	270 Horsepower.............V8-390	2 Barrel	4.05 x 3.78	390	9.5	270 @ 4400	390 @ 2600	35–60
	280 Horsepower (1969).......V8-390	2 Barrel	4.05 x 3.78	390	10.5	280 @ 4400	403 @ 2600	35–60
	320 Horsepower.............V8-429	2 Barrel	4.36 x 3.59	429	10.5	320 @ 4400	460 @ 2200	35–60
	360 Horsepower.............V8-429	4 Barrel	4.36 x 3.59	429	11.0	360 @ 4600	476 @ 2800	35–60
1971	240 Horsepower.............V8-351	2 Barrel	4.00 x 3.50	351	9.0	240 @ 4600	350 @ 2600	35–60
	260 Horsepower.............V8-400	2 Barrel	4.00 x 4.00	400	9.0	260 @ 4400	400 @ 2200	35–60
	320 Horsepower.............V8-429	2 Barrel	4.36 x 3.59	429	10.5	320 @ 4400	460 @ 2200	35–60
	360 Horsepower.............V8-429	4 Barrel	4.36 x 3.59	429	10.5	360 @ 4600	480 @ 2800	35–60
1972	153 Horsepower①...........V8-351	2 Barrel	4.00 x 3.50	351	8.3	153 @ 3800	266 @ 2000	35–60
	163 Horsepower①...........V8-351	2 Barrel	4.00 x 3.50	351	8.6	163 @ 3800	277 @ 2000	35–60
	172 Horsepower①...........V8-400	2 Barrel	4.00 x 4.00	400	8.4	172 @ 4000	298 @ 2200	35–60
	208 Horsepower①...........V8-429	4 Barrel	4.36 x 3.59	429	8.5	208 @ 4400	322 @ 2800	35–60
	200 Horsepower①...........V8-460	4 Barrel	4.36 x 3.85	460	8.5	200 @ 4400	326 @ 2800	35–75
1973	157 Horsepower①...........V8-351	2 Barrel	4.00 x 3.50	351	8.0	157 @ 4000	246 @ 2400	45–75
	161 Horsepower①...........V8-351	2 Barrel	4.00 x 3.50	351	8.0	161 @ 4000	254 @ 2400	45–75
	167 Horsepower①...........V8-400	2 Barrel	4.00 x 4.00	400	8.0	167 @ 3600	312 @ 2200	45–75
	171 Horsepower①...........V8-400	2 Barrel	4.00 x 4.00	400	8.0	171 @ 3600	314 @ 2000	45–75
	198 Horsepower①...........V8-429	4 Barrel	4.36 x 3.59	429	8.0	198 @ 4400	320 @ 2800	35–65
	202 Horsepower①...........V8-429	4 Barrel	4.36 x 3.59	429	8.0	202 @ 4400	320 @ 2800	35–65
	198 Horsepower①...........V8-460	4 Barrel	4.36 x 3.85	460	8.0	198 @ 4400	328 @ 2800	35–65
	202 Horsepower①...........V8-460	4 Barrel	4.36 x 3.85	460	8.0	202 @ 4400	330 @ 2800	35–65

Continued

GENERAL ENGINE SPECIFICATIONS—Continued

Year	Engine	Carburetor	Bore and Stroke	Piston Displacement, Cubic Inches	Compression Ratio	Maximum Brake H.P. @ R.P.M.	Maximum Torque Lbs. Ft. @ R.P.M.	Normal Oil Pressure Pounds
MERCURY—Continued								
1974	170 Horsepower①..........V8-400	2 Barrel	4.00 x 4.00	400	8.0	170 @ 3400	330 @ 2000	45–75
	195 Horsepower①..........V8-460	4 Barrel	4.36 x 3.85	460	8.0	195 @ 3800	335 @ 2600	35–65
	275 Horsepower①..........V8-460	4 Barrel	4.36 x 3.85	460	8.8	275 @ 4400	395 @ 2800	35–65
1975	148 Horsepower①..........V8-351	2 Barrel	4.00 x 3.50	351	8.0	148 @ 3800	243 @ 2400	50–75
	150 Horsepower①..........V8-351	2 Barrel	4.00 x 3.50	351	8.0	150 @ 3800	244 @ 2800	50–75
	144 Horsepower①..........V8-400	2 Barrel	4.00 x 4.00	400	8.0	144 @ 3600	255 @ 2200	50–75
	158 Horsepower①..........V8-400	2 Barrel	4.00 x 4.00	400	8.0	158 @ 3800	276 @ 2000	50–75
	218 Horsepower①..........V8-460	4 Barrel	4.36 x 3.85	460	8.0	218 @ 4000	369 @ 2600	40–65
1976	Horsepower①..........V8-400	2 Barrel	4.00 x 4.00	400	8.0	—	—	45–65
	Horsepower①..........V8-400	2 Barrel	4.36 x 3.85	460	8.0	—	—	45–65

①—Ratings are NET—as installed in the vehicle.

TUNE UP SPECIFICATIONS

The following specifications are published from the latest information available. This data should be used only in the absence of a decal affixed in the engine compartment.

OLD CAR SPECIFICATIONS: For 1946-68 Tune Up Specifications see main index.

★When using a timing light, disconnect vacuum hose or tube at distributor and plug opening in tube or hose so idle speed will not be affected.

●When checking compression, lowest cylinder must be within 75% of the highest.

▲Before removing wires from distributor cap, determine location of the No. 1 wire in cap, as distributor position may have been altered from that shown at the end of this chart.

Year	Spark Plug		Distributor		Ignition Timing★			Carb. Adjustments					
								Hot Idle Speed		Air Fuel Ratio		Idle "CO" %	
	Type ⑤	Gap Inch	Point Gap Inch	Dwell Angle Deg.	Firing Order Fig. ▲	Timing BTDC ①	Mark Fig.	Std. Trans.	Auto. Trans.②	Std. Trans.	Auto. Trans.	Std. Trans.	Auto. Trans.
1969													
6-240	BF-42	.034	.027	35–40	C	6°	D	775/500⑭	500D	—	13.9 to 1	—	—
V8-302 Std. Tr.	BF-42	.034	.021	24–29	B	6°	F	650	—	14.0 to 1	—	1.8	—
V8-302 Auto. Tr.	BF-42	.034	.017	26–31	B	6°	F	—	550D	—	13.8 to 1	—	2.2
V8-351	BF-42	.034	.017	26–31	A	6°	F	650	550D	14.0 to 1	14.0 to 1	1.8	1.6
V8-390	BF-42	.034	.017	26–31	B	6°	E	650	550D	14.0 to 1	14.4 to 1	1.6	1.1
V8-390⑮	BF-42	.034	.021	24–29	B	6°	E	—	550D	14.0 to 1	14.4 to 1	1.8	0.5
V8-429	BF-42	.034	.017	26–31	B	6°	E	—	550D	14.0 to 1	14.4 to 1	1.8	1.2
1970													
6-240	BF-42	.035	⑰	⑰	C	6°	D	775/500⑭	500④	14.45 to 1	14.70 to 1	—	—
V8-302	BF-42	.035	.021	24–29	B	6°	F	775/500⑭	575④	14.0 to 1	13.8 to 1	0.34	0.44
V8-351	BF-42	.035	.021	24–29	A	6°	F	775/500⑭	575④	⑩	⑩	0.30	0.14
V8-390 Std. Tr.	BF-42	.035	.021	24–29	B	6°	E	775/500⑭	—	14.1 to 1	—	0.12	—
V8-390 Auto. Tr.	BF-42	.035	.017	26–31	B	6°	E	—	600/500⑭	—	14.4 to 1	—	0.12
V8-429⑦	BF-42	.035	.021	24–29	B	4°	E	—	600/500⑭	13.5 to 1	13.5 to 1	0.20	0.20
V8-429⑥	BF-42	.035	⑬	⑬	B	4°	E	700	600	14.5 to 1	14.5 to 1	0.75	0.75

Continued

TUNE UP SPECIFICATIONS—Continued

The following specifications are published from the latest information available. This data should be used only in the absence of a decal affixed in the engine compartment.

OLD CAR SPECIFICATIONS: For 1946-68 Tune Up Specifications see main index.

★When using a timing light, disconnect vacuum hose or tube at distributor and plug opening in tube or hose so idle speed will not be affected.

●When checking compression, lowest cylinder must be within 75% of the highest.

▲Before removing wires from distributor cap, determine location of the No. 1 wire in cap, as distributor position may have been altered from that shown at the end of this chart.

Year	Spark Plug		Distributor		Ignition Timing★			Carb. Adjustments					
								Hot Idle Speed		Air Fuel Ratio		Idle "CO" %	
	Type ⑤	Gap Inch	Point Gap Inch	Dwell Angle Deg.	Firing Order Fig. ▲	Timing BTDC ①	Mark Fig.	Std. Trans.	Auto. Trans.②	Std. Trans.	Auto. Trans.	Std. Trans.	Auto. Trans.
1971													
6-240	BRF-42	.034	.027	35–40	C	6°	G	800/500⑭	500	14.5 to 1	14.5 to 1	1.0	0.5
V8-302 L/Air Cond.	BRF-42	.034	.021	24–29	B	6°	H	800/500⑭	575	12.5 to 1	12.5 to 1	0.3	0.4
V8-302 w/Air Cond.	BRF-42	.034	.021	24–29	B	6°	H	800/500⑭	600/500⑭	12.5 to 1	12.5 to 1	0.3	0.4
V8-351 L/Air Cond.⑪	BRF-42	.034	.021	24–29	A	6°	H	700/500⑭	575	13.50 to 1	13.90 to 1	1.1	0.5
V8-351 w/Air Cond.⑪	BRF-42	.034	.021	24–29	A	6°	H	700/500⑭	600/500⑭	13.50 to 1	13.90 to 1	1.1	0.5
V8-351⑫	ARF-42	.034	.021	24–29	A	6°	H	775/500⑭	600/500⑭	14.30 to 1	14.40 to 1	0.2	0.2
V8-390	BRF-42	.034	.021	24–29	B	6°	H	—	600/500⑭	—	—	—	—
V8-400	ARF-42	.034	.021	24–29	A	③	H	—	600/500⑭	14.5 to 1	14.5 to 1	—	0.7
V8-429⑦ L/Air Cond.	BRF-42	.034	.021	24–29	B	6°	H	—	590	14.5 to 1	14.5 to 1	—	0.2
V8-429⑦ w/Air Cond.	BRF-42	.034	.021	24–29	B	6°	H	—	600/500⑭	14.5 to 1	14.5 to 1	—	0.2
V8-429⑥	BRF-42	.034	.021	24–29	B	4°	H	700	600	14.50 to 1	14.50 to 1	0.2	0.2
1972													
6-240	BRF-42	.034	.027	35–39	C	6°	G	—	500	—	—	—	0.5
V8-302 w/Air Cond.	BRF-42	.034	.017	26–30	B	6°	H	800/500	600/500⑭	—	—	—	0.19
V8-302 L/Air Cond.	BRF-42	.034	.017	26–30	B	6°	H	—	575	—	—	—	0.19
V8-351 w/Air Cond.⑪	BRF-42	.034	.017	26–30	A	6°	H	—	600/500⑭	—	—	—	0.15
V8-351 L/Air Cond.⑪	BRF-42	.034	.017	26–30	A	6°	H	—	575	—	—	—	0.15
V8-351⑫	ARF-42	.034	.017	26–30	A	6°	H	—	700/500⑭	—	—	—	0.50
V8-351 Calif.⑫	ARF-42	.034	.017	26–30	A	6°	H	—	625/500⑭	—	—	—	0.50
V8-400	ARF-42	.034	.017	26–30	A	⑧	H	—	625/500⑭	—	—	—	0.10
V8-429	BRF-42	.034	.017	26–30	B	10°	H	—	600/500⑭	—	—	—	0.03
1973													
V8-351⑪	BRF-42	.034	.017	26–30	A	6°	H	—	600	—	—	—	0.5⑨
V8-351⑪	ARF-42	.034	.017	26–30	A	10°	H	—	600	—	—	0.5⑨	0.5⑨
V8-400	ARF-42	.034	.017	26–31	A	6°	H	—	625	—	—	—	0.5⑨
V8-429	ARF-42	.034	.017	26–30	B	10°	H	—	600	—	—	—	0.5⑨
V8-460	ARF-42	.034	.017	26–30	B	10°	H	—	650	—	—	—	0.4⑨
1974													
V8-351⑪	BRF-42	.044	.017	26–30	A	6°	H	—	600	—	—	—	—
V8-351⑪⑱	BRF-42	.044	—	—	A	6°	H	—	600	—	—	—	—
V8-351⑫	ARF-42	.044	.017	26–30	A	14°	H	—	650	—	—	—	—
V8-351⑫⑱	ARF-42	.044	—	—	A	14°	H	—	650	—	—	—	—
V8-400⑱	ARF-42	.044	—	—	A	12°	H	—	625	—	—	—	—
V8-460⑱	ARF-52	.054	—	—	B	⑯	H	—	⑯	—	—	—	.25
1975													
V8-351	ARF-42	.044	—	—	A	14°	H	—	700D	—	—	—	—
V8-400	ARF-42	.044	—	—	A	12°	H	—	625D	—	—	—	—
V8-460	ARF-52	.044	—	—	B	14°	H	—	650D	—	—	—	—
1976													
V8-351	ARF-42	.044	—	—	A	8°	H	—	650D	—	—	—	—
V8-400	ARF-42	.044	—	—	A	10°	H	—	⑲	—	—	—	—
V8-460	ARF-52	.044	—	—	B	10°	H	—	650D	—	—	—	—

Continued

TUNE UP SPECIFICATIONS—Continued

①—BTDC: Before top dead center.
②—D: Drive. N: Neutral.
③—California vehicles 6°; all others 10°.
④—With headlights on and A/C off.
⑤—Autolite.
⑥—Four barrel carburetor.
⑦—Two barrel carburetor.
⑧—California vehicles 6°; all others 8°.
⑨—For Calif. 0.2%.
⑩—With 2 Bar. Carb.—14.0 to 1, with 4 Bar. Carb.—14.5 to 1.
⑪—Windsor engine.
⑫—Cleveland engine.
⑬—Dual diaphragm dist., .021 gap, 24°-29° dwell. Single diaphragm .017 gap, 26°-31° dwell.
⑭—Higher figure is with throttle modulator energized.
⑮—Premium fuel.
⑯—Exc. Police Interceptor 14° at 650, Police Interceptor 10° at 700.
⑰—Dual diaphragm dist., .027 gap, 35°-40° dwell. Single diaphragm dist., .025 gap, 37°-42° dwell.
⑱—Breakerless distributor.
⑲—Exc. Calif.; 650, Calif.; 625.

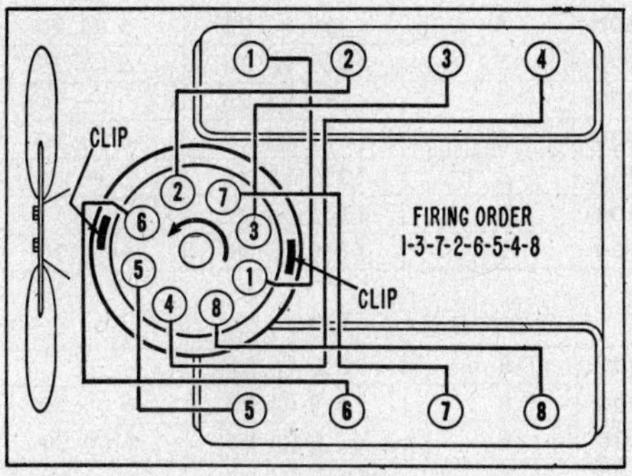

Fig. A

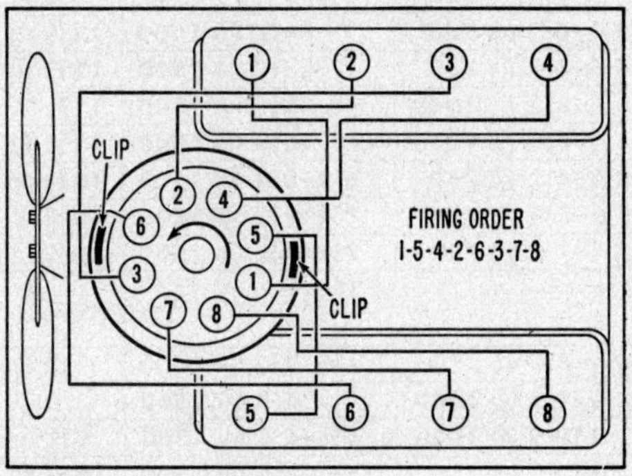

Fig. B

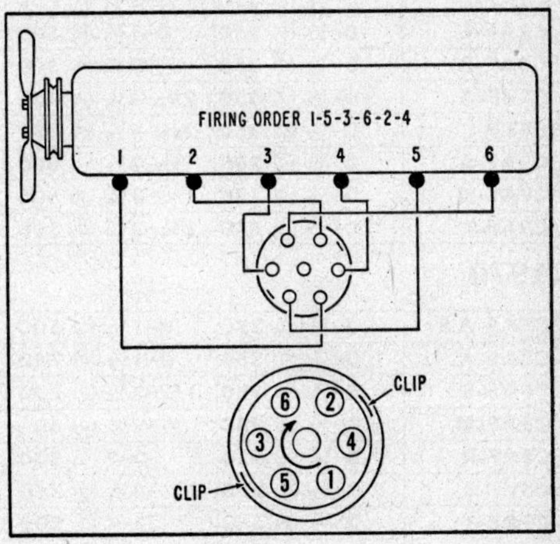

Fig. C

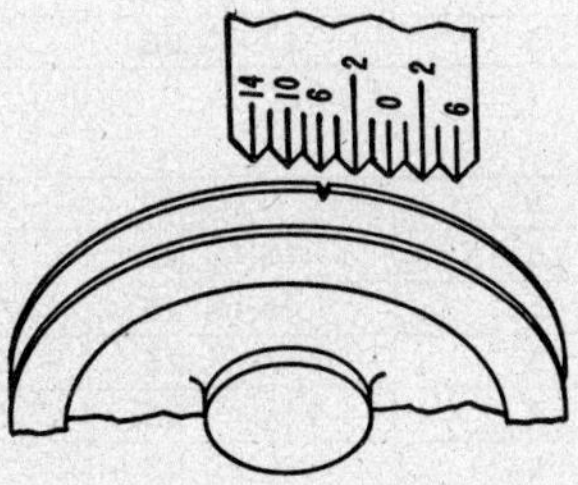

Fig. D

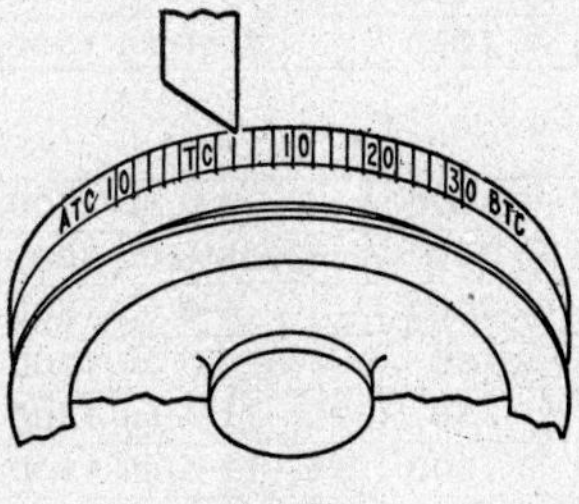

Fig. E

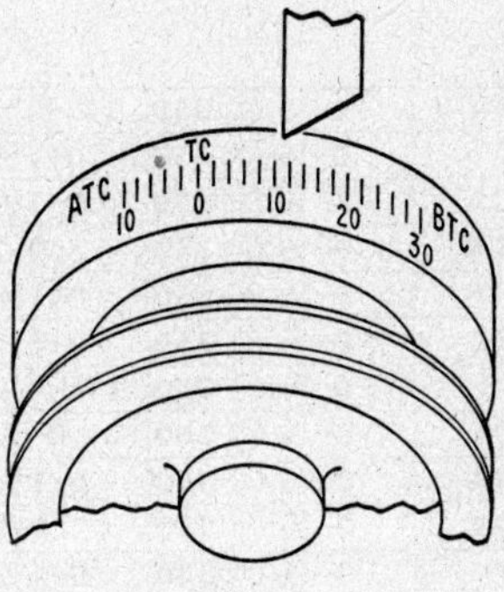

Fig. F

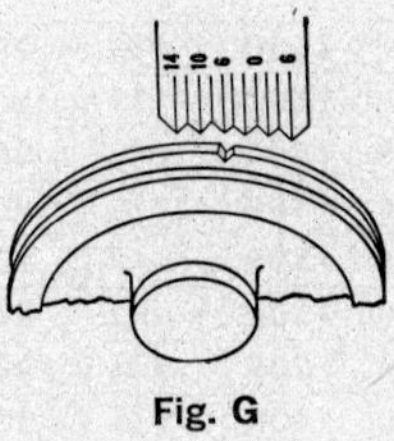

Fig. G

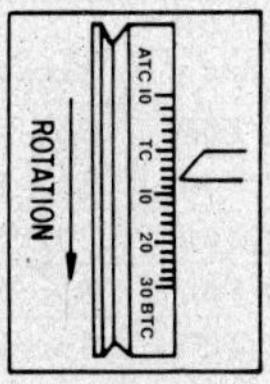

Fig. H

DISTRIBUTOR SPECIFICATIONS

★Note: If unit is checked on vehicle, double the RPM and degrees to get crankshaft figures.
Breaker arm spring tension—17–21.

Distributor Part No.①	Centrifugal Advance Degrees @ RPM of Distributor					Vacuum Advance		Distributor Retard
	Advance Starts	Intermediate Advance			Full Advance	Inches of Vacuum to Start Plunger	Max. Adv. Dist. Deg. @ Vacuum	Max. Retard Dist. Deg. @ Vacuum
1969								
C7AF-AA	0–½ @ 350	0–1½ @ 500	4¼–6¼ @ 750	7½–9¾ @ 1500	12 @ 2000	5	11½ @ 20	—
C8AF-A	0–½ @ 350	0–1½ @ 750	2¼–4½ @ 1000	6–8 @ 1500	10 @ 2000	5	8 @ 25	6 @ 20
C8AF-B	0–½ @ 350	1¾–3¾ @ 750	2¾–5 @ 1000	4½–6¾ @ 1500	8¾ @ 2000	5	9 @ 25	6 @ 20
C8AF-E	0–½ @ 350	0–1¾ @ 500	4½–6¾ @ 750	8–10 @ 1500	12½ @ 2000	5	11 @ 25	6 @ 20
C8AF-R	0–½ @ 350	0–2 @ 500	4¾–6¾ @ 750	7¾–10 @ 1500	12¼ @ 2000	5	12½ @ 25	6 @ 20
C8VF-A	0–½ @ 350	2½–4½ @ 750	6½–8½ @ 1000	8–10¼ @ 1500	11½ @ 2000	5	12 @ 25	—
C8VF-C	0–½ @ 350	2¼–4¼ @ 750	6½–8½ @ 1000	8–10¼ @ 1500	11½ @ 2000	5	12 @ 25	6 @ 20
C9AF-J	0–½ @ 350	½–2½ @ 500	2¼–4¼ @ 750	5¾–8¼ @ 1500	10¾ @ 2000	5	12½ @ 20	—
C9AF-N	0–½ @ 350	1¾–3¾ @ 500	5½–7½ @ 750	7¼–9½ @ 1500	10¾ @ 2000	5	11½ @ 20	—
C9AF-R	0–½ @ 350	1¾–3¾ @ 500	5½–7½ @ 750	7¼–9½ @ 1500	10¾ @ 2000	5	11½ @ 20	6 @ 20
1970								
C7AF-AA	0–½ @ 350	0–1½ @ 500	4¼–6¼ @ 750	7½–9¾ @ 1500	12 @ 2000	5	11½ @ 25	—
C8AF-A	0–½ @ 350	0–1½ @ 750	2¼–4½ @ 1000	6–8 @ 1500	10 @ 2000	5	8 @ 25	6 @ 20
C8AF-B	0–½ @ 350	1¾–3¾ @ 750	2¾–5 @ 1000	4½–6¾ @ 1500	8¾ @ 2000	5	9 @ 25	6 @ 20
C8AF-M	0–½ @ 350	½–2½ @ 500	2¼–4¼ @ 750	5¾–8¾ @ 1500	10¾ @ 1500	5	12½ @ 25	6 @ 20
C8AF-R	0–½ @ 350	0–2 @ 500	4¾–6¾ @ 750	7¾–10 @ 1500	12¼ @ 2000	5	12½ @ 25	6 @ 20
C8VF-C	0–½ @ 350	2¼–4½ @ 750	6½–8½ @ 1000	9–10¼ @ 1500	11½ @ 2000	5	12 @ 25	6 @ 20
C9AF-Y	0–½ @ 350	2–4 @ 500	6¾–8¾ @ 750	8½–10¾ @ 1500	12 @ 2000	5	8½ @ 25	6 @ 20
D0AF-AC	0–½ @ 350	0–2 @ 500	4¼–6¼ @ 750	7½–9¾ @ 1500	12¼ @ 2000	5	9½ @ 25	6 @ 20
D0AF-H	0–½ @ 350	0–2 @ 500	4¼–6¼ @ 750	7½–9¾ @ 1500	12¼ @ 2000	5	8 @ 25	6 @ 20
D0AF-M	0–½ @ 350	2¼–4½ @ 750	6½–8½ @ 1000	9–10¼ @ 1500	11½ @ 2000	5	12 @ 25	6 @ 20
D0AF-T	0–½ @ 350	0–1¾ @ 750	6–8 @ 1000	7¼–9¼ @ 1500	10¾ @ 2000	5	11½ @ 25	3½ @ 20
D0AF-Y	0–½ @ 350	0–1¾ @ 750	6–8 @ 1000	8–10¼ @ 1500	12½ @ 2000	5	11 @ 25	6 @ 20
D0AF-Z	0–½ @ 350	2¼–4¼ @ 750	6½–8½ @ 1000	8–10 @ 1500	11½ @ 2000	5	11 @ 25	—
1971								
C8AF-B	0–½ @ 350	1¾–3¾ @ 750	2¾–5 @ 1000	4½–6¾ @ 1500	8¾ @ 2000	5	9 @ 25	7 @ 20
C8VF-C	0–½ @ 350	0–1½ @ 500	2¼–4½ @ 750	6½–8½ @ 1000	11½ @ 2000	5	12 @ 25	7 @ 20
D1AF-CA	0–½ @ 350	0–1½ @ 500	1–3¼ @ 1000	6–8 @ 1500	10 @ 2000	5	8 @ 25	7 @ 20
D1AF-GA	0–½ @ 350	0–2 @ 500	4½–6¼ @ 750	7½–9¾ @ 1500	12 @ 2000	5	7 @ 25	7 @ 20
D1AF-HA	0–½ @ 350	0–1½ @ 500	4¾–7 @ 750	9½–11¾ @ 1500	14 @ 2000	5	9½ @ 25	7 @ 20
D1AF-KB	0–½ @ 350	0–1½ @ 750	5¼–8 @ 1000	9¾–11¾ @ 1500	14½ @ 2000	5	9½ @ 25	7 @ 20
D1AF-LB	0–½ @ 350	1¼–4½ @ 500	7¾–9¾ @ 750	11½–13½ @ 1500	16¼ @ 2000	5	9 @ 25	4 @ 20
D1MF-FA	0–½ @ 350	0–½ @ 500	2¼–4¼ @ 750	8–10¼ @ 1500	11½ @ 2000	5	11 @ 25	7 @ 20
D1OF-GA	0–½ @ 350	0–1½ @ 500	3¼–5¼ @ 750	8¼–10½ @ 1500	12½ @ 2000	5	11 @ 25	4 @ 20
D0AF-AE	0–½ @ 350	0–2 @ 750	5¼–8 @ 1000	7¼–9¼ @ 1500	10¾ @ 2000	5	2½ @ 25	7 @ 20
D0AF-Y	0–½ @ 350	0–1¾ @ 750	6–8 @ 1000	8–10 @ 1500	12½ @ 2000	5	11 @ 25	7 @ 20
D0OF-U	0–½ @ 350	1¼–3¼ @ 500	5¼–7¼ @ 750	8¼–10½ @ 1500	12¾ @ 2000	5	12½ @ 25	—
1972								
D2AF-BA	0–1 @ 500	1½–4 @ 750	3–5 @ 1000	4½–7 @ 1500	8½ @ 2000	5	9½ @ 20	7 @ 20
D2AF-CA	0–½ @ 500	0–1½ @ 750	4½–6½ @ 1000	7–9½ @ 1500	11 @ 2000	5	5½ @ 20	7 @ 20
D2AF-KA	1–3 @ 500	5½–7½ @ 750	6½–8½ @ 1000	8½–10½ @ 1500	12½ @ 2000	5	13 @ 20	7 @ 20
D2AF-PA	0–1 @ 500	1½–4 @ 750	6½–8½ @ 1000	8½–11 @ 1500	13½ @ 2000	5	9½ @ 20	7 @ 20
D2AF-RA	½–2½ @ 500	2–4 @ 750	3½–5½ @ 1000	6½–9 @ 1500	12 @ 2000	5	13½ @ 20	4 @ 20
D2AF-SA	½–2½ @ 500	2–4 @ 750	3½–5½ @ 1000	6½–9 @ 1500	12 @ 2000	5	13½ @ 20	4 @ 20
D2MF-EA	0–1 @ 500	2–4 @ 750	4½–6½ @ 1000	6–8 @ 1500	9½ @ 2000	5	11½ @ 20	4 @ 20
D2MF-FA	0–1 @ 500	2–4 @ 750	4½–6½ @ 1000	6–8 @ 1500	9½ @ 2000	5	11½ @ 20	4 @ 20
D2SF-EA	0–1 @ 500	0–1½ @ 750	2½–4½ @ 1000	4½–6½ @ 1500	8½ @ 2000	5	—	—
D2ZF-AA	0–1½ @ 500	2–4 @ 750	4½–6½ @ 1000	8–10 @ 1500	12½ @ 2000	5	12½ @ 20	7 @ 20
D2ZF-CA	0–1 @ 500	3½–5½ @ 750	6½–8½ @ 1000	8½–10½ @ 1500	12½ @ 2000	5	10 @ 20	4 @ 20

Continued

DISTRIBUTOR SPECIFICATIONS—Continued

★Note: If unit is checked on vehicle, double the RPM and degrees to get crankshaft figures.

Breaker arm spring tension—17–21.

Distributor Part No.①	Centrifugal Advance Degrees @ RPM of Distributor					Vacuum Advance		Distributor Retard
	Advance Starts	Intermediate Advance			Full Advance	Inches of Vacuum to Start Plunger	Max. Adv. Dist. Deg. @ Vacuum	Max. Retard Dist. Deg. @ Vacuum
1973								
D3AF-AA	0–1½ @ 500	4–6 @ 750	6–8 @ 1000	10–12½ @ 1500	16½ @ 2000	5	13 @ 20	—
D3ZF-CA	0–1 @ 500	3–5½ @ 750	4½–6½ @ 1000	6½–9 @ 1500	11½ @ 2000	5	13 @ 20	—
D3ZF-GA	0–1 @ 500	0–2½ @ 750	2–4 @ 1000	5½–7½ @ 1500	11 @ 2000	5	12½ @ 20	—
D3AF-BA	0–1½ @ 500	3½–5½ @ 750	5½–7½ @ 1000	8½–11 @ 1500	14½ @ 2000	5	13 @ 20	—
D3MF-GA	0–1 @ 500	0–2 @ 750	1½–3½ @ 1000	4½–7 @ 1500	10½ @ 2000	5	13½ @ 20	—
D3MF-DA	0–1 @ 500	1½–4 @ 750	5½–7½ @ 1000	7½–9 @ 1500	10½ @ 2000	5	11 @ 20	—
D3SF-BA	0–1 @ 500	½–2½ @ 750	5½–8 @ 1000	10–12 @ 1500	13½ @ 2000	5	11 @ 20	—
D3VF-CA	0–1 @ 500	0–1½ @ 750	4–6 @ 1000	6–8 @ 1500	10 @ 2000	5	11 @ 20	—
D3MF-BA	0–1 @ 500	0–1½ @ 750	3½–5½ @ 1000	7½–10 @ 1500	11 @ 2000	5	7 @ 20	—
D3VF-AA	0–1 @ 500	0–1½ @ 750	3–5 @ 1000	4–6 @ 1500	7½ @ 2000	5	11 @ 20	—
D3VF-BA	0–1 @ 500	0–2½ @ 750	5–7 @ 1000	7½–10 @ 1500	12 @ 2000	5	7 @ 20	—
1974								
D3AF-AA	0–1½ @ 500	4–6 @ 750	6–8¼ @ 1000	10–12½ @ 1500	15¼ @ 2000	5	13¼ @ 20	—
D3ZF-GA	0–1 @ 500	0–2½ @ 750	2¼–4¼ @ 1000	5½–7½ @ 1500	9½ @ 2000	5	12½ @ 20	—
D4OE-CA	0–½ @ 500	0–1½ @ 750	2–4 @ 1000	5½–7½ @ 1500	9¼ @ 2000	5	13½ @ 20	—
D4VE-CA	0–½ @ 500	0–½ @ 750	3–5 @ 1000	7–9 @ 1500	9½ @ 2000	5	11¼ @ 20	—
1975								
D4AE-AA	0–2 @ 520	4½–7½ @ 800	10–12½ @ 1500	—	15¼ @ 2000	4	12¾ @ 17	—
D4OE-EA	0–2 @ 520	4½–7½ @ 800	10–12½ @ 1500	—	15¼ @ 2000	4	12¾ @ 17	—
D4VE-CA	−1½–+½ @ 800	3–5 @ 1000	—	—	19 @ 2000	4	19¾ @ 14	—
1976								
D5AE-KA	0–3 @ 550	2½–4½ @ 650	5¼–8 @ 1500	—	10¼ @ 2250	5	15¼ @ 15	—
D5OE-FA	0–3 @ 500	4½–6¼ @ 650	6½–8¼ @ 1500	—	11½ @ 2500	5	13¼ @ 15	—
D6AE	0–2 @ 500	4–6 @ 700	7¼–9½ @ 1500	—	14 @ 2500	3	15¼ @ 11½	—
D6AE-AA	0–2 @ 500	4–6 @ 700	7¼–9½ @ 1500	—	14 @ 2500	3	15¼ @ 11½	—
D6AE-BA	0–3 @ 525	3½–5½ @ 650	8½–10¾ @ 1500	—	14¼ @ 2500	5	13¼ @ 23	—

①—Basic part No. 12127.

VALVE SPECIFICATIONS

Year	Engine Model	Valve Lash		Valve Angles		Valve Spring Installed Height	Valve Spring Pressure Lbs. @ In.	Stem Clearance		Stem Diameter	
		Int.	Exh.	Seat	Face			Intake	Exhaust	Intake	Exhaust
1969	6-240	1 Turn⑤		45	44	1 9/16	80 @ 1.70	.001–.0027	.001–.0027	.3416–.3423	.3416–.3423
	8-302	1 Turn⑤		45	44	1 5/8	75 @ 1.66	.001–.0027	.0015–.0032	.3416–.3423	.3411–.3418
	8-351	1 Turn⑤		45	44	1 25/32	83 @ 1.79	.001–.0027	.0015–.0032	.3416–.3423	.3411–.3418
	8-390	.100–.200④		45	44	1 7/8	90 @ 1.82	.001–.0027	.0015–.0032	.3711–.3718	.3706–.3713
	8-429	1 Turn⑤		45	44	1 13/16	80 @ 1.81	.001–.0027	.001–.0027	.3416–.3423	.3416–.3423
1970	6-240	1 Turn⑤		45	44	1 9/16	80 @ 1.70	.001–.0027	.001–.0027	.3416–.3423	.3416–.3423
	8-302	.067–.167⑤		45	44	1 5/8	75 @ 1.66	.001–.0027	.0015–.0032	.3416–.3423	.3411–.3418
	8-351	.083–.183⑤		45	44	1 25/32	83 @ 1.79	.001–.0027	.0015–.0032	.3416–.3423	.3411–.3418
	8-390	.100–.200④		①	44	1 7/8	90 @ 1.82	.001–.0027	.0015–.0032	.3711–.3718	.3706–.3713
	8-429	.075–.175④		45	44	1 51/64	80 @ 1.81	.001–.0027	.001–.0027	.3416–.3423	.3416–.3423
1971–72	6-240	1 Turn⑤		45	44	1 9/16	80 @ 1.70	.001–.0027	.001–.0027	.3416–.3423	.3416–.3423
	8-302	.090–.190⑤		45	44	1 5/8	75 @ 1.66	.001–.0027	.0015–.0032	.3416–.3423	.3411–.3418
	8-351③	.100–.200⑤		45	44	1 25/32	83 @ 1.79	.001–.0027	.001–.0027	.3416–.3423	.3411–.3418
	8-351⑥	.100–.200⑤		45	44	1 13/16	80 @ 1.82	.001–.0027	.0015–.0032	.3416–.3423	.3411–.3418
	8-390	.100–.200④		45	44	1 7/8	90 @ 1.82	.001–.0027	.0015–.0032	.3711–.3718	.3706–.3713
	8-400	.100–.200⑤		45	44	1 13/16	80 @ 1.82	.001–.0027	.0015–.0032	.3416–.3423	.3411–.3418
	8-429	⑦④		45	44	1 51/64	80 @ 1.81	.001–.0027	.001–.0027	.3416–.3423	.3416–.3423
1973	8-351③	.106–.156⑤		45	44	1 25/32	79 @ 1.79	.0010–.0027	.0015–.0032	.3416–.3423	.3411–.3418
	8-351⑥	.100–.150⑤		45	44	1 53/64	84 @ 1.82	.0010–.0027	.0015–.0032	.3416–.3423	.3411–.3418
	8-400	.100–.150⑤		45	44	1 53/64	84 @ 1.82	.0010–.0027	.0015–.0032	.3416–.3423	.3411–.3418
	8-429	.075–.125④		45	44	1 13/16	79 @ 1.78	.0010–.0027	.0010–.0027	.3416–.3423	.3416–.3423
	8-460	.105–.155④		45	44	1 13/16	79 @ 1.78	.0010–.0027	.0010–.0027	.3416–.3423	.3416–.3423
1974	8-351③	.106–.156④		45	44	1 51/64	75 @ 1.79	.0010–.0027	.0015–.0032	.3416–.3423	.3411–.3418
	8-351⑥	.100–.150④		45	44	1 13/16	80 @ 1.82	.0010–.0027	.0015–.0032	.3416–.3423	.3411–.3418
	8-400	.100–.150④		45	44	1 13/16	80 @ 1.82	.0010–.0027	.0015–.0032	.3416–.3423	.3411–.3418
	8-460	.075–.125④		45	44	1 13/16	80 @ 1.81	.0010–.0027	.0010–.0027	.3416–.3423	.3416–.3423
1975	V8-351③	.131–.181④		45	45	1.82	⑧	.001–.0027	.0015–.0032	.3416–.3423	.3411–.3418
	V8-351②	.125–.175④		45	45	1 53/64	226 @ 1.39	.001–.0029	.0015–.0032	.3416–.3423	.3411–.3418
	V8-400	.100–.200④		45	45	1.82	⑧	.001–.0027	.0015–.0032	.3416–.3423	.3411–.3418
	V8-460	.100–.150④		45	45	1.81	253 @ 1.33	.001–.0027	.0015–.0032	.3416–.3423	.3416–.3423
1976	V8-351	—		45	45	1.82	226 @ 1.39	.0010–.0027	.0015–.0032	.3416–.3423	.3411–.3418
	V8-400	.075–.175④		45	45	1.82	226 @ 1.39	.0010–.0027	.0015–.0032	.3416–.3423	.3411–.3418
	V8-460	.075–.175④		45	45	1.81	253 @ 1.33	.0010–.0027	.0010–.0027	.3416–.3423	.3416–.3423

①—Intake 30°, exhaust 45°.
②—Modified engine.
③—Windsor engine.
④—Clearance specified is obtainable at valve stem tip with lifter collapsed. See "Valves, Adjust" text.
⑤—See text under Valves, Adjust for procedure.
⑥—Cleveland engine.
⑦—1971, .155; 1972, .075–.125.
⑧—Intake 210 @ 1.39, Exhaust 226 @ 1.39.

PISTONS, PINS, RINGS, CRANKSHAFT & BEARINGS

Year	Engine Model	Piston Clearance	Ring End Gap①		Wrist-pin Diameter	Rod Bearings		Main Bearings			
			Comp.	Oil		Shaft Diameter	Bearing Clearance	Shaft Diameter	Bearing Clearance	Thrust on Bear. No.	Shaft End Play
1969–73	6-240	.0014–.0022	.010	.015	.9121	2.1228–2.1236	.0008–.0015	2.3982–2.3990	.0005–.0015	5	.004–.008
	8-302	.0018–.0026	.010	.015	.9121	2.1228–2.1236	.0008–.0015	2.2482–2.2490	.0005–.0015	3	.004–.008
	8-351②	.0018–.0026	.010	.015	.912	2.3103–2.3110	.0008–.0015	2.9994–3.0002	.0013–.0025	3	.004–.008
	8-351③	.0014–.0022	.010	.015	.912	2.3103–2.3110	.0008–.0026	2.7484–2.7492	.0009–.0026	3	.004–.008
	8-390	.0015–.0023	.010	.015	.975	2.4380–2.4388	.0008–.0015	2.7484–2.7492	.0013–.0025	3	.004–.010
	8-400	.0014–.0022	.010	.015	.975	2.3103–2.3111	.0008–.0026	2.9994–3.0002	.0009–.0026	3	.004–.010
	8-427	.0030–.0038	.018	.015	.975	2.4380–2.4388	.0008–.0015	2.7484–2.7492	.0005–.0015	3	.004–.010
	8-428	.0015–.0023	.010	.015	.975	2.4380–2.4388	.0008–.0015	2.7484–2.7492	.0005–.0015	3	.004–.010
	8-429	.0014–.0022	.010	.015	1.041	2.4992–2.5000	.0008–.0015	2.9994–3.0002	.0005–.0015	3	.004–.008
	8-460	.0014–.0022	.010	.015	1.040	2.4992–2.5000	.0008–.0015	2.9994–3.0002	.0005–.0015	3	.004–.008
1974	8-351②	.0018–.0026	.010	.015	.912	2.3103–2.3111	.0008–.0026	2.9994–3.0002	.0008–.0026	3	.004–.008
	8-351③	.0014–.0022	.010	.015	.912	2.3103–2.3111	.0008–.0026	2.7482–2.7492	.0009–.0026	3	.004–.008
	8-400	.0014–.0022	.010	.015	.975	2.3103–2.3111	.0008–.0026	2.9994–3.0002	.0009–.0026	3	.004–.008
	8-460	.0022–.0030	.010	.015	1.040	2.4992–2.5000	.0008–.0026	2.9994–3.0002	④	3	.004–.008
1975–76	V8-351	.0014–.0022	.010	.015	.9752	2.3107	.0008–.0026	2.9988	.0009–.0026	3	.004–.008
	V8-400	.0014–.0022	.010	.015	.9752	2.3107	.0008–.0026	2.9988	.0009–.0026	3	.004–.008
	V8-460	.0013–.0023	.010	.015	1.0401	2.4992–2.500	.0008–.0028	2.9998	⑤	3	.004–.008

①—Fit rings in tapered bores for clearance listed in tightest portion of ring travel.
②—Windsor engine.
③—Cleveland engine.
④—#1, .0004–.0020; others, .0012–.0028.
⑤—No. 1—.0004–.0022; #2, 3, 4, 5—.0009–.0027.

ENGINE TIGHTENING SPECIFICATIONS★

★Torque specifications are for clean and lightly lubricated threads only. Dry or dirty threads produce increased friction which prevents accurate measurement of tightness.

Year	Engine Model	Spark Plugs Ft. Lbs.	Cylinder Head Bolts Ft. Lbs.	Intake Manifold Ft. Lbs.	Exhaust Manifold Ft. Lbs.	Rocker Arm Shaft Bracket Ft. Lbs.	Rocker Arm Cover Ft. Lbs.	Connecting Rod Cap Bolts Ft. Lbs.	Main Bearing Cap Bolts Ft. Lbs.	Flywheel to Crank-shaft Ft. Lbs.	Vibration Damper or Pulley Ft. Lbs.
1969–72	6-240	15–20	70–75	23–28	23–28	—	7–9	40–45	60–70	75–85	130–150
1969–72	V8-302	15–20	65–72	23–25	12–16	17–23①	3–5	19–24	60–70	75–85	70–90
1969–72	V8-351	15–20	95–100	23–25	18–24	17–23①	3–5	40–45	95–105	75–85	70–90
1969–71	V8-390	15–20	80–90	32–35	18–24	40–45	4–7	40–45	95–105	75–85	70–90
1971–72	V8-400	15–20	95–105	②	12–16	18–25⑧	3–5	40–45	95–105	75–85	70–90
1969–72	V8-429	15–20	130–140	25–30	28–33	⑦	5–6	40–45	95–105	75–85	70–90
1973	V8-351③	15–20	105–112	23–25	18–24	17–23①	3–5	40–45	95–105	75–85	100–130
	V8-351④	15–20	95–105	②	12–22	18–25⑧	3–5	40–45	⑤	75–85	70–90
	V8-400	15–20	95–105	②	12–16	18–25⑧	3–5	40–45	⑤	75–85	70–90
	V8-429, 460	15–20	130–140	25–30	28–33	18–25⑧	5–6	40–45	95–105	75–85	70–90
1974	V8-351③	15–20	105–112	19–27	18–24	17–23①	3–5	40–45	95–105	75–85	100–130
	V8-351④	15–20	95–105	⑥	12–22	18–25⑧	3–5	40–45	⑤	75–85	70–90
	V8-400	15–20	95–105	⑥	12–16	18–25⑧	3–5	40–45	⑤	75–85	70–90
	V8-460	15–20	130–140	22–32	28–33	18–25⑧	5–6	40–45	95–105	75–85	70–90
1975–76	V8-351	10–15	95–105	②	18–24	18–25⑧	3–5	40–45	95–105	75–85	70–90
	V8-400	10–15	95–105	②	18–24	18–25⑧	3–5	40–45	95–105	75–85	70–90
	V8-460	10–15	130–140	25–30	28–33	18–25⑧	5–6	40–45	95–105	75–85	70–90

①—Rocker arm stud nut.
②—5/16″ bolts 21–25 ft. lbs. 3/8″ bolts 27–33 ft. lbs. 1/4″ bolts 6–9 ft. lbs.
③—Windsor engine.
④—Cleveland engine.
⑤—1/2″ bolts 95–105 ft. lbs. 3/8″ bolts 35–45 ft. lbs.
⑥—5/16″ bolts, 22–32, 3/8″ bolts, 17–25, 1/4″ bolts, 6–9.
⑦—1969–70 rocker arm stud to cylinder head; 65–75 ft. lbs. 1971–72 fulcrum bolt to cylinder head; 18–25 ft. lbs.
⑧—Fulcrum bolt to cylinder head.

WHEEL ALIGNMENT SPECIFICATIONS

OLD CAR SPECIFICATIONS: For 1946-68 Wheel Alignment Specifications see main index.

Year	Model	Caster Angle, Degrees		Camber Angle, Degrees				Toe-In. Inch	Toe-Out on Turns, Deg.①	
		Limits	Desired	Limits		Desired			Outer Wheel	Inner Wheel
				Left	Right	Left	Right			
1969	All	0 to +2	+1	−¼ to +1¼	−¼ to +1¼	+½	+½	3/16	②	20
1970–72	All	0 to +2	+1	−¼ to +1¼	−¼ to +1¼	+½	+½	3/16	19⅛	20
1973	All	0 to +4	+2	−1 to +1	−1 to +1	0	0	3/16	18¾	20
1974	All	0 to +4	+2	−½ to +1½	−¾ to +1¼	+½	+¼	3/16	18¾	20
1975	All	+1¼ to +2¾	+2	−¼ to +1¼	−½ to +1	+½	+¼	3/16	18.72	20
1976	All	+1¼ to +2¾	+2	+½ to +¾	+¼ to +¾	+¾	+¾	¼	18.69	20

①—Incorrect toe-out, when other adjustments are correct, indicates bent steering arms.
②—Manual steering 18.96°; Power steering 19.14°.

STARTING MOTOR SPECIFICATIONS

Year	Model	Ident. No.	Rotation	Brush Spring Tension Ounces	No Load Test			Torque Test		
					Amperes	Volts	R.P.M.	Amperes	Volts	Torque Ft. Lbs.
1969	6-240, 8-302, 351	C7AF-B	C	40	70	12	9500	670	5.0	15.5
	6-240, 8-302, 351	C9ZF-A	C	40	70	12	9500	670	5.0	15.5
	8-390	C7AF-C	C	40	70	12	9500	670	5.0	15.5
	8-429	C9AF-A	C	40	70	12	10000	700	5.0	15.5
1970	6-240, 8-302, 351	C7AF-B,F	C	40	70	12	—	670	5.0	15.5
	6-240, 8-302, 351	D0AF-B,C	C	40	70	12	—	670	5.0	15.5
	8-390	D0AF-A	C	40	70	12	—	670	5.0	15.5
	8-390	D0OF-A	C	40	70	12	—	670	5.0	15.5
	8-390	C9AF-B	C	40	70	12	—	670	5.0	15.5
	8-429	C9AF-A	C	40	70	12	—	700	5.0	15.5
1971	6-240, 8-302, 351	D0AF-B, C	C	40	70	12	—	670	5.0	15.5
	8-390	D0AF-A	C	40	70	12	—	670	5.0	15.5
	8-400	D1AF-AA	C	40	70	12	—	670	5.0	15.5
	8-429	C9AF-A	C	40	70	12	—	—	—	—
1972	8-429	D2AF-EA	C	40	70	12	—	—	—	—
1972–73	6-240, 8-302, 351	D2AF-CA	C	40	70	12	—	670	5.0	15.5
	8-400	D2AF-AA	C	40	70	12	—	670	5.0	15.5
1973	8-429, 460	D4OF-AA, BA	C	40	70	12	—	—	—	—
1974	8-351	D2AF-CA	C	80	80	12	—	670	5.0	15.5
	8-400	D4AF-AA	C	80	80	12	—	670	5.0	15.5
	8-460	D4OF-AA, BA	C	80	90	12	—	—	—	—
1975	8-351	D2AF-JA	C	80	80	12	—	670	5.0	15.5
	8-400	D5AF-EA	C	80	80	12	—	670	5.0	15.5
	8-460	D5AF-AB	C	80	90	12	—	—	—	—
1976	8-351	D5AF-FA	C	—	—	12	—	—	—	—
	8-400	D5AF-EA	C	—	—	12	—	—	—	—
	8-460	D6AF-AA	C	—	—	12	—	—	—	—

ALTERNATOR & REGULATOR SPECIFICATIONS

Year	Make or Model	Current Rating		Field Current @ 75°F.		Voltage Regulator				Field Relay	
		Amperes	Volts	Amperes	Volts	Part No. (10316)	Voltage @ 75°F.	Contact Gap	Armature Air Gap	Armature Air Gap	Closing Voltage @ 75°F.
1969	C6AF-10300-A	55	15	2.8–3.3	12	C3SZ-B	13.5–15.3	②	②	②	2.3–4.2
	C6AF-10300-B	42	15	2.8–3.3	12	C3SZ-B	13.5–15.3	②	②	②	2.3–4.2
	C6AF-10300-F	55	15	2.8–3.3	12	C3SZ-B	13.5–15.3	②	②	②	2.3–4.2
	C6AF-10300-G	55	15	2.8–3.3	12	C3SZ-B	13.5–15.3	②	②	②	2.3–4.2
	C7AF-10300-A	65	15	2.9	12	C3SZ-B	14.3–15.1	.045–.052	.010–.015	.015	3.0–4.0
	C9AF-10300-A	42	15	2.8–3.3	12	C3SZ-B	13.5–15.3	②	②	②	2.3–4.2
1970–72	D0ZF-10300-B	38	15	2.4	12	C3SZ-B	13.5–15.3	②	②	②	2.0–4.2
	D0AF-10300-C	42	15	2.9	12	C3SZ-B	13.5–15.3	②	②	②	2.0–4.2
	D0AF-10300-F	42	15	2.9	12	C3SZ-B	13.5–15.3	②	②	②	2.0–4.2
	D0AF-10300-G	42	15	2.9	12	C3SZ-B	13.5–15.3	②	②	②	2.0–4.2
	D0AF-10300-E	55	15	2.9	12	C3SZ-B	13.5–15.3	②	②	②	2.0–4.2
	D0AF-10300-H	55	15	2.9	12	C3SZ-B	13.5–15.3	②	②	②	2.0–4.2
	D0ZF-10300-A	55	15	2.9	12	C3SZ-B	13.6–15.1	.018–.020	.042–.052	.011–.018	6.2–7.2
	D0ZF-10300-C	55	15	2.9	12	C3SZ-B	13.6–15.1	.018–.020	.042–.052	.011–.018	6.2–7.2
	D0SF-10300-A	55	15	2.9	12	C3SZ-B	13.6–15.1	.018–.020	.042–.052	.011–.018	6.2–7.2
	D1ZF-10300-AA	55	15	2.9	12	C3SZ-B	13.5–15.3	②	②	②	2.0–4.2
	D1AF-10300-AA	61	15	2.9	12	C3SZ-B	13.5–15.3	②	②	②	2.0–4.2
	D1AF-10300-BA	65	15	2.9	12	C3SZ-B	13.5–15.3	②	②	②	2.0–4.2
	D0AF-10300-A③	65	15	2.9	12	—	—	—	—	—	—
	C5TF-10300-K	65	15	2.9	12	—	—	—	—	—	—
1973	Orange①	42	15	2.9	12	D3TZ-A	13.5–15.3	—	—	—	2½–4
	Red①	55	15	2.9	12	D3TZ-A	13.5–15.3	—	—	—	2½–4
	Green①	61	15	2.9	12	D3TZ-A	13.5–15.3	—	—	—	2½–4
	All	70	15	2.9	12	D3TZ-A	13.5–15.3	—	—	—	2½–4
	All	90	15	2.9	12	D3TZ-A	13.5–15.3	—	—	—	2½–4
1974	Purple①	38	15	2.9	12	D4TZ-A	—	—	—	—	—
	Orange①	42	15	2.9	12	D4TZ-A	—	—	—	—	—
	Red①	55	15	2.9	12	D4TZ-A	—	—	—	—	—
	Green①	61	15	2.9	12	D4TZ-A	—	—	—	—	—
	All	70	15	2.9	12	D4TZ-A	—	—	—	—	—
	All	90	15	2.9	12	D4TZ-A	—	—	—	—	—
1975	D3VF-10300-AB	90	15	2.9	12	D4TF-AA	13.5–15.3	—	—	—	2.5–4
	D30F-10300-AA	70	15	2.9	12	D4TF-AA	13.5–15.3	—	—	—	2.5–4
	D30F-10300-BA	55	15	2.9	12	D4AF-AA	13.5–15.3	—	—	—	2.5–4
	D30F-10300-EA	61	15	2.9	12	D4AF-AA	13.5–15.3	—	—	—	2.5–4
	D30F-10300-FA	42	15	2.9	12	D4AF-AA	13.5–15.3	—	—	—	2.5–4
	D32F-10300-BA	61	15	2.9	12	D4AF-AA	13.5–15.3	—	—	—	2.5–4
	D4LF-10300-AA	55	15	2.9	12	D4AF-AA	13.5–15.3	—	—	—	2.5–4
	D4LF-10300-CA	55	15	2.9	12	D4AF-AA	13.5–15.3	—	—	—	2.5–4
	D40F-10300-AA	42	15	2.9	12	D4AF-AA	13.5–15.3	—	—	—	2.5–4
	D40F-10300-BA	55	15	2.9	12	D4AF-AA	13.5–15.3	—	—	—	2.5–4
	D40F-10300-DA	61	15	2.9	12	D4AF-AA	13.5–15.3	—	—	—	2.5–4
	D40F-10300-EA	70	15	2.9	12	D4TF-AA	13.5–15.3	—	—	—	2.5–4
	D42F-10300-BA	61	15	2.9	12	D4AF-AA	13.5–15.3	—	—	—	2.5–4
1976	D5AF-10300-AA	40	15	2.9	12	D4AF-AA	13.5–15.3	—	—	—	2.5–4
	All	70	15	2.9	12	D4TF-AA	13.5–15.3	—	—	—	2.5–4
	D50F-10300-BA	90	15	2.9	12	D4TF-AA	13.5–15.3	—	—	—	2.5–4

①—Identification tag. ②—Not adjustable. ③—Integral regulator.

REAR AXLE SPECIFICATIONS

Year	Model	Carrier Type	Ring Gear & Pinion Backlash Inch	Nominal Pinion Locating Shim, Inch	Pinion Bearing Preload				Differential Bearing Preload	Pinion Nut Torque Ft.-Lbs.①
					New Bearings With Seal Inch-Lbs.	Used Bearings With Seal Inch-Lbs.	New Bearings Less Seal Inch-Lbs.	Used Bearings Less Seal Inch-Lbs.		
1969-73	④	Integral	.008–.012	.030	22–32	8–14	—	—	.008–.012⑥	140
	Others	Removable	.008–.012	.015	22–32⑤	8–14	—	—	.008–.012③	⑦
1974–75	—	Integral	.008–.012	.030	17–27	8–14	—	—	.008–.012⑥	140
	—	Removable	.008–.012	.015	17–27⑤	8–14	—	—	.008–.012③	⑧

①—If torque cannot be obtained, install new spacer.
③—Case spread with new bearings; with used bearings .005–.008".
④—8-302 engine with two-barrel carburetor and 6-240 engine.
⑤—Solid spacer 13–33 inch-lbs.
⑥—Case spread with new bearing; with used bearings .006–.010".
⑦—Collapsible spacer 175 ft. lbs.; solid spacer 200 ft. lbs.
⑧—Collapsible spacer 170 ft. lbs.; solid spacer 200 ft. lbs.

BRAKE SPECIFICATIONS

Year	Model	Brake Drum Inside Diameter	Wheel Cylinder Bore Diameter			Master Cylinder Bore Diameter		
			Front Disc Brake	Front Drum Brake	Rear Brake	With Disc Brakes	With Drum Brakes	With Power Brakes
1969–72	All Passenger Cars	11.03	2.755	1.125	.938	1.00	1.00	1.00
	All Wagons	11.03	2.755	1.094	.938	1.00	1.00	1.00
1973	All Passenger Cars	11.03	3.100	1.125	1.00	1.00	1.00	1.00
	All Wagons	11.03	3.100	0.875	0.875	1.00	1.00	1.00
1974	All	11.03	3.100	—	1.00	1.00	1.00	1.00
1975–76	All①	11.03	3.1	—	1	1	—	1
	All②	—	3.1	—	2.6	1⅛	—	1⅛

①—Except 4 wheel disc brakes with hydro boost.
②—With 4 wheel disc brakes with hydro boost.

COOLING SYSTEM & CAPACITY DATA

Year	Model or Engine	Cooling Capacity, Qts.			Radiator Cap Relief Pressure, Lbs.		Thermo. Opening Temp. ①	Fuel Tank Gals.	Engine Oil Refill Qts. ②	Transmission Oil			Rear Axle Oil Pints
		No Heater	With Heater	With A/C	With A/C	No A/C				3 Speed Pints	4 Speed Pints	Auto. Trans. Qts. ③	
FORD													
1969	6-240	12	13	13	12–15	12–15	195	24⑥	4	3½	4	⑬	4½
	8-302	14	15	15	12–15	12–15	195	24⑥	4	3½	4	⑬	4½
	8-390	19½	20½	20½	12–15	12–15	195	24⑥	4	3½	4	⑫	5
	8-429	19½	20½	20½	12–15	12–15	195	24⑥	4	3½	4	12¾	5
1970	6-240	13½	14½	14½	12–15	12–15	195	24⑩	4	3½	4	⑬	4½
	8-302	13½	14½	14½	12–15	12–15	195	24⑩	4	3½	4	⑬	4½
	8-351	15½	16½	16½	12–15	12–15	195	24⑩	4	3½	4	⑬	5
	8-390	19	20	20	12–15	12–15	195	24⑩	4	3½	4	⑬	5
	8-429	17½	18½	18½	12–15	12–15	195	24⑩	4	3½	4	12¾	5
1971	6-240	13	14	14	12–15	12–15	195	23⑩	4	3½	4	⑬	⑤
	8-302	14	15	15	12–15	12–15	195	23⑩	4	3½	4	⑬	⑤
	8-351	15¼	16¼	16¼	12–15	12–15	195	23⑩	4	3½	4	11	5
	8-390	19	20	20	12–15	12–15	195	23⑩	4	3½	4	12¾	5

Continued

COOLING SYSTEM & CAPACITY DATA—Continued

Year	Model or Engine	Cooling Capacity, Qts.			Radiator Cap Relief Pressure, Lbs.		Thermo. Opening Temp. ①	Fuel Tank Gals.	Engine Oil Refill Qts. ②	Transmission Oil			Rear Axle Oil Pints
		No Heater	With Heater	With A/C	With A/C	No A/C				3 Speed Pints	4 Speed Pints	Auto. Trans. Qts. ③	
FORD—Continued													
1971	8-400	16½	17½	18¼	12–15	12–15	195	23⑩	4	3½	4	12¾	5
	8-429	18	19	19	12–15	12–15	195	23⑩	4	3½	4	12¾	5
1972	6-240	13¼	14¼	14¼	12–15	12–15	195	22④	4	—	—	⑬	⑤
	8-302	14¼	15¼	15¼	12–15	12–15	195	22④	4	—	—	⑬	⑤
	8-351	15½	16½	16½	12–15	12–15	195	22④	4	—	—	⑬	5
	8-400	16¾	17¾	18¼	12–15	12–15	195	22④	4	—	—	12¾	5
	8-429	17¾	18¾	20	12–15	12–15	195	22④	4	—	—	12¾	5
1973	8-351⑧	14½	15½	16¼	12–15	12–15	195	22④	4	—	—	10¼	4
	8-351⑨	15½	16½	17	12–15	12–15	195	22④	4	—	—	10¼	4
	8-400	17	18	18	12–15	12–15	195	22④	4	—	—	12½	⑤
	8-429	18½	19½	19½	12–15	12–15	195	22④	4	—	—	12½	5
1974	8-351⑧	15.3	16½	17.2	12–16	12–16	191	22④	4	—	—	⑪	⑤
	8-351⑨	15.3	16½	16½	12–16	12–16	191	22④	4	—	—	⑭	⑤
	8-400	17	18	18½	12–16	12–16	191	22④	4	—	—	⑭	⑤
	8-460	18.4	19½	19½	12–16	12–16	191	22④	4	—	—	12½	⑤
1975	8-351	—	17.1	17.6	12–16	12–16	191	24.2④	4	—	—	⑪	⑤
	8-400	—	17.1	17.6	12–16	12–16	191	24.2④	4	—	—	⑬	⑤
	8-460	—	18.5	18.5⑮	12–16	12–16	191	24.2④	4⑯	—	—	⑬	⑤
1976	8-351	16.3	17.1	17.6	12–16	12–16	191	24.2④	4	—	—	⑪	⑤
	8-400	16.3	17.1	17.6	12–16	12–16	191	24.2④	4	—	—	⑬	⑤
	8-460	—	18.5	18.5⑮	12–16	12–16	191	24.2④	4⑯	—	—	⑬	⑤
MERCURY													
1969	All	19½	20½	20½	12–15	12–15	195	24⑥	4	3½	—	12¾	5
1970	8-390	19	20	20	12–15	12–15	195	24⑩	4	3½	—	12¾	5
	8-429	17½	18½	18½	12–15	12–15	195	24⑩	4	3½	—	12¾	5
1971	8-351	15¼	16¼	16¼	12–15	12–15	195	23⑩	4	3½	—	11	5
	8-400	16½	17½	18¼	12–15	12–15	195	23⑩	4	3½	—	12¾	5
	8-429	18	19	19	12–15	12–15	195	23⑩	4	3½	—	12¾	5
1972	8-351	15¼	16¼	16¾	12–15	12–15	195	22½⑦	4	—	—	11	5
	8-400	16.6	17.6	18.3	12–15	12–15	195	22½⑦	4	—	—	12¾	5
	8-429	17.8	18.8	19½	12–15	12–15	195	22½⑦	4	—	—	12¾	5
1973	8-351⑧	14½	15½	16¼	12–15	12–15	195	22④	4	—	—	10¼	⑤
	8-351⑨	15½	16½	17	12–15	12–15	195	22④	4	—	—	10¼	⑤
	8-400	17	18	18	12–15	12–15	195	22④	4	—	—	12½	⑤
	8-429, 460	18½	19½	19½	12–15	12–15	195	22④	4	—	—	12½	⑤
1974	8-351⑧	15.3	16½	17	12–16	12–16	191	22④	4	—	—	10½	⑤
	8-400	17	18	18½	12–16	12–16	191	22④	4	—	—	⑬	⑤
	8-460	18.4	19½	19½	12–16	12–16	191	22④	4	—	—	12½	⑤
1975	8-400	—	17.1	17.6	12–16	12–16	191	24.2④	4	—	—	⑬	⑤
	8-460	—	18.5	18.5⑮	12–16	12–16	191	24.2④	4⑯	—	—	⑬	⑤
1976	8-400	16.3	17.1	17.6	12–16	12–16	191	24.2④	4	—	—	⑬	⑤
	8-460	17.6	18.5	18.5⑮	12–16	12–16	191	24.2④	4⑯	—	—	⑬	⑤

①—With permanent type anti-freeze.
②—Add one quart with filter change.
③—Approximate. Make final check with dipstick.
④—Station Wagons 21 gals. Add 8 gals. with auxiliary tank.
⑤—WER axles 4, all others 5.
⑥—Station Wagons 20 gals.
⑦—Station Wagons 21½ gallons.
⑧—Cleveland engine.
⑨—Windsor engine.
⑩—Station Wagon 22 gals.
⑪—C4 10½ qts., FMX 11 qts.
⑫—Three spd. 11 qts., C6 13 qts.
⑬—FMX 11 qts., C6 12¼ qts. C4, 10¼ qts.
⑭—FMX 11 qts., C6 12½ qts., CW 11⅛ qts.
⑮—Medium duty 19 qts., Heavy duty and Police 20 qts.
⑯—Police Models 6½ qts.

Electrical Section

DISTRIBUTOR, REPLACE

Removal

1. On vehicles with conventional ignition systems, disconnect primary wire from ignition coil. On vehicles with electronic ignition, disconnect harness connector from engine wiring.
2. Disconnect vacuum control line and remove distributor cap.
3. Scribe a mark on the distributor body indicating the position of the rotor, and scribe another mark on the body and engine block indicating position of distributor body in block. These marks can be used as guides when installing distributor in a correctly timed engine.
4. Remove hold down screw or screws and lift distributor out of block. *Do not crank engine while distributor is removed or the initial timing operation will have to be performed.*

Installation

If the crankshaft has not been disturbed, install the distributor, using the scribed marks previously made on the distributor body and engine block as guides.

If the crankshaft has been rotated while the distributor was removed from the engine, it will be necessary to retime the engine. Crank the engine to bring No. 1 piston on top dead center of its compression stroke. Align the timing mark on the vibration damper or pulley with the timing pointer (see *Tune Up*) chart). Install the distributor so that the rotor points to the No. 1 spark plug wire terminal in the distributor cap.

NOTE: On all overhead valve engines, make sure the oil pump intermediate shaft properly engages the distributor shaft. It may be necessary to crank the engine with the starter, after the distributor drive gear is properly engaged, in order to engage the oil pump intermediate shaft.

STARTER REPLACE

SERVICE BULLETIN

STARTER PROBLEMS: If the starter is noisy or if it locks up, before condemning the starter, loosen the three mounting bolts enough to hand fit the starter properly into the pilot plate. Then tighten the mounting bolts, starting with the top bolt.

1969-76 Ford & Mercury

1. Disconnect battery ground cable and disconnect cable at starter cable.
2. Remove screws and take off starter.
3. Reverse procedure to install and torque bolts to 15-20 ft. lbs.

IGNITION LOCK, REPLACE

1970-76

1. Disconnect battery ground cable.
2. *Units with Fixed Steering Columns:* Remove steering wheel and trim pad. Insert a .090 inch wire pin in the hole located inside the column halfway down the lock cylinder housing, Fig. 1. *Units with Tilt Steering Columns:* Insert wire pin in the hole located on the outside of the flange casting next to the emergency flasher button, Fig. 1.
3. Place the gear shift lever in *PARK* (with automatic trans) or *REVERSE* (with manual trans) position, and turn the lock cylinder with the ignition key to *RUN* position.
4. Depress the wire pin while pulling up on the lock cylinder to remove. Remove the wire pin.
5. To install insert the lock cylinder into housing in the flange casting, and turn the key to *OFF* position. Be certain that the cylinder is fully inserted before turning to the *OFF* position. This action will extend the cylinder retaining pin into the cylinder housing.
6. Turn the key to check for correct operation in all positions.
7. Install the steering wheel and trim pad on fixed column units.
8. Connect the battery ground cable.

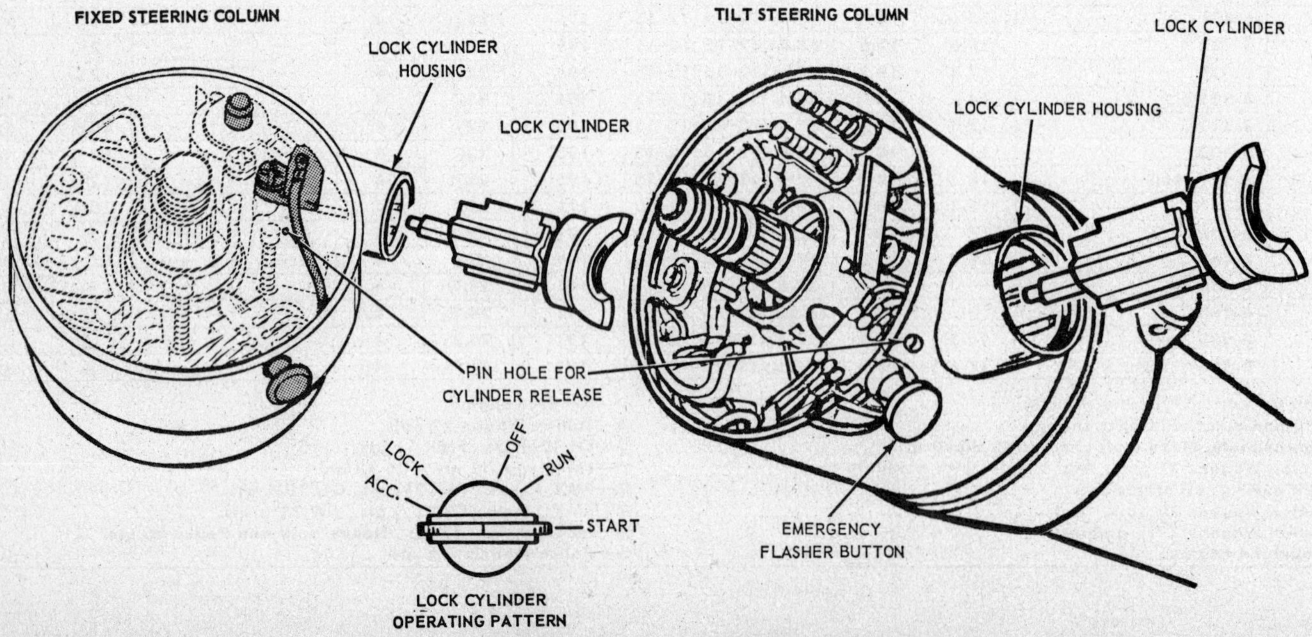

Fig. 1 Ignition Lock. 1970-76

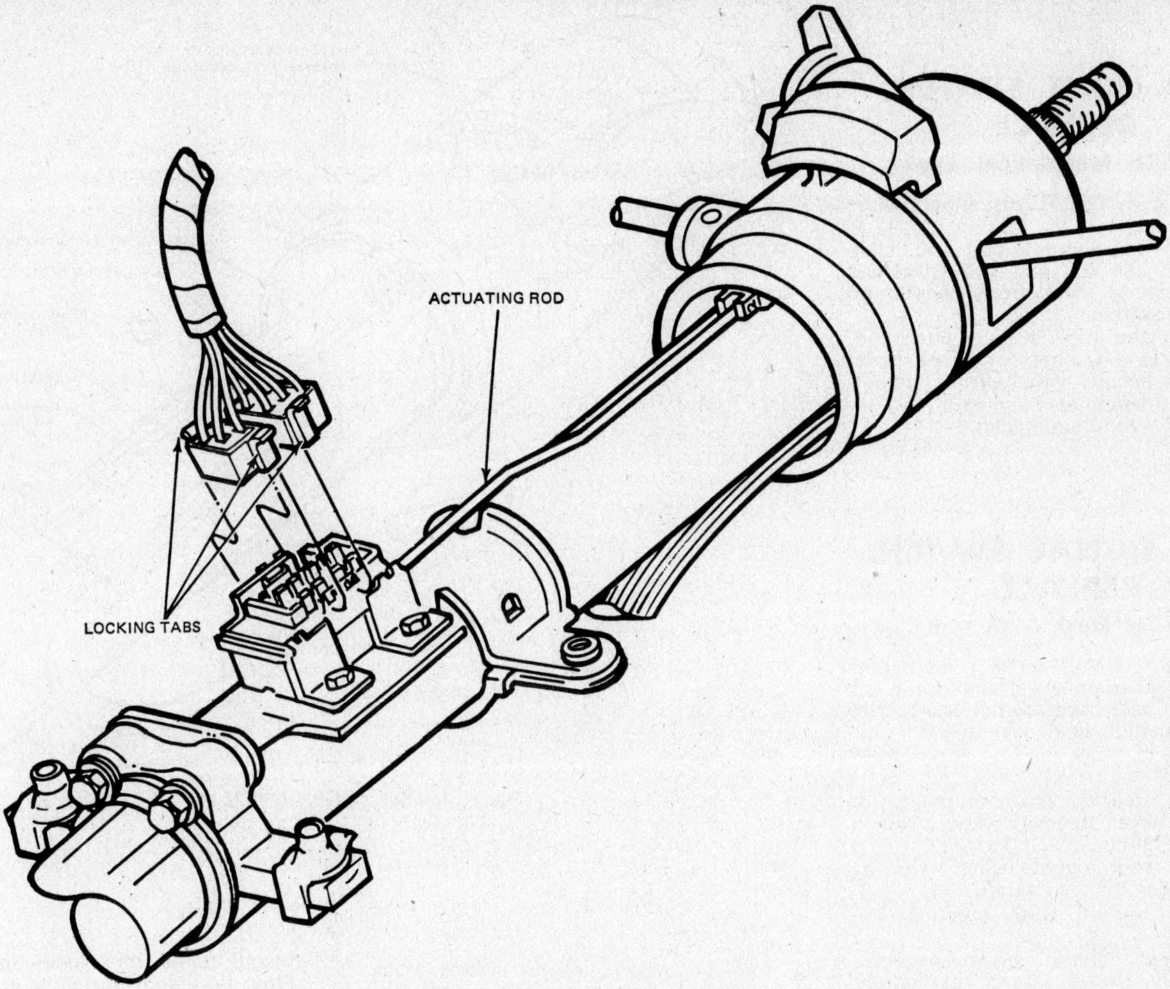

Fig. 2 Ignition switch. 1972-76 All

IGNITION SWITCH, REPLACE

1970-76 Ford & Mercury

1. Remove steering column shroud and lower steering column from brake support bracket.
2. Disconnect battery cable.
3. Disconnect switch wiring and remove two switch retaining nuts. Disconnect switch from actuator and remove switch, Figs. 2 and 3.
4. Move shift lever to Park position on automatic transmissions and Reverse on standard transmission units. Place ignition key in Lock position and remove the key.

NOTE: New replacement switches are pinned in the Lock position by a plastic shipping pin inserted in a locking hole in the switch. For an existing switch, pull plunger out as far as it will go then back one detent to Lock position and insert a 3/32" drill in locking hole to retain switch in Lock position.

5. With locking pin in place, install switch on steering column, determine mid position of actuator lash and tighten retaining bolts.
6. Remove locking pin.

1969 Ford & Mercury

1. Disconnect battery ground cable.
2. Insert ignition key in switch. Turn key to accessory position and insert a wire pin in hole on ignition switch. Slightly depress pin while turning counterclockwise past the accessory position; this will release lock cylinder from switch. Pull lock cylinder from switch with the key, Fig. 4.
3. Remove bezel nut retaining switch to instrument panel and lower switch.
4. Depress tabs securing multiple connector from switch and remove switch.
5. Reverse procedure to install.

LIGHT SWITCH, REPLACE

All Except 1970 Mercury

1. Disconnect battery ground cable.
2. On 1973-76 models, remove wiper switch knob.
3. With headlight switch "On" on all models, depress release button on switch housing and remove knob and shaft, Fig. 5.
4. On 1969 Mercury, remove wiring harness bracket screws from switch and remove bracket.
5. On all models remove bezel nut from light switch and on 1973-76 models, remove lower finish panel.
6. On 1972-76 models, remove three mounting plate screws.
7. Disconnect electrical connector. If equipped with concealed headlamps, disconnect vacuum hoses from switch and remove switch.

1970 Mercury

1. Disconnect battery ground cable.
2. From behind instrument panel, disconnect wiring plug and vacuum hoses if so equipped from switch.
3. Remove four retaining screws and pull switch out from behind instrument panel.

STOP LIGHT SWITCH, REPLACE

1969-76 Mechanical Type

1. Referring to Fig. 6, disconnect wires at connector.
2. Remove hairpin retainer and slide switch, push rod and nylon washers and bushing away from pedal, and remove switch.
3. Position the new switch, push rod, bushing and washers on brake pedal pin and secure with hairpin retainer.
4. Connect wires at connector and install wires in retaining clip.

TURN SIGNAL SWITCH, REPLACE

1970-76 Ford & Mercury

1. Remove retaining screw from underside of steering wheel spoke and lift off the pad horn switch/trim cover and medallion as an assembly.
2. Disconnect horn switch wires from terminals.
3. Remove steering wheel retaining nut and remove steering wheel with a suitable puller.
4. Remove turn signal switch lever by unscrewing it from column.
5. Remove shroud from steering column.
6. Disconnect column wiring connector plug and remove screws that secure switch to column.
7. On tilt column models, remove wires and terminals from column wiring plug.

NOTE: *Record the color code and position of each wire before removing it from plug. A hole provided in the flange casting on fixed column models makes it unnecessary to separate wires from plug. The plug with wires installed can be guided through the hole.*

8. Remove plastic cover sleeve from wiring harness and remove the switch from top of column.
   On vehicles equipped with speed control, transfer the ground brush located in the turn signal cancelling cam to the new switch assembly.

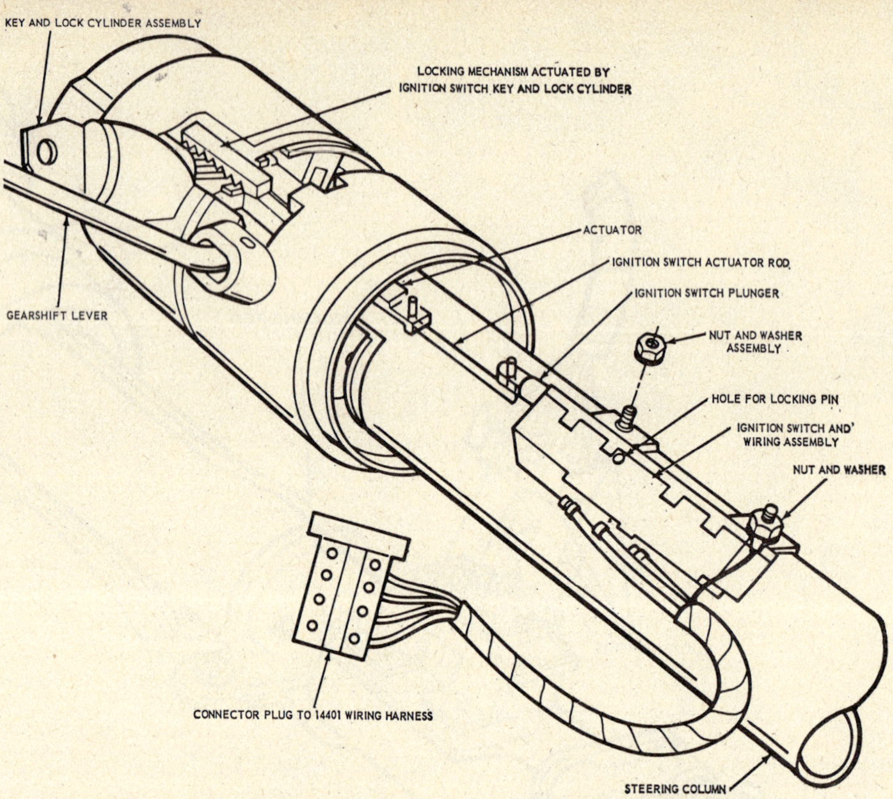

Fig. 3 Ignition switch. 1970-71 All

1969 Ford & Mercury

Removal

1. Disconnect battery ground cable.
2. Remove horn button and steering wheel.
3. Remove turn indicator handle.
4. Unscrew and remove turn indicator switch from steering column tube.
5. Disconnect connector blocks at the column.
6. From lower portion of column, remove cover from wiring and tie a cord to the wire ends. Remove switch from top of column, feeding wiring and cord up the column.

Installation

1. Attach the wire ends of a new turn indicator switch to the cord and feed wires down through steering column. Remove cord and install cover.
2. Install connector blocks in column. Plug in electrical leads and secure wiring in retaining clip.
3. Install switch to steering column tube and install indicator handle.
4. Install steering wheel and horn button, and connect battery cable.

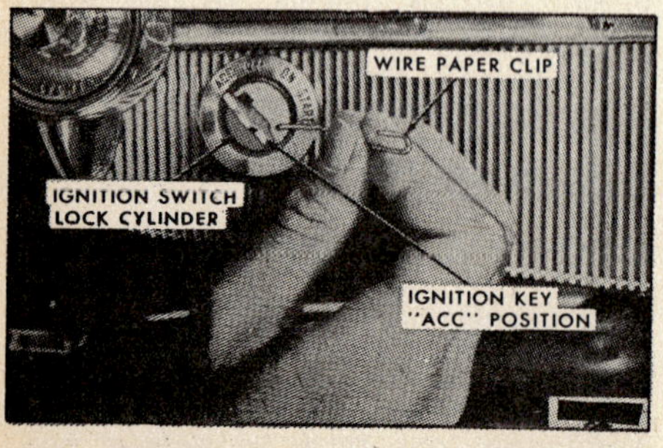

Fig. 4 Ignition switch lock. 1969 Ford and Mercury

NEUTRAL SAFETY SWITCH

1970-76 Ford & Mercury

Column Shift

The neutral safety switch has been

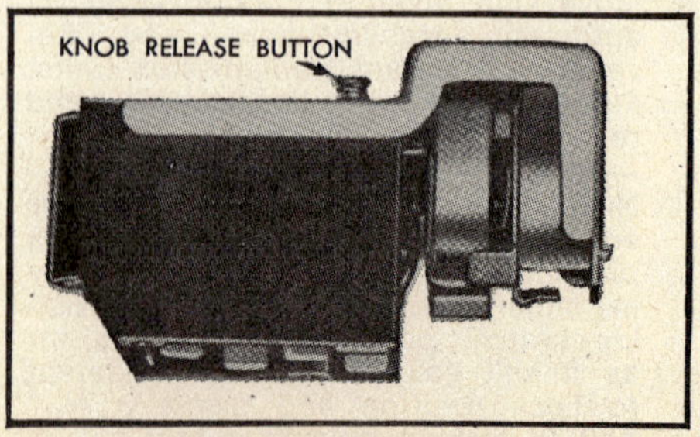

Fig. 5 Light switch. Ford and Mercury. Typical

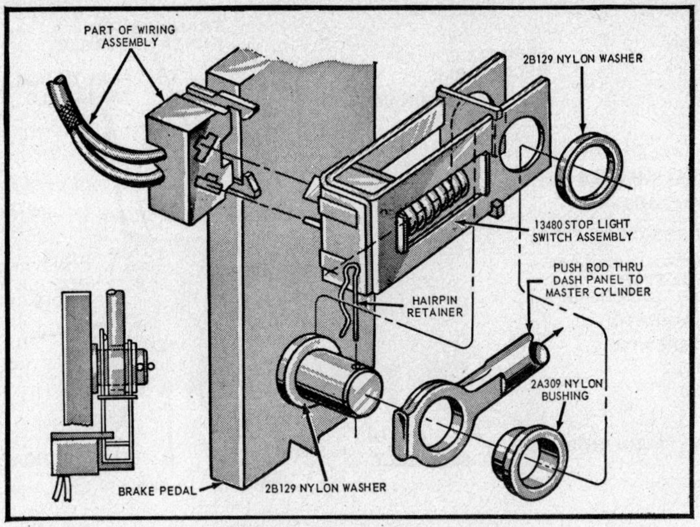

Fig. 6 Mechanical stop light switch. 1969-76 Ford and Mercury

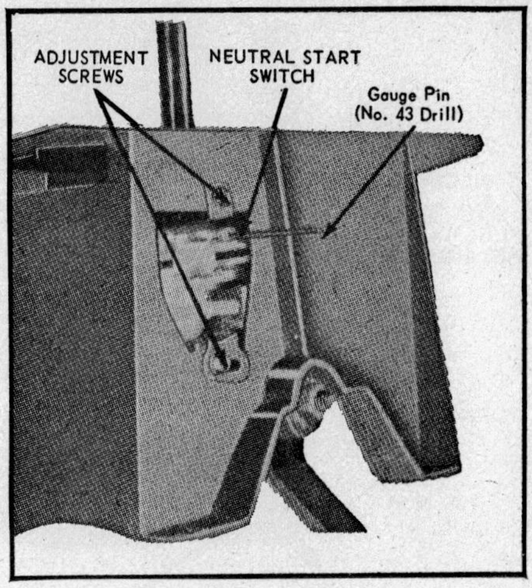

Fig. 7 Neutral safety switch (console shift). 1969 with C4 and C6 transmissions

eliminated and is replaced by a series of steps designed into the steering column selector lever hub casting.

Console Shift

1. Remove four screws and plates securing selector lever handle to the lever. Remove handle and detent control.
2. Remove two screws from rear of console top panel. Pull panel back to unhook it from front of console and remove panel.
3. Loosen two switch attaching screws.
4. Move selector lever back and forth until gauge pin (No. 43 drill) can be fully inserted into gauge pin holes.
5. Place selector lever firmly against the stop of the neutral detent position.
6. Slide switch forward or rearward as required until the switch lever contacts the selector lever actuator. If an adjustment cannot be made, loosen the actuating lever attaching screw and adjust the lever.
7. Tighten switch attaching screws and if actuator lever was adjusted tighten the actuator lever bolt.
8. Turn ignition key to ACC position and place selector lever in the reverse position and check operation of back-up lights. Turn key off.
9. Place console top panel on console and install screws.
10. Position selector lever detent control and handle on selector lever and secure with plates and screws.

1969 Ford & Mercury

Column Shift

To adjust the switch proceed as follows:

1. Place transmission selector lever against the stop of the neutral detent.
2. Loosen two retaining screws on the steering column.
3. With the selector lever against neutral stop, rotate the switch until a start in neutral position is obtained. Then tighten the two screws.
4. With the switch properly adjusted in neutral, place the selector lever in the "1" position and push the park reset button, located on the right side of the switch, to the left until it stops.

NOTE: The park reset must be performed whenever the switch has been adjusted.

1969 C4 & C6 Units

Console Shift, Fig. 7

1. Remove selector lever handle.
2. Unfasten and position console to one side.
3. Loosen adjusting screws.
4. Move selector lever back and forth until gauge pin (#43 drill) can be fully inserted into gauge pin holes.
5. Place transmission selector lever firmly against stop of neutral detent position.
6. Slide combination neutral start and back-up light switch forward or rearward as required until switch actuating lever contacts selector lever.
7. Tighten switch attaching screws, remove gauge pin and check for starting in Park position.

HORN SOUNDER, REPLACE

Rim-Blow Type

The rubber insert and copper strip assembly is not replaceable on all 1970-71 vehicles and the 1972-76 vehicles with speed control. If a new insert assembly is required the entire steering wheel must be replaced.

1. Remove the pad from the steering wheel (three screws).
2. Remove medallion from the pad.
3. After removing the steering wheel nut the wheel can be removed from the shaft with a wheel puller.

Except Rim-Blow Type

1. Disconnect battery ground cable.
2. Remove steering column pad (2 screws).
3. Push down and turn horn ring and remove ring and spring.
4. Reverse procedure to install.

INSTRUMENT CLUSTER

1973-76 Ford & Mercury

1. Disconnect the battery ground cable.
2. Remove two steering column cover screws and remove the cover.
3. Remove two instrument cluster trim cover attaching screws and remove cover, Fig. 7A.
4. Reach behind the cluster and disconnect the cluster feed plug from its receptacle.
5. Disconnect the speedometer cable.
6. Remove the screw attaching the transmission indicator cable to the steering column.
7. Remove the four cluster attaching screws and remove the cluster assembly.

1971-72 Ford

Instrument Cluster

1. Disconnect battery ground cable.
2. Remove instrument panel pad.
3. Disconnect speedometer cable and all electrical connections to the instrument cluster.
4. Remove four cluster attaching screws (two at bottom and two at top), and lift the cluster from the instrument panel.

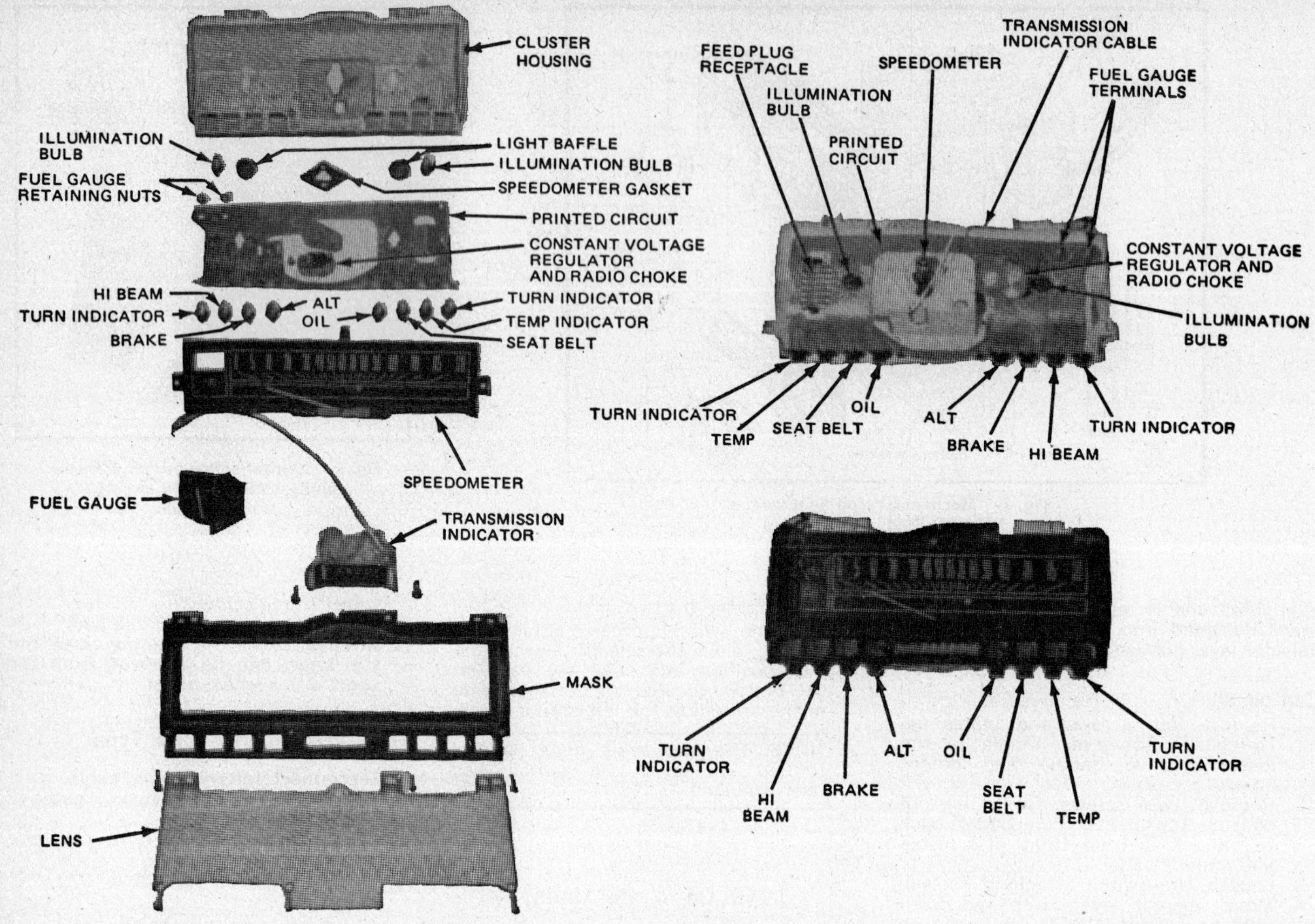

Fig. 7A Instrument cluster. 1973-76 Ford & Mercury

Fuel Gauge & Speedometer

1. Remove instrument cluster as described above.
2. Remove lights, wiper-washer and clock reset knobs.
3. Remove eight screws attaching lens, shield and mask to cluster.
4. Remove two screws attaching speedometer to cluster.
5. To remove fuel gauge, remove two terminal nuts.

1971-72 Mercury

Instrument Cluster

1. Disconnect battery ground cable.
2. Remove instrument panel pad.
3. Disconnect speedometer cable and all wires from back of cluster.
4. Remove four screws from lower edge of cluster and remove cluster.

Fuel Gauge & Speedometer

1. Remove cluster as described above.
2. Remove seven screws attaching the cluster housing and lens and inner mask and the cluster mask.
3. To remove speedometer, remove two screws from rear of cluster housing.
4. To remove fuel gauge, remove terminal nuts and gauge attaching screw.

1969-70 Ford

Fuel Gauge & Speedometer

1. From passenger side of instrument panel, remove lighter element and all of the control knobs, and then remove the cluster trim cover (10 screws).
2. Remove the attaching screws and the lens.
3. Remove attaching screws and lift the mask from the speedometer and fuel gauge.
4. Remove spring washers and nuts (one at each side) and the attaching screw at the bottom, then remove fuel gauge from the studs.
5. Remove attaching screws and pull speedometer head out of cluster far enough to reach the cable disconnect.
6. To disengage the cable, press on the flat surface of the disconnect and at the same time pull the cable from the head.
7. Remove the speedometer.
8. Reverse the procedure to install.

Instrument Cluster

1. Remove the instrument panel pad (20 screws).
2. From behind cluster disconnect all electrical connections to cluster.
3. Disconnect heater, air conditioner and speedometer cables.
4. Remove lighter element and all control knobs from passenger side of cluster.
5. Remove cluster trim cover (10 screws).
6. Remove eight mounting screws and withdraw cluster from panel.
7. Reverse procedure to install.

1969-70 Mercury

Instrument Cluster

1. Remove wiper knob and bezel, cigar lighter element and the finish panel.
2. Remove instrument panel pad.
3. Remove wiper nut and the bracket (3 screws) from left end of pad support.

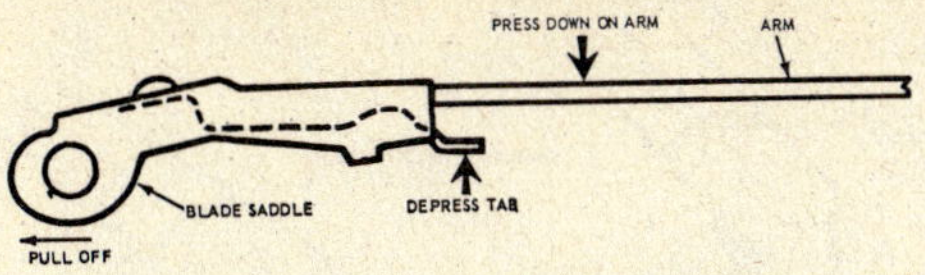

Fig. 8 Trico bayonet type blade

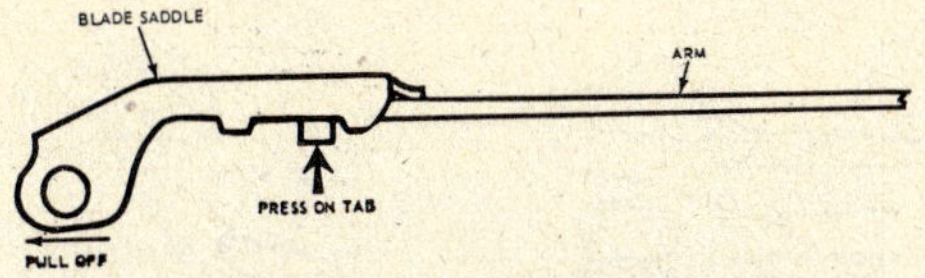

Fig. 9 Anco bayonet type blade

4. Remove lighter socket and bracket from right end of pad support.
5. Remove five pad support-to-instrument panel screws and three lower left panel-to-instrument panel screws and then remove pad support and lower panel as an assembly.
6. From behind cluster, disconnect all electrical connections to cluster and speedometer cable.
7. Remove the cluster assembly from panel (6 screws).
8. Reverse procedure to install.

Fuel Gauge & Speedometer

After removal of cluster as described above, the speedometer and fuel gauge can be removed from the cluster.

W/S WIPER BLADES

NOTE: Trico and Anco blades are used and both come in two types Bayonet type and Side Saddle Pin type.

Bayonet Type: To remove the *Trico* type press down on the arm to unlock the top stud. Depress the tab on the saddle, Fig. 8, and pull the blade from the arm. To remove the *Anco* type press inward on the tab, Fig. 9, and pull the blade from the arm. To install a new blade assembly slide the blade saddle over the end of the wiper arm so that the locking stud snaps into place.

Side Saddle Pin Type: To remove the pin type (Trico or Anco) insert an appropiate tool into the spring release opening of the blade saddle, depress the spring clip and pull the blade from the arm, Fig. 10. To install, push the blade saddle on to the pin so that the spring clip engages the pin. To replace the rubber element in a *Trico* blade squeeze the latch lock release and pull the element out of the lever jaws, Fig. 11, Remove *Anco* element by depressing the latch pin, Fig. 11, and sliding the element out of the yoke jaws. To install insert the element through the yoke or lever jaws. Be sure the element is engaged at all points.

W/S WIPER ARMS

Swing the arm and blade assembly away from the windshield and insert a 3/32" inch pin through the pin hole as shown in Fig. 12. Swinging the assembly away from the windshield will release the spring loaded attaching clip in the arm from the pivot shaft. Inserting the pin will hold it in the released position. The arm can now be pulled off the pivot shaft. *Do not* pry off with a screw driver. Leave the pin in the arm until after installation. A new replacement arm comes with a pin already installed to hold it in released position. Make sure the pivot shaft is in park position when installing new assembly. Push the arm onto the pivot shaft. Lock the arm to the pivot shaft by removing the pin.

W/S WIPER MOTOR

1971-76 Ford & Mercury

1. Disconnect battery ground cable.
2. Remove wiper arm and blade assemblies from pivot shafts.
3. Remove left cowl screen (four screws) for access.
4. Disconnect linkage drive arm from the motor output crankpin by removing retaining clip.
5. From engine side of dash, remove two wire connectors from motor.
6. Remove three bolts that retain motor to dash and remove motor. If output arm catches on dash during removal, handturn the arm clockwise so it will clear opening in the dash. Before installing motor, be sure output arm is in park position.

1969-70 Ford & Mercury

1. Remove wiper arm and blade assemblies from pivot shafts.
2. Remove cowl top grille (10 screws).
3. Disconnect linkage drive arm from motor output arm crankpin by removing clip.
4. Disconnect wire push-on connectors from motor.
5. From engine side of dash, remove bolts retaining motor to dash and remove motor. If output arm catches on dash during removal, hand turn the arm clockwise so it will clear opening in dash.
6. When installing motor, align output arm with opening in dash and turn the arm clockwise as necessary.

W/S WIPER SWITCH

1973-76 All

1. Disconnect battery ground cable.
2. Remove wiper and headlight switch knobs.
3. Remove headlight switch bezel, trim panel retaining screw and trim panel.
4. Remove wiper switch retaining screws, pull switch rearward, disconnect connector and remove switch.

1972 All

1. Disconnect battery ground cable.
2. Remove instrument panel pad.
3. On Mercury models, remove nut and bezel.

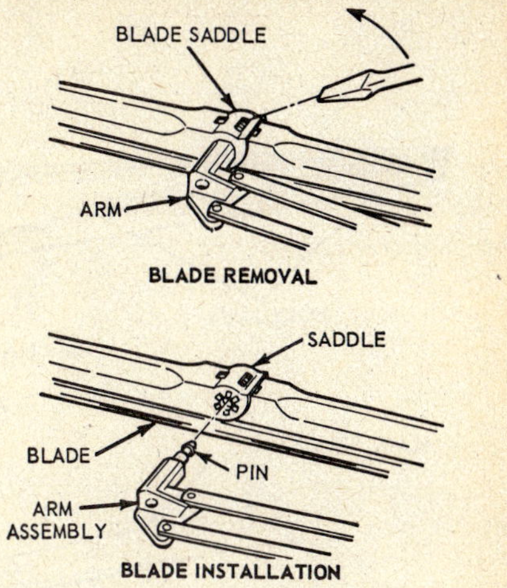

Fig. 10 Pin type blade

4. Remove three mounting plate retaining screws and remove switch and mounting plate.

1971 Ford & Mercury

1. Disconnect battery ground cable.
2. Remove instrument panel pad and instrument cluster.
3. Remove three switch attaching screws and remove switch.

1970 Ford

1. Disconnect battery ground cable and remove two piece cover from steering column (2 screws).
2. To allow removal of cluster trim cover, remove radio, wiper, washer, interval, heater and defogger switch knobs and lighter element.
3. Remove screw that retains the PRND21 dial cable to column and loosen set screw that retains the cable pin in the shaft housing.
4. Remove retaining screws and cluster trim cover assembly.
5. Remove the two screws that retain switch to cluster, lower the switch and disconnect multiple connector and hoses.

1969 Ford

1. To allow removal of cluster trim cover, remove the radio, wiper, washer, interval and heater switch knobs and the lighter element.
2. Remove retaining screws and cluster trim cover assembly.
3. Remove screws that retain the switch to the cluster, lower switch and disconnect multiple connector and vacuum hose from switch.

1969-70 Mercury

1. Remove wiper knob and bezel from switch shaft.
2. Remove nut that retains switch to bracket.
3. Lower switch from behind panel and

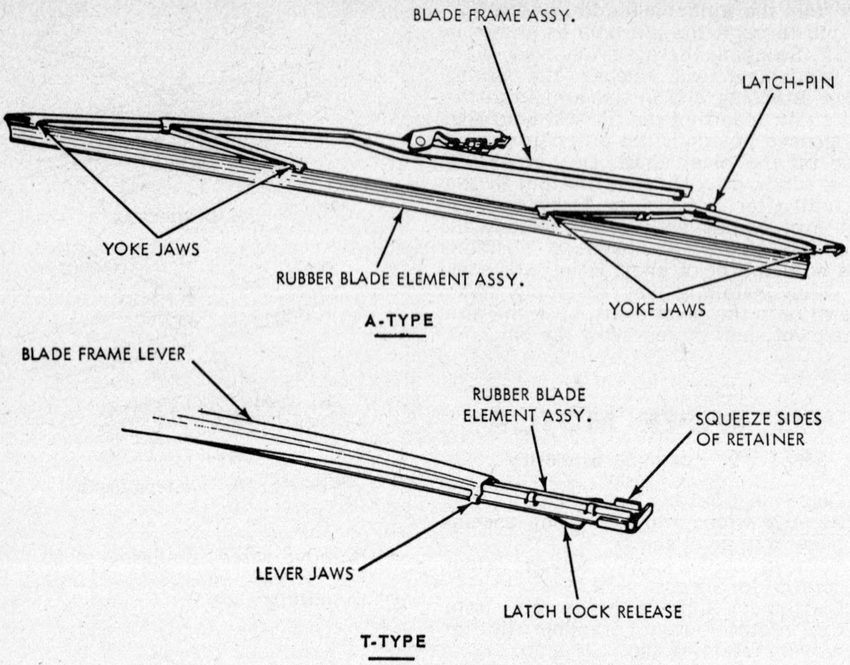

Fig. 11 Wiper blade element

disconnect multiple connector and vacuum hoses from the switch.

W/S WIPER TRANSMISSION

1971-76 Ford & Mercury

1. Disconnect battery and remove wiper arm and blade assemblies from pivot shafts.
2. Remove cowl screens for access to linkage.
3. Disconnect the left linkage arm from the drive arm by removing the clip.
4. Remove the three bolts retaining the left pivot shaft assembly to the cowl.
5. Remove the left arm and pivot shaft assembly through the cowl opening.
6. Disconnect linkage drive arm from motor crankpin by removing the clip.
7. Remove three bolts that connect drive arm pivot shaft assembly to the cowl and remove the pivot shaft drive arm and right arm as an assembly.

1969-70 Ford & Mercury

1. Remove wiper arm and blades from pivot shafts.
2. Remove cowl top grille.
3. The right pivot shaft and link on Mercury has to be removed before the left assembly or drive arm can be removed. Remove three retaining screws at right pivot shaft and disconnect the right link from the plate on the inner side of the dash panel by removing the clip. Lift the pivot assembly out of cowl opening.
4. Remove three retaining screws at the left pivot shaft and disconnect the left link from the plate and lift out of cowl opening.
5. Reverse the procedure to install.

RADIO, REPLACE

NOTE: When installing radio, be sure to adjust antenna trimmer for peak performance.

1973-76 Ford & Mercury

1. Disconnect battery ground cable.
2. Remove radio knobs and instrument panel bezel.
3. Remove mounting plate retaining screws and pull radio disengaging it from rear bracket.
4. Disconnect antenna and electrical wires and remove radio.

1971-72 Ford & Mercury

1. Disconnect battery ground cable.
2. Remove radio and fader knobs.
3. Remove radio bezel.
4. Remove upper and lower rear radio support bolts and brackets.
5. Disconnect power, antenna and speaker.
6. Remove two nuts retaining radio to instrument panel and remove radio.

1969-70 Ford

1. Remove the trim cover from instrument panel.
2. Remove three nuts retaining front of radio to cluster.
3. Remove screw retaining the rear support to the bottom of the radio.
4. Pull radio from cluster and disconnect wires from chassis.

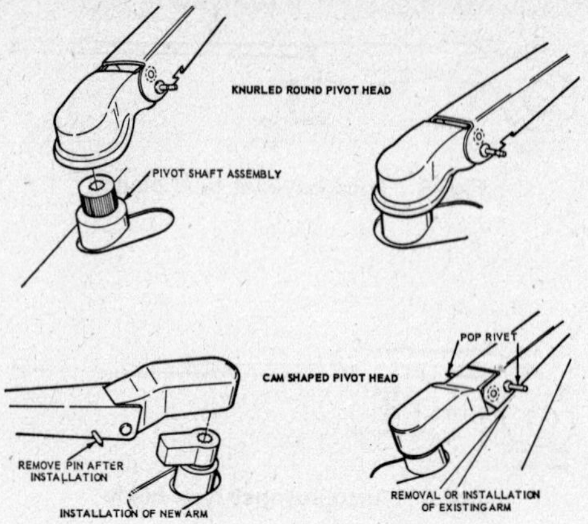

Fig. 12 Wiper arms

1969-70 Mercury

1. Disconnect battery ground cable.
2. Remove control knobs from radio.
3. Disconnect all wires from radio and remove nut attaching the rear support to the back of the radio.
4. Remove nut attaching the front edge of radio to instrument panel and remove radio.
5. Remove nut from radio control shaft.

HEATER CORE REMOVAL

1973-76 Ford & Mercury with Air Cond.

1. Drain cooling system.
2. Disconnect heater hoses from core tubes.
3. Remove heater core cover plate and gasket.
4. Pull heater core and mounting gasket up out of case.
5. Remove core mounting gasket. Remove heater core.

1969-76 Ford & Mercury less Air Cond.

1. Drain cooling system and disconnect heater hoses from the core.
2. Remove the core cover and gasket and remove the heater core.

1969-72 Ford & Mercury with Air Cond.

1. Drain cooling system.
2. Remove carburetor air cleaner.
3. Remove vacuum manifold from dash panel.
4. Disconnect heater hoses, remove core cover and remove the core.

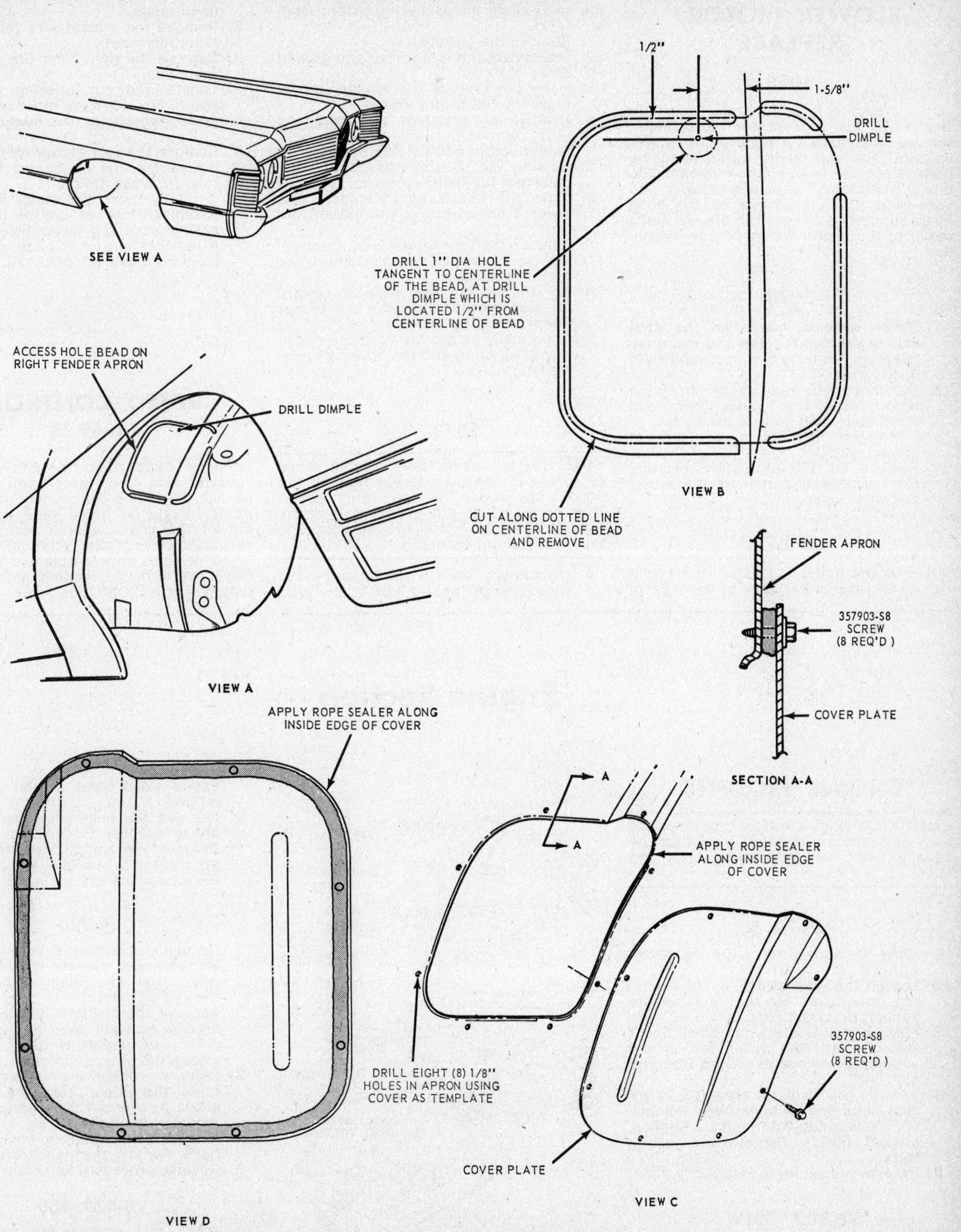

Fig. 13 Blower motor access. 1973-76 Ford & Mercury

BLOWER MOTOR, REPLACE

1969

The motor is located under the right fender. To gain access for replacement, remove the battery, right front wheel, vacuum tank and fender apron attaching bolts. Move the fender apron inboard. Remove the blower attaching screws and vent hose, and pry upward on the hood hinge support to remove the blower. Apply sealer to the blower flange before installation.

1970

1. Scribe location marks on the right side of the hood, hinge and mounting panel and remove the hood and right hood hinge.
2. Disconnect the right fender from the apron, cowl side panel, front end sheet metal and remove the fender.
3. Disconnect the blower motor lead and ground wires.
4. Remove the blower motor mounting plate screws and remove the motor and wheel assembly.

1971-72

The blower motor is located in the right side of the evaporator case to the right of the hood hinge under the right front fender.

1. Remove the battery.
2. Disconnect the motor lead and ground wires.
3. Raise the front of the vehicle and remove the right front wheel.
4. Remove the vacuum tank retaining nuts, remove the tank from the right fender apron, and set it to one side.
5. Remove the 15 fender apron retaining bolts and the fender-to-apron brace.
6. Move the fender apron inboard and lower it for access to the blower motor.
7. Remove the four blower motor mounting plate screws, and disconnect the motor vent hose.
8. Pry the hood hinge support upward for clearance and remove the blower motor and wheel assembly.
9. Apply body sealer between the motor mounting plate and the housing on installation.

1973-76

NOTE: The blower motor is located in the right side of the case to the right of the hood hinge under the right front fender. It is necessary to cut an opening in the right front fender apron to gain access to the blower motor assembly.

1. Disconnect the blower motor lead wire (orange) at the rear of the right hood hinge.
2. Remove the ground wire (black) from the upper cowl.
3. Remove the right front tire and wheel assembly.
4. Locate and cut opening in fender apron, Fig. 13. Care must be taken to avoid damage to the heater case by drill push-through or over-travel.
5. Remove the four blower motor mounting screws and disconnect the cooler tube from the motor.
6. Carefully move the motor and wheel assembly forward out of the heater case through the access hole.
7. A replacement cover plate is available from Ford part no. #18A475.

SPEED CONTROLS

1969-76

Adjust bead chain to obtain .06-.25" actuator arm free travel when engine is at hot idle. The adjustment should be made to take as much slack as possible out of the chain without restricting the carburetor lever from returning to idle.

On vehicles with a solenoid anti-diesel valve, perform adjustment with ignition switch in the "ON" position.

Engine Section

ENGINE MOUNTS

CAUTION: Whenever self-locking mounting bolts and nuts are removed, they must be replaced with new self-locking bolts and nuts.

1969-72 Six

1. Raise the hood. Then raise the vehicle with a hoist.
2. Loosen the nuts attaching the insulator assemblies to the intermediate support brackets, Fig. 1.
3. Place a jack and a wood block under the oil pan and raise the engine enough to remove its weight from the supports.
4. Remove the insulator assembly to engine bolts and lock washers. Remove the insulator to intermediate support bracket nut(s). Remove the insulator(s).
5. Reverse procedure to install.

V8-302, 351W

1. Remove the nut and through bolt attaching the insulator to the support bracket, Fig. 2.
2. Raise the engine slightly with a jack and a wood block placed under the oil pan.
3. Remove the engine insulator assembly to cylinder block attaching bolts. Remove the engine insulator assembly and the heat shield, if so equipped.
4. Reverse procedure to install.

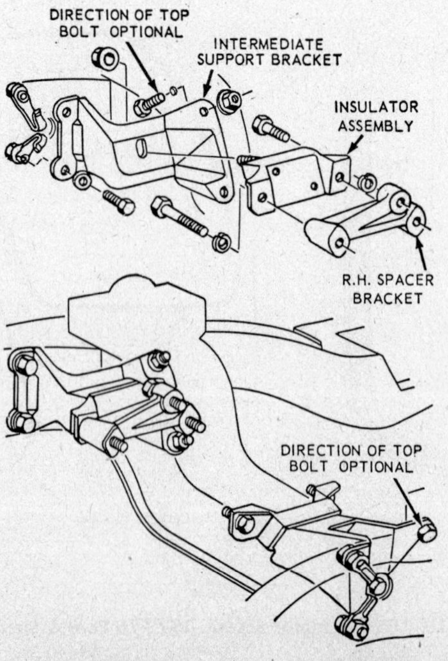

Fig. 1 Engine mount. 1969-72 Six

V8-390

1. On vehicles equipped with automatic transmission, remove the oil cooler tubes from the retaining bracket on the cylinder block.
2. Remove the insulator to intermediate support bracket lock nut, Fig. 3. If only one support is being removed, loosen the other support.
3. Using a jack and a wood block placed under the oil pan, raise the engine to allow just enough clearance for removal of the insulator(s).
4. Remove the insulator to engine locking bolts. Remove the insulator(s).
5. Reverse procedure to install.

V8-429, 460

1. Block rear wheels and set parking brake. Raise front of vehicle with floor jack and install safety stands.
2. Position jack under the front area of the oil pan. Place a wood block be-

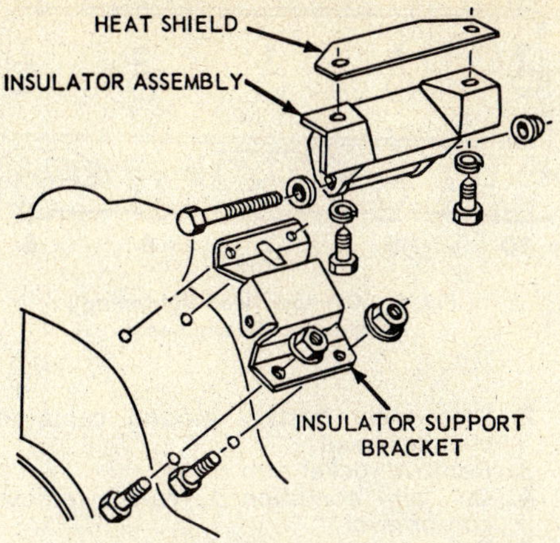

Fig. 2 Engine mount. V8-302, 351W

tween the jack and oil pan. Raise the jack just enough to support the engine.

3. Remove the nut through bolt that attaches the front support insulator to the lower support bracket, Figs. 4, 5.
4. Remove the bolts attaching the support insulator and heat shield to the cylinder block. Replace the insulator on one side before proceding to the other insulator.
5. Reverse procedure to install.

V8-351C, 351M, 400

1. Remove the fan shroud attaching bolts. Remove the transmission oil cooler lines from the retaining bracket on the block.
2. Remove the through bolt and lock nut attaching the insulator support bracket, Fig. 6. Remove the bolt and nut on the opposite mount to prevent distortion of the insulator.
3. Raise the engine slightly with a jack

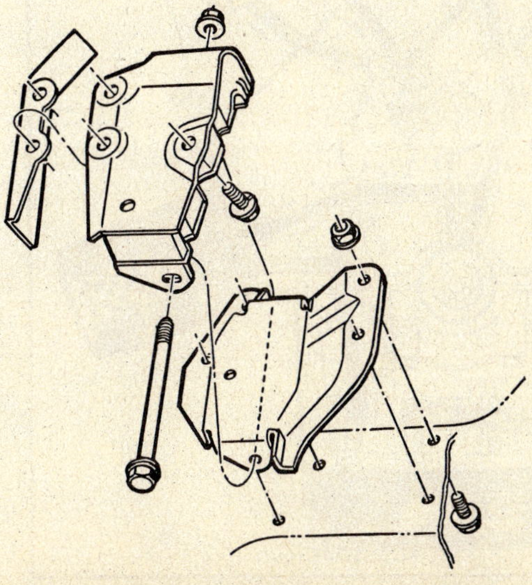

Fig. 5 Engine mount. 1972-76 V8-429, 460

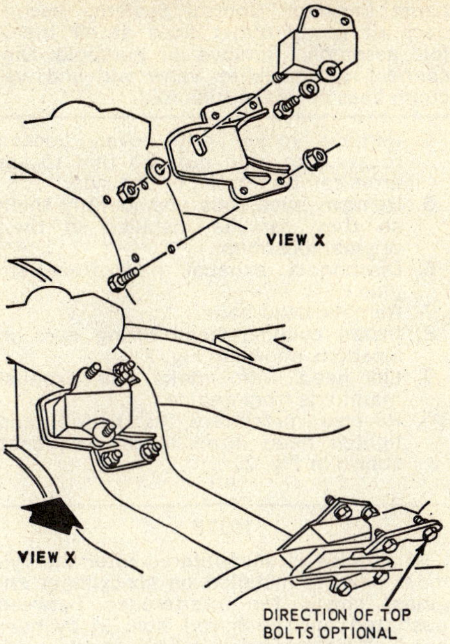

Fig. 3 Engine mount. V8-390

and a wood block placed under the oil pan.

4. Remove the engine insulator assembly to cylinder block attaching bolts and lockwashers.
5. Remove the engine insulator assembly and heat shield, if equipped.
6. Reverse procedure to install.

ENGINE, REPLACE

NOTE: Because of engine compartment tolerances, the engine should not be removed and installed with the transmission attached.

1. Disconnect battery ground cable, drain cooling system and crankcase and remove hood and air cleaner assembly.
2. Disconnect or remove all Thermactor components that may interfere with engine removal.
3. Disconnect hoses and oil cooler lines from radiator, then remove radiator, fan shroud and fan.
4. Remove all drive belts.
5. Disconnect power steering pump and alternator and position units out of the way.
6. If equipped with air conditioning, isolate and remove compressor.
7. Disconnect all hoses, lines and wiring from engine. Make certain to remove ground wires from block and right cylinder head.
8. Disconnect fuel line from pump and plug line.
9. Disconnect accelerator cable or linkage, then disconnect downshift linkage (if used).
10. Raise and properly support vehicle, then disconnect exhaust system from engine and remove starter. On six

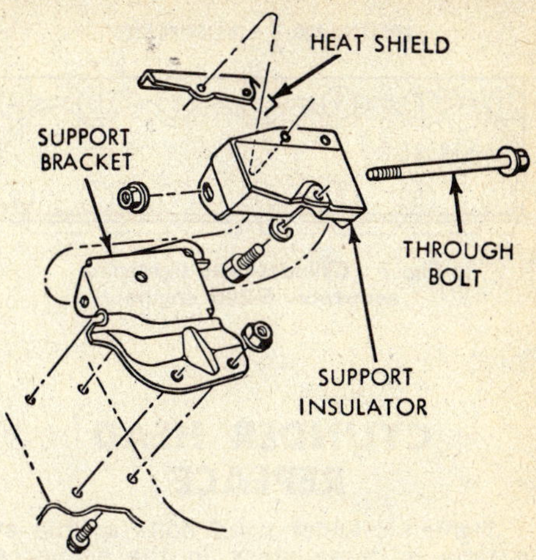

Fig. 4 Engine mount. 1969-71 V8-429

cylinder engines, remove transmission oil filler tube bracket.

11. Remove engine front support through bolts.
12. If equipped with automatic transmission, remove converter cover, converter to flywheel bolts, and downshift rod.
13. If equipped with manual transmission, remove clutch linkage from engine block.
14. Remove the four lower engine to clutch housing or converter housing bolts, then lower vehicle and remove the two upper engine to clutch housing or converter housing bolts.
15. Position jack under transmission, then using a suitable hoist, carefully remove engine from vehicle.
16. Reverse procedure to install.

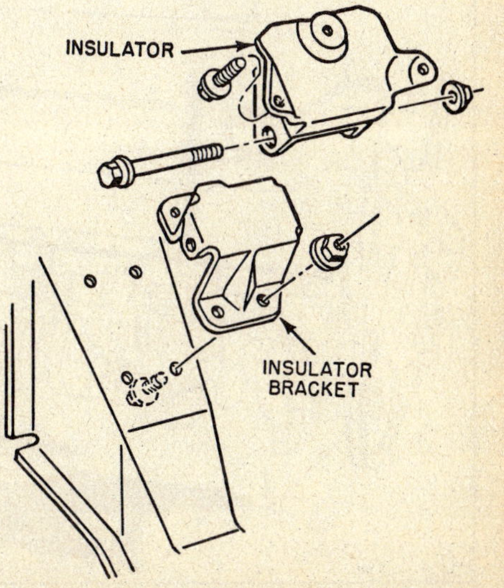

Fig. 6 Engine mount. V8-351C, 351M, 400

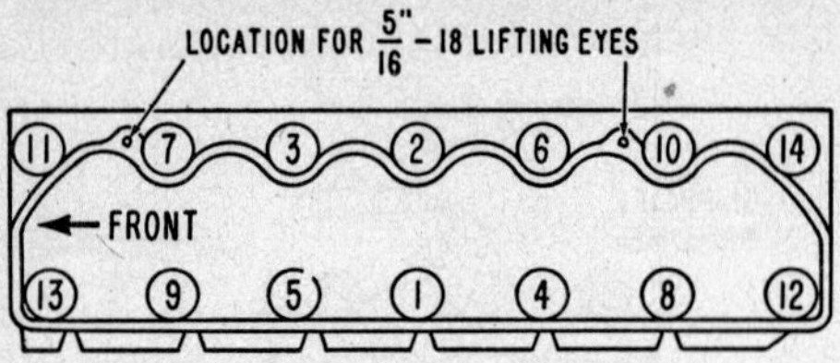

Fig. 7 Cylinder head tightening sequence. 6-240 engine

CYLINDER HEAD REPLACE

Tighten cylinder head bolts a little at a time in three steps in the sequence shown in the illustrations. Final tightening should be to the torque specifications listed in the *Engine Tightening* table. After bolts have been tightened to specifications, *they should not be disturbed.*

In instances where cylinder head gasket leakage is hard to control, aluminum paint can be applied to the gasket as a sealer.

Spray one coat of the aluminum paint on both sides of the gasket and allow the paint to dry. Then spray a second coat on both sides and, while the paint is still wet, install the gasket. Torque the head and manifold bolts to specifications to complete the job.

1969-72 6-240 Engine

1. Drain cooling system and remove air cleaner.
2. Disconnect hoses, tubing, wires and linkage attached to head.
3. Grasp crankcase ventilation regulator valve and pull it from rocker arm cover. Disconnect crankcase vent hose from inlet tube on intake manifold and remove hose and valve.

NOTE: If equipped with Thermactor Exhaust Emission Control System, disconnect air pump outlet hose at air manifold assembly. Remove air manifold. Disconnect anti-backfire valve air and vacuum lines at intake manifold.

4. Remove rocker arm cover. Loosen rocker arm stud nuts so that rocker arms can be rotated to one side.
5. Remove push rods and identify them so they can be installed in their original locations.
6. Disconnect exhaust pipe from engine.
7. Remove head bolts.
8. Install cylinder head lifting eyes in locations shown in Fig. 7.
9. Lift head with intake and exhaust manifolds from engine.
10. Reverse procedure to install and tighten head down in the sequence shown in Fig. 7.

NOTE

If a noise is encountered after the cylinder head is installed on six-cylinder engines, check for interference between push rod and push rod hole in cylinder head. If interference exists check push rod for straightness. If push rod is straight, then the cylinder head has not been aligned properly during its installation. The use of cylinder head guide pins (one at each end) on cylinder head installation will insure correct head alignment.

1969-76 V8-302, 351, 400

1. Remove intake manifold and carburetor as an assembly.

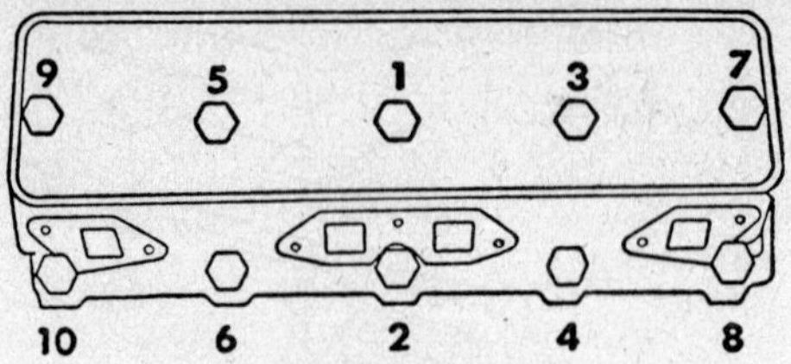

Fig. 8 Cylinder head tightening sequence. V8 engines

2. Disconnect battery ground cable at cylinder head.
3. Remove rocker arm cover.
4. On air conditioned cars, remove compressor.
5. On car with power steering, disconnect pump bracket from left cylinder head and remove drive belt. Wire power steering pump out of the way and in position that will prevent oil from draining out.

NOTE: If left cylinder head is being removed on an engine equipped with Thermactor Exhaust Emission Control System, disconnect hose from air manifold on left head. If a right head is to be removed, remove air pump and bracket and disconnect hose on right head.

6. Remove generator or alternator.
7. Disconnect exhaust manifold at exhaust pipes.
8. Loosen rocker arm stud nuts or bolts so that rocker arms can be rotated to the side.

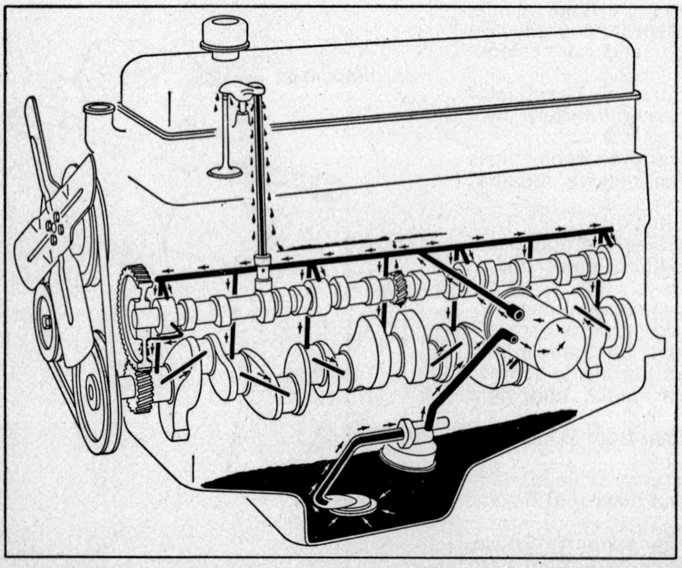

Engine oiling system. 6-240 engine

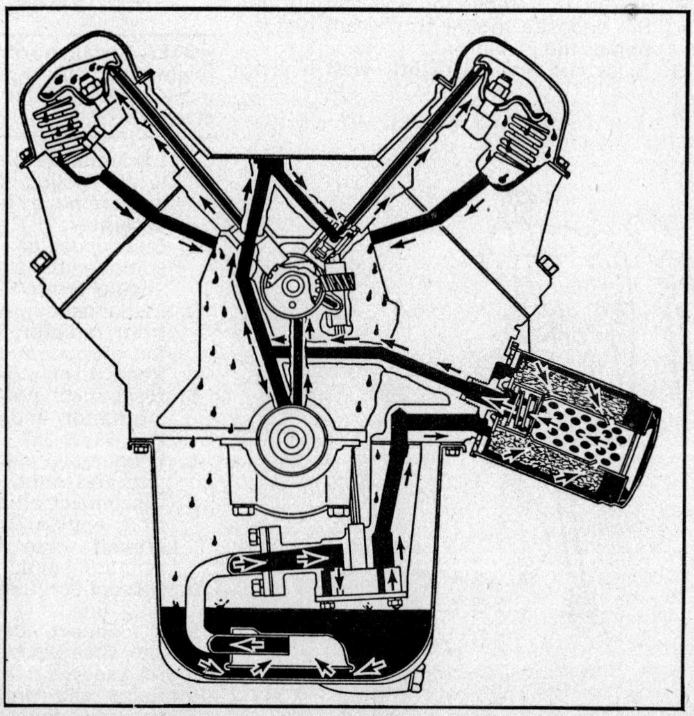

Engine oiling system. V8-302, 351, 400

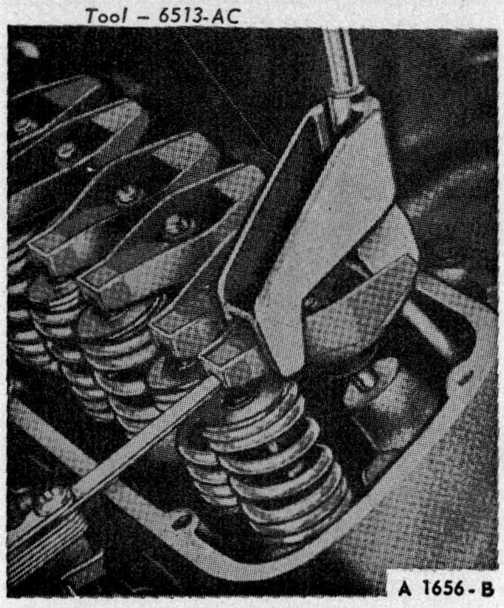

Fig. 9 Checking valve clearance on models with hydraulic lifters

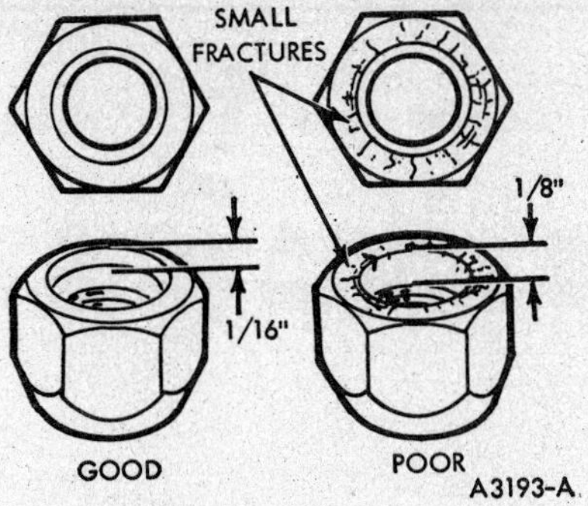

Fig. 10 Inspection of rocker arm stud nut. V8-302, 351W

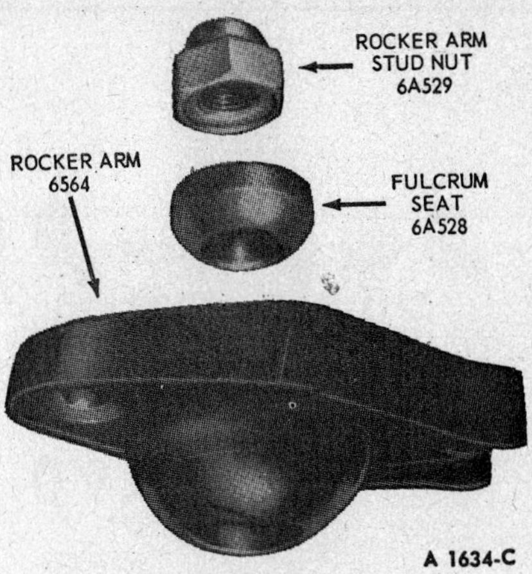

Fig. 11 Valve rocker arm parts. V8-302, 351W

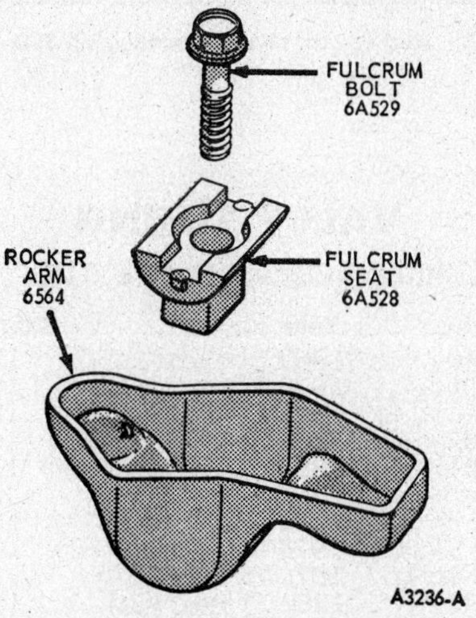

Fig. 12 Rocker arm and related parts. V8-351C, 351M & 1971-76 V8-400, 429, 460

9. Remove push rods, keeping them in sequence so they may be returned to their original locations.
10. Unfasten and remove cylinder head.
11. Reverse removal procedure to install the head. Tighten cylinder head down in the sequence shown in Fig. 8.

1969-76 V8-390, 429, 460

1. Remove intake manifold and carburetor as an assembly.
2. Disconnect muffler inlet pipe at exhaust manifold.
3. Loosen air conditioner compressor belt if so equipped.
4. Loosen alternator retaining bolts and remove bolt retaining alternator bracket to right head.
5. If air conditioned, isolate compressor at service valves and hoses from compressor. Remove nuts retaining compressor bracket to water pump. Remove bolts retaining compressor to upper mounting bracket and lay compressor out of way. Remove compressor upper bracket from head.
6. If not air conditioned, remove bolts retaining power steering reservoir bracket to left head and position reservoir out of way.
7. Remove rocker arm covers and rocker arms. Remove push rods in sequence so they can be installed in their positions.
8. Remove head retaining bolts and lift head with exhaust manifold.

NOTE: If necessary to break gasket

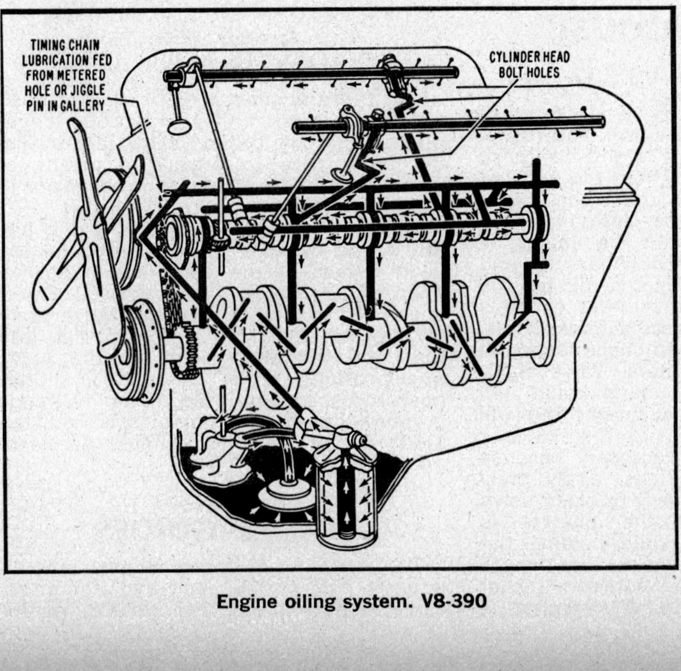

Engine oiling system. V8-390

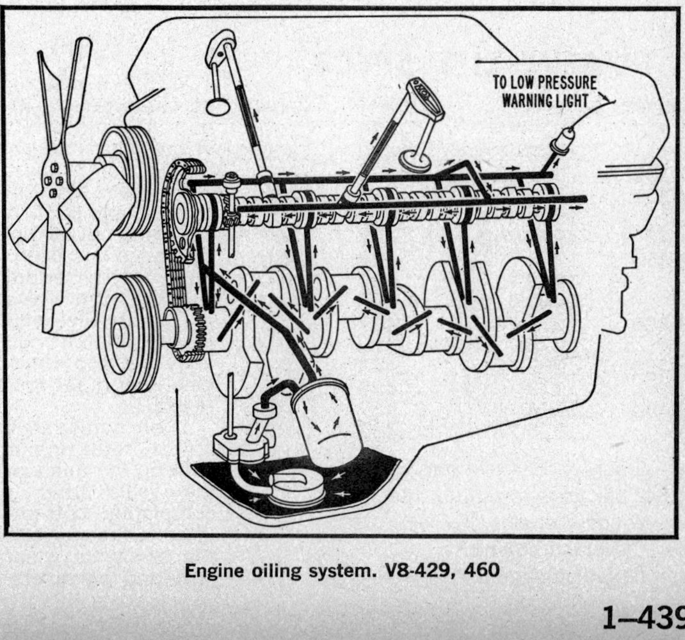

Engine oiling system. V8-429, 460

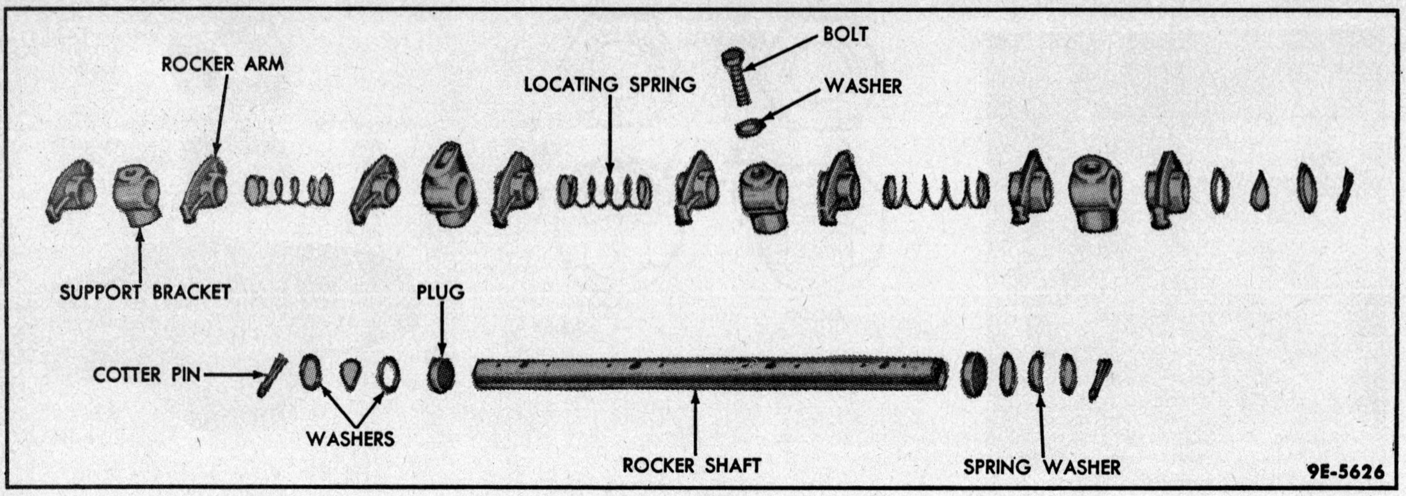

Fig. 13 Rocker arm shaft assembly. V8-390 engine

seal, pry at forward corners of cylinder heads against casting bosses provided on cylinder block. Avoid damaging machined surfaces on head and gasket.

9. Install cylinder heads in reverse order of removal and torque bolts in sequence shown in Fig. 8.

VALVE ARRANGEMENT

Front to Rear

6-240	E-I-E-I-E-I-E-I-E-I-E-I
302, 351, 429 Right	I-E-I-E-I-E-I-E
302, 351, 429 Left	E-I-E-I-E-I-E-I
361, 390	E-I-E-I-I-E-I-E
400, 460 Right	I-E-I-E-I-E-I-E
400, 460 Left	E-I-E-I-E-I-E-I

VALVE LIFT SPECS.

Engine	Year	Intake	Exhaust
6-240	1969-71	.376	.400
	1972	.400	.400
8-302	1969-72	.368	.381
8-351	1969-75[5]	.418	.448
	1970-74[6]	.407	.407
	1975-76[7]	.407	.407
8-390	1969-71[1]	.427	.430
	1969[2]	.481	.490
	1969[4]	.440	.440
8-400	1971-74	.427	.433
	1975	.428	.432
8-429	1969-72	.443	.486
	1973	.437	.480
8-460	1974	.437	.480
	1975-76	.437	.451

[1]—2 bar. carb. [2]—4 bar. carb.
[4]—2 bar. carb.-premium fuel.
[5]—Windsor engine.
[6]—Cleveland engine.
[7]—Modified engine.

VALVE TIMING

Intake Opens Before TDC

Engine	Year	Degrees
6-240	1969	12
	1970-71	10
	1972	18
8-302	1969-72	16
8-351	1969-72[1]	11
	1971-73[2]	12
	1974-75[1]	15
	1974[2]	19½
	1975-76[3]	19½
8-390	1969-71 Reg. Fuel	13
	1969 Prem. Fuel	16
8-400	1971-76	17
8-429	1969-73	16
8-460	1974-76	8

[1]—Windsor engine.
[2]—Cleveland engine.
[3]—Modified engine.

VALVES, ADJUST

6-240, 1969 V8-302

With the piston at top dead center of its compression stroke, loosen the rocker arm stud nut until there is end clearance in the push rod. Then tighten the nut just to the point where all end clearance is eliminated. This may be determined by moving the push rod with the fingers as the stud nut is tightened. When the end clearance has been eliminated, tighten the stud nut the additional number of turns listed in the *Valve Specifications Table*.

Operate the engine and check for rough engine idle or noisy lifters. Valve clearance set too tight will cause rough engine idle; if set too loose, noisy lifters will result.

If an adjustment is necessary because of the foregoing conditions, apply pressure on the push rod slowly to bleed down the valve lifter until the plunger is completely bottomed. While holding the lifter in the fully collapsed position, check the available clearance between rocker arm and valve stem tip. If clearance is not within specifications, turn the rocker arm stud nut clockwise to decrease the clearance and counterclockwise to increase clearance.

V8-302, 351W & 1969-429

These engines use a new positive stop rocker arm stud. If clearance between valve stem and rocker arm, with lifter collapsed, as in Fig. 9, is not as shown in *Valve Specifications* table a .060″ shorter or a .060″ longer push rod is available to compensate for dimensional changes in valve train. If clearance is less than specified, install an undersized push rod. If clearance is greater install an oversize push rod.

Valve clearance should be correct when stud nut is tightened until it contacts the stop and torqued to 18-22 ft-lbs. Inspect stud nuts to be sure they are in acceptable condition for re-use, Fig. 10.

V8-351C, 351M, 390, 400, 429 & 460 With Non-Adjustable Hydraulic Lifters

For these engines a .060″ shorter push rod (color coded white) or a .060″ longer push rod (color coded yellow) are available for service to provide a means of compensating for dimensional changes in the valve mechanism.

To check the clearance, bring the piston of the cylinder being checked on top dead center of the compression stroke. Then with hydraulic lifter collapsed, Fig. 9 check the clearance between valve stem and rocker arm. If the clearance is less than the minimum listed in the *Valve Specifications* table, the .060″ shorter push rod should be used. If the clearance is more than the maximum specified, the .060″ longer push rod should be used.

VALVE GUIDES

Valve guides in these engines are an integral part of the head and, therefore, cannot be removed. For service, guides

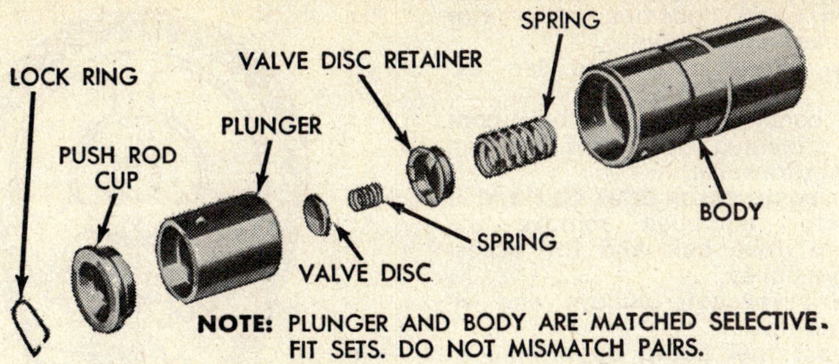

Fig. 14 Hydraulic valve lifter disassembled (typical)

Fig. 15 Removing valve lifter. 6-240 engine

can be reamed oversize to accommodate one of three service valves with oversize stems (.003", .015" and .030").

Check the valve stem clearance of each valve (after cleaning) in its respective valve guide. If the clearance exceeds the service limits of .004" on the intake or .005" on the exhaust, ream the valve guides to accommodate the next oversize diameter valve.

ROCKER ARMS & SHAFTS

V8-390

To disassemble the rocker arms, remove cotter pins from each end of the shaft and remove the flat washers and spring washers. Slide rocker arms, springs and supports off shaft, being sure to identify all parts so they can be assembled in the same position.

If it is necessary to remove the plugs from each end of the shaft, drill or pierce one plug, then insert a steel rod through the plug and knock out the plug on the opposite end. Working from the open end, knock out the remaining plug.

Assemble the rocker arms and related parts as indicated by Fig. 13.

ROCKER ARM STUD

6-240, 1969 V8-302

If necessary to replace a rocker arm stud, a kit is available which contains a stud remover, a stud installer, and two reamers, one .006" and the other 015".

Rocker arm studs that are broken or have damaged threads may be replaced with standard studs. Loose studs in the head may be replaced with .006", .010" or .015" oversize studs which are available for service. *The standard studs have no identification marks, whereas the .006" oversize stud has two grooves around the pilot end of the stud. The .015" oversize stud has a step produced by the increased diameter of the stud approximately 1$^5/_{32}$" from the pilot end.*

When going from a standard size stud to a .010 or .015" oversize stud, always use a ,006" reamer before finish reaming with a .010 or .015" reamer.

If a stud is broken off flush with the stud boss, use an easy-out to remove the broken stud, following the instructions of the tool manufacturer.

V8-302 & V8-351W

A new type positive stop rocker arm stud and nut eliminates the need for adjusting valve lash, Fig. 11.

Installation

1. Position the piston of the cylinder to be worked on at TDC compression stroke.
2. Locate stud properly with tool T69P-6049D. Make sure tool bottoms on the head.
3. Lubricate rocker arm components and place rocker arm and fulcrum on the stud.
4. Thread nut onto the stud until it contacts the shoulder then tighten nut to 18-22 ft-lbs.

V8-351C, 351M & 1971-76 V8-400, 429, 460

The rocker arm is supported by a fulcrum bolt which fits through the fulcrum seat and threads into the cylinder head. To disassemble, remove the bolt, oil deflector, fulcrum seat and rocker arm, Fig. 12.

1969-70 V8-429

The rocker arm support studs are threaded into the cylinder head and can be replaced as follows:

1. Position the piston of the cylinder being worked on the TDC compression stroke.
2. Thread the rocker arm stud into the head until the shoulder contacts the head. Tighten to the torque specified in the *Engine Tightening Table.*
3. Lubricate rocker arm components and position the arm and the fulcrum on the stud.
4. Thread the nut until it contacts the shoulder, then tighten to 18-22 ft. lbs.

HYDRAULIC VALVE LIFTERS

The internal parts of each hydraulic valve lifter assembly are a matched set. If these are mixed, improper valve operation may result. Therefore, disassemble, inspect and test each assembly separately to prevent mixing the parts.

Fig. 14 illustrates the type of hydraulic lifter used. See the *Trouble Shooting Chapter* under the heading *Engine Noises* for causes of hydraulic valve lifter noise.

Hydraulic Lifters, Replace

6-240 Engine

1. Remove valve rocker arm cover.
2. Remove valve push rod cover.
3. Loosen rocker arm stud nuts until rocker arms can be disengaged from push rods.
4. Remove push rods, keeping them in a rack so they may be installed in their original location.
5. Remove valve lifters, using the tool shown in Fig. 15 (or equivalent). Place lifters in a rack so they may be installed in their original location.

V8-302, 351, 400, 429, 460

1. Remove intake manifold and related parts.
2. Remove rocker arm covers.
3. Loosen rocker arm stud nuts or bolts and rotate rocker arms to the side.
4. Lift out push rods, keeping them in sequence in a rack so they may be installed in their original location.
5. Using a magnet rod, remove valve lifters and place them in sequence in a rack so they may be installed in their original location.

V8-390

1. Remove intake manifold.
2. Remove rocker arms and shafts.
3. Remove push rods, keeping them in a rack in sequence so they may be installed in their original location.
4. Remove valve lifters with a magnet rod and place them in a rack in sequence so they may be installed in their original location.
5. Reverse procedure to install.

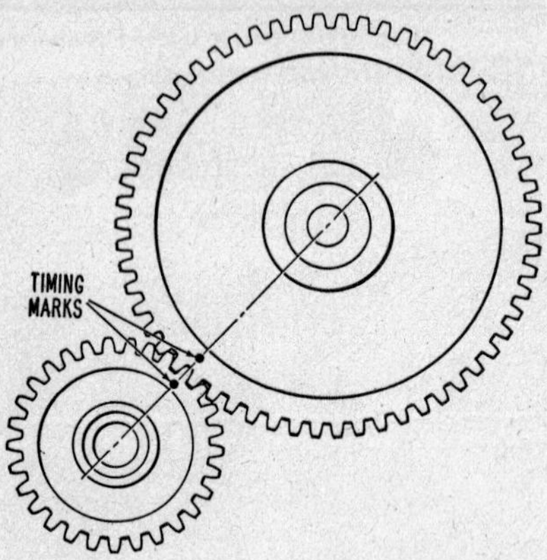

Fig. 16 Timing marks aligned for correct valve timing. 6-240 engine

TIMING CASE COVER

NOTE: To replace the seal in the timing gear cover, it is necessary to remove the cover as outlined below.

1969-76 V8-302, 351, 400

1. Drain cooling system and oil pan.
2. Disconnect lower radiator hose from water pump.
3. Disconnect heater hose from water pump and slide water pump bypass hose clamp toward pump.
4. Unfasten and position alternator and bracket out of way.
5. If equipped with power steering or air conditioning, remove the drive belts.
6. Remove the fan, spacer, pulley and drive belt.
7. Remove crankshaft pulley and vibration damper.
8. Disconnect fuel pump outlet line from pump and remove pump retaining bolts and lay pump to one side with flex line attached.
9. Remove oil dipstick and the oil pan to front cover attaching bolts.
10. Unfasten and remove the cylinder front cover and water pump as an assembly.
11. Reverse the foregoing to install.

1969-71 V8-390

1. Drain cooling system and oil pan.
2. Disconnect battery ground cable.
3. Disconnect transmission cooler lines from radiator if so equipped and disconnect water hoses from radiator and remove radiator.
4. Remove heater hose from water pump and remove hose from choke housing clamp.
5. Slide water pump bypass hose clamp toward engine.
6. If equipped with power steering, remove pump bracket bolts and position pump to left side to prevent fluid from draining.
7. If air conditioned, remove compressor mounting bolts and lay compressor out of way.
8. Unfasten and position alternator and bracket out of way.
9. Remove water pump and fan assembly.
10. If air conditioned, unfasten condenser attaching bolts and position condenser forward.
11. Remove compressor drive belt and if Thermactor equipped remove the air pump drive belt and the accessory drive pulley.
12. Remove crankshaft pulleys and vibration damper.
13. Disconnect fuel pump outlet line at pump and remove pump attaching bolts and lay pump to one side with flex line attached.
14. Remove crankshaft sleeve.
15. Unfasten and remove the front cover.
16. Reverse the foregoing to install.

V8-429, 460

1. Drain cooling system and oil pan.
2. Remove bolts attaching fan to water pump and remove screws attaching radiator shroud to radiator.
3. Remove fan assembly and radiator shroud.
4. Disconnect radiator hoses at engine and cooler lines at radiator and remove radiator upper support and radiator assembly.
5. Loosen alternator and remove drive belt with water pump pulley.
6. If air conditioned, loosen idler pulley and remove compressor support.
7. Remove vibration damper from crankshaft.
8. Disconnect power steering lines at pump and unfasten and remove the pump.
9. Loosen bypass hose at pump and disconnect heater return tube at pump.
10. Disconnect and plug fuel inlet line at pump and disconnect fuel outlet line from pump. Unfasten and remove fuel pump.
11. Unfasten and remove cylinder front cover.
12. Reverse procedure to install.

1969-72 6-240

1. Drain cooling system and oil pan.
2. Disconnect automatic transmission oil cooler lines from radiator.
3. Remove radiator.
4. Remove power steering drive belt. Unfasten and position power steering pump and bracket out of the way.
5. Remove fan, spacer, belt and pulley.
6. Remove vibration damper.
7. Remove oil pan, oil pump screen and inlet tube.
8. Unfasten and remove front cover.
9. Reverse procedure to install.

TIMING GEARS

1969-72 6-240 Engine

CAUTION

When the camshaft and crankshaft lose their timing relationship through removal of the timing gears, interference may occur between crankshaft and cam lobes. Therefore, to prevent possible damage to the camshaft lobes, do not rotate the camshaft or crankshaft in the engine without the timing gears installed.

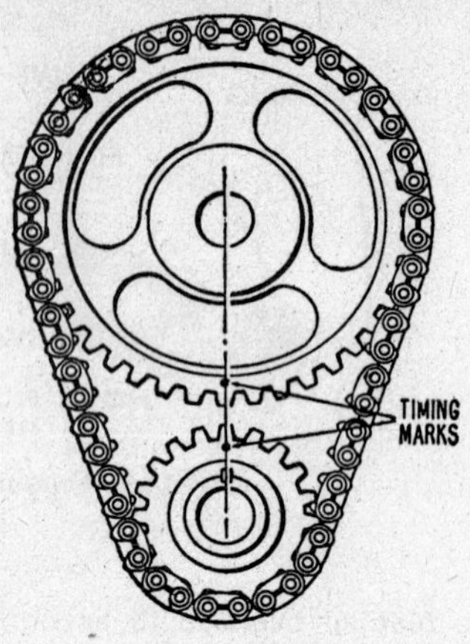

Fig. 17 Timing marks aligned for correct valve timing. V8s

1. To remove gears, remove cylinder front cover and camshaft.
2. Remove oil slinger from crankshaft.
3. Use a puller to remove crankshaft gear.
4. Press gear off camshaft and remove thrust plate, spacer and key.
5. Reverse procedure to install, being sure timing marks are aligned as shown in Fig. 16.

NOTE: Be sure the camshaft gear and spacer are tight against the shoulder on camshaft and that the thrust plate can be moved freely.

TIMING CHAIN

V8-302, 351, 400

After removing the cover as outlined above, remove the crankshaft front oil slinger. Crank the engine until the timing marks are aligned as shown in Fig. 17. Remove crankshaft sprocket retaining bolt and washer. Slide both sprockets and chain forward and remove them as an assembly.

Reverse the order of the foregoing procedure to install the chain and sprockets, being sure the timing marks are aligned.

V8-390, 429, 460

1. To remove the chain, first take off the front cover as outlined previously.
2. Crank engine until timing marks on camshaft sprocket is adjacent to timing mark on crankshaft sprocket, Fig. 17.
3. Remove camshaft sprocket cap screw and fuel pump eccentric.
4. Slide both sprockets and chain forward and remove as an assembly.

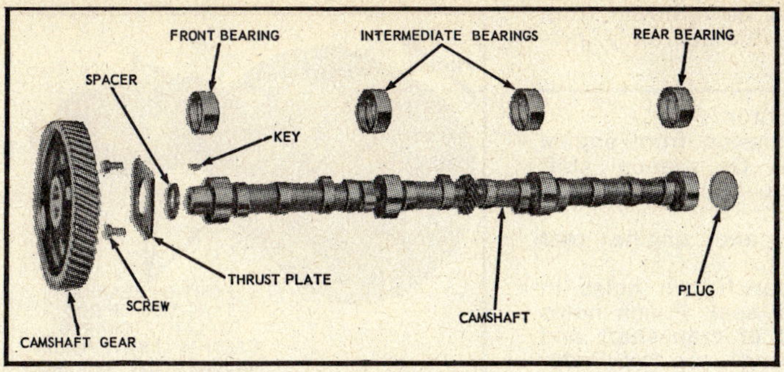

Fig. 18 Camshaft and related parts. 6-240 engine

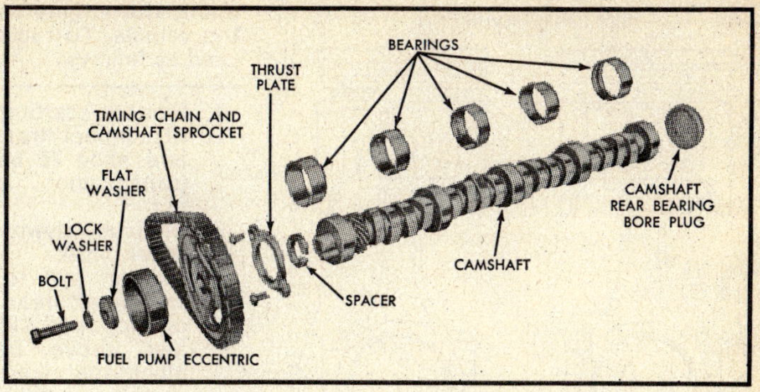

Fig. 19 Camshaft and related parts. V8-302

5. Reverse foregoing procedure to install the chain, being sure to align the timing marks as shown.

CAMSHAFT, REPLACE

1969-72 6-240 Engine

1. To remove camshaft, remove radiator and grille.
2. Remove rocker arm cover. Loosen rocker arm stud nuts and move rocker arms to one side and take out push rods. Place push rods in a rack so they can be installed in their original location.
3. Remove valve push rod cover and take out valve lifters. Place valve lifters in a rack so they may be installed in their original location.
4. Remove cylinder front cover.
5. Turn crankshaft to align timing marks as shown in Fig. 16.
6. Carefully remove camshaft with gear attached, Fig. 18.
7. Reverse procedure to install.

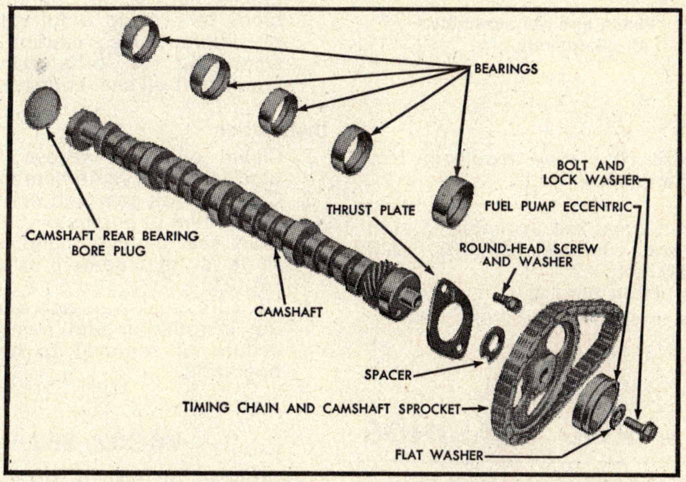

Fig. 20 Camshaft and related parts. V8-390, 400, 429, 460
NOTE: Spacer not used on all models

V8-302, 351, 400

1. To remove camshaft, remove cylinder front cover and timing chain.
2. Remove distributor cap and spark plug wires, then remove distributor.
3. Disconnect automatic transmission oil cooler lines from radiator and remove radiator.
4. Remove intake manifold and carburetor as an assembly.
5. Remove rocker arm covers.
6. Loosen rocker arm stud nuts or bolts and rotate rocker arms to one side.
7. Remove push rods, keeping them in sequence in a rack so they may be installed in their original location.
8. Using a magnet, remove valve lifters and place them in a rack in sequence so they may be installed in their original location.
9. Remove camshaft thrust plate, Fig. 19, and carefully pull camshaft from engine, using care to avoid damaging camshaft bearings.
10. Reverse procedure to install.

V8-390, 429, 460

1. Remove timing chain cover, chain, sprockets and intake manifold.
2. Remove grille and distributor.
3. Remove rocker arm assembly.
4. Remove push rods.
5. Position an inspection light through push rod opening and into valve push rod valley. Remove valve lifters with a magnet through push rod openings. *It may be necessary in some cases to transfer the lifter over to an adjoining push rod opening in order to remove it.*
6. Remove oil pan.
7. Slide camshaft out of engine, Fig. 20.

CAMSHAFT BEARINGS

When necessary to replace camshaft bearings, the engine will have to be removed from the vehicle and the plug at the rear of the cylinder block will have to be removed in order to utilize the special camshaft bearing removing and installing tools required to do this job. If properly installed, camshaft bearings require no reaming—nor should this type bearing be reamed or altered in any manner in an attempt to fit bearings.

PISTON & ROD, ASSEMBLE

All V8's

Assemble the pistons to the rods as shown in Fig. 21.

All Sixes

Piston heads are marked for location on the forward side, Fig. 22. Rods and caps are numbered on the same side as the piston they serve.

PISTONS, PINS & RINGS

SERVICE BULLETIN

Piston and Pin Replacement: When servicing engines using press fit piston pins, the piston and pin must be replaced as an assembly if either does not meet specifications. These components are not serviced separately for the principle reason that excess clearances are usually caused by piston wear rather than pin wear. Elimination of excessive clearance by

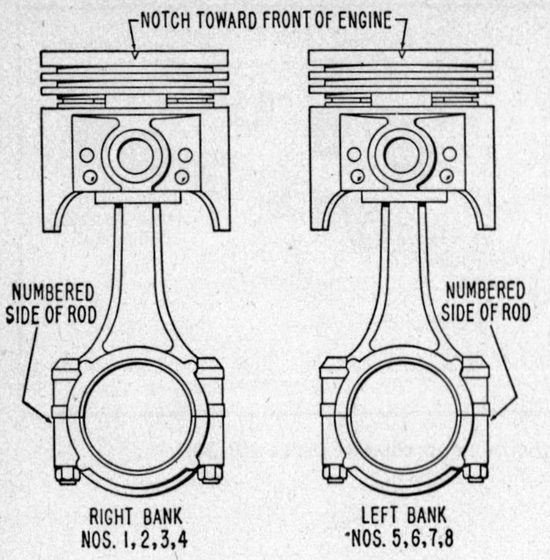

Fig. 21 Piston and rod assembly. All V8 engines

using oversize pins may result in fracture of the connecting rod.

Pistons and rings are available in standard sizes and oversizes of .003, .020, .030 and .040 inch.

Oversizes piston pins of .001 and .002" are available on 6-240 and V8-390 only.

MAIN & ROD BEARINGS

Main and rod bearings are available in standard sizes and the following undersizes:

V8-302, 390, 400, 429, 460: .002, .010, .020, .030, .040".

All other engines: .002, .010, .020, .030".

NOTE

Main and rod bearings are a selective fit. Do not file or lap bearing caps or use bearing shims to obtain proper bearing clearance. Selective fit bearings are available for service in standard sizes only. Standard bearings are divided into two sizes and are identified by a daub of red or blue paint. Red marked bearings increase the clearance; blue marked bearings decrease clearance. When replacing standard bearings with new bearings, it is good practice first to try to obtain the proper clearance with two blue bearing halves.

CRANKSHAFT OIL SEAL

1969-72 6-240 Engine

NOTE: If crankshaft rear oil seal replacement is the only operation being performed, it can be done in the vehicle. If the oil seal is being replaced in conjunction with a rear main bearing replacement, the engine must be removed from the vehicle. To replace the seal only, proceed as follows:

1. Remove starting motor.
2. Disconnect transmission from engine and slide it back. On manual shift transmission, remove clutch assembly.
3. Remove flywheel and engine rear cover plate.
4. Use an awl to punch two holes in crankshaft rear oil seal. Punch holes on opposite sides of crankshaft and just above bearing cap-to-cylinder block split line. Insert a sheet metal screw in each hole.
5. Use two large screwdrivers or pry bars and pry against both screws at the same time to remove seal. It may be necessary to place small blocks of wood against cylinder block to provide a fulcrum point for pry bars. Use caution to avoid scratching or otherwise damaging crankshaft oil seal surfaces.

Installation

1. Clean oil seal recess in cylinder block and rear main bearing cap.
2. Coat new oil seal and crankshaft with a light film of engine oil.
3. Start seal in recess and install it until it is fully seated in seal recess, Fig. 23.
4. Be sure seal was not damaged during installation and reverse the procedure of removal to complete the operation.

V8-302, 351

A braided oil seal is pressed into the upper and lower grooves behind the rear main bearing. Directly in front of this seal is an oil slinger which deflects the oil back into the oil pan. Should the braided seal require replacement, the installation of the lower half is accomplished as follows:

With the bearing cap and lower bearing half removed, install a new seal so that both ends protrude above the cap. Tap the seal down into position or roll it snugly in its groove with a smooth rounded tool. Then cut off the protruding end of the seal with sharp knife or razor blade.

V8-302, 351 NOTE

The crankshaft rear seal is in contact with the outer surface of the flywheel flange rather than on the main bearing journal. In this design, the front face of the flange is exposed to crankcase splash. Therefore, to eliminate the possibility of oil leaks through the threaded holes in the flange, the flywheel attaching capscrews must be coated with an oil resistant sealer.

V8-302, 351, 390, 400, 429, 460

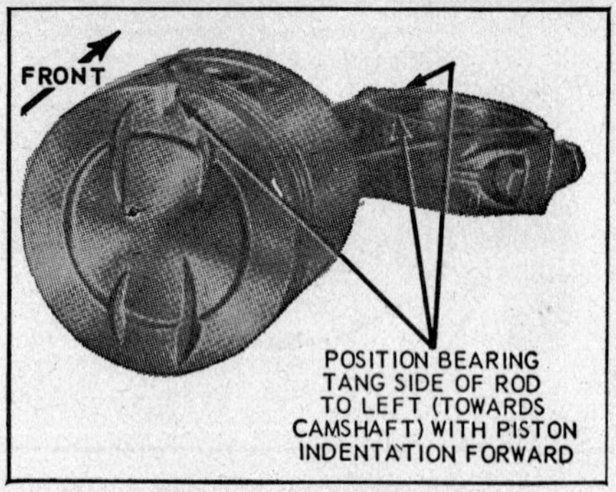

Fig. 22 Piston and rod assembly. 1969-72 6-240 engine

A new rubber split-lip rear crankshaft oil seal is released for service. This seal can be installed without removal of the crankshaft and also eliminates the necessity of seal installation tools.

1. Remove oil pan.
2. Remove rear main bearing cap.
3. Loosen remaining bearing caps, allowing crankshaft to drop down about 1/32".
4. Remove old seals from both cylinder block and rear main bearing cap. Use a brass rod to drift upper half of seal from cylinder block groove. Rotate crankshaft while drifting to facilitate removal.
5. Carefully clean seal groove in block with a brush and solvent. Also clean seal groove in bearing cap. Remove the oil seal retaining pin from the bearing cap if so equipped. *The pin is not used with the split-lip seal.*
6. Dip seal halves in clean engine oil.
7. Carefully install upper seal half in its groove with undercut side of seal toward front of engine, Fig. 24, by rotating it on shaft journal of crankshaft until approximately 3/8" protrudes below the parting surface. *Be sure no rubber has been shaved from outside diameter of seal by bottom edge of groove.*
8. Retighten main bearing caps and torque to specifications.
9. Install lower seal in main bearing cap with undercut side of seal toward front of engine, and allow seal to protrude about 3/8" above parting surface to mate with upper seal upon cap installation.
10. Apply suitable sealer to parting faces of cap and block. Install cap and torque to specifications.

NOTE: If difficulty is encountered in installing the upper half of the seal in position, lightly lap (sandpaper) the side of the seal opposite the lip side using a medium grit paper. After sanding, the seal must be washed in solvent, then dipped in clean engine oil prior to installation.

SERVICE BULLETIN

A revised crankshaft rear oil seal has been released for service. This new seal may be received when ordering an oil pan gasket kit and is installed in the same manner as described previously, Fig. 25.

OIL PAN, REPLACE

1969-72 6-240 Engine

1. Drain crankcase and cooling system.
2. Remove radiator.
3. Disconnect flexible fuel line at fuel pump.
4. With automatic transmission disconnect kickdown rod at bellcrank.
5. With manual shift transmission, disconnect clutch linkage.
6. Raise car and remove starter.
7. Remove engine front support retaining nuts.
8. Raise transmission, remove rear support insulator and lower transmission to crossmember.
9. Raise engine and place 3" thick wood blocks between both front support insulators and intermediate support brackets.
10. Remove oil pan bolts and oil pump retaining bolts.
11. Remove oil pump from block and lay it in bottom of oil pan.
12. Rotate crankshaft as required to remove oil pan.
13. Remove inlet tube and screen from pump.

Installation

1. After cleaning, install inlet tube and screen on oil pump, using a new gasket.
2. Clean gasket surfaces of oil pump, pan and block.
3. Remove rear main bearing cap-to-oil pan seal and cylinder front cover-to-oil pan seal. Clean seal grooves.

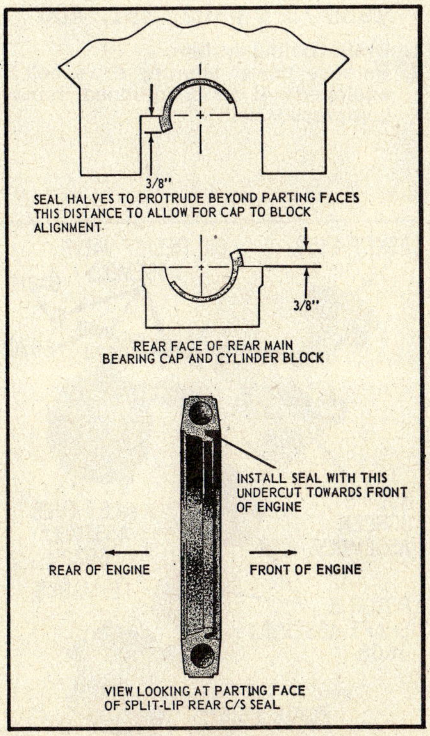

Fig. 24 Split-lip rear crankshaft seal installation on V8-351, 390, 400, 429, 460

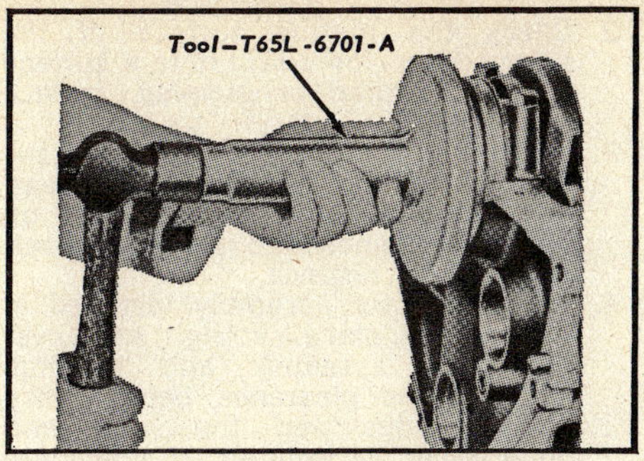

Fig. 23 Crankshaft rear oil seal installation. 6-240 engine

4. Apply oil-resistant sealer in cavities between bearing cap and cylinder block, Fig. 26.
5. Install new side gaskets on oil pan with oil-resistant sealer.
6. Position a new cylinder front cover seal on oil pan.
7. Place oil pump assembly in pan. Position pan under engine. Install oil pump with new gasket on cylinder block. Install oil pan and tighten screws securely.
8. Complete the operation by reversing the removal procedure.

1969 V8-302, 351

1. Drain crankcase and remove dipstick.
2. Unfasten and position oil pan on No. 2 crossmember.
3. Remove one of the inlet tube retaining bolts and loosen the other. This allows inlet tube to be positioned out of the way when removing oil pan.
4. Crank engine as required to obtain clearance and remove oil pan.
5. Remove oil pump inlet tube and screen.

Installation

1. Position oil inlet tube and loosely install one retaining bolt.
2. Clean gasket surfaces of block and pan.
3. Coat block surface and oil pan gasket with sealer. Position pan gaskets on cylinder block.
4. Position front seal on cylinder front cover, being sure tabs on seal are over oil pan gasket.
5. Position rear oil pan seal on rear main bearing cap, being sure tabs on seal are over oil pan gasket.
6. Place pan on No. 2 crossmember and install other inlet tube retaining bolt.
7. Complete installation in reverse order of removal.

1970-76 V8-302, 351C, 351W, 351M, 400

1. Drain oil pan and remove dip stick.
2. Remove fan shroud bolts and place shroud over fan.
3. On V8-351C, 351M, 400 engines, remove starter.
4. On all engines, remove engine front mount to chassis bolts, raise engine and install wood blocks between mounts and chassis, then lower engine onto the wood blocks.
5. On V8-302, 351W engines, disconnect stabilizer bar from lower control arms and position stabilizer bar and control arms to permit oil pan removal. Also, on all models with automatic transmission, position oil cooler lines aside for pan removal.
6. Remove oil pan bolts and oil pan.

Installation

1. Coat all gasket surfaces with a suitable sealer and install gaskets on engine.
2. On V8-351C, 351M, 400 engines, install pan seal on cylinder front cover and pan rear seal on rear main bearing cap.
3. On all engines, install oil pan. Torque $\frac{1}{4}$ inch bolts to 7-9 ft. lbs.; V8-302, 351W, $\frac{5}{16}$ inch bolts to 9-11 ft. lbs.; V8-351C, 351M, 400, $\frac{5}{16}$ inch bolts to 11-13 ft. lbs.
4. On V8-302, 351W engines, connect stabilizer bar to lower control arms and torque bolts to 6-12 ft. lbs.
5. On all engines, raise engine, remove wood blocks, lower engine and install front mount to chassis bolts.
6. Install starter, if removed, fan shroud and dip stick.

1969-71 V8-390 & 1969-76 429, 460

1. Drain oil pan.
2. On V8-429, 460 engines, remove starter.
3. On 1969-71 engines, disconnect stabilizer bar at connecting links and pull ends down. On 1972-76 engines disconnect sway bar and pull forward on struts.

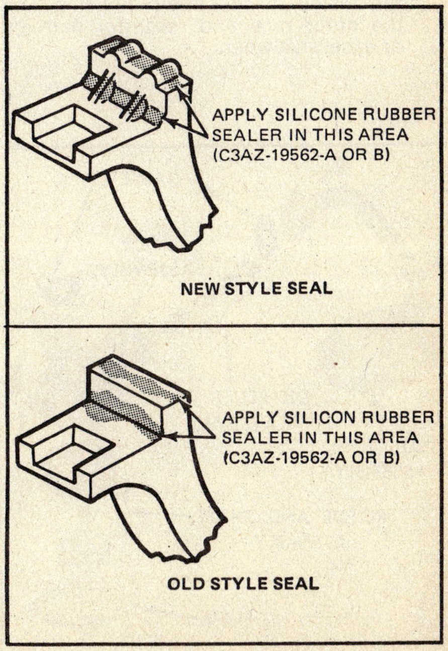

Fig. 25 Crankshaft rear oil seals

Fig. 26 Oil pan rear seal installation. 6-240 engine

4. Remove fan shroud bolts, place shroud on fan and remove oil filter.

NOTE: *To allow clearance for removal of oil pan, remove the front engine mount nuts. Then position floor jack under front leading edge of oil pan (use wood block between pan and jack). Raise engine about 1¼" and insert a 1" block of wood between insulators and frame crossmember. Then remove floor jack.*

5. Remove oil pan screws and lower pan to crossmember.
6. Crank engine to obtain necessary clearance between crankshaft counterweight and rear of oil pan. Then remove pan.

OIL PUMP REPAIRS

Figs. 27, 28, 29, 30

1. With all parts clean and dry, check the inside of the pump housing and the outer race and rotor for damage or excessive wear.
2. Check the mating surface of the pump cover for wear. If this surface is worn, scored or grooved, replace the cover.
3. Measure the clearance between the outer race and housing. This clearance should be .006-.013 inch on 1969-73 engines and .001-.013 inch on 1974-76 engines.
4. With the rotor assembly installed in the housing, place a straight edge over the rotor assembly and housing. Measure the clearance between the straight edge and the rotor and outer race. Recommended limits are .0011-.0041". *The outer race, shaft and rotor are furnished only as an assembly.*
5. Check the drive shaft-to-housing bearing clearance by measuring the O.D. of the shaft and the I.D. of the housing bearing. The recommended clearance limits are .0015-.0029".
6. Inspect the relief valve spring for a collapsed or worn condition.
7. Check the relief valve piston for scores and free operation in the bore. The specified piston clearance is .0015-.0029".

BELT TENSION DATA

	New Lbs.	Used Lbs.
1969-76 All	140	110

WATER PUMP, REPLACE

1969-72 Six

1. Drain cooling system.
2. Loosen and remove alternator, power steering and air conditioning belts.
3. Disconnect radiator lower hose and heater hose at pump.
4. Remove fan, spacer, pulley and belt.

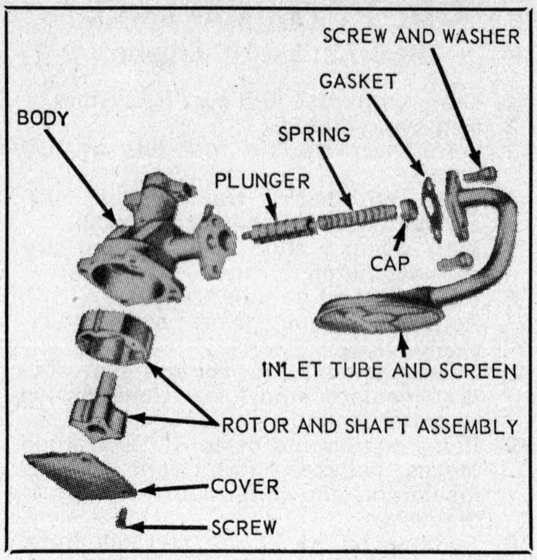

Fig. 27 Oil pump. 6-240 engine

5. Unfasten and remove the water pump.

1969-76 V8-351, 400, 429, 460

1. Drain cooling system, remove fan shroud, fan and all belts.
2. Disconnect alternator bracket and position bracket out of way.
3. Remove power steering pump, A/C compressor and position units out of way.
4. Remove A/C compressor bracket and disconnect hoses from water pump.
5. Remove water pump and remove separator plates from pump.

1969-73 V8-302, 351, 400

1. Drain cooling system.
2. Remove power steering drive belt (if equipped). If air conditioned, remove compressor belt.

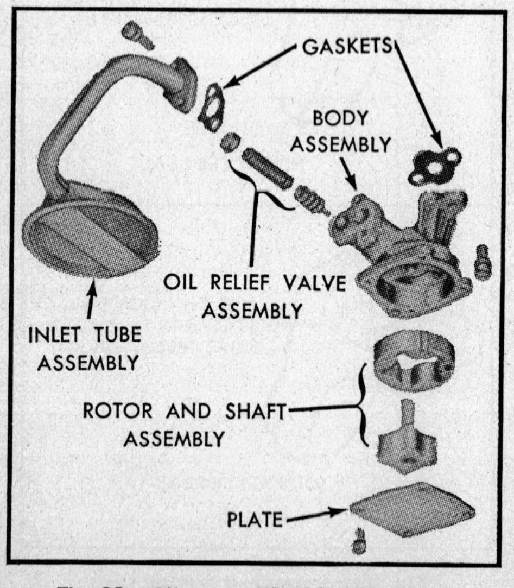

Fig. 28 Oil pump. V8-302, 351, 351W

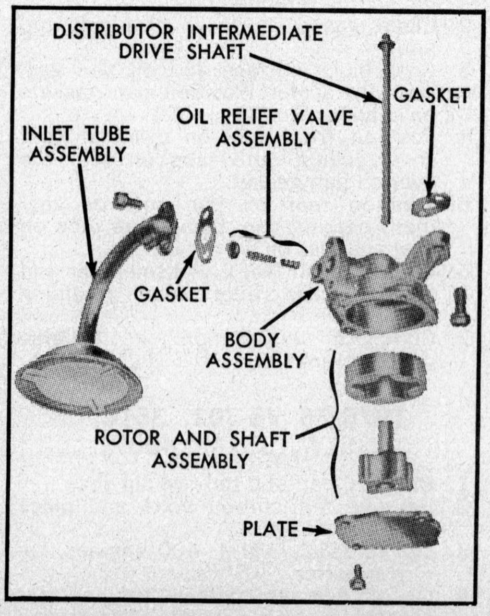

Fig. 29 Oil pump. V8-390, 429, 460

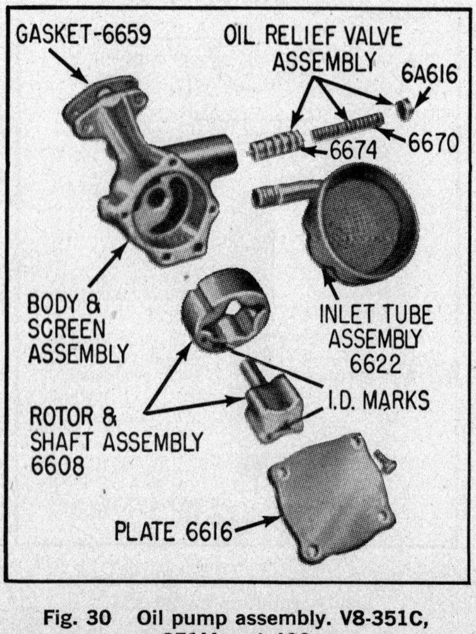

Fig. 30 Oil pump assembly. V8-351C, 351M and 400

3. Disconnect radiator lower hose and heater hose at water pump.
4. Remove drive belt, fan, spacer or fan drive clutch and pulley.
5. Unfasten and remove water pump from cylinder front cover.

1969-73 V8-390

1. Drain cooling system.
2. Remove power steering drive belt (if equipped). If air conditioned, remove compressor drive belt.
3. Disconnect radiator lower hose and heater hose at water pump.
4. Remove radiator upper support and fan guard.
5. Remove fan belt or belts, fan, spacer or fan drive clutch and pulley.
6. Unfasten and remove water pump.

FUEL PUMP PRESSURE

Year	Engine	Pressure Lbs.
1969-72	6-240	4-6
1969	All V8	4½-6½
1970-71	V8 exc. 302	5-7
1970-71	V8-302	4-6
1972-75	V8 exc. 1974-75 460	5-7
1974	V8-460 Calif.	5.7-7.7
1974	V8-460 exc. Calif.	5-7
1975-76	V8-460	5.7-7.7

FUEL PUMP, REPLACE

1. Remove all gasket material from the pump and block gasket surfaces. Apply sealer to both sides of new gasket.
2. Position gasket on pump flange and hold pump in position against its mounting surface. Make sure rocker arm is riding on camshaft eccentric.
3. Press pump tight against its mounting. Install retaining screws and tighten them alternately.
4. Connect fuel lines. Then operate engine and check for leaks.

NOTE: Before installing the pump, it is good practice to crank the engine so that the nose of the camshaft eccentric is out of the way of the fuel pump rocker arm when the pump is installed. In this way there will be the lease amount of tension on the rocker arm, thereby easing the installation of the pump.

Clutch and Transmission Section

NOTE: 1976 linkage adjustment information is in this section. Repair procedures on both automatic and manual shift transmissions are covered elsewhere in this manual. Procedures for removing automatic transmissions as well as linkage adjustments on 1969-75 models are included in the automatic transmission chapters. See Chapter Index.

CLUTCH PEDAL, ADJUST

1969-71

1. Disconnect clutch release lever spring from release lever.
2. Loosen release lever rod locknut and adjusting nut.
3. Move clutch release lever rearward until release bearing lightly touches pressure plate release fingers.
4. Adjust adapter length until adapter seats in release lever pocket.
5. Insert a .194″ feeler against the back face of rod adapter and tighten the adjusting nut finger tight against the feeler gauge.
6. Tighten locknut against adjusting nut being careful not to disturb adjustment.
7. Remove feeler gauge.
8. Install release lever spring and check pedal free travel. Travel should be ⅞″ to 1⅛″.

CLUTCH, REPLACE

1969-71

1. Remove transmission as outlined.
2. On cars with aluminum clutch housing, remove starter. Unfasten housing from engine and move housing back just far enough to clear pressure plate, then move it to the right to free the pivot from clutch equalizer bar. Be careful not to lose bushing or disturb linkage or assist spring.
3. Remove flywheel housing cover (cast iron housings only).
4. Remove release lever return spring. Then slide release bearing and hub off release lever (cast iron housings only).
5. Loosen pressure plate attaching bolts gradually and evenly to release spring tension. If same clutch is being installed, first mark clutch cover and flywheel so installation may be made in same position.
6. Remove clutch and driven plate.

MANUAL SHIFT TRANS.

Transmission Replace

Three Speed Units

1. Raise car and drain lube from transmission.
2. Mark drive shaft so that it may be installed in the same relative posi-

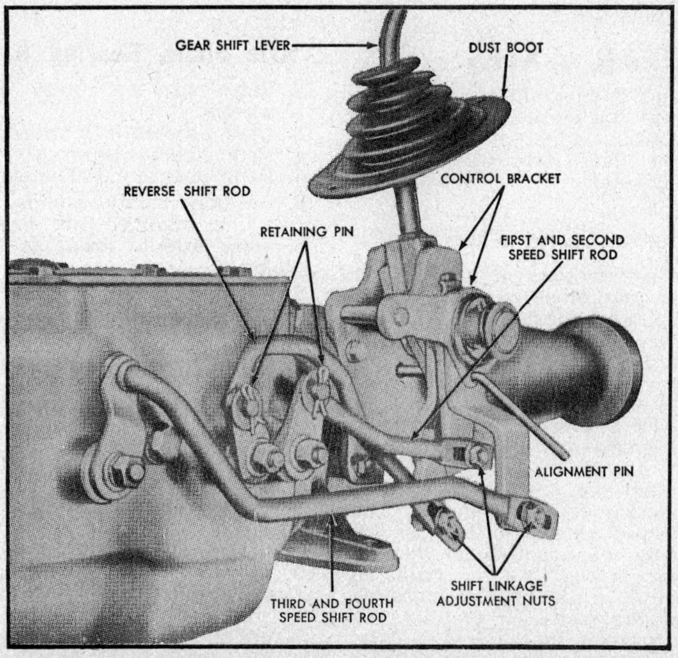

Fig. 1 Shift linkage. 1969-71 four speed transmission

tion.
3. Disconnect drive shaft from U-joint flange and slide it out of extension housing.
4. Disconnect speedometer cable.
5. Disconnect shift rods from levers at transmission.
6. Disconnect parking brake cable at equalizer.
7. Unfasten extension from rear support.
8. Raise engine high enough to remove weight from frame crossmember.
9. Support transmission with a jack and detach it from flywheel housing.
10. Slide transmission back and out of car.
11. Reverse removal procedure to install transmission.

Four Speed Units

1. Remove transmission gearshift lever boot retainer. Working under boot, remove shift lever retaining bolts and remove lever. The remaining shift linkage may be left on the transmission during removal.
2. Raise car and disconnect drive shaft from rear U-joint flange and remove drive shaft.
3. Disconnect speedometer cable.
4. Disconnect parking brake cable at equalizer bar and raise rear of engine.
5. Unfasten extension housing from engine rear support.
6. Raise transmission slightly with a jack. Disconnect and remove crossmember and engine rear support as a unit.
7. Unfasten transmission from clutch housing and install guide pins in the two lower holes.
8. Remove transmission. If necessary, lower engine to gain enough clearance for removal of transmission.
9. Reverse removal procedure to install.

SHIFT LINKAGE, ADJUST

1969-71 3 Speed Units

1. Place selector lever in neutral.
2. Loosen the two shift rod adjusting nuts.
3. Insert a $^3/_{16}$" drill shank through the low-reverse and 2-3 shift levers at steering column. It may be necessary to align levers to insert drill.
4. Tighten adjusting nuts.
5. Remove drill gauge.
6. Start engine and move shift lever to each position to be sure it operates freely.

1969-71 4 Speed Units

1. Loosen three shift linkage adjustment nuts.
2. Install a $^1/_4$" diameter alignment tool through control bracket and levers as shown in Fig. 1.

NOTE: An alignment tool can be made from $^1/_4$" diameter drill rod bent to an "L" shape. The extensions should be $1^1/_2$" and $3^3/_4$" from elbow. Short end of tool should be inserted into control bracket and linkage holes until it bottoms.

3. Tighten three linkage adjusting nuts and then remove alignment tool.
4. Check shift lever for smooth crossover.

NOTE: On 1970-71 models, a transmission lock rod is incorporated in the shift linkage. This link must be adjusted AFTER the shift linkage is properly adjusted.

1. Place shift lever in Neutral and loosen lock rod adjustment nut.
2. Align the hole in the steering column socket casting with the column alignment mark and insert a .180" dia. gauge rod. The column casting must not rotate with the gauge rod installed.
3. Tighten rod lock nut and check for proper operation.

1976 AUTO. TRANS. LINKAGE, ADJUST

Linkage adjustment procedures are the same as those described for 1974-75 models described in the "Automatic Transmissions" section of this manual.

Rear Axle, Propeller Shaft & Brakes

REAR AXLES

Figs. 1 and 3 illustrate the rear axle assemblies used on these cars. When necessary to overhaul either of these units, refer to the *Rear Axle Specifications* table in this chapter.

Integral Carrier Type

The gear set consists of a ring gear and an overhung drive pinion which is supported by two opposed tapered roller bearings, Fig. 1. The differential case is a one-piece design with openings allowing assembly of the internal parts and lubricant flow. The differential pinion shaft is retained with a threaded bolt (lock) assembled to the case.

The roller type wheel bearings have no inner race, and the rollers directly contact the bearing journals of the axle shafts. The axle shafts do not use an inner and outer bearing retainer. Rather, they are held in the axle by means of C-locks, Fig. 2. These C-locks also fit into a machined recess in the differential side gears within the differential case. There is no retainer bolt access hole in the axle shaft flange.

Axle Shaft, Bearing & Oil Seal

1. Raise car on hoist and remove wheels.
2. Drain differential lubricant.
3. Remove brake drums.
4. Remove differential housing cover.
5. Position safety stands under rear frame member and lower hoist to allow axle to lower as far as possible.
6. Working through differential case opening, remove pinion shaft lock bolt and pinion shaft.
7. Push axle shaft(s) inward toward center of axle housing and remove C-lock(s) from housing, Fig. 2.
8. Remove axle shaft, using extreme care to avoid contact of shaft seal lip with any portion of axle shaft except seal journal.
9. Use a hook-type puller to remove seal and bearing, Fig. 7.
10. Reverse procedure to install, using suitable driving tools, Fig. 8, to install seal and bearing. New seals are pre-packed with lubricant and do not require oil soaking before installation.

Removable Carrier Type

In these axles, Fig. 3, the drive pinion is straddle-mounted by two opposed tapered roller bearings which support the pinion shaft in front of the drive pinion gear, and straight roller bearing that supports the pinion shaft at the rear of the pinion gear. The drive pinion is assembled in a pinion retainer that is bolted to the differential carrier. The tapered roller bearings are preloaded by a collapsible spacer between the bearings. The pinion is positioned by a shim or shims located between the drive pinion retainer and the differential carrier.

The differential is supported in the carrier by two tapered roller side bearings. These bearings are preloaded by two threaded ring nuts or sleeves between the bearings and pedestals. The differential assembly is positioned for proper ring gear and pinion backlash by varying the adjustment of these ring nuts. The differential case houses two side gears in mesh with two pinions mounted on a pinion shaft which is held in place by a pin. The side gears and pinions are backed by thrust washers. With high performance engines, an optional rear axle having a four-pinion differential is also used.

The axle shafts are of unequal length, the left shaft being shorter than the right. The axle shafts are mounted on sealed ball bearings or tapered roller bearings which are pressed on the shafts.

Service Bulletin

All Ford Built Rear Axles: Recent manufacturing changes have eliminated the need for marking rear axle drive pinions for individual variations from nominal shim thicknesses. In the past, these pinion markings, with the aid of a shim selection table, were used as a guide to select correct shim thicknesses when a gear set or carrier assembly replacement was performed.

With the elimination of pinion markings, use of the shim selection table is no longer possible and the methods outlined below must be used.

1. Measure the thickness of the original pinion depth shim removed from the axle. Use the same thickness upon installation of the replacement carrier or drive pinion. If any further shim change is necessary, it will be indicated in the tooth pattern check.
2. If the original shim is lost, substitute a nominal shim for the original and use the tooth pattern check to determine if further shim changes are required.

Axle Shaft, Bearing & Seal

1. Remove wheel assembly.
2. Remove brake drum or caliper.
3. Working through hole provided in axle shaft flange, Fig. 4, remove nuts that secure bearing retainer.
4. Pull axle shaft out of housing. If bearing is a tight fit in axle housing, use a slide hammer-type puller, Fig. 5. *Brake carrier plate must not be dislodged. Install one nut to hold plate in place after axle shaft is removed.*

NOTE: On 1974-76 models, remove brake carrier plate.

5. If axle shaft bearing is to be replaced, drill a 1/4 inch hole not more than 5/16 inch deep in retainer ring surface and then nick it deeply in several places and slide bearing retain off, Fig. 6.
6. Press bearing from axle shaft.
7. Inspect machined surfaces of axle shaft and housing for rough spots that would affect the sealing action of the oil seal. Carefully remove any burrs or rough spots.
8. Press new bearing on shaft until it seats firmly against shoulder on shaft.
9. Press inner bearing retainer on shaft until it seats firmly against bearing.
10. If oil seal is to be replaced, use a hook-type tool to pull it out of the housing, Fig. 7. Wipe a small amount of oil resistant sealer on outer edge of seal before it is installed, Fig. 8.

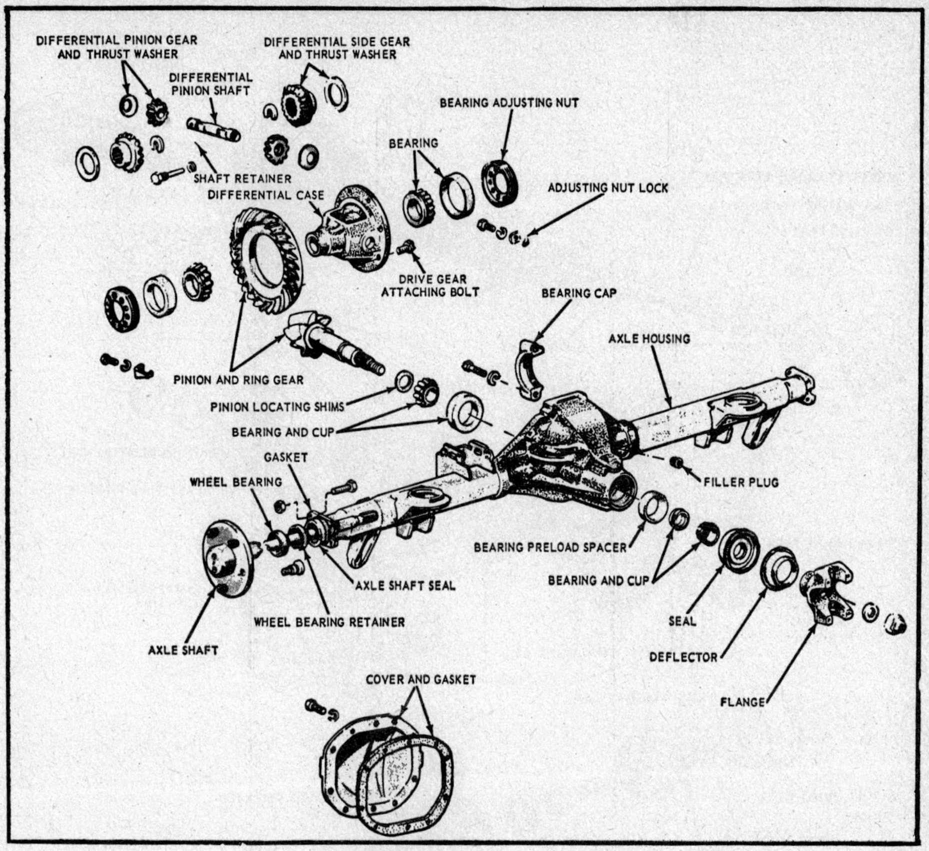

Fig. 1 Integral type rear axle assembly

Installation

NOTE: On models equipped with rear disc brakes, ensure disc brake adapters are installed in the correct position and are identified with a stamping.

1. On 1974-76 models, place a new gasket on each side of brake carrier plate and slide axle shaft into housing.
2. On all models start the splines into the differential side gear and push the shaft in until bearing bottoms in housing.
3. Install retainer and tighten nuts to 30-40 ft. lbs.
4. Install brake drum and wheel.

Fig. 2 Axle shaft C-locks. 1969-76 Ford integral type axle

PROPELLER SHAFT

Remove & Replace

1. Disconnect rear U-joint from drive pinion flange.
2. Pull dirve shaft toward rear of car until front U-joint yoke clears transmission extension housing and output shaft.
3. Install a suitable tool, such as a seal driver, in seal to prevent lube from leaking from transmission.
4. Before installing, check U-joints for freedom of movement. If a bind has resulted from misalignment after overhauling the U-joints, tap the ears of the drive shaft sharply to relieve the bind.
5. If rubber seal installed on end of transmission extension housing is damaged, install a new seal.
6. On a manual shift transmission, lubricate yoke spline with conventional transmission lubricant. On an automatic transmission, lubricate yoke spline with special spline lubricant. *This spline is sealed so that trans-*

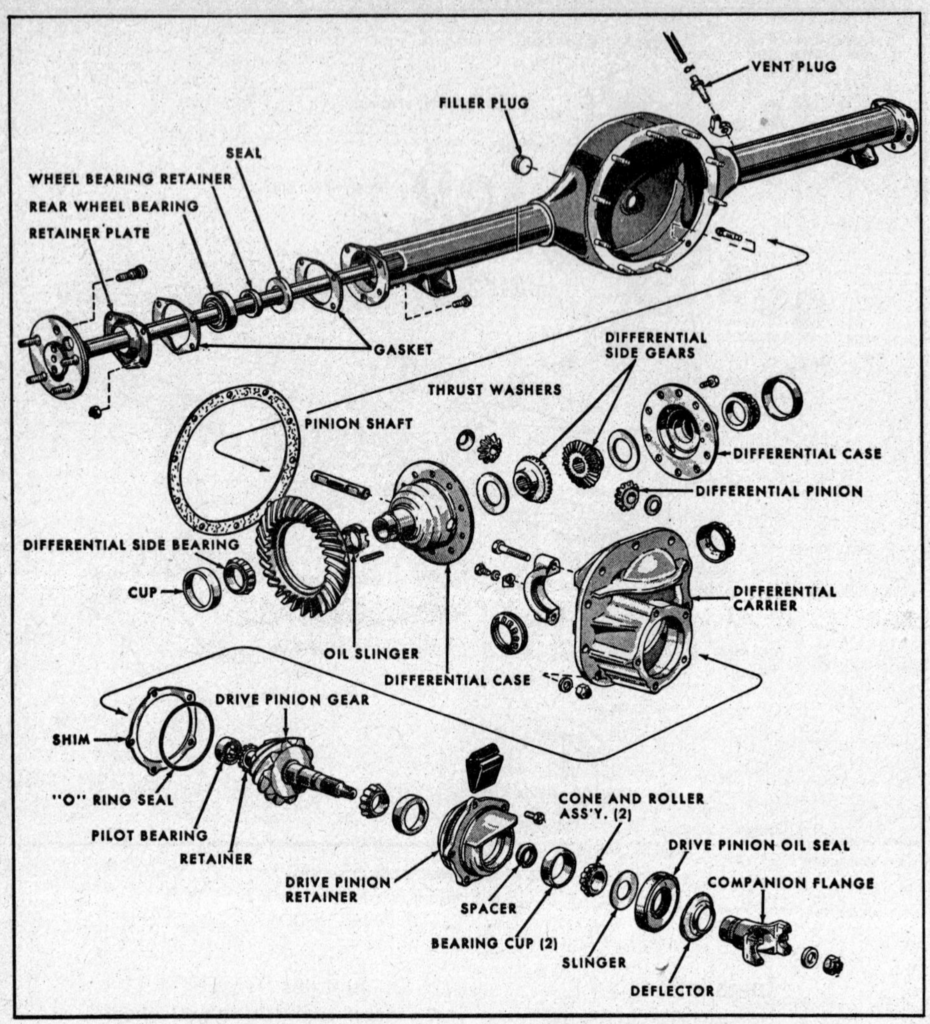

Fig. 3 Removable carrier type of rear axle assembly. On some high performance engines a four-pinion differential is also used

Fig. 4 Removing nuts from rear bearing retainer

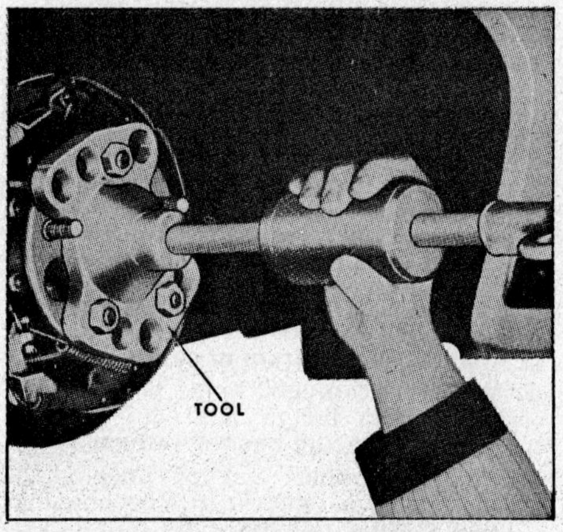

Fig. 5 Removing axle shaft with slide hammer-type puller

mission fluid does not "wash" away spline lubricant.

7. Install yoke on transmission output shaft.
8. Install U-bolts and nuts which attach U-joint to pinion flange. Tighten U-bolts evenly to prevent binding U-joint bearings.

BRAKE ADJUSTMENTS

1. Use the brake shoe adjustment gauge shown in Fig. 9 to obtain the drum inside diameter as shown.

Fig. 6 Splitting bearing inner retainer for bearing removal

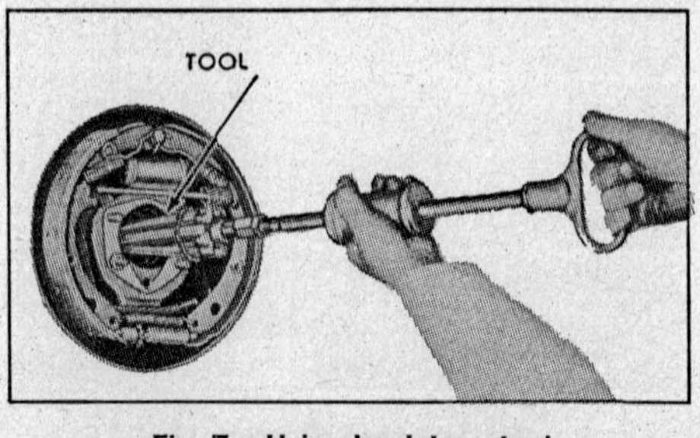

Fig. 7 Using hook-type tool to remove oil seal

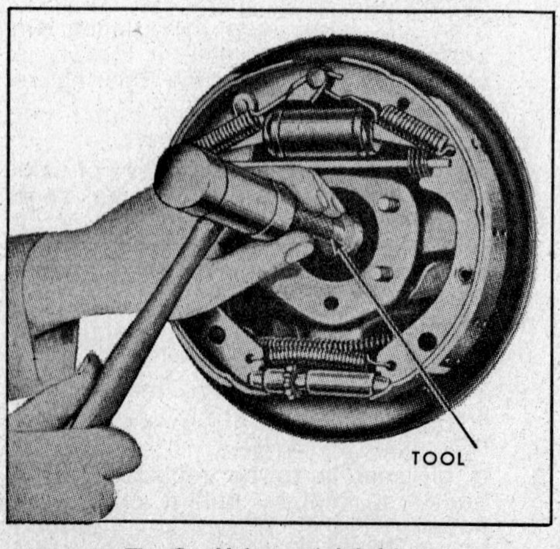

Fig. 8 Using special driver to install oil seal

Tighten the adjusting knob on the gauge to hold this setting.

2. Place the opposite side of the gauge over the brake shoes and adjust the shoes by turning the adjuster screw until the gauge just slides over the linings. Rotate the gauge around the lining surface to assure proper lining diameter adjustment and clearance.
3. Install brake drum and wheel. Final adjustment is accomplished by making several firm reverse stops, using the brake pedal.

Self-Adjusting Brakes

These brakes, Fig. 10, have self-adjusting shoe mechanisms that assure correct lining-to-drum clearances at all times. The automatic adjusters operate only when the brakes are applied as the car is moving rearward or when the car comes to an uphill stop.

Although the brakes are self-adjusting, an initial adjustment is necessary after the brake shoes have been relined or replaced, or when the length of the star wheel adjuster has been changed during some other service operation.

Frequent usage of an automatic transmission forward range to halt reverse vehicle motion may prevent the automatic adjusters from functioning, thereby inducing low pedal heights. Should low pedal heights be encountered, it is recommended that numerous forward and reverse stops be performed with a firm pedal effort until satisfactory pedal height is obtained.

NOTE

If a low pedal height condition cannot be corrected by making numerous reverse stops (provided the hydraulic system is free of air), it indicates that the self-adjusting mechanism is not functioning. Therefore, it will be necessary

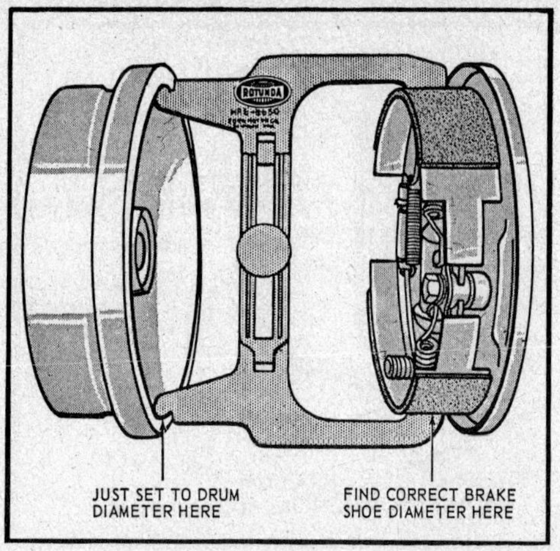

Fig. 9 Brake adjustment gauge

to remove the brake drums, clean, free up and lubricate the adjusting mechanism. Then adjust the brakes, being sure the parking brake is fully released.

Initial Adjustment

1. Remove adjusting hole cover from brake backing plate and, from the backing plate side, turn the adjusting screw upward with a screwdriver or other suitable tool to expand the shoes until a slight drag is felt when the drums are rotated.
2. Remove the drum.
3. While holding the adjusting lever out of engagement with the adjusting screw, Fig. 11, back off the adjusting screw about one full turn with the fingers.

NOTE: *If finger movement will not turn the screw, free it up. If this is not done, the adjusting lever will not turn the screw during vehicle operation. Lubricate the screw with oil and coat with wheel bearing grease. Any other adjustment procedure may cause damage to the adjusting screw with consequent self-adjuster problems.*

4. Install wheel and drum, and adjusting hole cover. Adjust brakes on remaining wheels in the same manner.
5. If pedal height is not satisfactory, drive the vehicle and make sufficient reverse stops with a firm pedal effort until proper pedal height is obtained.

PARKING BRAKE, ADJUST

Check parking brake cables when brakes are fully released. If cables are loose, adjust as follows:

Ford & Mercury

1969-76 Rear Drum Brakes

1. Make sure parking brake is released.
2. Place transmission in neutral and raise the vehicle.
3. Tighten the adjusting nut against the cable equalizer to cause rear brakes to drag.
4. Then loosen the adjusting nut until the rear wheels are fully released. There should be no drag.
5. Lower vehicle and check operation.

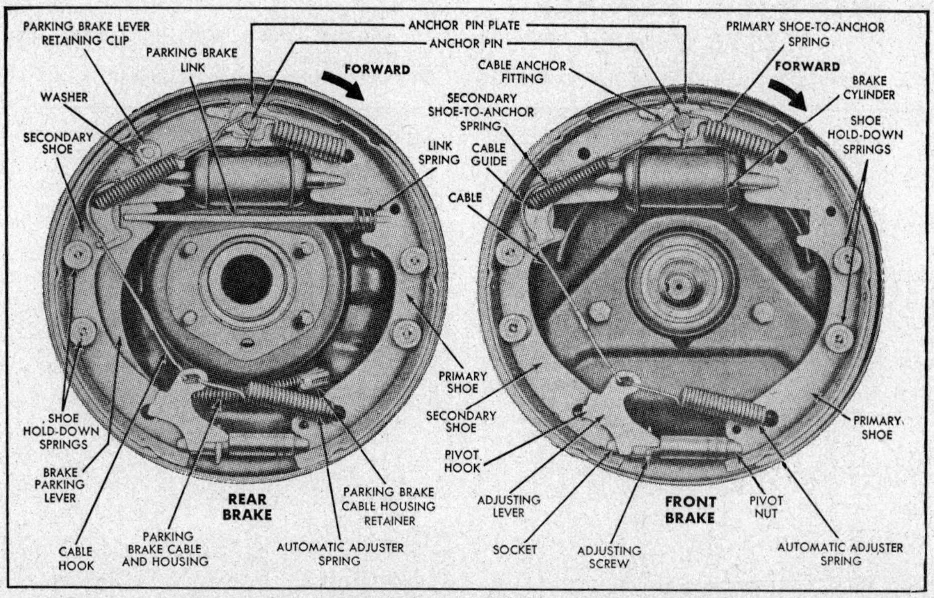

Fig. 10 Right front and rear drum brakes

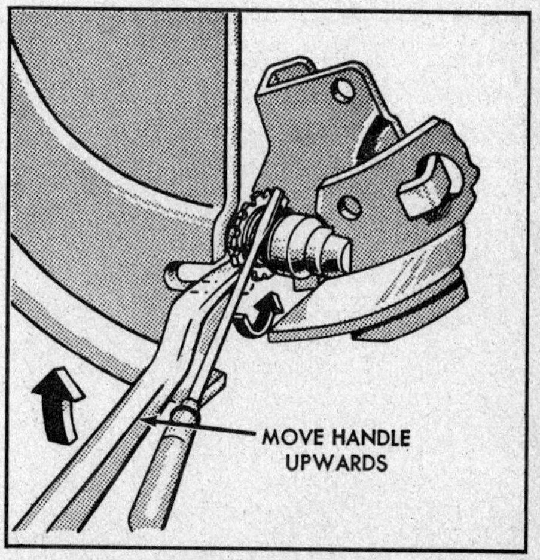

Fig. 11 Backing off brake adjustment by disengaging adjusting lever with screwdriver

1975-76 Rear Disc Brakes

1. Fully release parking brake, then place transmission in neutral and support vehicle at rear axle.
2. Tighten adjuster nut until levers on calipers just begin to move, then loosen adjuster nut until levers just return to stop position.
3. Apply and release parking brake. Check levers on caliper to determine if they are fully returned by attempting to pull lever rearward. If lever moves, the adjustment is too tight and must be readjusted.

VACUUM RELEASE PARKING BRAKE

The vacuum power unit will release the parking brakes automatically when the shift lever is moved into any drive position with the engine running. The brakes will not release automatically, however, when the shift lever is in neutral or park position with the engine running, or in any position with the engine off.

The power unit piston rod is attached to the release lever. Since the release lever pivots against the pawl, a slight movement of the release lever will disengage the pawl from the ratchet, allowing the brakes to release. The release lever pivots on a rivet pin in the pedal mount.

As shown in Fig. 14, hoses connect the power unit and the engine manifold to a vacuum release valve in the transmission neutral safety switch. Moving the transmission selector lever into any drive position with the engine running will open the release valve to connect engine manifold vacuum to one side of the actuating piston in the power unit. The pressure differential thus created will cause the piston and link to pull the release lever.

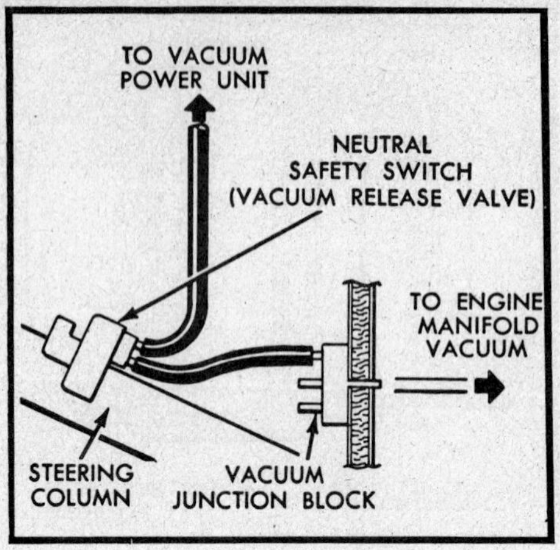

Fig. 14 Connections for automatic parking brake release. Typical

MASTER CYLINDER, REPLACE

1969-76 Ford & Mercury

Less Power Brakes

1. Working from inside of car beneath instrument panel, disconnect master cylinder push rod from brake pedal.
2. Disconnect stop light switch wires at connector. Remove hairpin retainer. Slide stop light switch off brake pedal pin just enough to clear end of pin, then lift switch straight upward from pin.
3. Slide master cylinder push rod, nylon washers and bushings from pedal pin.
4. Remove brake tubes from both outlet ports of master cylinder.
5. Unfasten and remove master cylinder from dash panel.

With Power Brakes

1. Disconnect brake lines from master cylinder.
2. Remove nuts retaining master cylinder to brake booster.
3. Remove master cylinder.

HYDRO-BOOST BRAKE BOOSTER, REPLACE

1975-76 Mercury

1. Working from inside of car under instrument panel, disconnect booster push rod link from brake pedal. To do this, proceed as follows:
2. Disconnect stop light switch wires at connector. Remove hairpin retainer. Slide switch off brake pedal pin just far enough for switch outer hole to clear pin. Then lift switch straight upward from pin. Slide master cylinder push rod and nylon washers and bushing off brake pedal pin.
3. Open hood and disconnect brake line at master cylinder outlet fitting.
4. Disconnect the pressure, steering gear and return lines, then plug lines and ports.
5. Remove Hydro-Boost to dash panel nuts and remove assembly from panel, sliding push rod link from engine side of dash panel.
6. Reverse procedure to install.

POWER BRAKE UNIT REPLACE

1969-76 Ford & Mercury

1. Working from inside of car under instrument panel, disconnect booster push rod link from brake pedal. To do this, proceed as follows:
2. Disconnect stop light switch wires at connector. Remove hairpin retainer. Slide switch off brake pedal pin just far enough for switch outer hole to clear pin. Then lift switch straight upward from pin. Slide master cylinder push rod and nylon washers and bushing off brake pedal pin.
3. Open hood and disconnect brake line at master cylinder outlet fitting.
4. Disconnect vacuum hose from booster unit. If equipped with automatic transmission disconnect transmission vacuum unit hose.
5. Remove four attaching nuts and remove booster and bracket from dash panel, slide push rod link out from engine side of dash panel. Remove four spacers.
6. Remove push rod link boot from dash panel.
7. Reverse procedure to install.

Rear Suspension

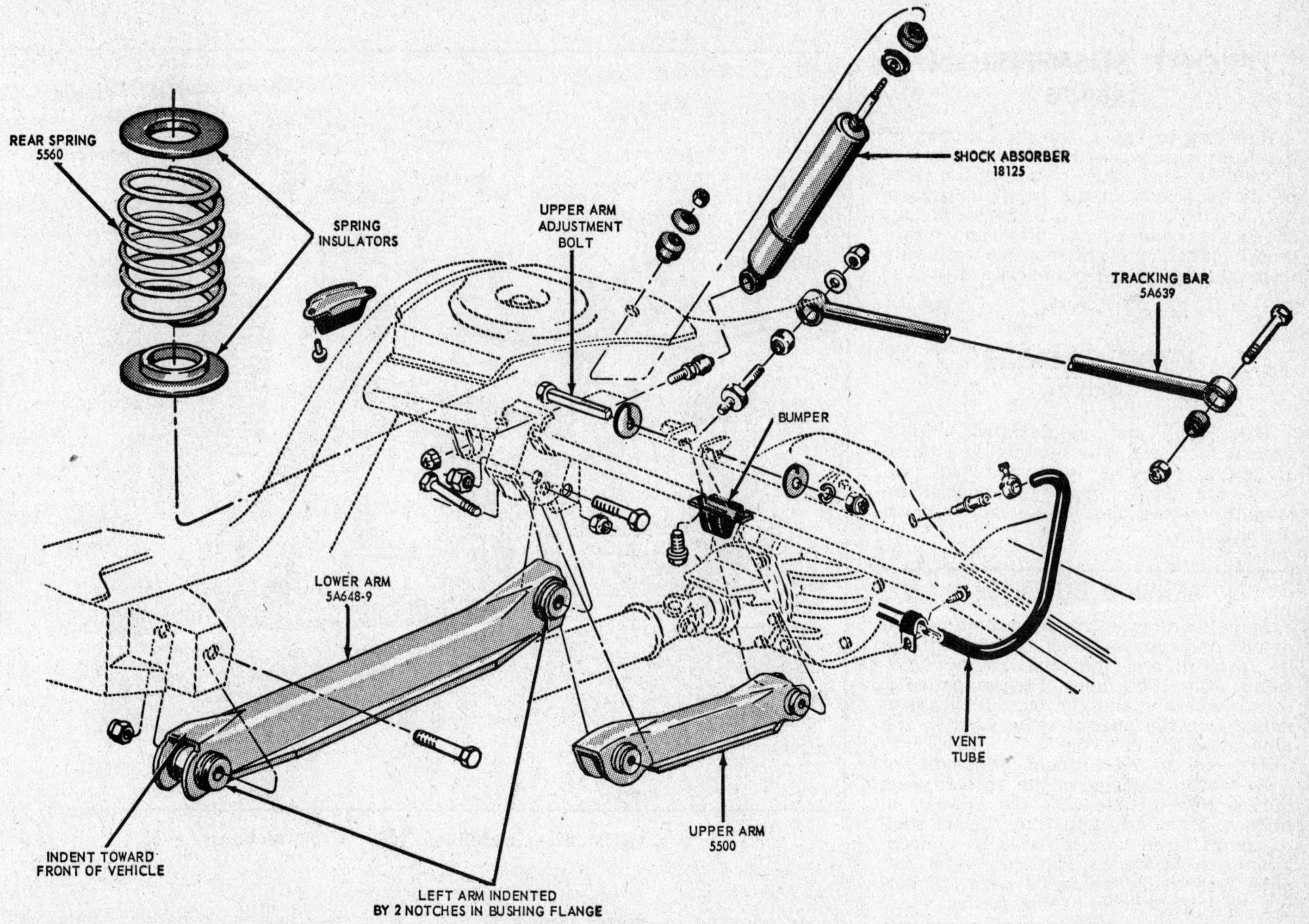

Fig. 1 Rear suspension (typical)

SHOCK ABSORBER, REPLACE

1. With the rear axle supported properly disconnect shock absorber at upper mounting and compress it to clear hole in spring seat.
2. Disconnect shock absorber from stud on axle bracket.
3. Reverse procedure to install.

COIL SPRING, REPLACE

1. Raise rear of vehicle and support at frame. Support rear axle with a suitable jack.
2. Disconnect shock absorbers at lower mountings.
3. On 1973-76 models, disconnect brake line from rear brake hose and remove hose to bracket clip.
4. Lower axle to remove springs.
5. Reverse procedure to install. Install an insulator between upper and lower seats and the spring.

CONTROL ARMS, REPLACE

NOTE: On 1972-76 models, lower arms must be replaced in pairs.

1. Raise rear of vehicle and support at frame. Support axle with a suitable jack.
2. Disconnect track bar from frame mounting bracket.
3. Lower axle and install a second jack under differential pinion nose.
4. Disconnect control arm from axle bracket. On upper arms, disconnect arm from crossmember and on lower arms, disconnect arm from frame attachment bracket.
5. Reverse procedure to install.

TRACK BAR, REPLACE

1. Remove cover from track bar axle attachment and disconnect track bar from mounting stud.
2. Disconnect track bar from frame side rail.
3. Reverse procedure to install.

Front End and Steering Section

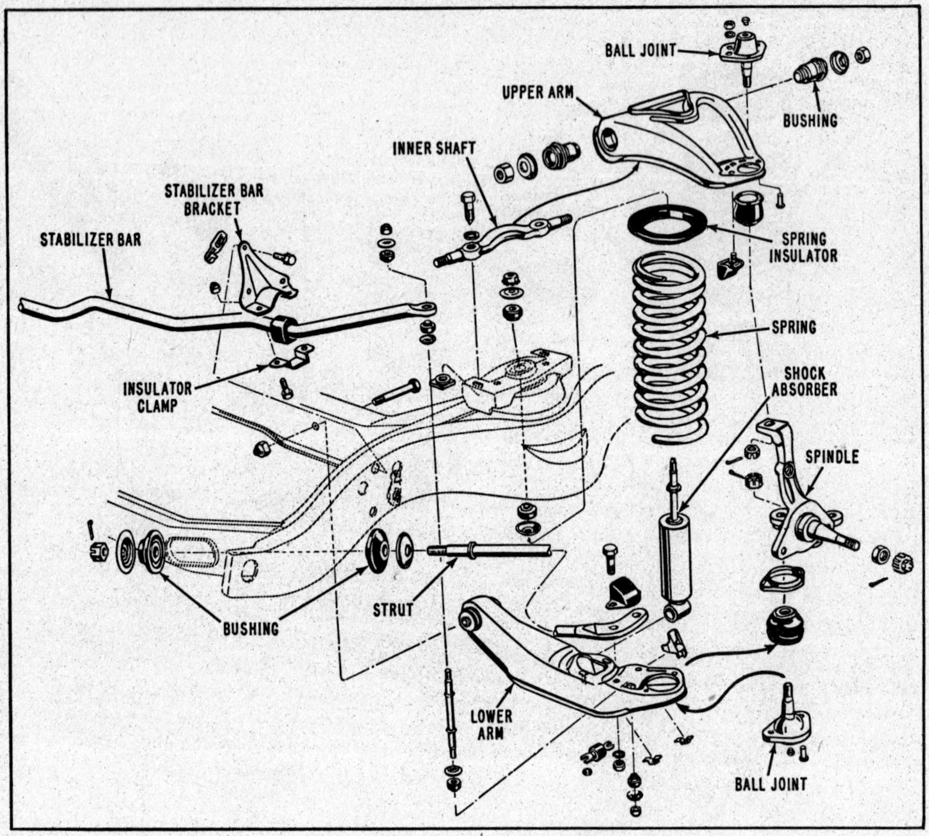

Fig. 1 Front suspension. 1969-76 Ford & Mercury

FRONT SUSPENSION
1969-76

Referring to Fig. 1, the construction of the front suspension differs from earlier models in that the lower control arm pivots on a bolt in the front crossmember. The struts, which are connected between the lower control arms and frame crossmember, prevent the control arms from moving forward and backward.

LUBRICATION
1969-76

Ball joints are prelubricated with a special lubricant. The lubricating interval is 36,000 miles. At these intervals, remove the plugs, apply the special lubricant, remove the fittings and replace the plugs.

SERVICE BULLETIN

Some uninformed service people recommend that conventional grease fittings be installed and that the car be lubricated every 1000 miles. This is completely unnecessary and, in fact, may cause damage to the special seals used in the lubrication points.

The use of conventional lubricants not only can do damage to the special seals but is incompatible with the special lubricant. Moreover, after the special sealing plugs have been replaced by conventional grease fittings, dirt and water can enter and cause excessive wear, rendering the units unfit for further service.

SERVICE BULLETIN, 1969-76

BALL JOINT LUBE: The ball joint seals on these models have been redesigned to provide improved sealing and longer life. The new seals can be damaged and the sealing characteristics destroyed if excessive lubricant is used. Specifications call for the addition of only 10 grams (level teaspoon) of lubricant to the ball joints at 36,000 miles intervals. The initial application of 10 grams of lubricant insures forcing grease into the bearing area and still allows for three subsequent lubrications of 10 grams each without ballooning the seals and resultant premature failure.

For the above reasons the ball joint seals on new cars might appear to be collapsed and give the mistaken impression that additional lubricant is required. This is not the case and under no circumstances should more than 10 grams of lubricant be added to the ball joints at the 36,000 mile intervals.

SERVICE BULLETIN

1969-76 STEERING LINKAGE LUBE: The steering linkage on these models should be lubricated at intervals of 36,000 miles. Normal breathing of socket joints permits moisture condensation within the joint. Moisture inside the joint assembly will cause no appreciable damage and the joint will function normally. However, if the moisture is concentrated in the bearing grease grooves and is frozen at the time of attempted lubrication, grease cannot flow and pressure greasing may damage the joint.

Do not attempt to lubricate the steering linkage on these vehicles if it has set in temperatures lower than 20 deg. above zero F. The vehicle should be allowed to warm up in a heated garage for 30 minutes or until the joints accept lubrication.

IMPORTANT: A torch must not be used to heat joints because this quantity of heat will melt the nylon bearing within the joint.

WHEEL ALIGNMENT

SERVICE BULLETIN

WHEEL BALANCING DIFFERS: On cars with disc brakes, dynamic balancing of the wheel-and-tire assembly on the car should not be attempted without first pulling back the shoe and lining assemblies from the rotor. If this is not done, brake drag may burn out the motor on the wheel spinner.

The drag can be eliminated by removing the wheel, taking out the two bolts holding the caliper splash shield, and detaching the shield. Then push the pistons into their cylinder bores by applying steady pressure on the shoes on each side of the rotor for at least a minute. If necessary, use waterpump pliers to apply the pressure.

After the pistons have been retracted, reinstall the splash shield and wheel. The wheel-and-tire assembly can then be dynamically balanced in the usual way. After the balancing job has been completed, be sure to pump the brake pedal several times until the shoes are seated and a firm brake pedal is obtained.

1969-76 Ford & Mercury

Caster and camber can be adjusted by loosening the bolts that attach the upper suspension arm to the shaft at the frame side rail, and moving the arm assembly

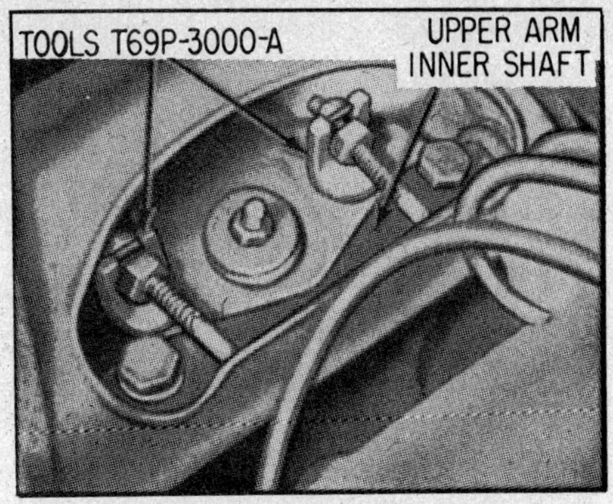

Fig. 2 Adjusting caster and camber. 1969-76 Ford and Mercury

in or out in the elongated bolt holes, Fig. 2. Since any movement of the arm affects both caster and camber, both factors should be balanced against one another when making the adjustment.

NOTE: Install the tool with the pins in the frame holes and the hooks over the upper arm inner shaft. Tighten the hook nuts snug before loosening the upper arm inner shaft attaching bolts.

Caster, Adjust

1. Tighten the tool front hook nut or loosen the rear hook nut as required to increase caster to the desired angle.
2. To decrease caster, tighten the rear hook nut or loosen the front hook nut as required.

NOTE: The caster angle can be checked without tightening the inner shaft retaining bolts.

3. Check the camber angle to be sure it did not change during the caster adjustment and adjust if necessary.
4. Tighten the upper arm inner shaft retaining bolts and remove tool.

Camber, Adjust

1. Loosen both inner shaft retaining bolts.
2. Tighten or loosen the hook nuts as necessary to increase or decrease camber.
3. Recheck caster and readjust if necessary.

TOE-IN, ADJUST

Position the front wheels in their straight-ahead position. Then turn both tie rod adjusting sleeves an equal amount until the desired toe-in setting is obtained.

WHEEL BEARINGS, ADJUST

1. With wheel rotating, tighten adjusting nut to 17-25 ft. lbs.
2. Back off adjusting nut ½ turn and retighten nut to 10-15 inch lbs.
3. Place nut lock on nut so that castellations on lock are aligned with cotter pin hole in spindle and install cotter pin.
4. Check front wheel rotation, if it rotates noisily or rough, clean, inspect or replace wheel bearings as necessary.

WHEEL BEARINGS, REPLACE

(Disc Brakes)

1. Raise car and remove front wheels.
2. Remove caliper mounting bolts.

NOTE: It is not necessary to disconnect the brake lines for this operation.

3. Slide caliper off of disc, inserting a spacer between the shoes to hold them in their bores after the caliper is removed. Position caliper assembly out of the way.

NOTE: Do not allow caliper to hang by brake hose.

4. Remove hub and disc assembly. Grease retainer and inner bearing can now be removed.

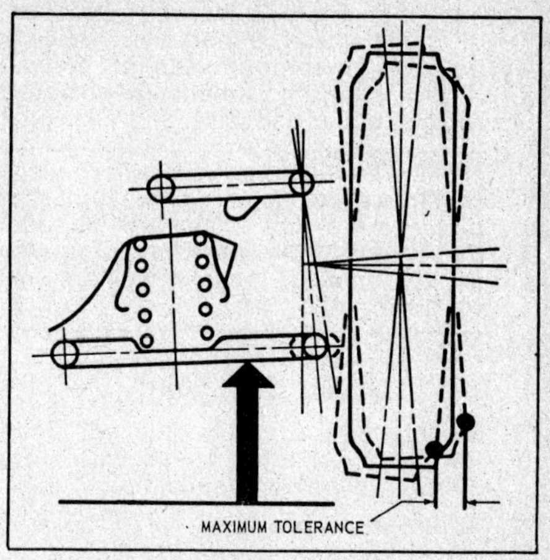

Fig. 3 Measuring lower ball joint radial play, which should not exceed ¼″

CHECKING BALL JOINTS FOR WEAR

Upper Ball Joint

1. Raise car on floor jacks placed beneath lower control arms.
2. Grasp lower edge of tire and move wheel in and out.
3. As wheel is being moved in and out, observe upper end of spindle and upper arm.

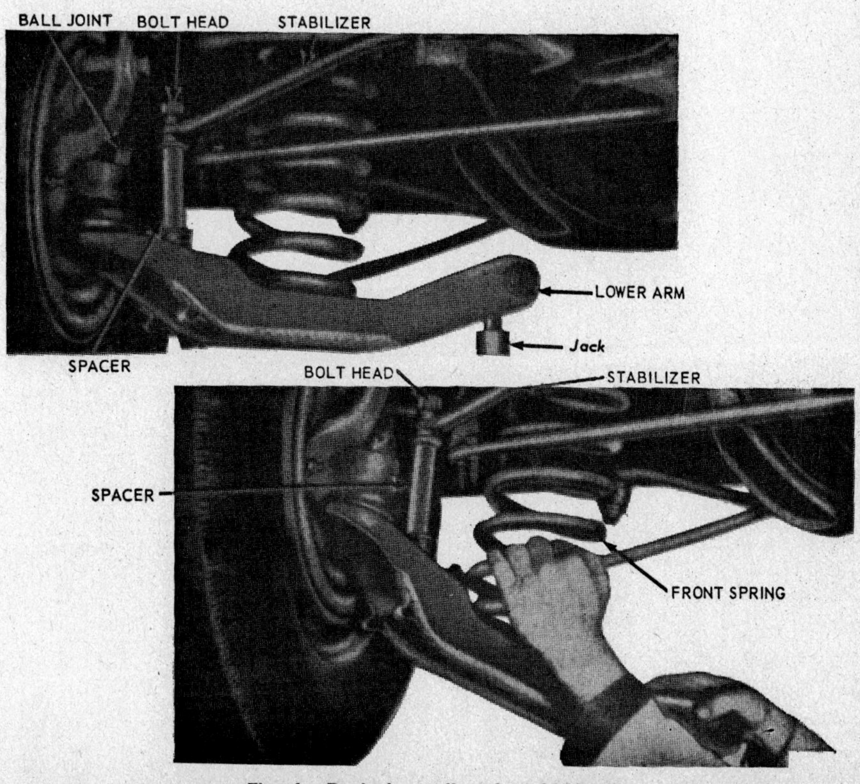

Fig. 4 Replacing coil spring. 1969-76

4. Any movement between upper end of spindle and upper arm indicates ball joint wear and loss of preload. If any such movement is observed, replace upper ball joint.

NOTE: During the foregoing check, the lower ball joint will be unloaded and may move. Disregard all such movement of the lower ball joint. Also, do not mistake loose wheel bearings for a worn ball joint.

Lower Ball Joint

1. Raise car on jacks placed under lower control arms as shown in Fig. 3. This will unload ball joints.
2. Adjust wheel bearings.
3. Attach a dial indicator to lower control arm and position so that its plunger rests against the inner side of the wheel rim adjacent to the lower ball joint.
4. Grasp tire at top and bottom and slowly move it in and out as shown in Fig. 3.
5. If reading on dial indicator exceeds ¼", replace lower ball joint.

SHOCK ABSORBER, REPLACE

To remove a shock absorber, unfasten it at the top and bottom and lower it through the opening in the lower control arm.

COIL SPRING, REPLACE

1. Raise vehicle and support front end of frame with jack stands.
2. Disconnect shock absorber from lower arm, collapse shock absorber into the spring and place a jack under the lower arm for support, Fig. 4.
3. Remove strut and rebound bumper bolts and disconnect lower end of sway bar stud from lower arm.
4. Remove the nut and bolt retaining the inner end of the lower arm to the crossmember.
5. Carefully lower jack to relieve spring pressure on lower arm, then remove the spring.

BALL JOINTS, REPLACE

NOTE: Ford Motor Company recommends that new ball joints should not be installed on used control arms and that the control arm be replaced if ball joint replacement is required. However, aftermarket ball joint repair kits which do not require control arm replacement, are available and can be installed using the following procedure.

The ball joints are riveted to the upper and lower control arms. The ball joints can be replaced on the car by removing the rivets and retaining the new ball joint to the control arm with the attaching bolts, nuts and washers furnished with the ball joint kit.

When removing a ball joint, use a suitable pressing tool to force the ball joint out of the spindle.

STEERING GEAR, REPLACE

1969-71 Ford & Mercury

1. Remove flex coupling retaining bolts.
2. Remove pitman arm to sector shaft nut and lock washer.
3. Using suitable puller, remove pitman arm from sector shaft.

NOTE: On vehicles equipped with manual transmission, it may be necessary to disconnect the clutch linkage, and on vehicles equipped with V8 engines, it may be necessary to lower the exhaust system.

4. Remove steering gear retaining bolts and remove gear.

POWER STEERING GEAR

1969-76 Ford & Mercury

1. Disconnect pressure and return lines from steering gear. Plug lines and ports in gear to prevent entry of dirt.
2. Remove two bolts that secure flex coupling to steering gear and to column.
3. Raise car and remove sector shaft nut.
4. Use a puller to remove pitman arm.
5. If car has a standard transmission, remove clutch release lever retracting spring to provide clearance for gear removal.
6. Support steering gear, then remove attaching bolts.
7. Work steering gear free of flex coupling and remove it from car.
8. Reverse procedure to install.

FORD & MERCURY

Compact & Intermediate Models

NOTE: The 1974-76 Ford Mustang II and Pinto and Mercury Bobcat are located elsewhere in this manual.

OLD CAR SPECIFICATIONS: For 1946-68 Tune Up and Wheel Alignment Specifications see main index.

INDEX OF SERVICE OPERATIONS

ENGINE & SERIAL NUMBER LOCATION: Vehicle warranty plate on rear face of left front door.

ENGINE IDENTIFICATION: Engine code is last letter in serial number on vehicle warranty plate.

Year	Engine	Engine Code
1969	6-170	U
	6-200	T
	6-250	L
	V8-302①	F
	V8-351①	H
	V8-351②	M
	V8-390②	S
	V8-427②	W
	V8-428	Q
	V8-428⑧	R
1970	6-170	U
	6-200	T
	6-250	L
	V8-302	F
	V8-302 "Boss"	G
	V8-351①	H
	V8-351②	M
	V8-428 CJ	Q
	V8-428 CJ⑧	R
	V8-429	N
	V8-429 CJ	C
	V8-429 "Boss"	Z
1971	6-170	U
	6-200	T
	6-250	L
	V8-302	F
	V8-302 H.O.	G
	V8-351①	H
	V8-351②	M
	V8-351②	Q (GT)
	V8-429 CJ	C
	V8-429 SCJ	J

Year	Engine	Engine Code
1972	6-170	U
	6-200	T
	6-250	L
	V8-302	F
	V8-351①	H
	V8-351②	Q
	V8-351⑥	R
	V8-400	S
	V8-429	N
1973-74	6-200	T
	6-250	L
	8-302	F
	8-351①	H
	8-351②	Q
	8-351⑥	R
	8-400	S
	8-429	N
	8-460	A
	8-460⑨	C
1975-76	6-200	T
	6-250	L
	8-302	F
	8-351	H
	8-400	S
	8-460	A
	8-460⑨	C

①—Two barrel carburetor.
②—Four barrel carburetor.
④—With transistorized ignition.
⑤—Premium fuel.
⑥—High Performance.
⑦—Special.
⑧—Ram Air.
⑨—Police Interceptor.

GRILLE IDENTIFICATION

1969 Montego and Cyclone

1969 Fairlane, Cobra

1969 Torino GT

1969, 1970 (Early) Falcon

1969 Cougar

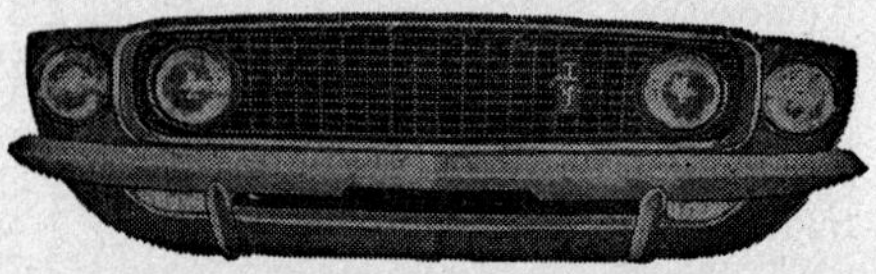

1969 Mustang

1970 Torino/Fairlane & 1970 (Late) Falcon

1970 Torino Cobra

1970 Torino GT

GRILLE IDENTIFICATION—Continued

1970 Cougar

1970 Cyclone

1970 Montego

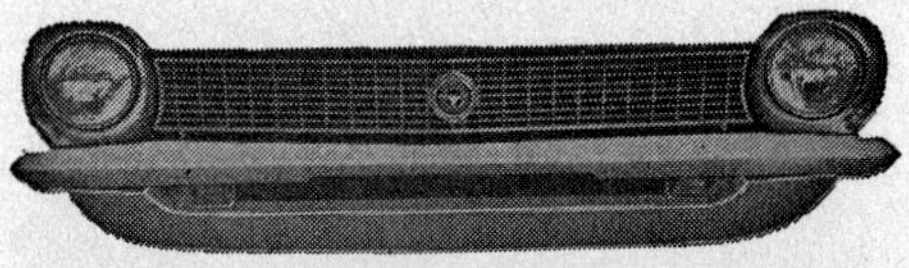
1970-72 Maverick

1970 Mustang

1970 Mustang Mach 1

1971-72 Maverick Grabber

1971 Cyclone

1971 Montego

1971-72 Cougar

1971-72 Comet

1971-72 Cougar XR7

1971-72 Mustang

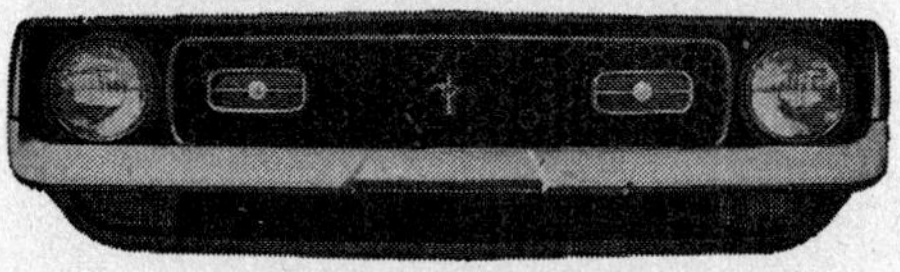
1971-72 Mustang Mach 1

1971 Torino

1972 Torino

1972 Gran Torino

1972 Montego

1972 Montego GT

1973 Comet

1973 Cougar

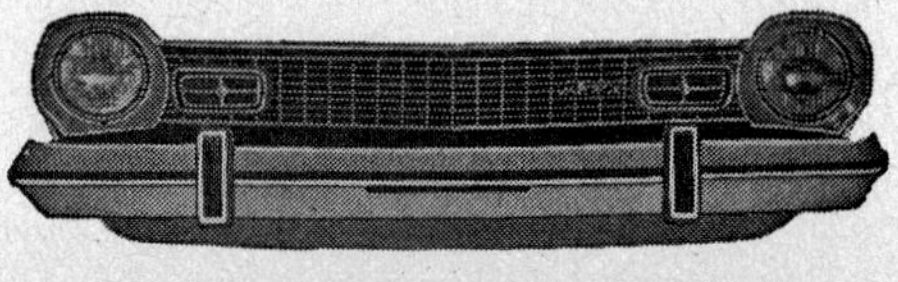
1973 Maverick

1973 Montego

1973 Mustang

GRILLE IDENTIFICATION—Continued

1973 Gran Torino

1973 Torino

1974 Comet

1974 Cougar

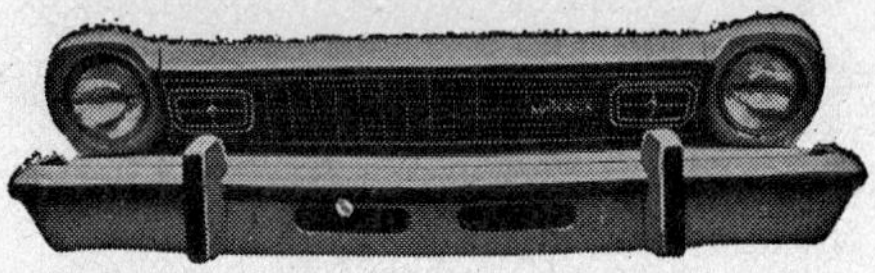

1974-75 Maverick

1974 Montego

1974 Tcrino

1974-75 Gran Torino

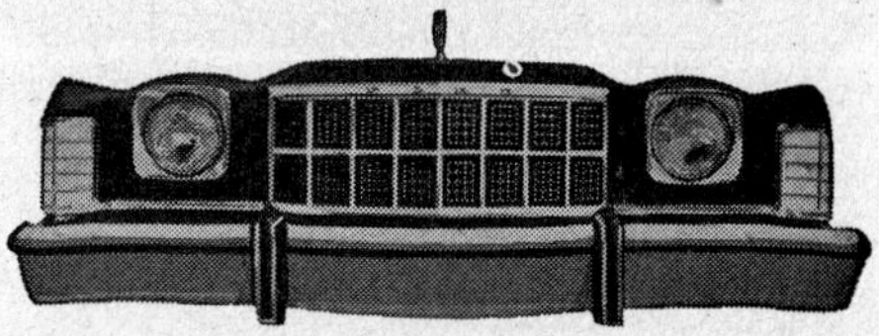

1974-76 Elite

1975-76 Granada

1975 Comet

1975-76 Cougar

1975-76 Montego

1975-76 Monarch

1976 Maverick

1976 Comet

1976 Torino

GENERAL ENGINE SPECIFICATIONS

Year	Engine	Carburetor	Bore and Stroke	Piston Displacement, Cubic Inches	Compression Ratio	Maximum Brake H.P. @ R.P.M.	Maximum Torque Lbs. Ft. @ R.P.M.	Normal Oil Pressure Pounds
1969	105 Horsepower..............6-170	1 Barrel	3.50 x 2.94	170	9.1	105 @ 4400	158 @ 2400	35–60
	120 Horsepower..............6-200	1 Barrel	3.68 x 3.13	200	8.8	120 @ 4400	190 @ 2400	35–60
	155 Horsepower..............6-250	1 Barrel	3.68 x 3.91	250	9.0①	155 @ 4000	240 @ 1600	35–60
	210 Horsepower............V8-302	2 Barrel	4.00 x 3.00	302	9.5	210 @ 4400	295 @ 2400	35–60
	290 Horsepower "H.O."......V8-302	4 Barrel	4.00 x 3.00	302	10.5	290 @ 5800	290 @ 4300	35–60
	250 Horsepower............V8-351	2 Barrel	4.00 x 3.50	351	9.5	250 @ 4600	355 @ 2600	35–60
	290 Horsepower............V8-351	4 Barrel	4.00 x 3.50	351	10.7	290 @ 4800	385 @ 3200	35–60
	320 Horsepower............V8-390	4 Barrel	4.05 x 3.78	390	10.5	320 @ 4800	427 @ 3200	35–60
	390 Horsepower............V8-427	4 Barrel	4.23 x 3.78	427	10.9	390 @ 5600	460 @ 3200	35–60
	335 Horsepower............V8-428	4 Barrel	4.13 x 3.98	428	10.6	335 @ 5200	440 @ 3400	35–60
	Ram Air...............V8-428	4 Barrel	4.13 x 3.98	428	10.6	—	—	35–60
1970	105 Horsepower..............6-170	1 Barrel	3.50 x 2.94	170	9.1	105 @ 4400	158 @ 2400	35–60
	120 Horsepower..............6-200	1 Barrel	3.68 x 3.13	200	8.8	120 @ 4400	190 @ 2400	35–60
	155 Horsepower..............6-250	1 Barrel	3.68 x 3.91	250	9.1	155 @ 4000	240 @ 1600	35–60
	210 Horsepower............V8-302	2 Barrel	4.00 x 3.00	302	9.5	210 @ 4400	295 @ 2400	35–60
	290 Horsepower "BOSS"......V8-302	4 Barrel	4.00 x 3.00	302	10.5	290 @ 5800	290 @ 4300	35–60
	250 Horsepower............V8-351	2 Barrel	4.00 x 3.50	351	9.5	250 @ 4600	355 @ 2600	35–60
	300 Horsepower............V8-351	4 Barrel	4.00 x 3.50	351	11.0	300 @ 5400	380 @ 3400	35–60
	335 Horsepower "CJ"........V8-428	4 Barrel	4.13 x 3.98	428	10.6	335 @ 5200	440 @ 3400	35–60
	Ram Air...............V8-428	4 Barrel	4.13 x 3.98	428	10.6	—	—	35–60
	360 Horsepower............V8-429	4 Barrel	4.36 x 3.59	429	11.0	360 @ 4600	476 @ 2800	35–60
	345 Horsepower "CJ"........V8-429	4 Barrel	4.36 x 3.59	429	11.5	345 @ 5800	450 @ 3400	35–60
	375 Horsepower "Boss"......V8-429	4 Barrel	4.36 x 3.59	429	10.5	375 @ 5200	450 @ 3400	20–60
1971	100 Horsepower..............6-170	1 Barrel	3.50 x 2.94	170	8.7	100 @ 4200	148 @ 2600	35–60
	115 Horsepower..............6-200	1 Barrel	3.68 x 3.13	200	8.7	115 @ 4000	180 @ 2200	35–60
	145 Horsepower..............6-250	1 Barrel	3.68 x 3.91	250	9.0	145 @ 4000	232 @ 1600	35–60
	210 Horsepower............V8-302	2 Barrel	4.00 x 3.00	302	9.0	210 @ 4600	296 @ 2600	35–60
	290 Horsepower H.O.........V8-302	4 Barrel	4.00 x 3.00	302	9.4	290 @ 5800	290 @ 4300	35–60
	240 Horsepower............V8-351	2 Barrel	4.00 x 3.50	351	9.0	240 @ 4600	350 @ 2600	35–60
	285 Horsepower............V8-351	4 Barrel	4.00 x 3.50	351	10.7	285 @ 5400	370 @ 3400	35–60
	330 Horsepower............V8-351	4 Barrel	4.00 x 3.50	351	11.7	330 @ 5400	370 @ 4000	35–60
	370 Horsepower "CJ"........V8-429	4 Barrel	4.36 x 3.59	429	11.3	370 @ 5400	450 @ 3400	35–60
	375 Horsepower "SCJ".......V8-429	4 Barrel	4.36 x 3.59	429	11.5	375 @ 5600	450 @ 3400	35–60
1972	82 Horsepower②.............6-170	1 Barrel	3.50 x 2.94	170	8.3	82 @ 4400	129 @ 1800	35–60
	91 Horsepower②.............6-200	1 Barrel	3.68 x 3.13	200	8.3	91 @ 4000	154 @ 2200	35–60
	95 Horsepower②.............6-250	1 Barrel	3.68 x 3.91	250	8.0	95 @ 3600	181 @ 1600	35–60
	98 Horsepower②.............6-250	1 Barrel	3.68 x 3.91	250	8.0	98 @ 3600	183 @ 1600	35–60
	99 Horsepower②.............6-250	1 Barrel	3.68 x 3.91	250	8.0	99 @ 3600	184 @ 1600	35–60
	140 Horsepower②...........V8-302	2 Barrel	4.00 x 3.00	302	8.5	140 @ 4000	230 @ 2200	35–60
	141 Horsepower②...........V8-302	2 Barrel	4.00 x 3.00	302	8.5	141 @ 4000	242 @ 2000	35–60
	143 Horsepower②...........V8-302	2 Barrel	4.00 x 3.00	302	8.5	143 @ 4200	242 @ 2000	35–60
	161 Horsepower②...........V8-351	2 Barrel	4.00 x 3.50	351	8.6	161 @ 4000	276 @ 2000	35–60
	164 Horsepower②...........V8-351	2 Barrel	4.00 x 3.50	351	8.6	164 @ 4000	276 @ 2000	35–85
	177 Horsepower②...........V8-351	2 Barrel	4.00 x 3.50	351	8.6	177 @ 4000	284 @ 2000	35–85
	248 Horsepower② "CJ".......V8-351	4 Barrel	4.00 x 3.50	351	8.6	248 @ 5400	299 @ 3800	50–70
	262 Horsepower②...........V8-351	4 Barrel	4.00 x 3.50	351	8.6	262 @ 5400	299 @ 3600	35–85
	266 Horsepower②...........V8-351	4 Barrel	4.00 x 3.50	351	8.6	266 @ 5400	301 @ 3600	35–85
	168 Horsepower②...........V8-400	2 Barrel	4.00 x 4.00	400	8.4	168 @ 4200	297 @ 2200	35–85
	205 Horsepower②...........V8-429	4 Barrel	4.36 x 3.59	429	8.5	205 @ 4400	322 @ 2600	35–60
1973	84 Horsepower..............6-200	1 Barrel	3.68 x 3.13	200	8.3	84 @ 3600	151 @ 1800	30–50
	88 Horsepower②.............6-250	1 Barrel	3.68 x 3.91	250	8.0	88 @ 3200	196 @ 1600	35–60
	92 Horsepower②.............6-250	1 Barrel	3.68 x 3.91	250	8.0	92 @ 3200	197 @ 1600	35–60
	95 Horsepower②.............6-250	1 Barrel	3.68 x 3.91	250	8.0	95 @ 3200	199 @ 1600	35–60

Continued

GENERAL ENGINE SPECIFICATIONS—Continued

Year	Engine	Carburetor	Bore and Stroke	Piston Displacement, Cubic Inches	Compression Ratio	Maximum Brake H.P. @ R.P.M.	Maximum Torque Lbs. Ft. @ R.P.M.	Normal Oil Pressure Pounds
1973	135 Horsepower②............V8-302	2 Barrel	4.00 x 3.00	302	8.0	135 @ 4200	228 @ 2200	35–60
	136 Horsepower②............V8-302	2 Barrel	4.00 x 3.00	302	8.0	136 @ 4200	232 @ 2200	35–60
	137 Horsepower②............V8-302	2 Barrel	4.00 x 3.00	302	8.0	137 @ 4200	230 @ 2200	35–60
	138 Horsepower②............V8-302	2 Barrel	4.00 x 3.00	302	8.0	138 @ 4200	234 @ 2200	35–60
	154 Horsepower②............V8-351	2 Barrel	4.00 x 3.50	351	8.0	154 @ 4000	246 @ 2400	35–60
	156 Horsepower②............V8-351	2 Barrel	4.00 x 3.50	351	8.0	156 @ 3800	260 @ 2400	35–60
	159 Horsepower②............V8-351	2 Barrel	4.00 x 3.50	351	8.0	159 @ 4000	250 @ 2400	35–60
	246 Horsepower②............V8-351	4 Barrel	4.00 x 3.50	351	8.0	246 @ 5400	312 @ 3600	35–60
	264 Horsepower②............V8-351	4 Barrel	4.00 x 3.50	351	8.0	264 @ 5400	314 @ 3600	35–60
	163 Horsepower②............V8-400	2 Barrel	4.00 x 4.00	400	8.0	163 @ 3800	300 @ 2000	35–60
	168 Horsepower②............V8-400	2 Barrel	4.00 x 4.00	400	8.0	168 @ 3800	310 @ 2000	35–60
	197 Horsepower②............V8-429	4 Barrel	4.36 x 3.59	429	8.0	197 @ 4400	320 @ 2600	35–60
	201 Horsepower②............V8-429	4 Barrel	4.36 x 3.59	429	8.0	201 @ 4400	322 @ 2600	35–60
	274 Horsepower..............V8-460	4 Barrel	4.36 x 3.85	460	8.8	274 @ 4600	392 @ 2800	35–65
1974	84 Horsepower②..............6-200	1 Barrel	3.68 x 3.13	200	8.0	84 @ 3800	150 @ 1800	30–50
	91 Horsepower②..............6-250	1 Barrel	3.68 x 3.91	250	8.0	91 @ 3200	190 @ 1600	40–60
	140 Horsepower②............V8-302	2 Barrel	4.00 x 3.00	302	8.0	140 @ 3800	230 @ 2600	40–60
	162 Horsepower②............V8-351	2 Barrel	4.00 x 3.50	351	8.0	162 @ 4000	275 @ 2200	40–65
	163 Horsepower②............V8-351	2 Barrel	4.00 x 3.50	351	8.0	163 @ 4200	278 @ 2000	45–75
	255 Horsepower②............V8-351	4 Barrel	4.00 x 3.50	351	8.0	255 @ 5600	290 @ 3400	45–75
	170 Horsepower②............V8-400	2 Barrel	4.00 x 4.00	400	8.0	170 @ 3400	330 @ 2000	45–75
	195 Horsepower②............V8-460	4 Barrel	4.36 x 3.85	460	8.0	195 @ 3800	335 @ 2600	35–65
	220 Horsepower②③..........V8-460	4 Barrel	4.36 x 3.85	460	8.0	220 @ 4000	355 @ 2600	35–65
	260 Horsepower②............V8-460	4 Barrel	4.36 x 3.85	460	8.0	260 @ 4400	355 @ 2700	35–65
1975	Horsepower②..............6-200	1 Barrel	3.68 x 3.13	200	8.3	—	—	30–50
	70 Horsepower②⑤............6-250	1 Barrel	3.68 x 3.91	250	8.0	70 @ 2800	175 @ 1400	40–60
	72 Horsepower②④............6-250	1 Barrel	3.68 x 3.91	250	8.0	72 @ 2900	180 @ 1400	40–60
	122 Horsepower②⑥..........V8-302	2 Barrel	4.00 x 3.00	302	8.0	122 @ 3800⑥	208 @ 1800⑦	40–60
	129 Horsepower②⑧..........V8-302	2 Barrel	4.00 x 3.00	302	8.0	129 @ 3800⑧	220 @ 1800⑨	40–60
	148 Horsepower②④.........V8-351M	2 Barrel	4.00 x 3.50	351	8.0	148 @ 3800	243 @ 2400	50–75
	150 Horsepower②⑤.........V8-351M	2 Barrel	4.00 x 3.50	351	8.0	150 @ 3800	244 @ 2800	50–75
	143 Horsepower②⑩........ V8-351W	2 Barrel	4.00 x 3.50	351	8.1	143 @ 3600⑩	255 @ 2200⑪	40–65
	154 Horsepower②⑫.........V8-351W	2 Barrel	4.00 x 3.50	351	8.1	154 @ 3800⑫	268 @ 2200	40–65
	144 Horsepower②⑤..........V8-400	2 Barrel	4.00 x 4.00	400	8.0	144 @ 3600	255 @ 2200	50–65
	158 Horsepower②④..........V8-400	2 Barrel	4.00 x 4.00	400	8.0	158 @ 4000	276 @ 2000	50–65
	216 Horsepower②④..........V8-460	4 Barrel	4.36 x 3.85	460	8.0	216 @ 4000	366 @ 2600	40–65
	217 Horsepower②⑤..........V8-460	4 Barrel	4.36 x 3.85	460	8.0	217 @ 4000	365 @ 2600	40–65
1976	Horsepower②..............6-200	1 Barrel	3.682 x 3.126	200	—	—	—	30–50
	Horsepower②..............6-250	1 Barrel	3.682 x 3.91	250	—	—	—	40–60
	Horsepower②............V8-302	2 Barrel	4.00 x 3.00	302	—	—	—	40–60
	Horsepower②............V8-351	2 Barrel	4.00 x 3.50	351	—	—	—	45–65⑬
	Horsepower②............V8-400	2 Barrel	4.00 x 4.00	400	—	—	—	35–65
	Horsepower②............V8-460	4 Barrel	4.36 x 3.85	460	—	—	—	35–65

①—Mustang "E" Model, 9.5.
②—Ratings are NET—as installed in the vehicle.
③—Cougar XR-7 requires A/C in California.
④—Except California.
⑤—California.
⑥—Comet & Maverick, California vehicles rated at 115 H.P. @ 3600 RPM.
⑦—California vehicles rated at 203 @ 1400 RPM.
⑧—Granada & Monarch, California vehicles rated at 115 H.P. @ 3600 RPM.
⑨—California vehicles rated at 203 @ 1800 RPM.
⑩—Granada, Monarch, Torino & Elite, California vehicles rated at 153 H.P. @ 3400 RPM.
⑪—California vehicles rated at 270 @ 2400 RPM.
⑫—Cougar & Montego, not available in California.
⑬—V8-351M: 45–75 pounds.

TUNE UP SPECIFICATIONS

The following specifications are published from the latest information available. This data should be used only in the absence of a decal affixed in the engine compartment.

OLD CAR SPECIFICATIONS: For 1946-68 Tune Up Specifications see main index.

★When using a timing light, disconnect vacuum hose or tube at distributor and plug opening in hose or tube so idle speed will not be affected.

●When checking compression, lowest cylinder must be within 75% of the highest.

▲Before removing wires from distributor cap, determine location of the No. 1 wire in cap, as distributor position may have been altered from that shown at the end of this chart.

Year	Spark Plug		Distributor		Ignition Timing★			Carb. Adjustments					
								Hot Idle Speed③		Air Fuel Ratio		Idle "CO" %	
	Type	Gap Inch	Point Gap Inch	Dwell Angle Deg.	Firing Order Fig. ▲	Timing BTDC ①	Mark Fig.	Std. Trans.	Auto. Trans.②	Std. Trans.	Auto. Trans.	Std. Trans.	Auto. Trans.
1969													
6-170	BF-82	.034	.027	35–40	H	6°	G	750	550D	13.80 to 1	13.60 to 1	2.2	2.6
6-200	BF-82	.034	.027	35–40	H	6°	G	750	550D	14.50 to 1	14.00 to 1	0.9	1.8
6-250 Less Air Cond.	BF-82⑱	.034	.025	37–42	H	6°	G	700	550D	14.0 to 1	13.5 to 1	1.8	2.8
6-250 With Air Cond.	BF-82⑱	.034	.025	37–42	H	6°	G	700/500⑥	550/450D⑥	14.0 to 1	13.5 to 1	1.8	2.8
V8-302 Std. Trans.	BF-42	.034	.021	24–29	E	6°	C	650	—	14.0 to 1	—	1.8	—
V8-302 Auto. Trans.	BF-42	.034	.017	26–31	E	6°	C	—	550D	—	13.8 to 1	—	2.2
V8-302 "H.O."	AF-32	.030	.020	30–33	E	16°	C	800	—	—	—	—	—
V8-351 2 B. Carb.	BF-42	.034	.017	26–31	F	6°	C	650	550D	14.0 to 1	14.0 to 1	1.8	1.8
V8-351 4 B. Carb.	BF-32	.034	.017	26–31	F	6°	C	650	550D	14.0 to 1	14.3 to 1	1.8	1.2
V8-390 Std. Trans.	BF-42	.034	.017	26–31	E	6°	B	700	—	㉕	—	㉗	—
V8-390 Auto. Trans.	BF-32	.034	.017	26–31	E	6°	B	—	550D	—	㉖	—	㉔
V8-427	BF-32	.034	.017	26–31	E	6°	C	—	600D	—	—	—	—
V8-428	BF-32	.034	.017	26–31	E	6°	B	700	650D	14.1 to 1	14.1 to 1	1.6	1.6
1970													
6-170	BF-82	.035	.027	35–40	H	6°	G	750	550	14.45 to 1	14.45 to 1	1.0	1.0
6-200	BF-82	.035	.027	35–40	H	6°	G	750	550	14.45 to 1	14.20 to 1	1.0	1.5
6-250	BF-82	.035	.027	35–40	H	6°	G	800/500⑥	500	14.20 to 1	14.20 to 1	1.5	1.5
V8-302	BF-42	.035	.021	24–29	E	6°	C	800/500⑥	600/500D⑥	14.0 to 1	13.80 to 1	0.34	0.44
V8-302 "BOSS"	AF-32	.034	.020	30–33	E	16°	C	800/500⑥	—	13.50 to 1	—	2.80	—
V8-351⑦⑯	AF-42	.034	⑪	⑪	F	6°	C	700/500⑥	600/500D⑥	14.0 to 1	14.0 to 1	0.30	0.14
V8-351⑦⑰ Std. Tr.	BF-42	.034	⑪	⑪	F	6°	C	700/500⑥	—	14.0 to 1	—	0.30	—
V8-351⑦⑰ Auto. Tr.	BF-42	.034	⑪	⑪	F	10°	C	700/500⑥	—	—	14.0 to 1	—	0.14
V8-351⑧	AF-32	.034	⑪	⑪	F	6°	C	800/500⑥	600/500D⑥	14.44 to 1	14.45 to 1	0.90	0.75
V8-428	BF-32	.034	⑪	⑪	E	6°	B	800/500⑥	600/500D⑥	13.8 to 1	14.3 to 1	2.15	0.95
V8-429 Auto. Trans.⑧	BF-42	.034	⑪	⑪	E	4°	C	700/500⑥	700/500⑥	—	14.45 to 1	—	0.75
V8-429 "CJ"	AF-32	.034	⑪	⑪	E	10°	C	700	600	14.4 to 1	14.45 to 1	0.35	0.35
V8-429 "BOSS"	AF-32	.034	.020	30–33	E	10°	D	700	—	14.2 to 1	—	2.3	—
1971													
6-170	BRF-82	.034	.027	35–40	H	6°	A	750③	—	14.5 to 1	14.5 to 1	1.0	1.0
6-200	BRF-82	.034	.027	35–40	H	6°	A	750③	550D③	14.5 to 1	14.2 to 1	1.0	1.5
6-250	BRF-82	.034	⑤	⑤	H	6°	A	750/500⑥	600/500D⑥	14.2 to 1	14.2 to 1	1.5	1.5
V8-302	BRF-42	.034	.021	24–29	E	6°	I	800/500⑥	600/500D⑥	14.5 to 1	14.5 to 1	0.3	0.4
V8-302 H.O.	AF-32	.034	.020	30–33	E	16°	I	800/500⑥	—	14.5 to 1	14.5 to 1	0.3	0.4
V8-351⑦	ARF-42	.034	⑪	⑪	F	6°	I	700/500⑥	600D	14.5 to 1	14.5 to 1	0.2	0.2
V8-351⑧	ARF-42	.034	.021	24–29	F	6°	I	800/500⑥	600D	14.5 to 1	—	1.1	—
V8-429 "CJ"	AF-32	.034	.021	24–29	E	10°	C	700/500⑥	650/500D⑥	14.5 to 1	14.5 to 1	0.3	0.3
V8-429 "SCJ"	AF-32	.034	.021	24–29	E	10°	C	650/500⑥	700/500D⑥	14.5 to 1	14.5 to 1	0.3	0.3

Continued

TUNE UP SPECIFICATIONS—Continued

The following specifications are published from the latest information available. This data should be used only in the absence of a decal affixed in the engine compartment.

OLD CAR SPECIFICATIONS: For 1946-68 Tune Up Specifications see main index.

★When using a timing light, disconnect vacuum hose or tube at distributor and plug opening in hose or tube so idle speed will not be affected.

●When checking compression, lowest cylinder must be within 75% of the highest.

▲Before removing wires from distributor cap, determine location of the No. 1 wire in cap, as distributor position may have been altered from that shown at the end of this chart.

Year	Spark Plug		Distributor		Ignition Timing★			Carb. Adjustments					
								Hot Idle Speed③		Air Fuel Ratio		Idle "CO" %	
	Type	Gap Inch	Point Gap Inch	Dwell Angle Deg.	Firing Order Fig. ▲	Timing BTDC ①	Mark Fig.	Std. Trans.	Auto. Trans.②	Std. Trans.	Auto. Trans.	Std. Trans.	Auto. Trans.
1972													
6-170	BRF-82	.034	.027	35–39	H	6°	A	750	—	—	—	1.2	1.2
6-200 Less Air Cond.	BRF-82	.034	.027	35–39	H	6°	A	750	550D	—	—	0.8	—
6-200 With Air Cond.	BRF-82	.034	.027	35–39	H	6°	A	800/500⑥	600/500D⑥	—	—	—	1.2
6-250	BRF-82	.034	.027	35–39	H	6°	A	550	550D	—	—	1.0	1.0
6-250	BRF-82	.034	.027	35–39	H	6°	A	750/500⑥	600/500D⑥	—	—	1.0	1.0
V8-302	BRF-42	.034	.017	26–30	E	6°	I	575	575D	—	—	—	0.19
V8-302	BRF-42	.034	.017	26–30	E	6°	I	800/500⑥	600/500D⑥	—	—	—	0.19
V8-351⑦	ARF-42	.034	.017	26–30	F	6°	I	750/500⑥	575/500D⑥	—	—	0.50	0.50
V8-351⑦ Calif.	ARF-42	.034	.017	26–30	F	6°	I	—	625/500D⑥	—	—	0.50	0.50
V8-351⑧ Std. Tr.	ARF-42	.034	.020	26–30	F	16°	I	1000/500⑥	—	—	—	0.5	—
V8-351⑧ Auto. Tr.	ARF-42	.034	.017	26–30	F	16°⑬	I	—	700/500D⑥	—	—	—	0.5
V8-351⑧ Auto. Tr. Calif.	ARF-42	.034	.017	26–30	F	16°	I	—	800/500D⑥	—	—	—	0.5
V8-351 H.O.	ARF-42	.034	.020	26–30	F	10°	I	1000/500⑥	—	—	—	—	—
V8-400	ARF-42	.034	.017	26–30	F	⑫	I	—	625/500D⑥	—	—	—	0.10
V8-429	BRF-42	.034	.017	26–30	E	10°	I	—	600/500D⑥	—	—	0.3	0.3
1973													
6-200	BRF-82	.034	.027	35–39	H	6°	C	800	650D	—	—	—	—
6-250	BRF-82	.034	.025	33–39	H	6°	A	750	600D	—	—	0.5	0.5
8-302	BRF-42	.034	.017	24–30	E	6°	I	800	650D	—	—	0.5⑲	0.5⑲
8-351⑦㉓	BRF-42	.034	.017	24–30	F	6°	I	—	625D	—	—	—	0.5⑲
8-351⑦㉒	ARF-42	.034	.017	24–30	F	10°	I	—	625D	—	—	0.5⑲	0.5⑲
8-351⑧	ARF-42	.034	㉑	⑨	F	⑳	I	900	700D	—	—	0.6⑲	0.5⑲
8-400	ARF-42	.034	.017	24–30	F	6°	I	—	625D	—	—	—	0.5⑲
8-429	ARF-42	.034	.017	24–30	E	18°	I	—	600D	—	—	—	0.5⑲
8-460	ARF-42	.034	.017	24–30	E	14°	I	—	600D	—	—	—	0.4⑲
1974													
6-200	BRF-82	.034	.025	33	H	6°	A	750	550D	—	—	—	—
6-200⑭	BRF-82	.034	—	—	H	6°	A	750	550D	—	—	—	—
6-250	BRF-82	.034	.025	33	H	6°	A	600	600D	—	—	—	—
6-250⑭	BRF-82	.034	—	—	H	6°	A	600	600D	—	—	—	—
8-302	BRF-42	.034	.017	26–30	E	6°	I	850	575D	—	—	—	.5
8-302⑭	BRF-42	.034	—	—	E	6°	I	850	575D	—	—	—	.5
8-351⑦㉓	BRF-42	.034	.017	26–30	F	6°	I	—	600D	—	—	—	.4
8-351⑦㉓⑭	BRF-42	.034	—	—	F	6°	I	—	600D	—	—	—	.4
8-351⑦㉒	ARF-42	.044	.017	26–30	F	10°	I	—	650D	—	—	—	.5
8-351⑦㉒⑭	ARF-42	.044	—	—	F	10°	I	—	650D	—	—	—	.5
8-351⑧	ARF-42	.034	.017㉑	26–31	F	⑳	I	900	800D	—	—	—	—
8-351⑧⑭	ARF-42	.034	—	—	F	⑳	I	900	800D	—	—	—	—
8-400⑭	ARF-42	.044	—	—	F	⑩	I	—	625D	—	—	—	—
8-460⑭	ARF-52	.054	—	—	E	⑮	I	—	650D	—	—	—	.25
1975													
6-200⑭	BRF-82	.044	—	—	H	6°	A	750	600D	—	—	—	—
6-250⑭	BRF-82	.044	—	—	J	6°	A	750	600D	—	—	—	—
V8-302⑭	ARF-42	.044	—	—	E	6°	I	900	650D	—	—	—	—

Continued

TUNE UP SPECIFICATIONS—Continued

The following specifications are published from the latest information available. This data should be used only in the absence of a decal affixed in the engine compartment.

OLD CAR SPECIFICATIONS: For 1946–68 Tune Up Specifications see main index.

★When using a timing light, disconnect vacuum hose or tube at distributor and plug opening in hose or tube so idle speed will not be affected.

●When checking compression, lowest cylinder must be within 75% of the highest.

▲Before removing wires from distributor cap, determine location of the No. 1 wire in cap, as distributor position may have been altered from that shown at the end of this chart.

Year	Spark Plug		Distributor		Ignition Timing★			Carb. Adjustments					
								Hot Idle Speed③		Air Fuel Ratio		Idle "CO"%	
	Type	Gap Inch	Point Gap Inch	Dwell Angle Deg.	Firing Order Fig. ▲	Timing BTDC ①	Mark Fig.	Std. Trans.	Auto. Trans.②	Std. Trans.	Auto. Trans.	Std. Trans.	Auto. Trans.
1975—Continued													
V8-351④⑭	ARF-42	.044	—	—	F	12°	I	—	650D	—	—	—	—
V8-351⑭㉓	ARF-42	.044	—	—	F	12°	I	—	650D	—	—	—	—
V8-400⑭	ARF-42	.044	—	—	F	12°	I	—	650D	—	—	—	—
V8-460⑭	ARF-52	.044	—	—	E	14°	I	—	650D	—	—	—	—
1976													
6-200⑭	BRF-82	.044	—	—	H	6°	A	800	650D	—	—	—	—
6-250⑭㉚	BRF-82	.044	—	—	J	㉙	A	850	600D	—	—	—	—
6-250⑭㉛	BRF-82	.044	—	—	J	㉜	A	850	600D	—	—	—	—
8-302⑭㉛㉝㉞	ARF-42	.044	—	—	E	4°	I	—	700D	—	—	—	—
8-302⑭㉚㉝	ARF-42	.044	—	—	E	㉙	I	750	650D	—	—	—	—
8-302⑭㉚㉟	ARF-42	.044	—	—	E	㊱	I	750	650D	—	—	—	—
8-302⑭㉛㉞㉟	ARF-42	.044	—	—	E	8°	I	—	700D	—	—	—	—
8-351⑭㉓	ARF-42	.044	—	—	F	10°	I	—	650D	—	—	—	—
8-351⑭④	ARF-42	.044	—	—	F	8°	I	—	650D	—	—	—	—
8-400⑭㉚	ARF-42	.044	—	—	F	㊲	I	—	650D	—	—	—	—
8-400⑭㉛㊳	ARF-42	.044	—	—	F	10°	I	—	625D	—	—	—	—
8-400⑭㉛㉙	ARF-42	.044	—	—	F	12°	I	—	625D	—	—	—	—
8-460⑭	ARF-52	.044	—	—	E	㊵㉙	I	—	650D	—	—	—	—

①—BTDC: Before top dead center.
②—D: Drive.
③—For A/C add 50 R.P.M. Set with headlamps on high beam and A/C off.
④—Modified Engine.
⑤—Dual diaphragm .027 gap, 35–40 dwell. Single diaphragm .025 gap, 37–42 dwell.
⑥—Higher figure is with throttle solenoid energized.
⑦—With two barrel carburetor.
⑧—With four barrel carburetor.
⑨—Manual trans. 32–35, Auto. trans. 24–30.
⑩—Cougar 12°, all others 6°.
⑪—Dual diaphragm .021 gap, 24°–29° dwell. Single diaphragm .017 gap, 26°–31° dwell.
⑫—Exc. Calif. 8°, Calif. 6°.
⑬—Cougar with 12 inch converter 10° BTDC.
⑭—Breakerless distributor.
⑮—Cougar 14°, all others 10°.
⑯—Fairlane & Montego.
⑰—Mustang & Cougar
⑱—Mustang "E" BF-92.
⑲—For California 0.2%.
⑳—Manual trans. 16° BTDC, Auto. trans. 18° BTDC.
㉑—Manual trans. .020, Auto. trans. .017.
㉒—Cleveland engine.
㉓—Windsor engine.
㉔—2 Bar. Carb. 1.1, 4 Bar. Carb. 0.5.
㉕—2 Bar. Carb. 14.1 to 1, 4 Bar Carb. 14.0 to 1.
㉖—2 Bar. Carb. 14.4 to 1, 4 Bar. Carb. 14.8 to 1.
㉗—2 Bar. Carb. 1.6, 4 Bar. Carb. 1.8.
㉘—California and Police Interceptor; 14°. All others; 10°.
㉙—With auto. trans.; 6°. With manual trans.; 4°.
㉚—Except Calif.
㉛—California.
㉜—With auto. trans.; 8°. With manual trans.; 6°.
㉝—Granada and Monarch.
㉞—With auto. trans.
㉟—Comet and Maverick.
㊱—With auto. trans.; 8°. With manual trans.; 4°.
㊲—Cougar; 10°. Elite, Montego, Torino; 12°.
㊳—Cougar.
㊴—Elite, Torino and Montego.
㊵—In drive with service and parking brake applied.

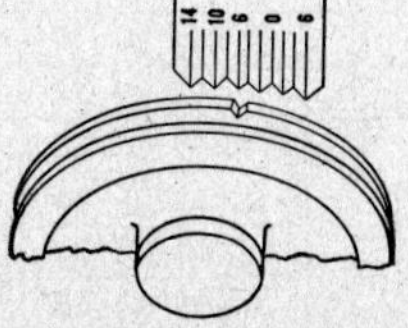

Fig. A

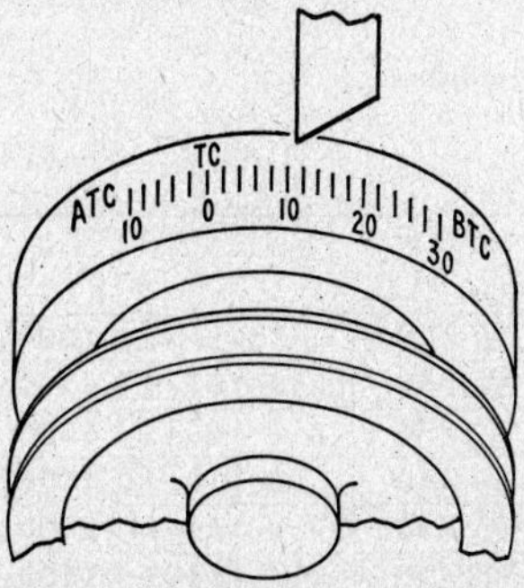

Fig. B

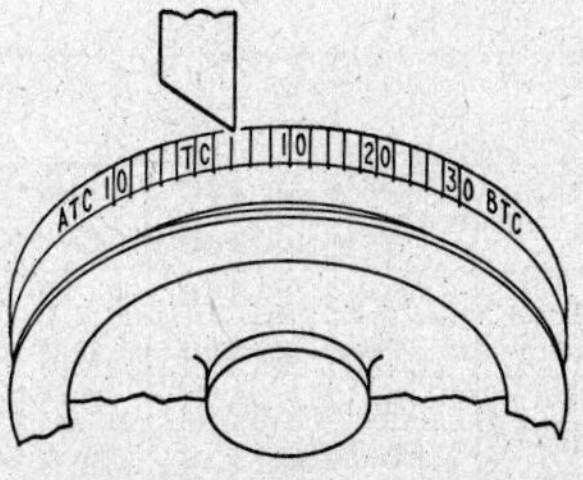

Fig. C

Continued

TUNE UP NOTES—Continued

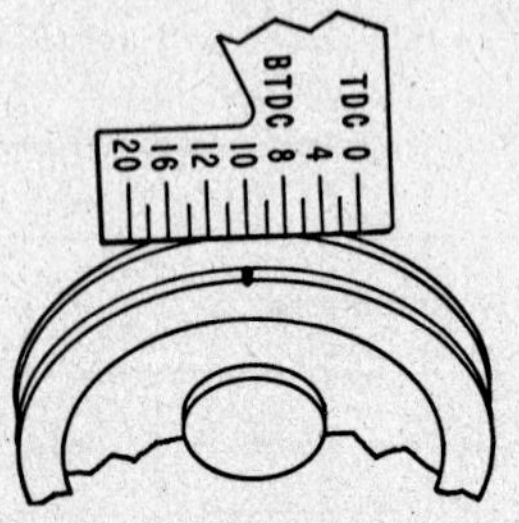

Fig. D

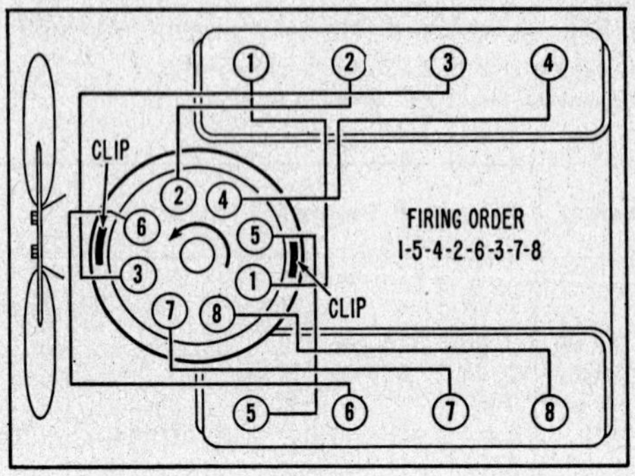

Fig. E

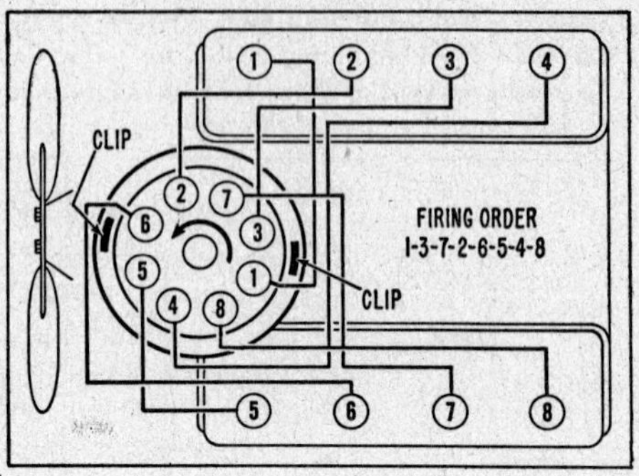

Fig. F

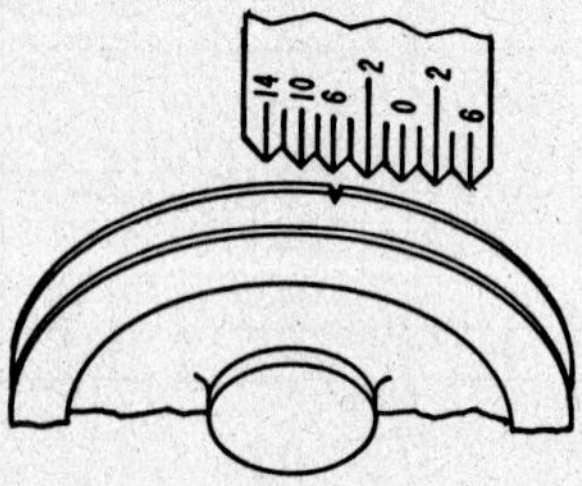

Fig. G

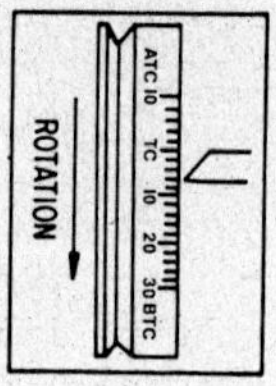

Fig. I

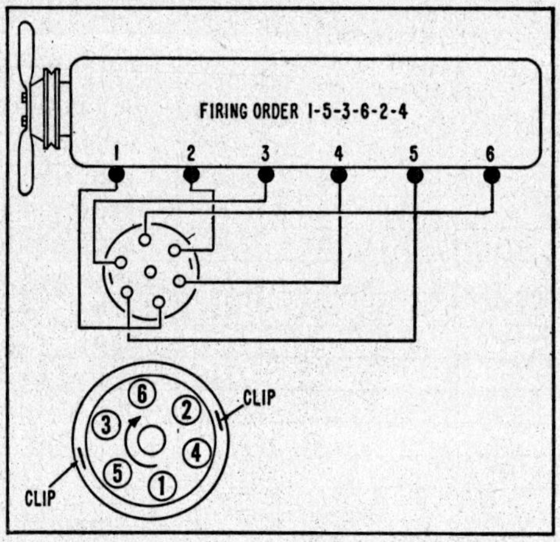

Fig. H

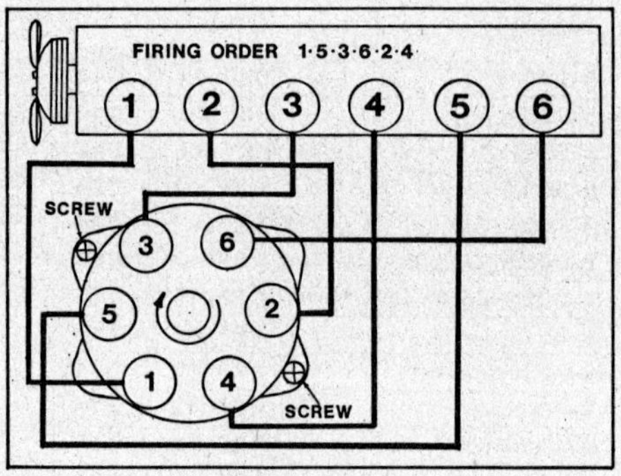

Fig. J

DISTRIBUTOR SPECIFICATIONS

★**Note: If unit is checked on vehicle, double the RPM and degrees to get crankshaft figures.**

Breaker arm spring tension—17–21.

Distributor Part No.①	Centrifugal Advance Degrees @ RPM of Distributor: Advance Starts	Intermediate Advance			Full Advance	Vacuum Advance: Inches of Vacuum to Start Plunger	Max. Adv. Dist. Deg. @ Vacuum	Distributor Retard: Max. Retard Dist. Deg. @ Vacuum
1969								
C7AF-AC	0–½ @ 350	0–2 @ 500	4¾–6¾ @ 750	7¾–10 @ 1500	12¼ @ 2000	5	12½ @ 20	—
C8AF-E	0–½ @ 350	0–1¾ @ 500	4½–6¾ @ 750	8–10¼ @ 1500	12½ @ 2000	5	11 @ 25	6 @ 20
C8DF-C	0–½ @ 350	0–2 @ 500	3¼–5¼ @ 750	8¼–10½ @ 1500	12 @ 2000	5	12½ @ 25	6 @ 20
C8DF-D	0–½ @ 350	0–1¼ @ 500	2¾–4¾ @ 750	9–11¼ @ 1500	14 @ 2000	5	11 @ 25	6 @ 20
C8DF-J	0–½ @ 350	0–1½ @ 500	3¼–5¼ @ 750	9½–11¾ @ 1500	14¾ @ 2000	5	11 @ 25	6 @ 20
C8OF-H	0–½ @ 350	¾–2¾ @ 500	7½–9½ @ 750	9½–11¾ @ 1500	13½ @ 2000	5	9½ @ 25	3½ @ 20
C8OF-J	0–½ @ 350	¾–2¾ @ 500	7½–9½ @ 750	9¾–12 @ 1500	13½ @ 2000	5	11 @ 20	—

Continued

DISTRIBUTOR SPECIFICATIONS—Continued

★Note: If unit is checked on vehicle, double the RPM and degrees to get crankshaft figures.

Breaker arm spring tension—17–21.

Distributor Part No.①	Centrifugal Advance Degrees @ RPM of Distributor					Vacuum Advance		Distributor Retard
	Advance Starts	Intermediate Advance			Full Advance	Inches of Vacuum to Start Plunger	Max. Adv. Dist. Deg. @ Vacuum	Max. Retard Dist. Deg. @ Vacuum
1969—Continued								
C9AF-K	0–½ @ 350	2¾–4¾ @ 750	5¾–7¾ @ 1000	7¾–10 @ 1500	12¼ @ 2000	5	11½ @ 25	—
C9AF-N	0–½ @ 350	1¾–3¾ @ 500	5½–7½ @ 750	7¼–9½ @ 1500	10¾ @ 2000	5	11½ @ 20	6 @ 20
C9DF-B	0–½ @ 350	0–1½ @ 500	4½–6½ @ 750	9–11¼ @ 1500	13¾ @ 2000	5	8 @ 25	—
C9OF-M	0–½ @ 350	0–2 @ 500	4¼–6¼ @ 750	7½–9¾ @ 1500	12¼ @ 2000	5	8 @ 20	—
C9OF-N	0–½ @ 350	0–2 @ 500	3¾–5¾ @ 750	7½–9¾ @ 1500	11 @ 2000	5	10½ @ 25	—
C9OF-R	0–½ @ 350	0–2 @ 500	4½–6½ @ 750	7¼–9½ @ 1500	11½ @ 2000	5	8½ @ 20	—
C9OF-T	0–½ @ 350	¼–2½ @ 500	4¾–6¾ @ 750	8¾–11 @ 1500	12 @ 2000	5	12½ @ 20	—
C9OF-U	0–½ @ 350	1¼–3¼ @ 500	4½–6½ @ 750	7–9¼ @ 1500	11¾ @ 2000	5	12½ @ 20	—
1970								
C8AF-A	0–½ @ 350	0–1½ @ 750	2¼–4½ @ 1000	6–8 @ 1500	10 @ 2000	5	8 @ 25	6 @ 20
C8AF-B	0–½ @ 350	1¾–3¾ @ 750	2¾–5 @ 1000	4½–6¾ @ 1500	8¾ @ 2000	5	9 @ 25	6 @ 20
C9AF-Y	0–½ @ 350	2–4 @ 500	6¾–8¾ @ 750	8½–10¾ @ 1500	12 @ 2000	5	8½ @ 25	6 @ 20
C9DF-B	0–½ @ 350	0–1½ @ 500	4½–6½ @ 750	9–11¼ @ 1500	14 @ 2050	5	8 @ 25	6 @ 20
C9ZF-D	0–½ @ 350	0–1½ @ 500	2¾–4¾ @ 750	8¼–10½ @ 1500	11 @ 2000	5	10 @ 25	6 @ 20
C9ZF-E	0–½ @ 350	0–1¾ @ 750	2½–4¾ @ 1000	7¾–10 @ 1500	11 @ 2000	5	7 @ 25	6 @ 20
D0AF-AC	0–½ @ 350	0–2 @ 500	4¼–6¼ @ 750	7½–9¾ @ 1500	14 @ 2375	5	9½ @ 25	6 @ 20
D0AF-H	0–½ @ 350	0–2 @ 500	4¼–6¼ @ 750	7½–9¾ @ 1500	14 @ 2375	5	8 @ 25	6 @ 20
D0AF-T	0–½ @ 350	0–1¾ @ 750	6–8 @ 1000	7¼–9¼ @ 1500	14 @ 3150	5	11½ @ 25	3½ @ 20
D0AF-Y	0–½ @ 350	0–1¾ @ 750	6–8 @ 1000	8–10¼ @ 1500	14 @ 2325	5	11 @ 25	6 @ 20
D0AF-Z	0–½ @ 350	2¼–4¼ @ 750	6½–8½ @ 1000	8–10¼ @ 1500	14 @ 2950	5	11 @ 25	—
D0DF-C	0–½ @ 350	½–2½ @ 500	4¾–6¾ @ 750	10–12¼ @ 1500	16 @ 2800	5	7 @ 25	6 @ 20
D0DF-E	0–½ @ 350	¼–2½ @ 750	7–9 @ 1000	9¾–12 @ 1500	19 @ 2650	5	11 @ 25	6 @ 20
D0OF-A	0–½ @ 350	½–2½ @ 500	6–8 @ 750	10–12¼ @ 1500	16 @ 2175	5	9 @ 25	6 @ 20
D0OF-AB	0–½ @ 350	0–1¾ @ 500	4¾–6¾ @ 750	8¼–10½ @ 1500	12 @ 2000	5	8½ @ 25	6 @ 20
D0OF-T	0–½ @ 350	0–1½ @ 500	2–4 @ 750	4½–6½ @ 1000	14 @ 2300	5	12½ @ 25	6 @ 20
D0OF-U	0–½ @ 350	1¼–3¼ @ 500	5¼–7¼ @ 750	8¼–10½ @ 1500	14 @ 2275	5	12½ @ 25	—
D0OF-V	0–½ @ 350	¾–2¾ @ 750	3¾–5¾ @ 1000	8½–10¾ @ 1500	14 @ 2300	5	10½ @ 25	6 @ 20
D0OF-Z	0–½ @ 350	3–5 @ 750	5½–7½ @ 1000	6¾–9 @ 1500	10¼ @ 2000	5	10½ @ 25	3½ @ 20
D0ZF-C	0–½ @ 350	5–7 @ 750	8¾–10¼ @ 1000	9¾–12 @ 1500	14 @ 2150	5	11 @ 25	3½ @ 20
D0ZF-G	0–½ @ 350	4¼–6¼ @ 750	6¾–8¾ @ 1000	8¾–11 @ 1500	14 @ 2075	5	9½ @ 25	—
1971								
C9DF-B	0–½ @ 350	4½–6½ @ 750	7–9 @ 1000	9–11¼ @ 1500	14 @ 2050	5	8 @ 25	7 @ 20
D0AF-Y	0–½ @ 350	0–1¾ @ 750	6–8 @ 1000	8–10¼ @ 1500	14 @ 2325	5	11 @ 25	7 @ 20
D0DF-C	0–½ @ 350	½–2½ @ 500	9–11 @ 1000	10–12¼ @ 1500	16 @ 2800	5	7 @ 25	7 @ 20
D0OF-AA	0–½ @ 350	0–1½ @ 500	5–7¼ @ 750	10¼–12½ @ 1500	16 @ 2200	5	8½ @ 25	7 @ 20
D0OF-AC	0–½ @ 350	0–1¾ @ 750	5½–8 @ 1000	7¼–9¼ @ 1500	13 @ 2500	5	5 @ 25	7 @ 20
D0OF-G	0–½ @ 350	3–5 @ 750	5½–7½ @ 1000	6¾–8¾ @ 1500	13 @ 2500	5	10½ @ 25	7 @ 20
D0OF-T	0–½ @ 350	2–4 @ 750	4½–6½ @ 1000	7¾–10 @ 1500	14 @ 2300	5	12½ @ 25	7 @ 20
D0OF-U	0–½ @ 350	1¼–3¼ @ 500	6¼–8¼ @ 1000	8¼–10½ @ 1500	14 @ 2275	5	12½ @ 25	—
D0OF-V	0–½ @ 350	¾–2¾ @ 750	3¾–5¾ @ 1000	8½–10¾ @ 1500	14 @ 2300	5	10½ @ 25	7 @ 20
D1AF-NA	0–½ @ 350	3–7 @ 750	8–10 @ 1000	10½–12½ @ 1500	14 @ 2500	5	8½ @ 25	7 @ 20
D1DF-BB	0–½ @ 350	0–¾ @ 500	3–5 @ 750	9½–11¾ @ 1500	14 @ 2500	5	7½ @ 25	7 @ 20
D1DF-EA	0–½ @ 350	0–1¾ @ 750	5½–8 @ 1000	7¼–9¼ @ 1500	13 @ 2500	5	11½ @ 25	7 @ 20
D1DF-FA	0–½ @ 350	0–2 @ 500	3¼–5 @ 750	9–11 @ 1500	14 @ 2500	5	8 @ 25	7 @ 20
D1DF-GA	0–½ @ 350	0–¾ @ 750	3¼–5¾ @ 1000	9–11¼ @ 1500	14 @ 2500	5	7 @ 25	7 @ 20
D1OF-AB	0–½ @ 350	0–1½ @ 500	3½–5¼ @ 750	7–9¼ @ 1500	14 @ 2500	5	7½ @ 25	—
D1OF-BB	0–½ @ 350	0–1 @ 750	3½–5½ @ 1000	7–9¼ @ 1500	14 @ 2500	5	7½ @ 25	—

Continued

DISTRIBUTOR SPECIFICATIONS—Continued

★Note: If unit is checked on vehicle, double the RPM and degrees to get crankshaft figures.

Breaker arm spring tension—17–21.

Distributor Part No.①	Centrifugal Advance Degrees @ RPM of Distributor					Vacuum Advance		Distributor Retard
	Advance Starts	Intermediate Advance			Full Advance	Inches of Vacuum to Start Plunger	Max. Adv. Dist. Deg. @ Vacuum	Max. Retard Dist. Deg. @ Vacuum
1971—Continued								
D1OF-CA	0–½ @ 350	0–1½ @ 500	3¼–5¼ @ 750	7–9¼ @ 1500	14 @ 2500	5	10 @ 25	7 @ 20
D1OF-GA	0–½ @ 350	3¼–5¼ @ 750	6½–8¾ @ 1000	8¼–10½ @ 1500	13 @ 2500	5	11 @ 25	4 @ 20
D1OF-LA	0–½ @ 350	½–3¼ @ 750	5¼–7¼ @ 1000	6¾–9 @ 1500	18 @ 2500	5	10½ @ 25	7 @ 20
D1OF-PA	0–½ @ 350	0–1½ @ 500	4½–6½ @ 750	10½–12¾ @ 1500	14 @ 2500	5	11 @ 25	—
D1ZF-AA	0–½ @ 350	0–1½ @ 750	2¾–4¾ @ 1000	8¾–10 @ 1500	11 @ 2500	5	2½ @ 25	7 @ 20
D1ZF-EA	0–½ @ 350	1–3 @ 750	4–6 @ 1000	9½–12 @ 1500	15 @ 2500	5	10½ @ 25	7 @ 20
D1ZF-FA	0–½ @ 350	1¼–3½ @ 750	4¼–6¼ @ 1000	10–12 @ 1500	16½ @ 2500	5	8½ @ 25	—
1972								
D2AF-RA	½–2½ @ 500	2–4 @ 750	3½–5½ @ 1000	6½–9 @ 1500	12 @ 2000	5	13½ @ 20	4 @ 20
D2AF-SA	½–2½ @ 500	2–4 @ 750	3½–5½ @ 1000	6½–9 @ 1500	12 @ 2000	5	13½ @ 20	4 @ 20
D2AF-CA	0–½ @ 500	0–1½ @ 750	4½–6½ @ 1000	7–9½ @ 1500	11 @ 2000	5	5½ @ 20	7 @ 20
D2AF-KA	1–3 @ 500	5½–7½ @ 750	6½–8½ @ 1000	8½–10½ @ 1500	12½ @ 2000	5	13 @ 20	7 @ 20
D2DF-AA	0–2 @ 500	4½–6½ @ 750	7–9 @ 1000	9–11½ @ 1500	13½ @ 2000	5	7½ @ 20	7 @ 20
D2DF-BA	0–2 @ 500	4½–6½ @ 750	7–9 @ 1000	9–11½ @ 1500	13½ @ 2000	5	7½ @ 20	7 @ 20
D2DF-CA	0–1 @ 500	3–5 @ 750	7½–9½ @ 1000	9½–12 @ 1500	12½ @ 2000	5	7½ @ 20	7 @ 20
D2DF-EA	0–1 @ 500	3–5 @ 750	7½–9½ @ 1000	9½–12 @ 1500	12½ @ 2000	5	7½ @ 20	7 @ 20
D2DF-DA	½–2½ @ 500	5–7 @ 750	9–11 @ 1000	10½–12½ @ 1500	13½ @ 2000	5	7½ @ 20	4 @ 20
D2DF-FA	0–1 @ 500	0–1½ @ 750	4½–7 @ 1000	10½–12½ @ 1500	13½ @ 2000	5	7½ @ 20	4 @ 20
D22F-LA	0–½ @ 500	0–1½ @ 750	2½–4½ @ 1000	8–10 @ 1500	13½ @ 2000	5	12½ @ 20	4 @ 20
D22F-HA	0–1 @ 500	1½–3½ @ 750	4½–6½ @ 1000	10–12 @ 1500	14½ @ 2000	5	9½ @ 20	—
D2MF-EA	0–1 @ 500	2–4 @ 750	4½–6½ @ 1000	6–8 @ 1500	9½ @ 2000	5	11½ @ 20	4 @ 20
D2MF-FA	0–1 @ 500	2–4 @ 750	4½–6½ @ 1000	6–8 @ 1500	9½ @ 2000	5	11½ @ 20	4 @ 20
D2OF-PA	0–2½ @ 500	6–8 @ 750	7½–9½ @ 1000	10–12½ @ 1500	15 @ 2000	5	9½ @ 20	7 @ 20
D2OF-DA	0–1½ @ 500	5–7 @ 750	5½–7½ @ 1000	7–9 @ 1500	10½ @ 2000	5	9½ @ 20	7 @ 20
D2OF-EA	0–1 @ 500	1½–3½ @ 750	4–6 @ 1000	7–9 @ 1500	10½ @ 2000	5	9½ @ 20	4 @ 20
D2OF-HA	0–1 @ 500	0–2 @ 750	5–7 @ 1000	7½–9½ @ 1500	11 @ 2000	5	5½ @ 20	7 @ 20
D2OF-JA	0–1 @ 500	0–2 @ 750	5–7 @ 1000	7½–9½ @ 1500	11 @ 2000	5	5½ @ 20	7 @ 20
D2OF-RA	0–½ @ 500	0–1½ @ 750	4½–6½ @ 1000	7–9½ @ 1500	11 @ 2000	5	5½ @ 20	7 @ 20
D2ZF-EA	0–½ @ 500	0–1 @ 750	1–3 @ 1000	5½–8 @ 1500	13 @ 2000	5	5½ @ 20	7 @ 20
D2ZF-AA	0–1½ @ 500	2¼–4¼ @ 750	4½–6½ @ 1000	8–10 @ 1500	12½ @ 2000	5	12½ @ 20	7 @ 20
D2ZF-CA	0–1 @ 500	3½–5½ @ 750	6½–8½ @ 1000	8½–10½ @ 1500	12½ @ 2000	5	10 @ 20	4 @ 20
D2ZF-FA	0–½ @ 500	0–1 @ 750	1–3 @ 1000	5½–8 @ 1500	13 @ 2000	5	5½ @ 20	7 @ 20
D2ZF-GC	0–½ @ 500	0–1 @ 750	0–2 @ 1000	4½–6½ @ 1500	11 @ 2000	5	9½ @ 20	—
D2ZF-HA	0–1 @ 500	1½–3½ @ 750	4½–6½ @ 1000	10–12 @ 1500	14½ @ 2000	5	9½ @ 20	—
D2ZF-JA	0–1 @ 500	0–1½ @ 750	2½–4½ @ 1000	7½–9½ @ 1500	13 @ 2000	5	11½ @ 20	7 @ 20
1973								
D3AF-AA	0–1½ @ 500	4–6 @ 750	6–8 @ 1000	10–12½ @ 1500	16½ @ 2000	5	13 @ 20	—
D3AF-BA	0–1½ @ 500	3½–5½ @ 750	5½–7½ @ 1000	8½–11 @ 1500	14½ @ 2000	5	13 @ 20	—
D3MF-BA	0–1 @ 500	0–1½ @ 750	3½–5½ @ 1000	7½–10 @ 1500	11 @ 2000	5	7 @ 20	—
D3MF-DA	0–1 @ 500	1½–4 @ 750	5½–7½ @ 1000	7½–9 @ 1500	10½ @ 2000	5	11 @ 20	—
D3MF-GA	0–1 @ 500	0–2 @ 750	1½–3½ @ 1000	4½–7 @ 1500	10½ @ 2000	5	13½ @ 20	—
D3OF-AA	0–1½ @ 500	6½–8½ @ 750	7–9 @ 1000	8–10½ @ 1500	11½ @ 2000	5	9 @ 20	4 @ 20
D3OF-BA	0–1½ @ 500	4–6 @ 750	4½–6½ @ 1000	6–8 @ 1500	10 @ 2000	5	9 @ 20	4 @ 20
D3OF-DA	0–1 @ 500	0–1½ @ 750	4–6 @ 1000	7–9½ @ 1500	11 @ 2000	5	5 @ 20	7 @ 20
D3OF-EA	0–1 @ 500	0–1½ @ 750	4–6 @ 1000	7–9½ @ 1500	11 @ 2000	5	5 @ 20	4 @ 20
D3OF-FA	0–1 @ 500	0–1½ @ 750	1–3 @ 1000	5½–8 @ 1500	13 @ 2000	5	5 @ 20	7 @ 20
D3OF-GA	0–1 @ 500	0–1½ @ 750	1–3½ @ 1000	6–8½ @ 1500	13½ @ 2000	5	5 @ 20	—
D3OF-HA	0–1 @ 500	0–2½ @ 750	2–4½ @ 1000	5½–8½ @ 1500	12 @ 2000	5	5 @ 20	—

Continued

DISTRIBUTOR SPECIFICATIONS—Continued

★Note: If unit is checked on vehicle, double the RPM and degrees to get crankshaft figures.

Breaker arm spring tension—17–21.

Distributor Part No.①	Centrifugal Advance Degrees @ RPM of Distributor					Vacuum Advance		Distributor Retard
	Advance Starts	Intermediate Advance			Full Advance	Inches of Vacuum to Start Plunger	Max. Adv. Dist. Deg. @ Vacuum	Max. Retard Dist. Deg. @ Vacuum
1973—Continued								
D3OF-JA	0–1 @ 500	0–2½ @ 750	2–4½ @ 1000	5½–8½ @ 1500	12 @ 2000	5	5 @ 20	—
D3OF-LA	0–1½ @ 500	4½–6½ @ 750	6–8 @ 1000	8–10½ @ 1500	12 @ 2000	5	9 @ 20	4 @ 20
D3OF-PA	0–1 @ 500	0–2½ @ 750	2–4½ @ 1000	6–8½ @ 1500	12½ @ 2000	5	5 @ 20	4 @ 20
D3OF-RA	0–1½ @ 500	3½–5½ @ 750	5½–7½ @ 1000	6½–8½ @ 1500	10 @ 2000	5	9½ @ 20	4 @ 20
D3SF-BA	0–1 @ 500	½–2½ @ 750	5½–8 @ 1000	10–12 @ 1500	13½ @ 2000	5	11 @ 20	—
D3UF-GA	0–1 @ 500	0–2½ @ 750	2–4½ @ 1000	5½–8½ @ 1500	12 @ 2000	5	5 @ 20	4 @ 20
D3VF-CA	0–1 @ 500	0–1½ @ 750	4–6 @ 1000	6–8 @ 1500	10 @ 2000	5	11 @ 20	—
D3ZF-CA	0–1 @ 500	3½–5½ @ 750	4½–6½ @ 1000	6½–9 @ 1500	11½ @ 2000	5	13 @ 20	—
D3ZF-GA	0–1 @ 500	0–2½ @ 750	2½–4 @ 1000	5½–7½ @ 1500	11 @ 2000	5	12½ @ 20	—
1974								
D3AF-AA	0–1½ @ 500	4–6 @ 750	6–8¼ @ 1000	10–12½ @ 1500	16½ @ 2000	5	13¼ @ 20	—
D3BF-DA	0–1¼ @ 500	2½–4½ @ 750	5½–7½ @ 1000	8½–10½ @ 1500	12 @ 2000	5	11¼ @ 20	5–7 @ 20
D3DF-FA	0–1½ @ 500	1½–3½ @ 750	4½–6½ @ 1000	6½–9 @ 1500	9¼ @ 2000	5	9¼ @ 20	2–4 @ 20
D3DF-HA	0–1½ @ 500	3½–5½ @ 750	7½–9½ @ 1000	9½–11½ @ 1500	11¼ @ 2000	5	7¼ @ 20	—
D3DF-KA	0–1½ @ 500	4½–6½ @ 750	7½–9½ @ 1000	8½–11 @ 1500	11¼ @ 2000	5	7¼ @ 20	2–4 @ 20
D3OF-FA	0–1 @ 500	0–1½ @ 750	1–3 @ 1000	5½–8½ @ 1500	13 @ 2000	5	5¼ @ 20	5–7 @ 20
D3OF-GA	0–1 @ 500	0–1½ @ 750	1–3½ @ 1000	6½–8½ @ 1500	13½ @ 2000	5	5¼ @ 20	—
D3OF-HB	0–1½ @ 500	½–3 @ 750	2½–5 @ 1000	6½–9 @ 1500	13½ @ 2000	5	11¼ @ 20	—
D30F-RA	0–1½ @ 500	3½–5½ @ 750	5½–7½ @ 1000	6½–8½ @ 1500	10 @ 2000	5	9¼ @ 20	2–4 @ 20
D3UF-EA	0–1¼ @ 500	2½–4½ @ 750	5½–7½ @ 1000	9–11 @ 1500	14½ @ 2000	5	9¼ @ 20	5–7 @ 20
D3ZF-GA	0–1 @ 500	0–2½ @ 750	2¼–4¼ @ 1000	5½–7½ @ 1500	11 @ 2000	5	12½ @ 20	—
D4AE-AA	0–½ @ 500	0–2 @ 750	2½–4½ @ 1000	5½–7½ @ 1500	11 @ 2000	5	12½ @ 20	—
D4AE-HA	0–1½ @ 500	5–7 @ 750	6–8 @ 1000	8–10½ @ 1500	12½ @ 2000	5	9½ @ 20	5–7 @ 20
D4DE-FA	0–1½ @ 500	4½–6½ @ 750	7½–9½ @ 1000	9–11 @ 1500	12½ @ 2000	5	7¼ @ 20	2–4 @ 20
D4DE-LA	0–1½ @ 500	4–6 @ 750	4½–6½ @ 1000	6–8 @ 1500	9½ @ 2000	5	9 @ 20	2–4 @ 20
D4DE-MA	0–½ @ 500	½–2½ @ 750	3½–5½ @ 1000	7½–9½ @ 1500	13½ @ 2000	5	11½ @ 20	—
D4DE-NA	0–1½ @ 500	3½–5½ @ 750	5½–7½ @ 1000	6½–8½ @ 1500	10 @ 2000	5	9¼ @ 20	2–4 @ 20
D4DE-RA	0–1½ @ 500	3½–5½ @ 750	5½–7½ @ 1000	6½–8½ @ 1500	10½ @ 2000	5	9¼ @ 20	2–4 @ 20
D4OE-CA	0–½ @ 500	0–1½ @ 750	2–4 @ 1000	5½–7½ @ 1500	10½ @ 2000	5	13½ @ 20	—
D4VE-CA	0–½ @ 500	0–½ @ 750	3–5 @ 1000	7–9 @ 1500	10½ @ 2000	5	11¼ @ 20	—
1975								
D3ZF-GA	0–2 @ 775	2–4 @ 975	5.4–7.7 @ 1500	—	10¾ @ 2500	6	13¼ @ 24	—
D4OE-EA	0–2 @ 520	4½–7½ @ 800	10–12½ @ 1500	—	17¾ @ 2000	4	13 @ 17	—
D4VE-CA	0–2 @ 825	4–6 @ 1025	7–9¼ @ 1500	8⅓–11¼ @ 2000	12½ @ 2500	4	11¾ @ 14.6	—
D5AE-BA	0–2½ @ 975	—	3¾–6¼ @ 1500	—	11¼ @ 2150	4	13¼ @ 13	—
D5AE-DA	0–2½ @ 975	—	3¾–6¼ @ 1500	—	11¼ @ 2150	4.3	15¼ @ 13	—
D5AE-EA	0–2½ @ 975	—	3¾–6¼ @ 1500	—	11¼ @ 2150	3	13¼ @ 11	—
D5DE-HA	0–2 @ 550	3½–5½ @ 1000	7–9¼ @ 1500	10½–13 @ 2000	16½ @ 2500	4½	13¼ @ 19	—
1976								
D4VE-CA	0–4 @ 825	4–6 @ 1025	7–9¼ @ 1500	—	12½ @ 2500	6.3	11¼ @ 14.6	—
D5AE-BA	0–2 @ 900	—	3¾–6 @ 1500	—	11½ @ 2150	4	13¼ @ 13	—
D5AE-DA	0–2½ @ 975	—	3¾–6¼ @ 1500	—	10¾ @ 2150	4	15¼ @ 13	—
D5AE-EA	0–2½ @ 975	—	3¾–6¼ @ 1500	—	10¾ @ 2150	3	13¼ @ 11	—
D5DE-ACA	—	—	—	—	—	—	—	—
D5DE-AFA	—	—	—	—	—	—	—	—

Continued

DISTRIBUTOR SPECIFICATIONS—Continued

★Note: If unit is checked on vehicle, double the RPM and degrees to get crankshaft figures.

Breaker arm spring tension—17–21.

Distributor Part No.①	Centrifugal Advance Degrees @ RPM of Distributor: Advance Starts	Intermediate Advance			Full Advance	Vacuum Advance: Inches of Vacuum to Start Plunger	Max. Adv. Dist. Deg. @ Vacuum	Distributor Retard: Max. Retard Dist. Deg. @ Vacuum
1976—Continued								
D5DE-AGA	—	—	—	—	—	—	—	—
D5DE-FA	—	—	—	—	—	—	—	—
D5DE-NA	—	—	—	—	—	—	—	—
D5DE-YA	—	—	—	—	—	—	—	—
D6BE-BA	—	—	—	—	—	—	—	—
D6DE-AA	—	—	—	—	—	—	—	—
D6DE-BA	—	—	—	—	—	—	—	—
D6DE-CA	—	—	—	—	—	—	—	—
D6DE-EA	—	—	—	—	—	—	—	—
D6DE-GA	—	—	—	—	—	—	—	—
D6OE-AA	0–2 @ 550	3½–5½ @ 1000	7–9 @ 1500	10½–13 @ 2000	16½ @ 2500	4.5	13¼ @ 19	—

①—Basic part No. 12127.

VALVE SPECIFICATIONS

Year	Engine	Valve Lash: Int.	Valve Lash: Exh.	Valve Angles: Seat	Valve Angles: Face	Valve Spring Installed Height	Valve Spring Pressure Lbs. @ In.	Stem Clearance: Intake	Stem Clearance: Exhaust	Stem Diameter, Standard: Intake	Stem Diameter, Standard: Exhaust
1969	6-170	.066–.117④		45	44	1 9/16	150 @ 1.22	.0008–.0025	.001–.0027	.3100–.3107	.3098–.3105
	6-200, 250	.095–.195④		45	44	1 9/16	150 @ 1.22	.0008–.0025	.001–.0027	.3100–.3107	.3098–.3105
	8-302	1 Turn①		45	44	1 5/8	180 @ 1.23	.001–.0027	.0015–.0032	.3416–.3423	.3411–.3418
	8-302 "H.O."	.025H	.025H	45	44	1 13/16	315 @ 1.32	.001–.0027	.0015–.0032	.3416–.3423	.3411–.3418
	8-351	1 Turn①		45	44	1 25/32	215 @ 1.34	.001–.0027	.0015–.0032	.3416–.3423	.3411–.3418
	8-390	.100–.200④		45	44	1 7/8	220 @ 1.38	.001–.0027	.0015–.0032	.3711–.3718	.3706–.3713
	8-427	.100–.200④		⑦	⑫	1 13/16	270 @ 1.32	.001–.0027	.0015–.0032	.3711–.3718	.3706–.3713
	8-428	.100–.200④		45	44	1 13/16	270 @ 1.32	.001–.0027	.0015–.0032	.3711–.3718	.3706–.3713
1970	6-170	.066–.117④		45	44	1 9/16	150 @ 1.22	.0008–.0025	.001–.0027	.3100–.3107	.3098–.3105
	6-200, 250	.095–.195④		45	44	1 9/16	150 @ 1.22	.0008–.0025	.001–.0027	.3100–.3107	.3098–.3105
	8-302	.067–.167④		45	44	1 5/8	180 @ 1.23	.001–.0027	.0015–.0032	.3416–.3423	.3411–.3418
	8-302 "BOSS"	.025H	.025H	45	44	1 13/16	315 @ 1.32	.001–.0027	.0015–.0032	.3416–.3423	.3411–.3418
	8-351⑧	.083–.183④		45	44	1 25/32	215 @ 1.34	.001–.0027	.0015–.0032	.3416–.3423	.3411–.3418
	8-351⑨⑩	.100–.200④		45	44	1 13/16	210 @ 1.42	.001–.0027	.0015–.0032	.3416–.3423	.3411–.3418
	8-351⑨⑪	.100–.200④		45	44	1 13/16	285 @ 1.32	.001–.0027	.0015–.0032	.3416–.3423	.3411–.3418
	8-428 "CJ"	.100–.200④		45	44	1 13/16	270 @ 1.32	.001–.0027	.0015–.0032	.3711–.3718	.3706–.3713
	8-429 "CJ"	.019H	.019H	⑦	⑫	1 13/16	306 @ 1.36	.001–.0024	.002–.0034	.3416–.3423	.3416–.3418
	8-429 "SCJ"	.019H	.019H	⑦	⑫	1 13/32	306 @ 1.36	.001–.0024	.002–.0034	.3711–.3718	.3701–.3708
	8-429 "BOSS"	.013C	.013C	⑦	⑫	1 13/32	315 @ 1.32	.001–.0024	.002–.0034	.3711–.3718	.3701–.3708
1971	6-170	.066–.117④		45	44	1 9/16	150 @ 1.22	.0008–.0025	.001–.0027	.3100–.3107	.3098–.3105
	6-200	.079–.144④		45	44	1 9/16	150 @ 1.22	.0008–.0025	.001–.0027	.3100–.3107	.3098–.3105
	6-250	.095–.145④		45	44	1 9/16	150 @ 1.22	.0008–.0025	.001–.0027	.3100–.3107	.3098–.3105
	8-302	.090–.190④		45	44	1 5/8	180 @ 1.23	.001–.0027	.001–.0027	.3416–.3423	.3411–.3418
	8-302 H.O.	.025H	.025H	45	44	1 13/16	315 @ 1.32	.001–.0027	.001–.0027	.3416–.3423	.3411–.3418
	8-351⑧	.100–.200④		45	44	1 25/32	215 @ 1.34	.001–.0027	.001–.0027	.3416–.3423	.3411–.3418

Continued

VALVE SPECIFICATIONS—Continued

Year	Engine Model	Valve Lash		Valve Angles		Valve Spring Installed Height	Valve Spring Pressure Lbs. @ In.	Stem Clearance		Stem Diameter Standard	
		Int.	Exh.	Seat	Face			Intake	Exhaust	Intake	Exhaust
1971	8-351⑨⑩	.100–.200④		45	44	1 13/16	210 @ 1.42	.001–.0027	.001–.0027	.3416–.3423	.3411–.3418
	8-351⑨⑪	.100–.200④		45	44	1 13/16	285 @ 1.32	.001–.0027	.001–.0027	.3416–.3423	.3411–.3418
	8-429	.075–.155④		45	44	1 13/32	315 @ 1.32	.001–.0027	.001–.0027	.3416–.3423	.3416–.3423
1972	6-170	.100–.150④		45	44	1 19/32	150 @ 1.22	.0008–.0025	.001–.0027	.3100–.3107	.3098–.3105
	6-200	.100–.150④		45	44	1 19/32	150 @ 1.22	.0008–.0025	.001–.0027	.3100–.3107	.3098–.3105
	6-250	.100–.150④		45	44	1 37/64	150 @ 1.22	.0008–.0025	.001–.0027	.3100–.3107	.3098–.3105
	6-250 Calif.	.100–.150④		45	44	1 37/64	146 @ 1.20	.0008–.0025	.001–.0027	.3100–.3107	.3098–.3105
	8-302	.090–.140④		45	44	1 11/16	200 @ 1.31	.001–.0027	.0015–.0032	.3416–.3423	.3411–.3418
	8-351⑩	.106–.156④		45	44	1 13/16	210 @ 1.42	.001–.0027	.0015–.0032	.3416–.3423	.3411–.3418
	8-351⑪	.100–.150④		45	44	1 13/16	277 @ 1.34	.001–.0027	.0015–.0032	.3416–.3423	.3411–.3418
	8-351 H.O.	.025H	.025H	45	44	1 5/8	315 @ 1.32	.001–.0027	.0015–.0032	.3416–.3423	.3411–.3418
	8-400	.100–.150④		45	44	1 13/16	226 @ 1.39	.001–.0027	.0015–.0032	.3416–.3423	.3411–.3418
	8-429	.075–.125④		45	44	1 13/16	229 @ 1.33	.001–.0027	.0015–.0032	.3416–.3423	.3411–.3418
1973	6-200	.079–.129④		45	45	1 19/32	150 @ 1.22	.0008–.0025	.001–.0027	.3100–.3107	.3098–.3105
	6-250	.095–.145④		45	44	1 37/64	149 @ 1.20	.0008–.0025	.001–.0027	.3100–.3107	.3098–.3105
	8-302	.090–.140④		45	44	1 11/16	200 @ 1.31	.001–.0027	.0015–.0032	.3416–.3423	.3411–.3418
	8-351⑧	.106–.156④		45	44	1 25/32	200 @ 1.34	.001–.0027	.0015–.0032	.3416–.3423	.3411–.3418
	8-351⑨⑩	.100–.150④		45	44	1 53/64	210 @ 1.42	.001–.0027	.0015–.0032	.3416–.3423	.3411–.3418
	8-351⑨⑪	.100–.150④		45	44	1 53/64	285 @ 1.32	.001–.0027	.0015–.0032	.3416–.3423	.3411–.3418
	8-400	.100–.150④		45	44	1 53/64	226 @ 1.39	.001–.0027	.0015–.0032	.3416–.3423	.3411–.3418
	8-429	.075–.125④		45	44	1 13/16	169 @ 1.39	.001–.0027	.001–.0027	.3416–.3423	.3416–.3423
	8-460	.105–.155④		45	44	1 25/32	169 @ 1.39	.0010–.0027	.0010–.0027	.3416–.3423	.3416–.3423
1974	6-200	.079–.129④		45	44	1 37/64	146 @ 1.20	.0008–.0025	.0010–.0027	.3100–.3107	.3098–.3105
	6-250	.095–.195④		45	44	1 37/64	146 @ 1.20	.0008–.0025	.0010–.0027	.3100–.3107	.3098–.3105
	8-302	.090–.140④		45	44	⑥	⑭	.0010–.0027	.0015–.0032	.3416–.3423	.3411–.3418
	8-351⑨⑩	.100–.150④		45	44	1 53/64	226 @ 1.42	.0010–.0027	.0015–.0032	.3416–.3423	.3411–.3418
	8-351⑧⑩	.106–.156④		45	44	1 51/64	200 @ 1.34	.0010–.0027	.0015–.0032	.3416–.3423	.3411–.3418
	8-351⑪	.100–.150④		45	44	1 53/64	280 @ 1.34	.0010–.0027	.0015–.0032	.3416–.3423	.3411–.3418
	8-400	.100–.150④		45	44	1 53/64	226 @ 1.39	.0010–.0027	.0015–.0032	.3416–.3423	.3411–.3418
	8-460	.075–.125④		45	44	1 53/64	253 @ 1.33	.0010–.0027	.0010–.0027	.3416–.3423	.3416–.3423
1975	6-200	.079–.209		45	44	1 19/32	150 @ 1.22	.0008–.0025	.001–.0027	.3100–.3107	.3098–.3105
	6-250	.112–.176④		45	44	③	150 @ 1.18	.0008–.0025	.001–.0027	.3100–.3107	.3098–.3105
	V8-302	.115–.165④		45	44	⑬	⑭	.001–.0027	.0015–.0032	.3416–.3423	.3411–.3418
	V8-351⑧	.131–.181④		45	44	1 51/64	200 @ 1.34	.001–.0027	.0015–.0032	.3416–.3423	.3411–.3418
	V8-351⑤, 400	.100–.200④		45	44	1 53/64	226 @ 1.39	.001–.0027	.0015–.0032	.3416–.3423	.3411–.3418
	V8-460	.100–.150④		45	44	1 53/64	315 @ 1.32	.001–.0027	.001–.0027	.3416–.3423	.3416–.3423
1976	6-200	.079–.209		45	44	1 37/64	②	.0008–.0025	.0010–.0027	.3100–.3107	.3098–.3105
	6-250	.079–.209		45	44	1 37/64	②	.0008–.0025	.0010–.0027	.3100–.3107	.3098–.3105
	8-302	.090–.190		45	44	1 11/16	⑮	.0010–.0027	.0015–.0032	.3416–.3423	.3411–.3418
	8-351⑧	.106–.206		45	44	1 51/64	200 @ 1.34	.0010–.0027	.0015–.0032	.3416–.3423	.3411–.3418
	8-351⑤, 400	.100–.200		45	44	1 13/16	226 @ 1.39	.0010–.0027	.0015–.0032	.3416–.3423	.3411–.3418
	8-460	.075–.175		45	44	1 13/16	229 @ 1.33	.0010–.0027	.0010–.0027	.3416–.3423	.3416–.3423

①—Tighten rocker arm adjusting screw to eliminate all push rod end clearance, then tighten screw the number of turns listed.

②—Intake 156 @ 1.20, exhaust 148 @ 1.23.

③—Intake 1 19/32, exhaust 1 9/16.

④—Clearance is obtained at valve stem tip with hydraulic lifter collapsed. If clearance is less than the minimum install an undersize push rod; if clearance is greater than the maximum install an oversize push rod.

⑤—Modified engine.

⑥—Intake 1 11/16, exhaust 1 9/16.

⑦—Intake 30°, Exhaust 45°.

⑧—Windsor engine.

⑨—Cleveland engine.

⑩—2 barrel carb.

⑪—4 barrel carb.

⑫—Intake 29°, exhaust 44°.

⑬—Intake 1 45/64, exhaust 1 39/64.

⑭—Intake 200 @ 1.31, exhaust 200 @ 1.22.

⑮—Intake 200 @ 1.31, exhaust 200 @ 1.20.

STARTING MOTOR SPECIFICATIONS

Year	Engine Model	Part No.	Brush Spring Tension, Ounces	No Load Test			Torque Test		
				Amperes	Volts	R.P.M.	Amperes	Volts	Torque Lbs. Ft.
1969–72	6-170, 200	C6OZ-A	40	70	12	8500	460	6	9
	6-170	C2DZ-A	40	70	12	8500	460	6	9
	6-200	C3OZ-C	40	70	12	9500	670	5	15½
	6-250	C2OZ-A	40	70	12	9500	670	5	15½
	8-302, 351	C2OZ-A	40	70	12	9500	670	5	15½
	8-302, 351	C5TZ-A	40	70	12	9500	670	5	15½
	8-390, 427	C3OZ-C	40	70	12	9500	670	5	15½
	8-428	C8AZ-A	40	70	12	11000	700	5	15½
	8-429	C8VY-C	40	70	12	10000	700	5	15½
1970	6-170	D0DF-A	40	70	12	—	460	5	9
	6-170, 200	C7DF-A	40	70	12	—	460	5	9
	6-170, 200	C7ZF-A	40	70	12	—	670	5	15½
	6-200	C7AF-B	40	70	12	—	670	5	15½
	6-200	C7OF-A	40	70	12	—	670	5	15½
	6-200	D0OF-A	40	70	12	—	670	5	15½
	6-200	D0ZF-B	40	70	12	—	670	5	15½
	6-250	D0ZF-A	40	70	12	—	670	5	15½
	6-250	C9ZF-A	40	70	12	—	670	5	15½
	6-250, 8-302, 351	D0AF-B	40	70	12	—	670	5	15½
	8-302, 351	C7AF-B	40	70	12	—	670	5	15½
	8-302, 351	C7AF-F	40	70	12	—	670	5	15½
	8-302, 351	D0AF-C	40	70	12	—	670	5	15½
	8-428	C8AF-A	40	70	12	—	670	5	15½
	8-428	D0TF-A	40	70	12	—	670	5	15½
1971	6-170, 200	D0ZF-B	40	70	12	—	670	5	15½
	6-200	D0OF-A	40	70	12	—	670	5	15½
	6-250, 8-302	D0ZF-A	40	70	12	—	670	5	15½
	6-250, 8-302, 351	D0AF-B	40	70	12	—	670	5	15½
	8-429	C9AF-A	40	70	12	—	—	—	—
1972	6-170, 200	D2ZF-BA	40	70	12	—	670	5	15½
	6-200	D2OF-AA	40	70	12	—	670	5	15½
	6-250, 302	D2ZF-AA	40	70	12	—	670	5	15½
	6-250, 8-302, 351	D2AF-CA	40	70	12	—	670	5	15½
	8-302, 351	D2ZF-DA	40	70	12	—	670	5	15½
	8-400	D2AF-AA	40	70	12	—	670	5	15½
	8-429	D2AF-EA	40	70	12	—	—	—	—
1973	6-200	D2ZF-BA	40	70	12	—	670	5	15½
	6-200	D2OF-AA, BA	40	70	12	—	670	5	15½
	6-250, 8-302, 351	D2AF-CA	40	70	12	—	670	5	15½
	8-302, 351	D2AF-DA	40	70	12	—	670	5	15½
	8-400	D2AF-AA	40	70	12	—	670	5	15½
	8-429, 460	D4OF-AA, BA	40	70	12	—	—	—	—
1974	6-200	D2ZF-BA, CA	80	90	12	—	670	5	15.5
	6-200	D2OF-AA, BA	80	90	12	—	670	5	15.5
	6-250, 8-302, 351	D2AF-CA	80	90	12	—	670	5	15.5
	8-302, 351	D4AF-AA, BA	80	90	12	—	670	5	15.5
	8-302, 351	D5TF-DA	80	80	12	—	670	5	15.5
	8-400	D2AF-AA	80	80	12	—	670	5	15.5
	8-460	D4OF-AA	80	90	12	—	—	—	—
1975	6-200	D2OF-BA	80	80	12	—	670	5	15.5
	6-200	D2ZF-CA	80	80	12	—	670	5	15.5
	6-250, 8-302 351①	D5OF-BA	80	80	12	—	670	5	15.5

Continued

STARTING MOTOR SPECIFICATIONS—Continued

Year	Engine Model	Part No.	Brush Spring Tension, Ounces	No Load Test			Torque Test		
				Amperes	Volts	R.P.M.	Amperes	Volts	Torque Lbs. Ft.
1975	8-351②	D2AF-JA	80	80	12	—	670	5	15.5
	8-400	D5AF-EA	80	80	12	—	670	5	15.5
	8-460	D5AF-AB	80	90	12	—	—	—	—
1976	6-200	D2ZF-BA	—	—	12	—	—	—	—
	6-250, 8-302, 351	D5OF-AA	—	—	12	—	—	—	—
	8-351, 400	D2AF-AA	—	—	12	—	—	—	—
	8-351②, 400	D5AF-EA	—	—	12	—	—	—	—
	8-460	D5AF-B	—	—	12	—	—	—	—
	8-460	D6AF-AA	—	—	12	—	—	—	—

PISTONS, PINS, RINGS, CRANKSHAFT & BEARINGS

Year	Engine	Piston Clearance	Ring End Gap①		Wrist-pin Diameter	Rod Bearings		Main Bearings			
			Comp.	Oil		Shaft Diameter	Bearing Clearance	Shaft Diameter	Bearing Clearance	Thrust on Bear. No.	Shaft End Play
1969	6-170, 200	.0014–.0020	.010	.015	.912	2.1232–2.1240	.0008–.0015	2.2482–2.2490	.0005–.0015	④	.004–.008
	6-250	.0014–.0020	.010	.015	.912	2.1232–2.1240	.0008–.0015	2.3982–2.3990	.0005–.0015	5	.004–.008
	8-302	.0018–.0026	.010	.015	.912	2.1228–2.1236	.0008–.0015	2.2482–2.2490	.0005–.0015	3	.004–.008
	8-302 "H.O."	.0034–.0042	.010	.015	.912	2.1222–2.1230	.001–.0028	2.2482–2.2490	.0005–.0015	3	.004–.008
	8-351	.0018–.0026	.010	.015	.912	2.3103–2.3110	.0008–.0015	2.9994–3.0002	.0013–.0025	3	.004–.008
	8-390	.0015–.0023	.010	.015	.975	2.4380–2.4388	.0008–.0015	2.7484–2.7492	.001–.002	3	.004–.010
	8-427, 428	.0030–.0038	.018	.015	.975	2.4380–2.4388	.002–.003	2.7484–2.7492	.001–.002	3	.004–.010
1970–71	6-170, 200	.0014–.0020	.010	.015	.912	2.1232–2.1240	.0008–.0026	2.2482–2.2490	.0005–.0025	④	.004–.010
	6-250	.0014–.0020	.010	.015	.912	2.1232–2.1240	.0008–.0026	2.3982–2.3900	.0005–.0025	5	.004–.008
	8-302	.0018–.0026	.010	.015	.912	2.1228–2.1236	.0008–.0026	2.2482–2.2400	.0005–.0025	3	.004–.008
	8-302 "BOSS"	.0034–.0042	.010	.015	.912	2.1222–2.1230	.001–.0028	2.2482–2.2490	.0005–.0015	3	.004–.008
	8-351②	.0018–.0026	.010	.015	.912	2.3103–2.3110	.0008–.0026	2.9994–3.0002	.0013–.0025	3	.004–.008
	8-351③	.0014–.0022	.010	.015	.912	2.3103–2.3110	.0008–.0026	2.7484–2.7492	.0013–.0029	3	.004–.008
	8-428	.003–.0038	.018	.015	.975	2.4380–2.4388	.002–.003	2.7484–2.7492	.001–.002	3	.004–.010
	8-429 "CJ"	.003–.0038	.010	.010	1.040	2.4992–2.5000	.001–.0015	2.9994–3.0002	.0005–.0025	3	.004–.008
	8-429 "BOSS"	.003–.0038	.010	.010	1.040	2.4992–2.5000	.0008–.0026	2.9994–3.0002	.0009–.0025	3	.004–.008
1972	6-170	.0013–.0026	.008	.015	.912	2.1232–2.1240	.0002–.0024	2.2482–2.2490	.0007–.0026	3	.004–.008
	6-200	.0014–.0020	.008	.015	.912	2.1232–2.1240	.0002–.0024	2.2482–2.2490	.0007–.0026	5	.004–.008
	6-250	.0013–.0021	.008	.015	.912	2.1232–2.1240	.0008–.0024	2.3982–2.3900	.0005–.0022	5	.004–.008
	8-302	.0018–.0026	.010	.015	.912	2.1232–2.1240	.0008–.0026	2.2482–2.2490	.0005–.0024	3	.004–.008
	8-351⑤	.0014–.0022	.010	.015	.912	2.3103–2.3110	.0008–.0026	2.7484–2.7492	.0009–.0026	3	.004–.010
	8-351⑥	.0018–.0026	.010	.015	.912	2.3103–2.3110	.0011–.0026	2.7484–2.7492	.0011–.0028	3	.004–.010
	8-400	.0014–.0022	.010	.015	.975	2.3103–2.3110	.0008–.0026	2.9994–3.0002	.0009–.0026	3	.004–.008
	8-429	.0014–.0022	.010	.015	.975	2.4992–2.5000	.0008–.0028	2.9994–3.0002	⑦	3	.004–.008
1973	6-200	.0014–.0020	.008	.015	.912	2.1232–2.1240	.0002–.0024	2.2482–2.2490	.0007–.0026	5	.004–.008
	6-250	.0013–.0021	.008	.015	.912	2.1232–2.1240	.0008–.0024	2.3982–2.3990	.0008–.0024	5	.004–.008
	8-302	.0018–.0026	.010	.015	.912	2.1228–2.1236	.0008–.0026	2.2482–2.2490	.0005–.0024	3	.004–.008
	8-351⑤②	.0018–.0026	.010	.015	.912	2.3103–2.3111	.0008–.0026	2.9994–3.0002	.0008–.0026	3	.004–.008
	8-351⑤③	.0014–.0022	.010	.015	.912	2.3103–2.3111	.0008–.0026	2.7484–2.7492	.0009–.0026	3	.004–.010
	8-351⑥	.0034–.0042	.010	.015	.912	2.3103–2.3111	.0011–.0026	2.7484–2.7492	.0011–.0028	3	.004–.010
	8-351⑥⑧	.0034–.0042	.010	.015	.912	2.3103–2.3111	.0011–.0026	2.7484–2.7492	.0011–.0028	3	.004–.010
	8-400	.0014–.0022	.010	.015	.912	2.3103–2.3111	.0008–.0026	2.9994–3.0002	.0009–.0026	3	.004–.008
	8-429	.0014–.0022	.010	.015	.975	2.4992–2.5000	.0008–.0028	2.9994–3.0002	⑦	3	.004–.008
	8-460	.0014–.0022	.010	.015	1.040	2.4992–2.5000	.0008–.0028	2.9994–3.0002	.0005–.0025	3	.004–.008

Continued

PISTONS, PINS, RINGS, CRANKSHAFT & BEARINGS—Continued

Year	Engine	Piston Clearance	Ring End Gap①		Wrist-pin Diam-eter	Rod Bearings		Main Bearings			
			Comp.	Oil		Shaft Diameter	Bearing Clearance	Shaft Diameter	Bearing Clearahce	Thrust on Bear. No.	Shaft End Play
1974	6-200	.0013–.0021	.008	.015	.912	2.1232–2.1240	.0008–.0024	2.2482–2.2490	.0007–.0026	5	.004–.008
	6-250	.0013–.0021	.008	.015	.912	2.1232–2.1240	.0008–.0024	2.3982–2.3990	.0005–.0022	5	.004–.008
	8-302	.0018–.0026	.010	.015	.912	2.1228–2.1236	.0008–.0026	2.2482–2.2490	⑩	3	.004–.008
	8-351⑤②	.0018–.0026	.010	.015	.912	2.3103–2.3111	.0008–.0026	2.9994–3.0002	.0008–.0026	3	.004–.008
	8-351⑤③	.0014–.0022	.010	.015	.912	3.3103–2.3111	.0008–.0026	2.7484–2.7492	.0009–.0026	3	.004–.008
	8-351⑥③	.0018–.0026	.010	.015	.912	2.3103–2.3111	.0011–.0026	2.7484–2.7492	.0011–.0028	3	.004–.008
	8-400	.0014–.0022	.010	.015	.975	2.3103–2.3111	.0008–.0026	2.9994–3.0002	.0009–.0026	3	.004–.008
	8-460	.0022–.0030	.010	.015	1.040	2.4992–2.5000	.0008–.0028	2.9994–3.0002	⑦	3	.004–.008
1975	6-200	.0013–.0021	.008	.015	.9121	2.136	.0008–.0026	2.2486	.0008–.0024	5	.004–.008
	6-250	.0013–.0021	.008	.015	.9121	2.136	.0008–.0024	2.3986	.0008–.0024	5	.004–.008
	8-302	.0018–.0026	.010	.015	.9121	2.136	.0008–.0026	2.2486	.0005–.0024	3	.004–.008
	8-351②	.0018–.0026	.010	.015	.9121	2.3207	.0008–.0026	2.9998	.0008–.0026	3	.004–.008
	8-351⑨, 400	.0014–.0022	.010	.015	.9751	2.3207	.0008–.0026	2.9998	.0009–.0026	3	.004–.008
	8-460	.0014–.0022	.010	.015	1.040	2.4992–2.5000	.0008–.0028	2.9998	⑦	3	.004–.008
1976	6-200	.0013–.0021	.008	.015	.9121	2.1236	.0008–.0026	2.2486	.0007–.0024	5	.004–.008
	6-250	.0013–.0021	.008	.015	.9121	2.1236	.0008–.0024	2.3986	.0005–.0022	5	.004–.008
	8-302	.0018–.0026	.010	.015	.9121	2.1236	.0008–.0026	2.2486	.0005–.0024	3	.004–.008
	8-351②	.0014–.0022	.010	.015	.9121	2.3207	.0008–.0026	2.9998	.0008–.0026	3	.004–.008
	8-351⑨, 400	.0014–.0022	.010	.015	.9752	2.3107	.0008–.0026	2.9998	.0009–.0026	3	.004–.008
	8-460	.0014–.0022	.010	.015	1.0401	2.4992–2.500	.0008–.0028	2.9998	⑪	3	.004–.008

①—Fit rings in tapered bores for clearance listed in tightest portion of ring travel.
②—Windsor engine.
③—Cleveland engine.
④—No. 3 in 6–170, No. 5 on 6–200.
⑤—Two barrel carburetor.
⑥—Four barrel carburetor.
⑦—No. 1, .0004–.0020; others, .0012–.0028.
⑧—Hi-Perf engine.
⑨—Modified engine.
⑩—No. 1, .0001–.0020, No. 2, 3, 4, 5, .0005–.0024.
⑪—No. 1, .0004–.0020, No. 2,3,4,5, .0009–.0026.

ENGINE TIGHTENING SPECIFICATIONS★

★Torque specifications are for clean and lightly lubricated threads only. Dry or dirty threads produce increased friction which prevents accurate measurement of tightness.

Year	Engine	Spark Plugs Ft. Lbs.	Cylinder Head Bolts Ft. Lbs.	Intake Manifold Ft. Lbs.	Exhaust Manifold Ft. Lbs.	Rocker Arm Shaft Bracket Ft. Lbs.	Rocker Arm Cover Ft. Lbs.	Connecting Rod Cap Bolts Ft. Lbs.	Main Bearing Cap Bolts Ft. Lbs.	Flywheel to Crank-shaft Ft. Lbs.	Vibration Damper or Pulley Ft. Lbs.
1969–72	6 Cyl.	15–20	70–75	—	13–18	30–35	3–5	③	60–70	75–85	85–100
	8-302④	15–20	65–72	23–25	12–16	17–23⑨⑮	3–5	19–24	60–70	75–85	100–130
	8-302⑤	5–10	65–72	⑥	12–16	17–23⑨⑮	3–5	40–45	60–70⑦	75–85	70–90
	8-351	15–20	95–100	23–25	18–24	⑯	3–5	40–45	95–105	75–85	70–90
	8-390	15–20	80–90	32–35	18–24	40–45	4–7	40–45	95–105	75–85	70–90
	8-400	10–15	95–105	⑥	⑥	17–23⑨⑮	3–5	40–45	95–105	75–85	70–90
	8-427	15–20	90–100	32–35	18–24	40–45	4–7	53–58	95–105	75–85	70–90
	8-428	15–20	80–90	32–35	18–24	40–45	4–7	53–58	95–105	75–85	70–90
	8-429	15–20⑧	130–140	25–30	28–33	②	2½–4	40–45	95–105	75–85	75–90
1973	6 Cyl.	15–20	70–75	—	13–18	30–35	3–5	21–26	60–70	75–85	85–100
	8-302	15–20	65–72	23–25	12–16	17–23⑨	3–5	19–24	60–70	75–85	70–90
	8-351⑩	15–20	105–112	23–25	18–24	17–23⑨	3–5	40–45	95–105	75–85	100–130
	8-351⑪	15–20	95–105	⑫	12–22	18–25⑮	3–5	40–45	⑬	75–85	70–90
	8-400	10–15	95–105	⑫	12–16	18–25⑮	3–5	40–45	⑬	75–85	70–90
	8-429	15–20	130–140	25–30	28–33	18–25⑮	5–6	40–45	95–105	75–85	70–90

Continued

ENGINE TIGHTENING SPECIFICATIONS—Continued

Year	Engine	Spark Plugs Ft. Lbs.	Cylinder Head Bolts Ft. Lbs.	Intake Manifold Ft. Lbs.	Exhaust Manifold Ft. Lbs.	Rocker Arm Shaft Bracket Ft. Lbs.	Rocker Arm Cover Ft. Lbs.	Connecting Rod Cap Bolts Ft. Lbs.	Main Bearing Cap Bolts Ft. Lbs.	Flywheel to Crank-shaft Ft. Lbs.	Vibration Damper or Pulley Ft. Lbs.
1974	6 Cyl.	15–20	70–75	—	13–18	30–35	3–5	(3)	60–70	75–85	85–100
	8-302	15–20	65–72	19–27	12–16	17–23(9)	3–5	19–24	60–70	75–85	70–90
	8-351(10)	15–20	105–112	19–27	18–24	17–23(9)	3–5	40–45	95–105	75–85	100–130
	8-351(11)	15–20	95–105	(14)	12–22	18–25(15)	3–5	40–45	(13)	75–85	70–90
	8-400	10–15	95–105	(14)	12–16	18–25(15)	3–5	40–45	(13)	75–85	70–90
	8-460	15–20	130–140	22–32	28–33	18–25(15)	5–6	40–45	95–105	75–85	70–90
1975–76	6-200	15–25	70–75	30–35	13–18	30–35	3–5	21–26	60–70	75–85	85–100
	6-250	15–25	70–75	18–24	13–18	30–35	3–5	21–26	60–70	75–85	85–100
	V8-302	10–15	65–72	23–25	18–24	17–23(9)	3–5	19–24	60–70	75–85	70–90
	V8-351(10)	10–15	105–112	23–25	18–24	17–23(9)	3–5	40–45	95–105	75–85	70–90
	V8-351(1)	10–15	95–105	(14)	18–24	18–25	3–5	40–45	95–105	75–85	70–90
	V8-400	10–15	95–105	(14)	18–24	18–25	3–5	40–45	95–105	75–85	70–90
	V8-460	10–15	130–140	25–30	28–33	18–25	5–6	40–45	95–105	75–85	70–90

(1)—Modified engine.
(2)—1969–70 rocker arm stud to cylinder head; 65–75 ft. lbs. 1971–72 fulcrum bolt to cylinder head; 18–25 ft. lbs.
(3)—6-170, 200 is 19–24; 6-250 is 21–26.
(4)—Except "H.O."
(5)—"H.O."
(6)—5/16 Bolts, 23–25, 3/8 Bolts, 28–32, 1/4 Bolt, 6–9
(7)—Outer bolts on caps 2, 3 & 4—35–40.
(8)—"CJ", "SCJ", "BOSS" use 5-10 ft.-lbs.
(9)—Rocker arm stud nut.
(10)—Windsor engine.
(11)—Cleveland engine.
(12)—5/16" bolts, 21–25, 3/8" bolts, 27–33, 1/4" bolt, 6–9.
(13)—1/2" bolts, 95–105, 3/8" bolts, 35–45.
(14)—5/16" bolts, 21–25, 3/8" bolts, 27–33.
(15)—Fulcrum bolt to cylinder head.
(16)—Windsor, rocker arm stud nut; 17–23. Cleveland, fulcrum bolt to cylinder head; 18–25.

ALTERNATOR & REGULATOR SPECIFICATIONS

Year	Make or Model	Current Rating(1)		Field Current @ 75°F.		Voltage Regulator(2)				Field Relay	
		Amperes	Volts	Amperes	Volts	Basic Part No. (10316)	Voltage @ 75°F.	Contact Gap	Armature Air Gap	Armature Air Gap	Closing Voltage @ 75°F.
1969	C5TF-10300-K	38	15	2.4	12	C3SZ-B	13.5–15.3	(4)	(4)	(4)	2.3–4.2
	C6AF-10300-B	42	15	2.8–3.3	12	C3SZ-B	13.5–15.3	(4)	(4)	(4)	2.3–4.2
	C6AF-10300-F	55	15	2.8–3.3	12	C3SZ-B	13.5–15.3	(4)	(4)	(4)	2.3–4.2
	C6AF-10300-G	55	15	2.8–3.3	12	C3SZ-B	13.5–15.3	(4)	(4)	(4)	2.3–4.2
	C6DF-10300-A	38	15	2.4	12	C3SZ-B	13.5–15.3	(4)	(4)	(4)	2.3–4.2
	C7AF-10300-A	65	15	2.9	12	C3SZ-B	13.5–15.3	(4)	(4)	(4)	2.3–4.2
	C9AF-10300-A	42	15	2.8–3.3	12	C3SZ-B	13.5–15.3	(4)	(4)	(4)	2.3–4.2
	C9AF-10300-B	55	15	2.8–3.3	12	C3SZ-B	13.5–15.3	(4)	(4)	(4)	2.3–4.2
	C9AF-10300-C	42	15	2.8–3.3	12	C3SZ-B	13.5–15.3	(4)	(4)	(4)	2.3–4.2
	C9SF-10300-A	55	15	2.8–3.3	12	C3SZ-B	13.5–15.3	(4)	(4)	(4)	2.3–4.2
	C9ZF-10300-B	55	15	2.8–3.3	12	C3SZ-B	13.5–15.3	(4)	(4)	(4)	2.3–4.2
	C9ZF-10300-C	55	15	2.8–3.3	12	C3SZ-B	13.5–15.3	(4)	(4)	(4)	2.3–4.2
1970–72	D0ZF-10300-B	38	15	2.4	12	C3SZ-B	13.5–15.3	(4)	(4)	(4)	2.0–4.2
	D0AF-10300-C	42	15	2.9	12	C3SZ-B	13.5–15.3	(4)	(4)	(4)	2.0–4.2
	D0AF-10300-F	42	15	2.9	12	C3SZ-B	13.5–15.3	(4)	(4)	(4)	2.0–4.2
	D0AF-10300-G	42	15	2.9	12	C3SZ-B	13.5–15.3	(4)	(4)	(4)	2.0–4.2
	D0AF-10300-E	55	15	2.9	12	C3SZ-B	13.5–15.3	(4)	(4)	(4)	2.0–4.2
	D0AF-10300-H	55	15	2.9	12	C3SZ-B	13.5–15.3	(4)	(4)	(4)	2.0–4.2
	D0ZF-10300-A	55	15	2.9	12	C3SZ-B	13.6–15.1	.018–.020	.042–.052	.011–.013	6.2–7.2
	D0ZF-10300-C	55	15	2.9	12	C3SZ-B	13.6–15.1	.018–.020	.042–.052	.011–.013	6.2–7.2
	D0SF-10300-A	55	15	2.9	12	C3SZ-B	13.6–15.1	.018–.020	.042–.052	.011–.013	6.2–7.2
	D0LF-10300A(6)	55	15	2.9	12	C8SZ-A	—	—	—	—	—

Continued

ALTERNATOR & REGULATOR SPECIFICATIONS—Continued

Year	Make or Model	Current Rating①		Field Current @ 75°F.		Basic Part No. (10316)	Voltage Regulator②			Field Relay	
		Amperes	Volts	Amperes	Volts		Voltage @ 75°F.	Contact Gap	Armature Air Gap	Armature Air Gap	Closing Voltage @ 75°F.
1970–72	D1ZF-10300AA	55	15	2.9	12	C3SZ-B	13.5–15.3	④	④	④	2.0–4.2
	D1AF-10300AA	61	15	2.9	12	C3SZ-B	13.5–15.3	④	④	④	2.0–4.2
	D0AF-10300A⑤	65	15	2.9	12	C3SZ B	—	—	—	—	—
1973	Purple	38	15	2.9	12	D3TZ-A	13.5–15.3	④	④	④	2.5–4.0
	Orange	42	15	2.9	12	D3TZ-A	13.5–15.3	④	④	④	2.5–4.0
	Red	55	15	2.9	12	D3TZ-A	—	④	④	④	2.5–4.0
	Green	61	15	2.9	12	D3TZ-A	13.5–15.3	④	④	④	2.5–4.0
	Green	70	15	2.9	12	D3TZ-A	13.5–15.3	④	④	④	2.5–4.0
1974–75	Purple	38	15	2.9	12	D4TZ-A	—	—	—	—	—
	Orange	42	15	2.9	12	D4TZ-A	—	—	—	—	—
	Red	55	15	2.9	12	D4TZ-A	—	—	—	—	—
	Green	61	15	2.9	12	D4TZ-A	—	—	—	—	—
	All	70	15	2.9	12	D4TZ-A	—	—	—	—	—
	All	90	15	2.9	12	D4TZ-A	—	—	—	—	—
1976	D3ZF-10300-AA	38	15	2.9	12	D4AF-AA	13.5–15.3	—	—	—	2.5–4.0
	D5AF-10300-AA	40	15	2.9	12	D4AF-AA	13.5–15.3	—	—	—	2.5–4.0
	D30F-10300-FA	42	15	2.9	12	D4AF-AA	13.5–15.3	—	—	—	2.5–4.0
	—	70	15	2.9	12	D4TF-AA	13.5–15.3	—	—	—	2.5–4.0
	D50F-10300-BA	90	15	2.9	12	D4TF-AA	13.5–15.3	—	—	—	2.5–4.0

①—Current rating stamped on housing. ②—Voltage regulation stamped on cover. ④—Not adjustable.
⑤—Integral regulator alternator.

WHEEL ALIGNMENT SPECIFICATIONS

OLD CAR SPECIFICATIONS: For 1946-68 Wheel Alignment Specifications see main index.

Year	Model	Caster Angle, Degrees		Camber Angle, Degrees				Toe-In. Inch	Toe-Out on Turns, Deg	
		Limits	Desired	Limits		Desired			Outer Wheel	Inner Wheel
				Left	Right	Left	Right			
1969	Falcon	−1¾ to +¼	−¾	−½ to +1	−½ to +1	+¼	+¼	3/16	18.1③	20
	Fairlane	−1¾ to +¼	−¾	−½ to +1	−½ to +1	+¼	+¼	3/16	18.1③	20
	Montego	−1¾ to +¼	−¾	−½ to +1	−½ to +1	+¼	+¼	3/16	18.1③	20
	Mustang	−¾ to +1¼	+¼	+¼ to +1¾	+¼ to +1¾	+¾	+¾	3/16	18.68	20
	Cougar	−¾ to +1¼	+¼	+¼ to +1¾	+¼ to +1¾	+¾	+¾	3/16	18.68	20
1970	Maverick	−1½ to +½	−½	−½ to +1	−½ to +1	+¼	+¼	¼	18.48	20
	Falcon	−1¾ to +¾	+¾	−½ to +1	−½ to +1	+¼	+¼	3/16	17.32④	20
	Torino	−1¾ to +¾	+¾	−½ to +1	−½ to +1	+¼	+¼	3/16	17.32④	20
	Montego	−1¾ to +¾	+¾	−½ to +1	−½ to +1	+¼	+¼	3/16	17.32④	20
	Mustang	−1 to +1	Zero	+¼ to +1¾	+¼ to +1¾	+1	+1	3/16	16.68	20
	Cougar	−1 to +1	Zero	+¼ to +1¾	+¼ to +1¾	+1	+1	3/16	16.68	20
1971	Maverick	−2½ to +1½	−½	−¾ to +1¼	−¾ to +1¼	+¼	+¼	3/16	18.48	20
	Torino	−2¾ to +1¼	+¾	−¾ to +1¼	−¾ to +1¼	+¼	+¼	3/16	17.32④	20
	Mustang	−2 to +2	Zero	−½ to +1½	−½ to +1½	+¾	+¾	3/16	18.68	20
	Comet	−2½ to +1½	−½	−¾ to +1¼	−¾ to +1¼	+¼	+¼	3/16	18.48	20
	Montego	−2¾ to +1¼	+¾	−¾ to +1¼	−¾ to +1¼	+¼	+¼	3/16	17.32④	20
	Cougar	−2 to +2	Zero	−½ to +1½	−½ to +1½	+¾	+¾	3/16	18.68	20

Continued

WHEEL ALIGNMENT SPECIFICATIONS—Continued

OLD CAR SPECIFICATIONS: For 1946-68 Wheel Alignment Specifications see main index.

Year	Model	Caster Angle, Degrees		Camber Angle, Degrees				Toe-In. Inch	Toe-Out on Turns, Deg.	
				Limits		Desired				
		Limits	Desired	Left	Right	Left	Right		Outer Wheel	Inner Wheel
1972	Maverick	−2½ to +1½	−½	−¾ to +1¼	−¾ to +1¼	+¼	+¼	3/16	18.21(5)	20
	Torino	−1¼ to +2¾	+¾	−¼ to +1¾	−¼ to +1¾	+¾	+¾	3/16	17.73	20
	Mustang	−2 to +2	Zero	−½ to +1½	−½ to +1½	+½	+½	3/16	17.72	20
	Comet	−2½ to +1½	−½	−¾ to +1¼	−¾ to +1¼	+¼	+¼	3/16	18.21(5)	20
	Montego	−1¼ to +2¾	+¾	−¼ to +1¼	−¼ to +1¾	+¾	+¾	3/16	17.73	20
	Cougar	−2 to +2	Zero	−½ to +1½	−½ to +1½	+½	+½	3/16	17.72	20
1973	Maverick	−2½ to +1½	−½	−¾ to +1¼	−¾ to +1¼	+¼	+¼	¼	18.21(5)	20
	Torino	−¾ to +2¼	+¾	−¼ to +1¾	−¼ to +1¾	+¾	+¾	⅜	17.73	20
	Mustang	−2 to +2	0	−½ to +1½	−½ to +1½	+½	+½	¼	17.72	20
	Comet	−2½ to +1½	−½	−¾ to +1¼	−¾ to +1¼	+¼	+¼	¼	18.21(5)	20
	Montego	−¾ to +2¼	+¾	−¼ to +1¾	−¼ to +1¾	+¾	+¾	⅜	17.73	20
	Cougar	−2 to +2	0	−½ to +1½	−½ to +1½	+½	+½	¼	17.72	20
1974	Maverick	−2½ to +1½	−½	−¾ to +1¼	−¾ to +1¼	+¼	+¼	3/16	18.16(6)	20
	Torino	+½ to +3½	+2	−⅜ to +1⅝	−⅞ to +1⅛	+⅝	+⅛	⅛	18.11	20
	Comet	−2½ to +1½	−½	−¾ to +1¼	−¾ to +1¼	+¼	+¼	3/16	18.16(6)	20
	Montego	+½ to +3½	+2	−⅜ to +1⅝	−⅞ to +1⅛	+⅝	+⅛	⅛	18.11	20
	Cougar	+½ to +3½	+2	−⅜ to +1⅝	−⅞ to +1⅛	+⅝	+⅛	⅛	18.84	20
1975	Comet	+¼ to −1¼	−½	−½ to +1	−½ to +1	+¼	+¼	⅛	18.36(1)	20
	Cougar	+3¼ to +4¾	+4	−¼ to +1¼	−½ to +1	+½	+¼	⅛	18.06	20
	Elite	+3¼ to +4¾	+4	−¼ to +1¼	−½ to +1	+½	+¼	⅛	18.06	20
	Granada	+¼ to −1¼	−½	−½ to +1	−½ to +1	+¼	+¼	⅛	18.43(2)	20
	Maverick	+¼ to −1¼	−½	−½ to +1	−½ to +1	+¼	+¼	⅛	18.36(1)	20
	Monarch	+¼ to −1¼	−½	−½ to +1	−½ to +1	+¼	+¼	⅛	18.43(2)	20
	Montego	+3¼ to +4¾	+4	−¼ to +1¼	−½ to +1	+½	+¼	⅛	18.06	20
	Torino	+3¼ to +4¾	+4	−¼ to +1¼	−½ to +1	+½	+¼	⅛	18.06	20
1976	Comet	+¼ to −1¼	−½	−½ to +1	−½ to +1	+¼	+¼	⅛	—	—
	Cougar	+3¼ to +4¾	+4	−¼ to +1¼	−½ to +1	+½	+¼	⅛	—	—
	Elite	+3¼ to +4¾	+4	−¼ to +1¼	−½ to +1	+½	+¼	⅛	—	—
	Granada	+¼ to −1¼	−½	−½ to +1	−½ to +1	+¼	+¼	⅛	—	—
	Maverick	+¼ to −1¼	−½	−½ to +1	−½ to +1	+¼	+¼	⅛	—	—
	Monarch	+¼ to −1¼	−½	−½ to +1	−½ to +1	+¼	+¼	⅛	—	—
	Montego	+3¼ to +4¾	+4	−¼ to +1¼	−½ to +1	+½	+¼	⅛	17.73	20
	Torino	+3¼ to +4¾	+4	−¼ to +1¼	−½ to +1	+½	+¼	⅛	17.73	20

(1)—Power steering 18.13°.
(2)—Power steering 18.20°.
(3)—Power steering 17⅞°.
(4)—Power steering 17.81°.
(5)—Manual Steering 18.44°.
(6)—Manual Steering 18.39°.

COOLING SYSTEM & CAPACITY DATA

Year	Model or Engine	Cooling Capacity, Qts. No Heater	Cooling Capacity, Qts. With Heater	Cooling Capacity, Qts. With A/C	Radiator Cap Relief Pressure, Lbs. With A/C	Radiator Cap Relief Pressure, Lbs. No A/C	Thermo. Opening Temp. ①	Fuel Tank Gals.	Engine Oil Refill Qts. ②	Transmission Oil 3 Speed Pints	Transmission Oil 4 Speed Pints	Transmission Oil Auto. Trans. Qts. ⑨	Rear Axle Oil Pints
1969	6-170, 200	8½	9½	9½	12–15	12–15	190	⑥	3½	3½	4	8	2½
	6-250	9	10	10	12–15	12–15	190	⑥	3½	3½	4	9	4
	8-302	14	15	15	12–15	12–15	190	⑥	4	3½	4	9	4
	8-351	14	15	15	12–15	12–15	190	⑥	4	3½	4	11	5
	8-390	19½	20½	20½	12–15	12–15	190	⑥	4	3½	4	12¾	5
	8-427	19½	20½	20½	12–15	12–15	190	⑥	5	3½	4	12¾	5
	8-428	19½	20½	20½	12–15	12–15	190	⑥	4	3½	4	12¾	5
1970	6-170, 200	9	10	10	12–15	12–15	190	⑫	3½	3½	4	8	2½
	6-250	10½	11½	11½	12–15	12–15	190	⑫	3½	3½	4	9	4
	8-302	14½	15½	15½	12–15	12–15	190	⑫	4	3½	4	9	4
	8-351	15½	16½	16½	12–15	12–15	190	⑫	4	3½	4	⑬	5
	8-428	19	20	20	12–15	12–15	190	⑫	4	3½	4	12¾	5
	8-429	17½	18½	18½	12–15	12–15	190	⑫	4	3½	4	12¾	5
	8-429 "CJ"	18½	19½	19½	12–15	12–15	190	⑫	4	3½	4	12¾	5
	8-429 "BOSS"⑩	18½	19½	19½	12–15	12–15	190	⑫	4	—	4	—	5
	8-429 "BOSS"⑪	18½	19½	19½	12–15	12–15	190	⑫	6	—	4	—	5
1971	6-170	8¼	9¼	9¼	12–15	12–15	190	16	3½	3½	—	8	4
	6-200, 250③	7½	8½	8½	12–15	12–15	190	16	3½	3½	—	8	4
	6-250	7½	8½	8½	12–15	12–15	190	⑭	3½	3½	—	9	4
	8-302	14	15	15	12–15	12–15	190	⑭	4	3½	4	9	4
	8-351	15¼	16¼	16¼	12–15	12–15	190	⑭	4	3½	4	⑬	5
	8-429	18½	19½	19½	12–15	12–15	190	⑭	6	3½	4	12¾	5
1972	Maverick 6-170	8¼	9¼	9¼	12–15	12–15	190	15	3½	3½	—	8	4
	Comet 6-170	8	9	9	12–15	12–15	190	15	3½	3½	—	8	4
	Maverick 6-200	8	9	9	12–15	12–15	190	15	3½	3½	—	8	4
	Comet 6-200	7¾	8¾	9	12–15	12–15	190	15	3½	3½	—	8	4
	Maverick 6-250	9	10	10	12–15	12–15	190	15	3½	3½	—	8	4
	Comet 6-250	8¾	9¾	9¾	12–15	12–15	190	15	3½	3½	—	8	4
	Torino 6-250	10½	11½	12	12–15	12–15	190	⑮	3½	3½	—	9	4
	Montego 6-250	10½	11½	11½	12–15	12–15	190	⑯	3½	3½	—	9	4
	Mustang 6-250	10¼	11¼	11¼	12–15	12–15	190	19½	3½	3½	—	9	4
	8-302③	12½	13½	14¼	12–15	12–15	190	15	4	3½	—	9	4
	Torino 8-302	14¼	15¼	16¼	12–15	12–15	190	⑮	4	3½	4	9	4
	Montego 8-302	14¼	15¼	15¼	12–15	12–15	190	⑯	4	3½	4	9	4
	Mustang 8-302	14¼	15¼	15½	12–15	12–15	190	19½	4	3½	4	9	4
	Torino 8-351	14½	15½	15¾	12–15	12–15	190	⑮	4	3½	4	11	5
	Montego 8-351	14½	15½	15⅞	12–15	12–15	190	⑯	4	3½	4	11	5
	Mustang 8-351	14¾	15¾	15¾	12–15	12–15	190	19½	4⑦	3½	4	11	5
	Cougar 8-351	14¾	15¾	15¾	12–15	12–15	190	16½	4	3½	4	⑧	5
	Torino 8-400	16¾	17¾	17¾	12–15	12–15	190	⑮	4	—	4	12¾	5
	Montego 8-400	16¾	17¾	17¾	12–15	12–15	190	⑯	4	—	4	12¾	5
	Torino 8-429	17¾	18¾	20	12–15	12–15	190	⑮	4	—	4	12¾	5
	Montego 8-429	17⅞	18⅞	18⅞	12–15	12–15	190	⑯	4	—	—	12¾	5
1973	Maverick 6-200	8	9	9	12–16	12–16	190	15	4	3½	—	8	4
	Maverick 6-250	8¾	9¾	9¾	12–15	12–15	190	15	3½	3½	—	8	4
	Comet 6-250	8¾	9¾	9¾	12–15	12–15	190	15	3½	3½	—	8	4
	Torino 6-250	10½	11½	11½	12–15	12–15	190	⑰	3½	3½	4	9	4
	Montego 6-250	10½	11½	11½	12–15	12–15	190	⑰	3½	3½	4	9	4
	Mustang 6-250	10¼	11¼	11¼	12–15	12–15	190	19½	3½	—	4	9	4
	8-302③	12½	13½	14¼	12–15	12–15	190	15	4	3½	—	9	4
	Torino 8-302	14¼	15¼	15¾	12–15	12–15	190	⑰	4	3½	4	9	4
	Montego 8-302	14¼	15¼	15¾	12–15	12–15	190	⑰	4	3½	4	9	4
	Mustang 8-302	14¼	15¼	15¼	12–15	12–15	190	19½	4	—	4	9	4

Continued

COOLING SYSTEM & CAPACITY DATA—Continued

Year	Model or Engine	Cooling Capacity, Qts.			Radiator Cap Relief Pressure, Lbs.		Thermo. Opening Temp. ①	Fuel Tank Gals.	Engine Oil Refill Qts. ②	Transmission Oil			Rear Axle Oil Pints
		No Heater	With Heater	With A/C	With A/C	No A/C				3 Speed Pints	4 Speed Pints	Auto. Trans. Qts. ⑨	
1973	Torino 8-351	14¾	15¾	16¾	12–15	12–15	190	⑰	4	3½	4	⑲	5
	Montego 8-351	14¾	15¾	16¾	12–15	12–15	190	⑰	4	3½	4	10½	5
	Mustang 8-351	14¾	15¾	15¾	12–15	12–15	190	19½	4	—	4	⑳	5
	Cougar 8-351	14¾	15¾	15¾	12–15	12–15	190	19½	4	—	4	⑳	5
	Montego 8-400	16¾	17¾	17¾	12–15	12–15	190	⑰	4	—	4	12½	5
	Torino 8-400	16¾	17¾	17¾	12–15	12–15	190	⑰	4	—	4	⑲	5
	Montego 8-429	17¾	18¾	18¾	12–15	12–15	190	⑰	4	—	4	12¾	5
	Torino 8-429	17¾	18¾	18¾	12–15	12–15	190	⑰	4	—	4	12½	5
	8-460⑩	18.5	19.5	19.5	12–16	12–16	190	⑰	6	—	—	13	5
1974	6-200③	8.0	9.0	9.0	12–16	12–16	191	15	4	3½	—	8	4
	6-250③	8.7	9½	9½	12–16	12–16	191	15	4	3½	—	9	4
	6-250⑩	—	11.9	—	12–16	12–16	191	26½	4	3½	—	10¼	4
	8-302③	12.4	13½	14¼	12–16	12–16	191	15	4	3½	—	9	4
	8-302⑩	14.2	15½	15½	12–16	12–16	191	④	4	3½	—	⑤	4
	8-351C⑩	14.8	16	16½	12–16	12–16	191	④	4	—	—	10¼	4
	Cougar 8-351C	14.9	16	16½	12–16	12–16	191	26½	4	—	—	㉓	5
	8-351 W⑩	15.3	16½	16½	12–16	12–16	191	④	4	—	—	㉓	4
	Cougar 8-351 W	15.3	16½	16½	12–16	12–16	191	26½	4	—	—	㉓	5
	8-351 "CJ"⑩	14.7	16½	16½	12–16	12–16	191	④	4	—	4	㉓	4
	Cougar 8-351 "CJ"	14.9	16½	16½	12–16	12–16	191	26½	4	—	—	㉓	5
	8-400⑩	16.7	17¾	18.3	12–16	12–16	191	④	4	—	—	㉓	5
	8-460⑩	18.4	19	19½	12–16	12–16	191	④	4	—	—	㉓	5
	Cougar 400	—	17½	18½	12–16	12–16	191	26½	4	—	—	㉓	5
	Cougar 460	—	19	19½	12–16	12–16	191	26½	6	—	—	㉓	5
1975	6-200③	8	9	9	12–16	12–16	191	19.2	4	3½	—	7.8	4
	6-250③	8	9.7	9.7	12–16	12–16	191	19.2	4	3½	—	8.8	4
	6-250⑱	—	10.5	10.7	12–16	12–16	191	19.2	4	3½	—	㉓	4
	8-302③	—	13.5	14.1	12–16	12–16	191	19.2	4	3½	—	8.8	4
	8-302⑱	—	14.4	14.6	12–16	12–16	191	19.2	4	3½	—	㉓	4
	8-351W⑱	—	15.7	16.7	12–16	12–16	191	19.2	4	3½	—	㉓	4
	8-351W㉒	—	15.9	16.2	12–16	12–16	191	④	4	—	—	㉓	5
	8-351M㉒	—	17.1	18	12–16	12–16	191	④	4	—	—	㉓	5
	8-400㉒	—	17.1	17.5	12–16	12–16	191	④	4	—	—	㉓	5
	8-460㉒	—	19.2	19.7	12–16	12–16	191	④	6	—	—	㉓	5
1976	6-200③	8.2	9	9	12–16	12–16	191	19.2	4	3½	—	8	4½
	6-200⑱	9.1	9.9	9.9	12–16	12–16	191	19.2	4	3½	—	8½	㉔
	6-250③	8.9	9.7	9.7	12–16	12–16	191	19.2	4	3½	—	9	4½
	6-250⑱	9.7	10.5	10.7	12–16	12–16	191	19.2	4	3½	—	8½	㉔
	8-302③	12.7	13.5	14.1	12–16	12–16	191	19.2	4	3½	—	9	4½
	8-302⑱	14.9	15.7	16.7	12–16	12–16	191	19.2	4	3½	—	8½	㉔
	8-351W⑱	14.9	15.7	16.7	12–16	12–16	191	19.2	4	3½	—	10	4
	8-351W㉒	15.1	15.9	16.2	12–16	12–16	191	④	4	—	—	㉑	5
	8-351㉕	16.3	17.1	17.5	12–16	12–16	191	26½	4	—	—	㉑	5
	8-351M㉒	16.3	17.1	17.5	12–16	12–16	191	④	4	—	—	㉑	5
	8-400	16.3	17.1	17.5	12–16	12–16	191	④	4	—	—	12½	5
	8-460	18.4	19.2	19.2	12–16	12–16	191	④	6	—	—	12½	5

①—Use with permanent type anti-freeze.
②—Add 1 qt. with filter change.
③—Maverick and Comet.
④—Sta. Wagons 21 gals.; others 26½ gals.
⑤—C4 9 quarts, FMX 11 quarts.
⑥—Falcon cars and Mustang 16, Cougar 17, Fairlane, Montego and Falcon Wagons 20.
⑦—H.O., 5.
⑧—12″ converter, 12¾; others, 11.
⑨—Approximate. Make final check with dipstick.
⑩—Torino & Montego.
⑪—Mustang & Cougar.
⑫—Early Falcon cars & Maverick 16, Mustang, Cougar, Fairlane, Late Falcon & Montego cars 22, Fairlane, Falcon & Montego wagons 19. Subtract 2 gals. for California cars.
⑬—C4 10¼, FMX 11, C6 12¾.
⑭—Torino, Montego, Mustang & Cougar cars, 20. Station wagons, 18.
⑮—Sedans, 22½; wagons, 20½.
⑯—Sedans, 23; wagons, 21.
⑰—Sta. Wagons 20½ gals.; others 22½ gals.
⑱—Granada and Monarch.
⑲—12½″ converter, 12½; others, 10½.
⑳—With 2 Bar. Carb. 11; with 4 Bar. Carb. & 10¼″ converter, 10¾; others 12¾.
㉑—C4 10 quarts, FMX 11 quarts, C6 12½ quarts.
㉒—Torino, Montego and Cougar.
㉓C4—10¼ qts., C6—10½ qts., FMX 11 qts.
㉔—With 8 inch ring gear; 4. With 9″ ring gear 5.
㉕—Ford Elite.

BRAKE SPECIFICATIONS

Year	Model	Brake Drum Inside Diameter	Wheel Cylinder Bore Diameter: Front Disc Brake	Wheel Cylinder Bore Diameter: Front Drum Brake	Wheel Cylinder Bore Diameter: Rear Brake	Master Cylinder Bore Diameter: With Disc Brakes	Master Cylinder Bore Diameter: With Drum Brakes	Master Cylinder Bore Diameter: With Power Brakes
1969–70	Falcon & Maverick 6 Cyl. Cars②	9.00	2.381	1.062	.844	.9375	1.00	.9375
	Falcon V8 Cars	10.00	2.381	1.094	.906	.9375	1.00	.9375
	Falcon Wagons	10.00	2.381	1.125	.938	.9375	1.00	.9375
	Mustang 6 Cyl.	9.00	2.381	1.062	.844	1.00	1.00	1.00
	Mustang & Cougar 8-302	10.00	2.381	1.125	.875	1.00	1.00	1.00
	Mustang & Cougar 8-351, 390	10.00	2.381	1.094	.813	1.00	1.00	1.00
	Montego & Fairlane 6-250, 8-302①③	10.00	2.381	1.125	.906	.9375	1.00	.9375
	Montego & Fairlane 8-351, 390	10.00	2.381	1.094	.875	.9375	1.00	.9375
	Montego & Fairlane Conv. (Exc. 8-351, 390)	10.00	2.381	1.094	.906	.9375	1.00	.9375
	Montego & Fairlane Wagons (Exc. 8-390)	10.00	2.381	1.094	.938	.9375	1.00	.9375
1971	Maverick & Comet 6 Cyl.	9.00	—	1.062	.844	—	1.00	.9375
	Maverick & Comet V8	10.00	—	1.125	.875	—	1.00	.9375
	Torino & Montego Cars	10.00	2.381	1.125	.906	.9375	1.00	.9375
	Torino & Montego Wagons	10.00	2.381	1.125	.968	.9375	1.00	.9375
	Mustang & Cougar 6-250, 8-302	10.00	2.381	1.125	.875	.9375	1.00	1.00
	Mustang & Cougar 8-351, 429	10.00	2.381	1.125	.906	.9375	1.00	1.00
1972	Maverick & Comet 6 Cyl.	9.00	—	1.062	.844	—	1.00	.9375
	Maverick & Comet V8	10.00	—	1.125	.875	—	1.00	.9375
	Torino & Montego	10.00	3.100	—	1.00	1.00	1.00	1.00
	Mustang & Cougar 6-250, 8-302	10.00	2.381	1.125	.875	1.00	1.00	1.00
	Mustang & Cougar 8-351	10.00	2.381	1.125	.906	1.00	1.00	1.00
1973	Maverick & Comet	10.00	—	1.125	.875	—	1.00	—
	Montego & Torino	10.00	3.100	—	1.00	1.00	1.00	1.00
	Mustang & Cougar	10.00	2.380	1.125	.875	.9375	1.00	1.00
1974	Maverick & Comet	10.00	2.600	1.125	.875	.9375	.9375	—
	Montego & Torino	④	3.100	—	1.00	1.00	1.00	1.00
	Cougar	9.00	3.100	—	.875	1.00	.9375	.9375
1975	Maverick & Comet	10.00	2.6	1.125	.843	.938	.938	.938
	Granada & Monarch	10.00	2.6	—	.938	.938	—	.938
	Torino, Montego, Cougar, Elite	10.00	3.1	—	1.00	1.00	—	1.00
1976	Maverick & Comet	10	2.6	—	.938	.938	—	.938
	Granada & Monarch	10	2.6	—	⑤	.938	—	⑥
	Torino, Montego, Cougar, Elite	11	3.1	—	.94	1.00	—	1.00

①—Except convertible.
②—Includes only early 1970 Falcons.
③—Includes late 1970 Falcons.
④—Exc. Sta. Wagons 10.00, Sta. Wagons 11.0.
⑤—Drum .938″, Disc 2.125″.
⑥—Power rear drum .938″; Power rear disc 1.00″.

REAR AXLE SPECIFICATIONS

Year	Model	Carrier Type	Ring Gear & Pinion Backlash Inch	Nominal Pinion Locating Shim, Inch	Pinion Bearing Preload: New Bearings With Seal Inch-Lbs.	Pinion Bearing Preload: Used Bearings With Seal Inch-Lbs.	Pinion Bearing Preload: New Bearings Less Seal Inch-Lbs.	Pinion Bearing Preload: Used Bearings Less Seal Inch-Lbs.	Differential Bearing Preload	Pinion Nut Torque Ft.-Lbs.(1)
1969		Integral	.008–.012	.017	17–27	6–12	—	—	.008–.012(3)	140
		Removable	.008–.012	(4)	(5)	8–14	—	—	.008–.012(6)	175
1970–71		Integral	.008–.012	.030	22–32	8–14	—	—	.008–.012(2)	140
		Removable	.008–.012	(4)	20–30	8–14	—	—	.008–.012(6)	175
1972		Removable	.008–.012	.015(7)	20–30(8)	8–14	—	—	.008–.012(6)	(9)
1973		Removable	.008–.012	.015(7)	17–27(8)	8–14	—	—	.008–.012(6)	(9)
1974		Removable	.008–.012	.015(7)	17–27(8)	8–14(10)	—	—	.008–.012(11)	(9)
1975–76		Integral	.008–.012	.030	17–27	8–14	—	—	.008–.012	140
		Removable	.008–.012	.015(7)	17–27(8)	8–14	—	—	.008–.012(11)	(9)

(1)—If torque cannot be obtained, install new spacer.
(2)—Case spread with new bearings. With used bearings .006–.010″.
(3)—Case spread with new bearings. With used bearings .003–.005″.
(4)—With 7¾″ and 8″ ring gear .022″. With 8¾″ and 9″ ring gear .015 .
(5)—With 7¾″ and 8″ ring gear 17–32 inch-lbs. With 8¾″ and 9″ ring gear 22–32 inch-lbs.
(6)—Case spread with new bearings. With used bearings .005–.008″.
(7)—With 8″ ring gear .022″.
(8)—Solid spacer 13–33 inch-lbs.
(9)—With collapsible spacer 170 ft. lbs., with solid spacer 200 ft. lbs.
(10)—With 6¾″ ring gear 6–12 inch-lbs.
(11)—Case spread with new bearings. With used bearings for 6¾″ ring gear .003–.005, 8″ and 9″ ring gear .005–.008.

Electrical Section

DISTRIBUTOR, REPLACE

1. On vehicles with conventional ignition systems, disconnect primary wire from ignition coil. On vehicles with breakerless ignition systems, disconnect distributor wiring from connector from main wiring harness.
2. Disconnect vacuum advance lines and remove distributor cap.
3. Scribe a mark on the distributor body indicating the position of the rotor, and scribe another mark on the body and engine block indicating position of distributor body in block. These marks can be used as guides when installing distributor in a correctly timed engine.
4. Remove hold down screw or screws and lift distributor out of block. *Do not crank engine while distributor is removed or the initial timing operation will have to be performed.*

Installation

If the crankshaft has not been disturbed, install the distributor, using the scribed marks previously made on the distributor body and engine block as guides.

If the crankshaft has been rotated while the distributor was removed from the engine, it will be necessary to retime the engine. Crank the engine to bring No. 1 piston on top dead center of its compression stroke. Align the timing mark on the crankshaft pulley with the timing pointer (see *Tune Up* chart). Install the distributor so that the rotor points to the No. 1 spark plug wire terminal in the distributor cap.

Make sure the oil pump intermediate shaft properly engages the distributor shaft. It may be necessary to crank the engine with the starter, after the distributor drive gear is properly engaged, in order to engage the oil pump intermediate shaft.

STARTER, REPLACE

1969-76

1. Raise car on hoist.
2. Disconnect cable at starter terminal.
3. On all models, unfasten and remove starter.

NOTE: On engines equipped with a solenoid actuated starter, turn wheel to full right and remove the idler arm to frame bolts. On some 1973-76 models, it may be necessary to turn wheels aside to aid starter removal. Also, on 1969-70 Montego with V8 engine and power steering and 1970 models equipped with a V8-428 engine, disconnect and lower the idler arm.

NOISY STARTER OR STARTER LOCKUP: If either of these situations occur, loosen the three mounting bolts enough to hand fit the starter properly into pilot plate. Then tighten starter mounting bolts, starting with top bolt. Starter should not be replaced until it has been proven noisy after proper alignment has been established by the above method.

IGNITION LOCK, REPLACE

1970-76

1. Disconnect the battery ground cable.
2. *Units With Fixed Steering Columns:* Remove steering wheel and trim pad. Insert a .090 inch wire pin into the hole inside the column halfway down the lock cylinder housing, Fig. 1. *Units with Tilt Steering Columns:* Insert wire pin in the hole located on the outside of the flange casting next to the emergency flasher button, Fig. 1.
3. Place the gear shift lever in *Park* (with automatic trans) or *Reverse* (with manual trans) position, and turn the lock cylinder with the ignition key to *Run* position.
4. Depress the wire pin while pulling up on the lock cylinder to remove. Remove the wire pin.
5. To install insert the lock cylinder into housing in the flange casting, and turn the key to *Off* position. Be certain that the cylinder is fully inserted before turning to the *Off* position. This action will extend the cylinder retaining pin into the cylinder housing.
6. Turn the key to check for correct operation in all positions.
7. Install the steering wheel and trim pad on fixed column units.
8. Connect the battery ground cable.

IGNITION SWITCH, REPLACE

1970-76 All (Except Early 1970 Maverick)

1. Remove steering column shroud and lower steering column from brake support bracket.
2. Disconnect battery cable.
3. Disconnect switch wiring and remove two switch retaining nuts. Disconnect

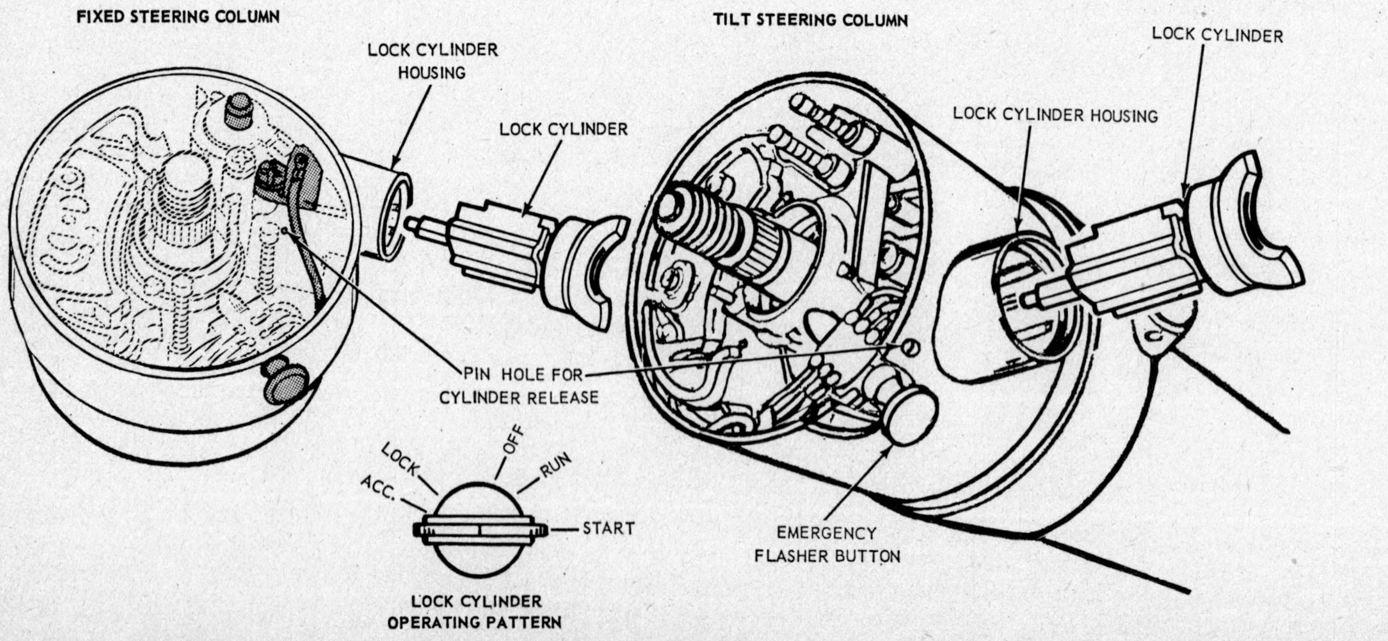

Fig. 1 Ignition lock

switch from actuator and remove switch, Figs. 2 and 3.

4. Move shift lever to Park position on automatic transmissions and Reverse on standard transmission units. Place ignition key in Lock position and remove the key.

NOTE: New replacement switches are pinned in the Lock position by a plastic shipping pin inserted in a locking hole in the switch. For an existing switch, pull plunger out as far as it will go then back one detent to Lock position and insert a $3/32''$ drill in locking hole to retain switch in Lock position.

5. With locking pin in place, install switch on steering column, determine mid position of actuator lash and tighten retaining bolts.
6. Remove locking pin.

1969 All & Early 1970 Maverick

1. Disconnect battery ground cable.

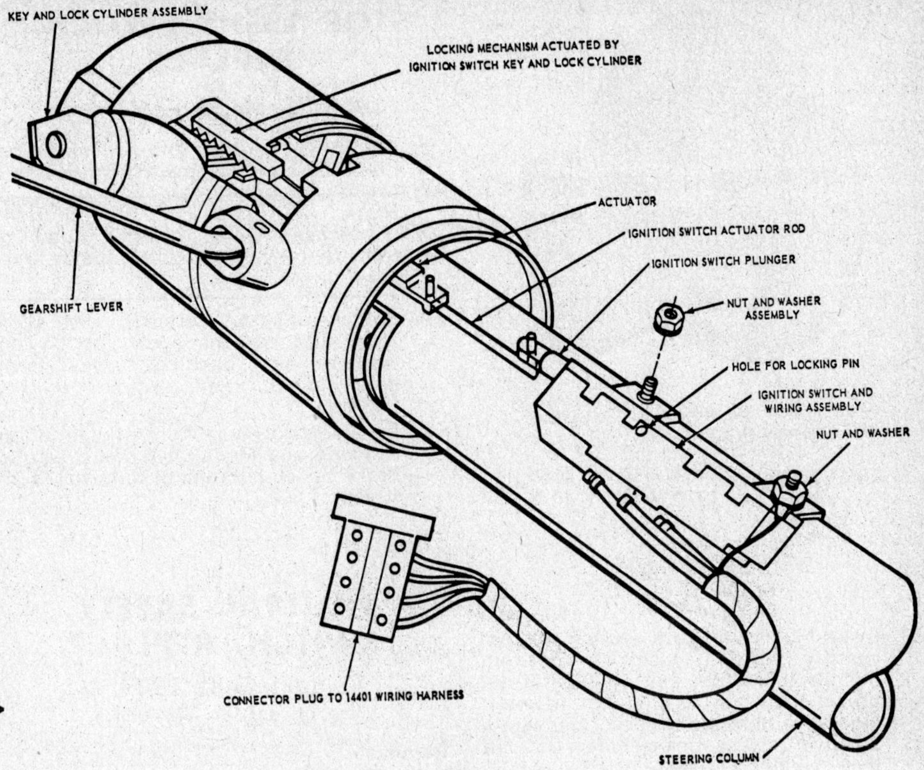

Fig. 2 Ignition switch installation. 1970-71 Montego & Torino, 1970-73 Mustang, 1970-75 Comet, Cougar & late 1970, 1971-75 Maverick ➡

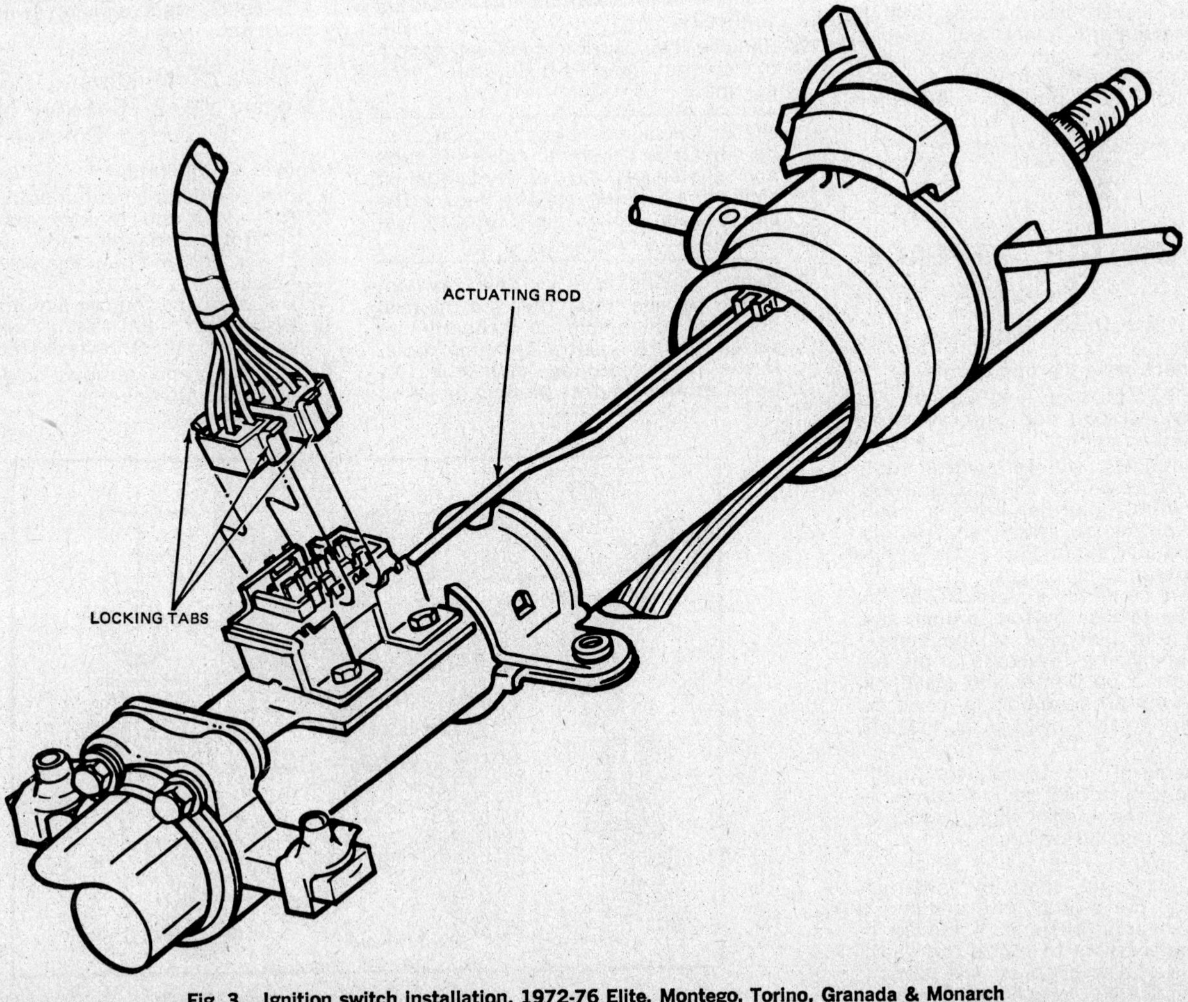

Fig. 3 Ignition switch installation. 1972-76 Elite, Montego, Torino, Granada & Monarch

Fig. 4 Ignition switch removal. 1969 All & Early 1970 Maverick

2. Insert key and turn switch to accessory position. Insert a wire pin in hole in switch, Fig. 4. Slightly depress pin while turning key counterclockwise past the accessory position. This will release lock cylinder. Pull out lock cylinder with key.
3. Remove bezel nut. Lower switch from instrument panel and remove accessory wire nut. Depress tabs securing multiple connector to rear of switch. Pull multiple connector from switch and remove switch.
4. Reverse procedure to install.

LIGHT SWITCH, REPLACE

1969-76 All

1. Disconnect battery ground cable.
2. On 1969-71 Mustang and Cougar, remove two screws and lower parking brake and air control.
3. On all models remove control knob and shaft assembly, by placing knob in full on position, then pressing knob release button on switch and pulling out knob and shaft, Fig. 5. On 1972-73 Mustang and Cougar, it will be necessary to use a screwdriver to press the release button through the hole in the underside of the instrument panel. To gain access to the release button on Comet and Maverick models with air conditioning it will be necessary to first disconnect the left A/C duct.
4. Remove bezel nut. Disconnect multiple plug connector, vacuum hoses if vehicle is equipped with headlight doors and remove switch.
5. Reverse procedure to install. However, install knob and shaft by inserting shaft into the switch until a distinct click is heard. In some instances it may be necessary to rotate the shaft slightly until it engages the switch carrier.

STOP LIGHT SWITCH, REPLACE

1969-76 Mechanical Type

1. Disconnect battery ground cable and disconnect wires at switch connector.

NOTE: On Granada and Monarch models, loosen brake booster nuts 1/4 turn, thereby eliminating binding during removal.

2. Remove hairpin retainer and slide stop light switch, push rod, nylon washers and bushings away from brake pedal, and remove switch, Fig. 6.
3. Reverse procedure to install and on Granada and Monarch models, torque brake booster retaining nuts to 13-25 ft. lbs.

NEUTRAL SAFETY SWITCH, REPLACE

Column Shift 1969 & Early 1970 Maverick

Removal

1. Disconnect the switch wires at plug connector.
2. Remove two screws securing switch to column and lift switch from column.

NOTE: Check the switch actuator to be sure it is secure to the shift tube and seated as far forward against shift tube bearing as possible. If the actuator is broken or damaged, replace it.

3. Before installing a new switch, check to see that the red neutral position gauge pin is properly inserted in the neutral pinning hole. If the pin is missing, align the two holes at the neutral pinning hole on top of the switch and install a No. 43 drill.
4. While holding selector lever against the stop in neutral position, place switch on column and install attaching screws.
5. Remove the gauge pin and connect the wires to the switch.

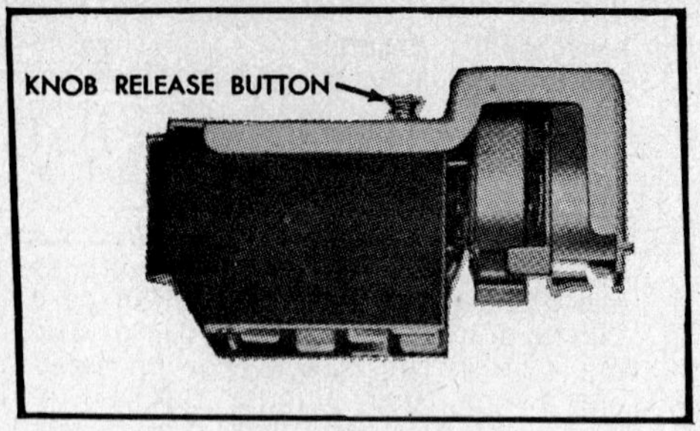

Fig. 5 Light switches. Typical

Adjustment

1. With selector lever against neutral stop, loosen two switch retaining screws.
2. Rotate switch until a start is obtained and tighten switch screws.
3. Place selector lever in "1" position and push the park reset button, located on right side of switch, to the left until it stops.

1969-73 Mustang, 1969-76 Cougar, 1972-76 Elite, Montego & Torino Console

Removal & Adjustment

1. Place selector lever in neutral.
2. Raise car and remove manual lever control rod attaching nut.
3. Lower car and remove selector lever handle.
4. Unfasten and remove dial housing.
5. Disconnect dial light and switch wires at connectors at dash panel.
6. Unfasten and remove selector lever and housing assembly.

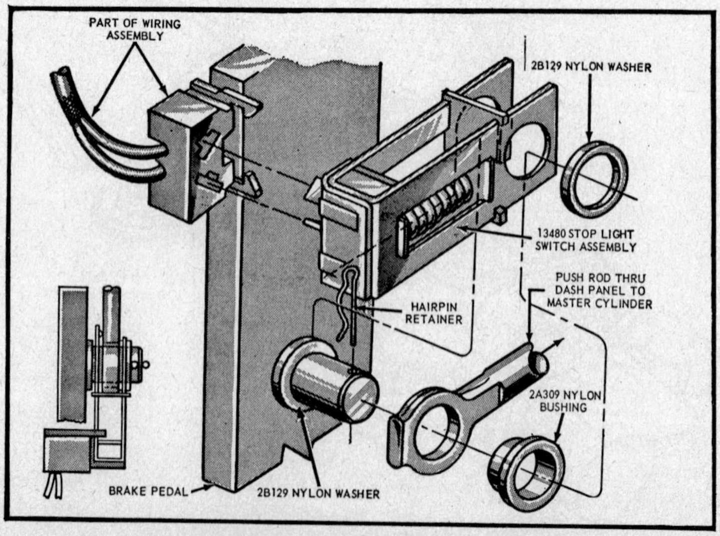

Fig. 6 Mechanical stop light switch. 1969-76

7. Remove pointer back up shield screws and remove the shield.
8. Remove the two switch screws, push the switch harness plug inward and remove the switch and harness assembly.
9. When installing switch, hold it with wires facing down and move the actuator lever all the way to the left. Then return it to neutral positon, Fig. 7.
10. Position the harness and secure the switch to the housing.
11. Install pointer back up shield.
12. Position selector lever and housing assembly on console and fasten.
13. Connect dial light and switch wires.
14. Install dial housing and selector lever handle.
15. Raise car and attach manual lever control rod.
16. Lower car and check operation of switch in Park position.

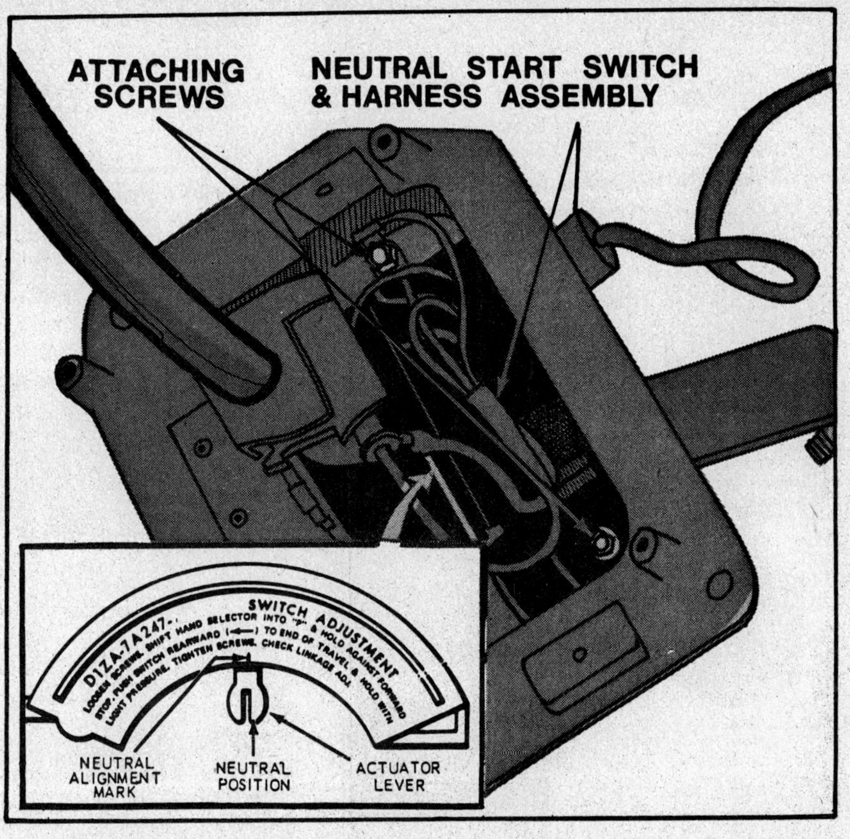

Fig. 7 Neutral switch. 1969-73 Mustang, 1969-76 Cougar, 1972-76 Elite, Montego & Torino with console

1969-70 Montego, Comet, Fairlane & 1970-71 Montego & Torino

Console Shift, Fig. 8

1. Remove handle from selector lever.
2. Remove trim panel from top of console.
3. Remove cover and dial indicator as a unit.
4. Unfasten and remove selector lever retainer from housing (6 screws).
5. Unfasten switch from lever housing (2 screws). Disconnect wires at plug connector and remove switch.
6. With selector lever in neutral, move lever back and forth until gauge pin (#43 drill) can be fully inserted in gauge pin holes, Fig. 8.
7. Place transmission selector lever firmly against stop of neutral detent position.
8. Slide combination neutral start and back-up light switch forward or rearward as required until switch actuating lever contacts selector lever.
9. Tighten switch screws and remove gauge pin.
10. Complete installation in reverse order of removal.

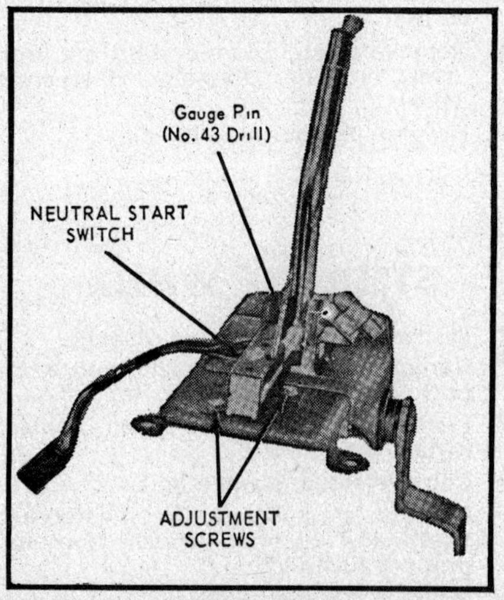

Fig. 8 Neutral safety switch (console shift). 1969-71 Comet, Fairlane and Montego

1971-76 Maverick, Comet

Transmission Mounted Switch, Fig. 9

1. Remove downshift linkage rod from transmission downshift lever.
2. Apply penetrating oil to downshift lever shaft and nut; then remove downshift outer lever.
3. Remove switch attaching bolts.
4. Disconnect multiple wire connector and remove switch from transmission.
5. Install new switch.
6. With transmission manual lever in neutral, rotate switch and install gauge pin (#43 drill) into gauge pin holes.
7. Tighten switch attaching bolts and remove gauge pin.
8. Complete the installation in reverse order of removal.

TURN SIGNAL SWITCH, REPLACE

1971-76

1. Remove retaining screw from underside of steering wheel spokes and lift off pad horn switch/trim cover and medallion as an assembly.
2. Disconnect horn switch wires from terminals.
3. Remove steering wheel retaining nut and remove steering wheel with suitable puller.

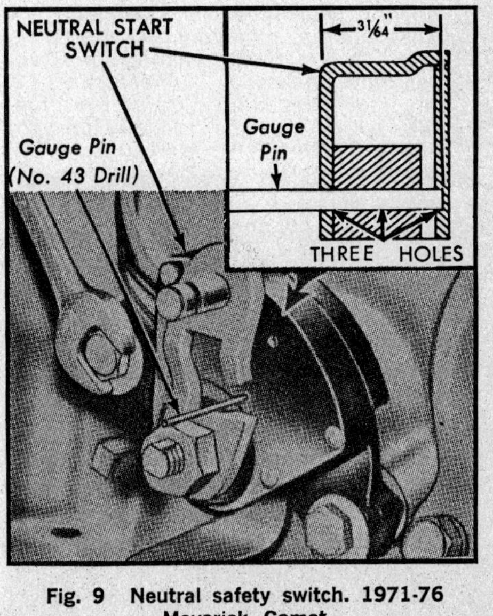

Fig. 9 Neutral safety switch. 1971-76 Maverick, Comet

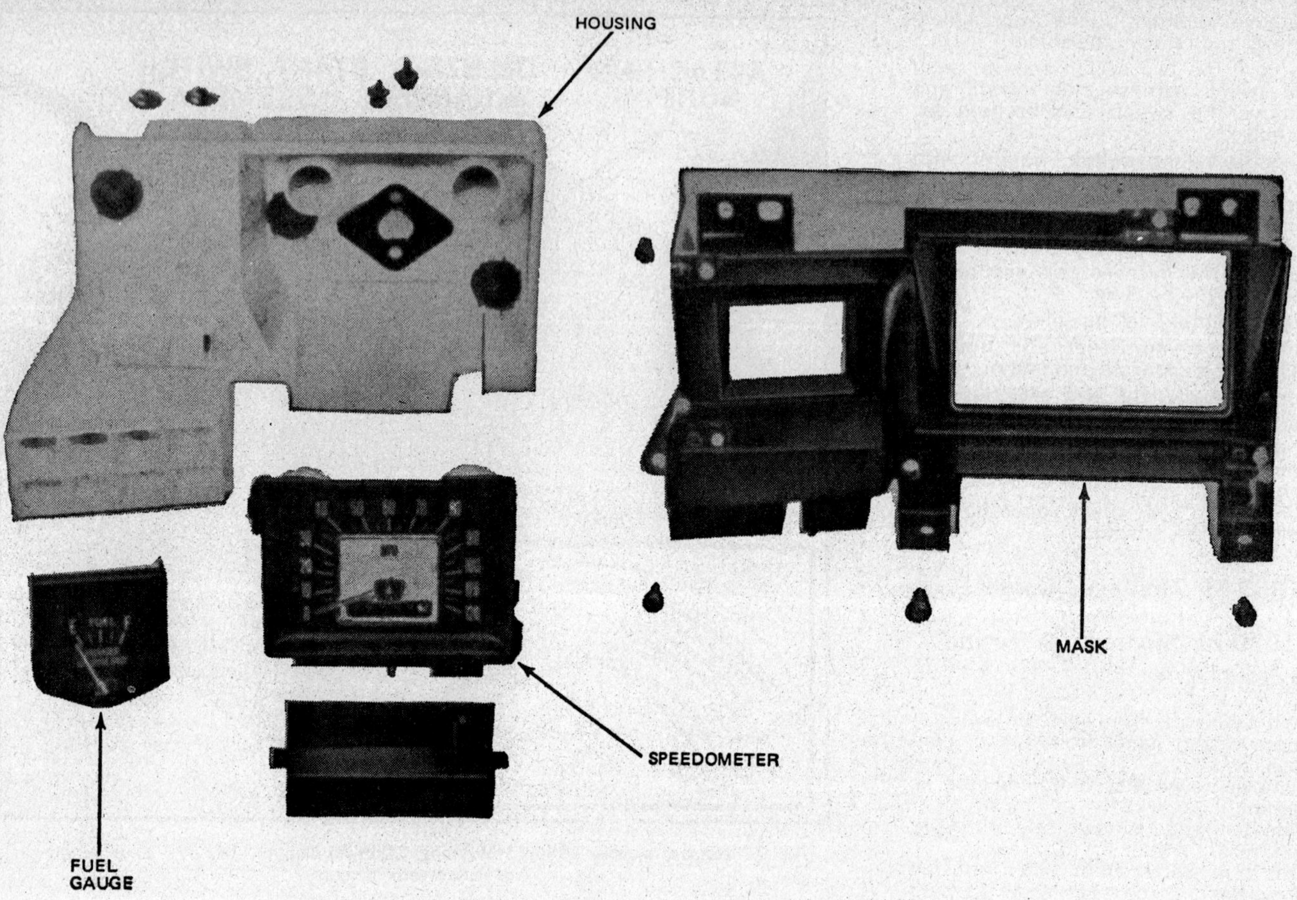

Fig. 10 Instrument cluster. 1975-76 Granada & Monarch

4. Remove turn signal switch lever by unscrewing it from steering column.
5. Remove shroud from under steering column.
6. Disconnect steering column wiring connector plugs and remove screws that secure switch to column.
7. On tilt column, remove wires and terminals from column plug. *NOTE: Record color code and position of wires before removing. A hole provided in the flange casting on fixed columns makes it unnecessary to separate wires from plug as the plug with wires can be guided through hole.*
8. Remove plastic cover sleeve from wiring harness and remove switch from top of column.

1969-70

The emergency warning flasher switch and the turn signal flasher switch are integral parts of the same switch assembly. To remove, proceed as follows:

1. Disconnect battery ground cable.
2. Remove steering wheel hub.
3. Remove horn button (3 screws).
4. Remove steering wheel.
5. Remove turn signal switch lever and emergency flasher control knob. If so equipped, disconnect set speed switch wiring connector.
6. Remove steering column upper collar (2 screws).
7. Disconnect turn signal switch wiring multiple connector near bottom of steering column. It may be necessary to lower hand brake control and left air vent control on some Mustang and Cougar models to provide access to turn signal wiring connector.
8. Remove wires and terminals from connector blocks. This can be done by depressing the tab on the wire terminal with an awl or with an empty ball point pen refill cartridge, then pull wire and terminal from connector block. *Record color code and location of each wire before removing it from connector block.*
9. Tape wires together and attach a piece of heavy cord to the wires to help pull them through steering column during installation.
10. Remove plastic cover from over wires.
11. Push lower steering column collar down. Remove wiring retainer clip.
12. Remove switch from steering column (2 screws) and pull switch and wiring out of column.
13. Reverse procedure to install.

HORN SOUNDER

The horn button or ring used on some models may be removed by twisting the button or ring counter-clockwise. The horn switch on some models is part of the trim cover assembly and if defective, the trim cover must be replaced. The horn sounder on 1970-76 rim-blow steering wheels equipped with speed control cannot be replaced and if defective, the steering wheel assembly must be replaced. On 1972-76 rim-blow steering wheels without speed control, the horn sounder (plastic strip and copper insert) may be replaced using the following procedure:

1. Remove trim pad and disconnect lead wires.
2. Remove plastic cover and lift out horn insert on inner diameter of steering wheel.
3. Reverse procedure to install.

STEERING WHEEL

1. Disconnect battery ground cable.
2. Remove steering wheel trim pad, horn button or ring.
3. Disconnect horn and speed control wiring, if equipped.
4. Remove steering wheel nut.
5. Mark relationship between steering shaft and steering wheel hub for proper reinstallation.
6. Remove steering wheel with a suitable puller.
7. Reverse procedure to install.

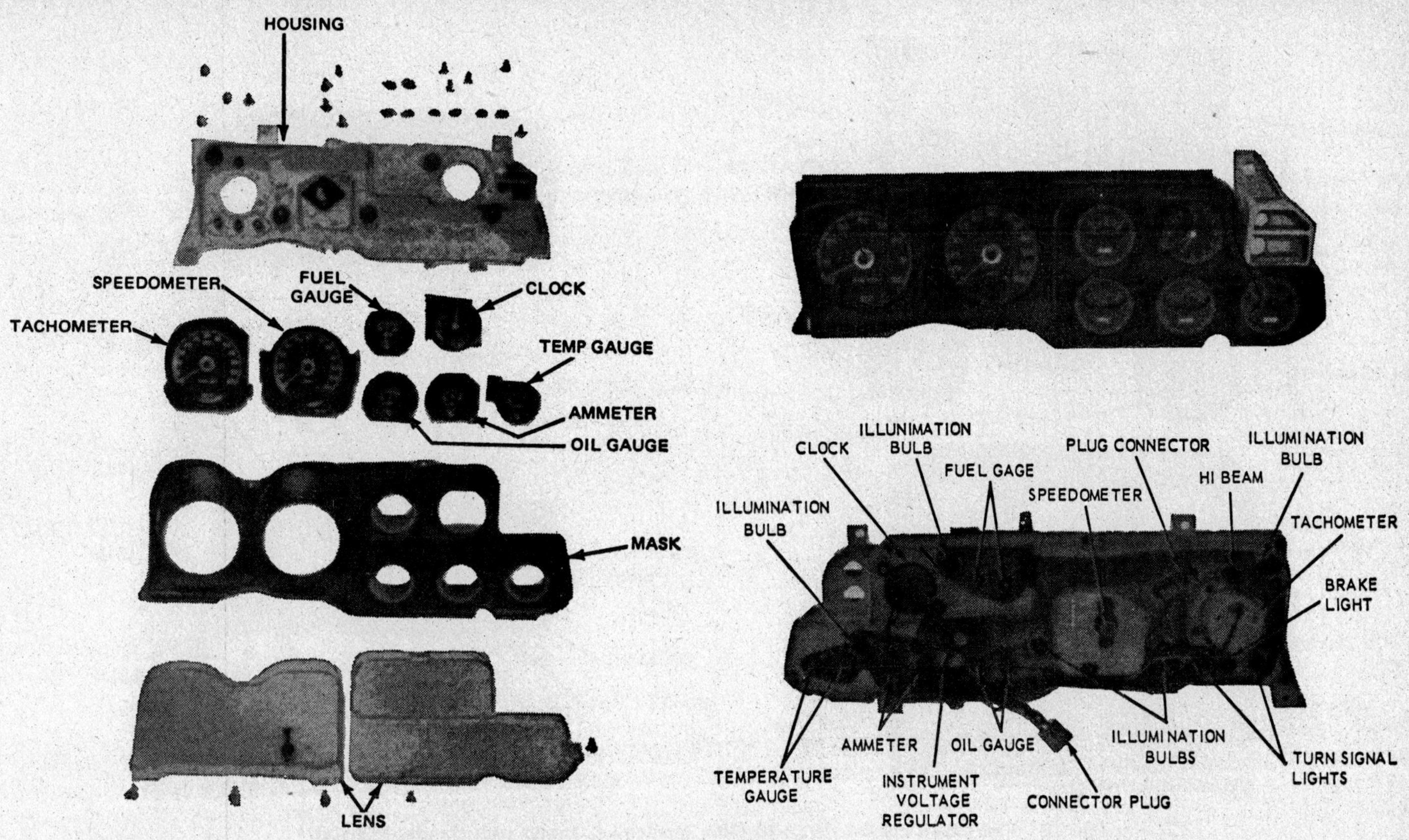

Fig. 11 Instrument cluster. 1974-76 Cougar; 1972-76 Elite, Montego & Torino with performance cluster

INSTRUMENT CLUSTER

1975-76 Granada & Monarch

1. Disconnect battery ground cable.
2. Remove windshield wiper switch knob, then the headlamp switch knob and shaft assembly and bezel.
3. Remove front and lower cluster cover screws, then the steering column shroud, Fig. 10.
4. Using a right angle screwdriver, pry along edges of finish panel, thereby removing stud from retainers, and remove finish panel.
5. Disconnect automatic transmission indicator cable and speedo cable.
6. Remove cluster to instrument panel screws, pull cluster out and disconnect feeder plug from printed circuit, then remove cluster.

1974-76 Cougar XR7 & 1972-76 Elite, Montego, Torino

1. Disconnect battery ground cable.
2. Remove upper and lower retaining screws from cluster trim cover and remove the cover, Figs. 11 and 12.
3. Remove two upper and two lower screws retaining cluster to the panel.
4. Pull cluster away from panel and disconnect speedo cable.
5. Disconnect cluster feed plug from receptable in printed circuit.
6. If equipped, remove Park, Belts and Fuel Economy lights from receptacles.
7. On Elite, Montego and Torino performance cluster, disconnect clock and tachometer wire loom at connector.
8. On Cougar models, disconnect overlay harness connector.
9. Remove cluster.

1972-73 Cougar XR7

1. Disconnect battery ground cable.
2. Remove instrument panel pad, radio bezel retaining screws and the heater-A/C control screws.
3. Disconnect the cluster front cover.
4. Remove oil pressure and ammeter connectors.
5. Disconnect map light and the two bulb sockets from auxiliary gauge assembly.
6. Remove cluster front cover and disconnect speedometer cable.
7. Remove cluster to instrument panel screws.
8. Disconnect cluster feed plug from the printed circuit.
9. Disconnect tachometer and remove cluster.

1971-73 Mustang

1. Disconnect battery ground cable.
2. Remove instrument panel end finish panel and the cluster opening finish panel.
3. Remove four cluster attaching screws and pull cluster away from instrument panel.
4. Disconnect speedometer cable and wiring.

1971-73 Cougar

1. Disconnect battery ground cable.
2. Remove instrument panel pad.
3. Remove four screws attaching cluster to instrument panel, Fig. 13.
4. Disconnect speedometer cable and wiring and remove cluster.

1969-70 Mustang & Cougar

1. Disconnect battery ground cable.
2. Remove instrument panel pad for access to cluster mounting screws.
3. Remove six screws retaining cluster to panel and withdraw cluster slightly.
4. Disconnect plug to printed circuit and tachometer if so equipped.
5. Disconnect speedometer cable by pressing on knurled surface of plastic connector and pulling cable away from head.
6. The cluster can now be removed from the panel.
7. Reverse the foregoing to install.

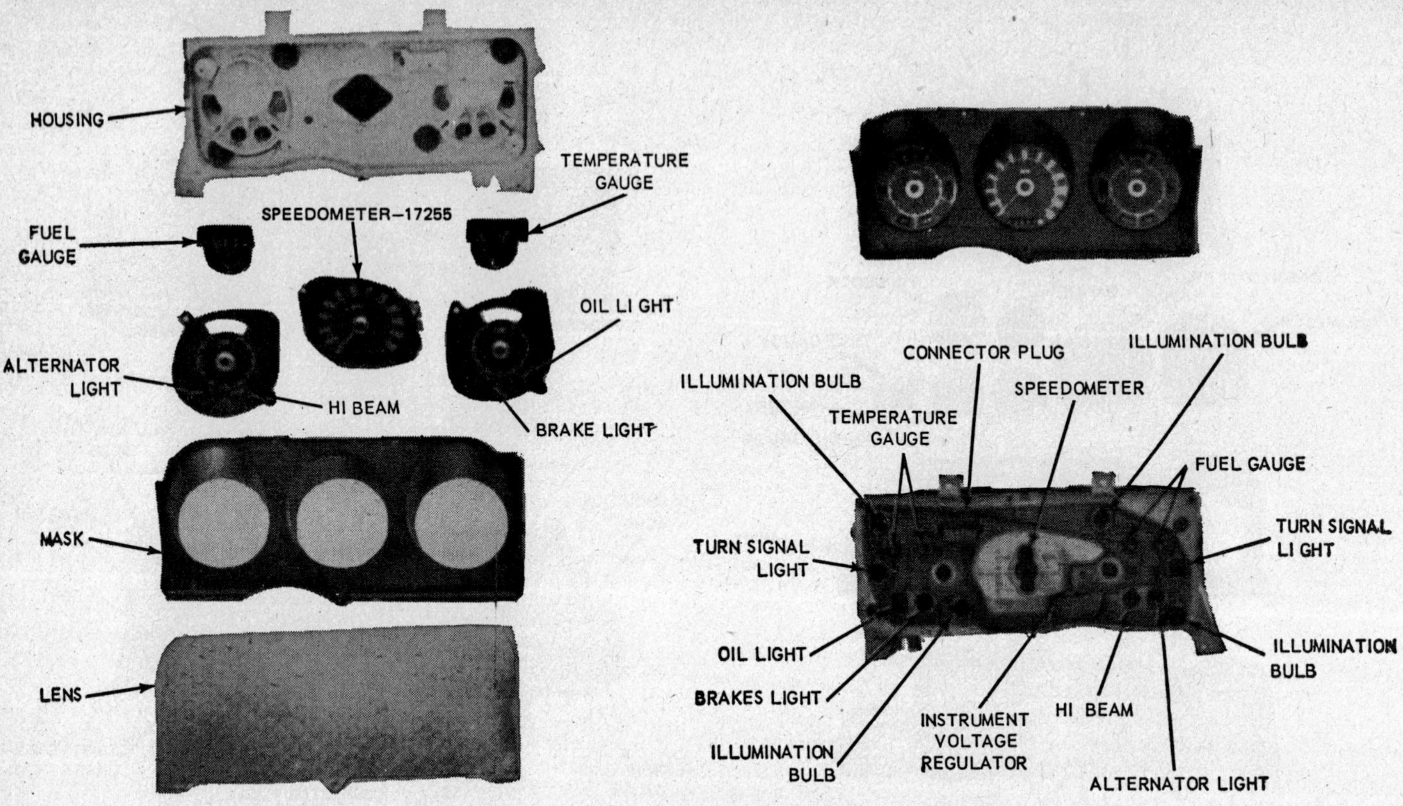

Fig. 12 Instrument cluster. 1972-76 Elite, Montego & Torino with standard cluster

1969-71 Fairlane/Torino Late 1970 Falcon

1. Disconnect battery ground cable.
2. Remove instrument panel cover.
3. Remove right instrument panel shield.
4. Unfasten cluster (5 screws) and position cluster out.
5. Disconnect speedometer cable, tachometer (if equipped) and multiple plug from printed circuit and remove cluster. Cluster components are now accessible for service.

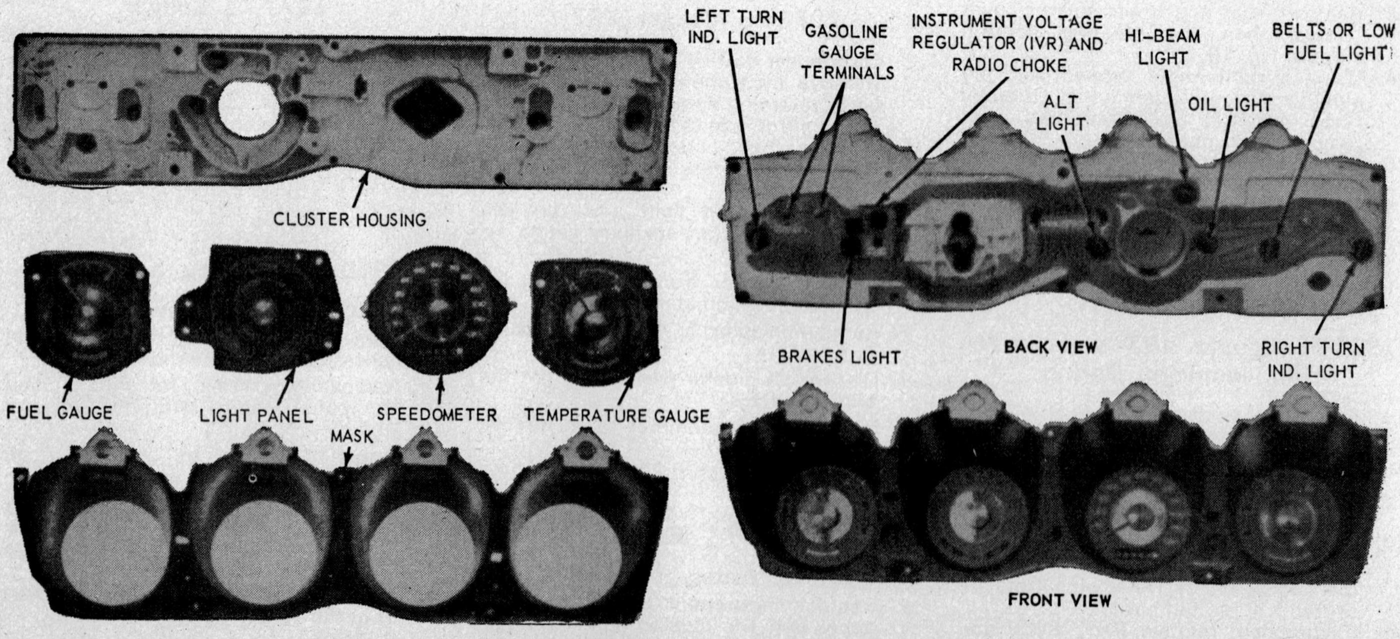

Fig. 13 Instrument cluster. 1971-73 Cougar (Typical)

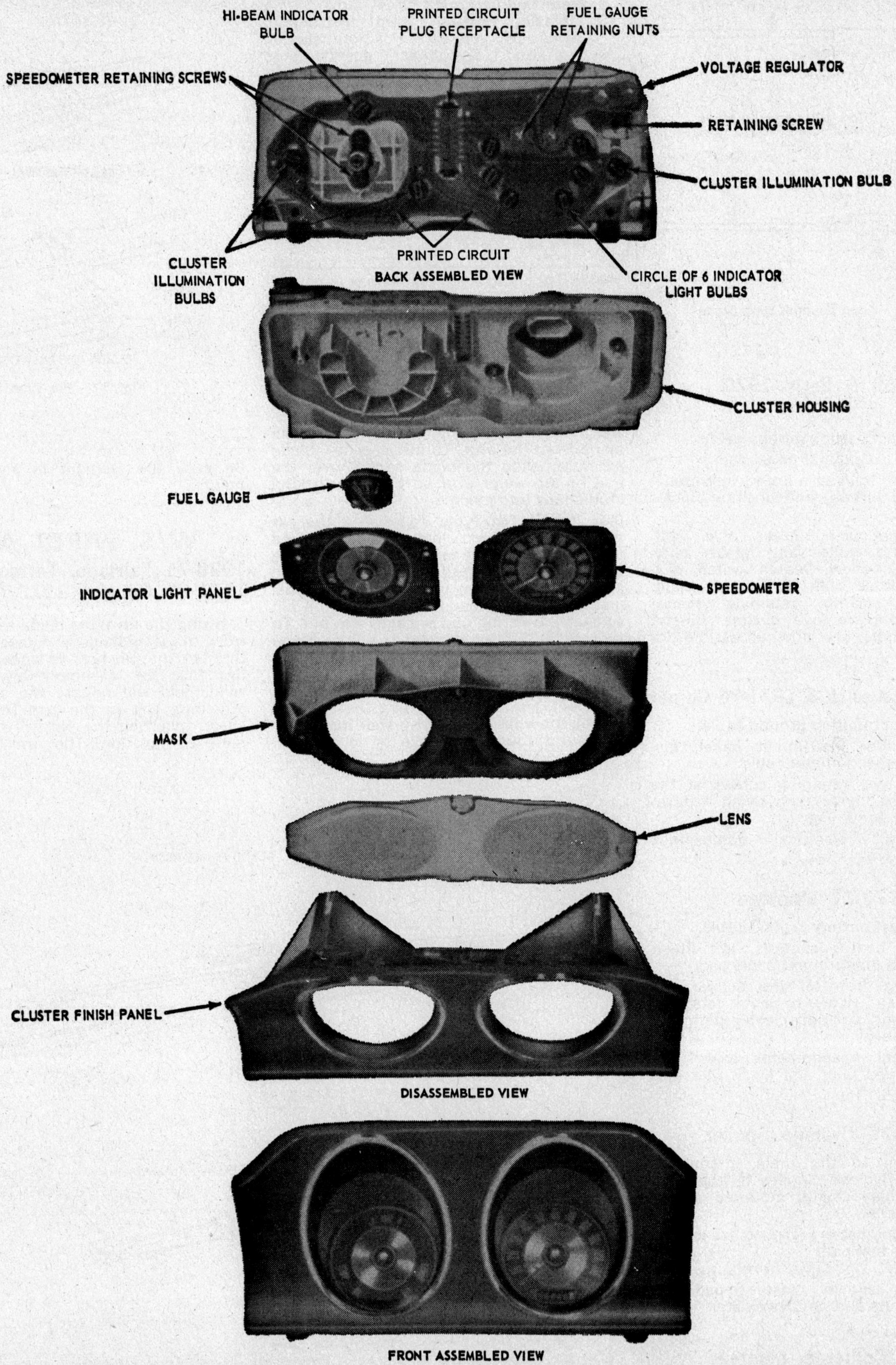

Fig. 14 Instrument cluster. 1971-76 Comet & Maverick

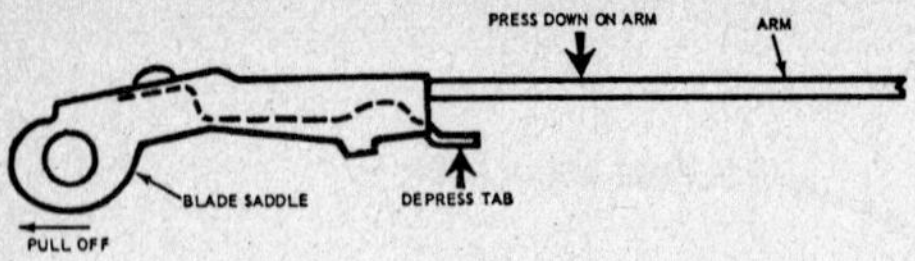

Fig. 15 Trico bayonet type blade

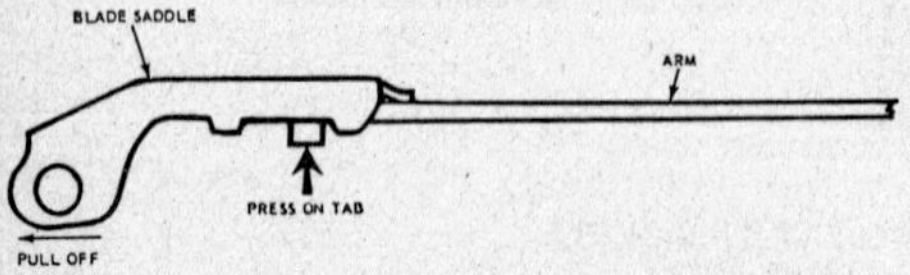

Fig. 16 Anco bayonet type blade

1969 & Early 1970 Falcon

1. Disconnect battery ground cable.
2. Remove instrument panel pad.
3. Unfasten cluster from instrument panel (5 screws) and position cluster out.
4. Disconnect speedometer cable, heater control cables and heater bulb. Also disconnect heater switch plug and multiple plug to printed circuit. Remove clamp retaining heater cables and remove cluster. Cluster components are now accessible for service.

1970-76 Maverick & 1971-76 Comet

1. Disconnect battery ground cable.
2. From under instrument panel, disconnect speedometer cable.
3. Remove two retaining screws at the top of the cluster and swing it down from the panel, Fig. 14.
4. Disconnect electrical connections and remove cluster.

1970-71 Montego

1. Disconnect battery ground cable.
2. Remove heater controls, left finish panels and instrument panel pad.
3. From the front of the cluster, remove four cluster-to-panel retaining screws and position cluster part way out of panel.
4. Disconnect speedometer cable and cluster feed plug and clock wire and remove cluster.

1970-71 Cyclone Spoiler

In addition to the main instrument cluster, the Cyclone Spoiler is equipped with an auxiliary cluster mounted to the right of the panel.

1. Disconnect battery ground cable and remove panel pad.
2. From the inner side of the pad assembly, remove cluster-to-pad retaining nuts and remove cluster.

1969 Comet & Montego

1. Disconnect battery ground cable.
2. Remove instrument panel pad.
3. Unfasten cluster from instrument panel (8 screws). Position cluster out and disconnect speedometer cable. Also disconnect multiple plug to cluster, multiple plug to convenience control lights (if equipped), heater control cables and switch.
4. Disconnect clock (if equipped) and remove cluster. Cluster components are now accessible for service.

W/S WIPER BLADES

NOTE: Trico and Anco blades are used and both come in two types Bayonet type and Side Saddle Pin type.

Bayonet Type: Remove the Trico type by pressing down on the arm, this will unlock the top stud. Depress the tab on the saddle, Fig. 15, and pull the blade from the arm. To remove the *Anco* type press inward on the tab, Fig. 16, and pull the blade from the arm. To install a new blade assembly slide the blade saddle over the end of the wiper arm so that the locking stud snaps into place.

Side Saddle Pin Type: To remove the pin type (Trico or Anco) insert an appropriate tool into the spring release opening of the blade saddle, depress the spring clip and pull the blade from the arm, Fig. 17. To install, push the blade saddle onto the pin so that the spring clip engages the pin. To replace the rubber element in a *Trico* blade squeeze the latch lock release and pull the element out of the lever jaws, Fig. 18. Remove *Anco* element by depressing the latch pin, Fig. 18, and sliding the element out of the yoke jaws. To install insert the element through the yoke or lever jaws.

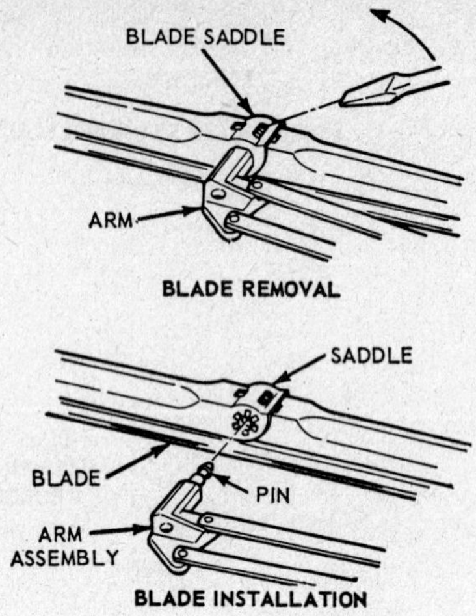

Fig. 17 Pin type blade

Be sure the element is engaged at all points.

W/S WIPER ARMS

1970-71 Fairlane, Torino, Montego; 1971-73 Cougar; 1972-73 Mustang

Swing the arm and blade assembly away from the windshield and insert a 3/32" pin through the pin hole as shown in Fig. 19. Swinging the assembly away from the windshield will release the spring loaded attaching clip in the arm from the pivot shaft. Inserting the pin will hold it in the released position. The arm can now be

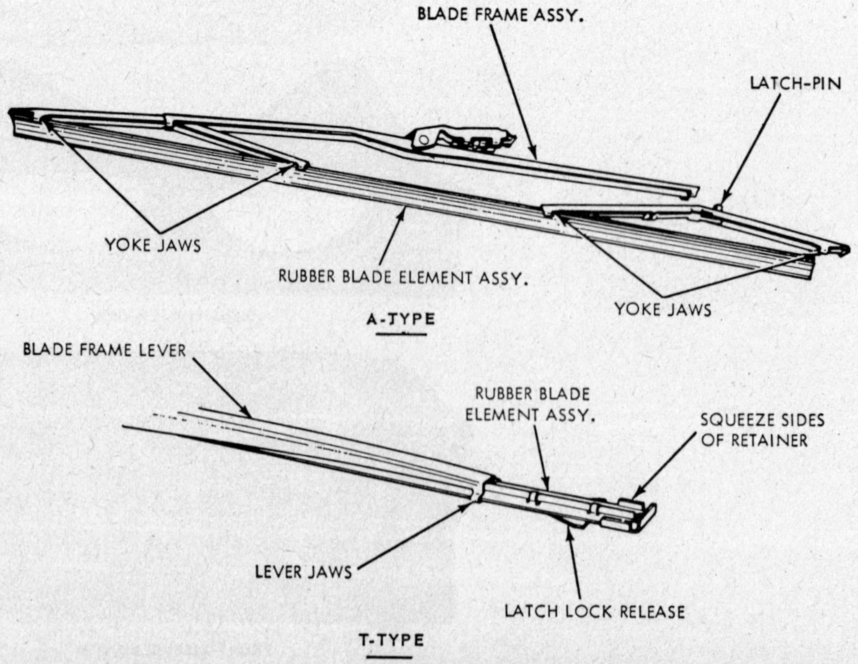

Fig. 18 Wiper blade element

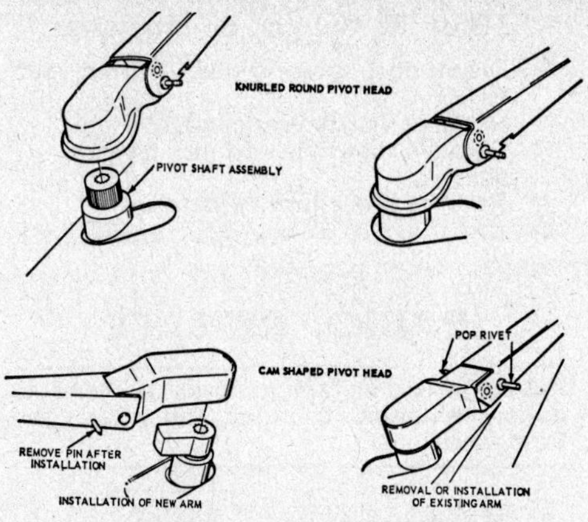

Fig. 19 Wiper arm (Exc. 1972-76 Elite, Montego, Torino & 1974-76 Cougar)

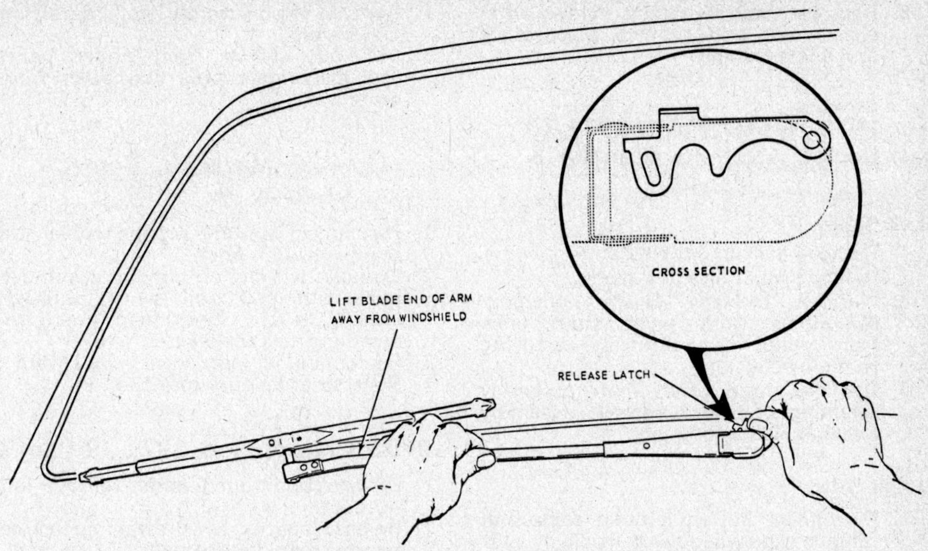

Fig. 20 Wiper arm. 1972-76 Elite, Montego, Torino & 1974-76 Cougar

pulled off the pivot shaft. *Do not pry* off with a screwdriver. Leave the pin in the arm until after installation. A new replacement arm comes with a pin already installed to hold it in released position. Make sure the pivot shaft is in parked position when installing new assembly. Push the arm onto the pivot shaft. Lock the arm to the pivot shaft by removing the pin.

1970 Falcon, Cougar, Mustang; 1970-76 Comet, Maverick, Granada & Monarch

NOTE: These models do not have a pin and hole arrangement to hold the attaching clip in the released position.

To remove, swing the arm and blade assembly away from the windshield to release the spring loaded attaching clip in the arm from the pivot. Hold in this position and pull the arm off the pivot shaft. To install, hold the arm and blade assembly in the swing out position and push the arm on to the pivot shaft. The arm will lock to the pivot when it is moved back against the windshield.

1972-76 Elite, Montego, Torino & 1974-76 Cougar

Raise the blade end of the arm off of the windshield and move the slide latch, Fig. 20, away from the pivot shaft. This action will unlock the wiper arm from the pivot shaft and hold the blade end of the arm off the glass at the same time. The wiper arm can now be pulled off of the pivot shaft without the aid of any tools.

W/S WIPER MOTOR

1972-76 Elite, Torino, Montego & 1974-76 Cougar

Procedure for removal is the same as for 1971 Ford & Mercury Full Size Models and can be found in that car chapter.

1971 Montego & 1971-73 Mustang & Cougar

1. Disconnect battery ground cable.
2. Disconnect wiper motor wiring connector.
3. Remove cowl top left vent screen.
4. Remove wiper link retaining clip from wiper motor arm.
5. Remove three motor retaining bolts and remove wiper and bracket.

1970-76 Maverick & 1971-76 Comet

1. Remove instrument cluster.
2. On air conditioned units, remove center connector and duct assembly.
3. Working through cluster opening, disconnect two pivot shaft links from motor drive arm by removing retaining clip.
4. Disconnect wiring and remove mounting bolts and remove motor through cluster opening.

1969-70 Cougar & Mustang

1. Remove wiper arm and blades.
2. Disconnect washer hose at "T" fitting (left side) on the cowl grille.
3. Remove cowl top grille.
4. Disconnect motor ground wire at forward edge of plenum chamber.
5. Disconnect motor harness at plug and push it back into plenum chamber.
6. Disconnect linkage drive arm from motor output arm crankpin by removing clip.
7. Remove three bolts that retain motor to bracket and rotate motor output arm 180° and remove the motor.

NOTE: Before installing motor, rotate arm 180° and before connecting linkage to motor, turn on ignition to ACC position to allow motor to go into park position.

1969-70 Montego, Comet, Fairlane, Falcon & 1970-71 Torino

1. Disconnect wiper motor wiring connector.
2. Remove wiper arms and blades.
3. Remove cowl top grille panel.
4. Remove wiper link clip from motor arm.
5. Unfasten and remove wiper motor and mounting bracket (4 bolts).
6. Reverse procedure to install.

W/S WIPER TRANSMISSION

1972-76 Elite, Torino, Montego & 1974-76 Cougar

1. Disconnect battery ground cable.
2. Remove wiper arm and blade assemblies.
3. Remove cowl screen. Screen snaps into cowl and the arm stop is integral with the screen.
4. Disconnect linkage drive arm from motor by removing retaining clip.
5. Remove pivot shaft retaining bolts and remove linkage and pivot shaft assemblies.

1971 Montego & 1971-73 Mustang & Cougar

1. Disconnect battery ground cable.
2. Remove wiper arm and blade assemblies from pivot shafts.
3. Remove cowl top left vent screen (four retaining drive pins).
4. Remove drive arm to pivot retaining clip.

5. Remove three retaining screws from each pivot and remove pivot shaft and link assembly.

1970-76 Maverick & 1971-76 Comet, 1975-76 Granada & Monarch

Left Side:

1. Remove instrument cluster.
2. Remove wiper arm and blade.
3. Working through cluster opening, disconnect both pivot shaft links from motor drive arm by removing retaining clip.
4. Remove three pivot shaft assembly retaining bolts and remove assembly through cluster opening.

Right Side:

1. Disconnect battery ground cable and remove wiper blade and arm.
2. On air conditioned units, remove right duct assembly.
3. From under the instrument panel, disconnect first left then right pivot shaft link from motor drive arm.
4. Reaching between utility shelf and instrument panel, remove pivot shaft retaining bolts and lower assembly out from under panel.

1969-70 Cougar & Mustang

1. Remove arm and blade assemblies from pivot shafts.
2. Disconnect washer hose at "T" fitting on cowl grille.
3. Remove cowl top grille.
4. Disconnect linkage drive arm from motor output arm crankpin by removing clip.
5. Disconnect right link from right arm and pivot shaft and remove the arm and pivot shaft assembly.
6. Unfasten and remove the left arm and pivot shaft and lift out to the right: the pivot shaft and arm, left link and linkage drive arm as one assembly.

NOTE: When installing the linkage, install the left pivot shaft and linkage first.

1969-70 Montego, Comet, Fairlane, Falcon & 1970-71 Torino

1. Remove wiper arms and blades.
2. Remove cowl top grille panel.
3. Remove drive arm to pivot clip and remove pivot shaft and link assembly.
4. Reverse procedure to install.

W/S WIPER SWITCH

1971-73 Montego, Mustang, Cougar 1974-76 Elite, Montego, Torino, Cougar

1. Disconnect battery ground cable.
2. Remove wiper switch knob, bezel nut and bezel.
3. Pull out switch from under panel and disconnect plug connector from switch.

1974-76 Maverick, Comet, Granada & Monarch

1. Disconnect battery ground cable and remove switch knob.
2. On units without air condition, remove bezel nut and pull switch through panel. On air conditioned units, remove instrument cluster.
3. Disconnect wiring and remove switch.
4. Reverse procedure to install.

1970-73 Maverick & 1971-73 Comet

1. On air conditioned units, remove left AC duct.
2. Release control knob retaining spring by pressing in through slot in knob and pull knob from shaft.
3. Remove switch retaining nut and lower switch from under instrument panel.
4. Disconnect switch wiring and remove switch.

1969-73 Fairlane & Torino Late 1970 Falcon

1. Disconnect battery ground cable.
2. Remove switch control knob and bezel nut.
3. Unplug and remove switch.
4. Reverse procedure to install.

1969 & Early 1970 Falcon

1. Disconnect ground cable from battery.
2. Remove switch knobs and unplug connectors at switch.
3. Unfasten (2 screws) and remove switch from under instrument panel.
4. Separate washer switch from wiper switch (2 screws).
5. Reverse procedure to install.

1969-70 Montego, Comet

Standard Wiper

1. Disconnect battery ground cable.
2. Remove switch knob, nut and bezel.
3. Pull switch out from instrument panel.
4. Disconnect plug connector from switch and remove switch.
5. Reverse procedure to install.

Intermittent Wipers

1. Disconnect ground cable from battery.
2. Remove set screws and remove control knob.
3. Remove bezel nut and bezel.
4. Lower switch assembly.
5. Remove three vacuum hoses from switch, disconnect plug connector and remove switch.
6. Reverse procedure to install, using color code provided on switch and hoses.

1969-70 Cougar & Mustang

1. Disconnect ground cable from battery.
2. Remove instrument cluster.
3. Remove switch from cluster (2 screws).
4. Reverse procedure to install.

RADIO, REPLACE

NOTE: When installing radio, be sure to adjust antenna trimmer for peak performance.

1975-76 Granada & Monarch

1. Disconnect battery ground cable.
2. Remove headlamp switch.
3. Remove knobs from heater and A/C control, windshield wiper switch and radio.
4. Remove instrument panel applique.
5. Disconnect antenna lead.
6. Remove radio bezel to instrument panel screws, then pull radio and bezel out to disconnect remaining electrical leads and remove radio from panel.
7. Remove rear support bracket and bezel from radio.
8. Reverse procedure to install.

1974-76 Comet & Maverick

1. Disconnect battery ground cable.
2. On 1975 models, remove seat belt interlock module, located below radio.
3. Disconnect power lead, speaker lead and antenna lead from radio.
4. Remove radio knobs, discs, control shaft nuts and washers.
5. Remove radio rear support attaching nut.
6. Remove radio from bezel and from rear support.
7. Reverse procedure to install.

1969-73 Cougar, Mustang, Maverick & 1971-73 Comet

1. Disconnect battery ground cable.
2. Pull control knobs, discs and sleeve from radio shafts.
3. Remove radio applique panel from dash.
4. Remove right and left finish panels.
5. Remove two mounting plate attaching screws.
6. Pull radio out of panel and disconnect wires.
7. Remove mounting plate and rear support from radio.
8. Reverse procedure to install.

1969-73 Fairlane, Falcon, Montego 1974-76 Elite, Montego, Torino, Cougar

1. Disconnect battery ground cable.
2. Pull off radio control knobs.
3. Remove radio support to instrument panel attaching screw.
4. Remove bezel nuts from radio control shafts, then lower radio and disconnect speaker, power and an-

tenna wire from radio.
5. Reverse procedure to install.

HEATER CORE REMOVAL

1975-76 Granada & Monarch

Less Air Conditioning

1. Drain cooling system and disconnect heater hoses from heater core.
2. Remove glove box and the right and floor air distribution ducts.
3. Disconnect heater control cables and wiring harness from resistor assembly.
4. Remove vent duct to upper cowl mounting bolt.
5. Remove heater case to dash panel nuts, then the heater case and vent duct assembly.
6. Remove core cover and seal, then slide core from case, Fig. 21.

With Air Conditioning

1. Disconnect battery ground cable and drain coolant system, then disconnect heater hoses from core at engine side of dash panel.

NOTE: Easier access may be obtained by first disconnecting suction hose and moving it out of the way.

2. Remove heat distribution duct from instrument panel, seat belt interlock module and glove box liner, then loosen right door sill scuff plate and remove right cowl side from trim panel.
3. Loosen instrument panel to right cowl side bolt and remove instrument panel brace bolt at lower rail under glove box.
4. Remove defroster nozzle by removing instrument panel crash pad, removing radio speaker or panel cowl brace, removing four nozzle to cowl bracket screws and lifting defroster nozzle upward through crash panel, Fig. 22.
5. Disconnect vacuum hoses from A/C-Defrost and A/C-Heat door motors and remove vacuum harness to plenum clip screw, then remove the two A/C-Heat mounting nuts and swing door rearward on crankarm.
6. Remove two plenum to left mounting bracket screws and remove the two screws and three clips securing plenum to evaporator case.
7. Swing bottom of plenum away from evaporator case and disengage S-clip on forward flange of plenum, then raise plenum to clear tabs on top of evaporator case.
8. Move plenum to left as far as possible (about 4 inches), pulling rearward on instrument panel to gain clearance.

NOTE: Use extreme care when pulling back on instrument panel to avoid cracking plastic panel. Also, there is very little clearance between plenum and wiper motor assembly.

9. Using tab molded into rear heater core seal, pull heater core to left, then as rear surface of heater core clears evaporator case, pull core rearward and downward to clear instrument panel.

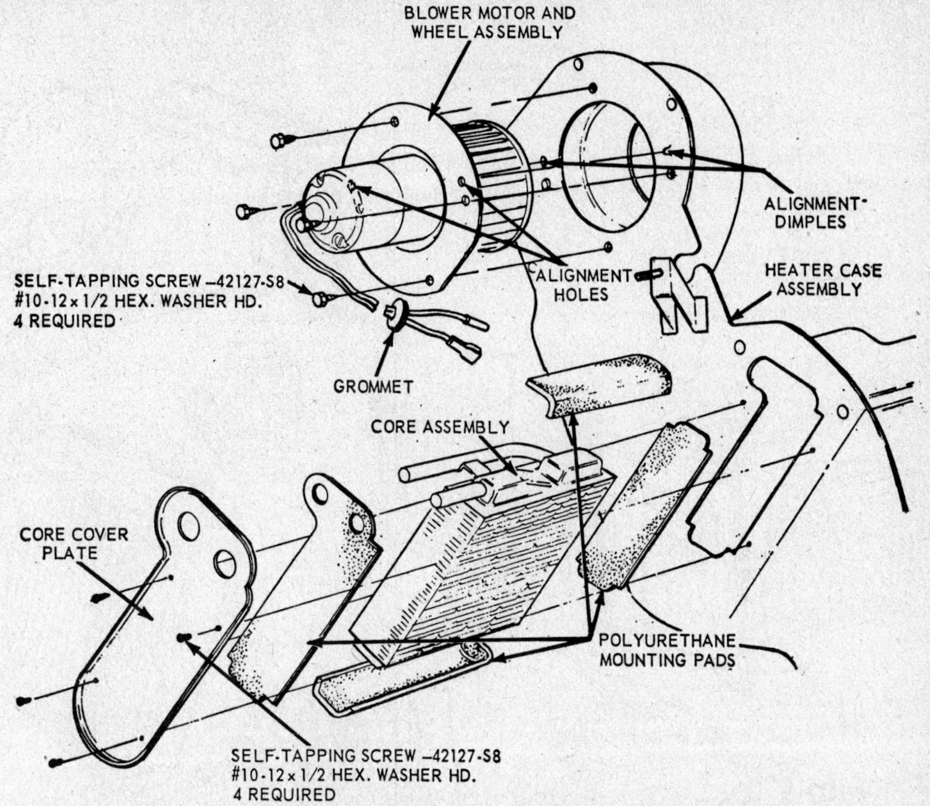

Fig. 21 Heater core & blower motor less air conditioning. 1969-70 Fairlane, Falcon & Maverick; 1969-76 Elite, Montego & Torino; 1971-73 Mustang; 1971-76 Cougar; 1975-76 Granada & Monarch

10. Reverse procedure to install, making certain that heater core tube to dash panel seal is in place between evaporator case and dash panel.

1972-76 Elite, Torino, Montego & 1974-76 Cougar

Less Air Conditioning

Removal of the core is done the same as for previous models and is described further on, Fig. 21.

With Air Conditioning

1. Drain engine coolant and disconnect heater hoses from core.
2. Remove glove box.
3. Remove heater air outlet register from plenum assembly (2 snap clips).
4. Remove temperature control cable assembly mounting screw, and disconnect the end of the cable from the blend door crank arm (1 spring nut).
5. Remove the blue and red vacuum hoses from the high-low door vacuum motor; the yellow hose from the panel-defrost door motor, and the brown hose at the inline tee connector to the temperature by-pass door motor.
6. Disconnect wiring connector from resistor.
7. Remove ten screws from around flange of plenum case and remove rear case half of the plenum, Fig. 23.
8. Remove mounting nut from heater core tube support bracket.
9. Remove heater core.

NOTE: During installation of heater core, be sure to apply body sealer around the case flanges to provide a positive seal. Also make sure core mounting gasket is properly installed.

1971-76 Comet & Maverick

Less Air Conditioning

1. Disconnect battery ground cable and drain engine cooling system.
2. Disconnect blower motor ground wire from fender apron at engine side of dash.
3. Disconnect heater hoses at engine block.
4. Remove heater assembly to dash panel mounting nuts.
5. On 1971-73 units, disconnect ignition switch and plate from package tray and remove package tray from instrument panel.
6. On 1974-76 units, remove glove compartment and on all models remove right cowl trim panel.
7. Remove cable retaining clips and the push nuts at door crank arms. Disconnect control cables from crank arms.
8. Remove defroster air duct from left side of heater assembly.
9. Disconnect motor lead from resistor assembly on bottom of heater.
10. Remove heater case to instrument panel support bracket mounting screw.
11. Pull heater hoses through dash panel and disconnect hoses from heater core.
12. Separate halves of heater and remove core, Fig. 24.

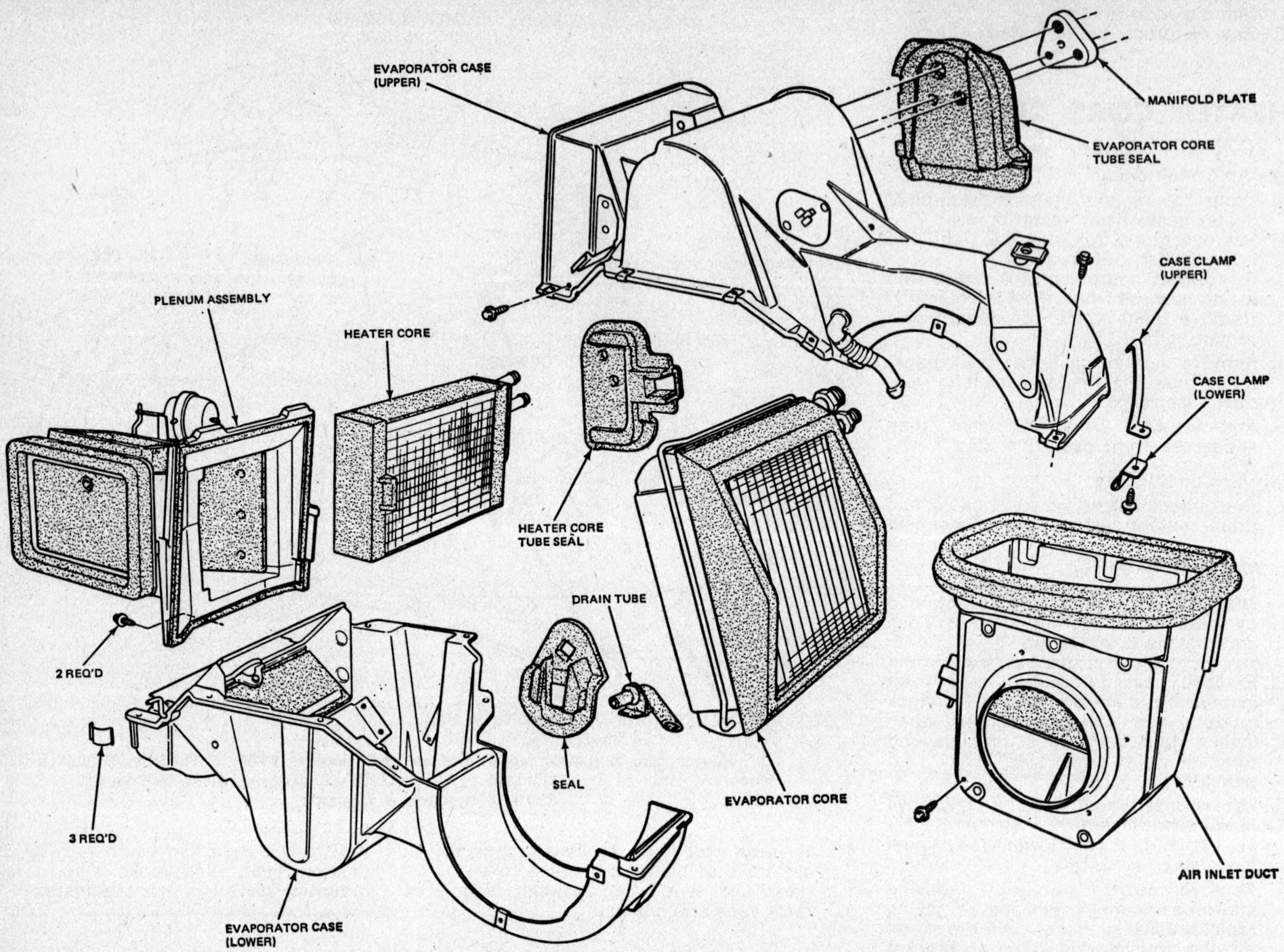

Fig. 22 Heater core. 1975-76 Granada & Monarch with air conditioning.

With Air Conditioning

1. Disconnect battery ground cable, remove air cleaner and drain cooling system.
2. Connect gauge set to compressor and discharge system.
3. Disconnect evaporator core tubes and tape ends to keep out foreign material and disconnect heater hoses from heater core tubes.
4. Remove A/C assembly-to-dash panel stud nuts.
5. On 1971-73 units, remove utility shelf from lower edge of instrument panel, the right cowl trim panel, utility bracket and remove radio.
6. On 1974 units, remove right cowl trim panel and remove glove compartment.
7. On all models disconnect right and left A/C air ducts.
8. Remove floor distribution duct from blower. Disconnect vacuum hoses from door motor and the vacuum harness multiple connector.
9. Disconnect temperature control cable from door crank arm on evaporator.
10. Remove screw that retains evaporator to cowl upper support and remove A/C rearward and away from dash panel.
11. Disconnect wire connectors from A/C thermostat switch on the evaporator housing and also the resistor on blower housing.
12. Disconnect blower motor ground wire from wiper mounting bracket and lower the blower and evaporator housing removing it from vehicle.
13. Separate halves of evaporator housing and remove water valve vacuum switch.
14. Remove temperature blend door shaft, frames and door from lower half of evaporator housing.
15. Lift heater core out of lower housing and remove pads from core.

1971-73 Mustang & Cougar

1. Drain coolant and remove heater hoses from core.
2. Remove glove box and right vent air duct assembly.
3. Disconnect control cables from heater case.
4. Remove heater case assembly (4 heater case-to-dash panel mounting stud nuts).
5. Remove heater core cover and pad and remove core from case.

1969-70 Cougar & Mustang

Less Air Conditioning

1. Disconnect battery ground cable and drain cooling system.
2. Remove instrument panel pad.
3. Remove glove box liner and door.
4. Remove air distribution duct from heater.
5. Disconnect control cables from heater.
6. Disconnect wires from blower motor resistor.
7. Remove right courtesy light from underside of dash, if so equipped.
8. Remove heater support to dash panel screw.
9. Disconnect vacuum hoses and remove power vent air duct.
10. Disconnect blower motor ground wire in engine compartment.
11. Disconnect heater hoses from heater at dash.
12. In engine compartment, remove heater retaining nuts.
13. Remove instrument panel-to-cowl panel attaching screws.
14. Remove instrument panel right side brace.
15. Pull heater assembly and right side of instrument panel rearward and

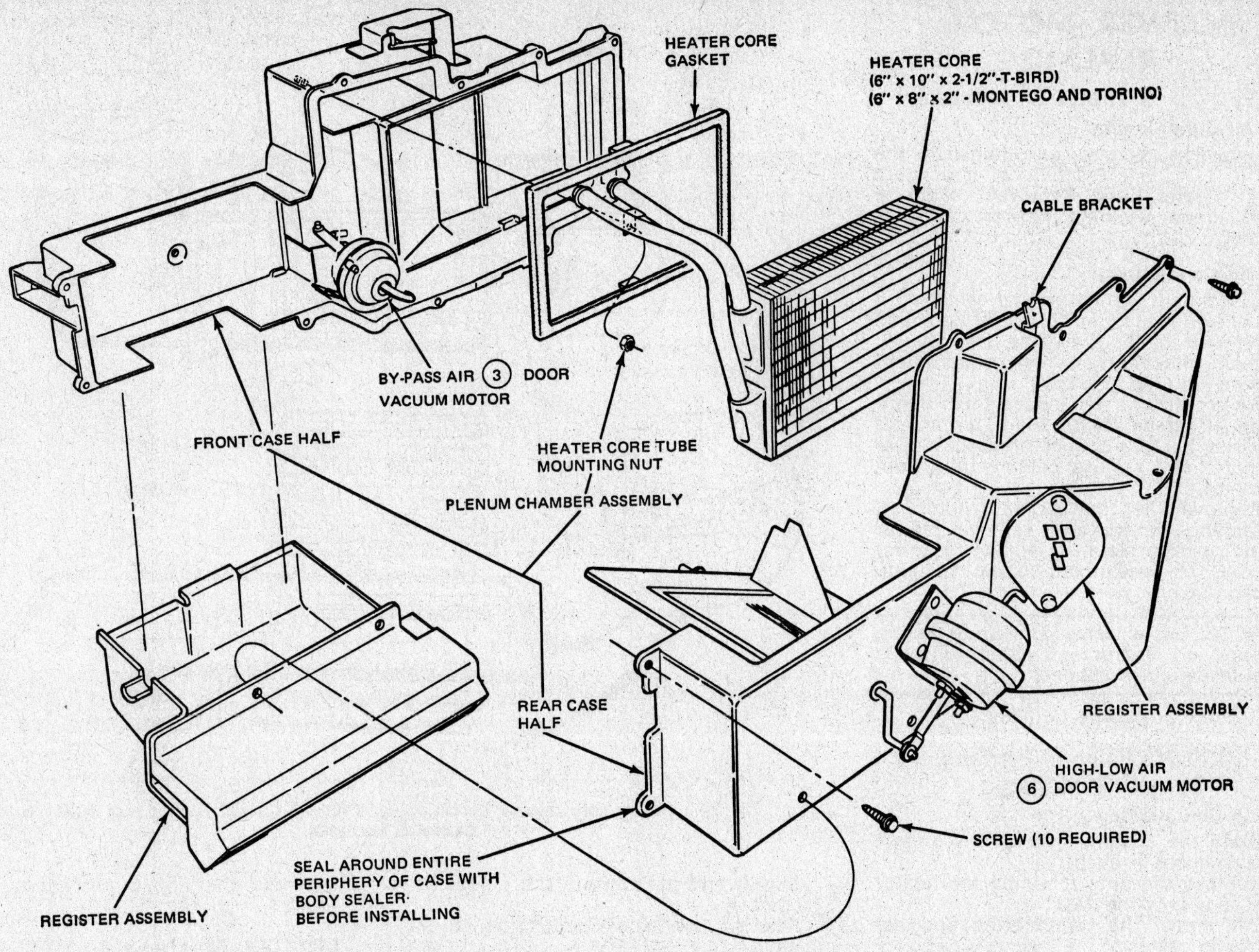

Fig. 23 Heater core with air conditioning. 1972-76 Montego & Torino; 1974-76 Cougar

remove heater assembly.
16. Remove air inlet seal from heater.
17. Separate halves of heater and remove core, Fig. 24.

With Air Conditioning

1. Disconnect battery ground cable and remove air cleaner.
2. Connect gauge set to compressor valves and isolate compressor.
3. Drain cooling system and remove heat shield from expansion valve.
4. Disconnect low pressure hose and service valve from compressor.
5. Disconnect high pressure hose at quick disconnect.
6. Remove straps retaining refrigerant hoses to the dash-to-fender apron supports.
7. Disconnect heater hoses from heater core.
8. Remove upper and lower seal retainers and remove hose seal.
9. From engine side of dash, remove evaporator housing mounting nuts and blower housing mounting nut.
10. Remove instrument panel pad.
11. Remove instrument cluster.
12. Disconnect vacuum hoses from reheat door and outside recirc door vacuum motors.
13. Disconnect vacuum hoses from water valve vacuum switch.
14. Disconnect control cable from temperature blend door.
15. Disconnect wires from thermostat switch.
16. Remove right and left air ducts from defrost plenum chamber.
17. Remove defrost plenum chamber.
18. Remove instrument panel right side brace.
19. Remove evaporator housing upper rear support bracket-to-cowl screw.
20. Remove blower housing-to-cowl screws.
21. Move blower housing to left away from evaporator housing.
22. Cover the carpet and pull drain tube from hole in floor.
23. Remove instrument panel-to-cowl screws from right side.
24. Remove instrument panel finish cover from around steering column.
25. Unfasten instrument panel from steering column support.
26. Remove instrument panel-to-cowl screws from left side.
27. Position instrument panel back and remove evaporator housing.
28. Separate halves of evaporator housing.
29. Remove water valve vacuum switch.
30. Remove temperature blend door shaft, frames and door from lower half of evaporator housing.
31. Remove heater core from evaporator lower housing and remove pads from core.
32. Reverse procedure to install.

1969-70 Fairlane, Falcon, Comet, Montego, Maverick 1971 Montego & Torino

1. Drain cooling system and disconnect both heater hoses at dash.
2. Unfasten heater from dash.
3. Disconnect temperature and defroster cables at heater.
4. Disconnect wires from resistor, and blower motor wires and clip retaining heater to defroster nozzle.
5. Remove glove box.
6. Remove bolt and nut retaining right air duct control to instrument panel.
7. Remove nuts retaining right air duct and remove duct.
8. Take heater assembly to bench. Then remove heater core cover and pad and lift out core.

BLOWER MOTOR, REPLACE

1975-76 Granada & Monarch

Less Air Conditioning

Remove heater case as outlined in the "1975-76 Granada & Monarch" procedure under "Heater Core, Removal," then remove screws securing blower motor to case, Fig. 21.

With Air Conditioning

1. Disconnect battery ground cable.
2. Loosen right door sill scuff plate and remove right cowl side trim panel, then remove right lower instrument panel to cowl side bolt.
3. Remove cowl to loosen instrument panel brace bolt, then disconnect wiring harness connectors from blower motor and remove cooling tube from motor.
4. Remove the four blower motor attaching screws and remove blower motor from scroll by pulling on lower edge of instrument panel to gain clearance.

NOTE: Use extreme care when pulling back on instrument panel to avoid cracking plastic panel.

5. Reverse procedure to install.

1969-70 Falcon, Fairlane, Comet & Montego

Less Air Conditioning

1. Drain the cooling system. Disconnect both hoses at dash.
2. Remove the nuts retaining the heater assembly to the dash.
3. Disconnect the temperature and defroster cables to the dash.
4. Disconnect the wires from the resistor, and disconnect the blower motor wires and the clip retaining the heater assembly to the defroster nozzle.
5. Remove the glove box.
6. Remove the bolt and nut retaining the right air duct control to the instrument panel. Remove the nuts retaining the right air duct and remove the duct assembly.
7. Remove the heater assembly to a bench. Remove the blower mounting screws and remove the motor and wheel assembly.
8. Reverse above procedure to install.

With Air Conditioning

1. Remove the glove box.
2. Remove the right hand fresh air duct. Disconnect the vacuum line from the actuator and position it out of the way.
3. Disconnect the plug from the resistor block and remove the resistor block.
4. Remove the blower motor cover and remove the motor and blower wheel.
5. Reverse above procedure to install.

1969-70 Cougar & Mustang

Less Air Conditioning

1. Follow the procedure to remove the heater core as described previously.
2. Remove the heater assembly from the vehicle.
3. Disconnect the blower motor wire from the resistor.
4. Remove the four blower and motor mounting plate nuts, and remove the blower and motor from the heater assembly.
5. Reverse the above procedure to install.

With Air Conditioning

1. Remove the accelerator pedal assembly.
2. Remove the lower air distribution duct.
3. Disconnect the blower motor cooling tube and wiring.
4. Remove the four blower mounting plate attaching screws, and remove the blower and motor.
5. Reverse the above procedure to install.

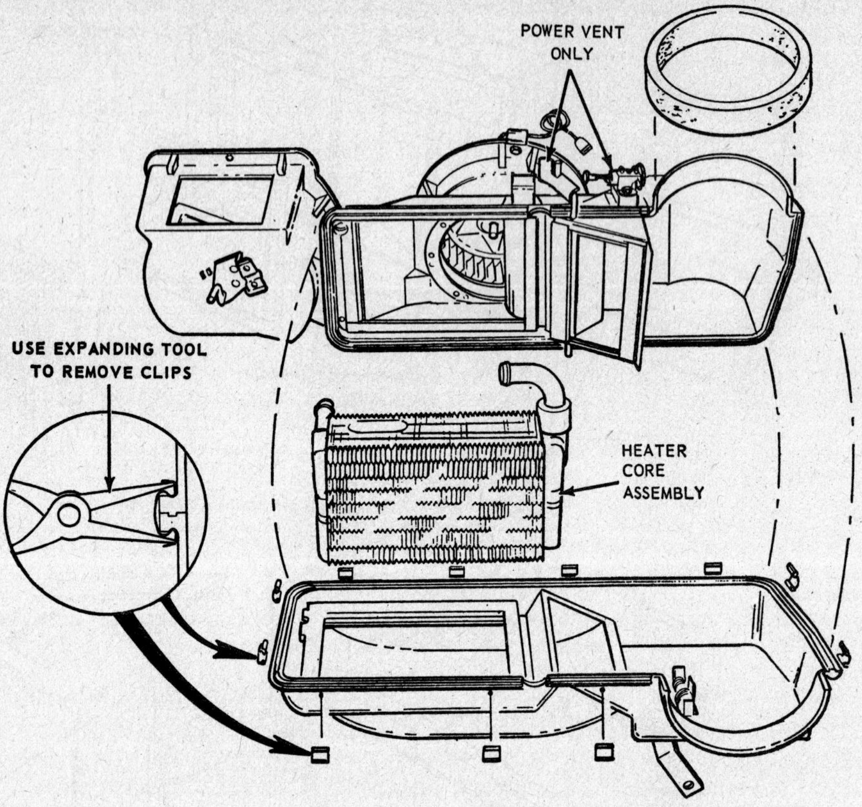

Fig. 24 Heater core less air conditioning. 1969-70 Cougar & Mustang; 1971-76 Comet & Maverick

1970 Maverick

Less Air Conditioning

Follow procedure shown for 1970 Falcon, Fairlane, Montego.

With Air Conditioning

1. Remove the radio. Disconnect the ignition switch from the package tray.
2. Remove the package tray and disconnect both left and right duct assemblies.
3. Remove the air distribution duct from the blower housing.
4. Remove the lower nut retaining the blower housing to the evaporator assembly. Rotate the blower housing until it disengages from the evaporator housing, and disconnect the vacuum hoses at the motor on the blower housing. Disconnect the resistor assembly and the blower ground wire and remove the blower.
5. Reverse the above procedure to install.

1971-73 Mustang & 1971-76 Cougar, Montego, Torino

Less Air Conditioning

1. Follow the procedure to remove the heater core as described previously.
2. Remove the heater assembly from the vehicle, and place it on a bench.
3. Remove the four mounting screws and remove the blower motor and wheel assembly from the blower, Fig. 21.
4. Reverse the above procedure to install.

With Air Conditioning: 1971-73 Mustang & Cougar

1. Remove the blower housing mounting bracket stud nut, from the engine side of the dash panel.
2. Remove the two blower housing-to-instrument panel support mounting screws.
3. Remove the floor air distribution duct (three plastic buttons).
4. Disconnect the blower motor ground wire (black) from the upper cowl.
5. Disconnect the blower motor lead wire (orange-black) from the resistor.
6. Rotate the blower motor housing to a diagonal position.
7. Remove the four blower motor mounting plate screws, and remove the blower motor and wheel as an assembly.

With Air Conditioning: 1971 Montego & Torino

1. Remove the glove box.
2. Remove the right hand fresh air duct.

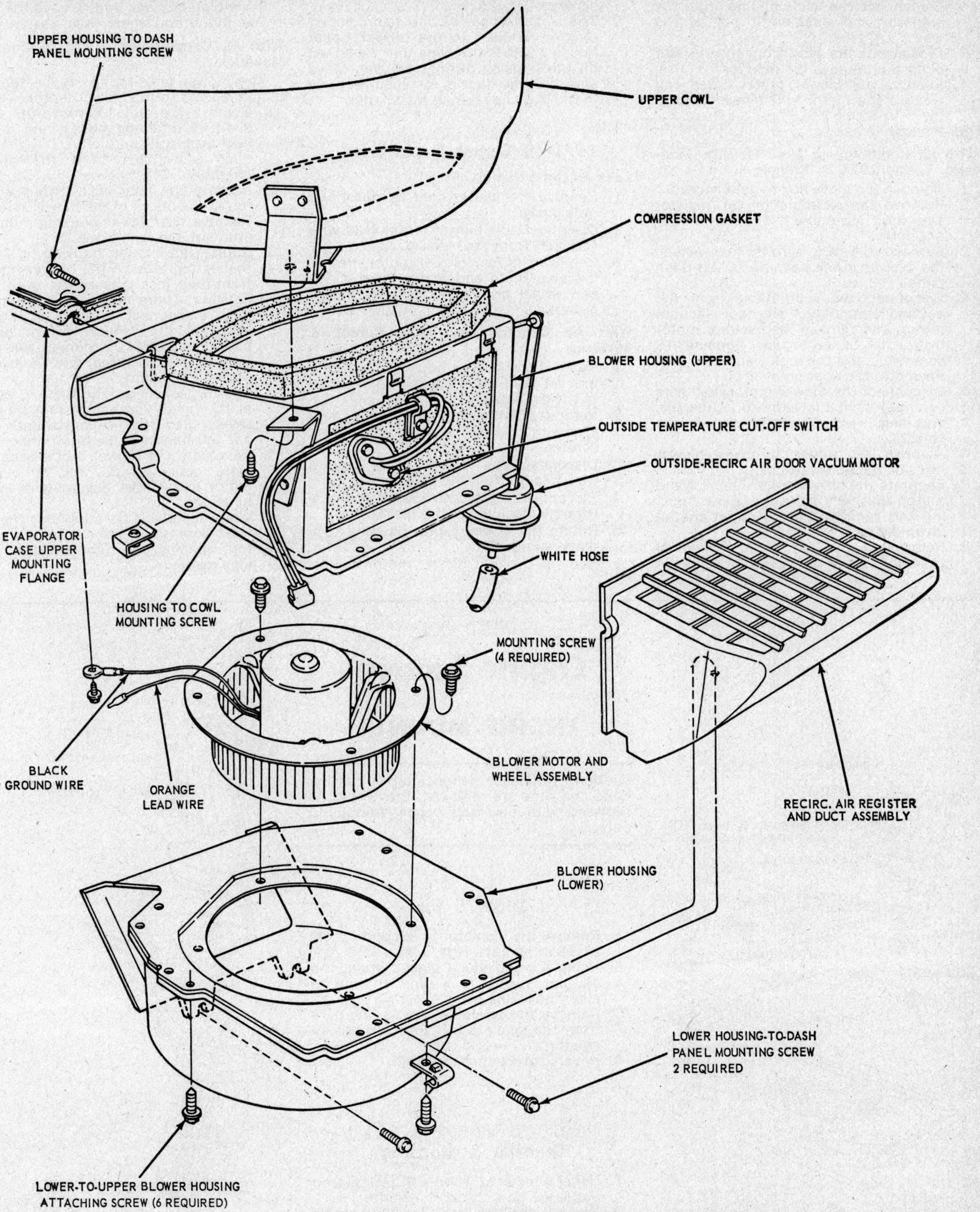

Fig. 25 Blower motor with air conditioning. 1972-76 Elite, Montego & Torino; 1974-76 Cougar

Disconnect the vacuum line from the actuator and position it out of the way.
3. Disconnect the plug from the resistor block and remove the resistor.
4. Remove the blower motor cover and remove the motor and blower assembly.
5. Reverse the above procedure to install.

With Air Conditioning: 1972-76 Elite, Montego, Torino 1974-76 Cougar

1. Remove the glove box to gain access.
2. Remove the recirculation air register and duct assembly from the blower assembly.
3. Remove the two screws that attach the blower lower housing to the dash panel.
4. Disconnect the white hose from the outside-recirculation air door vacuum motor, and remove the vacuum motor from the blower lower housing (2 screws). Leave the motor actuator connected to the door crank arm.
5. Disconnect the blower motor lead wire (orange) from the harness connector, and disconnect the motor ground wire (black).
6. Remove six upper to lower blower housing flange screws.
7. Separate the blower lower housing and motor assembly from the upper housing and remove it from under the instrument pad.
8. Remove the blower motor and wheel assembly from the lower housing (4 screws), Fig. 25.
9. The upper flange of the recirculating duct is retained to the blower upper housing with two S-clips that remained on the housing during removal. Make sure that the duct is installed properly in the two clips during installation.

1971-76 Comet & Maverick

Less Air Conditioning

1. Remove the heater core as described previously.
2. Disconnect the blower motor lead wire (orange) from the resistor.
3. Remove the four blower motor mounting plate nuts and remove the motor and wheel assembly from the heater assembly.

With Air Conditioning: 1971 Comet & Maverick

1. Remove the blower housing mounting bracket stud nut from the engine side of the dash panel.
2. Remove the two blower housing-to-instrument panel support mounting screws.
3. Disconnect the blower motor ground wire (black) from the upper cowl.
4. Disconnect the blower motor lead wire (orange-black) from the resistor.
5. Rotate the blower motor housing to a diagonal position.
6. Remove the four blower motor mounting plate screws, and remove the blower motor and wheel as an assembly.

With Air Conditioning: 1972-76 Comet & Maverick

The blower housing has to be removed to gain access to the blower motor.

1. Remove the radio. Remove the utility shelf (5 retaining bolts), and disconnect both right and left A/C register air duct assemblies from the plenum chamber.
2. Remove the floor air distribution duct from the bottom of the blower housing. Remove the blower housing mounting stud nut and lock plate.
3. Rotate the blower housing to unlock the slotted tabs on the blower housing from their lock pins on the evaporator housing. There are two tabs and two pins. Disconnect the red and yellow hoses at the vacuum motor on the blower housing. Disconnect the resistor and ground wires, and remove the blower housing.
4. Cut the gaskets around the A/C outlets at the break line before removing the seven clips that separate both right and left halves of the blower housing.
5. Remove three blower motor mounting plate retaining nuts, and remove the motor and wheel assembly from the housing.
6. When installing be sure that the A/C Heat-Door is positioned properly before clipping the right and left housings together.

Engine Section

ENGINE MOUNTS

Caution: Whenever self-locking mounting bolts and nuts are removed, they must be replaced with new self-locking bolts and nuts.

1969-76 Six

1. Remove the insulator to support bracket retaining nuts, Figs. 1 thru 7.
2. Using a wood block placed under the oil pan, raise the engine enough to clear the insulator.
3. Remove the retaining screws and nuts from the insulator(s). Remove the insulator(s).
4. Reverse procedure to install.

1975-76 V8-302 & 351 Granada & Monarch

1. Using a block of wood and jack placed under oil pan, support engine.
2. Remove through bolt attaching motor mount to crossmember, Fig. 8.
3. Remove engine mount to engine attaching bolts and remove engine mount and heat shield (if used).
4. Reverse procedure to install.

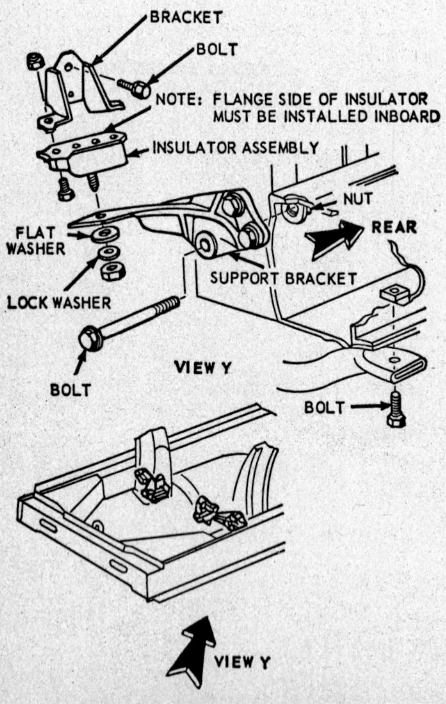

Fig. 1 Engine mounts. 1969-70 Mustang 6-170, 200

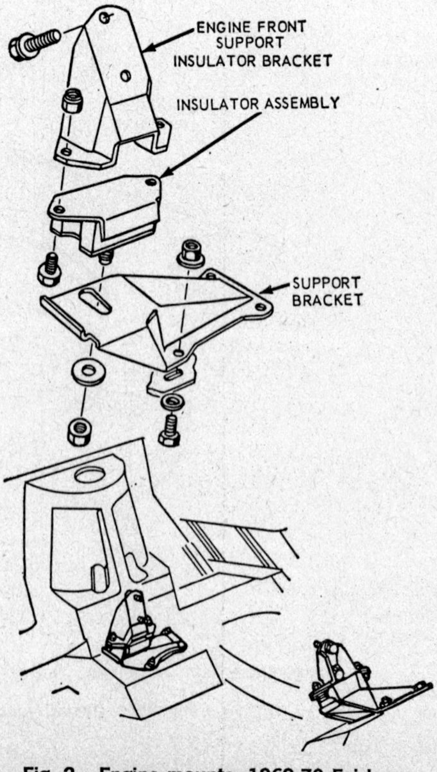

Fig. 2 Engine mounts. 1969-70 Fairlane, Falcon, Montego, Comet 6-170, 200

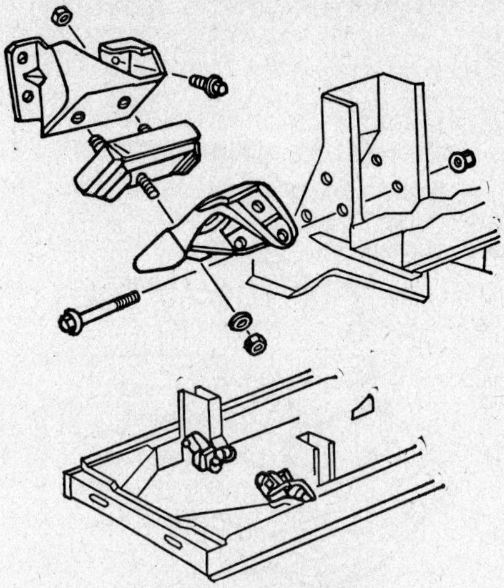

Fig. 3 Engine mounts. 1969-71 Mustang 6-250

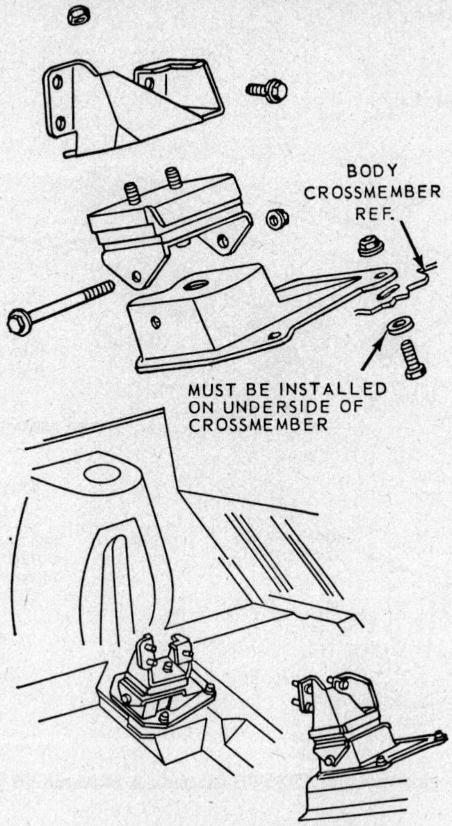

Fig. 4 Engine mounts. 1969-71 Fairlane, Falcon, Montego, Torino 6-250

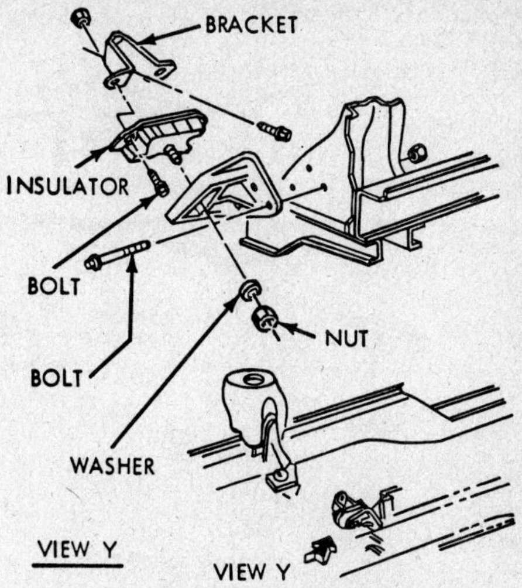

Fig. 5 Engine mounts. 1970-72 Comet, Maverick & All 1973-76 Six

1969-76 V8-302, 351, 400 Exc. Granada & Monarch

1. Raise the engine with a jack and wood block placed under the oil pan.
2. Remove bolts and washers attaching engine mount to engine, Fig. 9 thru 12.
3. Remove bolts and washers attaching engine mount brackets to side support.
4. Remove engine mount bracket assembly.
5. Detach engine mount from bracket by removing long bolts and nuts.

1969 V8-390, 428

1. On vehicles with automatic transmission, remove the oil cooler lines, and retaining bracket on the cylinder block.
2. Remove the insulator to intermediate support bracket nut and lock washer, Fig. 14. If only one support is being removed, loosen the other support.
3. Using a jack and a wood block placed under the oil pan, raise the engine to allow just enough clearance for the removal of the insulator(s).
4. Remove the insulator to engine bolts and lock washers. Remove the insulator.
5. Reverse procedure to install.

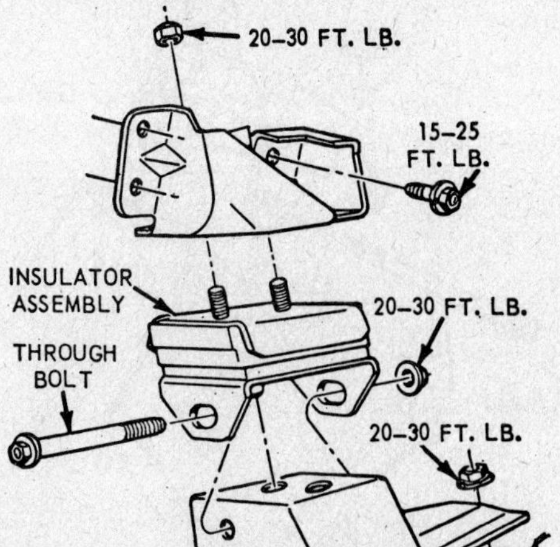

Fig. 6 Engine mounts. 1972 Mustang 6-250

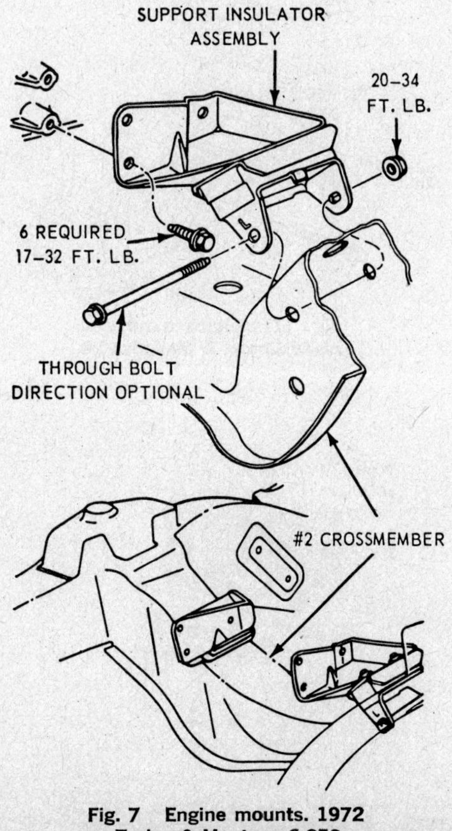

Fig. 7 Engine mounts. 1972 Torino & Montego 6-250

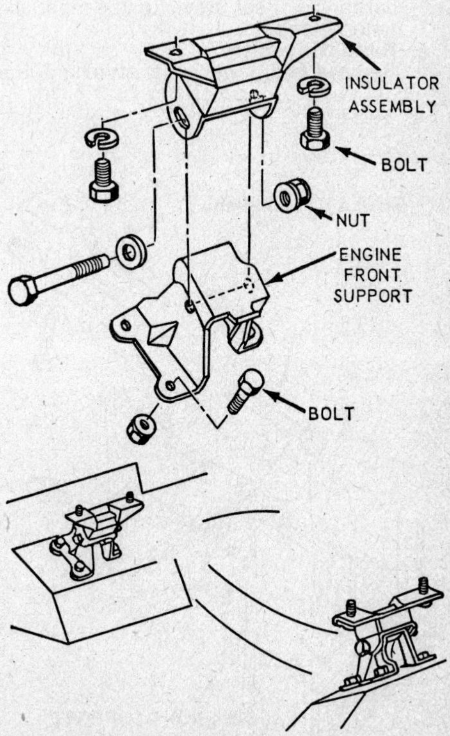

Fig. 9 Engine mounts. 1969-71 V8-302, 351 Fairlane/Torino Falcon, Montego & Comet

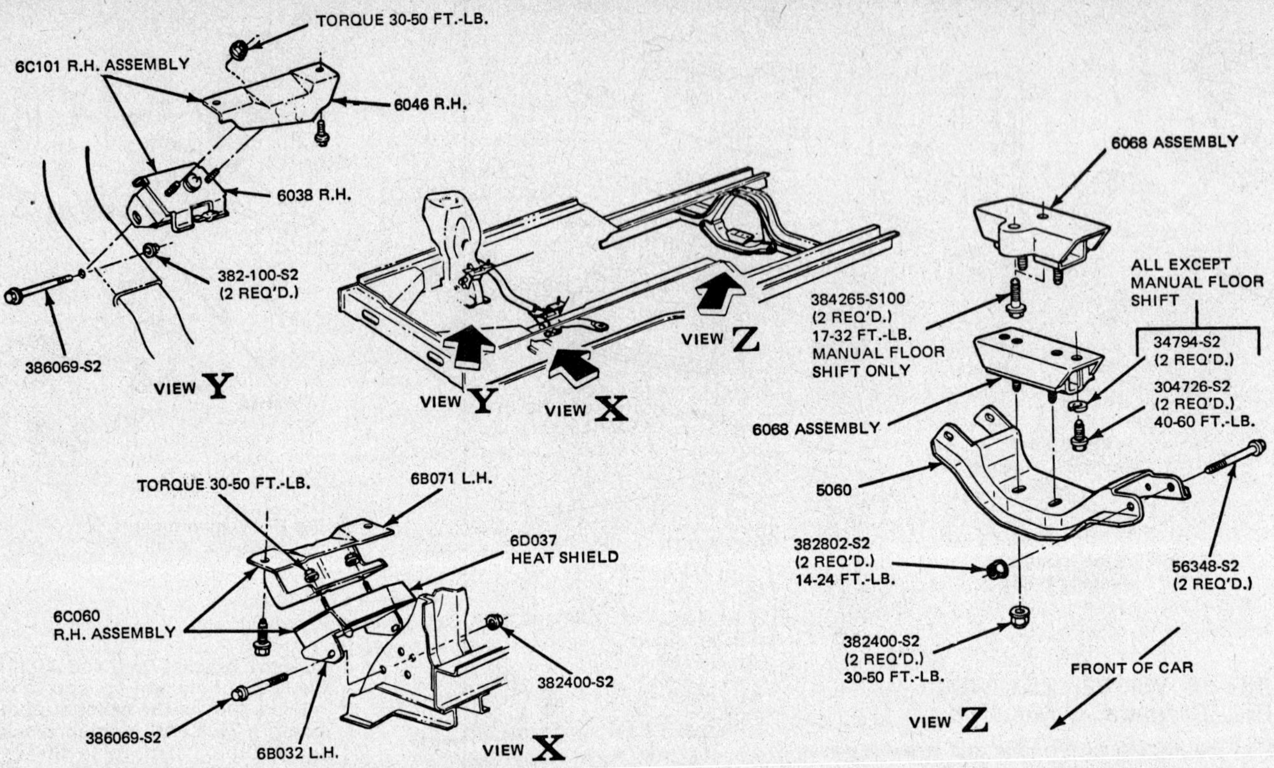

Fig. 8 Engine mounts. 1975-76 Granada & Monarch V8-302 & 351

1970 V8-428 Cobra Jet

1. Remove the air cleaner.
2. Remove the attaching bolts to radiator shroud and leave it loose to allow fan clearance when raising the engine.
3. Remove the upper nut attaching the carburetor heat stove to the right exhaust manifold.
4. Raise the vehicle.
5. Remove the lower nut attaching the carburetor heat stove to the right exhaust manifold. Rotate the stove up and around the front of the manifold and position it under the battery carrier, out of the way.
6. Disconnect the power steering oil cooler lines at the engine oil filter adapter.
7. Remove the power steering line retainer and position lines out of the way.
8. Place a floor jack under the engine.
9. Place a wood block between the oil pan and the jack and raise the engine slightly.

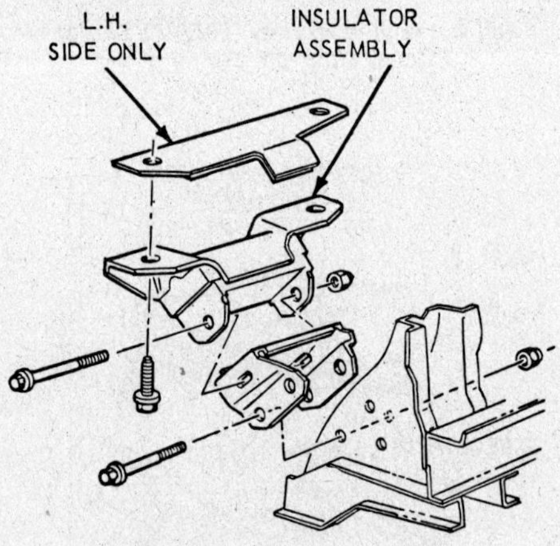

Fig. 11 Engine mounts. 1971-76 Comet & Maverick V8

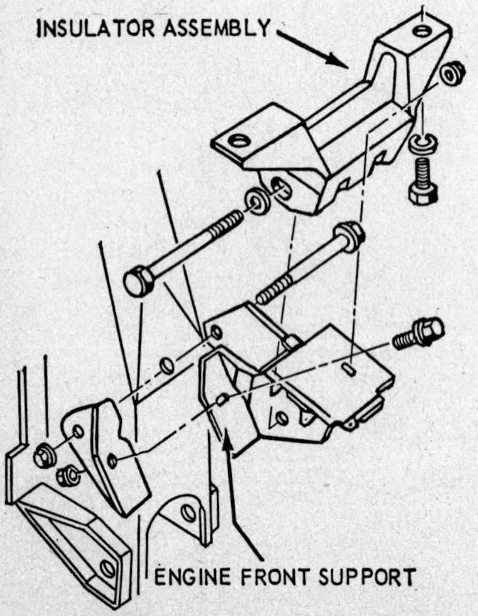

Fig. 10 Engine mounts. 1969-73 Cougar & Mustang V8-302, 351

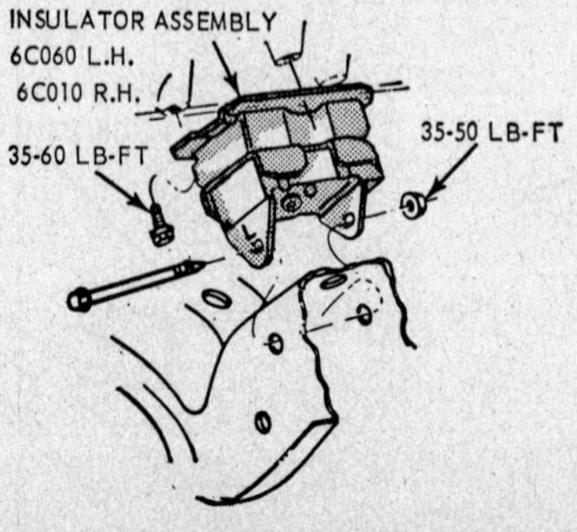

Fig. 11A Engine mounts. 1973-76 Elite, Torino & Montego V8-400

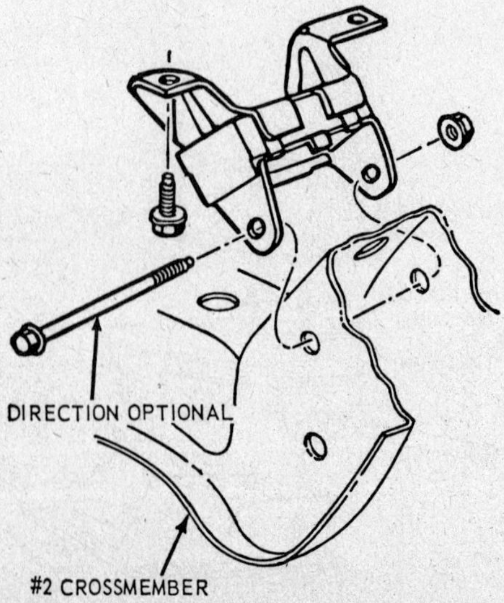

Fig. 12 Engine Mounts. 1972-76 Torino & Montego V8-302, 351 & 1974-76 Cougar V8-351C

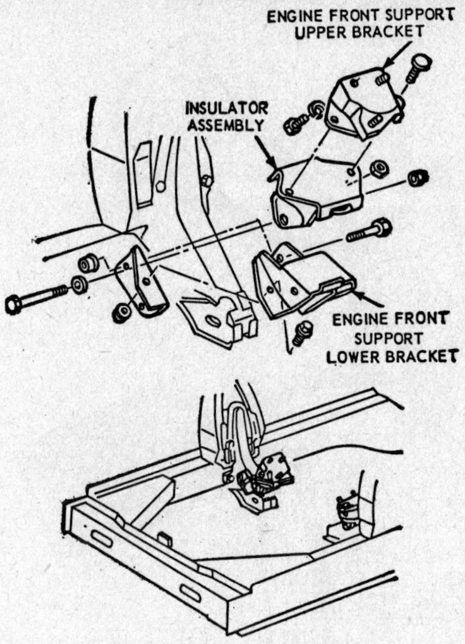

Fig. 13 Engine mounts. 1970 V8-428 "CJ"

10. Remove the engine support to chassis thru bolt and nut on either or both sides, Fig. 13.
11. Remove nuts and bolts attaching the engine support bracket to the chassis on either or both sides. Raise the engine.
12. Remove the nuts attaching the engine support insulators to the engine brackets and remove the insulators.

1970-71 V8-429

1. Block the rear wheels and set the parking brake. Raise the front of the vehicle with a floor jack and install safety stands.
2. Remove the two nuts attaching the upper insulator bracket to the insulator, Fig. 15.
3. Place a jack under the front area of the oil pan. Position a wood block between the jack and the oil pan. Raise the engine enough to take the load off the through-bolt.
4. Remove the through bolt and raise the engine to a point where the insulator can be removed from the lower bracket.
5. Remove the two nuts that attach the insulator to the insulator upper bracket and remove the insulator.

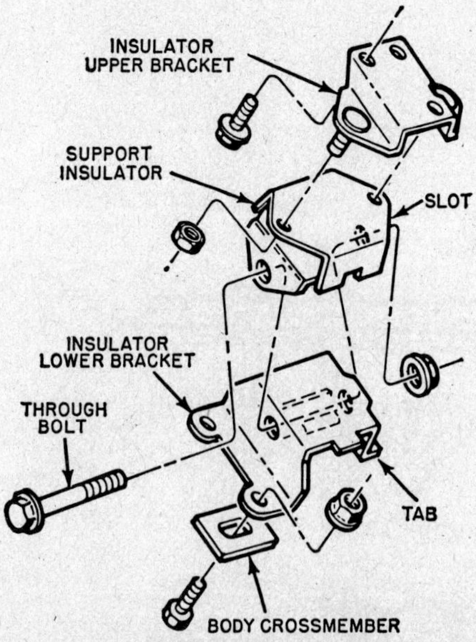

Fig. 15 Engine mounts. V8-429 1970 Fairlane & Montego, 1971 Cougar, Montego, Mustang & Torino

1972-76 V8-429, 460

1. Remove the fan shroud attaching screws and support the engine using a jack and block of wood under the oil pan.
2. Remove the through bolt and nut attaching the insulator to the frame crossmember.
3. Remove the insulator to upper bracket attaching nuts, Fig. 16.
4. Raise the engine enough to remove the insulator and heat shield if so equipped.
5. If necessary, the upper bracket can now be removed by removal of the three screws holding the bracket to the cylinder block.

ENGINE, REPLACE

1969-76 Six-Cylinder

NOTE: The engine is removed from the chassis, leaving the transmission in place. First disconnect and/or remove as required wires, tubes, hoses and linkage attached to engine. Then do the following:

1. Remove hood, radiator, fan and pulley.
2. On vehicles with air conditioning, remove compressor from mounting bracket and place aside.
3. On cars with manual shift transmission, remove clutch equalizer shaft and arm bracket.
4. Remove flywheel or converter housing-to-engine upper bolts through access holes in underbody.
5. Disconnect engine right and left mount at underbody bracket.
6. Remove flywheel or converter housing cover.
7. Remove flywheel or converter housing-to-engine lower bolts.
8. Support transmission and flywheel or converter housing with a jack.
9. Attach a lifting rig to engine and remove it from vehicle.

1969-76 V8-390, 427, 428, 429, 460

NOTE: The engine is removed from the vehicle, leaving the transmission in place. First disconnect and/or remove as required wires, tubes, hose and linkage attached to engine. Then do the following:

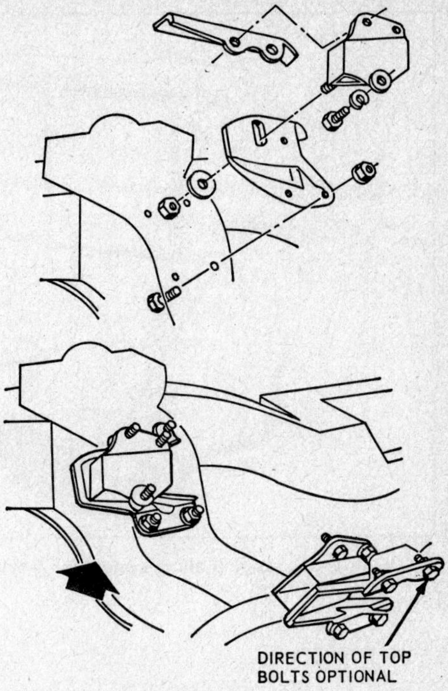

Fig. 14 Engine mounts. 1969 V8-390, 428 Cougar, Fairlane, Montego & Mustang

1. Drain cooling system and crankcase.
2. Remove hood and radiator.
3. If air conditioned, unfasten air compressor from its mounting and position it out of the way, leaving refrigerant lines attached.
4. Unfasten and wire power steering pump in a position that will prevent oil from draining out.

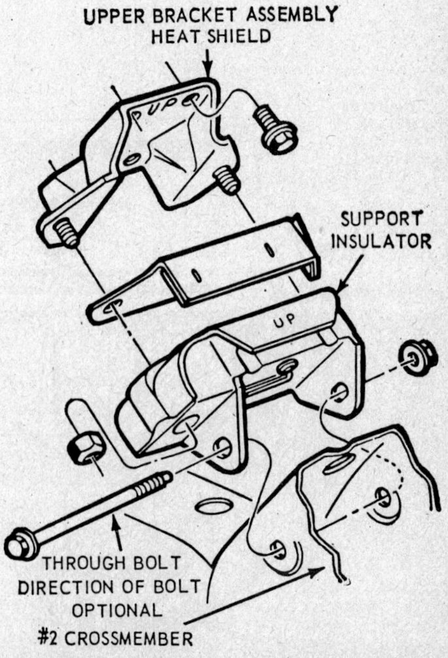

Fig. 16 Engine mounts. 1972-76 V8-429 460 Torino & Montego

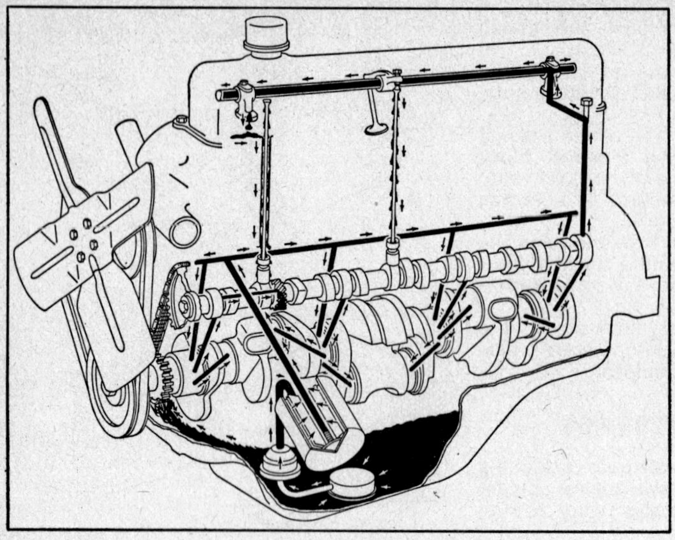

Engine oiling system for 6 cylinder engines

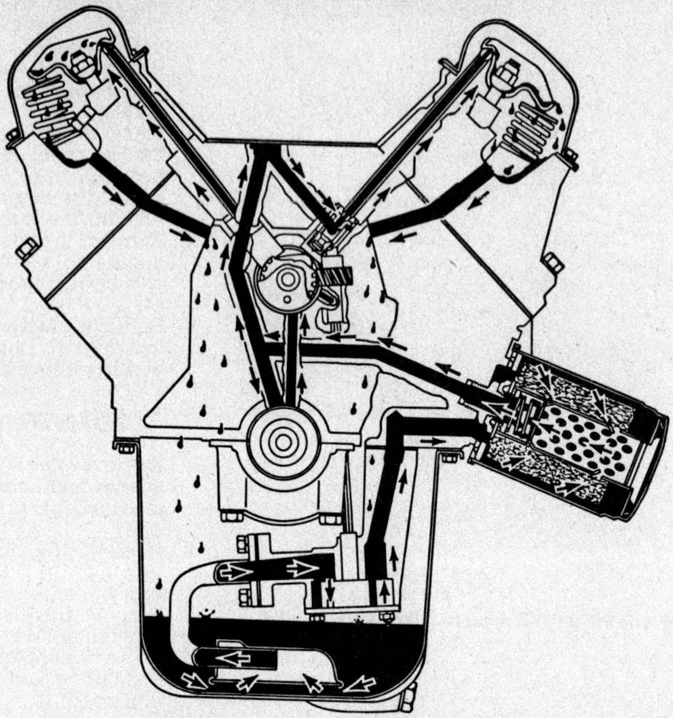

Engine oiling system. V8-302, 351, 400

5. Raise front of car.
6. Remove starter and transmission fluid filler tube bracket.
7. Remove engine intermediate support bracket-to-crossmember retaining nut on right and left engine front supports.
8. Remove converter housing cover.
9. Remove flywheel-to-converter nuts.
10. Secure converter to housing.
11. Remove converter housing-to-engine lower bolts.
12. Lower car and support transmission.
13. Remove converter housing upper bolts.
14. Lift engine out of chassis.

1969-76 V8-302, 351, 400

NOTE: The engine is removed from the vehicle, leaving the transmission in place. First disconnect and/or remove as required wires, tubes, hose and linkage attached to engine. Then do the following:

1. Drain cooling system and crankcase.
2. Remove oil filter, hood, radiator, fan and pulley.
3. Remove air cleaner and intake duct.
4. If air conditioned, unfasten and position air compressor out of the way.
5. Remove and position power steering pump to one side.
6. If equipped with Thermactor Exhaust Emission Control System, remove air pump air filter if it is not connected to engine.
7. Remove flywheel-to-converter housing upper bolts.
8. Raise front of car and remove starter.

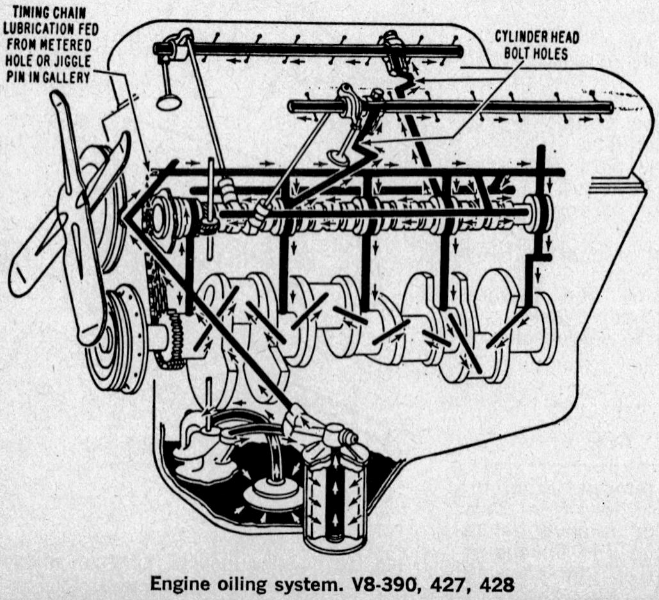

Engine oiling system. V8-390, 427, 428

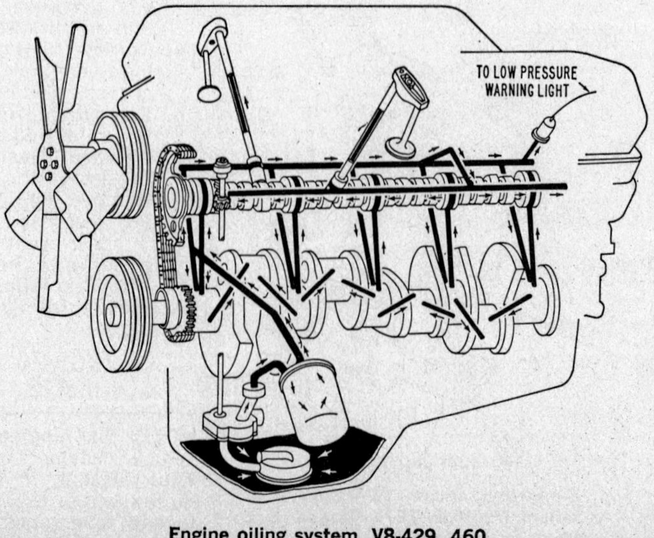

Engine oiling system. V8-429, 460

Fig. 17 Cylinder head tightening sequence. Six cylinder

9. Disconnect engine support insulators at brackets on frame underbody.
10. Remove remaining flywheel or converter housing-to-engine bolts.
11. Lower car, then support transmission.
12. Attach a lifting rig to engine and remove from vehicle.

CYLINDER HEAD, REPLACE

Tighten cylinder head bolts a little at a time in three steps in the sequence shown in the illustrations. Final tightening should be to the torque specifications listed in the *Engine Tightening* table. After tightening the bolts to specifications, *they should not be disturbed.*

1969-76 Six-Cylinder

1. Drain cooling system and remove air cleaner.
2. Unfasten exhaust pipe from manifold and pull it down.
3. Disconnect accelerator rod from carburetor.
4. Disconnect fuel inlet line at fuel filter hose, and distributor vacuum line at carburetor.
5. Disconnect coolant lines at carburetor spacer. Remove radiator upper hose at outlet housing.
6. Disconnect distributor vacuum line at distributor. Disconnect carburetor fuel inlet line at fuel pump. Remove lines as an assembly.
7. Disconnect spark plug wires at plugs and temperature sending unit wire at sending unit.
8. Remove crankcase ventilation system. Remove hoses from Thermactor system as necessary for accessability.
9. Remove valve rocker arm cover.
10. Remove rocker arm shaft assembly.
11. Remove valve push rods.
12. Remove remaining cylinder head bolts and lift off head.
13. Reverse procedure to install and tighten head bolts in the sequence shown in Fig. 17.

1969-76 V8-302, 351, 400

1. Remove intake manifold and carburetor as an assembly.
2. Disconnect battery ground cable at cylinder head.
3. If left head is being removed, remove A/C compressor (if equipped). Also remove and wire power steering pump out of the way. If equipped with Thermactor System, disconnect hose from air manifold on left cylinder head.
4. If right head is to be removed, remove alternator mounting bracket bolt and spacer, ignition coil and air cleaner inlet duct.
5. If right head is to be removed on an engine with Thermactor System, remove air pump from bracket. Disconnect hose from air manifold.
6. Disconnect exhaust manifolds at exhaust pipes.

NOTE: On V8-351, separate the exhaust manifolds from the heads first in order to gain access to the lower row of cylinder head attaching bolts.

7. Remove rocker arm covers. If equipped with Thermactor System, remove check valve from air manifold.
8. Loosen rocker arm nuts so rocker arms can be rotated to one side. Remove push rods.
9. Remove head bolts and lift head off block.
10. Reverse removal procedure to install and tighten head bolts in the sequence shown in Fig. 18.

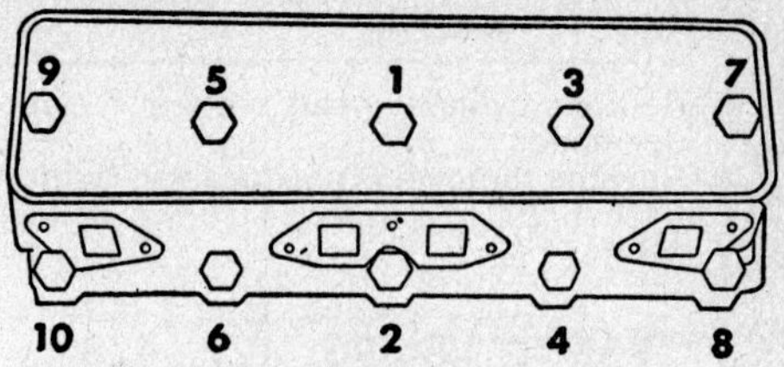

Fig. 18 Cylinder head tightening. V8 engines

1969-76 V8-429 & 460

1. Remove intake manifold and carburetor as an assembly and disconnect exhaust pipes from manifold.
2. Loosen A/C belt if equipped and remove alternator.
3. If equipped with A/C, isolate compressor at service valves and disconnect valves and hoses from compressor. Remove compressor attaching bolts and position unit out of way.
4. If not equipped with A/C, remove power steering retaining bolts and position unit out of way.
5. Remove valve covers. On 1969-70, back off adjusting nuts, turn rocker arms to one side and remove push rods. On 1971-76, remove rocker arm bolts, rocker arms, oil deflectors, fulcrums and push rods.

NOTE: Remove all parts in sequence so that they are installed in their

Fig. 19 Rocker arm stud removal. 6 cylinder and V8 engines with pressed-in rocker arm studs

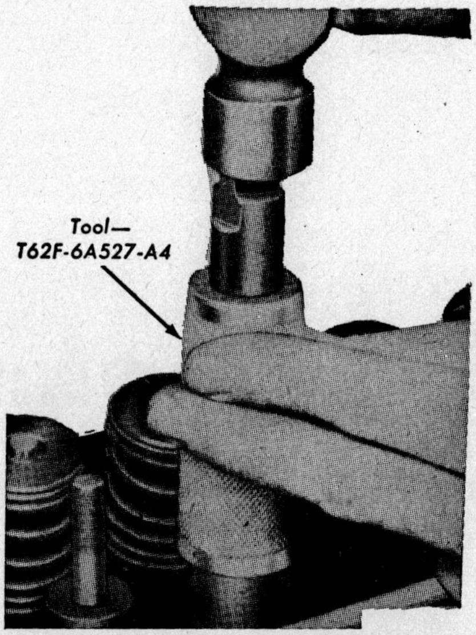

Fig. 20 Conventional type rocker arm stud installation. 6 cylinder and V8 engines with pressed-in studs

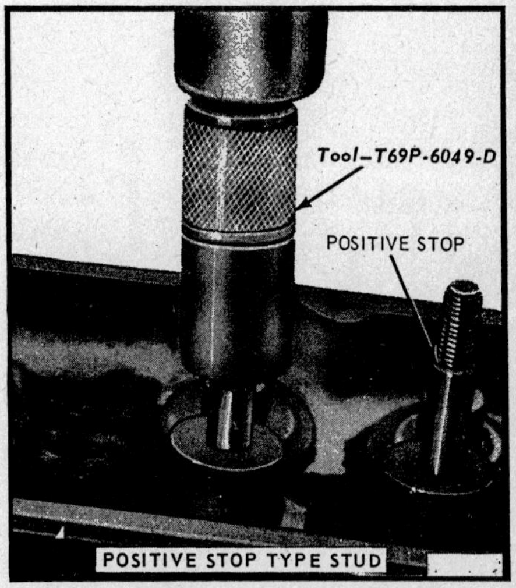

Fig. 21 Positive stop type rocker arm stud installation. 6 cylinder and V8 engines with pressed-in studs

original positions.

6. Remove cylinder head bolts and cylinder head.
7. Reverse removal procedure and tighten head bolts in sequence shown in Fig. 18.

1969-70 V8-390, 427, 428

1. If equipped with Thermactor System, disconnect air hoses as necessary for accessability, and position them out of the way.
2. Remove intake manifold and carburetor as an assembly.
3. Remove rocker arms and push rods. Keep all parts in sequence so they can be installed in the same locations.
4. Disconnect exhaust manifolds at exhaust pipes.
5. If left head is being removed, remove ignition coil and engine identification tag, and remove power steering pump mounting bolt from right cylinder head.
6. Remove head bolts and take off head.
7. Reverse procedure to install and tighten head bolts in the sequence shown in Fig. 18.

VALVE ARRANGEMENT

Front to Rear

Sixes	E-I-I-E-I-E-E-I-E-I-I-E
302 Right	I-E-I-E-I-E-I-E
302 Left	E-I-E-I-E-I-E-I
8-351, 400 Right	I-E-I-E-I-E-I-E
8-351, 400 Left	E-I-E-I-E-I-E-I
V8-390, 427, 428	E-I-E-I-I-E-I-E
V8-429, 460 Right	I-E-I-E-I-E-I-E
V8-429, 460 Left	E-I-E-I-E-I-E-I

VALVE LIFT SPECS.

Engine	Year	Intake	Exhaust
6-170	1969-72	.348	.348
6-200	1969-73	.348	.348
6-200	1974	.382	.350
6-200	1975	.372	.372
6-250	1969-72	.368	.368
6-250	1973	.348	.348
6-250	1974-75	.382	.350
8-302	1969-72	.368	.381
8-302	1973	.371	.382
8-302	1974-75	.371	.382
8-302 "H.O."	1969-71	.477	.477
8-351[4]	1969-75	.418	.448
8-351[5][1]	1970-74	.407	.407
8-351M[1][6]	1975	.407	.407
8-351[5][2]	1970-72	.427	.427
8-351[5][2]	1973-74	.481	.490
8-351[5][2][3]	1972-73	.490	.490
8-390[3]	1969	.440	.440
8-390[7]	1969	.481	.490
8-390	1969-71	.427	.430
8-400	1972	.422	.427
8-400	1971-74	.427	.433
8-400	1975	.428	.432
8-427	1969	.481	.490
8-428	1969-70	.481	.490
8-429 "CJ"	1970-71	.500	.500
8-429 "SCJ"	1970-71	.515	.515
8-429	1972	.442	.486
8-429	1973-74	.437	.480
8-460	1973-74	.437	.480
8-460	1975	.437	.451

[1]—Two bar. carb.
[2]—Four bar. carb.
[3]—Hi Perf. engine.
[4]—Windsor engine.
[5]—Cleveland engine.
[6]—Modified.
[7]—Police Interceptor.

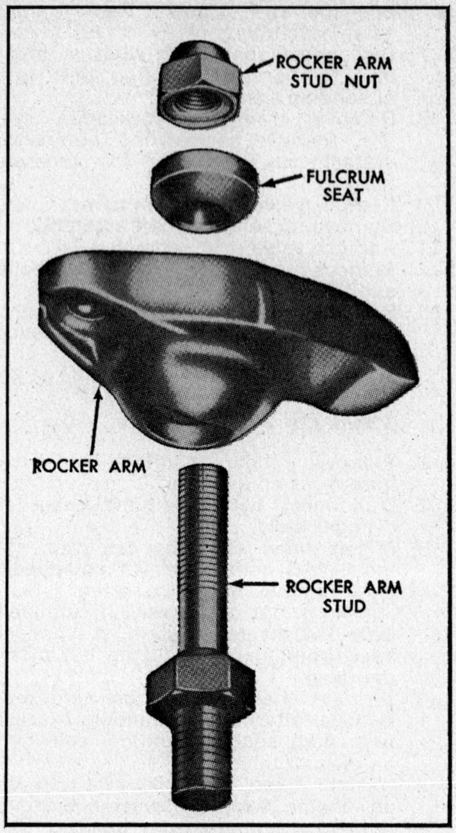

Fig. 22 Valve rocker arm and stud. High Performance V8-302 & 1969-70 V8-429 Except "BOSS"

VALVE TIMING

Intake Opens Before TDC

Engine	Year	Degrees
6-170	1969-72	9
6-200	1969-72	9
	1973	17
	1974-75	28
6-250	1969-72	10
	1972[6]	16

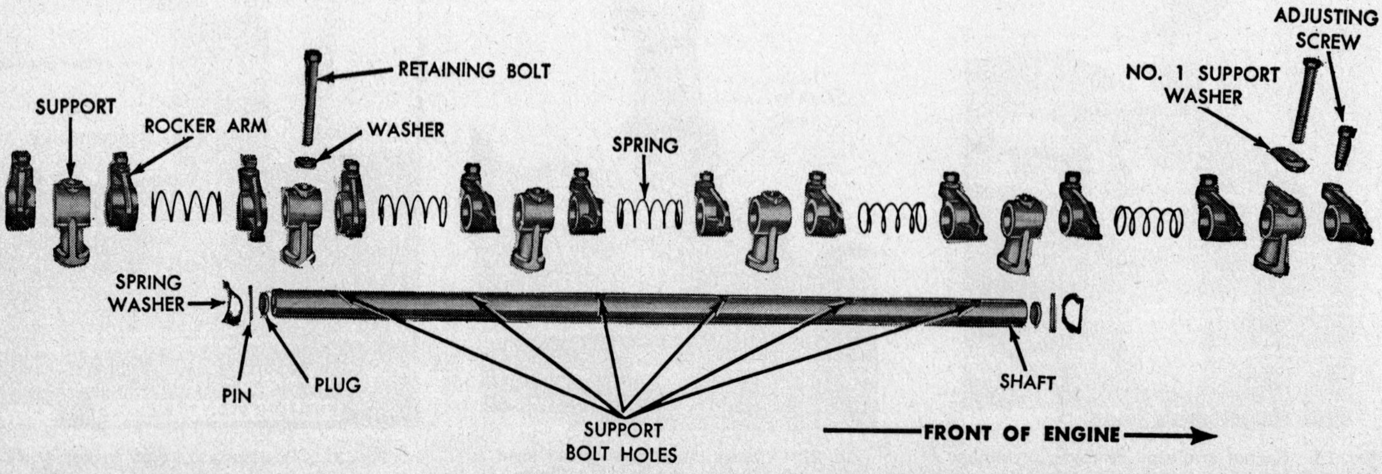

Fig. 23 Rocker arm shaft assembly. Six cylinder engines

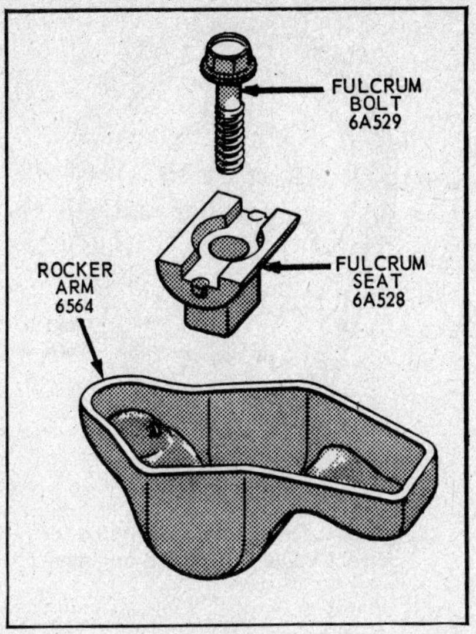

Fig. 24 Rocker assembly. 1970-76 V8-351C, 351M, 400 & 1971-76 V8-429, 460

6-250	1973	17
	1974-75	26
8-302	1969-72	16
8-302	1969	16
8-302 "BOSS"	1970	40
8-302 "H.O."	1971	34
8-302[7]	1973-74	16
8-302[4]	1973-75	20
8-351[5]	1969-71	11
	1974-75	15
8-351[3][1]	1970-73	12
	1974	19½
8-351[2][1]	1970	14
8-351[2][1]	1971-73	18
8-351[2]	1974	14
8-351[8]	1975	19.5
8-390	1969	16
8-400	1972-75	17
8-427	1969	18
8-428	1969-70	18
8-429	1972-73	8
8-429 "CJ"	1970-71	32
8-429 "SCJ"	1970-71	45.5
8-429 "BOSS"	1970	40
8-460	1973-75	8
8-460[9]	1974-75	18

[1]—Cleveland engine.
[2]—Four bar. carb.
[3]—Two bar. carb.
[4]—Auto. trans.
[5]—Windsor engine.
[6]—California engine.
[7]—Std. trans.
[8]—Modified.
[9]—Police Interceptor.

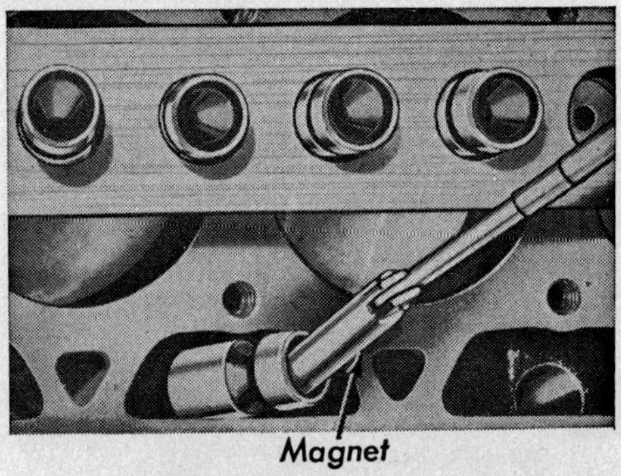

Fig. 25 Removing valve lifter with magnetic rod

VALVES, ADJUST

Six Cylinder

The procedure used to check the valve clearance is to rotate the crankshaft with an auxiliary starter switch until the No. 1 piston is near TDC at the end of the compression stroke. At this point the following valves can be checked:

No. 1 Intake	No. 3 Exhaust
No. 1 Exhaust	No. 4 Intake
No. 2 Intake	No. 5 Exhaust

After the clearance of these valves have been checked, rotate the crankshaft until the No. 6 piston is on TDC at the end of its compression stroke (1 revolution of the crankshaft) and check the following valves:

No. 2 Exhaust	No. 5 Intake
No. 3 Intake	No. 6 Intake
No. 4 Exhaust	No. 6 Exhaust

Non-Adjustable Rocker Arm:

1. Position cylinder to be checked at T.D.C.
2. Apply pressure on the push rod end of the rocker arm until the tappet plunger is completely bottomed.
3. Hold rocker arm in this position and check the clearance between rocker arm and valve stem.
4. If the clearance is not within limits, install the appropriate undersize or oversize push rod, which are available in .060″ undersize and .060″ oversize.

Mechanical Valve Lifters

Before the final lash adjustment is made, operate the engine for 30 minutes at a fast idle to stabilize engine temperatures. To set the lash accurately, use only a step-type feeler gauge. For example, to obtain the correct setting if the clearance is .019″, the .018″ portion of the gauge should slip between valve tip and rocker arm but the "no go" end (.020″) should not.

Early V8-302 w/Hydraulic Lifters

1. Turn crankshaft to position No. 1 piston at TDC of the compression stroke. With the piston in this position, adjust the following valves to obtain the clearances listed in the *Valve Specifications* table.

No. 1 Intake	No. 4 Exhaust
No. 1 Exhaust	No. 5 Exhaust
No. 2 Exhaust	No. 7 Intake
No. 3 Intake	No. 8 Intake

2. Turn crankshaft to position No. 6 piston on TDC of the compression stroke, and adjust the following valve clearances, using the same procedure as in Step 1.

No. 2 Intake	No. 6 Intake
No. 3 Exhaust	No. 6 Exhaust
No. 4 Intake	No. 7 Exhaust
No. 5 Intake	No. 8 Exhaust

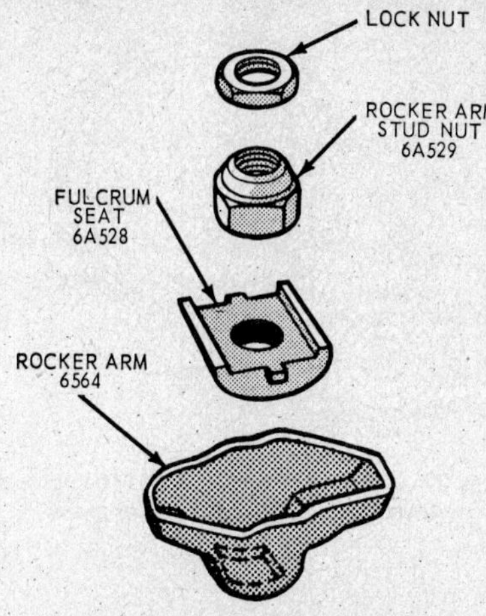

Fig. 24A Rocker assembly. 1969-71 V8-302 "BOSS"

V8-302 "BOSS", V8-429 "SCJ"

The valves in these engines use mechanical lifters which are adjusted in the conventional manner to the clearance listed in the *Valve Specifications* table.

Late V8-302 w/Hydraulic Lifters, V8-351 "Windsor" Engine

These engines incorporate a positive stop rocker arm stud and nut. Thread nut onto stud until it contacts the stud then tighten nut to 18-22 ft-lbs.

V8-351 "Cleveland" Engine V8-351M, 390, 400, 427, 428, 429, "CJ", 460

For these engines, a .060″ longer or a

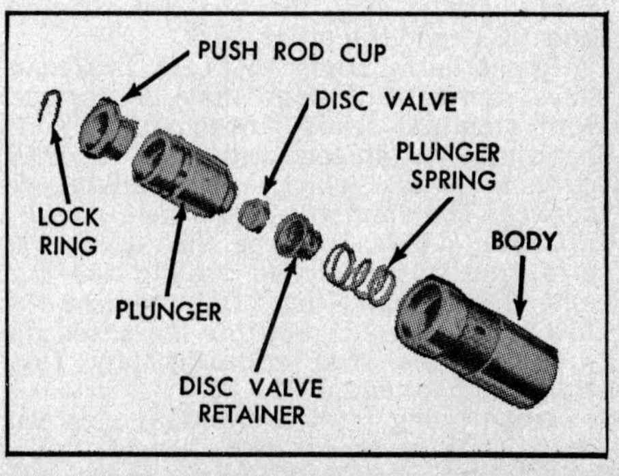

Fig. 26 Hydraulic valve lifter

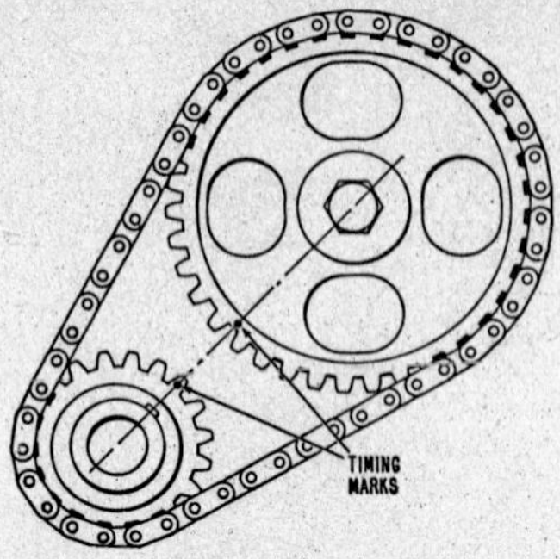

Fig. 27 **Timing marks aligned for correct valve timing. Six cylinder engines**

.060″ shorter push rod is available to provide a means of compensating for dimensional changes in the valve train.

To check the clearance, bring the piston of the cylinder being checked on top dead center of the compression stroke. Then with hydraulic lifter collapsed, check the clearance between valve stem and rocker arm. If the clearance is less than the minimum, the .060″ shorter push rod should be used. If clearance is more than the maximum the .060″ longer push rod should be used.

V8-429 "BOSS"

The valve clearance should be set only with the engine cold.

1. Remove rocker arm covers and position No. 1 piston at TDC of its compression stroke.
2. Torque rocker arm shaft nuts for No. 1 cylinder to specifications then adjust valve lash.
3. Rotate crankshaft 90° to bring No. 5 piston to TDC and repeat adjustment for that cylinder.
4. Proceed through the firing order (1-5-4-2-6-3-7-8) until all valves have been adjusted.

ROCKER ARM STUD

V8-302, 351W

If necessary to replace a rocker arm stud, a rocker arm stud kit is available and contains a stud remover, Fig. 19, a stud installer, Fig. 20, and two reamers, one .003″ and the other .015″.

Rocker arm studs that are broken or have damaged threads may be replaced with standard studs. Loose studs in the head may be replaced with .003″ or .015″ oversize studs which are available for service. *The standard studs have no identification marks, whereas the .003″ oversize stud has a groove around the pilot end of the stud. The .015″ oversize stud has a step produced by the increased diameter of the stud approximately 1 5/32″ from the pilot end.*

When going from a standard size stud to a .015″ oversize stud, always use a .003″ reamer before finish reaming with a .015″ reamer.

If a stud is broken off flush with the stud boss, use an easy-out to remove the broken stud, following the instructions of the tool manufacturer.

Late V8-302 w/Hydraulic Lifters & V8-351 "Windsor" Engine

A new type positive stop rocker arm stud and nut eliminates the need of adjusting valve lash.

Installation

1. Position the piston of the cylinder being worked on at TDC compression stroke.
2. Locate stud properly with tool T69P-6049D, Fig. 21. Make sure tool bottoms on the head.
3. Lubricate rocker arm components and place rocker arm and fulcrum on the stud.
4. Thread nut onto the stud until it contacts the shoulder, then tighten nut to 18-22 ft lbs.

High Performance V8-302 & 1969-71 V8-429 Except "BOSS"

Threaded rocker arm studs are used in these engines, Fig. 22. To remove stud, remove rocker arm cover, lock nut, if used, stud nut, fulcrum seat and rocker arm. Unscrew stud from cylinder head.

Installation:

1. Apply water resistant sealer to threads.
2. Install push rod guide if used, stud and torque 60-70 ft-lbs. for V8-289, 85 ft-lbs. for V8-302, and 65-75 for V8-429.
3. Lubricate and install rocker arm, fulcrum and stud nut.
4. With engine warmed up, adjust valve lash and install lock nut where used.

VALVE GUIDES

Valve guides consist of holes bored in the cylinder head. For service the guide holes can be reamed oversize to accommodate valves with oversize stems of .003, .015 and .030″.

ROCKER ARM SERVICE

6 Cylinder Engines

1. To disassemble, remove pin and spring washer from each end of rocker shaft, Fig. 23.
2. Slide rocker arms, springs and supports off the shaft, being sure to identify location of parts for reassembly.
3. If it is necessary to remove the plugs from the shaft ends, drill or pierce the plug on one end. Then use a steel rod to knock out the plug on the opposite end. Working from the open end, knock out the remaining plug.

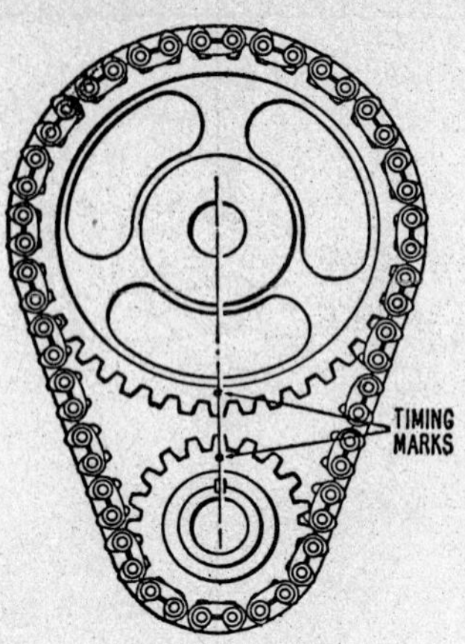

Fig. 28 **Timing marks aligned for correct valve timing. V8 engines**

Assemble

1. Lubricate all parts with engine oil. Apply Lubriplate to the rocker arm pads.
2. If plugs were removed from shaft ends, use a blunt tool or large diameter pin punch and install a plug (cup side out) in each end of shaft.
3. Install spring washer and pin on one end of shaft.
4. Install rocker arms, supports and springs in order shown in Fig. 23. *Be sure oil holes in shaft are facing downward.*
5. Complete the assembly by installing remaining spring washer and pin.

1970-76 V8-351C, 351M, 400 & 1972-74 V8-429, 460

These engines use stamped steel rocker arms retained by a fulcrum seat, Fig.

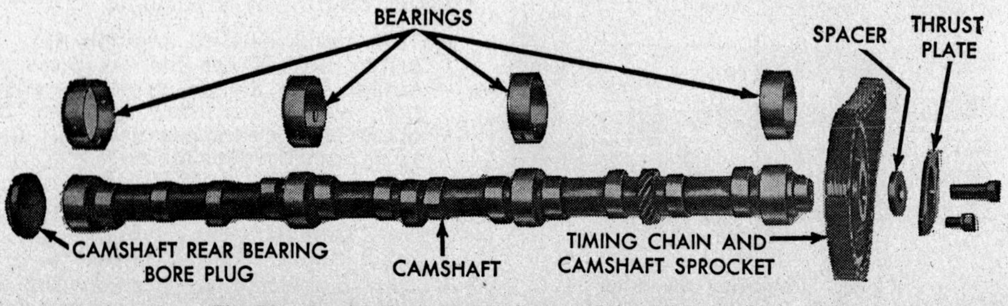

Fig. 29 **Camshaft and related parts. Six cylinder engines**

24. The fulcrum seat bolts directly to the cylinder head and guides the rocker arm.

1969-71 V8-302 "BOSS"

These engines use stamped steel rocker arms mounted on their own studs and retained by a fulcrum seat, stud nut and locknut, Fig. 24A. To remove rocker arm, remove valve cover, then remove locknut, fulcrum and seat and rocker arm.

V8-390, 427, 428, 429 Except "BOSS" & 460

See Ford-Mercury chapter.

V8-429 "BOSS"

Each rocker arm is mounted on its own shaft which in turn is mounted on a pedestal on the cylinder head. All valves and shafts are independent of each other.

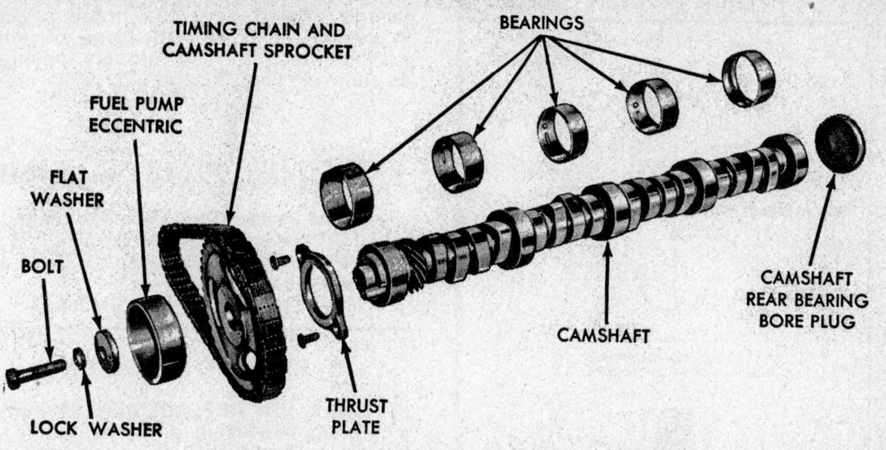

Fig. 30 **Camshaft and related parts. V8 engines**

VALVE LIFTERS, REPLACE

6 Cylinder Engines

When necessary to replace valve lifters, remove cylinder head and related parts as outlined previously. Then, using a magnet rod, Fig. 25, remove and install one lifter at a time to be sure they are placed in their original bores.

When installing, apply Lubriplate to each lifter foot and coat the remainder of lifter with oil before installation.

V8 Engines

1. Remove intake manifold.
2. Remove rocker arm covers. On engines with stud-mounted rocker arms, loosen stud nuts and rotate rocker arms to one side. On other engines, remove rocker arms and shafts.
3. Remove push rods in sequence so they can be installed in their original bores.
4. Using a magnet rod, Fig. 25, remove the lifters and place them in a numbered rack so they can be installed in their original bores. *If the lifters are stuck in their bores by excessive varnish, etc., it may be necessary to use a plier-type tool to remove them. Rotate the lifter back and forth to loosen it from the gum or varnish.*
5. The internal parts of each lifter are matched sets. Do not intermix parts. Keep the assemblies intact until they are to be cleaned, Fig. 26.

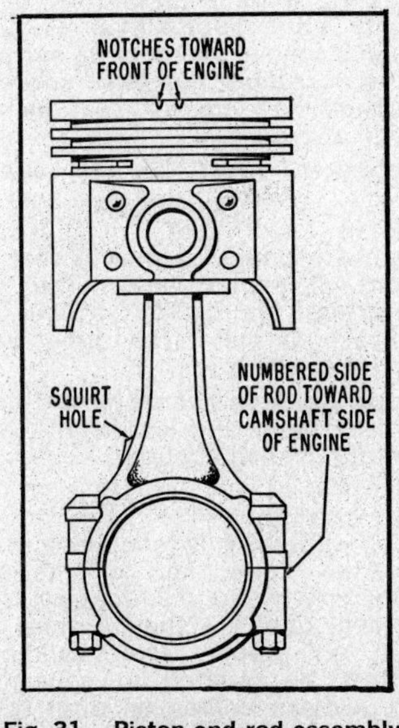

Fig. 31 **Piston and rod assembly. Six cylinder engines**

TIMING CASE COVER

NOTE: On all except V8-351 "Cleveland" engine; if necessary to replace cover oil seal, the cover and oil pan must be removed.

6 Cylinder Engines

1. Drain cooling system and crankcase.
2. Remove radiator, fan and pulley.
3. Use puller to remove damper.
4. Remove front cover and gasket.
5. Remove crankshaft oil slinger.
6. Drive seal out of cover with a pin punch and clean out recess in cover.
7. Coat a new seal with grease and drive it in until fully seated in recess. Check seal after installation to be sure spring is properly positioned in seal.
8. Reverse removal procedure to install.

V8 Engines

1. To remove cover, drain cooling system and crankcase. Remove air cleaner and disconnect battery ground cable.
2. Remove water hose as necessary.
3. Remove generator support bolt at water pump, and loosen generator mounting bolts.
4. Remove fan, spacer and pulley.
5. Remove power steering drive belt (if equipped). If air conditioned, remove compressor drive belt.
6. Remove crankshaft pulley and adapter.
7. Remove fuel pump and lay it to one side with flexible fuel line attached.
8. Remove oil level dipstick tube bracket and oil filler tube bracket.
9. Remove oil pan-to-front cover bolts.
10. Remove cover and water pump as an assembly.
11. Drive out cover seal with a pin punch. Clean out recess in cover.
12. Coat a new seal with grease and drive seal in until it is fully seated in recess. Check seal after installation to be sure spring is properly positioned in seal.
13. Reverse removal procedure to install cover.

TIMING CHAIN

After removing the cover as outlined

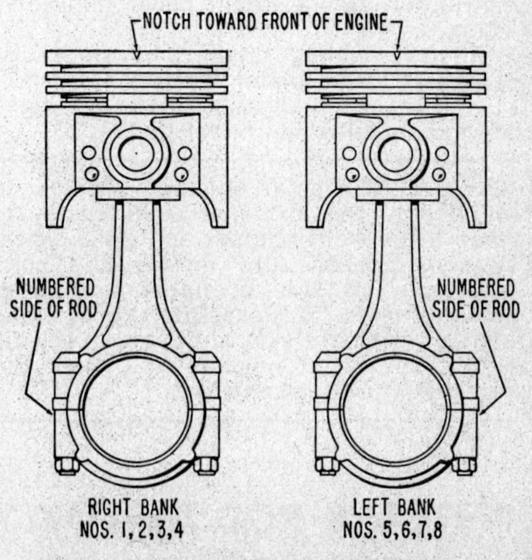

Fig. 32 **Piston and rod assembly. V8s**

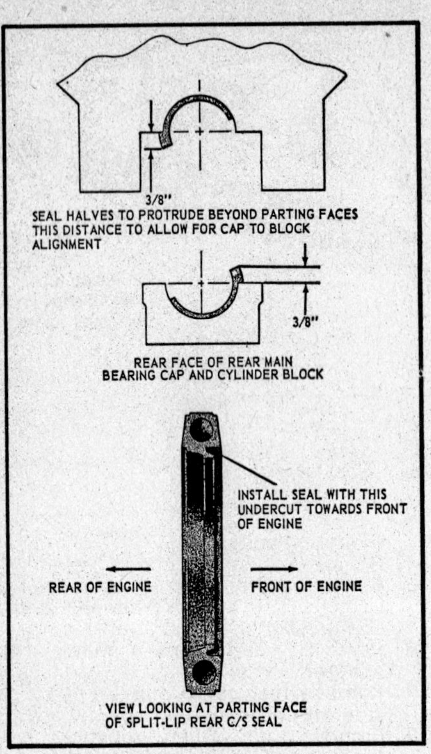

Fig. 33 New split-lip crankshaft rear seal installation

above, remove the crankshaft front oil slinger. Crank the engine until the timing marks are aligned as shown in Figs. 27 and 28. Remove camshaft sprocket retaining bolt and washer. Slide both sprockets and chain forward and remove them as an assembly.

Reverse the order of the foregoing procedure to install the chain and sprockets, being sure the timing marks are aligned.

CAMSHAFT, REPLACE

To remove camshaft, remove cylinder head and related parts, radiator, oil pan, distributor, timing chain cover and valve lifters.

Remove timing chain and sprockets, camshaft thrust plate, and pull camshaft out of engine. If thrust plate shows signs of wear, install a new one, Figs. 29, 30.

SERVICE NOTE: On Mustang models, in addition to the above, it is necessary to remove the front bumper and grille center support bracket. Also remove the bolts from the left side of upper and lower stone shields. If necessary, loosen the bolts on the right side of the stone shields and raise the stone shields out of the way to remove the camshaft.

PISTON & ROD ASSEMBLY

When installed, piston and rod assembly should have the notch or arrow in piston head toward front of engine with connecting rod numbers positioned as shown in Figs. 31 and 32.

PISTONS, PINS & RINGS

Pistons and rings are furnished in standard sizes and oversizes of .020, .030 and .040". On 6-200 engines, .060" oversizes are also available.

Oversize pins are not furnished.

SERVICE BULLETIN

PISTON & PIN REPLACEMENT: When servicing engines using press fit piston pins, the piston and pin must be replaced as an assembly if either does not meet specifications. These components are not serviced separately for the principle reason that excess clearances are usually caused by piston wear rather than pin wear. Elimination of excess clearance by using oversize pins may result in fracture of the connecting rod.

MAIN & ROD BEARINGS

NOTE: Some High Output engines are equipped with an oil baffle tray connected to the main bearing caps. This baffle must be removed to service main and rod bearings.

CRANKSHAFT OIL SEAL

Braided Type

To replace both upper and lower halves, engine and crankshaft must be removed. To replace lower half only, proceed as follows:

1. Remove oil pan and rear main bearing cap.
2. Remove seal and clean groove and cap and block mating surfaces. Preform new seal to approximate radius of cap.
3. Insert seal in cap, seating the center first and allowing seal to extend equally at both ends.
4. Using seating tool, firmly seat seal in groove and cut off ends flush with cap.
5. Apply sealer to cap at rear of the top mating surfaces. Do not apply sealer forward of the oil slinger groove.
6. Install cap and torque to specification.

Split Lip Type

1. Remove oil pan.
2. Remove rear bearing cap.
3. Loosen remaining bearing caps, allowing crankshaft to drop down about $\frac{1}{32}$".

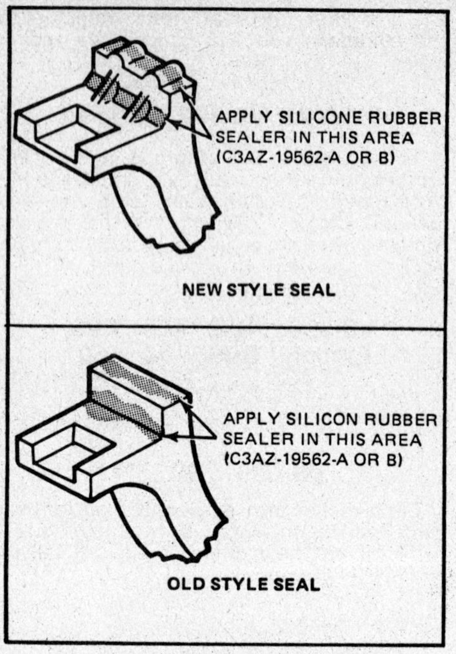

Fig. 34 Crankshaft oil seals

4. Remove old seals from both cylinder block and rear main bearing cap. Use a brass rod to drift upper half of seal from cylinder block groove. Rotate crankshaft while drifting to facilitate removal.
5. Carefully clean seal groove in block with a brush and solvent. Also clean seal groove in bearing cap. Remove the oil seal retaining pin from the bearing cap if so equipped. *The pin is not used with the split-lip seal.*
6. Dip seal halves in clean engine oil.
7. Carefully install upper seal half in its groove with undercut side of seal toward front of engine, Fig. 33, by rotating it on shaft journal of crankshaft until approximately $\frac{3}{8}$" protrudes below the parting surface. *Be sure no rubber has been shaved from outside diameter of seal by bottom edge of groove.*
8. Retighten main bearing caps and torque to specifications.
9. Install lower seal in main bearing cap with undercut side of seal toward front of engine, and allow seal to protrude about $\frac{3}{8}$" above parting surface to mate with upper seal upon cap installation.
10. Apply suitable sealer to parting faces of cap and block. Install cap and torque to specifications.

NOTE: If difficulty is encountered in installing the upper half of the seal in position, lightly lap (sandpaper) the side of the seal opposite the lip side using a medium grit paper. After sanding, the seal must be washed in solvent, then dipped in clean engine oil prior to installation.

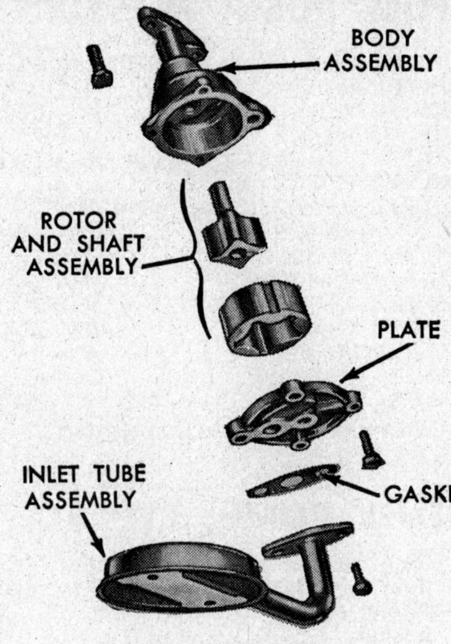

Fig. 36 Oil pump assembly.
Six cylinder engines

SERVICE BULLETIN

A new crankshaft rear oil seal has been released for service. This new seal may be received when ordering an oil pan gasket kit and is installed in the same manner as described above, Fig. 34.

OIL PAN, REPLACE

1969-76 V8-390, 427, 428, 429, 460

1. If equipped with air conditioning, remove fan shroud from radiator and position it over fan.
2. Disconnect stabilizer bar and pull ends down.

NOTE: To allow for clearance for removal of oil pan, remove engine front support insulator-to-intermediate support bracket nuts. Install a block of wood on a floor jack and position the jack under the front leading edge of the oil pan. Raise the engine about 1¼" and insert a 1" block of wood between insulators and frame crossmember; then remove floor jack.

3. Unfasten and lower oil pan to crossmember.
4. Crank engine to obtain necessary clearance between crankshaft counterweight and rear of oil pan. Remove upper bolt and loosen lower bolt on inlet tube.
5. Position inlet tube out of the way and remove oil pan.
6. Reverse procedure to install.

1969-76 6 Cyl. Exc. Granada & Monarch

1. Remove oil level dipstick and flywheel housing inspection cover.
2. On a Mustang, disconnect stabilizer bar and pull it downward out of the way. Remove one bolt and loosen the other on the No. 2 crossmember and lower it out of the way.
3. Remove oil pan and gasket.
4. Reverse procedure to install.

1975-76 Granada & Monarch 6 Cyl.

1. On automatic transmission vehicles, disconnect transmission oil cooler lines at radiator.
2. Remove both radiator top support bolts, then raise vehicle and drain crankcase.
3. Remove the four bolts and nuts retaining sway bar to chassis and allow bar to hang down.
4. Remove starter motor, then remove both engine mount retaining nuts to support brackets.
5. Loosen but do not remove both rear mount insulator to crossmember bolts.
6. Raise front of engine and place 1¼ inch wooden blocks between engine mount and chassis bracket.
7. Lower engine and raise transmission slightly, then remove oil pan bolts and lower oil pan to crossmember.
8. Position crankshaft rear throw in the up position, then position transmission oil cooler lines aside and remove oil pan.
9. Reverse procedure to install.

1970-73 Mustang & 1970-76 Cougar V8-351C, 351M, 400

1. Remove oil dipstick and drain crankcase.
2. Remove starter.
3. Remove sway bar retaining bolts and No. 2 crossmember (under engine)-to-chassis bolts.

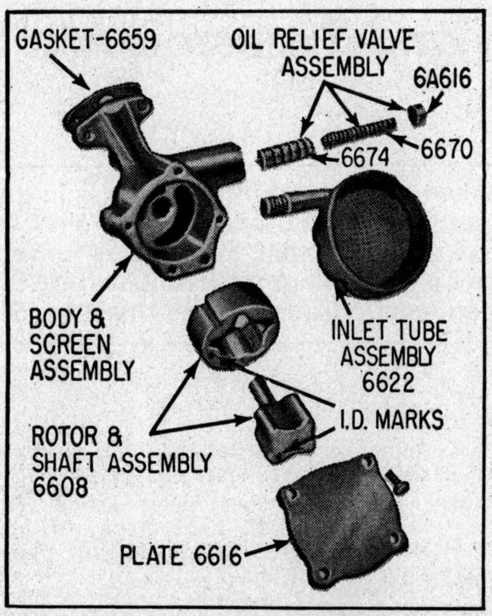

Fig. 37A Oil pump assembly.
V8-351C, 351M & 400

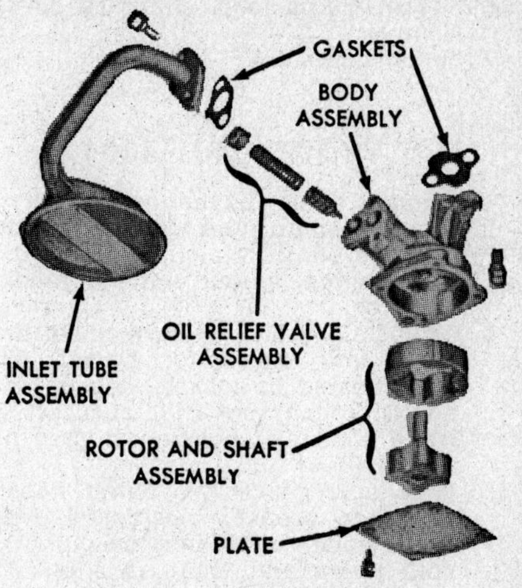

Fig. 37 Oil pump assembly.
V8-302, 351W

4. Remove oil pan bolts, turn crankshaft for maximum clearance and remove oil pan.

1970-76 Fairlane, Montego & Torino V8-351C, 351M & 400

1. Remove dipstick, remove fan shroud retaining bolts and position shroud over fan.
2. Raise car and drain crankcase.
3. Remove starter.
4. Disconnect and lower sway bar.
5. Remove front engine support through bolts, raise engine and place wood blocks between supports and brackets.
6. Remove oil pan bolts and lower oil pan.

1969-73 V8-302, 351 "Windsor" Engine

1. Drain crankcase and remove oil level dipstick.
2. Lower stabilizer bar. On a Mustang the idler arm will also have to be lowered.
3. Remove oil pan bolts and crank engine as required to obtain clearance for removal of pan.

OIL PUMP REPLACE

6 Cylinder Engines

1. Remove oil pan and related parts as directed above.
2. Unfasten and remove pump, gasket and intermediate drive shaft.
3. Prime pump by filling either the inlet or outlet port with engine oil. Rotate pump shaft to distribute oil within pump body.
4. Position intermediate drive shaft into distributor socket.
5. Position new gasket on pump housing. Insert intermediate drive shaft into oil pump.

6. Install pump and shaft as an assembly.
7. Install oil pan.

V8-302, 351, 400

1. Remove oil pan as outlined above.
2. Remove pump inlet tube and screen.
3. Remove pump retaining bolts and remove pump, gasket and intermediate shaft.
4. To install, position intermediate drive shaft into distributor socket. With shaft seated in socket, stop on shaft should touch roof of crankcase. Remove shaft and position stop as necessary.
5. With new gasket on pump housing and stop properly positioned, insert intermediate shaft into oil pump. Install pump and shaft as a unit. *Do not force pump into position if it will not seat readily. The drive shaft hex may be misaligned with distributor shaft. To align, rotate shaft into new position.*

V8-390, 427, 428, 429, 460

1. Remove oil pan as outlined above.
2. Remove oil pump screws, oil pump and intermediate shaft.
3. Remove inlet tube and screen from pump and discard gasket.
4. Prime pump by filling either inlet or outlet port with engine oil. Rotate pump shaft to distribute oil within pump body.
5. Position new gasket on pump housing.
6. Insert intermediate drive shaft into oil pump.
7. Install pump and shaft as a unit.
8. Complete installation in reverse order of removal.

OIL PUMP REPAIRS

V8-390, 427, 428, 429

See Ford-Mercury Full Size Car chapter for an illustration and service procedure for the oil pump used on this engine.

V8-302, 351, 400 & All Sixes

Referring to Figs. 36, 37 and 37A, disassemble pump. To remove the oil pressure relief valve, insert a self-threading sheet metal screw of the proper diameter into the oil pressure relief valve chamber cap and pull cap out of chamber. Remove spring and plunger.

The inner rotor and shaft and the outer race are serviced as an assembly. One part should not be replaced without replacing the other.

Install the pump cover and tighten to 6-9 ft. lbs. torque.

BELT TENSION DATA

	New	Used
1969-76 All	140	110

WATER PUMP, REPLACE

Drain cooling system and disconnect radiator lower hose and heater hose at water pump. Remove drive belt, fan and pulley (also spacer on V8's). Unfasten and remove pump.

FUEL PUMP PRESSURE

Year	Engine	Pressure Lbs.
1969-75	Six Cyl.	4½-6½
1969	V8	4½-6½
1970	V8-429	5-7
1970-71	V8-429 "CJ"	6½-8½ [1]
1970-71	V8-302 exc. "Boss"	4-6
1970-71	V8-302 "Boss", V8-428, V8-429 "Boss" "SCJ"	4½-6½
1970-75	V8-351C	5-7
1971	V8-429 "CJ"	4½-6½
1972-75	V8-302, 400, 429	5-7
1973-75	V8-460	5½-7½
1974-75	V8-302	4½-5½
1974-75	V8-351W	4-6
1974-75	V8-351C, 351M, 400	5½-6½

[1]—With fuel return line inoperative.

FUEL PUMP, REPLACE

1. Remove all gasket material from pump and block gasket surfaces. Apply sealer to both sides of new gasket.
2. Position gasket on pump flange and hold pump in position against its mounting surface. Make sure rocker arm is riding on camshaft eccentric.
3. Press pump tight against its mounting. Install retaining screws and tighten them alternately.
4. Connect fuel lines. Then operate engine and check for leaks.

NOTE: Before installing the pump, it is good practice to crank the engine so that the nose of the camshaft eccentric is out of the way of the fuel pump rocker arm when the pump is installed. In this way there will be the least amount of tension on the rocker arm, thereby easing installation of the pump.

Clutch and Transmission Section

NOTE: 1976 linkage adjustment information is in this section. Repair procedures on both automatic and manual shift transmissions are covered elsewhere in this manual. Procedures for removing automatic transmissions as well as linkage adjustments on 1969-75 models are included in the automatic transmission chapters. See Chapter Index.

CLUTCH PEDAL, ADJUST

1969-76

1. Disconnect clutch return spring from release lever.
2. Loosen release lever rod lock nut.
3. Move release lever rearward until release bearing lightly contacts clutch pressure plate release fingers.
4. Adjust adapter length until adapter seats in release lever pocket.
5. Insert the proper feeler gauge (see below) against back face of rod adapter, then tighten lock nut finger tight against feeler gauge.

1973-75 Exc. Below	.194″
1973-74 Comet & Maverick	.136″
1975 Comet & Maverick	.190″
1975 Granada & Monarch	.136″
1972	.194″
1969-71 except 8-390 engine	.136″
1969-71 8-390 engine	.178″

6. Remove feeler gauge. Hold lock nut in position and tighten adapter against nut.
7. Install release spring and check for free travel of pedal which should be ⅞ to 1⅛ inch for all except Mustang II and 1⅜ to 1⅝ inch for Mustang II.

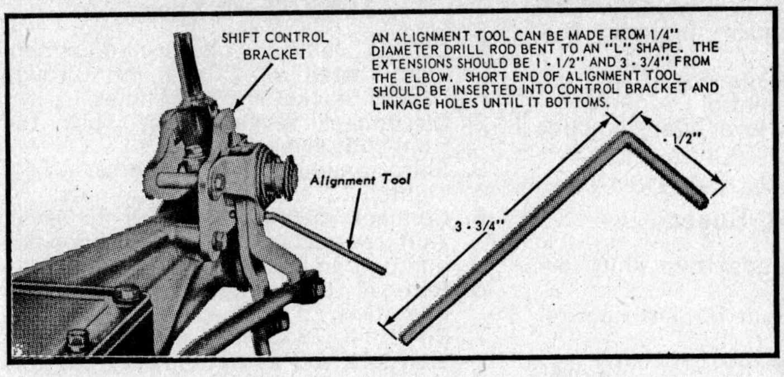

Fig. 1 Gearshift linkage. Typical 1969-76

SERVICE BULLETIN

CLUTCH ROD INTERFERENCE: High clutch pedal effort and/or binding on acceleration may exist on some 1969 Cougar and Montego V8-351 with 3 or 4 speed manual transmission due to the clutch release rod contacting the flywheel housing during acceleration. If evidence of this condition exists, grind off flywheel housing cover flush with clutch housing.

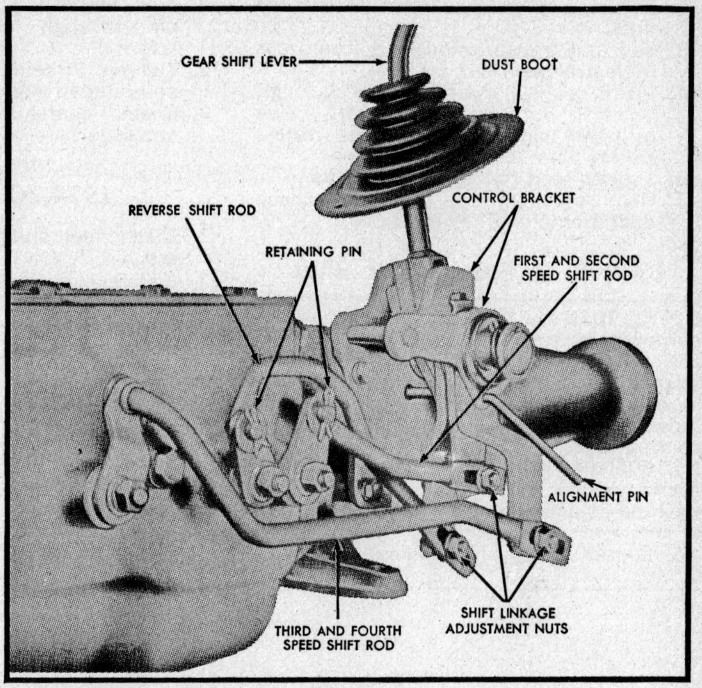

Fig. 2 Ford four speed shift linkage. Typical 1969-76

CLUTCH, REPLACE

1. Remove transmission as outlined further on.
2. Remove release lever retracting spring. Then slide release bearing and hub off release lever.
3. Remove starter motor.
4. Remove inspection cover (if used) and flywheel housing.
5. Loosen clutch cover attaching bolts a little at a time until spring tension is relieved. *If the same cover and pressure plate is to be installed, mark the cover and flywheel so that the pressure plate can be installed in the same position.*
6. Remove cover, pressure plate and clutch disc.
7. Remove clutch release lever from housing.
8. Reverse removal procedure to install the clutch and adjust the pedal as outlined previously.

THREE SPEED TRANS. REPLACE

1969-76

1. Raise and support vehicle on a hoist or safety stands.
2. Disconnect electrical connectors from transmission.
3. Mark drive shaft so that it may be installed in the same position then disconnect it from rear U joint and remove it from output shaft.

NOTE: Plug extension housing to prevent loss of lubricant.

4. Disconnect speedometer cable and shift linkages from transmission.
5. On 1969-72 Ford and Mercury models, disconnect parking brake cable from equalizer lever and remove lever from crossmember. Remove the extension housing to engine rear support bolts.
6. On 1971-72 intermediate models, raise transmission with suitable jack and remove engine rear support nuts. Remove the crossmember and rear support insulator.
7. Support engine and remove transmission.

NOTE: Do not depress clutch pedal with transmission removed.

FOUR SPEED TRANS. REPLACE

1969-76 Ford Design

1. Raise vehicle and support on hoist or safety stands.
2. Disconnect electrical connectors from transmission.
3. Mark driveshaft so that it may be installed in the same position then disconnect it from rear U joint and remove it from output shaft.
4. Disconnect speedometer cable from transmission and parking brake cable from equalizer lever and remove lever from crossmember.
5. Remove hairpin retainer from parking brake cable and remove cable from crossmember.
6. Remove the shift rods from the shift levers and remove the shift control.
7. Support engine with suitable jack and remove rear support bolts then raise engine enough to relieve weight from crossmember and remove crossmember bolts and crossmember.
8. Support transmission with suitable jack and remove transmission bolts and transmission.

NOTE: Do not depress clutch pedal with transmission removed.

GEARSHIFT LINKAGE

NOTE: *If the transmission shifts hard or will not engage, the gearshift levers may need adjusting at the cross-over. Move the shift lever through all positions to see that the cross-over operation is smooth. If not, adjust as follows:*

1972-76 Torino & Montego Column Shift

1. Working inside car, loosen clamp nut holding gearshift tube and shaft.
2. Loosen shift rod adjusting nuts at transmission.
3. Remove plastic covers over gearshift lever assembly and insert $\frac{1}{4}$" rod.
4. Set gearshift lever in Neutral in the 2-3 plane and torque clamp nut inside car to 12-18 ft. lbs.
5. Set transmission levers in Neutral and insert $\frac{1}{4}$" rod in locator holes in lower end of steering column.
6. Torque shift rod adjusting nuts using care to prevent motion between the rods and the studs on transmission levers.
7. Remove alignment pins and install covers.
8. Check operation to assure smooth crossover.

1969-71 Column Shift

1. Place shift lever in neutral.

2. Loosen two gearshift rod adjustment nuts.
3. See that transmission shift levers are in neutral position.
4. Insert a $\frac{3}{16}$" diameter rod through holes in both levers and both holes in lower casting. It may be necessary to align levers to insert tool.
5. Tighten shift rod adjustment nuts.
6. Remove alignment tool and check operation of shift levers.

1969-76 3 Speed Floor Shift

NOTE: 1970-76 floor shift linkages incorporate a transmission lock rod. This rod must be adjusted AFTER the shift linkage has been adjusted. With shift lever in Neutral and lock rod adjustment nut loose, align hole in steering column socket casting with alignment mark and insert a .180" dia. rod. The casting must not rotate with the rod in this position. Tighten lock rod adjustment nut.

1. Loosen three shift linkage adjusting nuts. Install a $\frac{1}{4}$" diameter alignment pin through control bracket and levers, Fig. 1.
2. Tighten three linkage adjusting nuts and remove alignment pin.
3. Check gearshift lever for smooth crossover.

1974-76 4 Speed

1. Disconnect shift rods from shift levers.
2. Install alignment pin in shift control, Fig. 2.
3. Turn reverse lever clockwise into gear to make sure forward gears are in neutral.
4. Attach forward speed shift rods on levers and rotate reverse lever counterclockwise until it stops then attach reverse rod on reverse lever.
5. Attach column lock rod and remove alignment pin.
6. Adjust transmission lock rod as outlined for 1970-76 three speed units.

1969-73 4 Speed

1. Loosen shift linkage adjustment nuts. Install a $\frac{1}{4}$" dia. rod through control bracket and lever holes.
2. Disconnect reverse gear shift rod from shift lever.
3. Shift transmission into reverse.
4. Tighten 1-2 and 3-4 shift rod nuts.
5. Connect reverse shift rod to lever, shift reverse lever to neutral position and tighten nut.
6. Remove alignment tool and check operation.
7. On 1970-73 units, adjust transmission lock rod as outlined for 1970-76 three speed units.

1976 AUTO. TRANS. LINKAGE, ADJUST

NOTE: All 1976 vehicles, use the same procedure as previous models and can be found elsewhere in this book.

Rear Axle, Propeller Shaft & Brakes

REAR AXLES

Figs. 1 and 2 illustrate the rear axle assemblies used on these cars. When necessary to overhaul either of these units, refer to the *Rear Axle Specifications* table in this chapter.

Conventional Integral Carrier Type, Fig. 1

In these axles, Fig. 1, the rear axle housing and differential carrier are cast into an integral assembly. The drive pinion assembly is mounted on two opposed tapered roller bearings. Spacers are used and are located between the front and rear bearing cones. Ring gear and pinion tooth contact is adjusted by shims between the rear bearing cone and pinion gear.

The differential carrier assembly is mounted on two opposed tapered roller bearings. The bearings are retained in the housing by removable caps. Differential bearing preload and drive gear backlash is adjusted by threaded ring nuts or sleeves located behind each differential bearing cup.

All service operations on the differential case assembly and the drive pinion assembly can be performed with the housing in the vehicle. The axle shafts and bearings can be pulled out of the housing ends. The differential assembly and then the drive pinion can be removed from the housing after the cover is removed from the rear face of the carrier casting.

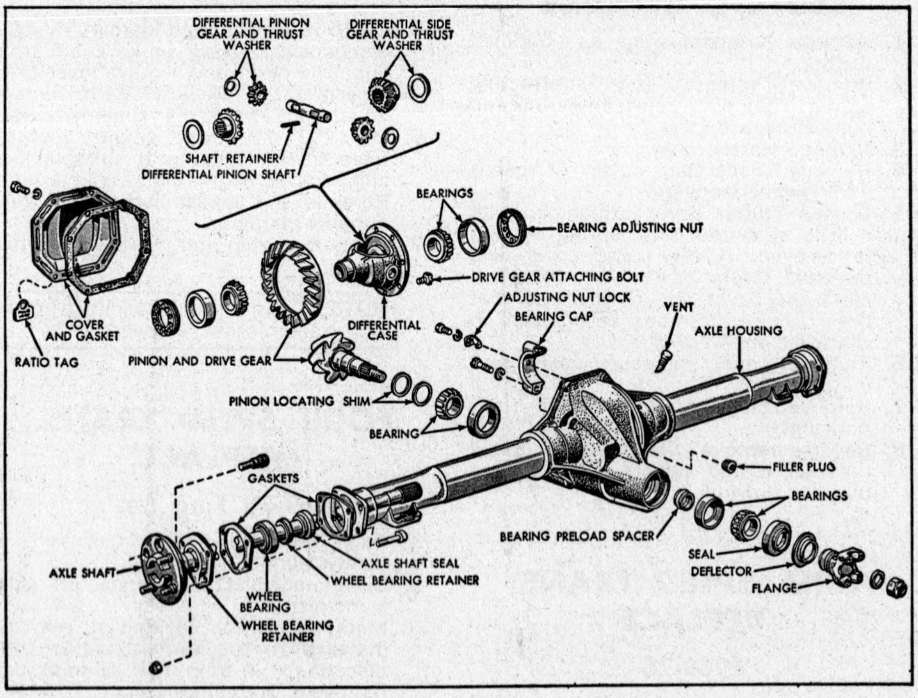

Fig. 1 Disassembled view of conventional integral carrier type rear axle assembly

Service Bulletin

All Ford Built Rear Axles: Recent manufacturing changes have eliminated the need for marking rear axle drive pinions for individual variations from nominal shim thicknesses. In the past, these pinion markings, with the aid of a shim selection table, were used as a guide to select correct shim thicknesses when a gear set or carrier assembly replacement was performed.

With the elimination of pinion markings, use of the shim selection table is no longer possible and the methods outlined below must be used.

1. Measure the thickness of the original pinion depth shim removed from the axle. Use the same thickness upon installation of the replacement carrier or drive pinion. If any further shim change is necessary, it will be

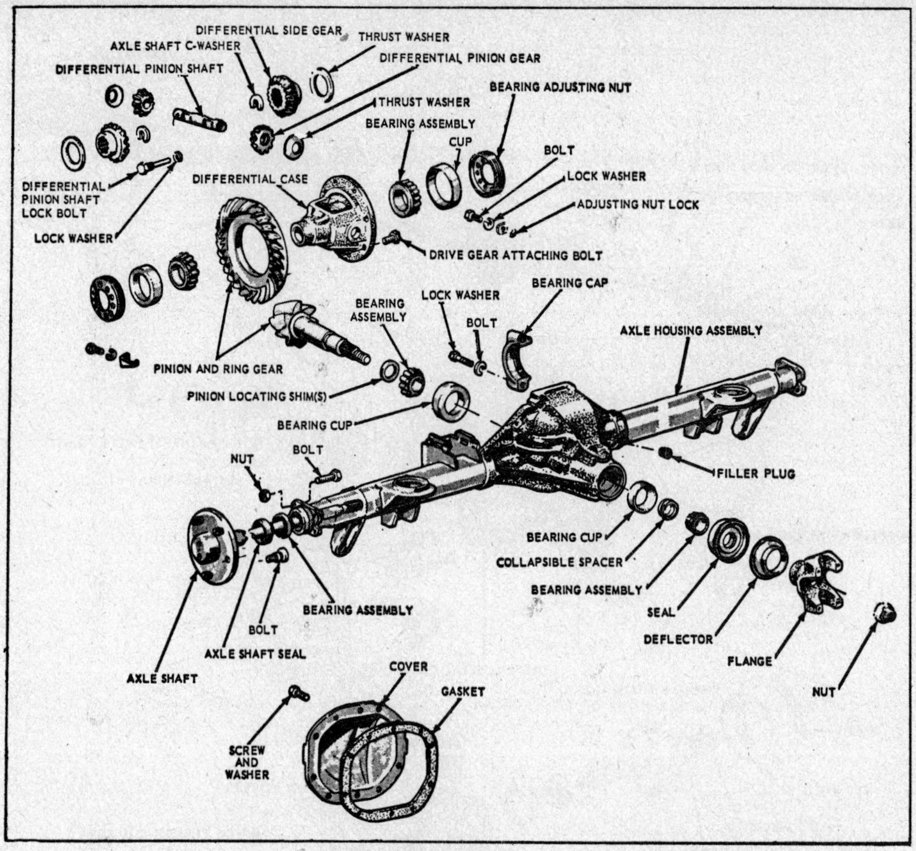

Fig. 2 Disassembled view of Ford WER integral carrier type rear axle assembly

Fig. 3 Axle shaft C-locks. Ford WER integral type axle

Fig. 5 Removing nuts from wheel bearing retainer

indicated in the tooth pattern check.

2. If the original shim is lost, substitute a nominal shim for the original and use the tooth pattern check to determine if further shim changes are required.

Ford WER Integral Carrier, Fig. 2

The gear set consists of a ring gear and an overhung drive pinion which is supported by two opposed tapered roller bearings. The differential case is a one-piece design with openings allowing assembly of the internal parts and lubricant flow. The differential pinion shaft is retained with a threaded bolt (lock) assembled to the case.

The roller type wheel bearings have no inner race, and the rollers directly contact the bearing journals of the axle shafts. The axle shafts do not use an inner and outer bearing retainer. Rather, they are held in the axle by means of C-locks, Fig. 3. These C-locks also fit into a machined recess in the differential side gears within the differential case. There is no retainer bolt access hole in the axle shaft flange.

Removable Carrier Type

In these axles, Fig. 4, the drive pinion is straddle-mounted by two opposed tapered roller bearings which support the pinion shaft in front of the drive pinion gear, and a straight roller bearing that

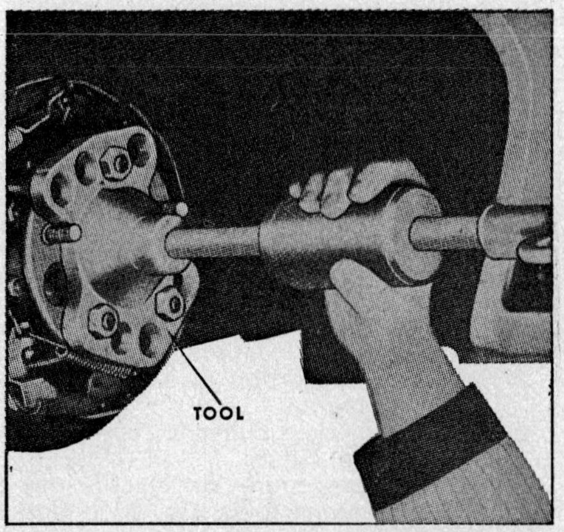

Fig. 6 Removing axle shaft with slide hammer-type puller

Fig. 7 Splitting bearing inner retainer for bearing removal

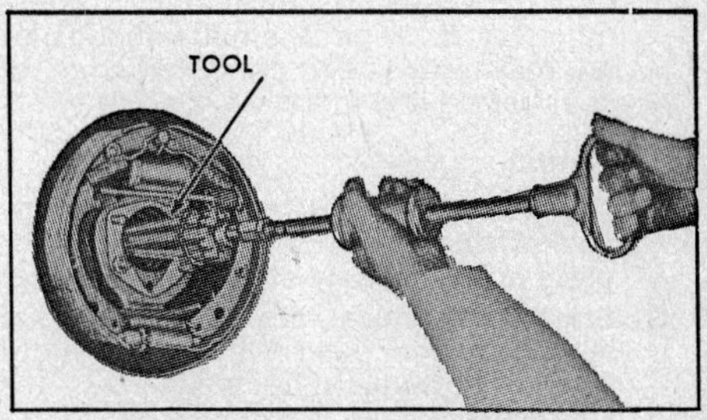

Fig. 8 Using hook-type tool to remove oil seal

supports the pinion shaft at the rear of the pinion gear. The drive pinion is assembled in a pinion retainer that is bolted to the differential carrier. The tapered roller bearings are preloaded by a collapsible spacer between the bearings. The pinion is positioned by a shim or shims located between the drive pinion retainer and the differential carrier.

The differential is supported in the carrier by two tapered roller side bearings. These bearings are preloaded by two threaded ring nuts or sleeves between the bearings and the pedestals. The differential assembly is positioned for proper ring gear and pinion backlash by varying the adjustment of these ring nuts. The differential case houses two side gears in mesh with two pinions mounted on a pinion shaft which are held in place by a pin. The side gears and pinions are backed by thrust washers.

The axle shafts are of unequal length, the left shaft being shorter than the right. The axle shafts are mounted in sealed ball bearings which are pressed on the shafts.

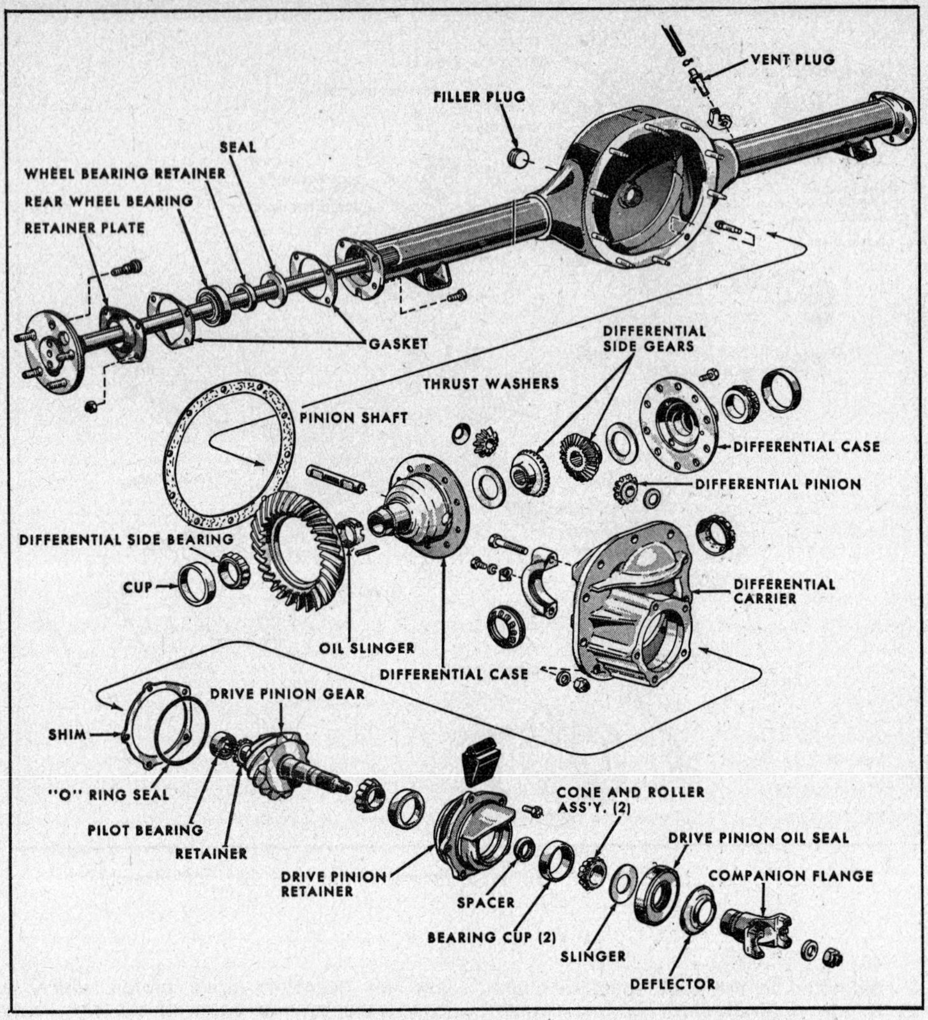

Fig. 4 Rear axle assembly with removable carrier

Axle Shaft, Replace

Conventional Integral & Removable Carrier Type

1. Remove wheel assembly.
2. Remove brake drum from flange.
3. Working through hole provided in axle shaft flange, Fig. 5, remove nuts that secure wheel bearing retainer.
4. Pull axle shaft out of housing. If bearing is a tight fit in axle housing use a slide hammer-type puller, Fig. 6. *Brake carrier plate must not be dislodged. Install one nut to hold the plate in place after axle shaft is removed.*

NOTE: On 1974-76 models, remove brake carrier plate.

5. If the axle shaft bearing is to be replaced, loosen the inner retainer by nicking it deeply with a chisel in several places, Fig. 7. The bearing will then slide off easily.
6. Press bearing from axle shaft.
7. Inspect machined surface of axle shaft and housing for rough spots that would affect sealing action of the oil seal. Carefully remove any burrs or rough spots.
8. Press new bearing on shaft until it seats firmly against shoulder on shaft.
9. Press inner bearing retainer on shaft until it seats firmly against bearing.
10. If oil seal is to be replaced, use a hook-type tool to pull it out of housing, Fig. 8. Wipe a small amount of oil resistant sealer on outer edge of seal before it is installed, Fig. 9.

Installation

1. On 1974-76 models, place a new gasket on each side of brake carrier plate and slide axle shaft into housing.
2. Start the splines into the differential side gear and push the shaft in until bearing bottoms in housing.
3. Install retainer and tighten nuts to 30-40 ft. lbs.
4. Install brake drum and wheel.

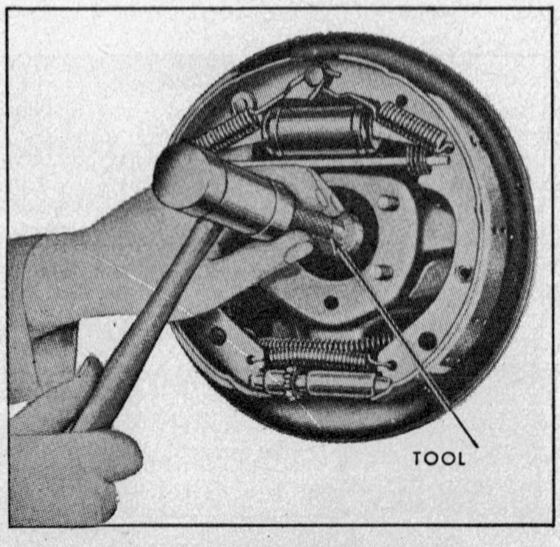

Fig. 9 Using special driver to install oil seal

Ford WER Integral Carrier Type

1. Raise car on hoist and remove wheels.
2. Drain differential lubricant.
3. Remove brake drums.
4. Remove differential housing cover.
5. Position safety stands under rear frame member and lower hoist to allow axle to lower as far as possible.
6. Working through differential case opening, remove pinion shaft lock bolt and pinion shaft.
7. Push axle shaft(s) inward toward center of axle housing and remove C-lock(s) from housing, Fig. 3.
8. Remove axle shaft, using extreme care to avoid contact of shaft seal lip with any portion of axle shaft except seal journal.
9. Use a hook-type puller to remove seal and bearing, Fig. 8.
10. Reverse procedure to install, using suitable driving tools, Fig. 9, to install seal and bearing. New seals are prepacked with lubricant and do not require oil soaking before installation.

PROPELLER SHAFT

Remove & Replace

1. Disconnect rear U-joint from drive pinion flange.
2. Pull drive shaft toward rear of car until front U-joint yoke clears transmission extension housing and output shaft.
3. Install a suitable tool, such as a seal driver, in seal to prevent lube from leaking from transmission.
4. Before installing, check U-joints for freedom of movement. If a bind has resulted from misalignment after overhauling the U-joints, tap the ears of the drive shaft sharply to relieve the bind.
5. If rubber seal installed on end of transmission extension housing is damaged, install a new seal.
6. On a manual shift transmission, lubricate yoke spline with conventional transmission grease. On an automatic transmission, lubricate yoke spline with special spline grease. *This spline is sealed so that transmission fluid does not "wash" away spline lubricant.*
7. Install yoke on transmission output shaft.
8. Install U-bolts and nuts which attach U-joint to pinion flange. Tighten U-bolts evenly to prevent binding U-joint bearings.

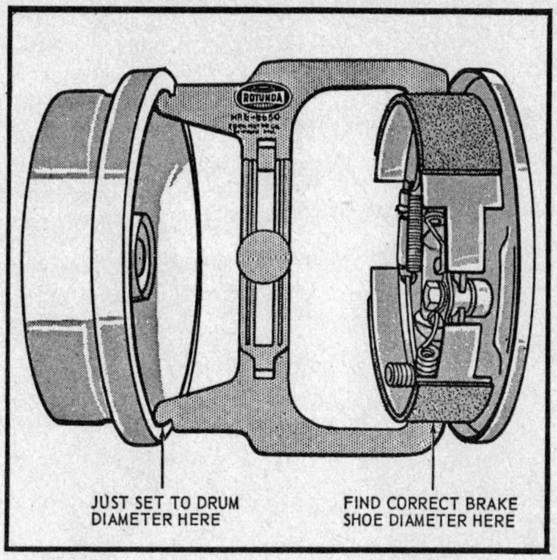

Fig. 10 Revised brake adjustment

BRAKE ADJUSTMENTS

SERVICE BULLETIN

REVISED BRAKE ADJUSTMENT PROCEDURE: Some models use a new front and rear brake backing plate which omits the adjusting slot for manual brake adjustment. The backing plates have a partially stamped knock-out slot for use ONLY when the brake drums cannot be removed in a normal manner. The open slot is then covered with a rubber plug as used in the past to prevent contamination of the brakes.

When servicing a vehicle requiring a brake adjustment, the metal knock-out plugs should NOT be removed. Rather the drums should be removed and brakes inspected for a malfunction.

Although the brakes are self-adjusting, an initial adjustment will be necessary after a brake repair, such as relining or replacement. The initial adjustment can be obtained by the new procedure which follows:

1. Use the brake shoe adjustment gauge shown in Fig. 10 to obtain the drum inside diameter as shown. Tighten the adjusting knob on the gauge to hold this setting.
2. Place the opposite side of the gauge over the brake shoes and adjust the shoes by turning the adjuster screw until the gauge just slides over the linings. Rotate the gauge around the lining surface to assure proper lining diameter adjustment and clearance.
3. Install brake drum and wheel. Final adjustment is accomplished by making several firm reverse stops, using the brake pedal.

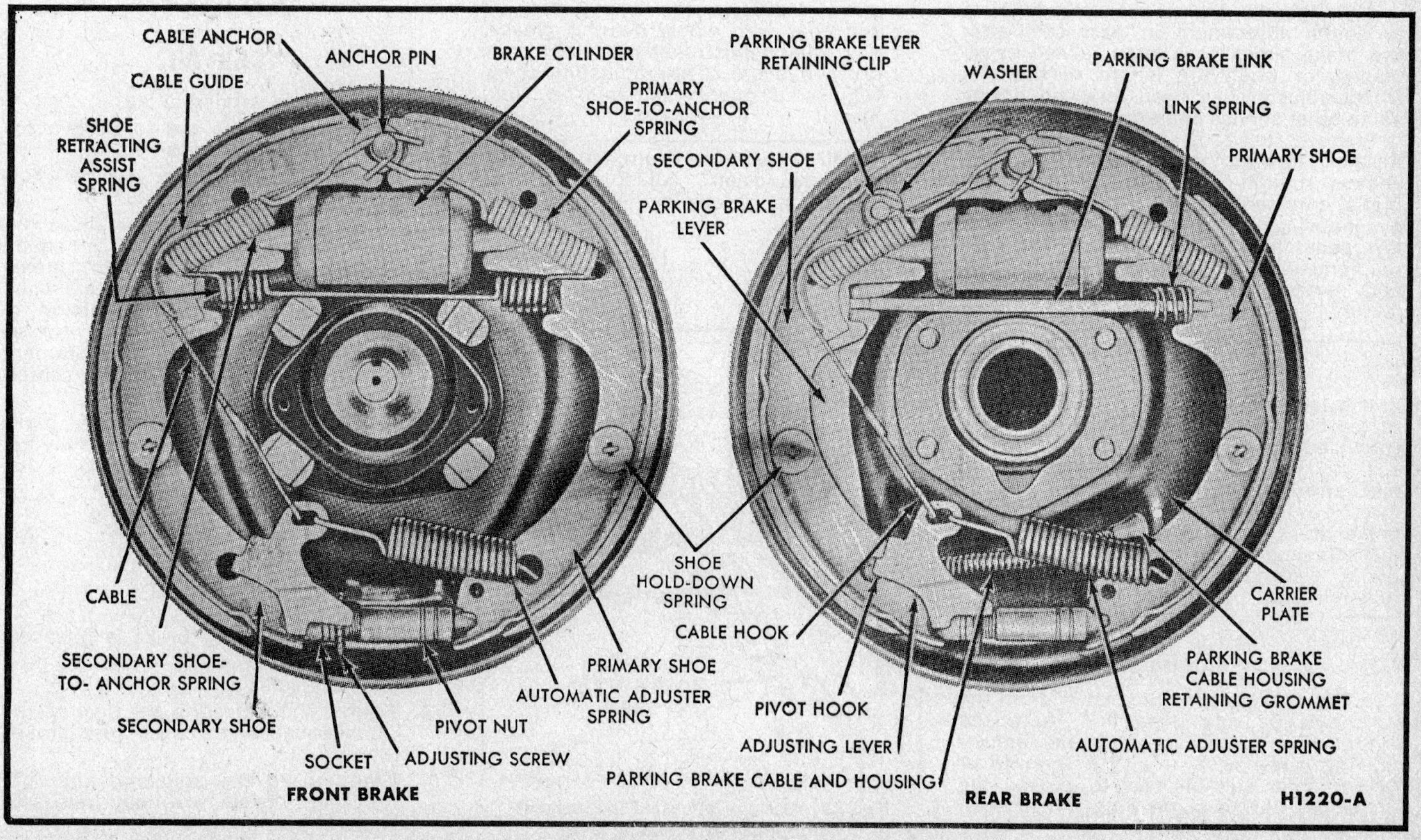

Fig. 11 Right front and rear brake mechanism. 1969-75 V8s

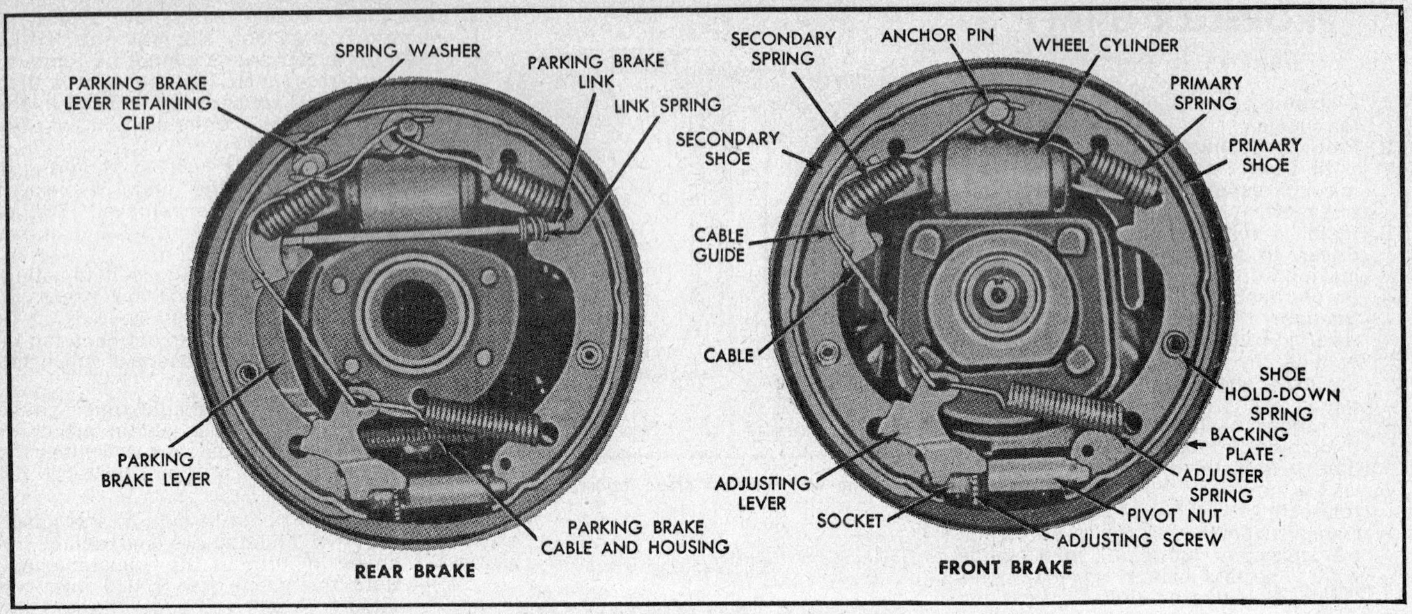

Fig. 12 Right front and rear brakes. 1969-75 Six-Cylinder models

Self-Adjusting Brakes

These brakes, Figs. 11 and 12, have self-adjusting shoe mechanisms that assure correct lining-to-drum clearances at all times. The automatic adjusters operate only when the brakes are applied as the car is moving rearward or when the car comes to an uphill stop.

Although the brakes are self-adjusting, an initial adjustment in necessary after the brake shoes have been relined or replaced, or when the length of the star wheel adjuster has been changed during some other service operation.

Frequent usage of an automatic transmission forward range to halt reverse vehicle motion may prevent the automatic adjusters from functioning, thereby inducing low pedal heights. Should low pedal heights be encountered, it is recommended that numerous forward and reverse stops be made until satisfactory pedal height is obtained.

NOTE

If a low pedal condition cannot be corrected by making numerous reverse stops (provided the hydraulic system is free of air) it indicates that the self-adjusting mechanism is not functioning. Therefore, it will be necessary to remove the brake drum, clean, free up and lubricate the adjusting mechanism. Then adjust the brake, being sure the parking brake is fully released.

Adjustment

1. Remove adjusting hole cover from the brake backing plate and, from the backing plate side, turn the adjusting screw upward with a screwdriver or other suitable tool to expand the shoes until a slight drag is felt when the drum is rotated.
2. Remove the drum.
3. While holding the adjusting lever out of engagement with the adjusting screw, Fig. 13, back off the adjusting screw about 3/4 turn with the fingers.

NOTE—*If finger movement will not turn the screw, free it up. If this is not done, the adjusting lever will not turn during subsequent vehicle operation. Lubricate the screw with oil and coat with wheel bearing grease. Any other adjustment procedure may cause damage to the adjusting screw with consequent self-adjuster problems.*

4. Install wheel and drum, and adjusting hole cover. Adjust brakes on remaining wheels in the same manner.

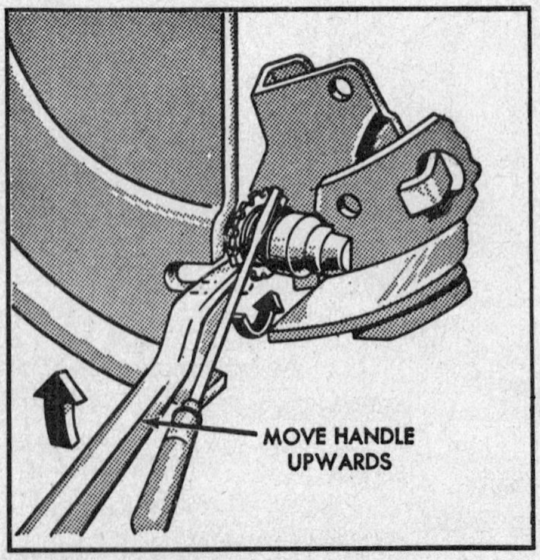

Fig. 13 Backing off brake adjustment by disengaging adjusting lever with screwdriver

5. If pedal height is not satisfactory, drive the vehicle and make sufficient reverse stops until proper pedal height is obtained.

PARKING BRAKE, ADJUST

1969 All

1. Fully release parking brake.
2. Pull brake handle out to third notch from fully released position.
3. Raise vehicle and remove wheel cover.
4. Turn locking adjustment nut forward against cable guide on equalizer until there is 100 ft-lbs breakaway torque at rear wheel when turning rear wheels in direction of forward rotation with a torque wrench. This torque measurement must be made relative to the centerline of the wheel.
5. Release parking brake and make sure brake shoes return to fully released position and no drag is felt when turning rear wheels.

1970-76 All

Rear Drum Brakes

1. Make sure parking brake is released.
2. Place transmission in neutral and raise the vehicle.
3. Tighten the adjusting nut against the cable equalizer to cause rear brakes to drag.
4. Then loosen the adjusting nut until the rear wheels are fully released. There should be no drag.
5. Lower vehicle and check operation.

1975-76

Rear Disc Brakes

1. Fully release parking brake, then place transmission in neutral and support vehicle at rear axle.
2. Tighten adjuster nut until levers on calipers just begin to move, then loosen adjuster nut until levers just return to stop position.
3. Apply and release parking brake. Check levers on caliper to determine if they are fully returned by attempting to pull lever rearward. If lever moves, the adjustment is too tight and must be readjusted.

MASTER CYLINDER, REPLACE

1969-76 Less Power Brakes

1. Working from inside vehicle below instrument panel, disconnect master cylinder push rod from brake pedal.
2. Disconnect stop light switch wires, remove hairpin retainer and slide stop light switch off brake pedal pin just far enough to clear end of pin. Then lift switch straight upward from pin.
3. Slide master cylinder push rod with nylon washers and bushings from brake pedal pin.
4. Remove brake tubes from outlet ports of master cylinder.
5. Remove lock nuts that secure master cylinder to dash panel and lift cylinder forward and upward from vehicle.
6. Reverse procedure to install.

1969-76 Power Brakes

1. Disconnect brake lines from master cylinder.
2. Remove nuts retaining master cylinder to brake booster.
3. Remove master cylinder.

POWER BRAKE UNIT, REPLACE

1969-75 Comet, Cougar, Montego & Fairlane & 1969-73 Mustang

1. Working under instrument panel, disconnect stop light switch wires at connector.
2. Remove hairpin type retainer. Slide stop light switch off brake pedal pin just far enough for the switch outer hole to clear the pin, then lower switch away from pin.
3. Slide master cylinder push rod link and nylon washers and bushing off brake pedal pin.
4. Disconnect brake line from master cylinder.
5. Disconnect vacuum hose from booster at check valve.
6. Unfasten and remove booster and bracket assembly from dash panel, sliding push rod link out from engine side of dash panel.

Rear Suspension

SHOCK ABSORBER, REPLACE

To replace shock absorber, support rear axle properly and disconnect shock absorber at upper and lower mountings. On all models except convertibles and 1971-73 Comet and Maverick, the upper nut is accessible from inside trunk. On 1969-71 Montego, Torino and all Cougar, Mustang Models access cover must be removed from trunk floor panel. On convertibles the upper attachment access cover is accessible by removing the rear seat.

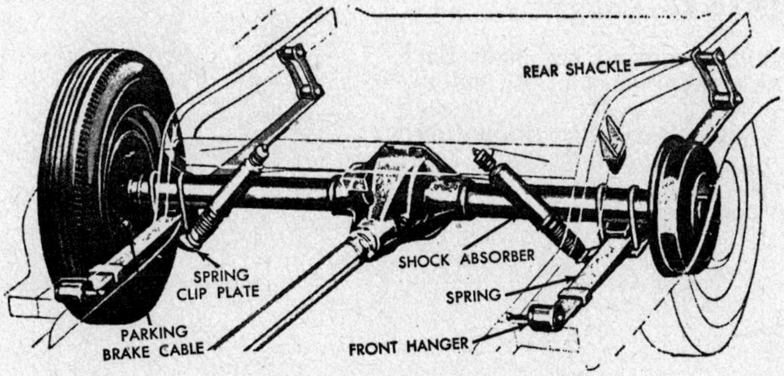

Fig. 1 Leaf spring suspension (typical)

LEAF SPRINGS & BUSHINGS REPLACE

All Exc. 1972-76 Elite, Montego, Torino & 1974-76 Cougar

1. Raise rear of vehicle and support at frame. Support axle with a suitable jack.
2. Disconnect shock absorbers from lower mountings.
3. Lower jack and remove spring plate "U" bolts and spring plate, Fig. 1.
4. Raise axle to remove weight from spring and disassemble rear shackle.
5. Remove spring front mount bolt.
6. Replace spring front eye bushings as necessary, Figs. 2 and 3.
7. Reverse procedure to install.

COIL SPRING, REPLACE

1972-76 Elite, Montego, Torino & 1974-76 Cougar

1. Raise rear of vehicle and support at frame. Support axle with a suitable jack.
2. Disconnect shock absorbers from lower mountings.
3. Lower axle to remove springs.
4. Reverse procedure to install. Install an insulator between upper and lower seats and spring.

CONTROL ARMS, REPLACE

1972-76 Elite, Montego, Torino & 1974-76 Cougar

NOTE: Upper and lower control arms must be replaced in pairs.

1. Raise rear of vehicle and support at frame. Support axle with a suitable jack. Also support differential pinion nose.
2. Disconnect control arm from axle

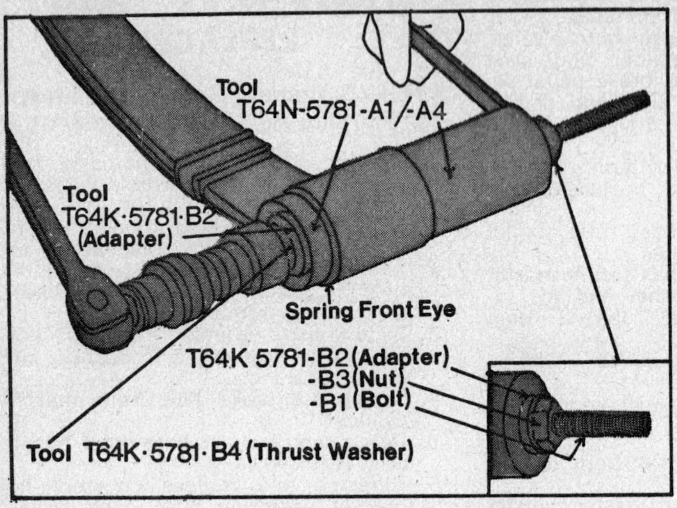

Fig. 2 Spring front bushing removal

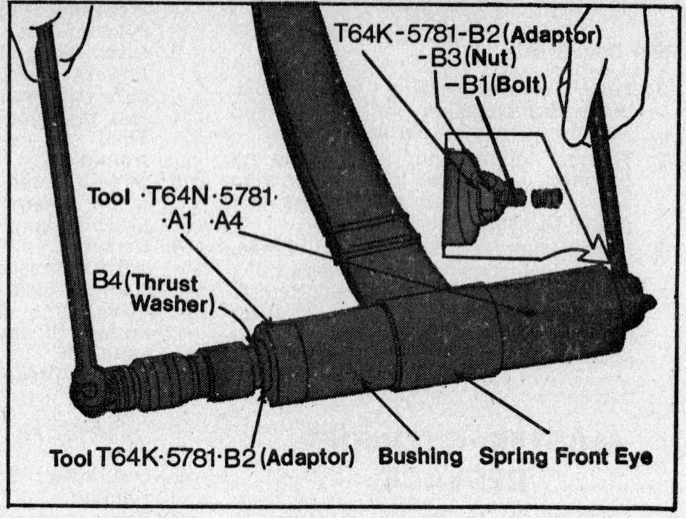

Fig. 3 Spring front bushing installation

bracket. On upper arms, disconnect arm from crossmember and on lower arms, disconnect arm from frame attachment bracket, Fig. 4.

3. Reverse procedure to install.

STABILIZER BAR, REPLACE

1972-76 Elite, Montego, Torino & 1974-76 Cougar

1. Remove bolts securing stabilizer bar to rear link assemblies on both sides, Fig. 5.
2. Remove nuts securing mounting bracket to lower mounting clamp and remove bar.
3. Reverse procedure to install.

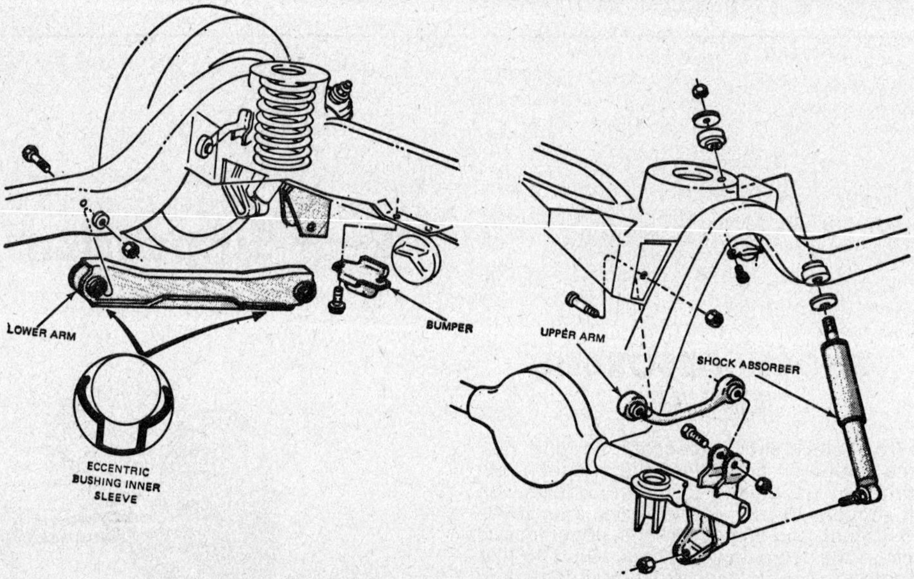

Fig. 4 Rear suspension, 1972-75 Elite, Montego & Torino & 1974-75 Cougar

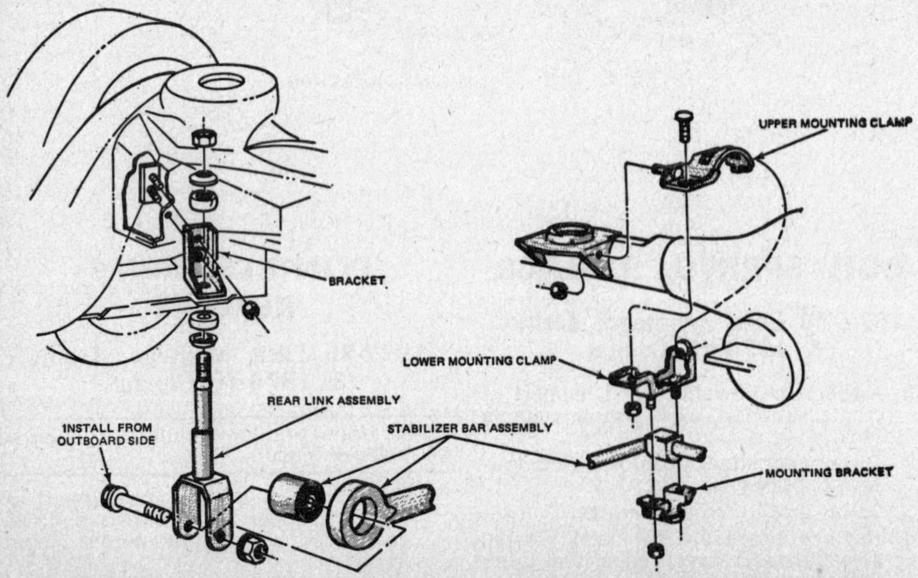

Fig. 5 Stabilizer bar installation 1972-75 Elite, Montego & Torino & 1974-75 Cougar

Front End & Steering Section

FRONT SUSPENSION

1972-76 Elite, Torino, Montego & 1974-76 Cougar

The front suspension, Fig. 1, has been redesigned with the coil spring mounted on the lower arm rather than on the upper arm as previous.

All Except 1972-76 Elite, Torino, Montego & 1974-76 Cougar

Referring to Fig. 2, each front wheel rotates on a spindle. The upper and lower ends of the spindle are attached to ball joints that are mounted to an upper and lower control arm. The upper arm pivots on a bushing and shaft assembly that is bolted to the underbody. The lower arm pivots on a bolt that is located in an underbody bracket.

A coil spring seats between the upper arm and the top of the spring housing. A double-acting shock absorber is bolted to the arm and the top of the spring housing.

Struts, which are connected between the lower control arms and the underbody, prevent the arms from moving fore and aft.

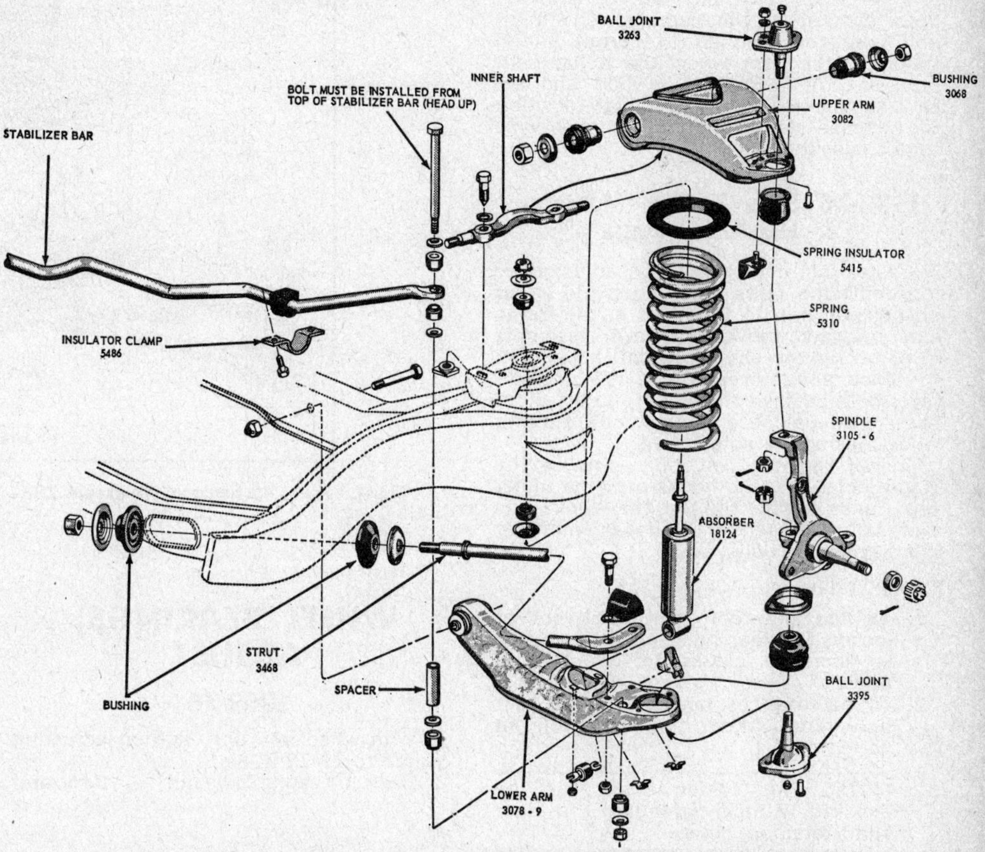

Fig. 1 Front suspension. 1972-76 Elite, Montego, Torino & 1974-76 Cougar

LUBRICATION

1969-76

These cars are equipped with an extended chassis lubrication feature which is made possible by a new type of special lubricant combined with special seals and bearing materials which extends the lubrication period to 36,000 miles.

SERVICE BULLETIN

Some uninformed service people recommended that conventional grease fittings be installed and that the car be lubricated every 1000 miles. This is completely unnecessary and, in fact, may cause damage to the special seals used in the lubrication points.

The use of conventional lubricants not only can do damage to the special seals but is incompatible with the special lubricant. Moreover, after the special sealing plugs have been replaced by conventional grease fittings, dirt and water can enter and cause excessive wear, rendering the units unfit for further service.

SERVICE BULLETIN

BALL JOINT LUBRICATION: The ball joint seals have been redesigned to provide improved sealing and longer life. The new seals can be damaged and sealing characteristics destroyed if excessive lubricant is used. Specifications call for the addition of only 10 grams (one level teaspoonful) of lubricant to the ball joint at each 36,000 mile interval. The initial application of 10 grams of lubricant insures forcing grease into the bearing area and still allows for three subsequent lubrications of 10 grams each without ballooning the seals and resultant premature failure.

For the above reasons the ball joint seals on new vehicles might appear to be collapsed and give the mistaken impression that additional lubricant is required. This is not the case and under no circumstances should more than 10 grams of grease be added to the ball joint at the 36,000 mile intervals.

SERVICE BULLETIN

STEERING LINKAGE LUBRICATION:

The steering linkage should be greased at 36,000 mile intervals. Normal breathing of the socket joints permits moisture condensation within the joint. Moisture inside the joint will cause no appreciable damage and the joint will function normally. However, if the moisture is concentrated in the bearing grease grooves and is frozen at the time of lubrication, grease cannot flow and pressure greasing may damage the joint.

Do not attempt to grease the steering linkage on these models if it has set in temperatures lower than 20 deg. above zero F. The car should be allowed to warm up in a heated garage for 30 minutes or until the joints accept lubrication.

IMPORTANT: A torch must not be used to heat joints because this quantity of heat will melt the nylon bearing within the joint.

WHEEL ALIGNMENT

All Except 1972-76 Elite, Torino, Montego & 1974-76 Cougar

As shown in Fig. 3, caster is controlled by the front suspension strut. To obtain positive caster, loosen the strut rear nut and tighten the strut front nut against the bushing. To obtain negative caster, loosen the strut front nut and tighten the strut rear nut against the bushing.

Camber is controlled by the eccentric cam located at the lower arm attachment to the side rail. To adjust camber, loosen the camber adjustment bolt nut at the rear of the body bracket. Spread the body bracket at the camber adjustment bolt area just enough to permit lateral travel of the arm when the adjustment bolt is turned. Rotate the bolt and eccentric clockwise from the high position to increase camber or counterclockwise to decrease it.

1972-76 Elite, Torino, Montego, & 1974-76 Cougar

Caster and camber can be adjusted by loosening the bolts that attach the upper suspension arm to the shaft at the frame side rail, and moving the arm assembly in or out in the elongated bolt holes, Fig. 4. Since any movement of the arm affects both caster and camber, both factors should be balanced against one another when making the adjustment.

Install the tool with the pins in the frame holes and the hooks over the upper arm inner shaft. Tighten the hook nuts snug before loosening the upper arm inner shaft attaching bolts, Fig. 4.

Caster, Adjust

1. Tighten the tool front hook nut or loosen the rear hook nut as required to increase caster to the desired angle.
2. To decrease caster, tighten the rear hook nut or loosen the front hook nut as required.

NOTE: The caster angle can be checked without tightening the inner shaft retaining bolts.

3. Check the camber angle to be sure it did not change during the caster adjustment and adjust if necessary.
4. Tighten the upper arm inner shaft retaining bolts and remove tool.

Camber, Adjust

1. Install as previously outlined.
2. Loosen both inner shaft retaining bolts.
3. Tighten or loosen the hook nuts as necessary to increase or decrease camber.
4. Recheck caster angle.

TOE-IN, ADJUST

Check the steering wheel spoke position when the front wheels are in the straight-ahead position. If the spokes are not in the normal position, they can be adjusted while toe-in is being adjusted.

1. Loosen clamp bolts on each tie rod end sleeve.
2. Adjust toe-in. If steering wheel spokes are in their normal position, lengthen or shorten both rods equally to obtain correct toe-in. If spokes are not in normal position, make necessary rod adjustments to obtain correct toe-in and steering wheel spoke alignment.

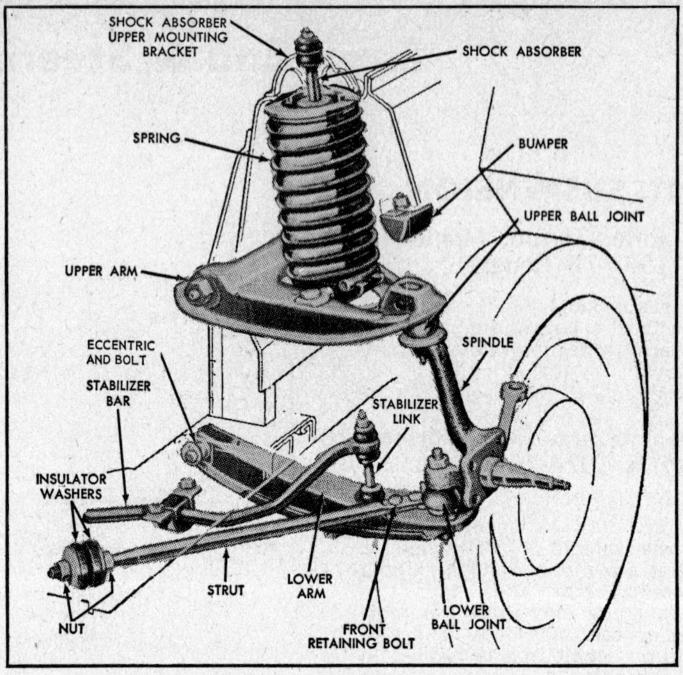

Fig. 2 Front suspension. All except 1972-76 Elite, Montego, Torino & 1974-76 Cougar

WHEEL BEARINGS, ADJUST

1969-76

1. With wheel rotating, tighten adjusting nut to 17-25 ft. lbs.
2. Back off adjusting nut ½ turn and retighten nut to 10-15 inch lbs.
3. Place nut lock on nut so that castellations on lock are aligned with cotter pin hole in spindle and install cotter pin, Fig. 5.
4. Check front wheel rotation, if it rotates noisily or rough, clean, inspect or replace wheel bearings as necessary.

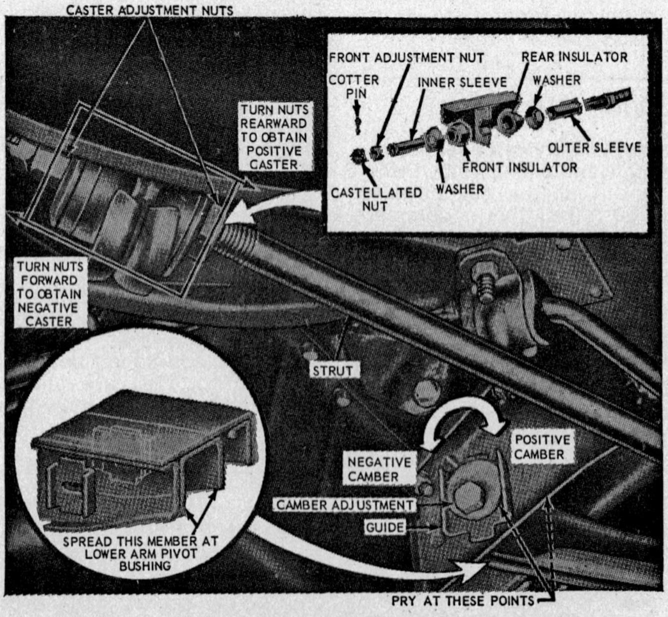

Fig. 3 Caster and camber adjustments. All except 1972-76 Elite, Torino and Montego & 1974-76 Cougar

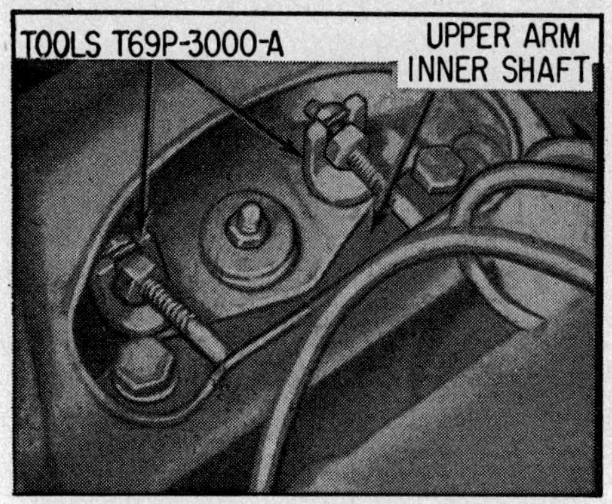

Fig. 4 Caster and Camber adjustment. 1972-76 Elite, Torino, Montego & 1974-76 Cougar

Fig. 5 Front wheel bearing adjustment

WHEEL BEARINGS, REPLACE

(Disc Brakes)

1. Raise car and remove front wheels.
2. Remove caliper mounting bolts.

NOTE: It is not necessary to disconnect the brake lines for this operation.

3. Slide caliper off of disc, inserting a clean spacer between the shoes to hold them in their bores after the caliper is removed. Position caliper out of the way.

NOTE: Do not allow caliper to hang by brake hose.

4. Remove hub and disc assembly. Grease retainer and inner bearing can now be removed.

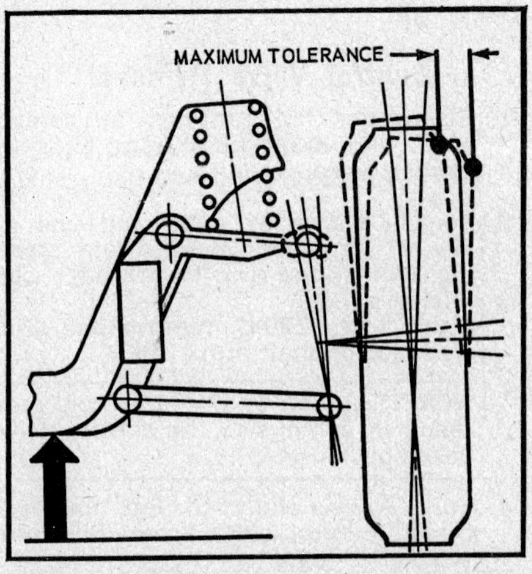

Fig. 6 Measuring upper ball joint for radial play. All except 1972-76 Elite, Montego, Torino & 1974-76 Cougar

CHECKING BALL JOINTS FOR WEAR

Upper Ball Joint

1972-76 Elite, Montego, Torino & 1974-76 Cougar

1. Raise vehicle and place floor jacks beneath lower control arms.
2. Grasp the lower edge of tire and move the wheel in and out.
3. While the wheel is being moved observe any movement between the upper end of the spindle and upper arm. If any movement is observed replace the ball joint.

Exc. 1972-76 Elite, Torino, Montego & 1974-76 Cougar

1. Raise car on frame contact hoist or by floor jacks placed beneath underbody until wheel falls to full down position as shown in Fig. 6. This will unload upper ball joint.
2. With front wheel bearings properly adjusted, attach a dial indicator to the upper control arm and position the indicator so that its plunger rests against the inner side of the wheel rim adjacent to the upper arm ball joint.
3. Grasp tire at top and bottom and slowly move it in and out, Fig. 6. Reading on dial will indicate the amount of radial play. If reading exceeds 1/4", replace the upper ball joint.

Lower Ball Joint

1972-76 Elite, Montego, Torino & 1974-76 Cougar

1. Raise vehicle and place floor jacks under the lower control arms, Fig. 7.
2. Adjust wheel bearings and place a dial indicator to the lower arm and position the indicator so that the plunger rests against the inner side of the wheel rim adjacent to the lower ball joint.
3. Grasp tire at top and bottom and move it slowly in and out. Reading on dial will indicate the amount of radial play. If reading exceeds 1/4 inch replace the ball joint.

Exc. 1972-76 Elite, Torino, Montego & 1974-76 Cougar

1. With car jacked up as directed above, grasp the lower edge of the tire and move it in and out.
2. As wheel is being moved in and out, observe lower end of spindle and lower arm.
3. Any movement between lower end of spindle and lower arm indicates ball joint wear and loss of preload. If such movement is observed, replace lower arm and/or ball joint.

NOTE: During the foregoing check, the ball joints will be unloaded and may move. Therefore disregard any movement of the upper ball joint when checking the lower ball joint and any movement of the lower ball joint when checking the upper ball joint. Also, do not mistake loose wheel bearings for a worn ball joint.

BALL JOINT, REPLACE

NOTE: Ford Motor Company recommends that new ball joints should not be installed on used control arms and that the control

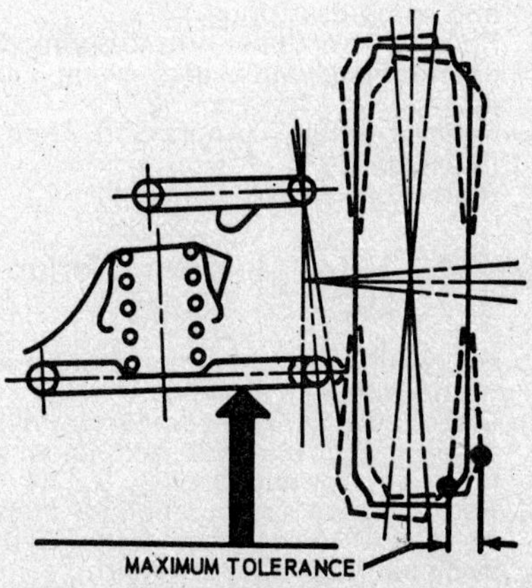

Fig. 7 Measuring lower ball joint for radial play. 1972-76 Elite, Montego, Torino & 1974-76 Cougar

arm be replaced if ball joint replacement is required. However, aftermarket ball joint repair kits which do not require control arm replacement, are available and can be installed using the following procedure.

The ball joints are riveted to the upper and lower control arms. The upper ball joint can be replaced by removing the rivets and retaining the new ball joint to the upper control arm with bolts, nuts and washers furnished with the ball joint repair kit. When removing an upper ball joint, use a suitable pressing tool to loosen the ball joint from the spindle.

SHOCK ABSORBER, REPLACE

Exc. 1972-76 Elite, Torino, Montego & 1974-76 Cougar

1. Raise hood and remove upper mounting bracket-to-spring tower retaining nuts.
2. Raise front of car and place safety stands under lower control arms.
3. Remove shock absorber lower retaining nuts and washers.
4. Lift shock absorber from spring tower.
5. Reverse procedure to install.

1972-76 Elite, Torino, Montego & 1974-76 Cougar

1. Remove upper mounting nut, washer and bushing from shock absorber.
2. Raise vehicle and install safety stands.
3. Remove the shock absorber lower retaining screws and remove shock absorber.

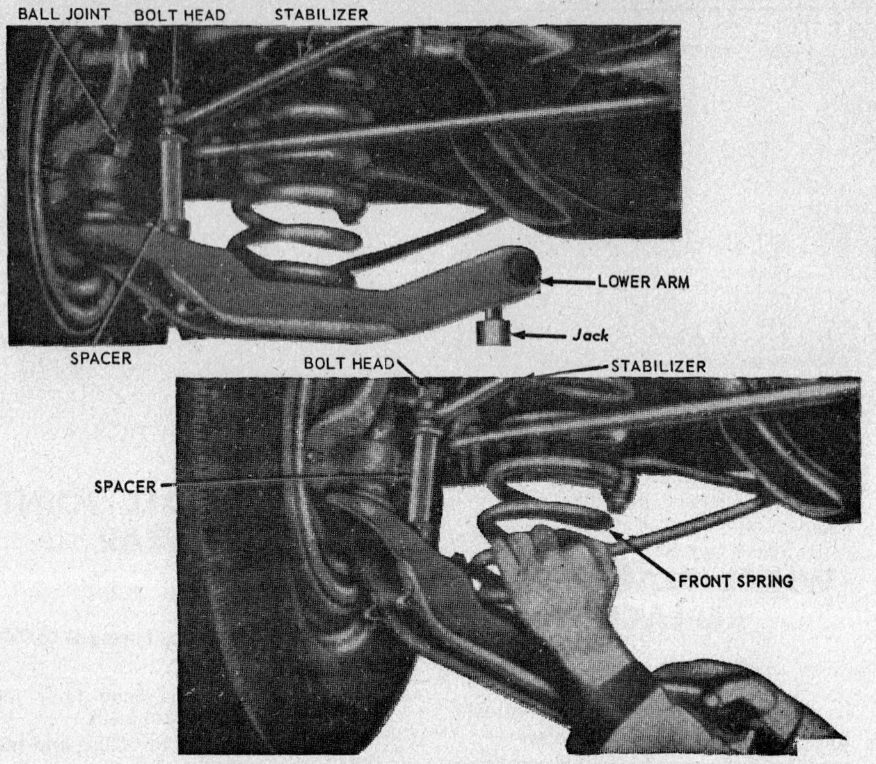

Fig. 8 Removing or installing front spring. 1972-76 Elite, Montego, Torino & 1974-76 Cougar

COIL SPRING, REPLACE

All Except 1972-76 Elite, Montego, Torino & 1974-76 Cougar

1. Remove shock absorber and upper mounting bracket as an assembly.
2. Raise car on hoist and install safety stands.
3. Remove wheel, hub and drum.
4. Install a suitable spring compressor and compress spring.
5. Remove two upper-arm-to-spring tower retaining nuts and swing upper arm outward from spring.
6. Release spring compressor. Then remove spring.
7. Reverse procedure to install.

1972-76 Elite, Montego, Torino & 1974-76 Cougar

1. Raise vehicle and support front end of frame with jack stands.
2. Disconnect shock absorber from lower arm and place a jack under the lower arm for support, Fig. 8.
3. Remove strut and rebound bumper bolts and disconnect lower end of sway bar stud from lower arm.
4. Remove the nut and bolt that retains the inner end of the lower arm to the crossmember.
5. Carefully lower jack relieving spring pressure on the lower arm, then remove the spring.

STEERING GEAR, REPLACE

1969-76 Integral Power Steering Gear

These models use an integral power steering gear which replaces the linkage power steering used previously. Remove the gear assembly as follows:

1. Disconnect pressure and return lines from gear and plug openings to prevent entry of dirt.
2. Remove two bolts that secure flex coupling to the gear and to the column.
3. Raise vehicle and remove pitman arm with suitable puller.
4. If vehicle is equipped with synchromesh transmission, remove clutch release lever retracting spring to provide clearance to remove gear.
5. Support gear and remove three gear attaching bolts.

1969-76 Manual Steering Gear

1. Remove flex coupling bolts.
2. Remove pitman arm nut and remove pitman arm from sector shaft using a puller.
3. On vehicles with manual transmission it may be necessary to disconnect the clutch linkage and on V8 models it may be necessary to lower the exhaust system.
4. Unfasten and remove steering gear.

POWER STEERING

Pump Pressure

The normal oil pressure against either steering stop with engine idling is 750 to 900 psi.

Control Valve Removal

1. Disconnect fluid fittings at control valve and drain fluid from lines by turning wheels to left and right several times.
2. Loosen clamp at right-hand end of sleeve. Remove roll pin from steering arm-to-idler arm rod through slot in sleeve.
3. Using tool 3290-C, remove ball stud from sector shaft arm.

NOTE: The use of any other tool may result in damage to the control valve assembly.

4. Turn wheels fully to left and unthread control valve from idler arm rod.

FORD MUSTANG II & PINTO MERCURY BOBCAT

NOTE: 1969-73 Mustang Models are covered elsewhere in this manual.

OLD CAR SPECIFICATIONS: For 1946-68 Tune Up and Wheel Alignment Specifications see main index.

INDEX OF SERVICE OPERATIONS

ENGINE & SERIAL NUMBER LOCATION: Vehicle warranty plate on rear face of left front door.

ENGINE IDENTIFICATION: Engine code is last letter in serial number on vehicle warranty plate.

Year	Engine	Engine Code
1971–73	4-98①	W
1971–74	4-122②	X
1974-76	4-140③	Y
1974-76	V6-171④	Z
1975-76	V8-302	F

①—1600 cc engine.
②—2000 cc engine.
③—2300 cc engine.
④—2800 cc engine.

GRILLE IDENTIFICATION

1971-73 Pinto

1974-75 Pinto

1974 Mustang II

1975 Mustang II
1976 Mustang II
Base Model

1975-76 Mercury Bobcat

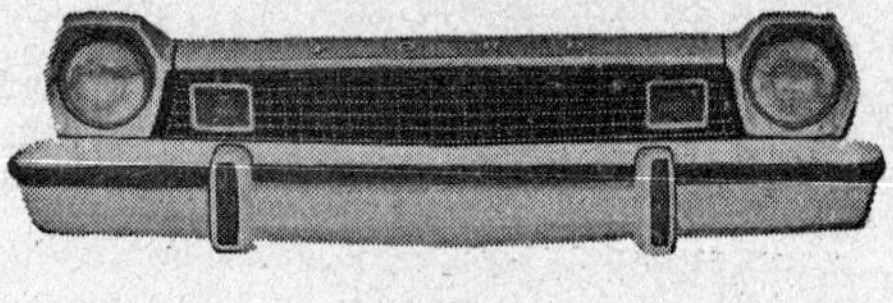

1976 Pinto

1976 Mustang II Stallion, Mach 1

BRAKE SPECIFICATIONS

Year	Model	Brake Drum Inside Diameter	Wheel Cylinder Bore Diameter			Master Cylinder Bore Diameter		
			Front Disc Brakes	Front Drum Brakes	Rear Brakes	With Disc Brakes	With Drum Brakes	With Power Brakes
1971–72	Pinto	9.0	2.125	1.00	.7187	.9375	.9375	—
1973	Pinto	9.0	2.127	1.00	.7187①	.9375	.9375	—
1974	Mustang II & Pinto	9.0	2.60	—	.875	.9375	—	.9375
1975–76	Mustang II, Pinto & Bobcat	9.0	2.60	—	.875	.9375	—	.9375

①—Sta. Wagon .875.

GENERAL ENGINE SPECIFICATIONS

Year	Engine	Carburetor	Bore and Stroke	Piston Displacement, Cubic Inches	Compression Ratio	Maximum Brake H.P. @ R.P.M.	Maximum Torque Lbs. Ft. @ R.P.M.	Normal Oil Pressure Pounds
1971	4-98①	1 Barrel	3.188 x 3.056	98①	8.4	75 @ 5000	96 @ 3000	35
	4-122②	2 Barrel	3.57 x 3.03	122②	9.0	100 @ 5600	120 @ 3600	50
1972	4-98①③	1 Barrel	3.188 x 3.056	98①	8.0	54 @ 4600	80 @ 2400	35–60
	4-122②③	2 Barrel	3.57 x 3.03	122②	8.2	86 @ 5400	103 @ 3200	35–60
1973	4-98①③	1 Barrel	3.188 x 3.056	98①	8.1	54 @ 4600	81 @ 2400	35
	4-122②③	2 Barrel	3.57 x 3.03	122②	8.2	④	98 @ 3800	50
1974	4-122②③	2 Barrel	3.575 x 3.029	122②	8.2	80 @ 5400	98 @ 3000	45–65
	4-140③⑤	2 Barrel	3.781 x 3.126	140⑤	8.4	82 @ 4600	113 @ 2600	40–60
	4-140③⑤	2 Barrel	3.781 x 3.126	140⑤	8.4	88 @ 5000	116 @ 2600	40–60
	V6-171③⑥	2 Barrel	3.66 x 2.70	171⑥	8.2	105 @ 4600	140 @ 3200	40–55
1975	4-140③⑤	2 Barrel	3.781 x 3.126	140⑤	8.4	85.5 @ 4800	113 @ 2600	40–60
	V6-171③⑥	2 Barrel	3.66 x 2.70	171⑥	8.2	110 @ 5000	135 @ 3200	40–60
	V8-302③⑦	2 Barrel	4.00 x 3.00	302	8.0	140 @ 3800	228 @ 2600	40–60
1976	4-140③⑤	2 Barrel	3.781 x 3.126	140⑤	—	—	—	—
	V6-171③⑥	2 Barrel	3.66 x 2.70	171⑥	—	—	—	—
	V8-302③⑦	2 Barrel	4.00 x 3.00	302	—	—	—	—

①—1600 cc engine.
②—2000 cc engine.
③—Net Rating—as installed in vehicle.
④—Exc. station wagon, 85 @ 5600; station wagon, 83 @5200.
⑤—2300 cc engine.
⑥—2800 cc engine.
⑦—Refer to the Ford & Mercury—Compact & Intermediate Chapter for Service procedures on this engine.

TUNE UP SPECIFICATIONS

The following specifications are published from the latest information available. This data should be used only in the absence of a decal affixed in the engine compartment.

OLD CAR SPECIFICATIONS: For 1946-68 Tune Up Specifications see main index.

★When using a timing light, disconnect vacuum hose or tube at distributor and plug opening in hose or tube so idle speed will not be affected.

●When checking compression, lowest cylinder must be within 75 percent of highest.

▲Before removing wires from distributor cap, determine location of the No. 1 wire in cap, as distributor position may have been altered from that shown at the end of this chart.

Year	Spark Plug		Distributor		Ignition Timing★			Carb. Adjustments					
								Hot Idle Speed③		Air Fuel Ratio		Idle "CO" %	
	Type	Gap Inch	Point Gap Inch	Dwell Angle Deg.	Firing Order Fig. ▲	Timing BTDC ①	Mark Fig.	Std. Trans.	Auto. Trans.②	Std. Trans.	Auto. Trans.	Std. Trans.	Auto. Trans.
1971													
4-98, 1600 cc	AGR-22	.030	.025	38–42	A	12°	C	800/500	—	—	—	1.2	—
4-122, 2000 cc	BRF-32	.025	.025	38–42	B	6°	D	750	650D	—	—	1.2	1.2
1972													
4-98, 1600 cc	AGR-22	.030	.025	36–40	A	12°	C	900/600	—	—	—	1.2	—
4-122, 2000 cc	BRF-42	.034	.025	36–40	B	6°④	D	750/500	650/500	—	—	1.2	1.2
1973													
4-98, 1600 cc	AGR-32	.034	.025	36–40	A	12°	C	900	—	11 to 1	—	1.5	—
4-122, 2000 cc	BRF-42	.030	.025	37–41	B	⑤	D	750	650	11 to 1	11 to 1	1.5	1.5
1974													
4-122, 2000 cc	BRF-42	.034	.025	35–41	B	6	D	750	650	⑥	⑥	—	—
4-140, 2300 cc	AGRF-52	.034	.027	35–41	G	6	E	850	750	11 to 1	11 to 1	.15	.15
V6-171, 2800 cc	AGR-42	.034	.027	35–41	H	12	F	750	650	12.7 to 1	12.7 to 1	.7	.4

Continued

TUNE UP SPECIFICATIONS—Continued

The following specifications are published from the latest information available. This data should be used only in the absence of a decal affixed in the engine compartment.

OLD CAR SPECIFICATIONS: For 1946-68 Tune Up Specifications see main index.

★When using a timing light, disconnect vacuum hose or tube at distributor and plug opening in hose or tube so idle speed will not be affected.

●When checking compression, lowest cylinder must be within 75 percent of highest.

▲Before removing wires from distributor cap, determine location of the No. 1 wire in cap, as distributor position may have been altered from that shown at the end of this chart.

Year	Spark Plug		Distributor		Ignition Timing★			Carb. Adjustments					
								Hot Idle Speed③		Air Fuel Ratio		Idle "CO" %	
	Type	Gap Inch	Point Gap Inch	Dwell Angle Deg.	Firing Order Fig. ▲	Timing BTDC ①	Mark Fig.	Std. Trans.	Auto. Trans.②	Std. Trans.	Auto. Trans.	Std. Trans.	Auto. Trans.
1975													
4-140, 2300 cc	AGRF-52	.034	—	—	G	⑦	E	850	750D	—	—	—	—
V6-171, 2800 cc	AGR-42	.034	—	—	H	⑧	F	850	700D	—	—	—	—
V8-302⑨	ARF-42	.044	—	—	I	6°	J	—	700D	—	—	—	—
1976													
4-140, 2300 cc	AGRF-52	.034	—	—	G	⑦	E	750	650D	—	—	—	—
V6-171, 2800 cc	AGR-42	.034	—	—	H	⑧	F	850	700D	—	—	—	—
V8-302⑨	ARF-42	.044	—	—	I	6°	J	—	700D	—	—	—	—

①—BTDC: Before top dead center.
②—D: Drive.
③—Headlamps on Hi Beam—Air Conditioner OFF. Where two speeds are listed, lower speed indicates solenoid disconnected.
④—California vehicles with Auto. Trans. 9°.
⑤—Auto. trans., 9°; man. trans., 6°.
⑥—11.7:1 Except Calif.; 10:1 Calif.
⑦—Exc. Calif. Auto. Trans., 6° BTDC; Calif. Auto. Trans., 10° BTDC.
⑧—Manual Trans., 6° BTDC; Exc. Calif. Auto. Trans., 10° BTDC; Calif. Auto. Trans., 8° BTDC.
⑨—Refer to the Ford & Mercury—Compact & Intermediate Chapter for service procedures on this engine.

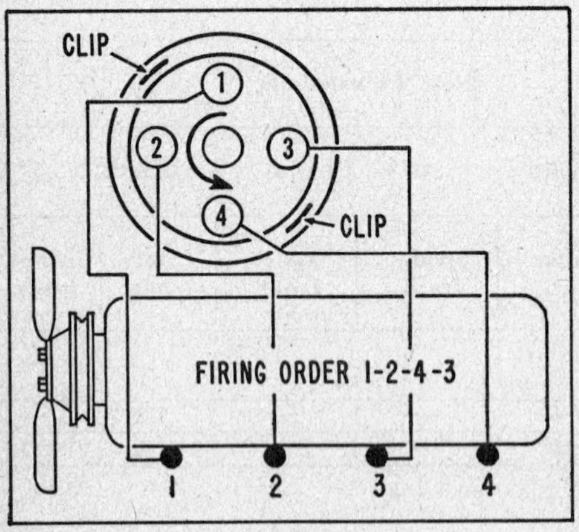

Fig. A

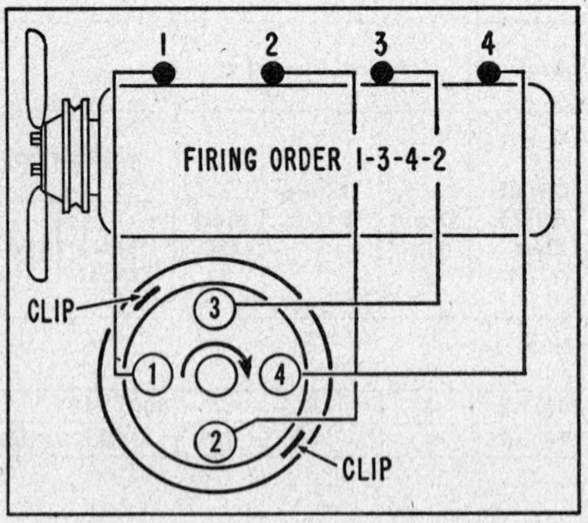

Fig. B

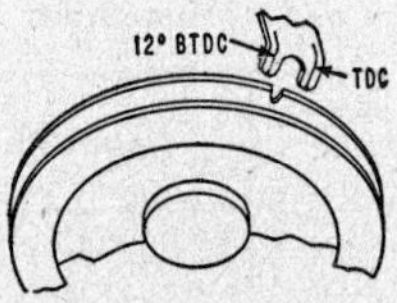

Fig. C

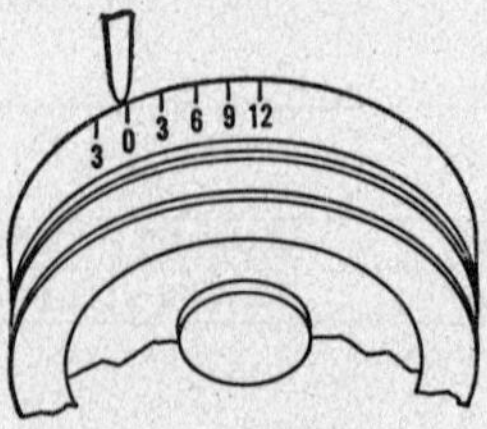

Fig. D

Continued

TUNE UP NOTES—Continued

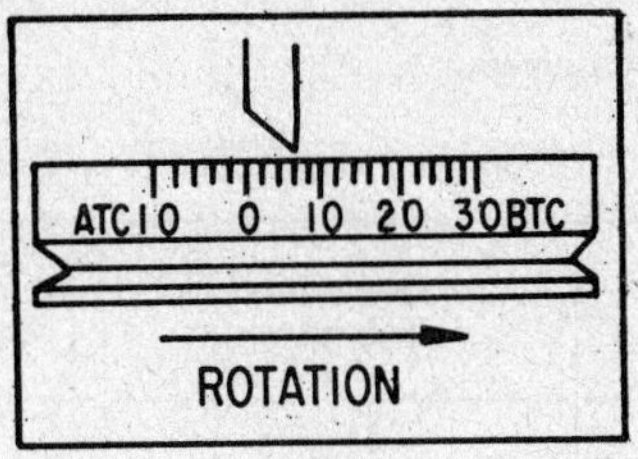

Fig. E

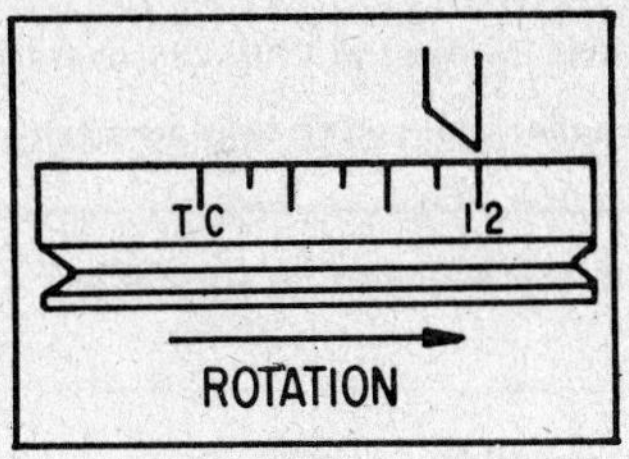

Fig. F

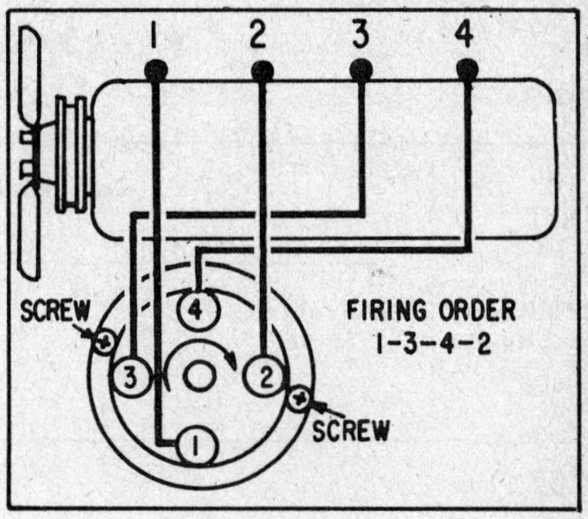

Fig. G

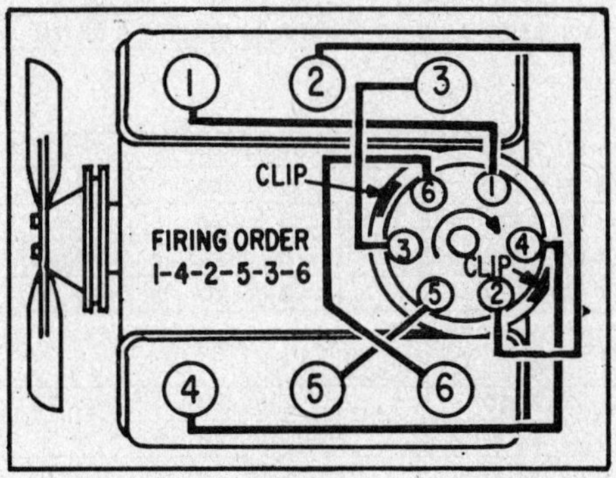

Fig. H

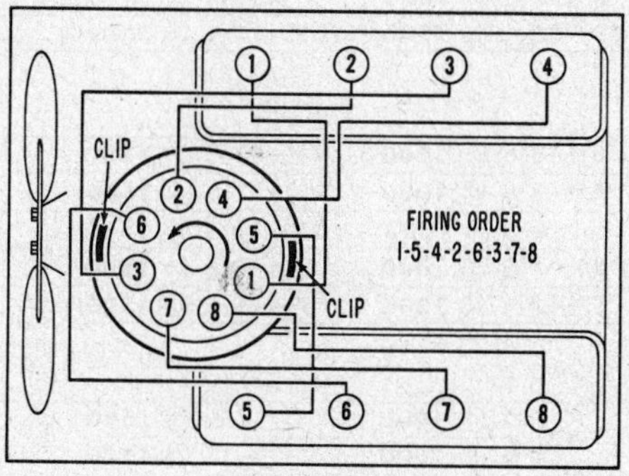

Fig. I

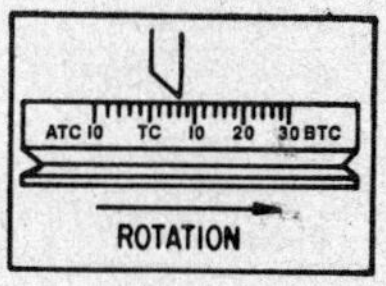

Fig. J

VALVE SPECIFICATIONS

Year	Engine	Valve Lash		Valve Angles		Valve Spring Installed Height	Valve Spring Pressure Lbs. @ In.	Stem Clearance		Stem Diameter, Standard	
		Int.	Exh.	Seat	Face			Intake	Exhaust	Intake	Exhaust
1971	4-98	.010H	.017H	45	45	1.263	118 @ .957	.0008–.0030	.0017–.0039	.3100	.3100
	4-122	.008①	.010①	45	45	1.417	144 @ 1.059	.0015–.0025	.0015–.0025	.3149	.3149
1972	4-98	.010H	.017H	44	45	1.263	122 @ .953	.0008–.0027	.0017–.0036	.3102	.3093
	4-122	.008①	.010①	44	45	1.418	176 @ 1.02	.0008–.0025	.0018–.0035	.3162	.3152
1973	4-98	.010H	.017H	45	44	1.266	118 @ .957	.0008–.0027	.0017–.0036	.3102	.3093
	4-122	.008①	.010①	45	44	1.418	176 @ 1.02	.0008–.0025	.0018–.0035	.3162	.3152
1974	4-122	.008①	.010①	45	44	1.406	176 @ 1.02	.0008–.0025	.0018–.0035	.3159–.3167	.3149–.3156
	4-140	.008C②	.010C②	45	44	1.56	210 @ 1.16	.0010–.0027	.0015–.0022	.3416–.3423	.3418–.3411
	6-171	.014H	.016H	45	44	1.593	144 @ 1.222	.0008–.0025	.0018–.0035	.3167–.3159	.3156–.3149
1975–76	4-140	.008C②	.010C②	45	44	1.56	189 @ 1.16	.0010–.0027	.0015–.0032	.3416–.3423	.3411–.3418
	V6-171	.014H	.016H	45	44	1.593	144 @ 1.222	.0008–.0025	.0018–.0035	.3158–.3167	.3149–.3156
	V8-302⑤	.115—.165⑥		45	45	③	④	.0017–.0027	.0015–.0032	.3416–.3423	.3411–.3418

①—Set Hot or Cold.
②—With hydraulic valve lash adjuster completely collapsed.
③—Intake 1.69; Exhaust 1.60.
④—Intake 200 @ 1.31; Exhaust 200 @ 1.22.
⑤—Refer to the Ford & Mercury—Compact & Intermediate Chapter for service procedures on this engine.
⑥—Clearance is obtained at valve stem tip with hydraulic lifter collapsed. If clearance is less than the minimum install an undersize push rod; if clearance is greater than the maximum install an oversize push rod.

DISTRIBUTOR SPECIFICATIONS

★If unit is checked on vehicle double the RPM and degrees to get crankshaft figures.

Breaker arm spring tension—17–21.

Distributor Part No.①	Centrifugal Advance Degrees @ RPM of Distributor					Vacuum Advance		Distributor Retard
	Advance Starts	Intermediate Advance			Full Advance	Inches of Vacuum To Start Plunger	Max. Adv. Dist. Deg. @ Vacuum	Max. Ret. Dist. Deg. @ Vacuum
1971								
DIFZ-A	0–½ @ 350	3¾–5¾ @ 750	6¼–8¼ @ 1000	7½–9½ @ 1500	11¾ @ 2500	5	7½ @ 25	4 @ 20
DIFZ-B	0–½ @ 400	1½–3½ @ 750	5½–7½ @ 1000	7¾–9¾ @ 1500	14 @ 2500	5	8 @ 25	7 @ 20
DIFZ-C	0–½ @ 350	3–5 @ 1000	8½–10½ @ 1500	10½–12½ @ 2000	12½ @ 2500	5	10 @ 25	4 @ 20
1972								
72HF-DC	½–2½ @ 500	5½–7½ @ 750	6½–8½ @ 1000	9½–11½ @ 1500	14 @ 2000	5	7½ @ 20	—
72HF-FA	½–2½ @ 500	5½–7½ @ 750	6½–8½ @ 1000	9½–11½ @ 1500	14 @ 2000	5	7½ @ 20	4 @ 20
72HF-SC	0–½ @ 500	0–1 @ 750	3–5 @ 1000	8½–10½ @ 1500	15½ @ 2000	5	7½ @ 20	—
72IF-TA	0–½ @ 500	1½–3½ @ 750	5½–7½ @ 1000	7½–9½ @ 1500	12 @ 2000	5	8 @ 20	7 @ 20
72IF-YA	0–½ @ 500	1½–3½ @ 750	5½–7½ @ 1000	7½–9½ @ 1500	12 @ 2000	5	8 @ 20	7 @ 20
1973								
72IF-AHA	0–½ @ 500	1–3 @ 750	5–7 @ 1000	7–9 @ 1500	11½ @ 2000	5	7 @ 20	7 @ 20
73HF-AA	½–2½ @ 500	5–7 @ 750	8½ @ 1000	9–11 @ 1500	14 @ 2000	5	7½ @ 20	—
73HF-BA	0–1 @ 500	1–3 @ 750	3–5 @ 1000	7½–9½ @ 1500	14 @ 2000	5	7½ @ 20	—
73HF-GA	0–½ @ 500	½–2½ @ 750	2½–4½ @ 1000	5½–7½ @ 1500	11 @ 2000	5	7½ @ 20	—
1974								
74HF-EA	1–3 @ 500	5–7 @ 750	6½–8½ @ 1000	9½–11½ @ 1500	12–14 @ 2000	5	6 @ 20	—
74HF-LA	0–1 @ 500	1–3 @ 750	4–6 @ 1000	7–9 @ 1500	9½–11½ @ 2000	5	6 @ 20	—
D4ZE-AA	0–1 @ 500	4–6 @ 750	4–6 @ 1000	9½–11½ @ 1500	11½–14 @ 2000	5	7½ @ 20	—
D4ZE-BA	0–1 @ 500	4–6 @ 750	7–9 @ 1000	9¼–11½ @ 1500	11½–14 @ 2000	5	7½ @ 20	—
D4ZF-DA	0–1 @ 500	1½–3½ @ 750	6–8 @ 1000	9½–11½ @ 1500	11½–14 @ 2000	5	4½ @ 20	—
74TF-LA	0–½ @ 500	1–3 @ 750	3½–5½ @ 1000	7½–9½ @ 1500	8–10 @ 2000	5	4 @ 20	7 @ 20
74TF-MA	0–½ @ 500	1–3 @ 750	3½–5½ @ 1000	7½–9½ @ 1500	8–10 @ 2000	5	4 @ 20	4 @ 20
D4ZE-KA	0–1 @ 500	4–6½ @ 750	7–9 @ 1000	9¼–11½ @ 1500	12¾ @ 2000	5	6 @ 20	—
74TF-SA	0–½ @ 500	1–3 @ 750	3½–5½ @ 1000	7½–9½ @ 1500	9 @ 2000	5	3 @ 20	—
1975								
D5DE-KA	0–2 @ 775	2¾–4¾ @ 950	4¾–7 @ 1500	6¾–9¼ @ 2000	10¾ @ 2350	4	11¼ @ 14½	3 @ 7
D5DE-NA	0–2 @ 550	2½–4½ @ 675	3¾–6 @ 1100	7–9½ @ 2000	11½ @ 2500	4	11¼ @ 14½	—
D5ZE-AA	0–2 @ 600	1–3 @ 650	2¾–5 @ 1000	5¼–7½ @ 1500	13 @ 2500	4.8	5¾ @ 7½	—
D52E-EA	0–2 @ 675	5¼–7¼ @ 1025	7½–9¾ @ 1500	10–12½ @ 2000	14 @ 2250	4	5¼ @ 7	—
D52E-FA	0–2 @ 725	5½–7½ @ 1125	7¾–10 @ 1500	10¾–13¼ @ 2000	14 @ 2150	4	5¼ @ 7½	3 @ 7½
75TF-EA	0–2 @ 650	2–4 @ 800	5–7 @ 1200	8–10 @ 1600	10 @ 2000	4¼	3 @ 6¾	6 @ 6¾
1976								
D52E-EA	0–2 @ 675	5¼–7¼ @ 1025	7½–9¾ @ 1500	10–12½ @ 2000	14 @ 2250	4	5¼ @ 7	—
D52E-FA	0–2 @ 725	5½–7½ @ 1125	7¾–10 @ 1500	10¾–13¼ @ 2000	14 @ 2150	4	5¼ @ 7½	4 @ 7½
D5ZE-AA	0–2 @ 600	1.3 @ 650	2¼–5 @ 1000	5¼–7½ @ 1500	10¼–13 @ 2500	4.8	5¼ @ 7½	—
75TF-EA	0–2 @ 650	2.4 @ 800	5–7 @ 1200	8–10 @ 1600	7½–10 @ 2000	4¼	3 @ 6¾	7 @ 10
75TF-NA	0–2 @ 650	2.4 @ 800	5¼–7¾ @ 1200	7–9½ @ 1500	9–12 @ 2500	4¼	7 @ 9½	—

①—Basic part No. 12127.

ALTERNATOR & REGULATOR SPECIFICATIONS

Year	Make or Model	Current Rating		Field Current @ 75°F.		Voltage Regulator				Field Relay	
		Amperes	Volts	Amperes	Volts	Make or Model	Voltage @ 75°F.	Contact Gap	Armature Air Gap	Armature Air Gap	Closing Voltage @ 75°F.
1971–72	D0ZF-B	38	15	2.4	12	D0AF-A	13.5–15.3	—	—	—	2.0–4.2
	D0AF-G	42	15	2.9	12	D0AF-A	13.5–15.3	—	—	—	2.0–4.2
1973	Orange①	42	15	2.9	12	D3TZ-A	13.5–15.3	—	—	—	2.5–4.0
	Purple①	38	15	2.4	12	D3TZ-A	13.5–15.3	—	—	—	2.5–4.0
	Green①	61	15	2.9	12	D3TZ-A	13.5–15.3	—	—	—	2.5–4.0
	All	70	15	2.9	12	D3TZ-A	13.5–15.3	—	—	—	2.5–4.0
1974–76	Purple①	38	15	2.9	12	D4AF-AA	13.5–15.3	—	—	—	2.5–4.0
	Green①	61	15	2.9	12	D4TF-AA	13.5–15.3	—	—	—	2.5–4.0
	All	70	15	2.9	12	D4TF-AA	13.5–15.3	—	—	—	2.5–4.0
	Orange①	42	15	2.9	12	D4AF-AA	13.5–15.3	—	—	—	2.5–4.0
	Red①	55	15	2.9	12	D4AF-AA	13.5–15.3	—	—	—	2.5–4.0
	All	40	15	2.9	12	D4AF-AA	13.5–15.3	—	—	—	2.5–4.0
	All	60	15	2.9	12	D4TF-AA	13.5–15.3	—	—	—	2.5–4.0

①—Color of identification tag.

STARTING MOTOR SPECIFICATIONS

Year	Engine Model	Ident. No.	Brush Spring Tension, Ounces	No Load Test			Torque Test		
				Amperes	Volts	R.P.M.	Amperes	Volts	Torque Lbs. Ft.
1971	All	D12F-AA	40 Min.	70	—	—	460	5	9
1971–72	All	D22F-AA	40 Min.	70	—	—	460	5	9
1972–73	All	D22F-AB	40 Min.	70	—	—	460	5	9
1973	All	D32F-AA	40 Min.	70	—	—	460	5	9
1974	4-122, 140	D42F-AA,AB,BA	40 Min.	70	—	—	460	5	9
	6-171	D4ZF-AA	40 Min.	70	—	—	460	5	9
1975–76	4-140	D42F-BA	40 Min.	70	—	—	460	5	9
	V6-171	D4ZF-BA	40 Min.	70	—	—	460	5	9
	V8-302	D5OF-BA	80 Min.	80	—	—	670	5	15.5

REAR AXLE SPECIFICATIONS

Year	Model	Carrier Type	Ring Gear & Pinion Backlash Inch	Nominal Pinion Locating Shim, Inch	Pinion Bearing Preload				Differential Bearing Preload	Pinion Nut Torque Ft.-Lbs.
					New Bearings With Seal Inch-Lbs.	Used Bearings With Seal Inch-Lbs.	New Bearings Less Seal Inch-Lbs.	Used Bearings Less Seal Inch-Lbs.		
1971	All	Integral	.008–.012	.030	17–32	6–12	—	—	.008–.012①	140
1972	All	Integral	.008–.012	.030	15–30	6–12	—	—	.004–.008①	140
1973	All	Integral	.008–.012	.030	19–27	6–12	—	—	.004–.008①	140
1974–76	All	Integral	.008–.012	.030	17–27	6–12	—	—	.008–.012①	140
	All	Removable	.008–.012	.022	17–27	8–14	—	—	.004–.008①	170

①—Case spread with new bearings; with used bearings, .003–.005″.

ENGINE TIGHTENING SPECIFICATIONS

★Torque specifications are for clean and lightly lubricated threads only. Dry or dirty threads produce increased friction which prevents accurate measurement of tightness.

Year	Engine	Spark Plugs Ft. Lbs.	Cylinder Head Bolts Ft. Lbs.	Intake Manifold Ft. Lbs.	Exhaust Manifold Ft. Lbs.	Rocker Arm Shaft Bracket Ft. Lbs.	Rocker Arm Cover Ft. Lbs.	Connecting Rod Cap Bolts Ft. Lbs.	Main Bearing Cap Bolts Ft. Lbs.	Flywheel to Crank-shaft Ft. Lbs.	Vibration Damper or Pulley Ft. Lbs.
1971–73	4-98	22–28	65–70	①	②	25–30	2½–3½	30–35	65–70	50–55	24–28
	4-122	14–20	65–80	12–15	15–18	③	4–6	29–34	65–75	47–51	39–43
1974–76	4-122	14–20	65–80	12–15	15–18	③	4–6	29–34	65–75	47–51	39–43
	4-140	10–15	80–90	14–21	16–23	—	4–7	30–36	80–90	54–64	80–114
	6-171	14–22	65–80	15–18	14–18	9–25④	3–5	21–25	65–75	47–51	92–103
1975–76	8-302⑤	15–20	65–72	17–25	12–16	17–23④	3–5	19–24	60–70	75–85	70–90

①—Studs 9-12 ft.-lbs., nuts 12-15 ft.-lbs., bolts 12-15 ft.-lbs.
②—Studs 9-12 ft.-lbs., nuts 15-18 ft.-lbs.
③—Rocker arm ball stud nut 32–36 ft.-lbs.
④—Rocker arm stud nut.
⑤—Refer to the Ford & Mercury—Compact & Intermediate Chapter for service procedures on this engine.

PISTONS, PINS, RINGS, CRANKSHAFT & BEARINGS

Year	Engine	Piston Clearance	Ring End Gap①		Wrist-pin Diam-eter	Rod Bearings		Main Bearings			
			Comp.	Oil		Shaft Diameter	Bearing Clearance	Shaft Diameter	Bearing Clearance	Thrust on Bear. No.	Shaft End Play
1971–72	4-98	②	.009	.009	.8119	1.9372	.0004–.0024	2.1257	.0004–.0024	3	.003–.011
	4-122	.001–.002	.0189	.016	.9475	2.0468	.0006–.0026	2.2441	.0006–.0019	3	.004–.008
1973	4-98	②	.009	.009	.8119	1.9372	.0004–.0024	2.1257	.0005–.0016	3	.003–.011
	4-122	0.001–0.002	.015	.016	.9448	2.0468	.0006–.0026	2.2436	.0006–.0016	3	.003–.011
1974	4-122	.0010–.0020	.015	.016	.9448	2.0468	.0006–.0026	2.2446	.0006–.0016	3	.003–.011
	4-140	.0013–.0021	.010	.015	.9120	2.0468	.0009–.0027	2.3996	.0008–.0015	3	.004–.008
	6-171	.0010–.0020	.015	.015	.9447	2.1268	.0006–.0022	2.2437	.0006–.0019	3	.003–.011
1975–76	4-140	.0023–.0039	.010	.015	.9120	2.0472	.0008–.0024	2.3990	.0008–.0026	3	.004–.008
	V6-171	.0020–.0036	.015	.016	.9456	2.1256	.0006–.0021	2.2437	.0006–.0019	3	.004–.008
	V8-302③	.0018–.0026	.010	.010	.9122	2.1236	.0008–.0026	2.2486	.0005–.0024	3	.004–.008

①—Fit rings in tapered bores for clearance listed in tightest portion of ring travel.
②—1971, .0016–.0022; 1972–73, No. 1, 2 & 3 bore, .0016–.0022, No. 4 bore, .0019–.0025.
③—Refer to the Ford & Mercury—Compact & Intermediate chapter for service procedures on this engine.

COOLING SYSTEM & CAPACITY DATA

Year	Model or Engine	Cooling Capacity, Qts.			Radiator Cap Relief Pressure, Lbs.		Thermo. Opening Temp.	Fuel Tank Gals.	Engine Oil Refill Qts. ①	Transmission Oil			Rear Axle Oil Pints
		No Heater	With Heater	With A/C	With A/C	No A/C				3 Speed Pints	4 Speed Pints	Auto. Trans. Qts.	
1971	4-98	6	6¾	—	—	12–15	186	11	3½②	—	2½	—	2.2
	4-122	6¾	7½	7½	12–15	12–15	186	11	5	—	2½	8	2.2
1972–73	4-98	6¾	7¾	—	—	12–15	186	11	3½②	—	2½	—	2¼
	4-122	7½	8½	8½	12–15	12–15	186	11	5	—	2½	8	2¼
1974	4-122	7½	8½	8½	12–16	12–16	186	③	5	—	2.8	8	3
	4-140	7½	8½	9½	12–16	12–16	186	③	5	—	⑤	8	3
	6-171	11½	12½	12½	12–16	12–16	186	13	5②	—	4.5	8	3
1975	4-140	—	8¾	9	12–16	12–16	191	④	4½②	—	⑤	⑥	⑦
	V6-171	—	12½	13¼	12–16	12–16	186	④	5②	—	⑤	⑥	⑦
	V8-302	—	16½	16½	12–16	12–16	191	16½	5①	—	—	—	4
1976	4-140	7.9	8.7	9	12–16	12–16	191	④	4½②	—	⑤	⑥	⑦
	V6-171	11.7	12½	13.2	12–16	12–16	186	④	5②	—	⑤	⑥	⑦
	V8-302	15½	16.3	16.3	12–16	12–16	191	16½	5①	—	—	⑨	4

①—Includes 1 qt. for filter.
②—Includes ½ qt. for filter.
③—Pinto: sedan, 13; wagon, 12.
④—Exc. Sta. Wag., 13 gals.; Sta. Wag. 14 gals. Add 3½ gals. with auxiliary fuel tank.
⑤—Pinto, 2.8 pts.; Mustang II, 3½ pts.
⑥—C3 trans., 8 qts.; C4 trans., 9 qts.
⑦—Pinto, 2.2 pts.; Mustang II, 3 pts.; units with 8 inch ring gear, 4 pts.

WHEEL ALIGNMENT SPECIFICATIONS

OLD CAR SPECIFICATIONS: For 1946-68 Wheel Alignment Specifications see main index.

Year	Model	Caster Angle, Degrees		Camber Angle, Degrees				Toe-In. Inch	Toe-Out on Turns, Deg.	
				Limits		Desired				
		Limits	Desired	Left	Right	Left	Right		Outer Wheel	Inner Wheel
1971–72	Exc. Wagon	+1 to +2	+1½	0 to +1½	0 to +1½	+¾	+¾	3/16	18.95	20
	Wagon	−½ to +3½	+1½	−¼ to +1¾	−¼ to +1¾	+¾	+¾	3/16	18.95	20
1973	Exc. Wagon	−1 to +3	+1	−¼ to +1¾	−¼ to +1¾	+¾	+¾	⅛	—	—
	Wagon	−1½ to +2½	+½	−¼ to +1¾	−¼ to +1¾	+¾	+¾	⅛	—	—
1974	Mustang II	−¼ to +1¾	+¾	−½ to +1½	−½ to +1½	+½	+½	⅛	18.84	20
	Pinto	−¾ to +3¼	+1¼	−¼ to +1¾	−¼ to +1¾	+¾	+¾	¼	18.84	20
1975–76	Bobcat & Pinto①	+½ to +2	+1¼	0 to +1½	0 to +1½	+¾	+¾	⅛–⅜	18.84	20
	Bobcat & Pinto②	+¾ to +2¼	+1½	0 to +1½	0 to +1½	+¾	+¾	⅛–⅜	18.84	20
	Mustang II	+⅛ to +1⅝	+⅞	−¼ to +1¼	−¼ to +1¼	+½	+½	0–⅛	18.84	20

①—Exc. sta. wag.
②—Sta. wag.

Electrical Section

DISTRIBUTOR, REPLACE

1600 cc Engine

1. Remove distributor cap and disconnect vacuum lines from distributor.
2. Scribe a mark on distributor body and the engine block indicating the position of the body in the block, and scribe another mark on the distributor body showing position of rotor.
3. Remove hold down bolt and remove distributor.

Exc. 1600 cc Engine

NOTE: On some 1974-76 2000 & 2300 cc engines, it may be necessary to position Thermactor pump aside to gain access to distributor.

1. Remove distributor cap and disconnect vacuum lines from distributor.
2. Rotate the engine until the timing notch in the pulley is at the specified BTC mark. Check that No. 1 cylinder is on compression stroke by removing oil filler cap. The cam lobe should be visible through the opening. The distributor rotor should now be pointing at the No. 1 cylinder mark on the distributor body.
3. Remove hold down bolt and remove distributor.

NOTE: *The hex shaft that drives the oil pump may stick in the distributor and be withdrawn from the pump.*

STARTER, REPLACE

1973-76

1. Disconnect battery ground cable.
2. Raise vehicle and remove four bolts from crossmember under bell housing.
3. Remove flex coupling clamping screw from steering gear, then the three bolts attaching steering gear to crossmember.
4. Disconnect steering gear from flex coupling and pull gear down to gain access to starter.
5. Disconnect starter cable and remove three attaching bolts and starter.

1971-72

1. Raise vehicle on hoist.
2. Disconnect starter cable from starter terminal.
3. Remove starter attaching bolts. Then move starter forward over the steering linkage and remove starter.

IGNITION LOCK

1971

1. Disconnect the battery ground cable.

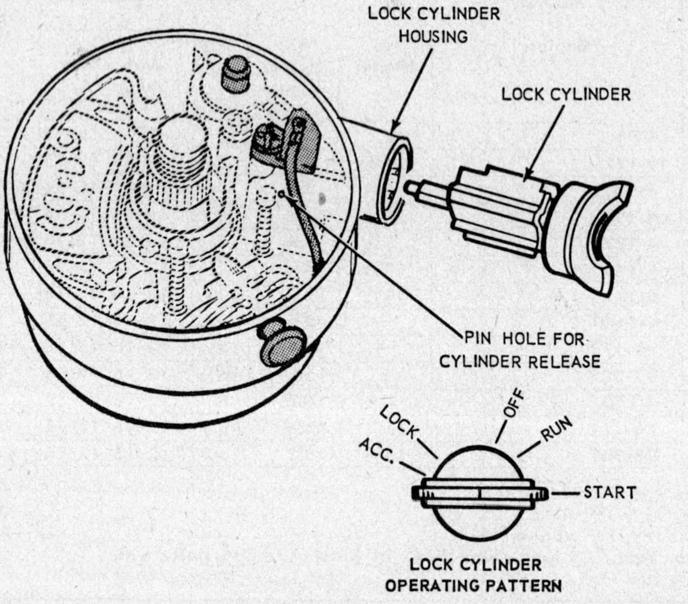

Fig. 1 Ignition lock

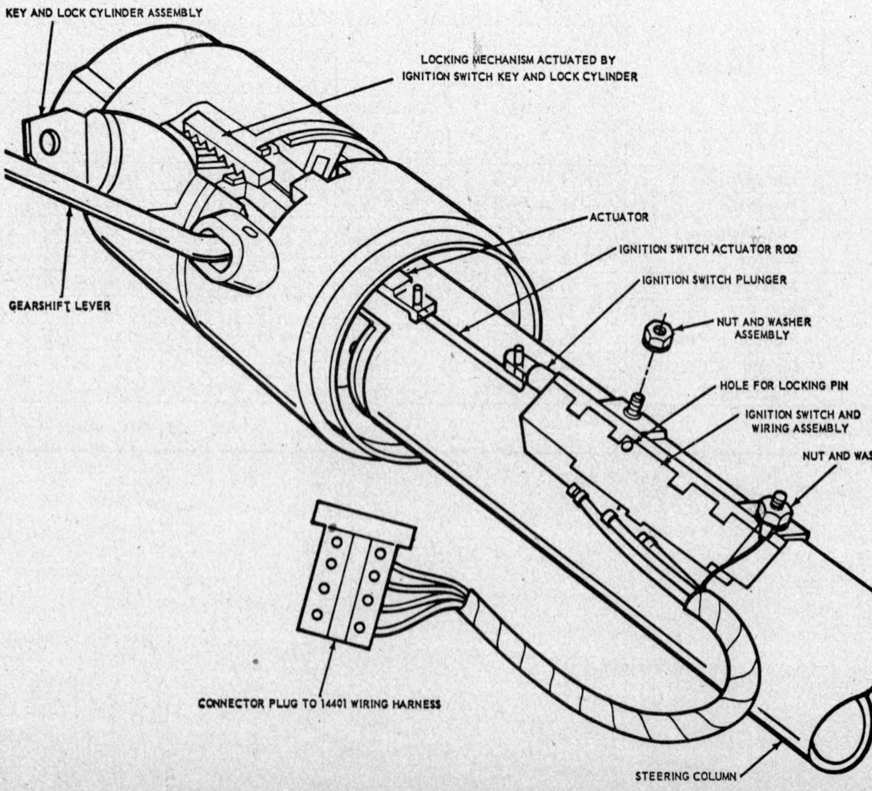

Fig. 1A Ignition switch installation. 1971-76 Bobcat, Pinto & 1974-76 Mustang II

2. Remove the steering wheel horn button or trim pad and steering wheel. Insert a wire pin in the lock cylinder hole located inside the column halfway down the housing, Fig. 1.
3. With the ignition key, turn the lock cylinder to the *ON* position.
4. Remove the lock cylinder by depressing the wire pin and pulling up on the cylinder; then, remove the wire pin.
5. To install insert the lock cylinder into the cylinder housing in the flange casting and turn the key to the *OFF* position. This action will extend the lock cylinder retaining pin into the flange.
6. With the key in the lock, rotate the cylinder to insure correct operation in all positions.
7. Replace the steering wheel and trim pad or horn button. Reconnect the battery cable.

1972-76

1. Disconnect the battery ground cable.
2. Remove the steering wheel trim pad and the steering wheel. Insert a wire pin in the hole located inside the column halfway down the lock cylinder housing, Fig. 1.
3. Place the gear shift lever in *PARK* (with auto. trans) or *REVERSE* (with manual trans) turn the lock cylinder with the ignition key to *RUN* position.
4. Depress the wire while pulling up on the lock cylinder to remove. Remove the wire pin.
5. To install insert the lock cylinder into the housing in the flange casting, and turn the key to *OFF* position. This action will extend the cylinder retaining pin into the cylinder housing.
6. Turn the key to check for correct operation in all positions.
7. Install the steering wheel and trim pad. Reconnect the battery ground cable.

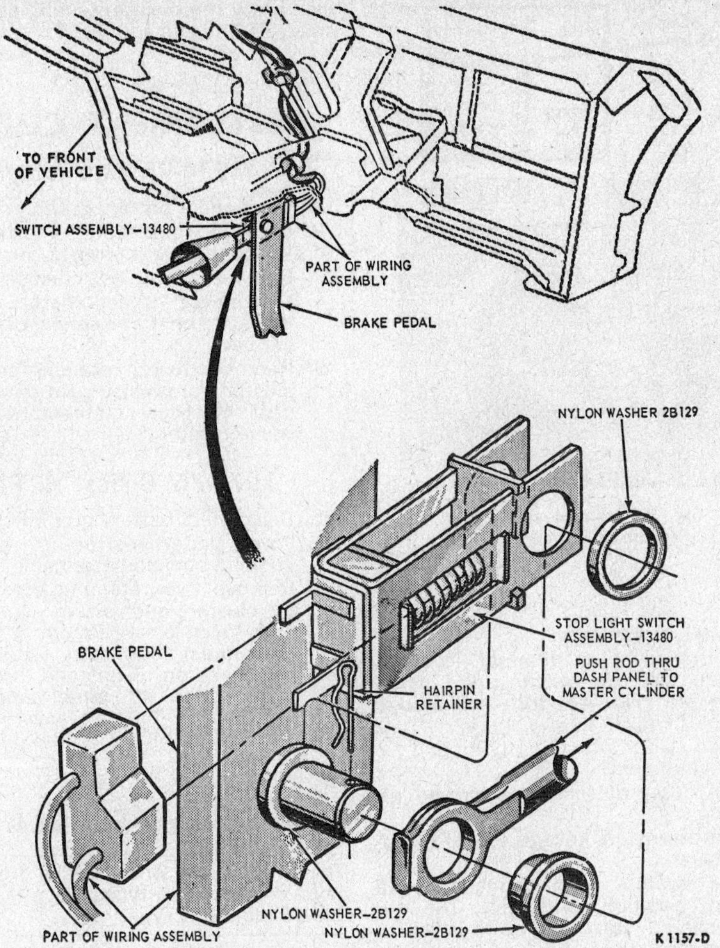

Fig. 2 Stoplight switch installation

IGNITION SWITCH, REPLACE

1. Remove shrouding from steering column and detach and lower steering column from brake support bracket.
2. Disconnect battery ground cable.
3. Disconnect switch wiring at plug, Fig. 1A.
4. Remove two nuts that retain switch to column.
5. Remove the pin that connects switch plunger directly to actuator and remove the switch.
6. To install switch, both the locking mechanism at top of column and the switch must be in LOCK position for correct adjustment.
7. Move shift lever into Park (with automatic transmission) or Reverse (with manual transmission), turn the key to LOCK position and remove the key.

NOTE: *New switches, when received, are already pinned in LOCK position by a plastic shipping pin inserted in a locking hole on top of switch.*

8. Position the hole in the end of switch plunger to the hole in the actuator and install the connecting pin.
9. Position switch on column and install retaining nuts, but do not tighten them.
10. Move switch up and down along column to locate the mid-position of rod lash and then tighten nuts.
11. Remove the plastic or substitute locking pin, connect battery and check switch for proper start in PARK or NEUTRAL.

LIGHT SWITCH, REPLACE

1974-76 Mustang II

1. Disconnect battery ground cable.
2. Depress shaft release button by inserting a screwdriver through hole in underside of instrument panel, then remove knob and shaft.
3. Remove bezel nut, lower switch, disconnect electrical connector and remove switch.

1971-76 Bobcat & Pinto

1. Disconnect battery ground cable.
2. Remove instrument cluster as described further on.
3. Remove headlight switch knob and shaft assembly and retaining nut.
4. Disconnect connector plug from switch and remove switch from cluster opening.

STOP LIGHT SWITCH REPLACE

1. Disconnect wires at connector.
2. Remove hairpin retainer, slide switch, push rod and nylon washers and bushing away from the pedal and remove the switch, Fig. 2.

NEUTRAL SAFETY SWITCH, REPLACE

1. Remove downshift linkage rod from transmission downshift lever.
2. Remove downshift outer lever retaining nut and lever.
3. Remove two switch attaching bolts.
4. Disconnect wire connector and remove switch.

Installation

1. Install switch on transmission and re-

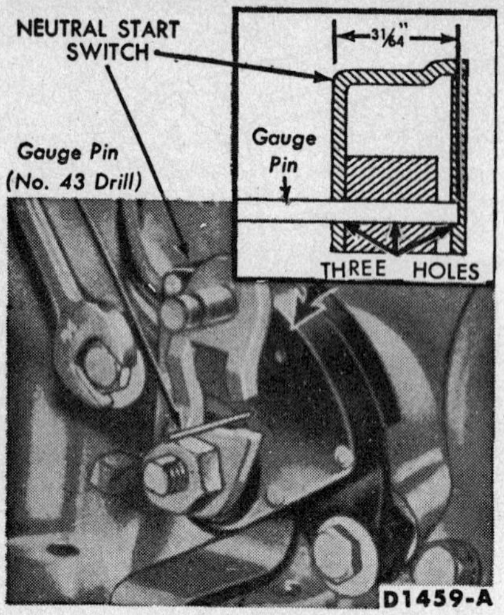

Fig. 3 Neutral safety switch adjustment

place attaching bolts.

2. With transmission manual lever in neutral, rotate switch and install gauge pin (No. 43 drill) into gauge pin hole, Fig. 3.
3. Tighten switch attaching bolts and remove gauge pin.
4. Install outer downshift lever and attaching nut.
5. Install downshift linkage rod to downshift lever.
6. Install switch wire connector and check operation of switch. The engine should start only with lever in Neutral or Park.

TURN SIGNAL SWITCH, REPLACE

1. Disconnect battery ground cable.
2. Remove horn button by pressing down and turning counterclockwise.
3. Remove steering wheel.
4. Remove turn signal switch lever by unscrewing it from column.
5. Snap off lower column shroud.
6. Remove two screws attaching upper column shroud and remove.
7. Disconnect main wiring harness from steering column connector.
8. Remove steering column wiring connector from metal retainer.
9. Remove plastic clip holding ignition and turn signal wire harnesses together to metal retainer.
10. Remove buzzer terminal from hard shell connector.
11. Remove one screw from buzzer terminal at column and three screws holding signal switch to casting.
12. Pull switch and wiring up out of column.

HORN SOUNDER

The horn sounder can be removed by depressing it and turning counter-clockwise or by removing retaining screws from under steering wheel spokes.

INSTRUMENT CLUSTER

1974-76 Mustang II

1. Disconnect battery ground cable.
2. Remove wiper switch and light switch knobs, then light switch bezel.
3. Remove instrument cluster trim cover.
4. Disconnect speedometer cable by pressing on flat section of quick disconnect.
5. Remove cluster retaining screws, pull cluster from instrument panel, disconnect electrical connectors, then remove cluster.

1971-76 Bobcat & Pinto

1. Disconnect battery ground cable.
2. From under instrument panel, disconnect speedometer cable.
3. Remove two retaining screws at top of cluster and swing cluster down away from panel, Fig. 4.
4. Disconnect connector from printed circuit, then disengage brackets on lower edge of panel and remove cluster.

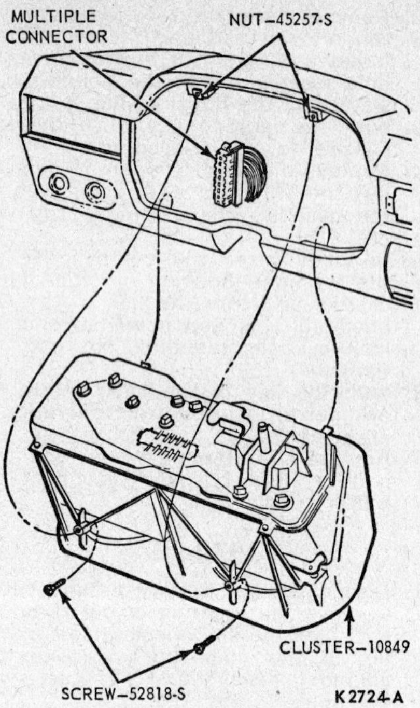

Fig. 4 Instrument cluster removal. 1971-76 Bobcat & Pinto

W/S WIPER BLADES

NOTE: Trico and Anco Blades are used and both come in two types Bayonet type and Side Saddle Pin Type.

Bayonet Type: To remove the *Trico* type press down on the arm to unlock the top stud. Depress the tab on the saddle, Fig. 5, and pull the blade from the arm. To remove the *Anco* type press inward on the tab, Fig. 6, and pull the blade from the arm. To install a new blade assembly slide the blade saddle over the end of the wiper arm so that the locking stud snaps into place.

Side Saddle Pin Type: To remove the pin type (Trico or Anco) insert an appropriate tool into the spring release opening of the blade saddle, depress the spring clip and pull the blade from the arm, Fig. 7. To install, push the blade saddle on to the pin so that the spring clip engages the pin. To replace the rubber element in a *Trico* blade squeeze the latch lock release and pull the element out of the lever jaws, Fig. 8. Remove *Anco* element by depressing the latch pin, Fig. 8, and sliding the element out of the yoke jaws. To install insert the element through the yoke or lever jaws. Be sure the element is engaged at all points.

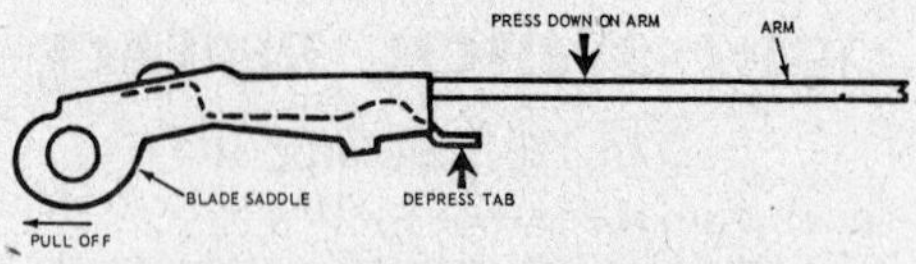

Fig. 5 Trico bayonet type blade

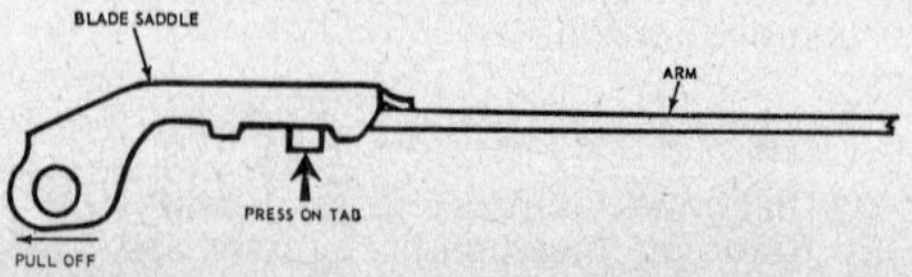

Fig. 6 Anco bayonet type blade

W/S WIPER ARMS

Swing the arm and blade assembly away from the windshield. While holding the assembly in this position pull the arm off the pivot shaft using an appropriate tool, Fig. 9. To install, hold the arm and blade in the swing out position and push the arm on to the pivot shaft.

W/S WIPER MOTOR

1. Loosen two nuts and disconnect wiper pivot shaft and link from the motor drive arm ball.
2. Remove three motor attaching screws and lower motor away from under the left side of the instrument panel.
3. Disconnect wiper motor wires and remove motor.

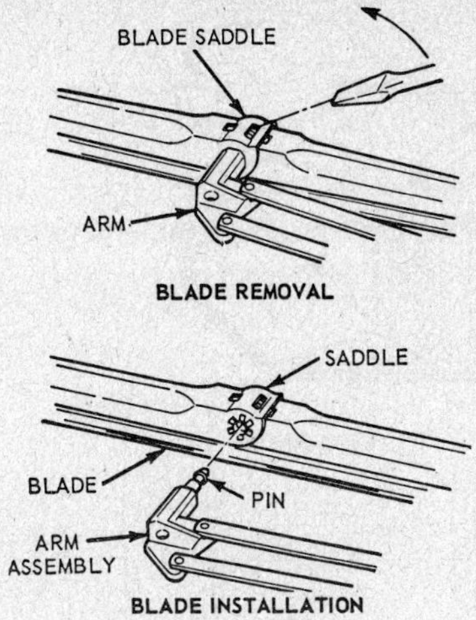

Fig. 7 Pin type blade

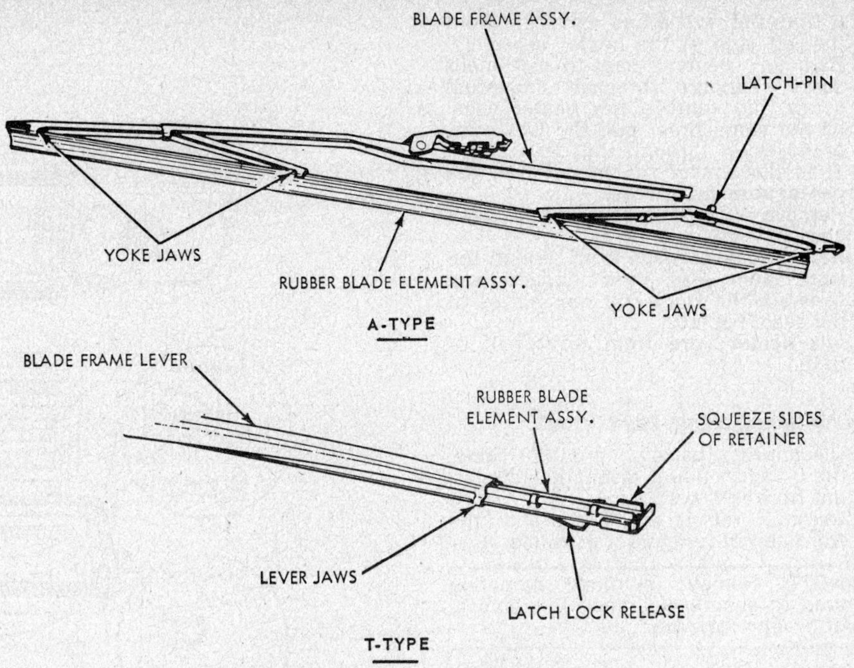

Fig. 8 Wiper blade element

W/S WIPER TRANSMISSION

NOTE: On 1973-76 Bobcat & Pinto equipped with air conditioning, remove blower motor to gain access to wiper transmission as described further on in this chapter.

1. Remove wiper arms and blades from pivot shafts.
2. Loosen two nuts retaining wiper pivot shaft and link assembly to the motor drive arm ball.
3. Remove three screws attaching each pivot shaft and remove assembly from under left side of instrument panel.

W/S WIPER SWITCH

1974-76 Mustang II

Dash Mounted Switch

1. Disconnect battery ground cable.
2. Remove switch knob and bezel nut, pull switch from panel, disconnect electrical connector and remove switch.

Column Mounted Switch

1. Disconnect battery ground cable.
2. Disconnect turn signal and wiper/washer switch wiring connector.
3. Remove lower instrument panel shield.
4. Pull wiring cover from bottom on column and remove cover.
5. With a screwdriver, disengage wiring shield tang and pry shield to remove.
6. Remove screw securing turn signal and wiper/washer switch assembly to column, then remove the assembly.

1971-76 Bobcat & Pinto

1. Remove instrument cluster as outlined previously.
2. Insert a thin bladed screwdriver into the slot in the switch knob and depress the spring. Then pull the knob from the switch shaft.
3. Remove wiper switch bezel nut. Then unplug wires and remove switch.

RADIO, REPLACE

NOTE: When installing radio, be sure to adjust antenna trimmer for peak performance.

1974-76 Mustang II

1. Disconnect battery ground cable.
2. Remove radio knobs, discs, shaft nuts and washers. On 1975-76 models, remove ash tray.

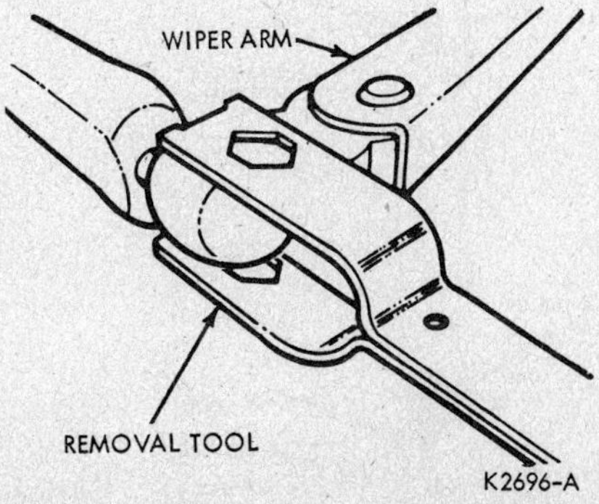

Fig. 9 Wiper arm

3. Remove radio rear support to instrument panel nut.
4. Lower radio, disconnect wiring and remove radio.

1971-76 Bobcat & Pinto

1. Disconnect battery ground cable.
2. On 1973-76 models remove instrument panel trim brace cover.
3. On all models, remove rear support to radio attaching bolt.
4. Remove four screws attaching the bezel to the instrument panel opening.
5. Pull radio out from instrument panel and disconnect speaker, power and antenna wires and remove radio.

HEATER CORE REMOVAL

Less Air Conditioning

1. Drain coolant and disconnect battery.
2. Disconnect blower motor wire at engine side of dash.
3. Disconnect heater hoses at engine block.
4. Remove four heater assembly-to-dash mounting nuts from the engine side of the dash.
5. Remove the glove box.
6. Disconnect control cables from heater. Remove mounting bracket clips and disconnect cables from door crank arms.
7. Remove radio as outlined previously.
8. Working inside car, remove snap rivet that attaches the forward side of the defroster air duct to the plenum chamber. Move the air duct back into the defroster nozzle to disengage it from the tabs on the plenum chamber. Now, tilt the forward edge of the duct up and forward to disengage

it from the nozzle and remove it from the left side of the heater assembly.

9. Remove heater case-to-instrument panel support bracket mounting screw and remove the heater case. At the same time, pull the two heater hoses in through the dash panel. Then disconnect the hoses from the heater core in the case.
10. Remove compression gasket from cowl air inlet.
11. Remove eleven clips from around the front and rear case flanges and separate the front and rear halves of the case, Fig. 10.
12. Lift heater core from front half of case.

With Air Conditioning, 1974-76 All

1. Disconnect battery ground cable, drain radiator and discharge refrigerant from A/C system.
2. Remove refrigerant lines and the front half of refrigerant manifold.

NOTE: Remove manifold mounting stud to ensure clearance for removal of evaporator case.

3. Disconnect heater hoses from core tubes and remove condensation drain hose in engine compartment.
4. Remove glove box.
5. Disconnect vacuum hoses from evaporator case and temperature control cable from blend door crank arm.
6. Remove heat distribution duct.

NOTE: On Mustang II, remove mode door vacuum motor.

7. On Mustang II, remove lower section of the air conditioning defrost plenum.
8. On Bobcat and Pinto, remove staples retaining fold down door on plenum, Fig. 11. Bend fold down door from locating tabs on plenum and remove adapter duct.

NOTE: During installation, position fold down door between locating tabs and tape in place with two pieces of black tape, Fig. 11.

9. On all models, remove blower motor and wheel from blower scroll.
10. Install a $\frac{1}{4}$-20 hex washer head screw in mounting tab of inlet duct to upper cowl bracket, holding inlet duct in place.

NOTE: Leave screw in position during installation of case assembly.

11. Remove inlet duct to evaporator case screws.

NOTE: One upper case to inlet duct screw is located under outside-recirc. motor mounting bracket.

12. Remove evaporator to cowl bracket screws and the evaporator to dash panel nuts in engine compartment. Rotate evaporator down and away from instrument panel and remove from under panel.
13. Remove upper to lower case screws and rubber seal from heater core tubes.
14. Remove upper half of evaporator case and move rubber seal on evaporator core forward and pull evaporator core from lower case.
15. Remove heater core upper straps, air deflector mounting screw, then remove air deflector and heater core.

With Air Conditioning, 1971-73 Pinto

1. Drain engine coolant, discharge refrigerant from A/C system and disconnect battery.
2. Disconnect heater hoses from heater core tubes.
3. Disconnect expansion valve from evaporator core tubes and apply tape over each core tube to keep out foreign material.
4. Remove three evaporator housing-to-dash panel mounting stud nuts.

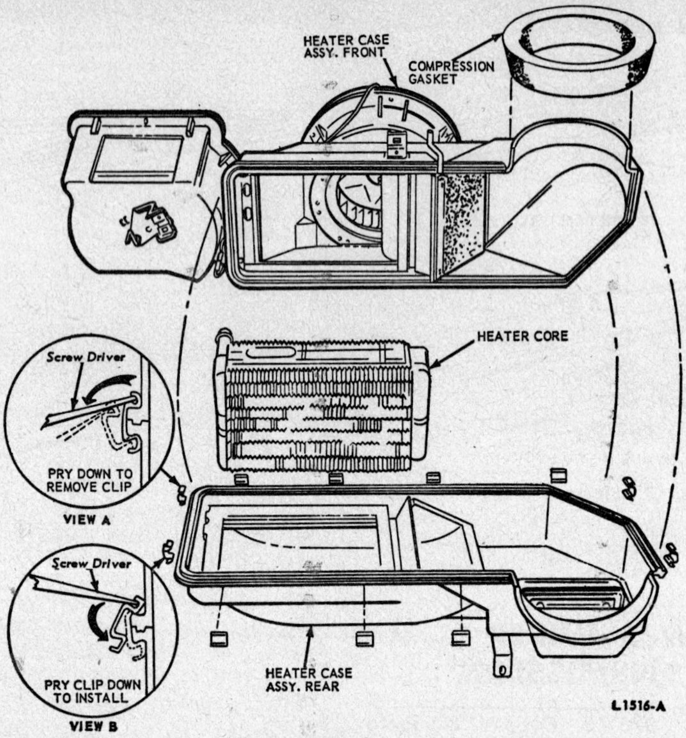

Fig. 10 Heater core removal

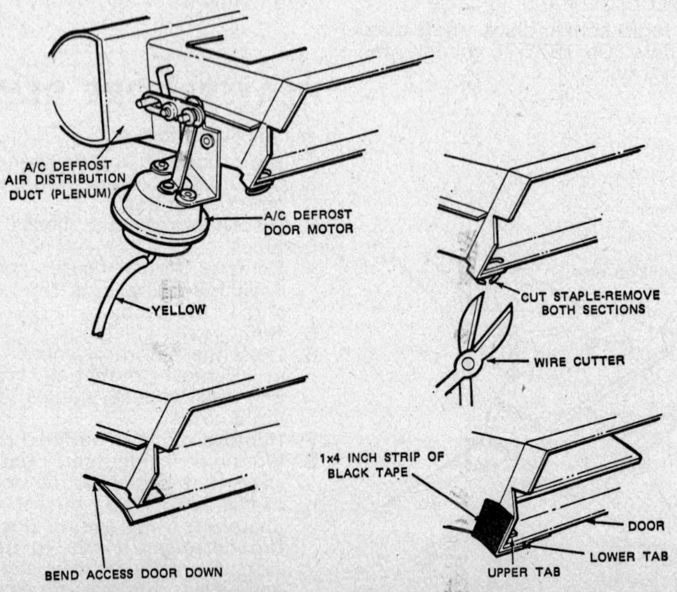

Fig. 11 A/C defrost air distribution duct fold down door. 1974-76 Bobcat & Pinto

5. Remove glove box and right air duct.
6. Disconnect blue vacuum hose from door motor and remove A/C defrost plenum chamber.
7. Disconnect red and yellow vacuum hoses from left door motor and the white hose from the right door motor.
8. Disconnect vacuum harness multiple connector from vacuum selector valve on the control assembly.
9. Disconnect temperature control cable from door crank arm on the evaporator housing, and disconnect purple and green vacuum hoses from the adjacent water valve vacuum switch.
10. Remove screw that retains evaporator housing to cowl upper support and remove the A/C assembly rearward and away from dash panel in order to clear mounting studs.
11. Remove tape from red and yellow vacuum hoses on top of blower housing. Disengage the green, white and purple vacuum hoses from clip on top of evaporator housing.
12. Disconnect wire connectors from A/C thermostat switch on evaporator housing and from resistor on blower housing. Disconnect blower ground wire.
13. Lower evaporator and blower housing and remove from vehicle.
14. Remove rubber dash panel seal from evaporator housing.
15. Remove eleven clips that hold two halves of housing together.
16. Remove A/C thermostat switch from upper housing.
17. Remove evaporator core from upper housing.
18. Remove temperature blend door upper frame.
19. Remove retaining clip from temperature door crank arm, slide crank arm out of door and lower housing and remove the door.
20. Remove temperature blend door lower frame.
21. Remove heater core from lower housing.

BLOWER MOTOR REMOVE

Less Air Conditioning

1. Follow procedure to remove the heater core as described previously.
2. Disconnect the blower motor lead wire (orange) from the resistor.
3. Remove the four blower mounting plate nuts and remove the motor and wheel assembly, Fig. 12.
4. Reverse procedure to install.

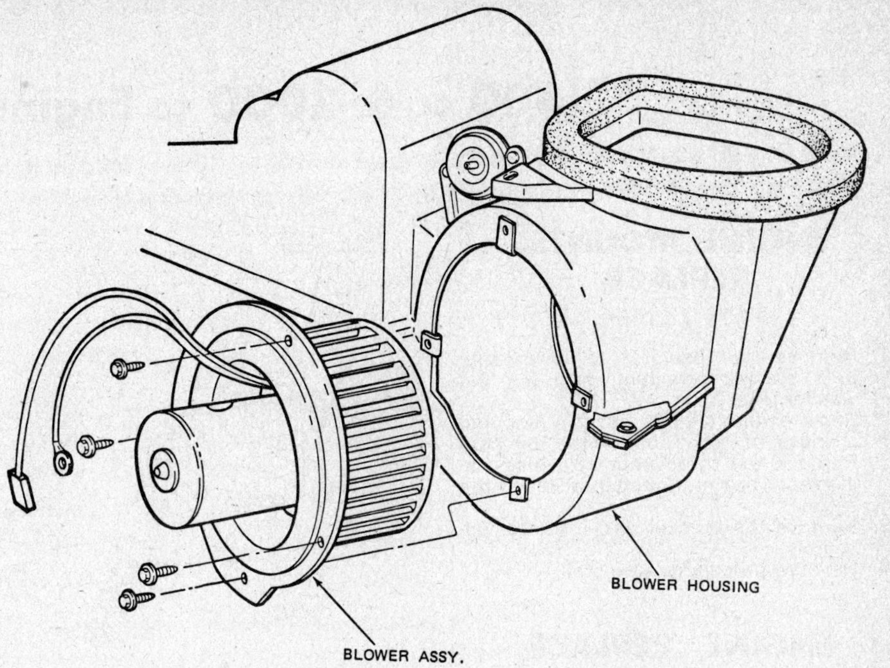

Fig. 12 Blower motor. 1971-76 All (Typical)

With Air Conditioning, 1974-76 All

1. Remove glove box.
2. Remove four screws retaining blower motor and wheel to blower scroll.
3. Remove blower motor and wheel assembly.

With Air Conditioning, 1971-73 Pinto

1. Remove the lower cover plate from the A/C De-Frost plenum by removing the two retaining screws.
2. Remove the air distribution duct from the blower housing.
3. Remove the blower housing mounting stud nut and lock plate.
4. Turn the blower housing clockwise to unlock the slotted tabs on the blower housing from their lock pins on the evaporator housing. There are two tabs and pins. Remove the blower housing.
5. Cut the gaskets around the A/C outlets at the break line.
6. Remove seven clips, and separate both halves of the blower housing.
7. Remove three blower motor mounting plate retaining nuts, and remove the motor and wheel assembly from the housing.
8. When installing, make sure that the A/C-Heat door is positioned properly before clipping the right and left housings together.

1600 cc & 2000 cc Engine Section

ENGINE MOUNTS REPLACE

1971-74 Pinto

1. Remove the insulator to lower support bracket attaching nut and flat washer, Fig. 1.
2. Raise engine slightly using a jack and a block of wood under the oil pan. Remove the three nuts attaching the insulator upper-support bracket to the engine.
3. Remove the bracket and insulator assembly.
4. Reverse procedure to install.

ENGINE, REPLACE

1600 cc Engine

1. Remove hood and disconnect battery lead and ground wire.
2. Drain coolant, disconnect radiator hoses at the engine and remove the radiator.
3. Disconnect hot air pipe at air cleaner and remove air cleaner.
4. Disconnect heater hoses from water pump and intake manifold.
5. Disconnect accelerator linkage from carburetor.
6. Disconnect temperature gauge and oil pressure gauge sender unit leads and alternator leads.
7. Disconnect exhaust pipe from manifold and remove hot air pipes from manifold, where applicable.
8. Disconnect fuel intake pipe from pump.
9. Disconnect distributor leads from coil, leads from spark plugs and remove distributor cap.
10. Jack up front of car and support with stands.
11. Remove starter motor, clutch housing lower bolts and remove the cover.
12. Remove stands and jack from under car.
13. Remove clutch housing to engine bolts.
14. Install suitable lifting bracket to engine and support engine with hoist.
15. Disconnect engine mounts from crossmember.
16. Support transmission and pull engine forward off main drive gear and lift from engine compartment.

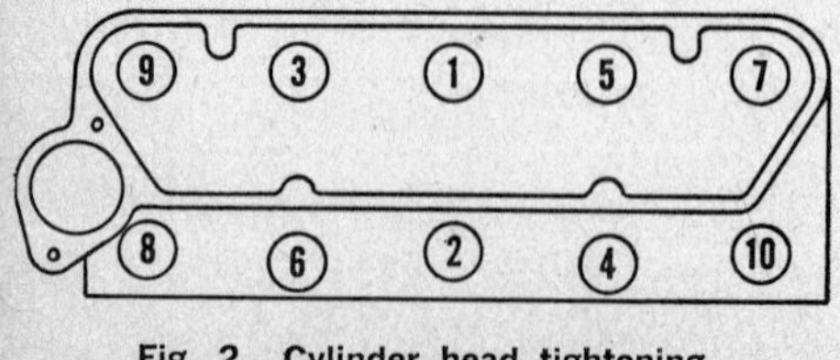

Fig. 2 Cylinder head tightening sequence. 1600 cc engine

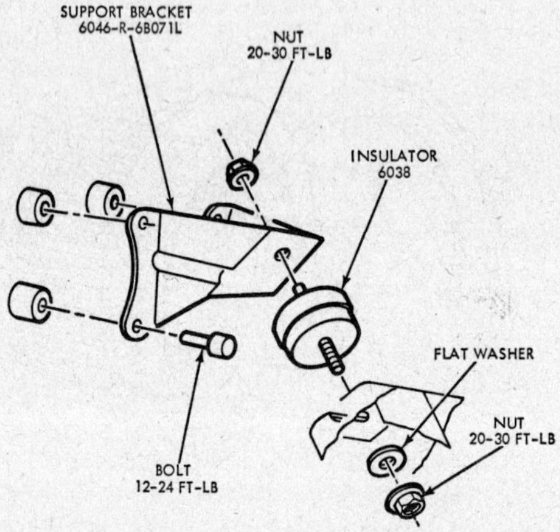

Fig. 1 Engine mount

2000 cc Engine

1. Drain coolant from radiator and oil from crankcase.
2. Raise hood and secure in vertical position.
3. Remove air cleaner and exhaust manifold shroud.
4. Disconnect battery ground cable.
5. Remove radiator hoses and remove radiator and fan.
6. Disconnect heater hoses from water pump and carburetor choke fitting.
7. Disconnect wires from alternator and starter and disconnect accelerator cable from carburetor. On A/C vehicles, remove compressor from bracket and position it out of way with lines attached.
8. Disconnect flex fuel line from tank line and plug tank line.
9. Disconnect primary wire at coil and disconnect oil pressure and temperature sending unit wires at sending units.
10. Remove starter and raise vehicle to remove the flywheel or converter housing upper attaching bolts.
11. Disconnect inlet pipe at exhaust manifold. Disconnect engine mounts at underbody bracket and remove flywheel or converter housing cover.
12. On vehicle with manual shift, remove flywheel housing lower attaching bolts.
13. On vehicle with automatic transmission, disconnect converter from flywheel and remove converter housing lower attaching bolts.
14. Lower vehicle and support transmission and flywheel or converter housing with a jack.
15. Attach engine lifting hooks to brackets and carefully lift engine out of engine compartment.

CYLINDER HEAD, REPLACE

1600 cc Engine

1. Disconnect hoses at air cleaner and remove air cleaner.
2. Disconnect fuel lines at pump and carburetor.
3. Drain coolant and disconnect spark plug leads.
4. Disconnect heater and vacuum hoses at intake manifold and at choke housing.
5. Disconnect lead from temperature gauge sender unit.
6. Detach exhaust pipe and move clear of cylinder head.
7. Disconnect throttle linkage and distributor vacuum pipe from carburetor.
8. Remove thermostat housing, pull to one side and remove thermostat.
9. Remove rocker cover and gasket.
10. Remove rocker arm shaft bolts evenly and lift off rocker shaft assembly.
11. Lift out push rods and keep them in their correct order for installation.
12. Remove cylinder head bolts and lift off cylinder head.
13. Reverse procedure to install and tighten bolts in sequence shown in Fig. 2.

2000 cc Engine

1. Drain cooling system and remove air cleaner and rocker arm cover.
2. Remove exhaust manifold and remove intake manifold, carburetor and decel valve as an assembly.
3. Remove camshaft drive belt cover.
4. Loosen drive belt tensioner and remove drive belt.
5. Remove water outlet elbow from head.
6. Remove cylinder head attaching bolts.

NOTE: The cylinder head retaining bolts have 12 point heads.

7. Lift head and camshaft assembly from engine.
8. Reverse procedure to install and tighten bolts in sequence shown in Fig. 2A.

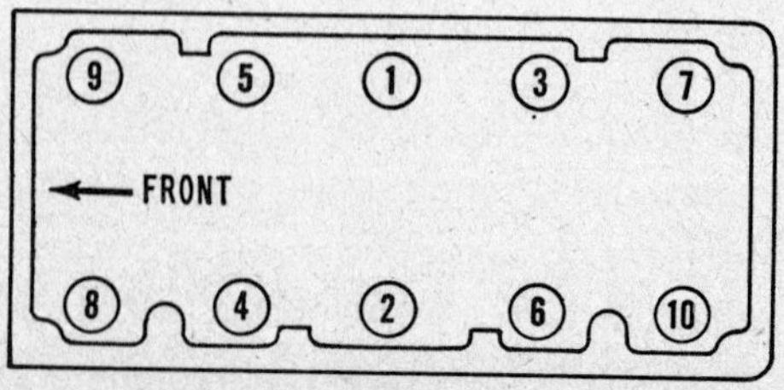

Fig. 2A Cylinder head tightening sequence. 2000 cc engine

1600 cc Engine — Set Hot

Valve Depressed	Valves to Adjust to .010	.017
no. 1	no. 3	no. 8
no. 2	no. 7	no. 5
no. 3	no. 6	no. 1
no. 5	no. 2	no. 4

Fig. 3 Valve adjustment table. 1600 cc engine

NOTE: Do not overtighten the valve cover retaining screws, also make certain the valve cover gasket is aligned properly. It is possible to block the oil return hole on the right side of the front camshaft bearing support if the above cautions are not adhered to. If the oil return hole becomes blocked, sufficient oil pressure can build up in the front camshaft bearing area and push the oil seal out of position.

VALVE ARRANGEMENT

Front to Rear

1600 cc Engine . . . . E-I-I-E-E-I-I-E
2000 cc Engine . . . . E-I-E-I-E-I-E-I

VALVE LIFT SPECS.

Engine	Year	Intake	Exhaust
1600 cc	1971	.2967	.3199
	1972-73	.3247	.3351
2000 cc	1971-74	.3993	.3993

VALVE TIMING

Intake Opens Before TDC

Engine	Year	Degrees
1600 cc	1971-73	17
2000 cc	1971	18
	1972-74	24

VALVES, ADJUST

1600 cc Engine

1. Start engine and allow it to idle for 20 minutes.
2. Turn engine off and remove air cleaner and wires from spark plugs.
3. Pull rubber grommet from bracket on rocker cover and lay wires over heater hose.
4. Remove screw retaining throttle cable to rocker cover.
5. Disconnect throttle cable link from ball stud and move cable out of way.
6. Pry linkage out of carburetor arm.
7. Unfasten and remove rocker cover being careful not to lose throttle rod retaining spring.
8. Rotate crankshaft clockwise by hand until number 1 valve is completely depressed and adjust valves 3 and 8. Valves are numbered from 1 to 8 (front to rear). Consult Fig. 3 and continue through the sequence shown until all valves are adjusted.

2000 cc Engine

1. If there is a clamp between heat shroud pipe and air cleaner duct and valve assembly, loosen the clamp.
2. Disconnect crankcase ventilation hose and carbon cannister hose at air cleaner.
3. Remove wing nuts and unsnap wire clips on air cleaner cover and lift air cleaner off carburetor.
4. Remove screws (11 mm) from rocker cover.
5. Remove spark plug wires from retainer and move out of way.
6. Remove rocker cover.
7. Rotate the crankshaft clockwise by hand until the high point of the number 1 cam lobe is pointing down. Check clearances on valves 6 and 7. Consult Fig. 4 and continue on through sequence until all valves are adjusted, Fig. 4A.

2000 cc Engine — Set at Any Temperature

Valve Depressed	Valves to Adjust to .008	.010
no. 1	no. 6	no. 7
no. 2	no. 8	no. 3
no. 3	no. 2	no. 5
no. 6	no. 4	no. 1

Fig. 4 Valve adjustment table. 2000 cc engine

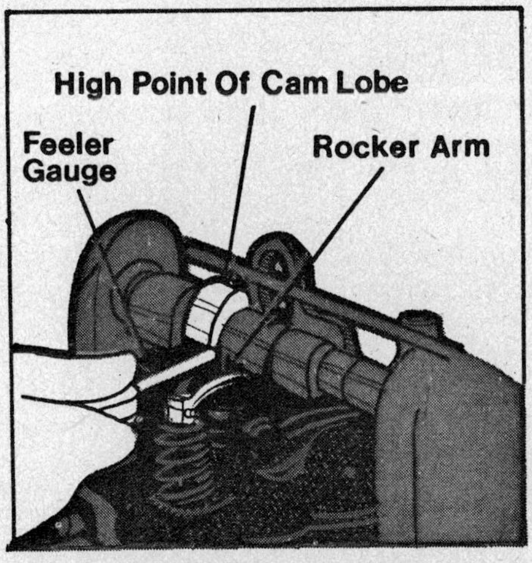

Fig. 4A Checking valve clearance. 2000 cc engine

VALVE GUIDES

Valve guides consist of holes bored in the cylinder head. For service, the guides can be reamed oversize to accommodate valves with oversize stems. Valves with oversize stems are available in .003", .015" and .030" oversizes on all engines except the 1974 2000 cc engine. Valves with oversize stems for the 1974 2000 cc engine are available in oversizes of .008", .016" and .032".

ROCKER ARM SERVICE

1600 cc Engine

1. Remove rocker arm cover.
2. Remove rocker shaft attaching bolts and lift off rocker shaft assembly.
3. Remove cotter pin from one end of shaft and slip flat washer, crimped washer and second flat washer off the shaft. The rocker arm shaft supports, rocker arms and springs can now be removed.
4. Remove plugs from shaft ends by drilling a hole in one plug. Insert a long rod through the drill plug and knock the opposite plug out of the shaft. Remove drilled plug in same manner.
5. To assemble, refit new plugs in ends of shaft. *The bolt hole in the rocker arm shaft support must be on the same side as the adjusting screw in the rocker arm. The rocker arms are right and left handed, the rocker pads being inclined towards the supports.* Install cotter pins with heads upwards and bend over the legs to secure.

2000 cc Engine

1. Remove air cleaner.
2. Remove rocker arm cover.
3. Rotate crankshaft as required to place the low side of the camshaft lobe next to the rocker arm that is being replaced.
4. Remove the rocker arm retaining spring.
5. Depress the valve spring with Tool T71P6565-A just enough to remove the rocker arm.

NOTE: Refer to "NOTE" under "Cylinder Head, Replace" before installing valve cover.

VALVES, GRIND

NOTE: The intake valves on the 1600 cc engine have aluminized faces. Under no conditions should the faces of aluminized valves be ground or lapped in as this will remove the coating and reduce the valves' wear and heat resistant properties. If valve faces are worn or pitted it will be necessary to install new valves and to resurface the valve seats or alternatively lap the seats using dummy valves. The exhaust valves may be lapped in or the faces ground as required.

VALVE LIFTERS, REPLACE

1600 cc Engine—The chilled cast iron tappets can only be removed from the engine after the camshaft has been removed. See "Camshaft, Replace" further on in this section.

Fig. 5 Removing crankshaft oil seal. 2000 cc engine

TIMING CASE COVER

1600 cc Engine

1. Drain coolant and disconnect radiator hoses at engine.
2. Remove radiator assembly.
3. Remove fan belt and then remove fan and water pump pulley.
4. Remove water pump.
5. Remove crankshaft pulley with suitable puller.
6. Remove the front cover. When replacing cover, install a new oil seal as this seal can only be replaced with the cover removed.

2000 cc Engine

It is not necessary to remove the front cover to replace the crankshaft oil seal. Proceed as follows:

1. Remove alternator belt.
2. Remove crankshaft pulley bolt and slide pulley off shaft.
3. Remove camshaft drive belt and slide sprocket and belt guide off the crankshaft. If sprocket cannot be slid off shaft use a puller.
4. Install tool T71P-6150A over end of crankshaft and remove seal, Fig. 5.
5. Install a new seal with tool T71P-6150B, Fig. 6.

TIMING CHAIN/BELT

1600 cc Engine

The timing chain can be replaced after removing the front cover as follows:

1. Remove the crankshaft oil slinger.
2. Remove the camshaft sprocket and disconnect timing chain. When replacing chain, be sure to align marks when the sprocket is fitted, Fig. 7.

2000 cc Engine

1. Place crankshaft on TDC.
2. Remove the three camshaft drive belt cover screws and remove cover.
3. Loosen camshaft drive belt tensioner adjustment bolt, Fig. 8 and force the tensioner toward the exhaust manifold side of engine to relax belt tension, then tighten the bolt.

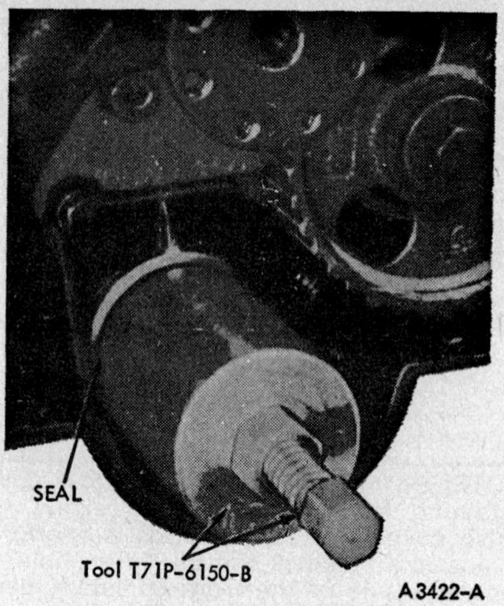

Fig. 6 Installing crankshaft oil seal. 2000 cc engine

4. Lift the belt off the sprockets.

NOTE: Do not rotate the crankshaft or the camshaft after the belt is removed. Rotating either one will impair valve timing.

5. To install, make sure timing marks are aligned as in Fig. 8 and place the belt over the sprockets.
6. Loosen the tensioner adjustment bolt to place tension on the belt.
7. Rotate the crankshaft two complete turns to place the timing marks in

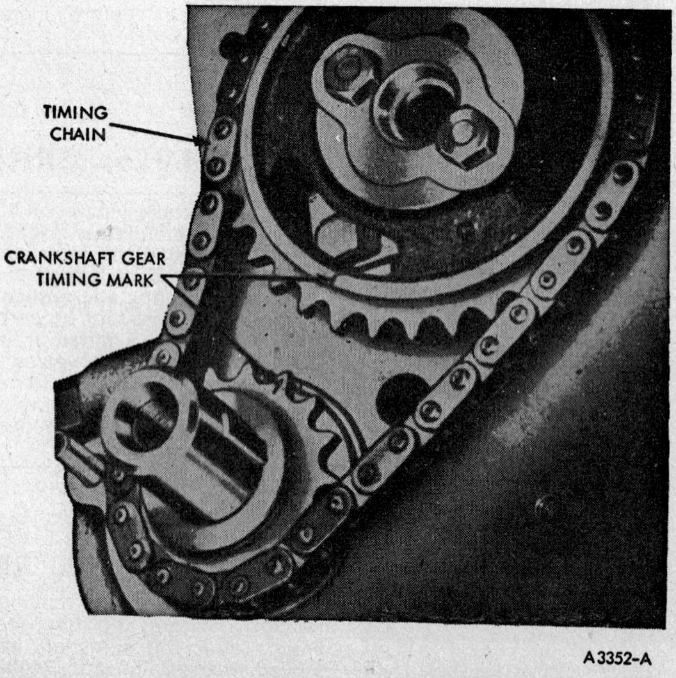

Fig. 7 Valve timing marks. 1600 cc engine

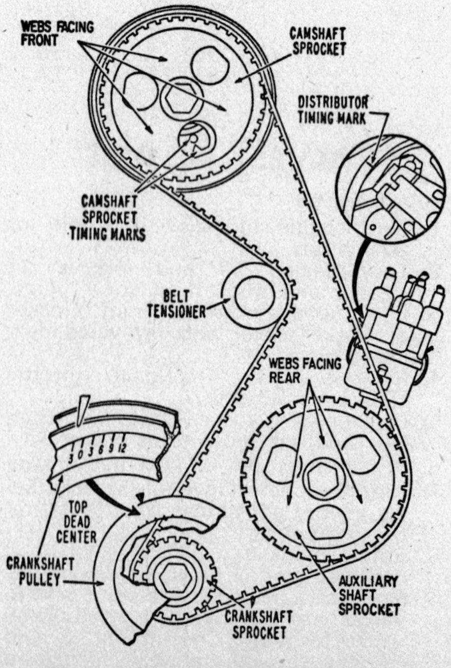

Fig. 8 Timing marks. 2000 cc engine

the proper position and to remove all slack from the belt. Torque the adjustment bolt and the pivot bolt.
8. Position camshaft drive belt cover and install screws.
9. Start engine and check ignition timing and adjust as required.

CAMSHAFT, REPLACE

1600 cc Engine

1. Remove engine as previously described.
2. With engine mounted on a stand, disconnect fuel line at pump.
3. Loosen alternator belt and remove belt.
4. Remove fan and water pump pulley.
5. Remove oil and fuel pumps from cylinder block.
6. Remove distributor.
7. Remove rocker arm cover and remove rocker arm shaft assembly.
8. Withdraw push rods from block and keep them in order for installation.
9. Invert engine on stand and remove the oil pan.
10. Remove dipstick, crankshaft pulley, front cover and oil slinger.
11. Remove timing chain tensioner and remove camshaft sprocket and chain.
12. With engine inverted, remove camshaft thrust plate and withdraw camshaft.

2000 cc Engine

After removal of cylinder head, proceed as follows:

1. Remove rocker arms.
2. Remove camshaft gear bolt and washer, and slide the gear and the belt guide plate off the shaft.
3. Remove camshaft thrust plate from rear of head and carefully slide camshaft from the rear of the head.

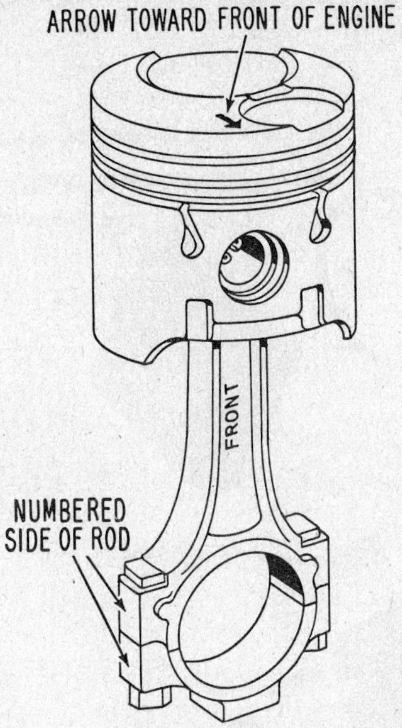

Fig. 9 Piston and rod. 1600 cc engine

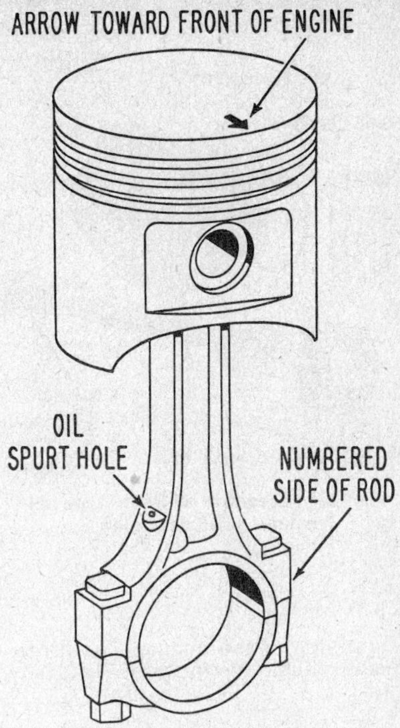

Fig. 10 Piston and rod. 2000 cc engine

PISTON & ROD, ASSEMBLE

1600 cc Engine

Assemble the rod to the piston with the Front mark on the rod on the same side of the assembly as the arrow in the piston crown, Fig. 9.

2000 cc Engine

Assemble the piston to the rod with the oil squirt hole in the rod positioned to

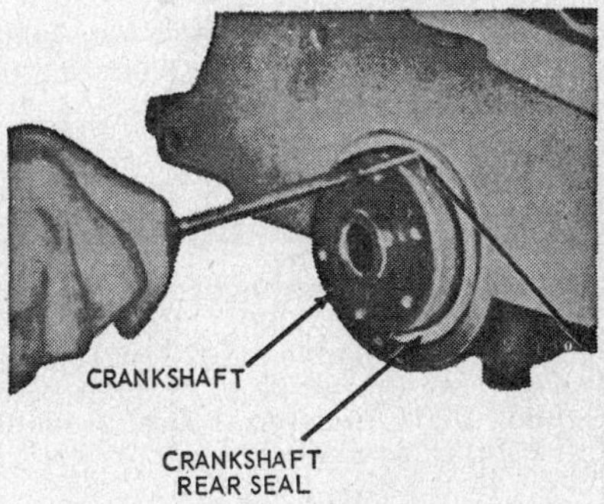

Fig. 12 Removing crankshaft rear oil seal. 2000 cc engine

Fig. 11 Installing crankshaft rear oil seal. 1600 cc engine

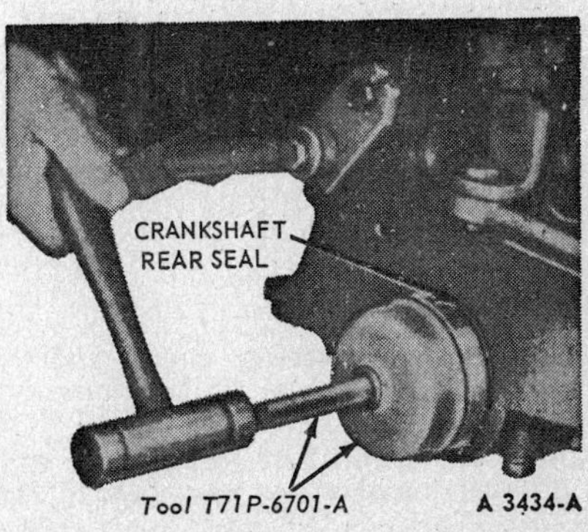

Fig. 13 Installing crankshaft rear oil seal. 2000 cc engine

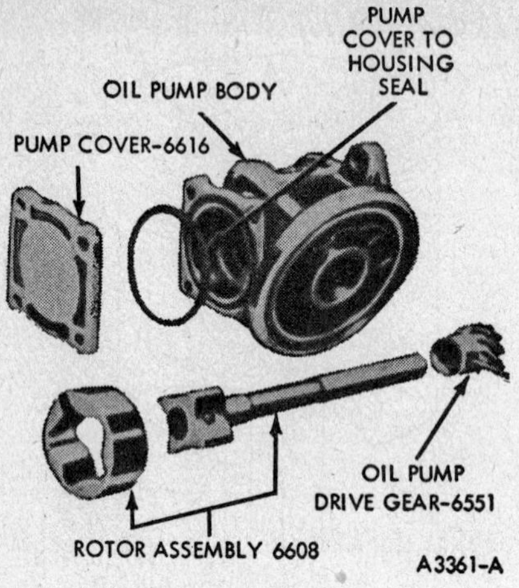

Fig. 14 Eccentric Bi-Rotor type oil pump. 1600 cc engine

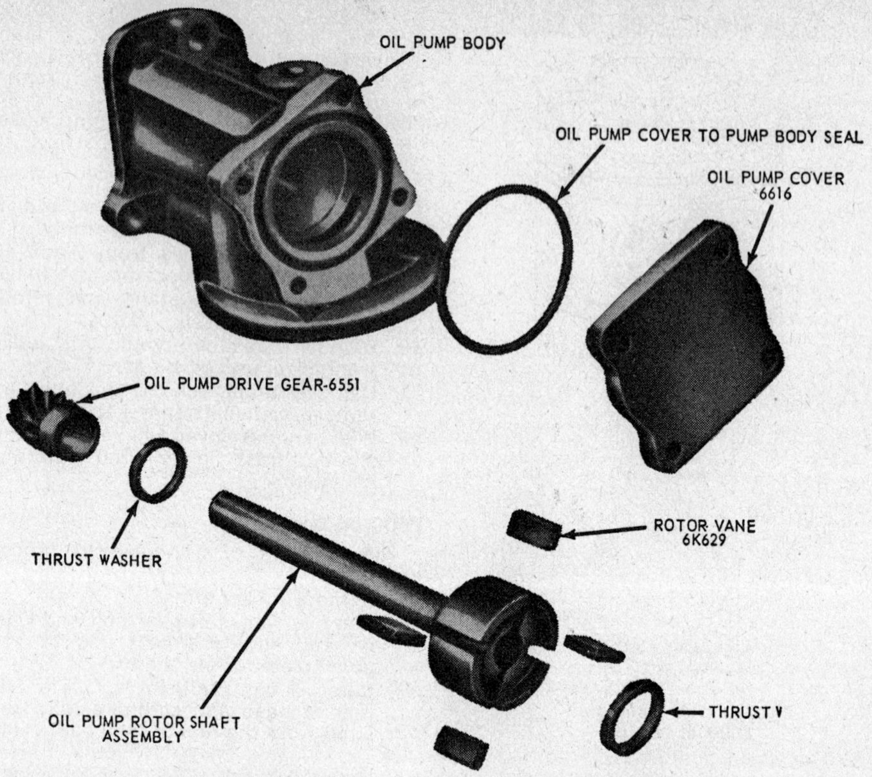

Fig. 15 Sliding vane type oil pump. 1600 cc engine

the right side of the engine. The arrow on the piston must face forward, Fig. 10.

PISTONS, PINS & RINGS

1600 cc Engine

Oversize pistons and rings are available in .0015, .0025 and .0030". Oversize pins are not available.

2000 cc Engine

Oversize pistons and rings are available in .0020 and .0040". Oversize pins are not available.

MAIN & ROD BEARINGS

Undersize main bearings are available in .010, .020 and .030". Rod bearing undersizes are available in .010, .020, .030 and .040" on the 1600 cc engine. Only .010 and .020" undersizes are available on the 2000 cc engine.

CRANKSHAFT OIL SEAL

1600 cc Engine

With the engine removed and mounted on a work stand, proceed as follows:

1. Remove the pressure plate bolts evenly and remove the pressure plate and clutch disc.
2. Remove the flywheel.
3. Remove the oil pan and gaskets.
4. Remove the rear oil seal carrier and use tool T70P-6165 to remove and install a new seal, Fig. 11.

2000 cc Engine

1. Remove transmission, clutch and flywheel or the automatic transmission, converter and flywheel.
2. Remove crankshaft rear seal with a sheet metal screw as shown in Fig. 12.
3. Install new seal with tool T71P-6701A as shown in Fig. 13.

OIL PAN, REPLACE

1600 cc Engine

1. Drain crankcase and remove oil dipstick.
2. Disconnect battery ground cable.
3. Disconnect throttle linkage from carburetor.
4. Disconnect steering cable from rack and pinion.
5. Disconnect rack and pinion from crossmember and move it forward to provide clearance for oil pan.
6. Remove three bolts and remove starter motor.
7. Remove left bottom bolt from lower rear cover and remove cover.
8. Unfasten and remove the oil pan.

2000 cc Engine

1. Drain crankcase and remove oil dipstick and flywheel inspection cover.
2. Disconnect steering cable from rack and pinion.
3. Disconnect rack and pinion from crossmember and move it forward to provide clearance.
4. Unfasten and remove the oil pan.

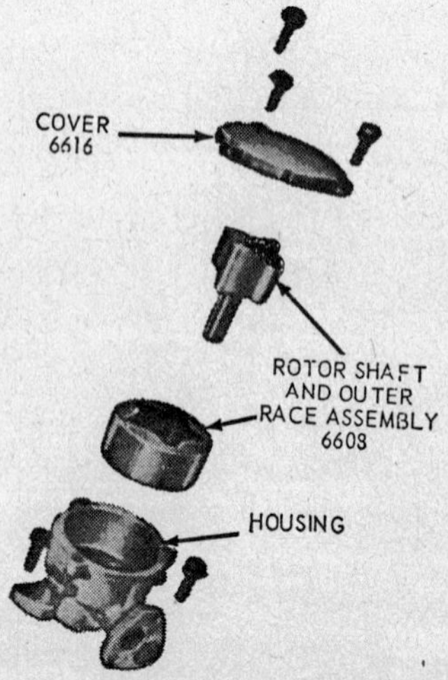

Fig. 16 Oil pump assembly. 2000 cc engine

OIL PUMP, REPLACE

The oil pump used on the 1600 cc engine may be one of two different types, an eccentric bi-rotor or a sliding vane type, Figs. 14 and 15. The pumps are directly interchangeable, differing only in internal design.

The oil pump on 2000 cc, Fig. 16, engines is easily removed after removal of the pan. It is not necessary to remove the pan on 1600 cc engines.

OIL PUMP REPAIRS

1. On the bi-rotor type pump, Figs. 14 and 16, remove filter body and element and extract sealing ring from the groove.
2. Remove end plate and withdraw O ring from groove in body.
3. Check clearance between lobes of

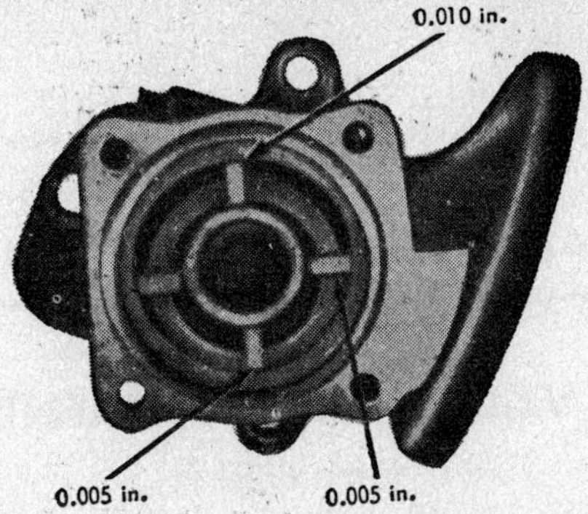

Fig. 17 Checking vane and rotor clearances. 1600 cc engine vane pump

inner and outer rotors. This should not exceed .006". Rotors are supplied only in a matched pair.

4. Check clearance between outer rotor and the housing. This should not exceed .010".
5. Place a straightedge across face of pump body. Clearance between face of rotors and straightedge should not exceed .005".
6. If necessary to replace rotor or drive shaft, remove outer rotor and then drive out retaining pin securing the skew gear to drive shaft and pull off the gear.
7. Withdraw inner rotor and drive shaft.

On the sliding vane type pump, Fig. 15, if clearances exceed those shown in Figs. 17 and 18, it will be necessary to rebuild or replace the pump.

BELT TENSION DATA

	New	Used
1971-74 All	140	110

WATER PUMP, REPLACE

1. Drain cooling system and disconnect heater hose and radiator lower hose from the pump.
2. Loosen alternator and remove the belt.
3. Remove the fan, spacer and pulley.
4. On 2000 cc engine, remove the camshaft drive belt cover.
5. Unfasten and remove the pump.

FUEL PUMP PRESSURE

Year	Engine	Pressure, Lbs.
1971-74	4 cyl	$3\frac{1}{2}$-$4\frac{1}{2}$

FUEL PUMP, REPLACE

1. Disconnect inlet and outlet lines at pump.
2. Unfasten and remove the fuel pump.
3. On 2000 cc engine, remove the actuator rod.

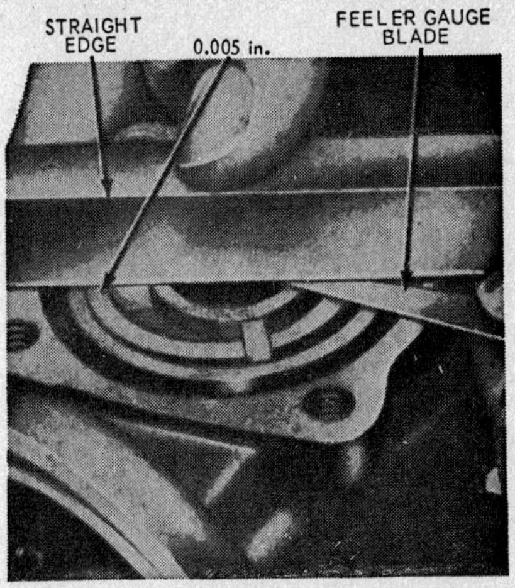

Fig. 18 Checking vane and rotor end-float. 1600 cc engine vane pump

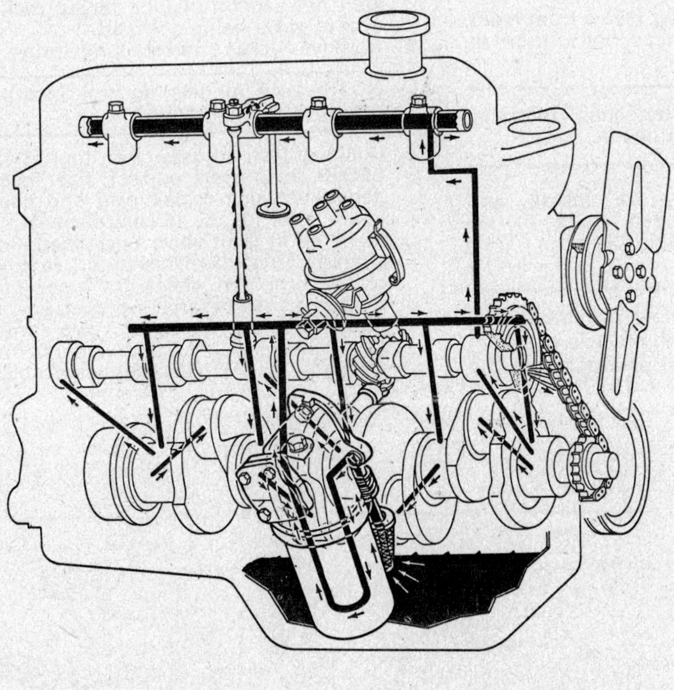

Engine oiling system. 1600 cc engine

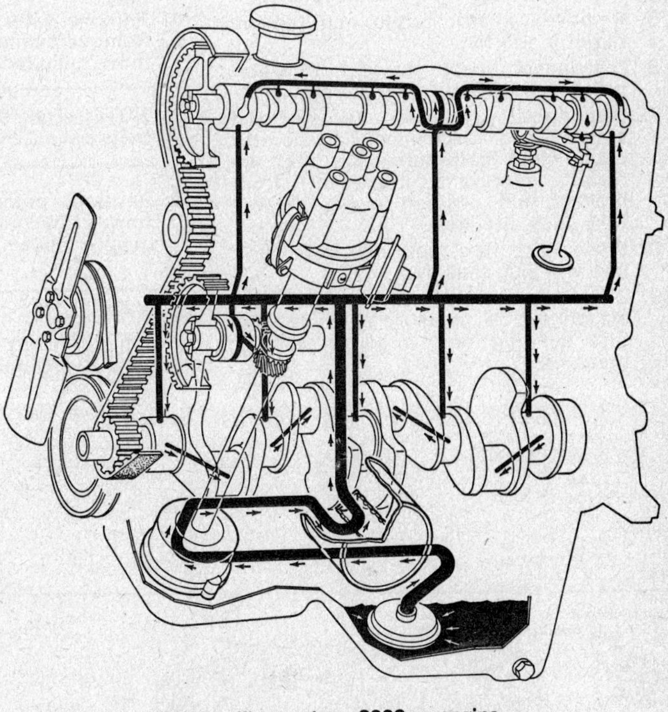

Engine oiling system. 2000 cc engine

2300 cc Engine Section

NOTE: This U.S. built engine is designed to metric specifications and therefore metric tooling will be required.

ENGINE MOUNTS, REPLACE

1974-76 Bobcat, Mustang II & Pinto

1. Remove fan shroud screws and support engine with a suitable jack and place a piece of wood under oil pan.
2. Remove insulator to support bracket through bolt, Figs. 1 and 2.
3. Remove support bracket mounting bolts, raise engine slightly, then remove support bracket.
4. Remove insulator to engine block bolts and insulator.
5. Reverse procedure to install.

ENGINE, REPLACE

1. Drain coolant from radiator and oil from crankcase.
2. Raise hood and secure in vertical position.
3. Remove air cleaner and exhaust manifold shroud.
4. Disconnect battery ground cable.
5. Remove radiator hoses and remove radiator and fan.
6. Disconnect heater hoses from water pump and carburetor choke fitting.
7. Disconnect wires from alternator and starter and disconnect accelerator cable from carburetor. On A/C vehicles, remove compressor from bracket and position it out of way with lines attached.
8. Disconnect flex fuel line from tank line and plug tank line.
9. Disconnect primary wire at coil and disconnect oil pressure and temperature sending unit wires at sending units.
10. Remove starter and raise vehicle to remove the flywheel or converter housing upper attaching bolts.
11. Disconnect inlet pipe at exhaust manifold. Disconnect engine mounts at underbody bracket and remove flywheel or converter housing cover.
12. On vehicle with manual shift, remove flywheel housing lower attaching bolts.
13. On vehicle with automatic transmission, disconnect converter from flywheel and remove converter housing lower attaching bolts.
14. Lower vehicle and support transmission and flywheel or converter housing with a jack.
15. Attach engine lifting hooks to brackets and carefully lift engine out of engine compartment.

CYLINDER HEAD, REPLACE

1. Drain cooling system and remove air cleaner and rocker arm cover.
2. Remove intake and exhaust manifolds and carburetor.
3. Remove timing case cover and drive belt.
4. Remove water outlet elbow from head.
5. Remove cylinder head bolts, then remove cylinder head.

NOTE: The cylinder head retaining bolts have 12 point heads.

6. Reverse procedure to install and torque cylinder head bolts in sequence shown in, Fig. 3.

CAUTION: When installing cylinder head, position camshaft at 5 o'clock position, Fig. 3, allowing minimal protrusion of the valves from the cylinder head.

VALVE ARRANGEMENT

Front to Rear

2300 cc Engine . . . . E-I-E-I-E-I-E-I

VALVE LIFT SPECS.

Engine	Year	Intake	Exhaust
2300 cc	1974-76	.400	.400

VALVE TIMING

Intake Opens Before TDC

Engine	Year	Degrees
2300 cc	1974-76	22

VALVES, ADJUST

The valve lash on this engine cannot be adjusted due to the use of hydraulic valve lash adjusters, Fig. 4. However, the valve train can be checked for wear as follows:

1. Crank engine to position camshaft with flat section of lobe facing rocker arm of valve being checked.
2. Remove rocker arm retaining spring.

NOTE: Late models do not incorporate the retaining spring.

3. Collapse lash adjuster with tool T74P-6565B and insert correct size feeler gauge between rocker arm and camshaft lobe, Fig. 5. If clearance is not as listed in the "Valve Specifications" chart in front of this chapter, remove rocker arm and check for wear and replace as necessary. If rocker arm is found satisfactory, check valve spring assembled height and adjust as needed. If valve spring assembled

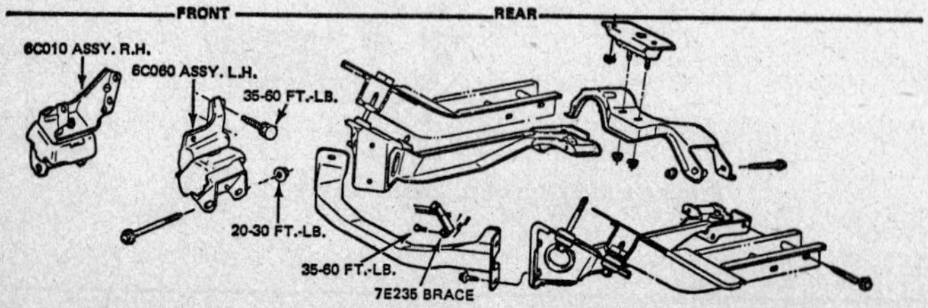

Fig. 1 Engine mounts installation. 1974-76 Mustang II

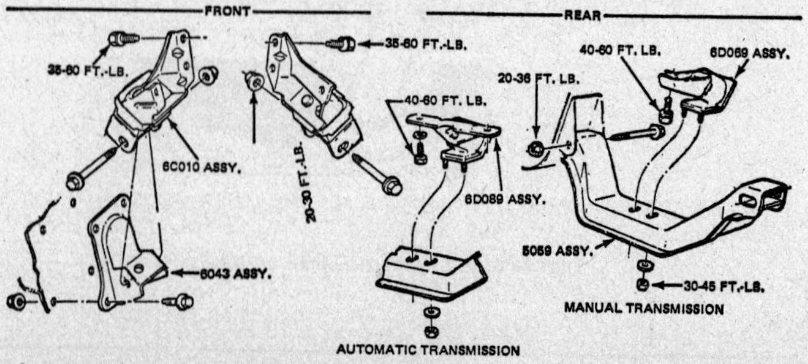

Fig. 2 Engine mount installation. 1974-76 Bobcat & Pinto

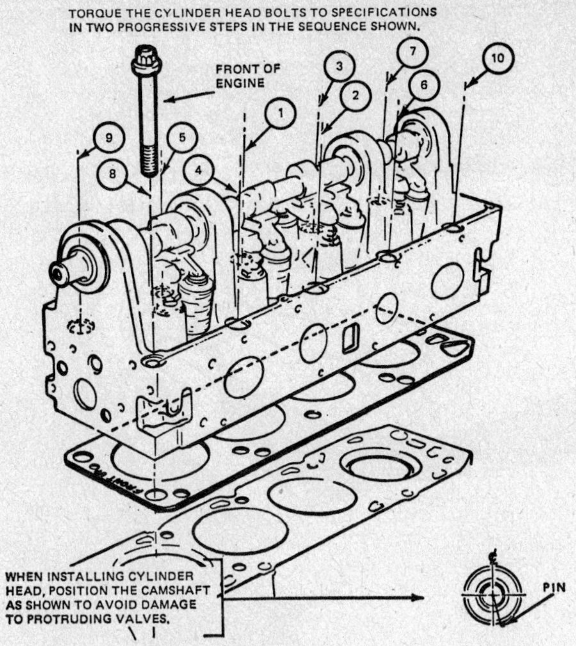

Fig. 3 Cylinder head installation. 2300 cc engine

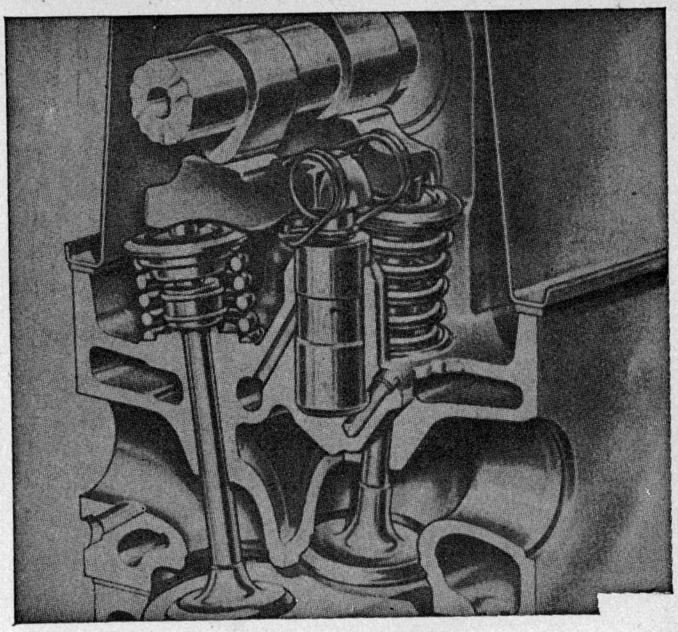

Fig. 4 Valve train installation. 2300 cc engine

height is within specifications listed in the front of this chapter, remove lash adjuster and clean or replace as necessary.

Fig. 6 Valve lash adjuster, Type I. 2300 cc engine

VALVE GUIDES

Valve guides consist of holes bored in the cylinder head. For service the guides can be reamed oversize to accommodate valves with oversize stems of .008, .016 and .032".

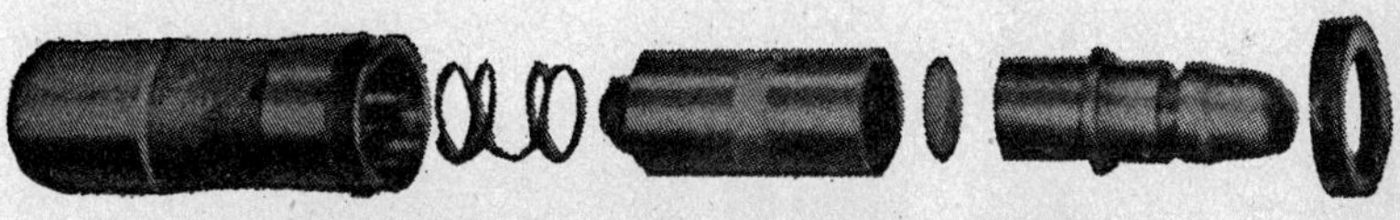

Fig. 7 Valve lash adjuster, Type II. 2300 cc engine

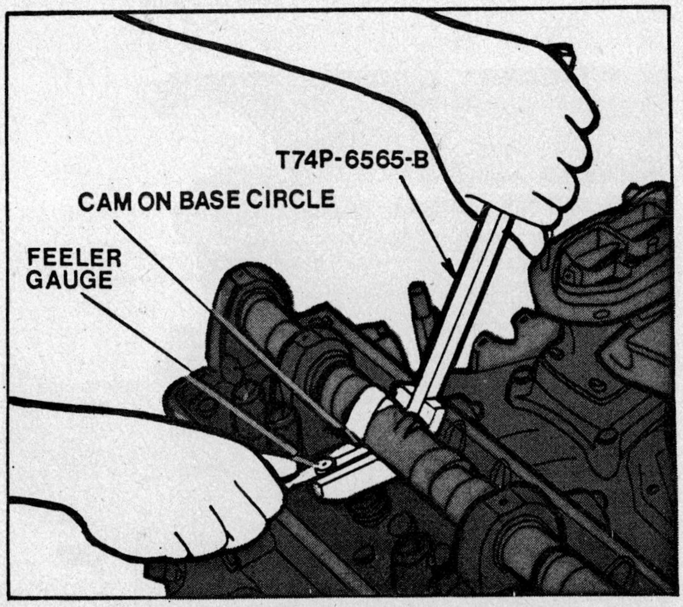

Fig. 5 Checking valve clearance. 2300 cc engine

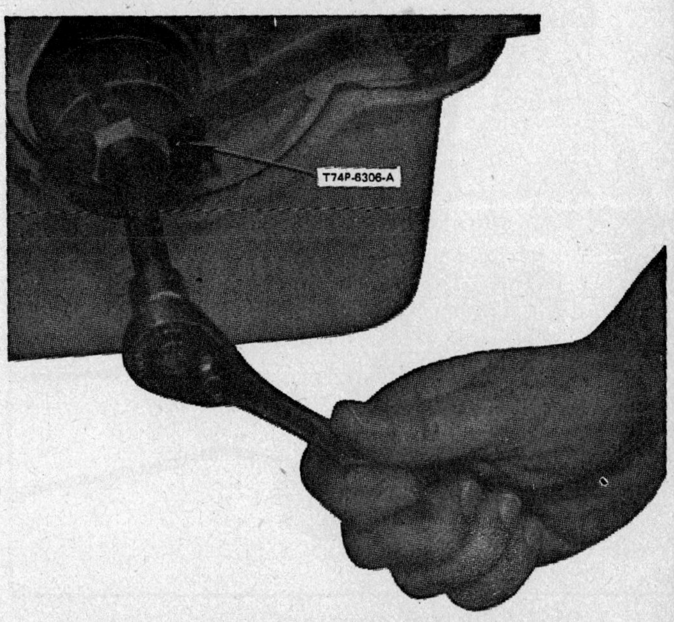

Fig. 8 Crankshaft sprocket removal. 2300 cc engine

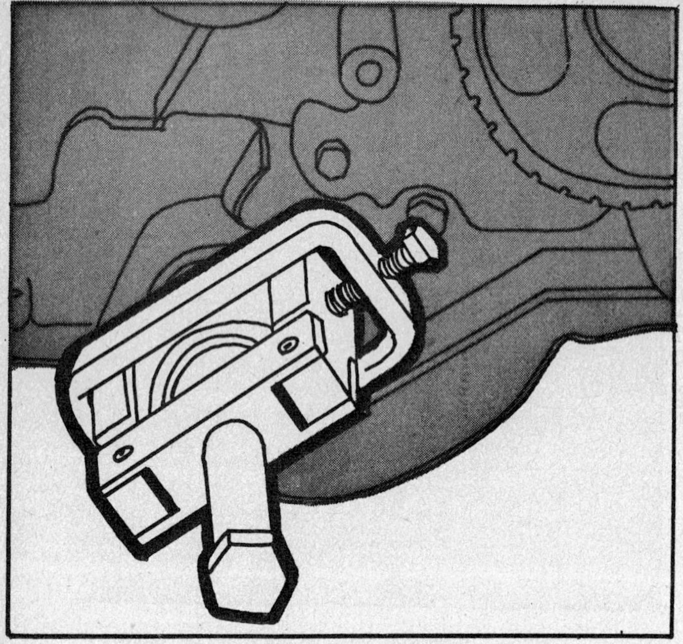

Fig. 9 Crankshaft front oil seal removal. 2300 cc engine

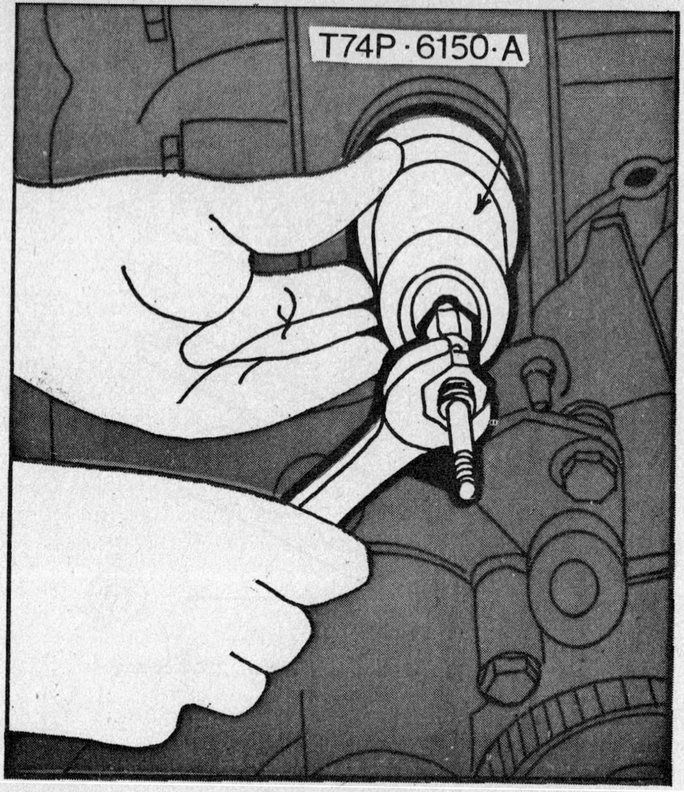

Fig. 10 Engine front seals installation. 2300 cc engine

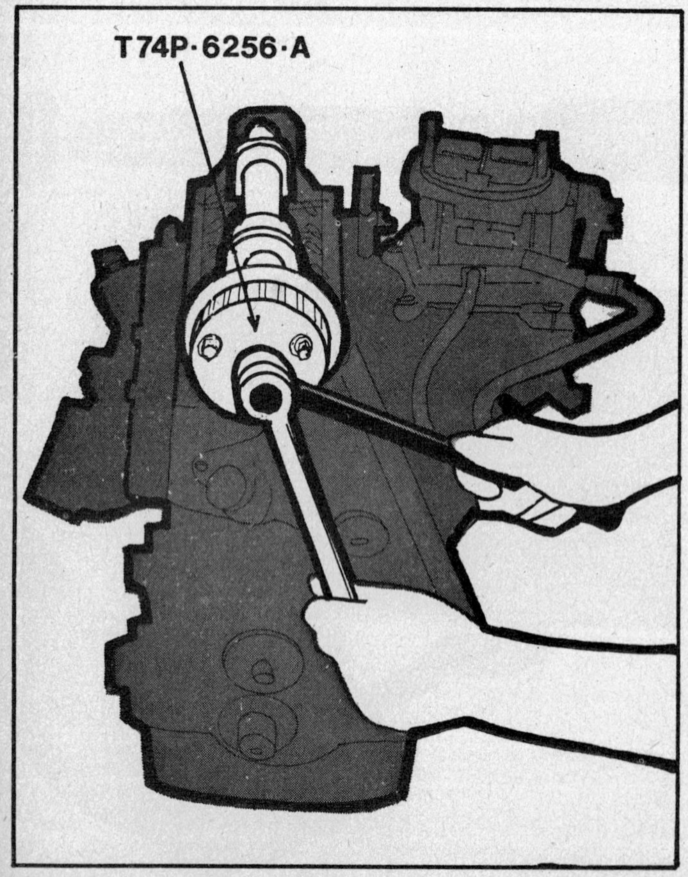

Fig. 12 Camshaft & auxiliary shaft sprockets removal. 2300 cc engine

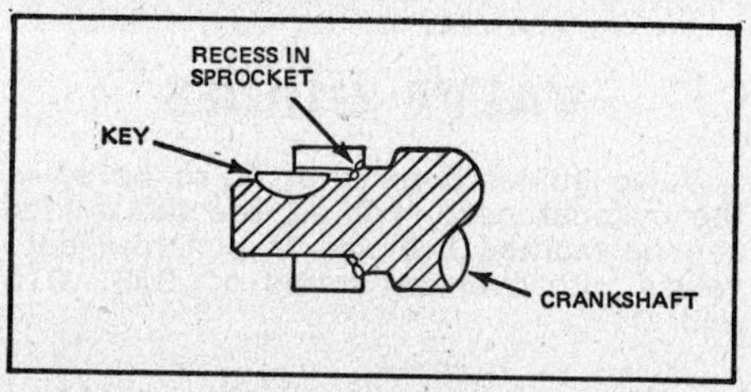

Fig. 11 Crankshaft sprocket installation. 2300 cc engine

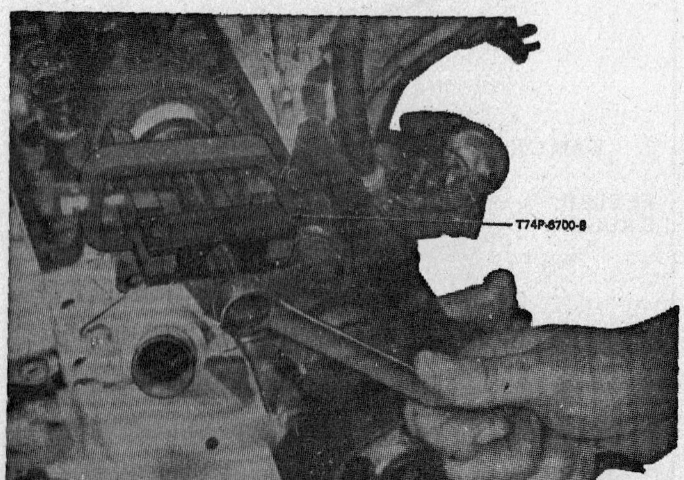

Fig. 13 Camshaft & auxiliary shaft seals removal. 2300 cc engine

ROCKER ARM SERVICE

1. Remove rocker arm cover.
2. Rotate camshaft until flat section of lobe faces rocker arm being removed.
3. With tool T74P-6565B, collapse lash adjuster and, if necessary, valve spring and slide rocker arm over lash adjuster.
4. Reverse procedure to install.

NOTE: Before rotating camshaft, ensure that lash adjuster is collapsed to prevent valve train damage.

LASH ADJUSTER, REPLACE

The hydraulic valve lash adjusters can be removed after rocker arm removal. There are two types of lash adjusters available, Type I, being the standard lash adjuster, Fig. 6, and Type II, having a .020 inch oversize outside diameter, Fig. 7.

FRONT ENGINE SEALS, REPLACE

To gain access to the front engine seals, remove the timing belt cover and proceed as follows:

Crankshaft Oil Seal

1. Without removing cylinder front cover, remove crankshaft sprocket with tool T74P-6306A, Fig. 8.
2. Remove crankshaft oil seal with tool T74P-6700B, Fig. 9.
3. Install a new crankshaft oil seal with tool T74P-6150A, Fig. 10.
4. Install crankshaft sprocket with recess facing engine block, Fig. 11.

Camshaft & Auxiliary Shaft Oil Seals

1. Remove camshaft or auxiliary shaft sprocket with tool T74P-6256A, Fig. 12.
2. Remove oil seal with tool T74P-6700B, Fig. 13.
3. Install a new oil seal with tool T74P-6150A, Fig. 10.
4. Install camshaft or auxiliary shaft sprocket with tool T74P-6256A with center arbor removed.

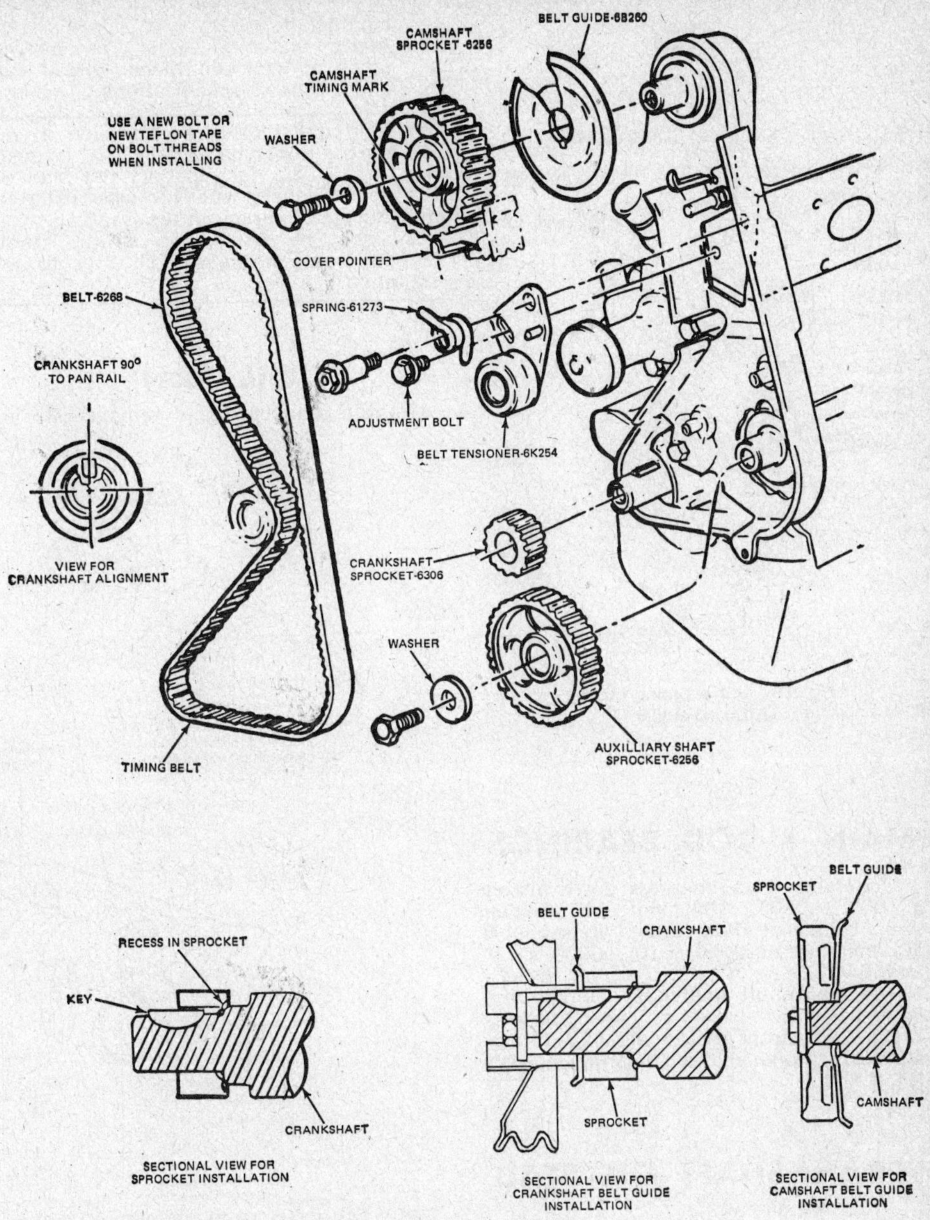

Fig. 14 Drive belt & sprockets installation. 2300 cc engine

TIMING BELT

1. Position crankshaft at TDC, No. 1 cylinder compression stroke.
2. Remove timing belt cover, loosen belt tensioner, and remove belt from sprockets, Fig. 14. Tighten tensioner bolt, holding tensioner in position.

NOTE: Do not rotate crankshaft or camshaft after belt is removed. Rotating either component will result in improper valve timing.

3. To install belt, ensure timing marks are aligned, Fig. 15, and place belt over sprockets.
4. Loosen tensioner bolt, allowing tensioner to move against belt.
5. Rotate crankshaft two complete turns, removing slack from belt. Torque tensioner adjustment and pivot bolts and check alignment of timing marks, Fig. 15.
6. Install timing belt cover.

CAMSHAFT, REPLACE

1. Remove rocker arm cover and rocker arms.
2. Remove timing belt cover, camshaft sprocket bolt and washer, then slide sprocket and belt guide off camshaft.
3. Remove camshaft retaining plate from rear of head, then remove camshaft from front of head.

PISTON & ROD, ASSEMBLE

Assemble the rod to the piston with the arrow on top of piston facing front of engine, Fig. 16.

PISTONS, PINS & RINGS

Oversize pistons are available in oversizes of .003″, .020″, .030″ and .040″. Oversize rings are available in .020″, .030″ and .040″ oversizes. Oversize pins are not available.

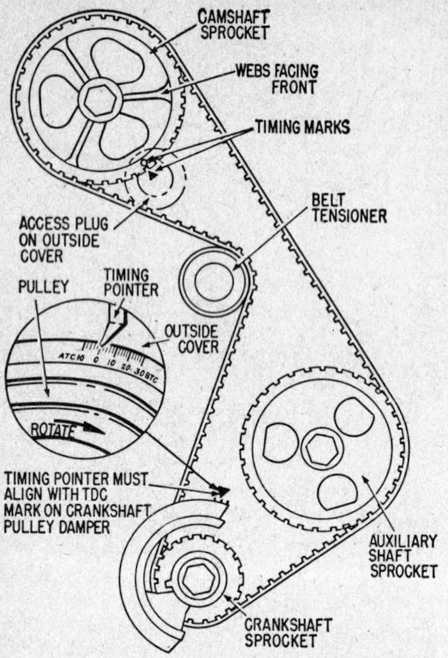

Fig. 15 Valve timing marks. 2300 cc engine

MAIN & ROD BEARINGS

Undersize main bearings are available in .002″, .020″, .030″ and .040″ undersizes. Undersize rod bearings are available in undersizes of .002″, .010″, .020″, .030″ and .040″.

The crankshaft and main bearings are installed with arrows on main bearing caps facing front of engine, Fig. 17. Install PCV baffle between bearing journals No. 3 and 4.

CRANKSHAFT OIL SEAL

1. Remove oil pan.
2. Remove rear main bearing cap.
3. Loosen remaining bearing caps, allowing crankshaft to drop down about 1/32″.
4. Install a sheet metal screw into seal and pull screw to remove seal.
5. Carefully clean seal groove in block with a brush and solvent. Also clean seal groove in bearing cap.
6. Dip seal halves in clean engine oil.
7. Carefully install upper seal half in its groove with locating tab toward rear of engine, Fig. 18, by rotating it on shaft journal of crankshaft until approximately 3/8″ protrudes below the parting surface. *Be sure no rubber has been shaved from outside diameter of seal by bottom edge of groove.*
8. Retighten main bearing caps and torque to specifications.
9. Install lower seal in main bearing cap with undercut side of seal toward front of engine, and allow seal to protrude about 3/8″ above parting surface to mate with upper seal upon cap installation.
10. Apply suitable sealer to parting faces of cap and block. Install cap and torque to specifications.

NOTE: If difficulty is encountered in installing the upper half of the seal in position, lightly lap (sandpaper) the side of the seal opposite the lip side using a medium grit paper. After sanding, the seal must be washed in solvent, then dipped in clean engine oil prior to installation.

OIL PAN

1. Drain crankcase and remove oil dip-

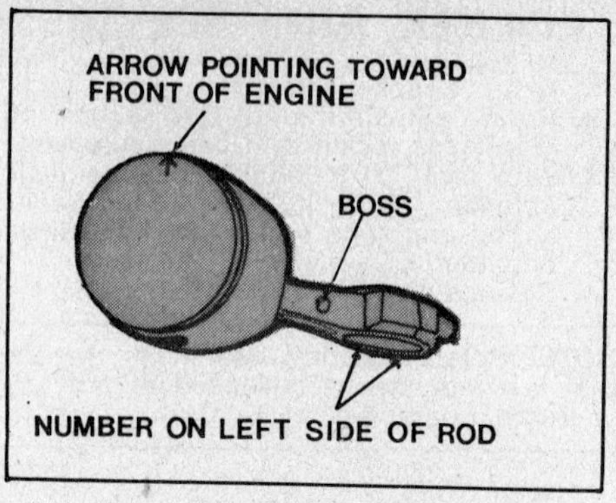

Fig. 16 Piston & rod. 2300 cc engine

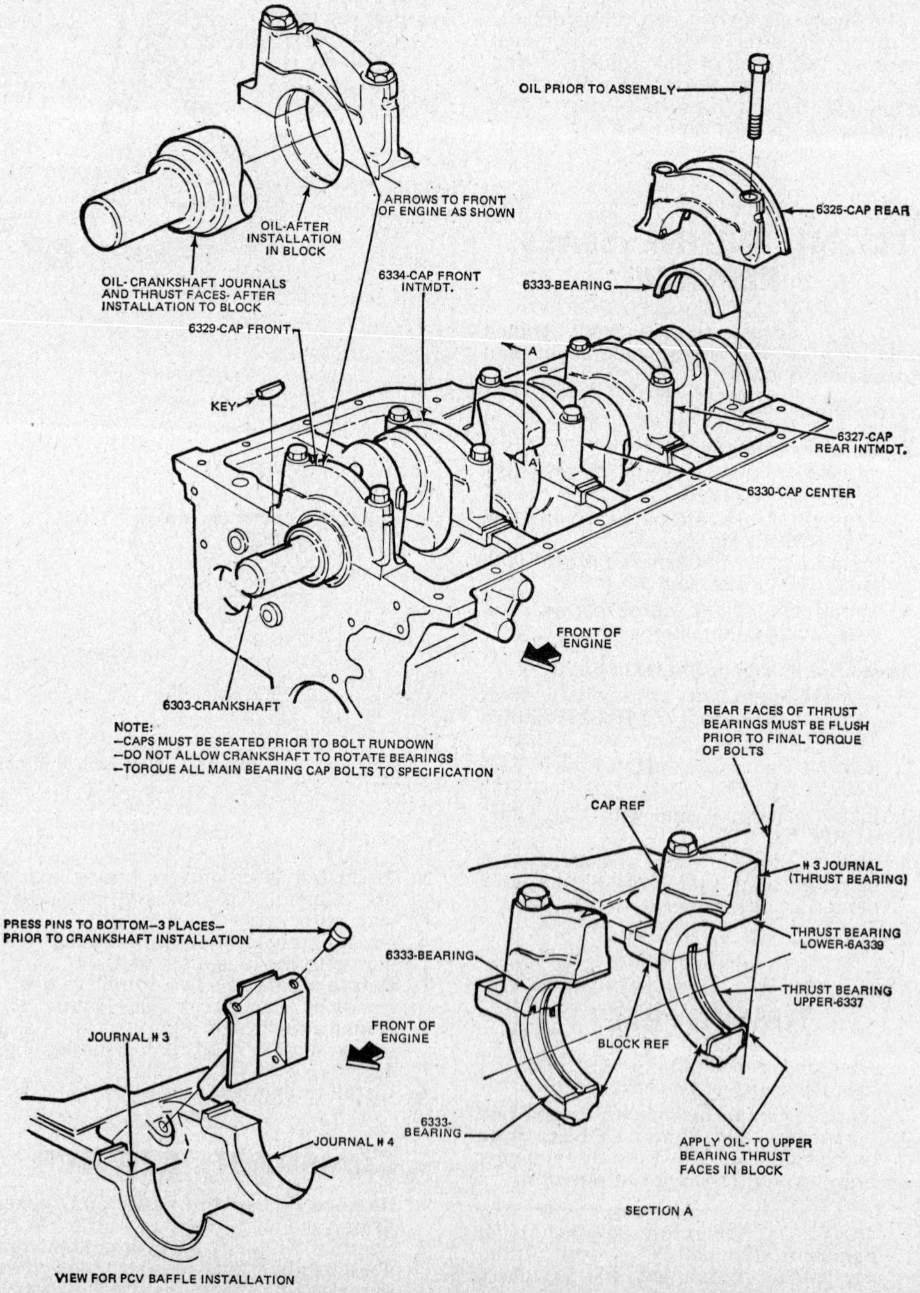

Fig. 17 Crankshaft & main bearing installation. 2300 cc engine

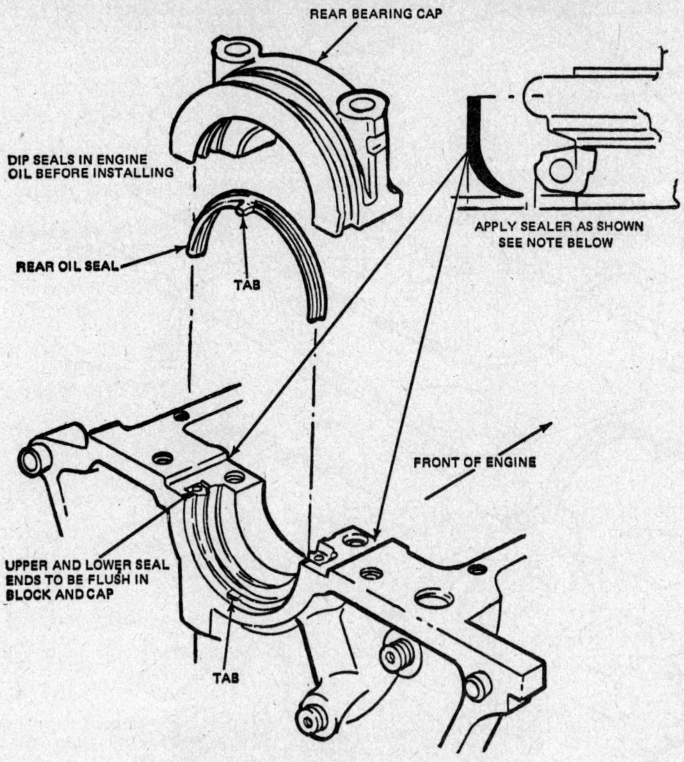

SEALER NOTE: CLEAN THE AREA WHERE SEALER IS TO BE APPLIED BEFORE INSTALLING THE SEALS. USE FORD SPOT REMOVER B7A-19521-A OR EQUIVALENT. AFTER THE SEALS ARE IN PLACE, APPLY A 1/16 INCH BEAD OF C3AZ-19562-A OR -B SEALER AS SHOWN. SEALER MUST NOT CONTACT SEALS.

Fig. 18 Crankshaft rear oil seal installation. 2300 cc engine

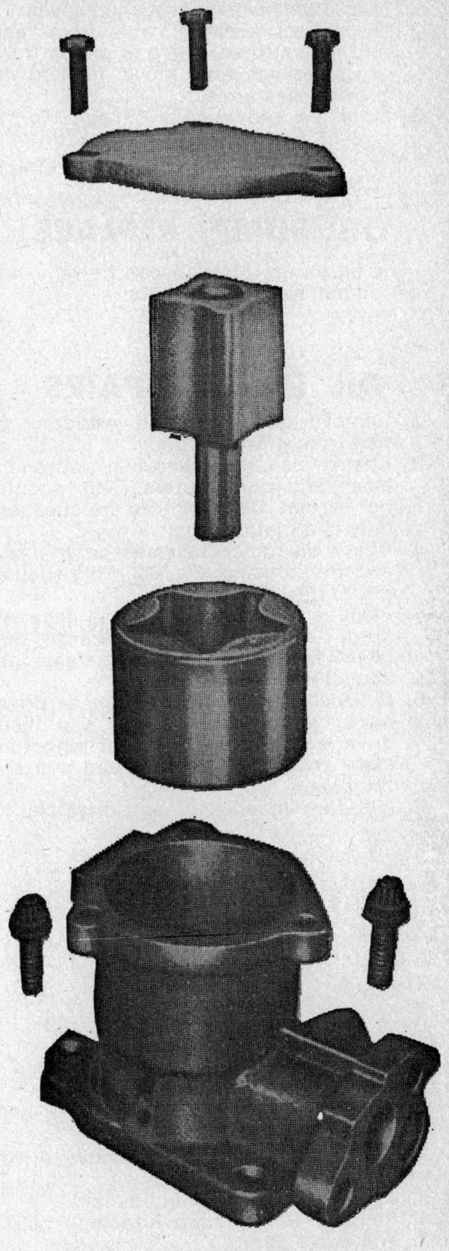

Fig. 20 Oil pump. 2300 cc engine

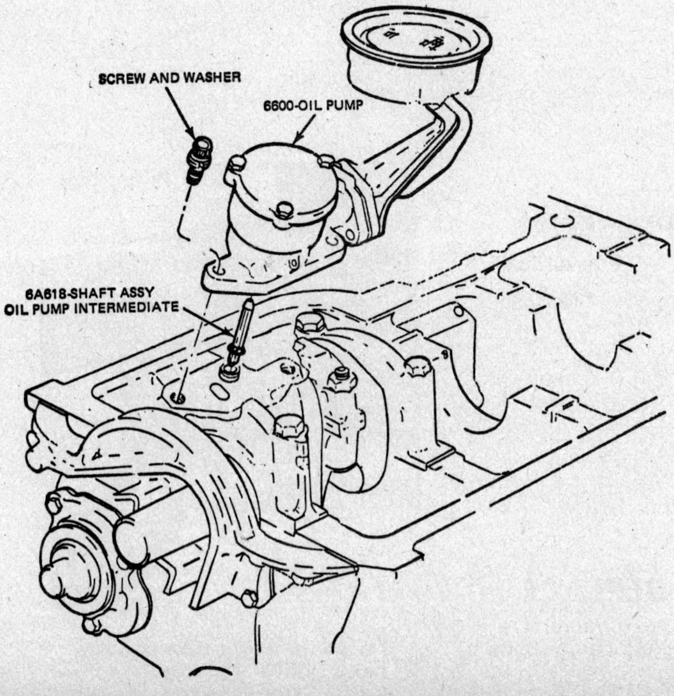

Fig. 21 Oil pump installation. 2300 cc engine

stick and flywheel inspection cover.
2. Disconnect steering cable from rack and pinion, then rack and pinion from crossmember and move forward to provide clearance.
3. Unfasten and remove oil pan. To install oil pan, refer to Fig. 19.

OIL PUMP, REPLACE

The oil pump, Fig. 20, can be removed after oil pan removal, Fig. 21.

OIL PUMP REPAIRS

1. Remove end plate and withdraw O ring from groove in body.
2. Check clearance between lobes of inner and outer rotors. This should not exceed .006″. Rotors are supplied only in a matched pair.
3. Check clearance between outer rotor and the housing, Fig. 22. This should not exceed .010″.
4. Place a straightedge across face of pump body, Fig. 23. Clearance between face of rotors and straightedge should not exceed .005″.
5. If necessary to replace rotor or drive shaft, remove outer rotor and then drive out retaining pin securing the skew gear to drive shaft and pull off the gear.
6. Withdraw inner rotor and drive shaft.

BELT TENSION DATA

	New Ft. Lbs.	Used Ft. Lbs.
1974-76 All	140	110

WATER PUMP, REPLACE

1. Drain cooling system and disconnect hoses from pump.
2. Loosen alternator and remove drive belt.
3. Remove fan, spacer and pulley.
4. Remove water pump attaching bolts and water pump.

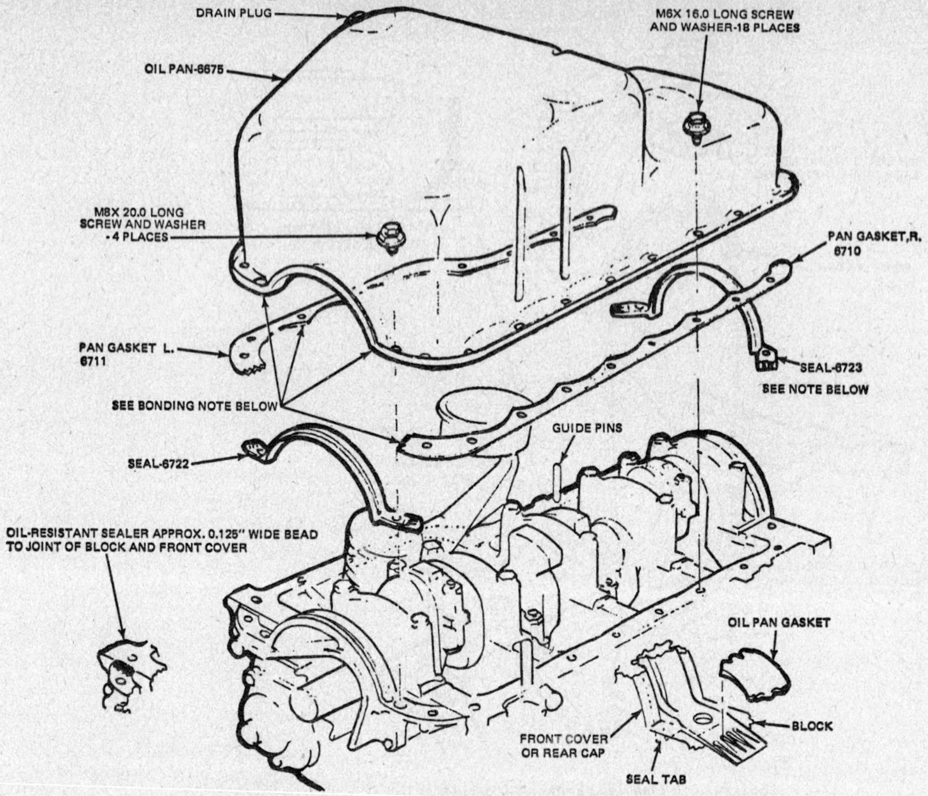

1. APPLY GASKET ADHESIVE EVENLY TO OIL PAN FLANGE AND TO PAN SIDE GASKETS. ALLOW ADHESIVE TO DRY PAST WET STAGE, THEN INSTALL GASKETS TO OIL PAN.
2. APPLY SEALER TO JOINT OF BLOCK AND FRONT COVER. INSTALL SEALS TO FRONT COVER AND REAR BEARING CAP AND PRESS SEAL TABS FIRMLY INTO BLOCK. BE SURE TO INSTALL THE REAR SEAL BEFORE THE REAR MAIN BEARING CAP SEALER HAS CURED.
3. POSITION 2 GUIDE PINS AND INSTALL THE OIL PAN. SECURE THE PAN WITH THE FOUR M8 BOLTS SHOWN ABOVE.
4. REMOVE THE GUIDE PINS AND INSTALL AND TORQUE THE EIGHTEEN M6 BOLTS, BEGINNING AT HOLE A AND WORKING CLOCKWISE AROUND THE PAN.

Fig. 19 Oil pan installation. 2300 cc engine

FUEL PUMP PRESSURE

Year	Engine	Pressure Lbs.
1974-76	2300 cc	3½-4½

FUEL PUMP, REPLACE

1. Disconnect fuel lines from pump.
2. Remove fuel pump attaching bolts and fuel pump.

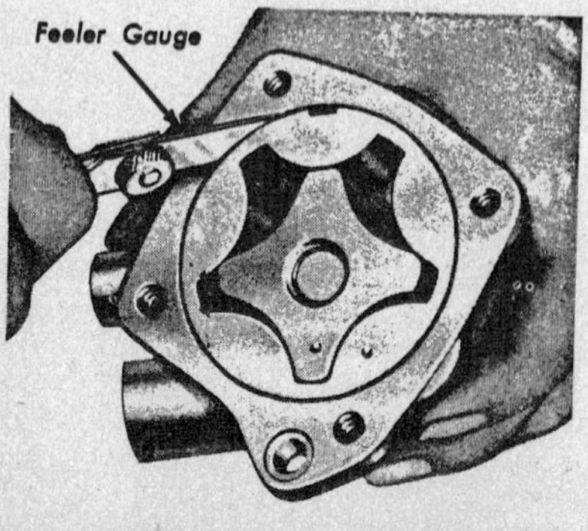

Fig. 22 Checking outer rotor to housing clearance. 2300 cc engine

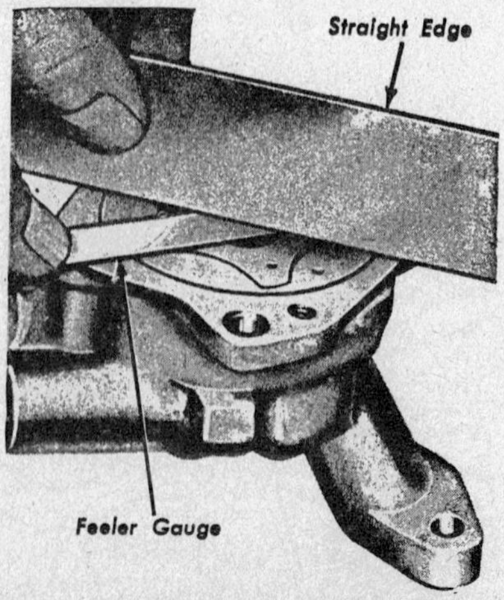

Fig. 23 Checking rotor end play. 2300 cc engine

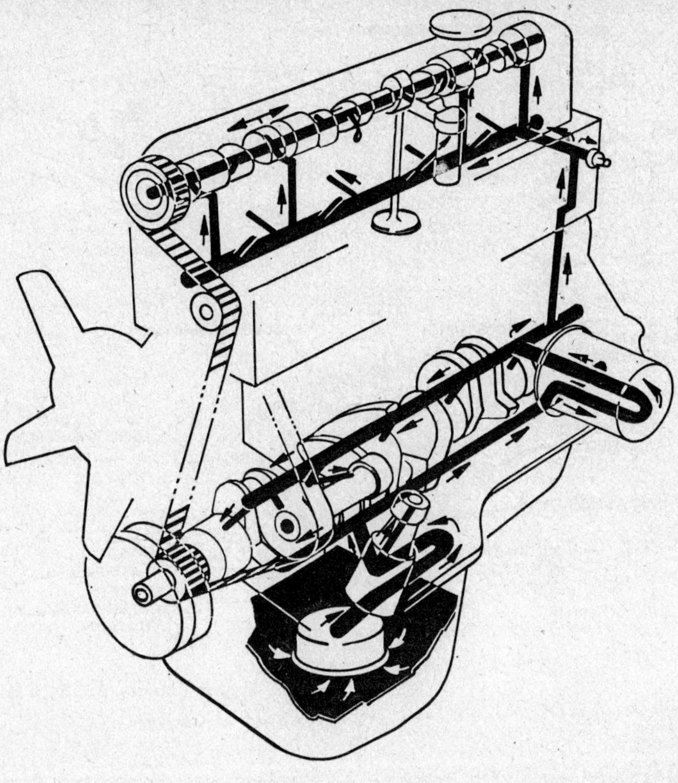

Engine oiling system. 2300 cc engine

2800 cc V6 Engine Section

ENGINE MOUNTS, REPLACE

1976 Bobcat & Pinto & 1974-76 Mustang II

1. Remove fan shroud screws and support engine with a suitable jack and a block of wood under the oil pan.
2. Remove insulator to insulator support bracket through bolt, support bracket to frame bolts, raise engine slightly, then remove support bracket, Figs. 1 & 1A.
3. Remove insulator assembly to engine block bolts, then remove insulator and heat shield.

ENGINE, REPLACE

1976 Bobcat & Pinto & 1974-76 Mustang II

1. Disconnect battery cables and remove hood.
2. Remove air cleaner and intake duct.
3. Drain cooling system, disconnect radiator hoses from radiator, then remove radiator. Disconnect heater hoses from engine block and water pump.

NOTE: Remove fan shroud and position shroud over fan before removing radiator.

4. Remove alternator and bracket.
5. Disconnect ground wires from engine block.
6. Disconnect fuel tank line from fuel pump and plug line.
7. Disconnect all linkage from engine and wires from ignition coil.

NOTE: If equipped with Thermactor system, remove or disconnect system components interfering with engine removal.

8. Raise vehicle and place on jack stands.
9. Disconnect exhaust pipes from exhaust manifold and remove starter.
10. Remove engine front support through bolts.
11. On vehicles equipped with automatic transmission, disconnect converter from flywheel, remove downshift rod, then remove converter housing to engine bolts and adapter plate to converter housing bolt.

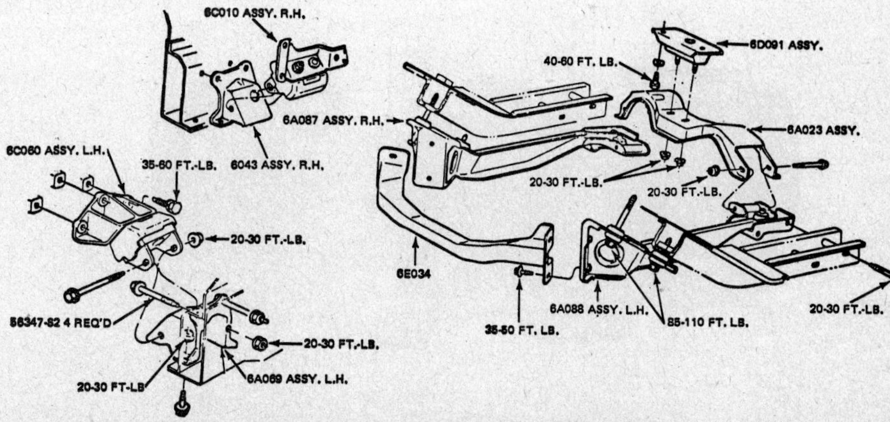

Fig. 1 Engine mounts. 1974-76 Mustang II

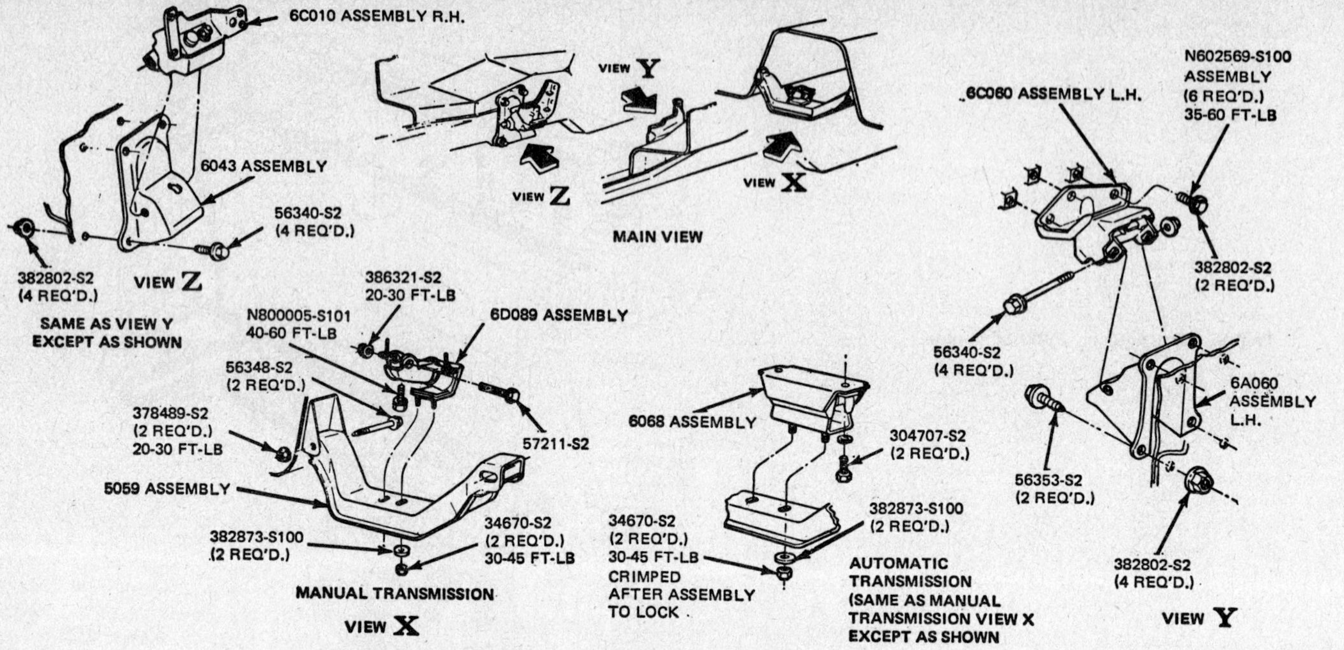

Fig. 1A Engine mounts. 1975-76 Bobcat & Pinto

12. On vehicles equipped with manual transmission, remove clutch linkage and bell housing to engine bolts.
13. On all models, lower vehicle and attach a suitable lifting sling to brackets on exhaust manifold.
14. Support transmission with a suitable jack, raise engine slightly and pull from transmission, then lift engine from engine compartment.

CYLINDER HEAD, REPLACE

1. Disconnect battery ground cable, disconnect linkage and drain coolant.
2. Remove distributor, coolant hoses, rocker arm covers, fuel line and filter, carburetor and intake manifold.
3. Remove rocker arm shaft, oil baffles and push rods.
4. Remove exhaust manifold.
5. Remove cylinder head bolts and cylinder head.
6. Reverse procedure to install and torque bolts in sequence shown in, Fig. 2.

VALVE ARRANGEMENT

Front to Rear

2800 cc engine—

Right .... I-E-I-E-E-I
Left .... I-E-E-I-E-I

VALVE LIFT SPECS.

Engine	Year	Intake	Exhaust
2800 cc	1974-76	.3730	.3730

VALVE TIMING

Intake Opens Before TDC

Engine	Year	Degrees
2800 cc	1974-76	20

VALVES, ADJUST

Cold Setting

1. Remove rocker arm covers.

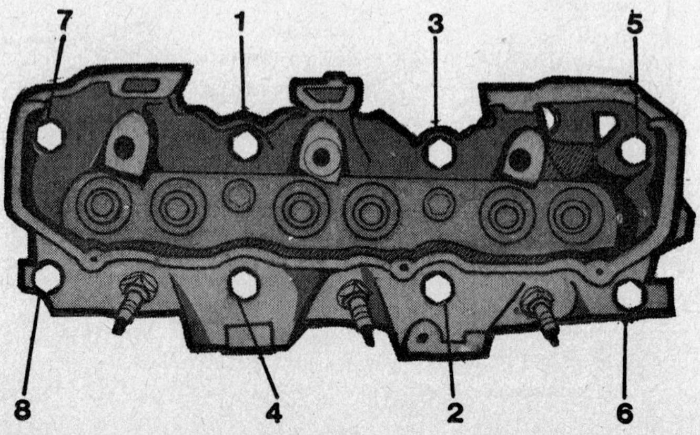

Fig. 2 Cylinder head tightening sequence. 2800 cc engine

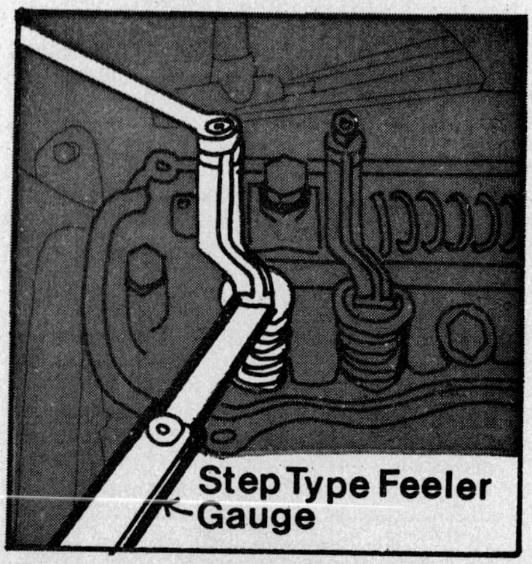

Fig. 3 Adjusting valve lash. 2800 cc engine

Fig. 4 Rocker arm replacement. 2800 cc engine

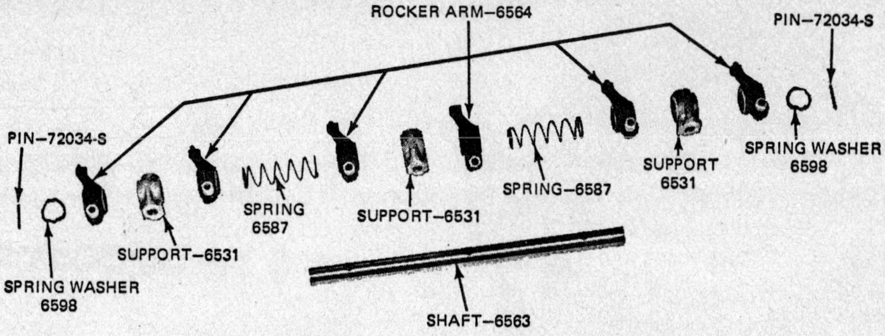

Fig. 5 Rocker arm shaft assembly. 2800 cc engine

2. Rotate crankshaft until No. 1 cylinder is at TDC, compression stroke, and set valve lash on No. 1 cylinder to specifications as listed in the "Valve Specifications" table.
3. Rotate crankshaft 120 degrees and set valve lash on No. 4 cylinder.
4. Rotate crankshaft an additional 120 degrees and set valve lash on No. 2 cylinder. The remaining valves are adjusted in the same manner as above, following the firing order found in the front of this chapter.

Hot Setting

With engine at operating temperature and idling, set valve lash with a step type feeler gauge, Fig. 3, to specifications as listed in the "Valve Specifications" table.

VALVE GUIDES

Valve guides consist of holes bored in the cylinder head. For service the guides can be reamed oversize to accommodate valves with oversize stems of .003, .015 and .030".

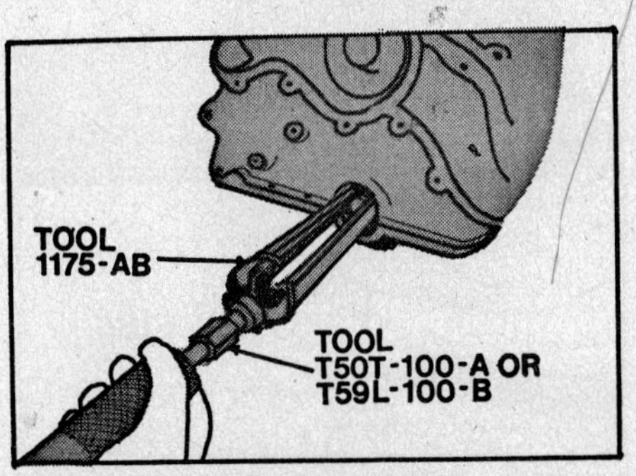

Fig. 6 Removing crankshaft front oil seal. 2800 cc engine

ROCKER ARM SERVICE

1. Disconnect throttle rod from carburetor and remove rocker arm cover.
2. Remove rocker arm shaft stand bolts, rocker arm shaft assembly and oil baffle, Fig. 4.
3. Remove cotter pin and spring washer from ends of rocker shaft and slide rocker arms, springs and shaft supports off shaft, marking components for proper reassembly, Fig. 5.
4. Remove plugs from shaft ends by drilling a hole in one plug, insert a long rod through drilled plug and knock the opposite plug from shaft. Remove the drilled plug in same manner.
5. With a blunt tool, install plugs in end of rocker shafts with cup side out.
6. Install spring washer and cotter pin on one end of shaft and install components in proper sequence as marked during disassembly.

NOTE: Oil holes in rocker shaft must face downward during installation.

VALVE LIFTERS, REPLACE

Remove cylinder head as outlined previously and using a magnet, remove lifters from their bores.

TIMING CASE COVER

1. Remove oil pan as described further on.
2. Drain coolant and remove radiator, then remove any other components as necessary to obtain clearance.
3. If equipped, disconnect A/C compressor and bracket and place aside.
4. Remove alternator, Thermactor pump, drive belts, fan, water pump, hoses and harmonic balancer.
5. Remove cover retaining bolts and remove cover.

CRANKSHAFT FRONT OIL SEAL

The crankshaft front oil seal may be serviced without removing the cylinder front cover as follows:

1. Drain coolant and remove radiator, crankshaft pulley and water pump drive belt.
2. Pull oil seal from front cover, Fig. 6.
3. Install new oil seal with tool T72C-6150, Fig. 7.
4. Install crankshaft pulley, water pump drive belt, radiator, then refill cooling system.

TIMING GEARS

1. Drain, then remove radiator and oil pan.
2. Remove cylinder front cover and water pump.
3. Align timing marks, Fig. 8.
4. Using a suitable gear puller, remove crankshaft gear and key.

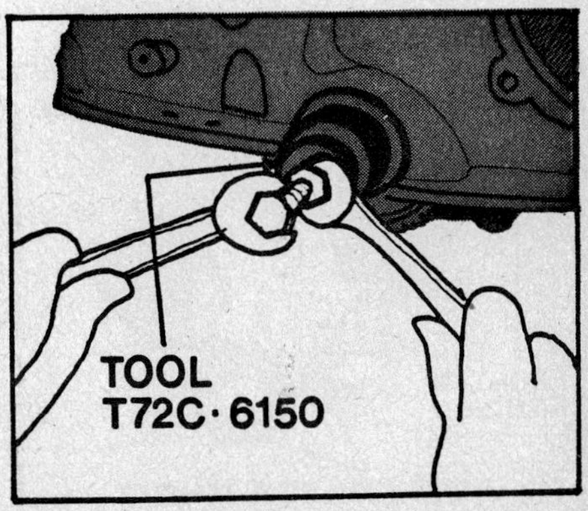

Fig. 7 Installing crankshaft front oil seal. 2800 cc engine

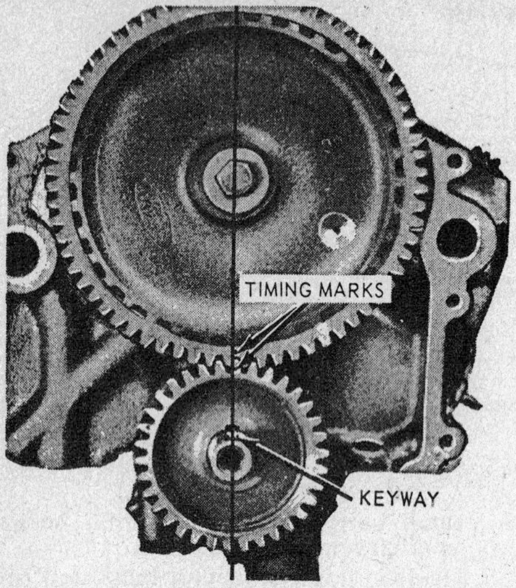

Fig. 8 Valve timing marks. 2800 cc engine

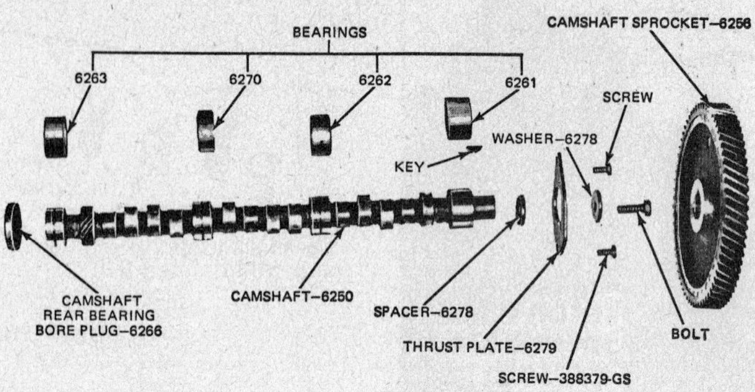

Fig. 10 Camshaft components. 2800 cc engine

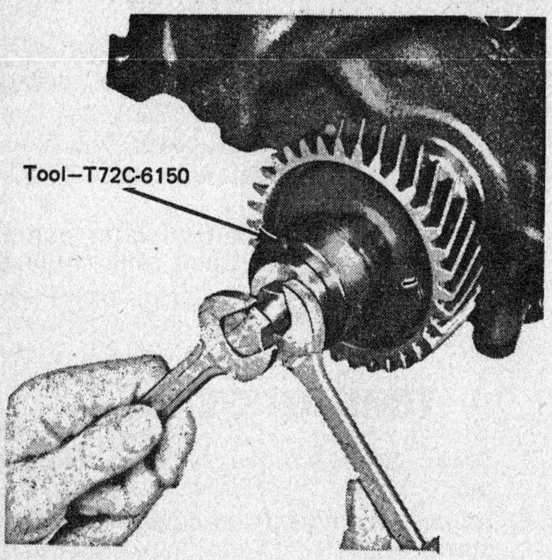

Fig. 9 Crankshaft gear installation. 2800 cc engine

5. Remove camshaft gear with a suitable gear puller.

NOTE: Do not rotate crankshaft or camshaft with gears removed as rotation of either component can result in improper valve timing.

---

6. Install key in camshaft, then press camshaft gear onto camshaft.
7. Install key in crankshaft, then press crankshaft gear onto crankshaft with tool T72C-6150, Fig. 9, and make sure timing marks are aligned, Fig. 8.
8. Install cylinder front cover, water pump, radiator and oil pan.
9. Refill cooling system and oil pan. Start engine and adjust ignition timing, if necessary.

CAMSHAFT, REPLACE

1. Drain coolant, then remove radiator, fan, water pump pulley and belt.
2. Remove distributor, alternator, Thermactor pump, fuel line, filter, carburetor and intake manifold.
3. Remove rocker arm covers, rocker arm assemblies, pushrods and lifters. Identify pushrods and lifters so they can be reinstalled in their original location.
4. Remove oil pan as described further on.
5. Remove timing case cover as described previously.
6. Remove camshaft gear retaining bolt and slide gear off shaft and remove camshaft thrust plate.
7. Carefully remove camshaft from engine using care to avoid damage to camshaft bearings.

PISTON & ROD ASSEMBLY

Assemble the piston to the rod with the notches facing front of engine and the numbered side of the rod toward left side of engine, Fig. 11.

PISTONS, PINS & RINGS

Oversize pistons and rings are available in .020″ and .040″ oversizes. Oversize pins are not available.

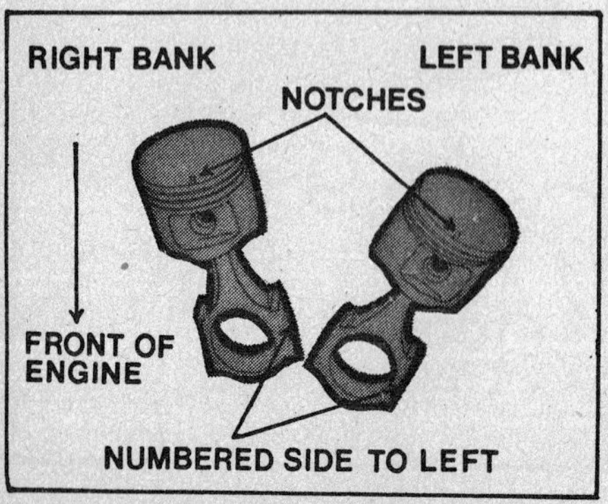

Fig. 11 Piston & rod. 2800 cc engine

Fig. 12 Removing crankshaft rear oil seal. 2800 cc engine

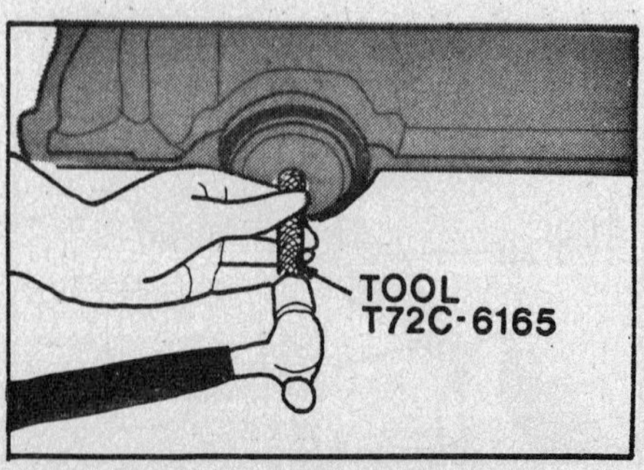

Fig. 13 Installing crankshaft rear oil seal. 2800 cc engine

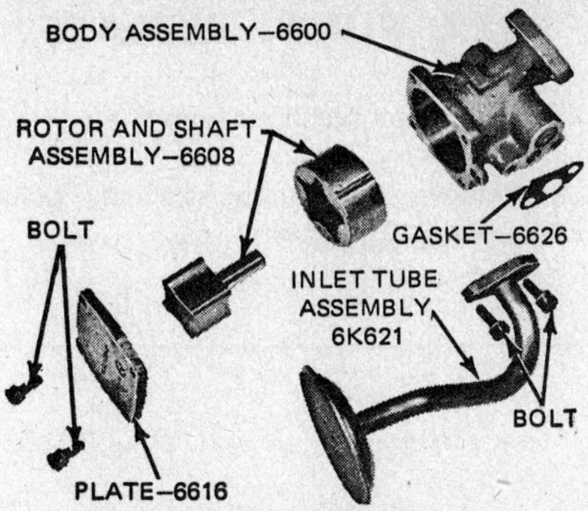

Fig. 14 Oil pump. 2800 cc engine

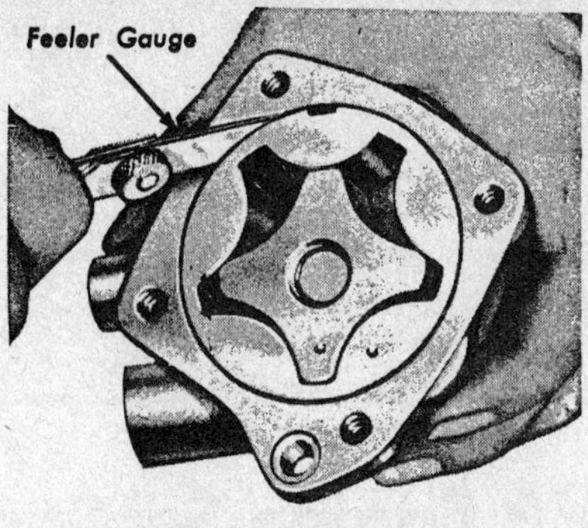

Fig. 15 Checking outer rotor to housing clearance. 2800 cc engine

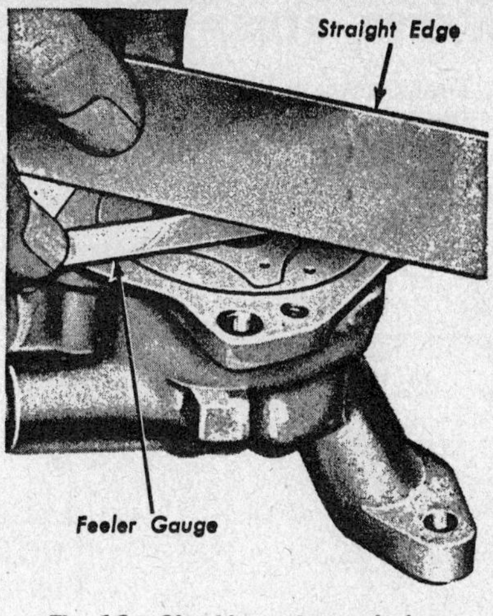

Fig. 16 Checking rotor end play. 2800 cc engine

CRANKSHAFT REAR OIL SEAL

1. Remove transmission assembly.
2. On automatic transmission vehicles, remove flywheel.
3. On manual transmission vehicles, remove clutch assembly, flywheel, clutch housing and rear plate.
4. Punch two holes on opposite sides of seal just above bearing cap to cylinder block split line and install a sheet metal screw in each hole. Using two large screwdrivers, pry evenly on both screws to remove seal, Fig. 12.

NOTE: Use care to avoid damaging the crankshaft oil seal surface.

5. Install new seal with tool T72C-6165, Fig. 13.

OIL PAN, REPLACE

1. Disconnect battery ground cable.
2. Remove oil level dipstick.
3. On 1975-76 vehicles, drain cooling system and disconnect upper and lower radiator hoses.
4. Remove fan shroud attaching bolts and place shroud over fan.
5. On 1974 vehicles, loosen alternator bracket and adjustment bolt.
6. Raise vehicle and drain crankcase, then remove splash shield and starter.
7. On 1975-76 vehicles, disconnect steering gear and power steering hoses (if equipped) and position gear aside, then disconnect sway bar and rotate to allow clearance.
8. Remove engine front support nuts, then raise engine and place wood blocks between engine front supports and chassis.
9. Remove converter or clutch housing cover.
10. Drain crankcase, then remove oil pan bolts and oil pan.

OIL PUMP, REPLACE

The oil pump, Fig. 14, can be removed after oil pan removal.

OIL PUMP REPAIRS

1. Remove end plate and withdraw O ring from groove in body.
2. Check clearance between lobes of inner and outer rotors. This should not exceed .006″. Rotors are supplied only in a matched pair.
3. Check clearance between outer rotor and the housing, Fig. 15. This should not exceed .010″.
4. Place a straightedge across face of pump body, Fig. 16. Clearance between face of rotors and straightedge should not exceed .005″.
5. If necessary to replace rotor or drive shaft, remove outer rotor and then drive out retaining pin securing the skew gear to drive shaft and pull off the gear.
6. Withdraw inner rotor and drive shaft.

BELT TENSION DATA

	New Ft. Lbs.	Used Ft. Lbs.
1974-76 All	140	110

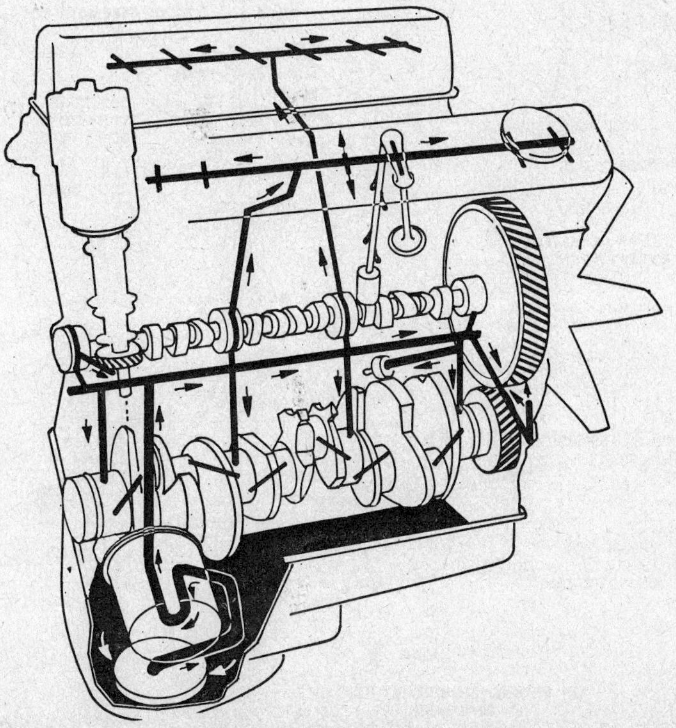
Engine oiling system. 2800 cc engine

WATER PUMP, REPLACE

1. Drain coolant and disconnect heater hose and lower radiator hose from pump.
2. Loosen alternator and remove drive belt.
3. Remove fan and pulley.
4. Remove water pump mounting bolts, water pump, water inlet housing and thermostat.

FUEL PUMP PRESSURE

Year	Engine	Pressure Lbs.
1974-76	2800 cc	3½-4½

FUEL PUMP, REPLACE

1. Disconnect fuel lines from pump.
2. Remove fuel pump attaching bolts and fuel pump.

V8 Engine Section

NOTE: Refer to the Ford & Mercury—Compact & Intermediate chapter for detailed service on this engine.

ENGINE MOUNTS REPLACE

1975-76 Mustang II

1. Remove fan shroud screws and support engine with a suitable jack and a block of wood under the oil pan.
2. Remove insulator to frame through bolt, Fig. 1.
3. Remove insulator to engine block attaching bolts.
4. Raise engine slightly, then remove insulator and heat shield, if equipped.
5. Reverse procedure to install.

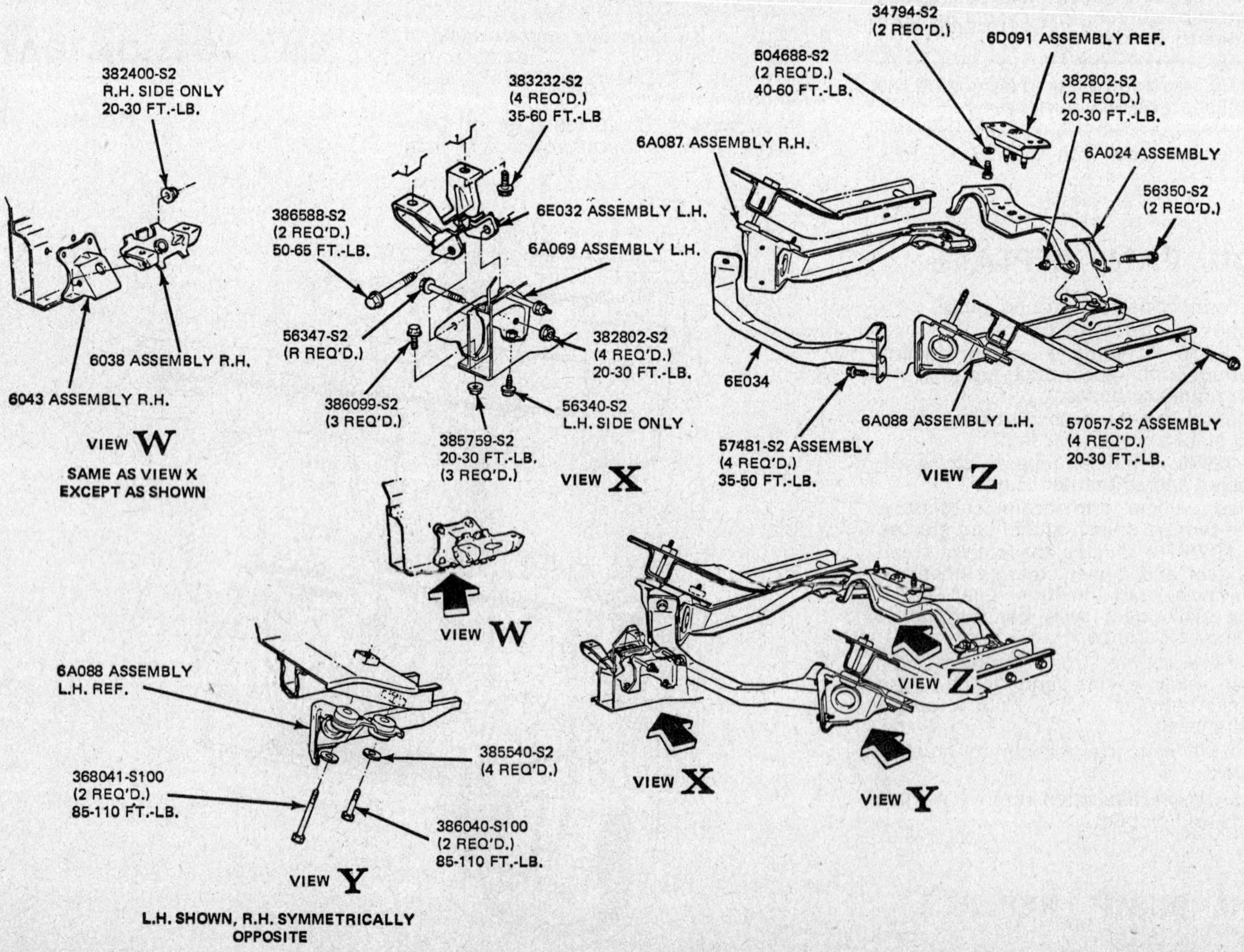

Fig. 1 Engine mounts. 1975-76 Mustang II

Clutch & Transmission Section

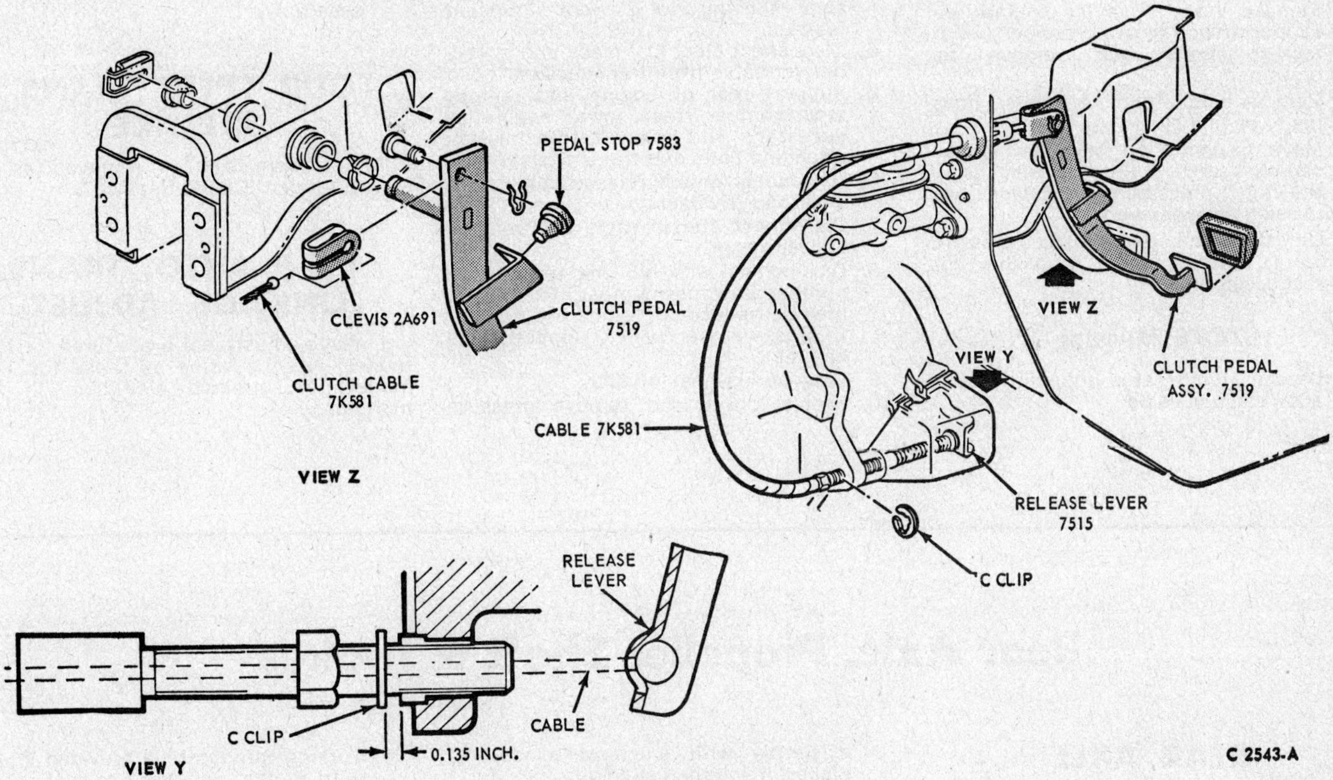

Fig. 1 Clutch linkage. Bobcat & Pinto

CLUTCH PEDAL, ADJUST

Early 1971

1. From under car, pull the flexible cable toward front of car until "C" clip can be removed from cable. Remove the clip.
2. Continue to pull cable toward front of car until all free movement of the release bearing is eliminated.
3. While holding cable in zero free movement position, place a .135" spacer against flywheel boss on the engine side and install the "C" clip into closest possible groove next to the spacer, Fig. 1.
4. Remove the spacer and release the cable.

Late 1971; 1972-74 Pinto

The C clip has been eliminated from the cable and is replaced by a locknut. To adjust, proceed as follows:

1. From under car, loosen cable locknuts and adjusting nut at flywheel housing.
2. Pull cable toward front of car until all free movement of the release lever is eliminated.
3. Holding cable in this position, place a 1/4" spacer block against the flywheel housing boss (on engine side). Run adjusting nut against the spacer finger tight.
4. Tighten the front locknut against the adjusting nut, being careful not to disturb the adjustment. Torque locknut to 40-60 ft. lbs.
5. With spacer still in place, tighten the rear locknut against the flywheel housing boss. Remove the spacer.

1974-76 Mustang II

1. Remove cable retaining clip at dash panel and remove cable retaining screw from fender apron.
2. Pull cable toward front of vehicle until nut can be rotated. Rotate nut from adjustment sleeve about 1/4 inch.
3. Release cable to neutralize the linkage and pull cable until free movement of release lever is eliminated.
4. Rotate adjusting nut toward adjustment sleeve until contact is made, then index into the next notch.
5. Install cable retaining clip, cable retaining bracket and retaining screw.

1975-76 Bobcat & Pinto

1. Loosen clutch cable lock nut at flywheel housing.
2. Pull cable toward front of vehicle so the nylon adjuster nut tabs are clear of the housing boss, then rotate nut toward vehicle front approximately 1/4 inch.
3. Release the cable, neutralizing the system, and pull cable forward again so release lever free movement is eliminated.
4. Rotate the adjusting nut until contact is made between index tab face and the housing, then index the tabs to engage the nearest housing groove.
5. Torque lock nut to 15 ft. lbs.

CLUTCH, REPLACE

1971-76 Bobcat & Pinto

1. Remove shift lever by removing knob and boot, then compress rubber spring and remove retaining snap ring. Bend shift lever lock tabs up and remove plastic dome nut from extension housing.
2. Raise vehicle on a hoist.
3. Disconnect drive shaft from U joint flange and slide drive shaft off transmission output shaft. Insert tool over output shaft to prevent loss of lubricant.
4. Disconnect speedometer cable from extension housing.
5. Disconnect lower end of clutch cable at release lever.

6. Remove starter motor.
7. Remove bolts securing engine rear plate to front lower part of flywheel housing.
8. Remove bolt attaching engine rear support. Also remove crossmember attaching bolts and remove the crossmember.
9. Remove bolts attaching flywheel housing to engine block.
10. Move transmission and flywheel assembly rearward until housing clears the clutch pressure plate. Lower transmission and remove.
11. Unfasten and remove the pressure plate, marking same to assure correct assembly.

1974-76 Mustang II

1. Loosen clutch cable adjusting nut to allow slack in cable.
2. Position gear shift lever in neutral, then remove the three lever attaching screws and remove lever.
3. Remove driveshaft then cover extension housing to prevent lubricant leakage.
4. Disconnect electrical leads and speedometer cable from transmission.
5. Support rear of engine and remove crossmember, then lower engine as necessary and remove transmission attaching bolts and transmission.
6. Disconnect clutch release cable from lever and flywheel housing.
7. Disconnect starter cable and remove starter motor.
8. On models with V6 engine, remove number 2A crossmember. This crossmember is located behind the number 2 crossmember which supports the engine.
9. Remove flywheel housing.
10. Evenly loosen and remove pressure plate attaching screws to prevent distortion of pressure plate. If pressure plate is to be reused, mark pressure plate and flywheel to assure correct assembly.

FOUR SPEED TRANS., REPLACE

The transmission is removed as described under "Clutch Replace".

1976 AUTO. TRANS. LINKAGE, ADJUST

Linkage adjustment procedures for 1976 models are the same as those for 1975 models as outlined elsewhere in this manual.

Rear Axle, Propeller Shaft & Brakes

REAR AXLE

Integral Type

This rear axle, Fig. 1, is an integral design hypoid with the centerline of the pinion set below the centerline of the ring gear. The semi-floating axle shafts are retained in the housing by ball bearings and bearing retainers at axle ends.

The differential is mounted on two opposed tapered roller bearings which are retained in the housing by removable caps. Differential bearing preload and drive gear backlash is adjusted by nuts located behind each differential bearing cup.

The drive pinion assembly is mounted on two opposed tapered roller bearings. Pinion bearing preload is adjusted by a collapsible spacer on the pinion shaft. Pinion and ring gear tooth contact is adjusted by shims between the rear bearing cone and pinion gear.

Axle Shaft, Bearing & Oil Seal

1. Remove wheel and tire from brake drum.
2. Remove Tinnerman nuts that secure brake drum to axle flange and remove brake drum.
3. Working through hole in each axle flange, remove nuts that secure wheel bearing retainer plate. Then pull the axle shaft assembly out of the housing being careful not to cut or rough up the seal.

NOTE: *The brake backing plate must not be dislodged. Replace one nut to hold the plate in place after shaft is removed.*

4. If wheel bearing is to be replaced, loosen inner retainer ring by nicking it deeply with a chisel in several places. It will then slide off.
5. Remove bearing from shaft.

Removable Carrier Type

In these axles, Fig. 2, the drive pinion is straddle-mounted by two opposed tapered roller bearings which support the pinion shaft in front of the drive pinion gear, and a straight roller bearing that supports the pinion shaft at the rear of the pinion gear. The drive pinion is assembled in a pinion retainer that is bolted to the differential carrier. The tapered roller bearings are preloaded by a collapsible spacer between the bearings. The pinion is positioned by a shim or shims located between the drive pinion retainer and the differential carrier.

The differential is supported in the carrier by two tapered roller side bearings. These bearings are preloaded by two threaded ring nuts or sleeves between the bearings and the pedestals. The differential assembly is positioned for proper ring gear and pinion backlash by varying the adjustment of these ring nuts. The differential case houses two side gears in mesh with two pinions mounted on a pinion shaft which held in place by a pin. The side gears and pinions are backed by thrust washers.

The axle shafts are of unequal length, the left shaft being shorter than the right. The axle shafts are mounted in sealed ball bearings which are pressed on the shafts.

Axle Shaft, Replace

1. Remove wheel assembly.
2. Remove brake drum from flange.
3. Working through hole provided in axle shaft flange, remove nuts that secure wheel bearing retainer.
4. Pull axle shaft out of housing. If bearing is a tight fit in axle housing use a slide hammer-type puller.

Brake carrier plate must not be dislodged. Install one nut to hold the plate in place after axle shaft is removed.

NOTE: On 1974-75 models, remove brake backing plate.

5. If the axle shaft bearing is to be replaced, loosen the inner retainer by nicking it deeply with a chisel in several places. The bearing will then slide off easily.
6. Press bearing from axle shaft.
7. Inspect machined surface of axle shaft and housing for rough spots that would affect sealing action of the oil seal. Carefully remove any burrs or rough spots.
8. Press new inner bearing retainer on shaft until it seats firmly against shoulder on shaft.
9. Press inner bearing retainer on shaft until it seats firmly against bearing.
10. If oil seal is to be replaced, use a hook-type tool to pull it out of housing. Wipe a small amount of oil resistant sealer on outer edge of seal before it is installed.

Installation

1. Place a new gasket on each side of brake carrier plate and slide axle shaft into housing. Start the splines into the differential side gear and

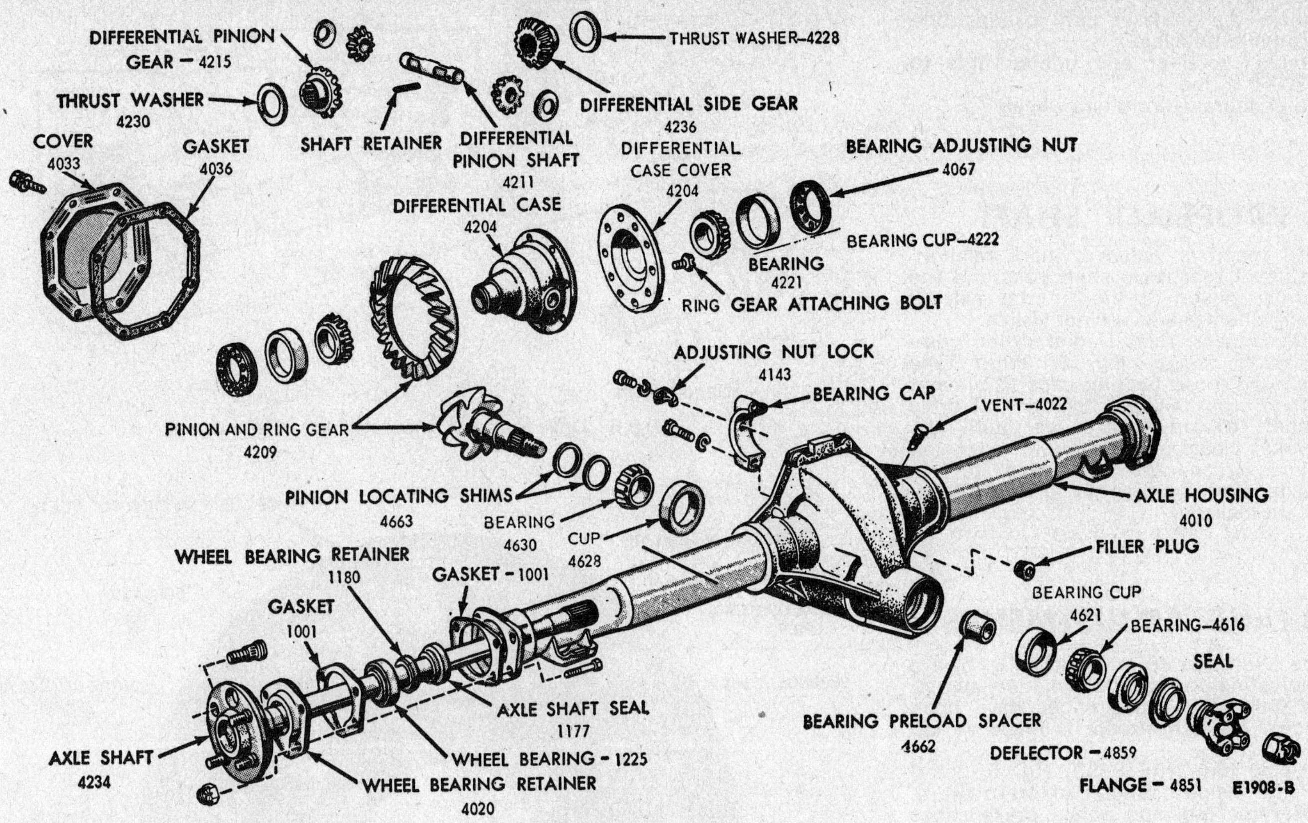

Fig. 1 Disassembled integral rear axle

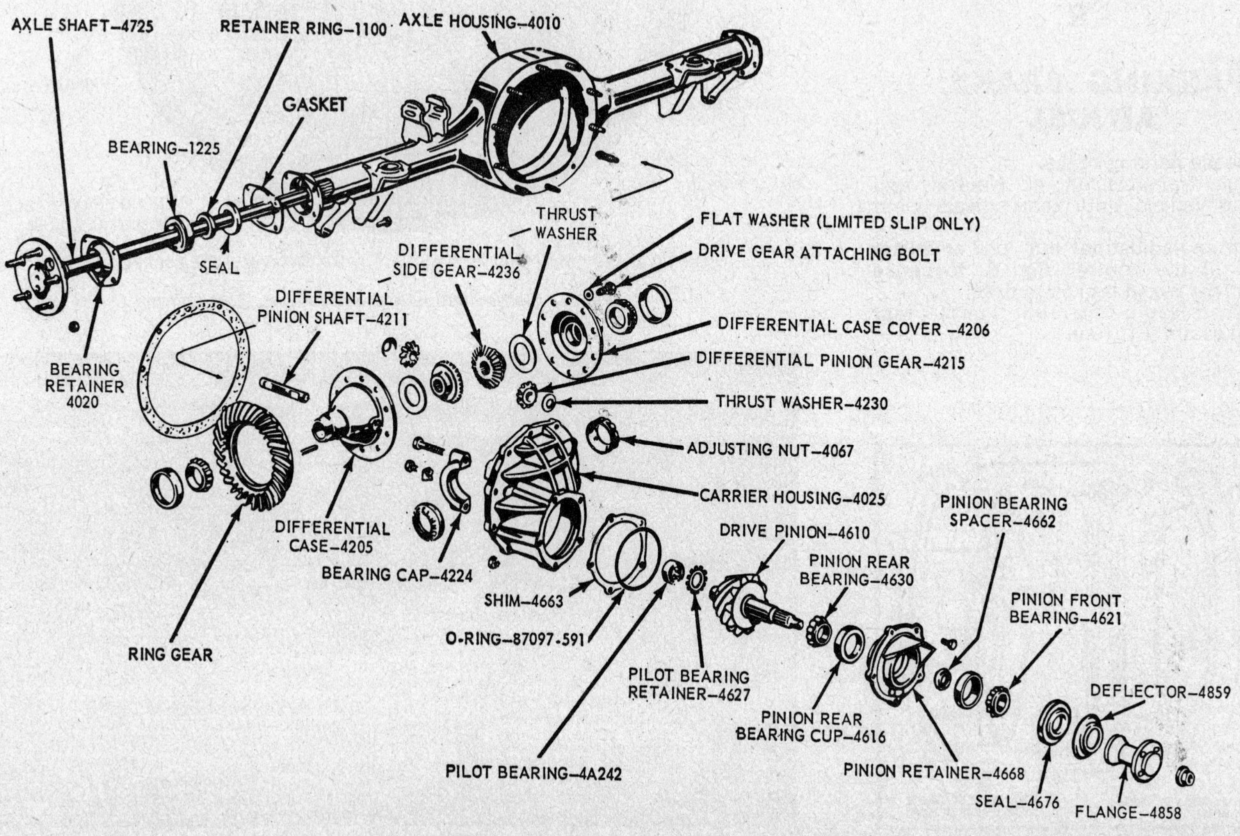

Fig. 2 Disassembled rear axle with removable carrier

push the shaft in until bearing bottoms in housing.
2. Install retainer and tighten nuts to 30-40 ft. lbs.
3. Install brake drum and wheel.

PROPELLER SHAFT

1. To maintain balance, mark relationship of rear drive shaft yoke and the drive pinion flange of the axle if alignment marks are not visible.
2. Disconnect rear U-joint from companion flange, Fig. 5. Wrap tape around loose bearing caps to prevent them from falling off spider. Pull drive shaft toward rear of car until slip yoke clears transmission extension housing and the seal. Install tool in extension housing to prevent lubricant leakage.

BRAKE ADJUSTMENTS

The hydraulic drum brakes, Fig. 6, are self-adjusting and require a manual adjustment only after brake shoes have been replaced. The adjustment is made as follows:

1. Using tool HRE 8650, Fig. 7, determine inside diameter of brake drum.
2. Reverse tool and adjust brake shoes to fit the gauge. Hold automatic adjusting lever out of engagement while rotating adjusting screw, to prevent burring slots in screw.

PARKING BRAKE, ADJUST

1. Release parking brake.
2. Place transmission in Neutral and raise vehicle until rear wheels clear floor.
3. Tighten adjusting nut on equalizer rod at the control, Fig. 8, to cause the rear wheel brakes to drag.
4. Loosen adjusting nut until rear brakes are just free.

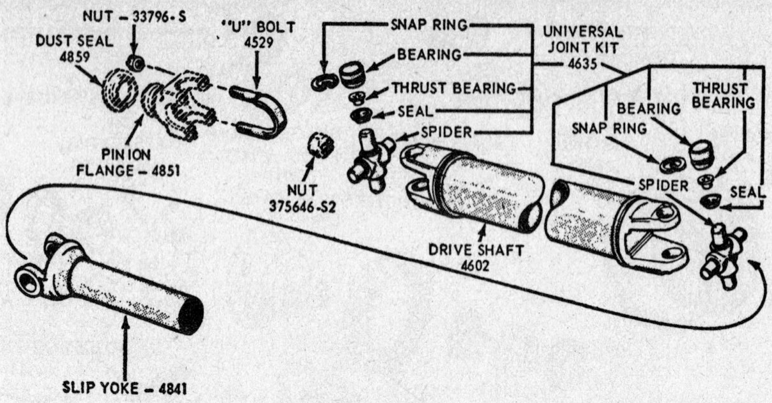

Fig. 5 Drive shaft and universal joints disassembled

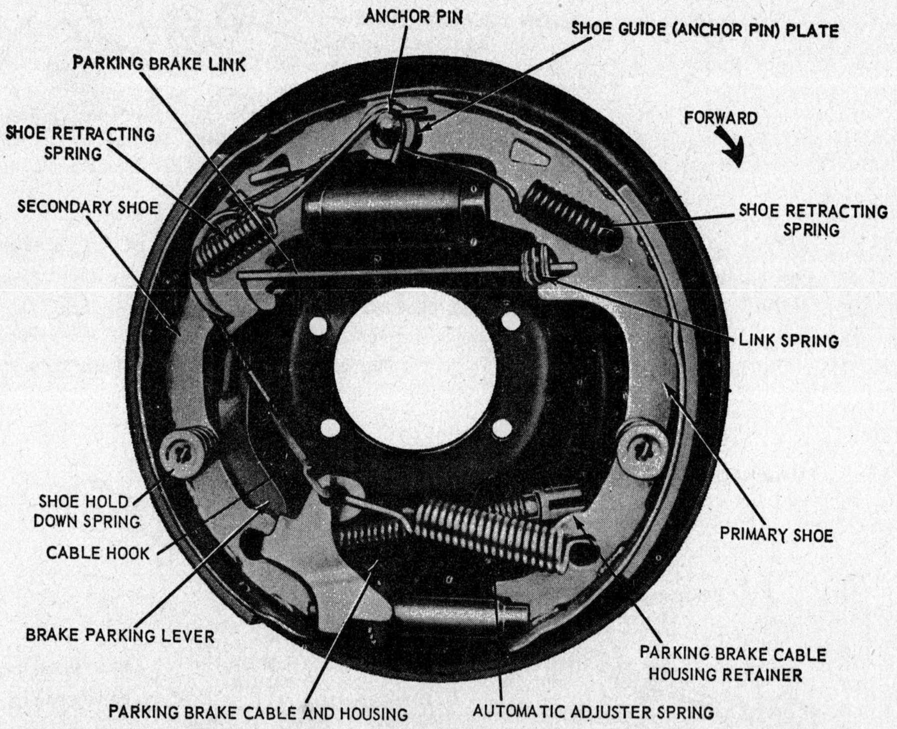

Fig. 6 Self-adjusting brake assembly. Rear shown

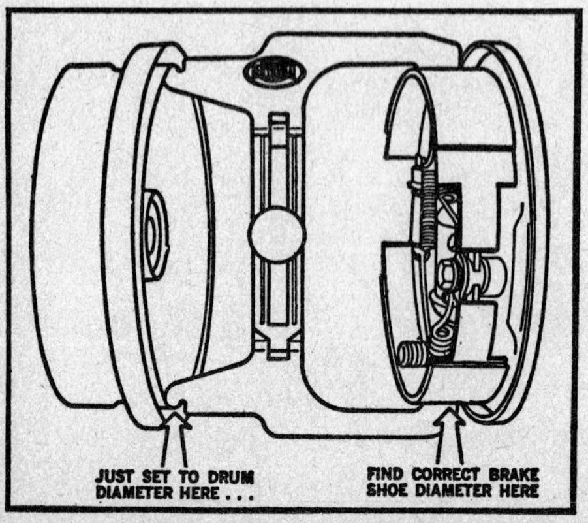

Fig. 7 Brake adjustment

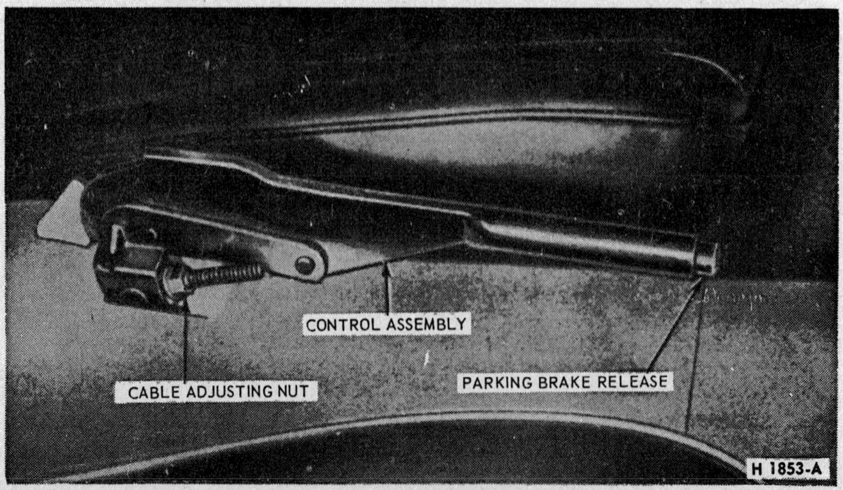

Fig. 8 Parking brake adjustment

MASTER CYLINDER, REPLACE

Exc. Power Brakes

1. Disconnect stoplight switch wires at connector. Remove spring retainer and slide stop light switch off brake pedal pin just far enough to clear end of pin, then lift switch straight upward from the pin.
2. Slide master cylinder push rod and nylon washers and bushings off brake pedal pin.
3. Remove brake tubes from master cylinder ports.
4. Unfasten and remove master cylinder forward and upward from vehicle.

Power Brakes

Disconnect brake tubes from master cylinder, then remove attaching nuts and slide master cylinder forward and upward from vehicle.

Rear Suspension

SHOCK ABSORBER, REPLACE

1. With rear axle supported properly disconnect shock absorber from lower mounting.
2. Remove bolts securing upper mounting bracket to underbody.
3. Remove bracket from shock absorber.
4. Reverse procedure to install.

LEAF SPRINGS & BUSHINGS, REPLACE

1. Raise rear of vehicle and support at frame. Support axle with a suitable jack.
2. Disconnect shock absorbers from lower mountings.
3. Lower jack and remove spring plate "U" bolts and spring plate, Fig. 1.
4. Raise axle to remove weight from spring and disassemble rear shackle.
5. Remove spring front mount bolt.
6. Replace spring front eye bushing as necessary, Figs. 2 and 3.
7. Reverse procedure to install.

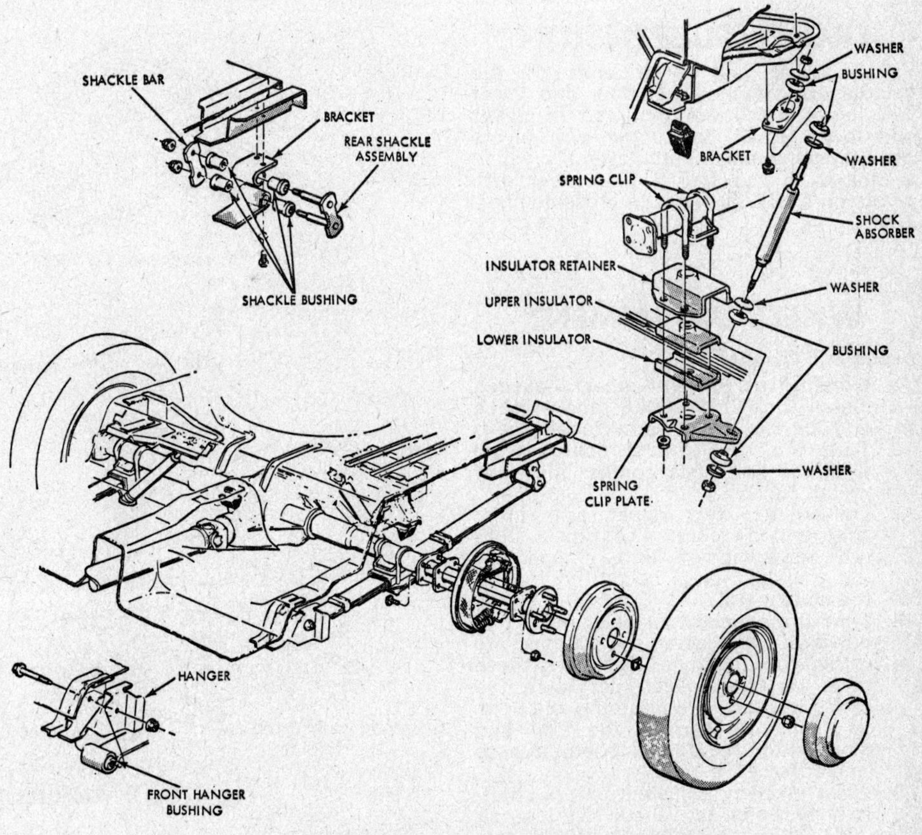

Fig. 1 Bobcat, Mustang II & Pinto rear suspension

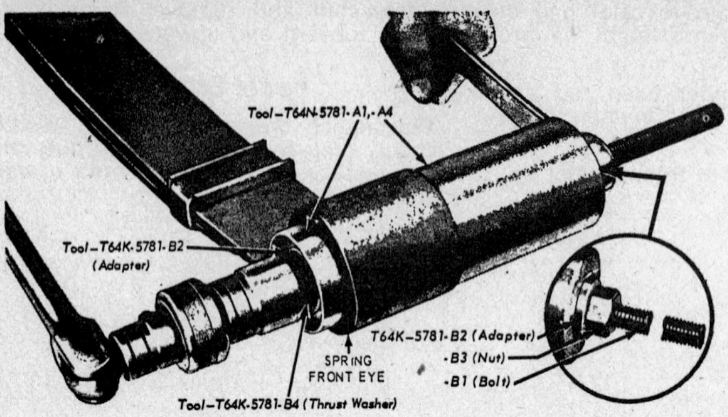

Fig. 2 Spring, front bushing removal

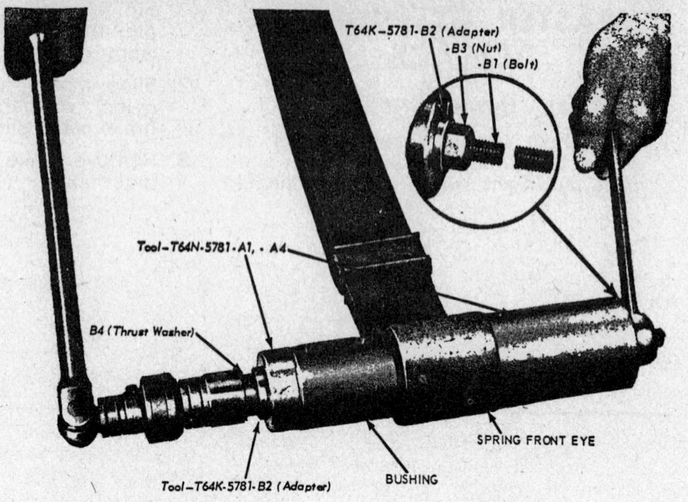

Fig. 3 Spring, front bushing installation

Front End & Steering Section

FRONT SUSPENSION

The upper and lower ends of the spindle are attached to upper and lower ball joints which are mounted in upper and lower arms. The upper arm pivots on a bushing and shaft assembly which is bolted to the frame. The lower arm pivots on a bolt in the front crossmember, Fig. 1.

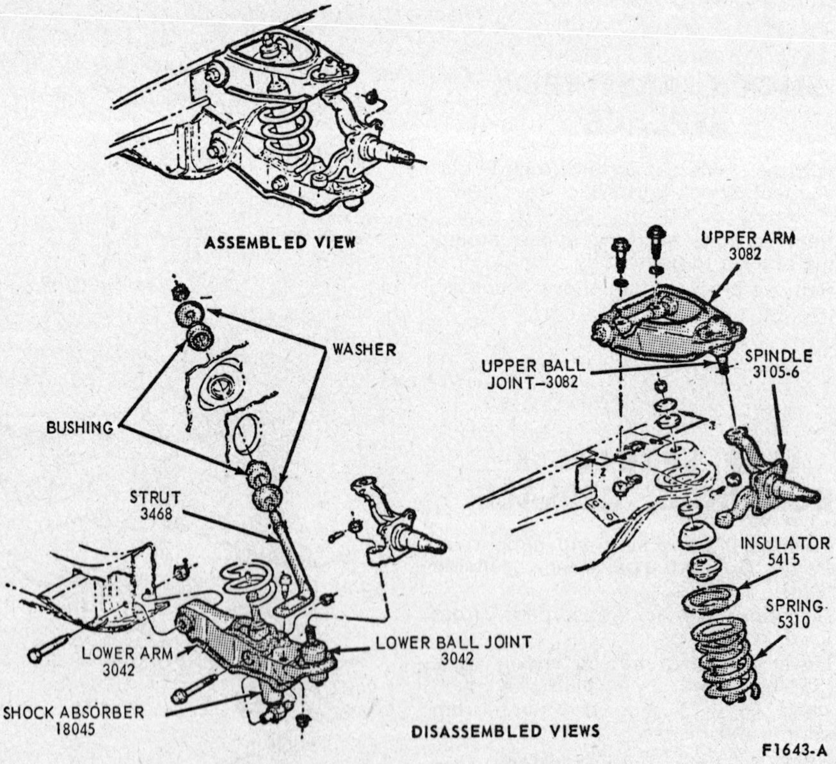

Fig. 1 Front suspension assembly. Typical

WHEEL ALIGNMENT

Caster and Camber

1. Working inside front wheel housing, install tool T71P-3000A, one at each end of the upper arm inner shaft. Turn the special tool bolts inward until the bolt ends contact the body metal, Fig. 2.
2. Loosen the two upper arm inner shaft-to-body bolts. The upper shaft will move inboard until stopped by the tool bolt ends solidly contacting the body metal.
3. Turn the special tool bolts inward or outward until caster and camber are within specifications. Tightening these bolts on the special tool force the arm outward; while loosening the bolts on the tools permits the arm and inner shaft to move inboard due to weight force.
4. When properly adjusted, torque shaft-to-body bolts to 75-105 ft. lbs. on 1971-74 models and 95-110 ft. lbs. on 1975-76 models, then remove the special tools.

TOE-IN, ADJUST

1. Check to see that steering shaft and steering wheel marks are in alignment and in the top position.
2. Loosen clamp screw on the tie rod bellows and free the seal on the rod to prevent twisting of the bellows, Fig. 3.
3. Loosen tie rod jam nut.
4. Use suitable pliers to turn the tie rod inner end to correct the adjustment to specifications. Do not use pliers on tie rod threads. Turning to reduce number of threads showing will increase toe-in. Turning in the opposite direction will reduce toe-in.

WHEEL BEARINGS, ADJUST

1. Raise vehicle until wheel and tire clear floor.
2. Remove wheel cover and dust cap from hub.
3. Remove cotter pin and lock nut.
4. While rotating wheel assembly, torque the adjusting nut to 17-25 ft. lbs. to seat the bearings.
5. Using a $1\frac{1}{8}$" box wrench, back off the adjusting nut one half turn. Retighten the nut to 10-15 in. lbs. with a torque wrench or finger tight.
6. Locate the nut lock on the adjusting nut so the castellations on the lock are aligned with the cotter pin hole in the spindle.
7. Install new cotter pin and replace dust cap and wheel cover.

WHEEL BEARINGS, REPLACE

(Disc Brakes)

1. Raise car and remove front wheels.
2. Use a $\frac{3}{4}$" wrench to loosen the large bolt at top of caliper assembly and washer at the front of the caliper. Loosen it until it can be turned with the fingers.
3. Remove the smaller bolt at bottom of caliper with a $\frac{5}{8}$" wrench.
4. Insert a strong piece of wire carefully through the upper opening in the caliper and fasten it. Position the free end of the wire over the suspension upper arm.
5. When removing caliper from disc the brake pads must be held apart. Do this by inserting a piece of wood or cardboard. While holding the caliper, remove the large bolt in front. Now carefully slide the caliper back and slightly upward to remove it. While doing this, insert the wood or cardboard between the brake pads.
6. Carefully move caliper back to sus-

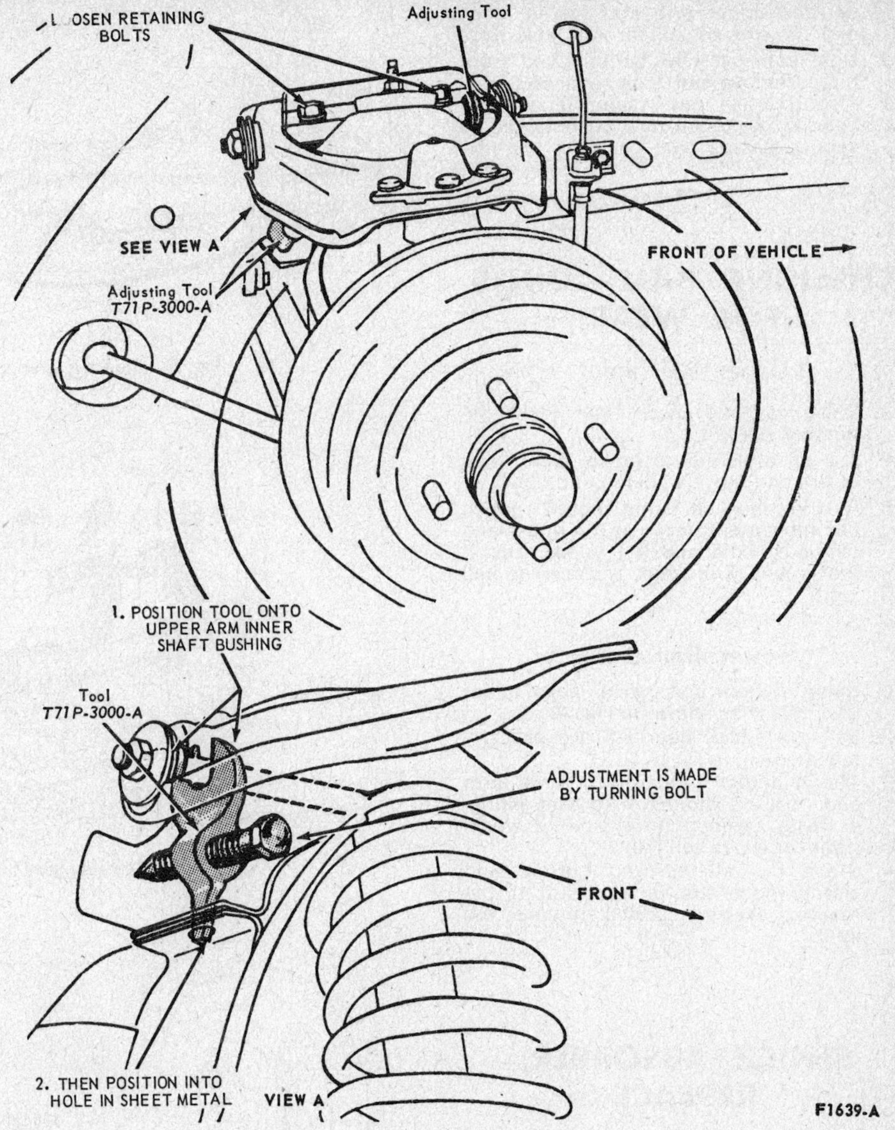

Fig. 2 Caster and camber adjustment

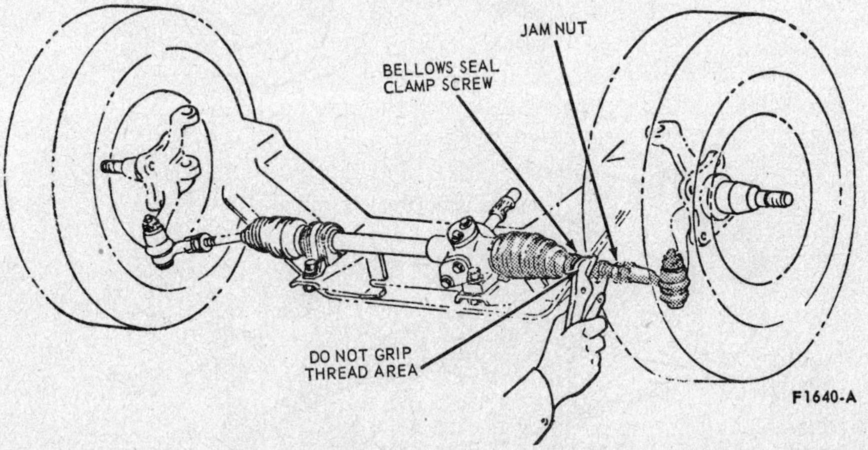

Fig. 3 Toe-in adjustment

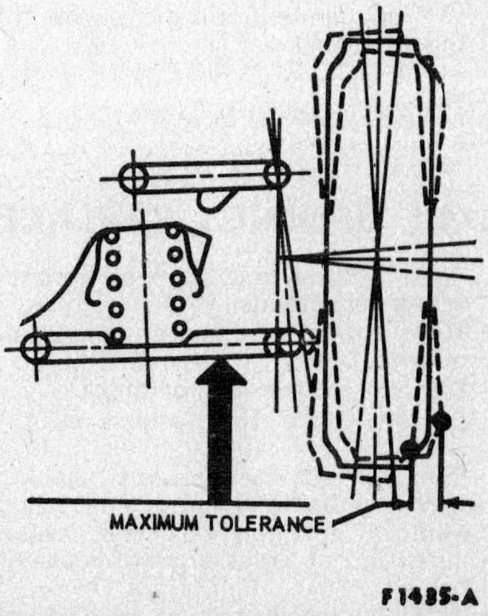

Fig. 4 Measuring lower ball joint play

pension upper arm and fasten loose end of wire so caliper will not drop.

7. Dust cap can now be removed from hub. Remove nut lock, etc. and rock disc to ease out washer and outer bearing. Disc can now be removed to service grease seal or inner bearing.

CHECKING BALL JOINTS FOR WEAR

Upper Ball Joint

1. Raise car and place floor jacks beneath lower arms.
2. Ask an assistant to grasp lower edge of tire and move wheel in and out.
3. As the wheel is being moved, notice any movement between the upper end of the spindle and the upper arm. If movement is present, replace the ball joint.

Lower Ball Joint

1. Raise vehicle and place jacks under lower arms as shown in Fig. 4.
2. Be sure wheel bearings are properly adjusted.
3. Attach a dial indicator to lower arm and position indicator so that plunger rests against inner side of wheel rim near lower ball joint.
4. Grasp tire at top and bottom and slowly move tire in and out. If the reading exceeds .250″, replace the joint.

SHOCK ABSORBER, REPLACE

1. Unfasten the upper end of the shock.
2. Raise vehicle and install safety stands.
3. Unfasten lower end of shock. It may be necessary to use a pry bar to free "T" shaped end of the shock from the lower end.

COIL SPRING, REPLACE

1. Raise vehicle and support front end with safety stands.
2. Place a jack under lower arm to support it.
3. Disconnect lower end of shock.
4. Remove bolts that attach strut to lower arm.
5. Remove nut that retains shock to crossmember and remove the shock.
6. Remove nut and bolt that secures inner end of lower arm to crossmember.
7. Carefully lower jack to relieve pressure from spring and remove spring.

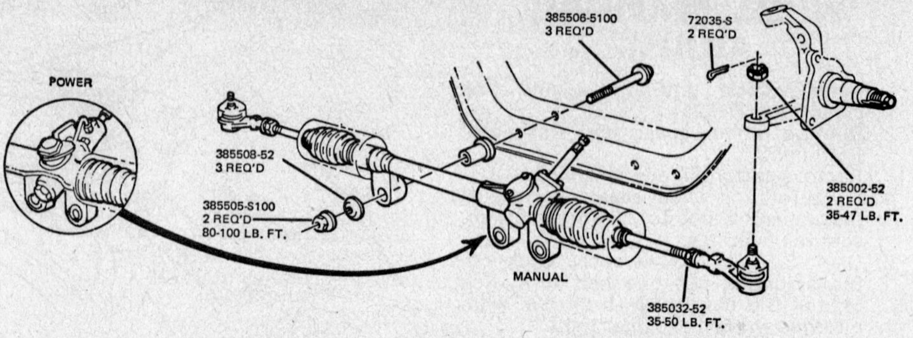

Fig. 5 Steering gear installation. 1974-76 Mustang II

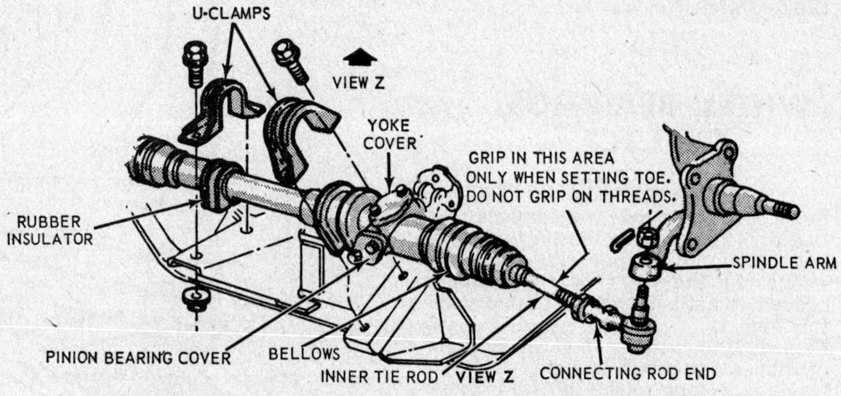

Fig. 6 Steering gear installation. 1973-76 Bobcat & Pinto

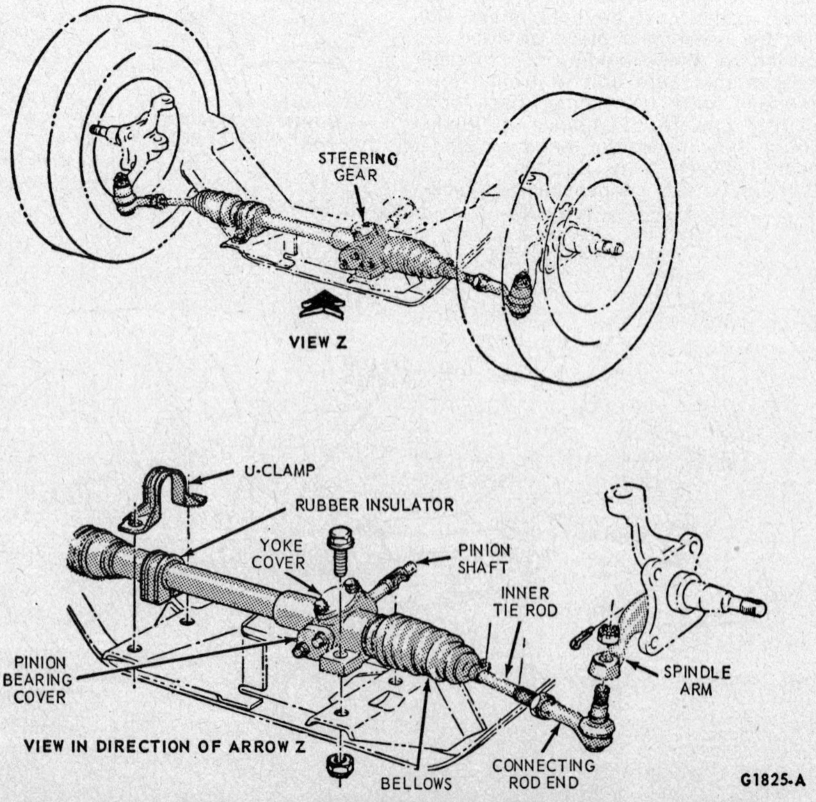

Fig. 7 Steering gear installation. 1971-72 Pinto

BALL JOINTS, REPLACE

NOTE: Ford Motor Company recommends that new ball joints should not be installed on used control arms and that the control arm be replaced if ball joint replacement is required. However, aftermarket ball joint repair kits which do not require control arm replacement, are available and can be installed using the following procedure.

The ball joints are riveted to the control arms. The ball joints can be replaced on the car by removing the rivets and replacing them with new attaching bolts, nuts and washers furnished with the kit.

When removing a ball joint, use a suitable pressing tool to force the ball joint out of the spindle.

STEERING GEAR, REPLACE

1973-76 Pinto, 1974-76 Mustang II & 1975-76 Bobcat

Manual Steering Gear

1. Disconnect battery ground cable, turn ignition "On" and raise vehicle.
2. Remove tie rod end retaining nuts and using ball joint separator (tool 3290C), separate tie rod ends from spindle arms, Figs. 5 and 6.
3. Remove pinion shaft to flexible coupling bolt and the bolts securing steering gear to crossmember.

NOTE: On 1975-76 Mustang II, the number 2A crossmember, located behind the front crossmember, must be removed to permit steering gear removal.

4. Turn front wheels, then remove steering gear from left side of vehicle.

Power Steering Gear

1. Disconnect battery ground cable.
2. Remove bolt retaining flexible coupling to input shaft.
3. Turn ignition key "ON" and raise vehicle.
4. Remove the two tie rod end retaining nuts, then separate studs from spindle arms, using a suitable tool.
5. Remove the number 2A crossmember located behind the front crossmember to allow removal of steering gear retaining bolts.
6. Support gear and remove attaching bolts, then lower gear enough to gain access to pressure and return lines, and remove bolt attaching the hose bracket to the gear.
7. Disconnect pressure and return lines and remove steering gear. Plug lines and ports to prevent entry of dirt.

1971-72 Pinto

1. Position steering wheel in straight ahead position and raise vehicle on a hoist.
2. Remove bolts retaining the flexible coupling to the pinion shaft, Fig. 7.
3. Remove four nuts and bolts securing the steering gear to the mounting pads on the crossmember.
4. Remove the U-clamp.
5. Remove cotterpins and loosen nuts securing connecting rod ends to the spindle arms.
6. Use ball joint separator (tool 3290C) to separate the connecting rod ends from the spindle arms.
7. Remove nuts from connecting rods and remove gear from vehicle left side.

FORD THUNDERBIRD

OLD CAR SPECIFICATIONS: For 1946-68 Tune Up and Wheel Alignment Specifications see main index.

Service procedures on the V8-400 are covered in the Ford & Mercury car chapters.

INDEX OF SERVICE OPERATIONS

ENGINE & SERIAL NUMBER LOCATION

Vehicle Warranty Plate On Left Front Door Pillar.

1969

1970

ENGINE IDENTIFICATION

*Serial number on vehicle Warranty Plate.

Year	Engine	Engine Code
1969–73	V8-429	N
1972	V8-400	S
1973–76	V8-460	A

1973

1974

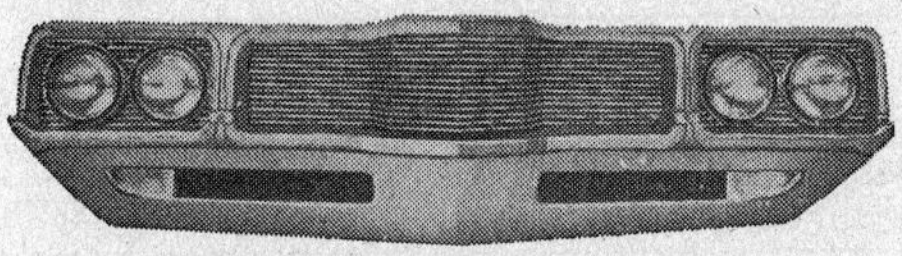

1971

1972

1975-76

GENERAL ENGINE SPECIFICATIONS

Year	Engine	Carburetor	Bore and Stroke	Piston Displacement, Cubic Inches	Compression Ratio	Maximum Brake H.P. @ R.P.M.	Maximum Torque Lbs. Ft. @ R.P.M.	Normal Oil Pressure Pounds
1969–70	360 Horsepower.............V8-429	4 Barrel	4.36 x 3.59	429	11.0	360 @ 4600	476 @ 2800	35–60
1971	360 Horsepower.............V8-429	4 Barrel	4.36 x 3.59	429	10.3	360 @ 4600	480 @ 2800	35–75
1972	212 Horsepower①...........V8-400	2 Barrel	4.00 x 4.00	400	8.4	172 @ 4000	298 @ 2200	35–75
	212 Horsepower①...........V8-429	4 Barrel	4.36 x 3.59	429	8.5	212 @ 4400	327 @ 2600	35–75
	224 Horsepower①...........V8-460	4 Barrel	4.36 x 3.85	460	8.5	224 @ 4400	357 @ 2800	35–60
1973	208 Horsepower① V8-429	4 Barrel	4.36 x 3.59	429	8.0	208 @ 4400	327 @ 2600	35–65
	208 Horsepower① V8-460	4 Barrel	4.36 x 3.85	460	8.0	208 @ 4400	338 @ 2800	35–65
1974	220 Horsepower① V8-460	4 Barrel	4.36 x 3.85	460	8.0	220 @ 4000	355 @ 2600	35–65
1975	216 Horsepower①...........V8-460	4 Barrel	4.36 x 3.85	460	8.0	216 @ 4000	366 @ 2600	40–65
1976	Horsepower①...........V8-460	4 Barrel	4.36 x 3.85	460	8.0	—	—	40–65

①—Ratings are NET—as installed in the vehicle.

TUNE UP SPECIFICATIONS

The following specifications are published from the latest information available. This data should be used only in the absence of a decal affixed in the engine compartment.

OLD CAR SPECIFICATIONS: For 1946-68 Tune Up Specifications see main index.

★When using a timing light, disconnect vacuum hose or tube at distributor and plug opening in hose or tube so idle speed will not be affected.

●When checking compression, lowest cylinder must be within 75 percent of highest.

▲Before removing wires from distributor cap, determine location of the No. 1 wire in cap, as distributor position may have been altered from that shown at the end of this chart.

Year	Spark Plug		Distributor		Ignition Timing★			Carb. Adjustments					
								Hot Idle Speed		Air Fuel Ratio		Idle "CO" %	
	Type	Gap Inch	Point Gap Inch	Dwell Angle Deg.	Firing Order Fig. ▲	Timing BTDC ①	Mark Fig.	Std. Trans.	Auto. Trans.②	Std. Trans.	Auto. Trans.	Std. Trans.	Auto. Trans.
1969													
8-429	BF-42	.034	.017	26–31	A	6°⑤	B	—	550D④	—	14.3 to 1	—	1.2
1970													
8-429	BRF-42	.034	.017	26–31	A	⑬	B	—	600D⑩	—	14.5 to 1	—	0.75
1971													
8-429	BRF-42	.034	.017	26–31	A	⑬	D	—	600D	—	14.5 to 1	—	.075
1972													
8-400	ARF-42	.034	.017	26–30	C	⑦	D	—	625D⑩	—	—	—	0.10
8-429	BRF-42	.034	.017	26–30	A	10°⑧	D	—	600D⑩	—	—	—	0.3
8-460	BRF-42	.034	.017	26–30	A	⑪	D	—	600D⑩	—	—	—	0.3
1973													
8-429	ARF-42	.034	.017	26–30	A	10°	D	—	600D⑩	—	—	—	0.5⑫
8-460	ARF-42	.034	.017	26–30	A	14°	D	—	600D⑩	—	—	—	0.4⑫
1974													
8-460	ARF-52	③	⑨	—	A	14°	D	—	650D⑥	—	—	—	—
1975													
8-460	ARF-52	③	⑨	—	A	14°	D	—	650D	—	—	—	—
1976													
8-460⑭	ARF-52	.044	⑨	—	A	8°⑮	D	—	650D⑯	—	—	—	—
8-460⑰	ARF-52	.044	⑨	—	A	14°⑮	D	—	650D⑯	—	—	—	—

①—BTDC: Before top dead center.
②—D: Drive. N: Neutral.
③—Exc. Calif., .054 inch; Calif. .044 inch.
④—With headlights and A/C on.
⑤—Whenever idle speed or ignition timing is adjusted, vacuum line to brake release mechanism must be disconnected and plugged to prevent parking brake from releasing when selector is moved to Drive.
⑥—With lights and A/C off.
⑦—Exc. Calif. 8° BTC, Calif. 6° BTC.
⑧—Built after Dec. 31, 1971, 14°.
⑨—Breakerless distributor.
⑩—With headlamp on Hi Beam—Air Conditioning OFF.
⑪—Exc. Calif. 10° BTC, Calif 6° BTC.
⑫—For Calif. 0.2.
⑬—Single diaphragm 14°, Dual diaphragm 4°.
⑭—Exc. Calif.
⑮—In Drive with service & parking brakes applied.
⑯—With A/C "On".
⑰—Calif.

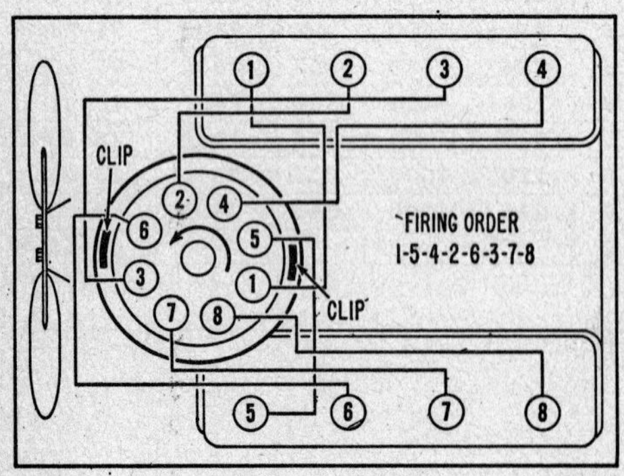

Fig. A

Fig. B

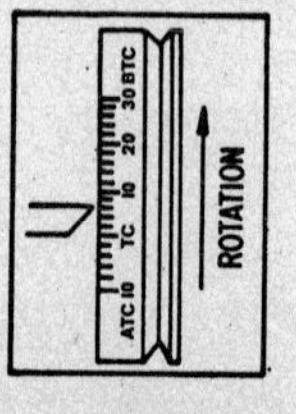

Fig. D

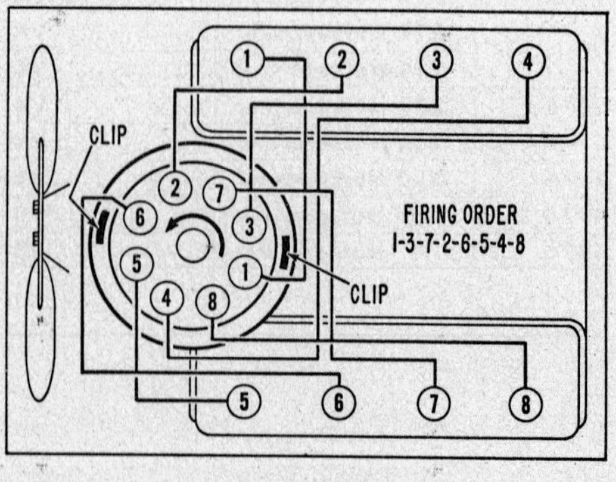

Fig. C

DISTRIBUTOR SPECIFICATIONS

★If unit is checked on vehicle, double the RPM and degrees to get crankshaft figures.
Breaker arm spring tension—17–21.

Distributor Part No.①	Centrifugal Advance Degrees @ RPM of Distributor					Vacuum Advance		Distributor Retard
	Advance Starts	Intermediate Advance			Full Advance	Inches of Vacuum to Start Plunger	Max. Adv. Dist. Deg. @ Vacuum	Max. Retard Dist. Deg. @ Vacuum
1969								
C8VF-A	0–½ @ 350	2½–4½ @ 750	6½–8½ @ 1000	8–10¼ @ 1500	14 @ 2975	5	12 @ 25	—
1970-71								
D0AF-Z	0–½ @ 350	2¼–4¼ @ 750	6½–8½ @ 1000	8–10¼ @ 1500	14 @ 2950	5	11 @ 25	—
1972								
D2AF-RA	½–2½ @ 500	2–4 @ 750	3½–5½ @ 1000	6½–9 @ 1500	12 @ 2000	5	13½ @ 20	4 @ 20
D2AF-SA	½–2½ @ 500	2–4 @ 750	3½–5½ @ 1000	6½–9 @ 1500	12 @ 2000	5	13½ @ 20	4 @ 20
D2MF-EA	0–1 @ 500	2–4 @ 750	4½–6½ @ 1000	6–8 @ 1500	9½ @ 2000	5	11½ @ 20	4 @ 20
D2MF-FA	0–1 @ 500	2–4 @ 750	4½–6½ @ 1000	6–8 @ 1500	9½ @ 2000	5	11½ @ 20	4 @ 20
D2SF-EA	0–1 @ 500	0–1½ @ 750	2½–4½ @ 1000	4½–6½ @ 1500	8½ @ 2000	5	—	—
D2VF-BA	0–2 @ 700	6–8 @ 1000	7½–10 @ 1500	—	12½ @ 2000	5	11½ @ 20	—
1973								
D3VF-AA	0–1 @ 500	0–1½ @ 750	3–5 @ 1000	4–6 @ 1500	7½ @ 2000	5	11½ @ 20	—
D3MF-DA	0–1 @ 500	1½–4 @ 750	5½–7½ @ 1000	7½–9 @ 1500	10½ @ 2000	5	11 @ 20	—
D3SF-BA	0–1 @ 500	½–2½ @ 750	5½–8 @ 1000	10–12 @ 1500	13½ @ 2000	5	11 @ 20	—
D3VF-CA	0–1 @ 500	0–1½ @ 750	4–6 @ 1000	6–8 @ 1500	10 @ 2000	5	11 @ 20	—
D3MF-BA	0–1 @ 500	0–1½ @ 750	3½–5½ @ 1000	7½–10 @ 1500	11 @ 2000	5	7 @ 20	—
D3VF-BA	0–1 @ 500	0–2½ @ 750	5–7 @ 1000	7½–10 @ 1500	12 @ 2000	5	7 @ 20	—
1974-75								
D4VE-CA	0–½ @ 500	0–½ @ 750	3–5 @ 1000	7–9 @ 1500	10½ @ 2000	5	11¼ @ 20	—
D5VE-BA	—	—	—	—	—	—	—	—
1976								
D6VE-AA	—	—	—	—	—	—	—	—
DVE-BA	—	—	—	—	—	—	—	—

①—Basic part No. 12127

ENGINE TIGHTENING SPECIFICATIONS

★Torque specifications are for clean and lightly lubricated threads only. Dry or dirty threads produce increased friction which prevents accurate measurement of tightness.

Year	Engine	Spark Plugs Ft. Lbs.	Cylinder Head Bolts Ft. Lbs.	Intake Manifold Ft. Lbs.	Exhaust Manifold Ft. Lbs.	Rocker Arm Shaft Bracket Ft. Lbs.	Rocker Arm Cover Ft. Lbs.	Connecting Rod Cap Bolts Ft. Lbs.	Main Bearing Cap Bolts Ft. Lbs.	Flywheel to Crank-shaft Ft. Lbs.	Vibration Damper or Pulley Ft. Lbs.
1969-70	V8-429	15–20	130–140	25–30	28–33	65–75①	5–6	40–45	95–105	75–85	75–90
1971	V8-429	15–20	130–140	25–30	28–33	75–85①	5–6	40–45	95–105	75–85	70–90
1972	V8-400	10–15	95–105	②	12–16	18–25①	3–5	40–45	95–105	—	70–90
1972-73	V8-429, 460	15–20	130–140	25–30	28–33	18–25①	5–6	40–45	95–105	75–85	70–90
1974-76	V8-460	10–15	130–140	25–30	28–33	18–25①	5–6	40–45	95–105	75–85	70–90

①—Rocker arm stud to cylinder head. ②—¼" bolt 6–9; 5/16" bolt 21–25; ⅜" bolt 27–33.

VALVE SPECIFICATIONS

Year	Engine Model	Valve Lash Int.	Valve Lash Exh.	Valve Angles Seat	Valve Angles Face	Valve Spring Installed Height	Valve Spring Pressure Lbs. @ In.	Stem Clearance Intake	Stem Clearance Exhaust	Stem Diameter Intake	Stem Diameter Exhaust
1969–70	8-429	.075–.125①		45	44	1 13/16	253 @ 1.33	.0010–.0027	.0010–.0027	.3416–.3423	.3416–.3423
1971	8-429	.105–.155①		45	45	1 13/16	229 @ 1.33	.0010–.0027	.0010–.0027	.3416–.3423	.3416–.3423
1972	8-400	.100–.150①		45	45	1 13/16	229 @ 1.33	.0010–.0027	.0015–.0032	.3416–.3423	.3411–.3418
	8-429	.075–.125①		45	45	1 13/16	229 @ 1.33	.0010–.0027	.0010–.0027	.3416–.3423	.3416–.3423
	8-460	.105–.155①		44.5	45.5	1 13/16	229 @ 1.33	.0010–.0027	.0010–.0027	.3416–.3423	.3416–.3423
1973	8-429	.075–.125①		45	44	1 13/16	169 @ 1.39	.0010–.0027	.0010–.0027	.3416–.3423	.3416–.3423
	8-460	.105–.155①		45	44	1 13/16	169 @ 1.39	.0010–.0027	.0010–.0027	.3416–.3423	.3416–.3423
1974	8-460	.075–.125①		45	44	1 13/16	252 @ 1.33	.0010–.0027	.0010–.0027	.3416–.3423	.3416–.3423
1975	8-460	.100–.150①		45	44	1 13/16	252 @ 1.33	.0010–.0027	.0010–.0027	.3416–.3423	.3416–.3423
1976	8-460	.075–.125①		45	44	1 13/16	228 @ 1.33	.0010–.0027	.0010–.0027	.3416–.3423	.3416–.3423

①—Clearance specified is obtained at valve stem with lifter collapsed. See "Valves, Adjust" text.

PISTONS, PINS, RINGS, CRANKSHAFT & BEARINGS

Year	Engine Model	Piston Clearance	Ring End Gap① Comp.	Ring End Gap① Oil	Wrist-pin Diameter	Rod Bearings Shaft Diameter	Rod Bearings Bearing Clearance	Main Bearings Shaft Diameter	Main Bearings Bearing Clearance	Main Bearings Thrust on Bear. No.	Main Bearings Shaft End Play
1969–73	8-429	.0014–.0022	.010	.015	1.0400	2.4992–.25000	.0008–.0028	2.9994–3.0002	.0005–.0015	3	.004–.008
1972	8-400	.0014–.0022	.010	.015	.9752	2.3103–2.3111	.0008–.0026	2.9994–3.0002	.0009–.0026	3	.004–.010
1972–73	8-460	.0014–.0022	.010	.015	1.0400	2.4992–2.5000	.0008–.0028	2.9994–3.0002	.0005–.0025	3	.004–.008
1974	8-460	.0014–.0022	.010	.015	1.0400	2.4992–2.5000	.0008–.0028	2.9994–3.0002	②	3	.004–.008
1975–76	8-460	.0014–.0022	.010	.015	1.0400	2.4992–2.5000	.0008–.0028	2.9994–3.0002	③	3	.004–.008

①—Fit rings in tapered boxes for clearance listed in tightest portion of ring travel.
②—#1, .0004–.0020, others .0012–.0028.
③—#1, .0004–.0020; others, .0009–.0026.

REAR AXLE SPECIFICATIONS

Year	Model	Carrier Type	Ring Gear & Pinion Backlash Inch	Nominal Pinion Locating Shim, Inch	Pinion Bearing Preload New Bearings With Seal Inch-Lbs.	Pinion Bearing Preload Used Bearings With Seal Inch-Lbs.	Pinion Bearing Preload New Bearings Less Seal Inch-Lbs.	Pinion Bearing Preload Used Bearings Less Seal Inch-Lbs.	Differential Bearing Preload	Pinion Nut Torque Ft.-Lbs.
1969–72	All	Removable	.008–.012	.015	22–32⑥	8–14⑥	—	—	.008–.012④	②⑤
1973	All	Removable	.008–.012	.015	17–27③	8–14③	—	—	.008–.012④	②⑤
1974-76	All	Removable	.008–.012	.015	17–27③	8–14③	—	—	.008–.012④	②⑦

②—If torque is not possible, install new spacer.
③—With collapsible spacer. With solid spacer 13–33 inch-lbs.
④—Case spread with new bearings. With used beaings .005–.008".
⑤—With 9" ring gear 175 ft.-lbs. With 9 3/8" ring gear 200 ft.-lbs.
⑥—With collapsible spacer. With solid spacer 15–35 inch-lbs. with seal in place.
⑦—With collapsible spacer 170 ft.-lbs., with solid spacer 200 ft.-lbs.

STARTING MOTOR SPECIFICATIONS

Year and Model	Ident.	Rotation ①	Brush Spring Tension, Ounces	No Load Test			Torque Test		
				Amperes	Volts	R.P.M.	Amperes	Volts	Torque Ft. Lbs.
1969 V8-429	C9AF-A	C	40	70	12	10000	700	5	15.5
1970–73 V8-429	D0VF-A	C	40	70	12	10000	700	5	15.5
1972–73 V8-460	D2VF-AA	C	40	70	12	10000	700	5	15.5
1974 V8-460	D4VF-AA	C	80	90	12	—	—	—	—
1974 V8-460	D5AF-AB	C	80	90	12	—	—	—	—
1975–76 V8-460	D5AF-AB	C	40	70	12	—	—	—	—

①—As viewed from the drive end. C—Clockwise.

ALTERNATOR & REGULATOR SPECIFICATIONS

Year	Model No. (10300)	Current Rating①		Field Current @ 75°F.		Voltage Regulator②				Field Relay	
		Amperes	Volts	Amperes	Volts	Part No. (10316)	Voltage @ 75°F.	Contact Gap	Armature Air Gap	Armature Air Gap	Closing Voltage @75°F.
1969	C9SF-B	65	15	2.9	12	C3SZ-B	13.5–15.3	③	③	③	2.0–4.2
1969–70	C8LF-B	55	15	2.9	12	C8SZ-A④	13.3–15.3	③	③	③	③
1970	D0AF-H	65	15	2.9	12	C3SZ-B	13.5–15.3	③	③	③	2.0–4.2
1971	D0SF-A	55	15	2.9	12	C3SZ-B	13.3–15.3	③	③	③	—
	D1AF-BA	65	15	2.9	12	C3SZ-B	13.5–15.3	③	③	③	2.0–4.2
1972	D2OF-EA	55	15	2.9	12	C3SZ-B	③	③	③	③	—
	D2AF-CA	65	15	2.9	12	D2TF-AA	13.5–15.3	③	③	③	2.5–4.0
	D2OF-AA	70	15	2.9	12	D3TF-AA	13.5–15.3	③	③	③	2.5–4.0
1973	All (Green)	61	15	2.9	12	D3TZ-A	13.5–15.3	③	③	③	2.5–4.0
	All	70	15	2.9	12	D3TZ-A	13.5–15.3	③	③	③	2.5–4.0
1974–75	All (Green)	61	15	2.9	12	D4AF-AA	13.5–15.3	③	③	③	2.5–4.0
	All	70	15	2.9	12	D4TF-AA	13.5–15.3	③	③	③	2.5–4.0
	All	90	15	2.9	12	—	13.5–15.3	③	③	③	2.5–4.0
1976	D5AF-EA	60	—	—	—	—	13.5–15.3	③	③	③	2.5–4.0
	All	61	—	—	—	D4AF-AA	13.5–15.3	③	③	③	2.5–4.0
	D5OF-AA	70	—	—	—	D4TF-AA	13.5–15.3	③	③	③	2.5–4.0

①—Stamped on housing.
②—Stamped on cover.
③—Not adjustable.
④—Integral regulator solid state.

WHEEL ALIGNMENT SPECIFICATIONS

OLD CAR SPECIFICATIONS: For 1955-68 Wheel Alignment Specifications see main index.

Year	Model	Caster Angle, Degrees		Camber Angle, Degrees				Toe-In. Inch	Toe-Out on Turns, Deg.	
				Limits		Desired				
		Limits	Desired	Left	Right	Left	Right		Outer Wheel	Inner Wheel
1969	All	0 to +2	+1	−¼ to +1¼	−¼ to +1¼	+½	+½	3/16	18¼	20
1970	All	0 to +2	+1	−¼ to +1¼	−¼ to +1¼	+½	+½	3/16	19¼	20
1971	All	−1 to +3	+1	−½ to +1½	−½ to +1½	+½	+½	3/16	19¼	20
1972	All	−1 to +3	+1	+¼ to +1¾	−¼ to +1¾	+¾	+¾	3/16	17.74	20
1973	All	−½ to +3½	+1½	−¼ to +1¾	−¼ to +1¾	+¾	+¾	3/16	17.74	20
1974	All	+½ to +3½	+2	0 to +2	−½ to +1½	+1	+½	3/16	18.07	20
1975	All	+3¼ to +4¾	+4	−¼ to +1¼	−½ to +1	+½	+¼	3/16	18.09	20
1976	All	+3¼ to +4¾	+4	+¼ to +1¼	−½ to +1	+½	+¼	3/16	18.09	20

BRAKE SPECIFICATIONS

Year	Model	Brake Drum Inside Diameter	Wheel Cylinder Bore Diameter			Master Cylinder Bore Diameter		
			Front Disc Brakes	Front Drum Brakes	Rear Brakes	With Disc Brakes	With Drum Brakes	With Power Brakes
1969–71	All	11.03	2¾	—	15/16	1.00	1.00	1.00
1972–74	All	11.03	3.100	—	1.00	1.00	—	1.00
1975	All	11.03	3.10	—	①	②	—	②
1976	All	11.03	3.10	—	①	1⅛	—	1⅛

①—Drum brakes; 1. Disc Brakes; 2.6.
②—Front wheel disc brakes only; 1. With 4-wheel disc brakes; 1⅛.

COOLING SYSTEM & CAPACITY DATA

Year	Model or Engine	Cooling Capacity, Qts.			Radiator Cap Relief Pressure, Lbs.		Thermo. Opening Temp. ①	Fuel Tank Gals.	Engine Oil Refill Qts. ②	Transmission Oil			Rear Axle Oil Pints
		No Heater	With Heater	With A/C	With A/C	No A/C				3 Speed Pints	4 Speed Pints	Auto. Trans. Qts. ③	
1969	V8-429	19½	20½	20½	12–15	12–15	195	24	4	—	—	12¾	5
1970	V8-429	17½	18½	18½	12–15	12–15	188	④	4	—	—	12¾	5
1971	V8-429	18½	19½	16½	12–15	12–15	188	23	4	—	—	12¾	5
1972	V8-400	16¾	17¾	18½	12–15	12–15	188	22½	4	—	—	12¾	5
	V8-429	17¾	18¾	19½	12–15	12–15	188	22½	4	—	—	12¾	5
	V8-460	19	20	20	12–15	12–15	188	22½	4	—	—	12½	5
1973	All	18½	19½	19½	12–15	12–15	188	22½	4	—	—	12½	5
1974	V8-460	—	19½	19½	12–16	12–16	191	26½	4	—	—	12½	5
1975–76	V8-460	—	19.3	19.3	12–16	12–16	191	26½	4	—	—	12½	5

①—For permanant type anti-freeze.
②—Add one quart with filter change.
③—Approximate. Make final check with dipstick.
④—California vehicles 22½; all others 24.

Electrical Section

DISTRIBUTOR, REPLACE

Removal

1. On vehicles with conventional ignition systems, disconnect primary wire from ignition coil. On vehicles with breakerless ignition systems, disconnect distributor wiring from connector from main wiring harness. On some models the work may be made easier if the accelerator pull back spring is disconnected.
2. Disconnect vacuum advance lines and remove distributor cap.
3. Scribe a mark on the distributor body indicating the position of the rotor, and scribe another mark on the body and engine block indicating position of distributor body in block. These marks can be used as guides when installing distributor in a correctly timed engine.
4. Remove hold down screw and screws and lift distributor out of block. *Do not crank engine while distributor is removed or the initial timing operation will have to be performed.*

Installation

If the crankshaft has not been disturbed, install the distributor, using the scribed marks previously made on the distributor body and engine block as guides.

If the crankshaft has been rotated while the distributor was removed from the engine, it will be necessary to retime the engine. Crank the engine to bring No. 1 piston on top dead center of its compression stroke. Align the timing mark on the vibration damper or pulley with the timing pointer (see *Tune Up* chart). Install the distributor so that the rotor points to the No. 1 spark plug wire terminal in the distributor cap.

NOTE: Make sure the oil pump intermediate shaft properly engages the distributor shaft. It may be necessary to crank the engine with the starter, after the distributor drive gear is properly engaged, in order to engage the oil pump intermediate shaft.

STARTER, REPLACE

1970-76

The starter does not use an auxiliary starter relay as did previous models therefore when replacing starter be sure to disconnect battery ground cable. Removal is otherwise the same as previous models.

1969

To provide working clearance, it may be necessary to turn wheels fully to the right and to disconnect the steering idler arm from the frame.

When installing the starter, be sure all mating surfaces are clean to insure a good electrical ground. To maintain proper alignment, hold the starter squarely against the mounting plate and fully inserted into the mounting hole while tightening the mounting bolts.

IGNITION LOCK, REPLACE

1970-76

1. Disconnect the battery ground cable.
2. **Units With Fixed Steering Columns:** Remove the steering wheel and trim pad. Insert a wire pin in the hole located inside the column halfway down the lock cylinder housing, Fig. 1. **Units With Tilt Steering Columns:** Insert wire pin in the hole located on the outside of the flange casting next to the emergency flasher button, Fig. 1.
3. Place the gear shift lever in *PARK* position, and turn the lock cylinder with the ignition key to *RUN* position.
4. Depress the wire pin while pulling up on the lock cylinder to remove. Remove the wire pin.
5. To install insert the lock cylinder into the housing in the flange casting, and turn the key to *OFF* position. Be certain that the cylinder is fully inserted before turning to the *OFF* position. This action will extend the cylinder retaining pin into the cylinder housing.
6. Turn the key to check for correct operation in all positions.
7. Install the steering wheel and trim pad on fixed column units.
8. Connect the battery ground cable.

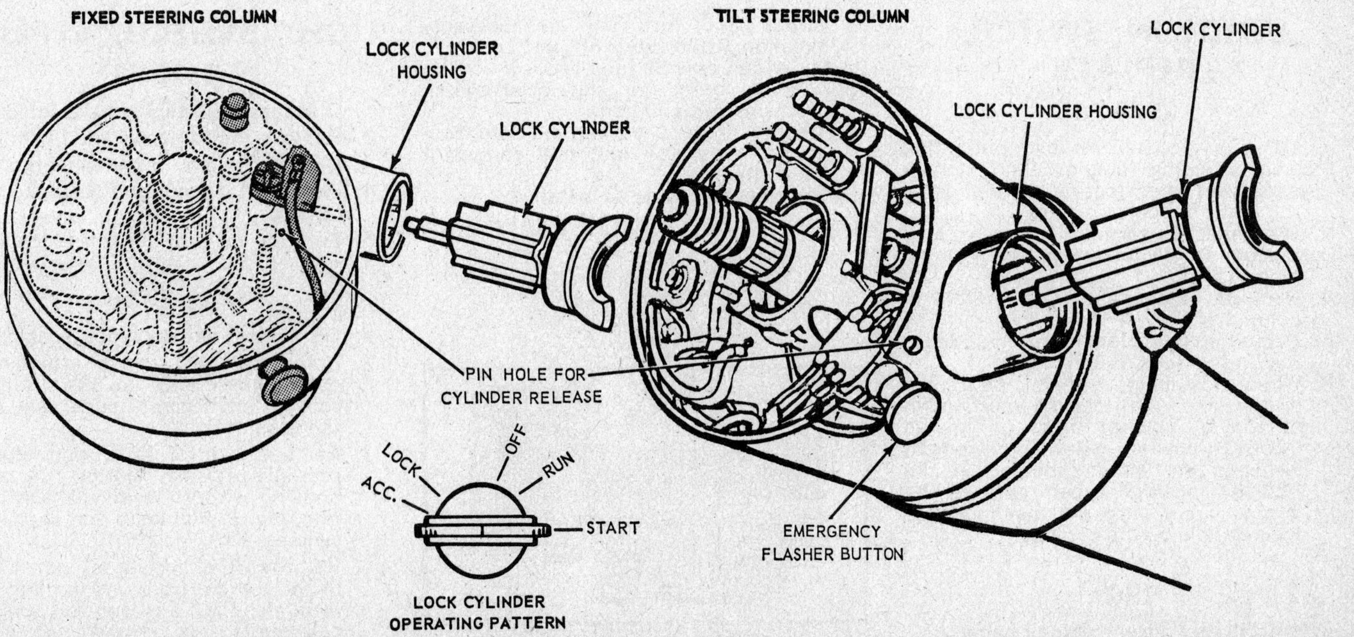

Fig. 1 Ignition lock, 1970-76

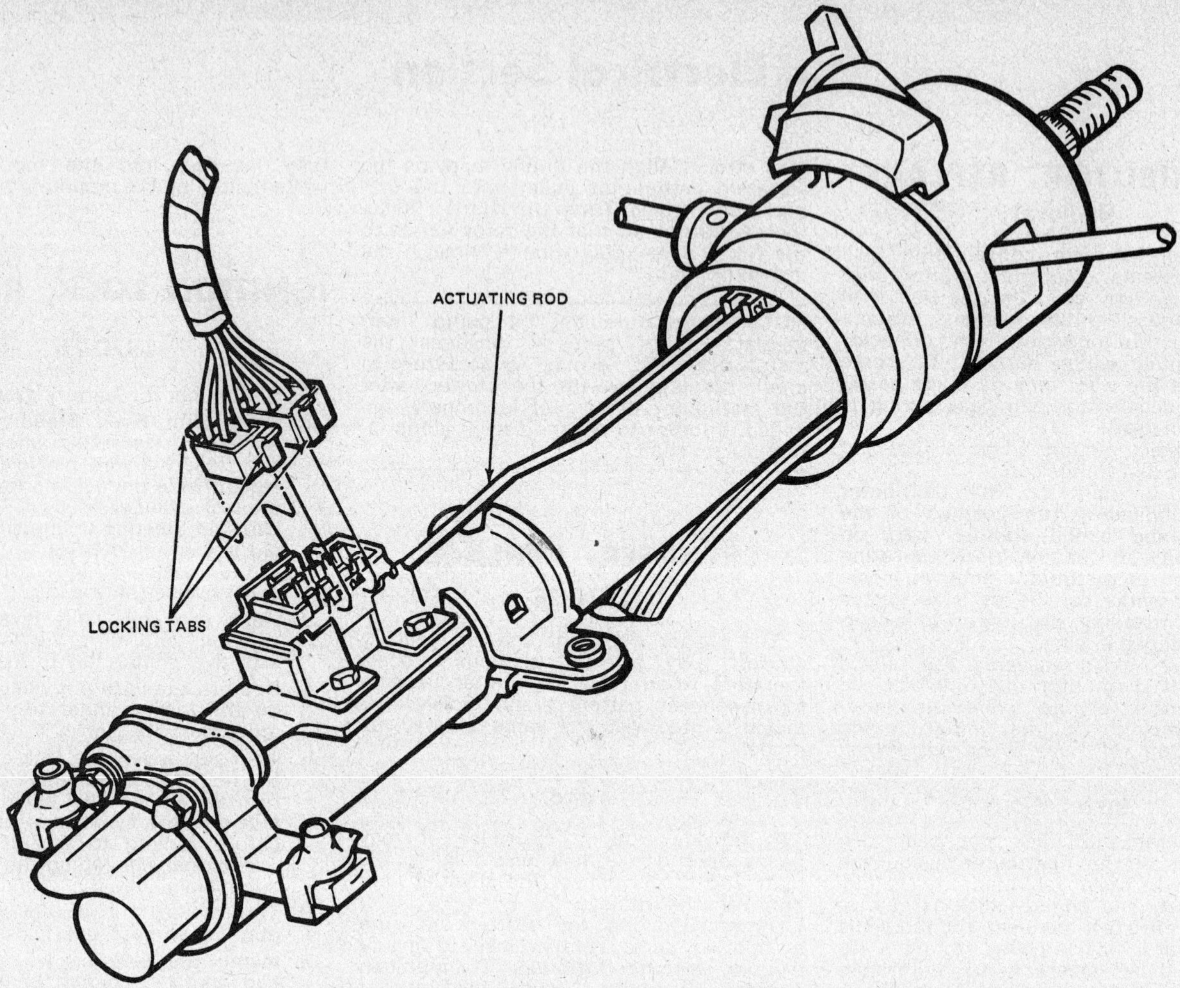

Fig. 2 Ignition switch installation. 1970-76 (typical)

IGNITION SWITCH, REPLACE

1970-76

1. To gain access, remove shrouding from steering column and detach and lower the steering column from the brake support bracket.
2. Disconnect battery cable.
3. Disconnect switch wiring at multiple plug.
4. Remove nuts that retain switch to column.
5. Detach switch plunger from actuator rod and remove switch.
6. When installing switch, both the switch and the locking mechanism at top of column must be in the "LOCK" position. New replacement switches are already pinned in the "LOCK" position when received by a plastic shipping pin inserted in a locking hole on top of switch.

1969

1. Disconnect battery ground cable.
2. Insert wire pin in hole in ignition switch and turn key to accessory position while pressing pin. Lock cylinder can now be removed.
3. Remove bezel nut and lower switch from instrument panel.
4. Depress tabs securing multiple connector to switch and pull connector off switch.
5. Reverse procedure to install.

Fig. 3 Light switch (typical)

LIGHT SWITCH, REPLACE

1972-76

1. Disconnect battery ground cable.
2. Remove instrument cluster trim panel.
3. Remove switch mounting plate.
4. Remove bezel nut and disconnect multiple connector.
5. Remove vacuum lines, if so equipped.

1969-71

1. Disconnect battery ground cable.
2. Remove three screws attaching air control assembly to the lower left side of instrument panel and lower the assembly.
3. Remove control knob and shaft by pressing release button on switch housing with knob in full "ON" position. Fig. 3. Pull knob and shaft out of switch.
4. Remove retaining bezel nut and remove switch from instrument panel through the air control assembly opening.
5. Reverse procedure to install.

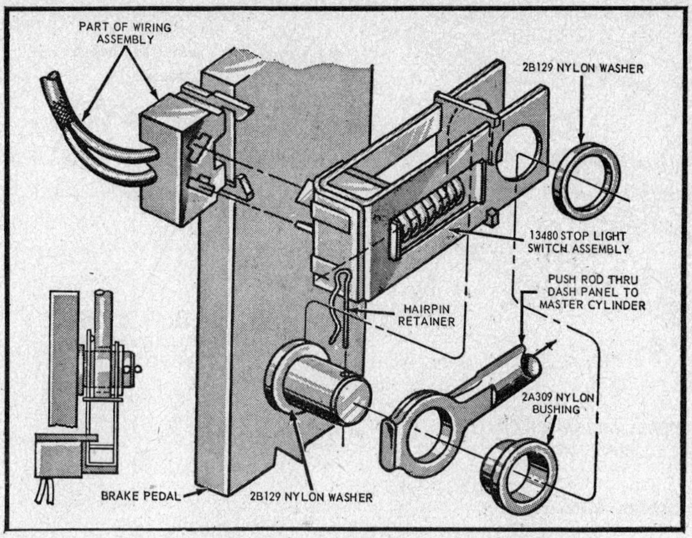

Fig. 4 Mechanical stop light switch. 1969-76

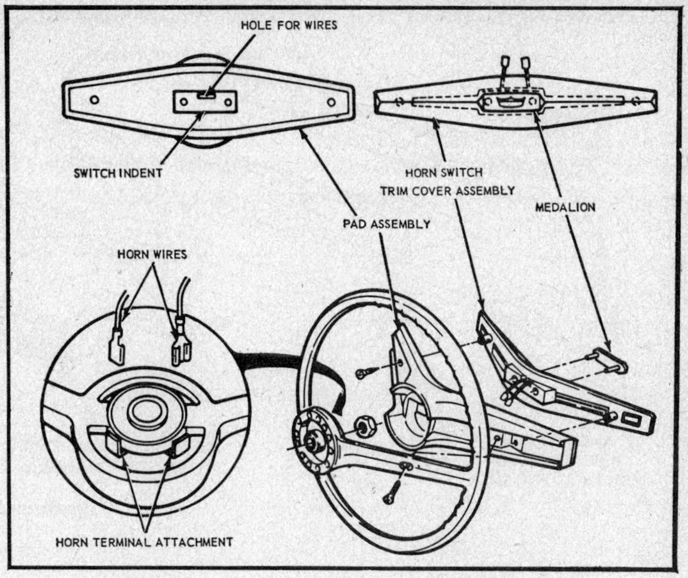

Fig. 5 Steering wheel and horn switch. 1970-76

STOP LIGHT SWITCH, REPLACE

1969-76

1. Disconnect wires at connector.
2. Remove hairpin retainer. Slide stop light switch, push rod and nylon washers and bushing away from pedal, and remove switch, Fig. 4.
3. Position switch, push rod, bushing and washers on brake pedal pin in the order shown, and install hairpin retainer.
4. Connect wires and install wires in clip.

NEUTRAL SAFETY SWITCH

1970-76

The neutral start switch has been eliminated from all column shift vehicles. A series of steps have been designed into the steering column selector lever hub casting which eliminates the need for a switch.

1969

Adjustment Procedure

1. Loosen retaining screws and with selector lever held lightly against the Neutral stop, rotate switch until a start is obtained. Tighten attaching screws to 20 in-lbs.
2. With the switch properly adjusted, place selector lever in the "1" position and push the Park reset button to the left until it stops. *The Park reset must be performed whenever the switch has been adjusted.*

HORN SOUNDER

1970-76

A two spoke steering wheel with a pressure sensitive horn switch built into the trim cover is used, Fig. 5. If the switch needs to be replaced, the entire horn switch/trim cover assembly will have to be replaced.

1969 Rim-Blow Type

The rubber insert and copper strip is not replaceable. If a new insert assembly is required, the entire steering wheel will have to be replaced.

1. Remove pad from steering wheel.
2. Remove medallion from pad.
3. After removing the retaining nut, the steering wheel can be removed with a suitable puller.

1969 Except Rim-Blow Type

1. Disconnect battery ground cable.
2. Remove center medallion from wheel.
3. Remove three screws from center of wheel and remove horn buttons.
4. Reverse procedure to install.

TURN SIGNAL SWITCH, REPLACE

1971-76

1. Remove retaining screw from underside of steering wheel spoke and lift off the pad horn switch/trim cover and medallion as an assembly. Disconnect horn switch wires from terminals.
2. Remove steering wheel retaining nut and steering wheel.
3. Remove turn signal switch lever by

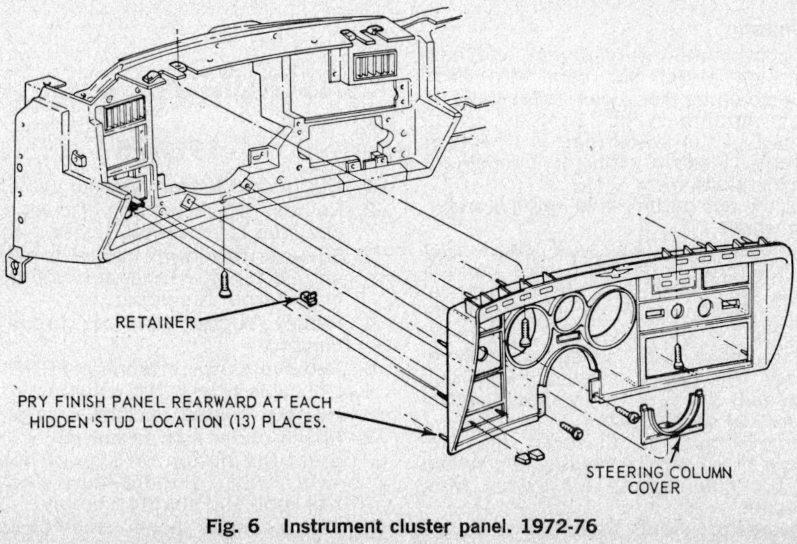

Fig. 6 Instrument cluster panel. 1972-76

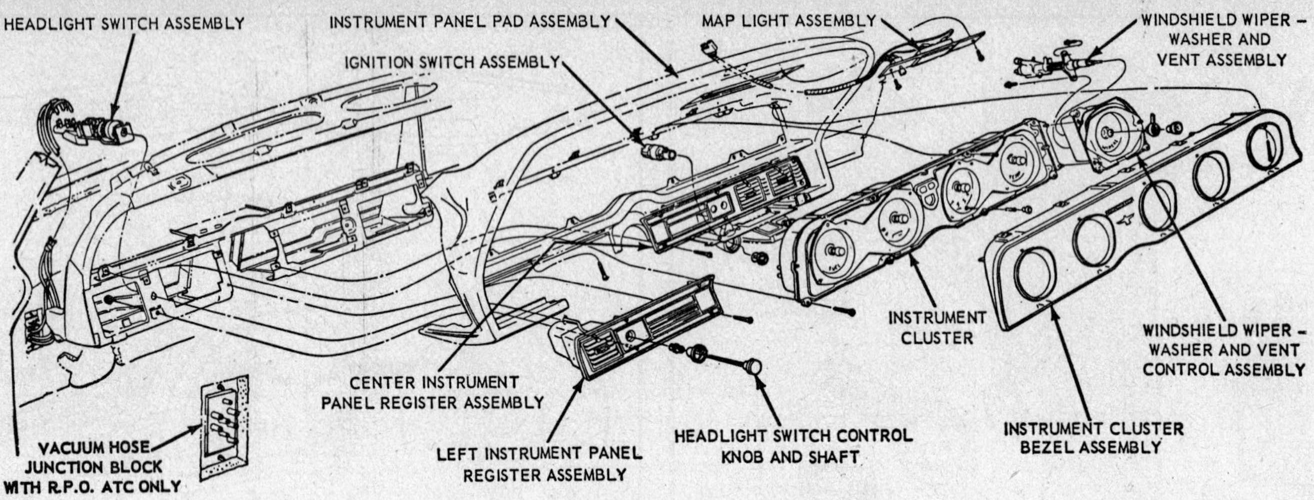

Fig. 7 Instrument cluster. 1969-71

unscrewing it from column.

4. Remove shroud from under steering column.
5. Disconnect steering column wiring connector plugs and remove screws that secure switch assembly to column.
6. On vehicles with tilt column, remove wires and terminals from steering column connector plug.

NOTE: *Record the color code and location of each wire before removing it from connector. A hole provided in the flange casting on fixed columns makes it unnecessary to separate the wires from the connector. The plug with wires installed can be guided through the hole.*

7. Remove plastic cover sleeve from wiring harness and remove the switch and wires from top of column.
8. Reverse procedure to install.

1969-70

Fixed Column

The combination turn signal and hazard warning switch for cars with fixed steering columns has been redesigned in that the warning switch knob is an integral part of the switch and is no longer replaceable. Electric wiring is bonded for ease of installation.

1. Disconnect battery and remove steering wheel.
2. Remove protective wire cover that runs along bottom of steering column tube and disconnect electrical plug, noting the color codes and location.
3. If equipped with speed control, remove the sleeve around the wiring and pull the first three speed control wires out of the column.
4. Remove turn signal lever, then remove three screws holding the switch to the column and lift switch from column.
5. Reverse procedure to install.

1969-70

Tilt Column

1. Disconnect ground cable at battery.
2. Remove steering wheel.
3. Unscrew turn indicator lever.
4. Remove emergency flasher control knob.
5. Remove upper tilt mechanism cover (2 screws).
6. Remove five screws from lower finish panel under steering column and remove plate.
7. Disconnect multiple connector beneath instrument panel and at base of steering column.
8. Disconnect connector block from bullet connectors by depressing tabs one at a time and remove plastic cover over wires.
9. Position lower collar down and remove plastic clip from side of steering column.
10. Remove two screws from switch and remove switch and wires from column.
11. Reverse procedure to install.

INSTRUMENT CLUSTER

1972-76

1. Disconnect battery ground cable.
2. Remove screw retaining lower cluster applique cover below steering column.
3. Squeeze the lower half of the column shroud together and separate the lower half from the upper.
4. Remove upper half of shroud from column.
5. Remove screw attaching PRNDL control cable wire to the column.
6. Remove heated backlite control knob.
7. Reach under instrument panel and depress the button on side of headlight switch while withdrawing switch control knob and shaft assembly.
8. Reach under panel and disconnect speedo cable.
9. Remove threaded headlight switch bezel.
10. Remove windshield wiper/washer control knob and bezel.
11. Remove cigar lighter from its receptacle.
12. Remove four screws retaining cluster front cover.
13. Insert a right angle standard tip screwdriver along edges of the finish panel withdrawing studs in sequence gradually around the periphery of the panel, Fig. 6.
14. Remove four screws retaining cluster to panel.
15. Pull cluster away from panel; disconnect cluster feed plug from its receptacle in the printed circuit.
16. Tilt cluster out, top first, and move the cluster toward center of car.

1969-71

1. Disconnect battery ground cable.
2. Remove instrument panel bezel, Fig. 7.
3. Remove clock adjusting knob, then push button retainers, mask and lens from instrument cluster.
4. Remove speedometer attaching screws and speedometer from cluster, then disconnect speedometer cable.
5. Remove cluster to instrument panel retaining screws.
6. Remove rear vent and wiper control assembly to instrument panel attaching screws.
7. Pull cluster from panel, disconnect electrical connectors from cluster and remove cluster.
8. Reverse procedure to install.

W/S WIPER BLADES

NOTE: Trico and Anco blades are used and both come in two types; Bayonet type and Side Saddle Pin type.

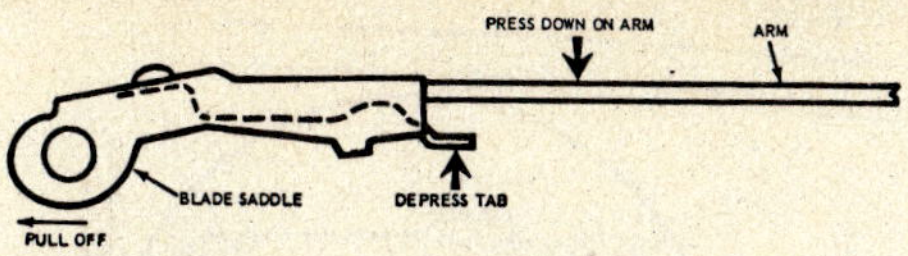

Fig. 8 Trico bayonet type blade

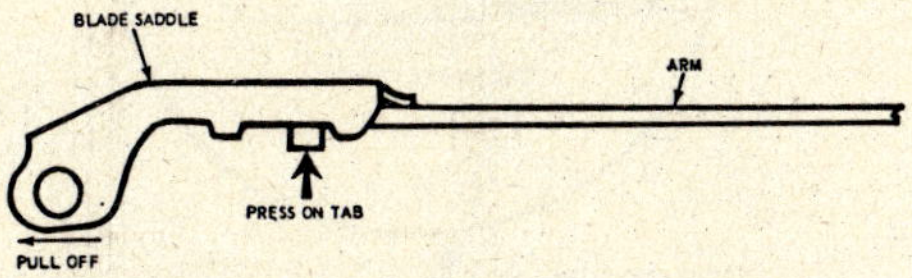

Fig. 8A Anco bayonet type blade

Bayonet Type: To remove the *Trico* type press down on the arm to unlock the top stud. Depress the tab on the saddle, Fig. 8, and pull the blade from the arm. To remove the *Anco* type press inward on the tab, Fig. 8A, and pull the blade from the arm. To install a new blade assembly slide the blade saddle over the end of the wiper arm so that the locking stud snaps into place.

Side Saddle Pin Type: To remove the pin type insert an appropriate tool into the spring release opening of the blade saddle, depress the spring clip and pull the blade from the arm, Fig. 9. To install, push the blade saddle on to the pin so that the spring clip engages the pin. To replace the rubber element in a *Trico* blade squeeze the latch lock release and pull the element out of the lever jaws, Fig. 10. Remove the *Anco* element by depressing the latch pin, Fig. 10, and sliding the element out of the yoke jaws. To install insert the element through the yoke or lever jaws. Be sure the element is engaged at all points.

W/S WIPER ARMS

1969-71

Swing the arm and blade assembly away from the windshield and insert a 3/32 inch pin through the pin hole as shown in Fig. 11. Swinging the assembly away from the windshield will release the spring loaded attaching clip in the arm from the pivot shaft. Inserting the pin will hold it in the released position. The arm can now be pulled off the pivot shaft. *Do not* pry off with a screwdriver. Leave the pin in the arm until after installation. A new arm comes with a pin already installed to hold it in the released position. Make sure the pivot shaft is in parked position when installing new assembly. Push the arm onto the pivot shaft. Lock the arm to the pivot shaft by removing the pin.

1972-76

Raise the blade end of the arm off the windshield and move the slide latch, Fig. 12, away from the pivot shaft. This action will unlock the wiper arm from the pivot shaft and hold the blade end of the arm off the glass at the same time. The wiper arm can now be pulled off the pivot shaft without the aid of any tools.

W/S WIPER MOTOR, REPLACE

1972-76

1. Disconnect battery ground cable.
2. Remove wiper arm and blade assemblies.
3. Remove left cowl screen for access through cowl opening.
4. Disconnect linkage drive arm from motor output arm crankpin by removing the retaining clip.
5. From engine side of dash, disconnect wire connectors from motor.
6. Remove bolts that retain motor to dash and remove motor. If output arm catches on dash during removal, handturn the arm clockwise so it will clear.

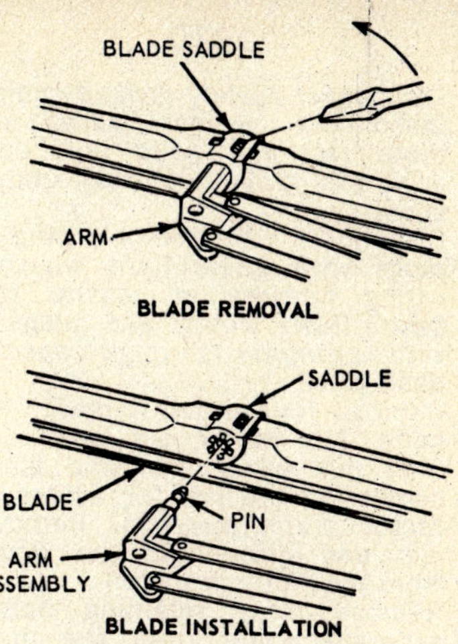

Fig. 9 Pin type blade

1971

1. Remove wiper arm and blade assemblies.
2. Remove air cleaner.
3. Remove retaining screws and remove cowl top panel. Disconnect washer hose.
4. Remove the clip retaining the left wiper link to the motor and disconnect the link.
5. Disconnect both hydraulic lines from motor. Be careful not to burn hands with hot hydraulic fluid.
6. Remove the three bolts retaining motor to cowl. Disconnect wiper control cable and remove motor through left cowl opening.

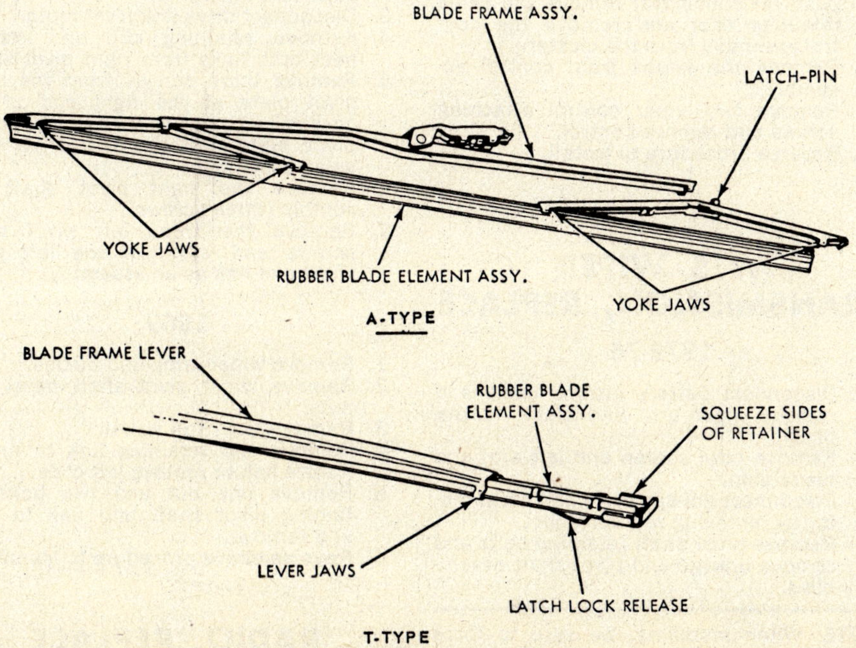

Fig. 10 Wiper blade element

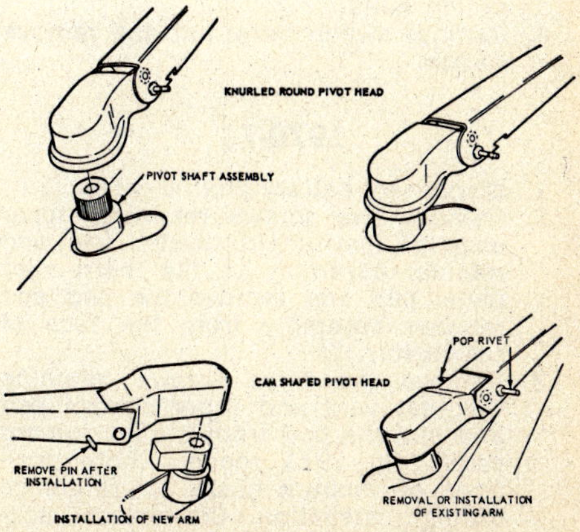

Fig. 11 Wiper arm. 1969-71

1970

1. Disconnect battery ground cable.
2. Disconnect washer hose, remove three retaining bolts and pull the cowl top grille out from under two slips.
3. Disconnect connector plug (two plugs with intermittent wiper) from wiring harness at engine side of dash. Push wiring and plugs along with grommet through opening in dash panel.
4. Remove four motor-to-cowl retaining bolts. Lift motor out and at same time pull wiper arm and blade assembly to the left for access to the motor crank pin clip. Remove the clip and disconnect the drive link from the motor crank pin.
5. Remove three retaining bolts and separate motor from the mounting plate and cover and wiring harness assembly.

1969

The cowl top panel is retained with eight screws and the two rear hood panel bumpers instead of 14 retaining screws. The wiper motor is attached to the upper cowl panel with three screws instead of two.

1. Remove wiper arm and blades.
2. Remove pivot shaft bezel.
3. Remove air cleaner.
4. Remove cowl top panel.
5. Remove seal plate from dash panel (2 screws).
6. Remove two clips retaining wiper links to motor. Rotate link to remove left clip.
7. Disconnect hydraulic line under hood.
8. Disconnect hydraulic line in cowl from motor.
9. Unfasten motor from mounting bracket (2 bolts). Disconnect control cable and remove motor.

W/S WIPER SWITCH, REPLACE

1972-76

1. Remove instrument cluster finish panel.
2. Remove wiper switch mounting plate.
3. Disconnect cigar lighter and wiper switch wires.
4. Remove switch bezel nut and remove switch.

1970-71

1. Disconnect battery ground cable.
2. Remove five screws retaining upper edge of instrument cluster pad and retainer assembly to the instrument panel pad and remove the pad and retainer assembly from the face of the cluster.
3. Remove the three screws retaining the rear vent and wiper control pod and pull the pod from the instrument panel. On 1971 models, note positions of vacuum hoses for reference during installation. Disconnect vacuum hoses and electrical connector and remove illumination bulbs.

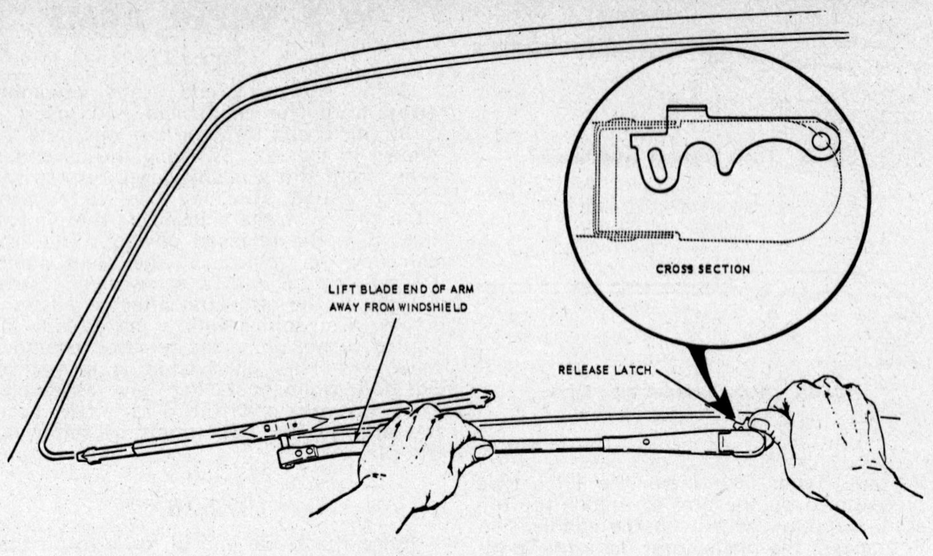

Fig. 12 Wiper arm. 1972-76

4. On 1971 models, remove clip retaining the control cable to the control and remove the control assembly.
5. Remove control knobs and remove the two screws retaining switch and remove the switch.

1969

1. Disconnect battery ground cable.
2. Remove five screws attaching the top of the cluster bezel to the instrument panel. Loosen five screws attaching bottom of the bezel to the instrument panel pad retainer and remove the bezel.
3. Remove three screws attaching the wiper and rear vent control to the instrument panel.
4. Disconnect vacuum hoses, lights and push on connector; remove the cable retaining clip; and remove the control assembly from the cluster.
5. Remove the knobs from control assembly.
6. Remove two wiper control attaching screws and remove control.
7. Reverse procedure to install.

W/S WIPER TRANSMISSION, REPLACE

1972-76

1. Disconnect battery ground cable and remove wiper arm and blade assemblies.
2. Remove cowl screen and left arm and blade stop.
3. Disconnect linkage drive arm from motor by removing retaining clip.
4. Remove pivot shaft retaining bolts and remove linkage and pivot shaft assemblies.

NOTE: When installing, be sure to force the linkage connecting clip into the locked position.

1971

1. Remove wiper arm and blade assemblies. Be sure to release tension arm retaining clip from the tension arm retaining stud on the left pivot assembly.
2. Remove retaining screws and remove the cowl vent grille. Disconnect windshield washer hoses.
3. Remove clip retaining the link assembly to the motor. Remove screws retaining pivot shaft and link assembly to the cowl and remove the pivot shaft and link assembly.

1970

1. Disconnect battery ground cable.
2. Disconnect washer hose and remove cowl vent top panel.
3. Disconnect drive link from motor.
4. Remove retaining clip and disconnect both links from right pivot shaft.
5. Remove three screws from the drive pivot plate at the right end of the cowl panel and withdraw the drive pivot plate and the two drive links as an assembly.
6. Remove the right pivot shaft assembly (three screws).
7. Remove the three left pivot shaft screws and withdraw the left pivot shaft and link as an assembly.

1969

1. Remove wiper arms and blades.
2. Remove wiper pivot shaft bezel and nuts.
3. Remove cowl top panel.
4. Remove clip retaining link to motor. Rotate link to remove left clips.
5. Remove one nut and two bolts retaining pivot shaft and link to cowl and remove.
6. Reverse above procedure to install.

RADIO, REPLACE

NOTE: When installing radio, be sure to

adjust antenna trimmer for peak performance.

1974-76

1. Disconnect battery ground cable.
2. Pull radio knobs, discs and nuts from shafts.
3. Remove twilight sentinel amplifier.
4. Remove air conditioning duct located under radio.
5. Remove radio rear support to panel screw and disconnect radio electrical leads, then remove radio.

1972-73

1. Disconnect battery ground cable.
2. Pull radio knobs off shafts.
3. Remove nut from radio control shafts.
4. Remove radio rear support attaching screw at instrument panel.
5. Disconnect power and speaker wires at connectors.
6. Disconnect antenna lead and remove radio.

1970-71

1. Disconnect battery ground cable.
2. Pull knobs off radio control shafts and remove cover plate located below steering column.
3. Remove the nut from the right radio control shaft.
4. Remove six screws and remove the trim applique from in front of radio.
5. Remove nut and washer from right radio control shaft.
6. Remove the screw attaching front left side of radio to instrument panel.
7. Remove radio rear support attaching screw.
8. Disconnect radio power wires at connectors.
9. Disconnect antenna lead-in cable and remove radio.

1969

1. Disconnect battery ground cable.
2. Pull knobs off radio control shafts.
3. Remove cover plate located below steering column.
4. Remove six screws and remove trim applique from in front of radio.
5. Remove nut and washer from right radio control shaft.
6. Remove screw attaching the front left side of radio to instrument panel.
7. Remove radio rear support attaching screw.
8. Disconnect radio power wires and speaker leads at connectors.
9. Disconnect antenna lead in cable and remove radio.
10. Reverse procedure to install.

BLOWER MOTOR REMOVE

1969-71

1. Disconnect the battery ground cable.
2. On **Air Cond Models:** Remove the courtesy light from the lower edge of the instrument panel. Remove the glove box liner.
3. Remove the right cowl side trim panel.
4. Remove the six duct mounting flange screws. Reach through the recirculating door opening and remove the vacuum motor hose from the vacuum motor and remove the duct assembly.
5. Disconnect the lead wire to the blower motor.
6. Remove one screw on the motor mounting plate. Rotate the motor mounting plate counterclockwise to un-lock the mounting plate from the case. Remove the motor and wheel assembly through the opening in the cowl side panel.
7. Reverse above to install.

1972-76

1. Disconnect the battery ground cable.
2. Remove the glove box for access.
3. Remove the recirc air register and duct assembly from the blower lower housing to the dash panel.
4. Disconnect the white hose from the outside-recirc air door vacuum motor and remove the vacuum motor from the blower lower housing. Leave the motor actuator connected to the door crank arm, Fig. 13.
5. Disconnect the blower lead wire (orange) from the harness connector, disconnect the motor ground wire (black).
6. Remove the six upper to lower blower housing flange screws.
7. Separate the blower lower housing and motor assembly from the upper housing and remove it from under the instrument panel.
8. Remove the blower motor and wheel assembly from the lower housing.
9. Reverse the above to install.

HEATER CORE REMOVAL

1972-76

1. Drain coolant and disconnect heater hoses from core.
2. Remove glove box.
3. Remove heater air outlet register from plenum assembly (2 snap clips).
4. Remove temperature control cable assembly mounting screw and disconnect end of cable from blend door crank arm (1 spring nut).
5. Remove the blue and red vacuum hoses from the high-low door vacuum motor; the yellow hose from the panel-defrost door motor, and the brown hose at the inline tee connector to the temperature by-pass door motor, Fig. 14.
6. Disconnect wiring connector from resistor.
7. Remove ten screws from around flange of plenum case and remove the rear case half of the plenum.
8. Remove mounting nut from heater core tube support bracket.
9. Remove core.

NOTE: When installing, apply body sealer between front and rear case halves and make sure core mounting gasket is properly installed.

1970-71

1. Remove hood and air cleaner and drain engine coolant.
2. Disconnect wiper hydraulic lines.
3. Disconnect heater hoses at heater core.
4. Disconnect vacuum hose at top of housing and remove oil pressure sender unit from back of engine.
5. Remove transmission dip stick and tube assembly.
6. Disconnect multiple connector leading to icing switch.
7. Remove evaporator housing front cover.
8. Remove heater core housing cover.
9. Remove heater core retaining bracket and remove the core.
10. Reverse the procedure to install.

1969

1. Remove hood and carburetor air cleaner.
2. Disconnect vacuum supply hose on top of heater case and pull it away from case mounting studs and disconnect hydraulic lines at wiper motor.
3. Remove transmission dipstick and tube.
4. Disconnect multiple connector leading to thermostat switch inside heater case cover and to resistor on front of cover.
5. Remove five nuts and two screws from heater core case cover and take off cover.
6. Slide heater core from case.

SPEED CONTROLS

1972-76

Adjust the bead chain to obtain .06-.25" actuator arm free travel when the engine is at hot idle, Fig. 15. The adjustment should be made to take as much slack as possible out of the chain without restricting the carburetor lever from returning to idle. On vehicles with a solenoid anti-diesel valve, perform this adjustment with the ignition switch in the ON position.

1970-71

Linkage Adjustment

Adjust the bead chain to obtain .06-.25" actuator arm free travel when the engine is at hot idle, Fig. 16.

1969

Bead Chain Adjustment

This is the only adjustment required for proper functioning of the system and assure a normal engine idle.

Adjust the chain to maintain ½" to 1" ball slack with engine at hot idle, Fig. 17.

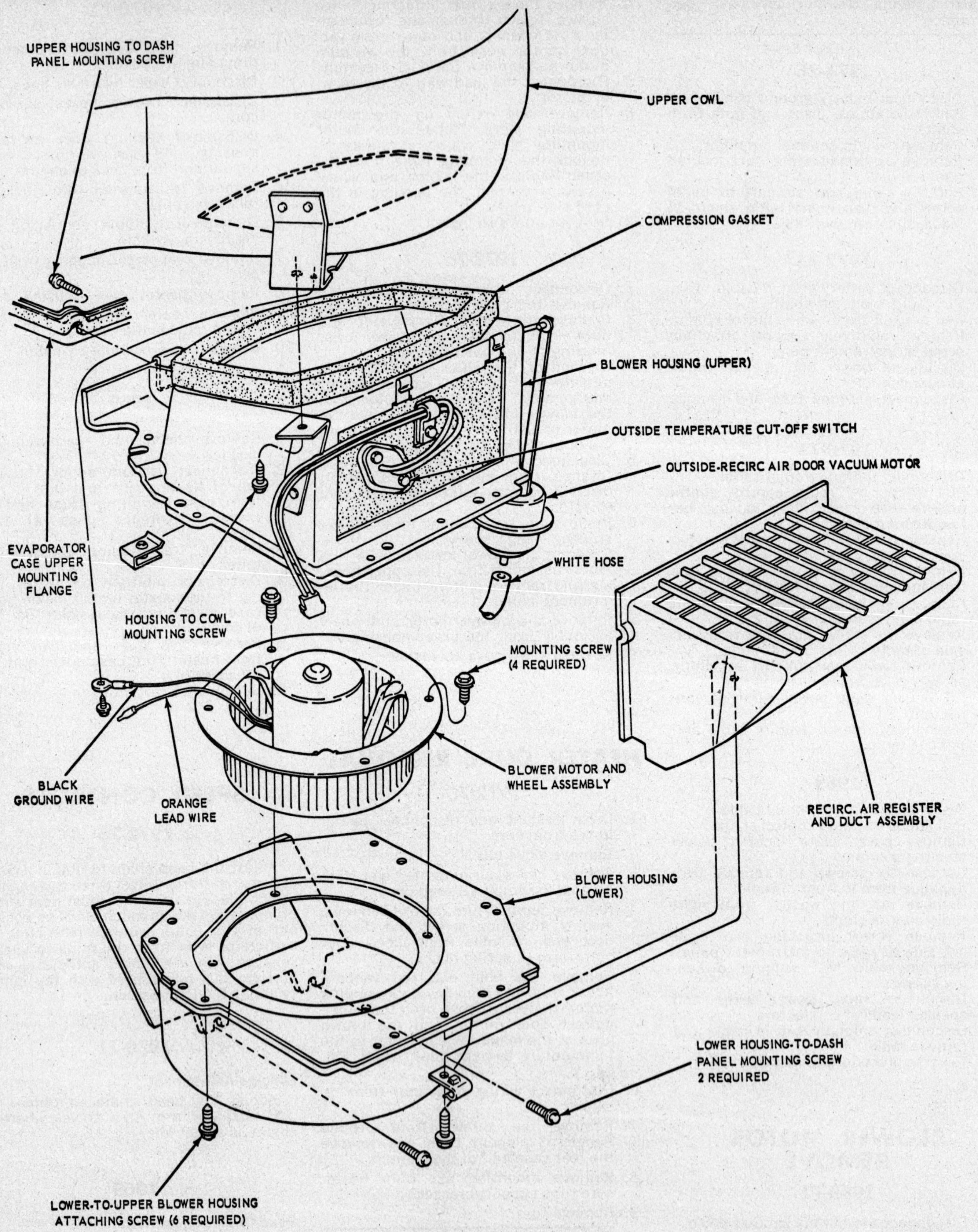

Fig. 13 Blower motor. 1972-76

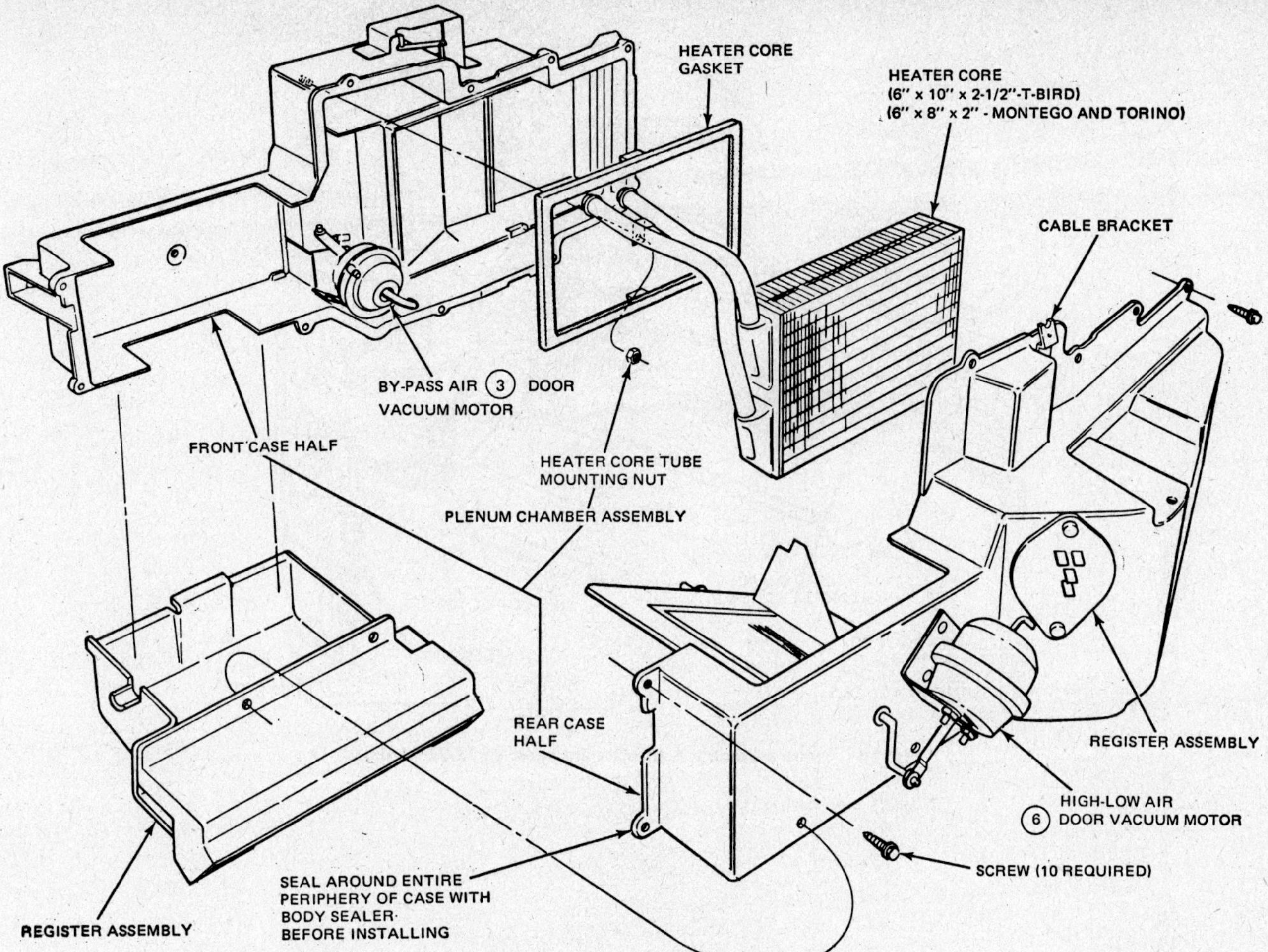

Fig. 14 Heater core. 1972-76

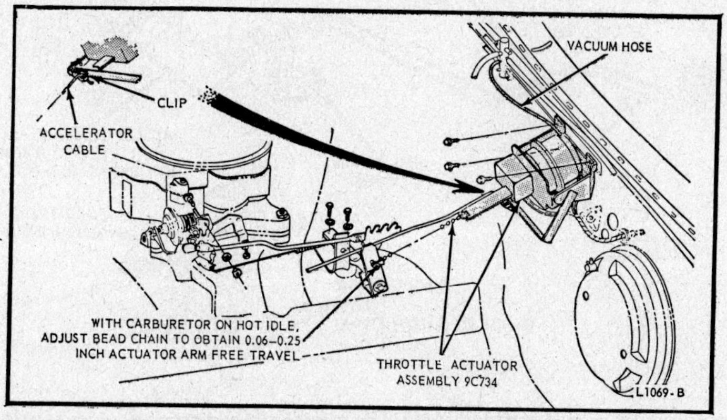

Fig. 16 Throttle actuator assembly. 1970-71

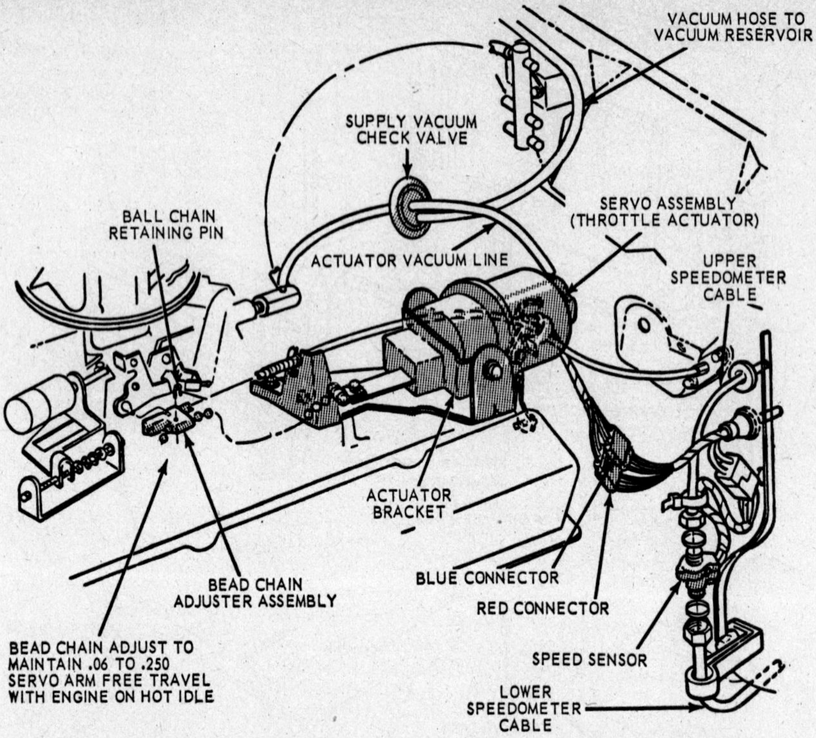

Fig. 15 Servo assembly & throttle linkages. 1972-76 (typical)

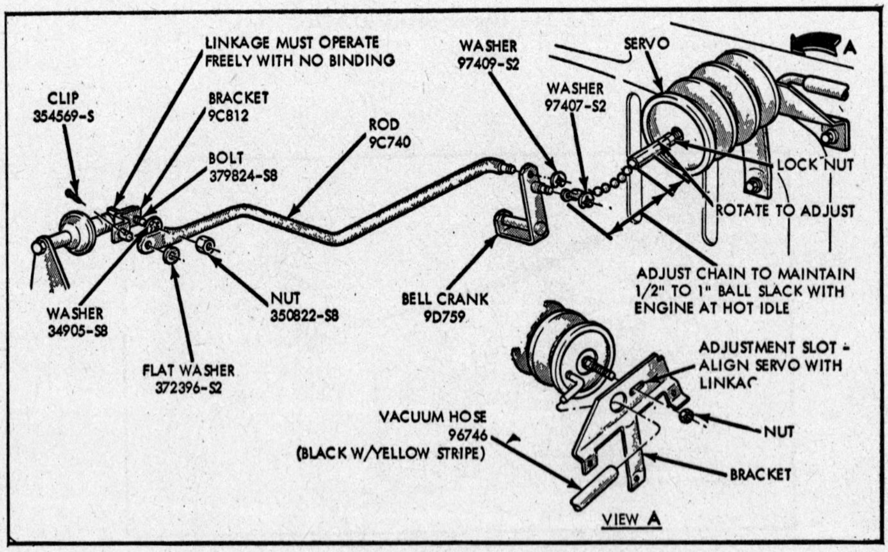

Fig. 17 Servo assembly and accelerator linkage. 1969

Engine Section

Service procedures on the V8-400 are covered in the Ford & Mercury car chapters

ENGINE MOUNTS REPLACE

1969-71

1. Disconnect battery and raise vehicle on hoist. Remove starter.
2. Remove the long bolt and nut that attaches the insulator to support bracket, Fig. 1.
3. Place jack and wood under the engine oil pan and raise engine about one inch.
4. Remove insulator to engine bolts and lock washers, and remove insulator.
5. Reverse above to install.

1972-76

1. Remove the fan shroud attaching screws and support the engine with a jack and a block of wood placed under the leading edge of the oil pan.
2. Remove the through bolt and nut attaching the insulator to the frame crossmember, Figs. 2 and 3.
3. Remove the insulator to upper bracket attaching nuts.
4. Raise the engine enough to remove the insulator and heat shield if so equipped.
5. Reverse above to install.

ENGINE, REPLACE

Because of engine compartment tolerances, the engine should not be removed and installed with the transmission attached.

1. Disconnect battery ground cable and drain cooling system and crankcase.

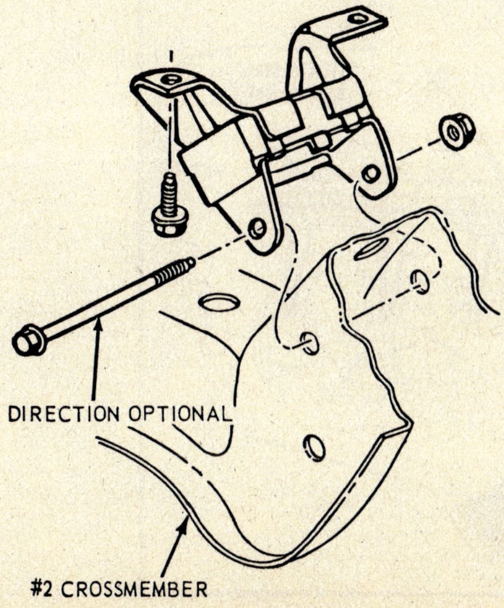

Fig. 2 Engine mount. 1972 V8-400

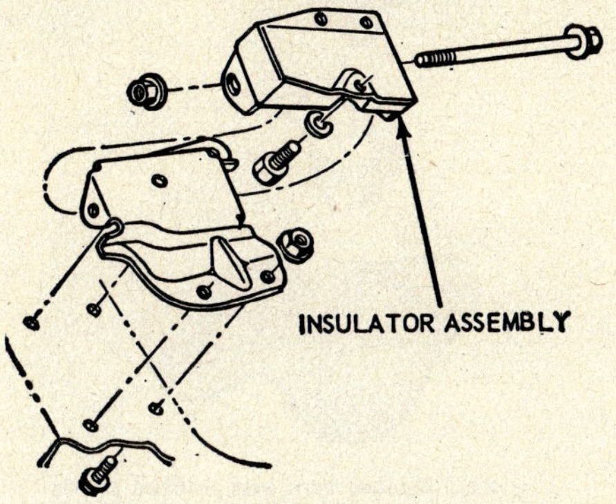

Fig. 1 Engine mount. 1969-71

2. Disconnect transmission oil cooler lines and remove radiator and air cleaner.
3. Remove hood.
4. Remove fuel and vacuum lines and all hoses, wires and linkage attached to engine.
5. Disconnect exhaust pipe from manifolds.
6. Remove starter and automatic transmission filler tube.
7. Remove converter or flywheel housing lower cover.
8. Support transmission with jack.
9. Unfasten converter or flywheel housing from engine.
10. Remove engine mounting bolts and lift engine out of chassis.

CYLINDER HEAD REPLACE

NOTES

Before installing cylinder head, wipe off engine block gasket surface and be certain no foreign material has fallen into cylinder bores, bolt holes or in the valve lifter area. It is good practice to clean out bolt holes with compressed air.

Some cylinder head gaskets are coated with a special lacquer to provide a good seal once the parts have warmed up. Do not use any additional sealer on such gaskets. If the gasket does not have this lacquer coating, apply suitable sealer to both sides.

Tighten cylinder head bolts at little at a time in three steps in the sequence shown in the illustrations. Final tightening should be to the torque specifications listed in the *Engine Tightening* table. After the bolts have been torqued to specifications, *they should not be disturbed.*

In instances where cylinder head gasket leakage is hard to control, aluminum paint can be applied to the gasket as a sealer.

Spray one coat of the aluminum paint on both sides of the gasket and allow the paint to dry. Then spray a second coat on both sides and, while the paint is still wet, install the gasket. Torque the head and manifold bolts to specifications to complete the job.

1969-76

1. Remove intake manifold and carburetor as an assembly.
2. Disconnect muffler inlet pipe at exhaust manifold.
3. Loosen air conditioner compressor belt if so equipped.
4. Loosen alternator retaining bolts and remove bolt retaining alternator bracket to right head.
5. If air conditioned, isolate compressor at service valves and hoses from compressor. Remove nuts retaining compressor bracket to water pump. Remove bolts retaining compressor to upper mounting bracket and lay compressor out of way. Remove compressor upper bracket from head.
6. If not air conditioned, remove bolts retaining power steering reservoir

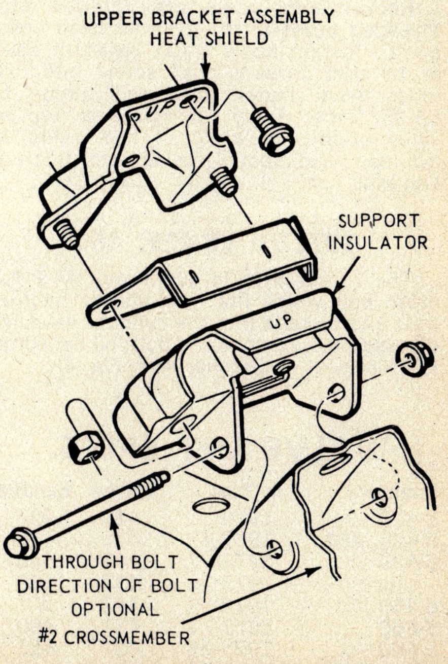

Fig. 3 Engine mount. 1972-76 V8-429, 460

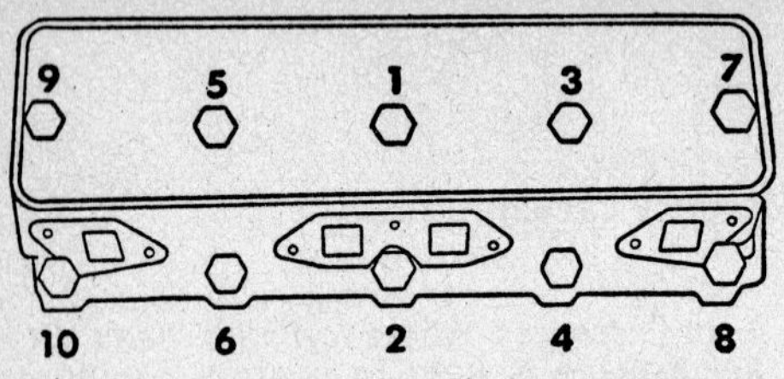

Fig. 4 Cylinder head tightening sequence

bracket to left head and position reservoir out of way.

7. Remove rocker arm covers and rocker arms. Remove push rods in sequence so they can be installed in their positions.
8. Remove head retaining bolts and lift head with exhaust manifold.

NOTE: If necessary to break gasket seal, pry at forward corners of cylinder heads against casting bosses provided on cylinder block. Avoid damaging machined surfaces on head and block.

---

9. Install cylinder heads in reverse order of removal and tighten bolts in the sequence shown in Fig. 4.

NOTE

The cylinder head gaskets are marked "Top" or "Front" stamped near the front end of the gasket. The gasket is properly installed when the word is at the forward end of the engine and water passage holes line up. This results in the sealing beads on the right head gasket being inverted with respect to the left head gasket.

ROCKER ARM STUDS

1969-70 V8-429, 460

Rocker arm studs are screwed into threaded bores in the cylinder head bosses. To install, apply water resistant sealer to stud threads that screw into cylinder head. Install stud and torque to 65-75 ft-lbs. Apply Lubriplate to top of valve stem and at push rod guide in cylinder head. Install rocker arm, fulcrum and stud nut.

1971-76 V8-429, 460

The rocker arm is supported by a fulcrum bolt which fits through the fulcrum seat and threads into the cylinder head. To disassemble, remove the bolt, oil deflector, fulcrum seat and rocker arm, Fig. 4A.

VALVE LIFT SPECS.

Engine	Year	Intake	Exhaust
8-400	1972	.422	.427
8-429, 460	1969-71	.443	.486
8-429	1972	.443	.486
	1973	.437	.480
8-460	1972	.443	[1]
8-460	1973-74	.437	.480
8-460	1975-76	.437	.451

[1] Exc. Calif., .486. Calif., .424

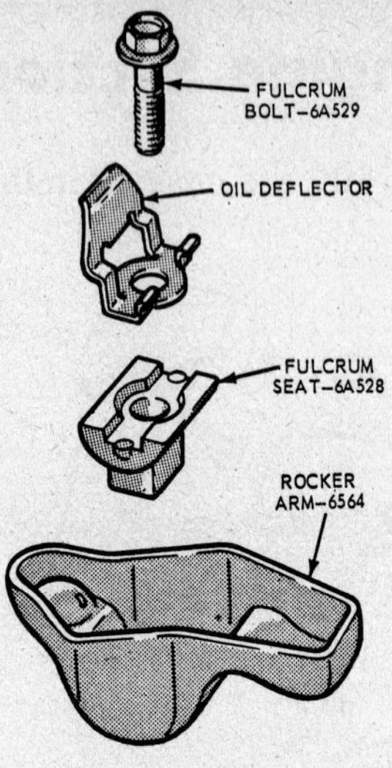

Fig. 4A Rocker arm and related parts. 1971-76 V8-429, 460

VALVE TIMING

Intake Opens Before TDC

Engine	Year	Degrees
8-400	1972	17
8-429	1969-71	16
	1972-73	8
8-460	1972-76	8

VALVE ARRANGEMENT

Front to Rear

V8-400, 429, 460 Right . . . . . I-E-I-E-I-E-I-E
V8-400, 429, 460 Left . . . . . . E-I-E-I-E-I-E-I

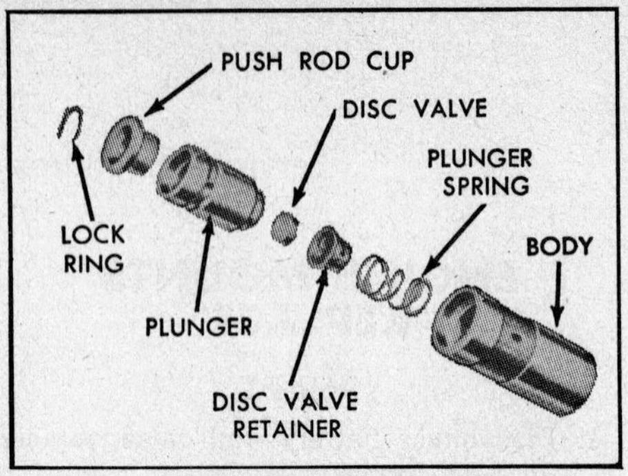

Fig. 5 Hydraulic valve lifter (typical)

VALVES, ADJUST

1969-76, V8-429, 460

To eliminate the need of adjusting valve lash, a positive stop rocker arm stud and nut is used on 1969-70 models, and a positive stop fulcrum bolt and seat is used on 1971-75 models, Fig. 4A.

It is very important that the correct push rod be used and all components be installed and torqued as follows.

NOTE: A .060" shorter push rod or a .060" longer push rod are available for service to provide a means of compensating for dimensional changes in the valve mechanism. Valve stem-to-rocker arm clearance should be as listed in the *Valve Specifications* table, with the hydraulic lifter completely collapsed. Repeated valve grind jobs will decrease this clearance to the point that if not compensated for the lifters will cease to function.

To check the clearance, bring the piston of the cylinder being checked on top dead center of the compression stroke. Then with

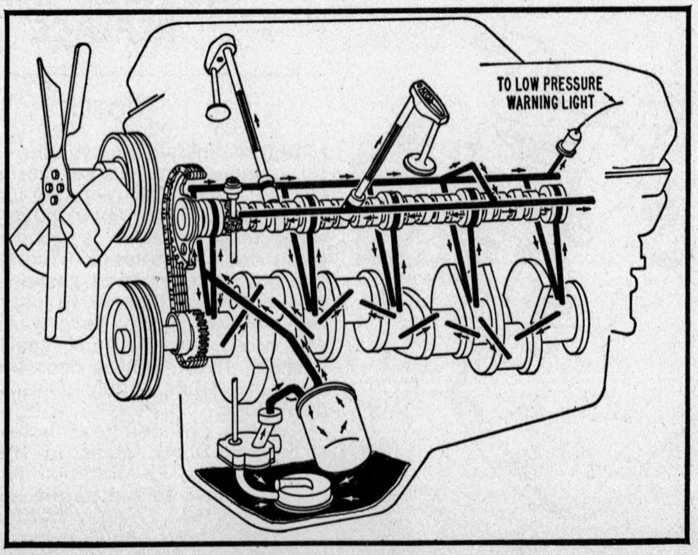

Engine lubrication system. V8-429, 460

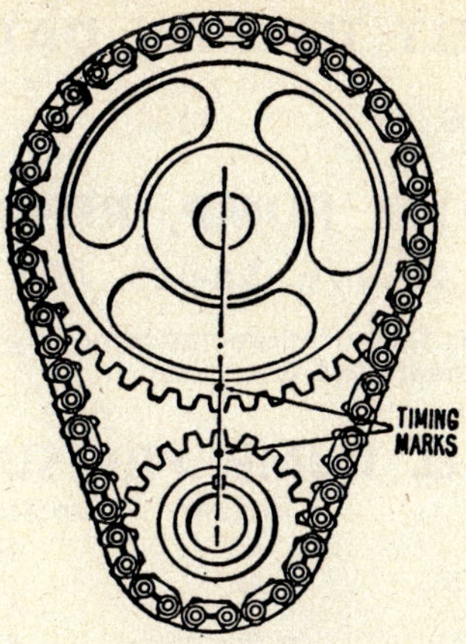

Fig. 6 Valve timing marks, V8-429, 460

hydraulic lifter collapsed, check the clearance between valve stem and rocker arm. If the clearance is less than the minimum, the .060″ shorter push rod should be used. If clearance is more than the maximum, the .060″ longer push rod should be used. (See *Valve Specifications* table).

1. Position the piston of the cylinder being worked on at TDC of its compression stroke.
2. On 1969-70 models, install positive stop stud and torque to 65-75 ft. lbs., then lubricate and install rocker arm and fulcrum seat on the stud. Thread nut onto the stud until it contacts the shoulder and torque to 18-22 ft. lbs.
3. On 1971-76 models, install rocker arm, fulcrum seat and oil deflector. Install fulcrum bolt and torque to 18-25 ft. lbs.

VALVE GUIDES

Valve guides in these engines are an integral part of the head and, therefore, cannot be removed. For service, guides can be reamed oversize to accommodate one of three service valves with oversize stems (.003″, .015″ and .030″).

Check the valve stem clearance of each valve (after cleaning) in its respective valve guide. If the clearance exceeds the service limits of .004″ on the intake or .005″ on the exhaust, ream the valve guides to accommodate the next oversize diameter valve.

HYDRAULIC VALVE LIFTERS

The internal parts of each hydraulic valve lifter assembly are a matched set. If these are mixed, improper valve operation may result. Therefore, disassemble, inspect and test each assembly separately to prevent mixing the parts, Fig. 5.

1. Remove the intake manifold.
2. Remove the rocker arms.
3. Remove the push rods, keeping them in a rack in sequence so that they may be installed in their original location.
4. Remove the valve lifters with a magnet rod and place them in a rack in sequence so that they may be installed in their original location.
5. Reverse procedure to install.

TIMING CASE COVER

NOTE: If it becomes necessary to replace the oil seal in the timing case cover the cover must be removed.

1969-76, V8-429, 460

1. Drain cooling system and crankcase.
2. Remove fan and shroud.
3. Remove radiator.
4. Remove drive belts and water pump pulley. Remove compressor support if so equipped.
5. Remove bolt and washer attaching crankshaft damper. Remove damper with suitable puller. Remove Woodruff key from crankshaft.
6. Remove power steering pump.
7. Remove fuel pump.
8. Remove front cover to cylinder block bolts. Cut the oil pan seal flush with cylinder block face prior to separating cover from cylinder block. Remove front cover and water pump as a unit.

TIMING CHAIN

1969-76

1. To remove the chain, first take off the front cover as outlined previously.
2. Crank engine until timing mark on camshaft sprocket is adjacent to timing mark on crankshaft sprocket, Fig. 6.
3. Remove camshaft sprocket cap screw and fuel pump eccentric.
4. Slide both sprockets and chain forward and remove as an assembly.
5. Reverse foregoing procedure to install the chain, being sure to align the timing marks as shown.

CAMSHAFT, REPLACE

1969-76, V8-429, 460

1. Remove timing cover, chain and sprockets as outlined previously.
2. Remove intake manifold and carburetor as an assembly.
3. Remove rocker arm covers. Back off rocker arm stud nuts or bolts, turn rocker arms sideways and remove push rods in sequence.
4. Remove valve lifters.
5. If air conditioned, remove condenser attaching bolts and carefully position and secure condenser on left fender.
6. Remove grille.

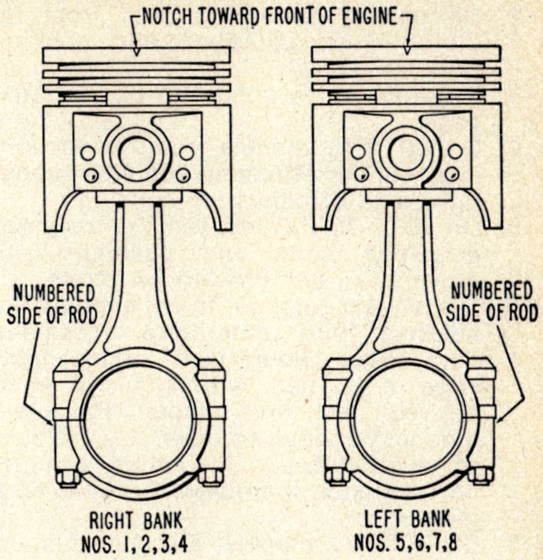

Fig. 7 Piston and rod assembly

7. Remove camshaft thrust plate retaining bolts and carefully remove camshaft from engine.
8. Reverse above procedure to install.

CAMSHAFT BEARINGS

When necessary to replace camshaft bearings, the engine will have to be removed from the vehicle and the plug at the rear of the cylinder block will have to be removed in order to utilize the special camshaft bearing removing and installing tools required to do this job. If properly installed, camshaft bearings require no reaming—nor should this type bearing be reamed or altered in any manner in an attempt to fit bearings.

PISTON & ROD, ASSEMBLE

All V8's

Assemble the pistons to the rods as shown in Fig. 7.

PISTONS, PINS & RINGS

Pistons are available in oversizes of .003, .020, .030, .040 and .060″.

Piston pins are available in oversizes of .001 and .002″.

Rings are available in oversizes of .002, .010, .020, .030 and .040″.

MAIN & ROD BEARINGS

Main and rod bearings are available in undersizes of .002, .010, .020, .030 and .040″.

OIL PAN, REPLACE

1969-76, V8-429, 460

1. Disconnect battery ground cable.

2. Disconnect radiator shroud from radiator if so equipped and position over fan.
3. Raise car on a hoist and drain crankcase.
4. On 1970-71 models, remove vacuum line retainer attaching bolt from upper right side of converter housing.
5. On 1972-76 models, remove sway bar attaching bolts and bushings and move sway bar forward on struts.
6. Disconnect engine front support insulators from underbody crossmember. Place floor jack under front edge of oil pan, with block of wood between jack and oil pan. Raise engine just enough to insert 1¼" blocks of wood between insulators and underbody side members. Remove floor jack.
7. On 1969-71 models, disconnect starter cable, then remove attaching bolts and position starter on steering linkage.
8. Remove oil filter.
9. On 1969-71 models, remove end attachments of stabilizer bar and rotate ends of bar down to raise center of bar.
10. Unbolt and remove oil pan ahead of underbody crossmember.
11. Reverse above procedure to install.

OIL PUMP REPAIRS

Rotor Type Pump, Fig. 8

1. With all parts clean and dry, check the inside of the pump housing and the outer race and rotor for damage or excessive wear.
2. Check the mating surface of the pump cover for wear. If this surface is worn, scored or grooved, replace the cover.
3. Measure the clearance between the outer race and housing. This clearance should be .006-013" for 1969-72 and .001-.013" for 1973-76.
4. With the rotor assembly installed in the housing, place a straight edge over the rotor assembly and housing. Measure the clearance between the straight edge and the rotor and outer race. Recommended limits are .001-.004". *The outer race, shaft and rotor are furnished only as an assembly.*
5. Check the drive shaft-to-housing bearing clearance by measuring the O.D. of the shaft and the I.D. of the housing bearing. The recommended clearance limits are .0015-.0029".
6. Inspect the relief valve spring for a collapsed or worn condition.
7. Check the relief valve piston for scores and free operation in the bore. The specified piston clearance is .0015-.0029".

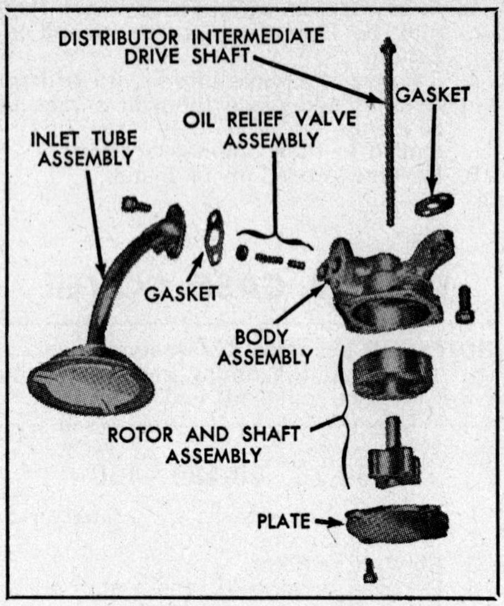

Fig. 8 Rotor type oil pump

BELT TENSION DATA

	New	Used
1969-76 All	140	110

WATER PUMP, REPLACE

1969-76, V8-429, 460

Refer to procedure for removing timing cover mentioned previously.

FUEL PUMP PRESSURE

Year	Engine	Pressure Lbs.
1969	All	4½-6½
1970-73	All	5-7
1974	[1]	5½-6½
1974	[2]	6½-7½
1975-76	All	5.7-7.7

[1] Exc. Calif.
[2] Calif.

FUEL PUMP, REPLACE

1. Remove all gasket material from the pump and block gasket surfaces. Apply sealer to both sides of new gasket.
2. Position gasket on pump flange and hold pump in position against its mounting surface. Make sure rocker arm is riding on camshaft eccentric.
3. Press pump tight against its mounting. Install retaining screws and tighten them alternately.
4. Connect fuel lines. Then operate engine and check for leaks.

SERVICE NOTE: Before installing the pump, it is good practice to crank the engine so that the nose of the camshaft eccentric is out of the way of the fuel pump rocker arm when the pump is installed. In this way there will be the least amount of tension on the rocker arm, thereby easing the installation of the pump.

Transmission Section

NOTE: 1976 linkage adjustment information is in this section. Repair procedures on both automatic and manual shift transmissions are covered elsewhere in this manual. Procedures for removing automatic transmissions as well as linkage adjustments on 1969-75 models are included in the automatic transmission chapters. See Chapter Index.

1976 AUTO TRANS. LINKAGE, ADJUST

Linkage adjustment procedures for 1976 models are the same as those for 1975 models as outlined elsewhere in this manual.

Rear Axle, Propeller Shaft & Brakes

REAR AXLES

Fig. 1 illustrates the rear axle assembly used on these cars. When necessary to overhaul the unit, refer to the *Rear Axle Specifications* table in this chapter.

Description

In these axles, Fig. 1, the drive pinion is straddle-mounted by two opposed tapered roller bearings which support the pinion shaft in front of the drive pinon gear, and straight roller bearing that supports the pinion shaft at the rear of the pinion gear. The drive pinion is assembled in a pinion retainer that is bolted to the differential carrier. The tapered roller bearings are preloaded by a collapsible spacer between the bearings. The pinion is positioned by a shim or shims located between the drive pinion retainer and the differential carrier.

The differential is supported in the carrier by two tapered roller side bearings. These bearings are preloaded by two threaded ring nuts or sleeves between the bearings and pedestals. The differential assembly is positioned for proper ring gear and pinion backlash by varying the adjustment of these ring nuts. The differential case houses two side gears in mesh with two pinions mounted on a pinion shaft which is held in place by a pin. The side gears and pinions are backed by thrust washers. With high performance engines, an optional rear axle having a four-pinion differential is also used.

The axle shafts are of unequal length, the left shaft being shorter than the right. The axle shafts are mounted in sealed ball bearings that are pressed on the shafts.

Diff. Carrier Assy.

Service Bulletin

All Ford Built Rear Axles: Recent manufacturing changes have eliminated the need for marking rear axle drive pinions for individual variations from nominal shim thicknesses. In the past, these pinion markings, with the aid of a shim selection table, were used as a guide to select correct shim thicknesses when a gear set or carrier assembly replacement was performed.

With the elimination of pinion markings, use of the shim selection table is no longer possible and the methods outlined below must be used.

1. Measure the thickness of the original pinion depth shim removed from the axle. Use the same thickness upon installation of the replacement carrier or drive pinion. If any further shim change is necessary, it will be indicated in the tooth pattern check.
2. If the original shim is lost, substitute a nominal shim for the original and use the tooth pattern check to determine if further shim changes are required.

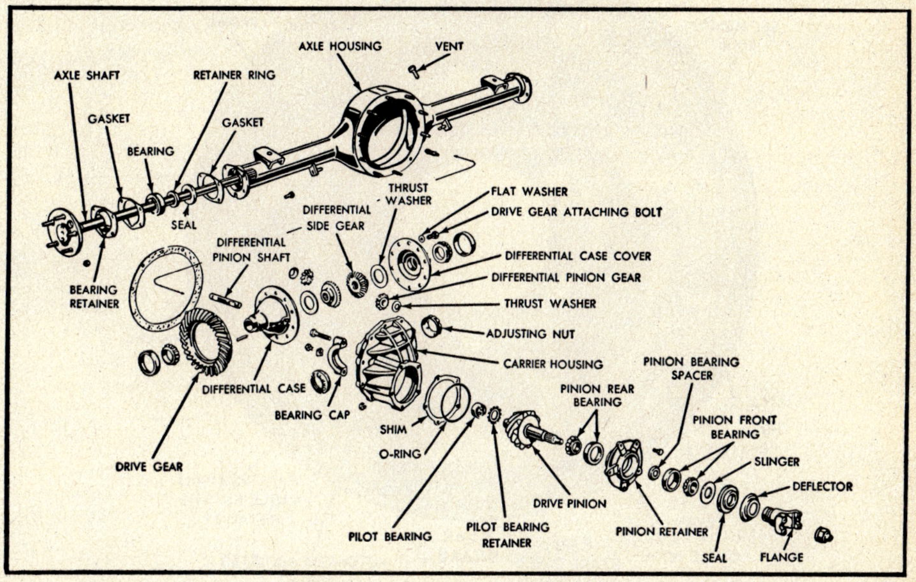

Fig. 1 Rear axle disassembled

Remove & Replace

In servicing the rear axles it is not necessary to remove the rear axle assembly for any normal repairs. The axle shafts and carrier assembly can easily be removed from the vehicle, leaving the axle housing in place.

1. Place a drain pan under the carrier and housing to catch the old grease when the carrier is separated from the housing.
2. Use a wire brush to clean dirt from the area around the carrier and housing mating surfaces. Then wipe the area clean with a cloth dampened in solvent.
3. Remove axle shafts and drive shaft as explained below.
4. Unfasten carrier from housing and lift out carrier.
5. Reverse removal procedure to install, using a new gasket between the carrier and housing.

AXLE SHAFTS

Removal

1. Remove wheel assembly.
2. Remove brake drum or caliper.
3. Working through hole provided in axle shaft flange, remove nuts that secure bearing retainer.
4. Pull axle shaft out of housing. If bearing is a tight fit in axle housing, use a slide hammer-type puller. *Brake carrier plate must not be dislodged. Install one nut to hold plate in place after axle shaft is removed.*
5. Loosen inner bearing retainer by nicking it deeply with a cold chisel in several places. On 1969-73 units with ball bearing axle bearings and 1974-76 units with tapered roller axle bearings, drill a 1/4 inch hole not more than 5/16 inch deep in retainer ring surface before using cold chisel. The bearing retainer will then slide off shaft easily.
6. Press bearing from axle shaft.
7. Inspect machined surfaces of axle shaft and housing for rough spots that would affect the sealing action of the oil seal. Carefully remove any burrs or rough spots.
8. Press new bearing on shaft until it seats firmly against shoulder on shaft.
9. Press inner bearing retainer on shaft until it seats firmly against bearing.
10. If oil seal is to be replaced, use a hook-type tool to pull it out of the housing. Wipe a small amount of oil resistant sealer on outer edge of seal before it is installed.

Installation

1. Place a new gasket on each side of brake carrier plate and slide axle shaft into housing. Start the splines into the differential side gear and push the shaft in until bearing bottoms in housing.
2. Install retainer and torque nuts to 50-75 ft. lbs. on 1969-72 units, 20-40 ft. lbs. on 1973 units and 35-55 ft. lbs. on 1974-76 units.
3. Install brake drum or caliper, then the wheel assembly.

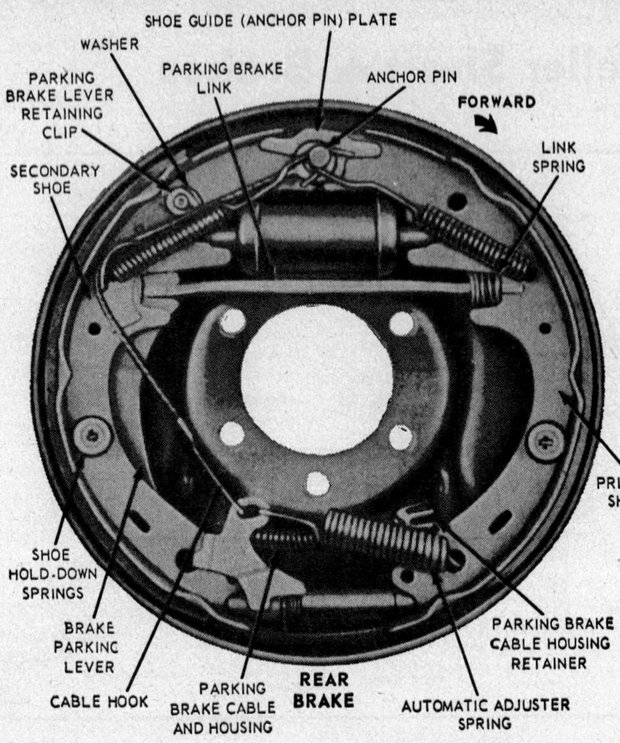

Fig. 2 Rear drum type brake on 1969-76

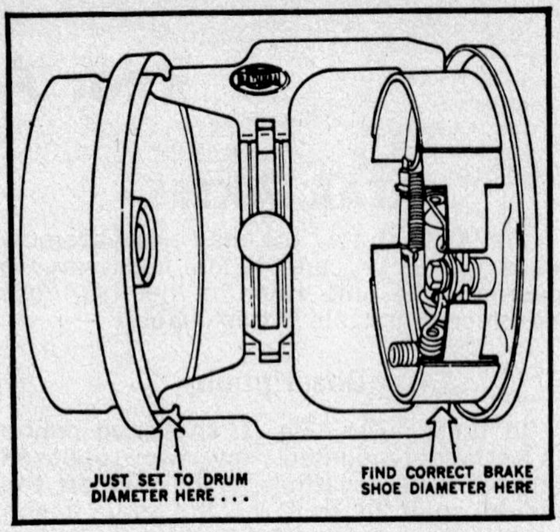

Fig. 3 Brake adjustment with gauge

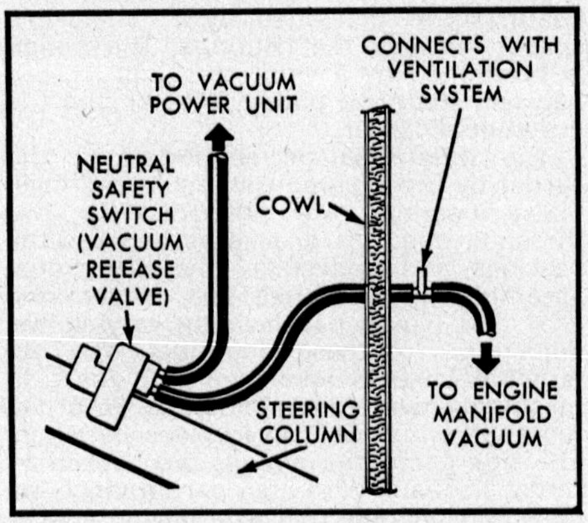

Fig. 4 Vacuum connections for automatic parking brake release. (typical)

PROPELLER SHAFT

Remove & Replace

1. Disconnect rear U-joint from drive pinion flange.
2. Pull drive shaft toward rear of car until front U-joint yoke clears transmission extension housing and output shaft.
3. Install a suitable tool, such as a seal driver, in seal to prevent lube from leaking from transmission.
4. Before installing, check U-joints for freedom of movement. If a bind has resulted from misalignment after overhauling the U-joints, tap the ears of the drive shaft sharply to relieve the bind.
5. If rubber seal installed on end of transmission extension housing is damaged, install a new seal.
6. On an automatic transmission, lubricate yoke spline with special spline lubricant. *This spline is sealed so that transmission fluid does not "wash" away spline lubricant.*
7. Install yoke on transmission output shaft.
8. Install U-bolts and nuts which attach U-joint to pinion flange. Tighten U-bolts evenly to prevent binding U-joint bearings.

BRAKE ADJUSTMENTS

1970-76

A new self centering pressure differential valve is used which no longer requires bleeding at the opposite end of the car to cause the brake warning light to go out. This is now accomplished as follows:

1. Turn on ignition switch.
2. Depress the brake pedal. This will automatically center the plunger in the switch causing the light to go out.

Self-Adjusting Brakes

These brakes, Fig. 2, have self-adjusting shoe mechanisms that assure correct lining-to-drum clearances at all times. The automatic adjusters operate only when the brakes are applied when the car is moving rearward or when it comes to an uphill stop.

Although the brakes are self-adjusting, an initial adjustment is necessary when the brake shoes have been relined or replaced, or when the length of the star wheel adjuster has been changed during some other service operation.

Frequent usage of an automatic transmission forward range to halt reverse vehicle motion may prevent the automatic adjusters from functioning, thereby inducing low pedal heights. Should low pedal heights be encountered, it is recommended that numerous forward and reverse stops be made until satisfactory pedal height is obtained.

Service Note

If a low pedal height condition cannot be corrected by making numerous reverse stops (provided the hydraulic system is free of air) it indicates that the self-adjusting mechanism is not functioning. Therefore, it will be necessary to remove the brake drum, clean, free up and lubricate the adjusting mechanism. Then adjust the brakes as follows, being sure the parking brake is fully released.

Adjustment

1. Remove adjusting hole cover from brake backing plate and, from the backing plate side, turn the adjusting screw upward with a screwdriver or other suitable tool to expand the shoes until a slight drag is felt when the drum is rotated.
2. Remove the drum.
3. While holding the adjusting lever out of engagement with the adjusting screw, back off the adjusting screw ¾ turn with the fingers.

NOTE—*If finger movement will not turn the screw, free it up. If this is not done, the adjusting lever will not turn the screw during subsequent vehicle operation. Lubricate the screw with oil and coat with wheel bearing grease. Any other adjustment procedure may cause damage to the adjusting screw with consequent self-adjuster problems.*

4. Install wheel and drum, and adjust-

ing hole cover. Adjust brakes on remaining wheels in the same manner.
5. If pedal height is not satisfactory, drive the vehicle and make sufficient reverse stops until proper pedal height is obtained.

Revised Brake Adjustment

The adjustment is made with the drums removed, using the brake gauge shown in Fig. 3. With the gauge, determine the inside diameter of the drum braking surface. Reverse the tool as shown and adjust the brake shoe diameter to fit the gauge. Hold the automatic adjusting lever out of engagement while rotating the adjusting screw to prevent burring the screw slots. Rotate the gauge around the brake shoes to be sure of the setting. After the brake drums and wheels have been installed, complete the adjustment by applying the brakes several times while backing the vehicle.

PARKING BRAKE, ADJUST

1969-76 Vacuum Release Unit

The vacuum power unit, Fig. 5, will release the parking brake automatically when the transmission selector lever is moved into any driving position with the engine running. The brakes will not release automatically, however, when the selector lever is in neutral or park position with the engine running, or in any other position with the engine off.

The lower end of the release handle extends out for alternate manual release in the event of vacuum power failure or for optional manual release at any time.

1970-76

Rear Drum Brakes

1. Make sure parking brake is released.
2. Place transmission in neutral and raise the vehicle.
3. Tighten the adjusting nut against the cable equalizer to cause rear brakes to drag.
4. Then loosen the adjusting nut until the rear wheels are fully released. There should be no drag.
5. Lower vehicle and check operation.

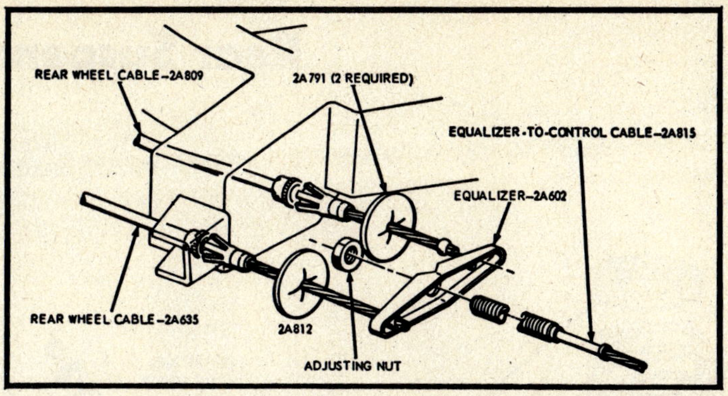

Fig. 5 Parking brake linkage. 1969-71

1975-76

Rear Disc Brakes

1. Fully release parking brake, then place transmission in neutral and support vehicle at rear axle.
2. Tighten adjuster nut until levers on calipers just begin to move, then loosen adjuster nut until levers just return to stop position.
3. Apply and release parking brake. Check levers on caliper to determine if they are fully returned by attempting to pull lever rearward. If lever moves, the adjustment is too tight and must be readjusted.

1969

1. Check the parking cables when the brakes are fully released. If the cables are loose, adjust as follows:
2. Fully release parking brake pedal by pushing down the manual release lever.
3. Depress the parking brake pedal approximately 1¼".
4. Raise the vehicle. With the transmission in neutral, turn the adjusting nut forward against the equalizer, Fig. 5, until a moderate drag is felt when turning the rear wheels (approximately 100 lbs. of force at the outside diameter of the tire is required to turn the rear wheels).
5. Release the parking brake and check to be sure the brake shoes return to the fully released position.
6. Depress the parking brake pedal until it is fully engaged.
7. Release the parking brake again and check as in step 5.
8. If the rear brakes do not fully release, check the cables for kinks or binds and free as required.
9. Depress parking brake pedal 2". Under normal conditions this will satisfactorily hold the car.
10. Release pedal again, then depress pedal ½"; the brakes should not drag.

POWER BRAKE UNIT, REPLACE

1. Disconnect vacuum hose at booster.
2. Remove three bolts and loosen one to allow brace between cowl and spring tower to be positioned inboard to obtain clearance.
3. Remove master cylinder from booster. It is not necessary to disconnect brake lines.
4. Working under instrument panel, disconnect booster push rod link from brake pedal as follows: 1) disconnect stop light switch wires at connector and remove hairpin clip, 2) slide stop light switch off pedal just far enough for switch outer hole to clear pin, then tilt switch straight upward from pin, 3) slide master cylinder push rod and nylon washer and bushing from brake pedal pin.
5. Unfasten and remove booster from dash panel, sliding push rod link out from engine side of dash panel.
6. Remove dust seal from push rod link and place it in slot of dash panel for installation.
7. Reverse procedure to install.

Rear Suspension

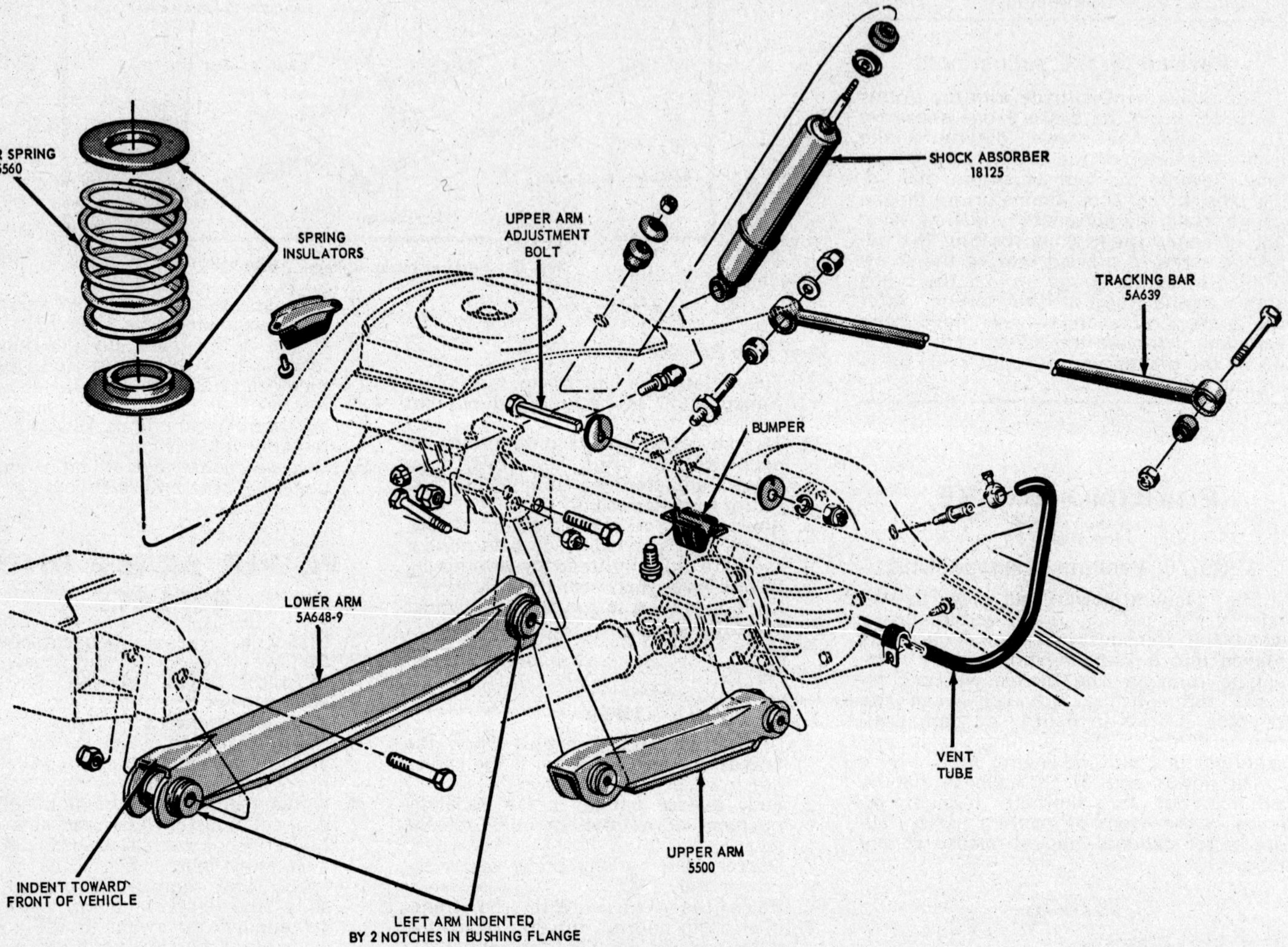

Fig. 1 Rear suspension, 1969-71 models (typical)

SHOCK ABSORBER, REPLACE

1. With rear axle supported properly disconnect shock absorber at upper mounting and compress it to clear hole in spring seat.
2. Disconnect shock absorber from stud on axle bracket.
3. Reverse procedure to install.

COIL SPRINGS, REPLACE

1. Raise rear of vehicle and support at frame. Support rear axle with a suitable jack.
2. Disconnect shock absorbers at lower mountings.
3. On 1973 models, disconnect brake line from rear brake hose and remove hose to bracket clip.
4. Lower axle to remove springs.
5. Reverse procedure to install. On 1969-71, install an insulator between upper

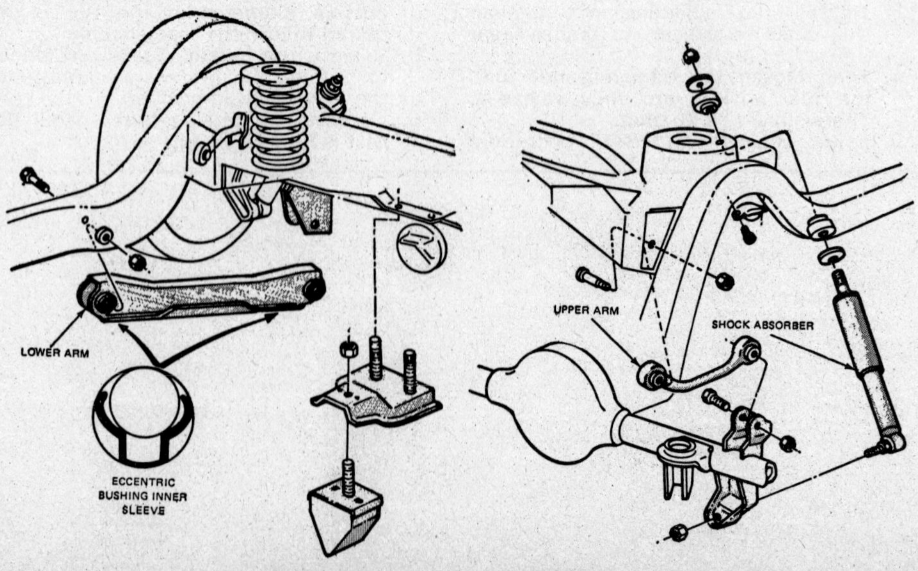

Fig. 2 Rear suspension, 1972-76 models

and lower seats and the spring, Fig. 1. On 1972-76, an insulator is installed only between upper seat and the spring, Fig. 2.

CONTROL ARMS, REPLACE

NOTE: On 1972-76 models, upper and lower control arms must be replaced in pairs.

1. Raise rear of vehicle and support at frame. Support axle with a suitable jack.
2. On 1969-71 models, disconnect track bar from frame mounting bracket, Fig. 1.
3. On all models lower axle and install a second jack under differential pinion nose.
4. Disconnect control arm from axle bracket. On upper arms, disconnect arm from crossmember and on lower arms, disconnect arm from frame attachment bracket.
5. Reverse procedure to install.

STABILIZER BAR, REPLACE

1972-76

1. Remove bolts securing stabilizer bar to rear link assemblies on both sides, Fig. 3.
2. Remove nuts securing mounting bracket to lower mounting clamp and remove bar.
3. Reverse procedure to install.

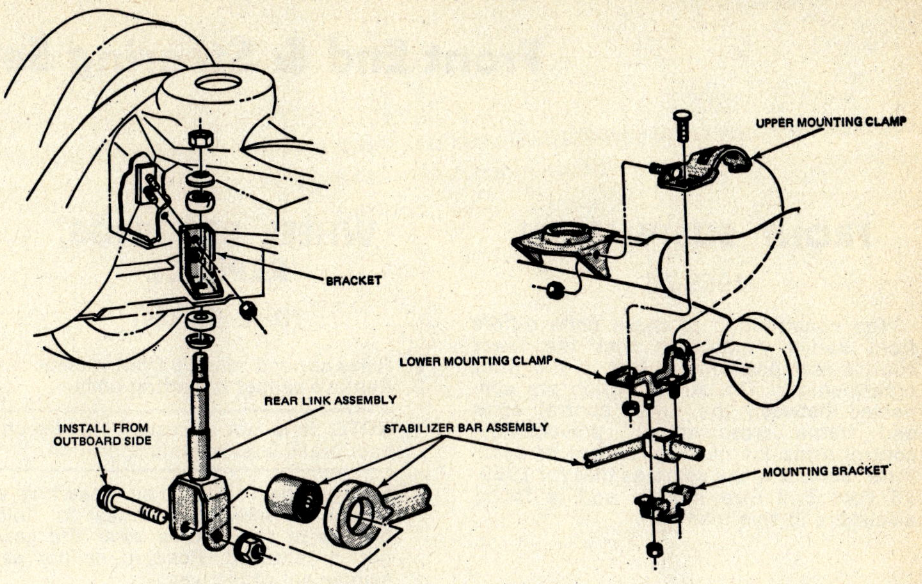

Fig. 3 Stabilizer bar installation, 1972-75 models

TRACK BAR & BUSHING, REPLACE

1969-71

1. Remove cover from track bar axle attachment, then disconnect track bar from mounting stud.
2. Disconnect track bar from frame side rail.
3. On 1970-71 models, replace frame bracket bushing, by collapsing the .030 inch ferrule and driving from bushing. Install bushing and new inner ferrule and compress bushing and washer to 2.16 inch and place inner ferrule over edge of inner washer.
4. Reverse procedure to install.

Front End & Steering Section

FRONT SUSPENSION

1969-76

The construction of these units differs from earlier models in that the lower control arm pivots on a bolt in the front crossmember. The struts, which are connected between the lower control arms and frame crossmember, prevent the control arms for moving forward or backward. Service is the same as that for 1969-75 Ford Full Size models and is found elsewhere in this manual.

LUBRICATION

Lubrication should be performed at 36,000-mile intervals at which time the special plugs should be removed and specially formulated grease applied with a hand-operated gun. This extended lubrication interval is made possible by a special type chassis lubricant combined with special seals and bearing materials. Under no circumstances should the special plugs be removed and fittings installed to accommodate conventional type grease as damage to the special seals may result.

WHEEL ALIGNMENT

1969-76

Wheel alignment is adjusted in the same manner as described for 1969-76 Ford Full Size models and is found elsewhere in this manual.

TOE-IN, ADJUST

Turn both tie rod adjusting sleeves an equal amount until toe-in is correct.

WHEEL BEARINGS, ADJUST

1969-76

1. With wheel rotating, tighten adjusting nut to 17-25 ft. lbs.
2. Back off adjusting nut ½ turn and retighten nut to 10-15 inch lbs.
3. Place nut lock on nut so that castellations on lock are aligned with cotter pin hole in spindle and install cotter pin.
4. Check front wheel rotation, if it rotates noisily or rough, clean, inspect or replace wheel bearings as necessary.

WHEEL BEARINGS, REPLACE

(Disc Brakes)

1. Raise car and remove front wheels.
2. Remove caliper mounting bolts.

NOTE: It is not necessary to disconnect brake lines for this operation.

3. Slide caliper off of disc, inserting a spacer between the shoes to hold pistons in their bores after the caliper is removed. Position caliper assembly out of the way.

NOTE: Do not allow caliper to hang by the brake line.

4. Remove hub and disc assembly. Grease retainer and inner bearing can now be removed.

CHECKING BALL JOINTS FOR WEAR

If loose ball joints are suspected, first be sure the front wheel bearings are properly adjusted and that the control arms are tight.

On 1969-75 models, the upper ball joint should be replaced if there is any noticeable looseness at the joint. Lower ball joint should be replaced if radial play exceeds .250", and if there is any noticeable axial play at the joint.

BALL JOINTS, REPLACE

NOTE: Ford Motor Company recommends that new ball joints should not be installed on used control arms and that the control arm be replaced if ball joint replacement is required. However, aftermarket ball joint repair kits which do not require control arm replacement, are available and can be installed using the following procedure.

The upper and lower ball joints are riveted to the control arms. On later models, the upper ball joint is pressed into the upper control arm whereas the lower ball joint is riveted to the lower control arm.

When replacing a riveted ball joint, remove the rivets and retain the ball joint

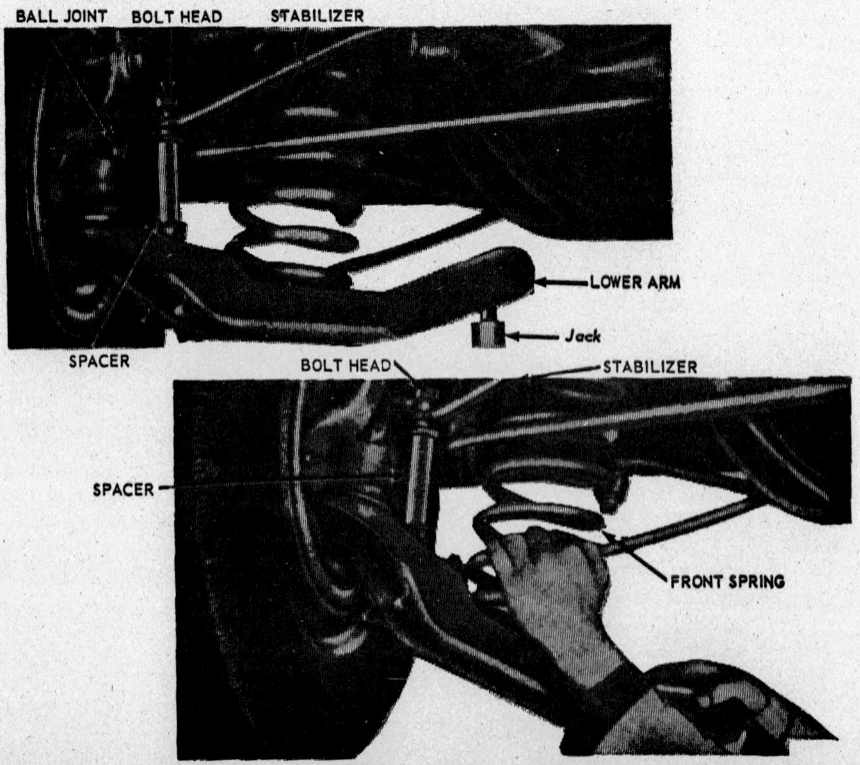

Fig. 1 Coil spring removal & installation

to its control arm with the bolts, nuts and washers furnished with the ball joint kit. Also, use a suitable pressing tool to force the ball joint studs out of the spindle.

SHOCK ABSORBER, REPLACE

1969-76

To remove a shock absorber, unfasten it from the frame at its upper end. Remove the two cap screws that retain the shock absorber mounting plate to the lower control arm and lower the shock absorber unit.

To install, reverse the removal procedure and tighten the two lower cap screws to 13-18 lbs. ft. torque.

COIL SPRING, REPLACE

1969-76

1. Raise vehicle and support front end of frame with jack stands.
2. Disconnect shock absorber from lower arm and place a jack under the lower arm for support, Fig. 1.
3. Remove strut and rebound bumper bolts and disconnect lower end of sway bar stud from lower arm.
4. Remove the nut and bolt that retains the inner end of the lower arm to the crossmember.
5. Carefully lower jack to relieve the spring pressure on the lower arm, then remove the spring.

STEERING GEAR, REPLACE

1969-76

1. Disconnect pressure and return lines from steering gear. Cap each line and plug each port to prevent entry of dirt.
2. Remove bolt that secures flex joint to steering gear.
3. Raise the vehicle and remove sector shaft nut and pitman arm.

NOTE: Do not damage seals.

4. Support steering gear and remove three retaining bolts. Remove flex coupling clamp bolt and work steering gear free of coupling and remove from vehicle.

LINCOLN CONTINENTAL

OLD CAR SPECIFICATIONS: For 1946-68 Tune Up and Wheel Alignment Specifications see main index.

INDEX OF SERVICE OPERATIONS

SERIAL & ENGINE NUMBER LOCATION

Vehicle Warranty Plate on Left Front Door Pillar

1969 Continental

1969-71 Mark III

1970 Continental

1971 Continental

1972 Continental

ENGINE IDENTIFICATION

*Serial number on vehicle Warranty Plate.

Engine code for 1969–76 is the last letter in the serial number.

Year	Engine	Engine Code*
1969–76	V8-460	A

1972 Mark IV

1973 Continental

1973 Mark IV

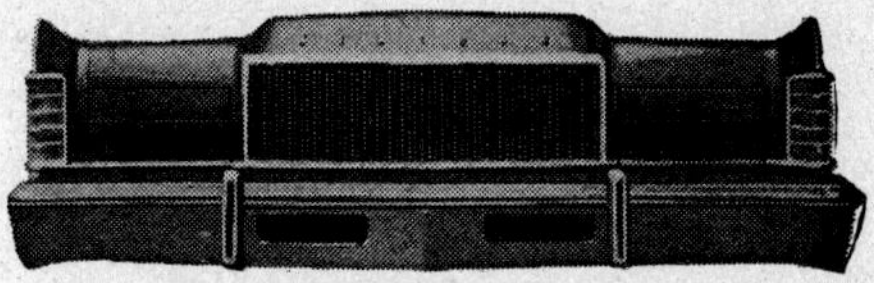
1974 Lincoln

1974 Mark IV

1975-76 Mark IV

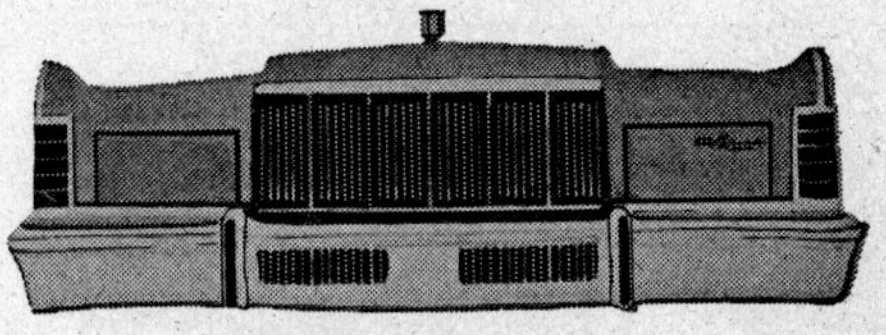
1975-76 Continental

GENERAL ENGINE SPECIFICATIONS

Year	Engine	Car-buretor	Bore and Stroke	Piston Dis-place-ment, Cubic Inches	Com-pres-sion Ratio	Maximum Brake H.P. @ R.P.M.	Maximum Torque Lbs. Ft. @ R.P.M.	Normal Oil Pressure Pounds
1969–70	365 Horsepower............V8-460	4 Barrel	4.3600 x 3.850	460	10.5	365 @ 4600	500 @ 2800	35–60
1971	365 Horsepower............V8-460	4 Barrel	4.3600 x 3.850	460	10.2	365 @ 4600	500 @ 2800	35–60
1972	212 Horsepower①...........V8-460	4 Barrel	4.3600 x 3.850	460	8.5	212 @ 4400	342 @ 2800	35–60
	224 Horsepower①...........V8-460	4 Barrel	4.3600 x 3.850	460	8.5	224 @ 4400	357 @ 2800	35–60
1973	208 Horsepower①...........V8-460	4 Barrel	4.3600 x 3.850	460	8.0	208 @ 4400	338 @ 2800	35–70
	219 Horsepower①...........V8-460	4 Barrel	4.3600 x 3.850	460	8.0	219 @ 4400	360 @ 2800	35–70
1974	215 Horsepower①...........V8-460	4 Barrel	4.362 x 3.850	460	8.0	215 @ 4000	350 @ 2600	35–65
	220 Horsepower①...........V8-460	4 Barrel	4.362 x 3.850	460	8.0	220 @ 4000	355 @ 2600	35–65
1975	253 Horsepower①...........V8-460	4 Barrel	4.362 x 3.850	460	8.0	253 @ 4400	386 @ 2600	40–65
1976	Horsepower①...............V8-460	4 Barrel	4.362 x 3.850	460	8.0	—	—	40–65

①—Ratings are NET—as installed in the vehicle.

TUNE UP SPECIFICATIONS

The following specifications are published from the latest information available. This data should be used only in the absence of a decal affixed in the engine compartment.

OLD CAR SPECIFICATIONS: For 1946-68 Wheel Alignment Specifications see main index.

★When using a timing light, disconnect vacuum hose or tube at distributor and plug opening in hose or tube so idle speed will not be affected.

●When checking compression, lowest cylinder must be within 75 percent of highest.

▲Before removing wires from distributor cap, determine location of No. 1 wire in cap, as distributor position may have been altered from that shown at the end of this chart.

Year	Spark Plug		Distributor		Ignition Timing★			Carb. Adjustments					
								Hot Idle Speed		Air Fuel Ratio		Idle CO %	
	Type	Gap Inch	Point Gap Inch	Dwell Angle Deg.	Firing Order Fig. ▲	Timing BTDC ①	Mark Fig.	Std. Trans.	Auto. Trans.②	Std. Trans.	Auto. Trans.	Std. Trans.	Auto. Trans.
1969													
V8-460③	BRF-42	.034	.017	26–31	C	10°⑥	B	—	550D⑤	—	14.3 to 1	—	1.2
1970													
V8-460	BRF-42	.034	.017	26–31	C	10°⑥	A	—	600D⑦	—	14.4 to 1	—	1.0
1971													
V8-460	BRF-42	.034	.017	26–31	C	5°⑥	D	—	600D⑦	—	14.2 to 1	—	0.9
1972													
V8-460 Exc. Calif.	BF-42	.034	.017	26–30	C	10°⑥	D	—	625D⑦	—	—	—	0.3
V8-460 Calif.	BF-42	.034	.017	26–30	C	6°⑥	D	—	625D⑦	—	—	—	0.3
1973													
V8-460	ARF-42	.034	.017	24–30	C	14°	D	—	600D	—	—	—	0.4⑧
1974													
V8-460	ARF-52	.054	⑨	—	C	14°	D	—	650D	—	—	—	—
1975													
V8-460	ARF-52	.044	⑨	—	C	14°	D	—	650D	—	—	—	—
1976													
V8-460⑩	ARF-52	.044	⑨	—	C	8°⑪	D	—	650D⑫	—	—	—	—
V8-460⑬	ARF-52	.044	⑨	—	C	14°⑪	D	—	650D⑫	—	—	—	—

①—BTDC-Before top dead center.
②—D-Drive. N-Neutral.
③—With IMCO system.
⑤—With headlights on and A/C "Full On".
⑥—Whenever idle speed or ignition timing is adjusted, vacuum line to brake release mechanism must be disconnected and plugged to prevent parking brake from releasing when selector is moved to Drive.
⑦—Headlamps on Hi Beam—Air Conditioner OFF.
⑧—For California 0.2.
⑨—Breakerless distributor.
⑩—Exc. Calif.
⑪—In Drive with service & parking brakes applied.
⑫—With A/C "On".
⑬—Calif.

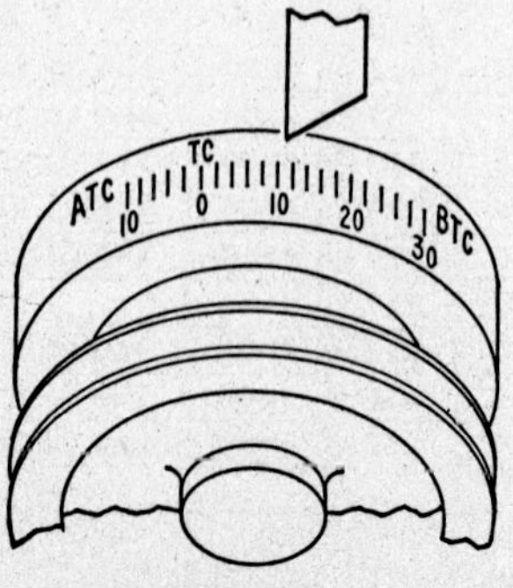

Fig. A

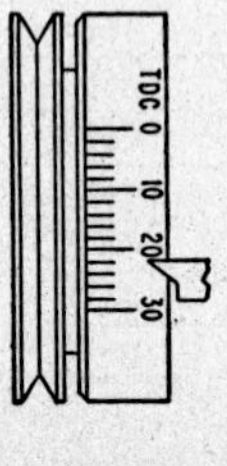

Fig. B

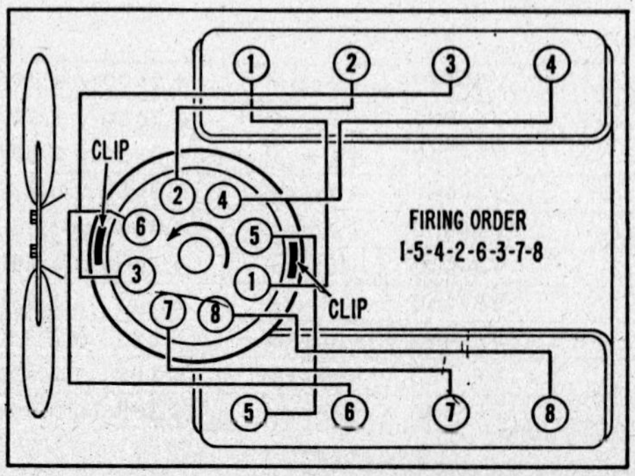

Fig. C

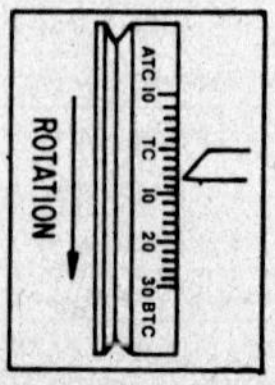

Fig. D

DISTRIBUTOR SPECIFICATIONS

★Note: If unit is checked on vehicle, double the RPM and degrees to get crankshaft figures.

Breaker arm spring tension—17–21.

Distributor Part No.①	Centrifugal Advance Degrees @ RPM of Distributor					Vacuum Advance		Distributor Retard
	Advance Starts	Intermediate Advance			Full Advance	Inches of Vacuum to Start Plunger	Max. Adv. Dist. Deg. @ Vacuum	Max. Retard Dist. Deg. @ Vacuum
1969								
C8VF-G	0–½ @ 350	¾–2¾ @ 750	4½–6½ @ 1000	6–8¼ @ 1500	11 @ 2575	5	12 @ 25	—
1970								
D0VF-B	0–½ @ 350	¾–2¾ @ 750	4½–6½ @ 1000	6–8¼ @ 1500	11 @ 2575	5	12 @ 25	—
1971								
D1VF-AA	0–½ @ 350	¾–2¾ @ 750	4½–6½ @ 1000	6–8¼ @ 1500	13 @ 2500	5	8½ @ 25	—
1972								
D2VF-AA	0–1 @ 500	1½–3½ @ 750	5½–7½ @ 1000	7–9 @ 1500	10½ @ 2000	5	11½ @ 20	—
D2VF-BA	0–1 @ 500	1–3 @ 750	6½–8½ @ 1000	7½–10 @ 1500	11½ @ 2000	5	12 @ 20	—
1973								
D3VF-AA	0–1 @ 500	0–1½ @ 750	3–5 @ 1000	4–6 @ 1500	7½ @ 2000	5	11 @ 20	—
D3VF-BA	0–1 @ 500	0–2½ @ 750	5–7 @ 1000	7½–10 @ 1500	12 @ 2000	5	7 @ 20	—
D3VF-AA	0–1 @ 500	0–1½ @ 750	4–6 @ 1000	6–8 @ 1500	10 @ 2000	5	11 @ 20	—
1974-75								
D4VE-CA	0–½ @ 500	0–½ @ 750	3–5 @ 1000	7–9 @ 1500	9½ @ 2000	5	11¼ @ 20	—
D5VE	—	—	—	—	—	—	—	—
D5VE	—	—	—	—	—	—	—	—
1976								
D6VE-AA	—	—	—	—	—	—	—	—
D6VE-BA	—	—	—	—	—	—	—	—

①—Basic part No. 12127.

VALVE SPECIFICATIONS

Year	Engine Model	Valve Lash		Valve Angles		Valve Spring Installed Height	Valve Spring Pressure Lbs. @ In.	Stem Clearance		Stem Diameter	
		Int.	Exh.	Seat	Face			Intake	Exhaust	Intake	Exhaust
1969–75	V8-460	①②		45	44	1 13/16	③	.0010–.0027	.0010–.0027	.3416–.3423	.3416–.3423
1976	V8-460	.075–.175②		45	45	1 13/16	228 @ 1.33	.0010–.0027	.0010–.0027	.3416–.3423	.3416–.3423

①—1969–70, .075–.125; 1971–73, .105–.155; 1974, .075–.125; 1975, .100–.150.

②—Clearance is obtained at valve stem tip with hydraulic lifter collapsed. If clearance is less than minimum, install an undersize push rod; if clearance is greater than maximum, install an oversize push rod.

③—1969–70, 253 @ 1.33; 1971–72, 229 @ 1.33; 1973, 169 @ 1.39; 1974–75, 252 @ 1.33.

STARTING MOTOR SPECIFICATIONS

Year	Car Model	Ident. No.	Brush Spring Tension Ounces	No Load Test			Torque Test		
				Amperes	Volts	R.P.M.	Amperes	Volts	Torque Lbs. Ft.
1969	All	C9AF-A	40	70	12	10000	700	5.0	15½
1970–71	All	D0VF-A	40	70	12	10000	700	5.0	15½
1972–73	All	D2VF-AA	40	70	12	10000	700	5.0	15½
1974	All	D4VF-AB	80	90	12	—	—	—	—
1975	All	D5AF-AB	40	70	12	—	—	—	—
1976	All	D6AF-AA	—	—	—	—	—	—	—

PISTONS, PINS, RINGS, CRANKSHAFT & BEARINGS

Year	Model	Piston Clearance	Ring End Gap①		Wrist-pin Diameter	Rod Bearings		Main Bearings			
			Comp.	Oil		Shaft Diameter	Bearing Clearance	Shaft Diameter	Bearing Clearance	Thrust on Bear. No.	Shaft End Play
1969–73	V8-460	.0014–.0022	.010	.015	1.040	2.4992–2.5000	.0008–.0015	2.9994–3.0002	.0005–.0015	3	.004–.008
1974	V8-460	.0022–.0030	.010	.015	1.040	2.4992–2.5000	.0008–.0028	2.9994–3.0002	②	3	.004–.008
1975–76	V8-460	.0014–.0022	.010	.015	1.040	2.4992–2.5000	.0008–.0028	2.9994–3.0002	③	3	.004–.008

①—Fit rings in tapered bores for clearance listed in tightest portion of ring travel.
②—#1 .0004–.002; others .0012–.0028.
③—#1, .0004–.0020; others, .0009–.0026.

BRAKE SPECIFICATIONS

Year	Model	Brake Drum Inside Diameter	Wheel Cylinder Bore Diameter			Master Cylinder Bore Diameter		
			Front Disc Brakes	Front Drum Brakes	Rear Brakes	With Disc Brakes	With Drum Brakes	With Power Brakes
1969	Lincoln	11.090	1.938	—	15/16	1	1	1
1969–70	Mark III	11.030	2.755	—	15/16	1	—	1
1970	Lincoln	11.090	2.755	—	15/16	1	—	1
1971	All	11.030	2.755	—	15/16	1	—	1
1972–74	Lincoln	11.030	3.100	—	1.000①	1	—	1
	Mark IV	11.030	3.100	—	1.000	1	—	1
1975	Lincoln	11.030	3.100	—	1.000②	1④	—	1④
	Mark IV	③	3.100	—	2.6	1⅛	—	1⅛
1976	Lincoln	11.030	3.100	—	1.000②	1⅛	—	1⅛
	Mark IV	③	3.100	—	2.6	1⅛	—	1⅛

①—1972 Lincoln .938.
②—On models with 4 wheel disc brakes, 2.6".
③—4 wheel disc brakes.
④—With 4 wheel disc brakes, 1⅛".

ALTERNATOR & REGULATOR SPECIFICATIONS

Year	Make or Model①	Current Rating②		Field Current @ 75°F.		Voltage Regulator③				Field Relay	
		Amperes	Volts	Amperes	Volts	Make or Model ⑥	Voltage @ 75°F.	Contact Gap	Armature Air Gap	Armature Air Gap	Closing Voltage @ 75°F.
1969–72	D0AZ-F	55	15	2.8–3.3	12	C3SZ-B⑤	—	—	—	—	—
1973	All	61	15	2.9	12	D3TZ-A	13.5–15.3	—	—	—	2.5–4.0
	All	70	15	2.9	12	D3TZ-A	13.5–15.3	—	—	—	2.5–4.0
	All	90	15	2.9	12	D3TZ-A	13.5–15.3	—	—	—	2.5–4.0
1974	All Red	55	15	2.9	12	D4TZ-A	—	—	—	—	—
	All Green	61	15	2.9	12	D4TZ-A	—	—	—	—	—
	All	70	15	2.9	12	D4TZ-A	—	—	—	—	—
	All	90	15	2.9	12	D4TZ-A	—	—	—	—	—
1975	Green	61	—	—	—	D4TF-AA	13.5–15.3	—	—	—	2.5–4.0
	All	70	—	—	—	D4TF-AA	13.5–15.3	—	—	—	2.5–4.0
	All	90	—	—	—	D3VF-AB	13.5–15.3	—	—	—	2.5–4.0
1976	D5AF-EA	60	—	—	—	D4AF-AA	13.5–15.3	—	—	—	—
	D5AF-GA	60	—	—	—	D4AF-AA	13.5–15.3	—	—	—	—
	D5OF-AA	70	—	—	—	D4TF-AA	13.5–15.3	—	—	—	—
	D5OF-BA	90	—	—	—	D4TF-AA	13.5–15.3	—	—	—	—

①—Basic No. 10300.
②—Stamped on housing.
③—Stamped on cover.
⑤—Integral regulator solid state.
⑥—Basic No. 10316.

ENGINE TIGHTENING SPECIFICATIONS*

★Torque specifications are for clean and lightly lubricated threads only. Dry or dirty threads produce increased friction which prevents accurate measurement of tightness.

Year	Spark Plugs Ft. Lbs.	Cylinder Head Bolts Ft. Lbs.	Intake Manifold Ft. Lbs.	Exhaust Manifold Ft. Lbs.	Rocker Arm Shaft Bracket Ft. Lbs.	Rocker Arm Cover Ft. Lbs.	Connecting Rod Cap Bolts Ft. Lbs.	Main Bearing Cap Bolts Ft. Lbs.	Flywheel to Crankshaft Ft. Lbs.	Vibration Damper or Pulley Ft. Lbs.
1969–73 V8-460	15–20	130–140	25–30	28–33	65–75①	5–6	40–45	95–105	75–85	75–90
1974-76 V8-460	15–20	130–140	22–32	28–33	18–25①	5–6	40–45	95–105	75–85	70–90

①—Rocker arm stud.

REAR AXLE SPECIFICATIONS

Year	Model	Carrier Type	Ring Gear & Pinion Backlash Inch	Nominal Pinion Locating Shim, Inch	Pinion Bearing Preload				Differential Bearing Preload	Pinion Nut Torque Ft.-Lbs.
					New Bearings With Seal Inch-Lbs.	Used Bearings With Seal Inch-Lbs.	New Bearings Less Seal Inch-Lbs.	Used Bearings Less Seal Inch-Lbs.		
1969	All	Removable	.008–.012	.015	15–35	—	12½–32½	—	.008–.012②	200①
1970–72	All	Removable	.008–.012	.015	8–14③	20–30③	—	—	.008–.012②	④
1973–76	All	Removable	.008–.012	.015	17–27③	8–14③	—	—	.008–.012②	④

①—If torque cannot be obtained, install new spacer.
②—Case spread with new bearings; with used bearings .005–.008″.
③—Bearing set with collapsible spacer; with solid spacer 13–33 inch-lbs.
④—With collapsible spacer 170 ft. lbs., with solid spacer 200 ft. lbs.

WHEEL ALIGNMENT SPECIFICATIONS

OLD CAR SPECIFICATIONS: For 1946-68 Wheel Alignment Specifications see main index.

Year	Model	Caster Angle, Degrees		Camber Angle, Degrees				Toe-In. Inch	Toe-Out on Turns, Deg.①	
		Limits	Desired	Limits		Desired			Outer Wheel	Inner Wheel
				Left	Right	Left	Right			
1969–70	Lincoln	−½ to −2½	−1½	−¼ to +1¼	−¼ to +1¼	+½	+½	⅛	17¾	20
1969–70	Mark III	0 to +2	+1	−¼ to +1¼	−¼ to +1¼	+½	+½	3/16	19¼	20
1971–74	Lincoln	−½ to +3½	+1½	−½ to +1½	−½ to +1½	+½	+½	⅛	②	20
1971	Mark III	−1 to +3	+1	−½ to +1½	−½ to +1½	+½	+½	3/16	19.28	20
1972	Mark IV	−1 to +3	+1	−¼ to +1¾	−¼ to +1¾	+¾	+¾	3/16	17.74	20
1973	Mark IV	−½ to +3½	+1½	−¼ to +1¾	−¼ to +1¾	+¾	+¾	3/16	17.74	20
1974	Mark IV	+¼ to +3¼	+1¾	−¼ to +1¾	−¾ to +1¼	+¾	+¼	3/16	18.07	20
1975	Lincoln	+1¼ to +2¾	+2	−¼ to +1¼	−½ to +1	+½	+¼	⅛	18.16	20
	Mark IV	+1¼ to +2¾	+2	−¼ to +1¼	−½ to +1	+½	+¼	3/16	18.09	20
1976	Lincoln	+1¼ to +2¾	+2	−¼ to +1¼	−½ to +1	+½	+¼	⅛	18.16	20
	Mark IV	+3¼ to +4¾	+4	−¼ to +1¼	−½ to +1	+½	+¼	3/16	18.09	20

①—Incorrect toe-out, when other adjustments are correct, indicates bent steering arms.

②—1971 18.70; 1972 18.43; 1973-74 18.16.

COOLING SYSTEM & CAPACITY DATA

Year	Model or Engine	Cooling Capacity, Qts.			Radiator Cap Relief Pressure, Lbs.		Thermo. Opening Temp. ①	Fuel Tank Gals.	Engine Oil Refill Qts. ②	Transmission Oil			Rear Axle Oil Pints
		No Heater	With Heater	With A/C	With A/C	No A/C				3 Speed Pints	4 Speed Pints	Auto. Trans. Qts. ③	
1969	V8-462	22½	23½	23½	12–15	12–15	185	25½	4	—	—	13½	5
	V8-460	—	22	22	12–15	12–15	185	25½	4	—	—	13½	5
1970	Lincoln	—	20½	—	12–15	12–15	188	24④	4	—	—	13	5
	Mark III	—	20½	—	12–15	12–15	188	24④	4	—	—	12¾	5
1971	Lincoln	—	19½	19½	12–15	12–15	188	23	4	—	—	13	5
	Mark III	—	19½	19½	12–15	12–15	188	23	4	—	—	12¾	5
1972–73	Lincoln	—	19½	19½	12–15	12–15	188	22	4	—	—	13⑤	5
	Mark IV	—	19½	19½	12–15	12–15	188	22½	4	—	—	12¾⑤	5
1974	Lincoln	—	20½	20½	12–16	12–16	191	22	4	—	—	12½	5
	Mark IV	—	20½	20½	12–16	12–16	191	26½	4	—	—	12½	5
1975	Lincoln	—	—	19¾	12–16	12–16	191	24¼	4	—	—	12½	5
	Mark IV	—	—	20	12–16	12–16	191	26½	4	—	—	12½	5
1976	Lincoln	—	—	19.7	12–16	12–16	191	24¼⑥	4	—	—	12	5
	Mark IV	—	—	19.8	12–16	12–16	191	26½	4	—	—	12	5

①—With permanent type anti-freeze.

②—Add one quart with filter change.

③—Approximate. Make final check with dipstick.

④—California vehicles 22½.

⑤—1973 Models 12½ quarts.

⑥—Add 8 gals. with auxiliary tank.

Electrical Section

DISTRIBUTOR, REPLACE

1. On vehicles with conventional ignition systems, disconnect primary wire from ignition coil. On vehicles with breakerless ignition systems, disconnect distributor wiring from connector from main wiring harness.
2. Disconnect vacuum advance lines and remove distributor cap.
3. Scibe a mark on distributor body and engine block indicating position of rotor in distributor and distributor in engine.
4. Remove distributor hold-down clamp and bolt and lift distributor from engine. Do not crank engine after distributor is removed, otherwise the distributor will have to be initially timed to the engine.
5. Install distributor in reverse order of removal. Then start engine and adjust timing.

STARTER, REPLACE

1969-76

1. Disconnect battery ground cable and support vehicle on hoist.
2. Disconnect cable and wires from solenoid.
3. On 1973-76 Mark IV models, loosen the two front brace retaining bolts, then remove all other brace retaining bolts and allow brace to hang.
4. Turn wheels fully to the right and remove the two bolts attaching the steering idler arm to the frame.
5. Remove starter attaching bolts and remove starter.

IGNITION LOCK REPLACE

1970-76

1. Disconnect the battery ground cable.
2. **Units With Fixed Steering Columns:** Remove the steering wheel trim pad. Insert a wire pin in the hole located inside the column halfway down the lock cylinder housing, Fig. 1. **Units With Tilt Steering Columns:** Insert wire pin in the hole located on the outside of the flange casting next to the emergency flasher button, Fig. 1.
3. Place the gear shift lever in *Park* position, and turn the lock cylinder with the ignition key to the *Run* position.
4. Depress the wire pin while pulling up on the lock cylinder to remove. Remove the wire pin.
5. To install insert the lock cylinder into the housing in the flange casting, and turn the key to the *Off* position. Be certain that the cylinder is fully inserted before turning to the *Off* position. This action will extend the cylinder retaining pin into the cylinder housing.
6. Turn the key to check for correct operation in all positions.
7. Install the steering wheel and trim pad on fixed column units.
8. Connect the battery ground cable.

IGNITION SWITCH, REPLACE

1970-76 All

1. Remove shrouding from steering column and detach and lower steering column from brake support bracket.
2. Disconnect battery cable.
3. Disconnect switch wiring at multiple plug, Figs. 2 and 2A.
4. Remove two nuts that retain switch to column.
5. Detach switch plunger from actuator rod and remove the switch.
6. When installing switch, both the locking mechanism at the top of the column and the switch must be in the "LOCK" position for correct adjustment. New replacement switches are already pinned in "LOCK" position by a plastic pin inserted in a locking hole.

1969 All

1. Disconnect battery ground cable.
2. Insert wire pin in hole in ignition switch and turn key to accessory position while pressing pin. Lock cylinder can now be removed, Fig. 3.
3. Remove bezel nut and lower switch from instrument panel.
4. Depress tabs securing multiple connector to switch and pull connector off switch.
5. Reverse procedure to install.

LIGHT SWITCH, REPLACE

1972-76 Mark IV

1. Disconnect battery ground cable.
2. Remove instrument cluster trim panel.
3. Remove the headlight switch mount-

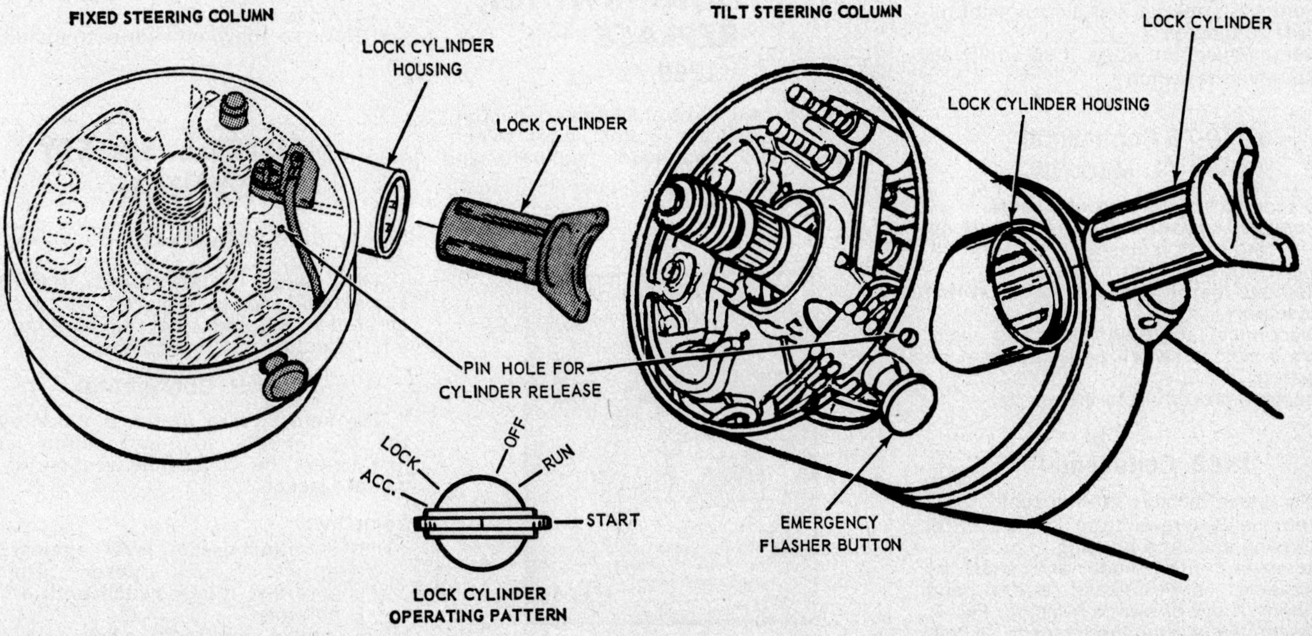

Fig. 1 Ignition lock cylinder

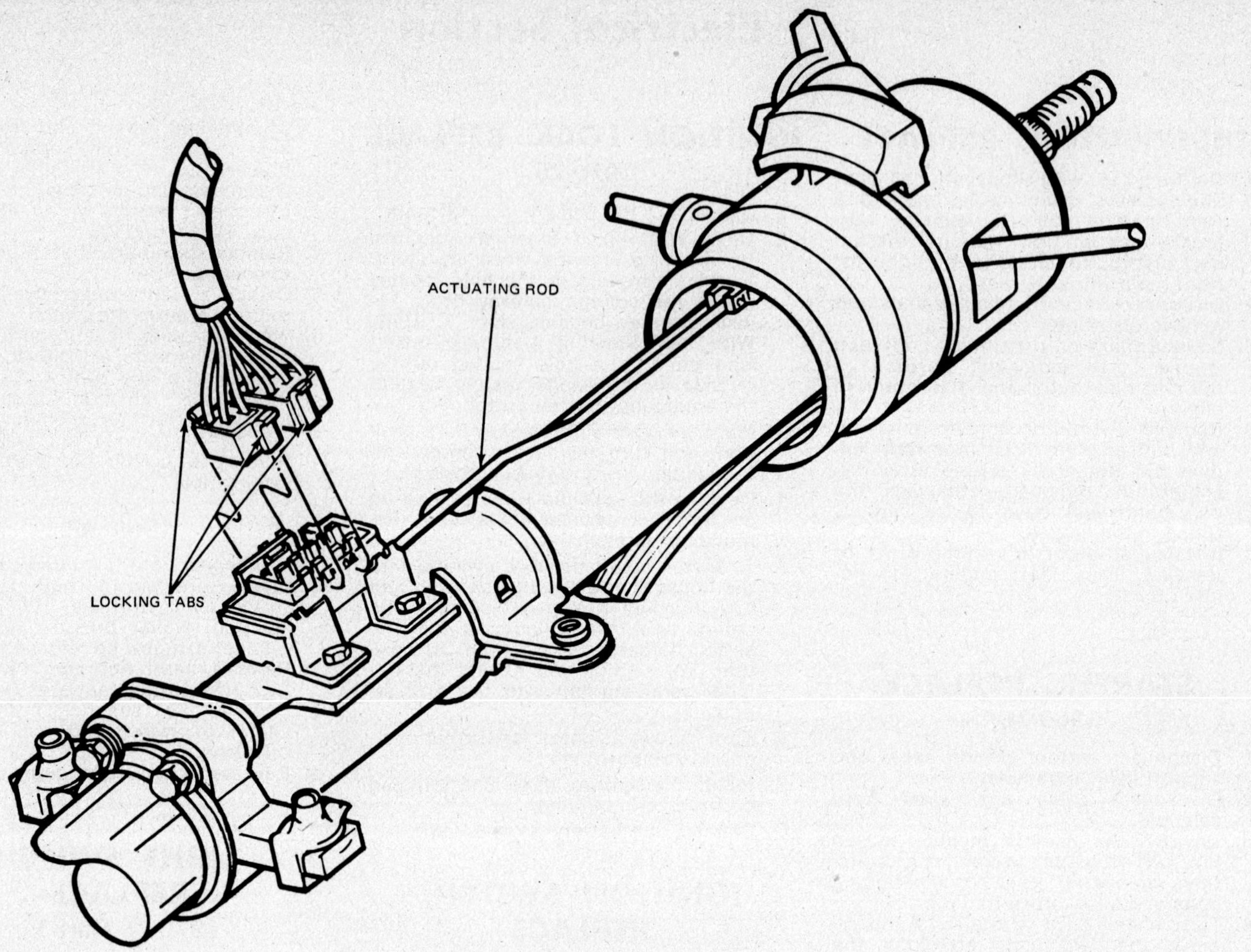

Fig. 2 Ignition switch installation. 1972-76 All

ing plate.
4. Remove bezel nut and disconnect multiple connector.
5. Remove vacuum lines, if so equipped.
6. Remove the switch.

1970-76 Continental
1969-71 Mark III

1. Disconnect battery ground cable.
2. Remove control knob and shaft by pressing knob release button and pulling it out of switch housing, Fig. 4.
3. Remove bezel nut and lower switch assembly.
4. Disconnect multiple plug and vacuum hoses at switch body and remove switch.
5. Reverse procedure to install.

1969 Continental

1. Disconnect battery ground cable.
2. Remove 8 screws from lower control housing and drop housing.
3. Remove control knob and shaft by pressing knob release button and pulling it out of switch housing, Fig. 4.
4. Disconnect wiring connector to switch.
5. Remove bezel nut and remove switch.
6. Reverse procedure to install.

STOP LIGHT SWITCH, REPLACE

1969-76

1. Disconnect wires at switch connector.
2. Remove hairpin retainer, slide switch, push rod and nylon washers and bushing away from brake pedal, and remove switch, Fig. 5.
3. Reverse above procedure to install.

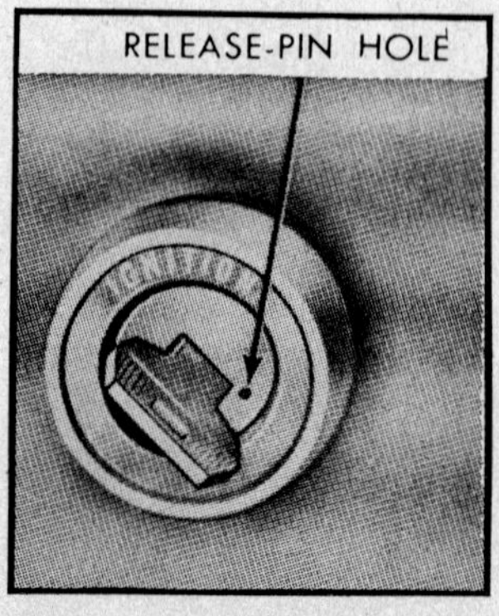

Fig. 3 Ignition lock cylinder release pin hole

NEUTRAL SAFETY SWITCH

1970-76

The neutral safety switch has been eliminated and is replaced by a series of steps designed into the steering column selector lever hub casting.

1969 Continental

The neutral start switch is mounted on the top of the steering column jacket just below the collapsible section of the column jacket.

Adjustment

1. With transmission lever against the stop in Neutral, loosen retaining screws and rotate switch until a start is obtained.
2. Tighten screws to 20 in-lbs.
3. Place selector in the "1" position and push the Park reset button counter-

clockwise until it stops. *The Park reset must be performed whenever the switch has been adjusted.*

1969 Mark III

The neutral start switch is mounted on the top of the steering column jacket just below the collapsible section of the column jacket.

To adjust the switch it must be removed from the column. Put the selector lever in neutral and set the parking brake. Then disconnect the electrical and vacuum connections, remove the two fastening screws and lift the switch straight up and out.

Adjustment

1. Hold the switch with the wire terminal facing you. Move the actuator lever all the way to the left but do not force as the switch will be damaged internally.
2. Insert a $3/32''$ drill shank in the hole in the tapered round boss.
3. Gently move the actuator lever to the right until it stops. This will move the Park circuit to its position of minimum travel, which must be done if the switch is to function properly upon installation.
4. Pull out the drill gauge and fit it in the hole on the top surface of the switch case to engage the switch internal carrier in the neutral position. Then reinsert the drill gauge.
5. With the selector lever held against the stop in the neutral detent position, set the switch in place on the column and fasten it with the two mounting screws.
6. Connect the electrical connector and any vacuum hose, and be sure to remove the drill gauge before operating the selector lever.

TURN SIGNAL SWITCH, REPLACE

1971-76

1. Remove retaining screw from underside of steering wheel spoke and lift off the pad horn switch/trim cover and medallion as an assembly.
2. Disconnect horn switch wires from terminals.
3. Remove steering wheel retaining nut and remove steering wheel using suitable puller.
4. Remove turn signal switch lever by unscrewing it from steering column.

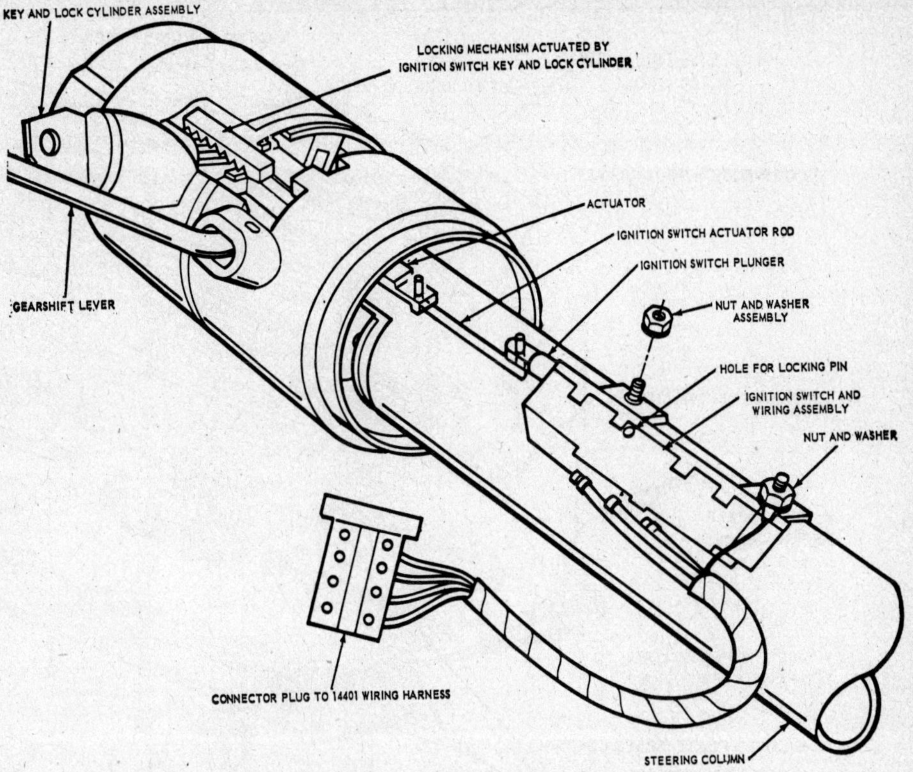

Fig. 2A Ignition switch installation. 1970-71 All

5. Remove shroud from under steering column.
6. Disconnect steering column wiring connector plugs and remove the screws that secure switch assembly to the column.
7. On vehicles with tilt column, remove wires and terminals from steering column wiring connector plug.

NOTE: *Record the color code and location of each wire before removing it from connector. A hole provided in the flange on fixed columns makes it unnecessary to separate the wires from the connector plug. The plug with wires installed can be guided through the hole.*

8. Remove the plastic cover sleeve from the wiring harness and remove the switch and wires from the top of the column.

Fig. 4 Light switch. 1969-76 (typical)

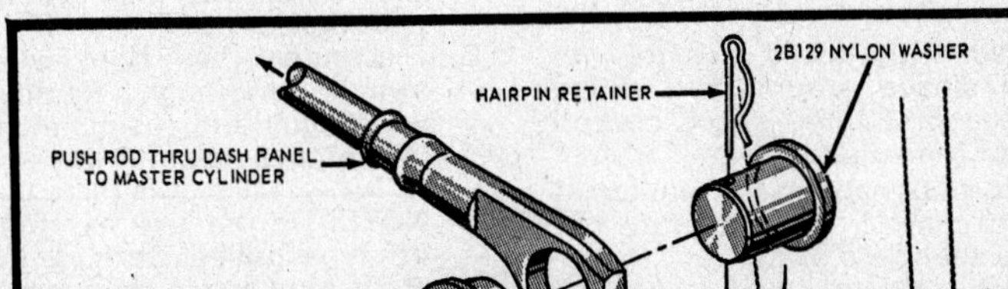

Fig. 5 Stop light switch. 1969-76

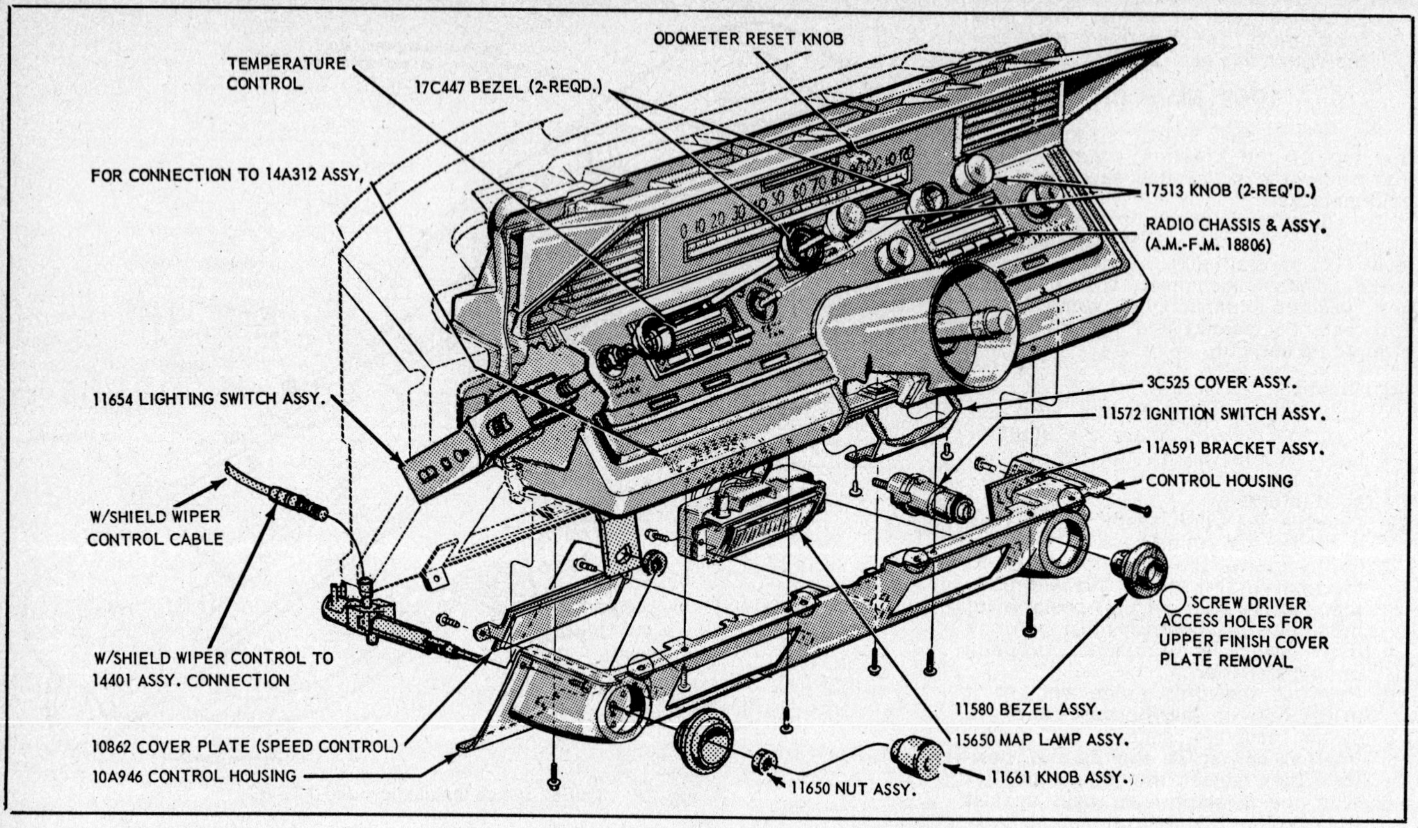

Fig. 6 Instrument cluster and controls. 1969 Continental

1969-70

1. Disconnect battery and remove steering wheel.
2. Remove protective wire cover that runs along bottom of steering column tube and disconnect electrical plug, noting color codes and location. If equipped with tilt steering column, tape wire ends together and attach a piece of heavy cord to the wires to help pull them through the column during installation.
3. If equipped with speed control, remove the sleeve around the wiring and pull the first three speed control wires out of the column.
4. Unscrew turn signal lever. If equipped with speed control, first separate the wire to the set-speed switch.
5. If equipped with tilt column, remove upper collar and push lower collar down. Remove wiring retaining clip.
6. Remove switch retaining screws and remove switch.
7. Reverse procedure to install.

HORN SOUNDER & STEERING WHEEL

1972-76 All

1. Disconnect battery ground cable.
2. Remove screws from behind wheel spoke holding crash pad to wheel. Lift pad and disconnect horn wires. Disconnect speed control wires if used and remove pad.
3. Remove steering wheel nut. Install a suitable puller and remove steering wheel.

1969-71 Mark III

1. Disconnect battery ground cable.
2. Remove the medallion from steering wheel spoke pad by gently prying out with a knife blade. Working from the underside of the spoke, remove the two screws that secure the crash pad. Remove pad.
3. Disconnect two horn wires at horn switch assembly. Remove steering wheel nut and, using a suitable puller, remove steering wheel.

NOTE: Do not use hammer or knock-off type puller. Striking the puller or shaft may cause damage to the bearings or collapsible column.

1969-71 Continental

1. Turn hub cap counterclockwise and lift it from steering wheel.
2. Remove hub mounting plate from top of wheel.
3. Remove nut and pull off wheel.

INSTRUMENT CLUSTER

1974-76 Continental

1. Disconnect battery ground cable.
2. Remove steering column trim shroud and instrument panel lower pad.
3. Disconnect lower instrument cluster electrical connector from printed circuit.
4. Remove PRND21 control cable to steering column screw.
5. Remove instrument cluster retaining screws, pull cluster from panel and disconnect electrical connector.
6. Remove instrument cluster.

1972-76 Mark IV

1. Disconnect battery ground cable.
2. Remove three screws attaching upper access cover to instrument panel pad.
3. Remove one screw retaining lower cluster applique cover below steering column.
4. Squeeze lower half of steering column shroud together and separate lower half from upper.
5. Remove upper half of shroud from column.
6. Remove one screw attaching PRNDL control cable to steering column.
7. Remove headed backlite control knob.
8. Reach under panel and depress the button on side of headlight switch while withdrawing switch control knob and shaft. Remove headlight switch bezel.
9. Reach under panel and disconnect speedo cable.
10. Remove wiper/washer control knob.
11. Remove threaded wiper/washer bezel.
12. Remove cigar lighter from its receptacle.
13. Remove four screws retaining cluster front cover.
14. Insert a right angle standard tip screwdriver along edges of finish panel with-

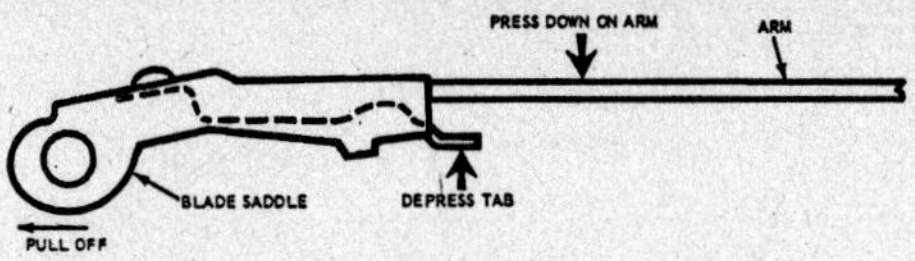

Fig. 7 Trico Bayonet wiper blade

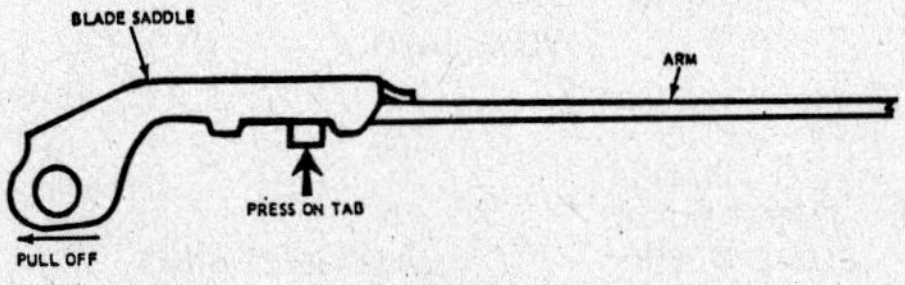

Fig. 8 Anco Bayonet wiper blade

drawing studs in sequence gradually around periphery of panel.

15. Remove two screws from cluster light baffle at cluster top.
16. Remove four screws retaining cluster to instrument panel.
17. Pull cluster away from panel and disconnect printed circuit feed plug.
18. Tilt cluster out, bottom first, and move cluster toward center of vehicle.

1970-73 Continental

1. Remove the instrument panel pad. The cluster trim cover does not have to be removed if only the cluster is being removed.
2. Reach under instrument panel and disconnect cluster printed circuit plug.
3. From the passenger side, remove the cluster-to-cluster housing retaining screws and swing the cluster away from housing.
4. From the underside of the cluster, unhook the pointer control cable from the PRND21 pointer lever.
5. Remove the cable retaining clip from cluster and remove the cluster.

NOTE: When replacing the cluster be sure to connect the control wire to the PRNDL pointer and attach the cable to the cluster with the retainer.

1969-71 Mark III

1. Disconnect battery ground cable.
2. Remove screws retaining upper edge of instrument cluster pad and retainer assembly to panel pad. Remove pad and retainer from face of cluster.
3. Remove clock knob and instrument cluster mask retainers and remove mask.
4. Remove three speedometer to cluster screws and pull speedometer from cluster. Disconnect two speedometer cable to cluster screws and clamps, release the tab of the plastic retainer and remove it from the cable. If equipped with speed control, the speedometer cable may be disconnected at the speed control unit instead.
5. Remove eight cluster to panel retaining screws and three screws retaining rear vent and wiper control pod. Pull cluster and pod out of the panel.
6. Disconnect the multiple connector and the low fuel warning and dual brake warning lights at the printed circuit and remove the cluster.

1969 Continental

Speedometer

1. Disconnect battery ground cable.
2. Remove air conditioning register casting screws, pull unit rearward and support on the ducts.
3. Remove cluster hood to crash pad screws, windshield molding screws and windshield molding.
4. Remove crash pad forward edge screws and nuts, disconnect speaker wire, then remove crash pad.
5. Remove radio knobs and bezels, heater, air conditioning and windshield wiper control knobs, then remove nuts and retainers from the control shafts.
6. Remove screws from lower control housing and lower the housing.
7. Remove screws securing finish cover to underside of instrument panel.
8. Remove odometer reset knob and screws securing upper cluster finish cover, then remove cover.
9. Remove right hand cluster air conditioning register screws and remove air hose.
10. Disconnect speedometer cable and speedometer retaining screws. Remove speedometer.

Fuel Gauge

1. Perform steps 1, 5, 6 and 7 outlined in the above procedure.
2. Remove fuel gauge retaining screws and electrical connectors, then remove fuel gauge.

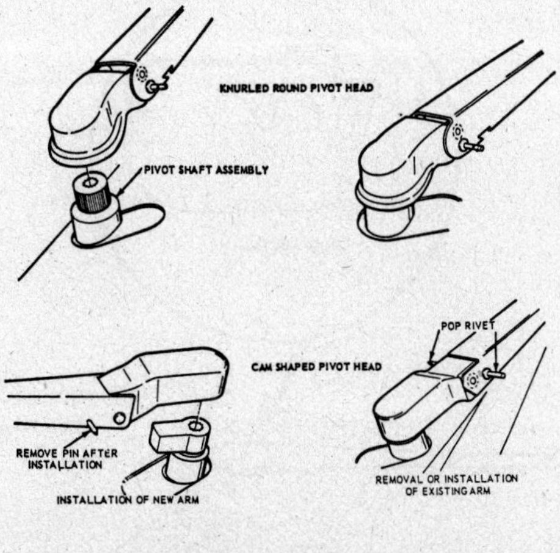

Fig. 11 Wiper arm. 1969-72 except Mark IV

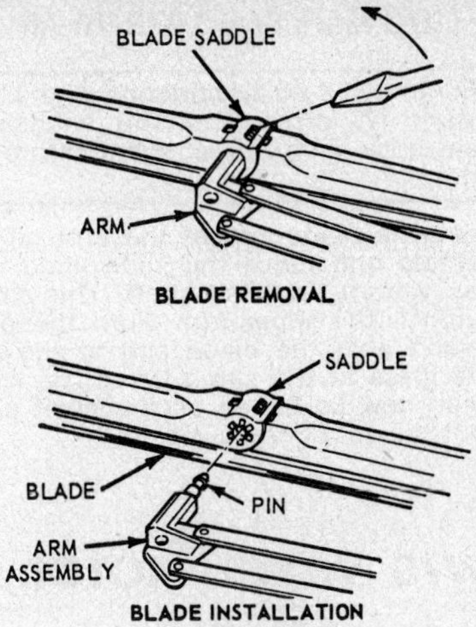

Fig. 9 Pin type wiper blade

W/S WIPER BLADES

NOTE: Trico and Anco blades are used and come in two types; Bayonet type and Side Saddle Pin type.

Bayonet Type: To remove the *Trico* type press down on the arm to unlock the top stud. Depress the tab on the saddle, Fig. 7, and pull the blade from the arm. To remove the *Anco* type press inward on the tab, Fig. 8, and pull the blade from the arm. To install a new blade assembly slide the blade saddle over the arm so that the locking stud snaps into place.

Side Saddle Pin Type: To remove the pin type insert an appropriate tool into the spring release opening of the blade saddle, depress the spring clip and pull the blade from the arm, Fig. 9. To install, push the blade saddle onto the pin so that the spring clip engages the pin. To replace the rubber element in a *Trico* blade squeeze the latch lock release and pull the element out of the lever jaws, Fig. 10. Remove the *Anco* element by depressing the latch pin, Fig. 10, and sliding the element out of the yoke jaws. To install, insert the element through the yoke or lever jaws. Be sure the element is engaged at all points.

W/S WIPER ARMS

1969-72 (Exc. Mark IV)

Swing the arm and blade assembly away from the windshield and insert a 3/32" inch pin through the pin hole as shown in Fig. 11. Swinging the assembly away from the windshield will release the spring loaded attaching clip in the arm from the pivot shaft. Inserting the pin will hold it in the released position. The arm can now be pulled off the pivot shaft. *Do not* pry off with a screwdriver. Leave the pin in the arm until after installation. A new arm comes with a pin already installed to hold it in the released position. Make sure the pivot shaft is in parked position when installing new assembly. Push the arm onto the pivot shaft. Lock the arm to the pivot shaft by removing the pin.

1972 Mark IV & 1973-76 All

NOTE: On 1973-76 Continental and 1974-76 Mark IV, disconnect the windshield washer hose from wiper arm before attempting to remove the arm.

Raise the blade end of the arm off the windshield and move the slide latch, Fig. 12, away from the pivot shaft. This action will unlock the wiper arm from the pivot shaft and hold the blade end of the arm off the glass at the same time. The wiper arm can now be pulled off the pivot shaft without the aid of any tools.

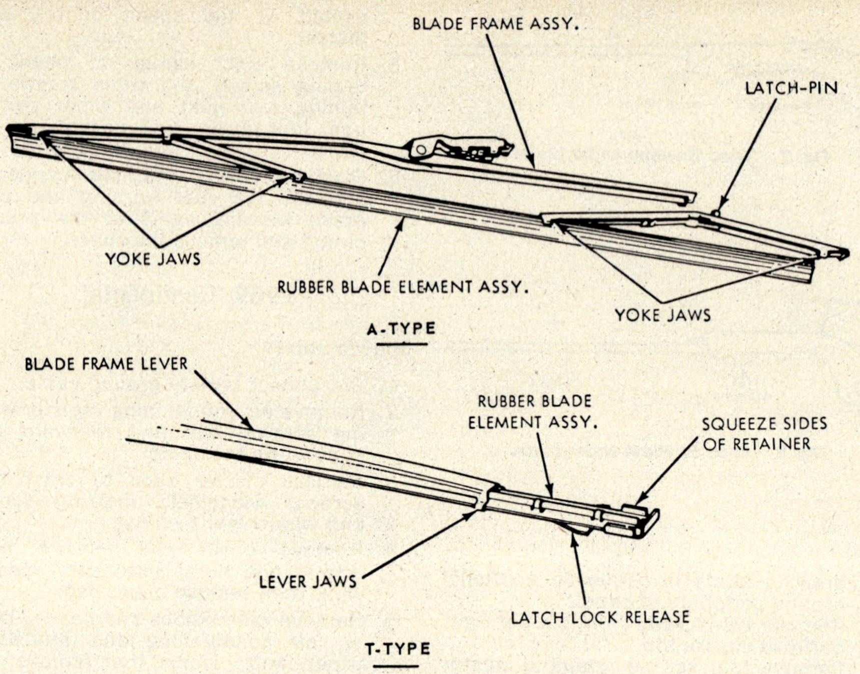

Fig. 10 Wiper blade element

W/S WIPER MOTOR, REPLACE

1971 Continental; 1972-76 All

1. Disconnect battery ground cable.
2. Remove wiper arm and blade assemblies.
3. Remove left cowl screen for access through cowl opening.
4. Disconnect linkage drive arm from motor output arm crankpin by removing the retaining clip.
5. From engine side of dash, disconnect wire connectors from motor.
6. Remove bolts that retain motor to dash and remove the motor. If the output arm catches on dash during removal, handturn the arm clockwise so it will clear the opening in dash.

NOTE: *Before installing motor be sure the output arm is in the Park position.*

1971 Mark III

1. Remove wiper arm and blade assemblies.
2. Remove the air cleaner.
3. Remove retaining screws and remove the cowl top panel. Disconnect washer hose.
4. Remove clip retaining left wiper link to motor and disconnect the link.
5. Disconnect hydraulic lines from motor being careful not to burn hands with hot hydraulic fluid.
6. Remove three bolts retaining motor to cowl and disconnect wiper control cable and remove through left cowl opening.

1970 Continental

For 1970, the hydraulic motor is replaced by a depressed park type electric motor. The motor is mounted on the engine side of the cowl panel under the left front fender. To remove, proceed as follows:

1. Disconnect battery ground cable.
2. Remove wiper arms and blades.
3. Remove cowl top grille and disconnect linkage drive arm from motor output crankpin.
4. Disconnect wires from the motor.
5. From engine side of dash, remove bolts retaining motor. If the output arm catches on the dash during removal, handturn the arm clockwise so it will clear the opening in the dash.

Late 1970 Mark III

All vehicles built after 9-22-69 will again use the hydraulic wiper system which was used on previous models. The major difference is the elimination of the surge chamber. Service is similar to 1969 models.

Early 1970 Mark III

The hydraulic motor used previously is replaced by an oscillating type, two speed electric motor. Removal is as follows:

1. Disconnect battery ground cable and washer hose.
2. Remove the cowl vent top panel.
3. Remove four bolts that retain the plastic cover and motor to the cowl panel.
4. Remove the plastic cover.
5. Disconnect the drive link from the

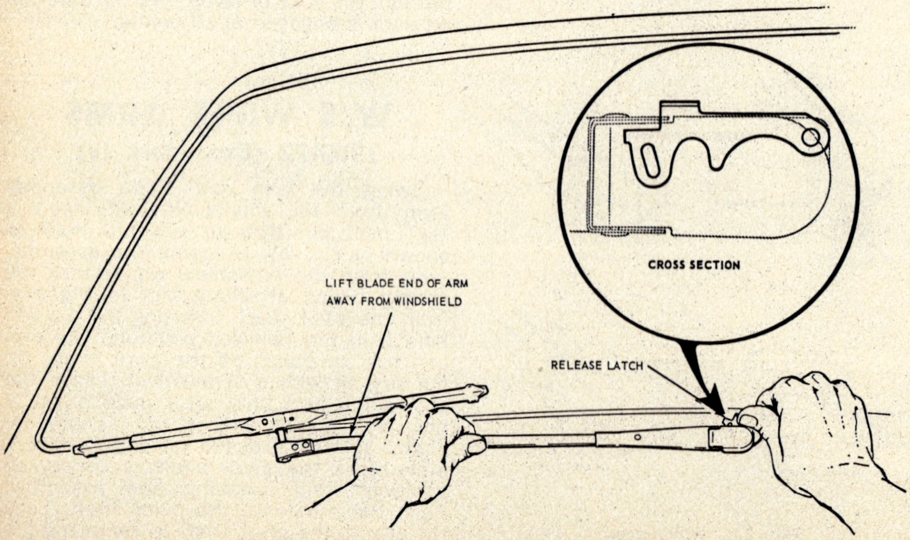

Fig. 12 Wiper arm. 1972 Mark IV & 1973-76 All

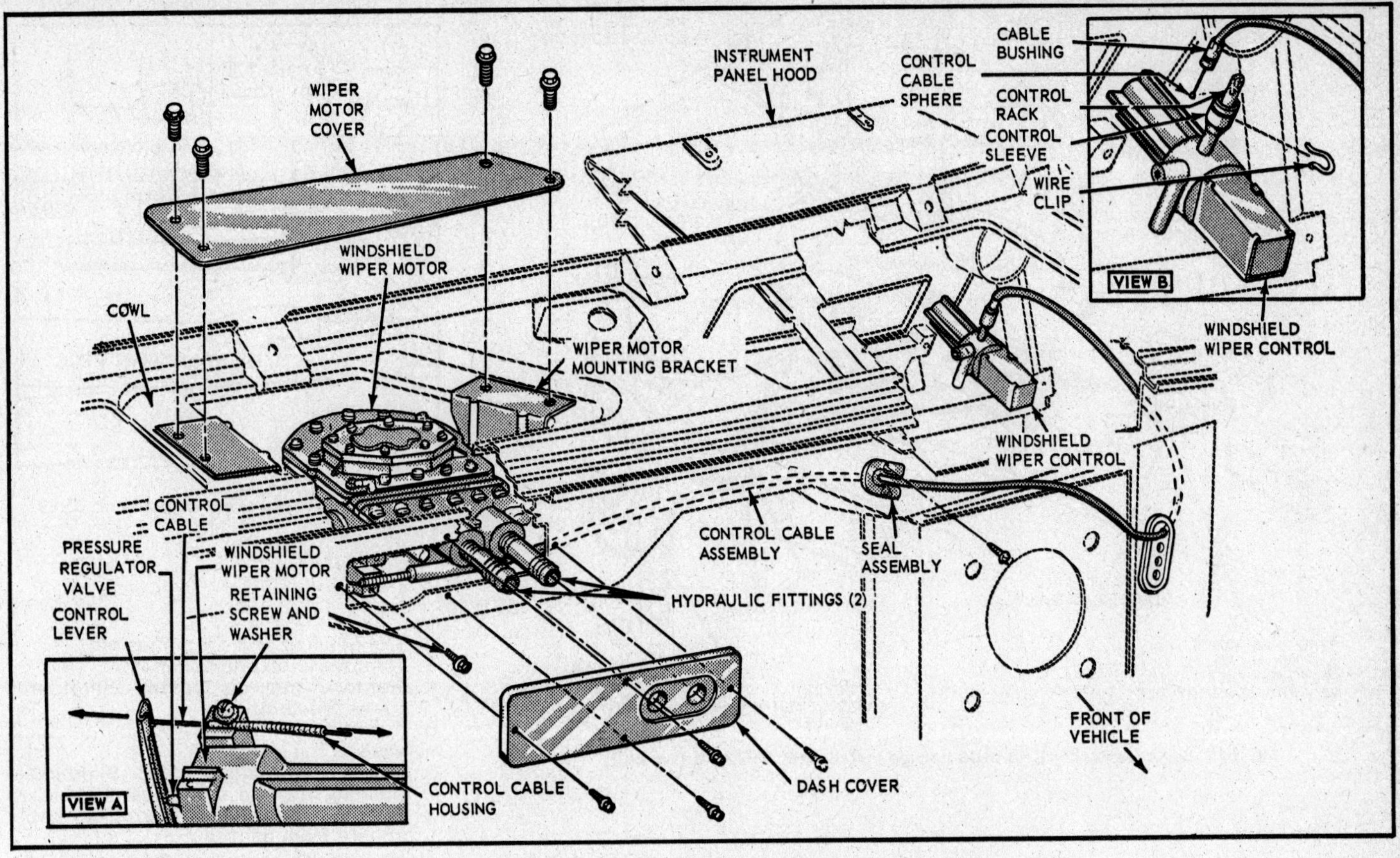

Fig. 13 Windshield wiper control and motor installation. 1969

motor arm.

6. Disconnect motor wiring at connector and remove the motor.

1969 Mark III

1. Remove wiper arm and blade assemblies.
2. Remove the air cleaner.
3. Remove retaining screws and remove cowl top panel. Disconnect washer hose.
4. Remove clip retaining left wiper link to motor and disconnect the link.
5. Disconnect hydraulic lines being careful not to burn hands with hot hydraulic fluid.
6. Remove three bolts retaining motor to cowl. Disconnect wiper control cable and remove motor through left cowl opening.

1969 Continental

1. Referring to Fig. 13, remove wiper arms and blades.
2. Remove cowl top grille, wiper nozzles and weatherstrip (6 screws).
3. Disconnect two hydraulic lines at wiper motor, *using care as oil may be hot.*

1969 NOTE

The windshield wiper is hydraulically operated. Hydraulic power for the motor is obtained from the power steering unit. Hydraulic fluid flows from the pump, through the steering gear to the wiper motor, and then to the fluid reservoir. During wiper operation, a part of the fluid is by-passed through the motor by a valve on the motor.

W/S WIPER TRANSMISSION

1972-76 Mark IV

1. Disconnect battery ground cable and remove wiper arm and blades.
2. Remove cowl screen and left arm and blade stop.
3. Disconnect linkage drive arm from the motor by removing retaining clip.
4. Remove pivot shaft retaining bolts and remove linkage and pivot shaft.

NOTE: When installing pivot shaft assemblies, be sure to force the linkage connecting clip into the locked position.

1971 Mark III

1. Remove wiper arm and blade assemblies. Be sure to release the tension arm retaining clip from the tension arm retaining stud on the left pivot assembly.
2. Remove retaining screws and remove the cowl vent grille. Disconnect washer hoses.
3. Remove clip retaining link assembly to the motor. Remove screws retaining pivot shaft and link assembly to the cowl and remove the pivot shaft and link assembly.

1970-76 Continental

1. Disconnect battery ground cable.
2. Remove wiper arm and blade assemblies.
3. Remove cowl screens for access to linkage.
4. Disconnect left linkage arm from the drive arm by removing the clip.
5. Remove three bolts retaining left pivot shaft assembly to the cowl and remove the left arm and pivot shaft assembly through cowl opening.
6. Disconnect linkage drive arm from motor crankpin by removing the clip.
7. Remove three bolts that connect the drive arm pivot assembly to the cowl and remove the pivot shaft drive arm and right arm as an assembly.

1970 Mark III

1. Disconnect battery ground cable.
2. Disconnect washer hose and remove cowl vent top panel.

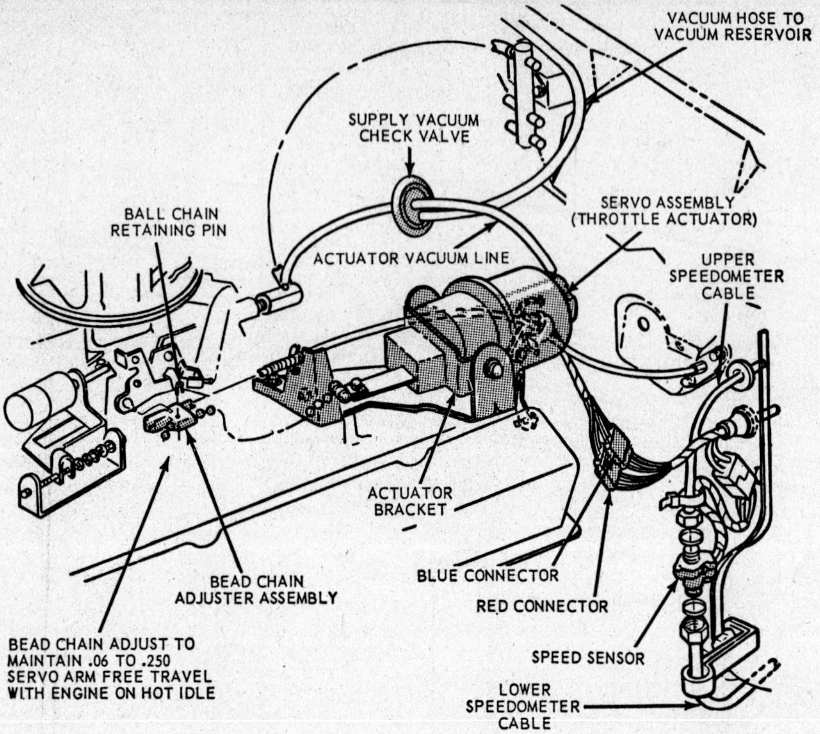

Fig. 14 Servo assembly & throttle linkage installation. 1972-76 (typical)

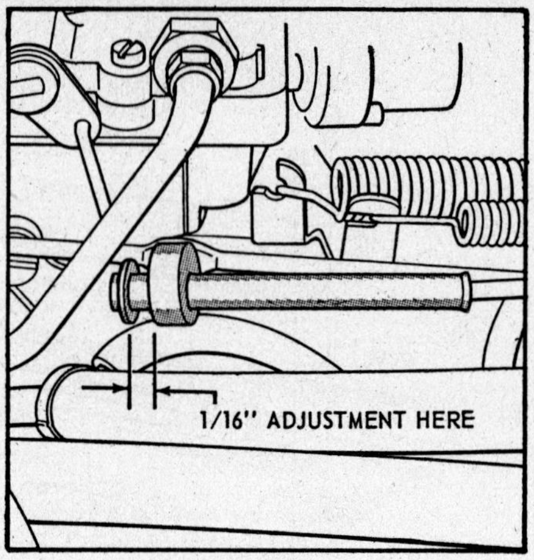

Fig. 15 Control cable adjustment. 1969

3. Disconnect drive link from motor arm and disconnect both links from the right pivot shaft assembly.
4. Remove three screws from the drive pivot plate at the right end of the cowl panel and withdraw the drive pivot plate and the two drive links as an assembly.
5. Remove the right pivot shaft (three screws).
6. Remove the three left pivot shaft screws and withdraw the left pivot shaft and link as an assembly.

1969 Mark III

1. Remove wiper arm and blade assemblies, being sure to release the tension arm retaining clip from the stud on the left pivot assembly.
2. Remove cowl top panel and disconnect washer hose.
3. Remove link retaining clip and pivot shaft assembly to cowl screws. Remove pivot shaft and link assembly.
4. Reverse procedure to install.

1969 Continental

1. Remove wiper arms and blades.
2. Remove cowl grille (6 screws) and disconnect washer hoses and weatherstrip.
3. Remove drive arm clip.
4. Remove three bolts at each pivot.
5. Remove pivots from connecting linkage and lift assembly out through left side of cowl.
6. Reverse procedure to install.

W/S WIPER SWITCH

1972-76 Mark IV

1. Disconnect battery ground cable.
2. Remove instrument cluster finish panel.
3. Remove switch mounting plate and disconnect cigar lighter and wiper switch wires.
4. Remove switch bezel nut and remove switch.

1970-76 Continental

1. Disconnect battery ground cable.
2. Pull the knob and remove retaining nut and gasket from switch shaft.
3. Lower switch from behind instrument panel and disconnect the multiple connector.

1969-71 Mark III

1. Disconnect battery ground cable.
2. Remove five retaining screws and remove instrument cluster pad and retainer from face of cluster.
3. Remove three retaining screws and pull wiper control pod from panel. Disconnect vacuum hoses and electrical connector and remove bulbs. Remove control cable retaining clip and remove control assembly.
4. Remove control knobs, two retaining screws and remove control unit.

1969 Continental

1. Disconnect battery ground cable.
2. Remove eight screws from lower control housing and drop housing.
3. Remove knob and bezel.
4. Remove nut on control shaft and lower the control.
5. Turn wiper control knob counterclockwise to the stop.
6. Remove wire clip, Fig. 13. Slide cable bushing up and out of control sleeve (view B). Remove control cable sphere from control rack.
7. Disconnect plug connector from wiring harness. Disconnect two vacuum hoses and remove wiper control.
8. Reverse procedure to install.

RADIO REMOVAL

NOTE: When installing radio, be sure to adjust antenna trimmer for peak performance.

1974-76 Mark IV

1. Disconnect battery ground cable.
2. Pull radio knobs and discs from shafts.
3. Remove twilight sentinel amplifier.
4. Remove air conditioning duct located under radio.
5. Remove radio rear support to panel screw, disconnect radio electrical leads and remove radio.

1973-76 Continental

1. Disconnect battery ground cable.
2. Remove radio knobs and discs, then map light assembly.
3. Remove steering column shroud, ash tray door pad and instrument cluster panel pad.
4. Remove center register applique, then disconnect cigar lighter and glove box light switch electrical connectors.
5. Remove radio bracket to instrument panel tab nut and radio mounting bracket to instrument panel screws.

6. Pull radio out, disconnect wiring and remove radio.

1972-73 Mark IV

1. Disconnect battery ground cable.
2. Pull radio control knobs off shafts.
3. Remove nut from both radio shafts.
4. Remove radio rear support attaching screw at instrument panel.
5. Disconnect radio power and speaker wires at connectors.
6. Disconnect antenna and remove radio.

1970-72 Continental

1. Disconnect battery ground cable.
2. Remove map light assembly.
3. Remove right and left inspection covers.
4. Remove lower instrument panel pad.
5. Remove the glove box. Open ash tray and let it hang open.
6. Remove glove box switch.
7. Through glove box opening remove two nuts retaining radio finish panel to the instrument panel.
8. Remove radio knobs.
9. Remove two screws at top of finish panel. Position panel out and disconnect cigar lighter and light from the right panel.
10. Through the glove box opening remove the nut from the lower right corner of the center finish panel.
11. Remove radio top support nut and the three mounting screws. Pull radio out, disconnect power leads and antenna cable and remove radio.

1969-71 Mark III

1. Disconnect battery ground cable.
2. Pull off control knobs.
3. Remove cover plate below steering column, and the nut from right radio control shaft.
4. Remove six screws and remove trim from front of radio.
5. Remove nut and washer from right control shaft, and screw attaching front left side of radio to instrument panel.
6. Remove rear support attaching screw, disconnect wires and remove radio.
7. Reverse procedure to install.

1969 Continental

1. Disconnect battery ground cable.
2. Remove eight screws in lower control housing and lower the housing.
3. Disconnect lead from speakers.
4. Disconnect power antenna lead.
5. Disconnect lead from foot-operated switch for AM/FM radios (if equipped).
6. Disconnect one two-way disconnect for pilot light and radio power.
7. Remove two knobs and bezels on selector shafts.
8. Remove two nuts and retainers on selector shafts.
9. Remove two screws attaching radio bracket to lower reinforcement on instrument panel.
10. Remove two nuts from selector shafts.
11. Disconnect antenna lead and remove radio.

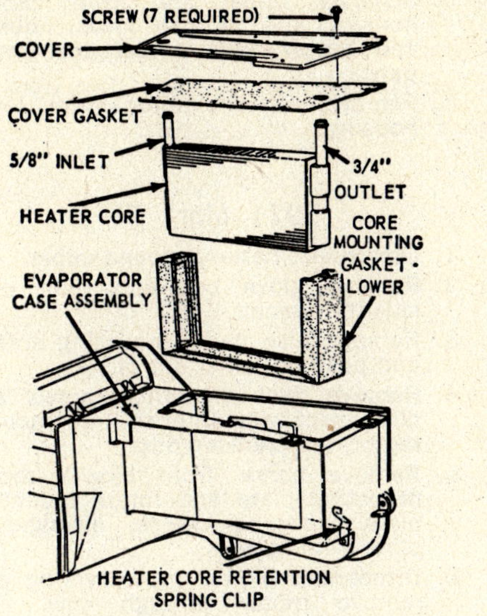

Fig. 16 Heater core removal. 1973-76 Continental

SPEED CONTROLS

1972-76

Adjust the bead chain to obtain .06-.25" actuator arm free travel when engine is at hot idle, Fig. 14. The adjustment should be made to take as much slack as possible out of the bead chain without restricting the carburetor lever from returning to idle. On vehicles with solenoid anti-diesel valve, perform the adjustment with the ignition switch in the ON position.

1970-71

A mercury cut-off switch has been mounted in the speed control regulator. The cut-off switch makes certain that the speed control is shut off when the vehicle is suddenly stopped. When installing a regulator, use a level to make certain the mercury will operate properly. It may be necessary to place washers under one of the mounting studs to make it level.

Linkage Adjustment

Adjust the servo chain to obtain a ½ to 1 ball slack when the engine is at hot idle.

1969 Mark III

Linkage Adjustment

Adjust the servo chain to obtain a ½ to 1 ball link slack when the engine is at hot idle.

1969 Continental

Control Cable, Adjust

With the carburetor set at hot idle, adjust the Bowden cable to provide a 1/16" clearance between the Bowden cable end "C" washer and the accelerator linkage sleeve, Fig. 15.

HEATER CORE REMOVAL

1973-76 Continental

1. Drain radiator and disconnect heater hoses from core.
2. Remove heater core cover and gasket, then lift heater core and lower mounting gasket from evaporator housing, Fig. 16.

1972-76 Mark IV

1. Drain radiator and disconnect heater hoses from core.
2. Remove glove box, then heater air outlet register from plenum assembly by disengaging the snap clips.
3. Disconnect temperature control cable from blend door crank arm.
4. Remove vacuum hoses from high-low door motor, panel-defrost door motor and the in-line tee connector to temperature by-pass door motor.
5. Disconnect resistor wiring and remove plenum case flange screws, then remove plenum case rear half.
6. Remove heater core tube support bracket nut and heater core.

1970-72 Continental

1. Drain the engine coolant.
2. Disconnect vacuum junction valve from dash panel and move valve and hoses away from case.
3. Disconnect speed control from dash, if so equipped, and move it away.
4. Disconnect multiple connector from blower resistor and the harness from the clip on the case.
5. Disconnect heater hoses from the case and the hose support clamp from the case. Move the hoses and the water valve away from the case.
6. Remove case cover-to-case flange screws and the wire harness clip.
7. Remove six case cover-to-back plate stud nuts.
8. Remove one upper case-to-dash panel mounting screw.
9. Remove two case-to-dash panel mounting stud nuts, one on the inboard mounting flange and one below the case on the lower flange.
10. Carefully move the heater core cover assembly forward to clear the mounting studs and lift it up and out of the car.
11. Remove the spring clips from the core tubes on the front of the core cover.
12. Remove three core end plate mounting screws and remove the plate. The heater core and gasket can now be removed from the cover.

1969-71 Mark III

1. Remove hood and air cleaner and drain radiator.

2. Disconnect wiper motor hydraulic lines, vacuum supply hose, icing switch multiple connector and heater hoses.
3. Remove oil pressure sending unit and transmission dip stick and tube.
4. Remove evaporator housing front cover and heater core housing cover.
5. Remove heater core retaining bracket and heater core.

1969 Continental

1. Remove air cleaner and harness clamp on top evaporator-heater case in engine compartment.
2. Take off actuator for temperature blending door, adjacent to evaporator-heater case.
3. Remove retaining screws and heater core cover plate and lift out core.

BLOWER MOTOR REMOVE

1972-76 Mark IV

1. Remove glove box, recirculating air register and duct assembly.
2. Remove blower lower housing to dash screws.
3. Disconnect vacuum hose from outside-recirculating air door motor and remove motor from blower lower housing, leaving motor actuator connected to door crank arm.
4. Disconnect blower motor wiring and remove blower motor housing flange screws.
5. Separate upper and lower blower housing and remove lower housing and motor from under instrument panel.
6. Remove blower motor from lower housing.

1971 Mark III

1. Disconnect battery ground cable.
2. Remove glove box and right cowl side trim panel.
3. Remove fuse panel mounting screws and position panel out of way.
4. Remove duct mounting screws and disconnect vacuum hose from vacuum motor, then remove duct.
5. Remove screw from blower motor mounting plate and rotate mounting plate counter-clockwise, disengaging plate from case.
6. Disconnect blower motor wiring and remove motor through cowl side panel opening.

1970 All & 1971-76 Continental

1. Remove the hood.
2. Remove the right hood hinge and right fender inner support brace as an assembly.
3. Disconnect the blower motor air cooling tube from the motor.
4. Disconnect the motor lead wire from the harness and the ground wire from the dash panel.
5. Disconnect the rear section of the right front fender apron from the fender around the wheel opening (7 screws) and remove the two lower fender-to-cowl mounting screws.
6. Separate the fender apron from the fender wheel opening so that the apron can be pushed downward away from the blower motor.
7. Remove the four blower motor mounting plate screws. Move the motor and wheel forward out of the blower scroll and remove the assembly through the opening while applying pressure to the fender apron to enlarge the opening at the hinge area.

1969

NOTE: The blower motor is located in the right front fender well.

1. Remove the right front fender splash shield which is held by six bolts.
2. Remove the four screws retaining the blower motor to the housing. Disconnect the electrical leads and remove the motor.
3. Reverse above to install.

Engine Section

ENGINE MOUNTS REPLACE

CAUTION: Whenever self-locking mounting bolts and nuts are removed, they must be replaced with new self-locking bolts and nuts.

1969 Lincoln; 1970-71 Mark III

1. Block the rear wheels and set the parking brake. Raise the front of the vehicle with a floor jack and install safety stands.
2. Remove the nuts and washers which secure the engine front support insulators to the underbody crossmember, Fig. 1.
3. Place a block of wood between a jack and the front edge of the oil pan. Raise the front of the engine about 2 inches to allow clearance for removal of the insulators.
4. Remove the bolts securing the support insulator to the cylinder block.

1970-76 Lincoln

1. Block the rear wheels and set the parking brake. Raise front of vehicle and install safety stands.
2. Remove the bolts that attach the support bracket to the cylinder block, Fig. 2.
3. Place a block of wood between a jack and the front edge of the oil pan. Raise the engine high enough to provide clearance to remove the support and not damage the radiator.
4. Remove the through bolts from the support insulator. Lift the insulator from the number 2 crossmember.
5. Remove the bracket-to-insulator attaching bolt and separate the units.

1972-76 Mark IV

1. Remove the fan shroud attaching screws and support the engine using a jack and a block of wood under the oil pan.
2. Remove the through bolt and nut attaching the insulator to the frame crossmember, Fig. 3.
3. Remove the insulator to upper bracket attaching nuts.
4. Raise the engine enough to remove the insulator and heat shield if so equipped.
5. If required, the upper bracket can now be removed by removal of the three screws holding the bracket to the cylinder block.

ENGINE, REPLACE

In addition to the usual items such as fuel lines, linkage and radiator hoses, the following operations must be performed:

1. Remove hood, radiator and air cleaner.
2. Remove power steering pump and starter.
3. Disconnect exhaust pipes from manifolds. On Mark III, disconnect idler arm at frame bracket.
4. Support transmission with a suitable jack.
5. Unfasten engine mountings and attach lifting rig to engine.
6. Remove bell housing to engine attaching bolts and carefully lift engine out of chassis. *Be sure to support transmission with front end tilted up to prevent converter from falling out.*

CYLINDER HEADS

1969-75 V8-460

NOTES

Before installing cylinder head, wipe off engine block gasket surface and be certain no foreign material has fallen into cylinder bores, bolt holes or in the

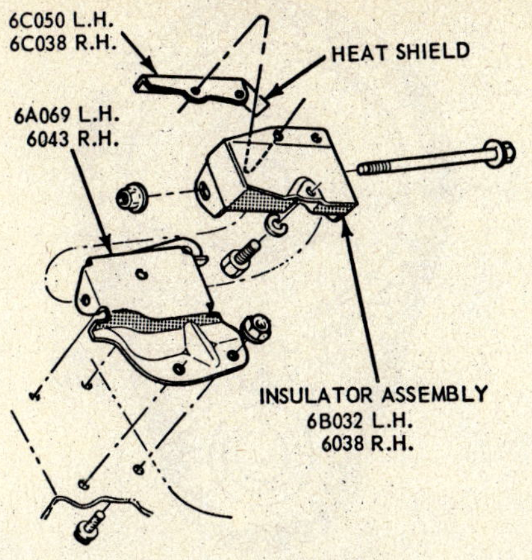

Fig. 1 Engine mount. 1969; 1970-71 Mark III

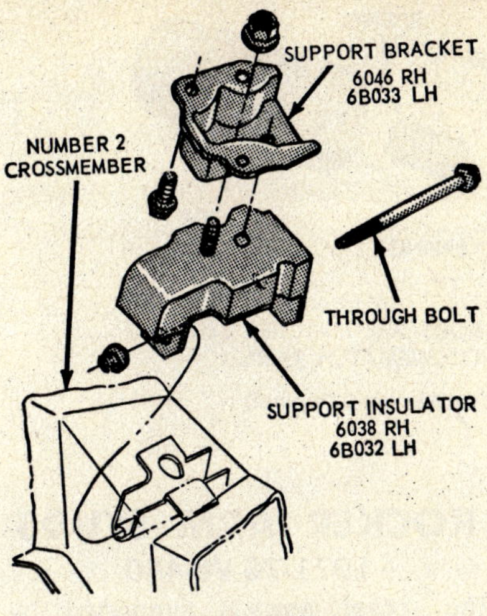

Fig. 2 Engine Mount. 1970-75 Lincoln

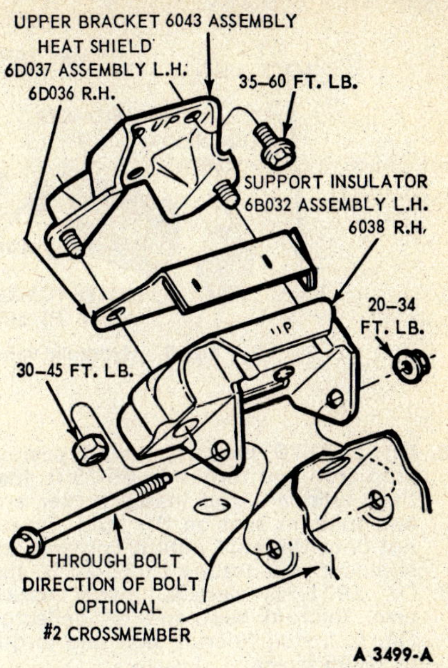

Fig. 3 Engine Mount. 1972-75 Mark IV

valve lifter area. It is good practice to clean out bolt holes with compressed air.

Some cylinder head gaskets are coated with a special lacquer to provide a good seal once the parts have warmed up. Do not use any additional sealer on such gaskets. If the gasket does not have this lacquer coating, apply suitable sealer to both sides.

Tighten cylinder head bolts at little at a time in three steps in the sequence shown in the illustrations. Final tightening should be to the torque specifications listed in the *Engine Tightening* table. After the bolts have been torqued to specifications, *they should not be disturbed.*

In instances where cylinder head gasket leakage is hard to control, aluminum paint can be applied to the gasket as a sealer.

Spray one coat of the aluminum paint on both sides of the gasket and allow the paint to dry. Then spray a second coat on both sides and, while the paint is still wet, install the gasket. Torque the head and manifold bolts to specifications to complete the job.

---

1. Remove intake manifold and carburetor as an assembly.
2. Disconnect muffler inlet pipe at exhaust manifold.
3. Loosen air conditioner compressor belt if so equipped.

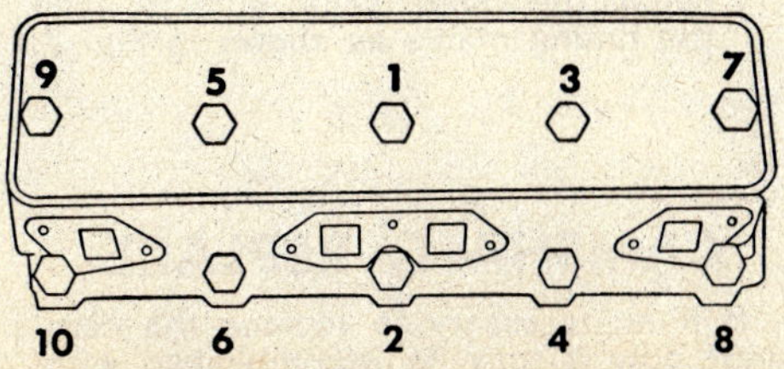

Fig. 4 Cylinder head tightening

4. Loosen alternator retaining bolts and remove bolt retaining alternator bracket to right head.
5. If air conditioned, isolate compressor at service valves and hoses from compressor. Remove nuts retaining compressor bracket to water pump. Remove bolts retaining compressor to upper mounting bracket and lay compressor out of way. Remove compressor upper bracket from head.
6. If not air conditioned, remove bolts retaining power steering reservoir bracket to left head and position reservoir out of way.
7. Remove rocker arm covers and rocker arms. Remove push rods in sequence so they can be installed in their positions.
8. Remove head retaining bolts and lift head with exhaust manifold.

NOTE: If necessary to break gasket seal, pry at forward corners of cylinder heads against casting bosses provided on cylinder block. Avoid damaging machined surfaces on head and block.

9. Install cylinder heads in reverse order of removal and torque bolts in the sequence shown in Fig. 4.

NOTE

The cylinder head gaskets are marked "Top" or "Front" stamped near the front end of the gasket. The gasket is properly installed when the word is at the forward end of the engine and water passage holes line up. This results in the sealing beads on the right head gasket being inverted with respect to the left head gasket.

VALVE CLEARANCE

1969-76 V8-460

To eliminate the need for valve lash adjustment, a positive stop rocker arm stud and nut is used on 1969-70 models, and a positive stop fulcrum bolt and seat is used on 1971-76 models, Fig. 5.

It is very important that the correct push rod be used and all components be installed and torqued as follows:

1. Position the piston of the cylinder being worked on at TDC of the compression stroke.

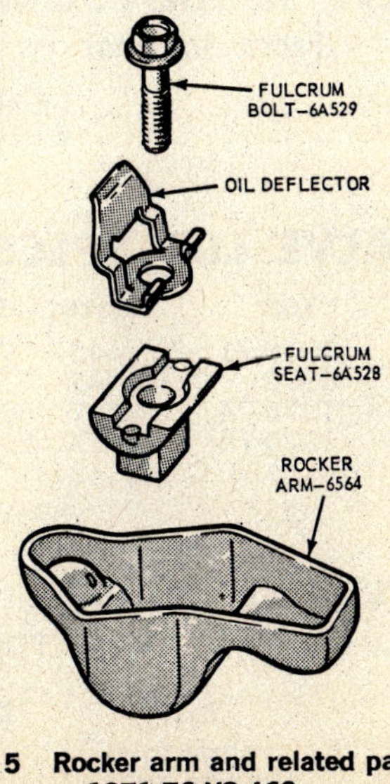

Fig. 5 Rocker arm and related parts. 1971-76 V8-460

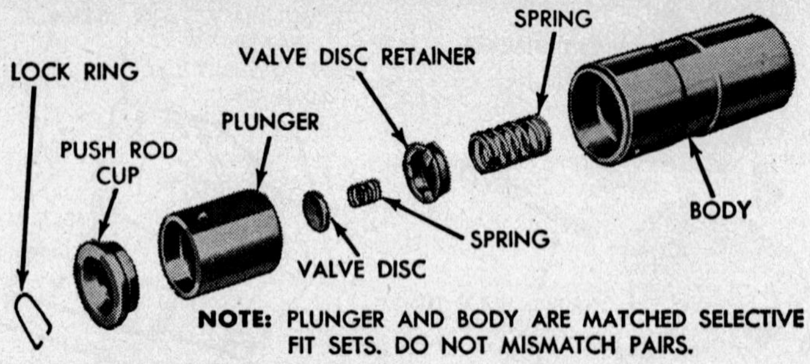

Fig. 6 Hydraulic valve lifter disassembled (typical)

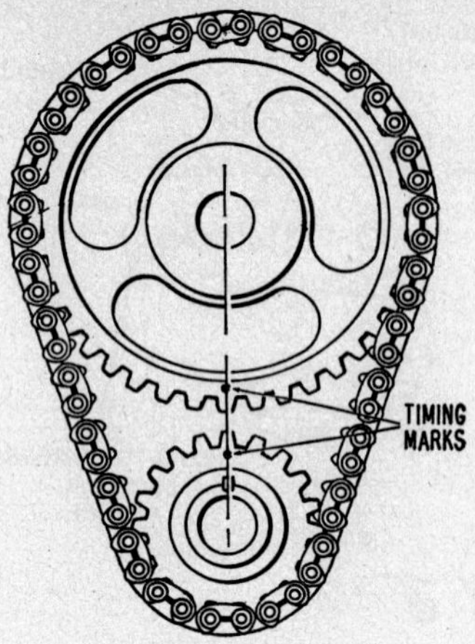

Fig. 7 Valve timing marks

2. On 1969-70 models, install positive stop stud and torque to 65-75 ft. lbs., then lubricate and install rocker arm and fulcrum seat on the stud. Thread nut onto the stud until it contacts the shoulder and torque to 18-22 ft. lbs.
3. On 1971-76 models, install rocker arm, fulcrum seat and oil deflector, Fig. 5. Install fulcrum bolt and torque to 18-25 ft. lbs.

NOTE: A .060" shorter push rod or a .060" longer push rod are available for service to provide a means of compensating for dimensional changes in the valve mechanism. Valve stem-to-rocker arm clearance should be as listed in the *Valve Specifications* table, with the hydraulic lifter completely collapsed. Repeated valve grind jobs will decrease this clearance to the point that if not compensated for the lifters will cease to function.

To check the clearance, bring the piston of the cylinder being checked on top dead center of the compression stroke. Then with hydraulic lifter collapsed, check the clearance between valve stem and rocker arm. If the clearance is less than the minimum, the .060" shorter push rod should be used. If clearance is more than the maximum, the .060" longer push rod should be used. (See *Valve Specifications* table).

VALVE ARRANGEMENT

Front to Rear

Right Bank . . . . I-E-I-E-I-E-I-E
Left Bank . . . . E-I-E-I-E-I-E-I

VALVE LIFT SPECS.

Engine	Year	Intake	Exhaust
V8-460	1969-71	.443	.486
	1972	.443	[1]
	1973-74	.437	.480
	1975-76	.437	.451

[1]Exc. Calif., .486. Calif., .424.

VALVE TIMING

Intake Opens Before TDC

Engine	Year	Degrees
V8-460	1969-71	16
	1972-76	8

ROCKER ARM STUDS

1971-76 V8-460

The rocker arm is supported by a fulcrum bolt which fits through the fulcrum seat and threads into the cylinder head. To disassemble, remove the bolt, oil deflector, fulcrum seat and rocker arm, Fig. 5.

1969-70 V8-460

Rocker arm studs are screwed into threaded bores in the cylinder head bosses. To install, apply water resistant sealer to stud threads that screw into cylinder head. Install stud and torque to 65-75 ft-lbs. Apply Lubriplate to top of valve stem and at push rod guide in cylinder head. Install rocker arm, fulcrum and stud nut.

VALVE GUIDES

Valve guides in these engines are an integral part of the head and, therefore, cannot be removed. For service, guides can be reamed oversize to accommodate one of three service valves with oversize stems (.003", .015" and .030").

Check the valve stem clearance of each valve (after cleaning) in its respective valve guide. If the clearance exceeds the service limits of .004" on the intake or .005" on the exhaust, ream the valve guides to accommodate the next oversize diameter valve.

HYDRAULIC VALVE LIFTERS

The internal parts of each hydraulic valve lifter assembly are a matched set. If these are mixed, improper valve operation may result. Therefore, disassemble, inspect and test each assembly separately to prevent mixing the parts.

Fig. 6, illustrates the type of hydraulic lifter used. See the *Trouble Shooting Chapter* under the heading *Engine Noises* for causes of hydraulic valve lifter noise.

TIMING CASE COVER

NOTE: If necessary to replace the cover oil seal the cover must first be removed.

V8-460

1. Drain cooling system and crankcase.
2. Remove fan and radiator.
3. Remove drive belts and water pump pulley. Remove compressor support if so equipped.
4. Remove bolt and washer retaining crankshaft damper and, using a suitable puller, remove damper. Remove Woodruff key from crankshaft.
5. Remove power steering pump.
6. Loosen by-pass hose and remove heater return tube at water pump.
7. Remove fuel pump.
8. Remove front cover to cylinder block bolts. Cut oil pan seal flush with cylinder block face prior to separating cover from the cylinder block. Remove front cover and water pump as a unit.

TIMING CHAIN

1. To remove the chain, first take off the timing chain cover as outlined previously.
2. Crank the engine until the timing mark on the camshaft sprocket is adjacent to the timing mark on the crankshaft sprocket, Fig. 7.
3. Remove cap screws, lock plate and fuel pump eccentric from front of camshaft.
4. Place a screwdriver behind the camshaft sprocket and carefully pry the sprocket and chain off the camshaft.
5. Reverse the foregoing procedure to install the chain, being sure to align the timing marks as shown in Fig. 7.

CAMSHAFT, REPLACE

If it is necessary to replace the camshaft only it may be accomplished without removing the engine from the

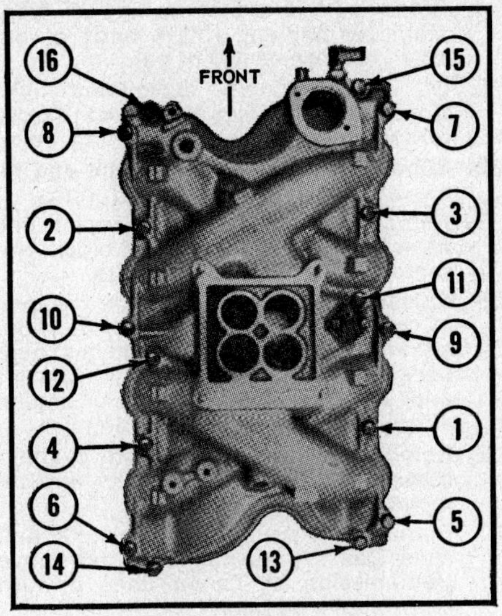

Intake manifold tightening sequence. V8-460

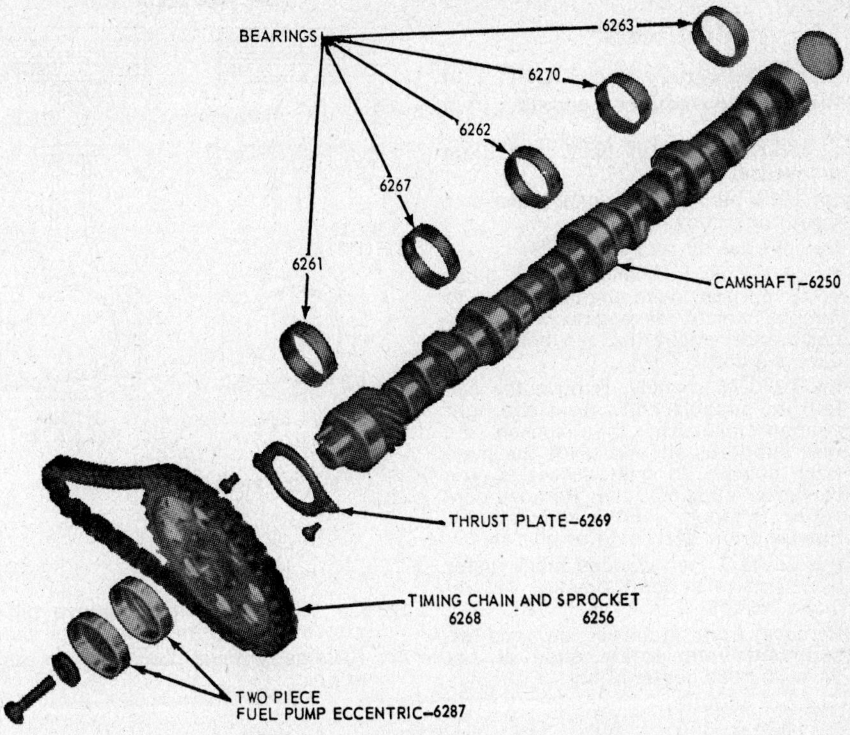

Fig. 8 Camshaft and related parts. V8-460

chassis. But if the camshaft bearings are to be replaced the engine will have to be removed. To remove the camshaft, proceed as follows:

V8-460

1. Drain crankcase and remove oil pan.
2. Remove timing cover, chain and sprockets as outlined previously.
3. Remove intake manifold and carburetor as an assembly.
4. Remove rocker arm covers. Back off all rocker arm stud nuts or bolts, turn rocker arms sideways and remove push rods in sequence.
5. Remove valve lifters.
6. If air conditioned, unbolt and lay condenser on left fender. Secure in this position.
7. Remove grille.
8. Remove camshaft thrust plate bolts and carefully remove camshaft from front of engine, Fig. 8.
9. Reverse above procedure to install.

PISTON & ROD, ASSEMBLE

If the old pistons are serviceable, make certain that they are installed on the rods from which they were removed. The assembly must be made as shown in **Fig. 9.**

PISTONS, RINGS & PINS

Pistons are available in oversizes of .003, .020, .030, .040 and .060".

Piston pins are available in oversizes of .001 and .002".

Rings are available in oversizes of .002, .010, .020, .030 and .040".

MAIN & ROD BEARINGS

Main and rod bearings are available in standard size and undersizes of .002, .010, .020, .030 and .040".

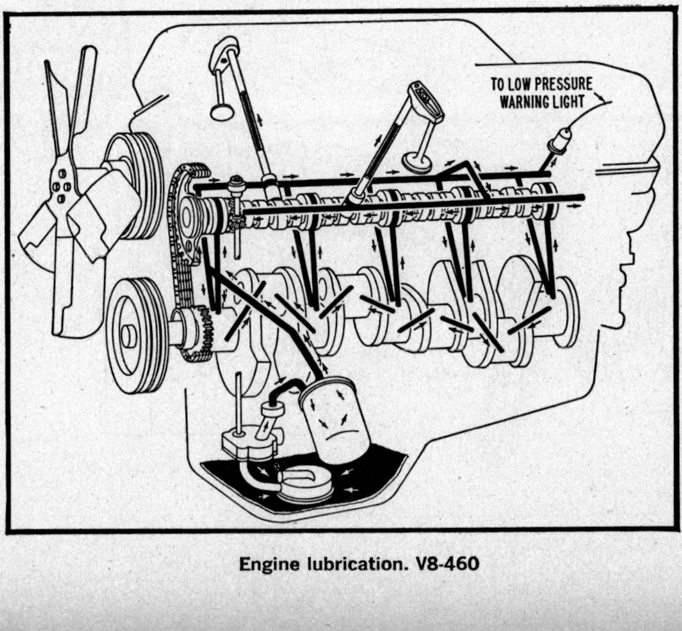

Engine lubrication. V8-460

OIL PAN, REPLACE

Continental

1. Disconnect battery ground cable.
2. Disconnect radiator shroud from radiator.
3. Support vehicle on hoist and drain crankcase.
4. On 1969 models, disconnect idler arm from underbody.
5. Remove starter retaining bolts.
6. Using a floor jack and wooden block under oil pan, raise engine enough to remove weight of engine from supports and remove through bolts from each support.
7. On 1970-76 models, remove the two forward support bolts from the right support insulator, then loosen the rear support bolt and pivot the insulator upward to gain access to the converter support bolts. Remove converter support bolts and remove bracket from each side of oil pan.
8. Position a 1 inch wooden block under each engine support bracket and remove jack.
9. Remove front stabilizer bar end attachments and rotate ends of bar down to raise center of bar.
10. Remove oil filter.
11. On 1969 models:
    a. Remove oil pan attaching bolts and lower oil pan to cross-member.
    b. Remove splash shield from right side of oil pan.
    c. Disconnect pressure line at power steering pump.
    d. Remove pump to front cover retaining bolts and rotate pump to clear oil pan.
    e. Remove oil pan.

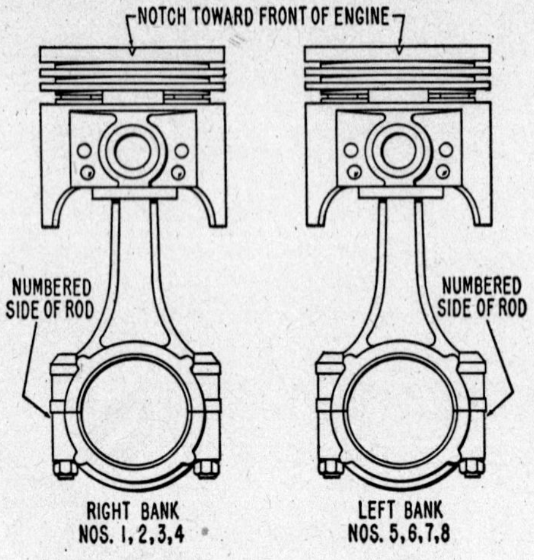

Fig. 9 Piston and rod assembly

12. On 1970-76 models, remove oil pan attaching bolts and remove oil pan.
13. Reverse removal procedure for installation.

Mark III & IV

1. Disconnect battery ground cable.
2. Disconnect transmission oil cooler lines from radiator, then remove radiator shroud attaching bolts and position shroud over fan.
3. Support vehicle on hoist, then drain crankcase and remove oil filter.
4. Remove end attachments of front stabilizer bar and rotate ends of bar down to raise center of bar.
5. On 1969 models, disconnect lines from power steering pump and remove power steering pump attaching bolts.
6. Support engine with floor jack and remove engine support through bolts.
7. Remove transmission oil cooler line attaching bolt from engine block and remove starter attaching bolts.
8. Raise engine enough to allow removal of right insulator heat shield, then position a wooden block 3 inches high between each engine support bracket and exhaust manifold. Lower engine allowing blocks to support engine.
9. Remove converter housing to cylinder block support attaching bolts and remove brackets.
10. Remove oil pan attaching bolts and lower pan to crossmember, then move transmission oil cooler lines upward and remove oil pan.
11. Reverse removal procedure for installation.

OIL PUMP

To remove the pump, drop the oil pan as outlined above. Then remove the two bolts that attach the pump to the crankcase and remove pump, gasket and the intermediate shaft.

To disassemble, remove the pump cover plate, Fig. 11, and lift out the rotor and shaft. Scrape the stake marks that hold the relief valve in the pump housing until the retainer can be removed. Then remove the retainer, spring and relief valve from the pump housing. Inspect the pump as follows:

1. With all parts clean and dry, check the inside of the pump housing and

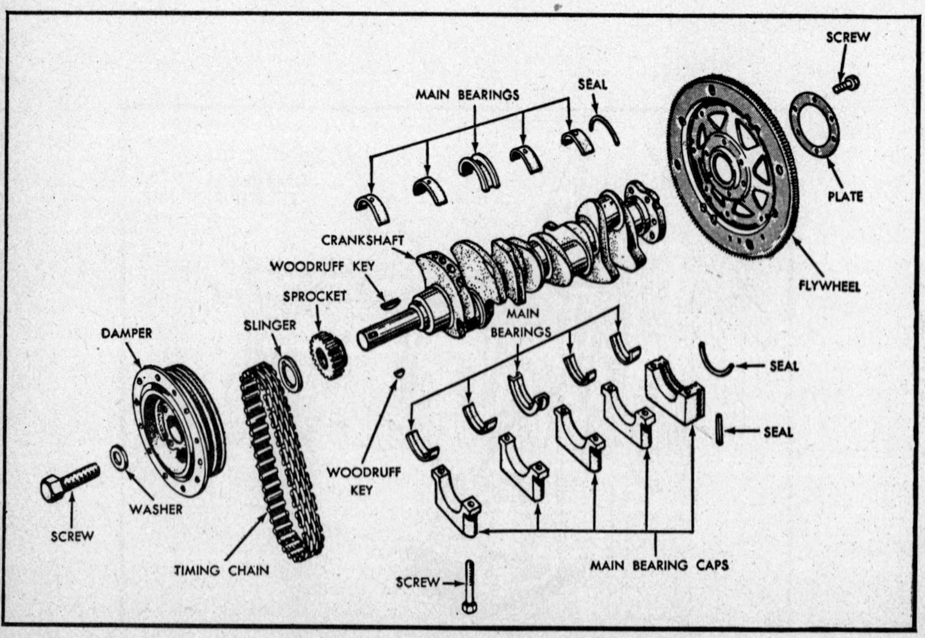

Fig. 10 Crankshaft and related parts

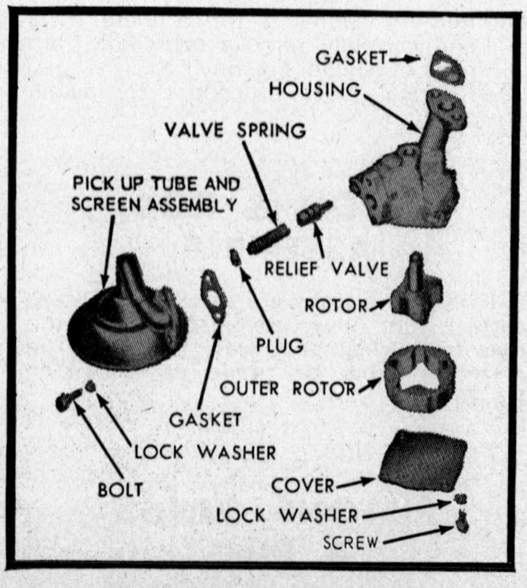

Fig. 11 Oil pump assembly

the outer race and rotor for damage or excessive wear.

2. Check the mating surface of the pump cover for wear. If this surface is worn, scored or grooved, replace the cover.
3. Measure the clearance between the outer race and housing. This clearance should be .006-.013″ for 1969-72 and .001-.013″ for 1973-75.
4. With the rotor assembly installed in the housing, place a straight edge over the rotor assembly and housing. Measure the clearance between the straight edge and the rotor and outer race. Recommended limits are .0016-.004″.
5. Check the drive shaft-to-housing bearing clearance by measuring the O.D. of the shaft and the I.D. of the housing bearing. The recommended clearance limits are .0015-.0029″.
6. Inspect the relief valve spring for a collapsed or worn condition.
7. Check the relief valve piston for scores and free operation in the bore. The specified clearance is .0015-.0029.

BELT TENSION DATA

	New	Used
1969-76 All	140	110

WATER PUMP, REPLACE

V8-460

1. Drain cooling system.
2. Remove shroud and fan.
3. Remove drive belts and water pump pulley.
4. Remove bolts attaching compressor bracket to water pump if so equipped.
5. Loosen alternator bracket and remove bolts attaching bracket to water pump.
6. Disconnect lower radiator hose, heater return tube and bypass hose at water pump.
7. Remove remaining water pump bolts and remove water pump.

FUEL PUMP PRESSURE

Year	Engine	Pressure, Lbs.
1969	V8-460	4½-6½
1970-73	V8-460	5-7
1974-75	V8-460[1]	5½-6½
1974-75	V8-460[2]	6.2-7.2
1976	V8-460[3]	6.2-7.2
1976	V8-460[4]	7.2-8.2

[1] Exc. Calif.
[2] Calif.
[3] Continental
[4] Mark IV

FUEL PUMP, REPLACE

Removal

1. Remove hose clamps and fuel line hoses from pump inlet and vapor discharge connections. Disconnect fuel filter from fuel line.
2. Loosen but do not remove pump attaching capscrews. Crank engine until fuel pump eccentric on camshaft is in a position which applies the least tension on fuel pump rocker arm. Then remove capscrews and pump.
3. If replacement of fuel pump push rod is required, remove access cover from cylinder front cover and remove push rod.

Installation

1. Remove old gasket material from pump mounting pad and pump flange.
2. Install adapter and vapor discharge valve in new pump. Install heat shield and tighten retaining nuts. If pump push rod was removed, install it on eccentric sleeve and install access cover to cylinder front cover.
3. Apply sealer to both sides of new gasket. Position gasket on pump flange and hold pump in position on cylinder front cover. Make sure rocker arm is riding on push rod.
4. Press pump tight against cylinder front cover. Install and tighten retaining screws.
5. Connect fuel lines. Then operate engine and check for leaks.

Transmission, Rear Axle, Propeller Shaft & Brakes

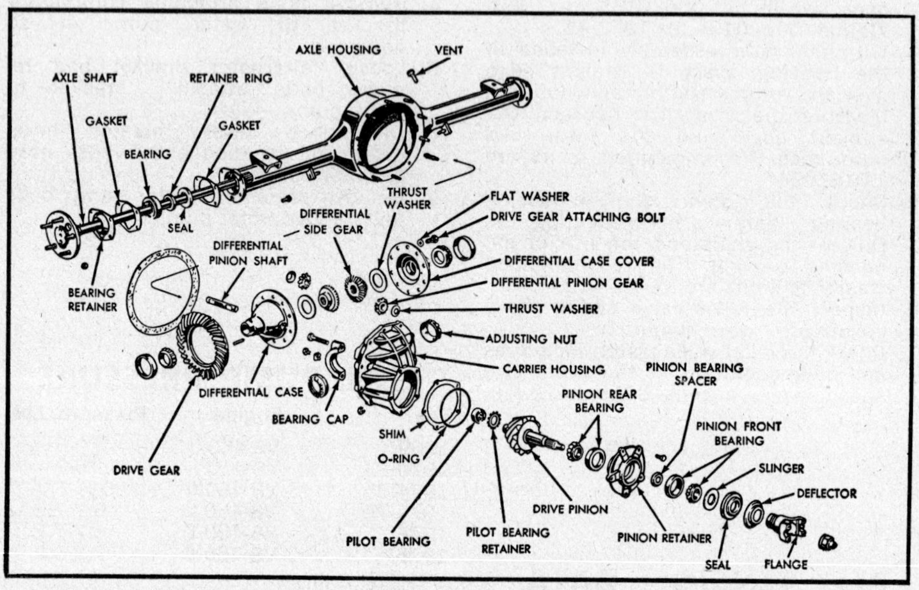

Fig. 1 Rear axle assembly. 1969-76

1976 AUTO. TRANS. LINKAGE ADJUST

The control linkage adjustment procedures are the same as described for the 1975 models in the rear of this manual.

REAR AXLES

Fig. 1 illustrates the rear axle assemblies used on these cars. When necessary to overhaul either of these units, refer to the *Rear Axle Specifications* table in this chapter.

1969-76 Removable Housing Type

In these axles, Fig. 1, the drive pinion is straddle-mounted by two opposed tapered roller bearings which support the pinion shaft in front of the drive pinion gear, and straight roller bearing that supports the pinion shaft at the rear of the pinion gear. The drive pinion is assembled in a pinion retainer that is bolted to the differential carrier. The tapered roller bearings are preloaded by a collapsible spacer between the bearings. The pinion is positioned by a shim or shims located between the drive pinion retainer and the differential carrier.

The differential is supported in the carrier by two tapered roller side bearings. These bearings are preloaded by two threaded ring nuts or sleeves between the bearings and pedestals. The differential assembly is positioned for proper ring gear and pinion backlash by varying the adjustment of these ring nuts. The differential case houses two side gears in mesh with two pinions mounted on a pinion shaft which is held in place by a pin. The side gears and pinions are backed by thrust washers. With high performance engines, an optional rear axle having a four-pinion differential is also used.

The axle shafts are of unequal length, the left shaft being shorter than the right. The axle shafts are mounted in sealed ball bearings that are pressed on the shafts.

Fig. 2 Removing nuts from rear bearing retainer

Service Bulletin

All Ford Built Rear Axles: Recent manufacturing changes have eliminated the need for marking rear axle drive pinions for individual variations from nominal shim thicknesses. In the past, these pinion markings, with the aid of a shim selection table, were used as a guide to select correct shim thicknesses when a gear set or carrier assembly replacement was performed.

With the elimination of pinion markings, use of the shim selection table is no longer possible and the methods outlined below must be used.

1. Measure the thickness of the original pinion depth shim removed from the axle. Use the same thickness upon installation of the replacement carrier or drive pinion. If any further shim change is necessary, it will be indicated in the tooth pattern check.
2. If the original shim is lost, substitute a nominal shim for the original and use the tooth pattern check to determine if further shim changes are required.

DIFFERENTIAL CARRIER, REPLACE

1969-76 Rear Axles

In servicing the rear axles shown in Fig. 1, it is not necessary to remove the rear axle assembly for any normal repairs. The axle shafts and carrier assembly can easily be removed from the ve-

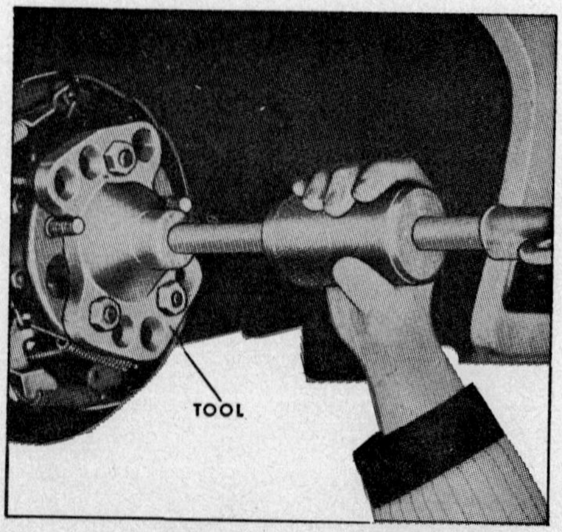

Fig. 3 Removing axle shaft with slide hammer-type puller

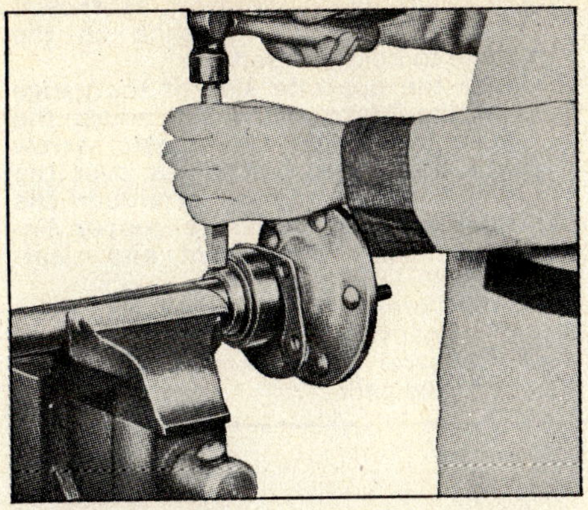

Fig. 4 Splitting bearing inner retainer for bearing removal

hicle, leaving the axle housing in place.

1. Place a drain pan under the carrier and housing to catch the old grease when the carrier is separated from the housing.
2. Use a wire brush to clean dirt from the area around the carrier and housing mating surfaces. Then wipe the area clean with a cloth dampened in solvent.
3. Remove axle shafts and drive shaft as explained below.
4. Unfasten carrier from housing and lift out carrier.
5. Reverse removal procedure to install, using a new gasket between the carrier and housing.

AXLE SHAFTS

1. Remove wheel assembly.
2. Remove brake drum from flange.
3. Working through hole provided in axle shaft flange, Fig. 2, remove nuts that secure wheel bearing retainer.
4. Pull axle shaft out of housing with a slide hammer-type puller, Fig. 3. *Brake carrier plate must not be dislodged. Install one nut to hold the plate in place after axle shaft is removed.*
5. If axle shaft bearing is to be replaced, loosen inner retainer by nicking it deeply with a chisel in several places, Fig. 4. The bearing will then slide off easily.
6. Press bearing from axle shaft.
7. Inspect machined surface of axle shaft and housing for rough spots that would affect the sealing action of the oil seal. Carefully remove any burrs or rough spots.
8. Press new bearing on shaft until it seats firmly against shoulder on shaft.
9. Press inner bearing retainer on shaft until it seats firmly against bearing.
10. If oil seal is to be replaced, use a hook-type tool to pull it out of the housing, Fig. 5. Wipe a small amount of oil resistant sealer on outer edge of seal before it is installed, Fig. 6.

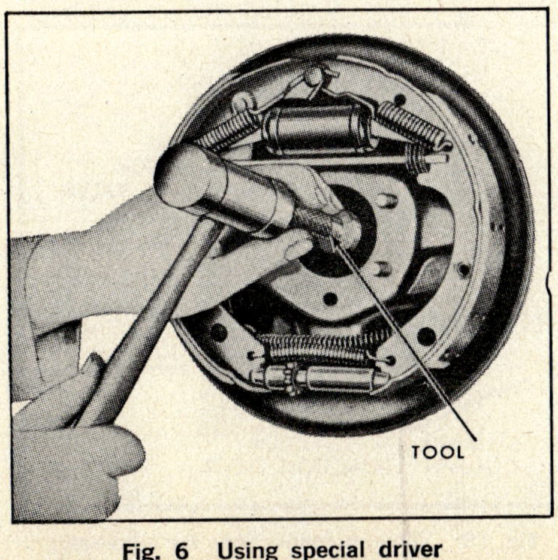

Fig. 6 Using special driver to install oil seal

Installation

1. Place a new gasket on each side of brake carrier plate and slide axle shaft into housing. Start splines into differential side gear and push the shaft in until bearing bottoms in housing.
2. Install retainer and tighten nuts to 30-40 ft. lbs.
3. Install brake drum and wheel.

PROPELLER SHAFT

To maintain proper drive line balance, mark the drive shaft, universal joints, slip yoke and companion flange before removing the shaft assembly so it can be

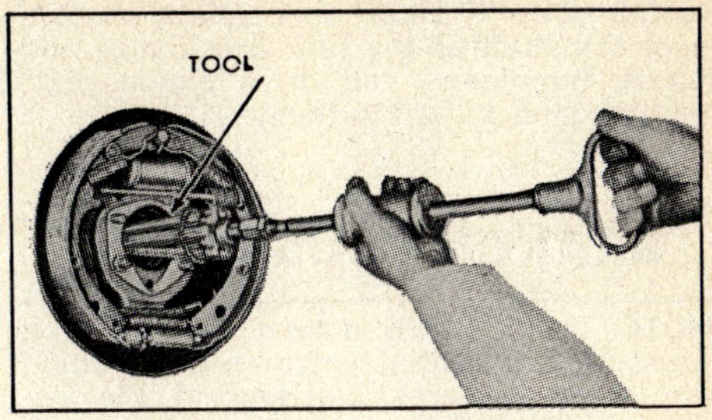

Fig. 5 Using hook-type tool to remove oil seal

reinstalled in its original position.

1. Remove cap-screws attaching slip yoke to front U-joint.
2. Push slip yoke forward on transmission output shaft and lower front of drive shaft.
3. Remove nuts, lockwashers and U-bolts attaching rear U-joint to differential drive pinion flange.
4. Remove shaft assembly.
5. Reverse removal procedure to install the assembly, and torque capscrews and U-bolts to 15-18 ft. lbs.

PROPELLER SHAFT BALANCE

If detailed parts of the drive shaft assembly have been replaced and shaft vibration is encountered after installation, disconnect the shaft at the slip yoke. Rotate the slip yoke and transmission output shaft 180°; then reconnect the shaft

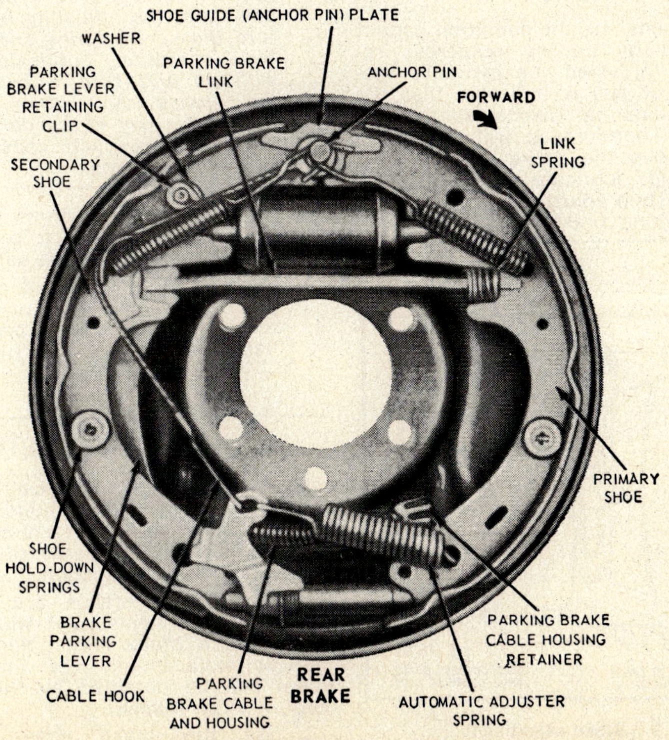

Fig. 7 Rear only on 1969-76

to the yoke. If vibration persists, disconnect the shaft at the rear axle flange and rotate the flange and drive pinion 180° and reconnect shaft to flange.

BRAKE ADJUSTMENTS

NOTE: For 1970-76 a new self-centering pressure differential valve is used which no longer requires bleeding at the opposite end of the car to cause the brake warning light to go out. To center the new valve, after any brake repair or bleeding, it is only necessary to turn the ignition switch on and depress the brake pedal. This action will center the piston and the light will go out.

These brakes, Fig. 7, have self-adjusting shoe mechanisms that assure correct lining-to-drum clearances at all times. The automatic adjusters operate only when the brakes are applied when the car is moving rearward or when the car comes to an uphill stop.

Although the brakes are self-adjusting, an initial adjustment is necessary when the brake shoes have been relined or replaced, or when the length of the star wheel adjuster has been changed during some other service operation.

Frequent usage of an automatic transmission forward range to halt reverse vehicle motion may prevent the automatic adjusters from functioning, thereby inducing low pedal heights. Should low pedal heights be encountered, it is recommended that numerous forward and reverse stops be made until satisfactory pedal height is obtained.

NOTE

If a low pedal height condition cannot be corrected by making numerous reverse stops (provided the hydraulic system is free of air) it indicates that the automatic adjusting mechanism is not functioning. Therefore, it will be necessary to remove the brake drum, clean, free up and lubricate the adjusting mechanism. Then adjust the brakes, being sure the parking brake is fully released.

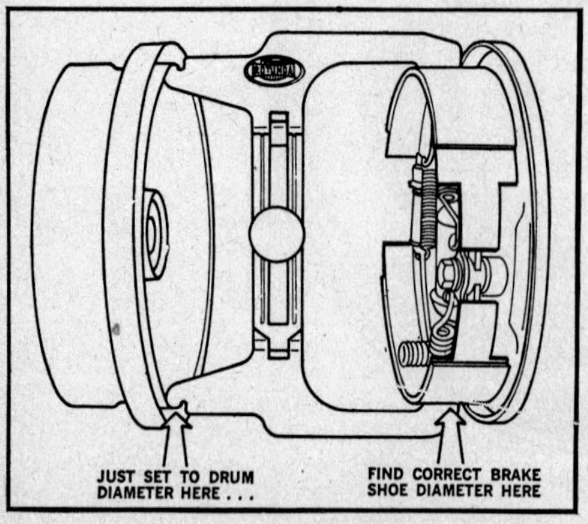

Fig. 9 Brake adjustment with gauge

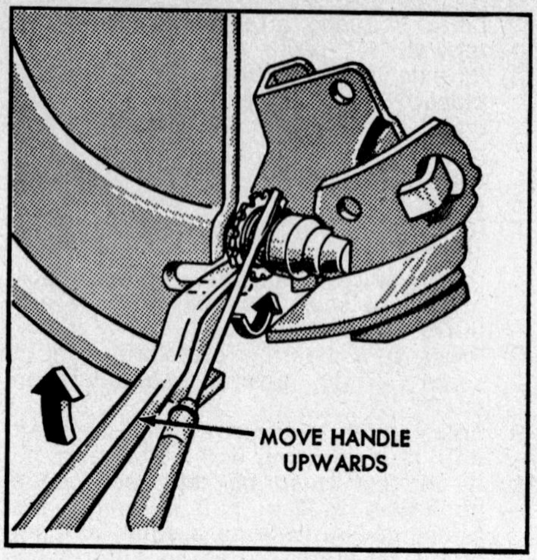

Fig. 8 Backing off brake adjustment by disengaging adjuster lever with screwdriver

Adjustment

1. Remove adjusting hole cover from brake backing plate and, from the backing plate side, turn adjusting screw upward with a screwdriver or other suitable tool to expand the shoes until a slight drag is felt when the drum is rotated.
2. Remove the drum.
3. While holding the adjusting lever out of engagement with the adjusting screw, Fig. 8, back off the adjusting screw ¾ turn with the fingers.

NOTE: *If finger movement will not turn the screw, free it up. If this is not done, the adjusting lever will not turn the screw during subsequent vehicle operation. Lubricate the screw with oil and coat with wheel bearing grease. Any other adjustment procedure may cause damage to the adjusting screw with consequent self-adjuster problems.*

4. Install wheel and drum, and adjusting hole cover. Adjust brakes on remaining wheels in the same manner.
5. If pedal height is not satisfactory, drive the vehicle and make sufficient reverse stops until proper pedal height is obtained.

SERVICE BULLETIN

When servicing a vehicle requiring a brake adjustment, the metal knock-out plugs should NOT be removed. Rather the drums should be removed and brakes inspected for a malfunction.

Although the brakes are self-adjusting, an initial adjustment will be necessary after a brake repair, such as relining or replacement. The initial adjustment can be obtained by the new procedure which follows:

1. Use the brake shoe adjustment gauge shown in Fig. 9 to obtain the drum inside diameter as shown. Tighten the adjusting knob on the gauge to hold this setting.
2. Place the opposite side of the gauge over the brake shoes and adjust the shoes by turning the adjuster screw until the gauge just slides over the linings. Rotate the gauge around the lining surface to assure proper lining diameter adjustment and clearance.
3. Install brake drum and wheel. Final adjustment is accomplished by making several firm reverse stops, using the brake pedal.

PARKING BRAKES, ADJUST

1975-76 Rear Disc Brakes

1. Fully release parking brake, then place transmission in neutral and support vehicle at rear axle.
2. Tighten adjuster nut until levers on calipers just begin to move, then loosen adjuster nut until levers just return to stop position.
3. Apply and release parking brake. Check levers on caliper to determine if they are fully returned by attempting to pull lever rearward. If lever moves, the adjustment is too tight and must be readjusted.

1970-76 Rear Drum Brakes

1. Make sure the parking brake is fully released.
2. Place transmission in neutral and raise the vehicle.
3. Tighten the adjusting nut against the cable equalizer to cause rear wheel brake drag. Then loosen the adjusting nut until the rear brakes are fully released. There should be no brake drag.
4. Lower the vehicle and check operation.

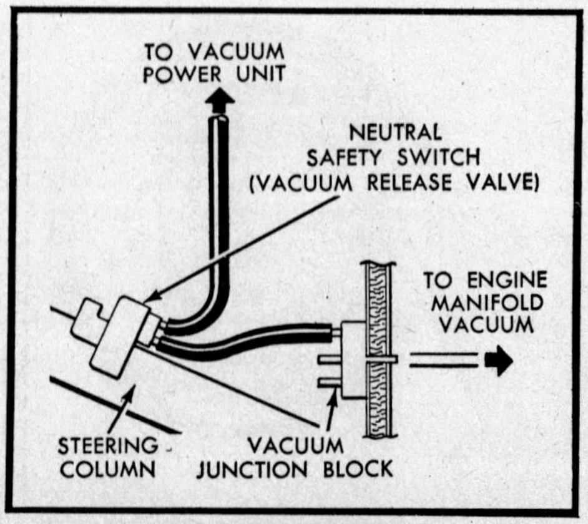

Fig. 10 Vacuum connections for automatic parking brake release. 1969-76

1969

1. Fully release parking brake.
2. Loosen adjusting nut on equalizer rod, then turn lock nut in front of equalizer several turns forward.
3. Depress parking brake slowly until initial locking position is obtained.
4. Turn adjusting nut forward against initial locking position is obtained. equalizer until about 100 lbs. of force at the outside diameter of the tire is required to turn the rear wheels.
5. Tighten lock nut against equalizer.
6. Release parking brake and check to make sure that there is no drag when rear wheels are turned.

1969-76 Vacuum Release Unit

The vacuum power unit, Fig. 10, will release the parking brake automatically when the transmission selector lever is moved into any driving position with the engine running. The brakes will not release automatically, however, when the selector lever is in neutral or park position with the engine running, or in any other position with the engine off.

The lower end of the release handle extends out for alternate manual release in the event of vacuum power failure or for optional manual release at any time.

POWER BRAKE UNIT, REPLACE

Vacuum Booster

1969-76

1. Disconnect battery ground cable.
2. Disconnect stop light switch wires and remove switch retaining pin. Slide switch off pin so switch outer hole clears pin, then remove switch. Slide booster push rod, nylon washers and bushing off brake pedal pin.
3. Remove master cylinder and position aside without damaging hydraulic lines.
4. Disconnect vacuum hose from booster and remove booster attaching nuts and booster.
5. Reverse procedure to install.

1975-76 Hydro-Boost

1. Disconnect stoplight switch wires at connector and remove hairpin retainer, then slide stoplight switch off brake pedal pin far enough for switch outer hole to clear pin and remove pin from switch.
2. Slide hydro-boost push rod and nylon washers and bushing off brake pedal pin.
3. Remove master cylinder and position to one side without disturbing hydraulic lines.

NOTE: It is not necessary to disconnect brake lines, but care should be taken not to deform lines.

4. Disconnect pressure, steering gear and return lines from booster, then plug lines and ports in hydro-boost to prevent entry of dirt.
5. Remove hydro-boost retaining nuts, and remove assembly sliding push rod link from engine side of dash panel.

Rear Suspension

SHOCK ABSORBER, REPLACE

1. With the rear axle supported properly disconnect shock absorber at upper mounting and compress it to clear hole in spring seat.
2. Disconnect shock absorber from stud on axle bracket.
3. Reverse procedure to install.

COIL SPRINGS, REPLACE

1. Raise rear of vehicle and support at frame. Support rear axle with a suitable jack.
2. Disconnect shock absorbers at lower mountings.
3. On 1973-76 Lincoln Continental models, disconnect brake hose at rear crossmember and remove hose to bracket clip.
4. Lower axle to remove springs.
5. Reverse procedure to install. On all models except 1972-76 Mark IV, install an insulator between upper and lower seats and the spring, Fig. 1. On 1972-76 Mark IV, an insulator is installed only between upper seat and the spring, Fig. 2.

CONTROL ARMS, REPLACE

NOTE: The 1972-76 Mark IV upper and lower control arms are replaced in pairs.

1. Raise rear of vehicle and support at frame. Support axle with a suitable jack.
2. On all models except 1972-76 Mark IV, disconnect track bar from frame mounting bracket.
3. Lower axle and install a second jack under differential pinion nose.
4. Disconnect control arm from axle bracket. On upper arms, disconnect arm from crossmember and on lower arms, disconnect arm from frame attachment bracket.
5. Reverse procedure to install.

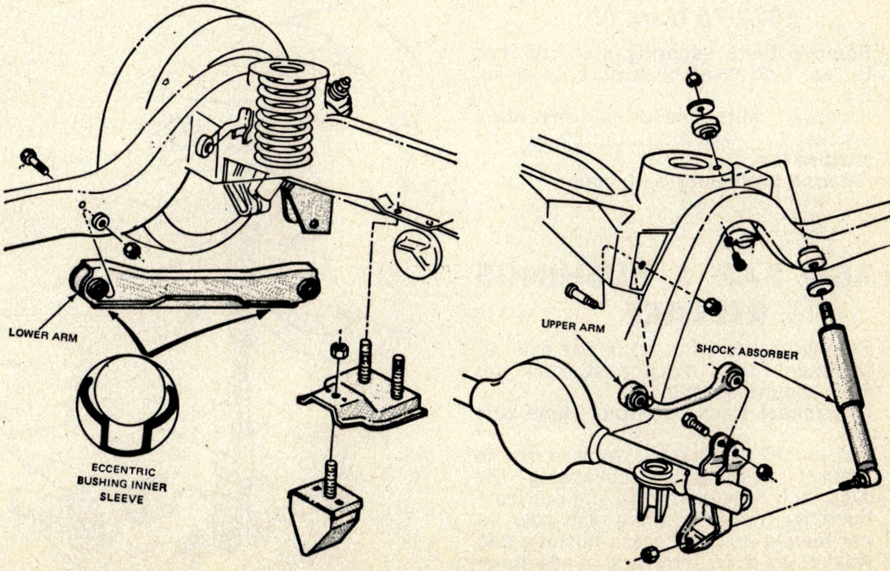

Fig. 2 Rear suspension. 1972-76 Mark IV

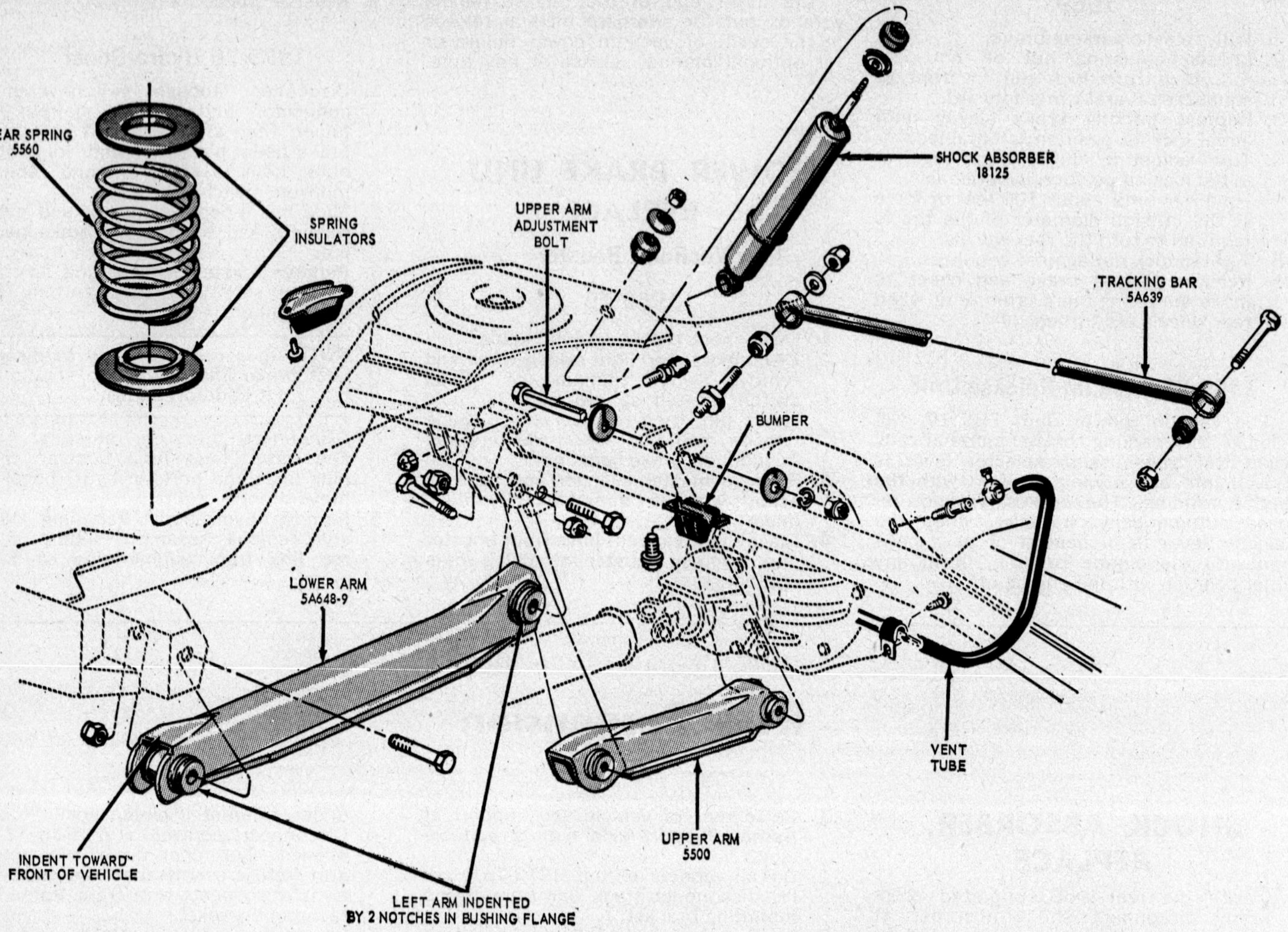

Fig. 1 Rear suspension, 1969-71 Mark III & 1969-76 Continental

STABILIZER BAR, REPLACE

1972-76 Mark IV

1. Remove bolts securing stabilizer bar to rear link assemblies on both sides, Fig. 3.
2. Remove nuts securing mounting bracket to lower mounting clamp and remove bar.
3. Reverse procedure to install.

TRACK BAR & BUSHINGS REPLACE

1. Remove cover from track bar axle attachment, then disconnect track bar from mounting stud.
2. Disconnect track bar from frame side rail.
3. On 1970-71 Mark III, replace frame bracket bushing by collapsing the .030 inch ferrule and driving out from bushing. Install bushing and new inner ferrule and compress bushing and washer to 2.16 inch and place inner ferrule over edge of inner washer.
4. Reverse procedure to install.

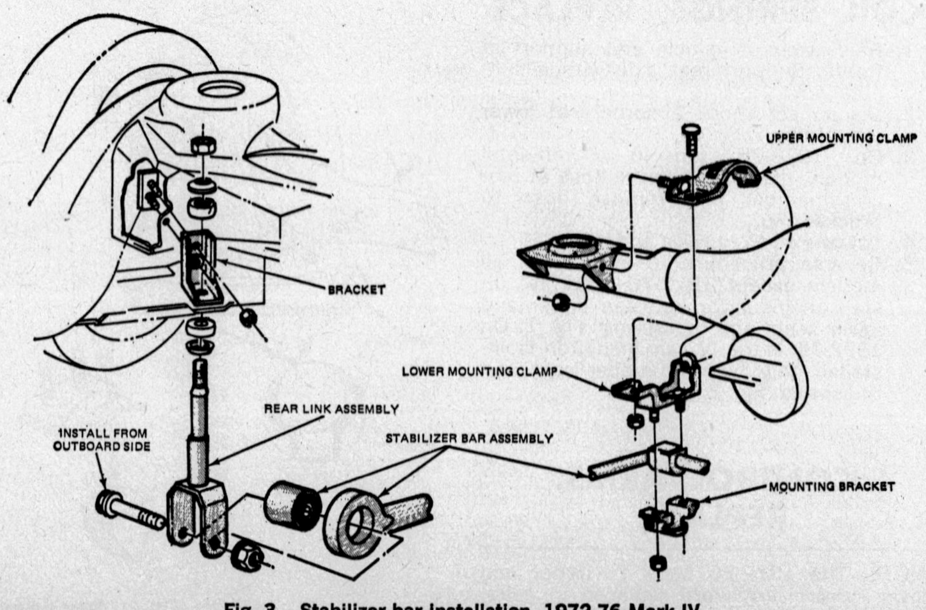

Fig. 3 Stabilizer bar installation. 1972-76 Mark IV

Front End and Steering Section

FRONT SUSPENSION

1969 Mark III & 1970-76 All

Referring to Fig. 1, each wheel rotates on a spindle. The upper and lower ends of the spindle are attached to upper and lower ball joints that are mounted to an upper and lower control arm. The upper control arm pivots on a shaft assembly that is bolted to the frame. The lower control arm pivots on a bolt in the front crossmember. The struts, which are connected between the lower control arms and frame crossmember, prevent the control arms from moving forward or backward.

1969 Continental

The front wheel suspension is a ball joint type utilizing coil springs and double acting shock absorbers. Fore and aft movement of each front wheel is controlled by a non-adjustable type stabilizing strut connected to the suspension lower control arm and to a point forward on the front crossmember, Fig. 2. A single rubber-cored bushing is used at the inner end of the lower control arm. Caster and camber are adjusted without the use of shims but rather by movement of the serrated upper control arm shaft.

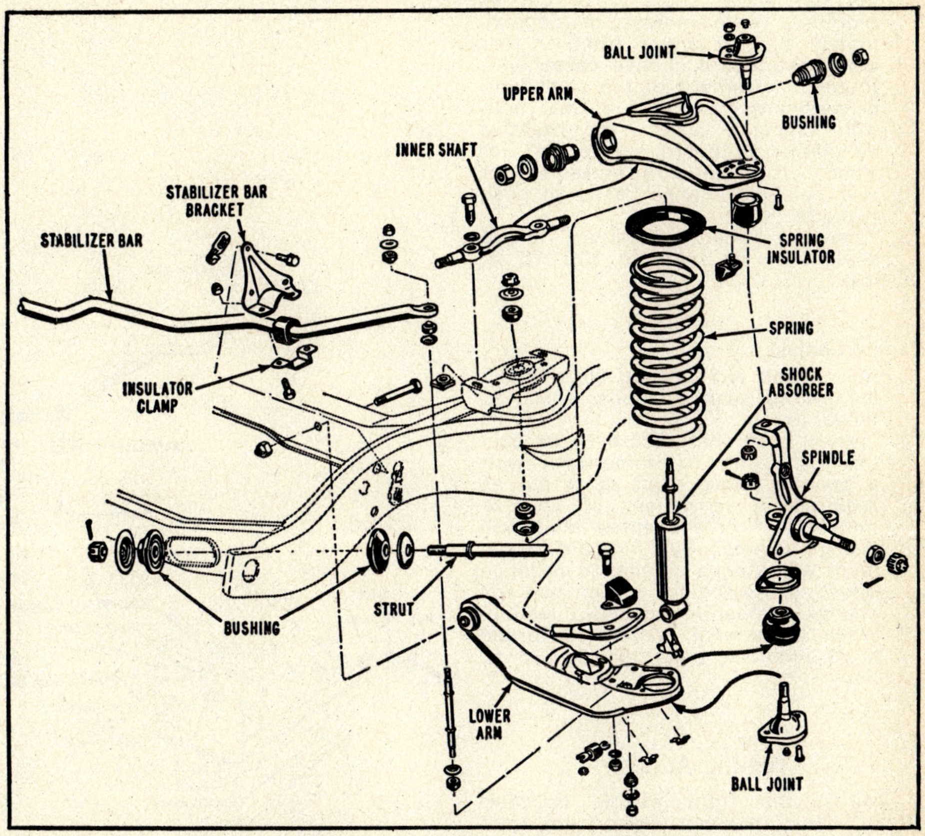

Fig. 1 Front suspension. 1969 Mark III & 1970-76 All

LUBRICATION

1969-76 STEERING LINKAGE: The steering linkage should be lubricated at 36,000 mile intervals. Normal breathing of socket joints permits moisture condensation within the joint. Moisture inside the joint assembly will cause no appreciable damage and the joint will function normally. However, if the moisture is concentrated in the bearing grease grooves and is frozen at the time of attempted lubrication, grease cannot flow and pressure greasing may damage the joint assembly.

Do not attempt to lubricate the steering linkage if it has set in temperatures lower than 20 deg. above zero F. The vehicle should be allowed to warm up in a heated garage for 30 minutes or until the joints accept lubrication.

IMPORTANT: A torch must not be used to heat joints because this quantity of heat will melt the nylon bearing within the joint.

Ball Joint Lubrication

Ball joints should be lubricated with special grease formulated just for this purpose every 36,000 miles on 1969.

Lubrication points are fitted with screw plugs. The plugs should be removed, grease fittings installed and, after applying the grease, remove the fittings and reinstall the plugs.

WHEEL ALIGNMENT

SERVICE BULLETIN

WHEEL BALANCING DIFFERS: On cars with disc brakes, dynamic balancing of the wheel-and-tire assembly on the car should not be attempted without first pulling back the shoe and lining assemblies from the rotor. If this is not done, brake drag may burn out the motor on the wheel spinner.

The drag can be eliminated by removing the wheel, taking out the two bolts holding the caliper splash shield, and detaching the shield. Then push the pistons into their cylinder bores by applying steady pressure on the shoes on each side of the rotor for at least a minute. If necessary, use waterpump pliers to apply the pressure.

After the pistons have been retracted, reinstall the splash shield and wheel. The wheel-and-tire assembly can then be dynamically balanced in the usual way. After the balancing job has been completed, be sure to pump the brake pedal several times until the shoes are seated and a firm brake pedal is obtained.

1969 Mark III & 1970-76 All

Caster and camber can be adjusted by loosening the bolts that attach the upper suspension arm to the shaft at the frame side rail, and moving the arm assembly in or out in the elongated bolt holes, Fig. 3. Since any movement of the arm affects both caster and camber, both factors should be balanced against one another when making the adjustment.

Caster, Adjust

1. To adjust caster, install the adjusting tool as shown in Fig. 3.
2. Loosen both upper arm inner shaft retaining bolts and move either front or rear of the shaft in or out as necessary to increase or decrease caster angle. Then tighten bolt to retain adjustment.

Camber, Adjust

1. Loosen both upper arm inner retaining bolts and move both front and rear ends of shaft inward or outward as necessary to increase or decrease camber angle.
2. Tighten bolts and recheck caster and readjust if necessary.

1969 Continental

Caster, Adjust

1. Raise hood and unsnap clips retaining top of rubber bushing shield to fender apron.
2. Loosen bolts that secure upper control arm shaft to frame and, with a pry bar, move shaft in or out as required. A movement of approximately 3/32″ at either front or rear bolt location will change caster ½°. Inboard movement of the front bolt, or outboard movement of rear bolt, will change caster in negative direction. Outboard movement of front bolt or inboard movement of rear bolt will change caster in positive direction.
3. When adjustment is correct, torque shaft retaining bolts to 100-125 ft. lbs.

Camber, Adjust

1. Raise hood and unsnap clips retaining top of bushing rubber shield to fender apron.
2. Loosen bolts that secure upper control arm shaft to frame and, with a pry bar, move shaft in or out as required. A movement of approximately 3/64″ of the entire shaft will change camber ¼°. Inboard movement will change camber in negative direction. Outboard movement will change camber in positive direction.
3. When adjustment is correct, torque shaft retaining bolts to 100-125 ft. lbs.

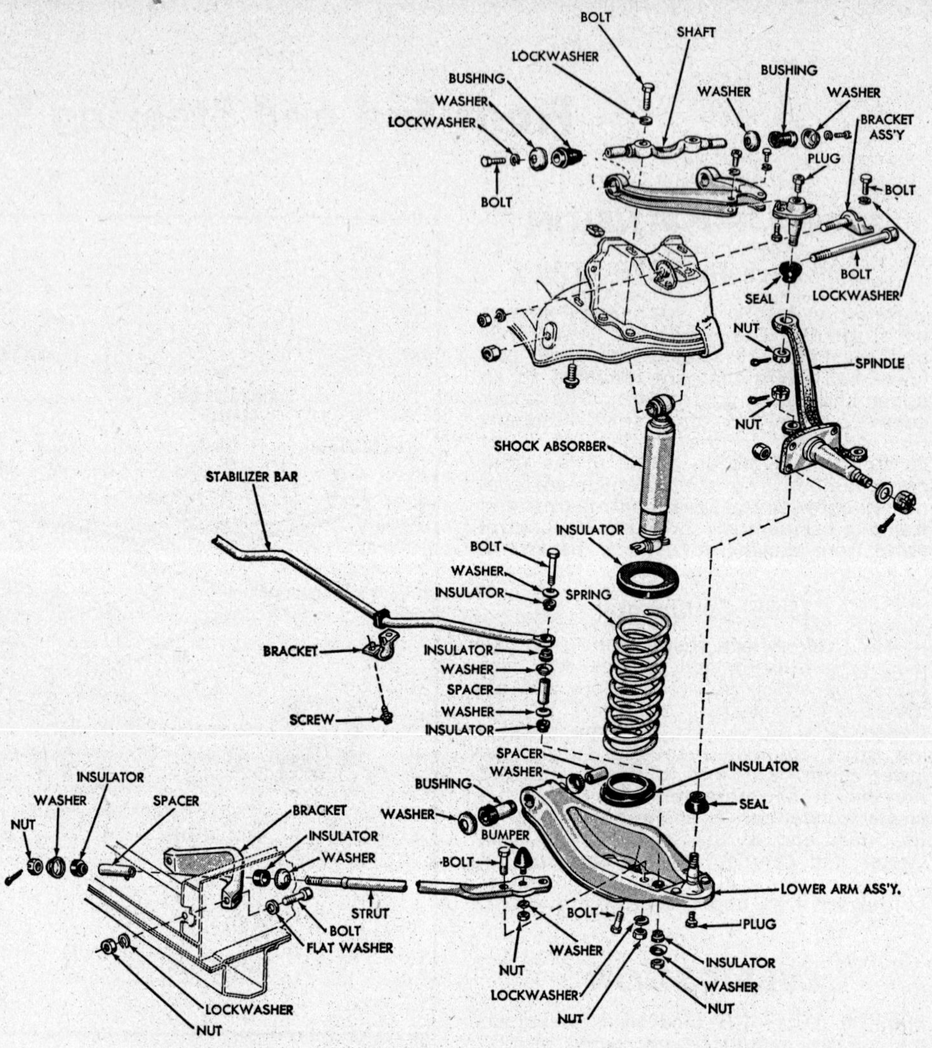

Fig. 2 Front suspension. 1969 Continental

Toe-In, Adjust

Position the front wheels in their straight-ahead position. Then turn both tie rod adjusting sleeves an equal amount until the desired toe-in setting is obtained.

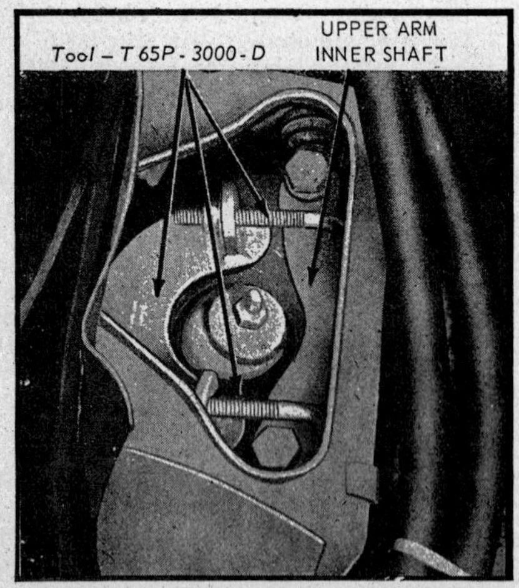

Fig. 3 Caster and camber adjusting tool. 1969 Mark III & 1970-76 All

WHEEL BEARINGS, ADJUST

1969-76

1. With wheel rotating, tighten adjusting nut to 17-25 ft. lbs.
2. Back off adjusting nut ½ turn and retighten nut to 10-15 inch lbs.
3. Place nut lock on nut so that castellations on lock are aligned with cotter pin hole in spindle and install cotter pin.
4. Check front wheel rotation, if it rotates noisily or rough, clean, inspect or replace wheel bearings as necessary.

WHEEL BEARINGS, REPLACE

(Disc Brakes)

1. Raise car and remove front wheels.

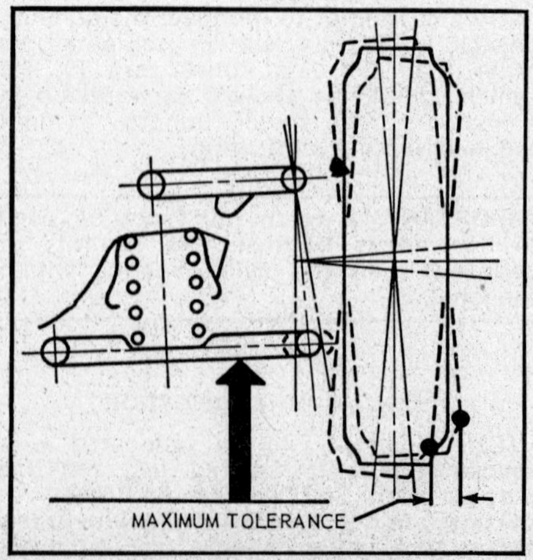

Fig. 4 Checking ball joints for wear

2. Remove caliper mounting bolts.

NOTE: It is not necessary to disconnect the brake line for this operation.

3. Slide caliper off of the disc, inserting a spacer between the shoes to hold them in their bores after the caliper is removed. Position caliper assembly out of the way.

NOTE: Do not allow caliper to hang by brake hose.

4. Remove hub and disc. Grease retainer and inner bearing can now be removed.

CHECKING BALL JOINTS FOR WEAR

Upper Ball Joint

1. Raise car on floor jacks placed beneath lower control arms.
2. Grasp lower edge of tire and move wheel in and out.
3. As wheel is being moved in and out, observe upper end of spindle and upper arm.
4. Any movement between upper end of spindle and upper arm indicates ball joint wear and loss of preload. If such movement is observed, replace upper ball joint.

NOTE: During the foregoing check, the lower ball joint will be unloaded and may move. Disregard all such movement of the lower joint. Also, do not mistake loose wheel bearings for a worn ball joint.

Lower Ball Joint

1. Raise car on jacks placed under lower control arms as shown in Fig. 4.
2. With a dial indicator attached to the lower arm, position indicator so that the plunger rests against inner side of wheel rim adjacent to lower ball joint.
3. Grasp tire at top and bottom and slowly move tire in and out. Note reading on dial, which is the radial play. If the reading exceeds ¼", replace lower ball joint.

BALL JOINTS, REPLACE

NOTE: Ford Motor Company recommends that new ball joints should not be installed on used control arms and that the control arm be replaced if ball joint replacement is required. However, aftermarket ball joint repair kits which do not require control arm replacement, are available and can be installed using the following procedure.

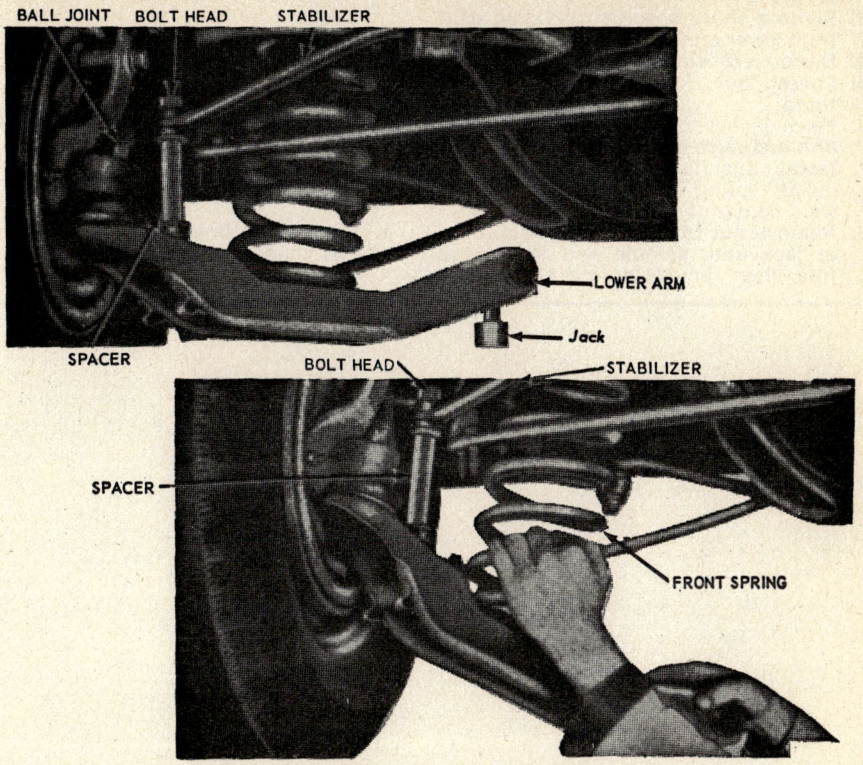

Fig. 5 Front spring replacement. 1969 Mark III & 1970-76 All

The upper ball joints on Continental models are bolted to the upper control arm, while on the Mark III, the ball joints are riveted to the arm. The lower ball joints are riveted to the lower control arm on all models.

When replacing a riveted joint, remove the rivets and retain the new joint in its control arm with the bolts, nuts and washers furnished with the ball joint kit.

Use a suitable pressing tool to force the ball joint from the spindle.

SHOCK ABSORBER, REPLACE

1969 Continental

1. Remove stud nut at upper eye of shock absorber. Remove upper eye stud bracket to crossmember bolt and remove stud bracket.
2. Unfasten shock absorber from lower control arm. Then lower and remove shock absorber.
3. Reverse above procedure to install.

1969 Mark III & 1970-76 All

1. Remove nut, washer and bushing from upper end of shock absorber.
2. Raise vehicle and support on stands.
3. Remove screws retaining shock absorber to lower control arm and remove shock absorber.
4. Reverse procedure to install.

COIL SPRING, REPLACE

1969 Mark III & 1970-76 All

1. Raise vehicle and support front end of frame with jack stands.
2. Disconnect shock absorber from lower arm and place a jack under the lower arm to support it, Fig. 5.
3. Remove strut and rebound bumper bolts and disconnect lower end of sway bar stud from lower arm.
4. Remove the nut and bolt retaining inner end of the lower arm to crossmember.
5. Carefully lower jack relieving spring pressure on the lower arm, then remove spring.

1969 Continental

1. Raise and support car as for replacing lower ball joint.
2. Remove brake assembly.

3. Remove shock absorber and drag strut from lower arm.
4. Disconnect stabilizer from lower arm.
5. Loosen nut from ball joint stud two turns.
6. Place jack under outer end of lower arm and raise arm several inches.
7. Install a suitable spring compressor inside spring with jaws of tool toward center of car.
8. Remove nut from ball joint stud. Lower jack until spindle and spring are free, then remove spring and insulators.
9. Reverse above procedure to install.

POWER STEERING UNIT, REPLACE

1969-76

1. Disconnect lines from steering gear and plug lines and ports to prevent entry of dirt.
2. Remove the two bolts securing flex coupling to steering gear and to column.
3. Raise vehicle and remove sector shaft nut and pitman arm.

NOTE: Do not damage the seals.

4. Support steering gear and remove three attaching bolts. Remove flex coupling clamp bolt and work steering gear free of coupling, then remove steering gear from vehicle.

OLDSMOBILE

(Exc. Starfire)

TORONADO: Service Procedures that apply to the Toronado only will be found starting on page 1-674.

OLD CAR SPECIFICATIONS: For 1946-68 Tune Up and Wheel Alignment Specifications see main index.

INDEX OF SERVICE OPERATIONS

PAGE NO.

ACCESSORIES

BRAKES

CLUTCH

COOLING SYSTEM

ELECTRICAL

ENGINE

ENGINE LUBRICATION

FUEL SYSTEM

PROPELLER SHAFT & U JOINTS (Conventional)

REAR AXLE & SUSPENSION

SPECIFICATIONS

STEERING GEAR

SUSPENSION, FRONT

TORONADO FRONT DRIVE

TRANSMISSIONS

TUNE UP

WINDSHIELD WIPER

VEHICLE IDENTIFICATION PLATE: 1969-1976 on left upper dash.

ENGINE NUMBER LOCATION

1969-71 & 1973-76 6-250: Right side of engine block directly to rear of distributor.
1969-74 V8s: Stamped on oil fill tube.

1975-76 Omega V8-350; on left bank cylinder head. V8-260, V8-455 & V8-350 Exc. Omega, stamped on oil filter tube.

ENGINE IDENTIFICATION CODE

YEAR	ENGINE	ENGINE PREFIX
1969	6-250	VA-B-E-F
	V8-350 2 Bar. Carb.	QI-A-B
	V8-350 4 Bar. Carb.	QV-N-P
	V8-350 2 Bar. Carb.	TL-B-D
	V8-350 4 Bar. Carb.	QX
	V8-400 4 Bar. Carb.	QW-R-S
	V8-400 4 Bar. Carb.	QU-T
	V8-455 2 Bar. Carb.	UJ-C-D
	V8-455 4 Bar. Carb.	UN-O
	V8-455 4 Bar. Carb.	UL
	V8-455 4 Bar. Carb.	US-T-V
	V8-455 4 Bar. Carb.	UW
1970	6-250	VB-F
	V8-350 2 Bar. Carb.	QI-A-J
	V8-350 2 Bar. Carb.	TL-D-C
	V8-350 4 Bar. Carb.	QV-N-P
	V8-350 4 Bar. Carb.	QX-D
	V8-455 2 Bar. Carb.	UJ-C-D
	V8-455 2 Bar. Carb.	TY-X
	V8-455 4 Bar. Carb.	TU-W-V
	V8-455 4 Bar. Carb.	TQ-P
	V8-455 4 Bar. Carb.	UN-O
	V8-455 4 Bar. Carb.	TS-T
	V8-455 4 Bar. Carb.	UL
	V8-455 4 Bar. Carb.	US-T-W-V
1971	6-250	VB-F
	V8-350 2 Bar. Carb.	QI-A-J
	V8-350 2 Bar. Carb.	TE-D-C
	V8-350 4 Bar. Carb.	QB-O-N-P
	V8-455 2 Bar. Carb.	UC-D-E
	V8-455 4 Bar. Carb.	TQ-P
	V8-455 4 Bar. Carb.	TU-N
	V8-455 4 Bar. Carb.	TS-B-T-L
	V8-455 4 Bar. Carb.	TW-V-A
	V8-455 4 Bar. Carb.	UN-O-S-T
1972	V8-350 2 Bar. Carb.	QA-B-C
	V8-350 2 Bar. Carb.	QN-O
	V8-350 4 Bar. Carb.	QD-E
	V8-350 4 Bar. Carb.	QJ-K-P-Q
	V8-455	UA-B-S-T
	V8-455	UL-N-O
	V8-455	UD-E-U-V
1973	6-250 Std. Tr.	CCC, CCD
	6-250 Auto Tr.	CCA, CCB
	V8-350 2 Bar. Carb.	QS, T, P
	V8-350 2 Bar. Carb.	QN, O, P
	V8-350 4 Bar. Carb.	QA, B, J, K
	V8-350 4 Bar. Carb.	QU, V, C
	V8-350 4 Bar. Carb.	QD, E, L
	V8-455	UD, B
	V8-455	UA, S, T
	V8-455	UU, V
1974	6-250 Std. Tr.	CCC, CCD
	6-250 Auto. Tr.	CCA, CCB
	V8-350 2 Bar. Carb.	QS, QT
	V8-350 4 Bar. Carb.	QB, QC, QL
	V8-350 4 Bar. Carb.	QO, QU, QW
	V8-350 4 Bar. Carb.	TB, TC, TL, TO
	V8-455 2 Bar. Carb.	UU, UW
	V8-455 4 Bar. Carb.	UA, UB, UC
	V8-455 4 Bar. Carb.	UD, UL, UO
	V8-455 4 Bar. Carb.	UP, UN, UR
	V8-455 4 Bar. Carb.	UV, UX, VP
	V8-455 4 Bar. Carb.	VA, VB, VC
	V8-455 4 Bar. Carb.	VD, VL, VO
1975	6-250 Std. Tr. [1]	CJU
	6-250 Auto. Tr. [1]	CJT
	6-250 Auto. Tr. [2]	CJL
	V8-260 Std. Tr. [1]	QA, QK
	V8-260 Std. Tr. a/c [1]	QD, QN
	V8-260 Std. Tr. [2]	TA, TK
	V8-260 Std. Tr. a/c [2]	TD, TN
	V8-260 Auto. Tr. [1]	QE, QP
	V8-260 Auto. Tr. a/c [1]	QJ, QQ
	V8-260 Auto. Tr. [2]	TE, TP
	V8-260 Auto. Tr. a/c [2]	TJ, TQ
	V8-350 Auto. Tr. [1]	RW, QL
	V8-350 Auto. Tr. a/c [1]	RX, QO, QX
	V8-350 Auto. Tr. [2]	RN, TL
	V8-350 Auto. Tr. a/c [2]	RO, TO, TX
	V8-400 Auto. Tr. [1]	YM, YT
	V8-455 Auto. Tr. [1]	UB, UE, UP
	V8-455 Auto. Tr. a/c [1]	UC, UD, UP
	V8-455 Auto. Tr. [2]	VB, VE, VP
	V8-455 Auto. Tr. a/c [2]	VC, VD, VP
	6-250 Std. [1]	CCD
	6-250 Auto. Tr. [1]	CCF
	6-250 Auto. Tr. [2]	CCC
	V8-260 Std. Tr. [1]	QA, QK
	V8-260 Std. Tr. a/c [1]	QD, QN
	V8-260 Auto. Tr. [1]	QB, QP
	V8-260 Auto. Tr. a/c [1]	QC, QT
	V8-260 Auto. Tr. [2]	TE, TP
	V8-260 Auto. Tr. a/c [2]	TJ, TT
	V8-350 2 Bar. Carb. Auto. Tr. [1]	PA
	V8-350 2 Bar. Carb. Auto. Tr. a/c [1]	PB
	V8-350 4 Bar. Carb. Auto. Tr. [1]	PE, Q2, Q6
	V8-350 4 Bar. Carb. Auto. Tr. a/c [1]	PF, Q3, Q5
	V8-4 Bar. Carb. Auto. Tr. [2]	PM, TL, TY, T6
	V8-350 4 Bar. Carb. Auto. Tr. a/c [2]	PN, TO, TW, TX
	V8-455 Auto. Tr. [1]	UB, UE, U5
	V8-455 Auto. Tr. a/c [1]	UC, UD, U3, U4
	V8-455 Auto. Tr. [2]	VB, VE, V5
	V8-455 Auto. Tr. a/c [2]	VD, V3, V4

[1]—Except California.
[2]—California.

GRILLE IDENTIFICATION

1969 F-85, Cutlass, Vista Cruiser

1969 "4-4-2"

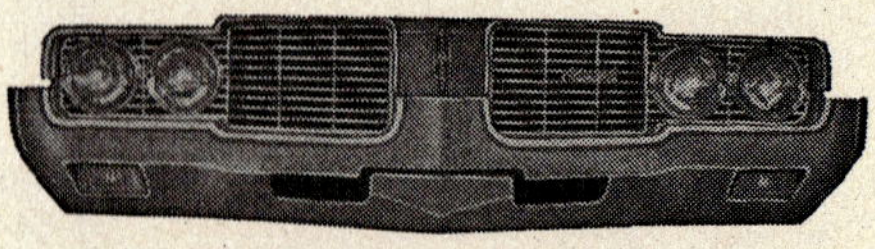
1969 Delta 88, Royale

1969 98

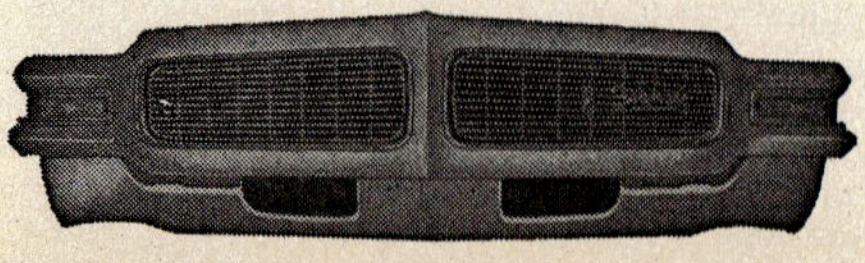
1969 Toronado

1970 F-85, Cutlass, Vista Cruiser

GRILLE IDENTIFICATION—Continued

1970 Cutlass Supreme

1970 "4-4-2"

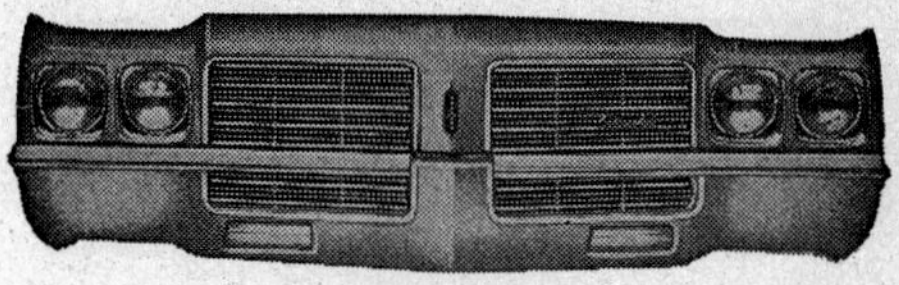

1970 Delta 88

1970 Delta Royale

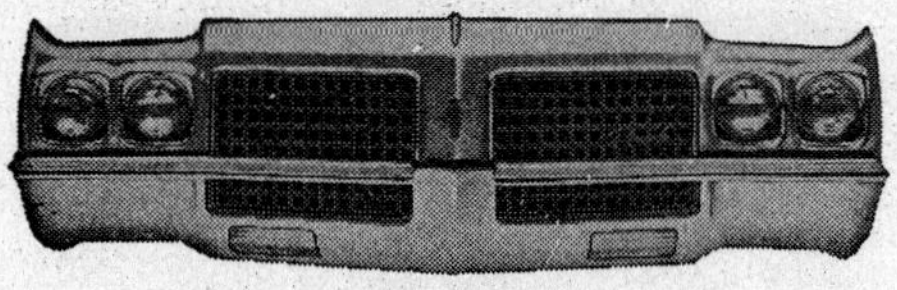

1970 98

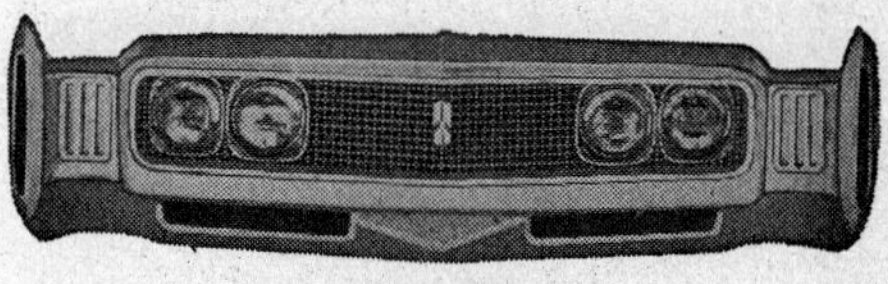

1970 Toronado

1971 F-85, Cutlass

1971 Cutlass Supreme

1971 Cutlass "S"

1971 "4-4-2"

1971 Delta Royale

1971 98

1971 Toronado

1972 Cutlass "S"

1972 Delta 88 & Royale

1972 98

1972 Toronado

1973 Omega

1973 Cutlass

1973 88

1973 98

GRILLE IDENTIFICATION—Continued

1973 Toronado

1974 Omega

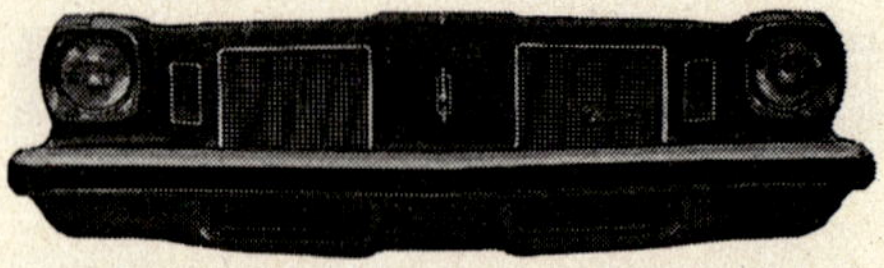
1974 Cutlass, Supreme & Vista Cruiser

1974 88

1974 98

1974 Toronado

1975 Omega & F-85

1975 Cutlass Supreme

1975 Cutlass Salon

1975 Cutlass "S"

1975 Delta 88

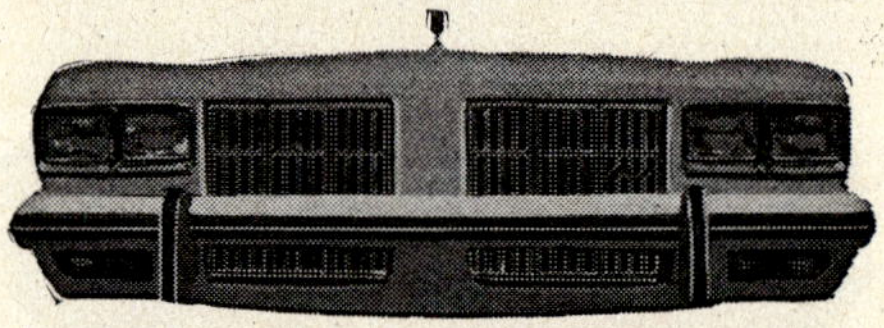
1975 98

1975 Toronado

1976 Omega

1976 Cutlass Supreme

1976 Cutlass "S"

1976 Delta 88

1976 Delta Royale

1976 98

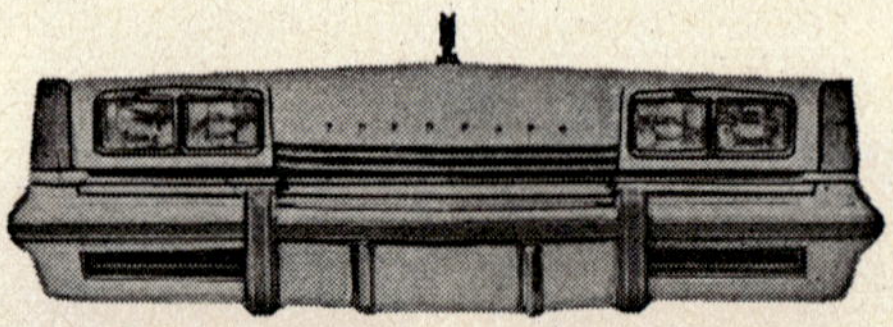
1976 Toronado

GENERAL ENGINE SPECIFICATIONS

Year	Engine	Carburetor	Bore and Stroke	Piston Displacement, Cubic Inches	Compression Ratio	Maximum Brake H.P. @ R.P.M.	Maximum Torque Lbs. Ft. @ R.P.M.	Normal Oil Pressure Pounds
1969	155 Horsepower............②6-250	1 Barrel	3.875 x 3.53	250	8.50	155 @ 4200	240 @ 2000	30-45
	250 Horsepower............V8-350	2 Barrel	4.057 x 3.385	350	9.00	250 @ 4400	355 @ 2600	30-45
	310 Horsepower............V8-350	4 Barrel	4.057 x 3.385	350	10.25	310 @ 4800	390 @ 3200	30-45
	310 Horsepower............V8-455	2 Barrel	4.125 x 4.250	455	9.00	310 @ 4200	490 @ 2400	30-45
	325 Horsepower............V8-350	4 Barrel	4.057 x 3.385	350	10.50	325 @ 5400	360 @ 3600	30-45
	325 Horsepower............V8-400	4 Barrel	3.870 x 4.250	400	10.50	325 @ 4600	440 @ 3000	35-50
	350 Horsepower............V8-400	4 Barrel	3.870 x 4.250	400	10.50	350 @ 4800	440 @ 3200	35-50
	360 Horsepower............V8-400	4 Barrel	3.870 x 4.250	400	10.50	360 @ 5400	440 @ 3600	35-50
	365 Horsepower............V8-455	4 Barrel	4.125 x 4.250	455	10.25	365 @ 4600	510 @ 3000	30-45
	375 Horsepower............V8-455	4 Barrel	4.125 x 4.250	455	10.25	375 @ 4600	510 @ 3000	30-45
	390 Horsepower............V8-455	4 Barrel	4.125 x 4.250	455	10.25	390 @ 5000	500 @ 3200	30-45
	400 Horsepower............V8-455	4 Barrel	4.125 x 4.250	455	10.25	400 @ 4800	500 @ 3200	30-45
1970	155 Horsepower............②6-250	1 Barrel	3.875 x 3.53	250	8.50	155 @ 4200	240 @ 2000	30-45
	250 Horsepower............V8-350	2 Barrel	4.057 x 3.385	350	9.00	250 @ 4400	355 @ 2600	30-45
	310 Horsepower............V8-350	4 Barrel	4.057 x 3.385	350	10.25	310 @ 4800	390 @ 3200	30-45
	310 Horsepower............V8-455	2 Barrel	4.125 x 4.250	455	9.00	310 @ 4200	490 @ 2400	30-45
	320 Horsepower............V8-455	2 Barrel	4.125 x 4.250	455	10.25	320 @ 4200	500 @ 2400	30-45
	325 Horsepower............V8-350	4 Barrel	4.057 x 3.385	350	10.50	325 @ 5400	360 @ 3600	30-45
	365 Horsepower............V8-455	4 Barrel	4.125 x 4.250	455	10.50	365 @ 5000	500 @ 3200	30-45
	370 Horsepower............V8-455	4 Barrel	4.125 x 4.250	455	10.50	370 @ 5200	500 @ 3600	30-45
	375 Horsepower............V8-455	4 Barrel	4.125 x 4.250	455	10.25	375 @ 4600	510 @ 3000	30-45
	390 Horsepower............V8-455	4 Barrel	4.125 x 4.250	455	10.25	390 @ 5000	500 @ 3200	30-45
	400 Horsepower............V8-455	4 Barrel	4.125 x 4.250	455	10.25	400 @ 3200	500 @ 3200	30-45
1971	110 Horsepower①...........②6-250	1 Barrel	3.875 x 3.53	250	8.10	110 @ 3800	185 @ 1600	30-45
	155 Horsepower①...........V8-350	2 Barrel	4.057 x 3.385	350	8.10	155 @ 4000	275 @ 2400	30-45
	180 Horsepower①...........V8-350	4 Barrel	4.057 x 3.385	350	8.10	180 @ 4000	275 @ 2400	30-45
	185 Horsepower①...........V8-455	2 Barrel	4.125 x 4.250	455	8.10	185 @ 3600	355 @ 2000	30-45
	225 Horsepower①...........V8-455	4 Barrel	4.125 x 4.250	455	8.10	225 @ 3600	360 @ 2600	30-45
	260 Horsepower①...........V8-455	4 Barrel	4.125 x 4.250	455	8.10	260 @ 4400	370 @ 3200	30-45
	265 Horsepower①...........V8-455	4 Barrel	4.125 x 4.250	455	8.10	265 @ 4200	375 @ 2800	30-45
1972	160 Horsepower①...........V8-350	2 Barrel	4.057 x 3.385	350	8.50	160 @ 4000	275 @ 2400	30-45
	175 Horsepower①...........V8-350	2 Barrel	4.057 x 3.385	350	8.50	175 @ 4000	295 @ 2600	30-45
	180 Horsepower①...........V8-350	4 Barrel	4.057 x 3.385	350	8.50	180 @ 4000	275 @ 2800	30-45
	200 Horsepower①...........V8-350	4 Barrel	4.057 x 3.385	350	8.50	200 @ 4400	300 @ 3200	30-45
	225 Horsepower①...........V8-455	4 Barrel	4.125 x 4.250	455	8.50	225 @ 3600	360 @ 2600	30-45
	250 Horsepower①...........V8-455	4 Barrel	4.125 x 4.250	455	8.50	250 @ 4200	370 @ 2800	30-45
	265 Horsepower①...........V8-455	4 Barrel	4.125 x 4.250	455	8.50	265 @ 4200	375 @ 2800	30-45
	270 Horsepower①...........V8-455	4 Barrel	4.125 x 4.250	455	8.50	270 @ 4400	370 @ 3200	30-45
	300 Horsepower①...........V8-455	4 Barrel	4.125 x 4.250	455	8.50	300 @ 4700	410 @ 3200	30-45
1973	100 Horsepower①...........②6-250	1 Barrel	3.875 x 3.53	250	8.25	100 @ 3600	175 @ 1600	40
	160 Horsepower①...........V8-350	2 Barrel	4.057 x 3.385	350	8.5	160 @ 3800	275 @ 2400	30-45
	180 Horsepower①...........V8-350	4 Barrel	4.057 x 3.385	350	8.5	180 @ 3800	275 @ 2800	30-45
	225 Horsepower①...........V8-455	4 Barrel	4.125 x 4.250	455	8.5	225 @ 3600	360 @ 2600	30-45
	250 Horsepower①...........V8-455	4 Barrel	4.125 x 4.250	455	8.5	250 @ 4000	375 @ 2800	30-45
	270 Horsepower①...........V8-455	4 Barrel	4.125 x 4.250	455	8.5	270 @ 4200	370 @ 3200	30-45
1974	100 Horsepower①...........②6-250	1 Barrel	3.87 x 3.53	250	8.5	100 @ 3600	175 @ 1800	40
	180 Horsepower①...........V8-350	4 Barrel	4.057 x 3.385	350	8.5	180 @ 3800	275 @ 2800	30-45
	200 Horsepower①...........V8-350	4 Barrel	4.057 x 3.385	350	8.5	200 @ 4200	300 @ 3200	30-45
	210 Horsepower①...........V8-455	4 Barrel	4.126 x 4.250	455	8.5	210 @ 3600	350 @ 2400	30-45
	230 Horsepower①...........V8-455	4 Barrel	4.126 x 4.250	455	8.5	230 @ 3800	370 @ 2800	30-45
	275 Horsepower①...........V8-455	4 Barrel	4.126 x 4.250	455	8.5	275 @ 4200	395 @ 3200	30-45
1975	105 Horsepower①...........6-250②	1 Barrel	3.87 x 3.53	250	8.25	105 @ 3800	185 @ 1200	36-41
	110 Horsepower①...........V8-260	2 Barrel	3.50 x 3.385	260	8.5	110 @ 3400	205 @ 1600	30-45
	165 Horsepower①.........V8-350③	4 Barrel	3.80 x 3.85	350	8.0	165 @ 3800	260 @ 2200	37
	170 Horsepower①...........V8-350	4 Barrel	4.057 x 3.385	350	8.5	170 @ 3800	275 @ 2400	30-45

Continued

GENERAL ENGINE SPECIFICATIONS—Continued

Year	Engine	Carburetor	Bore and Stroke	Piston Displacement, Cubic Inches	Compression Ratio	Maximum Brake H.P. @ R.P.M.	Maximum Torque Lbs. Ft. @ R.P.M.	Normal Oil Pressure Pounds
1975	190 Horsepower①..........V8-400④	4 Barrel	4.1212 x 3.75	400	7.6	190 @ 3400	350 @ 2000	55–60
	190 Horsepower①..........V8-455⑤	4 Barrel	4.126 x 4.25	455	8.5	190 @ 3600	350 @ 2400	30–45
	215 Horsepower①..........V8-455⑥	4 Barrel	4.126 x 4.25	455	8.5	215 @ 3600	370 @ 2400	30–45
1976	105 Horsepower①............6-250②	1 Barrel	3.87 x 3.53	250	8.5	105 @ 3800	185 @ 1200	36–41
	110 Horsepower①............V8-260	2 Barrel	3.50 x 3.385	260	8.0	110 @ 3400	205 @ 1600	30–45
	140 Horsepower①..........V8-350③	2 Barrel	3.80 x 3.85	350	8.0	140 @ 3200	280 @ 1800	37
	155 Horsepower①..........V8-350③	4 Barrel	3.80 x 3.85	350	8.0	155 @ 3400	280 @ 1800	37
	170 Horsepower①............V8-350	4 Barrel	4.057 x 3.385	350	8.5	170 @ 3800	275 @ 2400	30–45
	190 Horsepower①..........V8-455⑤	4 Barrel	4.126 x 4.25	455	8.5	190 @ 3400	350 @ 2000	30–45
	215 Horsepower①..........V8-455⑥	4 Barrel	4.126 x 4.25	455	8.5	215 @ 3600	370 @ 2400	30–45

①—All horsepower and torque ratings are net.
②—See Chevrolet Chapter for service procedure on this engine.
③—Omega only. See Buick Chapter for service procedures on this engine.
④—See Pontiac Chapter for service procedures on this engine.
⑤—Exc. Toronado.
⑥—Toronado.

TUNE UP SPECIFICATIONS

The following specifications are published from the latest information available. This data should be used only in the absence of a decal affixed in the engine compartment.

OLD CAR SPECIFICATIONS: For 1946-68 Tune Up Specifications see main index.

★When using a timing light, disconnect vacuum hose or tube at distributor and plug opening in hose or tube so idle speed will not be affected.

●When checking compression, lowest cylinder must be within 80 percent of highest.

▲Before removing wires from distributor cap, determine location of the No. 1 wire in cap, as distributor position may have been altered from that shown at the end of this chart.

Year	Spark Plug		Distributor		Ignition Timing★			Carb. Adjustments					
								Hot Idle Speed		Air Fuel Ratio		Idle "CO" %	
	Type	Gap Inch	Point Gap Inch	Dwell Angle Deg.	Firing Order Fig. ▲	Timing BTDC ①	Mark Fig.	Std. Trans.	Auto. Trans. ②	Std. Trans.	Auto. Trans.	Std. Trans.	Auto. Trans.
1969													
6-250 Std. Trans.⑭	R46N	.035	③	31–34	N	TDC	B	775⑪	—	—	—	—	—
6-250 Auto. Trans.⑭	R46N	.035	③	31–34	N	4°	B	—	625D⑪	—	—	—	—
8-350, 250 H.P.	R46S	.030	.016	30	F	6°	A	675⑪	600D⑪	—	—	—	—
8-350, 310 H.P.	R45S	.030	.016	30	F	8°	A	675⑪	575D⑪	—	—	—	—
8-350, 325 H.P.	R43S	.030	.016	30	F	12°	A	675⑪	575D⑪	—	—	—	—
8-400, 325 H.P.	R44S	.030	.016	30	F	8°⑨⑮	A	—	575D⑪	—	—	—	—
8-400, 350 H.P.	R44S	.030	.016	30	F	2°⑨	A	750⑪	—	—	—	—	—
8-400, 360 H.P.	R43S	.030	.016	30	F	14°⑫	A	750⑪	650D⑪	—	—	—	—
8-455, 310 H.P.	R45S	.030	.016	30	F	6°⑨	A	675⑪	600D⑪	—	—	—	—
8-455, 365 H.P.	R44S	.030	.016	30	F	8°⑨	A	—	575D⑪	—	—	—	—
8-455, 375 H.P.	R44S	.030	.016	30	F	8°⑨	A	—	575D⑪	—	—	—	—
8-455, 400 H.P.	R44S	.030	.016	30	F	10°⑨	A	—	575D⑪	—	—	—	—
1970													
6-250 Std. Trans.⑭	R46T	.035	③	31–34	N	TDC	B	750	—	—	—	—	—
6-250 Auto. Trans.⑭	R46T	.035	③	31–34	N	4°	B	—	600D	—	—	—	—
8-350, 250 H.P.⑤	R46S	.030	.016	30	F	10°⑯	A	750	575D	—	—	—	—
8-350, 250 H.P.⑥	R46S	.030	.016	30	F	8°⑯	A	675	575D	—	—	—	—

Continued

TUNE UP SPECIFICATIONS—Continued

The following specifications are published from the latest information available. This data should be used only in the absence of a decal affixed in the engine compartment.

OLD CAR SPECIFICATIONS: For 1946-68 Tune Up Specifications see main index.

★When using a timing light, disconnect vacuum hose or tube at distributor and plug opening in hose or tube so idle speed will not be affected.

●When checking compression, lowest cylinder must be within 80 percent of highest.

▲Before removing wires from distributor cap, determine location of the No. 1 wire in cap, as distributor position may have been altered from that shown at the end of this chart.

Year	Spark Plug		Distributor		Ignition Timing★			Carb. Adjustments					
								Hot Idle Speed		Air Fuel Ratio		Idle "CO" %	
	Type	Gap Inch	Point Gap Inch	Dwell Angle Deg.	Firing Order Fig. ▲	Timing BTDC ①	Mark Fig.	Std. Trans.	Auto. Trans. ②	Std. Trans.	Auto. Trans.	Std. Trans.	Auto. Trans.
1970—Continued													
8-350, 310 H.P.	R45S	.030	.016	30	F	10°⑯	A	650	575D	—	—	—	—
8-350, 325 H.P.	R43S	.030	.016	30	F	14°⑯	A	750	625D	—	—	—	—
8-455, 310 H.P.	R46S	.030	.016	30	F	8°⑯	A	675	575D	—	—	—	—
8-455, 320 H.P.	R45S	.030	.016	30	F	8°⑯	A	—	575D	—	—	—	—
8-455, 365 H.P.	R44S	.030	.016	30	F	8°⑨	A	750	650D	—	—	—	—
8-455, 370 H.P.	R44S	.030	.016	30	F	8°⑨	A	750	650D	—	—	—	—
8-455, 375 H.P.	R44S	.030	.016	30	F	12°⑯	A	750	600D	—	—	—	—
8-455, 390 H.P.	R45S	.030	.016	30	F	8°⑯	A	—	600D	—	—	—	—
8-455, 390 H.P.④	R44S	.030	.016	30	F	12°⑯	A	—	600D	—	—	—	—
8-455, 400 H.P.	R44S	.030	.016	30	F	12°⑯	A	—	600D	—	—	—	—
1971													
6-250⑭	R46TS	.035	⑧	31–34	N	4°	C	550	500D	—	—	1.0	1.0
8-350, 155 H.P.	R46S	.040	.016	30	F	10°⑯	A	750	600D	—	—	0.6	0.6
8-350, 180 H.P. S.Tr.	R45S	.040	.016	30	F	10°⑯	A	750	—	—	—	0.3	—
8-350, 180 H.P. A.Tr.	R46S	.040	.016	30	F	12°⑯	A	—	600D	—	—	—	0.3
8-350, 260 H.P.	R45S	.040	.016	30	F	10°⑯	A	750	600D	—	—	0.3	0.3
8-455, 185 H.P.	R46S	.040	.016	30	F	8°⑯	A	750	600D	—	—	0.6	0.6
8-455, 225 H.P.	R46S	.040	.016	30	F	8°⑯	A	—	600D	—	—	0.6	0.6
8-455, 260 H.P.④S.Tr.	R45S	.040	.016	30	F	12°⑯	A	750	—	—	—	0.3	—
8-455, 260 H.P.④A.Tr.	R45S	.040	.016	30	F	10°⑯	A	—	600D	—	—	—	0.3
8-455, 265 H.P.	R46S	.040	.016	30	F	10°⑯	A	—	600D	—	—	0.3	0.3
1972													
8-350 2 Bar. Carb.	R46S	.040	.016	30	F	8°⑯	A	750	650D	—	—	0.3	0.3
8-350 4 B. Carb. St. Tr.	R45S	.040	.016	30	F	8°⑯	A	750	—	—	—	0.3	—
8-350 4 B. Carb. A. Tr.	R46S	.040	.016	30	F	12°⑯	A	—	600D	—	—	—	0.3
8-455, 250, 270 H.P.	R46S	.040	.016	30	F	8°⑯	A	—	600D	—	—	0.3	0.3
8-455, 300 H.P.	R45S	.040	.016	30	F	10°	A	1000	650D	—	—	0.3	0.3
8-455 Others—St. Tr.	R45S	.040	.016	30	F	10°⑯	A	750	—	—	—	0.3	—
8-455 Others—A. Tr.	R46S	.040	.016	30	F	8°⑯	A	—	650D	—	—	—	0.3
1973													
6-250⑭	R46TS	.035	⑧	31–34	N	6°	C	700	600D	—	—	0.3	0.3
8-350 2 Bar.⑤	R46S	.040	.016	30	F	14°⑯	A	—	700D	—	—	—	0.3
8-350 2 Bar.⑥	R46S	.040	.016	30	F	12°⑯	A	—	700D	—	—	—	0.3
8-350 4 Bar. Std. Tr.	R45S	.040	.016	30	F	8°⑯	A	1100	—	—	—	0.3	—
8-350 4 Bar. Auto. Tr.	R46S	.040	.016	30	F	12°⑯⑮	A	—	650D	—	—	—	0.3
8-455⑤	R45S	.040	.016	30	F	10°⑯	A	1000	—	—	—	0.3	—
8-455⑥	R46S	.040	.016	30	F	8°⑯	A	—	650D	—	—	—	0.3

Continued

TUNE UP SPECIFICATIONS—Continued

The following specifications are published from the latest information available. This data should be used only in the absence of a decal affixed in the engine compartment.

OLD CAR SPECIFICATIONS: For 1946-68 Tune Up Specifications see main index.

★ When using a timing light, disconnect vacuum hose or tube at distributor and plug opening in hose or tube so idle speed will not be affected.

● When checking compression, lowest cylinder must be within 70 percent of highest.

▲ Before removing wires from distributor cap, determine location of the No. 1 wire in cap, as distributor position may have been altered from that shown at the end of this chart.

Year	Spark Plug		Distributor		Ignition Timing★			Carb. Adjustments					
								Hot Idle Speed		Air Fuel Ratio		Idle "CO" %	
	Type	Gap Inch	Point Gap Inch	Dwell Angle Deg.	Firing Order Fig. ▲	Timing BTDC (1)	Mark Fig.	Std. Trans.	Auto. Trans.(2)	Std. Trans.	Auto. Trans.	Std. Trans.	Auto. Trans.
1974													
6-250(14)	R46TS	.035	.019	31–34	N	8°(7)	C	850	600D	—	—	0.3	0.3
8-350 4 Bar.	R46S	.040	.019	30	(20)	12°(16)	H	—	650D	—	—	0.2	0.2
8-455, 275 H.P.	R45S	.040	.019	30	(20)	14°(16)	H	—	650D	—	—	0.2	0.2
8-455 4 Bar.	R46S	.040	.019	30	(20)	8°(16)	H	—	650D	—	—	0.2	0.2
8-455 4 Bar.(8)	R46SX	.080	—	—	(20)	8°(16)	H	—	650D	—	—	0.2	0.2
8-455 4 Bar.(13)	R46S	.040	.019	30	(20)	10°(16)	H	—	650D	—	—	0.2	0.2
8-455 4 Bar.(8)(13)	R46SX	.080	—	—	(20)	10°(16)	H	—	650D	—	—	0.2	0.2
1975													
6-250(14)	R46TX	.060	—	—	I	10°	C	850	(19)	—	—	—	—
V8-260	R46SX	.080	—	—	J	(17)	G	750	650D	—	—	—	—
V8-350(10)	R45TSX	.060	—	—	K	12°	D	—	600D	—	—	—	—
V8-350	R46SX	.080	—	—	J	20°(16)	G	—	650D	—	—	—	—
V8-400(18)	R45TSX	.060	—	—	L	16°(16)	E	—	650D	—	—	—	—
V8-455	R46SX	.080	—	—	J	16°(16)	G	—	650D	—	—	—	—
V8-455(13)	R46SX	.080	—	—	J	12°(16)	G	—	650D	—	—	—	—
1976													
6-250(14)(21)	R46TS	.035	—	—	I	6°	C	850	—	—	—	—	—
6-250(14)(22)	R46TS	.035	—	—	I	10°	C	—	(19)	—	—	—	—
V8-260(21)	R46SX	.080	—	—	J	16°(16)	G	750	—	—	—	—	—
V8-260(22)(23)	R46SX	.080	—	—	J	18°(16)	G	—	650D	—	—	—	—
V8-260(5)(24)	R46SX	.080	—	—	J	16°(16)	G	—	650D	—	—	—	—
V8-260(24)(25)	R46SX	.080	—	—	J	14°(16)	G	—	650D	—	—	—	—
V8-350	R46SX	.080	—	—	J	20°(16)	G	—	(19)	—	—	—	—
V8-350(10)	R46TSX	.060	—	—	K	12°	D	—	600D	—	—	—	—
V8-455	R46SX	.080	—	—	J	16°(16)	G	—	650D	—	—	—	—
V8-455(13)(23)	R46SX	.080	—	—	J	14°(16)	G	—	650D	—	—	—	—
V8-455(13)(24)	R46SX	.080	—	—	J	12°(16)	G	—	650D	—	—	—	—

(1)—BTDC: Before top dead center.
(2)—D: Drive. N: Neutral. Add 50 R.P.M. to slow idle speed for air conditioned cars with A/C off.
(3)—New points .019", used points .016".
(4)—Air Induction.
(5)—Cutlass.
(6)—Full size cars.
(7)—At 600 rpm with auto. trans. and 850 rpm with manual trans.
(8)—With High Energy Ignition system.
(9)—At 850 R.P.M.
(10)—Omega only. See Buick Chapter for service procedures on this engine.
(11)—With A/C "OFF" and idle compensator held closed.
(12)—At 1250 R.P.M.
(13)—Toronado.
(14)—See Chevrolet Chapter for service procedures on this engine.
(15)—Vista-Cruiser 10°.
(16)—At 1100 R.P.M.
(17)—Exc. Calif., 16° BTDC; Calif., 18° BTDC. At 1100 R.P.M.
(18)—See Pontiac Chapter for service procedures on this engine.
(19)—Exc. Calif., 550D; Calif., 600D.
(20)—Exc. H.E.I., Fig. C; H.E.I., Fig. G.
(21)—Manual trans.
(22)—Auto. trans.
(23)—Exc. California.
(24)—California.
(25)—Omega.

Continued

TUNE UP NOTES—Continued

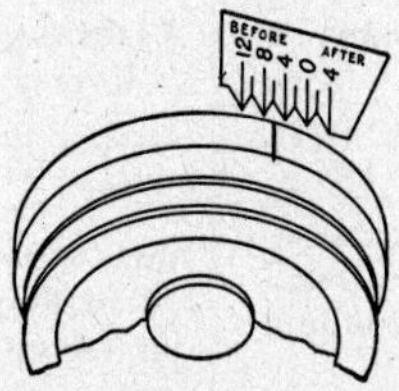

Fig. A

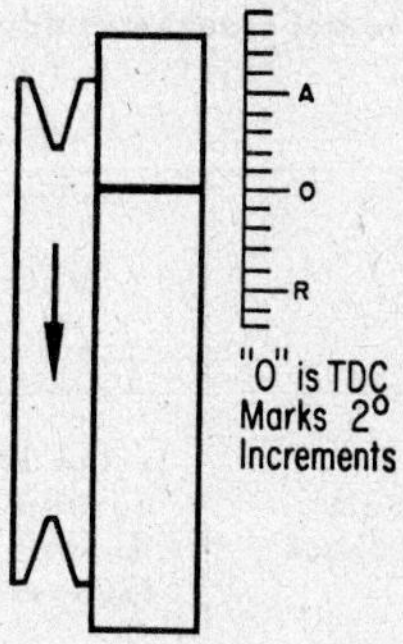

Fig. B

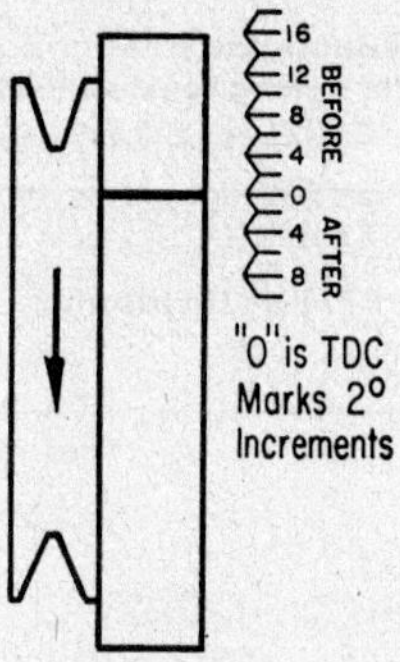

Fig. C

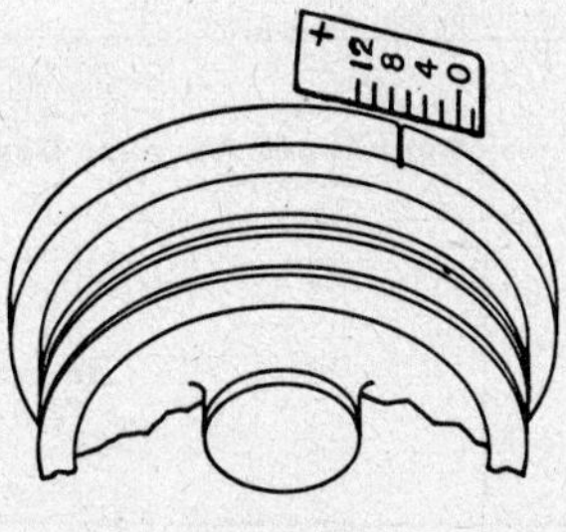

Fig. D

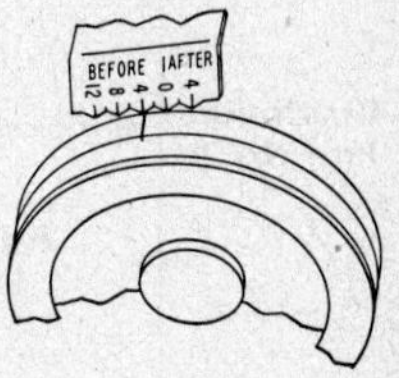

Fig. E

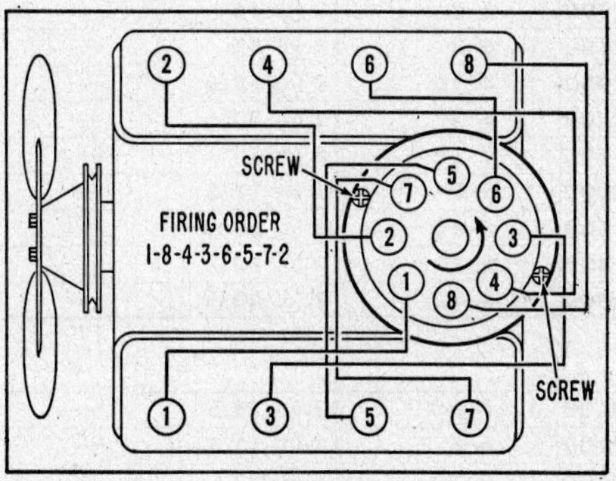

Fig. F

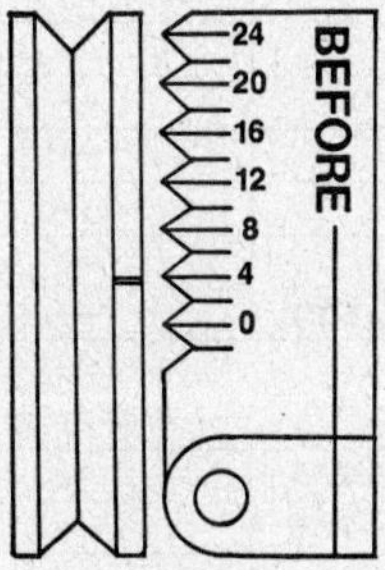

Fig. G

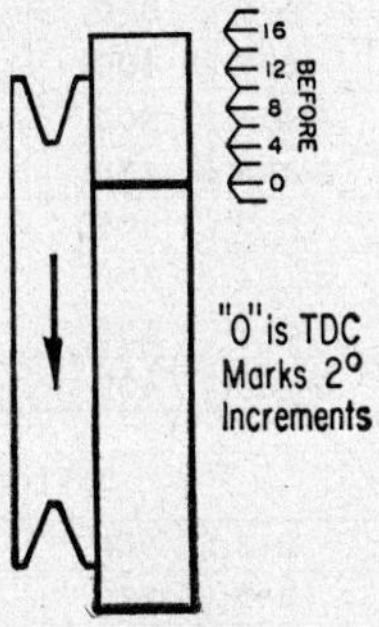

Fig. H

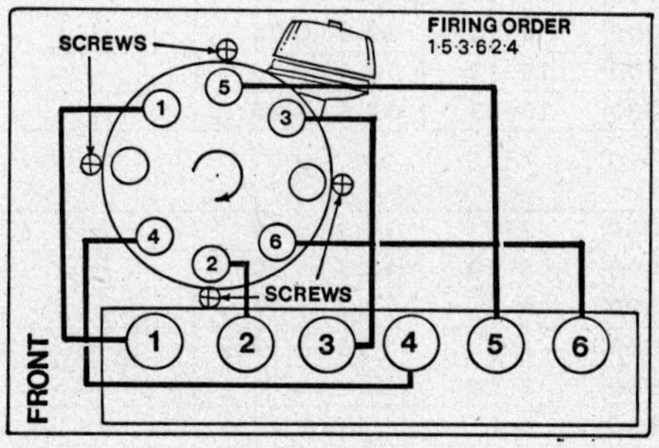

Fig. I

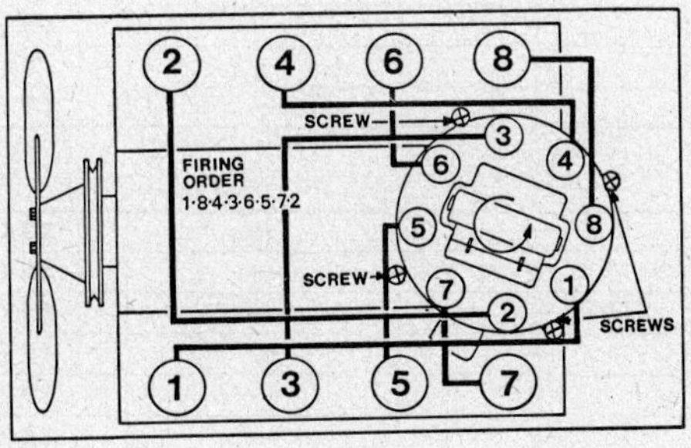

Fig. J

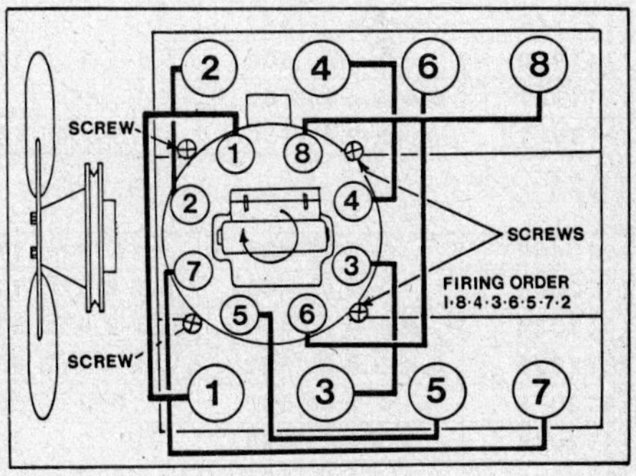

Fig. K

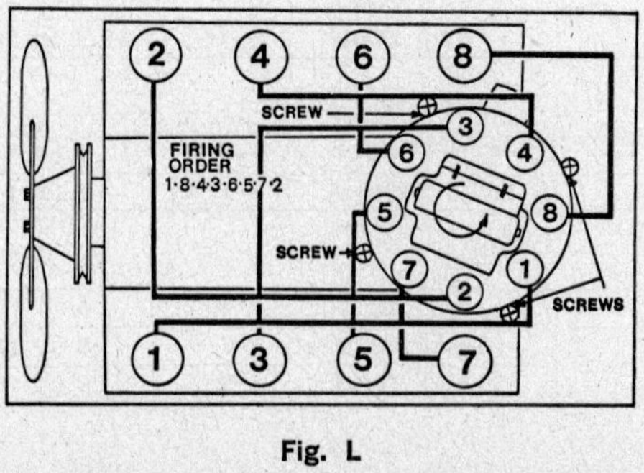

Fig. L

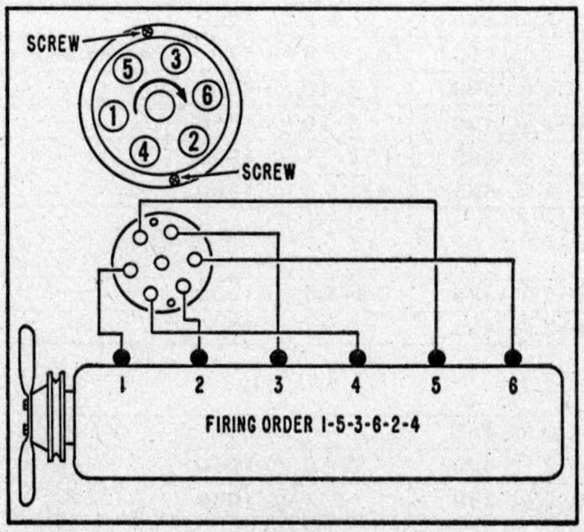

Fig. N

DISTRIBUTOR SPECIFICATIONS

★Note: If unit is checked on vehicle, double the RPM and degrees to get crankshaft figures.
Breaker arm spring tension—19–23.

Distributor Part No.①	Centrifugal Advance Degrees @ RPM of Distributor					Vacuum Advance		Distributor Retard
	Advance Starts	Intermediate Advance			Full Advance	Inches of Vacuum to Start Plunger	Max. Adv. Dist. Deg. @ Vacuum	Max. Retard Dist. Deg. @ Vacuum
1969								
1110463	0–3 @ 500	2.2–5.2 @ 580	9.5–11.5 @ 975	—	17 @ 2100	6–8	13 @ 16	—
1110464	0–2 @ 510	1.5–3.5 @ 600	7.5–9.5 @ 975	—	15 @ 2100	6–8	13 @ 16	—
1111930	0–2 @ 400	6–8 @ 900	—	—	12 @ 2000	8–10	9 @ 18½	—
1111932	.4–2.4 @ 500	8–10 @ 900	—	—	12 @ 2000	8–10	12 @ 20½	—
1111933	.5–2.5 @ 450	7–9 @ 1000	—	—	11 @ 1900	10–12	8 @ 18	—
1111934	0–2 @ 400	6.5–8.5 @ 1200	—	—	15 @ 2000	6–8	12 @ 17½	—
1111935	.4–2.4 @ 700	—	—	—	8 @ 1500	8–10	12 @ 20½	—
1111936	0–2 @ 550	5–7 @ 1000	—	—	11 @ 1800	8–10	12 @ 20½	—
1111961	0–1.8 @ 400	1.7–4 @ 500	2.2–4.2 @ 520	—	16 @ 2000	8–10	12 @ 20½	—
1970								
1110463	0–3 @ 500	2.2–5.2 @ 580	9.5–11.5 @ 975	—	17 @ 2100	6–8	13 @ 15.5	—
1110464	0–2 @ 510	1.5–3.5 @ 600	7.5–9.5 @ 975	—	15 @ 2100	6–8	13 @ 15.5	—
1111975	0–2 @ 400	6–8 @ 900	—	—	12 @ 2000	8–10	10.8 @ 23	—
1111976	0–2 @ 405	7¼–9¾ @ 1025	—	—	16 @ 2000	8–10	13.8 @ 25	—
1111977	0.4–2.4 @ 500	8–10 @ 900	—	—	12 @ 2000	10–13	12.8 @ 25	—
1111979	0–3.5 @ 375	7–10.3 @ 575	8.6–10.6 @ 625	—	16 @ 1500	10–13	12.8 @ 25	—
1111980	0–2 @ 400	6.5–8.5 @ 1200	—	—	15 @ 2000	8–10	13.8 @ 25	—
1111981	0.4–2.4 @ 700	—	—	—	8 @ 1500	10–13	12.8 @ 25	—
1111982	0–2 @ 550	5–7 @ 1000	—	—	11 @ 1800	10–13	12.8 @ 25	—
1971								
1110489	0–2 @ 635	6–8 @ 1150	—	—	13 @ 2050	7–9	10.5 @ 18	—
1112033	0–2 @ 400	6.5–8.5 @ 1200	—	—	15 @ 2000	8–10	12.5 @ 20½	—
1112034	0.8–4 @ 450	4.5–7.8 @ 570	6.2–8.2 @ 620	—	14 @ 1500	6–8	13 @ 17½	—
1112036	0.4–2.4 @ 500	8–10 @ 900	—	—	12 @ 2000	6–8	13 @ 17½	—
1112078	0–2 @ 500	0–3 @ 550	—	—	9 @ 1950	8–10	—	—
1112079	0–2 @ 385	10–12 @ 1000	—	—	18 @ 2050	6–8	13 @ 17½	—
1112085	0.6–2.6 @ 600	6–8 @ 1025	—	—	14 @ 2000	6–8	13 @ 17½	—
1972								
1112033	0–2 @ 540	5–7 @ 1000	—	—	11 @ 1800	8–10	12.5 @ 20½	—
1112034	0.8–4 @ 450	4.5–7.8 @ 570	6.2–8.2 @ 620	—	14 @ 1500	6–8	13 @ 17½	—
1112036	0.4–2.4 @ 500	8–10 @ 900	—	—	12 @ 2000	6–8	13 @ 17½	—
1112085	0–2 @ 400	8–10 @ 1050	—	—	16 @ 2000	6–8	13 @ 17½	—
1112106	0–2 @ 485	8–11 @ 1000	—	—	16 @ 2000	6–8	13 @ 17½	—
1112172	0–2 @ 575	3.5–5.5 @ 1000	—	—	9 @ 1700	8–10	12.5 @ 20½	—
1973								
1112198	0–2 @ 575	3.5–5.5 @ 1000	—	—	9 @ 1700	7–9	9 @ 14	—
1112222	0–2 @ 485	8–11 @ 1000	—	—	16 @ 2000	5–7	8 @ 12	—
1973-74								
1110499	0 @ 465	1 @ 635	7 @ 1150	—	12 @ 2050	6–8	6 @ 15½	—
1112195	0–2 @ 400	8–10 @ 1050	—	—	16 @ 2000	5–7	8 @ 12	—
1112197	0–2 @ 540	5–7 @ 1000	—	—	11 @ 1800	7–9	9 @ 16.6	—

Continued

DISTRIBUTOR SPECIFICATIONS—Continued

★Note: If unit is checked on vehicle, double the RPM and degrees to get crankshaft figures.
Breaker arm spring tension—19–23.

Distributor Part No.①	Centrifugal Advance Degrees @ RPM of Distributor					Vacuum Advance		Distributor Retard
	Advance Starts	Intermediate Advance			Full Advance	Inches of Vacuum to Start Plunger	Max. Adv. Dist. Deg. @ Vacuum	Max. Retard Dist. Deg. @ Vacuum
1973-74—Continued								
1112225	0–2 @ 380	5–7 @ 550	11–13 @ 1050	—	19 @ 2000	5–7	8 @ 12	—
1112226	0–2 @ 400	8–10 @ 1050	—	—	16 @ 2000	3½–4½	10 @ 13	—
1112506	0–2 @ 540	5–7 @ 1000	—	—	11 @ 1800	7–9	9 @ 16.6	—
1974								
1112531	0–2 @ 540	5–7 @ 1000	—	—	11 @ 1800	6	9 @ 10	—
1112532	0–2 @ 540	5–7 @ 1000	—	—	11 @ 1800	4	10 @ 13	—
1112550	0 @ 375	7 @ 600	—	—	13 @ 1500	12	8 @ 18	—
1112825	0–2 @ 575	3.5–5.5 @ 1000	—	—	9 @ 1700	6	9 @ 10	—
1112827	0–2 @ 575	3.5–5.5 @ 1000	—	—	9 @ 1700	7	12 @ 17	—
1112828	0–2 @ 400	8–10 @ 1050	—	—	16 @ 2000	3½–4½	10 @ 13	—
1112829	0–2 @ 575	3.5–5.5 @ 1000	—	—	9 @ 1700	4	10 @ 13	—
1112830	0–2 @ 575	3.5–5.5 @ 1000	—	—	9 @ 1700	7	12 @ 17	—
1975								
1110650	0 @ 550	3½ @ 1150	—	—	8 @ 2100	3–5	8¼ @ 12½	—
1112863	0 @ 550	3½ @ 1150	—	—	8 @ 2100	3–5	8¾ @ 12½	—
1112896	0 @ 550	3 @ 1150	—	—	6 @ 2250	6½–8½	7 @ 11½	—
1112928	0 @ 600	2 @ 700	—	—	8 @ 2200	6–8	12½ @ 12	—
1112936	0 @ 500	—	—	—	9½ @ 2000	6½	12 @ 16	—
1112937	0 @ 500	—	—	—	6½ @ 1800	8	9 @ 13	—
1112951	0 @ 325	9½ @ 1200	—	—	14 @ 2200	4	12 @ 15	—
1112952	0 @ 500	3½ @ 1050	—	—	7 @ 1800	8	9 @ 13	—
1112953	0 @ 500	—	—	—	9½ @ 2000	8	9 @ 16	—
1112956	0 @ 325	9½ @ 1200	—	—	14 @ 2200	—	—	—
1112958	0 @ 600	2 @ 700	—	—	8 @ 2200	5	12½ @ 11	—
1976								
1110666	0 @ 500	3½ @ 800	—	—	10 @ 2100	4	12 @ 15	—
1112863	0 @ 550	3½ @ 1150	—	—	8 @ 2100	3–5	13 @ 15.5	—
1112863	0 @ 550	3½ @ 1150	—	—	8 @ 2100	4	9 @ 12	—
1112936	0 @ 500	—	—	—	9½ @ 2000	6.5	12 @ 16	—
1112937	0 @ 500	—	—	—	6½ @ 1800	8	9 @ 13	—
1112952	0 @ 450	4½ @ 1000	—	—	8 @ 1700	8	9 @ 13	—
1112953	0 @ 500	—	—	—	9½ @ 2000	8	9 @ 16	—
1112956	0 @ 325	9½ @ 1200	—	—	14 @ 2200	—	—	—
1112988	0 @ 500	—	—	—	6½ @ 1800	8	9 @ 13	—
1112991	0 @ 713	—	—	—	10 @ 2213	7.5	10 @ 13.3	—
1112994	0 @ 325	9½ @ 1200	—	—	14 @ 2200	4.5	12 @ 10.5	—
1112995	0 @ 455	8½ @ 1188	—	—	13 @ 2233	4	15 @ 11	—

①—Stamped on distributor housing plate.

VALVE SPECIFICATIONS

Year	Model	Valve Lash		Valve Angles		Valve Spring Installed Height	Valve Spring Pressure Lbs. @ In.	Stem Clearance		Stem Diameter	
		Int.	Exh.	Seat	Face			Intake	Exhaust	Intake	Exhaust
1969	6-250⑧	1 Turn③		46	45	1.66	186 @ 1.27	.001–.0027	.0015–.0032	.3410–.3417	.3410–.3417
	V8-350	Hydraulic⑥		45	46	1.67	187 @ 1.27	.001–.0027	.0015–.0032	.3425–.3432	.3420–.3427
	V8-400 4 B. C.	Hydraulic⑥		④	⑤	1.67	187 @ 1.27	.001–.0027	.0015–.0032	.3425–.3432	.3420–.3427
	V8-400 2 B. C.	Hydraulic⑥		45	46	1.67	187 @ 1.27	.001–.0027	.0015–.0032	.3425–.3432	.3420–.3427
	V8-455	Hydraulic⑥		45	46	1.67	187 @ 1.27	.001–.0027	.0015–.0032	.3425–.3432	.3420–.3427
	Toronado	Hydraulic⑥		⑤	④	1.67	187 @ 1.27	.001–.0027	.0015–.0032	.3425–.3432	.3420–.3427
1970	6-250⑧	1 Turn③		46	45	1.66	186 @ 1.27	.001–.0027	.0015–.0032	.3410–.3417	.3410–.3417
	V8-350	Hydraulic⑥		④	⑤	1.67	187 @ 1.27	.001–.0027	.0015–.0032	.3425–.3432	.3420–.3427
	V8-455⑨	Hydraulic⑥		45	46	1.67	187 @ 1.27	.001–.0027	.0015–.0032	.3425–.3432	.3420–.3427
	V8-455⑩	Hydraulic⑥		④	⑤	1.67	187 @ 1.27	.001–.0027	.0015–.0032	.3425–.3432	.3420–.3427
	Toronado	Hydraulic⑥		④	⑤	1.67	187 @ 1.27	.001–.0027	.0015–.0032	.3425–.3432	.3420–.3427
1971	6-250⑧	1 Turn③		46	45	1.66	186 @ 1.27	.001–.0027	.0015–.0032	.3410–.3417	.3410–.3417
	V8-350	Hydraulic⑥		45	46	1.67	187 @ 1.27	.001–.0027	.0015–.0032	.3425–.3432	.3420–.3427
	V8-455⑨	Hydraulic⑥		45	46	1.67	187 @ 1.27	.001–.0027	.0015–.0032	.3425–.3432	.3420–.3427
	V8-455⑩	Hydraulic⑥		④	⑤	1.67	187 @ 1.27	.001–.0027	.0015–.0032	.3425–.3432	.3420–.3427
1972	V8-350①	Hydraulic⑥		⑪	⑫	1.67	187 @ 1.27	.001–.0027	.0015–.0032	.3425–.3432	.3420–.3427
	V8-350⑦	Hydraulic⑥		45	46	1.67	198 @ 1.23	.001–.0027	.0015–.0032	.3425–.3432	.3420–.3427
	V8-455 "98"	Hydraulic⑥		⑪	⑫	1.67	187 @ 1.27	.001–.0027	.0015–.0032	.3425–.3432	.3420–.3427
	V8-455 Toronado	Hydraulic⑥		45	46	1.67	196 @ 1.23	.001–.0027	.0015–.0032	.3425–.3432	.3420–.3427
	V8-455 Others	Hydraulic⑥		④	⑤	1.67	206 @ 1.19	.001–.0032	.0015–.0027	.3425–.3432	.3420–.3427
1973–74	6-250⑧	1 Turn③		46	45	1.66	186 @ 1.27	.001–.0027	.0015–.0032	.3410–.3417	.3410–.3417
	V8-350	Hydraulic⑥		②	⑬	1.67	187 @ 1.27	.001–.0027	.0015–.0032	.3425–.3432	.3420–.3427
	V8-455	Hydraulic⑥		②	⑬	1.67	187 @ 1.27	.001–.0027	.0015–.0032	.3425–.3432	.3420–.3427
1975–76	6-250⑧	1 Turn③		46	45	1.66	186 @ 1.27	.001–.0027	.0015–.0032	.3410–.3417	.3410–.3417
	V8-260	Hydraulic⑥		②	⑬	1.67	187 @ 1.27	.001–.0027	.0015–.0032	.3425–.3432	.3420–.3427
	V8-350⑭	Hydraulic⑥		45	45	1.727	177 @ 1.45	.0015–.0035	.0015–.0032	.3720–.3730	.3723–.3730
	V8-350	Hydraulic⑥		②	⑬	1.67	187 @ 1.27	.001–.0027	.0015–.0032	.3425–.3432	.3420–.3427
	V8-400 2 B. C.⑮	Hydraulic⑥		45	45	—	—	.0016–.0033	.0021–.0038	.3412–.3419	.3407–.3414
	V8-400 4 B. C.⑮	Hydraulic⑥		④	⑯	—	—	.0016–.0033	.0021–.0038	.3412–.3419	.3407–.3414
	V8-455	Hydraulic⑥		②	⑬	1.67	187 @ 1.27	.001–.0027	.0015–.0032	.3425–.3432	.3420–.3427

①—Except California.
②—Intake 45°, exhaust 31°.
③—Tighten rocker arm adjusting screw to eliminate all push rod end clearance. Then tighten screw the number of turns listed.
④—Intake 30°, exhaust 45°.
⑤—Intake 30°, exhaust 46°.
⑥—No adjustment.
⑦—California.
⑧—See Chevrolet Chapter for service procedures on this engine.
⑨—Except "4-4-2" models.
⑩—"4-4-2" models.
⑪—Intake 45°, exhaust 30°.
⑫—Intake 46°, exhaust 30°.
⑬—Intake 44°, exhaust 30°.
⑭—Omega only. See Buick Chapter for service procedures.
⑮—See Pontiac Chapter for service procedures.
⑯—Intake 29°; Exhaust 44°.

ALTERNATOR & REGULATOR SPECIFICATIONS

Year	Alternator					Regulator						
	Model	Rated Hot Output Amps.	Field Current 12 Volts @ 80° F.	Output @ 14 Volts		Model	Field Relay			Voltage Regulator		
				2000 R.P.M. Amps.	5000 R.P.M. Amps.		Air Gap In.	Point Gap In.	Closing Voltage	Air Gap In.	Point Gap In.	Voltage @ 125° F.
1969	1100734	42	2.2–2.6	28	40	1119515	.015	.030	6.3–8.3	.060	.014	13.5–14.4
	1100767	37	2.2–2.6	25	35	1119515	.015	.030	6.3–8.3	.060	.014	13.5–14.4
	1100777	55	2.2–2.6	32	50	1119515	.015	.030	6.3–8.3	.060	.014	13.5–14.4
	1100853	37	4.0–4.5	—	32	—	—	—	—	—	—	—

Continued

ALTERNATOR & REGULATOR SPECIFICATIONS—Continued

Year	Alternator					Regulator						
	Model	Rated Hot Output Amps.	Field Current 12 Volts @ 80° F.	Output @ 14 Volts		Model	Field Relay			Voltage Regulator		
				2000 R.P.M. Amps.	5000 R.P.M. Amps.		Air Gap In.	Point Gap In.	Closing Voltage	Air Gap In.	Point Gap In.	Voltage @ 125° F.
1970	1100777	55	2.2–2.6	32	50	1119515	.015	.030	6.3–8.3	.060	.014	13.5–14.4
	1100878	42	2.2–2.6	28	40	1119515	.015	.030	6.3–8.3	.060	.014	13.5–14.4
	1100879	37	2.2–2.6	25	35	1119515	.015	.030	6.3–8.3	.060	.014	13.5–14.4
	1100880	37	4.0–4.5	—	32	—	—	—	—	—	—	—
	1100886	63	2.8–3.2	35	59	1119515	.015	.030	6.3–8.3	.060	.014	13.5–14.4
	1100888	37	2.2–2.6	25	35	1119515	.015	.030	6.3–8.3	.060	.014	13.5–14.4
	1100890	55	4.0–4.5	—	50	—	—	—	—	—	—	—
	1100891	55	2.2–2.6	32	50	1119515	.015	.030	6.3–8.3	.060	.014	13.5–14.4
	1100892	55	2.2–2.6	32	50	1119515	.015	.030	6.3–8.3	.060	.014	13.5–14.4
	1100893	55	2.2–2.6	32	50	1119515	.015	.030	6.3–8.3	.060	.014	13.5–14.4
	1100907	61	2.2–2.6	33	58	1119515	.015	.030	6.3–8.3	.060	.014	13.5–14.4
1971	1100553	63	4.0–4.5	—	—	1119519	—	—	—	—	—	13.5–14.4
	1100566	37	2.2–2.6	—	—	1119515	—	—	—	—	—	13.5–14.4
	1100567	42	2.2–2.6	—	—	1119515	—	—	—	—	—	13.5–14.4
	1100568	55	2.2–2.6	—	—	1119515	—	—	—	—	—	13.5–14.4
	1100569	55	2.2–2.6	—	—	1119515	—	—	—	—	—	13.5–14.4
	1100570	61	2.2–2.6	—	—	1119515	—	—	—	—	—	13.5–14.4
	1100888	37	2.2–2.6	—	—	1119515	—	—	—	—	—	13.5–14.4
	1100934	37	4.0–4.5	—	—	—	—	—	—	—	—	—
	1100935	55	4.0–4.5	—	—	—	—	—	—	—	—	—
1972	1100573	42	4.0–4.5	—	—	—	—	—	—	—	—	—
	1100597	61	4.0–4.5	—	—	—	—	—	—	—	—	—
	1102435	42	2.2–2.6	—	—	1119515	—	—	—	—	—	13.5–14.4
	1102437	55	2.2–2.6	—	—	1119515	—	—	—	—	—	13.5–14.4
	1102439	55	2.2–2.6	—	—	1119515	—	—	—	—	—	13.5–14.4
	1102440	37	2.2–2.6	—	—	1119515	—	—	—	—	—	13.5–14.4
	1102463	61	2.2–2.6	—	—	1119515	—	—	—	—	—	13.5–14.4
1973-74	1100497	37	4–4.5	—	36①	Integral	—	—	—	—	—	13.8–14.8
	1100573	42	4–4.5	—	37	Integral	—	—	—	—	—	—
	1100934	37	4–4.5	—	32	Integral	—	—	—	—	—	—
	1102367	55	—	—	—	Integral	—	—	—	—	—	—
	1102368	61	—	—	—	Integral	—	—	—	—	—	—
1975–76	1102399	37	4–4.5	—	37②	Integral	—	—	—	—	—	—
	1102481	37	4–4.5	—	37②	Integral	—	—	—	—	—	—
	1102483	37	4–4.5	—	37②	Integral	—	—	—	—	—	—
	1102488	57	—	—	—	Integral	—	—	—	—	—	—
	1102493	42	4–4.5	—	38	Integral	—	—	—	—	—	—
	1102549	61	—	—	—	Integral	—	—	—	—	—	—
	1102550	63	—	—	—	Integral	—	—	—	—	—	—
1976	1101016	80	4–4.9	—	76	Integral	—	—	—	—	—	—
	1102388	37	4–4.5	—	33	Integral	—	—	—	—	—	13.6–14.2
	1102394	37	4–4.5	—	33	Integral	—	—	—	—	—	—
	1102491	37	4–4.5	—	33	Integral	—	—	—	—	—	13.8–14.8①
	1102841	42	—	—	—	Integral	—	—	—	—	—	—
	1102842	63	—	—	—	Integral	—	—	—	—	—	—
	1102843	61	4–4.5	—	57	Integral	—	—	—	—	—	—
	1102844	63	—	—	—	Integral	—	—	—	—	—	—

①—At 7000 RPM.
②—At 5500 RPM.
③—At 85° F.

COOLING SYSTEM & CAPACITY DATA

Year	Model or Engine	Cooling Capacity, Qts. With Heater	Cooling Capacity, Qts. With A/C	Radiator Cap Relief Pressure, Lbs. With A/C	Radiator Cap Relief Pressure, Lbs. No A/C	Thermo. Opening Temp. ①	Fuel Tank Gals.	Engine Oil Refill Qts. ②	Transmission Oil 3 Speed Pints	Transmission Oil 4 Speed Pints	Transmission Oil 5 Speed Pints	Transmission Oil Auto. Trans. Qts. ⑫	Rear Axle Oil Pints
1969	6-250	12.2	12.2	15	15	195	20	4	3½	—	—	⑩	3.69
	8-350	15.2	15.7	15	15	195	20⑬	4	3½	4.90	—	⑩	3.69
	8-400, 4-4-2	16.2	17.2	15	15	195	20	4	3½	4.90	—	⑮	3.69
	Delta, 98	17.5	18	15	15	195	25	4	4.9	—	—	⑩	5.32⑭
	Toronado	18	18.5	15	15	195	24	5	—	—	—	⑮	4
1970	6-250	12	12.0	15	15	195	20	4	3½	—	—	③	3¾
	8-350⑯	15	15.5	15	15	195	20⑬	4	3½	2¼	—	③	3¾
	8-350⑰	16.5	16.5	15	15	195	25	4	3½	2¼	—	⑩	3¾
	8-455⑯	16	16.5	15	15	195	20⑬	4	3½	2¼	—	⑩	3¾
	8-455⑰	17.5	18	15	15	195	25	4	4½	2¼	—	⑩	5⅓
	Toronado	18	18.5	15	15	195	24	5	—	—	—	⑩	4
1971	6-250	13	13	15	15	195	19⑱	4	3½	—	—	③	4¼
	8-350⑯	16	17	15	15	195	19⑱	4	⑲	2¼	—	⑩	4¼
	8-350⑰	16	17	15	15	195	24	4	3½	—	—	⑩	⑳
	8-455⑯	17	18	15	15	195	19	4	⑲	2¼	—	⑩	4¼
	8-455⑰	17	18	15	15	195	24	4	3½	—	—	⑩	⑳
	Toronado	18	19	15	15	195	24	5	—	—	—	⑩	4
1972	8-350⑯	15.2	15.7	15	15	195	19⑬	4	3½	3½	—	⑩	⑳
	8-350⑰	16.2	16.7	15	15	195	24	4	—	—	—	⑩	4¼
	8-455⑯	15.2	15.7	15	15	195	19⑬	4	3½	3½	—	⑩	4¼
	8-455⑰	17	17.5	15	15	195	24	4	—	—	—	⑩	5½
	Toronado	19.5	20	15	15	195	25	5	—	—	—	⑩	4
1973	6-250	12½	12½	15	15	195	21	4	3½	—	—	⑩	4¼
	8-350 Omega	15½	16½	15	15	195	21	4	3½	—	—	⑩	4¼
	8-350 Cutlass	16	16	15	15	195	22	4	3½	2¼	—	⑩	4¼
	8-350 "88"	16¼	16¼	15	15	195	26	4	—	—	—	⑩	5½
	8-455 Cutlass	17	18	15	15	195	22	4	—	2¼	—	⑩	4¼
	8-455⑰	17	17½	15	15	195	26㉑	4	—	—	—	⑩	5½
	Toronado	19½	20	15	15	195	26	4	—	—	—	㉒	4
1974	6-250	15½	—	15	15	195	21	4	3.5	—	—	⑩	4¼
	8-350 Omega	18½	19½	15	15	195	21	4	3.5	—	—	⑩	4¼
	8-350 Cutlass	20④	20④	15	15	195	22	4	—	—	—	⑩	4¼
	8-350⑰	21⑤	21⑤	15	15	195	26	4	—	—	—	⑩	4¼
	8-350 Sta. Wagon	20④	20④	15	15	195	22	4	—	—	—	⑩	5½
	8-455 Cutlass	21	21½⑥	15	15	195	22	4	—	—	—	⑩	4¼
	8-455⑰	21	21½⑥	15	15	195	26	4	—	—	—	⑩	5½
	8-455 Sta. Wagon	21	21½⑥	15	15	195	22	4	—	—	—	⑩	5½
	Toronado	21	21½	15	15	195	26	5	—	—	—	㉒	4
1975	6-250 Omega	15½	19½	15	15	195	21	4	3½	—	—	⑩	⑨
	6-250 Cutlass	17	17	15	15	195	22	4	3½	—	—	⑩	⑨
	8-260 Omega	18½	19½	15	15	195	21	4	3½	—	—	⑩	⑨
	8-260 Cutlass	23½	23½④	15	15	195	22	4	3½	—	—	⑩	⑨
	8-350 Omega	18½	19½	15	15	195	21	4	—	—	—	⑩	⑨
	8-350⑦	20	20④	15	15	195	26⑧	4	—	—	—	⑩	⑨
	8-400	21½	22⑥	15	15	195	26⑧	4	—	—	—	⑩	⑨
	8-455	21	21½⑥	15	15	195	26⑧	4	—	—	—	⑩	⑨
	Toronado	21	21½	15	15	195	26	5	—	—	—	㉒	4
1976	6-250 Omega	15½	16½	15	15	195	21	4	3½	—	—	⑩	㉓
	6-250 Cutlass	17	—	15	15	195	22	4	3½	—	—	⑩	㉔
	8-260 Omega	23	23½	15	15	195	21	4	3½	—	3½	⑩	㉓
	8-260 Cutlass	23½	23½④	15	15	195	22	4	3½	—	3½	⑩	㉔
	8-350 Omega	21½	22	15	15	195	21	4	—	—	—	⑩	㉓

Continued

COOLING SYSTEM & CAPACITY DATA—Continued

Year	Model or Engine	Cooling Capacity, Qts.		Radiator Cap Relief Pressure, Lbs.		Thermo. Opening Temp. (1)	Fuel Tank Gals.	Engine Oil Refill Qts. (2)	Transmission Oil				Rear Axle Oil Pints
		With Heater	With A/C	With A/C	No A/C				3 Speed Pints	4 Speed Pints	5 Speed Pints	Auto. Trans. Qts. (12)	
1976	8-350(7)	20	20(4)	15	15	195	26(8)	4	—	—	—	(10)	(24)
	8-455	21	21½(6)	15	15	195	26(8)	4	—	—	—	(10)	(24)
	Toronado	21½	21½	15	15	195	26	5	—	—	—	(22)	4

(1)—For Be.manent type anti-freexe.
(2)—Add one quart with filter change.
(3)—Oil pan only 2 qts. After overhaul 10 qts.
(4)—With heavy duty cooling system add 2½ qts.
(5)—With heavy duty cooling system add 1½ qts.
(6)—With heavy duty cooling system add 2 qts.
(7)—ntermediate and full size.
(8)—Intermediate and station wagons 22 gallons.
(9)—8½″ ring gear 4¼ pts., 8⅞″ ring gear 5½ pts., 9⅜ ring gear 5½ pts.
(10)—Oil pan only 3 qts. After overhaul 10 qts.
(12)—Approximate; make final check with dip-stick.
(13)—Vista-Cruiser 23 gallons.
(14)—With Jetaway, 3.69 pts.
(15)—Refill 4 qts.
(16)—Intermediate cars.
(17)—Full size cars.
(18)—Vista-Cruiser 22 gallons.
(19)—Standard unit 3½, heavy duty 4½.
(20)—With 10 bolt cover 4¼; with 12 bolt cover 5½.
(21)—Custom Cruiser 22 gallons.
(22)—Oil pan only 4 qts. after overhaul 12 qts.
(23)—7½″ ring gear, 3½ pts.; 8½″ ring gear, 4¼ pts.; 8⅞″ ring gear, 5½ pts.
(24)—8½″ ring gear, 4¼ pts.; 8⅞″ ring gear, 5½ pts.

ENGINE TIGHTENING SPECIFICATIONS★

★Torque specifications are for clean and lightly lubricated threads only. Dry or dirty threads produce increased friction which prevents accurate measurement of tightness.

Year	Engine Model	Spark Plugs Ft. Lbs.	Cylinder Head Bolts Ft. Lbs.	Intake Manifold Ft. Lbs.	Exhaust Manifold Ft. Lbs.	Rocker Arm Shaft Bracket Ft. Lbs.	Rocker Arm Cover Ft. Lbs.	Connecting Rod Cap Bolts Ft. Lbs.	Main Bearing Cap Bolts Ft. Lbs.	Flywheel to Crank-shaft Ft. Lbs.	Vibration Damper or Pulley Ft. Lbs.
1969–74	6-250(4)	25(1)	95	(5)	(5)	—	5	35	65	60	60(13)
	V8-350	35	80(6)	35–40(7)	25	25(2)	7	42	70–80(8)	(9)	160(3)
	V8-400	35	80(6)	35–40(7)	25	25(2)	7	42	100–120	(9)	160(3)
	V8-455	35	80(6)	35–40(7)	25	25(2)	7	42	100–120	(9)	160(3)
1975–76	6-250(4)	15	95	—	(5)	—	5	35	65	60	60
	V8-260	25	85(7)	40(7)	25	25(2)	7	42	80(12)	(14)	310
	V8-350(10)	15	80	45	28	30	4	40	115	60	140(3)
	V8-350	25	85(7)	40(7)	25	25(2)	7	42	80(12)	(14)	310
	V8-400(11)	15	95	40	30	20(5)	8	43	100(12)	95	—
	V8-455	25	85(7)	40(7)	25	25(2)	7	42	120	(14)	310

(1)—1973–74, 15 ft. lbs.
(2)—Rocker arm pivot bolt to head.
(3)—Minimum.
(4)—See Chevrolet Chapter for service procedures on this engine.
(5)—Outer clamp 20 ft.-lbs., all others 30 ft.-lbs.
(6)—1972–74, 85 ft. lbs.
(7)—Clean and dip entire bolt in engine oil before tightening.
(8)—Rear 100–120 ft.-lbs.
(9)—Auto. trans. 60, std. trans. 80.
(10)—Omega only. See Buick chapter for service.
(11)—See Pontiac Chapter for service.
(12)—Rear 120 ft.-lbs.
(13)—1970, pressed on.
(14)—Auto. trans., 60 ft. lbs.; manual trans., 90 ft. lbs.

PISTONS, PINS, RINGS, CRANKSHAFT & BEARINGS

Year	Model	Piston Clearance	Ring End Gap①		Wrist-pin Diameter	Rod Bearings		Main Bearings			
			Comp.	Oil		Shaft Diameter	Bearing Clearance	Shaft Diameter	Bearing Clearance	Thrust on Bear. No.	Shaft End Play
1969–70	6-250⑧	.0005–.0016	.010	.015	.9271	1.999–2.000	.0007–.0027	2.2983–2.2988	.0003–.0029	7	.002–.006
	V8-350	.001–.002	.010	.015	.9805	2.1238–2.1248	.0004–.0033	③	.0005–.0021⑤	3	.004–.008
	V8-400 (1969)	.001–.002	.010	.015	.9805	2.4988–2.4998	.0004–.0033	2.9993–3.0003	.0005–.0021⑥	3	.004–.008
	V8-455	.001–.002	.013	.015	.9805	2.4988–2.4998	.0004–.0033	2.9993–3.0003	.0005–.0021⑥	3	.004–.008
1971	6-250⑧	.0005–.0016	.010	.015	.9271	1.999–2.000	.0007–.0027	2.2983–2.2988	.0003–.0029	7	.002–.006
	V8-350	.001–.002	.010	.015	.9805	2.1238–2.1248	.0004–.0033	③	.0005–.0021⑤	3	.004–.008
	V8-455	.001–.002	.013	.015	.9805	2.4988–2.4998	.002–.011	2.9993–3.0003	.0005–.0021⑥	3	.004–.008
1972	V8-350	.001–.002	.010	.015	.9805	2.1238–2.1248	.0004–.0033	③	.0005–.0021⑤	3	.004–.008
	V8-455	.001–.002	.010	.015	.9805	2.4988–2.4998	.002–.011	2.9993–3.0003	.0005–.0021⑥	3	.004–.008
1973-74	6-250⑧	.0005–.0016	.010	.015	.9271	1.999–2.000	.0007–.0027	2.3004	.0003–.0029	7	.002–.006
	V8-350	.001–.002	.010	.015	.9805	2.1238–2.1248	.0004–.0033	③	.0005–.0021⑤	3	.004–.008
	V8-455	.001–.002	.010	.015	.9805	2.4988–2.4998	.0004–.0033	2.9993–3.0003	.0005–.0021⑥	3	.004–.008
1975–76	6-250⑧	.0005–.0015	.010	.015	.9271	1.999–2.000	.0007–.0027	2.2983–2.2993	.0003–.0029	7	.002–.006
	V8-260	.001–.002	.010	.015	.9805	2.1238–2.1248	.0004–.0033	③	.0005–.0021⑤	3	.004–.008
	V8-350④	.0008–.0014	.013	.015	.9392	1.991–2.000	.0005–.0026	2.9995	.004–.0015	3	.002–.006
	V8-350	.001–.002	.010	.015	.9805	2.1238–2.1248	.0004–.0033	③	.0005–.0021⑤	3	.004–.008
	V8-400⑦	.0029–.0037	②	.035	.9802	2.25	.0005–.0026	3.00	.0002–.0017	4	.003–.009
	V8-455	.001–.002	.010	.015	.9805	2.4988–2.4998	.0004–.0033	2.9993–3.0003	.0005–.0021⑥	3	.004–.008

①—Fit rings in tapered bores for clearance listed in tightest portion of ring travel.
②—Top—.019, #2—.015.
③—No. 1: 2.4988–2.4998; Nos. 2, 3, 4, 5: 2.4985–2.4995.
④—Omega only. See Buick Chapter for service procedures.
⑤—Rear .0015–.0031.
⑥—Rear .002–.0034.
⑦—See Pontiac Chapter for service procedures.
⑧—See Chevrolet Chapter for service procedures on this engine.

STARTING MOTOR SPECIFICATIONS

Year	Model	Starter Number	Brush Spring Tension Oz.①	Free Speed Test			Resistance Test③	
				Amps.	Volts	R.P.M.	Amps.	Volts
1969	8-455 "98"	1108333	35	70–99②	10.6	7800–12000	—	—
	8-350, 400, 455	1108348	35	70–105②	10.6	3800–6200	480–540②	3.0
	8-350 2 B. Carb.	1108349	35	65–100②	10.6	3600–5100	300–360	3.5
	Toronado	1108352	35	70–105②	10.6	3800–6200	480–540②	3.0
	6-250	1108365	35	49–87②	10.6	6200–10700	—	—
1970–71	6-250	1108365	35	49–87②	10.6	6200–10700	—	—
	8-350	1108386	35	55–80②	9	3500–6000	—	—
	8-350, 455	1108387	35	45–80②	9	4000–6500	—	—
	8-455	1108389	35	65–95②	9	7500–10500	—	—
	Toronado	1108352	35	70–105②	10.6	3800–6200	480–540②	3.0
1972–73	6-250	1108365	35	49–87②	10.6	6200–10700	—	—
	8-350	1108386	35	55–80②	9	3500–6000	—	—
	8-455	1108387	35	45–80②	9	4000–6500	—	—
	Toronado	1108352	35	70–105②	10.6	3800–6200	480–540②	3.0
1974	6-250	1108365	35	50–80②	9	5510–10500	—	—
	8-350	1108516	35	—	—	—	—	—
	8-455	1108517	35	—	—	—	—	—
	Toronado	1108518	35	—	—	—	—	—
1975	6-250	1108365	35	50–80	9	5500–10500	—	—
	6-250	1108774⑤	35	50–80	9	5500–10500	—	—
	8-260	1108765	35	55–80	9	3500–6000	—	—

Continued

STARTING MOTOR SPECIFICATIONS—Continued

Year	Model	Starter Number	Brush Spring Tension Oz.①	Free Speed Test			Resistance Test③	
				Amps.	Volts	R.P.M.	Amps.	Volts
1975	8-350④	1108762	35	55–80	9	3500–6000	—	—
	8-350	1108765	35	55–80	9	3500–6000	—	—
	8-400	1108758	35	65–95	9	7500–10500	—	—
	8-455	1108766	—	—	—	—	—	—
1976	6-250	1108774	35	50–80②	9	5500–10500	—	—
	6-250	1108778	—	—	—	—	—	—
	8-260, 350	1108765	35	55–80	9	3500–6000	—	—
	8-350④	1108762	35	55–80	9	3500–6000	—	—
	8-455	1108766	—	—	—	—	—	—
	8-455	1108794	—	—	—	—	—	—
	Toronado	1108795	—	—	—	—	—	—

①—Minimum. ②—Includes solenoid. ④—Omega only. ⑤—Has "R" terminal removed.
③—Check capacity of motor by using a 500 ampere meter and a carbon pile rheostat to control voltage. Apply volts listed across motor with armature locked. Current should be as listed.

WHEEL ALIGNMENT SPECIFICATIONS

OLD CAR SPECIFICATIONS: For 1946-68 Wheel Alignment Specifications see main index.

Year	Model	Caster Angle, Degrees		Camber Angle, Degrees				Toe-In. Inch	Toe-Out on Turns, Deg.①	
		Limits	Desired	Limits Left	Limits Right	Desired Left	Desired Right		Outer Wheel	Inner Wheel
1969	Intermediates	−½ to −2	−1¼	−¼ to +½	−¼ to +½	+⅛	+⅛	⅛–3⁄16	18.6	20
	F.S.C.②③	−½ to −1½	−1¼	−¼ to +½	−¼ to +½	+⅛	+⅛	⅛–3⁄16	18.3	20
	F.S.C.②④	−½ to −1½	−¾	−¼ to +½	−¼ to +½	+⅛	+⅛	⅛–3⁄16	18.7	20
	Toronado	−1½ to −2½	−2	−½ to +¼	−½ to +¼	+⅛	+⅛	0–1⁄16	18.1	20
1970	Intermediates	−½ to −2	−1¼	−¼ to +½	−¼ to +½	+⅛	+⅛	⅛–3⁄16	18.6	20
	F.S.C.②③	−½ to −1½	−1¼	−¼ to +½	−¼ to +½	+⅛	+⅛	⅛–3⁄16	18.3	20
	F.S.C.②④	−½ to −1½	−¾	−¼ to +½	−¼ to +½	+⅛	+⅛	⅛–3⁄16	17.7	20
	Toronado	−1½ to −2½	−2	−¼ to +½	−½ to +½	+⅛	+⅛	0–1⁄16	18.2	20
1971–72	Intermediates	−¾ to −2¼	−1¼	−¼ to +¾	−¾ to +¼	+¼	−¼	0	—	—
	F.S.C.②	+½ to +1½	+1	−¼ to +¾	−¾ to +¼	+¼	−¼	0	—	—
	Toronado	−1½ to −2½	−2	−¼ to +¾	−¾ to +¼	+¼	−¼	0	—	—
1973	Omega	0 to +1	+½	−¼ to +¾	−¼ to +¾	+¼	+¼	⅛–¼	—	—
	Cutlass③	−½ to −1½	−1	+½ to +1½	0 to +1	+1	+½	0–⅛	—	—
	Cutlass④	−½ to +½	0	+½ to +1½	0 to +1	+1	+½	0–⅛	—	—
	88, 98	+½ to +1½	+1	+½ to +1½	0 to +1	+1	+½	1⁄16–3⁄16	—	—
	Toronado	−1½ to −2½	−2	−¼ to +¾	−¾ to +¼	+¼	−¼	0	—	—
1974	Omega	0 to +1	+½	−¼ to +¾	−¼ to +¾	+¼	+¼	⅛–¼	—	—
	Cutlass Exc. Sal.⑤	−½ to +½	0	+½ to +1½	0 to +1	+1	+½	0–⅛	—	—
	Cutlass Salon⑤	+1½ to +2½	+2	+½ to +1½	0 to +1	+1	+½	0–⅛	—	—
	88, 98⑤	+½ to +1½	+1	+½ to +1½	0 to +1	+1	+½	−1⁄16 to +3⁄16	—	—
	Toronado⑤	−1½ to −2½	−2	−¼ to +¾	−¾ to +¼	+¼	−¼	0	—	—
1975	Omega③	−1½ to −½	−1	+¼ to +1¼	+¼ to +1¼	+¾	+¾	0 to ⅛	—	—
	Omega④	+½ to +1½	+1	+¼ to +1¼	+¼ to +1¼	+¾	+¾	0 to ⅛	—	—
	Cutlass⑤	+1½ to +2½	+2	+½ to +1½	0 to +1	+1	+½	0 to ⅛	—	—
	88, 98⑤	+1 to +2	+1½	+½ to +1½	0 to +1	+1	+½	0 to ⅛	—	—
	Sta. Wagons⑤	+1 to +2	+1½	+½ to +1½	0 to +1	+1	+½	0 to ⅛	—	—
	Toronado⑤	−1 to +1	0	−¼ to +¾	−¾ to +¼	+¼	−¼	0 to 1⁄16	—	—
1976	Omega③	−½ to −1½	−1	+¼ to +1¼	+¼ to +1¼	+¾	+¾	0 to ⅛	—	—
	Omega④	+½ to +1½	+1	+¼ to +1¼	+¼ to +1¼	+¾	+¾	0 to ⅛	—	—
	Cutlass⑤	+½ to +2½	+2	+½ to +1½	0 to +1	+1	+½	0 to ⅛	—	—
	88, 98⑤	+1 to +2	+1½	+½ to +1½	0 to +1	+1	+½	0 to ⅛	—	—
	Toronado⑤	−1 to +1	0	−¼ to +¾	−¾ to +¼	+¼	−¼	0 to 1⁄16	—	—

①—Incorrect toe-out, when other adjustments are correct, indicates bent steering arms.
②—F.S.C.-Full size car.
③—Manual Steering.
④—Power Steering.
⑤—Left and right side "camber" should be different at least ¼° and no more than ¾° with the left side having the greater (+) reading.

BRAKE SPECIFICATIONS

Year	Model	Brake Drum Inside Diameter	Wheel Cylinder Bore Diameter			Master Cylinder Bore Diameter		
			Disc Brake	Front Drum Brake	Rear Drum Brake	Disc Brakes	Drum Brakes	Power Brakes
1969	Intermediate Cars	9 1/2	2 15/16	1 1/8	15/16①	1 1/8	1	1
	Intermediate Wagons	9 1/2	2 15/16	1 1/8	15/16①	1 1/8	1	1
	Delta, 98	11	2 15/16	1 3/16	1	1	1	1
	Toronado	11	2 15/16	1 1/8	7/8	1 1/16	1	1
1970	Intermediate Cars	9 1/2	2 15/16	1 1/8	7/8	1 1/8	1	②
	Intermediate Wagons	9 1/2	2 15/16	1 5/8	1	1 1/8	1	②
	Delta, 98	11	2 15/16	1 3/16	15/16	1 1/8	1	②
	Toronado	11	2 15/16	—	15/16	1 1/8	—	1 1/8
1971–72	Intermediate Cars	9 1/2	2 15/16	1 1/8	7/8	1 1/8	1	1
	Intermediate Wagons	9 1/2	2 15/16	—	1	1 1/8	—	1
	Delta, 98	11	2 15/16	—	15/16	1 1/8	—	1
	Custom Cruiser	12	2 15/16	—	1	1 1/8	—	1
	Toronado	11	2 15/16	—	15/16	1 1/8	—	1
1973–74	Omega	9 1/2	2 15/16	1 1/8	7/8	1 1/8	1	1
	Cutlass	9 1/2	2 15/16	—	7/8	1 1/8	—	1 1/8
	Vista Cruiser	11	2 15/16	—	15/16	1 1/8	—	1 1/8
	Custom Cruiser	12	2 15/16	—	1	1 1/8	—	1 1/8
	88, 98 & Toronado	11	2 15/16	—	15/16	1 1/8	—	1 1/8
1975	Omega, Cutlass	9 1/2	2 15/16	—	7/8	1 1/8	—	1 1/8
	Vista Cruiser	11	2 15/16	—	7/8	1 1/8	—	1 1/8
	Custom Cruiser	12	2 15/16	—	1	1 1/8	—	1 1/8
	88, 98, Toronado	11	2 15/16	—	15/16	1 1/8	—	1 1/8
1976	Omega	9 1/2	2 15/16	—	7/8	1 1/8	—	1 1/8
	Cutlass	11	2 15/16	—	15/16	1 1/8	—	1 1/8
	Vista Cruiser	11	2 15/16	—	15/16	1 1/8	—	1 1/8
	Custom Cruiser	12	2 15/16	—	1	1 1/8	—	1 1/8
	88, 98, Toronado	11	2 15/16	—	15/16	1 1/8	—	1 1/8

①—13/16″ with front disc brakes.

②—With drum brakes 1″; with disc brakes 1 1/8″.

REAR AXLE SPECIFICATIONS

Year	Model	Carrier Type	Ring Gear & Pinion Backlash		Pinion Bearing Preload			Differential Bearing Preload		
			Method	Adjustment	Method	New Bearings Inch-Lbs.	Used Bearings Inch-Lbs.	Method	New Bearings Inch-Lbs.	Used Bearings Inch-Lbs.
1969–73	Toronado	Removable	Shims	.005–.009	Shims	2–15	2–5④	Shims	15–20①	5–7①
	Intermediates	Integral	Shims	.005–.008	②	20–30	5–15	Shims	.010	.010
	Full Size	Integral	Shims	.005–.009	Spacer	24–32	8–12	Shims	20–30	10–20
1974–75	Toronado	Removable	Shims	.005–.009	Shims	2–15	2–5	Shims	10–15①	5–7①
	Others	Integral	Shims	.005–.009	Spacer	24–32	8–12	Shims	—	—

①—Over pinion bearing preload.

②—Tighten pinion shaft nut with inch-pound torque wrench.

③—1969–73, 2–15 inch-lbs.

④—1969–73, 2–5 inch-lbs.

Electrical Section

DISTRIBUTOR, REPLACE

1. Disconnect primary wire from distributor and disconnect pipe from vacuum control unit.

NOTE: On H.E.I. systems, disconnect feed and module connectors from distributor cap.

2. Remove distributor cap.
3. Crank engine until distributor rotor is in position to fire No. 1 cylinder and the timing mark (see *Tune Up Chart*) is aligned with the timing indicator.
4. Remove distributor clamp and lift the distributor out of the crankcase.

Installation, All Models

1. Check to make sure that the timing mark is aligned with the timing indicator with No. 1 piston on the compression stroke in position to fire.
2. Place a new seal or gasket on distributor housing.
3. Rotate distributor shaft until rotor is in position to fire No. 1 cylinder.
4. Rotate oil pump shaft with screwdriver to align slot in shaft with tongue on lower end of distributor shaft.
5. Install distributor in crankcase.
6. Install distributor clamp and bolt with lockwasher, leaving bolt just loose enough to permit movement of distributor.
7. On conventional distributors, rotate distributor housing until the breaker points just open and tighten distributor clamp bolt.
8. Connect pipe to vacuum control and primary wire to terminal stud.
9. Install distributor cap. If spark plug wires are disconnected from cap make certain that wires are connected in accordance with firing orders.
10. Check and set ignition timing.

NOTE: When using a timing light to adjust ignition timing, the connection should be made at the No. 1 spark plug. Forcing foreign objects through the boot at the No. 1 terminal of the distributor cap will damage the boot and could cause engine misfiring.

IGNITION LOCK REPLACE

1969-76

CAUTION: On vehicles equipped with an Air Cushion Restraint system, turn ignition switch to "Lock," disconnect battery ground cable and tape end, thereby deactivating system.

1. Follow the procedure to remove the turn signal switch as described further on.

Fig. 1 Ignition lock remove. 1969-76

2. Position the lock assembly in *"ACCESSORY"* position for 1969-70 models and in *"RUN"* position for 1971-76 models. Insert a long thin screwdriver into the slot as shown in Fig. 1 and pull outward on lock assembly to remove.

IGNITION LOCK DISASSEMBLY

1969-76

The ignition lock assembly is not repairable due to staking of the adapter lock to the lock cylinder. Therefore a new lock assembly service package must be obtained for any failure.

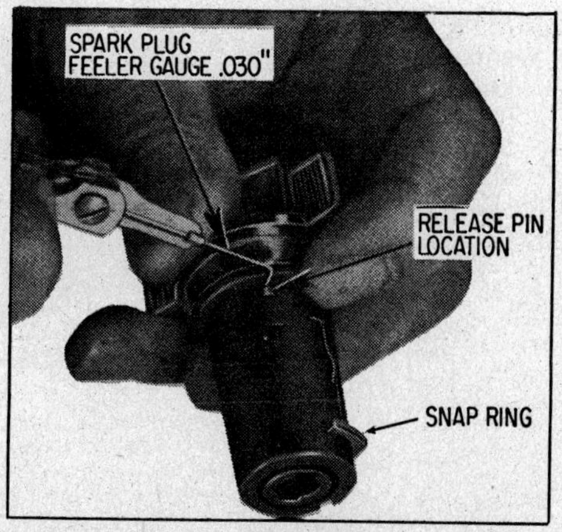

Fig. 2 Ignition lock disassembly. 1969-76

1. To separate the lock from the cylinder turn the ignition key to the "ACCESSORY" position, Fig. 2 (fully counterclockwise looking at key).
2. Depress the brass pin down and turn the lock cylinder clockwise; cylinder will pop out but will still be retained about ¼". Remove the key from the lock.
3. Tap sleeve lightly on bench with snap ring down to move the warning buzzer cam inboard. Cam must be inboard before lock cylinder can be removed.
4. Remove lock cylinder, lock & wave washer from sleeve.

STARTER, REPLACE

1969 V8-400

1. Disconnect battery and hoist car.
2. Disconnect clutch return spring at clutch release yoke.
3. Disconnect exhaust pipe from left exhaust manifold.
4. Loosen upper and remove lower starter-to-engine brace bolt.
5. Remove two starter-to-block bolts.
6. Move starter forward and downward, then disconnect wires from three starter terminals.
7. Rotate starter counterclockwise while pulling forward and downward on front of starter and remove starter. *On 4-4-2 with automatic transmission, adequate clearance can be obtained bv removing flywheel housing cover.*

1969-75 Except V8-400

1. Disconnect battery.
2. Noting position of wires, disconnect starter wiring.
3. Remove upper support attaching bolt.
4. Remove flywheel housing cover (4 screws).
5. Remove starter (2 bolts).

NOTE: If equipped with dual exhaust, the left-hand exhaust pipe may have to be disconnected to provide clearance.

IGNITION SWITCH, REPLACE

1969-76

CAUTION: On vehicles equipped with an Air Cushion Restraint system, turn ignition switch to "Lock," disconnect battery ground cable and tape end, thereby deactivating system.

1. Disconnect battery ground cable.

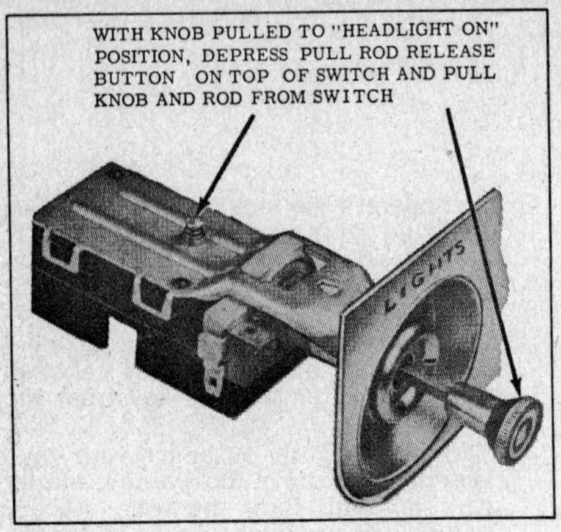

Fig. 3 Light switch with release button (typical)

2. Turn ignition lock to:
   a. "Lock" for all 1969 models.
   b. "Run" for all 1970 models.
   c. "Off-Unlock" for all 1971-76 models with fixed steering columns.
   d. "Accessory" for all 1971-76 models with tilt or tilt and telescope columns.
3. Remove cover attaching bolts, loosen toe pan clamp bolts and remove trim cap from lower part of panel.
4. Remove bracket retaining nuts and lower steering column to the seat.
5. Disconnect and remove switch.
6. Be sure that lock is in same position as when switch was removed then install switch onto actuator and column. On 1969-71 models, insert a .090 inch pin through positioning hole in switch.
7. Connect wiring and reinstall column.

LIGHT SWITCH, REPLACE

1974-76 Full Size

CAUTION: On vehicles equipped with an Air Cushion Restraint system, turn ignition switch to "Lock," disconnect battery ground cable and tape end, thereby deactivating system.

1. Disconnect battery ground cable.
2. Without disconnecting vacuum hoses or electrical connectors, remove heater or air conditioning control.
3. Remove switch escutcheon and pull switch through heater or air conditioning control opening in cluster and disconnect electrical connector.
4. Reverse procedure to install.

1973-76 Cutlass

Less Air Conditioning

1. Disconnect battery ground cable.
2. Remove steering column trim cover.
3. Pull switch to "On" and push in button on switch rear, then pull switch knob and shaft from switch.
4. Remove switch escutcheon, pull switch from panel and disconnect electrical connector.
5. Reverse procedure to install.

With Air Conditioning

1. Disconnect battery ground cable.
2. Remove steering column trim cover and disconnect A/C outlet hose.
3. Disconnect parking brake release cable.
4. Remove screws securing left hand control panel and pull control panel from instrument panel.
5. Disconnect switch electrical connector.
6. Pull switch to "On" and push in button on switch rear, then pull switch knob and shaft from switch.
7. Remove switch escutcheon and switch.
8. Reverse procedure to install.

1969-73 All & 1974-76 Omega

1. On 1969 intermediate models with A/C, remove left-hand outlet duct.
2. Disconnect multiple connector from switch.
3. Pull knob out to headlight ON position, then depress spring-loaded button on switch body and pull knob out of switch assembly, Fig. 3.
4. Remove switch escutcheon.
5. Remove switch from rear of panel.
6. On 1969 Toronado, disconnect vacuum hoses and note color coding of each.

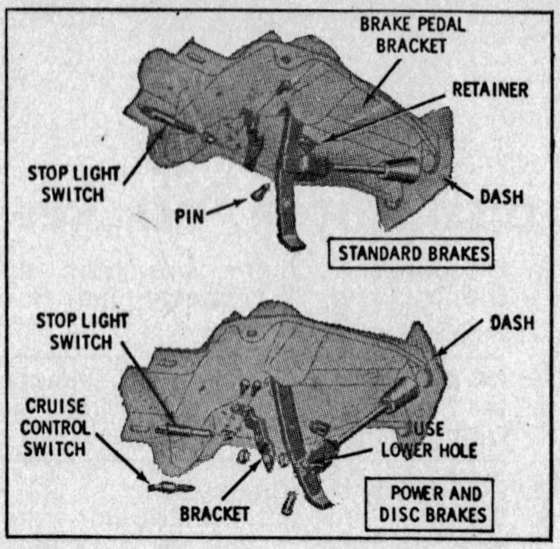

Fig. 4 Brake switch installation. 1969-76

STOP LIGHT SWITCH

1969-76

The stop light switch is attached to the brake pedal bracket and is actuated by the brake pedal arm, Fig. 4. When installing the switch, insert switch into tubular clip until switch body seats on tube clip. Pull brake pedal rearward until it contacts brake pedal stop. This moves the switch in the tubular clip providing proper adjustment.

CLUTCH START SWITCH

1969-76

All cars equipped with a manual trans-

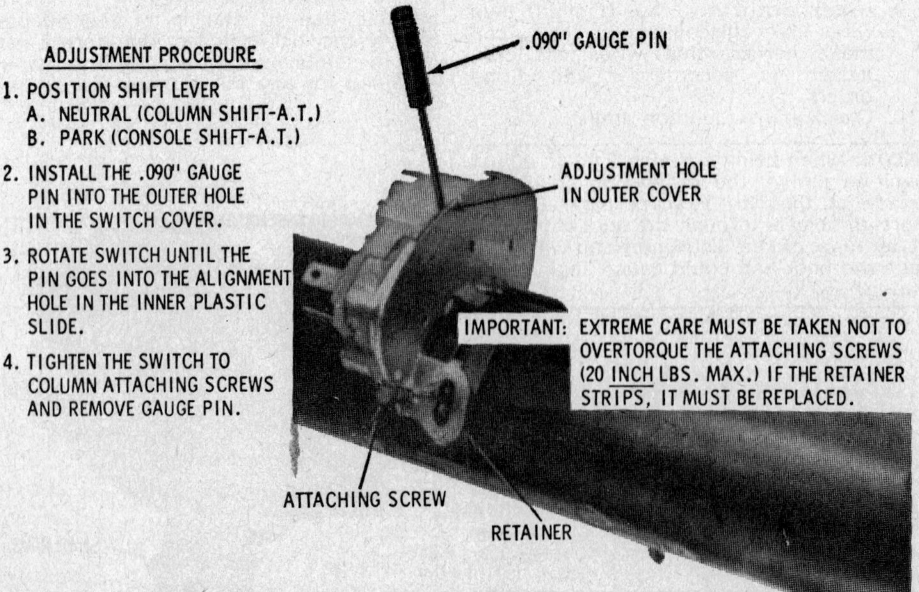

Fig. 5 Neutral safety switch adjustment. 1973-76 Auto. Trans.

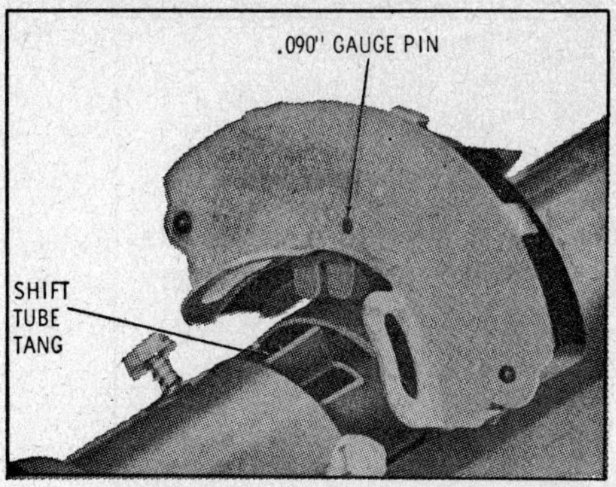

Fig. 6 Neutral safety switch adjustment. 1969-76 exc. 1973-76 auto. trans. (typical)

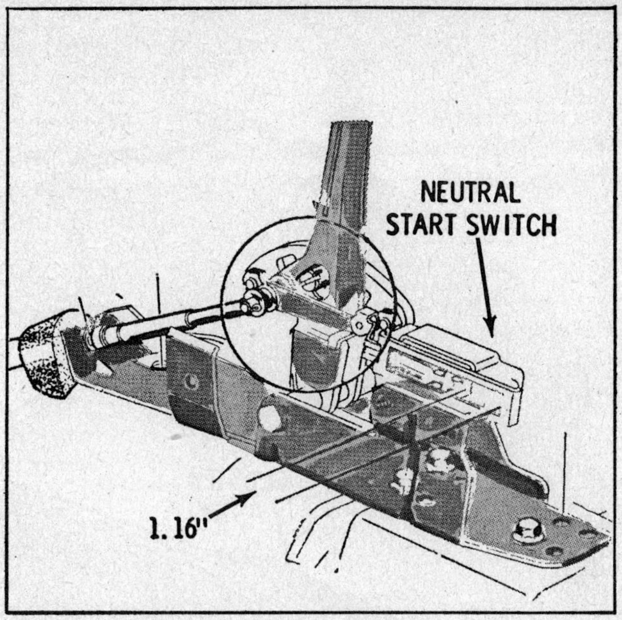

Fig. 7 Neutral start switch (console type). 1969-70

mission use a clutch start switch which is mounted on the pedal bracket. The switch closes when the clutch is depressed and completes solenoid connection. When installing switch, no adjustment is necessary.

NEUTRAL START & BACK-UP LIGHT SWITCH

The neutral safety switch is mounted on the steering column or inside the console on 1969-70 floor shift models with automatic transmission.

Checking

1. Apply parking brake firmly.
2. Position selector lever into "D" range and turn ignition switch to "Start".
3. While holding switch on "Start", slowly move selector lever toward "N" position until engine cranks and starts.
4. Without moving selector lever after engine starts, depress accelerator pedal slightly to determine whether or not transmission is in gear. If switch is properly adjusted, transmission will not be in gear.

NOTE: If equipped with back-up lights, lights should operate with ignition on and selector lever in reverse.

1971-76

For 1973-75 vehicles equipped with automatic transmission, refer to Fig. 5. For all other models, follow procedure outlined below:

1. Place selector in "Neutral" for column shift or "Park" for console shift vehicles.
2. Install a .090 inch gauge pin into outer hole in switch cover, Fig. 6.
3. On all except 1972-76 manual transmission vehicles, rotate switch until pin fits into alignment hole in inner plastic slide.
4. On all models, tighten switch to column screws and remove gauge pin.

1969-70

Adjustment (Except Console)

1. Place transmission in drive detent with switch installed and drive lug engaged in shift tube.
2. Adjust switch so gauge pin hole will align between contact support and switch bracket permitting insertion of gauge pin, Fig. 6.

Adjustment (Console Type)

1. Remove console.
2. On 1969 units, place shift lever against neutral stop. On 1970, place lever in Park.
3. On 1969 units, adjust switch to dimension shown in Fig. 7. On 1970, dimension is $2^1/_{16}$".

TURN SIGNAL SWITCH

1975-76 Tilt & Telescope

1. Disconnect battery ground cable and remove steering wheel.
2. Remove instrument panel lower trim panel, then disconnect turn signal harness connector. Remove connector from packet mounting bracket and wrap tape around connector and wires to prevent wires from snagging when removing switch.
3. Remove four bolts securing column bracket assembly to mast jacket.
4. Disconnect shift indicator.
5. Hold column in position and remove two nuts securing column bracket assembly. Then, remove bracket and turn signal wiring connector. Loosely re-install bracket to hold column in place.
6. Remove rubber bumper and plastic retainer.
7. Using a suitable compressor, depress lock plate far enough to remove "C" ring from shaft.
8. Remove lock plate and carrier assembly, then the upper bearing spring.
9. Place turn signal lever in right turn position and unscrew lever. Lift tilt lever and position in center position.
10. Push in hazard warning knob, then remove screw and hazard warning knob.
11. Remove turn signal switch attaching screws, then pull switch and wiring from top of column.

1969-76 Exc. 1975-76 Tilt & Telescope

CAUTION: On vehicles equipped with an Air Cushion Restraint system, turn ignition switch to "Lock," disconnect battery ground cable and tape end, thereby deactivating system.

1. Disconnect battery ground cable.
2. Remove steering wheel.
3. Remove cover screws and cover.
4. Using suitable compressor, depress lock plate far enough to remove the "C" ring from shaft.

NOTE: On Tilt & Telescope, compressor must be positioned on large lips of cancelling cam.

5. Remove lock plate, cancelling cam, spring and signal lever.
6. Depress hazard warning knob then unscrew knob and remove.
7. Position lever in right turn position and remove three switch attaching screws.
8. Remove panel lower trim cap, disconnect switch harness and remove bolts attaching bracket to column jacket.
9. Disconnect shift indicator if equipped.
10. Remove two nuts holding column in position, remove bracket and wire pro-

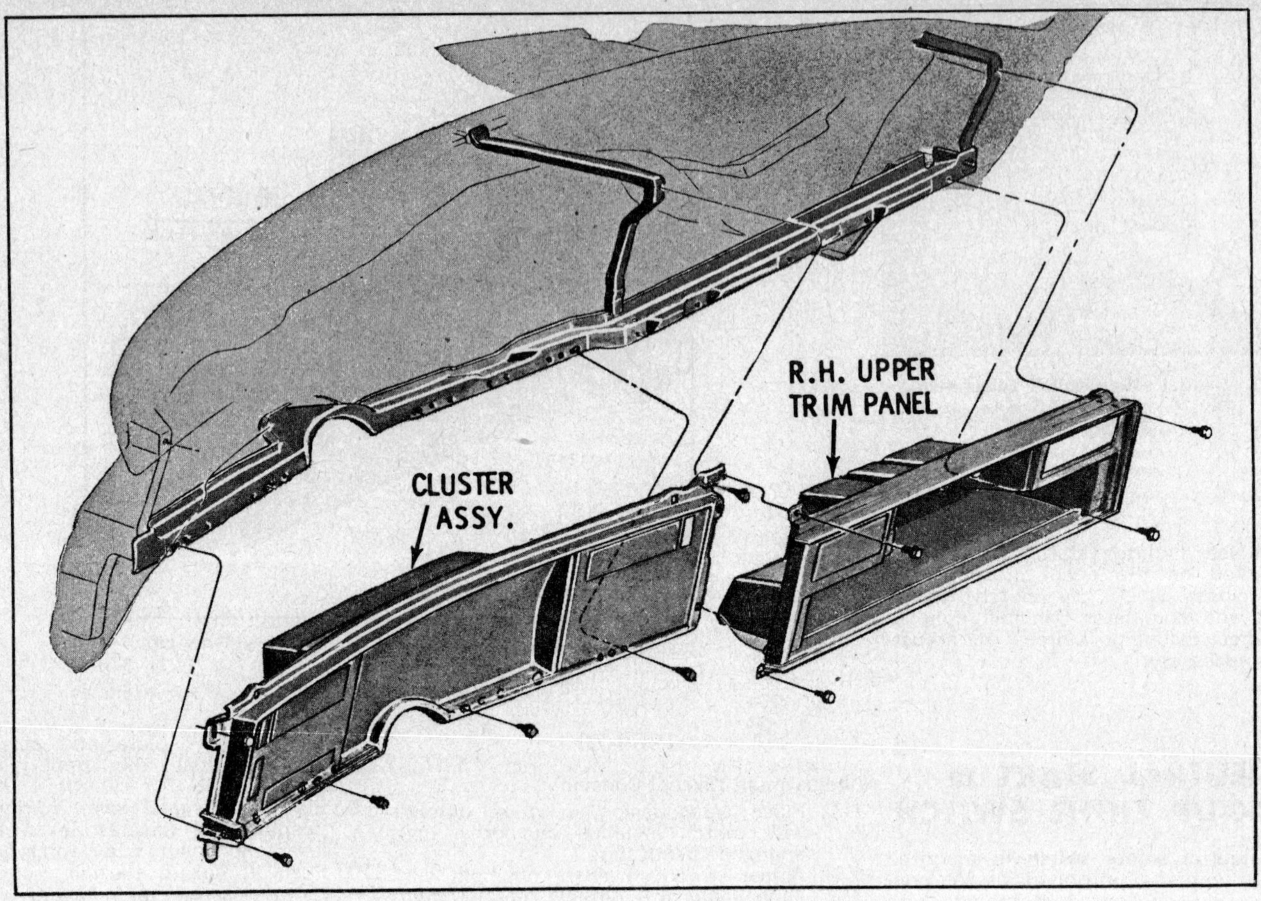

Fig. 8 Instrument cluster. 1974-76 Full Size models

tector while holding column in position then loosely install bracket to hold column in place.

11. Tape switch wires at connector keeping wires flat, then carefully remove wires and switch.

HORN SOUNDER & STEERING WHEEL

1969-76 All

CAUTION: On vehicles equipped with an Air Cushion Restraint System, turn ignition switch to "Lock," disconnect battery ground cable and tape end, thereby deactivating system. Also, on these vehicles, it is necessary to remove the drivers cushion module before removing steering wheel. With tool J-24628-2, remove the module to steering wheel screws, lift module and disconnect horn wire. Then, with tool J-24628-3, disconnect module wire connector from slip ring.

1. Disconnect battery ground cable.
2. For tilt and travel steering column proceed as follows:
   a. On 1969-71 models, pull up to remove pad assembly. On 1972-76 models, remove pad assembly retaining screws, disconnect connector and remove pad assembly.
   b. Move locking lever counterclockwise until full release is obtained. Scribe a mark on the plate assembly where the two screws attach plate assembly to locking lever (for ease of installation) and remove the two screws.
   c. Unscrew and remove plate assembly.

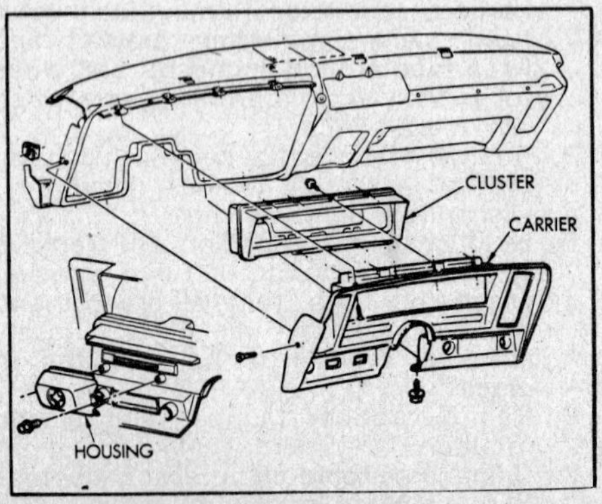

Fig. 9 Instrument cluster. 1973-76 Omega

3. For standard wheel, pull up on horn cap retainer assembly and on 1972-76 models, disconnect horn contacts.
4. For Deluxe and Rim Blow wheel, on 1969-71 models, pull down and out on pad assembly to remove. On 1972-76 models, remove three screws from pad assembly and disconnect connectors.
5. On sport wheel, pull up on and remove emblem and horn contact assembly from wheel.
6. On Wood Grain wheel, carefully pry horn cap assembly from wheel.
7. On all models, remove steering wheel nut and using a suitable puller, remove steering wheel.

NOTE: On 1975 models, remove snap ring before steering wheel nut.

8. Reverse procedure to install.

INSTRUMENT CLUSTER

1974-76 Full Size

CAUTION: On vehicles equipped with an Air Cushion Restraint system, turn ignition switch to "Lock," disconnect battery ground cable and tape end, thereby de-

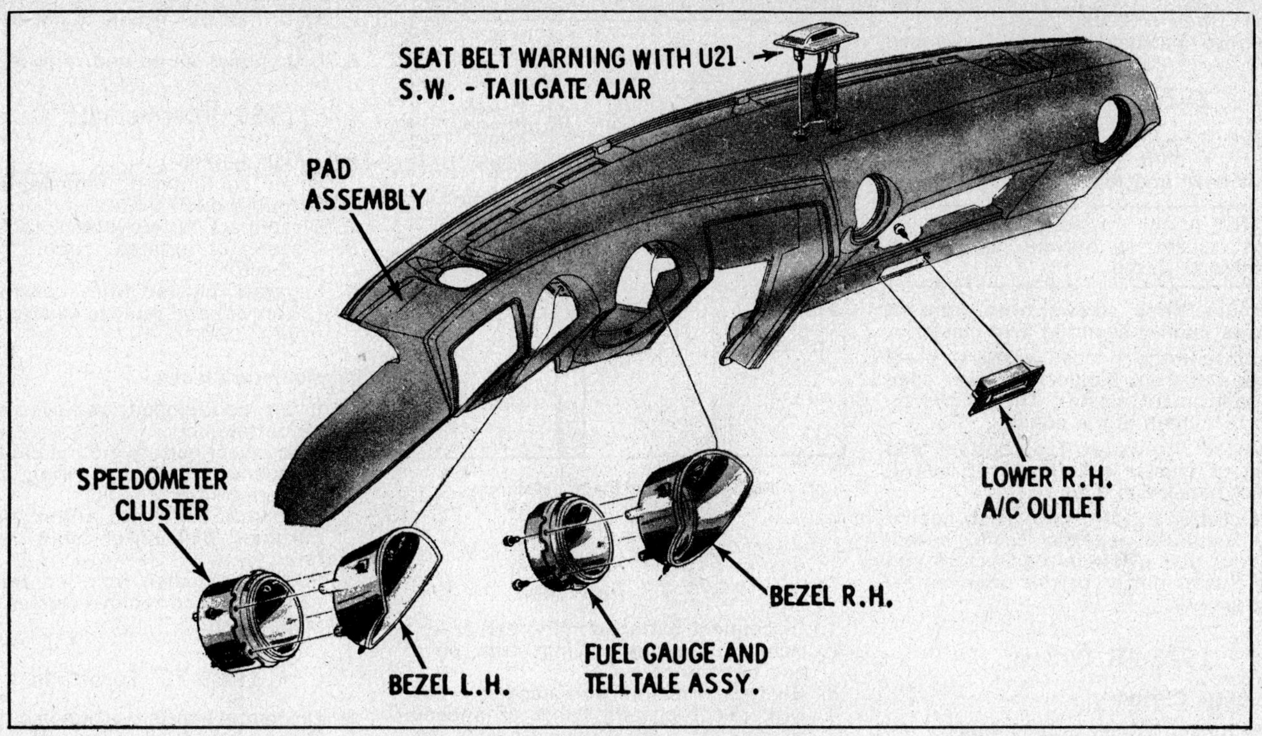

Fig. 10 Instrument cluster. 1973-76 Cutlass

activating system.

1. Disconnect battery ground cable.
2. Disconnect electrical connections at flood lamp or map lamp and remove lamp.
3. Carefully pry speakers from clips, disconnect electrical connections and remove speakers.
4. Remove instrument panel pad screws through speaker openings.
5. Remove screws from lower left outside edge of instrument panel pad, instrument cluster above speedometer, lower right hand corner of glove box and glove box top edge.
6. Pull at center edge of instrument panel pad, releasing clips holding instrument panel pad to windshield edge and remove instrument panel pad.
7. Disconnect the right hand instrument panel cover from lower trim panel and remove glove box door.

NOTE: Cover is secured by one screw and five studs pushed through clips mounted in lower panel.

8. If equipped with air conditioning, disconnect upper right hand air hose from duct.
9. On all models, disconnect electrical connectors from clock, trunk release and glove box lamp.
10. Remove the upper trim panel.
11. Disconnect the left hand instrument panel cover, slide steering column cover up the column and pull the left hand instrument panel cover from lower trim panel.

NOTE: Cover is secured by six studs pushed through clips mounted in lower panel.

12. If equipped with air conditioning, disconnect upper left hand air hose from duct.
13. On all models, disconnect temperature and defroster cables from heater case and remove radio to radio support nut.
14. Remove instrument cluster attaching screws, Fig. 8, pull cluster outward and disconnect speedometer cable.
15. Disconnect electrical connectors from instrument cluster and the vacuum harness from air conditioning or heater control.
16. Disconnect radio electrical connectors, pull fiber optic element from washer fluid indicator lens and remove screw from windshield wiper switch ground.
17. Remove three screws securing wiring harness to lower corners of instrument cluster, then disconnect harness

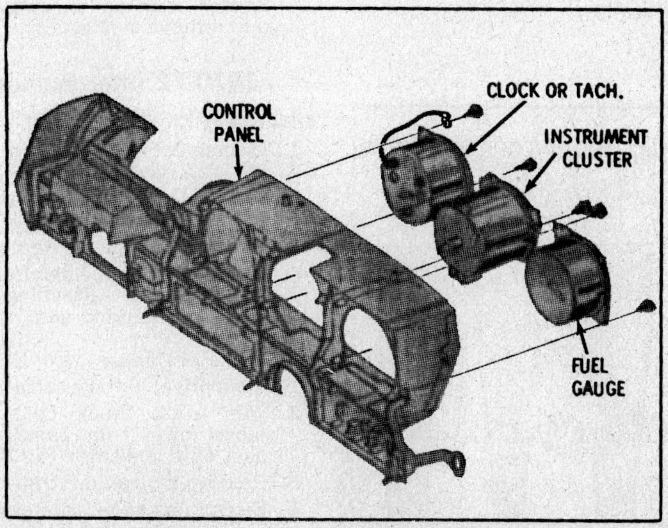

Fig. 11 Instrument cluster. 1970-72 Cutlass

from three retaining clips.

18. Remove instrument cluster assembly.

1973-76 Omega

1. Disconnect battery ground cable.
2. Remove shift indicator needle from shift bowl and lower steering column.

NOTE: Apply protective material to mast jacket to prevent damage to painted surfaces.

3. Remove three screws from front of heater control securing it to cluster.
4. Remove radio knobs, washers, bezel nuts and front support at lower edge of instrument cluster. This allows radio to remain in the panel.
5. Remove screws at top, bottom and side of cluster securing it to instrument panel, Fig. 9.
6. Tilt cluster forward and reach behind to disconnect speedo cable, speedminder and electrical connectors and lift cluster out of carrier after removing screws.

1973-76 Cutlass

Speedometer Cluster

1. Disconnect battery ground cable.
2. With automatic transmission column shift, remove lower trim cover (below steering column) then disconnect shift indicator clip on shift bowl.
3. On all models, lower steering column.
4. Disconnect speedo cable from speedometer.
5. Remove three cluster attaching screws and pull cluster out carefully so shift indicator needle is not damaged, Fig. 10. Disconnect wiring connector and remove cluster.

Fuel Gauge & Telltale Assembly

1. Perform steps 1 through 3 as outlined under "Speedometer Cluster" procedure.
2. Remove three assembly attaching screws and pull assembly out of pad, Fig. 10.
3. Disconnect wiring connectors and remove assembly.

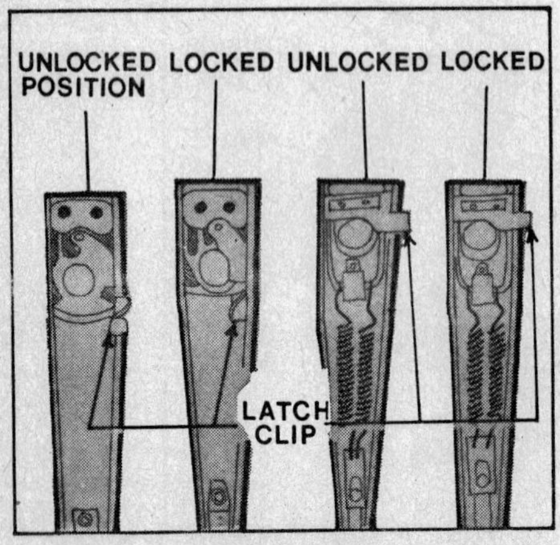

Fig. 13 Wiper arms

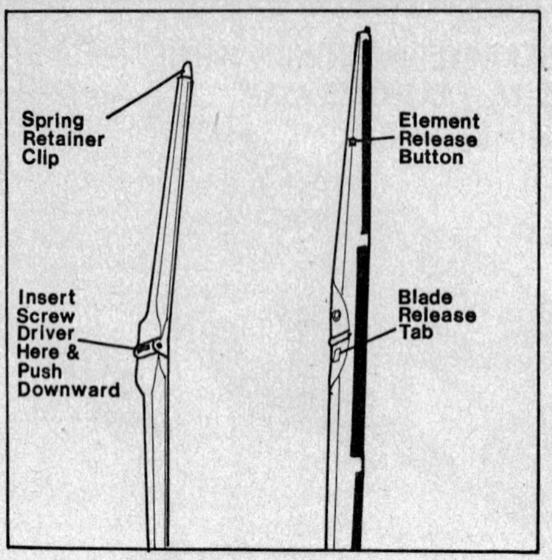

Fig. 12 Wiper blade retainers

1971-73 Full Size

1. Disconnect battery ground cable.
2. Remove steering column trim cover and left trim panel.
3. Remove shift indicator needle.
4. Disconnect speedo cable at speedo. If equipped with Cruise Control, disconnect cable at regulator and push about 6" of cable through toe pan.
5. Turn ignition to RUN position and place shift lever in Drive.
6. Remove two screws from sides of cluster.
7. Remove two screws (outboard) from top of cluster.
8. Pull out on cluster until wiring connector can be disconnected.
9. Pull cluster from dash.

1969-70 Full Size (Except Toronado)

1. Disconnect negative battery cable.
2. Disconnect printed circuit multiple connector.
3. Disconnect speedometer cable.
4. Remove pad and bezel assembly.
5. Remove three screws attaching cluster housing to instrument panel and remove cluster.

1970-72 Intermediate Cars

Fuel Gauge Cluster

1. Disconnect battery cable.
2. Remove instrument trim panel.
3. Remove three cluster attaching screws, pull cluster out and disconnect wiring, Fig. 11. Remove cluster.
4. When installing cluster, be sure that two of the attaching screws go through ground straps.

Speedometer Cluster

1. Disconnect battery cable.
2. With auto. trans. column shift, remove lower trim panel then disconnect shift indicator clip on shift bowl.
3. Remove instrument trim panel.
4. Disconnect speedometer cable.
5. Remove three cluster attaching screws and pull cluster out carefully so that shift indicator needle is not damaged, Fig. 11.
6. Disconnect wiring and remove cluster.

1969 Intermediate Cars

Fuel Gauge Cluster

1. If air conditioned, remove left-hand air outlet duct.
2. Disconnect battery ground cable.
3. Disconnect printed circuit multiple connector.
4. Separate cluster from control panel (2 screws) and remove cluster through front of pad.

Speedometer Cluster

1. If air conditioned, remove left-hand air outlet duct.
2. Disconnect battery ground cable.
3. Disconnect multiple wiring connector from printed circuit.
4. If column shift with automatic transmission, disconnect shift indicator link.
5. Separate cluster from control panel (2 screws) and remove cluster through front of pad.

1969-70 Toronado

1. Disconnect battery ground cable.
2. Disconnect Cruise Control cable at regulator (if equipped).
3. Disconnect speedometer cable at transmission.
4. Remove screws from right-hand lower pad and filler.
5. Unfasten steering column clamp (2 nuts) and retain wedge for use when installing.

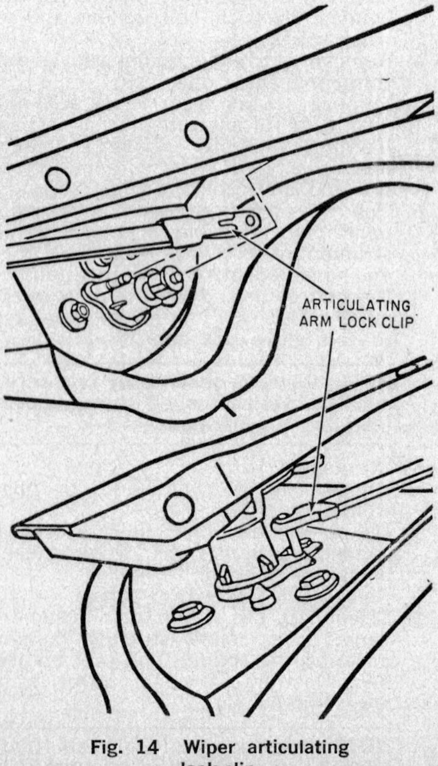

Fig. 14 Wiper articulating lock clip

6. If equipped with tilt and travel, tilt down to lowest position and let steering wheel rest on seat cushion. With a standard column, use a 4″ x 4″ block of wood between steering wheel and seat. This will prevent accidental disconnecting of hoses and wiring when tilting control panel out to perform service repairs.
7. Loosen two control housing bracket bolts about ¼″.
8. Disconnect radio lead-in.
9. Remove two control housing upper screws and tilt control panel outward from the top.
10. Disconnect printed circuit connectors.
11. Remove radio if speedometer is to be removed.
12. Separate cover from cluster (8 screws).
13. Cluster components are now accessible for service.

W/S WIPER BLADES

Two methods are used to retain wiper blades to wiper arms, Fig. 12. One method uses a press type release tab. When the release tab is depressed the blade assembly can be slid off the wiper arm pin. The other method uses a coil spring retainer. A screwdriver must be inserted on top of the spring and the spring pushed downward. The blade assembly can then be slid off the wiper arm pin. Two methods are also used to retain the blade element in the blade assembly, Fig. 12. One method uses a press type release button. When the button is depressed, the two piece blade assembly can be slid off the blade element. The other method uses a spring type retainer clip in the end of the blade element. When the retainer clip is squeezed together, the blade element can be slid out of the blade assembly.

NOTE: To be sure of correct installation, the element release button, or the spring element retaining clip should be at the end of the wiper blade assembly nearest the wiper transmission.

W/S WIPER ARMS

Models w/Rectangular Motor

1. Wiper motor must be in park position.
2. Use suitable tool to minimize the possibility of windshield or paint finish damage during arm removal.
3. Remove arm by prying up with tool to disengage arm from serrated transmission shaft.
4. To install arm to transmission rotate the required distance and direction so that blades rest in proper position.

Models w/Round Motor

1. Wiper motor must be in park position.
2. Raise hood to gain access to wiper arm.
3. *On Intermediate Models:* Lift arm off transmission shaft. On left arm, slide articulating arm lock clip, Fig. 13, away from transmission pivot pin and lift arm off pin. *On Full Size Models:* Lift arm and slide latch clip, Fig. 14, out from under wiper arm.
4. Release wiper arm and lift arm assembly off transmission shaft.

W/S WIPER MOTOR, REPLACE

1971-76

1. Raise hood and remove cowl screen or grille.
2. Reach through cowl opening and loosen transmission drive link attaching nuts to motor crankarm.
3. Disconnect wiring and washer hoses.
4. Disconnect transmission drive link from motor arm.
5. Remove motor attaching screws.
6. Remove motor while guiding crankarm through opening.

1969-70 Intermediate Cars

1. Disconnect wiring and washer hoses.
2. Remove wiper blade and arm.
3. Remove the motor attaching screws.
4. Lift rear edge of vent screen and loosen nuts attaching transmission to motor crank arm. Loosen bolts only until the crankarm will slide out of transmission socket.

1969-70 Full Size Cars

1. Disconnect wiring and washer hoses.
2. Remove three motor attaching screws.
3. Remove access hole plug.
4. Reaching through access hole, loosen two transmission crank arm nuts.
5. Hold motor with one hand and with the other move wiper arm halfway through its travel to center crank arm in dash hole.
6. Remove motor while guiding crank arm through hole.

W/S WIPER TRANSMISSION, REPLACE

1969-76

Rectangular Motor

1. Remove wiper arms and blades.
2. Raise hood and remove cowl vent screen or grille.
3. Disconnect wiring from motor.
4. Loosen, do not remove, transmission drive link to motor crankarm attaching nuts and disconnect drive link from crankarm.
5. Remove right and left transmission to body attaching screws and guide transmission and linkage out through cowl opening.

Round Motor

1. Raise hood and remove cowl vent screen.
2. On Intermediate models, remove right and left wiper arm and blade assemblies. On Full Size cars, remove arm and blade only from transmission to be removed.
3. Loosen, do not remove, attaching nuts securing transmission drive link to motor crankarm.

NOTE: On Full Size cars, if only the left transmission is to be removed, it will not be necessary to lossen attaching nuts securing the right transmission drive link to the motor.

4. Disconnect drive link from motor crankarm.
5. On Intermediate models, remove right and left transmission to body attaching screws. On Full Size cars, remove the attaching screws securing only the transmission to be removed.
6. Remove transmission and linkage by guiding it through opening.

W/S WIPER SWITCH

1974-76 Full Size

CAUTION: On vehicles equipped with an Air Cushion Restraint system, turn ignition switch to "Lock," disconnect battery ground cable and tape end, thereby deactivating system.

1. Disconnect battery ground cable.
2. Remove air conditioning or heater control without disconnecting vacuum harness or electrical connectors.
3. Remove headlamp switch escutcheon and pull headlamp switch through air conditioning or heater control opening without removing electrical connector.
4. Remove windshield wiper switch knob and two switch attaching nuts through air conditioning or heater control opening, then pull switch through opening, disconnect electrical connector and remove switch.

1973 Full Size

1. Disconnect battery ground cable.
2. Remove left hand flood lamp lens and let lamp assembly hang.
3. Remove screws from steering column, left side of trim cap.
4. Remove lower left hand trim panel screws, then disconnect the temperature cable on bottom of air conditioning or heater control.
5. Remove control panel attaching screws and control panel, guiding wiring and hoses through opening.
6. Remove electrical connector from wiper switch, wiper switch knob, switch retaining screws and switch.

1973-76 Cutlass

1. Disconnect battery ground cable.
2. Remove steering column trim cover screws and if equipped with air conditioning, disconnect outlet hose.

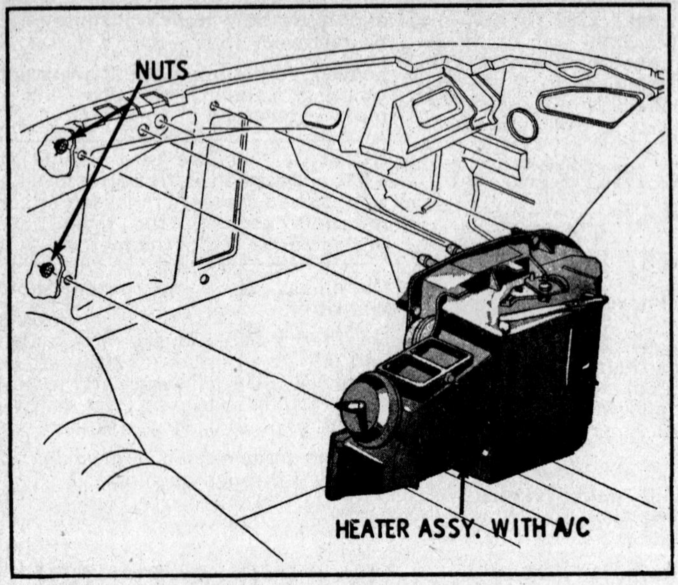

Fig. 15 Heater core. 1971-76 Full Size models & 1973-76 Cutlass (Typical)

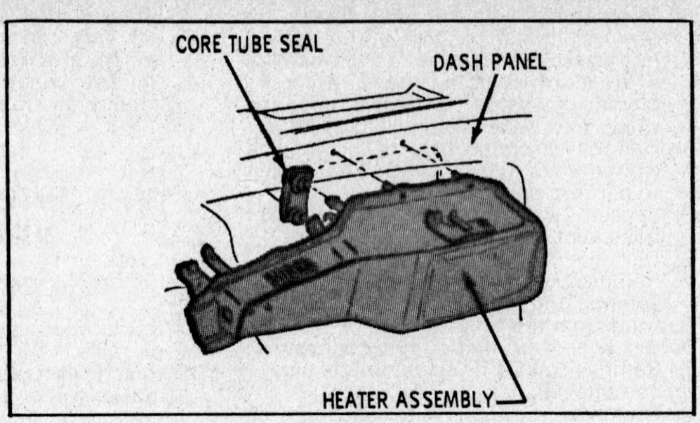

Fig. 16 Heater core. 1973-76 Omega. (Typical)

3. Disconnect parking brake cable.
4. Remove left hand control panel screws, pull control panel from instrument panel and disconnect wiring.
5. Remove electrical connector from wiper switch, wiper switch knob, switch retaining screws and switch.

1973-76 Omega

1. Disconnect battery ground cable.
2. Remove switch electrical connector, switch retaining screws and switch from behind instrument panel.

1971-72 Full Size Cars

1. Remove control panel.
2. Remove wiring connector from switch.
3. Pull wiper knob from front of switch.
4. Remove switch attaching screws and remove switch.

1971-72 Intermediate Cars

1. Disconnect battery ground cable.
2. Remove instrument trim panel.
3. Remove three fuel gauge cluster attaching screws and pull out cluster so cluster wiring can be disconnected. Remove cluster.
4. Disconnect washer/wiper connector from switch.
5. Remove knob from switch.
6. Remove switch attaching screws and remove switch.

1969-70 Except Toronado

1. On intermediate models with air conditioning, remove left-hand air outlet duct.
2. Pull wiper knob from switch.
3. Remove switch escutcheon.
4. Remove wiper and washer switch from rear of control panel.
5. Disconnect switch wiring.

1969-70 Toronado

1. Disconnect battery ground cable.
2. Disconnect wiring at switch.
3. Unfasten and remove switch from rear of control panel.

RADIO REMOVAL

NOTE: When installing radio, be sure to adjust antenna trimmer for peak performance.

1974-76 Full Size

CAUTION: On vehicles equipped with an Air Cushion Restraint system, turn ignition switch to "Lock," disconnect battery ground cable and tape end, thereby deactivating system.

1. Disconnect battery ground cable.
2. Disconnect all wiring from radio.
3. Disconnect throttle cable, then remove throttle lever and reinforcement.
4. Remove radio support bracket to tie-bar screw.
5. Remove radio knobs and two nuts securing radio to instrument cluster.
6. Lower radio and remove from behind instrument panel.

1973-76 Omega

1. Disconnect battery ground cable.
2. Remove ash tray and housing as necessary.
3. Remove knobs, controls, washers, trim plate and nuts from radio.
4. Remove hoses from center A/C distribution duct.
5. Disconnect all leads to radio.
6. Remove screws or nuts from radio rear mounting bracket and remove radio.

1973-76 Cutlass

1. Disconnect battery ground cable.
2. Remove four screws from steering column trim cover and remove cover.
3. Remove knobs from radio by pulling outward on knobs.
4. Remove nuts from front of radio.
5. Remove four screws holding R.H. Control Panel to dash and gently pull panel outward and up.
6. Remove screw from radio support bracket.
7. Remove four screws holding ash tray housing to tie bar and remove housing assembly.
8. Disconnect all leads and remove radio.

1971-73 Full Size

1. Loosen ground strap at lower valance panel support.
2. Remove radio support bracket nut at rear of radio.
3. Disconnect three flood lamps from bezel for clearance.
4. Remove four screws that hold bezel in instrument panel.
5. Pull bezel forward as far as possible.
6. Disconnect electrical connectors at rear of radio.
7. Remove radio knobs, attaching nuts and escutcheons.
8. Remove radio from rear of bezel.

1971-72 Intermediate Cars & 1969-70 Except Toronado

1. If Air Conditioned, remove manifold.

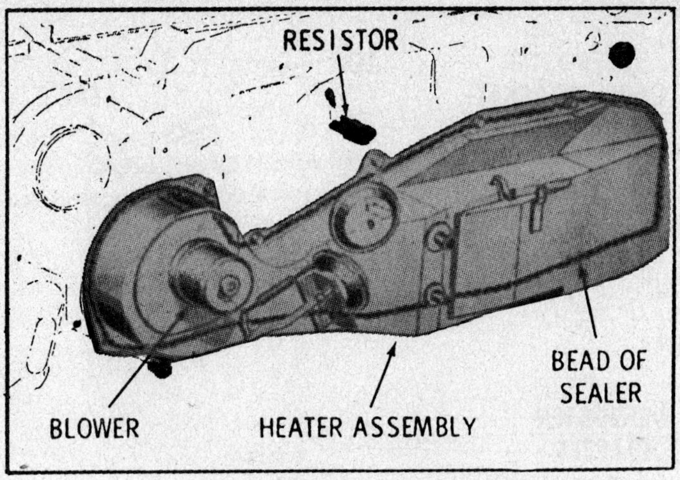

Fig. 17 Heater core & blower motor. 1969-70 Toronado (Typical)

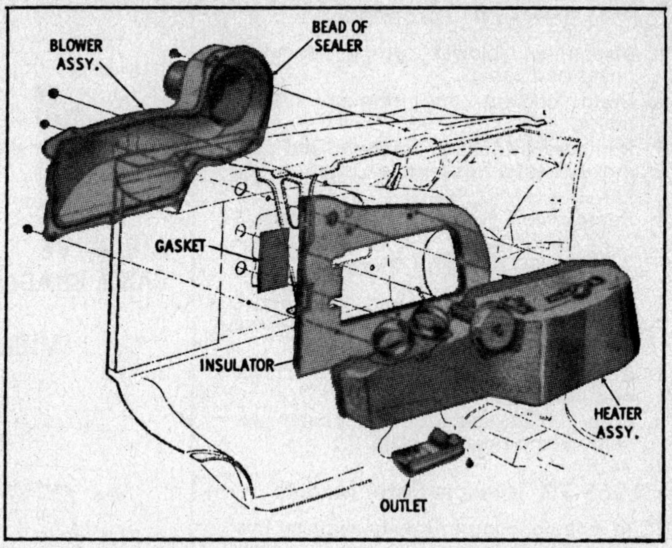

Fig. 18 Heater core & blower motor. 1969-70 Full Size models exc. Toronado (Typical)

2. Disconnect all wiring and antenna.
3. Remove knobs and attaching nuts from front of dash.
4. Remove attaching bolt (bracket to control panel) and remove radio from rear of panel.

1969-70 Toronado

1. Remove control knobs and nuts from front of dash.
2. Remove control panel.
3. Disconnect radio feed wire.
4. Disconnect dial lamp.
5. Remove attaching nut from support to radio.
6. Remove radio from rear of control panel.

HEATER CORE REMOVAL

1974-76 Full Size

CAUTION: On vehicles equipped with an Air Cushion Restraint system, turn ignition switch to "lock," disconnect battery ground cable and tape end, thereby deactivating system.

1. Disconnect battery ground cable and drain radiator.
2. Disconnect heater hoses and plug hoses and core openings to prevent coolant loss.
3. Remove heater case to dash panel attaching nuts, Fig. 15.
4. Remove instrument panel trim cover and the heater case to cowl bolts.
5. Remove lower air duct.
6. Remove instrument panel pad, then electrical connectors from glovebox light and clock.
7. Remove right hand upper trim panel and on models equipped with A/C, remove manifold from heater case.
8. On all models, remove defroster duct from heater case and disconnect lower dash trim panel.
9. Disconnect temperature and defroster cables, then the vacuum hose from heater case. Remove heater case from dash, then core from case.

1973-76 Intermediate Models

Less Air Conditioning

Cutlass

1. Disconnect battery ground cable and drain radiator.
2. Disconnect heater hoses and plug hoses and core openings to prevent coolant loss.
3. Remove heater case attaching nuts and screws, disconnect control cables and remove case from dash, then core from case, Fig. 15.

Omega

1. Disconnect battery ground cable and drain radiator.
2. Disconnect heater hoses and plug hoses and core openings to prevent coolant loss.
3. Remove heater case retaining nuts from engine side of dash, Fig. 16.
4. Remove glove box and door, then pull heater case from dash.
5. Disconnect blower resistor electrical connector and control cables from case, then remove heater case from dash and core from case.

With Air Conditioning

Cutlass

1. Disconnect battery ground cable and drain radiator.
2. Remove glove box and the center A/C manifold.
3. Disconnect vacuum hoses and temperature cable from heater case.
4. Disconnect heater hoses and plug hoses and core openings to prevent coolant loss.
5. Remove heater case to dash panel attaching bolts, then heater case from dash and core from case, Fig. 15.

Omega

1. Disconnect battery and drain coolant.
2. Disconnect upper heater hose from core.
3. Remove right front fender skirt bolts and lower skirt to gain access to lower heater hose clamp. Disconnect lower hose and remove lower right hand heater core and case attaching nut.
4. Remove glove box and door.
5. Remove recirculation vacuum diaphragm at right kick panel.
6. Remove heater outlet (at bottom of heater case).
7. Remove cold air distributor duct from heater case.
8. Remove heater case extension screws and separate extension from case.
9. Disconnect cables and wiring and remove case and core assembly, Fig. 16.
10. Separate core from case.

1971-73 Full Size

1. Disconnect battery ground cable.
2. Drain radiator below heater level, disconnect heater hoses.
3. Remove four heater case attaching nuts to dash panel, Fig. 15.
4. Disconnect temperature cable, defroster cable and vacuum hose from heater case.
5. Remove defroster duct to case attaching screw.
6. Disconnect right half of right trim panel.
7. Remove heater case from inside car.
8. Remove heater core from case.

1969-70 Toronado

1. Disconnect blower wiring, vacuum hoses and cables.
2. Drain coolant and remove heater hoses.
3. Remove attaching screws and remove heater assembly from cowl, Fig. 17.
4. Remove core from heater assembly.

1969-70 Full Size

1. Remove glove box and disconnect wiring, vacuum lines and defroster hoses from heater case.
2. Remove blower attaching screws and nuts, Fig. 18.
3. From inside car, remove heater assembly and take out core.

1969-72 Intermediate Models

1. In engine compartment, remove five attaching nuts from blower. The lower outboard nut is removed by drilling a 3/4" hole through fender filler panel at dimple or, if no dimple is provided, disconnect the right fender at the bottom and block it away from the body. Remove nut through this opening, Fig. 19.
2. Disconnect resistor wiring and three control cables.
3. Remove heater case assembly from under dash. Exposed core can now be removed.

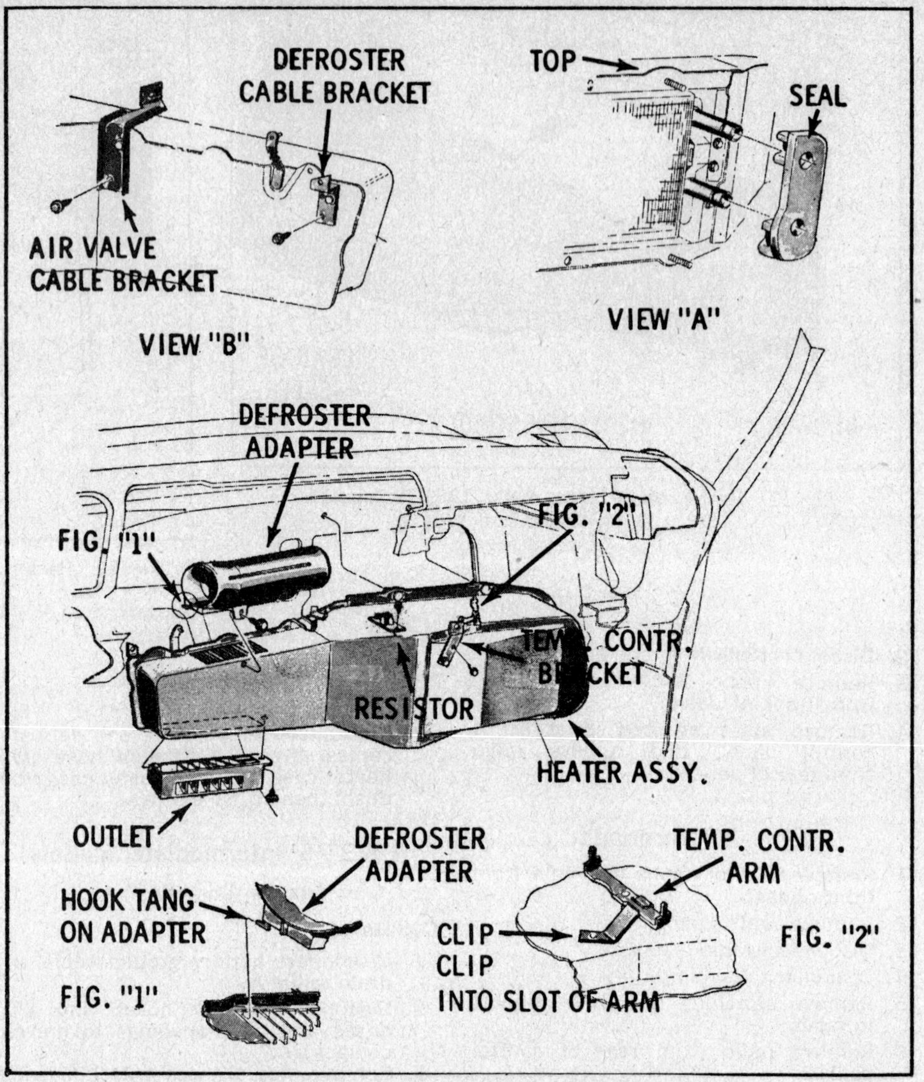

Fig. 19 Heater core. 1969-72 Cutlass (Typical)

BLOWER MOTOR REMOVE

1975-76 Full Size Models

1. Disconnect battery ground cable.
2. Raise vehicle and remove right front wheel.
3. Referring to Fig. 20, cut along inside of bead.
4. Fold flap over to expose blower motor, Fig. 21.
5. Remove blower motor attaching screws and the blower motor through opening.

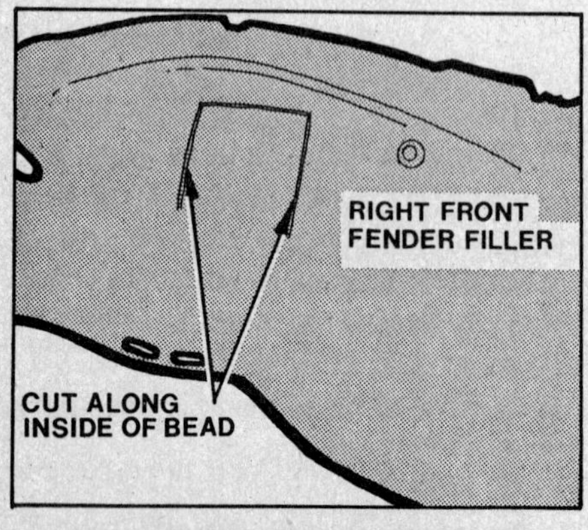

Fig. 20 Blower motor access hole location. 1975-76 Full size

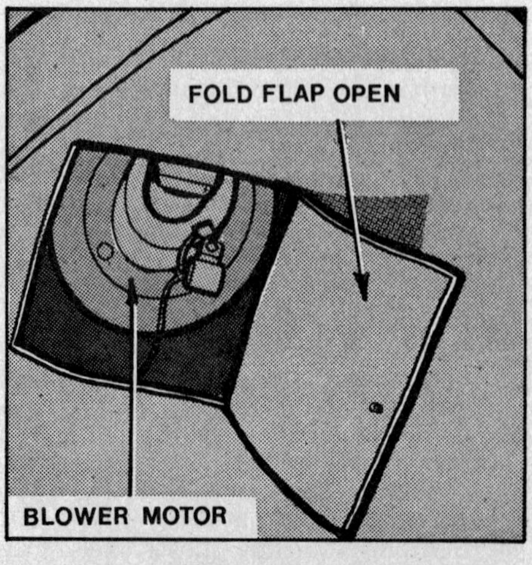

Fig. 21 Removing blower motor. 1975-76 Full size

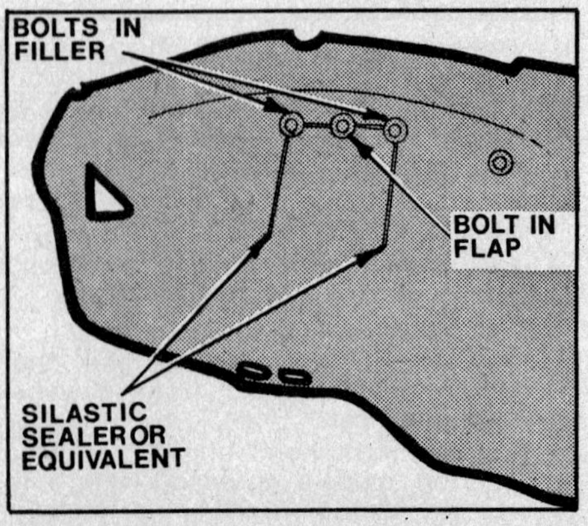

Fig. 22 Sealing blower motor access hole. 1975-76 Full size

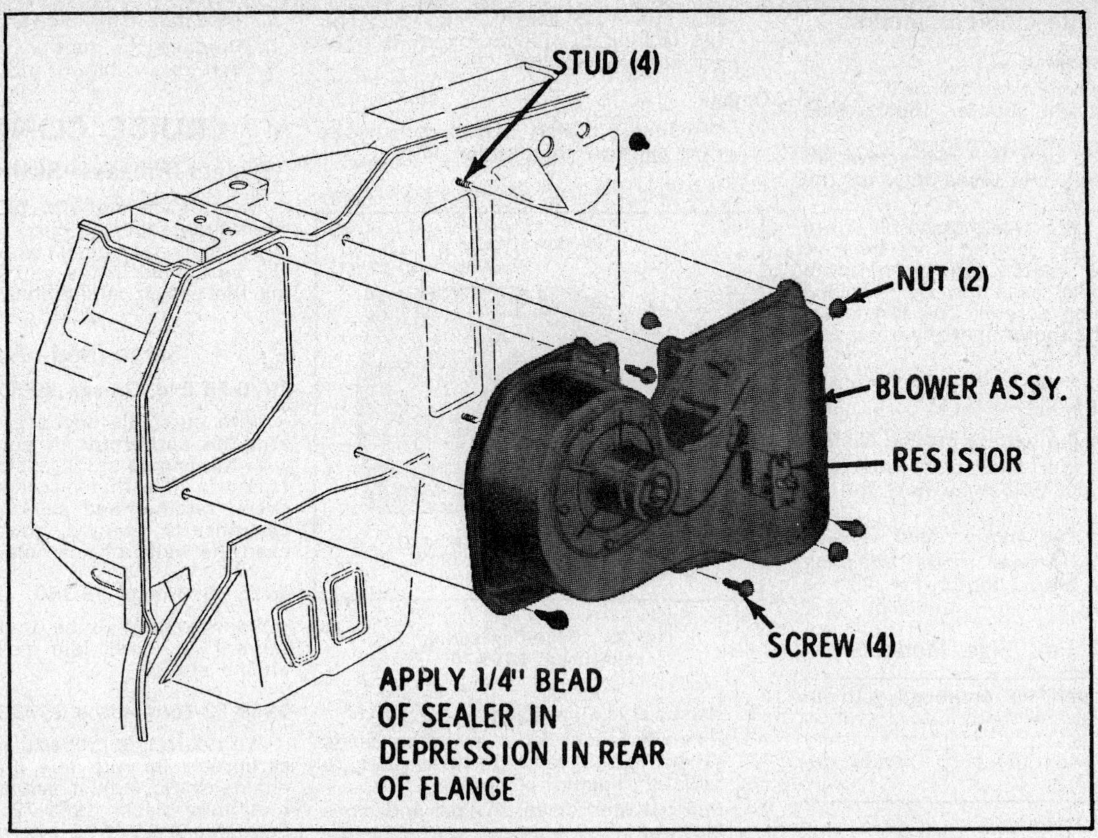

Fig. 23 Blower motor. 1969-76 Cutlass & 1971-76 Full Size models (Typical)

NOTE: After blower motor installation, the access hole must be sealed as follows:

a. Drill a hole in flap and install a $^5/_{16}$ inch self tapping bolt with a large flat washer, Fig. 22.

b. Close flap and drill two holes in inner filler panel and install two $^5/_{16}$ inch self tapping bolts with large flat washers, Fig. 22.

c. With a suitable sealer, seal edge of flap. Do not use weatherstrip adhesive.

1969-70 Full Size Models

Less Air Conditioning

1. Disconnect blower feed wire.
2. Remove upper sheet metal screw holding the heater inlet to dash by using approximately three feet of $^3/_8$" extensions and a $^7/_{16}$" socket through the opening between the fender filler plate and the fender forward of the right front wheel. Guide the socket onto the screw and remove.
3. Remove the rest of the nuts and screws attaching the heater assembly, Figs. 17 and 18.
4. Push the heater case studs back until studs do not protrude through dash.
5. Remove one fender to dash panel attaching screw just over the blower assembly case.
6. Remove the two bolts from the bottom rear of fender.
7. Push down on inner fender panel and remove blower assembly.
8. Remove blower motor attaching screws and remove blower motor.

With Air Conditioning

1. Disconnect the battery ground cable.
2. Disconnect the blower feed wire.
3. Remove blower motor attaching screws. Remove motor.

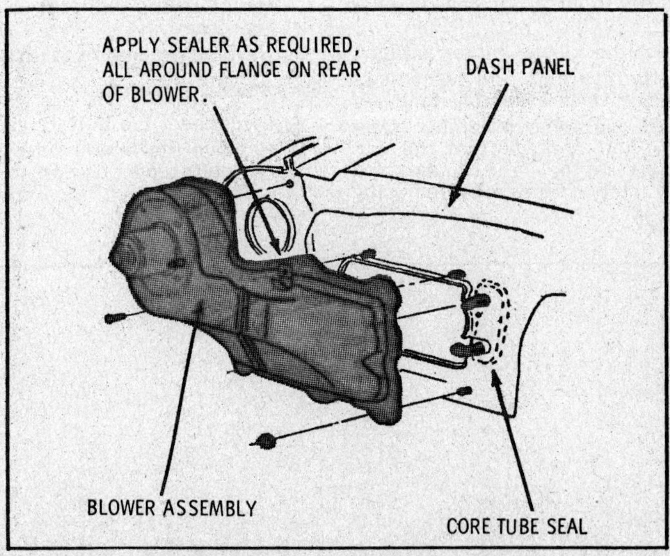

Fig. 24 Bower motor. 1973-76 Omega (Typical)

1969-70 Intermediate Models

Less Air Conditioning

1. Remove the right front wheel.
2. Disconnect the blower motor feed wire.
3. Remove five nuts and two screws attaching the inlet assembly to the dash, Fig. 23.
4. Disengage the inlet assembly from the studs and remove from the car. The blower motor can be removed from the inlet assembly by removing the attaching screws. The fan is secured to the motor shaft by a nut and lockwasher.

With Air Conditioning

1. Disconnect the battery ground cable.
2. Remove fender filler plate attaching bolts and position it forward and inboard.
3. Disconnect the blower feed wire.
4. Remove the blower motor attaching screws and blower motor, Fig. 23.

1971-74 Full Size Models

CAUTION: On vehicles equipped with an Air Cushion Restraint system, turn ignition switch to "Lock," disconnect battery ground cable and tape end, thereby deactivating system.

1. Disconnect battery.
2. Remove the right front wheel.
3. Remove canister or battery.
4. Remove the three filler plates to radiator support screws.
5. Remove the filler plate to wheelhouse attaching screws. Remove filler plate.
6. Remove the blower motor attaching screws and connector, Fig. 20.
7. Remove the blower motor.

1971-76 Intermediate Models

Less Air Conditioning

Cutlass

1. Remove the right front fender filler panel.
2. Disconnect the blower motor wiring.
3. Remove the five nuts and two screws securing the inlet assembly to dash.
4. Disengage the inlet assembly from the studs and remove from the car. The blower motor can be removed from the inlet assembly by removing the attaching screws, Fig. 23. The fan is held to the motor shaft by a nut and lock-washer.

Omega

1. Disconnect battery and detach hoses from clips on right fender skirt.

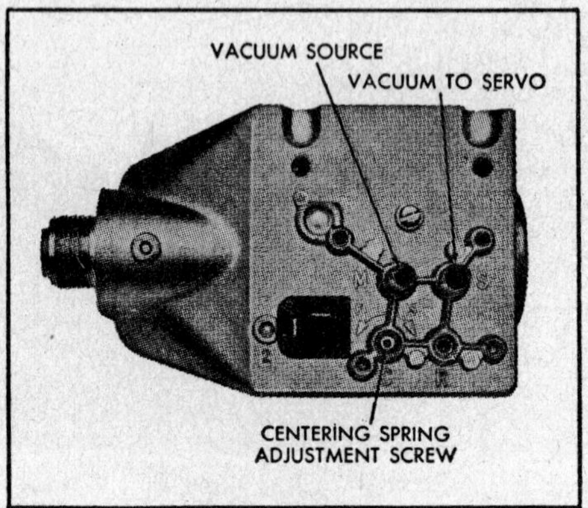

Fig. 25 Centering spring adjustment. 1969-70

2. Raise vehicle on hoist.
3. Remove fender skirt attaching bolts except those retaining the skirt to radiator support.
4. Pull out then down on skirt and place block of wood between skirt and fender to allow clearance for blower motor removal.
5. Disconnect blower motor cooling tube and electrical connections at blower motor.
6. Remove blower motor attaching screws and remove blower motor, Fig. 24. Gently pry motor flange if sealer acts as an adhesive.

With Air Conditioning

1973-75 Cutlass

1. Disconnect battery ground cable and blower motor feed wire.
2. Remove screws securing blower motor to dash, then the blower motor, Fig. 23.
3. Reverse procedure to install.

Exc. 1973-76 Cutlass

1. Disconnect the battery cable.
2. Remove the fender filler plate attaching bolts and position it forward and inboard.
3. Disconnect the blower feed wire.
4. Remove the blower motor attaching screws and blower motor, Fig. 21.

CRUISE CONTROLS

System Release Switches, Adjust

Insert switches into tubular clip until the switch seats on clip. Then, pull brake pedal rearward against stop. The switches will be moved in the clip, thereby providing the proper adjustment.

Servo Rod, Adjust

1975-76 Exc. Omega V8-350

With curb idle speed properly adjusted and the carburetor in curb idle position with the engine static, install servo rod retained in hole that provides clearance between retainer and servo bushing. Some clearance is required, however, not to exceed the width of one hole.

1975-76 Omega V8-350

Adjust chain to be taut with the carburetor in curb idle position and the engine static.

1969-72 Toronado & 1973-74 all

With curb idle properly adjusted and the carburetor in curb idle position with the engine static, adjust servo rod to provide minimum slack. 1969-72 Full Size Exc. Toronado & 1971-72 Intermediates

With curb idle speed properly adjusted and the carburetor in the curb idle position with the engine static, adjust servo rod to provide .020-.004 inch clearance between bellcrank and carburetor throttle lever.

Chain Linkage, Adjust

1969-70 Intermediates

No adjustment is required if the proper bead chain has been installed.

Centering Spring, Adjust

1. If speed control system holds speed three or more mph higher than selected speed, turn centering spring adjusting screw (C) toward (S) 1/32" or less, Fig. 25.
2. If speed control system holds speed three or more mph below selected speed, turn centering spring adjusting screw (C) toward (F) 1/32" or less. *Do not move adjustment screw (R).*

Engine Section

IMPORTANT: See the Toronado supplement for procedures on removing the engine and transmission, and method of servicing the front suspension, drive axles and final drive (differential) immediately following this chapter.

See Chevrolet Chapter for Service Procedures on 6-250 Engine.
See Buick Chapter for Service Procedures on 1975-76 Omega V8-350 Engine.
See Pontiac Chapter for Service Procedures on 1975 V8-400 Engine.

ENGINE MOUNTS

1973-76 Six

1. Remove engine mount nut and through bolt.
2. Raise engine and remove mount from frame.
3. Install new mount on frame, lower engine and install through bolt.

1969-71 Six

1. Remove nut, washer, spacer then engine mount through bolt, Fig. 1.
2. Raise engine to release weight from mount.
3. Remove mount, stop bracket and frame bracket assembly from crossmember, then remove stop bracket and mount from frame bracket.
4. Install stop bracket and new mount on frame bracket, then install assembly on crossmember.
5. Lower engine, install through bolt and tighten mount bolts.

1969-76 V8

Removal or replacement of a motor mount can be accomplished by supporting the weight of the engine at the area of the mount to be replaced, Figs. 1A thru 1G.

ENGINE, REPLACE

1969-76 V8s

1. Mark hood hinge before removing to aid in proper alignment upon reassembly.
2. Drain radiator and disconnect battery.
3. Disconnect radiator hoses, heater hoses, vacuum hoses, power steering pump hoses (if necessary), starter cable at junction block, engine to body ground strap, fuel hose from fuel line, wiring and accelerator linkage.
4. Remove fan blade and pulley, coil and upper radiator support.
5. Raise car.
6. Disconnect exhaust pipes at manifolds.

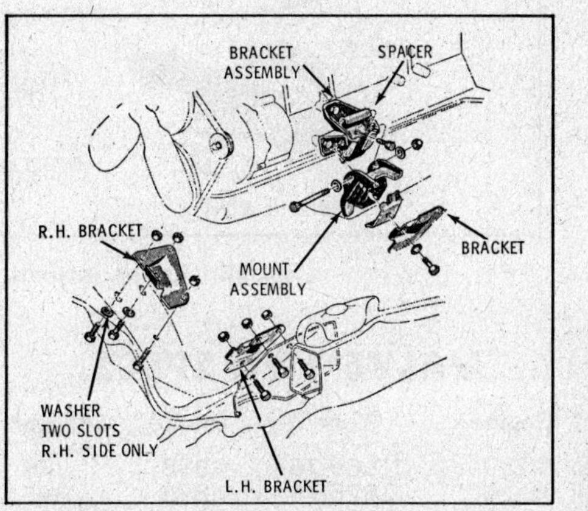

Fig. 1 Engine mounts. 1969-71 Six

7. Remove torque converter cover and the three bolts securing converter to flywheel.
8. Remove engine mount bolts, then three transmission to engine bolts on right side.
9. Remove starter with wiring attached and position aside.
10. Lower vehicle and support engine with suitable lifting equipment.
11. Support transmission with a suitable jack and remove three left hand transmission to engine bolts.
12. Remove engine from vehicle.

NOTE

On V8 engines, whenever installation of a front engine mounting becomes necessary, the cap screws fastening the mounting to the frame or bracket should first be screwed finger tight, then tightened alternately, one at a time. *Do not tighten one cap screw in position independently of the other.* This is extremely important since the lower portion of the assembly would not seat evenly in the upper portion. The front mounting must be properly positioned and tightened, otherwise the mounting will not properly function as an insulator.

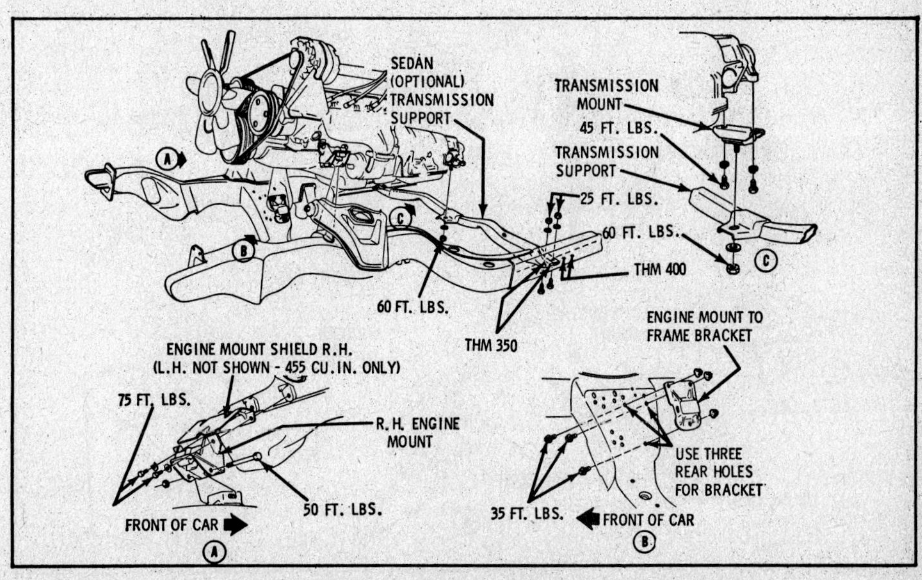

Fig. 1A Engine mounts. 1973-75 Cutlass (Typical)

CYLINDER HEAD, REPLACE

Some cylinder head gaskets are coated with a special lacquer to provide a good seal once the parts have warmed up. Do not use any additional sealer on such gaskets. If the gasket does not have this lacquer coating, apply suitable sealer to both sides.

Tighten cylinder head bolts a little at a time in three steps in the sequence shown in the illustrations. Final tightening should be to the torque specifications listed in the *Engine Tightening* table.

V8-260, 350, 400, 455

1. Drain radiator and cylinder block.
2. Disconnect spark plug wires and remove intake manifold.
3. Disconnect exhaust crossover pipe for left side and/or crossover pipe and exhaust pipe for right side.
4. Remove valve cover (loosen or remove any accessory brackets that interfere).
5. Remove ground strap from right cylinder head.
6. Remove exhaust manifold, rocker arms and push rods from head to be removed.
7. Reverse removal procedure to install head, and tighten bolts in the sequence shown in Fig. 2.

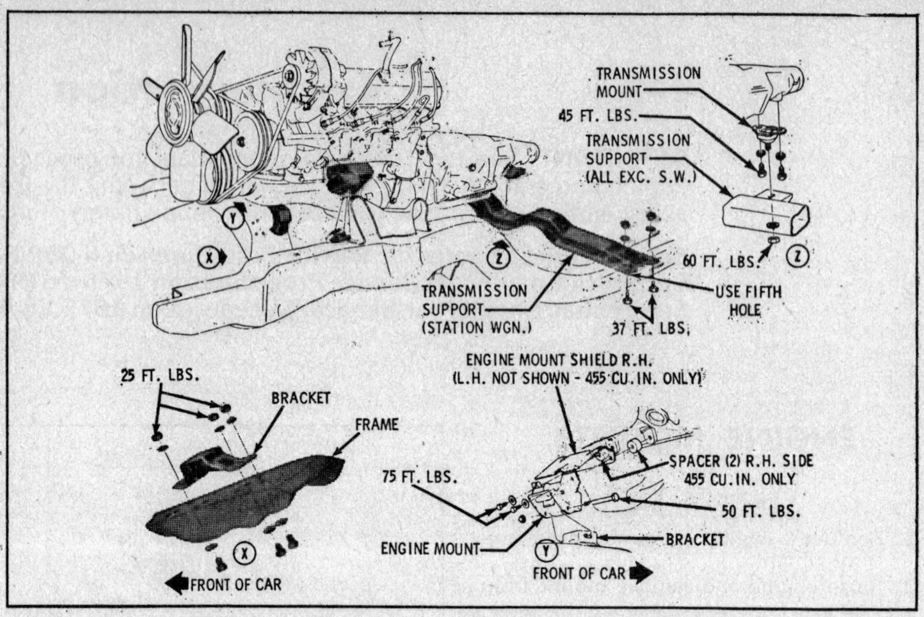

Fig. 1B Engine mounts. 1973-75 88 & 98 (Typical)

VALVE ARRANGEMENT

Front to Rear

6-250 .... E-I-I-E-E-I-I-E-E-I-I-E
V8-260, 350, 400, 455 .... I-E-I-E-E-I-E-I

VALVE LIFT SPECS.

Engine	Year	Intake	Exhaust
6-250	1969-76	.388	.388
V8-260	1975-76	.395	.395
V8-350	1969-70	.435	.435
	1971-72 [2] [5]	.400	.400
	1971-72 [4]	.472	.472
	1972 [10]	.440	.440
	1973-74 [1]	.440	.440
	1973-76 [2]	.400	.400
V8-350 [11]	1975-76	.382	.382
V8-400	1969 [4]	.472	.472
	1969 [5]	.430	.432
V8-400 [12]	1975	.410	.410
V8-455	1969-70 [5]	.435	.435
	1971 [1]	.435	.435
	1971 [9]	.472	.472
	1972	.435	.435
	1973-74 [3]	.474	.472
	1973-76	.435	.435

[1]—4 bar. carb. [2]—2 bar. carb.
[3]—Cutlass
[4]—4 bar. carb., std. trans.
[5]—4 bar. carb., auto. trans.
[6]—310, 365 H.P. [7]—375 H.P.
[8]—385 H.P.
[9]—"4-4-2".
[10]—California
[11]—Omega only. See Buick Chapter for service procedures.
[12]—See Pontiac Chapter for service procedures.

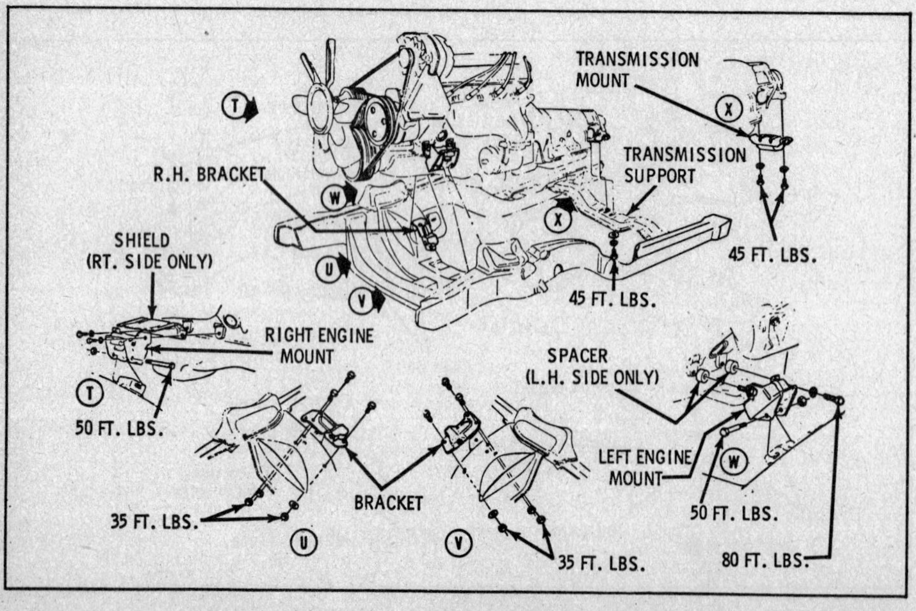

Fig. 1C Engine Mounts. 1973-75 Omega & F85 (Typical)

VALVE TIMING

Intake Opens Before TDC

Engine	Year	Degrees
6-250	1969-76	16
V8-260	1975-76	14
V8-350	1969-70	16
	1971 [5] [6]	14

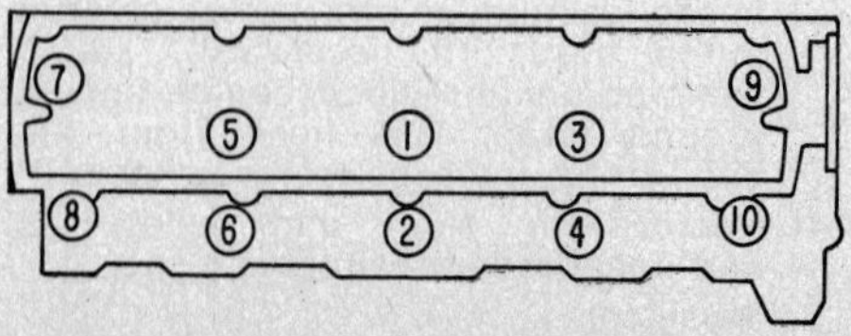

Fig. 2 Cylinder head tightening sequence. V8-260, 350, 400, 455

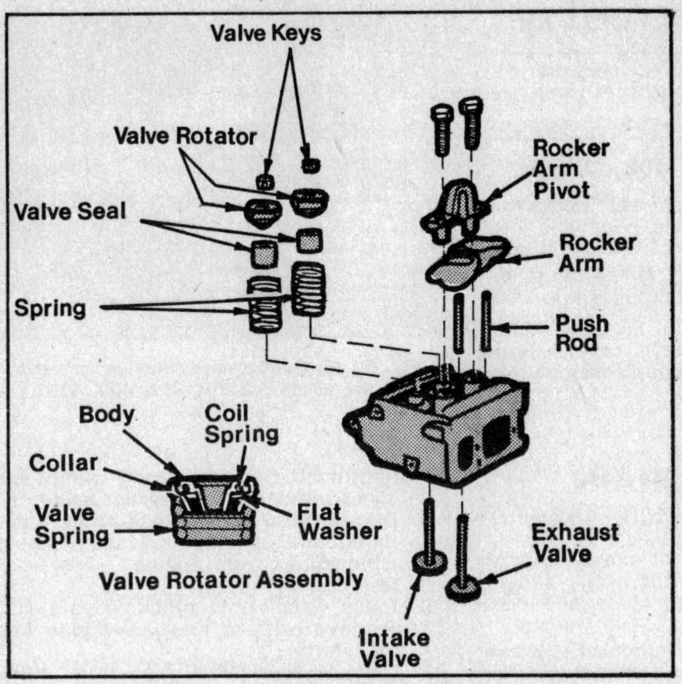

Fig. 3 Cylinder head exploded. 1970-76 V8

Fig. 4 Removing rocker arm. 1969 V8-350, 400, 455

5. Reverse removal procedure to install cover, noting data in Fig. 9.

	1971[7]	30
	1972[5][7]	16
	1972[6]	30
	1972[8]	22
	1973-74	22
	1973-74[5]	16
	1975-76	16
V8-350[9]	1975-76	19
V8-400	1969[1]	30
	1969[2]	21
V8-400[10]	1975	30
V8-455	1969-70[4]	20
	1971-76	20
	1971 Toronado	22
	1971 "4-4-2"[1]	30
	1971 "4-4-2"[2]	24
	1973-74[3]	28

[1]—Std. trans. [2]—Auto. trans.
[3]—Cutlass
[4]—GT & 400 H.P.—24.
[5]—2 bar. carb.
[6]—4 bar. carb., std. trans.
[7]—4 bar. carb., auto. trans.
[8]—California
[9]—Omega only. See Buick Chapter for service procedures.
[10]—See Pontiac Chapter for service procedures.

ROCKER ARMS

NOTE

On 1969 V8's, valve spring retainers are used in place of the valve rotators used on 1970-76 V8's, Fig. 3. The rotator operates on a sprag clutch principle utilizing the collapsing action of a coil spring to give rotation to the rotor body which turns the valve.

1970-76 V8-260, 350, 455

1. Remove valve cover.
2. Remove flanged bolts, rocker arm pivot and rocker arms, Fig. 3.
3. When installing rocker arm assemblies, lubricate wear surfaces with suitable lubricant. Torque flanged bolts to 25 ft. lbs.

1969 V8-350, 400, 455

1. To remove rocker arm assemblies, first remove valve cover.
2. Hold stud to prevent it from turning and remove rocker arm stud nuts, washer, lock plates, pivots and rocker arms, Fig. 4.

NOTE: Remove each set (one set per cylinder) as a unit.

Installation

1. Torque studs to 35 ft-lbs. Later engines use pivots and the pivot bolts are torqued to 25 ft-lbs.
2. Position a set of rocker arms (for one cylinder) on proper studs.
3. Install pivots and lock plates, being sure tangs on lock plates seat properly in the pivots. Coat wear points with lubricant.
4. Install flat washers and nuts, being sure both valve lifters are in closed valve position when nuts are tightened.

CAUTION: Never exceed specified torque of 25 ft-lbs on nuts. Check to be sure clearance exists between push rod and push rod hole in cylinder head. If no clearance exists, lock plate must be replaced.

VALVES, REMOVE

NOTE

Whenever a new valve is installed or after grinding valves, it will be necessary to measure valve stem height using the Special Tool shown in Figs. 5 and 6.

On 1975-76 V8-260, 350 and 455 engines, there should be a minimum clearance of .015 inch between gauge surface and the valve stem, Fig. 5. On 1969-74 V8-260, 350 engines, there should be .035 inch clearance between gauge surface and valve stem, Fig. 5. On 1969-74 V8-400, 455 engines, the minimum clearance is .005 inch.

On 1970-76 engines, check valve rotator height, Fig. 6. If valve stem tip extends less than .005 inch above rotator, replace the valve. On 1969 engines, check the valve retainer height, Fig. 6. If valve stem tip extends less than .030 inch above retainer, replace the valve.

Lacking this tool the only alternative is to lay flat feeler gauges on the retainer and check the distance between the retainer and valve stem tip.

VALVE GUIDES

V8-260, 350, 400, 455

Valve stem guides are not replaceable, due to being cast in place. If valve guide bores are worn excessively, they can be reamed oversize.

If a standard valve guide bore is being reamed, use a .003″ or .005″ oversize reamer. For the .010″ oversize valve guide bore, use a .013″ oversize reamer. If too

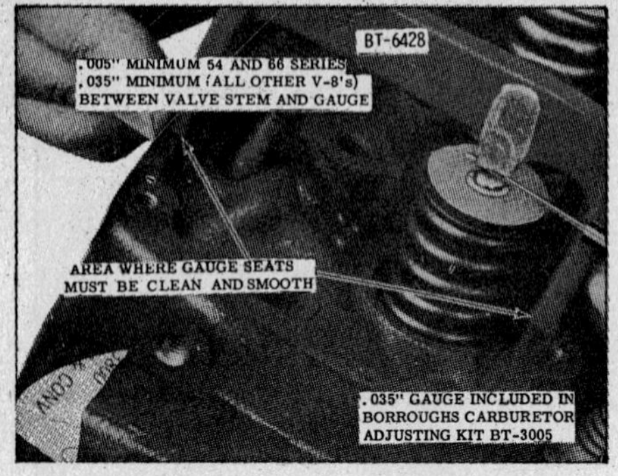

Fig. 5 Measuring valve stem height. V8-260, 350, 400, 455

large a reamer is used and the spiraling is removed, it is possible that the valve will not receive the proper lubrication.

NOTE: Occasionally a valve guide will be oversize as manufactured. These are marked on the cylinder head as shown in Fig. 7. If no markings are present, the guide bores are standard. If oversize markings are present, any valve replacement will require an oversize valve. Service valves are available in standard diameters as well as .003", .005", .010" and .013" oversize.

VALVE LIFTERS

Valve lifters are available in standard size and an oversize of .010 inch. An "O" is etched on the side of the .010 inch oversize lifter for identification. Also, the cylinder block near the valve lifter bore is marked with an "O". Ensure valve lifters are re-installed in original bores.

Plungers are not interchangeable because they are selectively fitted to the bodies at the factory.

If plunger and body appear satisfactory blow off with air to remove all particles of dirt. Install the plunger in the body without other parts and check for free movement. A simple test is to be sure that the plunger will drop of its own weight in the body, Fig. 8.

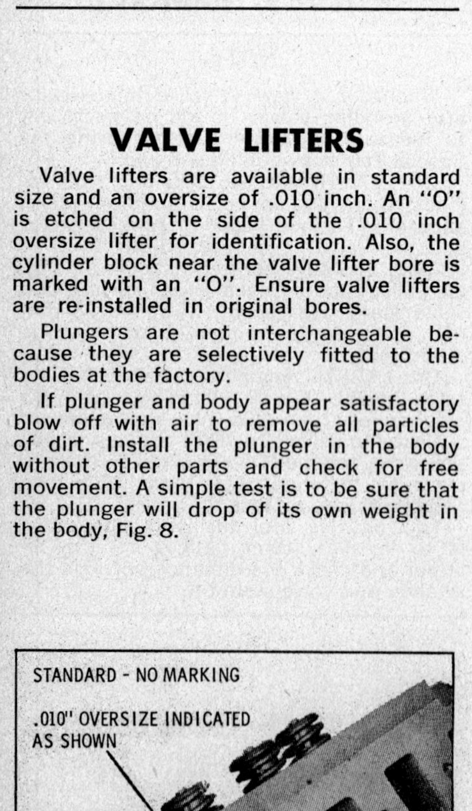

Fig. 7 Valve guide bore marking. V8-260, 350, 400, 455

TIMING CASE COVER

NOTE: When it becomes necessary to replace the cover oil seal, the cover need not be removed.

V8-260, 350, 400, 455

1. Drain cooling system and disconnect heater hose, by-pass hose and both radiator hoses.
2. Remove all belts, fan and pulley, crankshaft pulley and pulley hub.
3. Remove oil pan.
4. Unfasten and remove cover, timing pointer and water pump assembly.

TIMING CHAIN

V8-260, 350, 400, 455

1. After removing front cover, remove fuel pump eccentric, oil slinger, crankshaft sprocket, chain and camshaft sprocket.
2. Install camshaft sprocket, crankshaft sprocket and timing chain together, aligning timing marks as shown in Fig. 10.
3. Install fuel pump eccentric with flat side rearward, Fig. 11. Then install oil slinger and replace front cover.

CAMSHAFT, REPLACE

V8-260, 350, 400, 455

1. Disconnect battery ground cable and drain radiator.
2. Remove upper radiator baffle and disconnect upper radiator hose from water outlet.
3. Disconnect transmission oil cooler lines at radiator.
4. Remove raditor fan shroud, then the radiator.
5. Disconnect fuel lines from fuel pump.
6. Remove air cleaner and disconnect throttle cable.
7. Remove all drive belts and position alternator, power steering pump and air conditioning compressor aside.
8. Disconnect by-pass hose from water pump and all electrical and vacuum connections from engine.
9. Remove distributor.
10. Raise vehicle and drain oil pan.
11. Remove exhaust cross-over pipe and the starter.
12. Disconnect exhaust pipe from manifold.
13. Install engine support bar.
14. Remove engine mount to bracket bolts, raise engine and remove engine mounts.
15. Remove flywheel cover and engine oil pan.
16. Place wood blocks between exhaust manifolds and cross-member to support engine, then remove engine support bar.
17. Remove crankshaft pulley and balancer, then the engine front cover.
18. Lower vehicle and remove valve covers, intake manifold, rocker arms, push rods and valve lifters.

NOTE: Note position of valve train components to ensure installation in original location.

---

19. If equipped with A/C, discharge refrigerant and remove condenser.
20. Remove fuel pump eccentric, camshaft sprocket, oil slinger and timing chain.

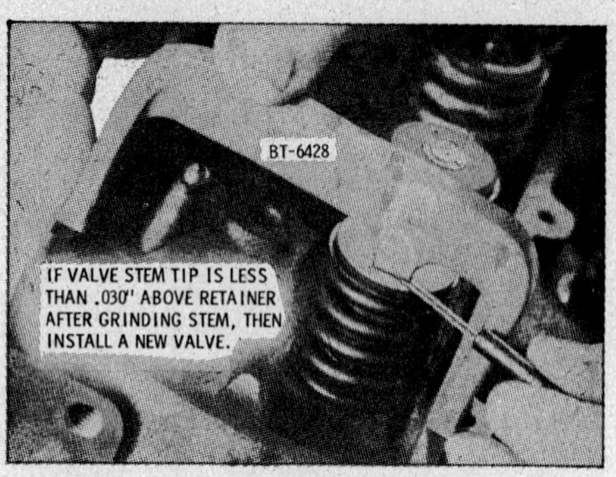

Fig. 6 Measuring valve retaining or valve rotor height. V8-260, 350, 400, 455

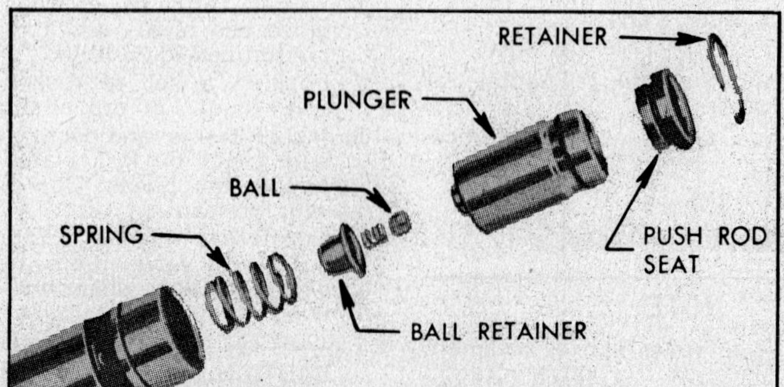

Fig. 8 Hydraulic valve lifter (typical)

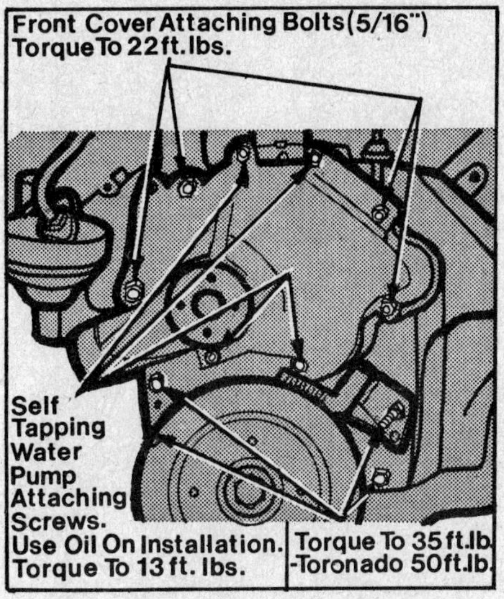

Fig. 9 Engine front cover bolts. V8-350, 400, 455

Fig. 10 Timing chain position. V8-350, 400, 455

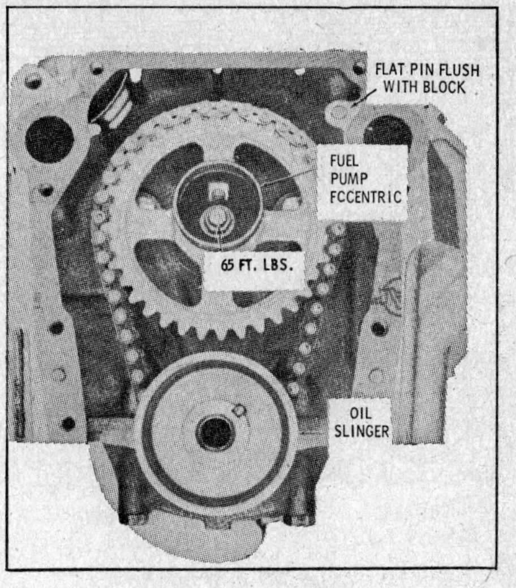

Fig. 11 Fuel pump eccentric. V8-350, 400, 455

21. Slide camshaft from front of engine.
22. Reverse procedure to install.

NOTE: *To insure proper camshaft installation, and to provide initial lubrication, it is extremely important that the camshaft be coated with GM Concentrate (Part No. 582099).*

PISTON & ROD, ASSEMBLE

Lubricate the piston pin hole and piston pin to facilitate installation of pin, then position the connecting rod with its respective piston as shown in Figs. 12 through 17.

PISTONS, RINGS & PINS

Pistons are available in standard sizes and oversizes of .010 and .030".

Rings are available in standard sizes and oversizes of .010 and .030".

MAIN & ROD BEARINGS

Main bearings are available in standard sizes and undersizes of .0005, .001, .0015, .002, .010 and .020".

Rod bearings are available in standard sizes and undersizes of .001, .002, .005, .010, .012 and .020 inch.

NOTE

Main bearing clearances not within specifications must be corrected by the use of selective upper and lower shells. Fig. 18 illustrates the undersize identification marking on the bearing tang.

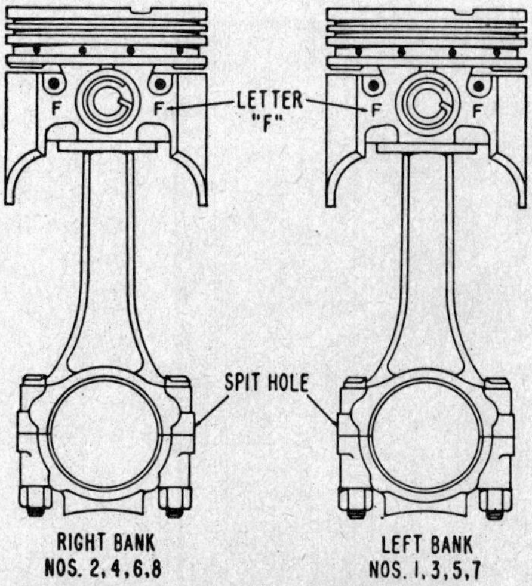

Fig. 12 Assembly of piston to rod. 1973-76 V8-260, 350

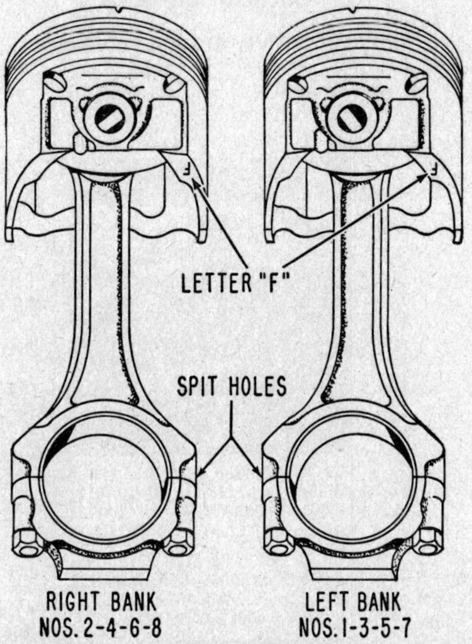

Fig. 13 Assembly of piston to rod. 1969 V8-455 low compression & 1970 V8-455 high compression

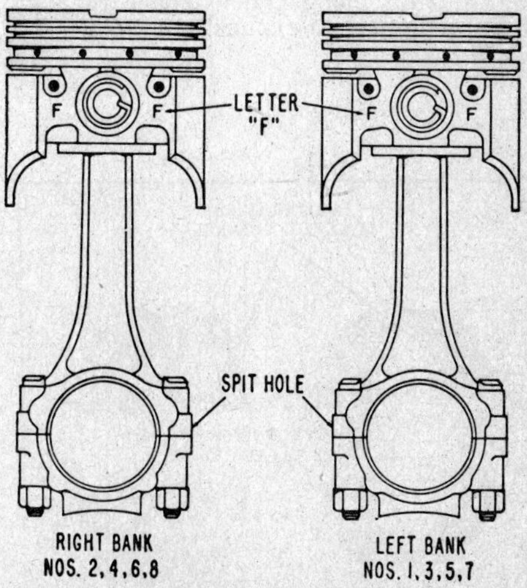

Fig. 14 Assembly of piston to rod. 1969-72 V8-350 exc. 1969-70 high performance

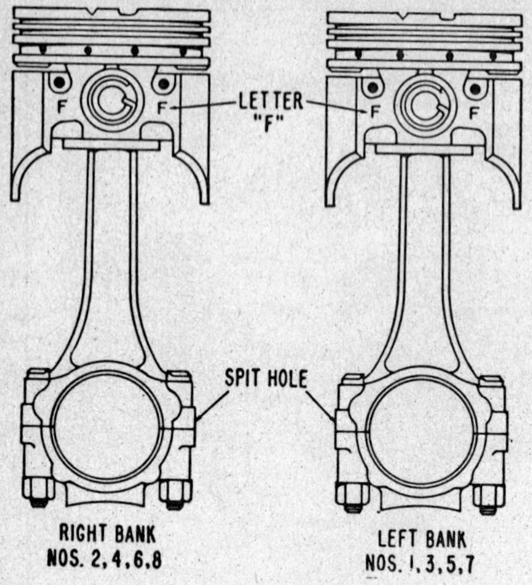

Fig. 15 Assembly of piston to rod. 1969-70 V8-350 high performance

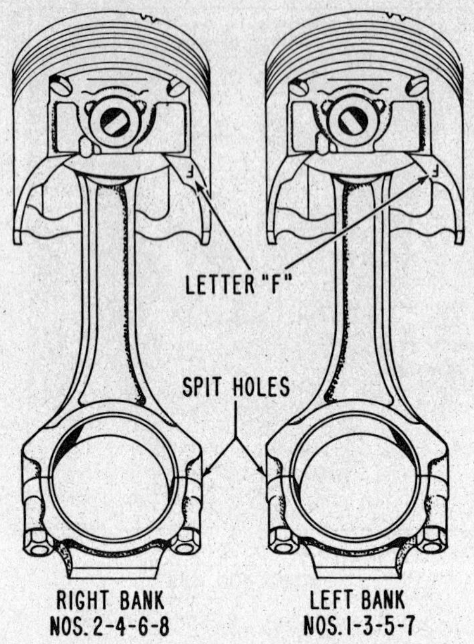

Fig. 16 Assembly of piston to rod. 1969 V8-400

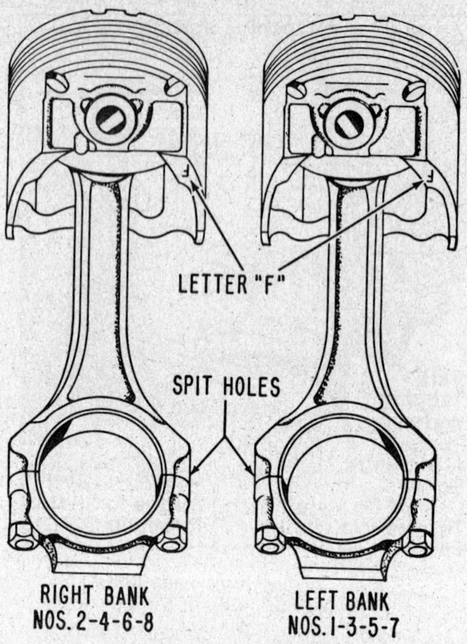

Fig. 17 Assembly of piston to rod. 1969 V8-455 high compression & all 1971-76 V8-455

OIL PAN

1973-76 V8

1. Remove distributor cap and align rotor with No. 1 firing position.
2. Disconnect ground cable, remove dip stick and drain oil pan.
3. Remove upper radiator support and fan shroud attaching screws.
4. Remove flywheel cover and starter.
5. Disconnect exhaust pipes and crossover pipe on single exhaust models.
6. Disconnect engine mounts and raise engine.
7. Remove oil pan bolts and oil pan.

1969-72 V8 (Except Cutlass V8-400 & 455)

1. Remove oil level dipstick.
2. Raise car and drain oil.
3. On F-85 models, disconnect exhaust pipe from right exhaust manifold.
4. On Olds F.S.C models, lower steering relay rod by disconnecting idler arm or pitman arm.
5. Disconnect engine mounts and raise front of engine as far as possible.

CAUTION: Be sure distributor does not contact cowl and fan blades do not contact fan shroud.

6. Bring No. 1 piston up on top center.
7. Remove crossover pipe and starter.
8. Unfasten and remove oil pan.

1969 Cutlass V8-400

1. Disconnect battery ground cable.
2. Remove upper radiator baffle.
3. Remove drive shaft, transmission crossmember, transmission and flywheel.
4. Disconnect left exhaust pipe and starter.
5. Disconnect right-hand engine mount.

NOTE: It is necessary to raise both the front and rear of engine and tilt it to the left, using the left engine mount as a pivot.

6. Use a suitable support bar and insert a 1" block between tool and right rear corner of oil pan to obtain additional lift on right side.

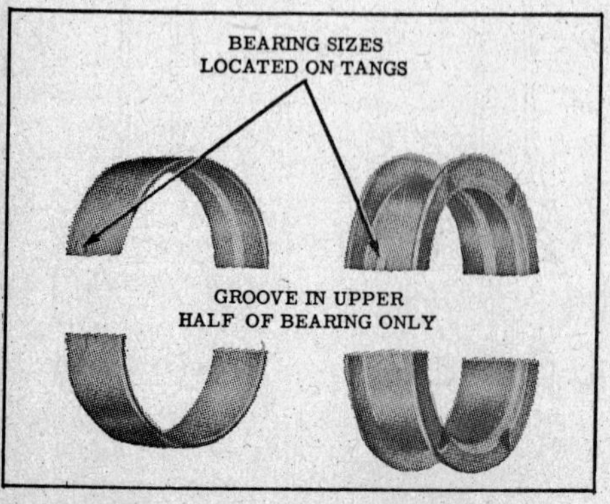

Fig. 18 Main bearing size location. V8-260, 350, 400, 455

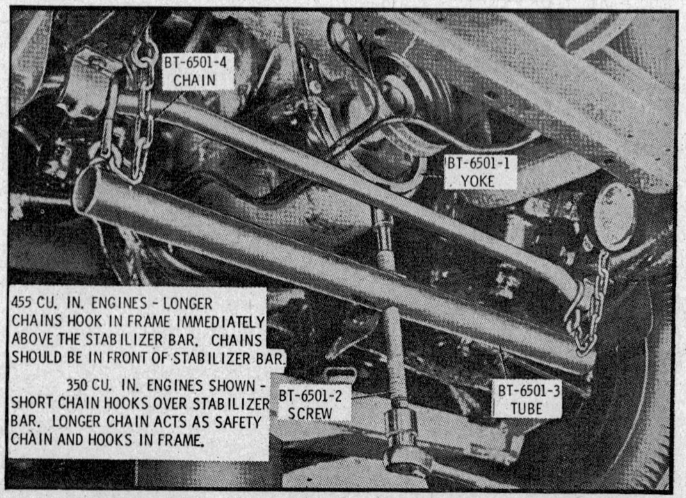

Fig. 19 Engine support bar. Cutlass & Omega (Typical)

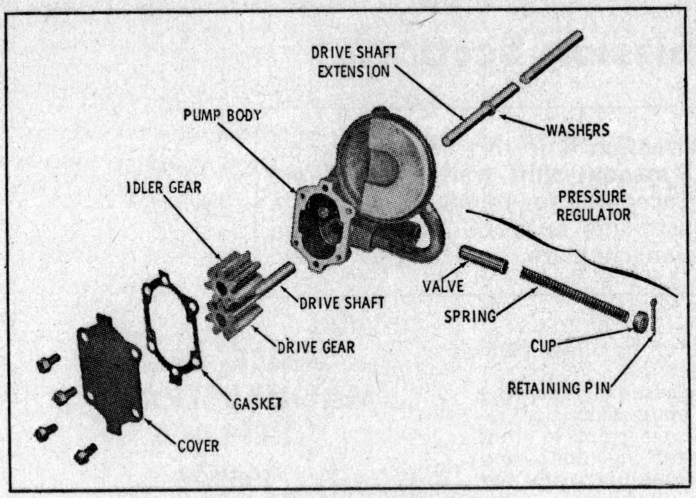

Fig. 20 Oil pump disassembled. V8-260, 350, 400, 455 (Typical)

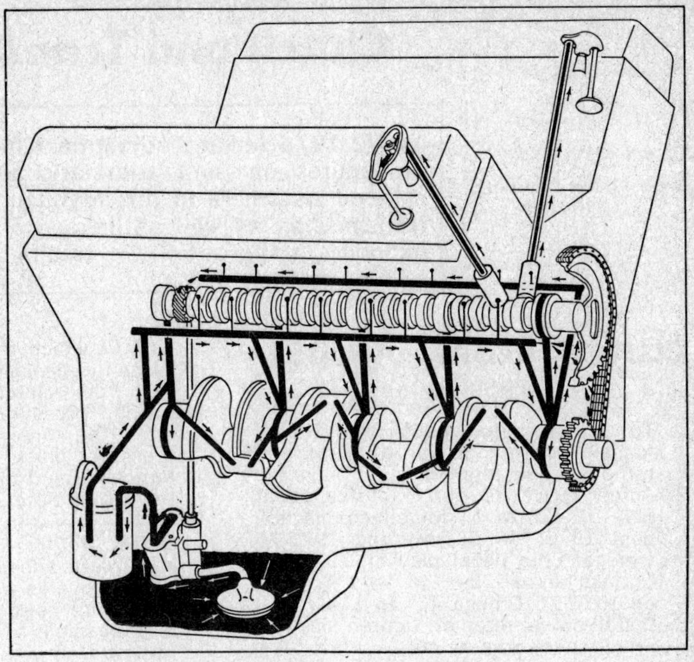

Engine lubrication. V8-260, 350, 400, 455

7. Raise front of engine.

NOTE: Remove lower two right-hand engine cover bolts to permit the lifting tool to clear water pump outlet. Do not bolt tool to block since engine must rock when being raised. Tool No. BT-6501-3 is available for this operation. Position fork of tool to lift pulley hub.

8. Raise right side of engine about 2½" and insert a block between upper and lower right-hand mounts.
9. Remove support bar tool from rear of engine.
10. Unfasten and remove pan.

NOTE: Before removing pan, have No. 1 and 2 connecting rod journals at the 4 or 7 o'clock position as viewed from front of engine.

11. Reverse procedure to install.

1969-72 Cutlass V8-455

1. Disconnect battery ground cable.
2. Disconnect fan shroud.
3. Raise car and drain oil.
4. Remove drive shaft and disconnect exhaust pipe and starter.
5. Install engine support bar, Fig. 19, and remove inspection cover. Disconnect modulator line, speedometer cable, oil cooler lines, solenoid wire and linkage.
6. Remove transmission crossmember, transmission and flywheel.
7. Remove right engine mount, raise engine 2" and install wedge block.
8. Loosen left engine mount-to-block bolts enough to permit removal of oil pan bolts.
9. Remove oil pan.

OIL PUMP REPAIRS

V8-260, 350, 400, 455

1. Remove oil pan and pump baffle. Remove attaching screws and remove pump and drive shaft extension.
2. To service the pump, refer to Fig. 20.
3. To install, insert the drive shaft extension through the opening in the block until the shaft mates into the distributor drive gear. Position pump onto rear main bearing cap and torque the attaching bolts to 35 ft-lbs.
4. Install oil pump baffle and pan.

BELT TENSION DATA

	New Lbs.	Used Lbs.
1969-76—		
V8 All	110-140	70
6 Cyl—		
Air Condition	135-145	95
Air Pump—		
1969	70-80	55
1970-75	120-130	75
Fan and Power Steering	120-130	75

WATER PUMP, REPLACE

V8-260, 350, 400, 455

1. Drain cooling system and remove heater and lower hoses from pump.
2. Loosen pulley belts and remove fan and pulley. On air conditioned cars, remove clutch fan assembly and pulley.
3. Unfasten and remove pump from front cover.

FUEL PUMP PRESSURE

Year	Engine	Pressure, Lbs.
1970-76	Six	4-5
1969	Six	3-5
1971-76	V8	5½-6½
1970	V8	5-6
1969	V8	5-7

FUEL PUMP, REPLACE

1. Remove all gasket material from the pump and block gasket surfaces. Apply sealer to both sides of new gasket.
2. Position gasket on pump flange and hold pump in position against its mounting surface. Make sure rocker arm is riding on camshaft eccentric.
3. Press pump tight against its mounting. Install retaining screws and tighten them alternately.
4. Connect fuel lines. Then operate engine and check for leaks.

SERVICE NOTE: Before installing the pump, it is good practice to crank the engine so that the nose of the camshaft eccentric is out of the way of the fuel pump rocker arm when the pump is installed. In this way there will be the least amount of tension on the rocker arm, thereby easing the installation of the pump.

Clutch and Transmission Section

NOTE: 1976 linkage adjustment information is in this section. Repair procedures on both automatic and manual shift transmissions are covered elsewhere in this manual. Procedures for removing automatic transmission as well as linkage adjustments on 1969-75 models are included in the automatic transmission chapters. See Chapter Index.

CLUTCH PEDAL, ADJUST

1969-76

1. To adjust clutch pedal free travel, loosen lock nut on adjusting rod and remove swivel retainer.
2. Adjust swivel to obtain a free pedal play of 3/4" to 1" for all models except 1973, 76 Cutlass and 1973-75 Omega. Free pedal play on 1973, 76 Cutlass should be 3/4" to 1 1/4" and on 1973-76 Omega 7/8" to 1 1/2".
3. Install swivel retainer, tighten lock nut and recheck adjustment.

CLUTCH, REPLACE

1971-72 Two Plate Clutch

1. Support engine and remove transmission.
2. Remove flywheel housing leaving starter attached to engine. Release yoke and ball stud will remain with housing.

NOTE: *Before removing clutch assembly, note location of "O" marks on edges of clutch cover, front pressure plate and flywheel. If marks are not visible, scribe or prick punch the parts so they can be reassembled in same position.*

3. Remove clutch cover to flywheel attaching bolts and remove the cover assembly with the front pressure plate attached, Fig. 1. The front driven plate can now be removed. Observe the driven plate before removing to be sure which side of the driven plate is the flywheel side.
4. With the clutch assembly on the bench, pressure plate side up, remove the three cover straps to front pressure plate attaching bolts and remove the front pressure plate and the driven plate.

NOTE: *Before lifting the front pressure plate and driven plate from the cover, observe which side of the pressure plate goes towards the flywheel. It is possible to assemble both parts reversed. A washer or shim (.027") is used between each strap and the front pressure plate.*

5. Reverse removal procedure to install referring to Fig. 2.

1969-76 Exc. Two Plate Clutch

1. Remove transmission.
2. Disconnect clutch release spring and clutch rod.
3. Remove clutch release bearing.
4. Remove flywheel housing, leaving starter attached to engine. Release yoke and ball stud will remain in housing.
5. Scribe mark on clutch cover to flywheel for correct assembly.
6. Unfasten and remove clutch cover and disc.

NOTE: Loosen pressure plate bolts alternately, one turn at a time.

7. Reverse removal procedure to install clutch and adjust clutch pedal free play.

THREE SPEED MANUAL TRANS., REPLACE

1969-76

1. Disconnect throttle linkage from cowl bracket to prevent damage.

NOTE: On 1975-76 Omega models with floor shift, remove shift lever from shifter assembly.

2. Raise car and remove drive shaft.
3. Disconncet shift rods from shift levers and the TCS switch wiring, if equipped.
4. Support rear of engine.
5. Remove cross support bar-to-rear transmission mount attaching bolts.
6. Disconnect parking brake cables from cross support and remove cross support bar.
7. If equipped with dual exhaust it may be necessary to disconnect left-hand exhaust pipe at exhaust manifold to provide clearance.

NOTE: On 1975-76 models, remove catalytic converter support bracket.

8. Disconnect speedometer cable.
9. Remove transmission upper attaching bolts and install aligning studs in the bolt holes.
10. Remove lower bolts and remove transmission.
11. Reverse procedure to install.

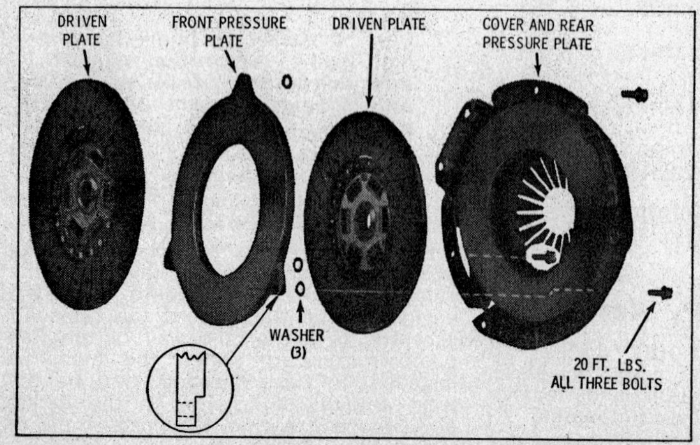

Fig. 1 Exploded view of two-plate clutch

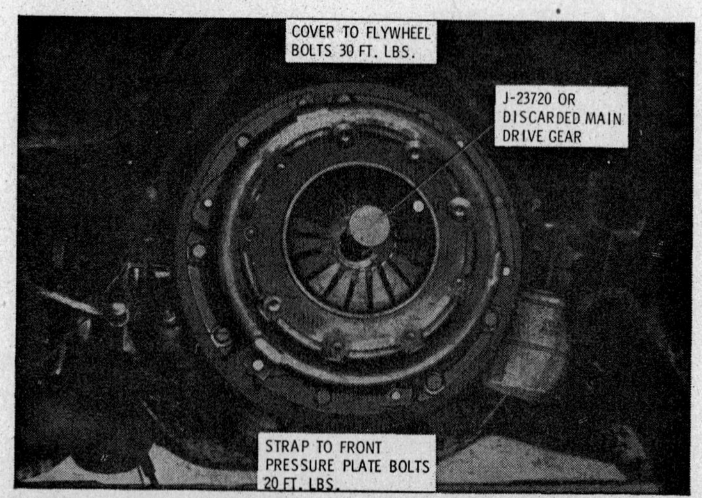

Fig. 2 Two-plate clutch installation

3 SPEED SHIFT LINKAGE, ADJUST

1969-75 Column Shift

1. Place transmission in reverse and raise vehicle.
2. Loosen swivel bolts on shift rods at transmission, ensuring rods are free to move in swivels.
3. Hold 1st-reverse column relay level in position, push up on shift rod until detent in column is felt, then torque swivel bolt to 20 ft. lbs. on 1969-74 models, 26 ft. lbs. on 1975-76 Omega and 23 ft. lbs. on 1975-76 Cutlass.
4. Place transmission in Neutral, insert $^3/_{16}$ inch rod through column relay lever into alignment hole and torque swivel nut to 20 ft. lbs. on 1969-74 models, 26 ft. lbs. on 1975-76 Omega and 23 ft. lbs. in 1975-76 Cutlass.
5. Lower vehicle and check shift operation.
6. Place transmission in reverse, turn ignition switch to lock position and ensure that the key can be removed, steering wheel will not turn and transmission will not shift out of reverse.
7. Place ignition switch in run position, shift transmission to second gear, ensure that steering wheel will turn and key can not be removed from ignition switch.

1973-75 Floor Shift

1. Place shift lever in Neutral and raise vehicle.
2. Loosen shift rod swivel nuts and disconnect rods from shifter, then insert a $^1/_4$ inch pin into shifter, Fig. 3.
3. Adjust swivel to obtain free pin length, tighten swivel nuts and connect shift rods to shifter.
4. With shift lever in Reverse and ignition key in Lock, loosen equalizer clamp and lightly pull back drive rod against stop. Torque equalizer clamp screw to 23 ft. lbs.
5. To check for proper operation:
   a. Place ignition key in Lock and shift lever in Reverse and ensure that key can be removed and steering wheel is locked.
   b. Place ignition key in Off and shift lever in Neutral and ensure that key cannot be removed and steering wheel is unlocked.

1969-71 3 SPEED HEAVY DUTY TRANS., REPLACE

1. Disconnect throttle linkage from cowl bracket to prevent damage.
2. Raise car and remove drive shaft.
3. Disconnect shift rods from shift levers (column shift).
4. With floor shift, remove shift lever knob and disconnect back-up lamp wiring (if equipped).
5. Disconnect equalizer shaft at side of transmission.
6. Support rear of engine.
7. Disconnect parking brake cable.
8. Remove cross support bar.
9. It may be necessary to disconnect left exhaust pipe from exhaust manifold to provide clearance if equipped with dual exhaust.
10. Remove transmission upper bolts and install aligning studs in the bolt holes.
11. Remove lower bolts and slide transmission out of car.
12. Reverse procedure to install.

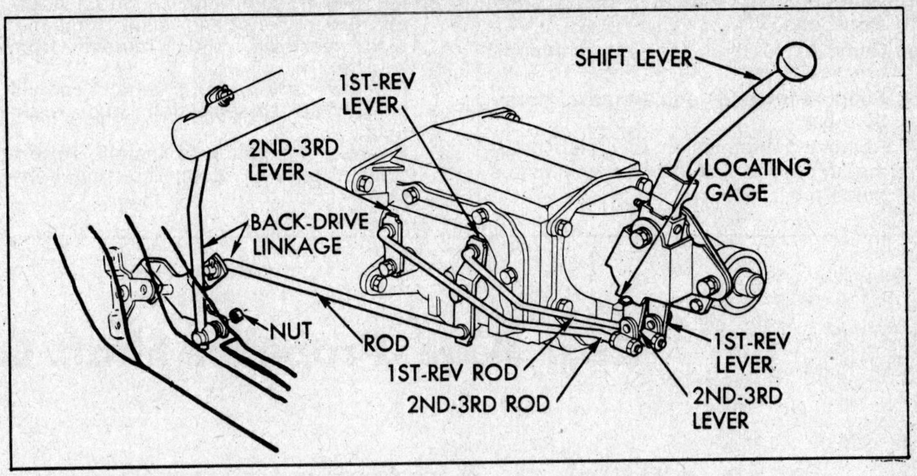

Fig. 3 Shift linkage. 1973-75 Omega floor shift

Shift Linkage, Adjust

On 1969-71 models, detent must be adjusted as follows: Place transmission in Reverse and loosen swivel bolt on back drive rod at equalizer. Push up lightly on back drive rod until stop is felt and tighten bolt to 20 ft. lbs. Continue adjustment.

1. Position transmission in neutral.
2. Loosen clamp screws on shift rods.
3. Use a suitable aligning rod and position it in slot provided in mast jacket so that it enters holes in low-reverse lever and the interlock pawl.
4. With floor shift, a $^1/_4$″ diameter pin is inserted through shift lever bracket and levers.
5. With transmission levers in neutral, tighten clamp screws to 23 ft-lbs. On 1969-71 models, place transmission in Reverse and ignition in "Lock" position. Check to be sure key can be removed, steering wheel will not turn and transmission will not shift out of Reverse.
6. Remove alignment rod. Make sure neutral positions between low-reverse and second-third are exactly in line. If not, readjust one rod to bring them in line.

4 SPEED TRANS., REPLACE

1969-73

1. Raise vehicle and drain transmission.
2. Remove propeller shaft.
3. Disconnect shift rods from levers.
4. Support engine using support bar No. BT-6424-1. Remove transmission mount to cross member bolts and disconnect parking brake cables from cross member, then remove cross-bar.

NOTE: On vehicles with dual exhaust, it maybe necessary to disconnect the left exhaust pipe at manifold.

5. Disconnect speedometer cable, remove shift lever to extension housing attaching bolts.

NOTE: On some models it may be necessary to remove shift lever.

6. Remove transmission to clutch housing upper attaching bolts, install guide studs.

NOTE: Guide studs are used to prevent the clutch plate from becoming distorted.

7. Remove transmission to clutch housing lower attaching bolts and carefully remove transmission.
8. Reverse procedure to install.

4 SPEED SHIFT LINKAGE, ADJUST

1969-73

1. On 1969-73, place transmission in Reverse and loosen swivel bolt on back drive rod at equalizer. Push up lightly on back drive rod until stop is felt and tighten bolt to 20 ft. lbs.
2. All models: Place transmission in Neutral and loosen shift rods.
3. Using a suitable rod, align levers in Neutral.
4. Adjust swivels to obtain a "free pin" fit at levers. Tighten swivel bolts to 20 ft. lbs.
5. On 1969-73 models, place transmission in Reverse and ignition in "Lock" position. Check to be sure key can be removed, steering wheel will not turn and transmission will not shift out of Reverse.

5 SPEED TRANS., REPLACE

1976

1. Remove Shifter assembly.
2. Raise vehicle and remove propeller shaft.
3. Disconnect speedometer cable from transmission.

4. Remove crossmember to transmission bolts.
5. Remove catalytic converter support bracket.
6. Support engine and remove crossmember.
7. Remove transmission to clutch housing upper retaining bolts and install guide pins.
8. Remove transmission to clutch housing lower bolts and slide transmission rearward, then remove from vehicle.
9. Remove back-up lamp switch and fill plug. Tilt transmission and drain fluid.
10. Reverse procedure to install. Torque transmission to clutch housing bolts to 60 ft. lbs.

1976 AUTO. TRANS. LINKAGE, ADJUST

Adjustment procedures for the 1975 models are essentially the same as those for the 1974 units as outlined elsewhere in this manual.

Rear Axle, Propeller Shaft & Brakes

REAR AXLE

Figs. 1 and 2 illustrate the rear axle assemblies used on conventional models. When necessary to overhaul any of these units, refer to the *Rear Axle Specifications* table in this chapter.

Integral Carrier 1969-76 Type "B" & "O"

As shown in Fig. 1, the drive pinion is mounted on two tapered roller bearings that are preloaded by two selected spacers. The drive pinion is positioned by shims located between a shoulder on the pinion and the rear bearing. The front bearing is held in place by a large nut.

The differential is supported in the carrier by two tapered roller side bearings. These are preloaded by inserting shims between the bearings and the pedestals. The differential assembly is positioned for ring gear and pinion backlash by varying these shims.

Major service work on the differential carrier assembly may be performed with the unit in the car provided a drive-on or twin-post hoist is available. If neither of these hoists are available, the rear axle assembly should be removed from the vehicle. The reason for this is that the axle tubes are pressed into the differential carrier housing and welded.

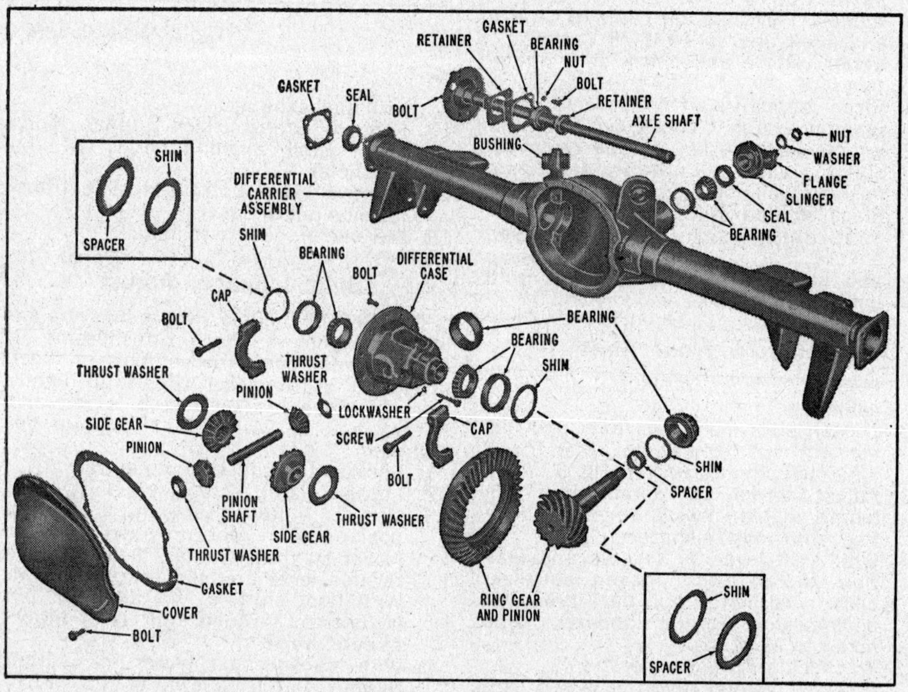

Fig. 1 Integral carrier type rear axle. 1969-76 Type B & O axle

1969-76 Type C, G, K, M & P

In these rear axles, Fig. 2, the rear axle housing and differential carrier are cast into an integral assembly. The drive pinion assembly is mounted in two opposed tapered roller bearings. The pinion bearings are preloaded by a spacer behind the front bearing. The pinion is positioned by a washer between the head of the pinion and the rear bearing.

The differential is supported in the carrier by two tapered roller side bearings. These bearings are preloaded by spacers located between the bearings and carrier housing. The differential assembly is positioned for proper ring gear and pinion backlash by varying these spacers. The differential case houses two side gears in mesh with two pinions mounted on a pinion shaft which is held in place by a lock pin. The side gears and pinions are backed by thrust washers.

Remove & Replace

Construction of the axle assembly is such that service operations may be performed with the housing installed in the vehicle or with the housing removed and installed in a holding fixture. The following procedure is necessary only when the housing requires replacement.

1. Raise car and remove rear wheels, drums and axle shafts.
2. Disconnect brake line from wheel cylinders.
3. Unfasten and support backing plates with wire hooks to frame kickup.
4. Disconnect shocks at housing.
5. Position jack stands under frame rear torque boxes, then lower axle housing to stands.
6. Remove springs.
7. Remove propeller shaft and support front of axle housing at companion flange to prevent assembly from rotating when the control arms are disconnected.
8. Remove control arm bolts at axle housing.
9. Remove support at companion flange and lower axle housing.
10. Remove assembly to bench and transfer parts to new axle housing.
11. Reverse procedure to install.

AXLE SHAFT, REPLACE

1969-76 Type C, G, K, M & P

1. Raise vehicle and remove wheel and brake drum.
2. Clean all dirt from area of carrier cover.
3. Drain lubricant from carrier by removing cover.

4. Remove differential pinion shaft lock screw and shaft.
5. Push flanged end of axle shaft toward center of vehicle and remove "C" lock from button end of shaft.
6. Remove axle shaft from housing, being careful not to damage oil seal.
7. Reverse procedure to install.

1969-76 Type B & O

Removal

1. Remove wheel and brake drum.
2. Remove axle bearing retainer (4 nuts).

NOTE: On 1976 8½ inch axles, new retainers are used and can be identified by a raised area around the axle shaft opening. This raised area provides proper seating of the seal and bearing.

3. Pull axle shaft from housing. If bearing is a tight fit in housing, use a slide hammer-type puller. Do not drag shaft over seal as this may damage seal.
4. Attach one axle bearing retainer nut to hold brake backing plate in position.
5. Before installing axle shaft, examine oil seal. The seals have feathered edges which form a tight seal around the shaft. If these edges are damaged in any way, seal must be replaced. Examine seal surface on shaft; if it is not smooth, dress it down with very fine emery cloth.
6. Reverse removal procedure to install axle shaft, being sure to grease outside of axle bearing, seal surface on axle shaft and bore of axle housing with differential lubricant. Place new gasket and bearing retainer over studs, install nuts and tighten them 45 to 60 ft. lbs.

Installation

Bearings should be replaced if found to be rough or have greater than .020" end play. Remove bearing only when new bearing is to be installed; once removed it must not be reused.

1. With axle shaft removed from housing, split bearing retainer with a chisel, Fig. 3.
2. Press bearing off shaft.
3. Press new bearing on shaft up against shoulder on shaft.
4. Press retainer on shaft up against bearing.
5. Reverse removal procedure to install axle shaft.

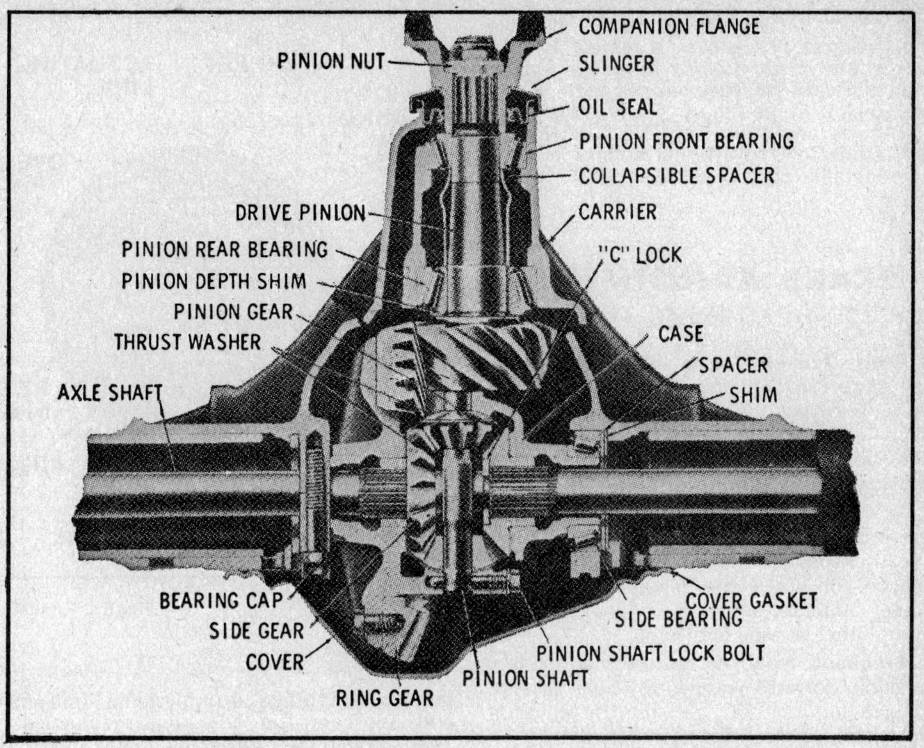

Fig. 2 Integral carrier type differential. 1969-76 Types C, G, K, M & P axle

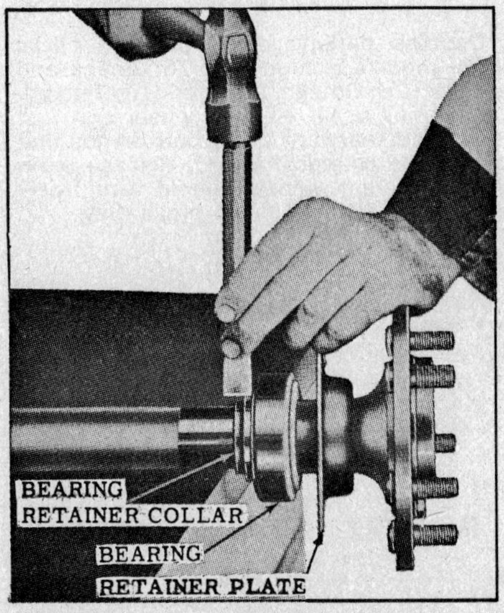

Fig. 3 Removing axle shaft bearing retainer

PROPELLER SHAFT
1969-76

The propeller shaft is of one or two piece construction with a single or double U-joint securing the shaft to the companion flange, Fig. 4. The front or rear yokes on some 1969-72 models with automatic transmissions are bonded in rubber to the inside of the shaft tube and is not removable for service.

1. Mark propeller shaft and companion flange so they can be installed in the same position.
2. On 1969-70 models remove U bolt nuts, on 1971-76 Cutlass, intermediate and full size station wagons and 1973-76 Omega remove strap bolts. On 1971-76 full size models except station wagon, remove flange yoke to companion flange attaching bolts. Use a piece of tape or wire to hold universal joint bearing caps in place.

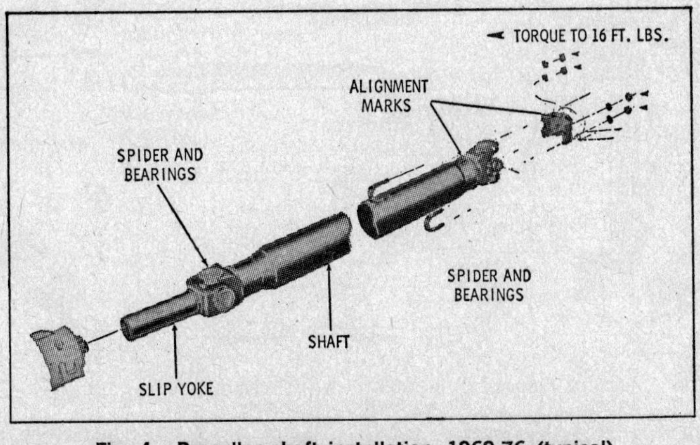

Fig. 4 Propeller shaft installation. 1969-76 (typical)

3. Lower rear of shaft and side rearward.
4. Reverse procedure to install. If drive shaft yokes do not have vent holes, lubricate internal splines with engine oil. If drive shaft yokes have vent holes apply lubricant No. 1050169 or equivalent to internal splines prior to installation.

BRAKE ADJUSTMENTS

1969-76

These brakes, Fig. 5, have self adjusting shoe mechanisms that assure correct lining-to-drum clearances at all times. The automatic adjusters operate only when the brakes are applied as the car is moving rearward.

Although the brakes are self-adjusting, an initial adjustment is necessary after the brake shoes have been relined or replaced, or when the length of the star wheel adjuster has been changed during some other service operation.

Frequent usage of an automatic transmission forward range to halt reverse vehicle motion may prevent the automatic adjusters from functioning, thereby inducing low pedal heights. Should low pedal heights be encountered, it is recommended that numerous forward and reverse stops be made until satisfactory pedal height is obtained.

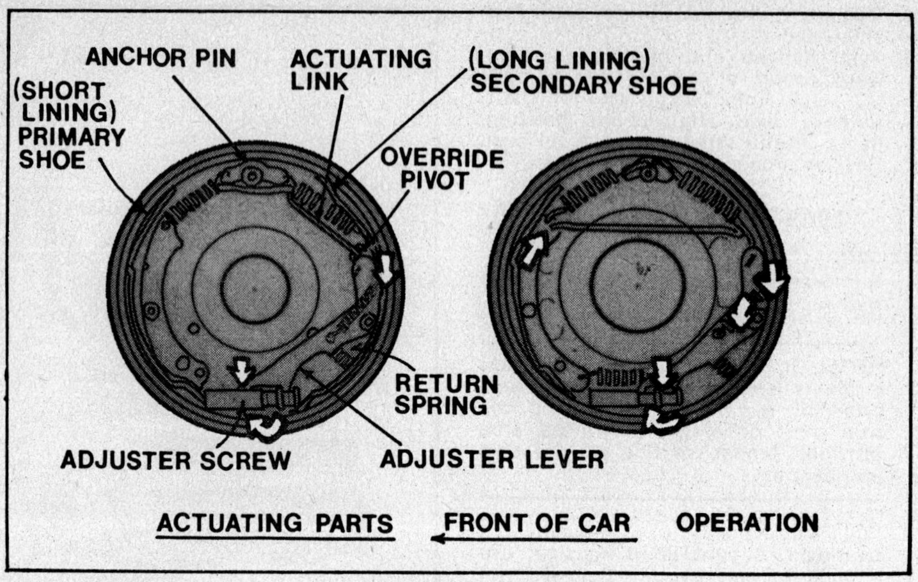

Fig. 5 Typical self-adjusting brake

NOTE

If a low pedal height condition cannot be corrected by making numerous reverse stops (provided the hydraulic system is free of air) it indicates that the self-adjusting mechanism is not functioning. Therefore, it will be necessary to remove the brake drum, clean, free up and lubricate the adjusting mechanism. Then adjust the brakes as follows, being sure the parking brake is fully released.

Adjustment

NOTE: Inasmuch as there is no way to adjust these brakes with the drums installed, the following procedure is mandatory after new linings are installed or if it becomes necessary to change the length of the brake shoe adjusting screw.

1. With brake drums removed, position the caliper shown in Fig. 6 to the inside diameter of the drum and tighten the clamp screw.
2. Next position brake shoe end of the caliper tool over the brake shoes as shown in Fig. 7.
3. Rotate the gauge slightly around the shoes to insure that the gauge contacts the linings at the largest diameter.
4. Adjust brake shoes until the gauge is a snug fit on the linings at the point of largest lining diameter.

NOTE: If it is necessary to back off the brake shoe adjustment, it will be necessary to hold the adjuster lever away from the adjuster screw, Fig. 8.

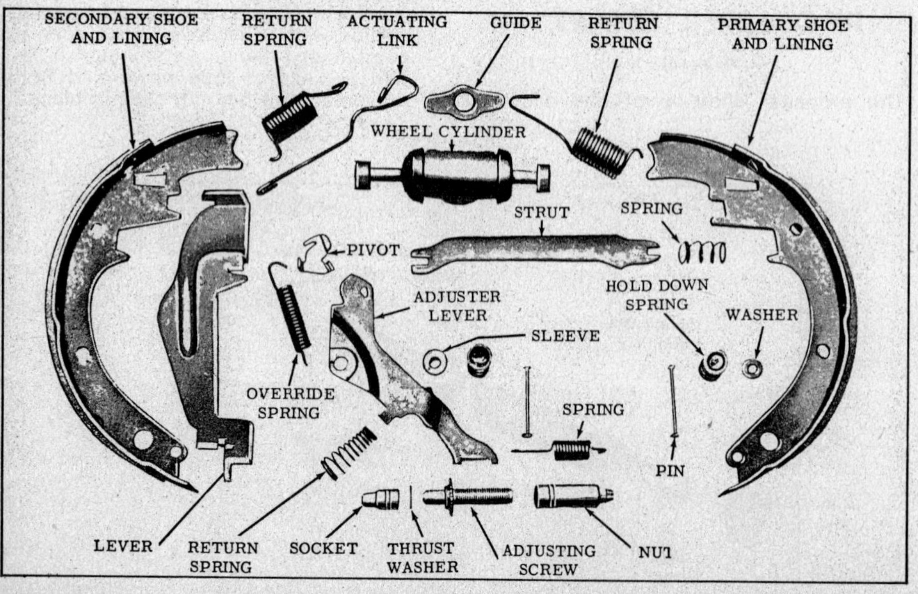

Fig. 5A Drum brake assembly exploded

PARKING BRAKE, ADJUST

1972-76

Depress parking brake pedal 3 clicks on all models except 1975-76 Omega and on 1975-76 Omega 2 clicks. Tighten adjusting nut until left rear wheel can just be rotated rearward using both hands and can not be rotated forward. Release parking brake, rear wheels should turn freely in either direction with no brake drag.

1969-71

With parking brake fully released, adjust rear cables by first tightening the brake equalizer adjusting nut until a heavy resistance is felt when rotating rear wheels forward. Then loosen equalizer adjusting nut 7 full turns.

POWER BRAKE UNIT, REPLACE

1969-74

1. Disconnect vacuum hose from vacu-

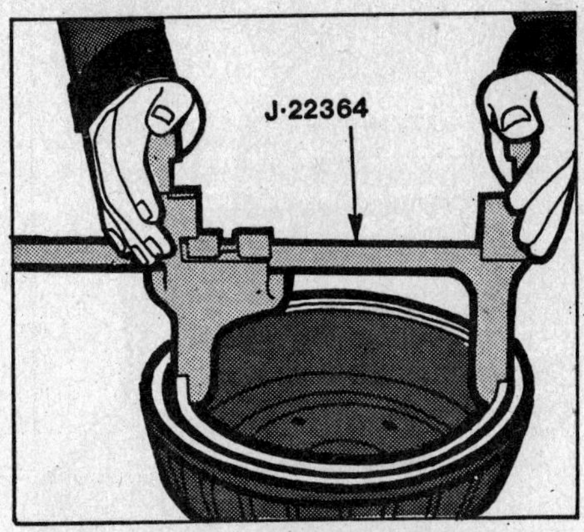

Fig. 6 Brake shoe gauge measuring inside diameter of brake drum. 1969-76

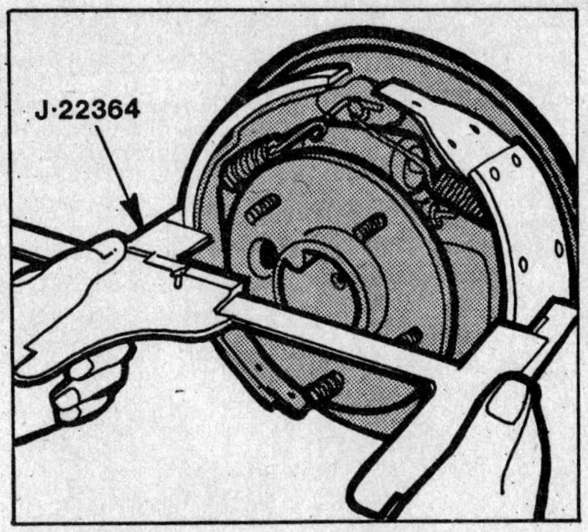

Fig. 7 Brake shoe gauge measuring outside diameter of brake shoes. 1969-76

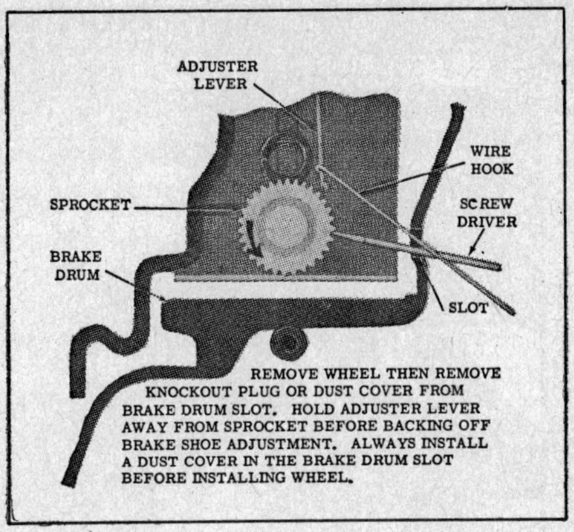

Fig. 8 Backing off brake shoe adjustment

um cylinder and cover openings to prevent entrance of dirt.
2. Disconnect pipes from master cylinder outlets and cover openings in master cylinder and end of pipes to prevent entrance of dirt.
3. Loosen inboard master cylinder attaching nut to disengage metering valve (disc brakes only).
4. Disconnect air valve rod from brake pedal.
5. Unfasten and remove power brake unit.
6. Reverse procedure to install.

BRAKE MASTER CYLINDER, REPLACE

1969-76

The standard brake master cylinder on Olds F.S.C. can be removed without disconnecting the push rod and clevis. On all cars equipped with power brakes, the master cylinder can be removed without removing the vacuum cylinder from the car.

1. Be sure area around master cylinder is clean, then disconnect the hydraulic lines at master cylinder. Plug or tape end of line to prevent entrance of dirt or loss of brake fluid.
2. On F-85 and Omega models, remove push rod-to-brake pedal clevis pin.
3. On all models, unfasten (4 bolts) and remove master cylinder.

Rear Suspension

SHOCK ABSORBER, REPLACE

With rear axle properly supported, disconnect shock absorber from upper and lower mountings.

LEAF SPRINGS & BUSHINGS, REPLACE

1973-76 Omega

1. Support vehicle at frame and support rear axle, relieving tension from spring.
2. Disconnect shock absorbers from lower mountings and loosen spring front mounting bolt.
3. Remove spring retainer bracket to underbody screws, lower rear axle and remove retainer bracket.
4. Disconnect parking brake cable from spring plate bracket.
5. Remove "U" bolts and spring plate, Fig. 1.
6. Support spring, remove spring front mounting bolt and rear shackle bolts.
7. Replace spring eye bushings and rear shackle frame bushings as necessary, Figs. 2 and 3.
8. Reverse procedure to install.

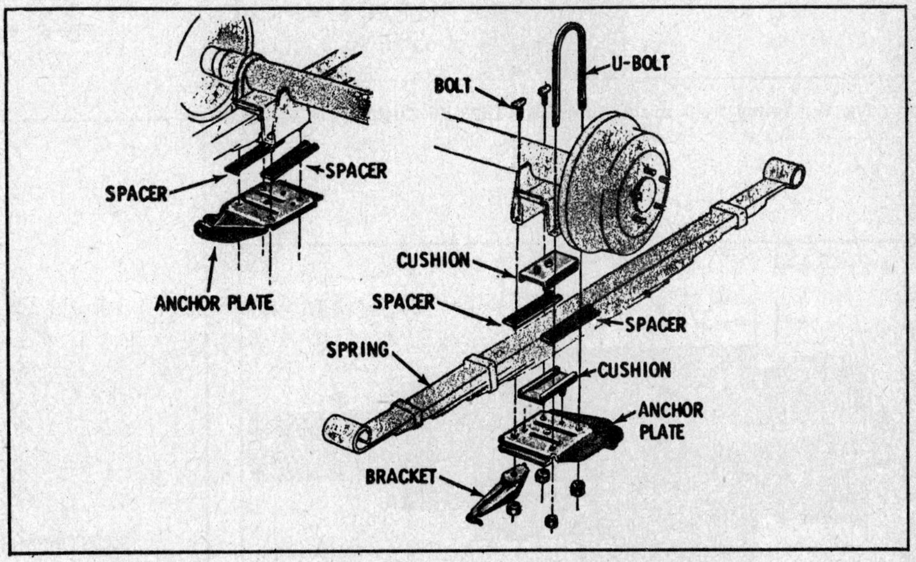

Fig. 1 Rear suspension. 1973-76 Omega

1971-76 Custom Cruiser

1. Support vehicle at frame and support rear axle, relieving tension from spring.
2. Disconnect shock absorbers from lower mountings and if removing right hand leaf spring, loosen tailpipe and

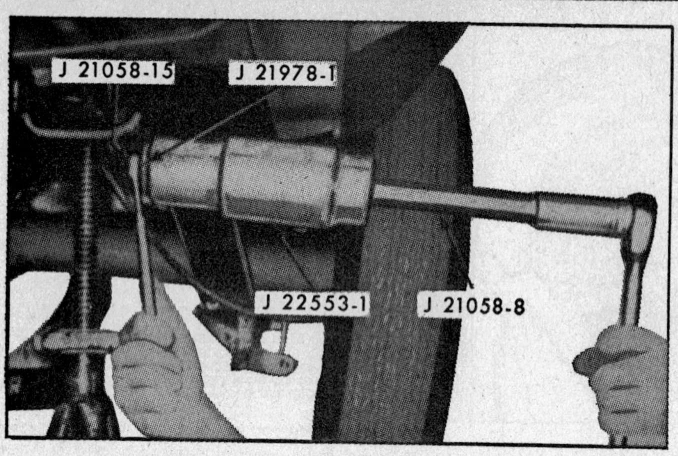

Fig. 2 Leaf spring bushings removal. 1973-76 Omega

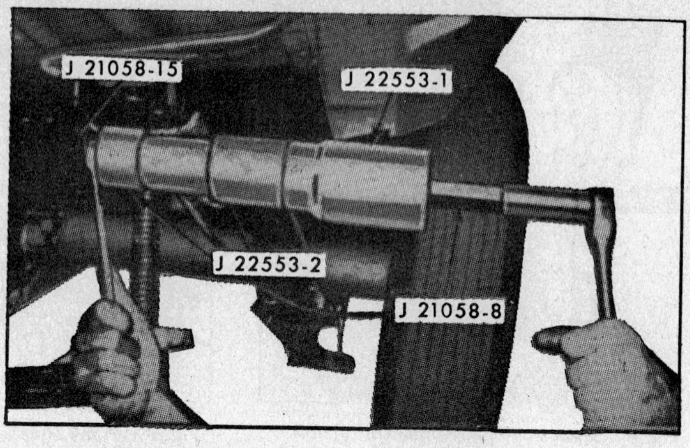

Fig. 3 Leaf spring bushings installation. 1973-76 Omega

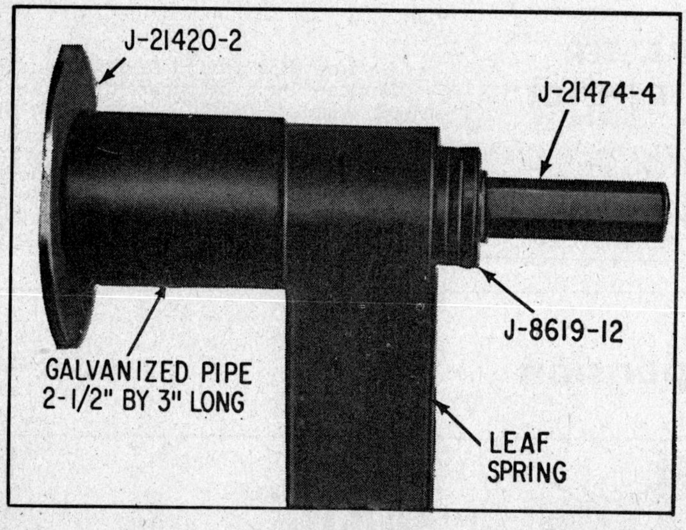

Fig. 5 Spring front bushing removal. 1971-76 Custom Cruiser

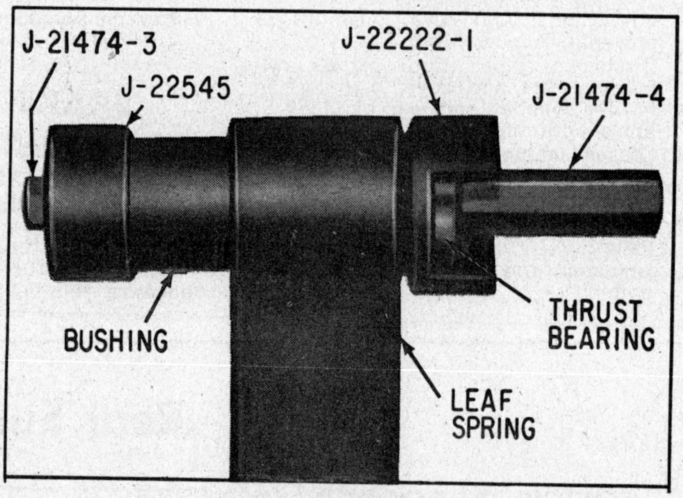

Fig. 6 Spring front bushing partial installation. 1971-76 Custom Cruiser

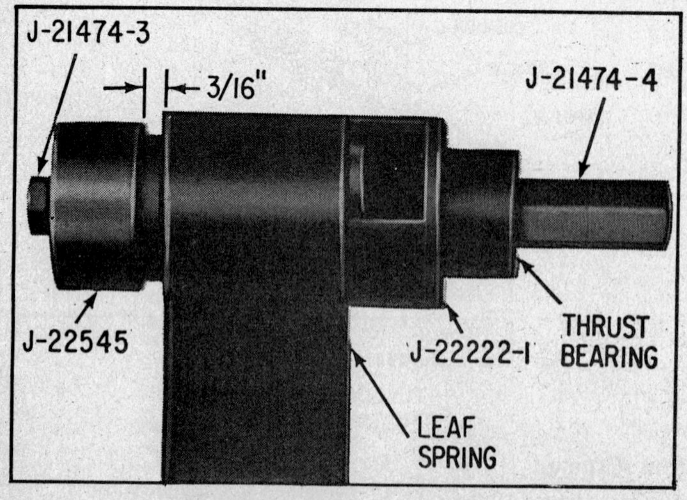

Fig. 7 Spring front bushing complete installation. 1971-76 Custom Cruiser

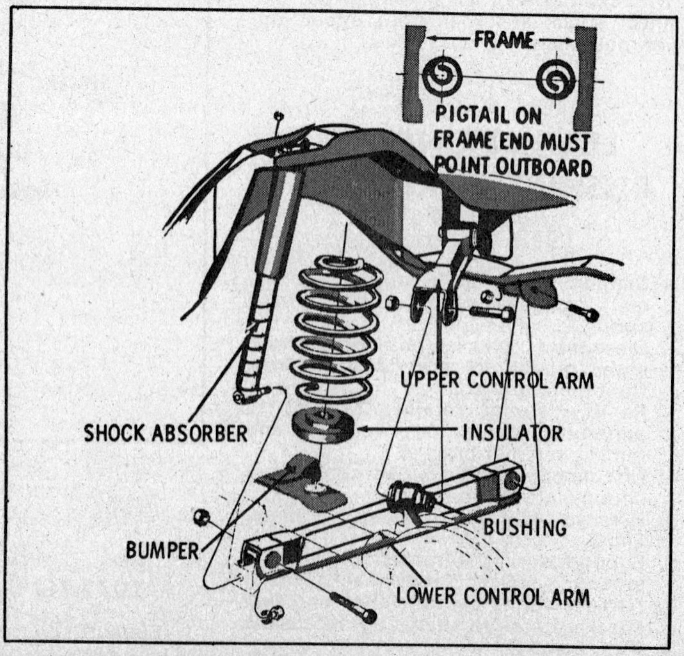

Fig. 8 Rear suspension. 1969-70 full size models

resonator assembly.
3. Remove rear shackle bolts, "U" bolts and spring plate, Fig. 4.
4. Support spring and remove spring front mount bolt.
5. Replace spring front bushing as necessary, Figs. 5, 6 and 7.
6. Reverse procedure to install.

COIL SPRING, REPLACE

1974-76

1. Raise rear of vehicle and support at frame and rear axle.
2. Disconnect shock absorber from lower mounting.
3. Compress coil spring using a suitable spring compressor.
4. Lower rear axle. Use caution not to damage or stretch brake hose.
5. Remove coil spring. It may be necessary to further compress spring and remove wheel.
6. Reverse procedure to install. Ensure spring is properly indexed, Fig. 9.

1969-73

1. Support vehicle at frame and support rear axle with a suitable jack.
2. Disconnect shock absorbers from lower mountings.
3. On 1972-73 full size models, disconnect brake line junction block from axle housing and remove brake lines from axle housing retainers. Disconnect upper control arms from axle brackets.
4. On all models lower rear axle to remove springs.
5. Reverse procedure to install insuring springs are indexed properly, Figs. 8 to 10.

CONTROL ARMS & BUSHINGS, REPLACE

NOTE: Replace control arms one at a time to prevent axle assembly misalignment, making installation difficult.

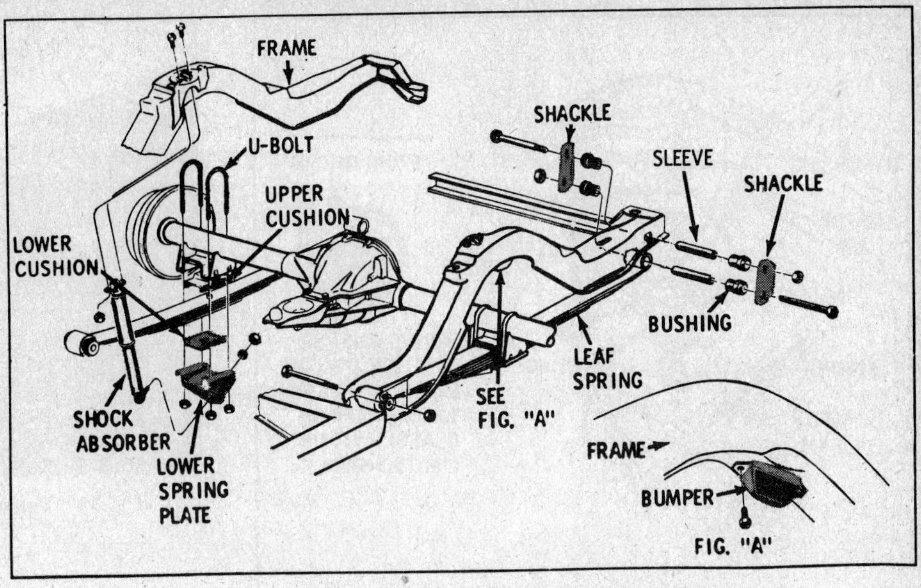

Fig. 4 Rear suspension. 1971-76 Custom Cruiser

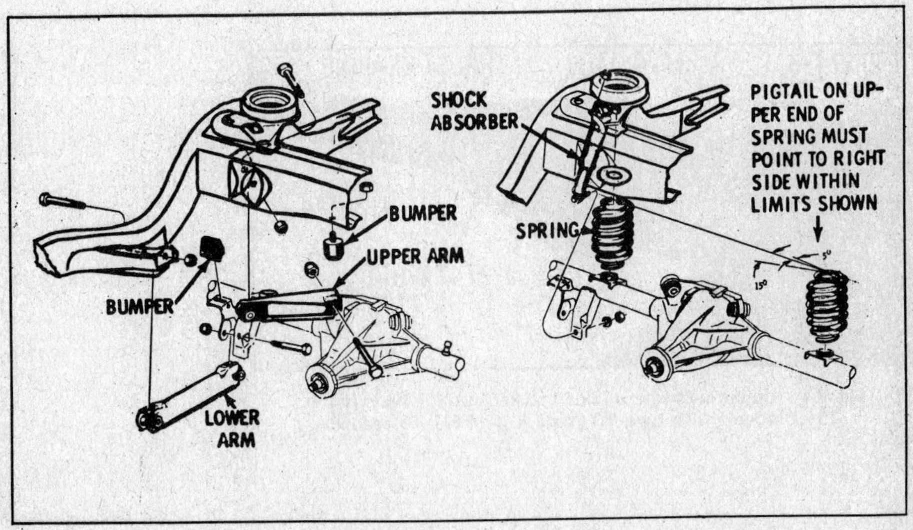

Fig. 9 Rear suspension. 1971-76 full size models & 1973-76 Cutlass

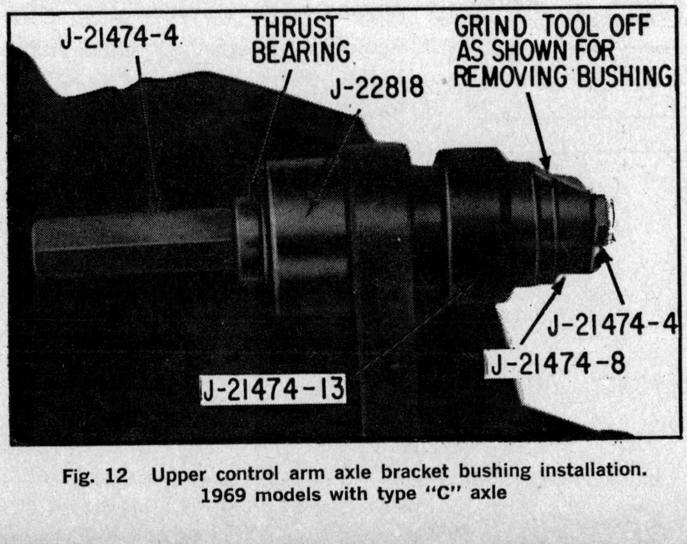

Fig. 12 Upper control arm axle bracket bushing installation. 1969 models with type "C" axle

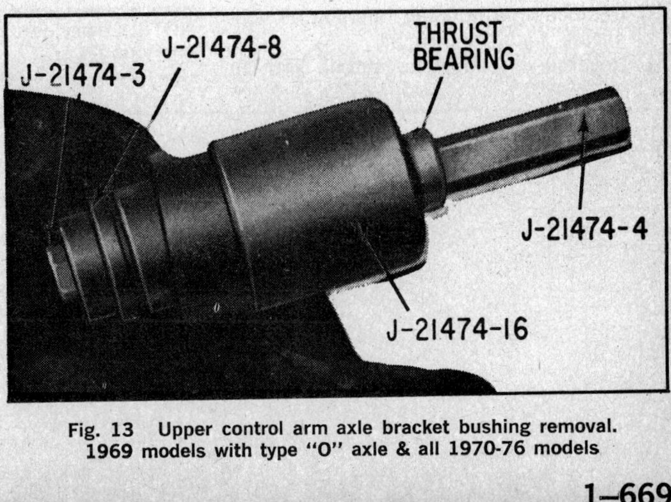

Fig. 13 Upper control arm axle bracket bushing removal. 1969 models with type "O" axle & all 1970-76 models

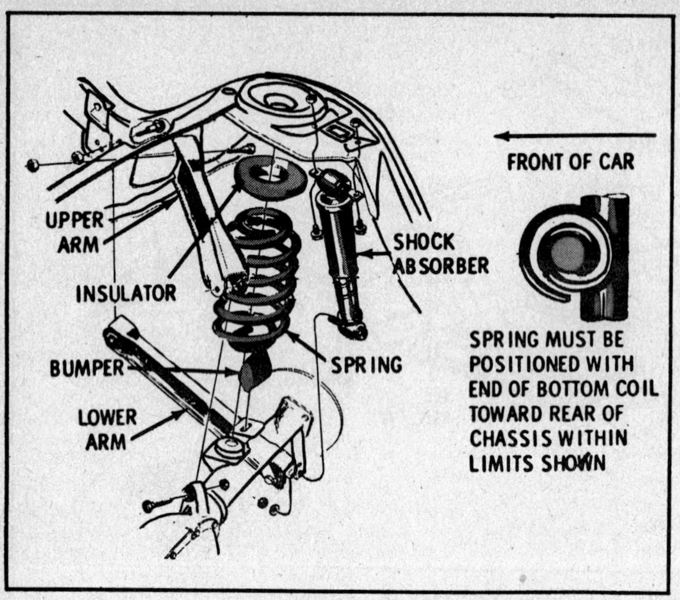

Fig. 10 Rear suspension. 1969-72 intermediate models

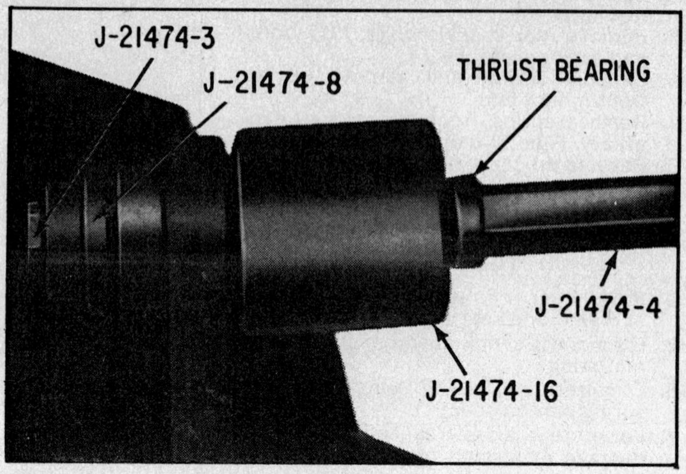

Fig. 11 Upper control arm axle bracket bushing removal. 1969 models with type "C" axle

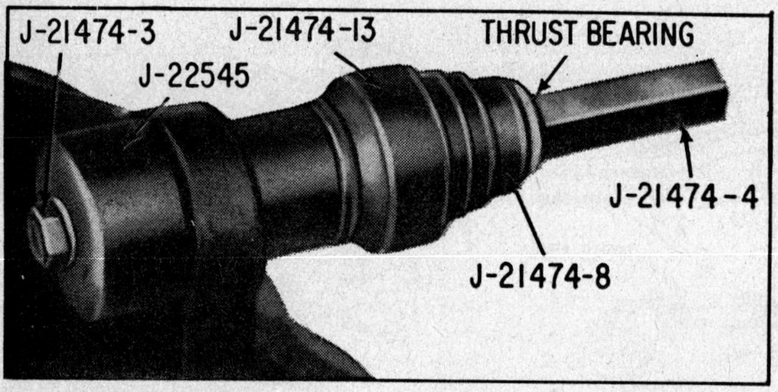

Fig. 14 Upper control arm axle bracket bushing installation. 1969 models with type "O" axle & all 1970-76 models

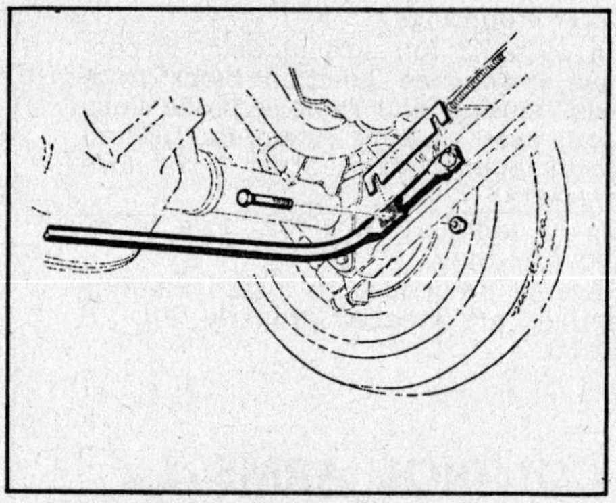

Fig. 15 Stabilizer bar installation (typical)

Upper Control Arms

1. Support vehicle at frame and rear axle.
2. Remove control arm front and rear mount bolts.
3. Replace axle housing bushing as necessary, Figs. 11 thru 14.
4. Reverse procedure to install. Tighten control arm bolts with vehicle at curb height.

Lower Control Arms

Follow "Upper Control Arms" procedure for replacement of lower control arms.

NOTE: Lower control arm bushings are not serviceable.

STABILIZER BAR, REPLACE

1. Support vehicle at rear axle.
2. Remove bolts attaching stabilizer bar to the lower control arms, Fig. 15.
3. Reverse procedure to install.

Front End and Steering Section

FRONT SUSPENSION

As shown in Figs. 1 and 2, the front suspension is of the conventional "A" frame design with ball joints. Double acting shock absorbers are mounted within the coil springs. Caster and camber are controlled by shims.

LUBRICATION

Front suspension lubrication is recommended every 12 months or 12,000 on 1969-70 models, 4 months or 6,000 miles on 1971-74 models and 6 months or 7,500 miles on 1975-76 models, whichever occurs first. The ball joints are fitted with plugs which must be removed and grease fittings installed. After applying the approved type of grease, remove the fittings and reinstall the plugs.

WHEEL ALIGNMENT

1969-76 Full Size Cars

Camber and caster are adjusted by shims placed between the upper pivot shafts and the frame. In order to remove or install shims, *do not remove weight from front wheels.* Loosen pivot shaft-to-

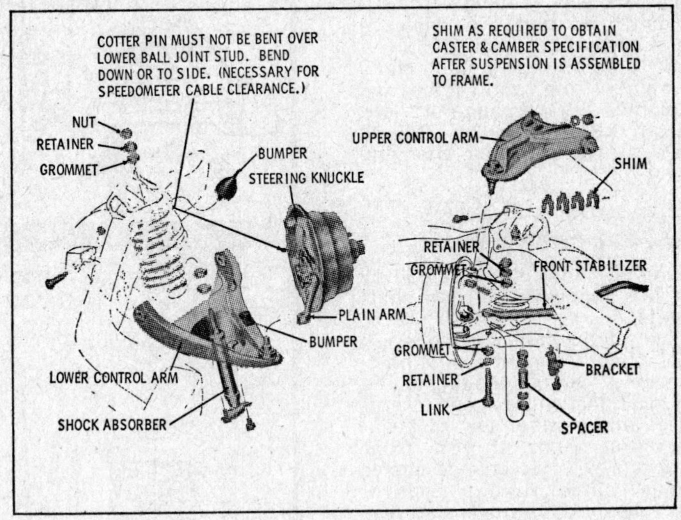

Fig. 1 Front suspension. 1969-76 Full Size Car (Typical)

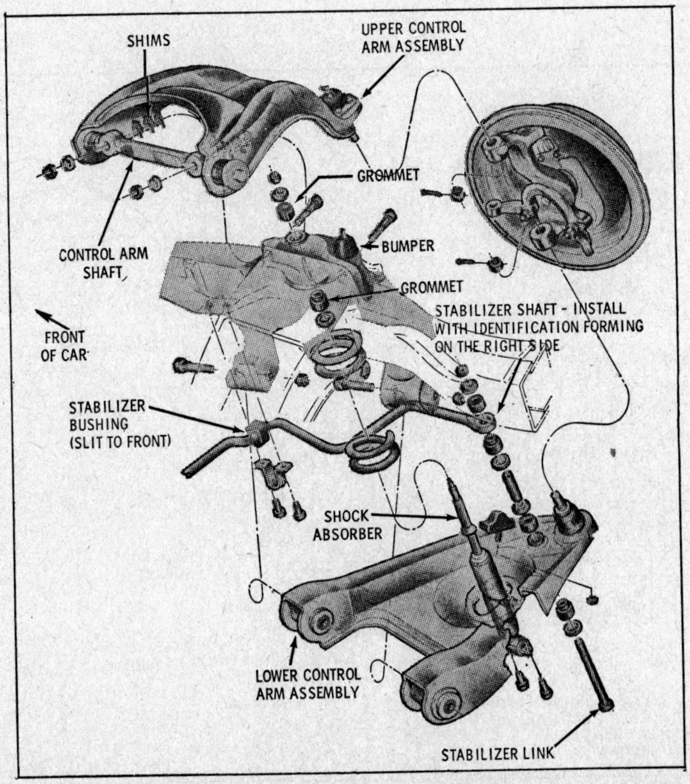

Fig. 2 Front suspension. 1969-76 Intermediate Models (Typical)

Shim Thickness	One shim added to or subtracted from BOTH BOLTS will change camber	One shim added to or subtracted from FRONT BOLT ONLY will change caster
.030"	1/8°	1/8°
.060"	5/16°	7/16°
.120"	5/8°	7/8°

Fig. 3 Wheel alignment shim data. 1969-72 Full size models

Shim Thickness	One shim added to or subtracted from BOTH BOLTS will change CAMBER	One shim added to or subtracted from FRONT BOLT ONLY will change CASTER
.020"	1/8°	3/16°
.030"	3/16°	1/4°
.060"	3/8°	1/2°
120"	3/4°	1°

Fig. 4 Wheel alignment shim data. 1969-72 Inter.

frame bolts. To gain access to these bolts, loosen top and rear fasteners on fender filler plate aprons.

To decrease positive caster, add shim at the front bolt. To increase positive caster, remove shim at the front bolt.

To increase camber, remove shims at both front and rear bolt. To decrease camber, add shims at both bolts.

By adding or subtracting an equal amount of shims from both front and rear bolts, camber will change without affecting caster adjustment. On 1969-72 models, refer to Fig. 3 when changing shims.

1969-76 Intermediates

Caster and camber is adjusted by shimming at the upper control arm shaft attaching points.

Adding shims at the front locations will change caster toward negative with practically no change in camber. Adding shims at the rear locations will change caster toward positive and camber toward negative. Adding equal shims at both front and rear locations will not change caster but will change camber toward negative.

To adjust, loosen both front and rear bolts to free shims for removal or addition. On 1969-72 models, refer to Fig. 4 when changing shims.

TOE-IN, ADJUST

To adjust the toe-in, loosen the clamps at both ends of the adjustable tubes at each tie rod. Then turn the tubes an equal amount until the toe-in is correct. Turning the tubes in the direction the wheels revolve when the car moves forward decreases the toe-in and vice-versa. When the adjustment is complete, tighten all clamp screws.

The steering knuckle and steering arm "rock" or tilt as front wheel rises and falls. Therefore, it is vitally important to position the bottom face of the tie rod end parallel with the machined surface at the outer end of the steering arm when tie rod length is adjusted. Severe damage and possible failure can result unless this precaution is taken. The tie rod sleeve clamps must be straight down to provide clearance.

WHEEL BEARINGS, ADJUST

1969-76

1. While rotating hub assembly, tighten nut to 30 ft-lbs to insure all parts are properly seated.
2. Back off nut ½ turn.
3. Retighten nut finger tight and install retaining ring or cotter key if possible. If unable to install retaining ring or cotter key, back off nut (not to exceed 1/24 of a turn) until tabs on clip align with serrations in nut.

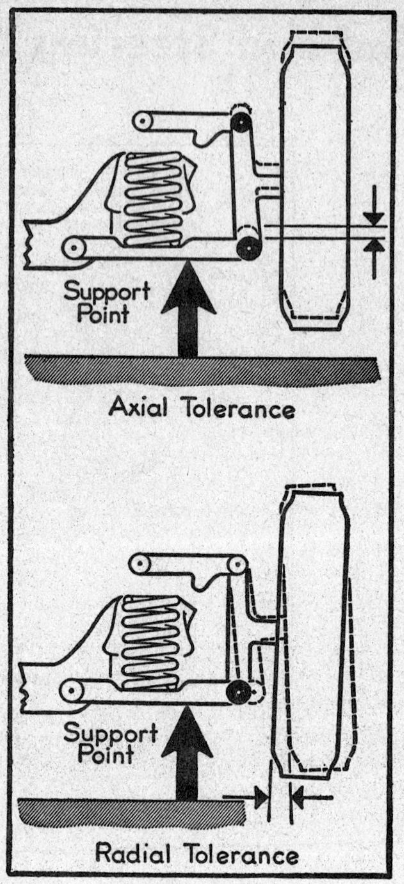

Fig. 5 Checking ball joints for wear

WHEEL BEARINGS, REPLACE

(Disc Brakes)
1969-76

1. Raise car and remove front wheels.
2. Remove brake pads and caliper assembly but do not disconnect brake line. Suspend caliper from a wire loop or hook to avoid strain on the brake hose.
3. Remove grease cap, cotter pin and nut. Pull off hub and disc assembly. Grease retainer and inner bearing can now be removed.

CHECKING BALL JOINTS FOR WEAR

If loose ball joints are suspected, first be sure the front wheel bearings are properly adjusted and that the control arms are tight. Then check ball joints for wear as follows:

Referring to Fig. 5, raise wheel with a jack placed under the lower control arm as shown. Then test by moving the wheel up and down to check axial play, and rocking it at the top and bottom to measure radial play.

1973-76 Omega

1. Upper ball joint should be replaced if looseness exceeds .125".
2. Lower ball joint should be replaced

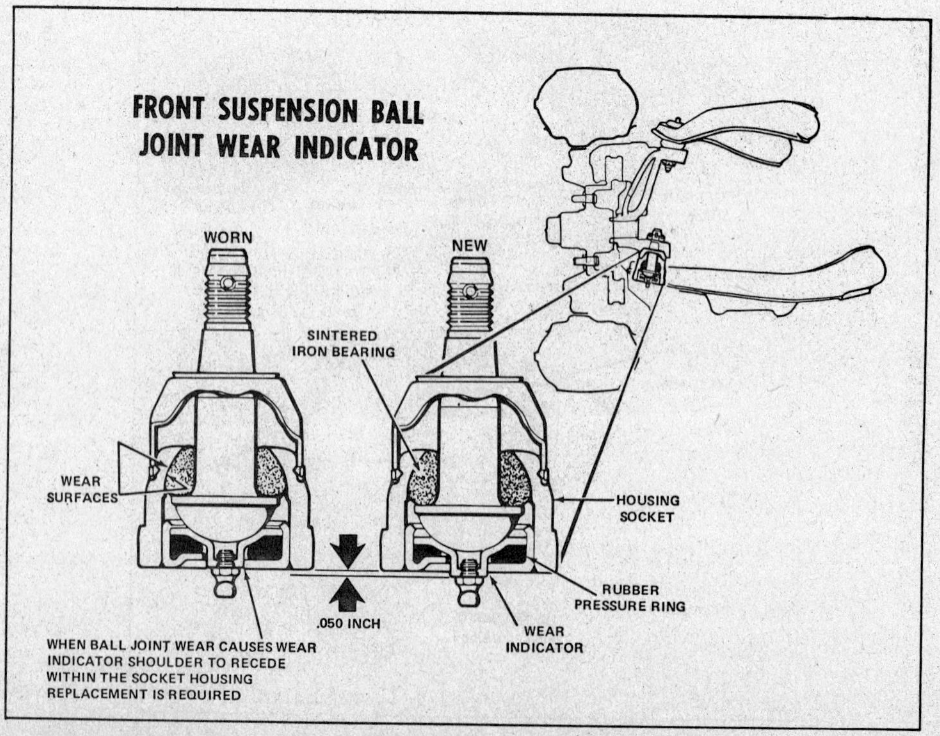

Fig. 6 Ball joint wear indicator. 1973 full size cars & 1974 Exc. Omega & 1975-76 All

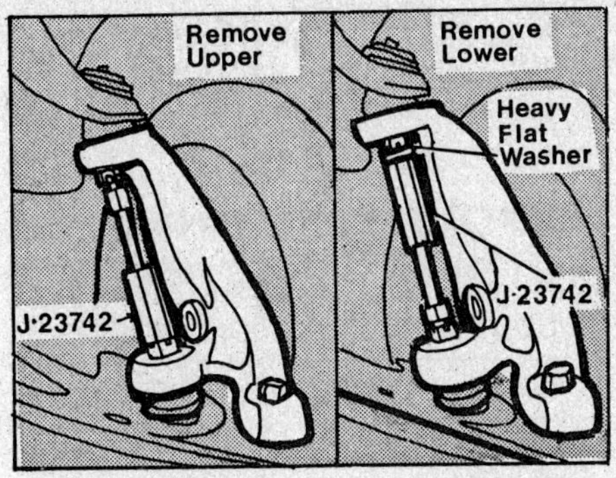

Fig. 7 Removing upper & lower ball joint stud from knuckle

if looseness exceeds 1/16".

NOTE: Beginning with 1975 models, a wear indicator is built into the lower ball joint, Fig. 6.

1969-76 Cutlass

1. Upper ball joint should be replaced if looseness exceeds .125".
2. Lower ball joint should be replaced if looseness exceeds .025" on 1973 models or .125" on previous models.

NOTE: Beginning with 1974 models, a wear indicator is built into the lower ball joint, Fig. 6.

1969-76 Full Size Cars

1. Upper ball joint should be replaced if looseness exceeds .125".
2. Lower ball joint should be replaced if looseness exceeds .125" on 1969-72 models.

NOTE: Beginning with 1973 models, a wear indicator is built into the lower ball joint. Refer to Fig. 6.

BALL JOINTS, REPLACE

On some models the ball joints are riveted to the control arms. All service ball joints, however, are provided with bolt, nut and washer assemblies for replacement purposes.

Some ball joints are pressed into the control arms, in which case they may be pressed out and new ones installed.

Lower Ball Joint

1. Raise vehicle and support at frame. Remove wheel.
2. Support lower control arm with a suitable jack.
3. Remove cotter pin and loosen ball joint stud nut 2-3 turns. Then, remove ball joint stud from knuckle, Fig. 7.
4. Remove stud nut and lower the control arm. Position knuckle assembly aside.
5. Pry ball joint seal retainer from joint and remove seal.
6. Press ball joint from lower control arm with suitable tools.
7. Reverse procedure to install. Torque ball joint stud nut to the following specifications:

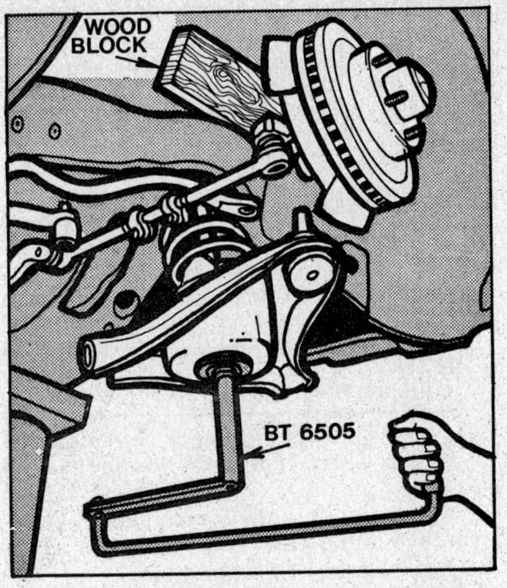

Fig. 8 Replacing coil spring. 1974-76

Year	Model	Ft. Lbs.
1969	Cutlass	40
1969-70	88, 98	70
1970-72	Cutlass	70
1971-74	88, 98	90
1973-74	Omega	80
1975-76	Cutlass, Omega	95
	88, 98	105

Upper Ball Joint

1. Raise vehicle and support at frame. Remove wheel.
2. Support lower control arm with suitable jack or jack stand.
3. Remove cotter pin and loosen upper ball joint stud nut 2-3 turns. Then, remove ball joint stud from knuckle, Fig. 7.
4. Remove stud nut and support knuckle assembly to prevent damage to brake hose.
5. Using a 1/8 inch twist drill, drill the 4 ball joint rivets approximately 1/4 inch. Then, drill off rivets heads using a 1/2 inch twist drill.
6. Punch out rivets and remove lower ball joint.
7. Install new ball joint in lower control arm and torque attaching bolts to 8 ft. lbs.
8. Reverse procedure to assemble. Torque ball joint stud nut to the following specifications:

Year	Model	Ft. Lbs.
1969	88, 98	40
1969-73	Cutlass	40
1970	88, 98	70
1971-73	88, 98	40
1974	Omega	50
	Cutlass, 88 & 98	60
1975-76	All	70

SHOCK ABSORBER REPLACE

1. Remove upper attaching nut, retainer and grommet from shock absorber.
2. Remove two bolts and washers attaching shock absorber to lower control arm and remove shock absorber.
3. To install, position grommet and retainer over shock and slide shock up through spring and frame. Install and

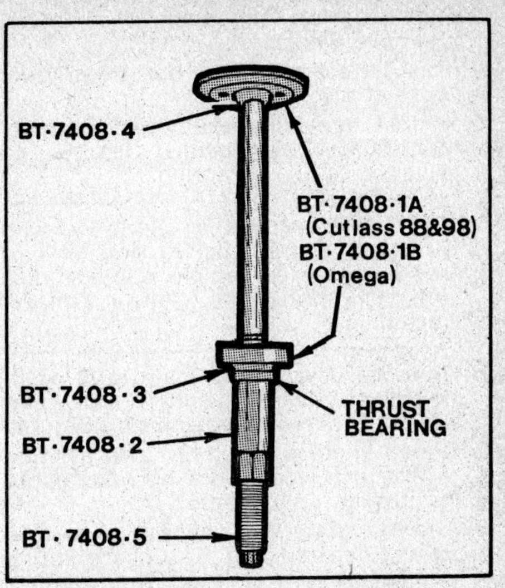

Fig. 9 Replacing coil spring 1969-73

tighten attaching nut and lower capscrews.

COIL SPRING, REPLACE

IMPORTANT: Left and right coil springs should not be interchanged. Spring part number is stamped on outer side of end coil.

1974-76

1. Place transmission in Neutral.
2. Disconnect shock absorber from upper mounting.
3. Raise vehicle and support at frame. Remove wheel.
4. Disconnect stabilizer bar from lower control arm.
5. Remove shock absorber.
6. Install lower plate BT-7408-1A or 1B, Fig. 8, with pivot ball seat facing downward into spring coils. Rotate plate to fully it in lower control arm spring seat.
7. Install upper plate BT-7408-1A or 1B, Fig. 8, with pivot ball seat facing upward into spring coils. Insert ball nut BT-7408-4 through spring coils and onto upper plate.
8. Install rod BT-7408-5 through shock absorber opening in lower control arm and through the upper and lower plates. Depress lock pin on shaft and thread into upper ball nut BT-7408-4. Ensure lock pin is fully extended above ball nut upper surface.
9. With ball nut tang engaged in slot in upper plate, rotate upper plate until it contacts upper spring seat.
10. Install lower pivot ball, thrust bearing and nut on rod and rotate nut until coil spring is compressed enough to be free in the seat.
11. Remove lower control arm pivot bolts. Move control arm rearward and remove coil spring.
12. Reverse procedure to install.

1969-73

1. Remove shock absorber.
2. Install long bolt of tool BT-6505 through upper shock absorber mounting hole.

3. Raise vehicle and support at frame. Remove wheel.
4. Disconnect stabilizer bar from lower control arm.
5. Install remaining components of tool BT-6505. Turn handle, Fig. 9, to slightly compress spring.

CAUTION: Ensure that clip is installed in upper end of tool to prevent personal injury or vehicle damage during coil spring and control arm removal.

6. Remove lower ball joint stud from knuckle as outlined under "Ball Joint, Replace", Lower Ball Joint procedure.
7. Raise knuckle assembly and position a 2x4 inch wood block between upper control arm and frame.
8. Slowly rotate tool handle to relieve spring tension, Fig. 9. Ensure spring tension is fully relieved, then remove tool handle.
9. Remove coil spring and lower control arm as an assembly.
10. Reverse procedure to install. When installing lower control arm assembly, install ball joint stud in knuckle before installing control arm pivot bolts.

MANUAL STEERING GEAR, REPLACE

1. Remove two flex coupling flange nuts.

NOTE: On 1974-76 models, remove coupling shield.

2. Hoist and support car with stands under outer ends of lower control arms.
3. Remove nut and use a puller to remove pitman arm.
4. Remove gear-to frame bolts.
5. Position steering linkage out of the way and withdraw gear assembly from under car.
6. Reverse procedure to install unit.

POWER STEERING GEAR, REPLACE

1. Remove coupling flange hub bolt.

NOTE: On 1974-76 models, remove coupling shield.

2. Disconnect hoses from pump and cap pump and hose fittings.
3. Remove pitman arm nut and, using a suitable puller, remove pitman arm.
4. Remove gear-to-frame bolts. Permit lower shaft to slide free of coupling flange, then remove gear with hoses attached.

TORONADO SUPPLEMENT

Items covered in this section apply to the Toronado only. For service procedures and specifications not covered here, refer to the conventional Oldsmobile section of this chapter.

Engine & Transmission Section

ENGINE, REPLACE

1. Drain radiator and remove hood, marking hinge as a guide for reassembly.
2. On 1972-76 models, if equipped with venturi shroud, unhook strap, remove seal to venturi ring clips and move towards radiator.
3. On all models, remove air cleaner and hot air pipe.
4. Disconnect engine ground strap, upper and lower radiator hoses, transmission oil cooler lines and heater hoses.

Fig. 1 Final drive supporting tool

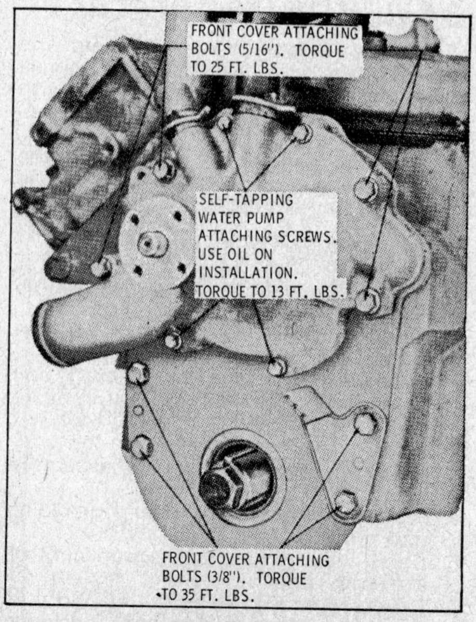

Fig. 3 Engine front cover bolts

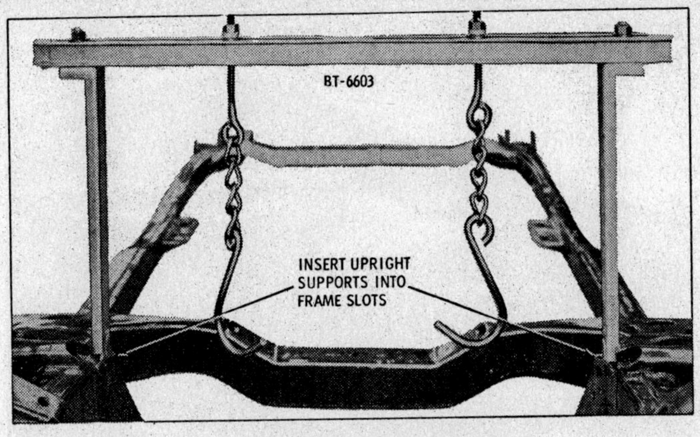

Fig. 2 Installing support bars

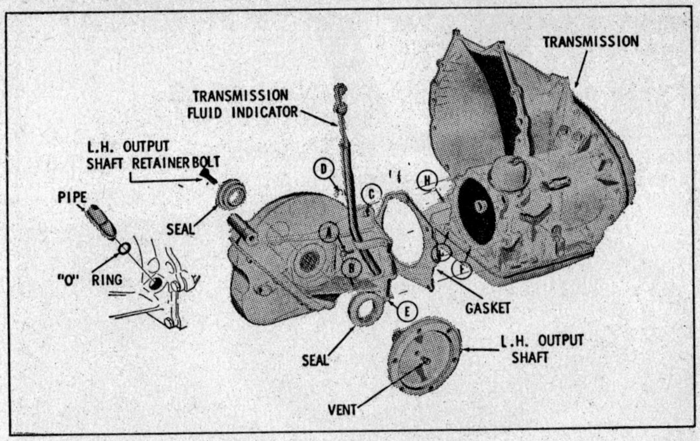

Fig. 4 Transmission attachment

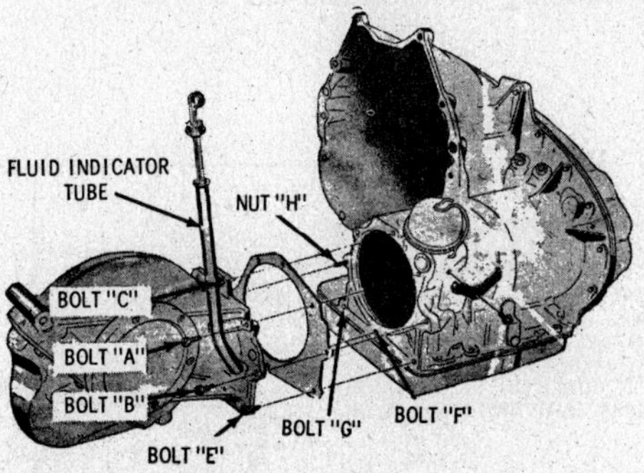

Fig. 4A Transmission attachment. 1970-76

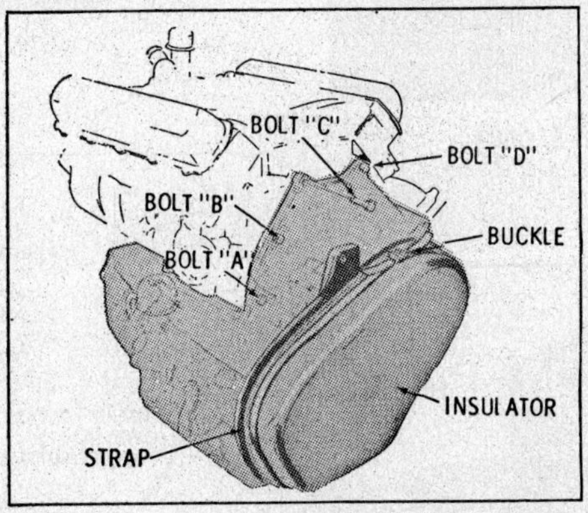

Fig. 5 Transmission-to-engine attachment. 1969

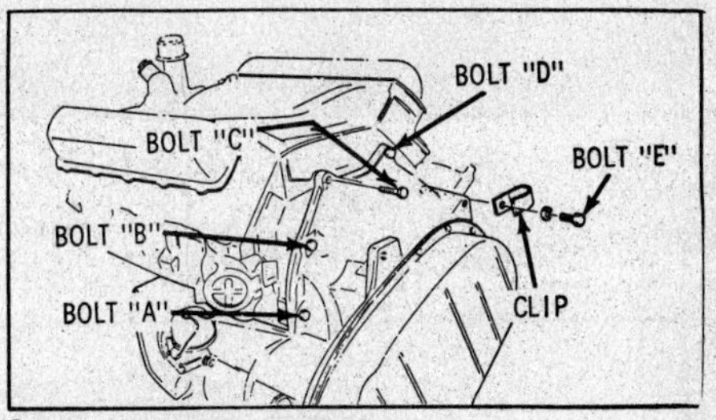

Fig. 5A Transmission to engine attachment. 1970-76

5. Remove upper radiator baffle, radiator, fan shroud and venturi components, if equipped.
6. Disconnect power steering pump bracket from engine without disconnecting oil lines. Position pump and bracket assembly aside.
7. Disconnect A/C compressor bracket from engine without disconnecting refrigerant lines. Position compressor and bracket assembly aside.
8. Disconnect fuel lines at fuel pump, throttle cable, vacuum hoses and electrical connections.
9. Disconnect exhaust pipes from manifolds.
10. Raise vehicle.
11. Loosen upper left hand flywheel cover bolt.
12. Remove starter, then the remaining flywheel cover bolts and pivot cover from upper left hand bolt slot.

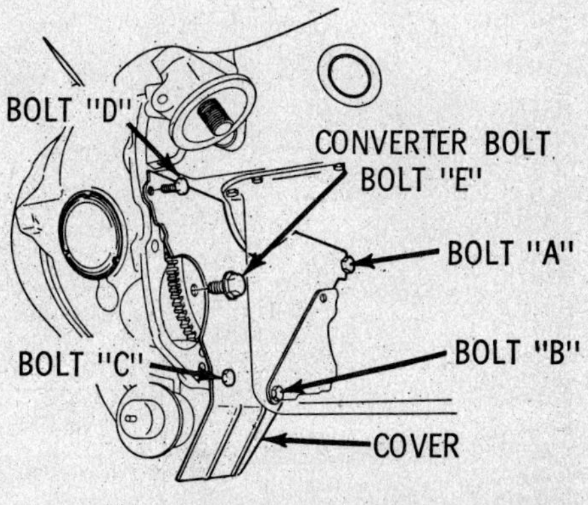

Fig. 5B Converter attachment. 1971-76

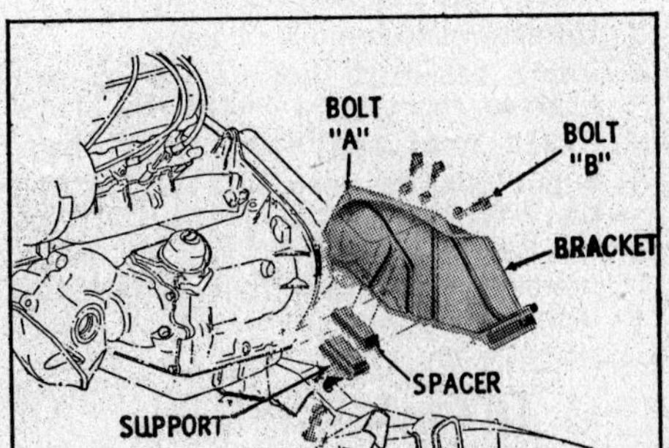

Fig. 6 Transmission-to-engine attachment

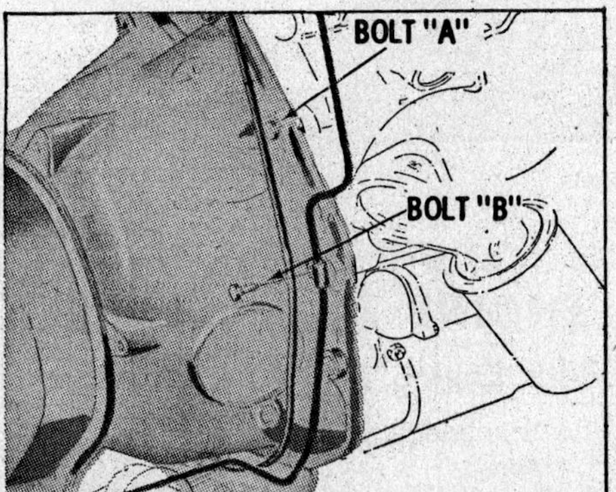

Fig. 7 Engine mount attachment

13. Mark relationship between converter and flywheel, then remove converter to flywheel bolts.
14. Remove engine front mounting attaching nuts.
15. Remove right hand output shaft support bracket bolts. Mark position of bracket to assist assembly.
16. From vehicle left side, remove final drive to engine through bolt and bracket.
17. Remove lower right hand transmission to engine attaching bolts.

Fig. 8 Removing or installing retaining rings

Fig. 9 Removing or installing sprockets and link assembly

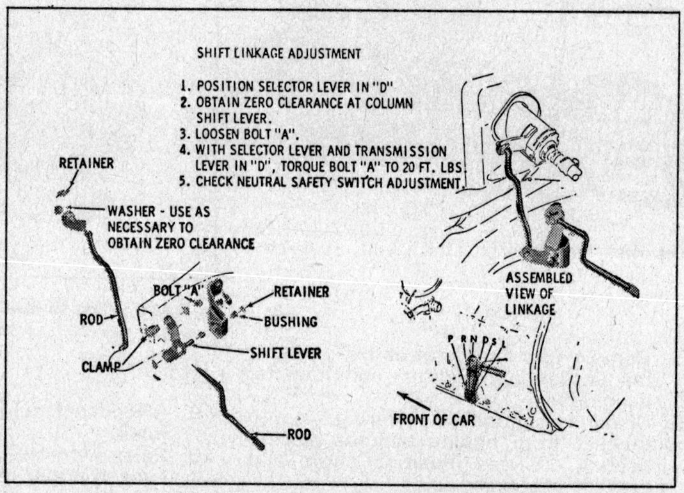

Fig. 11 Shift linkage adjustment

Fig. 10 Removing tight sprockets

18. Attach final drive support chain, Fig. 1, and lower vehicle.
19. Install suitable engine lifting equipment and remove remaining engine to transmission bolts.

NOTE: It may be necessary to raise or lower the transmission using a suitable jack with a wood block placed between the jack and transmission.

20. Remove engine from vehicle and place on support stand, Fig. 2.
21. Reverse procedure to install.

ENGINE FRONT COVER

With Engine & Oil Pan Removed

1. Disconnect by-pass hose from water pump.
2. Remove cover-to-block bolts and remove cover, timing pointer and water pump.
3. Install cover and torque as shown in Fig. 3.

Oil Pan

1. Remove engine as outlined.
2. Remove dipstick, drain oil and remove mount from front cover.
3. Unfasten and remove oil pan.
4. Apply sealer to both sides of pan gaskets (cork) and install on block.
5. Install front and rear rubber seals.
6. Wipe lube on seal area and install pan. Torque 5/16" bolts to 15 ft-lbs and 1/4" bolts to 10 ft-lbs.
7. Install mount on front cover and install engine.

TRANSMISSION

Less Final Drive

Removal

1. Disconnect battery.
2. Disconnect oil cooler lines at trans-

mission and radiator and speedometer cable at governor. Then, remove governor assembly and cover hole to prevent entry of dirt.
3. Install a suitable engine support, Fig. 2.
4. On 1969 models remove nut D and bolts A, B and C, Fig. 4. On 1970-75 models remove three upper final drive to transmission attaching bolts, Fig. 4A.
5. Remove bolts, Figs. 5 and 5A.
6. On 1969-70 models remove flywheel cover plate bolt A. On 1971-75 loosen, DO NOT REMOVE bolt A, Fig. 5B.
7. Hoist car and remove starter.
8. Remove bolts, B, C and D and remove flywheel cover. On 1971-75 models remove cover by pivoting the cover out of the upper bolt (A) slot, Fig. 5B.
9. Remove flywheel to converter bolts E, Fig. 5B. Rotate flywheel until bolts are removed.
10. Disconnect vacuum modulator line and detent switch wiring.
11. Install transmission lift.
12. Remove shift linkage.
13. Remove bolts E, F, G and nut H, Figs. 4 and 4A.
14. Remove bolts A and B, Fig. 7.
15. Remove two upper engine mount bracket-to-transmission bolts.
16. Remove four bracket-to-engine mount bolts.
17. Slide transmission rearward and down. Remove two R. H. output shaft bracket to engine bolts. Final drive will now swing down to allow clearance to remove transmission.
18. Attach converter holding strap to housing to prevent converter from falling out when transmission is being removed.
19. After transmission is removed from vehicle, the link assembly cover insulator can be removed or installed.

Link Belt or Sprockets

Removal

1. Remove sprocket housing cover attaching bolts and cover.
2. Remove sprocket bearing retaining snap rings from retaining grooves in support housing located under drive and driven sprockets, Fig. 8.

NOTE: Do not remove the snap rings from beneath the sprockets, leave them in a loose position between the sprockets and the bearing assemblies.

3. Remove drive and driven sprockets, link belt, bearings and shafts simultaneously by alternately pulling upwards and driven support housing, Fig. 9.

NOTE: If sprockets and link belt are difficult to remove, place a small piece of masonite, or similar material between the sprocket and a short pry bar. Alternately pry upward under each sprocket. Do not pry on links or aluminum case, Fig. 10.

4. Remove link belt from drive and driven sprockets.

Installation

1. Place link belt around the drive and driven sprockets so that the links engage the teeth of the sprockets, colored guide link which has etched numerals facing link cover.
2. Simultaneously place link belt, drive and driven sprockets into support housing, Fig. 8.
3. Using a plastic mallet, gently seat the sprocket bearing assemblies into the support housings.
4. Install sprocket assembly to support housing snap rings, Fig. 9.
5. Install new case to cover and plate assembly sprocket housing gasket.

NOTE: Important: One sprocket cover housing attaching bolt is 1/4 inch longer. This bolt must be installed in the tapped hole located directly over the cooler fittings on the transmission case.

6. Install sprocket housing cover and plate assembly and eighteen attaching bolts. Torque bolts to 8 ft. lbs.

Installation of Transmission

When installing the transmission, the motor mount bracket must be positioned loosely on the link assembly cover until the transmission is in place; then reverse removal procedure. Torque bolts to ft-lbs as follows:

Engine-to-converter housing	30
Engine bracket-to-transmission	55
Engine bracket-to-rubber mount	55
Oil cooler lines to transmission—	
1969-72	30
1973-76	20
Final drive-to-transmission	25
Converter-to-flywheel	35

Adjust shift linkage as directed in Fig. 11.

Drive Axles & Final Drive Section

DRIVE AXLES

Description

Drive axles are a complete flexible assembly and consist of an axle shaft and an inner and outer constant velocity joint, Fig. 1. The right axle shaft has a torsional damper mounted in the center. The inner constant velocity joint has complete flexibility plus inward and outward movement. The outer constant velocity joint has complete flexibility only.

NOTE: Whenever any operations call for disconnecting, connecting, removal or installation of the drive axles, care must be used to prevent damage to constant velocity joint seals. Seals may be wrapped with floor mat rubber or old inner tube, etc. Make sure rubber protective covers that are used are removed before car is started or driven.

DRIVE AXLE, REPLACE

Right Side Unit

Removal

1. Hoist car under lower control arms.
2. Remove axle nut, Fig. 3.
3. Remove oil filter element.
4. Remove inner constant velocity (C. V.) joint attaching bolts.
5. Push inner C.V. joint outward enough to disengage from R.H. final drive output shaft and move rearward.
6. Remove R.H. output shaft bracket bolts to engine and final drive.
7. Remove R.H. final drive output shaft.
8. Remove drive axle assembly.

NOTE: Care must be used to see that C.V. joints do not turn to full extremes and that seals are not damaged against shock absorber or stabilizer bar.

Installation

1. Place R.H. drive axle into lower control arm and enter outer race splines into knuckle.
2. Lubricate final drive output shaft seal with approved seal grease.
3. Install R.H. output shaft into final drive and attach support bolts to engine and brace. Torque to 50 ft-lbs.
4. Move R.H. drive axle toward front of car and align with R.H. output shaft. Install attaching bolts and torque to 65 ft-lbs.
5. Install oil filter element.
6. Install washer and nut on drive axle. Torque nut to 150 ft. lbs. on 1969-73 models, 140 ft-lbs. on 1974 models and 200 ft-lbs. on 1975-76 models. Install cotter pin.

Left Side Unit

Removal

1. Hoist car under lower control arms.
2. Remove wheel and drum.

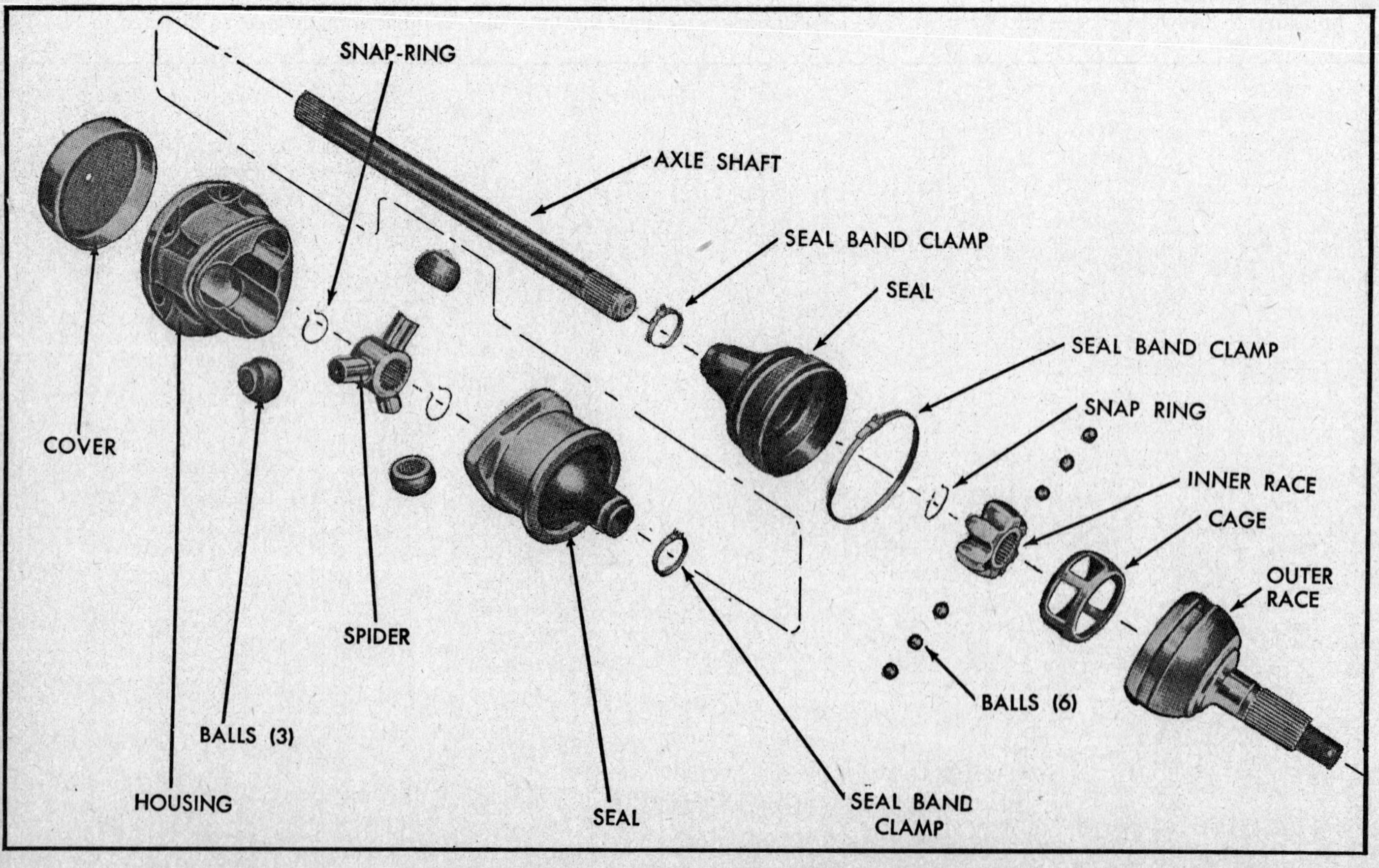

Fig. 1 Drive axle assembly. 1969-76

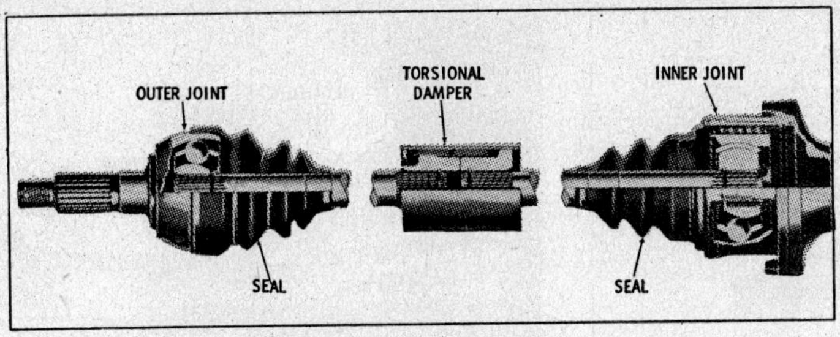

Fig. 2 Drive axle disassembled (right side)

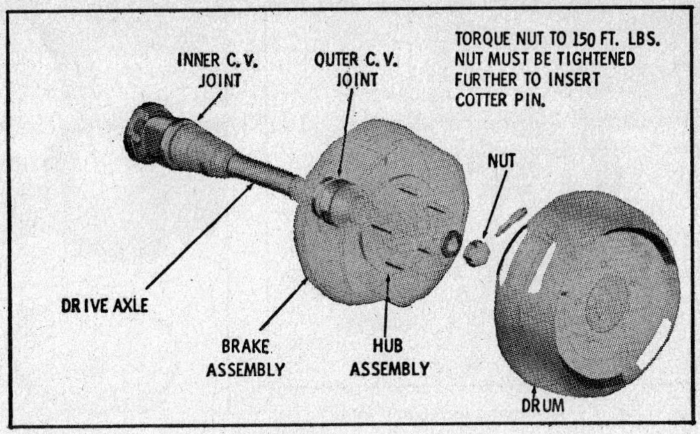

Fig. 3 Drive axle installed

3. Remove drive axle nut.
4. Position access slot in hub so that each attaching bolt (4) can be removed, Fig. 4. It will be necessary to push aside adjuster lever to remove one of the bolts.
5. On 1969 models, remove hub assembly, Fig. 4. It will be necessary to push aside adjuster lever for clearance.
6. Remove tie-rod end nut.
7. Using a hammer and brass drift, drive on knuckle until tie-rod end stud is free.
8. Remove bolts from drive axle and L.H. output shaft, Fig. 5.
9. Remove upper control arm ball joint nut. Using hammer and brass drift, drive on knuckle until upper ball joint stud is free, Fig. 6.
10. Remove ball joint, Fig. 7, being careful not to damage drive axle seal.
11. Remove knuckle and support so that brake hose is not damaged.
12. Carefully guide drive axle out.

NOTE: Care must be used to see that C.V. joints do not turn to full extremes and that seals are not damaged against shock absorber or stabilizer bar.

Installation

1. Guide L.H. drive axle onto lower control arm in position on block, Fig. 5.
2. Insert lower ball joint stud into knuckle and attach nut (do not tighten).
3. Center L.H. drive axle in opening of knuckle and insert upper ball joint stud.
4. Place brake hose clip over upper ball joint stud and install nut (do not tighten).
5. Insert tie-rod end stud into knuckle and attach nut. Torque to 45 ft-lbs and insert cotter pin.
6. Lubricate hub bearing OD with E.P. grease and install. Torque to 65 ft-lbs.
7. Align inner C.V. joint with output shaft and install attaching nuts. Torque to 65 ft-lbs.
8. Torque upper and lower ball joint nuts to 40 ft-lbs and insert cotter pins.

NOTE: Upper ball joint cotter pin must be crimped toward upper control arm to prevent interference with outer C.V. joint seal.

9. Install drive axle washer and nut. Torque nut to 150 ft. lbs. on 1969-73 models, 140 ft-lbs. on 1974 models and 200 ft. lbs. on 1975-76 models. Install cotter pin.
10. Install drum and wheel.
11. Lower car and check wheel alignment.

C.V. JOINT SERVICE

NOTE: The C.V. joints are to be replaced as a unit and are only disassembled for repacking and replacement of seals.

Outer C.V. Joint 1969-76

Disassemble

1. Insert axle in vise, clamping on mid-portion only.
2. Remove inner and outer seal clamps, Fig. 8.
3. Slide seal down axle shaft to gain access to C.V. joint.
4. Referring to Fig. 9, spread retaining ring until C.V. joint can be removed from axle spline.
5. Remove retaining ring, Fig. 10.
6. Slide seal from axle shaft.
7. Remove grease from C.V. joint.
8. Holding C.V. joint with one hand, tilt cage and inner race so that one ball can be removed. Continue until all six balls are removed, Fig. 11.
9. Turn cage 90° and with large slot in cage aligned with land in inner race, lift out, Fig. 12.
10. Turn inner race 90° in line with large hole in case, lift land on inner race up through large hole in cage and turn up and out to separate parts, Fig. 13.

Inspection

Wash all metal parts in cleaning sol-

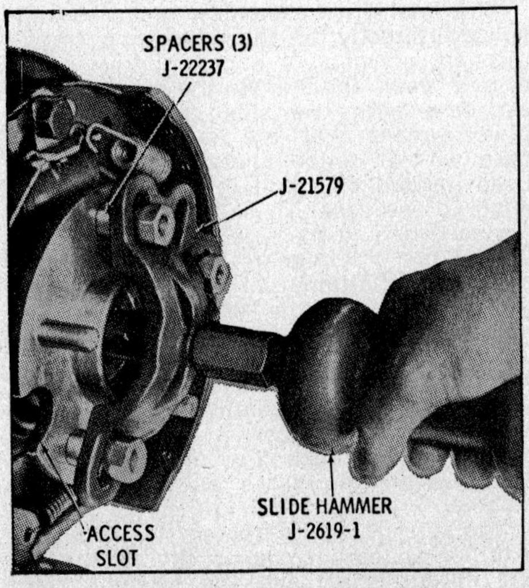

Fig. 4 Removing hub. 1969

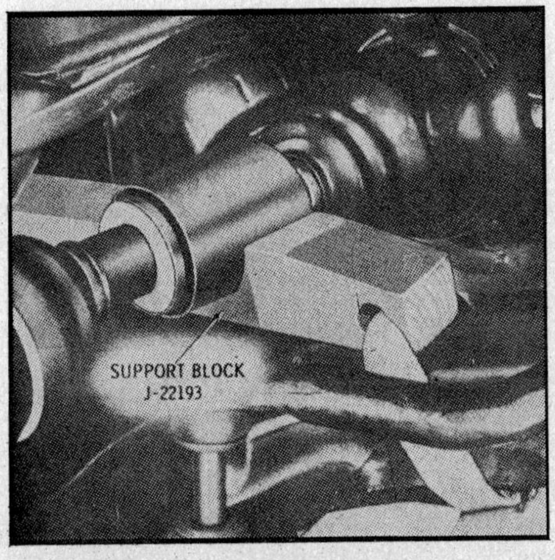

Fig. 5 Installing support block

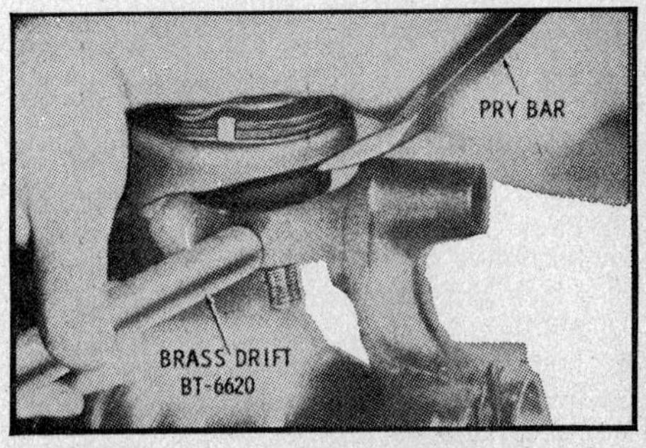

Fig. 6 Removing upper ball joint

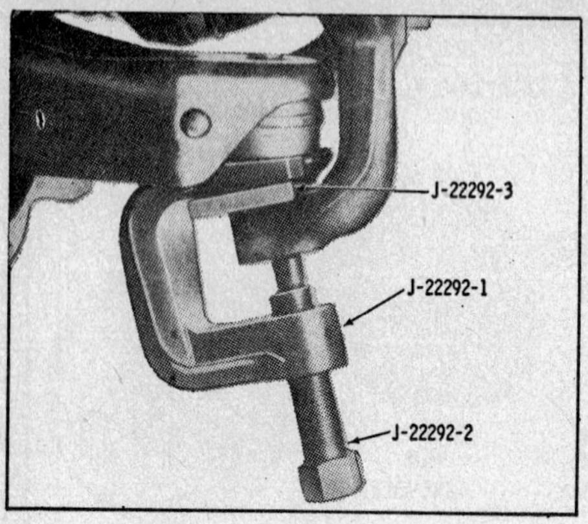

Fig. 7 Removing lower ball joint

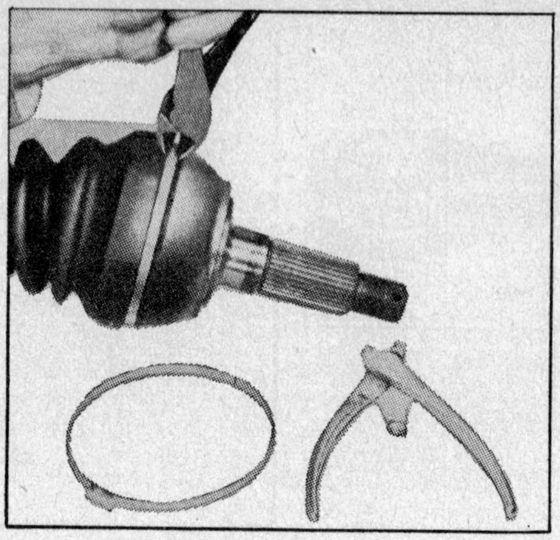

Fig. 8 Cutting seal clip

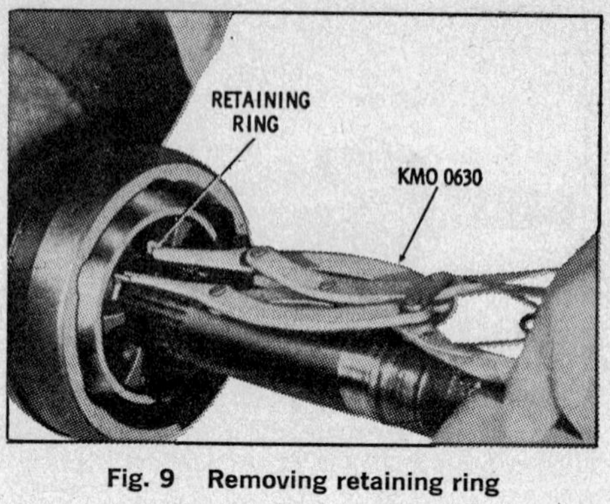

Fig. 9 Removing retaining ring

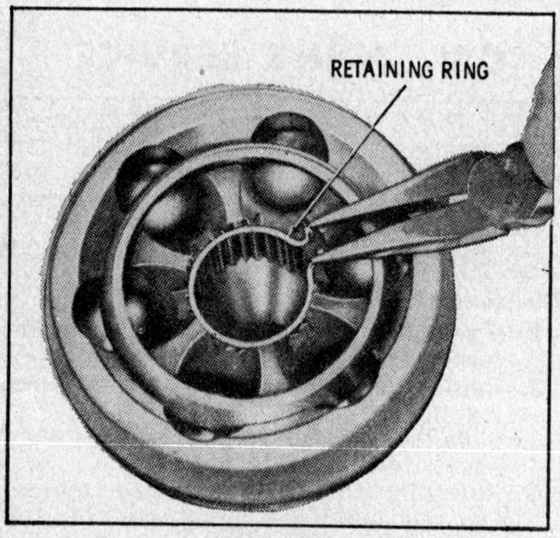

Fig. 10 Removing or installing retaining ring

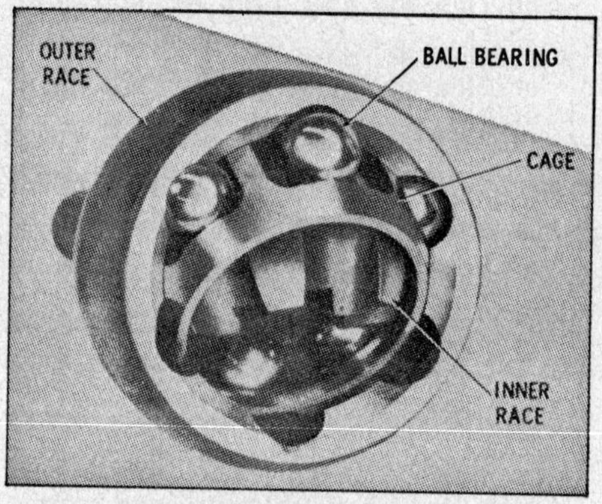

Fig. 11 Removing balls from outer race

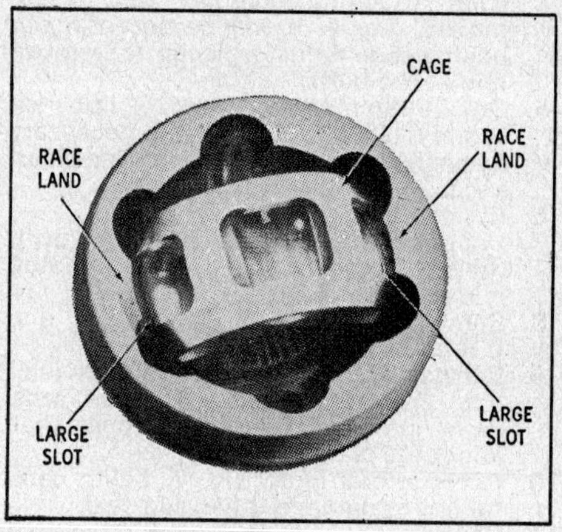

Fig. 12 Positioning cage for removal

vent and dry with compressed air. Rubber seal should be replaced whenever joint is disassembled for service. Inspect all metal parts for nicks, cracks, breaks or scores. If any defects are found the joint assembly will have to be replaced as a unit.

Reassemble

1. Insert land of inner race into large hole in cage and pivot to install in cage, Fig. 13.
2. Align inner race as shown in Fig. 12 and pivot inner race 90° to align in outer race as shown in Fig. 14.
3. Insert balls one at a time until all six are installed. Inner race and cage will have to be tilted as shown in Fig. 14 so that each ball can be inserted.
4. Pack joint full of approved lubricant. Pack inside of seal with approved lubricant until folds of seal are full.
5. Place small keystone clamp on axle shaft.
6. Install seal on axle shaft.
7. Install retaining ring into inner race, Fig. 10.
8. Insert axle shaft into splines of outer C.V. joint until retaining ring secures shaft.
9. Position seal in slot of outer race.
10. Install large keystone clamp over seal and secure, Fig. 15. Then install small keystone clamp over seal and secure, Fig. 16.

Inner C.V. Joint 1969-75

1. Clamp mid portion of axle shaft in vise and remove small seal clamp.
2. Remove large end of seal from C.V. joint by prying out peened spots and driving off with hammer and chisel, Fig. 17.
3. Carefully slide seal down axle shaft.
4. Carefully lift housing from spider assembly and remove "O" ring from housing outer surface.

NOTE: Place a rubber band over ends of spider to retain the three balls and needle bearings.

5. Remove retaining ring from end of axle.
6. Remove spider assembly from axle.
7. Remove inner retaining ring, seal and cover, Fig. 18.
8. Remove balls from spider, being careful not to lose any needles.
9. Reverse procedure to assemble, being sure to stake housing in six evenly spaced places after reassembly.

FINAL DRIVE, 1969-76

Description, Fig. 19

The final drive assembly, mounted and splined directly to the automatic transmission, consists of a pinion drive gear, a ring gear (bolted to the case), case assembly with two side gears and two pinion gears that are retained to the case with a pinion shaft. A lock pin is used instead of a bolt to lock the pinion shaft to the case. There are thrust washers used behind the side gears and shims behind the pinion gears the same as in a conventional differential. The left side gear is different than the right side gear in that it has a threaded retainer plate to which the left output shaft bolts. The two side bearings are the same and the preload shims are identical for the right and left side. The carrier is identical in external appearance and mounts to the transmission the same as in the past models.

The output shafts remain identical in external appearance as in the past. The left output shaft has the retainer bolt going through the shaft to the side gear.

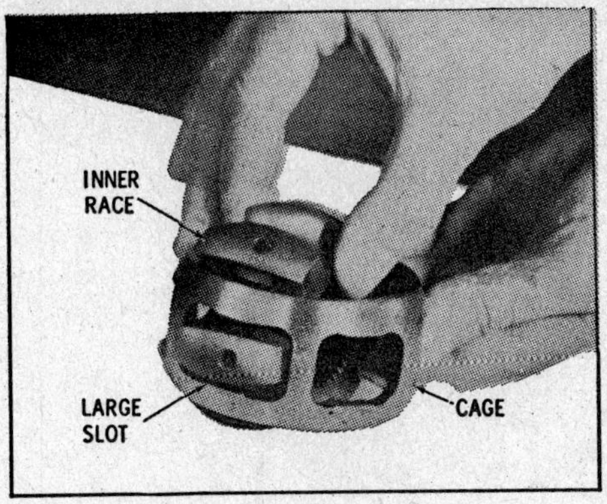

Fig. 13 Removing inner race from ball cage

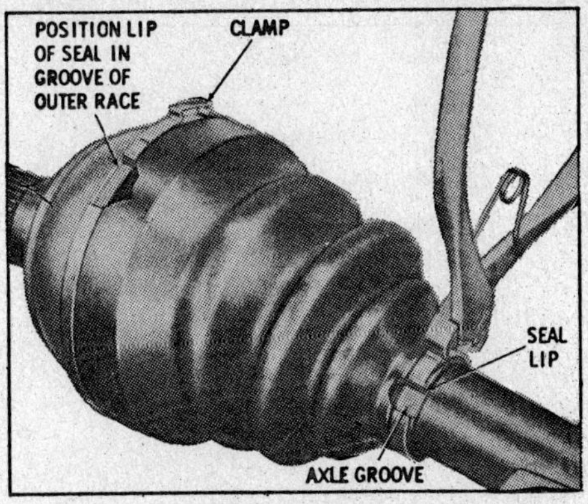

Fig. 16 Installing keystone clamp (small)

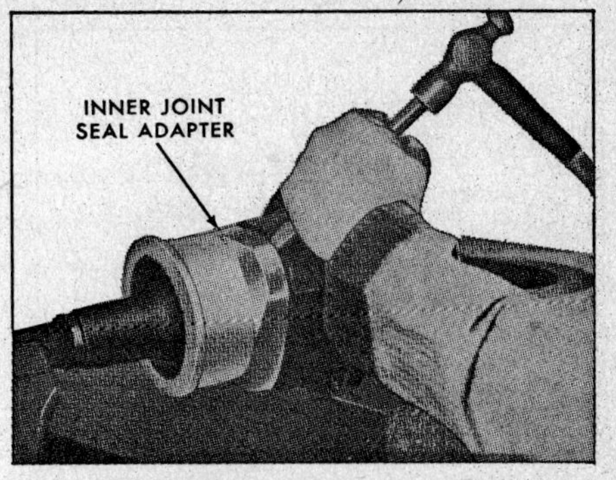

Fig. 17 Removing inner C.V. joint seal

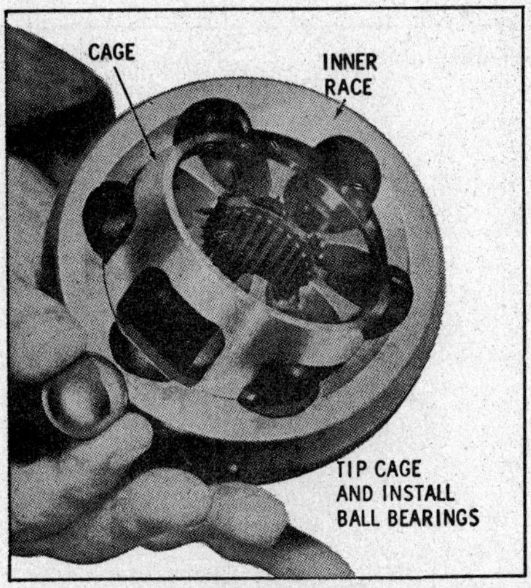

Fig. 14 Installing balls in outer race

OUTPUT SHAFT & SEALS

R.H. Shaft, Bearing & Seal

Removal

1. Disconnect battery. Hoist car.
2. Remove engine oil filter element.
3. Disconnect R.H. drive axle.
4. Disconnect support from engine and brace.
5. Remove output shaft assembly.

Installation Figs. 21 and 22

1. If removed, assemble bearing and related parts. Position assembly in a press and install bearing until seated against shoulder on shaft. Pack area between bearing and retainer with wheel bearing grease, then install slinger. Install seal if removed.
2. Install remaining parts removed in reverse order of removal.

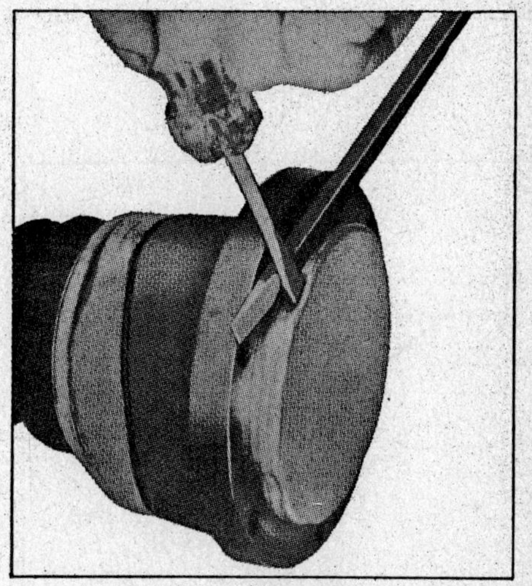

Fig. 18 Removing housing cover

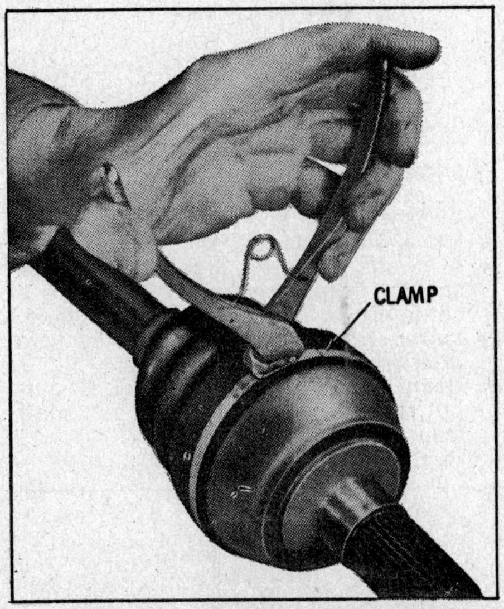

Fig. 15 Installing keystone clamp (large)

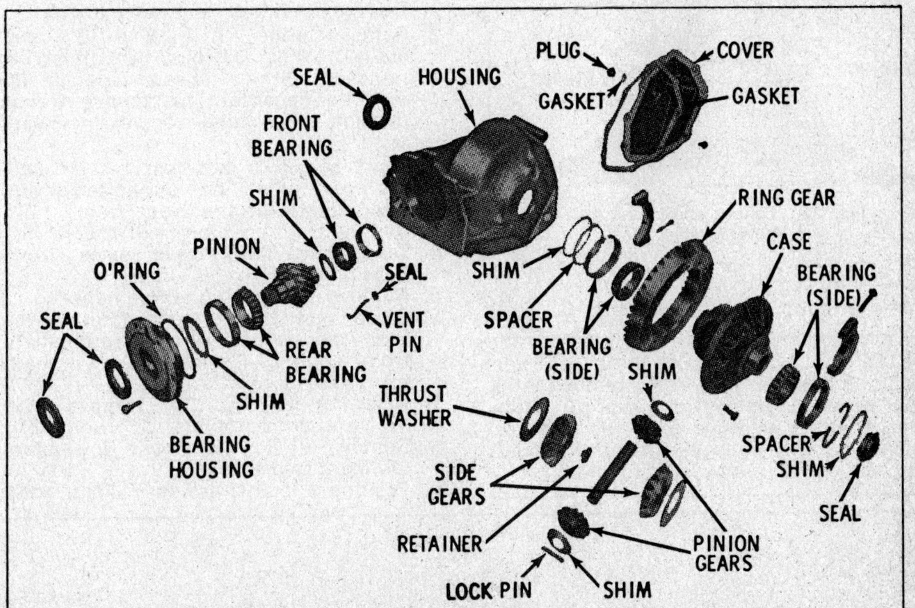

Fig. 19 Final drive disassembled. 1969-76

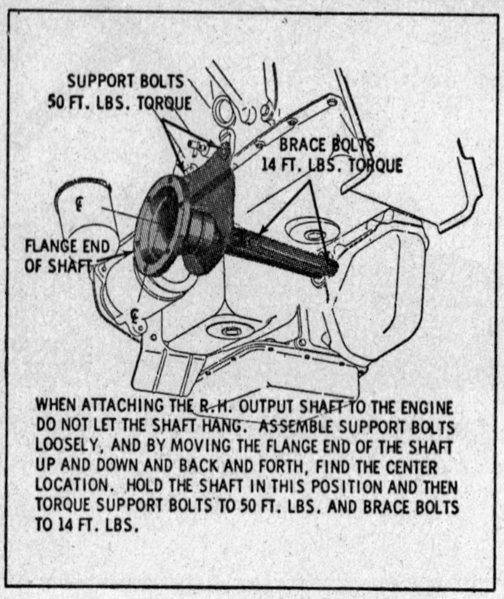

Fig. 21 Aligning right output shaft

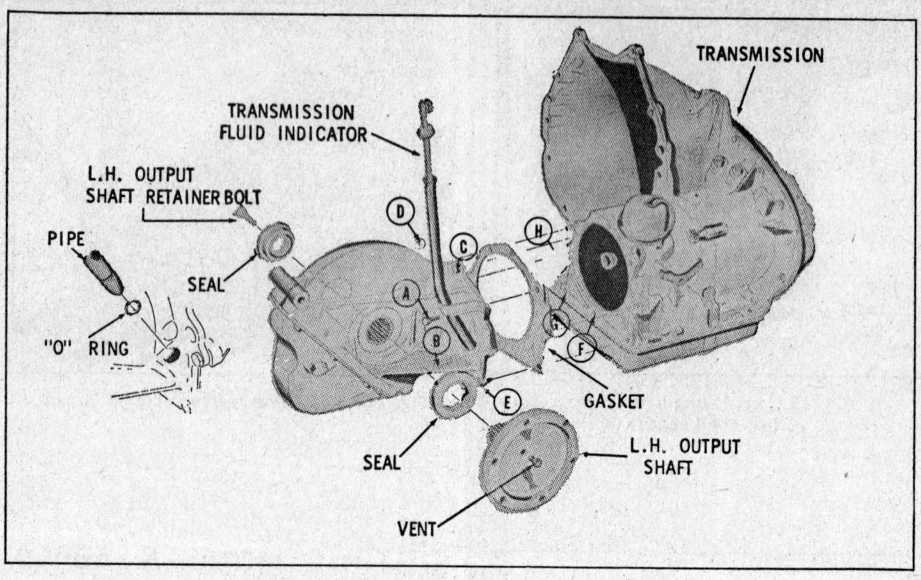

Fig. 23 Final drive attachment. 1969-76

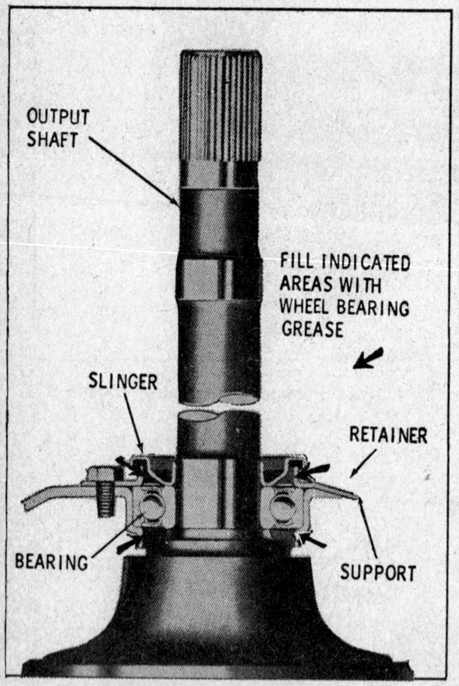

Fig. 22 Output shaft assembled (right side)

L.H. Output Shaft, Bearing & Seal

NOTE: The L.H. output shaft can normally be removed only after removing the final drive assembly from the car. However, if the L.H. drive axle has been removed for any reason, the output shaft and seal can be removed as follows:

1. Remove R.H. output shaft as outlined above.
2. Remove L.H. output shaft retaining bolt and remove shaft.
3. Apply approved lubricant to the seal, then insert output shaft into final drive, indexing splines of shaft with splines on final drive.
4. Install and torque L.H. output shaft retaining bolt to 45 ft-lbs.
5. Install R.H. output shaft in reverse order of removal.

FINAL DRIVE, REPLACE

1. Disconnect battery and raise hood.
2. Remove bolts "A", "B", "C" and nut "D", Fig. 23. Nut "D" must be removed with a special wrench. It may be necessary to remove transmission filler tube to obtain clearance.
3. Hoist car. If a two-post lift is used the car must be supported with floor stands at the front frame rails and the front post lowered.
4. Disconnect both drive axles from output shafts.
5. Remove engine oil filter element.
6. Disconnect brace from final drive, then disconnect R.H. output shaft from engine. Remove output shaft from final drive.
7. Referring to Fig. 24, remove bolt "X" and loosen bolts "Y" and "Z".
8. Remove final drive cover and allow lubricant to drain.
9. Position transmission lift with adapter for final drive. Install an anchor bolt through final drive housing and lift pad.
10. Referring to Fig. 23, remove bolts "E", "F", "G" and nut "H".
11. Move transmission lift toward front of car to disengage final drive splines from transmission. Provide a container to catch transmission fluid.
12. Lower transmission lift and remove final drive from lift.
13. Using a 9/16" socket, remove L.H. output shaft retaining bolt and pull shaft from final drive.
14. Reverse procedure to install.

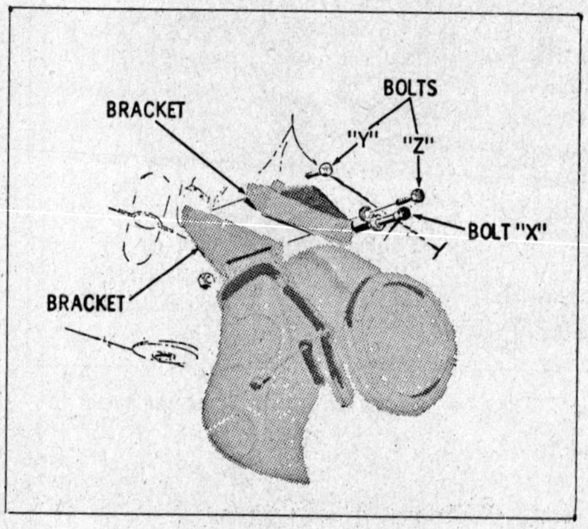

Fig. 24 Disconnecting final drive from engine

Rear Axle & Suspension

REAR AXLE

1969-70 models use a beam type rear axle, while 1971-76 models use the hollow tube type axle. The rear wheel spindles are a press fit and bolted to the rear axle assembly, Figs. 1 and 2. As shown, tapered roller bearings are used in the rear wheels. These bearings do not require regularly scheduled repacking. When major brake service work is to be performed, however, it is recommended that the bearings be cleaned and repacked.

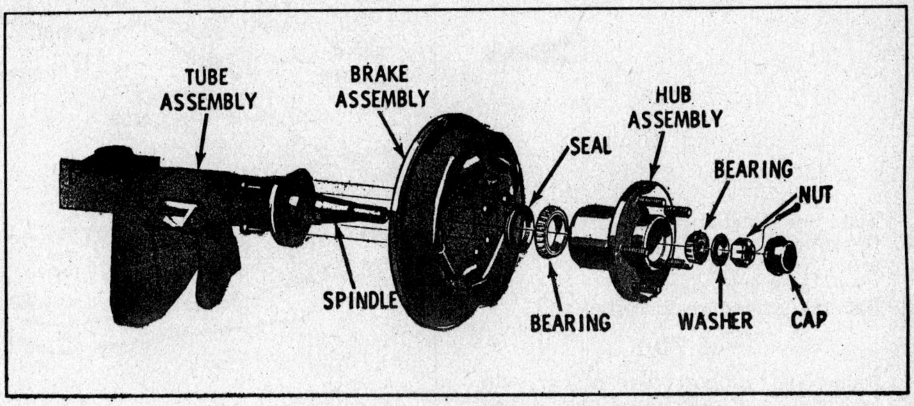

Fig. 1 Rear wheel hub & spindle less True-Track brake

Wheel Bearing Adjustment

Adjustment of the rear wheel bearings should be made while revolving the wheel at least three times the speed of the nut rotation when taking torque readings.

1. Check to make sure that hub is completely seated on wheel spindle.
2. While rotating wheel, tighten spindle nut to 30 ft-lbs. Make certain all parts are properly seated and that threads are free.
3. Back nut off ½ turn, then retighten nut to 2 ft. lbs. and install cotter pin.
4. If cotter pin cannot be installed in either of the two holes in the spindle, with the nut at 2 ft. lbs., back nut off until cotter pin can be installed.
5. The rear hub must be rotated at least three revolutions during tightening of spindle nut. The final adjustment to be 2 ft. lbs. to provide .004″ bearing end play.
6. Peen end of cotter pin snug against side of nut. If it can be moved with a finger, vibration may cause it to wear and break.

Wheel Spindle, Replace

1. Raise and support rear of car and remove hub.
2. Disconnect brake line at wheel cylinder.
3. Unfasten and remove brake backing plate and position out of the way.
4. Remove bolts securing spindle to axle.
5. On 1971-76 models, use a suitable puller to remove spindle from axle tube.
6. On 1969-70 models:
   a. Support rear axle with a suitable jack and remove four nuts from center spring clamp. Then, lower axle to provide access to spindle.
   b. Remove lower spring insulators from rear axle.
   c. Drive spindle from rear axle.

Installation

1. On 1971-76 models, use a suitable slide hammer to drive spindle into axle tube until fully seated.
2. On 1969-70 models:
   a. Place spindle on axle with keyway facing upward, then install four backing plate to spindle nuts and bolts.
   b. Alternately tighten nuts until spindle is fully seated, then remove the attaching nuts and bolts.
   c. Place lower spring insulator on rear axle.
   d. Position axle to spring clamp. Ensure alignment pin locates into axle, lower insulator is positioned properly and center spring clamp bolts enter axle mounting holes.
   e. Install and torque center spring clamp nuts to 30 ft-lbs.
3. Install new gasket on wheel spindle.
4. Install brake backing plate and tighten nuts to 40 ft-lbs.
5. Connect brake line to wheel cylinder, tightening fitting to 14 ft-lbs.
6. Install rear hub.

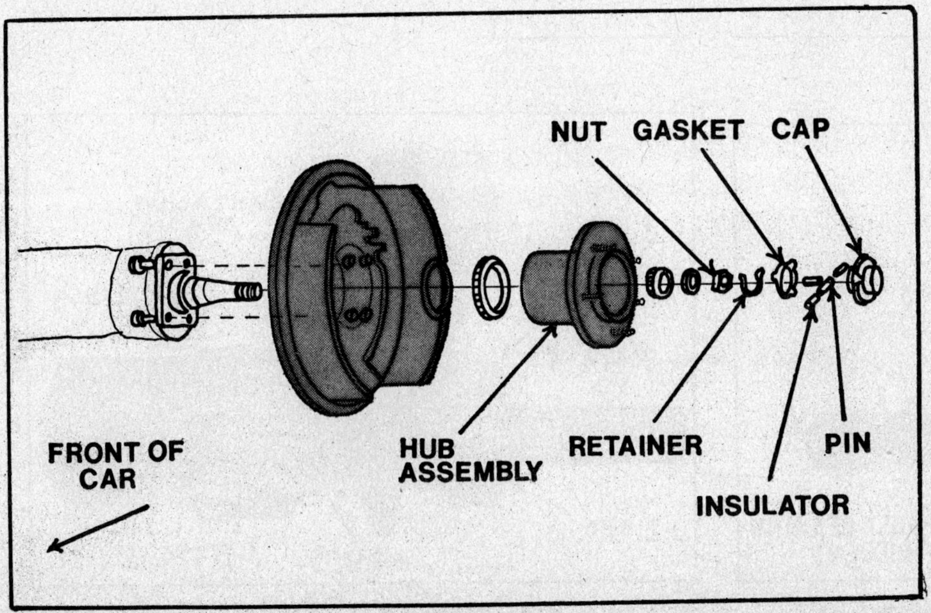

Fig. 2 Rear wheel hub & spindle with True-Track brake

Rear Axle, Replace 1971-76

1. Perform steps 1 through 5, outlined under 1969-70 "Rear Axle, Replace" procedure. Support axle with a suitable jack.
2. If equipped with True Track brakes, disconnect wiring connector from underbody connector.
3. Disconnect shock absorbers from lower mountings.
4. Lower axle and remove coil spring.
5. Remove brake backing plate attaching

Fig. 3 Rear suspension. 1969-70

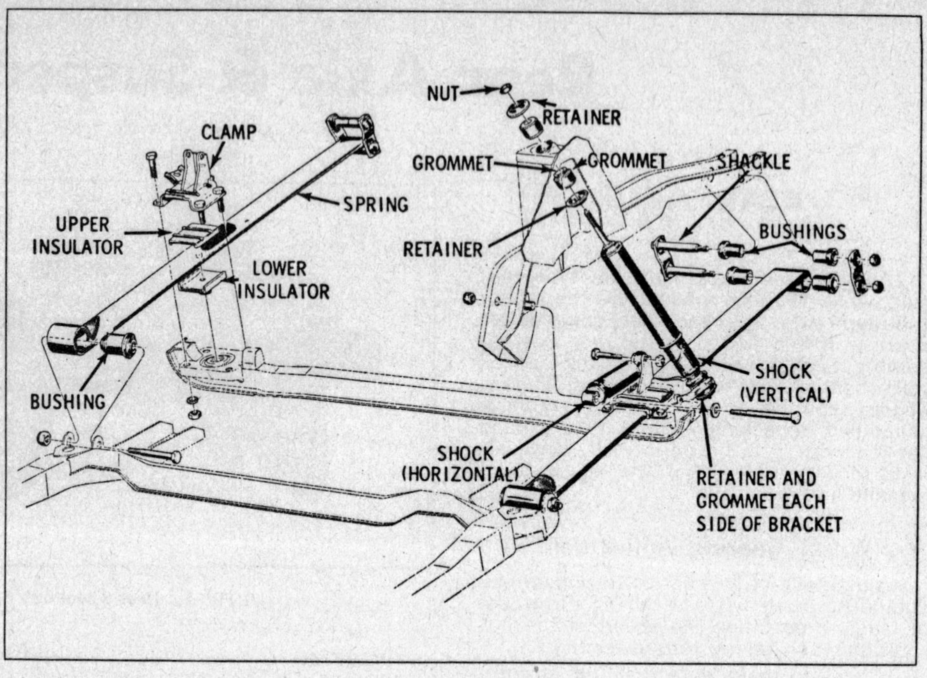

bolts and position plates aside.
6. Disconnect upper and lower control arms from axle.
7. Lower axle from vehicle.
8. Reverse procedure to install.

1969-70

1. Raise and support rear of car with jack stands at rear frame pads ahead of rear wheel opening.
2. Remove rear wheels and hubs.
3. Disconnect brake lines at wheel cylinders.

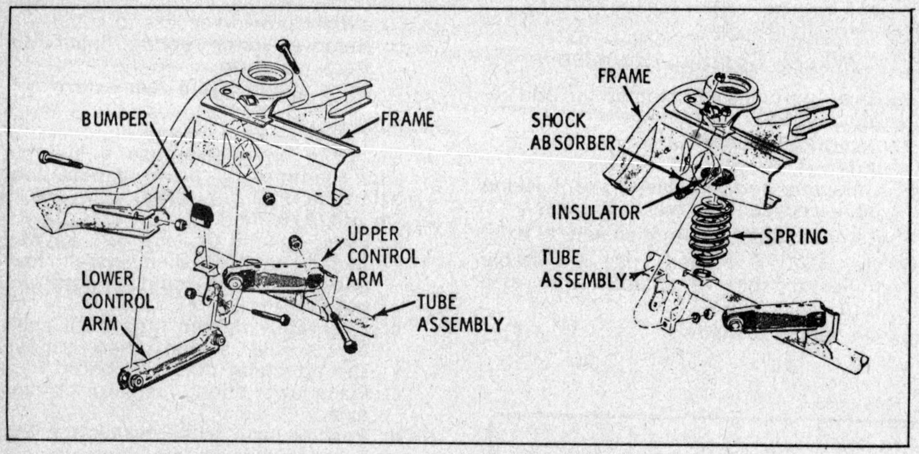

Fig. 4 Rear suspension. 1971-76

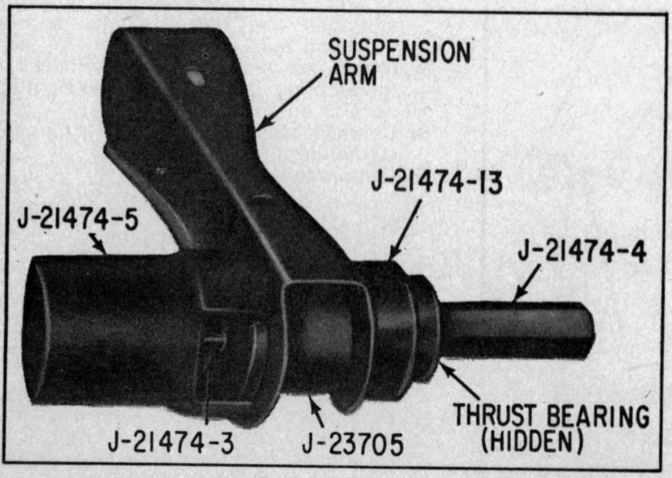

Fig. 5 All front bushings & lower control arm rear bushing removal. 1971 models

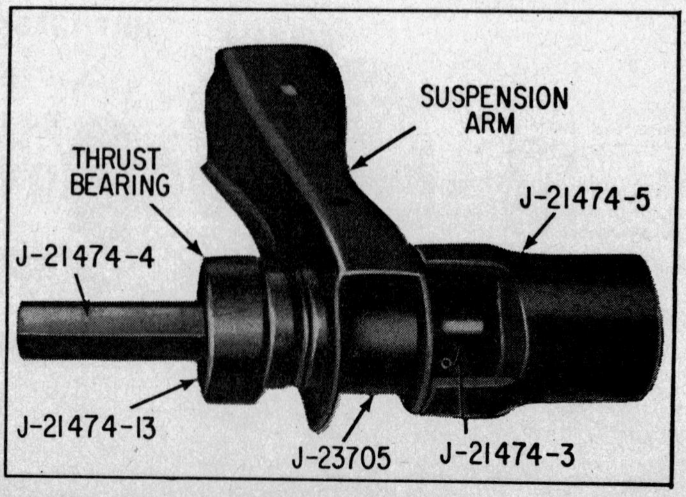

Fig. 6 All front bushings & lower control arm rear bushing installation. 1971 models

4. Disconnect parking brake cable at equalizer.
5. Disconnect rubber brake hose at underbody connector.
6. Disconnect overtravel lever link from bracket on rear axle.
7. Remove spring guides retaining parking brake cable to center spring clamp.
8. If rear axle is being replaced, remove brake backing plates.
9. Supporting rear axle at center with a jack, remove eight nuts (4 each side) from center spring clamp assemblies.
10. Lower rear axle with jack and remove from car.
11. Remove lower spring insulators from rear axle.
12. If rear axle is being replaced, remove bolt securing brake line junction fitting to axle. Remove brake line, overtravel lever link bracket, and drive spindles from axle.
13. Reverse procedure to install.

SHOCK ABSORBER, REPLACE

1. With rear axle properly supported, disconnect shock absorber from upper mounting.

NOTE: On 1969-70 models, the upper mounting is accessible from trunk by removing an access plug.

2. Disconnect shock absorber from lower mounting.
3. Reverse procedure to install.

NOTE: 1969-70 models incorporate horizontal shock absorbers in addition to the conventional type, Fig. 3. To replace, remove front and rear attaching bolts and lift unit from vehicle.

LEAF SPRINGS & BUSHINGS, REPLACE

1969-70

1. Support vehicle at frame and support rear axle, relieving tension from spring.
2. Remove nut from spring front mounting bolt.
3. Remove rear shackle nuts and shackle outer link.
4. Remove spring center clamp attaching bolts and lift clamp assembly up, shock absorber will hold clamp in position.
5. Disconnect resonator bracket from frame and allow resonator to hang free.
6. Lower rear axle until it clears spring, support spring, remove rear shackle bolts and spring front mounting bolt.
7. Replace spring eye bushing or rear shackle bushings as necessary. To replace eye bushing, wedge a small screwdriver or chisel between edge of spring and end of eye, expanding the eye. Drive bushing from eye with a hammer, with eye still expanded drive new bushing into eye.
8. Reverse procedure to install.

COIL SPRINGS, REPLACE

1971-76

1. Support vehicle at frame.
2. With rear axle properly supported, disconnect shock absorbers from lower mountings.
3. Carefully lower rear axle and remove springs.
4. Reverse procedure to install. On 1972-75 models, springs must be properly indexed, Fig. 4.

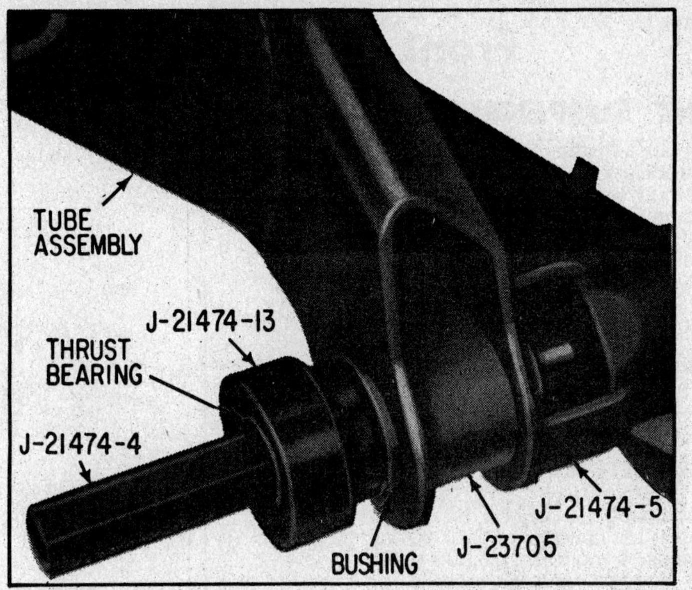

Fig. 7 Upper control arm bushing removal. 1971-76 models

CONTROL ARMS & BUSHINGS, REPLACE

1971-76

NOTE: Replace one control arm at a time to prevent rear axle misalignment, making installation difficult.

Upper Control Arms

1. Support vehicle at frame and rear axle.
2. Remove control arm front and rear mounting bolts.
3. Replace control arm bushings as necessary, Figs. 5, 6 and 7.

NOTE: When installing upper control arm rear bushing, reverse tool # J-21474-13.

4. Reverse procedure to install. Tighten control arm bolts with vehicle at curb height.

Lower Control Arms

Follow "Upper Control Arms" procedure for replacement of lower control arms.

NOTE: Lower control arm bushings are serviceable only on 1971 models, Figs. 5 and 6.

Front Suspension & Steering Section

FRONT SUSPENSION

The front suspension consists of control arms, stabilizer bar, shock absorbers and a right and left torsion bar, Figs. 1 and 2. Torsion bars are used instead of conventional coil springs. The front end of the torsion bar is attached to the lower control arm. The rear of the torsion bar is mounted into an adjustable arm at the torsion bar cross member. The riding height of the car is controlled by this adjustment.

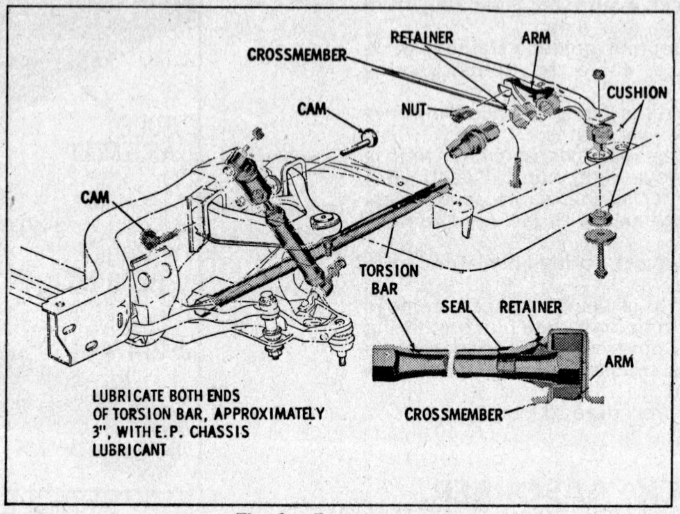

Fig. 1 Front suspension

LUBRICATION

Front suspension lubrication is recommended every 12 months or 12,000 on 1969-70 models, 4 months or 6,000 miles on 1971-74 models and 6 months or 7,500 miles on 1975-76 models, whichever occurs first.

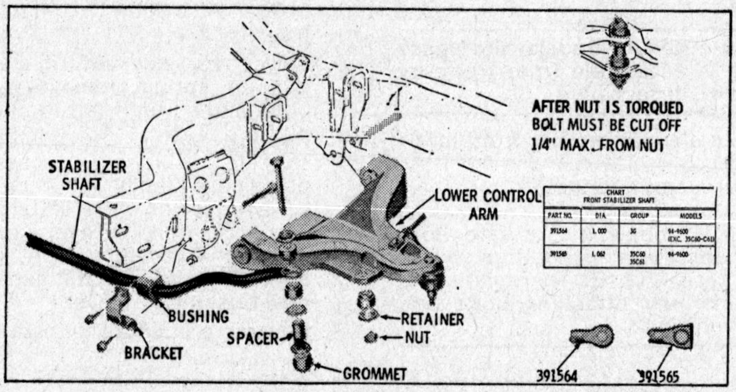

Fig. 2 Front suspension

WHEEL ALIGNMENT

NOTE: When checking wheel alignment the car must be on a level surface, gas tank full or a compensating weight added and front seat all the way to the rear. All doors must be closed and no passengers or additional weight should be in the car or trunk.

Caster and camber can be adjusted from under hood or under car. If under hood method is used, adjustments must be rechecked due to the change in weight distribution. After checking vehicle standing height adjust caster and camber as follows:

1. Loosen front and rear adjusting cam nuts, Fig. 4.

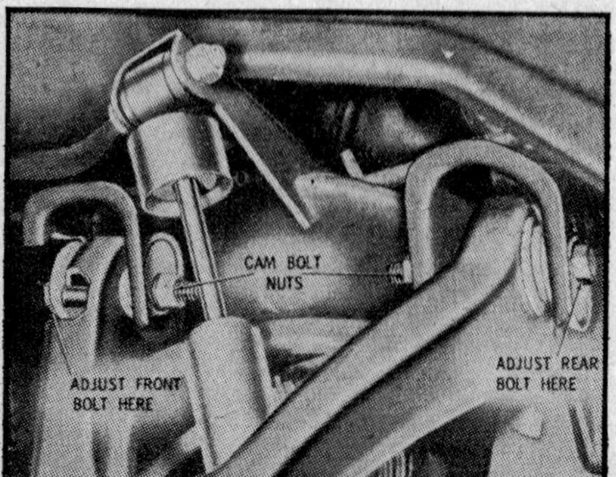

Fig. 4 Front wheel alignment cams

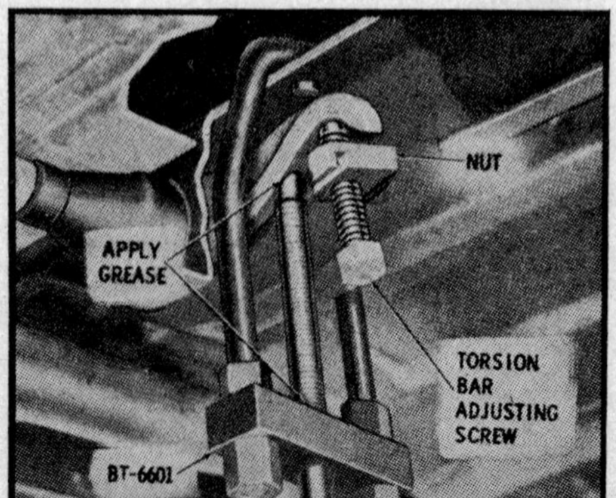

Fig. 5 Removing torsion bar

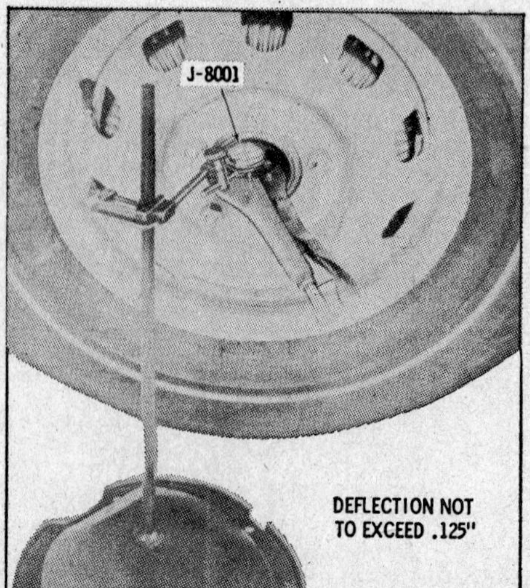

Fig. 6 Ball joint vertical check. 1969-75

Fig. 7 Pry bar installation

2. Rotate front cam to correct for ½ of incorrect camber reading.
3. Rotate rear cam to bring camber reading to 0°.
4. Tighten front and rear cam nuts, then check caster.

NOTE: If caster is within specifications, proceed to step 7.

5. If caster is to be adjusted, lossen front and rear cam nuts.
6. Rotate front cam bolt so that camber will change ¼ of desired caster change.

NOTE: A change of 1° in camber reading will change caster reading about 2°. To correct for excessive negative caster rotate front cam bolt to increase positive camber. To correct for excessive positive caster, rotate front cam bolt to increase negative camber.

7. Set camber to specifications by rotating rear cam bolt.
8. Torque front and rear cam nuts to 95 ft. lbs. on 1969-74 models and 110 ft. lbs. on 1975-76 models.

NOTE: When torquing cam nuts ensure cam bolt does not move, any movement of cam bolt will effect wheel alignment.

NOTE: If a problem exists where you should run out of cam to gain the correct reading, first turn front cam bolt so high part of cam is pointing up. Then turn rear cam bolt so high part of cam is pointing down. This is a location to start from and a correct setting can be obtained with the foregoing procedure.

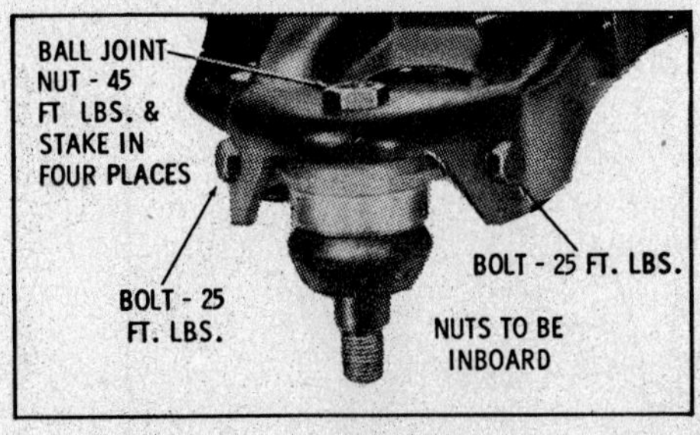

Fig. 9 Installing service ball joint

TOE-IN, ADJUST

1. Center steering wheel.
2. Loosen tie-rod nuts and adjust to proper setting.
3. Torque tie-rod clamp bolts to 20 ft-lbs. on 1969-74 models and 24 ft-lbs. on 1975-76 models. Position clamps so openings are facing upward.

TORSION BAR

Removal

1. Hoist car and place floor stands under front frame horns.
2. Slide seat at rear of torsion bar forward.
3. Use a tool of the type shown in Fig. 5.
4. Turn torsion bar adjusting bolt counterclockwise, counting the number of turns necessary to remove. Record this number for installing.
5. Remove adjusting bolt and nut.
6. Turn center screw of tool until torsion bar is completely relaxed.
7. Place block of wood on hoist (6″x6″x8″) and raise under lower control arm until drive axle is horizontal.
8. Remove stabilizer bolt and related parts. Discard bolt.
9. Place a daub of paint on bottom side of torsion bar.
10. Slide torsion bar forward until it bottoms in lower control arm (adjusting arm will drop out). *Do not mar, scratch or in any way damage torsion bar as replacement will be necessary if such conditions exist.*
11. Remove crossmember bolt from side torsion bar is being replaced.
12. Remove center bolt from tool.
13. Raise crossmember and twist rearward until torsion bar clears member.
14. Slowly raise lower control arm until maximum height is attained.

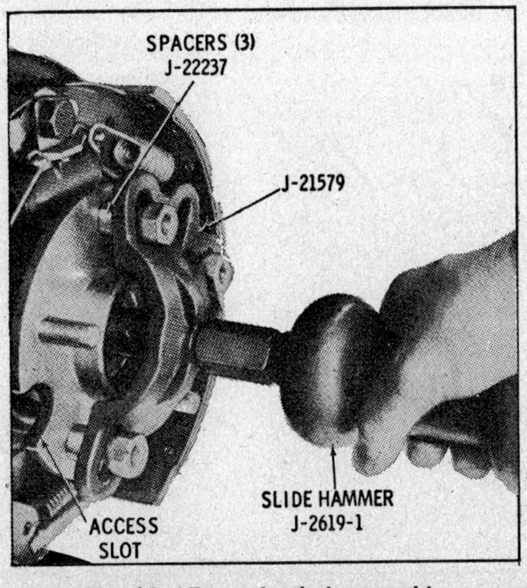

Fig. 10 Removing hub assembly

Fig. 8 Ball joint horizontal check 1969-71

15. Raise center crossmember until contact is made with floor pan.
16. Pull rearward on torsion bar with HANDS ONLY until bar is out of lower control arm. *It may be necessary to use air blowing into nut of lower control arm to relieve vacuum caused by grease.*

Installation

NOTE: Check rubber seal for damage; replace if necessary. Check retainer for excessive wear and replace if necessary. A new retainer is required on replacement of torsion bar. Stake as shown in Fig. 1.

1. Grease both ends of torsion bar for about 3″ with E.P. chassis lube.
2. With daub of paint in same location as when removed, insert bar into lower control arm nut and push forward until bar bottoms.
3. Pry crossmember back and align bar with hole in crossmember.

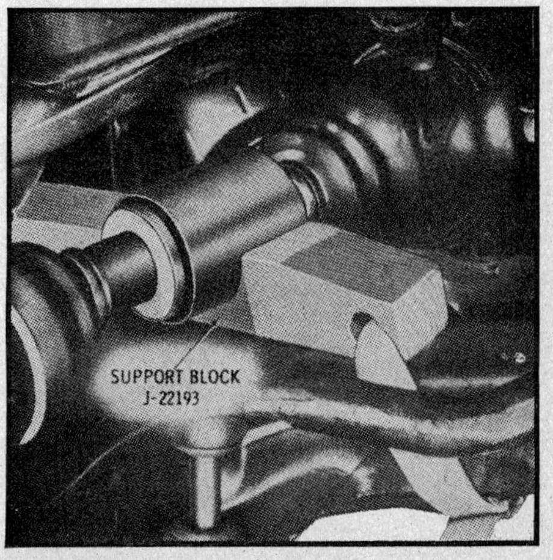

Fig. 11 Drive axle support block

4. Lower crossmember and install center bolt in tool, Fig. 5.
5. Lower front lower control arm until drive axle is horizontal.
6. Install torsion bar arm and pull bar rearward until fully seated in arm.
7. Install crossmember bolt through rubber mounting and torque to 40 ft-lbs.
8. Lower front lower control arm and remove wood block.
9. Install new bolt in stabilizer bar to lower control arm. Torque to 14 ft-lbs. *Cut off bottom of bolt so that ¼" is remaining below nut.*
10. Using the tool shown in Fig. 5 or its equivalent, tighten torsion bar arm to install lock plate under arm and through crossmember.
11. Grease threads of torsion bar adjuster with chassis lube and turn into nut the same number of turns required to remove.
12. Remove tool. Lower car to floor.
13. Check riding height. Correct by turning torsion bar adjusting bolt as required.

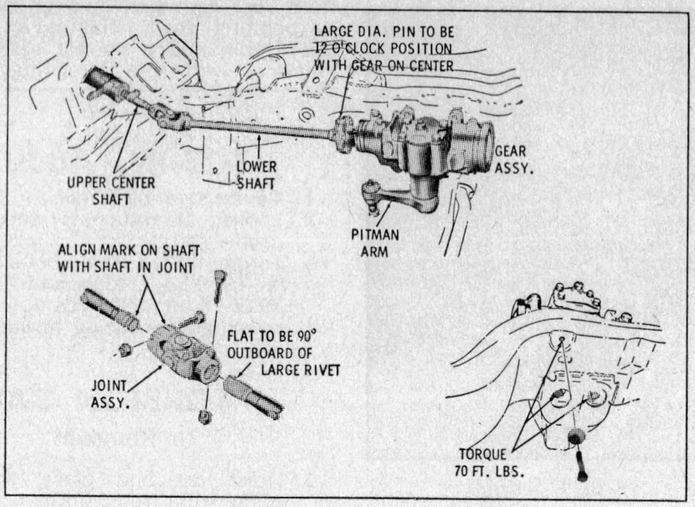

Fig. 12 Steering gear and shaft

BALL JOINT CHECKS

Vertical Check

1969-76

1. Raise car and place jack stands under both lower control arms as near as possible to lower ball joints.
2. Install dial indicator, Fig. 6.
3. Place a pry bar as shown in Fig. 7 and push down on bar. Use care to see that drive axle seal is not damaged. Reading must not exceed .125".

Horizontal Check

1969-71

1. With car placed on floor stands, install dial indicator, Fig. 8.
2. Grasp front wheel and rock top and bottom of tire. Read dial gauge, then reverse the push-pull procedure.
3. Horizontal deflection on dial gauge should not exceed .125" at wheel rim. This procedure checks both upper and lower ball joints.

BALL JOINTS, REPLACE

Lower Ball Joint

1. Remove knuckle.
2. Hacksaw three rivet heads off.
3. Using a 7/32" drill, drill side rivets 3/16" deep.
4. Drive center rivet until joint is out of control arm.
5. Install service ball joint into control arm and torque bolts and nuts as shown in Fig. 9.

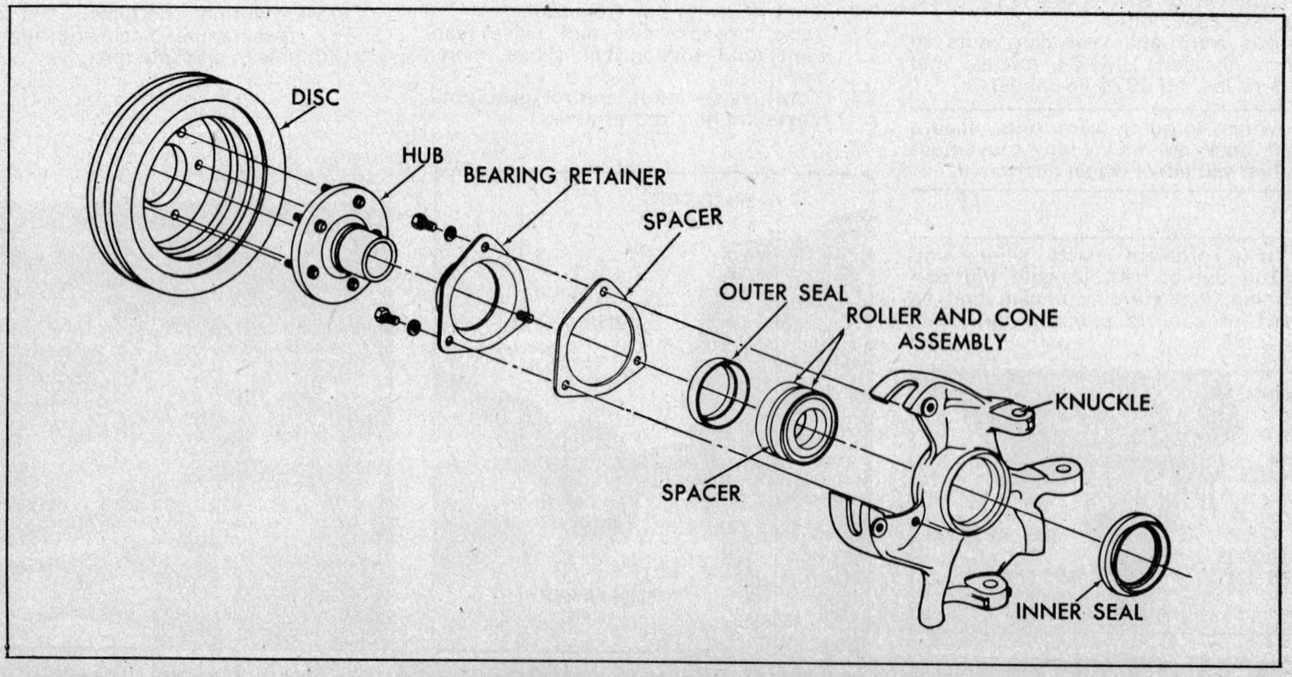

Fig. 13 Wheel bearing & steering knuckle assembly

6. Replace knuckle.

Upper Ball Joint

1. Raise car and support on floor stands.
2. Remove wheel and drum.
3. Remove hub assembly, Fig. 10.
4. Remove upper ball joint nut and brake line hose clip from ball joint.
5. Remove anchor bolt.
6. Lift brake plate outward over end of axle shaft and support so brake hose is not damaged.
7. Place support block as shown, Fig. 11.
8. Using a brass drift and hammer loosen and remove ball joint.
9. Install new ball joint and replace parts in reverse order of removal.

WHEEL BEARING & STEERING KNUCKLE, REPLACE

1. Raise and support vehicle under lower control arms.
2. Remove drive axle nut and washer and remove wheel and tire assembly.
3. Remove brake hose clip from ball joint and replace nut, then remove brake caliper off disc, and using a length of wire support caliper on suspension.

NOTE: Do not allow caliper to hang from brake hose as this could cause damage and premature failure of hose.

4. Mark hub and disc assembly for alignment during assembly and remove disc, then strike steering knuckle in area of upper ball joint until upper ball joint is loose.

CAUTION: Use extreme care to prevent striking and damaging brake hose or ball joint seal.

5. Place a short length of rubber hose over lower control arm torsion bar connector to avoid damage to inboard tri-pot joint seal when hub and knuckle are removed.
6. Using appropriate puller, disconnect tie rod end, upper and lower ball joints and remove steering knuckle and hub assembly, Fig. 12.

SHOCK ABSORBER, REPLACE

1. Remove upper shock attaching bolt.
2. Remove lower attaching nut and guide shock through upper control arm.
3. Reverse procedure to install.

STEERING GEAR, REPLACE

1. Referring to Fig. 12, remove coupling flange hub bolt.

NOTE: On 1974-76 models, remove coupling shield.

2. Disconnect hoses from power steering pump and cap pump and hose fittings. On cars equipped with a cooler disconnect return hose from cooler inlet pipe.
3. Raise vehicle and disconnect pitman arm from relay rod.
4. Unfasten gear from frame, permit lower shaft to slide free of coupling flange, then remove gear with hoses attached.

NOTE: Before installing steering gear, apply a sodium soap fibre grease to gear mounting pads to prevent squeaks between gear housing and frame. Make sure alignment pin on gear housing enters hole provided in frame side rail. Make sure there is a minimum of .040" clearance between coupling hub and steering gear upper seal.

Install coupling hub bolt and torque to 18 ft-lbs. on 1969-71 models and 30 ft-lbs. on 1972-76 models. Before torquing gear retaining bolts, position gear so the flexible coupling will not be distorted. Torque gear retaining bolts to 70 ft-lbs. Torque pitman arm to relay rod nut to 45 ft-lbs. on 1969-74 models and 85 ft-lbs. on 1975-76 models.

5. After hoses are connected to pump, add power steering fluid as necessary to bring the fluid level to the full mark. Run engine at idle for 30 seconds, then run at a fast idle for a minute before turning steering wheel. With engine running, turn steering wheel through its full travel two or three times to bleed air from system. Recheck oil level and add oil if necessary.

PONTIAC

(Exc. Astre & Sunbird)

OLD CAR SPECIFICATIONS: For 1946-68 Tune Up and Wheel Alignment Specifications see main index.

INDEX OF SERVICE OPERATIONS

PAGE NO.

ACCESSORIES

BRAKES

CLUTCH

COOLING SYSTEM

ELECTRICAL

ENGINE, OVERHEAD CAMSHAFT

PAGE NO.

ENGINE, Conventional

ENGINE LUBRICATION

FUEL SYSTEM

PROPELLER SHAFT & U JOINTS

REAR AXLE & SUSPENSION

SPECIFICATIONS

PAGE NO.

STEERING GEAR

SUSPENSION, FRONT

TRANSMISSIONS

TUNE UP

WINDSHIELD WIPER

SERIAL NUMBER LOCATION

1969-76: On plate fastened to upper left instrument panel area, visible through windshield.

ENGINE IDENTIFICATION

The V8 engine code is located beneath the production engine number on a machined pad on the right-hand bank of the engine block.

The 6-cylinder engine code for 1969 vehicles is stamped on the cylinder head-to-block contact surface behind oil filler tube and on pad at front right-hand side of cylinder block at rear of distributor for 1970-76 vehicles.

1969

CODE	TRANS.	ENGINE
ZC	④	6-250⑨
ZF	⑥	6-250⑨
ZH	④	6-250②
ZL	⑥	6-250②
WM	④	V8-350①
WN	④	V8-350②
WU	④	V8-350①
WV	④	V8-350②
XB	⑥	V8-350①
XC	⑥	V8-350②
XS	⑥	V8-350①
XU	⑥	V8-350②
WC	④	V8-350①
WP	④	V8-350①
XL	⑥	V8-350①
XR	⑥	V8-350①
YE	⑥	V8-350①
YN	⑥	V8-350①
YJ	⑥	V8-350①
WH	④	V8-400②
WQ	④	V8-400②
WS	④	V8-400②
WT	④	V8-400②
WZ	④	V8-400②
WW	④	V8-400②
XM	⑥	V8-400①
XN	⑥	V8-400②
XP	⑥	V8-400②
XX	⑥	V8-400①
YS	⑥	V8-400②
YT	⑥	V8-400②
YW	⑥	V8-400②
YZ	⑥	V8-400②
WD	④	V8-400①
WA	④	V8-400①
WB	④	V8-400①
WE	④	V8-400①
YA	⑥	V8-400①
YB	⑥	V8-400①
YC	⑥	V8-400①
YD	⑥	V8-400①
WX	④	V8-400②
XH	⑥	V8-400②
XZ	⑥	V8-400②
YF	⑥	V8-400①
WG	④	V8-428②
WJ	④	V8-428②
WF	④	V8-428②
WL	④	V8-428②
XE	⑥	V8-428②
XK	⑥	V8-428②
XJ	⑥	V8-428②
XF	⑥	V8-428②
XG	⑥	V8-428②
YL	⑥	V8-428②
YH	⑥	V8-428②
YK	⑥	V8-428②

1970

CODE	TRANS.	ENGINE
CG	⑥	6-250
RF	④	6-250
ZB	④	6-250
ZG	⑥	6-250
W7	④	V8-350①
X7	⑥	V8-350①
WU	④	V8-350①
YU	⑥	V8-350①
WE	④	V8-400①
YB	⑥	V8-400①
YD	⑥	V8-400①
XZ	⑥	V8-400②
XV	⑥	V8-400②
YZ	⑥	V8-400②
XX	⑥	V8-400①
XP	⑥	V8-400②
WS	④	V8-400②
WW	④	V8-400②
WT	④	V8-400②
WS	⑤	V8-400②
XH	⑥	V8-400②
YB	⑥	V8-400①
WX	④	V8-400②
YS	⑥	V8-400②
WH	⑤	V8-400⑧
XN	⑥	V8-400⑧
WG	④	V8-455②
YH	⑥	V8-455②
XF	⑥	V8-455②
WA	④	V8-455②
YA	⑥	V8-455②

1971

CODE	TRANS.	ENGINE
ZB	④	6-250
ZG	⑥	6-250
CAA	④	6-250
CAB	⑥	6-250
CCA	④	V8-307①
CCC	⑥	V8-307①
WN	④	V8-350①
YP	⑤	V8-350①
WP	⑥	V8-350①
YN	⑥	V8-350①
WR	④	V8-350①
WU	⑤	V8-350①
YU	⑥	V8-350①
YX	⑥	V8-400①
WX	④	V8-400①
XR	⑥	V8-350①
WT	④	V8-400②
WK	⑤	V8-400②
WS	④	V8-400①
XX	⑥	V8-400①
YS	⑥	V8-400②
WL	④	V8-455③
WC	⑤	V8-455③
WJ	④	V8-455②
YC	⑥	V8-455②
WG	④	V8-455①
YG	⑥	V8-455①
YE	⑥	V8-455③
YA	⑥	V8-455②

1972

CODE	TRANS.	ENGINE
W6	④	6-250
Y6	⑥	6-250
CBG	④	6-250
CBJ	⑥	6-250
CBA	④	6-250
CBC	⑥	6-250
CKG	④	V8-307
CKH	⑥	V8-307
CAY	④	V8-307
CAZ	⑥	V8-307
YV	⑥	V8-350
YR	⑥	V8-350
WR	④	V8-350
YX	⑥	V8-400①
WS	④	V8-400②
WK	④	V8-400②
YS	⑥	V8-400②
ZX	⑥	V8-400①
YH	⑥	V8-455①
YC	⑥	V8-455②
YA	⑥	V8-455②
WM	④	V8-455③
YB	⑥	V8-455③
ZH	⑥	V8-455①

1973

CODE	TRANS.	ENGINE
CCC	④	6-250
CCD	④	6-250
CCA	⑥	6-250
CCB	⑥	6-250
XR	④	V8-350
Y2	⑥	V8-350
YR	⑥	V8-350
Y7	⑥	V8-350
YV	⑥	V8-350
XV	④	V8-350
ZR	⑥	V8-350
ZV	⑥	V8-350
YL	⑥	V8-350
WA	⑥	V8-350①
WC	⑥	V8-350①
WD	⑥	V8-350①
WF	⑥	V8-350①
WL	⑥	V8-350①

Continued

ENGINE IDENTIFICATION—Continued

CODE	TRANS.	ENGINE
WN	⑥	V8-350①
XC	⑥	V8-350①
XF	⑥	V8-350①
X2	⑥	V8-350①
YW	⑥	V8-350①
ZB	④	V8-350①
ZD	④	V8-350①
W5	⑥	V8-400①
XH	⑥	V8-400①
XI	⑥	V8-400①
XK	⑥	V8-400②
XN	⑥	V8-400②
XX	⑥	V8-400②
XZ	⑥	V8-400②
X3	⑥	V8-400①
X4	⑥	V8-400①
X5	⑥	V8-400②
YF	④	V8-400②
YG	④	V8-400②
Y6	④	V8-400②
YP	⑥	V8-400①
Y4	⑥	V8-400①
YX	⑥	V8-400①
Y1	⑥	V8-400①
WK	⑤	V8-400②
YS	⑥	V8-400②
WS	④	V8-400②
WP	⑤	V8-400②
Y3	⑥	V8-400②
YN	⑥	V8-400②
ZS	⑥	V8-400②
YY	⑥	V8-400②
ZX	⑥	V8-400①
ZN	⑥	V8-400②
ZK	⑥	V8-400①
YZ	⑥	V8-400①
YT	⑥	V8-400②
WW	⑤	V8-455②
YC	⑥	V8-455②
YA	⑥	V8-455②
WT	⑤	V8-455②
W8	⑤	V8-455⑧
Y8	⑥	V8-455⑧
ZC	⑥	V8-455②
ZA	⑥	V8-455②
YK	⑥	V8-455②
YD	⑥	V8-455②
XA	⑥	V8-455②
XD	⑥	V8-455⑧
XE	⑥	V8-455②
XJ	⑥	V8-455②
XL	⑥	V8-455②
XM	⑥	V8-455②
XO	⑥	V8-455②
XT	⑥	V8-455②
XY	⑥	V8-455②
X7	⑥	V8-455②
ZE	④	V8-455②
ZJ	④	V8-455⑧
ZZ	④	V8-455②

1974

CODE	TRANS.	ENGINE
CCR	④	6-250⑨
CCX	⑥	6-250⑨
CCW	⑥	6-250⑨
AA	⑥	V8-350①⑩
WB	④	V8-350①
WA	④	V8-350①
YB	⑥	V8-350①
YA	⑥	V8-350①
YC	⑥	V8-350①
YS	⑥	V8-350②⑩
AD	⑥	V8-400①⑩
ZB	⑥	V8-350①
ZA	⑥	V8-350①
WP	④	V8-350②
WN	④	V8-350②
YP	⑥	V8-350②
YN	⑥	V8-350②
ZP	⑥	V8-350②
AH	⑥	V8-400①⑩
AT	⑥	V8-400②⑩
A3	⑥	V8-400②⑩
YH	⑥	V8-400①
YJ	⑥	V8-400①
ZH	⑥	V8-400①
WR	④	V8-400②⑩
WT	④	V8-040②
YF	⑥	V8-400①⑩
YK	⑥	V8-400①⑩
YL	⑥	V8-400②⑩
YM	⑥	V8-400②⑩
YZ	⑥	V8-400②⑩
Y3	④	V8-400②⑩
YT	⑥	V8-400②
ZD	⑥	V8-400①⑩
ZJ	⑥	V8-400①⑩
ZK	⑥	V8-400①⑩
ZS	⑥	V8-400②⑩
ZT	⑥	V8-400②
AW	⑥	V8-455②⑩
A4	⑥	V8-455②⑩
YR	⑥	V8-455②⑩
YW	⑥	V8-455②⑩
YX	⑥	V8-455②⑩
YY	⑥	V8-455②⑩
Y4	⑥	V8-455②⑩
Y6	⑥	V8-455②⑩
Y9	⑥	V8-455②⑩
YU	⑥	V8-455②
ZU	⑥	V8-455②
ZW	⑥	V8-455②⑩
ZX	⑥	V8-455②⑩
Z4	⑥	V8-455②⑩
Z6	⑥	V8-455②⑩
W8	④	V8-455⑧
Y8	⑥	V8-455⑧

1975

CODE	TRANS.	ENGINE
JU	④	6-250⑨
JT	⑥	6-250⑨
JL	⑥	6-250⑨
QA	④	V8-260①
QD	④	V8-260①
QE	⑥	V8-260①
QJ	⑥	V8-260①
TE	⑥	V8-260①
TJ	⑥	V8-260①
RS	⑥	V8-350①
RI	⑥	V8-350①
YA	⑥	V8-350①
YB	⑥	V8-350①
RW	⑥	V8-350②
RX	⑥	V8-350②
RN	⑥	V8-350②
RO	⑥	V8-350②
WN	⑥	V8-350②
YN	④	V8-350②
ZP	⑥	V8-350②
YH	⑥	V8-400①
YT	⑥	V8-400②
ZT	⑥	V8-400②
YM	⑥	V8-400②
WT	④	V8-400②
YS	⑥	V8-400②
YW	⑥	V8-455②
YU	⑥	V8-455②
ZU	⑥	V8-455②
ZW	⑥	V8-455②

①—Two barrel carburetor.
②—Four barrel carburetor.
③—High output engine.
④—Manual trans.
⑤—Four speed manual trans.
⑥—Automatic trans.
⑧—Super Duty engine.
⑨—One barrel carburetor.
⑩—High Energy Ignition System (H.E.I.).

GRILLE IDENTIFICATION

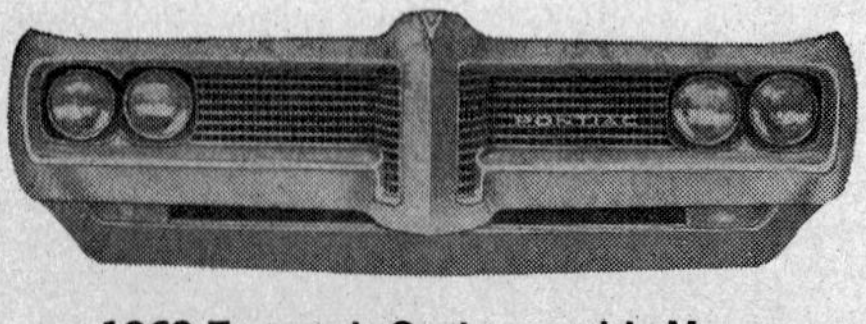
1969 Tempest, Custom and LeMans

1969 GTO

1969 Firebird

GRILLE IDENTIFICATION—Continued

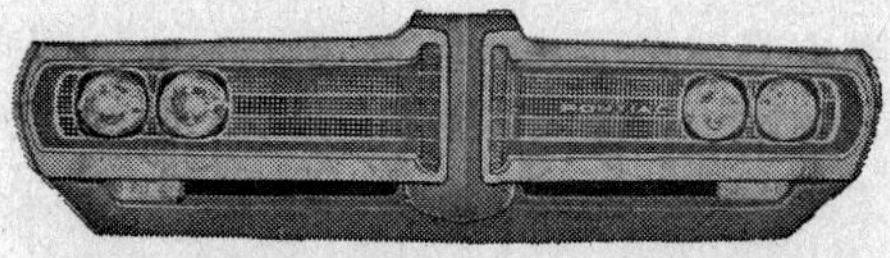
1969 Catalina, Executive, Ventura

1969 Bonneville

1969 Grand Prix

1970 GTO

1970 Tempest and LeMans

1970-71 Firebird

1970 Catalina, Executive

1970 Bonneville

1970 Grand Prix

1971 T-37 and LeMans

1971 GTO

1971 Catalina

1971 Grand Prix

1971 Bonneville

1971 Grand Ville

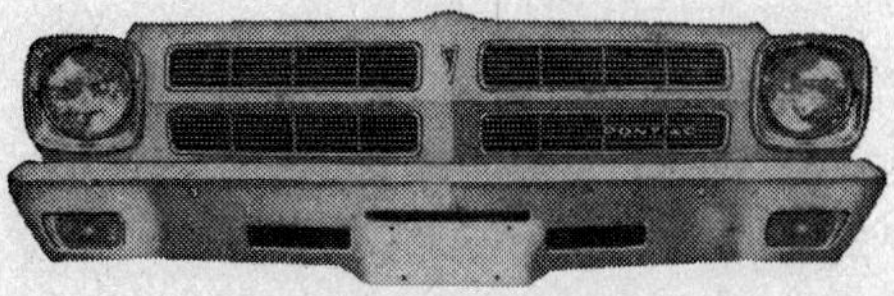
1971-72 Ventura II

1972 LeMans

1972 Catalina

1972 Grand Ville

1972 Bonneville

1972 Grand Prix

GRILLE IDENTIFICATION—Continued

1973 Ventura

1973 Luxury LeMans

1973 GTO

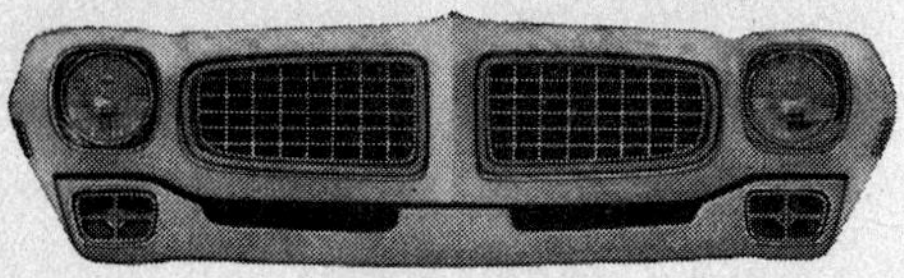
1973 Firebird

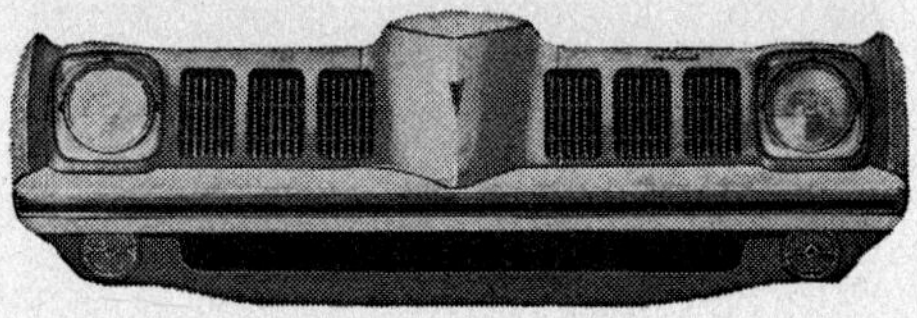
1973 Grand Am

1973 Catalina

1973 Grand Ville

1973 Grand Prix

1974 Ventura Custom

1974 GTO

1974 LeMans Sport

1974 Luxury LeMans

1974 Firebird

1974 Grand Am

1974 Catalina

1974 Bonneville

1974 Grand Prix

1974 Grand Ville

1975 Ventura

1975 LeMans

1975 Grand LeMans

GRILLE IDENTIFICATION—Continued

1975 Firebird

1975 Grand Am

1975 Catalina

1975 Bonneville

1975 Grand Prix

1975 Grand Ville & Grand Safari

1976 Ventura

1976 LeMans

1976 Grand LeMans

1976 Firebird

1976 Catalina

1976 Grand Prix

1976 Bonneville & Grand Safari

GENERAL ENGINE SPECIFICATIONS

Year	Engine	Carburetor	Bore and Stroke	Piston Displacement, Cubic Inches	Compression Ratio	Maximum Brake H.P. @ R.P.M.	Maximum Torque Lbs. Ft. @ R.P.M.	Normal Oil Pressure Pounds
1969	175 Horsepower...............6-250	1 Barrel	3.8750 x 3.52	250	9.00	175 @ 4800	240 @ 2600	26–36
	215 Horsepower...............6-250	4 Barrel	3.8750 x 3.52	250	10.50	215 @ 5200	255 @ 3800	26–36
	230 Horsepower...............6-250	4 Barrel	3.8750 x 3.52	250	10.50	230 @ 5400	260 @ 3600	26–36
	265 Horsepower.............V8-350	2 Barrel	3.8750 x 3.75	350	9.20	265 @ 4600	355 @ 2800	30–40
	325 Horsepower.............V8-350	4 Barrel	3.8750 x 3.75	350	10.50	325 @ 5100	380 @ 3200	55–60
	330 Horsepower.............V8-350	4 Barrel	3.8750 x 3.75	350	10.50	330 @ 5100	380 @ 3200	55–60
	265 Horsepower.............V8-400	2 Barrel	4.1200 x 3.75	400	8.60	265 @ 4600	397 @ 2400	30–40

Continued

GENERAL ENGINE SPECIFICATIONS—Continued

Year	Engine	Carburetor	Bore and Stroke	Piston Displacement, Cubic Inches	Compression Ratio	Maximum Brake H.P. @ R.P.M.	Maximum Torque Lbs. Ft. @ R.P.M.	Normal Oil Pressure Pounds
1969	290 Horsepower............V8-400	2 Barrel	4.1200 x 3.75	400	10.50	290 @ 4600	428 @ 2500	30–40
	330 Horsepower............V8-400	4 Barrel	4.1200 x 3.75	400	10.75	330 @ 4800	430 @ 3300	30–40
	335 Horsepower............V8-400	4 Barrel	4.1200 x 3.75	400	10.75	335 @ 5000	430 @ 3400	30–40
	345 Horsepower............V8-400	4 Barrel	4.1200 x 3.75	400	10.75	345 @ 5400	430 @ 3700	30–40
	350 Horsepower............V8-400	4 Barrel	4.1200 x 3.75	400	10.50	350 @ 5000	445 @ 3000	55–60
	350 Horsepower............V8-400	4 Barrel	4.1200 x 3.75	400	10.75	350 @ 5000	445 @ 3000	55–60
	366 Horsepower............V8-400	4 Barrel	4.1200 x 3.75	400	10.75	366 @ 5100	445 @ 3600	30–40
	370 Horsepower............V8-400	4 Barrel	4.1200 x 3.75	400	10.75	370 @ 5500	445 @ 3900	30–40
	360 Horsepower............V8-428	4 Barrel	4.1200 x 4.00	428	10.50	360 @ 4600	472 @ 3200	30–40
	370 Horsepower............V8-428	4 Barrel	4.1200 x 4.00	428	10.50	370 @ 4800	472 @ 3200	55–60
	390 Horsepower............V8-428	4 Barrel	4.1200 x 4.00	428	10.75	390 @ 5200	465 @ 3400	55–60
1970	155 Horsepower............①6-250	1 Barrel	3.875 x 3.53	250	8.50	155 @ 4200	235 @ 1600	30–45
	255 Horsepower............V8-350	2 Barrel	3.8750 x 3.75	350	8.80	255 @ 4600	355 @ 2800	30–40
	265 Horsepower............V8-400	2 Barrel	4.1200 x 3.75	400	8.80	265 @ 4600	397 @ 2400	30–40
	290 Horsepower............V8-400	2 Barrel	4.1200 x 3.75	400	10.00	290 @ 4600	428 @ 2500	30–40
	330 Horsepower............V8-400	4 Barrel	4.1200 x 3.75	400	10.00	330 @ 4800	445 @ 2900	30–40
	350 Horsepower............V8-400	4 Barrel	4.1200 x 3.75	400	10.25	350 @ 5000	445 @ 3000	55–60
	366 Horsepower............V8-400	4 Barrel	4.1200 x 3.75	400	10.50	366 @ 5100	445 @ 3600	30–40
	370 Horsepower............V8-400	4 Barrel	4.1200 x 3.75	400	10.50	370 @ 5500	445 @ 3900	30–40
	360 Horsepower............V8-455	4 Barrel	4.1510 x 4.21	455	10.00	360 @ 4300	500 @ 2700	30–40
	370 Horsepower............V8-455	4 Barrel	4.1510 x 4.21	455	10.25	370 @ 4600	500 @ 3100	30–40
1971	145 Horsepower............①6-250	1 Barrel	3.875 x 3.53	250	8.50	145 @ 4200	230 @ 1600	30–45
	200 Horsepower............②V8-307	2 Barrel	3.875 x 3.25	307	8.50	200 @ 4600	300 @ 2400	30–45
	250 Horsepower..............8-350	2 Barrel	3.8750 x 3.75	350	8.0	250 @ 4400	350 @ 2400	30–40
	265 Horsepower..............8-400	2 Barrel	4.1200 x 3.75	400	8.2	265 @ 4400	400 @ 2400	30–40
	300 Horsepower..............8-400	4 Barrel	4.1200 x 3.75	400	8.2	300 @ 4800	400 @ 3600	30–40
	280 Horsepower..............8-455	2 Barrel	4.1510 x 4.21	455	8.2	280 @ 4400	455 @ 2000	30–40
	325 Horsepower..............8-455	4 Barrel	4.1510 x 4.21	455	8.2	325 @ 4400	455 @ 3200	30–40
	335 Horsepower..............8-455	4 Barrel	4.1510 x 4.21	455	8.4	335 @ 4800	480 @ 3600	30–40
1972	110 Horsepower③..........①6-250	1 Barrel	3.875 x 3.53	250	8.50	110 @ 3800	185 @ 1600	30–45
	130 Horsepower③..........②8-307	2 Barrel	3.875 x 3.25	307	8.50	130 @ 4400	230 @ 2400	30–45
	160 Horsepower③............8-350	2 Barrel	3.875 x 3.75	350	8.00	160 @ 4400	270 @ 2000	30–40
	175 Horsepower③............8-350	2 Barrel	3.875 x 3.75	350	8.00	175 @ 4400	275 @ 2000	30–40
	175 Horsepower③............8-400	2 Barrel	4.1200 x 3.75	400	8.2	175 @ 4000	310 @ 2400	30–40
	200 Horsepower③............8-400	2 Barrel	4.1200 x 3.75	400	8.2	200 @ 4000	325 @ 2400	30–40
	200 Horsepower③............8-400	4 Barrel	4.1200 x 3.75	400	8.2	200 @ 4000	295 @ 2800	30–40
	250 Horsepower③............8-400	4 Barrel	4.1200 x 3.75	400	8.2	250 @ 4400	325 @ 3200	30–40
	185 Horsepower③............8-455	2 Barrel	4.1510 x 4.21	455	8.2	185 @ 4000	350 @ 2000	30–40
	200 Horsepower③............8-455	2 Barrel	4.1510 x 4.21	455	8.2	200 @ 4000	370 @ 2000	30–40
	220 Horsepower③............8-455	4 Barrel	4.1510 x 4.21	455	8.2	220 @ 3600	350 @ 2400	30–40
	230 Horsepower③............8-455	4 Barrel	4.1510 x 4.21	455	8.2	230 @ 4400	360 @ 2800	30–40
	250 Horsepower③............8-455	4 Barrel	4.1510 x 4.21	455	8.2	250 @ 3600	370 @ 2400	30–40
	250 Horsepower③............8-455	4 Barrel	4.1510 x 4.21	455	8.2	250 @ 3600	375 @ 2400	30–40
	300 Horsepower③............8-455	4 Barrel	4.1510 x 4.21	455	8.4	300 @ 4000	415 @ 3200	30–40
1973	100 Horsepower③..........①6-250	1 Barrel	3.87 x 3.53	250	8.2	100 @ 3600	175 @ 1600	30–45
	150 Horsepower③............8-350	2 Barrel	3.88 x 3.75	350	7.6	150 @ 4000	270 @ 2000	55–60
	175 Horsepower③..........④8-350	2 Barrel	3.88 x 3.75	350	7.6	175 @ 4400	280 @ 2400	55–60
	170 Horsepower③............8-400	2 Barrel	4.12 x 3.75	400	8.0	170 @ 3600	320 @ 2000	55–60
	185 Horsepower③..........④8-400	2 Barrel	4.12 x 3.75	400	8.0	185 @ 4000	320 @ 2400	55–60
	200 Horsepower③............8-400	4 Barrel	4.12 x 3.75	400	8.0	200 @ 4000	310 @ 2400	55–60
	230 Horsepower③..........④8-400	4 Barrel	4.12 x 3.75	400	8.0	230 @ 4400	325 @ 3200	55–60
	215 Horsepower③............8-455	4 Barrel	4.1510 x 4.21	455	8.0	215 @ 3600	350 @ 2400	55–60
	250 Horsepower③..........④8-455	4 Barrel	4.1510 x 4.21	455	8.0	250 @ 4000	370 @ 2800	55–60
	290 Horsepower③..........④8-455	4 Barrel	4.1510 x 4.21	455	8.4	290 @ 4000	395 @ 3200	75–80
	310 Horsepower③..........④8-455	4 Barrel	4,1510 x 4.21	455	8.4	310 @ 4000	390 @ 3600	75–80

Continued

GENERAL ENGINE SPECIFICATIONS—Continued

Year	Engine	Carburetor	Bore and Stroke	Piston Displacement, Cubic Inches	Compression Ratio	Maximum Brake H.P. @ R.P.M.	Maximum Torque Ft. Lbs. @ R.P.M.	Normal Oil Pressure Pounds
1974	100 Horsepower③............①6-250	1 Barrel	3.88 x 3.53	250	8.2	100 @ 3600	175 @ 1600	30–45
	155 Horsepower③.............8-350	2 Barrel	3.88 x 3.75	350	7.6	155 @ 3600	275 @ 2400	55–60
	170 Horsepower③............④8-350	2 Barrel	3.88 x 3.75	350	7.6	170 @ 4000	290 @ 2400	55–60
	170 Horsepower③.............8-350	4 Barrel	3.88 x 3.75	350	7.6	170 @ 4000	280 @ 2000	55–60
	200 Horsepower③............④8-350	4 Barrel	3.88 x 3.75	350	7.6	200 @ 4400	295 @ 2800	55–60
	175 Horsepower③.............8-400	2 Barrel	4.12 x 3.75	400	8.0	175 @ 3600	315 @ 2000	55–60
	190 Horsepower③............④8-400	2 Barrel	4.12 x 3.75	400	8.0	190 @ 4000	330 @ 2400	55–60
	200 Horsepower③.............8-400	4 Barrel	4.12 x 3.75	400	8.0	200 @ 4000	320 @ 2400	55–60
	225 Horsepower③............④8-400	4 Barrel	4.12 x 3.75	400	8.0	225 @ 4000	330 @ 2800	55–60
	215 Horsepower③.............8-455	4 Barrel	4.15 x 4.21	455	8.0	215 @ 3600	355 @ 2400	55–60
	250 Horsepower③............④8-455	4 Barrel	4.15 x 4.21	455	8.0	250 @ 4000	380 @ 2800	55–60
	290 Horsepower③............④8-455	4 Barrel	4.15 x 4.21	455	8.4	290 @ 4000	395 @ 3200	75–80
1975	105 Horsepower③............6-250①	1 Barrel	3.87 x 3.53	250	8.25	105 @ 3800	185 @ 1200	36–41
	110 Horsepower③..........V8-260⑤	2 Barrel	3.50 x 3.385	260	8.5	110 @ 3400	205 @ 1600	30–45
	155 Horsepower③............V8-350	2 Barrel	3.88 x 3.75	350	7.6	155 @ 4000	—	55–60
	175 Horsepower③............V8-350	4 Barrel	3.88 x 3.75	350	7.6	175 @ 4000	—	55–60
	145 Horsepower③..........V8-350⑥	2 Barrel	3.80 x 3.85	350	8.0	145 @ 3200	270 @ 3000	37
	165 Horsepower③..........V8-350⑥	4 Barrel	3.80 x 3.85	350	8.0	165 @ 3800	260 @ 2200	37
	170 Horsepower③............V8-400	2 Barrel	4.12 x 3.75	400	7.6	170 @ 4000	—	55–60
	185 Horsepower③............V8-400	4 Barrel	4.12 x 3.75	400	7.6	185 @ 3600	—	55–60
	200 Horsepower③............V8-455	4 Barrel	4.15 x 4.21	455	7.6	200 @ 3500	—	55–60
1976	110 Horsepower③............6-250①	1 Barrel	3.87 x 3.53	250	8.3	110 @ 3600	185 @ 1200	36–41
	110 Horsepower............V8-260⑤	2 Barrel	3.50 x 3.385	260	7.5	110 @ 3400	205 @ 1600	30–45
	160 Horsepower③............V8-350	2 Barrel	3.8762 x 3.75	350	7.6	160 @ 4000	280 @ 2000	35–40
	175 Horsepower③............V8-350	4 Barrel	3.8762 x 3.75	350	7.6	165 @ 4000	260 @ 2400	35–40
	140 Horsepower③..........V8-350⑥	2 Barrel	3.80 x 3.85	350	8.0	140 @ 3200	280 @ 1600	37
	155 Horsepower③..........V8-350⑥	4 Barrel	3.80 x 3.85	350	8.0	155 @ 3400	280 @ 1800	37
	170 Horsepower③............V8-400	2 Barrel	4.1212 x 3.75	400	7.6	170 @ 4000	310 @ 1600	35–40
	185 Horsepower③............V8-400	4 Barrel	4.1212 x 3.75	400	7.6	185 @ 3600	310 @ 1600	35–40
	200 Horsepower③............V8-455	4 Barrel	4.1522 x 4.21	455	7.6	200 @ 3500	330 @ 2000	55–60

①—For service on this engine, see Six Cylinder in Chevrolet Chapter.
②—For service on this engine, see Eight Cylinder in Chevrolet Chapter.
③—Ratings are NET—as installed in the vehicle.
④—With dual exhausts.
⑤—See Oldsmobile Chapter for service procedures.
⑥—Ventura only. See Buick Chapter for service procedures.

TUNE UP SPECIFICATIONS

The following specifications are published from the latest information available. This data should be used only in the absence of a decal affixed in the engine compartment.

OLD CAR SPECIFICATIONS: For 1946-68 Tune Up Specifications see main index.

★When using a timing light, disconnect vacuum hose or tube at distributor and plug opening in hose or tube so idle speed will not be affected.

●When checking compression, lowest cylinder must be within 80% of the highest.

▲Before removing wires from distributor cap, determine location of the No. 1 wire in cap, as distributor position may have been altered from that shown at the end of this chart.

Year	Spark Plug		Distributor		Ignition Timing★			Carb. Adjustments					
								Hot Idle Speed③		Air Fuel Ratio		Idle CO %	
	Type	Gap Inch	Point Gap Inch	Dwell Angle Deg.	Firing Order Fig. ▲	Timing BTDC ①	Mark Fig.	Std. Trans.	Auto. Trans.②	Std. Trans.	Auto. Trans.	Std. Trans.	Auto. Trans.
1969													
6-250 1 Bar. Carb.	R44NS	.035	.016	31–34	A	TDC	F	700/500	600/500D⑦	—	—	—	—
6-250 4 Bar. Carb.	R44NS	.035	.016	31–34	A	5°	F	850/600	600/500D⑦	—	—	—	—
8-350 2 Bar. Carb.	R45S	.035	.016	30	D	9°	C	850	650D⑦	—	—	—	—
8-350 4 Bar. Carb.	R45S	.035	.016	30	D	9°	C	1000	650D⑦	—	—	—	—
8-400 2 Bar. Carb.	R45S	.035	.016	30	D	9°	C	850	650D⑦	—	—	—	—
8-400 4 Bar. Carb.	⑪	.035	.016	30	D	9°	C	1000	650D⑦	—	—	—	—
8-400 Ram Air	⑪	.035	.016	30	D	15°	C	1000/650	650/500D⑦	—	—	—	—
8-428	R44S	.035	.016	30	D	9°	C	1000	650D⑦	—	—	—	—
1970													
6-250⑭ Std. Tr.	R46T	.035	.019	31–34	E	TDC	G	750/400	—	—	—	—	—
6-250⑭ Auto. Tr.	R46T	.035	.019	31–34	E	4°	G	—	600/400D	—	—	—	—
8-350 2 Bar. Carb.	R46S	.035	.016	30	D	9°	C	800	650D	—	—	—	—
8-400 2 Bar. Carb.	R46S	.035	.016	30	D	9°	C	800	650D	—	—	—	—
8-400 4 Bar. Carb.	R45S	.035	.016	30	D	9°	C	950	650D	—	—	—	—
8-400 Ram Air	R44S	.035	.016	30	D	15°	C	1000/650	750/500D	—	—	—	—
8-455	R45S⑫	.035	.016	30	D	9°	C	950	650D	—	—	—	—
1971													
6-250⑭	R46TS	.035	.019	32½	E	4°	H	850/550	650/500D	—	—	1.0	1.0
8-307⑭ Std. Tr.	R45TS	.035	.019	30	B	4°	H	550	—	—	—	1.0	—
8-307⑭ Auto. Tr.	R45TS	.035	.019	30	B	8°	H	—	550D	—	—	—	1.0
8-350 2 Bar. Carb.	R47S	.035	.016	30	D	12°	I	800	600D	—	—	1.0	1.0
8-400 2 Bar. Carb.	R47S	.035	.016	30	D	8°	I	—	600D	—	—	1.0	1.0
8-400 4 Bar. Carb.	R46S	.035	.016	30	D	12°	I	1000/600	700D	—	—	1.0	1.0
8-455	R46S	.035	.016	30	D	12°	I	—	650D	—	—	1.0	1.0
8-455 H.O.	R46S	.035	.016	30	D	12°	I	1000/600	700D	—	—	1.0	—
1972													
6-250⑭	R46T	.035	.019	32½	E	4°	H	850/450	650/450D	—	—	—	—
8-307⑭ Std. Tr.	R44T	.035	.016	30	B	4°	H	900/450	—	—	—	—	—
8-307⑭ Auto. Tr.	R44T	.035	.016	30	B	8°	H	—	600/450D	—	—	—	—
8-350	R46TS	.035	.016	30	D	8°	I	800	—	—	—	—	—
8-350	R46TS	.035	.016	30	D	10°	I	—	625D	—	—	—	—
8-400 2 Bar. Carb.	R46TS	.035	.016	30	D	10°	I	—	625D	—	—	—	—
8-400 4 Bar. Carb.	R45TS	.035	.016	30	D	10°	I	1000/600	700/500D	—	—	—	—
8-455 2 Bar. Carb.	R45TS	.035	.016	30	D	10°	I	—	625D	—	—	—	—
8-455 4 Bar. Carb.	R45TS	.035	.016	30	D	10°	I	—	650/500D	—	—	—	—
8-455 H.O. Std. Tr.	R45TS	.035	.016	30	D	8°	I	1000/600	—	—	—	—	—
8-455 H.O. Auto. Tr.	R45TS	.035	.016	30	D	10°	I	—	700/500D	—	—	—	—

Continued

TUNE UP SPECIFICATIONS—Continued

The following specifications are published from the latest information available. This data should be used only in the absence of a decal affixed in the engine compartment.

OLD CAR SPECIFICATIONS: For 1946-68 Tune Up Specifications see main index.

★When using a timing light, disconnect vacuum hose or tube at distributor and plug opening in hose or tube so idle speed will not be affected.

●When checking compression, lowest cylinder must be within 80% of the highest.

▲Before removing wires from distributor cap, determine location of the No. 1 wire in cap, as distributor position may have been altered from that shown at the end of this chart.

Year	Spark Plug		Distributor		Ignition Timing★			Carb. Adjustments					
								Hot Idle Speed③		Air Fuel Ratio		Idle CO %	
	Type	Gap Inch	Point Gap Inch	Dwell Angle Deg.	Firing Order Fig. ▲	Timing BTDC ①	Mark Fig.	Std. Trans.	Auto. Trans.②	Std. Trans.	Auto. Trans.	Std. Trans.	Auto. Trans.
1973													
6-250⑭	R46T	.035	⑮	32½	E	6°	H	700/450	600D	—	—	—	—
V8-350 Std. Tra.	R46TS	.040	.016	30	D	10°	I	900/600	—	—	—	0.2	—
V8-350 Auto. Tra.	R46TS	.040	.016	30	D	12°	I	—	650D	—	—	—	0.2
V8-400 2 Bar. Carb.	R46TS	.040	.016	30	D	⑯	I	—	650D	—	—	0.2	0.2
V8-400 4 Bar. Carb.	R45TS	.040	.016	30	D	⑯	I	1000/600	650D	—	—	0.2	0.2
V8-455	R45TS	.040	.016	30	D	⑯	I	1000	650D	—	—	0.2	0.2
V8-455 S.D.	R44TS	.040	.016	30	D	⑯	I	1000/600	750/500D	—	—	0.2	0.2
1974													
6-250⑭	R46T	.035	⑮	32½	E	6°	H	850	600D	—	—	0.2	0.2
V8-350 2 Bar. Carb.	R46TS	.040	⑮	30	D	⑯	I	900	650D	—	—	0.2	0.2
V8-350 2 Bar. Carb.④	R46TS	.040	⑮	30	D	10°	I	—	625D	—	—	—	0.2
V8-350 4 Bar. Carb.	R46TS	.040	⑮	30	D	⑯	I	1000	650D	—	—	0.2	0.2
V8-350 4 Bar. Carb.④	R46TS	.040	⑮	30	D	10°	I	—	625D	—	—	—	0.2
V8-400 2 Bar. Carb.	R46TS	.040	⑮	30	⑩	⑯	I	—	650D	—	—	—	0.2
V8-400 2 Bar. Carb.④	R46TS	.040	⑮	30	⑩	10°	I	—	625D	—	—	—	0.2
V8-400 4 Bar. Carb.	R45TS	.040	⑮	30	⑩	⑯	I	1000	650D	—	—	0.2	0.2
V8-400 4 Bar. Carb.④	R45TS	.040	⑮	30	⑩	10°	I	—	625D	—	—	—	0.2
V8-455	R45TS	.040	⑮	30	⑩	⑯	I	—	650D	—	—	—	0.2
V8-455④	R45TS	.040	⑮	30	⑩	10°	I	—	625D	—	—	—	0.2
V8-455 S.D.	R44TS	.040	⑮	30	⑩	12°	I	1000	750D	—	—	0.2	0.2
1975													
6-250⑭	R46TX	.060	—	—	L	10°	H	850	⑳	—	—	—	—
V8-260⑤㉑	R46SX	.080	—	—	M	16°⑨	J	750	650D	—	—	—	—
V8-260⑤㉒	R46SX	.080	—	—	M	18°⑨	J	750	650D	—	—	—	—
V8-350⑥	R45TSX	.060	—	—	N	12°	K	800	600D	—	—	—	—
V8-350㉑	R46TSX	.060	—	—	O	12°	I	775	㉓	—	—	—	—
V8-350㉒	R46TSX	.060	—	—	O	16°	I	775	㉓	—	—	—	—
V8-400㉑	㉔	.060	—	—	O	12°	I	775	㉕	—	—	—	—
V8-400㉒	㉔	.060	—	—	O	16°	I	775	㉕	—	—	—	—
V8-455㉖	R45TSX	.060	—	—	O	10°	I	—	650D	—	—	—	—
V8-455④	R45TSX	.060	—	—	O	16°	I	—	675D	—	—	—	—
1976													
6-250⑭	R46T	.035	—	—	L	㉗	H	850	㉘	—	—	—	—
V8-260⑤㉙	R46SX	.080	—	—	M	16°⑨	J	750	—	—	—	—	—
V8-260⑤㉚	R46SX	.080	—	—	M	㉛	J	—	㉘	—	—	—	—

Continued

TUNE UP SPECIFICATION—Continued

The following specifications are published from the latest information available. This data should be used only in the absence of a decal affixed in the engine compartment.

OLD CAR SPECIFICATIONS: For 1946–68 Tune Up Specifications see main index.

★When using a timing light, disconnect vacuum hose or tube at distributor and plug opening in tube or hose so idle speed will not be affected.

●When checking compression, lowest cylinder must be within 80% of the highest.

▲Before removing wires from distributor cap, determine location of the No. 1 wire in cap, as distributor position may have been altered from that shown at the end of this chart.

Year	Spark Plug		Distributor		Ignition Timing★			Carb. Adjustments					
								Hot Idle Speed③		Air Fuel Ratio		Idle CO %	
	Type	Gap Inch	Point Gap Inch	Dwell Angle Deg.	Firing Order Fig. ▲	Timing BTDC ①	Mark Fig.	Std. Trans.	Auto. Trans.②	Std. Trans.	Auto. Trans.	Std. Trans.	Auto. Trans.
1976—Continued													
V8-350⑥	R45TSX	.060	—	—	N	12°	K	—	600D	—	—	—	—
V8-350	R46TSX	.060	—	—	O	16°	I	—	550D	—	—	—	—
V8-400 2 Bar. Carb.	R46TSX	.060	—	—	O	16°	I	—	550D	—	—	—	—
V8-400 4 Bar. Carb.	R45TSX	.060	—	—	O	⑧	I	775	575D	—	—	—	—
V8-455㉑	R45TSX	.060	—	—	O	12°	I	775	600D	—	—	—	—
V8-455㉒	R45TSX	.060	—	—	O	16°	I	—	550D	—	—	—	—

①—BTDC: Before top dead center.
②—D: Drive. N: Neutral.
③—Where two figures are given, the higher is with solenoid active.
④—California.
⑤—See Oldsmobile Chapter for service procedure.
⑥—Ventura only. See Buick Chapter for service procedures.
⑦—With A/C off.
⑧—Std. Trans. 12°, Auto. Trans. 16°.
⑨—At 1100 RPM.
⑩—Exc. H.E.I., Fig. D; H.E.I., Fig. M.
⑪—GTO uses R44S; all others use R45S.
⑫—Use R44S on Ram Air option.
⑬—All except Firebird, Gran Prix and California.
⑭—For service on this engine, see Chevrolet Chapter.
⑮—New points .019″, used points .016″.
⑯—Std. trans. 10° BTDC. Auto trans. 12° BTDC.
⑰—Firebird and Grand Prix except California.
⑱—Std. Trans. 12°, Auto. Trans. 10°.
⑲—2 BBl. Carb. 600 RPM, 4 BBl. Carb. 650 RPM.
⑳—Exc. Calif., 550 RPM; Calif., 600 RPM.
㉑—Std. trans. & all California models.
㉒—Auto. trans. exc. California.
㉓—2 bbl. carb., 600 RPM; 4 bbl. carb.—Exc. Calif., 650 RPM; Calif., 625 RPM.
㉔—2 bbl. carb., R46TSX; 4 bbl. carb., R45TSX.
㉕—2 bbl. carb., 650 RPM; 4 bbl. carb.—Exc. Calif. & Grand Safari sta. wag., 650 RPM; Grand Safari sta. wag. exc. Calif., 625 RPM; All Calif. models, 600 RPM.
㉖—Exc. California.
㉗—Std. trans., 6° BTDC; auto. trans., 10° BTDC.
㉘—Exc. Calif., 550D; Calif., 600D.
㉙—Std. trans.
㉚—Auto. trans.
㉛—Exc. Calif., 18° BTDC; Calif., 14° BTDC. At 1100 RPM.

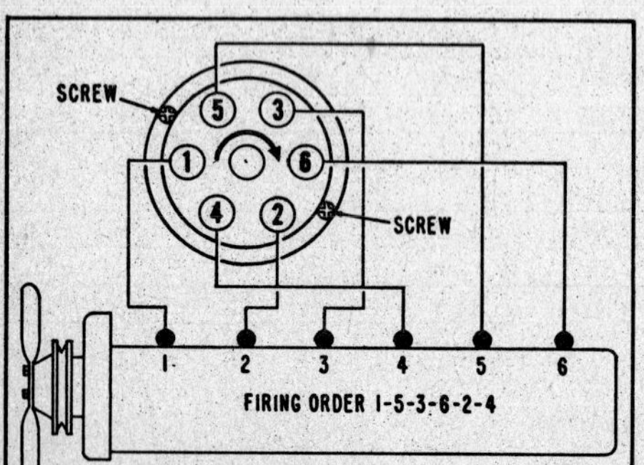

Fig. A

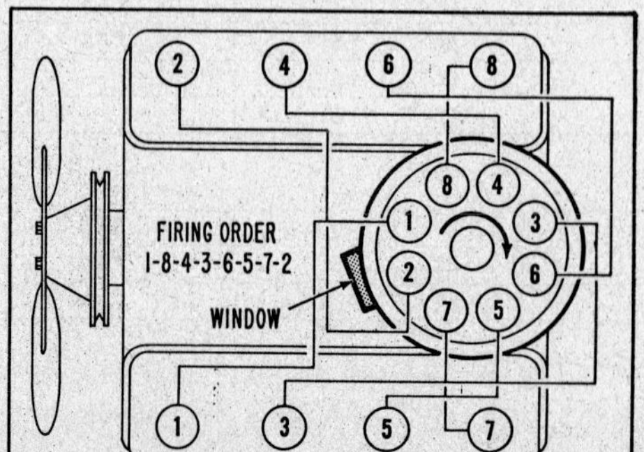

Fig. B

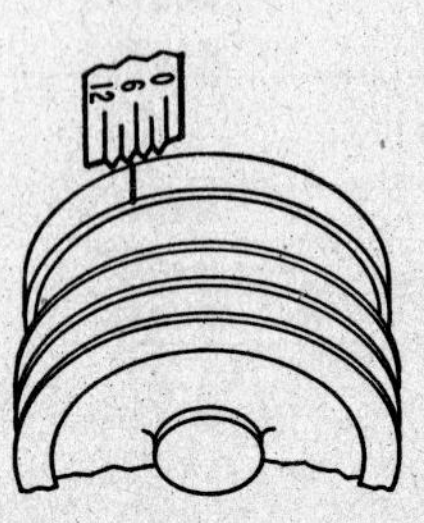

Fig. C

Continued

TUNE UP SPECIFICATIONS—Continued

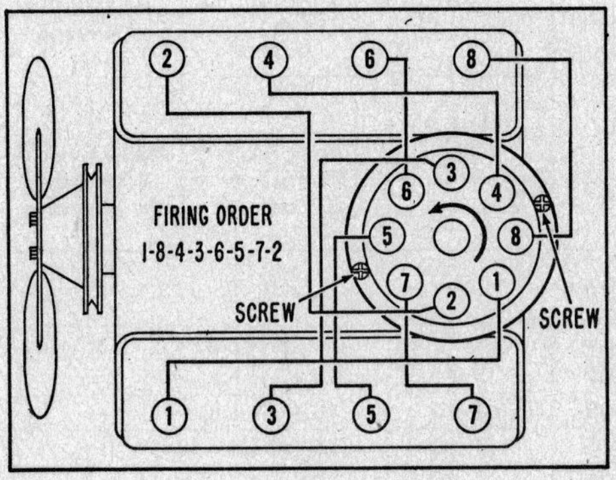

Fig. D

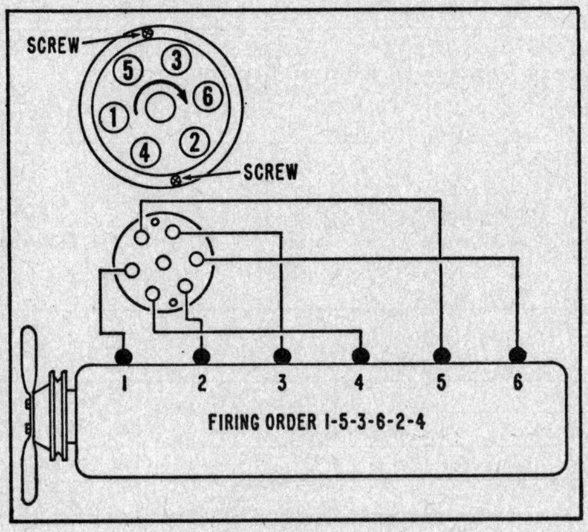

Fig. E

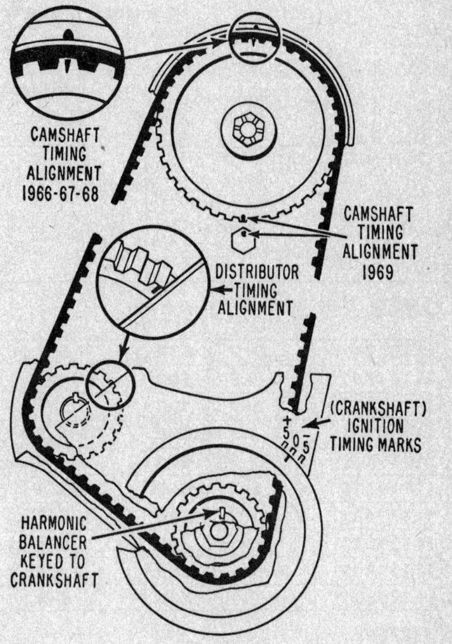

Fig. F

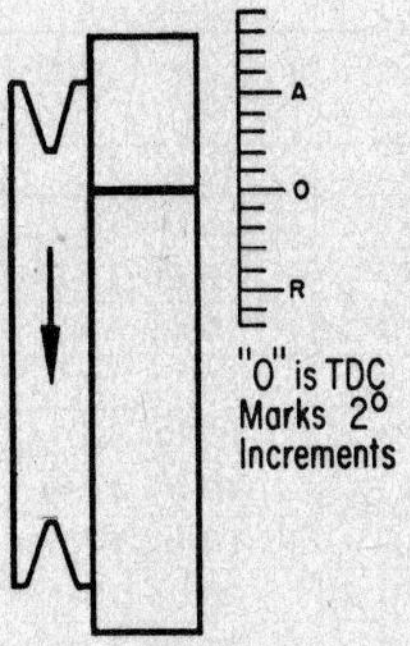

Fig. G

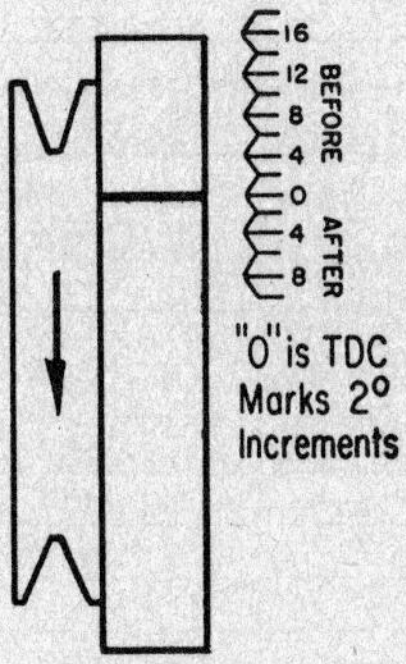

Fig. H

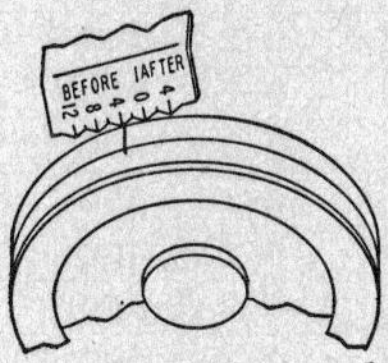

Fig. I

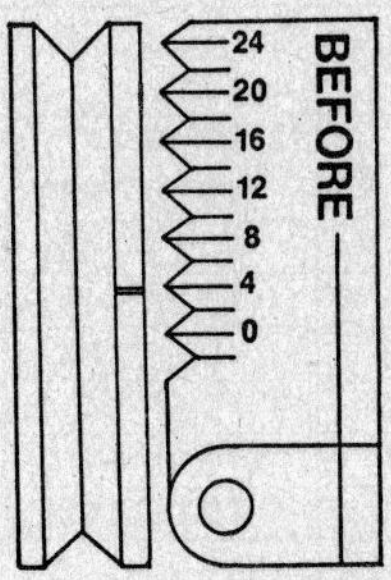

Fig. J

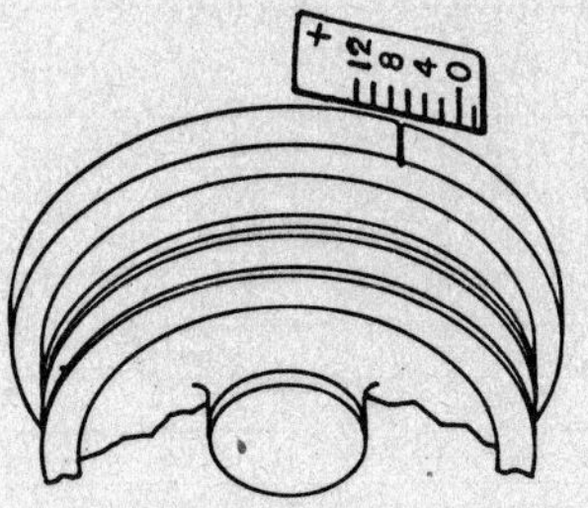

Fig. K

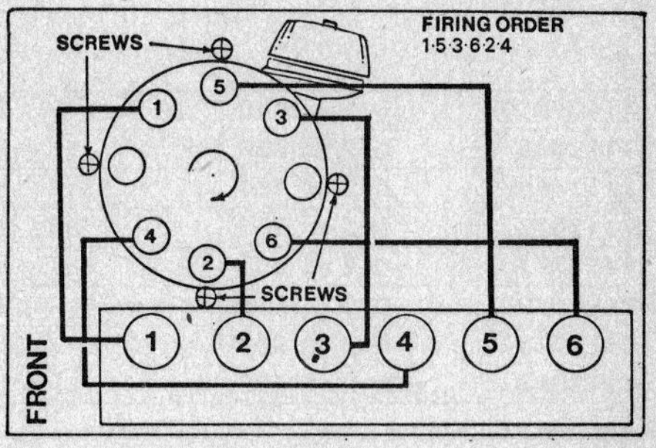

Fig. L

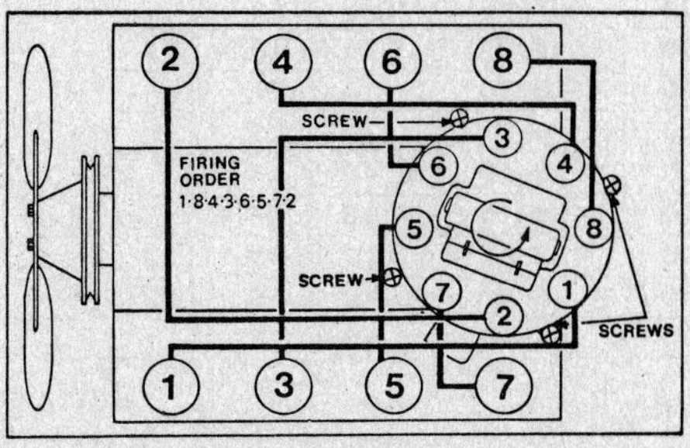

Fig. M

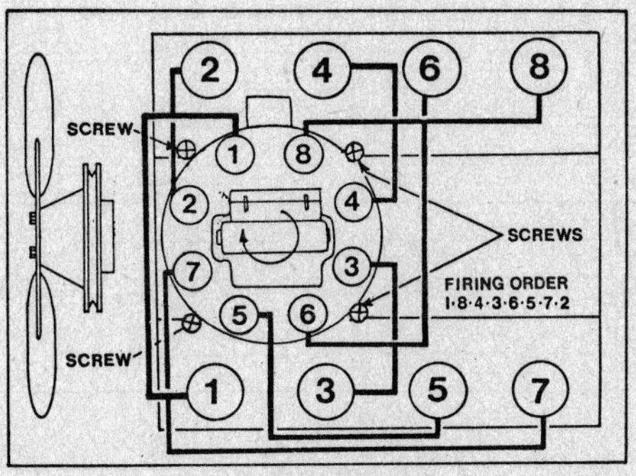

Fig. N

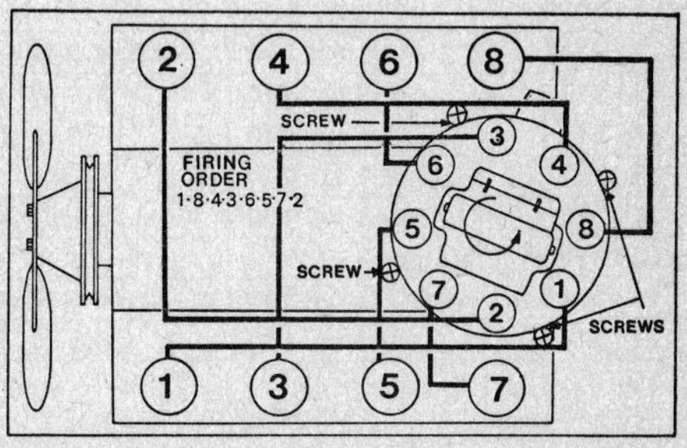

Fig. O

DISTRIBUTOR SPECIFICATIONS

★Note: If unit is checked on vehicle, double the RPM and degrees to get crankshaft figures.

Distributor Part No.(1)	Centrifugal Advance Degrees @ RPM of Distributor					Vacuum Advance		Distributor Retard
	Advance Starts	Intermediate Advance			Full Advance	Inches of Vacuum to Start Plunger	Max. Adv. Dist. Deg. @ Vacuum	Max. Retard Dist. Deg. @ Vacuum
1969								
1110474(4)	0–2 @ 550	6–8 @ 875	—	—	13.5 @ 2375	5–7	7.5 @ 11.5	—
1110475(4)	0–4.5 @ 500	9–11 @ 1050	—	—	15 @ 2200	5–7	7.5 @ 11.5	—
1111253(4)	0–2 @ 500	5–7 @ 1000	—	—	11 @ 2300	10–12	10 @ 18.8	—
1111940(4)	0.5–2.5 @ 500	7.5–9.5 @ 950	—	—	17 @ 2275	8–10	10 @ 17	—
1111941(5)	0–2 @ 660	5–7 @ 1100	—	—	12 @ 2300	8–10	10 @ 17	—
1111942(4)	1–3 @ 550	6.5–8.5 @ 975	—	—	13 @ 2400	6–8	10 @ 14¾	—
1111946(5)	0–2 @ 500	5–7 @ 1000	—	—	11 @ 2300	8–10	10 @ 17	—
1111952(5)	1–3 @ 700	5–7 @ 1000	—	—	11 @ 2300	10–12	10 @ 18.8	—
1111959(4)	0–2.2 @ 550	4.5–6.5 @ 1200	—	—	12 @ 2300	8–10	10 @ 17	—
1111960(4)	1–3.3 @ 700	6–8 @ 1000	—	—	12 @ 2300	8–10	10 @ 17	—
1970								
1110463(4)	0–2 @ 450	11.5 @ 975	—	—	16 @ 2100	6–8	11½ @ 17	—
1110464(4)	0–2 @ 450	8.5 @ 975	—	—	14 @ 2100	6–8	11½ @ 17	—
1111105(6)	0–2 @ 800	—	—	—	8 @ 2200	8–10	10 @ 17	—
1111148(6)	0–2 @ 400	6 @ 950	—	—	13 @ 2300	8–10	10 @ 17	—
1111176(6)	0–2 @ 550	6 @ 950	—	—	13 @ 2350	8–10	10 @ 17	—
1112007(4)	0–2 @ 400	6 @ 950	—	—	13 @ 2300	8–10	10 @ 17	—
1112008(4)	0–2 @ 550	6 @ 950	—	—	13 @ 2350	6–8	10 @ 14¾	—
1112009(6)	0–2 @ 400	5.5 @ 2200	—	—	11 @ 2300	8–10	10 @ 17	—
1112010(6)	0–2 @ 550	5.5 @ 1100	—	—	11 @ 2300	8–10	10 @ 17	—
1112011(6)	0–2 @ 600	6 @ 1050	—	—	14 @ 3050	8–10	10 @ 17	—
1112012(6)	0–2 @ 400	—	—	—	8 @ 2200	8–10	10 @ 17	—
1112013(6)	0–2 @ 600	6 @ 1050	—	—	14 @ 3050	8–10	10 @ 17	—
1112024(6)	0–2 @ 550	5.5 @ 1150	—	—	11 @ 2300	8–10	10 @ 17	—
1971								
1110489(4)	0–2 @ 550	7 @ 1150	—	—	12 @ 2050	7–9	11½ @ 17	—
1112068	0–2 @ 800	6 @ 1000	—	—	14 @ 2300	6–8	10 @ 14¾	—
1112069	0–2 @ 800	5½ @ 1000	—	—	11 @ 2300	6–8	10 @ 14¾	—
1112070	0–2 @ 575	3 @ 825	—	—	11 @ 2300	6–8	10 @ 14¾	—
1112071	0–2 @ 800	—	—	—	12 @ 2300	8–10	10 @ 17	—
1112072	0–2 @ 825	3 @ 825	—	—	11 @ 2300	8–10	10 @ 17	—
1112073	0–2 @ 500	4 @ 750	—	—	13 @ 2225	6–8	10 @ 14¾	—
1112083	0–2 @ 600	6 @ 1050	—	—	11 @ 2300	6–8	10 @ 14¾	—
1972								
1110489(4)	0–2 @ 650	7 @ 1150	—	—	12 @ 2050	7–9	11½ @ 17	—
1112005	0–2 @ 600	6 @ 1100	—	—	12 @ 2250	7–9	10.3 @ 17.8	—
1112039	0–2 @ 650	—	—	—	11 @ 2100	7–9	10½ @ 18	—
1112118	0–1 @ 800	2.5–4.5 @ 1000	—	—	10 @ 2300	6–8	10 @ 15	—
1112119	0–1 @ 800	3.5 @ 1000	—	—	13 @ 2300	8–10	10 @ 17	—
1112121	0–1 @ 700	4–6 @ 1200	—	—	13 @ 2300	6–8	10 @ 15	—
1112122	0–1 @ 800	—	—	—	13 @ 2300	8–10	10 @ 17	—
1112127	0–1 @ 700	3–5 @ 1000	—	—	11 @ 2300	8–10	10 @ 17	—
1112133	0–1 @ 575	4.5–6.5 @ 1000	—	—	15 @ 2300	6–8	10 @ 15	—
1112140	0–1 @ 800	4.5–6.5 @ 1000	—	—	12 @ 2300	6–8	10 @ 15	—
1112145	0–1 @ 700	3.5 @ 1000	—	—	11 @ 2300	8–10	10 @ 17	—

Continued

DISTRIBUTOR SPECIFICATIONS—Continued

★Note: If unit is checked on vehicle, double the RPM and degrees to get crankshaft figures.

Distributor Part No.①	Centrifugal Advance Degrees @ RPM of Distributor					Vacuum Advance		Distributor Retard
	Advance Starts	Intermediate Advance			Full Advance	Inches of Vacuum to Start Plunger	Max. Adv. Dist. Deg. @ Vacuum	Max. Retard Dist. Deg. @ Vacuum
1973								
1110499	0 @ 650	7 @ 1150	—	—	12 @ 2050	7	11½ @ 15	—
1112191	0 @ 575	3.8 @ 850	—	—	9 @ 1950	10	12½ @ 16	—
1112199	0 @ 615	4 @ 790	—	—	12 @ 1875	8	12½ @ 15	—
1112201	0 @ 600	6.5 @ 1050	—	—	12 @ 1790	7	12½ @ 12	—
1112202	0 @ 700	8 @ 1050	—	—	13 @ 1725	7	12½ @ 12	—
1112203②	0 @ 600	4 @ 750	—	—	9 @ 2000	10	12½ @ 16	—
1112216	0 @ 600	6½ @ 1050	—	—	12 @ 1790	3	10 @ 9.6	—
1112218	0 @ 600	8.5 @ 950	—	—	10 @ 1650	6	10 @ 12	—
1112220	0 @ 575	3.8 @ 850	—	—	9 @ 1950	5	12.5 @ 11	—
1112224	0 @ 610	4 @ 790	—	—	12 @ 1875	5	12.5 @ 11	—
1112232	0 @ 600	5 @ 850	—	—	11 @ 2300	5	12.5 @ 11	—
1112233②	0 @ 600	5 @ 850	—	—	11 @ 2300	9	12.5 @ 14	—
1112507②	0 @ 600	4 @ 750	—	—	9 @ 2000	5	12.5 @ 11	—
1112510	0 @ 500	4 @ 700	—	—	12 @ 1900	7	12.5 @ 12	—
1112511	0 @ 500	4 @ 700	—	—	12 @ 1900	8	12.5 @ 15	—
1112810②	0 @ 600	4 @ 750	—	—	9 @ 2000	9	10 @ 16	—
1112811	0 @ 575	3.8 @ 850	—	—	9 @ 1950	7	12.5 @ 12	—
1973-74								
1112205	0 @ 600	5 @ 850	—	—	11 @ 2300	9	12.5 @ 14	—
1112231	0 @ 600	5 @ 850	—	—	11 @ 2300	9	12.5 @ 13.5	—
1112804	0 @ 600	6.5 @ 1065	—	—	12 @ 1790	7	10 @ 14	—
1112805	0 @ 600	4 @ 775	—	—	12 @ 1875	9	10 @ 16	—
1112806	0 @ 650	8 @ 1050	—	—	13 @ 1725	7	10 @ 14	—
1112807	0 @ 575	4.5 @ 850	—	—	9 @ 1950	9	10 @ 16	—
1112808	0 @ 600	6.5 @ 1075	—	—	12 @ 1790	5	12.5 @ 11	—
1112809	0 @ 600	4 @ 790	—	—	12 @ 1875	7	12.5 @ 12	—
1112812②	0 @ 600	5 @ 850	—	—	11 @ 2300	9	10 @ 16	—
1112813	0 @ 600	5 @ 850	—	—	11 @ 2300	9	10 @ 16	—
1112814	0 @ 600	5 @ 850	—	—	11 @ 2300	7	12.5 @ 11.5	—
1974								
1110499	0 @ 550	7 @ 1150	—	—	12 @ 2050	7	12 @ 15	—
1112210	0 @ 600	4 @ 850	—	—	9 @ 1950	9	10 @ 16	—
1112212	0 @ 600	5 @ 850	—	—	11 @ 2300	9	10 @ 16	—
1112213	0 @ 600	5 @ 850	—	—	11 @ 2300	7	12½ @ 12	—
1122341③	0 @ 600	6.5 @ 1065	—	—	12 @ 1790	7	10 @ 14	—
1112235③	0 @ 600	6.5 @ 1075	—	—	13 @ 1725	5	12½ @ 11	—
1112236③	0 @ 650	8 @ 1050	—	—	13 @ 1725	7	10 @ 14	—
1112237③	0 @ 600	4 @ 775	—	—	12 @ 1875	9	10 @ 16	—
1112238③	0 @ 600	4 @ 790	—	—	12 @ 1875	7	12 @ 12	—
1112239③	0 @ 600	5 @ 850	—	—	11 @ 2300	9	12½ @ 14	—
1112240③	0 @ 600	5 @ 850	—	—	11 @ 2300	7	12½ @ 11½	—
1112243③	0 @ 600	5 @ 850	—	—	11 @ 2300	9	12½ @ 13½	—
1112512③	0 @ 600	5 @ 850	—	—	11 @ 2300	9	10 @ 16	—
1112513③	0 @ 575	4.5 @ 850	—	—	9 @ 1950	9	10 @ 16	—
1112546	0 @ 600	4 @ 800	—	—	12 @ 1800	9	10 @ 16	—
1112547	0 @ 600	4 @ 800	—	—	12 @ 1800	7	12½ @ 12	—
1112810②	0 @ 600	4 @ 800	—	—	9 @ 2000	9	10 @ 16	—

Continued

DISTRIBUTOR SPECIFICATIONS—Continued

★Note: If unit is checked on vehicle double the RPM and degrees to get crankshaft figures.

Distributor Part. No.①	Centrifugal Advance Degrees @ RPM of Distributor					Vacuum Advance		Distributor Retard
	Advance Starts	Intermediate Advance			Full Advance	Inches of Vacuum to Start Plunger	Max. Adv. Dist. Deg. @ Vacuum	Max. Retard Dist. Deg. @ Vacuum
1974—Continued								
1112821	0 @ 500	4 @ 700	—	—	12 @ 1900	7	10 @ 14	—
1112822③	0 @ 500	4 @ 700	—	—	12 @ 1900	7	10 @ 14	—
1112833	0 @ 600	3 @ 825	—	—	12 @ 2000	9	10 @ 16	—
1112834③	0 @ 600	3 @ 825	—	—	12 @ 2000	9	10 @ 16	—
1112856③	0 @ 610	5 @ 800	—	—	12 @ 1900	—	10 @ 14	—
1112857	0 @ 610	5 @ 800	—	—	12 @ 1900	—	10 @ 14	—
1112859	0 @ 575	4 @ 850	—	—	9 @ 1950	7	10 @ 14.5	—
1112860③	0 @ 575	4 @ 850	—	—	9 @ 1950	7	10 @ 14.5	—
1112871	0 @ 600	5 @ 875	—	—	11 @ 2300	9	12½ @ 16	—
1112876	0 @ 600	3 @ 775	—	—	12 @ 900	9	10 @ 16	—
1112878	0 @ 600	4 @ 800	—	—	9 @ 1950	7	10 @ 14	—
1975								
1112495⑦	0 @ 550	7 @ 1100	—	—	9 @ 2200	7	12½ @ 12	—
1112498⑦	0 @ 600	2 @ 700	—	—	8½ @ 1800	6	12 @ 13	—
1112500⑦	0 @ 600	3 @ 800	—	—	8 @ 2000	7	12½ @ 12	—
1112863⑦	0 @ 362	4½ @ 1300	—	—	8 @ 2100	4	9 @ 12	—
1112896⑦	0 @ 525	3 @ 1050	—	—	6 @ 2250	7	8 @ 12	—
1112918⑦	0 @ 500	5½ @ 1000	—	—	7 @ 2200	7	10 @ 11	—
1112928⑦	0 @ 600	2 @ 700	—	—	8 @ 2200	7	12½ @ 12	—
1112929⑦	0 @ 500	4 @ 700	—	—	10 @ 2200	7	10 @ 14	—
1112930⑦	0 @ 700	3½ @ 1000	—	—	5 @ 2200	7	12½ @ 12	—
1112946⑦	0 @ 500	4 @ 700	—	—	10½ @ 1800	7	12 @ 15	—
1112947⑦	0 @ 600	2 @ 700	—	—	10 @ 1900	8	10 @ 15	—
1112948⑦	0 @ 500	4 @ 700	—	—	10 @ 2200	8	10 @ 15	—
1112949⑦	0 @ 500	5½ @ 1000	—	—	7 @ 2200	7	10 @ 14	—
1112950⑦	0 @ 600	2 @ 700	—	—	10 @ 1700	7	10 @ 11	—
1112951⑦	0 @ 325	9½ @ 1200	—	—	14 @ 2200	4	12 @ 15	—
1112956⑦	0 @ 325	9½ @ 1200	—	—	14 @ 2200	—	—	—
1112500⑦	0 @ 600	3 @ 800	—	—	10 @ 2200	7	12½ @ 12	—
1976								
1103201⑦	0 @ 600	2 @ 700	—	—	10 @ 1900	6	12 @ 13	—
1103205⑦	0 @ 600	2 @ 700	—	—	8 @ 2200	7	10 @ 11	—
1103206⑦	0 @ 600	2 @ 700	—	—	8½ @ 1800	7	10 @ 11	—
1110666⑦	0 @ 500	3½ @ 800	—	—	10 @ 2100	4	12 @ 15	—
1112495⑦	0 @ 550	7 @ 1100	—	—	4 @ 2200	7	12½ @ 12	—
1112497⑦	0 @ 600	2 @ 700	—	—	10 @ 1900	5	12½ @ 11	—
1112500⑦	0 @ 600	3 @ 800	—	—	10 @ 2200	7	12½ @ 12	—
1112863⑦	0 @ 363	4½ @ 1300	—	—	8 @ 2200	4	9 @ 12	—
1112923⑦	0 @ 500	5½ @ 1100	—	—	7 @ 2200	7	12½ @ 12	—
1112928⑦	0 @ 600	2 @ 700	—	—	8 @ 2200	7	12½ @ 12	—
1112930⑦	0 @ 700	3½ @ 1000	—	—	5 @ 2200	7	12½ @ 12	—
1112950⑦	0 @ 600	2 @ 700	—	—	10 @ 1900	7	10 @ 11	—
1112956⑦	0 @ 325	9½ @ 1200	—	—	14 @ 2200	—	—	—
1112958⑦	0 @ 600	2 @ 700	—	—	8 @ 2200	5	12½ @ 11	—
1112960⑦	0 @ 500	5½ @ 1000	—	—	7 @ 2200	9	10 @ 16	—
1112991⑦	0 @ 713	—	—	—	10 @ 2213	7	10 @ 13	—
1112992⑦	0 @ 450	—	—	—	10 @ 2200	8	7 @ 11	—
1112994⑦	0 @ 325	9½ @ 1200	—	—	14 @ 2200	5	12 @ 11	—
1112995⑦	0 @ 450	8½ @ 1200	—	—	13 @ 2225	4	15 @ 11	—

①—Stamped on distributor housing plate.
④—Breaker arm spring tension—19–23.
②—Transistorized Unitized distributor.
⑤—Breaker arm spring tension—28–31.
③—Uni-Set ignition points.
⑥—Breaker arm spring tension—28–32.
⑦—High Energy Ignition.

STARTING MOTOR SPECIFICATIONS

Year	Model	Starter Number	Brush Spring Tension Oz①	Free Speed Test			Resistance Test③	
				Amps. ①	Volts	R.P.M. ①	Amps. ①	Volts
1969	6-230, 6-250	1107499	35	49–76②	10.6	6200–9600	270–310	4.3
	V8s	1107293	35	—	—	—	—	—
	V8s	1107355	35	70–99②	10.6	6800–9400	410–480②	3.0
	6-250 Auto. Trans.	1107594	35	50②	9	5500–10500	—	—
	6-250 Firebird	1108329	35	50②	9	5500–10500	—	—
	V8-400 Firebird	1108335	35	65②	9	7500–10500	—	—
	V8-400 Ram Air Eng.	1108353	35	65②	9	7500–10500	—	—
	V8-350 Firebird	1108328	35	55②	9	3500–6000	—	—
1970	6-250	1108439	35	50②	9	5500–10500	—	—
	V8-350	1108434	35	55②	9	3500–6000	—	—
	V8-400, 455	1108435	35	65②	9	7500–10500	—	—
1971-72	6-250	1108365	35	50②	9	5500–10500	—	—
	V8-307	1108367	35	50–80②	9	5500–10500	—	—
	V8-350	1108445	35	55②	9	3500–6000	—	—
	V8-400, 455	1108446	35	65②	9	7500–10500	—	—
	V8-455	1108436	35	65②	9	7500–10500	—	—
1973–74	6-250	1108365	35	50②	9	5500–10500	—	—
	V8-350	1108224	35	—	—	—	—	—
	V8-400, 455	1108225	35	—	—	—	—	—
	V8-455④	1108226	35	65②	9	7500–10500	—	—
1975	6-250	1108365	35	50②	9	5500–10500	—	—
	6-250	1108745	35	—	—	—	—	—
	V8-260	1108516	35	55–80	9	3500–6000	—	—
	V8-350⑤	1108762	35	—	—	—	—	—
	V8-350	1108758	35	65–95	9	7500–10500	—	—
	V8-400, 455	1108759	35	—	—	—	—	—
1976	6-250 Std. Trans.	1108365	35	50②	9	5500–10500	—	—
	6-250 Auto. Trans.	1108745	35	—	—	—	—	—
	V8-260⑥	1108516	35	55–80	9	3500–6000	—	—
	V8-350⑤	1108762	35	55–80	9	3500–6000	—	—
	V8-350	1108758	35	65–95	9	7500–10500	—	—
	V8-400, 455	1108759	35	—	—	—	—	—

①—Minimum. ②—Includes solenoid.
③—Check capacity of motor by using a 500 ampere meter and a carbon pile rheostat to control voltage. Apply volts listed across motor with armature locked. Current should be as listed.
④—With super-duty engine.
⑤—Ventura models only. See Buick Chapter for services procedures.
⑥—See Oldsmobile Chapter for service procedures.

VALVE SPECIFICATIONS

Year	Model	Valve Lash		Valve Angles		Valve Spring Installed Height	Valve Spring Pressure Lbs. @ In.	Stem Clearance		Stem Diameter	
		Int.	Exh.	Seat	Face			Intake	Exhaust	Intake	Exhaust
1969	6-250, 1 B.C.	Hydraulic⑦		46	45	1.629	170 @ 1.23	.0016–.0033	.0021–.0038	.3407–.3414	.3407–.3414
	6-250, 4 B.C.	Hydraulic⑦		46	45	1.629	122 @ 1.10	.0016–.0033	.0021–.0038	.3407–.3414	.3407–.3414
	8-350, 2 B.C.	Hydraulic⑦		45	⑧	1.582④	140 @ 1.13	.0016–.0033	.0021–.0038	.3407–.3414	.3407–.3414
	8-350, 4 B.C.	Hydraulic⑦		45	⑧	1.591④	140 @ 1.13	.0016–.0033	.0021–.0038	.3407–.3414	.3407–.3414
	8-400, 2 B.C.	Hydraulic⑦		⑨	⑩	1.582④	140 @ 1.13	.0016–.0033	.0021–.0038	.3407–.3414	.3407–.3414
	8-400, 4 B.C.	Hydraulic⑦		⑨	⑩	1.561④	185 @ 1.29	.0016–.0033	.0021–.0038	.3407–.3414	.3407–.3414
	8-400 Ram Air	Hydraulic⑦		⑨	⑩	1.591④	185 @ 1.29	.0016–.0033	.0021–.0038	.3407–.3414	.3407–.3414
	8-428	Hydraulic⑦		⑨	⑩	1.582④	140 @ 1.13	.0016–.0033	.0021–.0038	.3407–.3414	.3407–.3414

Continued

VALVE SPECIFICATIONS—Continued

Year	Model	Valve Lash		Valve Angles		Valve Spring Installed Height	Valve Spring Pressure Lbs. @ In.	Stem Clearance		Stem Diameter	
		Int.	Exh.	Seat	Face			Intake	Exhaust	Intake	Exhaust
1970	6-250⑤	1 Turn②		46	45	1.66	186 @ 1.27	.0010–.0027	.0010–.0027	.3410–.3417	.3410–.3417
	8-350	Hydraulic⑦		45	44	1.582④	130 @ 1.70	.0016–.0033	.0021–.0038	.3412–.3419	.3412–.3419
	8-400, 265 H.P.	Hydraulic⑦		45	44	1.582④	④⑥	.0016–.0033	.0021–.0038	.3412–.3419	.3407–.3414
	8-400, 290 H.P.	Hydraulic⑦		45	44	1.582④	④⑥	.0016–.0033	.0021–.0038	.3412–.3419	.3407–.3414
	8-400, 330 H.P.	Hydraulic⑦		45	44	1.582④	④⑪	.0016–.0033	.0021–.0038	.3412–.3419	.3407–.3414
	8-400, 350 H.P.	Hydraulic⑦		⑨	⑬	1.591④	132 @ 1.181	.0016–.0033	.0021–.0038	.3412–.3419	.3407–.3414
	8-400 Ram Air	Hydraulic⑦		⑨	⑬	1.591④	132 @ 1.178	.0016–.0033	.0021–.0038	.3412–.3419	.3407–.3414
	8-400 Ram Air IV	Hydraulic⑦		⑨	⑬	1.818④	222 @ 1.291	.0016–.0033	.0021–.0038	.3412–.3419	.3407–.3414
	8-455	Hydraulic⑦		⑨	⑬	1.561④	④⑫	.0016–.0033	.0021–.0038	.3412–.3419	.3407–.3414
1971	6-250⑤	1 Turn②		46	45	1.66	186 @ 1.27	.0010–.0027	.0015–.0032	.3410–.3417	.3410–.3417
	8-307①	1 Turn②		46	45	1.70	200 @ 1.25	.0010–.0027	.0010–.0027	.3410–.3417	.3410–.3417
	8-350	Hydraulic⑦		⑨	⑬	—	—	.0016–.0033	.0021–.0038	.3412–.3419	.3412–.3419
	8-400	Hydraulic⑦		⑨	⑬	—	—	.0016–.0033	.0021–.0038	.3412–.3419	.3407–.3414
	8-400 Ram Air	Hydraulic⑦		⑨	⑬	—	—	.0016–.0033	.0021–.0038	.3412–.3419	.3407–.3414
	8-455	Hydraulic⑦		⑨	⑬	—	—	.0016–.0033	.0021–.0038	.3412–.3419	.3407–.3414
1972	6-250⑤	1 Turn②		46	45	1.66	186 @ 1.27	.0010–.0027	.0015–.0032	.3410–.3417	.3410–.3417
	8-307①	1 Turn②		46	45	1.70	200 @ 1.25	.0010–.0027	.0010–.0027	.3410–.3417	.3410–.3417
	8-350	Hydraulic⑦		⑭	⑮	1.66	186 @ 1.27	.0010–.0027	.0010–.0027	.3410–.3417	.3410–.3417
	8-400 2 Bar. Carb.	Hydraulic⑦		45	44	1.59④	126 @ 1.21④	.0016–.0033	.0021–.0038	.3412–.3419	.3407–.3414
	8-400 4 Bar. Carb.	Hydraulic⑦		⑧	⑧	1.56④	135 @ 1.15④	.0016–.0033	.0021–.0038	.3412–.3419	.3407–.3414
	8-455	Hydraulic⑦		⑧	⑧	1.56④	137 @ 1.14④	.0016–.0033	.0021–.0038	.3412–.3419	.3412–.3419
1973-74	6-250	1 Turn②		46	45	1.66	186 @ 1.27	.0010–.0027	.0015–.0032	.3410–.3417	.3410–.3417
	8-350	Hydraulic⑦		⑯	⑬	—	—	.0016–.0033	.0021–.0038	.3400	.3400
	8-400 2 Bar. Carb.	Hydraulic⑦		⑯	⑬	—	—	.0016–.0033	.0021–.0038	.3400	.3400
	8-400 4 Bar. Carb.	Hydraulic⑦		⑯	⑬	—	—	.0016–.0033	.0021–.0038	.3400	.3400
	8-455	Hydraulic⑦		⑯	⑬	—	—	.0016–.0033	.0021–.0038	.3400	.3400
1975	6-250⑤	1 Turn②		46	45	1.66	186 @ 1.27	.0010–.0027	.0010–.0027	.3410–.3417	.3410–.3417
	V8-260⑰	Hydraulic㉑		⑱	⑲	1.67	187 @ 1.27	.0010–.0027	.0015–.0032	.3425–.3432	.3420–.3427
	V8-350⑳	Hydraulic㉑		45	45	1.727	㉒	.0015–.0035	.0015–.0032	.3720–.3730	.3723–.3730
	V8-350	Hydraulic⑦		⑯	⑬	1.727	180 @ 1.34	.0016–.0033	.0021–.0038	.3412–.3419	.3407–.3414
	V8-400	Hydraulic⑦		⑯	⑬	1.54	135 @ 1.13	.0016–.0033	.0021–.0038	.3412–.3419	.3407–.3414
	V8-455	Hydraulic⑦		⑯	⑬	1.57	135 @ 1.16	.0016–.0033	.0021–.0038	.3412–.3419	.3407–.3414
1976	6-250⑤	Hydraulic②		46	45	1.66	186 @ 1.27	.0010–.0027	.0010–.0027	.3410–.3417	.3410–.3417
	V8-260⑰	Hydraulic㉑		46	45	1.67	187 @ 1.27	.0010–.0027	.0015–.0032	.3425–.3432	.3420–.3427
	V8-350⑳	Hydraulic㉑		45	45	1.727	㉓	.0015–.0035	.0015–.0032	.372–.373	.3723–.3730
	V8-350	Hydraulic⑦		⑯	⑬	1.56	㉔	.0016–.0033	.0021–.0038	.3412–.3419	.3407–.3414
	V8-400 2 Bar. Carb.	Hydraulic⑦		⑯	⑬	1.54	㉕	.0016–.0033	.0021–.0038	.3412–.3419	.3407–.3414
	V8-400 4 Bar. Carb.	Hydraulic⑦		⑯	⑬	1.54	㉖	.0016–.0033	.0021–.0038	.3412–.3419	.3407–.3414
	V8-455	Hydraulic⑦		⑯	⑬	1.57	㉗	.0016–.0033	.0021–.0038	.3412–.3419	.3407–.3414

①—For service on this engine, see Chevrolet Chapter.
②—With valve fully closed, turn rocker arm nut down until all play in push rod is eliminated, then tighten nut one additional turn.
③—Intake 30°, exhaust 45°.
④—Outer spring.
⑤—For service on this engine, see Six Cylinder in Chevrolet Chapter.
⑥—Intake 126 @ 1.206, exhaust 134 @ 1.170.
⑦—No adjustment. On V8's, rocker arms are correctly positioned when ball retainer nuts are tightened to 20 ft.-lbs.
⑧—Intake 29°, exhaust 44°.
⑨—Small valve engines, intake and exhaust 45°. Large valve engines intake 30° and exhaust 45°.
⑩—Small valve engines, intake 46° and exhaust 44°. Large valve engines, intake 29° and exhaust 44°.
⑪—Intake 133 @ 1.172, exhaust 134 @ 1.168.
⑫—Intake 137 @ 1.151, exhaust 137 @ 1.148.
⑬—Small valve engines, intake 44° and exhaust 44°. Large valve engines, intake 29° and exhaust 44°.
⑭—Small valve engines, 46°. Large valve engines, 45°.
⑮—Small valve engines, intake 46° and exhaust 45°. Large valve engines, intake and exhaust 44°.
⑯—Small valve engines, intake 45° and exhaust 45°. Large valve engines, intake 30° and exhaust 45°.
⑰—See Oldsmobile Chapter for service procedures.
⑱—Intake 45°, exhaust 59°.
⑲—Intake 46°, exhaust 60°.
⑳—Ventura only. See Buick Chapter for service procedures.
㉑—No adjustment.
㉒—Intake, 180 @ 1.34; exhaust, 177 @ 1.45.
㉓—Intake, 180 @ 1.34; Exhaust, 175 @ 1.34.
㉔—Intake, 131.7 @ 1.18; Exhaust, 137.8 @ 1.15.
㉕—Intake, 134.5 @ 1.16; Exhaust, 135.1 @ 1.13.
㉖—Intake, 135.1 @ 1.13; Exhaust, 140.8 @ 1.12.
㉗—Intake, 135.8 @ 1.16; Exhaust, 141.6 @ 1.15.

PISTONS, PINS, RINGS, CRANKSHAFT & BEARINGS

Year	Model	Piston Skirt Clearance	Ring End Gap①		Wrist-pin Diameter	Rod Bearings		Main Bearings			
			Comp.	Oil		Shaft Diameter	Bearing Clearance	Shaft Diameter	Bearing Clearance	Thrust on Bear. No.	Shaft End Play
1969	6-250	.0022–.0038	.015	.035	.9272	2.00	.0005–.0028	2.30	.0003–.0020	7	.002–.006
	V8-350	.0025–.0031	.020	.035	.9802	2.25	.0005–.0025	3.00	.0002–.0020	4	.003–.009
	V8-400	.0025–.0031⑤	.020	.035	.9802	2.25	.0005–.0025	3.00	.0002–.0020	4	.003–.009
	V8-428	.0030–.0036	.020	.035	.9802	2.25	.0005–.0025	3.25	.0002–.0020	4	.003–.009
1970–72	6-250④	.0005–.0015	.010	.015	.9272	2.00	.0007–.0027	2.30	.0003–.0029	7	.002–.006
	V8-307⑧	.0005–.0011	.010	.015	.927	2.099–2.100	.0013–.0035	⑨	②	5	.002–.006
	V8-350	.0025–.0033	.019	.035	.9802	2.25	.0005–.0025	3.00	.0002–.0017	4	.0035–.0085
	V8-400	.0025–.0033	⑥	.035	.9802	2.25	.0005–.0025	3.00	.0002–.0017	4	.0035–.0085
	V8-455	.0025–.0033	⑦	.035	.9802	2.25	.0005–.0026	3.25	.0005–.0021	4	.0035–.0085
1973-74	6-250	.0005–.0015	.010	.015	.9272	2.00	.0007–.0027	2.30	.0003–.0029	7	.002–.006
	V8-350	.0029–.0037	⑥	.035	.9802	2.25	.0005–.0025	3.00	.0002–.0017	4	.003–.009
	V8-400	.0029–.0037	⑥	.035	.9802	2.25	.0005–.0025	3.00	.0002–.0017	4	.003–.009
	V8-455	⑩	⑦	.035	.9802	2.25	⑪	3.25	⑫	4	.003–.009
1975	6-250⑧	.0005–.0015	.010	.015	.9272	2.000	.0007–.0027	2.30	.0003–.0029	7	.002–.006
	V8-260⑬	.001–.002	.010	.015	.9805	2.1238–2.1248	.0005–.0026	⑮	⑯	3	.004–.008
	V8-350⑭	.0008–.0020	.010	.015	.9393	2.000	.0005–.0026	3.000	.0004–.0015	3	.003–.009
	V8-350	.0029–.0037	⑥	.035	.9802	2.25	.0005–.0025	3.00	.0002–.0017	4	.003–.009
	V8-400	.0029–.0037	⑥	.035	.9802	2.25	.0005–.0025	3.00	.0002–.0017	4	.003–.009
	V8-455	.0021–.0029	⑦	.035	.9802	2.25	.0005–.0025	3.25	.0005–.0021	4	.003–.009
1976	6-250⑧	.0005–.0015	.010	.035	.9272	2.00	.0007–.0027	2.30	.0003–.0029	7	.002–.006
	V8-260⑬	.00075–.00175⑰	.010	.015	.9805	2.124	.0005–.0026	2.50	⑯	3	.004–.008
	V8-350⑭	.0008–.0014	.013	.015	.9396	2.00	.0005–.0026	2.9995	.0004–.0015	3	.003–.009
	V8-350	.0029–.0037	.019	.035	.9802	2.25	.0005–.0025	3.00	.0002–.0017	4	.0035–.0085
	V8-400	.0029–.0037	⑥	.035	.9802	2.25	.0005–.0025	3.00	.0002–.0017	4	.0035–.0085
	V8-455	.0021–.0029	⑦	.035	.9802	2.25	.0005–.0025	3.25	.0005–.0021	4	.0035–.0085

①—Fit rings in tapered bores for clearance listed in tightest portion of ring travel.
②—No. 1: .0008–.002; No. 2, 3, 4, .0011–.0023; No. 5: .0017–.0033.
④—For service on this engine, see Six Cylinder in Chevrolet Chapter.
⑤—V8-400 Ram Air IV use .0055–.0061″.
⑥—Top ring .019″, second ring .015″.
⑦—Top ring .021″, second ring .015″.
⑧—For service on this engine, see Chevrolet Chapter.
⑨—Front: 2.4502, Rear: 2.4507, others: 2.4505.
⑩—1973 exc. super-duty .0025″–.0033″, super-duty; .0064″–.0072″; 1974 exc. super-duty .0021″–.0029″, super-duty .0064″–.0072″.
⑪—Exc. super-duty .0005″–.0025″, super-duty .0015″–.0031″.
⑫—Exc. super-duty .0005″–.0021″, super-duty .0010″–.0026″.
⑬—See Oldsmobile Chapter for service procedures.
⑭—Ventura only. See Buick Chapter for service procedures.
⑮—No. 1—2.4988–2.4998; No. 2, 3, 4, 5; 2.4985–2.4995.
⑯—No. 1, 2, 3, 4 .0005–.0021; No. 5—.0015–.0031.
⑰—Measured .77 below piston pin centerline.

BRAKE SPECIFICATIONS

Year	Model	Brake Drum Inside Diameter	Wheel Cylinder Bore Diameter			Master Cylinder Bore Diameter		
			Disc Brake	Front Drum Brake	Rear Drum Brake	Disc Brakes	Drum Brakes	Power Brakes
1969–73	①	9½	2 15/16	1⅛	⅞	1⅛	1	1
	Pontiac	11②	2 15/16	1⅛	15/16	1⅛	1	1
1974	①	9½③	2 15/16	1⅛	1⅛④	1⅛	1	1
	Pontiac	11⑤	2 15/16	—	15/16	1⅛	1	1
1975	Ventura	9½	2 15/16	—	⅞	1	—	1⅛
	⑥	9½③	2 15/16	—	⅞⑦	1	—	1⅛
	Grand Prix	11	2 15/16	—	15/16	—	—	1⅛
	Pontiac	11⑤	2 15/16	—	15/16⑧	—	—	1⅛
1976	Firebird & Ventura	9½	2 15/16	—	15/16	1	—	1⅛
	LeMans	11	2 15/16	—	15/16	15/16	—	1⅛
	Grand Prix	11	2 15/16	—	15/16	—	—	1⅛
	Pontiac	11⑤	2 15/16	—	1	—	—	1⅛

①—Intermediates, Ventura II, Firebird and Grand Prix.
②—1971-73 Wagon, 12″.
③—Wagon, 11″.
④—Wagon, 15/16″.
⑤—Wagon 12″.
⑥—Firebird and intermediates.
⑦—Wagon 15/16″.
⑧—Wagon 1″.

ALTERNATOR & REGULATOR SPECIFICATIONS

Year	Alternator					Regulator						
	Model	Rated Hot Output Amps.	Field Current 12 Volts @ 80° F.	Output @ 14 Volts 2000 R.P.M. Amps.	Output @ 14 Volts 5000 R.P.M. Amps.	Model	Field Relay Air Gap In.	Field Relay Point Gap In.	Field Relay Closing Voltage	Voltage Regulator Air Gap In.	Voltage Regulator Point Gap In.	Voltage Regulator Voltage @ 125° F.
1969–70	1100699	42	2.2–2.6	28	40	1119515	.015	.030	1.5–3.2	.067	.014	13.5–14.4
	1100700	55	2.2–2.6	32	50	1119515	.015	.030	1.5–3.2	.067	.014	13.5–14.4
	1100702	60	4.0–4.5	36	58	1116368	—	—	—	—	—	13.4–14.1
	1100703	60	4.0–4.5	36	58	1116368	—	—	—	—	—	13.4–14.1
	1100704	37	2.2–2.6	25	35	1119515	.015	.030	1.5–3.2	.067	.014	13.5–14.4
	1100736	37	2.2–2.6	25	35	1119515	.015	.030	1.5–3.2	.067	.014	13.5–14.4
	1100737	55	2.2–2.6	32	50	1119511	—	—	—	.067	.015	13.5–14.4
	1100738	55	2.2–2.6	32	50	1119511	—	—	—	.067	.015	13.5–14.4
	1100739	42	2.2–2.6	28	40	1119511	—	—	—	.067	.015	13.5–14.4
	1100740	60	4.0–4.5	36	58	1116370	—	—	—	—	—	13.7–14.3
	1100745	55	2.2–2.6	32	50	1119515	.015	.030	1.5–3.2	.067	.014	13.5–14.4
	1100747	55	2.2–2.6	32	50	1116370	—	—	—	—	—	13.7–14.3
	1100758	66	4.1–4.5	38	62	1116370	—	—	—	—	—	13.7–14.3
	1100760	55	2.2–2.6	32	50	1119515	.015	.030	1.5–3.2	.067	.014	13.5–14.4
	1100761	37	2.2–2.6	25	35	1119515	.015	.030	1.5–3.2	.067	.014	13.5–14.4
	1100762	37	2.2–2.6	25	35	1119515	.015	.030	1.5–3.2	.067	.014	13.5–14.4
	1100763	66	4.1–4.5	38	62	1116370	—	—	—	—	—	13.7–14.3
	1100800	55	2.2–2.6	—	—	—	—	—	—	—	—	—
	1100801	42	—	—	—	—	—	—	—	—	—	—
	1100830	55	2.2–2.6	32	50	1119515	.015	.030	1.5–3.2	.067	.014	13.5–14.4
	1100832	37	2.2–2.6	25	35	1119515	.015	.030	1.5–3.2	.067	.014	13.5–14.4
	1100905	37	2.2–2.6	25	35	1119515	.015	.030	1.5–3.2	.067	.014	13.5–14.4
1971	1100550	37	4.0–4.5	—	32	—	—	—	—	—	—	—
	1100920	55	4.0–4.5	—	50	—	—	—	—	—	—	—
	1100927	37	4.0–4.5	—	32	—	—	—	—	—	—	—
	1100928	55	4.0–4.5	—	50	—	—	—	—	—	—	—
	1101015	80	4.0–4.5	—	74	—	—	—	—	—	—	—
	1100566	37	2.2–2.6	25	35	1119515	.015	.030	1.5–3.2	.067	.014	13.8–14.8
	1100836	37	2.2–2.6	25	35	1119515	.015	.030	1.5–3.2	.067	.014	13.8–14.8
	1100843	61	2.2–2.6	33	58	1119515	.015	.030	1.5–3.2	.067	.014	13.8–14.8
1972	1100927	37	4.0–4.5	—	32	—	—	—	—	—	—	—
	1100928	55	4.0–4.5	—	50	—	—	—	—	—	—	—
	1101015	80	4.0–4.5	—	74	—	—	—	—	—	—	—
	1100566	37	—	—	35	1119515	.015	.030	1.5–3.2	.067	.014	13.8–14.8
	1102440	37	—	—	32	1119515	.015	.030	1.5–3.2	.067	.014	13.8–14.8
	1102463	61	—	—	55	1119515	.015	.030	1.5–3.2	.067	.014	13.8–14.8
1973–74	1100497	37	4.0–4.9	—	36①	Integral	—	—	—	—	—	—
	1100927	37	4.0–4.5	—	32	Integral	—	—	—	—	—	—
	1100928	55	4.0–4.5	—	50	Integral	—	—	—	—	—	—
	1101015	80	4.0–4.5	—	74	Integral	—	—	—	—	—	—
1975	1100497	37	4.0–4.5	—	33	Integral	—	—	—	—	—	—
	1101016	80	4.0–4.9	—	76	Integral	—	—	—	—	—	—
	1101027	80	4.0–4.9	—	76	Integral	—	—	—	—	—	—
	1102347	61	4.0–4.5	—	57	Integral	—	—	—	—	—	—
	1102384	37	4.0–4.5	—	33	Integral	—	—	—	—	—	—
	1102385	55	4.0–4.5	—	51	Integral	—	—	—	—	—	—
	1102388	37	4.0–4.5	—	33	Integral	—	—	—	—	—	—
	1102390	55	4.0–4.5	—	51	Integral	—	—	—	—	—	—
	1102394	37	4.0–4.5	—	33	Integral	—	—	—	—	—	—
	1102399	37	4.0–4.5	51	53	Integral	—	—	—	—	—	—
	1102457	55	4.0–4.5	—	51	Integral	—	—	—	—	—	—
	1102481	37	4.0–4.5	—	33	Integral	—	—	—	—	—	—

Continued

ALTERNATOR & REGULATOR SPECIFICATIONS—Continued

Year	Alternator					Regulator						
	Model	Rated Hot Output Amps.	Field Current 12 Volts @ 80° F.	Output @ 14 Volts		Model	Field Relay			Voltage Regulator		
				2000 R.P.M. Amps.	5000 R.P.M. Amps.		Air Gap In.	Point Gap In.	Closing Voltage	Air Gap In.	Point Gap In.	Voltage @ 125° F.
1975	1102482	55	4.0–4.5	—	51	Integral	—	—	—	—	—	—
	1102483	37	4.0–4.5	—	33	Integral	—	—	—	—	—	—
	1102486	61	4.0–4.5	—	57	Integral	—	—	—	—	—	—
	1102488	55	4.0–4.5	—	51	Integral	—	—	—	—	—	—
	1102491	37	4.0–4.5	—	33	Integral	—	—	—	—	—	—
	1102840	55	4.0–4.5	—	51	Integral	—	—	—	—	—	—
1976	1101016	80	4–4.9	—	76	Integral	—	—	—	—	—	—
	1102384	37	4–4.5	—	33	Integral	—	—	—	—	—	—
	1102385	55	4–4.5	—	51	Integral	—	—	—	—	—	—
	1102388	37	4–4.5	—	33	Integral	—	—	—	—	—	—
	1102390	55	4–4.5	—	51	Integral	—	—	—	—	—	—
	1102394	37	4–4.5	—	33	Integral	—	—	—	—	—	—
	1102486	61	4–4.5	—	57	Integral	—	—	—	—	—	—
	1102491	37	4–4.5	—	33	Integral	—	—	—	—	—	—
	1102840	55	4–4.5	—	51	Integral	—	—	—	—	—	—

①—At 7000 R.P.M.

ENGINE TIGHTENING SPECIFICATIONS★

★Torque specifications are for clean and lightly lubricated threads only. Dry or dirty threads produce increased friction which prevents accurate measurement of tightness.

Year	Model	Spark Plugs Ft. Lbs.	Cylinder Head Bolts Ft. Lbs.	Intake Manifold Ft. Lbs.	Exhaust Manifold Ft. Lbs.	Rocker Arm Ft. Lbs.	Rocker Arm Cover Ft. Lbs.	Connecting Rod Cap Bolts Ft. Lbs.	Main Bearing Cap Bolts Ft. Lbs.	Flywheel to Crank-shaft Ft. Lbs.	Vibration Damper or Pulley Ft. Lbs.
1969	6 Cyl.	20	95	30	30	—	15	33	100	60	160
	V8's	20	95	40	30	20	5	43	100③	95	160
1970–72	6 Cyl.④	15	95	②	25	—	55⑤	35	65	60	—
	V8-307⑦	15	65	30	20⑥	—	45⑤	45	75	60	60
	V8's	25	95	40	30	20	8	43	100③	95	160
1973–74	6 Cyl.④	15	95	②	25	—	55⑤	35	65	60	—
	V8's	15	95	40	30	20	8	①	100③	95	160
1975	6 Cyl.④	15	95	35	30	—	55⑤	35	65	60	60
	V8-260⑧	25	85	40	15–20	25	7	42	80③	⑩	200–310
	V8-350⑨	15	80	45	28	30⑪	4	40	115	60	140 Min.
	V8's	15	95	40	30	20	8	43	100③	95	160

①—Exc. V8-455 Super Duty 43 ft. lbs. On V8-455 Super Duty torque in 10 ft. lbs. steps until 60 lbs. measure for .006″–.008″ bolt elongation from original length, if less then .006″ proceed with 5 ft. lb. steps as necessary to obtain proper elongation.
②—Outer 20, all others 30.
③—Rear 120.
④—For service on this engine, see Six Cylinder in Chevrolet Chapter.
⑤—Inch pounds.
⑥—Inside bolts 30 ft. lbs.
⑦—For service on this engine, see Chevrolet Chapter.
⑧—For service on this engine, see Oldsmobile Chapter.
⑨—Ventura only. For service on this engine, see Buick Chapter.
⑩—Exc. manual trans. 60 ft. lbs.; manual trans. 90 ft. lbs.
⑪—Rocker arm shaft to cylinder head.

REAR AXLE SPECIFICATIONS

Year	Model	Carrier Type	Ring Gear & Pinion Backlash		Pinion Bearing Preload			Differential Bearing Preload		
			Method	Adjustment	Method	New Bearings Inch-Lbs.	Used Bearings Inch-Lbs.	Method	New Bearings Inch-Lbs.	Used Bearings Inch-Lbs.
1969–72	④	Integral	Shims	.005–.009	③	20–30	14–20	Shims	②	②
1970–72	①	Integral①	Shims	.005–.008	Spacer	20–25	10–15	Shims	②	②
1971–72	Ventura II	Integral①	Shims	.005–.008	Spacer	25	10	Shims	②	②
1973	All	Integral	Shims	.005–.009	③	20–25	10–15	Shims	②	②
1974–75	All	Integral	Shims	.005–.009	③	24–32	8–12	Shims	②	②

①—Type C differential (use "C" washers to retain axle shafts).
②—Slip fit plus .008″ tight.
③—Tighten pinion shaft nut with inch-pound torque wrench.
④—Exc. type "C" differential.

WHEEL ALIGNMENT SPECIFICATIONS

OLD CAR SPECIFICATIONS: For 1946-68 Wheel Alignment Specifications see main index.

Year	Model	Caster Angle, Degrees		Camber Angle, Degrees				Toe-In, Inch	Toe-Out on Turns, Deg.	
		Limits	Desired	Limits Left	Limits Right	Desired Left	Desired Right		Outer Wheel	Inner Wheel
1969–70	Firebird (1969)	0 to +1	+½	0 to +½	0 to +½	+¼	+¼	⅛–¼	18	20
	Firebird (1970)	−½ to +½	Zero	+½ to +1½	+½ to +1½	+1	+1	⅛–3/16	18	20
	Intermediates	−1 to −2	−1½①	0 to +½	0 to +½	+¼	+¼	0–⅛	18	20
	Full Size Models	−1 to −2	−1½	0 to +½	0 to +½	+¼	+¼	0–⅛	18	20
1971-72	Firebird	−½ to +½	Zero	+½ to +1½	+½ to +1½	+1	+1	⅛–¼	18	20
	Ventura II	0 to +1	+½	−¼ to +¾	−¼ to +¾	+½	+½	⅛ to ¼	—	20
	②	−1 to −2	−1½	−½ to +½	−½ to +½	Zero	Zero	1/16–3/16	18	20
	③	+½ to +1½	+1	+¼ to +1¼	+¼ to +1¼	+¾	+¾	⅛–¼	18	20
1973–74	Firebird	−½ to +½	Zero	+½ to +1½	+½ to +1½	+1	+1	⅛–¼	—	—
	Ventura	0 to +1	+½	−¼ to +¾	−¼ to +¾	+¼	+¼	⅛–¼	—	—
	Grand Prix	+2½ to +3½	+3	+½ to +1½	0 to +1	+1	+½	0–⅛	—	—
	Intermediates	④	⑤	+½ to +1½	0 to +1	+1	+½	0–⅛	—	—
	③	+½ to +1½	+1	+½ to +1½	0 to +1	+1	+½	0–⅛	—	—
1975	Ventura⑥	−½ to −1½	−1	+¼ to +1¼	+¼ to +1¼	+¾	+¾	0–⅛	—	—
	Ventura⑦	+½ to +1½	+1	+¼ to +1¼	+¼ to +1¼	+¾	+¾	0–⅛	—	—
	Firebird	−½ to +½	Zero	+½ to +1½	+½ to +1½	+1	+1	0–⅛	—	—
	Intermediate⑥	+½ to +1½	+1	+½ to +1½	0 to +1	+1	+½	0–⅛	—	—
	Intermediate⑦	+1½ to +2½	+2	+½ to +1½	0 to +1	+1	+½	0–⅛	—	—
	Grand Prix	+2½ to +3½	+3	+½ to +1½	0 to +1	+1	+½	0–⅛	—	—
	Pontiac	+½ to +2½	+1½	+½ to +1½	0 to +1	+1	+½	0–⅛	—	—
1976	Ventura⑥	−½ to −1½	−1	+¼ to +1¼	+¼ to +1¼	+¾	+¾	0–⅛	—	—
	Ventura⑦	+½ to +1½	+1	+¼ to +1¼	+¼ to +1¼	+¾	+¾	0–⅛	—	—
	Firebird	−½ to +½	Zero	+½ to +1½	+½ to +1½	+1	+1	0–⅛	—	—
	Intermediate⑥	+½ to +1½	+1	+½ to +1½	0 to +1	+1	+½	0–⅛	—	—
	Intermediate⑦	+1½ to +2½	+2	+½ to +1½	0 to +1	+1	+½	0–⅛	—	—
	Grand Prix	+2½ to +3½	+3	+½ to +1½	0 to +1	+1	+½	0–⅛	—	—
	Pontiac	+1 to +2	+1½	+½ to +1½	0 to +1	+1	+½	0–⅛	—	—

①—Station Wagons, −2°.
②—Intermediates and Grand Prix.
③—Catalina, Grandville and Bonneville.
④—Manual steering −1½ to −½, Power steering −½ to +½.
⑤—Manual steering −1°, Power steering 0°.
⑥—Manual Steering.
⑦—Power steering.

COOLING SYSTEM & CAPACITY DATA

Year	Model or Engine	Cooling Capacity, Qts.		Radiator Cap Relief Pressure, Lbs.		Thermo. Opening Temp. (1)	Fuel Tank Gals.	Engine Oil Refill Qts. (2)	Transmission Oil				Rear Axle Oil Pints
		With Heater	With A/C	With A/C	No A/C				3 Speed Pints	4 Speed Pints	5 Speed Pints	Auto. Trans. Qts. (12)	
1969	6-250	12	12¼	14–17	14–17	190	21½(15)	4½(16)	3½	3½	—	(17)	3
	8-350 Tempest	20	21¼	14–17	14–17	190	21½(15)	5	3½	3½	—	(17)	3
	8-400 Tempest	18¼	19¾	14–17	14–17	190	21½(15)	5	3½	3½	—	(17)	3
	8-350 Firebird	19½	20¼	14–17	14–17	190	18½	5	3½	3½	—	(17)	3
	8-400 Firebird	18½	18¾	14–17	14–17	190	18½	5	3½	3½	—	(17)	3
	8-400 Pontiac	18	18	14–17	14–17	190	26½(7)	5	2.8	—	—	(17)	4½
	8-428 Pontiac	17¼	17¼	14–17	14–17	190	26½(7)	5	2.8	—	—	(17)	4½
	8-400 Grand Prix	18¾	21	14–17	14–17	190	21½	5	2.8	2.5	—	(17)	3
	8-428 Grand Prix	17½	17½	14–17	14–17	190	21½	5	2.8	2.5	—	(17)	3
1970	6-250	11.3	13	14–17	14–17	195	21.5	4	3.5	2½	—	3¼(19)	3
	V8-350 Tempest	19.6	19.6	14–17	14–17	190	21.5	5	(18)	2½	—	3¼(19)	3
	V8-350 Pontiac	19.6	19.6	14–17	14–17	190	26(7)	5	(18)	—	—	3¼(19)	4½
	V8-400 G.T.O.	18.3	18.3	14–17	14–17	190	21.5	5	2.8	2½	—	3¼(19)	3(20)
	V8-350 Firebird	19½	20¼	14–17	14–17	190	19½(3)	5	(4)	2½	—	(5)	3¾
	V8-400 Firebird	18½	18¾	14–17	14–17	190	19½(3)	5	(4)	2½	—	(5)	3¾
	V8-400 Pontiac	18	18	14–17	14–17	190	26(7)	5	2.8	—	—	3¼(19)	4½
	V8-400 Gr'd Prix	18.7	21.1	14–17	14–17	190	21.5	5	2.8	2½	—	3¼(19)	3(20)
	V8-455 Gr'd Prix	17.5	19.9	14–17	14–17	190	21.5	5	2.8	2½	—	3¼(19)	3(20)
	V8-455 Pontiac	17.2	17.2	14–17	14–17	190	26(7)	5	2.8	—	—	3¼(19)	4½
1971	6-250 Firebird	12	—	14–17	14–17	195	17	4	3.5	3.5	—	3(19)	4¼
	6-250 Tempest	13	12.4	14–17	14–17	195	19(6)	4	3.5	—	—	3(19)	3
	6-250 Ventura II	12.4	—	14–17	14–17	195	16	4	3.5	—	—	3(19)	3¾
	8-307 Ventura II	15.5	16.5	14–17	14–17	195	16	4	3.5	—	—	3(19)	3¾
	8-350 Firebird	19.4	20.3	14–17	14–17	195	19	5	3.5	3.5	—	3(19)	4¼
	8-350 Tempest	20.2	20.9	14–17	14–17	195	19(6)	5	3.5(21)	2.5	—	3(19)	3
	8-350 Pontiac	20.2	21	14–17	14–17	195	23.6(22)	5	2.8	—	—	3(19)	3
	8-400 Firebird	18.6	18.7	14–17	14–17	195	17	5	2.8	2.5	—	3¾(19)	4¼
	8-400 Tempest	18.6	20.8	14–17	14–17	195	19(6)	5	2.8	2.5	—	3¾(19)	3
	8-400 Pontiac	18.6	19.6	14–17	14–17	195	23.6(22)	5	2.8	2.5	—	3¾(19)	3
	8-400 Gr'd Prix	18.7	19.7	14–17	14–17	195	23.5	5	2.8	2.5	—	3¾(19)	3(20)
	8-455 Firebird	17.9	18.7	14–17	14–17	195	17	5	2.8	2.5	—	3¾(19)	4¼
	8-455 Tempest	18.6	20.8	14–17	14–17	195	19(6)	5	2.8	2.5	—	3¾(19)	3
	8-455 Pontiac	17.9	19	14–17	14–17	195	23.6(22)	5	—	2.5	—	3¾(19)	3
	8-455 Gr'd Prix	18.7	19.7	14–17	14–17	195	23.5	5	—	—	—	3¾(19)	3(20)
1972	6-250 Ventura II	12	16	14–17	14–17	195	16	4	3½	—	—	(5)	3¾
	6-250 Firebird	12	—	14–17	14–17	195	17	4	3½	—	—	(5)	4¼
	6-250 Le Mans	13	12.4	14–17	14–17	195	20(23)	4	3½	—	—	(5)	5(20)
	8-307 Ventura II	15	16	14–17	14–17	195	16	4	3	—	—	(5)	3¾
	8-350 Ventura II	19.4	20.3	14–17	14–17	195	16	5	—	—	—	(5)	3¾
	8-350 Firebird	19.4	20.3	14–17	14–17	195	17	5	2.8	2.5	—	6	4¼
	8-350 LeMans	20.2	20.9	14–17	14–17	195	20(23)	5	3½	2.5	—	6	3(20)
	8-400 Pontiac	18.6	19.6	14–17	14–17	195	25(23)	5	—	—	—	7½	5½
	8-400 Firebird	18.6	18.7	14–17	14–17	195	17	5	—	2.5	—	7½	4¼
	8-400 LeMans	18.6	20.8	14–17	14–17	196	20(23)	5	2.8	2.5	—	7½	3(20)
	8-400 Gr'd Prix	18.7	19.7	14–17	14–17	195	26	5	—	—	—	7½	3(20)
	8-455 Pontiac	17.9	19.0	14–17	14–17	195	25(23)	5	—	—	—	7½	5½
	8-455 Firebird	17.9	18.9	14–17	14–17	195	17	5	—	2.5	—	7½	4¼
	8-455 LeMans	17.9	18.9	14–17	14–17	195	20(23)	5	—	2.5	—	7½	3(20)
1973	6-250 Ventura II	12	—	14–17	14–17	195	21.5	4	3½	—	—	1½(8)	3¾
	6-250 Firebird	12.4	—	14–17	14–17	195	18	4	3½	—	—	4(8)	4¼
	6-250 Le Mans	12.4	13.4	14–17	14–17	195	22	4	3½	—	—	4(8)	3(28)
	8-350 Ventura II	19.5	20.5	14–17	14–17	195	21.5	4	3½	3½	—	4(27)	3¾
	8-350 Firebird	19.5	20.5	14–17	14–17	195	18	5	3½	3½	—	4(27)	4¼
	8-350 Le Mans	20.2	21.4	14–17	14–17	195	22	5	3½(26)	3½	—	4(27)	3(28)

Continued

COOLING SYSTEM & CAPACITY DATA—Continued

Year	Model or Engine	Cooling Capacity, Qts.		Radiator Cap Relief Pressure, Lbs.		Thermo. Opening Temp. ①	Fuel Tank Gals.	Engine Oil Refill Qts. ②	Transmission Oil				Rear Axle Oil Pints
		With Heater	With A/C	With A/C	No A/C				3 Speed Pints	4 Speed Pints	5 Speed Pints	Auto. Trans. Qts. ⑫	
1973	8-400 Pontiac	18.6	19.4	14–17	14–17	195	26⑭	5	—	—	—	3¾⑧	5½㉙
	8-400 Firebird	18.6	19.2	14–17	14–17	195	18	5	—	2½	—	3¾⑧	4¼
	8-400 Le Mans	18.6	19.8	14–17	14–17	195	22㉔	5	2¾	2½	—	3¾⑧	3㉘
	8-400 Grand Prix	18.6	19.2	14–17	14–17	195	25	5	—	—	—	3¾⑧	3㉘
	8-455 Pontiac	18.0	18.4	14–17	14–17	195	26⑭	5	—	—	—	3¾⑧	5½㉙
	8-455 Firebird	18.0	19.0	14–17	14–17	195	18	5	—	2½	—	3¾⑧	4¼
	8-455 Le Mans	18.0	19.0	14–17	14–17	195	22㉔	5	—	2½	—	3¾⑧	3㉘
1974	6-250 Ventura	13.1	—	14–17	14–17	195	21.5	4	3½	—	—	4½㉖	3¾
	6-250 Firebird	13.5	—	14–17	14–17	195	21.5	4	3½	—	—	4½㉖	4¼
	6-250 Le Mans	14.7	—	14–17	14–17	195	22	4	3½	—	—	4½㉖	3㉘
	8-350 Ventura	20.0	20.0	14–17	14–17	195	21.5	5	3½	2½	—	4½㉖	3¾
	8-350 Firebird	22.0	22.9	14–17	14–17	195	21.5	5	3½	2½	—	4½㉖	4¼
	8-350 Le Mans	21.3	23.6	14–17	14–17	195	22	5	3½㉕	2½	—	4½㉖	3㉘
	8-350 Pontiac	21.6	22.4	14–17	14–17	195	25.8	5	—	—	—	4½⑪	4¼㉙
	8-400 Firebird	21.9	22.9	14–17	14–17	195	21.5	5	—	2½	—	4½⑪	4¼
	8-400 Le Mans	21.3	22.8	14–17	14–17	195	22㉔	5	—	2½	—	4½⑪	3㉘
	8-400 Grand Prix	21.6	24.0	14–17	14–17	195	25	5	—	—	—	4½⑪	3㉚
	8-400 Pontiac	21.6	22.4	14–17	14–17	195	25.8	5	—	—	—	4½⑪	4¼㉙
	8-455 Firebird	20.3	21.3	14–17	14–17	195	21.5	5	—	2½	—	4½⑪	4¼
	8-455 Le Mans	21.1	21.6	14–17	14–17	195	22㉔	5	—	—	—	4½⑪	3㉘
	8-455 Grand Prix	20.2	22.2	14–17	14–17	195	25	5	—	—	—	4½⑪	3㉚
	8-455 Pontiac	19.8	22.3	14–17	14–17	195	25.8	5	—	—	—	4½⑪	4¼㉙
1975	6-250 Ventura	13.1	—	14–17	14–17	195	20.5	4	3½	—	—	2½㉗	3¾
	6-250 Firebird	13.1	—	14–17	14–17	195	21.5	4	3½	—	—	4⑧	4¼
	6-250 LeMans	14.7	—	14–17	14–17	195	21	4	3½	—	—	4⑧	3㉘
	V8-260 Ventura	22.4	22.9	14–17	14–17	195	20.5	4	3½	2½	—	2½㉗	3¾
	V8-350 Ventura	20	20	14–17	14–17	195	20.5	4	—	2½	—	2½㉗	3¾
	V8-350 Firebird	22	23.3	14–17	14–17	195	21.5	5	3½	2½	—	4㉗	4¼
	V8-350 LeMans	21.3	23.6	14–17	14–17	195	21	5	—	—	—	4⑧	3㉘
	V8-400 Firebird	21.3	⑨	14–17	14–17	195	21.5	5	—	2½	—	4㉗	4¼
	V8-400 LeMans	21.3	⑩	14–17	14–17	195	21⑬	5	—	—	—	3¾⑧	3㉘
	V8-400 Grand Prix	21.5	24	14–17	14–17	195	25	5	—	—	—	3¾⑧	3㉘
	V8-400 Pontiac	21.6	22.4	14–17	14–17	195	25.8	5	—	—	—	3¾⑧	5½
	V8-455 LeMans	19.9	21.6	14–17	14–17	195	21⑬	5	—	—	—	3¾⑧	3㉘
	V8-455 Grand Prix	20.2	22.5	14–17	14–17	195	25	5	—	—	—	3¾⑧	3㉘
	V8-455 Pontiac	19.8	22.3	14–17	14–17	195	25.8⑭	5	—	—	—	3¾⑧	5½
1976	6-250 Ventura	13.0	—	14–17	14–17	195	20.5	4	3.5	—	—	2½㉗	3¾
	6-250 Firebird	13.5	—	14–17	14–17	195	21.5	4	3.5	—	—	4⑧	4¼
	6-250 LeMans	15.0	—	14–17	14–17	195	21	4	3.5	—	—	4⑧	3㉘
	V8-260 Ventura	20.6	21.3	14–17	14–17	195	20.5	4	3.5	—	3.0	2½㉗	3¾
	V8-260 LeMans	23.5	26.0	14–17	14–17	195	21	4	3.5	—	3.0	4⑧	3㉘
	V8-350 Ventura	17.3	18.0	14–17	14–17	195	20.5	4	—	—	—	2½㉗	3¾
	V8-350 Firebird	21.2	21.6	14–17	14–17	195	21.5	5	—	—	—	4㉗	4¼
	V8-350 LeMans	21.4	22.0	14–17	14–17	195	21	5	—	—	—	4⑧	3㉘
	V8-350 Grand Prix	21.6	22.0	14–17	14–17	195	25	5	—	—	—	3¾⑧	3㉘
	V8-400 Firebird	21.6	23.3	14–17	14–17	195	21.5	5	—	2.5	—	4㉗	4¼
	V8-400 Le Mans	21.4	22.0	14–17	14–17	195	21⑭	5	—	—	—	3¾⑧	3㉘
	V8-400 Grand Prix	22.2	22.2	14–17	14–17	195	25	5	—	—	—	3¾⑧	3㉘
	V8-400 Pontiac	21.6	22.4	14–17	14–17	195	25.8⑭	5	—	—	—	3¾⑧	5½

COOLING SYSTEM & CAPACITY DATA—Continued

Year	Model or Engine	Cooling Capacity, Qts.		Radiator Cap Relief Pressure, Lbs.		Thermo. Opening Temp. ①	Fuel Tank Gals.	Engine Oil Refill Qts. ②	Transmission Oil				Rear Axle Oil Pints
		With Heater	With A/C	With A/C	No A/C				3 Speed Pints	4 Speed Pints	5 Speed Pints	Auto. Trans. Qts. ⑫	
1976	V8-455 Firebird	23.3	23.3	14–17	14–17	195	21.5	5	—	2.5	—	4㉗	4¼
	V8-455 LeMans	21.6	21.6	14–17	14–17	195	21⑭	5	—	—	—	3¾⑧	3㉘
	V8-455 Grand Prix	22.2	22.2	14–17	14–17	195	25	5	—	—	—	3¾⑧	3㉘
	V8-455 Pontiac	22.1	22.1	14–17	14–17	195	25.8⑭	5	—	—	—	3¾⑧	5½

①—With alcohol-type anti-freeze use a 160° unit.
②—Add one quart with filter change.
③—With Evaporative Control System 17.
④—Saginaw 3.5, Muncie 4.
⑤—Two speed unit; oil pan 1½ qts., complete refill 9½ qts. Turbo Hydra Matic 350; oil pan 2½ qts., complete refill 5 qts. Turbo Hydra Matic 400; oil pan 3¾ qts., complete refill 9½ qts.
⑥—Station Wagons 21.5.
⑦—Station Wagons 24 gals.
⑧—Oil pan only. After overhaul 9½ qts.
⑨—2 BBl. Carb. 22.5 qts., 4 BBl. Carb. 23.3 qts.
⑩—2 BBl. Carb. 21.9 qts., 4 BBl. Carb. 23.6 qts.
⑪—Oil pan only. After overhaul 12 qts.
⑫—Approximate. Make final check with dip-stick.
⑬—Grand Am 25 gals., Sta. Wagon 22 gals.
⑭—Station Wagons 22 gals.
⑮—Station Wagons 20 gals.
⑯—Add ½ qt. with filter change.
⑰—Two speed unit; oil pan 2½ qts. and complete refill 7½ qts. Three speed unit ("J" Prefix); oil pan 3 qts. and complete refill 10 qts. Three speed unit ("P" Prefix); oil pan 3.7 qts. and complete refill 9½ qts.
⑱—Standard unit 3.5 pints, heavy duty 2.8 pints.
⑲—Oil pan only.
⑳—5 pts. with 8¾" ring gear.
㉑—Heavy duty 2.8.
㉒—Station Wagons 22.5.
㉓—Station Wagon 23.
㉔—Grand Am 25 gals.
㉕—Muncie 2¾ pints.
㉖—Oil pan only. After overhaul 11 qts.
㉗—Oil pan only. After overhaul 10½ qts.
㉘—"C" Type 4.9 pints.
㉙—"C" Type 4¼ pints.
㉚—"C" Type 3.9 pints.
㉛—Station Wagon 5.31 pts.
㉜—8½" ring gear 4¼ pts.

Electrical Section

DISTRIBUTOR, REPLACE

1. Disconnect distributor-to-coil primary wire.

NOTE: On H.E.I. systems, disconnect feed and module connectors from distributor cap.

2. Remove distributor cap.
3. Crank engine so rotor is in position to fire No. 1 cylinder and timing mark on vibration damper is indexed with pointer.
4. Remove vacuum line from distributor.
5. Remove distributor clamp.
6. Lift distributor from engine.

Installation

1. Check to see that engine is at firing position for No. 1 cylinder.
2. Install new gasket on block.
3. Install distributor so vacuum unit faces right side of engine and rotor points toward contact in cap for No. 1 cylinder.
4. Install distributor clamp, leaving screw loose enough to allow distributor to be turned for adjustment.
5. Attach vacuum line to distributor.
6. Install wires in distributor cap.
7. Attach distributor primary wire.
8. Adjust point gap, replace cap and set ignition timing.

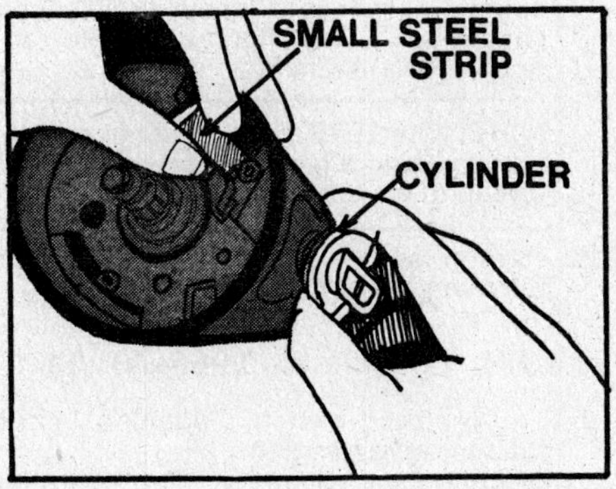

Fig. 1 Ignition lock replace: 1969-76

NOTE: When using a timing light to adjust ignition timing, the connection should be made at the No. 1 spark plug. Forcing foreign objects through the boot at the No. 1 terminal of the distributor cap will damage the boot and could cause engine misfiring.

STARTER, REPLACE

1969-76 Six-Cyl.

1. Disconnect ground cable at battery.
2. Disconnect cable and wiring harness leads from starter solenoid.
3. Unfasten and remove starter.

1969-76 V8s

1. Disconnect cable from battery.
2. Raise front of car and pull battery cable and solenoid wire loom down so they hang free of surrounding parts.
3. Unfasten and remove starter with cable and solenoid wire loom.
4. Remove wires from solenoid and cable from clamp or solenoid bracket.
5. Reverse procedure to install.

IGNITION LOCK REPLACE

1969-76

1. Follow the procedure to remove the turn signal switch as described further on.
2. Remove the lock cylinder in "RUN" position by inserting a thin tool (screwdriver or knife blade), Fig. 1, into the slot next to the switch mounting screw boss (right hand slot) and depress spring latch at bottom of slot, which releases lock. Remove lock by pulling out of housing.

NOTE: If this is the first time the lock cylinder is being removed, the slot will be covered by a thin casting "flash" which is easily broken when inserting thin tool.

IGNITION SWITCH, REPLACE

1969-76

1. Disconnect battery and loosen toe pan screws.
2. On 1971-76 models, remove column to instrument panel trim plates.
3. Lower steering column from instrument panel.

NOTE: On some models it may be necessary to remove the upper column mounting bracket if it hinders the servicing of switch.

4. Disconnect switch wires and remove switch.
5. To replace switch move key lock to OFF-LOCK position.
6. Move actuator rod hole in switch to OFF-LOCK position, Fig. 2.
7. Install switch with rod in hole.

LIGHT SWITCH, REPLACE

1969-76

1. Remove battery cable, pull knob to "On" position, depress latch button and remove knob and shaft.
2. Remove retaining nut.
3. Remove wire connector and remove switch.
4. On vacuum-operated headlight models, remove vacuum connector.
5. Reverse procedure to install.

STOP LIGHT SWITCH, REPLACE

1969-76

The stop light switch has a slip fit in the mounting sleeve which permits positive adjustment by pulling the brake pedal up firmly against the stop. The pedal arm forces the switch body to slip in the mounting sleeve bushing to position the switch properly.

1. Disconnect wires from switch and remove switch from bracket.
2. Position switch in bracket and push in to maximum distance. Brake pedal arm moves switch to correct distance on rebound. Check if pedal is in full return position by lifting slightly by hand.
3. Connect switch wires by inserting plug on switch.

NEUTRAL SAFETY SWITCH

1974-76

1. Loosen switch retaining screws and place selector lever in "Reverse" on manual transmission models or "Neutral" on automatic transmission models.

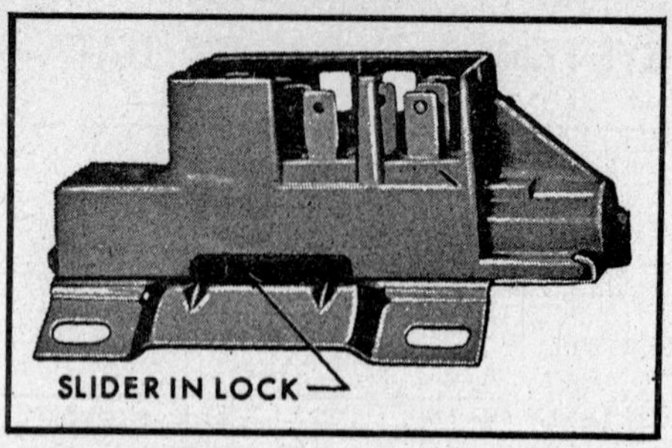

Fig. 2 Ignition switch. 1969-76 (typical)

2. Rotate switch until alignment hole in outer cover aligns with hole in inner contact carrier. Insert a .089 inch gauge pin into alignment hole on manual transmission models. On models with automatic transmission, insert a .096 inch guage pin on 1974 models and .082 inch guage pin on 1975-76 models, into alignment hole on top of switch. On all models, guage pin is inserted approximately ⅜ inch.
3. Tighten switch retaining screws and remove gauge pin.

1971 Column & 1972-73 All

1. Loosen switch retaining screws and place selector lever in "Drive" for 1971 column shift and 1972 console shift models, or "Neutral" for 1972 column shift and all 1973 models.
2. Rotate switch until alignment hole in outer cover aligns with hole in inner contact carrier and insert a .092 inch gauge pin into alignment hole approximately ⅜ inch.

NOTE: On 1973 automatic transmission models, alignment hole is located on top of switch.

3. Tighten switch retaining screws and remove gauge pin.

1971 Console & 1969-70 All

1. To replace switch, position shift tube in drive position.
2. Insert switch drive tang in shifter tube slot and assemble switch to column jacket.
3. To adjust switch, position and hold shift lever in low range while inserting blade of reset gauge J23056-2 into the RESET slot of switch. Be sure the tool is inserted about 9/16".
4. Insert blade of gauge J22701-5 into ADJUST slot of switch and move shift lever to park position and remove gauge.
5. Check starter in all ranges.

CLUTCH START SWITCH

1969-76

All cars equipped with a manual transmission use a clutch start switch which is mounted on the pedal bracket. The switch closes when the clutch is depressed and completes solenoid connection. When installing switch, no adjustment is necessary.

TURN SIGNAL SWITCH

1969-76

NOTE: *On tilt column, the column must first be lowered from panel.*

1. Remove steering wheel using puller.

CAUTION: *Do not hammer on end of shaft as hammering could collapse shaft or loosen plastic injections which maintain column rigidity.*

2. Remove three cover screws and lift cover off shaft.

NOTE: Screw retainers will be lost if screws are removed completely from cover.

3. Depress lock plate and pry round wire lock ring out of shaft groove.

NOTE: On 1976 models, remove lock plate cover with a suitable screwdriver.

4. Slide upper bearing preload spring and turn signal cancelling cam off shaft.
5. Slide thrust washer off shaft and remove turn signal lever.
6. Push hazard warning switch in and unscrew knob.
7. Pull turn signal wiring connector out of bracket on jacket and disconnect.
8. Remove three turn signal switch screws and pull switch straight up.

NOTE: Place tape around upper part of connector and wires to prevent snagging when switch is being removed.

HORN SOUNDER & STEERING WHEEL

1969-76

1. Lift ornament out of wheel hub.
2. Remove nut and washer from shaft.
3. Remove horn bar (deluxe wheel) or extension and switch assembly (standard wheel).
4. Use a suitable puller to remove wheel.
5. Reverse procedure to install, making sure wheel is in straight ahead position.

INSTRUMENT CLUSTER

1975-76 All Exc. GTO & Ventura

1. Disconnect battery ground cable.
2. Remove upper and lower instrument panel trim plates.
3. Remove shift indicator cable.
4. On Firebird, loosen steering column nuts and lower column approximately ½ inch.
5. On all models, remove cluster retaining screws, pull cluster outward and disconnect speedometer cable and printed circuit connector, then remove cluster.

1973 GTO & LeMans

1. Disconnect battery ground cable.
2. Remove ash tray and lower instrument panel trimplate.
3. Remove radio knobs and upper trim panel plate.
4. Lower steering column.
5. Remove cluster screws, disconnect speedometer cable circuit connector.
6. Remove cluster.

1973 Grand Prix & Grand Am

1. Disconnect battery ground cable.
2. Remove upper and lower trim panel plates.
3. Lower steering column.
4. Remove cluster screws, disconnect speedometer cable and remove cluster.

1971-76 Ventura & 1974 GTO

1. Disconnect battery and remove steering column cover trim.
2. Remove three screws retaining heater or A/C control panel to instrument panel carrier.
3. Remove radio control knobs, bezels and nuts.
4. Remove screws at top, bottom and side of carrier securing it to instrument panel pad.
5. Disconnect shift quadrant indicator cable at shaft bowl (if automatic), remove two steering column to panel nuts.
6. Remove toe plate cover and five toe plate to cowl screws, lower column from panel and protect it with shop towels or tape.
7. Remove ground wire screw from under left side of panel pad above kick pad and disconnect speedo cable from under dash.
8. Tilt carrier and cluster rearward, disconnect printed circuit and cluster ground connectors and rest assembly on top of column.
9. Remove screws from cluster to carrier assembly and remove cluster.

1971-73 Catalina, Grand Ville, Bonneville

1. Disconnect battery and remove upper instrument panel trim plate.
2. Remove transmission shift indicator (if automatic) and speedometer cluster bezel.
3. Remove cluster retaining screws, pull rearward, disconnect speedometer cable and printed circuit connector.

1971-72 T-37, LeMans & GTO

1. Disconnect battery and remove lower A/C duct if equipped.
2. Remove lower instrument panel trim and glove box.
3. Lower steering column.
4. Disconnect speedometer cable and heater cable at heater case.
5. Remove three instrument panel screws at gauges.
6. Remove three right upper instrument panel nuts.
7. Remove lower instrument panel bolts at right and left ends and at steering column.
8. Position crash pad outward on column.
9. Disconnect printed circuit.
10. Remove instrument panel harness retaining screws.
11. Remove cluster retaining screws and remove cluster.

1971-72 Grand Prix

1. Disconnect battery and remove lower A/C duct if equipped.
2. Remove pillar post mouldings and filler plate at windshield.
3. Remove lower instrument panel trim at steering column.
4. Disconnect speedometer cable and radio antenna lead.
5. Lower steering column.
6. Remove upper air outlet vents and instrument panel attaching screws at steering column, console, ends and upper center.
7. Remove lower defroster duct screw at heater case.
8. Position instrument panel pad outward on column.
9. Disconnect printed circuit.
10. Remove wire harness retaining screws and cluster ground screw.
11. Remove cluster mounting screws and cluster.

1970-73 Firebird

1. Disconnect battery cable and upper instrument panel trim plate.
2. Remove lower instrument panel trim and bracket at steering column.
3. Loosen two steering column nuts to lower column.
4. Remove cluster screws, pull rearward, disconnect speedometer cable and printed circuit connector.

1969-70 Grand Prix

1. Disconnect battery and remove glove box.
2. Disconnect speedo cable and wire connectors at headlight switch, wipers, turn signal, ignition switch, printed circuit, heater and air conditioner panel.
3. Remove lower column trim and disconnect air control cable at heater case.
4. Remove screws at instrument locations and instrument panel to column support.
5. Remove screws at outboard lower ends of panel.
6. Remove screws retaining right and left side of instrument panel to body. To do this, remove upper vent nozzles by inserting two thin blade screwdrivers on right and left side of nozzle to disengage retaining clips.
7. Remove upper instrument panel cover plate. Studs are pushed into retainers, pry up carefully to remove.
8. Remove two upper panel retaining screws and upper speaker support bracket screw.
9. Remove console to instrument panel screws.
10. Loosen toe plate screws.
11. Remove column to lower panel retaining nuts and lower column.
12. Pull entire instrument panel rearward on left far enough to gain access to cluster.
13. Remove ground strap screws and panel harness retaining screws.
14. Remove cluster retaining screws and remove cluster.

1969-70 Pontiac

1. Disconnect battery.
2. Remove lower instrument panel cover and remove radio.
3. Remove heater or air conditioning control panel.
4. Disconnect speedometer, printed circuit connector, cigar lighter and instrument panel harness from rear of cluster.
5. Remove four cluster retaining screws.
6. Position ground straps so that cluster may be pulled from studs and removed towards center of car.

1969-70 Tempest

1. Disconnect battery and remove glove box.
2. Disconnect speed cable and wire connectors at light switch, wiper, turn signal, ignition switch, printed circuit, heater and air conditioning control panel.
3. Remove lower column trim and disconnect air control cable at heater case.
4. Remove three screws at instrument locations.
5. Remove two screws at instrument panel to column support, one screw at each outboard lower end of panel and three nuts above glove box (inner).
6. Loosen two right hand toe plate screws and two screws clamping right toe plate to left toe plate.
7. Remove column to instrument panel bracket nuts and lower column.
8. Pull entire instrument panel rearward far enough to gain access to cluster.
9. Remove ground strap screws and two panel harness retaining screws.
10. Remove cluster retaining screws and remove cluster.

1969 Firebird

1. Disconnect battery and remove lower instrument panel cover.
2. Remove ash tray bracket screws and radio retaining nuts and glove box.
3. Disconnect heater control cables and wire connectors.
4. Disconnect speedo cable and remove upper L. H. vent duct connector.
5. Disconnect headlamp switch shaft.
6. Remove screws across top and bottom of instrument plate and nut on right side (stud through steel por-

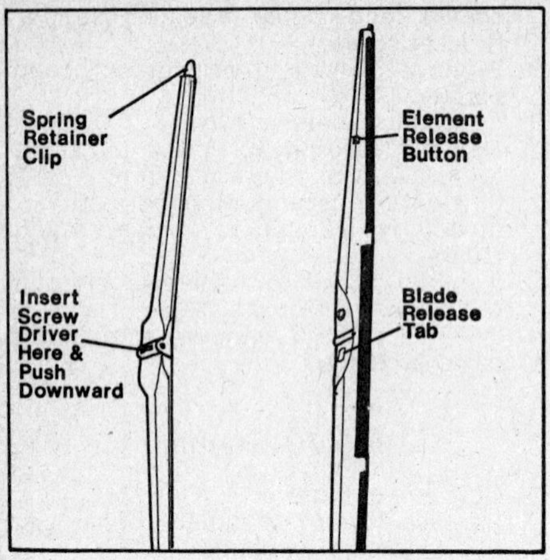

Fig. 3 W/S wiper blades

tion of dash).

7. Drop steering column by loosening toe plate screws and removing lower column support nuts.
8. Protect top of column and pull panel rearward to rest on column.
9. Disconnect printed circuit accessory, wiper and cigar lighter.
10. Remove ground strap, cluster retaining screws and carefully remove cluster.

W/S WIPER BLADES

Two methods are used to retain wiper blades to wiper arms, Fig. 3. One method uses a press type release tab. When the release tab is depressed the blade assembly can be slid off the wiper arm pin. The other method uses a coil spring retainer. A screwdriver must be inserted on top of the spring and the spring pushed downward. The blade assembly can then be slid off the wiper arm pin. Two methods are also used to retain the blade element in the blade assembly, Fig. 3. One method uses a press type release button. When the button is depressed, the two piece blade assembly can be slid off the blade element. The other method uses a spring type retainer clip in the end of the blade element. When the retainer clip is squeezed together, the blade element can be slid out of the blade assembly.

NOTE: To be sure of correct installation, the element release button, or the spring element retaining clip should be at the end of the wiper blade assembly nearest the wiper transmission.

W/S WIPER ARMS

With Rectangular Motor

1. Wiper motor must be in park position.
2. Use suitable tool to minimize the possibility of windshield or paint finish damage during arm removal.
3. Remove arm by prying up with tool to disengage arm from serrated transmission shaft.
4. To install arm to transmission rotate the required distance and direction so that blades rest in proper position.

With Round Motor

1. Wiper motor must be in park position.
2. Raise hood to gain access to wiper arm.
3. **On Intermediate Models:** Lift arm off transmission shaft. On left arm, slide articulating arm lock clip, Fig. 4, away from transmission pivot pin and lift arm off pin. **On Full Size Models:** Lift arm and slide latch clip, Fig. 5, out from under wiper arm.
4. Release wiper arm and lift arm assembly off transmission shaft.

W/S WIPER MOTOR, REPLACE

1971-76

1. Raise hood and remove cowl screen or grille.
2. Disconnect wiring and washer hoses.
3. Reaching through opening, loosen transmission drive link to crankarm attaching nuts.
4. Remove drive link from motor crankarm.
5. Remove three motor attaching screws and remove motor while guiding crankarm through opening.

1970 Firebird

1. Remove two cowl screen attaching screws from center of cowl screen.
2. Pry cowl screen up at eight integral clips and remove.

NOTE: Install screw as an alternate method of retention in hole provided in screen adjacent to clips.

3. Reaching through cowl opening, loosen two transmission crankarm attaching nuts.
4. Remove transmission crankarm.
5. Disconnect wiring and washer hoses.
6. Remove three motor attaching screws and remove motor while guiding crankarm through hole.

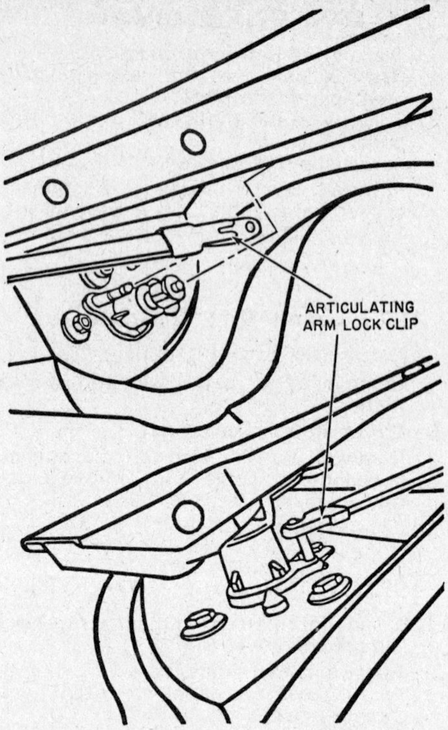

Fig. 4 W/S wiper articulating arm

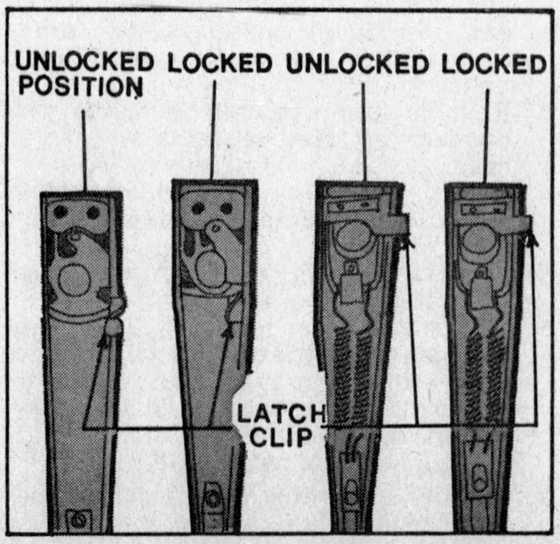

Fig. 5 W/S wiper arms

1969 All; 1970 Exc. Firebird

1. Remove hoses and wires connected to wiper.
2. On Pontiac, remove retainer securing wiper crank to transmission linkage. This can be done by removing plastic plug from left side of upper shroud, above wiper motor.
3. On Tempest and Grand Prix, remove screen and loosen clamp securing wiper crank to linkage.
4. Remove screws securing wiper assembly to dash.
5. On Firebird, carefully pull wiper assembly away from firewall until retainer securing wiper crank to transmission can be removed.
6. Remove wiper motor.

W/S WIPER TRANSMISSION, REPLACE

1971-76

With Rectangular Motor

1. Remove wiper arms and blades.
2. Raise hood and remove cowl vent screen or grille.
3. Disconnect wiring from motor.
4. Loosen, do not remove, transmission drive link to motor crankarm attaching nuts and disconnect drive link from crankarm.
5. Remove right and left transmission to body attaching screws and guide transmission and linkage assembly out through opening.

NOTE: When installing, motor must be in Park position.

With Round Motor

1. Raise hood and remove cowl vent screen.
2. On Intermediates, remove right and left wiper arm and blade assemblies. On Full Size cars, remove the arm and blade only from the transmission to be removed.
3. Loosen, do not remove, attaching nuts securing transmission drive link to motor crankarm.

NOTE: On Full Size cars, if only the left transmission is to be removed, it will not be necessary to loosen nuts securing the right assembly.

4. Disconnect transmission drive link from motor crankarm.
5. On Intermediate models, remove right and left transmission to body attaching screws. On Full Size cars, remove the attaching screw securing the transmission to be removed.
6. Remove transmission and linkage assembly by guiding it through opening.

NOTE: When installing, motor must be in Park position.

1969-70

1. Remove wiper arms and blades.
2. Remove fresh air intake screen or grille.
3. Remove wiper transmission screws.
4. On Pontiac models, remove center support screws.
5. On Depressed Park Optional System models, remove retainer securing linkage that attaches to wiper motor crank. On Pontiac models, this can be done by removing plastic plug from left side of upper shroud, directly above wiper motor.
6. On Tempest models, remove retainer securing right wiper transmission to linkage that attaches to wiper motor crank.
7. Remove transmission and linkage.
8. Reverse procedure to install.

W/S WIPER SWITCH

1973-76 Exc. Ventura & 1974 GTO

1. Disconnect battery ground cable. Remove upper and lower instrument panel trimplates.
2. Remove switch mounting plate and disconnect connector.
3. Remove switch retaining screws and remove switch.

1971-72 Catalina, Grand Ville & Bonneville

1. Disconnect battery and remove upper and lower instrument panel trim plates.
2. Remove speedometer bezel and cluster.
3. Disconnect headlamp, windshield wiper and accessory switch connectors.
4. Remove instrument panel bezel assembly screws, pull rearward and disconnect vent control cable if equipped.
5. Remove wiper switch from backside of bezel assembly.

1971-72 T-37, LeMans, GTO & Grand Prix

1. Disconnect battery and remove lower A/C duct if equipped.
2. Disconnect wire connector.
3. Remove ground strap screw.
4. Unfasten and remove switch.

1971-76 Ventura & 1974 GTO

1. Disconnect battery ground cable.
2. From under dash, disconnect wiring from switch.
3. Remove three screws retaining switch to lower instrument panel and remove switch from panel.

1970-72 Firebird

1. Remove upper and lower instrument panel trim plates.
2. Disconnect wire and unfasten and remove switch.

1969-70 Pontiac, Tempest & Grand Prix

1. Disconnect battery and wire from switch.
2. Remove retaining nuts.
3. Position ground straps and remove switch.

1969 Firebird

1. Disconnect battery.
2. Remove upper vent duct.
3. Remove switch retaining screws and reposition ground straps.
4. Remove switch.

RADIO, REPLACE

NOTE: When installing radio, be sure to adjust antenna trimmer for peak performance.

1973-76 LeMans & 1973 GTO

1. Disconnect battery and remove radio knobs and bezels.
2. Remove upper and lower trimplates.
3. Remove radio retaining screws.
4. Remove radio from opening, disconnect connections and antenna lead-in while radio is being pulled out.
5. If new unit is being installed, remove bushing from old unit and install on replacement unit.

1973-76 Grand Am, Grand Prix & 1975-76 Grand LeMans

1. Disconnect battery ground cable.
2. Remove radio knobs and bezels, then the hex nut from right hand tuning shaft.
3. Remove radio trim plate.
4. Remove radio and radio bracket to instrument panel retaining screws.

NOTE: Radio and bracket must be removed as an assembly.

5. Remove radio and bracket from instrument panel, disconnect wiring harness and antenna lead, then separate bracket from radio.

1971-76 Catalina, Grand Ville & Bonneville

1. Disconnect battery and remove radio knobs and hex nuts.
2. Remove upper and lower instrument panel trim plates and front lower radio bracket.
3. Remove glove box and disconnect all connections to radio.
4. Loosen screw holding radio brace to side of radio and slide radio toward front seat.

1971-72 T-37, LeMans, GTO & Grand Prix

1. Disconnect battery and remove lower A/C duct if equipped.
2. Remove radio control knobs and hex nuts.
3. Remove radio support bracket bolt.
4. Disconnect all leads to radio and remove radio.

1971-76 Ventura & 1974 GTO

1. Disconnect battery ground cable.
2. Remove radio knobs, bezels, nuts and side braces screw and disconnect wiring and antenna.
3. Remove radio from under dash.

1970-76 Firebird

1. Disconnect battery.

2. Remove glove box and door and right lower A/C duct if equipped.
3. Remove radio knobs and hex nuts and trim plate.
4. Disconnect all leads to radio.
5. Remove radio bracket and radio from passenger side of instrument panel.

1969 All; 1970 Exc. Firebird

1. Disconnect antenna and wires from radio.
2. Remove knobs, springs, nuts and bezels from control shafts.
3. Remove screws securing radio to dash or to brace and carefully remove radio.

HEATER CORE REMOVAL

1969-76 Without Air Cond.

All except Ventura & 1974 GTO

1. Drain radiator and disconnect heater hoses from core.
2. Remove retaining nuts from core case studs on engine side of dash.
3. Inside car, remove glove box and door on Firebird and heater outlet from case on Firebird.
4. Remove defroster duct retaining screw from heater case and pull entire heater assembly from firewall.
5. Disconnect control cables and wiring and remove assembly. On 1969-72 Grand Prix, disconnect vacuum hoses.
6. Remove core tube seal and core retaining strips and remove core.

Ventura & 1974 GTO

1. Drain radiator and disconnect heater hoses from core.
2. Disconnect battery ground cable.
3. Remove retaining nuts from core case studs on engine side of dash.
4. On 1974-76 models, remove glove box and door.
5. On all models, drill out lower right hand heater case stud with 1/4 inch drill from inside vehicle.
6. Pull entire heater core and case assembly from firewall.
7. Disconnect cables and wiring and remove assembly from car.
8. Remove core tube seal and core retaining strips and remove core.

1969-76 With Air Cond.

1972-76 Catalina, Grand Ville & Bonneville

1. Drain radiator and disconnect heater hoses from core.
2. Remove three nuts and one screw retaining core and case to dash.
3. Remove glove box and upper and lower instrument panel trim plates.
4. Remove radio.
5. Remove cold air duct and heater outlet duct.
6. Remove defroster duct to heater case screw.
7. Disconnect A/C temperature cable at heater case.
8. Disconnect vacuum hoses from diaphragms on heater case and remove core and case assembly.
9. Remove core from case (three screws).

1973-76 LeMans, Grand Am, Grand Prix & 1973 GTO

1. Drain radiator and disconnect heater hoses from core.
2. Remove glove box, cold air duct and heater outlet and defroster duct screw.
3. Disconnect heater core case from dash. Remove blower motor resistor to gain access to case upper retaining nut.
4. Move case assembly rearward, freeing case studs from cowl and remove assembly.
5. Disconnect temperature cable and vacuum hoses from case and remove screws securing core inside of case.

1969-70 All, 1971 Full Size & 1971-72 LeMans, Grand Prix

1. Drain radiator and disconnect heater hoses from core.
2. Remove lower duct and outlet assembly.
3. Remove glove box.
4. Remove defroster duct attaching screw.
5. Remove screws retaining case to dash.

NOTE: On some 1969 models it may be necessary to drill a hole in the fender skirt to remove lower case attaching nut.

6. Move core and case assembly rearward to free attaching studs from cowl and remove assembly.
7. Disconnect cables and wiring.
8. Mark heater cam and bracket assembly in three places to insure proper installation.
9. Remove heater cam and bracket assembly.
10. Remove front case to rear case attaching screws.
11. Separate front and rear case.
12. Remove screws retaining core attaching bands and remove core.

1971-76 Firebird

1. Drain radiator.
2. Remove glove box and door.
3. Remove cold air duct (lower right hand duct) and remove left and center lower A/C ducts.
4. Jack right front area of car and place on safety stand.
5. Remove rocker panel trim on right side and remove screws holding forward portion of rocker panel trim attaching bracket.
6. Remove three lower fender bolts at rear of fender.
7. Remove four fender to skirt bolts at rear of wheel opening.
8. Remove two fender skirt bolts near blower motor area.
9. Pry rear portion of fender out at bottom to gain access to hose clamp on lower core hose and disconnect hose.
10. Disconnect water pump to core hose at core.
11. Remove heater case retaining nuts under hood at dash.
12. Remove two heater case retaining bolts (inside car).
13. Remove console, if equipped. If equipped with tape player, remove console with tape player intact. If equipped with tape player and no console, remove tape player.
14. Disconnect temperature cable at heater case.
15. Remove heater outlet duct.
16. Remove lower defroster duct screw at heater case.
17. Remove right kick panel.
18. Remove heater core and case.
19. Disconnect vacuum hoses from heater case.
20. Remove core from case.

1971-76 Ventura & 1974 GTO

1. Disconnect battery and drain coolant.
2. Disconnect upper heater hose from core.
3. Remove right front fender skirt bolts and lower skirt to gain access to lower heater hose clamp. Disconnect lower hose and remove lower right hand heater core and case attaching nut.
4. Remove glove box and door.
5. Remove recirculation vacuum diaphragm at right kick panel.
6. Remove heater outlet (at bottom of heater case).
7. Remove cold air distributor duct from heater case.
8. Remove heater case extension screws and separate extension from case.
9. Disconnect cables and wiring and remove case and core assembly.
10. Separate core from case.

BLOWER MOTOR REMOVE

1970-76

1970-76 Full Size & Firebird, 1970-72 LeMans & 1970-72 Grand Prix With A/C

1. Raise car and remove right front wheel.
2. Cut access hole approx. 3/4 of the way along the outline stamped on the right hand fender skirt. Bend cut section of skirt outward for access to blower motor.

NOTE: Some 1970-71 Grand Prix models do not have an outline stamped on the right hand fender skirt. On these models a template must be used.

3. Disconnect blower motor feed wire.

1970-72 Grand Prix Less A/C, 1973 GTO, 1973-76 Grand Prix & LeMans

1. Disconnect blower motor feed wire.
2. Remove blower motor retaining screws.
3. Remove blower motor.

1971-76 Ventura & 1974 GTO

1. Disconnect battery and detach hoses from clips on right fender skirt.
2. Raise vehicle on hoist.

3. Remove fender skirt attaching bolts except those retaining the skirt to radiator support.
4. Pull out then down on skirt and place block of wood between skirt and fender to allow clearance for blower motor removal.
5. Disconnect blower motor cooling tube and electrical connections at blower motor.
6. Remove blower motor attaching screws and remove blower motor. Gently pry motor flange if sealer acts as an adhesive.

1969

Less Air Conditioning

1. On all models except Grand Prix, remove battery and tray and on Firebird, unclip hoses.
2. On all models except Grand Prix, remove fender skirt.
3. On all models, disconnect blower motor feed wire.
4. Remove blower motor retaining screws. Remove blower motor.

With Air Conditioning

1. Unclip hoses from fender skirt.
2. Remove rocker moulding.
3. Loosen lower rear fender retaining screws to allow bottom of fender to move.
4. Remove fender skirt. On Firebird first remove battery and tray then remove fender and skirt as an assembly.
5. Remove blower motor retaining screws. Remove motor feed wire and cooling tube.
6. Remove blower motor.

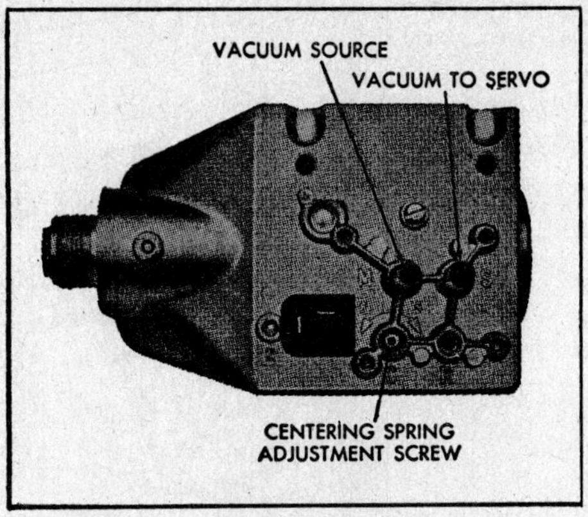

Fig. 6 Centering spring adjustment. 1969-76 Cruise Control

SPEED CONTROLS

1969-76 Cruise Control

Brake Release Switch, Adjust

Apply brake pedal and push both switches forward as far as possible. Pull pedal forcibly rearward to adjust switches.

Chain Linkage, Adjust 1970-76

This adjustment is no longer necessary because of the use of cable-connection to the carburetor.

Chain Linkage, Adjust 1969

1. Start engine and set carburetor to hot idle position.
2. Thread bead chain through hole in carburetor lever extension.
3. Adjust bead chain at extension to provide minimum slack and assemble clip to extension by straddling extension.
4. A minimum of two beads must extend outside of clip after adjustment. Cut off excess chain.

Centering Spring, Adjust

If speed control holds speed three or more mph higher than selected speed, turn centering screw (C) clockwise 1/8 turn or less, Fig. 6.

If speed control holds speed three or more mph below selected speed, turn centering adjustment screw (C) counterclockwise 1/8 turn or less. *Do not move adjustment screw* (R).

Conventional Engine Section

For service on 1975-76 V8-350 Ventura see Buick Chapter

See page 1-727 for the Overhead Camshaft Engine

For service on 6-250 & V8-307 see Chevrolet Chapter

For service on V8-260 see Oldsmobile Chapter

ENGINE MOUNTS REPLACE

1970-76 Six

1. Raise engine to release weight from mount to be replaced.
2. Remove insulator to engine bracket through bolt(s), Figs. 1 and 2.
3. Raise engine approximately 1" above front insulators.
4. Remove front insulator by unscrewing frame bracket to insulator bolt and lifting insulator from bolt.

1971-76 V8

1. Disconnect battery ground cable.
2. Raise engine to release weight off front mounts.
3. Remove bolts fastening engine insulators to engine.
4. Raise engine just clear of insulator.
5. Remove insulator.

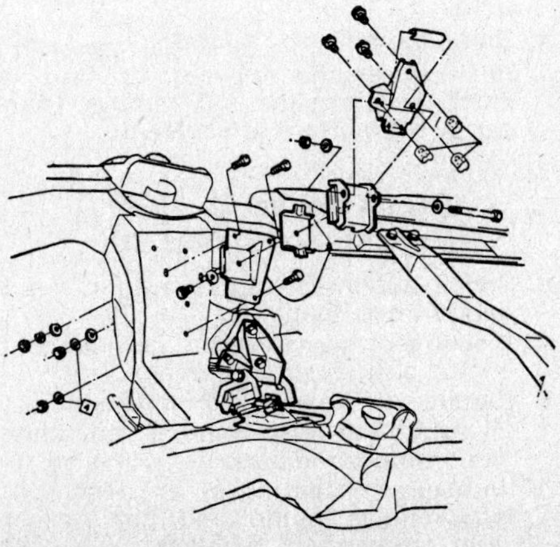

Fig. 1 1970-72 Tempest & Firebird 6 cyl.

1971-76 V8

1. Disconnect battery ground cable.
2. Raise engine to release weight off front mounts.
3. On **Intermediate Models** remove stabilizer bracket to frame bolts. **On Grand Prix Models** remove the idler arm to frame bolts and disconnect pitman arm from shaft.
4. Remove bolts securing insulators to frame and engine.
5. Remove insulators.

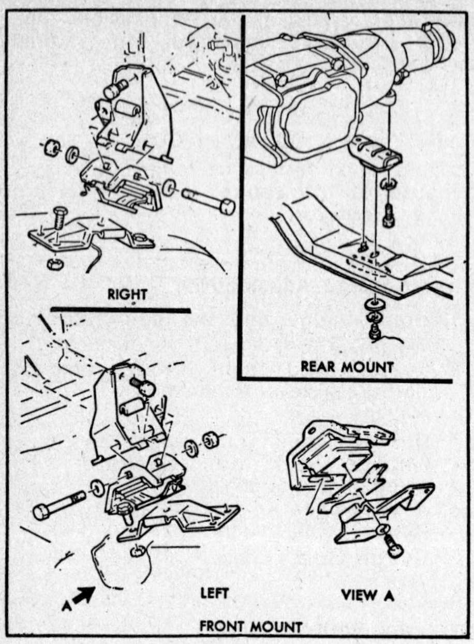

Fig. 2 Engine mounts. 1971-72 Ventura 6 cyl.

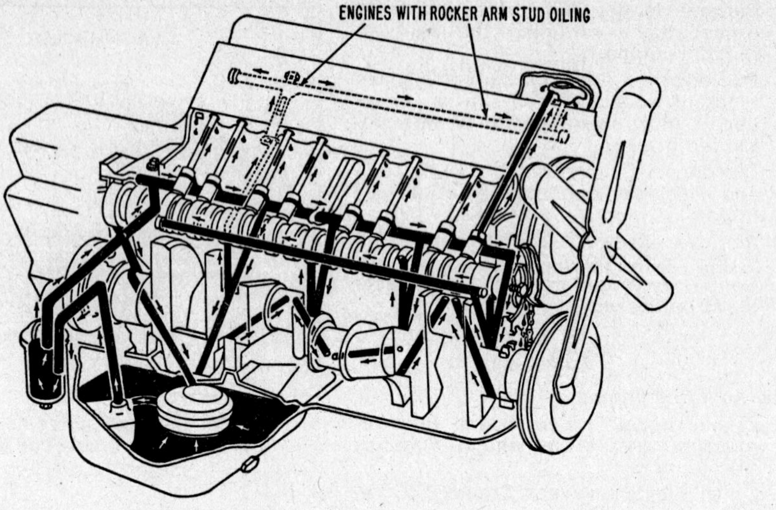

Fig. 3 Engine oiling system. V8s

ENGINE, REPLACE

1970-76

1. Disconnect battery cables at battery and drain cooling system.
2. Scribe alignment marks on hood and remove hood from hinges.
3. Disconnect all wiring, ground straps, fuel lines and vacuum hoses from engine.

NOTE: On 1973-76 V8's, remove thermal feed switch, located on rear of left cylinder head on all models except Ventura. On Ventura models, switch is located on the right cylinder head.

---

4. Remove air cleaner and upper radiator shield assembly.
5. Disconnect radiator hoses and heater hoses at engine.
6. Remove fan and disconnect accelerator linkage.
7. If equipped with power steering or air conditioning, remove pump and/or compressor from mountings and set aside. Do not disconnect hoses.
8. On V8, disconnect transmission vacuum modulator line and power brake vacuum line at carburetor and fold back out of way.
9. Raise vehicle on hoist and drain crankcase.
10. Disconnect exhaust pipe from manifold and remove starter.
11. If equipped with automatic transmission, remove converter cover and three converter retaining bolts and slide converter to rear.
12. With manual transmission, disconnect clutch linkage and remove clutch cross shaft.
13. Remove four lower bell housing bolts.
14. Disconnect transmission filler tube support and starter wire harness shield from cylinder head.
15. Remove two front motor mount to frame bracket bolts.
16. Lower vehicle and using a jack and block of wood, support transmission.
17. Support weight of engine with suitable lifting device.
18. Remove two remaining bell housing bolts.
19. Raise transmission slightly.
20. Position engine forward to free it from transmission and remove from car by tilting front of engine up.

Fig. 4 Slots filed in rocker arm stud

1969 Pontiac & 1969 Tempest & Firebird V8

1. Scribe alignment marks around hood and remove hood.
2. Disconnect engine wiring harness and engine-to-body ground straps.
3. Remove air cleaner and fan shield.
4. If equipped with manual transmission, remove radiator.
5. With power steering or air conditioning, remove pump and compressor from mounting brackets and set aside. Do not disconnect hoses.
6. Remove engine fan and pulley.
7. Disconnect accelerator control linkage and linkage support bracket.
8. If equipped, disconnect automatic transmission vacuum modulator line and power brake vacuum line at carburetor and fold back out of the way.

NOTE: On Firebird with A/C, remove wiper motor.

---

9. Raise vehicle and drain crankcase.
10. Disconnect fuel lines and exhaust pipes.
11. Disconnect starter wires.
12. With automatic transmission, remove converter cover, converter retaining bolts and slide converter to rear.
13. With manual transmission, disconnect clutch linkage, remove clutch cross shaft, starter and lower flywheel cover.
14. Remove lower bell housing bolts.
15. Remove front engine mounts at frame.
16. Lower vehicle.
17. Using jack and block of wood, support transmission.

Fig. 5 Tools positioned to remove rocker arm stud

18. Remove remaining bell housing bolts.
19. Raise transmission slightly. Then, using suitable lifting equipment, remove engine.
20. Reverse procedure to install.

CYLINDER HEAD

NOTE: Pontiac recommends no numerical sequence for tightening cylinder heads. However, they may be tightened by starting at the center bolts and then working alternately from side to side and outward toward the ends.

V8 Engines

1. Remove intake manifold, push rod cover and rocker arm cover.
2. Loosen all rocker arm nuts and move rocker arms off push rods.
3. Remove push rods, keeping them in order so they may be installed in their original locations.
4. Detach exhaust crossover pipe from manifolds.
5. Remove battery ground strap and engine ground strap on left head or engine ground strap.
6. Unfasten and remove head with exhaust manifold attached.

NOTE: To remove left cylinder head on 1972-76 V8-455 HO, SD engines, it is necessary to remove the exhaust manifold and the inner panel of the carburetor heat stove from the cylinder head.

CAUTION: *Use extreme care when handling heads as the rocker arm studs are hardened and may crack if struck.*

NOTE: *If left head is being removed, it will be necessary to raise head off dowel pins, move it forward and "jockey" it in order to clear power steering and power brake equipment if so equipped.*

7. Reverse removal procedure to install the heads.

ROCKER ARM STUDS

1969-76 V8 With Screw in Stud

1. Remove rocker arm cover.
2. Remove rocker arm and nut.
3. Using a deep socket, remove rocker stud.
4. Install new stud and tighten to 50 ft. lbs.
5. Install rocker arm and tighten nut to 20 ft. lbs.
6. Install rocker cover using new gasket.

1969-73 V8 With Pressed in Stud

NOTE: On these engines, rocker arm studs can now be replaced without removing the cylinder head. This is accomplished by the use of a new .005" oversize stud and a new reamer (J-22126), which is available from the Service Tool Division of Kent-Moore Inc.

1. Remove rocker cover, rocker arm nut and rocker arm.
2. Place oily rags around stud and in engine openings.
3. File two slots on opposite sides of stud approximately 3/32 to 1/8 inch deep. Top of slots should be 1/4 to 3/8 inch below threaded portion of stud, Fig. 4.
4. Position washer at bottom of stud, install stud remover J-8934, spacer J-8934 and a threaded 7/8 inch standard nut on stud, remove stud, Fig. 5.
5. Carefully ream stud hole with reamer J-22126. Hole must be reamed with pilot shaft end of reamer first.
6. Clean stud hole and surrounding area, apply white lead and oil to rocker arm stud, then position stud in installer J-23342.
7. Carefully drive stud in until installer bottoms on stud boss.
8. Install rocker arm and rocker cover.

VALVE ARRANGEMENT

Front to Rear

Six Cyl. . . . . E-I-I-E-E-I-I-E-E-I-I-E
V8s . . . . E-I-I-E-E-I-I-E

VALVE LIFT SPECS.

Year	Engine	Intake	Exhaust
1969	6-250[7]	.400	.400
	6-250[2]	.438	.438
	8-350[1]	.376	.412
	8-350[2][5]	.414	.413
	8-350[2][6]	.410	.413
	8-400[1][3]	.375	.410
	8-400[1][4][5]	.410	.414
	8-400[1][4][6]	.376	.412
	8-400[2][5]	.413	.413
	8-400[2][6]	.410	.413
	8-400[5]	.413	.413
	8-400[6]	.414	.413
	8-400[8]	.516	.516
	8-428[5]	.410	.413
	8-428[6]	.410	.414
	8-428[9][5]	.414	.413
	8-428[9][6]	.410	.413
1970	6-250	.388	.388
	8-350	.376	.412
	8-400[1]	.376	.412
	8-400[2]	.410	.414
	8-400[12]	.527	.527
	8-455[5]	.414	.413
	8-455[6]	.410	.413
1971	6-250	.388	.388
	8-350[5]	.376	.412
	8-350[6]	.410	.414
	8-400[1]	.376	.412
	8-400[2]	.410	.413
	8-455[1]	.376	.412
	8-455[2]	.410	.413
	8-455[9]	.414	.413
1972	6-250[13]	.388	.388
	6-250[14]	.388	.405
	8-307	.390	.409
	8-350[15]	.374	.407
	8-350[16]	.404	.408
	8-400[1][15]	.374	.407
	8-400[1][16]	.404	.408
	8-400[2]	.403	.406
	8-455[1][15]	.404	.408
	8-455[1][16]	.403	.406
	8-455[2]	.403	.406
	8-455[9]	.408	.406
1973	6-250[15]	.388	.388
	6-250[16]	.388	.405
	8-350	.374	.407
	8-400[5]	.403	.406
	8-400[6][2]	.404	.408
	8-400[6][1]	.374	.407
	8-455[11]	.470	.470
1974	6-250[15]	.388	.388
	6-250[16]	.388	.405
	8-350	.374	.407
	8-350[5]	.403	.406
	8-400[6][2]	.404	.408
	8-400[6][1]	.374	.407
	8-455[10]	.403	.406
	8-455[11]	.406	.406
1975-76	6-250	.388	.405
	8-260	.395	.400
	8-350[17]	.3818	.3984
	8-350	.377	.413
	8-400[1]	.377	.415
	8-400[2]	.410	.415
	8-455	.410	.414

[1]—2 bar. carb. [2]—4 bar. carb.
[3]—Reg. fuel. [4]—Premium fuel.
[5]—Std. trans. [6]—Auto. trans.
[7]—One bar. carb. [8]—Ram air eng.
[9]—Hi perf. eng.
[10]—Except Super Duty.
[11]—Super Duty Engine.
[12]—Ram Air IV engine.
[13]—Without Air Injector Pump.
[14]—With Air Injector Pump.
[15]—Exc. California cars.
[16]—California cars.
[17]—1975-76 Ventura only.

VALVE TIMING

Intake Opens Before TDC

Engine	Year	Degrees
6-250	1969[7]	14
	1969[2][5]	22
	1969[2][6]	14
	1970-76	16
8-260	1975-76	14
8-307	1971-72	28
8-350	1969[1]	22
	1969[2][5]	31
	1969[2][6]	23
	1970	22
	1971[5]	26
	1971[6]	30

	1972[1][12]	26
	1972[1][13]	30
	1973-75	26
8-350[14]	1975	19
8-350[14]	1976	13.5
8-350[1]	1976	22
8-350[2]	1976	26
8-400	1969-70[4]	23
	1969-70[4]	30
	1969-70[1][3]	22
	1969-70[8]	42
	1971[1]	26
	1971[2]	23
	1972[1][12]	26
	1972[1][13]	30
	1972[2]	23
	1973[2][5]	23
	1973-74[2][6]	30
	1973-74[1][6]	38
	1975-76[1]	26
	1975-76[2]	30
8-428	1969[5]	23
	1969[5]	23
	1969[6]	30
	1969[9][5]	31
	1969[9][6]	23
8-455	1970[5]	31
	1970[6]	23
	1970[9]	23
	1971[1]	30
	1971[2]	23
	1971[9]	31
	1972[1][12]	30
	1972[1][13]	23
	1972[2]	23
	1972[9]	31
	1973[6][1]	26
	1973-74[10]	23
	1973-74[6][2]	30
	1974[11]	38
	1975-76	23

[1]—2 bar. carb. [2]—4 bar. carb.
[3]—Regular fuel. [4]—Prem. fuel.
[5]—Std. trans. [6]—Auto. trans.
[7]—1 bar. Carb.
[8]—Ram Air IV engine.
[9]—Hi perf. eng.
[10]—Except Super Duty.
[11]—Super Duty Engine.
[12]—Except California cars.
[13]—California cars.
[14]—1975-76 Ventura only.

VALVE GUIDES

Valve guides are cast integral with the cylinder head. Valves with oversize stems are available in .001", .003" and .005" larger than standard.

Oversize reamers are required to enlarge valve guide holes to fit the oversize stems. For best results when installing .005" oversize valve stem use a .003" oversize reamer first and then ream to .005" oversize. Always reface the valve and valve seat after reaming valve guide. Valves are marked .001, .003 or .005 with colored ink.

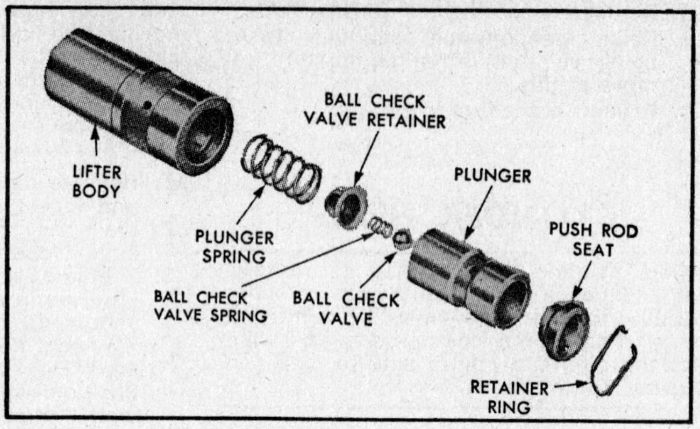

Fig. 6 Hydraulic valve lifter. Some lifters do not have the ball check valve spring shown

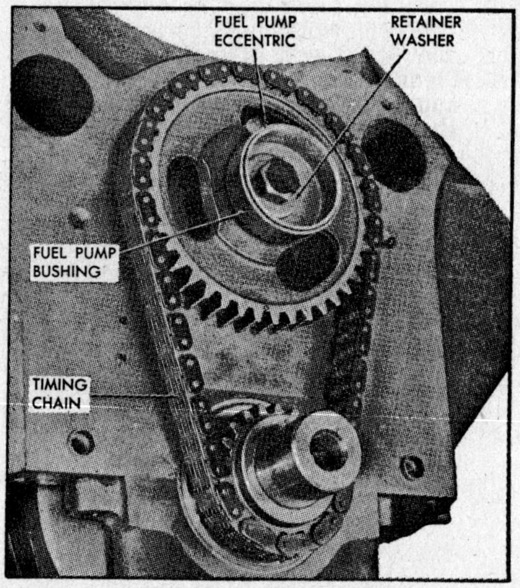

Fig. 7 Front of engine with timing case cover removed. V8 engines

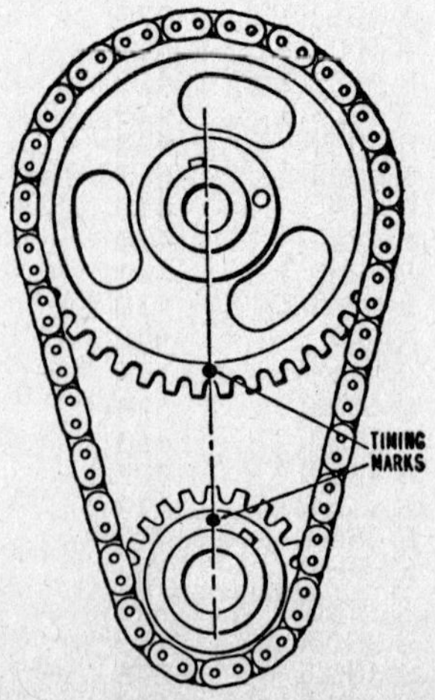

Fig. 8 Valve timing marks. V8 engines

VALVE LIFTERS

Remove intake manifold, push rod cover and rocker arm cover. Loosen rocker arm ball nut and move rocker arm off push rod. Remove push rod.

Remove lifter or lifters. If more than one lifter is to be replaced, be sure to identify them (push rods as well) so they will be installed in the same position, Fig. 6.

TIMING COVER

NOTE: If necessary to replace the cover oil seal it can be accomplished without removing the timing chain cover.

1969-76 V8s

1. Drain cooling system.
2. Loosen alternator adjusting bolts.
3. Remove fan and accessory drive belts.
4. Remove fan and pulley.
5. Remove water pump.
6. Disconnect radiator hoses.
7. Remove fuel pump.
8. Remove vibration damper.
9. Remove front four oil pan-to-timing chain cover screws, Fig. 14.
10. Remove cover attaching screws.
11. Pull cover forward to clear studs and remove.

TIMING CHAIN

V8 Engines

1. Remove timing chain cover, making certain O-ring seal and hollow dowels are retained for installation at assembly.
2. Remove fuel pump eccentric, bushing and timing chain cover oil seal, Fig. 7.
3. Align timing marks to simplify proper positioning of sprockets during assembly, Fig. 8.

NOTE: The valve timing marks, Fig. 8, does not indicate TDC, compression stroke for No. 1 cylinder for use during distributor installation. When installing the distributor, rotate engine until No. 1 cylinder is on compression stroke and the camshaft timing mark is 180° from the valve timing position shown in Fig. 8.

4. Slide off chain and sprockets.
5. Install new chain and sprockets, making sure timing marks are aligned exactly on a straight line passing through the shaft centers, Fig. 8. Camshaft should extend through sprocket so that hole in fuel pump eccentric will locate on shaft.
6. Install fuel pump eccentric and bushing, indexing tab on eccentric with keyway cutout in sprocket. Install retainer bolt with washer and tighten securely.
7. Making sure hollow dowels are in place in block, place timing chain cover gasket over studs and dowels.
8. Install cover, making sure O-ring seal is in place.

CAMSHAFT

All V8s

The camshaft and camshaft bearings can be replaced with the engine installed in the car or with engine removed and disassembled for overhaul. However, to replace the rear camshaft bearing without removing and completely disassembling the engine, the propeller shaft, transmission and clutch housing must first be removed. The procedure for removing the camshaft is as follows:

1. Drain cooling system and remove air cleaner.
2. Disconnect radiator and heater hoses, distributor vacuum hose and spark plug wires.
3. Disconnect carburetor linkage, fuel lines and wire connector from temperature sending unit.
4. Remove hood latch brace.
5. Remove radiator, fan and pulleys.
6. On air conditioned cars, remove alternator and its mounting bracket.
7. Remove crankcase ventilator hose or outlet pipe.
8. Remove distributor.
9. Remove rocker arm covers.
10. Remove intake manifold. *Make certain "O" ring seal between intake manifold and timing chain cover is retained and installed during assembly.*
11. Remove push rod cover.
12. Loosen rocker arm ball retaining nuts so that rocker arms can be disengaged from push rods and turned sideways.
13. Remove push rods and hydraulic lifters, keeping them in proper sequence so that they may be returned to their original locations.
14. Remove vibration damper.
15. Remove fuel pump.
16. Remove timing chain cover.
17. Remove fuel pump eccentric and fuel pump bushing.
18. Remove chain and sprockets.
19. Remove camshaft thrust plate and carefully pull camshaft from engine. *Clearance for camshaft removal is very limited and, in cases where engine mounts are worn excessively, it may be necessary to raise the front of the engine to permit removal.*

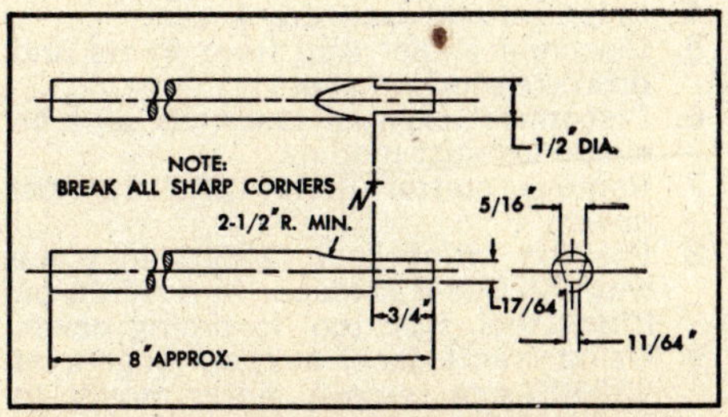

Fig. 10 Rear main bearing oil seal tool. V8-350, 400

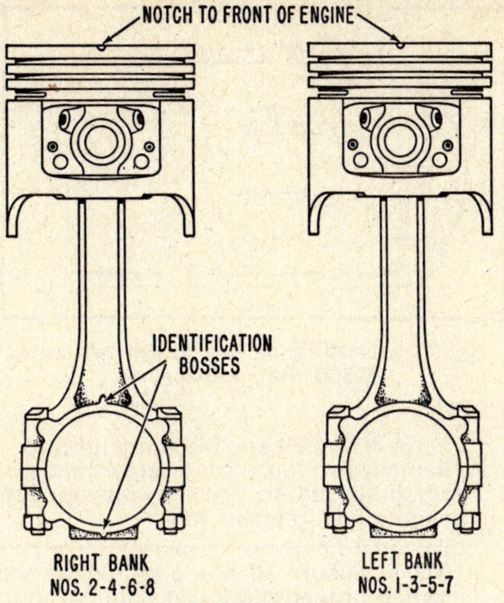

Fig. 9 Piston and rod assembly. 1969-76 V8s (Oil spurt hole toward camshaft)

PISTON & ROD, ASSEMBLE

Assemble pistons and rods as indicated in Fig. 9.

NOTE: The connecting rod squirt hole has been eliminated from all early production 1973 engines. The early and late connecting rods are completely interchangeable without any complications starting with the 1971 production engines.

PISTONS, PINS & RINGS

Pistons and rings are available in standard sizes and oversizes of .005, .010, .020 and .030 inch.

Piston pins are available in oversizes of .001 and .003".

MAIN & ROD BEARINGS

Main bearings are available in standard sizes and undersizes of .001 and .003".

Rod bearings are available in standard sizes and undersizes of .001 and .003".

CRANKSHAFT OIL SEAL

1969-76 V8

1. Remove oil pan, oil pump and pump drive shaft.
2. Remove oil baffle and cylinder block-to-oil baffle tube.
3. Remove rear main bearing cap.
4. Use tool shown in Figs. 10 & 11 made from brass bar stock to pack upper seal as follows:
   a. Insert tool against one end of oil seal in cylinder block and drive seal gently into groove until tool bottoms.
   b. Remove tool and repeat at other end of seal in cylinder block.
5. Clean block and bearing cap parting line thoroughly.
6. Form a new seal in cap.
7. Remove newly formed seal from cap and cut four pieces about 3/8" long from this seal.
8. Work two 3/8" pieces into each end of the gaps which have been made at the end of seal in cylinder block. Without cutting off the ends, work these seal pieces in until flush with parting line, being sure that no fibers are protruding over the metal adjacent to the groove.
9. Form another new seal in the cap.
10. Assemble the cap to the block and torque to specifications.
11. Remove cap and inspect parting line to insure that no seal material has been compressed between the block and cap.
12. Apply a 1/16" bead of sealer from the center of the seal to the external cork groove.
13. Reassemble the cap and torque to specifications.

OIL PAN

1970-76 6-250

1. Remove radiator upper shield.
2. Disconnect battery ground cable, raise vehicle and drain crankcase.
3. Disconnect exhaust pipe at manifold.

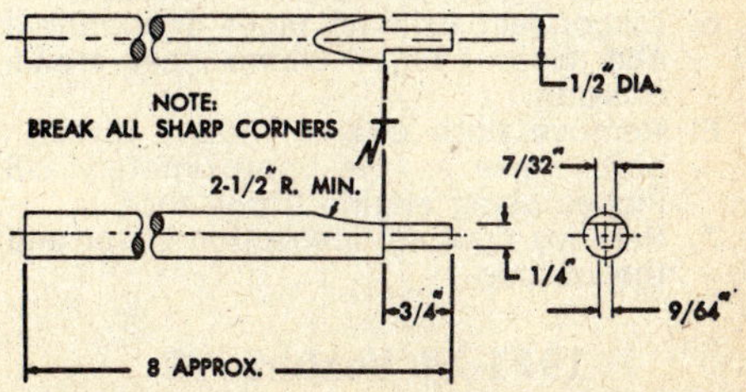

Fig. 11 Rear main bearing oil seal tool. V8-455

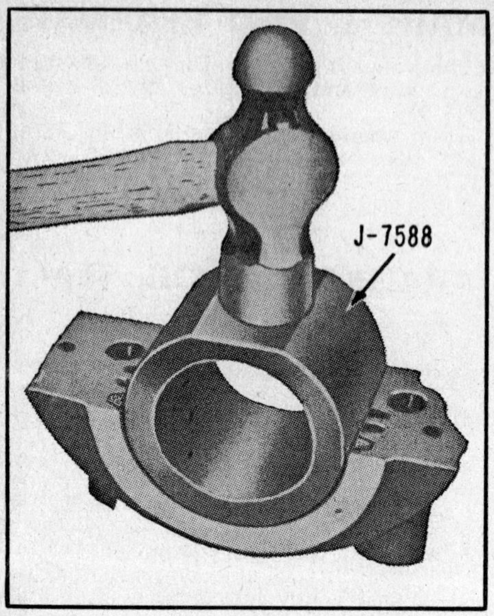

Fig. 12. Installing rear main bearing oil seal. V8-350, 400, 455 engines

4. Remove starter, then the flywheel housing inspection cover.
5. Install engine lifting tool and remove both engine mount through bolts.
6. Remove engine mount from right side.
7. Remove oil pan attaching bolts and raise engine until pan can be removed.

1975-76 Ventura V8-260

1. Remove distributor cap and crank engine until rotor is in position to fire number 1 cylinder.
2. Disconnect battery ground cable and remove dip stick.
3. Remove radiator upper support and fan shroud attaching screws.
4. Raise vehicle and drain crankcase.
5. Remove flywheel housing inspection cover, then the starter.
6. Disconnect exhaust pipes at manifold.
7. Remove both engine mount through bolts, raise engine as far as possible using engine lifting tool, then remove oil pan attaching bolts and oil pan.

1975-76 Ventura V8-350

1. Disconnect battery ground cable.
2. Loosen A/C compressor adjusting bolt and tilt compressor inward.
3. Remove fan.
4. Raise vehicle and drain crankcase.
5. Disconnect exhaust pipes at manifold and from hanger clamp near transmission.
6. Remove both engine mount through bolts, raise engine approximately 2-3 inches using engine lifting tool.
7. Remove flywheel inspection cover and the oil pan.

1971-74 Ventura V8

1. Disconnect battery ground cable.
2. If equipped with power steering, remove drive belt and tilt pump upward.
3. Remove two fan shroud screws and position shroud so it will swing up with engine and remove fan.

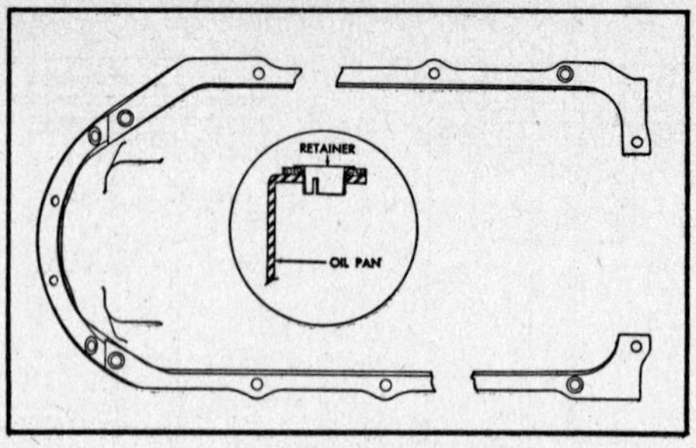

Fig. 13 Installing oil pan gasket retainers. V8-350, 400, 455 engines

NOTE: Ensure all hoses and wires are routed properly to avoid binding when engine is raised.

4. On 1973-76 models, remove thermal override switch from right cylinder head.
5. Raise car and drain oil.
6. Disconnect exhaust pipe to manifold bolts and let exhaust pipes hang down.
7. Remove flywheel dust cover.
8. Remove starter with wiring attached and position out of way.
9. Remove both frame bracket to engine mount thru bolts.
10. Attach a suitable engine lifting tool to engine.
11. Remove oil pan bolts.
12. Raise engine until pan can be removed.

1970-76 V8 Except Ventura

NOTE: On 1970-71 Tempest, T-37 and LeMans models with manual transmission, it will be necessary to remove the engine before removing the oil pan. On all other models, proceed as follows:

1. Disconnect battery cable at battery and remove fan.
2. On 1972-76 models equipped with power steering, remove pump adjusting bolt and drive belt, then tilt pump upward.
3. Remove fan shroud attaching screws and position shroud so it will swing up with engine.
4. On 1973-76 models, remove thermal switch from left cylinder head.
5. Make sure all hoses and wiring are routed properly to avoid bind when engine is raised.
6. Raise vehicle and drain crankcase.
7. On 1971-76 Firebird and Grand Prix, disconnect steering idler arm at frame and pitman arm from shaft. On 1970 full size Pontiac, disconnect steering idler arm support from frame.
8. Disconnect exhaust pipes from manifolds.
9. Remove starter assembly (set to one side with wires attached), starter motor bracket and flywheel inspection cover.
10. On 1970-76 T-37, LeMans and Grand Am and 1970-73 GTO, remove stabilizer shaft to frame bracket attaching bolts to insure free movement of lifting device. It may also be necessary to loosen the fuel pump to timing cover bolts for clearance.
11. On all models, attach lifting tool, loosen oil pan bolts and raise engine until oil pan can be removed.

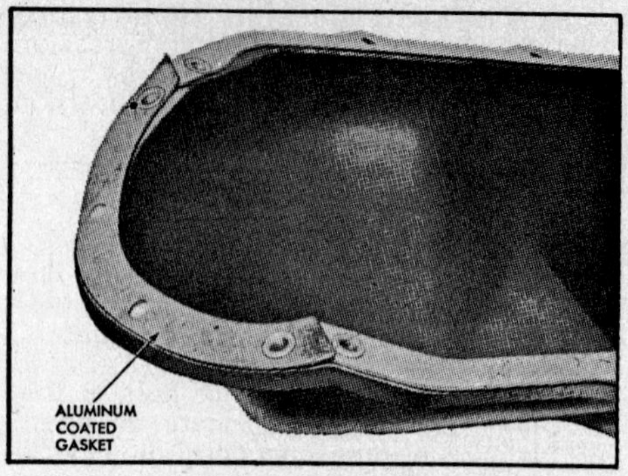

Fig. 14 Front oil pan gasket overlapping side gaskets. V8-350, 400, 455 engines

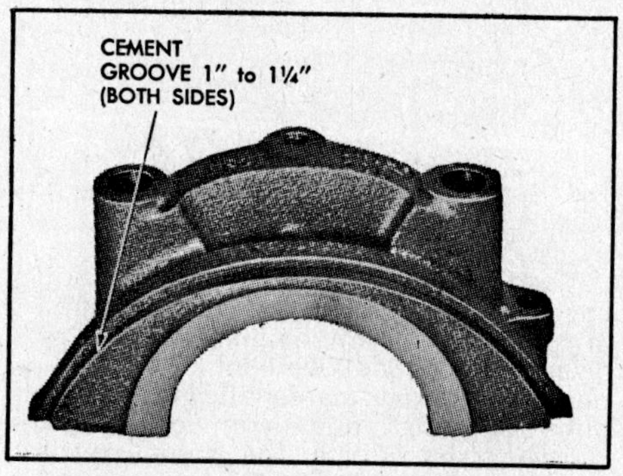

Fig. 15 Rear oil pan gasket positioned in bearing cap. V8-350, 400, 455 engines

NOTE: On Firebird and Grand Prix models it may be necessary to loosen steering gear to frame attaching bolts to install lifting tool.

1969 Tempest V8

1. Disconnect battery positive cable.
2. Remove fan blade assembly.
3. Inspect all hoses and wiring for proper routing to avoid excessive binding when engine is raised 4½".
4. Raise car and drain crankcase.
5. Disconnect idler arm from frame and drop steering linkage.
6. Disconnect exhaust crossover pipe on single exhaust models.
7. Remove starter motor and flywheel cover.
8. Position lifting tool J22603 in place with J22603-8 crossbar in position on lifting tool. Bolt tool to timing cover.
9. Using frame jack, support engine at J22603 and remove motor mount to frame bolts.
10. Raise engine about 4½" and remove oil pan.

1969 Firebird V8

1. Disconnect positive battery cable.
2. Disconnect fan shroud from radiator.
3. Remove air cleaner and distributor cap.
4. Make sure all hoses and wiring are free so as not to bind when engine is raised 4½″ and moved forward 1½″.
5. Raise car and drain crankcase.
6. Disconnect idler arm from frame.
7. Disconnect crossover pipe. If equipped with dual exhausts, disconnect exhaust pipes from manifolds.
8. Remove starter and flywheel cover.
9. Attach lifting tool J22603 to engine and place crossbar J22603-8 in position on tool.
10. Using frame jack or transmission jack, support engine at tool J22603 and remove motor mounts from engine.
11. Remove rear transmission to crossmember bolts.
12. Raise engine 4½″ and at the same time move engine forward 1½″.
13. Remove oil pan.

1969 Pontiac

1. Disconnect battery ground cable.
2. Remove fan shield.
3. On A/C cars, remove fan and pulley.
4. Disconnect engine ground cables.
5. On A/C cars, remove compressor from mounting brackets and position to one side.

NOTE: At this time, inspect all water hoses and wiring harnesses for proper routing to avoid excessive bind when engine is raised about 4½″.

6. Raise car and drain crankcase.
7. Disconnect steering idler arm from frame.
8. Remove exhaust crossover pipe. If equipped with dual exhausts disconnect exhaust pipes from manifolds.
9. Remove starter and flywheel cover.
10. Remove front engine mount-to-frame bolts.

On 1969 models, special Engine Lifting Tool (J-22603) is available. Bolt tool to timing chain cover with bolts provided with tool. Using frame jack or automatic transmission jack, support engine at Tool J-22603 and remove engine mounts. Loosen rear transmission mount. It may be necessary to remove this mount and rest rear of transmission on cross-member to obtain necessary clearance. Remove oil pan bolts and raise engine *straight up* until transmission is against floor pan. Remove pan by first rotating clockwise (facing forward) to clear oil pump.

NOTE: If work other than oil pan gasket replacement is to be performed, support engine with suitable blocks of wood and remove engine support.

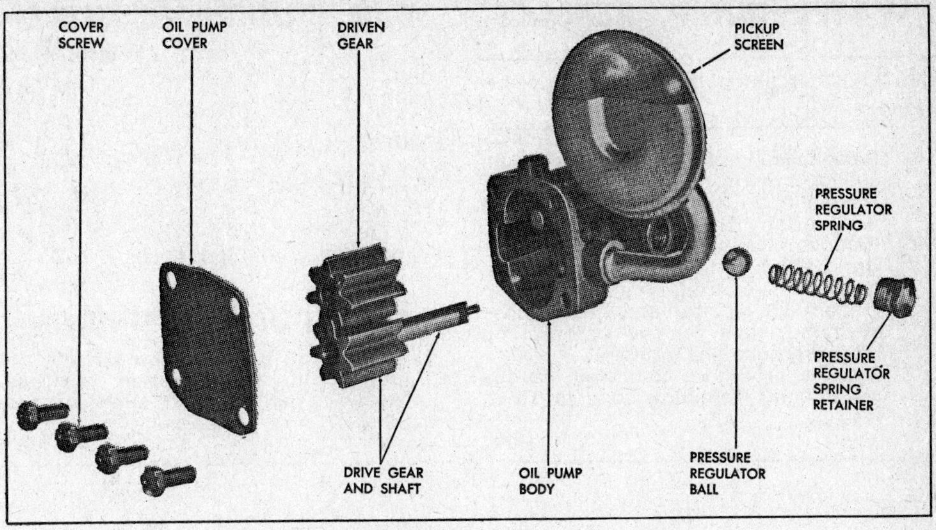

Fig. 16 Oil pump disassembled. V8 Pontiac engines

OIL PUMP

V8 Engines

Remove oil pan. While holding pump in place, remove attaching screws. Lower the pump away from the block with one hand while removing the oil pump drive shaft with the other.

Remove oil screen and pressure regulator parts. Detach cover from pump body and take out gears, Fig. 16.

Examine all parts for damage and assemble. Do not attempt to change oil pressure by varying length of pressure regulator spring.

Position drive shaft in distributor and oil pump drive gear. Place pump in position in the block, indexing the drive shaft with pump drive gear shaft. Install attaching screws with lock washers and tighten securely.

Removal and installation of pump does not affect distributor timing since the oil pump and distributor drive gear are mounted on the distributor shaft.

BELT TENSION DATA

	New Lbs.	Used Lbs.
1969—		
Air Condition	145	105
Generator to Power Steering	60	50
Power Steering Pump	145	105
Water Pump to Generator	120	75
1970—		
Six Cylinder—		
Air Condition	130-150	90-110
Power Steering	100-125	60-85
Water Pump to Generator	100-125	60-85
V8-Models—		
Air Condition	135-150	100-105
Power Steering	135-150	100-105
Water Pump to Generator	110-125	70-75
1971-72—		
Six Cylinder—		
Air Condition	135-145	85-95
Generator	120-130	70-80
Power Steering	120-130	70-80
V8-Models—		
Air Condition	135-150	100-105
Generator—		
With A/C	135-150	100-105
Without A/C	110-125	70-75
Power Steering	120-130	70-80
1973—		
V8 Models—		
Air Condition	135-150	100-105
Generator—		
With A/C	135-150	100-105
Without A/C	110-125	70-75
Power Steering	135-150	100-105
1973-76—		
Six Cylinder—		
Air Pump	100-125	70-90
Alternator	100-125	70-90
Power Steering	100-130	70-90
1974—		
V8 Models—		
Air Condition	135-165	100-105
Air Pump	110-140	70-75
Generator—		
With A/C	110-140	70-75
Without A/C	135-165	100-105
Power Steering	135-165	100-105
1975-76—		
V8 Models—		
Air Conditioning		
Exc. V8-260	135-165	100-105
V8-260	110-140	70
Air Pump		
Exc. Ventura V8-350	110-140	70
Ventura V8-350	65-80	50-70
Generator		
Exc. 80 amp.	110-140	70
80 amp.	135-165	100-105
Power Steering		
Exc. V8-260	135-165	100-105
V8-260	110-140	70

WATER PUMP, REPLACE

NOTE: Water pump is serviced as an assembly only.

1. Disconnect battery cable and drain cooling system.
2. Loosen alternator adjusting bolt and remove fan belt.
3. Remove fan and pulley.
4. On 1972-76 models, remove alternator front bracket.
5. Disconnect radiator and heater hose at pump, then remove water pump retaining bolts and pump.
6. Reverse procedure to install. Torque water pump retaining bolts to 15 ft. lbs.

FUEL PUMP PRESSURE

Year	Engine	Pressure Lbs.
1969	Six	4-5½
1969	V8	5-6½
1970-71	Six	4-5
1970-71	V8	5-6½ [1]
1972-74	Six	3-5
1972-74	V8	3-6½
1975-76	Six	4-5
1975-76	V8	5-6½

[1] With A/C and all 4 Bar. Carbs. 6½-8

FUEL PUMP, REPLACE

1. Remove all gasket material from the pump and block gasket surfaces. Apply sealer on both sides of new gasket.
2. Position gasket on pump flange and hold pump in position against its mounting surface. Make sure rocker arm is riding on camshaft eccentric.
3. Press pump tight against its mounting. Install retaining screws and tighten them alternately.
4. Connect fuel lines. Then operate engine and check for leaks.

SERVICE NOTE: Before installing the pump, it is good practice to crank the engine so that the nose of the camshaft eccentric is out of the way of the fuel pump rocker arm when the pump is installed. In this way there will be the least amount of tension on the rocker arm, thereby easing the installation of the pump.

Overhead Camshaft Engine Section

INDEX OF SERVICE OPERATIONS

ENGINE MOUNTS REPLACE

1969 Six

1. Support engine using suitable engine lifting tool.
2. Remove insulator to engine bracket through bolts, Figs. 1 and 2.
3. Raise engine approximately 1" above front insulators.
4. Remove front insulator by removing insulator to block bolt.
5. Remove insulator.

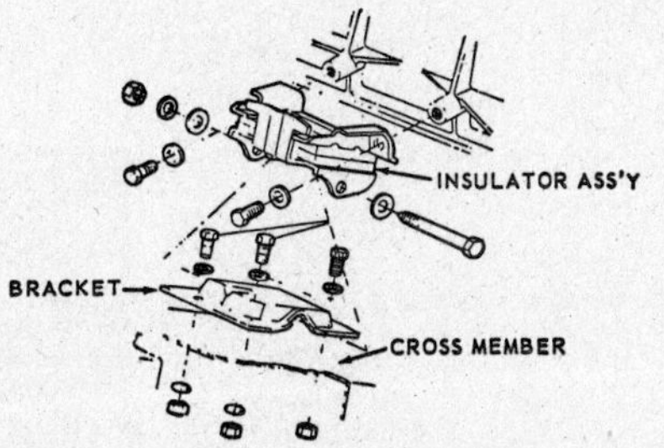

Fig. 1 Engine mounts. 1969 Tempest

ENGINE, REPLACE

1. Disconnect cables at battery.
2. Drain cooling system.
3. Scribe alignment marks on hood around hood hinges and remove hood.
4. Disconnect engine wiring harness and engine ground straps.
5. Remove air cleaner and fan shield.
6. Disconnect all hoses at engine.
7. With manual transmission, remove radiator.
8. If equipped with power steering or air conditioning, remove pump and compressor from mounting brackets and set aside (do not disconnect hoses).
9. Remove engine fan and pulley.
10. Disconnect accelerator linkage.
11. Disconnect transmission vacuum modulator line and power brake vacuum line at carburetor and fold back out of the way.
12. Raise vehicle and drain crankcase.
13. Disconnect fuel lines at fuel pump, exhaust pipes from manifold and starter wires.
14. If equipped with automatic transmission, remove converter cover and 3 converter retaining bolts and slide converter to rear.
15. If equipped with manual transmission, disconnect clutch linkage and remove clutch shaft. Remove four lower bell housing bolts (two each side).
16. Disconnect transmission filler tube support and starter wire harness shield from cylinder head.
17. Remove two front engine mounts to-frame thru bolts.
18. Lower vehicle.
19. Support transmission with a jack and block of wood.
20. Remove two remaining bell housing bolts, raise transmission slightly and remove engine.
21. Reverse procedure to install. Do not lower engine completely while jack is supporting transmission.

TIMING BELT, ADJUST

1. Remove 3 screws on front of top cover. Lift up cover to disengage side clips. Remove retaining clips from cover.
2. Using the equipment shown in Fig. 3, set the pointer of the fixture on the zero mark. *This calibration must be performed prior to each use of the fixture to insure an accurate belt adjustment.*
3. Remove camshaft sprocket-to-camshaft bolt and install the fixture on the belt with the rollers on the outside (smooth) surface of belt. Thread the fixture mounting bolt into camshaft sprocket bolt location finger tight.
4. Squeeze indicator end (upper) of fixture and quickly release so that fixture assumes released or relaxed position.
5. With the tool installed as directed, adjust accessory drive housing, Fig. 3A, up or down as required to obtain a tension adjuster indicator centered in the green range with drive housing mounting bolts torqued to 15 ft-lbs.
6. Remove tension fixture and install sprocket retaining bolt, making sure bolt threads and washer are free of dirt. Install cover.

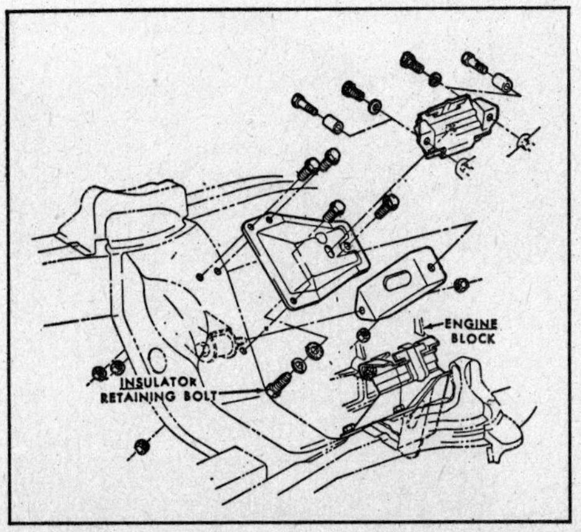

Fig. 2 Engine mounts. 1969 Firebird

CAMSHAFT SPROCKET OR SEAL

CAUTION: Do not use tools of any type, other than hands, to pry on timing belt during belt removal or replacement or during other service operations.

1. Remove timing belt top front cover. *For ease on reassembly, index three timing marks as shown in Fig. 3B.*
2. Loosen accessory drive housing bolts, Fig. 3A.
3. Remove belt from camshaft sprocket.
4. Remove sprocket.

NOTE: If necessary to replace camshaft seal, reinstall sprocket bolt. Thread a tool of the type shown, Fig. 4, into camshaft seal. Tighten center bolt on tool until seal is extracted. Install a suitable seal protector and pilot, Fig. 5, on end of camshaft. Slide seal over on tool, then drive it in place, Fig. 6.

5. Install sprocket, indexing pin on sprocket with hole in camshaft.
6. Install sprocket bolt finger tight.
7. Align timing marks and install belt.
8. Adjust timing and belt tension.

9. Torque sprocket bolt to 40 ft-lbs and install top cover.

ROCKER ARM COVER

1. Drain cooling system and disconnect radiator hose at fitting on cover.
2. Remove timing belt top cover.
3. Align timing marks and remove belt from camshaft sprocket.
4. Disconnect necessary fuel and vacuum lines.
5. Unfasten and remove rocker arm cover.
6. Reverse procedure to install.

CAMSHAFT

1. Remove camshaft sprocket and seal.
2. Remove rocker arm cover.
3. Using tools of the type shown in Fig. 7, drive camshaft from rocker cover.

CAUTION: *Do not allow camshaft to damage bearing surfaces of rocker cover.*

4. Remove parts shown from rear of camshaft, Fig. 8.
5. Remove water outlet fitting and thermostat from rocker cover.
6. Clean all parts and inspect for wear or damage. Minor nicks or scratches on edge of bearing surface can be corrected with a suitable scraper or file.

Installation

1. Install camshaft into rocker cover.
2. Install thrust washer, Fig. 9.
3. Install retaining bolt and washer and torque to 40 ft-lbs.
4. Using a tool of the type shown in Fig. 10, drive plug in so it is fully seated.

IMPORTANT: A camshaft bore plug not fully seated could result in excessive camshaft end play which is .003″ to .009″ when read at the sprocket end with a dial indicator.

5. Replace water outlet fitting and thermostat.

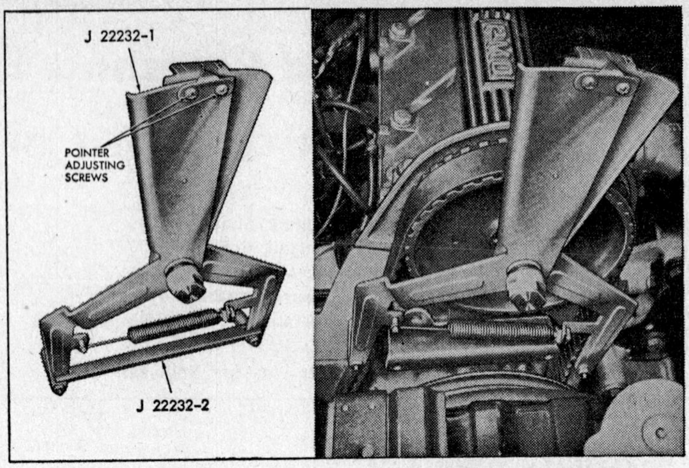

Fig. 3 Adjusting timing belt with tension fixture shown

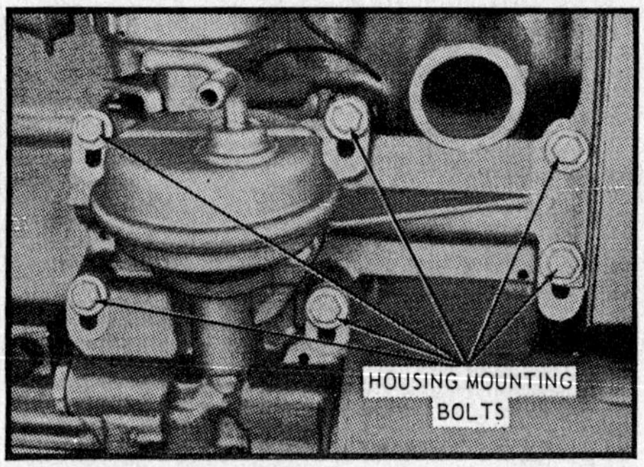

Fig. 3A Accessory drive housing mounting bolts

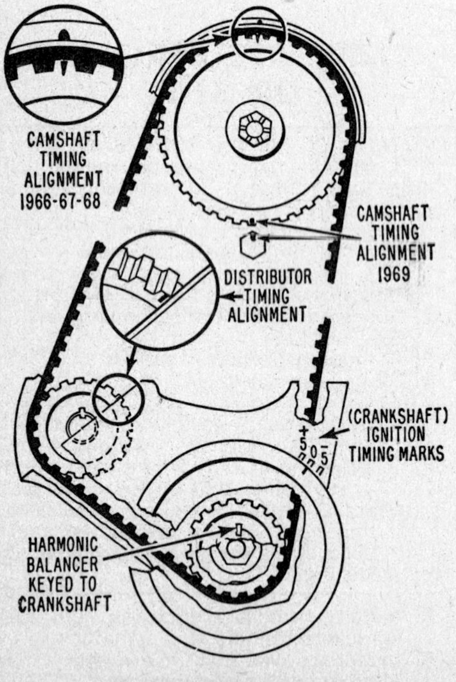

Fig. 3B Valve timing marks

Fig. 4 Removing camshaft seal

Fig. 5 Seal and protector tool installed

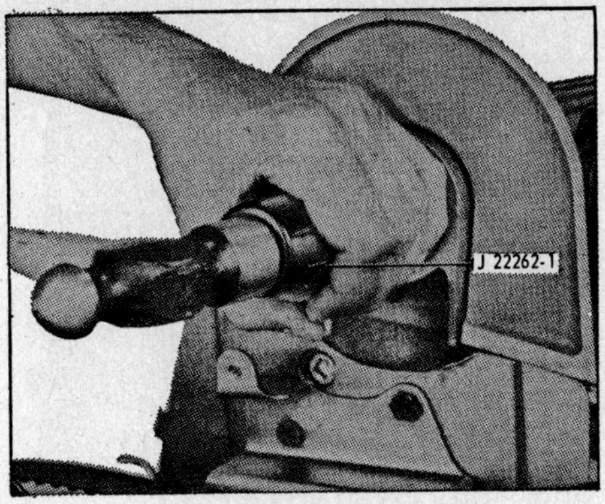

Fig. 6 Installing camshaft seal

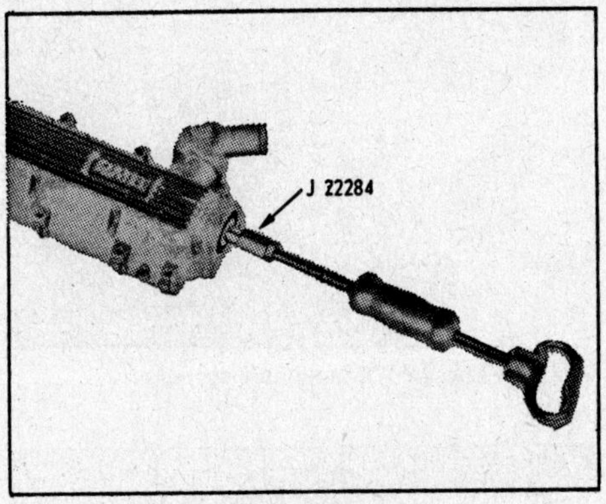

Fig. 7 Removing camshaft

ROCKER ARM OR VALVE LASH ADJUSTER

1. Remove rocker arm cover.
2. Remove rocker arm and hydraulic lash adjuster assembly and store so that each assembly can be installed in its original location.

NOTE: If a new lash adjuster is to be installed, it will be necessary to check the leak-down rate and prime the adjuster before installation.

3. Install retainer clip, Fig. 11.
4. Place each lash adjuster in its original location, and install rocker cover.

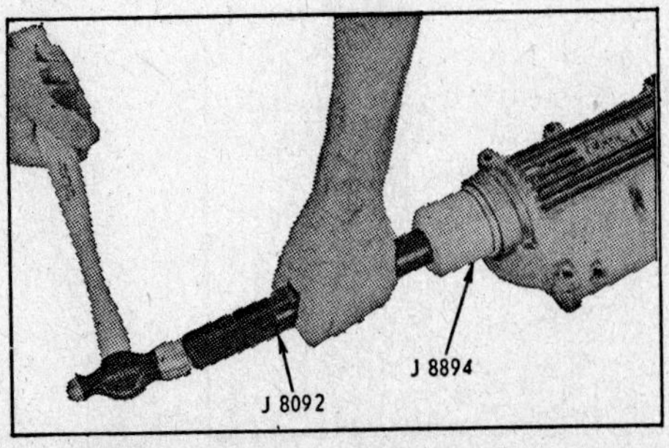

Fig. 10 Installing camshaft bore plug

Lash Adjuster Service

NOTE: Because of the important part hydraulic lash adjusters play in the operation of the engine, and the close tolerances to which they are manufactured, proper handling and, above all, cleanliness, cannot be overstressed when servicing these parts.

New adjusters are serviced as individual units, packaged with a plastic coating. Leave the coating on until ready to check leak-down rate. It is necessary to remove the oil from the new adjusters prior to checking leak-down rate since special oil is in new adjusters. Fill adjusters with SAE 10 oil before checking leak-down rate. If leak-down test equipment is not available, the supplier should check lash adjusters for you.

Removing Stuck Lash Adjuster

After removing the rocker arm, fill the vent hole adjacent to the stuck lifter with engine oil. Using a length of 3/16" diameter rod approximately 4" in length, insert it into the top of the vent hole and strike the end of the rod with a hammer. The hydraulic effect of the oil on the base of the lash adjuster will then break the adjuster free from the boss in the cylinder head. The use of pliers for this operation is not recommended since lifter damage will result.

VALVE SPRINGS & SEALS

Valve springs and seals can be replaced without removing the cylinder head if tools of the type shown in Fig. 14 are available.

1. Remove rocker arm cover.
2. Remove rocker arm of valve being serviced.
3. Remove spark plug from cylinder of valves to be serviced and install an air hose with an adapter that will screw into the spark plug hole.

CAUTION: When applying compressed air into the spark plug hole to hold the valve up, be sure the piston is all the way down in the cylinder, otherwise the fan will turn.

4. Install the hook end of the tool into the oil feed hole in lash adjuster bore. Hold In place and install fork end of valve spring compressor as shown. Compress valve spring, remove valve cup locks. Then remove tool, valve spring, cup shield and valve stem seal.

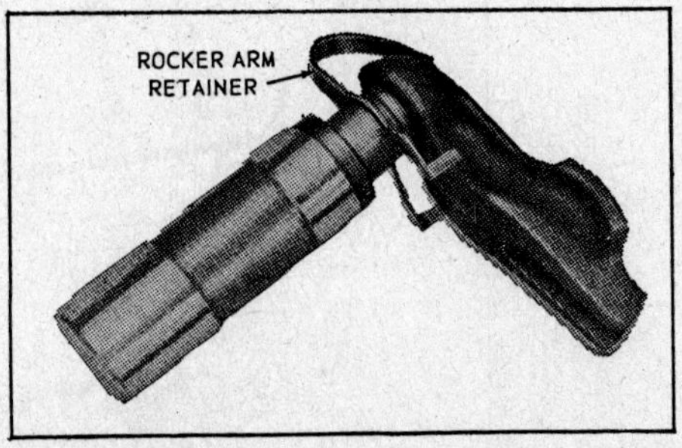

Fig. 11 Rocker arm retainer and lash adjuster

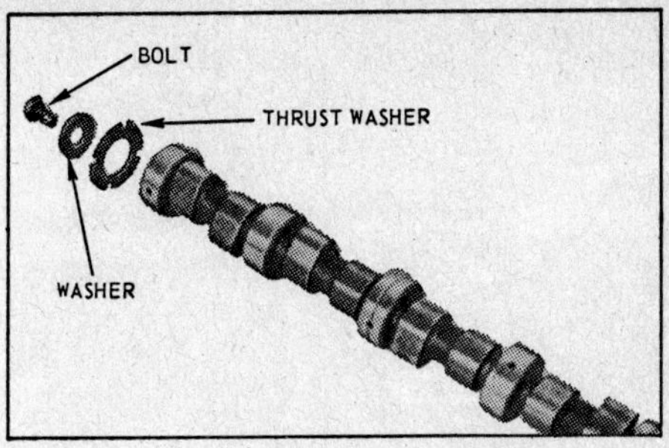

Fig. 8 Camshaft and related parts

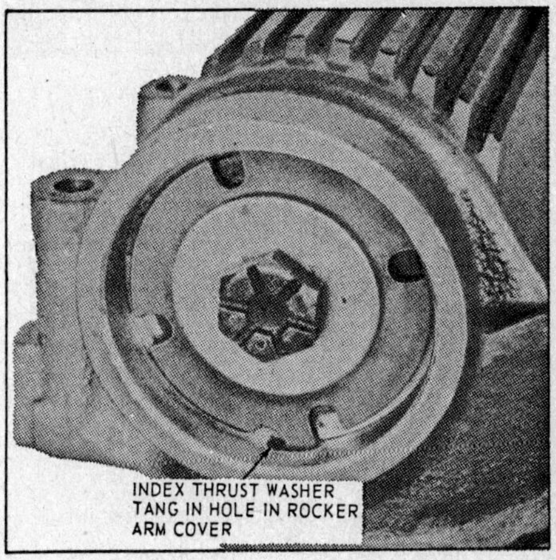

Fig. 9 Thrust washer installation

CYLINDER HEAD

Removal

1. Drain cooling system and remove air cleaner.

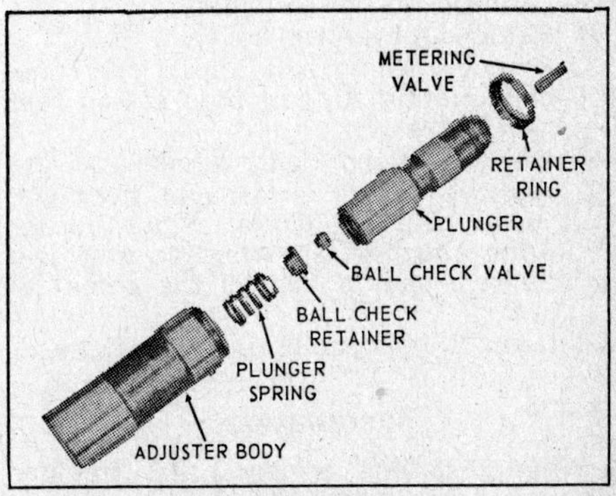

Fig. 12 Hydraulic lash adjuster disassembled

Fig. 14 Depressing valve spring

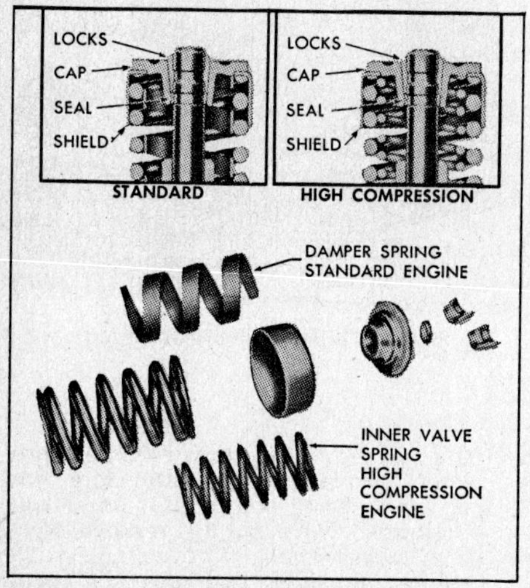

Fig. 15 Valves and related parts

2. Disconnect accelerator pedal cable at bellcrank on manifold and fuel and vacuum lines at carburetor.
3. Disconnect exhaust pipe at manifold flange, then remove manifold bolts and clamps. Remove manifolds and carburetor as an assembly.
4. Remove rocker arm cover.
5. Remove timing belt upper front cover mounting support bracket and rear lower cover.
6. Disconnect spark plug wires.
7. Remove rocker arms and hydraulic valve lash adjusters. *Store rocker arms and lash adjusters so they can be replaced in exactly the same location.*
8. Remove head bolts and lift off head.

Installation

When installing a new head, transfer all serviceable parts to new head using new seals on intake and exhaust valve stems and new manifold gaskets.

Head bolts are of two different lengths. When inserted in proper holes, all bolts will project an equal distance from head. Do not use sealer of any kind on threads.

Pontiac recommends no numerical sequence for tightening cylinder head bolts. Tighten them a little at a time, starting at the center bolts and working alternately from side to side outward to the end bolts.

VALVE SERVICE

Fig. 15 shows the valves and related parts.

Valves with oversize stems are available in .001″, .003″ and .005″ larger than standard. The same valve stem-to-guide clearance listed in the *Valve Specifications* table applies for oversize stems.

Carefully ream the valve guide. For best results when installing a .005″ oversize valve stem, use a .003″ oversize reamer first, then ream the .005″ oversize. Always reface the valve seat after reaming valve guide.

When installing valve springs, place the closed coil end toward the cylinder head.

CRANKCASE COVER SEAL

1. Remove upper timing cover.
2. Align timing marks, Fig. 3B.
3. Remove fan and water pump pulley.
4. Remove vibration damper.
5. Remove timing belt lower cover, Fig. 16.
6. Loosen accessory drive mounting bolts to provide slack in timing belt, Fig. 3A.
7. Remove timing belt.
8. Remove crankshaft timing belt flange and sprocket, Fig. 17.
9. Pry seal from front of crankcase.
10. Install new seal with lip of seal inward, using a suitable seal driver.
11. Reverse procedure to install, being sure to align the belt timing marks and adjust belt tension, Fig. 3.

Fig. 16 Lower front timing belt cover bolts

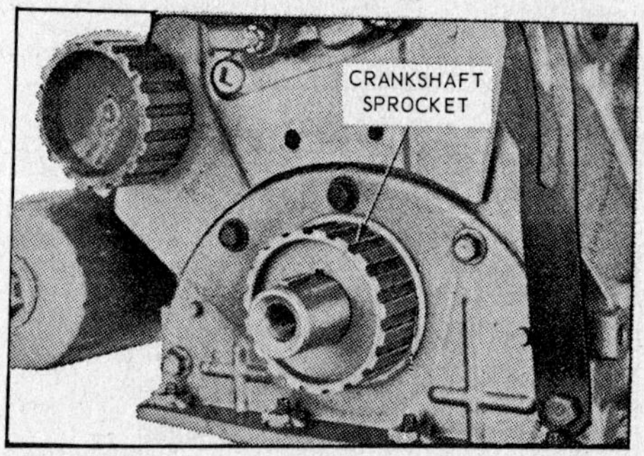

Fig. 17 Crankshaft sprocket

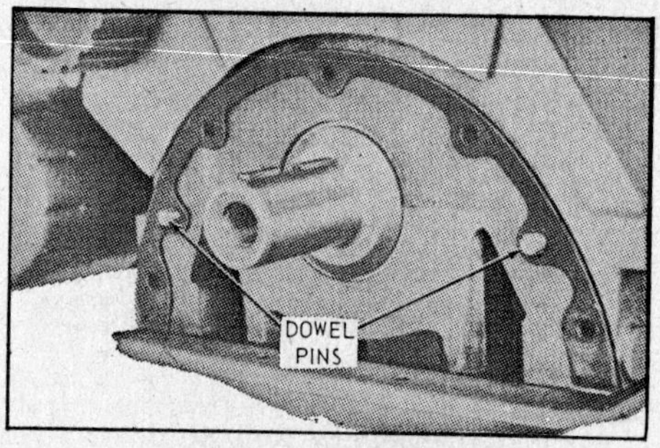

Fig. 18 Front crankcase with cover removed

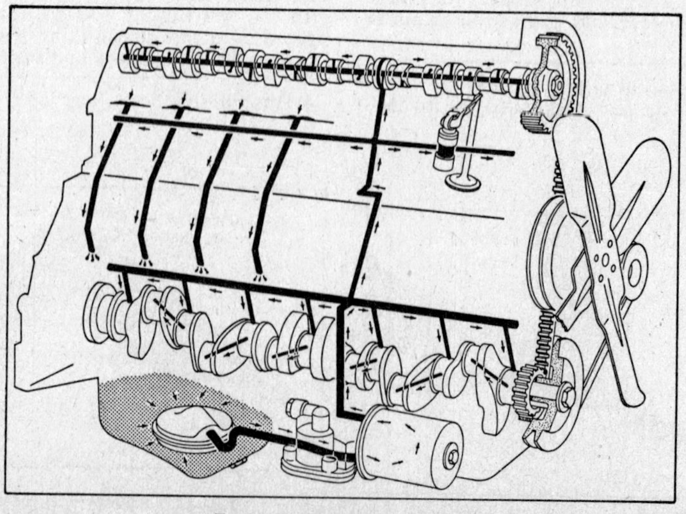

Engine oiling system. OHC Six

Fig. 20 Removing eccentric and distributor drive gear retaining pin

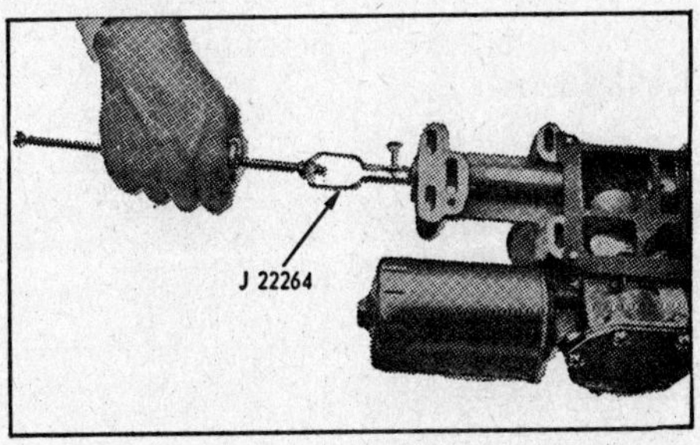

Fig. 21 Removing bearing and seal

FRONT CRANKCASE COVER

1. Remove crankshaft sprocket.
2. Remove 4 front oil pan-to-crankcase cover bolts.
3. Loosen remaining oil pan bolts as necessary to provide clearance between cover and pan. *It may be necessary to jar oil pan to gain necessary clearance.*
4. Remove 5 cover bolts.
5. Remove cover and gasket. Clean gasket surface, using care to see that gasket particles do not fall into oil pan.
6. Reverse procedure to install, Fig. 18.

ACCESSORY DRIVE HOUSING

Oil Pressure Regulator Valve

1. Remove cap, washer and spring from housing.
2. Using a magnet, remove valve from housing.
3. Reverse procedure to install.

Oil Pump

1. Remove pump cover and gasket.
2. Remove two gears.

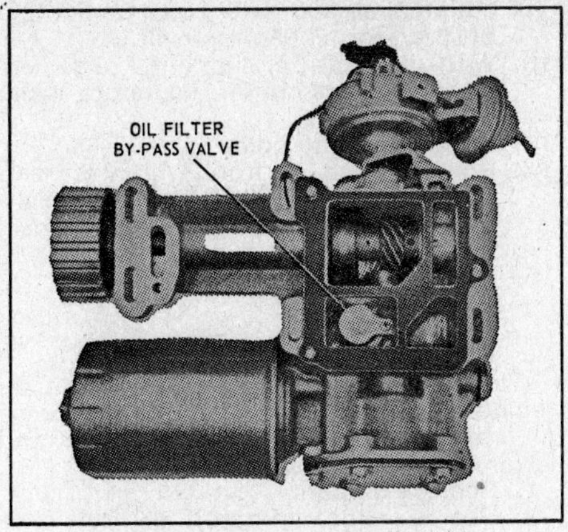

Fig. 19 Oil filter by-pass valve

3. Replace worn parts, install cover with new gasket and torque bolts to 20 ft-lbs.

Housing Assembly

1. Remove timing belt top cover.
2. Align timing marks, Fig. 3B.
3. Loosen 6 housing-to-cylinder block bolts, Fig. 3A.
4. Remove timing belt from camshaft sprocket and distributor drive.
5. Disconnect fuel lines from fuel pump.
6. Remove distributor cap, vacuum lines and wires from distributor.

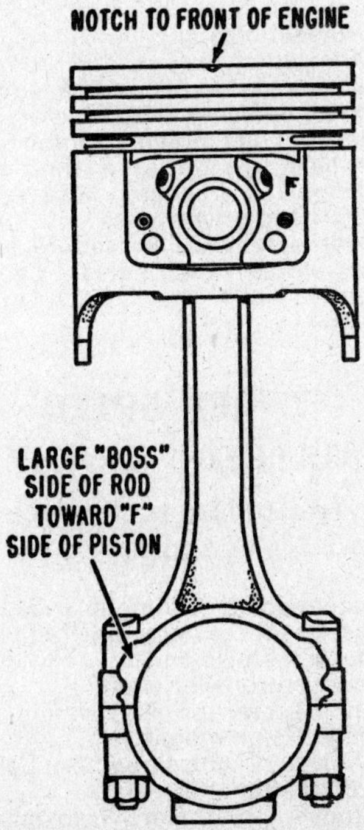

Fig. 24 Piston and rod assembly

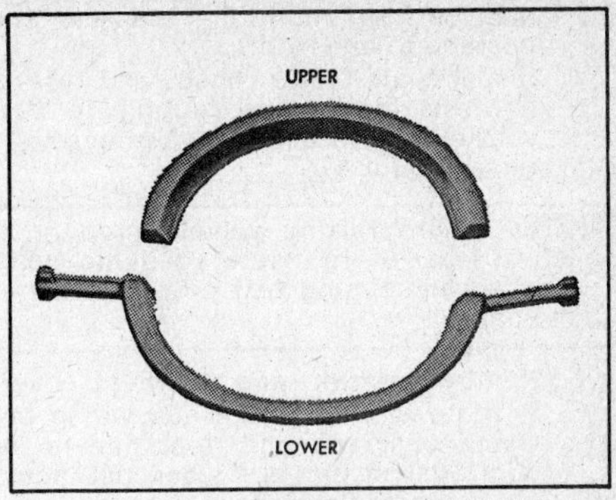

Fig. 22 Rear main bearing oil seal

Fig. 23 Removing rear main bearing cap

7. Remove housing assembly.
8. Reverse procedure to install, using new housing gasket. Be sure to align timing marks and adjust timing belt tension.

Oil Filter By-Pass Valve

1. Remove housing assembly.
2. Remove valve retaining screw, Fig. 19.
3. Remove by-pass valve.
4. Reverse procedure to install.

Sprocket, Seal, Fuel Pump, Distributor Drive

1. Remove housing assembly.
2. Look for and write down location of sprocket timing mark and direction of distributor rotor, then remove distributor.
3. Remove drive gear retaining pin, Fig. 20. Position shaft to allow adequate clearance in housing body for pin removal.
4. Remove shaft and sprocket.
5. If necessary to replace bearing or seal use a slide hammer and tool of the type shown in Fig. 21 to remove seal or bearing and seal together.
6. Reverse procedure to install.

OIL PAN

1. Disconnect battery cable.
2. Remove air cleaner.
3. On air conditioned cars, remove com-

pressor from mounting brackets and position to one side.
4. Inspect all water hose and wiring harness for proper routing to prevent excessive binding when engine is raised about 4½".

NOTE: Before raising vehicle prop hood open at least 6" to insure adequate clearance between timing belt cover and inner hood panel.

5. Remove starter and flywheel cover.
6. Reroute or disconnect any wiring between bellcrank and floor pan to insure against damage when bell housing contacts floor pan.
7. Loosen transmission insulator-to-crossmember bolts.
8. Remove right and left engine insulator-to-frame bracket thru bolts.
9. Rotate vibration damper until timing mark is at bottom so crankshaft counterweights will be in proper position for oil pan removal.
10. With suitable equipment, raise engine at front until insulators clear frame brackets.
11. Remove oil pan bolts.
12. Raise engine at front. Apply a rearward force on engine-transmission assembly until oil pan clears flywheel housing and remove pan.
13. Reverse procedure to install.

Rear Main Bearing Seal

The rear main bearing oil seal can be replaced (both halves) without removal of crankshaft. Always replace upper and lower seal as a unit, Fig. 22.

1. Remove oil pan.
2. Use a slide hammer of the type shown in Fig. 23 to remove bearing cap.
3. Pry seal from groove and bearing cap side seals.
4. Clean crankshaft surface.
5. Place new side seals in position and place bearing cap in vise and compress seals into place.
6. Insert new seal well lubricated with engine oil in bearing cap groove. Gradually push with a hammer handle until seal is rolled into place.
7. To replace upper half of seal, use a small hammer and brass pin punch to tap one end of oil seal until it protrudes far enough to be removed with pliers. Push new seal into place.
8. Install bearing cap and oil pan.

PISTONS & CRANKSHAFT BEARINGS

Service on pistons, rings, pins, main and rod bearings are conventional.

Assemble pistons to rods as shown in Fig. 24.

Clutch & Transmission Section

NOTE: 1976 linkage adjustment information is in this section. Repair procedures on both automatic and manual shift transmissions are covered elsewhere in this manual. Procedures for removing automatic transmissions as well as linkage adjustments on 1969-75 models are included in the automatic transmission chapters. See Chapter Index

CLUTCH PEDAL, ADJUST

1969-76 All Models

1. Unhook linkage return spring.
2. On 1975 models, loosen swivel lock nut, remove swivel from lower countershaft hole and position in countershaft upper gauge hole.
3. With clutch pedal against stop, loosen lock nut enough to allow adjusting rod to be turned out of swivel (V8) or push rod (6-cyl.) and rearward against clutch fork until clutch release bearing contacts pressure plate fingers lightly.
4. On 1969-74 models rotate adjustable rod into swivel or push rod 3½ turns and tighten lock nut.
5. On 1975 models rotate push rod until all lash is removed, remove swivel from gauge hole and position in countershaft lower hole, tighten lock nut.

NOTE: Use care when transfering push rod from gauge hole to lower hole to ensure push rod length does not change.

6. Install return spring and check for 1-1½ inch on 1969-74 models, ¾-1¼ inch on 1975 models of free pedal travel.

CLUTCH, REPLACE

1969-76 All

1. Disconnect battery-to-starter cable.
2. Remove transmission as outlined below.
3. Remove release bearing through rear opening in clutch housing.
4. Remove pedal return spring, starter, front flywheel housing shield and flywheel housing.
5. Mark clutch cover and flywheel to insure reassembly in the same position as balanced at the factory.
6. Loosen bolts holding clutch cover to flywheel a little at a time until all tension is relieved. Then remove clutch and driven disc.
7. Reverse removal procedure to install the clutch and adjust pedal free travel.

THREE SPEED MANUAL TRANS.

Transmission, Replace

1969-76

1. Disconnect speedometer cable.
2. Disconnect wiring and shift levers from transmission.
3. Remove propeller shaft.
4. Support rear of engine and remove transmission mount.
5. Remove 4 crossmember bolts and slide member rearward.
6. Remove 2 upper transmission-to-flywheel housing bolts, insert guide pins.
7. Remove two lower bolts.
8. Remove transmission.
9. Reverse procedure to install.

SHIFT LINKAGE 3 SPEED TRANS.

1972-76

Column Shift

1. Set gearshift lever in Reverse and lock ignition.
2. Loosen swivel clamp screw at 1st & Rev. transmission shifter lever (both levers on Ventura) and loosen screw to swivel clamp at cross shaft assembly.
3. Position 2nd & 3rd transmission shifter levers in Neutral and 1st & Rev. shifter lever in Reverse.
4. Tighten swivel clamp screw at 1st & Rev. shifter level to 20 ft. lbs, unlock steering column and shift into Neutral. *On Ventura models*, pull down slightly on 1st & Rev. gearshift control rod to remove any slack in column mechanism before tightening clamp screw.
5. Align steering column lower levers in Neutral (unlock steering column on Ventura) and insert a .186" dia. pin through hole in steering column lower levers.
6. Tighten swivel clamp screw at cross shaft to 20 ft. lbs. *On Ventura models* position 2 & 3 shifter lever in Neutral

and tighten swivel clamp nut at 2nd & 3rd shifter lever to 20 ft. lbs.

7. Remove gauge pin and check complete shift pattern.

Floor Shift (Exc. 1973-74 Muncie)

Follow procedure outlined for 1970-71 floor shift.

Floor Shift (1973-74 Muncie)

1. Place shift control lever in Neutral and insert a .250" pin into shifter assembly alignment holes. Be sure pin engages notch in far side of shifter housing.
2. Loosen swivel clamp screw to upper rod, loosen jam nuts at shifter assembly control rod trunnions, and disconnect 2nd & 3rd and 1st & Rev. trunnion and pin assemblies from both shifter levers.
3. With control rods disconnected, manually move shifter levers to check that transmission gears are in Neutral.
4. With gears and shifter assemblies in Neutral, adjust pin assemblies on their control rods so they align with and enter freely into their holes in shifter levers. Retain each with a plain washer and retainer clip and tighten jam nuts to 20 ft. lbs.
5. Remove gauge pin and check complete pattern for smooth operation.
6. Position shaft lever in Reverse, set steering column upper rod in Lock position and lock ignition. With transmission in Reverse, the ignition key must turn freely to its lock position.
7. Push up on upper rod to take up clearance in steering column lock mechanism and torque screw of adjusting swivel clamp to 20 ft. lbs.

1970-71

Column Shift

1. Set gearshift lever in Reverse position and lock ignition.
2. Loosen swivel clamp screws at transmission shifter lever (1st & Rev) and at cross shaft assembly.
3. Position front transmission shifter lever (2nd & 3rd) in Neutral position and rear transmission shifter lever (1st & Rev) in Reverse position.
4. Tighten swivel clamp screw at transmission 1st & Rev lever, unlock steering column and shift into Neutral.
5. Align gearshift lower control levers on column in Neutral position and insert a .185" pin through hole in levers.
6. Tighten swivel clamp screw at cross shaft to 20 ft. lbs., remove gauge and check operation.
7. Shift transmission into third speed (high gear) and adjust the T.C.S. switch so that the plunger is fully depressed against the 2nd & 3rd shifter lever.

Floor Shift

1. With steering column unlocked, position the shaft control lever into Neutral.
2. Loosen swivel clamp nut retaining the gearshift control rod to the idler lever.
3. Loosen trunnion jam nuts on 1st & Rev control rod, loosen jam nuts on 2nd & 3rd control rod and insert a .250" gauge pin into shifter assembly.
4. Manually position both transmission shifter levers in Neutral and torque jam nuts to 25 ft. lbs.
5. Remove gauge pin from shifter assembly and check operation.
6. Adjust backdrive by shifting into Reverse gear, set gearshift control rod in Lock position and lock the steering column.
7. Push up on gearshift control rod to take up clearance in column lock mechanism and torque adjusting swivel clamp nut to 20 ft. lbs.

1969 All

Column Shift

1. Align upper and lower gearshift levers on steering column in neutral position. Where provision has been made, insert a 3/16" gauge pin through holes in levers.
2. Loosen clamp screws at transmission control rods.
3. Position levers on transmission in neutral.
4. Tighten clamp screws to 20 ft-lbs.

Floor Shift

1. Position selector lever in neutral.
2. Loosen trunnion jam nut on transmission control rods.
3. Place transmission lever and bracket assembly in neutral. Where provision has been made, install a 1/4" gauge pin through hole in bracket.
4. Position transmission lever in neutral.
5. Tighten jam nuts to 30 ft-lbs.
6. Remove gauge pin.

4 SP. TRANS., REPLACE

1969-76

Follow procedure outlined for 1969-76 three-speed model.

SHIFT LINKAGE 4 SPEED TRANS.

1. Place selector lever in neutral.
2. Loosen trunnion nuts on transmission control rod.
3. Place transmission lever and bracket assembly in neutral and install a gauge pin through hole and slot provided in bracket.
4. Position lever on transmission in neutral.
5. Tighten trunnion nuts and remove gauge pin.
6. On 1969-75 units, position shift lever in Reverse, set steering column lever in Lock position and lock ignition. Push up on control rod to remove clearance and tighten nut of adjusting swivel to 20 ft.-lbs.

1976 AUTO. TRANS. LINKAGE, ADJUST

Adjustment procedures for the 1976 linkages are essentially the same as those for 1975 units which are outlined elsewhere in this manual.

Rear Axle, Propeller Shaft & Brake Section

REAR AXLES

Figs. 1 and 2 illustrate the rear axle assemblies used on 1969-76 conventional models. When necessary to overhaul either of these units, refer to the *Rear Axle Specifications* table in this chapter.

1969-76

In this rear axle, Fig. 1, the rear axle housing and differential carrier are cast into an integral assembly. The drive pinion assembly is mounted in two opposed tapered roller bearings. The pinion bearings are preloaded by a spacer behind the front bearing. The pinion is positioned by a shim between the head of the pinion and the rear bearing.

The differential is supported in the carrier by two tapered roller side bearings. These bearings are preloaded by shims located between the bearings and carrier housing. The differential assembly is positioned for proper ring gear and pinion backlash by varying these shims. The differential case houses two side gears in mesh with two pinions mounted on a pinion shaft which is held in place by a lock screw. The side gears and pinions are backed by thrust washers.

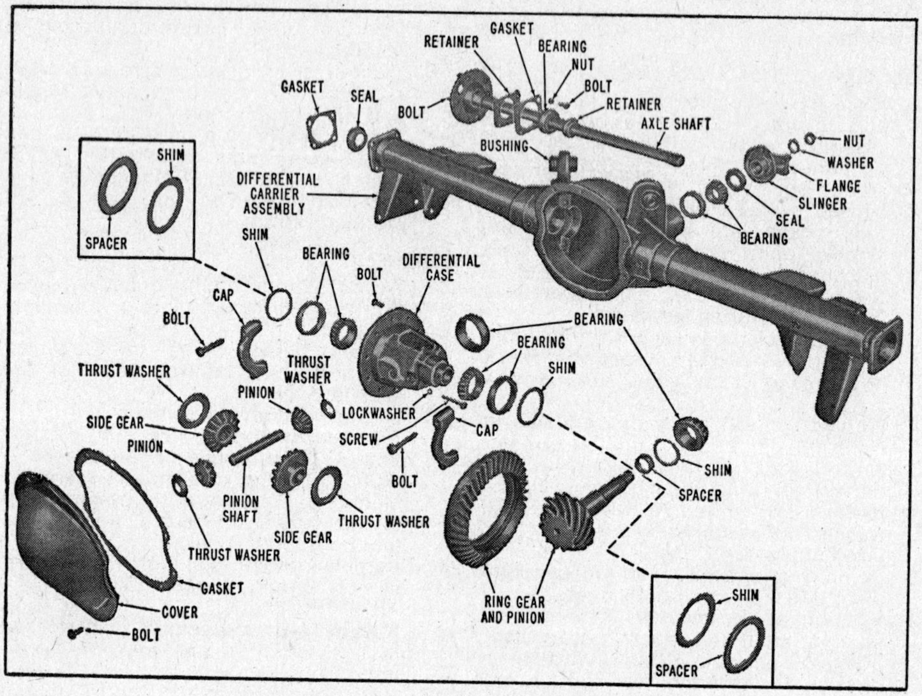

Fig. 1 Rear axle assembly exploded. All 1969 and 1970-76 type "B & O"

Diff. Carrier Assy

1969-76

Construction of the axle assembly is such that service operations may be performed with the housing installed in the vehicle or with the housing removed and installed in a holding fixture. The following procedure is necessary only when the housing requires replacement.

1. Raise car and place a floor jack under center of axle housing so it starts to raise rear axle assembly. Place jack stands solidly under frame members on both sides.
2. Disconnect rear U-joint from drive pinion flange and support propeller shaft out of the way.
3. Remove both axle shafts.
4. Support both brake backing plates out of the way.
5. Disconnect rear brake hose bracket by removing top cover bolt. Remove brake line from housing by bending back tabs.
6. Loosen remaining cover bolts, break loose cover about 1/8" and allow lube to drain.
7. Disconnect shock absorbers at axle housing.
8. On models with coil springs,
   a. Disconnect upper control arms at axle housing.
   b. Slowly lower jack until all spring tension is relieved and remove springs.
   c. Disconnect lower control arms and remove axle assembly.
9. On models with leaf springs,
   a. While supporting spring, remove rear shackle from spring.
   b. Lower spring and remove nut and bolt from spring front bushing, remove spring.
   c. Remove axle assembly.

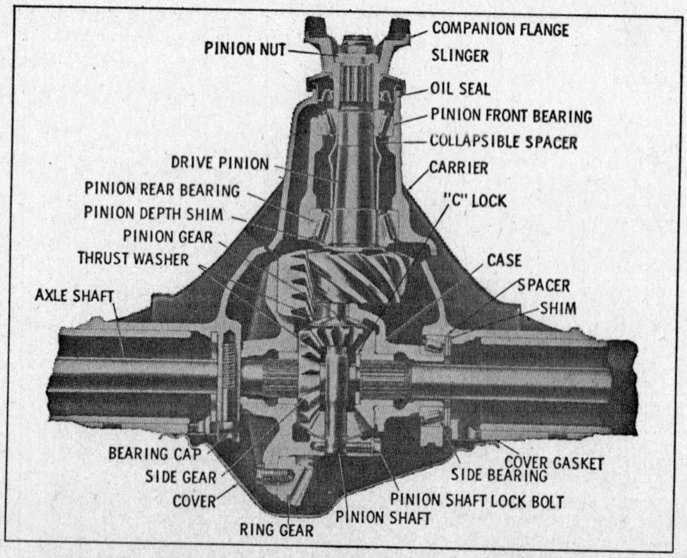

Fig. 2 1970-75 type "C", "G", "K" & "P" differential case

Fig. 3 Removing axle shaft with slide hammer-type propeller

Axle Shaft, Replace

NOTE: Design allows for axle shaft end play of .032" on 1969-73 models, .018" on 1974-76 "B" and "O" axles and .022" on 1974-76 "C", "G", "K" and "P" axles. This end play can be checked with the wheel and brake drum removed by measuring the difference between the end of the housing and the axle shaft flange while moving the axle shaft in and out by hand.

On 1973-76 "C" axles, excessive end play can be compensated for by using selective thickness "C" locks. On all other 1969-75 axles, excessive end play cannot be compensated for by inserting a shim inboard of the bearing in the housing since it ignores the end play of the bearing itself, and may result in improper seating of the gasket or backing plate against the housing. If end play is excessive, the axle shaft and bearing should be removed and the cause of the excessive end play corrected.

All 1969 & 1970-76 Type "B & O"

1. To remove, take off wheels and brake drums.
2. Remove nuts holding retainer plates and brake backing plates. Pull retainers clear of bolts and reinstall two lower nuts finger tight to hold backing plate in position.
3. Use a slide hammer-type puller to remove axle shaft, Fig. 3.

Axle Shaft Bearing

1. Press axle shaft bearing off shaft.
2. Press new bearing against shoulder on shaft.

CAUTION: Outer retainer plate which retains bearing in housing must be on axle shaft before bearing is installed. A new outer retainer gasket can be installed after bearing. Use care not to wedge outer retainer between bearing and shoulder of shaft. Do not press bearing and inner retainer on in one operation.

3. Press new inner retainer ring against bearing.

Axle Shaft Seal

1. Insert suitable tongs behind seal and pull straight out to remove seal.
2. Apply sealer to outside diameter of new seal.
3. Position seal over a suitable installer and drive straight into axle housing until tool bottoms on bearing shoulder in housing.

Axle Shaft, Install

1. Apply a coat of wheel bearing grease in bearing recess of housing. Also lightly lubricate the axle shaft with rear axle lube from the sealing surface to about 6" inboard.
2. Install *new* axle housing-to-brake backing plate gasket.
3. Install brake assembly with backing plate in proper position.
4. With a *new* outer retainer gasket in proper position, insert axle shaft until splines engage differential. *Do not allow shaft to drag on seal.*
5. Drive axle shaft into position.
6. Place new outer retainer gasket and retainer over studs and install nuts, torqueing them to 45-55 ft-lbs.
7. Install brake drums and wheels.

1970-76 Type "C", "G", "K" & "P" axle

1. Raise and support car leaving the rear wheels and differential suspended.
2. Remove rear wheels and brake drums.
3. Remove differential cover and drain lubricant.
4. Remove pinion shaft lock bolt and pinion shaft.
5. Push axle shaft inward to permit removal of "C" locks then remove axle shaft.
6. Install axle shaft bearing and seal remover and remove the bearing and seal.

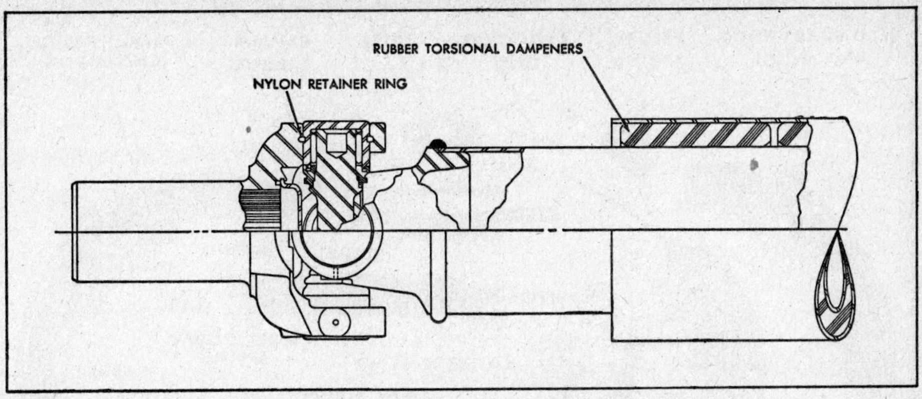

Fig. 4 Insulated propeller shaft. 1969 (typical)

PROPELLER SHAFT

1969-76

Two designs of propeller shaft are used, a solid type which is of one piece tubular steel construction, and a "rubber" type which incorporates rubber torsional dampers between two concentric tubes of steel.

Two methods are used to retain the U-joint bearings to the yoke: conventional snap rings are used at the companion flange of the differential: the remaining six U-joint bearings are held in place with nylon rings.

SERVICE NOTE: Because of the elastic properties of the nylon retainers, Fig. 4, it is not possible to drive the bearings out in the conventional manner. They must be pressed out, which shears the nylon retainers in half, rendering the bearings and journal unsuitable for further use. Therefore, when reassembling, a new bearing and journal assembly employing the conventional snap ring retainers must be used.

CAUTION: Do not attempt to replace the U-joints of propeller shafts from cars equipped with axles of 3.42 ratio and over. They are high speed balanced assemblies and replacement of composite U-joint will destroy this balance.

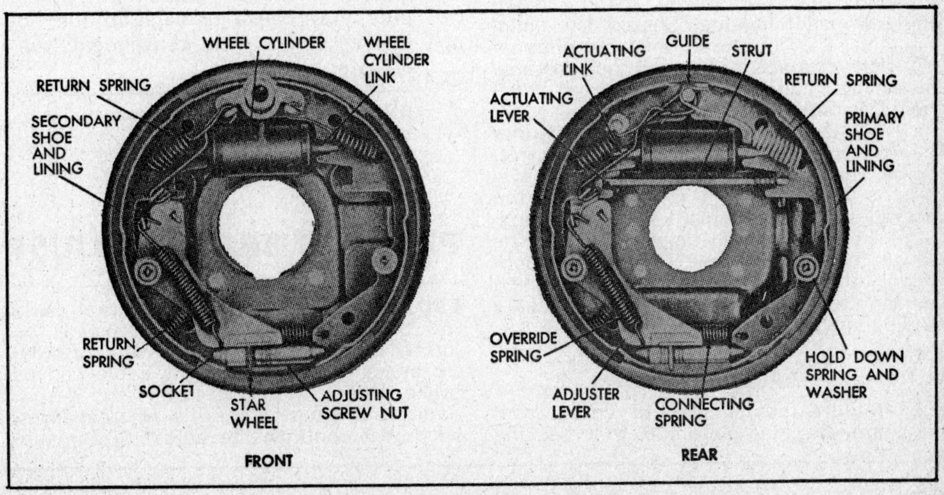

Fig. 5 Right front and rear brake. 1969-76 (typical)

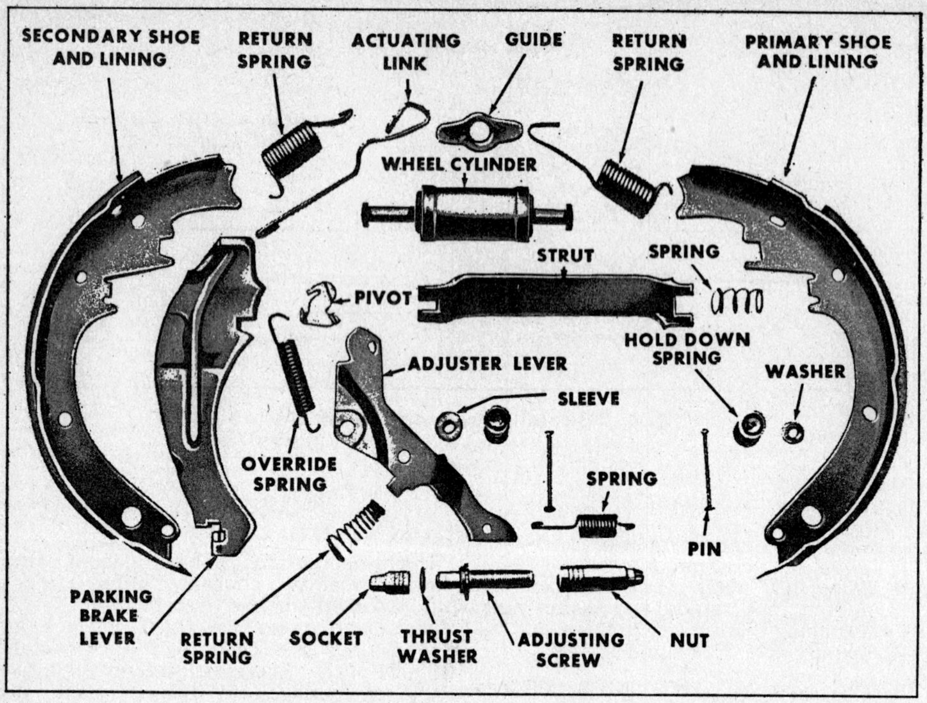

Fig. 5A Typical right rear self adjusting brake

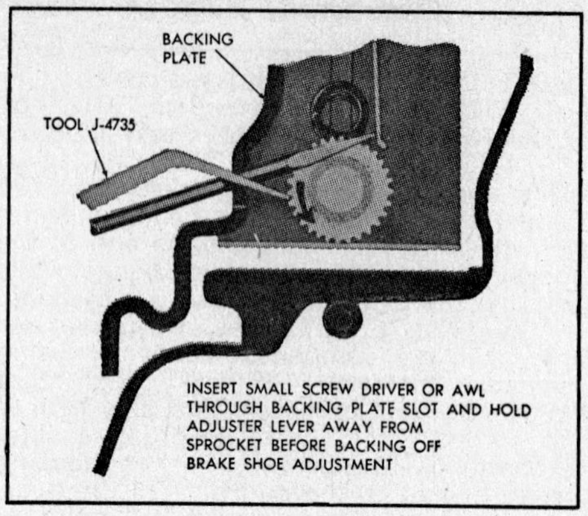

Fig. 6 Backing off adjusting screw. 1969-76

BRAKE ADJUSTMENTS

1969-76 Self-Adjusting Brakes

These brakes, Figs. 5, 5A have self-adjusting shoe mechanisms that assure correct lining-to-drum clearances at all times. The automatic adjusters operate only when the brakes are applied as the car is moving rearward.

Although the brakes are self-adjusting, an initial adjustment is necessary after the brake shoes have been relined or replaced, or when the length of the star wheel adjuster has been changed during some other service operation.

Frequent usage of an automatic transmission forward range to halt reverse vehicle motion may prevent the automatic adjusters from functioning, thereby inducing low pedal heights. Should low pedal heights be encountered, it is recommended that numerous forward and reverse stops be made until satisfactory pedal height is obtained.

If a low pedal height condition cannot be corrected by making numerous reverse stops (provided the hydraulic system is free of air) it indicates that the self-adjusting mechanism is not functioning. Therefore, it will be necessary to remove the drum, clean, free up and lubricate the adjusting mechanism. Then adjust the brakes as follows, being sure the parking brake is fully released.

Adjustment

1. Remove adjusting hole cover from brake backing plate and, from backing plate side, turn adjusting screw upward with a screwdriver or other suitable tool to expand the shoes until a slight drag is felt when the drum is rotated.
2. Remove brake drum.
3. While holding adjusting lever out of engagement with the adjusting screw, Fig. 6, back off the adjusting screw one full turn with the fingers.

NOTE: *If finger movement will not turn the screw, free it up. If this is not done, the adjusting lever will not turn the screw during subsequent vehicle operation. Lubricate the screw with oil and coat with wheel bearing grease. Any other adjustment procedure may cause damage to the adjusting screw with consequent self-adjuster problems.*

PARKING BRAKE, ADJUST

CAUTION: It is very important that parking brake cables are not adjusted too tightly causing brake drag. With automatic brake adjusters, a tight cable causes brake drag and also positions the secondary brake shoe, hence the adjuster lever, so that it continues to adjust to compensate for wear caused by the drag. The result is a cycle of wear and adjustment that can wear out linings very rapidly.

1969-76

1. Jack up both rear wheels.
2. Pull parking brake pedal five to seven notches from fully released position.
3. Loosen equalizer rear lock nut. Adjust forward nut until a light to moderate drag is felt when rear wheels are rotated.
4. Tighten lock nut. Fully released parking brake and rotate rear wheels to be sure there is no drag.

POWER BRAKE UNIT, REPLACE

1969-76 All Models

1. Disconnect vacuum hose at vacuum check valve. Plug hose and cover valve opening to exclude dust.
2. Disconnect pipe(s) from master cylinder hydraulic port and cover opening and pipe end to exclude dirt.
3. Remove clevis pin from brake pedal inside car.
4. Remove nuts from rear half housing and remove power cylinder assembly.
5. Reverse removal procedure to install.

BRAKE MASTER CYLINDER, REPLACE

1969-76

1. Disconnect brake lines from two outlets on master cylinder and tape end of lines to prevent entrance of dirt.
2. Disconnect master cylinder push rod from brake pedal.
3. Remove master cylinder from dash.

Rear Suspension

SHOCK ABSORBER, REPLACE

NOTE: If vehicle is equipped with Superlift shock absorbers, bleed system air pressure through service valve before disconnecting lines at shock absorber fittings.

1. With rear axle supported properly, disconnect shock absorber from lower mounting stud. Use a wrench to prevent mounting stud rotation.
2. Disconnect shock absorber from upper mounting.
3. Reverse procedure to install.

LEAF SPRING & BUSHINGS, REPLACE

1970-76 Firebird, 1971-76 Station Wagons, Ventura & 1974 GTO

1. Support vehicle at frame and support rear axle, removing tension from spring.
2. Disconnect shock absorbers at lower mountings.
3. Remove "U" bolts, spring plate and upper cushion pad.
4. Loosen rear shackle nuts and spring front mounting bolt, Fig. 1.
5. On all except station wagons, disconnect spring front mounting bracket from underbody, drop spring and remove bracket from spring. On station wagons, remove spring front mount bolt.
6. Support spring and remove lower rear shackle bolt.
7. To replace bushings, refer to Figs. 2 and 3.
8. Reverse procedure to install.

1969 Firebird

1. Perform steps 1 and 2 as outlined in above procedure.
2. Loosen spring front mount bolt.

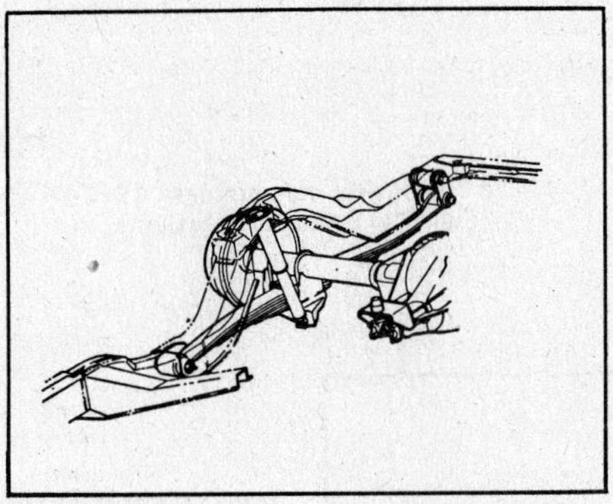

Fig. 1 Leaf spring suspension. 1969-76 Firebird, 1971-76 Station Wagons, Ventura & 1974 GTO (typical)

3. Remove spring retainer bracket to underbody screws, lower axle and remove bracket.
4. Disconnect parking brake cable from spring plate bracket and remove "U" bolts and spring plate.
5. Support spring, remove spring front mount bolt and rear shackle bolt from spring.
6. To replace bushings, refer to Figs. 2 and 3.
7. Reverse procedure to install.

Leaf Spring Service

NOTE: The main leaf may be serviced separately, however, if any of the smaller leaves require replacement, the entire assembly must be replaced.

1. Clamp spring in a vise, remove spring clips and center bolt, then carefully

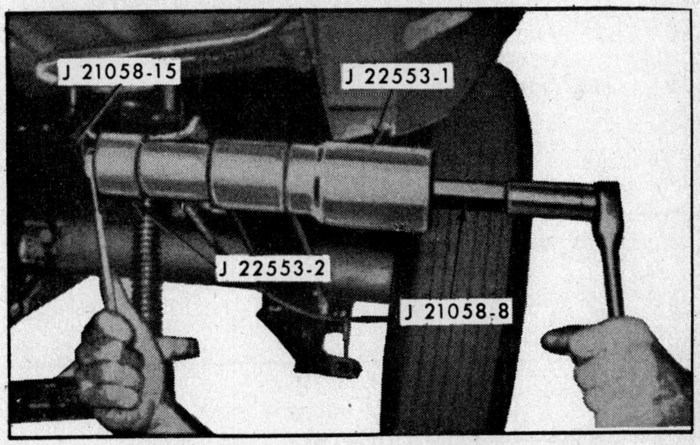

Fig. 3 Leaf spring front bushing installation

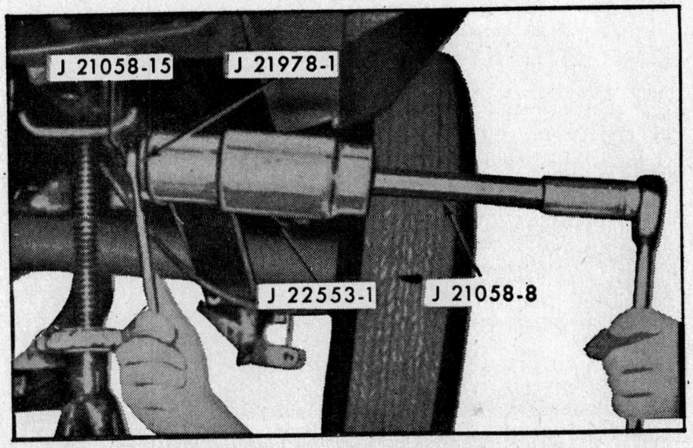

Fig. 2 Leaf spring front bushing removal

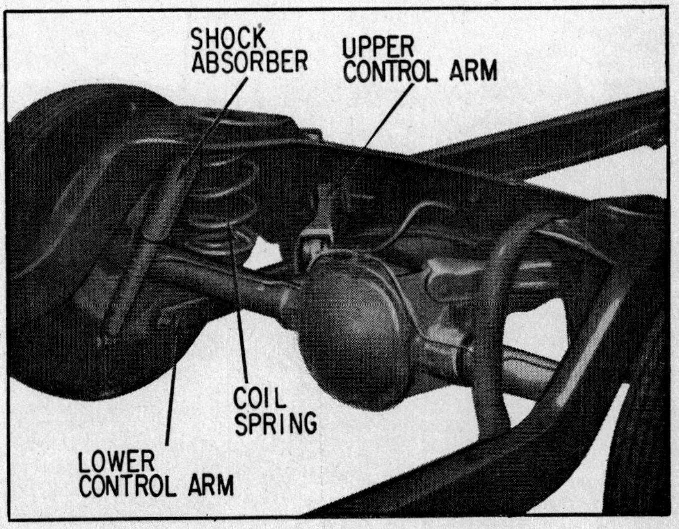

Fig. 4 Coil spring suspension (typical)

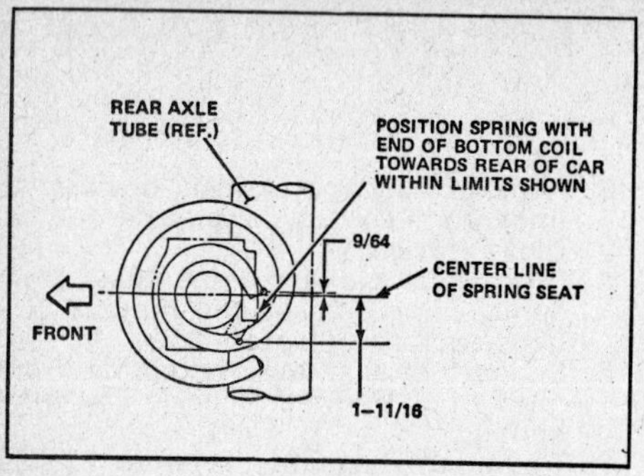

Fig. 5 Indexing coil springs. All 1970 models & 1971-76 Intermediates exc. Firebird, Ventura & 1974 GTO

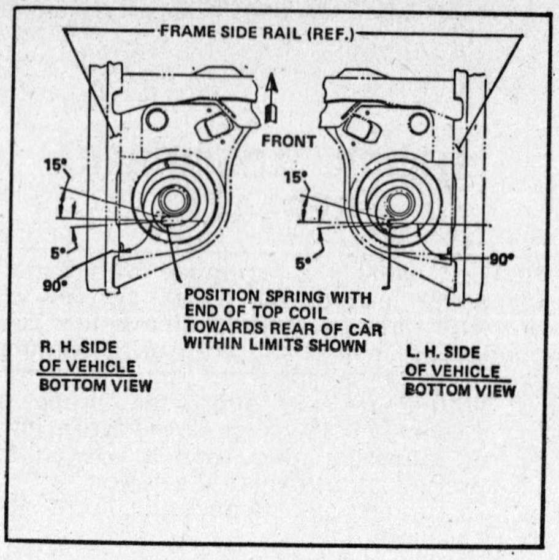

Fig. 6 Indexing coil springs. 1971-76 Full Size exc. station wagons

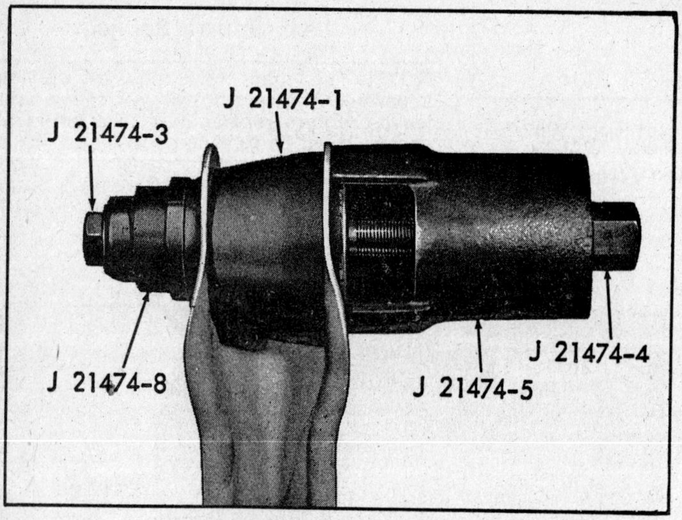

Fig. 7 All front bushings & lower control arm rear bushing removal

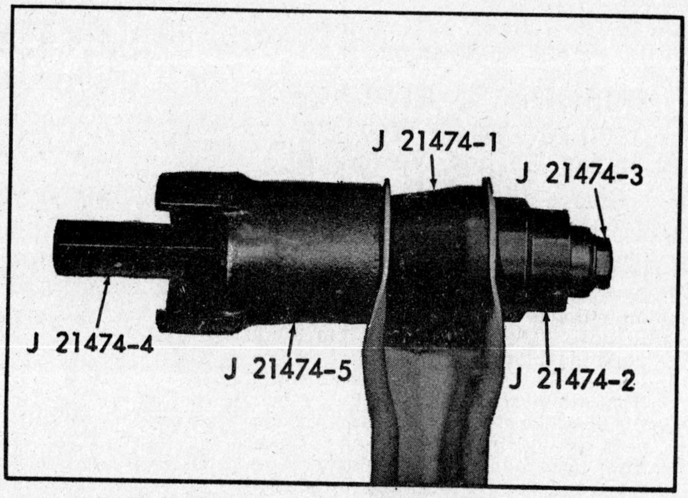

Fig. 8 All front bushings & lower control arm rear bushing installation

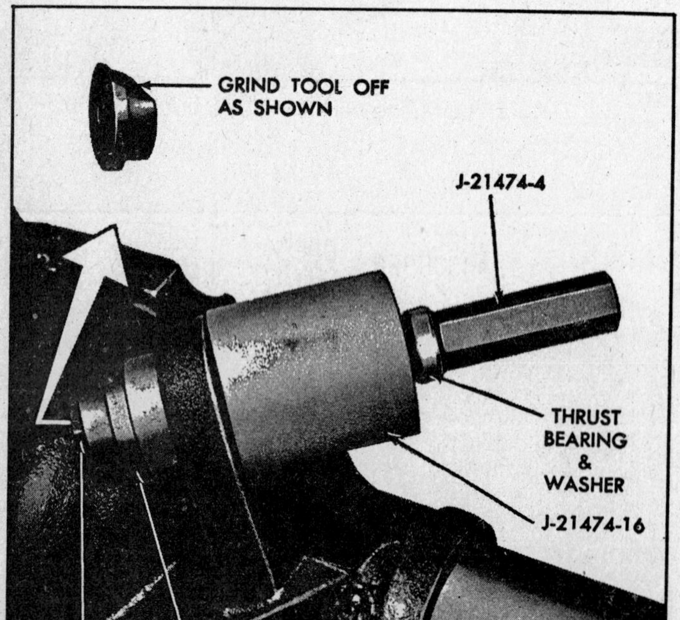

Fig. 9 Upper control arm rear bushing removal. 1970-76 models

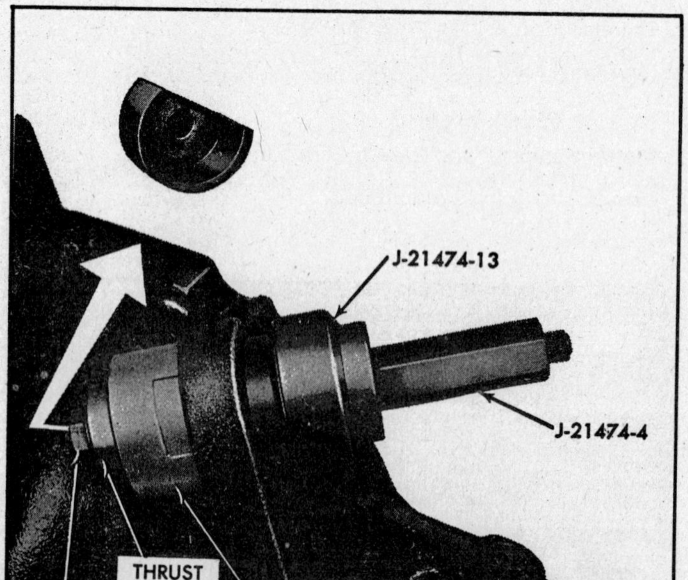

Fig. 10 Upper control arm rear bushing installation. 1970-76 models

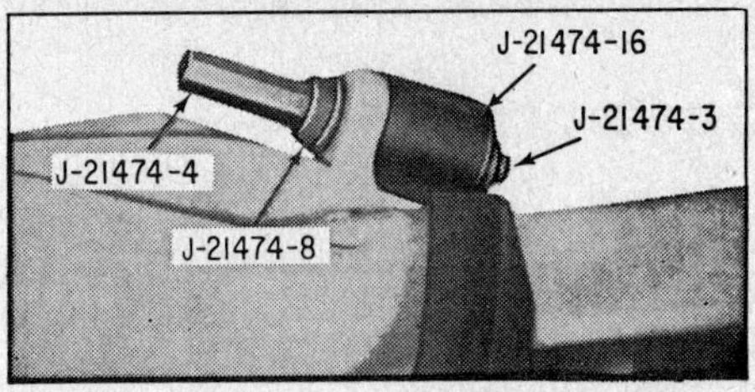

Fig. 11 Upper control arm rear bushing removal. 1969 models

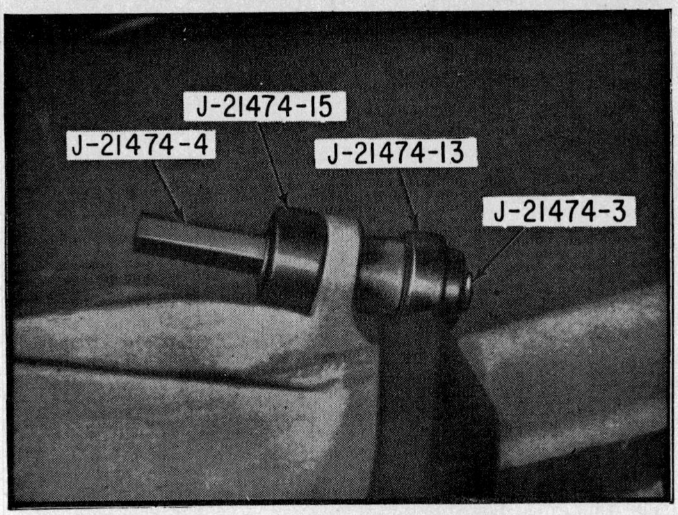

Fig. 12 Upper control arm rear bushing installation. 1969 models

open vise, allowing spring to expand.
2. Replace main leaf and use a drift to align center bolt holes, compress spring in a vise, remove drift and install new center bolt.
3. Align spring leaves and bend spring clips into position.

NOTE: Over tightening spring clips will bind spring action.

COIL SPRING, REPLACE

1. Support vehicle at frame and support rear axle with a suitable jack.
2. Disconnect brake line and axle housing connector from axle housing.
3. Disconnect shock absorbers from lower mountings, lower axle and remove springs, Fig. 4.
4. Replace upper spring seat insulator, if necessary.
5. Reverse procedure to install and properly index springs, Figs. 5 and 6.

CONTROL ARMS & BUSHINGS, REPLACE

NOTE: Replace control arms one at a time to prevent axle misalignment, making installation difficult.

Upper Control Arms

1. Support vehicle at frame and rear axle.
2. Remove control arm front and rear mounting bolts.
3. Replace bushings as necessary, Figs. 7 thru 12.
4. Reverse procedure to install. Tighten control arm bolts with vehicle at curb height.

Lower Control Arms

Follow "Upper Control Arms" procedure for replacement of lower control arms. Lower control arm bushings are serviceable, Figs. 7 and 8.

NOTE: On 1970-73 GTO, remove stabilizer bar, outlined under "Stabilizer Bar, Replace" 1970-73 GTO procedure, before removing lower control arm mounting bolts.

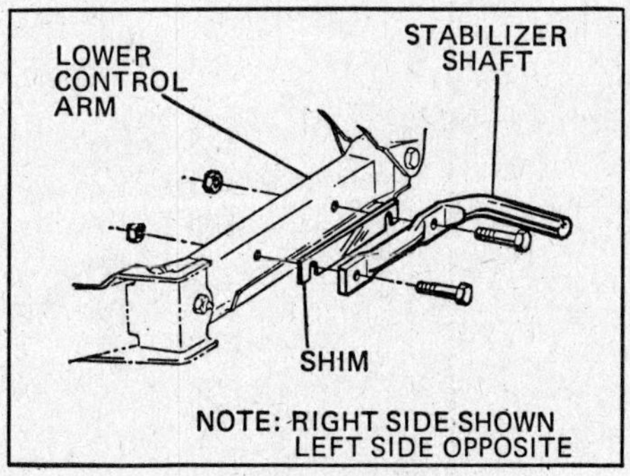

Fig. 13 Stabilizer bar installation. 1970-73 GTO

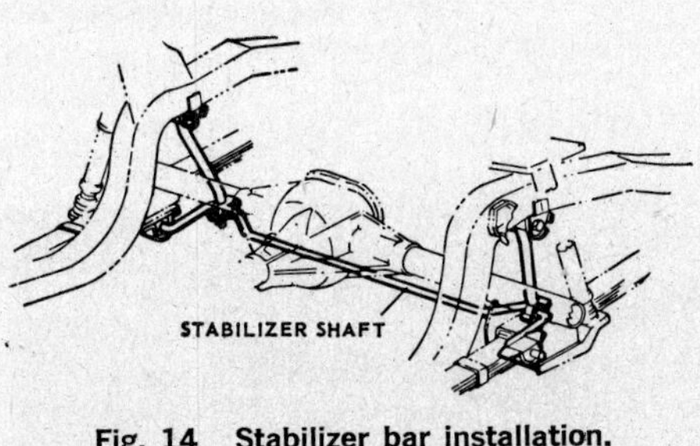

Fig. 14 Stabilizer bar installation. 1970-76 Firebird Formula, Trans Am, 1972-76 Ventura & 1974 GTO

STABILIZER BAR & BUSHINGS, REPLACE

1970-73 GTO

1. With vehicle supported at rear axle, remove bolts attaching stabilizer bar to lower control arms, Fig. 13.
2. Replace bushings as necessary.
3. Reverse procedure to install. Install shims, Fig. 13, if needed, equally on each side of stabilizer bar and tighten attaching bolts with vehicle at curb height.

1974 GTO, 1972-76 Ventura, 1970-76 Firebird Formula & Trans Am

1. With vehicle supported at rear axle, disconnect stabilizer bar from spring plate attachments, Fig. 14.
2. Disconnect stabilizer bar from upper support links and remove support links from frame brackets.
3. Replace bushings as necessary.
4. Reverse procedure to install. Tighten attaching bolts with vehicle at curb height.

Front Suspension & Steering Section

FRONT SUSPENSION
1969-76

The front suspension is of the conventional "A" frame design with coil springs and ball joints. The ball joints have a "fixed boot" grease seal for protection against the entry of dirt and water. The steering knuckles and spindles are of integral design and brake cylinders are rigidly attached to the knuckles with the backing plates serving principally as a support for brake shoes and as a protective cover.

On most 1971-76 models, an integral steering knuckle which is a combination steering knuckle, brake caliper support and steering arm is used. On other models, the steering knuckle is of the conventional type with a separate steering arm.

Rubber bushings at the inner ends of the upper control arms pivot on shafts attached to the car frame. Caster and camber adjustments are made with shims at this point, Fig. 1. Direct acting shock absorbers operate within the coil springs.

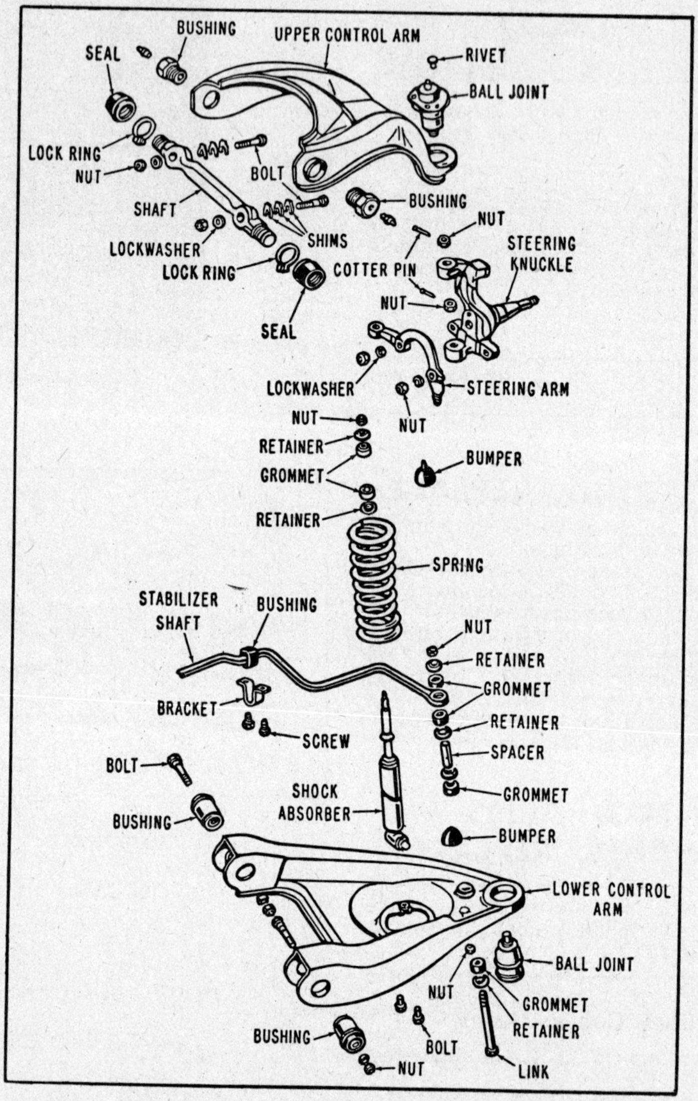

Fig. 1. Disassembled view of front suspension. 1969-76 (typical)

LUBRICATION

Under normal driving conditions, lubrication is required every 4 months or 6000 miles on 1969-74 models and 6 months or 7500 miles on 1975-76 models.

WHEEL ALIGNMENT
1969-76 All Models

Caster and camber adjustments are made by placing shims between the upper pivot shafts and frame, Fig. 2. Both adjustments can be made at the same time. In order to remove or install shims, raise car to remove weight from front wheel, then loosen control arm shaft-to-frame bolts.

1. To decrease positive caster add shims to front bolt.
2. To increase positive caster remove shims from front bolt.
3. To increase camber remove shims from both front and rear bolts.
4. To decrease camber add shims to both front and rear bolts.
5. Compensate for drift to right due to road camber by setting left angle ¼° greater than right.

NOTE: *By adding or subtracting an equal amount of shims from front and rear bolts, camber will be changed without affecting caster.*

6. After proper shim pack has been installed, torque pivot shaft mounting bolts to 50 ft. lbs. on 1969-72 Grand Prix and LeMans and 1971-74 Ventura, 75 ft. lbs. on 1969-71 full size models and Firebird and 80 ft. lbs. on 1972-76 full size models and Firebird, 1973-76 Grand Prix and LeMans and 1975-76 Ventura.

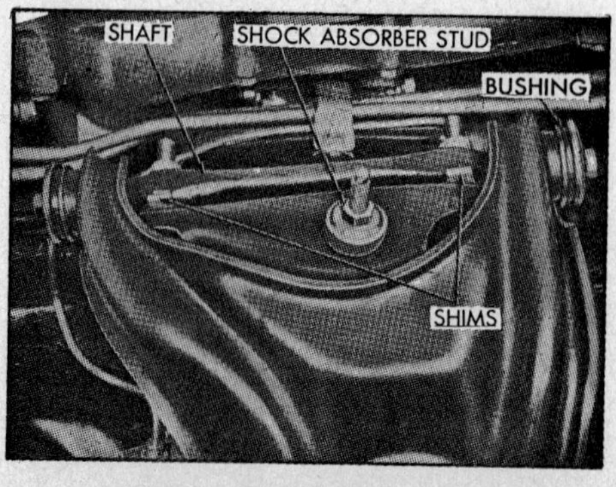

Fig. 2 Caster and camber shim location

TOE-IN, ADJUST

1. With wheels in straight ahead position, loosen tie rod clamp bolts and turn tubes an equal amount until toe-in is according to specifications. Turn right tie rod in direction of forward rotation of wheels to increase toe-in. Turn left tie rod in opposite direction to increase toe-in.
2. Tighten tie rod adjuster sleeve bolts to 14-20 ft-lbs, making sure bolts are to lower rear side of tie rod.

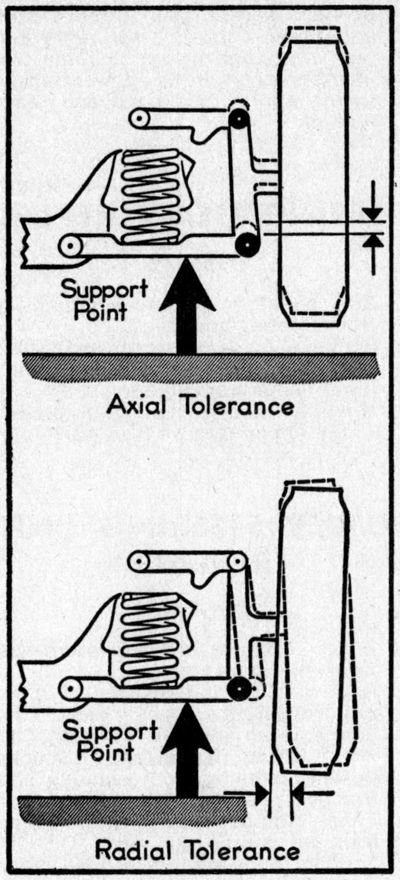

Fig. 3 Checking ball joints for wear

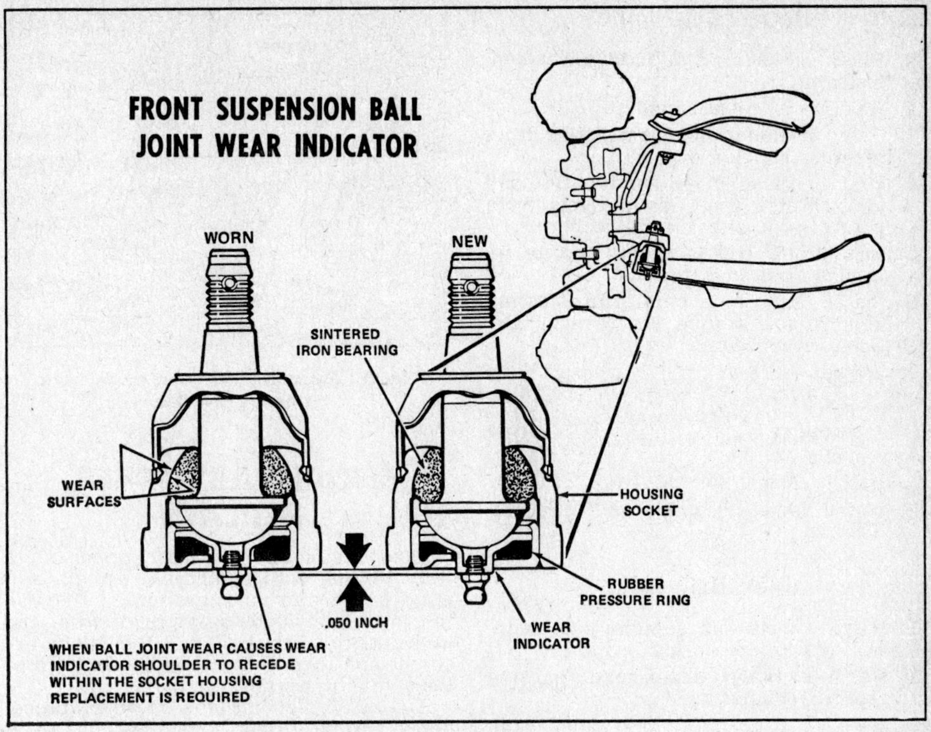

Fig. 4 Ball joint wear indicator. 1973 Catalina, Bonneville, Grandville & 1974 models exc. GTO & Ventura & all 1975-76

WHEEL BEARINGS, ADJUST

1974-76

1. While rotating wheel, torque spindle nut to 12 ft. lbs.
2. Back off spindle nut until just loose, then retighten by hand.
3. Loosen spindle nut until cotter pin can be inserted, however, do not loosen spindle nut more than ½ flat.
4. With bearing properly adjusted, there should be .001-.005 inch end play.

1969-73

1. While rotating wheel, snug-up spindle nut.
2. Back off spindle nut, then retighten by hand.
3. Loosen spindle nut until cotter pin can be inserted.

NOTE: Spindle nut must not be even finger tight.

---

4. With bearing properly adjusted, there should be .001-.005 inch end play on 1969-71 models and .001-.008 inch end play on 1972-73 models.

WHEEL BEARINGS, REPLACE

(Disc Brakes)

1. Raise car and remove front wheels.
2. Remove brake hose support to caliper mounting bracket screw.
3. Remove caliper to mounting bracket bolts.

NOTE: Do not place strain on brake hose.

---

4. Remove spindle nut, and disc and hub assembly. Grease retainer and inner bearing can now be removed.

CHECKING BALL JOINTS FOR WEAR

Before checking ball joints for wear, make sure the front wheel bearings are properly adjusted and that the control arms are tight.

Referring to Fig. 3, raise wheel with a jack placed under the lower control at the point shown. Then test by moving the wheel up and down to check axial play, and rocking it at the top and bottom to measure radial play.

1. Upper ball joint should be replaced if there is any noticeable looseness at this joint.
   If the ball joint is the type using a built in rubber pre-load cushion it will be necessary to remove the ball stud from the knuckle. Then replace the ball joint retaining nut on the ball stud. Using a socket and torque wrench, measure amount of torque required to turn the ball stud in its socket. If any torque is required, the ball joint is satisfactory. If no torque is required, the ball joint must be replaced.
2. Lower ball joint should be replaced if radial play exceeds .250".
3. Lower ball joint should be replaced if axial play between lower control arm and spindle exceeds the following:
   1969-74 . . . . .050"

NOTE: A visual wear indicator is built into the lower ball joint on the 1973 Bonneville, Catalina and Grand Ville and 1974 models except GTO and Ventura and all 1975-76 models, Fig. 4.

If the ball joint is the type using a built-in rubber pre-load cushion it will be necessary to remove the ball stud from the knuckle. Then replace the ball joint retaining nut on the ball stud. Using a socket and torque wrench, measure the amount of torque required to turn the ball stud in its socket. If any torque is required, the ball joint is satisfactory. If no torque is required, the ball joint must be replaced.

BALL JOINTS, REPLACE

On all models the upper ball joint is riveted to the control arm. All service ball joints, however, are provided with bolt, nut and washer assemblies for replacement purposes.

The lower ball joint is pressed into the control arm. They may be pressed out and new joints pressed in.

Upper Ball Joint

1. Raise vehicle and support lower control arm.
2. Remove wheel assembly.
3. Remove upper ball joint stud from steering knuckle, Fig. 5.
4. Drill or chisel rivet heads from ball joint rivets. Then, drive rivets from control arm with a suitable punch.
5. Install new ball joint and torque retaining bolts to 9 ft. lbs.
6. Install ball joint stud into steering knuckle and torque nut to specifications listed below:

Year	Ft. Lbs.
1975-76	50
1974	70
1972-73	40
1969-71	50

   Install cotter pin.
7. Install wheel assembly and lower vehicle.

Lower Ball Joint

1. Raise vehicle and support lower control arm under spring seats.
2. Remove brake drum and backing plate or caliper.
3. Remove lower ball joint stud from steering knuckle, Fig. 5.
4. With a screwdriver, pry ball joint seal and retainer from ball joint.
5. Press lower ball joint from control arm.
6. Press new ball joint into lower control arm.
7. Install ball joint stud into steering knuckle and torque nut to specifications listed below:

Year	Model	Ft. Lbs.
1972-76	Full Size	80
1972-76	Intermediates	70
1971	All	90
1970	Full Size	85
1969-70	Intermediates	85
1969	Full Size	50

   Install cotter pin.
8. Install brake backing plate and drum or caliper, then the wheel assembly.

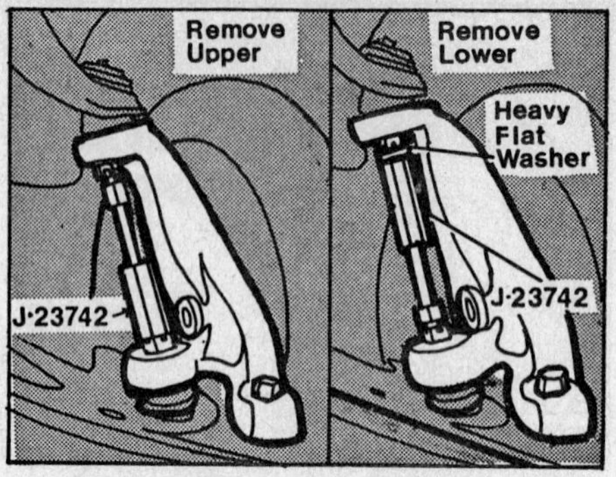

Fig. 5 Removing ball joint studs from steering knuckle.

SHOCK ABSORBER, REPLACE

Hold the shock absorber upper stem from turning with a suitable wrench and remove the nut and grommet. Remove the lower shock absorber pivot from the lower control arm and pull the shock absorber and mounting out at the bottom of the spring housing.

To install, reverse the removal procedure. Torque upper retaining nut to 8 ft. lbs. on 1969-70 models and 10 ft. lbs. on 1971-76 models. Torque shock absorber lower retaining bolts to 20 ft. lbs.

COIL SPRING, REPLACE

1969-76

1. Support vehicle at frame and remove wheel.
2. Remove shock absorber on 1969-70 vehicles, while on 1971-76 vehicles disconnect shock absorber from lower control arm and push shock absorber through hole in lower control arm up into spring.
3. Remove stabilizer link nut, link, spacer, grommets and retainer.
4. Support lower control arm with tool J-23028 bolted onto a suitable jack.
5. Install a safety chain through spring and lower control arm, remove the two lower control arm to frame crossmember pivot bolts. Lower jack, allowing spring to expand and remove spring.

STEERING GEAR, REPLACE

1969-76

1. Use puller to remove pitman arm from steering gear shaft.
2. Scribe a mark on worm shaft flange and steering shaft and disconnect lower flange from steering shaft
3. Unfasten gear housing from frame (3 bolts) and remove from car.

POWER STEERING UNIT, REPLACE

1969-76

1. Disconnect pressure and return hoses from gear housing.
2. Raise vehicle and disconnect pitman arm from shaft.
3. Scribe a mark on steering shaft worm shaft flange and disconnect flexible coupling from steering shaft.
4. Remove gear housing-to-frame bolts, noting number and location of gear-to-frame shims (if any). On 1969 Pontiac models a brake hose bracket must be removed prior to removing frame bolts.

NOTE: *Metal-to-metal contact between flanges of stub shaft and steering shaft will transmit and amplify gear noise to driver. Therefore, when installing the gear, align steering column jacket, shaft and steering gear so head of lower coupling bolt has ¼" clearance from flange on steering shaft. Adjust mast jacket up or down to avoid metal-to-metal contact.*

1946-68 Tune Up Data & Wheel Alignment

Year	Engine or Car Model	Spark Plug Gap	Distributor		Firing Order	Ignition Timing		Valve Lash		Wheel Alignment		
			Point Gap	Dwell Angle		BTDC or Mark	Mark Location	Intake	Exhaust	Caster Degrees	Camber Degrees	Toe-In Inch
AMERICAN MOTORS												
1958	Six L-Head	.035	.016	28–35	153624	3°	Damper	.016H	.018C	Zero	+¼	1/16–3/16
	Six O-Head	.035	.016	28–33	153624	5°	Damper	.012H	.016H	Zero	+¼	1/16–3/16
	V8-250	.035	.016	28–32	18436572①	5°	Damper	.012H	.014H	+½④	Zero	1/16–3/16
	V8-327	.035	.016	28–32	18436572①	5°	Damper	Zero	Zero	+½④	Zero	1/16–3/16
1959	Flat Head Six	.035	.016	28–35	153624	3°	Damper	.016C	.018C	+¼④	Zero	1/16–3/16
	OHV Six	.035	.016	28–35	153624	5°	Damper	.012H	.016H	+¼④	Zero	1/16–3/16
	V8-250	.035	.016	28–32	18436572①	TDC	Damper	.012H	.014H	+½②	Zero	1/16–3/16
	V8-327	.035	.016	28–32	18436572①	5°	Damper	Zero	Zero	+½②	Zero	1/16–3/16
1960	American⑤	.035	.020	37–41	153624	3°	Damper	.016C	.018C	+¼⑧	Zero	1/16–3/16
	American Custom⑥	.035	.016	28–35	153624	8°	Damper	.012H	.016H	+¼⑧	Zero	1/16–3/16
	American Custom⑦	.035	.016	28–35	153624	10°	Damper	.012H	.016H	+¼⑧	Zero	1/16–3/16
	Rambler 6	.035	.016	28–35	153624	5°	Damper	.012H	.016H	+¼②	Zero	1/16–3/16
	V8-250	.035	.016	28–32	18436572①	TDC	Damper	.012H	.014H	+½②	Zero	1/16–3/16
	V8-327-250 H.P.	.035	.016	28–32	18436572①	TDC	Damper	Zero	Zero	+½②	Zero	1/16–3/16
	V8-327-270 H.P.	.035	.016	28–32	18436572①	5°	Damper	Zero	Zero	+½②	Zero	1/16–3/16
1961	American⑤	.035	.020	37–41	153624	3°	Damper	.016C	.018C	+½⑧	Zero	1/16–3/16
	American Custom⑥	.035	.016	28–35	153624	8°	Damper	.012H	.016H	+½⑧	Zero	1/16–3/16
	American Custom⑦	.035	.016	28–35	153624	10°	Damper	.012H	.016H	+½⑧	Zero	1/16–3/16
	Rambler 6	.035	.016	28–35	153624	8°	Damper	.012H	.016H	+½④	Zero	1/16–3/16
	V8-250 Std. Trans.	.035	.016	28–32	18436572①	TDC	Damper	.012H	.014H	+½④	Zero	1/16–3/16
	V8-250 Auto. Trans.	.035	.016	28–32	18436572①	5°	Damper	.012H	.014H	+½④	Zero	1/16–3/16
	V8-327	.035	.016	28–32	18436572①	5°	Damper	Zero	Zero	+½④	Zero	1/16–3/16
1962	American L-Head 6	.035	.020	36–42	153624	3°	Damper	.016C	.018C	Zero⑦	⑩	⅛
	American OHV-6⑤	.035	.016	28–35	153624	8°	Damper	.012H⑧	.016H⑧	Zero⑦	⑩	⅛
	American OHV-6⑥	.035	.016	28–35	153624	10°	Damper	Zero⑨	Zero⑨	Zero⑦	⑩	⅛
	Rambler 6	.035	.016	28–35	153624	5°	Damper	.012H	.016H	Zero③	⑩	⅛
	V8-327 Std. Tr.	.035	.020	28–32	18436572①	5°	Damper	Zero	Zero	Zero③	⑩	⅛
	V8-327 Auto. Tr.	.035	.020	28–32	18436572①	3°	Damper	Zero	Zero	Zero③	⑩	⅛
1963	American L-Head 6	.035	.016	31–34	153624	3°	Damper	.016C	.018C	Zero⑦	⑩	⅛
	American OHV-6⑤	.035	.016	31–34	153624	8°	Damper	.012H⑫	.016H⑫	Zero⑦	⑩	⅛
	American OHV-6⑥	.035	.016	31–34	153624	10°	Damper	.012H⑫	.016H⑫	Zero⑦	⑩	⅛
	Rambler 6	.035	.016	31–34	153624	⑪	Damper	.012H⑫	.016H⑫	Zero③	⑩	⅛
	6-232	.035	.016	31–34	153624	5°	Damper	Zero	Zero	Zero③	⑩	⅛
	V8 Std. Trans.	.035	.016	28–32	18436572①	TDC	Damper	Zero	Zero	Zero③	⑩	⅛
	V8 Auto. Trans.	.035	.016	28–32	18436572①	5°	Damper	Zero	Zero	Zero③	⑩	⅛
1964	American L-Head 6	.035	.016	31–34	153624	3°	Damper	.016C	.018C	+¼⑬	Zero	1/16–3/16
	American OHV-6⑤	.035	.016	31–34	153624	8°	Damper	.012H⑫	.016H⑫	+¼⑬	Zero	1/16–3/16
	American OHV-6⑥	.035	.016	31–34	153624	10°	Damper	.012H⑫	.016H⑫	+¼⑬	Zero	1/16–3/16
	Rambler 6	.035	.016	31–34	153624	5°	Damper	.012H	.016H	+¼⑬	Zero	1/16–3/16
	6-232	.035	.016	31–34	153624	5°	Damper	Zero	Zero	+¼⑬	Zero	1/16–3/16
	V8-Std. Trans.	.035	.016	28–32	18436572①	TDC	Damper	Zero	Zero	+¼⑬	Zero	1/16–3/16
	V8-Auto. Trans.	.035	.016	28–32	18436572①	5°	Damper	Zero	Zero	+¼⑬	Zero	1/16–3/16
1965	American L-Head 6	.035	.016	31–34	153624	3°	Damper	.016C	.018C	+¼⑬	Zero	1/16–3/16
	American OHV-6⑤	.035	.016	31–34	153624	8°	Damper	.012H	.016H	+¼⑬	Zero	1/16–3/16
	American OHV-6⑥	.035	.016	31–34	153624	10°	Damper	.012H	.016H	+¼⑬	Zero	1/16–3/16
	6-199, 232	.035	.016	31–34	153624	5°	Damper	Zero	Zero	+¼⑬	Zero	1/16–3/16
	V8-287, 327	.035	.016	28–32	18436572①	5°	Damper	Zero	Zero	+¼⑬	Zero	1/16–3/16
1966	6-199	.035	.016	31–34	153624	10°⑭	Damper	Zero	Zero	+¼⑬	Zero	1/16–3/16
	6-232	.035	.016	31–34	153624	5°⑭	Damper	Zero	Zero	+¼⑬	Zero	1/16–3/16
	V8-287, 327	.035	.016	28–32	18436572①	5°⑭	Damper	Zero	Zero	+¼⑬	Zero	1/16–3/16
	V8-290	.035	.016	29–31	18436572①	TDC	Damper	Zero	Zero	+¼⑬	Zero	1/16–3/16
1967	6-199 Delco-Remy	.035	.016	31–34	153624	3°⑭	Damper	Hydraulic⑮		Zero③	Zero	⅛
	6-199 Prestolite	.035	.019	36–42	153624	3°⑭	Damper	Hydraulic⑮		Zero③	Zero	⅛
	6-232 Delco-Remy	.035	.016	31–34	153624	5°⑭	Damper	Hydraulic⑮		Zero③	Zero	⅛
	6-232 Prestolite	.035	.020	36–42	153624	5°⑭	Damper	Hydraulic⑮		Zero③	Zero	⅛
	V8's Delco-Remy	.035	.016	29–31	18436572①	TDC	Damper	Hydraulic⑮		Zero③	Zero	⅛
	V8's Prestolite	.035	.016	27–32	18436572①	TDC	Damper	Hydraulic⑮		Zero③	Zero	⅛

Continued

AMERICAN MOTORS—Continued

Year	Engine or Car Model	Spark Plug Gap	Distributor Point Gap	Distributor Dwell Angle	Firing Order	Ignition Timing BTDC or Mark	Ignition Timing Mark Location	Valve Lash Intake	Valve Lash Exhaust	Wheel Alignment Caster Degrees	Wheel Alignment Camber Degrees	Wheel Alignment Toe-In Inch
1968	6-199, 232 Std. Tr.	.035	.016	31–34	153624	TDC	Damper	Hydraulic(15)		(16)	Zero	1/8
	6-199, 232 Auto. Tr.	.035	.016	31–34	153624	5°	Damper	Hydraulic(15)		(16)	Zero	1/8
	V8-290, 343, 390	.035	.016	29–31	18436572	TDC	Damper	Hydraulic(15)		(16)	Zero	1/8

(1)—Cylinder numbering (front to rear): Left bank 1-3-5-7, right bank 2-4-6-8.
(2)—With power steering +3/4°.
(3)—With power steering +1°.
(4)—Deluxe and Super.
(5)—Standard transmission.
(6)—Automatic transmission.
(7)—Power steering +2.°
(8)—Cast iron block.
(9)—Aluminum block.
(10)—Left +1/4°, right 0.
(11)—Early 5°, late 8°.
(12)—With aluminum block hydraulic lifters are used and clearance is "Zero".
(13)—With power steering +1½.
(14)—California vehicle set at TDC.
(15)—No adjustment.
(16)—Rebel & Ambassador; −½, American, AMX & Javelin; w/Power Steering +3/4; w/Manual Steering; Zero.

BUICK

Year	Engine or Car Model	Spark Plug Gap	Distributor Point Gap	Distributor Dwell Angle	Firing Order	Ignition Timing BTDC or Mark	Ignition Timing Mark Location	Valve Lash Intake	Valve Lash Exhaust	Wheel Alignment Caster Degrees	Wheel Alignment Camber Degrees	Wheel Alignment Toe-In Inch
1946–49	All Models	.025	.016	21–30	16258374	"ADV"	Flywheel	.015H(1)	.015H(1)	+3/8	+3/8	1/32
1950–52	All Models	.025	.016	21–30	16258374	"ADV"	Flywheel	.015H(1)	.015H(1)	+3/4	+3/8	3/32
1953	Series 40	.025	.016	21–30	16258374	"ADV"	Flywheel	.015H	.015H	Zero	+3/8	3/32
	All Others	.032	.016	21–30	12784563(2)	"ADV"	Flywheel	Zero	Zero	Zero	+3/8	3/32
1954	All	.032	.016	21–30	12784563(2)	5°	Damper	Zero	Zero	−1/4	+3/8	1/32
1955	All Models	.032	.016	26–33	12784563(2)	5°	Damper	Zero	Zero	−1/8	+3/8	0–1/16
1956	All Models	.032	.016	26–33	12784563(2)	5°	Damper	Zero	Zero	−½	+3/8	1/16–1/8
1957	All Models	.032	(3)	30	12784563(2)	5°	Damper	Zero	Zero	−1 3/4	+½	1/16–1/8
1958	All Models	.032	(3)	30	12784563(2)	(4)	Damper	Zero	Zero	−½	+½	1/16–1/8
1959	364, 401, Std. Tr.	.032	.016(3)	30	12784563(2)	5°	Damper	Zero	Zero	−1½	+½	1/16–5/32
	364, 401 Auto. Tr.	.032	.016(3)	30	12784563(2)	12°	Damper	Zero	Zero	−1½	+½	1/16–5/32
1960	364, 401 Std. Tr.	.032	.016(3)	30	12784563(2)	5°	Damper	Zero	Zero	−2	+½	1/16–5/32
	364, 401 Auto. Tr.	.032	.016(3)	30	12784563(2)	12°	Damper	Zero	Zero	−2	+½	1/16–5/32
1961–62	364, 401	.032	.016(3)	30	12784563(2)	12°	Damper	Zero	Zero	−2	+1/3	3/16–5/32
	V6-198	.035	.016(3)	30	165432(5)	5°	Damper	Zero	Zero	−2	−5/8	7/32–5/16
	V8-215	.035	.016(3)	30	18436572(6)	5°	Damper	Zero	Zero	−2	−5/8	7/32–5/16
1963	401, 425 Std. Tr.	.035	.016(3)	30	12784563(2)	5°	Damper	Zero	Zero	−½	+1/3	7/32–5/16
	401, 425 Auto. Tr.	.035	.016(3)	30	12784563(2)	12°	Damper	Zero	Zero	−½	+1/3	7/32–5/16
	V6-198	.035	.016(3)	30	165432(5)	5°	Damper	Zero	Zero	−2	−5/8	7/32–5/16
	V8-215	.035	.016(3)	30	18436572(6)	5°	Damper	Zero	Zero	−2	−5/8	7/32–5/16
1964	V6-225	.035	.016(3)	30	165432(5)	5°	Damper	Zero	Zero	−½	+½	7/32–5/16
	V8-300 Special	.035	.016(3)	30	18436572(6)	2½°	Damper	Zero	Zero	−½	+½	7/32–5/16
	V8-300 LeSabre	.035	.016(3)	30	18436572(6)	2½°	Damper	Zero	Zero	−½	+1/4	7/32–5/16
	V8-401, 425	.035	.016(3)	30	12784563(2)	2½°(7)	Damper	Zero	Zero	−½	+1/4	7/32–5/16
1965	V6-225	.035	.016(3)	30	165432(5)	5°	Damper	Zero	Zero	−½	+½	1/8–1/4
	V8-300 Special	.035	.016(3)	30	18436572(6)	2½°	Damper	Zero	Zero	−½	+½	1/8–1/4
	V8-300 Le Sabre	.035	.016(3)	30	18436572(6)	2½°	Damper	Zero	Zero	+1	+½	7/32–5/16
	V8-401, 425	.035	.016(3)	30	12784563(2)	2½(7)	Damper	Zero	Zero	+1	+½	7/32–5/16
1966	V6-225	.035	.016(3)	30	165432(5)	5°	Damper	Zero	Zero	−½	+½	1/8–1/4
	V8-300, 340	.035	.016(3)	30	18436572(6)	2½°	Damper	Zero	Zero	(8)	+½	(9)
	V8-401, 425	.035	.016(3)	30	12784563(2)	2½(7)	Damper	Zero	Zero	(8)	+½	(9)
1967	V6-225 Exc. Cal.	.035	.016(3)	30	165432(5)	5°	Damper	Hydraulic(11)		(8)	(12)	(9)
	V6-225 Cal.(10)	.035	.016(3)	30	165432(5)	5°	Damper	Hydraulic(11)		(8)	(12)	(9)
	V8's Exc. Cal.	.035	.016(3)	30	18436572(6)	2½°	Damper	Hydraulic(11)		(8)	(12)	(9)
	V8's Cal.(10)	.035	.016(3)	30	18436572(6)	2½°	Damper	Hydraulic(11)		(8)	(12)	(9)
1968	6-250 Std. Tr.	.030	.019	32	153624	TDC	Damper	1 Turn(13)		−½	+½	1/8–1/4
	6-250 Auto. Tr.	.030	.019	32	153624	4°	Damper	1 Turn(13)		−½	+½	1/8–1/4
	V8-350	.030	.016	30	18436572(6)	TDC	Damper	Hydraulic(11)		(8)	(12)	(9)
	V8-400	.030	.016	30	18436572(6)	TDC	Damper	Hydraulic(11)		(8)	(12)	(9)
	V8-430	.030	.016	30	18436572(6)	TDC	Damper	Hydraulic(11)		(8)	(12)	(9)

(1)—With mechanical valve lifters. Zero lash with hydraulic lifters.
(2)—Cylinders numbering (front to rear): Right bank 1-3-5-7, left bank 2-4-6-8.
(3)—Turn adjusting screw clockwise until engine begins to misfire, then back off screw ½ turn.
(4)—Std. Trans. 5°, Auto. Trans. 12°.
(5)—Cylinders numbering (Front to rear): Right bank 2-4-6, Left bank 1-3-5.
(6)—Cylinders numbering (Front to rear): Right bank 2-4-6-8, Left bank 1-3-5-7.
(7)—12° for V8-425 with 2 carbs. and automatic transmission.
(8)—Full size cars +1; others −½.
(9)—Full size cars 7/32–5/16; others 1/8–1/4.
(10)—Air Injection Reactor (A.I.R.).
(11)—No adjustment.
(12)—Full size cars +1/4; others +½.
(13)—Turn rocker arm adjusting screw down until all lash is removed. Then turn screw down the additional number of turns specified.

Year	Engine or Car Model	Spark Plug Gap	Distributor: Point Gap	Distributor: Dwell Angle	Firing Order	Ignition Timing: BTDC or Mark	Ignition Timing: Mark Location	Valve Lash: Intake	Valve Lash: Exhaust	Wheel Alignment: Caster Degrees	Wheel Alignment: Camber Degrees	Wheel Alignment: Toe-In Inch
CADILLAC												
1946–47	All Models	.030	.016	21–30	18736542(1)	(2)	Damper	Zero	Zero	−2⅛	Zero	1/16
1948–51	All Models	.035	.016	21–30	18436572(1)	(3)	Damper	Zero	Zero	Zero	Zero	3/32
1952–53	All Models	.035	.0125	31	18436572(1)	(3)	Damper	Zero	Zero	Zero	Zero	3/32
1954	All Models	.035	.016	31	18436572(1)	(3)	Damper	Zero	Zero	Zero	Zero	3/32
1955	All Models	.035	.016	26–33	18436572(1)	"A"	Damper	Zero	Zero	−½	Zero	3/16–¼
1956–57	All Models	.035	(4)	30	18436572(1)	"A"	Damper	Zero	Zero	−½	Zero	3/16–¼
1958	All Models	.035	(4)	30	18436572(1)	5°	Damper	.025″ Min.(5)	.025″ Min.(5)	−½	Zero	3/16–¼(6)
1959–62	All Models	.035	(4)	30	18436572(1)	5°	Damper	.025″ Min.(5)	.025″ Min.(5)	(7)	(8)	3/16–¼(6)
1963–66	All Models	.035	(4)	30	18726543(1)	5°	Damper	Zero	Zero	−1	(9)	3/16–¼
1967	Eldorado	.035	(4)	30	18726543(1)	5°	Damper	Hydraulic(10)		Zero	Zero	0–⅛
	All Others	.035	(4)	30	18726543(1)	5°	Damper	Hydraulic(10)		−1	(9)	3/16–¼
1968	Eldorado	.035	(4)	30	15634278(11)	5°	Damper	Hydraulic(10)		Zero	Zero	0–⅛
	All others	.035	(4)	30	15634278(11)	5°	Damper	Hydraulic(10)		−1	(9)	3/16–¼

(1)—Cylinder numbering (front to rear): Left bank 1-3-5-7, right bank 2-4-6-8.
(2)—"IGA" or "IGN" mark.
(3)—"A" mark for premium fuel, "C" mark for regular fuel.
(4)—Turn adjusting screw clockwise until engine begins to misfire, then back out screw ½ turn.
(5)—Clearance between valve stem and rocker arm with hydraulic lifters collapsed and lifters on heel of cam.
(6)—Air suspension cars 1/16″.
(7)—1959–60 Ser. 60, 62 −½, Ser. 75 −1¼, 1961–62 All −1.
(8)—1959–60 Zero, 1961–62 −¼.
(9)—Left side Zero; right side −¼.
(10)—No adjustment.
(11)—Cylinder numbering (front to rear): Left bank 2-4-6-8, right bank 1-3-5-7.

Year	Engine or Car Model	Spark Plug Gap	Point Gap	Dwell Angle	Firing Order	BTDC or Mark	Mark Location	Intake	Exhaust	Caster Degrees	Camber Degrees	Toe-In Inch
CHECKER MOTORS												
1968	6-230 Std. Tr.	.035	(1)	31–34	153624	TDC	Damper	Hydraulic(2)		+2	+1(3)	1/16–⅛(4)
	6-230 Auto. Tr.	.035	(1)	31–34	153624	4°	Damper	Hydraulic(2)		+2	+1(3)	1/16–⅛(4)
	V8-307	.035	(1)	28–32	18436572(6)	2°	Damper	Hydraulic(2)		+2	+1(3)	1/16–⅛(4)
	V8-327	.035	(1)	28–32	18436572(6)	4°	Damper	Hydraulic(2)		+2	+1(3)	1/16–⅛(4)
	V8-327 (5)	.035	(1)	28–32	18436572(6)	8°	Damper	Hydraulic(2)		+2	+1(3)	1/16–⅛(4)

(1)—New points, .019″, used, .016″. On V8's, turn adjusting screw in (clockwise) until engine misfires, then back off ½ turn.
(2)—No adjustment.
(3)—Aerobus +1½.
(4)—Aerobus ⅛–3/16.
(5)—Aerobus.
(6)—Cylinder numbering (front to rear): Right bank 2-4-6-8; left bank 1-3-5-7.

Year	Engine or Car Model	Spark Plug Gap	Point Gap	Dwell Angle	Firing Order	BTDC or Mark	Mark Location	Intake	Exhaust	Caster Degrees	Camber Degrees	Toe-In Inch
CAMARO												
1967	6-230	.035	(2)	31–34	153624	4°	Damper	1 Turn(4)		+½	+¼	⅛–¼
	6-230(1)	.035	(2)	31–34	153624	4°	Damper	1 Turn(4)		+½	+¼	⅛–¼
	6-250	.035	(2)	31–34	153624	4°	Damper	1 Turn(4)		+½	+¼	⅛–¼
	6-250(1)	.035	(2)	31–34	153624	4°	Damper	1 Turn(4)		+½	+¼	⅛–¼
	8-302	.035	(2)	28–32	18436572(3)	6°	Damper	.030H	.030H	+½	+¼	⅛–¼
	8-327, 210 H.P.	.035	(2)	28–32	18436572(3)	2°	Damper	1 Turn(4)		+½	+¼	⅛–¼
	8-327, 210 H.P.(1)	.035	(2)	28–32	18436572(3)	2°	Damper	1 Turn(4)		+½	+¼	⅛–¼
	8-327, 275 H.P.	.035	(2)	28–32	18436572(3)	8°	Damper	1 Turn(4)		+½	+¼	⅛–¼
	8-327, 275 H.P.(1)	.035	(2)	28–32	18436572(3)	6°	Damper	1 Turn(4)		+½	+¼	⅛–¼
	8-327, 350 H.P.	.035	(2)	28–32	18436572(3)	10°	Damper	1 Turn(4)		+½	+¼	⅛–¼
	8-327, 350 H.P.(1)	.035	(2)	28–32	18436572(3)	10°	Damper	1 Turn(4)		+½	+¼	⅛–¼
	8-396, 375 H.P.	.035	(2)	28–32	18436572(3)	6°	Damper	.020H	.024H	+½	+¼	⅛–¼
	8-350	.035	(2)	28–32	18436572(3)	4°	Damper	1 Turn(4)		+½	+¼	⅛–¼
	8-350(1)	.035	(2)	28–32	18436572(3)	4°	Damper	1 Turn(4)		+½	+¼	⅛–¼
1968	6-230, 250 Std. Tr.	.035	(2)	31–34	153624	TDC	Damper	1 Turn(4)		+½	+½	⅛–¼
	6-230, 250 Auto. Tr.	.035	(2)	31–34	153624	4°	Damper	1 Turn(4)		+½	+½	⅛–¼
	8-302	.035	(2)	28–32	18436572(3)	4°	Damper	.030H	.030H	+½	+½	⅛–¼
	8-327, 210 H.P.(5)	.035	(2)	28–32	18436572(3)	2° ATC	Damper	1 Turn(4)		+½	+½	⅛–¼
	8-327, 210 H.P.(6)	.035	(2)	28–32	18436572(3)	2°	Damper	1 Turn(4)		+½	+½	⅛–¼
	8-327, 275 H.P.(5)	.035	(2)	28–32	18436572(3)	TDC	Damper	1 Turn(4)		+½	+½	⅛–¼
	8-327, 275 H.P.(6)	.035	(2)	28–32	18436572(3)	4°	Damper	1 Turn(4)		+½	+½	⅛–¼

Continued

CAMARO—Continued

Year	Engine or Car Model	Spark Plug Gap	Distributor: Point Gap	Distributor: Dwell Angle	Firing Order	Ignition Timing: BTDC or Mark	Ignition Timing: Mark Location	Valve Lash: Intake	Valve Lash: Exhaust	Wheel Alignment: Caster Degrees	Wheel Alignment: Camber Degrees	Wheel Alignment: Toe-In Inch
1968	8-350, 295 H.P.⑤	.035	②	28–32	18436572③	TDC	Damper	1 Turn④		+½	+½	⅛–¼
	8-350, 295 H.P.⑥	.035	②	28–32	18436572③	4°	Damper	1 Turn④		+½	+½	⅛–¼
	8-396, 325 H.P.	.035	②	28–32	18436572③	4°	Damper	1 Turn④		+½	+½	⅛–¼
	8-396, 375 H.P.	.035	②	28–32	18436572③	10°	Damper	.024H	.028H	+½	+½	⅛–¼

①—With air injection reactor system.
②—New points .019″, used .016″. On V-8s, turn adjusting screw in (clockwise) until engine misfires; then back off ½ turn.
③—From front to rear, left bank 1-3-5-7, right bank 2-4-6-8.
④—Turn rocker arm stud until all lash is eliminated, then tighten nut the additional turn listed.
⑤—With standard transmission.
⑥—With automatic transmission.

CHEVELLE

Year	Engine or Car Model	Spark Plug Gap	Distributor: Point Gap	Distributor: Dwell Angle	Firing Order	Ignition Timing: BTDC or Mark	Ignition Timing: Mark Location	Valve Lash: Intake	Valve Lash: Exhaust	Wheel Alignment: Caster Degrees	Wheel Alignment: Camber Degrees	Wheel Alignment: Toe-In Inch
1964	6-194	.035	①	31–34	153624	8°	Damper	1 Turn③		−1	+¾	1/16–3/16
	6-230	.035	①	31–34	153624	4°	Damper	1 Turn③		−1	+¾	1/16–3/16
	V8-283	.035	①	28–32	18436572②	4°	Damper	1 Turn③		−1	+¾	1/16–3/16
	V8-327, 250 H.P.	.035	①	28–32	18436572②	4°	Damper	1 Turn③		−1	+¾	1/16–3/16
	V8-327, 300 H.P.	.035	①	28–32	18436572②	8°	Damper	1 Turn③		−1	+¾	1/16–3/16
1965	6-194	.035	①	31–34	153624	8°	Damper	1 Turn③		④	+½	⅛–¼
	6-230	.035	①	31–34	153624	4°	Damper	1 Turn③		④	+½	⅛–¼
	V8-283, 195 H.P.	.035	①	28–32	18436572②	4°	Damper	1 Turn③		④	+½	⅛–¼
	V8-283, 220 H.P.	.035	①	28–32	18436572②	6°	Damper	1 Turn③		④	+½	⅛–¼
	V8-327, 250 H.P.	.035	①	28–32	18436572②	4°	Damper	1 Turn③		④	+½	⅛–¼
	V8-327, 300 H.P.	.035	①	28–32	18436572②	8°	Damper	1 Turn③		④	+½	⅛–¼
	V8-327, 350 H.P.	.035	①	28–32	18436572②	8°	Damper	1 Turn③		④	+½	⅛–¼
1966	6-194	.035	①	31–34	153624	⑥	Damper	1 Turn③		④	+½	⅛–¼
	6-230	.035	①	31–34	153624	4°	Damper	1 Turn③		④	+½	⅛–¼
	V8-283	.035	①	28–32	18436572②	4°	Damper	1 Turn③		④	+½	⅛–¼
	V8-327	.035	①	28–32	18436572②	⑦	Damper	1 Turn③		④	+½	⅛–¼
	V8-396	.035	①	28–32	18436572②	4°	Damper	1 Turn③		④	+½	⅛–¼
1967	6-230, 250	.035	⑧	31–34	153624	4°	Damper	1 Turn③		④	+½	⅛–¼
	6-230, 250⑤	.035	⑧	31–34	153624	4°	Damper	1 Turn③		④	+½	⅛–¼
	8-283	.035	⑧	28–32	18436572②	4°	Damper	1 Turn③		④	+½	⅛–¼
	8-283 Std. Tr.⑤	.035	⑧	28–32	18436572②	TDC	Damper	1 Turn③		④	+½	⅛–¼
	8-283 Auto. Tr.⑤	.035	⑧	28–32	18436572②	4°	Damper	1 Turn③		④	+½	⅛–¼
	8-327, 275 H.P.	.035	⑧	28–32	18436572②	8°	Damper	1 Turn③		④	+½	⅛–¼
	8-327, 275 H.P.⑤	.035	⑧	28–32	18436572②	6°	Damper	1 Turn③		④	+½	⅛–¼
	8-327, 325 H.P.	.035	⑧	28–32	18436572②	10°	Damper	1 Turn③		④	+½	⅛–¼
	8-327, 325 H.P.⑤	.035	⑧	28–32	18436572②	10°	Damper	1 Turn③		④	+½	⅛–¼
	8-396	.035	⑧	28–32	18436572②	4°	Damper	1 Turn③		④	+½	⅛–¼
	8-396⑤	.035	⑧	28–32	18436572②	4°	Damper	1 Turn③		④	+½	⅛–¼
1968	6-230, 250 Std. Tr.	.035	⑧	31–34	153624	TDC	Damper	1 Turn③		④	+½	⅛–¼
	6-230, 250 Auto. Tr.	.035	⑧	31–34	153624	4°	Damper	1 Turn③		④	+½	⅛–¼
	8-307	.035	⑧	28–32	18436572②	2°	Damper	1 Turn③		④	+½	⅛–¼
	8-327, 275 H.P.⑨	.035	⑧	28–32	18436572②	TDC	Damper	1 Turn③		④	+½	⅛–¼
	8-327, 275 H.P.⑩	.035	⑧	28–32	18436572②	4°	Damper	1 Turn③		④	+½	⅛–¼
	8-327, 325 H.P.	.035	⑧	28–32	18436572②	4°	Damper	1 Turn③		④	+½	⅛–¼
	8-396, 350 H.P.⑨	.035	⑧	28–32	18436572②	TDC	Damper	1 Turn③		④	+½	⅛–¼
	8-396, 350 H.P.⑩	.035	⑧	28–32	18436572②	4°	Damper	1 Turn③		④	+½	⅛–¼

①—New points .019″, used .016″.
②—Cylinder numbering (front to rear): Left bank 1-3-5-7, right bank 2-4-6-8.
③—Turn rocker arm adjusting screw down until all lash is removed. Then turn screw down the additional number of turns specified.
④—Super Sport —½; all others —1.
⑤—With Air Injection Reactor system (A.I.R.).
⑥—With manual transmission and Air Injection Reactor 3°; all others 8°.
⑦—With automatic transmission and Air Injection Reactor 2° ATDC; all others 8° BTDC.
⑧—New points, .019″, used .016″. On V8s, turn adjusting screw in (clockwise) until engine misfires; then back off ½ turn.
⑨—With standard transmission.
⑩—With automatic transmission.

Year	Engine or Car Model	Spark Plug Gap	Distributor		Firing Order	Ignition Timing		Valve Lash		Wheel Alignment		
			Point Gap	Dwell Angle		BTDC or Mark	Mark Location	Intake	Exhaust	Caster Degrees	Camber Degrees	Toe-In Inch
CHEVROLET												
1946–48	All Models	.040	.022	31–37	153624	Ball	Flywheel	.006H	.013H	Zero	−¼	1/32
1949–52	Std. Trans.	.035	.022	31–37	153624	Ball	Flywheel	.006H	.013H	+½	+½	1/16
1950–52	Powerglide	.035	.022	31–37	153624	Ball	Flywheel	1½ Turns②		+½	+½	1/16
1953	Std. Trans.	.035	.016	38–45	153624	①	Flywheel	.006H	.013H	+½	+½	⅛
	Powerglide	.035	.016	38–45	153624	①	Flywheel	1½ Turns②		+½	+½	⅛
1954	Std. Trans.	.035	.016	38–45	153624	Ball	Flywheel	.006H	.016H	+½	+½	⅛
	Powerglide	.035	.016	38–45	153624	Ball	Flywheel	1½ Turns②		+½	+½	⅛
1955	6 Std. Trans.	.035	③	28–35	153624	Ball	Flywheel	.006H	.013H	Zero	+½	⅛–3/16
	6 Powerglide	.035	③	28–35	153624	Ball	Flywheel	1½ Turns②		Zero	+½	⅛–3/16
	V8 Std. Trans.	.035	③	26–33	18436572④	4°	Damper	.008H	.016H	Zero	+½	⅛–3/16
	V8 Powerglide	.035	③	26–33	18436572④	4°	Damper	1 Turn②		Zero	+½	⅛–3/16
1956–57	Six	.035	③	28–35	153624	Ball	Flywheel	1½ Turns②		+1	+½	⅛–3/16
	V8	.035	③	26–33	18436572④	4°	Damper	1 Turn②		+1	+½	⅛–3/16
1958	Six	.035	③	28–35	153624	Ball	Flywheel	1½ Turns		Zero	+½	1/16–3/16
	V8	.035	③	30	18436572④	4°	Damper	1 Turn		Zero	+½	1/16–3/16
1959	6-235	.035	②	28–35	153624	Ball	Flywheel	1½ Turns		Zero	+½	1/16–3/16
	8-283	.035	③	30	18436572④	4°	Damper	1 Turn		Zero	+½	1/16–3/16
	8-283 Special Cam	.035	③	30	18436572④	4°	Damper	.012H	.018H	Zero	+½	1/16–3/16
	8-348	.035	③	30	18436572④	8°	Damper	1 Turn		Zero	+½	1/16–3/16
1960–61	6-235 Hydraulic	.035	③	28–35	153624	Ball	Flywheel	1½ Turns		Zero	+½	1/16–3/16
	6-235 Mechanical	.035	③	28–35	153624	Ball	Flywheel	.008H	.015H	Zero	+½	1/16–3/16
	8-283 170 H.P.	.035	③	30	18436572④	4°	Damper	1 Turn⑦		Zero	+½	1/16–3/16
	8-283 230 H.P.	.035	③	30	18436572④	4°	Damper	1 Turn⑦		Zero	+½	1/16–3/16
	8-283 Special Cam	.035	③	30	18436572④	4°	Damper	.012H	.018H	Zero	+½	1/16–3/16
	8-348 250 H.P.	.035	③	30	18436572④	8°	Damper	1 Turn⑦		Zero	+½	1/16–3/16
	8-348 280 H.P.	.035	⑤	⑥	18436572④	8°	Damper	1 Turn⑦		Zero	+½	1/16–3/16
	8-348 305 H.P.	.035	③	30	18436572④	8°	Damper	1 Turn⑦		Zero	+½	1/16–3/16
	8-348 320 H.P.	.035	③	30	18436572④	8°	Damper	.008H	.018H	Zero	+½	1/16–3/16
	8-348 335 H.P.	.035	⑤	⑥	18436572④	12°	Damper	.008H	.018H	Zero	+½	1/16–3/16
	8-348 340 H.P. (1961)	.035	③	30	18436572④	8°	Damper	.008H	.018H	Zero	+½	1/16–3/16
	8-348 350 H.P. (1961)	.035	⑤	⑥	18436572④	8°	Damper	.008H	.018H	Zero	+½	1/16–3/16
	8-409 360 H.P. (1961)	.035	③	30	18436572④	12°	Damper	.008H	.018H	Zero	+½	1/16–3/16
1962	6-235	.035	③	28–35	153624	⑧	Flywheel	1 Turn⑦		Zero	+½	⅛
	V8-283	.035	③	30	18436572④	4°	Damper	1 Turn⑦		Zero	+½	⅛
	V8-327	.035	③	30	18436572④	4°	Damper	1 Turn⑦		Zero	+½	⅛
	V8-327 Special Cam	.035	③	30	18436572④	8°	Damper	.008H	.018H	Caster	+½	⅛
	V8-409	.035	⑤	⑥	18436572④	12°	Damper	.008H	.018H	Zero	+½	⅛
1963	6-230	.035	③	31–34	153624	4°	Damper	1 Turn②		Zero	+½	1/16–3/16
	V8-283	.035	③	30	18436572④	4°	Damper	1 Turn②		Zero	+½	1/16–3/16
	V8-327 250 H.P.	.035	③	30	18436572④	4°	Damper	1 Turn②		Zero	+½	1/16–3/16
	V8-327 300 H.P.	.035	③	30	18436572④	8°	Damper	1 Turn②		Zero	+½	1/16–3/16
	V8-409 340 H.P.	.035	③	30	18436572④	6°	Damper	1 Turn②		Zero	+½	1/16–3/16
	V8-409 400 H.P.	.035	③	30	18436572④	12°	Damper	⑨		Zero	+½	1/16–3/16
	V8-409 425 H.P.	.035	③	30	18436572④	12°	Damper	⑨		Zero	+½	1/16–3/16
1964	6-230	.035	③	31–34	153624	4°	Damper	1 Turn②		Zero	+½	1/32–3/32
	V8-283	.035	③	30	18436572④	4°	Damper	1 Turn②		Zero	+½	1/32–3/32
	V8-327, 250 H.P.	.035	③	30	18436572④	4°	Damper	1 Turn②		Zero	+½	1/32–3/32
	V8-327, 300 H.P.	.035	③	30	18436572④	8°	Damper	1 Turn②		Zero	+½	1/32–3/32
	V8-409, 340 H.P.	.035	③	30	18436572④	6°	Damper	1 Turn②		Zero	+½	1/32–3/32
	V8-409, 400 H.P.	.035	③	30	18436572④	12°	Damper	⑨		Zero	+½	1/32–3/32
	V8-409, 425 H.P.	.035	③	30	18436572④	12°	Damper	⑨		Zero	+½	1/32–3/32
1965	6-230	.035	③	31–34	153624	4°	Damper	1 Turn②		+¾	+¼	⅛–¼
	V8-283	.035	③	30	18436572④	4°	Damper	1 Turn②		+¾	+¼	⅛–¼
	V8-327, 250 H.P.	.035	③	30	18436572④	4°	Damper	1 Turn②		+¾	+¼	⅛–¼
	V8-327, 300 H.P.	.035	③	30	18436572④	8°	Damper	1 Turn②		+¾	+¼	⅛–¼
	V8-396, 325 H.P.	.035	③	30	18436572④	4°⑩	Damper	1 Turn②		+¾	+¼	⅛–¼
	V8-396, 425 H.P.	.035	③	30	18436572④	10°	Damper	.020H	.024H	+¾	+¼	⅛–¼
	V8-409, 340 H.P.	.035	③	30	18436572④	6°	Damper	1 Turn②		+¾	+¼	⅛–¼
	V8-409, 400 H.P.	.035	③	30	18436572④	12°	Damper	.018H	.030H	+¾	+¼	⅛–¼

Continued

Year	Engine or Car Model	Spark Plug Gap	Distributor		Firing Order	Ignition Timing		Valve Lash		Wheel Alignment		
			Point Gap	Dwell Angle		BTDC or Mark	Mark Location	Intake	Exhaust	Caster Degrees	Camber Degrees	Toe-In Inch

CHEVROLET—Continued

Year	Engine or Car Model	Spark Plug Gap	Point Gap	Dwell Angle	Firing Order	BTDC or Mark	Mark Location	Intake	Exhaust	Caster Degrees	Camber Degrees	Toe-In Inch
1966	6-250	.035	③	31–34	153624	6°	Damper	1 Turn②		+¾	+¼	⅛–¼
	V8-283	.035	③	30	18436572④	4°	Damper	1 Turn②		+¾	+¼	⅛–¼
	V8-327	.035	③	30	18436572④	8°	Damper	1 Turn②		+¾	+¼	⅛–¼
	V8-327⑪	.035	③	30	18436572④	2°⑫	Damper	1 Turn②		+¾	+¼	⅛–¼
	V8-396	.035	③	30	18436572④	4°	Damper	1 Turn②		+¾	+¼	⅛–¼
	V8-427, 390 H.P.	.035	—	—	18436572④	4°	Damper	1 Turn②		+¾	+¼	⅛–¼
	V8-427, 425 H.P.	.035	—	—	18436572④	10°	Damper	.020H	.024H	+¾	+¼	⅛–¼
1967	6-250	.035	⑬	31–34	153624	4°	Damper	1 Turn②		+¾	+¼	⅛–¼
	6-250⑪	.035	⑬	31–34	153624	4°	Damper	1 Turn②		+¾	+¼	⅛–¼
	8-283	.035	⑬	28–32	18436572④	4°	Damper	1 Turn②		+¾	+¼	⅛–¼
	8-283 Std. Tr.⑪	.035	⑬	28–32	18436572④	TDC	Damper	1 Turn②		+¾	+¼	⅛–¼
	8-283 Auto. Tr.⑪	.035	⑬	28–32	18436572④	4°	Damper	1 Turn②		+¾	+¼	⅛–¼
	8-327, 275 H.P.	.035	⑬	28–32	18436572④	8°	Damper	1 Turn②		+¾	+¼	⅛–¼
	8-327, 275 H.P.⑪	.035	⑬	28–32	18436572④	6°	Damper	1 Turn②		+¾	+¼	⅛–¼
	8-396, 325 H.P.	.035	⑬	28–32	18436572④	4°	Damper	1 Turn②		+¾	+¼	⅛–¼
	8-396, 325 H.P.⑪	.035	⑬	28–32	18436572④	4°	Damper	1 Turn②		+¾	+¼	⅛–¼
	8-427, 385 H.P.	.035	⑬	28–32	18436572④	4°	Damper	1 Turn②		+¾	+¼	⅛–¼
	8-427, 385 H.P.⑪	.035	⑬	28–32	18436572④	4°	Damper	1 Turn②		+¾	+¼	⅛–¼
1968	6-250 Std. Tr.	.035	⑬	31–34	153624	TDC	Damper	1 Turn②		+¾	+¼	⅛–¼
	6-250 Auto. Tr.	.035	⑬	31–34	153624	4°	Damper	1 Turn②		+¾	+¼	⅛–¼
	8-307	.035	⑬	28–32	18436572④	2°	Damper	1 Turn②		+¾	+¼	⅛–¼
	8-327, 250 H.P.	.035	⑬	28–32	18436572④	4°	Damper	1 Turn②		+¾	+¼	⅛–¼
	8-327, 275 H.P.⑭	.035	⑬	28–32	18436572④	TDC	Damper	1 Turn②		+¾	+¼	⅛–¼
	8-327, 275 H.P.⑮	.035	⑬	28–32	18436572④	4°	Damper	1 Turn②		+¾	+¼	⅛–¼
	8-396, 325 H.P.	.035	⑬	28–32	18436572④	4°	Damper	1 Turn②		+¾	+¼	⅛–¼
	8-396, 375 H.P.	.035	⑬	28–32	18436572④	4°	Damper	.024H	.028H	+¾	+¼	⅛–¼
	8-427, 385 H.P.	.035	⑬	28–32	18436572④	4°	Damper	1 Turn②		+¾	+¼	⅛–¼

①—Before Engine No. LAA-283230 and LAQ-117209, set timing 7/32″ after timing ball. Starting with these engine numbers, set timing at the ball.
②—Turn rocker arm adjusting screw down until all lash is removed. Then turn screw down the additional number of turns specified.
③—New points .019″, used points .016″.
④—Cylinder numbering (front to rear): Left bank 1-3-5-7, right bank 2-4-6-8.
⑤—New Points .018″, used .015″.
⑥—Each set of points 28–30° total both sets 33–35°.
⑦—For long travel lifters 2 turns.
⑧—First short vertical line clockwise from ball in flywheel.
⑨—Early production: intake .012H, exhaust .020H. Engines date stamped TO-1118QA or QB: intake .018H, exhaust .030H.
⑩—Transistorized ignition 6° BTDC.
⑪—Automatic transmission and Air Injection Reactor system (A.I.R.).
⑫—After top dead center.
⑬—New points, .019″, used .016″. On V8s, turn adjusting screw in (clockwise) until engine misfires; then back off ½ turn.
⑭—With standard transmission.
⑮—With automatic transmission.

CHEVY II & NOVA

Year	Engine or Car Model	Spark Plug Gap	Point Gap	Dwell Angle	Firing Order	BTDC or Mark	Mark Location	Intake	Exhaust	Caster Degrees	Camber Degrees	Toe-In Inch
1962–63	4-153	.035	②	31–34	1342	4°	Damper	1 Turn④		+1	+½	¼
	6-194	.035	②	31–34	153624	8°	Damper	1 Turn④		+1	+½	¼
1964	4-153	.035	②	31–34	1342	4°	Damper	1 Turn④		+1	+½	3/16–5/16
	6-194	.035	②	31–34	153624	8°	Damper	1 Turn④		+1	+½	3/16–5/16
	6-230	.035	②	31–34	153624	4°	Damper	1 Turn④		+1	+½	3/16–5/16
	V8-283, 195 H.P.	.035	②	30	18436572③	4°	Damper	1 Turn④		+1	+½	3/16–5/16
	V8-283, 220 H.P.	.035	②	30	18436572③	8°	Damper	1 Turn④		+1	+½	3/16–5/16
1965	4-153	.035	②	31–34	1342	4°	Damper	1 Turn④		+1	+½	¼–⅜
	6-194	.035	②	31–34	153624	8°	Damper	1 Turn④		+1	+½	¼–⅜
	6-230	.035	②	31–34	153624	4°	Damper	1 Turn④		+1	+½	¼–⅜
	V8-283, 195 H.P.	.035	②	30	18436572③	4°	Damper	1 Turn④		+1	+½	¼–⅜
	V8-283, 220 H.P.	.035	②	30	18436572③	6°	Damper	1 Turn④		+1	+½	¼–⅜
	V8-327, 250 H.P.	.035	②	30	18436572③	4°	Damper	1 Turn④		+1	+½	¼–⅜
	V8-327, 300 H.P.	.035	②	30	18436572③	8°	Damper	1 Turn④		+1	+½	¼–⅜
1966	4-153	.035	②	31–34	1342	4°	Damper	1 Turn④		+1	+½	¼–⅜
	6-194	.035	②	31–34	153624	8°	Damper	1 Turn④		+1	+½	¼–⅜
	6-194⑤	.035	②	31–34	153624	3°	Damper	1 Turn④		+1	+½	¼–⅜

Continued

CHEVY II & NOVA—Continued

Year	Engine or Car Model	Spark Plug Gap	Distributor Point Gap	Distributor Dwell Angle	Firing Order	Ignition Timing BTDC or Mark	Ignition Timing Mark Location	Valve Lash Intake	Valve Lash Exhaust	Wheel Alignment Caster Degrees	Wheel Alignment Camber Degrees	Wheel Alignment Toe-In Inch
1966	6-230	.035	②	31–34	153624	4°	Damper	1 Turn④		+1	+½	¼–⅜
	V8-283	.035	②	30	18436572③	4°	Damper	1 Turn④		+1	+½	¼–⅜
	V8-327, 275 H.P.	.035	②	30	18436572③	8°	Damper	1 Turn④		+1	+½	¼–⅜
	V8-327, 275 H.P.⑥	.035	②	30	18436572③	2°⑦	Damper	1 Turn④		+1	+½	¼–⅜
	V8-327, 350 H.P.	.035	②	30	18436572③	10°	Damper	1 Turn④		+1	+½	¼–⅜
	V8-396	.035	②	30	18436572③	4°	Damper	1 Turn④		+1	+½	¼–⅜
1967	4-153	.035	⑨	31–34	1342	4°	Damper	1 Turn④		+1	+½	¼–⅜
	6-194	.035	⑨	31–34	153624	4°	Damper	1 Turn④		+1	+½	¼–⅜
	6-194 Std. Tr.⑧	.035	⑨	31–34	153624	2°	Damper	1 Turn④		+1	+½	¼–⅜
	6-194 Auto. Tr.⑧	.035	⑨	31–34	153624	4°	Damper	1 Turn④		+1	+½	¼–⅜
	6-250	.035	⑨	31–34	153624	4°	Damper	1 Turn④		+1	+½	¼–⅜
	6-250⑧	.035	⑨	31–34	153624	4°	Damper	1 Turn④		+1	+½	¼–⅜
	8-283	.035	⑨	28–32	18436572③	4°	Damper	1 Turn④		+1	+½	¼–⅜
	8-283 Std. Tr.⑧	.035	⑨	28–32	18436572③	TDC	Damper	1 Turn④		+1	+½	¼–⅜
	8-283 Auto. Tr.⑧	.035	⑨	28–32	18436572③	4°	Damper	1 Turn④		+1	+½	¼–⅜
	8-327, 275 H.P.	.035	⑨	28–32	18436572③	8°	Damper	1 Turn④		+1	+½	¼–⅜
	8-327, 275 H.P.⑧	.035	⑨	28–32	18436572③	6°	Damper	1 Turn④		+1	+½	¼–⅜
1968	4-153 Std. Tr.	.035	⑨	31–34	1342	TDC	Damper	1 Turn④		+½	+½	⅛–¼
	4-153 Auto. Tr.	.035	⑨	31–34	1342	4°	Damper	1 Turn④		+½	+½	⅛–¼
	6-230, 250 Std. Tr.	.035	⑨	31–34	153624	TDC	Damper	1 Turn④		+½	+½	⅛–¼
	6-230, 250 Auto. Tr.	.035	⑨	31–34	153624	4°	Damper	1 Turn④		+½	+½	⅛–¼
	8-307	.035	⑨	28–32	18436572③	2°	Damper	1 Turn④		+½	+½	⅛–¼
	8-327, 275 H.P.⑩	.035	⑨	28–32	18436572③	TDC	Damper	1 Turn④		+½	+½	⅛–¼
	8-327, 275 H.P.⑪	.035	⑨	28–32	18436572③	4°	Damper	1 Turn④		+½	+½	⅛–¼
	8-350, Std. Tr.	.035	⑨	28–32	18436572③	TDC	Damper	1 Turn④		+½	+½	⅛–¼
	8-350, Auto. Tr.	.035	⑨	28–32	18436572③	4°	Damper	1 Turn④		+½	+½	⅛–¼

①—For long travel lifters 2 turns.
②—New points .019″, used points .016″.
③—Cylinder numbering (front to rear): Left bank 1-3-5-7, right bank 2-4-6-8.
④—Turn rocker arm adjusting screw down until all lash is removed. Then turn screw down the additional number of turns specified.
⑤—Manual transmission and Air Injection Reactor system (A.I.R.).
⑥—Automatic transmission and Air Injection Reactor system (A.I.R.).
⑦—After top dead center.
⑧—With Air Injection Reactor (A.I.R.).
⑨—New points, .019″, used .016″. On V8s, turn adjusting screw in (clockwise) until engine misfires; then back off ½ turn.
⑩—With standard transmission.
⑪—With automatic transmission.

CHRYSLER

Year	Engine or Car Model	Spark Plug Gap	Distributor Point Gap	Distributor Dwell Angle	Firing Order	Ignition Timing BTDC or Mark	Ignition Timing Mark Location	Valve Lash Intake	Valve Lash Exhaust	Wheel Alignment Caster Degrees	Wheel Alignment Camber Degrees	Wheel Alignment Toe-In Inch
1946–48	Six	.025	.020	35–38	153624	②	Damper	.008H	.010H	Zero	+⅜	1/32
	Eight	.025	.017	37–30	16258374	②	Damper	.008H③	.010H③	Zero	+⅜	1/32
1949–50	Six	.035	.020	35–38	153624	"0"	Damper	.008H	.010H	−2	+⅜	1/32
	Eight	.035	.017	27–30	16258374	"0"	Damper	.008H③	.010H③	−2	+⅜	1/32
1951–52	Six	.035	.020	35–38	153624	②	Damper	.008H	.010H	−2	Zero	Zero
	V8	.035	.017	④	18436572⑤	"0"	Damper	Zero	Zero	−2	Zero	Zero
1953–54	Six	.035	.020	35–38	153624	"0"	Damper	.008H	.010H	−2	Zero	Zero
	V8	.035	.017	④	18436572⑤	⑥	Damper	Zero	Zero	−2	Zero	Zero
1955	300	.035	.017	④	18436572⑤	10°	Damper	.015H	.024H	⑨	⑩	0–1/16
	Others	.035	.017	④	18436572⑤	6°	Damper	Zero	Zero	⑨	⑩	0–1/16
1956	Windsor	.035	.017	31	18436572⑤	2°	Damper	Zero	Zero	⑨	⑩	⅛
	New Yorker	.035	.017	⑦	18436572⑤	4°	Damper	Zero	Zero	⑨	⑩	⅛
	300B	.035	.017	31	18436572⑤	8°	Damper	.015H	.024H	⑨	⑩	⅛
1957	300C	.035	.017	⑧	18436572⑤	4°	Damper	.015H	.024H	⑪	⑫	3/32–5/32
	New Yorker	.035	.017	⑧	18436572⑤	6°	Damper	Zero	Zero	⑪	⑫	3/32–5/32
	Others	.035	.017	29	18436572⑤	6°	Damper	Zero	Zero	⑪	⑫	3/32–5/32
1958	V8-354	.035	.017	27–32	18436572⑤	⑬	Damper	.060–.210⑭		⑮	⑫	3/32–5/32
	V8-392 New Yorker	.035	.017	⑧	18436572⑤	6°	Damper	.060–.210⑭		⑮	⑫	3/32–5/32
	V8-392 "300D"	.035	.017	⑧	18436572⑤	6°	Damper	.015H	.024H	⑮	⑫	3/32–5/32
1959	V8-383	.035	.017	27–32	18436572⑤	10°	Damper	.060–.210⑭		−¾	⑫	⅛
	V8-413	.035	.017	⑦	18436572⑤	10°	Damper	.060–.210⑭		−¾	⑫	⅛

Continued

Year	Engine or Car Model	Spark Plug Gap	Distributor		Firing Order	Ignition Timing		Valve Lash		Wheel Alignment		
			Point Gap	Dwell Angle		BTDC or Mark	Mark Location	Intake	Exhaust	Caster Degrees	Camber Degrees	Toe-In Inch
CHRYSLER—Continued												
1960	V8-383	.035	.017	27–32	18436572(5)	10°	Damper	.060–.210(14)		(11)	(16)	1/8
	V8-413 New Yorker	.035	.017	27–32	18436572(5)	10°	Damper	.060–.210(14)		(11)	(16)	1/8
	V8-413 "300F" S.T.	.035	.017	(7)	18436572(5)	10°	Damper	.060–.210(14)		(11)	(16)	1/8
	V8-413 "300F" A.T.	.035	.017	(7)	18436572(5)	5°	Damper	.060–.210(14)		(17)	(16)	1/8
1961	V8-361 Newport	.035	.017	27–32	18436572(5)	10°	Damper	.060–.210(14)		(17)	(18)	1/8
	V8-383 Windsor	.035	.017	27–32	18436572(5)	10°	Damper	.060–.210(14)		(17)	(18)	1/8
	V8-413 New Yorker	.035	.017	27–32	18436572(5)	10°	Damper	.060–.210(14)		(17)	(18)	1/8
	V8-413 "300G" S.T.	.035	.017	(7)	18436572(5)	10°	Damper	.060–.210(14)		(17)	(18)	1/8
	V8-413 "300G" A.T.	.035	.017	(7)	18436572(5)	5°	Damper	.060–.210(14)		(17)	(18)	1/8
1962	V8-361 Newport	.035	.017	27–32	18436572(5)	10°	Damper	.060–.210(14)		(17)	(18)	1/8
	V8-383 "300"	.035	.017	27–32	18436572(5)	10°	Damper	.060–.210(14)		(17)	(18)	1/8
	V8-413 New Yorker	.035	.017	27–32	18436572(5)	10°	Damper	.060–.210(14)		(17)	(18)	1/8
	V8-413 "300H"	.035	.017	(7)	18436572(5)	10°	Damper	.015H	.024H	(17)	(18)	1/8
1963	V8-361 Newport	.035	.017	27–32	18436572(5)	10°	Damper	.060–.210(14)		(17)	(18)	1/8
	V8-383 "300"	.035	.017	27–32	18436572(5)	10°	Damper	.060–.210(14)		(17)	(18)	1/8
	V8-413 "300"	.035	.017	(7)	18436572(5)	10°	Damper	.060–.210(14)		(17)	(18)	1/8
	V8-413 New Yorker	.035	.017	28–33	18436572(5)	10°	Damper	.060–.210(14)		(17)	(18)	1/8
	V8-413 "300J"	.035	.017	(7)	18436572(5)	12½°	Damper	.015H	.024H	(17)	(18)	1/8
1964	V8-361 Newport	.035	.017	28–33	18436572(5)	10°	Damper	.060–.210(14)		(17)	(18)	1/8
	V8-383 "300"	.035	.017	28–33	18436572(5)	10°	Damper	.060–.210(14)		(17)	(18)	1/8
	V8-413 "300"	.035	.017	(7)	18436572(5)	10°	Damper	.060–.210(14)		(17)	(18)	1/8
	V8-413 New Yorker	.035	.017	28–33	18436572(5)	10°	Damper	.060–.210(14)		(17)	(18)	1/8
	"300K" 1 Carb.	.035	.017	(7)	18436572(5)	10°	Damper	.060–.210(14)		(17)	(18)	1/8
	"300K" 2 Carbs.	.035	.017	(7)	18436572(5)	12½°	Damper	.017H	.028H	(17)	(18)	1/8
1965	V8-383	.035	.017	28–32	18436572(5)	10°	Damper	.060–.210(14)		(17)	(18)	1/8
	V8-413, 340 H.P.	.035	.017	28–32	18436572(5)	12½°	Damper	.060–.210(14)		(17)	(18)	1/8
	V8-413, 360 H.P.	.035	.017	(1)	18436572(5)	12½°	Damper	.060–.210(14)		(17)	(18)	1/8
1966	V8-383, 440	.035	.017	28–32	18436572(5)	12½°	Damper	.060–.210(14)		(17)	(18)	1/8
	V8-383, 440(19)	.035	.017	28–32	18436572(5)	5°	Damper	.060–.210(14)		(17)	(18)	1/8
1967	V8-383, 440(20)	.035	.017	28–32	18436572(5)	12½°	Damper	Hydraulic(23)		(17)	(18)	1/8
	V8-383, 440(21)	.035	.017	28–32	18436572(5)	5°	Damper	Hydraulic(23)		(17)	(18)	1/8
	V8-383, 440(22)	.035	.017	28–32	18436572(5)	TDC	Damper	Hydraulic(23)		(17)	(18)	1/8
1968	V8-383 Std. Tr.(24)	.035	.017	28–33	18436572(5)	TDC	Damper	Hydraulic(23)		(17)	(18)	1/8
	V8-383 Auto. Tr.(24)	.035	.017	28–33	18436572(5)	7½°	Damper	Hydraulic(23)		(17)	(18)	1/8
	V8-383 Std. Tr.(25)	.035	.017	28–33	18436572(5)	TDC	Damper	Hydraulic(23)		(17)	(18)	1/8
	V8-383 Auto. Tr.(25)	.035	.017	28–33	18436572(5)	5°	Damper	Hydraulic(23)		(17)	(18)	1/8
	V8-440 Std. Tr.	.035	.017	28–33	18436572(5)	TDC	Damper	Hydraulic(23)		(17)	(18)	1/8
	V8-440 Auto. Tr.	.035	.017	28–33	18436572(5)	7½°	Damper	Hydraulic(23)		(17)	(18)	1/8
	V8-440 Std. Tr.(26)	.035	.017	28–33	18436572(5)	5°	Damper	Hydraulic(23)		(17)	(18)	1/8
	V8-440 Auto. Tr.(26)	.035	.017	28–33	18436572(5)	TDC	Damper	Hydraulic(23)		(17)	(18)	1/8

(1)—Each set of points 27–31°; total dwell both sets 36–40°.
(2)—Second line after "O" mark.
(3)—Zero lash for hydraulic valve lifter jobs.
(4)—Each set of points 26–28°, total dwell both sets 32–36°.
(5)—Engine numbering (front to rear): Left bank 1-3-5-7, right bank 2-4-6-8.
(6)—Fourth line after "O" mark.
(7)—Each set of points 27–32°, total dwell both sets 34–40°.
(8)—Each set of points 29–32°, total dwell both sets 36–39°.
(9)—Manual steering −2°, power steering 0°.
(10)—Left side +½°, right side 0°.
(11)—Manual steering −¾°, power steering +¾°.
(12)—Left side +⅜°, right side 0°.
(13)—Windsor 8°, Saratoga 6°.
(14)—With lifter collapsed.
(15)—Manual steering −¾°, Power steering 0°.
(16)—Left side +⅜°, Right side +⅛°.
(17)—Manual Steering −½°, Power Steering +¾°.
(18)—Left side +½°, right side +¼°.
(19)—With CAP (Cleaner Air Package). Timing set after top dead center.
(20)—Without CAP (cleaner air package).
(21)—With CAP (cleaner air package) and Torqueflite.
(22)—With CAP (cleaner air package) and manual transmission.
(23)—No adjustment.
(24)—Two barrel carburetor.
(25)—Four barrel carburetor.
(26)—High performance engine.

Year	Engine or Car Model	Spark Plug Gap	Distributor: Point Gap	Distributor: Dwell Angle	Firing Order	Ignition Timing: BTDC or Mark	Ignition Timing: Mark Location	Valve Lash: Intake	Valve Lash: Exhaust	Wheel Alignment: Caster Degrees	Wheel Alignment: Camber Degrees	Wheel Alignment: Toe-In Inch
CORVAIR												
1960–61	All	.035	(1)	31–34	145236(6)	(2)	Pulley	1 Turn(3)		(4)	+1	(5)
1962–63	Turbo-Air, Std. Tr.	.035	(1)	31–34	145236(6)	4°	Damper	1 Turn(3)		+1¾	+½	¼–⅜
	Turbo-Air, Auto. Tr.	.035	(1)	31–34	145236(6)	13°	Damper	1 Turn(3)		+1¾	+½	¼–⅜
	Monza, Powerglide	.030	(1)	31–34	145236(6)	13°	Damper	1 Turn(3)		+1¾	+½	¼–⅜
	Super Turbo-Air	.030	(1)	31–34	145236(6)	13°	Damper	1 Turn(3)		+1¾	+½	¼–⅜
	Turbocharged Eng.	.030	(1)	31–34	145236(6)	24°	Damper	1 Turn(3)		+1¾	+½	¼–⅜
1964	95 H.P.-Std. Trans.	.035	(1)	31–34	145236(6)	6°	Damper	1 Turn(3)		+1¾	Zero	¼–⅜
	95 H.P.-Auto. Trans.	.035	(1)	31–34	145236(6)	14°	Damper	1 Turn(3)		+1¾	Zero	¼–⅜
	110 Horsepower	.030	(1)	31–34	145236(6)	14°	Damper	1 Turn(3)		+1¾	Zero	¼–⅜
	150 Horsepower	.030	(1)	31–34	145236(6)	24°	Damper	1 Turn(3)		+1¾	Zero	¼–⅜
1965–66	95 H.P.–Std. Trans.	.035	(1)	31–34	145236(6)	6°	Damper	1 Turn(3)		+3	+1	¼
	95 H.P.-Auto. Trans.	.035	(1)	31–34	145236(6)	14°	Damper	1 Turn(3)		+3	+1	¼
	110 Horsepower	.030	(1)	31–34	145236(6)	14°(8)	Damper	1 Turn(3)		+3	+1	¼
	4 Carb. Engine	.030	(1)	31–34	145236(6)	18°(7)	Damper	1 Turn(3)		+3	+1	¼
	Turbocharged Eng.	.030	(1)	31–34	145236(6)	24°	Damper	1 Turn(3)		+3	+1	¼
1967	95 H.P. Std. Tr.	.035	(1)	31–34	145236(6)	6°	Damper	1 Turn(3)		+1⅞	+1	3/16–5/16
	95 H.P. Auto. Tr.	.035	(1)	31–34	145236(6)	14°	Damper	1 Turn(3)		+1⅞	+1	3/16–5/16
	95 H.P.(9)	.035	(1)	31–34	145236(6)	TDC	Damper	1 Turn(3)		+1⅞	+1	3/16–5/16
	110 H.P.	.030	(1)	31–34	145236(6)	14°	Damper	1 Turn(3)		+1⅞	+1	3/16–5/16
	110 H.P. w P/G, A/C	.030	(1)	31–34	145236(6)	24°	Damper	1 Turn(3)		+1⅞	+1	3/16–5/16
	110 H.P.(9)	.030	(1)	31–34	145236(6)	4°	Damper	1 Turn(3)		+1⅞	+1	3/16–5/16
1968	95 H.P.	.035	(1)	31–34	145236(6)	(10)	Damper	1 Turn(3)		(12)	(13)	3/16–5/16
	110 H.P.	.030	(1)	31–34	145236(6)	(11)	Damper	1 Turn(3)		(12)	(13)	3/16–5/16
	140 H.P.	.030	(1)	31–34	145236(6)	4°	Damper	1 Turn(3)		(12)	(13)	3/16–5/16
1969	95 H.P. Std. Tr.	.035	(1)	31–34	145236(6)	6°	Damper	1 Turn(3)		(12)	(13)	3/16–5/16
	95 H.P. Auto. Tr.	.035	(1)	31–34	145236(6)	14°	Damper	1 Turn(3)		(12)	(13)	3/16–5/16
	110 H.P. Std. Tr.	.035	(1)	31–34	145236(6)	4°	Damper	1 Turn(3)		(12)	(13)	3/16–5/16
	110 H.P. Auto. Tr.	.035	(1)	31–34	145236(6)	12°	Damper	1 Turn(3)		(12)	(13)	3/16–5/16
	140 H.P.	.035	(1)	31–34	145236(6)	4°	Damper	1 Turn(3)		(12)	(13)	3/16–5/16

(1)—New points .019″, used .016″.
(2)—For Dist. No.'s (1110259, 260) 13°. Dist No.'s (1110256-7) 16°. All others 4°.
(3)—Tighten adjusting screw until all clearance between valve stem and rocker arm has been eliminated. Then tighten screw the additional number of turns listed to center the lifter plunger.
(4)—1960 +3, 1961 +2.
(5)—1960 ⅛–3/16, 1961 ¼–⅜.
(6)—Engine numbering (rear to front): Right bank 1-3-5, left bank 2-4-6.
(7)—With head gasket (Part No. 3891552) set at 14°.
(8)—1966 with A/C 24°.
(9)—With A.I.R. system.
(10)—Manual trans. 6°, automatic trans. 14°.
(11)—Manual trans. 4°, automatic trans. 12°.
(12)—+1½° to +2¼°.
(13)—+½ to +1½.

Year	Engine or Car Model	Spark Plug Gap	Distributor: Point Gap	Distributor: Dwell Angle	Firing Order	Ignition Timing: BTDC or Mark	Ignition Timing: Mark Location	Valve Lash: Intake	Valve Lash: Exhaust	Wheel Alignment: Caster Degrees	Wheel Alignment: Camber Degrees	Wheel Alignment: Toe-In Inch
CORVETTE												
1953–54	All Models	.035	.015	41–47	153624	Ball	Flywheel	.010H	.020H	+½	+½	⅛
1955	Six	.035	.015	41–47	153624	Ball	Flywheel	.006H	.013H	+½	+½	0–⅛
	V8	.035	.019	(1)	18436572(2)	4°	Damper	.008H	.018H	+½	+½	0–⅛
1956	All Models	.035	.019	(1)	18436572(2)	4°	Damper	.008H	.018H	+½	+½	0–⅛
1957	220 Horsepower	.035	.019	(1)	18436572(2)	4°	Damper	1 Turn(3)		+½	+½	0–⅛
	245 Horsepower	.035	.019	(1)	18436572(2)	12°	Damper	1 Turn(3)		+½	+½	0–⅛
	250 Horsepower	.035	.019	(1)	18436572(2)	13°	Damper	1 Turn(3)		+½	+½	0–⅛
	270 Horsepower	.035	.019	(1)	18436572(2)	12°	Damper	1 Turn(3)		+½	+½	0–⅛
	283 Horsepower	.035	.019	(1)	18436572(2)	6°	Damper	.012H	.018H	+½	+½	0–⅛
1958–59	230 Horsepower	.035	(4)	30	18436572(2)	4°	Damper	1 Turn(3)		(6)	(7)	0–⅛
	245 Horsepower	.035	(4)	30	18436572(2)	4°	Damper	1 Turn(3)		(6)	(7)	0–⅛
	250 Horsepower	.035	(5)	(1)	18436572(2)	4°	Damper	1 Turn(3)		(6)	(7)	0–⅛
	270 Horsepower	.035	(5)	(1)	18436572(2)	7°	Damper	.012H	.018H	(6)	(7)	0–⅛
	290 Horsepower	.035	(5)	(1)	18436572(2)	14°	Damper	.012H	.018H	(6)	(7)	0–⅛

Continued

CORVETTE—Continued

Year	Engine or Car Model	Spark Plug Gap	Distributor Point Gap	Distributor Dwell Angle	Firing Order	Ignition Timing BTDC or Mark	Ignition Timing Mark Location	Valve Lash Intake	Valve Lash Exhaust	Wheel Alignment Caster Degrees	Wheel Alignment Camber Degrees	Wheel Alignment Toe-In Inch
1960–61	230 Horsepower	.035	④	30	18436572②	4°	Damper	1 Turn③		+2	Zero	⑧
	245 Horsepower	.035	④	30	18436572②	12°	Damper	1 Turn③		+2	Zero	⑧
	270 Horsepower	.035	⑤	①	18436572②	12°	Damper	.012H	.018H	+2	Zero	⑧
	275 Horsepower	.035	⑤	①	18436572②	8°	Damper	.012H	.018H	+2	Zero	⑧
	315 Horsepower	.035	⑤	①	18436572②	18°	Damper	.012H	.018H	+2	Zero	⑧
1962	250 Horsepower	.035	④	30	18436572②	4°	Damper	1 Turn③		+2	Zero	¼
	300 Horsepower	.035	④	30	18436572②	8°	Damper	1 Turn③		+2	Zero	¼
	340 Horsepower	.035	⑤	①	18436572②	10°	Damper	.008H	.018H	+2	Zero	¼
	360 Horsepower	.035	⑤	①	18436572②	10°	Damper	.008H	.018H	+2	Zero	¼
1963	250 Horsepower	.035	⑤	30	18436572②	4°	Damper	1 Turn③		+½	+2½	¼
	300 Horsepower	.035	⑤	30	18436572②	8°	Damper	1 Turn③		+½	+2½	¼
	340 Horsepower	.035	⑤	30	18436572②	10°	Damper	.008H	.018H	+½	+2½	¼
	360 Horsepower	.035	⑤	30	18436572②	10°	Damper	.008H	.018H	+½	+2½	¼
1964	250 Horsepower	.035	⑤	30	18436572②	4°	Damper	1 Turn③		+½	+2½	3/16–5/16
	300 Horsepower	.035	⑤	30	18436572②	8°	Damper	1 Turn③		+½	+2½	3/16–5/16
	365 Horsepower	.035	⑤	30	18436572②	10°	Damper	.030H	.030H	+½	+2½	3/16–5/16
	375 Horsepower	.035	⑤	30	18436572②	10°	Damper	.030H	.030H	+½	+2½	3/16–5/16
1965	250 Horsepower	.035	⑤	30	18436572②	4°	Damper	1 Turn③		+1½	+¾	7/32–11/32
	300 Horsepower	.035	⑤	30	18436572②	8°	Damper	1 Turn③		+1½	+¾	7/32–11/32
	350 Horsepower	.035	⑤	30	18436572②	8°	Damper	1 Turn③		+1½	+¾	7/32–11/32
	365 Horsepower	.035	⑤	30	18436572②	12°	Damper	.030H	.030H	+1½	+¾	7/32–11/32
	375 Horsepower	.035	⑤	30	18436572②	12°	Damper	.030H	.030H	+1½	+¾	7/32–11/32
	425 Horsepower	.035	⑤	30	18436572②	10°	Damper	.020H	.024H	+1½	+¾	7/32–11/32
1966	300 Horsepower	.035	⑤	30	18436572②	6°	Damper	1 Turn③		+1	+¾	3/16–5/16
	300 Horsepower⑨	.035	⑤	30	18436572②	4°⑩	Damper	1 Turn③		+1	+¾	3/16–5/16
	350 Horsepower	.035	⑤	30	18436572②	10°	Damper	1 Turn③		+1	+¾	3/16–5/16
	390 Horsepower	.035	—	—	18436572②	4°	Damper	1 Turn③		+1	+¾	3/16–5/16
	425 Horsepower	.035	—	—	18436572②	8°	Damper	.020H	.024H	+1	+¾	3/16–5/16
1967	8-327, 300 H.P.	.035	⑬	28–32	18436572②	6°	Damper	1 Turn③		+1	+¾	3/16–5/16
	8-327, 300 H.P.⑨⑪	.035	⑬	28–32	18436572②	4° ATC	Damper	1 Turn③		+1	+¾	3/16–5/16
	8-327, 300 H.P.⑨⑫	.035	⑬	28–32	18436572②	4° ATC	Damper	1 Turn③		+1	+¾	3/16–5/16
	8-427	.035	⑬	28–32	18436572②	4°	Damper	1 Turn③		+1	+¾	3/16–5/16
	8-427⑨	.035	⑬	28–32	18436572②	4°	Damper	1 Turn③		+1	+¾	3/16–5/16
	8-427, 425 H.P.	.035	⑬	28–32	18436572②	12°⑭	Damper	.022H	.024H	+1	+¾	3/16–5/16
	8-427, 435 H.P.	.035	⑬	28–32	18436572②	5°	Damper	.024H	.028H	+1	+¾	3/16–5/16
	8-427, 435 H.P.⑨	.035	⑬	28–32	18436572②	5°	Damper	.024H	.028H	+1	+¾	3/16–5/16
1968	8-327, 300 H.P.	.035	⑬	28–32	18436572②	4°	Damper	1 Turn③		⑮	⑯	⑰
	8-327, 350 H.P.	.035	⑬	28–32	18436572②	4°	Damper	1 Turn③		⑮	⑯	⑰
	8-427, 390 H.P.	.035	⑬	28–32	18436572②	4°	Damper	1 Turn③		⑮	⑯	⑰
	8-427, 400 H.P.	.035	⑬	28–32	18436572②	4°	Damper	1 Turn③		⑮	⑯	⑰
	8-427, 425 H.P.	.035	⑬	28–32	18436572②	4°	Damper	.024H	.028H	⑮	⑯	⑰
	8-427, 430 H.P.	.035	⑬	28–32	18436572②	12°	Damper	.022H	.024H	⑮	⑯	⑰
	8-427, 435 H.P.	.035	⑬	28–32	18436572②	4°	Damper	.024H	.028H	⑮	⑯	⑰

①—Each set of points 28–30°, total dwell both sets 33–35°.

②—Engine numbering (front to rear): Left bank 1-3-5-7, right bank 2-4-6-8.

③—Turn rocker arm adjusting screw clockwise until all lash is eliminated, then turn screw the additional number of turns specified.

④—New points .018″, used .015″.

⑤—New points .019″, used .016″.

⑥—1958 +2¼°, 1959 +2°.

⑦—1958 +½°, 1959 Zero.

⑧—1960 0—⅛, 1961 ⅛+⅜°.

⑨—With Air Injection Reactor system (A.I.R.).

⑩—After top dead center.

⑪—Manual transmission.

⑫—Automatic transmission.

⑬—New points, .019″, used .016″. On V8s, turn adjusting screw in (clockwise) until engine misfires; then back off ½ turn.

⑭—At 800 R.P.M.

⑮—Manual steer. +1°, Power steer. +2¼°.

⑯—Manual steer., +¾°, Power steer. +¾°, Rear wheel align. —¾°.

⑰—Manual steer., 3/16 to 5/16, Power steer. 3/16 to 5/16, Rear wheel align. 1/32 to 3/32.

CROSLEY

Year	Engine or Car Model	Spark Plug Gap	Distributor Point Gap	Distributor Dwell Angle	Firing Order	Ignition Timing BTDC or Mark	Ignition Timing Mark Location	Valve Lash Intake	Valve Lash Exhaust	Wheel Alignment Caster Degrees	Wheel Alignment Camber Degrees	Wheel Alignment Toe-In Inch
1946–52	All Models	.025	.020	46	1-3-4-2	12°	Flywheel	.005C	.007C	10°	2°	3/64–1/16

DE SOTO

Year	Engine or Car Model	Spark Plug Gap	Distributor Point Gap	Distributor Dwell Angle	Firing Order	Ignition Timing BTDC or Mark	Ignition Timing Mark Location	Valve Lash Intake	Valve Lash Exhaust	Wheel Alignment Caster Degrees	Wheel Alignment Camber Degrees	Wheel Alignment Toe-In Inch
1946–48	All Models	.030	.020	35–38	153624	"O"	Damper	.008H	.010H	Zero	+3/8	1/32
1949–50	All Models	.035	.020	35–38	153624	(1)	Damper	.008H	.010H	−2	+3/8	1/32
1951	All Models	.035	.020	35–38	153624	(1)	Damper	.008H	.010H	−2	Zero	Zero
1952–53	Six	.035	.020	35–38	153624	(1)	Damper	.008H	.010H	−2	Zero	Zero
	V8	.035	.018	(2)	18436572(3)	(4)	Damper	Zero	Zero	−2	Zero	Zero
1954	Six	.035	.020	39	153624	(5)	Damper	.008H	.010H	−2	Zero	Zero
	V8	.035	.018	(2)	18436572(3)	(4)	Damper	Zero	Zero	−2	Zero	Zero
1955	2 Bar. Carb.	.035	.017	(2)	18436572(3)	4°	Damper	Zero	Zero	(6)	(7)	0–1/16
	4 Bar. Carb.	.035	.017	34	18436572(3)	10°	Damper	Zero	Zero	(6)	(7)	0–1/16
1956	V8-330 2 Bar. Carb.	.035	.017	27–32	18436572(3)	8°	Damper	Zero	Zero	(6)	(7)	3/32–5/32
	V8-330 4 Bar. Carb.	.035	.017	27–32	18436572(3)	4°	Damper	Zero	Zero	(6)	(7)	3/32–5/32
	V8-341	.035	.017	27–32	18436572(3)	6°	Damper	Zero	Zero	(6)	(7)	3/32–5/32
1957	V8-325	.035	.017	27–32	18436572(3)	6°	Damper	Zero	Zero	(8)	(9)	3/32–5/32
	V8-341	.035	.017	27–32	18436572(3)	6°	Damper	Zero	Zero	(8)	(9)	3/32–5/32
1958	V8-350, 361 1 Carb.	.035	.017	27–32	18436572(3)	6°	Damper	.060–.210(11)		(12)	(9)	3/32–5/32
	V8-361 2 Carbs.	.035	.017	(10)	18436572(3)	8°	Damper	.060–.210(11)		(12)	(9)	3/32–5/32
1959	V8-361, 383 1 Carb.	.035	.017	27–32	18436572(3)	10°	Damper	.060–.210(11)		(8)	(9)	1/8
	V8-383 Two Carbs.	.035	.017	(13)	18436572(3)	10°	Damper	.060–.210(11)		(8)	(9)	1/8
1960	V8-361, 383	.035	.017	27–32	18436572(3)	10°	Damper	.060–.210(11)		(8)	(7)	1/8
1961	V8-361	.035	.017	27–32	18436572(3)	10°	Damper	.060–.210(11)		(14)	(15)	1/8

(1)—Third line after "O" mark.
(2)—Each set of points 26–28°, total dwell both sets 32–36°.
(3)—Cylinder numbering (front to rear): Left bank 1-3-5-7, right bank 2-4-6-8.
(4)—Fourth line after "O" mark.
(5)—Second line after "O" mark.
(6)—Manual steering −2°, power steering 0°.
(7)—Left side +3/8°, right side +1/8°.
(8)—Manual steering −3/4°, power steering +3/4°.
(9)—Left side +3/8°, right side 0°.
(10)—Each set of points 29–32°; total dwell both sets 36–39°.
(11)—With lifter collapsed.
(12)—Manual steering −3/4°, power steering 0°.
(13)—Each set of points 27–32°; total dwell both sets 34–40°.
(14)—Manual steering −1/2°, power steering +3/4°.
(15)—Left side +1/2°, right side +1/4.°

DODGE

Year	Engine or Car Model	Spark Plug Gap	Distributor Point Gap	Distributor Dwell Angle	Firing Order	Ignition Timing BTDC or Mark	Ignition Timing Mark Location	Valve Lash Intake	Valve Lash Exhaust	Wheel Alignment Caster Degrees	Wheel Alignment Camber Degrees	Wheel Alignment Toe-In Inch
1946–48	All Models	.025	.020	35–38	153624	(1)	Damper	.008H	.010H	Zero	+3/8	1/16
1949	All Models	.025	.020	35–38	153624	(1)	Damper	.008H	.010H	Zero	+3/8	1/32
1950	All Models	.035	.020	35–38	153624	(1)	Damper	.008H	.010H	Zero	+3/8	1/32
1951–52	All Models	.035	.020	35–38	153624	(1)	Damper	.008H	.010H	Zero	Zero	Zero
1953–54	Six	.035	.020	36–42	153624	(1)	Damper	.010H	.010H	Zero	Zero	Zero
	V8	.035	.017	(2)	18436572(3)	4°	Pulley	Zero	Zero	Zero	Zero	Zero
1955	Six	.035	.020	39	153624	2°	Damper	.010H	.010H	(5)	(6)	0–1/16
	V8	.035	.017	(2)	18436572(3)	4°	Damper	Zero	Zero	(5)	(6)	0–1/16
1956	Six	.035	.020	39	153624	2°	Damper	.010H	.010H	(5)	(6)	3/32–5/32
	V8-270	.035	.017	29–32	18436572(3)	4°	Damper	Zero	Zero	(5)	(6)	3/32–5/32
	V8-315 2 Bar. Carb.	.035	.017	29–32	18436572(3)	6°	Damper	Zero	Zero	(5)	(6)	3/32–5/32
	V8-315 4 Bar. Carb.	.035	.017	29–32	18436572(3)	6°	Damper	.012H	.022H	(5)	(6)	3/32–5/32
1957	Six	.035	.020	39	153624	0°	Damper	.010H	.010H	(7)	(8)	3/32–5/32
	V8-325	.035	.017	29–32	18436572(3)	6°	Damper	Zero	Zero	(7)	(8)	3/32–5/32
	V8-354 One Carb.	.035	.017	29–32	18436572(3)	6°	Damper	Zero	Zero	(7)	(8)	3/32–5/32
	V8-354 Two Carbs.	.035	.017	(4)	18436572(3)	2°	Damper	.015H	.024H	(7)	(8)	3/32–5/32
1958	Six	.035	.020	39	153624	2°	Damper	.010H	.010H	(10)	(8)	3/32–5/32
	V8's	.035	.017	29–32	18436572(3)	6°	Damper	.060–.210(9)		(10)	(8)	3/32–5/32
1959	Six	.035	.020	39	153624	2½°	Damper	.010H	.010H	(7)	(8)	1/8
	V8	.035	.017	29–32	18436572(3)	10°	Damper	.060–.210(9)		(7)	(8)	1/8
1960	Six	.035	.020	36–42	153624	5°	Pulley	.010H	.020H	(7)	(11)	1/8
	V8	.035	.017	27–32	18436572(3)	(12)	Damper	(13)	(14)	(7)	(11)	1/8

Continued

Year	Engine or Car Model	Spark Plug Gap	Distributor Point Gap	Distributor Dwell Angle	Firing Order	Ignition Timing BTDC or Mark	Ignition Timing Mark Location	Valve Lash Intake	Valve Lash Exhaust	Wheel Alignment Caster Degrees	Wheel Alignment Camber Degrees	Wheel Alignment Toe-In Inch
DODGE—Continued												
1961	6-170, 225	.035	.020	40–45	153624	2½°	Pulley	.010H	.020H	⑮	⑯	⅛
	6-225 4 Bar. Carb.	.028	.020	40–45	153624	10°	Pulley	.010H	.020H	⑮	⑯	⅛
	V8-318 Std. Tr.	.035	.017	27–32	18436572③	5°	Damper	.010H	.018H	⑮	⑯	⅛
	V8-318 Auto. Tr.	.035	.017	27–32	18436572③	10°	Damper	.010H	.018H	⑮	⑯	⅛
	V8-361 2 Bar. Carb.	.035	.017	27–32	18436572③	10°	Damper	.060–.210⑨		⑮	⑯	⅛
	V8-361 4 Bar. Carb.	.035	.017	27–32	18436572③	10°	Damper	.016C	.028C	⑮	⑯	⅛
	V8-383	.035	.017	⑰	18436572③	7½°	Damper	.016C	.028C	⑮	⑯	⅛
1962	6 Cyl.	.035	.020	40–45	153624	2½°	Pulley	.010H	.020H	⑮	⑯	⅛
	8-318 Std. Tr.	.035	.017	27–32	18436572③	5°	Damper	.013H	.021H	⑮	⑯	⅛
	8-318 Auto. Tr.	.035	.017	27–32	18436572③	10°	Damper	.013H	.021H	⑮	⑯	⅛
	8-361 "880"	.035	.017	27–32	18436572③	10°	Damper	.060–.210⑨		⑮	⑯	⅛
	8-361 Dart, Polara	.035	.017	27–32	18436572③	10°	Damper	.060–.210⑨		⑮	⑯	⅛
	8-361 Hi-Perf.	.035	.017	⑰	18436572③	10°	Damper	.060–.210⑨		⑮	⑯	⅛
	8-383, 413 Hi-Perf.	.035	.017	⑰	18436572③	10°	Damper	.060–.210⑨		⑮	⑯	⅛
	8-413 Ram, Man.	.035	.017	⑰	18436572③	10°	Damper	.060–.210⑨		⑮	⑯	⅛
1963	6 Cyl.	.035	.020	40–45	153624	2½°	Damper	.010H	.020H	⑮	⑯	⅛
	8-318 Std. Trans.	.035	.017	28–33	18436572③	5°	Damper	.013H	.021H	⑮	⑯	⅛
	8-318 Auto. Trans.	.035	.017	28–33	18436572③	10°	Damper	.013H	.021H	⑮	⑯	⅛
	8-361, 383 2 B. Carb.	.035	.017	28–33	18436572③	10°	Damper	.060–.210⑨		⑮	⑯	⅛
	8-383 4 B. Carb.	.035	.017	⑰	18436572③	10°	Damper	.060–.210⑨		⑮	⑯	⅛
	8-426	.035	.017	⑰	18436572③	10°	Damper	.028C	.032C	⑮	⑯	⅛
1964	6 Cyl.	.035	.020	40–45	153624	2½°	Damper	.010H	.020H	⑮	⑯	⅛
	8-273 Std. Trans.	.035	.017	28–33	18436572③	5°	Damper	.013H	.021H	⑮	⑯	⅛
	8-273 Auto. Trans.	.035	.017	28–33	18436572③	10°	Damper	.013H	.021H	⑮	⑯	⅛
	8-318 Std. Trans.	.035	.017	28–33	18436572③	5°	Damper	.013H	.021H	⑮	⑯	⅛
	8-318 Auto. Trans.	.035	.017	28–33	18436572③	10°	Damper	.013H	.021H	⑮	⑯	⅛
	8-361, 383 2 B. Carb.	.035	.017	28–33	18436572③	10°	Damper	.060–.210⑨		⑮	⑯	⅛
	8-413	.035	.017	28–33	18436572③	10°	Damper	.060–.210⑨		⑮	⑯	⅛
	8-383, 426 4 B. Carb.	.035	.017	⑰	18436572③	10°	Damper	.060–.210⑨		⑮	⑯	⅛
	8-426 Hemi Charg.	.020	.017	⑱	18436572③	⑲	Damper	.028C	.032C	⑮	⑯	⅛
1965	6 Cyl.	.035	.020	40–45	153624	2½°	Damper	.010H	.020H	⑮	⑯	⅛
	8-273 Std. Trans.	.035	.017	28–33	18436572③	5°	Damper	.013H	.021H	⑮	⑯	⅛
	8-273 Auto. Trans.	.035	.017	28–33	18436572③	10°	Damper	.013H	.021H	⑮	⑯	⅛
	8-273 4 Bar. Carb.	.035	.017	28–33	18436572③	10°	Damper	.013H	.021H	⑮	⑯	⅛
	8-318 Std. Trans.	.035	.017	28–33	18436572③	5°	Damper	.013H	.021H	⑮	⑯	⅛
	8-318 Auto. Trans.	.035	.017	28–33	18436572③	10°	Damper	.013H	.021H	⑮	⑯	⅛
	8-361, 383 2 Bar. Carb.	.035	.017	28–32	18436572③	10°	Damper	.060–.210⑨		⑮	⑯	⅛
	8-383 4 Bar. Carb.	.035	.017	⑳	18436572③	10°	Damper	.060–.210⑨		⑮	⑯	⅛
	8-413	.035	.017	⑳	18436572③	12½°	Damper	.060–.210⑨		⑮	⑯	⅛
	8-426	.035	.017	⑳	18436572③	10°	Damper	.060–.210⑨		⑮	⑯	⅛
1966	6-170	.035	.020	40–45	153624	5°	Damper	.010H	.020H	⑮	⑯	⅛
	6-170㉑	.035	.020	40–45	153624	5°㉒	Damper	.010H	.020H	⑮	⑯	⅛
	6-225	.035	.020	40–45	153624	2½°	Damper	.010H	.020H	⑮	⑯	⅛
	6-225㉑	.035	.020	40–45	153624	5°㉒	Damper	.010H	.020H	⑮	⑯	⅛
	8-273 Std. Tr.	.035	.017	28–32	18436572③	5°	Damper	.013H	.021H	⑮	⑯	⅛
	8-273 Auto. Tr.	.035	.017	28–32	18436572③	10°	Damper	.013H	.021H	⑮	⑯	⅛
	8-273 2 Bar. Carb.㉑	.035	.017	28–32	18436572③	5°㉒	Damper	.013H	.021H	⑮	⑯	⅛
	8-273 4 Bar. Carb.	.035	.017	⑳	18436572③	10°	Damper	.013H	.021H	⑮	⑯	⅛
	8-273 4 Bar. Carb.㉑	.035	.017	⑳	18436572③	5°㉒	Damper	.013H	.021H	⑮	⑯	⅛
	8-318 Std. Tr.	.035	.017	28–32	18436572③	5°	Damper	.013H	.021H	⑮	⑯	⅛
	8-318 Auto. Tr.	.035	.017	28–32	18436572③	10°	Damper	.013H	.021H	⑮	⑯	⅛
	8-318 2 Bar. Carb.㉑	.035	.017	28–32	18436572③	4°㉒	Damper	.013H	.021H	⑮	⑯	⅛
	8-361, 383 2 B. C.	.035	.017	28–32	18436572③	12½°	Damper	.060–.210⑨		⑮	⑯	⅛
	8-361, 383 2 B. C.㉑	.035	.017	28–32	18436572③	5°㉒	Damper	.060–.210⑨		⑮	⑯	⅛
	8-383, 426, 440 4B.C.	.035	.017	28–32	18436572③	12½°	Damper	.060–.210⑨		⑮	⑯	⅛
	8-383,426,440 4B.C.㉑	.035	.017	28–32	18436572③	5°㉒	Damper	.060–.210⑨		⑮	⑯	⅛

Continued

Year	Engine or Car Model	Spark Plug Gap	Distributor		Firing Order	Ignition Timing		Valve Lash		Wheel Alignment		
			Point Gap	Dwell Angle		BTDC or Mark	Mark Location	Intake	Exhaust	Caster Degrees	Camber Degrees	Toe-In Inch
DODGE—Continued												
1967	6-170, 225㉓	.035	.020	40–45	153624	5°	Damper	.010H	.020H	⑮	⑯	1/8
	6-170㉑	.035	.020	40–45	153624	5°㉒	Damper	.010H	.020H	⑮	⑯	1/8
	6-225㉑	.035	.020	40–45	153624	TDC	Damper	.010H	.020H	⑮	⑯	1/8
	8-273 Std. Tr.㉓	.035	.017	28–32	18436572③	5°	Damper	.013H	.021H	⑮	⑯	1/8
	8-273 Auto. Tr.㉓	.035	.017	28–32	18436572③	10°	Damper	.013H	.021H	⑮	⑯	1/8
	8-273㉑	.035	.017	28–32	18436572③	5°㉒	Damper	.013H	.021H	⑮	⑯	1/8
	8-273, 4 B.C.㉓	.035	.017	⑳	18436572③	10°	Damper	.013H	.021H	⑮	⑯	1/8
	8-273, 4 B.C.㉑	.035	.017	⑳	18436572③	5°㉒	Damper	.013H	.021H	⑮	⑯	1/8
	8-318 Std. Tr.㉓	.035	.017	28–32	18436572③	5°	Damper	Hydraulic㉗		⑮	⑯	1/8
	8-318 Auto. Tr.㉑	.035	.017	28–32	18436572③	10°	Damper	Hydraulic㉗		⑮	⑯	1/8
	8-318㉑	.035	.017	28–32	18436572③	5°㉒	Damper	Hydraulic㉗		⑮	⑯	1/8
	8-383㉓	.035	.017	28–32	18436572③	12½°	Damper	Hydraulic㉗		⑮	⑯	1/8
	8-383㉑㉔	.035	.017	28–32	18436572③	5°	Damper	Hydraulic㉗		⑮	⑯	1/8
	8-383㉑㉕	.035	.017	28–32	18436572③	TDC	Damper	Hydraulic㉗		⑮	⑯	1/8
	8-440, 440 Hi Perf.㉓	.035	.017	28–32	18436572③	12½°	Damper	Hydraulic㉗		⑮	⑯	1/8
	8-440, 440 Hi Perf.㉑㉔	.035	.017	28–32	18436572③	5°	Damper	Hydraulic㉗		⑮	⑯	1/8
	8-440, 440 Hi Perf.㉑㉕	.035	.017	28–32	18436572③	TDC	Damper	Hydraulic㉗		⑮	⑯	1/8
	8-426 Hemi㉓	.035	.017	㉖	18436572③	12½°	Damper	.028C	.032C	⑮	⑯	1/8
	8-426 Hemi㉑	.035	.017	㉖	18436572③	TDC	Damper	.028C	.032C	⑮	⑯	1/8
1968	6-170 Std. Trans.	.035	.020	40–45	153624	5° ATC	Damper	.010H	.020H	⑮	⑯	1/8
	6-170 Auto. Tr.	.035	.020	40–45	153624	2½° ATC	Damper	.010H	.020H	⑮	⑯	1/8
	6-225	.035	.020	40–45	153624	TDC	Damper	.010H	.020H	⑮	⑯	1/8
	V8-273 Std. Trans.	.035	.017	28–33	18436572③	5° ATC	Damper	Hydraulic㉗		⑮	⑯	1/8
	V8-273 Auto. Tr.	.035	.017	28–33	18436572③	2½° ATC	Damper	Hydraulic㉗		⑮	⑯	1/8
	V8-318 Std. Trans.	.035	.017	28–33	18436572③	5° ATC	Damper	Hydraulic㉗		⑮	⑯	1/8
	V8-318 Auto. Tr.	.035	.017	28–33	18436572③	2½° ATC	Damper	Hydraulic㉗		⑮	⑯	1/8
	V8-340 Std. Trans.	.035	.017	⑳	18436572③	TDC	Damper	Hydraulic㉗		⑮	⑯	1/8
	V8-340 Auto. Tr.	.035	.017	⑳	18436572③	5°	Damper	Hydraulic㉗		⑮	⑯	1/8
	V8-383 Std. Tr.㉘	.035	.017	28–33	18436572③	TDC	Damper	Hydraulic㉗		⑮	⑯	1/8
	V8-383 Auto. Tr.㉘	.035	.017	28–33	18436572③	5°	Damper	Hydraulic㉗		⑮	⑯	1/8
	V8-383 Std. Tr.㉙	.035	.017	28–33	18436572③	TDC	Damper	Hydraulic㉗		⑮	⑯	1/8
	V8-383 Auto. Tr.㉙	.035	.017	28–33	18436572③	7½°	Damper	Hydraulic㉗		⑮	⑯	1/8
	V8-440 Std. Tr.㉚	.035	.017	㉖	18436572③	TDC	Damper	Hydraulic㉗		⑮	⑯	1/8
	V8-440 Auto. Tr.㉚	.035	.017	28–33	18436572③	5°	Damper	Hydraulic㉗		⑮	⑯	1/8
	V8-440 Std. Trans.	.035	.017	28–33	18436572③	TDC	Damper	Hydraulic㉗		⑮	⑯	1/8
	V8-440 Auto. Trans.	.035	.017	28–33	18436572③	7½°	Damper	Hydraulic㉗		⑮	⑯	1/8
	V8-426	.035	.017	㉖	18436572③	TDC	Damper	.028C	.032C	⑮	⑯	1/8

①—With Fluid Drive "O", others 2°.
②—Each set of points 26–28°, total dwell both sets 32–36°.
③—Cylinder numbering (front to rear): Left bank 1-3-5-7, right bank 2-4-6-8.
④—Each set of points 29–32°, total dwell both sets 36–39°.
⑤—Manual steering −2°, power steering 0°.
⑥—Left side +½°, right side 0°.
⑦—Manual steering −¾°, power steering +¾°.
⑧—Left side +⅜°, right side 0°.
⑨—With lifters collapsed.
⑩—Manual steering +¼, power steering +¾.
⑪—Left side +⅜°, right side +⅛°.
⑫—V8-383 Ram manifold 7½°, others 10°.
⑬—V8-318 .010H, V8-361, 383 .060–.210 w/lifters collapsed, V8-361, 383 Ram .016C.
⑭—V8-318 .018H, V8-361, 383 .060–.210 w/lifters collapsed, V8-361, 383 Ram .018C.
⑮—Manual steering −½, power steering +¾.
⑯—Left side +½°, right side +¼°.
⑰—Each set of points 27–32°. Total dwell both sets 34–40°.
⑱—Each set 27–31°, total dwell both sets 34–38°.
⑲—With N61Y plugs, 31° at 3000 R.P.M.; with N58R plugs, 34° at 3000 R.P.M.
⑳—Each set of points 27–31°; total dwell both sets 36–40°.
㉑—With Cleaner Air Package (CAP).
㉒—After top dead center.
㉓—Without CAP (cleaner air package).
㉔—Torqueflite.
㉕—Manual transmission.
㉖—Each set of points 27–32°; total dwell both sets 37–42°.
㉗—No adjustment.
㉘—Four barrel carb.
㉙—Two barrel carb.
㉚—High performance engine.

Year	Engine or Car Model	Spark Plug Gap	Distributor		Firing Order	Ignition Timing		Valve Lash		Wheel Alignment		
			Point Gap	Dwell Angle		BTDC or Mark	Mark Location	Intake	Exhaust	Caster Degrees	Camber Degrees	Toe-In Inch

EDSEL

Year	Engine or Car Model	Spark Plug Gap	Point Gap	Dwell Angle	Firing Order	BTDC or Mark	Mark Location	Intake	Exhaust	Caster Degrees	Camber Degrees	Toe-In Inch
1958	Std. Trans.	.034	.015	26–28	15426378①	3°	Damper	Hydraulic		+1②	+1③	⅛
	Auto. Trans.	.034	.015	26–28	15426378①	6°	Damper	Hydraulic		+1②	+1③	⅛
1959–60	6 Cyl. Std. Tr.	.034	.025	35–38	153624	4°	Damper	.019H	.019H	+½	+1¼	⅛
	6 Cyl. Auto. Tr.	.034	.025	35–38	153624	6°	Damper	.019H	.019H	+½	+1¼	⅛
	V8-292 Std. Tr.	.034	.015	26–28	15486372①	3°	Damper	.019H	.019H	+½	+1¼	⅛
	V8-292 Auto. Tr.	.034	.015	26–28	15486372①	6°	Damper	.019H	.019H	+½	+1¼	⅛
	Other V8s Std. Tr.	.034	.015	26–28	15426378①	3°	Damper	Hydraulic		+½	+1¼	⅛
	Other V8s Auto. Tr.	.034	.015	26–28	15426378①	6°	Damper	Hydraulic		+½	+1¼	⅛

①—Cylinder numbering (front to rear): Right bank 1-2-3-4, left bank 5-6-7-8.

②—Corsair and Citation −¾°.

③—Corsair and Citation +⅜°.

FORD & MERCURY—Compact & Intermediate Models

Year	Engine or Car Model	Spark Plug Gap	Point Gap	Dwell Angle	Firing Order	BTDC or Mark	Mark Location	Intake	Exhaust	Caster Degrees	Camber Degrees	Toe-In Inch
1960–61	6 Cyl. Std. Tr.	.034	.025	37–42	153624	4°	Pulley	.016H	.016H	+½	+½	¼–5/16
	6 Cyl. Auto. Tr.	.034	.025	37–42	153624	10°	Pulley	.016H	.016H	+½	+½	¼–5/16
1962	6 Cyl. Std. Tr.	.034	.025	37–42	153624	4°	Damper	.016H	.016H	①	+½	②
	6 Cyl. Auto. Tr.	.034	.025	37–42	153624	10°	Damper	.016H	.016H	①	+½	②
	V8-Std. Tr.	.034	.017	26–31	15426378③	4°	Damper	¾ Turn④		Zero	+½	⅛
	V8-Auto Tr.	.034	.017	26–31	15426378③	12°	Damper	¾ Turn④		Zero	+½	⅛
1963	6-144 Std. Tr.	.034	.025	37–42	153624	8°	Damper	.066–.216⑤		①	⑥	⑦
	6-144 Auto. Tr.	.034	.025	37–42	153624	12°	Damper	.066–.216⑤		①	⑥	⑦
	6-170 Std. Tr.	.034	.025	37–42	153624	6°	Damper	.066–.216⑤		①	⑥	⑦
	6-170 Auto. Tr.	.034	.025	37–42	153624	12°	Damper	.066–.216⑤		①	⑥	⑦
	V8-Std. Tr.	.034	.017	26–31	15426378③	6°	Damper	¾ Turn④		①	⑥	⑦
	V8-Auto. Tr.	.034	.017	26–31	15426378③	10°	Damper	¾ Turn④		①	⑥	⑦
1964	6-144 Std. Tr.	.034	.025	37–42	153624	8°	Damper	.066–.216⑤		+½	+½	⑧
	6-144 Auto. Tr.	.034	.025	37–42	153624	12°	Damper	.066–.216⑤		+½	+½	⑧
	6-170 Std. Tr.	.034	.025	37–42	153624	6°	Damper	.066–.216⑤		+½	+½	⑧
	6-170 Auto. Tr.	.034	.025	37–42	153624	12°	Damper	.066–.216⑤		+½	+½	⑧
	6-200 Std. Tr.	.034	.025	37–42	153624	6°	Damper	.066–.216⑤		+½	+½	⑧
	6-200 Auto. Tr.	.034	.025	37–42	153624	12°	Damper	.066–.216⑤		+½	+½	⑧
	V8-260 Std. Tr.	.034	.017	26–31	15426378③	6°	Damper	¾ Turn④		+½	+½	⑧
	V8-260 Auto. Tr.	.034	.017	26–31	15426378③	10°	Damper	¾ Turn④		+½	+½	⑧
	V8-289 Std. Tr.㉓	.034	.017	26–31	15426378③	6°	Damper	¾ Turn④		+½	+½	⑧
	V8-289 Auto. Tr.㉓	.034	.017	26–31	15426378③	10°	Damper	¾ Turn④		+½	+½	⑧
	V8-289 Hi. Perf.	.034	.020	33–36	15426378③	10°	Damper	.020H	.020H	+½	+½	⑧
1965	6-170 Std. Tr.	.034	.025	37–42	153624	6°	Damper	.066–.216⑤		⑩	⑪	⑫
	6-170 Auto. Tr.	.034	.025	37–42	153624	12°	Damper	.066–.216⑤		⑩	⑪	⑫
	6-200 Std. Tr.	.034	.025	37–42	153624	6°	Damper	.066–.216⑤⑨		⑩	⑪	⑫
	6-200 Auto. Tr.	.034	.025	37–42	153624	12°	Damper	.066–.216⑤⑨		⑩	⑪	⑫
	V8-260 Std. Tr.	.034	.017	26–31	15426378③	6°	Damper	¾ Turn④		⑩	⑪	⑫
	V8-260 Auto. Tr.	.034	.017	26–31	15426378③	10°	Damper	¾ Turn④		⑩	⑪	⑫
	V8-289㉓	.034	.017	26–31	15426378③	6°	Damper	¾ Turn④		⑩	⑪	⑫
	V8-289 Hi. Perf.	.030	.020	30–33	15426378③	12°	Damper	.018H	.018H	⑩	⑪	⑫
1966	6-170, 200 Std. Tr.	.034	.025	37–42	153624	6°	Damper	.067–.200⑤		⑬	⑭	¼
	6-170, 200 Auto. Tr.	.034	.025	37–42	153624	12°	Damper	.067–.200⑤		⑬	⑭	¼
	6-170, 200⑮	.034	.025	37–42	153624	TDC	Damper	.067–.200⑤		⑬	⑭	¼
	V8-289㉓	.034	.017	26–31	15426378③	6°	Damper	.067–.200⑤		Zero	⑭	¼
	V8-289⑮㉓	.034	.017	26–31	15426378③	TDC	Damper	.067–.200⑤		Zero	⑭	¼
	V8-289 Hi. Perf.	.030	.020	30–33	15426378③	12°	Damper	.018H	.018H	Zero	⑭	¼
	V8-390	.034	.017	26–31	15426378③	10°	Damper	.050–.150⑯		Zero	¼	¼
	V8-390⑮	.034	.017	26–31	15426378③	6°	Damper	.050–.150⑯		Zero	¼	¼
1967	6-170, 200 Std. Tr.	.034	.025	37–42	153624	6°	Damper	.066–.216⑤		⑰	⑱	⑲
	6-170, 200 Auto. Tr.	.034	.025	37–42	153624	12°	Damper	.066–.216⑤		⑰	⑱	⑲

Continued

Year	Engine or Car Model	Spark Plug Gap	Distributor		Firing Order	Ignition Timing		Valve Lash		Wheel Alignment		
			Point Gap	Dwell Angle		BTDC or Mark	Mark Location	Intake	Exhaust	Caster Degrees	Camber Degrees	Toe-In Inch
FORD & MERCURY—Compact & Intermediate Models—Continued												
1967	6-170⑮	.034	.025	37–42	153624	TDC	Damper	.066–.216⑤		⑰	⑱	⑲
	6-200⑮	.034	.025	37–42	153624	5°	Damper	.066–.216⑤		⑰	⑱	⑲
	V8-289㉓	.034	.017	26–31	15426378③	6°	Damper	¾ Turn④		⑰	⑱	⑲
	V8-289⑮㉓	.034	.017	26–31	15426378③	TDC	Damper	¾ Turn④		⑰	⑱	⑲
	V8-289 Hi. Perf.	.030	.020	30–33	15426378③	12°	Damper	.018H	.018H	⑰	⑱	⑲
	V8-289 Hi. Perf.⑮	.030	.020	30–33	15426378③	6°	Damper	.018H	.018H	⑰	⑱	⑲
	V8-390	.034	.017	26–31	15426378③	10°	Damper	.100–.200⑤		⑰	⑱	⑲
	V8-390⑮	.034	.017	26–31	15426378③	6°	Damper	.100–.200⑤		⑰	⑱	⑲
1968	6-170	.034	.027	37–42	153624	6°	Damper	.066–.166⑤		⑰	㉑	⑲
	6-200	.034	.027	37–42	153624	6°	Damper	.066–.166⑤		⑰	㉑	⑲
	8-289 2 B. Carb.	.034	.021	24–29	15426378③	6°	Damper	¾ Turn④		⑰	㉑	⑲
	8-289 4 B. Carb.㉒	.034	.020	30–33	15426378③	6°	Damper	.018H	.018H	⑰	㉑	⑲
	8-302 2 B. Carb.	.034	.021	24–29	15426378③	6°	Damper	¾ Turn④		⑰	㉑	⑲
	8-302 4 B. Carb.⑯	.034	.021	30–33	15426378③	6°	Damper	¾ Turn④		⑰	㉑	⑲
	8-302 4 B. Carb.⑳	.034	.017	26–31	15426378③	6°	Damper	¾ Turn④		⑰	㉑	⑲
	8-390 2 B. Carb.⑯	.034	.021	30–33	15426378③	6°	Damper	.100–.200⑤		⑰	㉑	⑲
	8-390 2 B. Carb.⑳	.034	.017	26–31	15426378③	6°	Damper	.100–.200⑤		⑰	㉑	⑲
	8-390 4 B. Carb.⑯	.034	.021	30–33	15426378③	6°	Damper	.100–.200⑤		⑰	㉑	⑲
	8-390 4 B. Carb.⑳	.034	.017	26–31	15426378③	6°	Damper	.100–.200⑤		⑰	㉑	⑲
	8-390 "GT"	.034	.016	26–31	15426378③	6°	Damper	.100–.200⑤		⑰	㉑	⑲
	8-427	.034	.017	26–31	15426378③	6°	Damper	.100–.200⑤		⑰	㉑	⑲
	8-428 Cobra Jet	.034	.017	26–31	15426378③	6°	Damper	.100–.200⑤		⑰	㉑	⑲

①—Comet and Falcon +½°, Fairlane and Meteor zero.
②—Comet and Falcon ¼″–5/16″, Fairlane and Meteor ⅛″.
③—Cylinder numbering (front to rear): right bank 1-2-3-4, left bank 5-6-7-8.
④—Tighten rocker arm adjusting screw to eliminate all push rod end clearance, then tighten screw the number of turns listed.
⑤—Clearance is obtained at valve stem tip with hydraulic lifter collapsed. If clearance is less than the minimum install an undersize push rod; if greater than the maximum install an oversize push rod.
⑥—Comet and Falcon +⅜°, Fairlane zero, Meteor +½°.
⑦—Comet and Falcon ¼″–5/16″, Fairlane ¼″, Meteor ⅛″.
⑧—Comet and Falcon ¼″–5/16″, Fairlane 3/16″–5/16″.
⑨—Engines built after 3-29-65 clearance should be .067–.200″.
⑩—Comet and Falcon +¾°, Fairlane zero, Mustang (6 cyl. +1°, V8 zero).
⑪—Fairlane +¼°, all others +½°.
⑫—Comet ¼″, Falcon 9/32″, Fairlane and Mustang 7/32″.
⑬—Mustang +1°, all others zero.
⑭—Mustang +½°, all others +¼°.
⑮—With Thermactor Emission System.
⑯—Engines built after 12-20-65, .050″–.200″.
⑰—Cougar and Mustang +¼°, all others −½°.
⑱—Cougar and Mustang +1°, Falcon and Comet +¼°, Fairlane +½°.
⑲—Cougar and Mustang 3/16″, all others ¼″.
⑳—With I.M.C.O. system.
㉑—Cougar and Mustang +1°, all others +¼°.
㉒—High performance.
㉓—All except high performance.

Year	Engine or Car Model	Spark Plug Gap	Distributor		Firing Order	Ignition Timing		Valve Lash		Wheel Alignment		
			Point Gap	Dwell Angle		BTDC or Mark	Mark Location	Intake	Exhaust	Caster Degrees	Camber Degrees	Toe-In Inch
FORD & MERCURY—Full Size Models												
1946–47	Six	.030	.015	36	153624	①	①	.014C	.014C	+6¾°	+⅝	1/16
	V8	.030	.015	36②	15486372③	①	①	.012C	.014C	+6¾°	+⅝	1/16
1948	Six	.030	.025	35	153624	Groove	Pulley	.014C	.014C	+6¾	+⅝	1/16
	V8	.030	.015	36②	15486372③	①	①	.012C	.014C	+6¾	+⅝	1/16
1949	Six	.030	.025	36	153624	Groove	Pulley	.010C	.014C	−¼	+½	⅛
	V8	.030	.015	28	15486372③	Groove	Pulley	④	④	−¼	+½	⅛
	V8⑤	.030	.015	28	15486372③	Groove	Pulley	.012C	.014C	Zero	+⅜	⅛
1950	Six	.030	.025	36	153624	Groove	Pulley	.010C	.014C	−¼	+½	3/16
	V8	.030	.015	28	15486372③	Groove	Pulley	.014C	.018C	−¼	+½	3/16
	V8⑤	.030	.015	28	15486372③	Groove	Pulley	.012C	.014C	Zero	+⅜	⅛
1951	Six	.030	.025	36	153624	Groove	Pulley	.010C	.014C	−¼	+½	3/16
	V8	.030	.015	28	15486372③	Groove	Pulley	.014C	.018C	⑰	⑱	⑲
1952–53	Six	.035	.025	36	153624	Groove	Damper	.015H	.015H	−½	+½	3/32
	V8	.030	.015	28	15486372③	Groove	Pulley	.014C	.018C	⑳	㉑	㉒
1954	Six	.035	.025	36	153624	3°	Damper	.015H	.015H	+½	+¾	3/32
	V8 Std. Trans.	.035	.015	28	15486372③	3°	Pulley	.019H	.019H	+½	+¾	3/32
	V8 Fordmatic	.035	.015	28	15486372③	6°	Pulley	.019H	.019H	+½	+¾	3/32
	V8⑤	.030	.015	28	15486372③	3°	Pulley	.018C	.018C	−¾	+⅜	⅛
1955	Six	.034	.025	35–38	153624	3°	Damper	.015H	.019H	+1	+¾	1/16–⅛
	V8 Std. Trans.	.034	.015	26–28	15486372③	3°	Damper	.019H	.019H	㉓	㉔	㉕
	V8 Fordmatic	.034	.015	26–28	15486372③	6°	Damper	.019H	.019H	㉓	㉔	㉕
1956	Six Std. Trans.	.034	.025	35–38	153624	4°	Damper	.019H	.019H	+1	+¾	1/16–⅛
	Six Fordmatic	.034	.025	35–38	153624	6°	Damper	.019H	.019H	+1	+¾	1/16–⅛
	V8 Std. Trans.	.034	.015	26–28	15486372③	3°	Damper	.019H	.019H	㉓	㉔	㉕
	V8 Fordmatic	.034	.015	26–28	15486372③	6°	Damper	.019H	.019H	㉓	㉔	㉕
1957	Six Std. Trans.	.034	.025	35–38	153624	4°	Damper	.019H	.019H	+1	+1	1/16–⅛
	Six Fordmatic	.034	.025	35–38	153624	6°	Damper	.019H	.019H	+1	+1	1/16–⅛
	V8 Std. Trans.	.034	.015	26–28	15486372③	3°	Damper	.019H	.019H	+1	+1	1/16–⅛
	V8 Fordmatic	.034	.015	26–28	15486372③	6°	Damper	.019H	.019H	+1	+1	1/16–⅛
	V8-312⑤	.034	.015	26–28	15486372③	6°	Damper	.019H	.019H	−¾	+⅜	3/16–5/16
	V8-368⑤	.034	.015	26–28	15486372③	8°	Damper	Zero	Zero	−¾	+⅜	3/16–5/16
1958	6-223 Std. Trans.	.034	.025	35–38	153624	4°	Damper	.019H	.019H	+½	+1	1/16–⅛
	6-223 Auto. Trans.	.034	.025	35–38	153624	6°	Damper	.019H	.019H	+½	+1	1/16–⅛
	V8-292 Std. Trans.	.034	.015	26–28	15486372③	3°	Damper	.019H	.019H	+½	+1	1/16–⅛
	V8-292 Auto. Trans.	.034	.015	26–28	15486372③	6°	Damper	.019H	.019H	+½	+1	1/16–⅛
	V8-332 Std. Trans.	.034	.015	26–28	15426378③	3°	Damper	.026H	.026H	+½	+1	1/16–⅛
	V8-332 Auto Trans.	.034	.015	26–28	15426378③	6°	Damper	.026H	.026H	+½	+1	1/16–⅛
	V8-312⑤	.034	.015	26–28	15486372③	㉖	Pulley	.019H	.019H	−¾	+⅜	3/16–5/16
	V8-383, 430⑤	.034	.015	26–28	15426378③	㉗	Damper	.078–.218⑩		−¾	+⅜	3/16–5/16
1959	6-223 Std. Trans.	.034	.025	37–42	153624	6°	Damper	.019H	.019H	+½	+1	1/32–⅛
	6-223 Auto. Trans.	.034	.025	37–42	153624	12°	Damper	.019H	.019H	+½	+1	1/32–⅛
	V8-292 Std. Trans.	.034	.017	26–31	15486372③	5°	Damper	.019H	.019H	+½	+1	1/32–⅛
	V8-292 Auto. Trans.	.034	.017	26–31	15486372③	12°	Damper	.019H	.019H	+½	+1	1/32–⅛
	V8-332 Std. Trans.	.034	.017	26–31	15426378③	5°	Damper	.062–.1875⑩		+½	+1	1/32–⅛
	V8-332 Auto. Trans.	.034	.017	26–31	15426378③	8°	Damper	.062–.1875⑩		+½	+1	1/32–⅛
	V8-352 Std. Trans.	.034	.017	26–31	15426378③	5°	Damper	.062–.1875⑩		+½	+1	1/32–⅛
	V8-352 Auto Trans.	.034	.017	26–31	15426378③	8°	Damper	.062–.1875⑩		+½	+1	1/32–⅛
	V8-312⑤	.034	.017	26–31	15486372③	㉖	Damper	.019H	.019H	−¾	+⅜	1/16–3/16
	V8-383, 430⑤	.034	.017	26–31	15426378③	8°	Damper	.078–.218⑩		−¾	+⅜	1/16–3/16
1960	6-233 Std. Trans.	.034	.025	37–42	153624	6°	Damper	.019H	.019H	+⅝	+¾	⅛–¼
	6-233 Auto. Trans.	.034	.025	37–42	153624	10°	Damper	.019H	.019H	+⅝	+¾	⅛–¼
	V8-292 Std. Trans.	.034	.017	26–31	15486372③	5°	Damper	.019H	.019H	+⅝	+¾	⅛–¼
	V8-292 Auto. Trans.	.034	.017	26–31	15486372③	12°	Damper	.019H	.019H	+⅝	+¾	⅛–¼
	V8-352 Std. Trans.	.034	.017	26–31	15426378③	5°	Damper	.078–.218⑩		+⅝	+¾	⅛–¼
	V8-352 Auto. Trans.	.034	.017	26–31	15426378③	8°	Damper	.078–.218⑩		+⅝	+¾	⅛–¼

Continued

Year	Engine or Car Model	Spark Plug Gap	Distributor		Firing Order	Ignition Timing		Valve Lash		Wheel Alignment		
			Point Gap	Dwell Angle		BTDC or Mark	Mark Location	Intake	Exhaust	Caster Degrees	Camber Degrees	Toe-In Inch
FORD & MERCURY—Full Size Models—Continued												
1960	V8-312⑤	.034	.017	26–31	15486372③	㉘	Damper	.019H	.019H	–¾	+⅜	1/16–3/16
	V8-383, 430⑤	.034	.017	26–31	15426378③	8°	Damper	.078–.218⑩		–¾	+⅜	1/16–3/16
1961	6-233 Std. Trans.	.034	.025	37–42	153624	6°	Damper	.019H	.019H	Zero	+⅜	⅛–¼
	6-233 Auto. Trans.	.034	.025	37–42	153624	12°	Damper	.019H	.019H	Zero	+⅜	⅛–¼
	V8-292 Std. Trans.	.034	.017	26–31	15486372③	5°	Damper	.019H	.019H	Zero	+⅜	⅛–¼
	V8-292 Auto. Trans.	.034	.017	26–31	15486372③	12°	Damper	.019H	.019H	Zero	+⅜	⅛–¼
	V8-352, 390 Std. Tr.	.034	.017	26–31	15426378③	5°	Damper	.078–.218⑩		Zero	+⅜	⅛–¼
	V8-352, 390 Auto. Tr.	.034	.017	26–31	15426378③	8°	Damper	.078–.218⑩		Zero	+⅜	⅛–¼
	V8-390, 375 H.P.	.034	.020	26–31	15426378③	14°	Damper	.020H	.020H	Zero	+⅜	⅛–¼
	V8-390, 401 H.P.	.034	.020	26–31	15426378③	14°	Damper	.020H	.020H	Zero	+⅜	⅛–¼
1962	6-223 Std. Trans.	.034	.025	37–42	153624	6°	Damper	.025H	.025H	Zero	+⅜	⅛–¼
	6-223 Auto. Trans.	.034	.025	37–42	153624	12°	Damper	.025H	.025H	Zero	+⅜	⅛–¼
	V8-292 Std. Trans.	.034	.017	26–31	15486372③	5°	Damper	.019H	.019H	Zero	+⅜	⅛–¼
	V8-292 Auto. Trans.	.034	.017	26–31	15486372③	12°	Damper	.019H	.019H	Zero	+⅜	⅛–¼
	V8-352, 390 Std. Tr.	.034	.017	26–31	15426378③	5°	Damper	.078–.218⑩		Zero	+⅜	⅛–¼
	V8-352, 390 Auto. Tr.	.034	.017	26–31	15426378③	8°	Damper	.078–.218⑩		Zero	+⅜	⅛–¼
	V8-390, 375 H.P.	.034	.020	26–31	15426378③	5°	Damper	.025H	.025H	Zero	+⅜	⅛–¼
	V8-390, 401 H.P.	.034	.020	26–31	15426378③	5°	Damper	.025H	.025H	Zero	+⅜	⅛–¼
	V8-406, 385 H.P.	.034	.020	26–31	15426378③	8°	Damper	.025H	.025H	Zero	+⅜	⅛–¼
	V8-406, 405 H.P.	.034	.020	26–31	15426378③	8°	Damper	.025H	.025H	Zero	+⅜	⅛–¼
1963	6-223 Std. Trans.	.034	.025	37–42	153624	4°	Damper	.025H⑧	.025H⑧	Zero	+⅜	⅛–¼
	6-223 Auto. Trans.	.034	.025	37–42	153624	10°	Damper	.025H⑧	.025H⑧	Zero	+⅜	⅛–¼
	V8-260 Std. Trans.	.034	.017	26–31	15426378③	6°	Damper	¾ Turn⑨		Zero	+⅜	⅛–¼
	V8-260 Auto. Trans.	.034	.017	26–31	15426378③	10°	Damper	¾ Turn⑨		Zero	+⅜	⅛–¼
	V8-289 Std. Trans.	.034	.017	26–31	15426378③	6°	Damper	¾ Turn⑨		Zero	+⅜	⅛–¼
	V8-289 Auto. Trans.	.034	.017	26–31	15426378③	10°	Damper	¾ Turn⑨		Zero	+⅜	⅛–¼
	V8-352 Std. Trans.	.034	.017	26–31	15426378③	5°	Damper	.083–.183⑩		Zero	+⅜	⅛–¼
	V8-352 Auto. Trans.	.034	.017	26–31	15426378③	8°	Damper	.083–.183⑩		Zero	+⅜	⅛–¼
	V8-390, Std. Trans.⑥	.034	.017	26–31	15426378③	5°	Damper	.083–.183⑩		Zero	+⅜	⅛–¼
	V8-390 Auto. Trans.⑥	.034	.017	26–31	15426378③	8°	Damper	.083–.183⑩		Zero	+⅜	⅛–¼
	V8-390⑦	.034	.017	26–31	15426378③	6°	Damper	.083–.183⑩		Zero	+⅜	⅛–¼
	V8-390, 330 H.P.	.034	.017	26–31	15426378③	10°	Damper	.025H	.025H	Zero	+⅜	⅛–¼
	V8-406, 427	.035	.017	26–31	15426378③	5°	Damper	.025H	.025H	Zero	+⅜	⅛–¼
	V8-406, 427 2 Carbs.	.034	.017	26–31	15426378③	8°	Damper	.025H	.025H	Zero	+⅜	⅛–¼
1964	6-223 Std. Trans.	.034	.025	37–42	153624	4°	Damper	.025H⑧	.025H⑧	Zero	+⅝	⅛–¼
	6-223 Auto. Trans.	.034	.025	37–42	153624	10°	Damper	.025H⑧	.025H⑧	Zero	+⅝	⅛–¼
	V8-289 Std. Trans.	.034	.017	26–31	15426378③	6°	Damper	¾ Turn⑨		Zero	+⅝	⅛–¼
	V8-289 Auto. Trans.	.034	.017	26–31	15426378③	10°	Damper	¾ Turn⑨		Zero	+⅝	⅛–¼
	V8-352 Std. Trans.	.034	.017	26–31	15426378③	6°	Damper	⑪		Zero	+⅝	⅛–¼
	V8-352 Auto. Trans.	.034	.017	26–31	15426378③	10°	Damper	⑪		Zero	+⅝	⅛–¼
	V8-390 Std. Trans.	.034	.017	26–31	15426378③	4°	Damper	⑪		Zero	+⅝	⅛–¼
	V8-390 Auto. Trans.	.034	.017	26–31	15426378③	6°	Damper	⑪		Zero	+⅝	⅛–¼
	V8-390 Std. Trans.⑫	.034	.017	26–31	15426378③	4°	Damper	.025H	.025H	Zero	+⅝	⅛–¼
	V8-390 Auto. Trans.⑫	.034	.017	26–31	15426378③	6°	Damper	.025H	.025H	Zero	+⅝	⅛–¼
	V8-427	.034	.017	26–31	15426378③	8°	Damper	.025H	.025H	Zero	+⅝	⅛–¼
1965	6-240 Std. Trans.	.034	.025	37–42	153624	6°	Damper	¾ Turn⑨		+1	+½	5/32
	6-240 Auto. Trans.	.034	.025	37–42	153624	8°	Damper	¾ Turn⑨		+1	+½	5/32
	V8-289	.034	.017	26–31	15426378③	6°	Damper	¾ Turn⑨		+1	+½	5/32
	V8-352	.034	.017	26–31	15426378③	6°	Damper	.050–.150⑩		+1	+½	5/32
	V8-390⑦⑤	.034	.017	26–31	15426378③	6°⑯	Damper	.050–.150⑩		+1	+½	3/16
	V8-390 Std. Trans.	.034	.017	26–31	15426378③	4°	Damper	.050–.150⑩		+1	+½	㉘
	V8-390 Auto. Trans.	.034	.017	26–31	15426378③	6°⑯	Damper	.050–.150⑩		+1	+½	㉘
	V8-390 Std. Trans.⑫	.034	.017	26–31	15426378③	4°	Damper	.025H	.025H	+1	+½	㉘
	V8-390 Auto. Trans.⑫	.034	.017	26–31	15426378③	6°⑯	Damper	.025H	.025H	+1	+½	㉘
	V8-427	.030	.020	⑬	15426378③	8°	Damper	.025H	.025H	+1	+½	㉘

Continued

Year	Engine or Car Model	Spark Plug Gap	Distributor		Firing Order	Ignition Timing		Valve Lash		Wheel Alignment		
			Point Gap	Dwell Angle		BTDC or Mark	Mark Location	Intake	Exhaust	Caster Degrees	Camber Degrees	Toe-In Inch
FORD & MERCURY—Full Size Models—Continued												
1966	6-240 Std. Trans.	.034	.025	37–42	153624	6°	Damper	¾ Turn⑨		+½	+¼	3/16
	6-240 Auto Trans.	.034	.025	37–42	153624	12°	Damper	¾ Turn⑨		+½	+¼	3/16
	6-240 Std. Trans.⑭	.034	.025	37–42	153624	TDC	Damper	¾ Turn⑨		+½	+¼	3/16
	6-240 Auto. Trans.⑭	.034	.025	37–42	153624	4°	Damper	¾ Turn⑨		+½	+¼	3/16
	V8-289	.034	.017	26–31	15426378③	6°	Damper	¾ Turn⑨		+½	+¼	3/16
	V8-289⑭	.034	.017	26–31	15426378③	TDC	Damper	¾ Turn⑨		+½	+¼	3/16
	V8-352, 390, 428	.034	.017	26–31	15426378③	10°⑯	Damper	.050–.150⑩⑮		+½	+¼	3/16
	V8-352, 390, 428⑭	.034	.017	26–31	15426378③	6°⑯	Damper	.050–.150⑩⑮		+½	+¼	3/16
	V8-410⑤	.034	.017	26–31	15426378③	10°⑯	Damper	.050–.150⑩⑮		+½	+¼	3/16
	V8-410⑤⑭	.034	.017	26–31	15426378③	6°⑯	Damper	.050–.150⑩⑮		+½	+¼	3/16
	V8-427	.030	.020	22–24	15426378③	10°	Damper	.025H	.025H	+½	+¼	3/16
1967	6-240 Std. Trans.	.034	.025	37–42	153624	6°	Damper	¾ Turn⑨		+1	+½	3/16
	6-240 Auto. Trans.	.034	.025	37–42	153624	10°	Damper	¾ Turn⑨		+1	+½	3/16
	6-240 Std. Trans.⑭	.034	.025	37–42	153624	TDC	Damper	¾ Turn⑨		+1	+½	3/16
	6-240 Auto. Trans.⑭	.034	.025	37–42	153624	4°	Damper	¾ Turn⑨		+1	+½	3/16
	V8-289	.034	.017	26–31	15426378③	6°	Damper	¾ Turn⑨		+1	+½	3/16
	V8-289⑭	.034	.017	26–31	15426378③	TDC	Damper	¾ Turn⑨		+1	+½	3/16
	V8-390,410, 428	.034	.017	26–31	15426378③	10°⑯	Damper	.100–.200⑩		+1	+½	3/16
	V8-390, 410, 428⑭	.034	.017	26–31	15426378③	6°⑯	Damper	.100–.200⑩		+1	+½	3/16
	V8-427	.030	.020	22–24	15426378③	8°	Damper	.025H	.025H	+1	+½	3/16
1968	6-240	.034	.027	35–40	153624	6°	Damper	¾ Turn⑨		+1	+½	3/16
	V8-302	.034	.021	24–29	15426378③	6°	Damper	¾ Turn⑨		+1	+½	3/16
	V8-390 Std. Trans.⑭	.034	.017	26–31	15426378③	6°	Damper	.100–.200⑩		+1	+½	3/16
	V8-390 Auto. Trans.㉙	.034	.021	26–31	15426378③	6°	Damper	.100–.200⑩		+1	+½	3/16
	V8-428 Std. Trans.⑭	.034	.017	26–31	15426378③	6°	Damper	.100–.200⑩		+1	+½	3/16
	V8-428 Auto. Trans.㉙	.034	.021	24–29	15426378③	6°	Damper	.100–.200⑩		+1	+½	3/16
	V8-427	.034	.017	26–31	15426378③	6°	Damper	.100–.200⑩		+1	+½	3/16

①—There are no timing marks. Spark can be advanced or retarded by adjusting the vacuum brake set screw on distributor housing.
②—Total dwell for both breakers.
③—Cylinder numbering (front to rear): Right bank 1-2-3-4, left bank 5-6-7-8.
④—Up to Serial No. 8BA-622468: intake .012C, exhaust .014C. Later cars, intake .014C, exhaust .018C.
⑤—Used on Mercury models.
⑥—Four barrel carburetor.
⑦—Two barrel carburetor.
⑧—"Silent Lash"—Clearance specified is obtained at valve stem tip with eccentric spring compressed.
⑨—Tighten rocker arm adjusting screw to eliminate all push rod end clearance, then tighten screw the number of turns listed.
⑩—Clearance specified is obtained at valve stem tip with hydraulic lifter collapsed.
⑪—Before 11-18-63 .083–.183″, from 11-18-63 .050–.150″ (see note ⑩).
⑫—High performance.
⑬—Conventional ignition 30–33°, transistor ignition 22–24°.
⑭—With Thermactor Emission System.
⑮—Engines built after 12-20-65, .050–.200″.
⑯—Whenever idle speed or ignition timing is adjusted, vacuum line to brake release mechanism (if equipped) must be disconnected and vacuum hole plugged to prevent parking brake from releasing when selector lever is moved to Drive.
⑰—Ford −¼°, Mercury zero.
⑱—Ford +½°, Mercury +⅜°.
⑲—Ford 3/16″, Mercury ⅛″.
⑳—Ford −½°, Mercury −¾°.
㉑—Ford +½°, Mercury +⅜°.
㉒—Ford 3/32″, Mercury ⅛″.
㉓—Ford +1°, Mercury −¾°.
㉔—Ford +¾°, Mercury +⅜°.
㉕—Ford 1/16″ to ⅛″, Mercury 3/32″ to 5/32″.
㉖—Standard transmission 3°, automatic transmission 6°.
㉗—V8-383-4°, V8-430-7°.
㉘—Ford 5/32″, Mercury 3/16″.
㉙—With I.M.C.O.

FORD THUNDERBIRD

Year	Engine or Car Model	Spark Plug Gap	Distributor		Firing Order	Ignition Timing		Valve Lash		Wheel Alignment		
			Point Gap	Dwell Angle		BTDC or Mark	Mark Location	Intake	Exhaust	Caster Degrees	Camber Degrees	Toe-In Inch
1955–56	All Models	.034	.015	26–28	15486372①	②	Damper	.019H	.019H	+1	+3/4	1/16–1/8
1957	All Models	.034	.015	26–28	15486372①	②	Damper	.019H	.019H	+1	+1	1/16–1/8
1958	All Models	.034	.015	26–28	15426378①	②	Damper	.062–.1875③		+1	+1	1/16–1/8
1959	All Models	.034	.017	26–31	15426378①	④	Damper	⑤		+1	+1	1/16–1/8
1960	All Models	.034	.017	26–31	15426378①	④	Damper	.078–.218③		+1	+1	1/16–1/8
1961	All Models	.034	.017	26–31	15426378①	8°	Damper	.078–.218③		+1/2	+1/2	1/16–1/8
1962	V8-390	.034	.017	26–31	15426378①	6°	Damper	.100–.200③		−1/2	+1/2	1/16–1/8
	V8-429	.035	.017	26–31	15426378①	6°	Damper	.075–.175③		−1/2	+1/2	1/16–1/8
1963	V8-390	.034	.017	26–31	15426378①	6°	Damper	.083–.183③		−1 1/2	+3/8	1/8–1/4
1964	V8-390	.034	.017⑥	26–31⑦	15426378①	6°	Damper	⑧		−1 1/2	+3/8	1/8–1/4
1965	V8-390	.034	.017⑥	26–31⑦	15426378①	6°⑨	Damper	.050–.150③		−1 1/2	+1/2	1/32–9/32
1966	L/Thermactor	.034	.017⑥	26–31⑦	15426378①	10°⑨	Damper	.050–.150③		−1 1/2	+1/2	3/16
	W/Thermactor	.034	.016⑥	26–31⑦	15426378①	6°⑨	Damper	.050–.150③		−1 1/2	+1/2	3/16
1967	L/Thermactor	.034	.017	26–31	15426378①	10°⑨	Damper	.100–.200③		+1/2	+1	3/16
	W/Thermactor	.034	.017	26–31	15426378①	6°⑨	Damper	.100–.200③		+1/2	+1	3/16
1968	8-390	.034	.017	26–31	15426378①	6°⑨	Damper	.100–.200③		+1	+1/2	3/16
	8-429	.034	.017	26–31	15426378①	6°⑨	Damper	.075–.125③		+1	+1/2	3/16

①—Cylinder numbering (front to rear): Right bank 1-2-3-4, left bank 5-6-7-8.
②—Standard transmission 3°, Fordomatic 6°.
③—Clearance specified is obtained at valve stem with lifter collapsed.
④—Standard transmission 5°, automatic transmission 8°.
⑤—V8-352 (.078–.218③), V8-430 (.126–.226③).
⑥—With transistor ignition .020″.
⑦—With transistor ignition 22–24°.
⑧—Engines built prior to Nov. 18, 1963 .083–.183″③; engines built from Nov. 18, 1963 .050–.150″③.
⑨—Whenever idle speed or ignition timing is adjusted, vacuum line to brake release mechanism must be disconnected and plugged to prevent parking brake from releasing when selector is moved to Drive.

HENRY J

Year	Engine or Car Model	Spark Plug Gap	Point Gap	Dwell Angle	Firing Order	BTDC or Mark	Mark Location	Intake	Exhaust	Caster Degrees	Camber Degrees	Toe-In Inch
1951–54	Four Cyl.	.030	.020	44–47	1342	5°	Flywheel	.014C	.014C	Zero	+1/2	1/4
	Six Cyl.	.030	.020	35–38	153624	TDC	Damper	.014C	.014C	Zero	+1/2	1/4

HUDSON

Year	Engine or Car Model	Spark Plug Gap	Point Gap	Dwell Angle	Firing Order	BTDC or Mark	Mark Location	Intake	Exhaust	Caster Degrees	Camber Degrees	Toe-In Inch
1946–47	Six Cyl.	.032	.020	35–38	153624	①	Flywheel	.010H	.012H	+1	+1	1/32
	Eight Cyl.	.032	.017	27–30	16258374	TDC	Flywheel	.006H	.008H	+1	+1	1/32
1948–49	Six Cyl.	.032	.020	35–38	153624	UDC 1-6	Flywheel	.010H	.012H	+1	+1	1/32
	Eight Cyl.	.032	.017	27–30	16258374	UDC 1-8	Flywheel	.006H	.008H	+1	+1	1/32
1950–52	Six Cyl.	.032	.020	35–38	153624	UDC 1-6	Flywheel	.008H	.010H	+1	+1	1/32
	Eight Cyl.	.032	.017	27–30	16258374	UDC 1-8	Flywheel	.008H	.010H	+1	+1	1/32
1953–54	Jets	.032	.020	39	153624	②	Damper	.010H	.012H	+1	+3/4	1/32
	Wasps, Hornet	.032	.020	39	153624	UDC 1-6	Flywheel	.008H	.010H	+1	+3/4	1/32
1955	Wasp	.032	.020	39	153624	Line	Damper	.010C	.014C	⑤	Zero	1/16–3/16
	Hornet 6	.032	.020	39	153624	UDC 1-6	Flywheel	.010C	.014C	⑤	Zero	1/16–3/16
	Hornet V8	.035	.017	38	18436572④	5°	Damper	Zero	Zero	⑤	Zero	1/16–3/16
1956	Wasp	.032	.020	39	153624	Line	Damper	.010C	.014C	⑤	Zero	1/16–3/16
	Hornet 6	.030	.020	39	153624	UDC 1-6	Flywheel	Zero	Zero	⑤	Zero	1/16–3/16
	Hornet V8	.035	.017	⑥	18436572④	5°	Damper	Zero	Zero	⑤	Zero	1/16–3/16
	Special	.035	.016	26–33	18436572④	5°	Damper	Zero	Zero	⑤	Zero	1/16–3/16
1957	All Models	.035	.016	28–32	18436572④	5°	Damper	Zero	Zero	⑤	Zero	1/16–3/16

①—1/2″ before DC mark.
②—Long line before "1 UDC" mark.
④—Cylinder numbering (front to rear): Left bank 1-3-5-7, right bank 2-4-6-8.
⑤—Manual steering +1/2°, power steering +1°.
⑥—Dual breaker units 38°, single breaker 31°.

IMPERIAL

Year	Engine or Car Model	Spark Plug Gap	Distributor Point Gap	Distributor Dwell Angle	Firing Order	Ignition Timing BTDC or Mark	Ignition Timing Mark Location	Valve Lash Intake	Valve Lash Exhaust	Wheel Alignment Caster Degrees	Wheel Alignment Camber Degrees	Wheel Alignment Toe-In Inch
1946–48	All Models	.025	.017	27–30	16258374	2°	Damper	.008H	.010H	Zero	+3/8	1/32
1949–50	All Models	.035	.017	27–30	16258374	"O"	Damper	.008H③	.010H③	−2	+3/8	1/32
1951–54	All Models	.035	.017	①	18436572②	"O"	Damper	Zero	Zero	−2	Zero	Zero
1955	All Models	.035	.017	①	18436572②	6°	Damper	Zero	Zero	⑥	⑦	0–1/16
1956	All Models	.035	.017	④	18436572②	4°	Damper	Zero	Zero	⑥	⑦	1/8
1957	All Models	.035	.017	⑤	18436572②	6°	Damper	Zero	Zero	⑧	⑨	3/32–5/32
1958	All Models	.035	.017	⑤	18436572②	6°	Damper	.060–.210⑩		⑪	⑨	3/32–5/32
1959	All Models	.035	.017	⑫	18436572②	10°	Damper	.060–.210⑩		−3/4	⑨	1/8
1960	All Models	.035	.017	27–32	18436572②	10°	Damper	.060–.210⑩		⑧	⑦	1/8
1961–63	All Models	.035	.017	27–32	18436572②	10°⑮	Damper	.060–.210⑩		⑬	⑭	1/8
1964	All Models	.035	.017	28–33	18436572②	10°⑮	Damper	.060–.210⑩		⑬	⑭	1/8
1965	All Models	.035	.017	28–32	18436572②	10°⑮	Damper	.060–.210⑩		⑬	⑭	1/8
1966	All Models	.035	.017	28–32	18436572②	12½°⑮	Damper	.060–.210⑩		⑬	⑭	1/8
	All Models⑯	.035	.017	28–32	18436572②	5°⑰⑮	Damper	.060–.210⑩		⑬	⑭	1/8
1967	V8-383⑱	.035	.017	28–32	18436572②	12½°	Damper	Hydraulic㉑		⑬	⑭	1/8
	V8-383⑯⑲	.035	.017	28–32	18436572②	5°	Damper	Hydraulic㉑		⑬	⑭	1/8
	V8-383 ⑯⑳	.035	.017	28–32	18436572②	TDC	Damper	Hydraulic㉑		⑬	⑭	1/8
	V8-440⑱	.035	.017	28–32	18436572②	12½°⑮	Damper	Hydraulic㉑		⑬	⑭	1/8
	V8-440⑯⑰	.035	.017	28–32	18436572②	5°⑮	Damper	Hydraulic㉑		⑬	⑭	1/8
1968	V8-440 Auto. Trans.	.035	.017	28–33	18436572②	7½°	Damper	Hydraulic㉑		⑬	⑭	1/8

①—Each set of points 26–28°, total dwell both sets 32–36°.
②—Cylinder numbering (front to rear): Left bank 1-3-5-7, right bank 2-4-6-8.
③—Zero lash with hydraulic valve lifters.
④—Each set of points 29–32°, total dwell both sets 32–36°.
⑤—Each set of points 29–32°, total dwell both sets 36–39°.
⑥—Manual steering −2°, power steering 0°.
⑦—Left side +3/8°, right side +1/8°.
⑧—Manual steering −3/4°, power steering +3/4°.
⑨—Left side +3/8°, right side 0°.
⑩—With lifter collapsed.
⑪—Manual steering −3/4°, power steering 0°.
⑫—Each set of points 27–32°. Total dwell both sets 34–40°.
⑬—Manual steering −1/2°, power steering +3/4°.
⑭—Left side +1/2°, right side +1/4°.
⑮—Whenever idle speed or ignition timing is adjusted, vacuum line to brake release mechanism must be disconnected and plugged to prevent parking brake from releasing when selector lever is moved to Drive. Set idle speed with A/C compressor operating.
⑯—With Cleaner Air Package (CAP).
⑰—After top dead center.
⑱—Without CAP (cleaner air package).
⑲—Torqueflite.
⑳—Manual transmission.
㉑—No adjustment.

KAISER-FRAZER

Year	Engine or Car Model	Spark Plug Gap	Distributor Point Gap	Distributor Dwell Angle	Firing Order	Ignition Timing BTDC or Mark	Ignition Timing Mark Location	Valve Lash Intake	Valve Lash Exhaust	Wheel Alignment Caster Degrees	Wheel Alignment Camber Degrees	Wheel Alignment Toe-In Inch
1947–48	All Models	.032	.020	35–38	153624	"O"①	Damper①	.014C②	.014C	Zero	+3/8	1/16
1949–50	All Models	.032	.020	35–38	153624	4°	Damper	.014C	.014C	Zero	+3/8	1/16
1951–53	All Models	.032	.020	35–38	153624	4°	Damper	.014C	.014C	Zero	+1/2	1/8
1954	All Models	.030	.016	38–45	153624	4°	Damper	.014C	.014C	Zero	+1/2	1/8

①—Early models TDC mark on flywheel.
②—Up to Eng. No. 10769, .010H.

LINCOLN CONTINENTAL

Year	Engine or Car Model	Spark Plug Gap	Distributor Point Gap	Distributor Dwell Angle	Firing Order	Ignition Timing BTDC or Mark	Ignition Timing Mark Location	Valve Lash Intake	Valve Lash Exhaust	Wheel Alignment Caster Degrees	Wheel Alignment Camber Degrees	Wheel Alignment Toe-In Inch
1946–48	All Models	.028	.015	36	①	③	③	Zero	Zero	+4	+7/8	3/32
1949–51	All Models	.030	.015	28	15486372②	Mark	Damper	Zero	Zero	Zero	+3/8	1/8
1952–54	All Models	.030	.015	28	15486372②	3°	Damper	Zero	Zero	−3/4	+3/8	1/8
1955	All Models	.034	.015	26–28	15486372②	5°	Damper	Zero	Zero	−3/4	+3/8	3/32–5/32
1956–57	All Models	.034	.015	26–28	15486372②	5°	Damper	Zero	Zero	−3/4	+3/8	1/8–3/16
1958	All Models	.034	.015	26–28	15426378②	6°	Damper	.078–.218④		−3/4	+3/8	1/8–3/16
1959–62	All Models	.034	.017	26–31	15426378②	⑤	Damper	.078–.218④		−3/4	+3/8	⑥
1963	All Models	.034	.017	26–31	15426378②	8°	Damper	.078–.178④		−1½	+1/2	1/8–1/4
1964	All Models	⑦	⑧	⑨	15426378②	8°	Damper	.078–.178④		−1½	+1/2	1/8–1/4
1965	All Models	⑦	⑧	⑨	15426378②	6°⑩	Damper	.050–.150④		−1½	+3/4	1/32–9/32
1966	All Models	⑦	⑧	⑨	15426378②	10°⑩	Damper	.050–.150④		−1½	+1/2	1/8
	All Models⑪	.034	.017	26–31	15426378②	10°⑩	Damper	.050–.150④		−1½	+1/2	1/8
1967	All Models	.034	.017	26–31	15426378②	10°⑩	Damper	.083–.183④		−1½	+1/2	1/8
	All Models⑪	.034	.017	26–31	15426378②	10°⑩	Damper	.083–.183④		−1½	+1/2	1/8

Continued

Year	Engine or Car Model	Spark Plug Gap	Distributor		Firing Order	Ignition Timing		Valve Lash		Wheel Alignment		
			Point Gap	Dwell Angle		BTDC or Mark	Mark Location	Intake	Exhaust	Caster Degrees	Camber Degrees	Toe-In Inch

LINCOLN CONTINENTAL—Continued

Year	Engine or Car Model	Spark Plug Gap	Point Gap	Dwell Angle	Firing Order	BTDC or Mark	Mark Location	Intake	Exhaust	Caster Degrees	Camber Degrees	Toe-In Inch
1968	V8-460⑫	.034	.017	26–31	15426378②	10°⑩	Damper	.083–.183④		−1½	+½	⅛
	V8-462⑫	.034	.017	26–31	15426378②	10°⑩	Damper	.075–.175⑬		−1½	+½	⅛

①—Cylinder numbering (front to rear): Left bank 1-3-5-7-9-11, right bank 2-4-6-8-10-12. Firing order 1-4-9-8-5-2-11-10-3-6-7-12.
②—Cylinder numbering (front to rear): Right bank 1-2-3-4, left bank 5-6-7-8.
③—There are no timing marks. Spark may be advanced or retarded by adjusting vacuum brake set screw on distributor housing.
④—With rocker arm rotated to collapse lifter, the clearance listed should exist between end of valve stem and rocker arm.
⑤—1959 6°, 1960–62 8°.
⑥—1959–60 ⅛–3⁄16, 1961–62 1⁄16–3⁄16.
⑦—Conventional ignition .034″, Transistor ignition .030″.
⑧—Conventional ignition .017″, Transistor ignition .020″.
⑨—Conventional ignition 26–31°, Transistor ignition 22–24°.
⑩—Whenever idle speed or ignition timing is adjusted, vacuum line to brake release mechanism must be disconnected and plugged to prevent parking brake from releasing when selector is moved to Drive.
⑪—With Thermactor Emission System.
⑫—With I.M.C.O.
⑬—Valve lifter adjustment is 1 turn down after contact. Check as described in note④.

NASH

Year	Engine or Car Model	Spark Plug Gap	Point Gap	Dwell Angle	Firing Order	BTDC or Mark	Mark Location	Intake	Exhaust	Caster Degrees	Camber Degrees	Toe-In Inch
1946–48	"600" 6	.025	.020	35–38	153624	IGN	Damper	.015H	.015H	+½	+½	5⁄32
	Ambassador	.025	.020	35–38	153624	IGN	Damper	.015H	.015H	−¼	+½	1⁄16
1949	"600" 6	.030	.022	31–37	153624	IGN	Damper	.015H	.015H	+¼	Zero	⅛
	Ambassador	.025	.022	31–37	153624	IGN	Damper	.015H	.015H	+¼	Zero	⅛
1950	Rambler	.030	.022	31–37	153624	TDC	Damper	.016C	.018C	+1	+½	¼
	Statesman	.030	.022	31–37	153624	TDC	Damper	.016C	.018C	+¼	Zero	⅛
	Ambassador	.030	.022	31–37	153624	TDC	Damper	.012H	.016H	+¼	Zero	⅛
1951	Rambler	.030	.022	31–37	153624	TDC	Damper	.016C	.018C	+1	+½	¼
	Statesman	.030	.022	31–37	153624	TDC	Damper	.016C	.018C	Zero	Zero	⅛
	Ambassador	.030	.022	31–37	153624	TDC	Damper	.012H	.016H	Zero	Zero	⅛
1952–54	Rambler	.030	.022	31–37	153624	4°③	Damper	.016C	.018C	+1	+½	¼
	Statesman	.030	.022	31–37	153624	4°③	Damper	.016C	.018C	+¼	Zero	⅛
	Ambassador	.030	.022	31–37	153624	TDC	Damper	.012H	.016H	+¼	Zero	⅛
1955	Statesman	.030	.022	31	153624	4°③	Damper	.016C	.018C	+½②	Zero	1⁄16–3⁄16
	Ambassador 6	.030	.016	38–45	153624	4°③	Damper	.012H	.016H	+½②	Zero	1⁄16–3⁄16
	Ambassador V8	.035	.017	38	18436572①	5°	Damper	Zero	Zero	+½②	Zero	1⁄16–3⁄16
1956	Statesman 6	.030	.016	28–35	153624	4°③	Damper	.012H	.016H	+½②	Zero	1⁄16–3⁄16
	Amb. Spec. V8	.035	.016	26–33	18436572①	5°	Damper	Zero	Zero	+½②	Zero	1⁄16–3⁄16
	Ambassador 6	.030	.020	39	153624	4°③	Damper	.012H	.016H	+½②	Zero	1⁄16–3⁄16
	Ambassador V8	.035	.017	31	18436572①	5°	Damper	Zero	Zero	+½②	Zero	1⁄16–3⁄16
1957	Ambassador V8	.035	.016	30	18436572①	5°	Damper	Zero	Zero	+½②	Zero	1⁄16–3⁄16

①—Cylinder numbering (front to rear): Left bank 1-3-5-7, right bank 2-4-6-8.
②—With power steering +¾°.
③—After top dead center.

OLDSMOBILE—All Intermediate & Full Size Models

Year	Engine or Car Model	Spark Plug Gap	Point Gap	Dwell Angle	Firing Order	BTDC or Mark	Mark Location	Intake	Exhaust	Caster Degrees	Camber Degrees	Toe-In Inch
1946–48	Six Cyl.	.040	.022	31–37	153624	TDC	Flywheel	.008H	.011H	−⅜	+¼	3⁄32
	Eight Cyl.	.030	.016	21–30	16258374	Ball	Flywheel	.008H	.011H	−⅜	+¼	3⁄32
1949–50	Six Cyl.	.040	.022	31–37	153624	TDC	Flywheel	.008H	.011H	−⅜	Zero	3⁄32
	V8	.030	.016	26–33	18736542①	②	Pulley	Zero	Zero	−⅜	Zero	3⁄32
1951	All Models	.030	.016	26–33	18736542①	Slot	Pulley	Zero	Zero	−⅜	Zero	3⁄32
1952–54	All Models	.030	.016	26–33	18736542①	Slot	Pulley	Zero	Zero	−⅜	+¼	3⁄32
1955–56	All Models	.030	.016	26–33	18736542①	④	Damper	Zero	Zero	−⅜	Zero	1⁄16–⅛
1957	All Models	.030	③	30	18736542①	④	Damper	Zero	Zero	+⅜	Zero	1⁄16–⅛
1958	All Models	.030	③	30	18736542①	5°	Damper	Zero	Zero	−½	+⅛	0–⅛
1959	V8-371, 394	.030	③	30	18736542①	5°	Damper	Zero	Zero	−½	+⅛	0–⅛
1960	With 2 Bar. Carb.	.030	③	30	18736542①	5°	Damper	Zero	Zero	−½	+⅛	0–⅛
	With 4 Bar. Carb.	.030	③	30	18736542①	5°	Damper	Zero	Zero	−½	+⅛	0–⅛
1961	F85 Std. Tr.-155 H.P.	.040	③	30	18436572①	5°	Damper	Zero	Zero	−1	+⅜	0–⅛
	F85 Auto. Tr.-155 H.P.	.040	③	30	18436572①	7½°	Damper	Zero	Zero	−1	+⅜	0–⅛
	F85 Std. Tr.-185 H.P.	.040	③	30	18436572①	5°	Damper	Zero	Zero	−1	+⅜	0–⅛
	F85 Auto. Tr.-185 H.P.	.040	③	30	18436572①	7½°	Damper	Zero	Zero	−1	+⅜	0–⅛
	F85-215 H.P.	.025	③	30	18436572①	10°	Damper	Zero	Zero	−1	+⅜	0–⅛

Continued

Year	Engine or Car Model	Spark Plug Gap	Distributor		Firing Order	Ignition Timing		Valve Lash		Wheel Alignment		
			Point Gap	Dwell Angle		BTDC or Mark	Mark Location	Intake	Exhaust	Caster Degrees	Camber Degrees	Toe-In Inch

OLDSMOBILE—All Intermediate & Full Size Models—Continued

Year	Engine or Car Model	Spark Plug Gap	Point Gap	Dwell Angle	Firing Order	BTDC or Mark	Mark Location	Intake	Exhaust	Caster Degrees	Camber Degrees	Toe-In Inch
1961	Olds. with 2 Bar. Carb.	.030	③	30	18736542①	5°	Damper	Zero	Zero	−½	+⅛	0–⅛
	Olds. with 4 Bar. Carb.	.030	③	30	18736542①	5°	Damper	Zero	Zero	−½	+⅛	0–⅛
1962	F85, 155 H.P. Std. Tr.	.030	③	30	18436572①	5°	Damper	Zero	Zero	−1¼	Zero	0–⅛
	F85, 155 H.P. Auto. Tr.	.030	③	30	18436572①	7½°	Damper	Zero	Zero	−1¼	Zero	0–⅛
	F85, 185 H.P. Std. Tr.	.030	③	30	18436572①	5°	Damper	Zero	Zero	−1¼	Zero	0–⅛
	F85, 185 H.P. Auto. Tr.	.030	③	30	18436572①	7½°	Damper	Zero	Zero	−1¼	Zero	0–⅛
	F85, 215 H.P.	.025	③	30	18436572①	10°	Damper	Zero	Zero	−1¼	Zero	0–⅛
	Olds 260 H.P.	.030	③	30	18736542①	2½°	Damper	Zero	Zero	−½	+⅛	0–⅛
	Olds All Others	.030	③	30	18736542①	5°	Damper	Zero	Zero	−½	+⅛	0–⅛
1963	F85, 155 H.P. Std. Tr.	.030	③	30	18436572①	5°	Damper	Zero	Zero	−1¼	Zero	0–⅛
	F85, 155 H.P. Auto. Tr.	.030	③	30	18436572①	7½°	Damper	Zero	Zero	−1¼	Zero	0–⅛
	F85, 185 H.P.	.025	③	30	18436572①	7½°	Damper	Zero	Zero	−1¼	Zero	0–⅛
	F85, 195 H.P.	.030	③	30	18436572①	7½°	Damper	Zero	Zero	−1¼	Zero	0–⅛
	F85, 215 H.P.	.025	③	30	18436572①	10°	Damper	Zero	Zero	−1¼	Zero	0–⅛
	Olds-260 H.P.	.030	③	30	18736542①	2½°	Damper	Zero	Zero	−½	+¼	0–⅛
	Olds-All Others	.030	③	30	18736542①	5°	Damper	Zero	Zero	−½	+¼	0–⅛
1964	F85, V6-225	.030	③	30	165432⑤	5°	Damper	Zero	Zero	−1	+⅛	1/16–⅛
	F85, V8-330	.030	③	30	18436572①	7½°	Damper	Zero	Zero	−1	+⅛	1/16–⅛
	Olds, V8-330	.030	③	30	18436572①	7½°	Damper	Zero	Zero	−½	+⅛	0–1/16
	V8-394 Std. Tr.	.030	③	30	18736542①	2½°	Damper	Zero	Zero	−½	+⅛	0–1/16
	V8-394 Auto. Tr.	.030	③	30	18736542①	5°	Damper	Zero	Zero	−½	+⅛	0–1/16
1965	V6-225	.030	③	30	165432⑤	5°	Damper	Zero	Zero	−1¼	+⅛	⅛–3/16
	V8-330	.030	③	30	18436572①	7½°	Damper	Zero	Zero	⑥	+⅛	⅛–3/16
	V8-400	.030	③	30	18436572①	7½°	Damper	Zero	Zero	⑥	+⅛	⅛–3/16
	V8-425, 2 Bar. Carb.	.030	③	30	18436572①	⑦	Damper	Zero	Zero	−1	+⅛	⅛–3/16
	V8-425, 4 Bar. Carb.	.030	③	30	18436572①	5°	Damper	Zero	Zero	−1	+⅛	⅛–3/16
1966	6-250	.035	.019	31–34	153624	6°	Damper	1 Turn⑧		−1¼	+⅛	⅛–3/16
	V8-330, 400	.030	③	30	18436572①	7½°	Damper	Zero	Zero	⑥	+⅛	⅛–3/16
	V8-425, 2 B.C.	.030	③	30	18436572①	⑦	Damper	Zero	Zero	⑥	+⅛	⅛–3/16
	V8-425, 4 B.C.	.030	③	30	18436572①	7½°	Damper	Zero	Zero	⑥	+⅛	⑨
1967	6-250	.035	⑩	31–34	153624	4°	Damper	1 Turn⑧		−1¼	+⅛	⅛–3/16
	V8-330, 400	.030	③	30	18436572①	7½°⑪	Damper	Hydraulic⑫		⑥	+⅛	⅛–3/16
	V8-425 2 B.C.	.030	③	30	18436572①	5°⑪	Damper	Hydraulic⑫		⑥	+⅛	⅛–3/16
	V8-425 4 B.C.	.030	③	30	18436572①	7½°⑪	Damper	Hydraulic⑫		⑥	+⅛	⑨
1968	6-250 Std. Trans.	.035	⑩	31–34	153624	TDC	Damper	1 Turn⑧		⑯	+⅛	⅛–3/16
	6-250 Auto. Trans.	.035	⑩	31–34	153624	4°	Damper	1 Turn⑧		⑯	+⅛	⅛–3/16
	8-350, 250 H.P.	.030	.016	30	18436572①	5°⑭	Damper	Hydraulic⑫		⑯	+⅛	⅛–3/16
	8-350, 300 H.P.	.030	.016	30	18436572①	7½°⑪	Damper	Hydraulic⑫		⑯	+⅛	⅛–3/16
	8-350, 310 H.P.	.030	.016	30	18436572①	7½°⑪	Damper	Hydraulic⑫		⑯	+⅛	⅛–3/16
	8-400, 290 H.P.	.030	.016	30	18436572①	5°⑪	Damper	Hydraulic⑫		⑯	+⅛	⅛–3/16
	8-400, 350 H.P.	.030	.016	30	18436572①	7½°⑪	Damper	Hydraulic⑫		⑯	+⅛	⅛–3/16
	8-400	.030	.016	30	18436572①	2½°⑪	Damper	Hydraulic⑫		⑯	+⅛	⅛–3/16
	8-400, 442⑬	.030	.016	30	18436572①	10°⑮	Damper	Hydraulic⑫		−1¼°	+⅛	⅛–3/16
	8-455, 310 H.P.	.030	.016	30	18436572①	5°⑪	Damper	Hydraulic⑫		⑯	+⅛	⅛–3/16
	8-455, 320 H.P.	.030	.016	30	18436572①	7½°⑪	Damper	Hydraulic⑫		⑯	+⅛	⅛–3/16
	8-455, 4 B. Carb.	.030	.016	30	18436572①	7½°⑪	Damper	Hydraulic⑫		⑯	+⅛	⅛–3/16
	Toronado⑬	.030	.016	30	18436572①	10°⑯	Damper	Hydraulic⑫		−2°	+⅛	0–1/16

①—Cylinder numbering (front to rear): Left bank 1-3-5-7, right bank 2-4-6-8.

②—Between two steel balls.

③—Turn adjusting screw in (clockwise) until engine begins to misfire, then back off screw ½ turn.

④—Cars with one notch, set at leading edge of notch. Cars with three notches, set at center one.

⑤—Cylinder numbering (front to rear): Left bank 1-3-5, right bank 2-4-6.

⑥—Intermediates set at −1¼, Full size car set at −1, Toronado set at −2.

⑦—Low compression engine set at 7½°; high compression engine set at 5°.

⑧—Tighten rocker arm adjusting screw to eliminate all pushrod end clearance. Then tighten screw the number of turns listed.

⑨—Toronado set at 0–1/16, all others set at ⅛–3/16.

⑩—New points .019″, used points .016″.

⑪—At 850 R.P.M.

⑫—No adjustment.

⑬—With A.I.R. system.

⑭—With a/c "off" and idle compensator held closed.

⑮—At 1250 rpms.

⑯—Intermediates, −1¼°, full size car manual steer, −1¼°. Full size car power steer, −¾°.

Year	Engine or Car Model	Spark Plug Gap	Distributor		Firing Order	Ignition Timing		Valve Lash		Wheel Alignment		
			Point Gap	Dwell Angle		BTDC or Mark	Mark Location	Intake	Exhaust	Caster Degrees	Camber Degrees	Toe-In Inch
PACKARD												
1946	Six Cyl.	.030	.020	35	153624	4°	Damper	.007H	.010H	−1	Zero	1/32
	Eight Cyl.	.030	.017	27–30	16258374	5°	Damper	.007H	.010H	−1	Zero	1/32
	Super Eight	.030	.017	27–30	16258374	4°	Damper	Zero	Zero	−2	Zero	1/32
1947	Six Cyl.	.030	.020	35	153624	6°	Damper	.007H	.010H	−1	Zero	1/32
	Eight Cyl.	.030	.017	27–30	16258374	7°	Damper	.007H	.010H	−1	Zero	1/32
	Super Eight	.030	.017	27–30	16258374	6°	Damper	Zero	Zero	−2	Zero	1/32
1948–50	Eight	.030	.017	27–30	16258374	6°	Damper	.007H	.010H	−1	Zero	1/32
	Super Eight	.030	.017	27–30	16258374	6°	Damper	.007H	.010H	−1	Zero	1/32
	Custom Eight	.030	.017	27–30	16258374	6°	Damper	Zero	Zero	−2	Zero	1/32
1951–53	200	.030	.017	27–30	16258374	6°	Damper	.007H	.010H	−1	Zero	1/32
	Others	.028	.017	26–33	16258374	6°	Damper	Zero	Zero	−1	Zero	1/32
1954	5400-1-2-11	.025	.017	27–30	16258374	6°	Damper	.007H	.010H	−1	Zero	1/32
	Others	.025	.016	26–33	16258374	TDC	Damper	Zero	Zero	−1	Zero	1/32
PLYMOUTH												
1946–50	All Models	.025	.020	35–38	153624	DC	Damper	.010H	.010H	0	+3/8	1/32
1951–53	All Models	.035	.020	35–38	153624	DC	Damper	.010H	.010H	0	0	0
1954	All Models	.035	.020	36–42	153624	2°	Pulley	.010H	.010H	0	0	0
1955	Six Cyl.	.035	.020	39	153624	2°	Damper	.010H	.010H	−1	④	1/8
	V8	.035	.017	①	18436572②	4°	Pulley	Zero	Zero	−1	④	1/8
1956	Six Cyl.	.035	.020	39	153624	2°	Damper	.010H	.012H	−1	④	1/8
	V8-270	.035	.017	31	18436572②	4°	Pulley	Zero	Zero	−1	④	1/8
	V8-277	.035	.017	31	18436572②	4°③	Damper	.012H	.020H	−1	④	1/8
	V8-318	.035	.017	31	18436572②	4°	Damper	.010H	.018H	−1	④	1/8
1957	Six Cyl.	.035	.020	39	153624	2°	Damper	.010H	.010H	⑧	⑨	1/8
	V8-277	.035	.017	29	18436572②	⑥	Pulley	.008H	.018H	⑧	⑨	1/8
	V8-301	.035	.017	29	18436572②	⑦	Pulley	.008H	.018H	⑧	⑨	1/8
	V8-318	.035	.017	⑤	18436572②	8°	Damper	.008H	.018H	⑧	⑨	1/8
1958	6-230	.035	.020	39	153624	2°	Damper	.010H	.010H	⑧	⑨	3/32–5/32
	V8-318 D't IBP-4003F	.035	.017	27–32	18436572②	10°	Damper	.010H	.018H	⑧	⑨	3/32–5/32
	V8-318 D't IBP-4003D	.035	.017	27–32	18436572②	10°	Damper	.010H	.018H	⑧	⑨	3/32–5/32
	V8-318 D't IBS-4003	.035	.017	⑤	18436572②	8°	Damper	.010H	.018H	⑧	⑨	3/32–5/32
	V8-350	.035	.017	⑤	18436572②	8°	Damper	.060–.210⑩		⑧	⑨	3/32–5/32
1959	6-230	.035	.020	39	153624	2½°	Damper	.010H	.010H	⑧	⑨	1/8
	8-318	.035	.017	27–32	18436572②	10°	Damper	.010H	.018H	⑧	⑨	1/8
	8-361	.035	.017	⑪	18436572②	7½°	Damper	.060–.210⑩		⑧	⑨	1/8
1960	6-170, 225 Valiant	.035	.020	40–45	153624	2½°	Pulley	.010H	.020H	⑬	⑭	1/8
	6-Cyl. 4 Bar. Carb.	.028	.020	40–45	153624	10°	Pulley	.010H	.020H	⑬	⑭	1/8
	6-225 Std. Tr. Ply.	.035	.020	36–42	153624	2½°	Pulley	.010H	.020H	⑬	⑭	1/8
	6-225 Auto. Tr. Ply.	.035	.020	36–42	153624	5°	Pulley	.010H	.020H	⑬	⑭	1/8
	8-318 Std. Trans.	.035	.017	27–32	18436572②	5°	Damper	.010H	.018	⑬	⑭	1/8
	8-318 Auto. Trans.	.035	.017	27–32	18436572②	10°	Damper	.010H	.018	⑬	⑭	1/8
	8-361, 383	.035	.017	27–32	18436572②	10°	Damper	.060–.210⑩		⑬	⑭	1/8
	8-383 Ram Manifold	.035	.017	27–32	18436572②	7½°	Damper	.016C	.028C	⑬	⑭	1/8
1961	6-170, 225	.035	.020	36–42	153624	2½°	Pulley	.010H	.020H	⑬	⑮	1/8
	8-318 Std. Trans.	.035	.017	27–32	18436572②	5°	Damper	.010H	.018H	⑬	⑮	1/8
	8-318 Auto. Trans.	.035	.017	27–32	18436572②	10°	Damper	.010H	.018H	⑬	⑮	1/8
	8-361	.035	.017	⑪	18436572②	10°	Damper	⑫	②	⑬	⑮	1/8
	8-383	.035	.017	⑪	18436572②	7½°	Damper	.016C	.028C	⑬	⑮	1/8
1962	6-170, 225	.035	.020	40–45	153624	2½°	Pulley	.010H	.020H	⑬	⑮	1/8
	8-318 Std. Tr.	.035	.017	27–32	18436572②	5°	Damper	.013H	.021H	⑬	⑮	1/8
	8-318 Auto. Tr.	.035	.017	27–32	18436572②	10°	Damper	.013H	.021H	⑬	⑮	1/8
	8-361	.035	.017	⑪	18436572②	10°	Damper	.060–.210⑩		⑬	⑮	1/8
1963	6-170, 225	.035	.020	40–45	153624	2½°	Pulley	.010H	.020H	⑬	⑮	1/8
	8-318 Std. Tr.	.035	.017	28–33	18436572②	5°	Damper	.013H	.021H	⑬	⑮	1/8
	8-318 Auto. Tr.	.035	.017	28–33	18436572②	10°	Damper	.013H	.021H	⑬	⑮	1/8

Continued

Year	Engine or Car Model	Spark Plug Gap	Distributor		Firing Order	Ignition Timing		Valve Lash		Wheel Alignment		
			Point Gap	Dwell Angle		BTDC or Mark	Mark Location	Intake	Exhaust	Caster Degrees	Camber Degrees	Toe-In Inch
PLYMOUTH—Continued												
1963	8-361, 383	.035	.017	28–33	18436572②	10°	Damper	.060–.210⑩		⑬	⑮	⅛
	8-383, 426 Hi-Perf.	.035	.017	⑪	18436572②	10°	Damper	.060–.210⑩		⑬	⑮	⅛
1964	6-170, 225	.035	.020	40–45	153624	2½°	Damper	.010H	.020H	⑬	⑮	⅛
	8-273 Std. Tr.	.035	.017	28–33	18436572②	5°	Damper	.013H	.021H	⑬	⑮	⅛
	8-273 Auto. Tr.	.035	.017	28–33	18436572②	10°	Damper	.013H	.021H	⑬	⑮	⅛
	8-318 Std. Tr.	.035	.017	28–33	18436572②	5°	Damper	.013H	.021H	⑬	⑮	⅛
	8-318 Auto. Tr.	.035	.017	28–33	18436572②	10°	Damper	.013H	.021H	⑬	⑮	⅛
	8-361	.035	.017	28–33	18436572②	10°	Damper	.060–.210⑩		⑬	⑮	⅛
	8-383, 426	.035	.017	⑪	18436572②	10°	Damper	.060–.210⑩		⑬	⑮	⅛
	8-426 Hemi-Charg.	.020	.017	⑯	18436572②	⑰	Damper	.028C	.032C	⑬	⑮	⅛
1965	6-170, 225	.035	.020	40–45	153624	2½°	Damper	.010H	.020H	⑬	⑮	⅛
	8-273 Std. Tr.	.035	.017	28–33	18436572②	5°	Damper	.013H	.021H	⑬	⑮	⅛
	8-273 Auto. Tr.	.035	.017	28–33	18436572②	10°	Damper	.013H	.021H	⑬	⑮	⅛
	8-273 4 Bar. Carb.	.035	.017	⑱	18436572②	10°	Damper	.013H	.021H	⑬	⑮	⅛
	8-318 Std. Tr.	.035	.017	28–33	18436572②	5°	Damper	.013H	.021H	⑬	⑮	⅛
	8-318 Auto. Tr.	.035	.017	28–33	18436572②	10°	Damper	.013H	.021H	⑬	⑮	⅛
	8-361, 383	.035	.017	28–32	18436572②	10°	Damper	.060–.210⑩		⑬	⑮	⅛
	8-426	.035	.017	⑱	18436572②	10°	Damper	.060–.210⑩		⑬	⑮	⅛
1966	6-170	.035	.020	40–45	153624	5°	Damper	.010H	.020H	⑬	⑮	⅛
	6-170⑲	.035	.020	40–45	153624	5°㉑	Damper	.010H	.020H	⑬	⑮	⅛
	6-225	.035	.020	40–45	153624	2½°	Damper	.010H	.020H	⑬	⑮	⅛
	6-225⑲	.035	.020	40–45	153624	5°㉑	Damper	.010H	.020H	⑬	⑮	⅛
	8-273 Std. Tr.	.035	.017	28–32	18436572②	5°	Damper	.013H	.021H	⑬	⑮	⅛
	8-273 Auto. Tr.	.035	.017	28–32	18436572②	10°	Damper	.013H	.021H	⑬	⑮	⅛
	8-273, 4 B.C.	.035	.017	28–32	18436572②	10°	Damper	.013H	.021H	⑬	⑮	⅛
	8-273⑲	.035	.017	28–32	18436572②	5°㉑	Damper	.013H	.021H	⑬	⑮	⅛
	8-318 Std. Tr.	.035	.017	28–32	18436572②	5°	Damper	.013H	.021H	⑬	⑮	⅛
	8-318 Auto. Tr.	.035	.017	28–32	18436572②	10°	Damper	.013H	.021H	⑬	⑮	⅛
	8-318⑲	.035	.017	28–32	18436572②	4°㉑	Damper	.013H	.021H	⑬	⑮	⅛
	8-361, 383 2 B.C.	.035	.017	28–32	18436572②	12½°	Damper	.060–.210⑩		⑬	⑮	⅛
	8-361, 383 2 B.C.⑲	.035	.017	28–32	18436572②	5°㉑	Damper	.060–.210⑩		⑬	⑮	⅛
	8-383, 440 4 B.C.	.035	.017	28–32	18436572②	12½°	Damper	.060–.210⑩		⑬	⑮	⅛
	8-383, 440 4 B.C.⑲	.035	.017	28–32	18436572②	5°㉑	Damper	.060–.210⑩		⑬	⑮	⅛
	8-426	.035	.017	⑳	18436572②	12½°	Damper	.060–.210⑩㉒		⑬	⑮	⅛
1967	6-170, 225㉗	.035	.020	40–45	153624	5°	Damper	.010H	.020H	⑬	⑮	⅛
	6-170⑲	.035	.020	40–45	153624	5°㉑	Damper	.010H	.020H	⑬	⑮	⅛
	6-225⑲	.035	.020	40–45	153624	TDC	Damper	.010H	.020H	⑬	⑮	⅛
	8-273, 318 Std. Tr.㉗	.035	.017	28–32	18436572②	5°	Damper	㉕		⑬	⑮	⅛
	8-273, 318 Auto. Tr.⑲	.035	.017	28–32	18436572②	10°	Damper	㉕		⑬	⑮	⅛
	8-273⑲	.035	.017	28–32	18436572②	5°㉑	Damper	.013H	.021H	⑬	⑮	⅛
	8-273, 4 B.C.㉗	.035	.017	⑱	18436572②	10°	Damper	.013H	.021H	⑬	⑮	⅛
	8-273, 4 B.C.⑲	.035	.017	⑱	18436572②	5°㉑	Damper	.013H	.021H	⑬	⑮	⅛
	8-318⑲	.035	.017	28–32	18436572②	5°㉑	Damper	Hydraulic㉖		⑬	⑮	⅛
	8-383, 440㉗	.035	.017	28–32	18436572②	12½°	Damper	Hydraulic㉖		⑬	⑮	⅛
	8-383, 440⑲㉓	.035	.017	28–32	18436572②	5°	Damper	Hydraulic㉖		⑬	⑮	⅛
	8-383, 440⑲㉔	.035	.017	28–32	18436572②	TDC	Damper	Hydraulic㉖		⑬	⑮	⅛
	8-426 Hemi㉗	.035	.017	⑳	18436572②	12½°	Damper	.028C	.032C	⑬	⑮	⅛
	8-426 Hemi⑲	.035	.017	⑳	18436572②	TDC	Damper	.028C	.032C	⑬	⑮	⅛
1968	6-170 Std. Trans.	.035	.020	40–45	153624	5° ATC	Damper	.010H	.020H	⑬	⑮	⅛
	6-170 Auto. Trans.	.035	.020	40–45	153624	2½° ATC	Damper	.010H	.020H	⑬	⑮	⅛
	6-225	.035	.020	40–45	153624	TDC	Damper	.010H	.020H	⑬	⑮	⅛
	V8-273 Std. Trans.	.035	.017	28–33	18436572②	5° ATC	Damper	Hydraulic㉖		⑬	⑮	⅛
	V8-273 Auto. Trans.	.035	.017	28–33	18436572②	2½° ATC	Damper	Hydraulic㉖		⑬	⑮	⅛
	V8-318 Std. Trans.	.035	.017	28–33	18436572②	5° ATC	Damper	Hydraulic㉖		⑬	⑮	⅛
	V8-318 Auto. Trans.	.035	.017	28–33	18436572②	2½° ATC	Damper	Hydraulic㉖		⑬	⑮	⅛
	V8-340 Std. Trans.	.035	.017	⑳	18436572②	TDC	Damper	Hydraulic㉖		⑬	⑮	⅛

Continued

PLYMOUTH—Continued

Year	Engine or Car Model	Spark Plug Gap	Distributor		Firing Order	Ignition Timing		Valve Lash		Wheel Alignment		
			Point Gap	Dwell Angle		BTDC or Mark	Mark Location	Intake	Exhaust	Caster Degrees	Camber Degrees	Toe-In Inch
1968	V8-340 Auto. Trans.	.035	.017	⑳	18436572②	5°	Damper	Hydraulic㉖		⑬	⑮	⅛
	V8-383 Std. Trans.㉘	.035	.017	28–33	18436572②	TDC	Damper	Hydraulic㉖		⑬	⑮	⅛
	V8-383 Auto. Trans.㉘	.035	.017	28–33	18436572②	5°	Damper	Hydraulic㉖		⑬	⑮	⅛
	V8-383 Std. Trans.㉙	.035	.017	28–33	18436572②	TDC	Damper	Hydraulic㉖		⑬	⑮	⅛
	V8-383 Auto. Trans.㉙	.035	.017	28–33	18436572②	7½°	Damper	Hydraulic㉖		⑬	⑮	⅛
	V8-440 Std. Trans.㉚	.035	.017	⑳	18436572②	TDC	Damper	Hydraulic㉖		⑬	⑮	⅛
	V8-440 Auto. Trans.㉚	.035	.017	28–33	18436572②	5°	Damper	Hydraulic㉖		⑬	⑮	⅛
	V8-440 Std. Trans.	.035	.017	28–33	18436572②	TDC	Damper	Hydraulic㉖		⑬	⑮	⅛
	V8-440 Auto. Trans.	.035	.017	28–33	18436572②	7½°	Damper	Hydraulic㉖		⑬	⑮	⅛
	V8-426 Hemi	.035	.017	⑳	18436572②	TDC	Damper	.028C	.032C	⑬	⑮	⅛

①—Each set of points 26–28°. Total dwell both sets 32–36°.
②—Cylinder numbering (front to rear): Left bank 1-3-5-7, right bank 2-4-6-8.
③—TDC on V8-277 engine with four barrel carburetor.
④—Left side +½°, right side 0°.
⑤—Each set of points 29–32°. Total dwell both sets 36–39°.
⑥—With IBJ distributor TDC, with IBP distributor 4°.
⑦—With IBP-4003A distributor 4°, with IBP-4003 10°.
⑧—With manual steering −¾°, power steering +¾°.
⑨—Left side +⅜°, right side 0°.
⑩—With lifter collapsed.
⑪—Each set of points 27–32°; Total dwell both sets 34–40°.
⑫—W/2 Bar. Carb. .060–.210 w/lifter collapsed. W–4 Bar. Carb. Intake .016C Exhaust .028C.
⑬—Manual steering −½°, power steering +¾°.
⑭—Left side +⅜°, right side +⅛°.
⑮—Left side +½°, right side +¼°.
⑯—Each set 27–31°. Total dwell both sets 34–38°.
⑰—With N61Y plugs, 31° at 3000 R.P.M. (with N58R plugs, 34° at 3000 R.P.M.).
⑱—Each set of points 27–31°; total dwell both sets 36–40°.
⑲—With Cleaner Air Package (CAP).
⑳—Each set of points 27–32°; total dwell both sets 37–42°.
㉑—After top dead center.
㉒—With mechanical lifters; intake .028C, exhaust .032C.
㉓—Torqueflite.
㉔—Manual transmission.
㉕—8-273 Valve Lash, intake .013H, exhaust .021H; 8-318, Hydraulic no adjustment.
㉖—No adjustment.
㉗—Without Cleaner Air Package (CAP).
㉘—With 4 bbl. carb.
㉙—With 2 bbl. carb.
㉚—With high performance engine.

PONTIAC—All Intermediate & Full Size Models

Year	Engine or Car Model	Spark Plug Gap	Point Gap	Dwell Angle	Firing Order	BTDC or Mark	Mark Location	Intake	Exhaust	Caster Degrees	Camber Degrees	Toe-In Inch
1946–48	Six Cyl.	.025	.022	31–37	153624	IGN	Flywheel	.012H	.012H	−¾	Zero	1/32
	Eight Cyl.	.025	.016	21–30	16258374	IGN	Flywheel	.012H	.012H	−¾	Zero	1/32
1949–52	Six Cyl.	.025	.022	31–37	153624	①	Damper	.012H	.012H	−¾	Zero	1/32
	Eight Cyl.	.025	.016	21–30	16258374	①	Damper	.012H	.012H	−¾	Zero	1/32
1953	Six Cyl.	.025	.016	31–37	153624	②	Damper	.012H	.012H	Zero	+½	1/32
	Eight Cyl.	.025	.016	21–30	16258374	①	Damper	.012H	.012H	Zero	+½	1/32
1954	Six Cyl.	.025	.016	38–45	153624	②	Damper	.012H	.012H	Zero	+½	1/32
	Eight Cyl.	.025	.016	21–30	16258374	①	Damper	.012H	.012H	Zero	+½	1/32
1955	All Models	.035	.016	26–33	18436572④	⑤	Damper	Zero	Zero	−1	+½	0–1/16
1956	V8 One Carb.	.035	.016	26–33	18436572④	⑤	Damper	Zero	Zero	−1	+½	0–1/16
	V8 Two Carbs.	.035	.016	26–33	18436572④	10°	Damper	Zero	Zero	−1	+½	0–1/16
1957	All Models	.035	③	30	18436572④	⑤	Damper	Zero	Zero	−1	+½	0–1/16
1958	All Models	.035	③	30	18436572④	6°	Damper	Zero	Zero	−½	+½	0–1/16
1959–60	All Models	.035	③	30	18436572④	6°	Damper	Hydraulic		−1½	+¼	0–1/16
1961	4-195	.035	.016	74–76	1342	6°	Damper	Hydraulic		−1⅔	+0° 8′	0–⅛
	V8-215	.032	016	28–32	18436572④	5°	Damper	Hydraulic		−1⅔	+0° 8′	0–⅛
	V8-389	.035	③	30	18436572④	6°	Damper	Hydraulic		−1⅔	+¼	0–⅛
1962	4-195 Dist. 1110282	.035	.016	74–76	1342	6°	Damper	Zero	Zero	−1⅔	+0° 8′	0–⅛
	4-195 Dist. 1110283	.035	.016	74–76	1342	6°	Damper	Zero	Zero	−1⅔	+0° 8′	0–⅛
	4-195 Dist. 1110284	.035	.016	31–34	1342	6°	Damper	Zero	Zero	−1⅔	+0° 8′	0–⅛
	4-195 Dist. 1110285	.035	.016	31–34	1342	6°	Damper	Zero	Zero	−1⅔	+0° 8′	0–⅛
	V8-215	.032	.016	28–32	18436572④	5°	Damper	Zero	Zero	−1⅔	+0° 8′	0–⅛
	V8-389	.035	③	30	18436572④	6°	Damper	Zero	Zero	−1½	+¼	0–⅛
1963	4-195	.035	.016	31–34	1342	6°	Damper	Zero	Zero	−1⅔	+0° 8′	0–⅛
	V8-326	.035	③	30	18436572④	6°	Damper	Zero	Zero	−1⅔	+0° 8′	0–⅛
	V8-389 One Carb.	.035	③	30	18436572④	6°	Damper	Zero	Zero	−1½	+¼	0–⅛
	V8-389 Tri-Carb.	.035	③	30	18436572④	6°	Damper	Zero	Zero	−1½	+¼	0–⅛
	V8-421 Tri-Carb.	.035	③	30	18436572④	6°	Damper	Zero	Zero	−1½	+¼	0–⅛

Continued

Year	Engine or Car Model	Spark Plug Gap	Distributor		Firing Order	Ignition Timing		Valve Lash		Wheel Alignment		
			Point Gap	Dwell Angle		BTDC or Mark	Mark Location	Intake	Exhaust	Caster Degrees	Camber Degrees	Toe-In Inch

PONTIAC—All Intermediate & Full Size Models—Continued

Year	Engine or Car Model	Spark Plug Gap	Point Gap	Dwell Angle	Firing Order	BTDC or Mark	Mark Location	Intake	Exhaust	Caster Degrees	Camber Degrees	Toe-In Inch
1964–65	6-215	.035	.016	31–34	153624	4°	Damper	1 Turn⑥		⑦	+¼	0–⅛
	V8-326	.035	③	30	18436572④	6°	Damper	Zero	Zero	⑦	+¼	0–⅛
	V8-389, 421	.035	③	30	18436572④	6°	Damper	Zero	Zero	⑦	+¼	0–⅛
1966	6-230	.035	.016	31–34	153624	5°	Damper	Zero	Zero	⑦	+¼	0–⅛
	6-230⑧	.035	.016	31–34	153624	5°⑨	Damper	Zero	Zero	⑦	+¼	0–⅛
	V8	.035	③	30	18436572④	6°	Damper	Zero	Zero	⑦	+¼	0–⅛
	V8⑧	.035	③	30	18436572④	4°	Damper	Zero	Zero	⑦	+¼	0–⅛
	Tri Carb. V8	.035	③	30	18436572④	6°	Damper	Zero	Zero	−1½	+¼	0–⅛
1967	6-230	.035	.016	31–34	153624	5°	Damper	Hydraulic⑩		−1½	⑪	0–1⅛
	6-230⑧	.035	.016	31–34	153624	TDC	Damper	Hydraulic⑩		−1½	⑪	0–1⅛
	V-8's	.035	③	30	18436572④	6°	Damper	Hydraulic⑩		−1½	⑪	0–1⅛
	V-8's⑧	.035	③	30	18436572④	6°	Damper	Hydraulic⑩		−1½	⑪	0–1⅛
1968	6-250 1 Bar. Carb.	.035	.016	31–34	153624	TDC	Damper	Hydraulic⑩		⑫	+¼	⑬
	6-250 4 Bar. Carb.	.035	.016	31–34	153624	5°	Damper	Hydraulic⑩		⑫	+¼	⑬
	8-350	.035	.016	28–32	18436572④	9°	Damper	Hydraulic⑩		⑫	+¼	⑬
	Firebird 400	.035	.016	28–32	18436572④	9°	Damper	Hydraulic⑩		⑫	+¼	⑬
	Firebird 400 H.O.	.035	.016	28–32	18436572④	9°	Damper	Hydraulic⑩		⑫	+¼	⑬
	Firebird Ram Air	.035	.016	28–32	18436572④	9°	Damper	Hydraulic⑩		⑫	+¼	⑬
	Pontiac 400, 2 B.C.	.035	.016	28–32	18436572④	9°	Damper	Hydraulic⑩		⑫	+¼	⑬
	Pontiac 400, 4 B.C.	.035	.016	28–32	18436572④	9°	Damper	Hydraulic⑩		⑫	+¼	⑬
	8-428	.035	.016	28–32	18436572④	9°	Damper	Hydraulic⑩		⑫	+¼	⑬

①—Early 1949, "IGN" mark. Late 1949 and 1950–54, first line to come under pointer is for standard heads, second line for high compression heads, and third line is TDC.
②—Center of three lines.
③—Turn adjusting screw in (clockwise) until engine misfires, then back off screw ½ turn.
④—Cylinder numbering (front to rear): Left bank 1-3-5-7, right bank 2-4-6-8.
⑤—First line to come under pointer while cranking engine.
⑥—With valve fully closed, turn rocker arm nut down until all play in push rod is eliminated, then tighten nut additional turns specified.
⑦—Tempest wagons −2, all other cars −1½.
⑧—With Air Injection Reactor system (A.I.R.).
⑨—After top dead center.
⑩—No adjustment. On V8's, rocker arms are correctly positioned when ball retainer nuts are tightened to 20 ft. lbs.
⑪—Intermediates +¼, full size +⅜.
⑫—Firebird +½°, Intermediates and full size −1½°.
⑬—Firebird ⅛"–¼", Intermediates and full size 0"–⅛".

STUDEBAKER

Year	Engine or Car Model	Spark Plug Gap	Point Gap	Dwell Angle	Firing Order	BTDC or Mark	Mark Location	Intake	Exhaust	Caster Degrees	Camber Degrees	Toe-In Inch
1946	Champion	.025	.020	35–38	153624	IGN	Damper	.016C	.016C	+1½	+½	3/32
1947–48	Champion	.025	.020	35–38	153624	IGN	Damper	.016C	.016C	+½	+½	3/32
	Commander	.025	.020	35–38	153624	IGN	Damper	.016C	.016C	+½	+½	3/32
1949	Champion	.025	.020	35–38	153624	IGN	Damper	.016C	.016C	+1	+½	3/32
	Commander	.025	.020	35–38	153624	IGN	Damper	.016C	.016C	−2½	+½	3/32
1950	Champion	.025	.020	35–38	153624	IGN	Damper	.016C	.016C	−1½	+½	3/32
	Commander	.025	.022	31–37	153624	IGN	Damper	.016C	.016C	−2	+½	3/32
1951–52	Champion	.025	.020	35–38	153624	IGN	Damper	.016C	.016C	−1¾	+½	3/32
	Commander	.035	.016	21–30	18436572①	IGN	Damper	.015H	.015H	−1¾	+½	3/32
1953–54	Champion	.025	.020	38–40	153624	IGN	Damper	.016C	.016C	−1¾	+½	3/32
	Commander	.035	.016	28–34	18436572①	IGN	Damper	.022H	.022H	−1¾	+½	3/32
1955	Six Cyl.	.030	.020	39	153624	IGN	Damper	.016C	.016C	−1¾	+½	1/16–⅛
	V8	.035	.013	28–34	18436572①	IGN	Damper	.026C	.026C	−1¾	+½	1/16–⅛
1956	Six Cyl.	.030	.020	39	153624	IGN	Damper	.016C	.016C	−1¾	+½	1/16–⅛
	Studebaker V8	.035	.016	26–33	18436572①	IGN	Damper	.026C	.026C	−1¾	+½	1/16–⅛
	Packard V8	.035	.017	31	18436572①	IGN	Damper	Zero	Zero	−1¾	+½	1/16–⅛
1957	Six Cyl.	.030	.020	39	153624	IGN	Damper	.016C	.016C	−1¾	+½	1/16–⅛
	V8	.035	.016	26–33	18436572①	IGN	Damper	.026C	.026C	−1¾	+½	1/16–⅛
	Golden Hawk	.035	.016	26–33	18436572①	IGN	Damper	.026C	.026C	−1¾	+½	1/16–⅛

Continued

Year	Engine or Car Model	Spark Plug Gap	Distributor		Firing Order	Ignition Timing		Valve Lash		Wheel Alignment		
			Point Gap	Dwell Angle		BTDC or Mark	Mark Location	Intake	Exhaust	Caster Degrees	Camber Degrees	Toe-In Inch
STUDEBAKER—Continued												
1958	6-186	.030	.020	38–40	153624	2°	Damper	.016C	.016C	−1¾②	+½③	1/16–1/8
	V8-259, 289	.035	.016	28–34	18436572①	4°	Damper	.026C	.026C	−1¾②	+½③	1/16–1/8
1959	Six Cyl.	.030	.020	38–40	153624	2°	Damper	.018C④	.018C④	−2¼②	+½③	1/16–1/8
	V8	.035	.016	28–34	18436572①	4°	Damper	.026C	.026C	−2¼②	+½③	1/16–1/8
1960	Six Cyl.	.030	.020	38–40	153624	2°	Damper	.018C	.018C	−2¼②	+½③	1/16–1/8
	V8	.035	.016	28–34	18436572①	4°	Damper	.026C	.026C	−2¼②	+½③	1/16–1/8
1961–62	Six Cyl.	.035	.020	37–41	153624	2°	Damper	.026C	.026C	−½②	+½③	¼④
	V8	.035	.016	28–32	18436572①	4°	Damper	.026C	.026C	−½②	+½③	¼④
1963–64	Six Cyl.	.035	.020	37–41	153624	2°	Damper	.026C	.026C	−½②	+½③	¼④
	V8 Std. Engine	.035	.017	27–31	18436572①	4°	Damper	.026C	.026C	−½②	+½③	¼④

①—Cylinder numbering (front to rear): Left bank 1-3-5-7, right bank 2-4-6-8.

②—Not more than ½° variation between wheels.

③—−½ more favored on drivers side.

④—With Power Steering ⅛″.

Imported Car Data

Year	Model	Spark Plug Gap, In.	Ignition: Breaker Gap, In.	Ignition: Dwell Angle Deg.	Ignition: Firing Order	Ignition: Timing & Mark Location	Valve Lash: Intake	Valve Lash: Ex-haust	Capacities: Engine Oil Qts.	Capacities: Coolant Qts. W/ Heater	Wheel Alignment: Caster	Wheel Alignment: Chamber	Wheel Alignment: Toe-in

AUSTIN-HEALEY

Year	Model	Spark Plug Gap, In.	Breaker Gap, In.	Dwell Angle Deg.	Firing Order	Timing & Mark Location	Intake	Ex-haust	Engine Oil Qts.	Coolant Qts. W/ Heater	Caster	Chamber	Toe-in
1969	Sprite	.024	.014	—	1342	4° BTC①	.012	.012	4	6	3°	3/4°	0—1/8
1969–70	America	.025	.015	—	1342	3° BTC①	.012	.012	5-1/2	4	②	③	④
1971	America	.025	.015	57–63	1342	Flywheel	.012	.012	6	4	5-1/2°	3/4°	④
1973	Marina	.035	.015	60	1342	12° BTDC⑤	.013H	.013H	3.8	5.3	+2°	+1/6°	1/16
1974	Marina	.035	.015	60	1342	12° BTDC⑤	.013H	.013H	3.8	5.3	+1-1/2°	0°	1/16
1975	Marina	.035	—	—	1342	10° BTDC⑤	.013	.013	4-1/2	5-1/2	+2°	—	1/16

①—Mark located on pulley. Timed at No. 1 cyl., front of engine.
②—+5-1/2° ± 1/2°.
③—+3/4 ± 1°.
④—+1/16″ toe-out.
⑤—Timed at No. 1 cyl., front of engine.

BMW

Year	Model	Spark Plug Gap, In.	Breaker Gap, In.	Dwell Angle Deg.	Firing Order	Timing & Mark Location	Intake	Ex-haust	Engine Oil Qts.	Coolant Qts. W/ Heater	Caster	Chamber	Toe-in
1969–75	1600, 2002	.035	.016	60	1342	①	.0069	.0069	4.3②	7.4	+3°	0°	.079
	2000 Ti	.035	.016	60	1342	③	.0069	.0069	4.3②	7.4	+3°	0°	.079
	6 Cyl.	.024	.014	35–41	153624	④	.011	.011	5.3②	12.2	+9-1/2°	0°	.04

①—Timed at No. 1 cyl., front of engine. Disconnect vacuum hose & align ball on flywheel with the pointer at 1400 RPM.
②—Less filter.
③—Timed at No. 1 cyl., front of engine. Disconnect vacuum hose & align ball on flywheel with the pointer at 2200 RPM.
④—Timed at No. 1 cyl., front of engine. Disconnect vacuum hose & align ball on flywheel with the pointer at 1700 RPM.

DATSUN

Year	Model	Spark Plug Gap, In.	Breaker Gap, In.	Dwell Angle Deg.	Firing Order	Timing & Mark Location	Intake	Ex-haust	Engine Oil Qts.	Coolant Qts. W/ Heater	Caster	Chamber	Toe-in
1969–70	SPL 311	.030	—	50–55	1342	10° BTC①	.017H	.017H	4	9	1°30′	1°26′	1/8
	SRL 311	.030	—	50	1342	10° BTC①	.008H	.012H	7	9	1°30′	1°25′	1/8
1971	PL 510 Sedan	.034	.020	49–55	1342	①②	.010	.012	4.5	6.8	+1-2/3°	+1°	5/16
	PL 510 Sta. Wag.	.034	.020	49–55	1342	①②	.010	.012	4.5	6.8	+2°	+1-1/6°	1/8
	PL 521	.034	.020	49–55	1342	①②	.010	.012	4.5	—	—	—	—
1971–72	LB 110	.034	.020	49–55	1342	5° BTDC①	.014	.014	3.3	5	—	—	—
	240Z Man. Tr.	.034	.020	35–41	153624	5° BTDC①	.010	.012	5	8	—	—	—
	240Z Auto. Tr.	.034	.020	33–39	153624	①③	.010	.012	5	8	—	—	—
1972	PL 510 Sedan	.034	.020	49–55	1342	①④	.010	.012	4.5	—	+1-2/3°	+1°	5/16
	PL 510 Sta. Wag.	.034	.020	49–55	1342	①④	.010	.012	4.5	—	+2°	+1-1/6°	1/8
	PL 521	.034	.020	49–55	1342	①④	.010	.012	4.5	—	—	—	—
1973	LB 110	.034	.020	49–55	1342	5° BTDC①	.014	.014	3.3	5	+1-1/12°	+1-1/12°	13/64
	PL 510 Series	.030	.020	49–55	1342	①⑤	.010	.012	4.5	7	+1-7/12°	+5/12°	11/64
	PL 610	.030	.020	49–55	1342	①⑤	.010	.012	4.5	7	+1-1/2°	+1-3/4°	19/64
	PL 620	.030	.020	49–55	1342	①⑤	.010	.012	4.5	6.5	+1-5/6°	+1-1/4°	3/16
	240Z Man Tr.	.034	.020	35–41	153624	7° BTDC①	.010	.012	5	8	+2-11/12°	+5/6°	9/64
	240Z Auto. Tr.	.034	.020	33–39	153624	①⑥	.010	.012	5	8	+2-11/12°	+5/6°	9/64
1974	B 210	.032	.020	49–55	1342	5° BTDC①	.014	.014	3.3	5.3	+1-3/4°	+1-1/6°	1/8
	PL 610	.030	.020	49–55	1342	12° BTDC①	.010	.012	4.5	7.3	+2°	+2°	1/2
	PL 620	.030	.020	49–55	1342	12° BTDC①	.010	.012	4.5	7	+1-1/2°	+1-1/4°	1/8
	PL 710	.030	.020	49–55	1342	12° BTDC①	.010	.012	4.5	6.9	+1-2/3°	+1-29/32°	1/2
	260Z	.033	.014⑦	—	153624	①⑧	.010	.012	5	10.7	+2-9/10°	+23/30°	9/64
1975	B210	.033	.020	49–55	1342	10° BTDC①	.014	.014	4.1	6.2	+1-3/4°	+1-1/6°	1/8
	610	.034	.020	49–55	1342	12° BTDC①	.010	.012	4.5	7.2	+2°	⑩	1/2
	710	.034	.020	49–55	1342	12° BTDC①	.010	.012	4.5	6.8	+1-11/20°	+2-1/6°	3/8
	620	.034	.020	49–55	1342	12° BTDC①	.010	.012	4.5	8.4	+1-5/6°	+1-1/4°	1/8
	280Z	.033	.014⑦	38–46⑧	153624	7° BTDC①	.010	.012	5	10.7	+2-4/5°	⑪	1/16

①—Mark located on pulley. Timed at No. 1 cyl., front of engine.
②—Primary points, 10° BTDC; secondary points, TDC.
③—Primary points, TDC; secondary points, 10° BTDC.
④—Primary points, 7° BTDC; secondary points, TDC.
⑤—Primary points, 12° BTDC; secondary points, 8° BTDC.
⑥—Primary points, 15° BTDC; secondary points, 5° BTDC.
⑦—Distributor air gap.
⑧—Man. trans., 8° BTDC; auto. trans., 8° BTDC retarded, 15° BTDC advanced.
⑨—Exc. wagon, +1-3/4°; wagon, +2-1/4°.
⑩—Exc. 2+2, +1-1/20°; 2+2, +1-1/10°.

Year	Model	Spark Plug Gap, In.	Ignition: Breaker Gap, In.	Ignition: Dwell Angle Deg.	Ignition: Firing Order	Ignition: Timing & Mark Location	Valve Lash: Intake	Valve Lash: Ex-haust	Capacities: Engine Oil Qts.	Capacities: Coolant Qts. W/ Heater	Wheel Alignment: Caster	Wheel Alignment: Camber	Wheel Alignment: Toe-in
DODGE COLT													
1971–73	All	.028	.020	52	1342	TDC①	.006H	.010H	3.7	7.2	+1-1/4°	+1°	②
1974	98	.030	.020	49–55	1342	①③	.006H	.010H	4.2	6.6	+7/8°	+7/8°	②
	122	.030	.020	49–55	1342	3° BTDC①	.006H	.010H	4.4	8.3	+7/8°	+7/8°	②
1975	98	.030	.020	49–55	1342	5° ATDC	.006H	.010H	4.2	6.3	+1-1/4°	+5/16°	②
	122	.030	.020	49–55	1342	5° ATDC	.006H	.010H	5	8.1	+1-1/4°	+5/16°	②

①—Mark located on pulley. Timed at No. 1 cyl., front of engine.
②—.08–.23 inch.
③—Exc. Calif. auto. trans., TDC; Calif. auto. trans., 3° BTDC.

Year	Model	Spark Plug Gap, In.	Ignition: Breaker Gap, In.	Ignition: Dwell Angle Deg.	Ignition: Firing Order	Ignition: Timing & Mark Location	Valve Lash: Intake	Valve Lash: Ex-haust	Capacities: Engine Oil Qts.	Capacities: Coolant Qts. W/ Heater	Wheel Alignment: Caster	Wheel Alignment: Camber	Wheel Alignment: Toe-in
FIAT													
1969	124 Sedan	.026	.018	60	1342	TDC①	.008C	.008C	4	6.3	+2-1/4°	+1/2°	1/4
1969–70	850 Sedan	.026	.018	57	1342	10° BTDC⑧	.006C	.006C	3.5	8	+9°	+2-1/64°	1/4
1970	124 Spider	.024	.017	55	1342	5° BTDC①	.008C	.020C	4.2	8	—	—	—
	850 Spider	.024	.017	55	1342	TDC⑧	.006C	.008C	4.2	8	—	—	—
1971	124 Sedan, Sta. Wag.	.022	.018	60③	1342	TDC①	.008C	.008C	4	8	+3°	+37/64°	5/16
	124 Spider, Sport Coupe	.022	.016	55③	1342	5° BTDC①	.018C	.020C	4.2	8	+3-1/2°	+1/2°	9/32
	850 Sedan	.022	.018	60②	1342	10° BTDC⑧	.006C	.006C	3.5	8	+10°	+1-1/6°	1/2
	850 Sport Coupe	.022	.016	55③	1342	TDC⑧	.006C	.008C	4.2	8	+10°	+2°	1/2
	850 Spider, Racer	.022	.016	55③	1342	TDC⑧	.006C	.008C	4.2	8	+10°	+2-1/6°	19/32
1972	124 Sedan, Sta. Wag.	.022	.017	60	1342	TDC①	.008C	.008C	4	8	+3°	+37/64°	17/64
	124 Cp., Spider	.022	.016	55	1342	TDC①	.018C	.020C	4.2	8	+3-1/2°	+37/64°	⑤
	128 Sedan, Sta. Wag.	.022	.016	55	1342	TDC⑨	.012C	.016C	4.5	7	+1-29/32°	+2°	−3/16
	850 Spider	.022	.016	55	1342	TDC⑧	.006C	.008C	4.2	8	+10°	④	⑥
1973	124 Coupe	.022	.016	55	1342	TDC①	.018C	.020C	4-1/4⑦	8	+3-5/64	+17/64°	9/32
	124 Spider	.022	.016	55	1342	TDC①	.018C	.020C	4-1/4⑦	8	+3-1/4°	+1/6°	15/64
	124 Special	.022	.018	60	1342	TDC①	.008C	.008C	4-1/4⑦	8	+2-2/3°	+1/4°	9/32
	128 Sedan, Cp.	.022	.016	55	1342	TDC⑨	.012C	.016C	4-1/2⑦	6-4/5	+1-2/3°	+1-2/3°	−3/16
	850 Spider	.022	.016	55	1342	TDC⑧	.006C	.008C	4-1/2⑦	8	+9°	+1-5/6°	35/64
1974	124	.022	.016	55	1342	TDC①	.018C	.020C	4⑦	8	+3-1/2°	+1/2°	1/8
	128 Exc. X1-9	.022	.016	55	1342	TDC⑨	.012C	.016C	4-1/2⑦	7	+2-1/4°	⑩	0
	X1-9	.022	.016	55	1342	TDC⑨	.012C	.016C	4-1/2⑦	11.6	+7°	−1/2°	5/32
1975	124 Coupe	.024	.016	55	1342	TDC①	.018C	.020C	4	8	+3°	+1/2°	9/32
	124 Spider	.024	.016	55	1342	TDC①	.018C	.020C	4	8	+3-1/6°	+1/2°	1/4
	128 Sedan, S.W.	.024	.016	55	1342	TDC①	.012C	.016C	4-1/2	7	+1-2/3°	⑪	⑫
	128 Coupe	.024	.016	55	1342	TDC⑨	.012C	.016C	4-1/2	7	+1-2/3°	⑬	⑭
	131	.024	.016	55	1342	TDC①	.018C	.020C	4	8	+3-3/4°	+11/12°	5/16
	X1-9	.024	.016	55	1342	TDC⑨	.012C	.016C	4-1/2	7	+7	⑮	⑯

①—Mark located on pulley. Timed at No. 1 cyl., front of engine.
②—At 700 RPM.
③—At 850 RPM.
④—850 Sedan, +1-1/2°; 850 Sport Coupe, +1-2/3°; 850 Spider, +1-5/6°.
⑤—124 Coupe, 17/64; 124 Spider, 15/64.
⑥—850 Sedan, 35/64; 850 Sport Coupe, 29/64; 850 Spider, 35/64.
⑦—With filter.
⑧—Mark located on pulley. Timed at No. 1 cyl., rear of engine.
⑨—Mark located on pulley. Timed at No. 1 cyl., right side of vehicle.
⑩—Exc. Sport L, +2°; Sport, +1°.
⑪—Front, +1-2/3°; rear, sedan, −1/6°, sta. wag., +2/3°.
⑫—Front, −3/64; rear, +5/64″.
⑬—Front, +1-1/3°; rear, −1-1/6°.
⑭—Front, .020″; rear, 5/32″.
⑮—Front, −1/2°; rear, −1-2/3°.
⑯—Front, 11/64″; rear, 13/64″.

FORD

Year	Model	Spark Plug Gap, In.	Ignition: Breaker Gap, In.	Ignition: Dwell Angle Deg.	Ignition: Firing Order	Ignition: Timing & Mark Location	Valve Lash: Intake	Valve Lash: Exhaust	Capacities: Engine Oil Qts.	Capacities: Coolant Qts. W/ Heater	Wheel Alignment: Caster	Wheel Alignment: Camber	Wheel Alignment: Toe-in
1969	Cortina 1600	.023	.025	39	1243	4° BTDC①③	.010H	.017H	4.2	6.9	−5/32°	+1-3/4°	5/32
	Cortina 1600 GT	.023	.025	39	1243	4° BTDC②③	.012H	.022H	4.2	6.9	−5/32°	+1-3/4°	5/32
1970	Cortina 1600	.023	.025	39	1243	12° BTDC①	.010H	.017H	4.2	6.1	−5/32°	+1-3/4°	5/32
	Cortina 1600 GT	.023	.025	39	1243	12° BTDC①	.012H	.022	4.2	6.1	−5/32°	+1-3/4°	5/32
1971	Capri 1600	.025	.025	39	1243	12° BTDC③	.010H	.017H	3.5	6.2	+1/2°	−1/2°	1/8
	Capri 2000	.025	.025	39	1342	6° BTDC③④	.008C	.018C	5	7.5	+1/2°	−1/2°	1/8
1972	Capri 1600	.025	.025	38	1243	③⑤	.010H	.017H	3.5	6.5	+1°	+1/2°	1/8
	Capri 2000	.025	.025	38	1342	6° BTDC③④	.008C	.018C	5	8	+1°	+1/2°	1/8
	Capri 2600	.034	.025	37	142536	12° BTDC③	.014C	.016C	5-1/4	8-1/4	+1°	+1/2°	1/8
	Pantera	.034	.020	28	⑦	6° BTDC⑥	Hydraulic		4-1/2	25-1/2	+3-1/2°	0°	1/8
1973	Capri 2000	.030	.025	39	1342	6° BTDC③	.008H	.010H	4	6-1/2	+1-3/4°	0°	1/8
	Capri 2600	.034	.025	39	142536	12° BTDC③	.014C	.016C	4	8-1/2	+1-3/4°	0°	1/8
1974	Capri 2000	.034	.025	35–41	1342	6° BTDC③	.008C	.010C	3-1/2	8.1	+1/2°	0°	1/8
	Capri 2800	.034	.024	35–41	142536	12° BTDC⑥	.014H	.016H	4-1/2	10.8	+1/2°	0°	1/8
	Pantera	.034	.020	30–33	⑦	16° BTDC⑥	Hydraulic		4	24	+2-3/4°	0°	5/32
1976	Capri 2300	.034	—	—	1342	10°	Hydraulic		4-1/2	8-1/2	+1-1/2°	+1-1/4°	5/32
	Capri 2800	.034	—	—	142536	⑧	.014C	.016C	4-1/2	8-1/2	+1-1/2°	+1-1/4°	5/32

①—10° if no emission pump is used.
②—8° if no emission pump is used.
③—Mark located on pulley. Timed at No. 1 cyl., front of engine.
④—Exc. Calif.; Calif. Auto only, 10° BTDC.
⑤—Carburetor No. 701W9510EA, 28mm, 12° BTDC; Carburetor No. 701W9510EB, 25mm, 6° BTDC.
⑥—Mark located on pulley. Timed at No. 1 cyl., right front of engine.
⑦—13726548.
⑧—Man. trans., 10° BTDC; auto. trans., 8° BTDC.

HONDA

Year	Model	Spark Plug Gap, In.	Ignition: Breaker Gap, In.	Ignition: Dwell Angle Deg.	Ignition: Firing Order	Ignition: Timing & Mark Location	Valve Lash: Intake	Valve Lash: Exhaust	Capacities: Engine Oil Qts.	Capacities: Coolant Qts. W/ Heater	Wheel Alignment: Caster	Wheel Alignment: Camber	Wheel Alignment: Toe-in
1971–72	600	.030	.014	—	12①	10° BTDC②	.004C	.004C	3.2	—	+1°	+1°	.078
1973–74	Civic	.029	.020	49–55	1342	TDC②	.006C	.006C	3.2	4.2	+1-3/4°	+1/2°	.04③
1975	C.V.C.C.	.030	.020	52	1342	④⑤	.006	.006	3.2	4.4	+1-3/4°	+1/2°	.04③
	Civic	.030	.020	49–55	1342	7° BTDC②	.005	.005	3.2	4.2	+1-3/4°	+1/2°	.04③

①—Simultaneous firing.
②—Located on pulley.
③—Toe-out.
④—Manual trans. T.D.C.; auto trans. 3° ATDC.
⑤—Located an flywheel.

JAGUAR

Year	Model	Spark Plug Gap, In.	Ignition: Breaker Gap, In.	Ignition: Dwell Angle Deg.	Ignition: Firing Order	Ignition: Timing & Mark Location	Valve Lash: Intake	Valve Lash: Exhaust	Capacities: Engine Oil Qts.	Capacities: Coolant Qts. W/ Heater	Wheel Alignment: Caster	Wheel Alignment: Camber	Wheel Alignment: Toe-in
1969	E Type	.025	.015	33–37	153624	10° BTDC①	.004	.006	9	16	2° ± 1/2	+1/4° ± 1/2	1/16–1/8
1970	XKE, XJ Sedan	.025	.015	32–38	153624	10° BTDC①	.004	.006	9	16	2° ± 1/2	1/4° ± 1/2	1/16–1/8
1971	XJ6	.025	.015	32–38	153624	10° BTDC①	.013	.013	9	10	2-1/4° ± 1/4	1/2° ± 1/4	1/16
	XKE	.025	.015	31–37	153624	10° BTDC①	.013	.013	9	10-1/4	1-3/4° ± 1/4	1/4°	1/16
1972	XJ6	.025	.015	35–41	153624	8° BTDC①	.004	.006	8-3/4	19-1/4	2°–2-1/2°	1/2° ± 1/4	1/16
	Ser-3 V-12	.025	.022	20–25	②	10° BTDC⑤	.014C	.014C	11-1/2	21.6	2-1/4°–2-3/4°	③	2°
1973	XJ-6	.025	.015	35	153624	8° BTDC①	.013	.013	8-3/4	19	+2-1/4°	+1/2°	3/32
	Ser-3 V-12	.025	.021④	—	②	10° BTDC⑤	.015	.015	11-1/2	21-1/2	+2-1/2°	0°	3/32
1974	Ser-3 V-12	.025	—	—	②	4° ATDC⑤	.013C	.013C	—	—	+21/2°	0°	1/16
1975	XJ6L	.025	—	—	153624	TDC	.013	.013	8-3/4	19.5	+2-1/4°	+1/2°	1/16–1/8
	XJ12	.035	—	—	②	10° BTDC	.013	.013	8-3/4	18	+2-1/4°	0°	1/16

①—Mark located on pulley. Timed at No. 1 cyl., rear of engine.
②—Firing order 1-7-5-11-3-9-6-12-2-8-4-10.
③—Front 0° ±1/4°. Rear 3/4° +1/4°.
④—Module gap.
⑤—Mark located on pulley. Timed at No. 1 cyl., right front of engine.

Year	Model	Spark Plug Gap, In.	Ignition: Breaker Gap, In.	Ignition: Dwell Angle Deg.	Ignition: Firing Order	Ignition: Timing & Mark Location	Valve Lash: Intake	Valve Lash: Ex-haust	Capacities: Engine Oil Qts.	Capacities: Coolant Qts. W/ Heater	Wheel Alignment: Caster	Wheel Alignment: Camber	Wheel Alignment: Toe-In
MAZDA													
1971–72	1200	.032	.018	58	1342	13° BTDC①	.010	.010	3.1	4.3	+3°	+55/60°	.12
	616, 808	.032	.020	52	1342	8° BTDC①	.012	.012	3.8	7.4	+1-1/20°	−1/4°	−.16 to +.08
	R100	.028	.018	58	12	②	—	—	—	7.5	+2°	+1°	.04 to .12
	RX-2, 3	.033	.018	58	12	②	—	—	4.7	8.5	+1-1/20°	−1/4°	−.16 to +.08
1973	618	.031	.020	49–55	1342	①③	.014H	.016H	4.1	8.0	+3/4°	+3/4°	1/8
	808	.031	.020	49–55	1342	①④	.012H	.012H	3.8	7.4	+1-1/4°	+2/3°	1/8
	RX-2	.024	.018	55–61	12	⑤⑥	—	—	4.8⑦	8.4	+1°	+3/5°	1/8
	RX-3	.024	.018	55–61	12	⑤⑥	—	—	4.8⑦	8.4	⑧	⑨	—
1974	808	.030	.018	49–55	1342	①⑩	.012	.012	3.9	7.4	+1-1/4°	+2/3°	1/8
	RX-2	.026	.018	58–62	12	⑥⑪	—	—	4.8	9.9	+1°	+1/2°	1/8
	RX-3	.026	.018	58–62	12	⑥⑪	—	—	4.8	10.2	+1-1/4°	+2/3°	1/8
	RX-4	.026	.018	58–62	12	⑥⑫	—	—	5.5	10.5	+2-1/30°	+1-1/20°	1/8
1975	808	.030	.020	49–55	1342	5° BTDC①	.012H	.012H	4	8	+1-2/3°	+1°	1/8
	RX-3	.025	.018	56–60	12	⑥⑬	—	—	5	10	⑭	+11/12°	1/8
	RX-4	.025	.018	56–60	12	⑥⑬	—	—	5	10	+2°	+1°	1/8

①—Mark located at crankshaft pulley. Timed at No. 1 cyl., front of engine.
②—Leading distributor, TDC; trailing distributor, 5° ATDC. Mark located at eccentric shaft pulley.
③—Man. Trans., 5° BTDC; Auto. Trans., 3° BTDC.
④—Exc. Sta. Wag., Auto. Trans., 5° BTDC; Sta. Wag. Auto. Trans., 8° BTDC.
⑤—Leading distributor, TDC; trailing distributor, 10° ATDC.
⑥—Timed at lower front (leading) spark plug. Mark located at eccentric shaft pulley.
⑦—Less filter.
⑧—Exc. Sta. Wag., +1/2°; Sta. Wag., +1-1/6°.
⑨—Exc. Sta. Wag., +11/12°; Sta. Wag., +1°.
⑩—Man. Trans., 5° BTDC; Auto. Trans., 8° BTDC.
⑪—Leading distributor, 5° ATDC; trailing distributor, 10° ATDC.
⑫—Leading distributor, 5° ATDC; trailing distributor, 15° ATDC.
⑬—Leading distributer, TDC; trailing distributer, 15° ATDC.
⑭—Sedan +1-2/3°; Coupe +1-5/6°; Sta. Wagon +1-7/12°.

Year	Model	Spark Plug Gap, In.	Ignition: Breaker Gap, In.	Ignition: Dwell Angle Deg.	Ignition: Firing Order	Ignition: Timing & Mark Location	Valve Lash: Intake	Valve Lash: Ex-haust	Capacities: Engine Oil Qts.	Capacities: Coolant Qts. W/ Heater	Wheel Alignment: Caster	Wheel Alignment: Camber	Wheel Alignment: Toe-In
MERCEDES-BENZ													
1969	230, 250/8	.024	.014	28–32	153624	TDC④	.003C	.007C	5-3/4	11	+2-1/2②	0°	13/64
1969–71	220/8	.024	.018	45–49	1342	5° ATDC④	.003C	.008C	4-1/4	11	+2-1/2②	0°	13/64
	280/8	.021⑤	.014	30–36⑥	153624	⑦④	.003C	⑧	5-3/4	12⑨	⑩	0°	⑪
1970–71	250/8	.024	.014	30–36	153624	4° ATDC④	.003C	.008C③	5-3/4	11	+2-1/2②	0°	13/64
1972	220	.024	.016	47–53	1342	5° ATDC④	.003C	.008C	5.3	11.2	4° ± 15′	⑫	⑮
	220D	—	—	—	1342	24° BTDC⑰	.004C	.016C	5.3	11.2	4° ± 15′	⑫	⑮
	250, 250CP	.024	.016	30–36	153624	4° ATDC④	.003C	.008C	6.8	10.5	4° ± 15′	⑫	⑮
	280 S.E. 2.8	.024	.016	30–36	153624	6° ATDC④	.003C	.008C	6.8	11.5	4° ± 15′	⑫	⑮
	280 SE/SEL 4.5	.024	.016	30–34	①	5° ATDC⑱	.003C	.008C	9.2	14.8	4° ± 15′	⑫	⑮
	300 SEL 4.5	.024	.016	30–34	①	5° ATDC⑱	.003C	.008C	9.2	14.8	4° ± 15′	⑬	⑮
	350 SL	.024	.016	30–34	①	5° ATDC⑱	.003C	.008C	9.2	15.8	3°15′ ± 20′	⑭	⑮
	600	.024	.016	34–38	①	5° ATDC⑲	.004C	.010C	7.4	24	2° ± 15′	⑭	⑯
1973	220	.024	—	47–53	1342	10° BTDC④	.003C	.008C	4-1/4	10.2	+3-2/3°	+1/4°	5/12°
	220D	—	—	—	1342	24° BTDC⑰	.004C	.016C	4-1/4	10.9	+3-2/3°	+1/4°	5/12°
	280SE/SEL	.024	—	30–34	①	4° ATDC⑱	.003C	.008C	6.2	11.5	⑳	+1/2°	1/3°
	300SEL	.024	—	30–34	①	5° ATDC⑱	.003C	.008C	8	14.7	+4°	+1/3°	1/3°
	450SE/SEL	.024	—	30–34	①	5° ATDC⑱	.003C	.008C	8	16	+10°	−1/6°	5/12°
	450SL/SLC	.024	—	30–34	①	5° ATDC⑱	.003C	.008C	8	16	+3-2/3°	0°	1/3°
1974	230	.024	—	47–53	1342	10° BTDC④	.004C	.008C	6.7	10.5	+3-2/3°	+1/4°	1/8°
	240D	—	—	—	1342	24° BTDC⑰	.004C	.016C	5.7	10.5	+3-2/3°	+1/4°	1/8°
	280, C	.024	—	34–40	153624	4° ATDC④	.004C	.010C	7	11.5	+3-2/3°	+1/4°	5/64°
	450SE/SEL	.024	—	30–34	①	5° ATDC⑱	.004C	.008C	9	16	+10°	−1/6°	1/8°
	450SL/SLC	.024	—	30–34	①	5° ATDC⑱	.004C	.008C	9	16	+3-2/3°	0°	5/64°

Continued

Year	Model	Spark Plug Gap, In.	Ignition				Valve Lash		Capacities		Wheel Alignment		
			Breaker Gap, In.	Dwell Angle Deg.	Firing Order	Timing & Mark Location	Intake	Ex-haust	Engine Oil Qts.	Coolant Qts. W/ Heater	Caster	Camber	Toe-In
MERCEDES—Continued													
1975	230	.024	—	47–53	1342	10° BTDC④	.004	.008	5.5	10.5	+3-2/3°	+1/4°	1/8
	240D	—	—	—	1342	24° BTDC⑰	.004	.012	6.5	10.5	+3-2/3°	+1/4°	1/8
	280, C	.024	—	34–40	153624	7° BTDC④	.004	.010	7	11.6	+2-2/3°	+1/4°	1/8
	280S	.024	—	34–40	153624	7° BTDC④	.004	.010	7	11.6	+10°	0°	1/8
	300D	—	—	—	12453	24° BTDC⑦	.004	.012	7.4	11.4	+3-2/3°	+1/4°	1/8
	450SE/SEL	.024	—	30–34	①	TDC⑱	.004	.008	8	15.9	+10°	+1/6°	1/8
	450SL/SLC	.024	—	30–34	①	TDC⑱	.004	.008	8	15.9	+3-2/3°	0°	5/64

①—Firing order 15486372.
②—With Power steer. +3-1/2.
③—1970 sedan—.007C.
④—Mark located on damper or pulley. Timed at No. 1 cyl., front of engine.
⑤—280 S & SEL .024.
⑥—1969 −28° −32°.
⑦—1969; 280S—@ 4000–4500 RPM, vacuum line disconnected 37° BTDC; 280 SE, SEL, SL 30° BTDC. 1970–71; except 280S, 8° ATDC; 280S, 4° ATDC.
⑧—1969 .007C; 1970–71 .008C.
⑨—280S—11-1/2 qts.
⑩—Except 280 SL—w/p. steer. +3-1/2; w/o p. steer. +2-1/2; 280 SL—w/p. steer. +4; w/o p. steer. +3-1/4.
⑪—Except 280 SL—13/64; 280 SL—5/64.
⑫—0°30′—20′.
⑬—0°20′—20′.
⑭—0° ±10′.
⑮—0°20′ ±10′.
⑯—0°45′ ±10′.
⑰—Injection timing.
⑱—Mark located on damper or pulley. Timed at No. 1 cyl., right front of engine.
⑲—Mark located on damper or pulley. Timed at No. 5 cyl., left front of engine.
⑳—With power steering, +4°; less power steering, +3-1/2°.

MG

Year	Model	Spark Plug Gap, In.	Breaker Gap, In.	Dwell Angle Deg.	Firing Order	Timing & Mark Location	Intake	Ex-haust	Engine Oil Qts.	Coolant Qts. W/ Heater	Caster	Camber	Toe-In
1969	MGC	.025	.015	—	153624	TDC①	.015	.015	7-1/2	11	5°	0°	0°
	MGC/GT	.025	.015	—	153624	TDC①	.015	.015	7-1/2	11	5°	0°	0°
1969–70	MG Midget	.025	.015	—	1342	4° BTDC①	.012H	.012H	3	6	3°	1°	1/16–1/8
	MGB/GT	.025	.015	—	1342	10° BTDC①	.015C	.015C	4-1/2	6	7-1/2°	1°	1/16–3/32
1971	MG Midget	.025	.015	57–63	1342	①②	.012	.012	4	3-1/2	3°	3/4°	0–1/8
	MGB/MGB. GT	.025	.015	57–63	1342	15° BTDC①	.015	.015	5	6	7°	1°	1/16–3/32
1972	MG Midget	.025	.015	57–63	1342	9° BTDC①	.012H	.012H	3.9	3-1/2	3°	3/4°	1/8
	MGB/MGB. GT	.025	.015	57–63	1342	16° BTDC①	.013H	.013H	4.8	6	7°	1°	1/16–3/32
1973–74	MG Midget	.025	.015	60	1342	9° BTDC①	.012H	.012H	3.9	3-1/2	+3°	+3/4°	1/16
	MGB/GT	.025	.015	60	1342	11° BTDC①	.013H	.013H	4.8	6	+7°	+1°	5/64
1975	MG Midget	.035	—	—	1342	2° ATDC①	.010	.010	4-1/2	5	+3°	+3/4°	0–1/8
	MGB	.035	—	—	1342	10° BTDC①	.013	.013	4-1/2	6	+7°	+1°	1/32–1/16

①—Mark located on pulley. Timed at No. 1 cyl., front of engine.
②—With emission controls, 10° BTDC; less emission controls, 13° BTDC.
③—Toe-out.

OPEL

Year	Model	Spark Plug Gap, In.	Breaker Gap, In.	Dwell Angle Deg.	Firing Order	Timing & Mark Location	Intake	Ex-haust	Engine Oil Qts.	Coolant Qts. W/ Heater	Caster	Camber	Toe-In
1969	1.1 Litre	.030	.018	50	1342	6-1/2° BTDC①	.006H	.010H	2-1/2	5-1/2	+3/4°④	+3/4°③	1/16
	1.9 Litre	.030	.018	50	1342	6-1/2° BTDC②	.012H	.012H	3	6	+3/4°④	+3/4°③	1/16
1970	1.1 Litre	.030	.018	50	1342	6-1/2° BTDC①	.006H	.010H	2-1/2	5-1/2	+2°	+1°	1/16
	1.9 Litre	.030	.018	50	1342	6-1/2° BTDC②	.012H	.012H	3	6	+2°	+1°	1/16
1971	1.1 Litre	.030	.018	50	1342	①⑤	.006H	.010H	2-3/4⑥	5	+2°	+1°	1/8
	1.9 Litre	.030	.018	50	1342	TDC②⑧	Hydraulic		3⑦	6⑨	⑩	+1°	1/8⑪
1972	1900	.030	.018	50	1342	TDC②⑧	Hydraulic		3/14	6	+4-1/2°	+1-1/2°	5/32
	GT	.030	.018	50	1342	TDC②⑧	Hydraulic		3-1/4	6	+3°	+1-1/2	1/16
1973	GT	.028	.016	50	1342	TDC②⑧	Hydraulic		2.9	5.8	+3°	+1°	1/16
1973–74	1900, Manta	.028	.016	50	1342	TDC②	Hydraulic		2.9	5.9	+4-½°④	−1°	5/32
1975	1900, Manta	.030	.016	47–53	1342	⑫	Hydraulic		4	7	+4-1/2°	−1/2°	1/8

①—Mark located on pulley. Timed at No. 1 cyl., front of engine.
②—Mark located on flywheel. Timed at No. 1 cyl., front of engine.
③—GT, +1/2°.
④—Left and right wheel angles must be within 1° of each other.
⑤—At 950 R.P.M. with Vac. Adv. & Retard units disconnected & plugged.
⑥—W/Filter, 3 Qts.
⑦—W/Filter, 3-1/4 Qts.
⑧—At 900 R.P.M. with Vac. Adv. and Retard units disconnected and plugged.
⑨—1900 Series, Std. Trans., 6 Qts.; Auto. Trans., 6.1 Qts.
⑩—1900 Series, +3-1/2°.; Opel, +2°; GT, +3°.
⑪—1900, 3/16.
⑫—Timed at No. 1 cyl., front engine. Disconnect & plug vacuum hose, align mark on pulley with mark on pointer.

IMPORTED CAR DATA—Continued

Year	Model	Spark Plug Gap, In.	Ignition				Valve Lash		Capacities		Wheel Alignment		
			Breaker Gap, In.	Dwell Angle Deg.	Firing Order	Timing & Mark Location	Intake	Ex-haust	Engine Oil Qts.	Coolant Qts. W/ Heater	Caster	Camber	Toe-in
PEUGEOT													
1969–70	504	.023	.016	55–59	1342	10° BTDC(2)	.004C	.010C	4.22	8.25	+2-2/3°	+5/8°	11/64
1970	404	.023	.016	55–59	1342	10° BTDC(2)	.004C	.010C	4.22	8.25	+2-2/3°	+5/8°	11/64
1971	304	.023	.016	57	1342	5°BTDC(3)	.004C	.010C	4.22	6.1	+1/2°	+1/2°	5/64
	504	.024	.016	57	1342	TDC(4)	.004C	.010C	4.22	8.5	+2-2/3°	+5/8°	7/64
1972	304	.024	.016	61	1342	5° BTDC(3)	.004C	(1)	4	7	+1/2°	+1/2°	5/64
	504	.025	.017	61	1342	5° BTDC(4)	.004C	.010C	5	8.25	+2-2/3°	+5/8°	5/64
1973–74	504	.024	.016	57	1342	5° BTDC(4)	.006C	.012C	4.2	8.25	+2-2/3°	+5/8°	1/8
1975	504 Gas	.028	.016	55–59	1342	5° BTDC(5)	(6)	.010	4.22	4.15	+2-2/3°	+19/32°	1/8
	504 Diesel	—	—	—	1342	5° BTDC(5)	.006	.010	4.22	5.28	+2-2/3°	+19/32°	1/8

(1)—Exc. Sta. Wag., .0004C; Sta. Wag., 10C.
(2)—Mark located on flywheel. Timed at No. 1 cyl., rear of engine.
(3)—Mark located on clutch housing. Timed at No. 1 cyl. Left side of vehicle.
(4)—Mark located on pulley. Timed at No. 1 cyl. rear of engine.
(5)—Mark located on pulley. Timed at cyl. No. 1 or No. 4 at 1200 RPM.
(6)—Cyl. 1 & 4, .008"; cyl. 2 & 3, .004".

Year	Model	Spark Plug Gap, In.	Breaker Gap, In.	Dwell Angle Deg.	Firing Order	Timing & Mark Location	Intake	Ex-haust	Engine Oil Qts.	Coolant Qts. W/ Heater	Caster	Camber	Toe-in
PLYMOUTH CRICKET													
1971–72	All	.025	.015	60	1342	7° BTDC(1)	.008	.016	3.4	7.7	+1-1/2°	+1-3/4°	5/64

(1)—Mark located on pulley. Timed at No. 1 cyl., front of engine.

Year	Model	Spark Plug Gap, In.	Breaker Gap, In.	Dwell Angle Deg.	Firing Order	Timing & Mark Location	Intake	Ex-haust	Engine Oil Qts.	Coolant Qts. W/ Heater	Caster	Camber	Toe-in
PORSCHE/PORSCHE AUDI													
1969	911 E & S	.014	.014	(2)	162435	(17)(23)	.004	.004	9-1/2	—	+6-3/4° ± 3/4°	(9)	(8)
	911T	.024	.014	(2)	162435	(17)(23)	.004	.004	9-1/2	—	+6-3/4° ± 3/4°	(9)	(8)
	912	.022	.016	47–53	1432	3° BTDC(5)	—	—	4.2	—	+6-3/4° ± 3/4°	(9)	(8)
1970–71	Audi S-90	.024	.016	47–53	1342	(12)(28)	.008	.008	4.2	8	1/6° ± 1/3°	(14)	0 to −.079"
	Audi 100LS	.024	.016	47–53	1342	(12)(28)	.016	.016	4.2	8	1/10° ± 1/3°	(15)	0 to −.08"
	911E, S & T	.024	.016	35–41	162435	(16)(23)	.004	.004	9-1/2	—	+6-3/4° ± 3/4°	(9)	(10)
	914	.028(18)	.016	44–50	1432	(5)(19)	.006	.006	(20)	—	6° ± 1/2°	(21)	(22)
	914/6	.024	.016	37–43	162435	(11)(27)	.004	.004	9-1/2	—	6° ± 1/2°	(21)	(22)
1972	911	.024	(1)	(2)	162435	(17)(23)	.004	.004	9-1/2(3)	—	6-5/6° ± 1/2°	(9)	0
	914	.028	.016	47-50	1432	5° BTDC(5)	.004	.004	3-3/4	—	6-1/2°	(21)	.33"
	Audi S-90	.024	.016	47–53	1342	(11)(28)	.008	.016	4-1/4	8	11/60° ± 1/3°	(14)	0 to −.08"
	Audi 100	.024	.016	47-53	1342	(12)(28)	.008	.016	4-1/4	8	1/10° ± 1/3°	(15)	0 to −.08"
1973	911E, S & T	.024	.016	(2)	162435	5° ATDC(23)	.004	.004	(6)	—	6-1/12° ± 1/4°	0° ± 1/6°	(24)
	914	.028	.016	44–50	1432	(5)(19)	.006	.006	3.7	—	6° ± 1/2°	(21)	(22)
	Audi 100	.035	.016	47–53	1342	8° ATDC(28)	.008	.016	4.2	8	1/15° ± 1/3°	(25)	(26)
1974	911, S	.022	.014	34–40	162435	5°ATDC(23)	.004	.004	10-1/2	—	+6-1/12°	+1/12°	0
	914	.028	.016	44–50	1432	(5)(19)	.006	.008	3.7	—	+6°	0°	(4)
	Audi 100LS	.038	.016	50	1342	8° ATDC(28)	.008	.016	4-1/4	8	+1/10°	+1/6°	(7)
	Audi Fox	.028	.016	50	1342	TDC(29)	.008	.016	3.1	6-1/2	+1/3°	+5/12°	(13)

Continued

Year	Model	Spark Plug Gap, In.	Ignition: Breaker Gap, In.	Ignition: Dwell Angle Deg.	Ignition: Firing Order	Ignition: Timing & Mark Location	Valve Lash: Intake	Valve Lash: Exhaust	Capacities: Engine Oil Qts.	Capacities: Coolant Qts. W/ Heater	Wheel Alignment: Caster	Wheel Alignment: Camber	Wheel Alignment: Toe-in

PORSCHE/PORSCHE AUDI—Continued

Year	Model	Spark Plug Gap, In.	Breaker Gap, In.	Dwell Angle Deg.	Firing Order	Timing & Mark Location	Intake	Exhaust	Engine Oil Qts.	Coolant Qts. W/ Heater	Caster	Camber	Toe-in
1975	911S	.022	.016	35–41	162435	5° ATDC(3)	.004	.004	11.6	—	+6-1/2°	+1/2°	0°
	914 1.8 Liter	.036	.016	44–50	1432	7-1/2° BTDC(5)	.006	.006	3.7	—	+6°	+1/3°	(4)
	914 2.0 Liter	.036	.016	44–50	1432	(5)(19)	.006	.008	3.7	—	+6°	+1/3°	(4)
	Audi 100LS	.036	.016	47–53	1342	6° ATDC(29)	.008	.016	4.7	—	+3/20°	+0°	1/3°
	Audi Fox	.028	.016	47–53	1342	3° ATDC(29)	.010	.018	3.7	6-1/2	+1/2°	0°	1/6°

(1)—Bosch .014. Marelli .016.
(2)—Bosch 35°–41°. Marelli 37°–43°.
(3)—Type 911S with oil cooled 10.6.
(4)—1/3° ±1/6°.
(5)—Mark located on pulley. Timed at No. 1 cyl., right front of engine.
(6)—Man. Trans.—911 E & T, 11 qts.; 911S, 13 qts.; Auto. Trans.—911 E & T, 13 qts.; 911S, 16 qts.
(7)—0.008 inch.
(8)—Front, .25; Rear, 0° ±1/6°.
(9)—Front, 0° ±1/3°; Rear —5/6 ±1/3°.
(10)—Front, 0°; Rear, 0° ±1/6°.
(11)—35° BTDC @ 6000 R.P.M. (Engine free or loaded).
(12)—27° BTDC @ 2500 R.P.M.
(13)—+5/12° ±1/4°.
(14)—Front, 1/4° ±1/3°; Rear, 1/2° ±1/4°.
(15)—Front, 11/60° ±1/3°; Rear, 0° ±1/2°.
(16)—911 E & S, see note (17). 911T, see note (11).
(17)—30° BTDC @ 6000 R.P.M. (Engine free or loaded).
(18)—For winter .020.
(19)—27° BTC @ 3500 R.P.M.
(20)—3.7 qts.; w/o oil filter 3.2 qts.
(21)—Front, 0° ±1/3°; Rear, —1/2° ±1/3°.
(22)—Front, 1/3° ±1/6°; Rear, 0° ±1/4°.
(23)—Mark located on pulley. Timed at No. 1 cyl., left rear of engine.
(24)—Front, 0°; Rear, 0° ±1/3°.
(25)—Front, 1/3° ±1/3°; Rear, —1/2° ±1/4°.
(26)—Toe out 0° to —1/3°.
(27)—Mark located on flywheel. Timed at No. 1 cyl., left rear of engine.
(28)—Mark located on pulley. Timed at No. 1 cyl., front of engine.
(29)—Mark located on flywheel. Timed at No. 1 cyl., front of engine.

RENAULT

Year	Model	Spark Plug Gap, In.	Breaker Gap, In.	Dwell Angle Deg.	Firing Order	Timing & Mark Location	Intake	Exhaust	Engine Oil Qts.	Coolant Qts. W/ Heater	Caster	Camber	Toe-in
1969	Renault 16	.025	.018	56	1342	TDC(1)	.008C	.010C	4-1/2(3)	7	+2-1/6°	+3/4°	(12)
	Renault 10	.025	.018	56	1342	TDC(1)	.006C	.008C	3(3)	7-1/2	+9°	+1-2/3°	(5)
1970–71	Renault 10	.025	.018	57	1342	3° ATDC(6)(1)	.006C	.008C	(3)(7)	7-1/2	+9°	+1-2/3°	(5)
	Renault 16	.024	.018	57	1342	TDC(1)(8)	.008C	.010C	4-1/2(3)	7	+2-2/3°	+3/4°	(9)
1972	Renault 12	.025	.016	57	1342	(2)(11)	.008C	.010C	4-1/2(3)	9.5	+4°	+1-1/2°	(12)
	Renault 15	.025	.016	57	1342	TDC(2)	.008C	.010C	4.2(3)	6	+4°	+1-1/2°	(12)
	Renault 16	.025	.016	57	1342	TDC(2)	.008C	.010C	4-1/2(3)	9.5	+4°	+1-1/2°	(12)
	Renault 17	.025	.016	57	1342	TDC(2)	.010C	.012C	4.7(3)	5.8	+4°	+1-1/2°	(12)
1973	Renault 12	.025	.016	57	1342	(2)(10)	.008C	.010C	4(3)	9.5	+4°	+1-1/2°	(12)
	Renault 15	.025	.016	57	1342	(2)(10)	.008C	.010C	4(3)	6	+4°	+1-1/2°	(12)
	Renault 17	.025	.016	57	1342	TDC(2)(4)	.010C	.012C	4.7(3)	5.8	+4°	+1-1/2°	(12)
1974	Renault 12	.026	.018	55	1342	10° BTDC(2)	.008C	.010C	4-1/2(3)	9-1/2	+4°	+1-1/2°	(12)
	Renault 15, 17TL	.026	.018	55	1342	10° BTDC(2)	.008C	.010C	4-1/2(3)	6	+4°	+1-1/2°	(12)
	Renault 17	.026	.018	55	1342	10° BTDC(2)	.010C	.012C	5(3)	5.8	+4°	+1-1/2°	(12)
1975	Renault 12	.026	.018	57	1342	10° BTDC(2)	.008	.010	4-1/2(3)	9-1/2	+2°	+1-1/2°	(12)
	Renault 15, 17TL	.026	.018	57	1342	10° BTDC(2)	.008	.010	4-1/2(3)	6	+2°	+1-1/2°	(12)
	Renault 17	.026	.018	57	1342	12° BTDC(2)	.010	.012	5(3)	5.8	+2°	+1-1/2°	(12)

(1)—Mark located at pulley. Timed at No. 1 cyl. toward front of vehicle.
(2)—Mark located at flywheel or converter. Timed at No. 1 cyl., rear of engine.
(3)—W/filter.
(4)—At 1000 R.P.M., vacuum hose off.
(5)—3/64 inch out to 5/64 inch in.
(6)—Notch on pulley 1/8″ past pointer (with hole) on timing chain cover.
(7)—Auto. Trans., 3; Man. Trans., 3-1/2.
(8)—Auto. Trans., with timing light, vacuum hose off, eng. idling below 800 R.P.M.
(9)—Toe-out, 0 to 1/8″ at 37mm (1-15/32″) steer. height.
(10)—Man. Trans., 5° BTDC; Auto. Trans., 3° BTDC.
(11)—Man. Trans., TDC; Auto. Trans., 6° BTDC.
(12)—Toe-out, 0 to 1/8″.

SAAB

Year	Model	Spark Plug Gap, In.	Breaker Gap, In.	Dwell Angle Deg.	Firing Order	Timing & Mark Location	Intake	Exhaust	Engine Oil Qts.	Coolant Qts. W/ Heater	Caster	Camber	Toe-in
1969	99	.028	—	(5)	1342	9° BTDC(6)	.009	.017	4(2)	9	+1-1/4°	+3/4°	0
1969–71	95, 96, 97	.026	—	50	1342	6° BTDC(1)	.014	.016	3-1/2(2)	7-1/2	+2°	(3)	(4)
1970–71	99C	.028	—	(5)	1342	9° BTDC(6)	.009	.017	4(2)	9	+1-1/4°	+3/4°	0
	99E	.028	—	(5)	1342	5° BTDC(6)	.009	.017	4(2)	9	+1-1/4°	+3/4°	0
1972	99C	.028	—	(5)	1342	9° BTDC(6)	.009	.017	4(2)	9	+3/4°	+3/4°	0
	99E	.028	—	(5)	1342	5° BTDC(6)	.009	.017	4(2)	9	+3/4°	+3/4°	0
1972–73	95, 96, 97	.026	—	50	1342	3° BTDC(1)	.014	.016	3-1/2(2)	7-1/2	+2°	(3)	(4)
1973	99C	.028	—	(5)	1342	14° BTDC(6)	.009	.017	4(2)	10	+3/4°	+3/4°	0
	99E	.028	—	(5)	1342	5° BTDC(6)	.009	.017	4(2)	10	+3/4°	+3/4°	0

Continued

SAAB—Continued

Year	Model	Spark Plug Gap, In.	Ignition: Breaker Gap, In.	Ignition: Dwell Angle Deg.	Ignition: Firing Order	Ignition: Timing & Mark Location	Valve Lash: Intake	Valve Lash: Exhaust	Capacities: Engine Oil Qts.	Capacities: Coolant Qts. W/ Heater	Wheel Alignment: Caster	Wheel Alignment: Camber	Wheel Alignment: Toe-in
1974	97	.026	.016	50	1342	3° BTDC①	.014	.016	3-1/2②	7-1/2	+2°	0°	.04
	99	.026	.014	⑤	1342	⑥⑦	.009	.017	4②	10	+3/4°	+3/4°	0
1975	99E	.026	—	50	1342	⑥⑧	.009	.017	4-1/2②	9	+3/4°	+3/4°	3/64

①—Mark located at pulley & front cover. Timed at No. 1 cyl., located right front.
②—Includes filter.
③—Models 95 & 96, +3/4°; Model 97, 0°.
④—Models 95 & 96, .08 inch; Model 97, .04 inch.
⑤—Carburetor equipped, 40°; fuel injected, 50°.
⑥—Mark located at flywheel. Timed at No. 1 cyl., located nearest firewall.
⑦—99C 14° BTDC; 99E exc. Calif. 12° BTDC; Calif. 4° BTDC.
⑧—Exc. Calif. 14° BTDC; Calif. 12° BTDC.

SIMCA

Year	Model	Spark Plug Gap, In.	Ignition: Breaker Gap, In.	Ignition: Dwell Angle Deg.	Ignition: Firing Order	Ignition: Timing & Mark Location	Valve Lash: Intake	Valve Lash: Exhaust	Capacities: Engine Oil Qts.	Capacities: Coolant Qts. W/ Heater	Wheel Alignment: Caster	Wheel Alignment: Camber	Wheel Alignment: Toe-in
1969	All	.024	.013	—	1342	4° BTDC①	.012H	.014H	—	—	+1-1/4	+1/4	1/36②
1970–71	All	.024	.013	—	1342	③①	.012H	.014H	—	—	+1-1/4	+1/4	1/36②

①—Mark located at pulley. Timed at No. 1 cyl., front of engine.
②—Toe-out.
③—Std. Trans. TDC; Auto Trans. 4° BTDC.

SUBARU

Year	Model	Spark Plug Gap, In.	Ignition: Breaker Gap, In.	Ignition: Dwell Angle Deg.	Ignition: Firing Order	Ignition: Timing & Mark Location	Valve Lash: Intake	Valve Lash: Exhaust	Capacities: Engine Oil Qts.	Capacities: Coolant Qts. W/ Heater	Wheel Alignment: Caster	Wheel Alignment: Camber	Wheel Alignment: Toe-in
1969–71	Exc. 1971 "G"	.029	.019	49–55	1324	TDC①	.009C	.010C	2.8	6.2	+2°	+1-5/6°	.2
1971	"G" Model	.029	.019	49–55	1324	TDC①	.012C	.012C	3.4	6.2	+2°	+1-5/6°	.2
1972	All	.029	.019	49–55	1324	6° BTDC①	.012C	.012C	3.4	6.2	+2°	+1-5/6°	.2
1973–74	All	.032	.019	49–55	1324	6° BTDC①	.012C	.012C	3.5	6.2	+3/4°	+1-1/3°	.08–.32
1975	All	.030	.020	49–55	1342	8° BTDC①	.012	.014	3.8	6.5	+3/4°	+1°	0–5/16

①—Mark located at flywheel. Timed at No. 1 cyl., right front of engine.

TOYOTA

Year	Model	Spark Plug Gap, In.	Ignition: Breaker Gap, In.	Ignition: Dwell Angle Deg.	Ignition: Firing Order	Ignition: Timing & Mark Location	Valve Lash: Intake	Valve Lash: Exhaust	Capacities: Engine Oil Qts.	Capacities: Coolant Qts. W/ Heater	Wheel Alignment: Caster	Wheel Alignment: Camber	Wheel Alignment: Toe-in
1969	Land Cruiser	.032	.018	—	153624	7° BTDC①	.008H	.014H	—	—	+1°	+1°	1/8
	Crown	.032	.018	—	153624	10° BTDC①	.004C	.007C	—	—	+1°	+1°	3/64
	Corolla	.032	.018	—	1432	5° BTDC①	.003C	.007C	4	5	+1/2°	+1/2°	5/64
	Corona Mark II	.030	.018	—	1342	②①	.008H	.014H	4-1/2	7.8	+1/6°	+1-1/4°	5/64
1969–70	Corona 1900③	.030	.018	50–54	1243	5° BTDC①	.008H	.014H	6.0	9.0	+1/12°	+1-1/6°	5/64
1970–71	Corona 1900④	.030	.018	50–54	1342	⑤①	.008H	.014H	5.3	8.3	+1/3°	+2/3°	13/64
	Corona MK II	.030	.018	50–54	1342	⑤①	.008H	.014H	5.3	⑥	+5/12°	+1-1/6°	9/32
	Crown	.030	.018	39–43	153624	TDC①	.007H	.010H	5.9	11.6	+1/2°	+1/2°	1/8
	Pickup	.030	.018	50–54	1342	⑤①	.008H	.014H	5.2	7.8	−1/3°	+1°	7/32
1970–73	Corolla 1200	.030	.018	50–54	1342	5° BTDC①	.008H	.012H	3.7	5.1	⑦	+5/6°	1/8
1970–74	Land Cruiser	.030	.018	39–43	153624	7° BTDC⑧	.008H	.014H	9.0	16.1	+1°	+1°	5/32
1971	Celica ST 1900	.030	.018	50–54	1342	⑤①	.008H	.014H	5.3	8.3	+1°	+1°	7/32
1971–74	Corolla 1600	.030	.018	50–54	1342	5° BTDC①	.008H	.013H	3.9	6.9	+1-2/3°	+1°	1/8
1972	Corona MK II	.030	.018	50–54	1342	7° BTDC①	.008H	.014H	5.3	8.7	+5/12°	+1-1/6°	1/4
	Crown	.030	.018	39–43	153624	5° BTDC①	.008H	.014H	5.9	11.5	+1/2°	⑨	3/64
	Mark II	.030	.018	39–43	153624	TDC①	.007H	.010H	5.5	11.4	⑩	+1-1/12°	5/32
1972–73	Carina 1600	.030	.018	50–54	1342	5° BTDC①	.008H	.013H	3.9	6.9	+1°	+1°	7/32
	Corona 1900	.030	.018	50–54	1342	7° BTDC①	.008H	.014H	5.3	⑪	⑫	⑬	⑭
	Pickup	.030	.018	50–54	1342	7° BTDC①	.008H	.014H	5.2	9.0	−1/3°	+1°	15/64
1972–74	Celica ST 2000	.030	.018	50–54	1342	7° BTDC①	.008H	.014H	5.3	8.3	+1°	+1°	13/64
1973–74	Mark II	.030	.018	39–43	153624	5° BTDC①	.007H	.010H	5.5	11.4	⑮	+1/2°	5/32
1974	Corolla 1200	.030	.018	50–54	1342	5° BTDC①	.008H	.014H	3.7	5.1	+2°	+5/6°	1/8
	Pickup	.030	.018	50–54	1342	7° BTDC①	.008H	.014H	5.2	9.0	+1/2°	+1	15/64

Continued

Year	Model	Spark Plug Gap, In.	Ignition				Valve Lash		Capacities		Wheel Alignment		
			Breaker Gap, In.	Dwell Angle Deg.	Firing Order	Timing & Mark Location	Intake	Ex-haust	Engine Oil Qts.	Coolant Qts. W/ Heater	Caster	Camber	Toe-In

TOYOTA—Continued

Year	Model	Spark Plug Gap, In.	Breaker Gap, In.	Dwell Angle Deg.	Firing Order	Timing & Mark Location	Intake	Exhaust	Engine Oil Qts.	Coolant Qts. W/ Heater	Caster	Camber	Toe-In
1975	Corolla	.030	.018	⑥	1342	⑰	.008	.013	4.6	8.2	⑱	⑳	3/64
	Corona	.030	.018	48–54	1342	8°	.008	.012	5.3	8.5	+5/6°	+7/12°	⑳
	Mark II	.030	.018	38–44	153624	㉑	.007	.010	5.9	12.3	㉒	+1-5/12°	7/32
	Celica	.030	.018	48–54	1342	8°	.008	.012	4.8	8.5	+1°	+1°	1/4
	Pickup	.030	.018	48–54	1342	8°	.008	.012	4.8	8.5	−1/2°	+1°	3/16
	Land Cruiser	.037	.018	38–44	153624	7°	.008	.014	8.5	17.4	—	—	—

①—Mark located at pulley. Timed at No. 1 cyl., front of engine.
②—Std. Trans., TDC; Auto. Trans., 5° BTDC.
③—Models RT 43 & 52.
④—Models RT 83 & 93.
⑤—TDC before 3/71; 10° BTDC after 3/71.
⑥—Model RT 78, 7.8 quarts; all others, 8.3 quarts.
⑦—Models KE 25 & 26, +2°; all others, +1-2/3°.
⑧—Mark located at flywheel. Timed at No. 1 cyl., front of engine.
⑨—Model MS 75, +5/12°; all others, +1/2°.
⑩—Model MX28, +2/3°; all others, +5/12°.
⑪—Models RT 104, 114 & 118, 8.7 quarts; all others, 8.4 quarts.
⑫—Models RT 104, 114 & 118, +5/6°; all others, +1/3°.
⑬—Models RT 104, 114 & 118, +7/12°; all others, +2/3°.
⑭—Models RT 104, 114 & 118, 3/64 inch; all others, 11/64 inch.
⑮—Model MX 29, +2/3°; all others, +5/12°.
⑯—Single point dist., 52°; dual point dist., primary, 57°, secondary, 52°.
⑰—Single point dist., 10° BTDC; dual point dist., primary, 12° BTDC, secondary, 19°–25° BTDC.
⑱—Exc. wagon, +1-5/6°; wagon, +1-1/12°.
⑲—2 dr. & 4 dr. sedan, +5/6°; 2 dr. coupe, +3/4°; wagon, +2/3°.
⑳—Exc. radial tires, 5/32″; radial tires, 13/16″.
㉑—Exc. Calif., 10° BTDC; Calif., 5° BTDC.
㉒—Exc. wagon, +5/12°; wagon, +2/3°.

TRIUMPH

Year	Model	Spark Plug Gap, In.	Breaker Gap, In.	Dwell Angle Deg.	Firing Order	Timing & Mark Location	Intake	Exhaust	Engine Oil Qts.	Coolant Qts. W/ Heater	Caster	Camber	Toe-In
1969–70	TR6	.025	.015	—	153624	4° ATC①	.010	.010	4.7	6.6	1°	④	1/16–1/8
	GT6+	.025	.020	—	153624	4° ATC①	.010	.010	4.7	6.6	1°	⑤	1/16–1/8
	Spitfire Mk III	.025	.015	—	1342	2° ATC①	.010C	.010C	4.2	4.8	+4°	+2°	0-1/16
1971	TR-6	.025	.015	32–38	153624	①	.010	.010	4.8	6.6	2-3/4°⑥	1/4°⑥	1/16
	GT-6X	.025	.015	40–42	153624	①	.010	.010	4.8	6.6	4°⑥	2°⑥	1/16
	Spitfire	.025	.015	40–42	1342	①	.010	.010	4.8	4.5	4-1/2°⑥	2° ⑥	1/16
1972	STAG	.025	.015	30–32	②	4° ATDC⑧	.008C	.010C	4.8	11	1°-3°	1	⑦
1972–73	GT-6 Mk-3	.025	.015	40–42	153624	4° ATDC①	.010C	.010C	4.8	6.6	3-1/2°	2-3/4°	1/32
1972–74	Spitfire	.025	.015	38–40	1342	2° ATDC①	.010C	.010C	4.2	4.8	+4°	+3°	1/16–1/8
	TR6	.025	.015	34–37	153624	4° ATDC①	.010C	.010C	4.8	6.6	+2-3/4°	+1/4°	⑦
1973	STAG	.025	.015	30–32	②	4° ATDC⑧	.009	.017	5.4③	11	+2°	+7/8°	1/16–1/8
1975	Spitfire	.025	—	—	1342	2° ATDC①	.010	.010	4-1/2	5	+4°	+3°	3/32
	TR6	.025	.015	34–48	153624	4° ATDC①	.010	.010	5	6-1/2	+2-3/4°	+1/4°	⑦

①—Mark located at pulley. Timed at No. 1 cyl., front of engine.
②—Firing order—1-2-7-8-4-5-6-3.
③—Includes filter.
④—+1/2° ± 1°.
⑤—2-3/4° ± 1°.
⑥—Plus or minus 1/2°.
⑦—Front 1/16″—1/8 Rear 0″ to 1/16.
⑧—Mark located at pulley. Timed at No. 2 cyl., right front of engine.

Year	Model	Spark Plug Gap, In.	Ignition				Valve Lash		Capacities		Wheel Alignment		
			Breaker Gap, In.	Dwell Angle Deg.	Firing Order	Timing & Mark Location	Intake	Ex-haust	Engine Oil Qts.	Coolant Qts. W/ Heater	Caster	Camber	Toe-in
VOLKSWAGEN													
1969	1500	.026	.016	①	1432	TDC②	.004C	.004C	2-1/3	—	+3	+1/2	1/8
	1600	.026	.016	①	1432	TDC②	③	④	2-1/3	—	+4	+1-1/3°	1/8
1970–73	1600⑤	.028	.016	47–53	1432	TDC②	.004C	.004C	2-1/3	—	+4	+1-1/3°	1/8
1970–74	1600⑥	.024	.018	44–50	1432	②⑦	.006C	.006C	2.6	—	⑧	⑨	1/2°
1971–73	1700⑩	.028	.017	—	1432	27° BTDC②	.004C	.004C	3-7/8	—	+1-1/6	+1-1/6°	1/8
1974	Dasher	.028	.015	47–53	1342	⑪	.008C	.016C	3.7	12.6	+1/6°	+5/12°	1/6°
	Type 4	.024	.016	44–50	1432	27° BTDC②	.006C	.006C	3.7	—	+1-3/4°	+1-1/6°	1/3°
1975	1600⑥	.028	.016	44–50	1432	②⑫	.006C	.006C	2.6	—	⑧	⑨	1/2°
	Dasher	.026	.016	44–50	1342	3° ATDC⑬	.010	.018	3.7	6.4	+1/6°	+5/12°	1/6°
	Rabbit	.028	.015	44–50	1342	3° ATDC⑬	.010	.018	3.7	6.8	+2°	+1/2°	1/6°
	Scirocco	.026	.015	44–50	1342	3° ATDC⑬	.010	.018	3.7	6.9	+2°	+1/2°	1/6°

①—Exc. VW-113-905-205-H Dist. 47–53. With VW-113-905-205-H Dist. 48-52.
②—Mark located at pulley. Timed at No. 1 cyl., right front of engine.
③—To 672748 (W/1 Carb.) .008. From 672749 (W/1 Carb.) .004. To 6727697 (w/2 Carbs.) .008. From 672698 (w/2 Carbs.) .004.
④—To 672748 (w/1 Carb.) .012. From 672749 (w/1 Carb.) .004. To 672697 (w/2 Carbs.) .012. From 672698 (w/2 Carbs.)
⑤—Type 3.
⑥—Type 1.
⑦—Engines built through July 1970, TDC; August 1970 to Spring 1973, 5° ATDC; From Spring 1973, 7½° BTDC.
⑧—Exc. Super Beetle, +3-1/3°; Super Beetle, +2°.
⑨—Exc. Super Beetle, +1/2°; Super Beetle, +1°.
⑩—Type 4.
⑪—30° BTDC at 3000 R.P.M. with vacuum hoses off. Timed at No. 1 cylinder, front of engine.
⑫—Man. trans., 5° BTDC; auto. trans., TDC.
⑬—Mark located on flywheel.

VOLVO

Year	Model	Spark Plug Gap, In.	Breaker Gap, In.	Dwell Angle Deg.	Firing Order	Timing & Mark Location	Intake	Ex-haust	Engine Oil Qts.	Coolant Qts. W/ Heater	Caster	Camber	Toe-in
1969	180S, B20B	.028	.016	60	1342	10° BTDC①	.021C	.021C	3-1/2②	9-1/2	+1/2°	+1/4°③	1/16
1969–71	140, B20B	.028	.016	60	1342	10° BTDC①	.021C	.021C	3-1/2②	9④	+1/2°	+1/4°③	1/16
	164, B30A	.028	.010	40	153624	10° BTDC①	.017C	.017C	5-1/2②	13	+1/2°	+1/4°③	1/16
1970–71	180E, B20E	.028	.016	60	1342	10° BTDC①	.017C	.017C	3-1/2②	9-1/2	+1/2°	+1/4°③	1/16
1972	140 Series	.028	.016	60	1342	10° BTDC①	⑤	⑤	3-1/2②	10	+1/2°	+1/4°	1/16
	164	.028	.016	40	153624	10° BTDC①	⑤	⑤	5-1/2②	13	+1/2°	+1/4°	1/16
	1800 E/ES	.028	.016	60	1342	10° BTDC①	⑤	⑤	3-1/2②	9-1/2	+1/2°	+1/4°	1/16
1973	140 Series	.030	.018	60	1342	10° BTDC①	.016C	.016C	3.4⑥	10.5	+1/2°	+1/4°	1/8
	164	.030	.010	40	153624	10° BTDC①	.021C	.021C	5.5⑦	13	+1-1/2°	+1/4°	1/8
	1800 ES	.030	.018	60	1342	10° BTDC①	.017C	.017C	3.4⑥	9.0	+2-1/4°	+1/4°	1/32
1974	140 Series	.028	.016	60	1342	10° BTDC①	.017C	.017C	4	10	+1-1/2°	+1/4°	1/8
	164	.028	.010	40	153624	10° BTDC①	.021C	.021C	5-1/2	13	+1-1/2°	+1/4°	1/8
1975	164	.030	—	—	153624	10° BTDC①	.021	.021	5.5⑦	11	+2°	+1/4°	1/8
	242, 244, 245	.030	—	—	1342	10° BTDC①	.017	.017	3.4⑥	10	+2-1/2°	+1-1/4°	⑧

①—Mark located at pulley. Timed at No. 1 cylinder, front of engine with vacuum line disconnected.
②—Add 1/2 qt. w/filter.
③—Models P444, 445, 544, 1100, 2100, +3/8°; all others, +1/4°.
④—100 H.P. B180 eng., add 1/2 qt.
⑤—B20B, B30 A & F engines, .021C; B20F engine, .017C. "F" engines are fuel injected.
⑥—W/filter, 4 qts.
⑦—W/filter, 6.3 qts.
⑧—Man. steer. 1/8 in.; power steer. 1/16".

Each purchaser of a new car receives a new car warranty certificate from the selling dealer. Before placing this certificate in the glove compartment, where it will often be forgotten, the car owner should carefully read it to the point where he fully understands what is required of him so as not to invalidate the terms of the warranty. Typical samples of the various new car warranties are shown.

It is also possible, a purchaser of a vehicle that has been used as a dealer demonstrator, will be given a warranty valid only for the balance or unused portion of the time or mileage left, whichever shall expire first. A person buying a used car may receive a 30 day guarantee or possibly even for longer periods. This would seem to imply that the selling dealer will guarantee this vehicle unconditionally for that expressed period of time. Or possibly, the guarantee is only a 50/50 guarantee which means the car owner would be asked to pay for one half the cost of the repairs and the selling dealer assumes the remaining one half the cost. Whatever the case, be sure you are fully aware of the terms and conditions of the warranty.

Since the passing of the 5 year or 50,000 mile warranty, whichever shall occur first, there has been a resurgence of the so called extended warranty. At the time of car purchase the dealer will explain that for a sum of say $15.00 or $25.00 you can have your warranty extended from the original 12 months or 12,000 miles to 5 years or 50,000 miles, whichever shall occur first. The provision that must be met usually states you must use a certain brand of lubricants and you must have prescribed maintenance servicing, for which you pay the usual charge. It might be necessary for you to return the car to the selling dealer for this servicing although it is generally permissable for the work to be done at another location providing you hold all invoices necessary for you to prove to the selling dealer that the prescribed work had been performed. It is advisable for you to have these conditions clearly defined so there will be no misunderstanding later when the car might be returned for service under this warranty.

Remember, there are many and varied kinds of warranties generally being offered. Know your warranty. But, at such time as a problem should occur and you feel it should be covered by your warranty, discuss this problem with a representative of the selling dealer first. Your next course of action would be to contact the vehicle manufacturers' Zone Office. If you still feel you haven't been satisfied you should then write to the vehicle manufacturers' Customer Relations Department outlining a full history of the problem. This process usually results in a better understanding and avoids the head to head battling which accomplishes very little in the way of straightening out the difficulty.

Warranty Facts Booklet

1974 New Car and Light Truck Warranty

(except Capri, Comet, Courier, Maverick, Mustang II and Pinto)

Ford* and the Selling Dealer jointly warrant for each 1974 model passenger car or light truck (P400 or lower series) sold by Ford that for the earliest of 12 months or 12,000 miles from either first use or retail delivery, the Selling Dealer will repair or replace free of charge any part except tires that is found to be defective in factory materials or workmanship under normal use in the United States or Canada.

All Ford and the Selling Dealer require is that you properly operate, maintain and care for your vehicle, and that you return for warranty service to your Selling Dealer's place of business or to any authorized Ford or Lincoln-Mercury dealer if you are traveling, have moved a long distance or need emergency repairs. Warranty repairs will be made with Ford Authorized Service or Remanufactured Parts.

To the extent allowed by law, THIS WARRANTY IS IN PLACE OF all other warranties, express or implied, including ANY IMPLIED WARRANTY OF MERCHANTABILITY OR FITNESS. Under this warranty, repair or replacement of parts is the only remedy.

* Ford Motor Company except in Canada where it is Ford Motor Company of Canada, Limited.

Please retain this Booklet in your vehicle.

Ford Customer Service Division

Fig. 1 A typical Ford Motor Co. warranty for 1974 models

CHRYSLER CORPORATION'S WARRANTY AND LIMITATION OF LIABILITY - FOR NEW 1974 MODEL PASSENGER CARS

WARRANTY PERIOD: Chrysler Corporation warrants this vehicle for 12 months or 12,000 miles, whichever occurs first, according to the following terms:

WHAT IS COVERED BY THE WARRANTY: Any part of this vehicle manufactured or supplied by Chrysler Corporation (except tires) found defective in material or workmanship.

WHO IS ELIGIBLE: All owners of this vehicle during the warranty period are covered by this warranty.

WARRANT START DATE: The warranty starts on the date of sale to the original retail purchaser or on the date this vehicle is originally placed in service, whichever occurs first.

OBTAINING WARRANTY SERVICE: Warranty service will be provided by the Selling Dealer at his place of business. The Selling Dealer will repair or replace, at Chrysler's option, the defective part without charge for parts or labor.

In the event the owner cannot return to the Seling Dealer (the Selling Dealer has ceased to do business as an authorized dealer or the owner is travelling, has moved or is living in a different locality and cannot return to the Selling Dealer), the owner may obtain warranty service at any authorized Chrysler Motors Corporation dealership.

WHAT IS NOT COVERED BY THE WARRANTY: This warranty will not apply to:

- Any vehicle on which the odometer mileage is altered;
- Normal maintenance services (as outlined in the Operator's Manual supplied with this vehicle) and the parts used in connection with such services;
- Repairs necessitated by accident, abuse, negligence or racing;
- Loss of use of the vehicle, loss of time, inconvenience or other consequential damages.

THIS WARRANTY IS IN LIEU OF ANY OTHER WARRANTIES OR CONDITIONS, INCLUDING MERCHANTABILITY OR FITNESS FOR A PARTICULAR PURPOSE. THE REMEDIES UNDER THIS WARRANTY ARE EXCLUSIVE AND NEITHER CHRYSLER CORPORATION NOR CHRYSLER MOTORS CORPORATION ASSUMES NOR AUTHORIZES ANYONE TO ASSUME FOR THEM ANY OTHER OBLIGATION.

• • •

TIRES—Tires originally installed on new passenger cars are warranted separately by the tire manufacturer and not by Chrysler.

Although the specific terms of these warranties will differ, generally tire manufacturers guarantee their tires for the life of the original tread against defects in material and workmanship and failure caused by normal road hazards such as blow-outs, fabric breaks, cuts, bruises and snags. These guarantees do not cover damage or failure caused by punctures, running flat, fire, wrecks, chain cuts, irregular wears, abuse, etc. Generally these guarantees provide that any tire determined to be defective or damaged within the terms of such guarantee will be repaired or replaced at the tire manufacturer's option. If replacement is made, the owner must pay the tire manufacturer's current adjustment base price plus transportation charges and taxes for such tire less a pro rata allowance based on the amount of the original tread remaining on the tire replaced. In some instances, the owner must also pay a service charge.

Warranty services pertaining to tires may be obtained from an authorized service station of the tire manufacturer or an authorized Chrysler Motors Corporation dealer providing such services. If necessary, an authorized Chrysler Motors Corporation dealer will assist an owner in requesting an adjustment.

The foregoing guarantees specifically provide that they do not cover consequential damage and that they are the only guarantees issued by the tire manufacturers.

BATTERIES—Original equipment batteries will be replaced for up to 36 months on the following basis. Should a battery prove defective during the first 12 months or 12,000 miles, whichever occurs first, it will be replaced without charge under the terms and conditions of the new vehicle warranty. Thereafter, and for up to 36 months from the date of original retail sale or original use of the vehicle, whichever occurred first, a defective battery will be replaced on a pro rata adjustment basis. The pro rata adjustment provides the owner credit towards the purchase of a new battery of equal or greater capacity. The number of months remaining between the time of failure and 36 months will determine the amount of credit. For additional information on the pro rata adjustment period, contact an authorized Chrysler Motors Corporation dealer.

Notice: A battery which is merely discharged is not considered defective.

Fig. 2 A typical Chrysler Corporation warranty for 1974 models

1974 PONTIAC NEW VEHICLE WARRANTY

WHAT IS WARRANTED AND FOR HOW LONG

Pontiac* (Pontiac Motor Division, General Motors Corporation) warrants to the owner of each 1974 model Pontiac passenger car that for a period of 12 months or 12,000 miles, whichever first occurs, it will repair any defective or malfunctioning part of the car —except tires which are warranted separately by the tire manufacturer. This warranty covers only repairs made necessary due to defects in material or workmanship, and needed service adjustments during the first 90 days of the warranty period.

The 12 month/12,000 mile warranty period shall begin on the date the car is delivered to the first retail purchaser or, if the car is first placed in service as a demonstrator or company car prior to sale at retail, on the date the car is first placed in such service.

WHAT IS NOT COVERED BY THE WARRANTY

This warranty does not cover:

1. Conditions resulting from misuse, negligence, alteration, accident, or lack of performance of required maintenance services;
2. The replacement of maintenance items (such as spark plugs, ignition points, positive crankcase ventilation valve, filters, brake and clutch linings) made in connection with normal maintenance services;
3. Loss of time, inconvenience, loss of use of the car or other consequential damages;
4. Any car on which the odometer mileage has been altered and the car's actual mileage cannot be readily determined; or
5. Any car registered and normally operated outside the United States or Canada. The warranty for these cars shall be that authorized for the country in which the car is registered and normally operated.

PONTIAC'S OBLIGATIONS

1. Repairs qualifying under this warranty will be performed by any authorized Pontiac dealer within a reasonable time following delivery of the car to the dealer's place of business.
2. During the first 90 days of the warranty period, any authorized Pontiac dealer will make any needed service adjustments.
3. Pontiac will pay the authorized Pontiac dealer for any repairs or 90-day service adjustments under the warranty.

* For vehicles sold in Canada, substitute the name General Motors of Canada Limited wherever the name Pontiac Motor Division appears in this folder.

OWNER'S OBLIGATIONS

1. After the first 90 days of the warranty period, needed service adjustments referred to in item 2 of "Pontiac's Obligations" are considered to be items of normal maintenance resulting from use and are to be paid for by the owner.
2. The car must be delivered to an authorized Pontiac dealer's place of business during regular business hours for the performance of warranty repairs or service adjustments.
3. The owner is responsible for maintenance services which may be performed at the owner's option by any repair outlet regularly performing such services.

• • • • •

This is the only express warranty applicable to 1974 model Pontiac passenger cars and Pontiac neither assumes nor authorizes anyone to assume for it any other obligation or liability in connection with such cars.

WHAT TO DO IF THERE IS A QUESTION REGARDING WARRANTY

The satisfaction and goodwill of owners of Pontiac products are of primary concern to Pontiac dealers and Pontiac Motor Division. In the event a warranty matter is not handled to your satisfaction, the following steps are suggested:

1. Discuss the problem with your Pontiac dealership management.
2. Contact the Pontiac Zone Office (General Motors Zone Office in Canada) closest to you as listed in the Pontiac General Information Owner's Manual.
3. Contact the Customer Services Manager* at the address below.

PONTIAC MOTOR DIVISION
GENERAL MOTORS CORPORATION
Pontiac, Michigan 48053

* In Canada contact the Owner Relations Manager at General Motors of Canada Limited, Oshawa, Ontario.

Fig. 3 A typical General Motors warranty for 1974 models

1975 BUICK NEW CAR WARRANTY

WHAT IS COVERED

Buick Motor Division, General Motors Corporation,* warrants to owners of 1975 Buick passenger cars which are registered and normally operated in the United States or Canada:

- The Buick dealer of the owner's choice will make any repairs on any part of the car, except tires, made necessary because of defects in material or workmanship for 12 months or 12,000 miles of use, whichever first occurs, from the date the car is delivered to the first retail purchaser or first placed in service as a demonstrator or company car, whichever is earlier, and will make any needed service adjustments during the first 90 days of use.
- Warranty repairs and needed service adjustments will be performed without charge to the owner by the Buick dealer at its place of business within a reasonable time after delivery of the car to the dealer.

WHAT IS NOT COVERED

- Repairs and service adjustments required because of misuse, negligence, alteration, accident or lack of reasonable and proper maintenance are not covered, nor are the replacement of maintenance items (such as spark plugs, positive crankcase ventilation valves, filters, brake and clutch linings) made in connection with normal maintenance services.
- Loss of time, inconvenience, loss of use of the car or other matters not specifically included are not covered.
- Any car registered and normally operated outside the United States or Canada. The warranty for these cars shall be that authorized for the country in which the car is registered and normally operated.

Buick Motor Division, General Motors Corporation, does not authorize any person to create for it any other obligation or liability in connection with these cars.

* For vehicles sold in Canada, substitute the name General Motors of Canada, Limited, wherever the name Buick Motor Division, General Motors Corporation, appears in this folder.

1975 BUICK EMISSION CONTROL SYSTEMS WARRANTY

Buick Motor Division, General Motors Corporation,* warrants to owners of 1975 Buick cars that the car (1) was designed, built, and equipped so as to conform at the time of sale with applicable regulations of the U.S. Federal Environmental Protection Agency, and (2) is free from defects in materials and workmanship at the time of sale which cause the car to fail to conform with applicable Federal Environmental Protection Agency regulations for a period of use of 50,000 miles or 5 years, whichever occurs first.

The 5-year/50,000-mile warranty period shall begin on the date the car is delivered to the first retail purchaser or, if the car is first placed in service as a demonstrator or company car prior to sale at retail, on the date the car is first placed in such service.

OWNER'S OBLIGATION

1. Perform emission system maintenance as required in Maintenance Schedule folder.
2. Use only unleaded gasoline.
3. Receipts covering the performance of regular maintenance should be retained in the event questions arise concerning maintenance. These receipts should be transferred to each subsequent owner of this car.

WHAT IS NOT COVERED BY THE WARRANTY

This warranty does not cover:

1. Conditions resulting from misuse, alteration, accident, failure to use unleaded gasoline or not performing maintenance services;
2. The replacement of maintenance parts (such as spark plugs, PCV valve, and filters) used in regular maintenance services;
3. Loss of time, inconvenience, loss of use of the car or other consequential damages;
4. Any car on which odometer mileage has been changed so that mileage cannot be readily determined.

Buick Motor Division, General Motors Corporation, does not authorize any person to create for it any other obligations or liability in connection with these systems. This warranty is in addition to the 1975 Buick New Car Warranty.

* For vehicles sold in Canada, substitute the name General Motors of Canada, Limited, wherever the name Buick Motor Division, General Motors Corporation, appears in this folder.

Fig. 4 A typical General Motors warranty for 1975 models

Flasher Locations

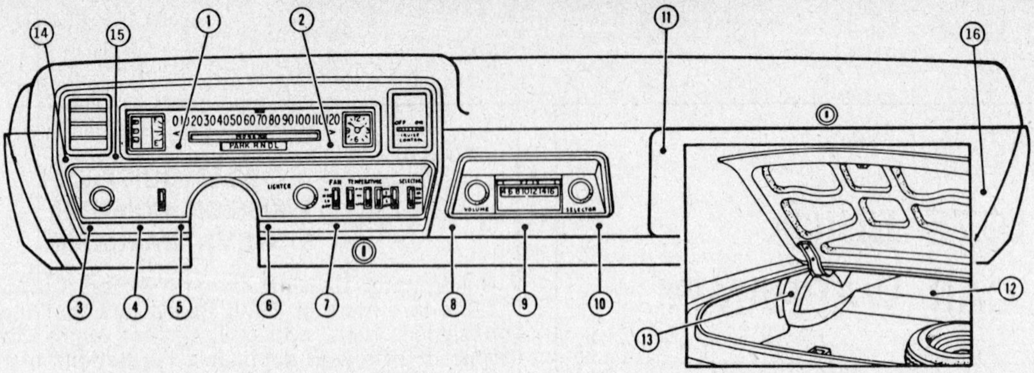

CAR	1969		1970		1971		1972		1973		1974		1975		1976	
	TSF	HWF	TSF	HWF	TSF	HWF	TSF	HWF	TSF	HWF	TSF	HWF	TSF	HWF	TSF	HWF
American Motors	2	3	2(14)	3	2(16)	3	3	3	3	3	3	3	3	4	3	4
Aspen & Volare	—	—	—	—	—	—	—	—	—	—	—	—	—	—	3	4
Astre & Vega	—	—	—	—	5	3	5	3	5	3	5	3	5	3	5	3
Buick, Special & Century	3	3	3	3	3	3	3	3	3	3	3	3	3	3	4	3
Buick Apollo & Skylark	—	—	—	—	—	—	—	—	10	3	10	3	10	3	10	3
Cadillac Exc. Seville	5	5	6	8	(15)	3	(15)	3	(15)	3	(15)	3	(15)	3	(15)	3
Cadillac Seville	—	—	—	—	—	—	—	—	—	—	—	—	—	—	4	3
Camaro	8	3	8	3	8	3	8	3	8	3	8	3	8	3	4	3
Chevelle & Monte Carlo	8	3	1	3	7	3	7	3	4	3	4	3	4	3	4	3
Chevette	—	—	—	—	—	—	—	—	—	—	—	—	—	—	15	3
Chevrolet	8	3	10	3	7	3	7	3	6	3	6	3	6	3	7	3
Chevrolet Nova	8	3	10	3	7	3	7	3	10	3	10	3	10	3	10	3
Chrysler	5	4	5	4	5	4	5	5	5	5	(18)	(19)	5	5	5	3
Comet & Montego	6	5	6	5	5	5	5	5	5	5	5(20)	5(20)	4	3	(25)	3
Cordoba & Charger SE	—	—	—	—	—	—	—	—	—	—	—	—	3	6	15	3(19)
Corvette	(8)	3	(8)	3	(8)	3	(8)	3	(8)	3	(8)	3	(8)	3	(8)	3
Cougar	8	10	8	10	11	6	8	10	8	10	5	5	5	5	3	3
Dart & Challenger	8	5	8	(9)	8	(9)	10	6	8	(3)	(18)	(19)	4	5	5	5
Dodge	(5)	(1)	5	6	(1)	—	(1)	5	8	6	(18)	(19)	5(22)	5(23)	(26)	3(23)
Fairlane, Torino & Elite	6	5	6	5	5	5	5	5	5	5	5	5	3	3	3	3
Falcon & Maverick	6	(10)	6	(10)	5	5	5	5	5	5	11	11	11	11	4	3
Firebird, Le Mans & Tempest	3	3	3	3	3	3	3	3	3	3	3	3	3	3	4	3
Ford	6	5	11	3	8	3	8	3	8	3	3	3	3	3	3	3
Ford Pinto & Mercury Bobcat	—	—	—	—	10	10	10	10	10	10	11	11	11	11	11	11
Granada & Monarch	—	—	—	—	—	—	—	—	—	—	—	—	8	4	10	4
Grand Am & Grand Le Mans	—	—	—	—	—	—	—	—	3	3	3	3	3	3	3	5
Imperial	13	13	5	4	5	4	5	5	5	6	(18)	(19)	5	5	—	—
Lincoln	11	11	3(12)	3(13)	6	6	3	3	3	3	7(21)	7(21)	11	3	10(21)	3
Mercury	4	6	11	3	8	3	8	3	8	3	3	3	3	3	3	3
Monza, Skyhawk, Starfire & Sunbird	—	—	—	—	—	—	—	—	—	—	—	—	5	3	14	14
Mustang & Mustang II	1	4	10	8	11	6	8	10	8	10	11	11	7(24)	7(24)	16	16
Oldsmobile	3	3	1	3	4	3	4	3	4	3	4	3	4	3	4	3
Oldsmobile Cutlass	3	3	3	3	4	3	4	3	4	3	4	3	4	3	4	3
Olds. Omega & Pont. Ventura	—	—	—	—	7	3	7	3	10	3	10	3	10	3	10	3
Oldsmobile Toronado	3	.3	4	3	4	3	4	3	4	3	4	3	4	3	4	3
Plymouth	(7)	(6)	(7)	6	(2)	—	(3)	5	8	6	(18)	(19)	5	5	(27)	3(28)
Pontiac	3	3	3	3	3	3	3	3	3	3	3	3	3	3	3	5
Thunderbird	13	13	(11)	11	11	6	3	3	3	3	3	3	3	3	3	3
Valiant & Barracuda	8	5	8	6	8	6	10	6	8	(17)	(18)	(19)	4	5	5	5

TSF: Turn Signal Flasher. **HWF: Hazard Warning Flasher.**

(1)—Location 10 on Coronet & Charger. Location 5 on Polara & Monaco.
(2)—Location 10 on Belvedere & Satellite. Location 5 on Fury & VIP.
(3)—Location 6 on Challenger. Location 5 on Dart.
(5)—Location 4 on Coronet & Charger. Location 8 on Polara & Monaco.
(6)—Location 6 on Belvedere & Satellite. Location 4 on Fury & VIP.
(7)—Location 4 on Belvedere & Satellite. Location 8 on Fury & VIP.
(8)—Extreme lower right corner of instrument panel.
(9)—Location 6 on Challenger. Location 10 on Dart.
(10)—Location 5 on Falcon. Behind ash tray on Maverick.
(11)—To right of glove box.
(12)—Mark III to right of glove box.
(13)—Mark III to left of glove box.
(14)—Location 3 on Hornet and Gremlin.
(15)—On the underside of steering column lower cover.
(16)—Location 3 on Hornet, Gremlin and Javelin.
(17)—Location 6 on Barracuda. Location 5 on Valiant.
(18)—Behind instrument panel, on right side of ash tray.
(19)—On right side of brake pedal support.
(20)—Location 11 on Comet.
(21)—Location 3 on Mark IV.
(22)—Location 3 on Coronet.
(23)—On right side of brake pedal support on Coronet models.
(24)—One flasher used for both systems.
(25)—Location 4 on Comet. Location 3 on Montego.
(26)—Location 15 on Coronet. Location 5 on Monaco.
(27)—Location 15 on Fury. Location 5 on Gran Fury.
(28)—On right side of brake support on Fury models.

TROUBLE SHOOTING

Index of Symptoms

ENGINE

ELECTRICAL

[1]—See Service Chapter.

Engine Troubles

STARTING A STALLED ENGINE

When an engine fails to start the chances are that 90 per cent of the cases will involve the ignition system and seldom the fuel system or other miscellaneous reasons. If a systematic procedure is followed the trouble can almost always be found without the use of special equipment.

To begin with, turn on the ignition switch and if the ammeter shows a slight discharge (or if the telltale lamp lights) it indicates that current is flowing. A glance at the gas gauge will indicate whether or not there is fuel in the tank.

Operate the starter and if the engine turns over freely, both the battery and starter are functioning properly. On the other hand, if the starter action is sluggish it may be due to a discharged or defective battery, loose, corroded or dirty battery terminals, mechanical failure in the starter, starter switch or starter drive. If the starter circuit is okay, skip this phase of the discussion and proceed to ignition.

Starter Circuit Checkout

To determine which part of the starter circuit is at fault, turn on the light switch and again operate the starter. Should the lights go out or become dim, the trouble is either in the battery, its connections or cables. A hydrometer test of the battery should indicate better than 1.250 specific gravity, while a voltmeter, placed across the positive and negative posts, should indicate about 12 volts. If either of these tests prove okay, clean and tighten the battery connections and cable terminals or replace any cable which seems doubtful.

If the lights remain bright when the starter is operated, the trouble is between the battery and the starter, or the starter switch is at fault, since it is evident that there is no electrical connection between these points. If these connections are clean and tight, it is safe to assume that the starter or starter switch is defective.

Neutral Safety Switch

If the ammeter shows a slight discharge (or if the telltale lamp lights) when the ignition is turned on, but the system goes dead when the starting circuit is closed, the neutral safety switch may be at fault. To check, bypass the switch with a suitable jumper. If the engine now starts, adjust or replace the switch.

CAUTION: With the safety switch bypassed, the car can be started in any gear. *Be sure the transmission is in neutral or park and the parking brake is applied.*

Primary Ignition Checkout

Let's assume that the battery and starter are doing their job, and that fuel is reaching the carburetor, but the car does not start, then the trouble must be somewhere in the ignition circuit. But first, before starting your diagnosis, it is advisable to give the whole system a visual inspection which might uncover obvious things such as broken or disconnected wires etc.

The best way to start tracking down ignition troubles is to begin with the primary circuit since this is where troubles show up most frequently. First remove the distributor cap and block the points open with a piece of cardboard, then turn on the ignition and with a test bulb or voltmeter check to see if there is current at the terminal on the distributor. If you do not get a reading at this point, the current is cut off somewhere in the connections leading back to the ignition switch or it may be that the condenser has an internal short to the ground. The latter possibility can be eliminated if you can restore current at the distributor terminal by disconnecting the condenser from the distributor plate so that its outside shell is not grounded. With the possibility of a bad condenser out of the way, work toward the ignition switch and test for current at each connection until you get to one where you get a reading. Between this connection and the distributor lies the trouble.

The foregoing steps in checking the primary circuit should include checking the ignition coil resistor for defects or loose connections. As this is done, bear in mind that while the starter cranks the engine, the resistor is by-passed by the starter switch on Ford and Delco-Remy systems (see Tune Up Chapter for details). This means that while the circuit through the resistor may be satisfactory, a broken connection or high resistance between the starter switch by-pass terminal and the coil would prevent starting. On the other hand, a satisfactory by-pass circuit might start the engine while the engine would stall immediately upon releasing the starter switch if there was a defect in the coil resistance circuit.

If, to begin with, the test equipment shows a current reading at the distributor terminal, it is safe to assume that the trouble is in the unit itself, most likely burned or dirty breaker points. A final positive test for defective breaker points can be made very simply by removing the cardboard from between the points, and positioning the distributor cam by turning the engine to where the points are closed. With the points closed there should be no current at the distributor terminal. If there is current, replace the points.

In an emergency, the points can be cleaned by using the sanded side of a match box, a knife blade, or the sharp edge of a screwdriver to scrape the scale from the contact faces. After cleaning the points, if a gauge is not available to set the gap, a quick adjustment can be made by using four layers of a piece of newspaper. The thickness of the paper is equivalent to about .020″, which is the approximate gap setting for most distributors. Of course, at the earliest opportunity, a precise point adjustment should be made.

If the procedure outlined under "Primary Ignition Checkout" does not uncover the trouble then it will be necessary to continue the tests into the secondary ignition circuit.

Secondary Ignition Checkout

First of all, remove the wire from one of the spark plugs, turn on the ignition and operate the starter. While the engine is cranking, hold the terminal of the spark plug wire about 1/4″ away from the engine or spark plug base. If the spark is strong and jumps the gap, the trouble is confined to either the spark plugs or lack of fuel. Before going any further, wipe the outside of the plugs to remove any dirt or dampness which would create an easy path for the current to flow, then try to start the engine again. It it still fails to start, remove one of the spark plugs and if it is wet around the base, it indicates that the fuel system is okay, so it naturally follows that the spark plugs are at fault. Remove all the plugs, clean them and set the gaps. An emergency adjustment of spark plug gaps can be made by folding a piece of newspaper into 6 or 7 layers. When changing the gap, always bend the side (ground) electrode and never the center one as there is danger of breaking the insulation.

Fuel System Checkout

If the spark plug that was removed showed no indication of dampness on its base, check the fuel system. A quick check can be made by simply removing the carburetor air cleaner and looking down into the carburetor. Open and close the throttle manually and if fuel is present in the carburetor, the throttle will operate the accelerating pump, causing it to push gasoline through the pump jet. If it does, check the choke valve. If the engine is cold, the choke valve should be closed. If the choke won't close, the engine can be started by covering the carburetor throat while the engine is cranking, provided, of course, that fuel is reaching the carburetor.

Check the operation of the fuel pump by disconnecting the fuel lines from the pump to the carburetor. Crank the engine and if the pump is working, fuel will pulsate out of the line. If not, either the pump isn't working or the line from the tank to the pump is clogged. Before blaming the pump, however, disconnect the line at the inlet side of the pump which leads to the tank and, while a companion listens at the tank, blow through the line. If a gurgling sound is heard back in the tank, the line is open and the trouble is in the pump. Remove the sediment bowl, if so equipped and clean the screen, then replace the bowl and screen, being sure that you have an air-tight fit. If the pump still refuses to function, it should be removed and repaired.

The foregoing discussion will, in most cases, uncover the cause of why an engine won't start. However, if further diagnosis is necessary, the following list will undoubtedly provide the answer.

CLOSED CRANKCASE VENTILATION

If the control valve becomes clogged with carbon or other foreign matter, the ventilation system will not operate and a slight pressure will build up in the crankcase which may cause oil leakage at the rear main bearing or by the piston rings. And should the valve fail to seat it will be impossible to make the engine idle satisfactorily.

SERVICE NOTE: If idle speed is slow, unstable, rolling, frequent stalling, breather backflow and oily engine compartment the ventilator valve may be completely plugged, or the valve may be stuck in the open position. A valve stuck in the closed position is indicated by breather backflow at heavy throttle and oily engine compartment. If the valve is stuck in the intermediate position it will be indicated by rough, fast idle and stalling.

The ventilation valve assembly should be cleaned every six months or 6000 miles (whichever comes first) and more frequently in service such as extensive engine idling during cold weather.

When the valve assembly is removed for cleaning, place a finger over the open end of the ventilator hose or tube and have the engine started. If the ventilator hose or tube and carburetor passages are open and operating normally, a strong suction will be felt and there will be a large change in engine idle quality when the end of the hose is uncovered. If these conditions are not observed, the carburetor passages and/or ventilator hose are plugged and must be cleaned. The carburetor should be removed from the engine and the ventilation passages cleaned by dipping the lower part of the carburetor in the cleaner. A pipe cleaner can be used to aid in cleaning passages.

ENGINE WON'T START

IMPORTANT—Alternator equipped cars cannot be push-started when the battery is completely dead because, unlike a generator, there is no residual magnetism in the rotor.

If the engine fires when the ignition switch is turned on but quits when the switch is released to its running position, it indicates that the ignition coil resistor has lost its continuity or there is a bad connection at the resistor terminals.

Due to Open Primary Ignition Circuit

1. Burned or oxidized ignition points.
2. Ignition coil resistance unit burned out or open.
3. Starting switch ignition coil resistance by-pass circuit open.
4. Ignition points not closing.
5. Breaker arm binding on pivot post, preventing closing of points.
6. Breaker arm spring weak or broken.
7. Breaker arm distorted or bent.
8. Dirty ignition points.
9. Primary lead connection loose at distributor or coil.
10. Primary windings in coil broken.
11. Open ignition switch circuit.

Due to Grounded Primary Ignition Circuit

A grounded coil primary winding, a grounded ignition switch, or a grounded switch-to-coil primary lead will cause excessive current flow and will usually cause wires to burn.

1. Ignition points not opening or closing due to improper adjustment.
2. Ignition points not opening due to worn rubbing block on breaker arm.
3. Faulty bushing in breaker arm.
4. Cracked or faulty insulator at distributor primary terminal.
5. Grounded condenser.
6. Distributor-to-coil lead grounded.
7. Primary coil winding grounded.

Due to Faulty Secondary Ignition Circuit

1. Corroded spark plug cable terminals.
2. Chafed or cracked cable insulation.
3. Ignition coil weak or inoperative.
4. Moisture on ignition coil, terminals, distributor cover, spark plug porcelains, or in distributor.
5. Improper type of spark plugs.
6. Cracked distributor cap or a burned carbon track from distributor cap center terminal to housing.
7. Improper installation of spark plug cables (not correct for firing order).
8. Spark plugs damaged, dirty or wet,

porcelains cracked, or gaps improperly spaced.
9. Rotor contact spring bent or broken.
10. Distributor rotor grounded.
11. Distributor cap center terminal (inner) broken or missing.
12. Broken or burned out radio suppressor in distributor cap.

Due to Battery

1. Battery run down.
2. Terminals loose or badly corroded.
3. Improper ground.
4. Battery cables frayed or undersize.

Due to Starter Motor

1. Not operating properly.
2. Congealed engine oil due to use of too heavy a grade of oil or to the formation of sludge.
3. Starter gear binding in flywheel gear.
4. Defective starter switch.
5. Faulty neutral safety switch on cars with automatic transmission.

Due to Excessive Fuel Supply (Flooding)

The engine is said to be flooded with fuel when a quantity of liquid fuel collects in the intake manifold, and perhaps also in the cylinders. This condition gives a mixture that is much too rich to ignite.

If the carburetor has a provision for opening the choke valve when the throttle is fully open, crank the engine with the throttle open until engine starts. It will start as soon as the extra fuel is pumped out.

If the choke valve is not designed to open when the throttle is fully opened, tie or block the choke valve open and crank the engine until it starts.

Flooding may also occur on the road. If the carburetor supplies too rich a mixture at full throttle, the intake manifold may be flooded with liquid fuel, with the result that when the engine is stopped, heat evaporates hte fuel and thus provides an over-rich incombustible mixture. The engine won't start until the rich mixture is pumped out by cranking.

1. Choke not operating property.
2. Automatic choke not properly set.
3. Carburetor unloaded linkage (if equipped) not properly set.
4. Float level set too high.
5. Dirty, worn or faulty needle valve and seat.
6. Float sticking or rubbing against side of fuel bowl.
7. Leak in float, allowing fuel to get inside.
8. Fuel pump pressure too great.

Due to Insufficient Fuel Supply

1. Carburetor inlet needle stuck in its seat, due to gum in fuel.
2. Float level too low.
3. Clogged inlet screen at carburetor.
4. Faulty fuel pump or one of insufficient capacity.
5. Fuel pump strainer clogged.
6. Faulty fuel pump bowl gasket.
7. Flexible line (if used) twisted, deteriorated or restricted.
8. Fuel line to tank clogged, kinked, restricted or leaking.
9. Vent in fuel tank filler cap clogged or restricted.
10. Worn fuel pump camshaft lobe.

HARD STARTING

When Engine is Hot

This condition is usually caused by an over-supply of fuel due to any of the items listed under *Engine Won't Start Due to Excessive Fuel Supply*. In rare cases, an ignition coil may lose its efficiency when it is hot and cause ignition failure.

When Engine is Cold

Many of the conditions enumerated under *Engine Won't Start* also may cause hard starting in cold weather. Of particular importance, however, are the following:

1. Choke setting too lean.
2. Fuel may have kerosene in it or water, or ice in bottom of tank.
3. Ice in fuel filter bowl.
4. Ice in fuel lines.
5. Engine is cranked too slowly or won't turn over because: (a) engine oil is too thick in sub-zero weather; (b) battery weak due to extremely low temperature.
6. Another possibility, although remote, is that the water pump is jammed with ice, which will interfere with cranking engine if fan belt is tight.

Due to Vapor Lock

The term vapor lock means the flow of fuel to the mixing chamber in the carburetor has been stopped (locked) by the formation of vaporized fuel pockets or bubbles caused by overheating the fuel by hot fuel pump, hot fuel lines or hot carburetor.

The more volatile the fuel the greater the tendency for it to vapor lock. Vapor lock is encouraged by high atmospheric temperature, hard driving, defective engine cooling and high altitude.

A mild case of vapor lock will cause missing and hard starting when engine is warm. Somewhat more severe vapor lock will stop the engine which cannot be started again until it has cooled off enough so that any vaporized fuel has condensed to a liquid.

SERVICE NOTE: Some cars equipped with air conditioning have a vapor bypass system. These cars have a special fuel filter which has a metering outlet in the top. Any vapor which forms is bled off and returned to the fuel tank through a separate line alongside the fuel supply line. This system greatly reduces the possibility of vapor lock. However, if vapor lock is suspected examine the bypass valve to see if it is functioning.

Due to Percolation

Percolation means simply that gasoline in the carburetor bowl is boiling over into the intake manifold. This condition is most apt to occur immediately after a hot engine is shut off. Most carburetors have a provision for relieving the vapor pressure of overheated fuel in the carburetor bowl by means of ports. If, however, percolation should take place, the engine may be started by allowing it to cool slightly and then holding the throttle wide open while cranking to clear the intake manifold of excess fuel.

After Long Storage

1. The more volatile components in the fuel have evaporated and those remaining are not sufficiently volatile to provide a combustible mixture.
2. Low or run-down battery.
3. Corrosion of engine parts mav result in so much friction that starter cannot crank engine at proper speed, if at all.
4. Pistons, etc. may be struck fast by gummy oil.
5. Engine valves may stick open due to gummy deposits.
6. There is the possibility that any small part essential to the running of the engine may be stuck due to gummy film or to corrosion.
7. Some of these troubles are most likely to occur in hot, humid climate and near salt water.

ENGINE STALLS

Many troubles which prevent smooth running at idle may cause stalling. The list includes almost everything that may cause hard starting or missing. Some of the more common causes are:

1. Engine idle speed set too low.
2. Large air leaks in intake manifold such as a disconnected windshield wiper vacuum line.
3. Ignition points need attention
4. Engine valves leaking.
5. Vapor lock.
6. Over-supply of fuel (flooding).
7. Valves set too tight.

If carburetor is equipped with a fast idle cam, which increases engine speed when the choke is in operation during the warm-up period, the engine may stall if the fast idle device fails to open the throttle due to sticking or need for adjustment.

On some cars equipped with a fluid coupling or torque converter, if the throttle is closed quickly the engine stalls. To avoid this trouble, most cars have a device which retards the speed of the throttle closing; this is called a throttle return check or dashpot and is usually mounted on the carburetor. It consists of a piston or diaphragm and a spring-closed check valve. If the linkage is out of adjustment or the check valve leaks, the engine will stall.

If the engine quits smoothly when car is in operation, the trouble is often caused by sudden lack of fuel due to:

1. Fuel tank empty.
2. Vapor lock.
3. Flooding.
4. Water in fuel.
5. Frozen fuel line.

Carburetor Icing

The carburetor discharges liquid fuel into the air stream in the form of an atomized spray which evaporates readily. The heat required to evaporate the gasoline is drawn from the entering air, thereby lowering its temperature. The cooler air chills the interior of the carburetor and may cause the moisture in the air to condense into droplets.

Under certain conditions of atmospheric temperature and humidity, the liberated moisture actually collects and freezes on the chilled carburetor surfaces, especially on the throttle plate and surrounding throttle body. When the throttle is almost completely closed for idling, this ice tends to bridge the gap between the throttle plate and throttle body, thereby cutting off the air supply and causing the engine to stall. Opening the throttle for restarting breaks the ice bridge but does not eliminate the possibility of further stalling until the engine and carburetor has warmed up.

For carburetor icing to occur, the outside air must be cool enough so that the refrigerating effect of fuel evaporation in the carburetor will lower the temperatures of the throttle plate and body below both the dew point of moist air and the freezing point of water. The air must also contain sufficient moisture for appreciable condensation of water to occur when it is chilled in the carburetor.

Generally speaking, carburetor icing occurs when winter grade gasoline (more volatile than summer grade) is used and when the atmospheric temperature ranges from 30° to 50° F. at relative humidities in excess of 65%.

Carburetor icing problems can be reduced by the use of anti-icing additives, such as alcohols, in the fuel. Some fuel refiners use anti-stalling additives in their gasolines which have proved effective in combating carburetor icing.

Another form of carburetor icing has been observed in some engines during high-speed driving on cool, moist days. When certain cars are driven steadily at 60 to 80 mph, the large quantities of cool air passing through the carburetor may result in gradual ice formation within the carburetor's venturi. Since this ice restricts the venturi passage, the resultant increased vacuum in the venturi tends to increase the rate of fuel flow. The fuel-air mixture thus becomes excessively rich, causing loss of power and high fuel consumption.

ENGINE STARTS BUT WON'T DRIVE CAR

1. Broken part in the drive line anywhere from clutch to rear axle shaft.
2. No oil or not enough oil in fluid coupling or torque converter.
3. Some defect in automatic transmission causes binding or dragging of clutches or slipping bands.
4. Engine develops only enough power to run itself due to: (a) extremely lean or rich mixture; (b) excessive engine friction; (c) throttle does not open; (d) very dirty air cleaner; (e) clogged exhaust system.
5. Oil in fluid coupling or torque converter is semi-solid due to zero temperature. This trouble is unlikely if the recommended oil is used.

ENGINE MISFIRES

At All Speeds

1. Fouled spark plug or broken porcelain.
2. Faulty spark plug cables.
3. Low battery voltage.
4. Low generator voltage.
5. Burned or pitted ignition points.
6. Incorrect ignition point gap.
7. Faulty condenser or coil.
8. Weak spark or no spark in one or more cylinders.
9. Faulty distributor cap or rotor.
10. Primary circuit restricted or open intermittently.
11. Primary circuit detoured by short intermittently.
12. Secondary circuit restricted or open intermittently.
13. Secondary circuit detoured by short intermittently.
14. Blown clyinder head gasket between cylinders. This can be noted when missing occurs in two adjacent cylinders.
15. Sticking valves.
16. Hydraulic tappet holds valve open slightly.
17. Broken valve spring.
18. Leak at intake manifold gaskets.
19. Mixture too rich or too lean.

At High Speed

1. Hot spark plugs. Change to colder type but note that a hot plug may be due to loose installation or lack of a plug gasket (if gasket is called for).
2. Ignition point gap much too wide.
3. Breaker arm binding or sticking.
4. Breaker arm spring weak.
5. Sticking engine valves.
6. Valve springs too weak to close valves promptly.
7. Valve springs broken.
8. Valve springs shimmy.
9. Intermittent delivery of fuel to carburetor so that momentarily the mixture is too weak for combustion.
10. Mild vapor lock.
11. Weak spark.
12. Exhaust manifold clogged with carbon.
13. Exhaust manifold, muffler or tail pipe restricted.
14. Improper ignition timing.
15. Centrifugal advance not functioning properly.
16. Manifold heater valve held closed.
17. Dirty carburetor air cleaner.
18. Choke valve not completely open.
19. Carburetor throttle lever loose on shaft.
20. Improper fuel pump operation.
21. Preignition.
22. Incorrect valve timing.

At Low or Idle Speeds

1. Faulty spark plugs.
2. Spark plugs gaps too narrow.
3. Dirty or corroded secondary circuit connections or faulty ignition cables.
4. Cracked or faulty distributor cap. Radial contacts in cap burned or worn.
5. Dirty air cleaner.
6. Leaky valves.
7. Ignition point gap too narrow.
8. Faulty carburetion due to: (a) float level too high or too low; (b) float valve leaking; (c) incorrect or loose jets; (d) restricted or partially clogged idle air passage or jet; (e) air leak occurring between upper and lower carburetor body; (f) air leak occurring around carburetor throttle shaft.
9. Air leaks in intake manifold or carburetor resulting from: (a) loose manifold connections or leaks occurring in vacuum lines; (b) loose manifold nuts or capscrews; (c) broken or damaged intake manifold or carburetor gaskets; (d) cracked manifold; (e) warped or damaged manifold contacting surface.
10. Slight leaks occurring at fuel pump check valves.
11. Air leak occurring around intake valve stem because of excessive valve stem-to-guide clearance.

When Car is Accelerated

If the engine misses when car is accelerated but does not miss when idling the reason is that the spark plugs stop firing because of increased compression pressure caused by:

1. Weak spark.
2. Plug gaps too wide.
3. Plug fouled or damp.
4. Plug porcelain below par.

Also see *Flat Spot*.

LACK OF POWER OR HIGH SPEED PERFORMANCE

It should be noted that the altitude at which the car is operated has a decided effect on performance. A car adjusted for normal altitudes will lack performance at high altitudes, whereas a car which operates normally at high altitudes may have a lean carburetor adjustment and show signs of preignition when operated at sea level.

1. Ignition timing incorrect.
2. Centrifugal governor advance not operating properly.
3. Vacuum advance not operating properly.
4. Ignition points burned, pitted, sticking or bouncing (due to weak breaker arm spring).
5. Faulty spark plugs.
6. Faulty ignition cables.
7. Faulty ignition coil.
8. Faulty carburetion.
9. Lack of engine compression.
10. Preignition.
11. Inoperative manifold heater valve (stuck closed).
12. Restricted carburetor inlet resulting from dirty air cleaner or choke valve not fully open.
13. Carburetor throttle lever loose on shaft.
14. Throttle linkage not properly adjusted.
15. Carburetor throttle valve not completely open.
16. Carburetor accelerating pump not functioning properly.
17. Improper fuel pump operation.
18. Partially restricted exhaust pipe, muffler or tail pipe.
19. Clutch slippage.
20. Excessive rolling resistance resulting from (a) dragging brakes, (b) tight wheel bearings, (c) misalignment of power transmitting units, (d) misalignment of rear axle, (e) underinflated tires.
21. Incorrect rear axle gear ratio.
22. Oversize tires.
23. Incorrect valve timing.
24. Inaccurate speedometer (gives impression of lack of performance).

ROUGH IDLE

The term "rough idle" means that the engine does not run smoothly when idling. The most likely cause is an over-rich mixture but any defect which produces uneven explosions or missing will cause a rough idle. The most common causes are:

1. Dirty idle jets and passages.
2. Improper idle mixture.
3. Dirty air cleaner.
4. Improper float level.
5. Choke set too rich.
6. Air leak into intake manifold.
7. Clogged idle jets.
8. Improper ignition point gap.
9. Improper spark plug gap.
10. Weak spark.
11. Leaky engine valve.
12. Sticking valve or rocker arm.
13. Broken valve spring.
14. Insufficient tappet clearance.
15. Improper fuel pump pressure.
16. Sticking breaker arm.
17. Hydraulic tappet holds valve open.
18. Fuel volatility too high or too low.

SPARK, KNOCK, PING, DETONATION

All three expressions mean the same thing. It is a sharp metallic knock caused by vibration of the cylinder head and block. The vibration is due to split-second high-pressure waves resulting from almost instantaneous abnormal combustion instead of the slower normal combustion.

The ping may be mild or loud. A mild ping does no harm but a severe ping will reduce power. A very severe ping may shatter spark plugs, break valves or crack pistons.

Pinging is most likely to occur on open throttle at low or moderate engine speed. Pinging is encouraged by:

1. Overheated engine.
2. Low octane fuel.
3. Too high compression.
4. Spark advanced too far.
5. Hot mixture due to hot engine or hot weather.
6. Heavy carbon deposit which increases the compression pressure.

Tendency to ping increases with mixture temperature including high atmospheric temperature; intake manifold heater valve "on" when engine is warm; hot cooling water; hot interior engine surfaces due to sluggish water circulation or water jackets clogged with rust or dirt especially around exhaust valves. Some of these troubles may be confined to one or two cylinders.

If an engine pings objectionably because of too low octane fuel, retard the spark setting but first be sure that the cooling system is in good condition, the mixture not too lean and the combustion chambers free of carbon deposit.

PRE-IGNITION

Pre-ignition means that the mixture is set on fire before the spark occurs, being ignited by a red hot spot in the combustion chamber such as an incandescent particle of carbon; a thin piece of protruding metal; an overheated spark plug, or a bright red hot exhaust valve. The result is reduction of power and overheating accompanied by pinging. The bright red hot exhaust valve may be due to a leak, to lack of tappet clearance, to valve sticking, or a weak or broken spring.

Pre-ignition may not be noticed if not severe. Severe pre-ignition results in severe pinging. The most common cause of pre-ignition is a badly overheated engine.

When the engine won't stop when the ignition is shut off, the cause is often due to red hot carbon particles resting on heavy carbon deposit in a very hot engine.

ENGINE KICKBACK

If ignition is set too far advanced, spark may occur too early when engine is cranked. The first (and only) explosion runs the engine backward. A kickback may jam the starter or break the starter drive housing.

BACKFIRE

Backfiring is a subdued explosion in the intake manifold. Causes are:

1. Lean mixture (often due to dirt or water in fuel).
2. Engine cold and choke too lean.
3. Leaky or sticking intake valve or weak or broken intake valve spring.
4. Leakage of current across distributor cap may cause backfire by enabling spark to occur in a cylinder which is on its intake stroke. Two mixed-up spark plug wires may also cause this trouble.
5. Popping back is synonymous with backfire.

MUFFLER EXPLOSION

1. Late ignition timing.
2. Late valve timing.
3. Burnt exhaust valve(s).
4. Weak or broken exhaust valve spring(s).
5. Tight exhaust valve(s).
6. Intermittent open circuit in primary (ammeter needle swings further away from zero when generator is charging).
7. Intermittent short in primary (ammeter swings toward zero when generator is charging).
8. Short in coil or secondary coil wire.
9. If just a couple of explosions are heard and then no more for a time (even for days) the trouble may be due to a gradually failing condenser.

AFTER-BURNING

A subdued put-putting at the exhaust tail pipe may be due to leaky exhaust valves which permit the mixture to finish combustion in the muffler. If exhaust pipe or muffler is red hot, better let it cool, as there is some danger of setting the car on fire. Most likely to occur when mixture is lean.

FLAT SPOT

If an engine does not respond promptly when the throttle is open quickly it (or the carburetor) is said to have a flat spot. This is usually caused by any of the following:

1. Accelerator pump piston (or diaphragm) leaks.
2. Accelerator pump valves leak.
3. Accelerator pump stroke too short.
4. Accelerator pump passages restricted.
5. Fuel volatility too low or too high.
6. Float level too low.
7. Fuel pump pressure too low.
8. The anti-percolating valve (on some carburetors) may open too soon when throttle is closed. If so, carburetor may have flat spot next time throttle is opened when engine is hot.
9. Fuel too hot due to hot engine and hot weather (see Vapor Lock).
10. If carburetor has a metering pin operated by throttle linkage and also a vacuum piston linked to the throttle to give a rich mixture at part throttle and moderate engine speed, a flat spot will be noted if the device fails to function properly because of stuck piston, vacuum leakage or restricted vacuum passages.
11. If carburetor has vacuum piston which provides richer mixture at part throttle and moderate engine speed by opening an additional passage or jet within carburetor, a flat spot will occur if fuel valves fail to work, or fuel passages are restricted, or if piston does not function because it is sticking, vacuum leakage or restricted vacuum passages.
12. Late ignition timing.

ENGINE FAILS TO REACH OPERATING TEMPERATURE

1. Defective thermostat.
2. Thermostat stuck open.
3. Thermostat removed from vehicle (during flushing cooling system and not replaced).
4. Defective temperature sending unit or dash unit.

ENGINE CONTINUES TO RUN AFTER IGNITION IS TURNED OFF

This condition, known as "dieseling," "run on," or "after running," is caused by improper idle speed and/or high temperature. Idle speed and engine temperature are affected by:

Carburetor Adjustment: High idle speed will increase the tendency to diesel because of the inertia of the engine crankshaft and flywheel. Too low an idle speed, particularly with a lean mixture, will result in an increase in engine temperature, especially if the engine is allowed to idle for long periods of time.

Ignition Timing: Because advanced ignition timing causes a corresponding increase in idle speed and retarded timing reduces idle speed, ignition timing influences the tendency to diesel in the same manner as Carburetor Adjustment.

Fuel Mixture: Enriching the idle fuel mixture decreases the tendency to diesel by causing the engine to run cooler.

Fuel Content: High octane fuels tend to reduce dieseling. Increased fuel content of lead alkyl increases the tendency to diesel. Phosphates and nickel fuel additives help prevent dieseling.

Spark Plugs: Plugs of too high a heat range for the engine in question can cause dieseling.

Throttle Plates: If the throttle plates are not properly aligned in the carburetor bore, a resulting leanness in fuel mixture occurs, contributing to dieseling.

Electrical System: Normally, during dieseling, ignition is self-supplied by a "hot spot," self-igniting fuel, etc. However, there is a possibility of the vehicle's electrical system supplying the necessary ignition. When the ignition switch is turned off, a small amount of current can flow from the generator into the primary of the ignition coil through the generator tell-tale light. This is particularly true when the warning light bulb has been changed for one of increased wattage.

NOTE: "Run on" is more prevalent in an engine when the ignition is turned off before the engine is allowed to return to idle. Therefore, it can be reduced by letting the engine return to idle before shutting off the ignition. "Run on" incidence can be reduced on automatic transmission units by turning off the engine when in gear.

A certain amount of "run on" can be expected from any gasoline engine regardless of make, size or configuration. (Diesel engines operate on this principle.) However, if the above suggestions are correctly employed, "run on" will be reduced to an unnoticeable level.

ENGINE OVERHEATS: WATER COOLED

Water is used to cool the engine and air is used to cool the water. Anything which prevents this water-air system from working properly will cause overheating. Oil or grease in the water will reduce the ability of the water to absorb heat from the block and to transfer heat in the water to the radiator. There are seven basic causes of overheating:

1. Water does not cool engine.
2. Air does not cool water.
3. Slow combustion.
4. Pre-ignition.
5. Detonation.
6. Excessive friction in engine or elsewhere in power transmitting units.
7. Excessive back pressure in exhaust system.

Water Too Hot

1. Slipping fan belt.
2. Not enough water in system.
3. Carburetor mixture too lean.
4. Clogged exhaust system.
5. Late ignition timing.
6. Centrifugal advance fails to advance spark as engine speed increases because weights stick or because of sticking elsewhere in mechanism.
7. Pre-ignition.
8. Detonation.
9. Water circulation impeded by installation of wrong head gasket.
10. Cylinder head gasket installed incorrectly, blocking off water holes.
11. Leaky cylinder head gasket permits exhaust gas to enter water. The gas bubbles interfere with the ability of the water to cool the engine.
12. Water circulation slowed down by rust, scale or dirt in water jackets.
13. Water distributing tube (when used) within cylinder block rusted out, dented or improperly installed so that not enough water reaches some cylinders, thus causing local overheating.
14. Local overheating at one cylinder (or more) due to heavy deposit of rust, scale or dirt in water jacket around cylinder or exhaust valve port.
15. Water circulation impeded by thermostat which fails to open fully or sticks closed.
16. Water temperature increased by thermostat which fails to open at correct temperature. Or the installation of a thermostat which opens at too high a temperature.
17. Any water hose which has rotted on inside, allowing loosened strips of rubber to impede water circulation.
18. The baffle in top tank may be bent in such a way as to interfere with free discharge of water from the hose.
19. Water passages in radiator are partially clogged with dirt, rust, corrosion or scale (mineral salts).
20. Exterior of radiator clogged with dirt, leaves or insects.
21. Rotting of water hose may weaken it so that pump suction causes it to collapse when engine is running fast, thus throttling the water flow.
22. If water pump seal leaks, air may be drawn into the water. Air bubbles in cooling water reduce the cooling ability of the water.
23. Water pump impeller loose on its shaft or impeller blades corroded.
24. Overheats due to alcohol type antifreeze during mild weather.

Water Leakage

Cylinder Head

1. Loose attaching bolts.
2. Dirty, corroded or burred surface prevents tight fit.
3. Warped surface does not fit tight against gasket.
4. Cracked due to freezing or excessive heat.
5. On overhead valve head, exhaust valve seats may be cracked, allowing water to leak into cylinders and crankcase.

Cylinder Block

1. Dirty, corroded or burred surface prevents tight fit.
2. Warped surface does not fit tight against gasket.
3. Cracked due to freezing or excessive heat.
4. If L-head design, excessive heat may crack exhaust valve seats, allowing water to leak into crankcase.
5. Block cracked due to use of cylinder head bolt which is too long.
6. Leaky expansion plugs or pipe plugs in water jacket.

Cylinder Head Gasket

1. Dirty, corroded or broken.
2. Loose because cylinder head bolts are loose.
3. Leaks because it cannot make tight contact between head and block.

Water Pump

1. Loose pump.
2. Faulty gasket.
3. Improper installation.
4. Warped pump body or dirty metal surfaces.
5. Hole or crack in pump body.
6. Worn seal.
7. Seal improperly installed.
8. Bent pump shaft.
9. Loose bearings or bushings or worn pump shaft.

Radiator

1. Leaks due to freezing or corrosion.
2. Strain due to improper attachment to car.
3. Fan striking radiator.
4. Drain plug or petcock leaks.
5. Radiator baffle bent so that water is directed into overflow pipe.
6. Clogged radiator causing water to pile up in upper tank which causes coolant to flow out overflow pipe.

Hose

1. Hose clamps loose.
2. Hose improperly installed.
3. Hose rotted through.

Heater: See that all heater connections are tight and that its radiator does not leak.

ENGINE OIL LEAKAGE

NOTE: If engine is equipped with a positive crankcase vent valve, check the valve for proper operation before checking cause of leak. A clogged crankcase vent valve can build up pressure in the crankcase which will cause seals and gaskets to leak.

1. Oil pan drain plug loose or gasket missing.
2. Crack or hole in oil pan.
3. Oil pan gasket leaks due to: (a) loose screws; (b) damaged gasket; (c) improperly installed gasket; (d) bent oil pan flange.
4. Timing case cover gasket leaks due to: (a) loose screws; (b) damaged gasket; (c) improperly installed gasket; (d) bent cover flange; (e) leakage at engine support plate.
5. Front crankshaft oil seal leaks due to: (a) worn oil seal; (b) seal not properly installed; (c) rough surface on crankshaft, or fan pulley or damper; (d) damper or pulley loose; (e) seal or cover not centered on crankshaft; (f) oil return passage to crankcase clogged up.
6. Rear main bearing oil seal leaks due to: (a) worn oil seal; (b) improper oil seal installation; (c) worn rear main bearing; (d) rough crankshaft.
7. Oil return passage to crankcase clogged.
8. Expansion plug in block at rear of camshaft leaks due to poor fit, careless installation, or corrosion.
9. Leakage at any external piping.
10. Plugs at ends of oil passages in cylinder block leak.
11. Oil filter leaks.
12. Leakage at distributor housing.
13. Valve cover leaks due to loose screws, defective gasket, improperly installed gasket or bent cover flange.
14. Rocker arm cover or push rod cover leaks because of loose screws, defective gasket, improper gasket instal-

lation or bent cover flange.
15. Pipe connections loose on oil gauge or oil filter lines.
16. Loose oil pump or faulty gasket (if pump is on outside of block).
17. Clogged breather and/or crankcase ventilating discharge pipe, permits increase in pressure within engine, thus causing oil to be forced out past any oil seals or gaskets.
18. If oil pressure relief valve is mounted on outside of block, leakage may occur if unit is loose or its gasket defective.

HIGH OIL CONSUMPTION

1. External oil leaks.
2. Leaky piston rings due to wear.
3. Leaky piston rings due to sticking caused by gummy deposit. Try to free up with suitable solvent poured in fuel tank. Blue smoke at tail pipe indicates badly leaking rings.
4. Worn pistons and cylinders.
5. Cylinder block distorted by tightening cylinder head bolts unevenly.
6. Excessive clearance between intake valve stems and guides allows oil mist to be sucked into cylinders.
7. Punctured vacuum pump diaphragm permits oil from crankcase to be sucked into intake manifold.
8. Worn main or rod bearings allow excessive leakage from bearings. Result is cylinder walls are flooded with oil.
9. Oil pressure too high due to faulty action of oil pressure relief valve, or clogged relief passage.
10. If pressure lubricated, loose piston pins may permit excessive leakage to cylinder walls.
11. Grade of oil used is too light. A poor quality oil may become far too thin when engine is hot. Hard driving on hot days will also consume more oil.
12. Clogged crankcase ventilator system.

OIL PRESSURE RELIEF VALVE LEAKS

1. Relief valve needs tighter adjustment.
2. Relief valve spring weak or broken.
3. Valve seat worn or distorted.
4. Plunger type valve face worn.
5. Plunger type valve stuck open.
6. Ball type valve damaged.
7. Pump discharge pipe or passages leak.

ENGINE OIL DILUTION

1. Oil contains foam caused by presence of water in oil. Water may be due to condensation within crankcase or to a leaky cylinder head gasket.
2. Extreme dilution of oil by fuel may add enough liquid to oil to mislead. In extreme cases, oil level may increase. Dilution is greatest when frequent stops are made in cold weather.

NO OIL PRESSURE

1. Oil pressure gauge defective.
2. Pipe to oil pressure gauge stopped up.
3. Not enough oil in crankcase.
4. Oil pump inoperative.
5. Oil pressure relief valve stuck open.
6. Oil passages on discharge side of pump stopped up.
7. Oil screen or passages on intake side of pump stopped up.

LOW OIL PRESSURE

1. Oil pressure gauge inaccurate.
2. Pipe to pressure gauge restricted.
3. Oil too thin due to dilution, poor quality, or too light a grade used.
4. Oil pressure relief valve adjustment too light.
5. Relief valve spring weak.
6. Oil pump gears worn.
7. Oil pump cover worn.
8. Oil pump body or cover loose.
9. Oil pump gasket damaged, improperly installed or too thick.
10. Air leak in oil intake pipe (if oil level is low).
11. Air leak in top of floating screen (if used).
12. Oil intake pipe or screen clogged with water, sludge, gummy oil, dirt or ice.
13. Oil leak in discharge pipe.
14. Loose connections in oil lines.
15. Worn main, rod or camshaft bearings.

HIGH OIL PRESSURE

1. Oil pressure gauge defective.
2. Oil too heavy.
3. Oil pressure relief valve adjustment too heavy.
4. Relief valve spring too stiff.
5. Oil pressure relief passage clogged.
6. Plunger type relief valve stuck by gummy oil or plunger is too tight a fit.
7. Main oil passages on pressure side of pump clogged.

ENGINE NOISES

Loose Main Bearing

A loose main bearing is indicated by a powerful but dull thud or knock when the engine is pulling. If all main bearings are loose a noticeable clatter will be audible.

The thud occurs regularly every other revolution. The knock can be confirmed by shorting spark plugs on cylinders adjacent to the bearing. Knock will disappear or be less when plugs are shorted. This test should be made at a fast idle equivalent to 15 mph in high gear. If bearing is not quite loose enough to produce a knock by itself, the bearing may knock if oil is too thin or if there is no oil at the bearing.

Loose Flywheel

A thud or click which is usually irregular. To test, idle the engine at about 20 mph and shut off the ignition. If thud is heard, the flywheel may be loose.

Loose Rod Bearing

A metallic knock which is usually loudest at about 30 mph with throttle closed. Knock can be reduced or even eliminated by shorting spark plug. If bearing is not loose enough to produce a knock by itself, the bearing may knock if oil is too thin or if there is no oil at the bearing.

Piston Pin

Piston pin, piston and connecting rod noises are difficult to tell apart.

A loose piston pin causes a sharp double knock which is usually heard when engine is idling. Severity of knock should increase when spark plug to this cylinder is short-circuited. However, on some engines the knock becomes more noticable at 25 to 35 mph on the road.

Piston pin rubs against cylinder wall, caused by lock screw being loose or snap ring broken.

Piston & Rings

1. Excessive clearance between pistons and cylinders (piston slap).
2. Out-of-round or tapered bores.
3. Top piston ring strikes ridge at top of cylinder bore.
4. Carbon deposit on top of piston strikes cylinder head.
5. Piston rubs against cylinder head gasket.
6. Broken piston ring.
7. Excessive side clearance of ring in groove.
8. Worn or broken piston ring lands.
9. Broken piston.

Valves

1. Valve click due to too much tappet clearance, hydraulic tappet not working properly, warped valve, sticking valve, binding rocker arm.
2. Insufficient oil to valve mechanism, especially overhead valves.
3. Worn or scored parts anywhere in valve mechanism.
4. Broken valve springs.
5. Weak valve springs.
6. Cocked valve springs.
7. Excessive tappet guide clearance.
8. Lower end of tappet scored, chipped, rough, worn or broken.
9. Very rough surface on cams.
10. Excessive valve stem-to-guide clearance.
11. Valve face not concentric with valve stem.
12. Valve seat face not concentric with valve stem.
13. Valve covers on overhead valve engines tightened excessively will amplify normal noise.

Hydraulic Lifters

The malfunctioning of a hydraulic valve lifter is amost always accompanied by a clicking or tapping noise. More or less hydraulic lifter noise may be expected when the engine is cold but if lifters are functioning properly the noise should disappear when the engine warms up.

If all or nearly all lifters are noisy, they may be stuck because of dirty or gummy oil.

If all lifters are noisy, oil pressure to them may be inadequate. Foaming oil may also cause this trouble. If oil foams there will be bubbles on the oil level dipstick. Foaming may be caused by water in the oil or by too high an oil level or by a very low oil level.

If the hydraulic plungers require an initial adjustment, they will be noisy if this adjustment is incorrect.

If one lifter is noisy the cause may be:

1. Plunger too tight in lifter body.
2. Weak or broken plunger spring.
3. Ball valve leaks.
4. Plunger worn.
5. Lock ring (if any) improperly installed or missing.
6. Lack of oil pressure to this plunger.

If ball valve leaks, clean plunger in special solvent such as acetone and reinstall. Too often, plungers are condemned as faulty when all they need is a thorough cleaning.

Gum and dirty oil are the most common causes of hydraulic valve lifter trouble. Engine oil must be free of dirt. Select a standard brand of engine oil and use no other. Mixing up one standard brand with another may cause gummy oil and sticking plungers. Do not use any special oils unless recommended by the car manufacturer and change oil filter or element as recommended intervals.

Timing Gears

1. Gears loose on hubs or shafts.
2. Gears misaligned.
3. Excessive gear backlash.
4. Eccentric gear, usually due to high key.
5. Teeth meshed too tight (new oversize gear).
6. Too much end play in camshaft or crankshaft.
7. Front crankshaft bearing clearance excessive.
8. Chipped tooth usually on camshaft gear.

Timing Chain

1. Chain loose due to wear.
2. Sprocket teeth worn.
3. Sprockets loose on hubs or shafts.
4. Sprockets misaligned.
5. Front camshaft bearing clearance excessive.
6. Front main bearing clearance excessive.
7. Loose vibration damper or drive pulley.

Loose Engine Mountings

Occasional thud with car in operation. Most likely to be noticed at the moment the throttle is opened or closed.

Excessive Crankshaft End Play

A rather sharp rap which occurs at idling speed but may be heard at higher speeds also. The noise should disappear when clutch is disengaged.

Water Pump

1. Water pump shaft pulley loose.
2. Impeller loose on shaft.
3. Too much end play in pump shaft.
4. Too much clearance between shaft and bearings.
5. Impeller blades rubbing against pump housing.
6. Impeller pin sheared off or impeller broken.
7. Rough bearing.
8. Pump seal too hard.

Fan Belt

1. Belt worn or burned.
2. Wrong belt. Does not fit pulley grooves properly.
3. Belt too tight. Squeaks.
4. Belt or pulley dirty or sticky with gummy oil.
5. Pulley bent, cracked or broken.
6. Belt pulleys misaligned.
7. Belt loose; squeaks when engine is accelerated.

Fan

1. Fan blades bent.
2. Fan blades loose on hub.
3. Fan out of balance when made.
4. Fan blades strike radiator.
5. Fan shaft end play excessive.
6. Fan shaft loose on its bearings.
7. Defective fan bearings.
8. Bearings need lubrication.

Engine Vibration

1. Unequal compression in cylinders.
2. Missing at high speed.
3. Unbalanced fan or loose fan blade.
4. Incorrect adjustment of engine mount or damaged mounts.
5. Loose engine mounts.
6. Engine support loose on frame or cylinder block.
7. Unbalanced or sprung crankshaft.
8. Excessive engine friction due to tight pistons, etc.
9. Defective vibration damper.

Fuel Pump Noise

Diagnosis of fuel pumps suspected as noisy requires that some form of sounding device be used. Judgment by ear alone is not sufficient, otherwise a fuel pump may be needlessly replaced in attempting to correct noise contributed by some other component. Use of a stethoscope, a long screwdriver, or a sounding rod is recommended to locate the area or component causing the noise. The sounding rod can easily be made from a length of copper tubing $1/4$ to $3/8$ inch in diameter.

If the noise has been isolated to the fuel pump, remove the pump and run the engine with the fuel remaining in the carburetor bowl. If the noise level does not change, the source of the noise is elsewhere and the original fuel pump should be reinstalled. On models using a fuel pump push rod, check for excessive wear and/or galling of the push rod.

Electrical Troubles

NOTE—Ignition troubles are included in the *Engine Troubles* section under the various operating difficulties these troubles could cause.

BATTERY REQUIRES FREQUENT RECHARGING

Insufficient Current Flow to Battery

1. Defective generator or alternator.
2. Incorrect voltage regulator setting.
3. Regulator contacts oxidized or burned.
4. Sulphated battery.
5. Corroded battery terminals.
6. Regulator not grounded.
7. Loose connections or grounds in lighting or ignition circuits.
8. Slipping fan belt.
9. Blown regulator fuse.
10. Wrong size generator drive pulley.
11. Shortened or open alternator rectifiers.
12. Grounded stator windings in alternator.

Excessive Starting Load Causing Abnormal Current Flow From Battery

1. Frequent use of starting motor.
2. Excessive use of starting motor due to difficulty in starting.
3. Faulty starting motor.
4. Excessive engine friction due to tight pistons, etc., or heavy engine oil.

Excessive Lighting Load

1. Car operation confined largely to night driving.
2. Tail and stop light wires reversed.
3. Stop light switch inoperative (closed at all times).
4. Unnecessary use of head lamps while parking.
5. Ground or short in lighting circuit.

Abnormal Accessory Load

1. Radio.
2. Heater.
3. Windshield defroster.
4. Cigar lighter.
5. Spotlights.

Internal Discharge of Battery

1. Plates badly sulphated.
2. Cell leak due to cracked jar or sealing compound.
3. Water level not maintained at proper height.
4. Plate separators ineffective.
5. Exterior of battery covered with corrosion and acid-soaked dirt which forms a path to ground for current.

Miscellaneous

Radio suppressor connected to generator or regulator field terminal.

STARTER WON'T ROTATE OR ROTATES SLOWLY

If lights become dim or go out when the starter switch is closed, the battery may be too weak to operate the starter. In this case, the engine may be started by pushing the car.

NOTE—Some cars cannot be started by pushing because their automatic transmissions have no rear oil pump to drive the engine through the transmission. In such cases, a fully charged battery should be installed or a "jumper" circuit should be used from another charged battery.

Cars equipped with alternators cannot be push started if the battery is completely dead because alternators retain no residual magnetism.

Due to Starter Circuit

1. Low battery. Lights grow very dim or go out when starter switch is closed.
2. Connections loose, dirty, corroded or broken at battery terminals, starter switch terminal, battery ground strap.
3. Short circuit across starter terminal.

Due to Starter Switch

1. Starter pedal (if any) stuck.
2. Starter switch stuck.
3. Pedal linkage fails to close starter switch (older cars).
4. Defective solenoid.
5. Neutral safety switch on cars with automatic transmissions out of adjustment or defective.
6. Starter switch makes poor contact due to dirt, corrosion, bent parts, weak contact spring.
7. Starter switch fails to close circuit because of sticking or broken contact parts.

Due to Armature & Field Circuits

1. Armature windings burned out, shorted, grounded or open-circuited.
2. Short circuit in armature winding or brush pigtail lead.
3. Broken wire in armature winding or brush pigtail lead.
4. Loose, dirty or corroded connections in armature circuit, including ground.
5. Field coils burned out, shorted or grounded.
6. Broken wire in field winding or broken lead.
7. Loose, dirty or corroded connections in field circuit.

Due to Commutator & Brushes

1. Brush pigtail leads loose or broken.
2. Starter brushes cracked crosswise (prevents flow of current).
3. Arm type brush holder sticks.
4. Brush sticks in sliding brush holder.
5. Bent brush holder misaligns brush and causes poor contact.
6. Starter brushes badly worn.
7. Brush leads shorted or have loose, dirty, corroded or broken connections.
8. Poor brush contact due to weak or broken springs.
9. Brushes coated with oil.
10. High mica between commutator segments prevents brush contact.
11. Commutator bars loose and/or solder melted.
12. Commutator dirty, corroded or burned.

Due to Engine Resistance

1. Piston sticking to cylinders in overheated engine.
2. Pistons struck to cylinders because of gummy oil.
3. Pistons binding in cylinders because of corrosion after long lay-up.
4. Jammed generator armature.
5. Combustion chamber full of water.
6. Solid ice in water pump.
7. Broken part in engine causes jamming.
8. Excessive engine friction, due to cold weather and too heavy oil.

Due to Improper Engine Repairs

1. New rings too tight.
2. New pistons too tight.
3. Main or rod bearings too tight.
4. New camshaft bearings too tight.

Due to Armature Binding

1. Loose field poles.
2. Armature shaft frozen in bearings.
3. Loose end plates.
4. Windings thrown out of armature slots.
5. Armature locked magnetically to field poles because of loose bearings or worn or bent armature shaft.
6. Bendix spring retaining screws loose (jammed against housing).
7. Cracked or distorted drive housing.
8. Starter misaligned.
9. Starter jams because of burred teeth on drive pinion or flywheel gear.
10. Starter pinion (sliding gear type) jams because of incorrect endwise clearance.

STARTER SPINS BUT WON'T ENGAGE FLYWHEEL GEAR

Bendix Type

1. Bendix pinion stuck on shaft due to dirty or gummy shaft or bent shaft.
2. Bendix spring broken.
3. Bendix spring bolt broken.
4. Pinion housing cracked.
5. Drive key sheared.
6. Pinion teeth broken off.
7. Starter ring gear has several teeth missing.
8. Armature shaft broken.

Sliding Gear Type

1. Weak or broken meshing spring.
2. Fault in sliding gear linkage.
3. Fault in solenoid.
4. Over-running clutch worn out or lubricant caked or gummy.
5. Drive key sheared.
6. Pinion teeth broken off.
7. Flywheel ring gear has several teeth missing.
8. Armature shaft broken, dirty or dry.
9. Wrong starter pinion clearance.

STARTER PINION JAMMED INTO FLYWHEEL GEAR

1. Burred teeth on pinion or ring gear.
2. Misalignment of starter or armature shaft.
3. If engine kicks back when being started, Bendix pinion may jam. Loosen starter to free pinion.

STARTER PINION DISENGAGES SLOWLY

Bendix Type

The most probable cause is a dirty Bendix drive shaft. Or the pinion may bind on its shaft due to a bent shaft or too tight a fit between pinion and splines.

When a Bendix Folo-Thru starter drive stays in mesh too long it is probably due to a sticking release pin which is designed to be released by centrifugal force at a certain engine rpm. In such an instance the drive should be replaced.

Sliding Gear Type

1. Pinion binds on its shaft due to too tight a fit or due to bent or burred shaft.
2. Pinion shaft sticky or dirty.
3. Sliding gear operating linkage sticking or binding.
4. Solenoid does not operate properly.

STARTER PINION WON'T RELEASE

Bendix Folo-Thru Drive

Failure to disengage would most probably be caused by a stuck release pin which is designed to be released by centrifugal force at a given engine rpm. If such is the case, replace the drive unit.

Sliding Gear Type

If solenoid operated, the solenoid may be defective. If pedal operated, the shift linkage may be binding or sticking. May also be caused by a defective starting switch on cars with key-starter switch or by improper starter pinion clearance.

STARTER NOISE

1. Loose pole pieces rubbing against armature.
2. Gear noise due to defective teeth.
3. Flywheel ring gear untrue.
4. Starter drive housing loose on flywheel housing.
5. Starter loose on drive housing.
6. Commutator end plate loose.
7. Armature shaft bent.

8. Worn armature shaft, bearings or bushings.
9. Drive pinion shaft bent.
10. Worn drive pinion shaft, bearings or bushings.
11. Misalignment caused by dirt or burrs on mating surfaces.

ALTERNATORS

Alternator Fails to Charge

1. Drive belt loose.
2. Brushes sticking.
3. Open charging circuit.
4. Open in stator winding circuit.
5. Faulty soldered connections at output terminal stud.
6. Rectifiers open circuited.

Low Unsteady Charging Rate

1. Drive belt loose.
2. High resistance at battery terminal posts.
3. Loose connections.
4. Poor ground between engine and body ground wire.
5. Resistance in charging circuit.
6. Open stator windings.

Low Output

1. Grounded stator.
2. Shorted rectifier.
3. Voltage regulator faulty.

Excessive Charging Rate

1. Voltage regulator faulty.
2. Open circuited rectifier.

Noisy Alternator

1. Misaligned belt or pulley, or loose pulley.
2. Shorted rectifier.
3. Worn bearings.
4. Rotor shaft sprung.

Regulator Points Oxidized

1. Poor ground connections.
2. Improper voltage regulator air gap setting.
3. Shorted field in alternator.
4. Voltage regulator setting too high.

Burned Points or Coil Windings in Regulator

1. Voltage regulator setting too high.

Voltage Regulator Points Stuck

1. Poor ground connections between alternator and regulator.

LOCATING ELECTRICAL SHORTS WITH TEST LAMP

Due to the complexity of locating electrical short circuits where several circuits are protected by the same fuse, fabricate a test lamp from the material shown in the accompanying illustration. By substituting the test lamp for the blown fuse the short circuit can be isolated.

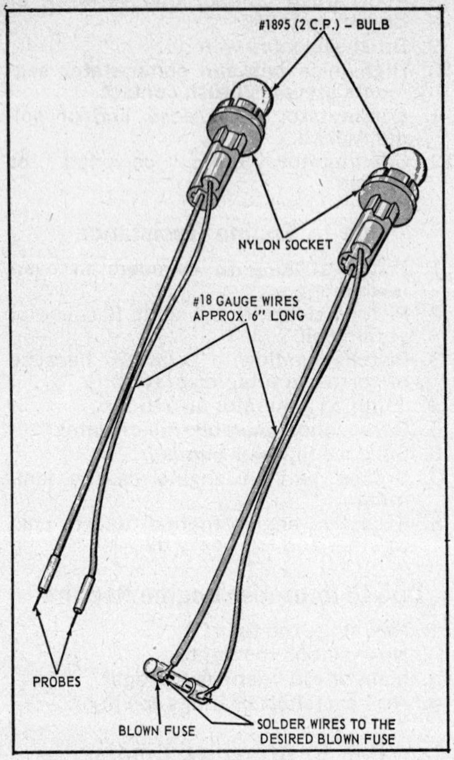

Test lamp for locating shorts

When the test lamp is inserted into the fuse panel, the bulb will light and continue to glow until the short circuit is removed. Determining which circuit is at fault can be accomplished by disconnecting the affected circuits one at a time until the test lamp goes out. Then trace the circuit to find the cause of the short (wire contacting sharp sheet metal edges, wire pinched between two metal objects, etc.).

For circuits that are not connected to the fuse panel but are protected by an in-line fuse cartridge, use a test lamp having two needle point probes in place of the blown fuse. Insert one probe through the insulation and into the wire on each side of the blown in-line fuse and follow the same testing procedure outlined above.

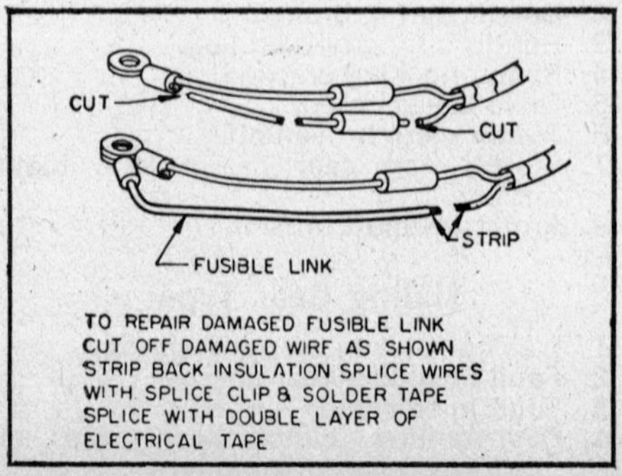

Repairing fusible links

FUSIBLE LINKS

Some cars starting with 1965 models have fusible links located between the battery and the lower ends of the main supply wires. These links are the weakest point in the electrical supply system for the entire car and, as such, will act as a fuse for every wire harness in the car. Every electrical accessory is still protected by a fuse or circuit breaker, of course, but fusible links have been added to protect the wiring harnesses *before* the fuses.

In the past, if a wire became grounded in the portion between the battery and the fuse block, a long section of the wire would burn out, making replacement of a complete wiring harness necessary. Now, with the fusible links, a short or ground in any unfused wire will cause only a short link to burn out. Because of its location, possibility of a fire, such as was sometimes caused by a burned-out wiring harness, is very remote.

A fusible link is simply a short section of wire that is several sizes smaller in gauge than the wire in the circuit which it protects. If a short or ground occurs the fusible link will melt before the insulation is damaged elsewhere in the circuit. Replace burned-out fusible link as directed in the illustration.

LIGHTS FLICKER

Circuit Breaker Vibrates

When the circuit breaker vibrates and causes lights to flicker it indicates a short in one of the lighting circuits, which may be traced as follows:

1. Pull switch successively to each lighting position. If circuit breaker vibrates in all positions except "off" the trouble should be found in the tail lamp and license lamp circuit, or instrument, map light, or clock light circuits.
2. If circuit breaker vibrates in parking lamp position only, look for a short in the parking lamp circuit.
3. If circuit breaker vibrates in headlamp position only, inspect headlamp wiring circuit and lamp assemblies. If both filaments in headlamps burn at the same time, check dimmer switch.

LAMPS FAIL TO BURN

1. Burned out bulb.
2. An open circuit in wiring.
3. A defective switch.
4. Burned out fuse.

LIGHTS FLARE UP WHEN ENGINE IS SPEEDED UP

This condition is caused by high voltage in the electrical system due to one or more of the following:

1. Electrolyte in battery low or weak.
2. High resistance in circuit between generator and battery due to loose or dirty connections.
3. Poor ground between generator and engine.
4. Voltage regulator adjusted too high.
5. Voltage regulator inoperative.

6. Ground or short in generator field circuit.

STOP LIGHT TROUBLES

1. If only one stop light fails to burn, check lamp bulb, socket and wiring.
2. If both stop lights fail to burn also check stop light switch and fuse.
3. If stop light burns when brake pedal is released, check stop light switch, brake pedal clearance and for dragging brakes.
4. If compensating port in brake master cylinder is plugged by foreign material, or is covered by the piston primary cut when brake pedal is released, high pressure will be maintained in hydraulic system and stop light switch will remain closed.

TURN SIGNAL TROUBLES

1. If signals are inoperative on both turns, look for a blown fuse or a defective flasher.
2. If stop lights burn, the fuse and rear signal lamp bulbs are okay.
3. An inoperative right signal light may be caused by a burned out bulb at the right indicator or a right signal lamp. The opposite applies for an inoperative left signal light.
4. If bulbs are okay, look for an open circuit or defective switch.
5. If indicator light on dash burns steady when lever is placed in a turn position, check for burned out bulb in park or stop light. If park and stop light bulbs are okay, check for faulty flasher.
6. If indicator light on dash does not burn when lever is in a turn position, check for burned out bulb or a faulty flasher.
7. If switch fails to cancel after completion of turn, remove steering wheel and check for worn or broken mechanism.

ELECTRIC CLOCKS

If clock does not run, check for blown "clock" fuse. If fuse is blown check for short in wiring. If fuse is not blown check for open circuit.

With an electric clock, the most frequent cause of clock fuse blowing is low voltage at the clock which will prevent a complete wind and allow clock contacts to remain closed. This may be caused by any of the following: discharged battery, corrosion on contact surface of battery terminals, loose connections at battery terminals, at junction block, at fuse clips, or at terminal connection of clock. Therefore, if in reconnecting battery or clock it is noted that the clock is not ticking, always check for blown fuse, or examine the circuits at the points indicated above to determine and correct the cause. See *Dash Gauge* chapter for Electric Clock data.

HAZARD WARNING FLASHER TROUBLES

To make a quick check of the system pull Hazard Warning switch to ON position. The rear turn signal bulbs should flash as well as the front turn signal bulbs, turn signal indicator bulbs and pilot bulb. All lights will burn continuously when the brake pedal is depressed; this is normal.

Pilot Bulb Fails to Flash

Check for burned out bulb and loose or defective ground wire. Replace bulb, repair ground wire or tighten ground wire screw. If this does not correct the condition, replace flasher switch and harness assembly. Then repeat quick check procedure.

All Bulbs Fail to Flash

1. Check for loose harness connections at Hazard Warning connectors and secure connectors if necessary.
2. Check for a burned out tail and stop light fuse and replace if necessary.
3. Check for a defective Hazard Warning flasher or switch. This may be done by removing the flasher and installing a known good flasher.
4. Pull switch to ON position. If flasher does not operate properly, replace flasher switch and harness assembly, installing old flasher. If system still does not operate properly, install new flasher along with new switch.

Some Bulbs Fail to Flash While Others are Operative

1. Turn ON ignition switch and turn OFF Hazard Warning switch.
2. Place turn signal lever first to right and then to left turn position. If turn signal circuits operate properly, the Hazard Warning switch and harness assembly should be replaced.
3. If the same bulbs fail to flash, the cause is most likely a burned-out bulb. In the case of turn signal indicator bulbs, a loose or defective ground wire can also cause this condition. Repair as necessary.

NOTE: If any turn signal bulb fails to flash when the turn signal circuit is actuated, the reduced current in the circuit will cause the remaining signals on the side of the car to burn steadily. If the Hazard Warning flasher is energized, however, all turn signal bulbs and indicator bulbs will flash except those that have a circuit defect. They will flash at a constant rate unless the battery is completely run down. This is because the Hazard Warning Flasher overrides the turn signal circuit flasher.

4. If the condition is still not resolved, disconnect the Hazard Warning connectors and again check the operation of the turn signal circuits. If the affected bulbs now flash, replace flasher switch and harness assembly.
5. If the condition is still not resolved, look for defects in the connectors to the affected bulb.
6. Repeat quick check test.

Clutch Troubles

CLUTCH DRAGS

Clutch drag means that when the clutch pedal is depressed fully the clutch disc is not completely released. In consequence it does not come to rest but continues to rotate, being dragged around by the rotation of the engine. Clutch dragging causes clashing of gears, especially when shifting from neutral to low or reverse.

1. Pedal cannot disengage clutch because of excessive free pedal travel. Pedal linkage should be adjusted so that the pedal shank is about 1" from the under side of the toe-board.
2. Worn clutch linkage.
3. Release levers need adjustment.
4. Clutch disc warped out of true.
5. High spots on clutch facing.
6. Broken or loose facings.
7. Loose rivet in facing.
8. Clutch disc hub binds on splined clutch shaft due to bent shaft, tight fit, burred splines or splines covered with gummy oil or dirt.
9. Clutch disc wobbles because of broken springs in hub.
10. Clutch disc hub out of true.
11. Clutch shaft bent.
12. Clutch shaft out of true because of worn bearings.
13. Transmission is not in alignment with flywheel housing.
14. Clutch pressure plate warped, thus throwing release levers out of adjustment.
15. Flange of clutch cover not in alignment with flywheel because of loose attaching screws, bent flange, dirt between flange and flywheel.
16. Grease on clutch facings.
17. Engine misaligned due to deteriorated or broken engine mounts.
18. Loose flywheel housing-to-engine attaching bolts.
19. Release fork pivot worn.

CLUTCH SLIPS

The clutch disc slips whenever the clutch pressure plate fails to hold it tight against the face of the flywheel. If clutch slippage is severe, the engine speed will rise above normal on full throttle in high gear. Slight but continuous slippage may go unnoticed until the clutch facings are ruined by excessive temperature caused by friction.

In a very high percentage of cases, clutch slippage is due to less than zero clearance between the shank of the pedal and the toe-board because of failure to have the pedal adjusted in time. The consequence is worn and burned clutch facings. Before the clutch starts slipping, the normal wear of the facings causes a gradual reduction in clutch pedal free play. When there is no free play of the pedal the clutch starts slipping.

Other causes of clutch slippage are:

1. Driving with foot resting on pedal.
2. Binding or sticking of pedal or its linkage.
3. Binding or sticking of clutch disc hub on clutch shaft.
4. Binding of release levers.
5. Release bearing sleeve sticks.
6. Weak or broken clutch pressure springs.
7. Worn clutch facings.
8. Facings covered with grease or oil.
9. Facings burned.
10. Release levers improperly adjusted.
11. Pressure plate sticks.

CLUTCH GRABS

A clutch is said to grab when it engages too abruptly. The usual causes are:

1. Loss of tension in cushioning plates in the rim of the steel clutch disc. These plates cause the clutch facings to bulge outward slightly. The resulting springy action of the facings aids in producing a smooth, gentle clutch engagement.
2. Use of wrong type of clutch facing.
3. Grease or oil on facings.
4. Clutch springs too stiff.
5. Momentary binding in clutch linkage while clutch is being engaged.
6. Exposed rivet heads due to excessively worn facings or loose rivets.

CLUTCH CHATTERS

If a clutch chatters while it is being engaged, the trouble is caused by rapid gripping and slipping. The usual causes are:

1. Somewhat sticky clutch friction surfaces due to engine or transmission oil leaking from defective seal.
2. Clutch friction surfaces damp or wet.
3. Weak clutch springs.
4. Slight binding in clutch linkage during engagement.
5. Slight binding of pressure plate during engagement.
6. Loose engine mounts.

CLUTCH PEDAL PULSATES

Clutch pedal pulsation has often been termed a nervous pedal. When a slight pressure is applied on the pedal, with the engine running, the pedal will vibrate or bounce with every revolution of the engine. As the pressure on the pedal is increased, the pulsation will cease.

1. Loose or improperly adjusted engine mounts.
2. Collar on clutch release sleeve does not run true due to a bent clutch shaft, or the clutch shaft misaligned because of misalignment between crankshaft and transmission.
3. Clutch release levers not adjusted to uniform height.

CLUTCH RATTLES

This condition will occur when the engine is idling with transmission in neutral.

1. Excessive clearance at pressure plate driving lugs.
2. Anti-rattle springs or retractor springs on release levers (or release bearing) weak, broken or disconnected.
3. Looseness in clutch pedal operating linkage.
4. Loose flywheel.

NOISE WHEN PEDAL IS DEPRESSED

1. Clutch release bearing worn, dirty, damaged, broken or inadequately lubricated.
2. Clutch shaft bearing or bushing in crankshaft worn, damaged, broken or inadequately lubricated.
3. Clutch shaft rear bearing at front end of transmission, worn, dirty or lacks lubricant.

NOISE WHEN PEDAL IS RELEASED

1. Misalignment of transmission with engine causing slight wobble of clutch disc hub—noticeable with engine idling or at low road speed.
2. Disc hub loose fit on splined clutch shaft.
3. Disc damper springs weak or broken.
4. No pedal play.
5. Weak or broken pedal return spring.
6. Weak or broken release sleeve spring.
7. Clutch linkage sticks.
8. Clutch pedal sticks.
9. Clutch release sleeve sticks.
10. Clutch release fork binds.
11. Bad clutch release bearing.
12. Loose flywheel.

BEARING NOISE

Clutch Release Bearing:—With engine idling, there is a high-pitched rubbing noise when foot rests on clutch pedal.

Clutch Pilot Bearing:—Fairly high-pitched noise when clutch pedal is fully depressed with engine idling.

Three Speed Transmission Troubles

Fully Synchronized Transmission

Car Application Listed in Three Speed Manual Shift Transmission Chapter

STICKING IN GEAR

1. Clutch not releasing completely.
2. Low lubricant level.
3. Corroded transmission levers.
4. Tight main drive gear pilot bushing.
5. Defective synchronizer sleeve or blocking ring.

FORWARD GEARS CLASH

1. Clutch not releasing completely.
2. Weak or broken springs in synchronizer assembly.
3. Worn blocking rings and/or cone surfaces.
4. Broken blocking ring.

NOISY IN FORWARD SPEEDS

1. Insufficient or incorrect lubricant.
2. Transmission misaligned or loose.
3. Main drive gear or bearings worn or damaged.
4. Countergear bearings worn or damaged.
5. Synchronizers worn or damaged.

NOISY IN REVERSE

1. Reverse idler gear or shaft worn or broken.
2. Reverse gear worn or broken.

HARD SHIFTING

1. Improper clutch or adjustment.
2. Worn or damaged shift linkage.
3. Incorrect lubricant.
4. Synchronizers worn or broken.

JUMPING OUT OF GEAR

1. Misadjusted, worn or loose shift linkage.
2. Transmission loose or misaligned.
3. Worn pilot bearing.
4. Excessive end play in main drive gear.
5. Weak detent cam spring.
6. Detent cam notches worn.
7. Worn clutch teeth on main drive gear or synchronizer sleeve.
8. Worn or broken synchronizer.
9. Bent output shaft.

Three Speed Transmission Troubles

General Motors Transmission with Second Speed Gear Located at Rear of Mainshaft

Car Application Listed in Three Speed Manual Shift Transmission Chapter

SLIPS OUT OF HIGH AND/OR 2ND GEAR

1. Transmission mounting bolts loose.
2. Control rods interfere with engine mounts or clutch release lever.
3. Control linkage does not work freely.
4. Gear does not fully engage.
5. Damaged mainshaft pilot bearing.
6. Clutch gear bearing retainer broken or loose.
7. Dirt between transmission case and clutch housing (front mounted), or between transmission case and differential carrier (rear mounted).
8. Misalignment of transmission.
9. Worn or broken synchronizer assembly.
10. Weak springs in transmission cover.

SLIPS OUT OF LOW AND/OR REVERSE

1. First and/or reverse gears damaged from operating at part engagement.
2. Improperly mated splines on inside of first and reverse gear and/or external spline on 2nd and 3rd synchronizer sleeve.
3. Improperly adjusted linkage.
4. Weak springs in transmission cover.

NOISY IN ALL GEARS

1. Not enough lubricant.
2. Worn countergear bearings.
3. Worn or damaged clutch gear and countershaft drive gear.
4. Damaged clutch gear or mainshaft ball bearings.
5. Damaged speedometer gears.

NOISY IN HIGH GEAR

1. Damaged clutch gear bearing.
2. Damaged mainshaft bearing.
3. Damaged speedometer gears.

NOISY IN NEUTRAL WITH ENGINE RUNNING

1. Damaged clutch gear bearing.
2. Damaged mainshaft pilot bearing roller.

NOISY IN ALL REDUCTION GEARS

1. Not enough lubricant.
2. Worn or damaged clutch gear or countershaft drive gear.

NOISY IN SECOND ONLY

1. Damaged or worn 2nd speed gears.
2. Worn or damaged countergear rear bearings.

NOISY IN LOW AND REVERSE ONLY

1. Worn or damaged 1st and reverse sliding gear.
2. Damaged or worn low and reverse countergear.

NOISY IN REVERSE ONLY

1. Worn or damaged reverse idler.
2. Worn reverse idler bushings.
3. Damaged or worn reverse countergear.

EXCESSIVE BACKLASH IN SECOND ONLY

1. Second gear thrust washer worn.
2. Mainshaft rear bearing improperly installed in case.
3. Worn countergear rear bearing.

EXCESSIVE BACKLASH IN REDUCTION GEARS

1. Worn countergear bushings.
2. Excessive end play in countergear.

LEAKS LUBRICANT

1. Too much lube in transmission.
2. Loose or broken clutch gear bearing retainer.
3. Clutch gear bearing retainer damaged.
4. Cover loose or gasket damaged.
5. Operating shaft seal leaks.
6. Idler shaft expansion plugs loose.
7. Countershaft loose in case.
8. Lack of sealant on bolts.

Three Speed Transmission Troubles

All Transmissions with Low-Reverse Gear Located at Rear of Mainshaft

Car Application Listed in Three Speed Manual Shift Transmission Chapter

NOISES

When diagnosing transmission noise note the gear position in which the noise occurs. Noise present in all gear positions may be due to worn or damaged constant mesh gears or bearings. Noise present in only one gear can usually be traced to the particular gear involved. Other causes of noise are as follows:

1. Misalignment due to loose mounting bolts.
2. Clutch housing misalignment.
3. Dirt or metal chips in lubricant.
4. Not enough lube in transmission.
5. Improper lubricant.

HARD SHIFTING

1. Clutch linkage out of adjustment.
2. Linkage improperly adjusted.
3. Linkage binding due to bent, worn or broken parts.
4. Gearshift tube binding due to misaligned steering gear housing.
5. Improper lube in transmission.
6. Damaged synchronizer assembly.

JUMPS OUT OF GEAR

1. Improper shift procedure.
2. Linkage parts worn, bent, broken or out of adjustment.
3. Excessive end play caused by wear in shift forks, sliding gear fork grooves, thrust washers, mainshaft and countershaft bearings, or clutch pilot bushing.
4. Misalignment or excessive clearance between sliding gear and mainshaft.
5. Damaged synchronizer.
6. Weak springs in transmission cover.

LEAKAGE

1. Overfilled transmission or using a lube that foams or expands while car is in operation.
2. Loose gearshift housing capscrews.
3. Damaged gaskets.
4. Transmission vent plugged.
5. Extension housing rear seal leaks.

Four Speed Transmission Troubles

NOISY IN ALL SPEEDS

1. Incorrect lubricant level.
2. Incorrect type lubricant.
3. Countergear bearings worn or damaged.
4. Countergear worn or damaged.
5. Clutch gear bearing worn or damaged.
6. Mainshaft bearing worn or damaged.
7. Clutch gear worn or damaged.
8. Transmission misaligned or loose.

NOISY IN 1st SPEED

1. First gear worn or damaged.
2. Countergear worn or damaged.
3. Countergear bearings worn or damaged.
4. Synchronizers worn or broken.
5. Countershaft worn or damaged.

NOISY IN 2nd SPEED

1. Second gear worn or damaged.
2. Countergear worn or damaged.
3. Countergear bearings worn or damaged.
4. Synchronizers worn or broken.
5. Countershaft worn or damaged.

NOISY IN 3rd SPEED

1. Third gear worn or damaged.
2. Countergear worn or damaged.
3. Countergear bearings worn or damaged.
4. Synchronizers worn or broken.
5. Countershaft worn or damaged.

NOISY IN 4th SPEED

1. Clutch shaft bearing worn or damaged.
2. Mainshaft bearing worn or damaged.
3. Synchronizers worn or broken.

NOISY IN REVERSE

1. Reverse idler gear or shaft worn or damaged.
2. Reverse sliding gear worn or damaged.
3. Shift linkage out of adjustment.
4. Shift linkage bent or damaged.
5. Shift linkage parts loose.
6. Shift levers, shafts or forks worn.

SHIFTS HARD

1. Clutch pedal free travel incorrect.
2. Clutch parts worn or damaged.
3. Shift linkage out of adjustment.
4. Shift linkage bent or damaged.
5. Shift linkage parts loose.
6. Shift levers, shafts or forks worn.
7. Lubricant type incorrect.
8. Lubricant level incorrect.

JUMPS OUT OF GEAR

1. Shift linkage out of adjustment.
2. Shift linkage bent or damaged.
3. Shift linkage parts loose.
4. Shift levers, shafts or forks worn
5. Shift cover loose or gasket damaged.
6. Transmission misaligned or loose.
7. Synchronizers worn or broken.
8. Clutch gear bearing retainer broken.
9. Clutch gear bearing worn or damaged.
10. Clutch pilot bearing worn or broken.
11. Mainshaft and/or pilot worn or damaged.
12. Mainshaft bearing worn or damaged.

LEAKS LUBRICANT

1. Lubricant level incorrect.
2. Lubricant type incorrect.
3. Vent plugged.
4. Clutch gear bearing retainer or gasket loose.
5. Clutch gear bearing retainer broken.
6. Shift cover loose or gasket damaged.
7. Shifter shaft seals leaking.
8. Shift cover bolts not sealed.
9. Countershaft loose in case bore.

Overdrive Troubles

DIAGNOSIS

Figs. 1, 2 and 3 illustrate the overdrive circuit diagrams in use. Since overdrive troubles may originate not only in the mechanical operation of the unit but also in the electrical circuit which controls that unit, always check the control system before disassembling the overdirve. If the trouble is not found after a thorough inspection of the control system, then the transmission and overdrive should be removed for examination.

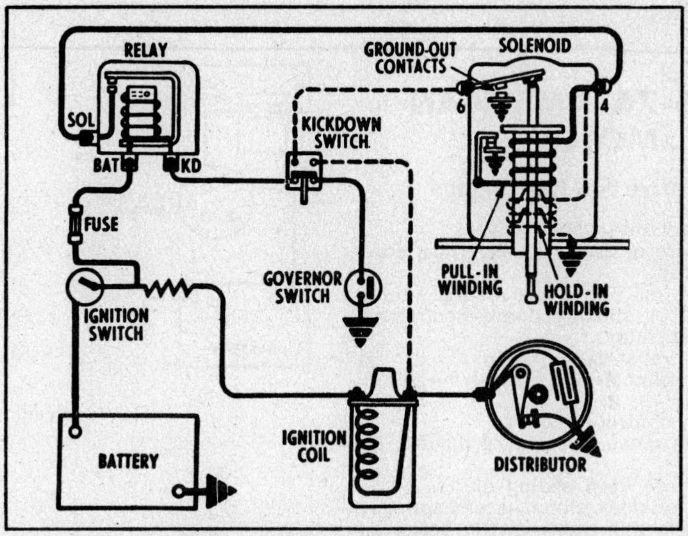

Fig. 1 Overdrive circuit diagram with relay. 1969 AMC

1969 AMERICAN MOTORS

Unsatisfactory Overdrive Operation

1. Blown fuse in governor-solenoid circuit.
2. Loose terminals on any of the connecting wires.
3. Incorrect terminal locations of connecting wires.
4. Circuits grounded by water, dirt or deformation.
5. Defective solenoid points.
6. Insufficient travel or unsatisfactory contacts in kickdown switch.
7. Excessive end play in governor shaft.
8. Improper adjustment of governor control springs.
9. Burned governor contact points.
10. Damage to governor cap and contacts.
11. Absence of rubber cover to exclude water and dirt.
12. Insufficiant travel of shift rod (adjust control cable).

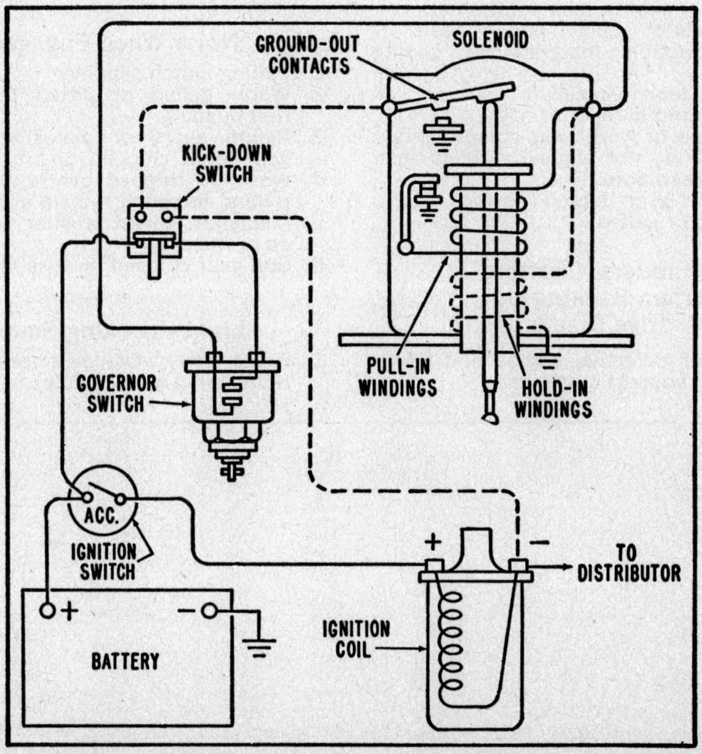

Fig. 2 Overdrive circuit diagram without relay. 1969 AMC

Overdrive Won't Drive Unless Locked Up Manually

1. Occasionally the unit may not drive the car forward in direct drive unless locked up by pulling the dash control. This may be caused by one or more broken rollers in the roller clutch, the remedy for which is to replace the entire set of rollers.
2. This condition may also be caused by sticking of the roller retainer upon the cam. This retainer must move freely to push the rollers into engaging position under the pressure of the two actuating springs.
3. Sometimes this condition is due to slight indentations, worn in the cam faces by the rollers spinning, remedied by replacing the cam.

Overdrive Does Not Engage or Lock-Up Does Not Release

1. Dash control improperly connected.
2. Transmission and overdrive improperly aligned.
3. Kickdown switch improperly adjusted.
4. Improper installation of solenoid.
5. Improper positioning of blocker ring.
6. Broken or slipping governor drive pinion.
7. Too much end play in mainshaft.

Overdrive Engages with Severe Jolt or Noise

Insufficient blocker ring friction may cause the ring to lose its grip on the hub of the sun gear control plate.

Free-Wheels At Speeds Over 30 MPH

If cam roller retainer spring tension is weak the unit will free-wheel at all times.

1975-76 AMERICAN MOTORS

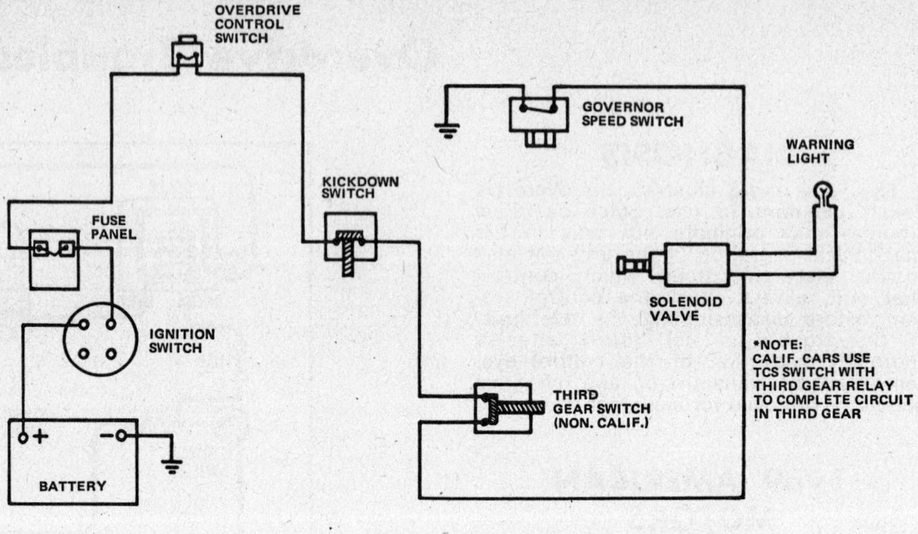

Fig. 3 Overdrive control circuit. 1975-76 American Motors

Overdrive Will Not Engage

1. Low lubricant level.
2. Open wire or switch in electrical control circuit.
3. Plugged oil pan or pressure filters, plugged or sticking pump non-return valve assembly.
4. Sticking relief valve piston.
5. Broken relief valve assembly or piston springs.
6. Plugged control orifice.
7. Solenoid valve sticking, grounded or open.
8. Sticking or worn sliding clutch.
9. Broken or weak clutch return springs.
10. Leaking clutch apply piston seals, or internal case leaks.
11. Damaged or worn gear components.
12. Pump body misaligned with oil feed slot in case bore.
13. Pump plunger of body worn excessively.

Overdrive Will Not Disengage

CAUTION: This condition requires immediate correction, as extensive damage may result if vehicle is moved in reverse.

1. Closed switch or short in electrical circuit.
2. Solenoid valve stuck or shorted.
3. Relief valve piston stuck or spring broke.
4. Control orifice plugged.
5. Sliding clutch sticking.
6. Damaged or seized gear components.

Slow Disengagement and/or Overdrive Freewheels On Overrun

1. Solenoid valve oil feed holes plugged or valve sticking.
2. Control orifice plugged.
3. Overrunning clutch worn or seized.
4. Worn or damaged sun gear or sliding clutch hub.
5. Worn or damaged friction material on sliding clutch.
6. Worn or damaged brake ring.
7. Relief valve piston sticking.

Overdrive Slips When Engaging

1. Low lubricant level.
2. Plugged oil pan or pressure filter.
3. Sticking or plugged non-return valve.
4. Broken spring on relief valve assembly.
5. Control orifice stuck.
6. Solenoid valve sticking.
7. Oil feed holes plugged.
8. Loose wire at solenoid terminal.
9. Worn clutch apply piston seals or worn friction material on sliding clutch.
10. Internal leak in case.
11. Worn pump plunger or body.
12. Damaged or worn gear components.
13. Pump body not aligned with oil feed slot in case bore.
14. Annulus gear clutch surface worn, burned, or galled.

Shudders, Chatters When Backing Up (Overdrive Disengaged)

1. Loose or defective engine and transmission support cushions.
2. Transmission clutch slipping, incorrectly adjusted.
3. Weak clutch return springs.
4. Worn or damaged friction material on sliding clutch.
5. Burned or galled annulus gear clutch surface.

No Kickdown

1. Kickdown switch not adjusted properly.
2. Kickdown switch stuck, shorted or broken.

Noisy When Engaged

1. Sliding clutch slipping.
2. Worn, galled or pitted bearings or rear bushing.
3. Pinion gears or mainshaft annulus gear teeth chipped or broken.
4. Worn or chipped overrunning clutch rollers, or clutch race in annulus gear.
5. Mainshaft thrust washer worn, broken or missing.
6. Sun gear chipped or missing.

Light Knocking Sound

1. Pump body installed improperly (flat not aligned with oil hole).

Rear Axle Troubles

Noise When Pulling Straight Ahead

1. Not enough oil.
2. Wrong grade of oil.
3. Poor quality oil.
4. Ring gear and pinion have excessive backlash.
5. Ring gear and pinion worn.
6. Pinion shaft bearings worn or loose.
7. Pinion shaft end play excessive.
8. Ring gear and pinion misaligned because of bent axle housing or distorted differential case.
9. Ring gear warped.
10. Differential bearings worn or loose.
11. Ring gear rivets or screws loose.
12. Ring gear and pinion not matched set.

Noise When Coasting In Gear

Any axle noise which is heard when the engine is pulling the car is likely to be heard when coasting although not as loud as when pulling.

If ring gear and pinion are meshed too tight, the noise will be greater when decelerating. The noise will disappear when the engine is pulling unless the gears are very tight.

Excessive end play of pinion shaft due to loose pinion nut or incorrect adjustment.

Intermittent Noise

1. Warped ring gear.
2. Loose ring gear rivets or screws.
3. Ring gear improperly installed on differential case due to dirt or burrs between the two.

Knocks or Clicks

1. Flat spot on ring gear or pinion tooth, or tooth chipped, or particle of metal lodged on tooth.
2. Flat spot in bearing.
3. Loose axle shaft key.
4. Loose splined shafts.
5. Mis-matched differential case halves.

Noise On Turns

1. Differential pinions or side gears chipped, scuffed or teeth broken.
2. Differential pinions binding on pinion shaft.
3. Differential pinions or side gears loose due to worn bushings or shaft.
4. Excessive backlash between pinions and side gears.
5. Excessive axle shaft end play.
6. Contacting surfaces between side gear and differential case burred, scored or otherwise damaged.

Oil Leak At Axle Ends

1. Oil level too high.
2. Oil too light or poor quality.
3. Axle shaft oil seals worn.
4. Axle shaft bearing retainer loose.
5. Cracked rear axle housing.
6. Vent (if any) clogged.

Oil Leak At Pinion Shaft

1. Oil level too high.
2. Oil too light or poor quality.
3. Pinion oil seal worn.
4. Pinion oil seal retainer distorted, loose in housing or improperly installed.
5. Oil return passage in carrier housing restricted.
6. Universal joint companion flange hub rough, scored or out of round.
7. Universal joint companion flange loose on pinion shaft.

Drum Brake Troubles

One Brake Drags

1. Brake line restricted.
2. Improperly adjusted or worn wheel bearing.
3. Distorted or improperly adjusted brake shoe.
4. Faulty retracting spring.
5. Drum out of round.
6. Loose backing plate.
7. Faulty wheel cylinder.
8. Dirty brake fluid.
9. Air in hydraulic system.
10. Insufficient shoe-to-backing plate lubrication.

All Brakes Drag

1. Mechanical resistance at pedal or shoes; damaged linkage.
2. Brake line restricted.
3. Distorted or improperly adjusted brake shoes.
4. Dirty brake fluid.
5. Faulty master cylinder.
6. Sticking booster control valve.

Hard Pedal

1. Mechanical resistance at pedal or shoes; damaged linkage.
2. Brake line restricted.
3. Distorted or improperly adjusted brake shoes.
4. Linings glazed or worn.
5. Oil or grease in lining.

Spongy Pedal

1. Leaks or insufficient fluid.
2. Air in hydraulic system.
3. Linings glazed or worn.
4. Faulty master cylinder.
5. Air in hydraulic system.
6. Self-adjusters not operating.
7. Cracked drum.

Car Pulls to One Side

1. Brake line restricted.
2. Improper tire pressure.
3. Improperly adjusted or worn wheel bearing.
4. Distorted or improperly adjusted brake shoes.
5. Faulty retracting spring.
6. Drum out of round.
7. Linings glazed or worn.
8. Oil or grease in lining.
9. Loose lining.
10. Faulty wheel cylinder.
11. Self-adjusters not operating.
12. Worn or binding front suspension parts.

One Wheel Locks

1. Distorted or improperly adjusted brake shoes.
2. Linings glazed or worn.
3. Oil or grease in lining.
4. Loose backing plate.
5. Faulty wheel cylinder.
6. Tire tread worn.

Brakes Chatter

1. Drum out of round.
2. Linings glazed or worn.
3. Oil or grease in lining.
4. Loose backing plate.
5. Loose lining.
6. Poor lining-to-drum contact.
7. Loose front suspension.

Excessive Pedal Travel

1. Leaks or insufficient fluid.
2. Distorted or improperly adjusted brake shoes.

Pedal Gradually Goes to Floor

1. Leaks or insufficient fluid.
2. Faulty master cylinder.

Brakes Uneven

1. Improper tire pressure.
2. Oil or grease in lining.
3. Scored drum.
4. Dirty brake fluid.

Shoe Click Release

1. Self-adjusters not operating.
2. Insufficient shoe-to-backing plate lubrication.
3. "Threads" left by drum turning tool pull shoes sideways.

Noisy or Grabbing Brakes

1. Distorted or improperly adjusted brake shoes.
2. Linings glazed or worn.
3. Oil or grease in lining.
4. Scored drum.
5. Dirty on drum-lining surface.
6. Faulty wheel cylinder.
7. Sticking booster control valve.

Brakes Do Not Apply

1. Leaks or insufficient fluid.
2. Linings glazed or worn.
3. Oil or grease in lining.
4. Dirty brake fluid.
5. Faulty master cylinder.
6. Air in hydraulic system.

Front End & Steering Troubles

Hard Steering

1. Low or uneven tire pressure.
2. Steering gear or connections adjusted too tight.
3. Insufficient or incorrect lubricant used.
4. Excessive caster.
5. Suspension arms bent or twisted.
6. Front spring sagged.
7. Frame bent or broken.
8. Steering knuckle bent.
9. Kingpin galled or frozen in bushing.
10. Excessive steering shaft coupling misalignment.

Excessive Play or Looseness In Steering

1. Steering gear connections adjusted too loose or worn.
2. Steering knuckle bushings worn.
3. Front wheel bearings incorrectly adjusted or worn.
4. Worn ball joints.
5. Worn or loose worm steering shaft bearings.
6. Worn control arm bushings.

Rattle or Chuckle in Steering Gear

1. Insufficient or improper lubricant in steering gear.
2. Excessive backlash in steering gear.
3. Worn or loose worm steering shaft bearings.
4. Pitman arm loose on shaft.

Erratic Steering On Application of Brakes

1. Oil or brake fluid on lining.
2. Brakes improperly adjusted.
3. Front spring weak.
4. Low or uneven tire pressure.
5. Insufficient or uneven caster.
6. Steering knuckle bent.

Car Pulls to One Side

1. Low or uneven tire pressure.
2. Incorrected or uneven caster or camber.
3. Wheel bearings adjusted too tight.
4. Uneven front car height.
5. Toe-in incorrect.
6. Oil or brake fluid on brake lining.
7. Brakes incorrectly or unevenly adjusted.
8. Steering knuckle or knuckle support bent.
9. Frame bent or broken.
10. Shock absorbers inoperative.
11. Rear wheels not tracking with front wheels.
12. Rear axle shifted (spring U bolts loose or center bolt sheared).
13. Broken or weak rear springs.

Scuffed Tires

1. Tire improperly inflated.
2. Toe-in incorrect.
3. Excessive wheel or tire run-out.
4. Steering knuckle bushings worn.
5. Uneven camber.
6. Incorrect toe-out on turns.
7. Suspension arm bent or twisted.
8. Steering knuckle bent.
9. Excessive speed on turns.

Cupped Tires

1. Improper toe-in.
2. Tires improperly inflated.
3. Wheels, tires or brake drums out of balance.
4. Dragging brakes.
5. Worn steering knuckle bushings.
6. Wheel bearings incorrectly adjusted or worn.
7. Uneven camber.
8. Steering knuckle bent.
9. Excessive mileage without rotating tires.

Front Wheel Shimmy

1. Low or uneven tire pressure.
2. Wheels, tires or brake drums out of balance.
3. Excessive wheel or tire run-out.
4. Shock absorbers inoperative.
5. Steering connections incorrectly adjusted or worn.
6. Steering gear incorrectly adjusted.
7. Front wheel bearings incorrectly adjusted or worn.
8. Incorrect or uneven caster.
9. Steering knuckle bushings worn.
10. Toe-in incorrect.
11. Steering knuckle bent.
12. Eccentric or bulged tires.
13. Stabilizer inoperative.
14. Worn ball joints.
15. Worn control arm bushings.

Front Wheel Tramp

1. Wheels, tires or brake drums out of balance.
2. Wheel or tire not concentric.
3. Shock absorbers inoperative.
4. Stabilizer inoperative.

Car Wanders

1. Low or uneven tire pressure.
2. Steering gear or connections adjusted too loose or worn.
3. Steering gear or connections adjusted too tight.
4. Steering knuckle bushings worn.
5. Improper toe-in.
6. Incorrect or uneven caster or camber.
7. Steering knuckle bent.
8. Kingpin bent.
9. Rear axle shifted (spring U bolts loose or center bolt sheared).
10. Stabilizer inoperative.
11. Kingpins or bushings tight.
12. Bind in lower or upper control arm shaft.
13. Bind in rear spring shackles or dry rear springs.
14. Excessive backlash in steering gear.

Road Shocks

1. High air pressure in tires.
2. Steering gear or connections incorrectly adjusted.
3. Excessive caster.
4. Shock absorbers inoperative.
5. Front springs weak or sagged.
6. Wrong type or size of tires used.
7. Steering knuckle bent.

Power Top, Window & Seat Troubles

Top Will Not Operate

1. Mechanical interference due to luggage or other objects.
2. Hold down strap not removed.
3. Top not free from windshield header studs.
4. Electrical shorts or loose connections in control switch circuit.
5. Dirty control switch contacts.
6. Inoperative power unit motor.
7. Hydraulic fluid low.
8. Power unit pump inoperative.
9. Stoppage in fluid pipes.
10. Faulty hydraulic control valve.
11. Broken port plate in hydraulic pump.

Top Operates in One Direction Only

1. Mechanical interference due to luggage or other objects.
2. Hold down strap not removed.
3. Top not free from windshield header studs.
4. Electrical shorts or loose connections in control switch circuit.
5. Dirty control switch contact.
6. Improperly adjusted control rod.
7. Hydraulic power cylinder faulty.
8. Stoppage in fluid pipes.
9. Faulty hydraulic control valve.

Window Won't Operate from Main Switch Only

1. Broken wire between relay and remote switch.
2. Defective switch in master switch group.
3. Break in wire where it enters door opening.

Window Won't Operate from Main or Door Switch

1. Burned out motor or relay.
2. Defective circuit breaker.
3. Break in battery feed wire from starter solenoid to circuit breaker.

Window Operates In One Direction Only from Main or Door Switch

1. Defective relay.
2. Defective switch.
3. Broken ground wires.
4. Burned out motor.
5. Broken control wire.

Circuit Breaker in Door Clicks On and Off Continuously and Window Won't Operate

1. Control wire grounded.
2. Defective switch.
3. Relay points stuck.

Main Or Door Switch Operates Window In Wrong Direction

1. Lead wires are not connected to proper terminals.

Window Operates Sluggishly

1. Binding window regulator.
2. Broken wires or loose connections.
3. Worn motor brushes.

All Windows Do Not Operate

1. Circuit breaker open in control circuit.
2. Circuit breaker open in power circuit.

Seat Regulators Inoperative

1. Circuit breaker open in control circuit.
2. Circuit breaker open in power circuit.

One Seat Regulator Inoperative

1. Defective wiring between relay and circuit breaker.
2. Defective motor.
3. Defective wiring between switch and circuit breaker.
4. Defective relay.

Seat Regulator Operates in One Direction Only

1. Defective wiring between switch and relay that applies to direction of travel desired.
2. Defective toggle switch.

Seat Regulator Operates Sluggishly

1. Binding mechanism.
2. Defective wiring.
3. Loose connectors or poor ground.
4. Worn or dirty brushes in motor.

Windshield Wiper Troubles

GENERAL INSPECTION

Before deciding that a windshield wiper needs servicing it might be well to consider some of the external factors which affect their operation.

It must be remembered that windshield wipers will operate more slowly when they do their work on dry glass. This is specially true on cars with curved windshields. You will also find that wiper blades may chatter or fail to travel a complete arc on dry glass. It is therefore obvious that any testing of windshield wiper operation should be done after the windshield has been sprayed with water.

Windshield wipers that chatter or do not wipe the glass clean under normal operating conditions (wet windshield) may need only replacement of the wiper arms or blades instead of more extensive service. This can be determined by visual inspection and most replacements can be made simply without the aid of any special tools.

Uneven movement of the wiper arms with respect to one another is usually caused by cables, pivots or cranks that are out of adjustment in the windshield wiper transmission system.

ELECTRIC TYPE

All passenger car electric windshield wiper circuits, regardless of manufacturer, include a control switch, a small shunt wound motor, and the wiring connecting these units to the battery. A circuit breaker or fuse may be mounted as a separate unit or incorporated in the control switch itself. A worm gear on the motor armature shaft drives one or two gears mounted on crankshafts for wiper operation.

A parking switch is mounted on the motor and actuated by a cam on one of the cranks. The parking switch, connected to the battery through a control switch, keeps the motor in operation for a brief period after the control switch has been shut off, allowing the wiper blades to return to the parked position.

Both single and two speed motors are used, the latter incorporating one or several resistors in the field circuit. The resistors may be located either in the parking switch housing or in the control switch.

In the following text you will find a list of the conditions you are likely to encounter when faced with a repair job on electric wipers. By consulting these possibilities you will simplify the job of locating the source of trouble. But before going further a few words of caution are in order: After you have made your diagnosis and are ready to make repairs, disconnect the battery to avoid damage under the dash or possible personal injury from accidental shorts. Also, on models which use off-glass parking windshield wipers, never remove or disassemble the motor while in "park" position.

Wipers Won't Operate

1. Discharged battery.
2. Blown fuse or faulty circuit breaker.
3. No power to control switch.
4. Faulty control switch.
5. Faulty parking switch.
6. Binding pivots, cranks or linkages.
7. Poor connection at switch.

8. No ground at motor.
9. Faulty motor.

Wipers Won't Park

1. Incorrect adjustment of parking switch lever.
2. Open circuit in lead feeding parking switch.
3. Faulty parking switch.
4. Faulty control switch.
5. No ground at control switch (variable speed wipers).
6. Motor crank and parking switch improperly assembled.
7. Cams in linkage reversed or binding (variable speed wipers).

Wipers Operate Slowly

1. Discharged battery.
2. Binding pivots, cranks or linkages.
3. Faulty motor windings.
4. High resistance connections or wiring.
5. High resistance in control switch contacts.
6. No ground at control switch (variable speed wipers).
7. Faulty resistance unit (if only high speed in affected).
8. Dirty commutator or sticking brushes.
9. Worn or damaged motor.

Multiple Speed Wipers Operate Only at Single Speed

1. Short or open in motor wiring harness.
2. Incorrect connections at control switch.
3. Faulty control switch.
4. Faulty resistance unit.
5. No ground at control switch.
6. Open shunt field in motor.

VACUUM TYPE

For satisfactory windshield wiper operation, it is necessary to have an adequate supply of vacuum. On some cars the vacuum is made available by tapping directly into the intake manifold. With this type of arrangement it is considered normal for the wipers to slow down or stop entirely while going up a hill or during acceleration, since under those conditions the manifold vacuum would drop below the 8″-10″ needed to operate the wipers. These conditions are almost completely eliminated on cars equipped with a vacuum booster pump. The purpose of this pump is to maintain enough vacuum to work the wipers under any driving condition.

Some of the conditions which prevent satisfactory windshield wiper operation are listed in the following text and may be used as a guide to help you locate the source of trouble. Always disconnect the battery when working under the dash.

Wipers Won't Operate

1. No vacuum supply to motor due to pinch, restriction or leak in the windshield wiper hose. A vacuum leak or a disconnected hose can easily be located because a hissing sound will be heard whenever the engine is running.
2. Faulty vacuum booster pump.
3. Wiper control switch inoperative or disconnected at motor.
4. Faulty wiper motor.
5. Frozen or binding pivots and linkages.
6. Linkages or cables improperly installed.

Wipers Operate Slowly

1. Low vacuum due to pinch or partial restriction in the wiper hose.
2. Loss of vacuum due to leaks at joints, fittings or in the wiper hose itself.
3. Faulty vacuum booster pump.
4. Faulty wiper motor.
5. Wiper control switch does not move operating valve on the motor to full "ON" position due to improper adjustment.
6. Air intake on motor (breather port) clogged.
7. Binding pivots, cranks, linkages or binding or frozen idler pulleys on cable tensioners.
8. Cables adjusted too tight.

Wipers Won't Park

1. Faulty parking valve on motor.
2. Wiper control switch out of adjustment.
3. Wiper arms not positioned properly on pivots.

PRESSURE WIPER

The windshield wiper is hydraulically operated. The hydraulic power for the motor is obtained from the power steering unit. Hydraulic fluid flows from the pump, through the steering gear to the wiper motor, and then to the fluid reservoir. During wiper operation, a part of the fluid is by-passed through the motor by a valve on the motor.

Checks and Adjustments: The only adjustment required is the control cable adjustment. To adjust, remove the seal plate mounting screws and position the plate and seal out of the way. Adjust the cable so that the control knob on instrument panel moves the valve control lever on motor from off to full on.

If the motor operates sluggishly, check the cable adjustment. If this is not the fault, check the hydraulic fluid pressure. If the power steering gear operates satisfactorily, it may be assumed that the fluid pressure is adequate. Check for binding wiper pivot shafts and arms. Repair or replace wiper motor and valves if necessary.

AIR CONDITIONING

NOTE: This chapter deals only with fundamentals and basic system tests. For complete air conditioning service details, Motor's Air Conditioner Service Manual is available. It includes unit replacement, servicing, diagnosis and testing, vacuum and wiring diagrams, aftermarket hang-on units, & truck/trailer refrigeration and all 1963-75 specifications.

CONTENTS

A/C System Testing

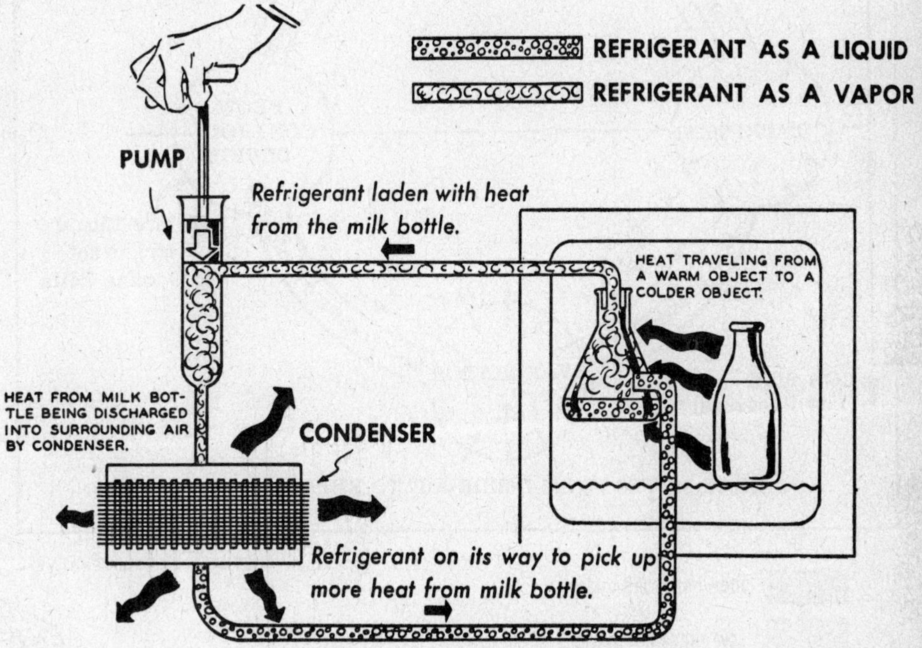

Diagram showing refrigerant cycle

FUNDAMENTALS OF REFRIGERATION

In order to understand how an air conditioning system works, we must have a knowledge of the fundamentals of refrigeration. And since refrigeration is the process of removing heat from things, an air conditioner removes heat from the surrounding air. In both cases, the process is simplified by one of Nature's laws—heat always moves from a warm object to a cold object.

When we put milk or vegetables into an ice box, they are warmer than the ice. Since heat always travels from warmer to cooler objects, the heat in the milk and vegetables naturally travels to the cold ice. Then, of course, as heat is removed from these articles, they begin to grow cooler—they have less heat than before.

Therefore, if refrigeration is the removal of heat, then anything can be made cooler by finding a method of absorbing heat from it.

In order to maintain cold temperatures, we have to have continuous refrigeration. This is accomplished in modern refrigerators and cooling systems (1) by using a refrigerant that will readily absorb heat and (2) by using the same refrigerant over and over.

Ice can refrigerate effectively *only when it is changing from a solid to a liquid.* A liquid can refrigerate *only when it is changing from a liquid to a vapor.* A vapor cannot refrigerate since the absorption of heat will not change it to any other state. The only thing to do with a vapor is to change it back to a liquid—by removing heat from it.

In other words, if the ice didn't melt, it could absorb a little heat from the objects and the air around it but not enough to do an effective job of cooling. But when it begins to change to a liquid (water), then the ice is absorbing heat rapidly and effectively.

The same applies to a liquid refrigerant. No matter how cold the liquid is, it won't actually "refrigerate" until it is absorbing heat fast enough to change its form into a vapor.

Operating Cycle of Car Air Conditioning System

1. The compressor compresses heat-laden, low-pressure vaporous refrigerant and discharges it into the condenser.
2. In the condenser, the vapor changes into a liquid as the heat is dissipated into the surrounding air.
3. From the condenser, the liquid is forced into the reservoir or receiver, which is simply a storage container the function of which is to ensure a supply of *liquid* Freon to the expansion valve. (The expansion valve functions in much the same manner as a hose nozzle).
4. From the reservoir, the liquid Freon passes through a combination strainer-drier, then through the expansion valve into the evaporator.
5. Being connected to the suction line of the compressor, the evaporator is a low-pressure region in which the refrigerant boils and reverts to a vapor, absorbing heat in the process and thereby cooling the evaporator coil and the air passing over it.
6. Upon leaving the evaporator, the vaporous refrigerant returns to the compressor inlet and completes the refrigeration cycle.

EXERCISE SYSTEM

An important fact most car owners ig-

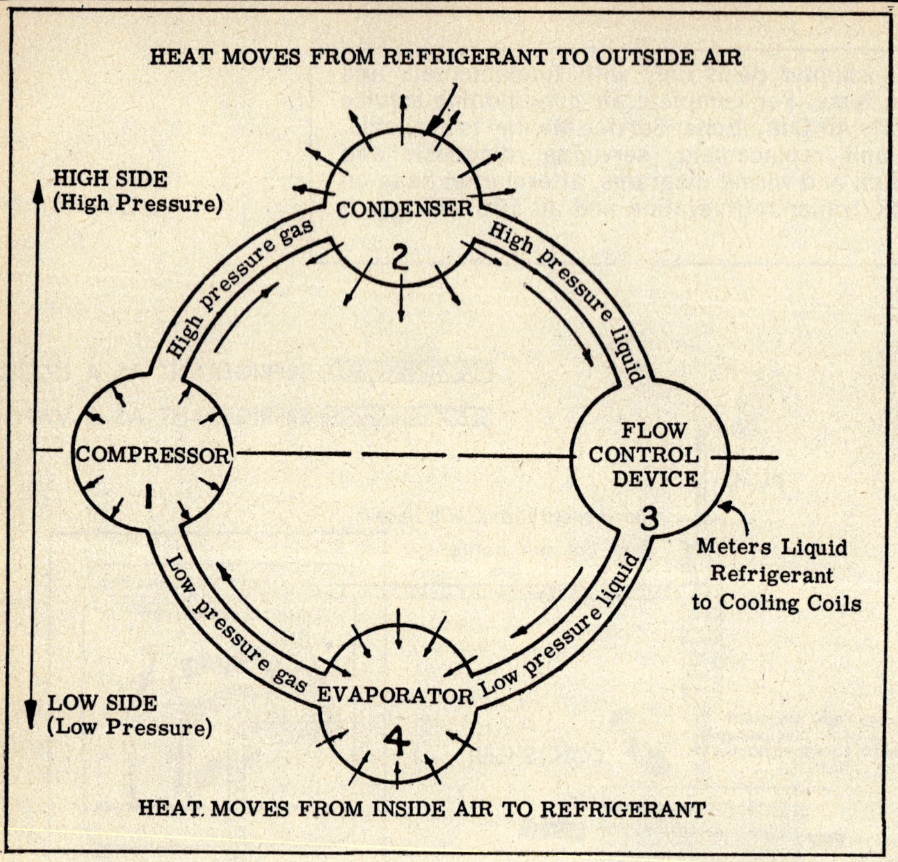

◀ Air conditioning cycle of operation

▼ How air conditioner works

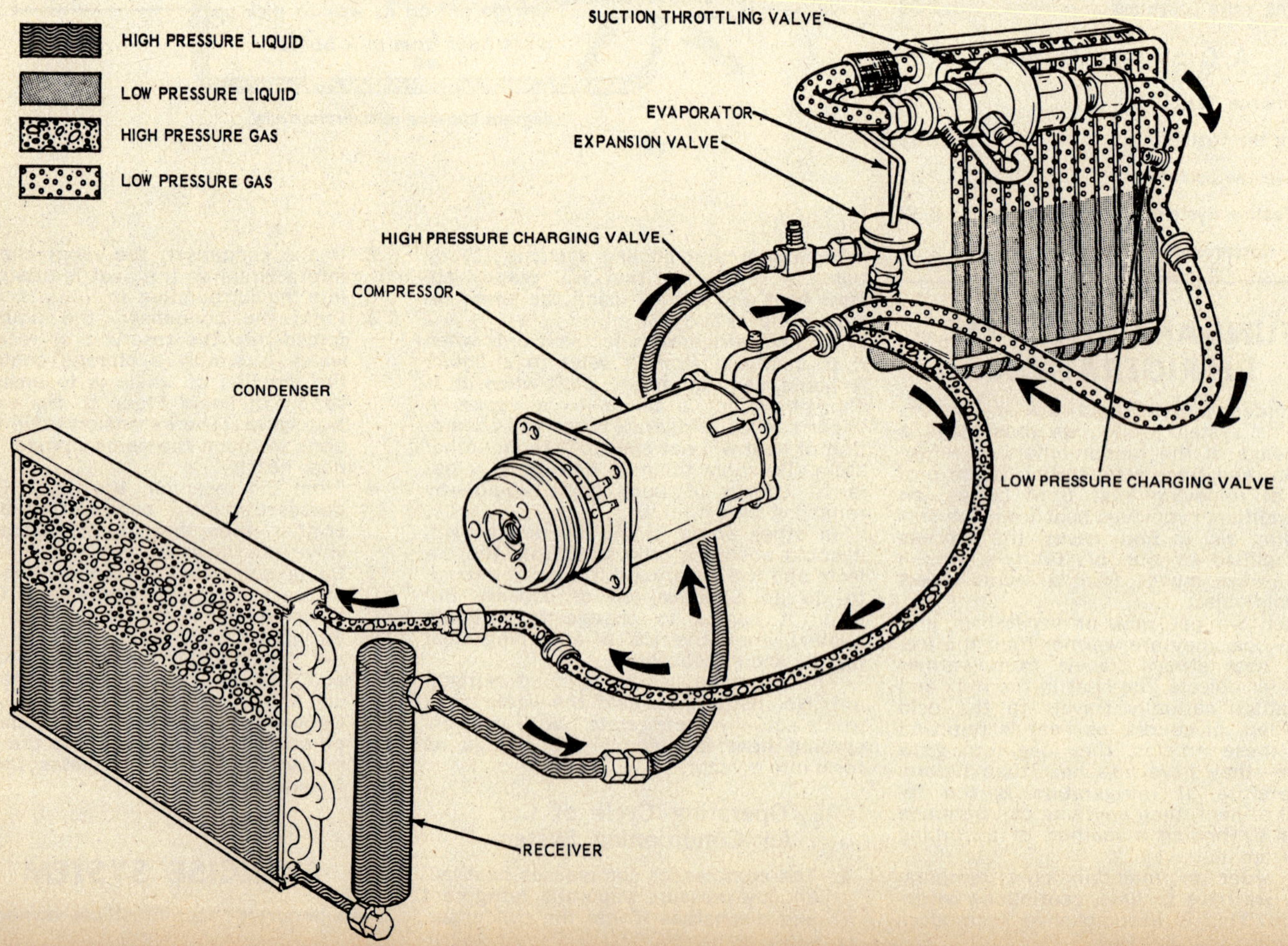

nore is that car coolers must be used periodically or they will go sour. Car manufacturers are emphatic on this point. They caution that, when the air conditioner is not used regularly, particularly during cold months, it should be turned on for a few minutes once every two or three weeks while the engine is running. This keeps the system in good operating condition.

Checking out the system for the effects of disuse before the onset of summer is one of the most important aspects of car cooler servicing.

First on the list is to clean out the condenser core, mounted in all cases at the front of the car's radiator. All obstructions, such as leaves, bugs, dirt, and the like, must be removed, as these will reduce heat transfer and impair the efficiency of the system. Make sure the space between the condenser and the radiator also is free of foreign matter.

Make certain the evaporator water drain is open. Certain systems have two evaporators, one in the engine compartment and one in the trunk. The evaporator is the device which cools and dehumidifies the air before it enters the car, and the place where refrigerant is changed from a liquid to a vapor. As the core cools the air, moisture condenses on it but the moisture is prevented from collecting in the evaporator by means of the water drain.

TROUBLE DIAGNOSIS

Insufficient or No Cooling

The first unit to check is the blower mechanism and its air-distribution system, since the cooling is produced by a combination of a cool evaporator and a blower to circulate the cooled air. If the blower is inoperative, check for a broken switch, a blown fuse, a loose connection, a broken wire or a defective motor activating the blower mechanism.

In the event the blower is working but the air output is low, look for loose wire connections or shorts, low battery charge, dirty or loose switch contacts or binding of the blower shaft or blades. Also check the air-distribution system for closed air-output valves or obstructions in the flexible hoses.

If the blower is circulating the air efficiently but there is no cooling, the trouble lies in the refrigerating circuit. Beginning at the source of power, check the compressor driving belt. Is it loose, slipping or broken? If the belt is tight and the pulley is turning but the compressor shaft is not, check the magnetic clutch for lack of current or a faulty coil.

A steady, resonant noise from the compressor is normal but any clicking or rattling sounds should be investigated. Using a standard set of twin gauges, check the compressor inlet and outlet pressures according to manufacturer's specifications. If the normally high outlet pressure is not up to specs, it is probably due to a low charge of coolant, excess moisture in the system, or a restriction or kink in the circulation lines.

Be sure that there is sufficient compressor oil of the proper viscosity. At the same time, make certain that there is no excess oil from the compressor flowing through the system, slowing coolant circulation and reducing compressor outlet pressure.

If the normally but inlet-suction pressure is high but the outlet pressure is normal, there is probably trouble either at the expansion valve, which controls the amount of fluid entering the evaporator, or with the sensor bulb that controls the expansion valve. Other causes would include excess moisture in the system or a defective suction throttle valve (if used) on the compressor.

Excessive vibration of the compressor indicates loose mounting bolts or an out-of-true belt pulley. A thumping noise from the compressor or a cool and sweating suction line into it, is usually caused by too much coolant in the system. Should there be no moisture in the system, evacuate the excess coolant into the atmosphere.

Check Condenser

If the condenser unit, located in front of the radiator, is suspected, the first thing to look for is clogged coils that limit the cooling capacity of the unit. Any bugs, leaves or debris caught in the coils should be removed by air pressure. In the event there is any icy or frosty spots apparent on the exterior of the condenser, they usually indicate a partial restriction inside the condenser at that point.

Check Sight Glass

The next spot to check in the refrigerating circuit is the line sight glass (if used). This small transparent device permits a visual check of the condition of the high-pressure liquid coolant. A cloudy, foamy appearance, or the presence of bubbles in the fluid, indicates air or excess moisture in the lines, or a low charge of coolant.

Check Receiver-Dehydrator

Sometimes fault will be found in the receiver-dehydrator unit, usually located alongside the condenser. This small storage unit for the coolant also contains an element for removing small amounts of moisture from the refrigerant and has a filtering element for catching small dirt particles.

If the system has been in operation for a considerable period of time, this dehydrating element may lose its moisture-absorbing ability. This condition causes the constant presence of small bubbles in the sight glass. It is also indicated when the high-pressure line into the dehydrator has a distinctly different temperature from that of the outlet line. This receiver-dryer component should be replaced as a unit if the above symptoms are observed, if all other parts of the system are in good operating condition, and the system is fully charged.

Check Evaporator

The proper functioning of the evaporator is always dependent on an efficient expansion valve at the intake end and an efficient sensor bulb and (if used) a suction throttle valve at the outlet end of the coils. In normal operation, this unit should be cold and sweating, but not iced. Icing calls for a check of the expansion valve, sensor bulb, or the suction throttle valve (if used).

Since there is a constant condensation of atmospheric moisture on the outside of the evaporator coils, make sure that the draining system is free and clean and able to dispose of the collected water through the car floor.

NOTE: Some of the larger, more expensive cars have a supplementary evaporator unit in the luggage compartment. Don't forget to check the operation of this unit.

Testing System for Leaks

Testing the system for leaks is one of the most important over-all phases of trouble-shooting an air conditioning system. There are several methods of performing this operation which is explained elsewhere in this chapter.

If there is a loss of refrigerant and no external leak can be found, gas might be leaking from a damaged piston diaphragm in the suction throttle valve (if used). In this case, vapor would be drawn into the engine intake manifold and be burned, undetected in the cylinders.

Refrigerant Dangerous

A final word of caution. The freon refrigerant used in car air-conditioners is also known as R-12 or F-12. It is colorless and odorless both as a gas and a liquid. Since it boils (vaporizes) at 21.7 def. F., it will usually be in a vapor state when being handled in a repair shop. But if a portion of the liquid coolant should come in contact with the hands or face, remember that its temperature momentarily will be at least 22 deg. below zero!

Protective goggles should be worn when opening any refrigerant lines. If liquid coolant does touch the eyes, bathe the eyes quickly in cold water. Then apply a bland disinfectant oil to the eyes. See an eye doctor.

TROUBLE SHOOTING GUIDE

NOTE: When a unit must be removed from the system for replacement or repairs, the dehydrator must be replaced also, and the system must be purged, evacuated and recharged to remove excess moisture.

System Produces No Cooling

Electrical

1. Blown fuse.
2. Disconnect or broken wire.
3. Disconnect or broken ground wire.
4. Clutch coil disconnected or burned out.
5. Switch contacts in thermostat (if used) burned excessively, or sensing element defective.
6. Blower motor burned out or disconnected.

Mechanical

1. Loose or broken drive belt.
2. Compressor completely or partially frozen.
3. Compressor reed valves inoperative.
4. Expansion valve stuck open.

Refrigeration

1. Broken refrigeration line.
2. Fusable plug blown (if used).
3. Leak in system.
4. Compressor shaft seal leaking.
5. Clogged screen or screens in receiver-

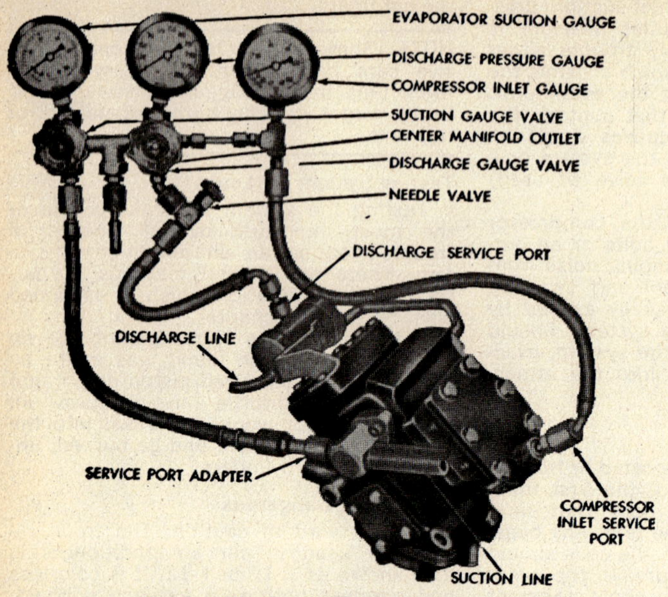

Fig. 1 Chrysler Air-Temp gauge set manifold connections (Typical)

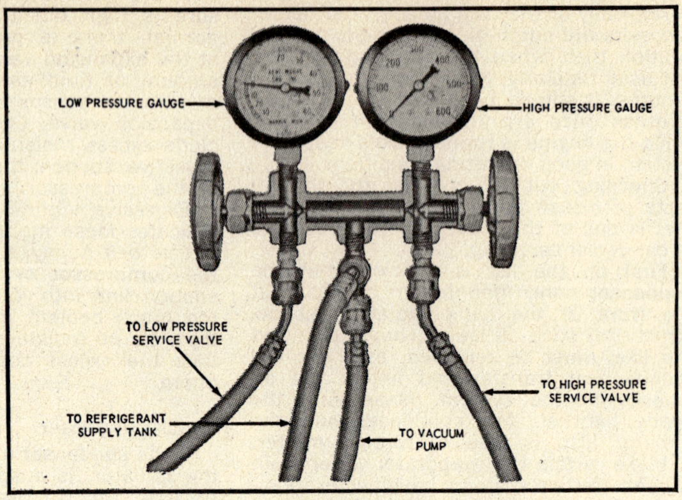

Fig. 2 Manifold gauge set for Frigidaire, Tecumseh and York compressors

dehydrator or expansion valve.
6. Plugged coil or hose.

Insufficient Cooling

Electrical

1. Blower motor operates sluggishly.

Mechanical

1. Compressor clutch slipping.
2. Obstructed blower discharge passage.
3. Clogged air intake filter.
4. Outside air vents open.
5. Insufficient air circulation over condenser coils; fins clogged with dirt, leaves or insects.
6. Evaporator clogged.
7. Evaporator regulator defective or improperly adjusted.

NOTE: The evaporator regulator includes the suction throttling valve (STV) used on Ford cars, the Pilot-Operated Absolute (POA) valve used on General Motors cars, and the Chrysler Evaporator Pressure Regulator (EPR) or Evaporator Temperature Regulator (ETR) valve. When any of these devices is used in a system, it can be assumed that the system contains no thermostat control.

Refrigeration

1. Insufficient refrigerant in system.
2. Clogged screen in expansion valve.
3. Expansion valve thermal bulb has lost its charge.
4. Clogged screen in receiver.
5. Excessive moisture in system.
6. Air in system.
7. Thermostat defective or improperly adjusted (if used).

System Cools Intermittently

Electrical

1. Defective circuit breaker, blower switch or blower motor.
2. Partial open, improper ground or loose connection in compressor clutch coil.

Mechanical

1. Compressor clutch slipping.

Refrigeration

1. Unit icing up; may be caused by excessive moisture in system, incorrect super-heat adjustment in expansion valve, or thermostat adjusted too low (if used).
2. Thermostat defective (if used).
3. Stuck hot gas by-pass valve, STV, POA or EPR valve.

OIL LEVEL CHECK

Air-Temp, Tecumseh & York Compressors

NOTE: The oil level of these compressors should be checked whenever refrigerant has been lost due to leakage or through normal system servicing.

1. Connect gauges into system, Figs. 1 and 2. Operate system for approximately 10 to 15 minutes with controls set for maximum cooling.

NOTE: On 1974 Fury, 1974-75 Chrysler, Imperial and Monaco, and 1975 Gran Fury models, disconnect and plug the water valve vacuum hose.

2. On all 1969-72 units and 1973 Tecumseh and York systems with manual service valves, isolate the compressor from system. On 1973 Tecumseh and York systems with Schrader type service valves, 1973 Air Temp units and all 1974 units, discharge the refrigerant system.
3. With proper eye protection, remove oil check plug slowly to bleed a small refrigerant charge which may be left in the compressor. Then check oil level, which should be within the limits found in the A/C Specification Table.

Frigidaire Compressors

NOTE: These compressors have no provision accurately to check oil level on the car. Periodic inspection of oil level has, therefore, been eliminated from installations using this compressor. Oil level check performed with the compressor removed from the car is to be made only under conditions of severe oil loss caused by a compressor oil leak, broken refrigerant hose, or a rupture from crash damage. The system is designed to hold the quantity of refrigeration oil listed in the *A/C Data Table*. The affinity of Freon 12 and refrigeration oil and the design of the compressor will prevent a full oil charge being contained in the compressor.

1. Purge refrigerant from compressor.
2. Remove compressor from car.
3. Remove oil from compressor.
4. Remove plug on pump.
5. Pour oil from compressor into a clean container calibrated in ounces.
6. Measure oil removed from compressor.
7. Install new oil into compressor. Install plug and tighten.
8. Install compressor on car. Repair damage to system as required to eliminate leaks. Evacuate system, using a suitable vacuum pump. Then charge system with new Freon 12.

Oil Distribution

Whenever a component of the A/C system is replaced, measured quantities of refrigeration oil should be added to the component to assure that the total oil charge in the system is correct before the unit is placed into operation.

The oil is poured directly into the replacement component. If an evaporator is installed, pour oil into the inlet pipe with the pipe held vertically so the oil will drain into the evaporator core. No additional oil is required if valves and hoses are replaced.

1. If the oil drained from the compressor is more than 4 fluid ounces, add to the new compressor the same amount of oil as drained from the replaced unit.

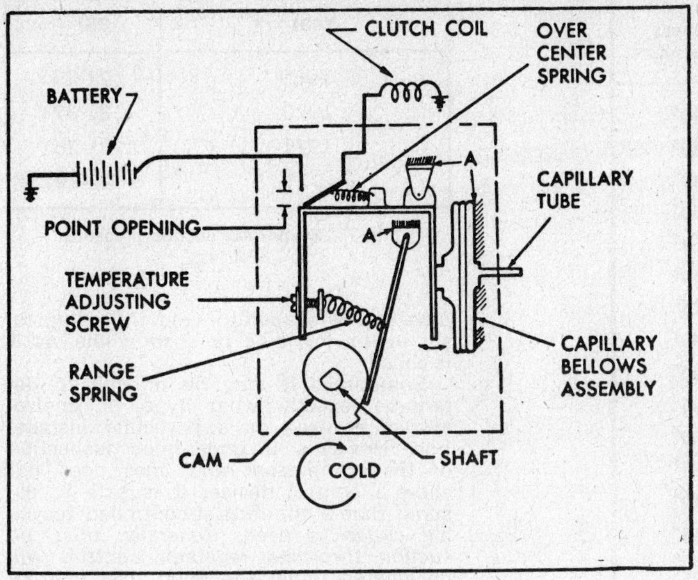

Fig. 3 Wiring diagram of a typical adjustable thermostatic switch

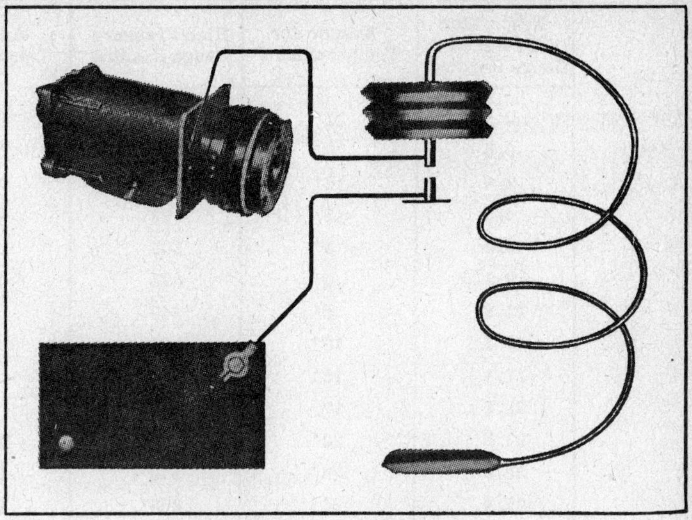

Fig. 4 Schematic layout of a thermostatic-controlled clutch circuit

2. If oil drained from the compressor is less than 4 fluid ounces, add 6 ounces of oil to the new compressor.
3. Oil should be added to replacement components as follows:
   Evaporator (front or rear) . . add 3 oz.
   Condenser . . . . add 1 oz.
   Receiver or VIR . . . . add 1 oz.
   Condenser-receiver assembly add 2 oz.

NOTE: If the system is flushed with sufficient quantity of a flushing agent that would remove oil from the system, install the full amount specified in the *A/C Data Table*. On a newly installed system, install the full capacity of oil prior to operation. On systems containing metal particles in the oil, replace or overhaul the compressor, replace the receiver-dehydrator and install a high capacity, low pressure drop filter in the liquid line to the filter to protect the expansion valve and new compressor from damage due to foreign particles.

THERMOSTAT CONTROL SWITCH

When an A/C system is not provided with a hot gas by-pass valve or a suction throttling valve (STV, POA or EPR), the pumping action of the compressor is stopped by a thermostatically-controlled switch, Figs. 3 and 4.

The opening and closing of the electrical contacts are controlled by a movement of a temperature sensitive diaphragm or bellows. The bellows has a capillary tube connected to it which is filled with Freon or CO_2. The capillary tube is positioned so that it may have either the cold air from the evaporator pass over it or it may be connected to the tail pipe of the evaporator. In either position, evaporator temperature will affect the temperature-sensitive compound in the capillary tube by causing it to contract as the evaporator becomes colder. The contraction of the gas will cause the bellows to contract. This action separates the electrical contacts and breaks the electrical circuit to the compressor clutch, which stops compressor operation.

The evaporator begins to warm which, in turn, causes the gas in the capillary tube to expand. The bellows will also begin to expand, moving the electrical contacts closer to each other. At a predetermined point, the expansion of the bellows will bring the contacts together, closing the circuit to the compressor clutch, energizing it and bringing the compressor into operation again. This cycling action will be repeated as long as air conditioning operation is required.

Thermostat, Adjust

Visual inspection must be made to determine if a thermostat is adjustable. All thermostats have a provision for regulating the range between opening and closing the contacts. Some models may have a removable fiber cover under which will be located the adjusting screw. If a set screw is not found here, it may be assumed that the thermostat is not adjustable.

All types, whether or not they have an internal adjustment, have a provision whereby the spring tension and contact spacing may be varied by the driver to regulate evaporator cooling. Temperature control is accomplished by rotating a cam which will increase or decrease spring tension on a pivoting contact hanger. Rotation is accomplished either by movement of a stem on which a knob is fastened or it may be accomplished by a cable arrangement. On thermostats that are adjustable, the procedure is as follows:

1. Connect gauges into the system and adjust A/C controls for maximum cooling.
2. Operate engine for 10 to 15 minutes to stabilize at 1500-1750 rpm.
3. Read high side gauge for full refrigerant charge, which will read high side pressure approximating the temperature-pressure relationship shown in Fig. 5.
4. Check sight glass for absence of bubbles.
5. Read low side gauge for thermostat operation. It should read from 26 psi to 14 psi after system has been stabilized for 10 to 15 minutes.

NOTE: The thermostat should disconnect the compressor clutch for evaporator defrost between high and low readings. If the thermostat will not recycle the clutch, move the temperature control toward the warmer position to check for thermostat contact opening.

6. Count number of psi required for warm-up until contacts close, which should indicate 26 to 32 psi rise between contact opening and reclosing. Check the thermostat at least 3 times for consistent operation.

NOTE: Thermostats are generally but not always located in the evaporator case. A certain amount of diligence will be required to locate the thermostat, especially with some models of luggage compartment units.

7. Remove parts as required to gain access to the thermostat. Open access door to the adjusting screw. Rotate adjusting screw counter-clockwise to lower contact opening, and clockwise to raise point opening pressure.

NOTE: Localities having high mean humidity will require higher contact opening pressure than localities with low mean humidity. Coastal areas with contact opening adjustment lower than 24-26 psi will result in evaporator freeze-up. So-called desert areas

Evaporator Pressure Gauge Reading	Evaporator Temperature F°	High Pressure Gauge Reading	Ambient Temperature
0	-21°	45	20°
0.6	-20°	55	30°
2.4	-15°	72	40°
4.5	-10°	86	50°
6.8	- 5°	105	60°
9.2	0°	126	70°
11.8	5°	140	75°
14.7	10°	160	80°
17.1	15°	185	90°
21.1	20°	195	95°
22.5	22°	220	100°
23.9	24°	240	105°
25.4	26°	260	110°
26.9	28°	275	115°
28.5	30°	290	120°
37.0	40°	305	125°
46.7	50°	325	130°
57.7	60°		
70.1	70°		
84.1	80°		
99.6	90°		
116.9	100°		
136.0	110°		
157.1	120°		
179.0	130°		

Fig. 5 Pressure-temperature relationship. Conditions equivalent to 30 mph or 1750 engine rpm

with very low humidity can easily tolerate contact opening adjustment of 14-16 psi without evaporator freeze-up.

8. Check operation of thermostat for newly adjusted cycle of operation. Replace thermostat if operating cycle is inconsistent or will not respond to adjustment.

CHRYSLER ETR CONTROL

The ETR System (Evaporator Temperature Regulator), used on 1969-73 Chrysler Corp. Auto Temp A/C systems, consists of an ETR solenoid valve and an ETR switch. The ETR system permits the evaporator to maintain a minimum air temperature without evaporator coil icing at low ambient temperatures. This control system replaces the EPR valve used in standard Chrysler Corp. A/C systems.

The ETR solenoid valve is actuated by the ETR switch which senses evaporator coil temperature through a sensing capillary placed on the evaporator coil fins. The ETR solenoid valve closes when energized, stopping the refrigerant flow to the compressor.

System Testing

1. Connect gauges into system, Fig. 1, with suction and discharge valves closed, then connect a test lamp in series with the ETR switch feed wire at the compressor.
2. Set thumbwheel temperature selector to 65, close windows and run engine at 1000 RPM, then depress Auto button.
3. After approximately five minutes, compare gauge readings and test lamp as indicated in Fig. 6. Discharge pressure may vary, however, a 25 to 30 pound pressure differential indicates actuation of the ETR switch.

Year	Test Lamp	Evap. Suction P.S.I.	Discharge Press. P.S.I.	Compressor Inlet P.S.I.
1969-70	Off	22	180	30
	On	30	150	0 or Below
1971-73	Off	25	180	24
	On	30	150	0 or Below

Fig. 6 Chrysler ETR Control test pressure chart

SUCTION THROTTLING REGULATORS

Suction throttling regulated systems include any type control used to regulate the flow of refrigerant from the evaporator to the compressor. This control will be located at some point between the tail pipe of the evaporator and the compressor in the low side or suction line, as it is called.

Year	PSI
1969	21-25
1970	21-27
1971-72	21-25
1973-75	22-29

Fig. 7 Evaporator suction pressure EPR test

Sometimes it may be difficult to determine exactly what type of control system is used on a particular installation. However, if under-hood inspection of the compressor and lines does not show a control device, it is safe to assume that a thermostat-controlled recycling clutch is used. *Generally, then, all suction throttling regulator controls will be located under the hood. The thermostat will not be used, and the clutch will be energized during total operation of this type system.*

Constant heat load, ram air, and compressor speed will allow the expansion valve to meter an even flow of refrigerant into the evaporator. Change any of these conditions, however, and the refrigerant flow through the expansion valve will change. Without some means provided to regulate refrigerant flow, the evaporator cooling will become excessive and freeze the moisture condensing on the coils. This results in evaporator icing or freeze-up and requires that the ice on the evaporator be melted before it can resume cooling action.

Installation of a suction throttling control device will compensate for these varying conditions. As the pressure and temperature within the evaporator drop, spring pressure forces a valve within the regulating device toward its seat, retarding the flow of refrigerant. This action increases the pressure back into the evaporator, raising both pressure and temperature. The temperature will not be allowed to drop low enough to freeze over the evaporator coils. A point of balance will soon be reached whereby the refrigerant flow past the valve in the regulator will reach a degree of pressure and temperature sufficient to maintain efficient cooling without evaporator freezing.

Chrysler Air-Temp EPR & EPR-2 Valve

The EPR valve (Evaporator Pressure Regulator) is calibrated to produce maximum cooling without causing frost or ice on the evaporator fins and tubing. If for

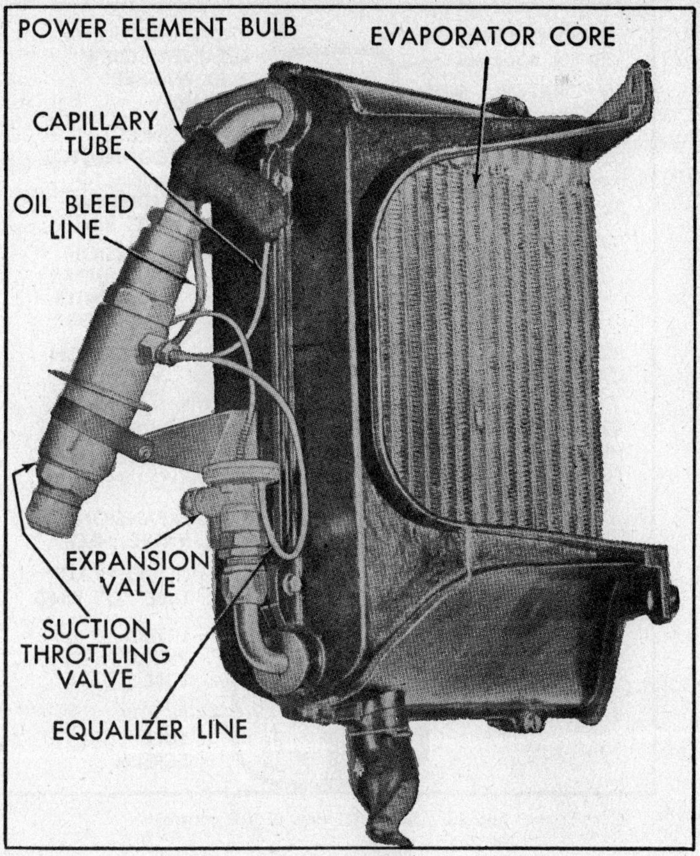

Fig. 8 Typical installation of GM POA valve and associated parts

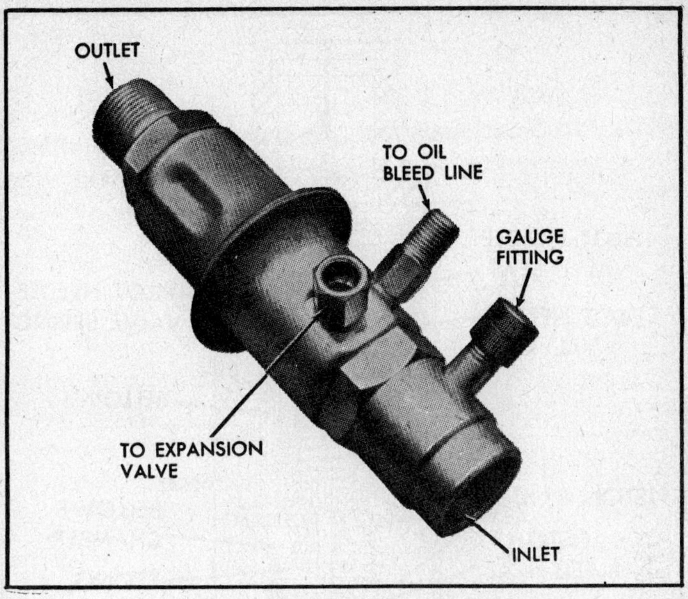

Fig. 9 Exterior view of POA valve

any reason the factory calibration has been disturbed, the EPR valve may restrict the flow of refrigerant at an evaporator pressure which is either too high for maximum performance or too low to prevent coil freeze-up. The following test determines whether or not the valve is functioning properly.

All 1974-75 units incorporate an EPR-2 valve which performs the same functions as the EPR valve described above. Servicing and testing procedures remain the same for the EPR-2 valve except unless noted otherwise.

EPR Valve Test

1970-75 Less H-Valve

1. Close windows and set A/C controls for "Max A/C", high blower and temperature control to full reheat position.
2. Run V8 engines at 800 RPM and 6 cylinder engines at 900 RPM.
3. Operate for a minimum of five minutes to obtain partial stabilization and sufficient reheat to load system.
4. Pressure at the discharge port should be 140 to 210 PSI at this time. If not, check pressure drop across EPR valve which should be 1 to 4 PSI on 1969-72 units or 1 to 6 PSI on 1973-75 units. If pressure drop is greater than the above limits, the EPR valve and nylon filter suction screen should be replaced.
5. If the EPR valve was found satisfactory, remove the expansion valve thermo bulb from suction line and immerse bulb into a container filled with ice and water. The compressor inlet pressure should now be 17 PSI or less on 1969-73 units and 15 PSI or less on 1974-75 units with the evaporator suction pressure as specified in Fig. 7. If the compressor inlet pressure is more than specified, the EPR valve and nylon filter suction screen is faulty and must be replaced.

1974-75 With H-Valve

NOTE: This test procedure cannot be used on these vehicles if equipped with the Auto Temp II A/C system since a special tester must be used to trouble-shoot this system.

1. Close windows and set A/C controls for "Max A/C", high blower and the temperature control to full reheat position.
2. Disconnect and plug water valve vacuum hose.
3. Run engine at 800 RPM and operate system for a minimum of five minutes to obtain partial stabilization and sufficient reheat to load system.
4. Pressure at the discharge port should now be 140-240 PSI. If not, check system for proper refrigerant charge. If system charge is satisfactory, check pressure drop across EPR-2 valve which should be 1 to 6 PSI. If pressure drop is greater than specified, replace the EPR-2 valve.
5. If system charge and EPR-2 valve were found satisfactory, place crushed dry ice on H-valve control head for a minimum of 30 seconds and observe evaporator suction pressure which should drop below 15 inches Hg. Remove dry ice and evaporator suction pressure should now rise to a minimum of 38 PSI and finally stabilize to 30 PSI. If system does not respond as described above, replace the H-Valve.
6. The EPR-2 valve should now be fully open with a 1 to 6 PSI pressure drop across the valve. If pressure drop is greater than specified, the EPR-2 valve is faulty or a restriction is present in the system.
7. If the EPR-2 valve was found satisfactory in Step 6, reconnect the water valve vacuum hose, open windows and operate system for 15 minutes. Then, set blower at low speed and run engine at 1100 RPM for five minutes to obtain stabilization. The compressor inlet pressure should be 15 PSI or less with the evaporator pressure at 22 to 29 PSI. If compressor inlet pressure is greater than specified, replace the EPR-2 valve.

Frigidaire POA Valve

The POA valve (Pilot Operated Absolute), Figs. 8, 9, 10, differs from the STV in that it has no vacuum diaphragm. The POA valve will not change calibration when the system is operated at a higher altitude due to the atmospheric pressure to which the STV was subjected because of its vacuum diaphragm.

NOTE: When it is found necessary to replace a POA valve, check the interior of the defective valve for corrosion or crystalization of salts. This should indicate moisture in the system. If this condition exists, the receiver-dehydrator should be replaced and the system evacuated for one hour.

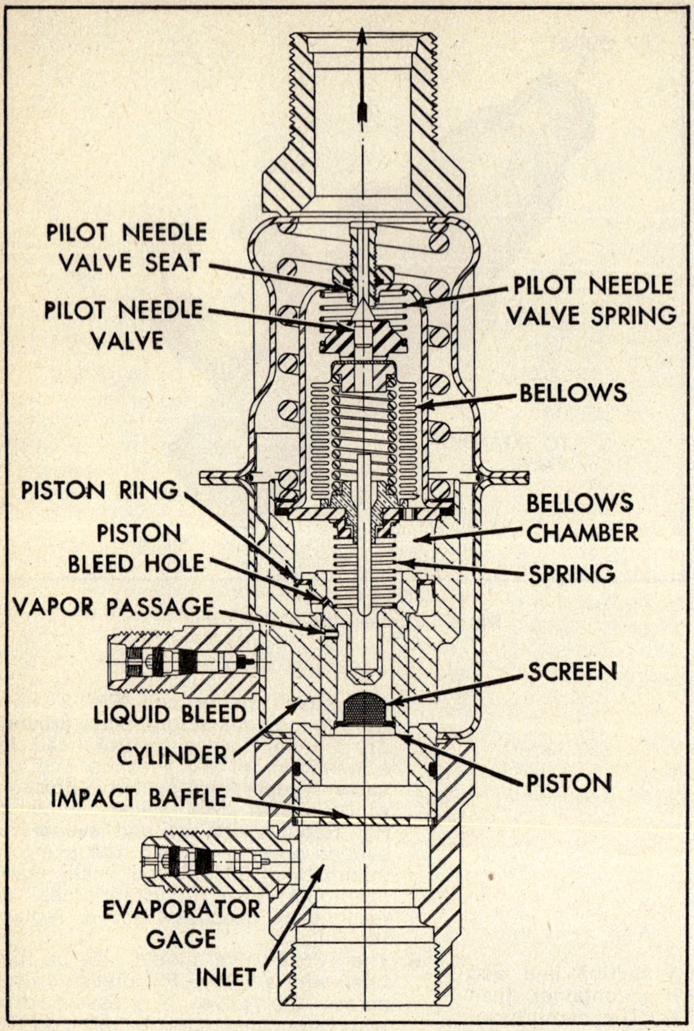

Fig. 10 Sectional view of POA valve

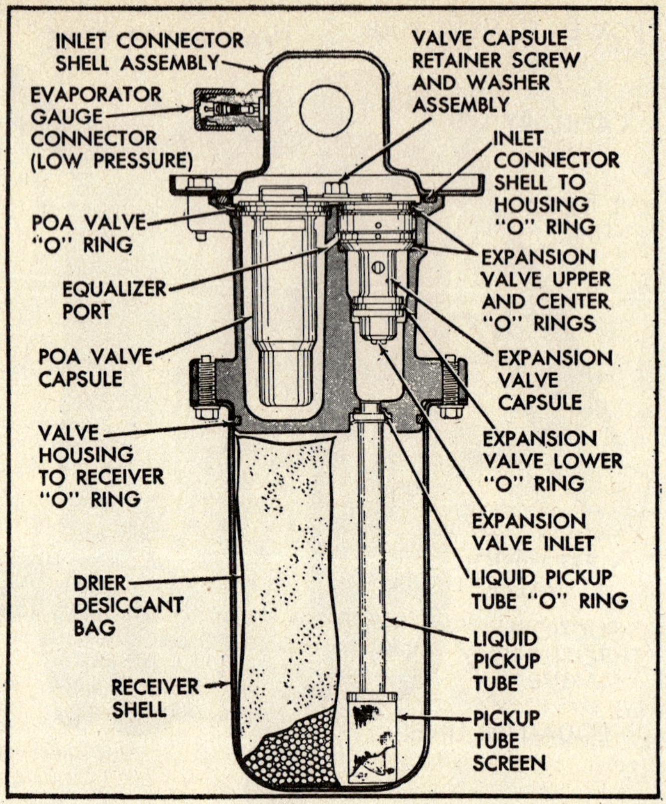

Fig. 11 Sectional view of VIR control

When leak-testing the POA valve, it is necessary to check only the hose coupling ends. When using a Propane Torch Leak Detector, no evidence of refrigerant should be present at the POA valve.

The POA valve will provide different gauge readings based on the altitude where the readings are being taken. Correct gauge reading at sea level is 29.5 psi. Gauge readings will be one-half psi higher for each additional 1000 feet of elevation.

The only test for POA valve operation is to check the suction pressure at the valve when making a performance test. If the valve gives improper gauge readings, it must be replaced since it is not repairable or adjustable.

VIR CONTROL

Valves-In Receiver

The VIR assembly, shown in Fig. 11, incorporates the thermostatic expansion valve, POA suction throttling valve, receiver-dehydrator and sight glass. The construction of this unit eliminates the need for an external equalizer line between the thermostatic expansion valve and POA suction throttling valve. Also, the thermobulb and capillary line has been eliminated.

The equalizer function is affected by a port in the wall separating the POA suction throttling valve and thermostatic expansion valve cavities in the VIR unit. The functions of the thermobulb and capillary line is now accomplished directly by the thermostatic expansion valve. The diaphragm end of the thermostatic expansion valve is exposed to refrigerant vapor entering the VIR from the outlet of the evaporator.

NOTE: A revised design VIR assembly is used in most 1975 GM vehicles. This assembly (VIREE) eliminates the equalizer port between the POA and thermostatic expansion valve cavities in the valve housing. The upper "O" ring is omitted from the thermostatic expansion valve, permitting pressure within the inlet connector shell to be applied to the thermostatic expansion valve equalizer port.

The sight glass is located at the inlet of the thermostatic expansion valve and the evaporator gage fitting is located in the inlet connector shell. The VIR unit is mounted adjacent to the evaporator.

The liquid bleed valve, located in the VIR valve housing, opens when a pressure differential greater than 10-20 PSI occurs between evaporator and suction pressures. The valve opening returns oil to the compressor under low refrigerant charge or low evaporator load conditions. The valve closes when the pressure differential is below 10 PSI to prevent loss of capacity.

NOTE: The filter screen, drier desiccant, thermostatic expansion valve and POA suction throttling valve are serviceable.

EXPANSION VALVES

Internally Equalized Type

Referring to Fig. 12, the refrigerant enters the inlet and screen as a high pressure liquid. The refrigerant flow is restricted by a metered orifice through which it must pass. As the refrigerant passes through this metered orifice, it changes from a high pressure liquid to a low pressure liquid or from the high side to the low side.

The refrigerant flow through the metered orifice is of the utmost importance. Anything that interferes with this flow seriously affects operation of the entire system.

A sufficient drop in temperature of the area cooled by the evaporator changes the heat transfer requirement of the evaporator. To continue the same flow of refrigerant into the evaporator will result in extreme cooling of the evaporator fins and coils to such an extent that they will freeze over with ice and stop air flow over them.

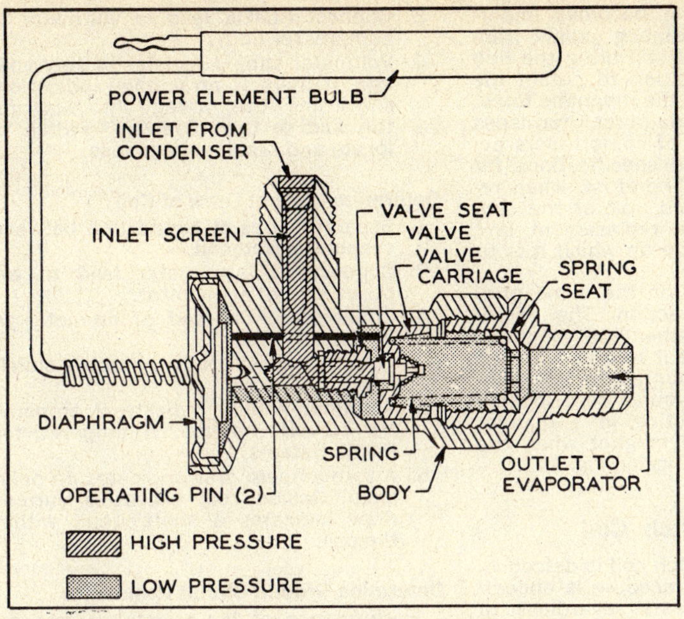

Fig. 12 Internally equalized expansion valve

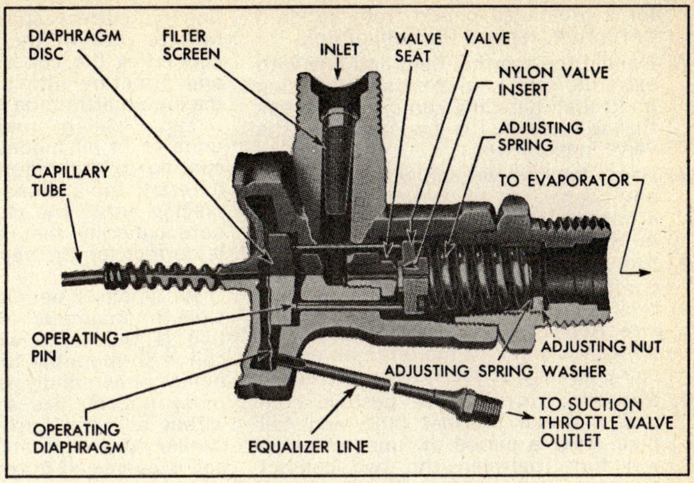

Fig. 13 Externally equalized expansion valve

The flow of refrigerant is controlled by varying the amount of refrigerant allowed to pass through the metered orifice. This is accomplished by internal balancing pressures moving a valve or a seat to control the refrigerant flow.

A thermal bulb connected to a diaphragm by a small line and filled with refrigerant gas such as Freon is secured firmly by a clamp to the evaporator tail pipe. The thermal bulb is sensitive to tail pipe temperatures. If the tail pipe becomes warm, the gas inside the bulb will begin to expand, exerting pressure against the diaphragm on the top plate that is connected to the seat or valve by a pin or pins (whichever construction is used). This expansion will then move the seat away from the orifice, allowing an increased refrigerant flow. As the tail pipe temperature drops, the pressure in the thermal bulb also drops, allowing the valve to restrict flow as required by the evaporator.

The pressure of the refrigerant entering the evaporator is fed back to the underside of the diaphragm through the internal equalizing passage. Expansion of the gas in the thermal bulb must overcome the internal balancing pressure before the valve will open to increase refrigerant flow.

A spring is installed against the valve and adjusted to a pre-determined setting at the time of manufacture. This is the super-heat spring, the purpose of which is to prevent slugging of the evaporator with excessive liquid. The adjusted tension of this spring is the determining factor in the opening and closing of the expansion valve. During opening or closing, the super-heat spring tension will retard or assist valve operation as required.

NOTE: For all practical purposes we should not be concerned with the adjustment of this spring. Tension is adjusted from 4 to 16 pounds as required for the unit on which it is to be installed. This original setting is sufficient for the life of the valve, and tampering with the adjustment will only lead to an unsatisfactory operating unit. Special equipment is required in most cases to accurately calibrate this adjustment.

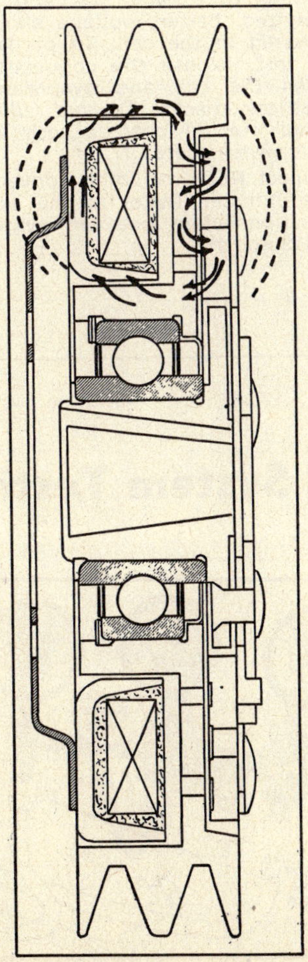

Fig. 14 Path of magnetic flow through compressor clutch

Externally Equalized Type

Operation of this type valve, Fig. 13, is the same as the internally equalized valve except that the evaporator pressure is fed against the underside of the diaphragm from the tail pipe of the evaporator by an equalizer line. This balances the temperature of the tail pipe through the expansion valve thermal bulb against evaporator pressure taken from the tail pipe.

Service Precautions

From the foregoing description of the thermostatic expansion valve, it can readily be seen that the valve is more sensitive to foreign materials than any other unit in the A/C system. Therefore, the following points should be observed to protect the valve while in service and while servicing the system.

1. The A/C system should be operated at least for short periods throughout the winter season to prevent the internal moving parts from corroding and sticking, which will result in the valve becoming inoperative and necessitating replacement. Periodic operation will also lubricate the entire system and especially the compressor seal, keeping it soft and in condition properly to seal the refrigerant.
2. Clean and replace all accessible screens when servicing.
3. Install a filter if system has excessive foreign materials present.
4. Follow correct evacuation procedure to remove all possible moisture from system.
5. Cap or cover any lines opened for service to prevent entrance of moisture or dirt.
6. Replace dehydrator as soon as excess moisture content is evident. Any system should have the dehydrator replaced at least upon opening the system the third time for service. Also, any time the system has been opened

for a prolonged period from accident or rupture, replace the dehydrator.

7. Handle the thermal bulb and line with extreme care, as excessive bending and rough handling can cause a break that will release the gas, rendering the valve inoperative.
8. Use a back-up wrench when removing any connection to prevent twisting of a line, which may result in weakening or breaking it.
9. Replace expansion valve only with a comparable valve. A numbering system is used to designate the orifice size. Replacing the valve with one with either too small or too large an orifice can seriously affect system operation.
10. Maintain corrosion-free positive contact between thermal bulb and tail pipe. Over a period of time corrosion will form between the two contact surfaces and insulate the two surfaces to such an extent that the operation of the expansion valve will be affected. The clamp may also work loose, preventing positive contact between the two and again affect operation of the expansion valve.

COMPRESSOR CLUTCHES

The clutch units found on automotive A/C systems are of two general types. The first type developed had the magnetic coil installed inside the pulley and rotated with it. This type is called the rotating coil type. The electrical current is carried to the coil by the use of brushes mounted to the compressor frame and contacting a slip ring mounted to the inside of the rotating pulley.

The second type has the magnetic coil mounted to the frame of the compressor and does not rotate. This is called the stationary coil type. This coil being stationary, correct spacing becomes important to prevent the rotating pulley from contacting the coil and still bring the hub and armature into position to obtain the maximum attraction of the magnetic force.

Each clutch manufacturer furnishes units to fit all models of compressors according to the required specifications for different applications. Therefore, when replacing either the clutch unit or the coil, note carefully that the replacement unit is correct for the vehicle on which it is to be installed.

All clutches operate on the same principle of magnetic attraction. The operation is the same whether the magnetic coil is so mounted that it rotates with the pulley or is solidly mounted to the compressor. Each has a wound core located within a metal cup acting in a manner similar to a horseshoe magnet when the coil is energized magnetically, Fig. 14.

Testing Clutch Coil

To determine if a clutch coil is defective, the following general procedure is enough to meet all requirements regardless of variations of installation. Meter connections given are for negative ground systems; reverse connections for positive ground systems.

Determine Voltage to Coil

1. With ignition switch on and clutch energized, battery voltage should be delivered to the coil. To prepare for the test, expose the connection between the coil and evaporator for electrical checks. *Do not allow exposed wire to contact body of car while switches are ON.*
2. Connect RED lead of a conventional Volt-Ammeter Tester to the exposed wire connection between evaporator and clutch coil.
3. Connect BLACK lead of voltmeter to compressor body.
4. Voltmeter should register battery voltage. If there is no voltage indication, check line fuse, voltage into evaporator, and to the "OFF-ON" switch to locate and repair voltage loss.

Determine Current Draw of Coil

1. Separate wires at connection between evaporator and coil.
2. Connect RED ammeter lead to exposed wire of evaporator.
3. Connect BLACK lead of ammeter to clutch coil wire.
4. Turn switches ON to energize clutch coil.
5. Ammeter should indicate 3 amperes for 12-volt systems; 5 amperes for 6-volt systems.
6. A zero ampere draw indicates an open circuit inside coil; excessive current draw indicates a short-circuit within the coil.

Determine Ground Circuit Resistance

Performance of the resistance test requires that the current draw of the coil be within specifications given above.

1. Connect evaporator and clutch coil wire.
2. Connect RED lead of voltmeter to compressor body.
3. Connect BLACK lead of voltmeter to battery post.
4. Turn switches ON to energize clutch coil.
5. Total resistance from compressor body to battery post cannot exceed .3 of a volt.
6. If resistance is excessive, clean all connections and metal-to-metal contacts, including engine-to-body and compressor-to-engine, to reduce resistance.

A/C System Testing

PERFORMANCE TEST

The system should be operated for at least 15 minutes to allow sufficient time for all parts to become completely stabilized. Determine if the system is fully charged by the use of test gauges and sight glass if one is installed on system. Head pressure will read from 180 psi to 220 psi or higher, depending upon ambient temperature and the type unit being tested. The sight glass should be free of bubbles if a glass is used in the system. Low side pressures should read approximately 15 psi to 30 psi, again depending on the ambient temperature and the unit being tested. It is not feasible to give a definite reading for all types of systems used, as the type control and component installation used on a particular system will directly influence the pressure readings on the high and low sides (see Fig. 5).

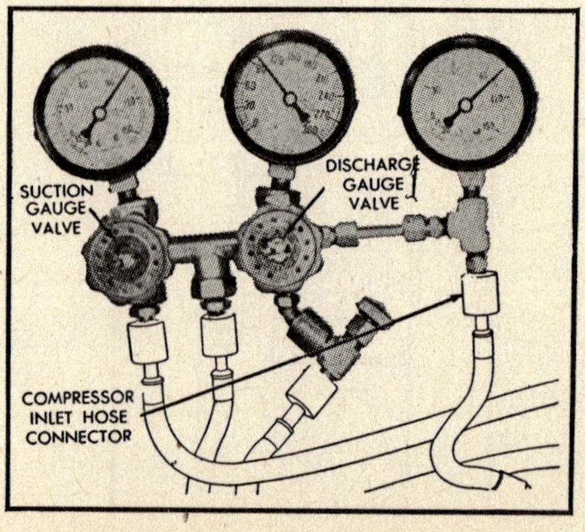

Fig. 15 Purge gauge hoses for Air-Temp compressors

The high side pressure will definitely be affected by the ambient or outside air temperature. A system that is operating normally will indicate a high side gauge reading between 150-170 psi with an 80° F ambient temperature. The same system will register 210-230 psi with an ambient temperature of 100°F. No two systems will register exactly the same, which requires that allowance for variations in head pressures must be considered. Following are the most important normal readings likely to be encountered during the season.

Ambient Temp.	High Side Pressure
80	150-170
90	175-195
95	185-205
100	210-230
105	230-250
110	250-270

Relative Temperature of High and Low Sides

The high side of the system should be uniformly hot to the touch throughout. A difference in temperature will indicate a partial blockage of liquid or gas at this point.

The low side of the system should be uniformly cool to the touch with no excessive sweating of the suction line or low side service valve. Excessive sweating or frosting of the low side service valve usually indicates an expansion valve is allowing an excessive amount of refrigerant into the evaporator. This condition will not necessarily be applicable to those units installed on General Motors vehicles that use the Suction Throttling valve. On these systems the line from the valve to the compressor will normally drop to a much lower reading than the evaporator pressure as the Suction Throttling Valve closes, resulting in the presence of moisture or frosting on this line. This is a normal reaction in this type system and is the result of the construction and operation of the STV or POA and the compressor, having approximately 35% more capacity than previous models of these vehicles that used the hot gas by-pass valve. These factors often cause the line from the STV or POA to the compressor to drop into a partial vacuum under normal operation.

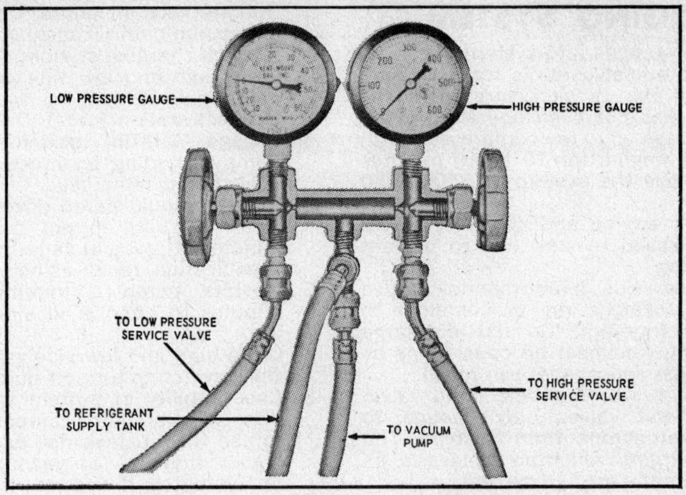

Fig. 16 Purge gauge hoses for Frigidaire, Tecumseh and York compressors

Evaporator Output

At this point, provided all other inspection tests have been performed, and components have been found to operate as they should, a rapid cooling down of the interior of the vehicle should result. The use of a thermometer is not necessary to determine evaporator output. Bringing all units to the correct operating specifications will insure that the evaporator performs as intended.

Temp. F.	Press. PSI	Temp. F.	Press. PSI	Temp. F.	Press. PSI	Temp. F.	Press. PSI	Temp. F.	Press. PSI
0	9.1	35	32.5	60	57.7	85	91.7	110	136.0
2	10.1	36	33.4	61	58.9	86	93.2	111	138.0
4	11.2	37	34.3	62	60.0	87	94.8	112	140.1
6	12.3	38	35.1	63	61.3	88	96.4	113	142.1
8	13.4	39	36.0	64	62.5	89	98.0	114	144.2
10	14.6	40	36.9	65	63.7	90	99.6	115	146.3
12	15.8	41	37.9	66	64.9	91	101.3	116	148.4
14	17.1	42	38.8	67	66.2	92	103.0	117	151.2
16	18.3	43	39.7	68	67.5	93	104.6	118	152.7
18	19.7	44	40.7	69	68.8	94	106.3	119	154.9
20	21.0	45	41.7	70	70.1	95	108.1	120	157.1
21	21.7	46	42.6	71	71.4	96	109.8	121	159.3
22	22.4	47	43.6	72	72.8	97	111.5	122	161.5
23	23.1	48	44.6	73	74.2	98	113.3	123	163.8
24	23.8	49	45.6	74	75.5	99	115.1	124	166.1
25	24.6	50	46.6	75	76.9	100	116.9	125	168.4
26	25.3	51	47.8	76	78.3	101	118.8	126	170.7
27	26.1	52	48.7	77	79.2	102	120.6	127	173.1
28	26.8	53	49.8	78	81.1	103	122.4	128	175.4
29	27.6	54	50.9	79	82.5	104	124.3	129	177.8
30	28.4	55	52.0	80	84.0	105	126.2	130	182.2
31	29.2	56	53.1	81	85.5	106	128.1	131	182.6
32	30.0	57	55.4	82	87.0	107	130.0	132	185.1
33	30.9	58	56.6	83	88.5	108	132.1	133	187.6
34	31.7	59	57.1	84	90.1	109	135.1	134	190.1

Fig. 17 Temperature-pressure relationship for F12 or R12 refrigerant

PURGING SYSTEM

1. Connect gauges into system, Figs. 15, 16, and adjust controls for maximum cooling. *This is necessary when the system has not been operating to return excess oil to the compressor.*
2. Operate engine for 10 to 15 minutes to stabilize the system at 1500-1750 rpm.
3. Shut off engine and controls. Adjust engine speed to slow idle to prevent "dieseling".
4. Open low side hand manifold valve slightly, using a rag or container to catch refrigerant. *Do not discharge the refrigerant near an open flame as a toxic gas (phosgene) can result.*
5. Open high side manifold hand valve. *Open hand valves only enough to bleed refrigerant from system. Too rapid purging will draw excessive oil from compressor and system.*
6. Close gauge manifold shut-off valves when refrigerant ceases to bleed from hose. Both gauges will read zero psi.

Evacuate System with Vacuum Pump

Vacuum pumps suitable for removing air and moisture from A/C systems are commercially available. A specification for system pump-down used here is 28 to 29½" vacuum. This reading can be attained at or near sea level only. For each 1000 feet of altitude this operation is being performed, the reading will be 1" vacuum higher. As an example, at 5000 feet elevation, only 23-24½" of vacuum can be obtained.

1. Connect vacuum pump to gauge manifold. With gauges connected into system, remove cap from vacuum hose connector. Install center hose from gauge manifold to vacuum pump connector. Mid-position high and low side compressor service valves (if used). Open high and low side gauge manifold hand valves.
2. Operate vacuum pump a minimum of 30 minutes for air and moisture removal. Watch compound gauge that system pumps down into a vacuum. System will reach 28-29½" vacuum in not over 5 minutes. If system does not pump down, check all connections and leak-test if necessary.
3. Close gauge manifold hand valves and shut off vacuum pump.
4. Check ability of system to hold vacuum. Watch compound gauge to see that gauge does not rise at a faster rate than 1" vacuum every 4 or 5 minutes. If compound gauge rises at too rapid a rate, install partial charge and leak-test. Then purge system as outlined above.
5. If system holds vacuum, charge system with refrigerant.

Evacuate System Using Charging Station

A vacuum pump is built into the charging station and is constructed to withstand repeated and prolonged use without damage. Complete moisture removal from the system is possible only with a vacuum pump constructed for the purpose.

1. Operate vacuum pump. Connect hose to vacuum pump if system was purged through charging station.
2. Open high and low side gauge valves of charging station.
3. Connect station into 110-volt current.
4. Engage "Off-On" switch to vacuum pump according to directions of specific station being used.
5. System should pump down into a 28-29½" vacuum in not more than 5 minutes. If system fails to meet this specification, repair as necessary.
6. Operate pump a minimum of 30 minutes to remove all air and moisture.
7. Close high and low side gauge valves. Open switch to turn off pump.
8. Check ability of system to hold vacuum by watching compound gauge to see that it does not rise at a rate higher than 1" of vacuum every 4 or 5 minutes: If rise rate is not within specifications, repair system as necessary. If rise rate is within specifications, charge system with refrigerant.

CHARGING THE SYSTEM

Using 15-Ounce Containers

To prevent waste and to aid in more accurate charging, refrigerant manufacturers package refrigerant in 15-ounce cans. The small containers are advantageous to small shops doing only a limited amount of A/C service work. The small containers are handled in the same manner as the larger drums except care must be taken not to overheat the cans because of the danger of explosion.

1. Install "Fitz-All" valve to container(s). The Fitz-All valve is available for single cans or three cans. Whichever is used, preliminary installation to the can(s) is the same.
2. Close shut-off valve of Fitz-All valve. Pierce can with mechanism which is part of Fitz-All valve.
3. With the system pumped-down, install charging hose to Fitz-All valve. Loosen charging hose at center connector on gauge manifold.
4. Crack Fitz-All shut-off valve to purge air from charging hose. Tighten charging hose connection on gauge manifold and close shut-off valve.
5. Complete charge of system. Do not overfill. Use A/C Data table to determine capacity of system being serviced. Close high side gauge manifold hand valve. Start engine and adjust throttle to 1500-1750 rpm. Adjust A/C controls for maximum cooling. Open low side gauge manifold hand valve to allow Freon to be drawn into system. If single containers are used, it will be necessary to replace each as it becomes empty. Watch sight glass until bubbles disappear.
6. To check charge in system, watch for bubbles in sight glass (if used). Read high pressure gauge indication. *Excessive head pressure with a normal low side pressure indicates an overcharge of Freon or air in system. Compressor may or may not be noisy.*
7. Listen for hissing noise in expansion valve as many systems have a hissing in the expansion valve until the system is fully charged.
8. Continue testing system and adjust controls for maximum efficiency.

Using Charging Station

Most stations contain a charging cylinder into which the exact amount of refrigerant required for the system being serviced may be placed while system pump-down is being performed. The refrigerant charging cylinder contained in the station is heated to the correct temperature to insure proper refrigerant flow to all parts of the system as a gas during the charging operation. Fig. 17. Following correct evacuation procedure as to length of time for complete moisture removal, the vacuum pump will so efficiently pump-down the system that opening the correct valves will completely charge the system from the high side, and the use of the compressor in the charging operation will not be required.

1. Prepare charging cylinder for filling: Open storage drum valve. Close all valves on station. Read storage tank gauge pressure. Rotate dial shroud on charging cylinder to correlate with pressure on gauge. Open cylinder fill valve.
2. Fill charging cylinder: Determine system capacity from A/C Data table. Intermittently open and close pressure relief valve. *When pressure relief valve opens, Freon will enter cylinder and boil. Closing the valve will increase pressure on the Freon, changing it to a liquid to stabilize the refrigerant in the sight glass. Fill to specified level in sight glass and close pressure relief valve.*
3. Charge system: With gauges connected into system, open refrigerant control valve. Open high pressure valve. Remove vacuum hose from pump and crack (barely open) low pressure valve. Allow refrigerant to escape through vacuum hose for approximately 3 seconds. Close high and low pressure valves, and close refrigerant control valve.
4. Continue testing system and adjust controls for maximum efficiency.

LEAK TEST SYSTEM

The propane torch Halide Leak Detector is the most widely used of the detection devices. Therefore, only the procedure for this device will be given. The procedure is the same for any electronic detector, except that the pick-up device registers the presence of refrigerant by a flashing light or high pitched squeal instead of changing the color of the flame. All other steps in preparing the system and leak testing are the same and can be followed as outlined below.

1. Stabilize system at 1500-1750 rpm. If system is empty of refrigerant, it will be necessary to install a partial charge before continuing. With gauges connected into system, adjust A/C controls for maximum cooling. Operate for 10 to 15 minutes, then shut off car engine.

2. Light leak detector. Open valve to a low flame that will not blow itself out. Warm up until copper element turns cherry red. Lower flame until flame tip is even or slightly below center of element. *(For electronic tester, follow preparation procedure as given in operating instructions.)*
3. Move leak detector pick-up under hoses, joints, seals, and any possible place for a leak to occur.

NOTE: Freon 12 refrigerant is heavier than air and will move downward. If concentration of refrigerant is located, move pick-up upward to locate leak. Do not inhale fumes produced by burning refrigerant.

4. Watch for color change of flame: Pale blue, no refrigerant; yellow, small amount of refrigerant; purplish-blue, large amount of refrigerant. Repair system as necessary if leaks are located.
5. Check sensitivity of reaction plate: Pass pick-up hose over empty can or crack open refrigerant container; flam should show violent reaction. If no color change, replace reaction plate, following instructions accompanying leak detector. *Too high a flame will result in short life to reaction plate and poor reaction and will soon burn out element.*
6. Charge system if repairs were necessary.

ISOLATE COMPRESSOR FROM SYSTEM

On systems having both a high side and low side service valve, the compressor may be isolated and refrigerant retained in the system while service work is being performed on the compressor or in the engine compartment of the car.

1. Stabilize system at 1500-1750 rpm: With gauges connected into system, adjust A/C controls for maximum cooling. Operate system for 10 to 15 minutes.
2. Isolate compressor: Slowly close low side service valve until low side gauge reads zero psi. Then shut off car engine.
3. Return car engine to idle speed to prevent "dieseling". Completely close low side service valve. Close high side service valve. Purge refrigerant from compressor by cracking low side hand manifold until both gauges read zero psi. *Purge refrigerant slowly to prevent pulling oil from compressor.*
4. Remove gauges from service valves and service valves from compressor. Then perform service work as required.
5. Place compressor in system: Install service valves to compressor, using new gaskets or O-ring seals, whichever are required. Purge air from compressor by cracking high side service valve for 3 seconds with high side hose connector capped and low side hose connector open.
6. Install gauges to service valve connectors and purge air from hoses.
7. Mid-position service valves, continue testing system and adjust controls for maximum performance.

Each make and model of A/C system has its own type of construction. To cover service work on components of all models is beyond the scope of this manual.

AIR CONDITIONING SPECIFICATIONS

Year	Model	Refrigerant Capacity, Lbs.		Refrigeration Oil				Temperature Control Device	Comp. Pulley to Clutch Plate Clearance
		With One Evaporator	With Two Evaporators	Viscosity	Capacity, Ounces: With One Evaporator	Capacity, Ounces: With Two Evaporators	Oil Level Inches		
AMERICAN MOTORS									
1969–71	Ambassador, Rebel	2¾	—	300	7	—	⑧	Thermo. Switch	—
	Matador	2¾	—	300	7	—	⑧	Thermo. Switch	—
	Others	2¼	—	300	7	—	⑧	Thermo. Switch	—
1972-73	Ambassador, Matador	2½	—	300	7	—	⑧	Thermo. Switch	—
	All Others	2	—	300	7	—	⑧	Thermo. Switch	—
1974–75	Ambassador, Matador	3¼	—	300	7	—	⑧	Thermo. Switch	—
	Others	2	—	300	7	—	⑧	Thermo. Switch	—
BUICK—FULL SIZE CARS									
1969–70	All	4¼	—	525	10½	—	①	POA Valve	.031–.062
1971–72	All	4½	—	525	10½	—	①	POA Valve	.031-.062
1973	All	4	—	525	10½	—	①	POA Valve	.031–.062
1974–75	All	3¾	—	525	⑮	—	①	POA Valve	⑯
BUICK—INTERMEDIATE CARS									
1969–72	All	3¾	—	525	10½	—	①	POA Valve	.031–.062
1973	All	4½	—	525	10½	—	①	POA Valve	.031–.062
1974	All	3¾	—	525	10½	—	①	POA Valve	.031–.062
1975	Skyhawk	3	—	525	⑮	—	①	POA Valve	⑯
	Apollo, Skylark	3½	—	525	⑮	—	①	POA Valve	⑯

AIR CONDITIONING SPECIFICATIONS—Continued

Year	Model	Refrigerant Capacity, Lbs.		Refrigeration Oil				Temperature Control Device	Comp. Pulley to Clutch Plate Clearance
		With One Evaporator	With Two Evaporators	Viscosity	Capacity, Ounces: With One Evaporator	Capacity, Ounces: With Two Evaporators	Oil Level Inches		
CADILLAC									
1969–73	All	4	5¼	525	10½	13½	①	POA Valve	.031–.057
1974–75	All	3¾	5	525	10½	13½	①	POA Valve	.031–.057
1976	Seville	3½	—	525	10½	—	①	POA Valve	.031–.057
CAMARO, CHEVROLET, CHEVELLE, NOVA & MONTE CARLO									
1969–74	Four Season	3¾	—	525	11	—	①	POA Valve	.022–.057
	Comfortron	3¾	—	525	11	—	①	POA Valve	.022–.057
	Universal	3	—	525	11	—	①	Thermo. Switch	.022–.057
	All Weather	2½	—	525	11	—	①	Thermo. Switch	.022–.057
1975	4 Cyl. Comp.	3¾	—	525	5½–6½	—	①	POA Valve	.020–.040
	6 Cyl. Comp.	3¾	—	525	11	—	①	POA Valve	.022–.057
CHEVROLET MONZA									
1975	All	2¾	—	525	11	—	①	POA Valve	.022–.057
CHEVROLET VEGA & PONTIAC ASTRE									
1971-72	All	2¾	—	525	11	—	①	POA Valve	.022-.057
1973–74	All	2¾	—	525	11	—	①	Thermo. Switch	.022–.057
1975	All	2¾	—	525	11	—	①	POA Valve	.022–.057
CHEVROLET CORVETTE									
1969–73	All	3¼	—	525	11	—	①	POA Valve	.022–.057
1974–75	All	3	—	525	11	—	①	POA Valve	.022–.057
CHECKER MOTORS									
1969–75	All	2¾	4½	525	10½	13½	①	Thermo. Switch	.022–.057
CHRYSLER & IMPERIAL									
1969–70	All	3¼	4¼	300	10–11	10–11	1⁹⁄₁₆–2⅜②	EPR Valve⑤	—
1971	All	3¼	4¼	300	10–11	10–11	1⅝–2⅜②	EPR Valve⑤	—
1972	All	3⅓	4⅓	300	10–11	10–11	1⅝–2⅜②	EPR Valve⑤	—
1973	All	3⅓	4⅓	300	10–12	10–12	1⅝–2⅜②	EPR Valve⑤	—
1974	All	5	—	300	10–12	—	1⅝–2⅜②	EPR-2 Valve	—
1975	Exc. Cordoba	4¼	—	300	10–12	—	1⅝–2⅜②	EPR-2 Valve	—
	Cordoba	4⅛	—	300	10–12	—	1⅝–2⅜②	EPR-2 Valve	—

AIR CONDITIONING SPECIFICATIONS—Continued

Year	Model	Refrigerant Capacity, Lbs.		Refrigeration Oil				Temperature Control Device	Comp. Pulley to Clutch Plate Clearance
		With One Evaporator	With Two Evaporators	Viscosity	Capacity, Ounces: With One Evaporator	Capacity, Ounces: With Two Evaporators	Oil Level Inches		
DODGE—FULL SIZE CARS									
1969–70	All	3¼	4¼	300	10–11	10–11	1$\frac{9}{16}$–2⅜②	EPR Valve⑤	—
1971	All	3¼	4¼	300	10–11	10–11	1⅝–2⅜②	EPR Valve⑤	—
1972	All	3⅓	4⅓	300	10–11	10–11	1⅝–2⅜②	EPR Valve⑤	—
1973	All	3⅓	4⅓	300	10–12	10–12	1⅝–2⅜②	EPR Valve⑤	—
1974	All	5	—	300	10–12	10–12	1⅝–2⅜②	EPR-2 Valve	—
1975	All	4¼	—	300	10–12	—	1⅝–2⅜②	EPR-2 Valve	—
DODGE—COMPACT & INTERMEDIATE CARS									
1969–70	Dart 6	2¾	—	300	10–11	—	1⅝–2⅛②	EPR Valve	—
	Coronet, Charger 6	3¼	—	300	10–11	—	2⅛–2¾②	EPR Valve	—
	All V-8	3¼	—	300	10–11	—	1$\frac{9}{16}$–2⅜②	EPR Valve	—
1971	6 Cylinder	2¾	—	300	10–11	—	1⅝–2⅛②	EPR Valve	—
	V8	3¼	—	300	10–11	—	1⅝–2⅜②	EPR Valve	—
1972	All	2⅞	—	300	10–11	—	⑥	EPR Valve	—
1973	All	2⅞	—	300	10–12	—	⑦	EPR Valve	—
1974	Exc. Dart	2¾	—	300	10–12	—	⑦	EPR-2 Valve	—
	Dart	3¼	—	300	10–12	—	⑦	EPR-2 Valve	—
1975	Exc. Dart	3¾	—	300	10–12	—	⑦	EPR-2 Valve	—
	Dart	3	—	300	10–12	—	⑦	EPR-2 Valve	—
FORD—FULL SIZE CARS									
1969	York Comp.	3	—	③	10	—	⅞–1$\frac{3}{16}$②	Thermo. Switch	—
	Tecumseh Comp.	3	—	③	11	—	⅞–1⅝②	Thermo. Switch	—
1970	York Comp.	3	—	③	10	—	②⑧	Thermo. Switch	—
	Tecumseh Comp.	3	—	③	11	—	②⑧	Thermo. Switch	—
1971–72	York Comp.	3¼	—	③	10	—	②⑧	Thermo. Switch	—
	Tecumseh Comp.	3¼	—	③	11	—	②⑧	Thermo. Switch	—
1973–74	York Comp.	4¼	—	③	10	—	②⑧	STV Valve	—
	Tecumseh Comp.	4¼	—	③	11	—	②⑧	STV Valve	—
1975	All	4¼	—	525	10½	—	①	STV Valve	—
FORD—COMPACT & INTERMEDIATE CARS (EXC. MUSTANG II)									
1969	Falcon, Fairlane	1⅞	—	③	10④	—	⅞–1⅝②④	Thermo. Switch	—
	Mustang	1¾	—	③	10④	—	⅞–1⅜②④	Thermo. Switch	—
1970–71	Exc. Mustang	1⅞	—	③	⑨	—	②⑧	Thermo. Switch	—
	Mustang	1¾	—	③	⑨	—	②⑧	Thermo. Switch	—
1972–75	York Comp.	⑩	—	③	10	—	②⑧	⑪	—
	Tecumseh Comp.	⑩	—	③	11	—	②⑧	⑪	—
FORD MUSTANG II									
1974–75	York Comp.	3¼	—	③	10	—	⑧	STV Valve	—
	Tecumseh Comp.	3¼	—	③	11	—	⑧	STV Valve	—

AIR CONDITIONING SPECIFICATIONS—Continued

Year	Model	Refrigerant Capacity, Lbs.		Refrigeration Oil				Temperature Control Device	Comp. Pulley to Clutch Plate Clearance
		With One Evaporator	With Two Evaporators	Viscosity	Capacity, Ounces: With One Evaporator	Capacity, Ounces: With Two Evaporators	Oil Level Inches		
FORD PINTO & MERCURY BOBCAT									
1971–73	York Comp.	1⅞	—	③	10	—	②⑧	Thermo. Switch	—
	Tecumseh Comp.	1⅞	—	③	11	—	②⑧	Thermo. Switch	—
1974–75	York Comp.	2¼	—	③	10	—	②⑧	STV Valve	—
	Tecumseh Comp.	2¼	—	③	11	—	②⑧	STV Valve	—
LINCOLN									
1969	Lincoln	3½	—	③	10④	—	⅞–1³⁄₁₆②④	Thermo. Switch	—
	Mark III	2¼	—	③	10④	—	⅞–1³⁄₁₆②④	Thermo. Switch	—
1970	Lincoln	4¼	—	③	⑨	—	⑧	Thermo. Switch	—
	Mark III	2¼	—	③	⑨	—	⑧	Thermo. Switch	—
1971	Lincoln	4¼	—	③	10④	—	⑧	Thermo. Switch	—
	Mark III	2¼	—	③	10	—	⑧	Thermo. Switch	—
	Six Cyl. Comp.	—	—	525	10½	—	—	STV Valve	.022–.057
1972–75	Lincoln	4½	—	525	10½	—	—	STV Valve	.022–.057
	Mark IV	4½	—	525	10½	—	—	STV Valve	.022–.057
MERCURY—FULL SIZE CARS									
1969	York Comp.	3	—	③	10	—	⅞–1³⁄₁₆②	Thermo. Switch	—
	Tecumseh Comp.	3	—	③	11	—	⅞–1⅝②	Thermo. Switch	—
1970	York Comp.	3	—	③	10	—	⑧	Thermo. Switch	—
	Tecumseh Comp.	3	—	③	11	—	⑧	Thermo. Switch	—
1971–72	York Comp.	3¼	—	③	10	—	⑧	Thermo. Switch	—
	Tecumseh Comp.	3¼	—	③	11	—	⑧	Thermo. Switch	—
1973–74	York Comp.	4¼	—	③	10	—	⑧	STV Valve	—
	Tecumseh Comp.	4¼	—	③	11	—	⑧	STV Valve	—
	V8-460 W/ATC	—	—	525	10½	—	①	STV Valve	.022–.057
1975	All	4¼	—	525	10½	—	①	STV Valve	.022–.057
MERCURY—COMPACT & INTERMEDIATE CARS									
1969	Montego	1⅞	—	③	10④	—	⅞–1³⁄₁₆②④	Thermo. Switch	—
	Cougar	1¾	—	③	10④	—	⅞–1³⁄₁₆②④	Thermo. Switch	—
1970	Montego	1⅞	—	③	⑨	—	⑧	Thermo. Switch	—
	Cougar	1¾	—	③	⑨	—	⑧	Thermo. Switch	—
1971–75	York Comp.	⑫	—	③	10	—	⑧	Thermo. Switch⑬	—
	Tecumseh Comp.	⑫	—	③	11	—	⑧	Thermo. Switch⑬	—
OLDSMOBILE—FULL SIZE CARS									
1969–72	All	4½	—	525	10½	—	①	POA Valve	.022–.057
1973–75	All	4	—	525	10½	—	①	POA Valve	.022–.057
OLDSMOBILE—INTERMEDIATE CARS									
1969–75	All	4	—	525	⑮	—	①	POA Valve	⑯

AIR CONDITIONING SPECIFICATIONS—Continued

Year	Model	Refrigerant Capacity, Lbs. — With One Evaporator	Refrigerant Capacity, Lbs. — With Two Evaporators	Refrigeration Oil — Viscosity	Refrigeration Oil — Capacity, Ounces — With One Evaporator	Refrigeration Oil — Capacity, Ounces — With Two Evaporators	Oil Level Inches	Temperature Control Device	Comp. Pulley to Clutch Plate Clearance
PLYMOUTH—FULL SIZE CARS									
1969	Six Cylinder	3¼	4¼	300	10–11	10–11	2⅛–2¾②	EPR Valve⑤	—
	V8	3¼	4¼	300	10–11	10–11	1⁹⁄₁₆–2⅜②	EPR Valve⑤	—
1970–71	Six Cylinder	3¼	4¼	300	10–11	10–11	1⅝–2⅛②	EPR Valve⑤	—
	V8	3¼	4¼	300	10–11	10–11	1⅝–2⅜②	EPR Valve⑤	—
1972	Six Cylinder	3¼	4¼	300	10–11	10–11	1¾–2⅜	EPR Valve⑤	—
	V8	3¼	4¼	300	10–11	10–11	1⅝–2⅜	EPR Valve⑤	—
1973	Six Cylinder	3¼	4¼	300	10–12	10–12	1¾–2⅜	EPR Valve⑤	—
	V8	3¼	4¼	300	10–12	10–12	1⅝–2⅜	EPR Valve⑤	—
1974	All	5	—	300	10–12	—	1⅝–2⅜	EPR-2 Valve	—
1975	All	4¼	—	300	10–12	—	1⅝–2⅜	EPR-2 Valve	—
PLYMOUTH—COMPACT & INTERMEDIATE CARS									
1969	Valiant, Barracuda 6	2¾	—	300	10–11	—	1⅝–2⅛②	EPR Valve	—
	Belvedere, Satellite 6	3¼	—	300	10–11	—	2⅛–2¾②	EPR Valve	—
	All V8	3¼	—	300	10–11	—	1⁹⁄₁₆–2⅜②	EPR Valve	—
1970–71	Barracuda, Satellite 6	3¼	4¼	300	10–11	—	1⅝–2⅛	EPR Valve⑤	—
	Valiant 6	2¾	—	300	10–11	—	1⅝–2⅛	EPR Valve⑤	—
	All V8	3¼	4¼	300	10–11	—	1⅝–2⅜	EPR Valve⑤	—
1972	Six Cylinder	2¾	—	300	10–11	—	1¾–2⅜	EPR Valve⑤	—
	V8	2¾	—	300	10–11	—	1⅝–2⅜	EPR Valve⑤	—
1973	All	2⅞	—	300	10–12	—	⑦	EPR Valve⑤	—
1974	Exc. Valiant	2¾	—	300	10–12	—	⑦	EPR-2 Valve	—
	Valiant	3¼	—	300	10–12	—	⑦	EPR-2 Valve	—
1975	Exc. Valiant	3¾	—	300	10–12	—	⑦	EPR-2 Valve	—
	Valiant	3	—	300	10–12	—	⑦	EPR-2 Valve	—
PONTIAC—FULL SIZE CARS									
1969–75	All	4⅛⑭	—	525	11	—	①	POA Valve	.022–.057
PONTIAC—INTERMEDIATE CARS EXC. ASTRE									
1969	Firebird	3¾	—	525	11	—	①	POA Valve	.022–.057
	All Others	4⅛	—	525	11	—	①	POA Valve	.022–.057
1970–75	All	4⅛⑭	—	525	11	—	①	POA Valve	.022–.057
THUNDERBIRD									
1969	All	2¼	—	③	10④	—	⅞–1³⁄₁₆②④	Thermo. Switch	—
1970	All	2¼	—	③	11④	—	⅞–1¹⁄₁₆②	Thermo. Switch	—
1971	All	2¼	—	③	10④	—	⅞–1⅜②	Thermo. Switch	—
1972–75	All	4½	—	525	10½	—	①	STV Valve	.022–.057

①—See text for procedure.
②—Dipstick reading with compressor installed.
③—Suniso 5G or Capella E.
④—For Tecumseh Comp., oil level is ⅞–1⅝. Capacity 11 oz.
⑤—Auto Temp uses ETR valve.
⑥—Six cylinder engines, 1⅝–2⅛; V8 engines, 1⅝–2⅜.
⑦—Six cylinder engines, 1¾–2⅜; V8 engines, 1⅝–2⅜.
⑧—York Comp.—Vertical mount., ⅞–1⅛; horizontal mount., 1³⁄₁₆–1³⁄₁₆. Tecumseh Comp.—Vertical mount., ⅞–1⅜; horizontal mount., ⅞–1⅝.
⑨—York Comp., 10 oz., Tecumseh Comp., 11 oz.
⑩—Maverick, 1⅞; Torino, 4½; 1972–73 Mustang, 1¾; Granada, 4.
⑪—Exc. Granada & Torino thermo. switch; Granada & Torino, STV valve.
⑫—Comet & 1971 Montego, 1⅞; 1971–72 Cougar, 1¾; 1972-75 Montego & 1973–75 Cougar, 4½; Monarch, 4.
⑬—Monarch, 1972–75 Montego & 1974–75 Cougar, STV valve.
⑭—Refer to compressor decal when servicing these systems.
⑮—Axial comp., 10½ oz.; radial comp., 5½–6½ oz.
⑯—Axial comp., .022–.057″; radial comp., .020–.040″.

CHARGING VALVE LOCATIONS

Year & Model	High Press.	Low Press.
AMERICAN MOTORS		
1969–75	Compressor	Compressor
BUICK-FULL SIZE		
1969 72	Compressor	STV(POA)①
1973–75	Compressor	VIR
BUICK INTERMEDIATE		
1969–73	Compressor	STV(POA)
1974—		
Exc. Apollo	Compressor	VIR
Apollo	Compressor	POA
1975	Compressor	VIR
CADILLAC		
1969–73②	Vapor Line	STV(POA)①
1973–75③	Vapor Line	VIR
CAMARO, CHEVELLE, CHEVROLET & MONTE CARLO		
1969–71—		
Exc. Below	Compressor	Compressor
Four Season	Exp. Valve	STV(POA)①
Comfortron	Exp. Valve	STV(POA)①
1972–73	Exp. Valve	POA①
1974–75	④	VIR
CHEVROLET VEGA & PONTIAC ASTRE		
1971	Exp. Valve	STV(POA)①
1972	Dis. Pres. Sw.	POA①
1973–75	Dis. Pres. Sw.	Accumulator
CHEVROLET MONZA		
1975	Compressor	VIR
CHEVY NOVA		
1969–71—		
Exc. Below	Compressor	Compressor
Four Season	Exp. Valve	STV(POA)①
1972–73	Exp. Valve	POA①
1974–75	④	VIR
CHECKER MOTORS		
1969–73	Exp. Valve	POA
1974–75	④	VIR
CHRYSLER & IMP.		
1969–75	Compressor	Compressor
CORVETTE		
1969–71	Compressor	STV(POA)①
1972	Compressor	Compressor
1973	Compressor	VIR
1974-75	④	VIR
DODGE (ALL)		
1969–75	Compressor	Compressor
FORD (ALL)		
1969–75	Compressor	Compressor
LINCOLN		
1969–71	Compressor	Compressor
1972–75	Compressor	Compressor
MERCURY (ALL)		
1969–75	Compressor	Compressor
OLDSMOBILE (ALL)		
1969–74②	Compressor	STV(POA)①
1973–75③	Compressor	VIR
PLYMOUTH (ALL)		
1969–75	Compressor	Compressor
PONTIAC (ALL)		
1969–74②	Compressor	STV(POA)①
1973–75③	Compressor	VIR
T-BIRD		
1969–71	Compressor	Compressor
1972–75	Compressor	Compressor

①—Located at cowl, right side of engine.
②—Except VIR system.
③—With VIR.
④—Axial comp., VIR inlet line; radial comp., located on compressor.

VARIABLE SPEED FANS

The fan drive clutch, Fig. 1, is a fluid coupling containing silicone oil. Fan speed is regulated by the torque-carrying capacity of the silicone oil. The more silicone oil in the coupling the greater the fan speed, and the less silicone oil the slower the fan speed.

Two types of fan drive clutches are in use. On one, Fig. 2, a bi-metallic strip and control piston on the front of the fluid coupling regulates the amount of silicone oil entering the coupling. The bi-metallic strip bows outward with a decrease in surrounding temperature and allows a piston to move outward. The piston opens a valve regulating the flow of silicone oil into the coupling from a reserve chamber. The silicone oil is returned to the reserve chamber through a bleed hole when the valve is closed.

On the other type of fan drive clutch, Fig. 3, a heat-sensitive, bi-metal spring connected to an opening plate brings about a similar result. Both units cause the fan speed to increase with a rise in temperature and to decrease as the temperature goes down.

In some cases a Flex-Fan is used instead of a Fan Drive Clutch. Flexible blades vary the volume of air being drawn through the radiator, automatically increasing the pitch at low engine speeds.

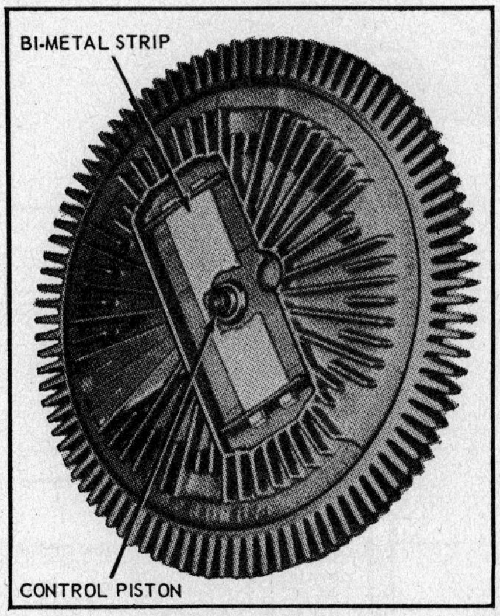

Fig. 2 Variable-speed fan with flat bi-metal thermostatic spring

Fig. 1 Typical variable-speed fan installed

Fan Drive Clutch Test

Run the engine at a fast idle speed (1000 rpm) until normal operating temperature is reached. This process can be speeded up by blocking off the front of the radiator with cardboard. Regardless of temperatures, the unit must be operated for at least five minutes immediately before being tested.

Stop the engine and, using a glove or a cloth to protect the hand, immediately check the effort required to turn the fan. If considerable effort is required, it can be assumed that the coupling is operating satisfactorily. If very little effort is required to turn the fan, it is an indication that the coupling is not operating properly and should be replaced.

Service Procedure

CAUTION: When it becomes necessary to remove a fan clutch of the silicone fluid type, Fig. 2, the assembly must be supported in the vertical (on car) position to prevent leaks of silicone fluid from the clutch mechanism. This loss of fluid will render the fan clutch inoperative.

The removal procedure for either type of fan clutch assembly is generally the same for all cars. Merely unfasten the unit from the water pump and remove the assembly from the car.

The type of unit shown in Fig. 2 may by partially disassembled for inspection and cleaning. Take off the capscrews that hold the assembly together and separate the fan from the drive clutch. Next remove the metal strip on the front by pushing one end of it toward the fan clutch body so it clears the retaining bracket. Then push the strip to the side so that its opposite end will spring out of place. Now remove the small control piston underneath it.

Check the piston for free movement of the coupling device. If the piston sticks, clean it with emery cloth. If the bi-metal strip is damaged, replace the entire unit. These strips are not interchangeable.

When reassembling, install the control piston so that the projection on the end of it will contact the metal strip. Then install the metal strip with any identification numerals or letters facing the clutch. After reassembly, clean the clutch drive with a cloth soaked in solvent. Avoid dipping the clutch assembly in any type of liquid. Install the assembly in the reverse order of removal.

The coil spring type of fan clutch cannot be disassembled, serviced or repaired. If it does not function properly it must be replaced with a new unit.

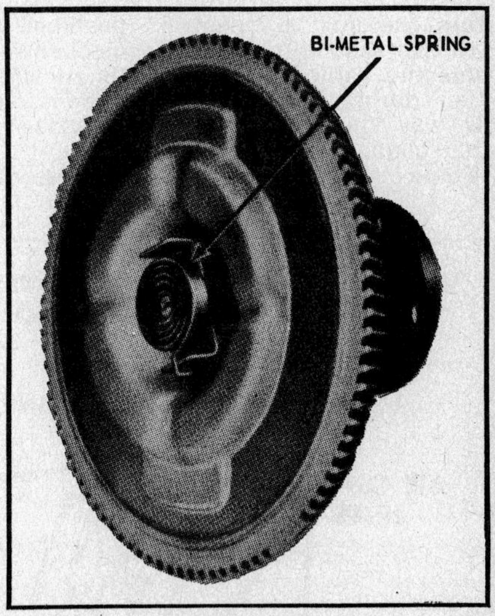

Fig. 3 Variable-speed fan with coiled bi-metal thermostatic spring

AUTOMATIC LEVEL CONTROL

SUPERLIFT SYSTEM

This system, Fig. 1, combines the Superlift shock absorber option with an air compressor and a height control valve. When the Superlift option is used alone, the shocks are filled with (or deflated of) compressed air at any gas station through a fill valve. When used in this system, however, the air pressure is supplied by the compressor and the amount of pressure added or removed is controlled by a sensing valve.

The compressor, a two stage type requiring no lubrication, is designed to operate off engine vacuum to replenish air used from the reservoir which is part of the compressor. As the compressor cycles, the reservoir air pressure gradually increases, causing a back pressure on the secondary stage piston, until it equals the engine vacuum pull against the diaphragm and the unit stops operating until reservoir pressure drops again.

A pressure regulator is attached to the output side of the compressor to limit reservoir outlet pressure to 125 psi. The rear standing height is controlled by a valve which is mounted to the frame, and senses changes in vehicle loading through a link attached to the suspension upper control arm. Changes in the position of the link cause the control valve to either admit or exhaust air from the shocks to return the link to "neutral" position. A 4 to 18 second time delay mechanism inside the control valve prevents transfer of air during normal ride conditions. In this way the system only responds to actual changes in vehicle loading.

Some later systems incorporate a vacuum regulator valve which consists of a relay valve and a deceleration valve. The relay valve is connected to the compressor and the P.C.V. line by rubber hose. The spark advance port of the regulator is connected into the distributor advance vacuum line and the intake manifold port is tapped into the rear of the intake manifold.

When the engine is at slow idle, the compressor will not operate due to insufficient vacuum, Fig. 3. As engine speed increases to fast idle or cruising speed, the increased vacuum is applied through a .020-.025″ orifice to the vacuum relay, overcoming valve spring tension and opening the relay valve. Vacuum from the P.C.V. line now acts upon the compressor and allows it to operate, Fig. 4.

During deceleration, vacuum at the rear of the intake manifold exceeds 17″ and vacuum at the spark advance port is negligible. Manifold vacuum overcomes deceleration valve spring tension, opening the valve and permitting vacuum to overcome the relay valve spring and open the relay valve to admit P.C.V. vacuum and operate the compressor, Fig. 5.

The Superlift, Fig. 6, is essentially a conventional shock absorber enclosed in an air chamber. A pliable nylon reinforced boot seals the dust tube (air dome) to the reservoir tube (air piston). The unit will extend when inflated and retract when deflated by the control valve. An 8 to 15 psi air pressure is maintained in the unit to minimize boot friction. This is accomplished by a check

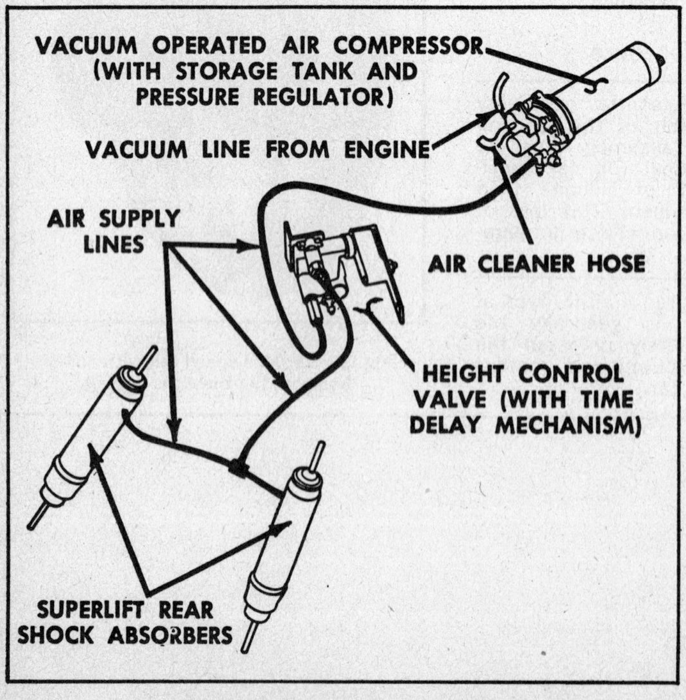

Fig. 1 Schematic diagram of Superlift Automatic Level Control System

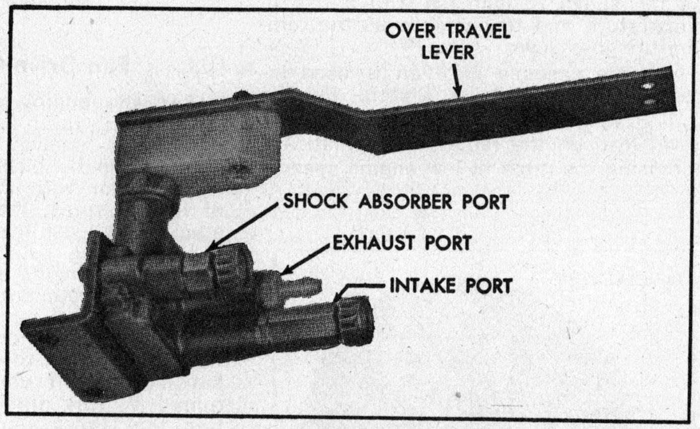

Fig. 2 Height control valve

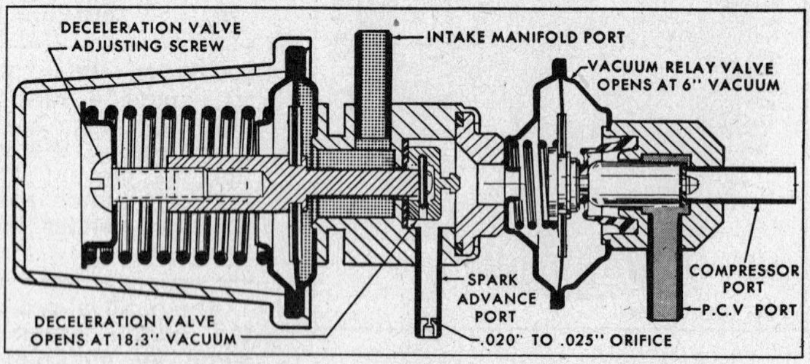

Fig. 3 Vacuum regulator valve in slow idle position

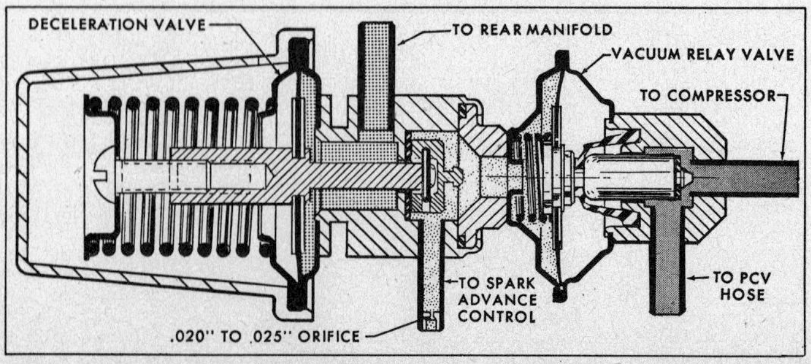

Fig. 4 Vacuum regulator valve in fast idle position

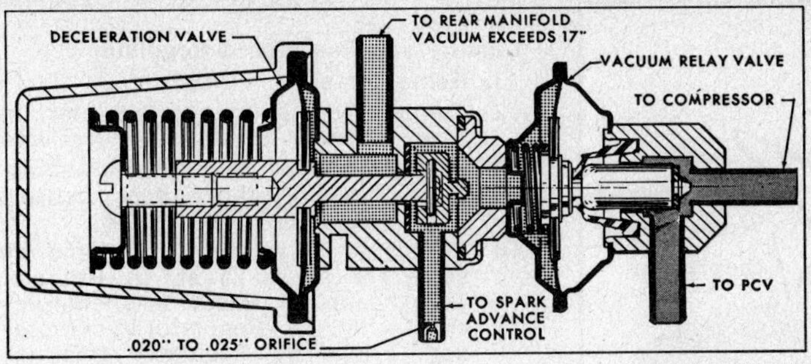

Fig. 5 Vacuum regulator valve in deceleration position

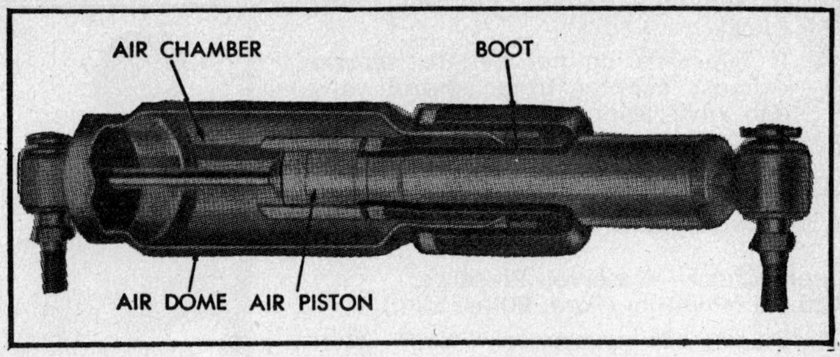

Fig. 6 Superlift shock absorber

valve in the exhaust fitting of the control valve.

TROUBLE SHOOTING GUIDE

Car Loaded, Will Not Rise

1. External damage or breakage.
2. Line leak.
3. Linkage to overtravel lever in wrong hole.
4. Control valve setting incorrect.
5. Defective component.

Car Loaded, Raises to Level, Then Leaks Down

1. Line leak.
2. Control valve exhaust leak.
3. Superlift leak.
4. Control valve leak.

Car Loaded, Raises Partially

1. Load excessive (over 500 lbs at axle) on cars with special springs.
2. Control valve setting incorrect.
3. Low supply pressure.

Car Unloaded, Rides too High, Will Not Come Down

1. Control valve setting incorrect.
2. Improper springs.
3. External damage or breakage.

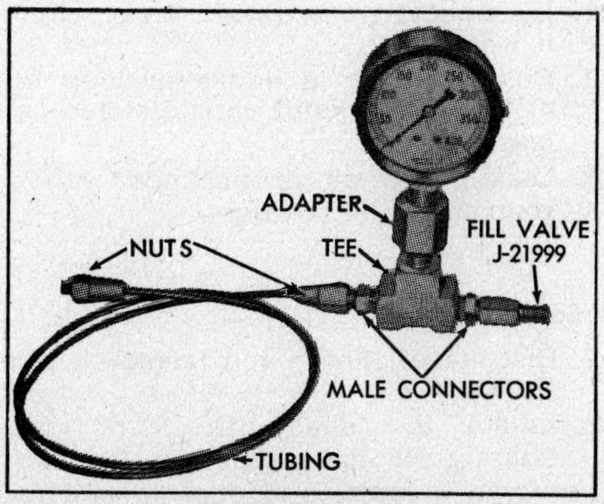

Fig. 7 Test gauge set
(Kent-Moore No. J-22124)

4. Linkage to overtravel in wrong hole.
5. Defective control valve.

Car Rises When Loaded but Leaks Down While Driving

1. Time delay mechanism not functioning properly.

CHECKS & ADJUSTMENTS

Quick Check of System

1. Record rear trim height of empty car (measure from center of rear bumper to ground).
2. Add weight equivalent to two-passenger load to rear of car. Car should begin to level in 4 to 15 seconds, and final position should be (plus or minus) ½" of measured dimension.
3. Remove weight. After 4 to 18 seconds car should begin to settle. Final unloaded position should be within approximately (plus or minus) ½" of the original measured dimension.

NOTE: To service the system it will be necessary to secure the gauge set shown in Fig. 7 or make one out of the materials illustrated.

Compressor Output Test

1. With all accessories off, run engine until engine settles to hot idle speed. Then turn off ignition.
2. Deflate system through service valve, then remove high pressure line at regulator adapter and connect test gauge.
3. Inflate reservoir to 70 psi through service valve.
4. Observe test gauge for evidence of compressor air leak.
5. If leaking, proceed to leak-test compressor reservoir and regulator. If not leaking, continue this test.
6. With engine running at hot idle speed, observe reservoir build-up for five minutes. Reservoir pressure should build up to a minimum of 90 psi.
7. If compressor fails to cycle, make sure vacuum and air intake lines are open and unobstructed before removing compressor for repair.
8. If build-up is too slow, repair compressor.
9. Satisfactory build-up indicates system problems to be in the control section. However, again observe the test gauge for evidence of an air leak and proceed accordingly.

Regulator Test & Adjustment General Motors

1. Performance test the regulator with a known good compressor on the car.
2. Deflate system through service valve, remove line at regulator and connect test gauge at regulator adapter.
3. Inflate reservoir through service valve to maximum pressure available. If less than 140 psi, start engine to build-up pressure.
4. Regulated pressure on test gauge should build up to 100-130 psi and hold steady within this range.
5. Recheck regulated pressure by momentarily depressing valve core on test gauge and observe reading.
6. If regulated pressure exceeds 130 psi, replace regulator as a unit.

Control Valve Test

Exhaust—Superlifts Inflated

1. Disconnect control valve lever from link.
2. Hold lever down in exhaust position for a period of 15-18 seconds except on 1973 Ford, Mercury which is 30 seconds and two minutes on 1974 models. If time allows, Superlifts may be totally deflated.

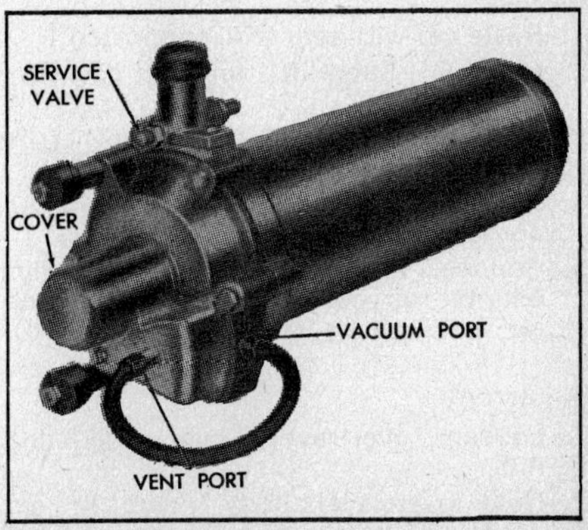

Fig. 8 Assembly leak test preparation

3. If Superlifts deflate, perform Intake Check.
4. If Superlifts do not deflate, remove exhaust adapter from control valve and hold lever down as in Step 2. Replace adapter, O-ring and filter if this deflates Superlifts.
5. Replace control valve if none of the above steps solve problem.

Intake Check—Reservoir Pressure, 125 psi Minimum (Ford, 90 psi Min.)

1. Disconnect overtravel lever from link.
2. Hold lever up in intake position for a period of 15-18 seconds except on 1973-75 Ford, Mercury which is a period of two minutes. If time allows, Superlifts may be totally inflated.
3. If Superlifts inflate and hold, proceed to Time Delay Test.
4. If Superlifts inflate and then leak down, perform leak test on lines and fittings and then on Superlifts and control valve. Repair or replace as required.

Time Delay Test Ford & GM Reservoir Pressure 125 PSI Minimum

1. Disconnect overtravel lever from link.
2. Disconnect lines at Superlift and intake ports.
3. Connect test gauge to intake valve port and apply air pressure (95 lbs.).
4. Move overtravel lever down approximately one inch from neutral position then quickly move lever up two inches. Air should begin to escape from the Superlift port in 4 to 30 seconds. Repeat test.
5. Remove test gauge and plug intake port with fill valve (female end).
6. Connect test gauge to the Superlift port and apply air pressure (95 lbs.).
7. Repeat Step 4. If either test is not within specifications, valve is defective or there has been a loss of silicone fluid.

Trim Adjustment On Car

Trim adjustment should be performed with a full gas tank or the equivalent in load at rate of 6 lbs. per gallon.

All Exc. 1972-74 Ford & Mercury

Preparation

1. Raise car with rear axle supported.
2. Remove Superlift line at control valve, Fig. 2.
3. Connect a Fill Valve Assembly (see Fig. 7).
4. Inflate Superlifts to 8 to 15 psi. Jounce car to neutralize suspension.
5. Connect test gauge to Superlift adapter on control valve and attach air pressure source (80 to 110 psi).

Adjustment

1. Loosen overtravel lever adjusting nut.
2. Hold overtravel body down in exhaust position until air escapes from exhaust valve port.

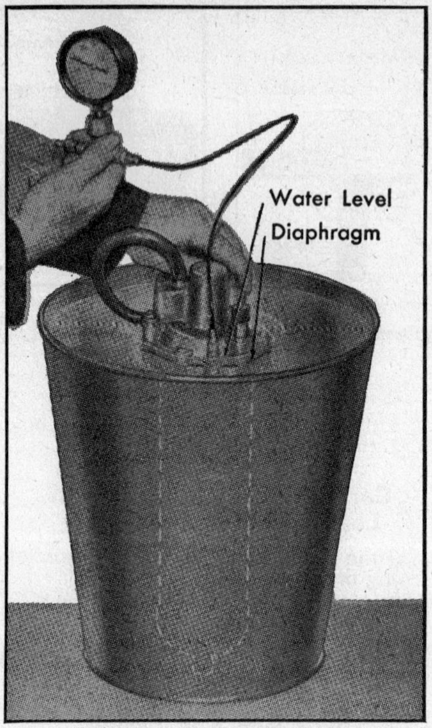

Fig. 9 Checking compressor, reservoir and regulator for leaks

3. Slowly move overtravel body and tighten nut at the point of minimum air bleed. With nut tight, a slight continuous air bleed should be noticeable.

Restore System

1. Remove test gauge and air pressure source from Superlift adapter.
2. Remove Fill Valve Assembly from Superlift line and reconnect line to control valve.
3. Lower car and inflate reservoir through service valve.

1972-75 Ford & Mercury

1. Support vehicle by the front and rear suspensions.
2. Disconnect control valve link from rear upper control arm and manually exhaust air from Superlifts.
3. Loosen control valve lever adjusting nut.
4. On 1972 models, place sufficient weight in rear of vehicle to allow for a trim height of 4.96 inches.
5. On all models, adjust lever arm until link enters control arm hole. Or if a downward movement of lever arm of $\frac{1}{16}$ inch on 1972 models and $\frac{1}{4}$ inch on 1973-75 models is required for a fit into the control arm hole.
6. Tighten adjusting nut and connect link to upper control arm.

Leak Tests

Compressor, Reservoir & Regulator

1. Remove assembly intact.
2. Connect test gauge to regulator. Inflate reservoir through service valve to 80-100 psi.
3. Route an 8″ rubber hose between vacuum and vent ports, Fig. 8.
4. Submerge in water, Fig. 9, and observe for air leaks at: a) Reservoir weld seam. b) Reservoir-to-compressor O-ring. c) Regulator-to-compressor O-ring. d) Regulator boot defective. e) Boot internal O-ring defective. f) Diaphragm between 1st and 2nd stage housing. g) Tightening thru bolts may correct leak. h) Cover gasket and retainer screw. A few bubbles here is not a leak. A continuous stream indicates defective compressor check valves. i) Service valve. j) Test gauge connections.
5. Correct any leaks detected by either tightening screws or replacing parts.

Control Valve

1. Remove control valve from car.
2. Clean exterior of valve thoroughly.
3. Connect test gauge and air pressure source to intake adapter and open air pressure (80-110 psi).
4. Submerge unit in water. No air should escape if overtravel lever is in "neutral" position. If bubbles escape from Superlift port, replace control valve.
5. Shut off air pressure and detach test gauge from air intake port. Plug intake port with Fill Valve Assembly.
6. Connect test gauge to Superlift port and open air pressure.
7. With overtravel lever in "neutral" position, no air should escape. If bubbles escape from exhaust port, replace control valve.
8. If air escapes around edge of cover plate, tighten screws or replace gasket.
9. Remove control valve from water. Actuate overtravel lever to expel any water from unit.
10. Shut off air pressure and remove line from Superlift port.

Lines and Fittings

1. Disconnect overtravel lever from link.
2. Hold lever up in intake position for maximum Superlift inflation and release.
3. Leak check all connections with a soap and water solution.

Superlifts

1. Disconnect lines and remove unit from car.
2. Inflate individually to 50-60 psi, utilizing Fill Valve (see Fig. 7). Submerge in water and observe unit for leaks.
3. Install Superlifts.

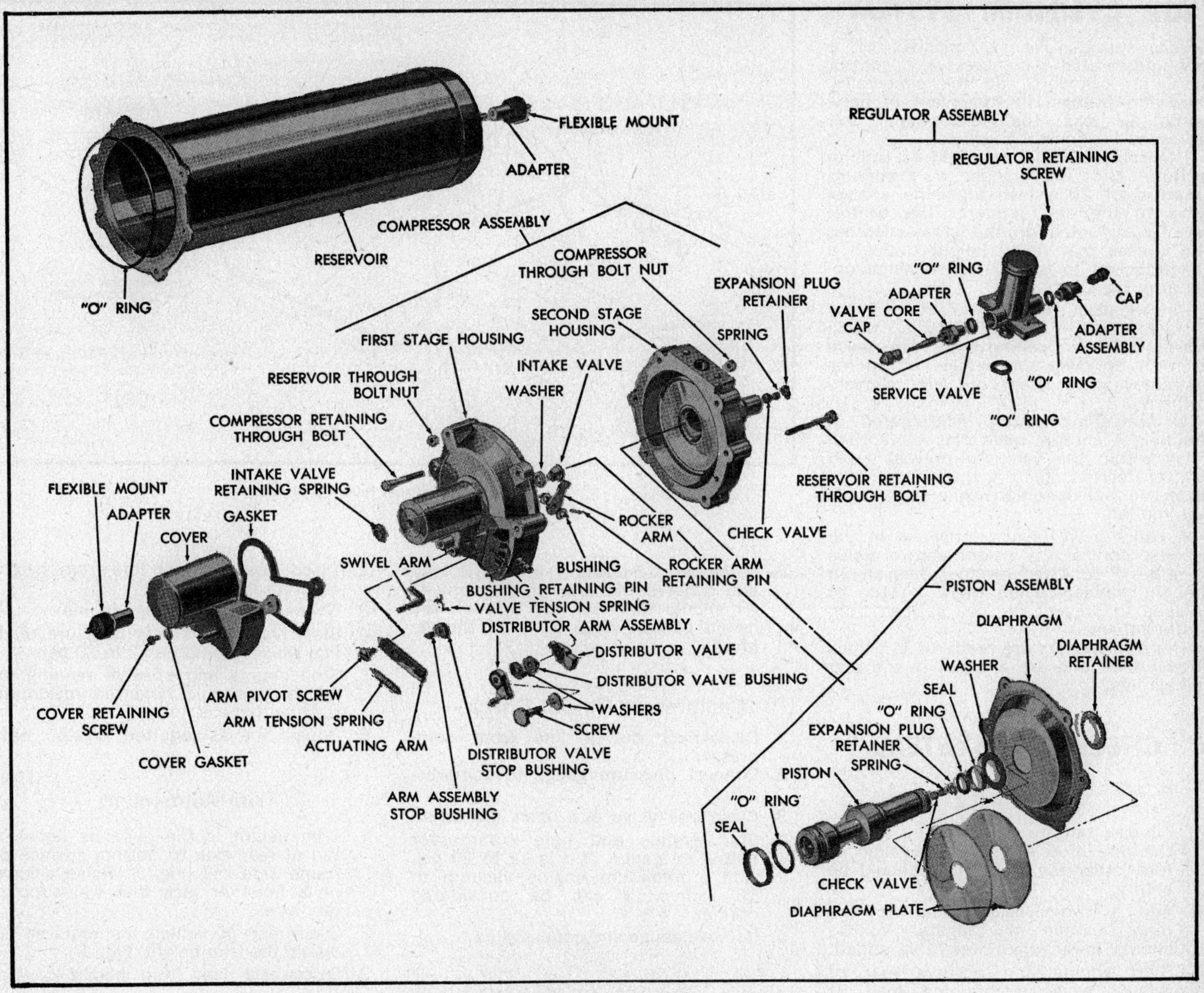

Fig. 10 **Exploded view of General Motors compressor, reservoir and regulator**

COMPRESSOR SERVICE

Removal

1. Disconnect vacuum and air intake hoses.
2. Remove air intake and vacuum hoses.
3. Unfasten compressor brackets from mounting points.
4. Remove compressor with brackets attached.
5. Deflate system, using service valve, and remove high pressure fitting at pressure regulator.
6. Remove brackets from compressor.

NOTE: The compressor is a precision-built mechanism, Fig. 10. If an overhaul is contemplated, all parts should be handled carefully. Take care to prevent entrance of dirt or foreign matter. Do not lubricate as unit is designed to run dry.

Installation

1. Attach brackets to compressor.
2. Install compressor to its mounting.
3. Attach air intake and vacuum hoses.
4. Secure air line to compressor pressure regulator.
5. Lower car.
6. Inflate reservoir to 140 psi through compressor service valve.
7. Be sure that vacuum and air intake lines are not rubbing against adjacent parts to prevent chafing.

AIR CYLINDER SYSTEM

This system, Fig. 1, consists of a vacuum-operated compressor, control valve, air cylinders and the connecting lines and fittings. In the event of accidental air loss, the conventional coil springs will support the vehicle.

The compressor is operated off engine vacuum and will supply a maximum pressure of 20 psi through the control valve to the air cylinders. The control valve is mounted on the cross member and senses rear riding height through a link attached to the rear suspension upper arm, Fig. 2.

The control valve is actuated by changes in the riding height, which moves the link, opening the proper valve to raise or lower the vehicle by adding or removing air from the air cylinders at the rear.

A dampening piston incorporated in the height control valve acts as a time delay within the valve to prevent rapid air transfer to and from the air cylinders under normal operation while the vehicle is in motion.

A check valve is also provided in the exhaust port of the height control valve to retain 2 to 4 psi residual pressure in the air cylinders when there is little or no load to prevent the coil springs from scuffing them.

The air cylinders are made of $\frac{1}{4}''$ thick butyl rubber and are encased in the rear coil springs.

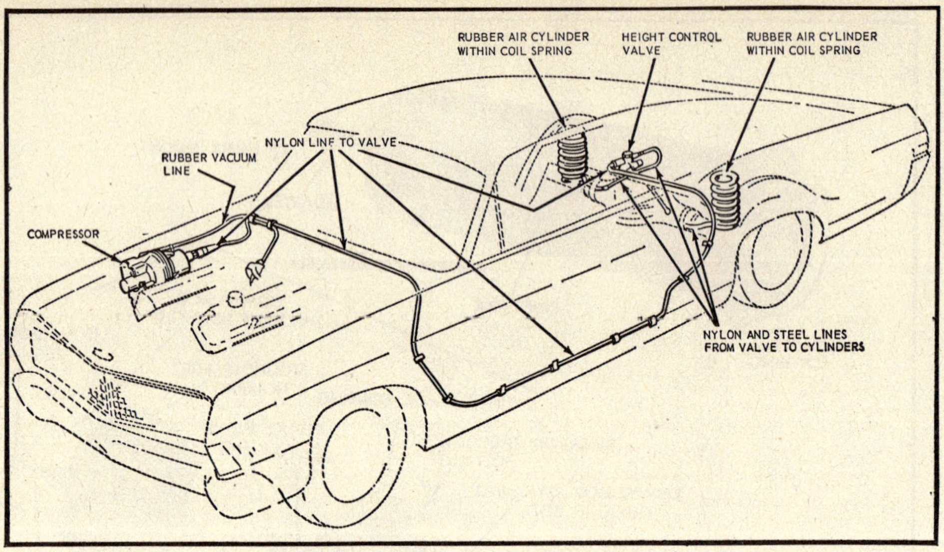

Fig. 1 Air cylinder leveling system

CHECKING SYSTEM

Quick Check

1. Fill fuel tank or simulate the load at the rate of 6 lbs for each gallon of fuel, otherwise the vehicle should be empty.
2. Add a two-passenger load to the rear bumper or tailgate. Vehicle should lower when weight is added.
3. Start engine and observe rear of vehicle while maintaining load at rear bumper or tailgate.
4. If vehicle does not raise when engine is started or within two minutes, the system is not operating properly.
5. If the vehicle raises, remove the load and observe the rear of the vehicle. Air should then exhaust through the height control valve to lower the vehicle.

Compressor Output Check

1. Disconnect output line from compressor.
2. Connect pressure gauge to compressor.
3. Close one valve and open the other.
4. Start engine and note compressor output on gauge. If it is 12 to 20 psi, with a minimum engine vacuum of 15 inches, it can be considered normal.
5. Remove gauge and connect line.

Compressor-to-Control Valve Line Leak Check

1. Disconnect inlet fitting from height control valve.
2. Disconnect air inlet line from height control valve.
3. Connect pressure gauge to line.
4. Start engine and note pressure reading, which should be 12 to 20 psi.
5. Stop engine and observe reading on gauge. It should hold the maximum reading if line is not leaking.
6. Repair line as required to stop leak.

Trim Adjustment

1. Trim height is the distance between top of rear axle to bottom surface of frame side rail, Fig. 3. Raise vehicle on a hoist or jack that will support rear axle.
2. Load rear of vehicle as required to obtain the trim height, Fig. 3.
3. Disconnect link from height control

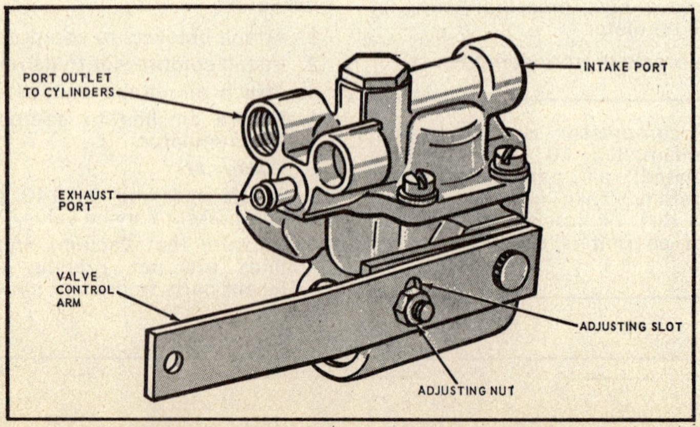

Fig. 2 Height control valve

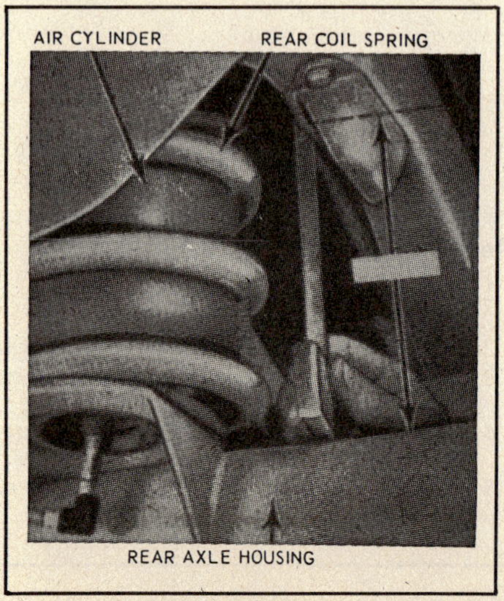

Fig. 3 Measuring trim adjustment; dimension should be 6.70″. Ford and Mercury

valve lever, Fig. 4.
4. Disconnect air line from inlet port of height control valve.
5. Connect test gauge to inlet port of height control valve.
6. Working from a 20 psi air source, hold height control valve lever in raised position for 60 seconds while inflating the system. Lower lever to the exhaust position. Hold lever in this position until air stops exhausting. This should leave approximately 2 to 4 psi in the air cylinders if the check valve is functioning properly.
7. Jounce vehicle to neutralize system.
8. Connect link to height control valve lever.
9. Readjust vehicle trim height to 6.70″ if not at this dimension.
10. Loosen height control valve lever adjusting nut and allow it to neutralize itself. After the lever is neutralized, tighten adjusting nut.
11. Remove gauge and added weight from vehicle.
12. Connect air line to height control valve and check for leaks.

Control Valve Check

Exhaust Check, Air Cylinders Inflated

1. Disconnect link from height control valve lever.
2. Hold lever down (exhaust position) until air cylinders deflate, or for a minimum of 30 seconds. Release arm and allow it to return to neutral position.
3. If cylinders deflate perform a trim adjustment.
4. If cylinders do not deflate, disconnect air cylinder line at height control valve to make sure that there is no restriction in line.
5. Connect air cylinder line to control valve and link to lever.

Intake Check, Compressor at Engine Idle

1. Disconnect link from height control valve lever.
2. Hold lever up (intake position) until cylinders inflate or for a minimum of 60 seconds, then release lever and allow it to return to neutral position.
3. If cylinders inflate and hold perform a trim adjustment.
4. If cylinders inflate then leak down, perform a leak test on lines or fittings then check air cylinders.
5. If cylinders do not inflate, disconnect compressor output line at the height control valve and check it for restrictions.
6. If lines and fittings are satisfactory, check height control valve for leaks.
7. Connect link to control valve lever.

Time Delay Check

1. Disconnect link from control valve lever.
2. Disconnect both air lines from height control valve, Fig. 4.
3. Connect test gauge to intake port of control valve. Attach a 20 psi air source to gauge.
4. Move control valve lever downward approximately one inch from the neutral position. One inch distance is measured at end of lever.

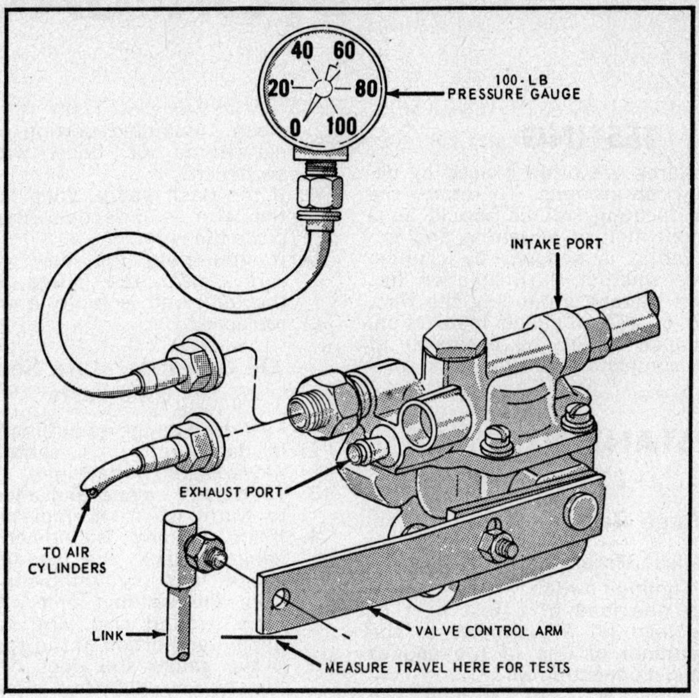

Fig. 4 Test gauge installation

5. Move lever upward two inches, at the same time start timing the number of seconds before air is expelled from air cylinder port. The time delay should be from 1 to 6 seconds. Repeat check to obtain an accurate reading. This operation is for air intake time only.
6. Connect air cylinder line to control valve outlet port.
7. Hold control valve lever in raised position and charge air cylinders with 15 to 20 psi air pressure from intake port of control valve.
8. Move lever to neutral position.
9. Move control valve lever upward approximately one inch from neutral position. One inch dimension is measured from end of lever. Quickly move lever downward two inches, at the same time start timing the number of seconds before air is expelled from exhaust port. The time delay should be from 1 to 6 seconds. Repeat check to obtain accurate reading.
10. If either delay is not within specifications, replace height control valve.

System Leak Test

1. Start engine and allow pressure to build up in system.
2. Apply a soap and water solution to all fittings and lines that are suspected to be leaking. If air bubbles appear at any fitting or section of line, make required repairs to eliminate the leak.

TROUBLE DIAGNOSIS

Vehicle Loaded, Will Not Raise

1. External damage or breakage.
2. Line or cylinder leak.
3. Pump inoperative or output inadequate.
4. Control valve setting incorrect.
5. Inadequate time delay.

Vehicle Loaded, Raises Partially

1. Load excessive (over 250 lbs) at axle.
2. Height control valve setting incorrect.
3. Low supply pressure.

Vehicle Unloaded, Rides too High, Won't Come Down

1. Control valve setting incorrect.
2. External damage or breakage.
3. Defective control valve.

Compressor Cycles Continuously

1. Line leak.
2. Air cylinder ruptured.
3. Inadequate time; may take five minutes to balance at idle.

Compressor Does Not Cycle

1. Vacuum hose off or leaking.
2. Pump internal failure.
3. Lines or hoses restricted.
4. Pump filter clogged.

Vehicle Loaded, Raises to Level and Then Leaks Down

1. Line leak.
2. Control valve exhaust leak.
3. Air cylinder leak.
4. Control valve exhaust leak.
5. If leak down while driving, check for control valve time delay less than one second.

DASH GAUGES

TESTING

Gauge failures are often caused by defective wiring or grounds. Therefore, the first step in locating trouble should be a thorough inspection of all wiring and terminals. If wiring is secured by clamps, check to see whether the insulation has been severed thereby grounding the wire. In the case of a fuel gauge installation, rust may cause failure by corrosion at the ground connection of the tank unit.

CONSTANT VOLTAGE TYPE

Voltage Regulator Test

Except American Motors

1. Turn on ignition switch.
2. Connect one lead of a test light or positive lead of a voltmeter to the feed terminal of one of the gauges without disconnecting wire.
3. If regulator is okay, voltage will oscillate.
4. If it does not, voltage regulator is defective or there is a short or ground between regulator and gauges.

CAUTION: Applying 12 volts to any part of the system except the regulator input terminal or grounding the regulator or system in any way except to connect test equipment may burn out one or more components. When replacing any part of the system, battery ground cable must be disconnected.

American Motors

1. Connect a 10 ohm resistor in series with each indicator wire.
2. Ground fuel and temperature gauge wires.
3. The fuel gauge should read Full to two needle widths above. The temperature gauge should read Hot to two needle widths above.

Dash Gauge Test

1. Turn off ignition switch.
2. Connect terminals of two series-connected flashlight batteries to the gauge terminals in question (fuel, oil or temperature).
3. The three volts of the batteries should cause the gauge to read approximately full scale.
4. If the gauge unit is inaccurate or does not indicate, replace it with a new unit.
5. If the gauge unit still is erratic in its operation, the sender unit or wire to the sender unit is defective.

Fuel Tank Gauge Test

1. Test the dash gauge as outlined above.
2. If dash gauge is satisfactory, remove flashlight batteries.
3. Then disconnect wire at tank unit and ground it momentarily to a clean, unpainted portion of the vehicle frame or body *with ignition switch on.*
4. If the dash gauge does not indicate, the wire is defective. Repair or replace the wire.
5. If grounding the new or repaired wire causes the gauge to indicate, the tank unit is faulty and should be replaced.

Oil & Temperature Sending Unit Tests

1. Test dash gauge as outlined above.
2. If dash gauge is satisfactory, remove flashlight batteries.
3. Then start engine and allow it to run to warm up to normal temperature.
4. If no reading is indicated on the gauge, check the sending unit-to-gauge wire by removing the wire from the sending unit and momentarily ground this wire to a clean, unpainted portion of the engine.
5. If the gauge still does not indicate, the wire is defective. Repair or replace the wire.
6. If grounding the new or repaired wire causes the dash gauge to indicate, the sending unit is faulty.

VARIABLE VOLTAGE TYPE

The procedure given herewith applies to AC, Auto-Lite and Stewart-Warner systems. Following is the method of quickly checking the gauge system to determine which component (sender or receiver) of a given system is defective.

Fuel Gauge Tank Unit Method

1. Use a spare gauge tank unit known to be correct.
2. To test whether the dash gauge in question (fuel, oil or temperature) is functioning, disconnect the wire at the gauge which leads to the sending unit.
3. Attach a wire lead from the dash gauge terminal to the terminal of the "test" tank gauge, Fig. 1.
4. Ground the test tank unit to an unpainted portion of the dash panel and move the float arm.
5. If the gauge operates correctly, the sending unit is defective and should be replaced.
6. If the gauge does not operate during this test, the dash gauge is defective and should be replaced.

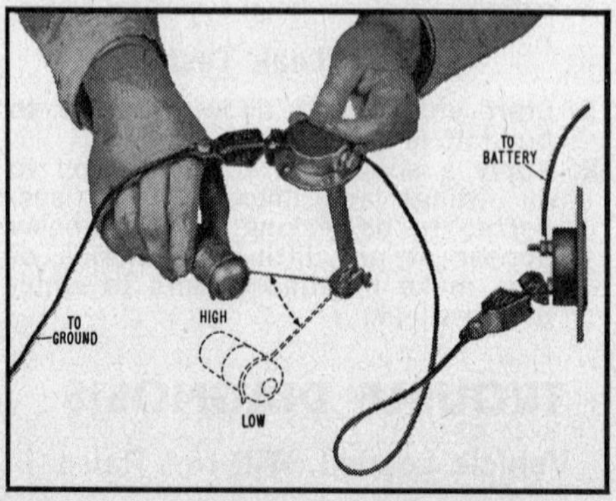

Fig. 1 Hook-up for testing dash gauge with a spare tank unit

AMMETERS

This instrument shows whether the battery is being charged by the generator or alternator or is being discharged by lights, radio, engine, etc. If a constant discharge is indicated on the ammeter, it is a signal that the battery is being run down. It is often a signal that the generator is out of order. Since both a charged battery and a working generator are very necessary—especially with vehicles equipped with many electricity-consuming devices such as heater, defroster, fog lights, radio, etc.—an inoperative ammeter should be given prompt attention.

The typical ammeter, Fig. 2, consists of a frame to which a permanent magnet is attached. The frame also supports an armature and pointer assembly.

When no current flows through the ammeter, the magnet holds the pointer armature so that the pointer stands at the center of the dial. When current passes in either direction through the ammeter, the resulting magnetic field attracts the armature away from the effect of the permanent magnet, thus giving a reading proportional to the strength of the current flowing.

Trouble Shooting

When the ammeter apparently fails to register correctly, there may be trouble in the wiring which connects the ammeter to the generator and battery or in the generator or battery themselves.

To check the connections, first tighten the two terminal posts on the back of the ammeter. Then, following each wire from the ammeter, tighten all connections on the ignition switch, battery and generator. Chafed, burned or broken insulation can be found by following each ammeter wire from end to end.

All wires with chafed, burned or broken insulation should be repaired or replaced. After this is done, and all connections are tightened, connect the battery cable and turn on the ignition switch. The needle should point slightly to the discharge (–) side.

Start the engine and speed it up a little above idling speed. The needle should then move to the charge side (+), and its movement should be smooth.

If the pointer does not behave correctly, the ammeter itself is out of order and a new one should be installed.

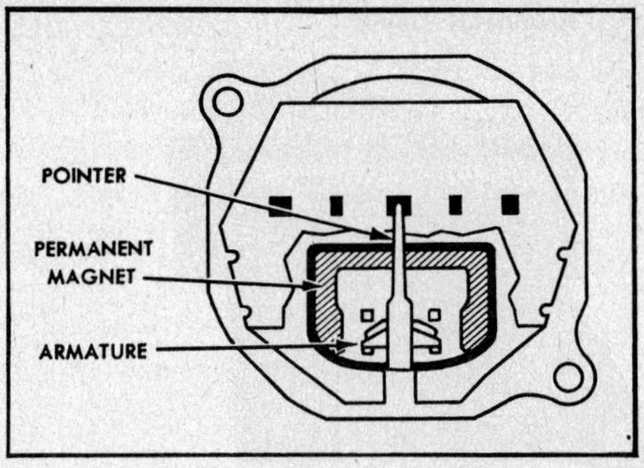

Fig. 2 Drawing of a typical automobile ammeter

ALTERNATOR INDICATOR LIGHT

A red alternator "no charge" light is used on many cars in lieu of an ammeter. This light flashes on if the battery is discharging and the alternator is not supplying current.

The light should glow when the ignition is turned on and before the engine is started. If the bulb does not light, either the bulb is burned out or the indicator light wiring has an open circuit. After the engine is started, the light should be out at all times.

If the light fails to go out when the engine is running, the drive belt may be loose or missing or the alternator or voltage regulator may be defective.

Light Circuit

A double contact voltage regulator together with a field relay is used on Delco-Remy and Ford alternators when used with the indicator light. The circuit is as follows:

With the ignition switch turned on (engine not running), current flow is through the ignition switch through the indicator light on the dash panel. From there it goes to a terminal of the regulator (marked "4" or "L" on Delco-Remy or "I" on Ford). The circuit continues through the lower contacts of the voltage regulator (held closed by a spring), out the "F" terminal of the regulator, in the "F" terminal of the alternator, through a brush and slip ring, through another brush and slip ring to ground.

After the engine is started, the voltage output of the alternator immediately closes the field relay. This causes battery voltage from the battery terminal of the regulator (marked "3" or "V" on Delco-Remy, "B" on Ford) to be present at the "4", "L" or "I" terminal. Since battery voltage is present on both sides of the indicator light, the light goes out.

If the generator light comes on with the engine running, the charging circuit should be tested as soon as possible to determine the cause of the trouble.

A typical light circuit is shown in Fig. 3.

OIL PRESSURE INDICATOR LIGHT

Many cars utilize a warning light on the instrument panel in place of the conventional dash indicating gauge to warn the driver when the oil pressure is dangerously low. The warning light is wired in series with the ignition switch and the engine unit—which is an oil pressure switch.

The oil pressure switch contains a diaphragm and a set of contacts. When the ignition switch is turned on, the warning light circuit is energized and the circuit is completed through the closed contacts in the pressure switch. When the engine is started, build-up of oil pressure compresses the diaphragm, opening the contacts, thereby breaking the circuit and putting out the light.

Trouble Shooting

The oil pressure warning light should go on when the ignition is turned on. If it does not light, disconnect the wire from the engine unit and ground the wire to the frame or cylinder block. Then if the warning light still does not go on with the ignition switch on, replace the bulb.

If the warning light goes on when the wire is grounded to the frame or cylinder block, the engine unit should be checked for being loose or poorly grounded. If the unit is found to be tight and properly grounded, it should be removed and a new one installed. (The presence of sealing compound on the threads of the engine unit will cause a poor ground).

If the warning light remains lit when it normally should be out, replace the engine unit before proceeding further to determine the cause for a low pressure indication.

The warning light sometimes will light up or will flicker when the engine is idling, even though the oil pressure is adequate. However, the light should go out when the engine is speeded up. There is no cause for alarm in such cases; it simply means that the pressure switch is not calibrated precisely correct.

TEMPERATURE INDICATOR LIGHTS

A temperature (bimetal) switch, located in cylinder head, controls the operation of a "Cold" temperature indicator light with a green lens and a "Hot" temperature indicator light with a red lens. When the cooling system water temperature is below approximately 110 defrees F., the temperature switch grounds the "Cold" indicator circuit and the green light goes on. When the green light goes out, the water temperature is high enough so that the heater can be turned on and be effective. *Note: The car should never be subjected to full throttle accelerations or high speeds until after the green light has gone out.*

If the engine cooling system is not

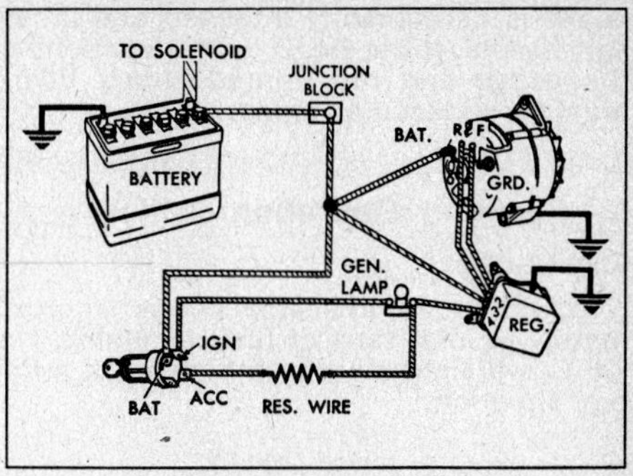

Fig. 3 Wiring diagram of a typical charge indicator light circuit

functioning properly and the water temperature should reach a point where the engine approaches an overheated condition, the red light will be turned on by the temperature switch.

NOTE: *As a test circuit to check whether the red bulb is functioning properly, a wire which is connected to the ground terminal of the ignition switch is tapped into its circuit. When the ignition is in the "Start" (engine cranking) position, the ground terminal is grounded inside the switch and the red bulb will be lit. When the engine is started and the ignition switch is in the "On" position, the test circuit is opened and the bulb is then controlled by the temperature switch.*

Trouble Shooting

If the red light is not lit when the engine is being cranked, check for a burned out bulb, an open in the light circuit, or a defective ignition switch.

If the red light is lit when the engine is running, check the wiring between light and switch for a ground, temperature switch defective, or overheated cooling system.

If the "Cold" light is not lit when ignition is on and engine cold, check for a burned out bulb, an open in the light circuit, or a defective temperature switch.

If the "Cold" light stays on after normal engine warm-up period, check for a ground between light and switch, defective temperature switch, or a defective cooling thermostat.

CHRYSLER GAUGE ALERT SYSTEM

L.E.D. (Light Emitting Diode)

The fuel, temperature and ammeter gauges are equipped with a L.E.D. (Light Emitting Diode) mounted in each of the gauge dials, Fig. 4. This diode will illuminate and alert the driver that the system the gauge is monitoring is malfunctioning. The electronic sensor circuit is mounted on the gauge housing. The printed circuit board is permanently attached and is not serviceable. If the L.E.D. is malfunctioning, the gauge and the printed circuit board must be replaced as an assembly.

Operation

Fuel Gauge

When gauge indicator shows approximately $\frac{1}{8}$ of a tank of fuel remaining, the L.E.D. will light alerting the driver of a low fuel situation.

Temperature Gauge

When gauge indicator shows engine temperature approximately 240 to 250 degrees F. the L.E.D. will light alerting the driver of an overheat condition.

Ammeter Gauge

This L.E.D. operates independently of the gauge indicator and monitors system voltage. The L.E.D. will alert the driver of three charging system potential malfunctions.

1. A discharging condition, caused by excessive electrical demand on charging system, (engine at idle rpm).
2. A weak or defective battery with ignition switch in the "ON" position, (before the ignition switch is moved to the "START" position).
3. A weak or defective battery with minimum demand on charging system, while vehicle is being used in stop and go driving (intermittent L.E.D. illumination occurring).

Testing

Fuel And Temperature L.E.D.

Use testor C-3826 for diagnosing systems.

Ammeter L.E.D.

NOTE: Only if battery and charging system are functioning properly can the following test be performed.

Turn ignition switch to the "ON" position and turn on headlights, windshield wipers and stoplights. This will cause excessive demand on charging system activating the L.E.D. immediately or within approximately one minute, if the L.E.D. does not light there is a malfunction in the system. If L.E.D. lights, run engine at approximately 2000 rpm, L.E.D. should stop emitting light, if the L.E.D. continues to emit light there is a malfunction in the system.

NOTE: In all cases of system malfunctions the complete gauge must be replaced.

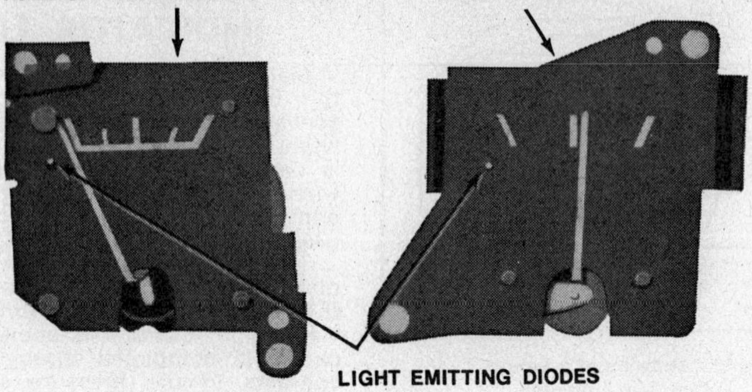

Fig. 4 Gauges incorporating the L.E.D. system

CHRYSLER FUEL PACER SYSTEM

The Fuel Pacer system, Figs. 5, 6 and 7, incorporates a vacuum switch, relay, vacuum hoses, diodes and the use of the left front fender mounted turn signal indicator lamp. When engine vacuum falls below $4\frac{1}{2}$ inches Hg., electrical contacts in the vacuum switch close, causing the indicator lamp to light, therefore indicating excessive fuel consumption by the engine.

The relay dims the indicator lamp during night operations by routing the current through a resistor, in turn, dimming the indicator lamp. The relay is activated when the parking lamps or headlamps are "On."

The Fuel Pacer system overrides the blinking action of the fender mounted left turn indicator lamp. However, an inline diode is incorporated in the circuit so the complete turn signal system is not affected. A second in-line diode is placed in the fender mounted indicator lamp feed wire, preventing feedback through the main harness and therefore avoiding some accessories from functioning when the hazard warning flashers are "On", with the ignition switch in the "Accessory" position.

Testing

Relay Tests

1. Turn ignition switch "On" and observe the left front fender mounted turn indicator lamp. If indicator lamp is lit, relay is satisfactory. If not, check for the following: defective wiring at the vacuum switch, relay and signal lamp, inoperative indicator bulb, open vacuum switch or a defective relay.
2. With ignition switch "On", turn headlamps "On". If indicator lampdims, relay is satisfactory. If not, check wiring at relay windings. If wiring is satisfactory, the relay is defective.

Vacuum Switch Tests

1. Disconnect the switch vacuum hose, turn ignition switch "On" and observe left front fender mounted turn signal indicator lamp. If lamp is lit, switch is satisfactory. If not, check for the following: defective wiring at vacuum switch, relay and signal lamp, inoperative indicator bulb, defective relay or a faulty vacuum switch.
2. Connect an external vacuum source to the vacuum switch, turn ignition "On", increase vacuum to switch and observe indicator lamp. If indicator lamp does not turn off at approximately $4\frac{1}{2}$ inches of vacuum, apply more vacuum until lamp goes out. If lamp goes out at this time, the switch must be adjusted. If lamp remains lit, regardless of amount of vacuum applied, the switch is defective, requiring replacement.

Vacuum Switch Adjustment

1. If 5.2 inches of vacuum or more was required to turn indicator lamp off,

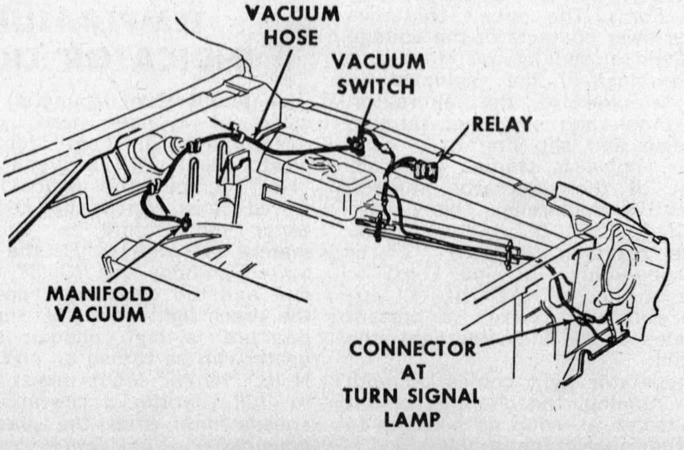

Fig. 5 Fuel pacer system. 1975 Dart & Valiant

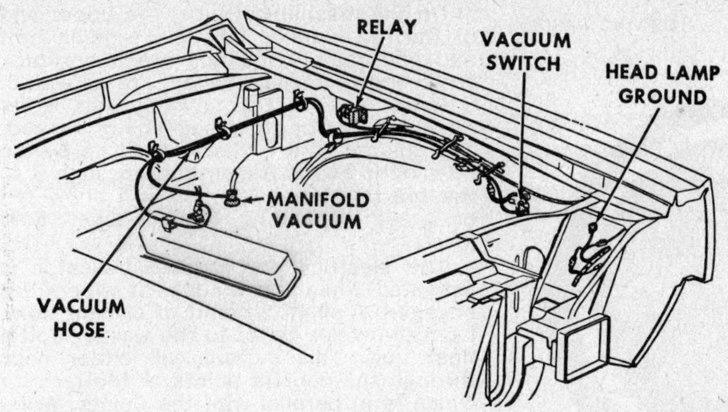

Fig. 6 Fuel pacer system. 1975 Charger SE, Cordoba, Coronet & Fury

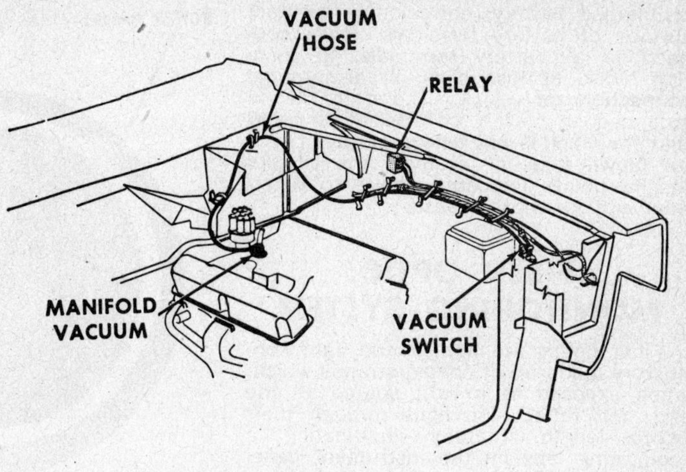

Fig. 7 Fuel pacer system. 1975 Chrysler, Gran Fury, Imperial & Monaco

turn adjusting screw counter-clockwise to decrease amount of vacuum required.

2. If 3.8 inches of vacuum or less was required to turn vacuum switch off, turn adjusting screw clockwise to increase amount of vacuum required.

NOTE: If more than two turns of adjusting screw (in either direction) was required to adjust switch, the switch is defective, requiring replacement.

SPEEDOMETERS

The following material covers only that service on speedometers which is feasible to perform by the average service man. Repairs on the units themselves are not included as they require special tools and extreme care when making repairs and adjustments and only an experienced speedometer mechanic should attempt such servicing.

The speedometer has two main parts—the indicating head and the speedometer drive cable. When the speedometer fails to indicate speed or mileage, the cable or housing is probably broken.

Speedometer Cable

Most cables are broken due to lack of lubrication or a sharp bend or kink in the housing.

A cable might break because the speedometer head mechanism binds. If such is the case, the speedometer head should be repaired or replaced before a new cable or housing is installed.

A "jumpy" pointer condition, together with a sort of scraping noise, is due, in most instances, to a dry or kinked speedometer cable. The kinked cable rubs on the housing and winds up, slowing down the pointer. The cable then unwinds and the pointer "jumps".

To check for kinks, remove the cable, lay it on a flat surface and twist one end with the fingers. If it turns over smoothly the cable is not kinked. But if part of the cable flops over as it is twisted, the cable is kinked and should be replaced.

Lubrication

The speedometer cable should be lubricated with special cable lubricant every 10,000 miles.

Fill the ferrule on the upper end of the housing with the cable lubricant. Insert the cable in the housing, starting at the upper end. Turn the cable around carefully while feeding it into the housing. Repeat filling the ferrule except for the last six inches of cable. Too much lubricant at this point may cause the lubricant to work into the indicating hand.

Installing Cable

During installation, if the cable sticks when inserted in the housing and will not go through, the housing is damaged inside or kinked. Be sure to check the housing from one end to the other. Straighten any sharp bends by relocating clamps or elbows. Replace housing if it is badly kinked or broken. Position the cable and housing so that they lead into the head as straight as possible.

Check the new cable for kinks before installing it. Use wide, sweeping, gradual curves where the cable comes out of the transmission and connects to the head so the cable will not be damaged during its installation.

If inspection indicates that the cable and housing are in good condition, yet pointer action is erratic, check the speedometer head for possible binding.

The speedometer drive pinion should also be checked. If the pinion is dry or its teeth are stripped, the speedometer may not register properly.

The transmission mainshaft nut must be tight or the speedometer drive gear may slip on the mainshaft and cause slow speed readings.

ELECTRIC CLOCKS

Regulation of electric clocks used on automobiles is accomplished automatically by merely resetting the time. If the clock is running fast, the action of turning the hands back to correct the time will automatically cause the clock to run slightly slower. If the clock is running slow, the action of turning the hands forward to correct the time will automatically cause the clock to run slightly faster (10 to 15 seconds a day).

Winding Clock When Connecting Battery or Clock Wiring

The clock requires special attention when reconnecting a battery that has been disconnected for any reason, a clock that has been disconnected, or when replacing a blown clock fuse. *It is very important that the initial wind be fully made.* The procedure is as follows:

1. Make sure that all other instruments and lights are turned off.
2. Connect positive cable to battery.
3. Before connecting the negative cable, press the terminal to its post on the battery. Immediately afterward strike the terminal against the battery post to see if there is a spark. If there is a spark, allow the clock to run down until it stops ticking, and repeat as above until there is no spark. Then immediately make the permanent connection before the clock can again run down. The clock will run down in approximately two minutes.
4. Reset clock after all connections have been made. *The foregoing procedure should also be followed when reconnecting the clock after it has been disconnected, or if it has stopped because of a blown fuse. Be sure to disconnect battery before installing a new fuse.*

Trouble Shooting

If clock does not run, check for blown "clock" fuse. If fuse is blown check for short in wiring. If fuse is not blown check for open circuit.

With an electric clock, the most frequent cause of clock fuse blowing is low voltage at the clock which will prevent a complete wind and allow clock contacts to remain closed. This may be caused by any of the following:

discharged battery, corrosion on contact surface of battery terminals, loose connections at battery terminals, at junction block, at fuse clips, or at terminal connection of clock. Therefore, if in reconnecting battery or clock it is noted that the clock is not ticking, always check for blown fuse, or examine the circuits at the points indicated above to determine and correct the cause.

FIBER OPTIC MONITORING SYSTEM

Fiber optics are non-electric light conductors made up of coated strands which, when exposed to a light source at one end, will reflect the light through their entire length, thereby illuminating a monitoring lens on the instrument panel or fender without the use of a bulb when the exterior lights are turned on.

LOW FUEL WARNING SYSTEM

There are two types of low fuel warning systems, the thermistor and switch types. The thermistor type incorporates an indicator light, low fuel relay and a thermocouple assembly which is attached to the fuel sender outlet tube in the fuel tank. The thermistor temperature is low when submerged in fuel and, in turn, high electrical resistance is evident. When the fuel level drops and the thermistor is exposed to the air, the thermistor heat will increase thus decreasing its resistance allowing current to energize the low fuel relay and activate the indicator light. The switch type consists of an indicator light and a low fuel warning switch located on the instrument panel.

The warning switch contacts are closed by the difference in voltage potential between the fuel gauge terminals. This voltage differential will activate the warning switch when the fuel tank is less than 1/4 full and, in turn, cause the indicator to light.

Trouble Shooting

Both systems incorporate an indicator light which may be checked in the same manner. With ignition switch turned to "ON", the indicator should light. If not, check bulb and all electrical connections. On General Motors switch type, replace warning switch if bulb and connections prove satisfactory. On Ford thermistor and switch type systems, perform additional tests outlined below.

On Ford thermistor system, using an ohmmeter and thermistor assembly out of the fuel, connect one lead to metal housing of sending unit and the other lead to thermistor terminal of sending unit. The ohmmeter should then indicate a resistance of 450 to 600 ohms. If resistance is not within above limits, replace fuel sender.

Improper operation of the warning switch will be indicated when the light remains "ON" when tank is more than 1/4 full. To test system, disconnect connector on warning switch and turn ignition "ON". Starting with the terminal on the end opposite the blank position, connect a jumper wire between the battery positive terminal and connector terminal. The indicator on instrument panel should light. If not, replace warning switch. Skip the next connector terminal and connect a test lamp between the battery positive terminal and connector terminal. The test lamp should light. If not, trace wire from ignition switch for an open circuit. Test the remaining connector terminals with the test lamp making connections between ground and terminal connectors. If lamp fails to light, trace particular wire for an incomplete circuit.

LOW WASHER FLUID INDICATOR

There are two types of low washer fluid indicating systems used on GM cars. They are the mechanical type and electrically controlled type. The mechanical type of a float and rod assembly, sending unit and a fiber optic. The electrically controlled type consists of a float, magnet, contact points and a resistor.

Fig. 8 Typical vacuum gauge

On the mechanical type, the upper end of the rod extends into the sending unit and has colored red and green portions. When the windshield wipers are activated, a lamp bulb in the sending unit lights either the red or green sections of the rod. The colored light is then picked up by the fiber optic and is transmitted through it to the tell tale lens. The lens will show red or green depending upon washer fluid level.

The electrically controlled indicator is activated when the windshield wipers are engaged. A slight amount of current flows from the wiper motor to the washer bottle float unit. This current will either pass through the contact points or the resistor which is in parallel with the points. When the washer fluid level is high, the magnet holds the contact points open. The current will now flow through the resistor where it is reduced so the indicator will not light. When the washer fluid level is low, the float drops and the magnet will separate from the cap assembly allowing the current to pass through the contact points and activate the indicator light.

Trouble Shooting

On the mechanical indicating system, if the tell tale lens fails to glow when the windshield wipers are activated, check lamp bulb in sending unit and see that fiber optic is not broken.

On the electrically controlled system, the first item to check is the indicator bulb. With the windshield wipers "ON", connect a jumper wire between the two terminals on the washer bottle cap. The indicator should then light. If not, replace bulb. If the bulb is found to be satisfactory, remove cap and float assembly from washer bottle. Float should be able to move to the bottom of the stem and the magnet should separate from the cap. If not, replace float and cap assembly.

VACUUM GAUGE

This gauge, Fig. 8, measures intake manifold vacuum. The intake manifold vacuum varies with engine operating conditions, carburetor adjustments, valve timing, ignition timing and general engine condition.

Since the optimum fuel economy is directly proportional to a properly functioning engine, a high vacuum reading on the gauge relates to fuel economy. For this reason some manufacturers call the vacuum gauge a "Fuel Economy Indicator." Most gauges have colored sectors the green sector being the "Economy" range and red the "Power" range. Therefore, the vehicle should be operated with gauge registering in the green sector or a high numerical number, Fig. 5, for maximum economy.

HEADLIGHT AIMING

All headlight adjustments should be made with a half-full tank plus or minus one gallon, with a person seated in the driver's seat and a person seated in the passenger seat, the car unloaded and the trunk empty except for the spare tire and jacking equipment, and recommended pressure in all tires. If necessary, compensate for unspecified gas capacity by adding or removing weight at the tank area at the rate of six pounds per gallon. Before each adjustment, bounce the car by pushing on the center of both the front and rear bumpers, then rock car side-to-side, to level the vehicle.

Headlights can be aimed mechanically or by means of a wall screen. When using a mechanical aimer, follow manufacturer's instructions. To align the No. 1 headlights by means of a wall screen, select a level portion of the shop floor. Lay out the floor and wall as shown in Fig. 1.

Establish the headlight horizontal centerline by subtracting 20 inches from the actual measured height of the headlight lens center from the floor and adding this difference to the 20-inch reference line obtained by sighting over the uprights to obtain dimension B (upper diagram Fig. 2). Draw a horizontal line 2 inches below, and parallel to the headlight horizontal centerline on the screen as measured on the vehicle (dimension A, upper diagram Fig. 2).

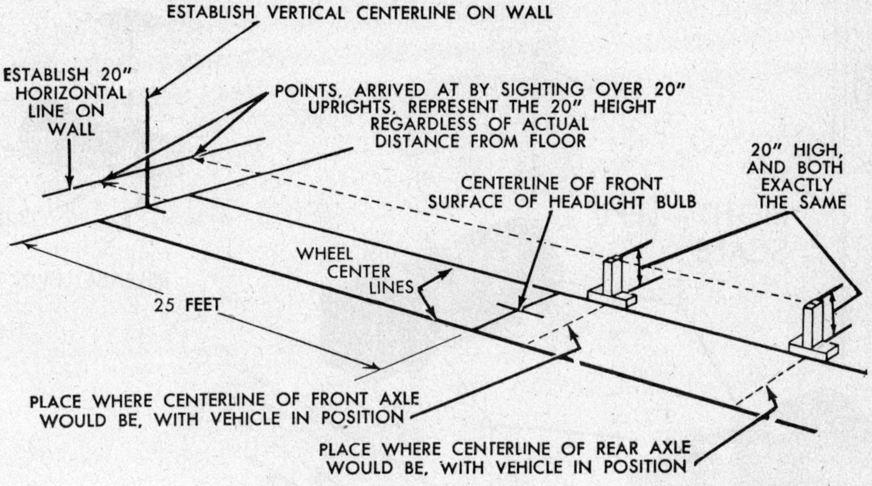

Fig. 1 Floor and wall layout

NO. 1 HEADLIGHT HIGH BEAM ADJUSTMENT

(Four Bulb Cars)

Adjust each No. 1 headlight beam as shown in Fig. 2 upper diagram. **Cover the No. 2 lights when making this adjustment.**

Some states may not approve of the 2-inch dimension for the No. 1 headlights. Check the applicable state law, as a 3-inch dimension may be required.

NO. 2 HEADLIGHT LOW BEAM ADJUSTMENT

(Four Bulb And All Two Bulb Cars)

To align the No. 2 headlights on four bulb cars or the headlights on two bulb cars, a different wall chart is used (lower diagram of Fig. 2). Dimension B for the No. 2 lights will be different than B for the No. 1 lights, but dimension A which is measured on the car will be the same as for the No. 1 lights. **Note that the line of adjustment of the No. 2 lights is the horizontal centerline of the No. 2 lights.** Turn the headlights to low beam and adjust each No. 2 light as shown in Fig. 2.

Each headlight can be adjusted by means of two screws located on the headlight adjusting ring. Always bring each beam into final position by turning the adjusting screws clockwise so that the headlights will be held against the tension springs when the operation is completed.

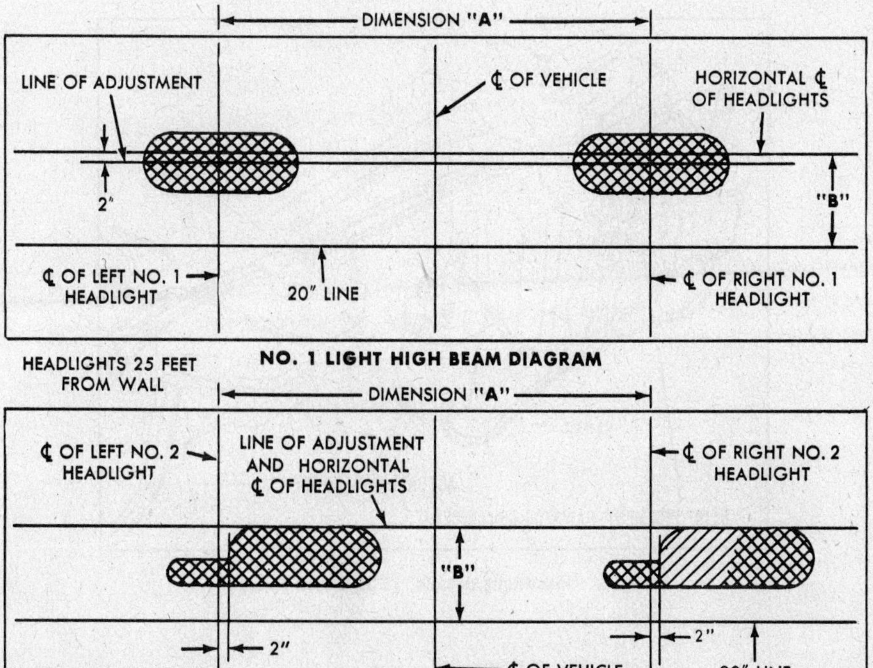

Fig. 2 Headlight wall screens

MECHANICAL AIMERS

NOTE: The rectangular headlamps used on some 1975 General Motors vehicles may be properly aimed using the mechanical aimer equipped with a special adapter that will fit these headlamps.

There are many different types of mechanical aiming equipment available one of which is shown in Figs. 3 and 4. Some may require a level floor whereas others have levelling devices which compensate for floor level variations.

When using this type of equipment, consult the equipment manufacturers instructions for its proper use.

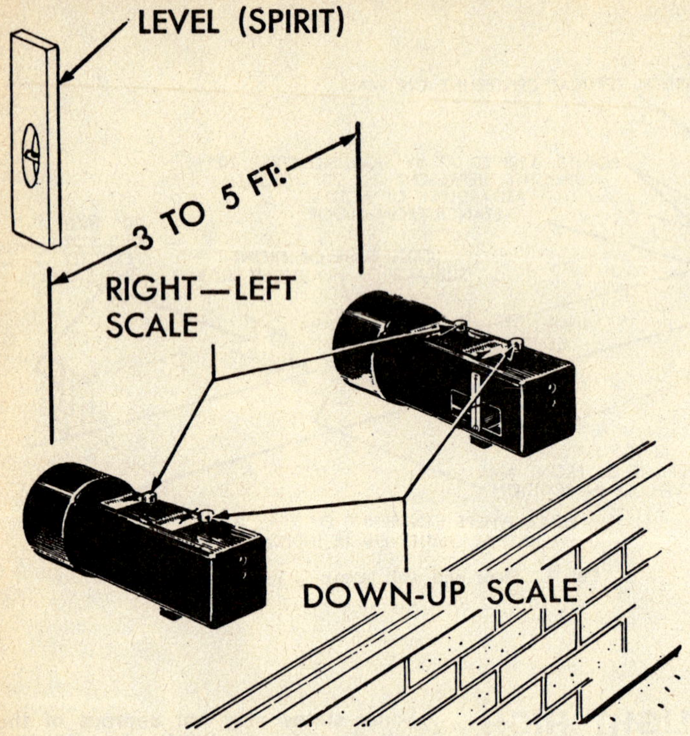

Fig. 3 Checking calibration of aimer

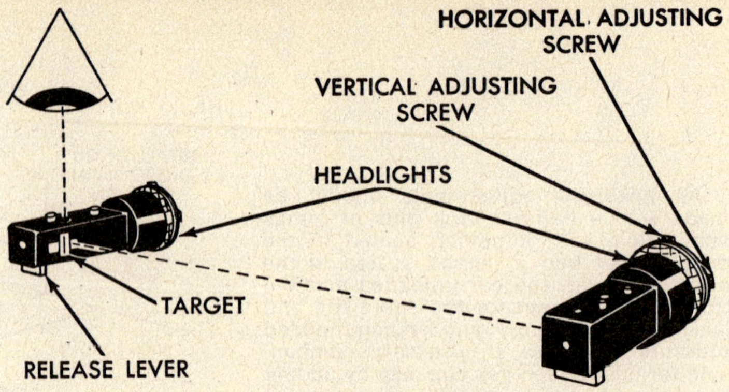

Fig. 4 Mounting and adjusting aimers

CONCEALED HEADLAMPS

BUICK RIVIERA

1969

These headlamps are vacuum controlled through a pair of vacuum actuators, one for each pair of headlamps, Fig. 1. Vacuum is taken at the intake manifold and vacuum storage tank and routed through the headlight switch to one side or the other of the actuator, depending on whether the lights are on or off.

When the lights are off, vacuum is supplied to the lower end of each actuator and the headlamp assemblies are rotated upward and held in this position by the actuators. When the lights are turned on, vacuum is supplied to the upper end of the actuators, causing the assemblies to rotate to the forward position as the electrical connection at the switch is closed to turn on the lights.

Also connected in the vacuum system is a vacuum relay. When the light switch is off, vacuum is applied to the diaphragm in the relay, moving the relay valve to connect the vacuum tank to the lower end of the actuators.

When the light switch is turned on, vacuum flow is cut off and normal air pressure acts on the relay to move the valve to connect the tank to the upper ends of the actuators. Thus, the headlights will remain in the forward position in case of a failure in the vacuum system.

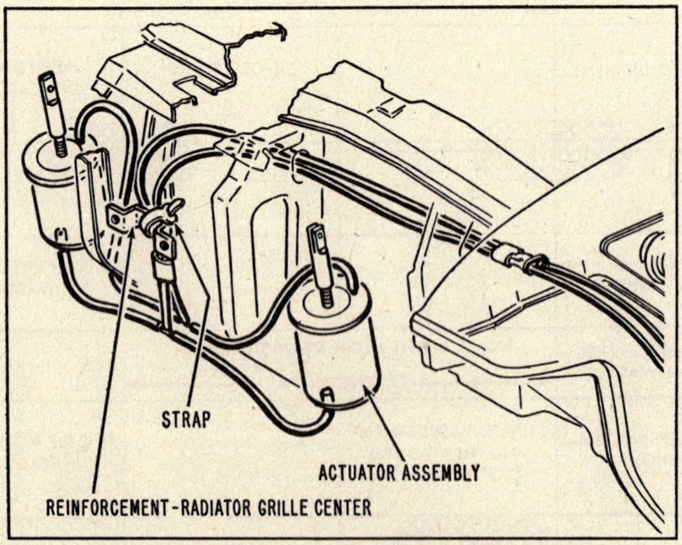

Fig. 1 Headlight details. 1969 Buick Riviera

CAMARO RALLY SPORT

1969

The headlamps are fixed and sections of the grille cover them when not in use. Each grille section, or door, is controlled by a vacuum actuator located behind it and connected to the door mechanically, Fig. 2.

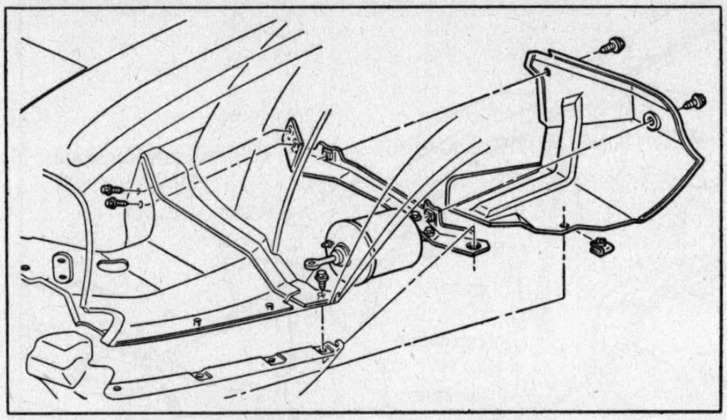

Fig. 2 Headlight details. 1969 Camaro

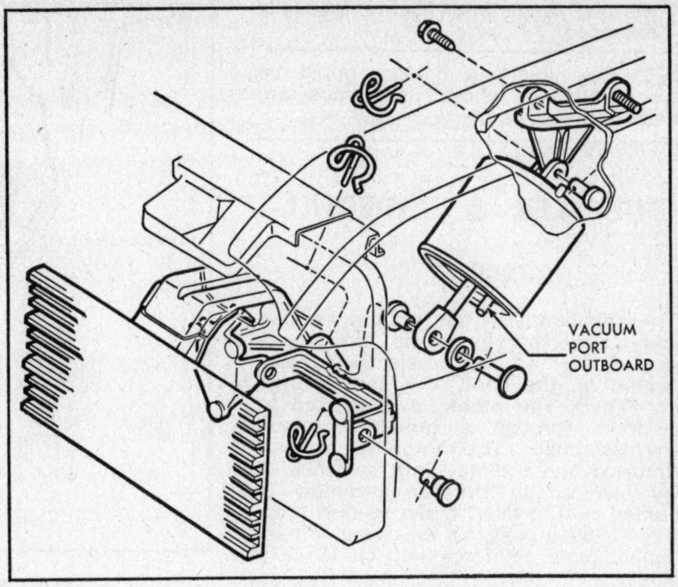

Fig. 3 Headlight details. 1969 Chevrolet

Vacuum from the vacuum storage tank is routed through a vacuum valve which is part of the headlight switch to the actuators to control the position of the doors.

A vacuum relay is also incorporated in the system so that the headlight doors will remain open in case of a failure in the vacuum system.

CHEVROLET

1969

This system uses fixed headlights with moveable, vacuum actuator operated headlight covers, Fig. 3, similar to the 1969 Camaro system.

CORVETTE

1969-75

These units are vacuum operated through separate actuators for each headlamp assembly, Fig. 4. The operation is the same as that for the 1969 Camaro Rally Sport except that the Corvette uses movable headlamp assemblies while the Camaro's lights are fixed. The vacuum actuators cause the headlamp "pods" to be raised up when in the on position and to be lowered flush with hood when turned off. Unlike the earlier Corvette system, the headlamps are lighted and brought into position by the same control switch. A vacuum valve on the headlamp switch routes vacuum to the proper side of the actuators according to the position of the light switch.

ALIGNMENT

1. "In-out", loosen screws fastening slotted bracket to underside of headlamp housing.
2. "Down", lamp cover top to opening; by turning hex head screw fastened to top of pivot link.
3. "Open," fully extended actuator with rod.
   a. Remove spring from actuator rod pin.
   b. Remove cotter pin from rod pin.
   c. Turn actuator rod until bushing hole aligns to forward end of slot in connecting link extended position, with engine idling for vacuum.
   d. Shut off engine, retract actuator rod and unscrew rod ½ turn to preload actuator rod in link.
4. "Up", (bezel to opening alignment). Loosen jam nut and turning bumper covered screw up or down to touch then up 1½ turns more. Micro switch on linkage must shut off

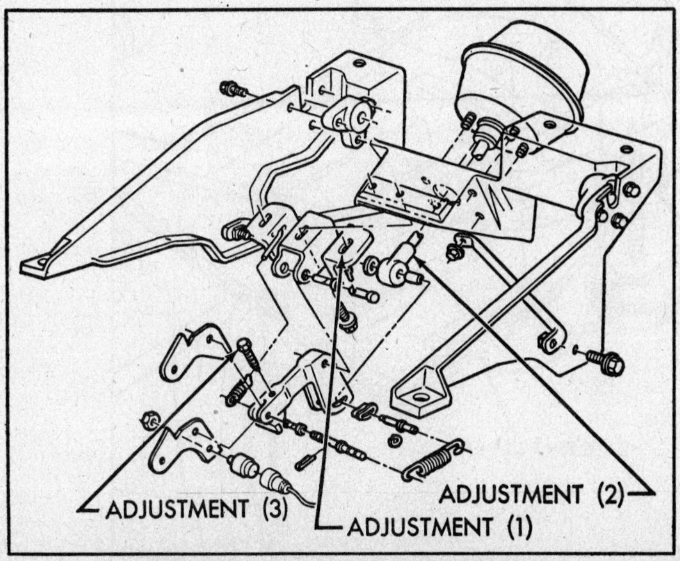

Fig. 4 Headlight details. 1969-75 Corvette

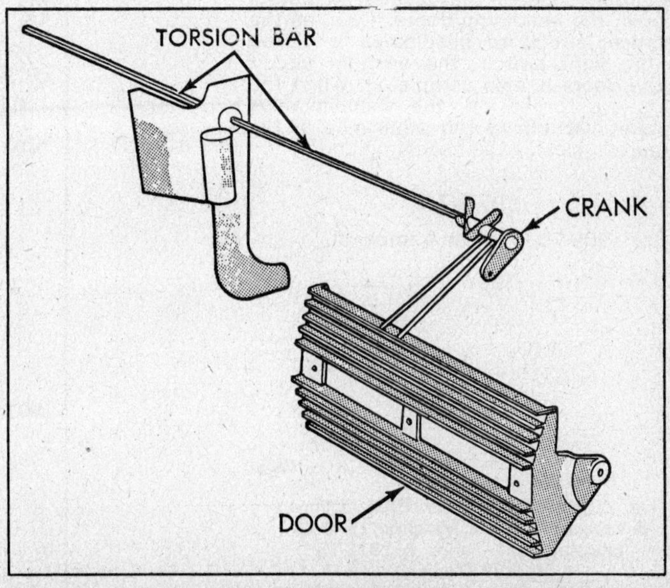

Fig. 5 Headlight details. 1969-75 Chrysler & Imperial (Typical)

warning lamp when lights are fully extended.

NOTE: The headlamp housing must be properly aligned before headlamps are aimed.

CHRYSLER & IMPERIAL

1969-75

The headlamps in this system are stationary and the headlamp doors are operated by a single electric motor mounted on the hood lock vertical support, Fig. 5. The motor is connected to the doors through a torsion bar and crank assemblies. The motor has a worm gear drive and internal limit switches. A relay and circuit breaker assembly is mounted on the hand brake release bracket on 1969 models or the lower left instrument panel reinforcement on 1970-75 models.

To manually open the headlamp doors, disconnect the electrical connectors from motor and rotate the hand wheel located at the lower end of the motor.

CAUTION: Rotating the wheel after the doors reach the end of travel will permanently damage the motor. Also, do not operate motor with headlamp doors disconnected as operating the motor without load will damage motor.

DODGE

1969

This system uses stationary headlamps and moveable headlamp doors. Each door is controlled by a separate vacuum actuator.

Vacuum is supplied from a vacuum tank to a vacuum control valve which is part of the ignition switch. When the headlamps are turned on, the vacuum valve routes vacuum through hoses to the proper side of the vacuum actuators to open the headlamp doors. Thus, as the electrical circuit to headlamps is closed by the light switch, the vacuum circuit to the doors is also completed. When the lights are turned off, the vacuum valve supplies vacuum to the other side of the actuators, closing the headlamp doors.

1970-75

See 1969-75 Chrysler & Imperial.

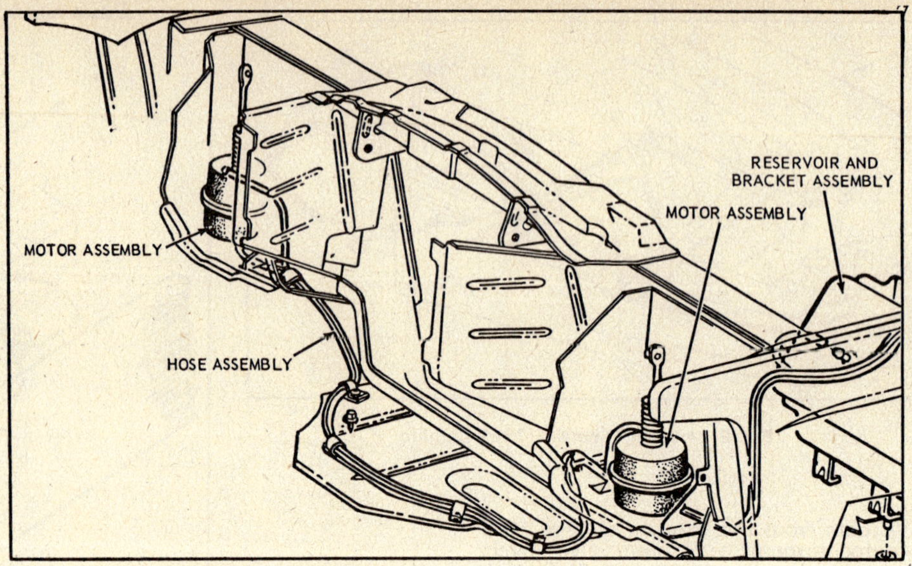

Fig. 6 Headlight details. 1969-71 Lincoln Mk III, 1969 Thunderbird, 1973-75 Mercury, 1974-75 Lincoln & 1975 Ford (Typical)

FORD MOTOR CO. CARS

On 1969 Thunderbird, 1969-71 Mark III, 1973-75 Mercury, 1974-75 Lincoln and 1975 Ford the headlamp covers are actuated by two vacuum motors, Fig. 6. On 1969-71 Ford and Cougar, 1969-72 Mercury, 1970-73 Lincoln, 1971 Torino and 1972-75 Mark IV, the headlamp covers are actuated by a single vacuum motor which is connected to them through a torsion bar, Fig. 7.

Vacuum is supplied to one side of the actuator from a vacuum storage tank through the vacuum control valve mounted on the back of the headlamp switch. When the headlamps are turned on, the vacuum valve routes vacuum to the "open" side of the actuator and vents the "close" side to atmospheric pressure. The reverse applies when the lights are turned off.

The covers are equipped with over

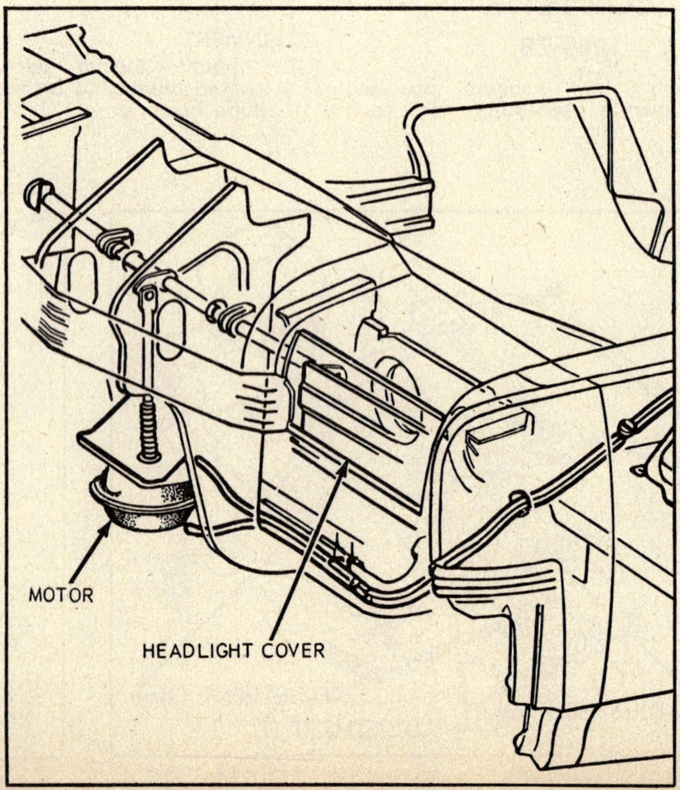

Fig. 7 Headlight details. 1969-71 Ford & Cougar, 1969-72 Mercury, 1970-73 Lincoln, 1971 Torino & 1972-75 Mark IV (Typical)

center springs which aid the actuators in holding the doors in position.

OLDSMOBILE TORONADO

1969

Headlamps are fixed and covers are moved by a single, centrally mounted vacuum actuator, Fig. 8. When the headlights are turned on, the vacuum valve on the light switch routes vacuum to the remote control valve which, in turn, routes it to the open side of the actuator and causes the headlamp covers to open. When the lights are turned off, vacuum is routed to the close side of the actuator.

A thermostatic vacuum switch is incorporated in the Outside Air Induction option so that if engine coolant temperature exceeds 220 degrees, vacuum is routed to the actuator and the headlamp covers open to provide more air flow to cool the engine.

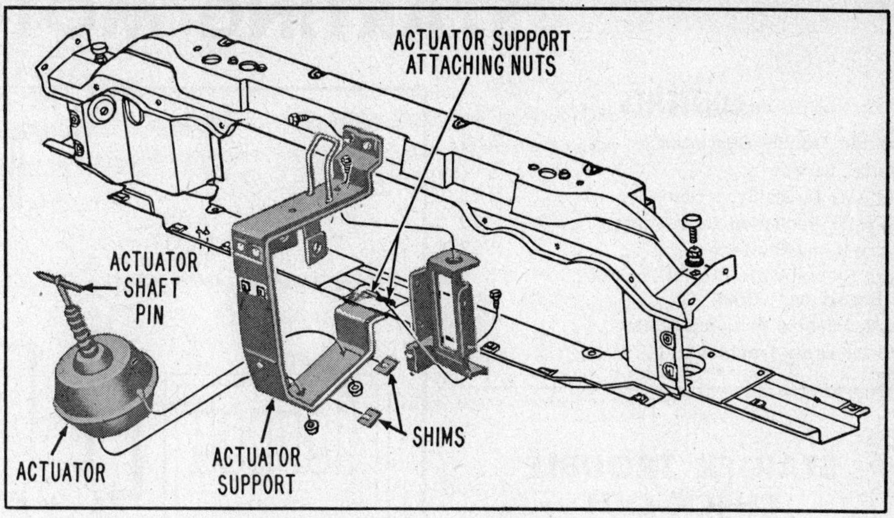

Fig. 8 Headlight details. 1969 Oldsmobile Toronado

PLYMOUTH

1970-72

See 1969-75 Chrysler & Imperial.

PONTIAC GTO

1969

This system uses fixed headlamps and moveable, vacuum controlled headlamp covers, Fig. 9. Each cover is controlled by a separate vacuum actuator which receives vacuum from a vacuum tank through a vacuum valve attached to the light switch. The covers are held in either the open or closed position by overcenter springs.

TROUBLE SHOOTING

Vacuum Type

Examine all hoses for splits, which occur most often around connections. Also, look for kinked or pinched hoses, a condition which often occurs when retaining clips are too tight, thus blocking off vacuum flow.

If inspection reveals that all the hoses are satisfactory, check each vacuum actuator by disconnecting the actuator hoses one at a time and hooking a vacuum gauge to the hose (s). With engine running, if the gauge indicates at least 14 inches of vacuum, the problem is either in the actuator, which must be replaced, or because of jammed covers or linkage.

If the gauge shows less than 14 inches of vacuum, check vacuum at the storage tank, distribution valve, check valve vacuum relay, if used, and at each hose connection.

Electrical Type

Connect a jumper wire directly from the battery to the motor (s). If the system operates, check the headlight or motor control switch. If the switch is eliminated as the cause of trouble, check the wiring; look for loose connections, broken wires or terminals.

If the system fails to operate with the jumper wire, remove the motor for repair or install a new or rebuilt unit.

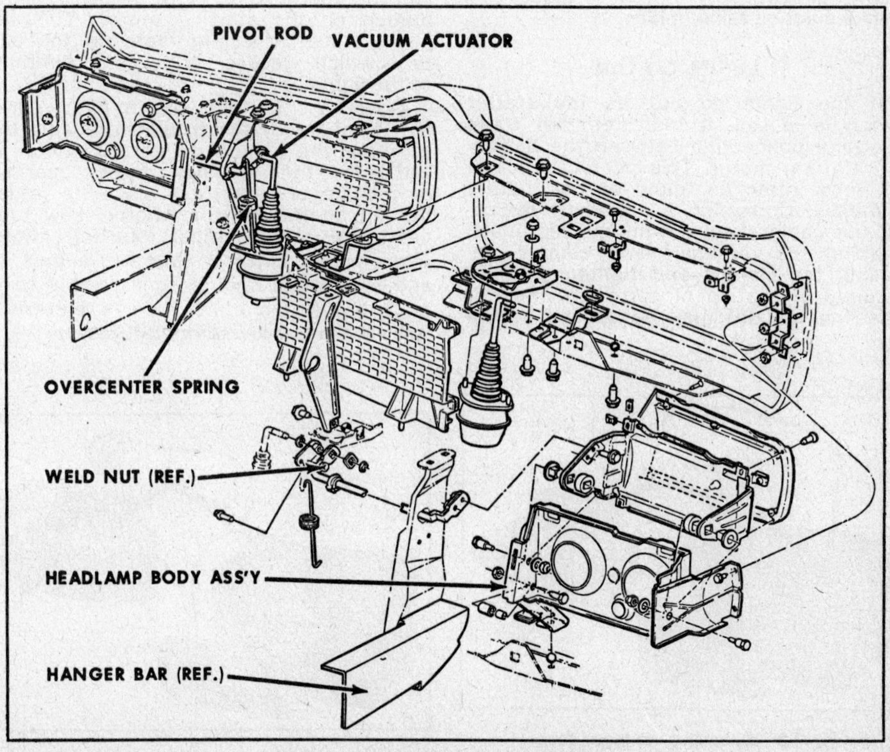

Fig. 9 Headlight details. 1969 GTO

STARTING MOTORS

CONTENTS

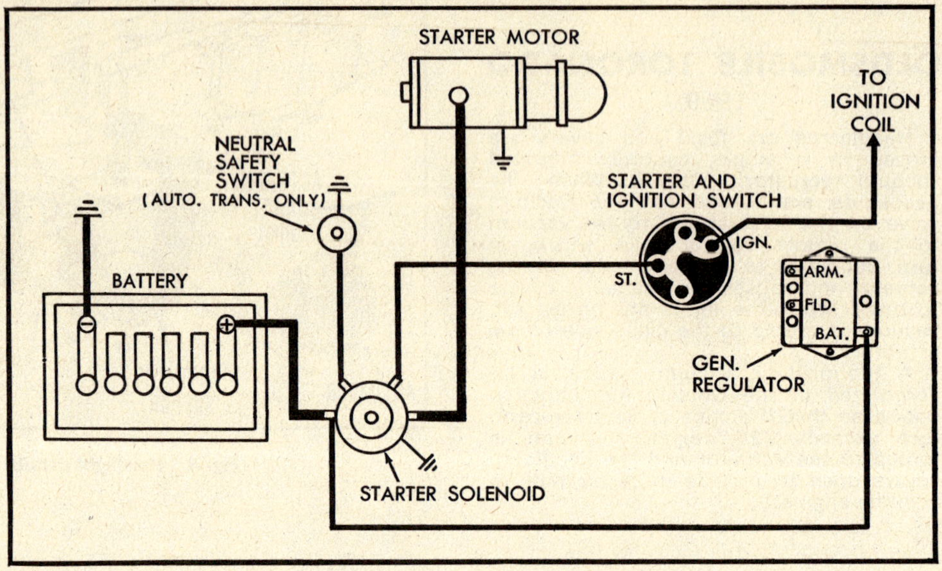

Fig. 1 Wiring diagram of a typical starting circuit

STARTER TROUBLE CHECK-OUT

When trouble develops in the starting motor circuit, and the starter cranks the engine slowly or not at all, several preliminary checks can be made to determine whether the trouble lies in the battery, in the starter, in the wiring between them, or elsewhere. Many conditions besides defects in the starter itself can result in poor cranking performance.

To make a quick check of the starter system, turn on the headlights. They should burn with normal brilliance. If they do not, the battery may be run down and it should be checked with a hydrometer.

If the battery is in a charged condition so that the lights burn brightly, operate the starting motor. Any one of three things will happen to the lights: (1) They will go out, (2) dim considerably or (3) stay bright without any cranking action taking place.

If Lights Go Out

If the lights go out as the starter switch is closed, it indicates that there is a poor connection between the battery and starting motor. This poor connection will most often be found at the battery terminals. Correction is made by removing the cable clamps from the terminals, cleaning the terminals and clamps, replacing the clamps and tightening them securely. A coating of corrosion inhibitor (vaseline will do) may be applied to the clamps and terminals to retard the formation of corrosion.

If Lights Dim

If the lights dim considerably as the starter switch is closed and the starter operates slowly or not at all, the battery may be run down, or there may be some mechanical condition in the engine or starting motor that is throwing a heavy burden on the starting motor. This imposes a high discharge rate on the battery which causes noticeable dimming of the lights.

Check the battery with a hydrometer. If it is charged, the trouble probably lies in either the engine or starting motor itself. In the engine, tight bearings or pistons or heavy oil place an added burden on the starting motor. Low temperatures also hamper starting motor performance since it thickens engine oil and makes the engine considerably harder to crank and start. Also, a battery is less efficient at low temperatures.

In the starting motor, a bent armature, loose pole shoe screws or worn bearings, any of which may allow the armature to drag, will reduce cranking performance and increase current draw.

In addition, more serious internal damage is sometimes found. Thrown armature windings or commutator bars, which sometimes occur on over-running clutch drive starting motors, are usually caused by excessive over-running after starting. This is the result of such conditions as the driver keeping the starting switch closed too long after the engine has started, the driver opening the throttle too wide in starting, or improper carburetor fast idle adjustment. Any of these subject the over-running clutch to extra strain so it tends to seize, spinning the armature at high speed with resulting armature damage.

Another cause may be engine backfire during cranking which may result, among other things, from ignition timing being too far advanced.

To avoid such failures, the driver should pause a few seconds after a false start to make sure the engine has come

Fig. 2 Checking voltage drop between vehicle frame and grounded battery terminal post

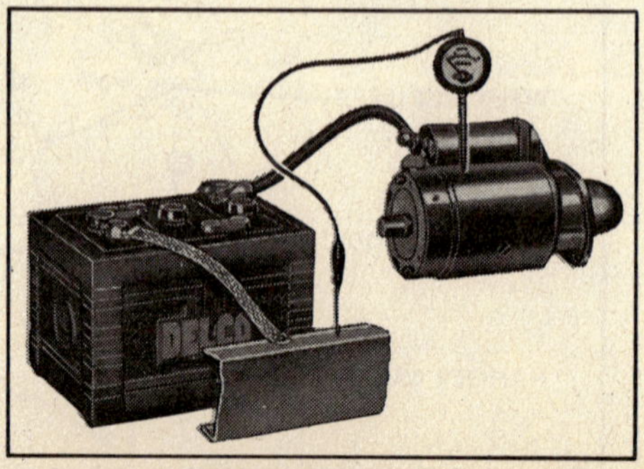

Fig. 3 Checking voltage drop between vehicle frame and starter field frame

Fig. 4 Checking voltage drop between ungrounded battery terminal post and battery terminal on solenoid

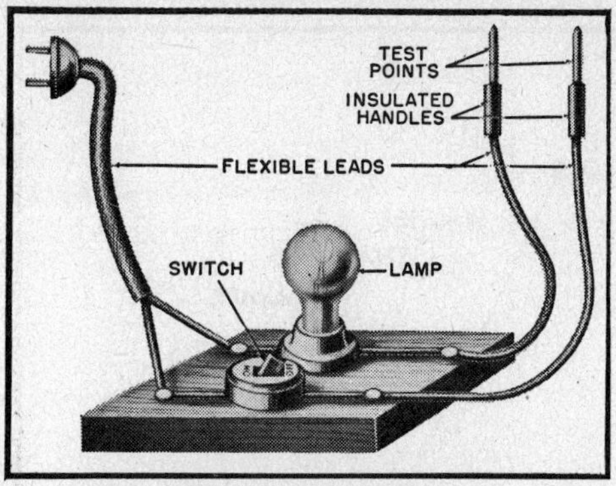

Fig. 5 A simple tester for use in making continuity and ground tests on armature and field windings

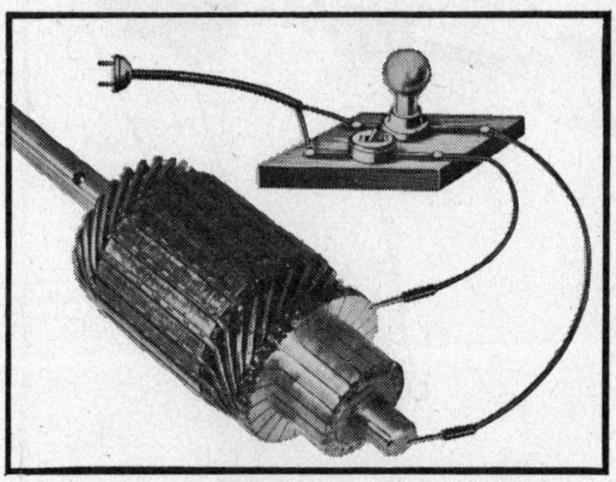

Fig. 6 Checking armature for grounds. If lamp lights armature is grounded and should be replaced

Fig. 7 Measuring commutator runout with dial indicator. Mount shaft in V blocks and rotate commutator. If runout exceeds .003″, commutator should be turned in a lathe to make it concentric

completely to rest before another start is attempted. In addition, the ignition timing should be reset if engine backfiring has caused the trouble.

Lights Stay Bright, No Cranking Action

This condition indicates an open circuit at some point, either in the starter itself, the starter switch or control circuit. The solenoid control circuit can be eliminated momentarily by placing a heavy jumper lead across the solenoid main terminals to see if the starter will operate. This connects the starter directly to the battery and, if it operates, it indicates that the control circuit is not functioning normally. The wiring and control units must be checked to locate the trouble, Fig. 1.

If the starter does not operate with the jumper attached, it will probably have to be removed from the engine so it can be examined in detail.

Checking Circuit With Voltmeter

Excessive resistance in the circuit between the battery and starter will reduce cranking performance. The resistance can be checked by using a voltmeter to measure voltage drop in the circuits while the starter is operated. There are three checks to be made:

1. Voltage drop between car frame and grounded battery terminal post (not cable clamp), Fig. 2.
2. Voltage drop between car frame and starting motor field frame, Fig. 3.
3. Voltage drop between insulated battery terminal post and starting motor terminal stud (or the battery terminal stud of the solenoid), Fig. 4.

Each of these should show no more than one-tenth (0.1) volt drop when the starting motor is cranking the engine. Do not use the starter for more than 30 seconds at a time to avoid overheating it.

If excessive voltage drop is found in any of these circuits, make correction by disconnecting the cables, cleaning the connections carefully, and then reconnecting the cables firmly in place. A coating of vaseline on the battery cables and terminal clamps will retard corrosion.

NOTE—On some cars, extra long battery cables may be required due to the location of the battery and starter. This may result in somewhat higher voltage drop than the above recommended 0.1 volt. The only means of determining the normal voltage drop in such cases is to check several of these vehicles. Then when the voltage drop is well above the normal figure for all cars checked, abnormal resistance will be indicated and correction can be made as already explained.

STARTING MOTOR SERVICE

To obtain full performance data on a starting motor or to determine the cause of abnormal operation, the starting motor should be submitted to a no-load and torque test. These tests are best performed on a starter bench tester with the starter mounted on it.

From a practical standpoint, however, a simple torque test may be made quickly with the starter in the car. Make sure the battery is fully charged and that the starter circuit wires and terminals are in good condition. Then operate the starter to see if the engine turns over normally. If it does not, the torque developed is below standard and the

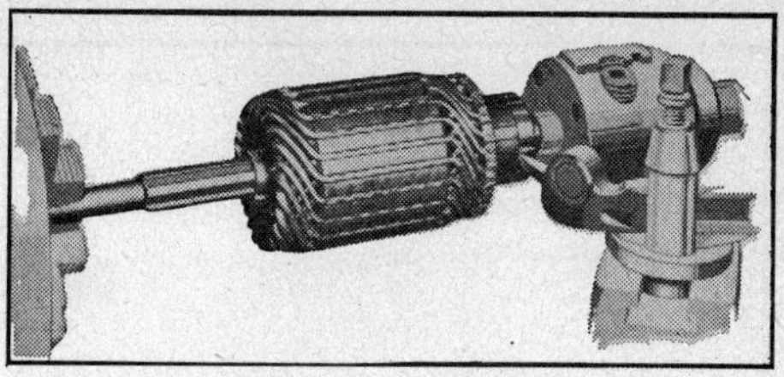

Fig. 8 Turning commutator in a lathe. Take light cuts until worn or bad spots are removed. Then remove burrs with No. 00 sandpaper

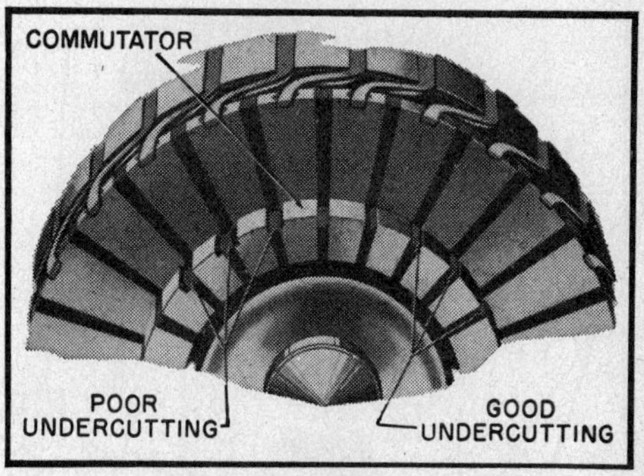

Fig. 9 Good undercutting should be .002″ wider than mica insulation, $\frac{1}{64}$″ deep and exactly centered so that there are no burrs on the mica. Do not undercut molded commutators

Fig. 10 Checking armature for short circuit. As armature is rotated by hand, steel strip (hacksaw blade) will vibrate if short circuit exists

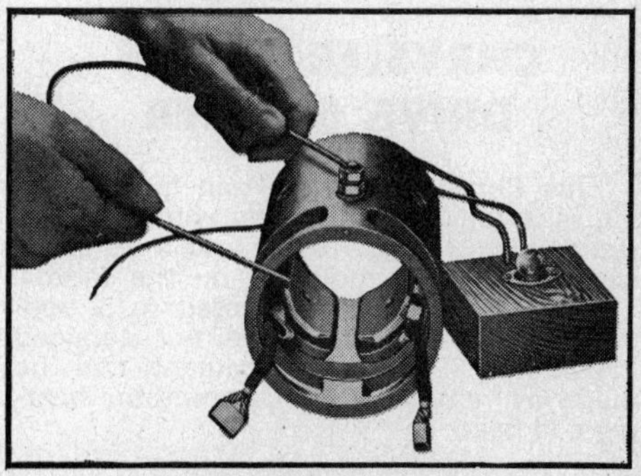

Fig. 11 Testing field coils for grounds. If a ground is present, lamp will light

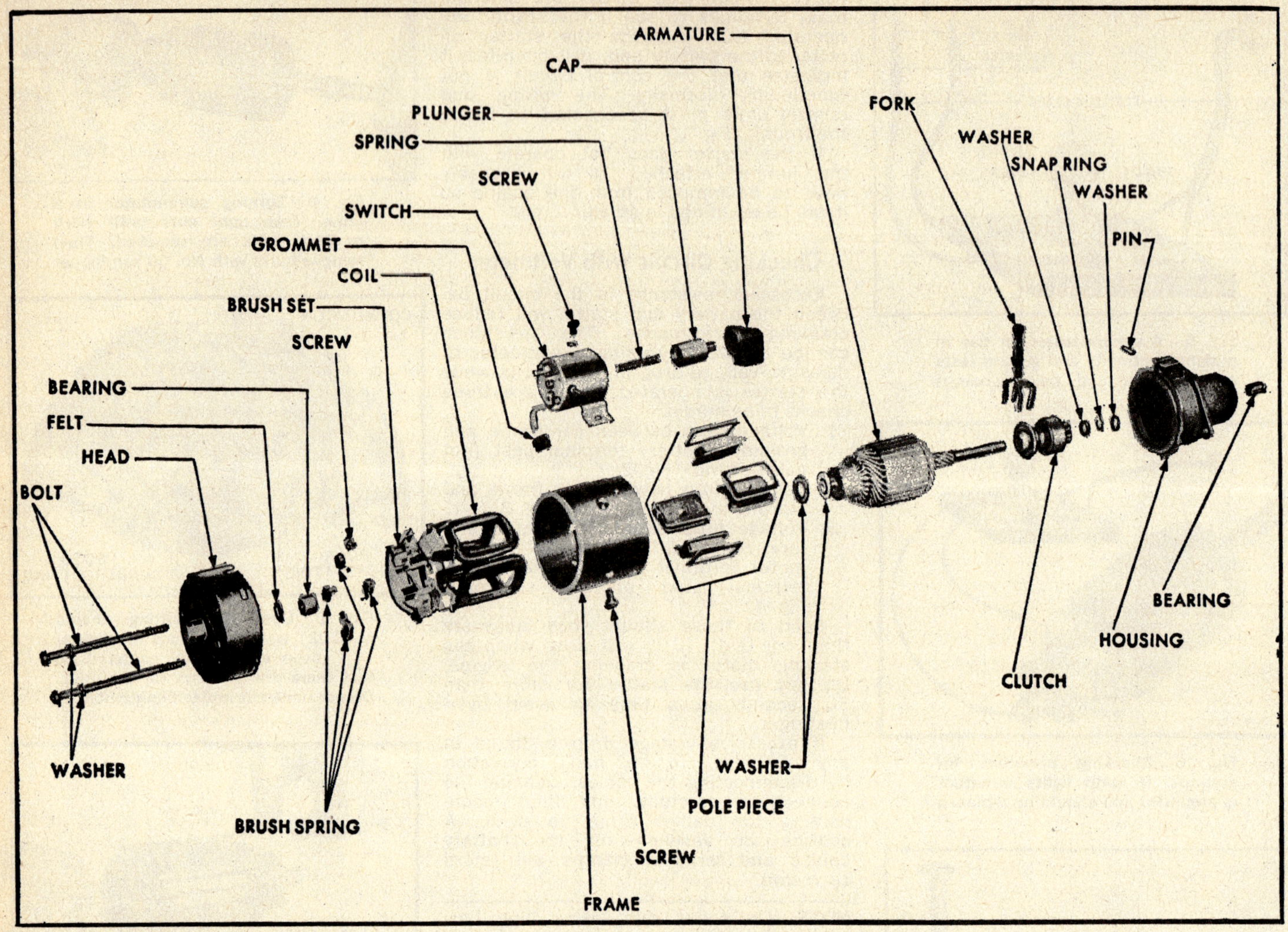

Fig. 12 Chrysler built direct drive starter

starter should be removed for further checking.

Remove the starter from the engine as outlined in the vehicle chapters, disassemble it as outlined further on and make the tests as suggested in Figs. 6 through 11.

CHRYSLER DIRECT DRIVE STARTER

This Chrysler built starting motor, Fig. 12, is a four coil assembly with an over-running clutch type drive and a solenoid shift-type switch mounted on the motor. The brush holders are riveted to a separate brush plate and are not serviced individually. Brush replacement can be made by removing the commutator bearing end head.

Disassembly

1. Remove through bolts and tap commutator end head from field frame.
2. Remove thrust washers from armature shaft.
3. Lift brush holder springs and remove brushes from holders.
4. Remove brush plate.
5. Disconnect field leads at solenoid connector.
6. Unfasten and remove solenoid and boot assembly.
7. Drive out over-running clutch shift fork pivot pin.
8. Remove drive end pinion housing and spacer washer.
9. Note position of shifter fork on starter and remove fork.
10. Slide over-running clutch pinion gear toward commutator end of armature. Drive stop retainer toward clutch pinion gear to expose snap ring and remove snap ring.
11. Slide clutch drive from armature shaft.
12. If necessary to replace the field coils, remove screw that holds ground brushes and raise brushes with the terminal and shunt wire up and away from field frame. Remove pole shoe screws and take out field coils.

Reassembly

1. Lubricate armature shaft and splines with SAE 10W or 30W rust preventive oil.
2. Install starter drive, stop collar (retainer), lock ring and spacer washer.
3. Install shifter fork over starter drive spring retainer washer with narrow leg of fork toward commutator. *If fork is not positioned properly, starter gear travel will be restricted, causing a lockup in the clutch mechanism.*
4. Install drive end (pinion) housing on armature shaft, indexing the shift fork with slot in drive end of housing.
5. Install shift fork pivot pin.
6. Install armature with clutch drive, shifter fork and pinion housing. Slide armature into field frame until pinion housing indexes with slot in field frame.
7. Install solenoid and boot assembly and tighten bolts securely.
8. Install ground brushes.
9. Connect field coil leads at solenoid connector.

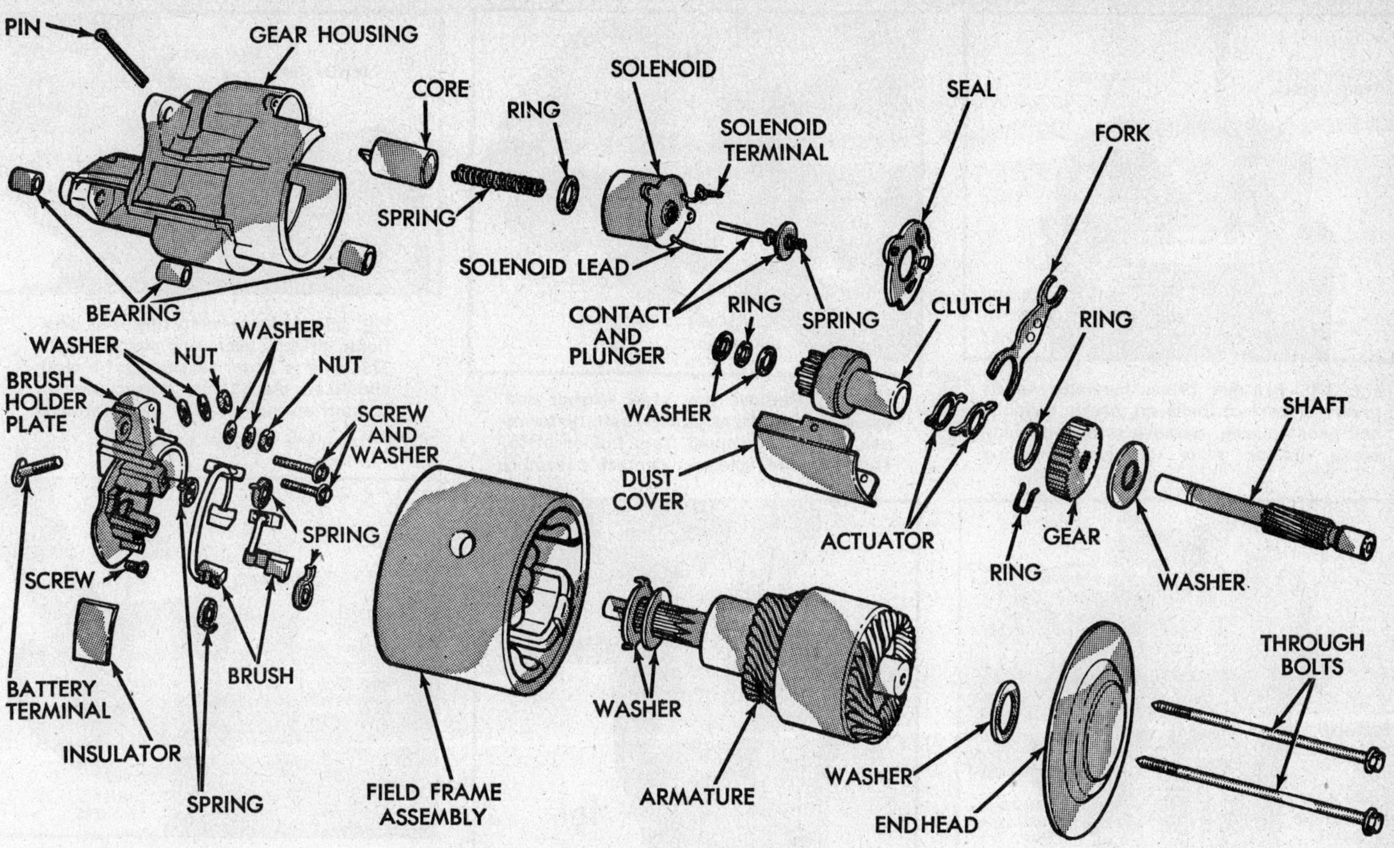

Fig. 14 Chrysler built reduction gear starting motor

10. Install brush holder ring, indexing tang of ring in hole of field frame.
11. Position brushes in brush holders. *Be sure field coil lead wires are properly enclosed behind brush holder ring and that they do not interfere with brush operation.*
12. Install thrust washer on commutator end of armature shaft to obtain .010" minimum end play.
13. Install commutator end head.
14. Install through bolts and tighten securely.

Adjusting Pinion Clearance

1. Place starter in vise with soft jaws and tighten vise enough to hold starter. *Place a wedge or screwdriver between bottom of solenoid and starter frame to eliminate all deflection in solenoid when making pinion clearance check.*
2. Push in on solenoid plunger link, Fig. 16 (not fork lever) until plunger bottoms.
3. Measure clearance between end of pinion and pin stop with plunger seated and pinion pushed toward commutator end. Clearance should be ⅛". Adjust by loosening solenoid attaching screws and move solenoid fore and aft as required.
4. Test starter operation for free running and install on engine.

CHRYSLER REDUCTION GEAR STARTER

This reduction gear starting motor,

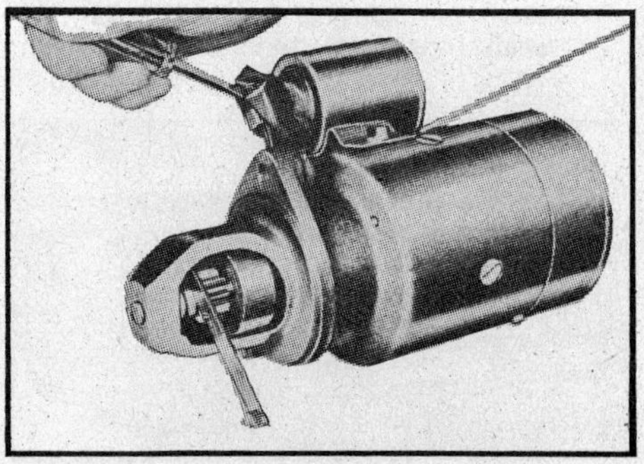

Fig. 13 Clearance between end of pinion and pin stop should be 1/8" with plunger seated and pinion pushed toward commutator end

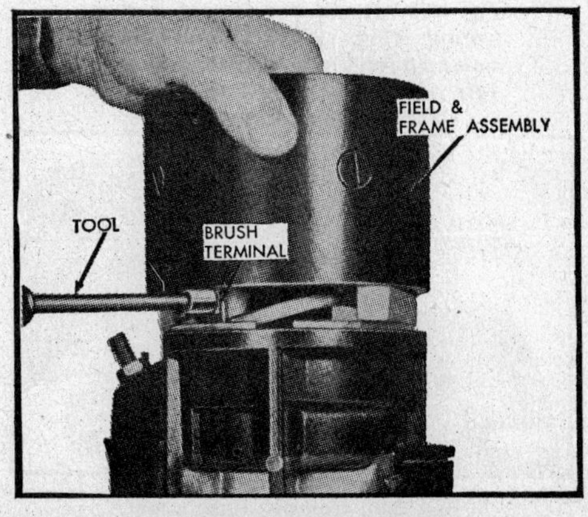

Fig. 15 Removing brush terminal screw

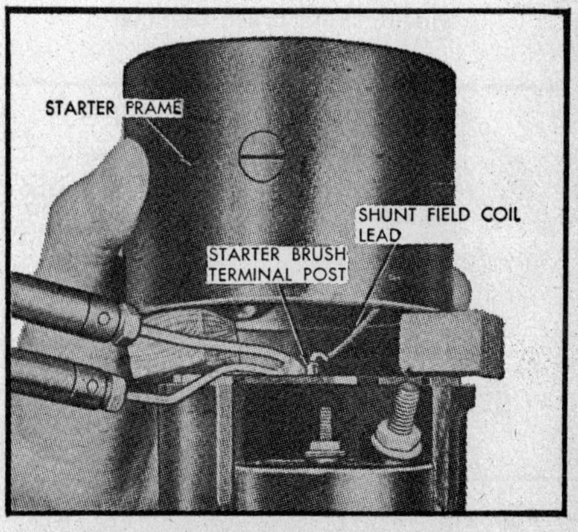

Fig. 16 Unsoldering shunt field coil lead from starter brush terminal

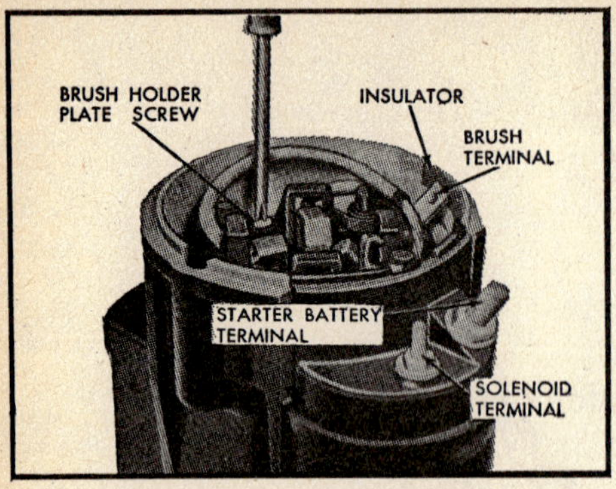

Fig. 17 Remove brush insulator which prevents contact between brush terminal and gear housing. Remove screw attaching brush holder plate to gear housing

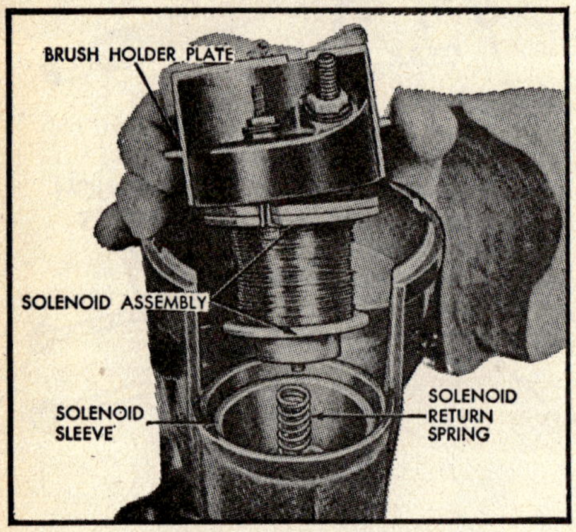

Fig. 18 Remove brush holder plate with brushes and solenoid as a unit

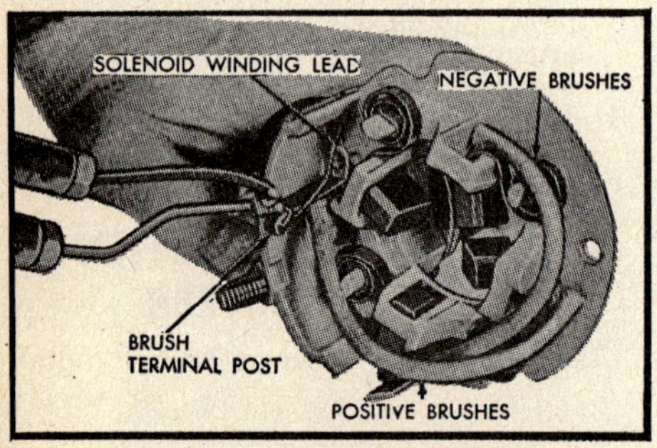

Fig. 19 Unsolder solenoid winding from brush terminal

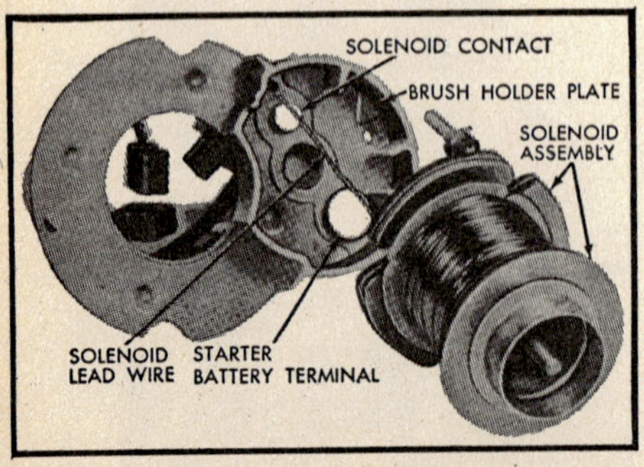

Fig. 20 Remove nut, steel washer and nylon washer from solenoid terminal. Separate brush holder plate from solenoid

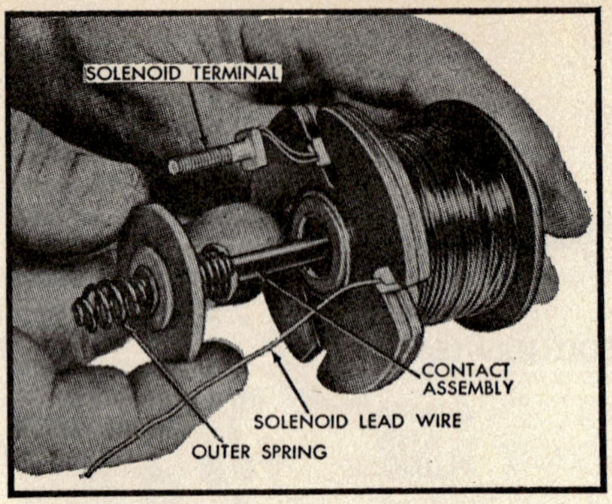

Fig. 21 Remove nut, steel washer and nylon washer from starter battery terminal. Remove terminal from holder plate. Then remove solenoid contact assembly

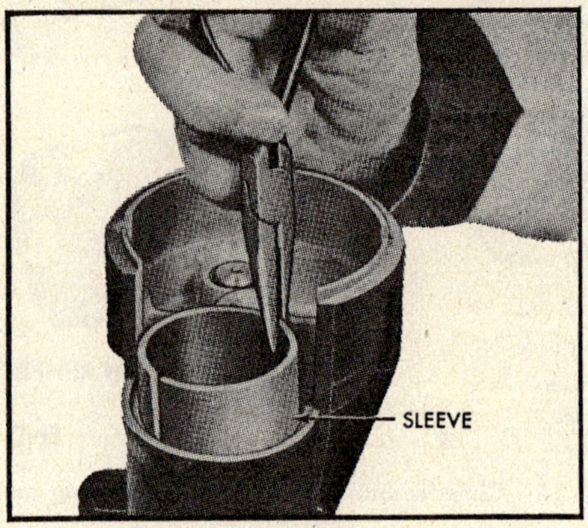

Fig. 22 Remove solenoid coil sleeve

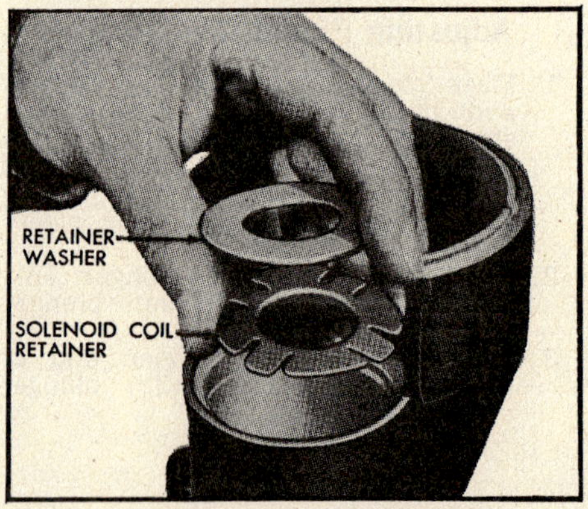

Fig. 23 Remove solenoid return spring (Fig. 21). Then remove solenoid coil retainer washer and retainer from solenoid housing

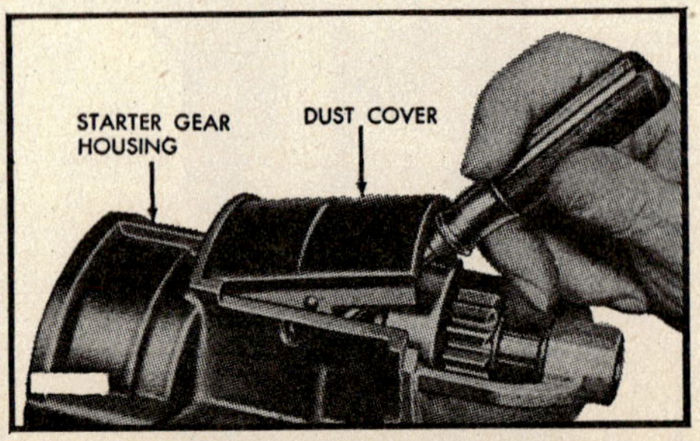

Fig. 24 Remove dust cover from gear housing

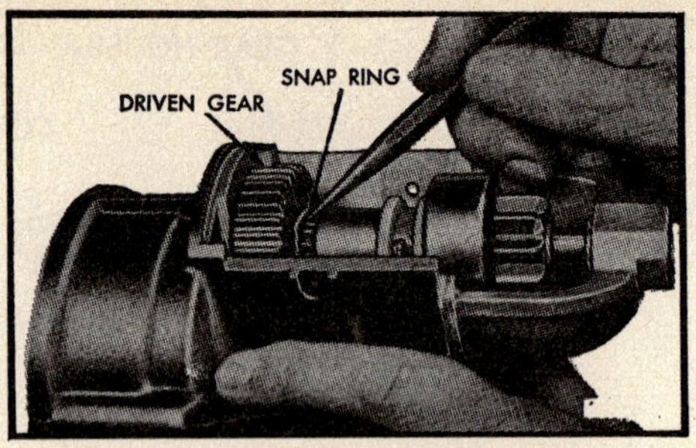

Fig. 25 Release snap ring that positions driven gear on pinion shaft. This ring is under tension and a cloth should be placed over ring to prevent it from springing away after removal

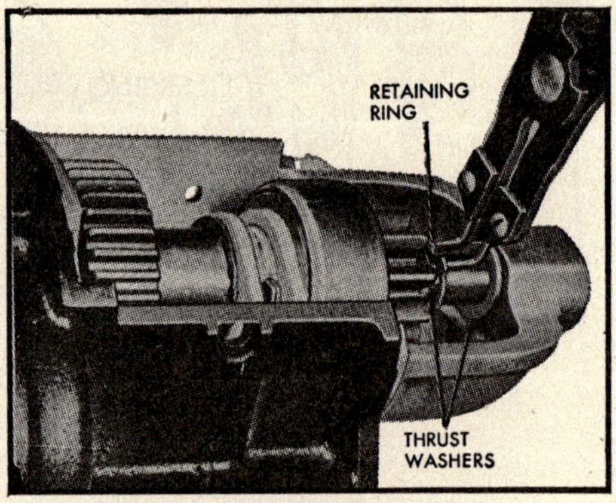

Fig. 26 Release retainer ring at front of pinion shaft. Do not spread ring any greater than the outside diameter of pinion shaft otherwise ring can be damaged

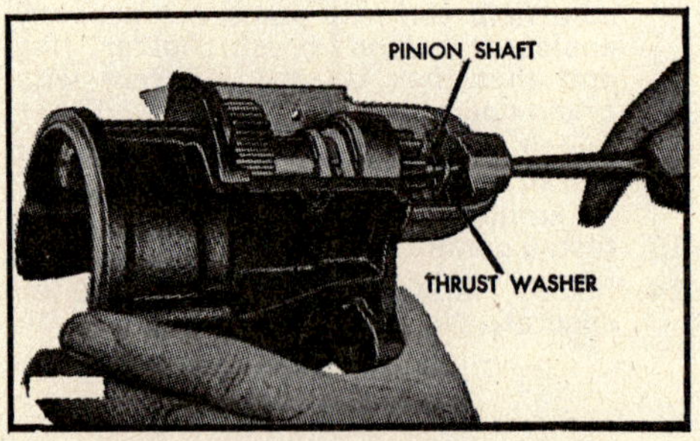

Fig. 27 Push pinion shaft toward rear of housing and remove snap ring and thrust washers

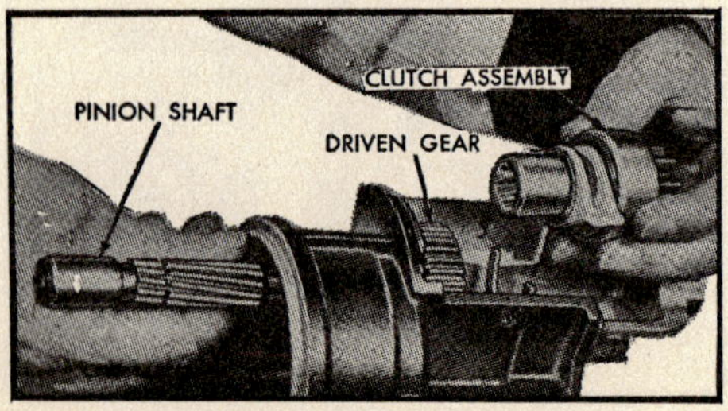

Fig. 28 Lift out clutch and pinion assembly with the two shifter fork nylon actuators

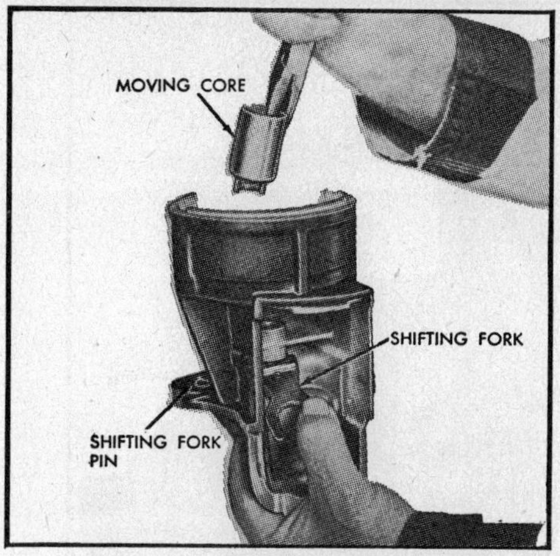

Fig. 29 Remove driven gear and friction washer. Then pull shift fork forward and remove solenoid moving core

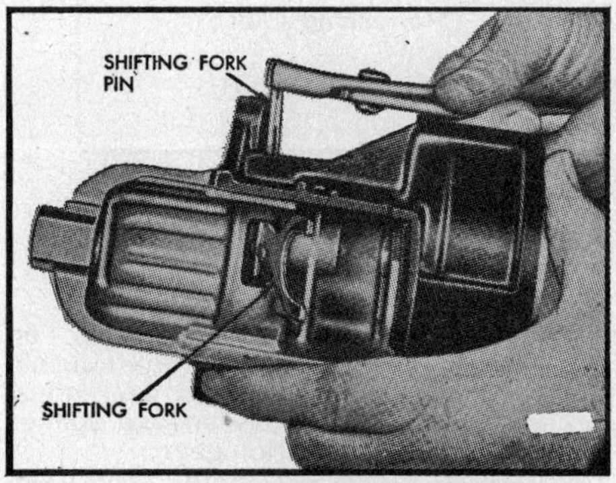

Fig. 30 Remove shift fork retainer pin and take out shift fork

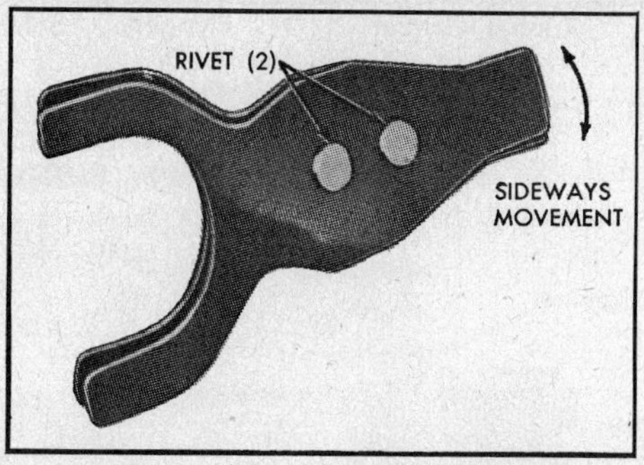

Fig. 31 Shifter fork assembly (Typical)

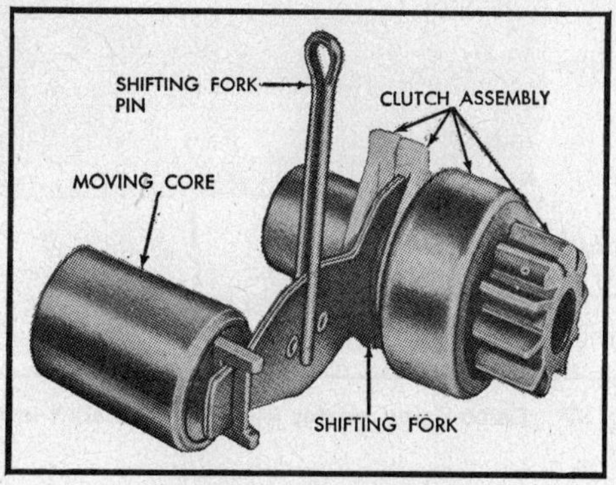

Fig. 32 Shifter fork and clutch arrangement

Fig. 33 Installing solenoid coil and sleeve

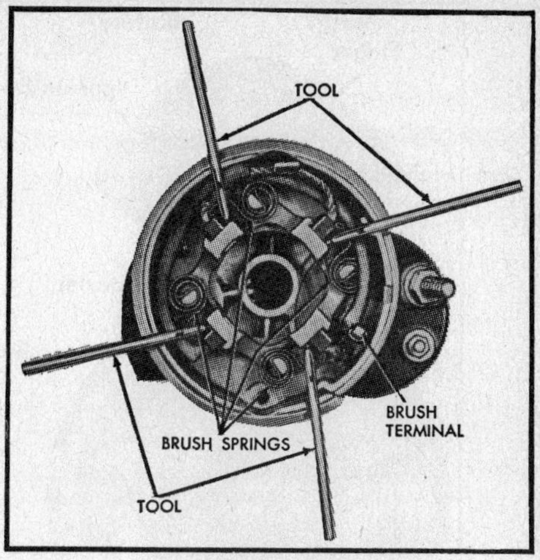

Fig. 34 Positioning of brushes with Tool Set C-3855

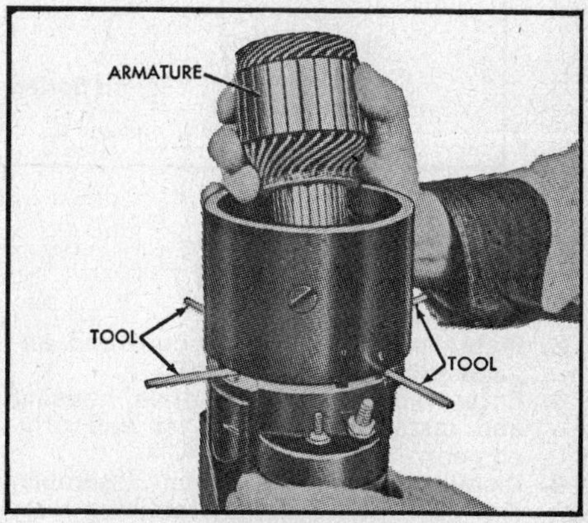

Fig. 35 Installing armature

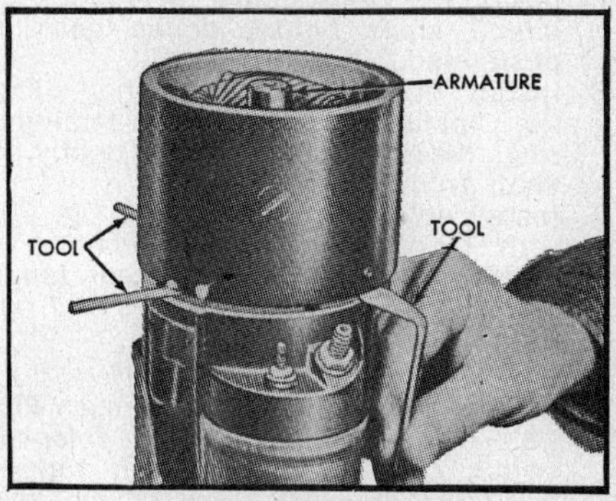

Fig. 36 Removing brush positioning tools

Fig. 14, has an armature-to-engine crankshaft ratio of 45 to 1; a 3½ to 1 reduction gear set is built into the motor assembly. The starter utilizes a solenoid shift. The housing of the solenoid is integral with the starter drive end housing.

Disassembly

1. Place gear housing of starter in a vise with soft jaws. *Use vise as a support fixture only; do not clamp.*
2. Remove through bolts and starter end head assembly.
3. Carefully pull armature up and out of gear housing, and starter frame and field assembly. Remove steel and fiber thrust washer. *The wire of the shunt field coil is soldered to the brush terminal. One pair of brushes are connected to this terminal. The other pair of brushes is attached to the series field coils by means of a terminal screw. Carefully pull the frame and field assembly up just enough to expose the terminal screw and the solder connection of the shunt field at the brush terminal. Place two wood blocks between starter frame and gear housing, Fig. 15, to facilitate removal of terminal screw and unsoldering of shunt field wire at brush terminal.*
4. Support brush terminal by placing a finger behind terminal and remove screw, Fig. 15.
5. Complete the disassembly procedure by referring to Figs. 16 through 30.

Reassembly

The shifter fork consists of two spring steel plates assembled with two rivets, Fig. 31. There should be about 1/16" side movement to insure proper pinion gear engagement. Lubricate between plates sparingly with SAE 10 engine oil.

1. Position shift fork in drive housing and install fork retaining pin, Fig. 30. *One tip of pin should be straight, the other tip should be bent at a 15 degree angle away from housing. Fork and pin should operate freely after bending tip of pin.*

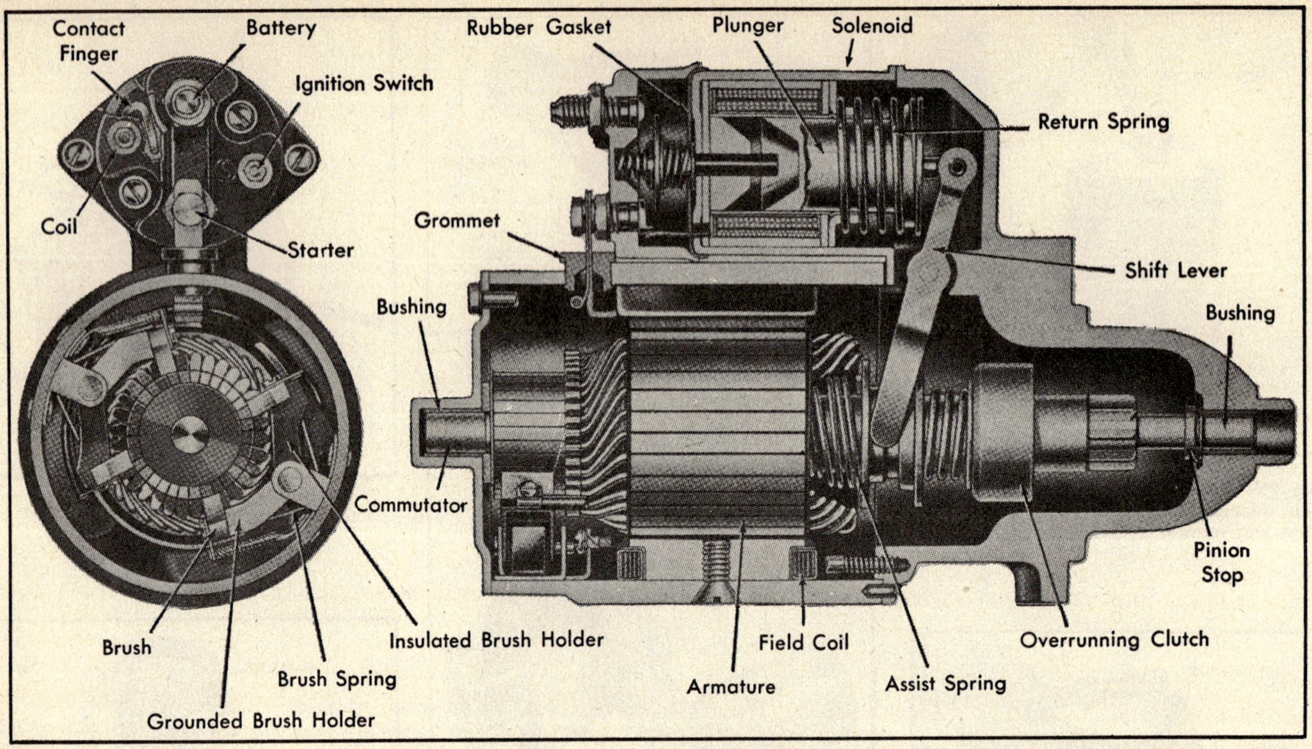

Fig. 37 Delco-Remy starter with enclosed shift lever

2. Install solenoid moving core and engage shifting fork, Fig. 29.
3. Enter pinion shaft in drive housing and install friction washer and driven gear.
4. Install clutch and pinion assembly, Fig. 28, thrust washer, retaining ring, and thrust washer.
5. Complete installation of pinion shaft, engaging fork with clutch actuators, Fig. 32. *Friction washer must be positioned on shoulder of splines of pinion shaft before driven gear is positioned.*
6. Install driven gear snap ring, Fig. 26. Install pinion shaft retaining ring, making sure ring fits tightly in shaft groove.
7. Install solenoid coil retainer, Fig. 23, with tangs down. *Space retainer in housing bore so that the four tangs rest on ridge in housing bore and not in the recesses.*
8. Install solenoid retainer washer.
9. Install solenoid return spring, Fig. 18. *Inspect condition of solenoid switch contacting washer. If top of washer is burned from arcing, disassemble switch and reverse washer.*
10. Install solenoid contact into solenoid, Fig. 21. Make sure contact spring is positioned in solenoid contact. *Inspect condition of contacts in brush holder plate. If contacts are badly burned, replace brush holder with brushes and contacts as an assembly.*
11. Enter solenoid lead wire through hole in brush holder, Fig. 20, and solenoid stud, insulating washer, flat washer and nut.
12. Solder solenoid lead wire to contact terminal, Fig. 19. Wrap wire securely around terminal and solder with a high temperature solder and resin flux.
13. Carefully enter solenoid coil and coil sleeve into bore of gear housing and position brush plate assembly into gear housing, Fig. 33. Align tongue of ground terminal with notch in brush holder.
14. After brush holder is bottomed in housing, install attaching screw, Fig. 17. Tighten screw and install flat insulating washer and hold in place with friction tape.
15. Position brushes with tools shown in Fig. 34 or equivalent.
16. Position field frame to exact position and resolder field coil lead, Fig. 16.
17. Install brush terminal screw, Fig. 15.

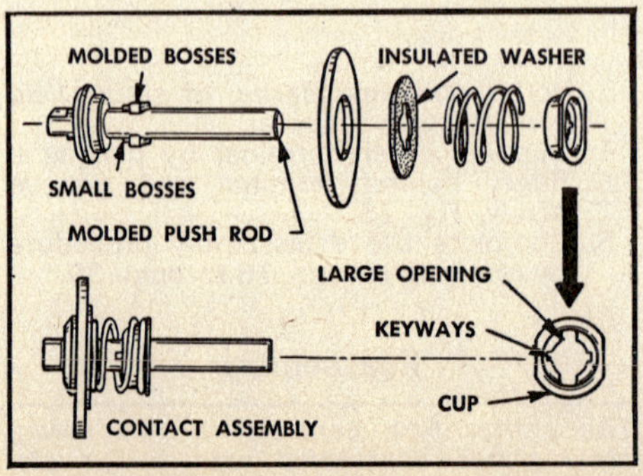

Fig. 38 Solenoid contact assembly

18. Install armature thrust washer on brush holder plate and enter armature into field frame and gear housing, Fig. 35. Carefully engage splines of shaft with reduction gear.
19. Remove brush positioning tools, Fig. 36. Install fiber and steel thrust washers on armature shaft.
20. Position starter end head, install screws and tighten securely.
21. Install gear housing dust cover. *Make sure dimples on cover are securely engaged in holes provided in gear housing.* Test starter for free running and install on engine.

DELCO-REMY STARTERS

This type starting motor, Fig. 37, has the solenoid shift lever mechanism and the solenoid plunger enclosed in the drive housing, thus protecting them from exposure to road dirt, icing conditions and splash. They have an extruded field frame and an overrunning clutch type of drive. The overrunning clutch is operated by a solenoid switch mounted to a flange on the drive housing.

Solenoid

The solenoid is attached to the drive end housing by two screws. The angle of the nose of the plunger provides a greater bearing area between the plunger and core tube. A molded push rod, Fig. 38, is assembled in the contact assembly. A shoulder molded on the push rod and a cup that can easily be assembled to the rod and locked into position over

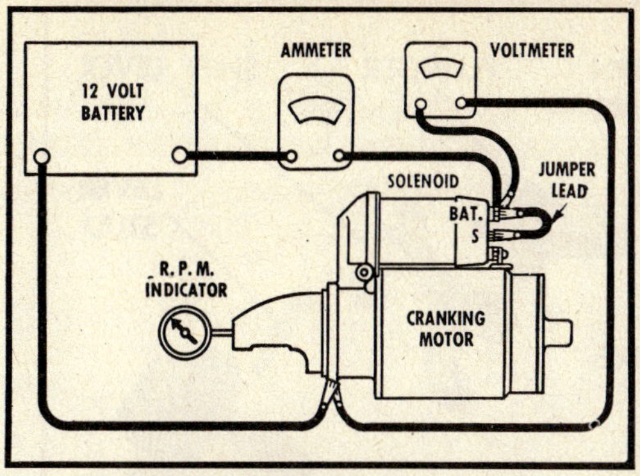

Fig. 39 Connections for checking free speed of motor

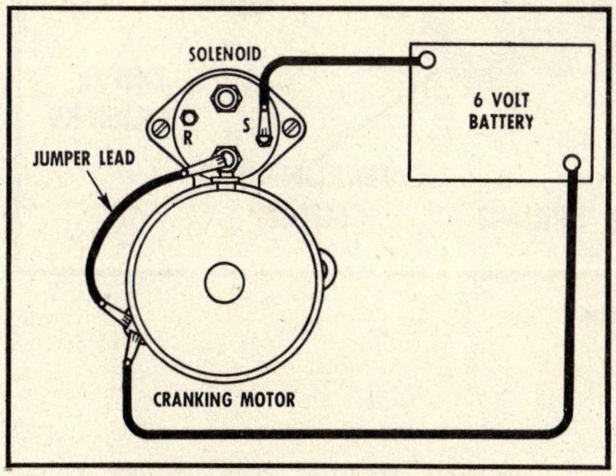

Fig. 40 Connections for checking pinion clearance

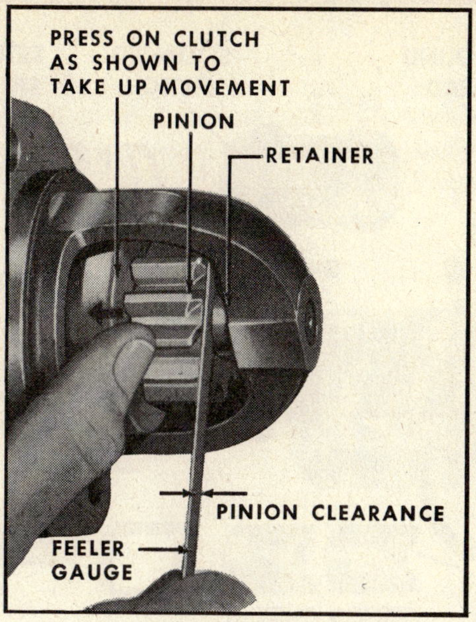

Fig. 41 Checking pinion clearance

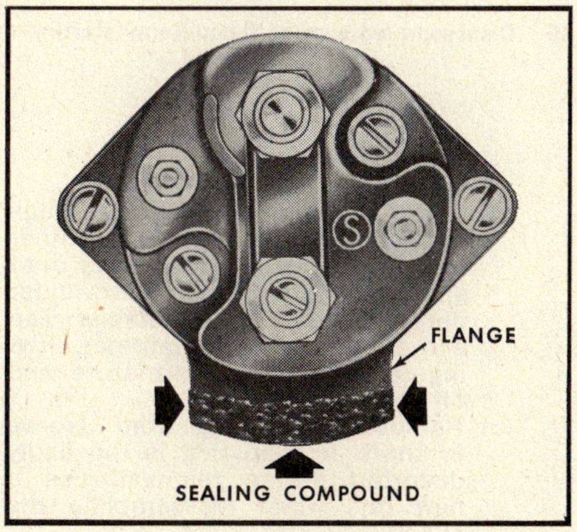

Fig. 42 Sealing solenoid housing to frame

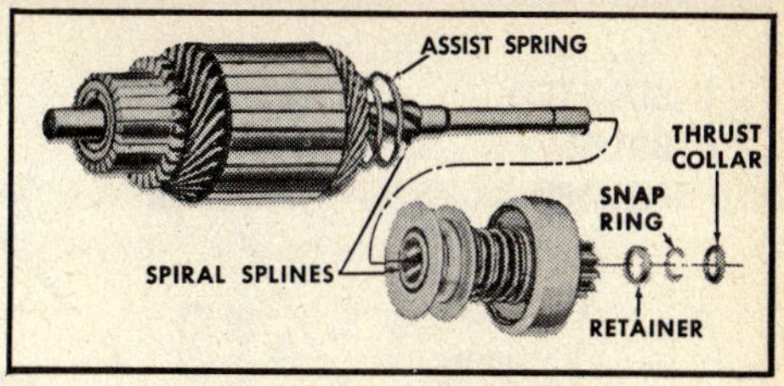

Fig. 43 View of armature and over-running clutch

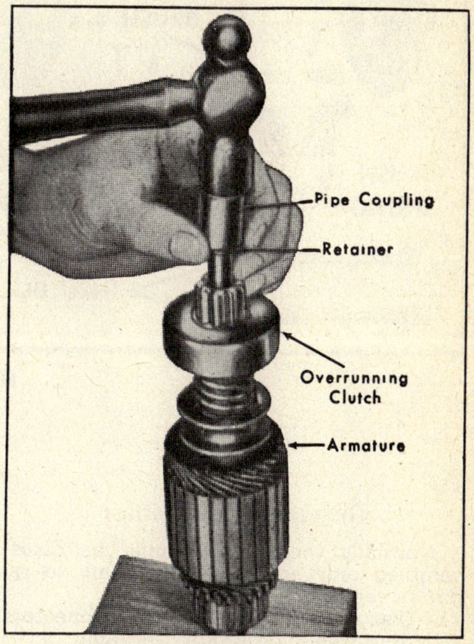

Fig. 44 Removing over-running clutch snap ring retainer

two molded bosses holds the contact assembly in place.

To disassemble the cup from the push rod, push in on the metal cup and rotate 1/4 turn so the molded bosses on the rod are in line with openings in the cup; then slide the metal cup off the rod.

To assemble the metal cup on the rod, locate the parts on the rod as shown and align the large openings in the cup with the molded bosses on the rod; then push in on the cup and rotate it 1/4 turn so the small bosses on the rod fall into the keyways of the cup.

Solenoid Terminals

The terminals of the solenoid are assembled in a molded cover. Some solenoids have an additional small terminal which is identified with the letter "R". To this terminal is attached a small metal finger which makes contact with a disc inside the solenoid when it is energized. On the vehicle, this terminal is connected to the battery side of the ignition coil. The purpose of this is to short out the ignition resistor during cranking and thereby provide high ignition coil output for starting the engine.

Maintenance

Most motors of this type have graphite and oil impregnated bronze bearings which ordinarily require no added lubrication except at times of overhaul when a few drops of light engine oil should be placed on each bearing before reassembly.

Motors provided with hinge cap oilers should have 8-10 drops of light engine oil every 5000 miles, or every 300 hours of operation. Since the motor and brushes cannot be inspected without disassembling the unit, there is no service that can be performed with the unit assembled on the vehicle.

Free Speed Test

With the circuit connected as shown in Fig. 39, use a tachometer to measure armature revolutions per minute. Failure of the motor to perform to specifications may be due to tight or dry bearings, or high resistance connections.

Pinion Clearance

There is no provision for adjusting pinion clearance on this type motor. When the shift lever mechanism is correctly assembled, the pinion clearance should fall within the limits of .010 to .140". When the clearance is not within these limits, it may indicate excessive wear of the solenoid linkage or shift lever yoke buttons.

Pinion clearance should be checked after the motor has been disassembled and reassembled. To check, make connections as shown in Fig. 40. **Caution:** *Do not connect the voltage source to the ignition coil terminal "R" of the solenoid. Do not use a 12-volt battery instead of the 6 volts specified as this will cause the motor to operate.* As a further precaution to prevent motoring, connect a heavy jumper lead from the solenoid motor terminal to ground.

After energizing the solenoid with the clutch shifted toward the pinion stop retainer, push the pinion back toward the commutator end as far as possible to take up any slack movement; then check the clearance with feeler gauge, Fig. 41.

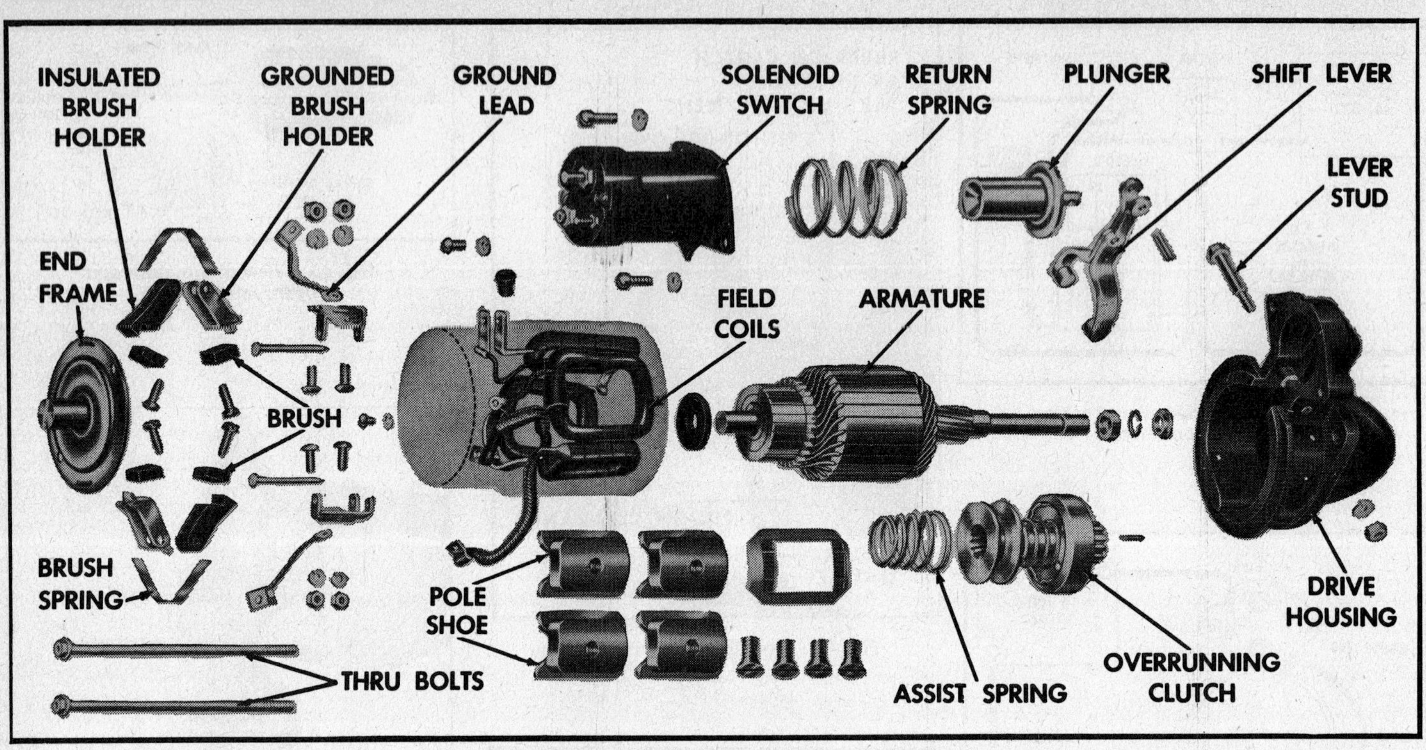

Fig. 45 Disassembled view of Delco-Remy starting motor

Disassembling Motor

Normally the motor should be disassembled only so far as necessary to repair or replace defective parts.

1. Disconnect field coil connectors from solenoid "motor" terminal.
2. Remove thru bolts.
3. Remove commutator end frame and field frame assembly.
4. Remove armature assembly from drive housing. On some models it may be necessary to remove solenoid and shift lever assembly from the drive housing before removing the armature assembly. *Important: When solenoid is installed, apply sealing compound between field frame and solenoid flange, Fig. 42.*
5. Remove overrunning clutch from armature shaft as follows:
   a) Slide thrust collar off end of armature shaft, Fig. 43.
   b) Slide a standard $\frac{1}{2}$" pipe coupling, a $\frac{5}{8}$" deep socket or other metal cylinder of suitable size onto shaft so end of coupling or cylinder butts against edge of retainer. Tap end of coupling with hammer, driving retainer toward armature and off snap ring, Fig. 44.
   c) Remove snap ring from groove in shaft. If snap ring is too badly distorted during removal, use a new one when reassembling the clutch.
   d) Slide retainer, clutch and assist spring from armature shaft.

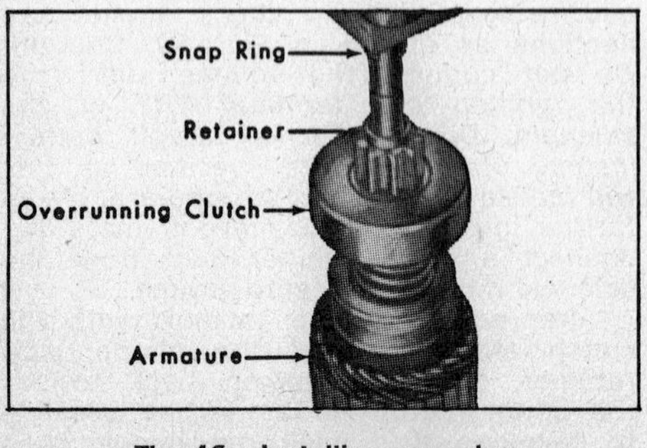

Fig. 46 Installing snap ring onto armature shaft

Reassembling Motor, Fig. 45

1. Lubricate drive end and splines of armature shaft with SAE 10 oil. *If heavier oil is used it may cause failure to mesh at low temperatures.*
2. Place "assist" spring on drive end of shaft next to armature, with small end against lamination stack.
3. Slide clutch assembly onto armature shaft with pinion outward.
4. Slide retainer onto shaft with cupped surface facing end of shaft.
5. Stand armature on end on wood surface with commutator down. Position snap ring on upper end of shaft and hold in place with a block of wood. Hit wood block with a hammer forcing snap ring over end of shaft, Fig. 46. Slide snap ring into groove, squeezing it to ensure a good fit in groove.
6. Assemble thrust collar on shaft with shoulder next to snap ring.
7. Position retainer and thrust collar next to snap ring. With clutch pressed against assist spring, for clearance next to retainer, use two pairs of pliers at the same time (one pair on either side of shaft) to grip retainer and thrust collar. Then squeeze until snap ring is forced into retainer, Fig. 47.
8. Place 4 or 5 drops of SAE 10 oil in drive housing bushing. Make sure thrust collar is in place against snap ring and retainer; then slide armature and clutch assembly into place in drive housing.
9. Attach solenoid and shift lever assembly to drive housing. Be sure lever buttons are located between sides of clutch collar.
10. Position field frame over armature, *applying sealing compound between frame and solenoid flange* (Fig. 42).

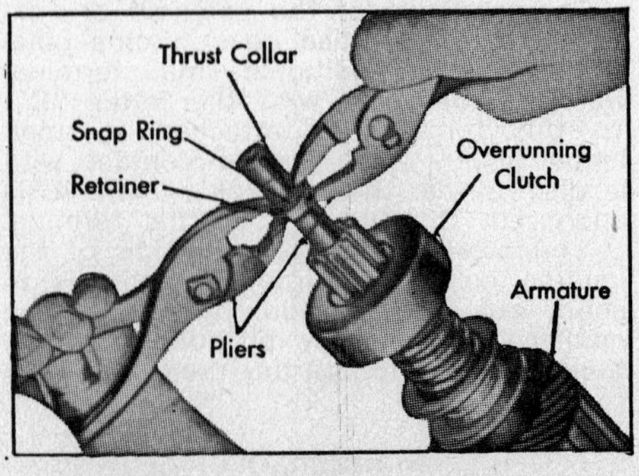

Fig. 47 Installing snap ring into retainer

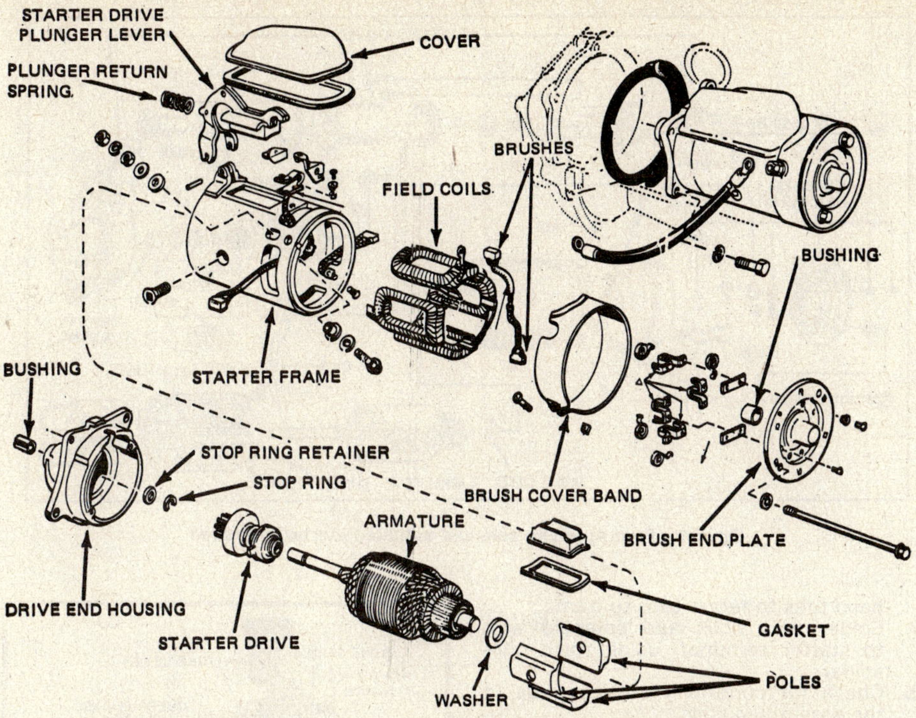

Fig. 48 Ford Auto-Lite starter with an integral positive engagement drive

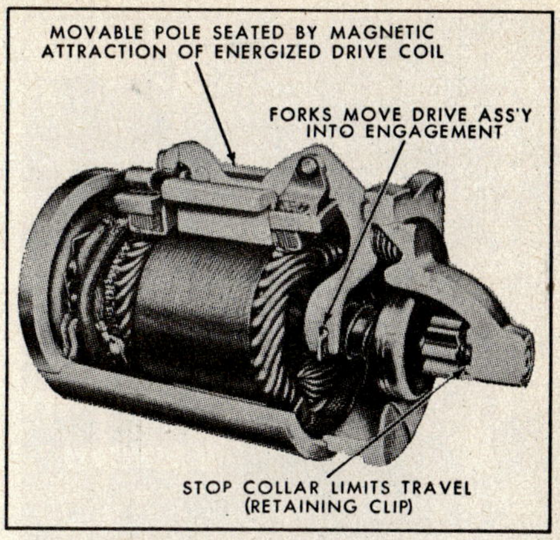

Fig. 49 Starter drive engaged

Position frame against drive housing, using care to prevent damage to brushes.

11. Place 4 or 5 drops of SAE 10 oil in bushing in commutator end frame. Make sure leather brake washer is on armature shaft; then slide commutator end frame onto shaft.
12. Install thru bolts and tighten securely.
13. Reconnect field coil connectors to solenoid "motor" terminal.

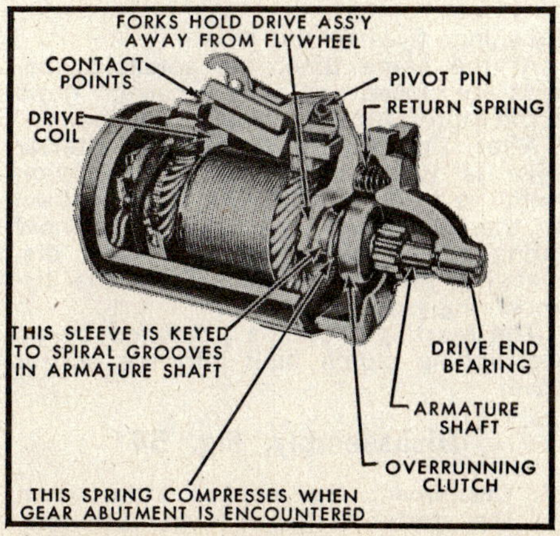

Fig. 50 Starter drive disengaged

FORD AUTO-LITE STARTER WITH INTEGRAL POSITIVE ENGAGEMENT DRIVE

This type starting motor, Fig. 48, is a four pole, series parallel unit with a positive engagement drive built into the starter. The drive mechanism is engaged with the flywheel by lever action before the motor is energized.

When the ignition switch is turned on to the start position, the starter relay is energized and supplies current to the motor. The current flows through one field coil and a set of contact points to ground. The magnetic field given off by the field coil pulls the movable pole, which is part of the lever, downward to its seat. When the pole is pulled down, the lever moves the drive assembly into the engine flywheel, Fig. 49.

When the movable pole is seated, it functions as a normal field pole and opens the contact points. With the points open, current flows through the starter field coils, energizing the starter. At the same time, current also flows through a holding coil to hold the movable pole in its seated position.

When the ignition switch is released from the start position, the starter relay opens the circuit to the starting motor. This allows the return spring to force the lever back, disengaging the drive from the flywheel and returning the movable pole to its normal position, Fig. 50.

Disassembly

It may not be necessary to disassemble the starter completely to accomplish repair or replacement of certain parts. Thus, before disassembling the motor, remove the cover band and starter drive actuating lever cover. Examine brushes to make sure they are free in their holders. Replace brushes if defective or worn beyond their useful limit. Check the tension of each brush spring with a pull scale. Spring Tension should not be less than 45 ounces. If disassembly is necessary, proceed as follows:

1. Remove cover band and starter drive actuating lever cover.
2. Remove through bolts, starter drive gear housing, drive gear retaining clip cup and starter drive actuating lever return spring. Some 1969-70 units incorporate needle bearings at the drive end. If needle bearings need not be replaced, insert a dummy shaft into the housing while removing the armature, Fig. 51. This will prevent loss of bearings.
3. Remove pivot pin retaining starter gear actuating lever and remove lever and armature.
4. Remove and discard spring clip retaining starter drive gear to end of armature shaft, and remove starter drive gear.

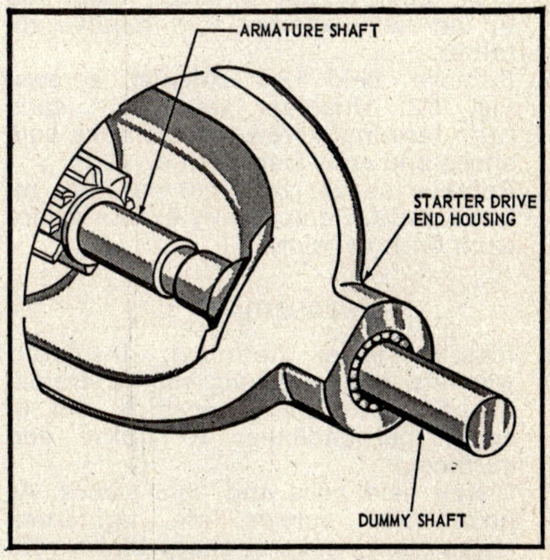

Fig. 51 Using dummy shaft to hold needle bearings in place while servicing 1969-70 starter motor

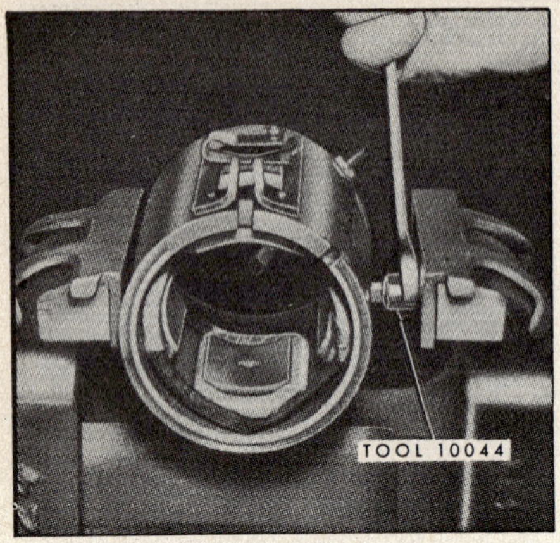

Fig. 52 Removing field coil pole shoe screws

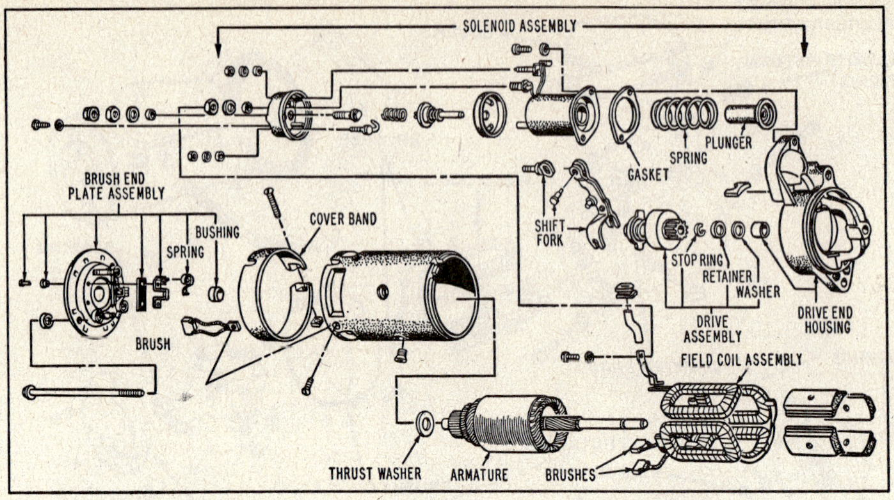

Fig. 54 Ford Auto-Lite solenoid actuated starter, exploded

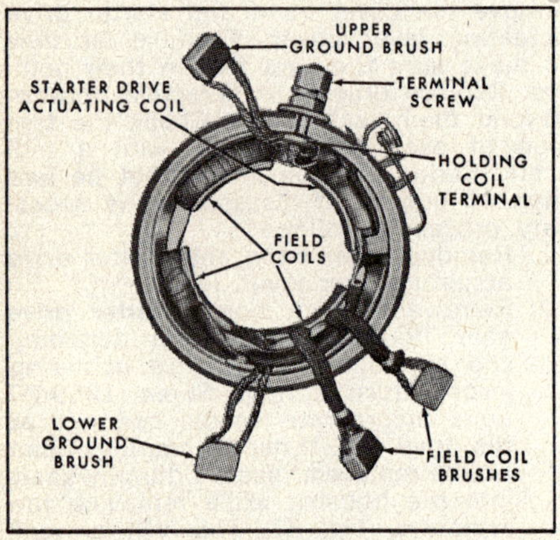

Fig. 53 Field coil assembly

5. Remove commutator brushes from brush holders and remove brush end plate.
6. Remove two screws retaining ground brushes to frame.
7. On the field coil that operates the drive gear actuating lever, bend tab up on field retainer and remove retainer.
8. Remove field coil retainer screws, Fig. 52. Unsolder field coil leads from terminal screw, and remove pole shoes and coils from frame.
9. Remove starter terminal nut and related parts. Remove any excess solder from terminal slot.

Reassembly

1. Install starter terminal, insulator, washers and retaining nut in frame, Fig. 53. Be sure to position slot in screw perpendicular to frame end surface.
2. Install field coils and pole pieces. As pole shoe screws are tightened, strike frame several sharp blows with a soft-faced hammer to seat and align pole shoes, then stake the screws.
3. Install solenoid coil retainer and bend tabs to retain tabs to frame.
4. Solder field coils and solenoid wire to starter terminal, using rosin core solder.
5. Check for continuity and grounds in the assembled coils.
6. Position solenoid coil ground terminal over ground screw hole nearest starter terminal.
7. Position ground brushes to starter frame and install retaining screws, Fig. 54.
8. Position starter brush end plate to frame with end plate boss in frame slot.
9. Install drive gear to armature shaft and install a new retaining spring clip.
10. Position fiber thrust washer on commutator end of armature shaft and install armature in frame.
11. Install starter drive actuating lever to frame and starter drive, and install pivot pin.
12. Position actuating lever return spring and drive gear housing to frame and install through bolts. Do not pinch brush leads between brush plate and frame.
13. Install brushes in holders, being sure to center brush springs on brushes.
14. Position drive gear actuating lever cover on starter and install brush cover band.

FORD AUTO-LITE SOLENOID ACTUATED STARTER

Description

The solenoid assembly, in this unit, is mounted to a flange on the starter drive housing which encloses the entire shift lever and solenoid plunger mechanism The solenoid incorporates a pull-in winding and a hold-in winding.

Operation

As the solenoid is energized, it shifts the starting motor pinion into mesh with the engine flywheel ring gear.

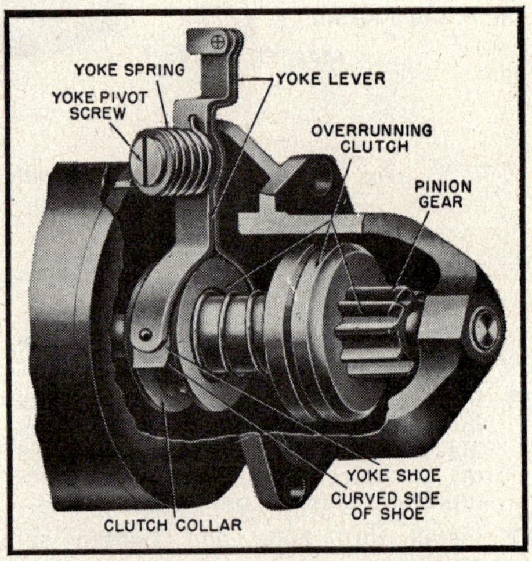

Fig. 55 Overrunning clutch drive. When assembling, make sure curved sides of yoke shoes are toward gear end of clutch. Reversed yoke shoes can cause improper meshing of pinion

At the same time, the solenoid contacts are closed and battery current flows to the motor, turning it and the engine.

After the engine starts, the starter drive is disengaged when the ignition switch is returned from the start position to the run position and the solenoid spring pushes the shift lever back, disengaging the starter drive from the flywheel ring gear.

The starting motor is protected by an overrunning clutch built into the starter drive.

Disassembly, Fig. 54

1. Disconnect the copper strap from the starter terminal of the solenoid, remove the retaining screws and remove solenoid.

2. Loosen retaining screw and slide brush cover band back on frame.
3. Remove commutator brushes from holders. Hold each spring away from the brush with a hook while sliding brush from holder.
4. Remove through bolts and separate end plates and frame.
5. Remove solenoid plunger and shift fork assembly.
6. Remove armature and drive assembly from frame. Remove drive stop ring and slide drive assembly from shaft. Remove fiber thrust washer from commutator end of shaft.
7. Remove drive stop ring retainer from shaft.

Reassembly

1. Install drive assembly on shaft and install new stop ring.
2. Install solenoid plunger and shift fork.
3. Place new retainer in drive housing and install armature and drive in housing. Be sure shift lever tangs properly engage drive assembly.
4. Install fiber washer on commutator end of shaft and position frame to drive housing, being sure to index frame and drive housing correctly.
5. Install brush plate assembly being sure to index it properly, install through bolts and tighten to 55-75 in lbs.
6. Install brushes by pulling each spring away from holder with a hook to allow entry of the brush. Center the brush springs on the brushes. Press insulated brush leads away from all other components to prevent possible shorts.
7. Install rubber gasket and solenoid.
8. Connect copper strap to starter terminal of solenoid.
9. Position cover band and tighten retaining screw.
10. Connect starter to battery and check operation.

STARTER DRIVE TROUBLES

Starter drive troubles are easy to diagnose and they usually cannot be confused with ordinary starter difficulties. If the starter does not turn over at all or if it drags, look for trouble in the starter or electrical supply system. Concentrate on the starter drive or ring gear if the starter is noisy, if it turns but does not engage the engine, or if the starter won't disengage after the engine is started. After the starter is removed, the trouble can usually be located quickly.

Worn or chipped ring gear or starter pinion are the usual causes of noisy operation. Before replacing either or both of these parts try to find out what caused the damage. With the Bendix type drive, incomplete engagement of the pinion with the ring gear is a common cause of tooth damage. The wrong pinion clearance on starter drives of the overrunning clutch type leads to poor meshing of the pinion and ring gear and too rapid tooth wear.

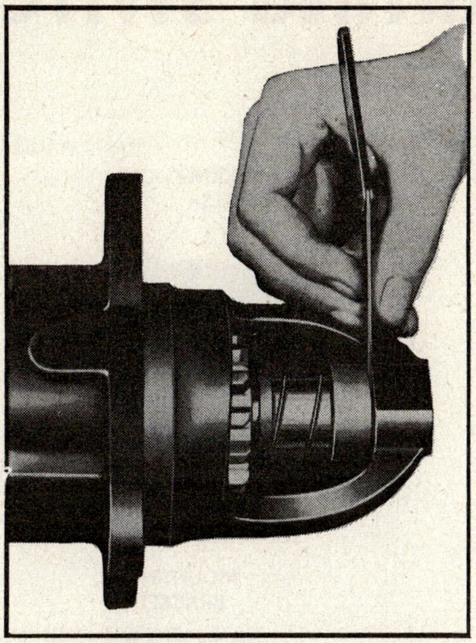

Fig. 56 Measuring overrunning clutch drive stop clearance. Do not compress anti-drift spring as this will give an incorrect clearance. If clearance is not present there is danger of the drive housing being broken as gear or collar slams back against it

A less common cause of noise with either type of drive is a bent starter armature shaft. When this shaft is bent, the pinion gear alternately binds and then only partly meshes with the ring gear. Most manufacturers specify a maximum of .003" radial run-out on the armature shaft.

When Clutch Drive Fails

The over-running clutch type drive seldom becomes so worn that it fails to engage since it is directly activated by a fork and lever, Fig. 55. The only thing that is likely to happen is that, once engaged, it will not turn the engine because the clutch itself is worn out. A much more frequent difficulty and one that rapidly wears ring gear and teeth is partial engagement. Proper meshing of the pinion is controlled by the end clearance between the pinion gear and the starter housing or pinion stop, if used.

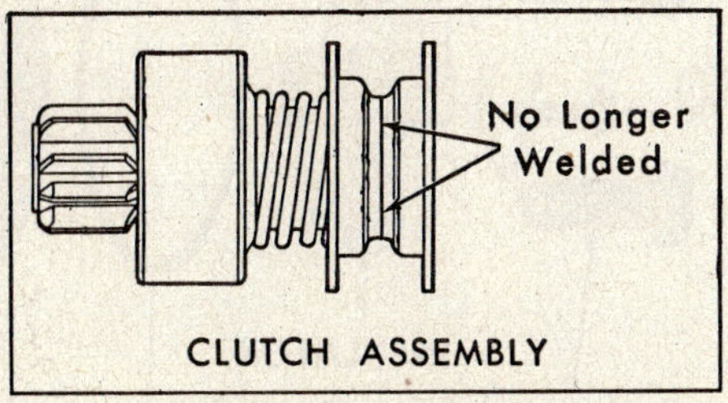

Fig. 57 New design overrunning clutch collar

The clearance is set with the starter off the car and with the drive in the engaged position. To check the clearance, supply current to the starter solenoid with the electrical connection between starter and solenoid removed. Supplying current to the solenoid but not the starter will prevent the starter from rotating during the test. Take out all slack by pushing lightly on the starter drive clutch housing while inserting a feeler gauge between pinion and housing or pinion stop, Fig. 56.

On late model cars, the solenoids are completely enclosed in the starter housing and the pinion clearance is not adjustable. If the clearance is not correct, the starter must be disassembled and checked for excessive wear of solenoid linkage, shift lever mechanism, or improper assembly of parts.

Failure of the over-running clutch drive to disengage is usually caused by binding between the armature shaft and the drive. If the drive, particularly the clutch, shows signs of overheating it indicates that it is not disengaging immediately after the engine starts. If the clutch is forced to over-run too long, it overheats and turns a bluish color. For the cause of the binding, look for rust or gum between the armature shaft and the drive, or for burred splines. Excess oil on the drive will lead to gumming, and inadequate air circulation in the flywheel housing will cause rust.

Over-running clutch drives cannot be overhauled in the field so they must be replaced. In cleaning, never soak them in a solvent because the solvent may enter the clutch and dissolve the sealed-in lubricant. Wipe them off lightly with kerosene and lubricate them sparingly with SAE 10 or 10W oil.

NOTE: Beginning in 1968 some Delco Remy starter drives used an overrunning clutch with a split collar. Side by side but not welded together as in previous models, Fig. 57. Do not mistake this split design as a defective collar.

When Bendix Drive Fails

When a Bendix type drive doesn't engage the cause usually is one of three things: either the drive spring is broken, one of the drive spring bolts has sheared off, or the screwshaft threads won't allow the pinion to travel toward the flywheel. In the first two cases, remove the drive by unscrewing the set screw under the last coil of the drive spring and replace the broken parts. Gummed or rusty screwshaft threads are fairly common causes of Bendix drive failure and are easily cleaned with a little kerosene or steel wool, depending on the trouble. Here again, as in the case of overrunning clutch drives, use light oil sparingly, and be sure the flywheel housing has adequate ventilation. There is usually a breather hole in the bottom of the flywheel housing which should be open.

The failure of a Bendix drive to disengage or to mesh properly is most often caused by gummed or rusty screwshaft threads. When this is not true, look for mechanical failure within the drive itself.

STARTING SWITCHES

MAGNETIC and SOLENOID SWITCHES are designed to perform mechanical jobs electromagnetically such as closing a heavy circuit or shifting the starter drive pinion with the engine flywheel ring gear for cranking. Switches of this type consist basically of contacts and a winding (or windings) around a hollow cylinder containing a movable core or plunger. When the winding (or windings) is energized by the battery through an external control circuit the plunger is pulled inward, producing the necessary mechanical movement.

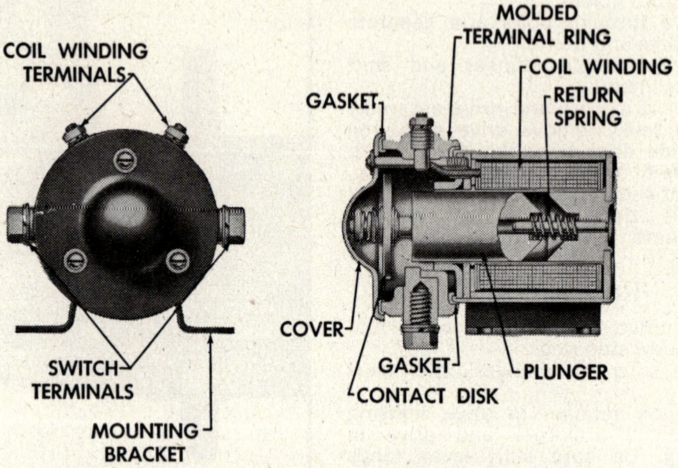

Fig. 1 End and sectional views of a sealed type magnetic switch which uses gaskets to seal the contact compartment

MAGNETIC SWITCHES

The switch shown in Fig. 1 is not designed for disassembly and must be replaced if defective.

In the Delco-Remy switch shown in Fig. 1 the terminals are assembled into a molded terminal ring which is held in place on the switch case by the cover and screws. Gaskets on both sides of the ring seals the contact compartment as a protection against moisture and dirt. The winding assembly is not removable from the case on this unit although the contact disk, plunger and plunger return spring can be removed after the cover is taken off.

NOTE

On vehicles with overrunning clutch starting motors, a magnetic switch is normally used to shift the drive pinion into mesh and to close the starter circuit. There are two variations of three-terminal switches and also one type with four terminals. Any of these switches may be manufactured with either a grounded or insulated base. When installing a switch that is not marked "grounded base" or "insulated base", it must be checked out as follows, using a battery and test lamp in series.

Three Terminal Switches

1. If the switch has a grounded base the test lamp will light when connected between the starter ("S") terminal or ignition ("I") terminal and switch mounting bracket.
2. If the switch has an insulated base the test lamp will light when connected between the "S" and "I" terminals and either one of the 5/16" threaded studs.
3. If the test lamp fails to light when connected between the "S" and "I" terminals and any external part of the switch, disassemble the switch to determine whether the base is grounded or insulated.

Four Terminal Switches

1. If the switch has a grounded base the test lamp will light when connected between the "S" terminal and switch mounting bracket.
2. If the switch has an insulated base the test lamp will light when connected between the "S" and "I" terminals.
3. If the test lamp fails to light when connected between the "S" terminal and any other external part of the switch, disassemble the switch to determine whether it has an insulated or grounded base.

SOLENOID SWITCHES

The solenoid switch on a cranking motor not only closes the circuit between the battery and the cranking motor but also shifts the drive pinion into mesh with the engine flywheel ring gear. This

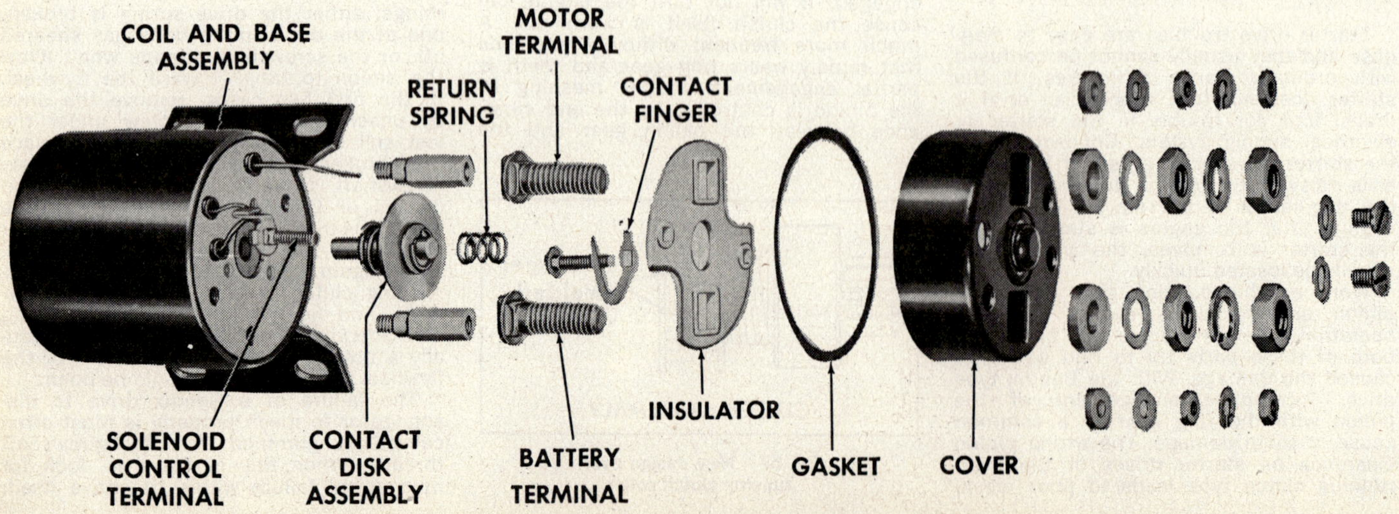

Fig. 2 Exploded view of a typical solenoid switch

is done by means of a linkage between the solenoid switch plunger and the shift lever on the cranking motor (see *Starting Motors* chapter).

Fig. 2 shows a solenoid switch used on vehicles with 12-volt systems. Like other solenoid switches, this type is energized by the battery through a separate starting switch. Note, however, that the switch includes an additional small terminal and contact finger. This terminal has no functional duty in relation to the switch, but is used to complete a special ignition circuit during the cranking cycle only. When the solenoid is in the cranking position, the finger touches the contact disk and provides a direct circuit between the battery and ignition coil.

When reassembling the switch the contact finger should be adjusted to touch the contact disk before the disk makes contact with the main switch terminals. There should be 1/16" to 3/32" clearance between the contact disk and the main terminals when the finger touches.

Fig. 3 is a wiring circuit of a typical solenoid switch. There are two windings in the solenoid; a pull-in winding (shown as dashes) and a hold-in winding (shown dotted). Both windings are energized when the external control switch is closed. They produce a magnetic field which pulls the plunger in so that the drive pinion is shifted into mesh, and the main contacts in the solenoid switch are closed to connect the battery directly to the cranking motor. Closing the main switch contacts shorts out the pull-in winding since this winding is connected across the main contacts. The magnetism produced by the hold-in winding is sufficient to hold the plunger in, and shorting out the pull-in winding reduces drain on the battery. When the control switch is opened, it disconnects the hold-in winding from the battery. When the hold-in winding is disconnected from the battery, the shift lever spring withdraws the plunger from the solenoid, opening the solenoid switch contacts and at the same time withdrawing the drive pinion from mesh. Proper operation of the switch depends on maintaining a definite balance between the magnetic strength of the pull-in and hold-in windings.

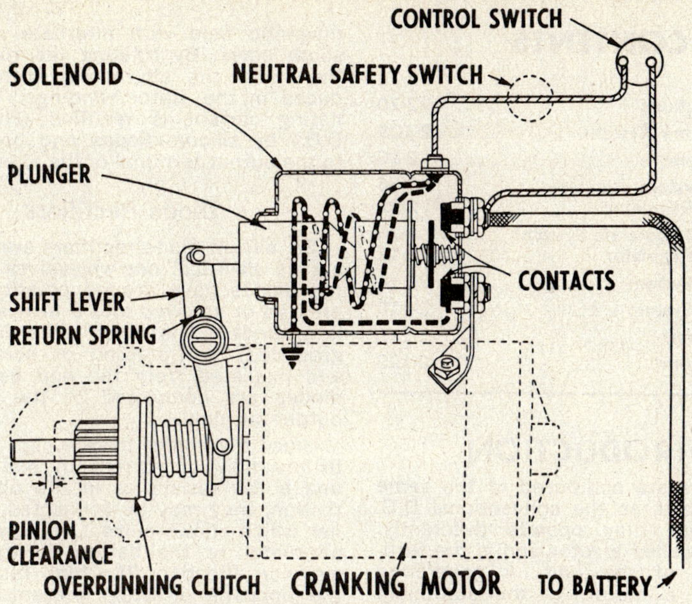

Fig. 3 Wiring circuit of a typical solenoid switch

This balance is established in the design by the size of the wire and the number of turns specified. *An open circuit in the hold-in winding or attempts to crank with a discharged battery will cause the switch to chatter.*

To disassemble the solenoid, remove nuts, washers and insulators from the switch terminal and battery terminal. Remove cover and take out the contact disk assembly.

ALTERNATOR SYSTEMS

CONTENTS

INTRODUCTION

Alternators are composed of the same functional parts as the conventional D.C. generator but they operate differently: The field is called a rotor and is the turning portion of the unit. A generating part, called a stator, is the stationary member, comparable to the armature in a D.C. generator. The regulator, similar to those used in a D.C. system, regulates the output of the alternator-rectifier system.

The power source of the system is the alternator. Current is transmitted from the field terminal of the regulator through a slip ring to the field coil and back to ground through another slip ring. The strength of the field regulates the output of the alternating current. This alternating current is then transmitted from the alternator to the rectifier where it is converted to direct current.

These alternators employ a three-phase stator winding in which the phase windings are electrically 120 degrees apart. The rotor consists of a field coil encased between interleaved sections producing a magnetic field with alternate north and south poles. By rotating the rotor inside the stator the alternating current is induced in the stator windings. This alternating current is rectified (changed to D.C.) by silicon diodes and brought out to the output terminal of the alternator.

Diode Rectifiers

Six silicon diode rectifiers are used and act as electrical one-way-valves. Three of the diodes have ground polarity and are pressed or screwed into a heat sink which is grounded. The other three diodes (ungrounded) are pressed or screwed into and insulated from the end head; these diodes are connected to the alternator output terminal.

Since the diodes have a high resistance to the flow of current in one direction and a low resistance in the opposite direction, they may be connected in a manner which allows current to flow from the alternator to the battery in the low resistance direction. The high resistance in the opposite direction prevents the flow of current from the battery to the alternator. Because of this feature no circuit breaker is required between the alternator and battery.

SERVICE PRECAUTIONS

1. Be certain that battery polarity is correct when servicing units. Reversed battery polarity will damage rectifiers and regulators.
2. If booster battery is used for starting, be sure to use correct polarity in hook up.
3. When a fast charger is used to charge a vehicle battery, the vehicle battery cables should be disconnected *unless the fast charger is equipped with a special Alternator Protector*, in which case the vehicle battery cables need not be disconnected. Also the fast charger should never be used to start a vehicle as damage to rectifiers will result.
4. Lead connections to the grounded rectifiers (negative) on Prestolite and Chrysler units should never be soldered as the excessive heat may damage the rectifiers.
5. Unless the system includes a load relay or field relay, grounding the alternator output terminal will damage the alternator and/or circuits. This is true even when the system is not in operation since no circuit breaker is used and the battery is applied to the alternator output terminal at all times. The field or load relay acts as a circuit breaker in that it is controlled by the ignition switch.
6. When adjusting the voltage regulator, do not short the adjusting tool to the regulator base as the regulator may be damaged. The tool should be insulated by taping or by installing a plastic sleeve.
7. Before making any "on vehicle" tests of the alternator or regulator, the battery should be checked and the circuit inspected for faulty wiring or insulation, loose or corroded connections and poor ground circuits.
8. Check alternator belt tension to be sure the belt is tight enough to prevent slipping under load.
9. The ignition switch should be off and the battery ground cable disconnected before making any test connections to prevent damage to the system.
10. The vehicle battery must be fully charged or a fully charged battery may be installed for test purposes.

Chrysler Alternators

Three types of alternators are available on Chrysler Corp. cars; one with an electro-mechanical voltage regulator, Figs. C1 and C1A, the insulated brush unit, Figs. C2 and C2A optional on some 1969 cars and the isolated field unit which is standard on later cars, Figs. C3 thru C4A. A 100 amp alternator is optional on some 1975 vehicles and is equipped with 12 silicone rectifiers, Fig. C5.

The difference between these units is in the area of the brushes. The electro-mechanical unit uses a ground brush, the insulated brush unit has a positive insulated brush with the system grounded through the regulator and the isolated field unit has two separate, insulated, field brushes.

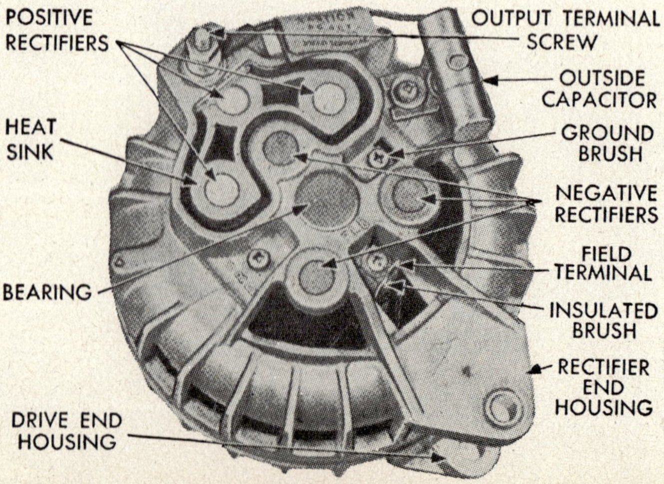

Fig. C1 Chrysler alternator, electro-mechanically regulated unit

Both the insulated brush and the isolated field units use a fully electronic voltage regulator which is a sealed, non-adjustable unit.

NOTE: The charging system on all 1975 Chrysler Corp. vehicles except Dart and Valiant incorporates a field-loads relay, Fig. C6. This protects the battery from overcharging and provides a more consistant voltage to the electrical accessories. This field-loads relay, located between the battery and voltage regulator, brings the regulator electrically closer to the battery, thereby reducing the voltage drop between these components and also makes the regulator more sensitive to battery requirements than the previous system.

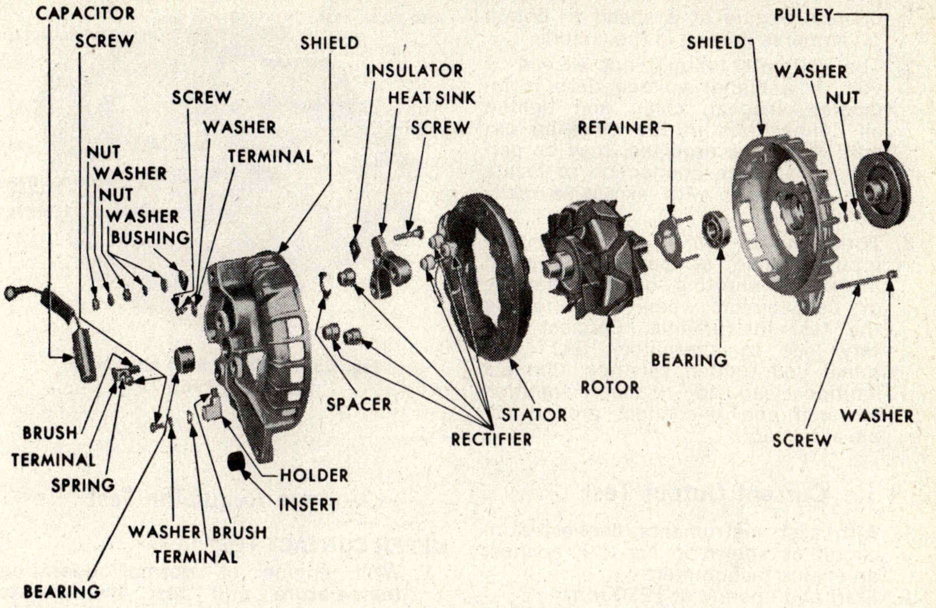

Fig. C1A Electro-mechanically regulated alternator disassembled

TESTING SYSTEM ON VEHICLE

ELECTRO-MECHANICAL UNIT

Field Circuit Resistance Test

1. Referring to Fig. C7, disconnect ignition wire at coil side of ballast resistor and connect a test ammeter and voltmeter in the circuit as shown. All lights and accessories should be turned off.
2. Turn ignition switch on and turn voltmeter selector switch to the low voltage scale and read the meter. The voltage should not exceed .55 volt. A reading in excess of .55 volt indicates high resistance in field circuit between battery and voltage regulator field terminal.
3. If high resistance is indicated, move negative voltmeter lead to each connection along the circuit to the battery. A sudden drop in voltage indicates a loose or corroded connection between that point and the last point tested. To test the terminals for tightness, attempt to move the terminal while observing the voltmeter. Any movement of the meter pointer indicates looseness.

NOTE: *Excessive resistance in the regulator wiring circuit will cause fluctuation in the ammeter.*

4. Turn ignition switch off, disconnect test instrument and reconnect ignition primary wire at the coil side of the ballast resistor.

Charging Circuit Resistance Test

With battery in good condition and fully charged, first disconnect the battery ground cable to avoid accidental shorting of the charging or field circuit when making the test connections shown in Fig. C8.

1. With the test instruments connected as shown and with battery ground cable re-connected, start and

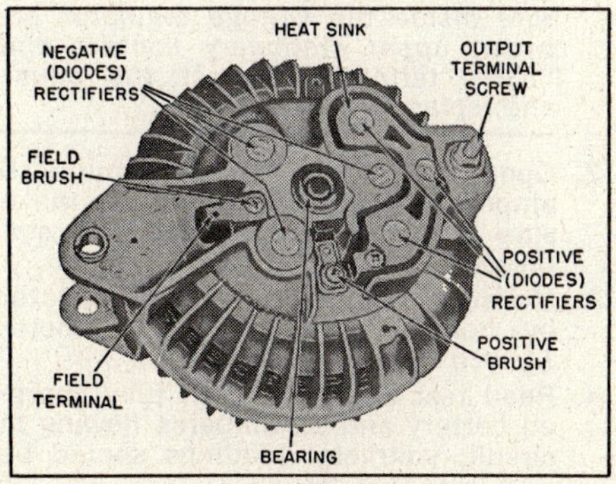

Fig. C2 Insulated brush alternator assembly

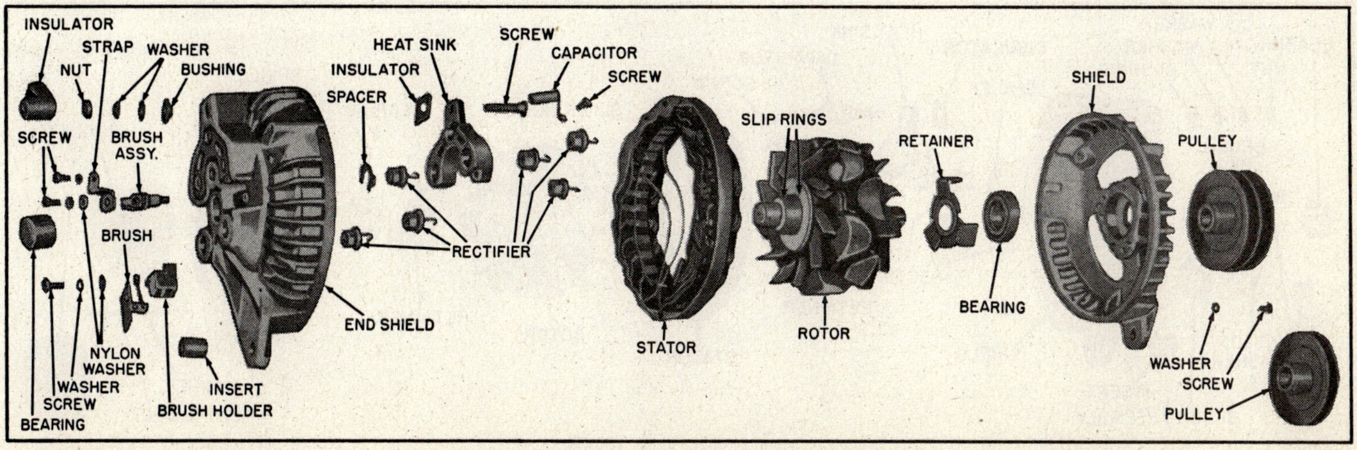

Fig. C2A Insulated brush alternator disassembled

operate engine at a speed to obtain 10 amperes flowing in the circuit.

2. The voltmeter should not exceed .3 volt. If a higher voltage drop is indicated, inspect, clean and tighten all connections in the charging circuit. A voltage drop test may be performed at each connection to locate the connection with excessive resistance.
3. Turn ignition switch off. Disconnect ground cable at battery to avoid accidental shorting of the charging or field circuit when disconnecting the test instruments. Connect battery lead to alternator "BAT" terminal and tighten securely. Connect ignition lead to regulator ignition terminal and re-connect ground cable at battery.

Current Output Test

1. With test instruments connected in circuit as shown in Fig. C9, connect an engine tachometer.
2. Start and operate at 1250 rpm.
3. Adjust carbon pile rheostat to obtain a reading of 15 volts on the test voltmeter.
4. Observe reading on test ammeter.
5. If the output is slightly less (5 to 7 amperes) than the rated output of the alternator, it may be an indication of an open-circuited diode or other internal alternator problem.
6. If the output is considerably lower than the rated output of the alternator, it may be an indication of a short-circuited diode or other internal alternator problem. In either case the alternator should be removed and tested.

NOTE: *Turn off the carbon pile rheostat immediately after observing reading on test ammeter.*

7. If the alternator current output tested satisfactorily, turn off the ignition switch and remove the jumper lead from the alternator field terminal and output terminal.

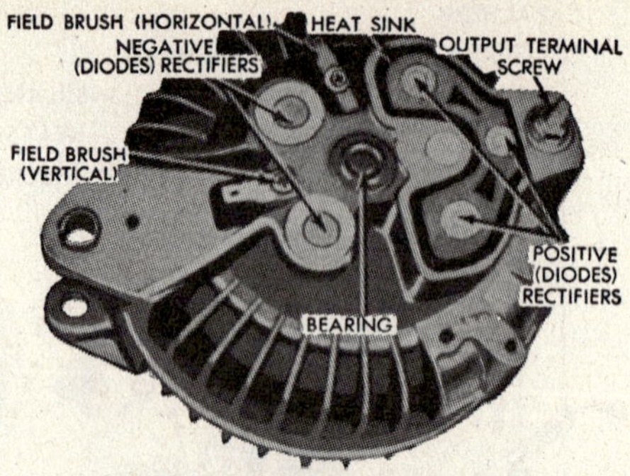

Fig. C3 Isolated field alternator assembly, 1969-71

Voltage Regulator Test

UPPER CONTACT TEST

1. With engine at normal operating temperature and test instruments connected as shown in Fig. C10, start and operate the engine at 1250 rpm.
   Adjust carbon pile to obtain a 15 ampere output as indicated on test ammeter.

NOTE: *No current reading on the ammeter would indicate either a low regulator setting or a blown fuse wire inside the voltage regulator between upper stationary contact and "IGN" terminal. Correct the cause and replace the fusible wire.*

2. Operate engine at 1250 rpm and a 15 ampere load for 15 minutes to make sure entire regulator system is stabilized.
3. Measure temperature at regulator by holding a reliable thermometer ¼ inch from regulator cover.
4. Read test ammeter. With fully charged battery and 15 amperes flowing in circuit, voltmeter readings should be within specifications.
5. If regulator operates within specifications, proceed to the lower contact voltage test. If not, remove cover and adjust voltage setting as outlined under "Regulator Adjustments".

LOWER CONTACT VOLTAGE TEST

1. Increase engine speed to 2200 rpm. Vary carbon pile to decrease current load to 7 amperes output as registered on test ammeter. The voltage should *increase* and amperage should decrease.

NOTE: *There will be a slightly higher voltage at higher engine speeds above 2200 rpm. However, this increased voltage must not exceed the voltage specified by more than .7 volt at any temperature range.*

2. If the regulator setting is outside the specified limits, the regulator must be removed to remove the cover.
3. To adjust the voltage setting, bend *the regulator lower spring hanger down to increase voltage, or up to* decrease voltage setting, Fig. C11. The regulator must be installed, correctly connected, and retested after each adjustment of the lower spring hanger.

NOTE: *If repeated readjustment is required, it is permissible to use a jumper wire to ground the regulator base to the fender splash shield for testing instead of reinstalling the regulator each time. However, it is important that the cover be reinstalled, the regulator connections correctly connected, and the regulator insulated to prevent grounding the regulator terminals or resistances. When testing, the regulator must be at the same attitude (or angle) as when installed on the vehicle.*

4. If the alternator and regulator tested satisfactorily, turn the igni-

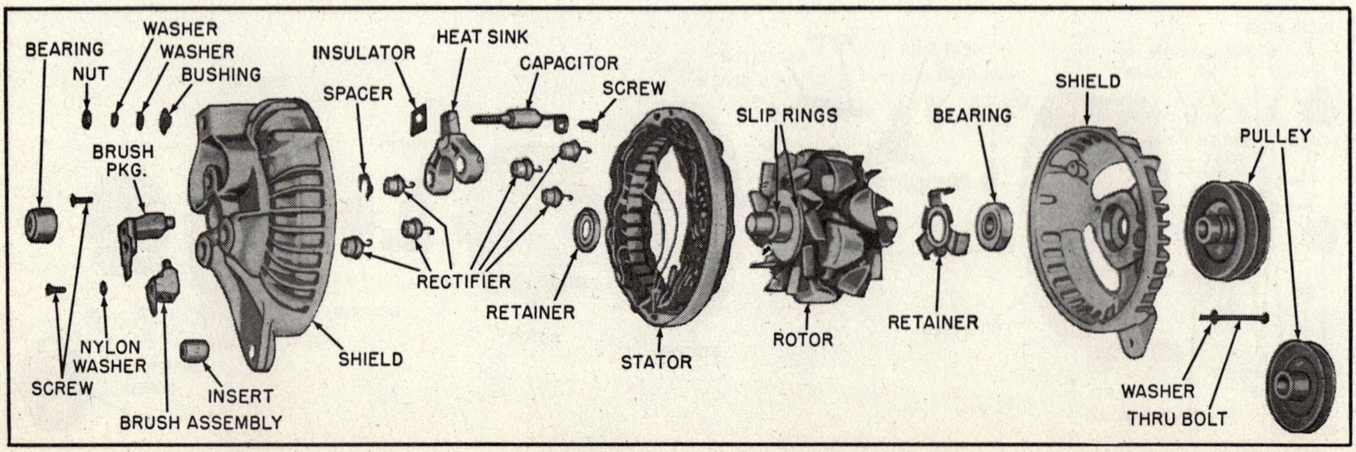

Fig. C3A Isolated field alternator disassembled. 1969-71

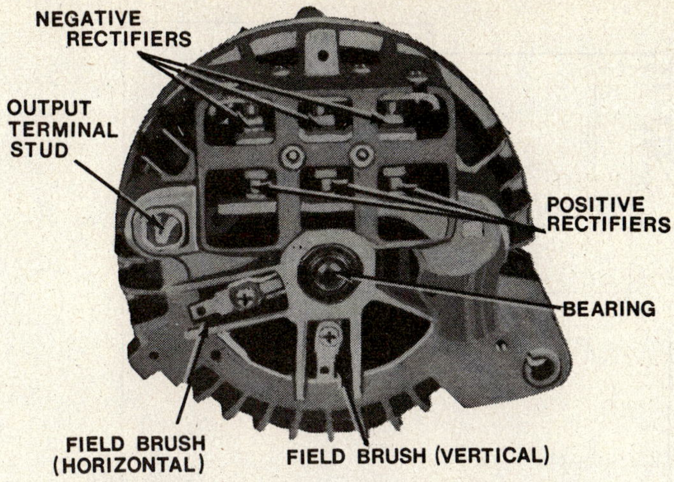

Fig. C4 Isolated field alternator assembly, 1972-75

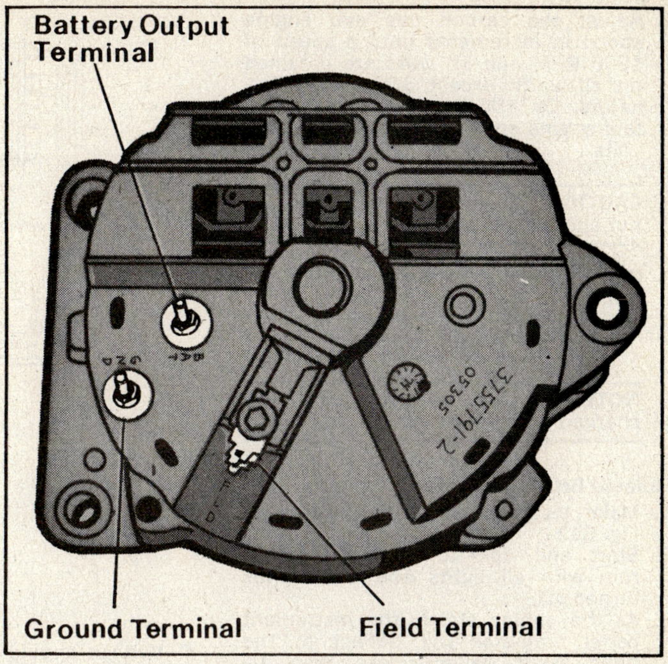

Fig. C5 100 amp alternator assembly. 1975

tion switch off. Disconnect battery ground cable, then the test instruments. Connect the leads to alternator and regulator. Finally reconnect battery ground cable.

Regulator Adjustments

If the regulator cannot be adjusted or voltage control, or if the regulator performance is erratic or malfunctions, it may be necessary to adjust the air gap and contact point gap.

1. Remove regulator from vehicle and take off cover.
2. Insert a .048" wire gauge between regulator armature and core, next to stop pin on spring hanger side, Fig. C12.
3. Press down on armature (not contact spring) until it contacts wire gauge. Upper contacts should just open.

NOTE: *A battery and test light connected in series to the "IGN" and "FLD" terminals may be used to determine accurately the contact opening. When the contacts open, the test light will do dim.*

4. Insert a .052" wire gauge between armature and core, next to stop pin on spring hanger side.
5. Press down on armature until it contacts wire gauge. The contacts should remain closed and test light should remain bright.
6. If adjustment is required, adjust air gap by loosening the screw and moving the stationary contact bracket. Make sure air gap is measured with attaching screw fully tightened. Re-measure the gap as directed above.
7. Remove wire gauge. Measure lower contact gap with feeler gauge. which should be .012 to .016". Adjust lower contact gap by bending lower stationary contact bracket.
8. Install regulator cover and then the regulator. Finally, make electrical adjustments as outlined above.

INSULATED BRUSH & ISOLATED FIELD ALTERNATORS

Charging Circuit Resistance Test

1. Disconnect battery ground cable. Disconnect "Batt" lead at the alternator.
2. Complete test connections as per Figs. C13, C14, C14A and C14B.

NOTE: Use a 0-100 amp ammeter.

3. Connect battery ground cable, start engine and operate at idle.
4. Adjust engine speed and carbon pile to obtain 20 amps in the circuit and check voltmeter reading. Reading should not exceed .7 volts. If a voltage drop is indicated, inspect, clean and tighten all connections in the circuit. A voltage drop test at each connection can be performed to isolate the trouble.

Current Qutput Test

1. Disconnect battery ground cable, complete test connections as per Figs. C15, C16, C16A and C16B and start engine and operate at idle. *Immediately after starting, reduce engine speed to idle.*

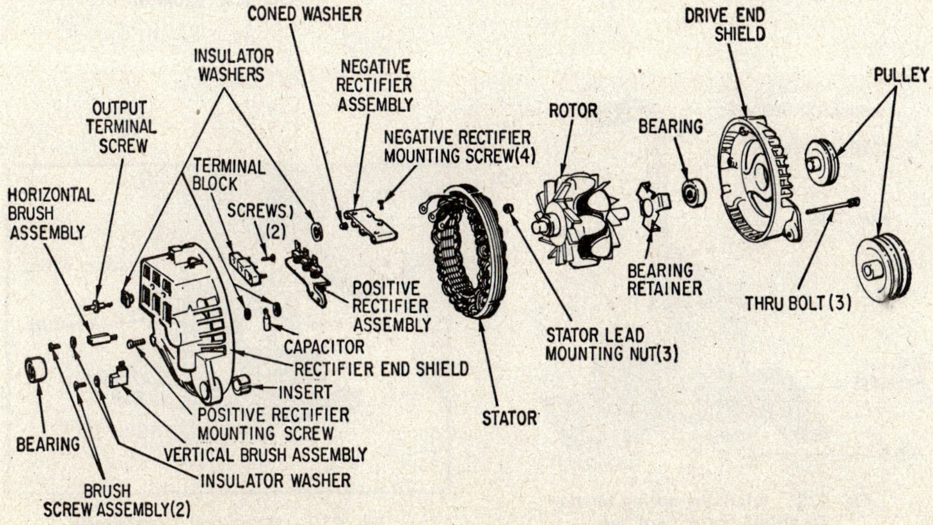

Fig. C4A Isolated field alternator disassembled. 1972-75

2. Adjust the carbon pile and engine speed in increments until a speed of 1250 RPM and 15 volts are obtained on all units except 100 amp alternators. On 100 amp alternators, obtain engine speed of 900 RPM and 13 volts.

CAUTION: While increasing speed, do not allow voltage to exceed 16 volts.

3. Check ammeter reading. Output current should be within specifications.

Voltage Regulator Test

NOTE: Battery must be fully charged for test to be accurate.

Insulated Brush Alternator

1. Make test connections as shown in Fig. C17.
2. Start and operate engine at 1250 rpm with all lights and accessories turned off.
3. As the engine starts, the instrument panel ammeter will deflect to the right. If it deflects less than ¼ scale, turn on high beams and heater blower. If deflection is greater than ¼ scale, turn on heater blower on high.

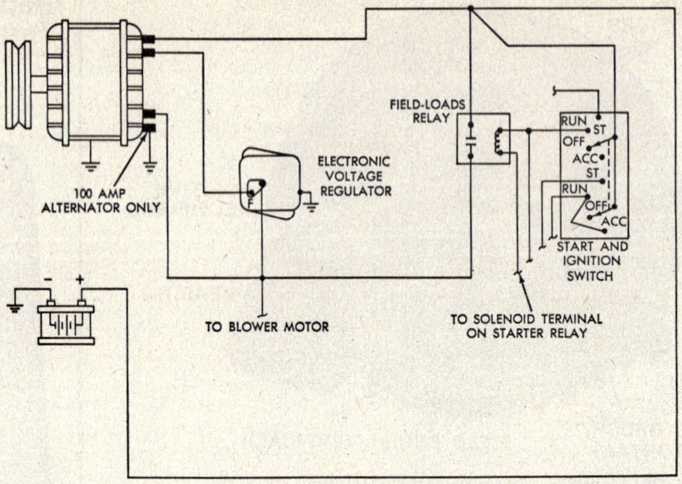

Fig. C6 Charging system with field-loads relay

4. Voltage should be as specified in Fig. C17A.
5. If voltage is not correct, and the alternator is satisfactory, turn off ignition and disconnect regulator connector. Check for battery voltage at black and green leads. Turn on ignition without starting engine and check for battery voltage at blue lead. If voltage is not present at these connections, wiring is at fault.
6. Check regulator for good ground.

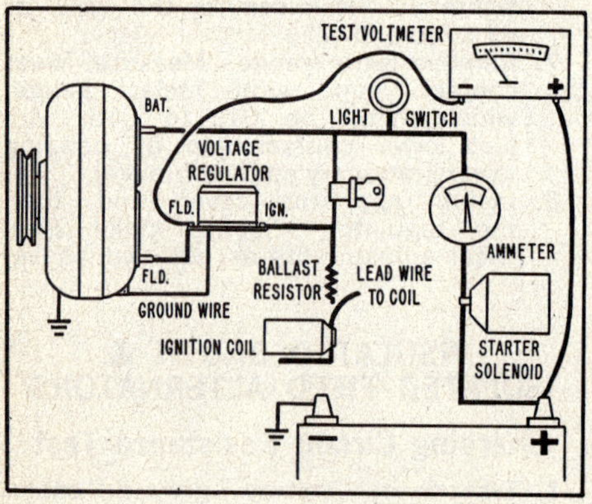

Fig. C7 Field resistance test, electro-mechanical regulator

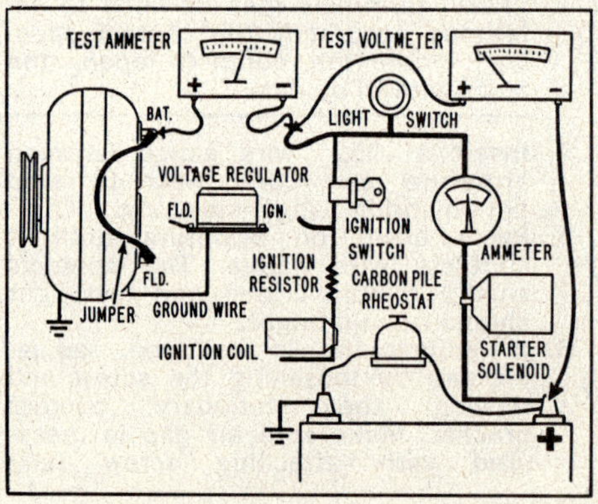

Fig. C8 Charging circuit resistance test, electro-mechanical regulator

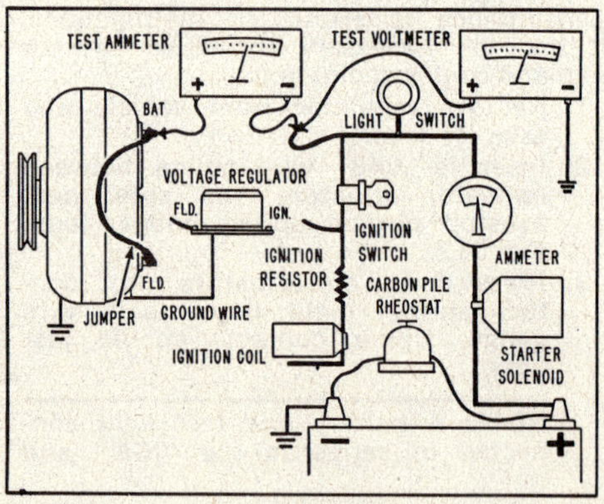

Fig. C9 Current output test, electro-mechanical regulator

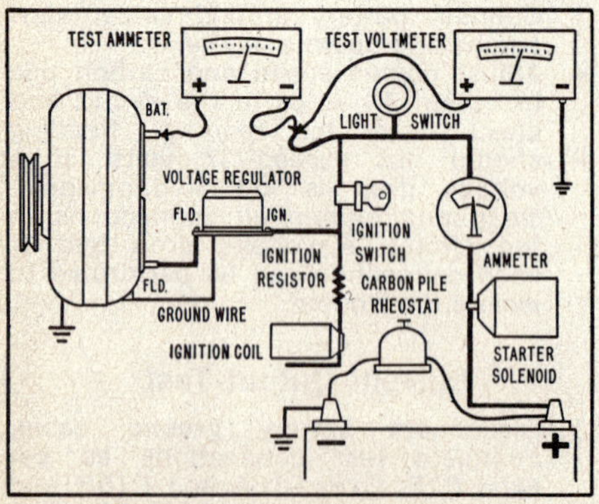

Fig. C10 Voltage regulator test, electro-mechanical regulator

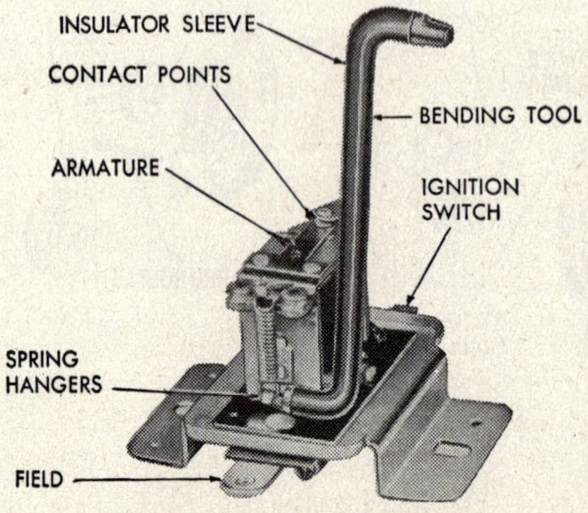

Fig. C11 Adjusting spring tension to obtain correct voltage, electro-mechanical units

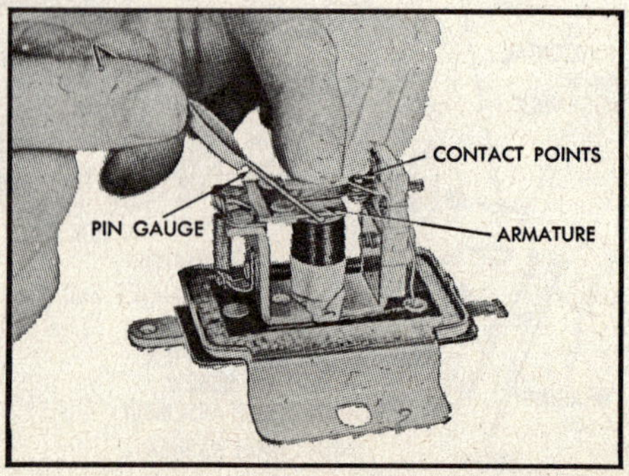

Fig. C12 Measuring armature air gap, electro-mechanical unit

7. If regulator voltage is .5 volt away from specification, it must be replaced.

NOTE: The field circuit is grounded through the regulator. A good ground is established through the use of cup shaped washers on the regulator mounting screws which cut through body paint. These washers must be reinstalled whenever the regulator is removed.

Isolated Field Alternator, 1973-75

1. Connect test equipment, Figs. C18 and C18A.
2. Start and run engine at 1250 RPM with all lights and accessories turned "Off". Voltage should be as specified in Fig. C17A.
3. It is normal for the vehicle ammeter to indicate an immediate charge, then gradually return to the normal position.
4. If voltage is below limits or is fluctuating, proceed with the following:
   A. Check voltage regulator for proper ground. The ground is obtained through the regulator case to mounting screws, then to the vehicle sheet metal.
   B. With ignition switch "Off", disconnect voltage regulator connector. Turn ignition "On" and check for battery voltage at the wiring harness terminal. Both green and the blue leads should have battery voltage.
   C. If voltage regulator was grounded properly and battery voltage was present at the green and blue leads, replace voltage regulator except on systems incorporating a field-loads relay. On systems with a field-loads relay, test the relay as outlined under "Field-Loads Relay Test". If relay tests satisfactory, replace voltage regulator. On all systems, repeat test.

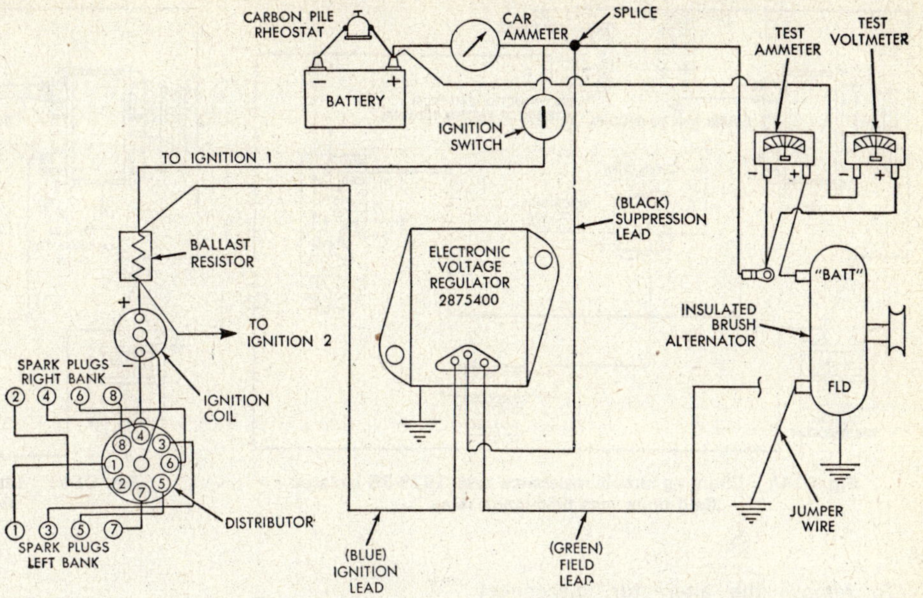

Fig. C13 Charging circuit resistance test, insulated brush units

5. If voltage is above limits, refer to Steps 4B and 4C.

Isolated Field Alternator, 1970-72

1. Complete test connections as per Fig. C18B.
2. Start and operate engine at 1250 rpm with all lights and accessories turned off. Voltage should be as specified in Fig. C17A.
3. It is normal for the car ammeter to show an immediate charge then gradually return to normal position.
4. If voltage is below limits, check for good voltage regulator ground, and voltage drop between regulator cover and body on low voltage scale of voltmeter.
5. Turn off ignition switch and disconnect regulator connector.
6. Turn on ingition switch but do not start engine. Check for battery voltage at the wiring harness terminal connected to the blue and green leads.

NOTE: Disconnect wiring harness from regulator when checking the leads.

7. Turn off ignition switch. If voltage is not present at either lead, the problem is in the vehicle wiring or alternator field circuit. *Use care to avoid bending the terminals with voltmeter probe.*
8. If Steps 4-7 tested satisfactorily, change voltage regulator and repeat Step 2. If voltage is slightly above limits, or is fluctuating, proceed as follows:
9. Check voltage regulator ground, check ground between vehicle body and engine.
10. Check ignition switch circuit between battery terminal of ignition switch and voltage regulator. If voltage is more than ½ volt above limits, replace the regulator.

Field-Loads Relay Test

1. Disconnect voltage regulator connector and turn ignition switch "On".
2. Using a voltmeter, check for battery voltage at terminals of disconnected connector.
3. If battery voltage is present, the field-loads relay is satisfactory.
4. If battery voltage is not measured at the terminals, check circuit wiring and connections. If wiring and connections are satisfactory, replace field-loads relay.

BENCH TESTS

If the alternator performance does not meet current output specification limits, it will have to be disassembled for further tests and servicing.

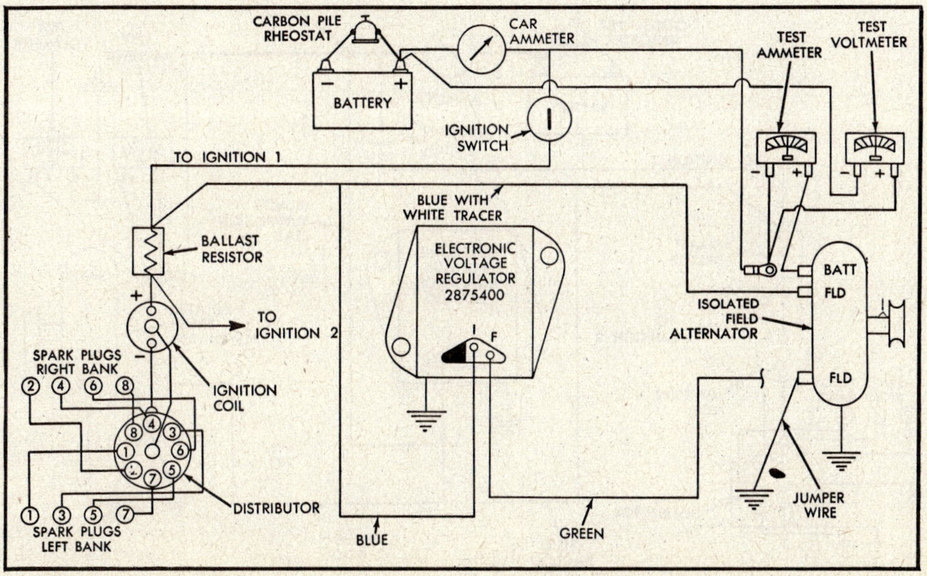

Fig. C14 Charging circuit resistance test 1970-72 isolated field units

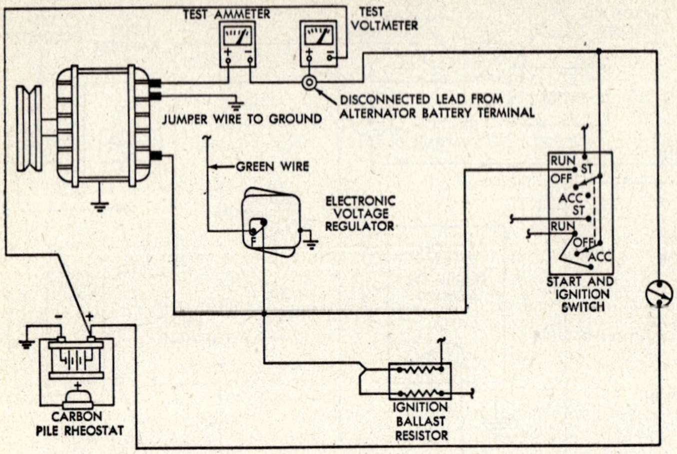

Fig. C14A Charging circuit resistance test. 1973-75 isolated field units less field-loads relay

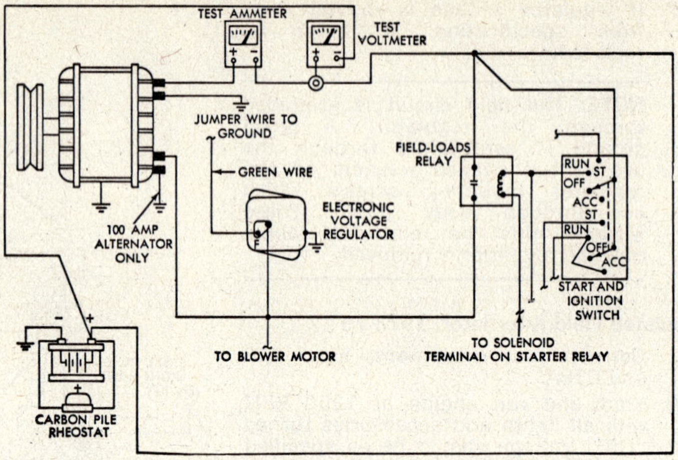

Fig. C14B Charging circuit resistance test. 1975 isolated field units with field-loads relay

To remove the alternator, disconnect the battery ground cable and the leads at the alternator. Then unfasten and remove the alternator from the vehicle.

Field Coil Draw

Electro-Mechanical Unit

1. Connect a test ammeter positive lead to the battery positive terminal of a fully charged battery.
2. Connect ammeter negative lead to the field terminal of the alternator.
3. Connect a jumper wire to negative terminal of battery, and ground it to the alternator end shield.
4. Slowly rotate alternator rotor by hand. Observe ammeter reading. The field coil draw should be 2.3 to 2.7 amperes at 12 volts.

Insulated Brush Alternator

1. Connect jumper wire between alternator "Batt" terminal and the positive terminal of a fully charged battery. Connect test ammeter positive lead to the alternator field terminal and negative lead to negative battery terminal.
2. Slowly rotate alternator rotor by hand and observe ammeter reading. Field coil draw should be 2.3 to 2.7 amperes at 12 volts.

A low rotor coil draw is an indication of a high resistance in the field coil circuit (brushes, slip rings or rotor coil). A higher rotor coil draw indicates a possible shorted rotor coil or a grounded rotor.

Isolated Field Alternator

1. Connect jumper wire between one alternator field terminal and the positive terminal of a fully charged battery.
2. Connect test ammeter positive lead to the other alternator field terminal and the ammeter negative lead to the negative battery terminal.
3. Slowly rotate alternator rotor by hand. Field current at 12 volts should be 2.3-2.7 amps on 1969-71 units, 2.5-3.1 amps on 1972-73 units, 2.5-3.7 amps on 1974-75 units except 100 amp alternators or 4.75-6.0 amps on 1975 100 amp alternators.
4. A low rotor coil draw is an indication of high resistance in the field coil circuit, (brushes, slip rings or rotor coil). A high rotor coil draw indicates shorted rotor coil or grounded rotor. No reading indicates an open rotor or defective brushes.

Testing Alternator Internal Field Circuit

Except Isolated Field Alternator

1. Remove ground brush (positive brush on insulated brush alternators). Touch one test prod from a 110 volt test lamp to the alternator insulated brush (field brush on insulated brush units) and the remaining test prod to the end shield. If rotor or insulated brush (field brush on insulated brush units) is not grounded, lamp will not light.
2. If lamp lights, remove insulated brush (field brush on insulated brush units) remove through bolts and separate end shields.
3. Touch one test prod to a slip ring and remaining prod to end shield. If lamp lights, rotor is grounded and must be replaced. If lamp does not brush (field brush on insulated brush units), the insulated brush is grounded.

Isolated Field Alternator, 1970-72

1. Touch one probe of a 110 volt test lamp to one of the alternator field brush terminals and the other to the end shield. If lamp lights, rotor assembly or a field brush is grounded.

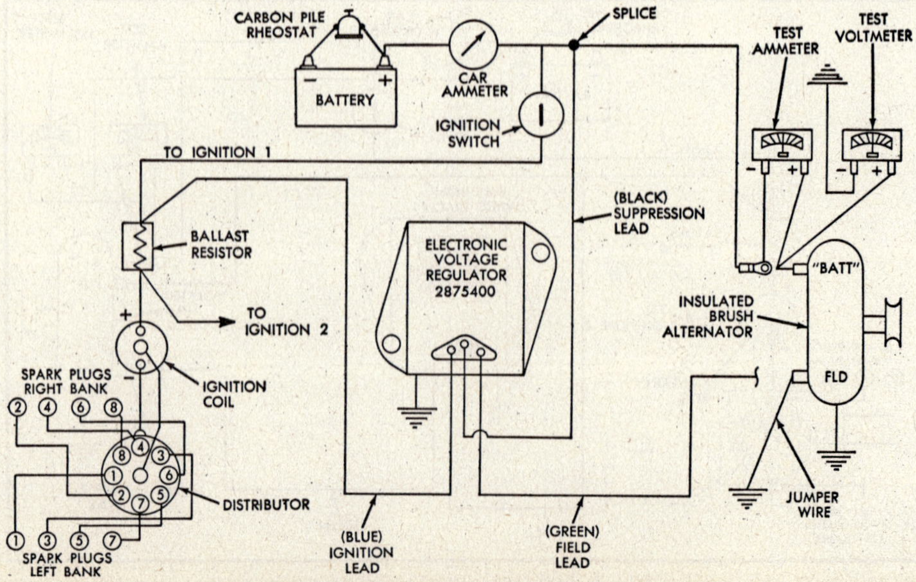

Fig. C15 Current output test, insulated brush unit

2. If lamp lights, remove field brush assemblies, remove through bolts and separate end shields.
3. Touch one test probe to a slip ring and the remaining probe to the end frame. If the lamp lights, rotor is grounded and must be replaced. If lamp does not light, cause is a grounded brush.

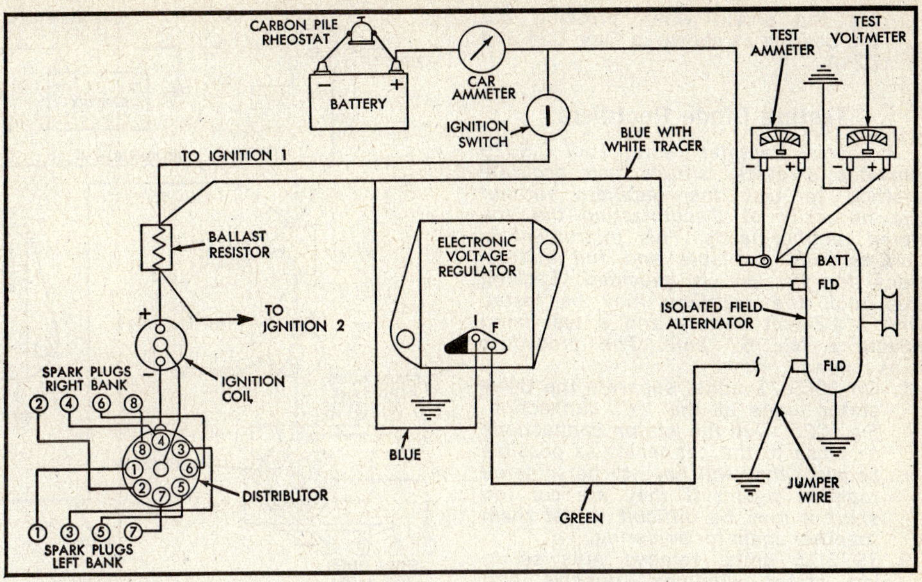

Fig. C16 Current output test, 1970-72 isolated field units

ALTERNATOR REPAIRS EXCEPT 100 AMP UNITS

Disassembly

To prevent possible damage to the brush assemblies, they should be removed before disassembling the alternator. The insulated (field) brush is mounted in a plastic holder which positions the brush against one of the slip rings. In the isolated field type alternator, both brushes are insulated and mounted in plastic holders. Disassembled views of all four types are shown in Figs. C1A, C2A, C3A and C4A.

1. On 1969-71 units, remove retaining screw lockwasher, insulated washer and field terminal. Carefully lift plastic holder containing the spring and brush from the end housing.
2. On insulated brush units, remove the positive brush strap screw at the heat sink, the brush assembly screw, insulating nylon washer and lift out brush assembly. On isolated field units, remove both brush screws, insulating nylon washers and remove brush assemblies. On electro-mechanical units, remove the ground brush retaining screw and lift the clip, spring and brush from end shield.

NOTE: The stator is laminated; do not burr it or the end shield.

3. Remove through bolts and pry between stator and drive end shield with a screwdriver. Carefully separate drive end shield, pulley and rotor from stator and diode rectifier shield, Fig. C19.
4. The pulley is an interference fit on the rotor shaft; therefore, a suitable puller must be used to remove it, Fig. C20.
5. Pry drive end bearing spring retainer from end shield with a screwdriver, Fig. C21.
6. Support end shield and tap rotor shaft with a plastic hammer to separate rotor from end shield.
7. The drive end ball bearing is an interference fit with the rotor shaft; therefore, a suitable puller must be used to remove it, Fig. C22.
8. On 1969-71 units, remove D.C. output terminal nuts and washers, terminal screw and capacitor, if used. Then remove heat sink and heat sink insulator, Fig. 23.
9. On 1972-75 units to remove rectifiers and heat sinks, loosen screws securing negative rectifier and heat sink assembly to end shield, remove the two outer screws and lift assembly from end shield. Remove nuts securing positive rectifier and heat sink assembly to insulated terminals in end shield. Then, remove capacitor ground screw and lift insulated washer, capacitor and positive rectifier and heat sink assembly from end shield.
10. The needle roller bearing in the rectifier end shield is a press fit. If it is necessary to remove the rectifier end frame needle bearing, protect the end shield by support-

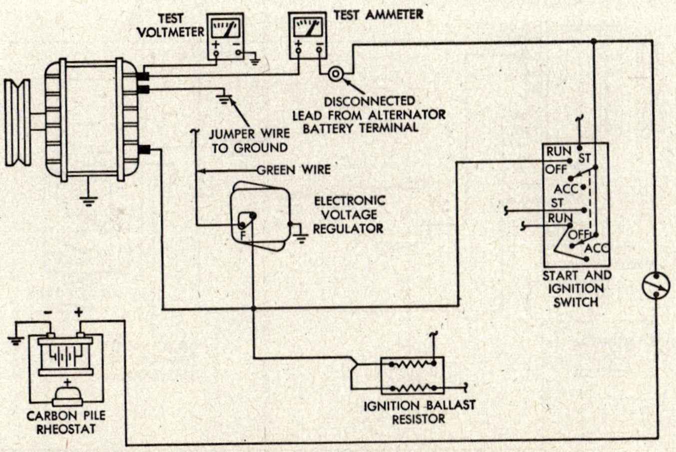

Fig. C16A Current output test. 1973-75 isolated field units less field-loads relay

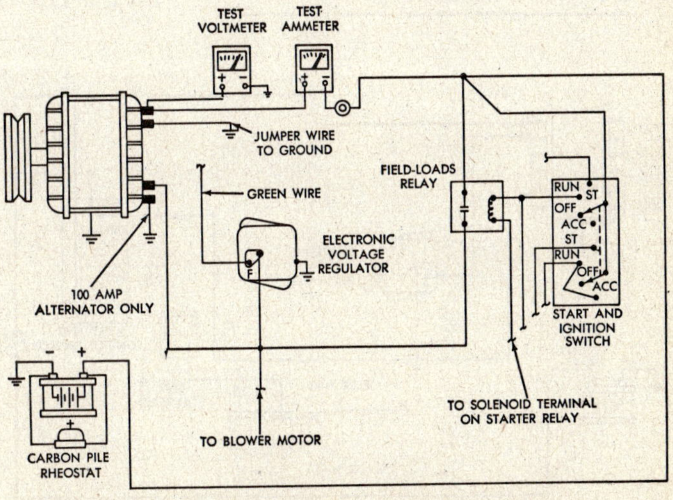

Fig. C16B Current output test. 1975 isolated field units with field-loads relay

ing the shield when pressing out the bearing as shown in Figs. C24 and C24A.

Testing Diode Rectifiers

A special Rectifier Tester Tool C-3829 provides a quick, simple and accurate method to test the rectifiers without the necessity of disconnecting the soldered rectifier leads. This instrument is commercially available and full instructions for its use is provided. Lacking this tool, the rectifiers may be tested with a 12 volt battery and a test lamp having a No. 67 bulb. The procedure is as follows:

1. On 1969-71 units, separate the three stator leads at the "Y" connection, Fig. C25. *Cut the stator connections as close to the connector as possible because they will have to be soldered together again. If they are cut too short it may be difficult to get them together again for soldering.*
2. 1972-75 units, remove nuts securing stator windings, positive and negative rectifier straps to terminal block. Remove stator winding terminals and pry stator from end shield.
3. On all units, connect one side of test lamp to positive battery post and the other side of the test lamp to a test probe. Connect another test probe to the negative battery post, Fig. C26.
4. On 1969-71 units, contact outer case of rectifier with one probe and other probe to rectifier center wire. On 1972-75 units, contact heat sink with one probe and strap on top of rectifier with the other probe.
5. On all units, reverse position of probes. If test lamp lights in one direction only, the rectifier is satisfactory. If test lamp lights in both directions, the rectifier is shorted. If test lamp lights in neither direction, the rectifier is open.

NOTE: *Possible cause of an open or a blown rectifier is a faulty capacitor or a battery that has been installed on reverse polarity. If the battery is installed properly and the rectifiers are open, test the capacitor capacity, which should be .50 microfarad plus or minus 20%.*

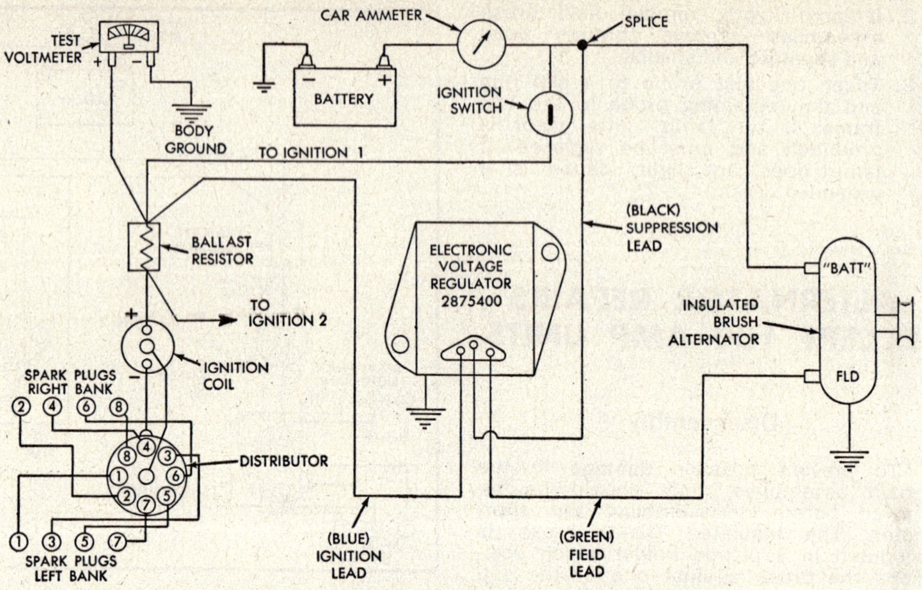

Fig. C17 Voltage regulator test, insulated brush unit

Ambient Temperature Near Regulator	−20°F	80°F	140°F	Above 140°F
1969–74	14.3–15.3	13.8–14.4	13.3–14.0	Less than 13.8
1975	14.9–15.9	13.9–14.6	13.3–13.9	Less than 13.6

Fig. C17A Voltage regulator test specifications

Testing Stator

1. On 1969-71 units, unsolder rectifiers from stator leads. On 1972-75 units, separate stator from end shields.
2. On 1969-71 units, insulate stator from rectifier shield with wood slats and using a 110 volt test lamp, Fig. C27, test stator for grounds by contacting one test probe to stator pole frame and the other to each stator lead. If test lamp lights, stator is grounded.
3. On 1972-75 units, using a 12 volt test lamp, Fig. C27A, test stator for grounds. Contact one test probe to

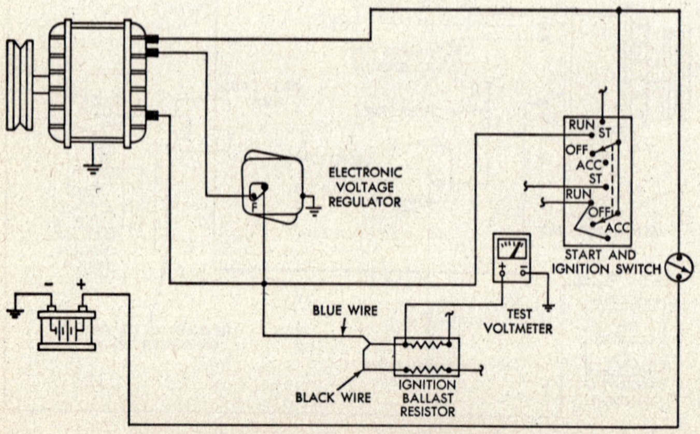

Fig. C18 Voltage regulator test. 1973-75 isolated field units less field-loads relay

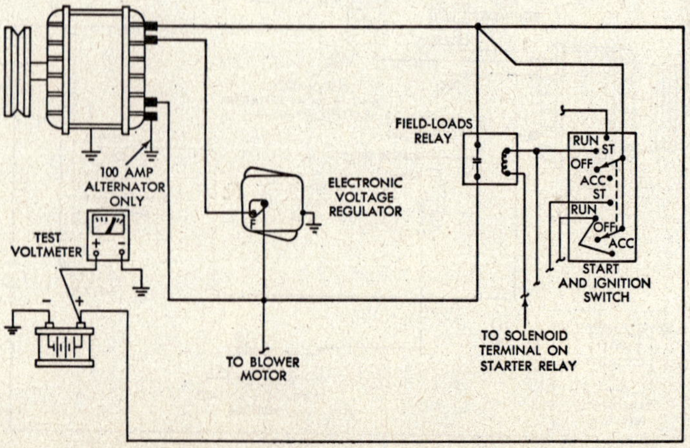

Fig. C18A Voltage regulator test. 1975 isolated field units with field-loads relay

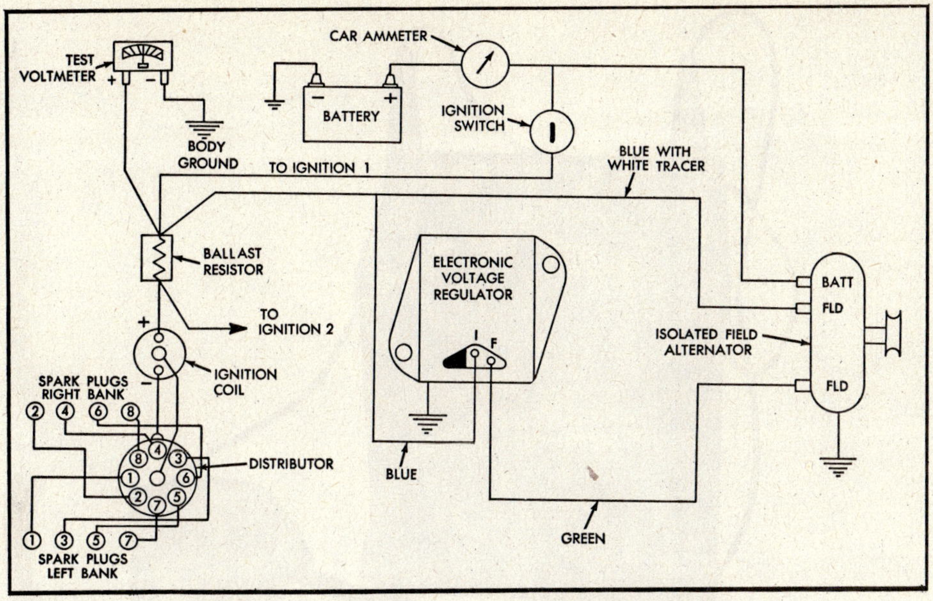

Fig. C18B Voltage regulator test, 1970-72 isolated field units

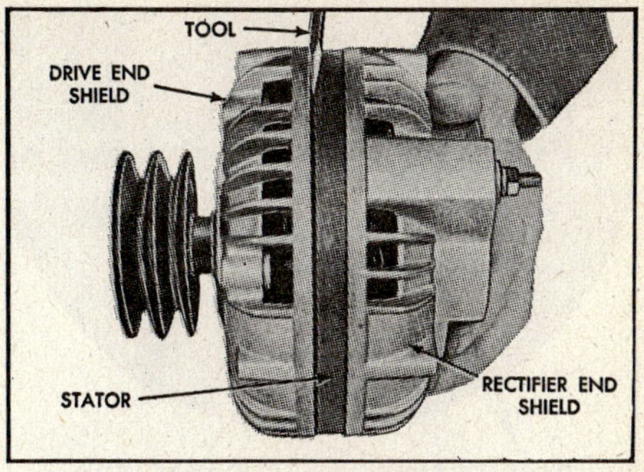

Fig. C19 Separating drive end shield from stator

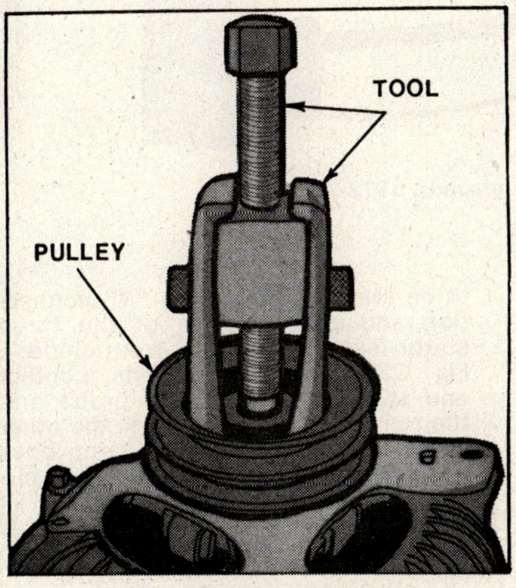

Fig. C20 Removing pulley

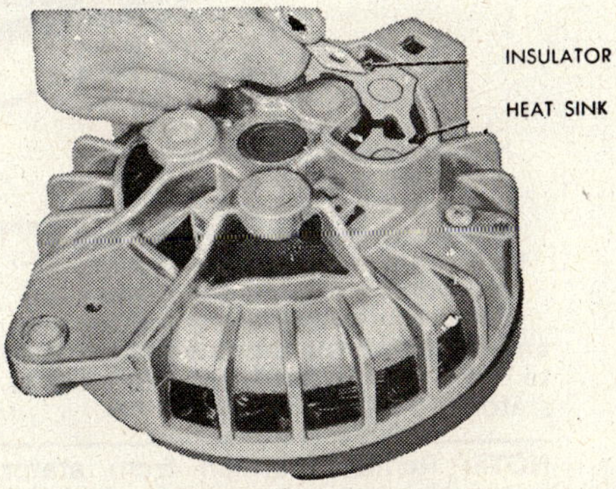

Fig. C23 Removing or installing heat sink insulator

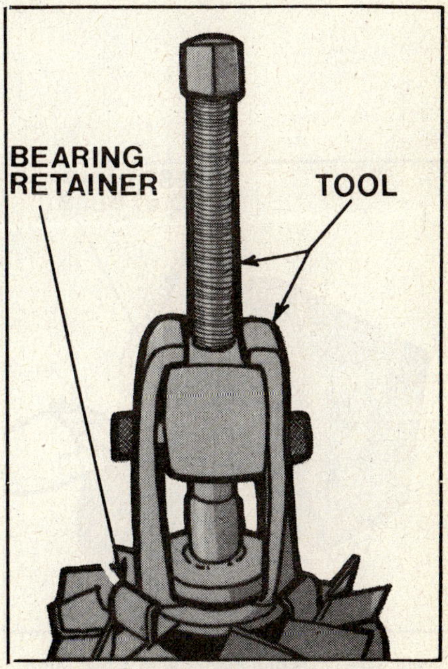

Fig. C22 Removing bearing from rotor shaft

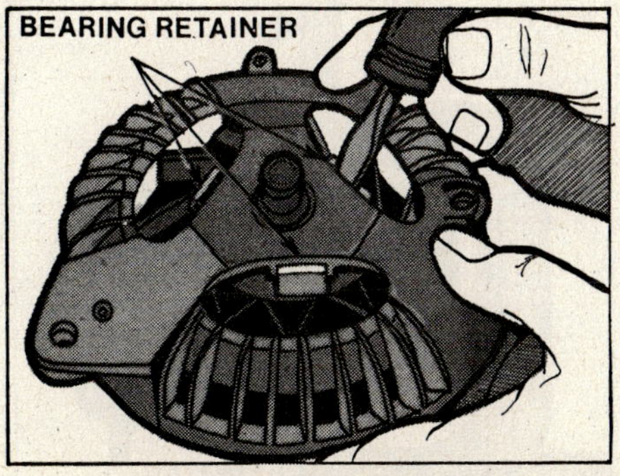

Fig. C21 Disengaging bearing retainer from end shield

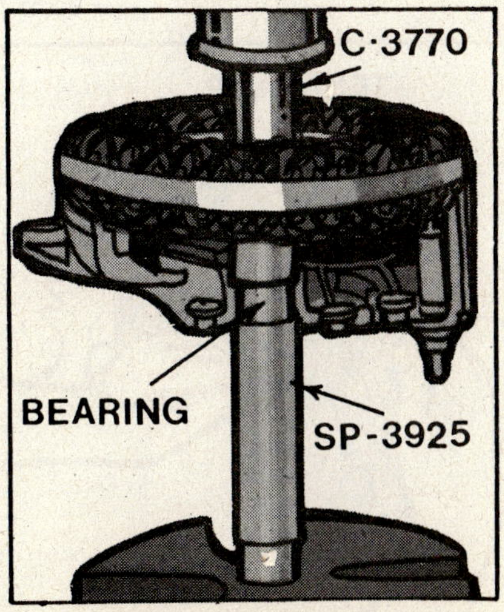

Fig. C24 Removing diode end shield bearing. 1969-71

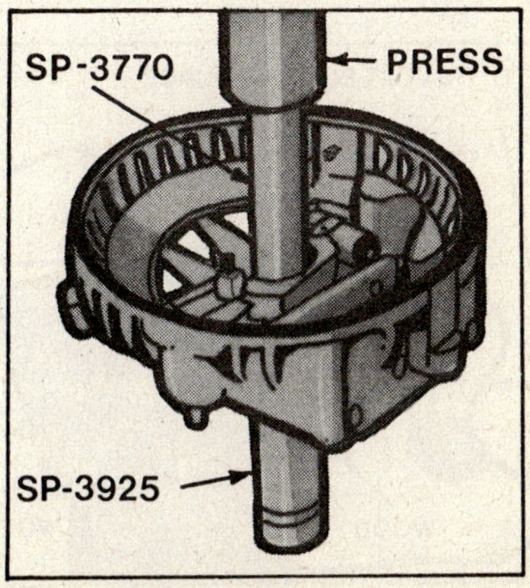

Fig. C24A Removing rectifier end shield bearing. 1972-75

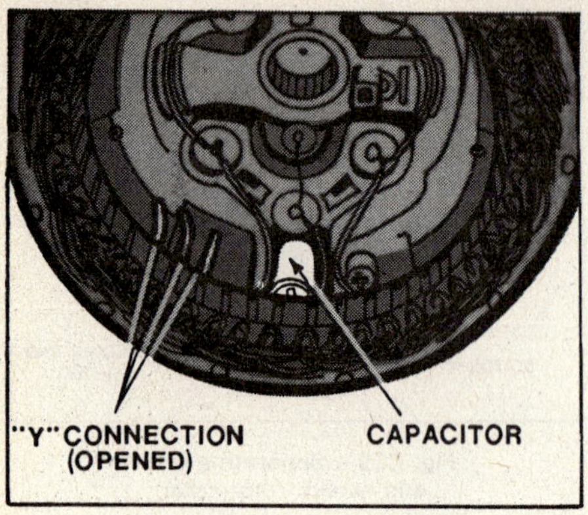

Fig. C25 Separating the three stator leads. 1969-71

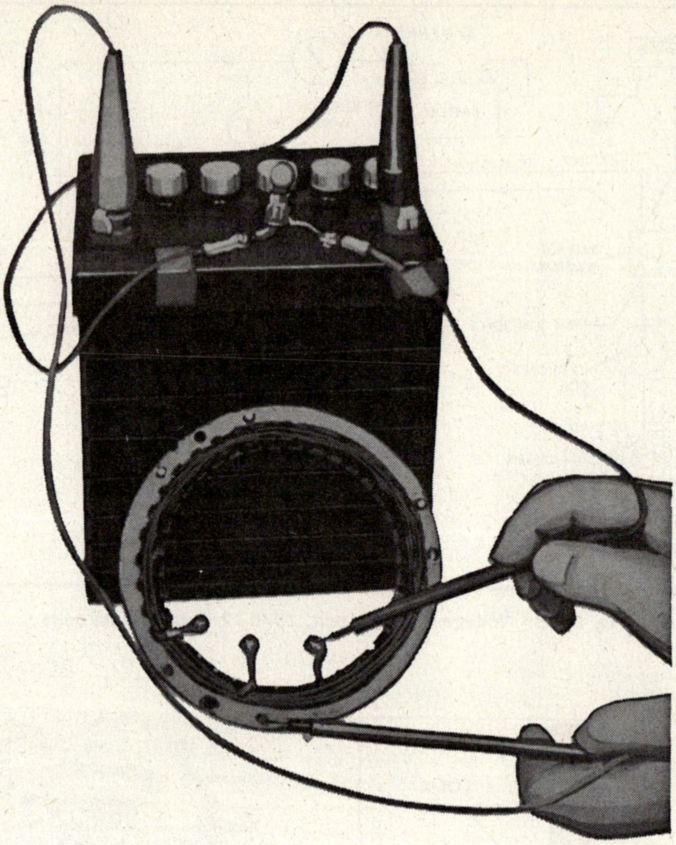

Fig. C27A Testing stator for grounds, 1972-75 units

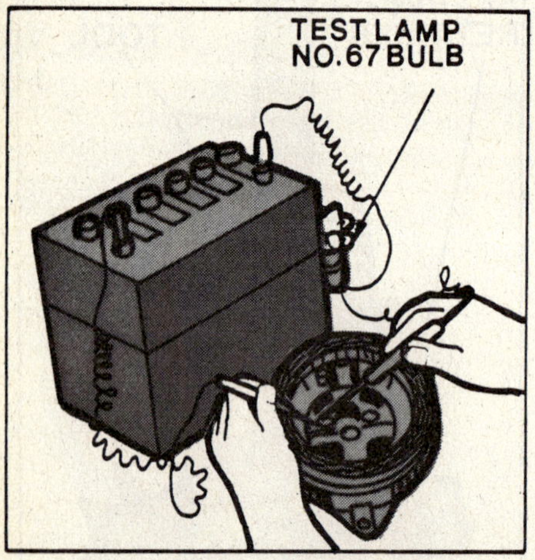

Fig. C26 Testing diodes with a test lamp

any pin on stator frame and the other to each stator lead. If lamp lights, stator is grounded.

NOTE: Remove varnish from stator frame pin to ensure proper electrical connection.

4. On all units, use a 12 volt test lamp to test stator for continuity. On 1969-71 units, connect one probe to all three stator leads at the "Y" connection and contact each of the three stator leads (disconnect from diodes), Fig. C28. On 1972-75 units, contact one stator lead with one probe and the remaining two leads with the other probe. On all units, if test lamp does not light, the stator has an open circuit.
5. Install new stator if one tested is defective.

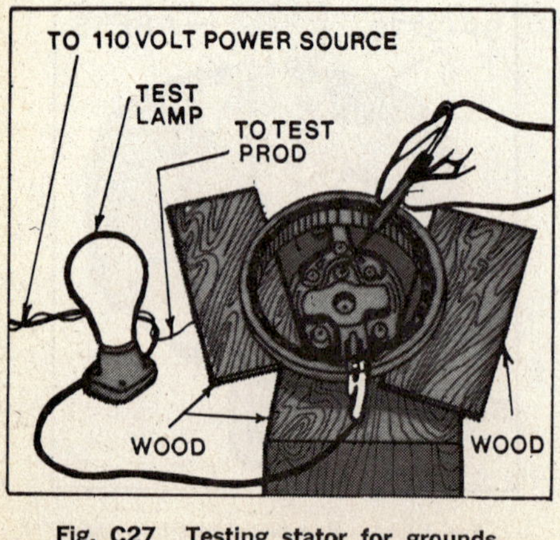

Fig. C27 Testing stator for grounds, 1969-71 units

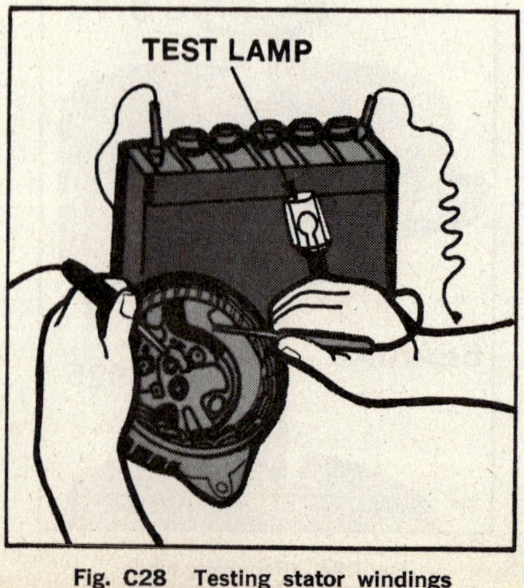

Fig. C28 Testing stator windings for continuity (typical)

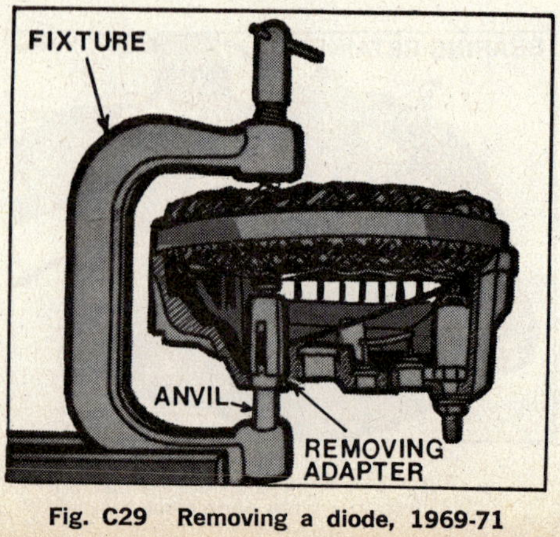

Fig. C29 Removing a diode, 1969-71

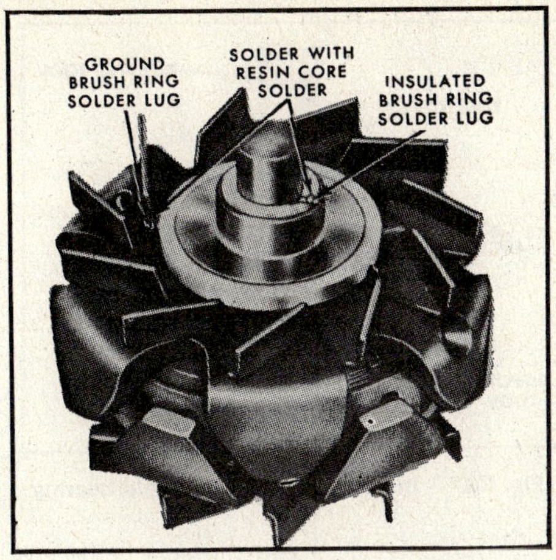

Fig. C30 Soldering points with slip ring installed. 1969-74

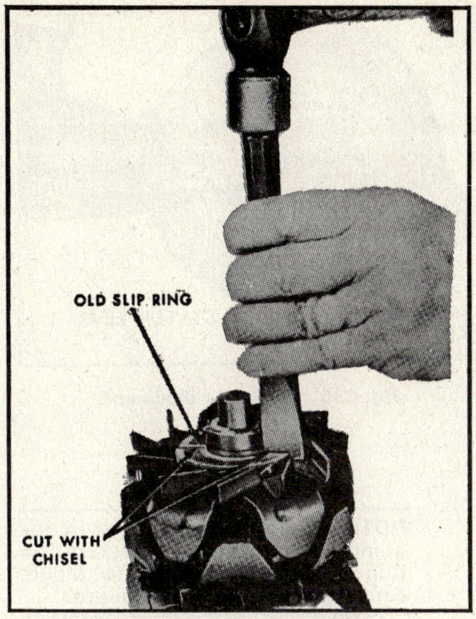

Fig. C31 Cutting old slip rings for removal. 1969-74

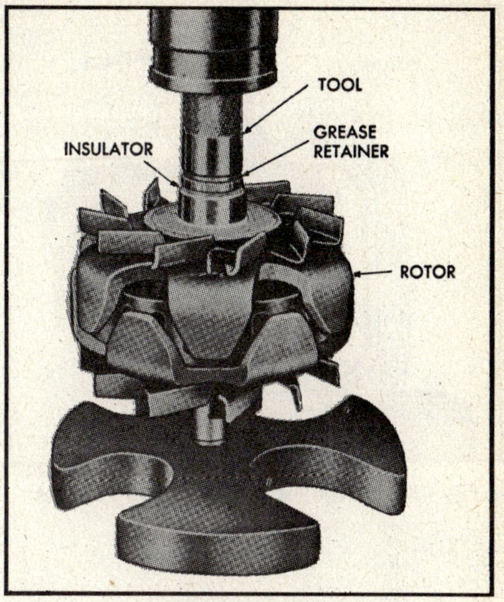

Fig. C34 Installing bearing grease retainer

Testing Rotor 1972-75

The rotor may be tested electrically for grounded, open or shorted field coils as follows:

Grounded Field Coil Test: Connect an ohmmeter between each slip ring and the rotor shaft. The ohmmeter should indicate infinite resistance. If reading is zero or higher, rotor is grounded.

Open or Shorted Field Coil Test: Connect an ohmmeter between the slip rings. If reading is below 3 ohms, field coil is shorted. A resistance of 3 to 4 ohms at room temperature indicates rotor is satisfactory. A reading of 4 to 6 ohms indicates the alternator was operated at a high underhood temperature, however, rotor is still satisfactory. Resistances above 6.5 ohms indicates a high resistance in the field coils and further testing or rotor replacement is required.

Removing Rectifiers 1969-71

1. Three diodes are pressed into the heat sink and three in the end shield. When removing the diodes, it is necessary to support the end shield and/or heat sink to prevent damage to these castings.
2. Install the tools shown in Fig. C29, making sure bore of tool completely surrounds diode.
3. Carefully apply pressure to remove diode from end shield.

Replacing Slip Rings 1969-75

NOTE: Slip rings on 1975 alternators are not serviced as a separate item. Rotor replacement is required when the slip rings are defective.

1. Cut through rotor grease retainer with a chisel and remove retainer and insulator.
2. Unsolder field coil leads at solder lugs, Fig. C30.
3. Cut through copper of both slip rings at opposite points with a chisel, Fig. C31.
4. Break insulator and remove old ring.
5. Clean away dirt and particles of old slip ring from rotor.
6. Scrape ends of field coil lead wires clean for good electrical contact.
7. Position field coil wires aside and place new slip ring on rotor shaft so slip ring lugs are properly positioned for field coil wire connections, Fig. C32.

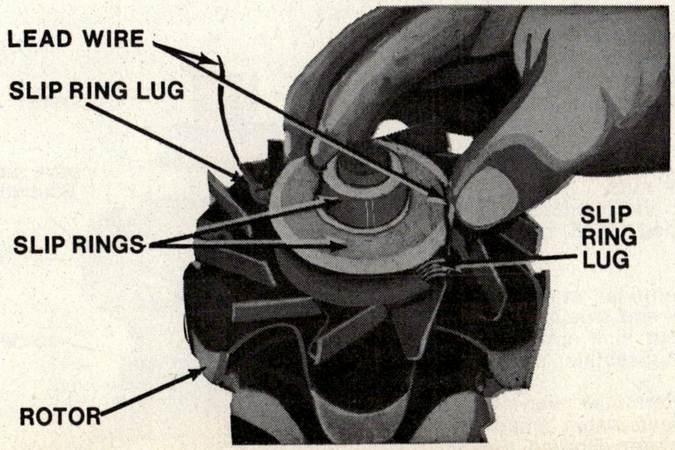

Fig. C32 Aligning slip ring with field lead wires. 1969-74

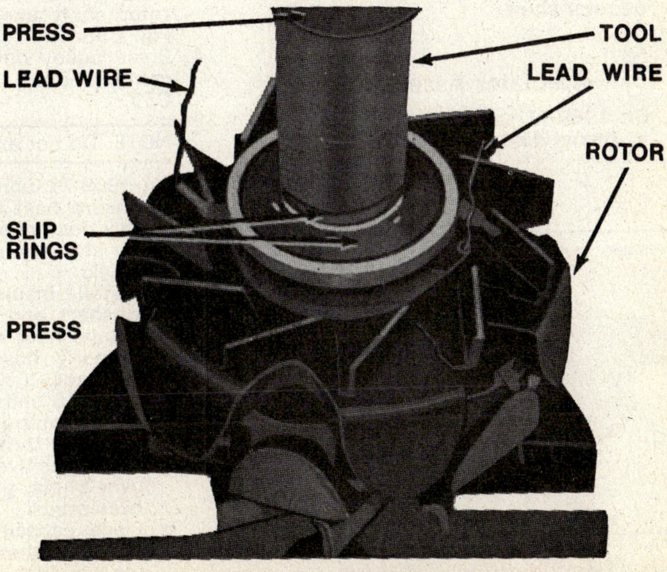

Fig. C33 Installing slip ring. 1969-74

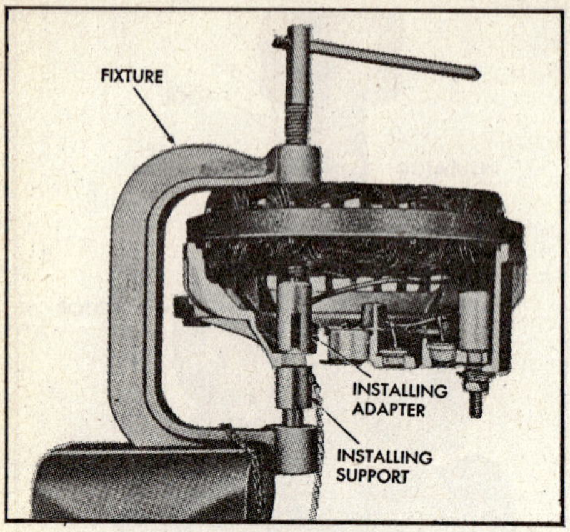

Fig. C35 Installing a diode, 1969-71 units

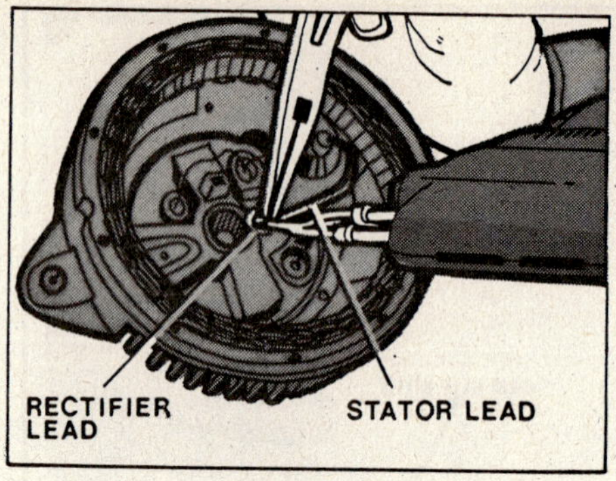

Fig. C36 Soldering diode and stator leads

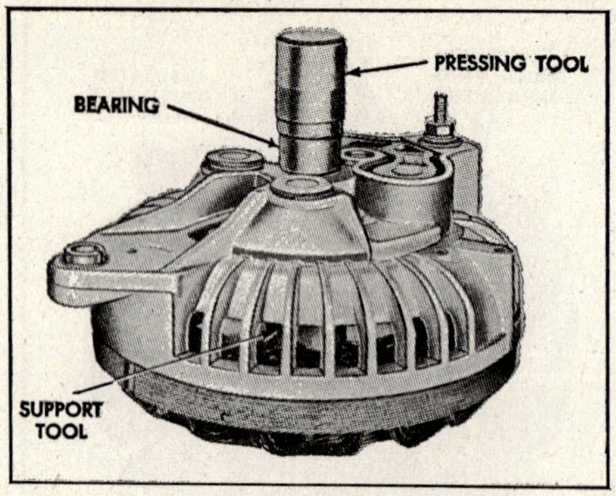

Fig. C37 Installing diode end shield bearing

8. Place installing tool over rotor shaft and with a suitable press, press slip ring onto shaft, Fig. C33.

NOTE: With slip ring bottomed against fan, the field lead wire should clear the access hole, fan and pole piece.

9. Tin field coil lead wires, then coil each lead wire around slip ring lug with first wrap against lug shoulder and winding outward. Solder leads with resin core solder.
10. Test slip rings for ground with a 110 volt test lamp by touching one test lead prod to rotor pole shoe and remaining prod to slip rings. The lamp should not light. If lamp lights, slip rings are shorted to ground, possibly due to a grounded insulated field lead when installing slip ring.
11. If rotor is not grounded, lightly clean slip ring surfaces with No. 00 sandpaper and assemble to alternator.
12. Position grease retainer gasket and retainer on rotor shaft and press retainer on shaft, Fig. C34. Retainer is properly positioned when inner bore of installer tool bottoms on rotor shaft.

Alternator Assemble

1. On 1969-71 units:
   a. Install diodes, Fig. C35.

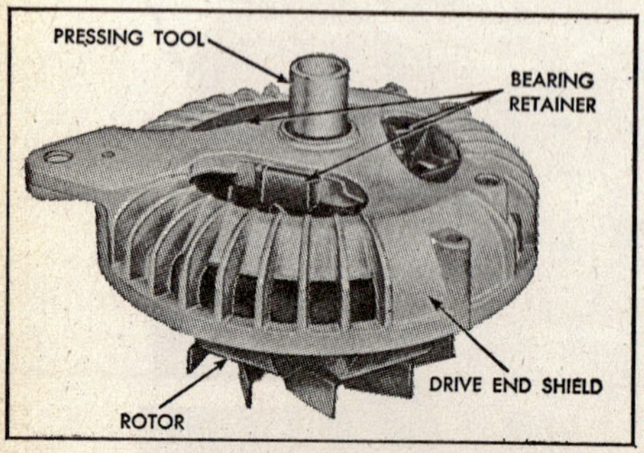

Fig. C38 Installing drive end shield and bearing

NOTE: Do not hammer diode in any manner since this will fracture the thin silicon wafer in the diode, causing complete diode failure.

   b. Clean leads and mate stator lead with diode wire loop, then bend loop snugly around stator lead to provide a good electrical and mechanical connection. Solder wires with resin core solder, holding diode lead wire with pliers just below joint, Fig. C31. Pliers will act as a heat sink and protect diode.

NOTE: After soldering, quickly cool soldered connection by touching a damp cloth against it. Also, this will aid in forming a solid joint.

   c. Push stator leads into slots in end shield and cement them to protect leads against possible interference with rotor fan (Cement is Mopar part No. 2299314). Test each diode to ensure proper installation.
2. On all units, install diode end shield bearing, Fig. C37.
3. Install drive end bearing in end shield with bearing retainer plate to hold bearing in position. Place assembly on rotor shaft and press into position, Fig. C38.
4. Press pulley onto rotor shaft until it contacts inner race of bearing, Fig. C39.

NOTE: Do not exceed 6800 lbs.

5. On 1969-71 units:
   a. Ensure heat sink insulator is properly positioned, then install capacitor stud through heat sink and end shield.
   b. Install insulating washers, lockwashers and lock nuts.
   c. Ensure heat sink and insulator are properly positioned, then tighten lock nut.
6. On 1972-75 units:
   a. Install output terminal stud and insulator through end shield. Then place positive heat sink assembly over studs, guiding rectifier straps over studs.
   b. Place capacitor terminal over capacitor end stud and install capacitor shoulder insulator. Ground the capacitor bracket to end shield with a metal screw. Install and tighten positive heat sink lockwashers and nuts.
   c. Slide negative rectifier and heat sink assembly into place, position straps on terminal block studs, then install and tighten attaching screws.
7. Position stator on diode end shield.
8. Position rotor end shield on stator and diode end shield.
9. Align through bolt holes in stator, diode end shield and drive end shield.
10. Compress stator and both end shields by hand and install through bolts, washers and nuts.
11. On 1969 units, install insulated (field) brush in diode end shield. Place bronze terminal on plastic holder with tab of terminal in recess of holder. Then place nylon washer on bronze terminal and install lockwasher and attaching screws.
12. On 1970-75 units, install field brushes into vertical and horizontal holders. Place an insulating washer on each field brush terminal and install lockwashers and attaching screws.

NOTE: Ensure brushes are not grounded.

13. Rotate pulley slowly by hand to be sure rotor fans do not touch diodes, capacitor lead and stator connections.
14. Install alternator and adjust drive belt.

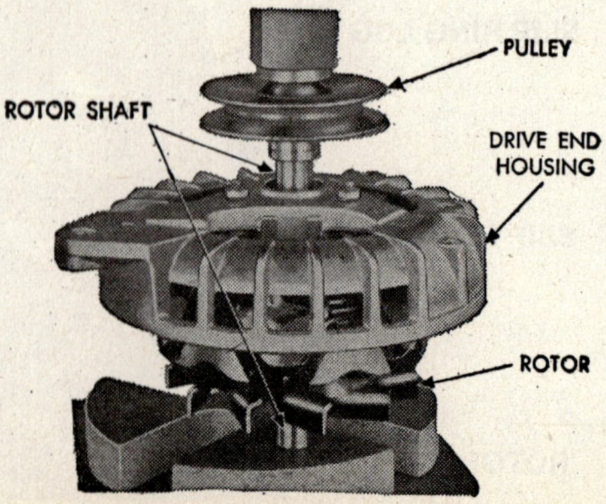

Fig. C39 Installing pulley

15. Connect leads to alternator.
16. Connect battery ground cable.
17. Start and operate engine and observe alternator operation.
18. If necessary, test current output and regulator voltage setting.

100 AMP ALTERNATOR

Disassembly & Testing

Separating End Shields

1. Remove brush holder screw and insulating washer, then lift brush holder from end shield.
2. Remove the through bolts, then using a screwdriver, pry between the stator and end shield in the slot provided to separate end shields, Fig. C40.

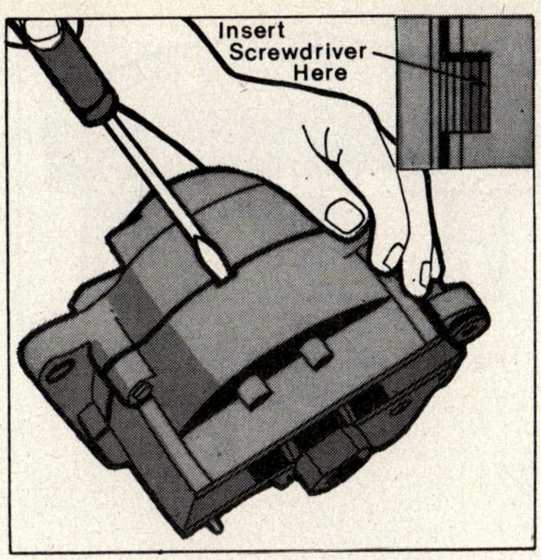

Fig. C40 Separating end shields. 100 amp units

Rectifier Testing

1. Remove stator winding leads to terminal block stud nuts, Fig. C41.
2. Lift stator winding leads and pry stator from end shield.
3. Using a 12 volt battery and a test lamp equipped with a #67 bulb, test rectifiers as follows:
   A. Connect one test probe to rectifier heat sink and the other test probe to the metal strap on top of rectifier, Fig. C42. Reverse the probes.
   B. If test lamp lights in one direction and does not light in the other, rectifier is satisfactory. If test lamp lights in both directions, rectifier is shorted. If test lamp does not light in either direction, rectifier is open.

Rectifier & Heat Sink Assembly Removal

1. Remove nut and insulator securing positive heat sink assembly to end shield stud.
2. Remove capacitor attaching screw.
3. Remove nut and insulator securing positive heat sink assembly stud to end shield, then remove positive heat sink assembly, Fig. C43, noting location of the insulators.
4. Remove screws securing negative heat sink assembly to end shield, then the negative heat sink assembly, Fig. C44.
5. Remove terminal block, then the capacitor and insulator.

Stator Testing

1. Contact one test lamp probe to outer diameter of stator frame and the other probe to each of the stator lead terminals, one at a time, Fig. C45.
2. If test lamp lights, the stator lead is grounded, requiring replacement.

NOTE: The stator windings are Delta Wound, therefore the windings cannot be tested for opens or shorts using a test lamp. If the stator is not grounded, and all other electrical circuits and alternator components test satisfactory, the stator may be open or shorted.

Pulley & Bearing Removal

1. Remove pulley with a suitable puller, Fig. C46.
2. Remove bearing retainer to drive end shield attaching screws.
3. Support end shield and using a mallet, tap rotor from end shield.
4. Remove bearing with a suitable puller, Fig. C47.
5. If necessary to remove needle roller bearing in rectifier end shield, use tool C-4330, Fig. C48.

Rotor Testing

Grounded Field Coil Test: Connect a test lamp between each slip ring and the rotor shaft, Fig. C49. If test lamp lights, the rotor is grounded, requiring replacement.

Open Field Coil Test: Connect a test lamp between the slip rings, Fig. C50. If test lamp does not light, the rotor is open, requiring replacement.

Shorted Field Coil Test: Connect an ohmmeter between the slip rings, Fig. C50. If reading is below 1.7 ohms, the rotor is shorted.

High Resistance Test: with an ohmmeter connected across the slip rings, reading should be between 1.7 and 2.1 ohms at 80° F. If not, replace rotor.

Assembly

1. Press grease retainer onto rotor shaft, Fig. C48 (Omit tool C-4330-3).
2. Place rectifier end shield bearing on base of tool C-4330-1, Fig. C48 (Omit tool C-4330-3), then place rectifier end shield on top of bearing. Using tool C-4330-2, press end shield onto bearing until end shield contacts press base.
3. Install drive end bearing and retainer in drive end shield.

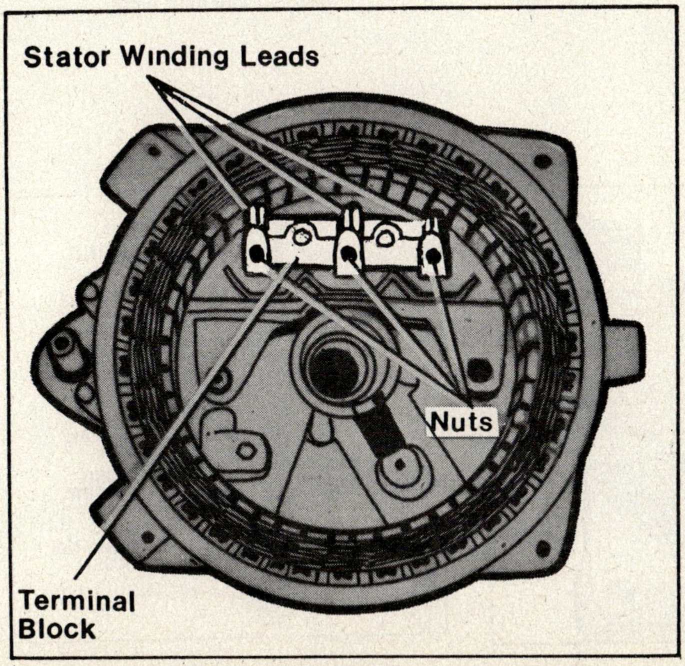

Fig. C41 Removing stator winding leads. 100 amp units

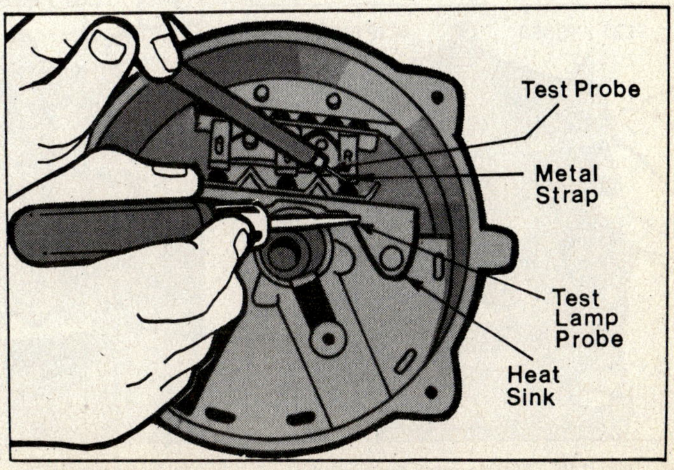

Fig. C42 Testing rectifiers with a test lamp. 100 amp units

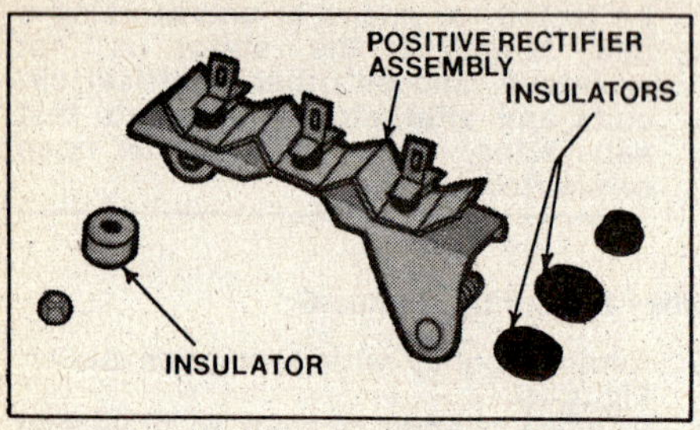

Fig. C43 Positive rectifier assembly. 100 amp units

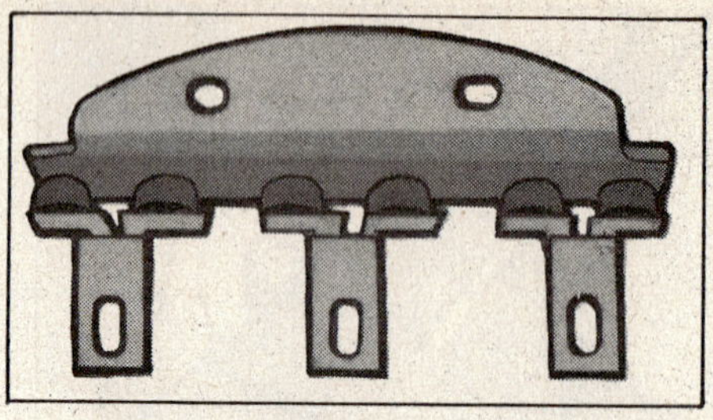

Fig. C44 Negative rectifier assembly. 100 amp units

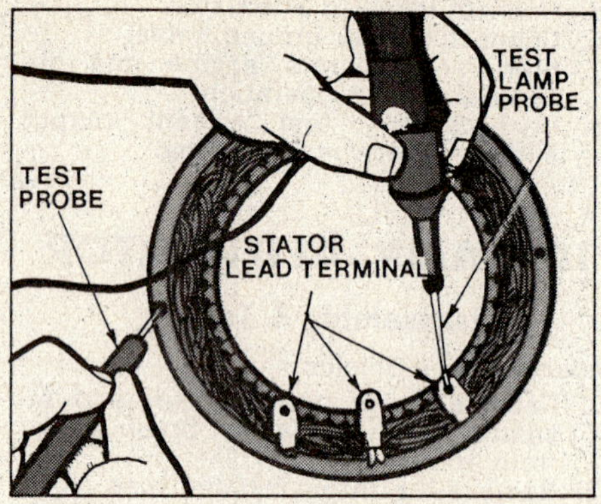

Fig. C45 Testing stator. 100 amp units

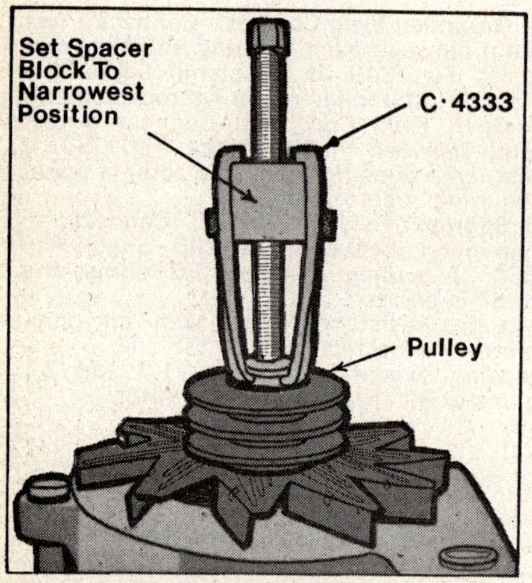

Fig. C46 Removing pulley. 100 amp units

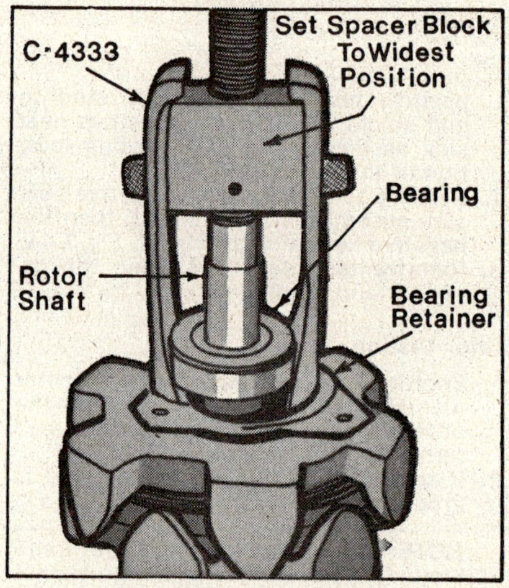

Fig. C47 Removing bearing from rotor shaft. 100 amp units

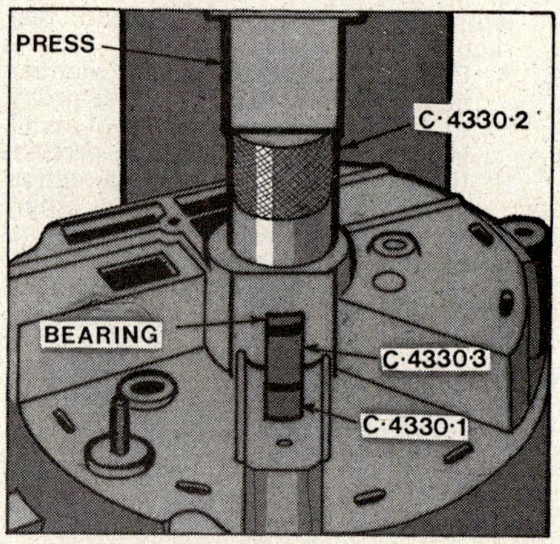

Fig. C48 Removing rectifier end shield bearing. 100 amp units

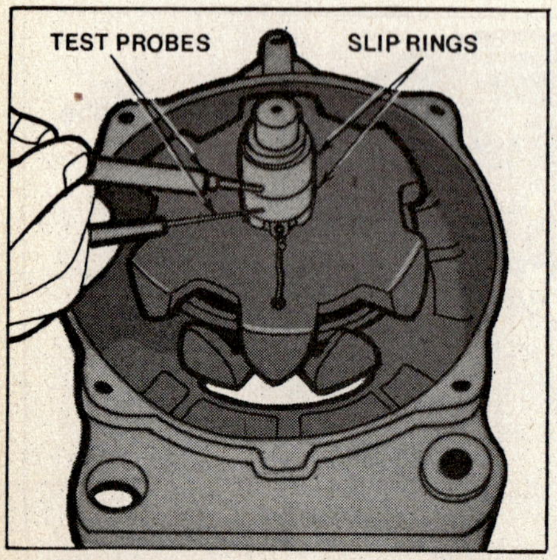

Fig. C49 Testing rotor for grounds. 100 amp units

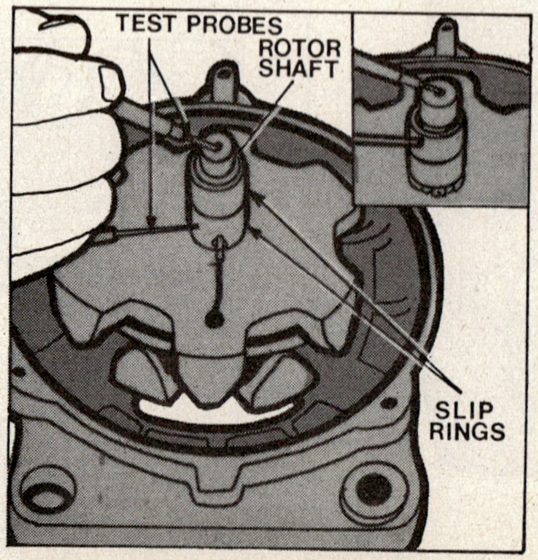

Fig. C50 Testing rotor for opens or shorts. 100 amp units

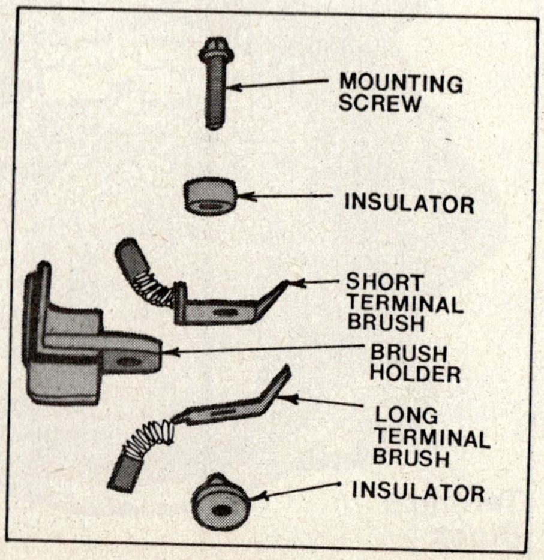

Fig. C51 Assembling field brushes. 100 amp units

4. Position bearing and drive end shield on rotor and while supporting base of rotor shaft, press end shield onto shaft. Ensure rotor spacer is in place before pressing bearing and end shield on shaft.
5. Press pulley onto rotor shaft until it contacts inner race of bearing.

NOTE: Do not exceed 6800 lbs.

6. Place insulator and capacitor on positive heat sink mounting stud, then install capacitor mounting screw.
7. Place terminal block into position in rectifier end shield and install mounting screws.
8. Place negative heat sink into position, ensuring metal straps are properly located over studs on terminal block, then install negative heat sink mounting screws.
9. Place insulator over positive heat sink stud and install positive heat sink assembly into position in end shield, ensuring metal straps are properly located over terminal block studs. From inside end shield, place insulator on positive heat sink mounting stud, then install mounting nut. From outside end shield, place insulator on positive heat sink stud, then install mounting nut.
10. Place stator over rectifier end shield and install terminals on terminal block. Then, press stator pins into end shield and install terminal nuts.

NOTE: Route leads to avoid contact with rotor or sharp edge of negative heat sink.

11. Place rotor and drive end shield assembly over stator and rectifier end shield assembly, aligning bolt holes. Compress stator and both end shields, then install and torque through bolts to 40-60 inch lbs.
12. Install field brushes into brush holder with long terminal on bottom and the short terminal on top, Fig. C51. Then, install insulators and mounting screw.
13. Place brush holder on end shield, ensure it is properly seated and tighten mounting screw.
14. Slowly rotate pulley to ensure rotor poles do not contact stator winding leads.

Delco-Remy "Delcotron" Alternator With External Regulator

TESTING SYSTEM IN VEHICLE

System Condition Test

1. Check alternator drive belt tension and adjust if necessary. Also, check charging system for proper electrical connections.
2. Connect a voltmeter between junction block relay and the ground at regulator base.

NOTE: Ensure that voltmeter clip does not contact a resistor or the terminal extension under regulator.

3. On models equipped with an indicator lamp, turn ignition switch to the "ON" position. Lamp should light. If not, check condition of bulb and wiring.
4. On models equipped with an ammeter, turn ignition switch to the "ACC" position with any one accessory turned on, such as the blower motor or radio. Ammeter should indicate a discharge condition. If not, check ammeter circuit.
5. On all models, start engine and observe indicator lamp or ammeter. The indicator lamp should go out or ammeter should move toward charge. If not, check respective circuit.
6. Turn headlamps to high beam and blower motor to high speed and set engine at 1500 RPM. If voltage is 12½ volts or more, proceed to next step. If voltage is less than 12½ volts, perform "Alternator Output Test" to follow. If output test is satisfactory, remove regulator cover and turn voltage adjusting screw to raise voltage to 12½ volts, Fig. D5. If 12½ volts cannot be obtained, replace regulator.
7. Connect a ¼ ohm, 25 watt resistor into charging circuit, Fig. D6.
8. Run engine at 1500 RPM or above for approximately 15 minutes. Cycle the regulator voltage control by disconnecting and reconnecting regulator connector. Note voltage.
9. If voltage reading is between 13.5 and 15.2 volts, regulator is satisfactory.
10. If voltage is not within above limits, remove regulator cover and adjust voltage to 14.2-14.6 volts. Replace regulator cover and run engine for approximately 10 minutes to obtain regulator internal operating temperature. Cycle the regulator and note voltage which should be between 13.5 and 15.2 volts. If not, replace regulator.

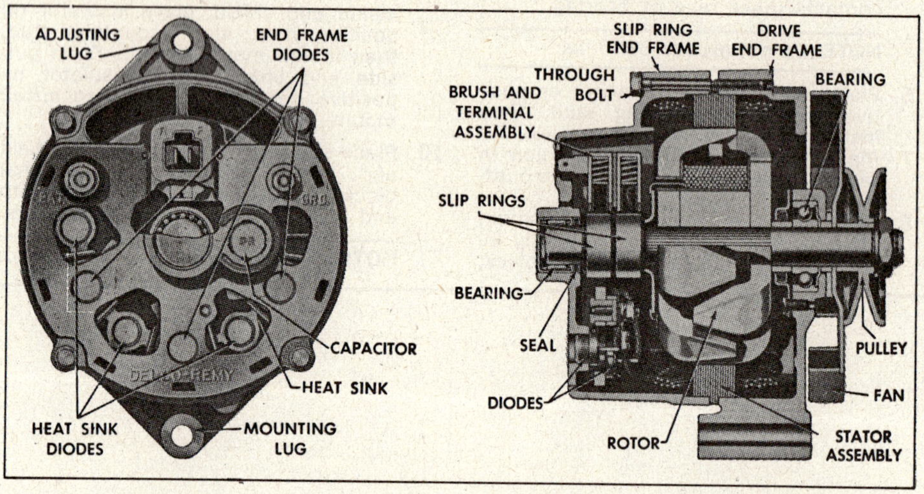

Fig. D1 Sectional end and side views of "Delcotron" alternator (typical)

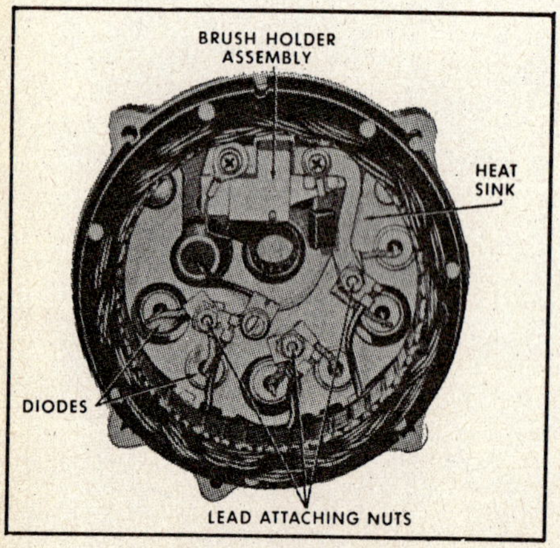

Fig. D2 Alternator internal lead connections, 1969-71

Alternator Output Test

1. Disconnect battery ground cable.
2. Connect an ammeter between alternator battery terminal and battery wire and a voltmeter between alternator battery terminal and a proper ground on the alternator, Fig. D7.
3. Disconnect the "F-R" terminal connector and connect a jumper wire between the "F" and alternator battery terminals, Fig. D7.
4. Reconnect the battery ground cable and connect a carbon pile rheostat across the battery terminals.
5. Start engine and slowly bring speed to 1500 RPM with carbon pile adjusted to hold 14 volts. Observe ammeter which should indicate alternator rated amperage. If not, service alternator.

Field Relay Test & Adjustment

1. Connect a voltmeter between regulator No. 2 terminal and the ground, Fig. D8.
2. Run engine at fast idle RPM and note voltage.
3. If zero voltage is indicated at regulator, check circuit between regulator No. 2 terminal and the alternator "R" terminal.
4. If closing voltage exceeds specifications as found in "Alternator & Regulator Specifications" in individual car chapters, adjust as follows:
   a. Connect a voltmeter and a 50 ohm variable resistor into circuit, Fig. D9.
   b. Turn resistor to "Open" position with ignition switch in the "Off" position.
   c. Adjust closing voltage by bending heel iron, Fig. D10.

Tailoring Voltage Regulator

It is important to remember that the voltage setting for one type of operating condition may not be satisfactory for a different type of operating condition. Vehicle under-hood temperatures, operating speeds, and night-time service all are factors which help determine the proper voltage setting. The proper setting is attained when the battery re-

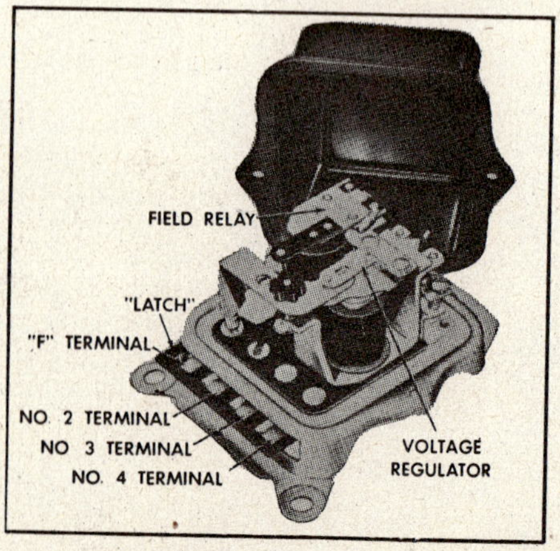

Fig. D3 Voltage regulator. The field relay is used only on vehicles having a charge indicator light instead of an ammeter

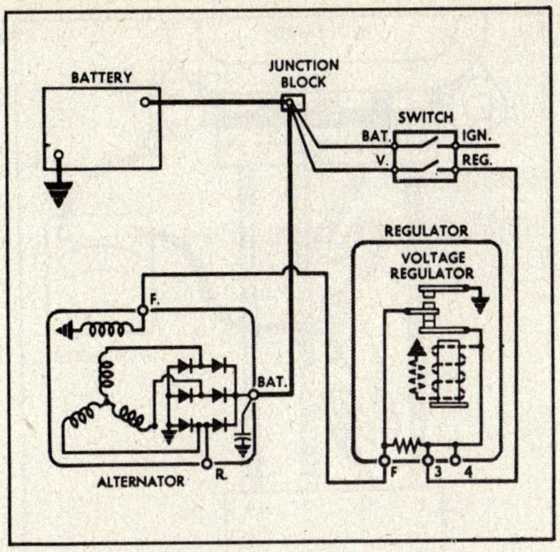

Fig. D4 Wiring diagram of the alternator charging circuit

mains fully charged with a minimum use of water.

If no circuit defects are found, yet the battery remains undercharged, raise the setting by .3 volt, and then check for an improved battery condition over a service period of reasonable length. If the battery remains overcharged, lower the setting by .3 volt, and then check for an improved battery condition. However, never adjust the voltage setting out of the limits specified in car chapters.

ALTERNATOR SERVICE

Alternator Removal

1. Disconnect battery negative cable.
2. Remove two leads at alternator.
3. Loosen adjusting bolts and remove drive belt.
4. Remove alternator retaining bolts and take off alternator.

Alternator Disassembly

1. Remove through bolts and loosen end frames by prying at bolt hole locations.
2. Scribe a reference mark across end frames to aid reassembly. Then, separate slip-ring end frame and stator assembly from drive end frame and rotor assembly.

NOTE: Brushes may drop onto rotor shaft when separating end frames, becoming contaminated with bearing lubricant. Clean brushes prior to installation.

3. Place a piece of tape over slip-ring end frame bearing to prevent entry of foreign material. Also, place tape on rotor shaft at slip-ring end.
4. Remove stator lead attaching nut and separate stator from end frame.
5. On 1972 units:
   a. Remove brush holder and terminal assemblies. Note that one of the three brush holder retaining screws is insulated.
   b. Remove rectifier bridge attaching screw, "Bat" terminal nut, capacitor lead and rectifier bridge from end frame.
   c. Remove capacitor from end frame.
6. On 1969-71 units:
   a. Remove screws, brushes and holder assembly.
   b. Remove "Bat" and "Grd" terminals, heat sink attaching screw and heat sink from end frame.
7. On all models, remove pulley retaining nut, slide washer, pulley and fan from rotor shaft.

NOTE: Use a $^5/_{16}$-inch Allen wrench inserted into shaft to hold shaft stationary while removing retaining nut.

8. Remove rotor, spacers, drive end bearing retainer plate and bearing from end frame.

Rotor Checks

1. To check for grounds, connect a 110 volt test lamp for either slip ring to the rotor shaft, Fig. D11. If the lamp lights the field winding is grounded.

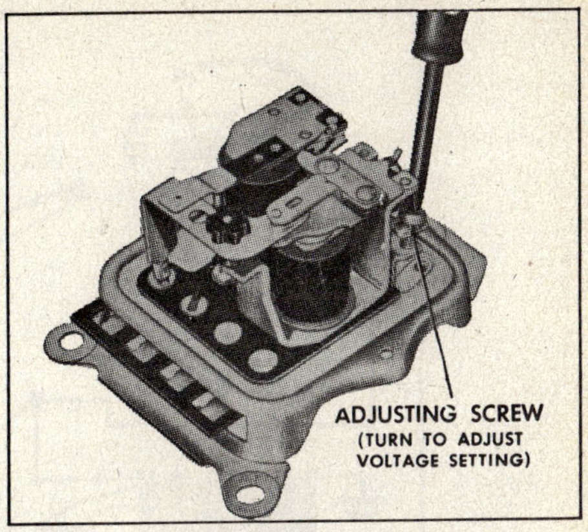

Fig. D5 Adjusting voltage regulator setting

2. To check for opens, connect the test lamp to each slip ring. If the lamp fails to light the winding is open.
3. To check for short circuits, connect a battery and ammeter in series with the two slip rings. The ammeter should indicate field current as specified in the Alternator and Regulator Specifications found in the car chapters.
4. An ammeter reading above the values given indicates shorted windings, and the rotor assembly should be replaced.

Stator Checks

1. To check the stator windings, remove all three stator lead attaching nuts and separate the stator from the end frame.
2. The stator winding may be checked with a 110 volt test lamp. If the lamp lights when connected from any stator lead to the frame, the windings are grounded. If the lamp fails to light when successively connected between each pair of stator leads, the windings are open, Fig. D12.
3. A short circuit in the stator windings is difficult to locate without laboratory test equipment due to the low resistance of the windings. However, if all other electrical checks are normal and the alternator fails to supply rated output, shorted stator windings are indicated.

Rectifier Bridge Check

Caution: Do not use a high voltage for this check, such as 110 volt test lamp.

1. Connect an ohmmeter to the grounded heat sink and one of the terminals. Then, reverse the probes, Fig. D13.
2. If readings in both directions are the same, replace the rectifier bridge assembly following procedure under "Alternator Disassembly", since diodes are not replaced individually.
3. Repeat check between grounded heat sink and other two terminals and between insulated heat sink and the three terminals.

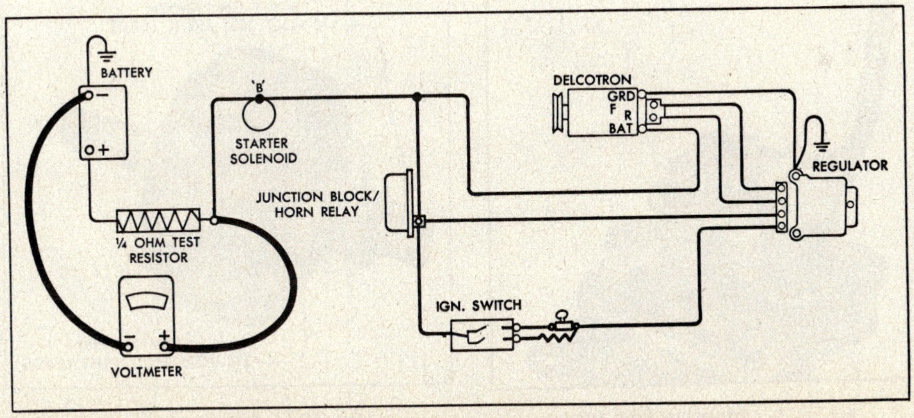

Fig. D6 Voltage setting test connections

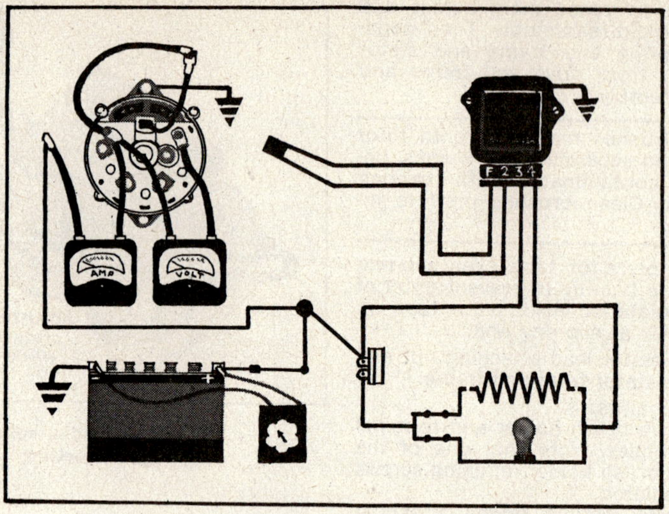

Fig. D7 Alternator output test connections

1969-71 Diode Checks

1. Each diode should be checked electrically for a shorted or open condition using a *test lamp of not more than 12 volts*, Fig. D14.
2. With the stator disconnected, connect the test lamp leads across each diode, first in one direction and then in the other.
3. If the lamp lights in both checks, or fails to light in both checks, the diode is defective.
4. When checking a good diode, the lamp will light in only one of the two directions.

1969-71 Diode Replacement

1. To remove a diode, place slip ring end frame in a vise with the remover equipment mounted as shown in Fig. D15. Tighten the vise to remove the defective diode.
2. To install a diode, place the new diode in the installer, Fig. D16. With the tools installed in the vise as shown, tighten the vise to install the new diode.

CAUTION: *Never attempt to remove or install a diode by striking it as the shock may damage the other diodes.*

Slip Ring Service

If the slip rings are dirty they may be cleaned with No. 400 silicon carbide paper and finish polished with crocus cloth. Spin the rotor in a lathe, or otherwise spin the rotor, and hold the polishing cloth against the slip rings until they are clean.

CAUTION: The rotor must be rotated in order that the slip rings will be cleaned evenly. Cleaning the slip rings by hand without spinning the rotor may result in flat spots on the slip rings, causing brush noise.

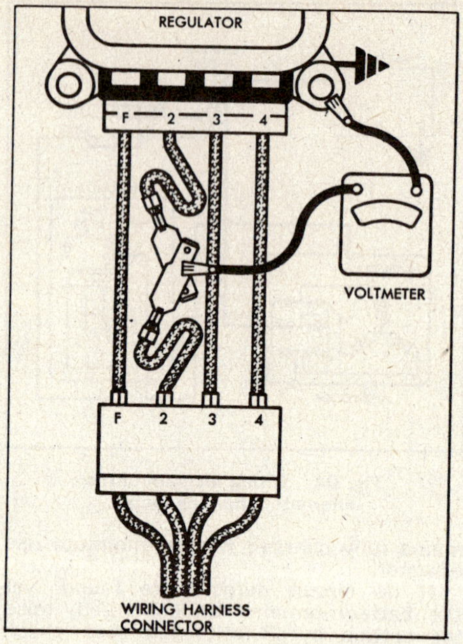

Fig. D8 Field relay test connections

Slip rings that are rough or out-of-round should be trued in a lathe to .002" maximum runout as indicated on a dial gauge. Remove only enough material to make the rings smooth and round. Finish polish with crocus cloth and blow away all dust.

Bearing Replacement

1. The bearing in the drive end frame can be removed by detaching the retainer plate screws and then pres-

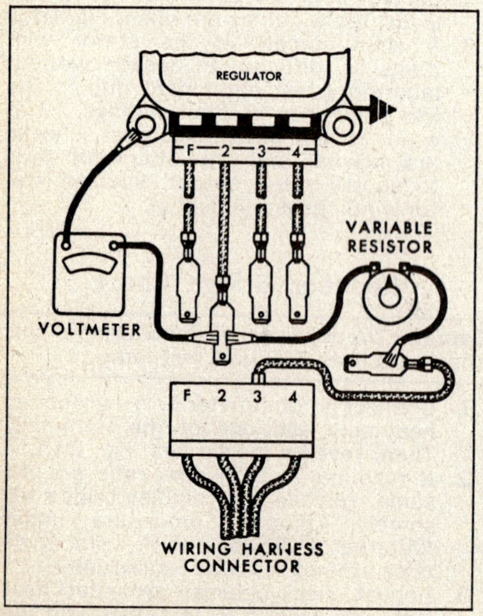

Fig. D9 Field relay closing voltage test connections

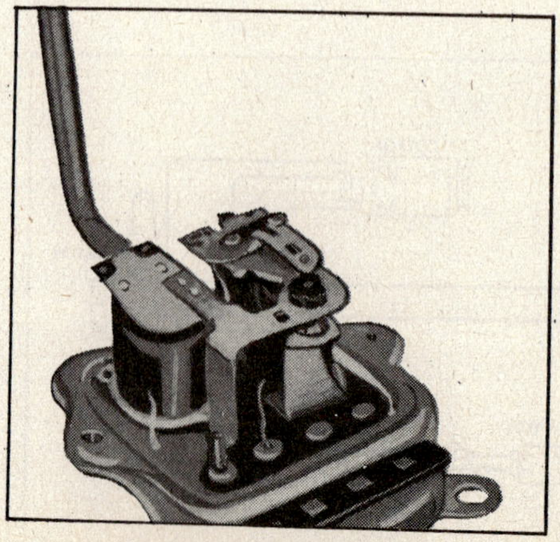

Fig. D10 Adjusting field relay closing voltage

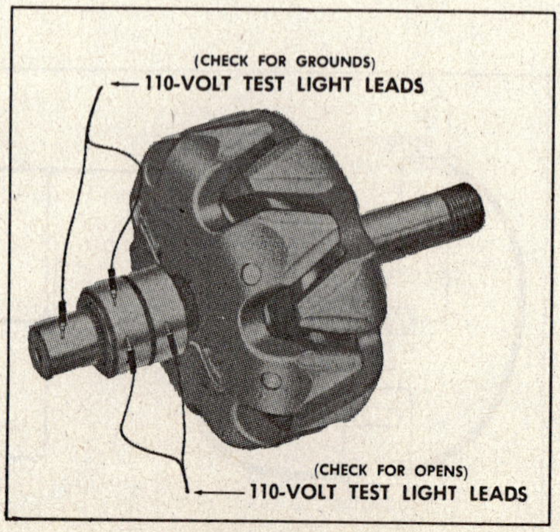

Fig. D11 Checking rotor for opens or grounds

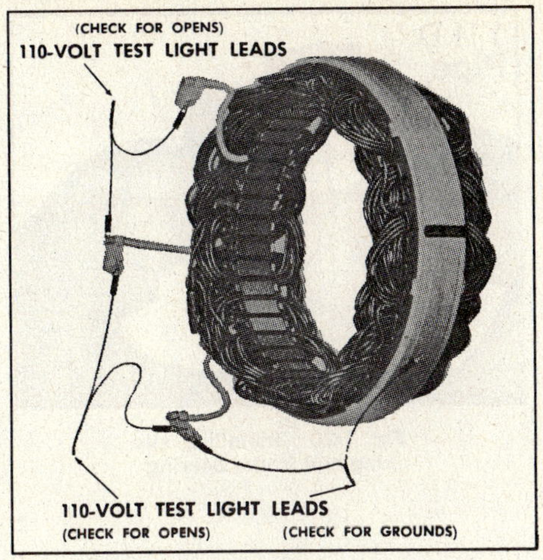

Fig. D12 Checking stator for opens or grounds

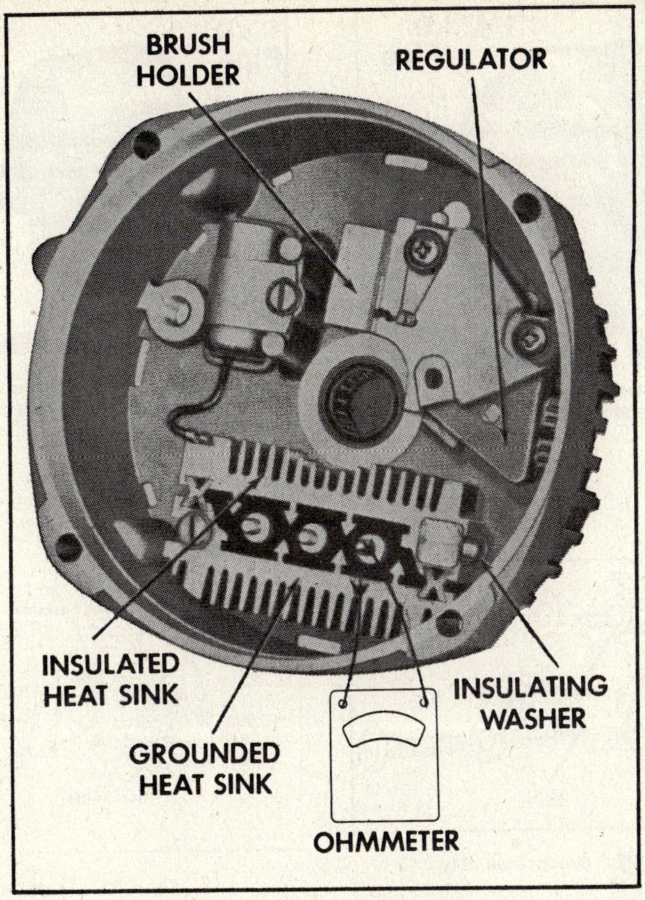

Fig. D13 Rectifier bridge check

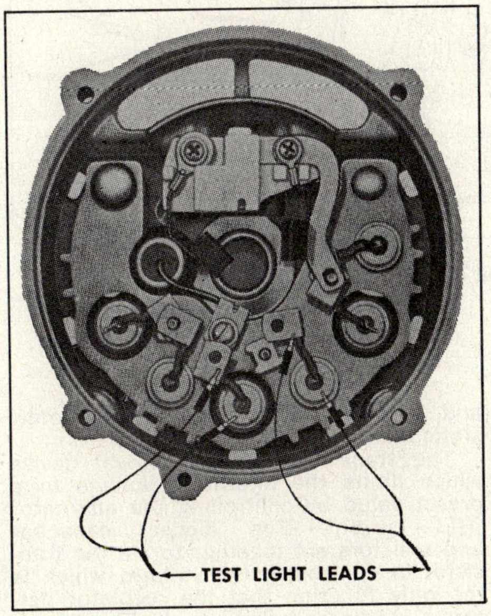

Fig. D14 Checking diodes for opens or shorts, 1969-71

sing the bearing from the end frame as shown in Fig. D17.

2. Press in new bearing with a tube or collar that just fits the outer race, Fig. D18. Install a new retainer plate if the felt seal in the existing plate is hardened or worn.
3. The bearing in the slip ring end frame can be removed by pressing with a tube or collar that just fits inside the end frame housing. Press from the outside of the housing towards the inside as shown in Fig. D19.
4. To install the new bearing, place a flat plate over the bearing and press in from the outside towards the inside of the frame until the bearing is flush with the outside of the end frame. Support the inside of the frame with the pipe shown to prevent breakage of the end frame, Fig. D20.
5. Saturate the felt seal with S.A.E. 20 oil and reassemble the felt seal and steel retainer.

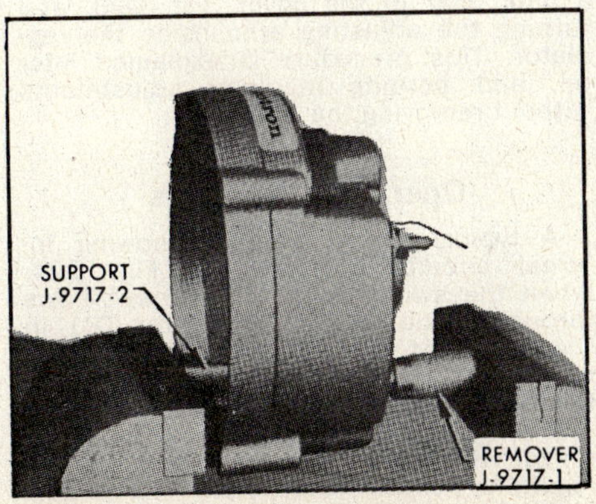

Fig. D15 Removing a diode, 1969-71

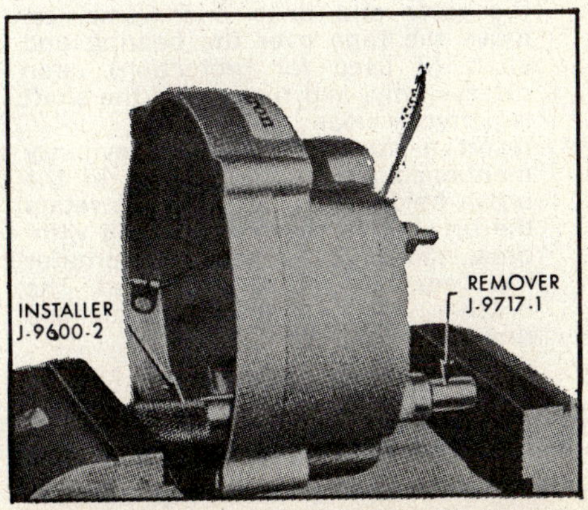

Fig. D16 Installing a diode, 1969-71

Fig. D17 Removing drive end frame bearing

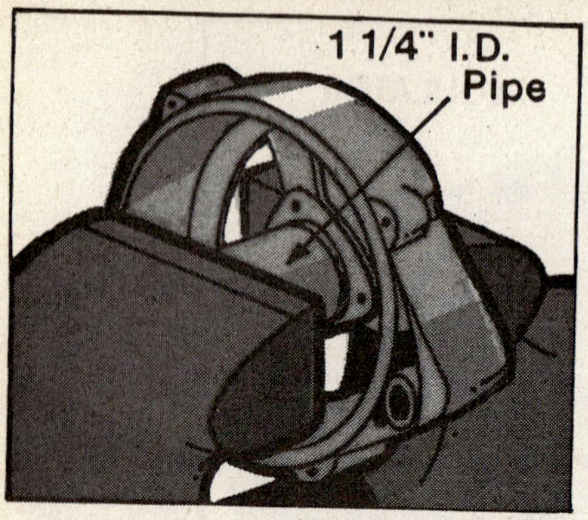

Fig. D18 Installing drive end frame bearing

Fig. D19 Removing slip ring end frame bearing

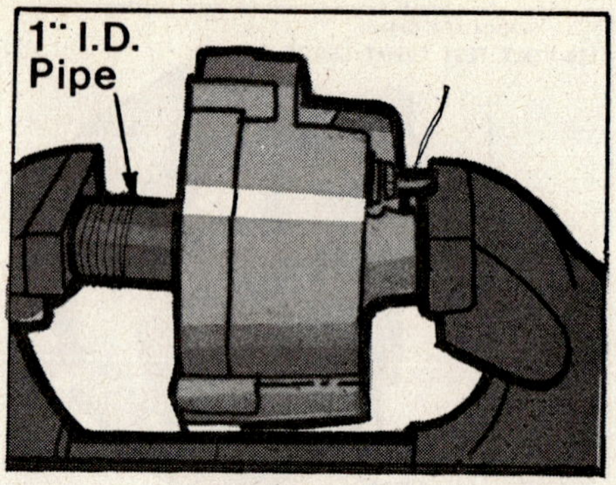

Fig. D20 Installing slip ring end frame bearing

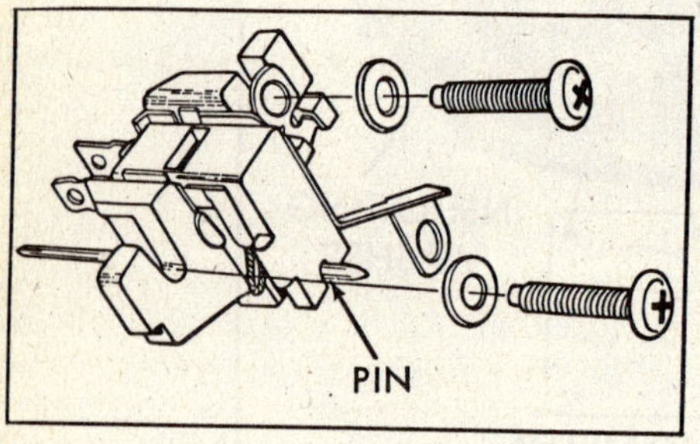

Fig. D21 Brush assembly

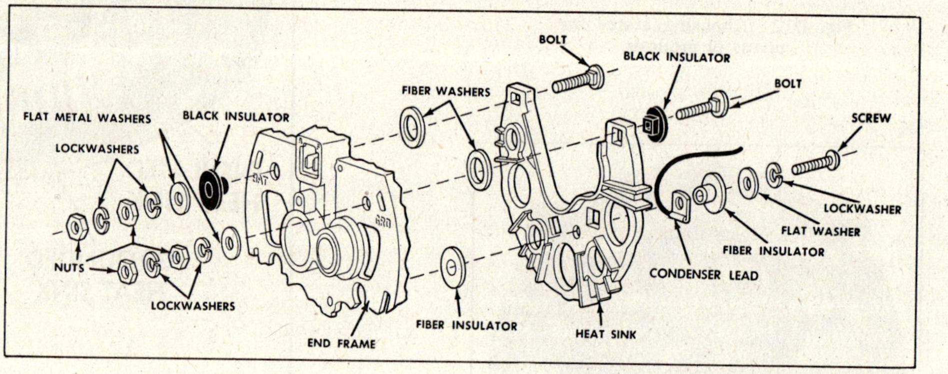

Fig. D22 Disassembled view of heat sink and related parts, 1969-71

Brush Replacement

1. When the slip ring end frame assembly is separated from the rotor and drive end frame, the brushes will fall down onto the shaft and come in contact with the lubricant. If the brushes are to be re-used, they must be thoroughly cleaned with a soft dry cloth immediately. Also, the shaft must be thoroughly cleaned before reassembly.
2. Inspect the brush springs for damage or corrosion. If there is any doubt as to the condition of the springs, they should be replaced.
3. To install new brushes, remove the brush holder assembly from the end frame by detaching the two screws.
4. Install the springs and brushes into the brush holder, and insert a straight wire or pin into the holes at the bottom of the holder to retain the brushes, Fig. D21.
5. Attach the brush holder assembly to the end frame, noting carefully the proper stack-up of the parts as shown. Allow the straight wire to protrude through the hole in the end frame.

1969-71 Heat Sink Replacement

1. Remove the "BAT" and "GRD" terminals from the end frame and the screw attaching the condenser lead to the heat sink.
2. During reassembly, note carefully the proper stack-up of parts as shown in Fig. D22.

Alternator Reassembly

1. Reassembly is the reverse of disassembly. Refer to Fig. D2 for connection of internal leads.
2. When installing the pulley, secure the rotor in a vise only tight enough to permit tightening the shaft nut to a torque of 40-50 ft. lbs. If excessive pressure is applied to the rotor, the assembly may become distorted.
3. To install the slip ring end frame to the rotor and drive end frame, remove the tape over the bearing and shaft (if used for protection) upon disassembly, and make sure the shaft is perfectly clean.
4. Insert a straight wire as previously mentioned through the holes in the brush holder and end frame to retain the brushes in the holder. Then withdraw the wire after the alternator has been completely assembled. The brushes will then drop onto the slip rings.

TRANSISTOR REGULATOR

The transistor regulator, Fig. D23, is an assembly composed principally of transistors, diodes, resistors, a capacitor, and a thermistor to form a completely static unit containing no moving parts.

The transistor is an electrical devise which limits the alternator voltage to a preset value by controlling the alternator field current. The diodes, capacitor and resistors act together to aid the transistor in controlling the voltage, which is the only function that the regulator performs in the charging circuit. The thermistor provides a temperature-compensated voltage setting.

The voltage at which the alternator operates is determined by the regulator adjustment. The regulator voltage setting can be adjusted externally by removing a pipe plug in the cover, Fig. D23, and turning the adjusting arm inside the regulator. This procedure is explained later on, and permits regulator adjustments without removing the cover.

Operating Principles

A typical wiring diagram showing internal circuits is shown in Fig. D24. When the switch is closed, current flows through diode D1 and transistor TR1 in the regulator to the alternator "F" terminal, and then through the alternator field winding to ground.

When alternator voltage reaches a preset value, the other components of the regulator cause transistor TR1 alternately to "turn-off" and "turn-on" the alternator field current. The regulator thus

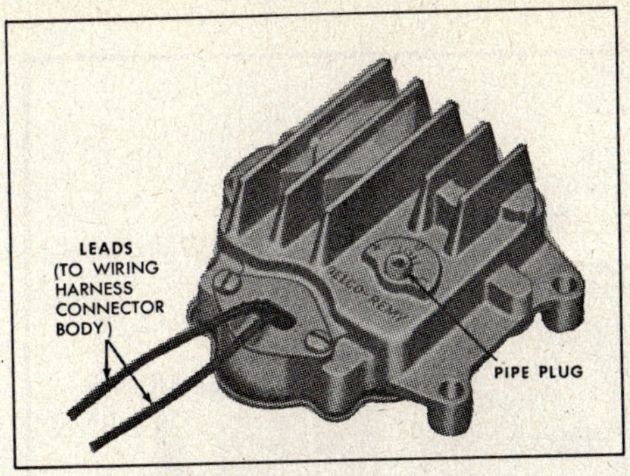

Fig. D23 **Transistor regulator**

operates automatically to limit the alternator voltage to a preset value.

Checking Circuit

1. Connect test ammeter and voltmeter in circuit as shown in Fig. D25, and connect a jumper wire from alternator "F" terminal to alternator "BAT" terminal.
2. Operate alternator at specified speed, turn on accessories as required to obtain specified voltage and observe output.
3. If current output is low, remove and check the alternator as outlined previously.
4. If the alternator failure was caused by a defective stator or diodes, the repaired alternator may be installed back on the vehicle and no further checks are needed.
5. If the alternator failure was caused by a defective field winding, the repaired alternator may be installed back on the vehicle and the following checks must be made to locate possible damage to regulator.
6. Referring to Fig. D25, remove jumper lead and reconnect wiring harness connector to alternator "F" terminal.
7. Turn on ignition switch but do not start engine.
8. Connect voltmeter positive lead to battery positive terminal and negative lead into regulator black lead connector body to make connection to regulator (Part 1, Fig. D26). Record voltage drop.
9. Connect voltmeter to negative terminal of battery and ground on regulator (Part 2, Fig. D26). Record voltage drop.
10. If addition of voltage readings is greater than .3 volt, check ignition switch for poor contacts and system wiring for high resistance. If voltage difference is less than .3 volt, proceed as follows, referring to Part 3, Fig. D26.
11. Connect voltmeter positive lead to regulator positive terminal and voltmeter negative lead to alternator "F" terminal. Slide voltmeter positive lead into regulator connector body (black lead terminal) to make connection. Record voltage.
12. If the voltage is .9 volt or less, replace regulator, as transistor is shorted. If voltage is 2.0 volts or greater, replace regulator as transistor is open. If voltage is between .9 and 2.0 volts, proceed as follows:
13. Operate engine at approximately 1500 rpm for 10 minutes with low beam headlights on. Referring to Part 4, Fig. D26, with engine running at 1500 rpm, record voltage reading from regulator positive terminal to ground by sliding voltmeter lead into regulator connector body black lead terminal to make connection.
14. Compare with Fig. D27. Ambient temperature is temperature of air measured 1/4" from regulator cover.
15. If voltage reading is within specifications, charging system is satisfactory but voltage setting may need to be changed to a different value to meet the requirements of driving conditions.
16. To do this, remove the pipe plug on regulator and insert a small screwdriver in adjustment slot. Turn counterclockwise for an undercharged battery one or two notches to increase setting.
17. For an overcharged battery, as evidenced by excessive water usage, turn clockwise one or two notches to decrease setting. For each notch moved, voltage setting will change by approximately .3 volt. Then check for an improved battery condition over a service period of reasonable length.
18. If voltage is not within specifications, check to see if the adjustment arm is in the center position. If the voltage reads out of specifications in the center position, replace the regulator.

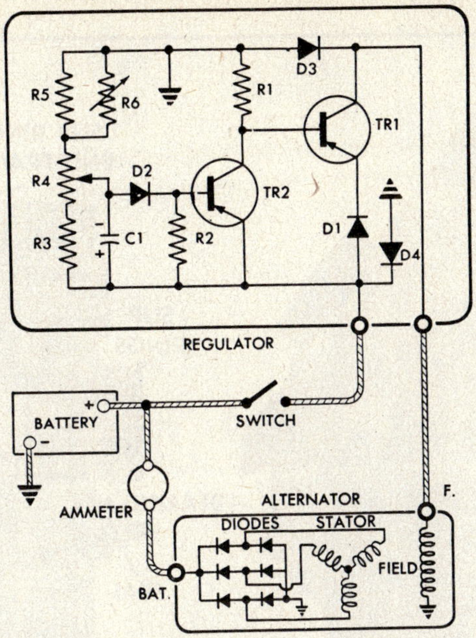

Fig. D24 **Wiring diagram of transistor regulator in charging circuit**

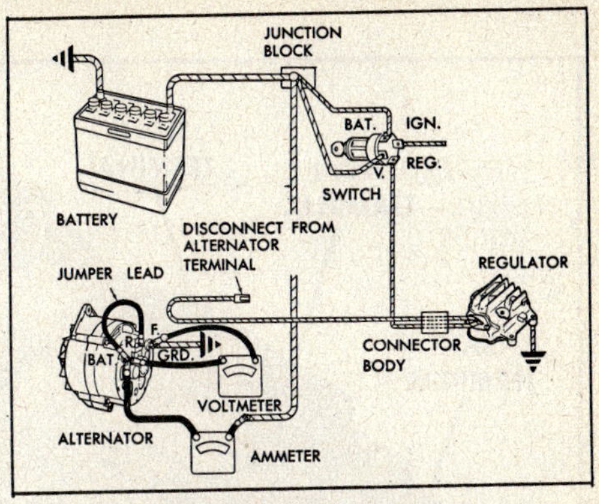

Fig. D25 **Checking charging circuit for undercharged battery condition**

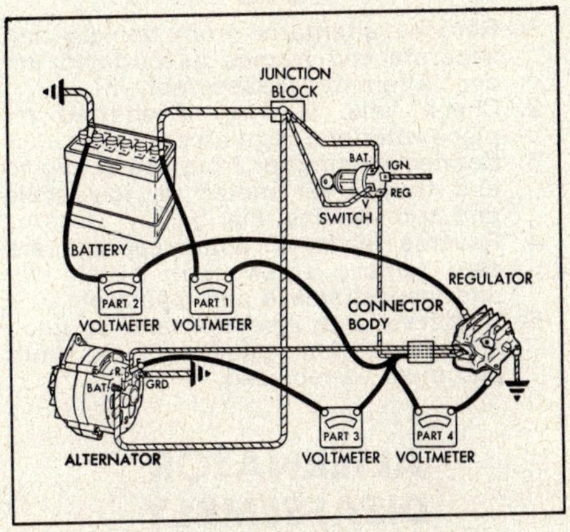

Fig. D26 **Checking charging circuit for overcharged battery condition**

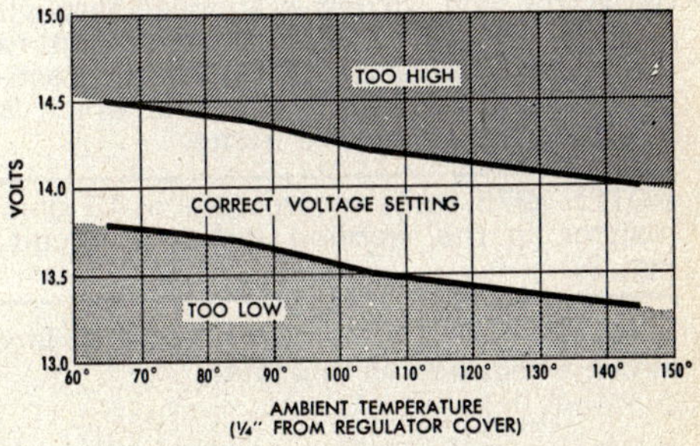

Fig. D27 **Temperature correction chart**

Delcotron Type SI Integral Charging System

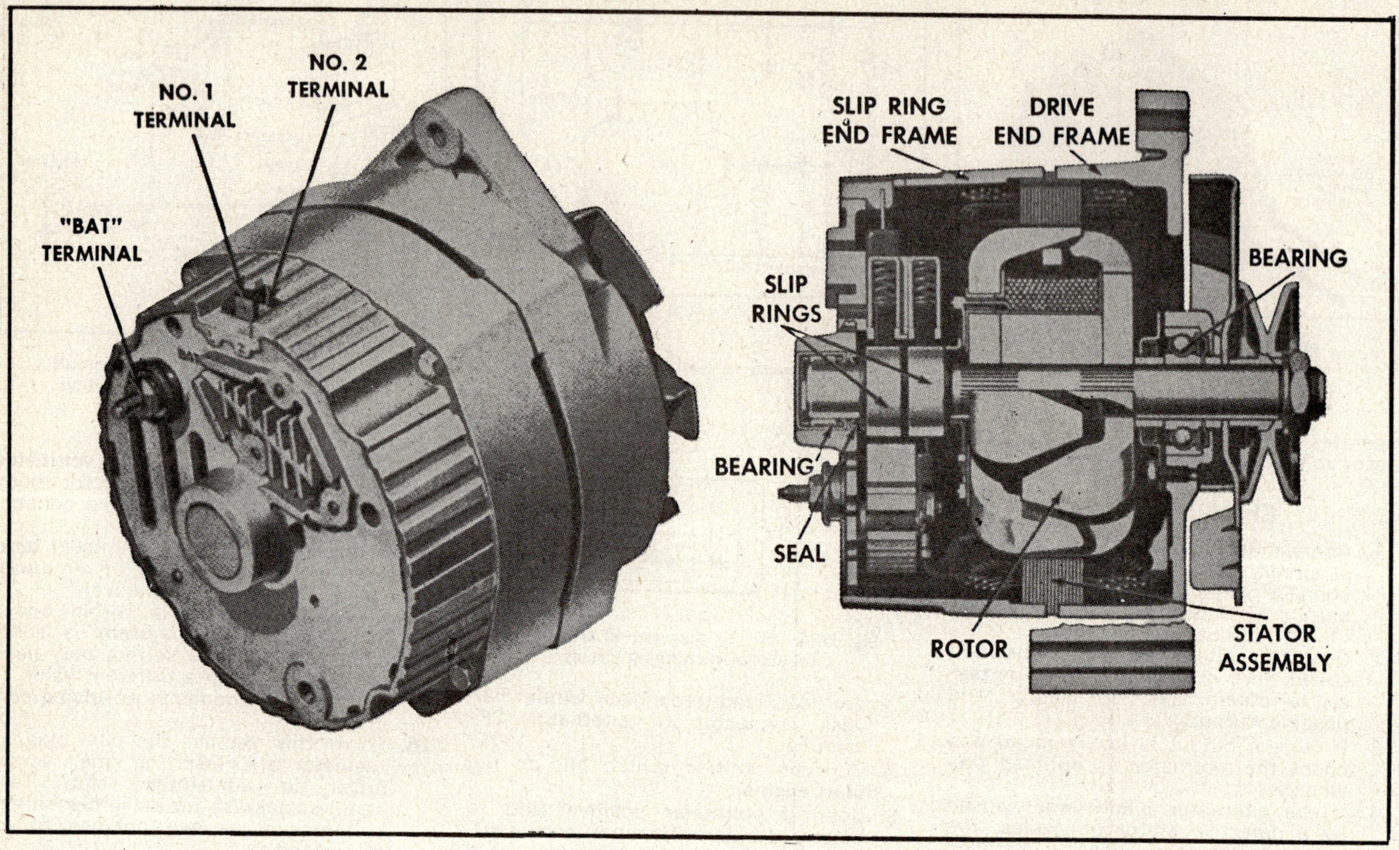

Fig. 1 Delcotron Type SI Integral Charging System

DESCRIPTION

This unit, Fig. 1, features a solid state regulator mounted inside the alternator slip ring end frame, along with the brush holder assembly. All regulator components are enclosed in a solid mold with no need or provision for adjustment of the regulator. A rectifier bridge, containing six diodes and connected to the stator windings, changes A.C. voltage to D.C. voltage which is available at the output terminal. Generator field current is supplied through a diode trio which is also connected to the stator windings. The diodes and rectifiers are protected by a capacitor which is also mounted in the end frame.

NOTE: 1975 units incorporate a 40 ohm resistor in the warning indicator circuit. Fig. 2.

No maintenance or adjustments of any kind are required on this unit.

TROUBLE SHOOTING

Undercharged Battery

SERVICE NOTE: If inadequate charging is encountered on 1974 Camaro's Z28 with optional gauge package the cause could be an omitted wire in the engine wiring harness. If wire in Fig. 3 is missing the alternator will not charge below 2,000 engine rpm. Add this wire to the existing harness and connect as shown in Fig. 3.

1. Disconnect battery ground cable.
2. Disconnect wire at "BAT" terminal of alternator, connect ammeter, positive lead to "BAT" terminal and negative lead to wire.
3. Connect battery ground cable.
4. Turn on all accessories, then connect a carbon pile regulator across battery.
5. Operate engine at moderate speed, adjust carbon pile regulator to obtain maximum current output.
6. If ammeter reading is within 10 percent of rated output, alternator is not at fault.

NOTE: Alternator rated output is stamped on alternator frame.

7. If ammeter reading is not within 10 percent of rated output, ground field winding by inserting screw driver in end frame hole, contacting tab. Fig. 4.

NOTE: Do not insert screw driver deeper than one inch, tab is usually located within ¾ inch of casing surface.

8. If reading is within 10 percent of rated output, regulator must be replaced. If reading is not within limits, check field winding, diode trio, rectifier bridge and stator.
9. Turn off all accessories and disconnect ammeter and carbon pile regulator.

Overcharging Battery

1. Remove alternator from vehicle and separate end frames as outlined under "Alternator Disassembly."
2. Check field winding, if shorted replace rotor and regulator.
3. Connect ohmmeter from brush clip to end frame, set meter on low scale and note reading. Fig. 5.
4. Reverse leads, if both readings are zero remove screw from brush clip and inspect sleeve and insulator.
5. If sleeve and insulator are in good condition, then regulator is at fault and must be replaced.

ALTERNATOR DISASSEMBLY

NOTE: When pressing bearings or seals from end frames, support frames from inside.

1. Scribe mark across end frames and stator ring so parts can be installed in same position.
2. Remove four through bolts, then using screw driver in stator slot pry end frames apart. Fig. 6.

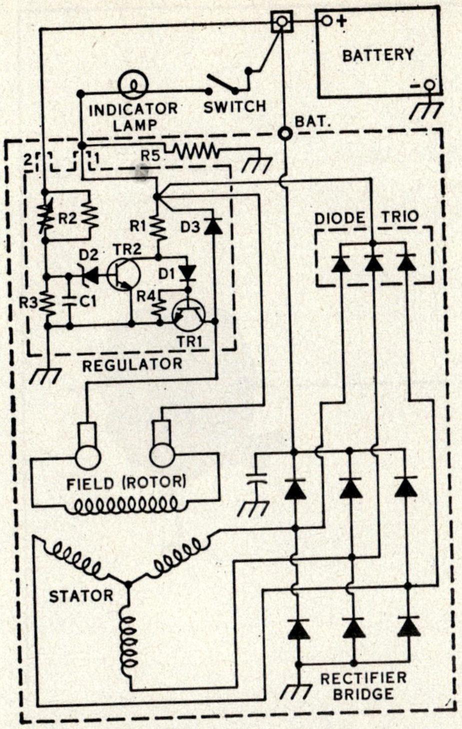

Fig. 2 Wiring diagram of charging circuit

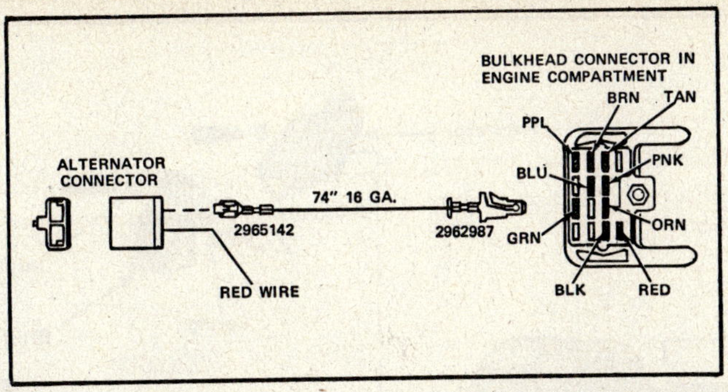

Fig. 3 Harness connections for omitted wire. 1974 Camaro Z-28

NOTE: Brushes may fall from holders and become contaminated with bearing grease, if so they must be cleaned prior to assembly.

3. Place tape over slip ring end frame bearing and shaft at slip ring end.
4. Remove nut, washer, pulley, fan and collar from rotor shaft, then slide drive end frame from shaft.
5. Remove bearing, retainer and seal from drive end frame.
6. Remove attaching bolts, then pry stator from slip ring end frame.
7. Remove capacitor, diode trio, rectifier bridge and battery terminal stud.
8. Remove resistor (if equipped), brush holder and regulator.
9. Remove bearing and seal from slip ring end frame.

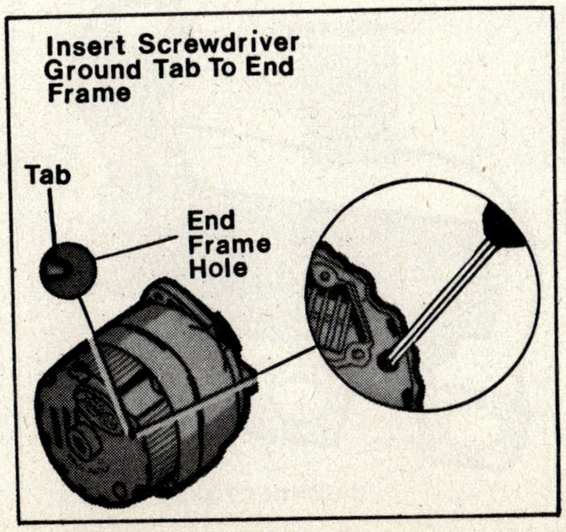

Fig. 4 Grounding field windings

BENCH TESTS

Rotor & Slip Ring Test

NOTE: Ohmmeter must be at low scale setting during this test.

1. Inspect rotor for wear or damage.
2. Touch ohmmeter leads to slip rings. Fig. 7.
3. If no reading is obtained an open circuit exist in windings.
4. If reading below 2.5 ohms is obtained winding is shorted.
5. If a reading above 3 ohms is obtained excessive resistance exists in windings.
6. Connect one ohmmeter lead to rotor shaft and touch slip rings with other lead, if any reading is obtained there is a ground in the circuit. Fig. 7.

NOTE: If any of the above problems are present the rotor assembly must be replaced.

Stator Winding Test

1. Inspect stator for discolored windings, loose connections and damage.
2. Connect an ohmmeter from stator lead to frame, if any reading is obtained windings are grounded. Fig. 8.
3. Connect ohmmeter between stator leads, if reading is high when connected between each pair of leads, an open circuit exists in windings.

NOTE: Shorted windings are difficult to locate without special equipment. If other test indicate normal, but rated alternator output cannot be obtained the windings are probably shorted.

Diode Trio

1. With diode unit removed, connect an ohmmeter to the single connector and to one of the three connectors. Fig. 9.

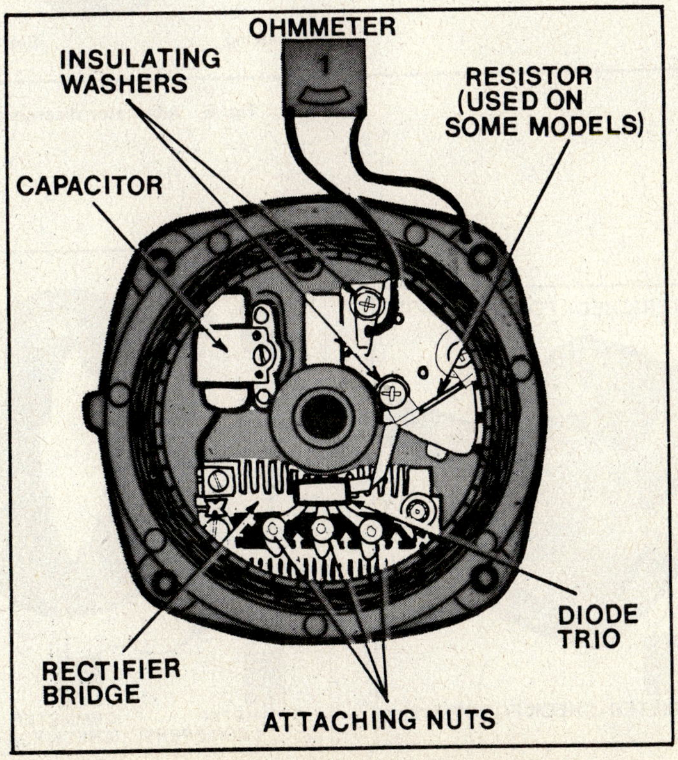

Fig. 5 Testing brush clip

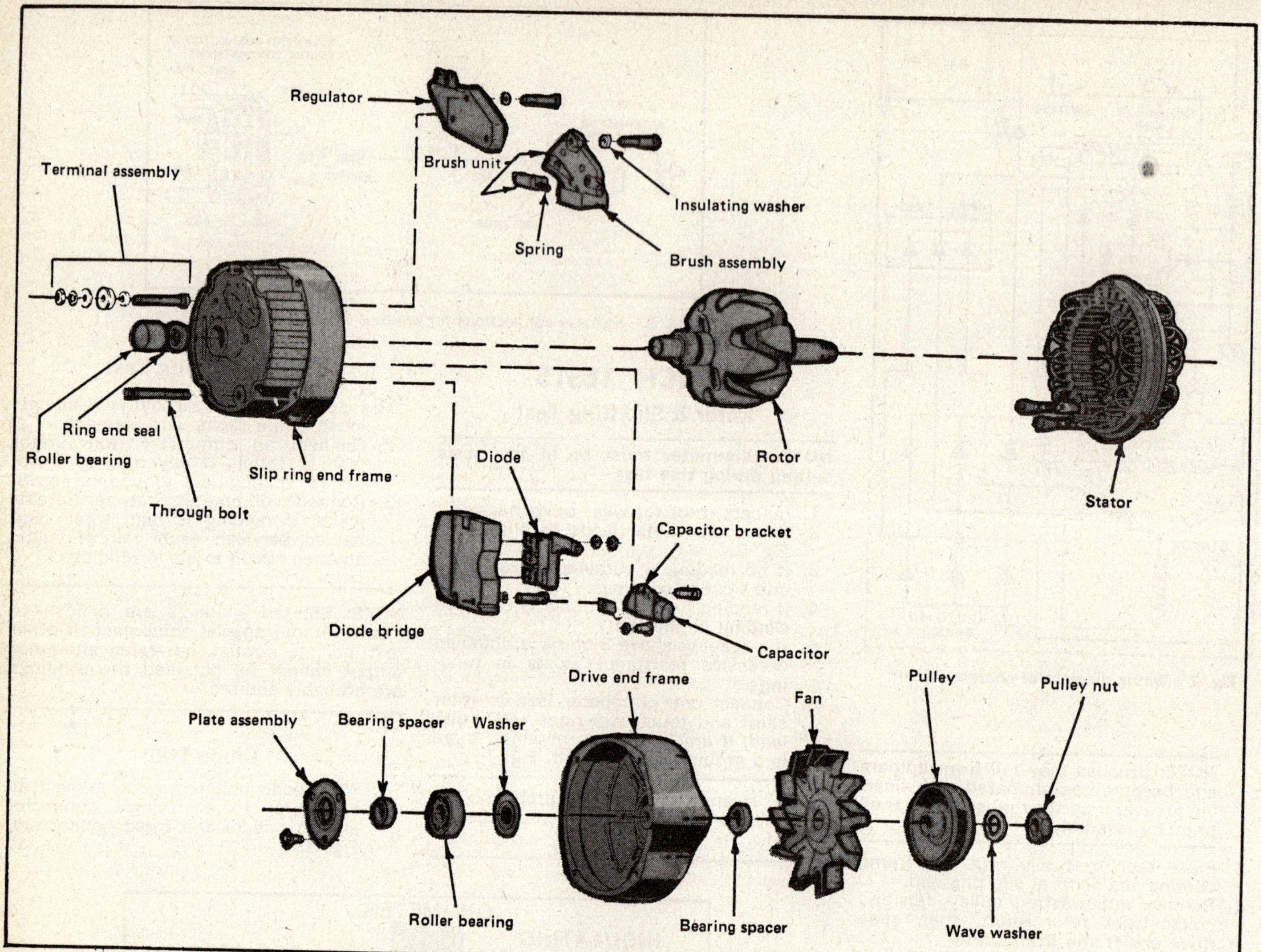

Fig. 6 Alternator disassembled

Fig. 7 Testing rotor & slip rings

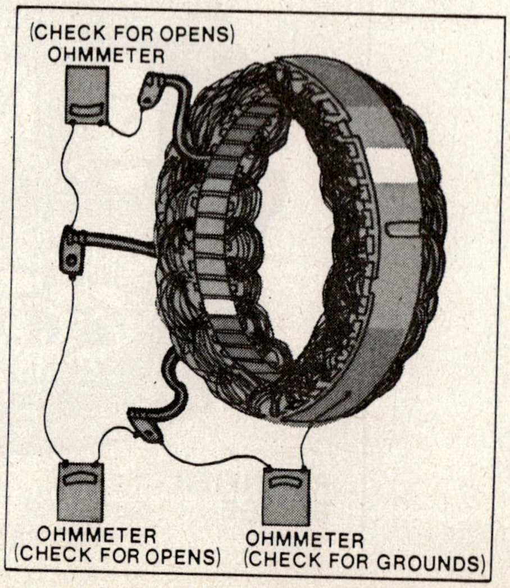

Fig. 8 Testing stator winding

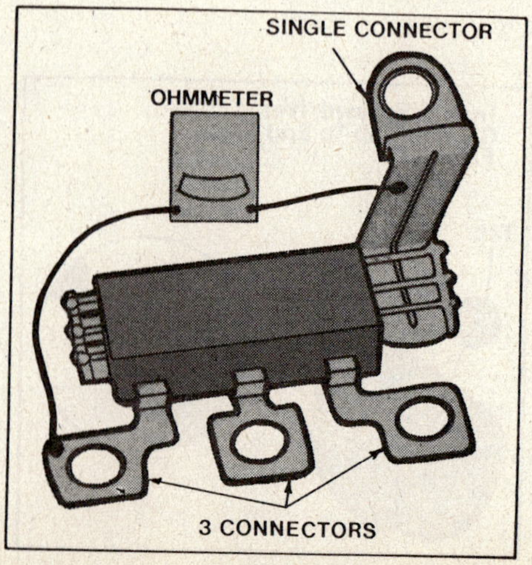

Fig. 9 Testing diode trio

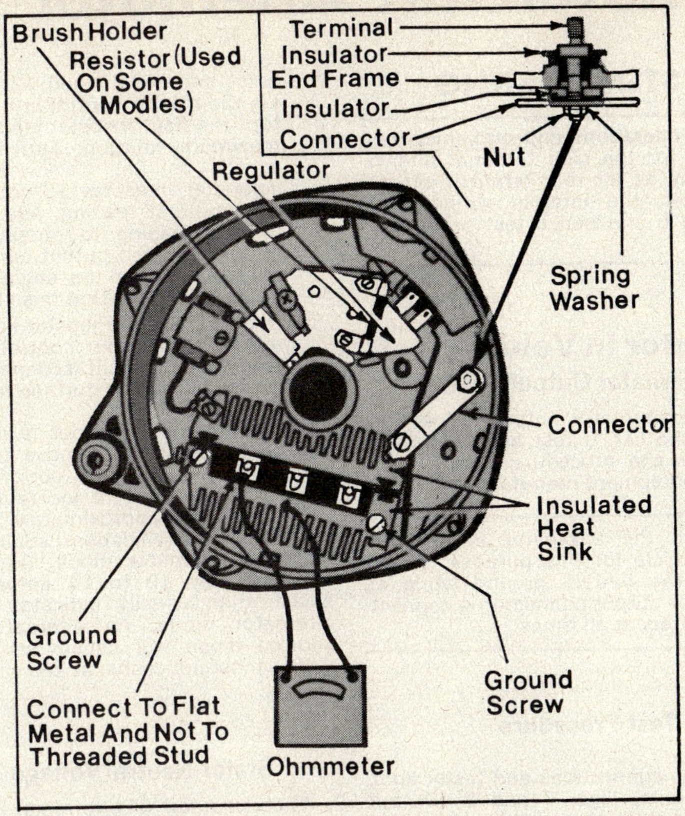

Fig. 10 Testing rectifier bridge diodes

2. Observe the reading. Reverse ohmmeter leads.
3. Reading should be high with one connection and low with the other. If both readings are the same, unit must be replaced.
4. Repeat between the single connector and each of the three connectors.

NOTE: There are two diode units differing in appearance. These are completely interchangeable.

The diode unit can be checked for a grounded brush lead while still installed in the end frame by connecting an ohmmeter from the brush lead clip to the end frame as in Steps 1 and 2 above. If both readings are zero, check for a grounded brush or brush lead.

Rectifer Bridge Test

1. Connect ohmmeter to the grounded heat sink and one of the three terminals. Fig. 10.
2. Observe the reading then reverse leads.
3. Reading should be high with one connection and low with the other. If both readings are the same, unit must be replaced.
4. Repeat test for each of the other terminals.

Voltage Regulator/Brush Lead Test

Connect an ohmmeter from the brush lead clip to the end frame, note reading, then reverse connections. If both readings are zero, either the brush lead clip is grounded or the regulator is defective.

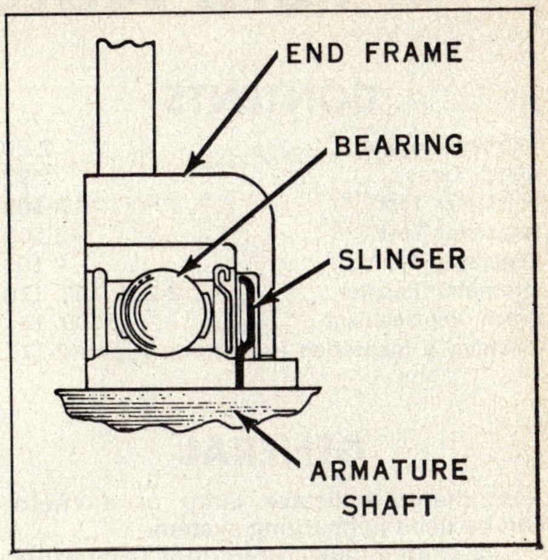

Fig. 11 Drive end frame bearing & slinger installed

ALTERNATOR ASSEMBLY

NOTE: When pressing bearings or seals, end frames must be supported from inside.

1. Lightly lubricate seal and position on slip ring end frame with lip facing toward rotor. Fig. 6.
2. Press seal part way into housing.
3. Position bearing and end plug on slip ring end frame, press bearing and plug in until flush with end frame.
4. Place regulator in end frame, install brushes and springs in brush holder, use pin to hold brushes in compressed position.
5. Install rectifier bridge and battery terminal stud.
6. Install diode trio, ensure current only flows one way through single connector.
7. Install capacitor.
8. Install stator, check the three leads for continuity, ensure stator is not grounded against case or holder.
9. Position slinger on drive end frame, then press ball bearing into end frame. Fig. 11.
10. Fill seal cavity ¼ full with special alternator lubricant, then install retainer.
11. Install rotor in drive end frame, then install collar, fan, pulley, washer and nut.
12. Align scribe marks on end frames and stator plate, install through bolts and remove brush retaining pins.

Ford Autolite/Motorcraft Alternator

CONTENTS

GENERAL

A charge indicator lamp or ammeter can be used in charging system.

If a charge indicator lamp is used in the charging system, Figs. 1A, 3A & 6, the system operation is as follows: when the ignition switch is turned ON, a small electrical current flows through the lamp filament (turning the lamp on) and through the alternator regulator to the alternator field. When the engine is started, the alternator field rotates and produces a voltage in the stator winding. When the voltage at the alternator stator terminal reaches about 3 volts, the regulator field relay closes. This puts the same voltage potential on both sides of the charge indicator lamp causing it to go out. When the field relay has closed, current passes through the regulator A terminal and is metered to the alternator field.

If an ammeter is used in the charging system, Figs. 2A, 4A & 5, the regulator I terminal and the alternator stator terminal are not used. When the ignition switch is turned ON, the field relay closes and electrical current passes through the regulator A terminal and is metered to the alternator field. When the engine is started, the alternator field rotates causing the alternator to operate.

NOTE: The ammeter indicates current flow into (charge) or out of (discharge) the vehicle battery.

SYSTEM TESTING

NOTE: The operations and on vehicle test procedures for the side terminal alternator are same as for rear terminal alternator. However, the internal wiring, Figs. 3, 4, 5 and 6 and bench test procedures differ.

Alternator in Vehicle Tests

Alternator Output Test

When the alternator output test is conducted off the car, a test bench must be used. Follow the procedure given by the test bench equipment manufacturer.

NOTE: When the alternator is removed from the vehicle for this purpose, always disconnect the battery ground cable as the alternator output connector is connected to the battery at all times.

Test Procedure

1. Make the connections and tester knob adjustments, Figs. 7 and 8 (Output Test). Be sure that the field rheostat knob is at the OFF position at the start of this test.
2. Close the battery adapter switch. Start the engine, then open the battery adapter switch.
3. Increase the engine speed to approximately 2000 rpm (use a tachometer following the manufacturers instructions). Turn off all lights and electrical accessories.
4. Turn the field rheostat clockwise until 15 volts is indicated on the voltmeter upper scale. Turn the master control clockwise until the voltmeter indicates between 11 and 12 volts. Holding the master control in this position, turn the field rheostat clockwise to its maximum rotation. Turn the master control counter clockwise until the voltmeter indicates 15 volts. Observe the ammeter reading. Add 2 amperes to this reading to obtain alternator output. If rated output cannot be obtained, increase the engine speed to 2900 rpm and repeat this step.
5. Return the field rheostat knob to OFF, release the master control knob, and stop the engine. Disconnect the test equipment, if no further tests are to be made.

If the alternator output is not O.K., it will be necessary to remove the alternator from the vehicle and perform the necessary bench tests to locate the defect.

An output of approximately 2 to 5 amperes below specification usually indicates an open alternator diode. An output of approximately 10 to 14 amperes below specification usually indicates a shorted alternator diode. An alternator with a shorted diode will usually whine, which will be most noticeable at idle speeds.

Stator Neutral Voltage Test

The alternator STA terminal is connected to the stator coil neutral or center point of the alternator windings, Figs. 1 thru 6. The voltage generated at this point is used to close the field relay in the charge indicator light system.

To test for the stator neutral voltage, disconnect the regulator connector plug from the regulator. Make the connections and tester knob adjustments, Fig. 9.

Start the engine and run it at 1000 rpm (use a tachometer). Turn off all lights and accessories. Rotate the field rheostat clockwise until at least 6 volts is indicated on the voltmeter upper scale. If 6 volts or more is not obtained, remove the alternator and perform the diode and sta-

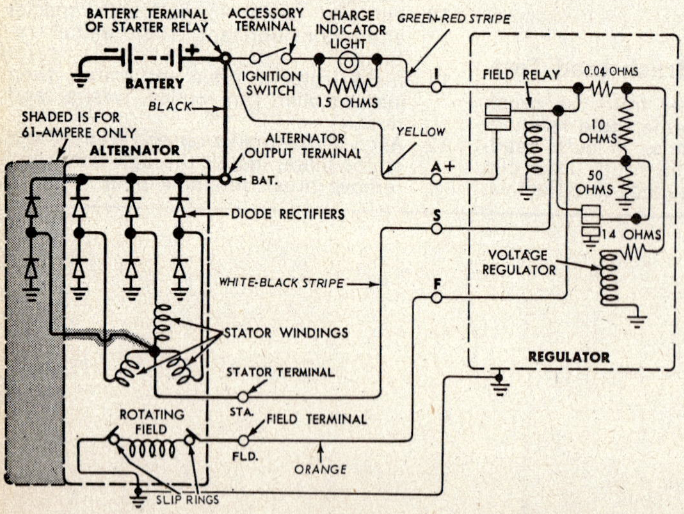

Fig. 1 Indicator light rear terminal alternator charging system less electric choke

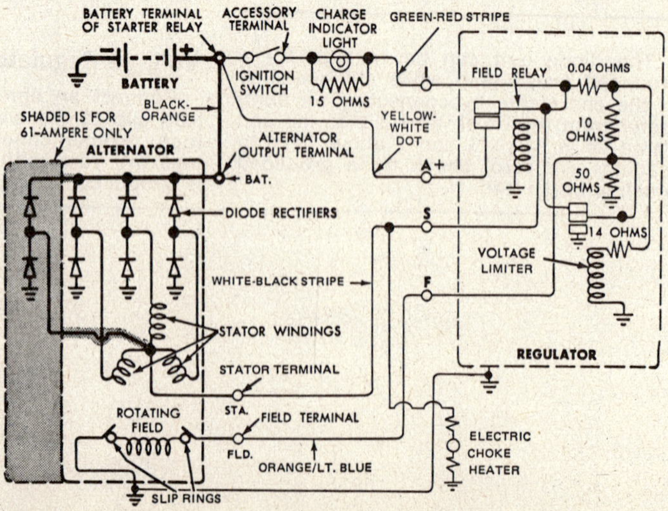

Fig. 1A Indicator light rear terminal alternator charging system with electric choke

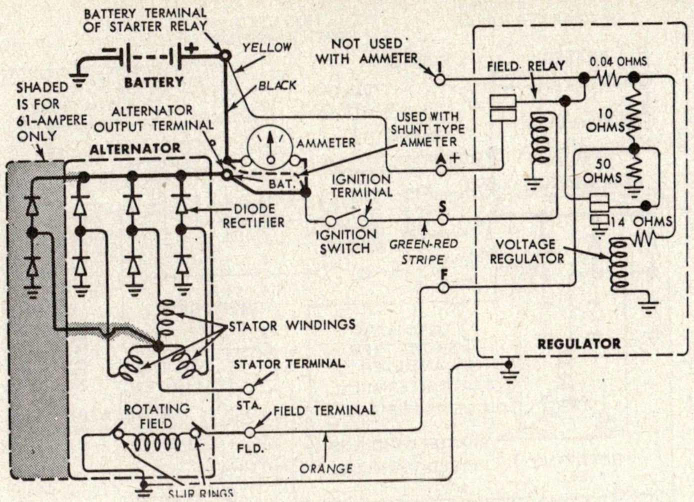

Fig. 2 Ammeter rear terminal alternator charging system less electric choke

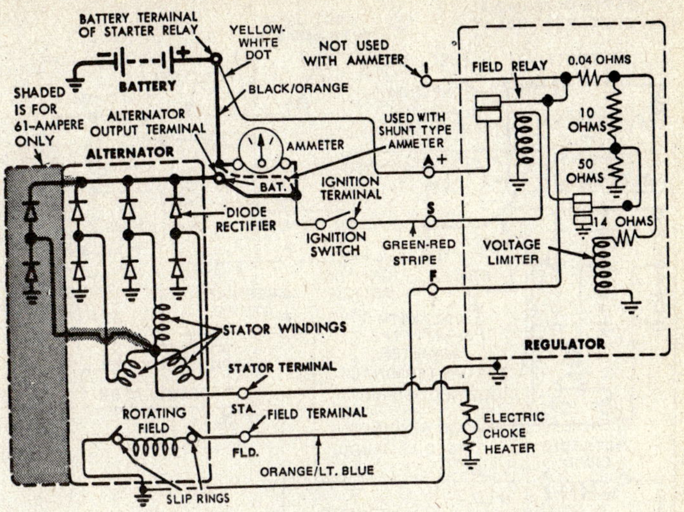

Fig. 2A Ammeter rear terminal alternator charging system with electric choke

tor tests to determine which part of the alternator is damaged.

Bench Tests

REAR TERMINAL ALTERNATOR

Field Open Or Short Circuit Test

The first part of this test will determine if the alternator portion of the field coil system, consisting of the field coil, the field coil slip rings and the field coil brush assembly is satisfactory. The second part of the test will indicate (in case of a field coil system malfunction), which of the above items is causing the malfunction.

Test Procedure

Make the connection, Fig 8. The current draw, as indicated by the ammeter, should be to specification as listed in car chapters. If there is little or no current flow, the field or brushes current flow considerably higher than that specified above indicates shorted or grounded field turns or brush leads touching. If the test shows that the field is shorted or open, determine if the field brush assembly or slip rings are at fault.

Disassemble front housing and rotor from the rear housing and stator, check the resistance of the rotor with ohmmeter. Set the ohmmeter multiply-by knob at 1 and calibrate the ohmmeter as indicated inside the ohmmeter cover.

Contact each ohmmeter probe to a slip ring. The resistance should be 3.5 to 5 ohms. A higher reading indicates a damaged slip ring soldered connection or a broken wire. A lower reading indicates a shorted wire or slip ring assembly.

Contact one ohmmeter probe to a slip

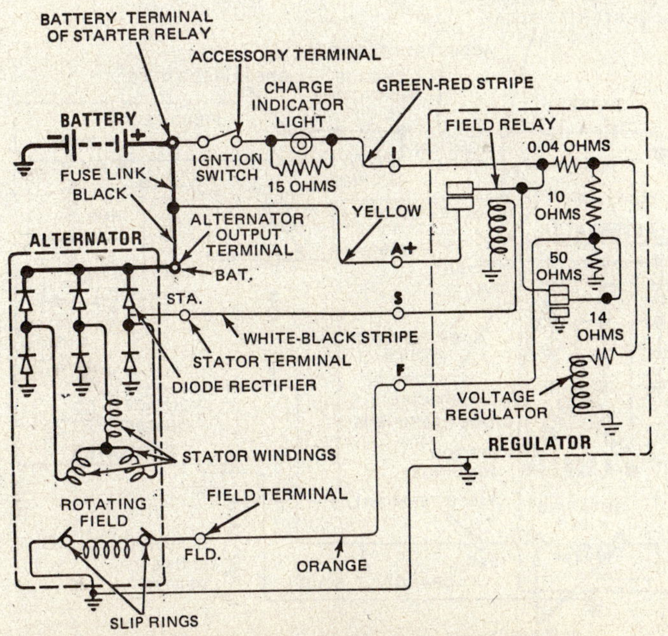

Fig. 3 Indicator light side terminal alternator charging system. 65 amp system

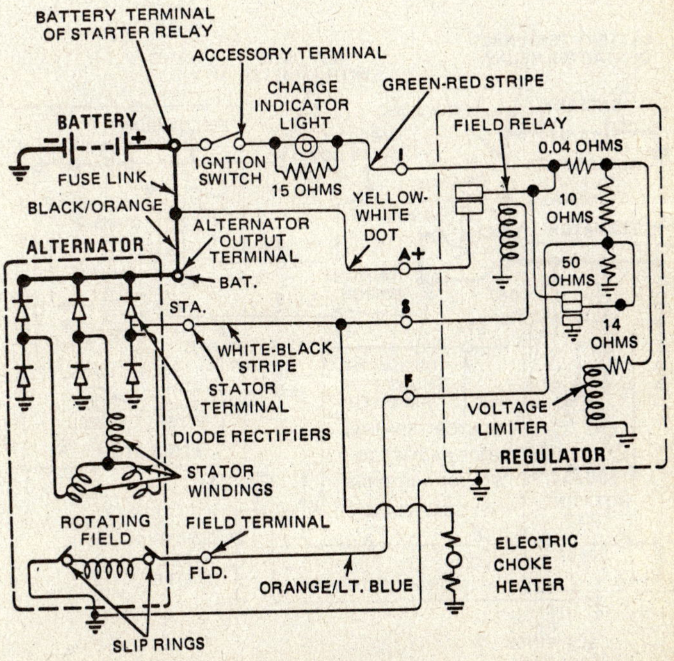

Fig. 3A Indicator light side terminal alternator charging system. 70 amp system

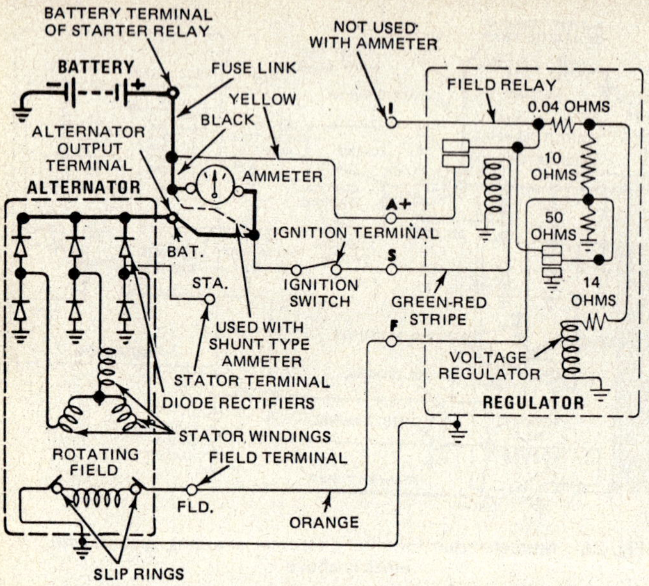

Fig. 4 Ammeter side terminal alternator charging system. 65 amp system

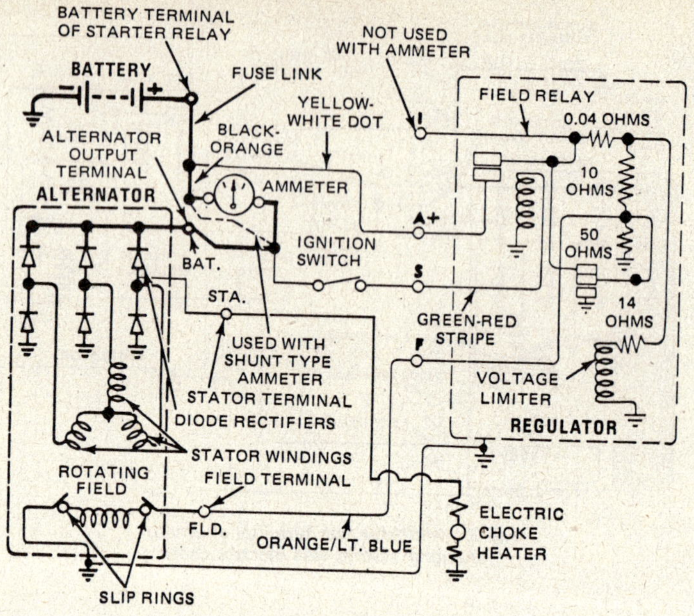

Fig. 4A Ammeter side terminal alternator charging system. 70 amp system

ring and the other probe to the rotor shaft. The resistance should be infinite. Any reading other than infinite indicates a short to ground. Inspect the slip ring soldered terminals to make certain that they are not bent and touching the shaft, or the excess solder is not grounding the rotor coil.

If the rotor checks indicate that it is in proper operating condition but the overall test, Fig. 8 indicates trouble, the brushes or brush assembly are the cause.

Diode Tests

Disassemble the alternator. Disconnect the rectifier assembly from the stator and connect leads, Fig. 10 or 11. To test one set of diodes, contact one probe to the diode plate and contact each of the three stator lead terminals with the other probe. Reverse the probes and repeat the test. Test the other set of diodes in the same way. On 61-ampere alternators, test the two additional diodes, Fig. 12.

All 6 tests (8 tests on 61-ampere alternator) should show a low reading of approximately 60 ohms in one direction and an infinite reading (no needle movement) with the probes reserved.

Open Or Grounded Stator Coil Tests

These tests are made to determine if the stator coil is operating properly. Disassemble the stator from the alternator and rectifier assembly.

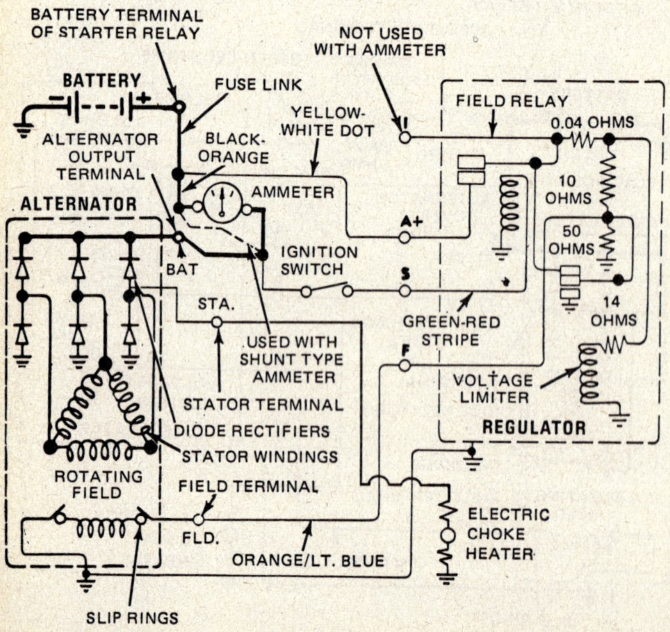

Fig. 5 Ammeter side terminal alternator. 90 amp system

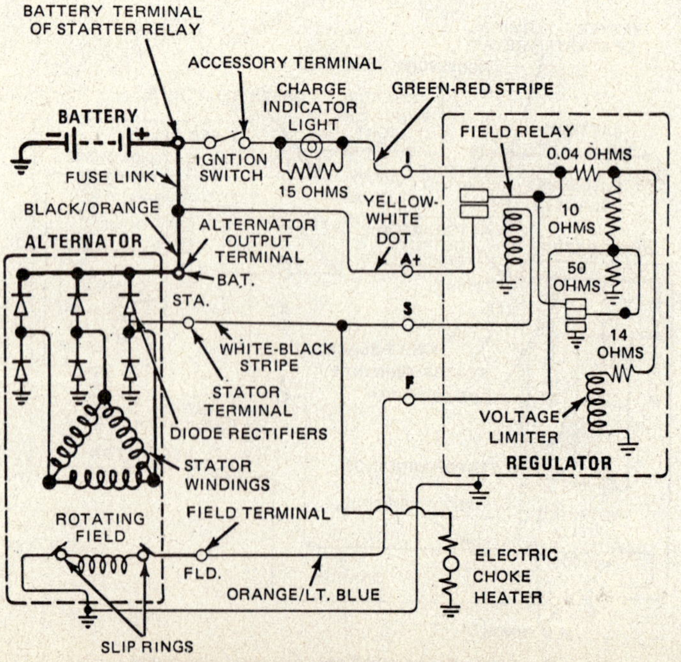

Fig. 6 Indicator light side terminal alternator. 90 amp system

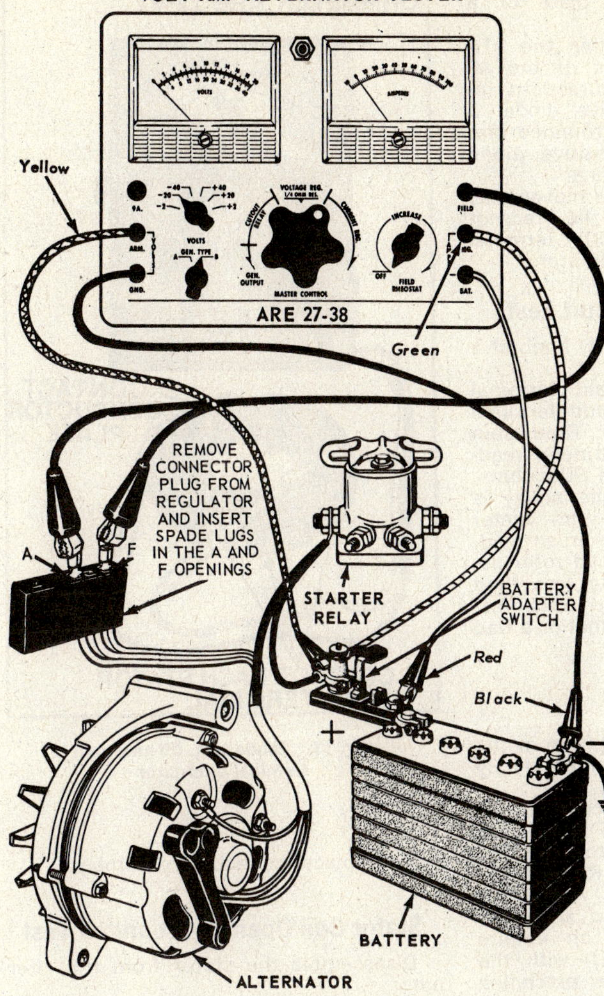

Fig. 7 Alternator output test

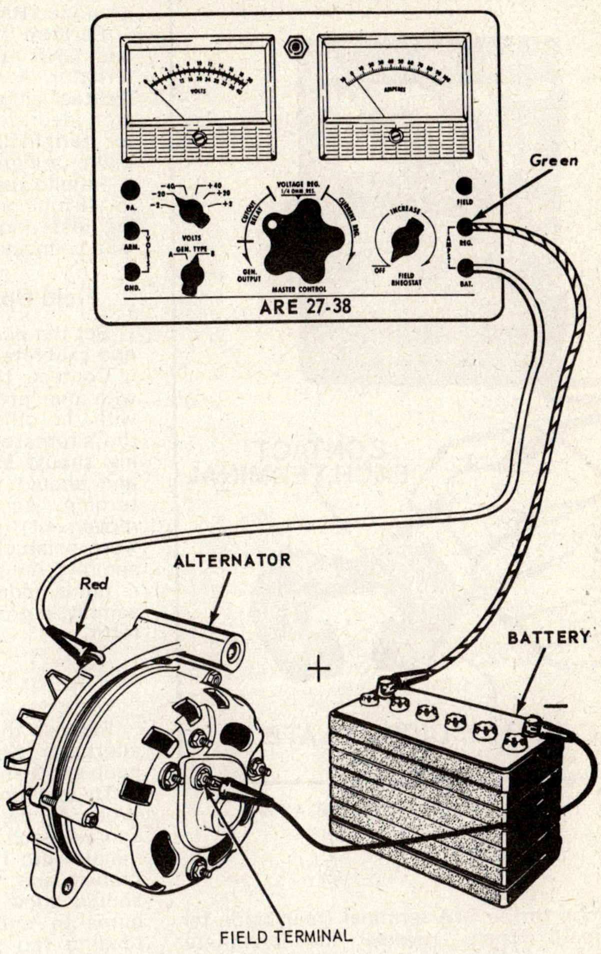

Fig. 8 Alternator field open or short circuit test

Open Stator Test

Set ohmmeter multiple-by knob at 1. Connect the ohmmeter probes between each pair of stator leads. If the ohmmeter does not show equal readings between each pair of stator leads, the stator is open and must be replaced.

Grounded Stator Test

Connect the ohmmeter probes to one of the stator leads and to the stator laminated core. Be sure that the probe makes a good electrical connection with the stator core. The metal should show an infinite reading (no meter movement). If the meter does not indicate an infinite reading (needle moves), the stator winding is shorted to the core and must be replaced. Repeat this test for each of the stator leads.

SIDE TERMINAL ALTERNATOR

Rectifier Short or Grounded and Stator Grounded Test

Set ohmmeter Multiply By knob at 10, and calibrate meter.

Contact one ohmmeter probe to the alternator BAT terminal, Fig. 13, the other

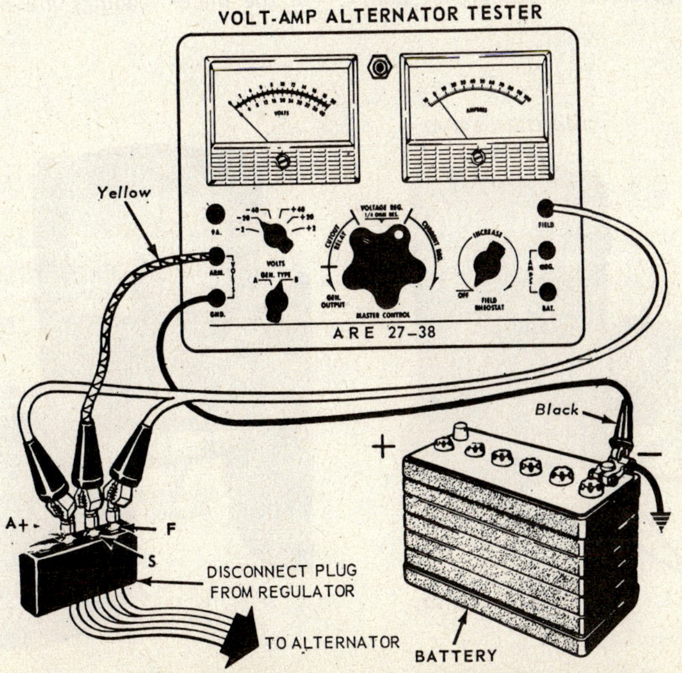

Fig. 9 Typical stator neutral voltage test

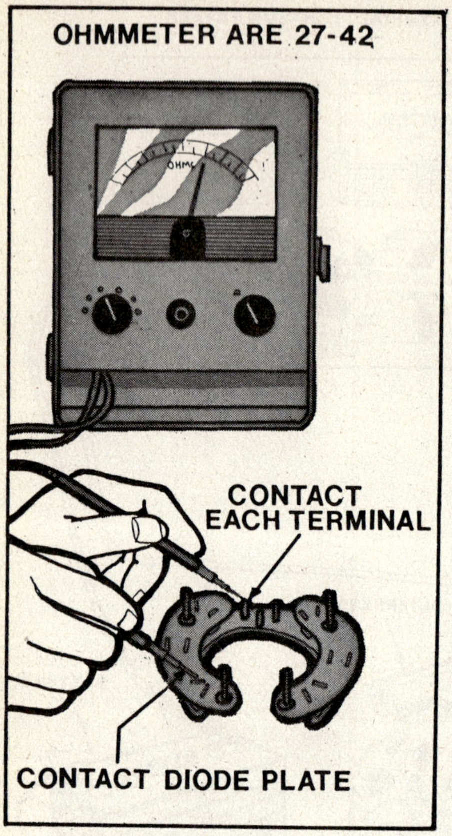

Fig. 10 Diode test. 38, 42 & 55 amp alternators

probe to the STA terminal (rear blade terminal). Then, reverse the ohmmeter probes and repeat the test. A reading of about 60 ohms should be obtained in one direction and no needle movement with the probes reversed. A reading in both directions indicates a bad positive diode, a grounded positive diode plate or a grounded BAT terminal.

Perform the same test using the STA and GND (ground) terminals of the alternator. A reading in both directions indicates either a bad negative diode, a grounded stator winding, a grounded stator terminal, a grounded positive diode plate, or a grounded BAT terminal.

Infinite readings (no needle movement) in all four probe positions in the preceeding tests indicates an open STA terminal lead connection inside the alternator.

Field Open or Short Circuit Test

Set the ohmmeter Multiply By knob at 1 and calibrate meter.

Contact the alternator field terminal with one probe and the ground terminal with the other probe, Fig. 14. Then, spin the alternator pulley. The ohmmeter reading should be between 4 and 200 ohms, and should fluctuate while the pulley is turning. An infinite reading (no meter movement) indicates an open brush lead, worn or stuck brushes, or a bad rotor assembly. An ohmmeter reading less than 4 ohms indicates a grounded brush assembly, a grounded field terminal or a bad rotor.

Diode Test

Remove the rectifier assembly from the alternator. Set the ohmmeter Multiply By knob at 10 and calibrate meter.

To test one set of diodes-contact one probe to the terminal bolt, Fig. 15 and contact each of the three stator lead terminals with the other probe. Reverse the probes and repeat the test. All diodes should show a low reading of about 60 ohms in one direction, and an infinite reading (no needle movement) with the probes reversed. Repeat the preceding tests for the other set of diodes except that the other terminal screw is used.

If the meter readings are not as specified, replace the rectifier assembly.

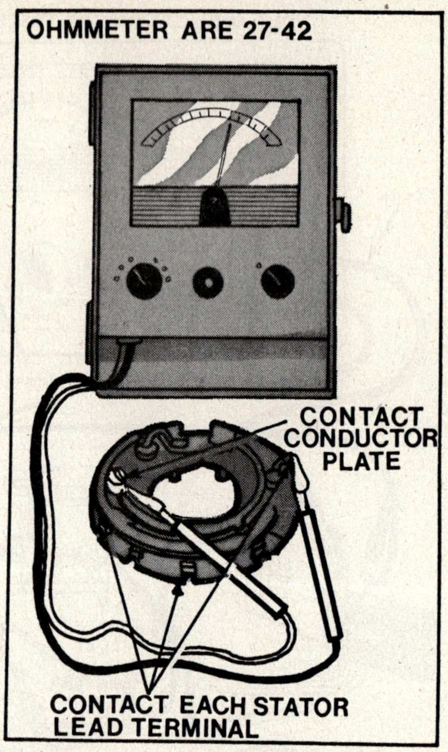

Fig. 11 Diode test. 65-amp rear terminal alternator

Stator Coil Open or Grounded Test

Disassemble the stator from the alternator.

Set ohmmeter Multiply By knob at 1, and calibrate meter. Connect the ohmmeter probes between each pair of stator

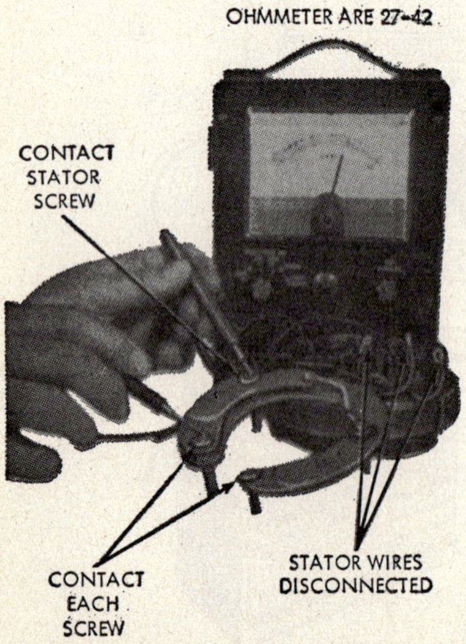

Fig. 12 Booster plate diode test. 61-amp alternator

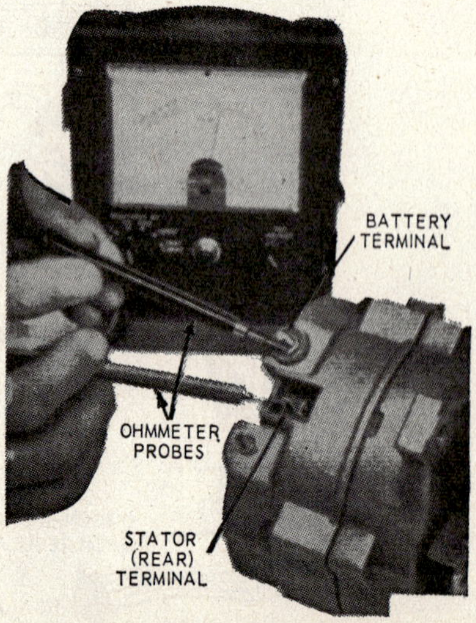

Fig. 13 Rectifier short or grounded and stator grounded test

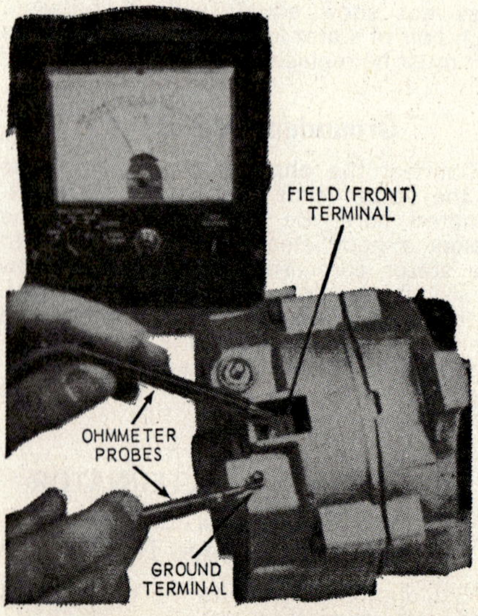

Fig. 14 Field open or short circuit test

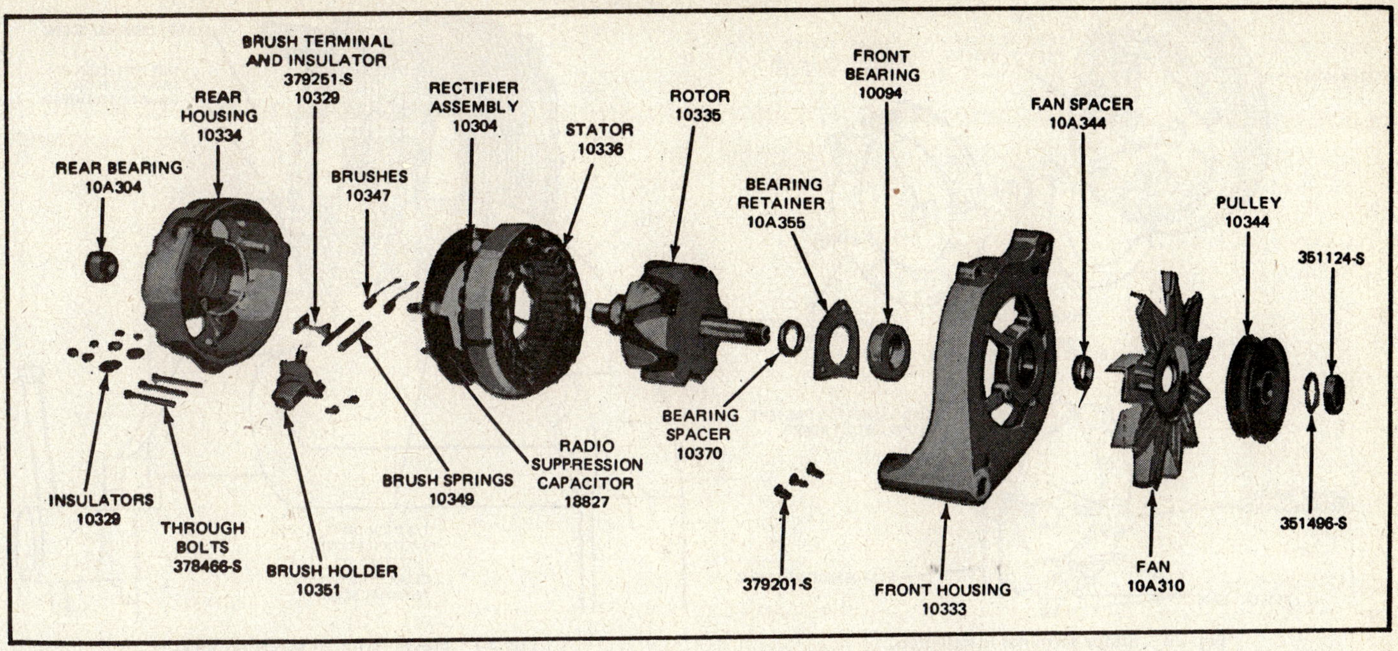

Fig. 27 Disassembled rear terminal alternator exc. 65 amp.

Assembly

NOTE: Refer to "Cleaning and Inspection" procedures before reassembly.

1. The rotor, stator and bearings must not be cleaned with solvent. Wipe these parts off with a clean cloth.
2. Press the front bearing in the front housing bearing boss (put pressure on the outer race only), and install the bearing retainer, Fig. 27.
3. If the stop-ring on the rotor drive shaft was damaged, install a new stop-ring. Push the new ring on the shaft and into the groove.

NOTE: *Do not open the ring with snap ring pliers as permanent damage will result.*

4. Position the rotor stop on the drive shaft with the recessed side against the stop-ring.
5. Position the front housing, fan spacer, fan, pulley and lock washer on the drive shaft and install the retaining nut. Torque the retaining nut, Fig. 31, to 60-100 ft lbs.
6. If the rear housing bearing was removed, support the housing on the inner boss and press in a new bearing flush with the outer end surface.
7. Place the brush springs, brushes, brush terminal and terminal insulator in the brush holder and hold the brushes in position by inserting a piece of stiff wire in the brush holder, Fig. 32.
8. Position the brush holder assembly in the rear housing and install the mounting screws. Position the brush leads in the brush holder, Fig. 33.
9. Wrap the three stator winding leads around the circuit board terminals and solder them. Position the stator neutral lead eyelet on the stator terminal screw and install the screw in the rectifier assembly, Fig. 34.
10. For a rectifier with the diodes exposed insert the special screws through the wire lug, dished washers and circuit board, Fig. 28. Turn them ¼ turn counterclockwise to lock them. For single circuit boards with built in diodes, insert the screws straight through the wire lug, insulating washer and rectifier into the insulator, Fig. 29.

NOTE: The dished washers are to be used only on the circuit board with exposed diodes, Fig. 28. If they are used on the single circuit board, a short circuit will occur. A flat insulating washer is to be used between the stator terminal and the board when a single circuit board is used, Fig. 29.

11. For a rectifier with a booster diode plate (61-Ampere Alternator only), proceed as follows:
    a. Position the stator wire terminal on the stator terminal screw and position the screw into the rectifier. Position the square insulator over the screw and into the square hole in the rectifier, Fig. 35.
    b. Rotate the terminal screw until it locks in position. Then, press the screw in finger tight.
    c. Position the stator wire, Fig. 36. Press the terminal screw into the rectifier and insulator, Fig. 37.
12. Position the radio noise suppression capacitor on the rectifier terminals. On the circuit board with exposed diodes, install the STA and DAT terminal insulators, Fig. 34. On the single circuit board, position the square stator-terminal insulator in the square hole in the rectifier assembly, Fig. 29.

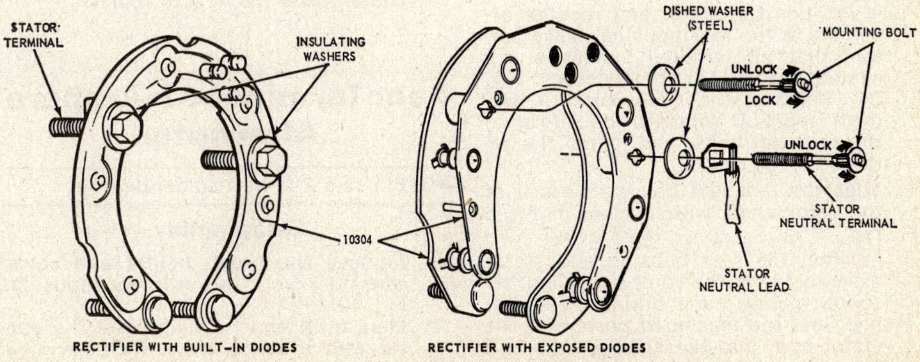

Fig. 28 Rectifier assembly

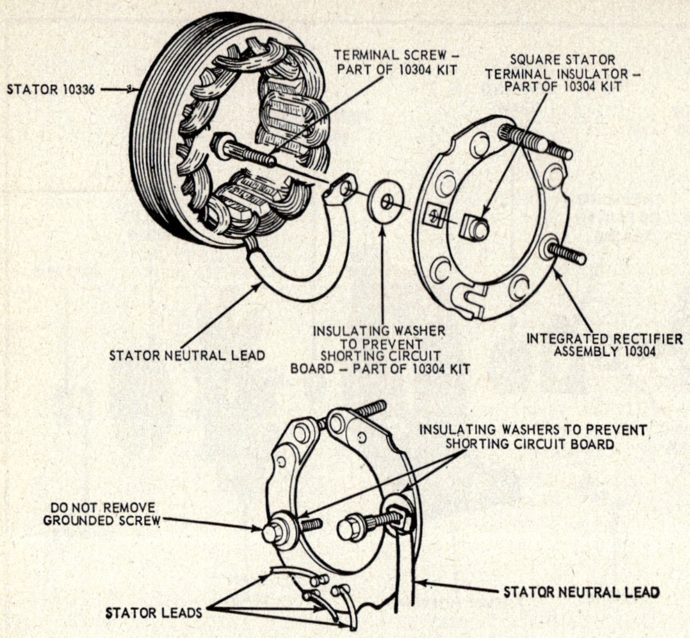

Fig. 29 Stator terminal installation. Integral rectifier circuit board

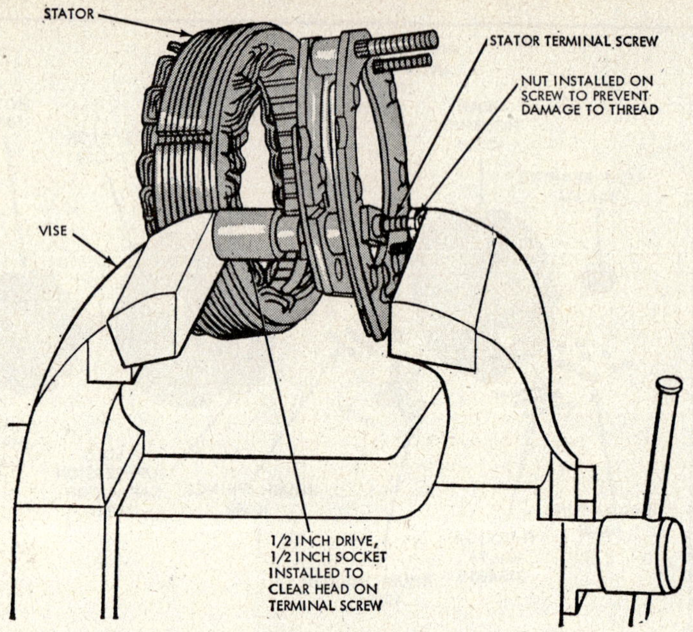

Fig. 30 Stator terminal screw removal. 61-amp alternator

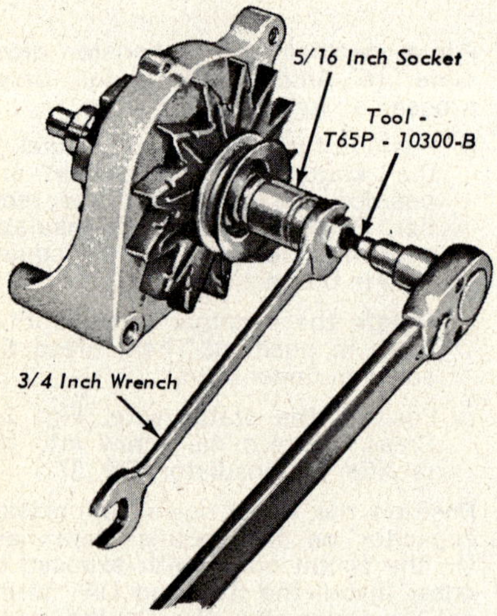

Fig. 31 Typical pulley removal

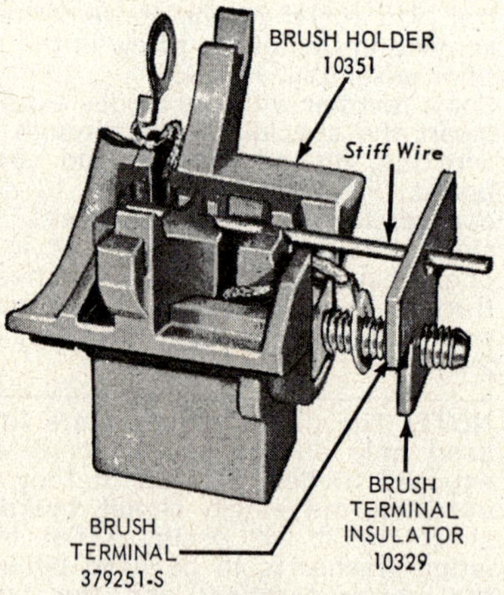

Fig. 32 Brush holder assembly

Fig. 33 Typical brush lead positions

Position the BAT terminal insulator, Fig. 38.

Position the stator and rectifier assembly in the rear housing. Make certain that all terminal insulators are seated properly in the recesses, Fig. 34. Position the STA (black), BAT (red) and FLD (orange) insulators on the terminal bolts, and install the retaining nuts, Fig. 39.

13. Wipe the rear end bearing surface of the rotor shaft with a clean lint-free rag.
14. Position the rear housing and stator assembly over the rotor and align the scribe marks made during disassembly. Seat the machined portion of the stator core into the step in both end housings. Install the housing through bolts. Remove the brush retracting wire, and put a daub of water-proof cement over the hole to seal it.

Rear Terminal 65-Ampere Alternator

NOTE: Use a 200 watt soldering iron.

Disassembly

1. Remove the brush holder and cover assembly from the rear end housing, Fig. 40.
2. Mark both end housings and the stator with a scribe mark for assembly.
3. Remove the three housing through bolts.
4. Separate the front housing and rotor from the stator and rear housing.
5. Remove the drive pulley nut, lockwasher, flat washer, pulley, fan, fan spacer and rotor from the front housing, Fig. 31.
6. Remove the three screws that hold the front bearing retainer, and remove the retainer. If the bearing is damaged or has lost its lubricant, support the housing close to the bearing boss and press out the bearing from the housing.
7. Remove all the nut and washer assemblies and insulators from the rear housing and remove the rear housing from the stator and rectifier assembly.
8. If replacement is necessary, press the bearing from the rear housing, supporting the housing on the inner boss.
9. Unsolder the three stator leads from

Fig. 34 Stator lead connections. Except 61-amp alternator

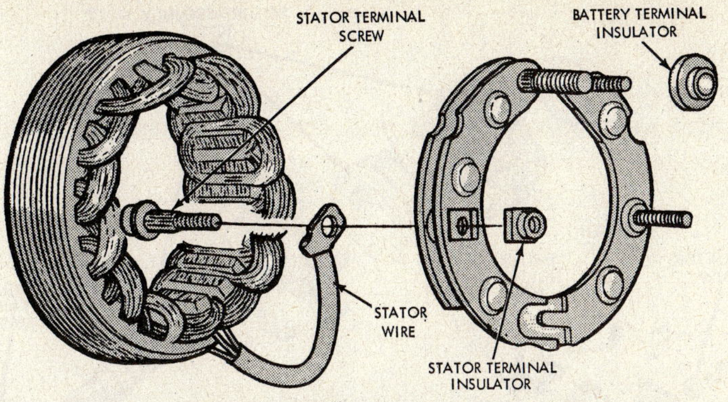

Fig. 35 Stator and rectifier assembly. 61-amp alternator

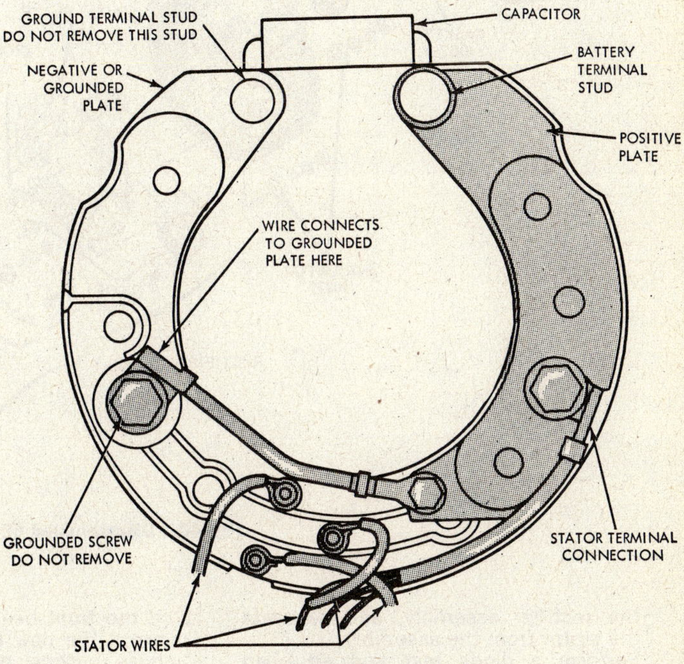

Fig. 36 Rectifier terminal locations. 61-amp alternator

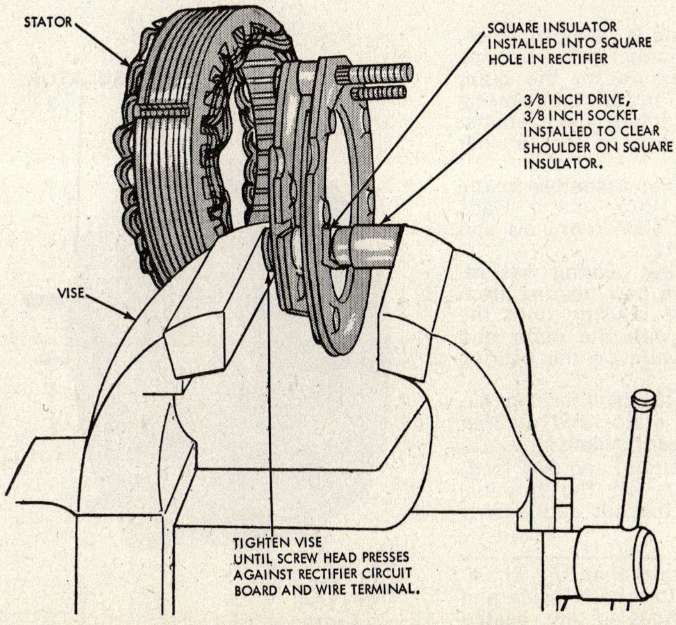

Fig. 37 Stator terminal screw installation. 61-amp alternator

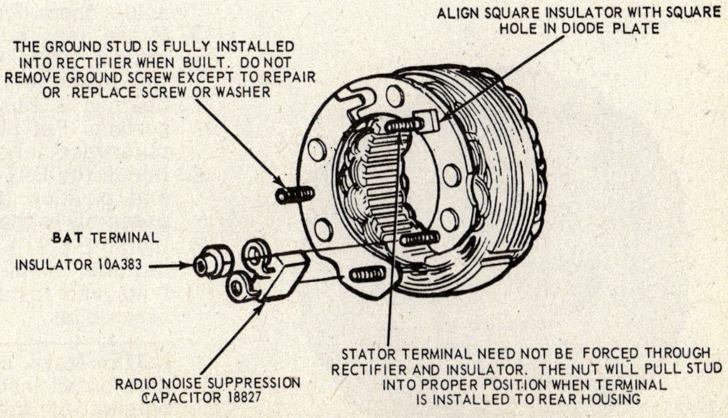

Fig. 38 Fiber-glass circuit board terminal insulators

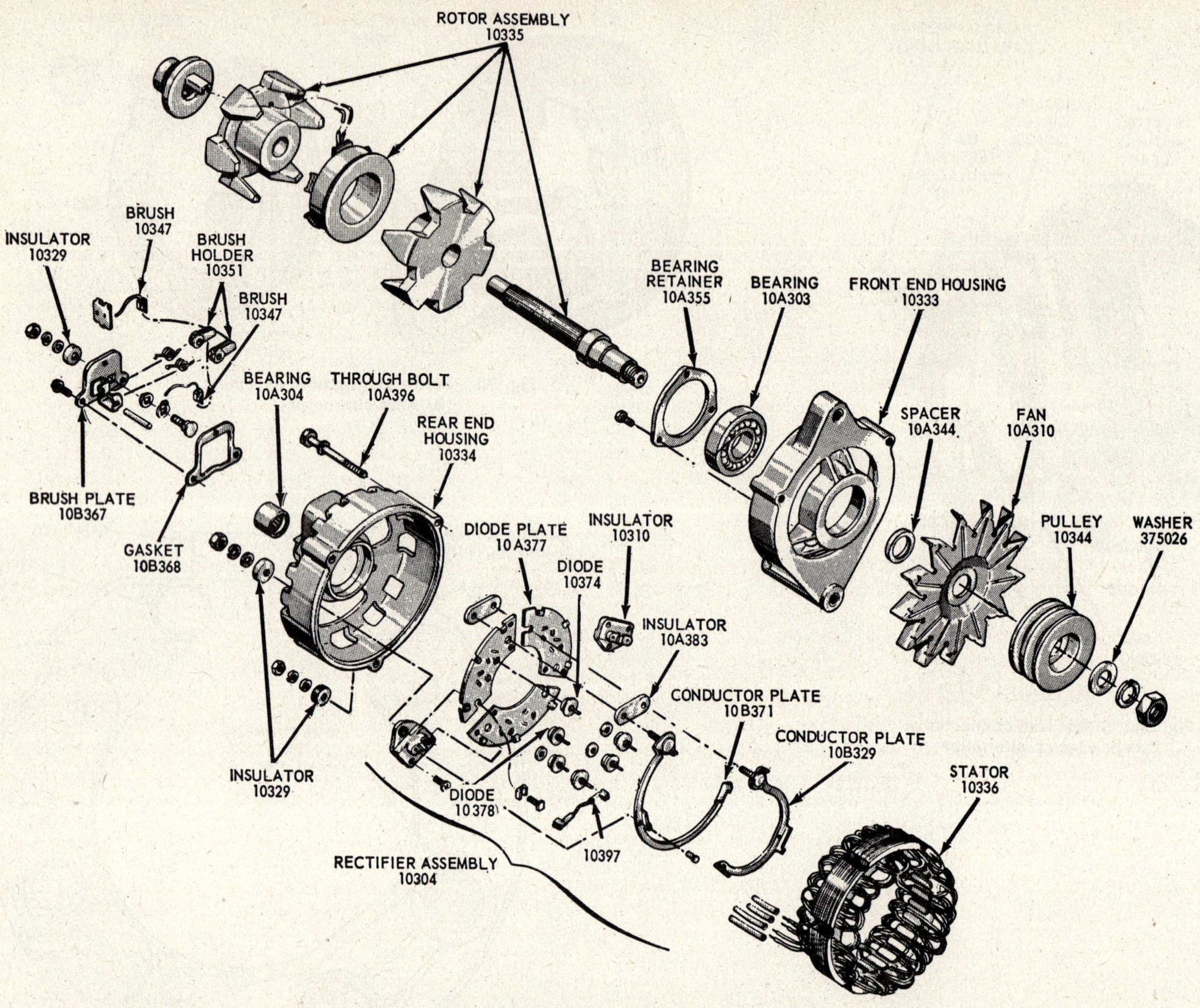

Fig. 40 Disassembled 65-amp rear terminal alternator

the rectifier assembly, and separate the stator from the assembly.

10. Perform a diode test and an open and grounded stator coil test.

Assembly

NOTE: Refer to "Cleaning and Inspection" procedures before reassembly.

Fig. 39 Alternator terminal locations

1. If the front bearing is being replaced, press the new bearing in the bearing boss putting pressure on the outer race only. Then, install the bearing retainer, and tighten the retainer screws until the tips of the retainer touch the housing.
2. Position the rectifier assembly to the stator, wrap the three stator leads around the diode plate terminals and solder them, Fig. 41.
3. If the rear housing bearing was removed, press in a new bearing from the inside of the housing until the bearing is flush with the outer end surface. Put pressure on the bearing outer race only.
4. Install the BAT-GRD insulator, Fig. 41, and position the stator and rectifier assembly in the rear housing.
5. Install the STA (purple) and BAT (red) terminal insulators on the terminal bolts and install the nut and washer assemblies.

NOTE: Make certain that the shoulders on all insulators both inside and outside of the housing are seated properly before tightening the nuts.

Fig. 41 Stator lead connections

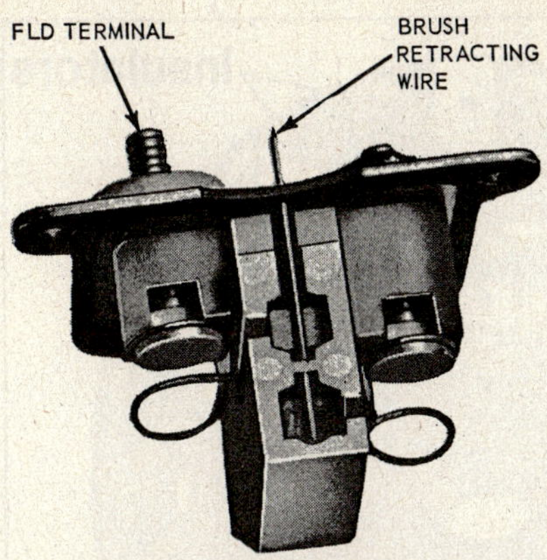

Fig. 42 Field brush assembly 65-amp alternator. 1970-72

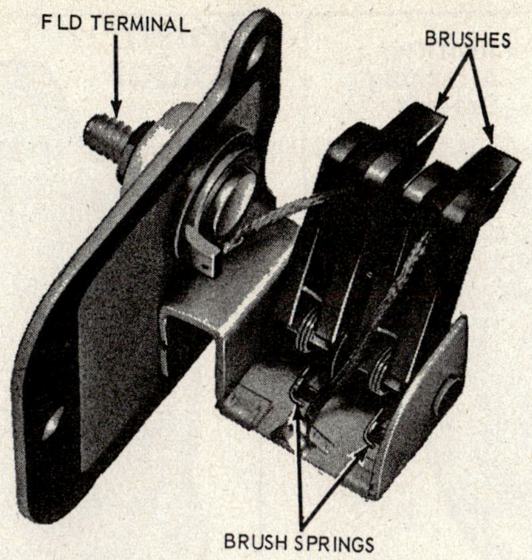

Fig. 43 Field brush assembly 65-amp alternator. 1969

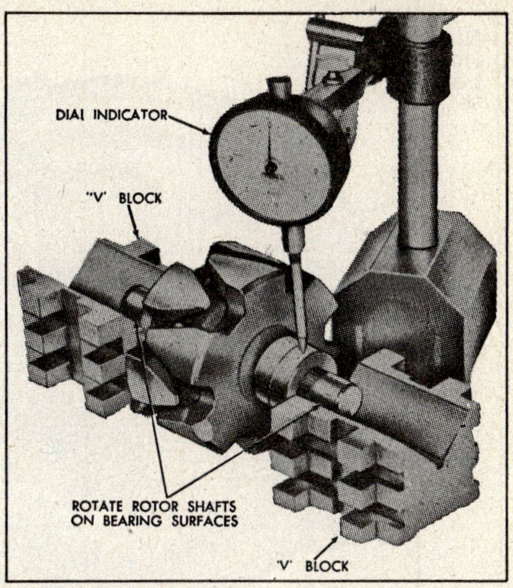

Fig. 44 Checking slip ring runout

6. Position the front housing over the rotor and install the fan spacer, fan, pulley, flat and lock washers and nut on the rotor shaft, Fig. 31.
7. Wipe the rear bearing surface of the rotor shaft with a clean lint free rag.
8. Position the rotor with the front housing into the stator and rear housing assembly, and align the scribe marks made during disassembly. Seat the machined portion of the stator core into the step in both housing, and install the through bolts.
9. On 1970-72 units hold the brushes in position by inserting a stiff wire in the brush holder, Fig. 42.
10. Position the brush holder assembly into the rear housing and install the three mounting screws. On 1970-73 units remove the brush retracting wire and put a daub of water-proof cement over the hole to seal hole.

NOTE: On 1969 units the brush holders rotate on a plastic rod. This rod must be properly seated in the recess in the rear housing.

Brush Replacement

1969 ALTERNATOR

1. Remove the brush holder and mounting plate assembly from the rear end housing.
2. Slide the FLD brush out of its holder and unsolder the brush lead from the terminal lug.
3. Slide the ground brush out of its holder. Unsolder the brush lead from the plate, and remove the brush.
4. Position the new ground brush to the mounting plate and solder the brush lead to the plate. Make sure that the brush springs are hooked over the lugs, Fig. 43. The entire brush and plate assembly is also available for service.
5. Solder the new FLD brush to the FLD stud lug, Fig. 43.
6. Install both brushes in their holders, Fig. 43.
7. Install the brush holder and mounting plate assembly to the rear end housing. Make certain that the plastic rod that the brush holders rotate on is seated in the recess in the rear end housing.

1970-72 ALTERNATOR

1. Remove the brush holder and cover assembly from the rear housing.
2. Remove the terminal bolts from the

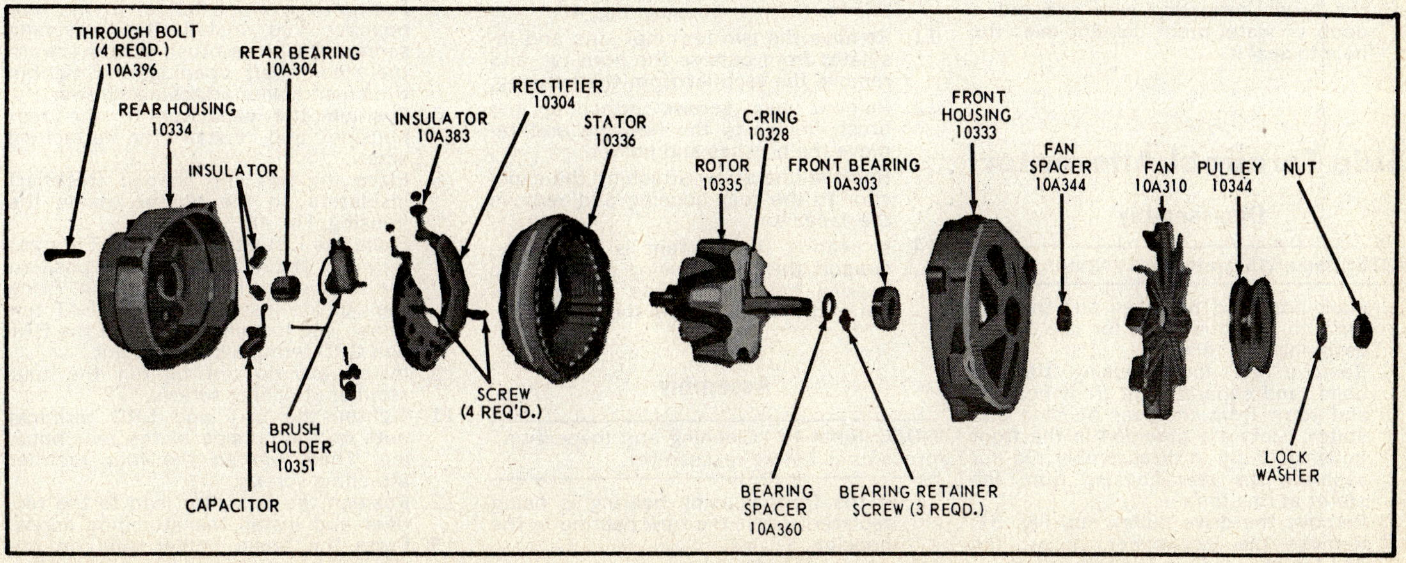

Fig. 45 Disassembled 65-amp side terminal alternator

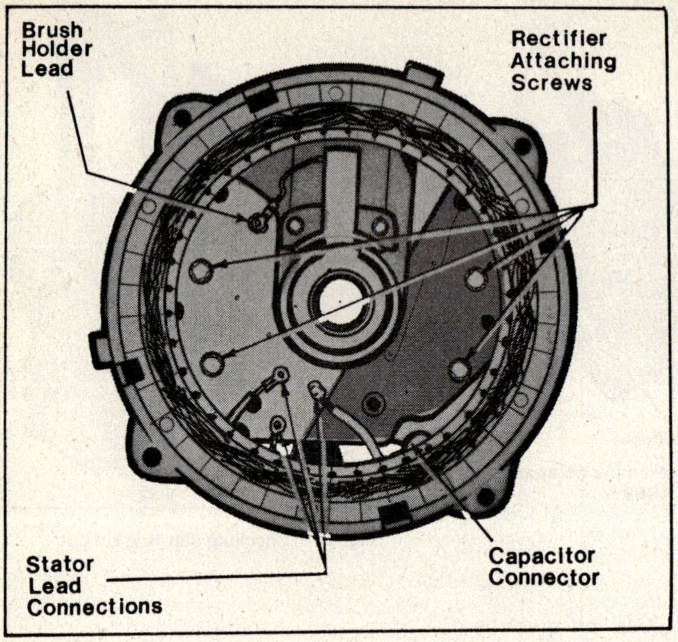

Fig. 46 Stator lead connections

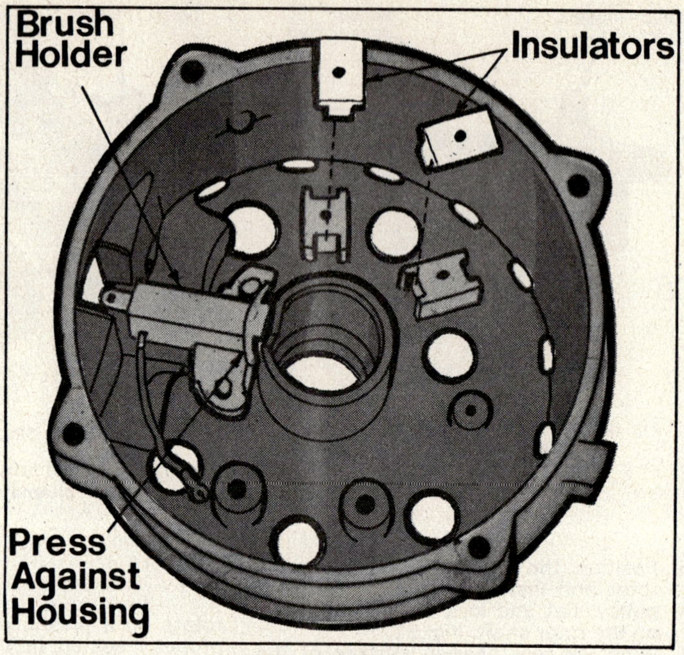

Fig. 48 Brush holder and rectifier insulators installed

brush holder and cover assembly, and remove the brush assemblies.

3. Position the new brush terminals on the terminal bolts and assemble the terminals, bolts, brush holder washers and nuts, Fig. 42. The insulating washer mounts under the FLD terminal nut. The entire brush and cover assembly is also available for service.
4. Depress the brush springs in the brush holder cavities and insert the brushes on top of the springs. Hold the brushes in position by inserting a stiff wire in the brush holder as shown in Fig. 42. Position the brush leads, Fig. 42.
5. Install the brush holder and cover assembly to the rear housing. Remove the brush retracting wire and put a daub of water-proof cement over the hole to seal it.

Side Terminal Alternator

Disassembly

NOTE: Use a 200 watt soldering iron.

1. Mark both end housings and the stator with a scribe mark for use during assembly, Fig. 45.
2. Remove the four housing through bolts, and separate the front housing and rotor from the rear housing and stator. Slots are provided in the front housing to aid in disassembly. *Do not separate the rear housing from the stator at this time.*
3. Remove the drive pulley nut, Fig. 31. Remove the lockwasher, pulley, fan and fan spacer from the rotor shaft.
4. Pull the rotor and shaft from the front housing, and remove the spacer from the rotor shaft, Fig. 45.
5. Remove three screws retaining the bearing to the front housing. If the bearing is damaged or has lost its lubricant, remove the bearing from the housing. To remove the bearing, support the housing close to the bearing boss and press the bearing from the housing.
6. Unsolder and disengage the three stator leads from the rectifier, Fig. 46.
7. Lift the stator from the rear housing.
8. Unsolder and disengage the brush holder lead from the rectifier.
9. Remove the screw attaching the capacitor lead to the rectifier.
10. Remove four screws attaching the rectifier to the rear housing, Fig. 46.
11. Remove the two terminal nuts and insulator from outside the housing, and remove the rectifier from the housing.
12. Remove two screws attaching the brush holder to the housing and remove the brushes and holder.
13. Remove one screw attaching the capacitor to the rear housing and remove the capacitor.
14. If bearing replacement is necessary, support the rear housing close to the bearing boss and press the bearing out of the housing from the inside.

Assembly

NOTE: Refer to "Cleaning and Inspection" procedures before reassembly.

1. If the front housing bearing is being replaced, press the new bearing in the housing.

NOTE: *Put pressure on the bearing outer race only.* Then, install the bearing retaining screws.

2. Place the inner spacer on the rotor shaft and insert the rotor shaft into the front housing and bearing.
3. Install the fan spacer, fan, pulley, lockwasher and nut on the rotor shaft, Fig. 31. Torque nut to 60-100 ft. lbs.
4. If the rear bearing is being replaced, press a new bearing in from inside the housing until it is flush with the boss outer surface.
5. Position the brush terminal on the brush holder, Fig. 47. Install the springs and brushes in the brush holder, and insert a piece of stiff wire to hold the brushes in place, Fig. 47.
6. Position the brush holder in the rear housing and install the attaching screws. Push the brush holder toward the rotor shaft opening and tighten the brush holder attaching screws.
7. Position the capacitor to the rear housing and install the attaching screw.
8. Place the two cup shaped (rectifier) insulators on the bosses inside the housing, Fig. 48.
9. Place the insulator on the BAT (large) terminal of the rectifier, and position the rectifier in the rear housing. Place the outside insulator on the BAT terminal, and install the nuts on the BAT and GRD terminals *finger tight*.
10. Install but do not tighten the four rectifier attaching screws.
11. Tighten the BAT and GRD terminal nuts on the outside of the rear housing. Then, tighten the four rectifier attaching screws.
12. Position the capacitor lead to the rectifier and install the attaching screw.
13. Press the brush holder lead on the rectifier pin and solder securely, Fig. 46.

14. Position the stator in the rear housing and align the scribe marks. Press the three stator leads on the rectifier pins and solder securely, Fig. 46.
15. Position the rotor and front housing into the stator and rear housing. Align the scribe marks and install the four through bolts. Tighten two opposing bolts and then the two remaining bolts.
16. Spin the fan and pulley to be sure nothing is binding within the alternator.
17. Remove the wire retracting the brushes, and place a daub of waterproof cement over the hole to seal it.

Brush Replacement

Removal

1. Mark both end housings and the stator with a scribe mark for use during assembly.
2. Remove the four housing through bolts, and separate the front housing and rotor from the rear housing and stator. Slots are provided in the front housing to aid in disassembly.

NOTE: Do not separate the rear housing and stator.

3. Unsolder and disengage the brush holder lead from the rectifier.
4. Remove the two brush holder attaching screws and lift the brush holder from the rear housing.
5. Remove the brushes from the brush holder.

Installation

1. Insert the brushes into the brush holder and position the terminal on the brush holder.
2. Depress the brushes and insert a 1½ inch piece of stiff wire, Fig. 47, to hold the brushes in the retracted position.
3. Position the brush holder to the rear housing, inserting the wire used to retract the brushes through the hole in the rear housing.
4. Install the brush holder attaching screws. Push the brush holder toward the rotor shaft opening and tighten the attaching screws.
5. Press the brush holder lead on the rectifier pin and solder securely.
6. Position the rotor and front housing into the stator and rear housing. Align the scribe marks and install the four through bolts. Tighten two opposing bolts and then the two remaining bolts.
7. Spin the fan and pulley to be sure nothing is binding within the alternator.
8. Remove the wire retracting the brushes, and place a daub of waterproof cement over the hole to seal it.

Rectifier Replacement

Removal

1. Mark both end housings and the stator with a scribe mark for use during assembly, Fig. 45.
2. Remove the four housing through bolts, and separate the front housing and rotor from the rear housing and stator. Slots are provided in the front housing to aid in disassembly.

NOTE: *Do not separate the rear housing and stator at this time.*

3. Unsolder and disengage the three stator leads from the rectifier, Fig. 46. Lift the stator from the rear housing.
4. Unsolder and disengage the brush holder lead from the rectifier.
5. Remove the screw attaching the capacitor lead to the rectifier.
6. Remove four screws attaching the rectifier to the rear housing, Fig. 46.
7. Remove two terminal nuts and insulator from outside the housing, and remove the rectifier from the housing.

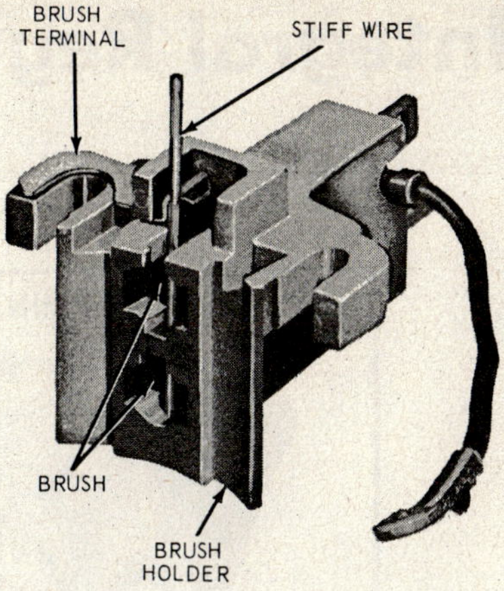

Fig. 47 Brush holder assembly

Installation

1. Insert a piece of wire through the hole in the rear housing to hold the brushes in the retracted position.
2. Place the two cup shaped (rectifier) insulators on the bosses inside the housing, Fig. 48.
3. Place the insulator on the BAT (large) terminal) of the rectifier, and position the rectifier in the rear housing. Place the outside insulator on the BAT terminal, and install the nuts on the BAT and GRD terminals finger tight.
4. Install but do not tighten the four rectifier attaching screws.
5. Tighten the BAT and GRD terminal nuts on the outside of the rear housing. Then, tighten the four rectifier attaching screws.
6. Position the capacitor lead to the rectifier and install the attaching screw.
7. Press the brush holder lead on the rectifier pin and solder securely, Fig. 46.
8. Position the stator in the rear housing and align the scribe marks. Press the three stator leads on the rectifier pins and solder securely, Fig. 46.
9. Position the rotor and front housing into the stator and rear housing. Align the scribe marks and install the four through bolts. Partially tighten all four through bolts. Then, tighten two opposing bolts and then the two remaining bolts.
10. Spin the fan and pulley to be sure nothing is binding within the alternator.
11. Remove the wire retracting the brushes in the brush holder, and place a daub of waterproof cement over the hole in the rear housing to seal it.

Cleaning and Inspection Procedures

1. The rotor, stator, and bearings must not be cleaned with solvent. Wipe these parts off with a clean cloth.
2. Rotate the front bearing on the drive end of the rotor drive shaft. Check for any scraping noise, looseness or roughness that will indicate that the bearing is excessively worn. Look for excessive lubricant leakage. If any of these conditions exist, replace the bearing.
3. Inspect the rotor shaft at the rear bearing surface for roughness or severe chatter marks. Replace the rotor assembly if the shaft is not smooth.
4. Place the rear end bearing on the slip-ring end of the shaft and rotate the bearing on the shaft. Make the same check for noise, looseness or roughness as was made for the front bearing. Inspect the rollers and cage for damage. Replace the bearing if these conditions exist, or if the lubricant is lost or contaminated.
5. Check the pulley and fan for excessive looseness on the rotor shaft. Replace any pulley or fan that is loose or bent out of shape. Check the rotor shaft for stripped or damaged threads. Inspect the hex hole in the end of the shaft for damage.
6. Check both the front and rear housing for cracks. Check the front housings for stripped threads in the mounting gear. Replace defective housings.
7. Check all wire leads on both the stator and rotor assemblies for loose soldered connections, and for burned insulation. Resolder poor connections. Replace parts that show burned insulation.
8. Check the slip rings for nicks and surface roughness. Check the slip rings for runout, Fig. 44. Nicks, scratches and slight slip ring runout may be removed by turning down the slip rings. Do not go beyond the minimum diameter limit. If the slip rings are badly damaged, the entire rotor will have to be replaced, as it is serviced as a complete assembly.
9. Replace any parts that are burned or cracked. Replace brushes and brush springs that are not to specification.

Ford Autolite/Motorcraft Alternator With Integral Regulator

DESCRIPTION

Alternator

The alternator used with this unit is basically the same as the standard Ford Autolite unit. Modifications have been made in the brush holder assembly, rear end frame and the stator assembly to accommodate the integral regulator. Except for tests outlined here, refer to preceding chapter for service procedures.

Integral Regulator

The integral regulator, Fig. 1, consists of an integrated, solid state circuit, made up of transistors, diodes and resistors, all connected by aluminum conductors and fabricated within a ⅛″ square silicon crystal. This is a one piece, non-adjustable unit which must be replaced if it malfunctions, or if it is not calibrated within specified limits of 13.5 to 15.3 volts between 50 and 125 degrees F.

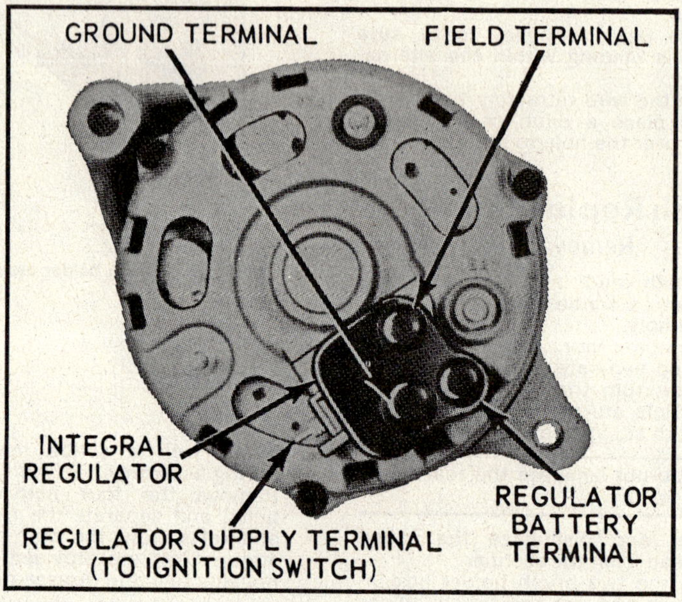

Fig. 1 Alternator with integral regulator

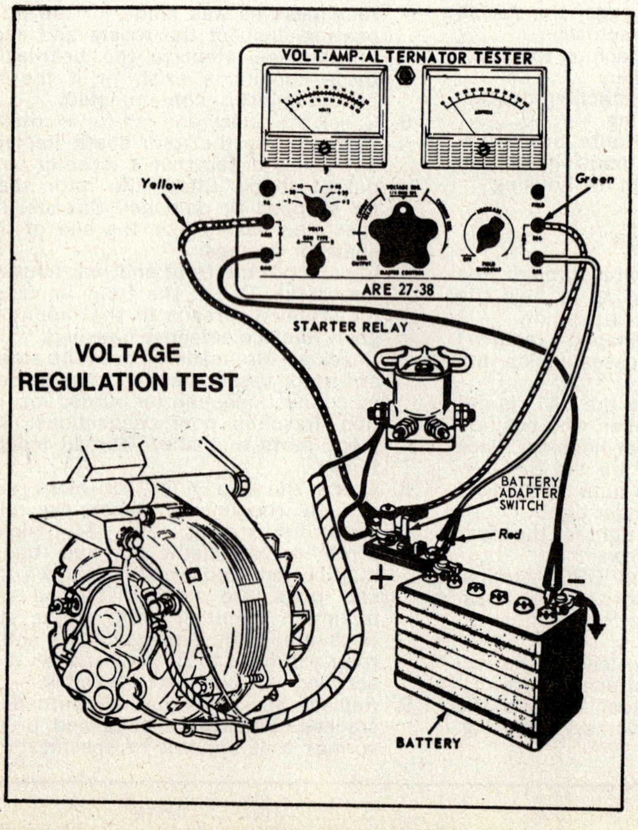

Fig. 2 Voltage regulator test

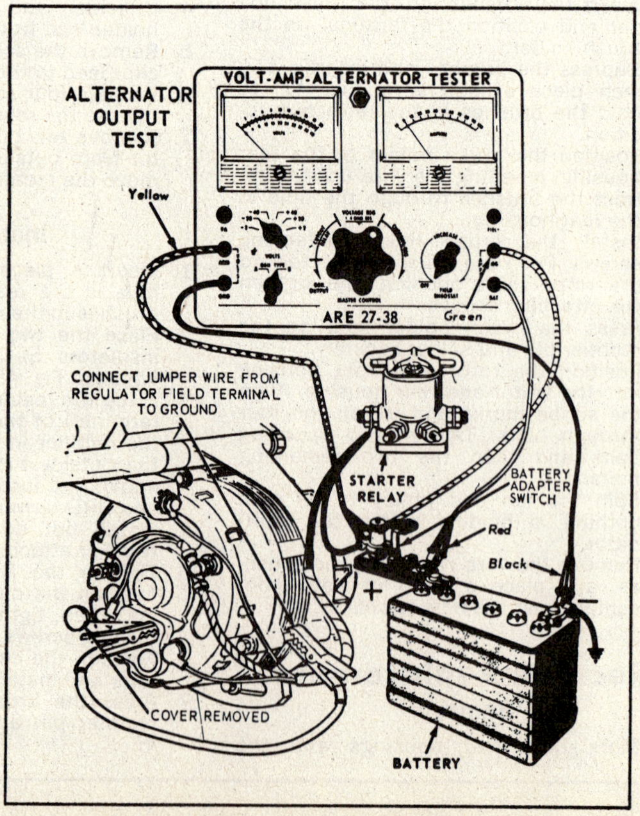

Fig. 3 Alternator output test

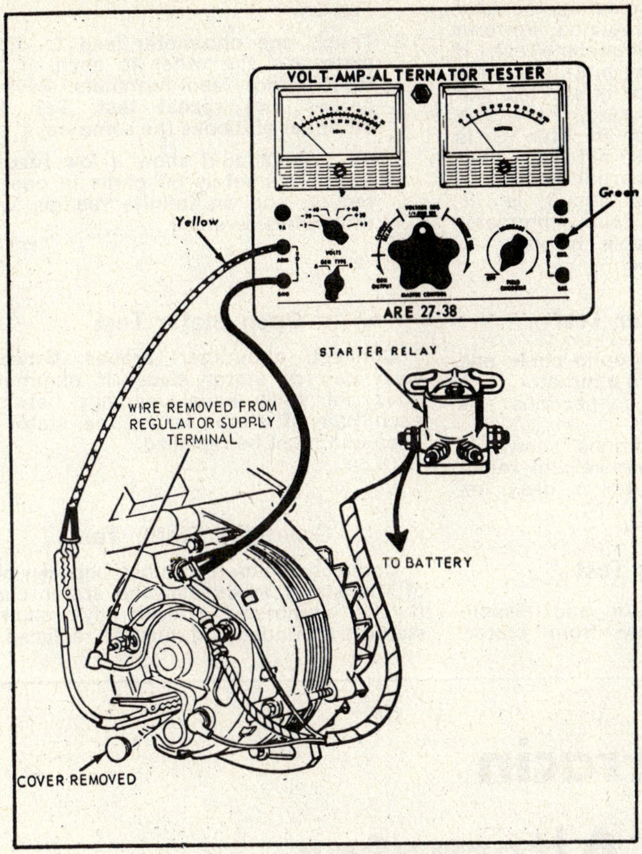

Fig. 4 Field voltmeter test

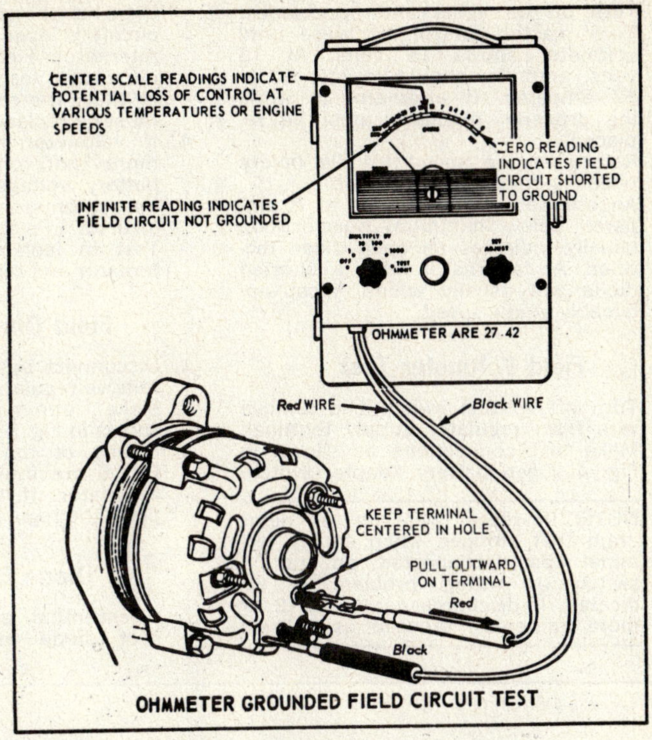

Fig. 5 Ohmmeter field circuit test

SYSTEM TESTS

NOTE: Because the voltage sensing circuit is permanently connected across the charging system, resulting in a small but harmless current drain, a voltmeter cannot be connected in series with the battery for diagnosis. The integral regulator is not defective and should not be replaced because of leakage indicated by a voltmeter connected in series with the battery or regulator.

Voltage Regulator Test

1. Using a fully charged battery, turn off all lights and accessories. Be sure ignition switch is off and make test connections as shown in Fig. 2.
2. Open battery adapter switch. Ammeter should show zero amperes. A discharge (2 amperes) indicates a malfunction in the alternator field coil or the regulator. Refer to Field Circuit Tests. If ammeter shows zero, proceed as follows:
3. With transmission in neutral or park and parking brake applied, place tester master control at the ¼ohm resistor position.
4. Close battery adapter switch and start engine. Be sure all lights and accessories are off and open battery adapter switch.
5. Operate engine at approximately 2000 rpm for 5 minutes, and check voltage. If voltage is between 13.3 and 15.3 volts, the regulator is functioning satisfactorily. If voltage does not rise above battery voltage, check regulator supply voltage. If voltage exceeds 15.3 volts, perform field circuit tests.

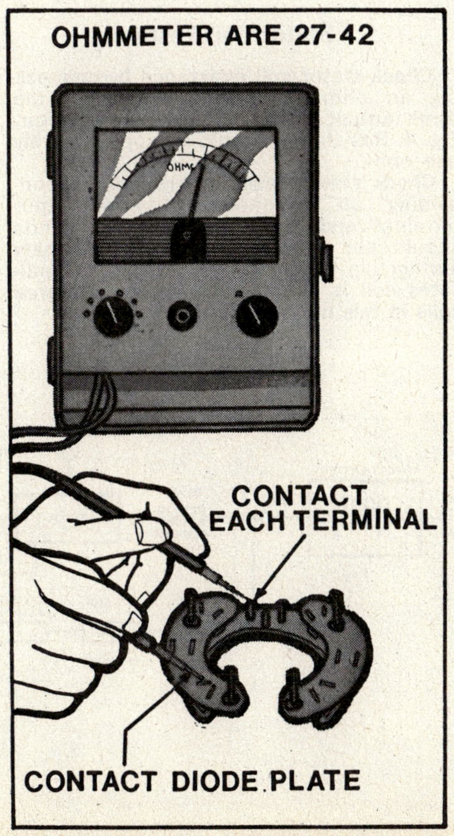

Fig. 6 Diode test

Supply Voltage Test

Check for voltage supply terminal of the alternator with a 12 volt test light or a voltmeter. If no voltage is indicated, the supply circuit is disconnected or broken. If voltage is present, perform Alternator Output Test.

Alternator Output Test Off Vehicle

When using test bench, refer to manufacturer's procedures and be sure to disconnect battery cable as the alternator output connector is always connected to the battery.

Alternator Output Test On Vehicle

NOTE: Under no circumstances should the regulator battery terminal be connected to the regulator field terminal. To do so will damage regulator.

1. With transmission in neutral or park and parking brake applied, make test connections as shown in Fig. 3.
2. Close battery adapter switch. Start engine and reopen adapter switch. Voltage reading must be maintained between 10 and 15 volts.

3. Increase engine speed to 2000 rpm. Turn off all lights and accessories.
4. Turn master control clockwise until voltmeter shows 15 volts. At 15 volts, ammeter should register 50 to 57 amperes. If alternator is working properly, regulator must be replaced.
5. Return engine speed to idle before releasing master control knob.
6. An alternator output of 2 to 8 amperes below minimum specification usually indicates an open diode rectifier. An alternator with a shorted diode will usually whine, most noticeably at idle speed.

Field Voltmeter Test

1. Turn off ignition switch and remove wire from regulator supply terminal.
2. Make test connections as shown in Fig. 4. Open battery adapter switch.

NOTE: if there was an ammeter drain that stopped when supply terminal was disconnected, an ignition switch or wiring problem is indicated. If discharge continues (2 or more amperes), proceed as follows:

3. Voltmeter should read 12 volts. If there is no voltage reading, the field circuit is open or grounded. Perform Alternator Field Ohmmeter Test. If ohmmeter tests show alternator field is okay, the regulator is shorted and must be replaced.
4. If voltmeter reading in Step 3 is more than one volt but less than battery voltage, a partial ground in the alternator field circuit is indicated. Perform Field Ohmmeter Test to isolate trouble between alternator and regulator.

Field Ohmmeter Test

1. Disconnect battery ground cable and remove regulator from alternator.
2. Make ohmmeter connections as shown in Fig. 5.
3. If any of the conditions shown in Fig. 5 are found, remove and repair alternator. If alternator is okay, replace the regulator.

Diode Bench Test

1. Disassemble alternator and disconnect diode assembly from stator. Make test connections as shown in Fig. 6.
2. Touch one ohmmeter lead to diode plate and the other to each of the three stator lead terminals. Reverse probes and repeat test. Test the other set of diodes the same way.
3. All tests should show a low reading of approximately 60 ohms in one direction and an infinite reading with the probes reversed.

Open Stator Test

Connect ohmmeter probes between each pair of stator leads. If ohmmeter does not show equal readings between each pair of stator leads, the stator is open and must be replaced.

Grounded Stator Test

Connect ohmmeter probes between one of the stator leads and the stator core. If the ohmmeter shows any reading, stator is grounded and must be replaced.

Ford Sierracin High Voltage Alternator

DESCRIPTION

This alternator, used with the Ford Sierracin system, Figs. 1 and 1A delivers approximately 120 volts to heat the front and rear windows on some Ford Co. vehicles. This system is completely isolated from the main vehicle system to prevent high voltage feed into the main system. Plastic shielded wiring is used between the alternator output terminals and the window heating elements, Fig. 2. Note that a high voltage warning tag is attached to the wiring at each junction.

ALTERNATOR TESTING

CAUTION: Since this is a 120 volt system and danger of electrical shock is present, all testing must be done with the field lead disconnected, using an ohmmeter and the engine stopped. Do not attempt to check alternator output voltage at the output terminal.

Field Coil Test

Connect an ohmmeter to the field terminal stud and the rear housing, avoiding contact with epoxy on housing since this will act as an insulator, Fig. 3. Field coil resistance is 2.8 ohms, however, reading will vary with brush contact resistance. A resistance between 3 and 250 ohms indicates the field circuit is satisfactory.

Stator Coil Tests

Check stator coil resistance by connecting an ohmmeter between two of the three output terminals in the connector, Fig. 4. Resistance should be approximately one ohm.

Check stator coils for grounds by connecting an ohmmeter to one output terminal and the rear housing. Hands should not contact probes or terminals during this test. A reading of infinity indicates coil is satisfactory. Check all three coils in this manner.

Field Circuit Voltage Test

Remove field terminal connector and connect a voltmeter between metal termination and rear housing. A voltage of 12 to 13 volts should be noted with the system control switch in the "On" position and zero volts in the "Off" position.

ALTERNATOR REPAIRS

Disassembly

1. Scribe reference marks between end

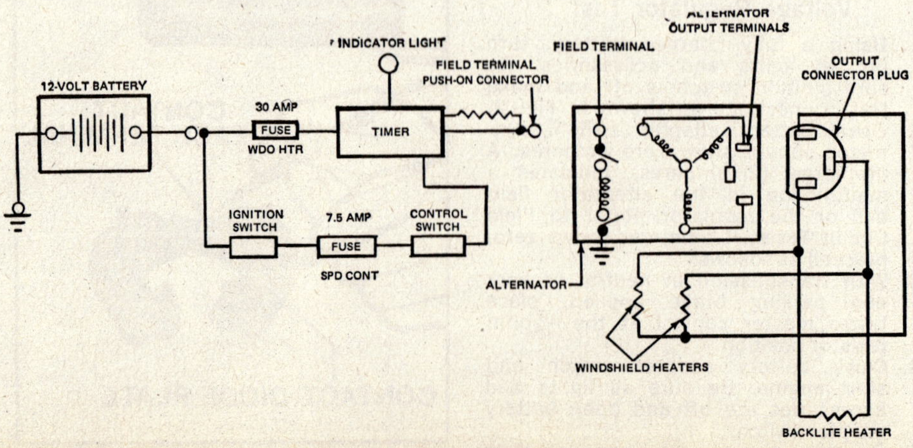

Fig. 1 High voltage alternator system circuit. 1974

housings to aid reassembly.
2. Remove through bolts and separate front end housing and rotor assembly from rear housing.
3. Remove outlet plug retaining nuts, then separate stator from rear end housing.
4. Remove brush holder attaching screws, brush holder, insulator and terminal.
5. If necessary, press out rear housing bearing while supporting housing on inner boss.
6. Remove drive pulley nut, lockwasher, pulley fan, fan spacer, front housing and rotor stop from rotor shaft.
7. If necessary, remove front bearing retainer and press bearing from housing while supporting housing on the bearing boss.

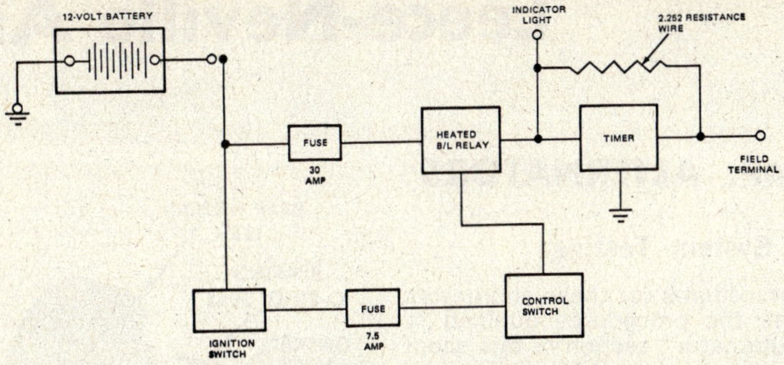

Fig. 1A High voltage system circuit. 1975

Assembly

1. If removed, press front bearing into housing putting pressure on outer race. Install bearing retainer.
2. To replace stop ring, if damaged, push new stop ring into groove in rotor shaft.

NOTE: Do not spread stop ring with snap ring pliers since permanent damage will result.

3. Position rotor stop on shaft with recessed side against stop ring, then install front housing, fan spacer, fan, pulley, lockwasher and retaining nut.
4. If removed, press rear housing bearing into bearing while supporting housing on inner boss. Bearing is properly installed when bearing is flush with outer end surface.
5. Assemble brush holder and insert a stiff wire to hold brushes in position. Place assembly in rear housing and install brush holder mounting screws and the output connector retaining nuts.
6. Place rear housing and stator assembly over front housing and rotor assembly, aligning reference marks, and seat the machined portion of stator core into the step in both end housings. Remove brush retracting wire.
7. Install alternator into vehicle and check for proper operation.

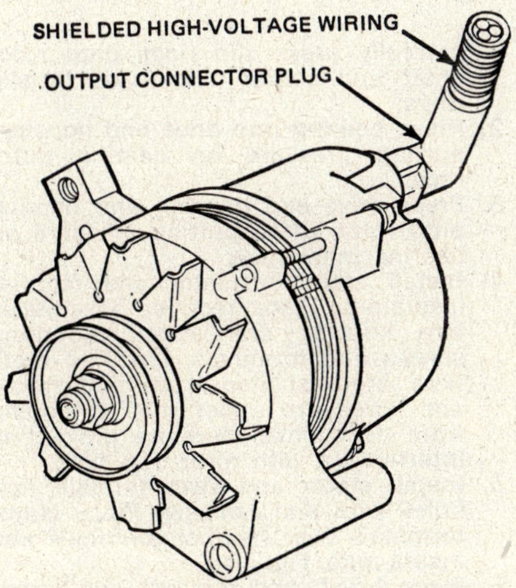

Fig. 2 High voltage alternator assembly

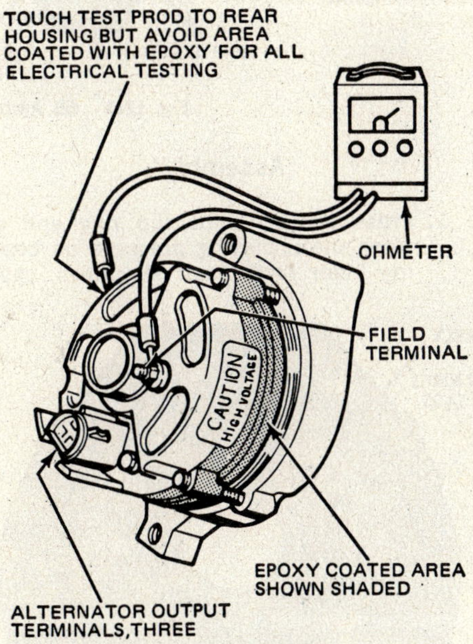

Fig. 3 Field coil test

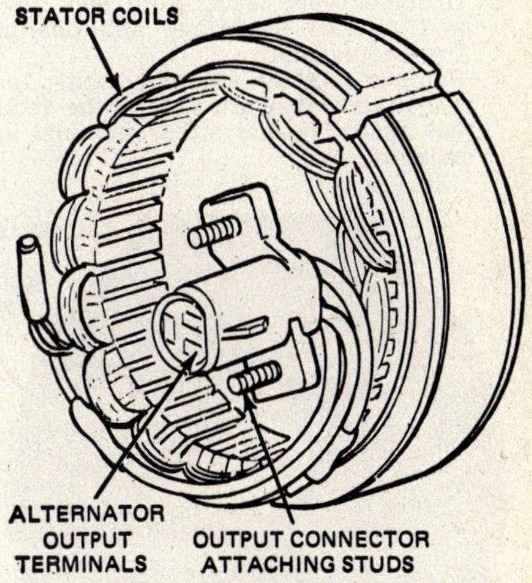

Fig. 4 Stator assembly

Leece-Neville Alternators

65 AMP. ALTERNATORS

System Testing

Testing procedures for these alternators are same as the procedures outlined in the "Ford Alternator" section of this chapter.

ALTERNATOR REPAIRS

Disassembly

1. Remove pulley nut, pulley, fan, shaft key and spacer.
2. Remove brushes and terminal insulator, then brush holder assembly from housing, Fig. LN1.
3. Remove through bolts and separate brush end housing and stator from alternator.
4. Remove AC terminal nuts and stator from end housing.
5. Remove rotor from drive end housing with a gear puller or a suitable press, then front bearing retainer and bearing from housing.
6. Unsolder field leads from slip rings and remove slip rings and bearing from rotor shaft.
7. Remove rectifier assembly bolts, terminals and insulators, then the rectifier assembly and stator terminal insulator.

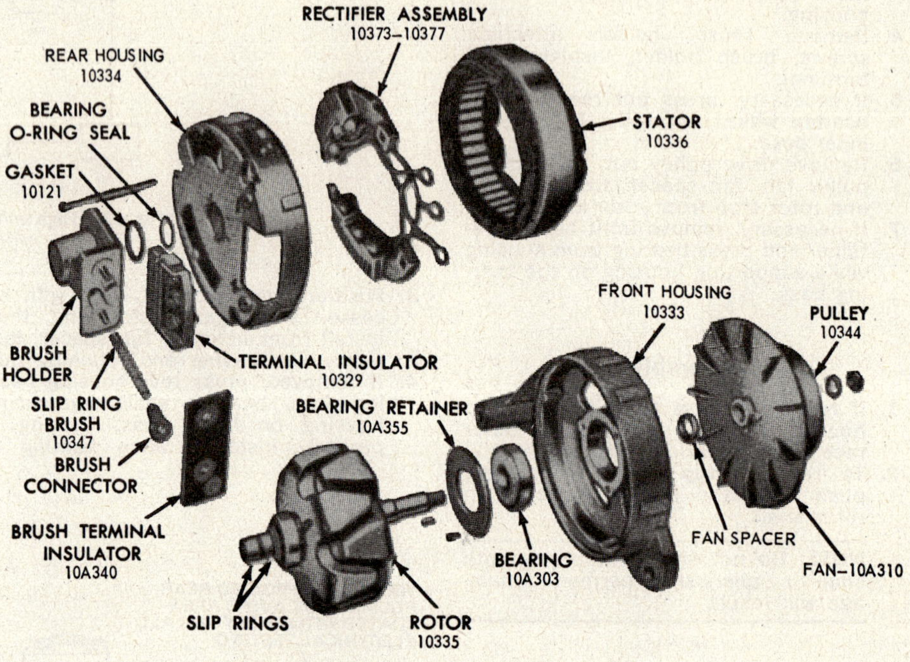

Fig. LN1 65 Amp alternator disassembled

Assembly

1. Press bearing onto slip ring end of rotor shaft, putting pressure on bearing inner race. Heat slip rings, then carefully press slip rings onto rotor shaft and solder field leads to slip rings.
2. Press bearing into drive end housing, putting pressure on bearing outer race.
3. Press drive end housing and bearing onto rotor shaft, putting pressure on bearing inner race.
4. Install stator insulator and rectifier insulators. Place rectifier assemblies into housing and install mounting screws and terminals, ensuring rectifiers are insulated from end frame and wires are under tabs extending from heat sinks therefore preventing interference with rotor, Fig. LN2.
5. Install stator and align through bolt holes with end housing. Place stator terminals over rectifier terminals and install nuts, Fig. LN2.
6. Place brush end housing and stator assembly over rotor and install and tighten through bolts.
7. Install brush holder with "O" ring between holder and frame, place brushes and springs into holder with extruded portion of brush connectors against terminal screw shoulders. Hold brush connectors in position and install terminal insulator.
8. Install spacer, shaft key, fan, puller and nut.

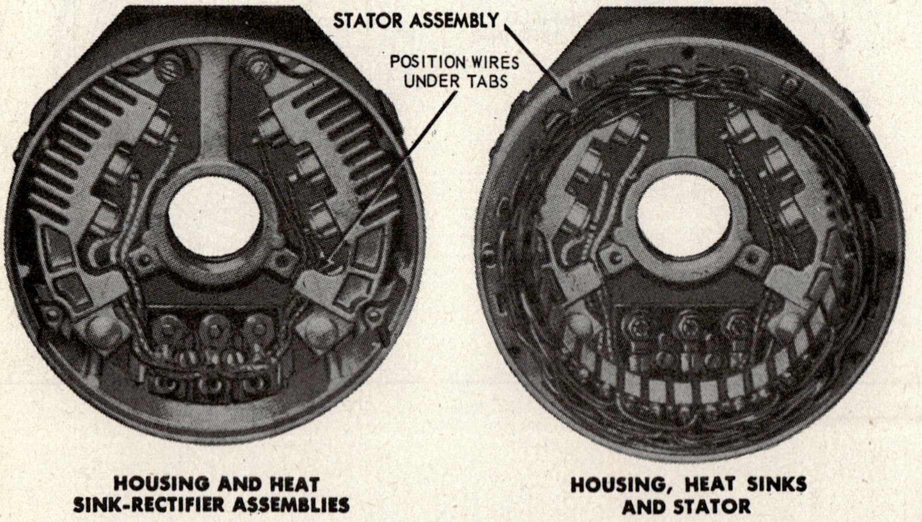

Fig. LN2 Slip ring end housing, rectifier & stator assembly, 65 amp alternator

Motorola Alternator

DESCRIPTION

1969-70

The electrical circuit of the alternator, Figs. 1 and 2, uses 6 silicon diodes in a full wave rectifier circuit. Since the diodes will pass current from the alternator to the battery or load but not in the reverse direction, the alternator does not use a circuit breaker. Fig. 3 shows the charging circuit.

The entire DC output of the system passes through the "Isolation Diode". This diode is mounted in a separate aluminum heat sink and is replaced as an assembly. The isolation diode is not essential for rectification. It is used to:

1. Provide an automatic solid state switch for illuminating the charge-discharge indicator light.
2. Automatically connect the voltage regulator to the alternator and battery when the alternator is operating.
3. Eliminate electrical leakage over the alternator insulators so that maximum leakage is less than one milliampere when the car is not in use.

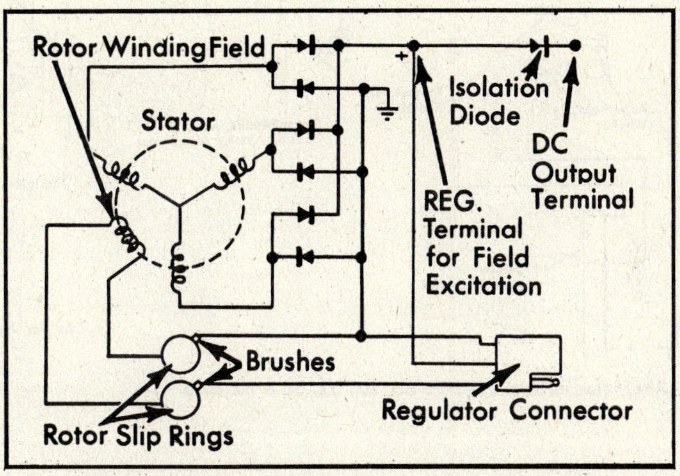

Fig. 1 Alternator circuit diagram. 1969-70 35 amp

1971-75

The electrical circuit, Fig. 4, of these units differ from previous units in that a field diode assembly is used. Also, on alternators used in 1973-75 American Motors cars equipped with a four barrel carburetor, an extra terminal is used on the rear housing which provides about seven volts of alternating current to the heating element of the electric assisted choke.

The field diode (diode trio) assembly incorporates three diodes mounted on a circuit board or on some 1974-75 units, a potted type diode trio is used, Fig. 4A. The input leads are connected to the stator windings in parallel with the positive diodes. The diode output leads are connected to a metal grommet in the circuit board and is secured to the insulated regulator terminal.

A portion of the alternating current and voltage developed in the stator windings is rectified by the field diode assembly. This voltage is sensed by the voltage regulator to provide current to the field windings. Fig. 5 shows the charging circuit.

Voltage Regulator

The voltage regulator is an electrical switching device sealed at the factory, requiring no adjustments. It senses the voltage appearing at the regulator terminal of the alternator and supplies the necessary field current for maintaining the system voltage at the output terminal.

TESTING SYSTEM IN VEHICLE

Alternator Output Test

1974-75

1. Connect a voltmeter to the battery, start engine and turn on headlamps.
2. Run engine for two minutes at 1000 RPM and observe voltmeter. If voltage remains above 13 volts, alternator and regulator are satisfactory.

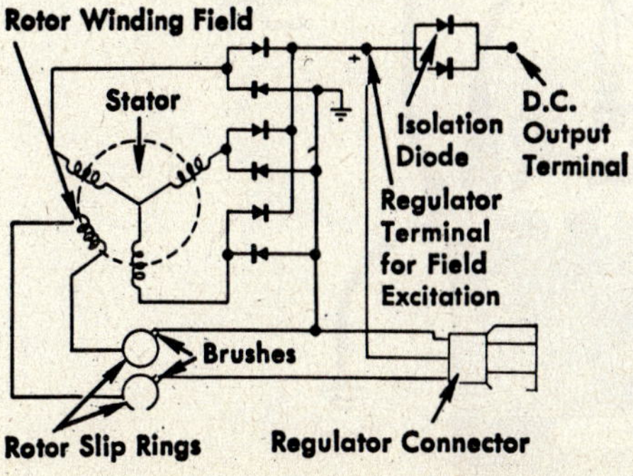

Fig. 2 Alternator circuit. 1969-70 55 amp

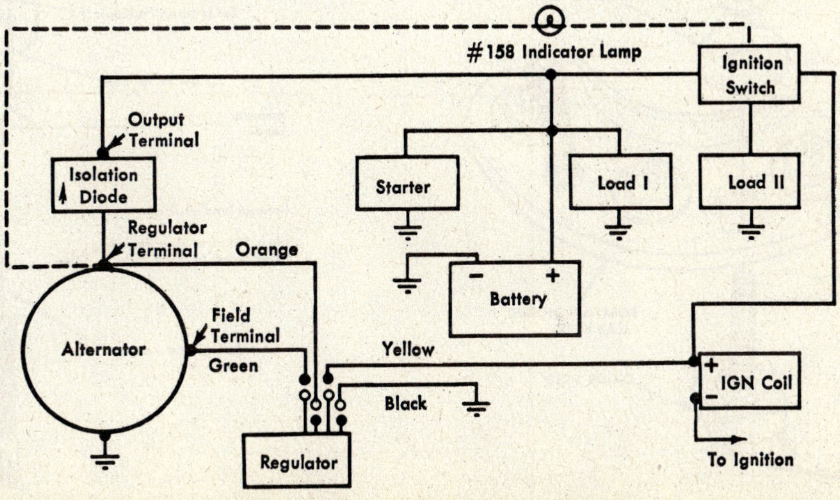

Fig. 3 Charging circuit diagram. 1969-70

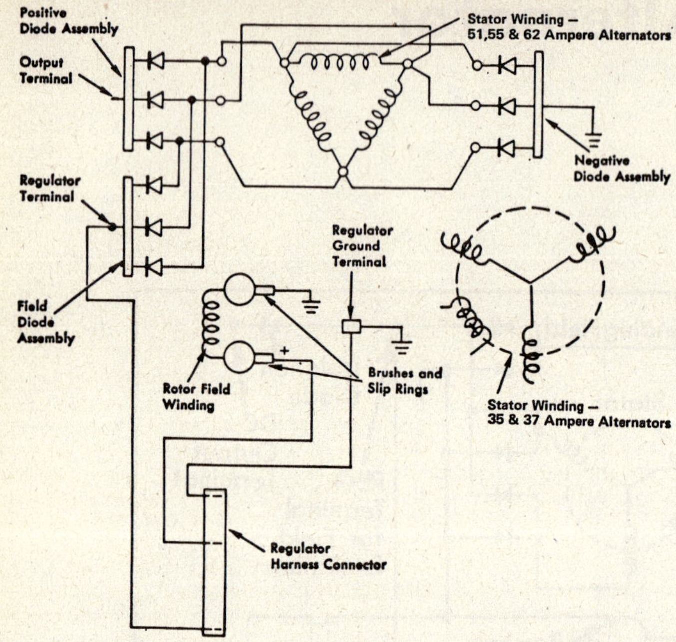

Fig. 4 Alternator circuit. 1971-75 35, 37, 51, 55 & 62 amp

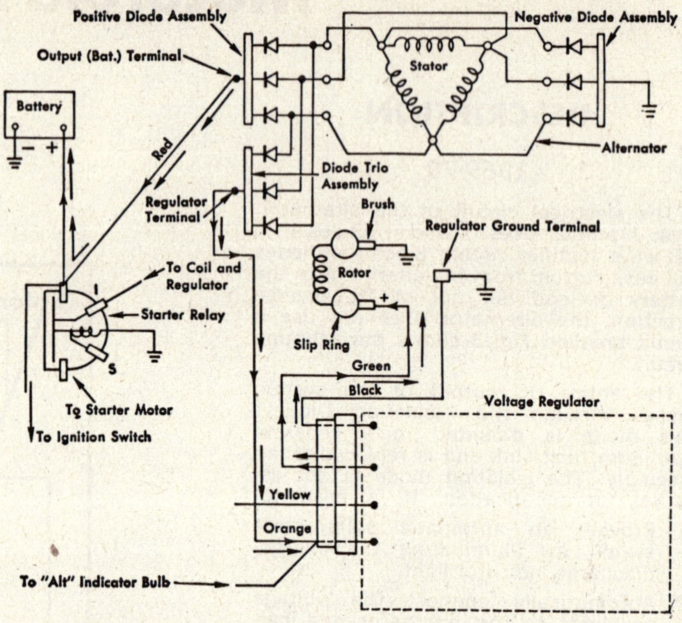

Fig. 5 Charging circuit diagram. 1971-75 (typical)

1969-73

1. Connect test leads as shown in Fig. 6.
2. Turn ground polarity switch to Negative and voltage selector to 16 volt position. Place load control knob in Direct position.
3. Close battery post adapter switch, Fig. 7, and start engine. Run engine at 2000 RPM and allow to warm up.
4. Turn load control knob until highest reading is noted on ammeter. This reading should be within 10 amps of the alternator rated output. If not, further proceed with this test as outlined below.
5. Disconnect field and ground wire connector at alternator.
6. Connect a jumper wire (J-21053) to field terminal and connect one test lead to jumper and the other to output terminal, Fig. 8.
7. Close battery post adapter switch and start engine. Open adapter switch and run engine at 2000 RPM.
8. Slowly rotate field control knob clockwise and note highest ammeter reading. Do not allow voltage to exceed 16 volts.
9. Stop engine and install a jumper wire

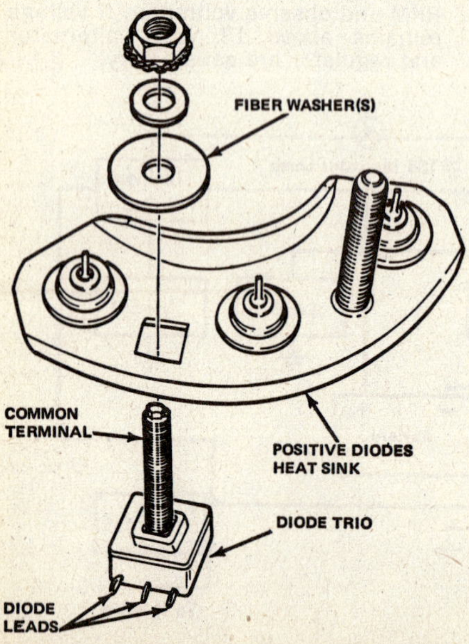

Fig. 4A Potted type diode trio. 1974-75

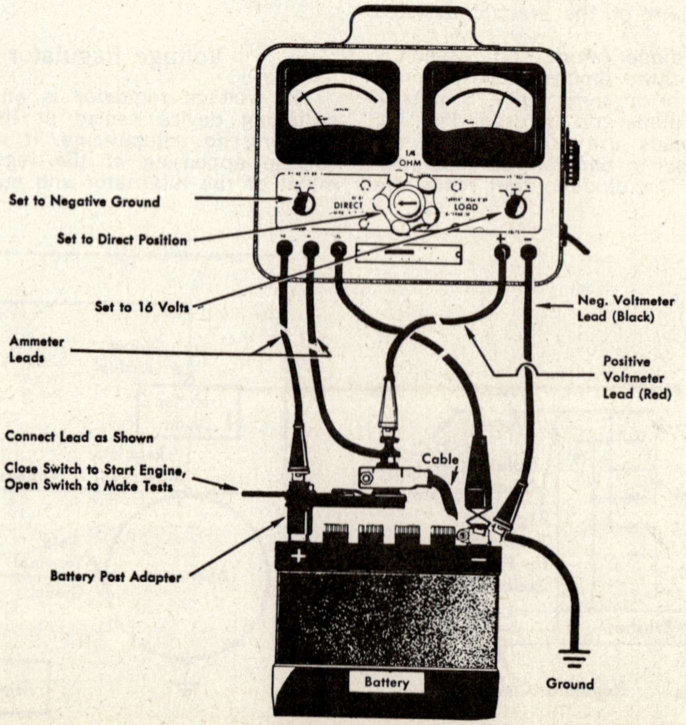

Fig. 6 Alternator test connections. 1969-73

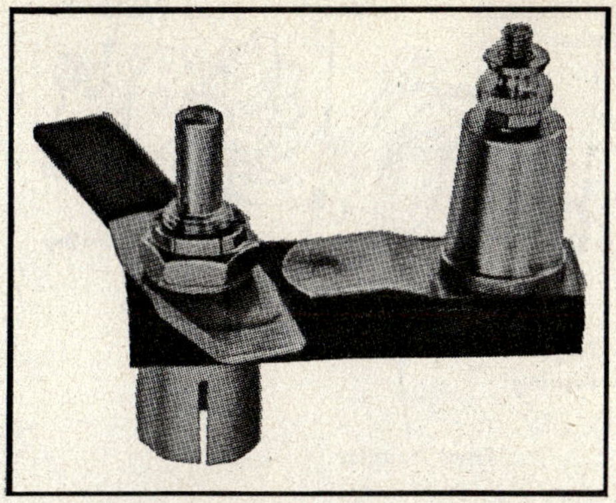

Fig. 7 Battery post adapter tool which provides a convenient method for connecting the ammeter leads of the volt-ammeter tester to the charging system

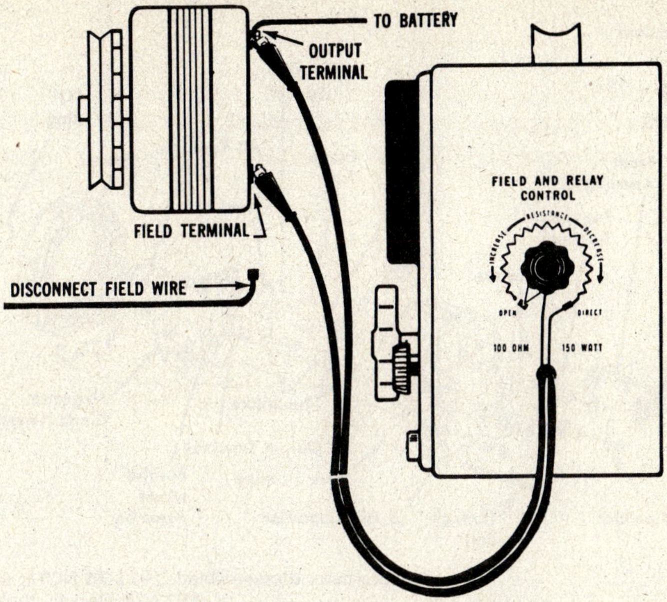

Fig. 8 Alternator output test with external field control

between ignition coil negative terminal and ground.

10. Turn ground polarity switch to Positive and ignition switch ON. Note ammeter reading, which should not exceed 5 amps, and add this reading to reading taken in Step 8. Total reading should be approximately the same as alternator rated output. If not, alternator must be repaired.

Isolation Diode Test

1969-70

If a commercial diode tester is used, follow the Test Equipment Manufacturer's instructions. If a commercial tester is not available, use a DC Test Lamp.

CAUTION: *Do not use a 120 volt test lamp as diodes will be damaged.*

1. Connect test lamp to output terminal and regulator terminal of isolation diode.
2. Reverse test probes.
3. The test lamp should light in one direction but should not light in the other direction.
4. If the test lamp lights in both directions the isolation diode is shorted.
5. If the test lamp does not light in either direction, isolation diode is open.

Rectifier Diode Tests

Any commercial in-circuit diode tester will suffice to make the test. Follow Test Equipment Manufacturer's instructions.

Check diodes individually after the diodes have been disconnected from the stator. A shorted stator coil or shorted insulating washers or sleeves on positive diodes would make diodes appear to be shorted.

A test lamp will not indicate an open condition unless all three diodes of either assembly are open. However, a shorted diode can be detected. This test is not 100% effective but can be used if so desired when an in-circuit diode tester is not available.

The test lamp should light in one direction but not in the other direction. If the test lamp lights in both directions, one or more of the diodes of the assembly being tested is shorted. If the test lamp does not light in either direction,

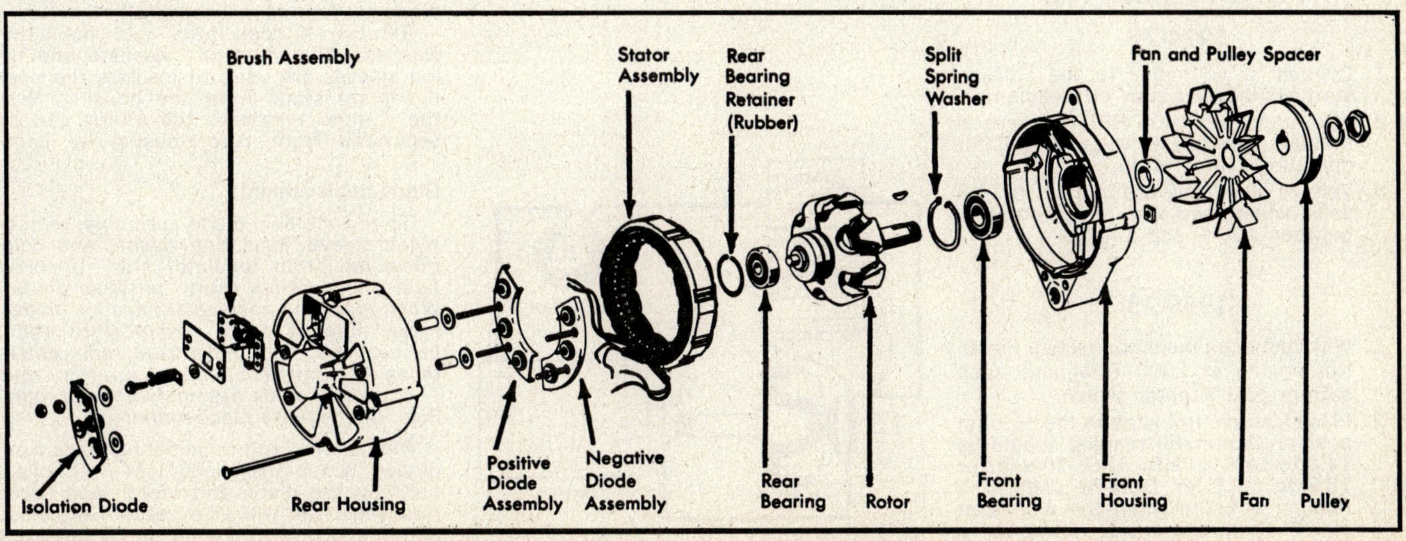

Fig. 9 Alternator disassembled. 1969-70

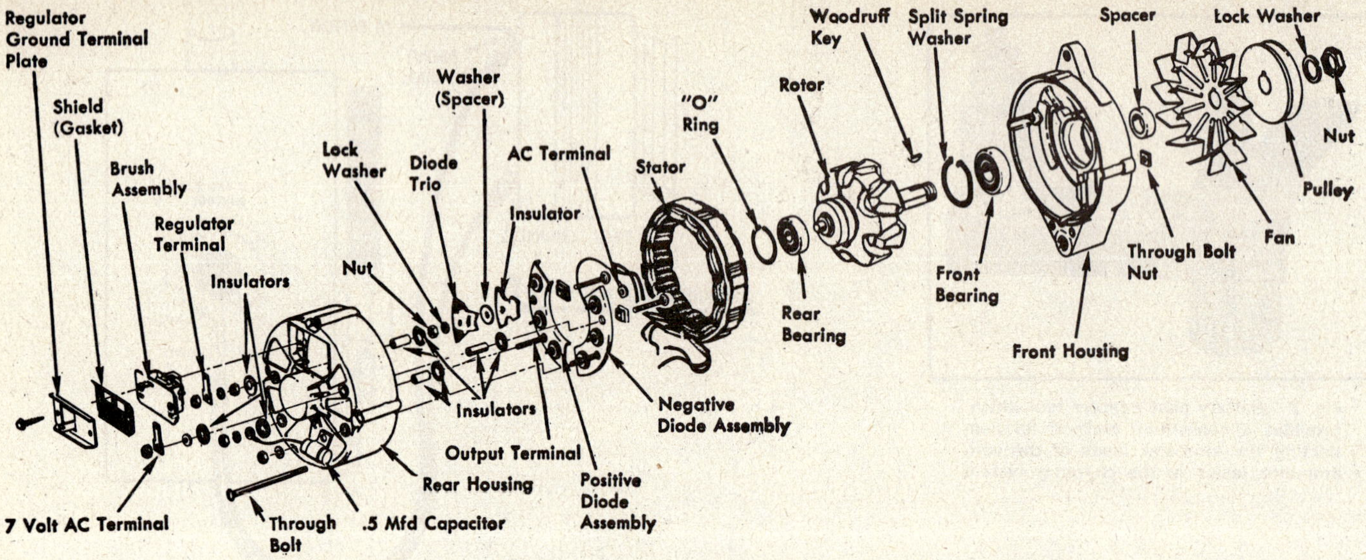

Fig. 10 Alternator disassembled. 1971-75 NOTE: model shown has AC terminal for 1973-75 electric choke

all three diodes in the assembly are open. Check diodes individually after disassembly to ascertain findings.

NOTE: *A shorted stator coil would appear as a shorted negative diode. Also check stator for shorts after disassembly.*

Field Diode (Diode Trio) Test

1. Using a voltmeter, connect positive lead to alternator output terminal and negative lead to regulator terminal.
2. Start and run engine at idle speed. The voltmeter should then read .6 volt or less. If reading is over .6 volt, replace field diode assembly.

Voltage Regulator Test

1974-75

1. Connect a voltmeter to the battery, start engine and turn on headlamps.
2. Run engine at 1000 RPM for several minutes to establish voltage regulator operating temperature.
3. Voltage should be within 13.1 to 14.3 volts when regulator temperature is between 100°F and 150°F.

1969-73

1. With test equipment connected, Fig. 6, run engine at 2000 RPM and open battery post adapter switch.
2. Place load control knob in the ¼ ohm position. Voltmeter reading should be 14.0 to 14.8 volts for 1968-70 units or 13.8 to 14.2 for 1971-73 units. The reading is taken when the unit is at 80° F. If voltage is not within above limits, replace voltage regulator.

ALTERNATOR REPAIRS

Disassembly, Figs. 9 and 10

Brush Assembly

The brush assembly can be removed in most cases with the alternator on the vehicle. The spring clip is bent back so that the field terminal plug can be removed. Remove the two self-tapping screws, field plug retainer spring and cover. Pull brush assembly straight out far enough to clear locating pins, then lift brush assembly out. The complete brush assembly is available for replacement.

Isolation Diode, 1969-70

Remove the 2 lock nuts securing the isolation diode to the rear housing and slide it off the studs. The diode is replaced as an assembly.

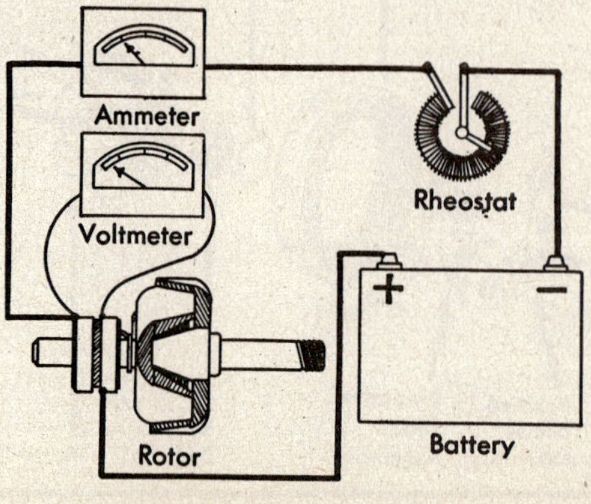

Fig. 11 Field coil test

Rear Housing

Remove the 4 through bolts and nuts. Carefully separate the rear housing and stator from the front housing by using 2 small screwdrivers and prying the stator from the front housing at 2 opposing slots where the "through bolts" are removed. Do not burr the stator core which would make assembly difficult.

Caution: *Do not insert screwdriver blade deeper than ¹⁄₁₆" to avoid damaging stator winding.*

Stator and Diode Assembly

Do not unsolder stator-to-diode wire junction. Remove stator and diode as an assembly. Avoid bending stator wire at junction holding positive and negative diode assembly from housing.

Remove 4 lock nuts and insulating washers. The insulating washers and nylon sleeves are used to insulate the positive plate studs from the housing. With the 4 nuts removed, the stator can be separated from rear housing by hand.

Diode Replacement

To replace field diode assembly, remove nylon sleeve, insulating washer and hold-down nut from regulator stud. Unsolder field diode wires from positive diodes. *When replacing positive or negative diodes, make note of diode assembly to stator connections, and make sure replacement diode assembly connections are the same. The positive diode assembly has red markings, the negative black markings.*

In soldering and unsoldering leads from diodes, grasp the diode lead with pliers between the diode and stator lead to be removed. This will give better heat dissipation and protect the diode. Do not exert excessive stress on diode lead.

Year	Output Amps	Field Current
1969-71	35	2.0-2.6
1969-71	55	1.8-2.4
1972-75	All	1.8-2.5

Fig. 12 Field current

ROTOR

The rotor should only require removal from the front housing if there is a defect in the field coil itself or in the front bearing. Front and rear bearings are permanently sealed, self-lubricating type. If the front housing must be removed from the rotor, use a two jaw puller to remove the pulley. The split spring washer must be loosened with snap ring pliers through the opening in the front housing. Remove the washer only after the housing is removed. The rotor and front bearing can be removed from the front housing by tapping the rotor shaft slightly.

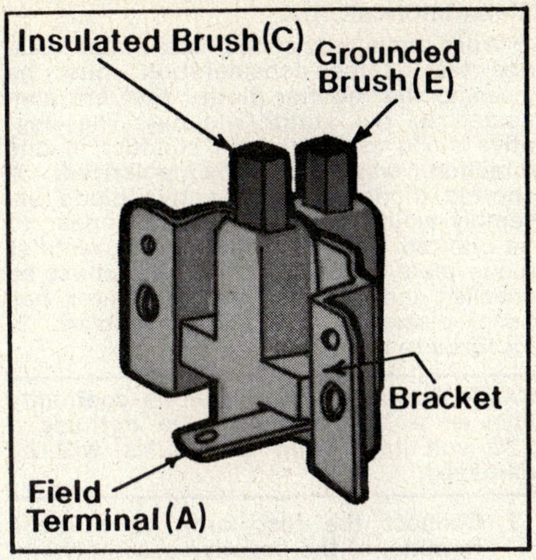

Fig. 13 Brush assembly test

NOTE: *Make certain that the split spring washer has been removed from its groove before attempting to remove the front housing from the bearing.*

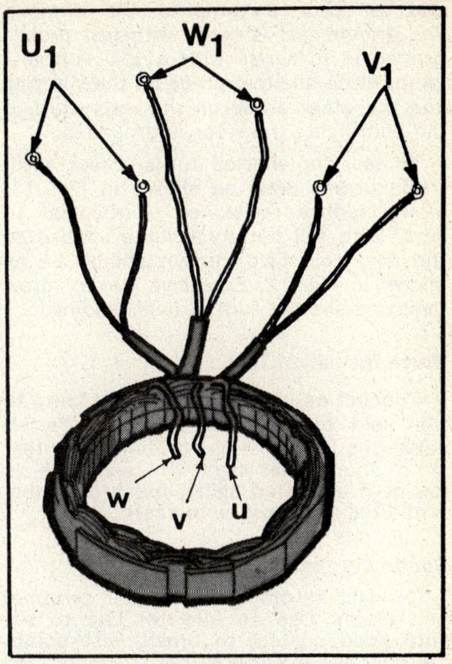

Fig. 16 Stator coil shorts and continuity tests

Alternator Bench Tests

Field Coil Test

The rotor should be tested for grounds

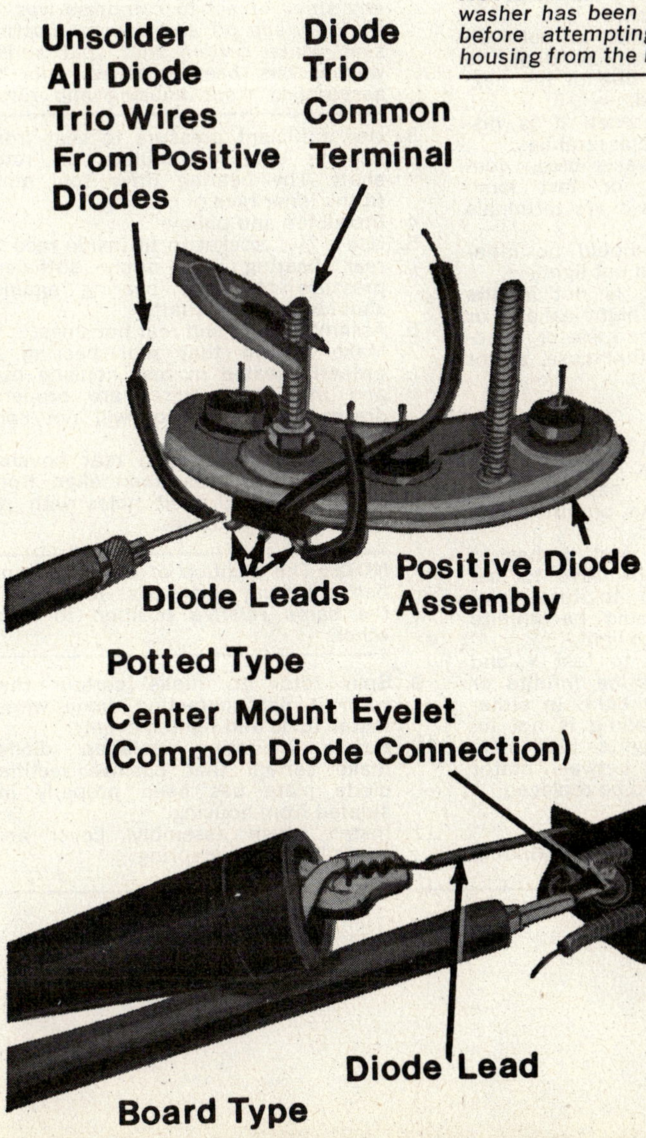

Fig. 14 Diode trio bench test. 1974-75 alternators

Fig. 15 Rectifier diodes bench test. 1974-75 alternators

and for shorted turns in the winding. The ground test is made with test probes connected in series with a 110 volt test lamp. Place one test probe on the slip ring and the other probe on the rotor core. If the bulb lights the rotor is grounded.

To test for shorted turns, check rotor field current draw as shown in Fig. 11. Slowly reduce resistance of rheostat to zero. With full battery voltage applied to the field coil, field current should be as shown in Fig. 12. Excessive current draw indicates shorted turn in field winding.

Brush Insulation Test

Connect an ohmmeter or a test lamp to the field terminal and bracket. Resistance should be high (infinite) or test lamp should not light. If resistance is low or if test lamp lights, brush assembly is shorted and must be replaced.

Continuity Test

Connect an ohmmeter to field terminal and brush. Use an alligator clip to assure good contact to brush, test points "A" and "C" in Fig. 13.

CAUTION: *Do not chip brush.*

Resistance reading should be zero. Move brush and brush lead wire to make certain that brush lead wire connections are not intermittent. Resistance reading should not vary when brush and lead wire are being moved around. Connect ohmmeter to bracket and grounded brush, test points "E" and "D", Fig. 13. Resistance reading should be zero.

1974-75 Diode Tests

Diode Trio: Unsolder diode trio leads and using a 12 volt meter which draws a one amp maximum load, connect test leads providing current path, Fig. 14. Hold this connection for at least two minutes, then reverse test leads immediately. Test each diode in this manner. If current flows in both directions, neither direction or flows intermittently, the diode trio must be replaced.

Rectifier Diodes: The same method of testing is used for the rectifier diodes as for the diode trio, Fig. 15. However, a 20 amp load is used when testing the rectifier diodes.

NOTE: A diode can be tested for opens or shorts using an ohmmeter. Reason for applying an electrical load to the diodes, in turn generating heat, is to detect intermittent diode failures.

Stator In-Circuit Test

When making the in-circuit stator leakage test, some consideration must be given to the rectifier diodes that are connected to the stator winding. The negative diode assembly will conduct in one direction when properly polarized. A shorted diode in the negative diode assembly would make the stator appear to be shorted. For this reason, the rectifier diode plate assembly and stator must be checked individually after alternator has been disassembled if the problem is localized to the stator.

CAUTION: *Use a special diode continuity light or a DC test lamp. Do not use a 120 volt test lamp as diodes will be damaged.*

1. Connect the test lamp to a diode terminal of the negative assembly and ground terminal.
2. Reverse test probes. The lamp should light in one direction but not in the other.
3. If the test lamp does not light in either direction, this indicates that all three rectifiers in the negative diode assembly are open.
4. If the test lamp lights in both directions, the stator winding is shorted to stator or one of the negative diodes is shorted.
5. Check stator again when it is disassembled from diode assemblies.
6. With alternator disassembled, connect an ohmmeter or test lamp probes to one of the diode terminals and to stator.
7. Resistance reading should be infinite or test lamp should not light.
8. If resistance reading is not infinite or test lamp lights, high leakage or a short exists between stator winding and stator. In either case, stator should be replaced.

Stator Coil Shorts Test

1. This test checks for shorts between stator coil windings. The winding junctions must be separated as shown in Fig. 16. An ohmmeter or test lamp may be used.
2. Connect one of the test probes to test point "U" and the other to test point "V" and then to test point "W". Resistance should be infinite or test lamp should not light.
3. Connect test probes to test V and W. Resistance should be infinite or test lamp should not light. In either test, if resistance reading is not infinite or test lamp lights, high leakage or a short exists between stator windings. Stator should be replaced.

Continuity Test

1. Measure resistance of each winding in stator between test points U and U1, V and V1, W and W1, Fig. 16. Resistance should be a fraction of an ohm (approximately .1 Ohm). An extremely accurate instrument would be necessary to ascertain shorted turns. Only an open condition can be detected with a commercial type ohmmeter.
2. If the alternator has been disassembled because of an electrical malfunction, replace stator only after all other components have been checked and found to be satisfactory.

Assemble Alternator

1. Clean bearing and inside of bearing hub of front housing. Support front housing and, using a suitable driver, apply sufficient pressure to outside race of bearing to seat bearing.
2. Insert split spring washer hub of front housing, seating washer into groove of hub.

NOTE: *Do not use a screwdriver or any small object to compress washer that can slip off and damage bearing seal. Make certain that split spring washer has been installed prior to assembling front housing and rotor.*

3. Use sufficient pressure to seat front bearing against shoulder on rotor shaft. The bearing drive tool must fit the inner race of bearing.
4. Install fan and pulley.
5. Use a 7/16" socket to fit inside race of rear bearing and apply sufficient pressure to drive bearing against shoulder of rotor shaft.
6. Assemble front and rear housings.
7. Make certain that rear bearing is properly seated in rear housing hub and that diode wires are properly dressed so that rotor will not contact diode wires.
8. Align stator slots with rear housing through bolt holes, then align front housing through bolt holes with respect to rear housing.

NOTE: *The position of the brush and belt adjusting screw boss must be in the same relative position to each other.*

9. Spin rotor to make certain that rotor is not contacting diode wires. Install bolts and tighten evenly.
10. Before mounting isolation diode, make certain that positive rectifier diode plate has been properly insulated from housing.
11. Install brush assembly, cover and field plug retainer spring.

Prestolite & American Motors Alternators

Alternator Output Test

1. Disconnect battery positive terminal.
2. Install battery post "knife switch" then complete test connections as per Fig. P1.
3. Close knife switch and start engine. Set engine speed at approximately 2000 rpm.
4. Open knife switch and adjust control knob clockwise to Load position until maximum ammeter reading is obtained. Add 5 amps to reading. Total reading should be the rated output of the alternator.

Regulator Test

1. Make test connections as per Fig. P1.
2. With engine speed at 2000 rpm, set control knob to ¼ ohm position.
3. Read voltmeter. The reading should be not less than 13 volts and not more than 15 volts.

NOTE: Voltage readings vary due to underhood temperatures.

4. Return engine to idle, stop engine. Ammeter should read zero, indicating that isolation diode is functioning properly.
5. If a current draw exists, the isolation diode must be tested for leakage.

Isolation Diode Test

1. Close knife switch. Disconnect auxiliary or regulator lead from alternator.
2. Connect the positive voltmeter lead to the auxiliary or regulator terminal of the alternator and negative lead to the case of the alternator.
3. Set voltmeter to the 2 volt scale.
4. With regulator connected to the alternator and the ignition switch and all accessories off, measure voltage at the auxiliary or regulator terminal.
5. Voltage should not exceed .1 volt. Voltage in excess of .1 volt indicates that the isolation diode is leaking and must be replaced.

External Field Control Test

1. With the engine stopped and all test connections completed as per Fig. P1, disconnect field terminal from alternator and connect field rheostat as per Fig. P2.
2. With knife switch open, rotate rheostat to full clockwise (direct) position. Ammeter should read approximately minus 2 amps. If not, alternator brushes or rotor field windings are defective. Rotate rheostat back to full counterclockwise (open) position.
3. Close knife switch and set engine speed at 2000 rpm. Open knife switch.
4. Rotate rheostat control until alternator specified output less 5 amps is obtained.
5. If output is low, alternator is defective. If output is satisfactory, regulator is at fault.

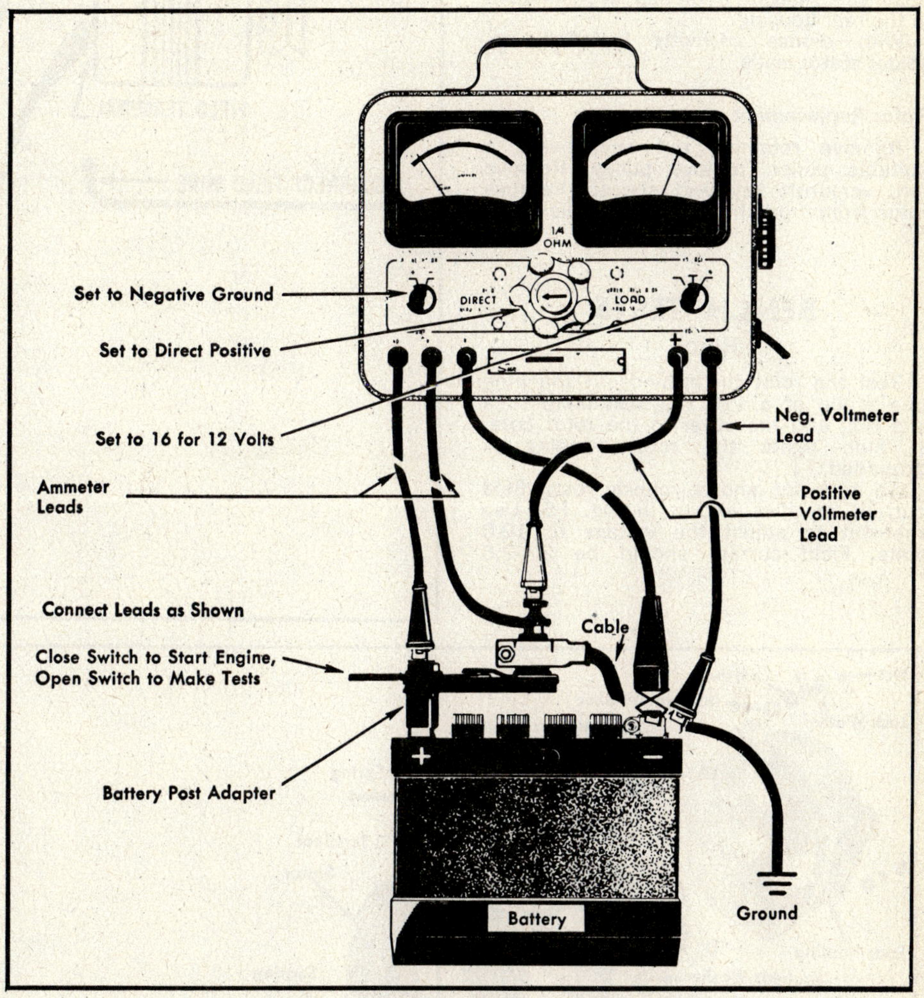

Fig. P1 Charging circuit test connections

ALTERNATOR REPAIRS

Disassembly, Fig. P3

Brush Assembly

The brush assembly can be removed with the alternator on the car. Remove the two mounting screws and the cover. Tip the brush holder away from the alternator and remove. The brushes can be replaced by removing the retaining screws.

Rear Housing and Stator

The rear housing and stator must be removed as a unit. Mark housing and stator before disassembling to insure correct assembly.

Remove four retaining screws and tap lightly on stator and rear housing with a plastic mallet to separate from front drive housing.

Diode Replacement

To replace negative or positive rectifier diodes or the isolation diode, the stator leads must be unsoldered. When soldering or unsoldering any diodes, use a pair of pliers and a small piece of water soaked cotton for heat dissipation to protect the diode.

With stator removed, remove mounting nuts from "Aux" and "Output" terminals and remove the heat sink. Use a suitable press type tool to remove and

Fig. P2 Alternator test with external field control

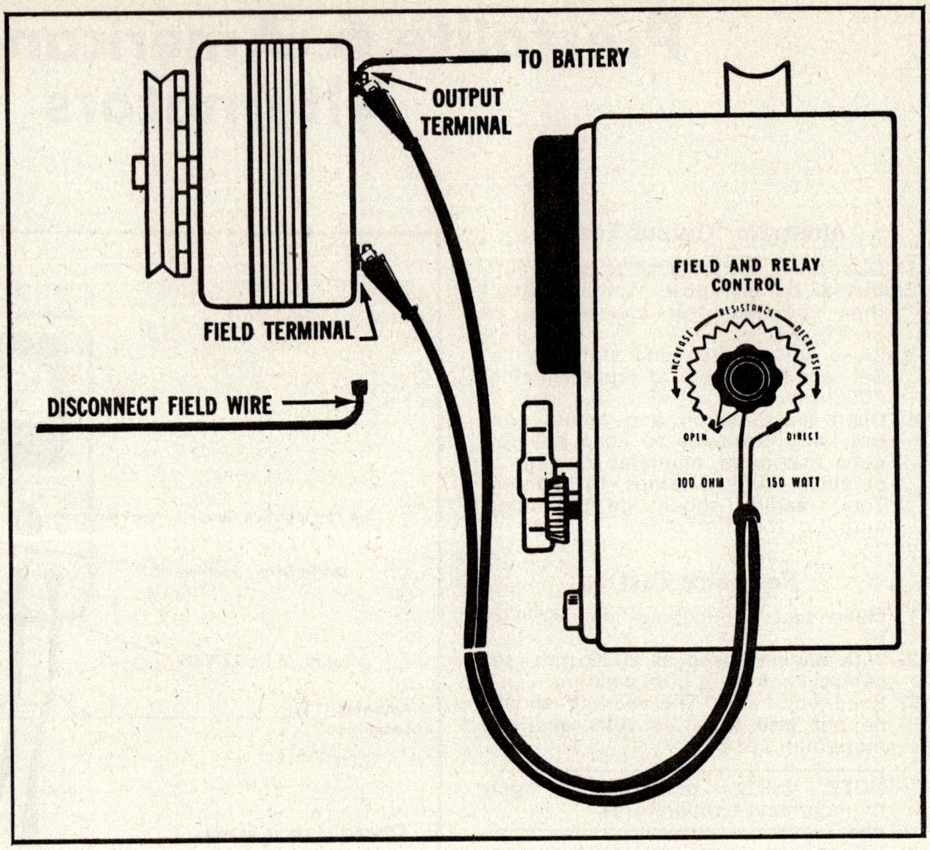

install diodes, Fig. P4. *Do not drive diodes in or out as damage will result.*

Positive diodes are installed in the heat sink; negative diodes are installed in the rear housing.

With diodes correctly installed, resolder stator leads.

Rotor Replacement

Remove retaining nut and, using a suitable puller, remove pulley. Remove fan, woodruff key and spacer. Remove rotor from front housing with puller, Fig. P5.

BENCH TESTS

Rotor

Test the rotor for grounds by touching one probe of a 110 volt test lamp to a slip ring and the other to the rotor core. If bulb lights the rotor winding is grounded.

To test for shorts, check rotor field current draw as shown in Fig. P6. Use rheostat to adjust the voltage to 10.0 volts. Field current should be 2.4-2.5

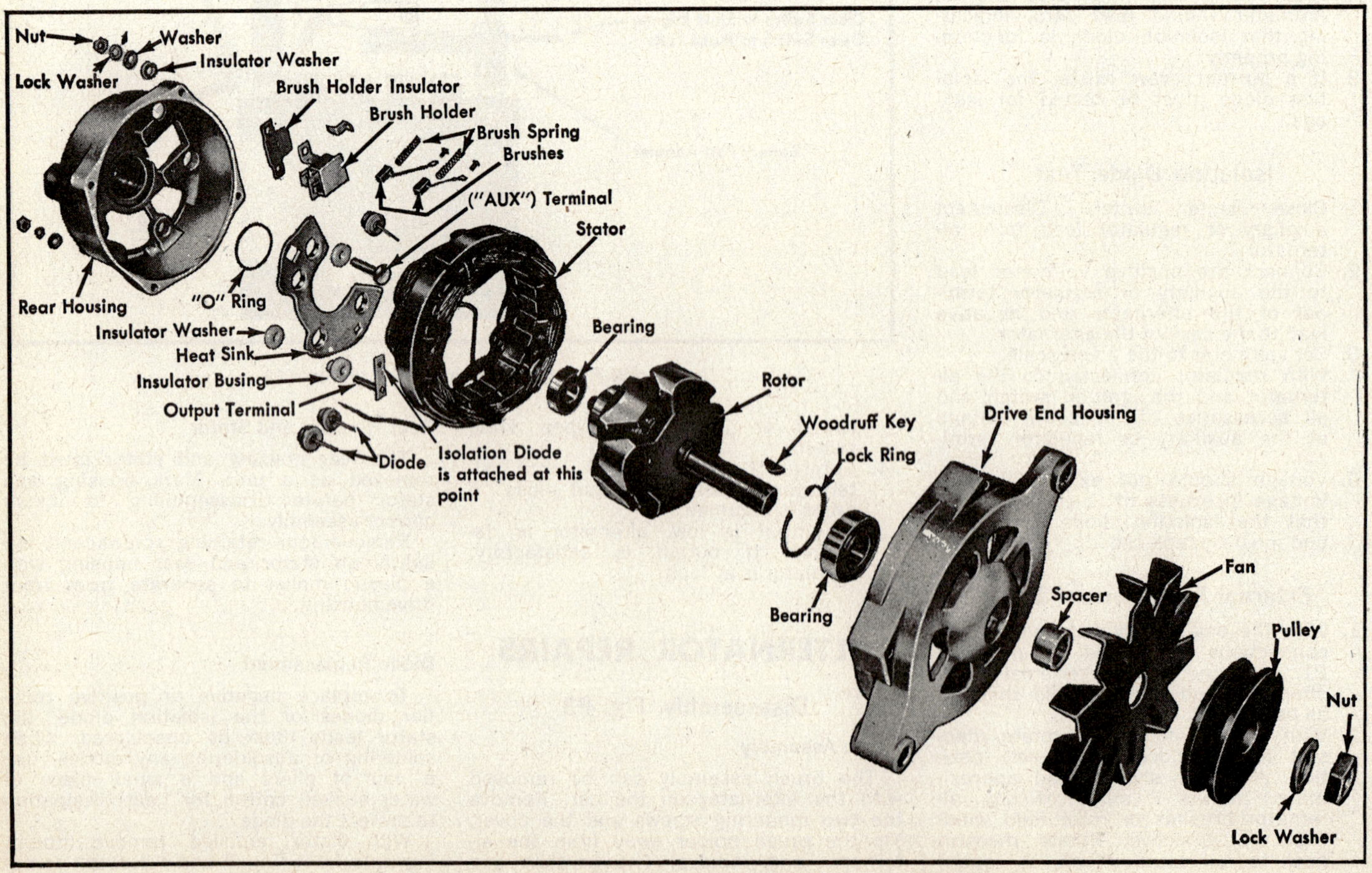

Fig. P3 Exploded view of Prestolite and American Motors alternator

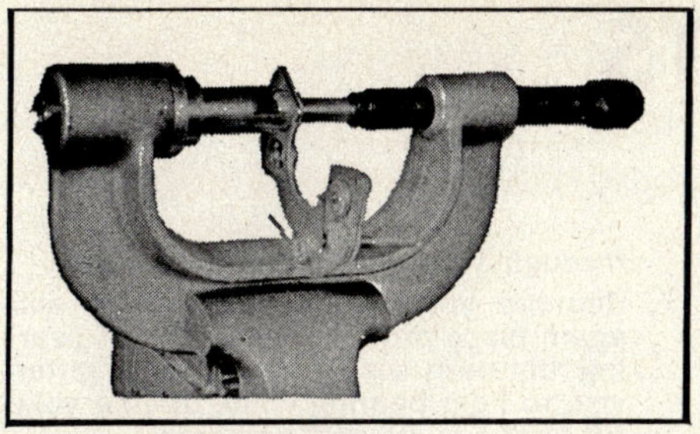

Fig. P4 Removing or installing diodes

amps. Battery must be fully charged to carry out this test.

Excessive draw indicates a shorted field winding.

Stator

To test for an open winding use a number 57 bulb in series with a 12 volt battery and test probes. Touch one test probe to the connection of the three stator windings and the other to each stator lead that is connected to the diodes. If the bulb fails to light, the winding is open.

Test for grounded winding with a 110 volt test lamp. For this test it is necessary to disconnect the diodes from the stator leads. Touch one lead to the stator core, the other to each of the three leads of the stator winding. If lamp lights, winding is grounded.

Shorted windings in the stator are difficult to detect, therefore, if rotor and diodes are not defective and the stator is not open or grounded, replace the stator.

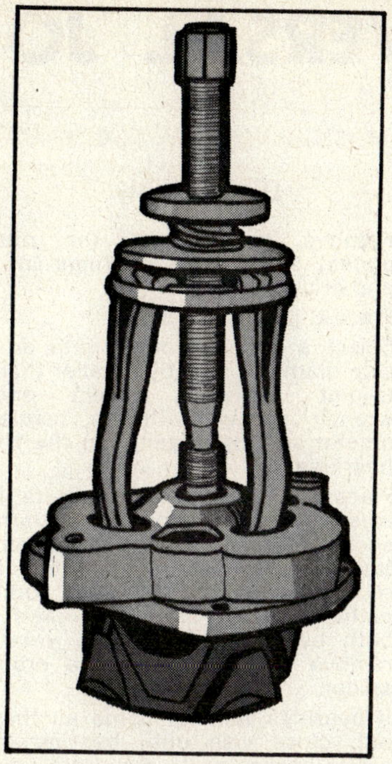

Fig. P5 Removing rotor from front end drive housing

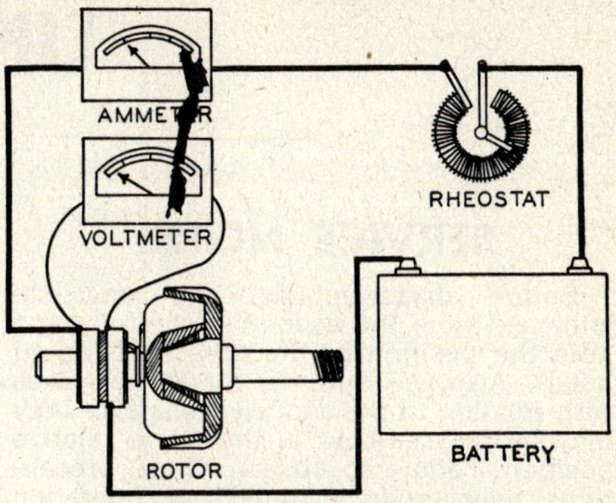

Fig. P6 Meter connections for testing rotor field current draw

Alternator, Assemble

1. Press front bearing into the front end drive housing, making sure the dust seal faces the rotor. Install bearing retainer stop ring.
2. Install spacer, fan and pulley. Install lockwasher and nut and tighten securely.
3. Install heat sink, negative diodes and stator. Solder all stator-to-diode connections that were previously disconnected.
4. Install rotor and front end drive housing to stator and rear housing.
5. Align marks made during disassembly on housings and install four retaining screws.
6. Install brush holder assembly and screws.
7. Check to be sure stator leads and brush holder assembly do not rub on the rotor and that the rotor turns freely when rotated by hand.

UNIVERSAL JOINTS

SERVICE NOTES

Before disassembling any universal joint, examine the assembly carefully and note the position of the grease fitting (if used). Also, be sure to mark the yokes with relation to the propeller shaft so they may be reassembled in the same relative position. Failure to observe these precautions may produce rough car operation which results in rapid wear and failure of parts, and place an unbalanced load on transmission, engine and real axle.

When universal joints are disassembled for lubrication or inspection, and the old parts are to be reinstalled, special care must be exercised to avoid damage to universal joint spider or cross and bearing cups.

NOTE: Some late model cars use an injected nylon retainer on the universal joint bearings. When service is necessary, pressing the bearings out will sheer the nylon retainer. Replacement with the conventional steel snap ring, type is then necessary.

CROSS & ROLLER TYPE

Figs. 1, 2 and 3 illustrate typical examples of universal joints of this type. They all operate on the same principle and similar service and replacement procedures may be applied to all.

Disassembly

1. Remove snap rings (or retainer plates) that retain bearings in yoke and drive shaft.
2. Place U-joint in a vise.
3. Select a wrench socket with an outside diameter slightly smaller than the U-joint bearings. Select another wrench socket with an inside diameter slightly larger than the U-joint bearings.
4. Place the sockets at opposite bearings in the yoke so that the smaller socket becomes a bearing pusher and the larger socket becomes a bearing receiver when the vise jaws come together, Fig. 4. Close vise jaws until both bearings are free of yoke and remove bearings from the cross or spider.
5. If bearings will not come all the way out, close vise until bearing in receiver socket protrudes from yoke as much as possible without using excessive force. Then remove from vise and place that portion of bearing which protrudes from yoke between vise jaws. Tighten vise to hold bearing and drive yoke off with a soft hammer.
6. To remove opposite bearing from yoke, replace in vise with pusher socket on exposed cross journal with receiver socket over bearing cup. Then tighten vise jaws to press bearing back through yoke into receiving socket.
7. Remove yoke from drive shaft and again place protruding portion of bearing between vise jaws. Then tighten vise to hold bearing while driving yoke off bearing with soft hammer.
8. Turn spider or cross ¼ turn and use the same procedure to press bearings out of drive shaft.

Reassembly

1. If old parts are to be reassembled, pack bearing cups with universal joint grease. *Do not fill cups completely or use excessive amounts as over-lubrication may damage seals during reassembly.* Use new seals.
2. If new parts are being installed, check new bearings for adequate grease before assembling.
3. With the pusher (smaller) socket, press one bearing part way into drive shaft. Position spider into the partially installed bearing. Place second bearing into drive shaft. Fasten drive shaft in vise so that bearings are in contact with faces of vise jaws, Fig. 5. *Some spiders are provided with locating lugs which must face toward drive shaft when installed, Fig. 6.*
4. Press bearings all the way into position and install snap rings or retainer plates.

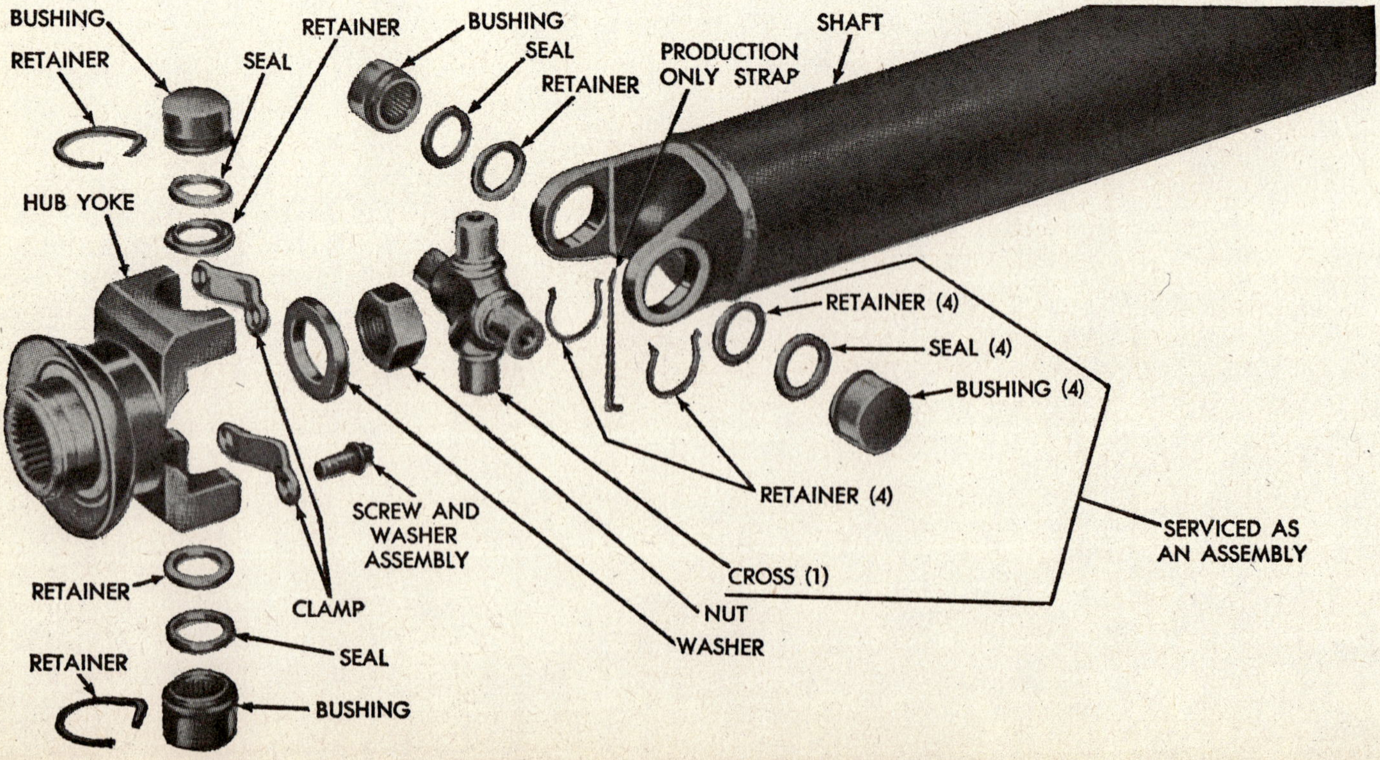

Fig. 1 Cross and roller type universal joint. Chrysler-built cars

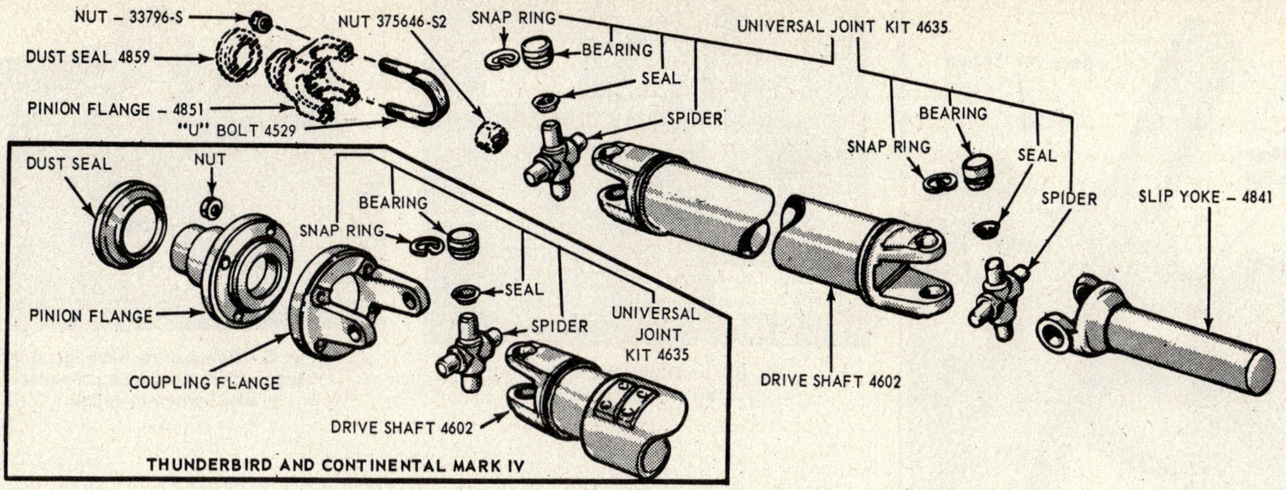

Fig. 2 Cross and roller universal joints and propeller shaft. Ford-built cars

5. Install bearings in yoke in same manner. When installation is completed, check U-joint for binding or roughness If free movement is impeded, correct the condition before installation in vehicle.

CONSTANT VELOCITY TYPE

This type of U-joint, Fig. 7, is composed of two conventional cross and roller joints connected with a special link yoke. Because the two joint angles are the same, even though the usual U-joint fluctuation is present within the unit, the acceleration of the front joint (within the yoke) is always neutralized by the deceleration of the rear joint (within the yoke) and vice versa. The end result is the front and rear propeller shafts always turn at a constant velocity.

General Motors

NOTE: Not all General Motors CV joints are serviceable. CV joints which are serviceable are 1974-75 Chevrolet, 1969-75 Buick, 1972-75 Cadillac and 1973-75 Oldsmobile and Pontiac. 1971-73 Chevrolet CV joints are not serviceable and require a complete propeller shaft replacement if the CV joints fail. However, beginning in September, 1972, the Chevrolet replacement propeller shaft is serviceable, Fig. 8.

For ease of handling and to prevent damage to the constant velocity U-joints, the front and rear propeller shafts must be separated at the slip joint before any service is attempted.

Disassemble Slip Joint

1. Pry lockwasher from flats on bearing locknut.

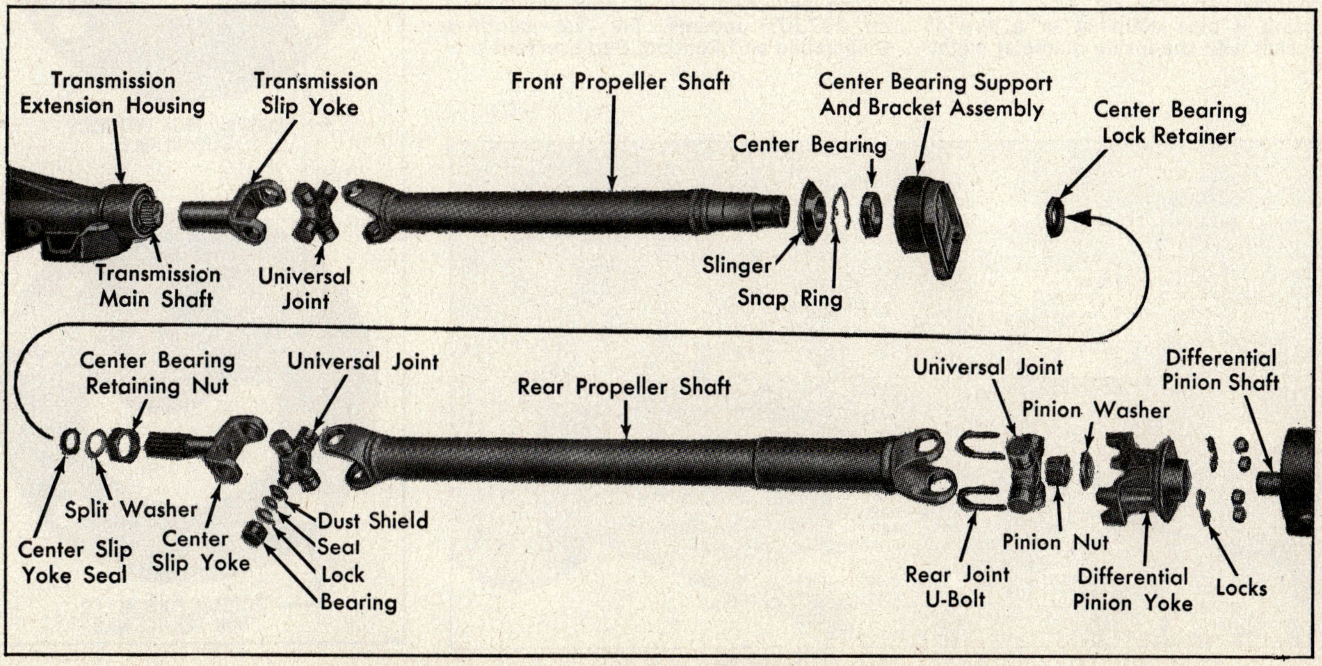

Fig. 3 Example of a two-piece propeller shaft with three cross and roller universal joints and center bearing support assembly. Cadillac

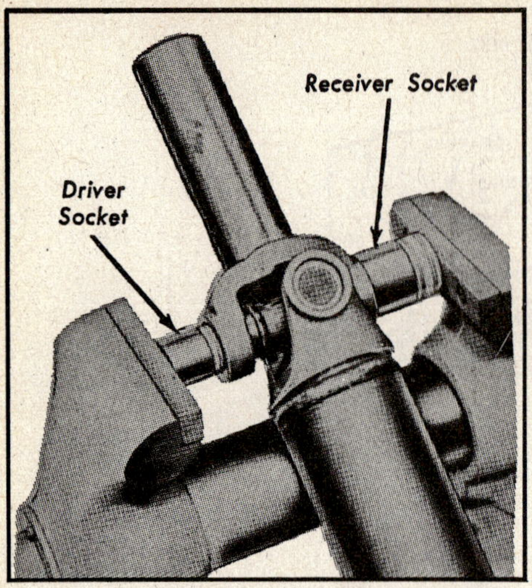

Fig. 4 Removing bearings from yoke using small and large wrench sockets as pusher and receiver tools, respectively

Fig. 5 Installing bearings into drive shaft yoke

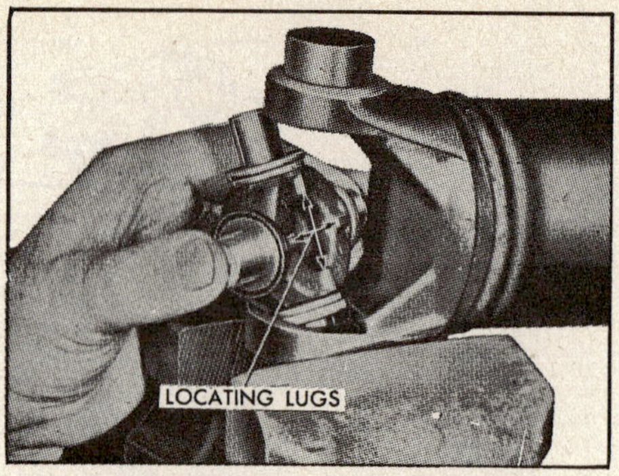

Fig. 6 Some units have locating lugs which must face propeller shaft when installed

2. Loosen locknut until free of threads and slide locknut and seal against constant velocity joint.
3. Slide rear propeller shaft from front propeller shaft, making sure that index spring wire in splines is not lost.

Disassemble Constant Velocity U-Joint

1. Mark yokes before disassembly to be sure reassembly is made in same relative position of components, Fig. 9.
2. Disassemble rear section of constant velocity U-joint first as follows:
3. Remove snap rings from bearings using a punch.
4. Place rear propeller shaft yoke in a vise. Shaft must be supported horizontally and link yoke must be free to move vertically, Fig. 10.
5. Using a pipe coupling or a wrench socket with the inside diameter slightly larger than outside diameter of bearing, Fig. 10, drive link yoke downward until about a 1/4" of bearing projects from yoke. *Do not attempt to drive yoke down farther than ball socket will allow easily.*
6. Rotate shaft 180 degrees and repeat Steps 3, 4 and 5.
7. Clamp 1/4" projecting portion of either bearing in vise and remove bearing by driving link yoke upward. Remove other bearing in same manner, Fig. 11.
8. Separate spider, shaft yoke and shaft from link yoke.
9. To remove bearings from shaft yoke, clamp spider in vise with its jaws bearing against ends of spider journals. Yoke must be free to move vertically between jaws of vise.
10. Using the same bearing remover tool as in Step 5, apply force on shaft yoke around bearing. Drive yoke downward until bearing is free of yoke.

Centering Ball Replacement

The centering ball is only serviceable on 1973-75 models. Fig. 12 identifies Oldsmobile and Pontiac. Replacement procedure is as follows:

1. With CV joint disassembled, position inner part of tool J-23677 on centering ball, Fig. 13.
2. Install outer cylinder of tool J-23677 over inner part, thread nut onto tool and pull centering ball off stud.
3. Place replacement ball on stud and using a suitable tool, drive ball onto stud until it seats firmly against

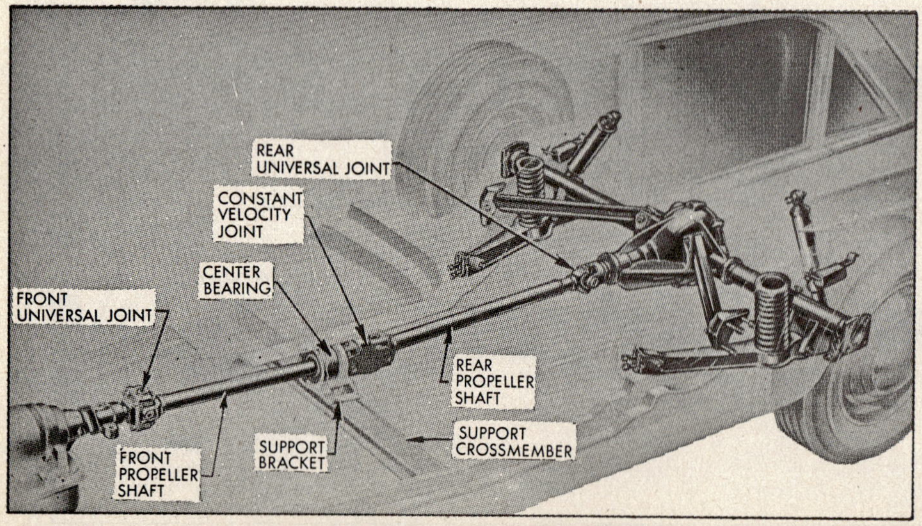

Fig. 7 Two piece propeller shaft with constant velocity universal joint. General Motors

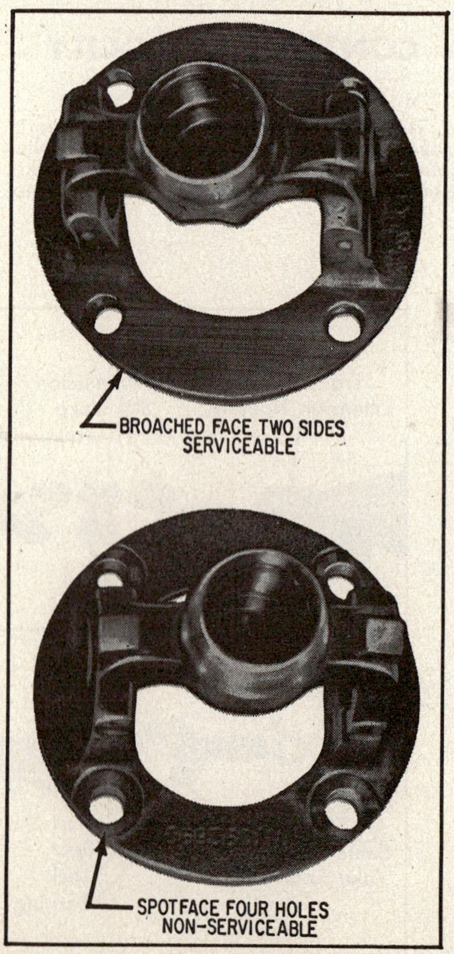

Fig. 8 Chevrolet serviceable and non-serviceable propeller shaft

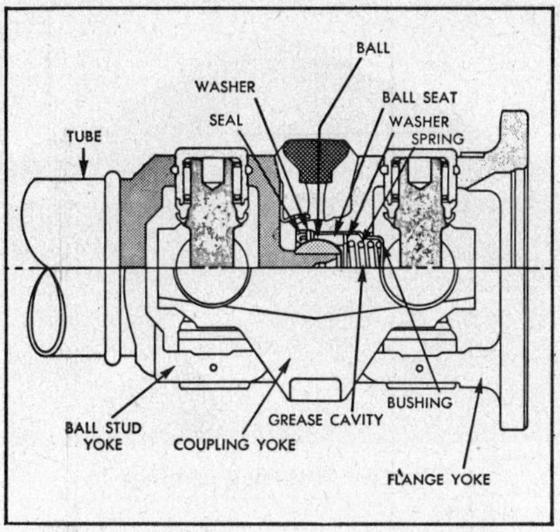

Fig. 9 Cross-section of typical GM constant velocity U-joint

shoulder at base of stud.

4. Assemble ball seats and related parts into ball cavity as shown in Fig. 14. All parts must be adequately lubricated with lubricant provided with kit.
5. Lubricate centering ball seal with approved lubricant and install with sealing lip tipping inward. Fill ball cavity with lubricant.

Reassemble Constant Velocity U-Joint—

All yokes must be carefully assembled using the marks made before disassembly for reference. Assemble front section of constant velocity joint first.

1. Position spider inside splined yoke. Install bearings by pressing between vise jaws. Make sure that spider journals enter bearings squarely to avoid damage, Fig. 15.
2. Fully install bearings and install snap rings.
3. Position splined yoke and spider inside link yoke and install bearings into link yoke in same manner as for splined yoke.
4. Position spider inside rear propeller shaft yoke and install bearings.
5. Lubricate ball and socket with a high grade of extreme pressure grease.
6. Position spider of rear propeller shaft assembly in link yoke.
7. Engage socket with ball of splined yoke assembly. *Make sure that all reference marks are properly aligned.*
8. Install bearings into link yoke in same manner as above while holding spring loaded ball and socket assembly together to make sure that spider journals enter bearings squarely.

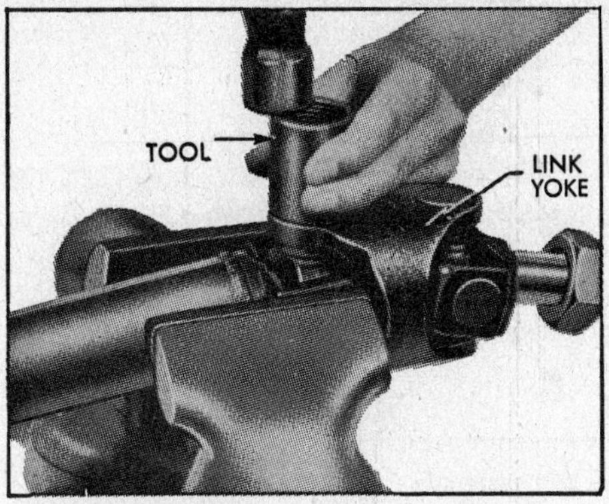

Fig. 10 Driving bearing from link yoke

Reassemble Slip Joint

1. Make sure locknut, seal and split washer are in place on smooth part of spline shaft. Also make sure that index spring wire is in place in splines and that spacer washer and large lockwasher are in place on rear end of front propeller shaft.
2. Align index spring with missing internal spline in rear end of propeller shaft and slide slip joint together, Fig. 16.
3. Install locknut and tighten securely. Bend in rim of lockwasher to engage flat of locknut firmly.

1969-71 Mark III, Thunderbird & 1969-75 Lincoln

These vehicles incorporate a double cardan type constant velocity joint at each end of the drive shaft, Fig. 17. Each double cardan has a center yoke (cage), a centering socket yoke and a stud yoke which is welded to each end of the tube assembly. The splines on the yoke and transmission output shaft permit drive shaft to move in and out as the axle moves up and down. All drive shaft assemblies are balanced and should be kept free of under coating.

Fig. 11 Removing bearing

Disassembly

1. Mark location of spiders, center yoke and centering socket yoke as related to stud yoke.

NOTE: The spiders must be assembled with bosses in their original position to provide proper clearance.

2. Remove snap rings that secure bearings in front of center yoke, then position tool, Fig. 18, and thread clockwise until bearing protrudes about ⅜-inch out of yoke.
3. Remove drive shaft from vise and tighten bearing in vise then tap the center yoke, Fig. 19, to free it from bearing.
4. Remove the two bearings from the spider, Fig. 20, then reposition the tool on the yoke and move the remaining bearing in the opposite direction so that it protrudes about ⅜ inch out the yoke.
5. Grip bearing in a vise, then drive the

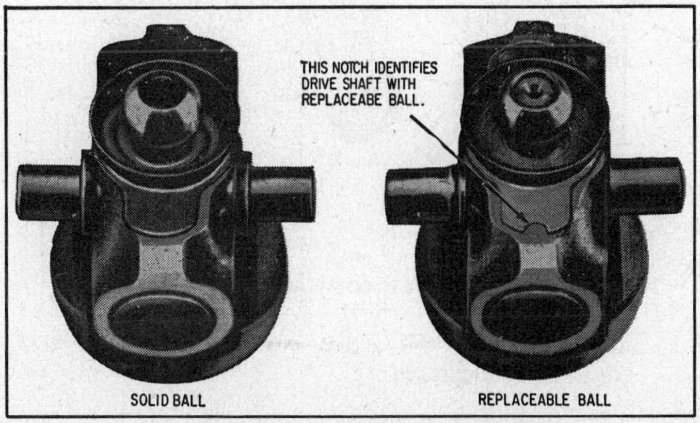

Fig. 12 Oldsmobile and Pontiac serviceable and non-serviceable centering balls

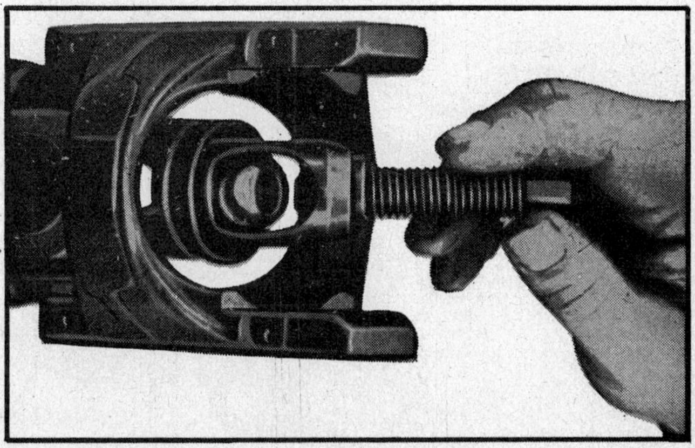

Fig. 13 Removing centering ball with tool J-23677

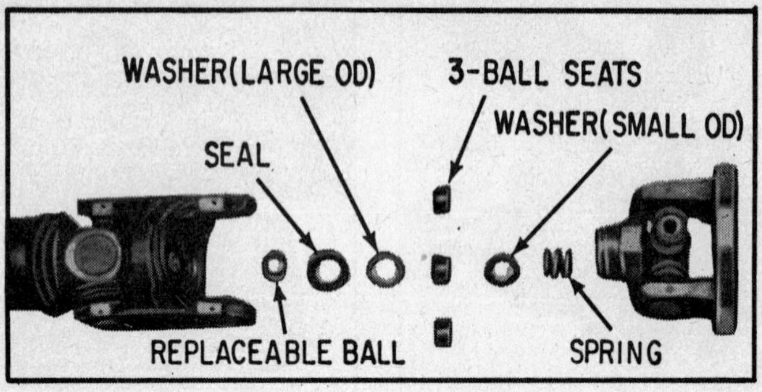

Fig. 14 Centering ball assembly

Fig. 15 Installing bearings

center yoke freeing it from bearing, Fig. 19, and remove spider from center yoke.

6. Pull centering socket yoke off center stud, Fig. 21, then remove rubber seal from centering ball stud.
7. Remove the snap rings from center and drive shaft yokes, then position tool on drive shaft yoke, Fig. 22, and press bearing outward until inside of center yoke almost contacts the slinger ring at the front of the drive shaft yoke.

NOTE: Pressing beyond this point can distort the slinger ring. Fig. 23 shows the interference point.

8. Clamp exposed end of bearing in a

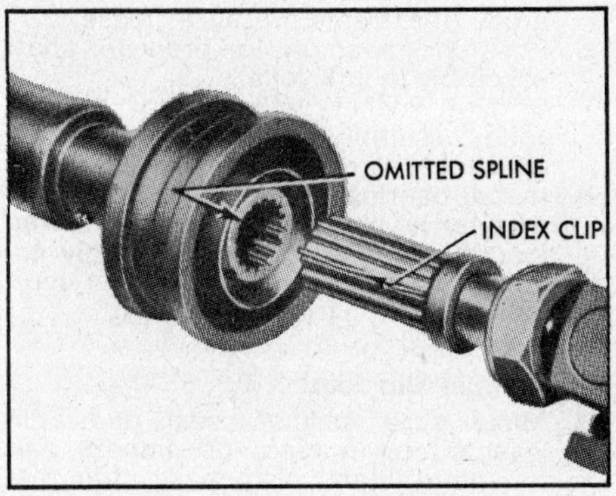

Fig. 16 Aligning index spring with missing internal spline on propeller shaft

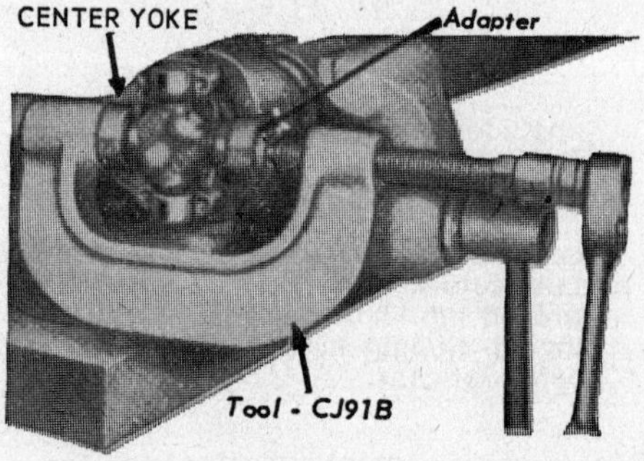

Fig. 18 Partially pressing bearing from center yoke

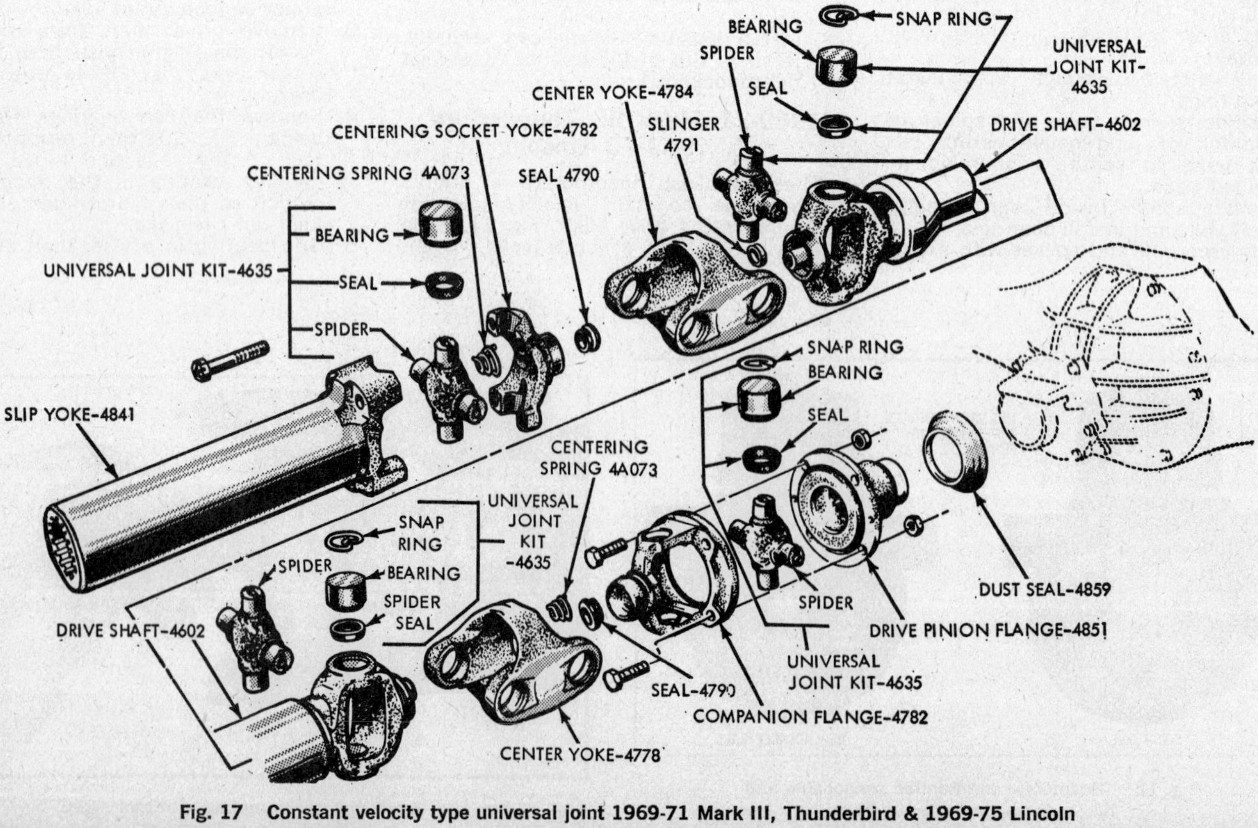

Fig. 17 Constant velocity type universal joint 1969-71 Mark III, Thunderbird & 1969-75 Lincoln

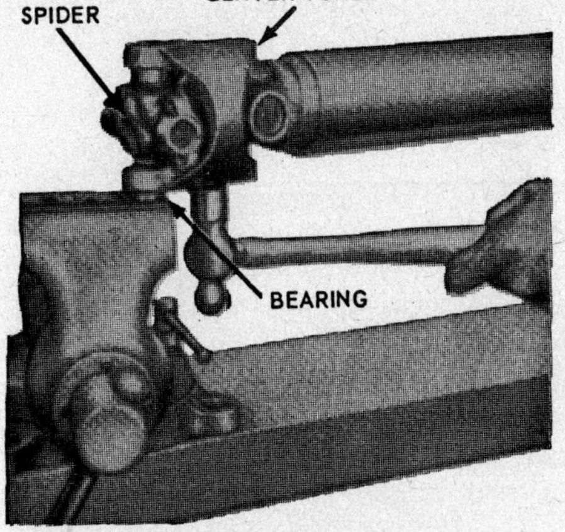

Fig. 19 Removing bearing from center yoke

vise and drive the center yoke with a soft face hammer freeing it from bearing, then reposition tool and

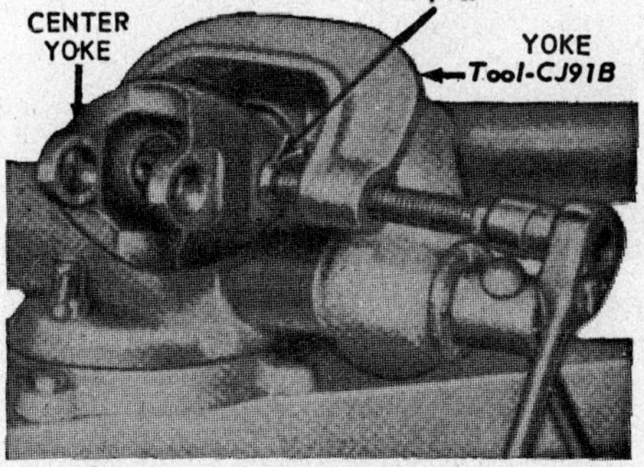

Fig. 22 Removing bearing from rear of center yoke

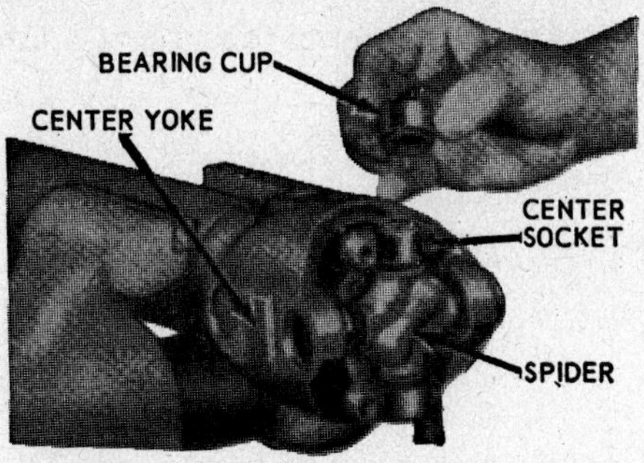

Fig. 20 Removing bearing cup from centering yoke

press on spider to remove opposite bearing.

9. Remove center yoke from spider and remove spider from drive shaft yoke in same manner.
10. Clean all serviceable parts in cleaning solvent. If using a repair kit, use all parts supplied in kit. If driveshaft is damaged, it should be replaced to insure a balanced assembly.

Assembly

1. Position spider in shaft yoke making sure that spider bosses (or lubrication plugs) will be in original position. Press in bearing cups and install snap rings.
2. Position center yoke over spider ends then press in bearing cups and install snap rings.
3. Install new seal on centering stud and position centering socket yoke on stud.
4. Place front spider in center yoke making sure that spider bosses (or lubrication plugs) are properly positioned. Press in bearing cups and install snap rings.
5. Applying pressure on centering socket, install remaining bearing cup.
6. If using a repair kit, remove plug from each spider and lubricate universal joints. Reinstall plug.

Fig. 21 Removing center socket yoke

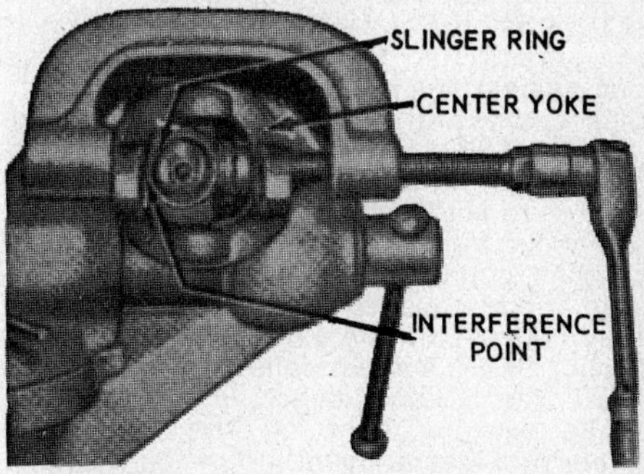

Fig. 23 Center yoke interference point

HYDRAULIC BRAKE SYSTEM

For Brake Adjustments, see Car Chapters

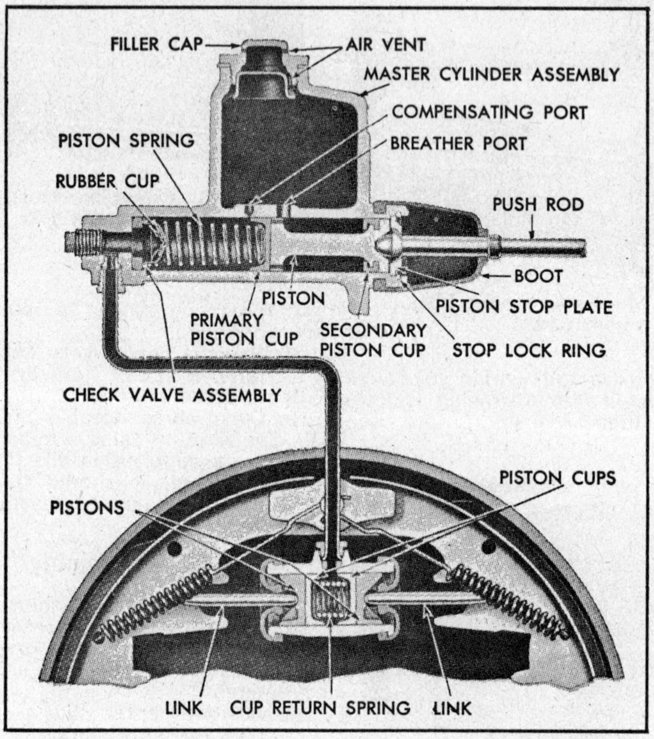

Fig. 1 Schematic diagram of a typical hydraulic brake system

DUAL MASTER CYLINDER SYSTEM

When the brake pedal is depressed, both the primary (front brake) and the secondary (rear brake) master cylinder pistons are moved simultaneously to exert hydraulic fluid pressure on their respective independent hydraulic system. The fluid displacement of the two master cylinders is proportioned to fulfill the requirements of each of the two independent hydraulic brake systems, Figs. 2 and 3.

If a failure of a rear (secondary) brake system should occur, initial brake pedal movement causes the unrestricted secondary piston to bottom in the master cylinder bore. Primary piston movement displaces hydraulic fluid in the primary section of the dual master cylinder to actuate the front brake system.

Should the front (primary) brake system fail, initial brake pedal movement causes the unrestricted primary piston to bottom out against the secondary piston. Continued downward movement of the brake pedal moves the secondary piston to displace hydraulic fluid in the rear brake system to actuate the rear brakes.

The increased pedal travel and the increased pedal effort required to compensate for the loss of the failed portion of the brake system provides a warning that a partial brake system failure has occurred When the ignition switch is turned on, a brake warning light on the instrument panel provides a visual indication that one of the dual brake systems has become inoperative.

Should a failure of either the front or rear brake hydraulic system occur, the hydraulic fluid pressure differential resulting from pressure loss of the failed brake system forces the valve toward the low pressure area to light the brake warning lamp.

Brake Warning Light Switches

There are three basic types of brake warning light switches as shown in Figs. 5, 6 and 7, and usually they form a common electrical circuit with the brake warning light.

When a pressure differential occurs between the front and rear brake systems, the valves will shuttle toward the side with the low pressure.

As shown in Fig. 5, movement of the differential valve forces the switch plunger upward over the tapered shoulder of the valve to close the switch contacts and light the dual brake warning lamp, signaling a brake system failure.

In Fig. 6 the valve assembly consists of two valves in a common bore that are spring loaded toward the centered position. The spring-loaded switch contact plunger rests on top of the valves in the centered position (right view). When a pressure differential occurs between the front and rear brake systems, the valves

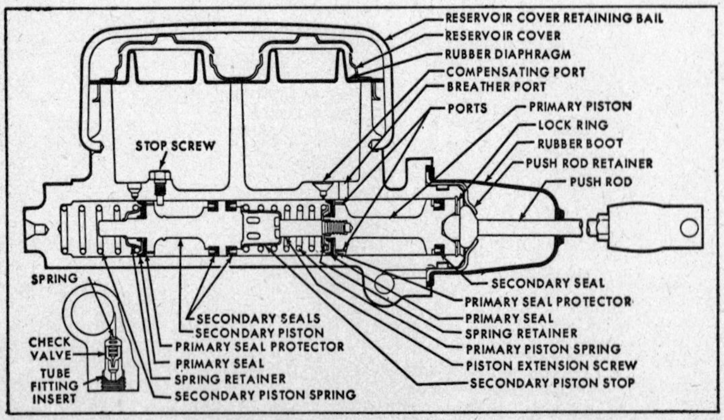

Fig. 2 Delco-Moraine dual master cylinder used with drum brakes (typical)

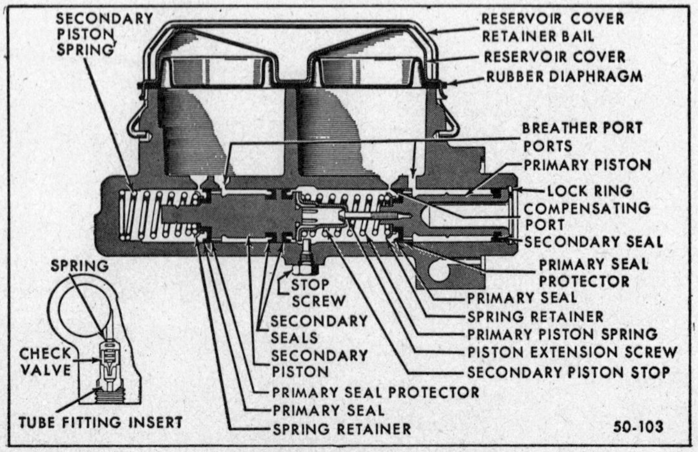

Fig. 3 Bendix dual master cylinder used with drum brakes (typical)

will shuttle toward the side with the low pressure. The spring-loaded switch plunger is "triggered" and the ground circuit for the warning light is completed, lighting the lamp (left view).

In Fig. 7, as pressure falls in one system, the other system's normal pressure forces the piston to the inoperative side, contacting the switch terminal, causing the warning light on the instrument panel to glow.

Testing Warning Light System

If the parking brake light is connected into the service brake warning light system, the brake warning light will flash only when the parking brake is applied with the ignition turned ON. The same light will also glow should one of the two service brake systems fail when the brake pedal is applied.

To test the system, turn the ignition ON and apply the parking brake. If the lamp fails to light, inspect for a burned out bulb, disconnected socket, a broken or disconnected wire at the switch.

Fig. 8 is an exterior view of one of these switches. They are usually mounted on the left frame side rail or on the brake pedal bracket.

To test the brake warning system, raise the car and open a wheel bleeder valve while a helper depresses the brake pedal and observes the warning light on the instrument panel. If the bulb fails to light, inspect for a burned out bulb, disconnected socket, or a broken or disconnected wire at the switch. If the bulb is not burned out, and the wire continuity is proven, replace the brake warning switch.

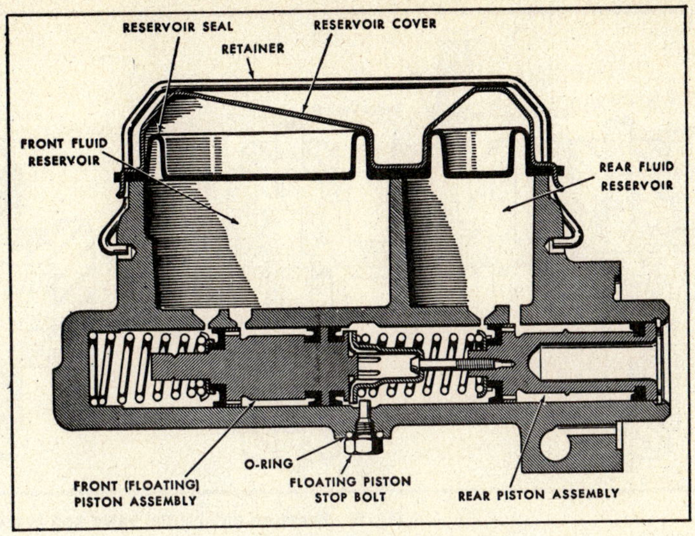

Fig. 4 Bendix dual master cylinder used with disc brakes (typical)

Combination Valve

The combination valve, Fig. 9 is a metering valve, failure warning switch, and a proportioner in one assembly and is used on disc brake applications. The metering valve delays front disc braking until the rear drum brake shoes contact the drum. The failure warning switch is actuated in event of front or rear brake system failure, in turn activating a dash warning lamp. The proportioner balances front to rear braking action during rapid deceleration.

Metering Valve

When the brakes are not applied, the metering valve permits the brake fluid to flow through the valve, thus allowing the fluid to expand and contract with temperature changes.

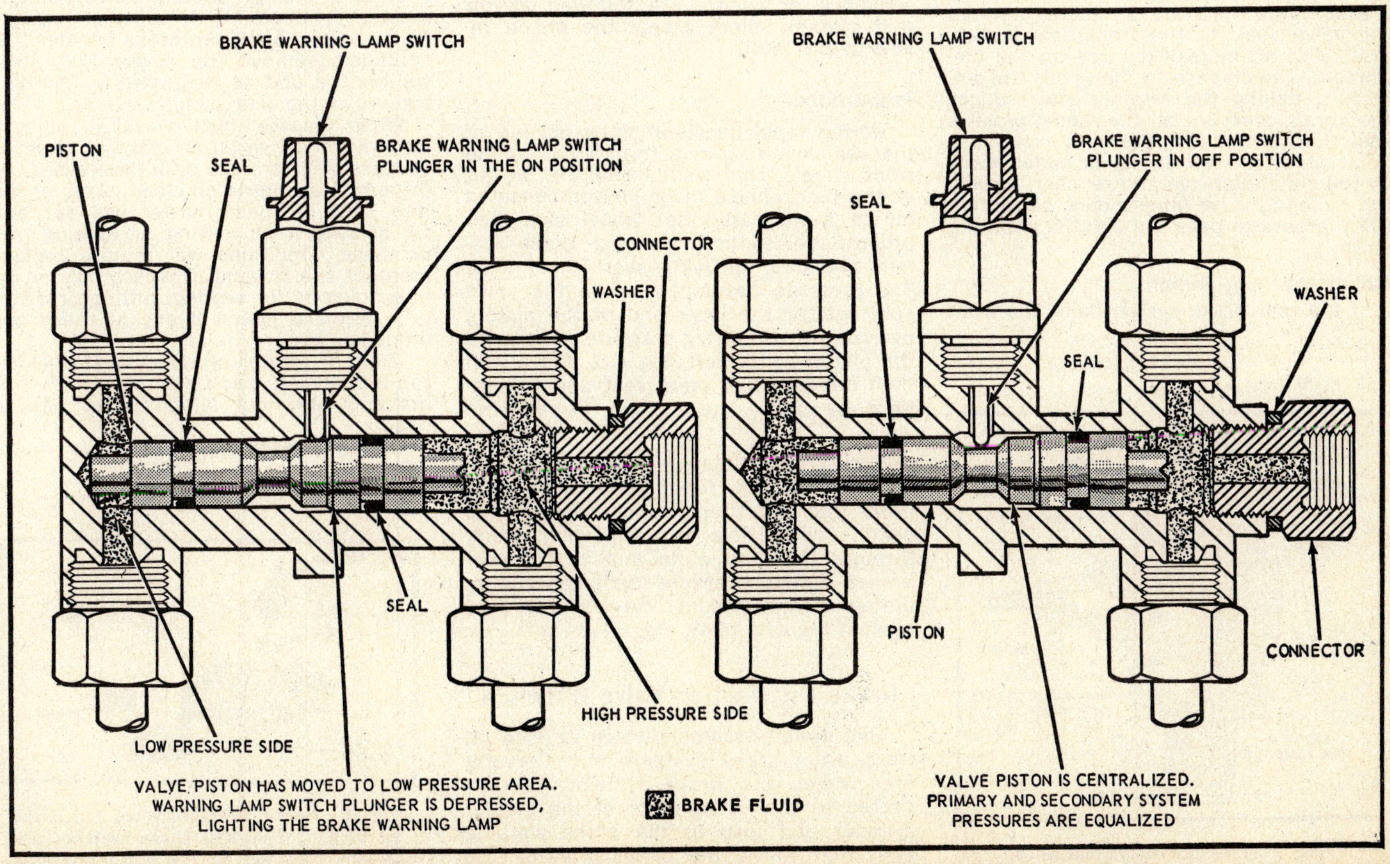

Fig. 5 Pressure differential valve and brake warning light switch

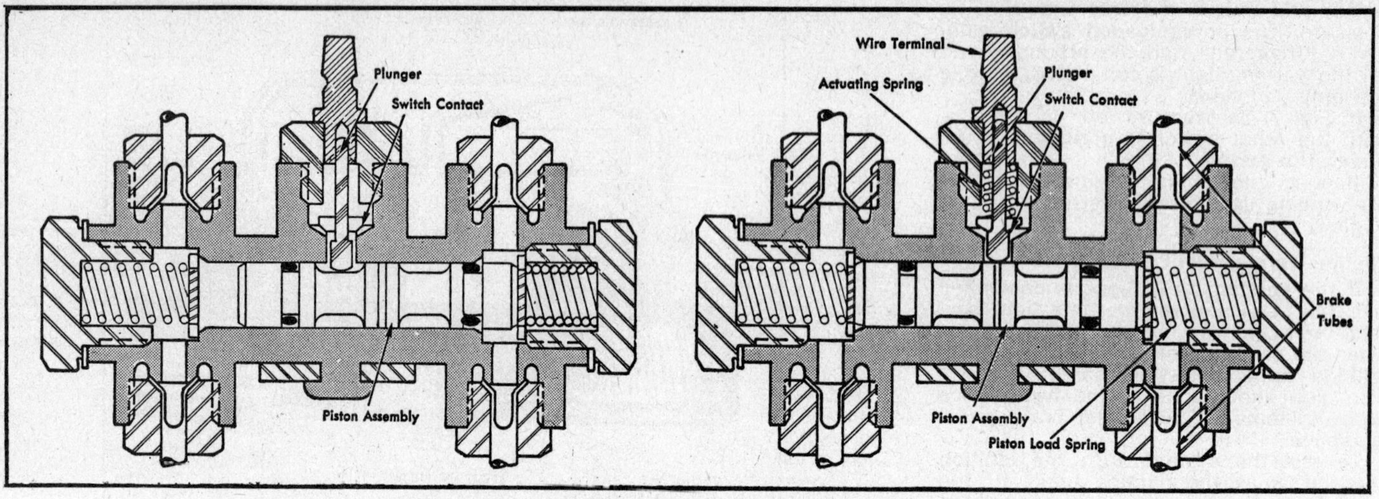

Fig. 6 Pressure differential valve and brake warning light switch

When the brakes are initially applied, the metering valve stem moves to the left, preventing fluid to flow through the valve to the front disc brakes. This is accomplished by the smooth end of the metering valve stem contacting the metering valve seal lip at 4 to 30 PSI, Fig. 10. The metering valve spring holds the retainer against the seal until a predetermined pressure is produced at the valve inlet port which overcomes the spring pressure and permits hydraulic pressure to actuate the front disc brakes, Fig. 11. The increased pressure into the valve is metered through the valve seal, to the front disc brakes, producing an increased force on the diaphragm. The diaphragm then pulls the pin, in turn pulling the retainer and reduces the spring pressure on the metering valve seal. Eventually, the pressure reaches a point at which the spring is pulled away by the diaphragm pin and retainer, leaving the metering valve unrestricted, permitting full pressure to pass through the metering valve.

Failure Warning Switch

If the rear brake system fails, the front system pressure forces the switch piston to the right, Fig. 12. The switch pin is then forced up into the switch, completing the electrical circuit and activates the dash warning lamp.

When repairs are made and pressure returns to the system, the piston moves to the left, resetting the switch. The detent on the piston requires approximately 100 to 450 PSI to permit full reset of the piston. In event of front brake system failure, the piston moves to the left and the same sequence of events are followed as for rear system failure except the piston resets to the right.

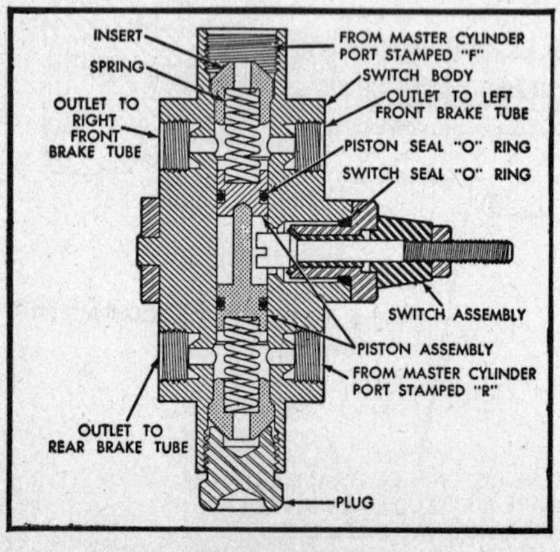

Fig. 7 Pressure differential valve and brake warning light switch

Proportioner

During rapid deceleration, a portion of the vehicle's weight is transferred to the front wheels. This resultant loss of weight at the rear wheels must be compensated for to avoid early rear wheel skid. The proportioner reduces the rear brake system pressure, delaying rear wheel skid. The pressure developed within the valve acts against the large end of the piston, overcoming the spring pressure and moves the piston to the left, Fig. 13. Ths piston then contacts the stem seat and proportions (restricts) line pressure through the valve.

During normal braking action, the proportioner is not functional. Brake fluid flows into the proportioner between the piston center hole and the valve stem, through the stop plate and to the rear brakes. Spring pressure loads the piston, during normal braking, causing it to rest against the stop plate, Fig. 14.

Brake Distribution Valve & Switch

This switch assembly which is used on front drum brake systems and Corvette four wheel disc brake systems, is connected to the outlet ports of the master cylinder and also to the brake warning light that warns the driver if either the front or rear brake system has failed.

When hydraulic pressure is equal in both front and rear brake systems, the switch remains centered, Fig. 15. If pressure fails in one of the systems, hydraulic pressure moves the piston toward the inoperative side, Fig. 16. The shoulder of the piston contacts the switch terminal, providing a ground and lighting the warning lamp.

Master Cylinder Service

Figs. 17 to 21 show an array of dual master cylinders. With cylinder removed from vehicle, and from brake booster if so equipped, remove the covers and disassemble the unit as suggested by the illustration of the unit being serviced.

When disassembled, wash all parts in alcohol *only*. Use an air hose to blow out all passages, orifices and valve holes. Air dry and place parts on clean paper or lint-free cloth. Inspect master cylinder bore for scoring, rust, pitting or etching. Any of these conditions will require replacement of the housing. Inspect master cylinder pistons for scoring, pitting or distortion. Replace piston if any of these conditions exist.

If either master cylinder housing or piston is replaced, clean new parts with alcohol and blow out all passages with air hose.

Examine reservoirs for foreign matter

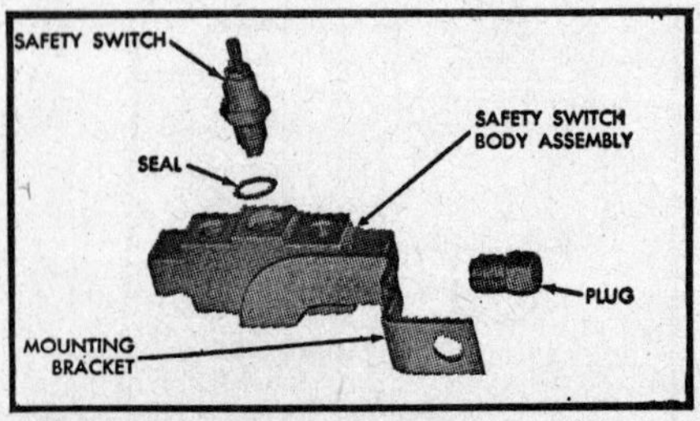

Fig. 8 Typical pressure valve and brake warning light switch. These switches are usually mounted on the left frame side rail

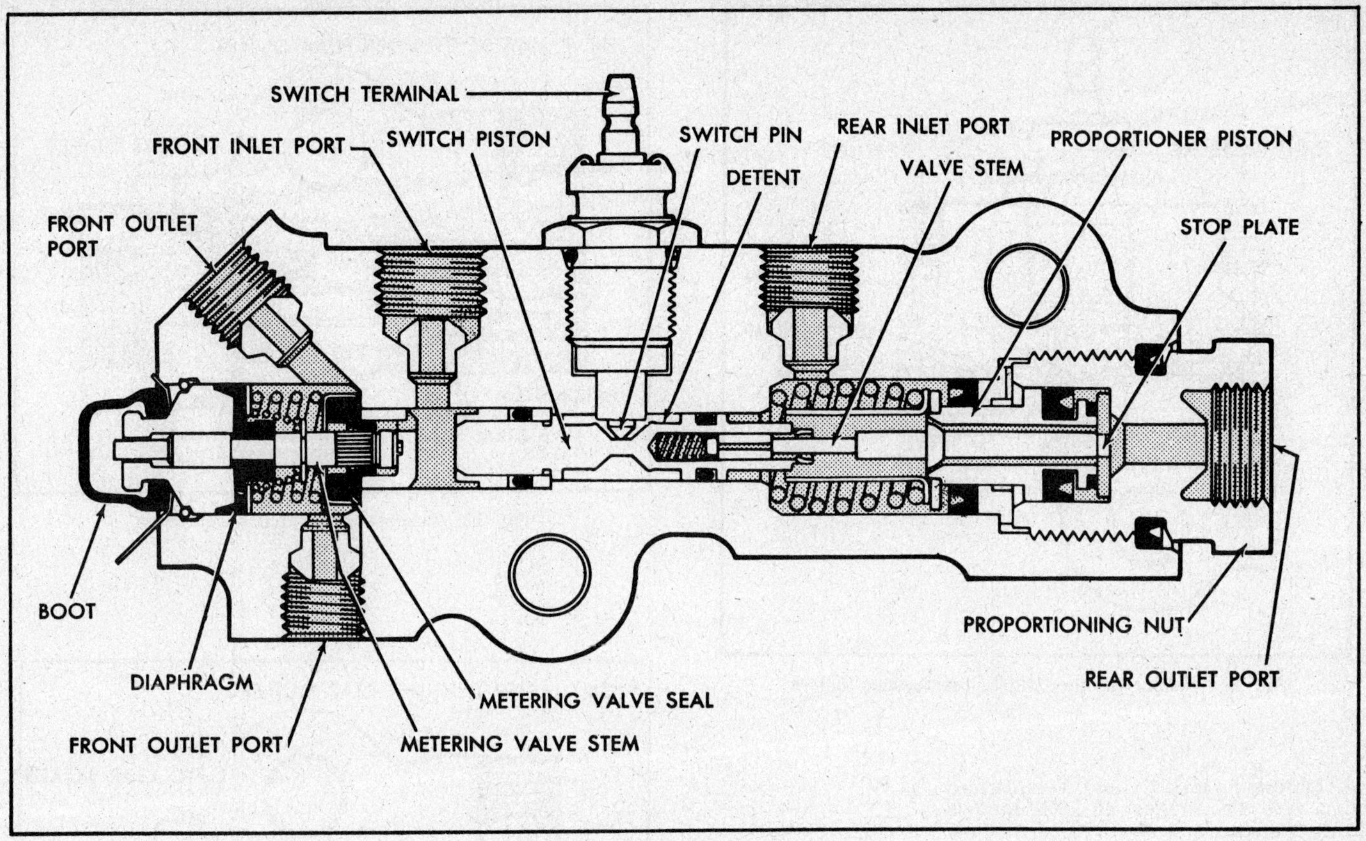

Fig. 9 Combination valve

and check all passages for restrictions. If there is any suspicion of contamination or evidence of corrosion, completely flush hydraulic system as outlined below.

When overhauling a master cylinder, use all parts contained in repair kit. Before starting reassembly, dip all cups, seals, pistons, springs, check valves and retainers in alcohol and place in a clean pan or on clean paper. *Wash hands with soap and water only to prevent contamination of rubber parts from oil, kerosene or gasoline.* During assembly, dip all parts in clean, heavy duty brake fluid.

Inspect through side outlet of dual master cylinder housing to make certain cup lips do not hand up on edge of hole or turn back, which would result in faulty operation. A piece of $^3/_{16}$" rod with an end rounded off will be helpful in guiding cups past hole.

BLEEDING BRAKES

NOTE: Chrysler Corp. recommends only pressure bleeding for all 1969-75 models. Ford Motor Co. recommends pressuring bleeding for 1969-71 Lincoln, 1970-71

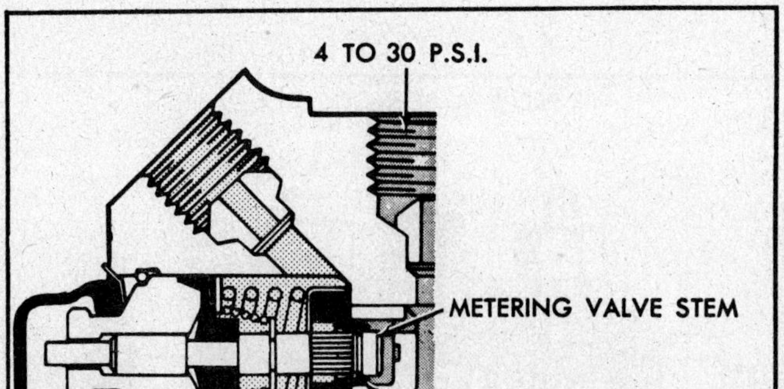

Fig. 10 Metering valve, initial braking

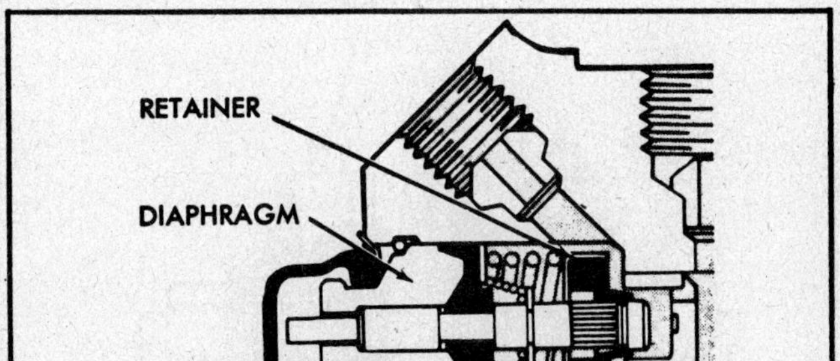

Fig. 11 Metering valve, continued braking

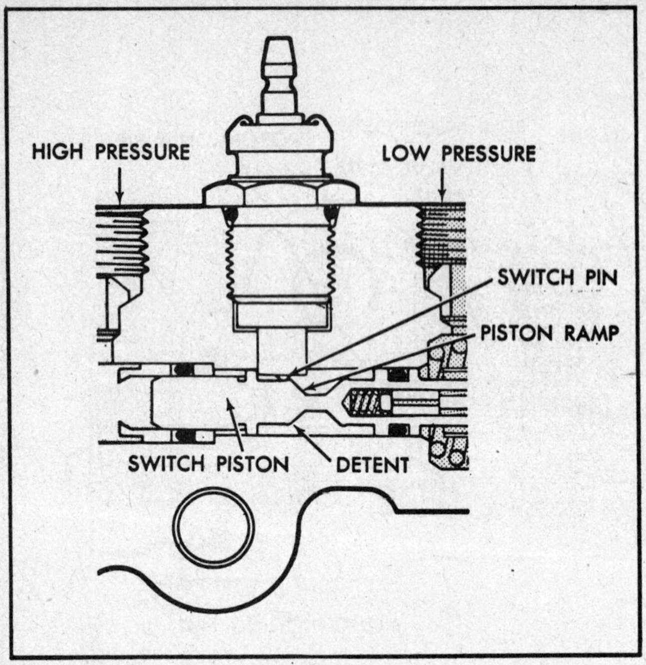

Fig. 12 Failure warning switch, rear system failure

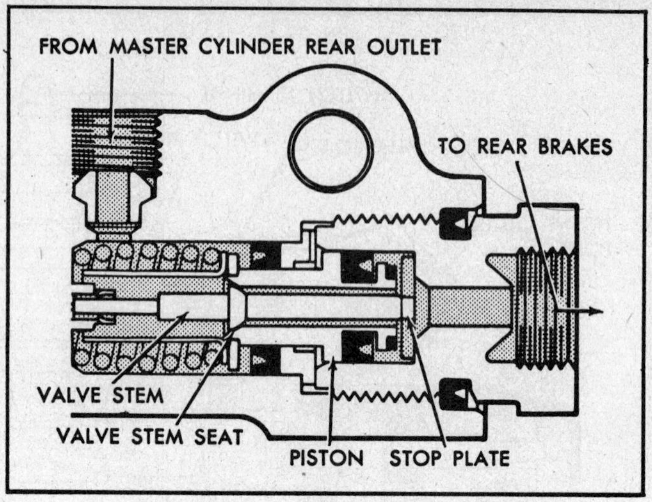

Fig. 13 Proportioner, rapid deceleration

Ford, Mercury, Mark III and Thunderbird models with disc brakes, all 1972 models with disc brakes and all 1973-75 models.

The bleeding operation itself is fairly well standardized. First step in all cases is cleaning the dirt from the filler cap before removing it from the master cylinder. This should be done thoroughly.

Pressure bleeding is fastest because the master cylinder doesn't have to be refilled several times, and the job can be done by one man. To prevent air from the pressure tank getting into the lines, do not shake the tank while air is being added to the tank or after it has been pressurized. Set the tank in the required location, bring the air hose to the tank, and

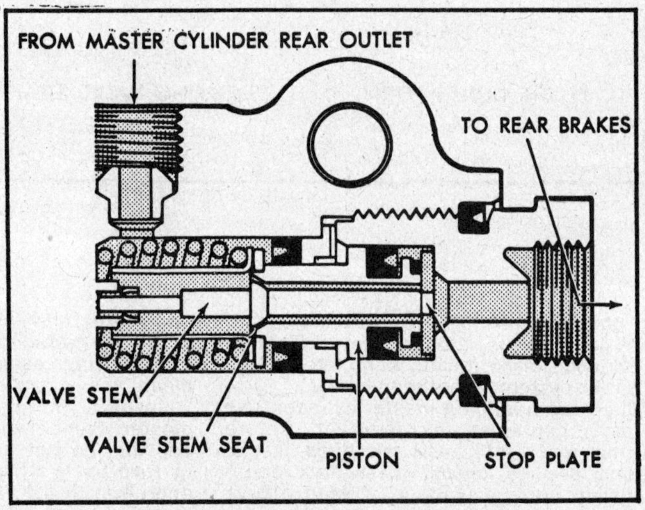

Fig. 14 Proportioner, normal braking

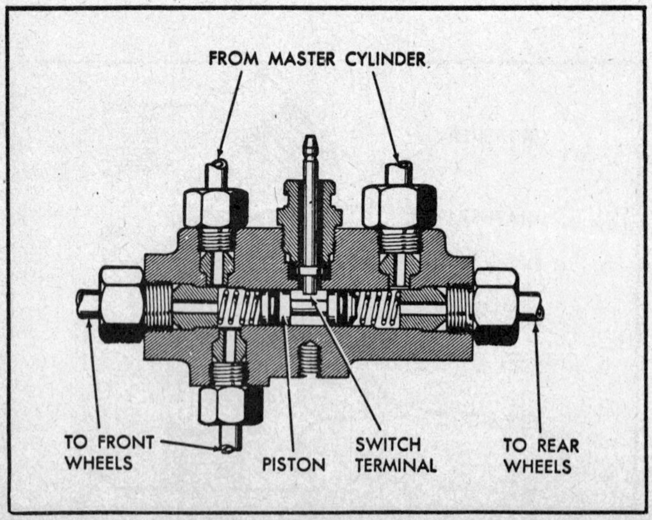

Fig. 15 Brake distribution switch (normal)

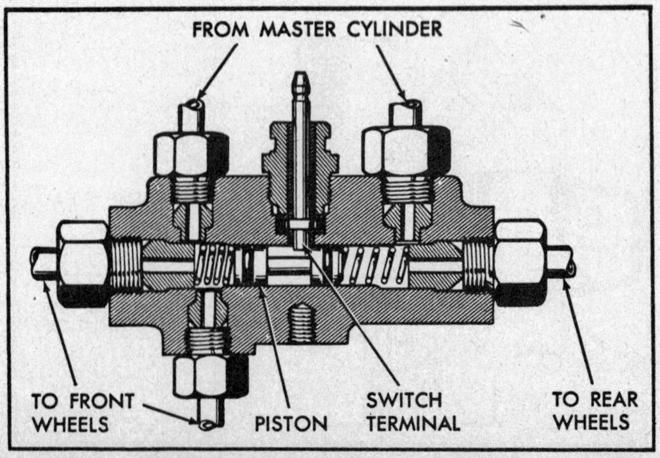

Fig. 16 Brake distribution switch (failed)

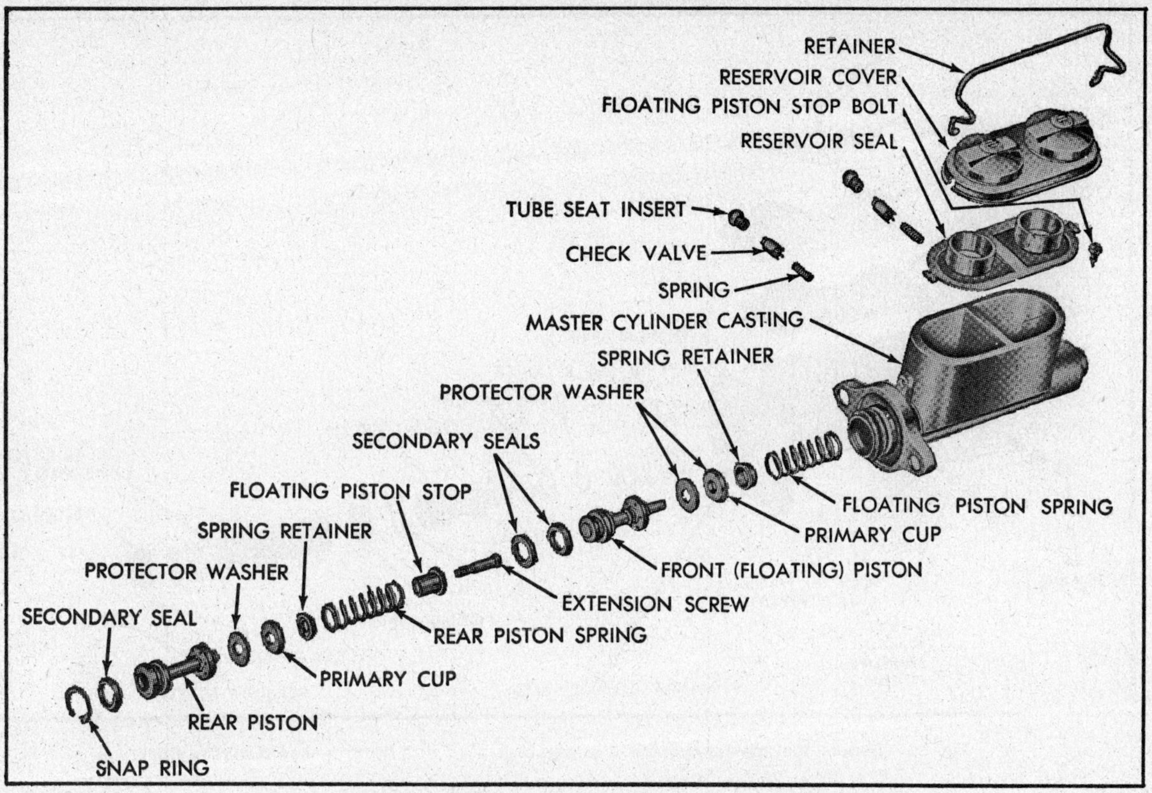

Fig. 17 Delco-Moraine dual master cylinder disassembled (GM cars)

do not move it during the bleeding operation. The tank should be kept at least one-third full.

NOTE: On vehicles equipped with disc brakes, the brake metering valve or combination valve must be held in its position, using the recommended tool, (C-4121 for Chrysler Corp., J-22742 for Ford Motor Co. and J-23709 for General Motors).

If air does get into the fluid, releasing the pressure will cause the bubbles to increase in size, rise to the top of the fluid, and escape. Pressure should not be greater than about 35 lb. per sq. in.

When bleeding without pressure, open the bleed valve three-quarters of a turn, depress the pedal a full stroke, then allow the pedal to return slowly to its released position. Some makers suggest that after the pedal has been depressed to the end of its stroke, the bleeder valve should be closed before the start of the return stroke. On cars with power brakes, first reduce the vacuum in the power unit to zero by pumping the brake pedal several times with the engine off before starting to bleed the system.

Pressure bleeding, of course, eliminates the need for pedal pumping. Chrysler Corp. suggests that, when pressure is used, the bleeder valve should be opened and closed intermittently at about four-second intervals. This gives a whirling action to the fluid in the wheel cylinder, and helps expel the air.

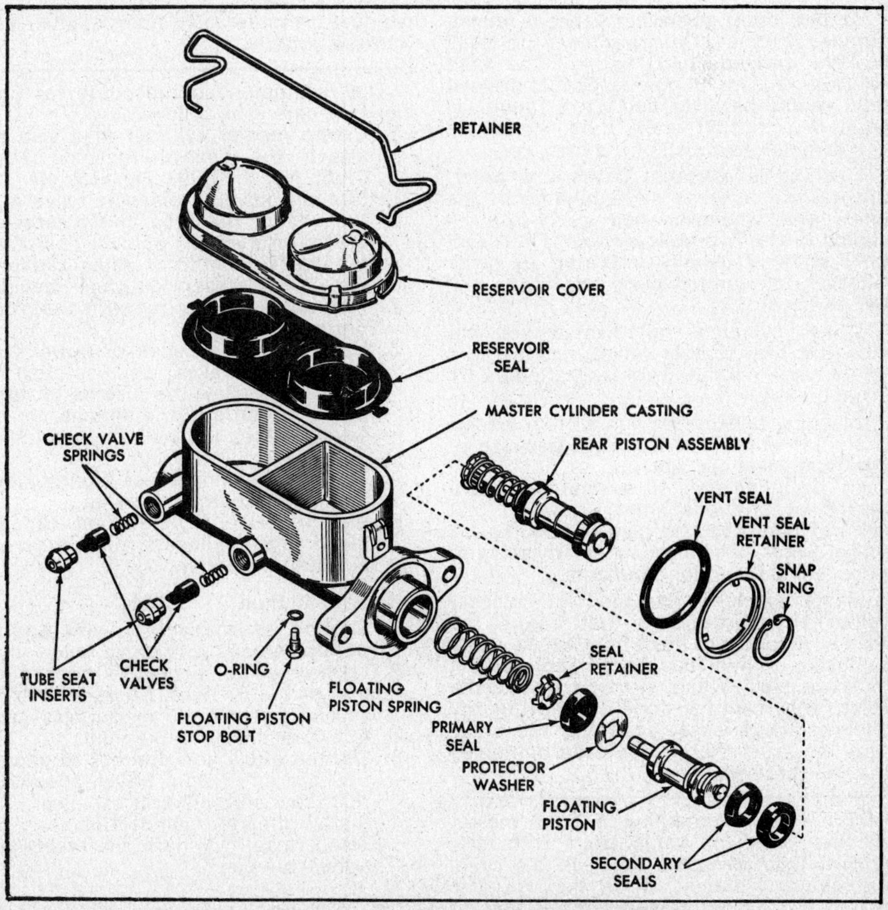

Fig. 18 Bendix dual master cylinder disassembled used with drum brakes (GM and American Motors)

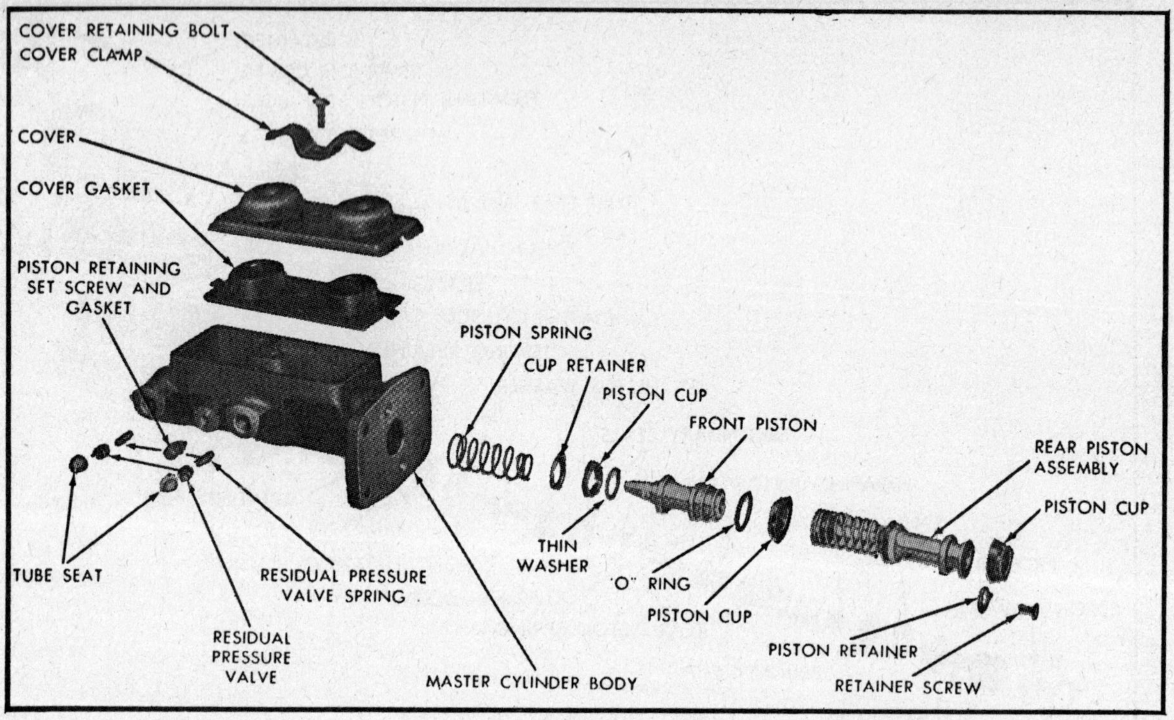

Fig. 19 Bendix dual master cylinder disassembled used with drum brakes (Chrysler line)

At one time, some car makers recommended that a clean container be used for the drained fluid, so that the fluid could be reused. All now agree that drained fluid should be discarded. Care should be taken not to spill brake fluid, since this can damage the finish of the car.

Flushing is essential if there is water, mineral oil or other contaminants in the lines, and whenever new parts are installed in the hydraulic system. Fluid contamination is usually indicated by swollen and deteriorated cups and other rubber parts.

Wheel cylinders on disc brakes are equipped with bleeder valves, and are bled in the same manner as wheel cylinders for drum brakes.

Bleeding is necessary on all four wheels if air has entered the system because of low fluid level, or the line or lines have been disconnected. If a line is disconnected at any one wheel cylinder, that cylinder only need be bled. Of course, on brake reline jobs, bleeding is advisable to remove any air or contaminants.

Master cylinders equipped with bleeder valves should be bled first before the wheel cylinders are bled. In all cases where a master cylinder has been overhauled, it must be bled. Where there is no bleeder valve, this can be done by leaving the line (or lines) loose, actuating the brake pedal to expel the air and then tightening the line (or lines).

NOTE: After overhauling a dual master cylinder used in conjunction with disc brakes, it is advisable to bleed the cylinder before installing it on the car. The reason for this recommendation is that air may be trapped between the master cylinder pistons because there is only one residual pressure valve (check valve) used in these units.

The recommended precedure for Chrysler Line cars is as follows:

1. Clamp master cylinder in a vise and attach the special Bleeding Tubes (Tool No. C-4029) Fig. 15. *Be sure that the residual pressure valve is on the end of the tube in the large capacity reservoir as shown. This keeps the brake fluid from being syphoned out of the reservoir while bleeding.*
2. Fill both reservoirs with approved brake fluid.
3. Using a wooden stick or dowel (cars with power brakes) depress push rod slowly and allow the pistons to return under pressure of the springs. Do this several times until all air bubbles are expelled.
4. Remove bleeding tubes from cylinder and install cover and gasket.
5. Install master cylinder on car and bleed wheel cylinders, preferably with a pressure bleeder.

Alternate Method

1. Support assembly in a vise and fill both reservoirs with brake fluid.
2. Loosely install a plug in each outlet port of the cylinder. Depress push rod several times until air bubbles cease to appear in the brake fluid.
3. Tighten plugs and attempt to depress the piston. Piston travel should be restricted after all air is expelled.
4. Install master cylinder on car and bleed wheel cylinders, preferably with a pressure bleeder.

Testing Dual Master Cylinders

Be sure that the master cylinder compensates in both ports. This can be done by applying the brake pedal lightly (engine running with power brakes), and observing for brake fluid squirting up in the reservoirs. This may only occur in the front chamber. To determine if the rear compensating port is open, pump up the brakes rapidly and hold the pedal down. Have an observer watch the fluid in the rear reservoir while the pedal is raised. A disturbance in the fluid indicates that the compensating port is open.

Wheel Bleeding Sequence

Difference of opinion as to whether the longest or shortest line should be bled first still exists. To be safe, use the sequence given below, recommended by the car manufacturers.

Chrysler Corp. cars .... RR-LR-RF-LF
Ford Company cars .... RR-LR-RF-LF
General Motors cars:
All except Chevrolet .... LF-RF-LR-RR
Chevrolet Division .... LR-RR-LF-RF
American Motors .... RR-LR-RF-LF

Dual Master Cylinder Bleeding Notes

Ford Motor Co. Cars

After the normal bleeding operation has been completed, note that the brake warning light will be ON because the pressure differential valve has moved off center, and must be returned to the central position. To do this, loosen the valve's inlet tube on the side opposite the wheel cylinder that was bled last. Apply the brake pedal slowly until the warning light goes out, and tighten fitting. Replace any

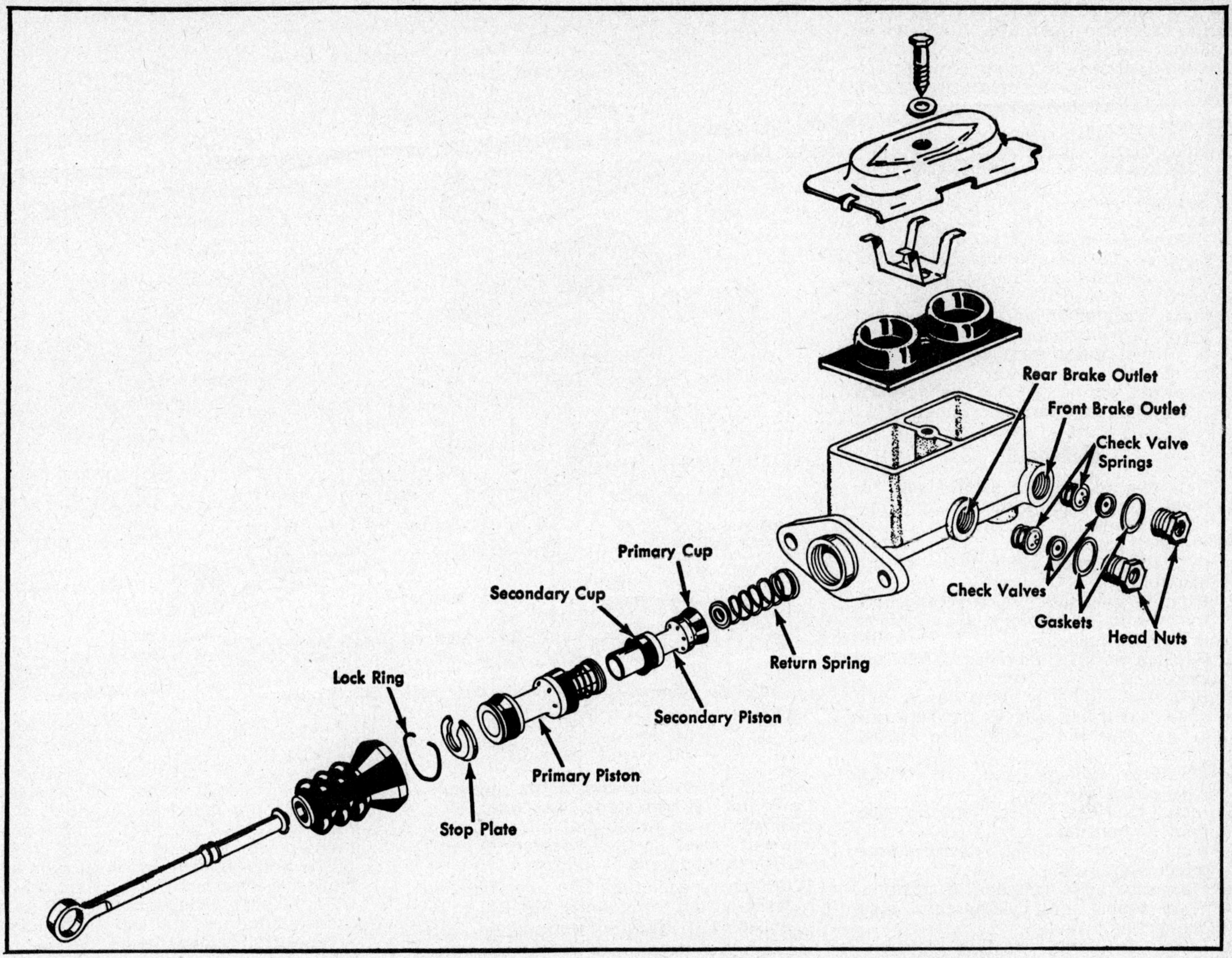

Fig. 20 Wagner dual master cylinder disassembled (Rambler American)

fluid that has leaked out during the operation.

NOTE: 1970-75 cars use a self-centering valve. After any bleeding operation, turn ignition switch to ACC or ON position and depress brake pedal. Valve will center itself.

General Motors Cars

On vehicles equipped with combined drum and disc brakes, the pressure differential valve must be held in its open position while bleeding the brakes using a pressure type bleeder. This is done by depressing the plunger and holding it in place using a special tool. When using pressure bleeding equipment follow the recommendations of the manufacturer.

Corvette

On Corvette models with disc brakes on all four wheels, there are two bleeder valves on each rear wheel disc brake. Remove the rear wheels to bleed. A single valve is used on front disc brakes.

American Motors

Before bleeding brakes, disconnect the switch terminal wire and remove nylon switch terminal, contact plunger actuating spring, and nylon plunger with contact.

In the event the valve has "triggered", the valve centering spring pressure may hold the switch plunger. If this happens, apply a slight amount of brake pedal pressure while releasing the plunger from the valve body.

After the bleeding operation, assemble the plunger spring and install valve with contact down. Install the nylon terminal and connect warning light wire to valve terminal. In the event brake fluid leaks from the center terminal body opening when the terminal is removed, replace the valve assembly.

Chrysler Corp.

Some Chrysler built cars with disc brakes are equipped with front disc brake pressure metering valve which is located on the left frame rail directly under the battery.

The purpose of the metering valve is to provide a better match of the front disc brakes with the rear drum brakes, resulting in improved braking balance *in light pedal applications.*

Gravity bleed and pedal methods are not affected by the presence of the metering valve. However, pressure bleeding is influenced by the metering valve.

Bleed pressure, which is normally about 35 psi, is high enough to cause the metering valve to close, which stops the flow of fluid to the front brakes. However, the valve can be held open manually by depressing the pressure release plunger (located at the bottom of the valve) in its uppermost position by hand or secured with masking tape while bleeding the brakes.

CAUTION: Under no conditions should a rigid clamp, wedge or block be used to secure the plunger as this can cause an internal failure in the valve. It should be noted that the pressure release plunger of the valve is already in its uppermost position when there is no pressure present.

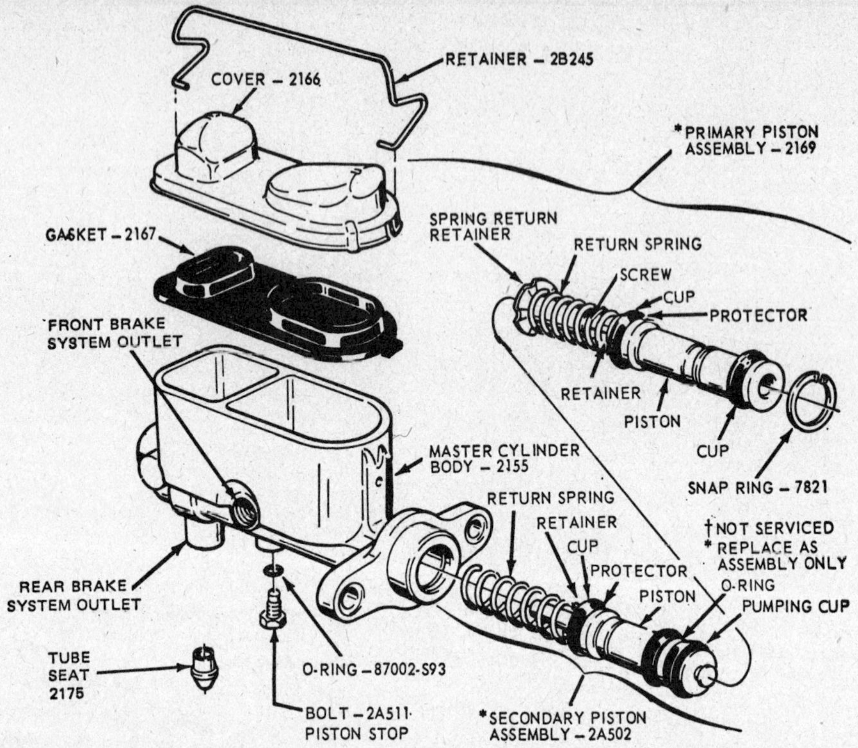

Fig. 21 Cutaway view of Bendix dual master cylinder used with disc brakes (typical)

WHEEL CYLINDERS

1. Remove wheel, drum and brake shoes.
2. Disconnect hydraulic line at wheel cylinder. *Do not pull metal line away from cylinder as the cylinder connection will bend metal line and make installation difficult. Line will separate from cylinder when cylinder is moved away from brake backing plate.*
3. Remove screws holding cylinder to brake plate and remove cylinder.

Overhaul

1. Referring to Fig. 22 as a guide, remove boots, pistons, springs and cups from cylinder.
2. Place all parts, except cylinder casting in alcohol. Wipe cylinder walls with alcohol.
3. Examine cylinder bore. A scored bore may be honed providing the diameter is not increased more than .005". Replace worn or damaged parts from the repair kit.
4. Before assembling, wash hands with soap and water only as oil, kerosene or gasoline will contaminate rubber parts.
5. Lubricate cylinder wall and rubber cups with brake fluid.
6. Install springs, cups, pistons and boots in housing.
7. Wipe end of hydraulic line to remove any foreign matter.
8. Place hydraulic cylinder in position. Enter tubing into cylinder and start connecting fitting.
9. Secure cylinder to backing plate and then complete tightening of tubing fitting.
10. Install brake shoes, drum and wheel.
11. Bleed system as outlined previously, and adjust brakes.

FLUSHING HYDRAULIC SYSTEM

It may sometime become necessary to flush out the system due to the presence of mineral oil, kerosene, gasoline, etc., which will cause swelling of rubber piston cups and valves so they become inoperative. The procedure is as follows:

1. Attach bleeder tube and open bleeder valve at left front wheel, Fig. 24.
2. Flush out system thoroughly with clean denatured alcohol, pumping the fluid from the master cylinder reservoir and out of the wheel cylinder bleeder valve.
3. Repeat Steps 1 and 2 at remaining wheel cylinders. To ensure thorough flushing, about ½ pint of alcohol should be bled through each wheel cylinder.
4. Replace all rubber parts in master and wheel cylinders. Thoroughly clean cylinders and pistons in alcohol before installing new parts.
5. After installing parts, fill system with recommended brake fluid and flush system of cleaning solution and then bleed brakes. In doing this, pump brake fluid from wheel cylinder bleeder valves until clear fluid flows from bleeder tube and then, if necessary, continue until no air bubbles emerge from bleeder tube.

HYDRAULIC TUBING

Steel tubing is used to conduct hydraulic pressure to the brakes. All fittings, tubing and hose should be inspected for rusted, damaged or defective flared seats. The tubing is equipped with a double flare or inverted seat to insure more positive seating in the fitting. To repair or reflare

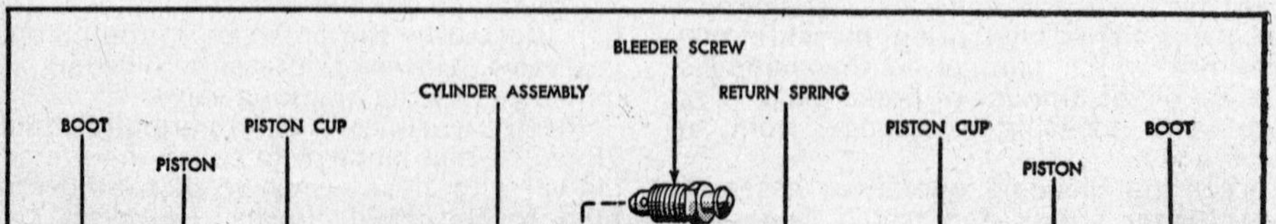

Fig. 22 Disassembled view of typical wheel cylinder

Fig. 23 Bleeding master cylinder used in conjunction with disc brakes

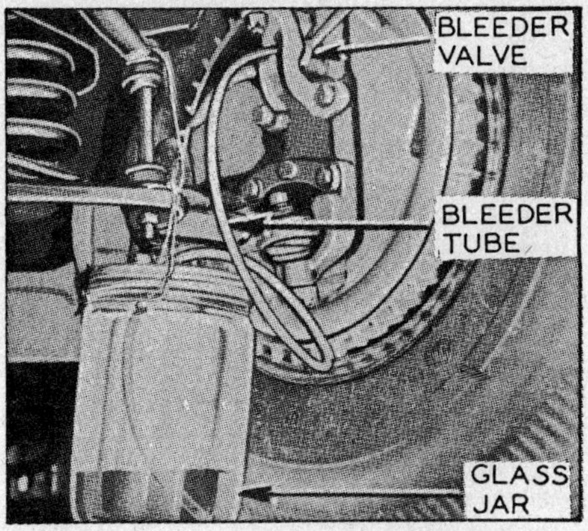

Fig. 24 Bleeding wheel cylinder

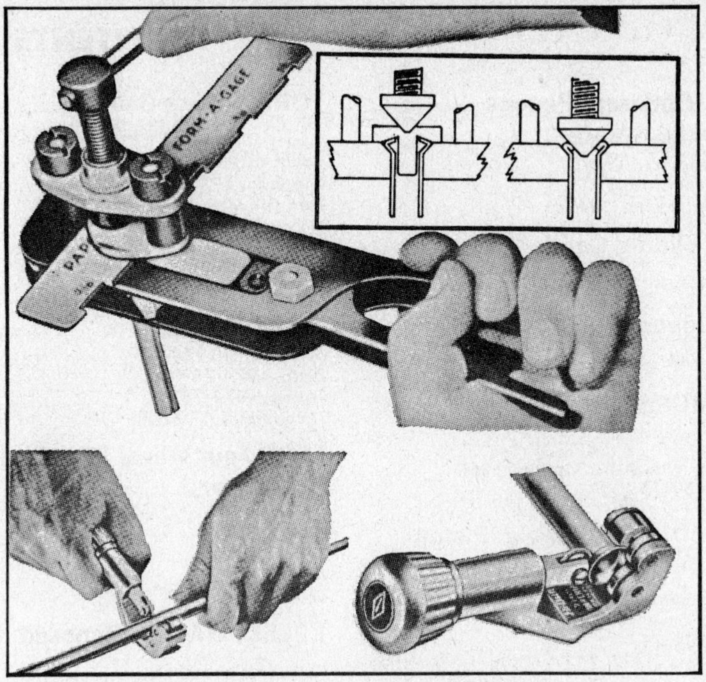

Fig. 25 Flaring hydraulic brake tubing

tubing, proceed as follows:

1. Using the tool shown in Fig. 25 or its equivalent, cut off the damaged seat or damaged tubing.
2. Ream out any burrs or rough edges showing on inside edges of tubing. This will make the ends of the tubing square and insure better seating of the flared end. *Before flaring tubing, place a compression nut on tubing.*
3. Open handles of flaring tool and rotate jaws of tool until mating jaws of tubing size are centered in the area between vertical posts.
4. Slowly close handles with tubing inserted in jaws but do not apply heavy pressure to handle as this will lock tubing in place.
5. Referring to Fig. 25, place gauge on edge over end of tubing and push tubing through jaws until end of tubing contacts recessed notch of gauge matching size of tubing.
6. Squeeze handles of flaring tool and lock tubing in place.
7. Place proper size plug of gauge down in end of tubing. Swing compression disc over gauge and center tapered flaring screw in recess in disc.
8. Lubricate taper of flaring or screw and screw in until plug gauge has seated in jaws of flaring tool. This action has started to invert the extended end of tubing.
9. Remove gauge and apply lubricant to tapered end of flaring screw and continue to screw down until tool is firmly seated in tubing.
10. Remove tubing from flaring tool and inspect the seat. If seat is cracked, cut off cracked end and repeat flaring operation.

DISC BRAKES

CONTENTS

TROUBLE SHOOTING

Excessive Pedal Travel

1. Shoe and lining knock back after violent cornering or rough road travel.
2. Piston and shoe and lining assembly not properly seated or positioned.
3. Air leak or insufficient fluid in system or caliper.
4. Loose wheel bearing adjustment.
5. Damaged or worn caliper piston seal.
6. Improper booster push rod adjustment.
7. Shoe out of flat more than .005".
8. Rear brake automatic adjusters inoperative.
9. Improperly ground rear brake shoe and lining assemblies.

Brake Roughness or Chatter; Pedal Pumping

1. Excessive lateral run-out of rotor.
2. Rotor excessively out of parallel.
3. Wheel bearings incorrectly adjusted.
4. Rear drums out of round.
5. Brake pad reversed.

Excessive Pedal Effort

1. Frozen or seized pistons.
2. Brake fluid, oil or grease on linings.
3. Shoe and lining worn below specifications.
4. Proportioning valve malfunction.
5. Booster inoperative.
6. Leaking booster vacuum check valve.
7. Restricted, dented, kinked or collapsed brake lines or hoses.
8. Partial system failure.
9. Fading due to incorrect brake lining.

Pull, Uneven or Grabbing Brakes

1. Frozen or seized pistons.
2. Brake fluid, oil or grease on linings.
3. Caliper out of alignment with rotor.
4. Loose caliper attachment.
5. Unequalized front tire pressure.
6. Unmatched tires on same axle.
7. Incorrect front end alignment.
8. Loose front end parts.
9. Lining protruding beyond end of shoe.
10. Restricted brake tubes or hoses.
11. Malfunctioning caliper assembly.
12. Inoperative combination valve.
13. Power brake unit malfunctioning.

Brake Rattle

1. Excessive clearance between shoe and caliper or between shoe and splash shield.
2. Shoe hold-down clips missing or improperly positioned.

Heavy Brake Drag

1. Frozen or seized pistons.
2. Operator riding brake pedal.
3. Incomplete brake pedal return due to linkage interference.
4. Faulty booster check valve holding pressure in hydraulic system.
5. Residual pressure in front brake hydraulic system.
6. Stuck master cylinder piston.
7. Restricted brake lines or hoses.
8. Incorrect parking brake adjustment.
9. Check valve in lines to front wheels.

Caliper Brake Fluid Leak

1. Damaged or worn caliper piston seal.
2. Scores in cylinder bore.
3. Corrosion build-up in cylinder bore or on piston surface.
4. Metal clip in seal groove.

No Braking Effect When Pedal is Depressed

1. Piston and shoe and lining assembly not properly seated or positioned.
2. Air leak or insufficient fluid in system or caliper.
3. Damaged or worn caliper piston seal.
4. Bleeder screw open.
5. Air in hydraulic system or improper bleeding.
6. Hydraulic leaks.
7. Master cylinder primary cup worn or damaged.
8. Master cylinder bore corroded, worn or scored.

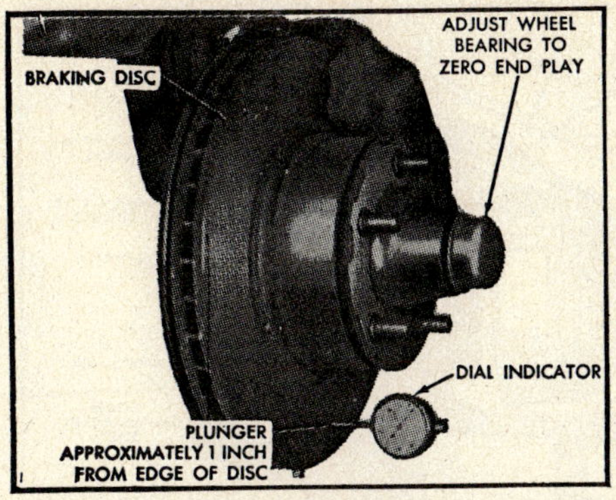

Fig. 1 Checking brake disc for runout

9. Caliper cylinder bore corroded, worn or scored.

Rear Brakes Locking On Application

On brake systems equipped with a proportioning or rear pressure regulator valve, should the valve malfunction rear brakes may receive excess pressure, resulting in wheel lock-up.

SERVICE PRECAUTIONS

Brake Lines & Linings

Remove one of the front wheels and inspect the brake disc, caliper and linings. (The wheel bearings should be inspected at this time and repacked if necessary).

Do not get any oil or grease on the linings. If the linings are worn to within .030" of the surface of the shoe, replace both sets of shoe and lining assemblies. It is recommended that both front wheel sets be replaced whenever a respective shoe and lining is worn or damaged. Inspect and, if necessary, replace rear brake linings also.

If the caliper is cracked or fluid leakage through the casting is evident, it must be replaced as a unit.

Shoe & Lining Wear

If visual inspection does not adequately determine the condition of the lining, a physical check will be necessary.

To check the amount of lining wear, remove a wheel from the car, the caliper from the steering knuckle, and the shoe and lining assemblies. Three thickness measurements should be taken (with a micrometer) across the middle section of the shoe and lining; one reading at each side and one reading in the center.

When a shoe and lining assembly has worn to a thickness of .180", it should be replaced. If shoes do not require replacement, reinstall them in their original inner and outer positions.

Brake Roughness

The most common cause of brake chatter on disc brakes is a variation in thickness of the disc. If roughness or vibration is encountered during highway operation or if pedal pumping is experienced at low speeds, the disc may have excessive thickness variation. To check for this condition, measure the disc at 12 points with a micrometer at a radius approximately one inch from edge of disc. If thickness measurements vary by more than .0005", the disc should be replaced with a new one.

Excessive lateral runout of braking disc may cause a "knocking back" of the pistons, possibly creating increased pedal travel and vibration when brakes are applied.

Before checking the runout, wheel bearings should be adjusted. The readjustment is very important and will be required at the completion of the test to prevent bearing failure. Be sure to make the adjustment according to the recommendations given under *Front Wheel Bearings, Adjust* in the car chapters.

Brake Disc Service

Servicing of disc brakes is extremely critical due to the close tolerances required in machining the brake disc to insure proper brake operation. In manufacturing brake discs, tolerances of the rubbing surfaces for flatness is .001" and usually for parallelism .0005". Lateral runout of the faces should not exceed .004 to .005" in most cases although the limit on Ford Company cars is only .002".

The maintenance of these close controls of the shape of the rubbing surfaces is necessary to prevent brake roughness. In addition, the surface finish must be non-directional and maintained at a micro inch finish. This close control of the rubbing surface finish is necessary to avoid pulls and erratic performance and promote long lining life and equal lining wear of both left and right brakes.

In light of the foregoing remarks, re-

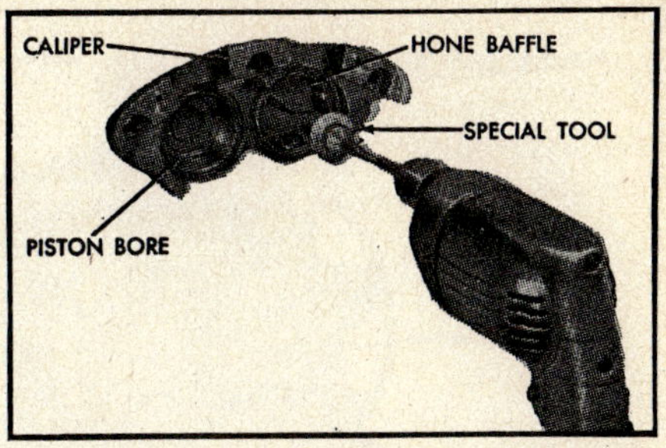

Fig. 2 Honing caliper piston bore

finishing of the rubbing surfaces should not be attempted unless precision equipment, capable of measuring in micro inches (millionths of an inch) is available.

To check runout of a disc, mount a dial indicator on a convenient part (steering knuckle, tie rod, disc brake caliper housing) so that the plunger of the dial indicator contacts the disc at a point one inch from the outer edge, Fig. 1. If the total indicated runout exceeds specifications, install a new disc.

General Precautions

1. Grease or any other foreign material must be kept off the caliper, surfaces of the disc and external surfaces of the hub, during service procedures. Handling the brake disc and caliper should be done in a way to avoid deformation of the disc and nicking or scratching brake linings.
2. If inspection reveals rubber piston

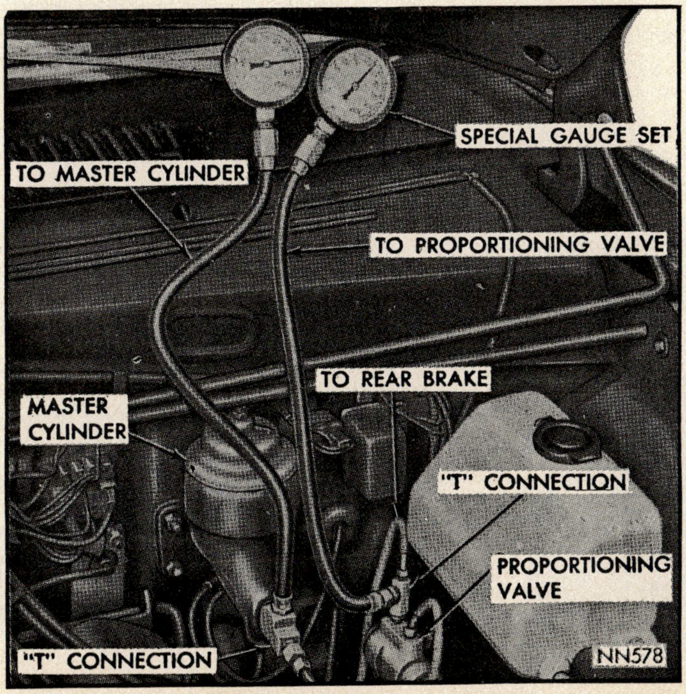

Fig. 3 Gauge hook-up for testing proportioning valve (typical)

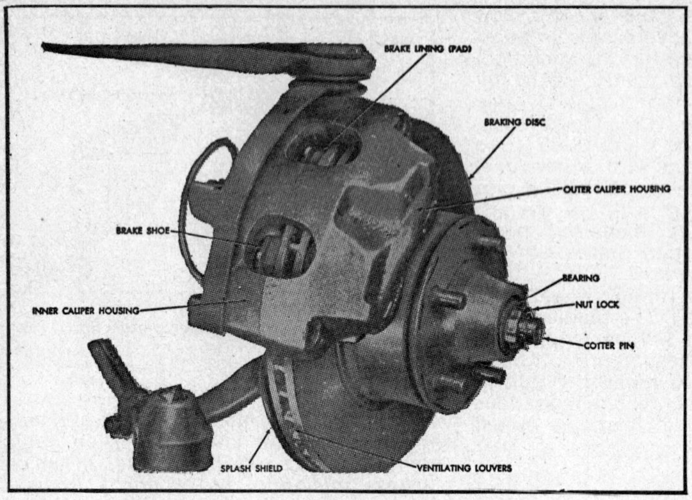

Fig. 4 Bendix opposed piston disc brake

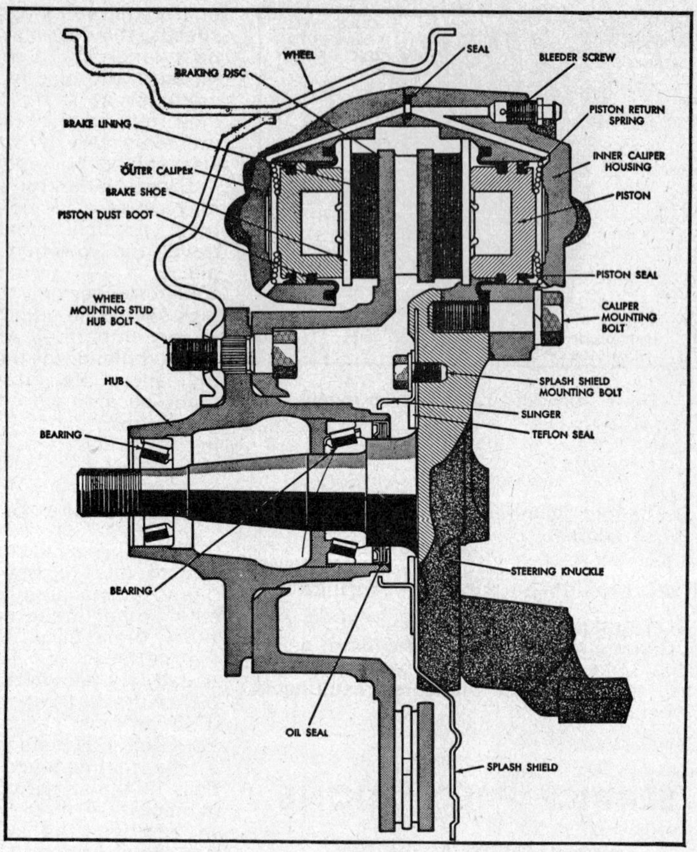

Fig. 5 Sectional view of Bendix opposed piston disc brake

seals are worn or damaged, they should be replaced immediately.

3. During removal and installation of a wheel assembly, exercise care so as not to interfere with or damage the caliper splash shield, the bleeder screw or the transfer tube.
4. Front wheel bearings should be adjusted to specifications.
5. Be sure vehicle is centered on hoist before servicing any of the front end components to avoid bending or damaging the disc splash shield on full right or left wheel turns.
6. Before the vehicle is moved after any brake service work, be sure to obtain a firm brake pedal.
7. The assembly bolts of the two caliper housings should not be disturbed unless the caliper requires service.

Inspection of Caliper

Should it become necessary to remove the caliper for installation of new parts, clean all parts in alcohol, wipe dry using lint-free cloths. Using an air hose, blow out drilled passages and bores. Check dust boots for punctures or tears. If punctures or tears are evident, new boots should be installed upon reassembly.

Inspect piston bores in both housings for scoring or pitting. Bores that show light scratches or corrosion can usually be cleaned with crocus cloth. However, bores that have deep scratches or scoring may be honed, provided the diameter of the bore is not increased more than .002". If the bore does not clean up within this specification, a new caliper housing should be installed (black stains on the bore walls are caused by piston seals and will do no harm).

When using a hone, Fig. 2, be sure to install the hone baffle before honing bore. The baffle is used to protect the hone stones from damage. Use extreme care in cleaning the caliper after honing. Remove all dust and grit by flushing the caliper with alcohol. Wipe dry with clean lint-less cloth and then clean a second time in the same manner.

Bleeding Disc Brakes

NOTE: Chrysler Corp. recommends only pressure bleeding for all 1969-75 models, while Ford Motor Co. recommends that 1969-71 Lincoln Continental, 1970-71 Ford, Mercury, Mark III and Thunderbird and all 1972-75 models, must be pressure bled.

The disc brake hydraulic system can be bled manually or with pressure bleeding equipment (except as noted above). On

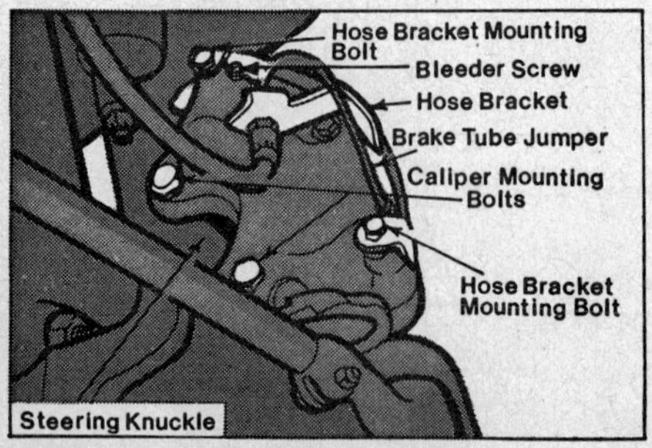

Fig. 6 Caliper mounting

Fig. 7 Removing or installing caliper

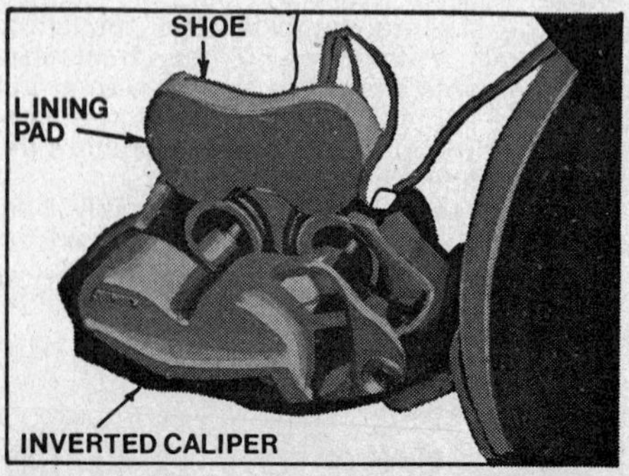

Fig. 8 Removing or installing brake shoe and lining

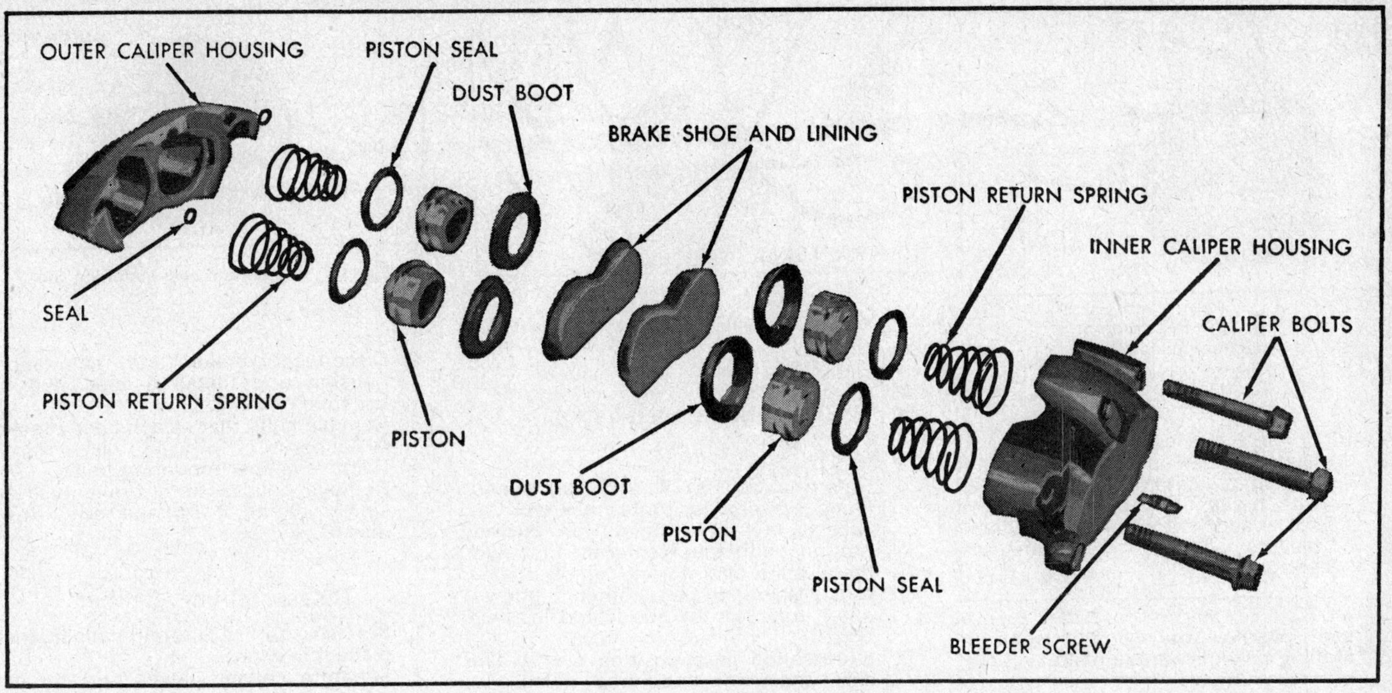

Fig. 9 Exploded view of caliper assembly

vehicles with disc brakes the brake pedal will require more pumping and frequent checking of fluid level in master cylinder during bleeding operation.

Never use brake fluid that has been drained from hydraulic system when bleeding the brakes. Be sure the disc brake pistons are returned to their normal positions and that the shoe and lining assemblies are properly seated. Before driving the vehicle, check brake operation to be sure that a firm pedal has been obtained.

Proportioning Valve

The proportioning valve (when used), Fig. 3, provides balanced braking action between front and rear brakes under a wide range of braking conditions. The valve regulates the hydraulic pressure applied to the rear wheel cylinders, thus limiting rear braking action when high pressures are required at the front brakes. In this manner, premature rear wheel skid is prevented.

Testing Proportioning Valve

When a premature rear wheel slide is obtained on a brake application, it usually is an indication that the fluid pressure to the rear wheels is above the 50% reduction ratio for the rear line pressure and that a malfunction has occured within the proportioning valve.

To test the valve, install gauge set shown in Fig. 3 in brake line between master cylinder and proportioning valve, and at output end of proportioning valve and brake line as shown. Be sure all joints are fluid tight.

Have a helper exert pressure on brake pedal (holding pressure). Obtain a reading on master cylinder output of approximately 700 psi. While pressure is being held as above, reading on valve outlet should be 550-610 psi. If the pressure readings do not meet these specifications, the valve should be removed and a new valve installed.

Fig. 10 Removing piston dust boot

BENDIX OPPOSED PISTONS

The front wheel disc brake, Fig. 4, consists of a fixed caliper (inner and outer housing), two friction pads (brake lining) molded to steel shoes, four pistons, piston return springs, piston seals and dust boots. The brake disc is made from high grade cast iron and has a series of air vent louvers to provide for cooling of the disc. The splash shield that is bolted to the spindle is used to prevent road contaminants from contacting the inboard side of the disc and lining surfaces, Fig. 5. The wheel itself provides for the outboard surface of the disc.

The brake disc is mounted on the front wheel hub by five bolts, and is straddled by the caliper which is attached to the steering knuckle by two bolts. Inserted between the pistons and the disc are the shoe and lining assemblies, which are held in position by parallel machined abutments within the caliper.

Brake Shoe Removal

1969-70 American Motors & 1969 Buick

1. Remove wheel assembly and disconnect hydraulic tubing from mounting bracket.
2. On 1969 Buick models:
   a. Remove upper and loosen lower caliper mounting bolt, then rotate upper end of caliper rearward.

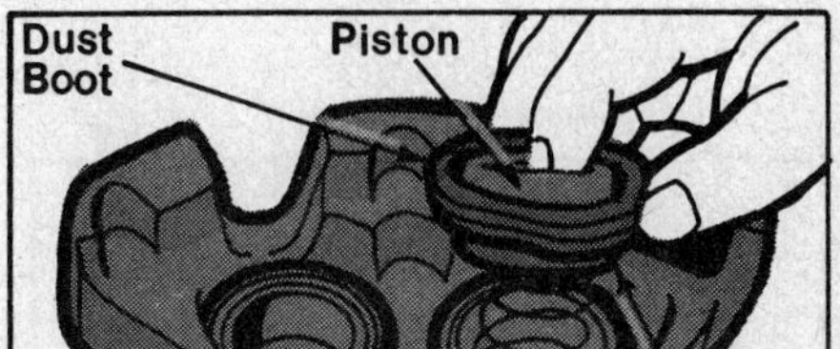

Fig. 11 Removing piston boot, seal and return spring

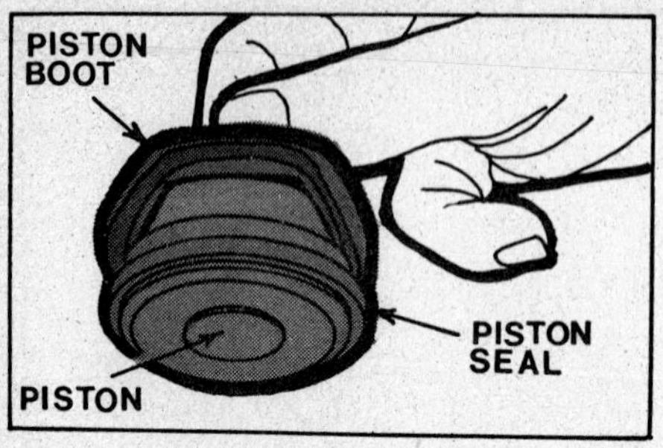

Fig. 12 Removing or installing piston boot

Fig. 13 Removing or installing piston seal

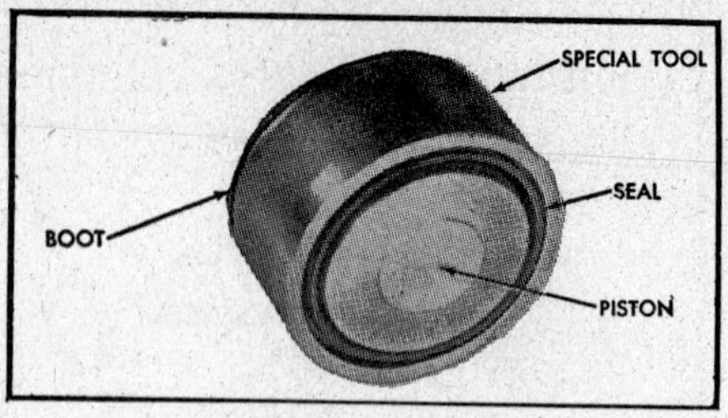

Fig. 14 Piston and seal installing tool

b. Connect one end of drain hose to bleeder screw and submerge other end in a container partially filled with brake fluid, then open bleed screw and bottom pistons. Tighten bleeder screw after pistons are bottomed.

NOTE: This step is necessary to prevent overflow of reservoir when installing new disc brake pads.

3. On 1969-70 American Motors:
   a. Loosen caliper mounting bolts then hold lower edge of caliper and remove lower bolt. Shake caliper until all shims fall out and mark these as "Lower".
   b. Press in on upper edge of caliper to hold shims on upper mounting bolt. Remove bolt and mark these shims as "Upper".

NOTE: Original shim thickness must be replaced in same locations when caliper is installed.

   c. Support caliper assembly using a length of wire.
4. Using two screwdrivers between shoe and pistons, press pistons into bottom of their cylinders and remove pad assembly from calipers.

1969 Belvedere & Coronet

1. Remove wheel assemblies.
2. Remove caliper-to-steering knuckle bolts, Fig. 6.
3. Remove caliper from disc by sliding it up and away from disc, Fig. 7.
4. Remove brake shoes and lining assemblies one at a time through bottom opening, Fig. 8.

Brake Shoe Installation

All Models

1. Referring to Fig. 8, slide shoe and lining assemblies into position in caliper, one at a time, with curved portion (with tabs) entering first and metal shoe against open ends of pistons. Using fingers, spread linings apart until pistons are seated in their bores.
2. Slide caliper into position over brake disc and align mounting holes. As caliper is being installed, be sure that lining slides easily along brake disc.
3. Install caliper mounting bolts and torque to 85 ft. lbs for American Motors, Belvedere and Coronet and to 70 ft. lbs for Buick models.

NOTE: On American Motors, the original shim thickness must be installed in the same location between the caliper mounting lugs and bracket. If exact shim thickness is unknown, the caliper must be adjusted as outlined under "Caliper Adjustment."

4. Make sure disc rotates freely and with minimum drag.
5. Install wheel assembly.

CAUTION: Road test vehicle and make several heavy 40 mph stops to wear off any foreign material on the brakes and to seat the linings. The vehicle may pull to one side if this is not done.

Removing Caliper

1. Remove wheels.
2. Disconnect brake line at caliper housing and install a pipe plug in the tube opening.
3. Remove bolts that attach hose bracket to caliper.
4. Remove caliper mounting bolts.
5. Remove caliper from brake disc by slowly sliding it up and away from disc.

Disassembling Caliper

1. Referring to Fig. 9, drain caliper, then place it in a vise.
2. Separate caliper halves and remove two crossover seals.
3. Using a screwdriver, pry exposed dust boot out of groove, Fig. 10. Be sure to hold piston compressed during this operation.
4. Remove piston, seal and dust boot from caliper, Fig. 11. Remove piston return spring.
5. Remove piston dust boot by grasping edge and pulling out of its groove, Fig. 12.
6. Remove fingers, roll piston seal out of its groove and discard seal, Fig. 13.
7. Remove remaining pistons in same manner.
8. Remove bleeder screw from inner caliper.

Assembling Caliper

1. Clamp caliper housing in vise and coat cylinder bores with silicone grease, then install piston return spring with large diameter down and seated in recess at bottom of bore.
2. Coat outside diameter and fill inner diameter of a new piston seal with silicone grease and work over piston land and down into position in groove,

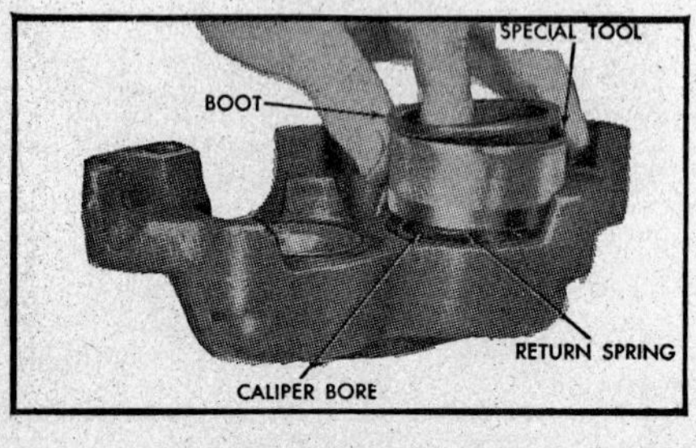

Fig. 15 Installing piston, seal and dust boot

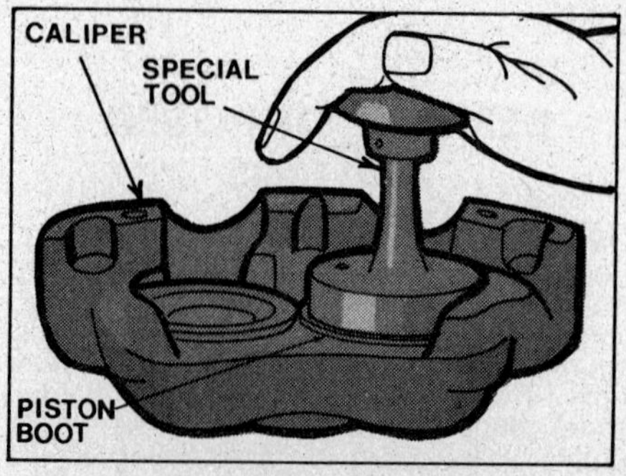

Fig. 16 Installing piston dust boot in caliper groove

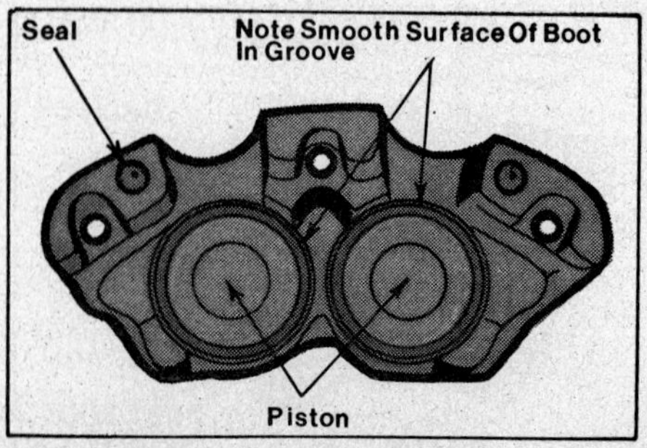

Fig. 17 Pistons and boots installed in caliper

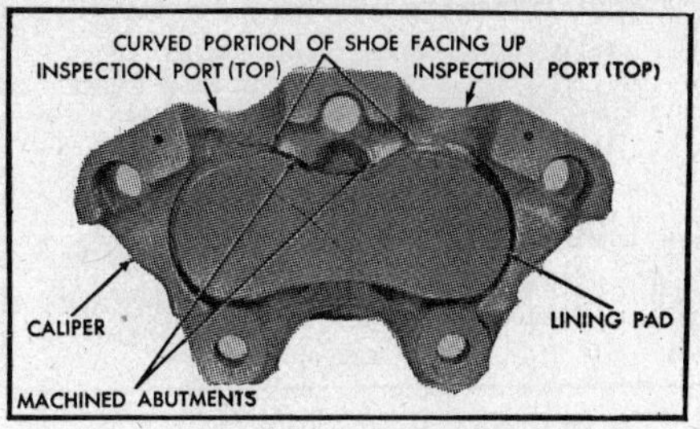

Fig. 18 Shoes and linings correctly positioned in caliper

using fingers only.

3. Install dust boot on piston with lip of seat toward piston.
4. Using a suitable tapered sleeve, Fig. 14, install piston, seal and boot in sleeve, with lip of seal towards taper. Push in on assembly until seal lip is even with knife edge of sleeve, Fig. 14.
5. Place installing tool over bore opening and return spring. Index with boot in caliper, Fig. 15. Press down on piston, sliding piston out of tool and into caliper bore until bottomed. Remove sleeve tool.
6. Position piston boot sealing lip over groove evenly. Using the tool shown in Fig. 16 or its equivalent, press down on tool, forcing boot sealing lip into caliper groove. Remove tool and install remaining pistons in same manner, Fig. 17.
7. Test pistons for smooth operation in their bores by depressing with fingers.
8. Lightly clamp outer caliper half in vise and install new crossover passage seals in position in recess of caliper mating surface, Fig. 17.
9. Place mating caliper half over one clamped in vise, install attaching bolts and torque to 120-140 ft-lbs.
10. Install brake shoes in caliper with curved portion (with tabs) entering first and metal shoe against open ends of pistons. Using fingers, spread linings apart until pistons are seated in their bores, Fig. 18.
11. Install bleeder screw and tighten lightly.

Installing Caliper

1. Check runout of brake disc with dial indicator as outlined previously.
2. Install caliper over disc and align mounting holes. Install mounting bolts and torque to 85 ft. lbs for American Motors and Plymouth Belvedere models and 70 ft. lbs for Buick models.

NOTE: As caliper is being lowered into place, be sure that linings ride freely over disc.

3. Connect brake line at caliper housing, then position brake hose bracket. Install bracket and attaching bolts and tighten securely.
4. Follow normal bleeding procedure, being sure all air bubbles have escaped.
5. Replenish brake fluid in master cylinder.
6. Install wheels and road test as suggested previously.

Caliper Adjustment 1969-70 American Motors

The caliper reference surfaces and friction faces of the disc must be within .010 inch on both sides of the disc, Fig. 19 (Dimension C). Use feeler gauges to check to see if both ends of the caliper are centered relative to the rubbing surfaces of the disc. If clearance is not within specifications, loosen the caliper mounting bolts and add or remove shims as necessary between the caliper mounting lugs and mounting bracket.

The caliper and disc must also be parallel so that dimension "D" Fig. 19, is equal within .005 inch at each end of the caliper. If clearance is not within specifications, add or remove shims as required on one or the other mounting lug bolt. After proper clearance has been obtained, torque mounting bolts to 85 ft. lbs.

BUDD OPPOSED PISTONS

The Budd Disc brake, Fig. 20, consists of a fixed caliper (inner and outer housing) two friction pads (brake lining) bonded to steel shoes, four pistons, piston return springs, piston seals, dust boots and retainers. The brake disc is made from high grade cast iron and has a series of air vent louvers to provide cooling for the brake assembly. The splash shield has a series of stamped vents so designed as to supply additional air for cooling.

Removing Lining

1. Remove wheels.

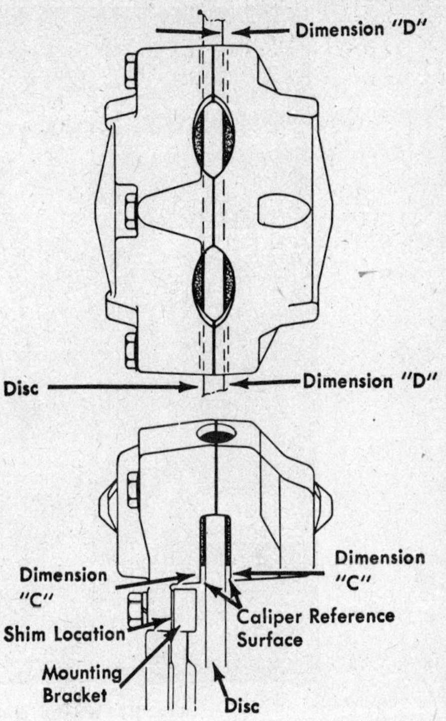

Fig. 19 Disc to caliper clearance checking points

2. Remove brake shoe anti-rattle spring.
3. Remove bolts that attach caliper assembly to steering knuckle and knuckle arm.
4. Slowly slide caliper up and away from brake shoe.
5. Carefully invert caliper and remove brake shoe assembly (one at a time).
6. Insert piston compressing tool C-3992

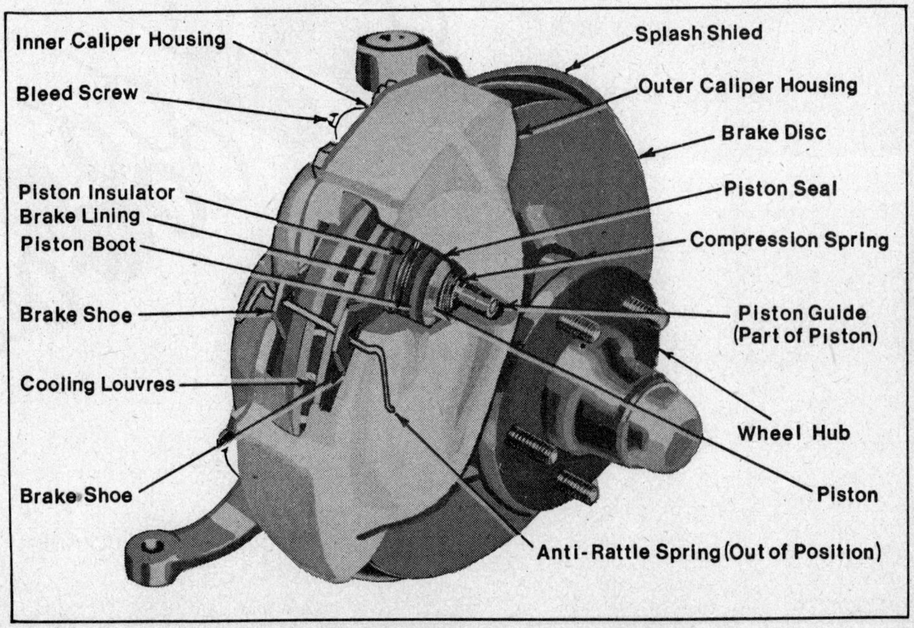

Fig. 20 Budd opposed piston disc brake assembly

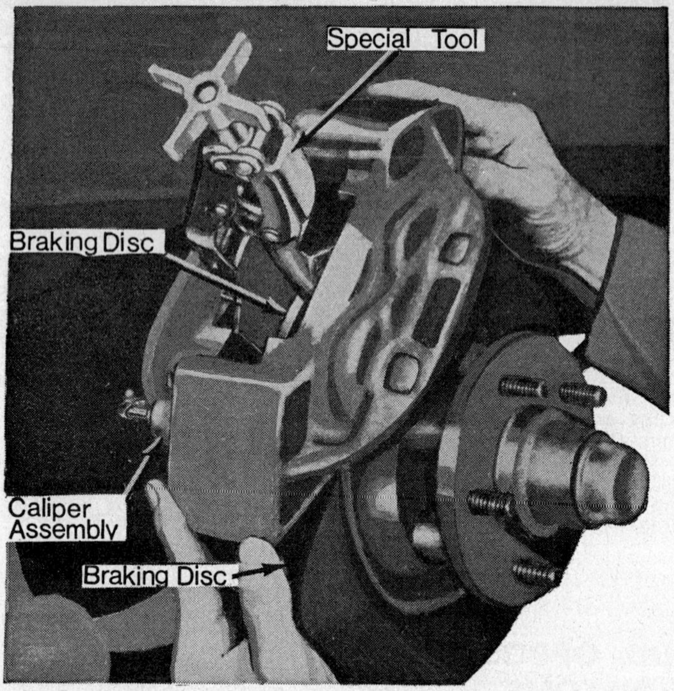

Fig. 21 Piston compression tool installed

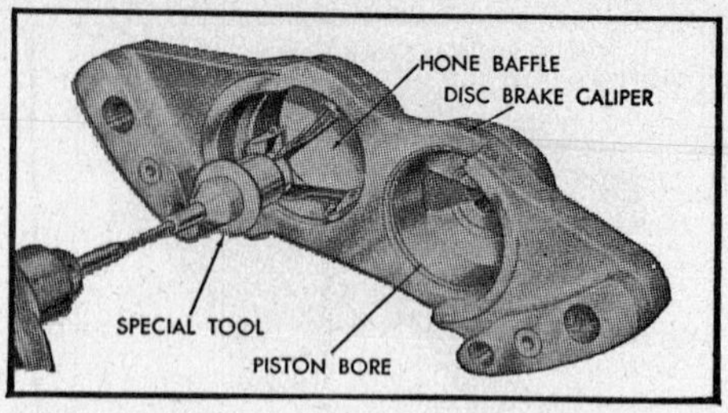

Fig. 23 Honing piston bore

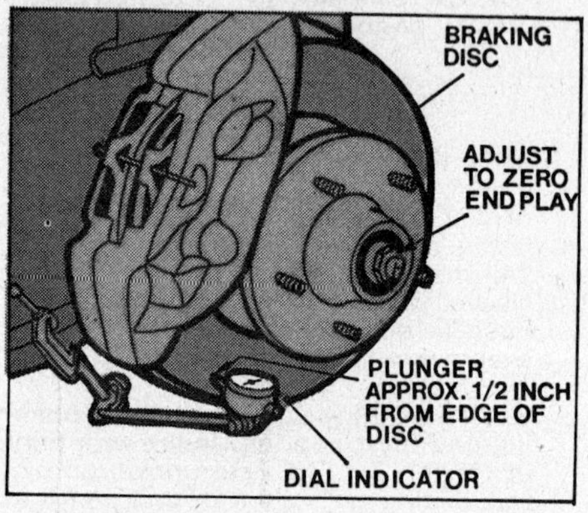

Fig. 24 Checking brake disc for runout

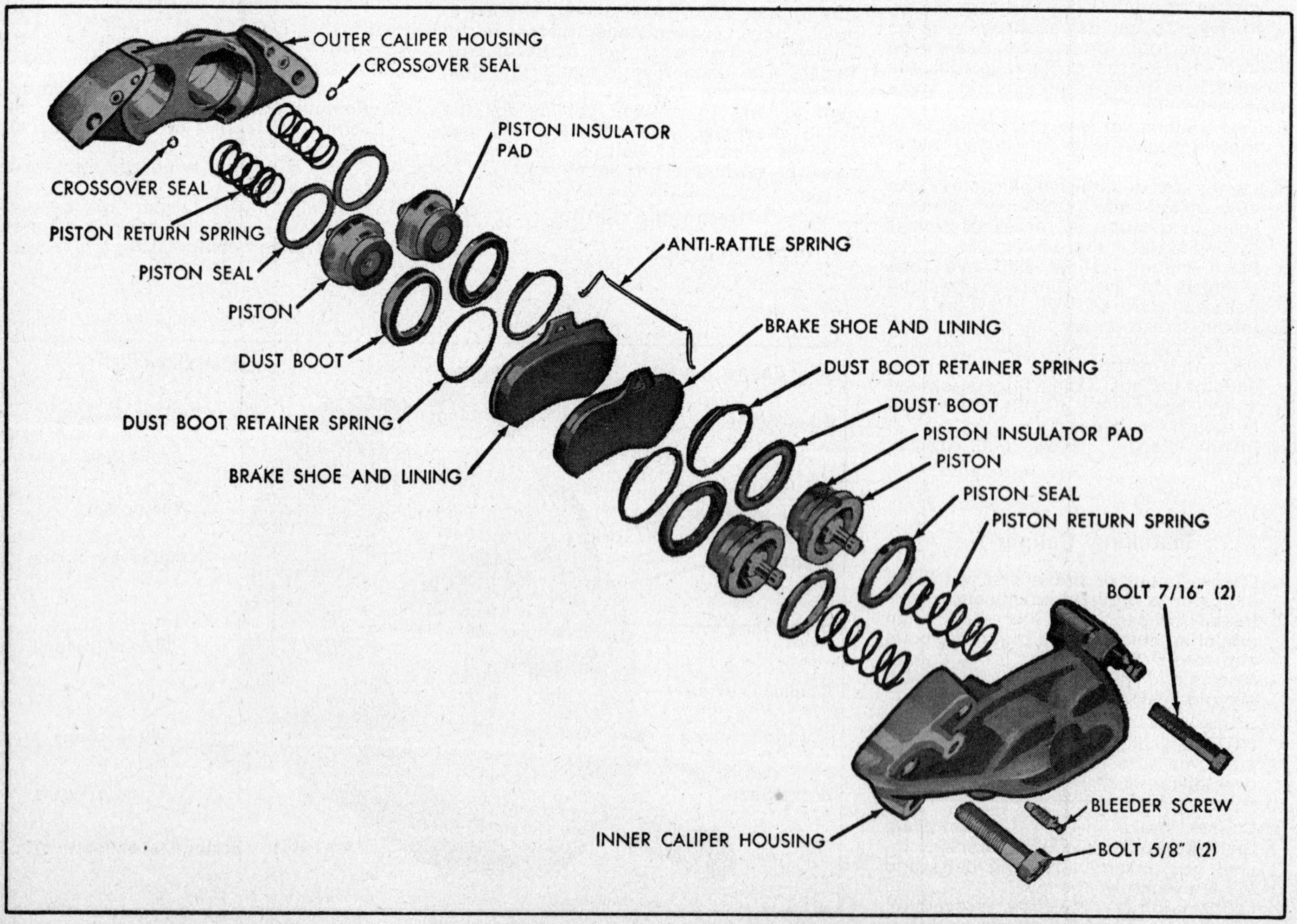

Fig. 22 Budd disc brake caliper disassembled

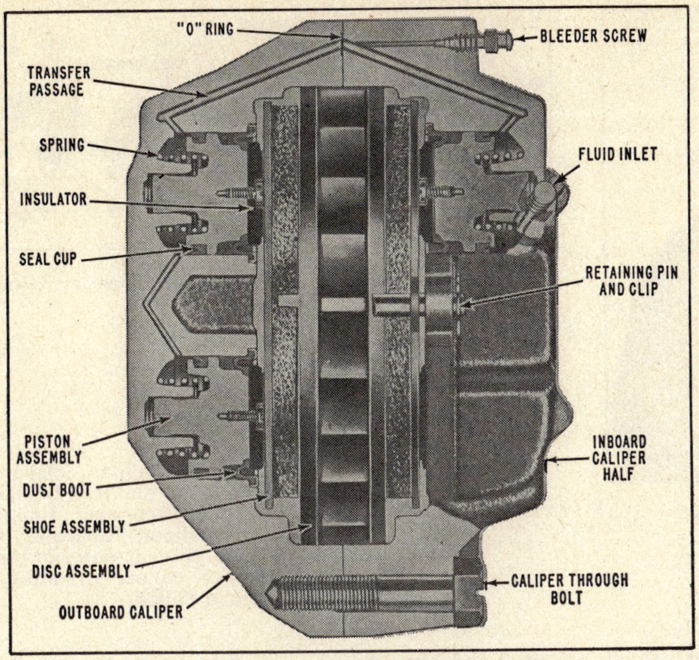

Fig. 25 Delco-Moraine disc brake assembly

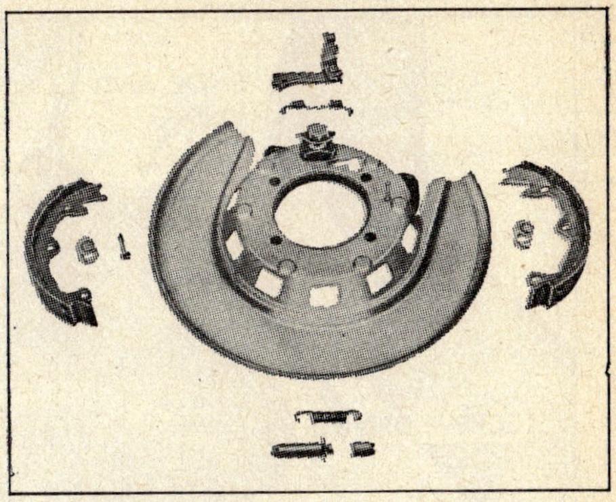

Fig. 26 Delco-Moraine parking brake components

between piston insulator pads and turn knob on tool until pistons are fully compressed, Fig. 21.

Installing Lining

1. Slide brake disc shoe and lining assembly into position in caliper (one at a time).
2. Position shoes, then install tool C-3992 between shoes. Adjust tool so that it holds shoes apart and that it will slip out easily when contacted by disc.
3. Install caliper mounting bolts and torque to 80-90 ft. lbs.

CAUTION: Road test vehicle and apply several heavy 40 mph stops to wear off any foreign material on brakes and to seat units. The vehicle may pull to one side or the other if this is not done. This condition will be more noticeable if only one wheel was worked on.

Removing Caliper

1. Remove wheels.
2. Remove anti-rattle clips, then disconnect brake line from caliper and install a plug.
3. Remove caliper mounting bolts, then slowly slide caliper assembly up and away from disc.
4. Remove lining pads, then install compressing tool C-3992 and turn knob until pistons are fully compressed, Fig. 21.

Disassembling Caliper

1. Referring to Fig. 22, remove piston compression tool and four bolts that hold two halves of caliper together.
2. Separate halves and remove two crossover seals.
3. Using a small screwdriver, pry out exposed end of piston dust boot retainer spring and uncoil from its groove to release dust boot.
4. Using same screwdriver, work dust boot out of groove. Be sure to hold piston compressed during this operation.
5. Remove piston, seal and dust boot from caliper. Remove piston return spring.
6. Remove piston dust boot by grasping edge and pulling out of its groove.
7. Pry piston seal out of its groove and discard.
8. Remove remaining three pistons in same manner.
9. Remove bleeder screw from inner caliper housing.

Cleaning & Inspection

1. Clean all parts in brake fluid and wipe dry. Using an air hose, blow out drilled passages and bores.
2. Check dust boots for punctures and tears. If punctures or tears are evident, new boots should be installed.
3. Inspect piston bores in both housings for scoring or pitting. Bores that show light scratches or corrosion can usually be cleaned with crocus cloth. However, bores that have deep scratches or scoring may be honed, Fig. 23, providing the diameter of bore is not increased more than .002". If the bore does not clean up within this specification, a new caliper housing must be installed. Black stains on bore walls are caused by the piston seals and will do no harm.
4. When honing, Fig. 23, be sure to install the hone baffle before honing bore. The baffle is used to protect hone stones from damage.
5. Use extreme care in cleaning the caliper after honing. Remove all dust and grit by flushing caliper with brake fluid. Wipe dry with clean, lintless cloth and then clean a second time in like manner.

Assembling Caliper

1. Referring to Fig. 22, clamp caliper in a vise, then install piston return spring. Be sure spring is seated in recess in bottom of bore.
2. Coat outside diameter and fill inner diameter of a new piston seal with silicone grease and work over piston land and down into position in groove, using fingers only.
3. Install dust boot on piston with lip of boot toward piston pad.
4. Install piston over return spring and press down until piston bottoms in bore.
5. Using a small, blunt screwdriver, work lip of boot into groove, around diameter of bore. Use care so as not to puncture boot during this operation, or a new boot will have to be installed.
6. Install retainer spring by inserting one end in position in groove and continue to install around diameter of bore until retainer is fully seated. Be sure boot is completely locked in position by retainer and that retainer is fully seated in groove.
7. Install remaining pistons in same manner, then test pistons for smooth operation in their bores by depressing with fingers.
8. Lightly clamp outer caliper half in vise and install new crossover seals, then place mating caliper half over one installed in vise.
9. Install attaching bolts and torque 7/16 inch bolts to 55 ft. lbs and 5/8 inch bolts to 150 ft. lbs.
10. Install bleeder screw.

Installing Caliper

1. Before installing caliper over disc, check disc for runout with a dial gauge as shown in Fig. 24. Lateral runout should not exceed .005"; if runout is excessive, remove disc and

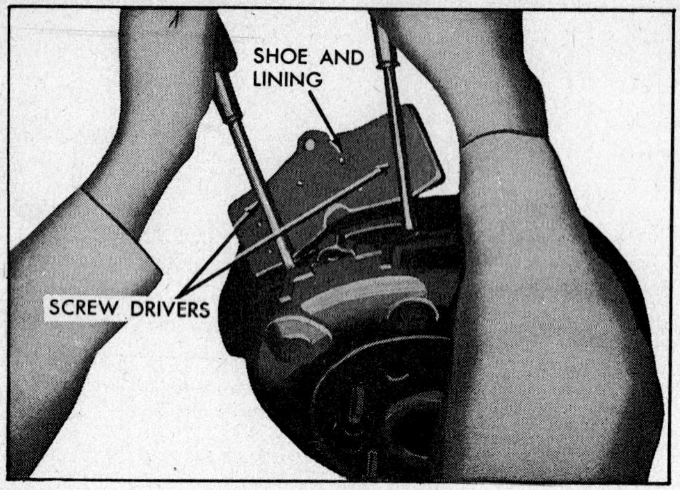

Fig. 27 Installing Delco Moraine disc brake shoes

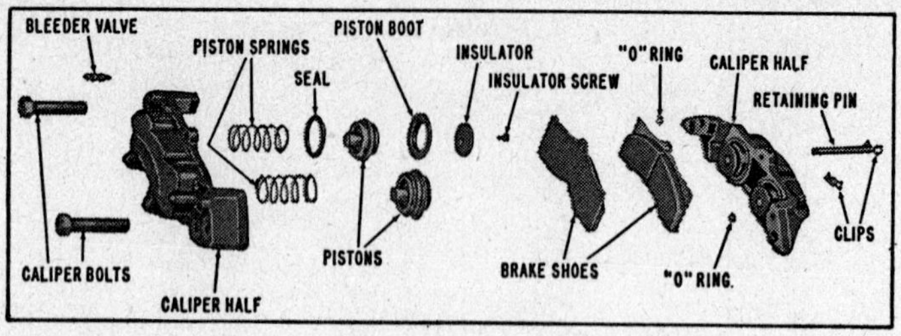

Fig. 28 Delco-Moraine disc brake caliper components

check its mounting surface on wheel hub. Runout of hub should not exceed .004".

2. Install caliper assembly over disc and align mounting holes. Install mounting bolts and torque to 80-90 ft. lbs.
3. Connect brake line at caliper housing and allow caliper to fill with brake fluid then close bleeder screw. Be sure all air bubbles have escaped when bleeding the caliper. Replenish brake fluid in master cylinder.
4. Install wheels and road test vehicle as outlined previously.

DELCO-MORAINE OPPOSED PISTONS

These brakes are used on all four wheels. The components of the disc brake system are shown in Fig. 25. The caliper assemblies replace the conventional wheel cylinder, brake shoes and linings, and the disc replaces the brake drum.

The caliper assembly contains four pistons, two acting on each shoe with one shoe on each side of the disc.

The brake disc is riveted to the hub flange at the front wheel and to the spindle flange at the rear wheel. The disc rotates through the caliper assembly, which is bolted to a support that is attached to the steering knuckle at the front wheel and the spindle support bolts at the rear wheel. The disc has cooling fins between the two shoe reacting surfaces. When a disc must be replaced, the rivets can be drilled out and then the wheel studs will be used for disc retension purposes.

A miniature set of brake shoes, mounted on a flange plate and shield assembly attached to the rear wheel spindle support bolts, are used for vehicle parking, Fig. 26.

Removing Lining

1. To prevent overflow, remove two thirds of brake fluid from master cylinder.
2. Support vehicle on hoist and remove wheel.
3. Remove cotter pin from inboard end of retaining pin. On 1969-70 models with heavy duty brakes, two retaining pins must be removed, one from each end of caliper.
4. Remove inboard and outboard shoe by pulling up.

Installing Lining

1. Install inboard and outboard shoe one at a time. Use two screwdrivers to push pistons back as shoes are inserted, Fig. 27.
2. Install retaining pin through outboard caliper half, outboard shoe, inboard shoe and inboard caliper half. Insert a new $^{3}/_{32}$ x $^{5}/_{8}$ inch plated cotter pin through retaining pin. On 1969-70 models with heavy duty brakes, install two retaining pins, one at each end of caliper.
3. Repeat above procedure at each wheel where shoes are to be replaced.
4. Refill master cylinder, then install wheel and lower vehicle.

CAUTION: Do not move vehicle until a firm brake pedal has been obtained.

Calipers

The caliper assembly, Fig. 28 comes in two halves assembled by strong bolts at the flange end. The two halves contain fluid crossover passages from one to the other, sealed with "O" rings.

The bleeder screw is threaded into a passage drilled to intersect the fluid crossover passage. The bleeder screws are located at the front of each caliper. There are two bleeder screws, one inboard, one outboard at the rear wheels, and one bleeder screw at the inboard side at the front wheel. It is necessary, therefore, to remove the rear wheel when bleeding the rear caliper.

Removing Caliper

1. Support vehicle on hoist and remove wheel.
2. On front caliper, disconnect brake hose from support bracket. On rear caliper, disconnect tubing from inboard caliper. Tape open tube or line end to prevent entry of dirt.
3. Remove caliper mounting bolts and remove caliper.

Disassembling Caliper

1. Remove brake hose from front caliper.
2. Remove cotter pin from retaining pin, then remove pin and shoe assembly from caliper.

NOTE: On 1969-70 models with heavy duty brakes, two retaining pins must be removed.

3. Remove caliper retaining bolts and separate caliper halves, then remove the two O-rings from fluid transfer cavities in ends of caliper halves.
4. Push piston into caliper as far as it will move, then insert a screwdriver under inner edge of steel ring in boot and using piston as a fulcrum, pry piston boot from its seat in caliper half.

CAUTION: Use care not to puncture seal when removing pistons from caliper.

5. Remove pistons and springs from caliper half, then remove boot and seal from piston.

Cleaning & Inspection

1. Clean all metal parts using clean brake fluid, removing all traces of dirt and grease.

CAUTION: Never use mineral base cleaning solvents as they can cause deterioration of rubber parts or make them soft and swollen.

2. Using air pressure, blow out all fluid passages in caliper halves, making sure that these passages are not obstructed.
3. Discard all rubber parts and replace with new service kit parts.
4. Inspect piston bores. They must be free of scores and pits. A damaged bore will cause leaks and unsatisfactory brake operation. If either caliper

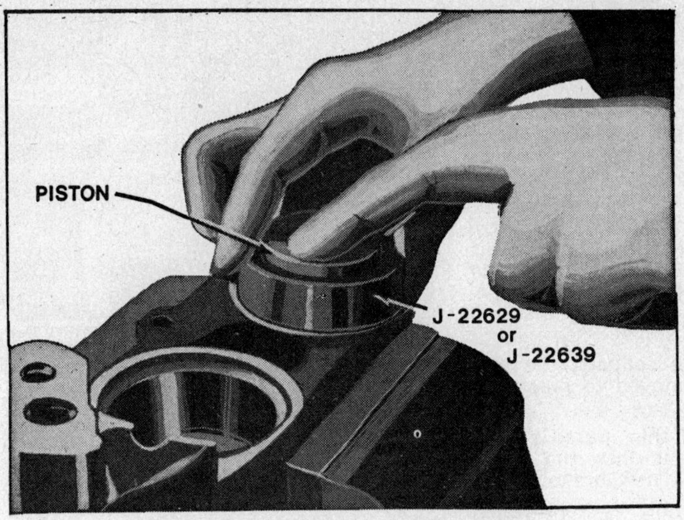

Fig. 29 Installing piston in caliper

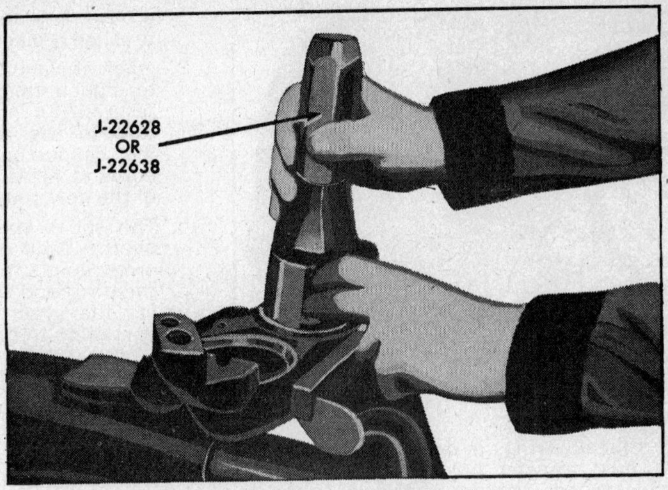

Fig. 30 Installing boot seal in caliper

half is damaged to the extent that polishing with fine crocus cloth will not restore to satisfactory condition, replace the caliper half.

5. Check fit of piston in bore using a feeler gauge. Clearance should be as follows:

   2$^1/_{16}$ inch bore .0045-.010
   1$^7/_8$ inch bore .0045-.010
   1$^3/_8$ inch bore .0035-.009

   If bore is not damaged and clearance exceeds specifications, only a new piston will be required.

Assembling Caliper

1. Install seal in piston groove which is closest to flat end of piston. The seal lip must face toward large end of piston.

NOTE: Make certain seal lips are in piston groove and do not extend over step in end of groove.

2. Place spring in piston bore, then lubricate seal with brake fluid.
3. Install piston assembly in bore using tool J-22591, 22629 or 22639, Fig. 29. Use care not to damage seal lip as piston is pressed past edge of bore.
4. Install piston boot in groove closest to concave end of piston with fold in boot facing toward end of piston with seal attached.
5. Make certain that piston slides smoothly into bore until end of piston is flush with end of bore. If not, recheck piston assembly and position of piston spring and seal.
6. Using boot seal installer tool J-22592, J-22628 or J-22638, Fig. 30, over piston, seat steel boot retaining ring evenly into counterbore.

NOTE: Boot retaining ring must be flush or below machined face of caliper. Any distortion or uneven seating could allow corrosive elements to enter bore.

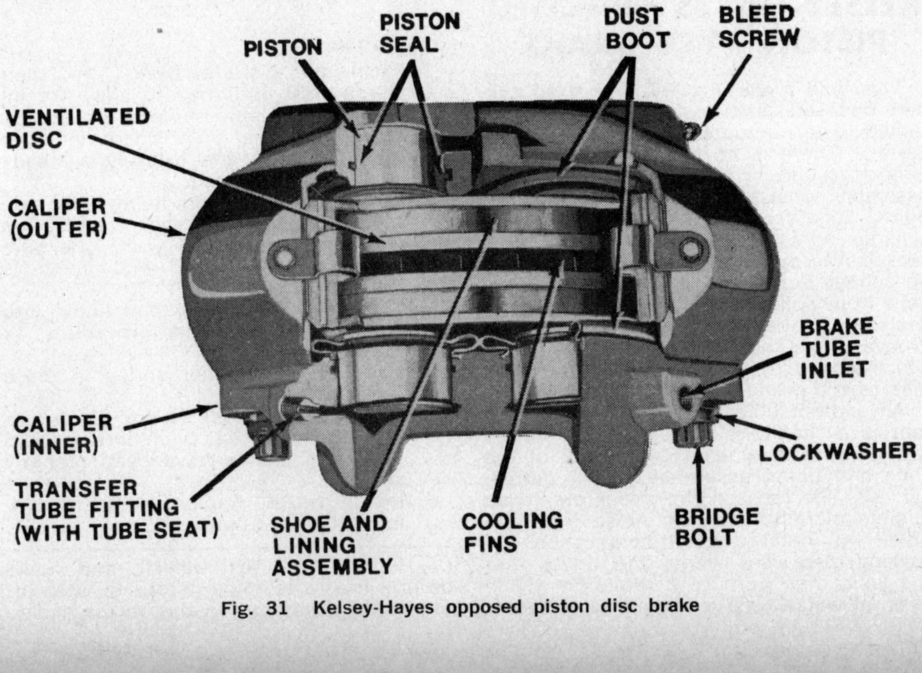

Fig. 31 Kelsey-Hayes opposed piston disc brake

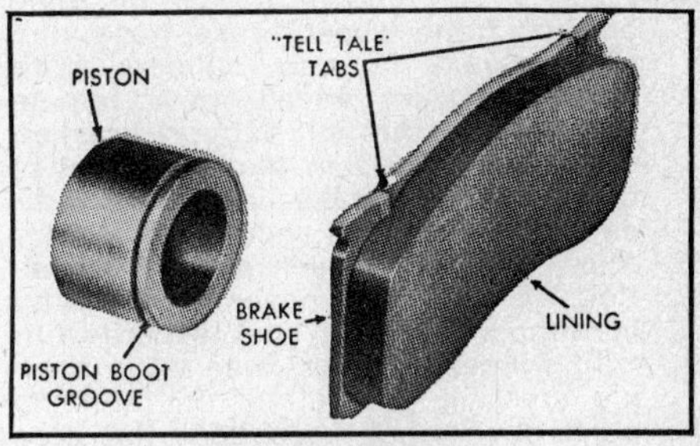

Fig. 32 Brake piston, shoe and lining assembly

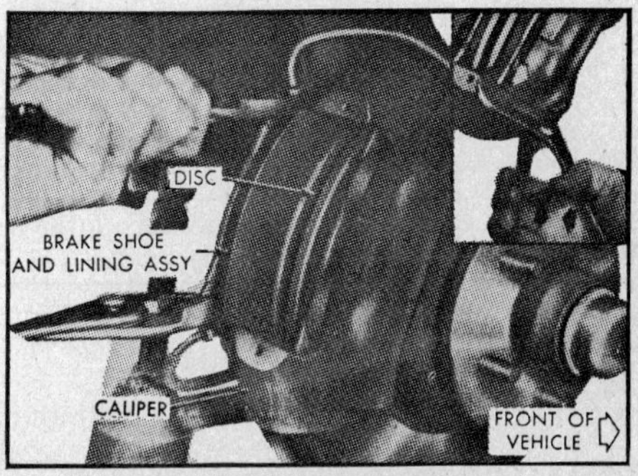

Fig. 33 Removing brake shoe and lining

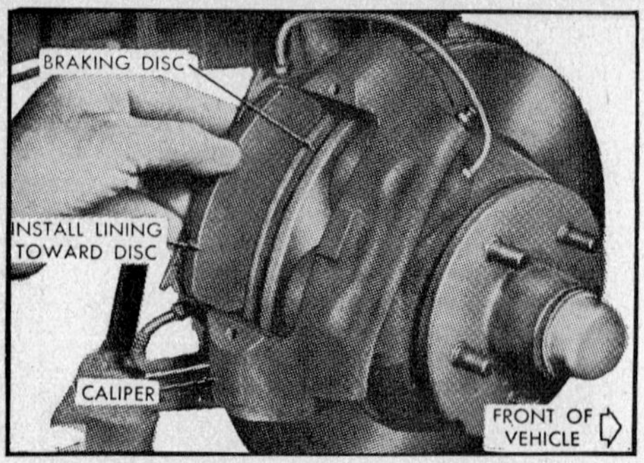

Fig. 34 Installing brake shoe and lining

7. Install O-rings in cavities around brake fluid transfer holes at both ends of outboard caliper halves. Lubricate caliper bolts with Delco Brake Lube #540032 (or equivalent) or clean brake fluid, then secure caliper halves together and torque caliper bolts to 130 ft. lbs.

Installing Caliper

1. Mount caliper over disc, then using two screwdrivers, depress pistons so that caliper can be lowered into place.

NOTE: Use care not to damage boots on edge of disc as caliper is installed.

2. Install mounting bolts and torque to 70 ft. lbs.

CAUTION: If reusing old shoe assemblies, be sure to install shoes in same location from which removed.

3. Install disc pads as outlined previously.
4. Place a new copper gasket on male end of front wheel brake hose and install brake hose in calipers. With wheels straight ahead, pass female end of hose through support bracket, then making certain that tube seat is clean, connect brake line tube nut to caliper and tighten securely.
5. Allowing hose to seek a normal position, without twist, insert hose fitting in support bracket and secure with "U" shaped retainer, then while turning steering geometry from stop to stop, check that hose does not contact other parts at anytime. If contact does occur, remove "U" shaped retainer and twist hose in a direction that will eliminate hose contact. Reinstall retainer and recheck for hose contact. If satisfactory, place steel tube connecter in hose fitting and tighten securely.
6. If rear caliper is being serviced, connect brake line to caliper.
7. Bleed brakes and install wheels.

CAUTION: Do not move vehicle until a firm pedal has been obtained.

Service Summary

1. There is no brake shoe adjustment on the disc brakes.
2. The groove in the brake shoe is an indicator of brake wear. When the groove is just about gone it is time for shoe replacement.
3. When replacing shoes it is necessary to siphon fluid from master cylinder reservoir to make room for fluid return to the reservoir when pushing the caliper pistons back into their bores to make room for the thickness of the new shoes.
4. The shoes have a directional arrow on the back of the shoe plate. This arrow points to the forward rotation of the disc, and the purpose is for aligning the grain of the lining material in relation to the disc.
5. When bleeding the calipers, the rear wheel must be removed to reach the outboard bleeder screw.
6. A retaining clip of thin metal is used to hold the pistons into the bores while installing the new brake shoes.
7. The caliper assembly is removable, after disconnecting the brake line, by removing the two mounting bolts and lifting the assembly off the disc.
8. The disc is riveted to the spindle flange in production. However, the rivets may be drilled out and the wheel studs and nuts are sufficient to hold the new disc in place when replacing the disc.
9. The rear wheel spindle must be removed to gain access to the parking brake shoes. It is necessary then to remove the caliper, the axle drive shaft, the spindle drive shaft yoke and remove the spindle and disc as an assembly from the wheel support. You now have access to the parking brake shoes the same as any other conventional bendix type brake shoe, Fig. 26.
10. If the car is equipped with the special optional knock-off hub assemblies, the adapters must be removed to gain access to the parking brake adjustment.

KELSEY-HAYES OPPOSED PISTON TYPE BRAKE

This type brake, Fig. 31, is a fixed caliper, opposed piston, non-energized, ventilated type, actuated by the hydraulic system. There is no lateral movement of either the disc or the caliper. The caliper assembly consists of two caliper housings bolted together. Each half contains two cylinder bores. Each cylinder contains a seal, piston and externally attached molded rubber dust boot to seal the cylinder bore from contamination. The pistons are sealed by rubber piston seals positioned in grooves machined in the cylinder bores which provide hydraulic sealing between pistons and cylinder bores.

An additional feature of this brake system is a "tell-tale" tab sounding device which indicates when replacement of the shoe and lining assemblies are required, Fig. 32. The tabs on the shoes create an audible metallic scraping noise from the brake by metal-to-metal contact on the braking disc. This warns the driver that the lining has worn to a minimum thickness, at which time it should be replaced.

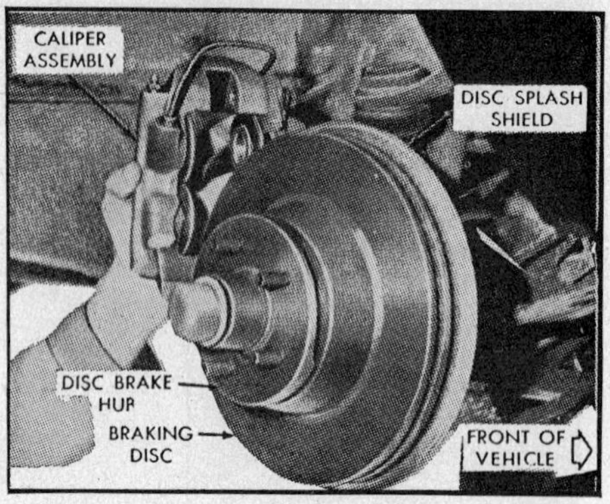

Fig. 35 Replacing brake disc caliper

Checking Running Clearance

To check the lining-to-disc running clearance, remove wheel and caliper splash shield. Insert a feeler gauge between lining and disc. Clearance ordinarily should be .003-.006". However, if the vehicle was stopped by a brake application just prior to checking the clearance, it is considered normal for the brakes to drag slightly.

Brake Shoe Replace

1. Raise vehicle on hoist or stands.
2. Remove wheel assembly.
3. Remove shoe retainer spring assemblies.
4. Using two pairs of pliers, grasp tabs on outer ends of shoes and remove shoe and lining by pulling outward, Fig. 33.

NOTE: Due to a ridge of rust that may have formed on the disc surface outside of lining contact area, it may be necessary to force the piston back slightly into its bore. This is done by forcing the shoe back with water pump pliers placed on corner of shoe and caliper housing as shown, Fig. 33.

Installation

1. Push all pistons back into their bores until bottomed to allow for installation of new shoes. This can be done by placing a flat-sided metal bar against piston and exerting a steady force until bottomed.
2. Slide new shoe and lining into caliper with ears of shoe resting on bridges of caliper, Fig. 34. Be sure shoe is fully seated and lining is facing disc.
3. Slide remaining shoe and lining into caliper, using same procedure as above.
4. Install shoe retainer spring in place on caliper.
5. Pump brake pedal several times until a firm pedal has been obtained and shoe and linings have been properly seated.
6. Install wheel. Replenish master cylinder fluid as required.

CAUTION: Road test vehicle and make several heavy 40 m.p.h. stops to wear off any foreign material on the brakes and to

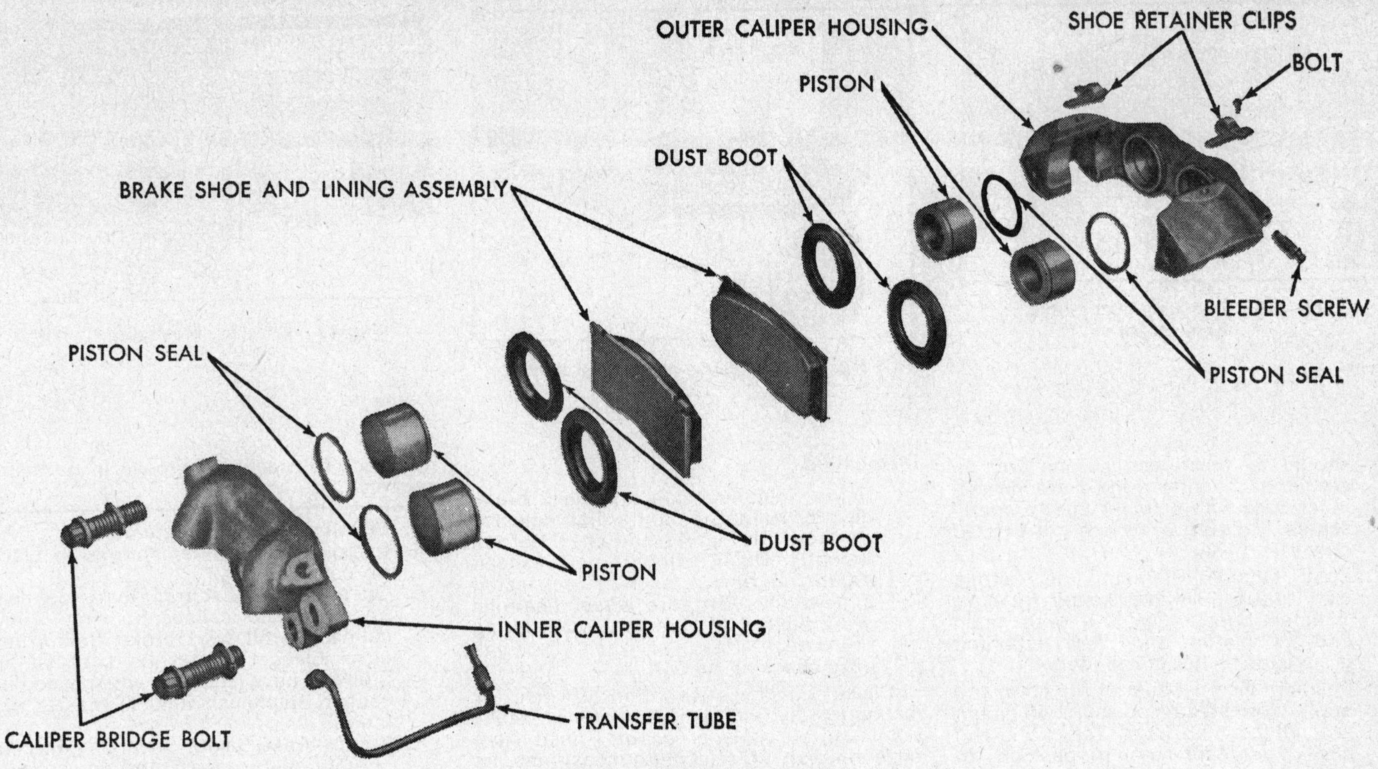

Fig. 36 Caliper disassembled

seat the units. The vehicle may pull to one side or the other if this is not done. It should not be necessary to bleed the system after replacing linings.

Servicing Caliper

Removal

1. Raise car on hoist or stands.
2. Remove wheel assembly.
3. Disconnect front brake flexible hose from brake tube at frame mounting bracket. Plug tube to prevent loss of fluid.
4. Remove bolts that attach caliper to steering knuckle.

NOTE: Should it become necessary to install a new flexible brake hose, scribe a mark on hose bracket on side where hose enters, and position of hose retaining clip underneath. When reassembling, be sure open end of retaining clip is facing out and away from caliper.

5. Slowly slide caliper up and away from brake disc, Fig. 35.

Disassembly

1. Referring to Fig. 36, remove shoe retainer clips and transfer tube from caliper.
2. Mount caliper in a vise with soft jaws and remove transfer tube.
3. Remove shoe and lining units.
4. Remove bridge bolts and separate caliper halves.
5. Peel dust boot out and away from caliper housing retainer and out of piston groove, Fig. 37. Remove remaining boots in same manner.
6. Using Tool C-3999, remove each piston, Fig. 38.

CAUTION: Care must be used so as not to scratch, burr or otherwise damage piston on outside diameter. To do so effects sealing qualities of piston. Draw piston straight out of its bore. If a piston becomes cocked, removal is more difficult and piston or bore may be damaged.

7. Using a small pointed wooden or plastic tool, remove piston seals from grooves in cylinder bore, Fig. 39. Discard old seals.

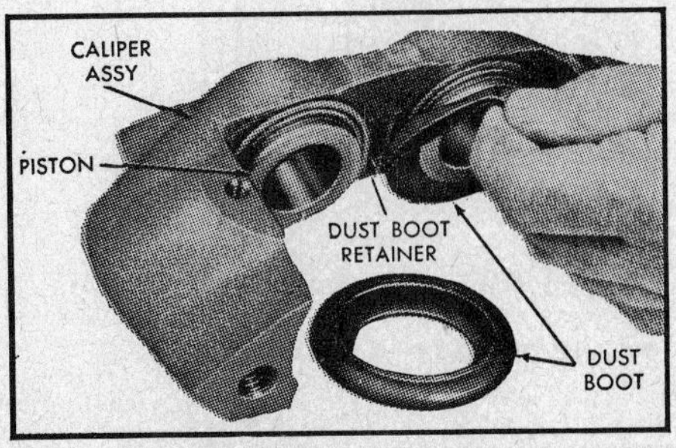

Fig. 37 Removing dust boot from piston and caliper

Cleaning & Inspection

Refer to "BUDD OPPOSED PISTONS" section for cleaning and inspection.

Assembling Caliper

1. Clamp inner caliper half in a vise (with protector jaws) by mounting lugs.
2. If it was necessary to install a new dust boot retainer ring, contact area on housing should be cleaned and Loctite Sealant Grade H (or equivalent) applied to retainer ring on surface where it seats in housing, then install retainer ring.
3. Dip new piston seals in brake fluid and install in caliper grooves. Seal

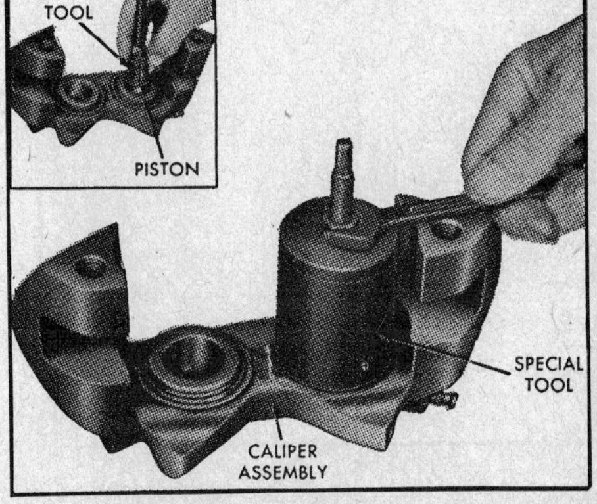

Fig. 38 Removing piston from caliper

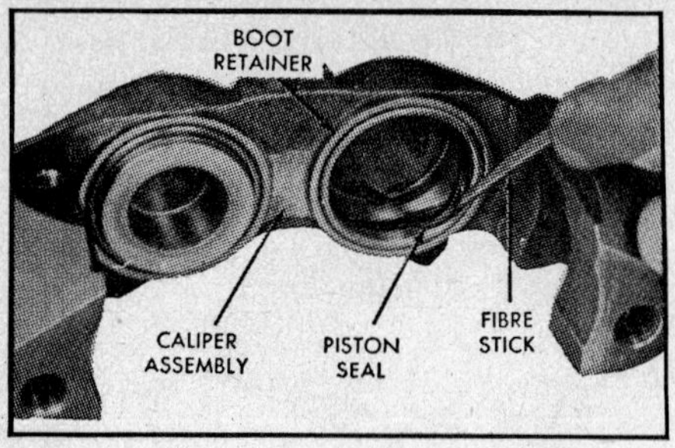

Fig. 39 Removing piston seals from caliper

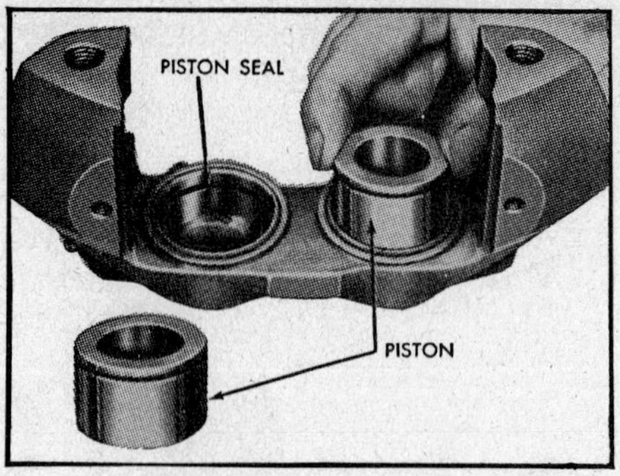

Fig. 40 Installing piston in caliper

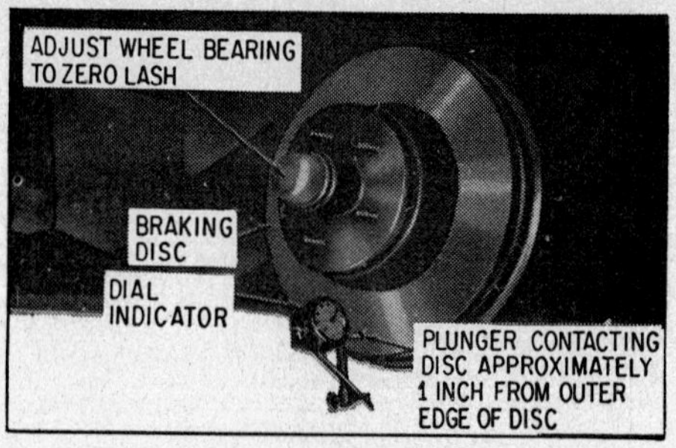

Fig. 41 Checking brake disc runout

should be positioned at one area in groove and gently worked around cylinder bore with a finger until properly seated. *Be sure seals are not twisted or rolled.*

4. Coat outside diameter of pistons with brake fluid and install them in cylinder bores, Fig. 40, with open end of piston and boot retaining groove facing out of cylinder.
5. Position piston squarely in bore and apply slow steady pressure. *If piston will not position itself, remove it and check seal for proper position in groove.*
6. Install new dust boot over caliper retaining ring and in piston groove. Install remaining boots in same manner.
7. Install caliper half on one clamped in vise. Assemble with bridge bolts and torque to 70-80 ft-lbs.

CAUTION: Under no circumstances should the bridge bolts be substituted or replaced by inferior bolts as this could cause caliper failure, resulting in an accident.

8. Install and tighten transfer tube.
9. Install bleeder screw loosely.

Installation

1. Before installing caliper, check brake disc for runout. Mount a dial indicator as shown in Fig. 41 and check lateral runout, which should not exceed .0025". If runout is excessive, install a new disc. *Be sure wheel bearings are adjusted to zero end play during this check. Readjust wheel bearings after checking runout.*

Check thickness variation of disc by measuring disc with a micrometer at twelve points at a radius of about 1 inch from edge of disc. If thickness measurements vary more than .0005 inch, the disc should be replaced.

2. Install caliper over disc. Install mounting bolts and torque as indicated:
   Dodge & Plymouth:
   1969-72 . . . . . 50-80 ft.-lbs.
   Lincoln Continental:
   1969 . . . . . . 100-140 ft.-lbs.

NOTE: A check should be made to be sure that brake disc runs squarely and centrally within caliper opening. There should be .090" to .120" clearance between outside diameter of disc and caliper. There should also be a minimum of .050" from either disc face to machined groove in outboard caliper.

3. Install shoe and lining units.
4. Install shoe retainer springs in place on caliper.
5. Open bleeder screw, then connect brake line to caliper housing. Allow caliper to fill with brake fluid, then close bleeder screw. Be sure all air bubbles have escaped when bleeding caliper. Replenish fluid in master cylinder.
6. Pump brake pedal several times to actuate piston seals and to position linings.
7. Road test vehicle as outlined previously.

KELSEY-HAYES FLOATING CALIPER

This type brake is a floating caliper, single piston, ventilated unit, actuated by the hydraulic system, Fig. 42. The caliper assembly, Fig. 43, is made up of a floating caliper assembly and an anchor plate. The anchor plate is bolted to the wheel spindle arm by two bolts. The caliper

Fig. 42 Single piston disc brake (typical)

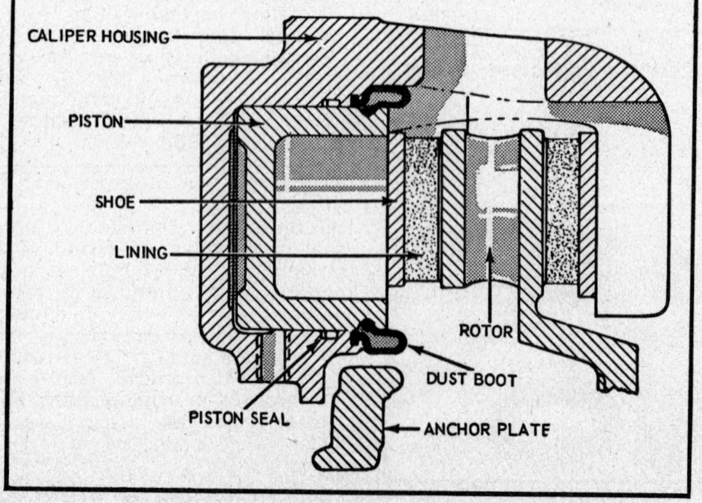

Fig. 43 Single piston disc brake caliper (typical)

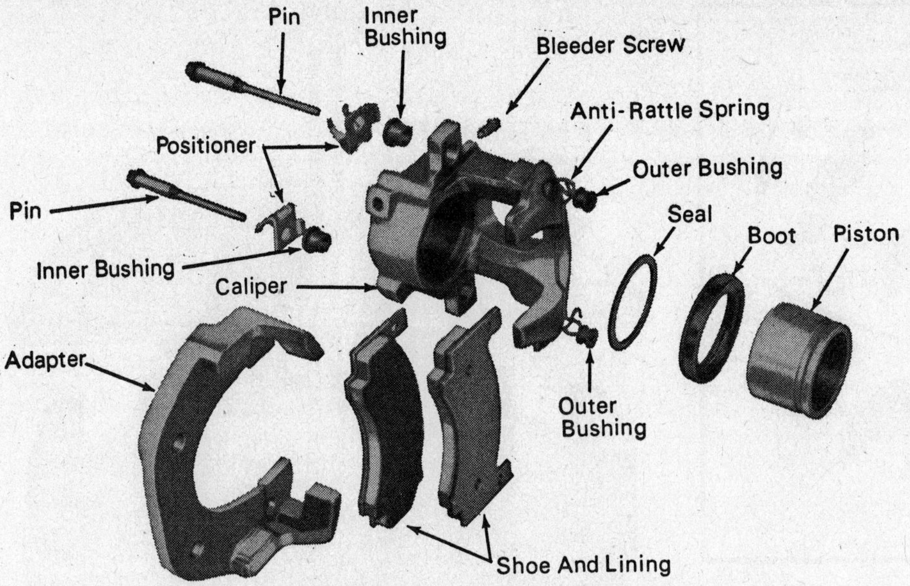

Fig. 44 Single piston disc brake caliper disassembled. (typical of American Motors and Chrysler Corp.)

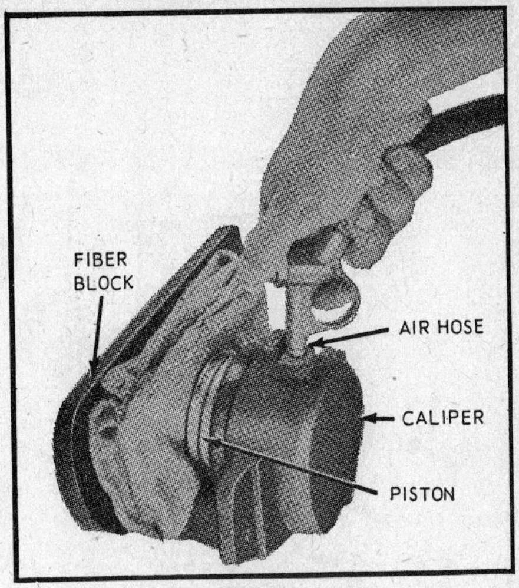

Fig. 45 Removing piston with air pressure

is attached to the anchor plate through two spring steel stabilizers. The caliper slides on two guide pins which also attach to the stabilizers. A single piston is used. The cylinder bore contains a piston with a molded rubber dust boot to seal the cylinder bore from contamination and also to return the piston to the released position when hydraulic pressure is released. Also a rubber piston seal is used to provide sealing between cylinder and piston.

Service Precautions

In addition to the precautions described at the beginning of this chapter, the following must be observed.

1. If the piston is removed for any reason the piston seal must be replaced.
2. During removal and installation of a wheel assembly, use care not to interfere with and damage the caliper splash shield or the bleeder screw fitting.
3. Be sure the vehicle is centered on the hoist before servicing any front end components to avoid bending or damaging the rotor splash shield on full right or left wheel turns.
4. The proportioning valve should not be disassembled or adjustments attempted on it.
5. The wheel and tire must be removed separately from the brake rotor.
6. The caliper assembly must be removed from the spindle prior to removal of shoe and lining assembly.
7. Do not attempt to clean or restore oil or grease soaked brake linings. When contaminated linings are found, linings must be replaced in complete axle sets.

American Motors & Chrysler Corp.

Removing Lining & Caliper

1. Support vehicle on hoist and remove wheel assembly.
2. Remove two thirds of brake fluid from reservoir that serves disc brakes.
3. If caliper is to be removed, disconnect front brake hose from tube from mounting bracket and plug brake tube to prevent loss of fluid.

NOTE: If pistons are to be removed, leave flex brake line connected to tube at mounting bracket.

4. Remove caliper guide pins, Fig. 44. On American Motors and 1969-72 Chrysler Corp. models, remove positioners and anti-rattle clips.
5. Slide outboard and inboard shoe assembly out of caliper and adaptor.
6. Remove inner and outer bushings from caliper.

Installing Lining & Caliper

1. Install new inner and outer bushings Fig. 44, then slide shoe and lining assembly into place in caliper and adaptor making certain that metal portion of shoe is fully in recess of caliper and adaptor.
2. Holding inboard shoe assembly in place, carefully slide caliper down into position in adaptor and over rotor. Align guide pin holes of adaptor and inboard and outboard shoes.
3. On American Motors and 1969-72 Chrysler Corp. models, install positioners over guide pins with open ends up and anti-rattle spring.
4. Press in on end of guide pin and thread pin into adaptor using extreme care not to cross threads. Torque guide pins from 30-35 ft. lbs. making sure that tabs of positioners are over machined surfaces of caliper.
5. If caliper was removed, connect brake hose to caliper and tighten securely, then open bleeder screw and allow caliper to fill with fluid. Make certain that all air bubbles have escaped when bleeding caliper.

CAUTION: Do not move vehicle until a firm brake pedal has been obtained.

Caliper Disassembly

1. Open bleeder screw and drain brake fluid from caliper then place caliper assembly in a soft jawed vice.

CAUTION: Do not overtighten vise as excessive pressure will cause bore distortion and binding of piston.

2. To remove piston place a cloth over piston and apply compressed air to fluid port in caliper, Fig. 45. Use extreme care to avoid damage to piston or bore. Allow dust boot to remain in caliper groove as piston is withdrawn.
3. Using a small wooden or plastic stick, work piston seal out of its groove and discard seal.

NOTE: Do not use screwdriver to remove piston seal as it could scratch bore or burr edges of seal groove.

4. Remove bleeder screw.

Cleaning & Inspection

1. Clean all parts with brake fluid and wipe dry, then blow out all drilled passages using compressor.
2. Inspect bore for scoring, pitting or corrosion. A deeply scored or corroded caliper should be replaced, although light scores and stains may be removed.

NOTE: If piston is pitted, scored or worn, replace piston.

3. Using crocus cloth, polish any discolored or stained area. Using finger pressure rotate crocus cloth in cylinder bore. Do not slide cloth in and out of bore under pressure or use any other form of abrasive or abrasive cloth. Black stains on bore wall are caused by piston seals and will do no harm.
4. Bores that have deep scratches or scores, should be honed providing that the diameter of the bore is not

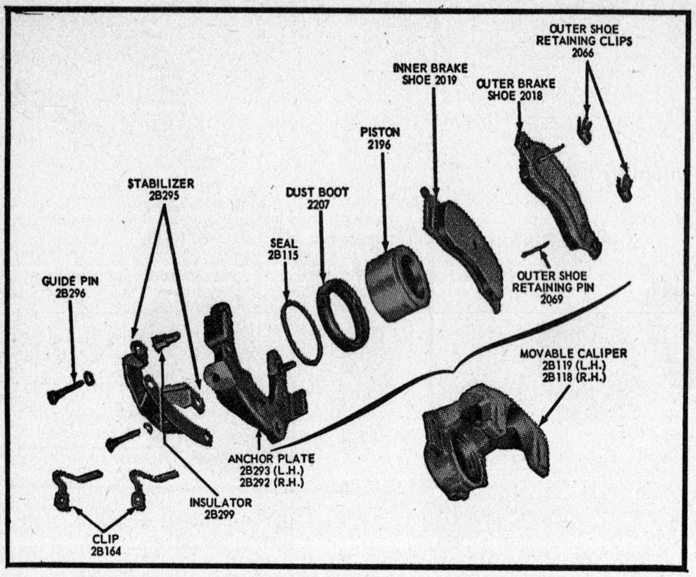

Fig. 46 Single piston disc brake caliper disassembled. (typical of Ford Motor Co.)

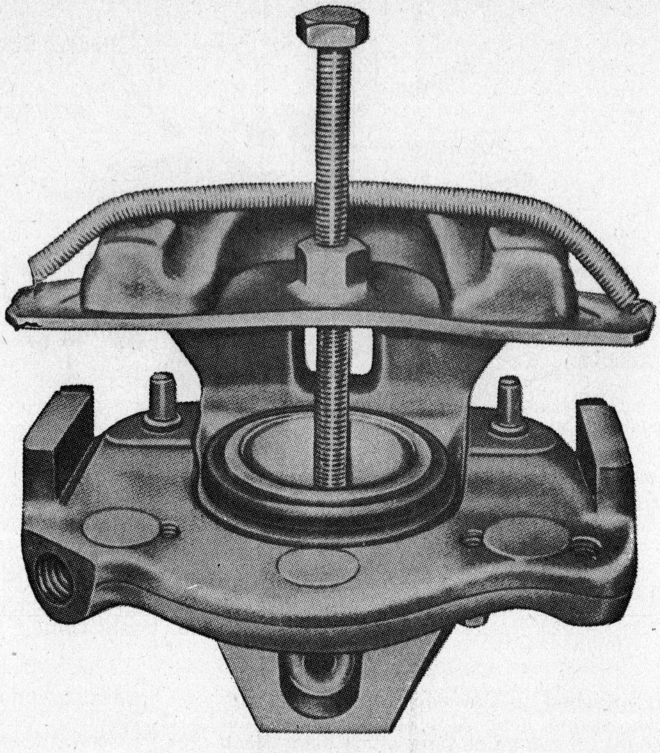

Fig. 47 Installing piston retracting tool

increased more than .002 inch.

5. Using a feeler gauge, check clearance of piston in bore. Clearance should be .002-.006 inch. If clearance exceeds this specification, replace caliper assembly.

Caliper Assembly

1. Dip new piston seal in clean brake fluid and install in bore groove. Seal should be positioned and gently worked around groove until properly seated.
2. Dip new dust boot in clean brake fluid and install in caliper by working into outer groove. Boot will seem larger than diameter of groove but will snap into place when properly seated in groove. Slide forefinger around inside of boot to be sure it is seated.
3. Dip piston in clean brake fluid, then with fingers spreading boot, work piston into boot and carefully down the bore until bottomed.

CAUTION: To avoid cocking, force must be uniformly applied on piston.

Ford Motor Co.

Removing Caliper

1. Support vehicle on hoist and remove wheel assembly being careful not to interfere with or damage caliper splash shield or bleeder screw.
2. Disconnect brake line from caliper and cap open end to prevent fluid loss.
3. On 1969 models, remove safety wire and the two caliper assembly to spindle attaching bolts.
4. On all except 1969 models, remove caliper locating pins and lower stabilizer attaching bolts and discard stabilizers.
5. Lift caliper assembly from rotor.

NOTE: If both caliper assemblies are removed, identify them as left and right as they are not interchangeable.

Installing Caliper

1. Install caliper assembly over rotor.
2. On 1969 models;
   a. Align mounting bolt holes with those in spindle.

NOTE: It may be necessary to push caliper piston into bore to obtain clearance between lining and rotor.

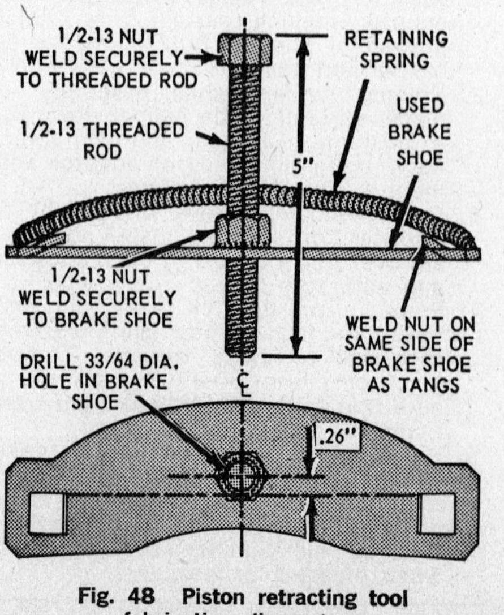

Fig. 48 Piston retracting tool fabrication dimensions

   b. Install caliper to spindle bolts and first torque upper bolt to 110-140 ft. lbs then torque lower bolt to 90-120 ft. lbs. for full size and 55-75 ft. lbs for intermediate models. Install safety wire and push ends against spindle.

3. On 1970-73 models;
   a. Place outer brake shoe against rotor surface during caliper installation, in anchor plate to prevent pinching piston boot between inner brake shoe and piston.
   b. Position new stablizer and install and torque locating pins to 25-35 ft. lbs.

NOTE: Make certain that locating pins are free of oil, grease or dirt.

   c. Install stabilizer to anchor plate screws and torque to 8-11 ft. lbs.

4. Using a new copper gasket on each side of hose fitting, install hose on caliper and torque attaching bolt to 17-25 ft. lbs.
5. Bleed brake system then centralize pressure differential valve as described:
   a. On 1969 models, turn ignition switch to "ON" or "ACC" positions and loosen brake line fitting at outlet port of differential valve on opposite side of brake system that was repaired or bled last, then depress brake pedal until brake warning light goes out and tighten brake line fitting.
   b. On 1970-73 models, turn ignition switch to "ON" or "ACC" and depress brake pedal until brake warning light goes out.
6. Apply brake pedal several times to

seat brake linings.

CAUTION: Do not move vehicle until a firm brake pedal has been obtained.

Removing Lining

1969 Models

1. Support vehicle on hoist and remove wheel assembly.
2. Remove caliper assembly as described previously. To aid in caliper removal, apply a steady inward pressure against inner shoe and lining assembly. Maintain pressure for at least a minute until piston is pushed into its bore.
3. Slide the outer two outer shoe retaining clips off retaining pins, Fig. 46, then remove retaining pins and remove shoe from stationary caliper.
4. Slide inner brake shoe outward until it is free of hold down springs, then remove brake shoe.
5. Remove caliper locating pins and stabilizer attaching bolts, then remove and discard stabilizers.
6. Remove locating pin insulators from anchor plate.

1970-72 Full Size Models

1. Remove half of brake fluid capacity from master cylinder.
2. Support vehicle on hoist and remove wheel assembly.
3. Remove inner shoe hold down clips, then place a small screwdriver under outer shoe retaining clip tang and lift away from pin groove and slide clip from shoe retaining pin. Remove outer shoe retaining clip and remove outer shoe.
4. Remove caliper locating pins and upper stabilizer to anchor plate bolt, then remove upper stabilizer to avoid interference with brake hose.
5. Lift caliper assembly from anchor plate and remove outer shoe and retaining pins from caliper assembly.
6. Support caliper from suspension with a wire and remove caliper locating pin insulators.
7. Remove inner shoe and lining assembly and inspect rotor.

1970-73 Intermediate

1. Support vehicle and remove wheel assembly being careful to avoid damage or interference with caliper splash shield or bleeder screw.
2. Remove caliper as described previously.
3. Remove inner brake shoe hold down clips and locating pin insulators from anchor plate and remove inboard shoe and lining assembly.
4. To remove outer brake shoe, place a small screwdriver under outer brake shoe retaining clip tang and lift away from pin groove and slide clip from brake shoe retaining pin. Remove outer brake shoe retaining clip and remove outer brake shoe.

Installing Lining

NOTE: When new shoe and lining assemblies are being installed to replace worn linings it will be necessary to push the piston all the way into the caliper bore. This will displace fluid from the caliper into the master cylinder reservoir. Check the primary (front) brake system reservoir level and remove fluid to approximately half full before replacing brake shoes. This will prevent overflow. Do not reuse the removal fluid.

1969 Models

1. Install new caliper locating pin insulators in anchor plate.
2. Position caliper in anchor plate, then install new stabilizers and loosely install caliper locating pins. Locating pins should be free of oil, grease or dirt.

NOTE: If caliper locating pins are rusted or corroded, they should be replaced.

3. Position outer brake shoe on caliper and install two retaining pins and clips.
4. Install inner brake shoe so that ears of shoe are on top of anchor plate bosses and under shoe hold down springs.
5. Position shoe and lining assemblies so that caliper can be placed over rotor. Rotate a hammer handle between linings to provide proper clearance.
6. Install caliper over rotor on spindle. Install the two caliper attaching bolts and first torque the upper bolt to 110-140 ft. lbs. then torque the lower bolt to 90-120 ft. lbs. for full size and 55-75 ft. lbs for intermediate models. Install safety wire, twist ends at least five turns and push wire ends against spindle.
7. With moderate pressure applied to brake pedal, torque stabilizer attaching screws to 8-11 ft. lbs. and locating pins to 25-35 ft. lbs.

1970-72 Full Size & 1970-73 Intermediate Models

1. Install inner brake shoe in anchor plate and new caliper locating pin insulators.
2. Install inner brake shoe hold down clips and torque retaining screws to 6-10 ft. lbs.
3. Install piston retracting tool in caliper with brake shoe lances positioned in slots in caliper outer legs and retract piston, Fig. 47. The piston retracting tool can be fabricated from a discarded outer brake shoe and threaded rod, Fig. 48. When using piston retracting tool, turn threaded rod one half turn at a time and pause to permit piston to move in seal. As piston nears bottom of travel, reduce time interval to insure bottoming of piston.
4. Install new outer brake shoe and lining assembly on caliper and install retaining pins and clips.
5. Install caliper as described previously.

Caliper Disassembly

1. Remove caliper assembly from vehicle.
2. Remove caliper locating pins from caliper assembly and remove anchor plate from caliper.
3. Slide the two outer shoe retaining clips off retaining pins, then remove retaining pins and outer brake shoe from caliper.
4. Slide inner brake shoe outward until it is free of hold down springs, then remove brake shoe.
5. To remove piston, apply air pressure to fluid port, Fig. 45. Place a cloth over piston before applying pressure to prevent damage to piston. If piston is seized and cannot be removed, tap lightly around piston while applying pressure.

NOTE: If piston is stuck, care should be taken when applying air as piston could develop considerable force.

6. Remove dust boot and piston seal from caliper.

Cleaning & Inspection

Clean all metal parts with alcohol, then using compressed air, clean out and dry all grooves and passages. Make certain that caliper bore and component parts are completely free of any foreign material.

Check cylinder bore and piston for damage or excessive wear. Replace piston if pitted, scored, or the chrome plating is worn off.

Caliper Assembly

1. Apply clean brake fluid to caliper piston seal and install it in cylinder bore. Be sure seal does not become twisted and that it is fully seated in groove.
2. Install a new dust boot by setting flange squarely in outer groove of caliper bore.
3. Coat piston with clean brake fluid and install it in cylinder bore. Spread dust boot over piston as it is installed, then seat dust boot in piston groove.
4. Position inner brake shoe so that ears of the shoe rest on top of anchor plate bosses and beneath hold down springs.
5. Install new caliper locating pin insulators in anchor plate and position caliper on anchor plate.
6. Apply alcohol to caliper locating pins and install them loosely in anchor plate.

NOTE: Make certain that guide pins are free of oil, grease or dirt.

7. Install caliper as previously outlined.

DELCO-MORAINE SINGLE PISTON

This single piston sliding caliper assembly, Fig. 49, incorporates a one piece housing with the inboard side of the housing bored for the piston. A seal within the housing bore provides a hydraulic seal between the piston and housing wall.

A spring steel scraper (wear sensor) is incorporated on each inboard shoe. When the shoe lining has worn to within .030 inch of the shoe, the sensor scrapes the rotor and emits an audible high frequency sound indicating that the linings should

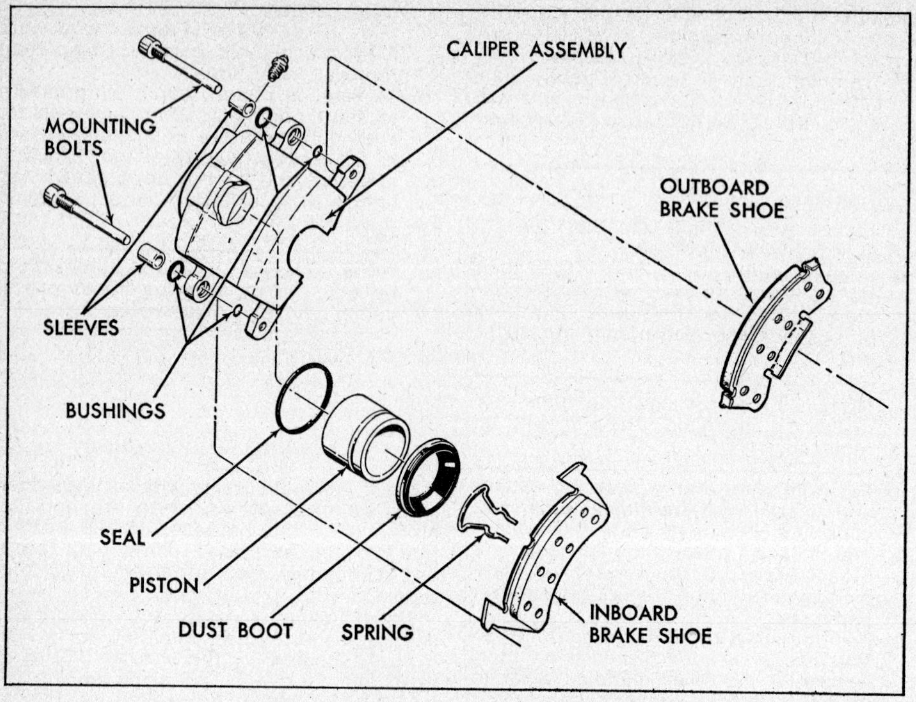

Fig. 49 Single piston caliper. Exploded

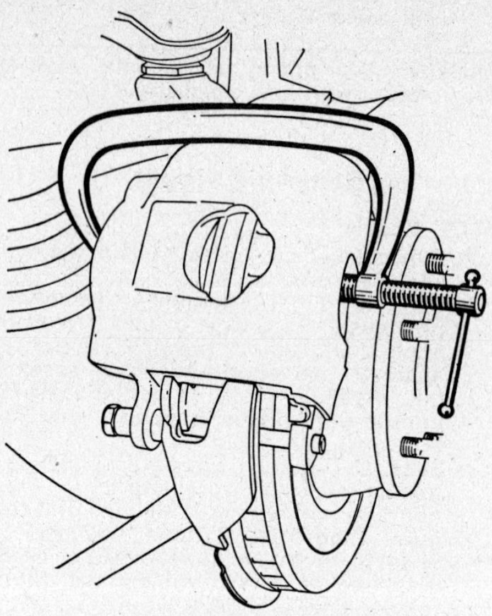

Fig. 50 Compressing piston and shoes with "C" clamp

be replaced.

The caliper assembly used on all models except Vega, Fig. 49, slides on its mounting bolts whereas the caliper on Vega models, the caliper assembly slides on mounting sleeves which are secured by two mounting pins. Upon brake application, fluid pressure against the piston forces the inboard shoe and lining assembly against the inboard side of the disc. This action causes the caliper assembly to slide until the outboard lining comes into contact with the disc. As pressure builds up, the linings are pressed against the disc with increased force.

ALL EXC. ASTRE, MONZA, SKYHAWK, STARFIRE & VEGA

Caliper Removal

1. Siphon enough brake fluid out of the master cylinder to bring fluid level to 1/3 full to avoid fluid overflow when the caliper piston is pushed back into its bore.

Fig. 51 Installing support spring

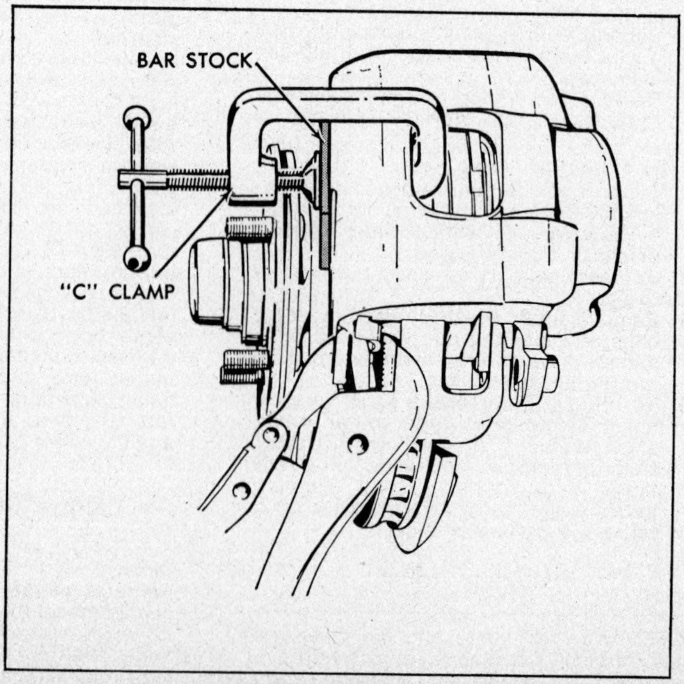

Fig. 52 Fitting shoe to caliper

2. Raise vehicle and remove front wheels.
3. Using a "C" clamp, as illustrated in Fig. 50, push piston back into its bore.
4. Remove two mounting bolts and lift caliper away from disc.

Brake Shoe Removal

1. Remove caliper assembly as outlined above.
2. Remove inboard shoe. Dislodge outboard shoe and position caliper on the front suspension so the brake hose will not support the weight of the caliper.
3. Remove shoe support spring from piston.
4. Remove two sleeves from inboard ears of the caliper.
5. Remove four rubber bushings from the grooves in each of the caliper ears.

Brake Shoe Installation

1. Lubricate new sleeves, rubber bushings, bushing grooves and mounting bolt ends with Delco Silicone Lube or its equivalent.
2. Install new bushings and sleeves in caliper ears.

NOTE: Position the sleeve so that the end toward the shoe is flush with the machined surface of the ear.

3. Install shoe support spring in piston cavity, Fig. 51.
4. Position inboard shoe in caliper so spring ends centrally contact shoe edge. Initially, this will place the shoe on an angle. Push upper edge of shoe down until shoe is flat against caliper. When properly seated, spring ends will not extend past shoe more than .100".
5. Position outboard shoe in caliper with shoe ears over caliper ears and tab at bottom of shoe engaged in caliper cutout.
6. With shoes installed, lift caliper and rest bottom edge of outboard lining on outer edge of brake disc to be sure there is no clearance between outboard shoe tab and caliper abutment.
7. On 1969-72 models, using a 1/4 x 1 x 2 1/2 inch metal bar to bridge caliper cutout, clamp outboard shoe to caliper with a "C" clamp. Bend both ears of outboard shoe over caliper until clearance between shoe ear and caliper (measured at both the edge and side of caliper) is .005 inch or less, Fig. 52.
8. Remove "C" clamp then install caliper and torque mounting bolts to 30-40 ft. lbs.
9. On 1973-75 vehicles, clinch upper ears of outboard shoe by positioning pliers with one jaw on top of upper ear and one jaw in notch on bottom shoe opposite ear, Fig. 53. Ears are to be flat against caliper housing with no radial clearance. If clearance exists, repeat clinching procedure.

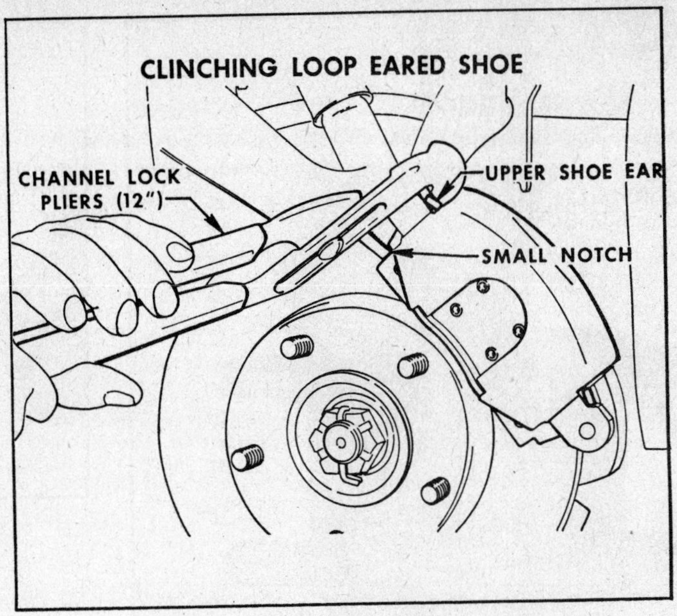

Fig. 53 Clinching loop eared brake shoe

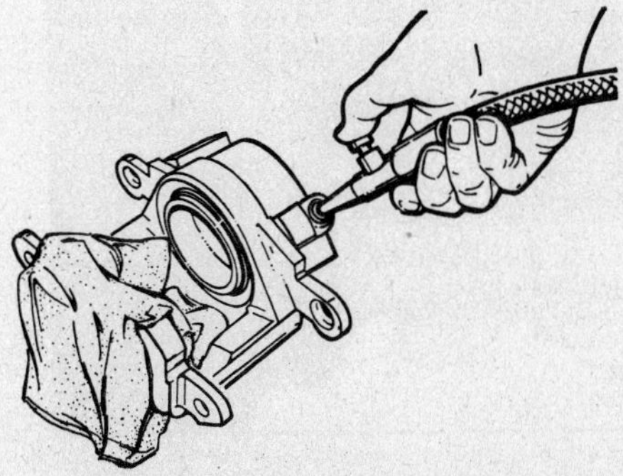

Fig. 54 Removing piston from caliper

Disassembling Caliper

1. Remove caliper as outlined above.
2. Disconnect hose from steel line, remove U shaped retainer and withdraw hose from frame support bracket.
3. After cleaning outside of caliper, remove brake hose and discard copper gasket.
4. Drain brake fluid from caliper.
5. Pad caliper interior with clean shop towels and use compressed air to remove piston, Fig. 54.

NOTE: Use just enough air pressure to ease piston out of bore. Do not blow piston out of bore.

CAUTION: Do not place fingers in front of piston in an attempt to catch or protect it when applying compressed air. This could result in serious injury.

6. Carefully pry dust boot out of bore.
7. Using a small piece of wood or plastic, remove piston seal from bore.

NOTE: Do not use a metal tool of any kind to remove seal as it may damage bore.

8. Remove bleeder valve.

Assembling Caliper

1. Lubricate caliper piston bore and new piston seal with clean brake fluid. Position seal in bore groove.
2. Lubricate piston with clean brake fluid and assemble a new boot into the groove in the piston so the fold faces the open end of the piston, Fig. 55.
3. Using care not to unseat the seal, insert piston into bore and force the piston to the bottom of the bore.
4. Position dust boot in caliper counterbore and install, Fig. 56.

NOTE: Check the boot installation to be sure the retaining ring moulded into the boot is not bent and

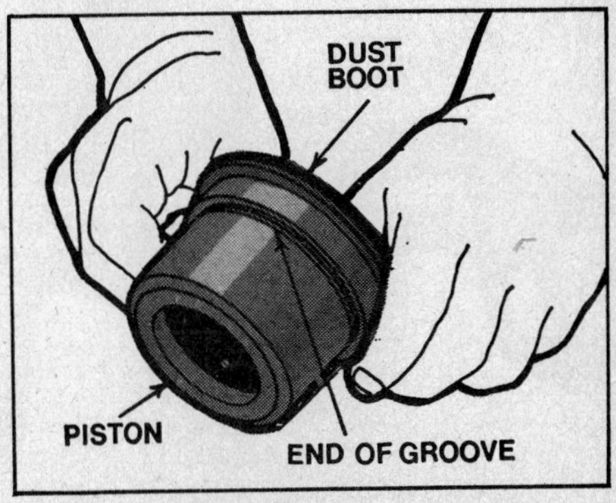

Fig. 55 Installing boot to piston

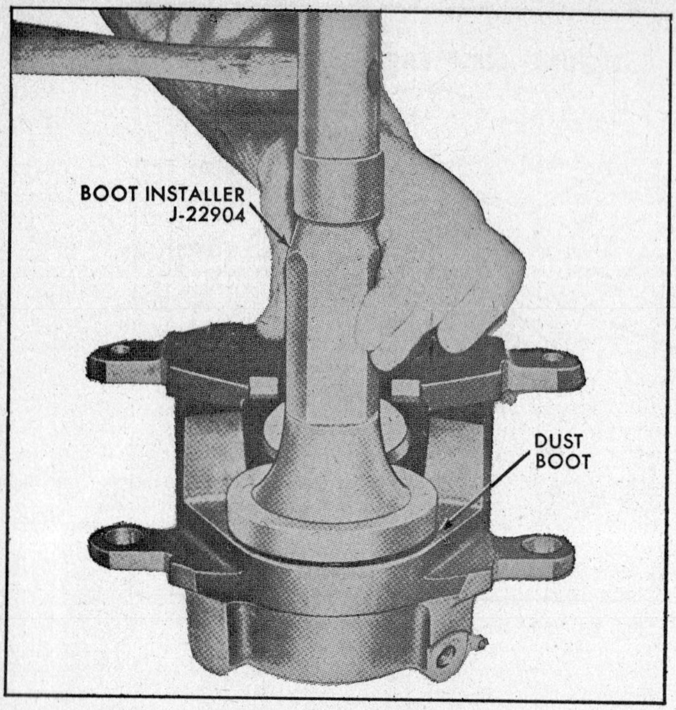

Fig. 56 Installing boot to caliper

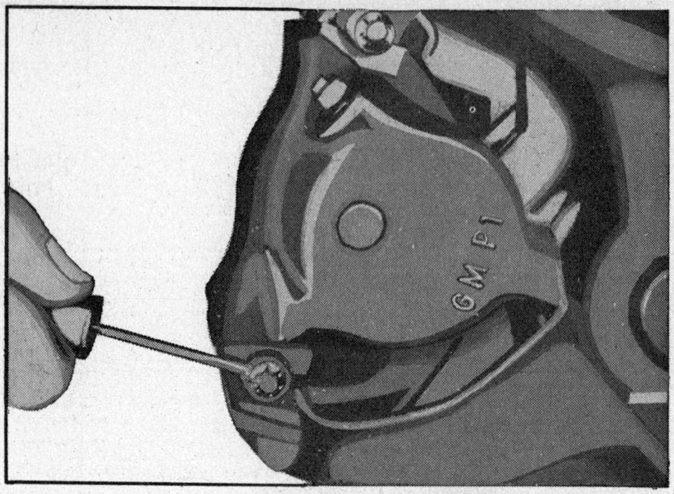

Fig. 57 Removing stamped nuts from mounting pins

that the boot is installed below the caliper face and evenly all around. If the boot is not fully installed, dirt and moisture may enter the bore and cause corrosion.

---

5. Install the brake hose in the caliper using a new copper gasket.
6. Install shoes and re-install caliper assembly.

Caliper Installation

1. Position caliper over disc, lining up holes in caliper with holes in mounting bracket. If brake hose was not disconnected during removal, be sure not to kink it during installation.
2. Start mounting bolts through sleeves in inboard caliper ears and the mounting bracket, making sure ends of bolts pass under ears on inboard shoe.

NOTE: Right and left calipers must not be interchanged.

---

3. Push mounting bolts through to engage holes in the outboard ears. Then thread mounting bolts into bracket.
4. Torque mounting bolts to 30-40 ft. lbs.
5. If brake hose was removed, reconnect it and bleed the calipers.
6. Replace front wheels, lower vehicle and add brake fluid to master cylinder to bring level to 1/4" from top.

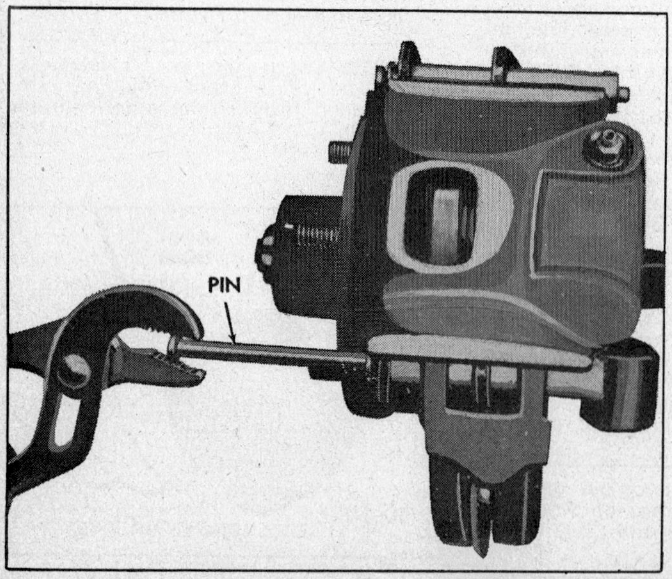

Fig. 58 Removing mounting pins

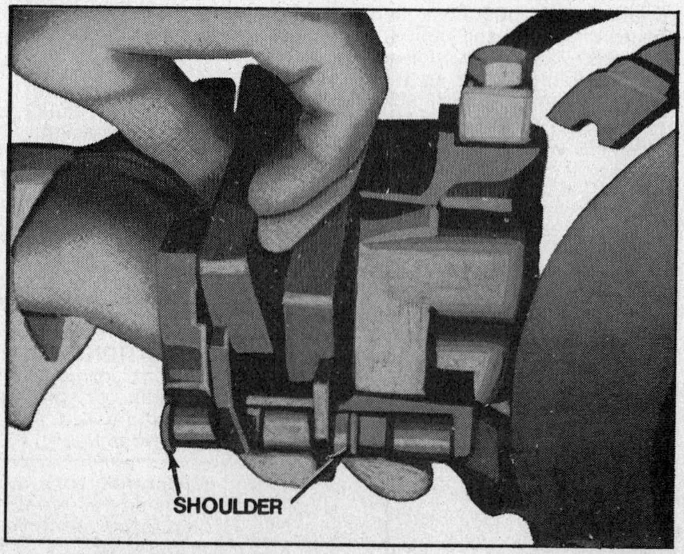

Fig. 59 Mounting sleeves and brake shoe installation

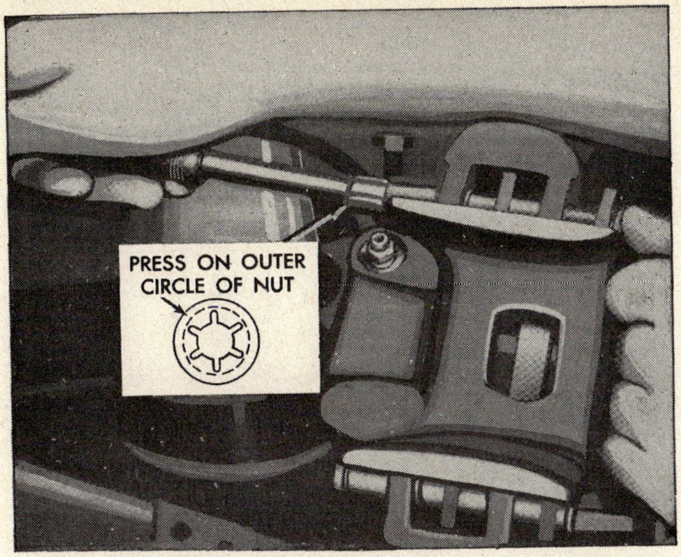

Fig. 60 Installing stamped nuts on mounting pins

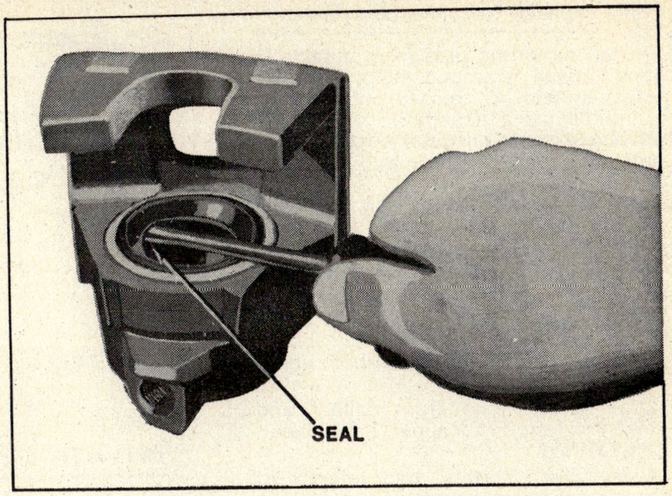

Fig. 61 Dust boot seal removal

NOTE: Before moving vehicle, pump brake pedal several times to be sure it is firm. Do not move vehicle until a firm pedal is obtained.

ASTRE, MONZA, SKYHAWK, STARFIRE & VEGA

Lining Removal

1. Support vehicle on hoist and remove wheel assembly.
2. Remove the two mounting pin stamped nuts, Fig. 57, and slide out the mounting pins, Fig. 58.
3. Lift caliper off disc and support caliper from suspension using wire.
4. Slide inboard and outboard shoes past mounting sleeve openings and remove mounting sleeves and bushing assemblies.
5. If caliper is to be removed, disconnect brake line.

Lining Installation

1. Install new sleeves with bushings on caliper grooves, Fig. 59.

NOTE: The "shouldered end" of sleeve must be installed toward outside.

2. Install inner shoe on caliper and slide shoe ears over sleeve, Fig. 59. Install the outer shoe in the same manner.

CAUTION: If pads are being re-used, they must be installed in same location as when removed.

3. Mount caliper on rotor. If brake line was disconnected, reconnect and torque bolt to 22 ft. lbs.

NOTE: To avoid overflow, it may be necessary to remove half of brake

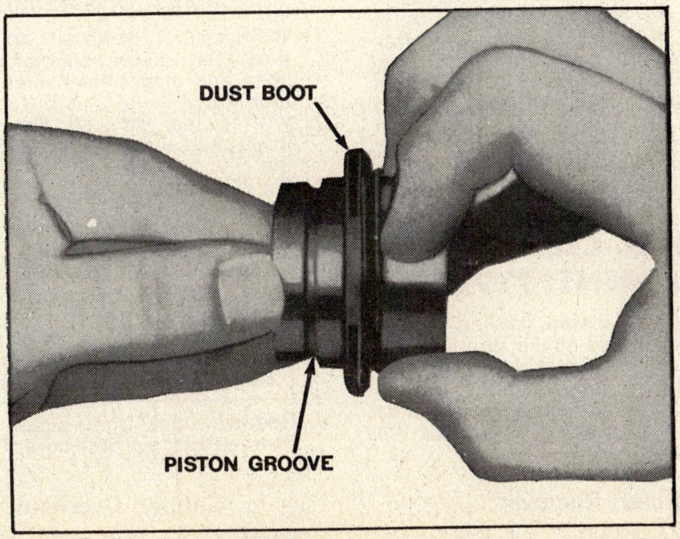

Fig. 62 Installing piston on boot

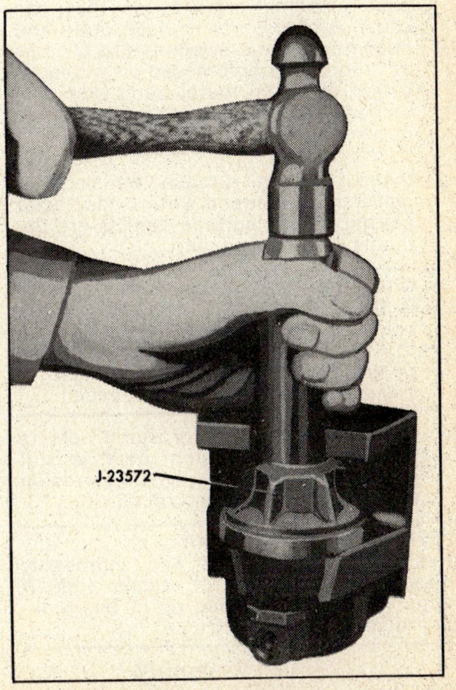

Fig. 63 Seating dust boot in caliper

fluid capacity from master cylinder.

4. Install mounting pins from outside in and install stamped nuts, Fig. 60. Nuts should be pressed on as far as possible using a suitable size socket that just seats on outer edge of nut.
5. Install wheel assembly and lower vehicle.
6. Add brake fluid to within 1/4 inch from top of master cylinder and test brake operation to insure a firm brake pedal before moving vehicle.

Caliper Disassembly

1. Remove caliper as described under "Lining Removal".
2. Drain brake fluid from caliper and clean exterior of caliper using clean brake fluid.
3. Using clean towels, pad interior of caliper and remove piston by applying just enough compressed air to fluid inlet port to ease piston out of bore.

CAUTION: Do not place fingers in front of piston in an attempt to catch or protect it when applying compressed air.

4. Carefully using a screwdriver so as not to scratch piston bore, pry dust boot out of piston bore, Fig. 61.
5. Using a piece of wood or plastic so as not to damage bore, remove piston seal from its groove in caliper bore.
6. Remove bleeder screw.

Cleaning & Inspection

1. Clean all metal parts in clean brake fluid, then using clean filtered air, dry parts and blow out all passages in caliper and bleeder valve.

NOTE: Always use clean brake fluid to clean caliper parts. Never use mineral base cleaning solvents as they can cause rubber parts to deteriorate and become soft and swollen, also the use of lubricated compressed air will leave a film of oil on metal parts that may damage rubber parts when they come in after reassembly.

2. Inspect piston surface for scoring, nicks, corrosion and worn or damaged plating. If any surface defects are detected, replace piston.

CAUTION: The piston outside surface is the primary sealing surface in the caliper. It is manufactured and plated to close tolerances, therefore refinishing by any means or the use of any abrasive is not recommended.

3. Check caliper bore for same defects as piston. The piston bore is not plated and stains or minor corrosion may be polished with crocus cloth.

CAUTION: Do not use emery cloth or any other form of abrasive and thoroughly clean caliper after use of crocus cloth. If caliper cannot be cleaned up in this manner, replace caliper.

Caliper Assembly

NOTE: The dust boot and piston seal are to be replaced each time that the caliper is disassembled.

1. Lubricate piston bore and new piston seal with clean brake fluid, then position seal in caliper bore groove.
2. Lubricate piston with clean brake fluid and assemble a new boot into groove in piston, Fig. 62.
3. Install piston into bore using care not to unseat seal, then force piston to bottom of bore.

NOTE: Approximately 50-100 pounds of force are required to push piston to bottom of bore.

4. Position dust boot in caliper counterbore and seat boot using tool shown in Fig. 63.

NOTE: Check boot installation to make sure that retaining ring moulded into boot is not bent and that boot is installed evenly all around. If boot is not fully installed, dirt and moisture may enter bore.

5. Install caliper as described under "Caliper Installation".

FORD CENTER ABUTMENT TYPE

This is a single piston, sliding caliper type brake but instead of the torque being taken by the caliper assembly, it is transmitted directly to the U-shaped anchor plate. The caliper housing performs only the necessary clamping action against the disc, Fig. 64.

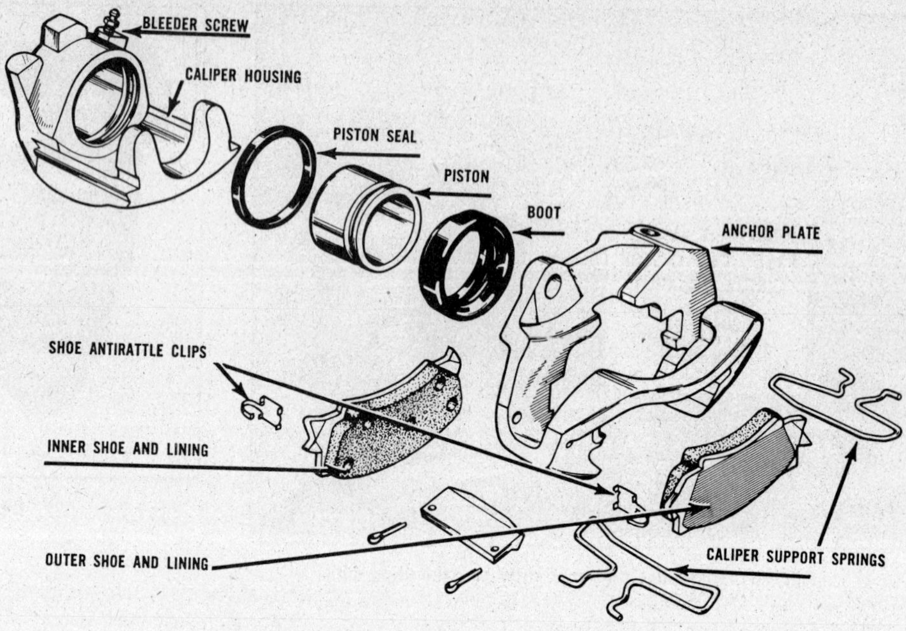

Fig. 64 Center abutment disc brake exploded

Caliper Removal

NOTE: Siphon a portion of the brake fluid from the larger master cylinder reservoir before servicing unit.

1. Remove wheel and tire and remove cotter pins from caliper support key.
2. Using a drift and a light hammer, remove caliper support key, being careful to avoid damaging key or machined surfaces, Fig. 65.
3. Rotate lower end of caliper housing toward the rear and upward and remove caliper from anchor plate. It is not necessary to disconnect hydraulic line for this operation.
4. Suspend caliper housing with wire or lay caliper on support strut. Do not let caliper hang by its own weight from hydraulic line.

Caliper Disassembly

1. With caliper removed as described previously, disconnect brake hose. Cap hose and plug caliper inlet to prevent fluid loss.
2. With caliper on work bench, remove inlet plug and drain fluid from housing.
3. Position caliper as in Fig. 66, and place shop cloths as shown.
4. Apply air pressure slowly to caliper inlet port to remove piston.

NOTE: If high pressure is applied quickly, piston may pop out and cause injury. A cocked or seized piston can be eased out by rapping sharply on piston end with a soft brass hammer.

5. Remove boot from piston and seal from caliper cylinder bore, Fig. 67.

Caliper Overhaul

1. Remove any rust or corrosion from machined surfaces of caliper housing.

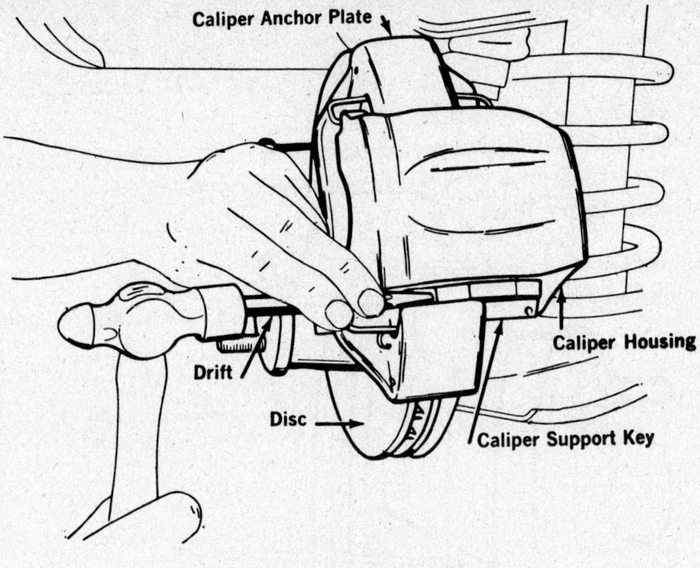

Fig. 65 Removing caliper support key

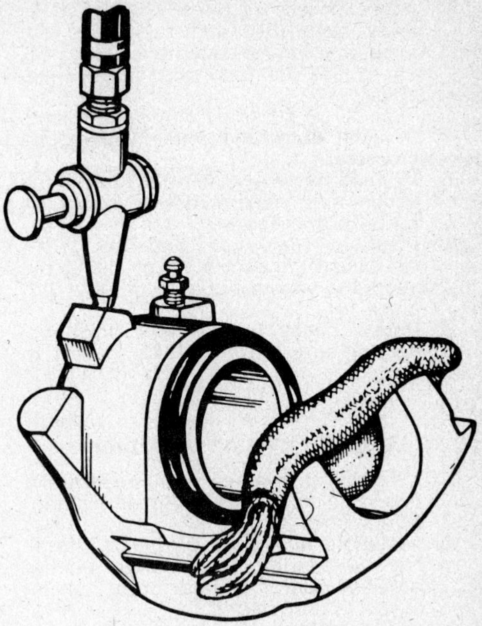

Fig. 66 Removing caliper piston with air pressure

Fig. 67 Caliper piston assembly

2. Clean housing and piston with isopropyl alcohol. Clean and dry out grooves and passages with compressed air. Make sure cylinder bore is free of all foreign material.
3. Check cylinder bore, seal groove and piston for wear or damage. Replace piston if pitted or worn.
4. Remove any rust or corrosion from machined surfaces of anchor plate still mounted on car and inspect plate for damage.
5. Inspect anti-rattle clips for damage. Clips should have four tabs, Fig. 67, and a looped type spring. Replace any damaged parts.

Brake Shoe and Lining, Replace

1. Remove caliper housing as outlined previously.
2. As shoe and lining assemblies are

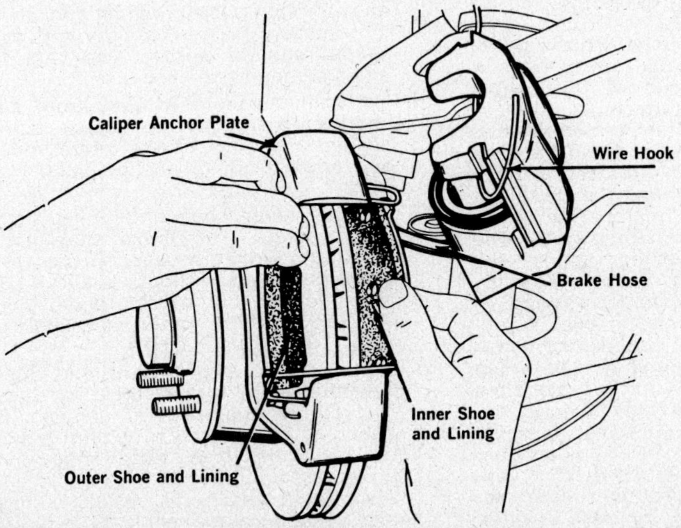

Fig. 68 Removing shoe and lining assemblies

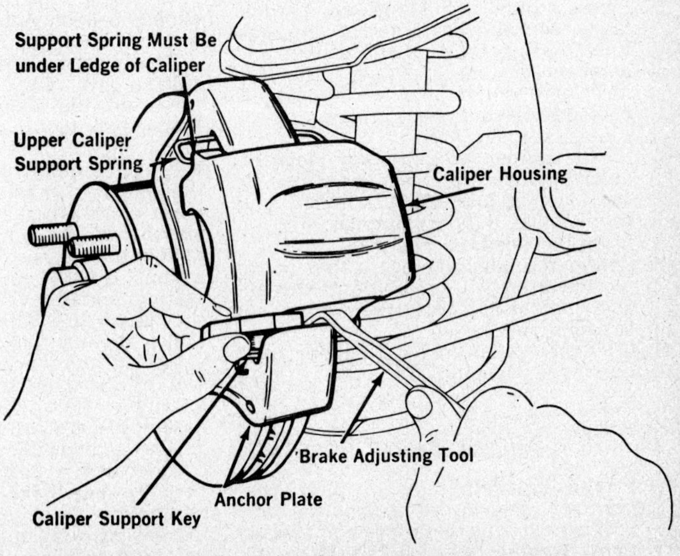

Fig. 69 Positioning caliper housing on anchor plate

now exposed, tilt upper edge of shoes away from disc and then take out shoe and lining assemblies, Fig. 68. Shoes are identical and interchangeable.

3. Take three thickness & measurements of each shoe and lining with a micrometer.
4. If there is less than .030" of lining above rivet heads, replace shoes on both front wheels.
5. Position new shoe and lining assemblies in anchor plate by tilting the shoe and sliding the bottom edge into place.
6. Rotate shoes into correct positions, making sure the lining side is next to disc.

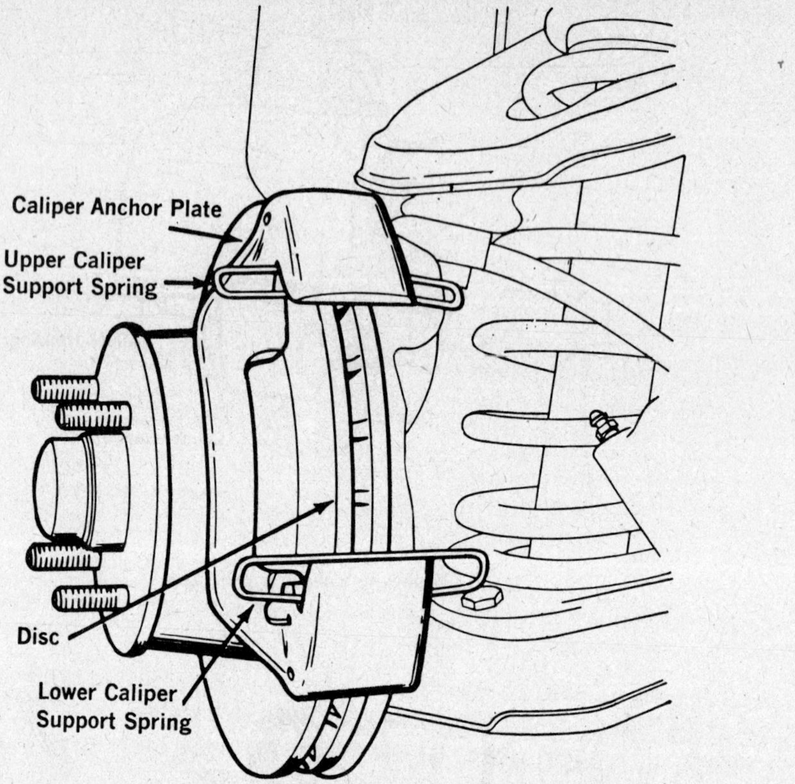

Fig. 70 Position of caliper support springs

Caliper Housing Assembly

1. Lubricate piston seal with clean brake fluid and position seal in its cylinder bore groove.
2. Assemble dust boot on caliper housing by seating boot flange in the outer groove of cylinder bore, making sure it is fully seated.
3. Coat piston with clean brake fluid and install in cylinder bore.
4. Spread dust boot over piston as piston is installed and then bottom piston in the bore. Seat dust boot in its piston groove.

Caliper Housing Installation

1. Assemble anti-rattle clips in anchor plate, making sure that tabs on clips are positioned properly and the loop type springs are positioned on the anchor plate side, away from the disc.
2. Position shoe and lining assemblies as described previously.
3. Place a thin coat of high-temperature lubricant on anchor plate and caliper surfaces that will be in contact after caliper is installed. Avoid getting lubricant on linings.
4. Position caliper in anchor plate, making sure that the top, trailing edge, of the caliper is properly positioned and the caliper support spring is under the projecting ledge of the caliper.
5. Insert a brake adjusting tool, or wide blade screwdriver, between the bottom, leading edge, of the caliper and adjacent anchor plate surface. Pry downward so that caliper housing is pressed upward and inward toward the spindle, Fig. 69.
6. Insert caliper support key between caliper housing and anchor plate. Be sure key is properly positioned and that caliper support springs are still in proper position, Fig. 70.
7. Center the support key so that cotter pin holes are on each side of the anchor plate and insert a new cotter pin in each hole.
8. Connect brake hose to caliper inlet port. Bleed system.

Anchor Plate

1. In the event the anchor plate was removed, position it on steering arm and install both mounting bolts finger tight. Tighten upper bolt first to 90-120 ft. lbs. and then tighten lower bolt to 55-75 ft. lbs.
2. Replace caliper housing as described previously.

FORD SLIDING CALIPER

The caliper assembly is made up of a sliding caliper housing assembly and an anchor plate, Fig. 71.

The anchor plate is bolted to the wheel spindle arm. Two angular machined surfaces on the upper end of the caliper housing contact mating machined surfaces of the anchor plate. A steel, plated key and a caliper support spring is fitted between the angular machined surfaces of the lower end of the caliper and the machined surface of the anchor plate. The key is held in position with a retaining screw. The caliper is held in position against the mating surfaces of the anchor plate by means of the caliper support spring. A brake shoe anti-rattle spring clip is provided on the anchor plate at the lower end of the inner brake shoe and lining assembly. The inner and outer brake shoe assemblies are not interchangeable.

The sliding caliper contains a single cylinder and a piston with a molded dust boot to seal the cylinder bore from contamination. A square section rubber piston seal is positioned in a groove in cylinder bore to provide sealing between cylinder and piston.

Caliper Removal

1. Raise car and support with safety stands. Block both rear wheels if a jack is used.
2. Remove wheel and tire assembly from hub.
3. Disconnect brake hose from caliper.
4. Remove retaining screw from caliper retaining key, Fig. 71.
5. Slide caliper retaining key and support spring either inward or outward from anchor plate. Use hammer and drift, if necessary, to remove the key and caliper support spring. Use care to avoid damaging the key.
6. Lift caliper assembly away from anchor plate by pushing caliper down against anchor plate and rotate upper end upward out of anchor plate, Fig. 72.
7. Remove inner shoe and lining from anchor plate. The brake shoe anti-rattle clip (inner shoe only) may become displaced at this time and if so, reposition it on anchor plate, Fig. 73. Tap lightly on outer shoe and lining to free it from caliper.
8. Clean caliper, anchor plate and rotor assemblies and inspect them for signs of fluid leakage, wear or damage. If either lining is worn to within 1/32" of any rivet head, both shoe and lining assemblies must be replaced. Also, if necessary to replace shoes and lining on one wheel, they must be replaced on both wheels to maintain equal brake action.

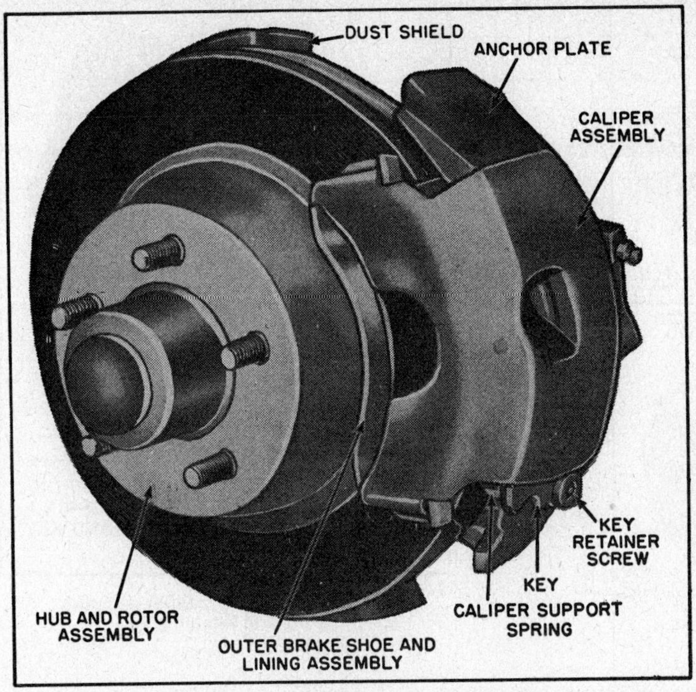

Fig. 71 Ford sliding caliper disc brake

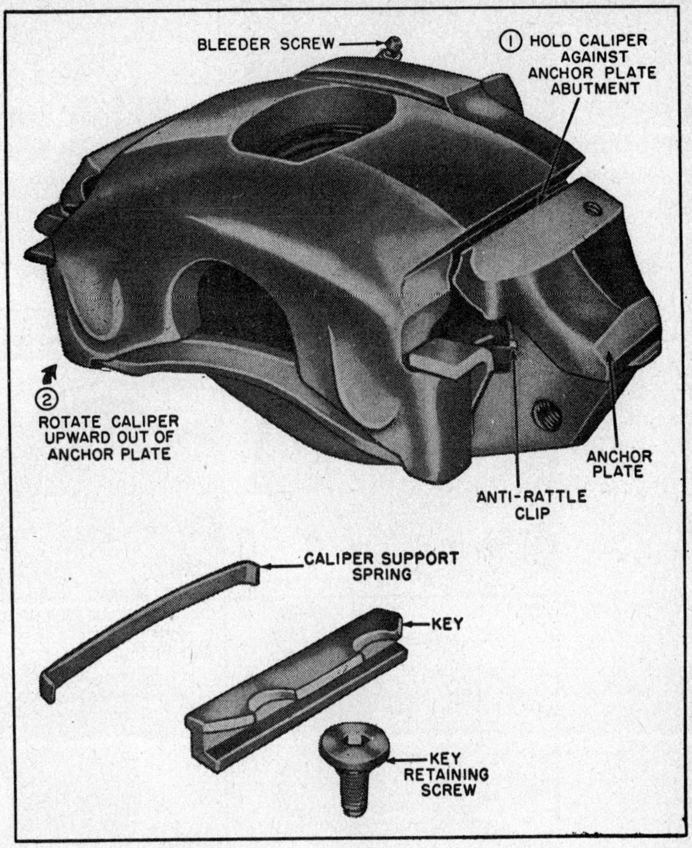

Fig. 72 Removing caliper assembly

Caliper Disassembly

1. With caliper removed as described previously, disconnect brake hose. Cap hose and plug caliper inlet to prevent fluid loss.
2. With caliper on work bench, remove inlet plug and drain fluid from housing.
3. Place a wooden block or an old brake pad into the caliper, then place a shop cloth between wooden block and piston.
4. Apply air pressure slowly to caliper inlet port to remove piston.

NOTE: If high pressure is applied quickly, piston may pop out and cause injury. A cocked or seized piston can be eased out by rapping sharply on piston end with a soft brass hammer.

5. Remove boot from piston and seal from caliper cylinder bore.

Caliper Housing Assembly

1. Lubricate piston seal with clean brake fluid and piston seal in its cylinder bore groove.
2. Assemble dust boot on caliper housing by seating boot flange in the outer groove of cylinder bore, making sure it is fully seated.
3. Coat piston with clean brake fluid and install in cylinder bore.
4. Spread dust boot over piston as piston is installed and then bottom piston in the bore. Seat dust boot in its piston groove.

Caliper Installation

1. If new shoe and lining assemblies are to be installed, use a 4″ C-clamp and a block of wood $1\frac{3}{4}$″ x 1″ and about $\frac{3}{4}$″ thick to seat the caliper piston in its bore. This must be done to provide clearance for the caliper to fit over new shoes when installed.
2. Be sure brake shoe anti-rattle clip is in place on lower inner brake shoe support on anchor plate with pigtail of clip toward inside of anchor plate. Position inner shoe and lining on anchor plate with lining toward rotor, Fig. 73.
3. Install outer shoe and lining with lower flange ends against the caliper leg abutments and the brake shoe upper flanges over the shoulders on caliper legs. The shoe upper flanges fit tightly against the shoulder machined surfaces. If the same brake shoes and linings are to be used, be sure they are installed in their original positions.
4. Remove C-clamp if used, from the caliper (the piston will remain seated

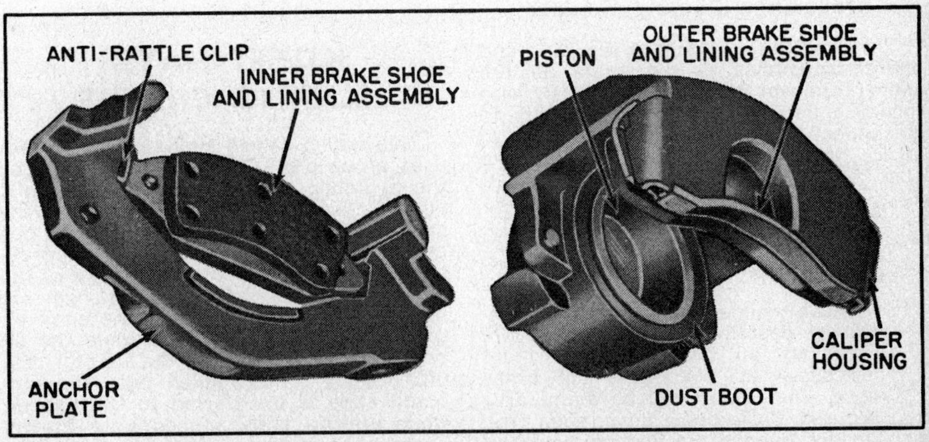

Fig. 73 Caliper and outer shoe removed from anchor plate

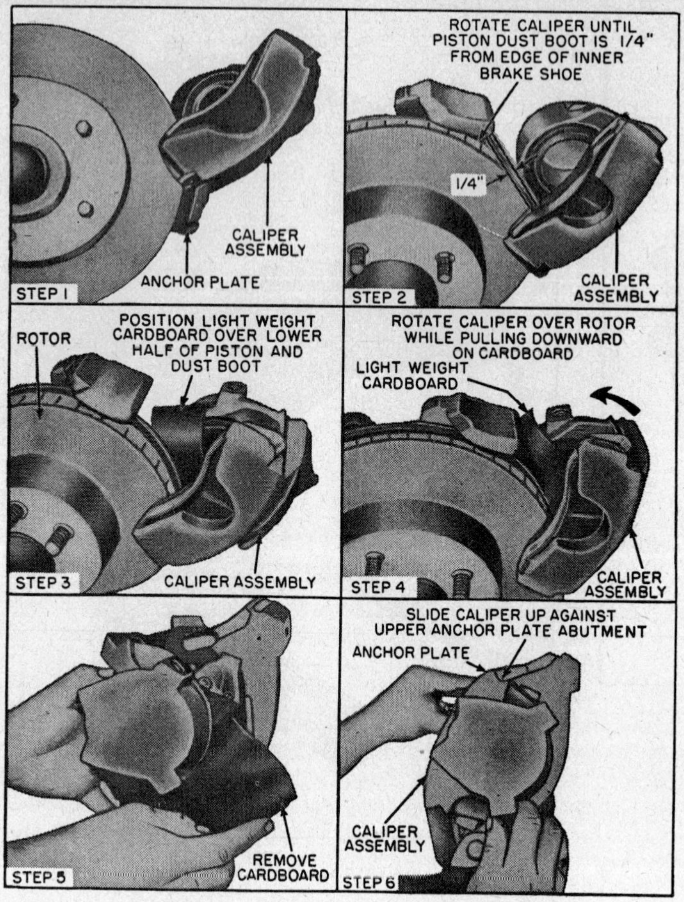

Fig. 74 Installing caliper assembly

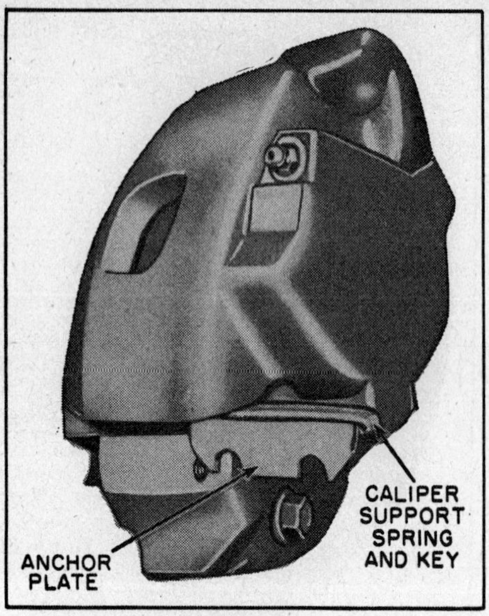

Fig. 75 Installing caliper support spring and retaining key

in its bore).
5. Position caliper housing lower V groove on anchor plate lower abutment surface, Fig. 74. Refer to Figs. 74 and 75 to complete assembly following steps shown. Connect brake hose, bleed brakes and replace wheel.
6. Install key retaining screw and torque to 12-20 ft. lbs.

Brake Shoe & Lining, Replace

The procedure to replace the shoe and lining assemblies is the same as the caliper removal discussed previously with the exception that it is not necessary to disconnect the brake hose. Use care to avoid twisting or stretching the brake hose.

Hub & Rotor Remove

1. Remove caliper and shoes as previously described. If no repairs are necessary on the caliper it is not necessary to disconnect the brake hose. The caliper can be temporarily secured to the upper suspension arm. Do not remove the anchor plate and be careful not to stretch or twist the brake hose.
2. Remove grease cap from wheel spindle and remove cotter pin and nut lock from wheel bearing adjustment nut.
3. Remove wheel bearing adjusting nut and grasp the hub and rotor and pull it out far enough to loosen the washer and outer wheel bearing. Then push it back in and remove the washer, outer wheel bearing and remove the hub and rotor.

KELSEY-HAYES SLIDING CALIPER

This sliding caliper single piston system uses a one piece hub and is actuated by the hydraulic system and disc assembly, Fig. 76. Alignment and positioning of the caliper is achieved by two machined guides or "ways" on the adaptor, while caliper retaining clips allow lateral movement of the caliper, Fig. 77. Outboard shoe flanges are used to position and locate the shoe on the caliper fingers, Fig. 78, while the inboard shoe is retained by the adaptor, Fig. 65. Braking force applied onto the outboard shoe is transferred to the caliper, while braking force applied onto the inboard shoe is transferred directly to the adaptor.

A square cut piston seal provides a hydraulic seal between the piston and the cylinder bore, Fig. 76. A dust boot with a wiping lip installed in a groove in the cylinder bore and piston, prevents contamination in the piston and cylinder bore area. Adjustment between the disc and the shoe is obtain automaticaly by the outward relocation of the piston as the inboard lining wears and inward movement of the caliper as the outboard lining wears.

Caliper Removal

1. Raise the vehicle and remove front wheel.
2. Remove caliper retaining clips and anti-rattle springs, Fig. 77.
3. Remove caliper from disc by slowly sliding caliper assembly out and away from disc.

NOTE: Use some means to support caliper. Do not let caliper hang from hydraulic line.

Brake Shoe Removal

1. Remove caliper assembly as outlined above.
2. Remove outboard shoe by prying between the shoe and the caliper fingers, Fig. 80, since flanges on outboard shoe retain caliper firmly.

NOTE: Caliper should be supported to avoid damage to the flexible brake hose.

3. Remove inboard brake shoe from the adaptor, Fig. 79.

Brake Shoe Installation

NOTE: Remove approximately ⅓ of the brake fluid out of the reservoir to prevent overflow when pistons are pushed back into the bore.

1. With care, push piston back into bore until bottomed.
2. Install new outboard shoe in recess of caliper.

NOTE: No free play should exist between brake shoe flanges and caliper fingers, Fig. 81.

If up and down movement of the shoe shows free play, shoe must be removed and flanges bent to provide a slight interference fit, Fig. 78. Reinstall shoe after modification, if shoe can not be finger snapped into place, use light "C" clamp pressure, Fig. 82.

3. Position inboard shoe with flanges inserted in adaptor "ways," Fig. 79.
4. Carefully slide caliper assembly into adaptor and over the disc while aligning caliper on machined "ways" of adaptor.

NOTE: Make sure dust boot is not pulled out from groove when piston and boot slide over the inboard shoe.

5. Install anti-rattle springs and retaining clips and torque retaining screws to 180 inch-pounds.

NOTE: The inboard shoe anti-rattle spring is to be installed on top of the retainer spring plate, Fig. 77.

Caliper Disassembly

1. With caliper and shoes removed as described previously, place caliper onto the upper control arm and slowly depress brake pedal, in turn hydraulically pushing piston out of bore.

NOTE: Pedal will fall when piston passes bore opening.

2. Support pedal below first inch of pedal travel to prevent excessive fluid loss.
3. To remove piston from the opposite caliper, disconnect flexible brake line at frame bracket, from vehicle side where piston has been removed previously and plug tube to prevent pressure loss. By depressing brake pedal this piston can also be hydraulically pushed out.
   Important: Air pressure should never be used to remove piston from bore.
4. Mount caliper in a vise equipped with protector jaws.

NOTE: Excessive vise pressure will distort caliper bore.

5. Remove the dust boot, Fig. 83.
6. Insert a suitable tool such as a small, pointed wooden or plastic object between the cylinder bore and the seal and work seal out of the groove in the piston bore.

NOTE: A metal tool such as a screwdriver should not be used since it can cause damage to the piston bore or burr the edges of the seal groove.

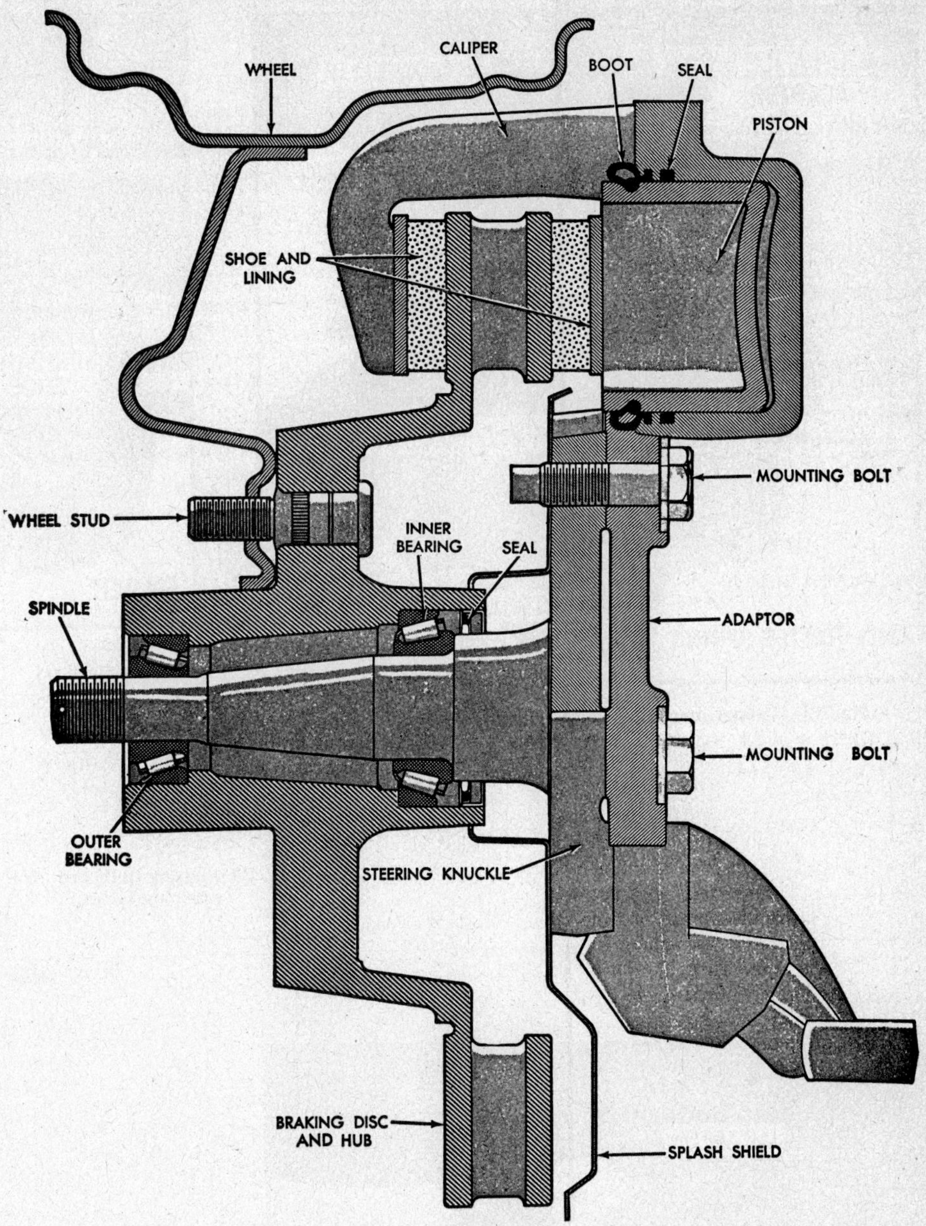

Fig. 76 Sectional view of Kelsey-Hayes sliding caliper front disc brake

Caliper Assembly

1. Before installing the new piston seal in groove of bore, dip seal in Ucon LB1145Y24 lubricant or equivalent. Work seal gently into the groove (using clean fingers) until seal is properly seated, make sure that seal is not twisted or rolled.

NOTE: Old seals should never be reused.

2. With new piston boot lubricated generously with Ucon LB1145Y24 or equivalent. Using finger pressure, install into caliper by pushing into outer groove of the caliper bore. When properly positioned in groove boot will snap into place. Double check to make sure boot is properly installed and seated by running finger around the inside of the boot.
3. Plug high pressure inlet to caliper and bleeder screw hole and coat piston with a generous amount of lubricant. Spread boot with finger and work piston into boot while pressing down on piston. As piston is depressed, entrapped air below piston will force boot around piston and into its groove.
4. Remove the plug and apply uniform force to the piston (avoid cocking piston) until piston bottoms in bore.
5. Install caliper and shoes as described under "Brake Shoe Installation."

Fig. 77 Caliper machined "ways" and assembly retention

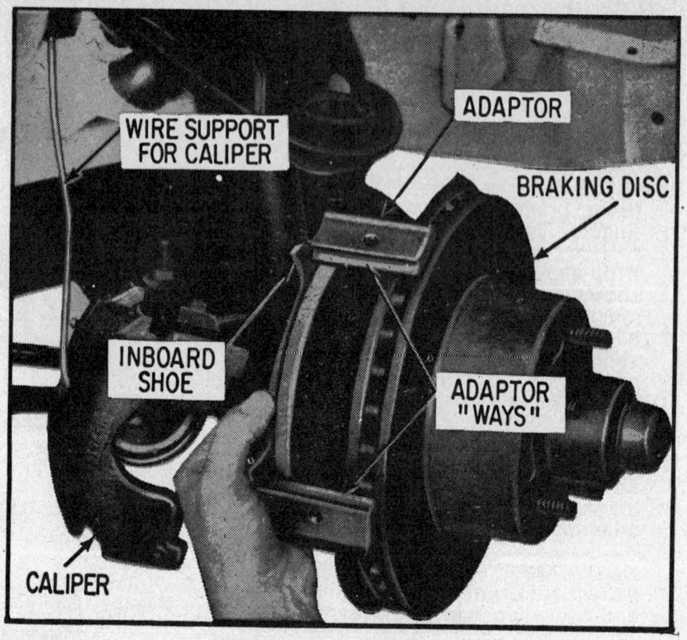

Fig. 79 Replacing inboard shoe

Fig. 78 Fitting outboard shoe retaining flange

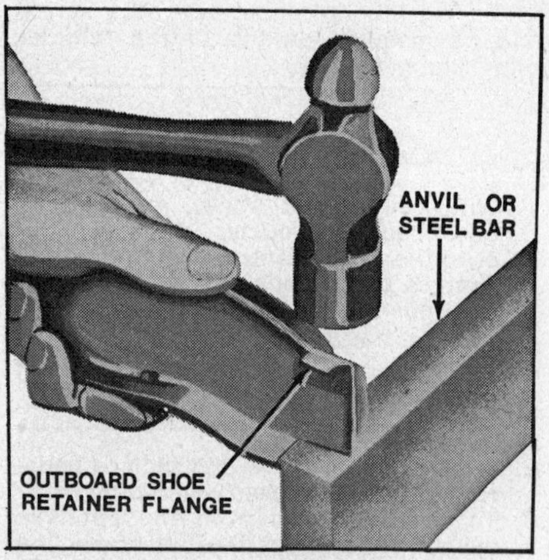

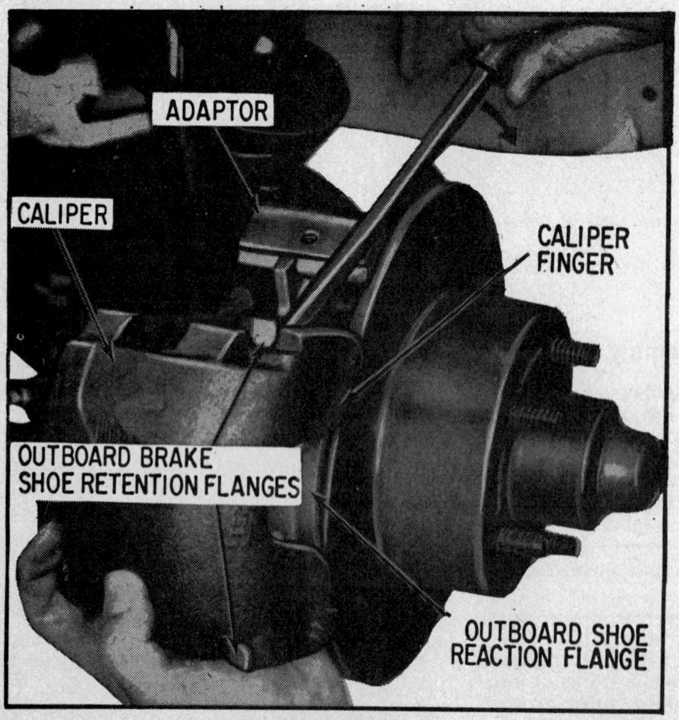

Fig. 80 Removing outboard shoe

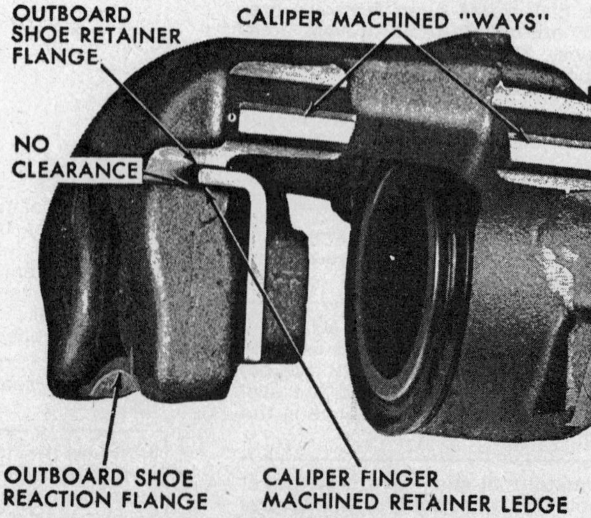

Fig. 81 Positioning outboard shoe onto caliper finger machined retainer ledge

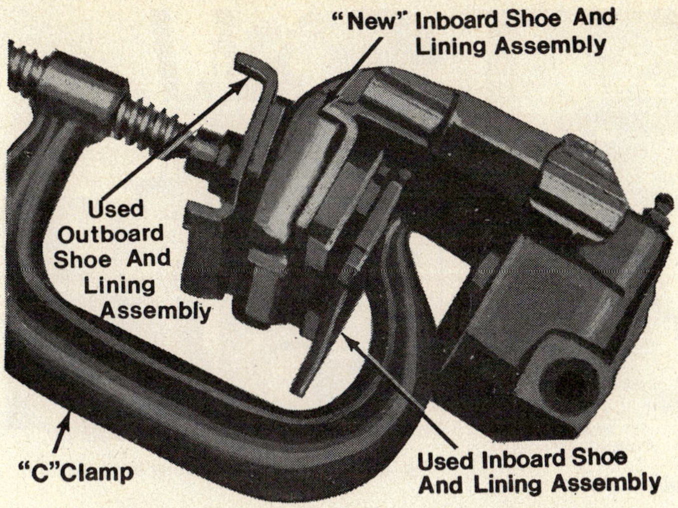

Fig. 82 Installing outboard shoe using "C" clamp

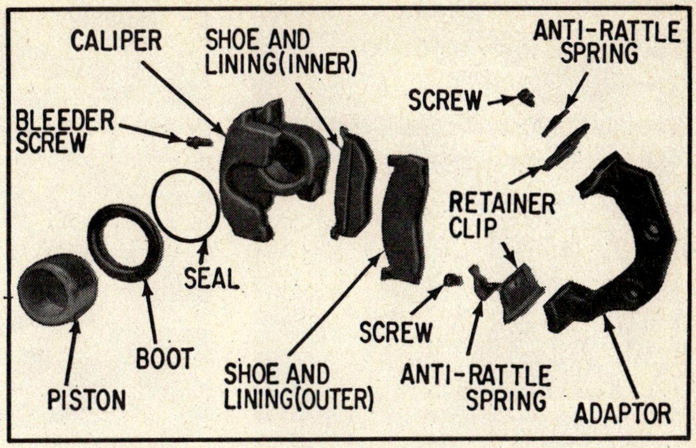

Fig. 83 Exploded view of a Kelsey-Hayes sliding caliper disc brake

REAR WHEEL DISC BRAKE & PARKING BRAKE SERVICE

1974-75 Imperial

The parking brake assembly of this system is mounted on the axle flange and intermediate disc brake adaptor. The rotor has an internal 7 inch drum surface for the internal expanding parking brake shoes, Fig. 84. Service to system should be performed as follows:

Disassembly

1. Support vehicle on a stand and remove rear wheel.
2. Remove caliper retaining screws and anti rattle spring assemblies, Fig. 77. Slide caliper from adaptor machined ways and position caliper on leaf spring for support, Fig. 85.
3. Remove inboard shoe from adaptor ways and remove parking brake adjusting access hole plug.
4. Using brake adjusting tool C-4223, release parking brake adjustment and remove disc/drum.
5. Remove lower shoe spring, Fig. 86. Spread brake shoes slightly and remove adjusting star wheel and upper shoe spring, Fig. 87.
6. Move brake shoes off support and remove brake shoe retainers, springs and nails and remove from support, Fig. 88.
7. Disconnect park brake cable from lever and snap ring retainer from shaft and remove lever, Fig. 89.
8. Remove inner snap ring from shaft and remove cam lever, cam and shaft, Fig. 90.

Cleaning & Inspection

1. Brush or wipe (dry) metal portions of the brake shoes. Inspect lining contact area, the lining should show contact marks across its entire area. Shoes showing contact on one side should be replaced.
2. Using a suitable solvent, clean brake support. Inspect for burrs and remove as necessary.
3. Clean cam lever, cam, shaft and operating lever. Inspect for damaged or distorted parts, including O-rings and replace as necessary. Replace any parts that do not permit freedom of operation. Apply lubricant to threads, sockets and washers.

Drum Refacing

Check drum diameter and runout. Drum diameter should not vary more than .002 inch. Drum runout should not exceed .006 inch. If drum runout or diameter exceed these limits, the drum should be refaced. Remove only as much metal as necessary, but do not remove more than .060 inch over the standard brake diameter.

NOTE: Maximum allowable diameter of 7.090 inches, Fig. 91, includes the .030 inch allowable drum wear.

Assembling Parking Brake Mechanism

1. Lubricate shaft with Mopar Lubricant #2932524 or equivalent and install in intermediate shaft adaptor.
2. Install cam and cam lever and snap ring on shaft, Fig. 90.
3. Install inner operating lever and snap ring on shaft and connect parking brake cable to lever, Fig. 89.
4. Lubricate shoe tab contact area (6 places) with Mopar Lubricant #2932524 or equivalent.
5. Place brake shoes on support and install retaining nails, springs and retainers, Fig. 88.
6. Install upper shoe spring and adjuster wheel, Fig. 87.

NOTE: Adjuster star wheel is forward on left side of vehicle and to the rear

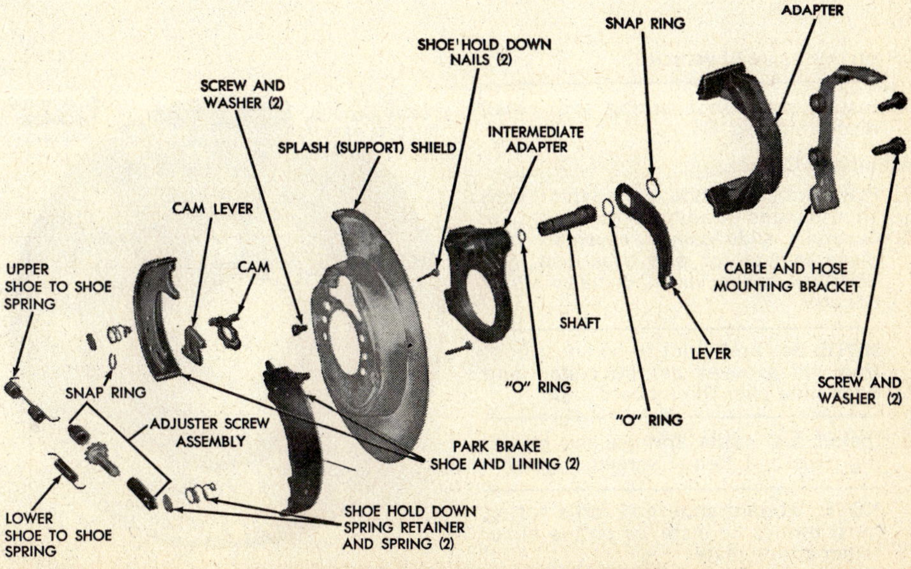

Fig. 84 Rear wheel parking brake assembly

Fig. 85 Replacing caliper assembly

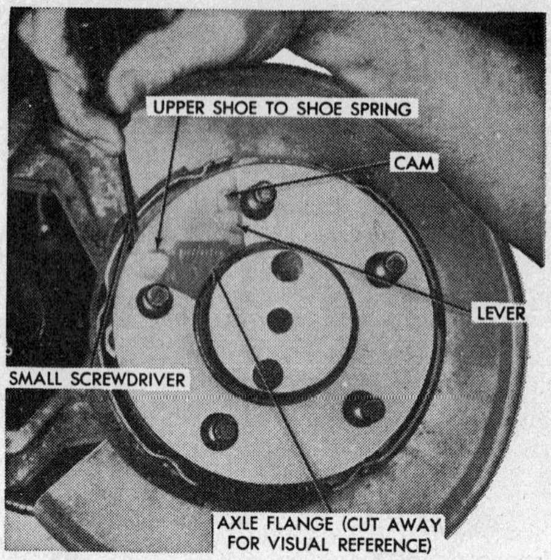

Fig. 87 Replacing upper shoe to shoe spring

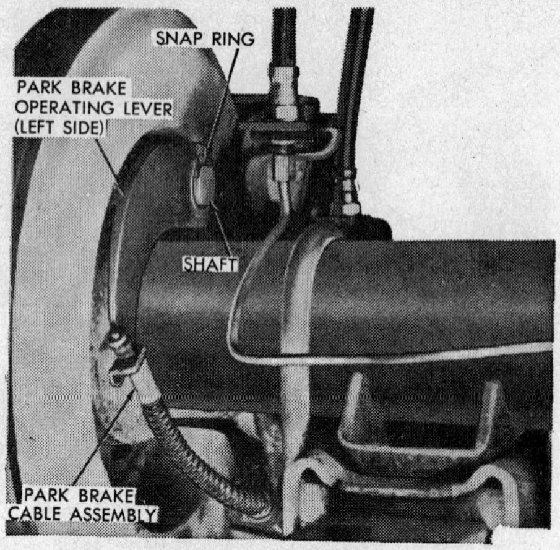

Fig. 89 Park brake operating level location

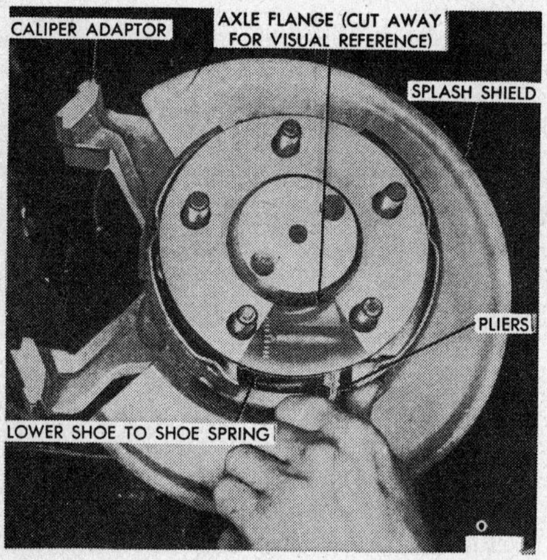

Fig. 86 Replacing lower shoe to shoe spring

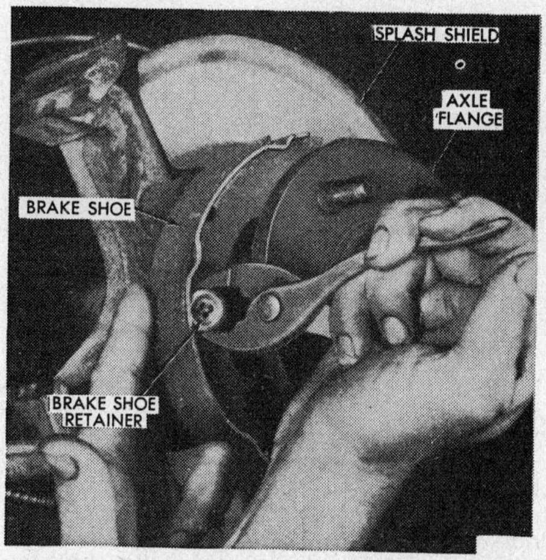

Fig. 88 Replacing brake shoe retainer, springs and nails

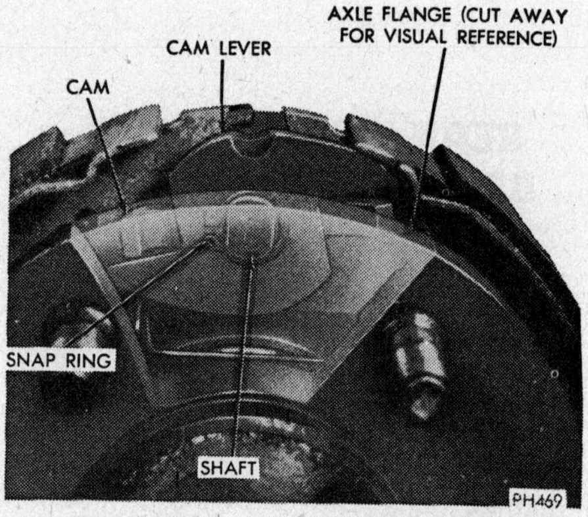

Fig. 90 Outer cam, cam lever, snap ring and shaft location

on right side of vehicle.

7. Install lower shoe spring and install disc drum, Fig. 86.

Assembling Caliper

1. Place inboard shoe on adaptor with shoe flanges in adaptor ways, Fig. 85.
2. Carefully slide caliper assembly into place in adaptor and over disc, Fig. 85. Align caliper on machined ways of adaptor.

NOTE: Be careful not to pull dust boot from its grooves as the piston and boot slide over the inboard shoe.

3. Install anti rattle springs and retaining clips and torque screws, Fig. 77.

NOTE: Inboard shoe anti rattle spring must always be installed on top of retainer spring plate.

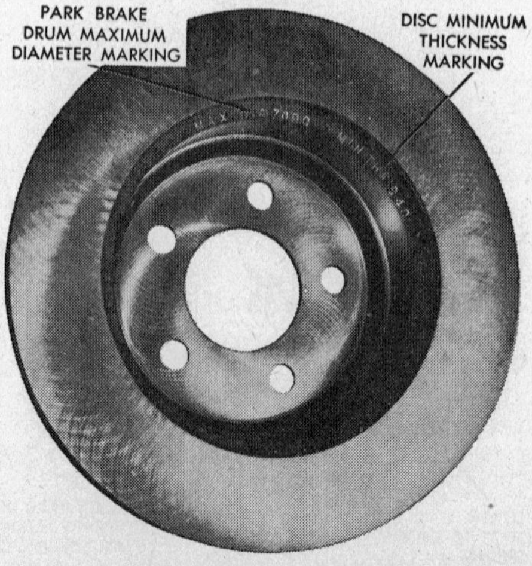

Fig. 91 Maximum drum diameter marking

FORD REAR WHEEL DISC BRAKE & PARKING BRAKE

Sliding caliper rear disc brakes are used on some 1975 models, Fig. 92. The caliper is basically the same as the larger front wheel caliper, however, a parking brake mechanism and a larger inner brake shoe anti-rattle spring has been added, Fig. 93. A hydraulically powered brake booster (Hydroboost) provides the power assist for this four wheel disc brake system.

The parking brake lever, located at the rear of the caliper, is actuated by a cable system similar to rear drum brake applications. When the parking brake is applied, the cable rotates the lever and operating shaft. Three steel balls, placed in pockets between the opposing heads of the operating shaft and thrust screw, roll between ramps formed in the pockets and force the thrust screw away from the operating shaft, in turn, driving the caliper piston and brake shoe assembly against the rotor. An automatic adjuster in the assembly compensates for lining wear and maintains proper clearance in the parking brake mechanism.

The cast iron rotors are ventilated by curved fins located between the braking surfaces and are designed to cause the rotor to act as an air pump when the vehicle is traveling forward. The rotors are not interchangeable and are identified by a Right or Left marking cast inside the hat section of the rotor. The rotor is secured to the axle flange in the same manner as a rear brake drum. A splash shield is bolted to a forged axle adapter to protect the inboard rotor surface.

Caliper Removal

NOTE: After performing any service work, obtain a firm brake pedal before moving vehicle.

1. Raise vehicle and support on safety stands, then remove tire and wheel assemblies.
2. Disconnect fitting on rear brake tube from hose end fitting at frame mounted bracket and plug end of brake tube to prevent loss of fluid and entry of dirt. Remove horseshoe retaining clip from hose fitting and disengage hose from bracket.
3. Disconnect parking cable from lever, Fig. 93, using care to avoid kinking or cutting cable or return spring, then remove retaining screw from caliper retaining key, Fig. 94.
4. Slide caliper retaining key and support spring from anchor plate, Fig. 94. If necessary, use a hammer and brass drift, being careful to avoid damaging key on sliding ways or hitting parking brake lever.

   NOTE: If caliper cannot be removed due to rust build-up on outer edge of rotor, scrape off loose scale, being careful not to damage braking surfaces. If rotor wear or scoring prevents removal of caliper, it will be necessary to loosen caliper end retainer ½ turn maximum, to allow piston to be forced back into its bore. To loosen end retainer, remove parking brake lever and mark or scribe end retainer and caliper housing to be sure that end retainer is not loosened more than ½ turn, then force piston back in its bore, Fig. 94, and move caliper back and forth to center rotor and remove caliper. If retainer must be loosened more than ½ turn, use caution, as the seal between the thrust screw and housing may be broken and brake fluid will enter parking brake mechanism chamber. In this case, the end retainer must be removed and the internal parts cleaned and lubricated.

5. Remove inner shoe and lining assembly from anchor plate, then tap lightly on outer shoe and lining assembly to free it from caliper. Mark each shoe for identification if they are to be reused.

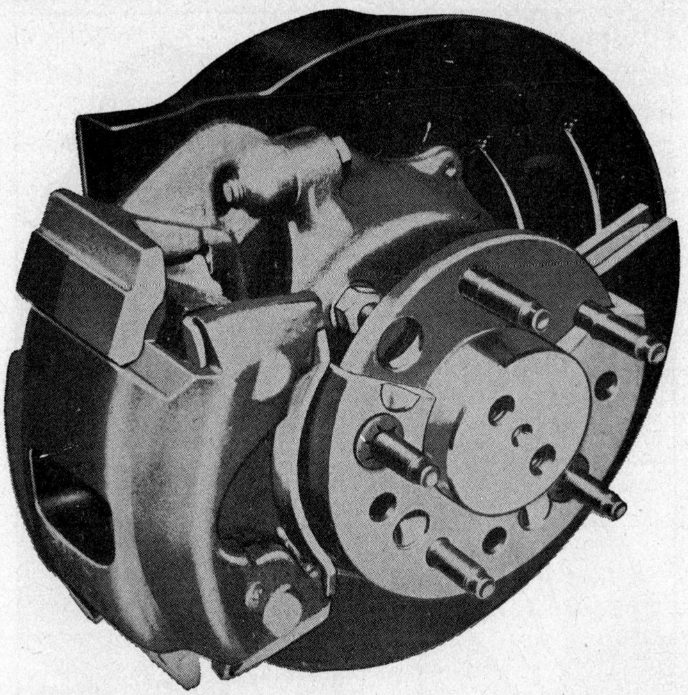

Fig. 92 Rear disc brake

Fig. 93 Caliper housing cutaway to show parking brake mechanism

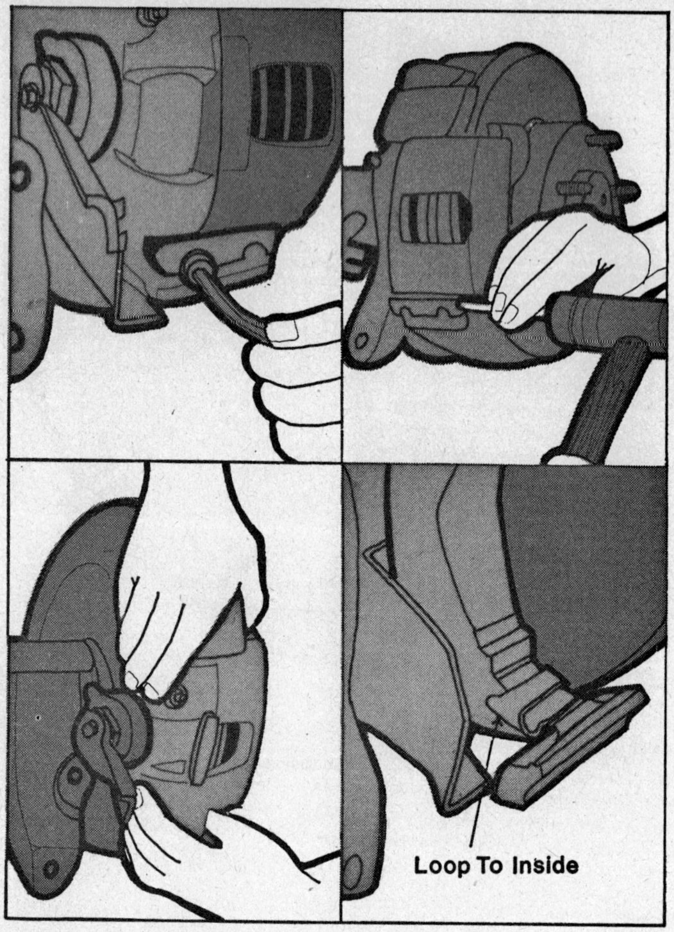

Fig. 94 Removing rear caliper assembly

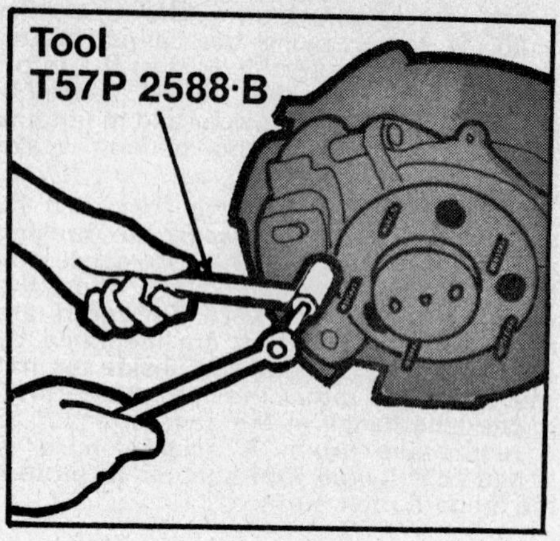

Fig. 95 Adjusting piston depth for lining installation

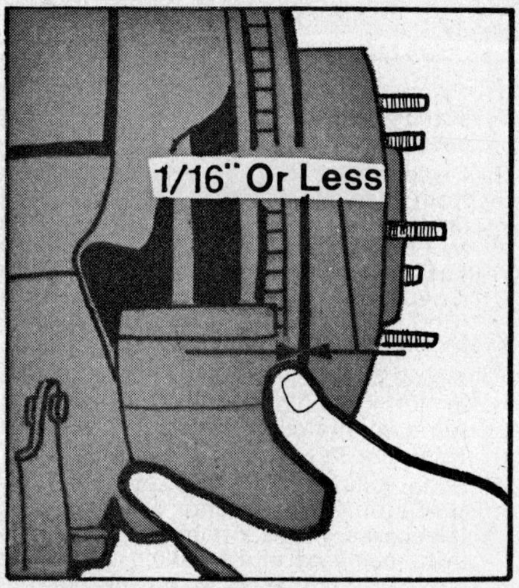

Fig. 96 Checking lining clearance

Cleaning & Inspection

Clean caliper, anchor plate and rotor assembly and inspect for signs of brake fluid leakage, excessive wear or damage. The caliper must be inspected for leakage both in piston boot area and operating shaft seal area. Lightly sand or wire brush any rust or corrosion from caliper and anchor plate sliding surfaces and inner brake shoe abutment surfaces in anchor plate. Inspect brake shoes for wear. If either lining is within 1/32 inch of any rivet head, replace both shoe and lining assemblies from both wheels in order to maintain equal brake action.

Caliper Installation

1. If end retainer has been loosened only ½ turn, reinstall caliper in anchor plate using key. Do not install shoe and lining assembly. Torque end retainer to 75-95 ft. lbs. and install parking brake actuating lever on its keyed spline. Lever arm must point down and rearward so that parking brake cable will pass freely under axle. Torque retainer screw to 16-22 ft. lbs.

NOTE: Parking brake lever must rotate freely after torquing retainer screw.

2. Remove caliper from anchor plate. If new shoe and lining assemblies are to be installed, the piston must be bottomed in caliper bore using tool T57P-2588-B to provide clearance. Remove rotor and install caliper without lining and shoe assemblies in anchor plate using key only. Install tool and while holding shaft, rotate tool handle counterclockwise until the tool seats firmly against piston, Fig. 95. Loosen handle about ¼ turn, and while holding handle rotate tool shaft clockwise until piston is fully bottomed in bore (piston will continue to turn even after it is bottomed). Turn tool handle until there is no further inward movement of piston and there is a firm seating force, then remove caliper from mounting plate and reinstall rotor.
3. Making certain that brake shoe anti rattle clip is in place in lower inner brake shoe support on anchor plate with loop of clip toward inside of anchor plate, Fig. 94, position inner brake shoe and lining assembly on anchor plate.
4. Install outer brake shoe with lower flange ends against caliper abutments and brake shoe upper flanges over shoulders on caliper legs. The shoe upper flanges fit tightly against machined shoulder surfaces.

NOTE: If old brake shoes and lining assemblies are re-used, be certain the shoes are installed in their original positions as marked for identification during removal.

5. Lubricate caliper and anchor sliding ways with M1C-167-A (LPS-ESA-100) grease, using care to prevent lubricant from getting on braking surfaces, then position caliper housing lower V-groove on anchor plate lower abutment surfaces.
6. Rotate caliper until it is completely over rotor, being careful not to damage piston dust boot, then pull caliper outboard until inner shoe and lining is firmly seated against rotor. Measure clearance between outer lining and rotor which should be 1/16 inch

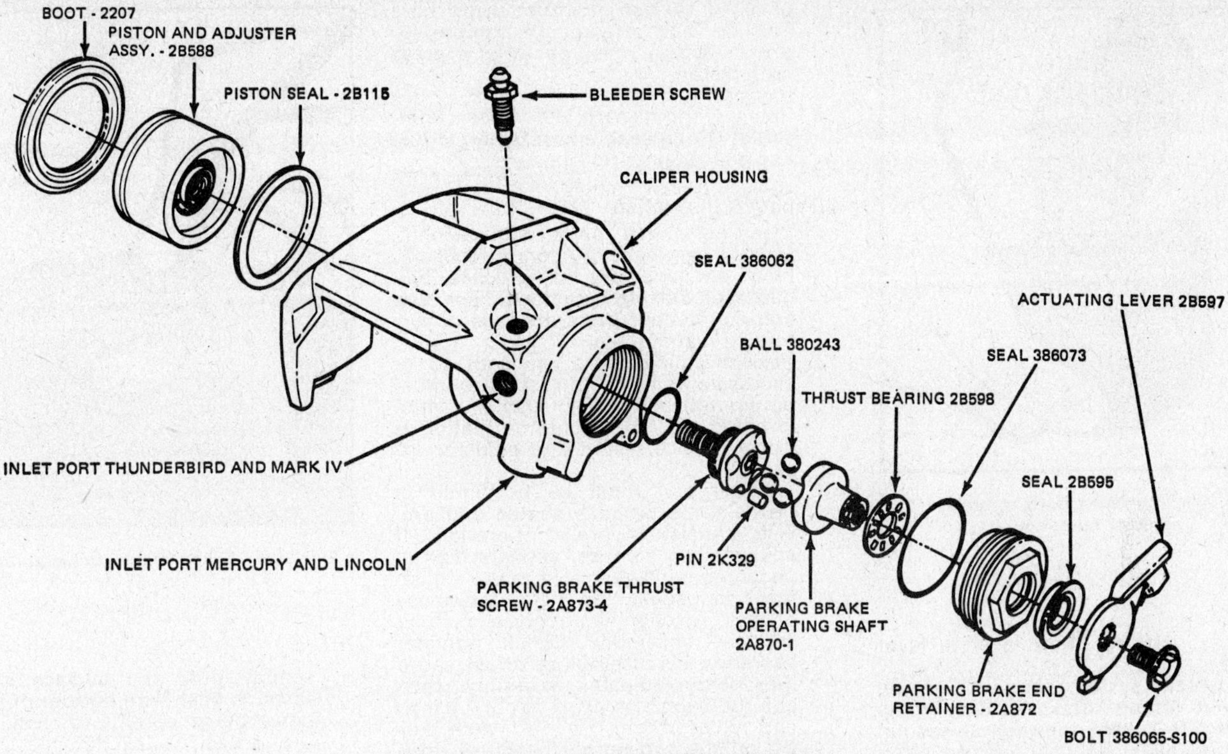

Fig. 97 Rear disc brake caliper assembly

or less, Fig. 96. If it is greater, remove caliper and move piston outward to narrow gap. Follow procedure in step 2 and note that 1/4 turn of the shaft counterclockwise, moves piston about 1/16 inch.

CAUTION: A clearance greater than 1/16 turn may allow adjuster to be pulled out of piston when service brake is applied, causing parking brake to fail to adjust. It will then be necessary to replace piston/adjuster assembly.

7. While holding caliper against anchor plate upper abutment surfaces, center caliper over lower anchor plate abutment, then position caliper support spring and key in slot and slide them into opening between lower end of caliper and lower anchor plate abutment until key semi-circular slot is centered over retaining screw threaded hole in anchor plate.
8. Install key retaining screw and torque to 12-16 ft. lbs., then reinstall brake hose on caliper. On Lincoln and Mercury models, place a new gasket on fitting and torque to 20-30 ft. lbs. On Thunderbird and Mark IV models, place a new gasket on each side of fitting outlet, then install attaching bolts through washers and fitting and torque bolts to 17-25 ft. lbs.
9. Position upper end of flexible hose in bracket and install retaining clip, then connect brake tube to hose and torque fitting to 10-15 ft. lbs.

NOTE: Do not twist or coil brake hose, the stripe on the hose must be kept straight.

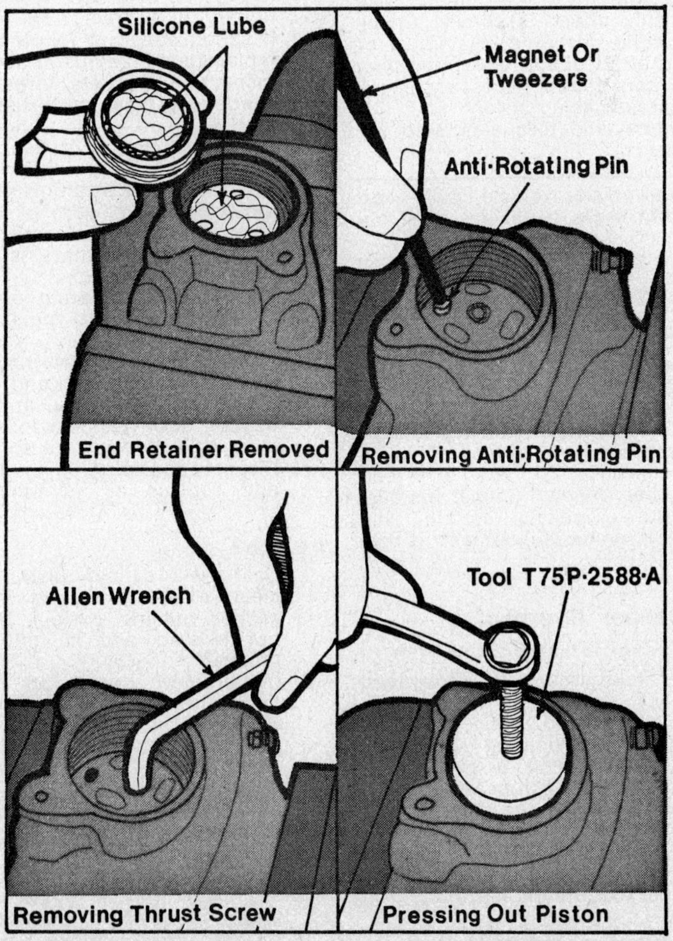

Fig. 98 Disassembling rear disc brake caliper

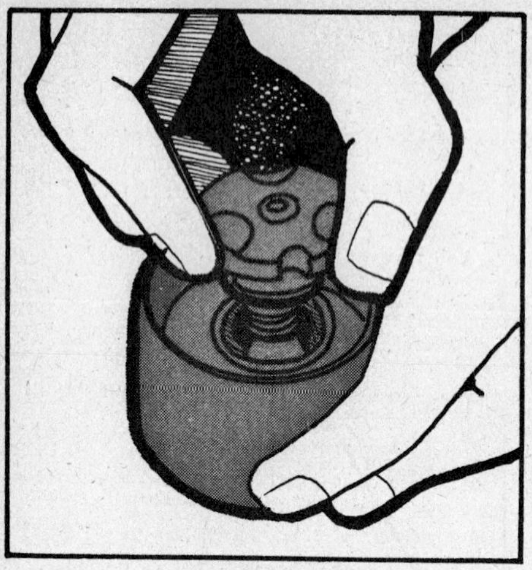

Fig. 99 Checking parking brake adjuster operation

10. Connect parking brake lever to lever on caliper.
11. Bleed brake system, then with engine running pump brake pedal lightly about 40 times allowing 1 second between pedal applications. An alternate with engine off is to pump brake pedal lightly about 10 times to discharge accumulator, then pump brake pedal firmly about 30 times. Check parking brake for excessive travel or very light effort, if so, repeat pumping brake pedal, and if necessary check parking brake cable tension.
12. Install wheel and torque nuts to 70-115 ft. lbs.

NOTE: Before moving vehicle, make certain that a firm brake pedal has been obtained.

Shoe & Lining Removal & Installation

To remove shoe and lining assemblies, follow "Caliper Removal" procedure and omit step 3 as it is not necessary to disconnect brake hose. After removing caliper, support it with a length of wire to avoid damaging brake hose. To install shoe and lining assemblies, follow "Caliper Installation" procedure, making certain that proper parking brake adjustment is obtained.

Caliper Overhaul

Disassemble

1. Remove caliper assembly as described previously.
2. Remove caliper end retainer, operating shaft, thrust bearing and balls, Fig. 97, then using a magnet or tweezers, remove thrust screw anti rotation pin, and remove thrust screw, Fig. 98.
3. Install tool T75P-2588-A through back of caliper housing and remove piston and carburetor assembly, Fig. 98.

CAUTION: Use care not to damage polished surface in thrust screw bore and do not attempt to remove or press adjuster can, as it is a press fit in piston.

4. Remove and discard piston seal, boot, thrust O-ring seal, end retainer, O-ring and end retainer lip seal.

Cleaning & Inspection

1. Clean all metal parts with alcohol, then using clean, dry compressed air, blow out and dry all grooves and passages making sure the caliper bore and component parts are free of any foreign material.
2. Inspect caliper bore for damage or excessive wear. The thrust screw must be smooth and free of pits. If piston is pitted, scored or chrome plating is worn, replace piston and adjuster assembly.
3. Adjuster can must be bottomed in piston to be properly seated and provide consistent brake operation. If adjuster can is loose, appears high in piston, is damaged, or if brake adjustment is usually too tight, too loose or not functioning, replace piston/adjuster assembly. Check adjuster operation by assembling thrust screw into piston/adjuster assembly, then pull the two parts apart about 1/4 inch and release them, Fig. 99. When pulling on the two parts, the brass drive ring must remain stationary causing the nut to rotate. When releasing the two parts, the nut must remain stationary and drive ring must rotate. If action does not follow this pattern, replace piston/adjuster assembly.
4. Inspect ball pockets, threads, grooves, bearing surfaces of thrust screw, operating shaft, balls and anti rotation pin for wear, brinnelling or pitting. Replace operating shaft, balls, thrust screw and anti rotation pin if any of these parts are worn or damaged. A polished appearance on the ball paths is acceptable if there is no sign of wear into the surface.
5. Inspect thrust bearing for corrosion, pitting or wear and replace as necessary.
6. Inspect end plug bearing surface for wear or brinnelling and replace as necessary. A polished appearance on bearing surface is acceptable if there is no sign of wear into surface.
7. Inspect operating lever for damage and replace as necessary.

Assemble

1. Coat new caliper piston seal with clean brake fluid and install it in caliper making certain that seal is not twisted and is fully seated in groove.
2. Install new dust boot by seating flange squarely in outer groove of caliper bore, then coat piston/adjuster assembly with clean brake fluid and install it in caliper bore. Spread dust boot over piston as it is installed and seat dust boot in piston groove.
3. Install caliper in vise, Fig. 100, and fill piston/adjuster assembly with clean brake fluid.
4. Coat new thrust screw O-ring with clean brake fluid and install it in thrust screw groove, then install thrust screw into piston adjuster assembly until top surface of thrust screw is flush with bottom of threaded bore, being careful to avoid cutting O-ring seal. Index notches on thrust screw and caliper housing and install anti rotation screw.

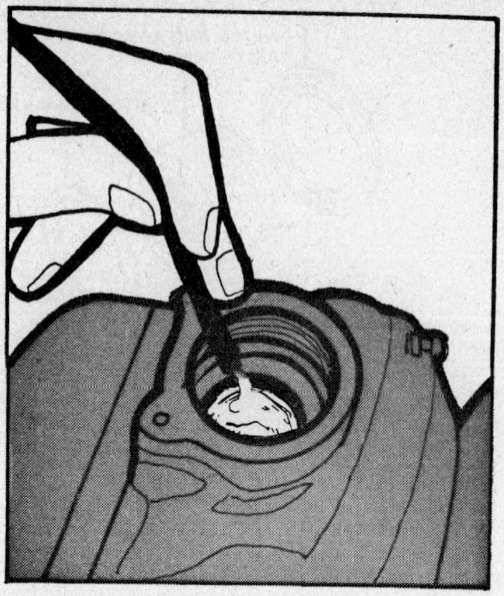

Fig. 100 Filling piston/adjuster assembly

5. Place a ball in each of three pockets of thrust screw and apply a liberal amount of silicone grease M1C-169-A on parking brake components, then install operating shaft on balls.
6. Coat thrust bearing with silicone grease and install it on operating shaft, then install a new lip seal and O-ring on end retainer.
7. Lightly coat O-ring seal and lip seal with silicone grease and install end retainer in caliper. Firmly hold operating shaft against internal mechanism while installing end retainer to prevent mislocation of balls. If lip seal moves out of position, reseat seal. Torque end retainer to 75-95 ft. lbs.

NOTE: Parking brake lever must rotate freely after torquing.

8. Install parking brake lever on keyed spline facing down and rearward. Torque retaining screw to 16-22 ft. lbs.
9. Bottom piston using tool T75P-2588-B, Fig. 101, and install caliper as described previously.

BENDIX SLIDING CALIPER

This sliding caliper disc brake assembly incorporates a hub and rotor assembly, caliper, brake shoes and linings, caliper anchor plate and a splash shield, Fig. 102.

Cooling fins are cast into the rotor between the two braking surfaces to ventilate and cool the rotor. The sliding caliper is positioned in, and slides on, the abutment surfaces on the leading and trailing edges of the caliper anchor plate. A caliper support key, located between the forward edge of the caliper and abutment surface, is secured with a retaining screw. A support spring is installed between the support key and caliper to maintain tension on the support key.

The caliper is a one-piece casting containing the piston, piston seal and dust seal, Fig. 103. The hydraulic seal between the caliper piston and piston bore is achieved by a square cut piston seal, located in a machined groove in the piston bore. The dust seal seats in a recess machined on the edge of the piston bore and into a groove in the caliper piston.

Caliper Removal

1. Siphon two-thirds of brake fluid from master cylinder reservoir serving front disc brakes.
2. Raise vehicle, support on jackstands and remove front wheels.
3. Bottom the caliper piston in bore. Insert a screwdriver between inboard shoe and piston, then pry piston back into bore. The piston can also be bottomed in the bore with a large "C" clamp.
4. Using a 1/4 inch allen wrench, remove support key retaining screw, Fig. 104.
5. Drive caliper support key and spring from anchor plate with a suitable drift and hammer, Fig. 105.
6. Lift caliper from anchor plate and off rotor, Fig. 106. Hang caliper from coil spring with wire. Do not allow caliper to hang from brake hose.
7. Remove inboard brake shoe from anchor plate, then the anti-rattle spring from the brake shoe, Fig. 107.
8. Remove outboard brake shoe from caliper, Fig. 108. It may be necessary to loosen the brake shoe with a hammer to permit shoe removal.

Caliper Disassembly

1. Drain brake fluid from caliper.
2. Position caliper with shop cloths, Fig. 109, and apply compressed air to fluid inlet port to ease piston from bore.

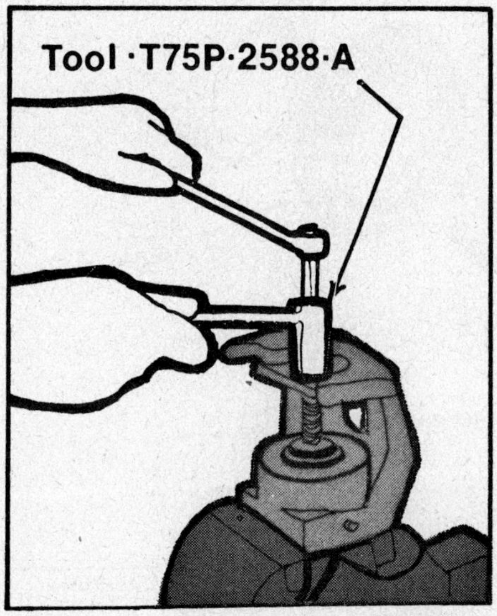

Fig. 101 Bottoming piston in caliper

NOTE: Do not attempt to catch piston or to protect it when applying compressed air since personal injury is possible.

3. Remove dust seal from piston, then the piston seal from bore, Fig. 103. Use wooden or plastic tool to remove piston seal since metal tools may damage piston.
4. Remove bleeder screw.

Caliper Assembly

1. Coat square cut piston seal with clean brake fluid, then install seal into piston bore. Work seal into groove with clean fingers.
2. Install and torque bleeder screw to 11 ft. lbs.
3. Lubricate dust seal and tool J-24387 with clean brake fluid, then place dust seal on tool, allowing 1/4 inch of tool to extend past small lip of dust seal, Fig. 110.
4. Place dust seal and tool over piston bore, then work large lip of dust seal into seal groove, Fig. 111. Ensure dust seal is fully seated.
5. Lubricate caliper piston and insert through tool. Center piston in bore and use a hammer handle to apply pressure to install piston halfway into bore, Fig. 111.
6. Remove tool J-24387 and seat small lip of dust seal in caliper piston groove, then bottom piston in bore.

Brake Shoe & Lining, Replace

The procedures to remove & install the brake shoe and lining assemblies are out-

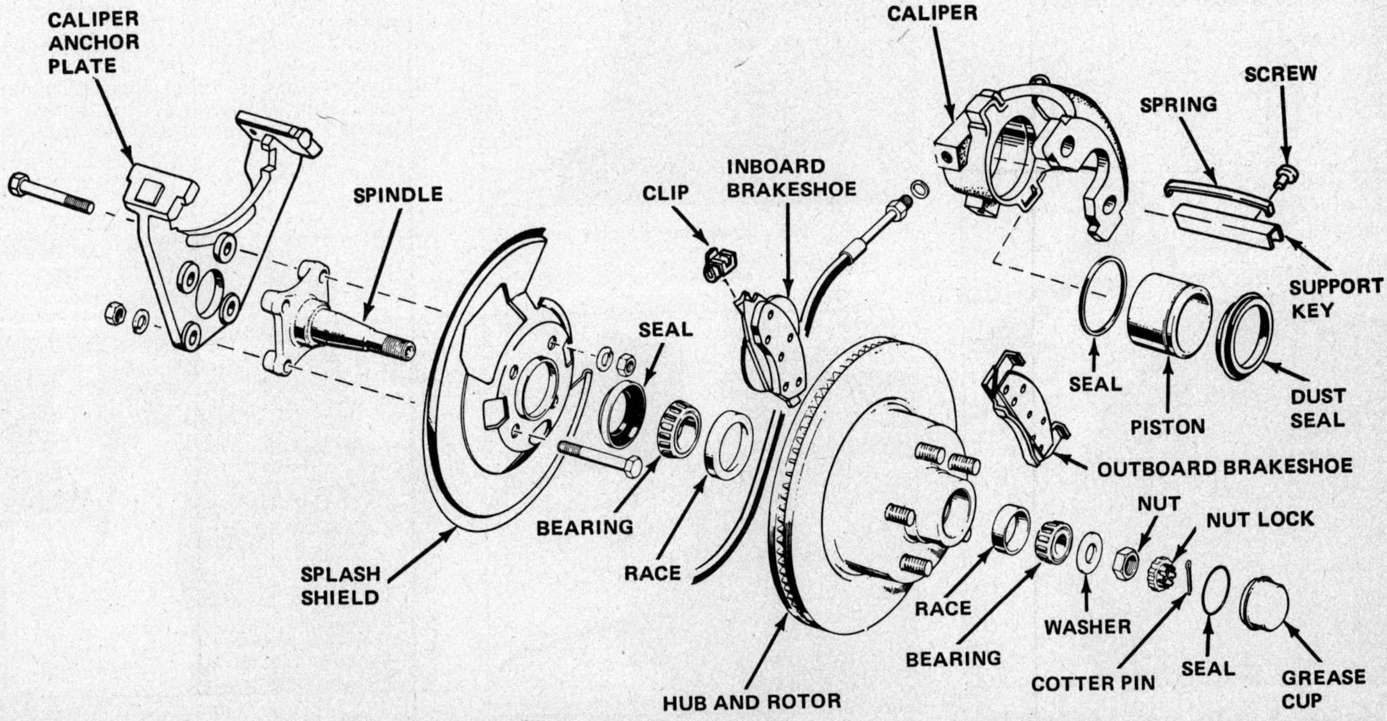

Fig. 102 Bendix sliding caliper disc brake

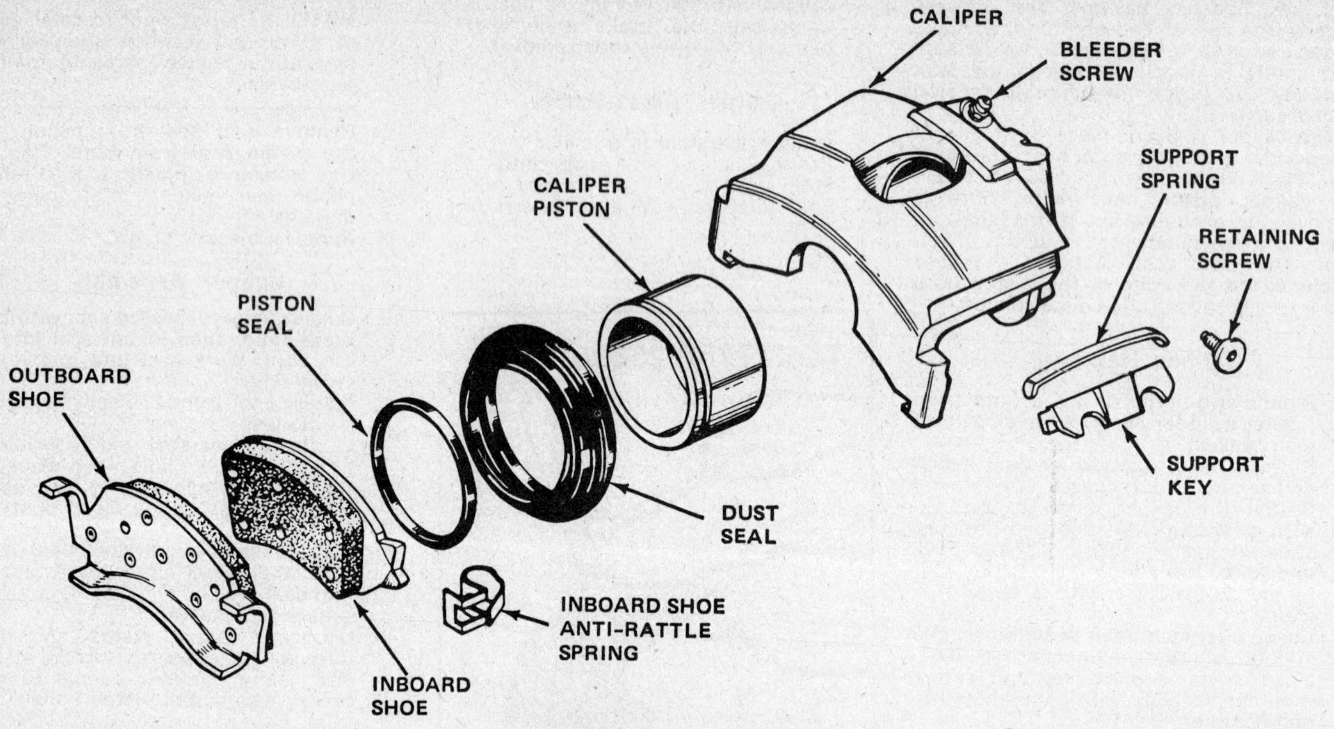

Fig. 103 Caliper assembly

lined under "Caliper Removal" and "Caliper Installation". It is not necessary to disconnect the brake hose, however, use caution not to twist or kink hose.

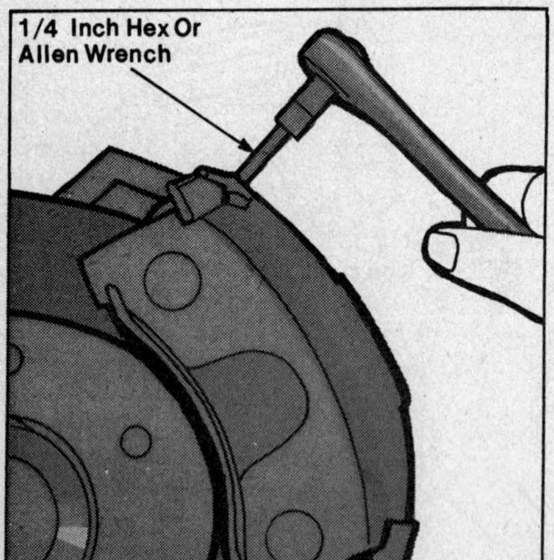

Fig. 104 Removing support key retaining screw

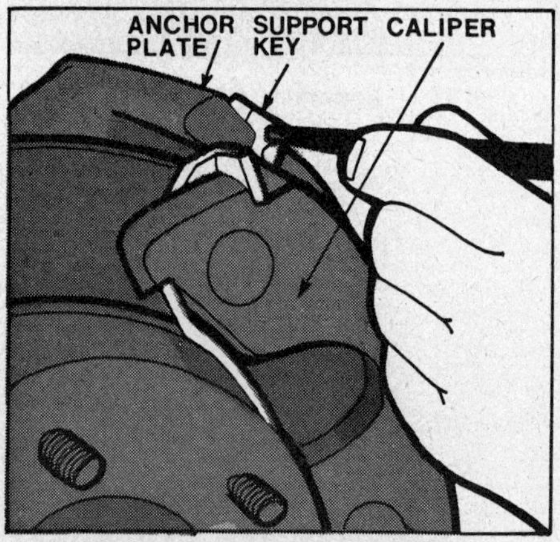

Fig. 105 Removing support key

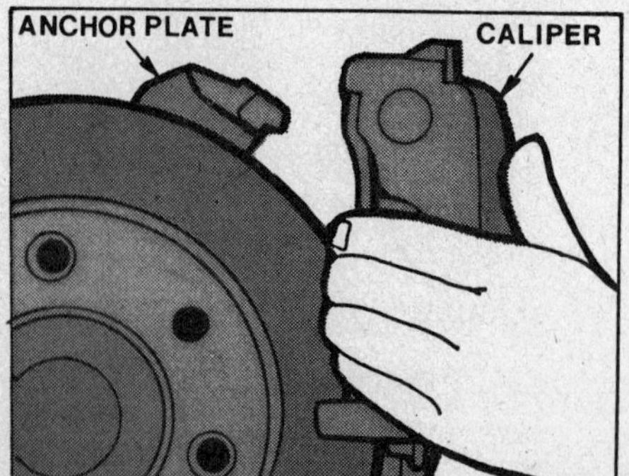

Fig. 106 Removing or installing caliper

Caliper Installation

1. Clean and lubricate abutment surfaces of caliper and the anchor plate with a suitable molydisulfide grease, Fig. 112.
2. Install inboard brake shoe anti-rattle spring on brake shoe rear flange, ensure looped section of clip is facing away from rotor, Fig. 13.
3. Install inboard brake shoe on caliper anchor plate, Fig. 107.
4. Install outboard brake shoe in caliper, Fig. 108. Ensure the shoe fiange is seated fully into outboard arms of

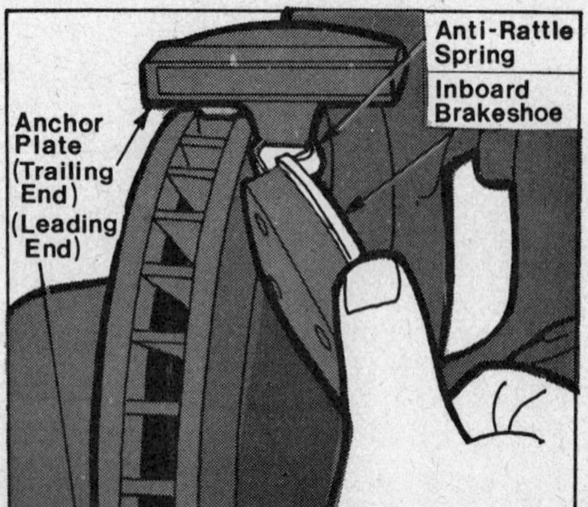

Fig. 107 Removing or installing inboard brake shoe

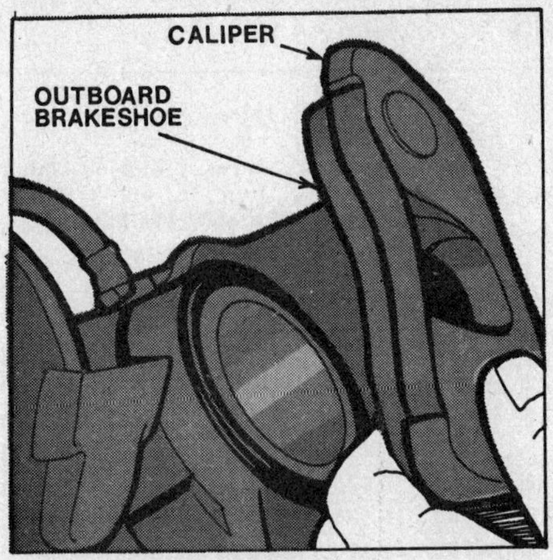

Fig. 108 Removing or installing outboard brake shoe

Fig. 109 Removing piston from caliper

caliper. It may be necessary to use a hammer to seat the shoe.

5. Place caliper assembly over rotor and position in caliper anchor plate. Ensure dust boot is not torn or mispositioned by inboard brake shoe during caliper installation.
6. Align caliper with anchor plate abutment surfaces, then insert support key and spring between abutment surfaces at the trailing end of caliper and anchor plate. With a hammer and brass drift, drive caliper support key and spring into position, then install and torque support key retaining screw to 15 ft. lbs.
7. Refill master cylinder to within one inch of rim. Press brake pedal several times to seat shoes.
8. Install front wheels and lower vehicle.

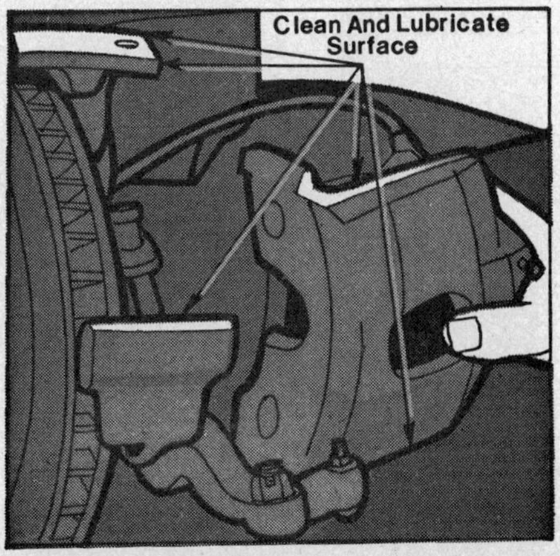

Fig. 112 Caliper & anchor plate abutment surfaces

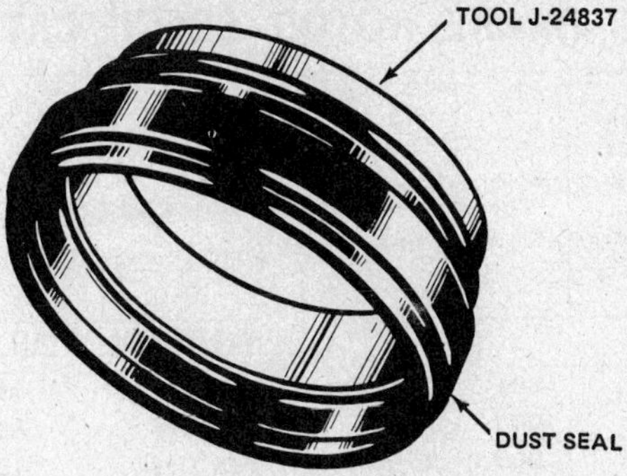

Fig. 110 Dust seal & installer tool assembly

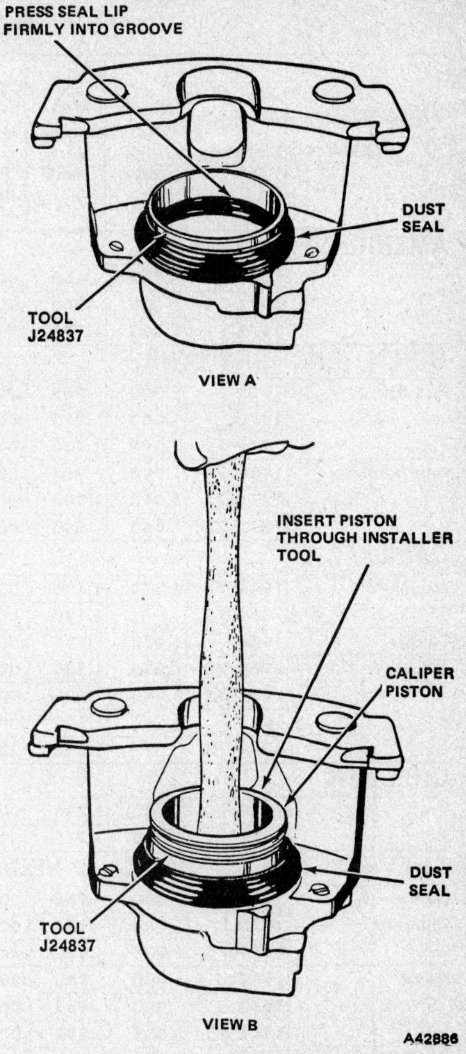

Fig. 111 Installing dust seal & caliper piston

Fig. 113 Installing inboard brake shoe anti-rattle spring

DISC BRAKE ROTOR SPECIFICATIONS

CAR	Year	Nominal Thickness	Minimum Thickness	Thickness Variation Parallelism	Run-out (T.I.R.)	Finish Micro-In.
AMERICAN MOTORS						
	1969–70	.500	.450	.0005	.005	15–80
	1971–74	1.000	.940	.0005	.005	15–80
	1975	1.190	1.130	.0005	.003	20–60
BUICK (EXC. SKYHAWK)						
Full Size	1969	1.000	.965	.0005	.005	30–50
	1970	1.000	.965	.0005	.004	30–50
	1971–75	1.290	1.230	.0005	.005	30–80
Intermediate	1969–70	1.000	.965	.0005	.004	30–50
	1971–72	1.040	.980	.0005	.004	30–80
	1973–75	1.040	.980	.0005	⑤	30–80
CADILLAC						
Eldorado	1969–73	1.210	1.190	.0005	.008	15–80
	1974–75	—	1.190	.0005	—	—
Seville	1976	1.030	.980	.0005	.005	20–60
All others	1969–70	1.250	1.230	.0007	.002	15–80
	1971–72	1.240	1.220	.0007	.002	15–80
	1973	1.240	1.220	.0007	.005	15–80
	1974–75	—	1.220	.0005	.005	—
CHECKER						
	1969–70	1.240	1.215	.0007	.004	20–60
	1971–75	1.290	1.215	.0007	.005	20–60
CHEVROLET (EXC. MONZA & VEGA)						
Camaro & Chevelle	1969	1.000	.965	.0005	.004	30–50
	1970	1.000	.965	.0005	.002	30–50
	1971–75	1.035	.980	.0005	.005	20–60
Nova	1969	1.000	.965	.0005	.004	30–50
	1970	1.000	.965	.0005	.002	30–50
	1971–75	1.035	.980	.0005	.005	20–60
Chevrolet	1969	1.250	1.215	.0005	.004	30–50
	1970	1.250	1.215	.0005	.002	30–50
	1971–72	1.250	1.230	.0005	.005	20–60
	1973–75	1.285	1.230	.0005	.005	20–60
Corvette	1969–70	1.250	1.215	.0005	.004	30–50
	1971–75	1.250	1.230	.0005	.005	20–60
CHEVROLET VEGA & PONTIAC ASTRE						
	1971–72	.500	.470	.0005	.002	20–60
	1973–75	.500	.455	.0005	.002	20–60
CHRYSLER, DODGE & PLYMOUTH						
Full Size Exc. Imperial	1969–70	1.250	1.200	.0005	.002	15–80
	1971–72	1.250	1.180	.0005	.002	15–80
	1973–75	1.250	1.180	.0005	.004	15–80
Imperial	1969	.877	—	—	.005	30–60
	1970	1.250	1.200	.0005	.002	15–80
	1971–72	1.250	1.180	.0005	.002	15–80
	1973–75	③	④	.0005	.004	15–80
Intermediate (Exc. Below)	1969	.886	.816	.0005	.005	40
	1970	1.000	.980	.0005	.002	15–80
	1971–72	1.000	.940	.0005	.002	15–80
	1973–75	1.000	.940	.0005	.004	15–80
Dart & Valiant	1969–72	.810	—	.0005	.002	15–80
	1973–75	1.000	.940	.0005	.004	15–80
Challenger & Barracuda	1969	.810	—	.0005	.002	15–80
	1970	1.000	.980	.0005	.002	15–80
	1971–72	1.000	.940	.0005	.002	15–80
	1973–75	1.000	.940	.0005	.004	15–80
FORD MOTOR COMPANY						
Ford & Mercury Full Size	1969	1.185	1.140	.0007	.002	15–80
	1970–72	1.180	1.120	.0007	.003	15–80
	1973–75	1.180⑦	1.120⑧	.0005②	.003⑨	15–80
Ford & Mercury Intermediate	1969	.940	.895	.0007	.002	15–80
	1970–73	.935	.875	.0007	.002	15–80
	1972⑥	1.180	1.120	.0007	.003	15–80
	1973–75①	1.180	1.120	.0005	.003	15–80
Comet & Maverick	1974–75	.870	.810	.0005	.003	15–80
Granada & Monarch	1975	.870⑦	.810⑧	.0005	.003④	15–80
Lincoln	1969	1.245	1.215	.0007	.002	15–80
	1970–72	1.180	1.120	.0007	.003	15–80
	1973–75	1.180	1.120	.0005	.003	15–80
	1974–75	1.180⑦	1.120⑧	.0025	.003⑨	15–80
Mark III, IV & Thunderbird	1969	1.185	1.140	.0007	.002	15–80
	1970–72	1.180	1.120	.0007	.003	15–80
	1973–75	1.180⑦	1.120⑧	.00025⑩	.003⑨	15–80
Mustang II,	1971–73	.750	.685	.0007	.003	15–80
Pinto & Bobcat	1974–75	.870	.810	.0005	.003	15–80
MONZA, SKYHAWK & STARFIRE						
	1975	.500	.455	.0005	.005	20–60
OLDSMOBILE (EXC. STARFIRE)						
Oldsmobile	1969	1.250	1.215	.0005	.004	30–50
	1970	1.250	1.215	.0007	.002	30–50
	1971–72	1.290	1.215	.0005	.002	30–50
	1973–75	1.290	1.215	.0005	.005	30–50
Olds F-85, Cutlass Omega	1969	1.000	.965	.0005	.004	30–50
	1970	1.035	.965	.0005	.004	30–50
	1971–75	1.040	.965	.0005	.004	30–50
Olds Toronado	1969	1.250	1.215	.0005	.004	30–50
	1970	1.250	1.215	.0005	.002	30–50
	1971–75	1.245	1.170	.0005	.002	30–50
PONTIAC (EXC. ASTRE)						
Pontiac Full Size	1969	1.240	1.195	.0007	.004	20–60
	1970–71	1.250	1.215	.0007	.004	20–60
	1972–75	1.285	1.215	.0007	.004	20–60
Pontiac Intermediate Exc. Firebird & Ventura	1969	1.000	.960	.0007	.004	20–60
	1970–71	1.085	.965	.0007	.004	20–60
	1972–75	1.035	.965	.0007	.004	20–60
Pontiac Firebird	1969	1.000	.960	.0007	.004	20–60
	1970	1.040	.965	.0005	.005	30–80
	1971	1.085	.965	.0007	.004	20–60
	1972–75	1.035	.965	.0007	.004	20–60
Ventura II	1971–73	1.035	.980	.0005	.004	20–60
	1974–75	1.035	.965	.0007	.004	20–60

①—Montego, Torino & 1974–75 Cougar.
②—1974–75 Mercury, Front disc. .0004, 1975 rear disc. .0005.
③—Front, 1.250; Rear, 1.000.
④—Front, 1.180; Rear, .940.
⑤—Exc. Apollo .004; Apollo .005.
⑥—Montego & Torino.
⑦—1975 rear disc. .945.
⑧—1975 rear disc. .895.
⑨—1975 rear disc. .004.
⑩—1975 rear disc. .0004.

ANTI-SKID BRAKE SYSTEMS

CHRYSLER SURE BRAKE

This system is designed to prevent any wheel from locking up during brake applications above a speed of about 5mph. The system reduces skid potential of a locked wheel and still maintains brake pressure for maximum stopping effort. The end result is to improve directional control and steerability of the vehicle, and in many cases to reduce the distance required to bring the vehicle to a stop.

The major components of the system are shown in Fig. 1. These include a mechanically driven speed sensor at each wheel, a logic controller located inside the right rear quarter panel and three pressure modulators; one modulator is under each front fender ahead of the wheels and one is in the engine compartment next to the radiator on the right side of the vehicle.

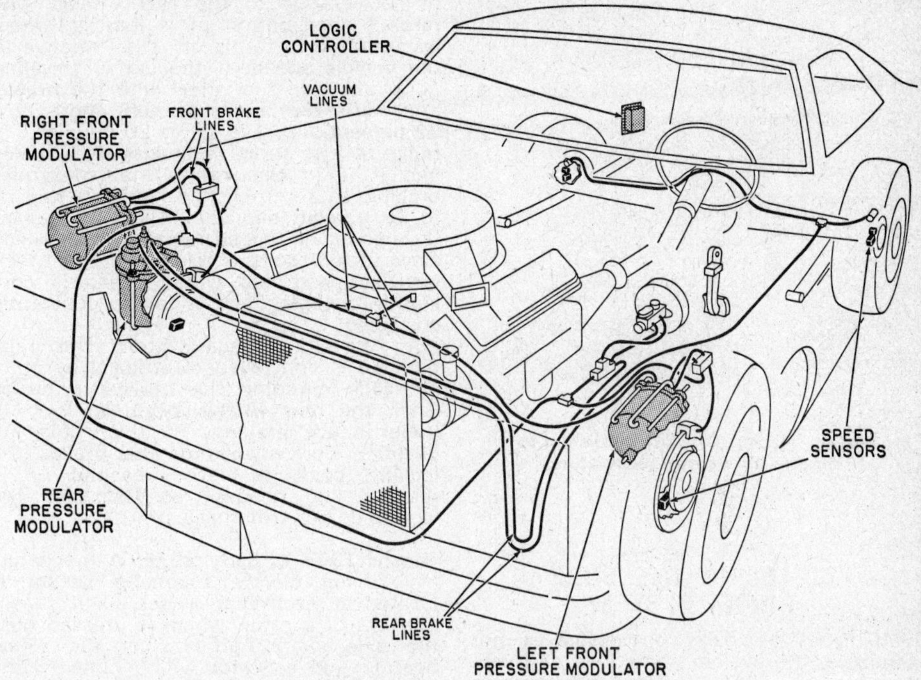

Fig. 1 Chrysler Sure Brake system components

OPERATION

Engine Running—Vehicle Not in Motion

The wheel sensors do not generate any signals for transmission to the logic controller when the vehicle is not in motion. Hence, the logic controller sends no commands to the pressure modulator.

Engine Running—Vehicle in Motion

With the vehicle in motion, alternating current voltage is generated at each wheel sensor and sent to the logic controller. The logic controller processes the signals received from the wheel sensors to sample the speed of each wheel. If the brakes are not applied or if they are applied lightly, the controller does not send any commands to the pressure modulator.

When the brakes are applied with greater force, the controller, based on wheel sensor signals, determines the rate at which each wheel is decelerating. If the rate is great and might produce wheel slippage or lockup, the controller sends a command to the modulator that controls the braking for the wheel or wheels concerned.

Exercise Cycle

If the engine is started with the brake pedal depressed, the pressure modulators go through two exercise cycles. When the ignition switch is turned from OFF to START all three modulators cycle once, and when the switch returns from START to ON the modulators cycle again. These cycles are to insure that the system is working properly. The cycles can be heard under some conditions, but should not be a cause for concern.

Warning System

The Sure Brake System includes a secondary system to warn the driver of certain types of failures in the system. The warning system uses the brake warning light. The warning light will be on under the following conditions:

1. It the pressure modulator is activated in the absence of a brake light signal.
2. If the controller sends a signal to open the air valve on a modulator in the absence of a signal to open the bypass valve.
3. If the electrical continuity of the air valve lead wire is broken.
4. If all speed sensor signals are not received at the controller, or if the signals are not properly converted in the controller at speeds above 15 mph.

NOTE: Due to the complexities of testing this system and the various voltage charts necessary, it is recommended the unit be returned to a dealer for service. For this reason, a procedure for disabling the Sure Brake System to allow the car to be operated safely, follows.

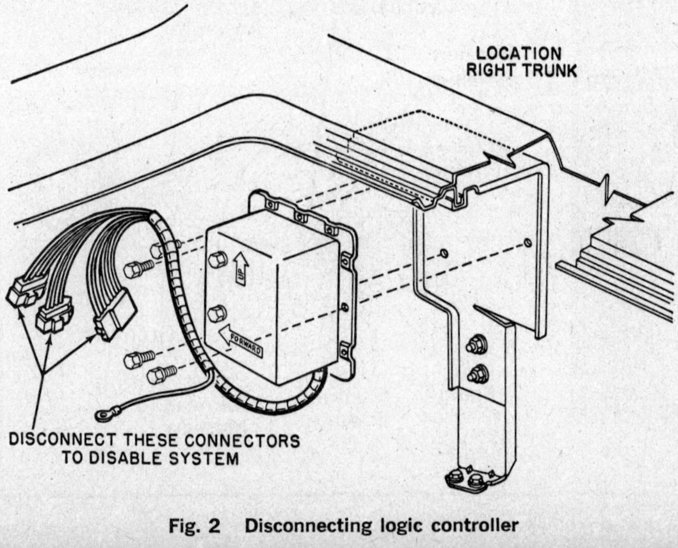

Fig. 2 Disconnecting logic controller

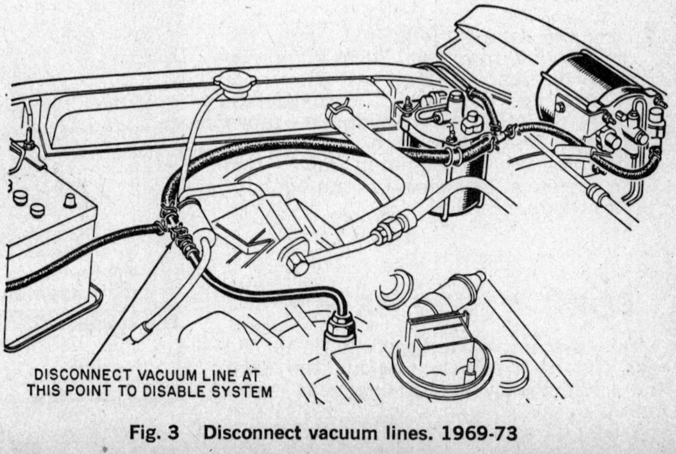

Fig. 3 Disconnect vacuum lines. 1969-73

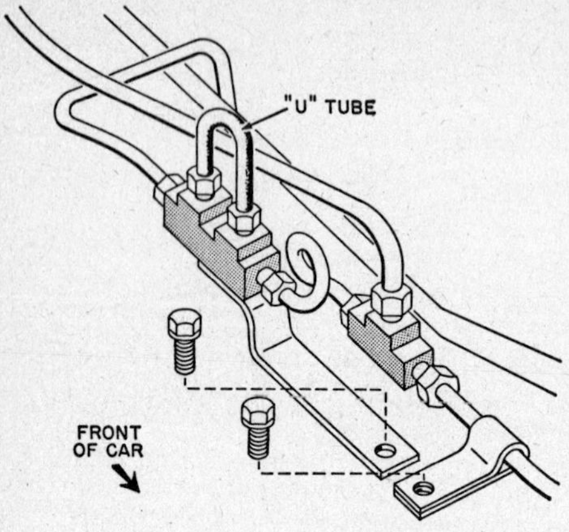

Fig. 4 Placing "U" tube in system to disable front wheel modulators

DISABLING THE SURE BRAKE SYSTEM

If necessary parts or service is not available, the system should be disabled. The brakes will then function the same as those on cars without this system. Proceed as follows:

Disconnect the logic controller (all three connectors), Fig. 2. On 1969-73 units, disconnect the vacuum supply hose at the "T" connection leading to the modulator, Fig. 3, and the line to the intake manifold should be plugged and the line to the modulators should be taped to keep out dirt.

An additional step may be necessary to disable the system. If a pressure modulator diaphragm plate sticks in the retracted or partially retracted position the basic hydraulic brake system will be affected. Test for this by placing car on a hoist and spin wheels by hand to insure they are free. With the brakes applied, attempt to spin wheels, release brakes and spin wheels again. If a wheel spins with the brakes applied or does not rotate with brakes released, the modulator for that wheel must be by-passed. This can be accomplished as follows:

1. On front wheel modulators, a "U" tube should be placed in the system after the modulator tubes have been removed, Fig. 4.

2. On rear wheel modulator, route the brake line from rear wheels directly into the rear of safety tee after modulator line has been disconnected, Fig. 5. Also, both hydraulic tubes should be disconnected at the brake warning switch metering valve. This completely by-passes the modulator for the rear brakes.

FORD "SURE TRACK"

This system is designed to keep the rear end of a vehicle tracking the front end correctly by controlling rear brake lockup during "panic stops." The system consists of sensors, valves, an actuator and a tiny computer to control the rate of deceleration of the rear wheels. The rate of deceleration of a rotating wheel is measured in terms of "slip" relative to the vehicle speed. If the car is traveling at 50mph and the wheel with the brakes applied is turning at 40mph, there is a 20 percent slip. The 10 to 20 percent slip range is just short of locking the brakes and it is in this range that maximum braking is accomplished. Ideally, the 10 to 20 percent slip should be maintained throughout the speed range as the vehicle slows down, so the relationship between vehicle speed and wheel speed is constantly changing and must be constantly corrected.

In the Sure Track system, this relationship is controlled electrically by automatically pumping the brakes in cycles when the rear wheels begin to lock up under heavy braking. By this rapid application and release of the brakes the locking point is never reached or is reached and released so fast that the wheels do not stop turning.

Sensor 1969 & Early 1970: A mechanically driven electro magnetic sensor is located at each rear wheel. Each sensor consists of a rotor, which is pressed onto the axle shaft outboard of the wheel bearing and a stator which bolts to the brake backing plate, Fig. 6. When the axle turns, teeth of the rotor pass the teeth of the stator, cutting magnetic lines of force and setting up an electrical current, in proportion to the speed of the rear wheels. Two wires from each of the sensors conduct impulses to the control module.

Sensor Late 1970 & 1971-75. A single sensor stator assembly, Figs. 7 and 8, is mounted on the rear axle drive pinion housing with the sensor rotor pressed onto the companion flange outboard of the grease seal. This unit operates in the same manner as the two wheel sensors used on the 1969 units to generate electric current. The AC voltage, generated in proportion to driveshaft speed, is conducted from the sensor to the control module.

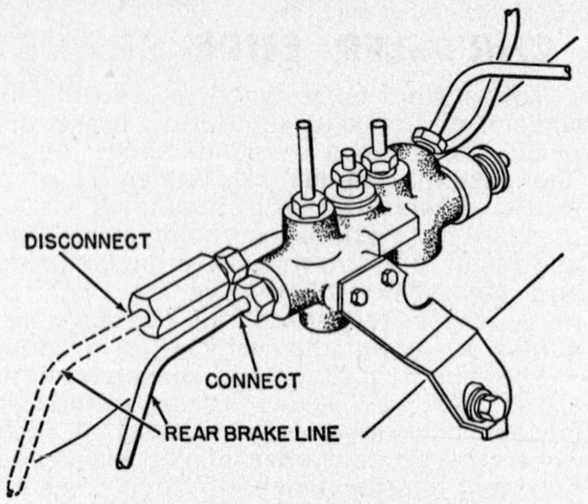

Fig. 5 Connections at safety tee to disable rear wheel modulators

Control Module: The control module is mounted under the glove box in the passenger compartment and operates on the sum of the signals from the sensor(s).

When the sum of the signals drops abruptly below a predetermined level due to rapid deceleration, the module sends an electrical signal to the actuator solenoid to release the rear brakes. Then, when rear wheel speed increases again, the module de-energizes the solenoid, allowing the brakes to reapply. This cycle continues until the driver releases the brake pedal past the full pressure point or the vehicle slows to less than 4 mph.

Actuator: The vacuum operated actuator, Fig. 9, is mounted on the vehicle right frame rail under the toe board. The actuator is divided into two chambers by a diaphragm with a spring positioned ahead of the diaphragm.

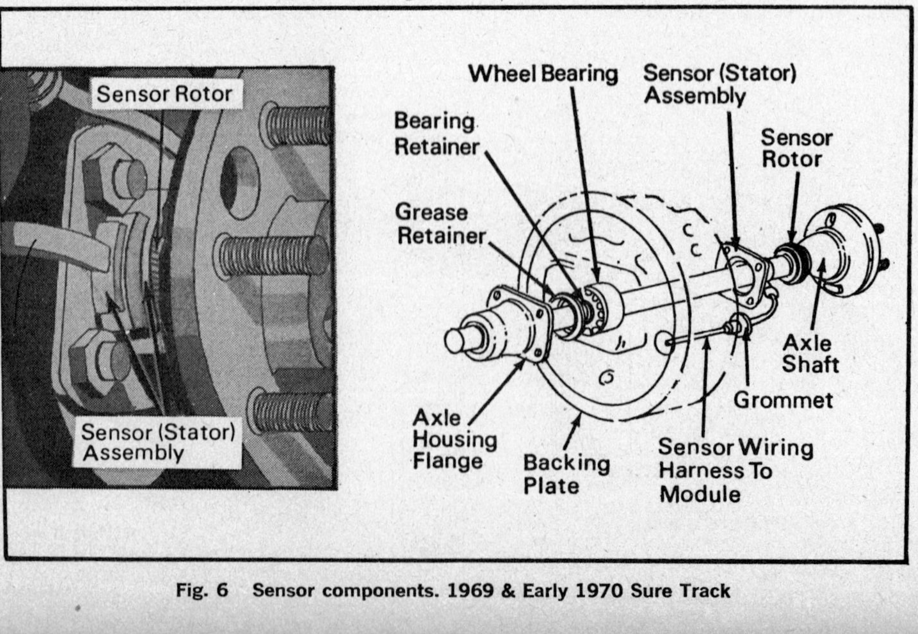

Fig. 6 Sensor components. 1969 & Early 1970 Sure Track

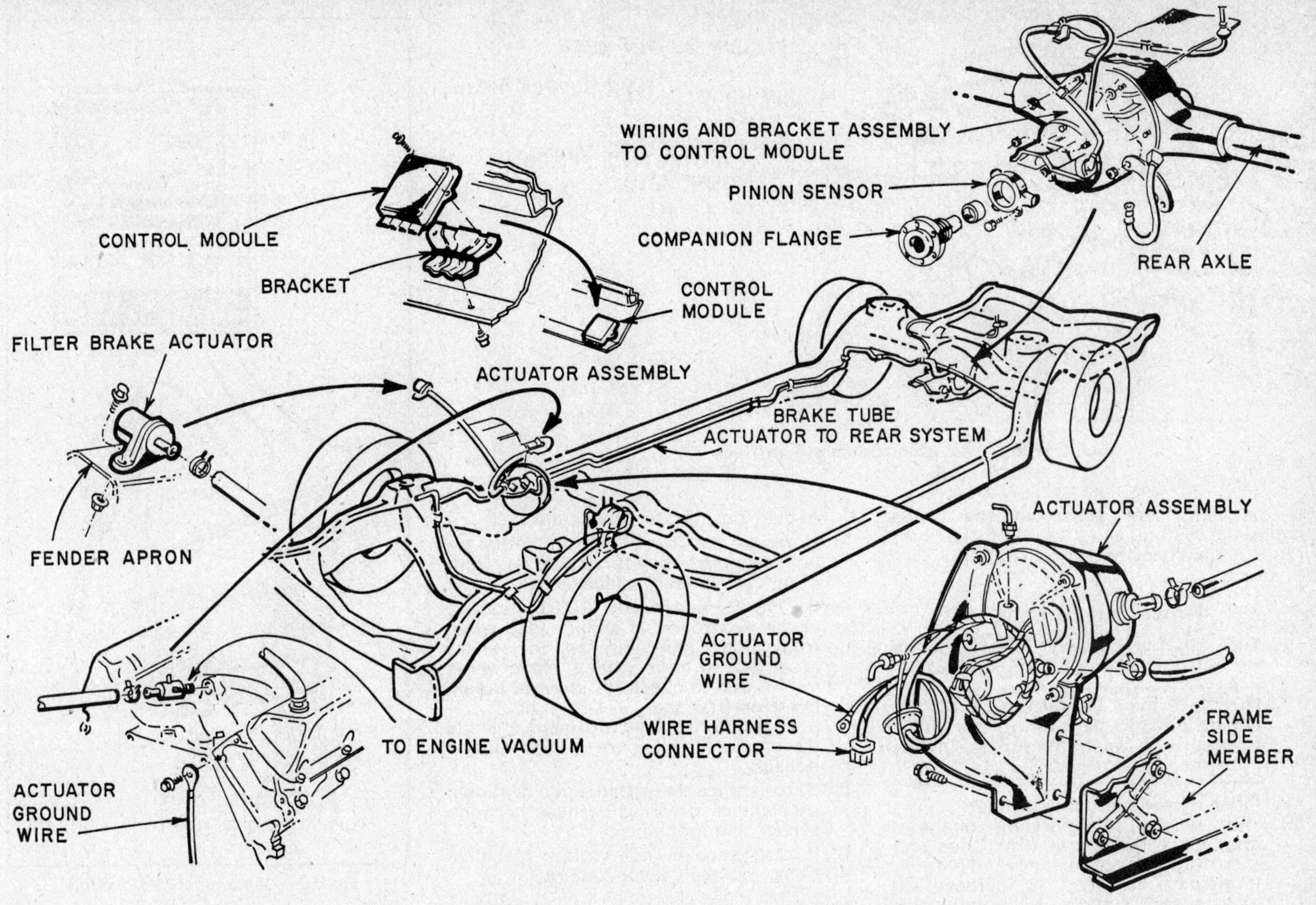

Fig. 7 Ford Sure Track components. Late 1970 & 1971-75 (typical)

In addition, the actuator contains a hydraulic cylinder, a solenoid and a time delay switch (failure switch), all functioning together to regulate pressure to the rear brakes upon control of the control module.

The actuator is connected to the brake hydraulic system between the pressure differential valve and the rear brake hydraulic system.

Electrical power is provided to the Sure Track system when the ignition switch is turned to any position except "Off" and "Accessory". The circuit is protected by a 3-amp fuse in the fuse box.

CAUTION: Do not use a fuse of higher rating than 3-amp to prevent damage to the control module. This fuse is identified by a red cover.

Vacuum for the operation of the actuator is supplied from the engine intake manifold. Atmospheric pressure is provided from an air filter mounted on the right hood hinge bracket. A vacuum check is included in the actuator.

Brake Warning Light: Illumination of the brake warning switch indicates a malfunction in the brake hydraulic system or in the Sure Track unit and requires immediate service of the brake system.

As a test, the warning light will go on when the ignition key is in the start position and will go out when the key is returned to the run position.

To determine if the trouble is in the Sure Track system or the brake hydraulic system, disconnect the switch plug from the hydraulic system differential switch.

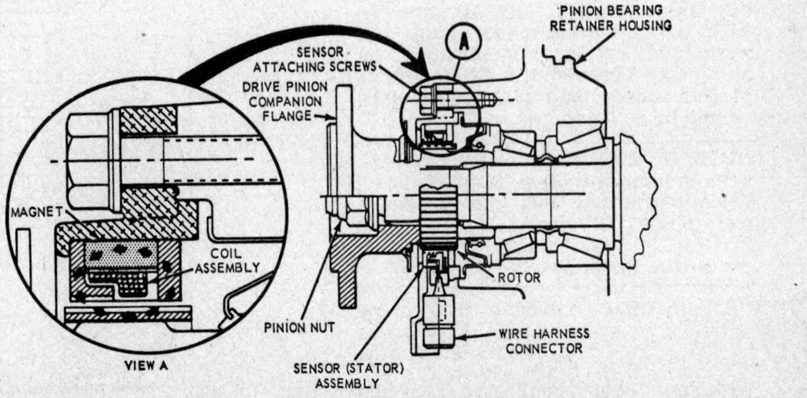

Fig. 8 Crossectional view of Sure Track sensor assembly. Late 1970 & 1971-75

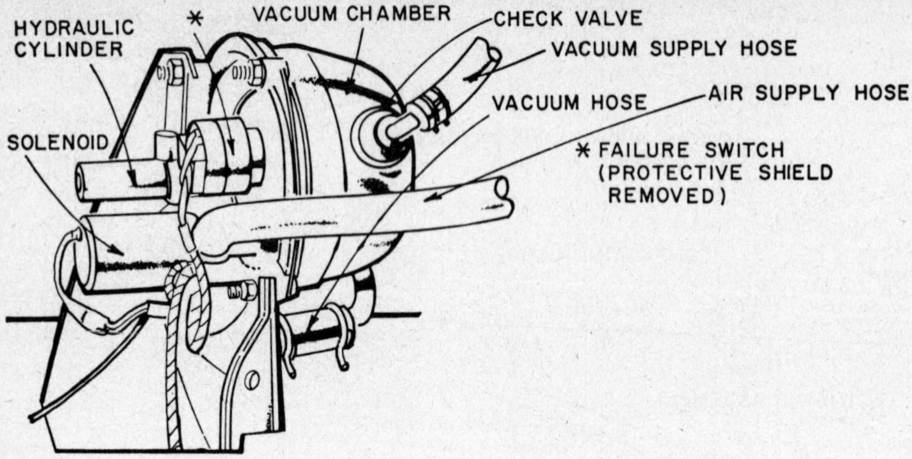

Fig. 9 Actuator assembly. Sure Track

If the light now goes out the brake hydraulic system is at fault. If the light remains on, check Sure Track system.

Functional Testing

1. Turn ignition key to "On" and listen for solenoid click and actuator cycle.

NOTE: It may be necessary to run engine prior to performing this check to assure a vacuum supply in the actuator.

---

2. Raise rear wheels to clear floor.
3. With engine at operating temperature, place transmission in Drive and accelerate to approximately 25 mph. It may be necessary to increase engine speed slightly to obtain sufficient cycling of the Sure Track system for easy observation of its operation.
4. Apply the brakes quickly and firmly. If the Sure Track is functioning properly, it will cycle five or six times or until the brake pedal is released.

NOTE: Both wheels must be turning for this test or the system will not operate.

Sensor Test

1. On 1969-74 models, disconnect plug "C" from module, Fig. 10, and on 1975 models, disconnect plug "A" from module, Fig. 10A. Connect an ohmmeter between the two contacts of the sensor lead plug. Resistance should be as listed below:

NOTE: On 1969 & Early 1970 models, a modulator with five plugs is used. When performing this check, disconnect plugs "C" and "E" and check for the specified resistance between the two contacts of either plug.

1969 & Early 1970	2000 (±500) ohms
Late 1970	1600-3200 ohms
1971-75	2400-3200 ohms

2. With ohmmeter connected between chassis ground and one of the two contacts of sensor lead plug, there should be infinite resistance.
3. If resistance is not within limits on both checks, isolate sensor(s) from other components at rear axle.
4. On 1969 and early 1970 models, check resistance of sensor lead wires from floor pan harness to sensor assembly at rear wheel. Move wires during test to check for internal break. On late 1970 and 1971-75 models, repeat resistance test with sensor leads disconnected from rear axle pinion housing.
5. If resistance is within specifications, problem is in wiring harness between sensor and module.
6. If resistance is not within specifications, replace sensor assembly.

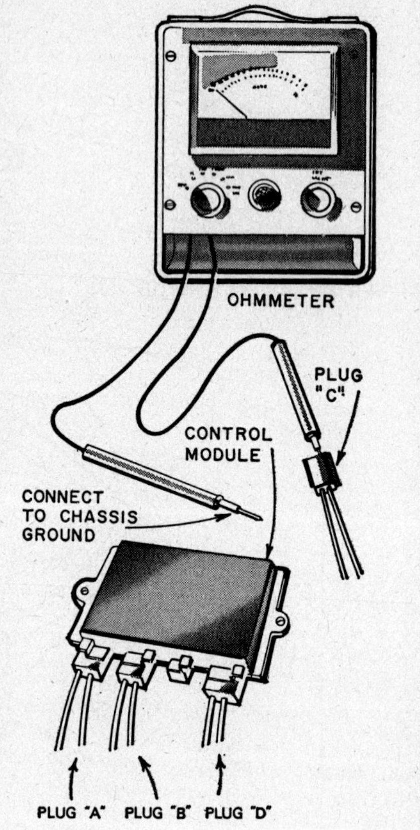

Fig. 10 Checking control module

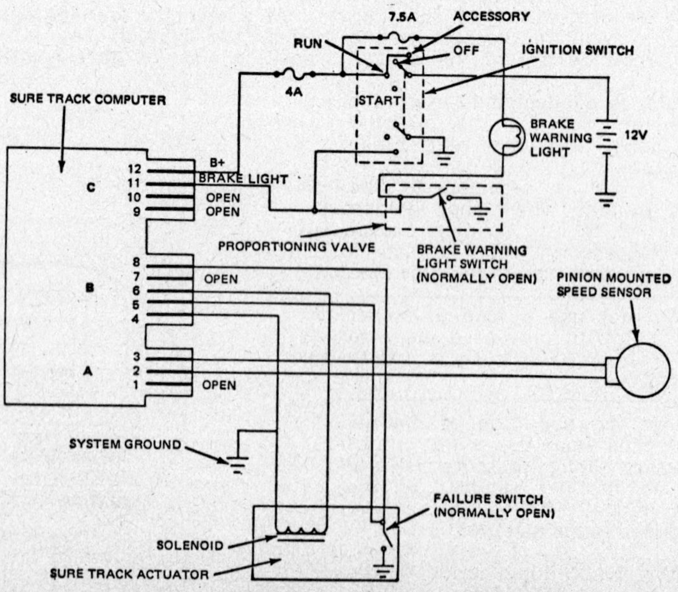

Fig. 10A Sure-Track brake system wiring. 1975

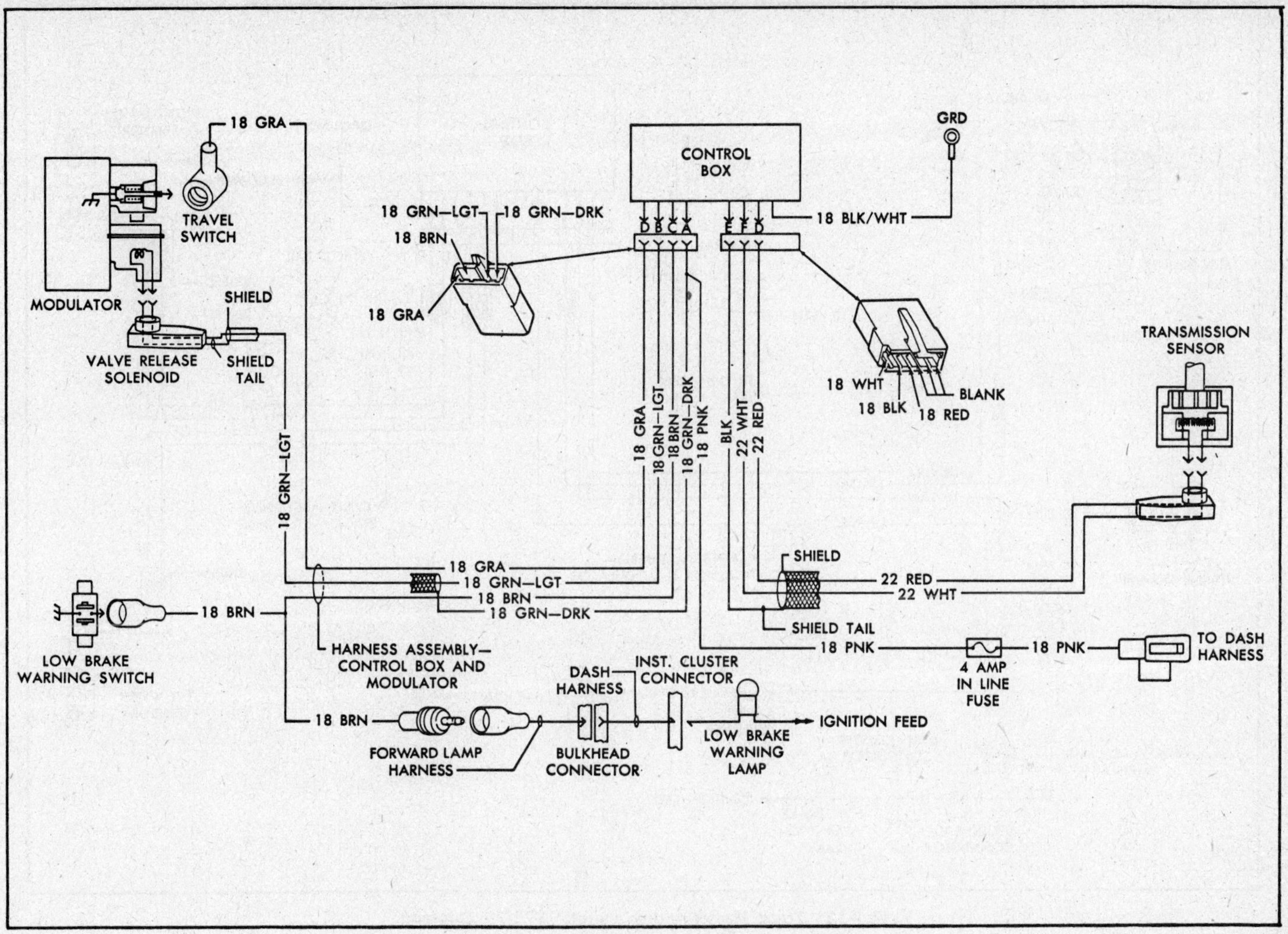

Fig. 11 Track Master wiring circuit. 1972-75 Cadillac

Sensor Output Voltage Test

1975 Models

1. Disconnect plug "A" from module, Fig. 10A, and connect an AC voltmeter between the two contacts.
2. Raise rear wheels off ground and run vehicle at 30 mph. Voltage reading should be above 5 volts AC.

Solenoid Test

1975 Models

1. Disconnect plug "B" from module, Fig. 10A, and connect a ohmmeter between pin 4 and pin 6. Resistance should be between 2 and 8 ohms.
2. If resistance is less than 2 ohms, check for grounded solenoid wiring.
3. If resistance is greater than 8 ohms, connect ohmmeter to solenoid terminals. If reading is between 2 and 8 ohms, check solenoid wiring harness for an open or for poor terminal contact at solenoid connector or at the intermediate connector on the engine compartment harness. If reading is not between 2 and 8 ohms, check solenoid wiring harness for an open or for poor terminal contact at solenoid connector or at the intermediate connector on the engine compartment harness. If reading is not between 2 and 8 ohms, replace the actuator assembly.

1974 Models

1. Remove plug "A" from control module, Fig. 10, then connect an ohmmeter between the two contacts of the solenoid (red) and system ground (black) lead plug. Resistance should be 4-8 ohms.
2. If resistance is lower than 4 ohms, inspect solenoid wiring for a grounded condition. If resistance is higher than 8 ohms, inspect solenoid wiring for an open condition.

NOTE: Refer to "System Ground Test" for connector check.

---

3. If wiring is satisfactory, replace actuator assembly.

System Ground Test

1. On 1969-74 models, disconnect plug "A" from module, Fig. 10, and on 1975 models, disconnect plug "B" from module, Fig. 10A. Connect an ohmmeter between the chassis ground and system ground (black plug lead). Resistance should be less than one ohm.
2. If resistance is greater than 1 ohm, check ground system wiring harness for loose or broken wires.
3. With ohmmeter still connected, move ground wires (black) at actuator harness to intermediate harness connector in and out. If meter fluctuates, it indicates a poor connection.
4. To repair a loose connection, separate connector and inspect system ground wiring connector terminals for an oversize condition. If terminal is oversize, use a pointed tool to force it together. Reinstall connectors and retest system for normal operation.
5. On 1969-72 models, connect an ohmmeter between system ground plug sleeve (black) and solenoid high (red), Fig. 10. If resistance is below 1000 ohms, the actuator coil or solenoid high (red) is shorted to ground.

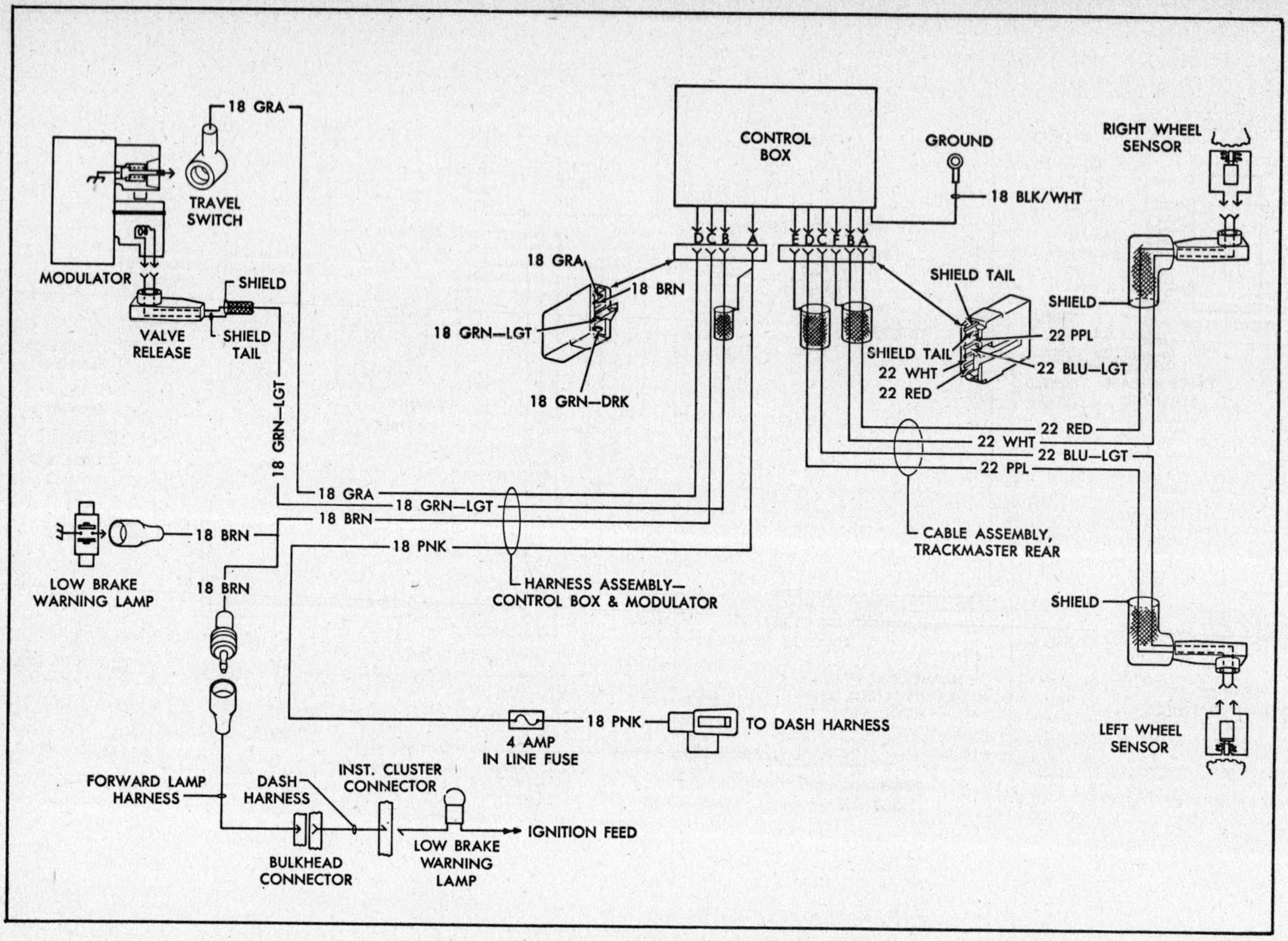

Fig. 12 Track Master wiring circuit. 1971-75 Eldorado

Failure Switch Test

1975

1. Disconnect plug "B" from module, Fig. 10A, and connect an ohmmeter between pin 5 and pin 8. Resistance should be less than one ohm.
2. Connect the ohmmeter between the chassis ground and one of the two failure switch terminals. Resistance should be infinite.
3. If resistance was not within specifications in Steps 1 and 2, check for proper connections at the intermediate connector in the engine compartment and the failure switch connector at the actuator. If connections are satisfactory, check for continuity in wiring harness between failure switch connector and plug "B" at module. If continuity is satisfactory, and the failure switch has not been shuttled due to brake release during Sure-Track cycling, replace the actuator.

GM TRACK MASTER & TRUE TRACK

Operation

This system Figs. 11, 12 and 13, incorporates a speed sensor mounted on each rear wheel, Fig. 14, on 1971-75 Eldorado and Toronado models and a transmission mounted speed sensor, Fig. 15 on 1972-75 Cadillac models. It combines a controller under the instrument panel and a modulator on the cowl in the engine compartment to prevent rear wheel lock-up.

The sensors produce an electrical signal that is proportional to wheel speed. This signal is monitored by the controller. A wheel that is not locking during brake application will gradually decelerate. When a wheel has started to lock, it will decelerate abruptly at a rate much greater than the car's deceleration. When the controller determines that the signal from the sensors changes too rapidly, and the car is approaching a skid, the controller energizes the modulator solenoid.

The modulator, Figs. 16 and 17, contains a diaphragm both sides of which are exposed to intake manifold vacuum as long as the solenoid is not energized. When the solenoid is energized, atmospheric pressure is allowed to reach the top side of the diaphragm, forcing it down against the support spring. The displacement piston is lowered along with the diaphragm. This seals off hydraulic pressure from the master cylinder to the rear wheel cylinders by closing the hydraulic check valve. The lowering of the displacement piston also increases the area in the hydraulic chamber immediately below the check valve. This increased area results in decreased hydraulic pressure and partial brake release.

With the brakes partially released, the wheels accelerate and, as car speed is approached, the controller senses the condition and de-energizes the modulator, allowing full brake application until a locking condition is again sensed.

The entire cycle of brake release and re-application occurs in about $1/3$ second. During a stop where brake pressure is

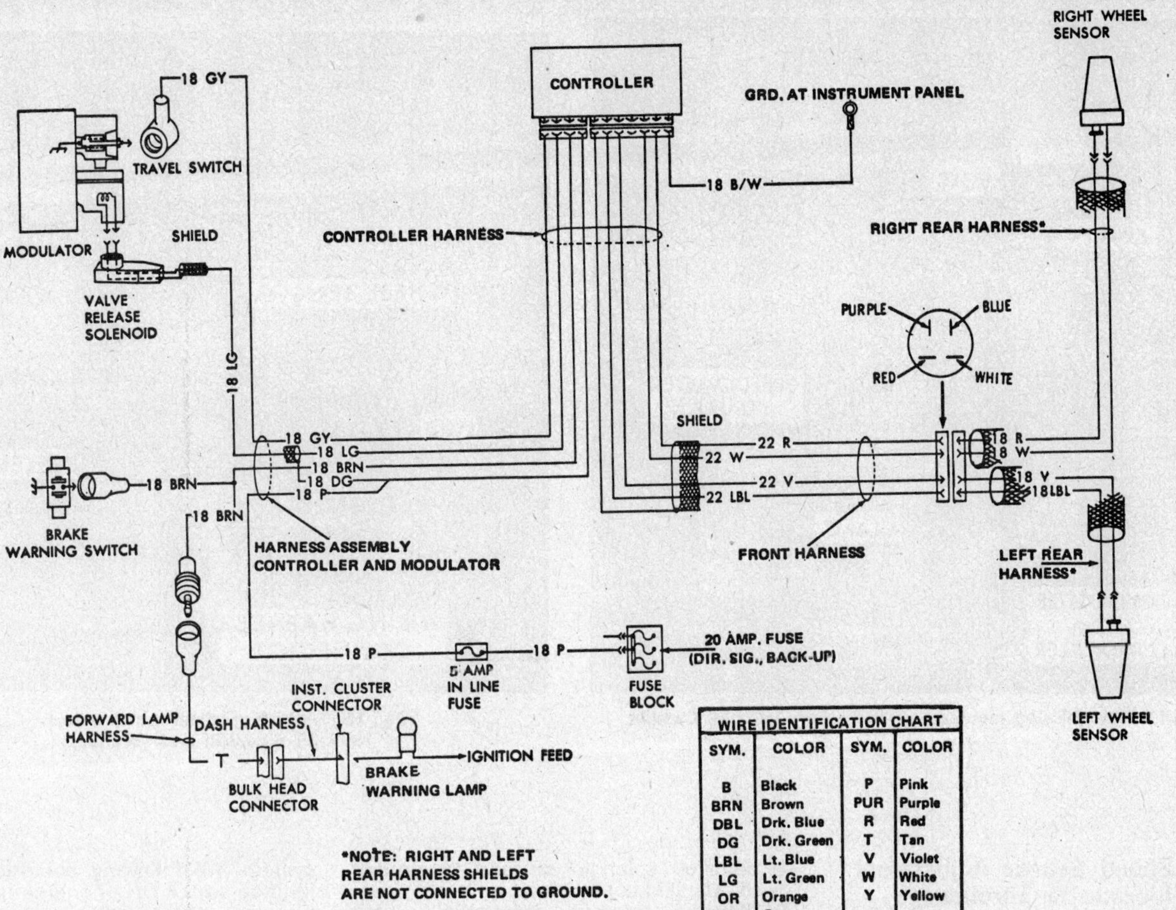

WIRE IDENTIFICATION CHART

SYM.	COLOR	SYM.	COLOR
B	Black	P	Pink
BRN	Brown	PUR	Purple
DBL	Drk. Blue	R	Red
DG	Drk. Green	T	Tan
LBL	Lt. Blue	V	Violet
LG	Lt. Green	W	White
OR	Orange	Y	Yellow
GY	Gray		

Fig. 13 True Track wiring circuit. 1971-75 Toronado

sufficient to cause lock-up, the cycling will continue until the car is slowed to approximately 5 mph or until the brakes are released by the driver.

The brake system warning light also operates in conjunction with the modulator travel switch in case of system malfunctions. When the switch is open for more than approximately four seconds, except during a skid controlled stop, the controller turns on the warning light. An exception to this is an open in the feed circuit from the ignition switch, which is indicated without the four second delay. When a malfunction is indicated, the lamp remains lighted until the ignition is turned off.

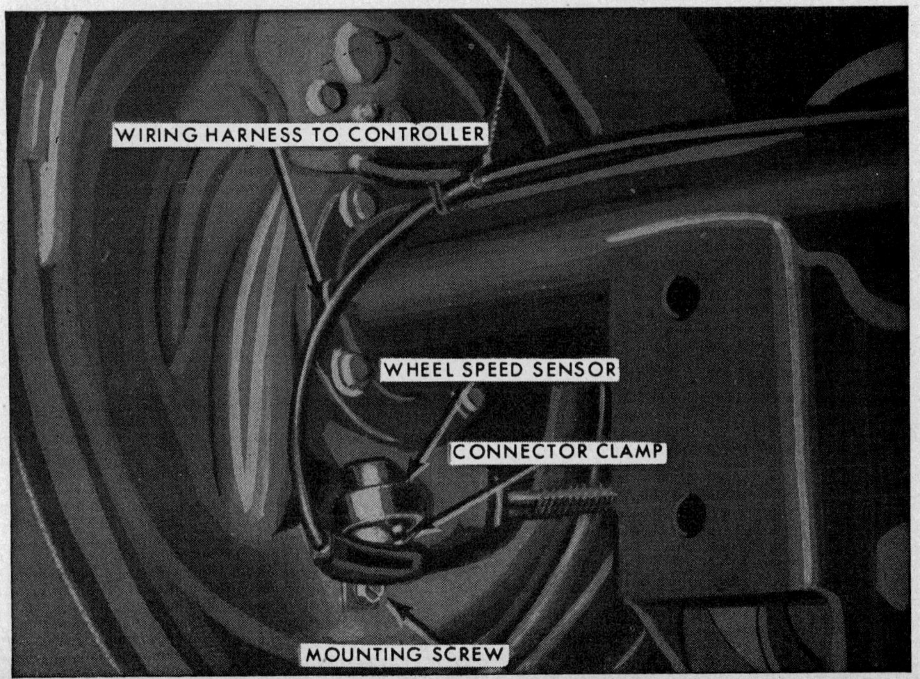

Fig. 14 Rear wheel speed sensor. 1971-75 Eldorado and Toronado

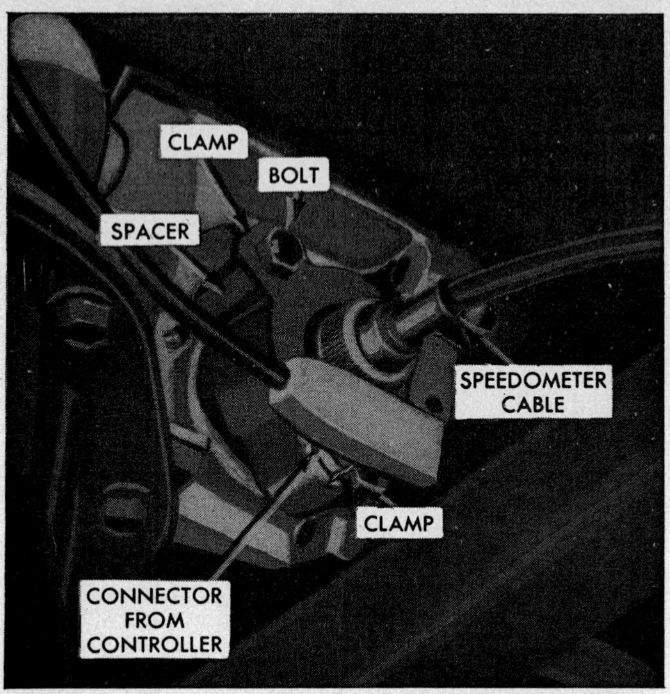

Fig. 15 Transmission mounted speed sensor. 1972-75 Cadillac

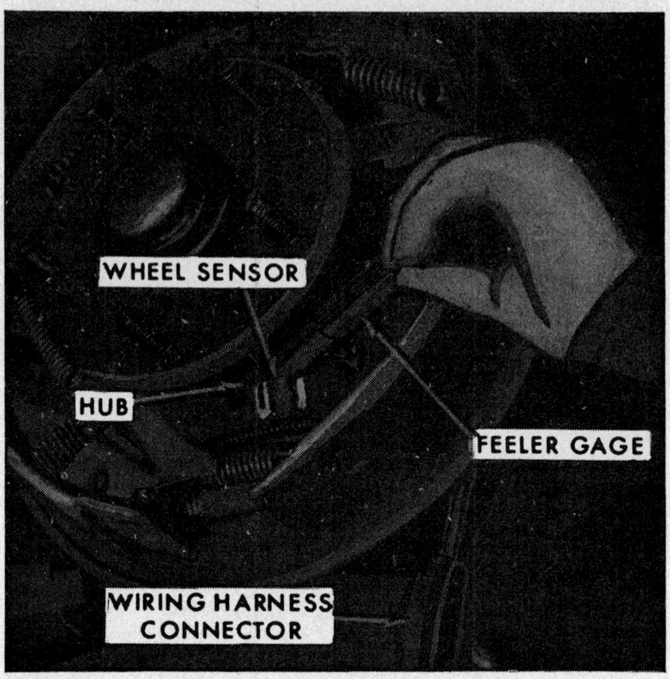

Fig. 18 Adjusting wheel speed sensor. 1971-75 Eldorado and Toronado

Wheel Speed Sensor Adjustment Eldorado & Toronado

NOTE: New sensors are furnished with a piece of tape on pick up end. This tape must be left on for adjustment.

1. Install sensor in backing plate so that taped end contacts teeth on hub.

   NOTE: If no tape is present on end of sensor, remove brake drum and a .010 inch feeler gauge between sensor and hub, Fig. 18.

2. While holding sensor against hub, install retaining screw and torque to 75 inch lbs.
3. Coat harness connector with wheel bearing grease and install connector and secure with clamp, Fig. 14.
4. Install brake drum, if removed.

Trouble Shooting

Brake Warning Light Does Not Light (Check When Ignition Switch Is In Start Position):

1. Burned out bulb.
2. Blown instrument fuse.
3. Open in bulb circuit.

Immediate Brake Light When Ignition Is Turned To On Position:

1. Leak in hydraulic system.
2. Blown skid control fuse.
3. Open in controller circuit.
4. Faulty controller.
5. Less than 12 volts at controller due to faulty connections or open in wiring circuit.
6. Shorted parking brake switch.
7. Shorted electrical lead to parking brake switch and or brake lamp.
8. Shorted solenoid or lead to ground blows skid control fuse.
9. Shorted pressure differential switch (System OK).

NOTE: For the following conditions, it is assumed that the brake lamp circuit operates normally and there are no hydraulic leaks or air trapped in the brake system. All following conditions are with ignition switch ON or engine running.

No Brake Light, System Inoperative, Exercise Cycle OK (Engine Running Or Ignition Switch On):

1. Wheel sensor not being driven due to missing dust cap or sensor quill

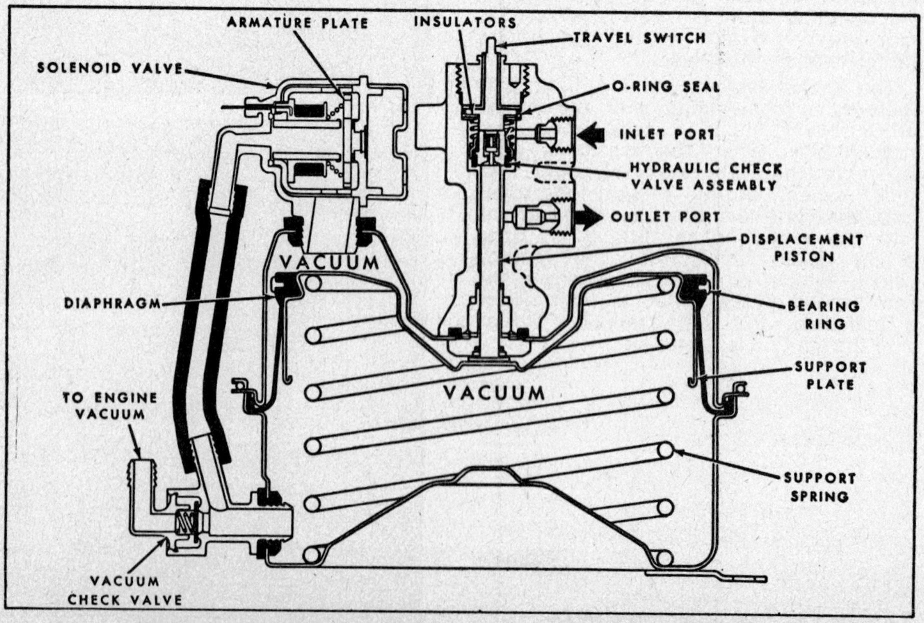

Fig. 16 Modulator assembly during normal braking. G.M. Computerized Skid Control

(Eldorado and Toronado).
2. Transmission sensor seized or not being driven (Cadillac).
3. Incorrect sensor adjustment (Eldorado and Toronado).
4. Both sensor leads shorted to each other (not shorted to ground).
5. Faulty controller.

No Brake Light, System Inoperative, No Exercise Cycle:

1. Loss of ground to controller.
2. Bad connection at controller or modulator.
3. Modulator seized in de-energized position.
4. Solenoid valve seized in de-energized position.
5. Faulty controller.

Brake Light Lights After 2-5 Second Delay, System Inoperative, Exercise Cycle O.K.

1. Speed sensor leads open.
2. Speed sensor leads shorted to ground.
3. Modulator travel switch open.
4. Faulty connection at travel switch.
5. Faulty controller.

Brake Light Lights After 2-5 Second Delay, System Inoperative, No Exercise Cycle:

1. Solenoid leads open.
2. Faulty controller.

False Releases While Car Is In Motion:

1. Frayed shield leads causing intermittent short.
2. Sensor drive shaft rubbing in wheel spindle due to improper adjustment (Eldorado and Toronado).
3. Missing or damaged insulators.
4. Sensor damper deteriorated.
5. Bad electrical connections.
6. Faulty controller.

False Releases While Vehicle Is Parked:

1. Bad electrical connections.
2. Faulty controller.

Does Not Cycle Down To 5 MPH During Maximum Braking:

1. Insufficient operating vacuum.
2. Faulty controller.

Brake Light On 2-5 Seconds After High Brake Pressure Is Applied:

1. Defective differential and proportioning valve.

Brake Light Comes On During a Stop, Or When Brake Is Depressed Firmly And Goes Off When Brake Pedal Is Released:

1. Hydraulic fluid leak.
2. Air in brake system.

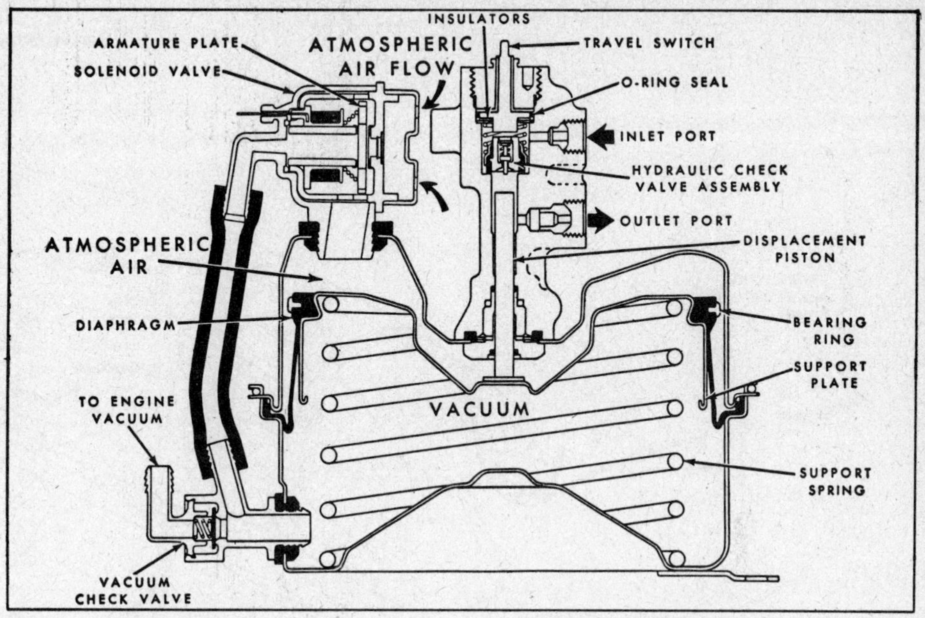

Fig. 17 Modulator assembly during release. G.M. Computerized Skid Control

MAX-TRAC SYSTEM

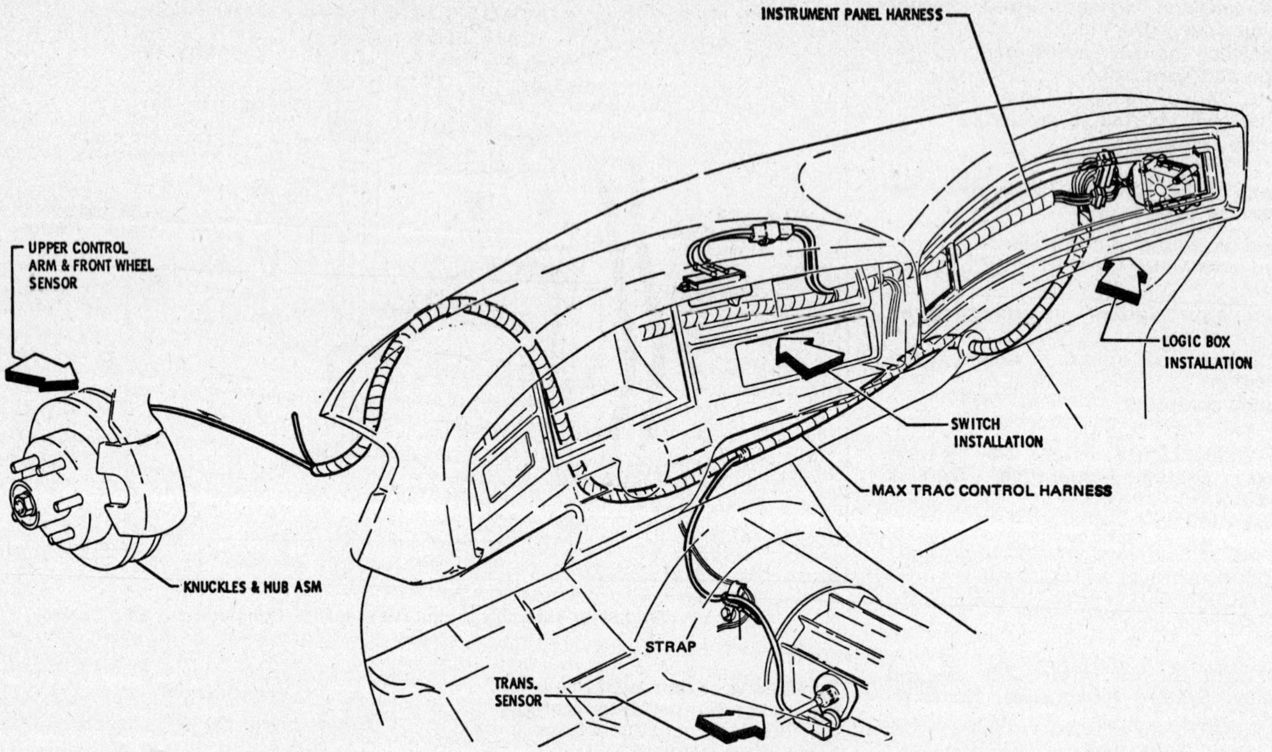

Fig. 1 Max-Trac system component location. Less Air Cushion Restraint system

DESCRIPTION

The Max-Trac system, Figs. 1 and 2, aids the driver in maintaining directional stability and maneuverability during acceleration and cornering. This system automatically limits engine power to a value required for maximum acceleration without excessive rear wheel slippage.

This system incorporates a transmission speed sensor, front wheel speed sensor and speed disc, electronic controller, "On-Off" switch and electrical harness. Also, the existing ignition switch, brake light switch, brake lights and ignition components are used.

Transmission Speed Sensor

This sensor is basically an AC generator which produces an AC voltage with the frequency proportional to the transmission output shaft speed. This speed sensor converts rear wheel speed into an electrical signal at a frequency of 25 Hz/mph which is then transmitted to the electronic controller through the electrical harness. The sensor is mounted in the transmission at the speedometer cable connection.

Front Wheel Speed Sensor & Speed Disc

This sensor is also an AC generator which produces an AC voltage with the frequency proportional to the speed of the disc, mounted on the hub of the left front wheel. The sensor converts front wheel speed into an electrical signal at a frequency of 6.25 Hz/mph which is then transmitted to the electronic controller through the electrical harness. An adjustable gap is provided between the sensor and speed disc.

Electronic Controller

This unit is a solid state electronic computer, mounted in the passenger compartment. The controller receives the driven (rear) and undriven (front) wheel speed signals and compares these two signals. When the difference exceeds a predetermined point, the unit generates a 30 Hz output signal with a duty cycle proportional to the difference which is then transmitted to the ignition system through the electrical harness.

On-Off Switch

The "On-Off" switch, located on the instrument panel, permits the driver to control the system. During normal driving conditions, it is recommended that the switch be left in the "On" position. However, under certain conditions, it is desirable to have the system inoperative, such as rocking the vehicle in snow or sand. If a malfunction occurs in the system, the vehicle may be operated with the switch in the "Off" position. In the "Off" position, the controller output signals are shorted, thereby preventing the signals being transmitted to the ignition system.

Diverter Valve Solenoid

Vehicles equipped with Max-Trac and AIR have a diverter valve solenoid mounted on the AIR pump to prevent engine back-fire when the MAX-Trac system is operating.

OPERATION

The system is activated when the ignition switch is "On", the "On-Off" switch is "On" and the brake are not applied. The vehicle can be operated in a normal manner without ignition interuption will occur if rear wheel slippage does not exceed a predetermined value. However, under certain conditions such as excessive acceleration or hard cornering, the normal load on the rear wheels is reduced, thereby allowing rear wheel slippage. The controller compares the speed signals from the two speed sensors, in turn generating an output signal which interupts the ignition. This results in a power reduction and a corresponding reduction in rear wheel slippage.

TROUBLE SHOOTING

System Inoperative At All Speeds

NOTE: Ensure "On-Off" switch is in the "On" position and the stop lamps are

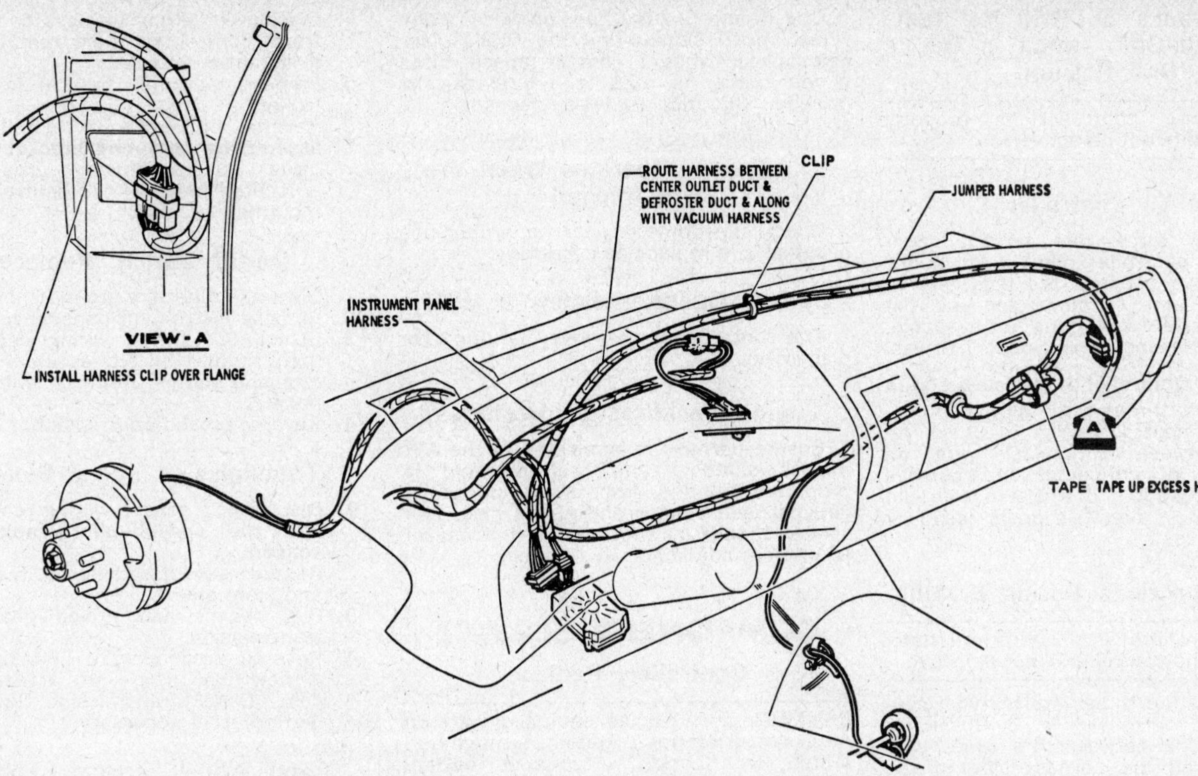

Fig. 2 Max-Trac system component location. With Air Cushion Restraint system

functioning. Also, make certain the controller is properly grounded or is not defective.

No Transmission Speed Sensor Signal

Disconnect the six wire terminal connector from controller, Fig. 3. Check resistance between red and white wires on harness side of connector. Resistance should be 1600-2400 ohms. If circuit is open, disconnect transmission sensor connector and check resistance across sensor terminals. If circuit is open, replace transmission sensor. If resistance is 1600-2400 ohms, replace wiring harness.

No Front Wheel Speed Sensor Signal

Disconnect six wire terminal connector from controller, Fig. 3. Check resistance between light green and white wires on harness side of connector. Resistance should be 1250-1750 ohms. If circuit is open, disconnect front wheel sensor connector and check resistance across sensor terminals. If circuit is open, replace front wheel speed sensor. If resistance is 1250-1750 ohms, replace wiring harness.

Low Front Wheel Speed Sensor

Check and adjust gap between front wheel speed sensor and speed disc. The felt spacer should contact disc. If the felt spacer is missing, adjust gap between end of sensor and speed disc to .05 inch.

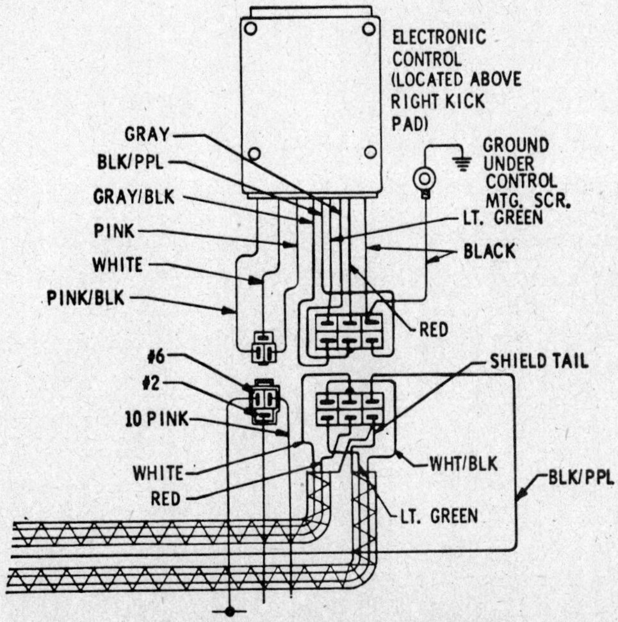

Fig. 3 Controller wiring (Typical)

Engine Cranks But Will Not Run With "On-Off" Switch In The "On" Position

NOTE: Ensure controller is properly grounded and is not defective.

No Current To Controller

Disconnect three wire terminal connector from controller, Fig. 3. Check for battery voltage between the pink wire at the harness side of connector and the ground. If battery voltage is not available, repair or replace instrument panel harness.

Open Circuit Controller & Resistance Wire

Disconnect three wire terminal connector from controller, Fig. 3. Check resistance between pink/black wire on sensor side of connector and positive terminal of ignition coil. Resistance should be 1.8 ohms. If circuit is open, repair or replace harness.

System Functions During Braking

NOTE: Ensure the stop lamps are functioning and the controller is not defective.

Disconnect the three wire terminal connector from controller, Fig. 3. Check for battery voltage between the white wire (#2 cavity) on the harness side of the connector and the ground with the brakes applied. If battery voltage is not available, repair or replace harness.

System Inoperative At Low Vehicle Speeds

NOTE: Ensure controller is not malfunctioning.

Check and adjust gap between front wheel speed sensor and the speed disc. Felt spacer should contact speed disc. If felt spacer is missing, adjust gap to provide .05 inch between sensor and speed disc.

System Operates When Not Needed

Check speedometer gear for proper application and replace as necessary.

System Operates Below 2 MPH

The controller is defective, requiring replacement.

Engine Backfires Or Hesitates

Ensure harness is connected to the AIR bypass solenoid and that solenoid is properly grounded. Also, check and adjust gap between sensor and speed disc. Felt spacer should contact disc. If felt spacer is missing, adjust gap to .05 inch.

COMPONENT SERVICE

Controller, Replace

CAUTION: On vehicles equipped with an Air Cushion Restraint System, turn ignition switch to "Lock", disconnect battery ground cable and tape end, thereby deactivating system.

1. On vehicles not equipped with an air cushion restraint system, remove glove box.
2. On vehicles equipped with an air restraint system, remove lower left cover assembly.
3. On all vehicles, remove controller and mounting bracket attaching screws.
4. Disconnect wiring harness connectors from controller, then remove controller from vehicle.
5. Remove mounting bracket from controller.
6. Reverse procedure to install. When assembling mounting bracket to controller, ensure controller ground wire is located between the controller and mounting bracket.

On-Off Switch, Replace

1. Disconnect battery ground cable.
2. Remove instrument panel cover.
3. Disconnect switch wiring connector, then remove switch escutcheon and retaining screws. Remove switch from panel.
4. Reverse procedure to install.

Transmission Speed Sensor

1. Disconnect sensor wiring connector and the speedometer cable from sensor.
2. Remove sensor retainer bolt, retainer and shim sleeve.
3. Pull sensor and driven gear from transmission.
4. Reverse procedure to install. Install a new "O" ring into transmission case. Check transmission fluid level and adjust if necessary.

Front Wheel Sensor, Replace

1. Remove cable attaching screw from rotor shield and disconnect wire connector from sensor.
2. Remove sensor retaining bolt and sensor from vehicle.
3. Reverse procedure to install. Felt spacer should contact speed disc. If felt spacer is missing, adjust gap between sensor and speed disc to .05 inch.

MANUAL STEERING GEARS

STEERING GEAR ADJUSTMENT SPECIFICATIONS

CAR MAKE	GEAR TYPE	WORM BEARING PRELOAD	CROSS SHAFT PRELOAD
AMERICAN MOTORS Exc. Pacer			
1969-73	Ball & Nut	④	⑤
1974-75	Ball & Nut	5 to 8①	9 to 16①
AMERICAN MOTORS PACER			
1975	Rack & Pinion	⑥	⑥
BUICK (EXC. APOLLO & SKYHAWK)			
1969	Ball & Nut	5 to 8①	9 to 16①
1970-72	Ball & Nut	2 to 7①	7 to 18①
1973-75	Ball & Nut	5 to 8①	9 to 16①
BUICK APOLLO			
1973	Ball & Nut	4 to 6①	9 to 15①
1974-75	Ball & Nut	5 to 8①	9 to 15①
CHECKER MOTORS			
1969-71	Ball & Nut	5 to 8①	9 to 16①
1972-75	Ball & Nut	4 to 6①	9 to 15①
CHEVROLET (EXC. CORVETTE, MONZA & VEGA)			
1969-71	Ball & Nut	5 to 8①	9 to 16①
1972-75	Ball & Nut	4 to 6①	9 to 15①
CHEVROLET VEGA & PONTIAC ASTRE			
1971-75	Ball & Nut	5 to 8①	9 to 16①
CHRYSLER			
1969-71	Ball & Nut	1 to 4①	8 to 11①
CORVETTE			
1969-71	Ball & Nut	5 to 8①	9 to 16①
1972-75	Ball & Nut	4 to 6①	9 to 15①
DODGE			
1969-75	Ball & Nut	1 to 4①	8 to 11①
FORD (EXC. MUSTANG II & PINTO)			
1969-73	Ball & Nut	4 to 5①	9 to 10①
1974-75	Ball & Nut	3 to 8①	10 to 16①
FORD MUSTANG II, PINTO & BOBCAT			
1971-75	Rack & Pinion	②	③
MERCURY (EXC. BOBCAT)			
1969-73	Ball & Nut	4 to 5①	9 to 10①
1974-75	Ball & Nut	3 to 8①	10 to 16①
MONZA, SKYHAWK & STARFIRE			
1975	Ball & Nut	5 to 8①	9 to 16①
OLDSMOBILE (EXC. STARFIRE)			
1969-71	Ball & Nut	4 to 7①	8 to 17①
1972-75	Ball & Nut	5 to 8①	9 to 16①
PLYMOUTH			
1969-75	Ball & Nut	1 to 4①	8 to 11①
PONTIAC (EXC. ASTRE)			
1969-70	Ball & Nut	7①	16①
1971	Ball & Nut	5 to 8①	4 to 10①
1972-75	Ball & Nut	5 to 8①	9 to 16①

①—Measured with inch-pound torque wrench attached to steering wheel nut.

②—For support yoke to rack adjustment, add selected shims to give a combined pack thickness of .005-.006 inch greater than gap between cover and housing flange.
For pinion bearing preload adjustment, add one .005 inch shim to pack.

③—Effort required to sustain input shaft rotation should be 7-15 in-lb, with gear filled with lubricant.

④—Exc. 1972-73 Gremlin, Hornet & Javelin, 2 to 6; 1972-73 Gremlin, Hornet & Javelin, 4 to 10.

⑤—Exc. 1972-73 Gremlin, Hornet & Javelin, 12 to 18; 1972-73 Gremlin, Hornet & Javelin, 14 to 18.

⑥—See text for adjustment procedure.

Recirculating Ball Worm & Nut Gear

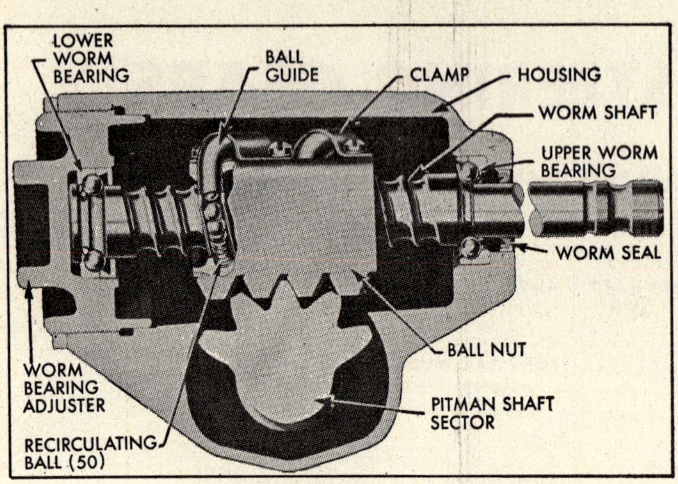

Fig. 1 Recirculating ball-worm and nut steering gear

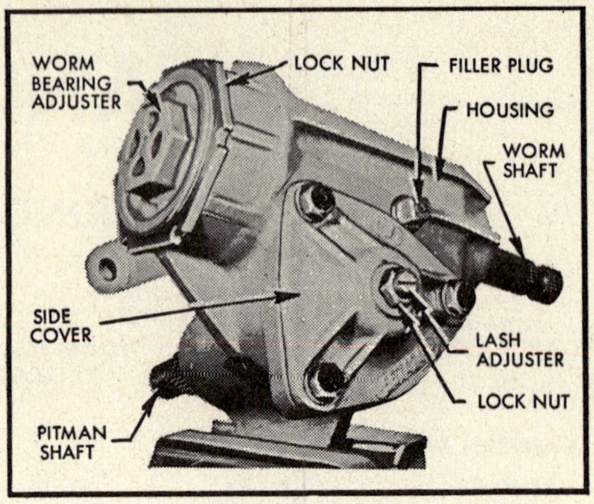

Fig. 2 Steering gear adjustments

DESCRIPTION

As shown in Fig. 1, the worm on the lower end of the steering shaft and the ball nut which is mounted on the worm have mating spiral grooves in which steel balls circulate to provide a low friction drive between worm and nut.

Two sets of balls are used, ranging in number from approximately 20 to 30 to a set, depending upon the size of the steering gear unit. Each set of balls operate independently of the other. The circuit through which each set of balls circulates includes the grooves in the worm and ball nut and a ball return guide attached to the outer surface of the nut.

When the wheel and steering shaft turn to the left, the ball nut is moved downward by the balls which roll between the worm and nut. As the balls reach the outer surface of the nut they enter the return guides which direct them across and down into the ball nut where they enter the circuit again.

When a right turn is made the ball nut moves upward and balls circulate in the reverse direction.

The teeth on the ball nut engage teeth on the sector which is forged integral with the pitman shaft. The teeth on the ball nut are made so that a "high point" or tighter fit exists between the ball nut and pitman shaft sector teeth when the front wheels are in the straight-ahead position. The teeth on the sector are tapered slightly so that a proper lash may be obtained by moving the pitman shaft endwise by means of a lash adjuster screw which extends through the gear housing side cover. The head of the lash adjuster and the selectively fitted shim fit snugly into a T-slot in the end of the pitman shaft so that the screw also controls end play of the shaft. The screw is locked by an external lock nut.

GEAR ADJUSTMENTS

There are two adjustments on the steering gear: worm bearing preload and pitman shaft overcenter preload, Fig. 2. **Important:** Never attempt to adjust the steering gear while it is connected to the steering linkage. The gear must be free of all outside load in order to properly make any steering gear adjustment.

Preliminary

1. Tighten steering gear mounting bolts.
2. Disconnect steering linkage from steering arm or gear.
3. Turn wheel slowly from one extreme to the other.
   Caution: Never turn the wheel hard against the stopping point in the gear as damage to the ball nut assembly may result.
4. Steering wheel should turn freely and smoothly throughout its entire range.
   Note: Roughness indicates faulty internal parts requiring disassembly of gear unit. Hard pull or binding indicates an excessively tight adjustment of the worm bearings, or excessive misalignment of the steering shaft. Any excessive misalignment must be corrected before the gear can be properly adjusted.

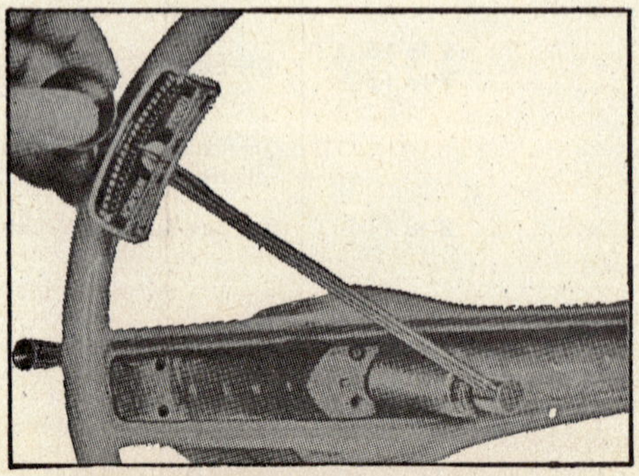

Fig. 3 Checking adjustments with inch-pound torque wrench

NOTE: Specifications in the Steering Gear Adjustment Specifications chart are given in inch pounds as measured with a torque wrench pulling on steering wheel nut, Fig. 3.

Checking Worm Bearing Preload

1. Turn steering wheel gently in one direction until it stops. This positions gear away from "high point" load.
2. Attach torque wrench and check the torque required to turn the wheel steadily in the range where lash exists between ball nut and pitman shaft sector.
3. If adjustment is not within specified limits, adjust worm bearing preload.

Adjust Worm Bearing Preload

1. Loosen worm bearing adjuster lock nut, using a drift, Fig. 2.
2. Turn bearing adjuster as required to bring the adjustment within specified limits.
3. Tighten lock nut and recheck preload.

Checking Pitman Shaft Over-Center Preload

1. Turn steering wheel from one extreme to the other while counting the total turns, then turn wheel back ½ the number of turns. This positions steering gear on "high point" where a preload should exist between ball nut and pitman shaft teeth.
2. Attach torque wrench and check torque or pull required to turn the wheel through the "high point" range.
3. If adjustment is not within the specified limits, adjust as follows:

Adjust Pitman Shaft Overcenter Preload

1. Loosen lock nut and turn pitman shaft lash adjuster screw as required to bring the adjustment within specified limits.
2. After tightening lock nut, rotate steering wheel back and forth through the "high point" and through the entire range to check for tight spots.

NOTE: If lash cannot be removed at the "high point", or if gear load varies greatly and feels rough, the gear should be removed for inspection of internal parts.

3. Attach linkage to steering gear when adjustments have been completed.

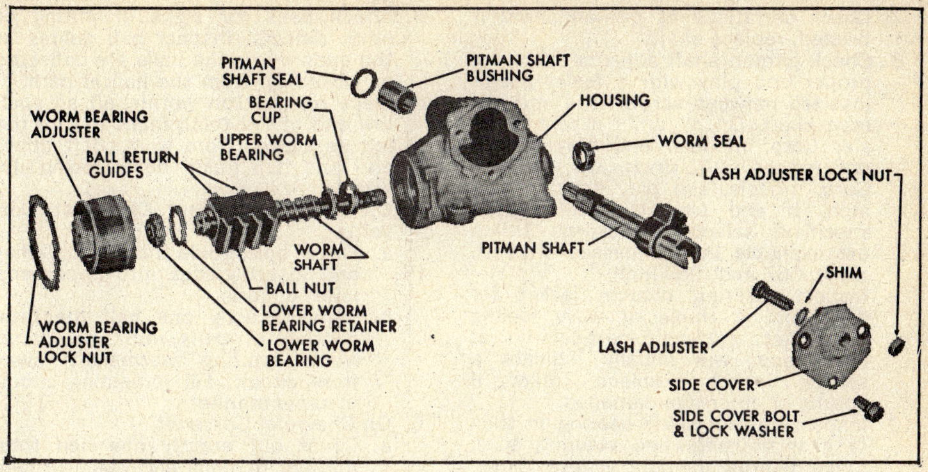

Fig. 4 Recirculating ball-worm and nut gear disassembled, exc. Chevrolet Corvette

STEERING GEAR REPAIRS

Disassembly

1. Clamp steering in a vise with wormshaft in horizontal position.
2. Rotate wormshaft from stop to stop, noting number of turns. Then, turn wormshaft back exactly halfway to center the gear. Loosen wormshaft lock nut.
3. Remove side cover attaching screws, lightly tap end of pitman shaft with a mallet and lift side cover and pitman shaft assembly from housing.

NOTE: If pitman shaft sector does not clear housing opening, rotate wormshaft until sector passes through opening.

4. Remove wormshaft lock nut and adjuster plug assembly, then wormshaft and ball nut from housing.
5. On all units except Chevrolet Corvette, remove upper wormshaft bearing from shaft, Fig. 4. Pry lower bearing retainer from adjuster plug and remove bearing.
6. On Chevrolet Corvette, remove lower wormshaft bearing from shaft, Fig. 5.
7. On all units, remove lash adjuster screw lock nut, turn adjuster screw clockwise through side cover, then slide screw and shim from pitman shaft slot.
8. Pry pitman shaft and wormshaft seals from housing.

Inspection of Parts

1. Clean and inspect all ball and roller bearings and races, including race in housing.
2. Inspect pitman shaft bushings in gear housing and end cover. Replace bushings in housing and replace end cover if bushings are worn excessively.
3. Replace pitman shaft and wormshaft seals to avoid leakage of lubricant.
4. Inspect steering shaft for wear or pits in bearing races, which would require replacement of shaft.
5. Check shaft for straightness.
6. Inspect teeth of ball nut and pitman shaft. If scored or excessively worn it is advisable to replace both parts to insure proper mating of teeth.

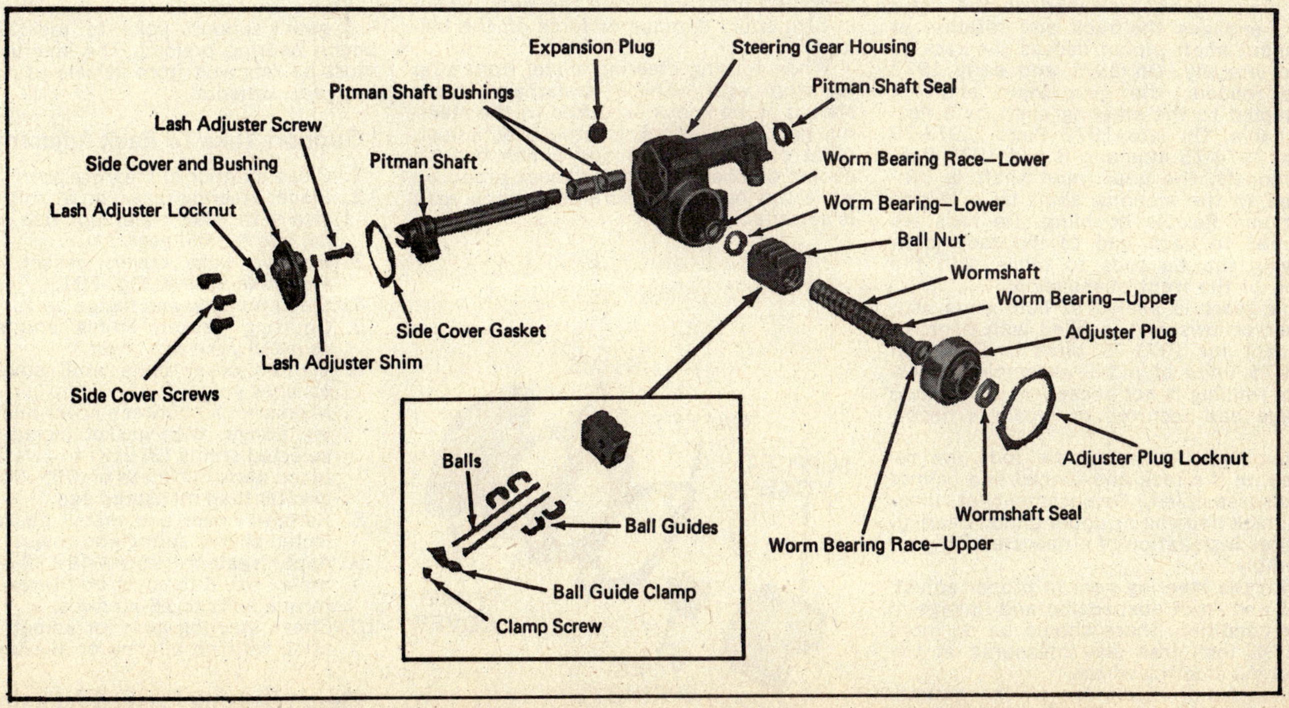

Fig. 5 Recirculating ball-worm and nut gear disassembled. Chevrolet Corvette

7. Check serrations of pitman shaft; if twisted, replace shaft.
8. Check pitman shaft adjuster screw for proper end play with a feeler gauge inserted between screw head and pitman shaft. On all units except Chrysler Corp. models, end play should not exceed .002 inch. On Chrysler Corp. models, end play limit is .004 inch. If end play is greater than specified, selective thickness shims are available in thicknesses of .063, .065, .067 and .069 inch.
9. Inspect steering column jacket for distortion. A ripple or wavy feeling of jacket surface, particularly at lower end, will usually indicate a sprung jacket. Replace jacket if sprung or otherwise damaged.
10. Inspect control shaft bearing in tube of gear housing, and steering shaft upper bearing in control lever housing support. Replace worn or damaged parts.

Ball Nut Servicing

NOTE: Servicing of the ball nut assembly is not necessary provided no binding or tightness is noted when rotated on the wormshaft.

1. Remove ball guide clamp and ball guides.
2. Turn ball nut upside down and rock wormshaft so balls drop from nut, into a clean container.
3. Check balls for signs of pitting or other defects. Inspect ball guides at the ends where the balls are deflected or picked up from the helical path.
4. Place ball nut on wormshaft so shallow end of nut teeth faces toward the left as viewed from wormshaft upper end and ball guide holes facing upward.
5. On all units except Chevrolet Corvette:
   a. Place ball guide halves together and insert guides into upper and lower circuits.
   b. Install exactly one half the total number of balls into one circuit while rotating wormshaft away from circuit. Fill remaining circuit in same manner.
6. On Chevrolet Corvette:
   a. Count out exactly one half total number of balls and place in separate container.
   b. Fill one circuit in ball nut while rotating wormshaft away from hole. With one half of a ball guide with groove facing upward, place remaining balls from circuit in groove. Place ball guide halves together and plug open ends with a suitable grease, then push guides into guide holes. Ball guides may be gently tapped into place, if necessary.
7. On all units, install ball guide clamp. Rotate wormshaft to check proper installation of balls.

Reassembly

NOTE: Lubricate all seals, bushings, bearings and gears with recommended lubricant prior to installation.

1. On all units except Chevrolet Corvette, place upper bearing over wormshaft and insert wormshaft assembly into housing, Fig. 4. Install bearing into wormshaft adjuster race and, guiding bearing onto wormshaft end, install adjuster and tighten until almost all lash is removed from wormshaft, then install lock nut.
2. On Chevrolet Corvette, place wormshaft lower bearing into housing race, Fig. 5. Install upper bearing and adjuster plug assembly onto wormshaft and insert assembly into housing, guiding wormshaft into lower bearing.
3. On all units, install lash adjuster and shim into pitman shaft slotted end.
4. Ensure ball nut is in center of travel and insert pitman shaft into housing with center tooth on sector meshing with center groove on ball nut.
5. Fill gear housing with recommended lubricant. Then, thread side cover onto lash adjuster screw using a screwdriver to turn screw counterclockwise through threaded opening in side cover until screw bottoms. Back off screw one half turn and install lock nut and cover retaining screws.

Ford Rack & Pinion

DESCRIPTION

A pinion gear, machined on the input shaft, engages the rack and rotation of the input shaft pinion causes the rack to move laterally. On 1971 and early 1972 Pinto models, the gear input shaft is connected to the steering shaft by a flexible cable. On late 1972 Pinto, 1973-75 Pinto, 1974-75 Mustang II and 1975 Bobcat models, the gear input shaft is connected to the steering shaft by a U-joint shaft and flexible coupling. Tie rods are attached to each end of the rack joint, allowing the tie rods to move with the action of the front suspension.

The gears is sealed at both ends with rubber bellows and is filled with 5 oz. of lubricant for 1971-73 units or 7 oz. on 1974-75 units at initial assembly. Checking or refilling is not necessary unless fluid leakage has occurred or repair is necessary.

Couplings connecting tie rods are retained on the rack and pinned and cannot be disassembled. Replacement of inner rods, rack housing or upper pinion bearing, requires installation of a new steering gear assembly.

With the steering gear in proper adjustment and front suspension and linkage in good condition, there should be no more than 3/8 inch free play measured at the rim of the steering wheel.

When front wheels are off the ground, do not turn steering wheel quickly or forcefully from stop to stop as this will build-up hydraulic pressure within the gear, which in turn could damage or force off the bellows.

When turning steering wheel from stop to stop with vehicle stationary, there should be no knock produced by the steering gear. If a knock is produced, adjustment of rack preload and pinion bearing should be checked. A faint knock produced while driving on an extremely rough road is acceptable.

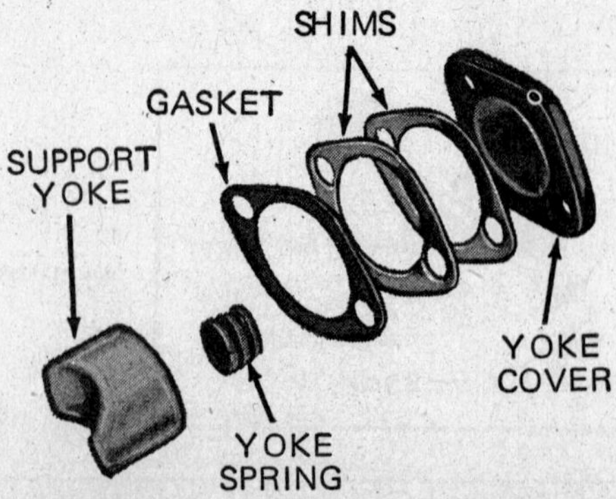

Fig. FR1 Support yoke arrangement

GEAR ADJUSTMENTS

There are two adjustments on the steering gear: support yoke to rack fit and pinion bearing preload. The steering gear must be removed from vehicle to perform both adjustments.

Support Yoke To Rack Adjustment

1. Clean exterior of steering gear.
2. Place steering gear in a soft jawed vise with yoke cover up. Use mounting pad to hold gear.
3. Remove yoke cover, gasket, shims and yoke spring, Fig. FR1.
4. Clean housing and flange area.
5. Omitting gasket, shims and spring, reinstall yoke and cover.
6. Tighten cover bolts until cover just touches yoke.
7. Measure gap between cover and housing flange. With gasket included add selected shims (3F518) to give a combined pack thickness of .005-.006 inch greater than measured gap.
8. Remove cover and install gasket, selected shims, spring and cover.
9. Apply sealant (ESW-M46-132-A) Permatex No. 3 to cover bolt threads and torque bolts to 15-20 ft-lbs.
10. Check steering gear for smooth operation with no binding or slackness.

Pinion Bearing Preload Adjustment

1. Clean exterior of steering gear.

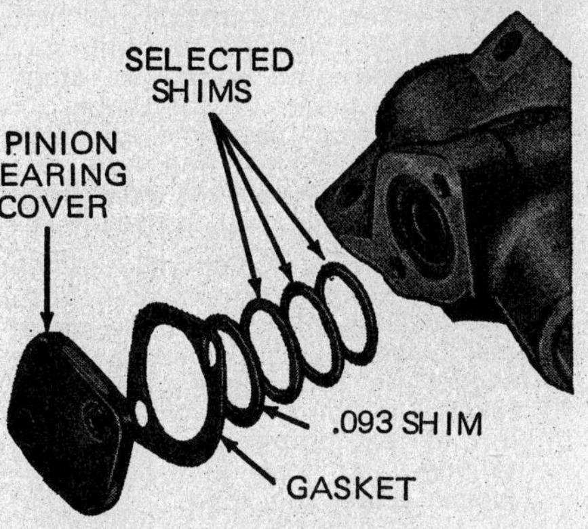

Fig. FR2 Pinion bearing cover and shim arrangement

2. Place steering gear in a soft jawed vise with pinion cover up. Use mounting pads to hold gear.
3. Loosen yoke cover bolts to relieve spring tension on rack.
4. Remove pinion cover and clean flange area.
5. Remove gasket and shims, Fig. FR2.
6. Install a new gasket and fit shims until shim pack is flush with gasket. Check with a straight edge using light pressure.
7. Assemble shim pack, with the thinnest selected shim first followed by the .093 inch shim and cover.
8. Install one .005 inch shim to preload bearings. The thickest shim (.093 inch) should be next to cover.
9. Apply sealant (ESW-M46-732-A) Permatex No. 3 to bolt threads and torque pinion bearing cover bolts and support yoke cover bolts to 15-20 ft-lbs.

STEERING GEAR REPAIRS

Bellows Seal (Either Side) Replace

Disassembly

1. Place steering gear in a soft jawed vise. Use mounting pads to hold gear.
2. Loosen both bellows retaining clips and remove bellows.
3. Keep gear body below level of bellows to avoid spilling the lubricant. Empty lubricant into suitable container.
4. With gear in a vertical position, move rack back and forth from stop to stop several times to remove remaining lubricant.

Assembly

1. Install steering gear in a soft jawed vise with end where bellows is to be installed on in an upright position.
2. Move rack so that the upper rod is fully extended.
3. Add 5 ounces of D2AZ-19580-B gear lubricant under bellows at inner tie rod joint into gear housing. Move rack back and forth to distribute the lubricant.

NOTE: Do not add more lubricant than specified, as pressurization of the gear could occur and could cause bellows to burst or be forced loose.

4. Using lubricant D2AZ-19580-B, lubricate tie rod in area of bellows contact.
5. Install bellows clips. To avoid interference with components of the vehicle, place bellows clips in the same plane as the pinion shaft and screw ends pointing upward.
6. Install screw end caps.

Input Shaft Seal

1. Clean input shaft and seal area. Use care not to scratch or damage pinion shaft.
2. Pry pinion seal from its bore and discard.
3. Using lubricant C1AZ-1959-B, lubricate seal and install seal over shaft.
4. Use a piece of tubing to engage outer flange of seal and press or tap seal into its bore until flange is flush with shoulder of bore.

NOTE: If outer edge of seal is not engaged at assembly, damage to the seal will result.

Rack Support Yoke, Spring, Gasket, Shims and Cover

Disassembly

1. Clean exterior of steering gear.
2. Place steering gear in a soft jawed vise with yoke cover up.
3. Remove yoke cover, shims, gasket and yoke spring.
4. Discard gasket.

Assembly

1. Clean cover and cover flange areas.
2. Assemble yoke and spring.
3. Install a new gasket. If yoke to rack spring tension adjustment is needed or new shims are to be installed, refer to Support Yoke to Rack.
4. Install shim pack and cover.
5. Apply sealant (ESW-M46-132-A) Permatex No. 3 to cover bolt threads and torque bolts to 15-20 ft-lbs.

Pinion Shaft Cover, Gasket, Shims, Lower Bearing, Shaft and Spacer

Disassembly

1. Clean exterior of steering gear.
2. Place gear in a soft jawed vise.
3. Loosen rack support cover bolts to relieve yoke spring tension on rack.
4. Move rack to either lock and note position of flat on input shaft.

NOTE: It is important that the flat be aligned in this position during assembly. If the pinion is not assembled in the same position, the steering wheel will not be in the straight position.

5. Remove pinion cover, shims and gasket. Discard gasket.
6. Remove pinion lower bearing and pinion shaft.
7. Remove spacer.
8. Inspect all parts for damage and replace as required.

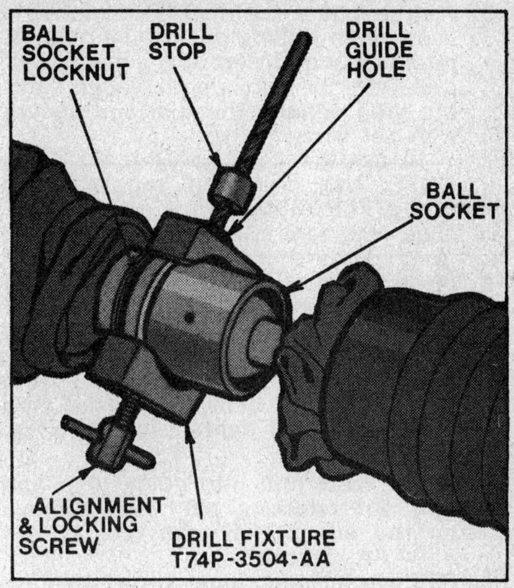

Fig. FR3 Ball socket retaining pin removal

Assembly

1. Clean pinion shaft cover flange. Clean all parts to be assembled in a suitable solvent.
2. Using lubricant D2AZ-19580-B, lubricate spacer, pinion shaft and bearing. Lubricate the inside diameter of the pinion seal with chassis lubricant.
3. Place steering gear in a soft jawed vise with pinion shaft cover flange up.
4. Assemble spacer into place against bearing face.
5. Move rack to one lock and assemble pinion. The flat on shaft must be in same position as noted before disassembly.
6. Assemble bearing, aligning shaft with inside diameter of bearing by rotating shaft and applying side pressure. Tap bearing into its bore until it is recessed below shoulder of bore.

NOTE: If bearing preload needs adjustment or a new shim pack is to be installed, refer to "Pinion Bearing Preload Adjustment" procedure.

7. Assemble shim pack with thinnest shim(s) first and the .093 inch next to the cover.
8. Install new gasket.
9. Apply sealant (ESW-M46-132-A) Permatex No. 3 to cover bolt threads and torque pinion bearing cover bolts and support yoke cover bolts to 15-20 ft-lbs.

Tie Rod Ends, Bellows & Ball Joint Sockets

Disassembly

1. With steering gear assembly secured in holding fixture, tool #T57L-500B, or a suitable vise, loosen tie rod jam nuts on outer ends adjacent to the tie rod ends, then remove the tie rod ends and jam nuts.
2. Remove bellows clamps and drain lubricant, then remove bellows.

3. A drill fixture tool #T74P-3504-AA, is used when drilling out the ball socket retaining pin. Insert a 5/32 inch twist drill into the tool depth gauge, butt the stop against the tool and secure with set screw.

NOTE: Use drill stop supplied with tool #T74P-3504-Z since none is supplied with tool #T74P-3504-AA.

4. Place protective cloths on each side of ball socket and install tool #T74P-3504-AA on ball socket with guide hole aligned with socket retaining pin. Place cone points of alignment and locking screws between locknut and ball socket and tighten securely, Fig. FR3.
5. Insert twist drill into guide hole and drill out retaining pin, then remove tie rod and socket with tool #T74P-3504-AA attached.
6. Remove locknut, inner thrust bearing and rack spring from recess in rack end.

Assembly.

1. Install new rack spring and assemble ball socket to tie rod, then install new inner thrust bearing in ball socket and lubricate assembly with lubricant, P/N D2AZ-19580-B.
2. Install new ball joint locknut on rack end and thread ball socket onto rack until tie rod movement becomes stiff and ball socket is tight.
3. Install ball joint and jam nut on tie rod end and using a spring scale, measure force required to move the tie rod in the socket which should be 4-6 pounds. Adjust ball housing to obtain specified reading.
4. Hold ball housing with tool #T74P-3504-CC and torque locknut against ball socket with a suitable tool to 35-40 foot lbs.
5. Remove tools and install drill fixture, too #T74P-3504-AA, on ball socket. Using procedure found in steps 3 through 5 under "Disassembly", drill new retaining pin hole on line of contact of locknut and ball socket. The new hole must not enter notches of locknut or old ball socket hole.
6. Insert new retaining pin into hole and tap until flush with ball housing and locknut surfaces. Stake metal of ball housing and locknut over pin to secure tie rod adjustment. Apply a reverse torque of 35 ft. lbs. on locknut to test proper installation of retaining pin. The joint should not loosen under this load.
7. Install bellows, breather tube and new bellows to housing clamps.
8. Install 7 ounces of lubricant, P/N D2AZ-19580-B, into bellows and secure bellows to tie rod with new clamps.

Saginaw Rack & Pinion

DESCRIPTION

This steering gear assembly, Fig. S1, incorporates an integral tube and housing, containing a pinion shaft and steering rack. The tube and housing are joined by a plastic injection-bonding process. The pinion shaft is supported in the housing by two thrust bearings and nylon bushings. The steering rack is supported at each end of the housing by nylon bushings. A preload spring maintains pinion bushing location and also, compensates for bushing wear.

The steering linkage incorporates two inner tie rods, two adjuster tube assemblies and two tie rod ends. The inner tie rods are secured to the steering rack by the tie rod housings. The adjuster tubes secure the inner tie rods to the tie rod ends, in turn connected to the steering arms.

Rubber boots protect the inner tie rods from damage. A breather tube, connected to each rubber tube, equalizes pressure within the boots during steering operations.

DISASSEMBLY

1. Clamp steering gear housing in a soft-jawed vise.
2. Cut boot clamp from steering housing end, then slide boot to expose rack teeth.
3. Rotate flexible coupling, moving rack toward housing end of gear assembly, to further expose rack teeth.
4. Inspect rack teeth. If teeth are damaged, replace steering gear assembly.

NOTE: If rack teeth have machining marks or appear excessively bright and shiny, steering gear replacement is not necessary since these are normal conditions.

5. Remove flexible coupling pinch bolt, then the flexible coupling from pinion shaft.
6. Remove adjusted plug locknut and adjuster plug, then lift pinion shaft, rotate counter-clock-wise and remove from housing, Fig. S2.
7. Inspect pinion shaft teeth. If teeth are damaged, replace steering gear assembly.
8. Remove contraction plug, lower pinion bushing and preload spring from housing.
9. Center rack in the tube and install pinion shaft and adjuster plug into housing.
10. Mark position of adjuster tubes on tie rods, loosen adjuster tube clamps and remove adjuster tubes and tie rod ends.
11. Mark position of breather tube on the steering gear assembly since boots and breather tube must be installed in original position to ensure proper sealing. Cut all boot clamps from boots, then remove boots.
12. Remove breather tube from mounting grommet hole, then the tube from gear assembly without mispositioning mounting clamp or grommet. If grommet replacement is necessary, rotate and pull grommet from housing.

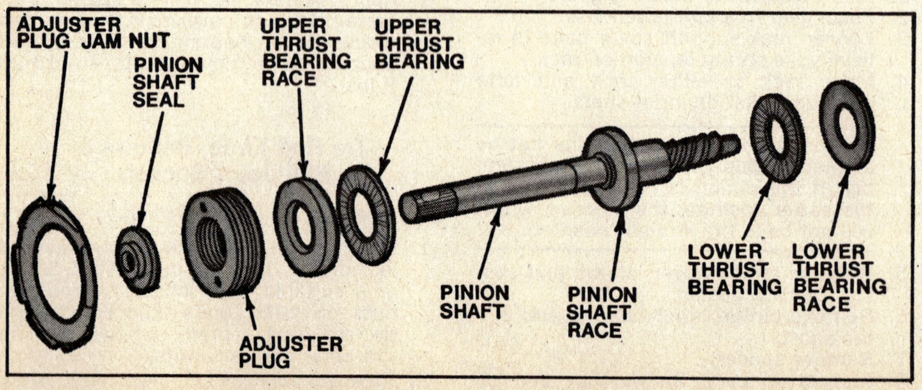

Fig. S2 Pinion shaft & adjuster plug assembly

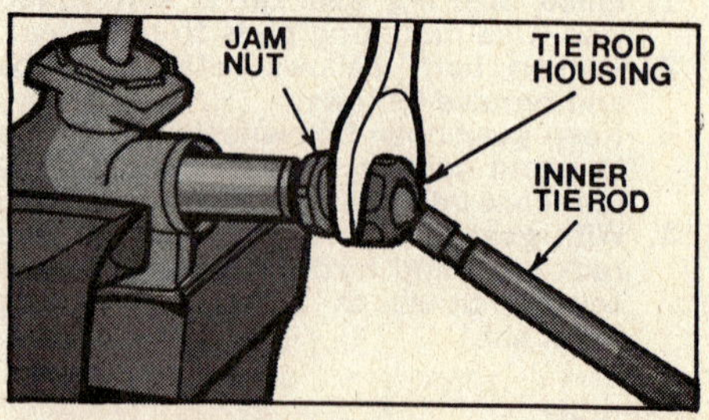

Fig. S3 Removing or installing inner tie rod

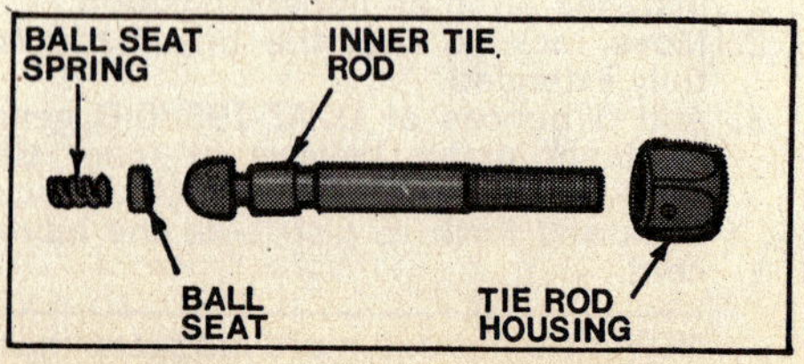

Fig. S4 Inner tie rod assembly

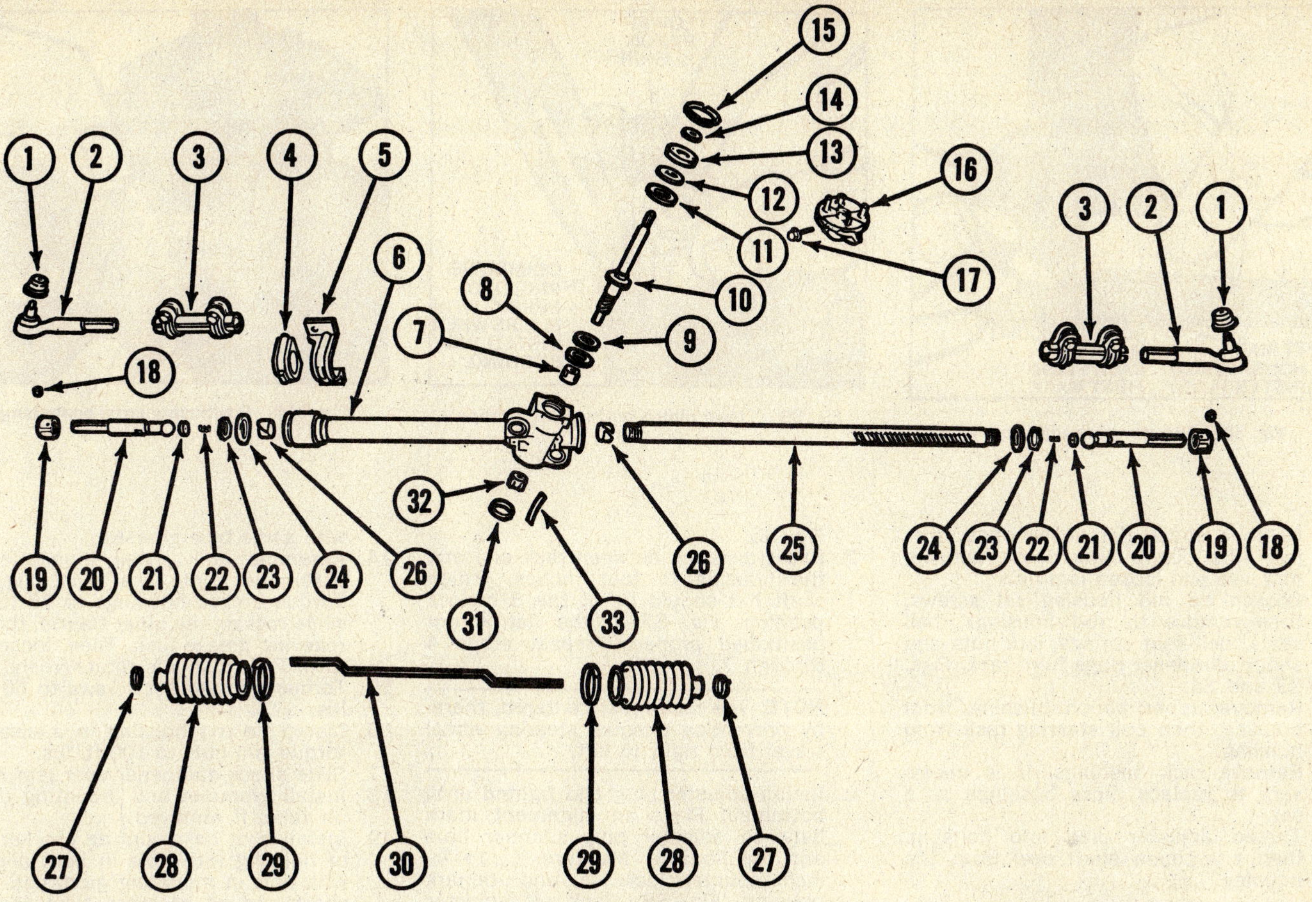

1. TIE ROD SEAL
2. TIE ROD END
3. ADJUSTER TUBE
4. MOUNTING GROMMET
5. MOUNTING CLAMP
6. TUBE AND HOUSING ASSEMBLY
7. UPPER PINION BUSHING
8. LOWER THRUST BEARING RACE
9. LOWER THRUST BEARING
10. PINION SHAFT
11. UPPER THRUST BEARING
12. UPPER THRUST BEARING RACE
13. ADJUSTER PLUG
14. PINION SHAFT SEAL
15. ADJUSTER PLUG LOCKNUT
16. FLEXIBLE COUPLING
17. PINCH BOLT
18. SET SCREW
19. TIE ROD HOUSING
20. INNER TIE ROD
21. BALL SEAT
22. BALL SEAT SPRING
23. JAM NUT
24. SHOCK DAMPENER RING
25. STEERING RACK
26. RACK BUSHING
27. BOOT RETAINER
28. BOOT
29. BOOT CLAMP
30. BREATHER TUBE
31. CONTRACTION PLUG
32. LOWER PINION BUSHING
33. PRELOAD SPRING

Fig. S1 Saginaw rack & pinion steering gear, disassembled

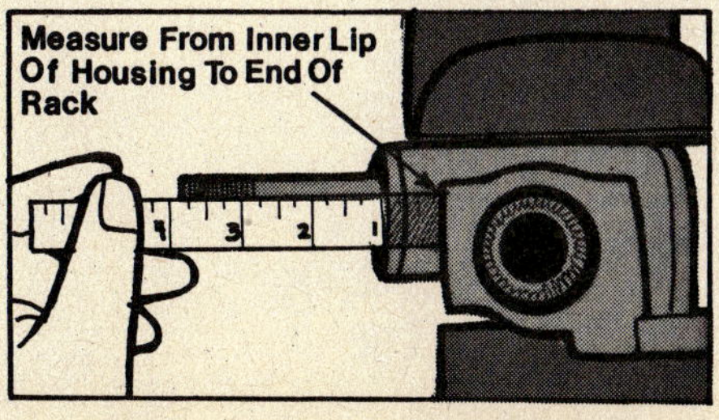

Fig. S5 Centering steering rack

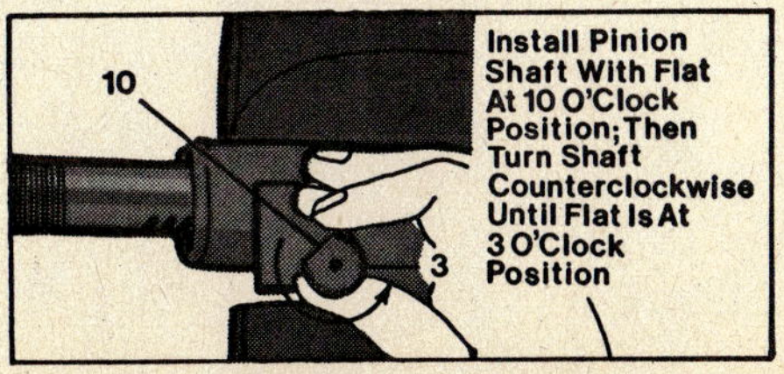

Fig. S6 Pinion shaft installation

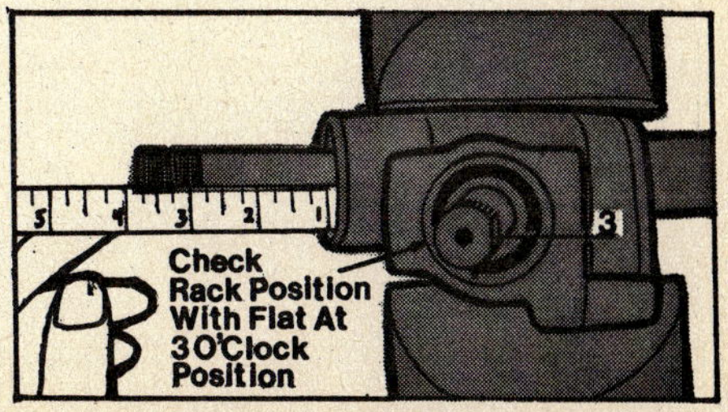

Fig. S7 Checking rack position

Fig. S8 Adjuster plug adjustment

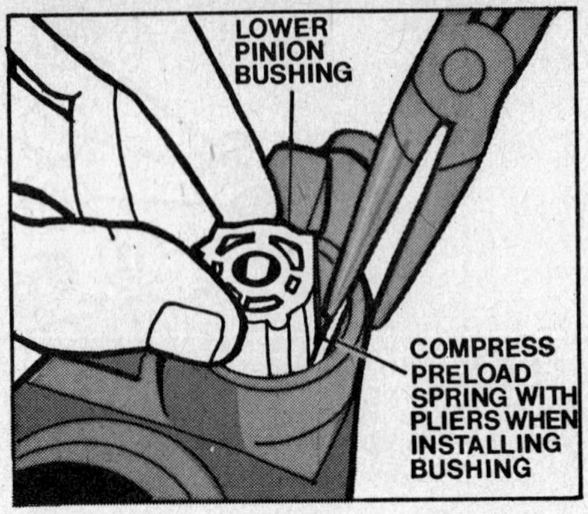

Fig. S9 Lower pinion bushing installation

Fig. S10 Positioning large boot clamp

13. Remove shock dampener rings from jam nuts, clamp inner tie rod housing in a vise and loosen jam nuts.
14. Loosen tie rod housing set screws, then remove tie rod housings, ball seats, bell seat springs, jam nuts and shock dampener rings from rack, Figs. S3 and S4.
15. Remove upper pinion bushing from housing, then pull steering rack from housing.
16. Remove rack bushings. It is necessary to replace these bushings as a set.
17. Thread adjuster plug into housing, then pry pinion shaft seal from the housing.

ASSEMBLY

1. Press new pinion shaft seal onto adjuster plug until flush with plug face.
2. Install rack bushings into housing.
3. Lubricate rack teeth with a suitable lithium based grease, then insert rack into housing. Install upper pinion bushing in housing.
4. Lubricate pinion shaft lower race and thrust bearing with a suitable lithium based grease, then install race and bearing into housing with race flanged edge facing upward.
5. Center steering rack in housing. Distance from rack end to housing inner lip should be four inches, Fig. S5.
6. Start pinion shaft into housing with flat on splined end of shaft at the 10 o'clock position. Rotate pinion shaft counter-clockwise and push downward until race bottoms on thrust bearing, Fig. S6.
7. Reset distance between rack end and the housing to four inches. Pinion shaft flat should be at the 3 o'clock position, Fig. S7. If the flat is not positioned properly, repeat steps 4 through 7.

NOTE: The rack must centered, thereby preventing unequal steering wheel travel from right to left.

8. Install adjuster plug and tighten until bottomed. Place an alignment mark between adjuster plug spanner hole and the housing. Measure 3/16 to 1/4 inch counter-clockwise and remark housing, Fig. S8. Back off adjuster plug to second housing mark. Install and torque adjuster plug lock nut to 50 ft. lbs.
9. Fill cavity around pinion shaft with a suitable lithium based grease. Do not overfill since pinion bushing and preload spring must be installed.
10. Install preload spring with the center hump bearing against the housing, allowing the spring to extend approximately 1/4 inch from housing end. Using needlenose pliers, hold top of preload spring against housing and install bushing into housing with chamfered end facing downward, Fig. S9.
11. Seat contraction plug in housing.
12. Install shock dampener rings on ends of steering rack with open ends facing outward, then install the jam nuts.
13. Lubricate all inner tie rod wear surfaces with a suitable lithium based grease. Also, pack tie rod housings with same type grease.
14. Assemble and install inner tie rod assemblies onto the rack, Fig. S4. Torque tie rod housings to 25 ft. lbs. while rocking the inner tie rod, thereby relieving grease lock. Then, loosen 1/8 turn. Tie rod should rotate freely.
15. Torque housing set screws to 60 inch lbs.
16. Clamp tie rod housing in a vise and torque jam nuts to 100 ft. lbs.
17. Slide shock dampener over jam nuts.
18. Install grommet and mounting clamp on tube, if removed.
19. Install boot on mounting bracket side of housing with hole in boot aligning with hole in mounting grommet. Slide short end of breather tube through the grommet and boot holes. The long end of tube should lie against the housing.
20. Install opposite side boot with hole in boot aligned with breather tube.
21. Slide outer collars of boots over inner tie rod undercuts. Install small boot clamps and compress with tool J-22610.
22. Install adjuster tubes and tie rod ends using alignment marks made during disassembly. A minimum of three threads must be visable on each end of adjuster tubes. Also, the number of threads per side should not differ by more than three threads.
23. Install flexible joint on pinion shaft and torque pinch bolt to 30 ft. lbs.
24. Install large boot clamps and compress with tool J-22610. Position clamp ear approximately 3/4 inch from breather tube before compressing clamp, Fig. S10.

THREE SPEED MANUAL SHIFT TRANSMISSIONS

See Car Chapters for procedures on removing the transmission and adjusting the gearshift linkage

APPLICATION INDEX

AMERICAN MOTORS

	Type No.	Page
1969-72 Heavy Duty	7	2-212
1969-75 Exc. Heavy Duty	1	2-209
1974-75 Heavy Duty	6	2-209

BUICK—Full Size

	Type No.	Page
1969-71 Heavy Duty	6	2-209
1969-71 Exc. Heavy Duty	8	2-214

BUICK—Intermediates

	Type No.	Page
1969 Floor Shift	6	2-209
1969-75 Exc. Heavy Duty	8	2-214
1970 Heavy Duty	2	2-201

CAMARO

	Type No.	Page
1969-75 Exc. Heavy Duty	8	2-214
1969-74 Heavy Duty	2	2-201

CHEVELLE & MONTE CARLO

	Type No.	Page
1969-75 Exc. Heavy Duty	8	2-214
1969-75 Heavy Duty	2	2-201

CHEVROLET

	Type No.	Page
1969-73 Exc. Heavy Duty	8	2-214
1969-73 Heavy Duty	2	2-201

CHEVROLET MONZA

	Type No.	Page
1975	8	2-214

CHEVROLET VEGA & PONTIAC ASTRE

	Type No.	Page
1971-72	9	2-216
1973-75	8	2-214

CHEVY NOVA

	Type No.	Page
1969-75 Exc. Heavy Duty	8	2-214
1969-75 Heavy Duty	2	2-201

CHRYSLER

	Type No.	Page
1969	4	2-206
1970-71	3	2-204

COMET & MONTEGO

	Type No.	Page
1969-75	6	2-209

CORVETTE

	Type No.	Page
1969 Exc. V8-396	8	2-214
1969 Heavy Duty	2	2-201

COUGAR

	Type No.	Page
1969-72	6	2-209

DODGE DART

	Type No.	Page
1969-74 Exc. Heavy Duty	5	2-206
1970-75 Heavy Duty	3	2-204
1975 Exc. Heavy Duty	6	2-209

DODGE

	Type No.	Page
1969-74 Exc. Heavy Duty	5	2-206
1969	4	2-206
1970-75 Heavy Duty	3	2-204

FAIRLANE & TORINO

	Type No.	Page
1969-74	6	2-209

FALCON & MAVERICK

	Type No.	Page
1969-75	6	2-209

FORD

	Type No.	Page
1969-72	6	2-209

GRANADA & MONARCH

	Type No.	Page
1975	6	2-209

MERCURY

	Type No.	Page
1969-71	6	2-209

MUSTANG

	Type No.	Page
1969-73	6	2-209

OLDSMOBILE—Full Size

	Type No.	Page
1969-71	6	2-209

OLDS—Intermediates

	Type No.	Page
1969 Floor Shift	6	2-209
1969-75	8	2-214
1970-72	2	2-201

PLYMOUTH

	Type No.	Page
1969-74 Exc. Heavy Duty	6	2-206
1969 Heavy Duty Unit	4	2-206
1970-75 Heavy Duty	3	2-204

PONTIAC—Full Size

	Type No.	Page
1969-71	6	2-209
1971 Grand Prix	2	2-201

PONTIAC—Intermediates

	Type No.	Page
1969-75 Exc. Heavy Duty	8	2-214
1969 Heavy Duty	6	2-209
1970-74 Heavy Duty	2	2-201

VALIANT

	Type No.	Page
1969-74 Exc. Heavy Duty	5	2-206
1970-75 Heavy Duty	3	2-204
1975 Exc. Heavy Duty	6	2-209

Type One

DISASSEMBLE TRANS.

1. Remove transmission cover and shift levers.
2. Remove front retainer and gasket and front bearing snap rings.
3. Align notch in clutch shaft with 3rd speed gear and use suitable puller to remove clutch shaft using care not to lose rollers.
4. Remove front bearing with puller.
5. Remove extension case and remove snap rings that retain speedometer drive gear.
6. Remove speedometer gear using care not to lose drive ball.
7. Remove rear bearing snap rings and remove rear bearing with a puller.
8. Move mainshaft to side and remove shift forks.
9. Place front synchronizer in 2nd speed position and remove mainshaft by tilting front of shaft up and lifting through top of case, Fig. 3.
10. If equipped, remove transmission controlled spark switch assembly.
11. Use a punch to remove roll pins from shift shafts and push shafts into case. Detent assembly may now be removed from case.
12. Using a brass drift drive reverse idler gear shaft out of rear of case and remove idler gear. Do not lose rollers.
13. To retain rollers in countershaft gear, use a dummy shaft to drive countershaft out of rear of case.
14. After disassembling mainshaft carefully inspect all bearings and gear.

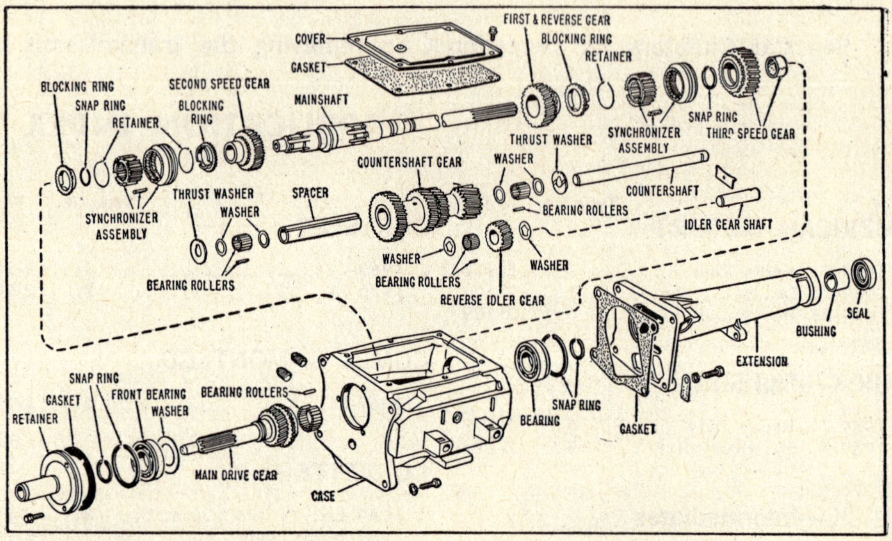

Fig. 1 Type 1 fully synchronized transmission exploded

Fig. 3 Removing mainshaft

REASSEMBLE TRANS.

Mainshaft Assemble

1. Place low speed gear and friction ring on shaft with friction ring hub facing toward rear of shaft.
2. Insert low speed synchro-gear into synchro-collar with deep end of gear facing the low speed gear.

NOTE: Synchro-plates and retainer ring are installed with large end of plates facing toward groove in synchro-hub.

3. Place synchro-clutch assembly on shaft with groove in synchro-collar facing toward low speed gear and

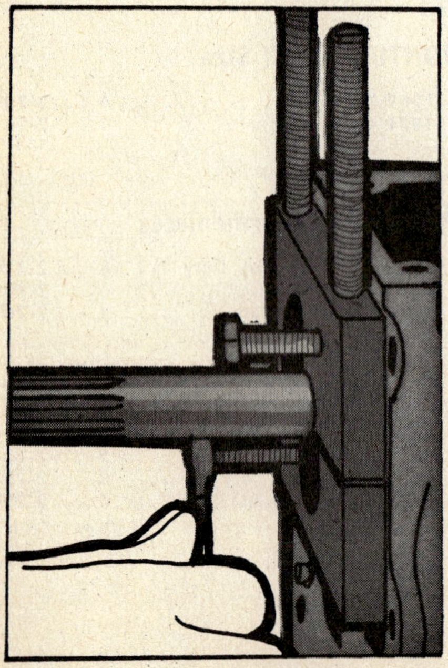

Fig. 2 Removing clutch shaft with puller

Fig. 4 Installing mainshaft with Pilot End Support J-22994

install low speed synchro-clutch snap ring.

NOTE: Snap rings are available in selective thicknesses. Clearance between first gear and collar on mainshaft must be .003-.012 inch.

4. Place second speed gear and friction ring on front of shaft with gear hub and friction ring facing toward front of shaft.
5. Install second speed synchro-gear into synchro-collar with deep end of gear facing toward rear of shaft.
6. Position synchro-clutch assembly with one synchro-plate in 12 o'clock position. Place either tang of retainer ring on synchro-plate at 12 o'clock position and install ring, feeding it in a clockwise direction. Install retainer ring on opposite side of assembly in same manner.
7. Place second speed synchro assembly on shaft with deep end facing toward rear of shaft and install second gear snap ring.

NOTE: Snap rings are available in selective thicknesses. Clearance between second gear and collar must be .003-.018 inch.

8. Install reverse gear onto mainshaft.

Countergear

1. Coat bore at each end of counter gear with grease to hold rollers in place.
2. Install dummy shaft in countergear and install spacer, washers and rollers.
3. Place countergear in transmission case and position thrust washers at each end so tabs align with slots in case.
4. Use a plastic mallet to install countershaft.

Reverse Idler Gear

1. Coat bore of idler gear with grease to retain rollers.
2. Install idler gear in case and position thrust washers.
3. Use plastic mallet to install idler gear shaft.

Shifter Shafts

1. Partially install shifter shafts in transmission case.
2. Align detent assembly with shifter shafts and case stud.
3. Push shift detent assembly and shift shafts into place and install roll pins.
4. If equipped, install transmission controlled spark switch.

Mainshaft Installation

1. Place front synchronizer in 2nd speed position and place mainshaft in case.
2. Move mainshaft to side and install shift forks by pulling detent lever up and placing forks in the shifting assembly.
3. Position mainshaft assembly in center of case and install Pilot End Support J-22994, Fig. 4.
4. Place rear bearing on mainshaft and drive bearing into position and install snap ring.
5. Install speedometer gear and snap ring.

Final Assembly

1. Install rollers in clutch shaft using grease to retain them.
2. Slide clutch shaft into position through front of case.
3. Install front bearing, snap rings, gasket and retainer.
4. Replace seal if necessary and install extension housing, shift levers and case cover.
5. Fill transmission with lubricant and check operation.

Type Two

DISASSEMBLE TRANS.

Case Components

1. Remove side cover assembly, gasket and shift forks.
2. Remove clutch gear retainer and gasket.
3. Remove clutch gear bearing to stem snap ring, then slide bearing off over clutch gear stem. The clutch gear bearing is a slip fit on the gear and into the case bore.
4. Remove rear extension to case bolts.
5. Rotate extension to left until groove in extension housing flange lines up with the reverse idler shaft. Drive reverse idler shaft out of gear and case, Fig. 2.
6. Remove entire clutch gear, mainshaft and extension assembly through case rear opening. Remove reverse idler gear from case.

Mainshaft Disassembly

1. Remove clutch gear from mainshaft, Fig. 3.
2. Expand extension snap ring and tap end of mainshaft to remove extension, Fig. 4.
3. Depress speedometer gear retaining clip and slide gear from mainshaft.
4. Remove rear bearing snap ring, support Reverse gear and press on rear of mainshaft to remove Reverse gear, thrust washer and rear bearing.
5. Remove 1st and Reverse sliding clutch hub snap ring from mainshaft.
6. Support 1st gear and press on rear of mainshaft to remove the clutch assembly, blocker ring and 1st gear.
7. Remove 2nd and 3rd speed sliding clutch hub snap ring.
8. Support 2nd gear and press on front of mainshaft to remove clutch assembly, 2nd speed blocker ring and 2nd speed gear.

Synchronizer Clutch Keys & Springs

NOTE: Clutch hubs and sleeves are a selected assembly and should be kept together as originally assembled, but the keys and springs may be replaced separately.

Before disassembling synchronizers, mark hub and sleeve so they can be matched upon reassembly.

Place the keys and springs in position so all three keys are engaged by both springs. The tanged end of each spring should be installed into different key cavities on either side. Slide sleeve onto hub, aligning marks made before disassembly.

NOTE: A groove around the outside of the hub identifies the end that must be opposite the fork slot in the sleeve.

Mainshaft Reassemble

With front of mainshaft up:

1. Install 2nd speed gear with clutching teeth upward; the rear face of the gear will butt against the shoulder on the shaft.
2. Install blocker ring with teeth downward over gear. All three blocker rings used in this unit are identical.
3. Install 2nd and 3rd synchronizer assembly with fork slot downward and press onto splines until it bottoms. Install snap ring.

NOTE: Be sure the notches of the blocker ring align with the keys of the synchronizer assembly.

With rear of mainshaft up:

4. Install 1st speed gear with clutching teeth upward. Install a blocker ring with teeth downward over synchronizer surface of the gear.
5. Install 1st and Reverse synchronizer assembly with fork slot up.

NOTE: Be sure the notches of the blocker ring align with the keys of the synchronizer assembly and that both synchronizer sleeves face the front of the mainshaft.

6. Install snap ring.
7. Install Reverse gear with clutching teeth downward. Install steel thrust washer.
8. Press rear bearing onto mainshaft with snap ring slot downward. Install snap ring.
9. Install speedometer drive gear and retaining clip.

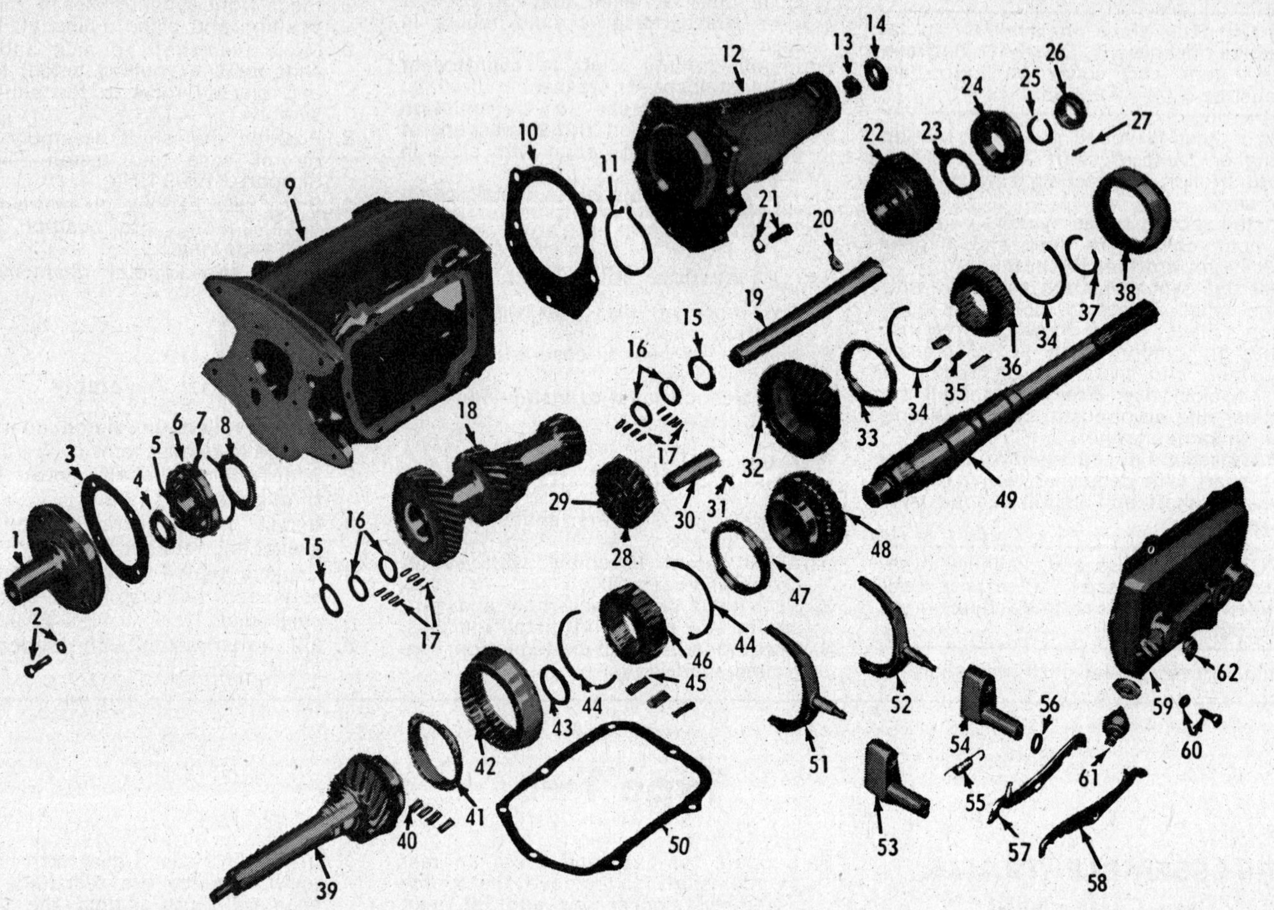

1. Bearing Retainer
2. Bolt and Lock Washer
3. Gasket
4. Oil Seal
5. Snap Ring (Bearing-to-Main Drive Gear)
6. Main Drive Gear Bearing
7. Snap Ring Bearing
8. Oil Slinger
9. Case
10. Gasket
11. Snap Ring (Rear Bearing-to-Extension)
12. Extension
13. Extension Bushing
14. Oil Seal
15. Thrust Washer
16. Bearing Washer
17. Needle Bearings
18. Countergear
19. Countershaft
20. Woodruff Key
21. Bolt (Extension-to-Case)
22. Reverse Gear
23. Thrust Washer
24. Rear Bearing
25. Snap Ring
26. Speedometer Drive Gear
27. Retainer Clip
28. Reverse Idler Gear
29. Reverse Idler Bushing
30. Reverse Idler Shaft
31. Woodruff Key
32. 1st Speed Gear
33. 1st Speed Blocker Ring
34. Synchronizer Key Spring
35. Synchronizer Keys
36. 1st and Reverse Synchronizer Hub Assembly
37. Snap Ring
38. 1st and Reverse Synchronizer Collar
39. Main Drive Gear
40. Pilot Bearings
41. 3rd Speed Blocker Ring
42. 2nd and 3rd Synchronizer Collar
43. Snap Ring
44. Synchronizer Key Spring
45. Synchronizer Keys
46. 2nd and 3rd Synchronizer Hub
47. 2nd Speed Blocker Ring
48. 2nd Speed Gear
49. Mainshaft
50. Gasket
51. 2nd and 3rd Shifter Fork
52. 1st and Reverse Shifter Fork
53. 2-3 Shifter Shaft Assembly
54. 1st and Reverse Shifter Shaft Assembly
55. Spring
56. O-Ring Seal
57. 1st and Reverse Detent Cam
58. 2nd and 3rd Detent Cam
59. Side Cover
60. Bolt and Lock Washer
61. TCS Switch and Gasket
62. Lip Seal

Fig. 1 Type two fully synchronized transmission. Exploded view

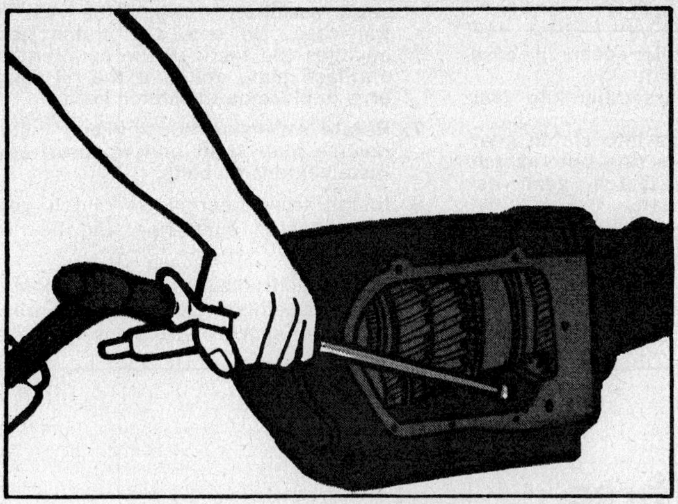

Fig. 2 Removing reverse idler shaft

Fig. 4 Removing extension housing snap ring

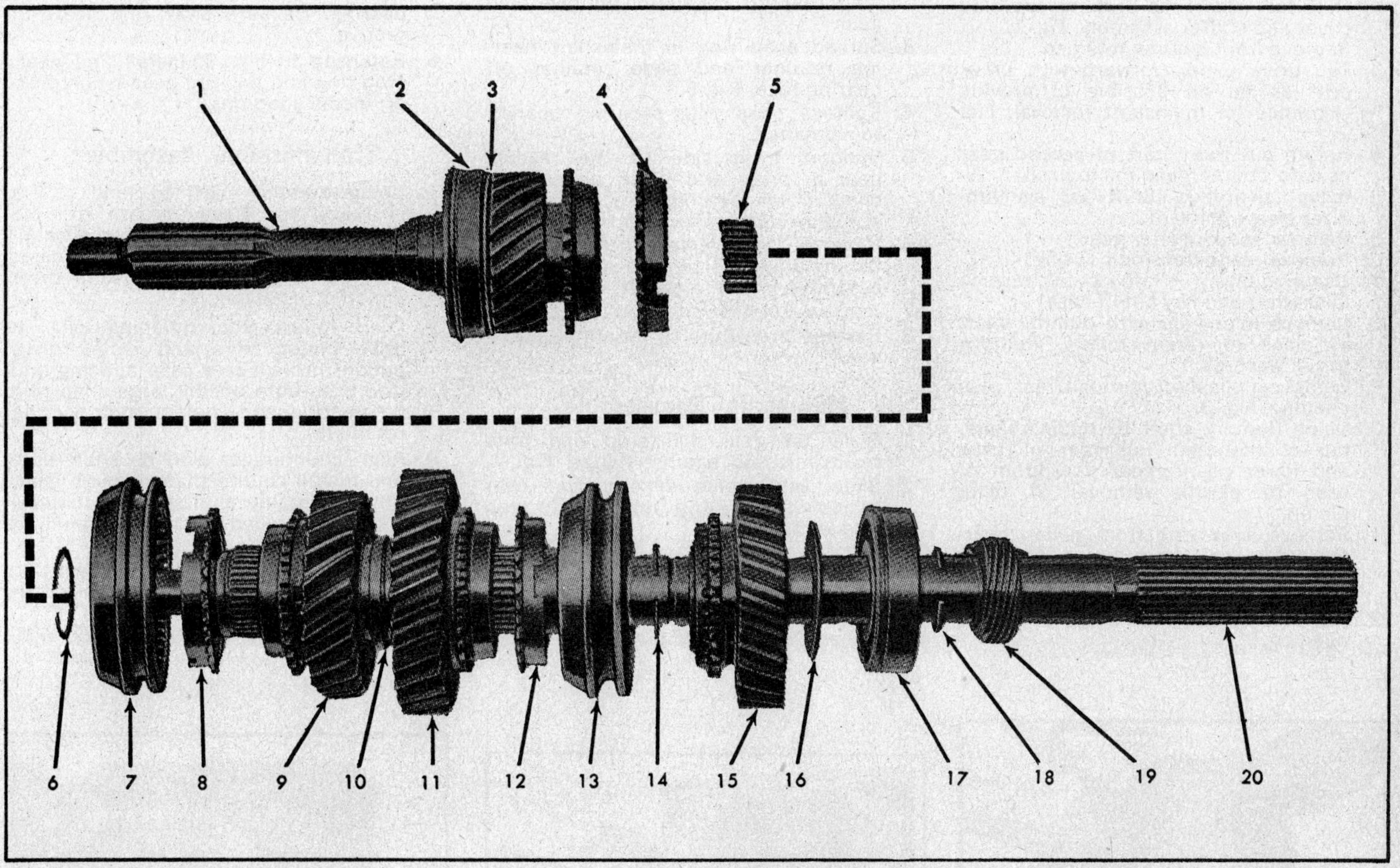

1. Clutch Gear
2. Clutch Gear Bearing
3. Oil Slinger
4. 3rd Speed Blocker Ring
5. Mainshaft Pilot Bearings (16)
6. Snap Ring
7. 2-3 Synchronizer Assembly
8. 2nd Speed Blocker Ring
9. 2nd Speed Gear
10. Shoulder (Part of Mainshaft)
11. 1st Speed Gear
12. 1st Speed Blocker Ring
13. 1st Speed Synchronizer Assembly
14. Snap Ring
15. Reverse Gear
16. Reverse Gear Thrust Washer
17. Rear Bearing
18. Snap Ring
19. Speedo Drive Gear
20. Mainshaft

Fig. 3 Mainshaft assembly

REASSEMBLE TRANS.

1. Load a double row of roller bearings and a thrust washer at each end of the countergear. Use heavy grease to hold them in place.
2. Place countergear through case rear opening with tanged thrust washer (tang away from gear) at each end and install countergear shaft and woodruff key from rear of case.

NOTE: Be sure countershaft picks up both thrust washers and that the tangs are aligned with their notches in the case.

3. Position reverse idler gear in case but do not install shaft.
4. Install mainshaft assembly into rear extension housing.
5. Load roller bearings into clutch gear bore, install blocker ring onto clutch gear and install clutch gear assembly onto mainshaft.

NOTE: Be sure notches in blocker ring align with keys in synchronizer assembly.

6. Using new gasket, install the mainshaft assembly through the rear of the case. Be sure the clutch gear engages the teeth of the countergear anti-lash plate and that the oil slinger is in place on the clutch gear.
7. Rotate extension housing and install reverse idler shaft and woodruff key. Install extension bolts.
8. Install front bearing on clutch gear stem, install snap ring and bearing retainer.
9. With synchronizers in Neutral, install cover assembly. Be sure forks align with their synchronizer sleeve grooves.

Type Three

DISASSEMBLE TRANS.

Case Components

1. Shift transmission into 2nd gear for shift fork clearance and remove side cover and shifter assembly, Fig. 1.
2. Remove front bearing retainer.
3. Tap drive pinion forward with brass drift as far as possible to provide clearance for mainshaft removal, Fig. 2.
4. Rotate cut away part of second gear next to countergear for mainshaft removal clearance. Shift 2-3 synchronizer sleeve forward.
5. Remove speedometer gear.
6. Remove rear extension housing.
7. Using dummy shaft, push reverse idler shaft and key out of case.
8. Remove idler gear with dummy shaft in place to retain rollers. Remove thrust washers.
9. Remove mainshaft through rear case opening, Fig. 3.
10. Using dummy shaft to retain rollers, tap countershaft out rear of case and lower countergear to bottom of case to permit removal of drive pinion.
11. Remove snap ring from pinion bearing outer race, drive pinion into case and remove through rear case opening, Fig. 4.
12. Remove countergear through rear case opening.

Mainshaft, Disassemble

1. Remove 2-3 synchronizer clutch gear retaining ring from front of mainshaft.
2. Slide 2-3 synchronizer and 2nd gear stop ring off of shaft. Remove 2nd gear.
3. Spread snap ring in mainshaft bearing retainer and slide retainer off bearing race, Fig. 5.
4. Remove snap ring securing bearing to mainshaft.
5. Support front side of the reverse gear in press and press bearing off shaft. *When bearing clears shaft, do not allow parts to drop through.*
6. Remove 1st-Reverse synchronizer retaining ring and remove synchronizer assembly from mainshaft, Fig. 6.
7. Remove 1st gear and stop ring.
8. Reverse procedure to assemble mainshaft.

Mainshaft, Reassemble

1. Slide 1st gear and stop ring onto mainshaft and against flange, Fig. 7.
2. Slide 1st-Reverse synchronizer over mainshaft, indexing hub slots to stop ring lugs.
3. Install clutch gear snap ring on mainshaft.
4. Install reverse gear and mainshaft bearing, support inner race of bearing and press shaft through to shoulder. *Be sure snap ring groove on outer race is forward.* Install bearing retaining ring on mainshaft.
5. Spread snap ring in mainshaft bearing retainer and slide retainer over bearing. Be sure snap ring seats in groove.
6. Referring to Fig. 1, install 2nd gear, stop ring and 2nd-3rd gear synchronizer. Install snap ring.

Transmission, Assemble

1. Using dummy shaft to hold rollers in place and heavy grease to hold thrust washers, carefully place countergear assembly in bottom of case. *Do not finish installation until drive pinion is installed.*
2. Load rollers and retaining ring in drive pinion bore and install drive pinion through rear case opening and into case bore. Install large snap ring on bearing and install front bearing retainer.
3. Align countergear with its shaft bore and install countershaft through gear, driving dummy shaft out as countershaft is installed. Install countershaft key.
4. Carefully tap drive pinion forward to provide mainshaft installation clearance.
5. With 2-3 synchronizer sleeve fully

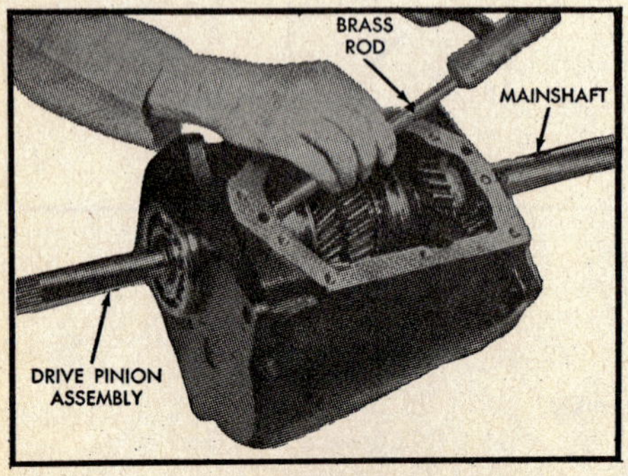

Fig. 2 Tapping drive pinion forward for mainshaft clearance

Fig. 3 Removing mainshaft assembly

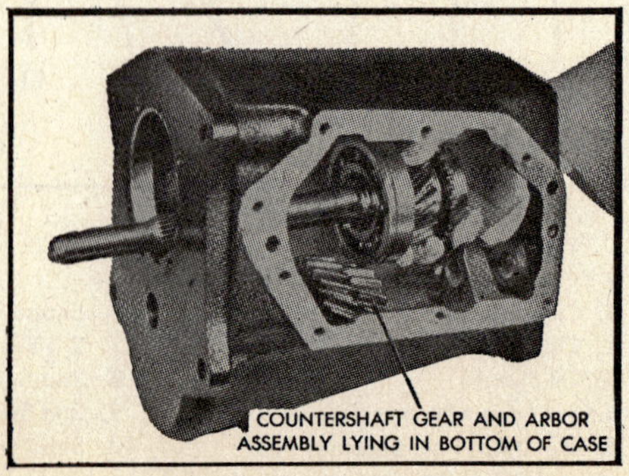

Fig. 4 Removing drive pinion

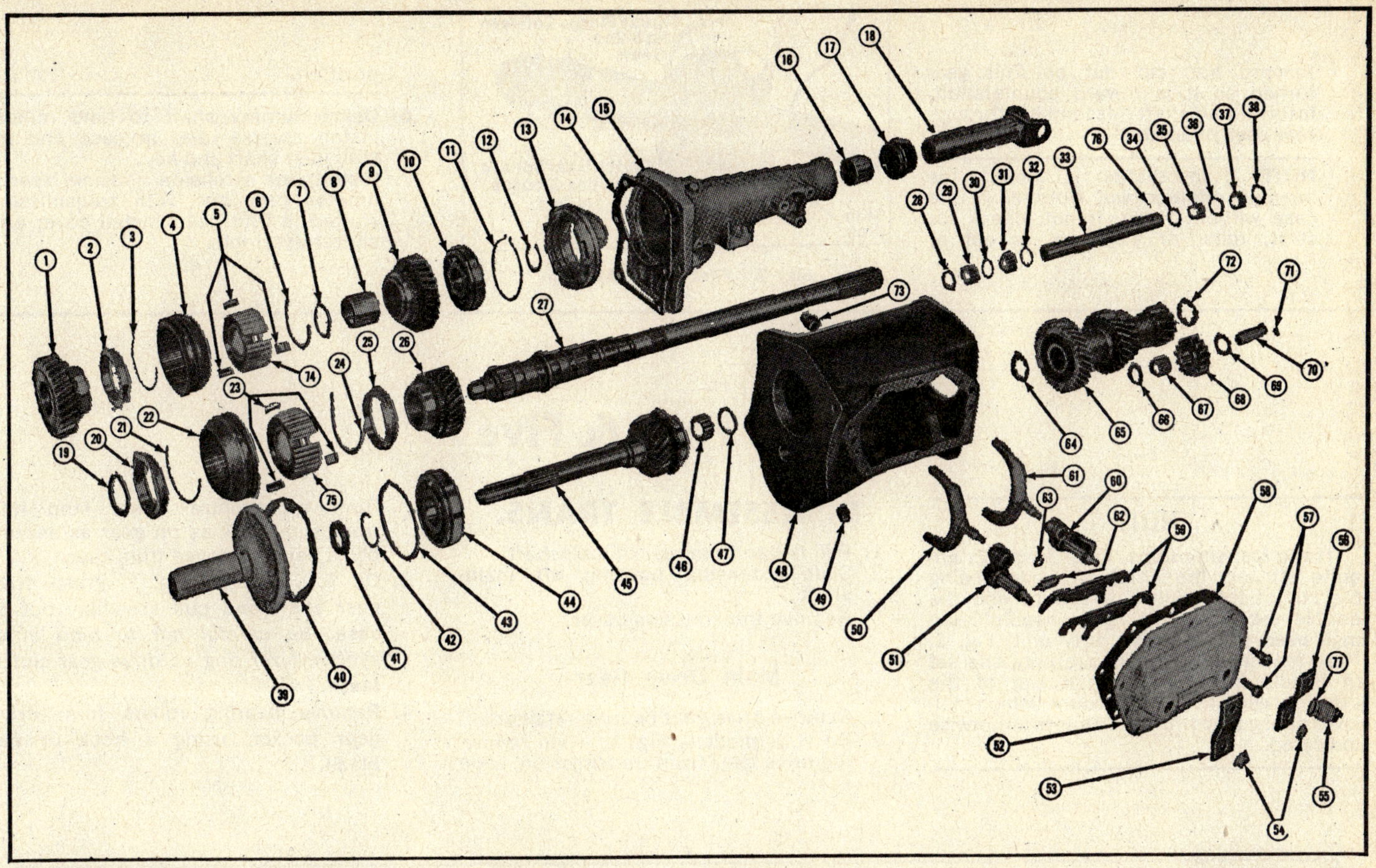

1. Gear, First	14. Gasket	27. Shaft, Output	40. Gasket	53. Lever	66. Washer
2. Ring	15. Extension	28. Washer	41. Seal	54. Nut Locking	67. Roller
3. Spring	16. Bushing	29. Roller	42. Snap Ring	55. Switch	68. Gear, Idler
4. Sleeve	17. Seal	30. Washer	43. Snap Ring	56. Lever	69. Washer
5. Struts (3)	18. Yoke	31. Roller	44. Bearing	57. Bolt	70. Shaft
6. Spring	19. Snap Ring	32. Washer	45. Pinion, Drive	58. Gasket	71. Key
7. Snap Ring	20. Ring	33. Countershaft	46. Roller	59. Lever, Interlock	72. Washer
8. Bushing	21. Spring	34. Washer	47. Snap Ring	60. Lever	73. Plug, Filler
9. Gear, Reverse	22. Sleeve	35. Roller	48. Case	61. Fork	74. Gear, Clutch
10. Bearing	23. Struts (3)	36. Washer	49. Plug, Drain	62. Spring	75. Gear, Clutch
11. Snap Ring	24. Spring	37. Roller	50. Fork	63. Snap Ring	76. Key
12. Snap Ring	25. Ring	38. Washer	51. Lever	64. Washer	77. Gasket
13. Retainer	26. Gear, Second	39. Retainer	52. Housing	65. Gear, Countershaft	

Fig. 1. Type 3 fully synchronized three speed transmission

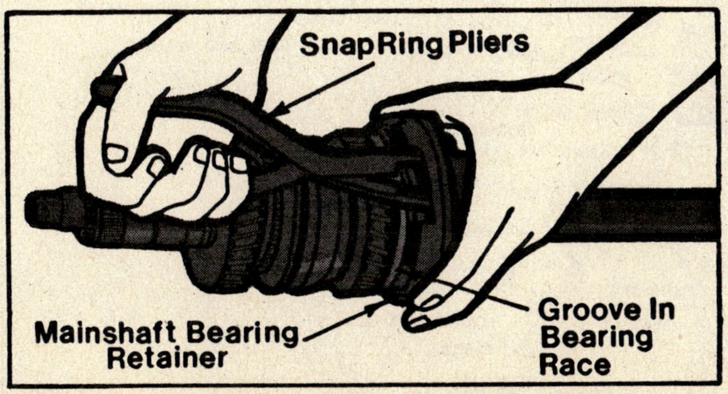

Fig. 5 Removing mainshaft bearing retainer

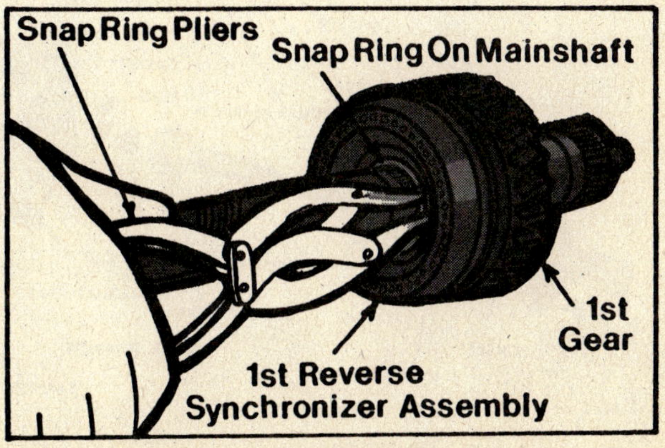

Fig. 6 Removing 1st-Reverse synchronizer snap ring

forward and cut out on 2nd gear turned so it is toward countershaft, insert mainshaft assembly through rear case opening.

NOTE: If installation is correct, the bearing retainer will bottom in the case without force. If not, check for strut, roller or stop ring is out of position.

6. Using dummy shaft to hold rollers, install reverse idler in case and install idler shaft and key.
7. Install rear extension housing, speedometer gear and with transmission shifted in 2nd gear, install cover and shifter assembly.

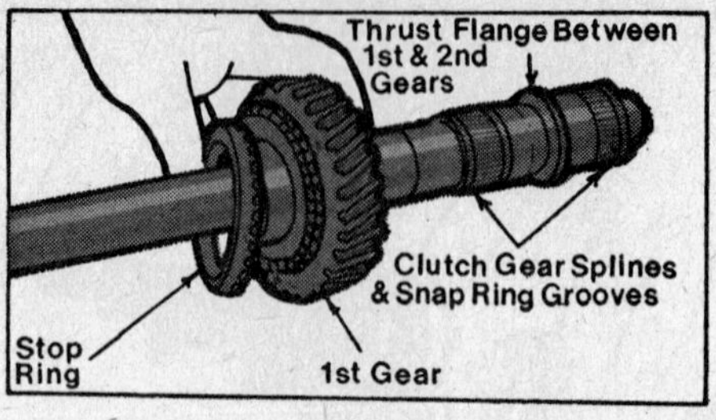

Fig. 7 First gear installation

Types Four & Five

NOTE

These transmissions, Figs. 1 and 2, are quite similar, the chief difference being the cut of the mainshaft splines and the double set of counter gear needle bearings used in the heavy duty unit, Fig. 1. Note that in Fig. 2 there are only one set of needle bearings at each end of the counter gear. The procedure which follows apply to both units unless otherwise indicated.

DISASSEMBLE TRANS.

1. Pull flange from rear of mainshaft.
2. Slide extension housing off mainshaft.
3. Remove transmission cover.

Main Drive Gear

1. Remove drive gear bearing retainer.
2. On H.D. models, Fig. 1, when removing drive gear from transmission, slide front synchronizer inner stop ring from short splines on gear as assembly is being removed from case.
3. On S.D. models, Fig. 2, grasp drive gear shaft and pull assembly out of case. *Be careful not to bind inner synchronizer ring on drive gear clutch teeth.*
4. Remove bearing rollers from drive gear pocket, using a hook or flat blade.

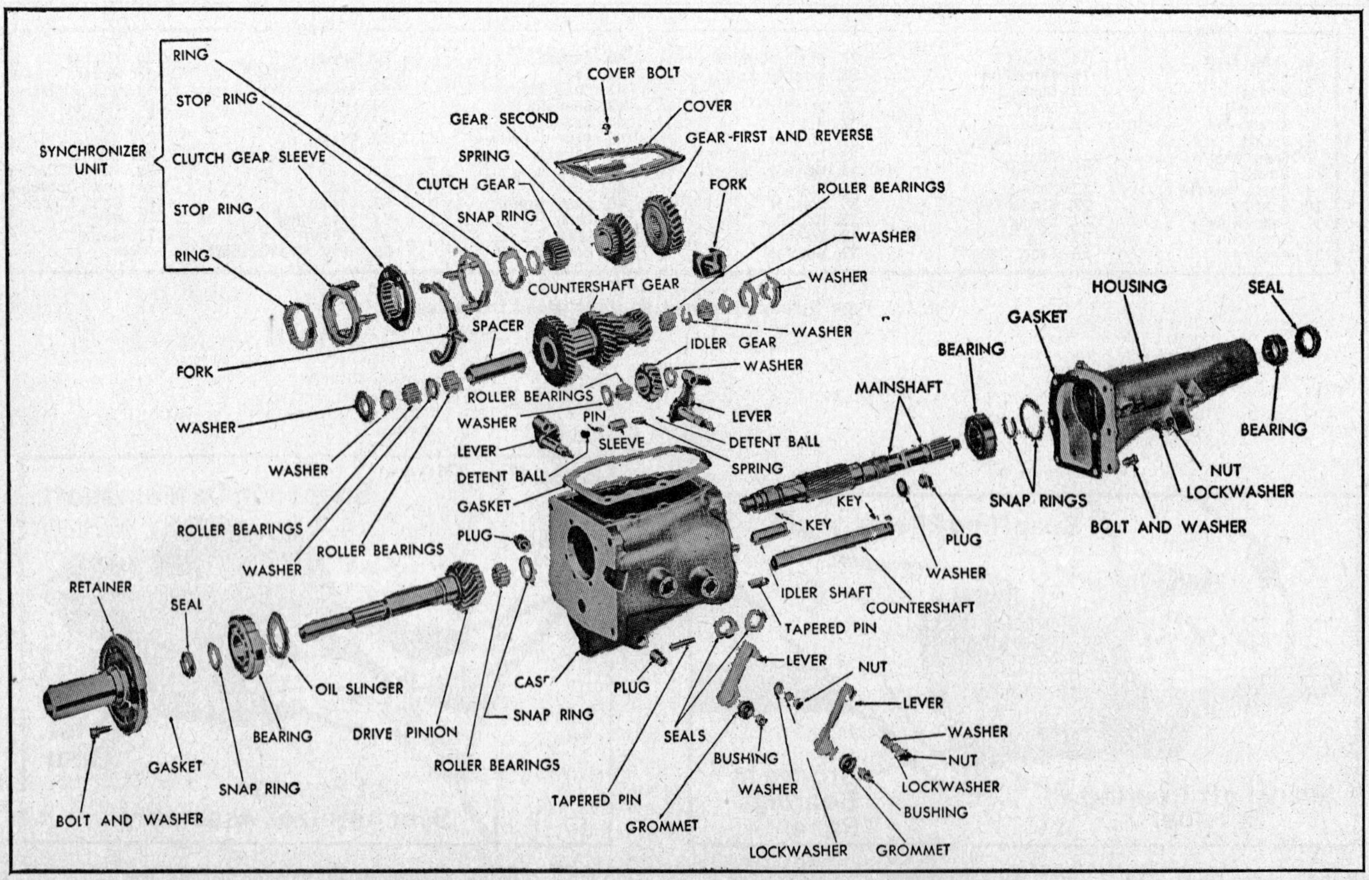

Fig. 1 Type 4 heavy duty transmission disassembled

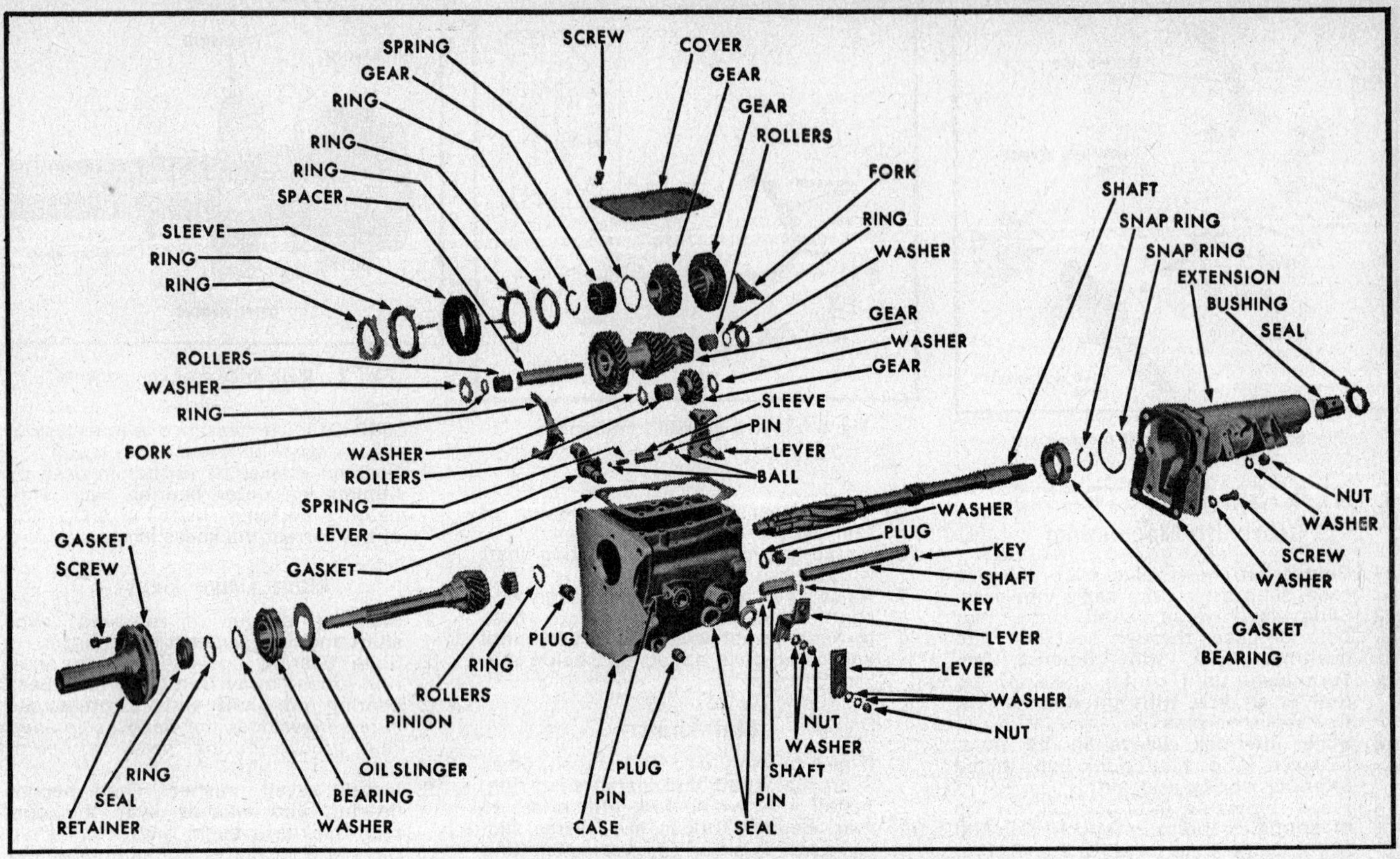

Fig. 2 Type 5 standard duty transmission disassembled. Note that low-reverse mainshaft gear shift fork groove goes to the front

Counter Gear

1. Use a suitable bearing loading tool to drive countershaft toward rear of case until key can be removed from countershaft. Then drive countershaft all the way out, keeping loading tool tight against end of countershaft to keep needle bearings in place.

Mainshaft

1. With transmission in reverse, remove outer center bearing snap ring, using a hook or flat blade, then partially remove mainshaft.
2. Cock mainshaft, then remove clutch sleeve, outer synchronizer rings, front inner ring and 2-3 shift fork, Figs. 3 and 4.
3. Remove clutch gear snap ring. Slide clutch gear off end of mainshaft.
4. Slide 2nd speed gear, stop ring and synchronizer spring off mainshaft.
5. Remove low-reverse sliding gear and shift fork as mainshaft is withdrawn from case.
6. Lift counter gear assembly from case.

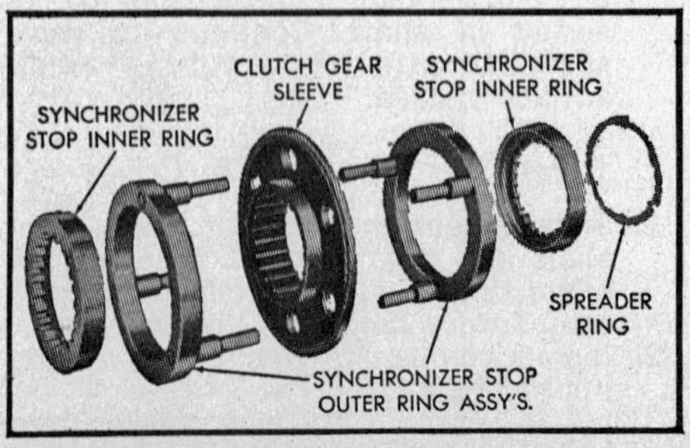

Fig. 3 Type 4 synchronizer unit

Reverse Idler Gear

1. Drive reverse idler shaft towards rear and out of case. Remove key from end of shaft.
2. Lift out idler gear, thrust washers and needle bearings from case.

Gearshift Mechanism

HEAVY DUTY UNIT, Fig. 5

1. Remove both lever shaft seals.
2. Drive tapered lock pins from lever shafts, driving from bottom toward top of transmission.
3. Remove lever shafts from transmission, being careful not to lose spring-loaded detent balls.
4. Remove interlock sleeve, spring, pin and balls.

STANDARD DUTY UNIT, Fig. 6

1. Remove operating levers from shafts.
2. Drive out tapered pin from either of two lever shafts, then withdraw shaft from inside of case. The detent balls are spring-loaded; as the shaft is being withdrawn, ball will drop to bottom of case.
3. Remove interlock sleeve, spring, pin and both balls from case. Then drive out remaining tapered pin and slide shaft out of transmission.
4. Drive out shaft seals and discard.

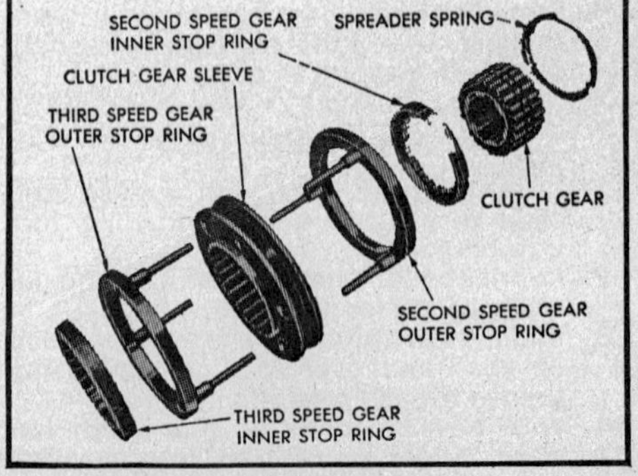

Fig. 4 Type 5 synchronizer unit

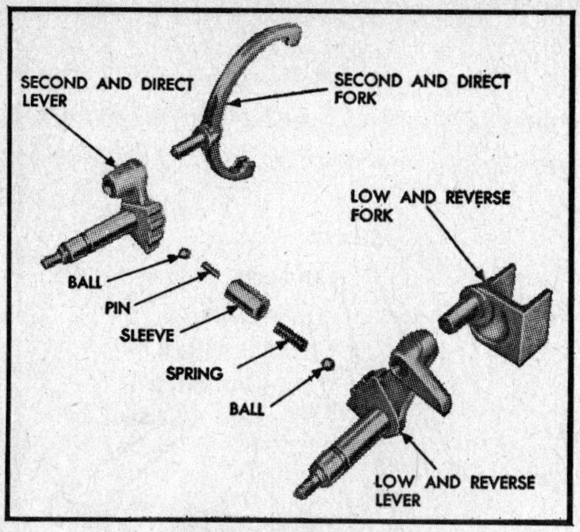

Fig. 5 Type 4 gearshift mechanism

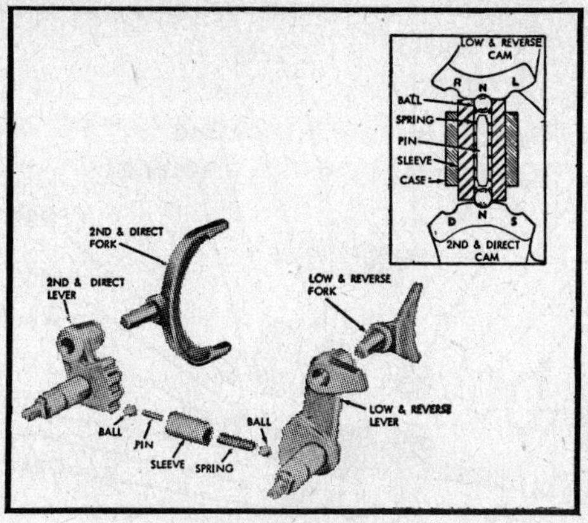

Fig. 6 Type 5 gearshift mechanism

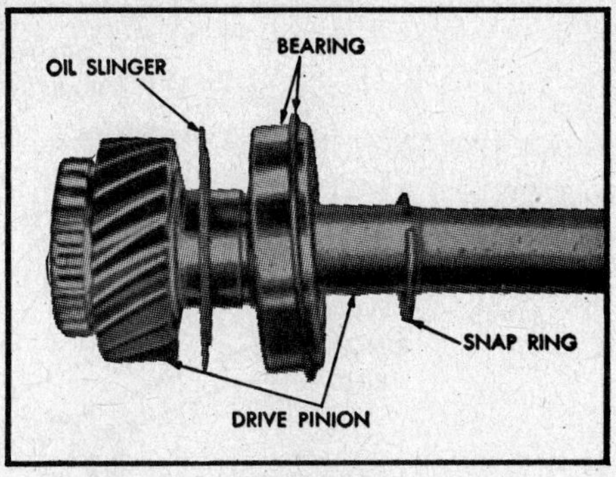

Fig. 7 Main drive gear components

REASSEMBLE TRANS.

Gearshift Mechanism

1. Center two new seals over holes in case, then drive both seals into case.
2. Slide low-reverse shaft into rear boss of case, through seal and into position. Lock with tapered pin. Turn lever until center (neutral) detent is in line with interlock bores. Use sealer on pins.
3. Slide interlock sleeve in its bore, followed by one interlock ball. Install interlock spring and pin.
4. Place remaining interlock ball on top of spring. Depress interlock ball and at the same time install 2-3 lever into fully seated position with center (neutral) detent aligned with detent ball. Secure shaft with tapered pin.

Counter Gear

1. Slide bearing spacer over bearing loading tool. Coat bore of gear with lubricant. Then slide tool and spacer into gear bore.
2. Lubricate needle bearings and install half the total of bearings at each end of gear around loading tool (Type 4 transmission uses a total of 88 bearings, Type 5 uses 44 total).
3. Install bearing retainer rings at each end of gear. Apply grease to hold bearings in place.
4. Thrust washers are available in two sizes, marked A and B. Make a selection to obtain .004-.012″ total end play of counter gear.
5. Install thrust washers at each end of assembly and over loading tool, using grease to hold washers in place.
6. Install counter gear assembly in case, making sure tabs on thrust washers slide into grooves in case.

Reverse Idler Gear

1. Coat bore of gear with grease and slide a suitable bearing loading tool into bore.
2. Lubricate 22 needle bearings and install around loading tool.
3. Install new thrust washer at each end of gear and over loading tool using grease to hold washers in place.
4. With bevelled ends of gear teeth forward, slide gear down into position in case.
5. Install gear shift in its opening at rear of case.
6. Install key in shaft, and position shaft with keyway.
7. Raise idler gear slightly to align with shaft, then drive shaft into case through thrust washer and gear until end of shaft is about 1/64″ below surface of case.

Mainshaft

1. If new bearing is to be installed, press it on mainshaft and install snap ring.
2. Install low-reverse fork with offset to rear. Engage fork in low-reverse sliding gear. Position in case by shifting into reverse.
3. Slide mainshaft into case and through low-reverse gear.
4. Install 2-3 shift fork with offset toward rear.
5. Install 2nd speed gear and spreader spring on mainshaft.

NOTE: *Synchronizer float should be .050″ to .090″ when measured between end of synchronizer outer ring pin and opposite synchronizer outer ring. This measurement must be made 180 degrees apart with equal gap on both pin ends for float determination. To be acceptable, the gauge should be a snug fit between pins and outer rings. In cases where float dimension is over .090″, synchronizer shims (part no. 2464724) should be installed to reduce float to .090″ or less. This shim is to be installed on the 2nd speed gear before the energizing spring is installed. In cases where float is below .050″, material should be removed from ends of all six synchronizer pins, using a magnetic grinder or other suitable equipment.*

6. Install 2nd gear inner stop ring and outer stop ring assembly. Engage synchronizer clutch sleeve with 2-3 shift fork.
7. Slide clutch gear over end of mainshaft and down against 2nd speed gear. Select a snap ring of the correct thickness and install. This snap ring eliminates end play and must be a snug fit.
8. Measure clearance between clutch gear and 2nd speed gear. Limits are .002-.011″. If clearance is in excess of .011″, "gear jump-out" may result.
9. Position mainshaft further in case by tapping on outer bearing race until bearing bottoms. Install a snap ring of the correct thickness in case.

Main Drive Gear

1. Slide oil slinger (if removed) over shaft and down against gear, Fig. 7.
2. Slide bearing over shaft with snap ring groove away from gear end. Seat bearing on shaft with a press. *Be sure slinger does not hang up in snap ring groove during pressing operation.*
3. Install keyed washer. Then secure bearing and washer with the correct thickness snap ring. Four snap rings are available to eliminate end play. If large snap ring around bearing was removed, install at this time.
4. Install rollers in drive gear pocket using grease to hold them in place, and retain with lock ring (14 rollers used with Fig. 2 unit, 15 with Fig. 1).
5. Install 3rd gear outer stop ring and inner stop ring. Guide drive gear through front of case and engage inner stop ring with clutch teeth. Then seat bearing. Bearing is fully seated when snap ring is in full contact with case.
6. Install new seal in drive gear bearing retainer. Slide bearing retainer (less gasket) down against case.
7. Hold retainer against case and measure clearance between case and retainer, using a feeler gauge. Select a gasket .003″ to .004″ thicker than the clearance to eliminate all end play in bearing.
8. Install gasket selected and reinstall bearing retainer. Install attaching bolts and tighten to 23 ft-lbs. torque.
9. Install the countershaft, driving the bearing loading tool forward and out of counter gear until key can be inserted in shaft. Continue to drive shaft into case until about 1/64″ below surface of case.

Extension Housing & Cover

1. Slide extension housing over mainshaft and down against case, at the same time guiding mainshaft into oil seal. Torque bolts to 50 ft-lbs.
2. Install companion flange and torque nut to 175 ft-lbs.
3. Install cover with new gasket and torque attaching bolts to 12 ft-lbs.

Type Six

DISASSEMBLE TRANS.

1. Remove transmission cover, Fig. 1.
2. Remove extension housing. To prevent mainshaft from following housing (with resultant loss of needle bearings) tap end of mainshaft while withdrawing housing.
3. Remove front bearing retainer.
4. Remove filler plug from right side of case. Then working through plug opening, drive roll pin out of case and countershaft with small punch, Fig. 2.
5. Hold counter gear with a hook and, with a dummy shaft, push countershaft out rear of case until counter gear can be lowered to bottom of case, Fig. 3.
6. Pull main drive gear forward until gear contacts case, then remove large snap ring.

NOTE: *On some models, it is necessary to move gear forward to provide clearance when removing mainshaft assembly. On other models, the drive gear is removed from front of case.*

7. Remove snap ring and slide speedometer drive gear off mainshaft. Remove lock ball from shaft.
8. Remove snap ring and remove mainshaft rear bearing from shaft and case, Fig. 4.
9. Place both shift levers in neutral (central) position.
10. Remove a set screw that retains detent springs and plugs in case. Remove one spring and plug. Fig. 5.
11. Remove low-reverse set screw and slide shift rail out through rear of case.
12. Rotate low-reverse shift fork upward and lift it from case.
13. Remove 2-3 set screw and rotate 2-3 shift rail 90 degrees with pliers.
14. Lift interlock plug from case with a magnet rod.
15. Tap inner end of 2-3 shift rail to remove expansion plug from front of case. Remove shift rail.
16. Remove 2-3 detent plug and spring from detent bore.
17. Rotate 2-3 shift fork upward and lift from case.
18. Lift mainshaft assembly out through top of case.
19. On some transmissions used on Ford Co. vehicles, push main drive gear into case until bearing is free of bore, then lift gear and bearing through top of case.
20. Working through front bearing opening, drive reverse idler gear shaft out through rear of case with a drift, Fig. 6.
21. Lift reverse idler gear and two thrust washers from case.
22. Lift counter gear and thrust washers from case, Fig. 7. Be careful not to allow dummy shaft and needle bearings to fall out of gear.

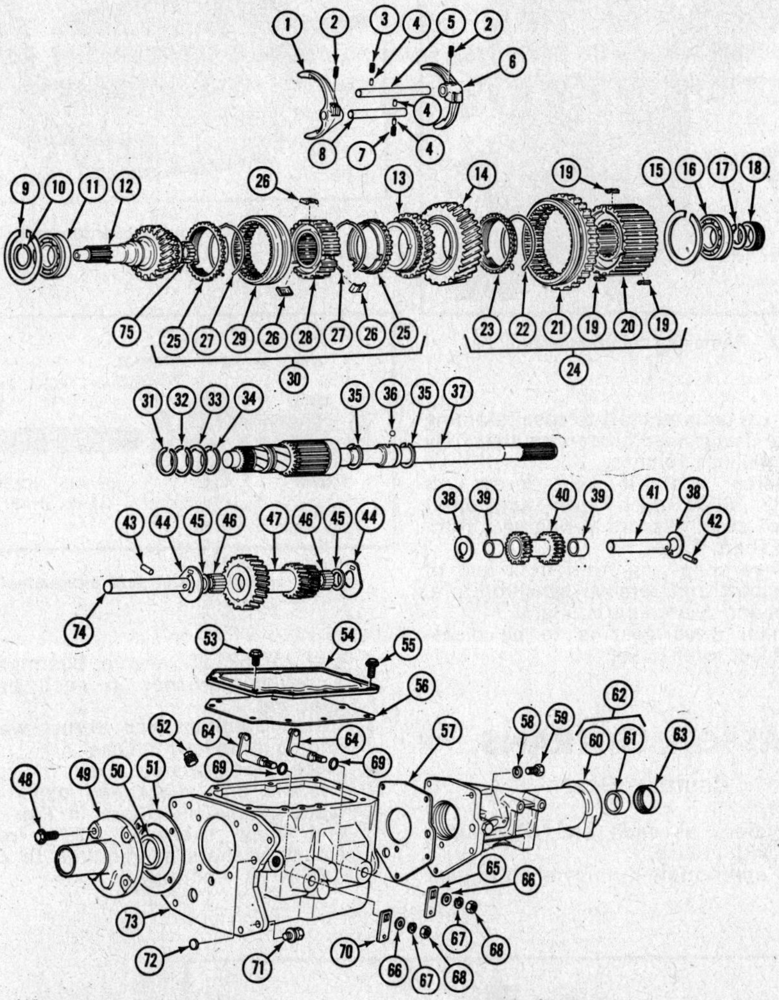

1. Fork, Second and Third Shifter
2. Setscrew, Shifter Fork (2)
3. Spring, Upper Detent (Long)
4. Plugs, Gear Shift Detent and Interlock (3)
5. Shift Rail, First and Reverse
6. Fork, First and Reverse Shifter
7. Spring, Lower Detent (Short)
8. Shift Rail, Second and Third
9. Ring, Locating (Snap) Front Bearing
10. Ring, Retaining (Snap) Front Bearing to Clutch Shaft
11. Bearing, Front
12. Shaft, Clutch
13. Gear, Second
14. Gear, First
15. Ring, Locating (Snap) Rear Bearing
16. Bearing, Rear
17. Ring, Retaining (Snap) Rear Bearing
18. Gear, Speedometer Drive
19. Insert, First and Reverse Synchronizer (3)
20. Hub, First and Reverse
21. Sleeve and Gear, First and Reverse
22. Spring, Insert, First and Reverse Synchronizer
23. Ring, Blocking, First and Reverse Synchronizer
24. Synchronizer Assembly, First and Reverse
25. Ring, Blocking, Second and Third Synchronizer (2)
26. Insert, Second and Third Synchronizer (3)
27. Spring, Insert, Second—Third Synchronizer (2)
28. Hub, Second and Third
29. Sleeve, Second and Third Synchronizer
30. Second—Third Synchronizer Assembly
31. Ring, Retaining (Snap) Output Shaft
32. Ring, Retaining (Snap) First Gear
33. Washer, Thrust (Tabbed) First Gear
34. Ring, Retaining (Snap) First-Reverse Hub
35. Ring, Retaining (Snap) (2) Rear Bearing and Speedometer Gear
36. Lock Ball, 1/4 Diameter—Speedometer Gear
37. Shaft, Output
38. Washer, Thrust, Reverse Idler Gear (2)
39. Bushing, Reverse Idler Gear (2)
40. Gear, Reverse Idler
41. Shaft, Reverse Idler Gear
42. Pin, Roll, Reverse Idler Gear Shaft
43. Pin, Roll, Countershaft
44. Washer, Thrust, Countershaft Gear (2)
45. Retainer, Countershaft Needle Bearing
46. Needle Bearing, Countershaft Gear (50)
47. Gear, Countershaft
48. Bolt, Front Bearing Cap
49. Cap, Front Bearing
50. Oil Seal, Front Bearing Cap
51. Gasket, Front Bearing Cap
52. Plug, Transmission Fill
53. Bolt, Top Cover
54. Top Cover, Case
55. Bolt, Top Cover
56. Gasket, Top Cover
57. Gasket, Extension Housing
58. Lockwasher, Extension Housing Bolt (5)
59. Bolt, Extension Housing
60. Extension Housing
61. Bushing, Extension Housing (Included with Housing)
62. Housing Assembly, Extension
63. Seal, Oil, Extension Housing
64. Shaft, Shifter Fork
65. Lever, First and Reverse Shifter
66. Flatwasher, Shifter Levers (2)
67. Lock, Washer Shifter Levers (2)
68. Nut, Hex, Shifter Levers (2)
69. O-Ring, Shifter Shaft (2)
70. Lever, Second and Third Shifter
71. Bolt, Transmission Mounting (4)
72. Plug, Expansion
73. Case, Transmission
74. Countershaft
75. Bearings, Clutch Shaft Roller

Fig. 1 Type 6 transmission disassembled

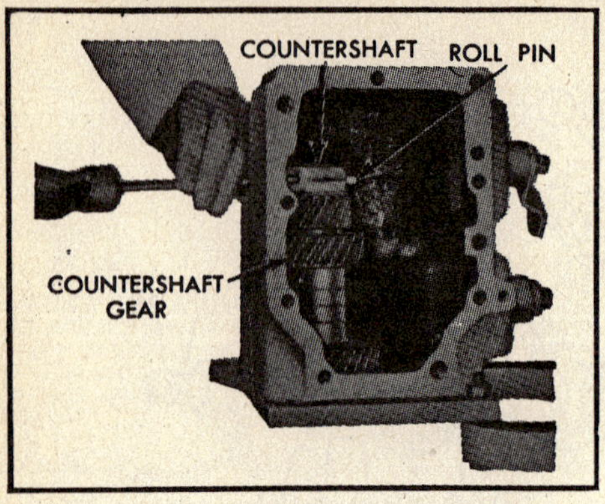

Fig. 2 **Removing countershaft roll pin**

Fig. 3 **Removing countershaft**

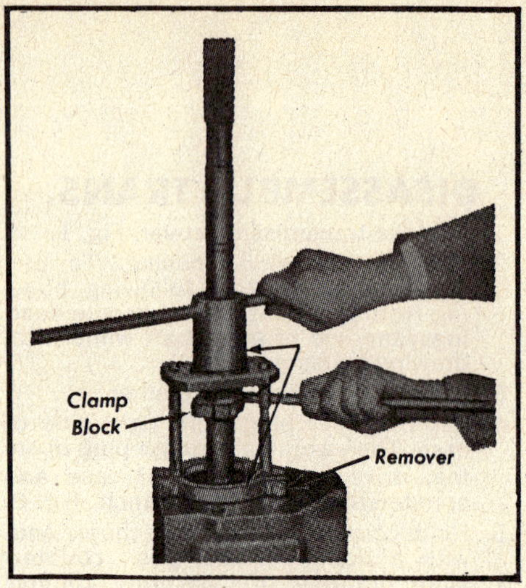

Fig. 4 **Removing mainshaft bearing**

23. Remove countershaft-to-case retaining pin and any needle bearings that may have fallen into case.
24. Unfasten and lift shift levers off shafts. Slide each lever and shaft out of case. Discard O-ring seal from each shaft, Fig. 8.
25. Remove snap ring from front end of mainshaft and remove synchronizers, ears and related parts, Fig. 9.
 ain drive gear is to be disassembled, refer to Fig. 10.

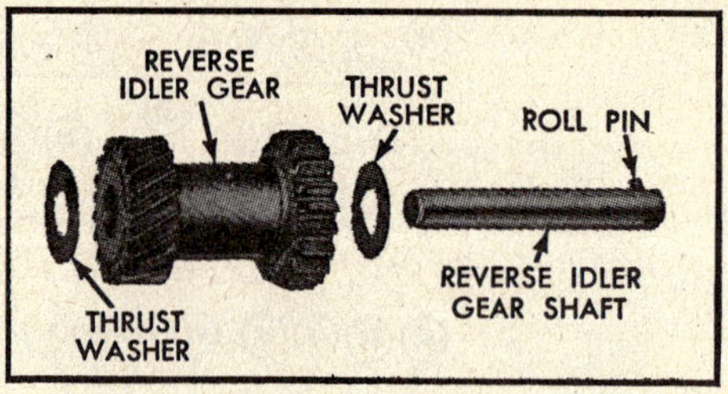

Fig. 6 **Reverse idler gear disassembled**

REASSEMBLE TRANS.

Counter Gear

1. Coat bore at each end of counter gear with grease.
2. Hold appropriate dummy shaft in gear and install 25 needle bearings and a retainer washer in each end of gear, Fig. 7.
3. Install counter gear, thrust washers and countershaft in case.
4. Place transmission case in vertical position and check end play with a feeler gauge as shown in Fig. 13. If end play exceeds .018", replace thrust washers as required to obtain .004-.018" end play.

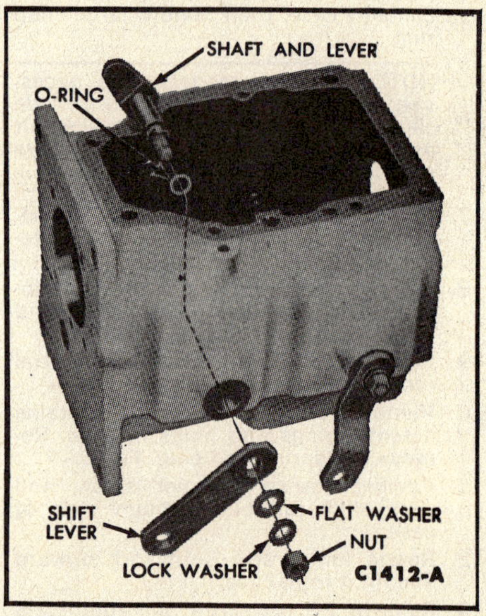

Fig. 8 **Shift lever and related parts**

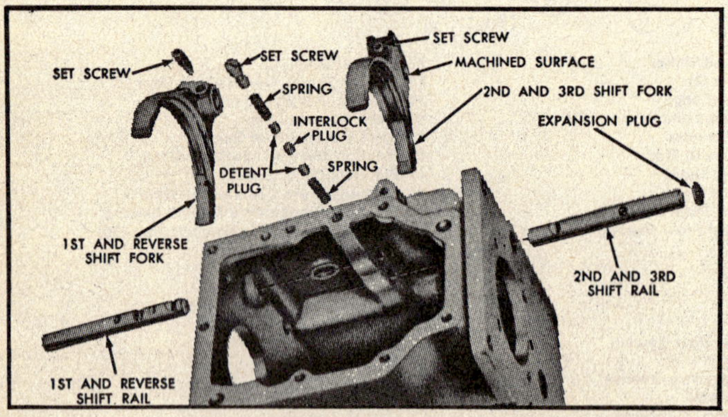

Fig. 5 **Shift rails and forks disassembled**

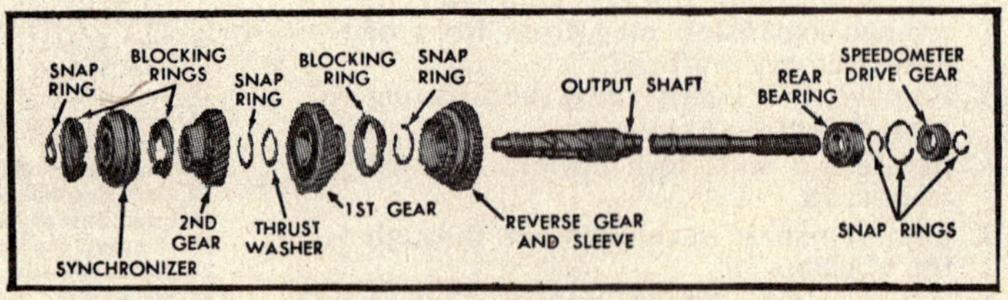

Fig. 9 **Mainshaft disassembled**

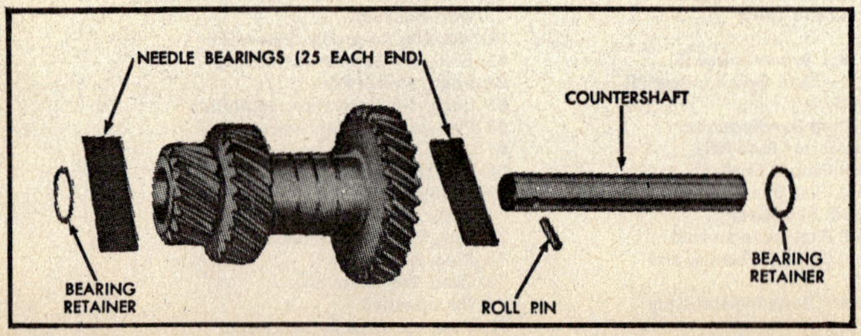

Fig. 7 **Counter gear disassembled**

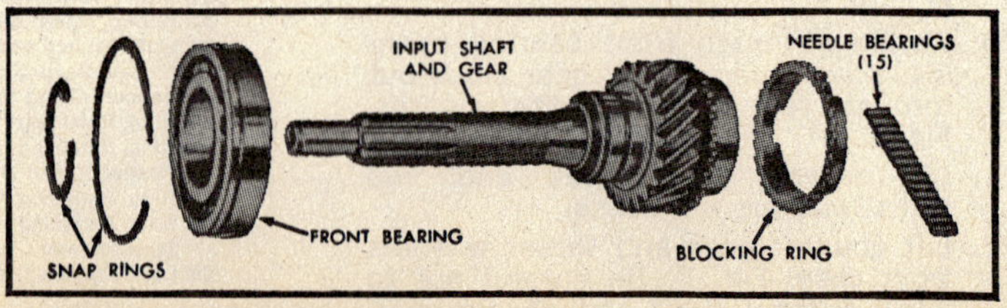

Fig. 10 **Main drive gear disassembled**

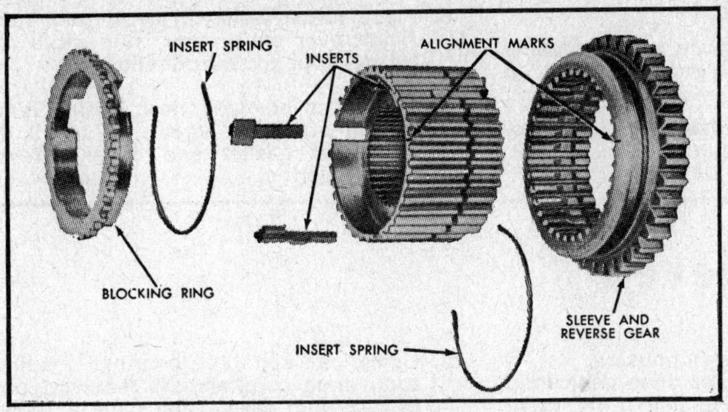

Fig. 11 Low-reverse synchronizer disassembled

5. Once end play has been established, remove countershaft with dummy shaft.
6. Allow counter gear assembly to remain in case.

Reverse Idler Gear

1. Install idler gear, thrust washers and shaft in case.
2. Make sure that thrust washer with flat side is at the web end, and that the spur gear is toward the rear of case, Fig. 6.
3. Check reverse idler gear end play in same manner and to the same clearance as the counter gear. If end play is within limits of .004-.018″ leave gear in case.

Low-Reverse Synchronizer

1. Install an insert spring, Fig. 11, in groove of low-reverse synchronizer hub. Make sure spring covers all insert grooves.
2. Start hub in sleeve, making sure that alignment marks are properly indexed.
3. Position three inserts in hub, making sure that small end is over spring and that shoulder is on inside of hub.
4. Slide sleeve and reverse gear onto hub until detent is engaged.
5. Install other insert spring in front of hub to hold inserts against hub.

Second-Third Synchronizer

1. Install one insert spring, Fig. 12, into a groove of synchronizer hub, making sure that all three insert slots are fully covered.
2. With alignment marks on hub and sleeve aligned, start hub into sleeve.
3. Place three inserts on top of retaining spring and push assembly together.
4. Install remaining insert spring so that spring ends cover same slots as do other spring. Do not stagger springs.
5. Place a synchronizer blocking ring in each end of sleeve.

Main Drive Line

1. Lubricate mainshaft splines and machined surfaces with transmission lube.
2. Slide low-reverse synchronizer, Fig. 9, onto mainshaft with teeth end of gear facing toward rear of shaft. Secure in place with snap ring.
3. Coat tapered machined surface of low gear with grease. Place blocking ring on greased surface.
4. Slide low gear onto mainshaft with blocking ring toward rear of shaft. Rotate gear as necessary to engage the three notches in blocking ring with synchronizer inserts. Secure low gear with thrust washer and snap ring.
5. Coat tapered machined surface of 2nd gear with grease and slide blocking ring onto it. Slide 2nd gear with blocking ring and 2-3 synchronizer onto mainshaft. Tapered machined surface of 2nd gear must be toward front of shaft. Make sure that notches in blocking ring engage synchronizer inserts. Secure synchronizer with snap ring.
6. Install new O-ring on each of two shift lever shafts, Fig. 8. Lubricate shafts with transmission lube and install them in case. Secure each lever on its shaft with a flat washer, lock washer and nut.
7. Coat bore of main drive gear shaft with thick coat of grease. Install 15 needle bearings in gear pocket, Fig. 10.

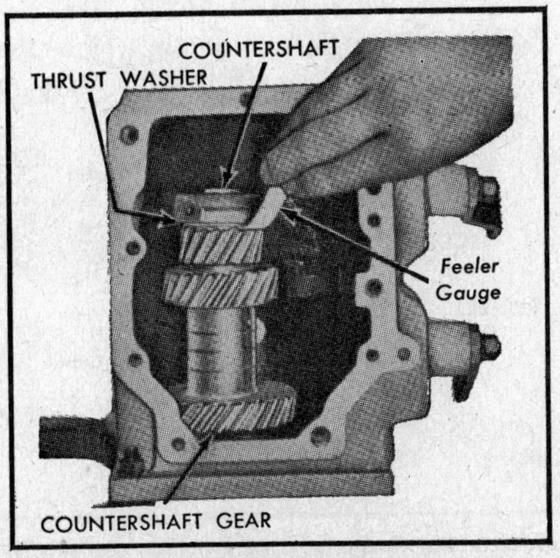

Fig. 13 Checking counter gear end play

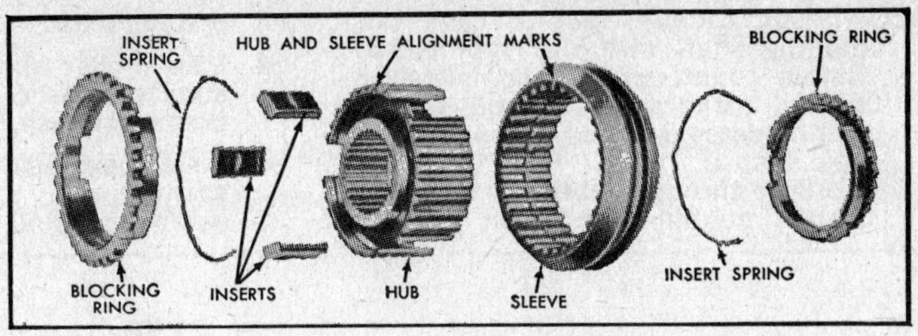

Fig. 12 Second-high synchronizer disassembled

8. Position drive gear assembly into case.
9. Place a detent spring and plug in case, Fig. 5. Place 2-3 shift fork in synchronizer groove. Rotate fork into position and install 2-3 shift rail. It will be necessary to depress detent plug to enter rail in bore. Move rail inward until detent plug engages center (neutral) notch. Secure fork to shaft with set screw.
10. Install interlock plug in case. If 2-3 shift rail is in neutral position, top of interlock will be slightly lower than surface of low-reverse shift rail bore.
11. Place low-reverse shift fork in groove of synchronoizer. Rotate fork into position and install low-reverse shift rail. Move rail inward until center (neutral) notch is aligned with detent bore. Secure fork to shaft with set screw.
12. Install remaining detent plug and spring. Secure spring with slotted head set screw. Turn set screw in until head is flush with case.
13. Install new expansion plug in case.
14. Install large snap ring on drive bearing. Work drive gear into case and onto mainshaft until snap ring is seated against case. Make sure that needle bearings do not fall out of place and that notches in blocking ring engage inserts in synchronizer.

NOTE: *The foregoing applies only to transmissions used with V8-352 and 390 engines. On all other models, the drive gear is installed through front of case with bearing in place on shaft.*

15. Position new front bearing retainer gasket on case. Place bearing retainer on case, making sure oil return groove is at the bottom. Install and tighten attaching screws to 19-25 ft-lbs.
16. Install large snap ring on mainshaft rear bearing. Place bearing on mainshaft with snap ring end toward rear of shaft. Press bearing into place and secure with snap ring.
17. Hold speedometer drive gear lock ball in detent and slide gear into place. Secure gear with snap ring.

Final Assembly

1. Place transmission in vertical position. Working through drain hole in bottom of case, align bore of counter gear and thrust washers with bore of case, using a screwdriver.

2. Working through rear of case, push dummy shaft out with countershaft. Before countershaft is completely inserted, make sure the hole that accommodates roll pin is aligned with hole in case.
3. Working through lubricant filler hole, install roll pin (Fig. 2) in case and countershaft.
4. Install filler and drain plugs, making sure magnetic plug is installed in bottom of case.
5. Install extension housing with new gasket and torque attaching cap screws to 42-50 ft-lbs.
6. Place transmission in gear, pour lubricant over entire gear train while rotating input and output shafts.
7. Install cover with new gasket and torque attaching screws to 14-19 ft-lbs. Coat gasket and cover screws with sealer.

Type Seven

DISASSEMBLE TRANS.

1. Referring to Fig. 1, remove cover.
2. Remove extension housing. *To prevent mainshaft from following housing (with resultant loss of needle bearings) tap end of mainshaft while withdrawing extension housing.*
3. Remove speedometer drive gear snap ring, gear and drive ball from mainshaft.
4. Remove idler gear shaft and countershaft retainer. If necessary tap front ends of both shafts to free retainer.
5. Using a suitable bearing loading tool, drive countershaft rearward out of gear and case. Then carefully lower counter gear to bottom of case, Fig. 2.
6. After removing main drive gear bearing retainer, remove drive gear and

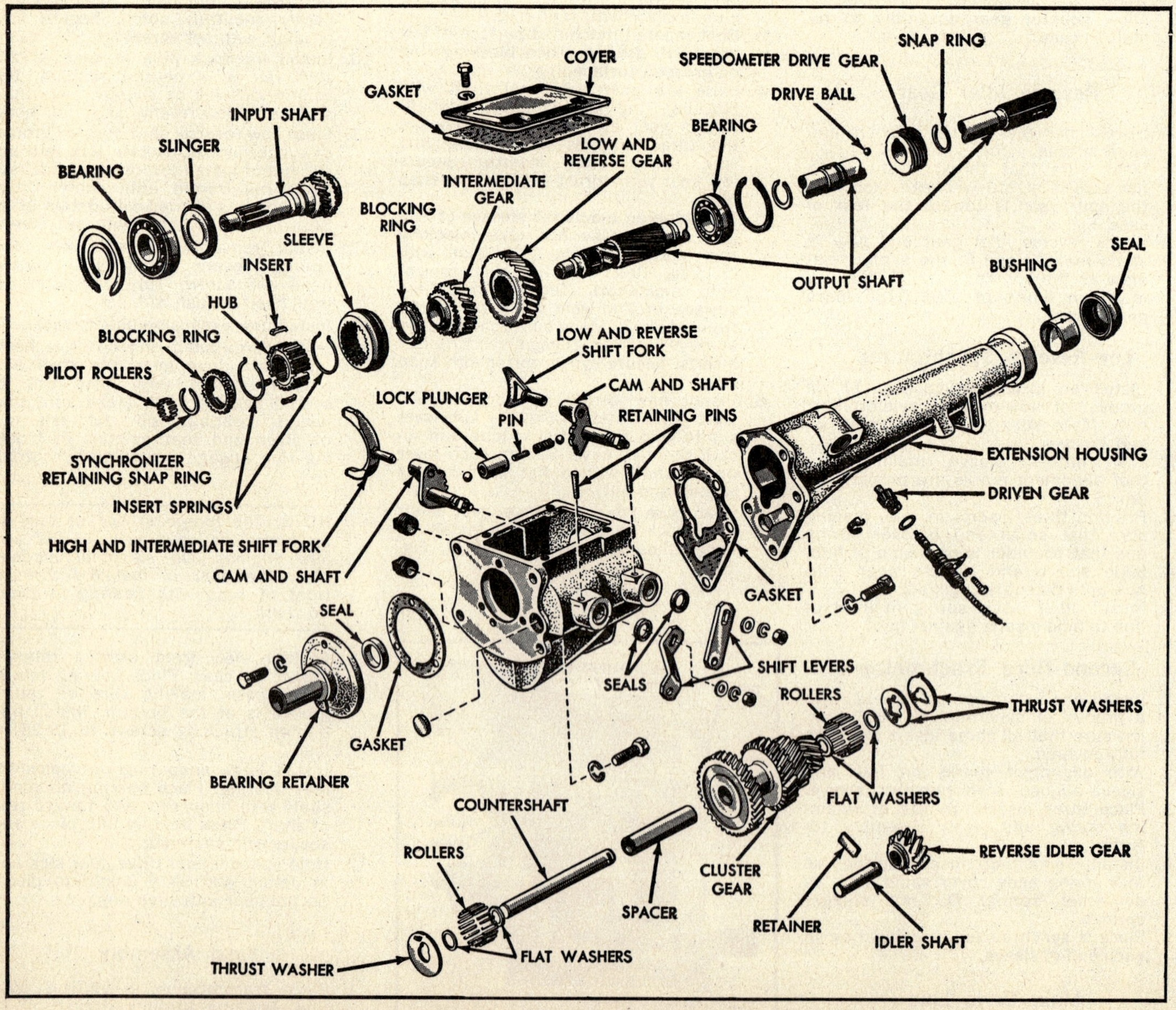

Fig. 1 Type 7 transmission. Note that longer end of synchronizer hub goes to the front. CAUTION: On some models the shift fork groove for low-reverse sliding gear goes to rear as shown; on other models fork goes to front. Check position after cover is removed

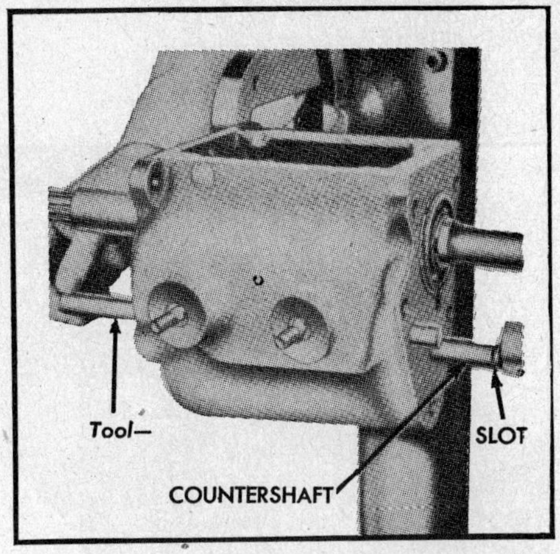

Fig. 2 Countershaft removal and installation

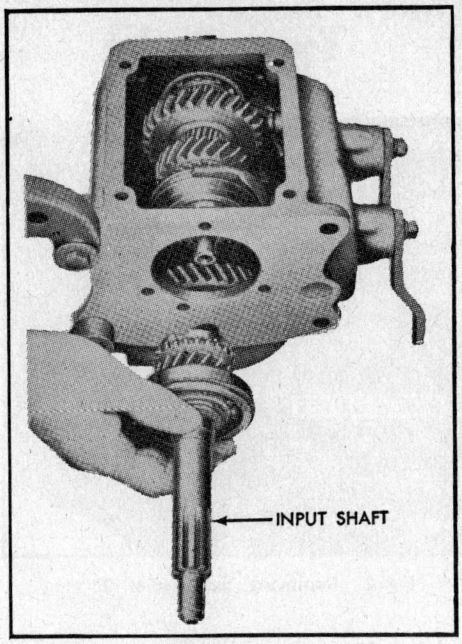

Fig. 3 Removing main drive gear

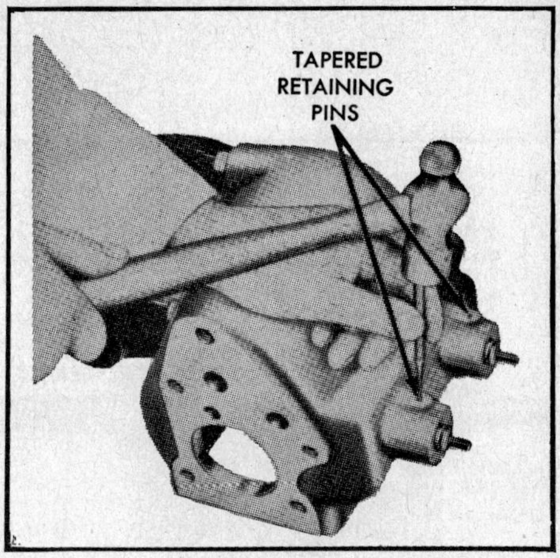

Fig. 4 Removing shift shaft retaining pins. Note that pins are driven out from the bottom

front synchronizer blocking ring from case, Fig. 3.

7. Remove synchronizer retaining snap ring from mainshaft. Then, while holding synchronizer together, pull mainshaft out of case. Lift synchronizer, gears and shift forks from case. *For reference on reassembly, note which end of synchronizer hub faces forward.*
8. Drive reverse idler shaft out of case and lift out idler gear, shaft and counter gear.
9. Remove shift levers.
10. From underside of case, drive out tapered pins, Fig. 4. Using a plastic hammer, drive 2-3 shift cam and shaft toward outside of case and separate balls and springs from plunger. Push out cam and shaft assemblies and remove plunger.
11. If necessary, shift lever oil seals may be removed with a slide hammer type hooked tool.
12. Remove snap ring and press main drive gear out of bearing and oil slinger.
13. Remove snap ring and press bearing from mainshaft.

REASSEMBLE TRANS.

1. Press main drive gear bearing and oil slinger on shaft, securing with snap ring.
2. Press mainshaft bearing on shaft and secure with snap ring.
3. Insert spacer and dummy shaft, Fig. 2, into counter gear. Position one flat washer at each end of spacer. Apply grease to needle bearings and assemble them around dummy shaft at each end of gear. Apply grease to other two flat washers and thrust washers and assemble at each end of counter gear. Note position of tangs on thrust washers, Fig. 1.
4. Position counter gear assembly in bottom of case with larger gear toward front.
5. Install low-reverse shift cam through case opening. Assemble spacer and spring in plunger. Hold plunger in position and install 2-3 cam and shaft in case opening, allowing balls to register in cam detents.
6. Align cam and shaft grooves with openings in shaft bosses, and install retaining pins, Fig. 4. Check cam action; bent pins may restrict movement.
7. Position reverse idle gear, and insert shaft (from rear) through case just far enough to hold gear.
8. After assembling three inserts in synchronizer hub and securing them with two spring retainers, insert hub into 2-3 synchronizer sleeve. Install one block ring in rear side of hub. Coat blocking rings with grease.
9. Using a coating of grease, assemble needle bearings in drive gear pocket and install front synchronizer blocking ring on drive gear shaft.
10. Install shift forks in shift lever shafts with large fork in 2-3 shaft. Web of low-reverse fork must be to rear of shaft center.
11. Start mainshaft through rear of case. Place low-reverse gear on shaft, followed by 2nd speed gear. Tilt mainshaft enough to allow rear shift fork to engage sliding gear groove.
12. *With longer hub forward,* slide synchronizer onto mainshaft and engage synchronizer sleeve in 2-3 shift fork.
13. Install synchronizer hub and snap ring.
14. Position main drive gear and front synchronizer blocking ring.
15. Place a new gasket on drive gear bearing retainer. If shaft oil seal was removed from retainer, install a new seal. Install bearing retainer, using sealer on bolts. Line up oil drain groove in retainer with oil hole in case.
16. Raise counter gear assembly and align dummy shaft with countershaft opening in case. Start countershaft into case from rear, and carefully drive shaft into position.
17. Install idler gear shaft and retainer.
18. Secure speedometer drive gear and ball with snap ring.
19. Using a new gasket, install extension housing.
20. Start new seal over each cam and shaft and drive in seals.
21. Fill transmission to proper level with lubricant. Then, using a new gasket and sealer on bolts, install transmission cover.

NOTE: *Cover gasket vent holes must be toward rear and cover vent hole must be toward front.*

Type Eight

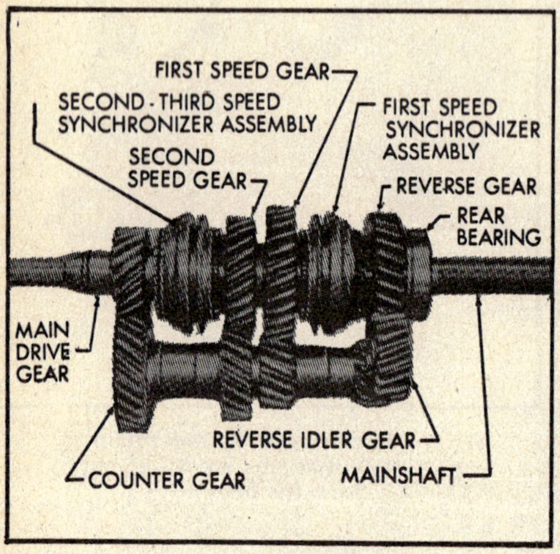

Fig. 1 Type 8 Saginaw three speed transmission

Fig. 2 Removing reverse idler "E" ring

Fig. 3. Removing rear bearing retainer after spreading snap ring as shown

Disassemble

1. Drain lubricant and remove damper assembly, if equipped.
2. Remove side cover and gasket.
3. Remove front bearing retainer.
4. Remove clutch gear bearing-to-stem snap ring then slide bearing off over clutch gear stem. The clutch gear bearing is a slip fit on the gear and into the case bore.
5. Remove extension-to-case bolts.
6. Remove reverse idler shaft-to-gear "E" ring, Fig. 2.
7. Remove clutch gear, mainshaft and extension assembly together through rear case opening.
8. Expand snap ring and remove rear bearing retainer and mainshaft from extension, Fig. 3.
9. Drive countershaft and woodruff key out through the rear of the case. Remove countergear and bearings.
10. Use a long drift and drive reverse

Fig. 4 Removing counter shaft, using aligning arbor to hold needle bearings in place

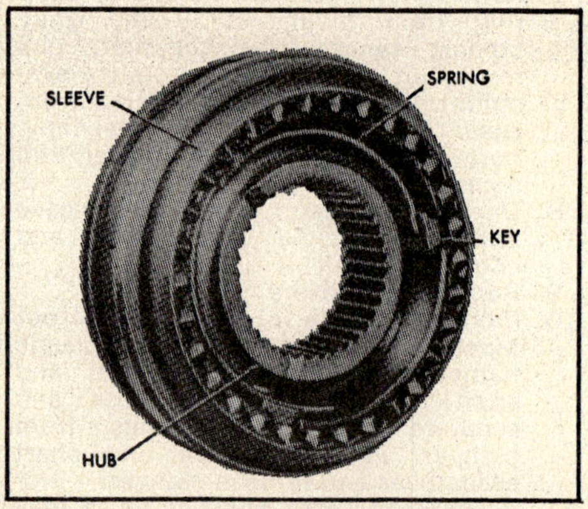

Fig. 6 Synchronizer assembly

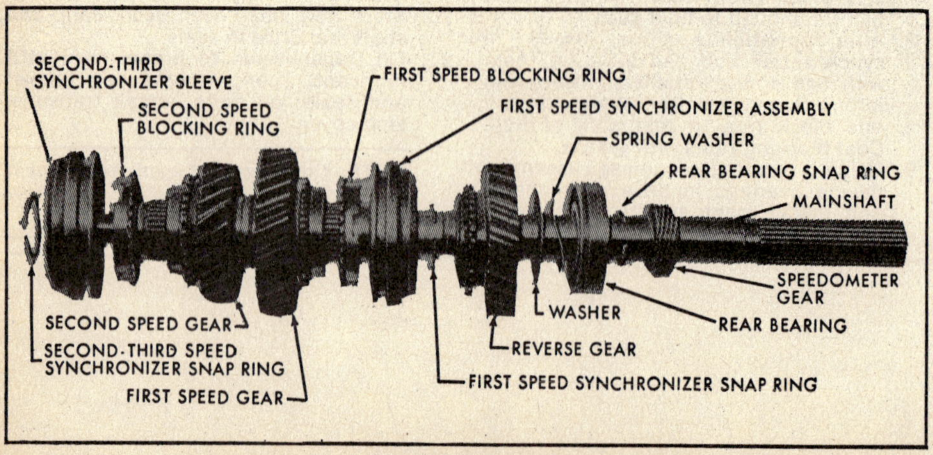

Fig. 5 Mainshaft and related parts assembled loosely to show location of parts

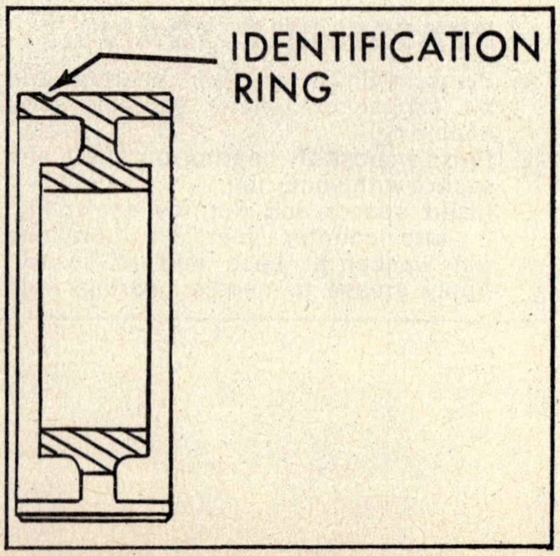

Fig. 7 Identification chamfer around synchronizer hub

Fig. 8 Installing second speed gear

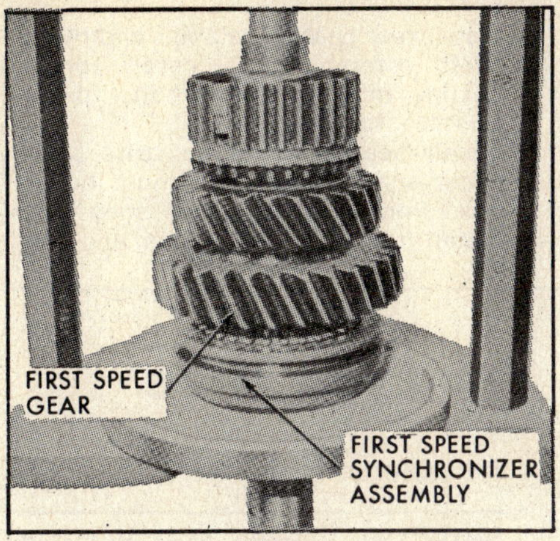

Fig. 9 Installing first speed gear

Fig. 10 Countergear with riveted anti-rattle gear. Except Astre, Monza & Vega

idler shaft and key through rear of case.

11. Remove reverse idler gear tanged thrust washer.

Disassemble Mainshaft

1. Remove clutch gear, roller bearings and blocking ring from mainshaft. Referring to Fig. 5, remove 2-3 synchronizer sleeve.
2. Remove speedometer gear.
3. Remove rear bearing snap ring, and press off rear bearing, spring washer, thrust washer and Reverse gear.
4. Remove 1st and Reverse sliding clutch hub snap ring and clutch assembly, 1st speed blocker ring and gear.

Synchronizers

NOTE: The synchronizer hubs and sliding sleeves are a selected assembly and should be kept together as originally assembled. The keys and springs may be replaced if worn or broken.

1. Mark hub and sleeve so they can be reassembled in same position.
2. Remove sleeve from synchronizer hub.
3. Remove keys and springs from hub.
4. Place three keys and two springs in position (one on each side of hub) so all three keys are engaged by both springs, Fig. 6. The tanged end of each synchronizer spring should be installed in different key cavities on either side of hub. Slide sleeve onto hub, aligning marks made before disassembly.

Fig. 11 Astre, Monza & Vega counter-gear

NOTE: A chamfer or groove around the outside of synchronizer hub identifies the end that must be opposite the fork slot in the sleeve, Fig. 7.

Assemble Mainshaft

With front of mainshaft up:

1. Install second speed gear with clutching teeth upward.
2. Install blocking ring with clutching teeth downward. All blocking rings in this unit are identical.
3. Press 2-3 synchronizer onto mainshaft, Fig. 8. *Be sure notches in blocking ring align with keys in synchronizer.*

With rear of mainshaft up:

4. Install first speed gear with clutching teeth upward.
5. Install blocking ring on gear with teeth downward.
6. Press first and reverse synchronizer onto mainshaft. *Be sure notches in blocking ring align with keys in synchronizer.* Install snap ring.
7. Install reverse gear, thrust washer, spring washer, rear bearing and bearing snap ring.
8. Install speedometer gear.

Assemble Transmission

1. Install countergear-to-case thrust washers. Install countergear into case from rear. Make certain wooruff key is in place. Note that anti rattle gear is riveted to countergear in four places and is not serviced separately, Fig. 10.

NOTE: Astre, Monza and Vega transmissions do not use an anti-rattle gear, Fig. 11.

2. Install reverse idler gear tanged steel thrust washer. Install idler gear, shaft and woodruff key. *Reverse idler gear snap ring will be installed after installation of mainshaft.*
3. Install rear bearing retainer. Spread snap ring in retainer to allow snap ring to drop around rear bearing. Press on end of mainshaft until snap ring engages groove in rear bearing.

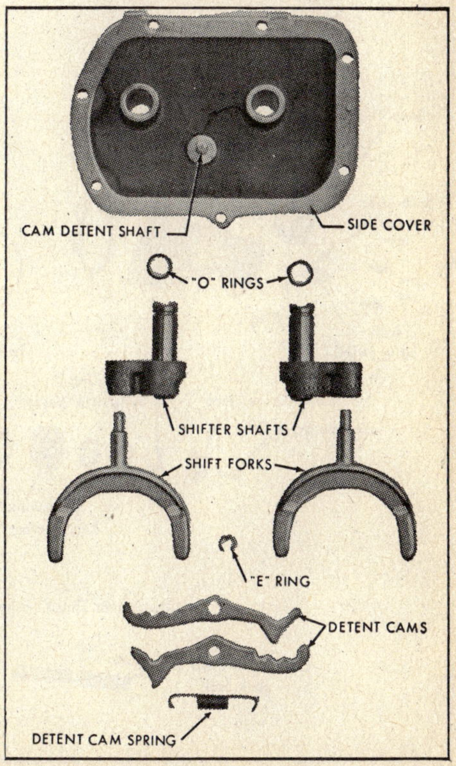

Fig. 12 Disassembled view of side cover

4. Install 14 needle rollers in main drive gear pocket, using grease to hold them in place. Assemble third speed blocking ring on main drive gear. Pilot main drive gear and blocking ring over front of mainshaft. Make certain notches in blocking ring align with keys in 2-3 synchronizer.
5. Install rear bearing retainer-to-case gasket, using heavy grease to hold gasket in place.
6. Install rear bearing retainer and mainshaft assembly into case. Torque bearing retainer-to-case bolts to 35-55 ft-lbs. torque.
7. Install bearing on main drive gear. Outer snap ring groove must be toward front of gear. Install snap ring.
8. Install front bearing retainer and gasket. Torque retainer bolts to 8-12 ft. lbs.
9. Install reverse idler gear "E" ring.
10. If repairs are required to the side cover, refer to Fig. 12.
11. Install side cover gasket. Place transmission gears in neutral and install side cover. Install attaching bolts and tighten evenly to avoid cover distortion.

Type Nine

DISASSEMBLE TRANS.

1. Refer to Fig. 1 and proceed as follows: Remove TCS and backup lamp switches and shift lever boot.
2. Remove cotter pins from each end of shift control rod, Fig. 2. Remove washers and control rod.
3. Remove serrated pin retaining selector lever to the boss on rear extension, Fig. 3.
4. Remove retaining rings, wave washer and selector ring from selector ring from selector shaft.

NOTE: Slide selector lever and shift idler lever shaft from intermediate shift lever assembly while simultaneously removing selector ring. Selector lever will just slip off selector ring.

5. Remove transmission case cover and invert transmission to drain oil.
6. Remove rear extension bolts and rotate extension until countergear shaft is exposed, Fig. 4.
7. From front of transmission and using tool J-23562, remove countergear shaft. Be sure that lock ball is not lost, Fig. 5. With special tool inserted lift countergear from case and remove thrust washers.
8. Using a ⅛" pin punch to remove lock pins, engage second gear to prevent 2-3 fork pin binding against case. Drive out lock pins from both shifter forks.

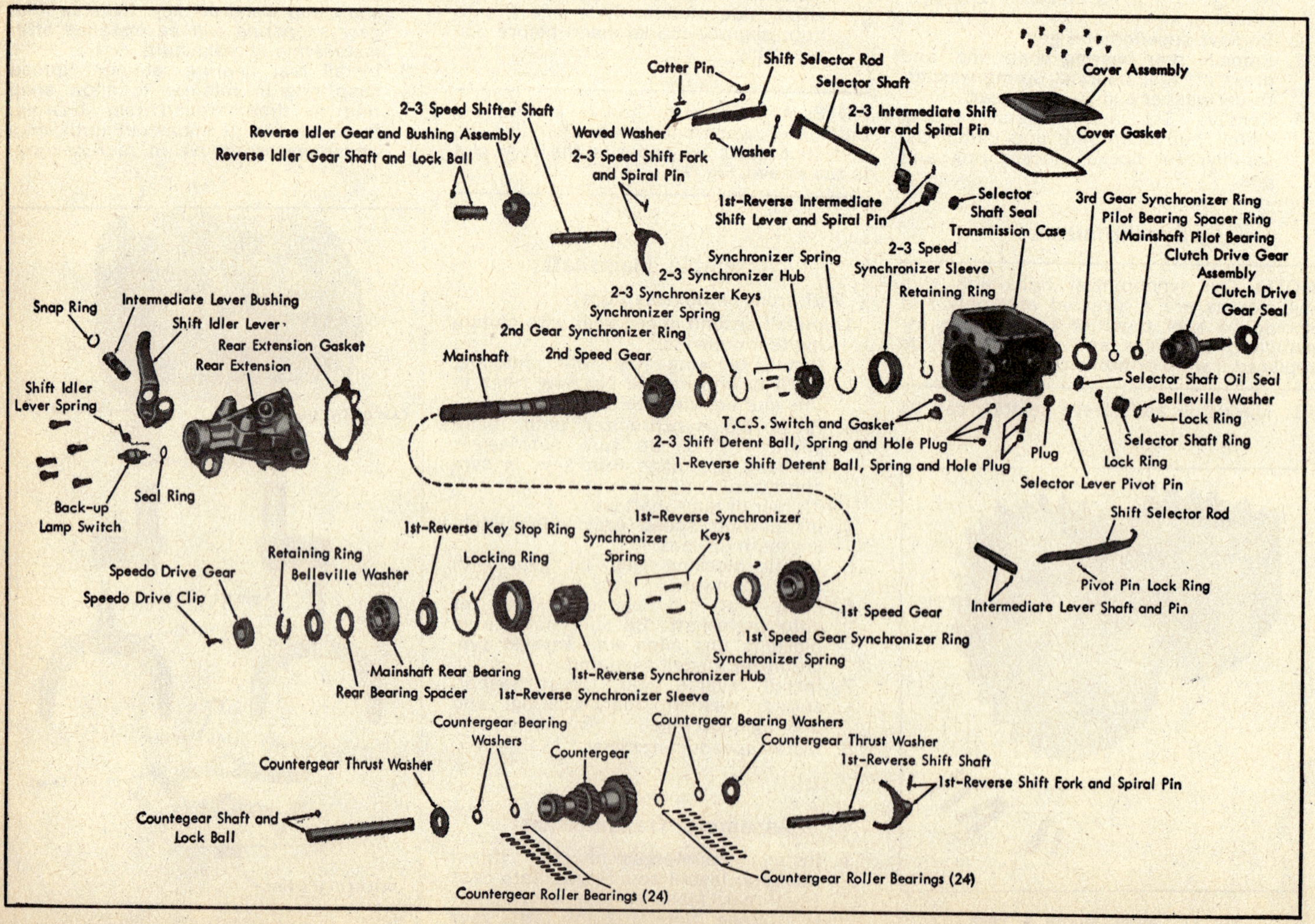

Fig. 1 Exploded view of Type 9 transmission

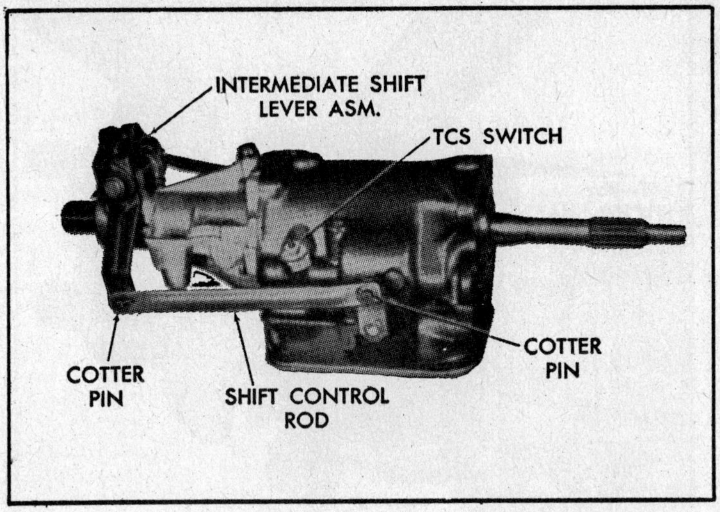

Fig. 2 Cotter pins securing shift control rod

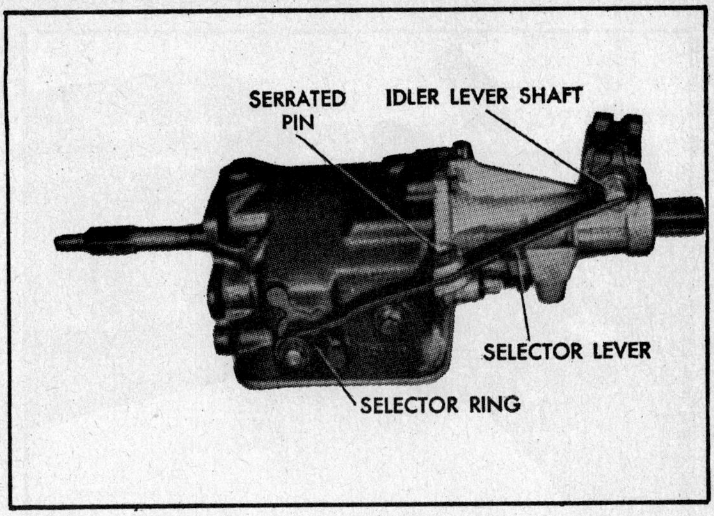

Fig. 3 Pin securing selector lever

NOTE: Before driving shifter shafts, place transmission in 3rd gear and make sure 2nd-3rd intermediate lever engages shifter shaft. This allows selector shaft and intermediate levers to pivot as shifter shaft is driven from case.

9. Insert a long narrow drift through bolt hole at rear of case and drive 2nd-3rd shifter shaft out front of case, Fig. 6. Remove fork from case.
10. Using a suitable drift, drive 1st-Reverse shifter shaft out front of case, Fig. 7 and remove fork.
11. Remove selector shaft intermediate lever lock pins, Fig. 8, and remove shaft and levers from case.
12. Remove snap ring from rear bearing retainer groove and slide rear extension from mainshaft assembly, Fig. 9.
13. Remove clutch drive gear from case. Position 1st-Reverse sliding gear to rear of hub shaft and remove mainshaft assembly, Fig. 10. Remove lock pins and detent balls from bottom of case.
14. Insert a punch into shaft rail detent holes and drive out hole plugs and springs.
15. Using tool J-22923, remove reverse idler shaft and gear from case, Fig. 11.

Mainshaft Disassembly

1. Remove snap ring in front of clutch hub, Fig. 12.

NOTE: The synchronizer hubs and sliding sleeves are a selected assembly and should be kept together as originally assembled.

2. Depress retaining clamp and slide speedo drive gear from shaft.
3. Remove snap ring, belleville washer and spacer from shaft, Fig. 13.
4. Support 1st gear and press mainshaft until bearing and synchronizers are free on shaft, Fig. 14. Remove all loose parts.
5. Support 2nd speed gear and press shaft from 2nd-3rd synchronizer and 2nd speed gear, Fig. 15.

Drive Gear Bearing

1. Remove snap ring retaining bearing on shaft.
2. Support outer race of bearing and press drive gear from bearing.
3. Position new bearing on shaft with slinger toward gear and press onto shaft using tool J-5390 or other suitable piece of pipe.
4. Install snap ring on shaft behind bearing, Fig. 16.

Mainshaft Assembly

IMPORTANT: The snychronizer hubs and

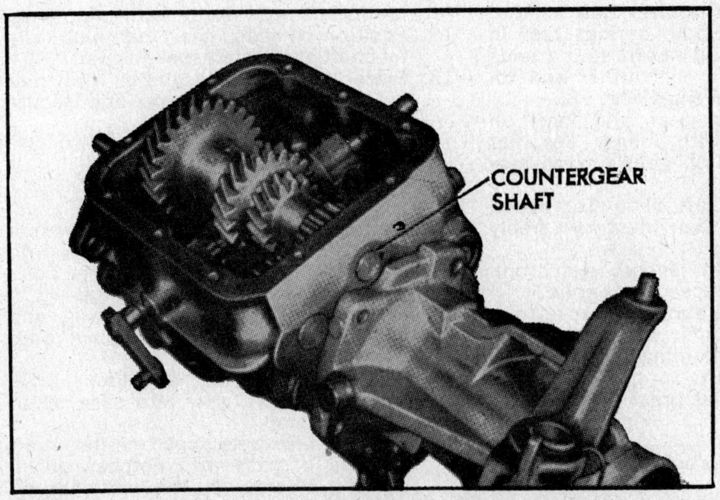

Fig. 4 Countergear shaft exposed for removal

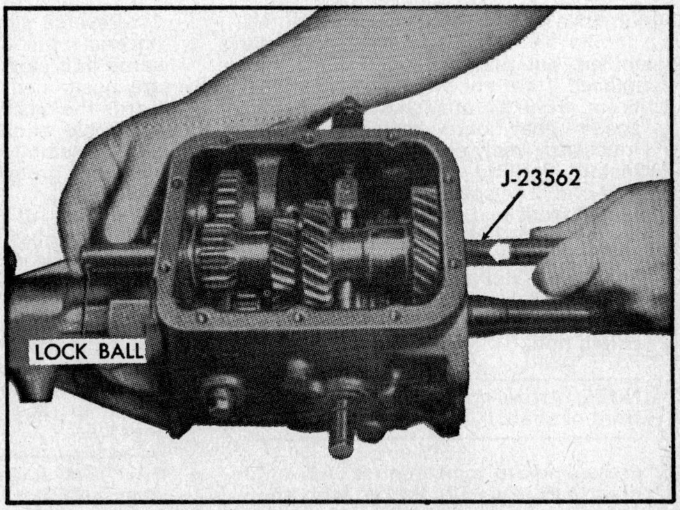

Fig. 5 Removing countergear shaft

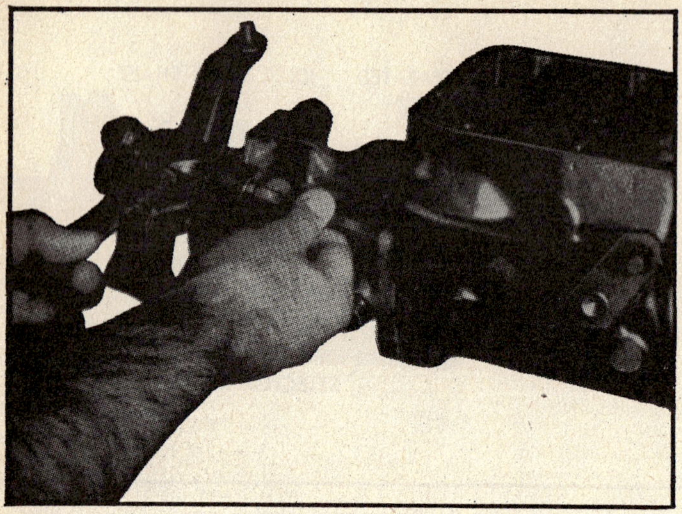

Fig. 6 Removing 2nd-3rd shifter shaft

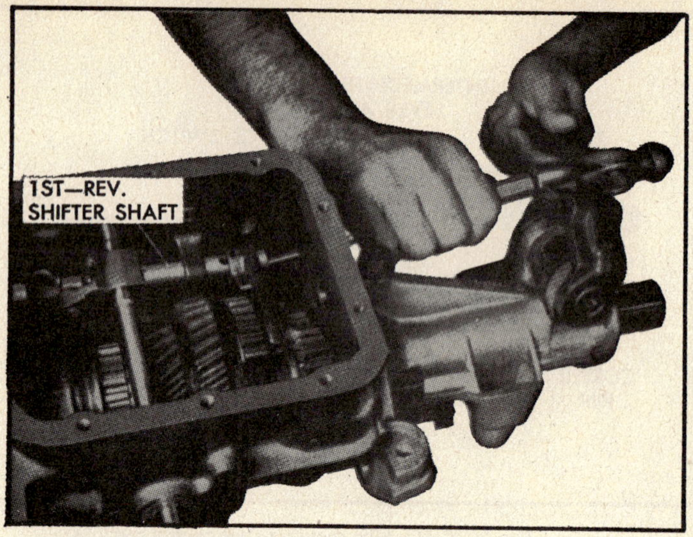

Fig. 7 Removing 1st-Reverse shifter shaft

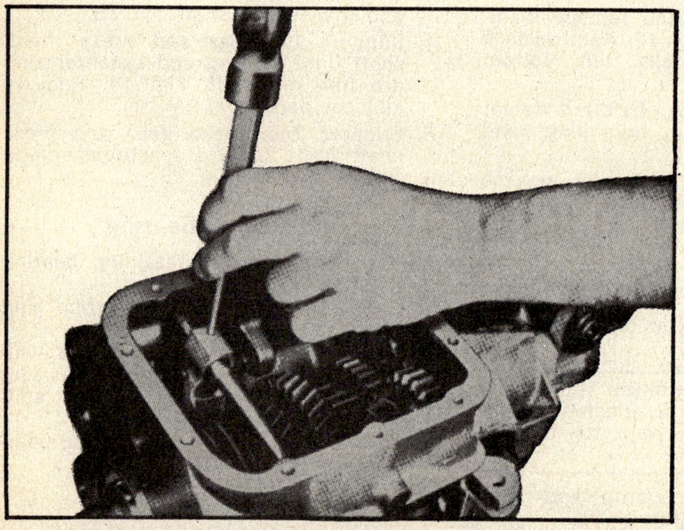

Fig. 8 Removing selector shaft intermediate lever pins

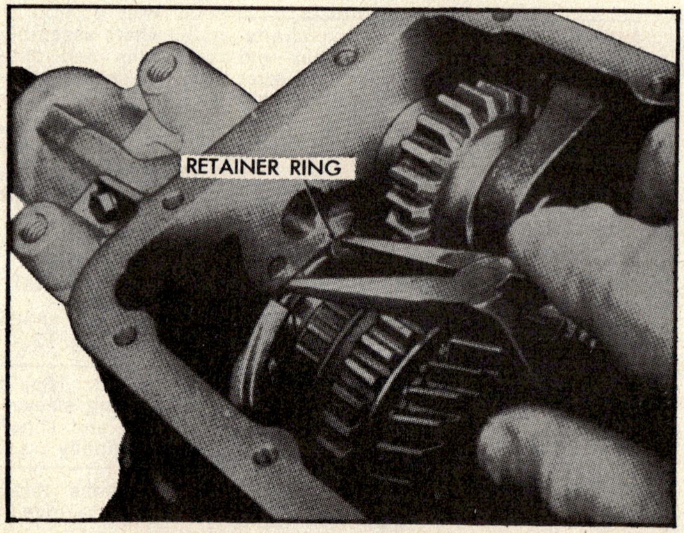

Fig. 9 Releasing rear extension retainer ring

sliding sleeves are a selected assembly and should be kept together as originally assembled, but the keys and springs may be replaced if worn or broken.

1. From front of mainshaft install 2nd speed gear onto mainshaft. Gear must turn freely on shaft.
2. Install 2nd-3rd speed synchronizer ring onto 2nd speed gear cone.
3. Install front and rear synchronizer key springs into 2nd-3rd synchronizer hub so that hooked spring ends rest in the same slot and raised ends are against the blocker rings.
4. Install sliding sleeve and keys on clutch hub.

NOTE: Arrow on keys must point to front of shaft.

5. Press 2nd-3rd synchronizer hub onto mainshaft, Fig. 17. Secure with snap ring.
6. Install both clutch key springs into 1st-Reverse synchronizer hub so that hooked ends of both springs rest in same hub slot and raised spring ends are positioned to each other and towards the blocker rings, Fig. 18.
7. Assemble sliding gear and keys on hub assembly with longer key flat and fork groove on gear toward rear of shaft.
8. From rear of shaft slide 1st speed gear onto shaft. Gear must turn freely on shaft, Fig. 19.
9. Place 1st-Reverse speed synchronizer ring onto 1st speed gear cone.
10. Slide 1st-Reverse synchronizer assembly onto mainshaft. Slide stop ring, rear extension retaining ring and rear bearing onto shaft. Support rear bearing inner race and press components together.

CAUTION: Align slots in synchronizer rings with keys.

11. Place spacer and belleville washer on mainshaft and secure with snap ring.
12. Position speedo gear retaining clip on shaft and install speedo gear.
13. Place mainshaft assembly into rear extension up to its stop and secure with retaining ring.

ASSEMBLE TRANS.

1. Install new gasket onto rear extension and slide mainshaft assembly into case. Install one or two retainer bolts to keep extension from rotating.
2. From the front, slide lock ring and pilot roller bearing assembly onto mainshaft.
3. Install blocker ring on clutch drive gear and install gear into case up to snap ring stop.
4. Insert 1st-Reverse speed shifter shaft at front of case with notches down, pushing it through the shifter fork, positioning fork shoulder toward front of case. Drive lock pin in place allow-

Fig. 10 Removing mainshaft assembly

Fig. 11 Removing reverse idler gear shaft

Fig. 12 Removing snap ring in front of clutch hub

Fig. 13 Removing rear bearing snap ring

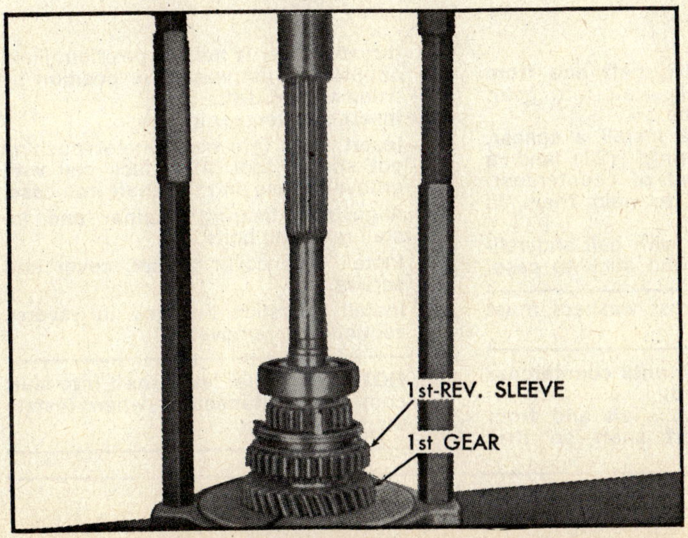

Fig. 14 Removing rear bearing from mainshaft

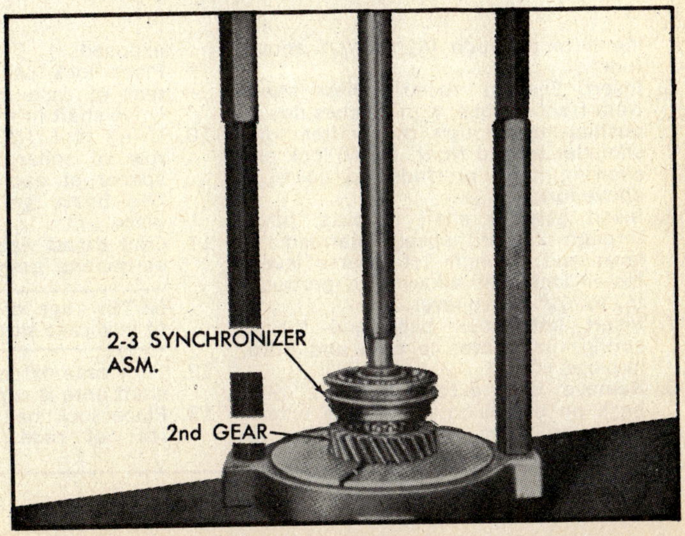

Fig. 15 Removing 2nd-3rd clutch hub from mainshaft

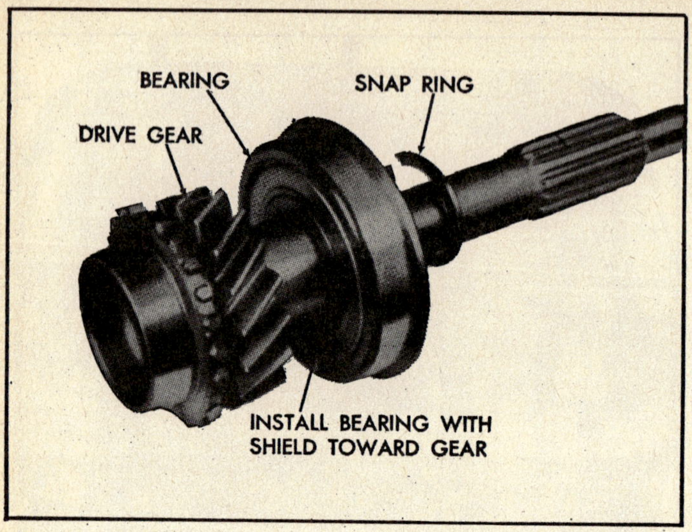

Fig. 16 Drive gear assembly

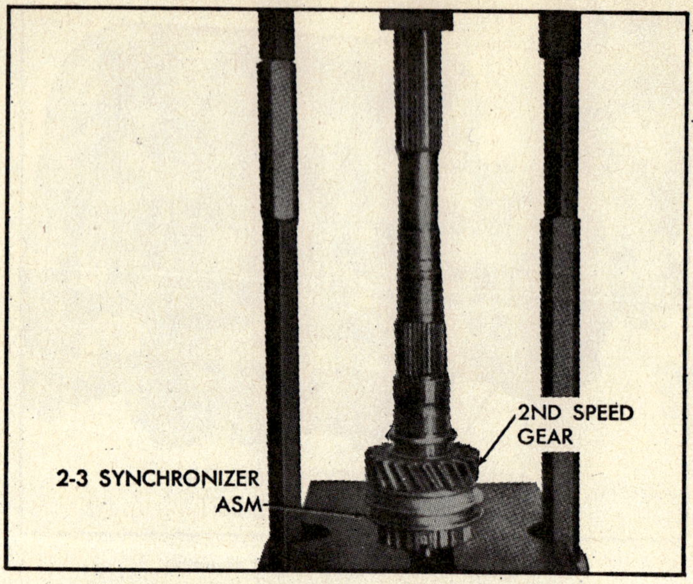

Fig. 17 Pressing 2nd-3rd synchronizer on mainshaft

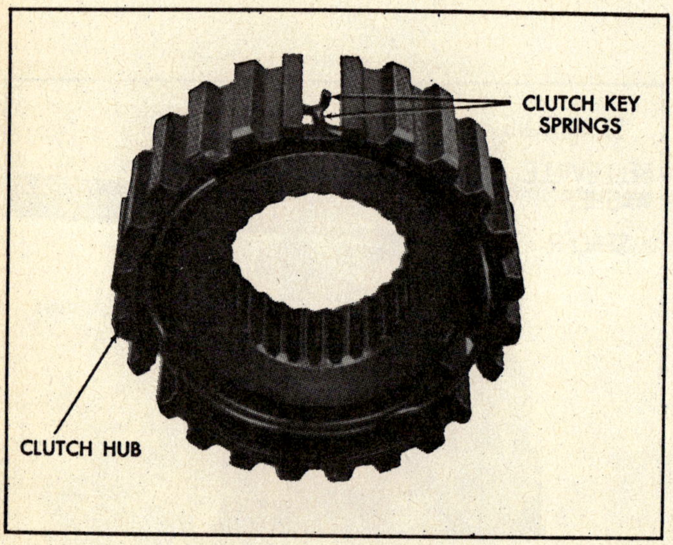

Fig. 18 Both clutch key springs installed

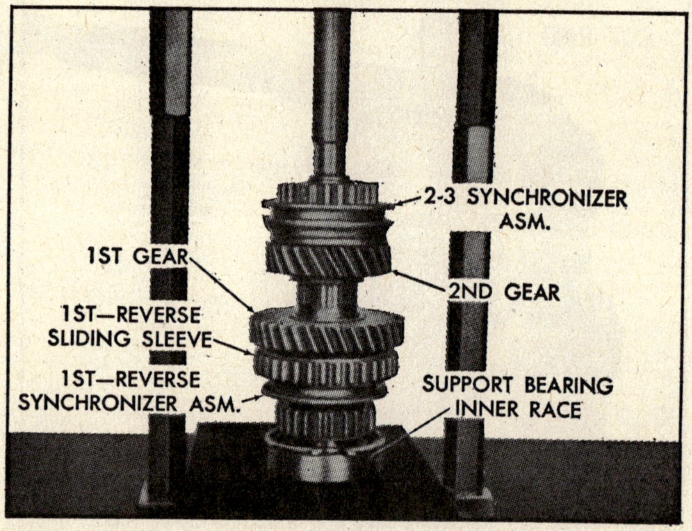

Fig. 19 Pressing mainshaft components together

ing it to protrude 1/16 to 5/64″ above fork.

5. Insert 2nd-3rd speed shifter shaft from front of case, with notches down, pushing it through the shifter fork shoulder toward front. Install lock pin allowing it to protrude 1/16 to 5/64″ above fork.
6. Insert selector shaft in case, push through 2nd-3rd speed intermediate lever and through 1st-Reverse lever. Install lock pins allowing to protrude 1/16 to 5/64″ above lever.
7. Insert both lock balls and thrust springs into bores in case and drive in plugs.
8. Remove rear extension bolts, pull back on extension and rotate extension until bore for reverse idler is exposed.
9. Place lock ball into shaft and from rear of case install shaft into gear. Drive shaft into place.
10. Using tool J-23562 install a spacer, row of roller bearings (24) and a spacer at each end of countergear. Use heavy grease to hold them in place.
11. Coat thrust washer with ball and roller bearing grease and stick to case.

NOTE: Lugs of thrust washers must fit into case slots.

12. Turn case extension until countergear shaft bore is exposed.
13. Place lock ball into shaft and from rear of case insert shaft so that thrust washer is held in position. Hold opposite thrust washer in position by using a short drift.
14. Insert countergear into case.
15. Insert shaft into countergear pushing out special tool. Align lock ball with groove in case and tap shaft into case.
16. Align rear bearing retainer and install retaining bolts.
17. Install case cover gasket, cover and screws.
18. Install gearshift linkages in reverse sequence to removal.

NOTE: Start idler lever shaft into shift control simultaneously when installing selector ring.

FOUR SPEED MANUAL SHIFT TRANSMISSIONS

See Car Chapters for Linkage Adjustments and Removal Procedures

APPLICATION INDEX

Type One Warner T-10

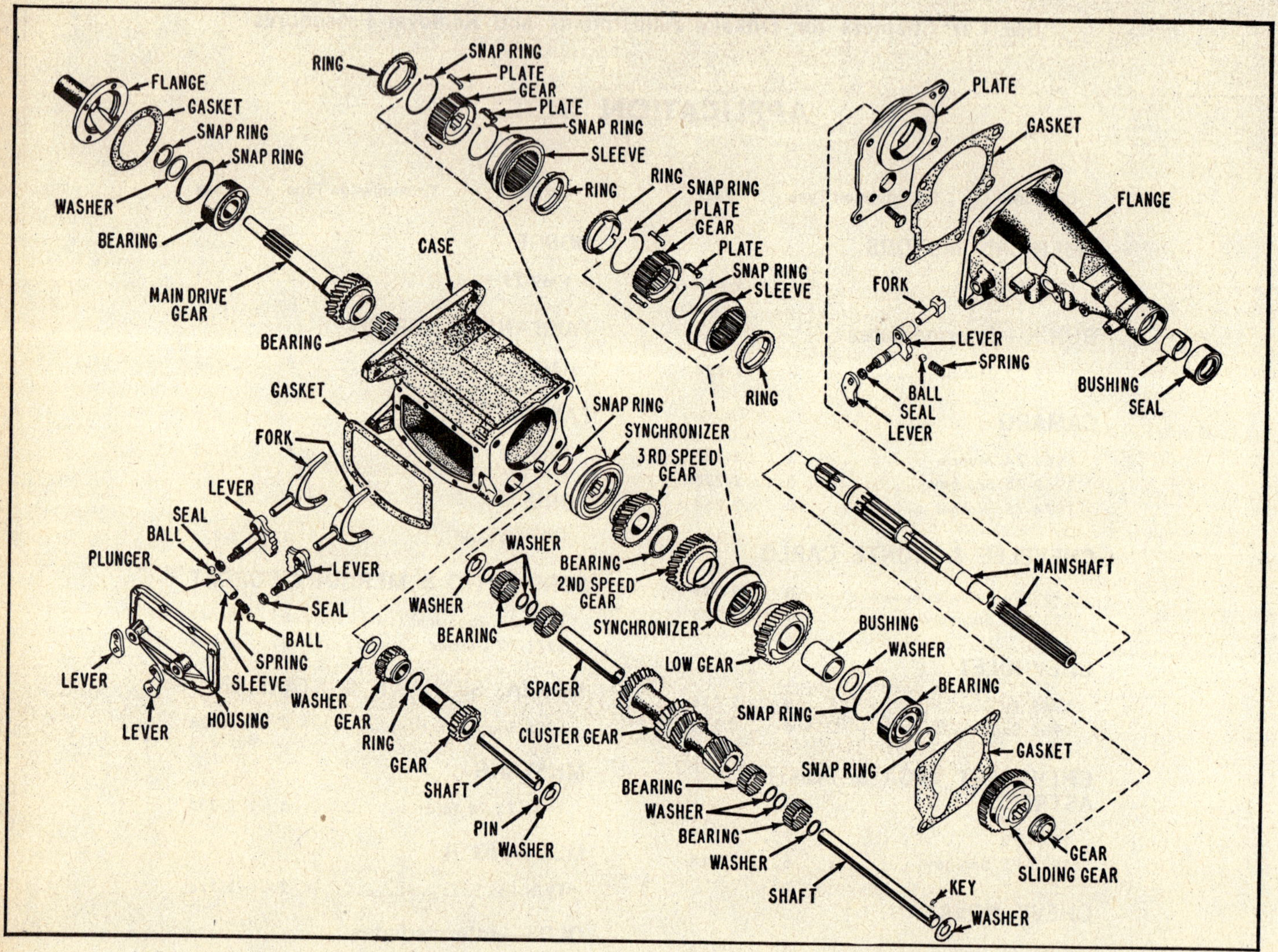

Fig. 1 Type 1 Warner T-10 four speed transmission. Some units do not have the mainshaft low gear bushing shown.

Side Cover, Remove

1. Disconnect control rods from levers.
2. Remove cover assembly from transmission and allow oil to drain.
3. Remove outer shift lever nuts and lockwashers and pull levers from shaft.
4. Carefully push shifter shafts into cover, allowing detent balls to fall free, then remove both shifter shafts.
5. Remove interlock sleeve, pin and poppet spring.

Side Cover, Install

1. Install interlock sleeve and one shifter shaft. Place detent ball into sleeve followed by poppet spring and interlock pin.
2. Start second shifter shaft into position and place second detent ball on poppet spring. Compress ball and spring with screwdriver and push shifter shaft fully in.
3. With transmission in neutral and shifter forks and levers in place, lower side cover into place. Install attaching bolts, using sealer on bolts to prevent leakage. Tighten bolts evenly.

Disassemble Transmission

1. Remove side cover as outlined above.
2. Remove front bearing retainer.
3. Drive lock pin from botton side of reverse shifter lever boss, Fig. 3, and pull shifter out about ⅛". This disengages reverse shift fork from reverse gear.
4. Unfasten extension housing from rear bearing retainer (adapter). Tap extension with soft hammer in rearward direction to start. When reverse idler shaft is out as far as it will go, move extension to left so reverse fork clears reverse gear, and remove extension and gasket.
5. Remove speedometer drive gear with a suitable puller.
6. The rear reverse idler gear, tanged thrust washer and reverse gear may now be removed, Figs. 4 and 5.
7. Remove self-locking bolts attaching rear bearing retainer (adapter) to transmission case. Then carefully remove entire mainshaft assembly.
8. Lift front reverse ider gear and thrust washer from case, Fig. 6.

Fig. 3 Removing reverse shifter lock pin

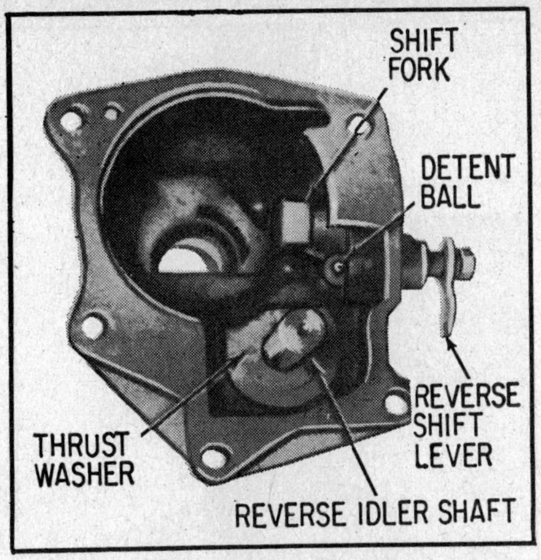

Fig. 4 Reverse idler shaft and reverse shifter

Fig. 5 Reverse gear rear idler removal

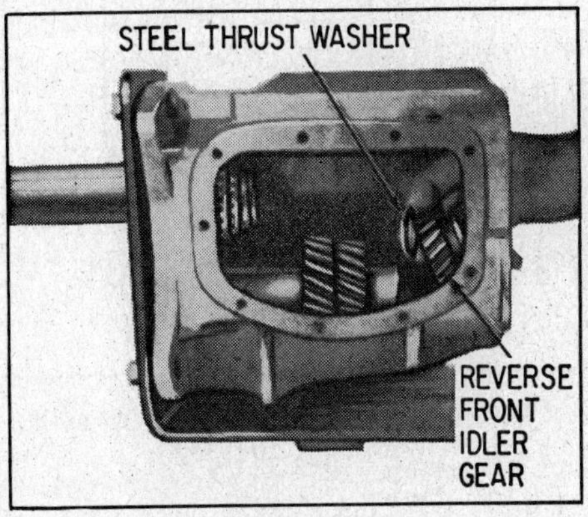

Fig. 6 Reverse gear front idler removal

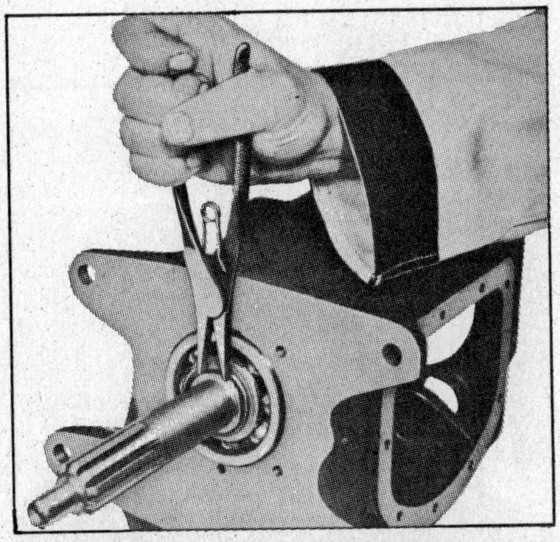
Fig. 7 Removing main drive gear snap ring

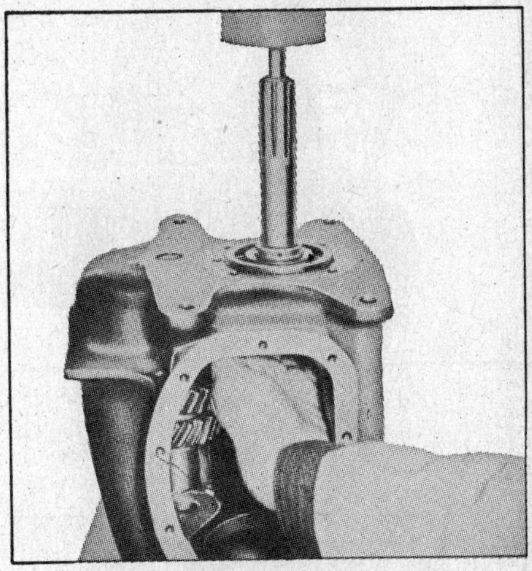
Fig. 8 Removing main drive gear from front bearing

9. Unload bearing rollers from main drive gear and remove 4th speed synchronizing ring.
10. Remove main drive gear snap ring, Fig. 7, and remove spacer washer.
11. With soft hammer, tap main drive gear down from front bearing, Fig. 8.
12. From inside case, tap out front bearing and snap ring.
13. From front of case, tap out countershaft, Fig. 9, using a dummy shaft as shown. Remove countergear and both tanged thrust washers.
14. Remove dummy shaft, all 80 rollers and six spacers from countergear.
15. Remove mainshaft front snap ring, Fig. 10, and slide 3-4 speed synchronizer, 3rd speed gear and synchronizing ring, 2-3 speed gear thrust washer (needle roller bearing), 2nd speed gear and synchronizing ring from front of mainshaft.
16. Spread rear bearing snap ring and press mainshaft out of retainer, Fig. 11.
17. Remove mainshaft rear snap ring. Support 1-2 speed synchronizer assembly Fig. 12, and press on rear of mainshaft to remove shaft from remaining parts on shaft.

Cleaning & Inspection

1. Wash transmission case inside and out with cleaning solvent and inspect for cracks. Inspect front face of case for burrs and, if present, dress them off with a fine cut mill file.
2. Wash front and rear bearings in cleaning solvent. Blow out bearings with compressed air. *Do not allow bearings to spin; turn them slowly by hand. Spinning bearings will damage race and balls.*
3. Make sure bearings are clean, then lubricate them with light engine oil and check them for roughness. Roughness may be determined by turning outer race by hand.
4. All main drive gear and countergear bearing rollers should be inspected closely and replaced if they show wear. Inspect countershaft and replace if necessary. Replace all worn spacers.
5. Inspect all gears and first speed gear bushing (or sleeve) and, if necessary, replace all that are worn or damaged.

Synchronizers

Clutch hubs and sliding sleeves are a selected assembly and should be kept together as originally assembled, but the three keys and two springs may be replaced if worn or broken.

Push hub from sliding sleeve. Keys will fall free and springs may be easily removed. To assemble, place the two springs in position (one on each side of hub) so tanged end of each spring falls into same keyway in hub. Place keys in position and, holding them in place, slide hub into sleeve.

Assemble-Mainshaft

1. From rear of mainshaft, assemble 1-

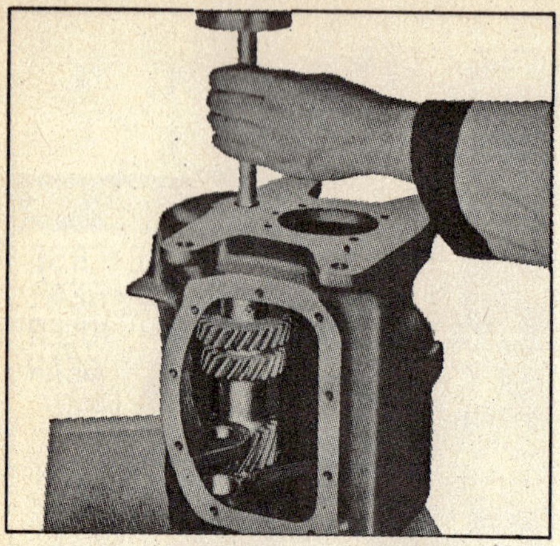

Fig. 9 Removing countershaft

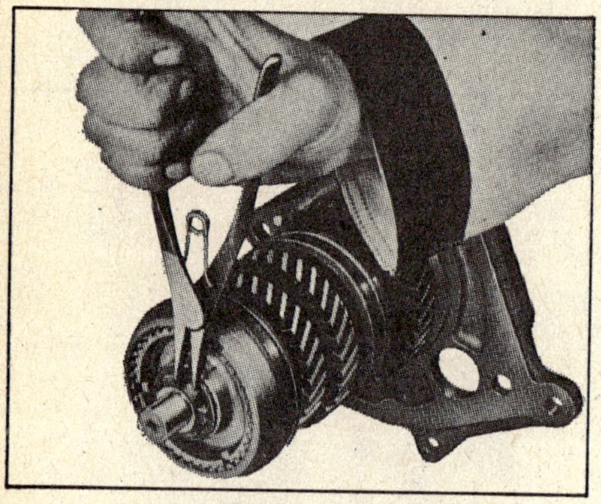

Fig. 10 Removing mainshaft front snap ring

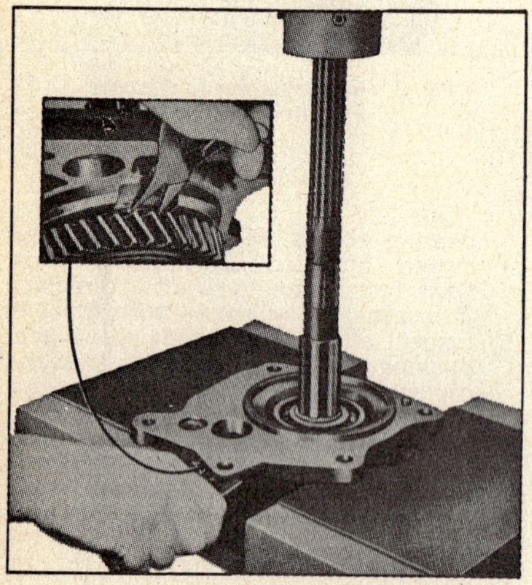

Fig. 11 Removing mainshaft from rear bearing retainer

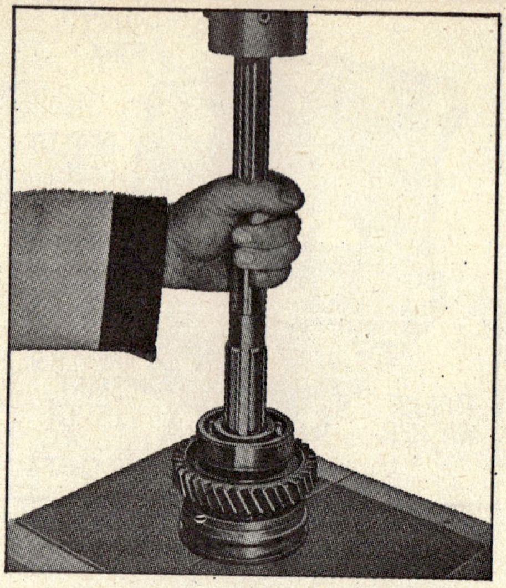

Fig. 12 Removing mainshaft from rear bearing and synchronizer

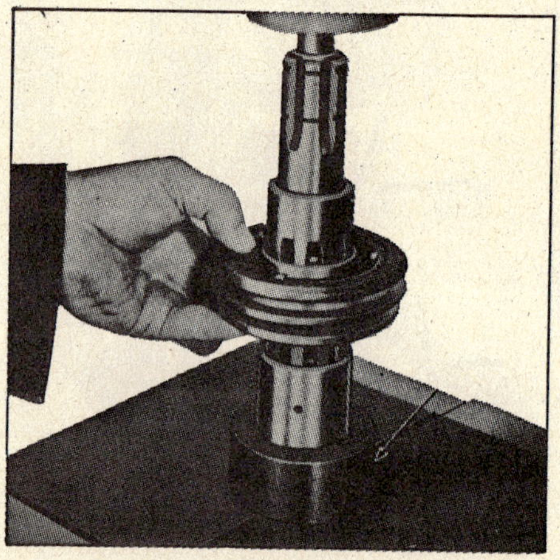

Fig. 13 Installing first speed gear bushing

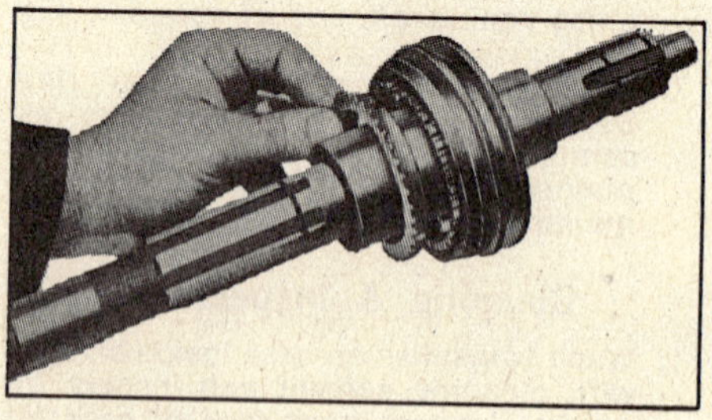

Fig. 14 Installing synchronizer ring

Fig. 15 Installing rear bearing retainer

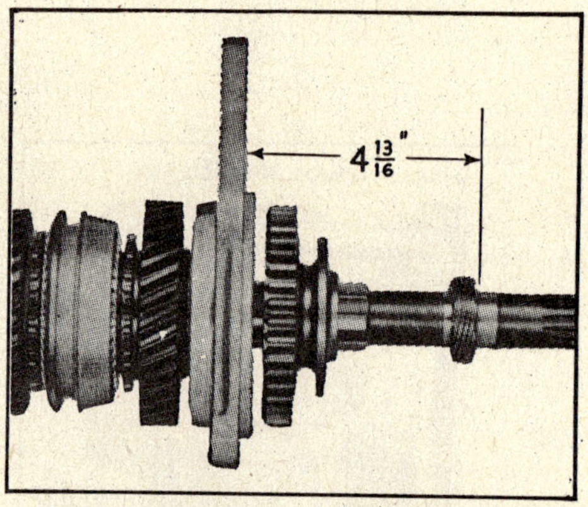

Fig. 16 Installing speedometer drive gear

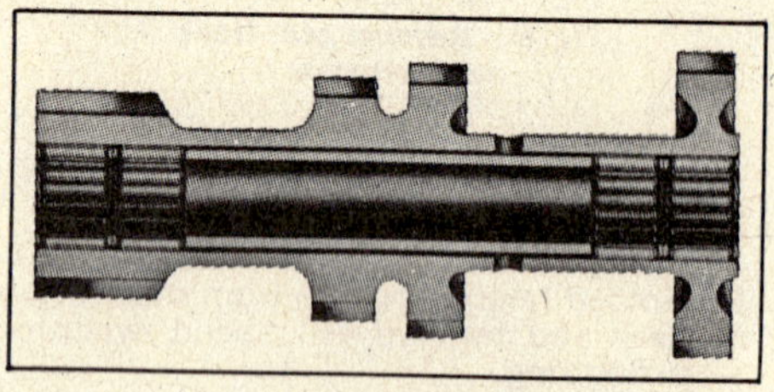

Fig. 17 Cross section of countergear assembly

2 speed synchronizer to mainshaft with sleeve taper toward rear. Press 1st gear bushing on shaft, Fig. 13.

2. Install 1st gear synchronizing ring so that notches in ring correspond to keys in hub, Fig. 14.
3. Install 1st gear with hub toward front, and 1st gear thrust washer. Make certain that grooves in washer are facing 1st gear.
4. Press on rear bearing with snap ring groove toward front of transmission.
5. Choose correct selective fit snap ring and install it in groove in mainshaft behind rear bearing. *Always use new snap ring and do not expand it further than necessary for assembly.*
6. From front of mainshaft, install 2nd gear synchronizing ring so notches in ring correspond to keys in hub.
7. Install 2nd gear with hub of gear toward back of transmission. Install 2-3 speed gear thrust washer (needle roller bearing).
8. Install 3rd gear with hub to front of transmission, and 3rd gear synchronizing ring with notches to front of transmission.
9. Install 3-4 synchronizer with sleeve taper toward front, making sure keys in hub correspond to notches

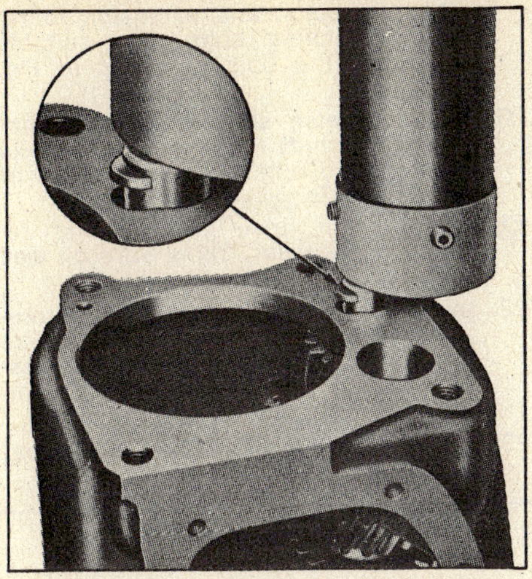

Fig. 18 Installing countershaft

Fig. 19 Checking countershaft end play

Fig. 20 Installing mainshaft assembly

with 3rd gear synchronizing ring.

10. Install snap ring in groove in mainshaft in front of 3-4 synchronizer. *With the correct size snap ring, 3rd speed gear will have .010-.015" end play with 3-4 synchronizer hub forward against snap ring.*
11. Install rear bearing retainer, Fig. 15. Spread selective fit snap ring in plate to allow snap ring to drop around rear bearing and press end of mainshaft until snap ring engages groove in rear bearing (use largest size snap ring that will fit into groove).
12. Install reverse gear with shift collar to rear.
13. Press speedometer drive gear onto mainshaft. Position speedometer gear as shown in Fig. 16.

Assemble Countergear

1. Install roller spacer in countergear.
2. Using heavy grease to retain rollers, install 20 rollers in either end of countergear, two spacers, 20 more rollers, then one spacer, Fig. 17.
3. Assemble rollers and spacers in the same manner in other end of countergear. Then insert dummy shaft in countergear to retain rollers.

Assemble Transmission

1. Rest case on its side with side cover opening toward you. Place countergear tanged thrust washers in place, retaining them with heavy grease, making sure that tangs are resting in notches in case.
2. Place countergear assembly in bottom of case, making sure that tanged thrust washers are not knocked out of place.
3. Press bearing onto main drive gear with snap ring groove to front.
4. Install spacer washer and selective fit snap ring in groove on gear stem.
5. Install main drive gear assembly through side cover opening and into position in transmission front bore. Tap lightly into place with a soft hammer, if necessary. Place snap ring in groove in front bearing.
6. With transmission resting on its front face, move countergear into mesh with main drive gear, making sure thrust washers remain in place. Install key in end of countershaft and, from front of case, tap or press shaft, Fig. 18, until end of shaft is flush with rear of case and dummy shaft is displaced.
7. Attach dial indicator as shown in Fig. 19 and check end play of countergear. End play must not be more than .025".
8. Install 14 rollers into main drive gear, using heavy grease to hold them in place. Place gasket in position on front face of rear bearing retainer, using heavy grease to hold it in position.
9. Install 4th gear synchronizing ring on main drive gear with clutch key notches toward rear of case.
10. Position reverse idler gear thrust washer (untanged) on machined face of ear cast in case for reverse idler shaft. Position front reverse idler gear on top of thrust washer, with hub facing toward rear of case.
11. Lower mainshaft assembly into case, making certain that notches on 4th gear synchronizing ring correspond to keys in synchronizer, Fig. 20.

Fig. 21 Self-locking bearing retainer-to-case bolt

Fig. 22 Installing extension housing on transmission case

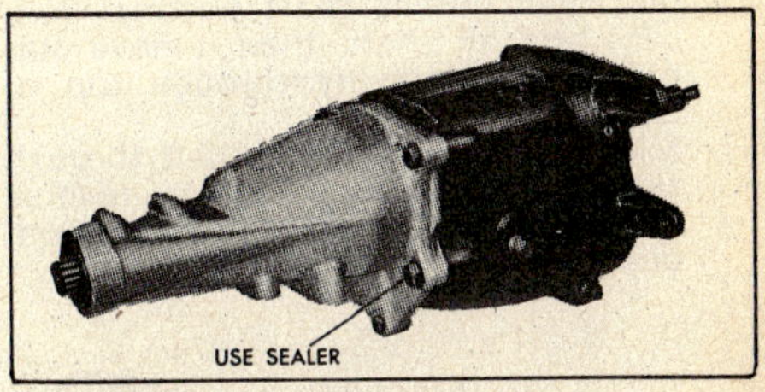

Fig. 23 Sealing lower right attaching bolt

12. Install self-locking bolt attaching rear bearing retainer to case, Fig. 21.
13. From rear of case, insert rear reverse idler gear, engaging splines with portion of gear within case.
14. Using heavy grease, place gasket into position on rear face of rear bearing retainer.
15. Install remaining tanged thrust washer into place on reverse idler shaft, being sure tang on washer is in notch in idler thrust face of extension.

16. Place two synchronizers in neutral position. *If locking-up of gears is encountered, a small amount of petrolatum may be applied to the 1st speed gear synchronizing ring, enabling it to turn freely on 1st speed gear hub.*

17. Pull reverse shifter shaft to left side of extension and rotate shaft to bring reverse shift fork to extreme forward position in extension. Line up front and rear reverse idler gears, making sure front thrust washer is in place.
18. Start extension into case. Fig. 22, by carefully inserting reverse idler shaft through reverse idler gears. Slowly push it on shifter shaft until shift fork engages reverse gear shift collar. When fork engages, rotate shifter shaft to move reverse gear rearward, permitting extension to slide onto transmission case.
19. Install extension and retainer to case attaching bolts, and extension to retainer attaching bolts. Use suitable sealer on the bolt indicated in Fig. 23.
20. Adjust reverse shifter shaft so that groove in shaft lines up with hole in boss and drive in lock pin from top of boss, Fig. 22.
21. Install main drive gear bearing retainer and gasket, being sure oil well lines up with oil outlet hole.
22. Install shift fork in each synchronizer sleeve. With both synchronizers in neutral, install side cover with gasket. Use suitable sealer when installing lower right cover bolt.
23. Install shifter levers, lock washers and nuts.

Type Two

Transmission Disassembly

1. Drain lubricant from transmission by removing the lower extension housing bolt.
2. Drive access plug from rear of extension housing and remove offset lever assembly, Fig. 2-1.
3. Remove extension housing from case.
4. Remove transmission cover, shifter fork and shift rod assembly.
5. Remove front bearing retainer and gasket.
6. Remove spring clip securing reverse lever assembly to pivot bolt, then remove pivot bolt and reverse level assembly.
7. Support countershaft gear with a wire hook. Drive a dummy countershaft through front of case until cluster gear drops to bottom of case, then remove counter shaft through rear of case and lower the cluster gear to bottom of case.
8. Remove input shaft, Fig. 2-2.
9. Remove speedometer drive gear snap ring, slide gear off output shaft and remove lock ball from output shaft.
10. Remove output shaft bearing retaining snap ring. Using the outer snap ring, pull output shaft bearing from case and off shaft, then remove output shaft assembly through top of case.
11. Slide reverse idler gear shaft through rear of case, then remove reverse gear and cluster gear from bottom of case.

Fig. 2-2 Removing input shaft. RAD four speed transmission

Cover Disassembly

1. Remove detent screw, spring and plunger, Fig. 2-1.
2. Pull shifter rod shaft rearward and rotate counter-clockwise. Remove manual selector and interlock to shaft spring pin, shifter shaft, manual selector and the interlock from cover.
3. Remove the first-second shift fork, then the third-fourth shift fork.
4. Reverse procedure for reassembly.

Output Shaft Disassembly

1. Scribe alignment marks on sychronizer and blocker ring for alignment during installation, remove snap ring from front of shaft and slide third and fourth speed synchronizer assembly, blocker rings and third speed gear off shaft, Fig. 2-1.
2. Remove next snap ring, then second speed gear thrust washer, second speed gear and blocker ring from shaft.

NOTE: The first and second speed synchronizer hub cannot be removed from output shaft.

3. Remove first gear thrust washer from rear of shaft, first gear spring pin retainer and slide first gear and blocker ring off shaft.

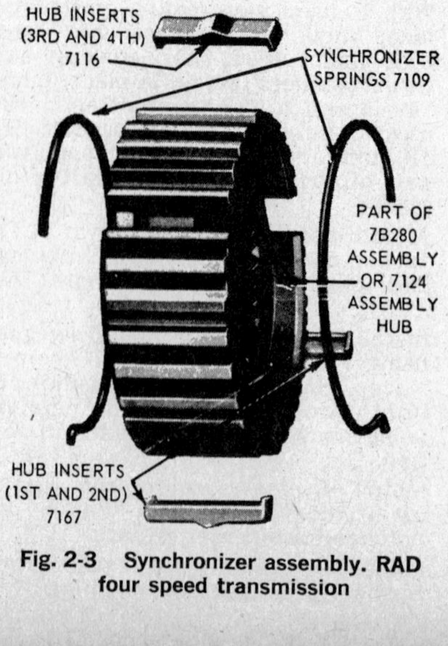

Fig. 2-3 Synchronizer assembly. RAD four speed transmission

Output Shaft Reassembly

1. Place blocker ring on cone of first gear and slide assembly onto output shaft, engaging notches in blocker ring with inserts in synchronizer, Fig. 2-3 and install first gear spring pin.
2. Place blocker ring on cone of second gear and slide assembly onto output shaft, engaging notches in blocker ring with inserts in synchronizer. Install second gear thrust washer and new snap ring.
3. Place blocker ring on cone of third gear and slide assembly onto output shaft, install third and fourth speed synchronizer, engaging notches on blocker ring with inserts on synchronizer. Install new snap ring.
4. Install first gear thrust washer and first gear spring pin retainer onto shaft.

NOTE: Oil grooves in thrust washer must be positioned against gear.

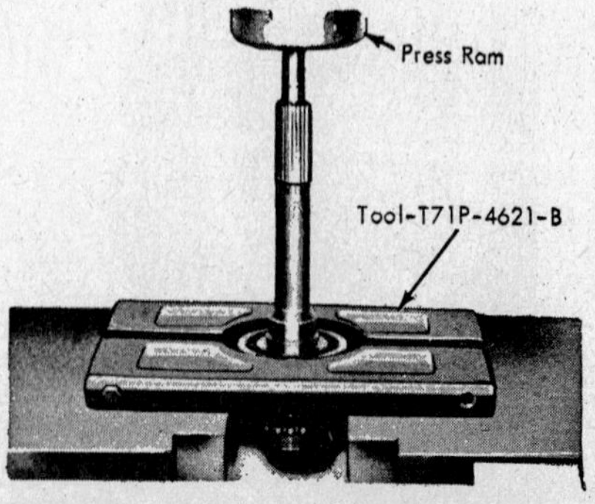

Fig. 2-4 Input shaft bearing removal. RAD four speed transmission

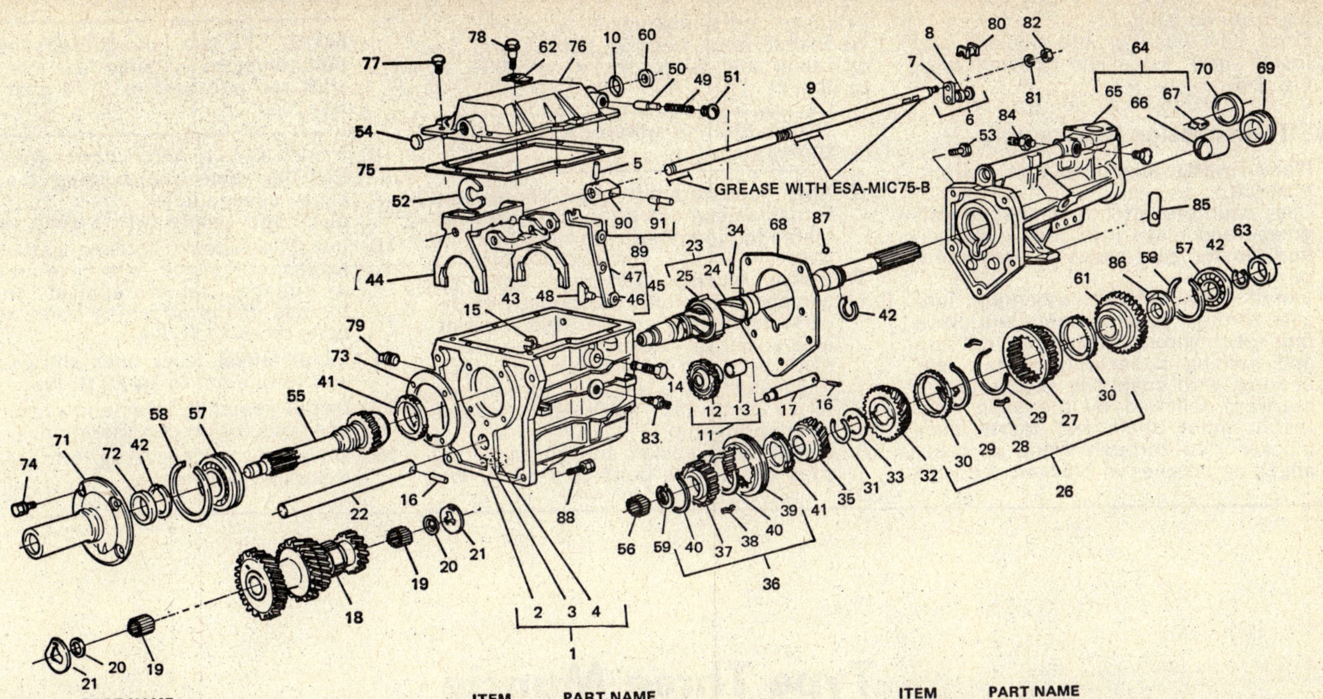

ITEM	PART NAME
1.	TRANS. CASE ASSY.
2.	CASE
3.	TRANS. CASE CHIP MAGNET
4.	SPRING NUT
5.	PIN
6.	LEVER ASSY.
7.	OFFSET LEVER
8.	PIN
9.	SHIFTER SHAFT
10.	O-RING SEAL
11.	REVERSE IDLER ASSY.
12.	REVERSE IDLER GEAR
13.	BUSHING
14.	REVERSE GEAR SELECTOR FORK PIVOT PIN
15.	RETAINING RING
16.	PIN
17.	REVERSE IDLER GEAR SHAFT
18.	COUNTERSHAFT GEAR
19.	COUNTERSHAFT ROLLER BEARINGS
20.	WASHER
21.	COUNTERSHAFT GEAR THRUST WASHERS
22.	COUNTERSHAFT
23.	OUTPUT SHAFT ASSY.
24.	OUTPUT SHAFT
25.	1ST & 2ND GEAR CLUSTER SYNCHRONIZER HUB
26.	OUTPUT SHAFT GEAR ASSY.
27.	REVERSE SLIDING GEAR
28.	SYNCHRONIZER HUB INSERT
29.	SYNCHRONIZER RETAINING SPRING
30.	SYNCHRONIZER BLOCKING RING
31.	2ND SPEED GEAR RETAINING SNAP RING
32.	2ND SPEED GEAR
33.	2ND SPEED GEAR THRUST WASHER
34.	PIN
35.	3RD SPEED GEAR
36.	3RD & 4TH SPEED SYNCHRONIZER ASSY.
37.	SYNCHRONIZER CLUTCH HUB
38.	SYNCHRONIZER HUB INSERTS
39.	SYNCHRONIZER SLEEVE
40.	SYNCHRONIZER RETAINING SPRINGS
41.	SYNCHRONIZER BLOCKING RING
42.	SNAP RING
43.	1ST & 2ND GEAR SHIFT FORK
44.	3RD & 4TH GEAR SHIFT FORK
45.	REVERSE GEAR SHIFT RELAY LEVER ASSY.
46.	RELAY LEVER RING
47.	RELAY LEVER
48.	REVERSE GEAR SHIFT FORK
49.	SHIFTER INTERLOCK SPRING
50.	MESHLOCK PLUNGER
51.	SCREW
52.	GEAR SELECTOR INTERLOCK PLATE
53.	EXTENSION HOUSING RETAINING BOLT
54.	WELSH PLUG
55.	INPUT SHAFT
56.	MAINSHAFT ROLLER BEARINGS
57.	BALL BEARING
58.	BEARING RETAINING SNAP RING
59.	RETAINING RING
60.	SHIFT SHAFT SEAL
61.	1ST SPEED GEAR
62.	SPARK CONTROL SWITCH WIRE RETAINING CLIP
63.	SPEEDOMETER DRIVE GEAR
64.	EXTENSION ASSY.
65.	EXTENSION HOUSING
66.	EXTENSION HOUSING BUSHING
67.	SHIFT LEVER REVERSE STOP
68.	EXTENSION HOUSING GASKET
69.	EXTENSION HOUSING OIL SEAL
70.	EXTENSION HOUSING PLUG
71.	INPUT SHAFT BEARING RETAINER
72.	INPUT SHAFT OIL SEAL
73.	INPUT SHAFT BEARING RETAINER GASKET
74.	INPUT SHAFT RETAINING BOLT
75.	CASE COVER GASKET
76.	CASE COVER
77.	CASE COVER RETAINING BOLT
78.	BOLT
79.	OIL FILLER PLUG
80.	GEAR SHIFT DAMPER BUSHING
81.	SPRING LOCK WASHER
82.	LEVER RETAINING NUT
83.	BACK-UP LAMP SWITCH
84.	SEAT BELT WARNING SENSOR SWITCH
85.	TRANSMISSION IDENTIFICATION TAG
86.	1ST GEAR THRUST WASHER
87.	BALL
88.	TRANS. RETAINING BOLT
89.	TRANS. CONTROL SELECTOR ARM ASSY.
90.	CONTROL SELECTOR ARM
91.	GEARSHIFT RETAINING PIN

Fig. 2-1 RAD four speed transmission disassembled.

Countershaft Gear Bearing, Replace

1. Remove dummy shaft, bearing retainer washers and needle bearings from gear.

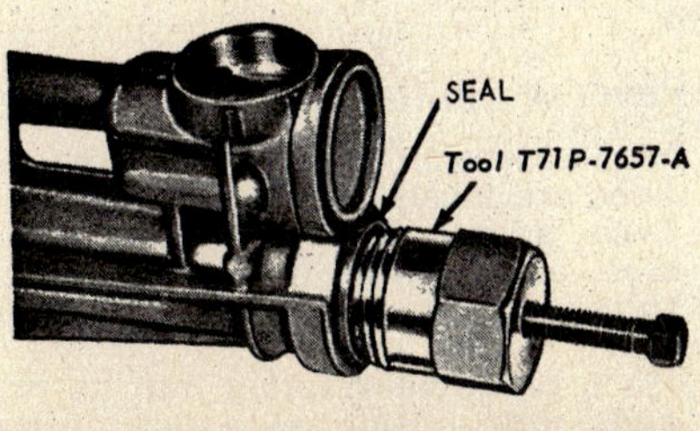

Fig. 2-5 Extension housing seal removal. RAD four speed transmission

2. Coat bore of gear with grease and holding dummy shaft in the gear, install needle bearings and retainer washers at each end of gear.

Input Shaft Bearing, Replace

1. Remove needle bearings from bore of input shaft.
2. Remove input shaft bearing snap ring and press bearing from shaft, Fig. 2-4.
3. Press new bearing onto shaft with snap ring groove facing front of shaft and install new snap ring.
4. Lightly coat bore of input shaft with grease and install needle bearings.

NOTE: A heavy coat of grease will plug bearing lubrication holes.

Extension Housing Bushing and Seal, Replace

1. Remove seal from extension housing, Fig. 2-5.
2. Using a suitable puller, remove bush-

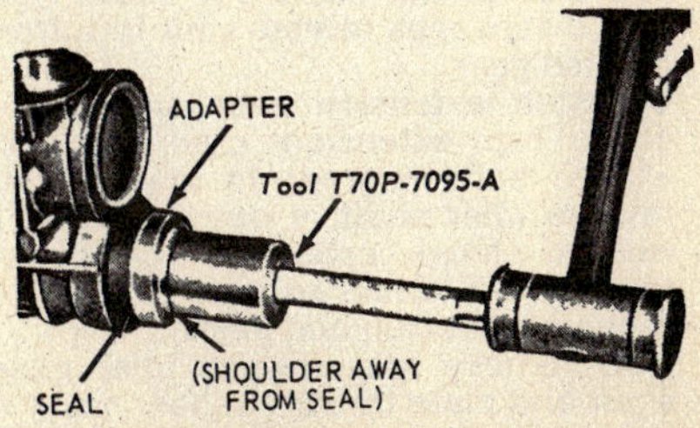

Fig. 2-6 Extension housing seal installation. RAD four speed transmission

ing from housing.
3. Drive new bushing into housing and install new extension housing seal, Fig. 2-6.

Transmission Reassembly

1. Place reverse idler gear and shaft in position.
2. Coat countershaft thrust washer with grease and place into case and position cluster gear assembly in bottom of case.
3. Install output shaft assembly into case through cover opening and place first gear thrust washer on roll pin and holding it securely. Install rear bearing, with snap ring groove facing rearward followed by the snap ring.
4. Install input shaft and fourth gear blocker ring through front of case, engaging notches in blocker ring with inserts on synchronizer.
5. Install front bearing gasket and retainer and torque bolts to 11-15 ft. lbs.

NOTE: Apply a suitable sealer to bolt threads.

6. Align countershaft gear bore and thrust washers with case bore and install countershaft through rear of case.
7. Install reverse idle gear lever assembly with fork positioned in reverse idler gear groove. Apply a suitable sealer to threads of reverse lever pivot bolt and install bolt into case. Align lever on pivot bolt, torque bolt to 15-25 ft. lbs., and install reverse lever spring clip.
8. Install cover gasket and cover, wiring clips and torque bolts to 7-10 ft. lbs.

NOTE: The two (shouldered) locating bolts must be installed first and place shift rail into first or third gear position.

9. Install speedometer drive gear lock ball into hole and holding the ball, slide speedometer drive gear into place and secure with a snap ring.
10. Install extension housing gasket and extension housing onto case and with a suitable sealer applied to the threads of the attaching bolts, torque bolts to 18-27 ft. lbs.
11. Install offset lever onto shifter shaft and torque nut to 14-20 ft. lbs.
12. Install gearshift lever and check gear positions for proper operation.
13. Drive new access plug into rear of extension housing.

Type Three Muncie

SIDE COVER, REPLACE

1. Referring to Fig. 3-2 disconnect control rods from levers.
2. Shift transmisson into 2nd speed before removing cover by moving 1-2 shifter lever into forward detent position.
3. Remove cover from case.
4. Reverse procedure to install the cover, being sure first to shift the transmission into 2nd gear. Make sure shift forks are aligned with their respective grooves in synchronizer sliding sleeves.

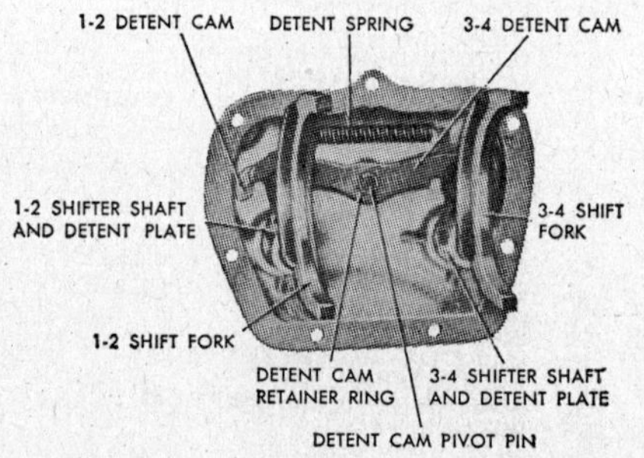

Fig. 3-2 Transmission side cover

DISASSEMBLE TRANSMISSION

1. Remove side cover.
2. Remove front bearing retainer.
3. Shift transmission into two gears at once to keep mainshaft from turning, then remove retainer nut from main drive gear.
4. With gears in neutral, drive lock pin from reverse shifter lever boss and pull shifter shaft out about ⅛". This disengages reverse shift fork from reverse gear.
5. Unfasten extension case from main case. Tap extension case rearward with a soft hammer to start. When reverse idler shaft is out as far as it will go, move extension to left so reverse fork clears reverse gear. Then remove extension and gasket.
6. Now remove rear reverse idler gear, shaft and plate thrust washer.
7. Use a suitable puller to remove speedometer gear from mainshaft, after which remove reverse gear.
8. Slide 3-4 synchronizer clutch sleeve to 4th gear position before trying to remove mainshaft assembly from case.
9. Remove rear bearing retainer and mainshaft assembly from case by tapping retainer with soft hammer.
10. Unload bearing rollers from main drive gear and remove 4th gear synchronizer blocking ring.
11. Lift front half of reverse idler gear and its tanged washer from case.
12. Press main drive gear down into case and remove. From inside of case, tap out front bearing and snap ring.
13. From front of case, press out countershaft. Then remove countergear and both tanged washers.
14. Remove mainshaft front snap ring and strip mainshaft of loose parts.
15. Spread rear bearing retainer snap ring and press mainshaft out of retainer.

NOTE: Early models of this transmission use a snap ring behind the 1-2 synchronizer hub to retain the hub in position while later units have a sleeve within the 1st speed gear. This sleeve acts as a spacer between the rear bearing and 1-2 synchronizer hub.

16. Remove mainshaft rear snap ring. Support 1st gear in a press and press against rear of shaft to remove it from rear bearing, 1st gear thrust washer, 1st gear and synchronizing ring.
17. Remove 1-2 synchronizer snap ring and remove 1-2 synchronizer unit, 2nd gear synchronizer ring and 2nd gear from shaft, Fig. 3-3.

UNIT REPAIRS

Reverse Idler

Because of the high degree of accuracy to which the reverse idler gear bushings are machined, the bushings are not serviced separately. Check bushings for excessive wear by using a narrow feeler gauge between shaft and bushing. Proper clearance is from .003 to .005".

Reverse Shifter Shaft & Seal

1. With extension case removed as outlined previously, remove shift fork.
2. Drive shifter shaft into case extension, allowing ball detent to drop into case. Remove shaft and ball detent spring.
3. Place ball detent spring into its hole and, from inside extension, install shifter shaft fully into its opening until detent plate is butted against inside

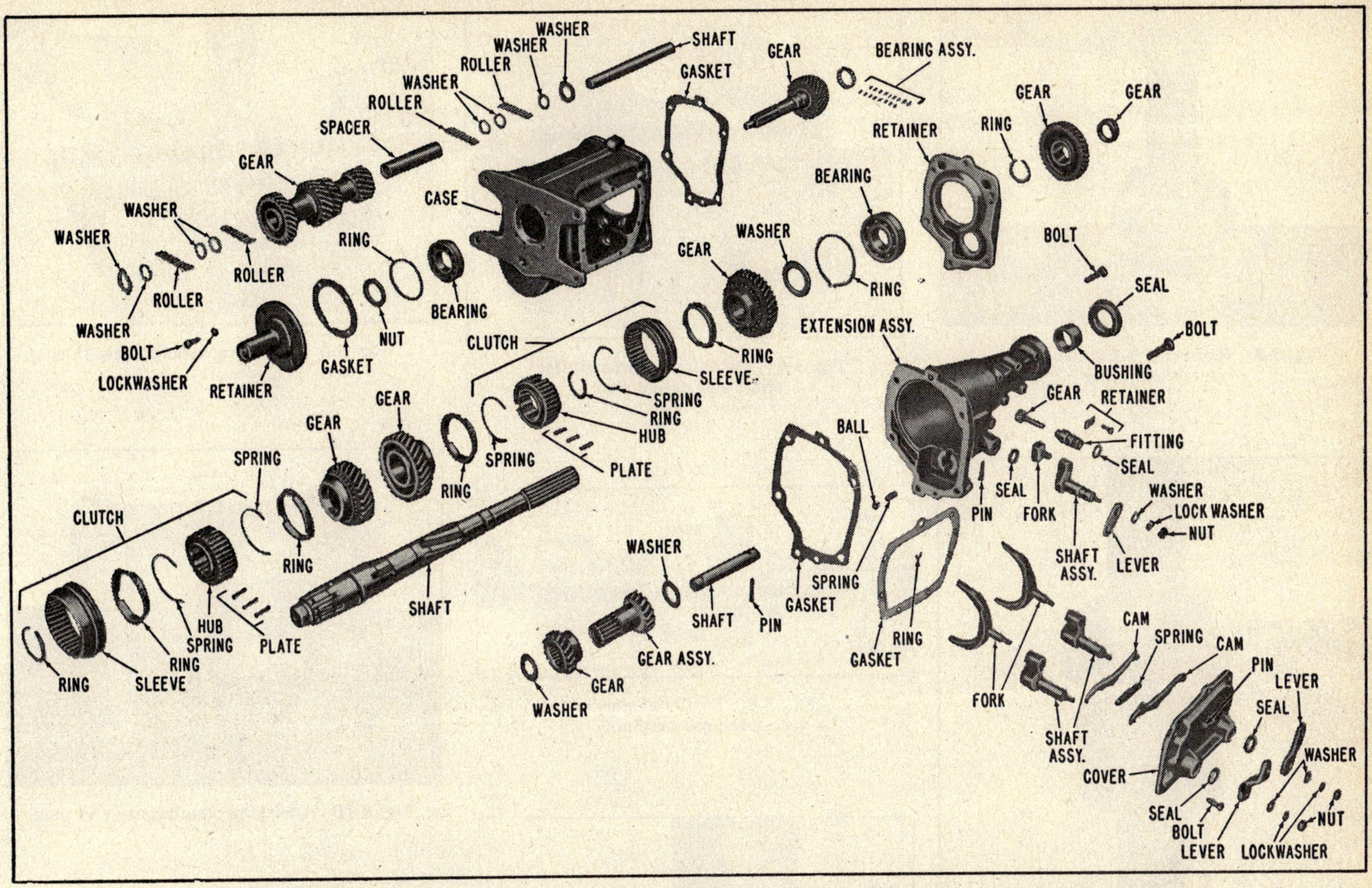

Fig. 3-1 Muncie four speed transmission. Early models use a snap ring (illustrated) behind the 1-2 synchronizer hub to retain the hub in position. Later units have a spacer sleeve within the 1st speed gear, (not illustrated)

of extension housing.

4. Place detent ball on spring, Fig. 3-4, and, holding ball down with thumb or a suitable tool, push shifter shaft back in, away from case until it is directly over ball and turn until ball drops into detent on shaft detent plate.
5. Install shift fork.

NOTE: Do not drive shifter shaft lock pin into place until extension has been installed on transmission case.

Extension Oil Seal or Bushing

If bushing in rear of extension requires replacement, remove oil seal and drive bushing into case extension. Drive new bushing in from the rear. Coat I.D. of bushing with transmission lubricant, then install new oil seal.

Clutch Keys & Springs

NOTE: The clutch hubs and sliding sleeves are a selected assembly and should be kept together as originally assembled. However, the three keys and two springs may be replaced if worn or broken.

1. To replace, push hub from sliding sleeve. Keys will fall free and springs easily removed.
2. Place the two springs in position (one on each side of hub) so all three keys are engaged by both springs. Place keys in position and, holding them in place, slide hub into sleeve.

Assemble Mainshaft

1. From rear of mainshaft, assemble 2nd gear with hub of gear toward rear of shaft.
2. Install 1-2 synchronizer clutch, with clutch sleeve taper to rear, together with a blocker ring on each side so their keyways line up with clutch keys, Fig. 3-5. On early units, install 1-2 synchronizer retainer snap ring (smaller of the two). On later units, install 1st gear spacer sleeve.
3. Install 1st gear (hub toward front) and 1st gear thrust washer.
4. Press rear bearing on mainshaft, Fig. 3-6, being sure to seat bearing firmly.
5. Choose correct selective fit snap ring and install it in groove of mainshaft behind rear bearing. (Snap rings of .084", .087", .090", .093", .096" are available.) With proper snap ring, maximum distance between ring and rear face of bearing will be from zero to .005".
6. Install 3rd gear (hub to front) and 3rd gear synchronizing ring (notches to front).
7. Install 3-4 clutch assembly with both sleeve taper and hub toward front, making sure keys in hub correspond to notches in 3rd gear synchronizing ring.
8. Install snap ring in mainshaft groove in front of 3-4 synchronizer clutch with ends of snap ring seated behind spline teeth.
9. Install rear bearing retainer. Spread snap ring in plate to allow ring to drop around rear bearing and press on end of mainshaft until snap ring engages groove in rear bearing.
10. Install reverse gear with shift collar to rear.
11. Press speedometer drive gear on mainshaft to distance shown in Fig. 3-7.

Assemble Countergear

1. Install roller spacer in countergear.
2. Using heavy grease to retain rollers, install 28 rollers in either end of countergear, two .050" spacers, 28 more rollers, then one .050" spacer.

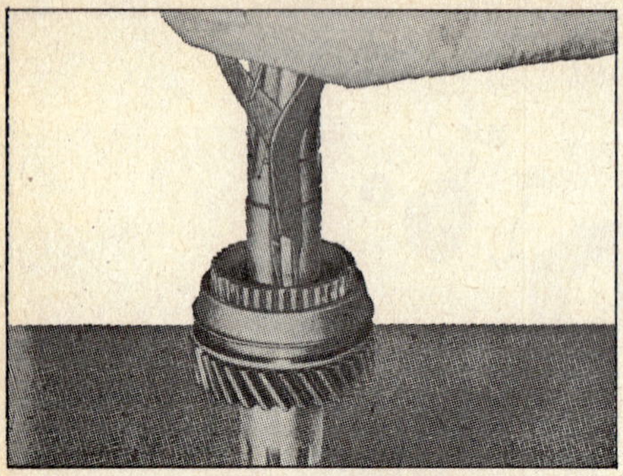

Fig. 3-3 Removing 1-2 synchronizer clutch snap ring

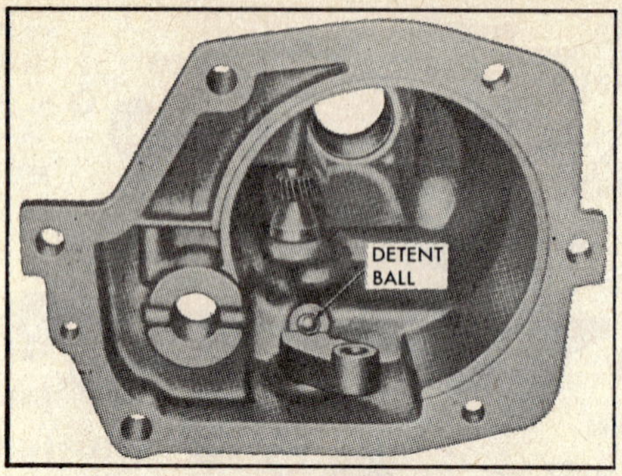

Fig. 3-4 Installing reverse shifter shaft and detent ball

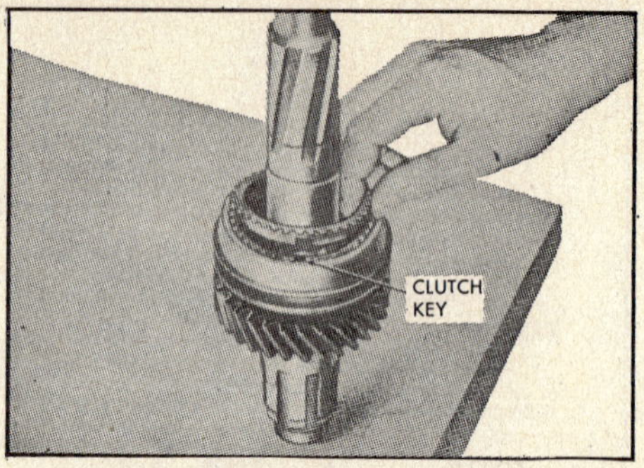

Fig. 3-5 Installing synchronizer ring

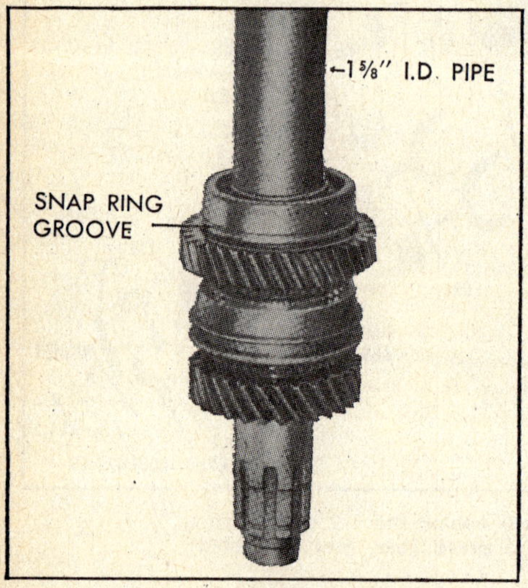

Fig. 3-6 Installing mainshaft rear bearing

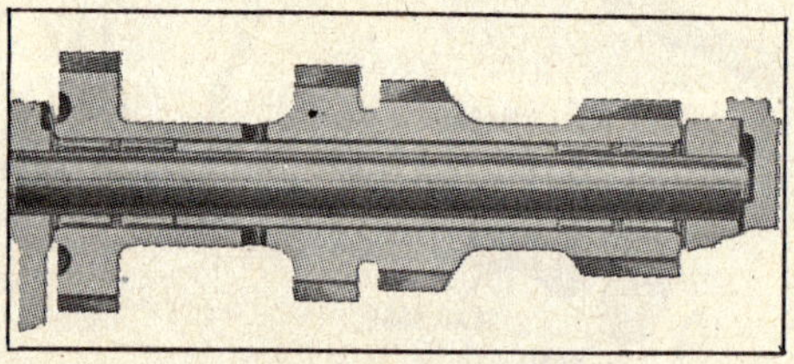

Fig. 3-8 Sectional view of countergear assembly

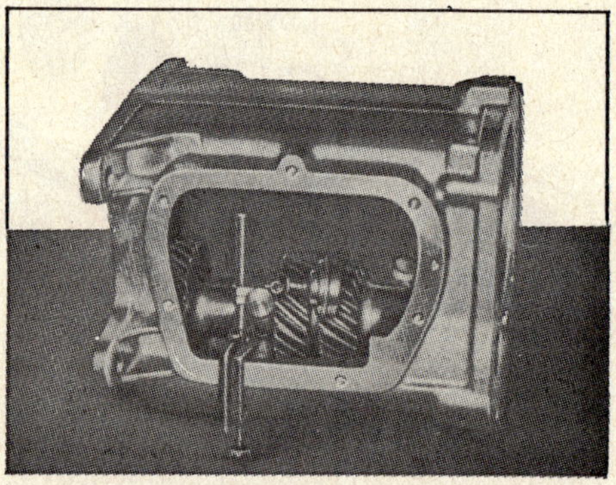

Fig. 3-10 Checking countergear end play

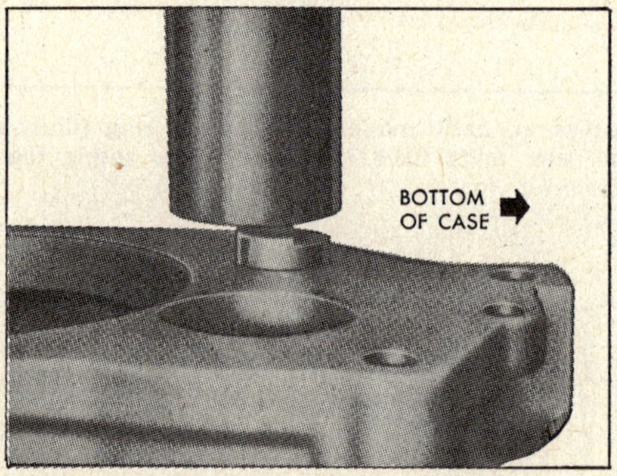

Fig. 3-9 Installing countershaft

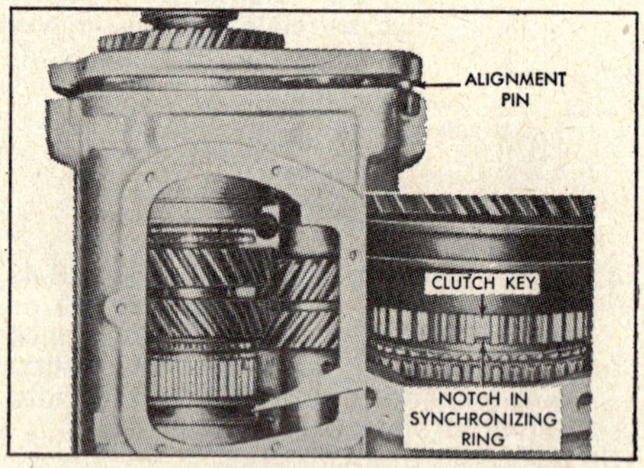

Fig. 3-11 Installing mainshaft assembly

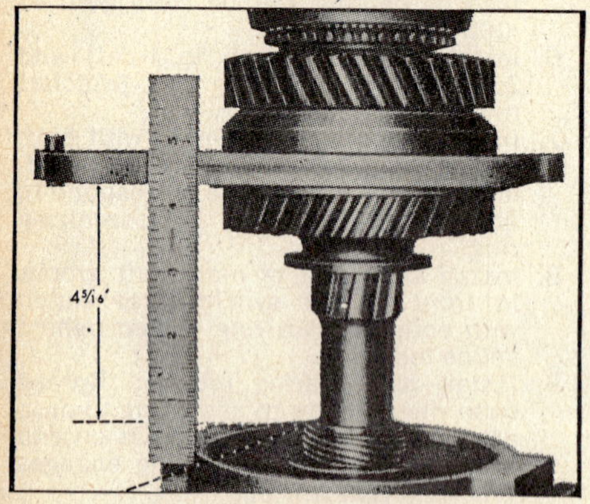

Fig. 3-7 Installing speedometer drive gear

Make same installation at other end of countergear, Fig. 3-8.

ASSEMBLE TRANS.

1. With transmission case on its side, put countergear tanged thrust washers in place, retaining them with heavy grease, and making sure tangs are resting in notches in case.
2. Set countergear in bottom of case.
3. Position transmission so it is resting on its front face.
4. Lubricate and insert countershaft through rear of case. Turn countershaft so flat on end of shaft is horizontal and facing bottom of case.
5. Align countergear with shaft in rear and hole in front of case. Press countershaft into case until flat on shaft is flush with rear of case, Fig. 3-9. Be sure thrust washers remain in place.
6. Attach a dial indicator as shown in Fig. 3-10 and check end play of countergear. If end play is greater than .025", new thrust washers must be installed.
7. Install 17 roller bearings in main drive gear, using heavy grease to hold bearings and cage in place.
8. Install main drive gear and pilot bearings through side cover opening and into transmission front bore.
9. Place gasket in position of front face of rear bearing retainer.
10. Install 4th gear synchronizing ring on main drive gear with notches toward rear.
11. Position tanged reverse idler gear thrust washer on machined face of ear cast in case for reverse idler shaft and hold with heavy grease. Position front reverse idler gear next to thrust washer, with hub facing toward rear.

NOTE: Before attempting to install mainshaft assembly in case, slide 3-4 synchronizer clutch sleeve forward into 4th speed detent position.

12. Lower mainshaft into case making certain notches in 4th speed synchronizing ring correspond to keys in clutch, Fig. 3-11.
13. Tap rear bearing into position.
14. Insert rear reverse idler gear.
15. Install remaining flat thrust washer on reverse idler shaft. If new idler shaft in being used, drive out roll pin and press it into new shaft.
16. Install reverse idler shaft, making sure to pick up rear tanged thrust washer. *Roll pin should be in a vertical position.*
17. Pull reverse shifter shaft to left side of extension and rotate shaft to bring reverse shift fork forward in extension (reverse detent position). Start extension onto transmission case, while slowly pushing in on shifter shaft to engage shift fork with reverse gear shift collar. Then pilot reverse idler shaft into extension housing, permitting extension to slide into transmission case.
18. Install extension attaching bolts. Torque upper three bolts to 15-25-ft-lbs, and lower three bolts to 25-35 ft-lbs.
19. Push or pull reverse shifter shaft to line up holes and drive in lock pin. Then install shifter lever.
20. Press bearing onto main drive gear (snap ring groove to front) and into case until several main drive gear retaining nut threads are exposed.
21. Lock transmission by shifting into two gears. Install main drive gear retaining nut. Be sure bearing fully seats against shoulder of gear. Torque nut to 40 ft-lbs and lock it in place by staking securely into shaft hole with center punch.
22. Install main drive gear bearing retainer, using sealer on bolts. Torque to 15-20 ft-lbs.
23. Shift 3-4 sliding sleeve into neutral and 1-2 sliding sleeve into 2nd gear position. Shift side cover 3-4 shift lever into neutral and 1-2 shift lever into 2nd gear.
24. Install side cover with gasket. Torque attaching bolts evenly to avoid cover distortion and torque to 15-20 ft-lbs.

Type Four Chrysler

DISASSEMBLE TRANS.

1. On early models, disconnect shift control rods from shift levers, shift transmission into two gears at once to prevent mainshaft from turning and remove flange from end of mainshaft.
2. All units, unfasten shift housing and with all levers in neutral position, Fig. 4-2, pull housing out and away from case. Work shift forks out of synchronizer sleeves and remove from case.
3. Remove main drive gear bearing retainer.
4. Unbolt extension housing from transmission case.
5. Slide 3-4 synchronizer slightly forward and slide mainshaft and extension housing out of case, Fig. 4-3.

Disassemble Mainshaft

1. Remove snap ring in front of 3-4 synchronizer and slide off synchronizer.
2. Slide off 3rd gear and stop ring.
3. While holding center bearing snap ring compressed, Fig. 4-4, pull mainshaft assembly and bearing out of extension housing.
4. Remove mainshaft rear bearing snap ring, insert steel plates on front side of 1st speed gear, and press gear and rear bearing from mainshaft. Remove 1st gear stop ring.
5. Remove snap ring and slide 1-2 synchronizer clutch unit from mainshaft.

NOTE: Fig. 4-6 shows mainshaft bearing surfaces. Inspect these surfaces for any condition that would warrant the use of a new mainshaft. Fig. 4-7 shows details of reverse gearing and related parts.

6. With a feeler gauge, Fig. 4-8, measure countergear end play by inserting gauge between thrust washer and gear. Measurement should not exceed .0155" to .028". If greater than specified, new thrust washers should be used upon reassembly.
7. Remove reverse lever detent parts.
8. Press or drive reverse slider gear shaft (from front to rear) far enough out of case to remove slider gear. Remove key from shaft and shaft from case, 4-9.
9. Push reverse lever shaft into case and remove as shown in Fig. 4-10. Lift out detent ball from bottom of case, and remove shift fork from shaft and detent plate.
10. Using a suitable arbor, drive countershaft out of case, allowing cluster gear to rest in bottom of case to permit removal of main drive gear.
11. Remove snap ring and remove main drive gear by driving it into case. Remove snap ring and press bearing from main drive gear.
12. Lift cluster gear from case.

REASSEMBLE TRANS.

Countergear

1. Using heavy grease, coat inside of gear bore at each end, then center bearing spacer. Insert arbor through gear and spacer.
2. Grease needle rollers and, at each end of gear, install 19 rollers, followed by a spacer ring, 19 more rollers and a spacer ring.
3. Coat thrust washers with grease and install them over arbor with tang side toward case boss.
4. Install countergear into case, Fig. 4-11. Allow assembly to rest in bottom of case until after main drive gear is installed.

Mainshaft

1. Slide 2nd gear over mainshaft, synchronizer cone toward rear, and into position against shoulder on shaft, Fig. 4-5.
2. Slide 1-2 clutch sleeve gear including 2nd gear stop ring over mainshaft with shift fork slot toward front, and down into position against 2nd gear. Be sure stop ring is indexed with shift plates and install snap ring.
3. Slide low gear stop ring over shaft and down into position and index with shift plates.
4. Slide 1st gear synchronizer (synchronizer cone toward clutch gear sleeve just installed) over mainshaft and down against clutch sleeve gear.
5. Install mainshaft bearing retainer ring followed by mainshaft bearing. Drive or press bearing down into position and secure with snap ring.
6. On earlier units, install extension housing bearing.
7. Install partially assembled mainshaft into extension housing far enough to engage retaining ring in slot in extension housing. Compress retaining ring and at same time rest mainshaft in extension housing (see Fig. 4-4). Be sure retaining ring is seated all around slot.
8. Slide 3rd gear on shaft, synchronizer cone toward front, followed by 3rd gear stop ring.
9. Install 3-4 synchronizer clutch assembly with shift fork slot toward rear. Be sure to index rear stop ring with clutch gear shift plates (see Fig. 4-4). Install snap ring.

Service Bulletin

Quiets Gear Rattle: A gear rattle, or neutral noise, having automatic transmission fluid in the case can be quieted by changing the lubricant. Drain the automatic transmission fluid and replace it with SAE-140 multipurpose gear lube. If the SAE-140 causes hard or stiff shifting during cold weather operation, change to SAE-80 or 90. This will lessen the effort required for shifting.

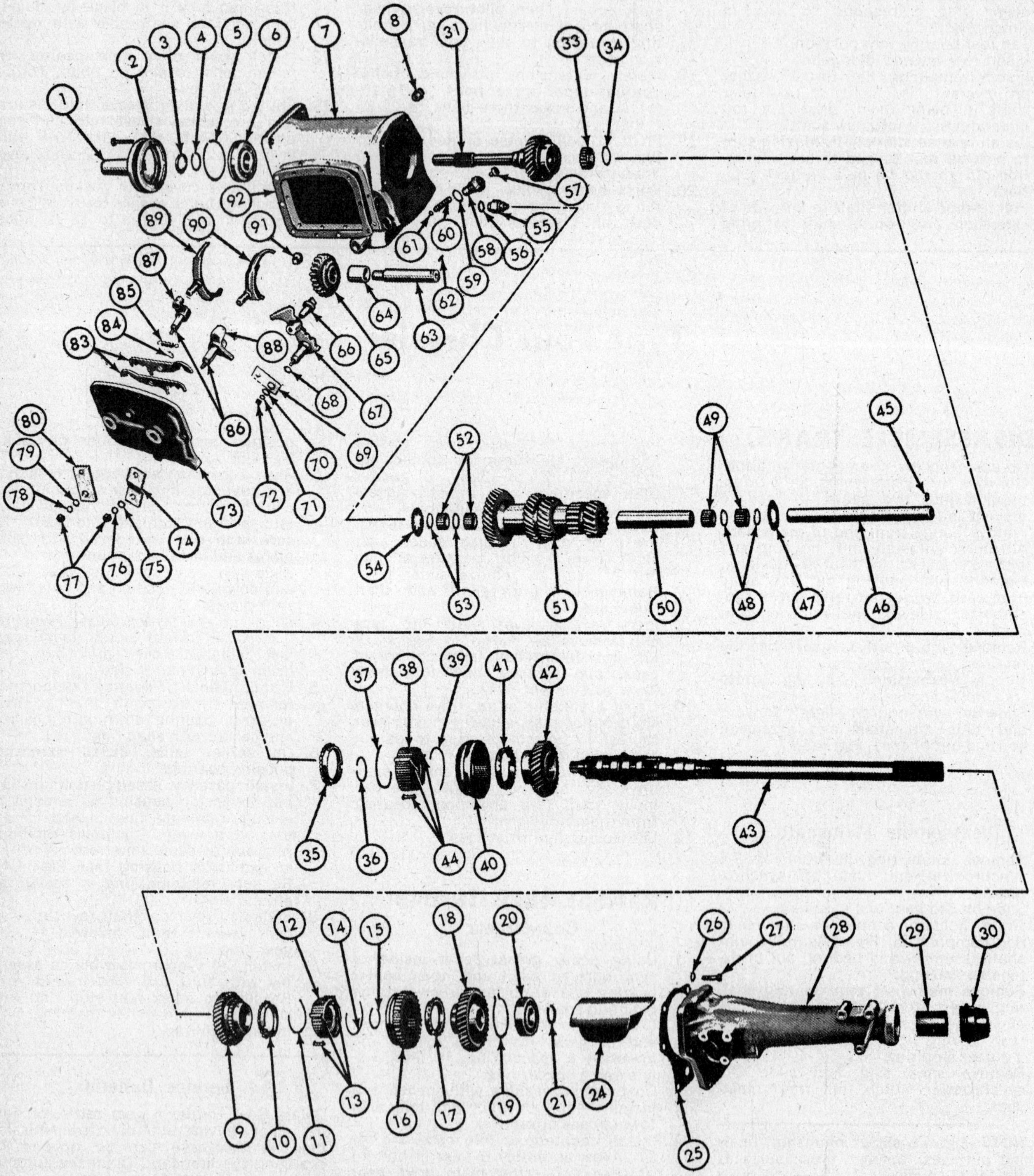
1
2
3
4
5
6
7
8
31
33
34
57
92
61
60
58
56
55
89
90
91
87
59
62
64
63
85
84
66
65
83
88
67
68
86
69
70
80
79
73
72
71
78
74
77
76
75
52
49
45
54
53
51
50
48
47
46
37
38
39
41
42
35
36
44
40
43
12
14
15
18
20
26
27
28
29
30
9
10
11
13
16
17
19
21
24
25

1. Bearing Retainer
2. Bearing Retainer Gasket
3. Bearing Retainer Oil Seal
4. Snap Ring, Bearing (Inner)
5. Snap Ring, Bearing (Outer)
6. Pinion Bearing
7. Transmission Case
8. Filler Plug
9. Gear, 2nd Speed
10. Snap Ring
11. Shift Strut Springs
12. Clutch Gear
13. Shift Struts (3)
14. Shift Strut Spring
15. Snap Ring
16. 1st and 2nd Clutch Sleeve Gear
17. Stop Ring
18. 1st Speed Gear
19. Bearing Retainer Ring
20. Rear Bearing
21. Snap Ring
24. Baffle
25. Gasket, Case to Extension Housing
26. Lockwasher
27. Bolt
28. Extension Housing
29. Mainshaft Yoke Bushing
30. Oil Seal
31. Main Drive Pinion
33. Needle Bearing Rollers
34. Snap Ring
35. Stop Ring
36. Snap Ring
37. Shift Strut Spring
38. Clutch Gear
39. Shift Strut Spring
40. Clutch Sleeve
41. Stop Ring
42. 3rd Speed Gear
43. Mainshaft (Output)
44. Shift Struts (3)
45. Woodruff Key
46. Countershaft
47. Thrustwasher, Gear (1)
48. Spacer Ring Needle Roller Bearing
49. Needle Bearing Rollers
50. Bearing Spacer
51. Countershaft Gear (Cluster)
52. Needle Bearing Rollers
53. Spacer Ring Needle Roller Bearing
54. Thrustwasher, Gear (1)
55. Backup Light Switch
56. Backup Light Switch Gasket
57. Plug
58. Retainer, Reverse Detent Ball Spring
59. Gasket
60. Spring, Reverse Detent Ball
61. Ball, Reverse Detent
62. Woodruff Key
63. Reverse Idler Gear Shaft
64. Bushing, Reverse Idler Gear
65. Gear, Reverse Idler
66. Fork, Reverse Shifter
67. Reverse Lever
68. Oil Seal, Reverse Lever Shaft
69. Reverse Operating Lever
70. Flatwasher
71. Lockwasher
72. Nut
73. Gearshift Control Housing
74. 1st and 2nd Operating Lever
75. Flatwasher
76. Lockwasher Lever
77. Nut, Lever
78. Lockwasher, Lever
79. Flatwasher, Lever
80. 3rd and 4th Operating Lever
83. Interlock Lever (2)
84. "E" Ring
85. Spring
86. Oil Seal (2)
87. 3rd and 4th Lever
88. 1st and 2nd Lever
89. 3rd and 4th Speed Fork
90. 1st and 2nd Speed Fork
91. Drain Plug
92. Gasket, Shift Control Housing

←

Fig. 4-1 Type Four Chrysler four-speed transmission.

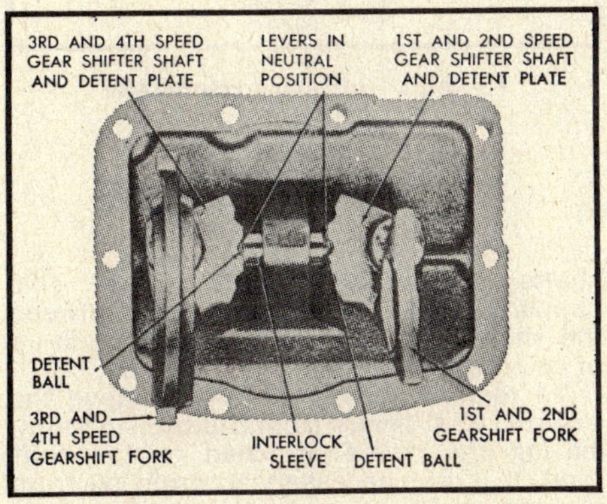

Fig. 4-2 Shift housing assembly

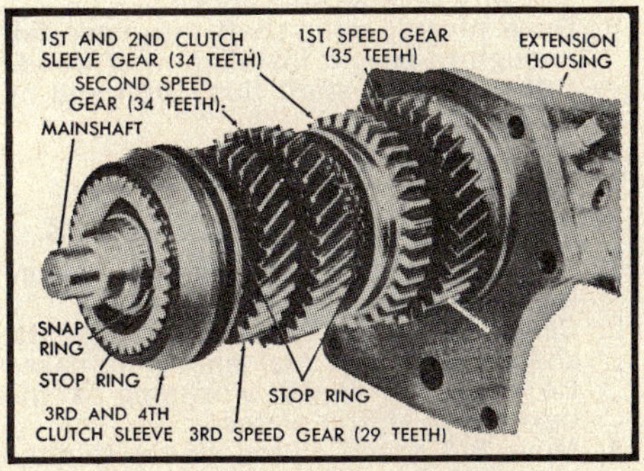

Fig. 4-3 Mainshaft assembly

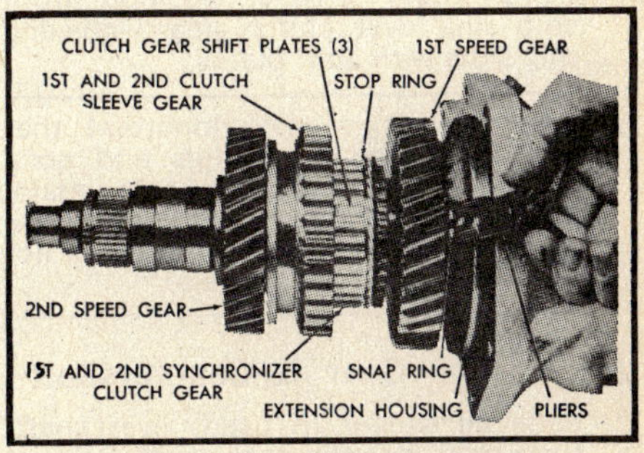

Fig. 4-4 Compressing center bearing snap ring so mainshaft can be pulled from bearing housing

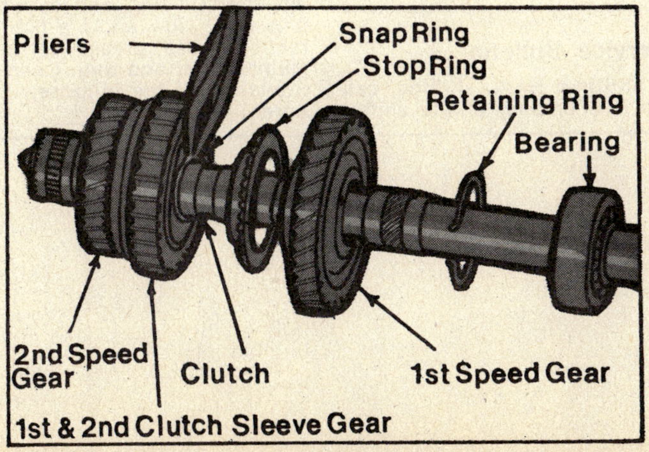

Fig. 4-5 Removing clutch gear snap ring

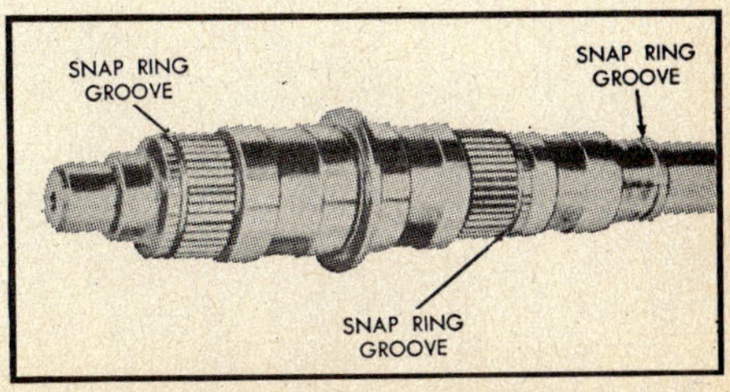

Fig. 4-6 Mainshaft bearing surfaces

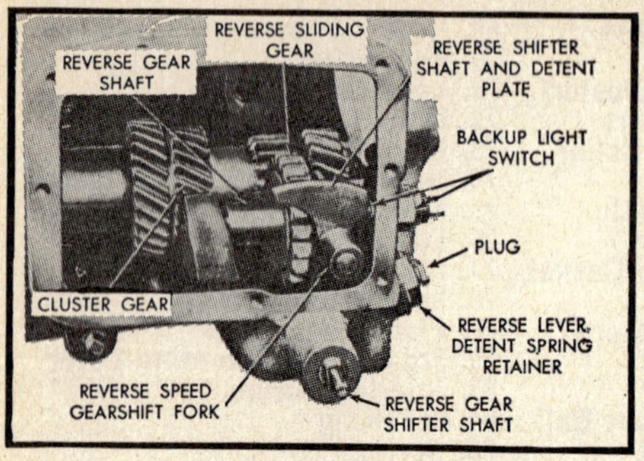

Fig. 4-7 Reverse gearing and countergear

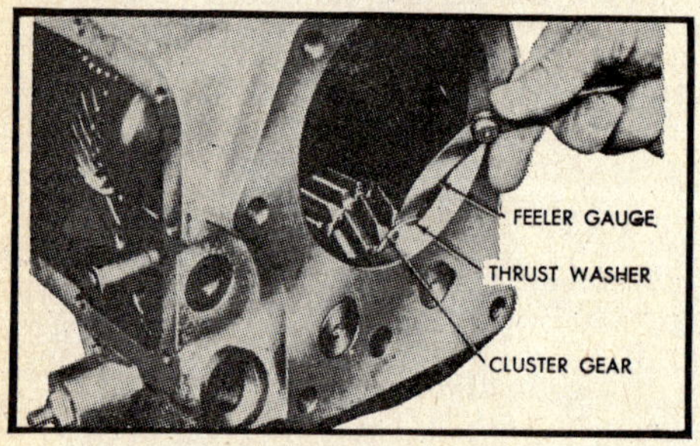

Fig. 4-8 Measuring countergear end play

10. Using heavy grease, position front stop ring over clutch gear, indexing ring slots with shift plates.

CAUTION: It is very important that indexing of all stop rings and positioning of gears and clutches on mainshaft be correct, or the mating of the extension housing to the case will not be possible without damage.

Main Drive Gear & Countershaft

1. Slide oil slinger (if used) over shaft, then press bearing onto shaft. Be sure outer snap ring groove is toward front. Seat bearing fully against shoulder of gear.
2. Install new inner snap ring into groove to retain bearing.
3. Install gear and bearing into case and position in front bore. Tap lightly into place. Install outer snap ring in bearing groove.
4. Start countershaft in its bore at rear of case. Raise countergear to mesh with main drive gear. Be sure thrust washer tangs are aligned with slots in case.

Fig. 4-9 Removing reverse slider gear

5. Drive or press countershaft into gear. Install woodruff key. Continue to press shaft into case until end of shaft is flush with rear face of case.
6. Measure countergear end play (see Fig. 4-8). If end play is greater than .028" install new thrust washers.

Reverse Gearing

1. Install reverse shaft detent and spring. Install spring retainer gasket and retainer (see Fig. 4-7).
2. Position reverse slider gear shaft in end of case and drive it far enough to position slider gear on protruding end of shaft with shift slot toward rear (see Fig. 4-9). At same time engage slot with reverse shift fork.
3. Drive reverse gear shaft into case far enough to permit installation of woodruff key. Drive shaft flush with end of case.

Mainshaft & Extension Housing

1. Grease both sides of extension housing gasket and stick it on case.
2. Center reverse slider gear on its shaft, then insert mainshaft into case. Be sure 3-4 speed stop ring is indexed with shifter plates.
3. Move 3-4 speed clutch sleeve slightly toward front and at same time align front of mainshaft with main drive gear. Push in on extension housing and bottom against case and housing.
4. Install bolts and tighten securely.
5. Move reverse slider gear ahead to neutral position and install shift housing.

Service Bulletin

Cures Hard Shifting: Hard shifting can result if there is a binding of the shifter shafts in the case or in the cover. This happens when moisture enters between the shaft and bore and causes a build-up of corrosion.

To restore normal shifting, remove the external shift levers from the transmission, leaving the levers attached to the shift rods. It is not necessary to drain oil from the unit.

Place a 3/4" diameter hole saw over the shifting shaft protruding from the case and cut a 1/8" deep counterbore in the case. After blowing out the cuttings, apply to the shaft a penetrating oil that will not damage the "O" ring, and work the shaft until it is free. Now apply a liberal amount of multi-purpose grease to the shaft and counterbore, and then insert an "O" ring, available for the purpose, in the counterbore.

Fig. 4-10 Removing reverse shift fork and lever

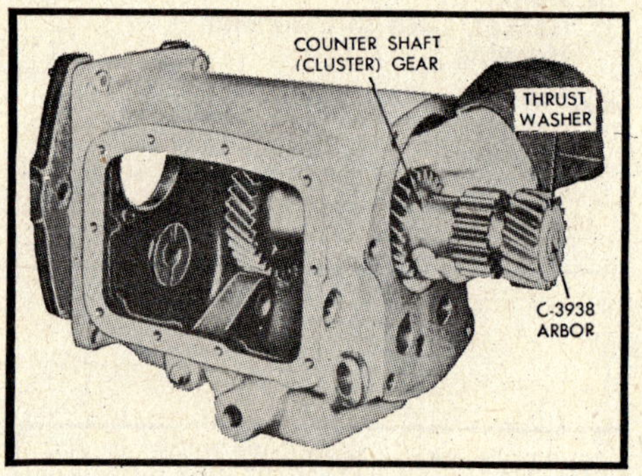

Fig. 4-11 Installing countergear

Type Five Ford

DISASSEMBLE TRANS.

1. Remove shift linkage and control bracket.
2. Remove transmission cover.
3. Remove extension housing.
4. Remove input shaft bearing retainer.
5. Support countergear with a wire hook. Then working from front of case, push countershaft out rear of case and lower countergear to bottom of case.
6. Place 1-2 shift lever and reverse shift lever in neutral. Place 3-4 shift lever in 3rd speed position.
7. Remove bolt that retains 3-4 shift rail detent spring and plug in left side of case, Fig. 5-2. Remove spring and plug with a magnet.
8. Remove detent mechanism set screw from top of case and take out detent spring and plug with a magnet.
9. Remove attaching screw from 3-4 speed shift fork. Tap on inner end of shift rail to unseat expansion plug from front of case. Then withdraw 3-4 shift rail from front of case. Do not lose interlock pin from shift rail.
10. Remove attaching screw from 1-2 shift fork. Slide 1-2 shift rail out of rear of case.
11. Remove interlock plug and detent plug from top of case with a magnet.
12. On early models, remove snap ring securing speedometer drive gear to output shaft. Slide gear off shaft and remove speedometer drive gear ball. On later units, depress tang on retaining clip and remove gear.
13. Remove snap ring that secures output shaft bearing to shaft and use a suitable puller to remove bearing.
14. Remove input shaft and bearing and blocking ring from front of case.
15. Move output shaft to right side of case to provide clearance for shift forks. Rotate forks as shown in Fig. 5-3 and lift them from case.
16. Support 1st gear to prevent it from sliding off shaft, then lift output shaft assembly from case, Fig. 5-4.
17. Remove reverse gear shift fork attaching screw. Rotate reverse shift rail 90 deg. as shown in Fig. 5-5. Slide shift rail out rear of case and lift fork from case.
18. Remove reverse detent plug and spring from case with a magnet.
19. Remove reverse idler gear shaft from case, Fig. 5-6.
20. Lift reverse idler gear and thrust washers from case.
21. Lift countergear and thrust washers from case.

Disassemble Output Shaft

1. Remove snap ring from front of output shaft. Slide 3-4 synchronizer blocking ring and 3rd gear off shaft, Fig. 5-7.
2. Remove next snap ring and 2nd gear thrust washer. Slide 2nd gear and blocking ring from shaft.
3. Remove next snap ring, thrust washer, 1st speed gear and blocking ring from rear of shaft. On later models, the 1-2 synchronizer hub is a press fit. On these units, use an arbor press to remove hub. On earlier models, slide hub off output shaft.

UNIT REPAIRS

Cam & Shaft Seals

NOTE: To facilitate reassembly, note position of cams and shafts assemblies and levers before removal from transmission case.

1. Referring to Fig. 5-8, remove three shift levers.
2. Remove three cams and shafts from inside of case.
3. Remove O-ring from each cam and shaft.
4. Dip new O-rings in gear lubricant and install them on cam and shafts.
5. Slide each cam and shaft into its respective bore in case.
6. Secure each shift lever.

Input Shaft Bearing

1. Referring to Fig. 5-9, remove snap ring that secures bearing to shaft.

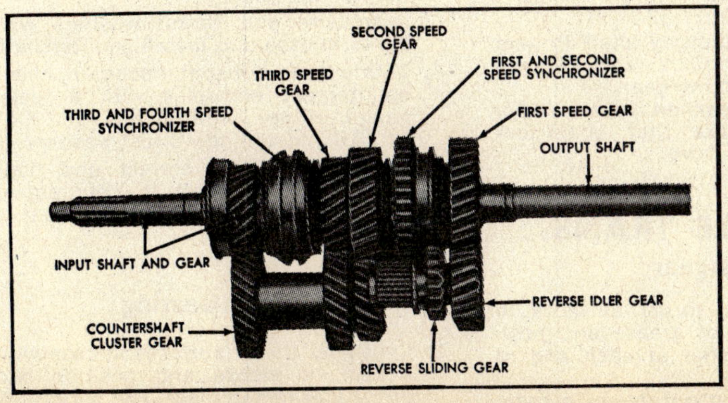

Fig. 5-1 Type Five Ford four speed transmission

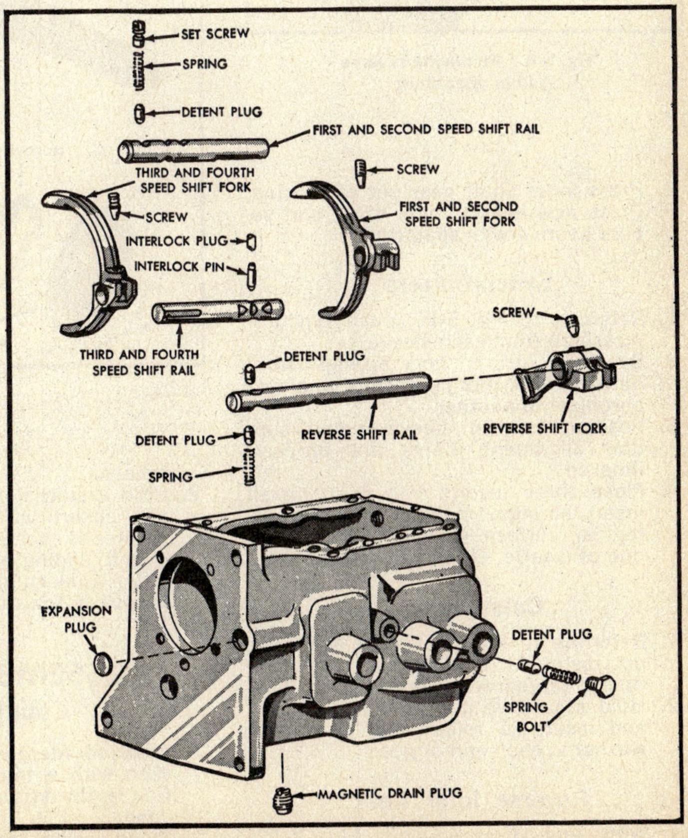

Fig. 5-2 Shift rails and forks disassembled

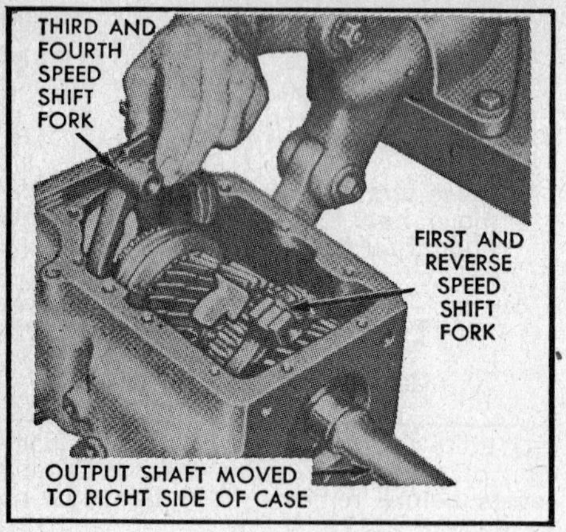

Fig. 5-3 Removing shift forks from case

Fig. 5-4 Removing output shaft assembly

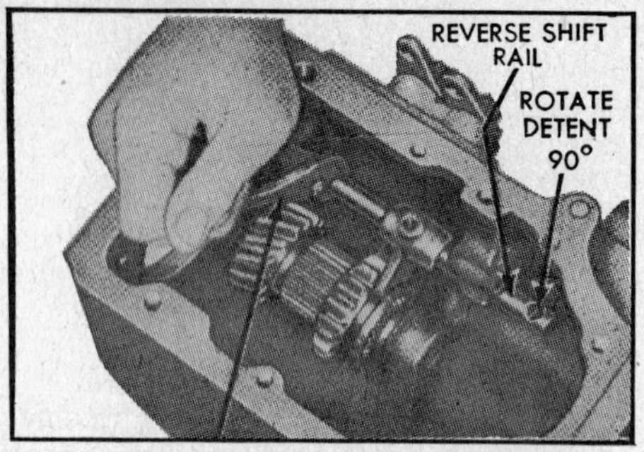

Fig. 5-5 Rotating reverse shift rail

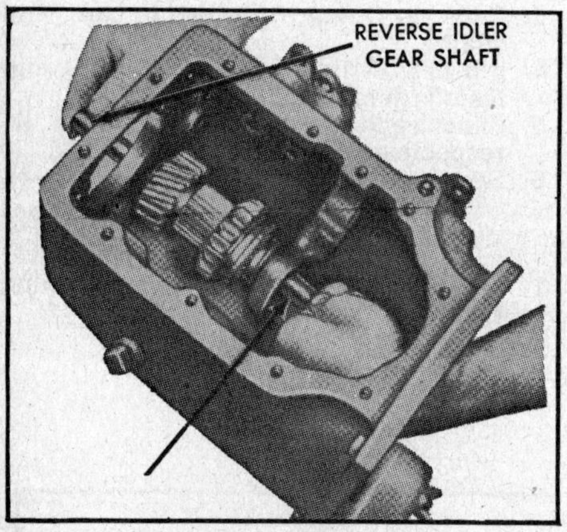

Fig. 5-6 Removing reverse idler gear shaft

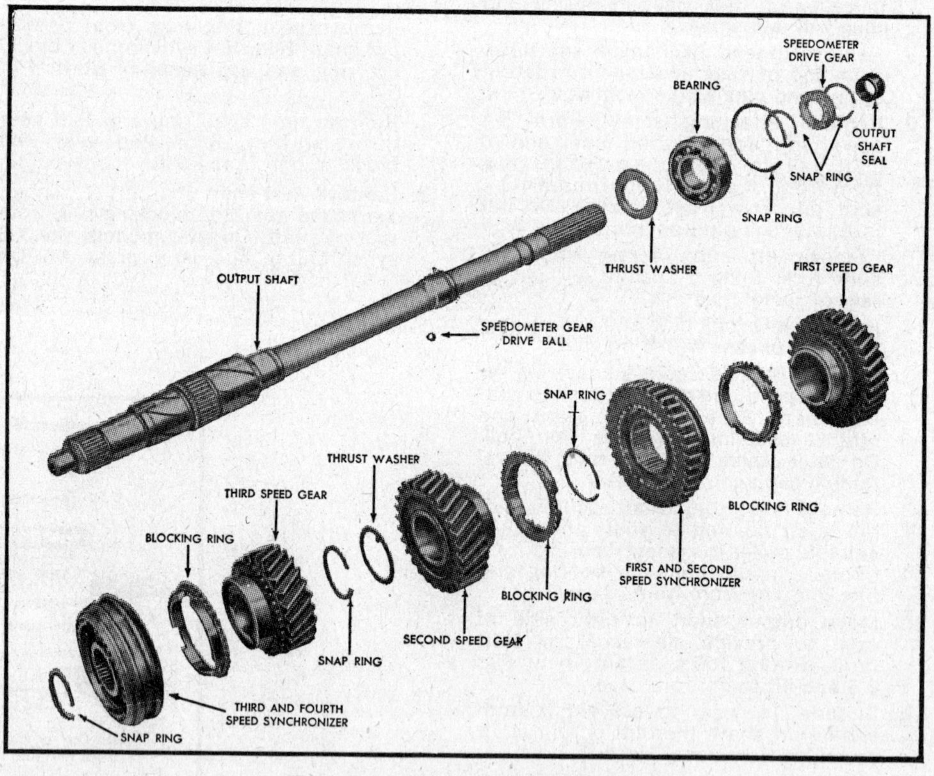

Fig. 5-7 Output shaft disassembled

2. Press input shaft gear out of bearing.
3. Press new bearing onto shaft and secure bearing with snap ring.

Synchronizers

1. Referring to Fig. 5-10, push synchronizer hub from each sleeve.
2. Separate inserts and springs from hubs. Do not mix parts from one synchronizer to another.
3. Position hub in sleeve, being sure that alignment marks are properly indexed.
4. Place three inserts into hub. Install insert springs, making sure that irregular surface (hump) is seated in one of inserts. Do not stagger springs.

Countergear

1. Referring to Fig. 5-11, and with unit disassembled, coat bore in each end of countergear with grease.
2. Hold a suitable dummy shaft in gear and insert 21 rollers and a retainer washer in each end of gear.

Reverse Idler Gear

1. With unit disassembled, Fig. 5-12, coat bore at each end of gear with grease.
2. Hold a suitable dummy shaft in gear and insert 22 rollers and retainer washer at each end of gear.
3. Install sliding gear on reverse idler gear, making sure that shift fork groove is toward front.

REASSEMBLE TRANS.

Countergear

1. Coat countergear thrust surfaces in case with a film of grease and position a thrust washer at each end of case.
2. Place assembled countergear in case.
3. With case in a vertical position, align gear bore and thrust washers with bores in case and install countershaft.
4. With case in horizontal position, check countergear end play with a feeler gauge. If not within limits of .004" to .018", install new thrust washers.
5. After establishing correct end play, install dummy shaft in countergear and allow gear to remain in bottom of case.

Reverse Gearing

1. Coat gear thrust surfaces in case with a film of grease and position two thrust washers in place, Fig. 5-12.
2. Position idler gear, sliding gear,

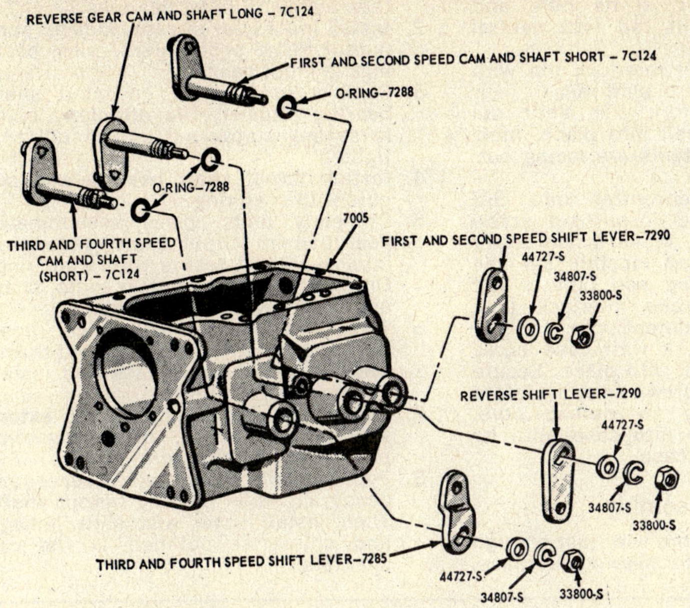

Fig. 5-8 **Cams and shafts assemblies and levers disassembled. To facilitate reassembly, note position of each part before disassembly**

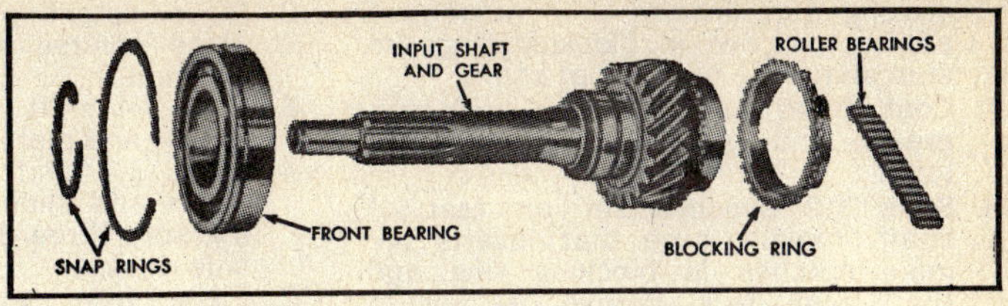

Fig. 5-9 **Input shaft gear disassembled**

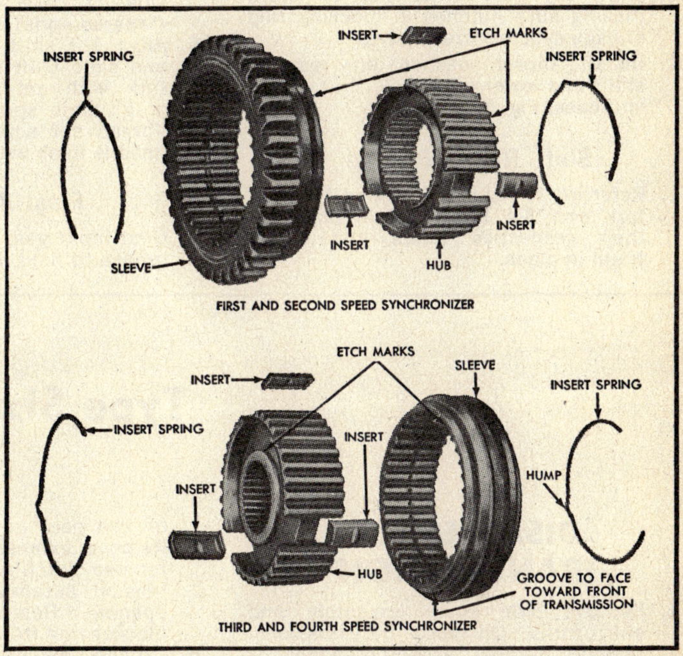

Fig. 5-10 **Synchronizers disassembled**

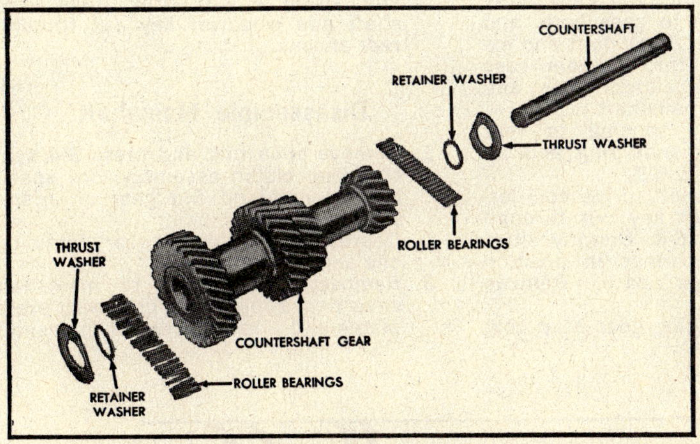

Fig. 5-11 **Countergear disassembled**

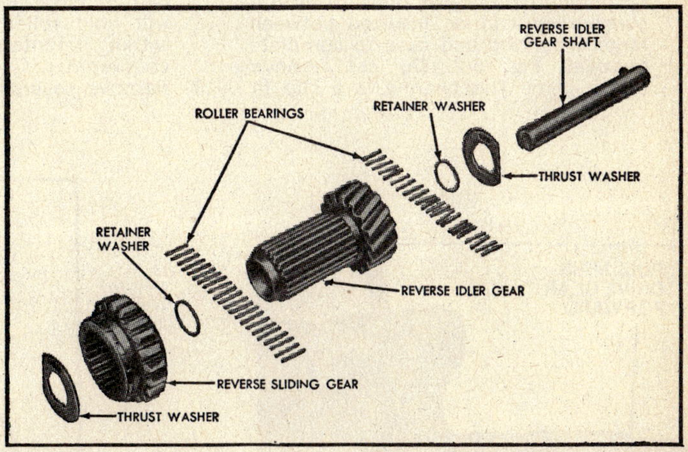

Fig. 5-12 **Reverse idler gear disassembled**

dummy shaft and roller bearings in place, making sure that shift fork groove in sliding gear is toward front of case.
3. Align gear bore and thrust washers with case bores and install reverse idler shaft.
4. Measure reverse idler gear end play with feeler gauge. If not within the limits of .004" to .018", install new thrust washers.
5. Position reverse gear shift rail detent spring and plug in case. Hold reverse shift fork in place on sliding gear and install shift rail from rear of case. Secure fork to rail with set screw.

Output Shaft

1. Referring to Fig. 5-7, install 1-2 synchronizer on shaft.
2. Slide 2nd gear onto front of shaft with synchronizer coned surface facing to rear.
3. Install 2nd gear thrust washer and snap ring.
4. Slide 3rd gear on shaft with synchronizer coned surface to front.
5. Coat coned surface of 3rd gear with grease and place blocking ring on gear.
6. Slide 3-4 synchronizer onto shaft

making sure inserts in synchronizer engage notches in blocking ring. Install snap ring on front of shaft.

7. Coat coned surface of 2nd gear with grease and place blocking ring on gear.
8. Slide 1-2 synchronizer on rear of shaft, making sure that inserts engage notches in blocking ring and that shift fork groove is toward rear.
9. Coat coned surface of 1st gear with grease and place blocking ring on it.
10. Slide 1st gear onto rear of shaft, making sure notches in blocking ring engage synchronizer inserts.
11. Install thrust washer on rear of shaft and lower output shaft assembly into case, Fig. 5-4.

Shift Rails & Forks

1. Referring to Fig. 5-2, position 1-2 shift fork and 3-4 shift fork in place on their respective gears and rotate them in place.
2. Place detent plug in its bore and place reverse shift rail into neutral position.
3. Coat 3-4 shift rail interlock pin with grease and plate it in shift rail.
4. Align 3-4 shift fork with shift rail bores and slide rail into place, making sure three detents are facing outside of case.
5. Place front synchronizer into 3rd gear position and install set screw in 3-4 shift fork. Move synchronizer to neutral position. Install 3-4 rail detent plug, spring and bolt in left side of case. Place interlock plug (tapered ends) in detent bore.
6. Align 1-2 shift fork with case bores and slide shift rail into place. Secure fork with set screw. Install detent plug and spring in detent bore. Thread set screw into case until its head is flush with case.

Final Assembly

1. Coat input gear bore with just enough grease to hold roller bearings in place, then install 15 roller bearings.
2. Install input gear in case, making sure output shaft pilot enters roller bearings in input gear.
3. Stick a new gasket on input shaft bearing retainer. Dip attaching bolts in sealing compound and install and tighten.
4. Install output shaft bearing and secure with snap ring.
5. On early units, place speedometer gear ball in output shaft, slide gear into place and secure with snap ring. On later units, slip gear onto shaft and secure with retaining clip.
6. With transmission in vertical position, align countergear bore and thrust washers with bore in case and install countershaft.
7. Use a new gasket and secure extension housing to case. Use sealing compound on attaching screws.
8. Pour specified gear lube over entire gear train while rotating output shaft. Then install cover and shift linkage and adjust as outlined in the car chapters.

Type Six Saginaw

DISASSEMBLE TRANSMISSION

1. Remove side cover assembly and shift forks. On Vega transmissions, remove damper assembly which is bolted to extension housing.
2. Remove clutch gear bearing retainer.
3. Remove clutch gear bearing to gear stem snap ring, then remove bearing by pulling clutch gear outward until a screwdriver can be inserted between large snap ring and case to complete removal, Fig. 6-2. Do not remove clutch gear. The bearing is a slip fit on the gear and into the case bore.
4. Remove extension to case bolts and remove clutch gear, mainshaft and extension assembly through rear case opening. Remove clutch gear and blocker ring from mainshaft.
5. Expand extension housing to rear mainshaft bearing snap ring and remove extension, Fig. 6-3.
6. Using a dummy shaft, drive countershaft and woodruff key out through rear of case, Fig. 6-4. Dummy shaft will hold roller bearings in position within countergear bore. Remove countergear.
7. Remove reverse idler gear stop ring and, using a long drift, drive idler shaft and woodruff key out through rear of case.

Disassemble Mainshaft

1. Remove snap ring and press 3-4 synchronizer clutch assembly, 3rd speed blocker ring and 3rd gear off mainshaft, Fig. 6-5.
2. Depress speedometer gear retaining clip and remove gear.
3. Remove rear bearing to mainshaft snap ring, support 1st gear with press plates and press on rear of mainshaft

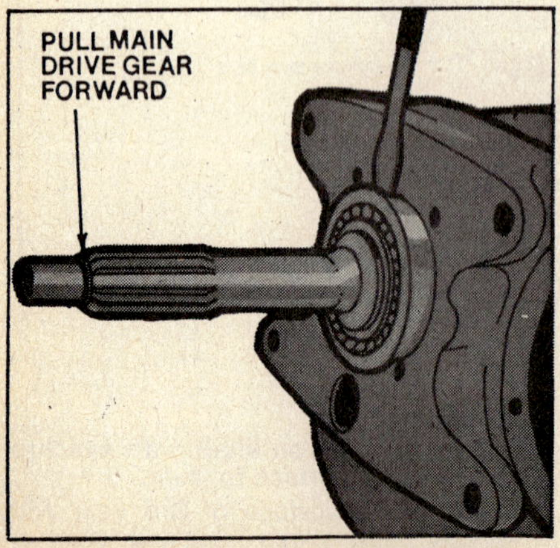

Fig. 6-2 Removing clutch gear bearing

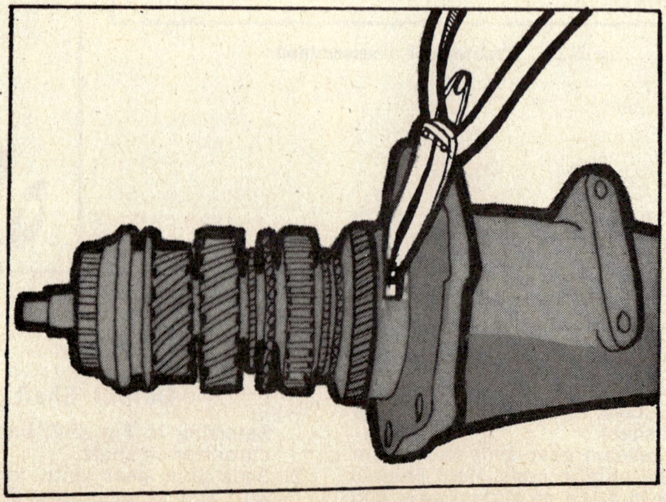

Fig. 6-3 Removing extension

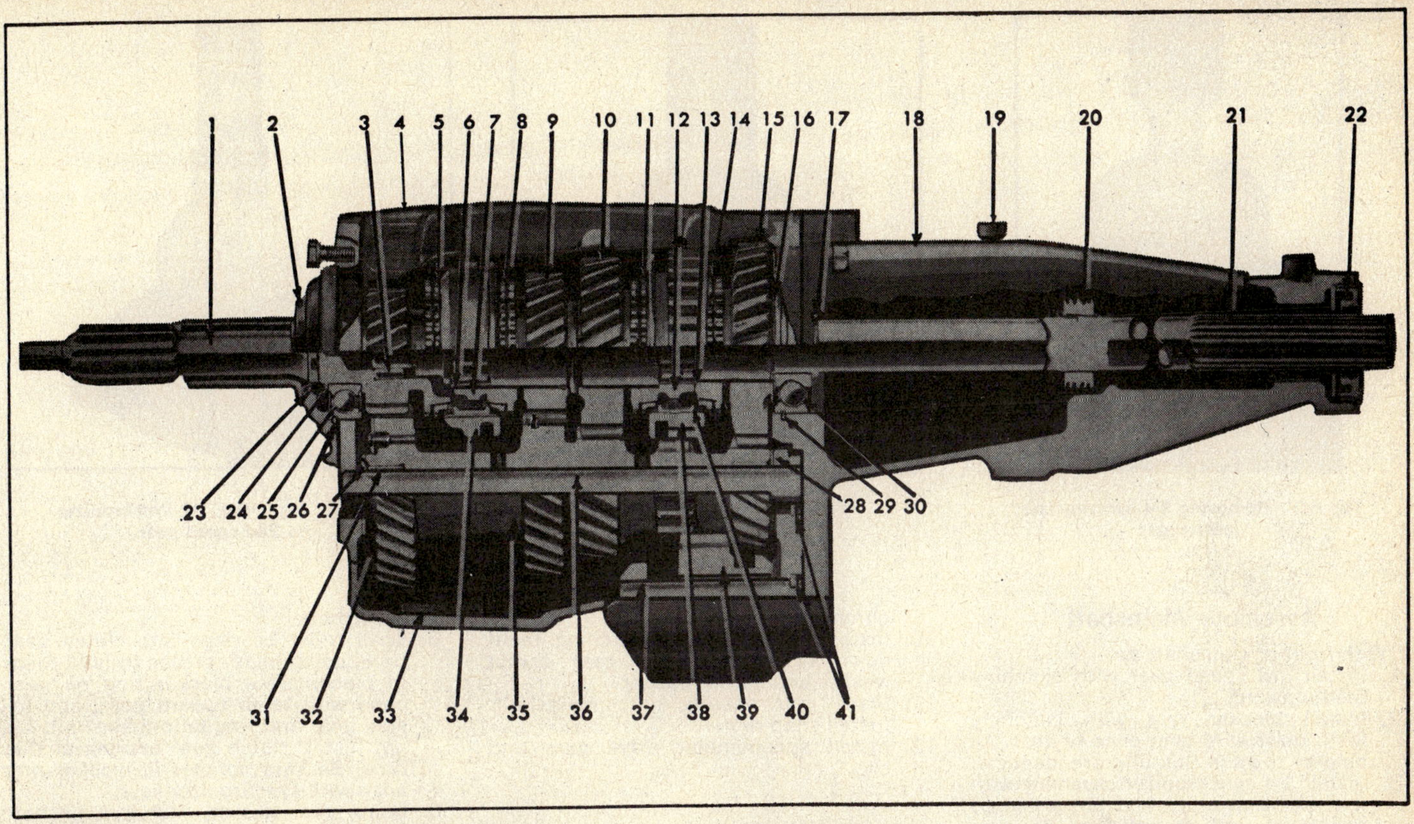

1. Clutch Gear
2. Bearing Retainer
3. Pilot Bearings
4. Case
5. 4th Speed Blocker Ring
6. 4-3 Snych. Snap Ring
7. 4-3 Synch. Hub
8. 3rd Speed Blocker Ring
9. 3rd Speed Gear
10. 2nd Speed Gear
11. 2nd Speed Blocker Ring
12. 1-2 Speed Synch. Hub
13. 1-2 Speed Synch. Snap Ring
14. 1st Speed Blocker Ring
15. First Gear
16. Reverse Gear Thrust and Spring Washers
17. Snap Ring-Bearing to Mainshaft
18. Extension
19. Vent
20. Speedometer Drive Gear and Clip
21. Mainshaft
22. Rear Oil Seal
23. Retainer Oil Seal
24. Snap Ring-Bearing to Gear
25. Clutch Gear Bearing
26. Snap Ring-Bearing to Case
27. Thrust Washer-Front
28. Thrust Washer-Rear
29. Snap Ring-Bearing to Extension
30. Rear Bearing
31. Countergear Roller Bearings
32. Anti-Lash Plate Assembly
33. Magnet
34. 4-3Snych. Sleeve
35. Countergear Assembly
36. Counter Shaft
37. Reverse Idler Shaft
38. 1-2 Speed Synch. Sleeve and Reverse Gear
39. Reverse Idler Gear (Sliding)
40. Clutch Key
41. Woodruff Key

Fig. 6-1 Type six Saginaw transmission

to remove 1st speed gear, thrust washer, spring washer and rear bearing, Fig. 6-6.

4. Remove 1-2 sliding clutch hub snap ring, support 2nd speed gear and press clutch assembly, 2nd speed blocker ring and gear from mainshaft, Fig. 6-7.

Clutch Keys & Springs

NOTE: The clutch hubs and sleeves are a selected assembly and should be kept together as originally assembled, but the keys and springs may be replaced separately.

1. Mark hub and sleeve so they can be matched upon reassembly.
2. Push hub from sliding sleeve and remove keys and springs.
3. Install the three keys and two springs so all three keys are engaged by both springs. The tanged end of each spring should be installed into different key cavities on either side. Slide sleeve onto hub, aligning marks made before disassembly.

NOTE: A groove around the outside of the hub identifies the end that must be opposite the fork slot in the sleeve when assembled. This groove indicates the end of the hub with a .070" greater recess depth.

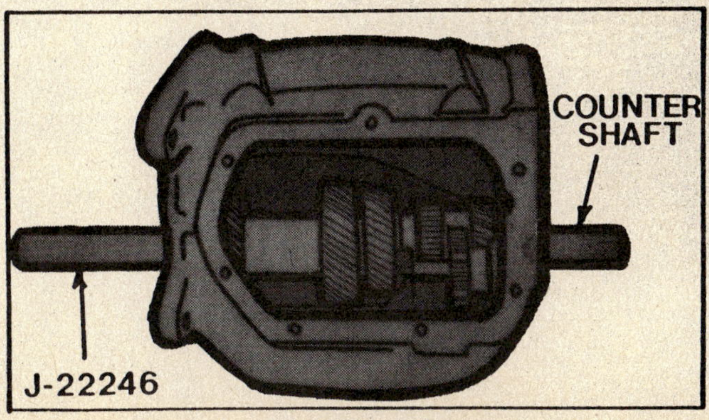

Fig. 6-4 Removing countershaft

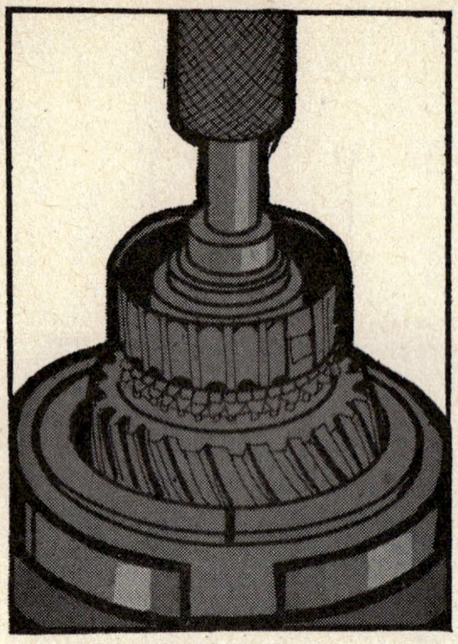

Fig. 6-5 Removing 3-4 synchronizer assembly

Fig. 6-6 Removing 1st gear and rear bearing

Fig. 6-7 Removing 1-2 synchronizer and 2nd speed gear

Assemble Mainshaft

With front of mainshaft up:

1. Install 3rd speed gear with clutching teeth upward.
2. Install blocker ring with clutching teeth downward over cone of gear. All blocker rings in this unit are identical.
3. Install 3-4 synchronizer assembly with fork slot downward and press it onto mainshaft until it bottoms. *Be sure the notches of the blocker ring align with the keys of the synchronizer assembly.*
4. Install synchronizer hub to mainshaft snap ring. Both synchronizer snap rings are identical.

With rear of mainshaft upward:

5. Install 2nd speed gear with clutching teeth upward.
6. Install a blocker ring with clutching teeth downward over cone of gear.
7. Press 1-2 synchronizer assembly onto mainshaft with fork slot downward. *Be sure notches in blocker ring align with keys of synchronizer assembly.*
8. Install synchronizer hub snap ring.
9. Install a blocker ring with notches downward so they align with the synchronizer keys.
10. Install 1st gear with clutching teeth downward. Install 1st gear thrust washer and spring washer.
11. Press rear bearing onto mainshaft. Install snap ring.
12. Install speedometer drive gear and clip.

ASSEMBLE TRANS.

1. Load a row of roller bearings and a thrust washer at each end of the countergear. Use heavy grease to hold them in place.
2. Install countergear through case rear opening with a tanged thrust washer at each end and install countershaft and woodruff key from rear of case. *Be sure countershaft picks up both thrust washers and that the tangs are aligned with their notches in the case.*
3. Install reverse idler gear, shaft and woodruff key from rear of case.
4. Expand extension housing snap ring and assemble extension housing over mainshaft.
5. Load roller bearings into clutch gear bore, using heavy grease to hold them in place, place blocker ring on gear cone with teeth toward gear, and install gear and ring onto mainshaft. Do not install clutch gear bearing at this time. *Be sure notches in blocker ring align with synchronizer keys.*
6. Using new gasket, install mainshaft and extension assembly through rear opening in case. Use sealing cement on bottom bolt.
7. Install large outer snap ring on clutch gear bearing and install bearing onto gear and into case bore. Install gear stem snap ring and bearing retainer.

NOTE: The retainer oil hole should be at the bottom.

8. With transmission in neutral, install cover assembly. *Be sure the shift forks are properly aligned in their grooves in the synchronizer sleeves before attempting to tighten cover bolts.*

Type Seven

DISASSEMBLE TRANS.

1. Remove TCS and backup lamp switches from case and extension, Fig. 1.
2. Remove cotter pins securing ends of shift control rod and remove rod, Fig. 2.
3. Remove serrated bolt retaining selector lever to boss on rear extension, Fig. 3.
4. Remove lock nut and selector ring.

NOTE: Slide lever and gearshift intermediate shaft from gearshift bracket while simultaneously removing selector ring. Selector lever will just slip off selector ring.

5. Remove snap ring from intermediate shift lever bushing and using a suitable brass drift, drive bushing from lever assembly.
6. Remove transmission case cover and discard gasket.
7. Remove detent cap, spring and ball, Fig. 4.
8. Invert transmission to drain oil.
9. Remove rear extension retaining bolts and rotate extension to expose countergear shaft, Fig. 5.
10. From front of transmission, using tool J-22911, drive out shaft. Be sure lock ball, Fig. 6, is not lost. With tool J-22911 inserted, take countergear out of case and remove thrust washers from case.
11. Drive out reverse intermediate shift lever pivot pin and remove intermediate lever, Fig. 7.

NOTE: Slide reverse shaft to rear of case so scallop in selector shaft will clear reverse shaft.

12. Shift transmission to neutral and push in on selector shaft. Turn selector shaft so lock pins are in vertical position. First drive lock pin out of 3rd-4th speed intermediate lever cam and then out of 1st-2nd speed lever cam, Fig. 8. Remove selector shaft.
13. With screwdriver, pry out selector shaft seal rings on transmission case.
14. Pull out both lock ball plugs using tool J-21715 with slide hammer J-7004, Fig. 9. Remove thrust springs and balls.
15. With transmission in 1st gear drive lock pins out of shifter forks and selector levers, Fig. 10. Remove 1st-2nd lever pin first.
16. From rear of transmission, drive out 1st-2nd shifter shaft using a brass drift, Fig. 11. Remove fork from sliding sleeve.
17. Tap 3rd-4th shifter shaft rearward until fork may be removed from shaft. Then drive out 3rd-4th shifter shaft through front of case.
18. Remove clutch drive gear from case.
19. Carefully slide rear extension and mainshaft assembly from case.
20. With tool J-22923 push out reverse idler gear shaft from front towards rear. Be sure lock ball is not lost, Fig. 12. Remove gear and shaft from case.
21. Using a brass drift and from the front of the transmission, drive out reverse shifter shaft. Remove shifter fork from case.

Mainshaft Disassemble

1. Remove snap ring from rear bearing retainer groove and remove mainshaft assembly from bearing retainer, Fig. 13.
2. Depress retaining clamp and slide speedo drive gear from shaft.
3. Remove loose parts such as needle bearing, spacer ring and synchronizer ring. The sliding sleeve, keys and front clutch key spring may also be removed.

NOTE: The snychronizer hubs and sliding sleeves are a selected assembly and should be kept together as originally assembled.

4. Remove snap ring in front of synchronizer hub, Fig. 14.
5. Remove snap ring, belleville washer and spacer from shaft, Fig. 15.
6. Support 2nd gear and press mainshaft until bearing and synchronizers are free on shaft. Remove all loose parts, Fig. 16.
7. Remove 3rd speed synchronizer hub snap ring. Support 3rd speed gear and press shaft from synchronizer assembly and 3rd speed gear, Fig. 17.

Mainshaft Assemble

1. From front of mainshaft install 3rd speed gear. Gear must turn freely on mainshaft.
2. Install 3rd speed synchronizer ring onto 3rd speed gear cone.
3. Install rear clutch key spring into 3rd-4th speed synchronizer hub so that hooked spring end rests in one of

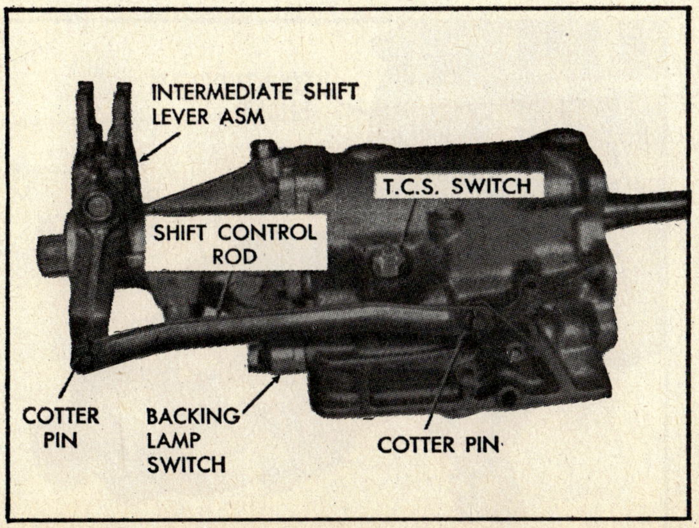

Fig. 2 Cotter pins securing shift control rod

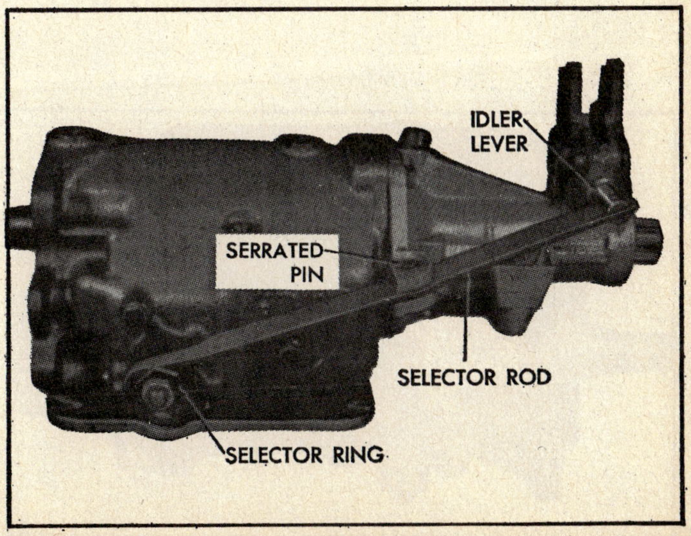

Fig. 3 Serrated pin location

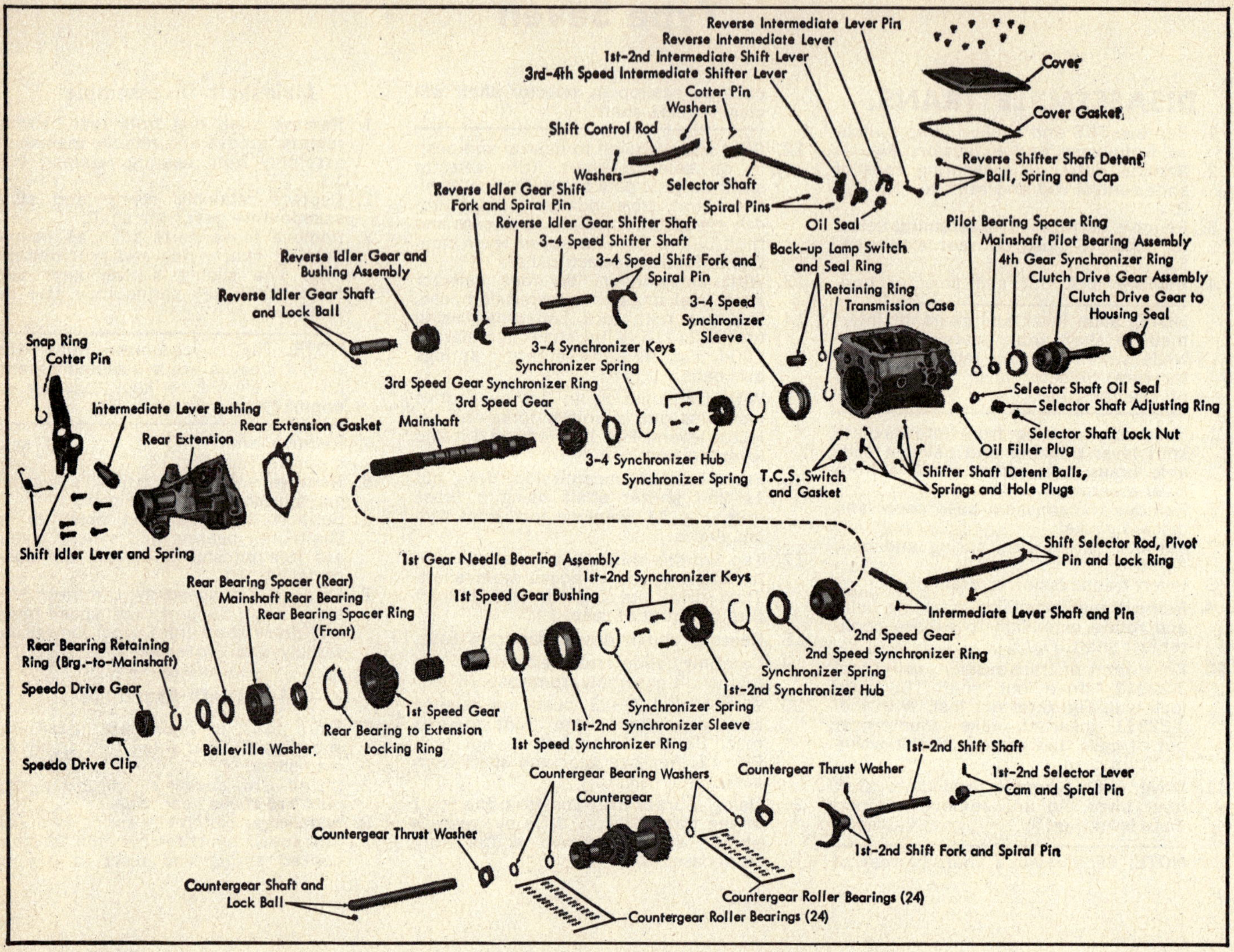

Fig. 1 Exploded view of Type 7 transmission

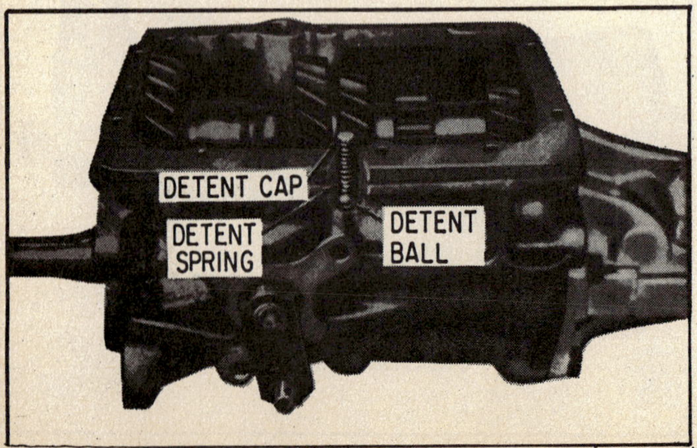

Fig. 4 Reverse detent cap, spring and ball

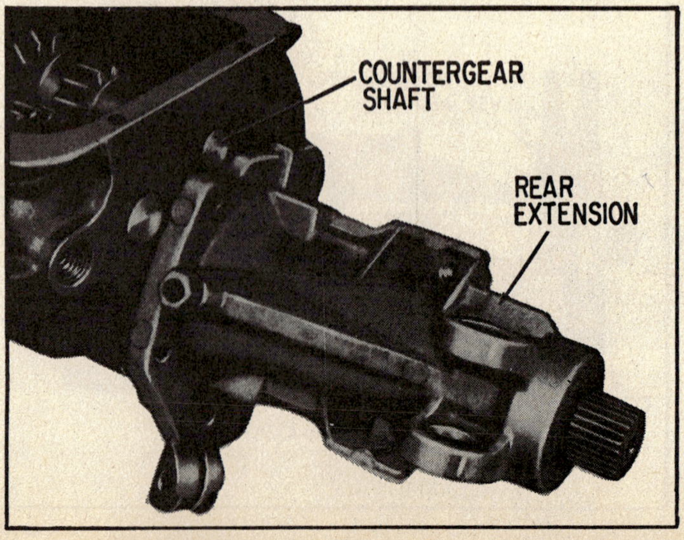

Fig. 5 Countergear shaft exposed for removal

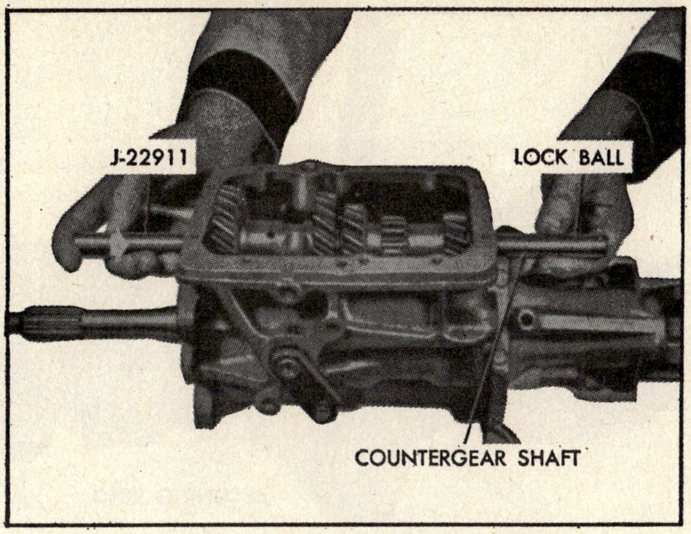

Fig. 6 Removing countergear shaft

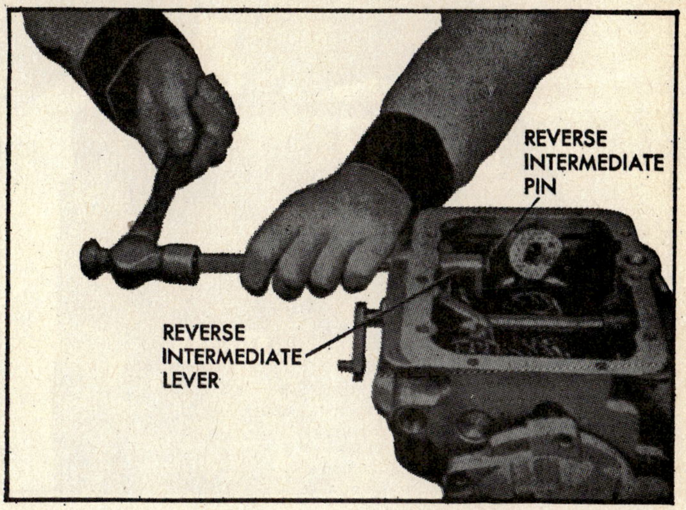

Fig. 7 Removing reverse intermediate lever pin

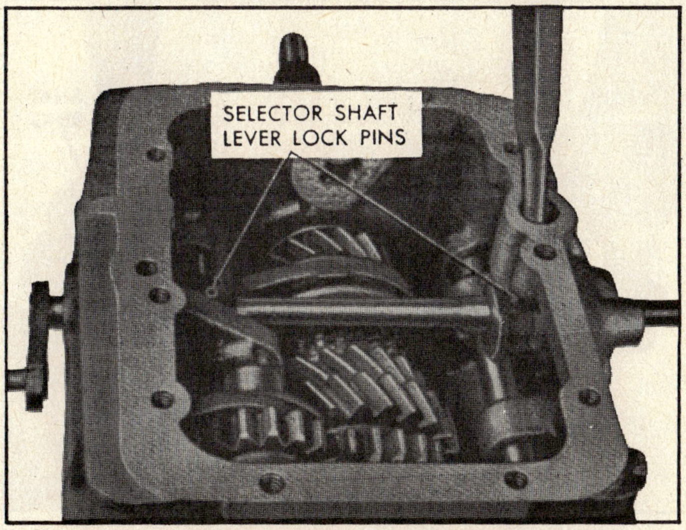

Fig. 8 Removing lock pins from selector shaft levers

Fig. 9 Removing detent ball lock plugs

Fig. 10 Position of lock pins

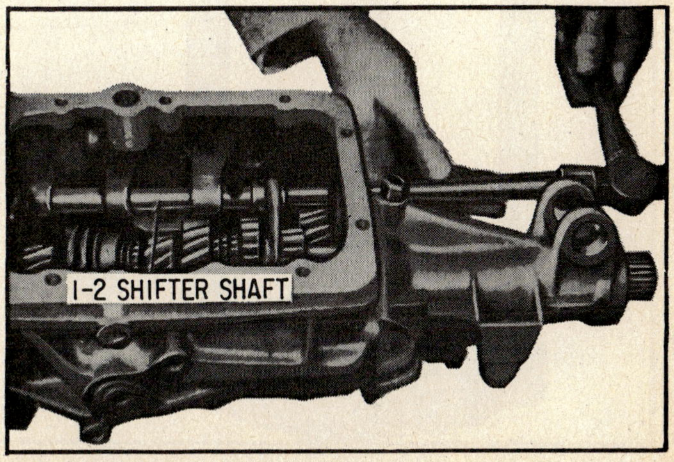

Fig. 11 Removing 1st-2nd shifter shaft

Fig. 12 Removing reverse idler gear shaft

Fig. 13 Removing snap ring from rear extension

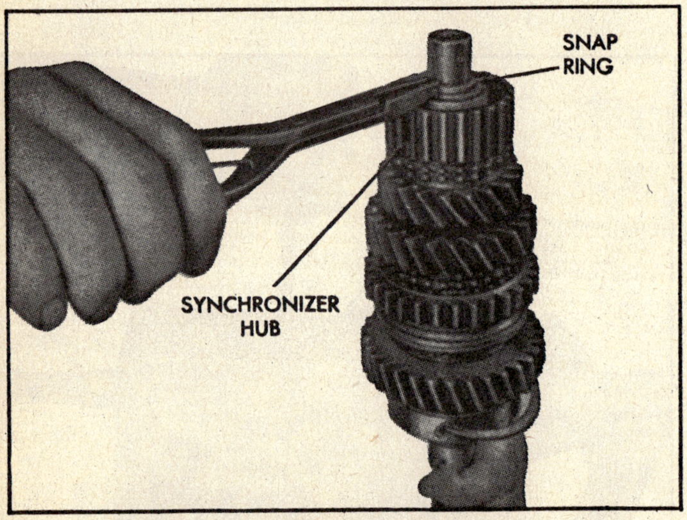

Fig. 14 Removing snap ring in front of synchronizer hub

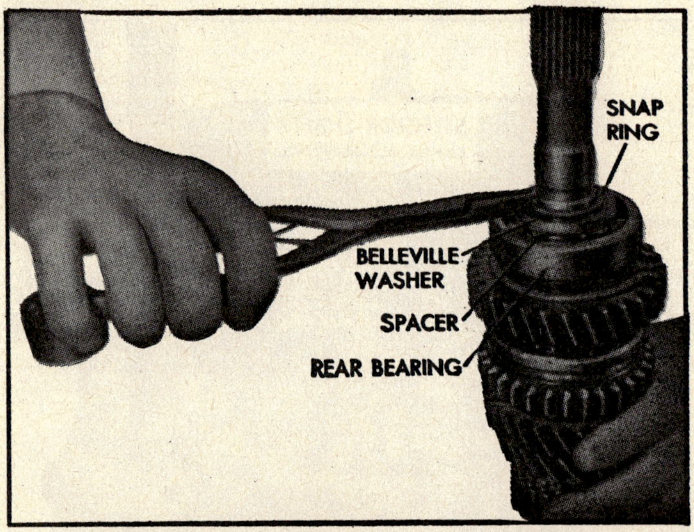

Fig. 15 Removing rear bearing snap ring

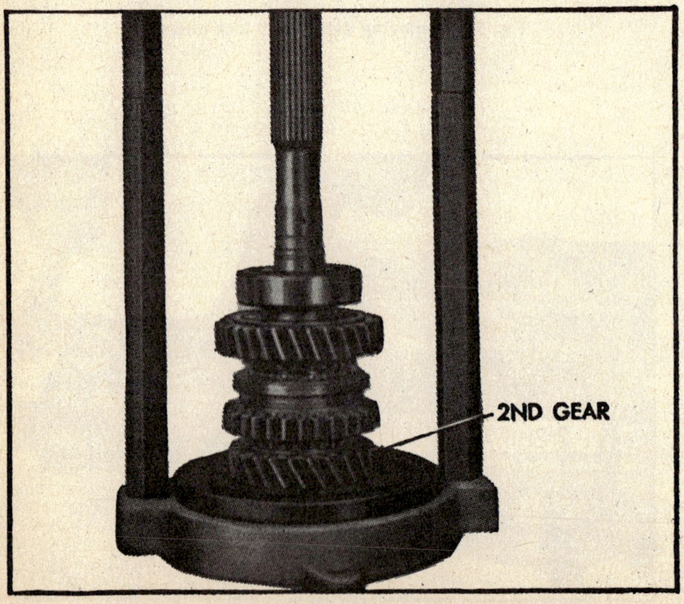

Fig. 16 Removing components from mainshaft

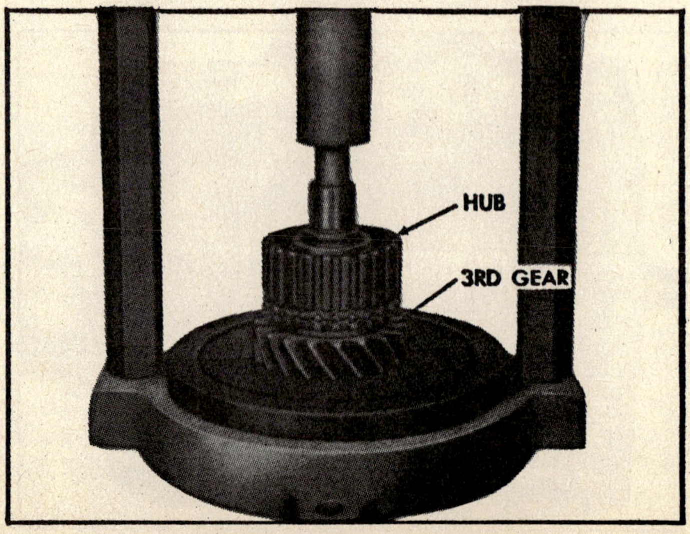

Fig. 17 Removing 3rd speed synchronizer hub

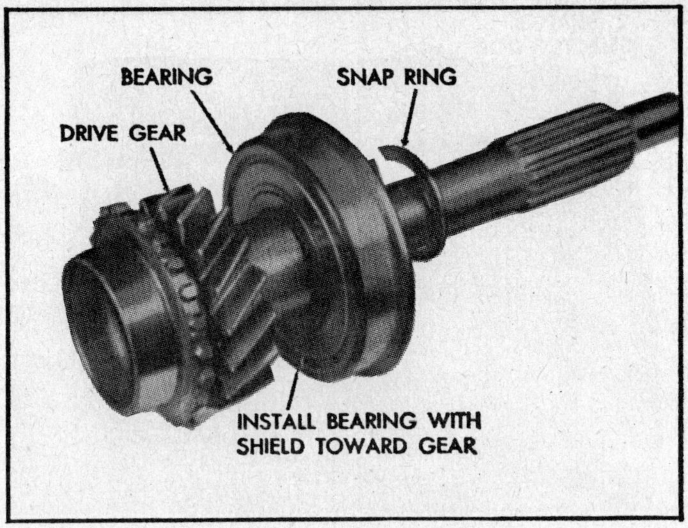

Fig. 18 Drive gear assembly

Fig. 19 Securing 3rd-4th synchronizer hub on mainshaft

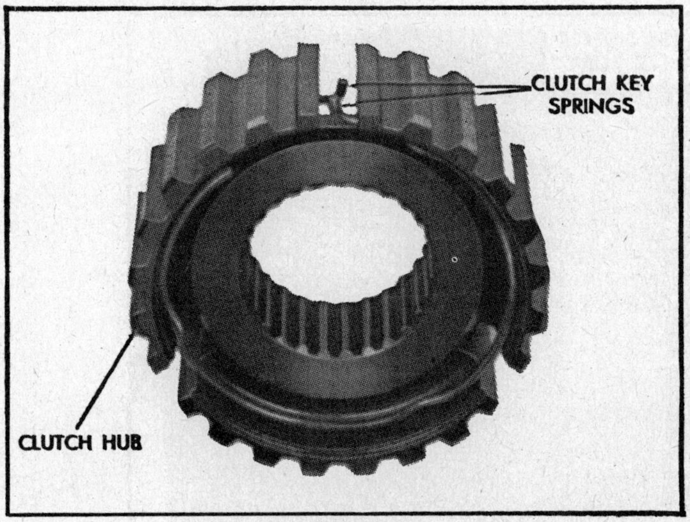

Fig. 20 Both synchronizer key springs installed

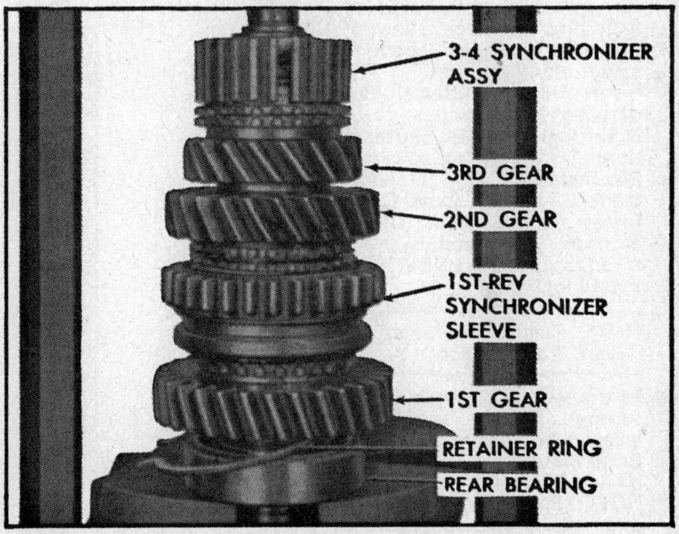

Fig. 21 Pressing mainshaft components

the slots and raised end is toward blocker ring.

4. Press 3rd-4th speed synchronizer hub on mainshaft.
5. Secure 3rd-4th speed synchronizer hub with snap ring, Fig. 19.
6. From rear of mainshaft slide 2nd speed gear onto mainshaft. Gear must turn freely on mainshaft.
7. Place 2nd speed synchronizer ring on-to to 2nd speed gear cone.
8. Install both synchronizer key springs into 1st-2nd synchronizer hub so that hooks of both springs rest in the same hub slot and other spring ends are positioned opposite to each other and toward blocker rings, Fig. 20. Install sliding gear and keys on hub.
9. Slide 1st-2nd speed synchronizer hub, needle bearing and inner sleeve onto mainshaft. Slide spacer, rear extension retaining ring and rear bearing onto shaft. Support rear bearing inner race and press components together, Fig. 21.

CAUTION: Align slots in synchronizer rings with synchronizer keys.

10. Place spacer and belleville washer on mainshaft and secure with snap ring.

NOTE: Concave side of belleville washer should face toward bearing.

11. Position speedo gear retaining clip on shaft and install gear.
12. Place mainshaft assembly into rear bearing retainer up to its stop. Secure with snap ring, Fig. 22.
13. Assemble 3rd-4th speed synchronizer on hub with raised end of key springs toward blocker ring.

NOTE: Arrow on keys point towards shifter fork groove (front of shaft).

ASSEMBLE TRANS.

1. With sealer, install new gasket onto rear extension.
2. Slide mainshaft assembly into case.
3. From front, slide spacer ring and needle bearing onto mainshaft.

NOTE: Coat needle bearing with ball and roller bearing grease.

4. Install synchronizer blocker ring on clutch drive gear and install gear into transmission case up to snap ring stop.
5. Insert 1st-2nd shifter shaft at front of case with notches down, pushing it first through "L" shaped selector dog. Dog should be positioned as shown in Fig. 23.

Then push 1st-2nd selector shaft through shifter fork, positioning shoulder toward front of case. Drive

Fig. 22 Installing rear bearing retainer ring

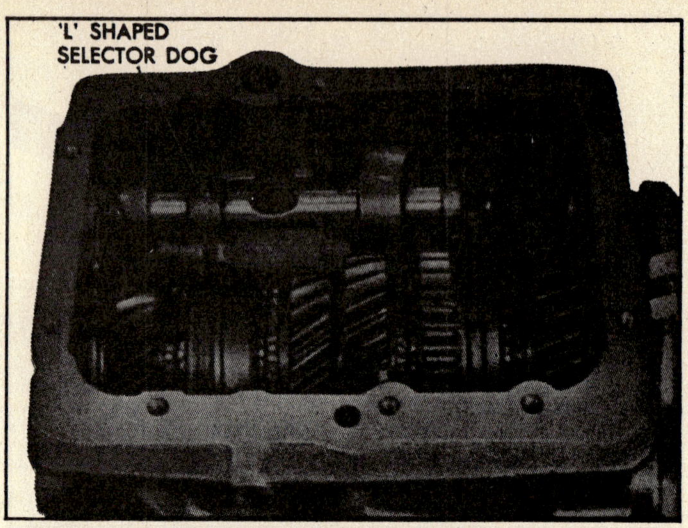

Fig. 23 1st-2nd shifter shaft installed

lock pins in place allowing them to protrude 1/16 to 5/64" above fork. Install selector dog pin first.

6. Insert 3rd-4th shifter shaft from front of case with notches down, pushing it through 3rd-4th shifter fork, positioning shouldering toward front, Fig. 24. Install lock pin allowing it to protrude 1/16 to 5/64" above fork.
7. Install reverse shifter shaft from rear of case, with notches up, pushing it through reverse shifter fork, Fig. 24. Install lock pin.

NOTE: Position shoulder of shift fork toward front of case.

8. Insert selector shaft in case, push through 3rd-4th speed intermediate lever, then through 1st-2nd speed intermediate lever. Install lock pins.

NOTE: Place transmission in neutral and rotate selector shaft to engage levers with shifter shafts.

9. Engage reverse speed intermediate lever with 3rd-4th speed intermediate lever and install pivot pin. Reverse speed intermediate lever end play on pin should be .004–.012".
10. Insert both lock balls and thrust springs into bores in transmission case and drive in plugs.
11. Turn extension case until bore for reverse idler shaft is exposed.
12. Place lock ball into shaft and from rear of case install shaft into gear.
13. Simultaneously position reverse idler gear and reverse shifter fork.

NOTE: Shifter fork groove of reverse idler gear and shoulder of shifter fork should be toward front of mainshaft.

Fig. 24 3rd-4th shifter shaft installed

14. Using tool J-22911, install a spacer, a roll of roller bearings (24) and a spacer at each end of countergear. Use heavy grease to hold them in place.
15. Coat thrust washers with ball and roller bearing grease and stick to case.

NOTE: Lugs of thrust washers must fit into case slots.

16. Turn case extension until countergear shaft bore is exposed.
17. Place lock ball into shaft and from rear of transmission insert shaft so thrust washer is held in position. Hold opposite thrust washer in position by using a short drift.
18. Insert countergear into case.
19. Insert shaft into countergear and push out special tool J-22911. Align lock ball with groove in case and drive shaft into case.
20. Align rear bearing retainer and torque bolts to case.
21. Install lock ball and thrust spring and spring cap into top transmission bore.
22. Install case cover gasket, cover and bolts.
23. Install gearshift linkage in reverse order of removal.

Type Eight

This section covers the English built unit which can be identified by having only 4 top cover bolts. The next section covers the German built unit which has 10 top cover bolts of 10 mm metric size.

DISASSEMBLE TRANS.

1. Remove clutch release bearing and lever from clutch housing.
2. Remove clutch housing to transmission bolts and remove housing.
3. Mount transmission in suitable holding fixture and remove top cover bolts.

NOTE: On late 1973 units, the detent shift spring and ball have been relocated, Fig. 1A. Also note that the fill plug location has been changed.

On 1973 units, remove TRS wire retaining clip and with a suitable magnet, remove detent shift spring and ball from case.

4. Using a suitable tool, pry blanking plug from extension housing.
5. On 1971-72 and early 1973 units, remove meshlock plunger set screw, spring and detent ball from side of case.
6. Using suitable punch, remove roll pin securing shift boss to rail. Be sure the pin can be punched through clear of any output shaft components. It may be necessary to position the synchronizer hub on the output shaft to suit.
7. Withdraw shift rail rearward, taking care not to let the shift boss and C-cam drop into case.
8. To remove shift forks, move 1st-2nd and 3rd-4th synchronizer hubs to their foremost position towards input shaft bearing.
9. Remove spring pin securing 3rd-4th shift fork to relay lever and remove fork.
10. Remove bolts securing extension housing to case.
11. Using a plastic faced mallet, tap the extension housing slightly rearward until it is possible to rotate it so countershaft aligns with cutaway in extension housing flange.
12. Tap countershaft rearward using a drift until it is just clear of front of case. Push countershaft out using a dummy countershaft. The countershaft gear will drop to bottom of case.
13. Remove extension housing and output shaft assembly. It is necessary to push 3rd-4th synchronizer sleeve forward to provide clearance between synchronizer and countershaft gear.

NOTE: Do not move synchronizer sleeve beyond 4th gear position or synchronizer inserts will fall out.

14. Unfasten bearing retainer from front of case and pry the oil seal from the retainer.

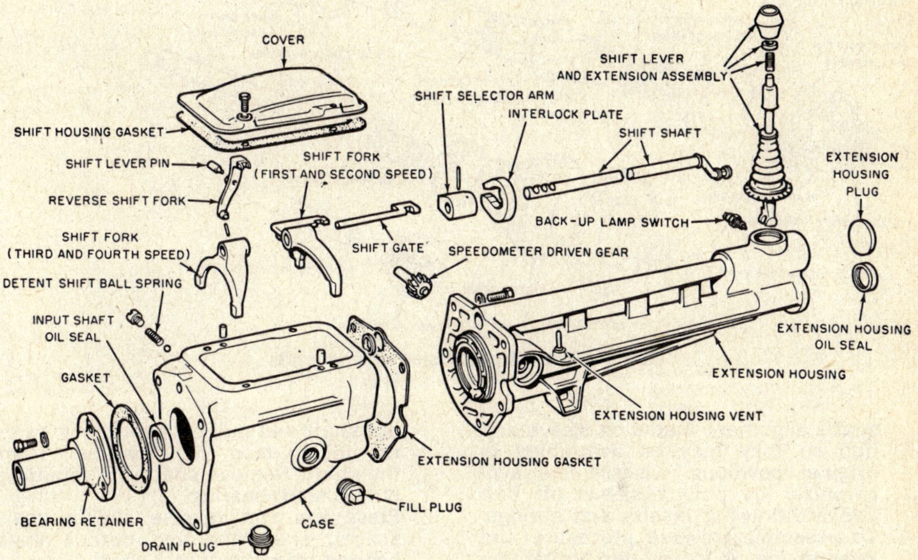

Fig. 1 Type 8 transmission case and related parts

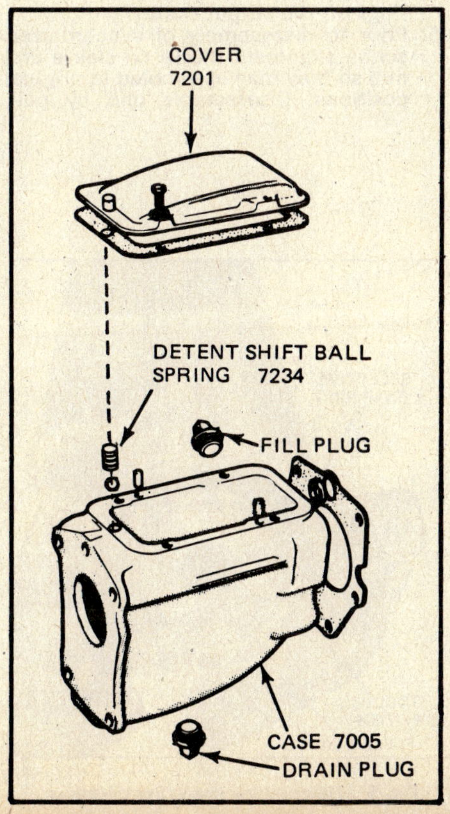

Fig. 1A Late 1973 detent shift ball, spring and fill plug location

15. Remove needle bearing from recess in end of input shaft gear.
16. Remove outer snap ring from around input shaft bearing. Using a suitable drift, tap the outer race inward evenly until bearing is free of case. Lift assembly out of transmission.
17. Remove countershaft gear and two thrust washers from case. In both ends of gear there are 20 needle rollers retained by a washer on each side of each set. Remove the rollers, washers and dummy shaft.
18. Withdraw reverse idler shaft, Fig. 3. Should these tools not be available, locate a nut, flat washer and a sleeve on a 5/16" 24UNF threaded bolt. Screw bolt into reverse idler shaft and tighten nut to withdraw shaft.
19. Slide reverse relay lever from fulcrum pin on case. Do not remove pin.

3rd-4th Synchronizer

1. Lift 4th gear blocking ring from front of 3rd-4th synchronizer.
2. Remove snap ring at forward end of output shaft and discard it. Press output shaft out of 3rd-4th synchronizer and 3rd gear while supporting the shaft so it will not drop.
3. Prior to disassembly of synchronizer

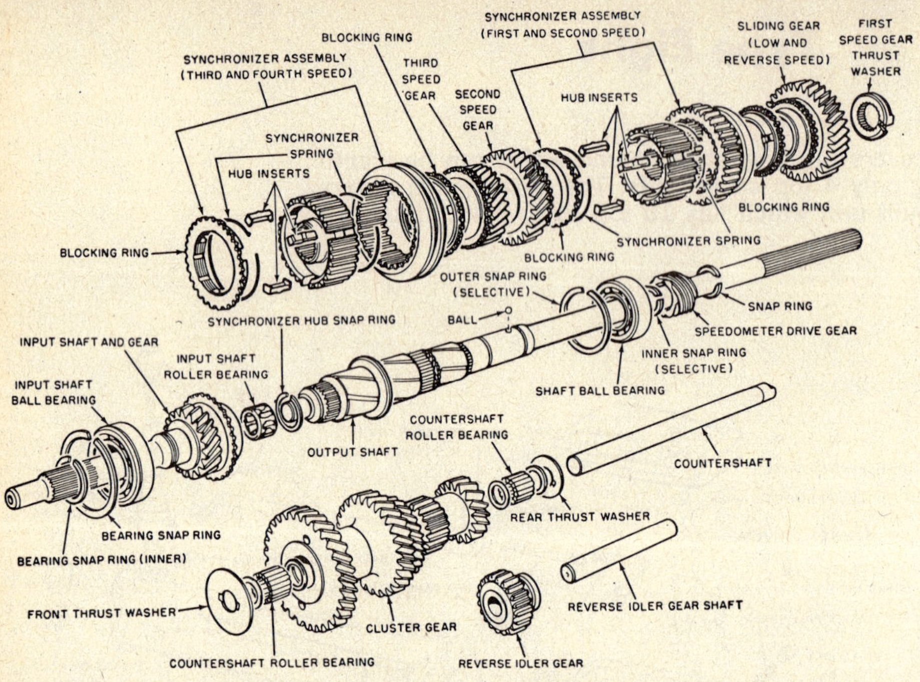

Fig. 2 Type 8 transmission internal parts

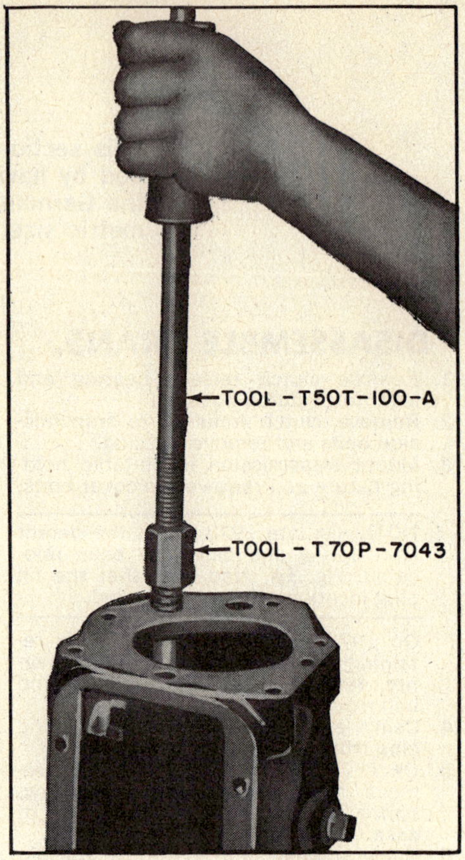

Fig. 3 Removing reverse idler shaft

scribe alignment marks on sleeve and hub so they may be assembled in original positions. Disassemble synchronizer by pulling sleeve off hub and withdrawing inserts and springs.

4. To assemble, reverse procedure and refer to Fig. 4 for proper positioning of synchronizer springs.

1st-2nd Synchronizer

1. Remove snap ring securing output shaft bearing to extension housing. Tap output shaft out of extension using a plastic faced mallet. Remove snap ring retaining speedo gear to output shaft and pull off gear being careful not to lose the drive ball from the shaft. Remove snap ring retaining output shaft bearing.
2. Press low and reverse sliding gear, spacer, snap ring and output shaft bearing from the output shaft.
3. Remove snap ring securing 1st-2nd synchronizer to output shaft.
4. Press 2nd gear and 1st-2nd synchronizer assembly complete with blocking rings off the output shaft.
5. Prior to disassembly of synchronizer, scribe alignment marks on sleeve and hub so they may assembled in original positions. Disassemble unit by pulling sleeve off hub and withdrawing the inserts and springs.
6. To assemble, reverse the procedure and refer to Fig. 4 for proper positioning of synchronizer springs.

Output Shaft

1. When installing output shaft assembly into extension, place master spacer tool T70P-7154 in output shaft bearing bore of extension, Fig. 5.
2. Determine thickness of the snap ring required to remove all end play from the master gauge as follows:
   a. Measure width of shaft bearing

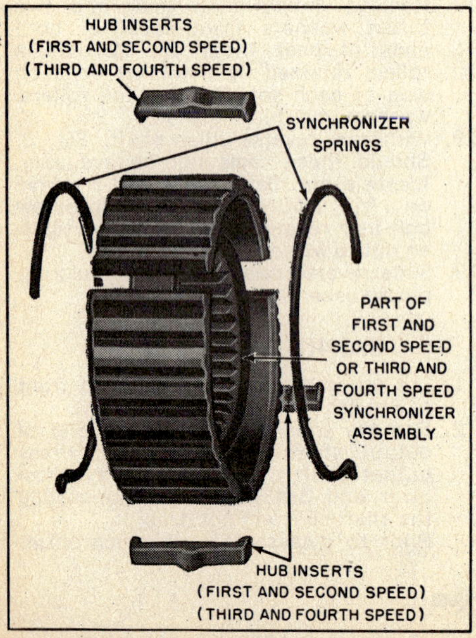

Fig. 4 Synchronizer spring rotation

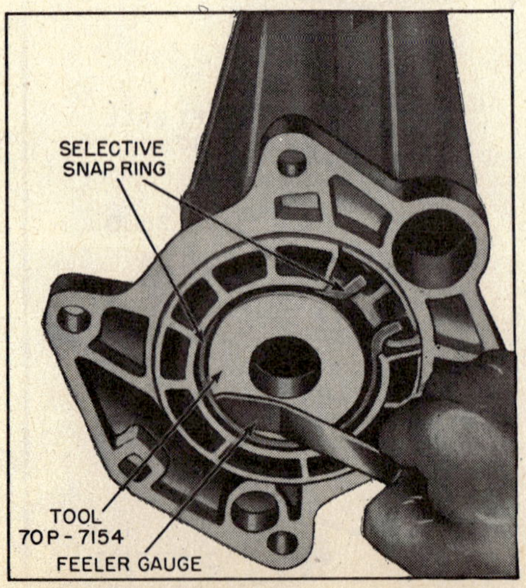

Fig. 5 Determining snap ring thickness

Part Number	Identification	Thickness
DORY-7030-A	Yellow	0.0726–0.0736
DORY-7030-B	Red	0.0715–0.0725
DORY-7030-C	Blue	0.0703–0.0713
DORY-7030-D	Violet	0.0691–0.0701
DORY-7030-E	Green	0.0679–0.0689
DORY-7030-F	Magenta	0.0677–0.0667
DORY-7030-G	Plain	0.0665–0.0655

Fig. 6 Output shaft bearing snap ring thickness chart

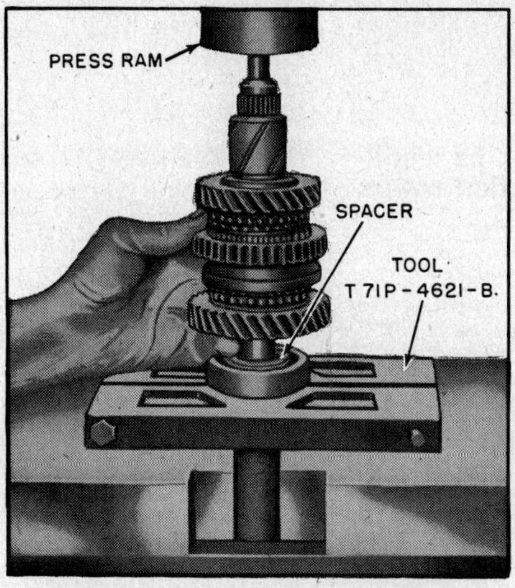

Fig. 7 Installing output shaft bearing

outer race with micrometer. The difference in thickness between master gauge and bearing outer race will determine thickness of selective ring to be used.

b. If bearing race thickness is more than that stamped on master gauge, the snap ring thickness must be decreased by that closest to available ring, Fig. 6.

c. If the thickness is less, the snap ring thickness must be increased.

3. Position the selected snap ring and the bearing on the output shaft.
4. Position tool T71P-4621B on the shaft, then place assembly into press. Press the bearing into place, Fig. 7, and secure with the thickest snap ring that will fit the groove in the output shaft.
5. Locate drive ball in output shaft detent and push speedo gear onto shaft so it just clears the snap ring groove in the shaft. Install a new snap ring to shaft to retain the gear.
6. Heat the front end of the extension housing, using a suitable hot plate or by placing in hot water. This will expand the extension housing so that the shaft can easily be installed.

NOTE: Do not use a welding torch.

7. Secure the output shaft bearing in the extension housing with the snap ring selected previously.

Input Shaft and Gear

1. Remove snap ring from input shaft and discard snap ring.
2. The bearing should not be removed unless it is noisey, rough, spalled or cracked. Position input shaft bearing, Fig. 8, and press bearing off shaft.
3. To assemble, press bearing on input shaft. The bearing must be placed on the shaft with the ring groove facing away from the gear. The tool used to replace the bearing insures that all the load is taken through the bearing inner race so the bearing will not be damaged in the pressing operation.
4. Install the thickest snap ring to secure the bearing to the shaft.

ASSEMBLE TRANS.

1. Slide reverse relay lever onto fulcrum pin on case.
2. Lubricate idler shaft and push it into case. Install reverse idler gear on shaft and locate reverse relay lever in groove in reverse idler gear.
3. Tap reverse idler shaft into position with copper mallet.
4. Slide dummy countershaft into countershaft gear. Install a retainer washer over dummy shaft and push it into the gear bore. Grease needle rollers and assemble 20 into the recess; install 2nd retaining washer and repeat procedure at other end of gear.
5. Grease thrust washers and locate them so their tab side seats into recess in case.
6. Position countershaft gear in bottom of case, taking care not to displace washers.
7. Assemble input shaft and gear into transmission. Using a copper drift on the bearing outer race, tap it into place until the snap ring groove appears on the outside of the case. Take care that the dog teeth on the input shaft gear are not damaged by the countershaft gear.

NOTE: The bearing is an interference fit in the case, it is important that the outer race is tapped. Do not tap on input shaft gear as the bearing will be damaged.

8. Install snap ring to periphery of the bearing.
9. Lubricate input shaft needle bearing and position it in recess of input shaft gear.
10. Place a new oil seal on input shaft retainer and install so lips of seal face transmission. Drive seal into retainer until it bottoms.
11. Lubricate front bearing retainer seal and the seal journal area on input shaft. Place a new gasket on retainer. Fabricate a plastic sleeve and slide it over input shaft splines to prevent damage to seal lip. Be sure that oil groove in retainer is in line with oil passage in case and that gasket does not cover this passage. Coat attaching screws with suitable sealer and install them. Remove plastic sleeve. Apply a light film of grease to release bearing surface on bearing retainer.
12. Prior to installing output shaft and extension assembly, lubricate input shaft gear cone and position 4th gear blocking ring on input shaft gear cone.
13. If necessary, install a new seal in shaft rail aperture in rear of case and use a standard socket to drive in the seal.
14. Thread cord or suitable plastic covered wire under countershaft gear at each end to facilitate lifting into position later.
15. Install new gasket on extension using a sealer.
16. Slide extension and output shaft assembly into position after pulling the 3rd-4th synchronizer sleeve forward to clear countershaft gear.

NOTE: Do not move sleeve beyond 4th gear position as synchronizer inserts will drop out. Be sure 4th gear blocking ring locates correctly.

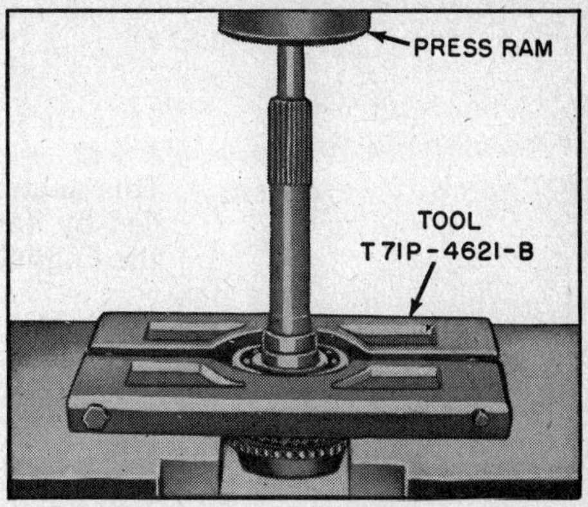

Fig. 8 Removing input shaft bearing

17. Align cutaway on extension housing with countershaft opening in rear face of transmission.
18. Carefully, with string or wire attached previously, lift countershaft gear into mesh with output shaft and input shaft gears. Take care that thrust in case at each end of gear are not displaced.
19. Check that countershaft gear bore aligns with apertures of countershaft. Push dummy shaft out of gear by inserting countershaft from rear. Finally, tap countershaft into position with suitable mallet. Be sure that lug on rear of countershaft is positioned horizontally so it will install into recess on extension housing flange. The front of the countershaft must be flush with front face of transmission case after installation.
20. Rotate extension so bolts align and push extension fully home onto the transmission. Secure housing to case using sealer on bolts.
21. Replace shift forks on relay lever and secure the 3rd and 4th fork to lever with new roll pin.
22. Position assembled shift forks on their synchronizer sleeves and move the synchronizer hubs into neutral position so that shift fork extension arms locate beneath the reverse idler shift arm mounted on side of case.
23. Grease shift rail oil seal in rear of case and slide the rail through the extension housing. Position the shift boss and the C-cam so that the cam locates the cutouts in the shift fork extension arms. Pass the rail through the boss and forks until the spring pin holes in the boss and rail align. Take care not to damage the shift rail seal.
24. Assemble detent ball and spring to their bore and install set screw using sealer.
25. Install roll pin to retain shift boss to shift rail.
26. Apply sealer to blanking plug and tap it into extension housing behind shift rail.
27. Using a new gasket and sealer, install top cover.
28. Remove transmission from holding fixture, install clutch housing, release lever and bearing. Refill with oil to proper level.

Type Nine

This section covers the German built unit which can be identified by having 10 top cover bolts. The previous section covers the English built unit which has only 4 top cover bolts.

DISASSEMBLE TRANS.

1. Remove clutch release bearing, lever and clutch housing.
2. Remove top cover bolts with 10mm wrench and drain lubricant.
3. Remove threaded plug, spring and shift rail detent plunger from case, Fig. 1.
4. Drive access plug from rear of case, Fig. 2, and drive the interlock plate retaining pin from case, Fig. 3. Lift interlock plate from case.
5. Remove roll pin from selector lever arm, Fig. 2.
6. Tap front end of shaft rail to displace plug at rear of extension housing.
7. Withdraw shift rail from extension and case, Fig. 4.
8. Lift selector arm and shift fork from case.
9. Remove extension housing bolts and tap extension housing with plastic mallet to loosen it from case so it may be rotated.
10. Rotate extension to align countershaft with cutaway in extension flange. Using a brass drift, drive countershaft rearward until it just clears front of case. Install a dummy shaft in the case and gear until the countershaft gear can be lowered to bottom of case, then remove the countershaft.
11. Lift extension housing and mainshaft from case as an assembly, Fig. 5.
12. Remove 10mm input shaft attaching bolts and remove the input shaft and bearing retainer from case as an assembly.
13. Remove reverse idler gear shaft from rear of case, Fig. 6. Remove reverse idler gear and on 74WT transmissions, reverse idler gear spacer, Fig. 6A.
14. Remove countershaft bearing retain-

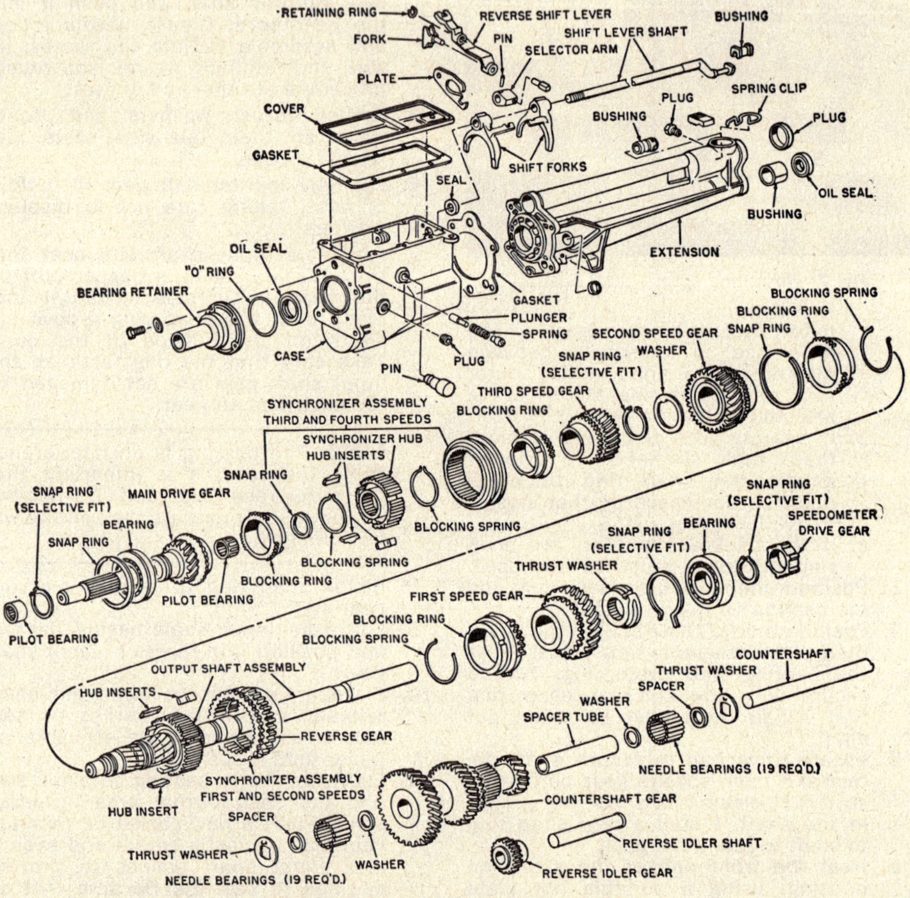

Exploded view of Type 9 transmission

Fig. 1 Shift rail detent plunger

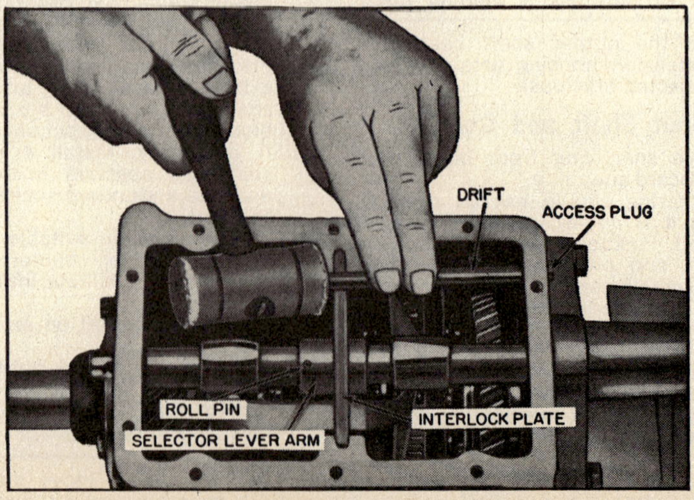

Fig. 2 Removing interlock plate access plug

ing washers, bearings (19 on each end for 74WG transmissions and 21 for 74WT models), dummy shaft and spacer from cuntershaft gear, Fig. 7.

NOTE: 74WG transmissions have one thrust washer located at each end of the countershaft whereas, the 74WT transmissions have two thrust washers.

15. Remove bearing retainer and pilot bearing from input shaft gear.

NOTE: Do not remove the ball bearing from input shaft unless replacement is necessary. If so, remove snap ring from input shaft, Fig. 8, and press the shaft out of the bearing, Fig. 9.

16. Pry input shaft seal out of bearing retainer, Fig. 8.
17. Lift 4th gear blocker ring, Fig. 10, from front of output shaft.
18. Remove snap ring from forward end of output shaft and discard it.
19. Position tool T69P-4621A behind 3rd speed gear. Place output shaft and extension assembly in a press and press output shaft out of 3rd-4th synchronizer and the 3rd gear while supporting the extension housing and output shaft to prevent it from dropping.
20. Remove snap ring and washer, then slide 2nd gear and blocker ring off output shaft and discard snap ring.
21. Disassemble synchronizer assembly by pulling sleeve off hub and removing inserts and springs.
22. Remove snap ring that retains output shaft bearing in extension housing.
23. Tap output shaft assembly out of extension with a plastic hammer.
24. Position tool T69P-4621A behind 1st gear, then place assembly in press as shown in Fig. 11.
25. The 1st-2nd speed synchronizer and hub are serviced only as an assembly and no attempt should be made to separate the hub from the shaft. The sleeve, springs and inserts may be removed from the hub.
26. Drive shift rail bushing from rear of

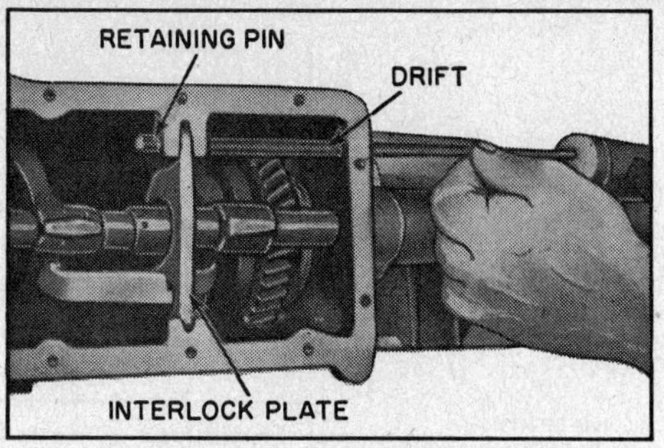

Fig. 3 Removing interlock plate retaining pin

extension with a 9/16" socket and extension. Do not remove bushing if serviceable.

27. Pry shift rail seal from rear of transmission case.
28. Remove remaining linkage from case.

ASSEMBLE TRANS.

1. Seat new shift rail seal in rear of case, Fig. 12.
2. If shift rail bushing was removed, drive a new one into place with a 9/16" socket and extension.
3. If 1st-2nd synchronizer was disassembled, slide the sleeve over the hub making sure the shift fork groove is toward front of shaft. The sleeve and hub are select fit and must be reassembled with etch marks in same location. Locate an insert in each of three slots cut in the hub. Install insert spring inside the sleeve beneath the inserts. The tab on end of spring must locate in the section of an insert. Fit the other spring to the opposite face of the synchronizer unit, being sure the spring tab locates in the same insert as the spring just installed and is in the same rotational direction. Looking down at the synchronizer, the tab end of one spring should be in line with the tab of the spring on the opposite side.

NOTE: Oil all parts at time of assembly.

4. Assemble a blocker ring on the 1st gear side of the 1st-2nd synchronizer. Apply grease to the cone surface of the 1st gear. Slide 1st gear onto output shaft so that cone surface engages blocker ring.
5. Position spacer on shaft making certain the large diameter is toward rear of staff.
6. Place master spacer tool T70P-7154 in the output shaft bearing bore of the extension, Fig. 13. Determine thickness of snap ring required to remove all end play from master gauge. Then measure width of the output shaft bearing outer race with a micrometer. The difference in thickness between the master gauge and the bearing outer race will determine thickness of selective snap ring. If the bearing race thickness is more than that stamped on the master gauge, the snap ring thickness must be de-

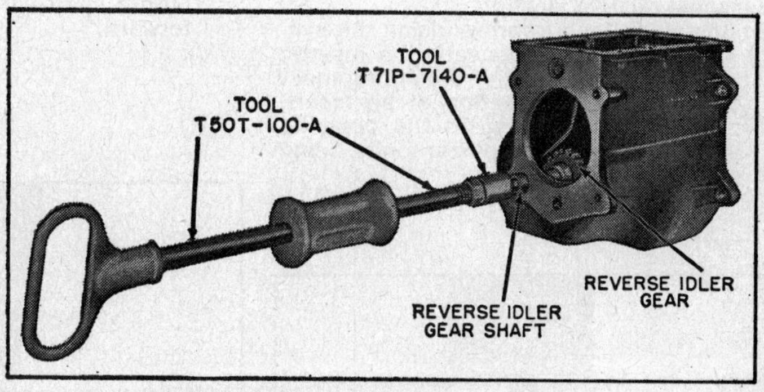

Fig. 6 Removing reverse idler gear shaft

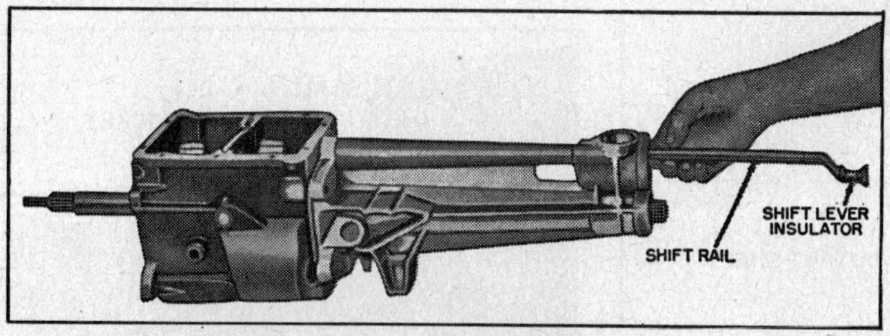

Fig. 4 Removing or installing shift rail

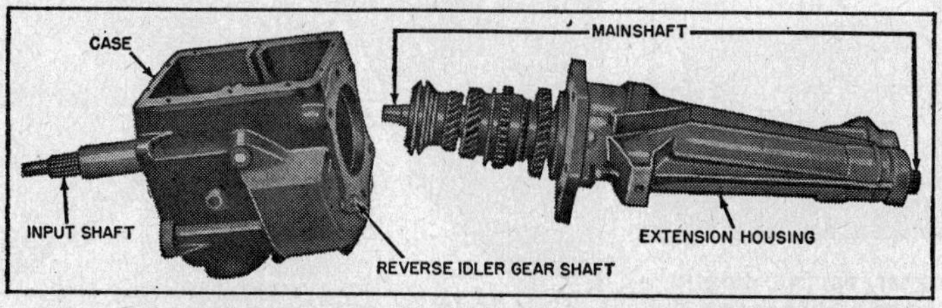

Fig. 5 Removing or installing extension and mainshaft

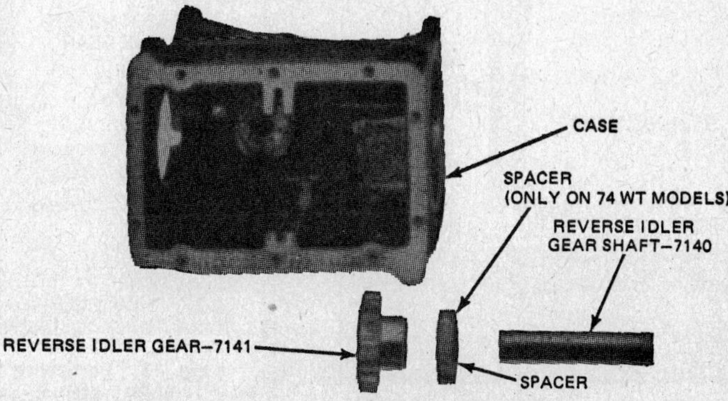

Fig. 6A Reverse idler gear spacer location

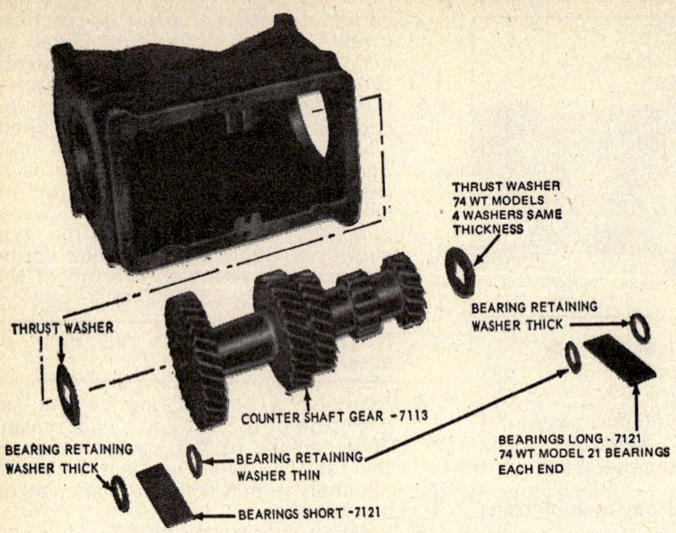

Fig. 7 Countershaft gear disassembled

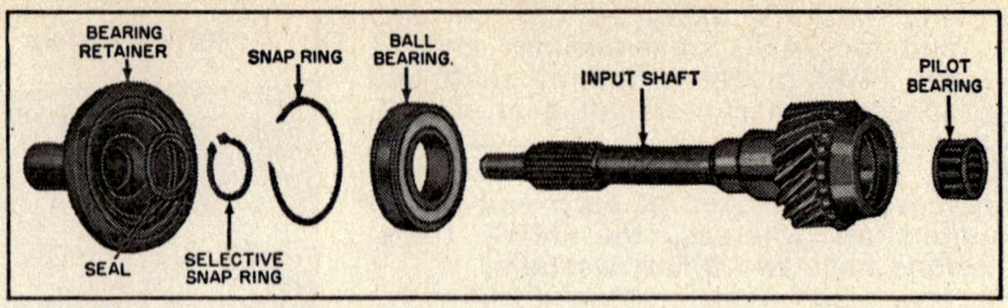

Fig. 8 Input shaft disassembled

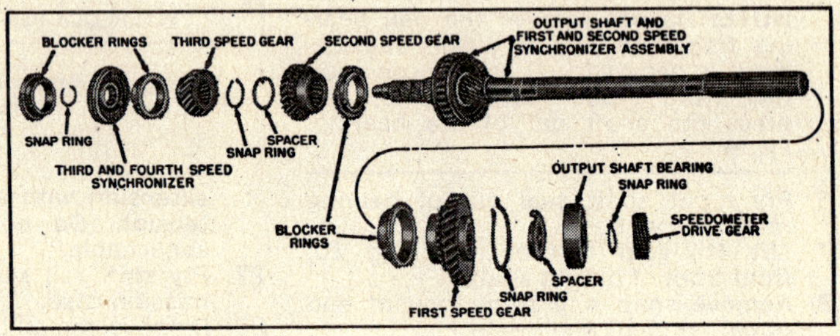

Fig. 10 Output shaft disassembled

creased to the closest available ring, Fig. 14. If the thickness is less, the snap ring thickness must be increased.

7. Position selected snap ring and the bearing on the output shaft. Position tool T69P-4621A on the shaft, then place assembly in a press and press bearing into place. Secure bearing with thickest snap ring that will fit the groove in the output shaft.
8. Slide the synchronizer sleeve over the hub and locate an insert in each of the three slots cut into sleeve. The sleeve and hub are select fit and must be reassembled with etch marks in same relative location.
9. Install an insert spring inside the synchronizer sleeve beneath the inserts. The tab on the end of the spring must locate in the U section of an insert. Fit the other spring to the opposite face of the synchronizer unit, being sure that the spring tab locates in the same insert as the spring just installed and is in the same rotational direction. Looking down at the unit, the tab end of one spring should be in line with the tab of the spring on the opposite side.
10. Position the 2nd gear and blocker ring on output shaft so that dog teeth face rearward. Install washer and snap ring. Position 3rd gear on output shaft so dog teeth face forward. Apply grease to cones of the gears and assemble the blocker ring on the 3rd gear cone.
11. Position the 3rd-4th synchronizer on output shaft with the hub boss facing forward.
12. Position tool T69P-4621A so that it butts against the boss on the synchronizer hub.
13. Place entire unit, extension end up, in a press and push the synchronizer unit onto the output shaft as far as possible.
14. Retain the 3rd-4th synchronizer to output shaft with snap ring. Pull up on the synchronizer assembly so the snap ring is tight in its groove.
15. Prior to assembling output shaft and extension housing into the case, apply lubricant to the cone of the gear and place the 4th gear blocker ring on the input shaft gear cone.
16. Press the speedo gear onto the shaft with tools shown in Fig. 15, until dowels just contact bearing outer race.

NOTE: The dowels must contact the

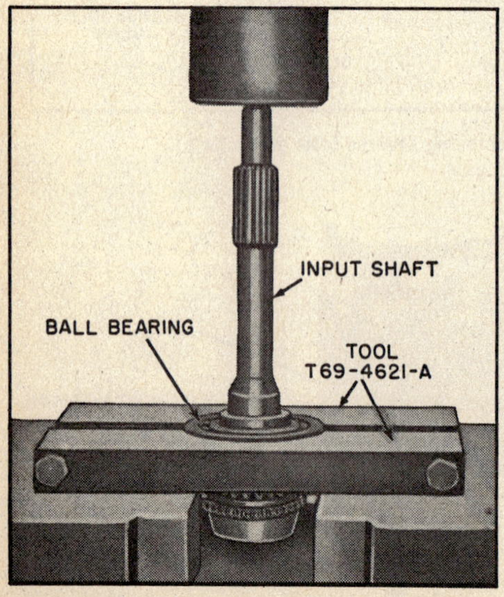

Fig. 9 Removing input shaft bearing

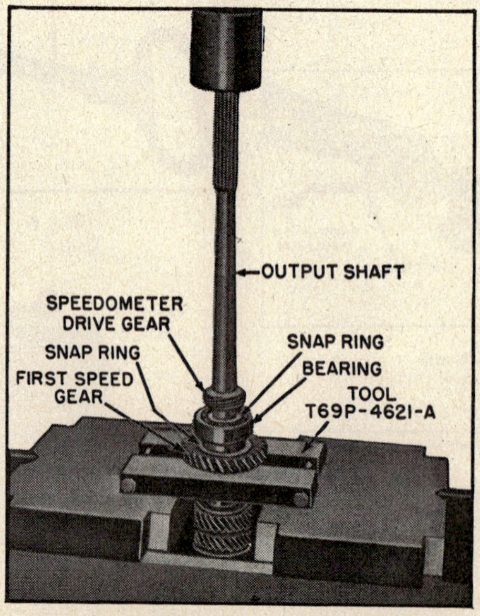

Fig. 11 Removing 1st gear, spacer, output shaft bearing, snap rings and speedo gear from output shaft

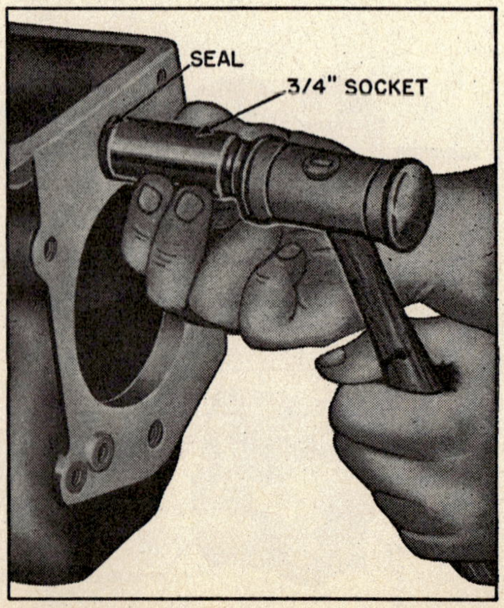

Fig. 12 Installing shift rail seal

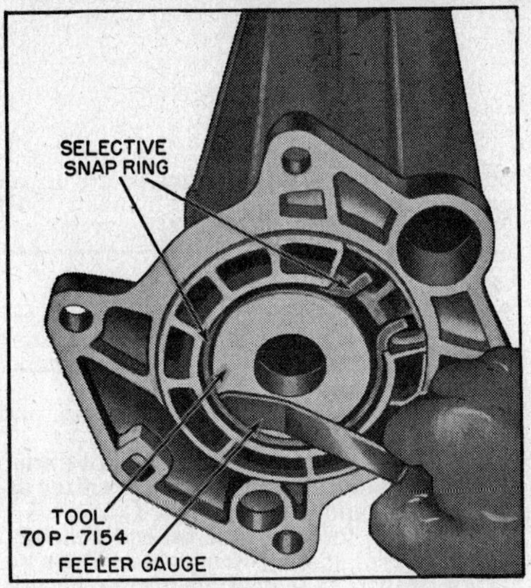

Fig. 13 Determining snap ring thickness

bearing outer race to properly locate the speedo gear on the shaft.

17. Coat the bearing bore of the extension housing with lubricant and install the output shaft in the housing. It may be necessary to tap the shaft with a plastic hammer while holding the two synchronizer sleeves firmly to prevent sleeve separation from the hubs. Secure it to the extension with the selective snap ring that was previously installed.
18. Press bearing on input shaft with snap ring groove toward front of shaft. Secure bearing to shaft with thickest selective snap ring.
19. Slide spacer and dummy shaft into countershaft gear. Position a thin bearing retaining washer at each end of dummy shaft. Coat the bearings with lubricant. Load long bearings into small end of countershaft gear and the short bearings into large end of gear. Note that the 74WG transmission has 19 bearings at each end and the 74WT transmission has 21 bearings at each end. Fit a thick retaining washer over each end of dummy shaft. Coat each thrust washer with lubricant and on 74WG models position one on each end of dummy shaft and on 74WT models, position two thrust washers on ends of dummy shaft. Make sure the tabs are in the same relative position so they may engage the slots in the case when the gear is lowered into place. Loop a piece of rope or wire around each end of the gear. Carefully install the countershaft gear with rope through rear end of case and lower the gear into place being careful not to disturb thrust washers and making sure that the tabs engage slots in case.
20. Apply lubricant to reverse idler gear shaft. If selector lever relay was removed, position it on the pivot pin. Secure lever on pin with spring clip. Hold gear in the lever with long hub toward rear of case. Slide reverse idler gear shaft into place and seat the shaft in the case with a copper hammer.
21. Install a new seal in input shaft bearing retainer.
22. Assemble input shaft to case using a new retainer O ring, Fig. 16. If necessary, tap outer race of bearing with copper hammer evenly until outer snap ring is seated against case.

NOTE: Do not tap on input shaft as this may damage the races/or bearings.

23. Carefully slide 3rd-4th speed synchronizer sleeve into 4th speed position (forward to provide clearance).
24. Position new gasket on extension housing.
25. Lubricate input shaft pilot bearing and install in shaft. Slide extension housing and output shaft into place being careful not to disturb 3rd-4th synchronizer.
26. Align cutaway in extension housing flange with countershaft bore in rear of case.
27. Lift countershaft gear into place with cord or wire, then slip countershaft into place making sure both thrust washers are in place. Make sure that the flat on the countershaft is toward top of case and in horizontal position, Fig. 17. Then tap it into case with brass hammer until front of shaft is flush with case.
28. Place shift forks in synchronizers sleeves. Position interlock lever and install new retaining pin. Lubricate shift rail oil seal and slide shift rail through extension housing, transmission case and the 1st and 2nd speed shift fork. Position selector arm on the rail, then slide the rail through the 3rd and 4th speed shift fork, then through the front of case until center detent is aligned with the detent plunger bore. Install new retaining pin in selector arm.

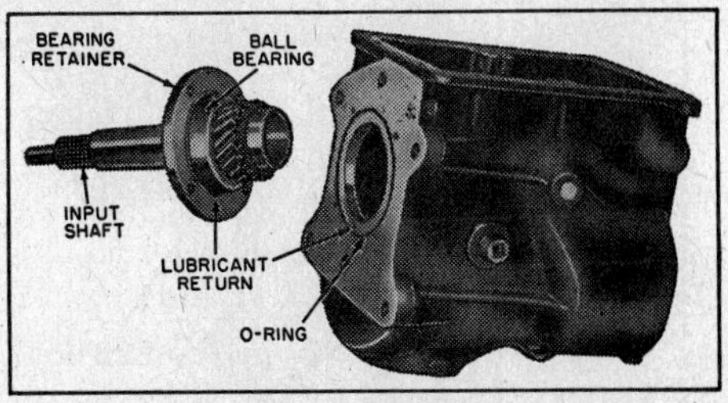

Fig. 16 Installing input shaft gear

Part No.	Thickness	Identification
D1FZ-7030-A	0.0679-	Color Coded–Copper
D1FZ-7030-B	0.0689-	Letter–W
D1FZ-7030-C	0.0699-	Letter–V
D1FZ-7030-D	0.0709-	Letter–U
D1FZ-7030-E	0.0719-	None
D1FZ-7030-F	0.0728-	Color Coded–Blue
D1FZ-7030-G	0.0738-	Color Coded–Black
D1FZ-7030-H	0.0748-	Color Coded–Brown

Fig. 14 Output shaft bearing snap ring thickness chart

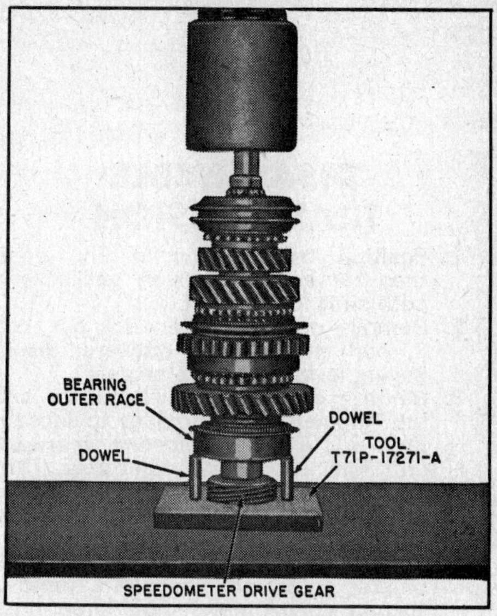

Fig. 15 Installing speedo drive gear

29. Install detent plunger, spring and plug with sealer.
30. Install new access plug in rear of case.
31. Rotate extension to align bolt holes, then install bolts loosely. Before tightening bolts make sure shift rail slides freely in bore.
32. Position new oil seal in input shaft retainer so tension spring and lip face the case. Drive seal into position.
33. Position new O ring in groove in face of case. Install input shaft bearing retainer to case. Be sure that oil passage in case is in line with oil groove in retainer, Fig. 16. Apply sealer to bolts and attach to case.
34. Reinstall clutch release bearing.
35. Apply sealer to new extension housing plug and install it.
36. Install top cover with vent toward rear. Make sure that sealer is applied to bolt over detent plunger bore.

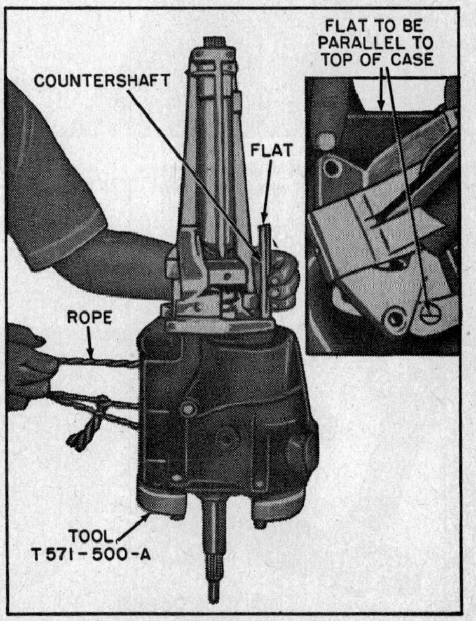

Fig. 17 Installing countershaft

Type 10 Warner

DISASSEMBLE TRANSMISSION

1. Position transmission in 2nd gear, then remove side cover, gasket and both shift forks. Fig. 1.
2. Remove reverse shifter lock pin, Fig. 2, pull shifter shaft outward, disengaging lever from reverse gear.
3. Remove extension housing bolts and tap housing with a mallet to loosen, then pull extension housing rearward until idler shaft clears idler gear. Turn housing to left thereby freeing reverse shift fork from collar and remove housing.
4. Remove snap ring, then slide speedometer gear from mainshaft, remove second speedometer gear snap ring. Fig. 3.
5. Remove reverse gear from mainshaft, then slide rear reverse idler gear from rear bearing retainer. Fig. 4.
6. Remove lock bolt from rear bearing retainer and drive locating pin from bearing retainer into case.
7. Rotate rear bearing retainer counterclockwise, providing access to rear of countershaft.
8. From front of case, using a dummy shaft, tap countershaft and key out the rear of case. Fig. 5.
9. Using a mallet, tap rear bearing retainer rearward, remove mainshaft through rear of transmission.
10. Remove bearing rollers and washer from main drive gear bore.
11. Remove front reverse idler gear and thrust washer from case.
12. Remove fourth gear synchronizing ring from main drive gear.
13. Remove main drive gear bearing retainer from case, remove snap ring and washer from main drive gear, then press gear from bearing, Fig. 6.
14. Tap drive gear bearing from case, remove countergear assembly and 2 washers, drive locating pin from case.

NOTE: It is not necessary to remove countershaft and main gear assemblies if only main drive gear bearing is to be replaced. This is accomplished by using tools J-22912 and J-8433-1. Fig. 7.

Disassemble Mainshaft

1. From rear of mainshaft, remove snap ring and slide rear bearing retainer from mainshaft. Fig. 8.
2. Remove rear bearing snap ring and washer.
3. From front of mainshaft, remove snap ring, then remove 3-4 synchronizer assembly and 3rd gear, Fig. 1.
4. Support forward face of 2nd gear and press mainshaft from rear then remove the rear bearing, thrust washer, 1st gear and bushing, 1-2 synchronizer and 2nd gear.

Disassemble Countergear

Remove dummy shaft, remove bearing rollers, spacers and tubular spacer, Fig. 9.

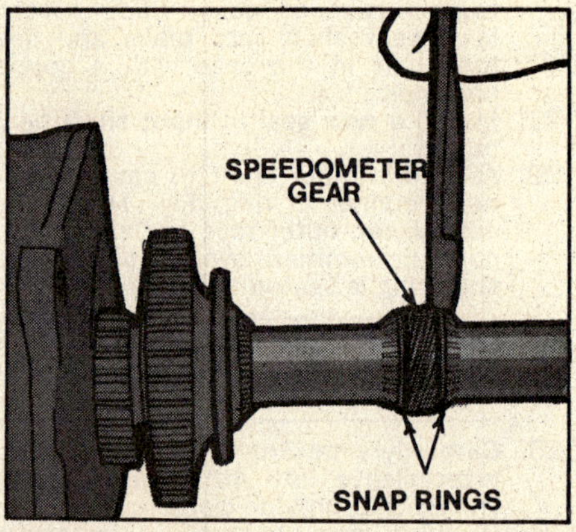

Fig. 3 Speedometer gear and snap rings

Fig. 4 Removing & installing reverse gear

Fig. 2 Removing reverse shifter shaft lock pin

Fig. 5 Removing countershaft

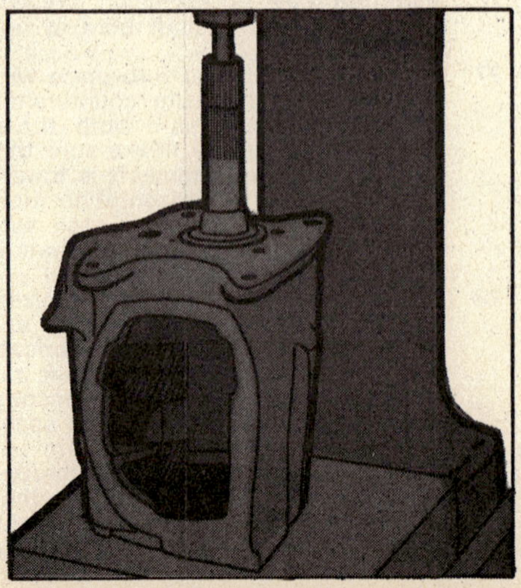
Fig. 6 Pressing drive gear from bearing

Fig. 7 Main drive gear bearing removal

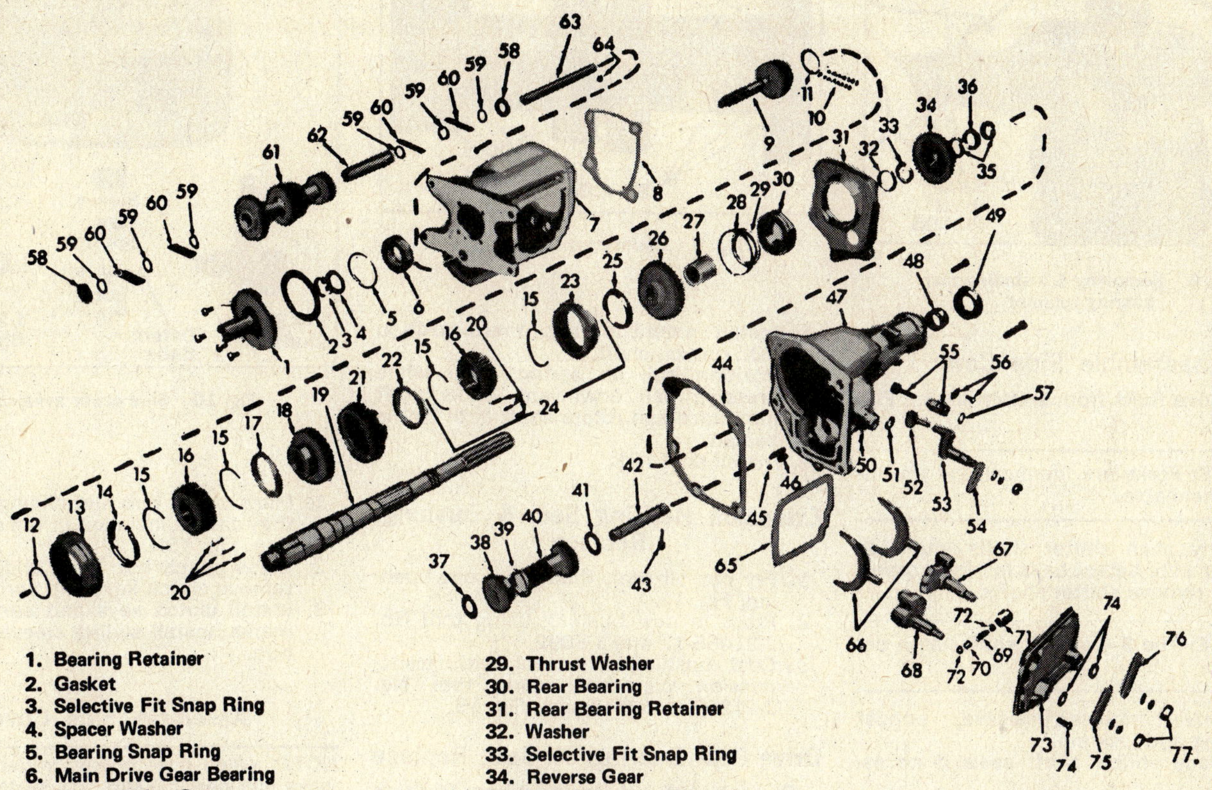

1. Bearing Retainer
2. Gasket
3. Selective Fit Snap Ring
4. Spacer Washer
5. Bearing Snap Ring
6. Main Drive Gear Bearing
7. Transmission Case
8. Rear Bearing Retainer Gasket
9. Main Drive Gear
10. Bearing Rollers (16)
11. Washer
12. Snap Ring
13. Third and Fourth Speed Clutch Sliding Sleeve
14. Fourth Speed Gear Synchronizing Ring
15. Clutch Key Spring
16. Clutch Hub
17. Third Speed Gear Synchronizing Ring
18. Third Speed Gear
19. Mainshaft
20. Clutch Keys (3)
21. Second Speed Gear
22. Second Speed Gear Synchronizing Ring
23. First and Second Speed Clutch Sliding Sleeve
24. First and Second Speed Clutch Assembly
25. First Speed Gear Synchronizing Ring
26. First Speed Gear
27. First Speed Gear Sleeve
28. Rear Bearing Snap Ring
29. Thrust Washer
30. Rear Bearing
31. Rear Bearing Retainer
32. Washer
33. Selective Fit Snap Ring
34. Reverse Gear
35. Snap Ring
36. Speedometer Drive Gear
37. Reverse Idler Front Thrust Washer (Flat)
38. Reverse Idler Gear (Front)
39. Snap Ring
40. Reverse Idler Gear (Rear)
41. Thrust Washer (Tanged)
42. Reverse Idler Shaft
43. Reverse Idler Shaft Lock Pin and Welch Plug
44. Rear Bearing Retainer To Case Extension Gasket
45. Reverse Shifter Shaft Detent Ball
46. Reverse Shifter Shaft Ball Detent Spring
47. Case Extension
48. Extension Bushing
49. Rear Oil Seal
50. Reverse Shifter Shaft Lock Pin
51. Reverse Shifter Shaft Lip Seal
52. Reverse Shift Fork
53. Reverse Shifter Shaft and Detent Plate
54. Reverse Shifter Lever
55. Speedometer Driven Gear and Fitting
56. Retainer and Bolt
57. "O" Ring Seal
58. Washer (Tanged)
59. Spacer (.050")
60. Bearing Rollers (28)
61. Countergear
62. Countergear Roller Spacer
63. Countershaft
64. Countershaft Woodruff Key
65. Gasket
66. Forward Speed Shift Forks
67. First and Second Speed Gear Shifter Shaft and Detent Plate
68. Third and Fourth Speed Gear Shifter Shaft and Detent Plate
69. Poppet Spring
70. Interlock Pin
71. Interlock Sleeve
72. Detent Balls
73. Transmission Side Cover
74. Lip Seals
75. Third and Fourth Speed Shifter Lever
76. First and Second Speed Shifter Lever
77. Lever Attaching Nuts

Fig. 1 Type 10 Warner four speed transmission

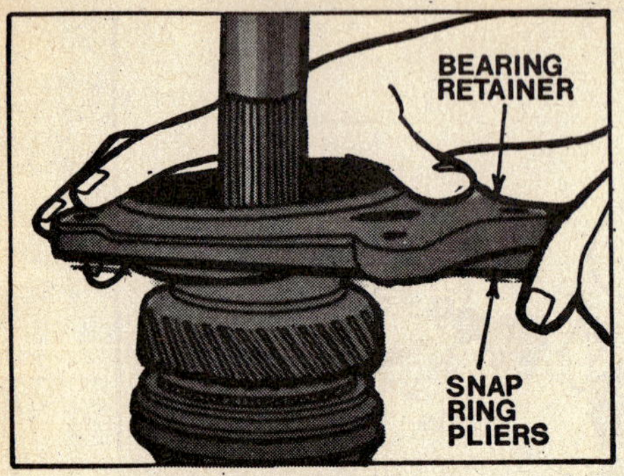

Fig. 8 Removing & installing rear bearing retainer

Disassemble Side Cover

1. Remove forks from shifter shafts. Fig. 10.

NOTE: Forks are identical and are interchangable.

2. Slowly push shifter shafts into side cover until detent balls fall from cover, then remove shifter shafts.

NOTE: The 3-4 shifter shaft has a detent.

3. Remove interlock sleeve, poppet spring and lock pin.
4. Replace shifter shaft seals if necessary.

Reverse Idler Shaft, Replace

1. Remove thrust washer, use punch to drive welch plug and pin into idler shaft, Fig. 11, until idler shaft can be removed.
2. Position idler shaft in extension housing.
3. Align hole in shaft with hole in extension housing, drive in pin and install welch plug with sealer.

Reverse Shifter & Seal, Replace

1. Remove shifter fork, tap shaft into extension housing until detent ball falls from housing, then remove shaft and detent spring.
2. Remove "O" ring and discard.

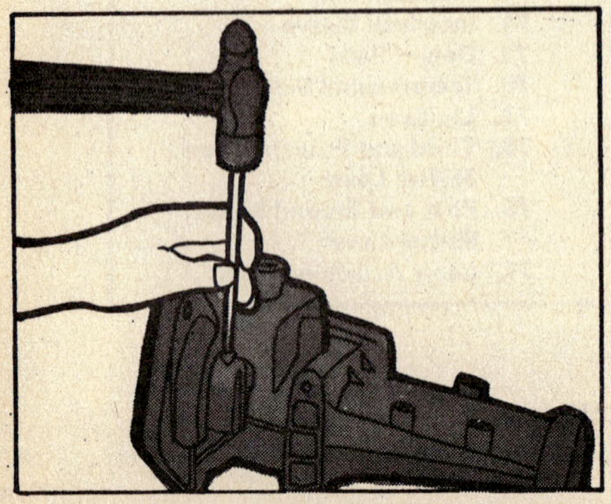

Fig. 11 Removing roll pin from reverse idler shaft

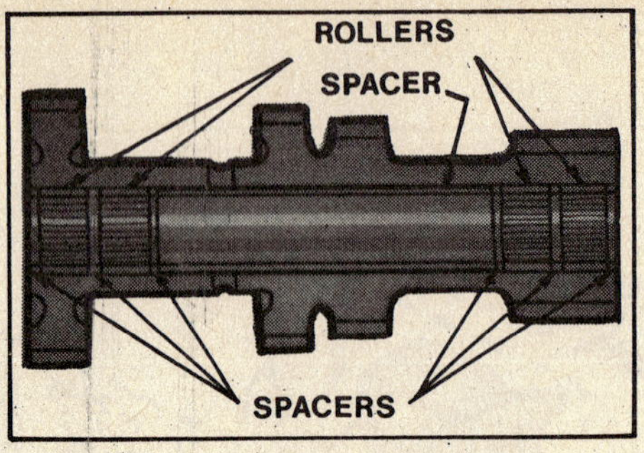

Fig. 9 Countergear assembly

3. Install detent spring, then position "O" ring on shaft.
4. Position ball on spring, then while pressing ball downward, slide shaft into place and rotate shaft until ball is seated.
5. Install shift fork.

Extension Housing Seal & Bushing, Replace

1. Pry out oil seal, then drive out bushing, Fig. 12.
2. Press in new bushing using tool No. J-21465-17 and J-8092.
3. Coat bushing and seal with transmission oil, then using tool No. J-21359 install oil seal. Fig. 13.

Drive Gear Bearing Oil Seal, Replace

Pry seal out with screw driver. Position seal in bore and tap in with flat plate, Fig. 14.

Synchronizers, Overhaul

NOTE: The synchronizer clutch hub and sliding sleeve are replaced as an assembly.

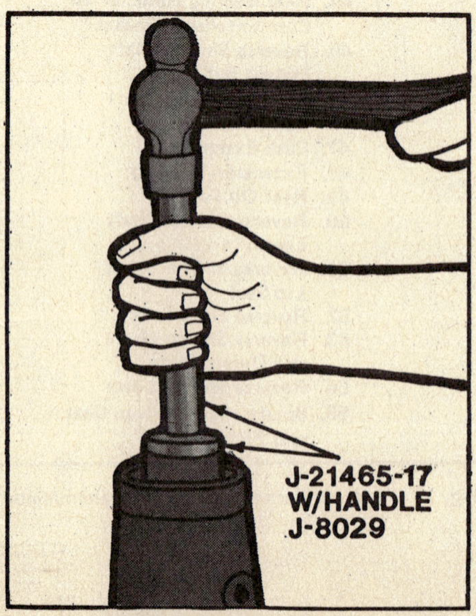

Fig. 12 Removing & installing extension housing bushing

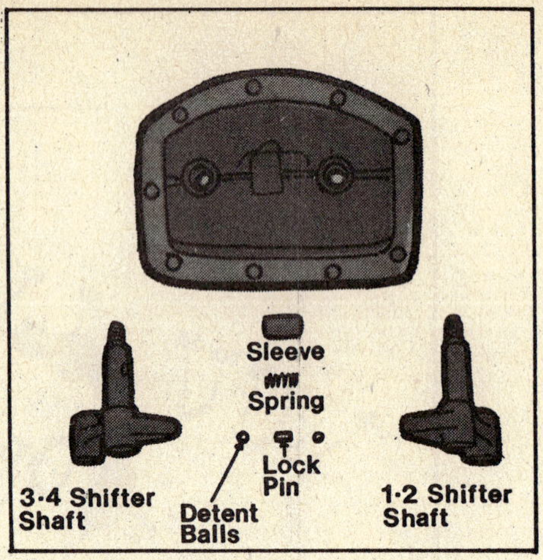

Fig. 10 Side cover assembly

1. Mark clutch hub and sliding sleeve so they can be installed in the same position, Fig. 15.
2. Press clutch hub from sliding sleeve, remove clutch keys and springs.
3. Install clutch keys and springs, align marks, install sliding sleeve on clutch hub.

Assemble Side Cover

NOTE: When assembling transmission always use new gaskets, oil seals and snap rings.

1. Install seals in shifter shaft bores, if removed.
2. Install 3-4 shifter shaft in bore, position in neutral, then install detent ball,

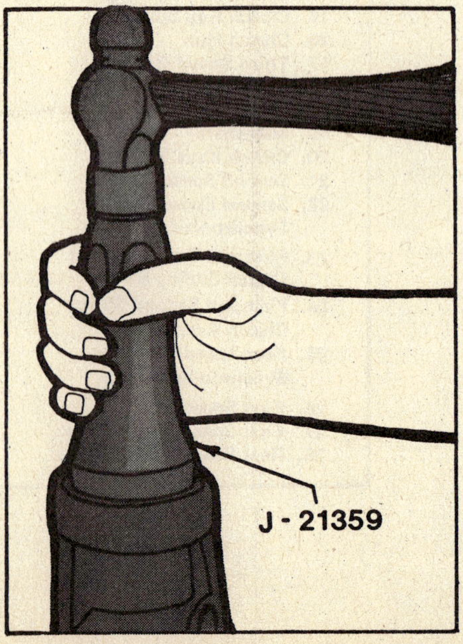

Fig. 13 Installing extension housing oil seal

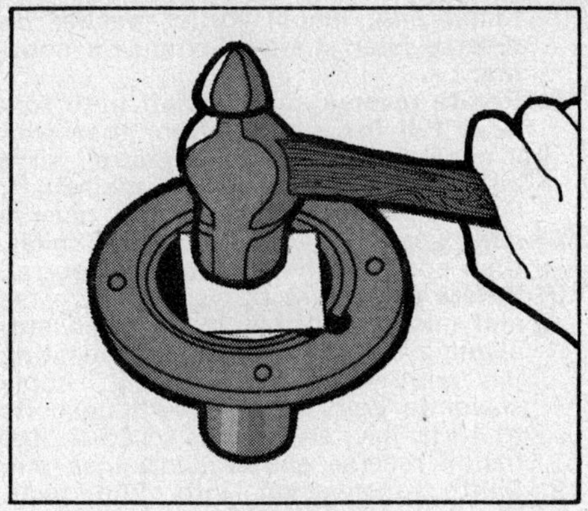

Fig. 14 Installing drive gear bearing retainer oil seal

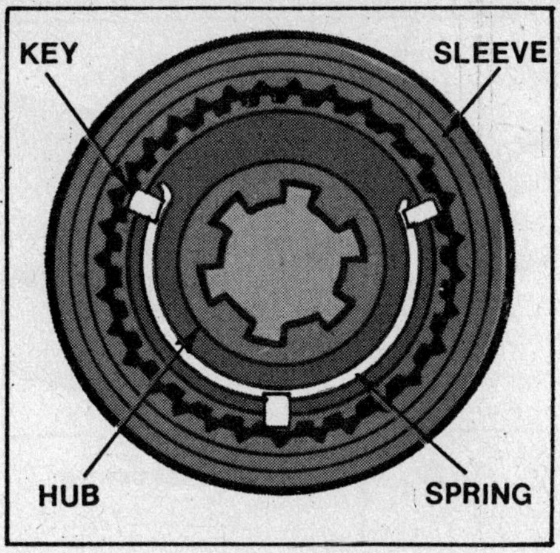

Fig. 15 Synchronizer assembly

Fig. 16 Countergear end play check

interlock sleeve, poppet spring and lock pin, Fig. 10.
3. Position detent ball in other end of interlock sleeve, depress ball and install 1-2 shifter shaft.
4. Position 1-2 shifter shaft in 2nd gear and install shift forks.

Countergear Assemble & Install

1. Position dummy shaft in countergear bore and install tubular spacer.
2. Install spacer in bore, then using heavy grease to hold rollers in place install 28 rollers. Fig. 9.
3. Install second spacer, then install 28 additional rollers and install third spacer.
4. Install rollers and spacers in other end of countergear using above procedure.
5. Install thrust washers in each end of transmission case with tangs inserted in case recesses, use grease to hold washers in place.
6. Insert countershaft through rear of case, in turn forcing dummy shaft out through front of case. Install woodruff key on countershaft, tap countershaft until flush with rear of case.

Countergear End Play Check

Attach dial indicator, Fig. 16, check end play. If end play is greater than .025 in. new selective thrust washers must be installed.

Assemble Mainshaft

1. From rear of mainshaft, install 2nd gear with hub facing rearward.
2. Position sychronizing rings on 1-2 synchronizer, then install synchronizer on mainshaft with hub facing forward.
3. Press 1st gear bushing on mainshaft until 2nd gear, 1-2 synchronizer and bushing are against mainshaft shoulder, Fig. 17.
4. Install 1st gear and inner race with gear hub facing front, then press rear bearing on mainshaft with snap ring groove facing front. Fig. 18.
5. Install spacer and selective thickness snap ring.

NOTE: Use the thickest snap ring that will fit in mainshaft groove.

6. Install 3rd gear with hub facing front and the 3rd gear synchronizing ring with notches facing front.
7. Install 3rd and 4th gear clutch assembly with the taper facing front.

NOTE: Ensure that keys in hub mate with notches in 3rd gear synchronizing ring.

8. Install selective thickness snap ring, use thickness snap ring that will fit mainshaft groove.
9. Install locating pin in rear bearing retainer, then place bearing retainer over rear of mainshaft, spread retainer snap ring and align with groove in rear bearing. Fig. 18.
10. Using heavy grease, install 16 rollers in rear bearing and washer in bearing bore. Fig. 19.
11. Position 4th gear synchronizer ring on 3-4 synchronizer, aligning ring notches with hub keys.
12. Install front drive gear on main shaft.

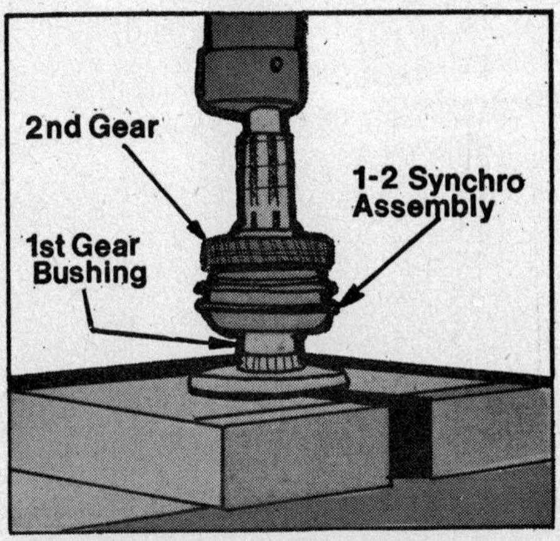

Fig. 17 Installing 2nd gear, 1-2 synchronizer assembly & 1st gear bushing

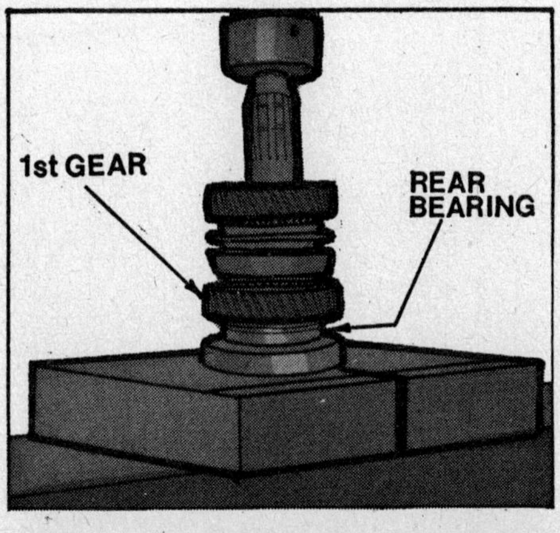

Fig. 18 Installing 1st gear & rear bearing

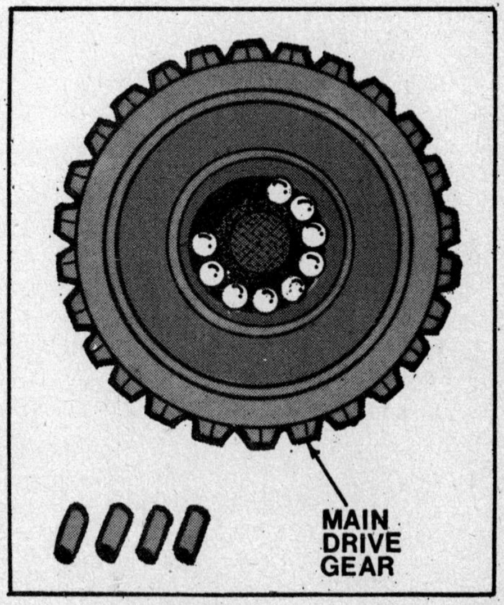

Fig. 19 Installing main drive bearing rollers

ASSEMBLE TRANSMISSION

1. Install drain plug in transmission case.
2. Install thrust washer and front section of reverse idler gear with hub facing rear and position gasket on rear of transmission case.
3. Position synchronizers in neutral and install mainshaft in case. Install retainer locating pin and torque retainer lock bolt to 25 ft. lbs.
4. Tap main drive gear bearing into case using suitable tool, pull drive gear shaft forward, and tap bearing again until bearing bottoms on main drive gear and the washer and snap ring can be installed on main gear drive shaft, Fig. 20.
5. Install main drive gear bearing retainer and gasket, apply sealer to bolts and torque to 18 ft. lbs.
6. Install rear reverse idler gear, engaging splines with front section.

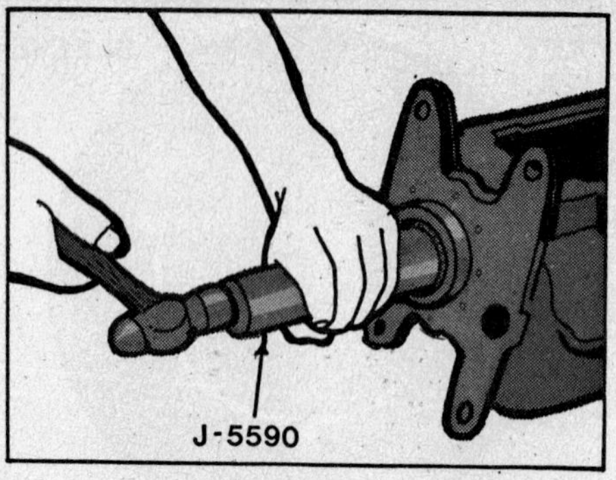

Fig. 20 Installing drive gear bearing

7. Install reverse sliding gear with hub facing rear, Fig. 4, then install the speedometer gear and two snap rings on mainshaft. Fig. 3.
8. Position gasket on rear bearing retainer and, install thrust washer on reverse idler shaft in extension housing.
9. Rotate reverse shifter shaft until fork is in full forward position, then with synchronizer sleeves in neutral, slide extension housing over mainshaft. Guide reverse idler shaft into reverse idler gears, engaging fork on collar, then position shifter shaft in neutral.
10. Rotate reverse shifter shaft clockwise and push extension housing forward, aligning rear bearing retainer locating pin with extension housing. Apply sealer to bolts and torque long bolts to 40 ft. lbs., short bolts to 25 ft. lbs.
11. Install reverse shifter shaft lock pin.
12. Shift transmission into 2nd gear. The front synchronizer sleeve should be in neutral and the rear sleeve in forward position to engage 2nd gear synchronizer teeth.
13. Position slide cover and gasket to transmission case and index shift forks to shift collars, then install and torque side cover bolts to 18 ft. lbs.

OVERDRIVE TRANSMISSIONS

See Trouble Shooting Chapter For Diagnosis Procedure On These Units

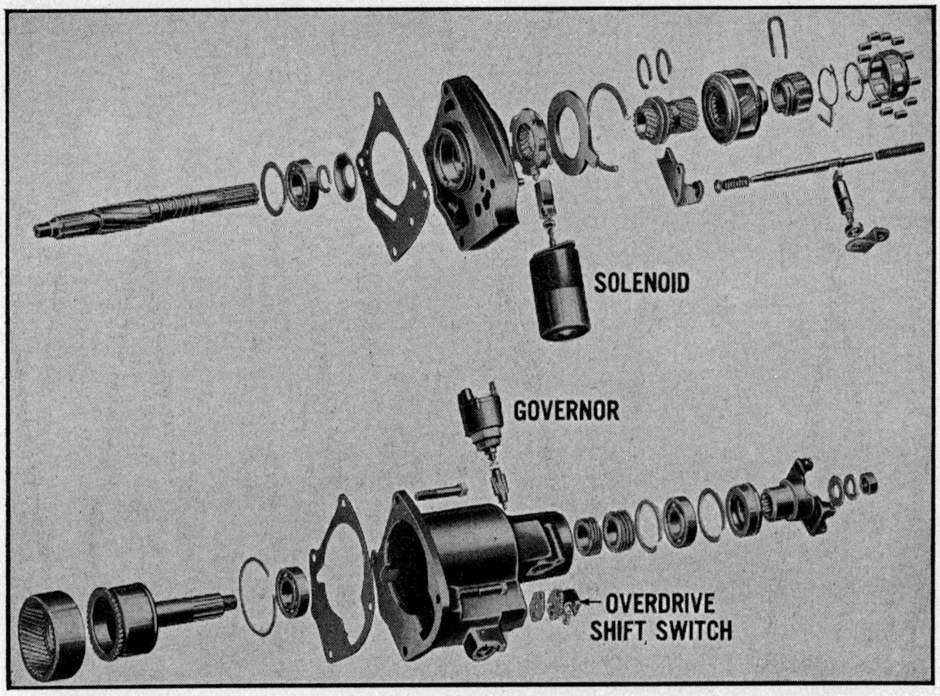

Fig. 2 Second version of full-electric overdrive with centrifugal governor. This unit is similar to the first version except the design of the second version is more compact. Some models do not use the shift (lockout) switch

1969 AMERICAN MOTORS

Overdrive units are essentially automatic two-speed planetary transmissions attached to the rear of conventional three-speed transmissions. As shown in Fig. 1, the heart of the over-drive is the planetary unit consisting of sun gear, planetary pinions and internal (ring) gear. In overdrive, the pinions are connected to the mainshaft, and revolve around the sun gear which holds against rotation. The internal gear, connected to the tailshaft, is thus forced to rotate at a speed greater than the mainshaft. The engagement of the gearset is controlled by coupling the internal gear to the tail shaft, or holding the sun gear stationary, or by a combination of the two methods.

By following the procedure shown pictorially in Figs. 3 through 12, no difficulty should be experienced in servicing these units.

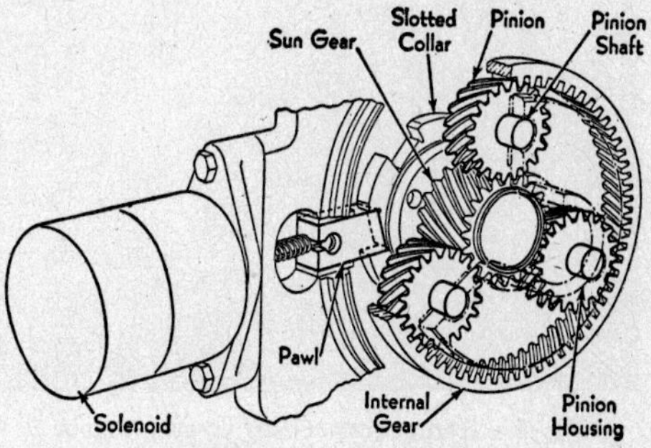

Fig. 1 Through the planetary unit shown, the overdrive provides a higher gear ratio, and when in operation, engine speed is approximately 30 per cent slower than when operating in conventional high gear

Fig. 3 Remove companion flange and governor. Also lockout switch if so equipped

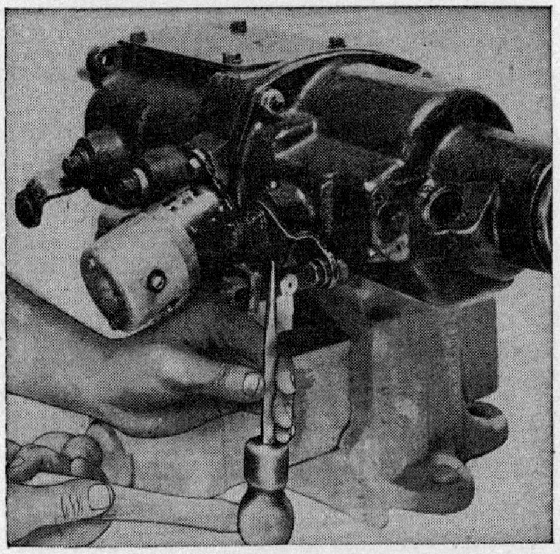

Fig. 4 After driving out locating pin, pull shift shaft as far as possible to disengage operating cam from shift rail. Remove overdrive housing. Tap end of shaft to prevent its coming off with housing and spilling free wheel rollers. Parts inside housing may then be removed.

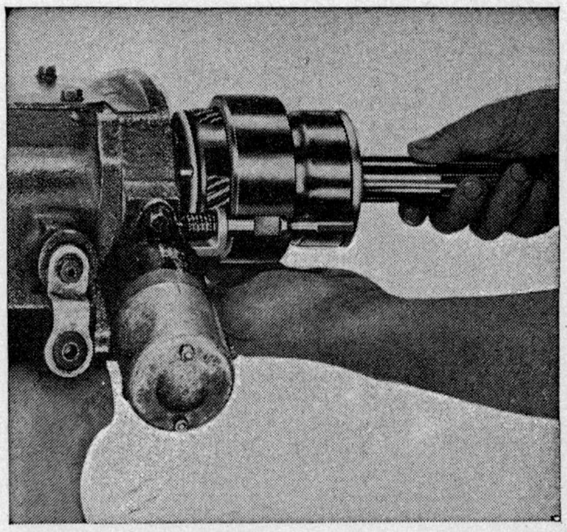

Fig. 5 Hold the adapter plate to the transmission case with one screw and remove the overdrive shaft, catching the free wheel rollers as shown. Removing snap ring permits ring gear to be taken off shaft.

Fig. 6 Remove retaining clip and take off free wheel unit and pinion cage

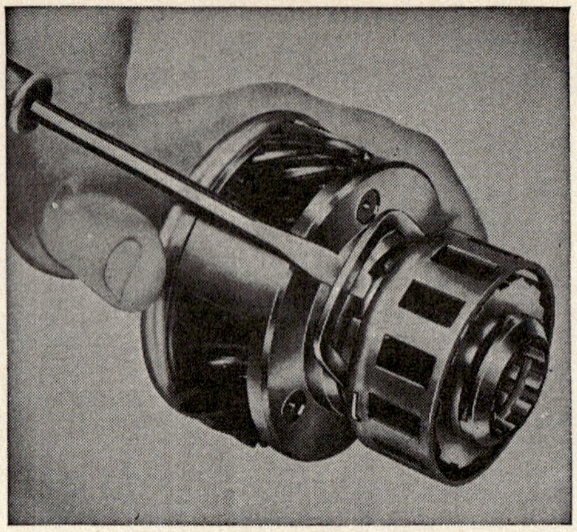

Fig. 7 Separate pinion cage from free wheel unit by removing retaining clip

Fig. 8 Remove overdrive sun gear and shift rail

Fig. 9 Remove attaching screws, rotate solenoid 1/4 turn and take off

Fig. 10 After releasing snap ring from adapter plate, remove sun gear cover plate, blocker and solenoid pawl

Fig. 11 At this point, if repairs are to be made on the transmission, remove the mainshaft, adapter plate, gears and synchronizer as a unit

Fig. 12 Reverse the order of disassembly to assemble the unit. After inserting the pawl with the notched side up as shown, install blocker assembly and cover plate, being sure blocker ring and pawl are properly positioned. Then install large snap ring in adapter plate

1975-76 AMERICAN MOTORS

The overdrive is an additional gear unit located between the transmission and propeller shaft, Fig. 14, and provides a higher overall gear ratio than normally provided by the drive gear and pinion. The overdrive permits highway cruising at a lower engine speed than in a normal third gear. This unit reduces engine-to-rear axle ratio approximately 25%, thus providing a transmission ratio in overdrive third gear of .75 to 1.00.

The overdrive is a hydraulically operated unit and its actuation is controlled by a solenoid valve. This solenoid valve is activated by a control switch mounted on the steering column. A governor speed switch, operated by the speedometer cable, determines cut-in and cut-out speeds for the unit. The overdrive may be engaged when the vehicle is above the governor cut-in speed of 38 MPH and in third gear only. A third gear switch installed in the transmission prevents overdrive engagement in any other gear.

NOTE: California vehicles require a TCS switch which is open when the transmission is in third gear. To complete the overdrive control circuit, but also prevent overdrive operation in any gear except third gear, a third gear relay is connected to the TCS switch. This relay performs the same function as the third gear switch and is closed only when the transmission is in third gear.

The overdrive is engaged by activating the control switch and until the switch is deactivated or the vehicle decelerates to below the governor cut-out speed of 32 MPH, the overdrive remains engaged. A kickdown switch is incorporated into the control circuit, Fig. 15. This switch, mounted on the carburetor base, is actuated by throttle linkage. When the throttle is fully opened, the switch is actuated and "Opens" the control circuit, in turn de-energizing the solenoid valve and disengaging the overdrive. When the throttle returns from the wide open position, the switch "Closes" the control circuit, re-energizing the solenoid and re-engaging the overdrive.

An indicator lamp, connected electrically to the solenoid valve, is provided to indicate overdrive engagement.

Hydraulic system pressure is developed by a cam operated plunger-type pump, Fig. 16, and is driven by the transmission output shaft through a drive key. The pump obtains oil from the air cooled sump, through the oil pan filter, and pumps it through the non-return valve and pressure filter to the clutch apply pistons, solenoid valve and relief valve assembly. A spring loaded piston within the relief valve assembly provides smooth engagement and disengagement under all operating conditions. When the overdrive is disengaged, a residual system pressure of 20 to 40 PSI is maintained. The system pressure with the overdrive engaged is 520 to 540 PSI and is maintained by the relief valve.

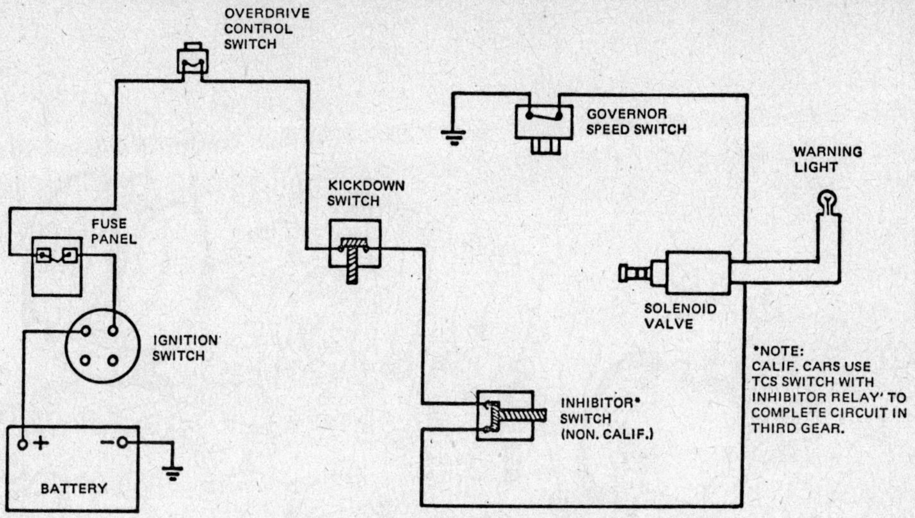

Fig. 14 Overdrive electrical control circuit

DISASSEMBLY

1. Remove solenoid valve, Fig. 13, with tool J-25304. Use only specified tool since solenoid damage may result.
2. Remove and discard nuts securing clutch piston apply bars to thrust bearing cover pins.
3. Separate main case from rear case. Note location of copper gaskets used on main case to rear case studs.
4. Remove clutch return springs, clutch brake ring and gaskets from main case. If clutch brake ring is stuck, tap lightly with a mallet.
5. Remove oil pan, gasket, filter and main case pressure plug from main case, Fig. 16, then pressure filter plug, filter and aluminum washer.
6. Remove pump body plug, non-return valve ball seat spring, check ball and valve seat.
7. Remove clutch apply pistons from bores, then "O" rings from pistons.

NOTE: Do not remove lubrication relief valve plug, spring or ball from main case.

8. Remove pump assembly as follows:
   a. Push pump body upward until it

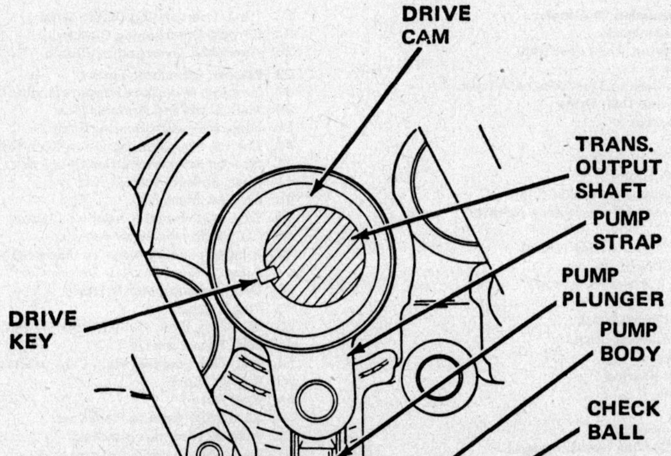

Fig. 15 Overdrive unit oil pump assembly

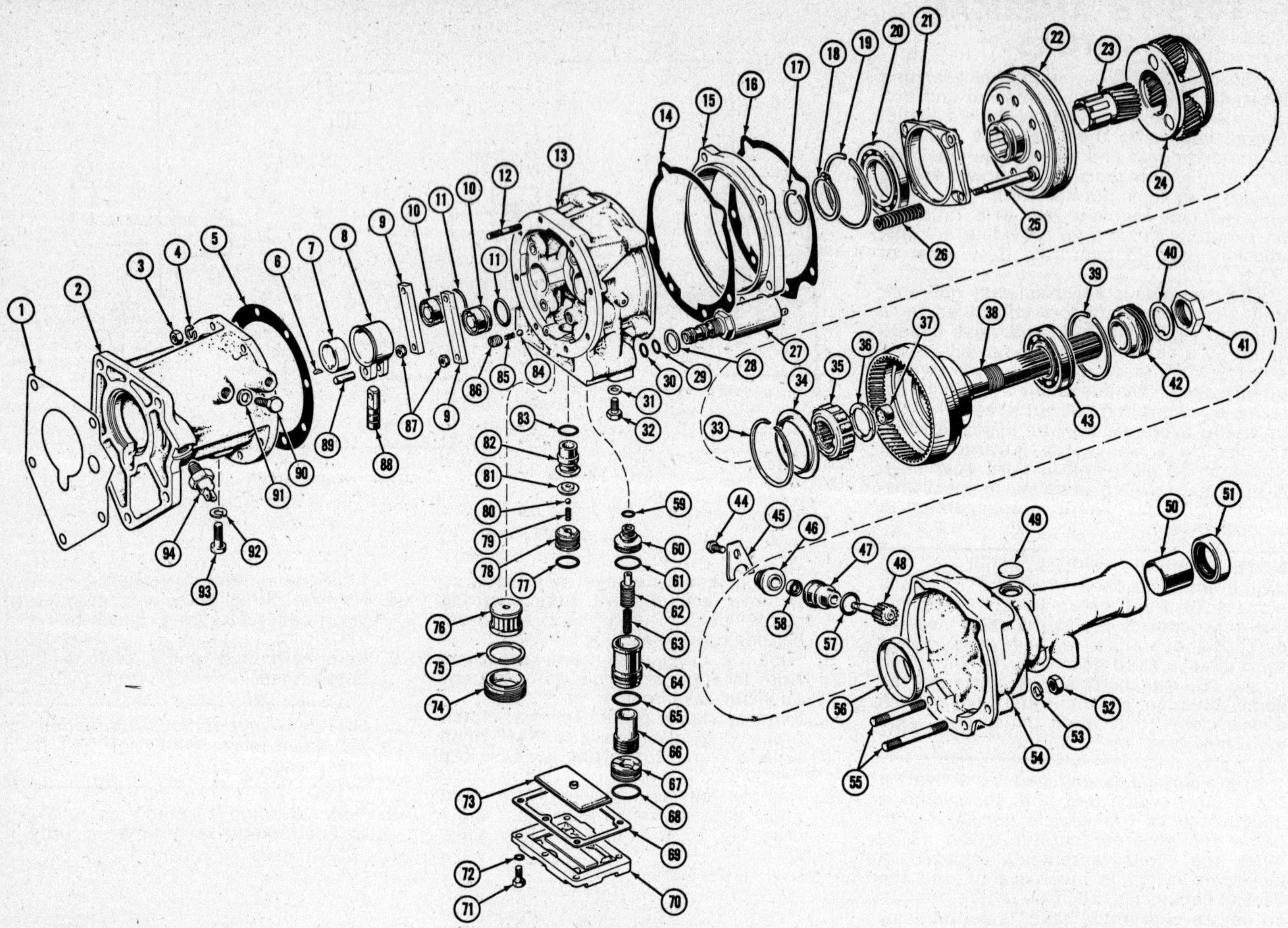

1. Gasket Transmission to Adapter
2. Adapter, Transmission
3. Nut, Self Locking, Main Case Stud
4. Washer, Lock
5. Gasket, Main Case to Transmission Adaptor
6. Key, Pump Strap Cam Drive
7. Cam, Pump Strap
8. Strap, Pump
9. Bar, Clutch Piston Apply
10. Piston, Clutch Apply
11. Seal, Clutch Apply Piston O-Ring
12. Stud, Main Case to Transmission Adapter
13. Main Case
14. Gasket, Clutch Brake Ring (front)
15. Brake Ring, Clutch
16. Gasket, Clutch Brake Ring (rear)
17. Ring, Sun Gear Snap
18. Ring Lock, Sliding Clutch
19. Ring, Thrust Bearing Snap
20. Bearing, Thrust
21. Cover, Thrust Bearing
22. Clutch, Sliding
23. Sun Gear
24. Assembly, Pinion Carrier
25. Bolt, Thrust Bearing Cover (4 reqd.)
26. Spring, Clutch Return (4 reqd.)
27. Solenoid Valve
28. Washer, Solenoid Valve
29. Seal, Solenoid Valve O-Ring
30. Seal, Solenoid Valve O-Ring
31. Gasket, Main Case Pressure Plug
32. Plug, Main Case Pressure
33. Ring, Overrunning Clutch Snap
34. Slinger, Overrunning Clutch Oil
35. Assembly, Overrunning Clutch
36. Washer, Mainshaft Thrust
37. Bushing, Mainshaft Support (Included in Mainshaft)
38. Main Shaft and Annulus Gear
39. Ring, Mainshaft Bearing Snap
40. Washer, Speedometer Drive Gear Tab
41. Nut, Speedometer Drive Gear Lock
42. Gear, Speedometer Drive
43. Bearing, Mainshaft
44. Bolt, Speedometer Adapter Clamp
45. Clamp, Speedometer Adapter
46. Adapter, Speedometer to Governor Speed Switch
47. Adapter, Speedometer Driven Gear
48. Gear, Speedometer Driven
49. Plug, Expansion
50. Bushing, Rear Case (Included in Case)
51. Seal, Rear Case Oil
52. Nut, Self Locking, Main Case to Rear Case Stud
53. Washer, Lock
54. Rear Case
55. Stud, Main Case to Rear Case
56. Washer, Disc (not removed: included in rear case)
57. Seal, Speedometer Adapter O-Ring
58. Seal, Speedometer Adaptor Oil
59. Seal, Relief Valve Body O-Ring (Inner)
60. Body, Relief Valve
61. Seal, Relief Valve Body O-Ring (Outer)
62. Assembly, Relief Valve and Spring
63. Spring, Relief Valve Residual Pressure
64. Sleeve, Relief Valve
65. Seal, Relief Valve Sleeve O-Ring
66. Piston, Relief Valve
67. Plug, Relief Valve Piston
68. Seal, Relief Valve Piston Plug O-Ring
69. Gasket, Oil Pan
70. Oil Pan
71. Bolt, Oil Pan
72. Washer, Lock
73. Filter, Oil Pan
74. Plug, Pressure Filter
75. Washer, Pressure Filter (Aluminum)
76. Filter, Pressure
77. Seal, Pump Body O-Ring
78. Plug, Pump Body
79. Spring, Non-return Valve Ball-seat
80. Ball, Non-return Valve Check
81. Seat, Non-return Valve
82. Body, Pump Plunger
83. Seal, Pump Plunger Body O-Ring
84. Ball, Lubrication Relief Valve Check
85. Spring, Lubrication Relief Valve
86. Plug, Lubrication Relief Valve
87. Nut, Self Locking, Clutch Piston Apply Bar
88. Plunger, Pump
89. Pin, Pump Plunger
90. Bolt, Gearshift Lever Retainer to Adapter
91. Washer, Lock
92. Washer, Lock
93. Bolt, Rear Support Cushion to Adapter
94. Switch, Back-up Light

Fig. 13 Overdrive unit, exploded view

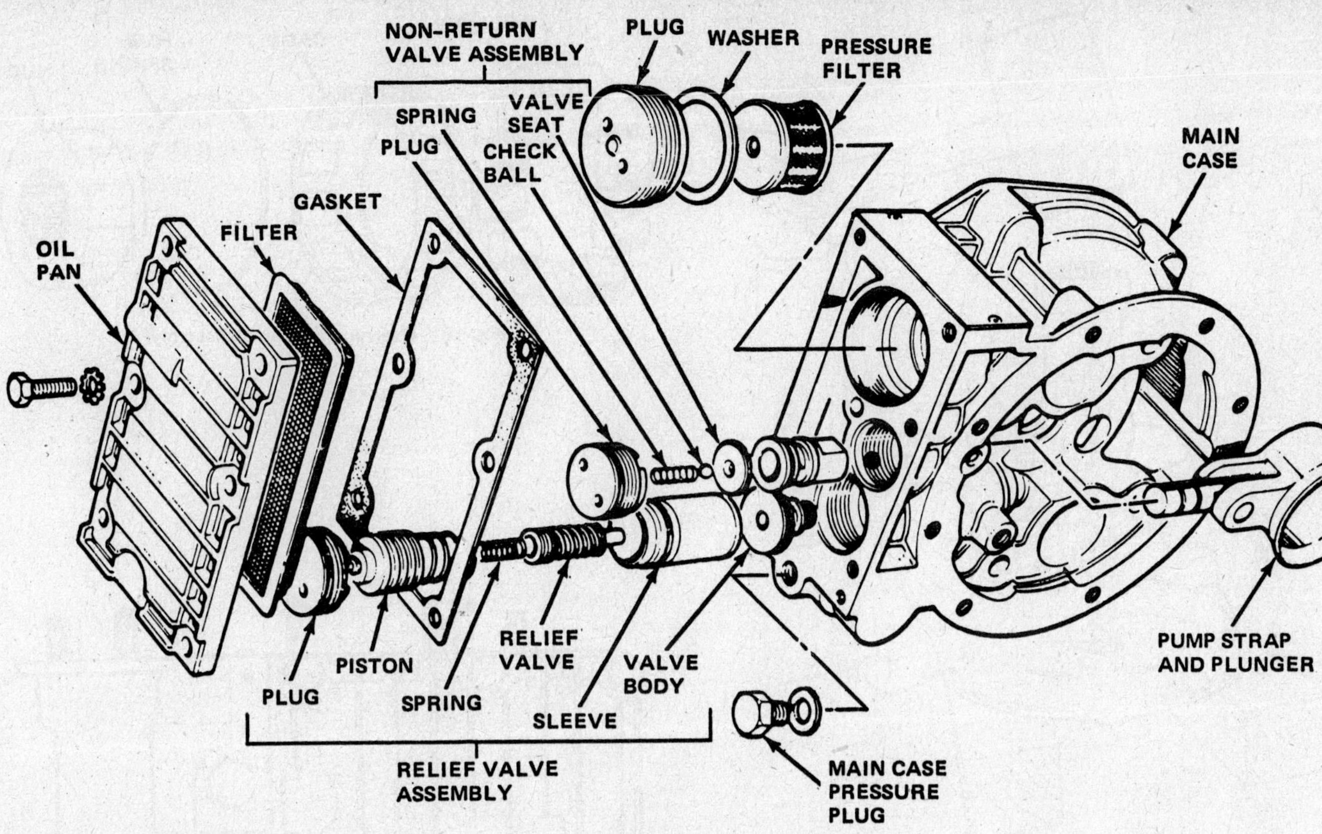

Fig. 16 Overdrive main case components

unseats from case bore and slide pump plunger from pump body, then remove body from case bore.

NOTE: The pump body has a flat machined on one side which aligns with oil feed hole and slot in main case bore. Note location of oil feed hole and slot for proper reassembly.

---

b. Remove drive cam and key from pump strap. Do not disassemble pump strap and plunger since they are serviced as an assembly.

9. Remove relief valve piston plug, piston and residual pressure spring.

NOTE: The residual pressure spring is the only loose spring in the relief valve assembly.

---

10. With a magnet or needlenose plier, remove relief valve and spring assembly. Do not remove calibrated spring from relief valve piston.
11. Using tool J-25307, remove relief valve sleeve and body as follows:
    a. Install hooked end of tool through relief valve body bore and hook tool over inner edge of body. Slide tool barrel downward, securing tool, and pull valve body and sleeve from case bore..
    b. Remove and discard "O" rings from valve body, sleeve and plug.
12. Remove sliding clutch, sun gear and thrust bearing cover from main shaft annulus gear in rear case, then remove pinion carrier from main shaft annulus gear.
13. Remove sun gear snap ring, sliding clutch ring lock and push sun gear from sliding clutch hub.
14. Install tool J-25315 into sliding clutch hub and tap end of tool to drive clutch hub from thrust bearing. Remove thrust bearing snap ring and press bearing from cover.
15. Remove overrunning clutch snap ring and brass oil slinger. Install tool J-25308 into main shaft annulus gear bore, reach through tool and pull overrunning clutch into tool and remove tool and overrunning clutch together. Remove main shaft thrust washer from main shaft annulus gear recess. Remove overrunning clutch from tool and disassemble.
16. Remove rear case expansion plug.
17. Place rear case on wooden blocks, expand main shaft bearing snap ring and tap main shaft out from rear case, Fig. 17.
18. Remove speedometer drive gear lock nut, tab washer and drive gear, then main shaft bearing.
19. Pry rear case oil seal from case and remove main shaft bearing snap ring from machined groove in rear case.

NOTE: Do not remove disc washer or rear bushing from rear case since they are not serviceable and are available as part of the rear case.

ASSEMBLY

1. Lubricate and install main shaft bearing on main shaft with the groove facing toward rear of main shaft.
2. Install speedometer drive gear with shoulder side of gear facing toward main shaft bearing, tab washer and drive gear lock nut on main shaft. Torque lock nut to 55 ft. lbs. while holding main shaft secure, then bend drive gear washer against lock nut in two places, securing lock nut.
3. Install main shaft bearing snap ring into rear case machined groove, place main shaft in upright position and lower rear case onto main shaft. Tap end of case to start main shaft bearing into counterbore of case, expand snap ring and tap case until bearing fully seats in counterbore and snap ring in bearing groove.
4. Lubricate and install rear case oil seal, then new rear case expansion plug.
5. Lubricate and install main shaft thrust washer into main shaft annulus gear recess.
6. Assembly overrunning clutch, Fig. 18, as follows:
    a. Insert longest hub spring hooked end into cage locating hole.
    b. Hold cage stationary and rotate

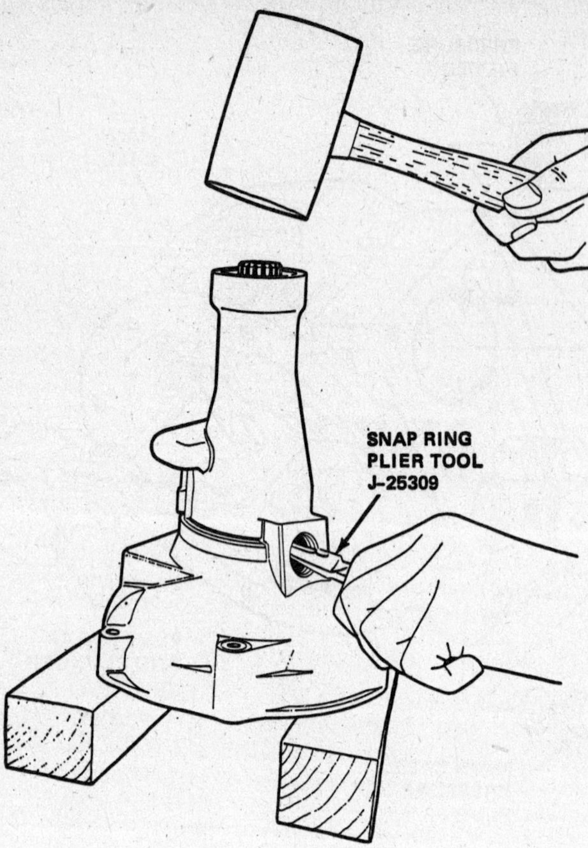

Fig. 17 Main shaft removal

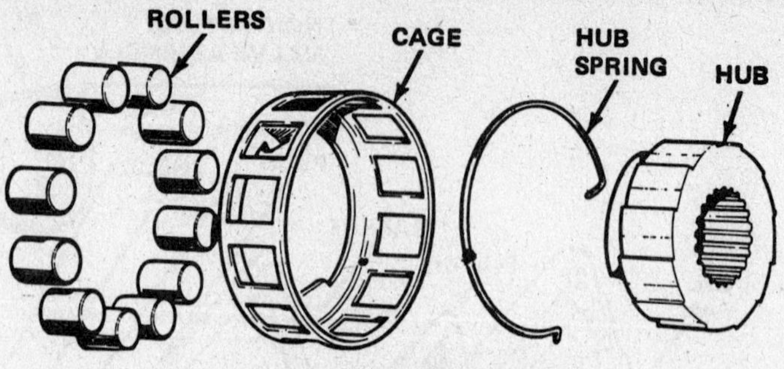

Fig. 18 Overrunning clutch assembly

hub against spring.

7. Place assembled cage and hub into tool J-25308 with cage open end facing outward. Install clutch rollers into cage slots through gate in tool while rotating cage in clockwise direction.
8. Lubricate and install overrunning clutch assembly in main shaft annulus gear bore. Remove tool J-25308 and install brass oil slinger with shoulder facing outward, then overrunning clutch snap ring, ensuring snap ring fully seats in groove.
9. Lubricate and install pinion carrier into main shaft annulus gear.
10. Press thrust bearing into thrust bearing cover, then install thrust bearing snap ring.
11. Position thrust bearing and cover on sliding clutch hub and tap cover to start bearing onto hub. Invert assembly and install tool J-25315 into clutch hub and drive hub into thrust bearing.
12. Install sun gear into sliding clutch hub, then sliding clutch lock ring with sharp edge facing upward and sun gear snap ring.
13. Install sliding clutch assembly onto main shaft annulus gear while engaging sun gear in pinion gears. Ensure sliding clutch fully seats on the annulus and the sun gear is fully engaged in pinion gears.

NOTE: To aid installation, rotate main

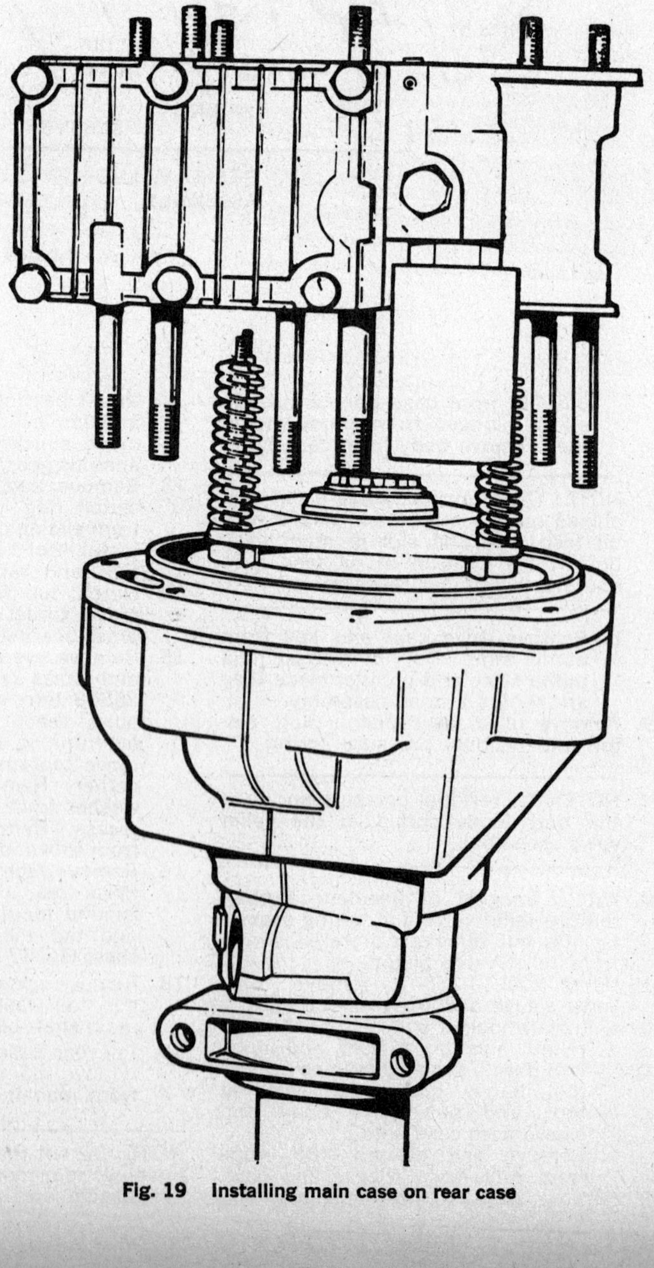

Fig. 19 Installing main case on rear case

shaft while engaging sun gear.

14. Lubricate and install clutch apply pistons with new "O" rings into main case bores with piston counterbored ends facing outward.
15. Install new "O" rings on relief valve body, sleeve and valve piston plug, then install relief valve assembly as follows:
    a. Install valve body into main case bore.
    b. Align valve sleeve oil hole with oil hole in bore and insert sleeve into bore with "O" ring end facing upward until sleeve and valve body are seated.
    c. Install relief valve and spring assembly into valve body, then residual pressure spring into the assembly.
    d. Install valve piston and relief valve piston plug and torque plug to 16 ft. lbs.
16. Install pressure filter into main case bore and install torque pressure filter plug with new aluminum washer to 16 ft. lbs.
17. Install new "O" rings on pump body and plug, then lubricate and install pump plunger, pump body and non-return valve components as follows:
    a. Align flat on pump body with oil hole in main case bore and install pump body halfway into bore.
    b. Insert pump plunger into pump body, then push pump body fully into main case.
    c. Place non-return valve seat on top of pump body with check ball seat facing upward, then place check ball onto valve seat.
    d. Place non-return valve spring into pump body plug and install assembly. Torque plug to 16 ft. lbs.
18. Install main case pressure plug, gasket, oil pan filter, oil pan gasket and oil pump cover on main case. Torque oil pan bolts to 6 ft. lbs. and pressure plug to 13 ft. lbs.
19. Place rear case assembly in a soft jawed vise and install new clutch return springs on thrust bearing cover bolts.
20. Install first clutch brake ring gasket on rear case, then clutch brake ring into case with tapered surface facing rearward. Install second new clutch brake ring gasket on brake ring. Ensure gaskets and brake ring are aligned with rear case stud holes.
21. Apply a light coat of suitable sealer to main case studs, place main case over rear case, aligning studs, and lower onto rear case, Fig. 19. Align thrust bearing cover bolts as main case is being lowered.
22. Install copper gaskets on the upper main case to rear case studs, four lockwashers on remaining studs and nuts and all studs. Torque nuts to 11 ft. lbs.
23. Install clutch apply bars on thrust bearing cover bolts and torque attaching nuts to 8 ft. lbs.
24. Install solenoid valve with tool J-25304.
25. Lubricate and install oil pump drive cam and drive key on transmission output shaft and secure with snap ring.

1975 CHRYSLER

This four speed overdrive transmission is basically the same transmission as the Chrysler four speed A-833 unit found in the "Four Speed Transmission" chapter, under "Type Four Chrysler". All service procedures for the overdrive transmission are the same as for the Chrysler Type Four unit.

The differences in these transmissions are as follows: second gear ratio changed from 1.92 to 1.67, third gear ratio changed from 1.40 to 1.00 (direct drive) and the fourth gear ratio changed from 1.00 (direct drive) to 0.73 (overdrive). Since the fourth gear ratio is less than 1 to 1, an overdrive condition is achieved.

GENERAL MOTORS 5-SPEED TRANSMISSION

Disassemble Trans.

1. Remove selector lever pivot.
2. Remove spring pin from shifter head and shift rail, Fig. 20.
3. Remove transmission case and extension housing to center support bolts, then slide case forward from transmission.
4. Slide extension housing rearward to remove. Insure the shifter head, shift rail or selector lever does not fall from housing. Also, retain needle rollers since the rollers must be installed in original race.
5. Remove rail selector pin and selector.
6. Press downward on speedometer gear retaining tab, then remove gear and retainer from output shaft.
7. Remove snap ring, thrust washer, first speed gear and blocking ring from output shaft, Fig. 21.
8. Remove snap ring from behind synchronizer hub, Fig. 22.
9. Move shift rail to position pawl, permitting removal of first-reverse shift link.
10. Slide first-reverse synchronizer, shift fork and rail rearward, then remove the reverse idler gear from idler shaft and reverse gear from output shaft.
11. Move interlock pawl to permit second-third speed shift link and shift fork removal. Then, position interlock pawl to permit fourth-fifth speed shift fork and link removal.
12. Remove center support from output shaft and cluster gear, Fig. 23.
13. Remove needle thrust race and bearing from output shaft, Fig. 24.
14. Remove cluster gear from remaining gears and the output shaft from input shaft.
15. Remove remaining output shaft components.

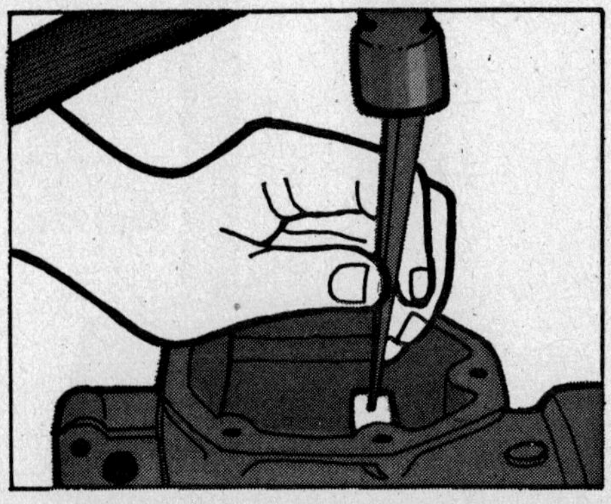

Fig. 20 Removing spring pin

Roller Bearings

Roller bearings in the center support, case and extension housing may be replaced if worn or damaged. All bearings can be removed with tool J-26225. Install countergear bearings in case, Fig. 25, and center support with tools J-8092 and J-26223. Install input shaft bearings in center support, Fig. 26, and case with tools J-8092, J-26222-1 and J-26222-2.

Fig. 21 Removing first gear snap ring

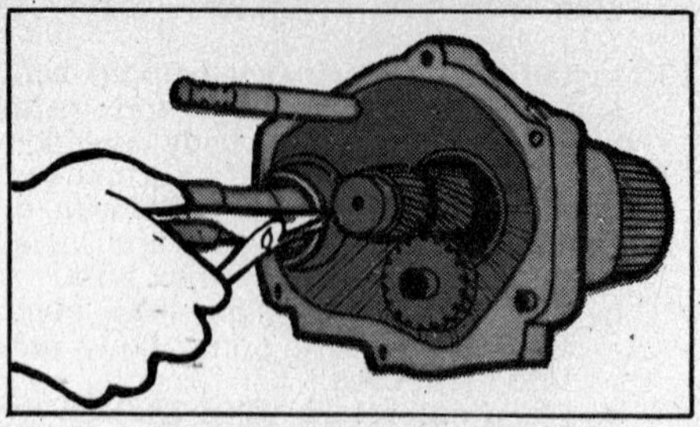

Fig. 22 Removing synchronizer hub snap ring

Fig. 23 Removing center support

Synchronizers

NOTE: The synchronizer hubs and sliding sleeves are a selected assembly and should be kept together as originally assembled. The keys and springs may be replaced if worn or broken.

1. Mark hub and sleeve so they can be reassembled in same position.
2. Remove sleeve from synchronizer hub.
3. Remove keys and springs from hub.
4. Place three keys and two springs in position (one on each side of hub) so all three keys are engaged by both springs. The tanged end of each synchronizer spring should be installed in different key cavities on either side of hub. Slide sleeve onto hub, aligning marks made before disassembly.

Clutch Bearing Retainer Oil Seal

To replace, pry seal from retainer and drive new seal into place with tool J-26224 or other similar tool.

Extension Housing Oil Seal & Bushing

1. Pry oil seal out and using tool J-5778 or J-21424-9, drive bushing from extension housing.
2. Drive new bushing into housing from the rear using same tools.
3. Lubricate bushing and new seal and drive new seal into bushing using tool J-5154 or J-21426.

Assemble Trans.

1. Install third gear with coned end facing upward on output shaft.
2. Install blocker rings with slots aligned with shift plates of the second-third synchronizer assembly, then slide the assembly over output shaft and position of third gear face.
3. Install a snap ring in shaft groove ahead of the synchronizer hub.
4. Install second gear with coned end facing toward blocker rings, then place a thrust washer on face of gear and a snap ring in shaft groove in front of thrust washer, Fig. 27.
5. Install fifth gear on output shaft, Fig. 28, then slide one needle spacer over shaft and into gear bore. Insert a row of 47 needle bearings and a second spacer into bearing bore, then a third spacer.

NOTE: Ensure that the correct needle bearings are used since other needle bearings used in this transmission are of similar length but are slightly smaller in diameter.

Fig. 27 Installing thrust washer snap ring

Fig. 24 Removing needle thrust race & bearing

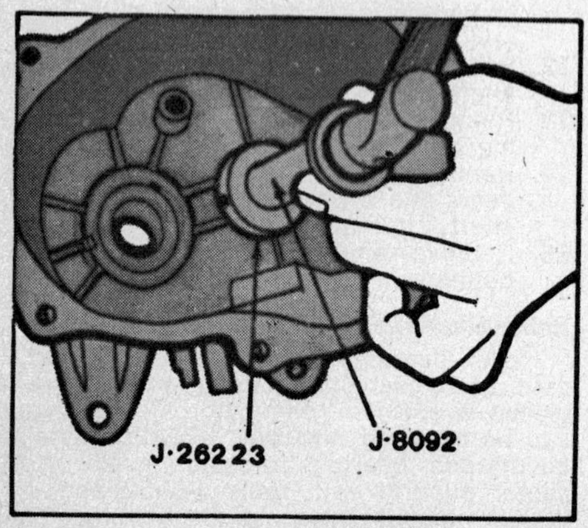

Fig. 25 Installing countergear bearing in case

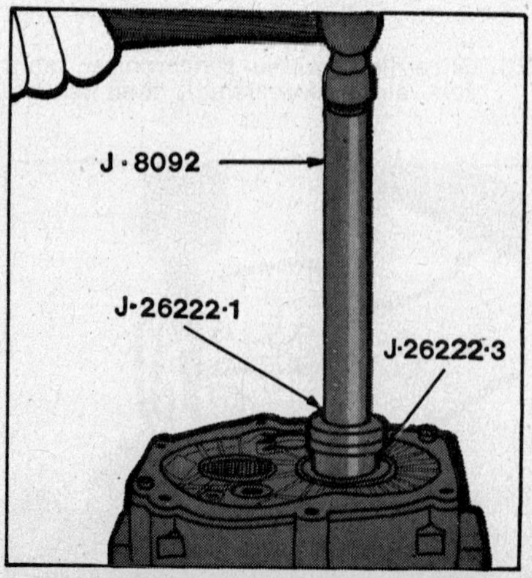

Fig. 26 Installing input shaft bearing in center support

Fig. 28 Fifth speed gear & bearings

Fig. 29 Input shaft & bearings

Fig. 30 Mating input & output shafts

6. Install blocker rings with slots aligned with shift plates of the fourth-fifth synchronizer assembly, then slide the assembly over output shaft and position on fifth gear face.
7. Install a snap ring in shaft groove ahead of synchronizer hub.
8. Install 19 needle rollers into second step of input shaft bore, using vaseline to hold rollers in position, Fig. 29. Then, place input shaft on output shaft, Fig. 30.
9. Mesh cluster gear with input shaft gear and gears assembled on output shaft.
10. Install a needle thrust washer and thrust plate over output shaft and against shaft shoulder.
11. Install interlock pawl into center support bore and secure with retaining plate, Fig. 31.
12. Install needle rollers into center support races and retain with vaseline. Then, assemble center support over output shaft and cluster gears.
13. Install reverse gear and bushing assembly over output shaft, contacting center support.
14. Install reverse idler gear and bushing on reverse idler shaft.
15. Install fourth and fifth link on shift fork. Position interlock pawl to allow shift link to be placed through outboard slot of center support (viewed from rear of transmission), Fig. 32. Engage shift fork in synchronizer collar.
16. Position interlock pawl to allow the second and third shift link to be placed through the middle slot of center support and installed on shift fork, Fig. 32. Engage shift fork in synchronizer collar.

Fig. 31 Interlock pawl & retaining plate installation

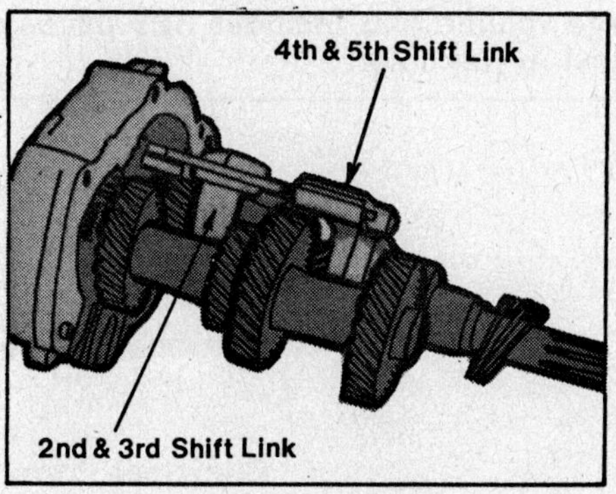

Fig. 32 Shift linkage

17. Position interlock pawl in a manner allowing first and reverse shift link to be placed through inboard slot in center support and install on shift fork.
18. Install selector lever arm over shift rail, aligning drilled hole near center of rail. Drive spring pin into selector arm and shift rail.
19. Install shift rail through shift fork with poppet notches toward rear of transmission, Fig. 33. Engage shift fork with first and reverse synchronizer. Slide synchronizer over output shaft as shift rail is installed through interlock pawl, second and third shift fork, and the fourth and fifth shift fork. Install a snap ring in shaft groove.
20. Install a blocker ring and the first gear over output shaft, behind first and reverse synchronizer assembly. Then, place a thrust washer and snap ring behind first gear.
21. Install speedometer gear retainer in output shaft hole, with retainer looped forward, then slide speedometer gear over output shaft until retainer snaps up to secure gear in position.

Fig. 33 Shift rail

Fig. 34 Mating extension housing & center support

22. Slide rail selector over shift rail with hole facing toward rear of transmission. Drive spring pin into selector and rail.
23. Secure selector lever to the shorter shift rail using a pin and two retaining clips.
24. Press a new oil seal into extension housing and install needle rollers into extension housing race, using vaseline to retain rollers.
25. Install selector lever and shift rail into extension housing hole, then assemble shifter head onto rail when rail is exposed in the housing opening.
26. Apply a bead of Silastic RTV 732 sealer to transmission case and extension housing faces.
27. Extend rail selector into extension housing, engaging rail selector as the extension housing is placed over output shaft and contacts the center support, Fig. 34.
28. Place lipped thrust race, needle thrust washer and flat thrust race over input shaft. Assemble transmission case and install bolts to retain extension housing and transmission case to the center support.
29. Drive spring pin into shifter head and shift rail.
30. Install transmission mesh lock plunger, poppet spring and threaded plug.
31. Install selector lever pivot and electrical switches, if removed, into extension housing.

AUTOMATIC TRANSMISSIONS

NOTE:—This chapter deals only with maintenance, adjustments and "in car" repairs. For major service work, Motor's Automatic Transmission Manual is available. Current edition is a 1040 page volume that includes 329 pages of oil circuit diagrams, most in full color.

INDEX

NOTE: For 1976 linkage adjustment information, see car chapters.

CHRYSLER UNITS

FORD UNITS

GENERAL MOTORS UNITS

MISCELLANEOUS

HOW TO PUSH AND TOW AUTOMATIC DRIVE CARS—Inside Back Cover

TURBO HYDRA-MATIC 375, 400

TRANSMISSION IDENTIFICATION

An identification plate is attached to the transmission. The plate indicates year of production, code letters, and serial number.

*Late production 1973 Pontiac transmissions have the number "2" between code letters.

BUICK		CODE
1969	V8-350 LeSabre	BU
	V8-400 Sportwagon	BW
	V8-400 G.S. 400	BA, BB
	V8-430 Wildcat, Electra	BC
	V8-430 Riviera	BT
1970	V8-455 G.S. 455	BA
	V8-455 G.S. 455 Stage 1	BB
	V8-455 Riviera	BT
	V8-455 All Others	BC
1971	V8-455 G.S. 455	BS
	V8-455 G.S. 455 Stage 1	BB, OW
	V8-455 Riviera	BT
	V8-455 All Others	BC
1972	V8-455 G. S. 455	BS
	V8-455 G. S. 455 Stage 1	BB
	V8-455 Centurion Hi Perf.	BU
	V8-455 Riviera G. S.; Perf. axles on all models	BT
	V8-455 All Others	BC
1973-74	V8-455 Century, Regal less Stage 1	BS
	V8-455 Century, Regal Stage 1	BB
	All Others— Hi Perf.	BT
	All Others— Exc. Hi. Perf.	BC
1975	V8-350 LeSabre	BK
	V8-400	OC
	V8-455	BC, BT

CADILLAC		
1969	75 and Commercial Chassis	AB
	All Others	AA
1970-75	All Models	AA
1976	Seville	AC

CHECKER MOTORS		
1973-75	6-250	NA
	V8-350	NB

CHEVROLET		
1969	V8-350 Chevrolet	CA
	V8-396 Chevrolet	CA
	V8-427, 390 H.P.	CB
	V8-396, 325 H.P.	CC
	V8-350, 250 H.P. Chevrolet	CD
	V8-327, 235 H.P. Chevrolet	CD
	V8-396, 350 H.P. Chevelle, Camaro, Nova	CE
	V8-427, 335 H.P.	CF
	V8-396, 265 H.P. Chevrolet	CG
	V8-427, 390 H.P. Chevrolet	CH
	V8-350, 300 H.P. Corvette	CK
	V8-427, 400 H.P. Corvette	CL
	V8-427, 335 H.P. Chevrolet	CQ

CHEVROLET—Continued		CODE
	V8-427, 425 H.P. Chevrolet	CY①
	V8-427, 430 H.P. Corvette	CY①
	V8-427, 435 H.P. Corvette	CY①
	V8-396, 375 H.P. Chevelle, Camaro, Nova	CY①
1970	V8-400 Chevrolet	CA
	V8-454 Chevrolet	CB
	V8-400 Chevelle, Monte Carlo	CD
	V8-400 Camaro, Chevrolet, Nova	CF
	V8-454 Police	CG
	V8-350 Corvette	CK
	V8-454 Camaro, Chevelle, Nova	CR
	V8-454 Chevelle, Monte Carlo, Corvette	CS
	V8-400 Camaro, Chevelle, Nova	CW
	V8-454 Camaro, Chevelle, Corvette	CY
1971	V8-400 Chevrolet	CA
	V8-454 Chevrolet	CB
	V8-400 Chevelle	CD
	V8-454 Chevelle	CF
	V8-350 Corvette	CK
	V8-454 Corvette	CS
	V8-454 Chevelle, Corvette Hi. Perf.	CY
1972	V8-350, 400 Chevrolet	CA
	V8-402 4 Bar. Carb. Chevelle	CD
	V8-454 Chevelle	CF
	V8-400 Camaro	CK
	V8-454 Chevrolet	CR
	V8-454 Corvette	CS
	V8-350 Camaro, Chevelle, Corvette	CY
1973	V8-400 2 Bar. Carb. Chevrolet	CA
	V8-350 2 Bar. Carb. Chevelle	CF
	V8-350 4 Bar. Carb. Chevelle	CG
	V8-350 Corvette 190 H.P.	CK
	V8-454 Chevrolet	CR
	V8-454 Corvette	CS
	V8-350 Corvette 250 H.P.	CY
1974	V8-400 Chevrolet	CA
	V8-400 2 Bar. Carb. Chevrolet	CB
	V8-400 Chevrolet	CG
	V8-350 Corvette 190 H.P.	CK
	V8-454 Chevrolet	CR
	V8-454 Corvette	CS
	V8-350 Corvette 250 H.P.	CZ
1975	V8-350 Police	CR
	V8-350 Corvette Exc. Sp. H/Perf.	CK
	V8-350 Corvette Sp. H/Perf.	CZ
	V8-400 Chevrolet with 3.08, 3.42 Axles Exc. Calif.	CA
	V8-400 Chevrolet with 2.56, 2.73 Axles Exc. Calif.	CB
	V8-454 Chevelle	CF
	V8-454 Chevrolet	CR

①High shift point.

OLDSMOBILE		CODE
1969	V8-350 Vista-Cruiser	OA
	V8-350 Delta 88 W/Cruise Control	OA
	V8-455 4 barrel carb. "98"	OB
	V8-455 2 barrel carb. Delta	OC
	V8-350 Delta 88	OF
	V8-400 4-4-2 Exc. Ram Air	OG
	V8-455 4 bar. carb. "98", Delta	OK
	V8-400 4 barrel carb. Vista-Cruiser	OP
	V8-455 Heavy Duty, Police	OL
	V8-455 2 barrel carb. Delta	OR
	V8-400 4-4-2 Ram Air	OW
1970	V8-350 Exc. Below	OF
	V8-350 W/Cruise Control	OA
	V8-455 2 bar. carb. Exc. Below	OC
	V8-455 Delta 88, 4 bar. carb.	OB
	V8-455 2 bar. carb. Cruise Control	OR
	V8-455 Cutlass Exc. SX	OD
	V8-455 Cutlass, 442 SX Exc. Ram Air	OG
	V8-455 Senior models W/Cruise Control & 4 bar. carb.	OK
	V8-455 Vista-Cruiser	OK
	V8-455 Police	OL
	V8-455 442 Ram Air	OW
1971	V8-455 Supreme 4 Bar. Carb.	OD
	V8-455 4-4-2	OG
	V8-455 4-4-2 W/Air Induction	OW
	V8-455 Vista Cruiser	OA
	V8-455 Vista Cruiser 2 Bar. Carb.	OR
	V8-455 Vista Cruiser 4 Bar. Carb.	OK
	V8-455 Senior Models 2 Bar. Carb.	OR
	V8-455 Senior Models Dual Exhaust	OK
	V8-455 Delta 88 Single Exhaust	OR
1972	V8-455 Cutlass, 4 Bar. Carb., Dual Exh.	OD
	V8-455 Cutlass, 4 Bar. Carb. (W30)	OW
	V8-350 Vista Cruiser	OA
	V8-350 Senior Models	OA
	V8-455 Senior Models	OR
	V8-455 Senior Models, Dual Exh.	OK
	V8-455 Vista Cruiser, Dual Exh.	OK
1973	V8-455 Cutlass, 4 Bar. Carb.	OD
	V8-455 Senior Models, Dual Exh.	OK
	V8-455 Senior Models	OR
1974	V8-350 88	OA
	V8-455 Cutlass Exc. Calif.	OD
	V8-455 Cutlass Calif.	OW
	V8-455 Cutlass 275 H.P.	OX
	V8-455 88 & 98 Single Exhaust Exc. Calif.	OR
	V8-455 88 & 98 Single Exh. Calif.	OL
	V8-455 88 & 98 Dual Exhaust	OK
	V8-455 Custom Cruiser Exc. Calif.	OK
	V8-455 Custom Cruiser Calif.	OE

Continued

TRANSMISSION IDENTIFICATION—CONTINUED

OLDSMOBILE—Continued		CODE
1975	V8-350 88 Exc. Calif.	OA
	V8-350 88 Calif.	OB
	V8-400 Custom Cruiser	OC
	V8-400 98	OF
	V8-455 Cutlass Exc. Calif.	OD
	V8-455 Cutlass Calif.	OW
	V8-455 Custom Cruiser Exc. Calif.	OK
	V8-455 Custom Cruiser Calif.	OE
	V8-455 88 & 98 Exc. Calif.	OR
	V8-455 88 & 98 Calif.	OL

PONTIAC		CODE
1969	V8-350 2 Barrel carb. (Tempest, Firebird)	PV
	V8-350 H.O. (Tempest, Firebird)	PS
	V8-400 2 barrel carb.	PB
	V8-400 2 barrel carb. (GTO, Grand Prix)	PT
	V8-400 4 barrel carb. (Grand Prix)	PW
	V8-400 4 barrel carb. (GTO, Firebird)	PX
	V8-400 Ram Air (GTO, Firebird)	PQ
	V8-428 H.O.	PC
	V8-428 Bonneville	PH
	V8-428 H.O. (Grand Prix)	PR
1970	V8-400 (Pontiac)	PB
	V8-400 Ram Air (GTO)	PD
	V8-400 (Tempest, Grand Prix)	PT
	V8-400 4 bar. carb. (Grand Prix)	PW
	V8-400 2 bar. carb. (Firebird)	PF
	V8-400 Sports Option	PX
	V8-400 Ram Air, HO	PQ
	V8-455 Exc. HO (Pontiac)	PA
	V8-455 HO (Pontiac)	PC
	V8-455 Heavy Duty (Pontiac)	PH
	V8-455 HO (Tempest, Grand Prix)	PR
1971	V8-455 (Pontiac)	PA
	V8-400 (Pontiac)	PB
	V8-455 4 Bar. Carb., W/Dual Exh. (Pontiac)	PC
	V8-400 (Pontiac)	PD
	V8-455 (Pontiac)	PF

PONTIAC—Continued		CODE
	V8-400 2 Bar. Carb. (Police & H.D.)	PH
	V8-455 4 Bar. Carb. (Police & H.D.)	PH
	V8-455 H.O. (LeMans, GTO, Firebird)	PQ
	V8-455 4 Bar. Carb. (LeMans, GTO, Firebird)	PR
	V8-400 2 Bar. Carb. (LeMans, GTO, Firebird)	PT
	V8-455 4 Bar. Carb. (LeMans, GTO, Firebird, Grand Prix)	PW
	V8-400 4 Bar. Carb. (LeMans, GTO, Firebird, Grand Prix)	PX
	V8-400 4 Bar. Carb. (LeMans, GTO, Firebird)	PY
1972	V8-400 Wagons	PA
	V8-455 2 Bar. Carb. (Pontiac)	PA
	V8-400 4 Bar. Carb.	PB
	V8-455 4 Bar. Carb. (Pontiac)	PC
	V8-400 2 Bar. Carb.	PD
	V8-400 4 Bar. Carb. (LeMans, Firebird)	PG
	V8-455 HO (LeMans, Firebird)	PQ
	V8-455 4 Bar. Carb. (LeMans, Grand Prix)	PR
	V8-400 2 Bar. Carb. (LeMans, Firebird)	PT
	V8-400 4 Bar. Carb. (Grand Prix)	PX
1973	V8-400 2 Bar. Carb. Wagons	PA
*	V8-400 4 Bar. Carb.	PA
	V8-400 4 Bar. Carb. (Pontiac)	PB
	V8-455 4 Bar. Carb. (Pontiac)	PC
	V8-400 2 Bar. Carb. (Exc. Wagon on Pontiac)	PD
	V8-350 2 Bar. Carb. (Pontiac)	PF
	V8-400 4 Bar. Carb. (Firebird, LeMans, Grand Am)	PT
	V8-455 4 Bar. Carb. (Police)	PH
	V8-455 SD (Firebird, LeMans, Grand Am)	PQ

PONTIAC—Continued		CODE
	V8-455 4 Bar. Carb. (LeMans, Grand Am, Grand Prix)	PR
	V8-400 2 Bar. Carb. (Firebird, LeMans, Grand Am)	PT
	V8-400 4 Bar. Carb. (Grand Prix)	PX
1974	V8-400 2 Bar. Carb. Sta. Wag.	PA
	V8-400 4 Bar. Carb. (Police & H.D.)	PA
	V8-400 4 Bar. Carb. (Pontiac)	PB
	V8-455 4 Bar. Carb. (Pontiac)	PC
	V8-400 2 Bar. Carb. (Pontiac Exc. Calif.)	PD
	V8-400 2 Bar. Carb. (Pontiac Calif.)	PF
	V8-400 4 Bar. Carb. (LeMans & Firebird, Exc. Calif.)	PG
	V8-400 4 Bar. Carb. (LeMans & Firebird, Calif.)	PW
	V8-455 4 Bar. Carb. (Police)	PH
	V8-455 SD (Firebird)	PQ
	V8-455 4 Bar. Carb. (LeMans & Grand Prix)	PR
	V8-400 2 Bar. Carb. (LeMans & Firebird, Exc. Calif.)	PT
	V8-400 2 Bar. Carb. (LeMans & Firebird, Calif.)	PL
	V8-400 4 Bar. Carb. (Grand Prix)	PX
	V8-455 4 Bar. Carb. (Firebird)	PZ
1975	V8-400 2 Bar. Carb. (Pontiac)	PD
	V8-400 2 Bar. Carb. (LeMans & Grand Prix)	PT
	V8-400 4 Bar. Carb. (Pontiac Exc. Calif.)	PB
	V8-400 4 Bar. Carb. (Pontiac, Calif.)	PF
	V8-400 4 Bar. Carb. (LeMans & Grand Prix Exc. Calif.)	PX
	V8-400 4 Bar. Carb. (LeMans & Grand Prix, Calif.)	PG
	V8-455 Pontiac Exc. Calif.	PC
	V8-455 Pontiac, Calif.	PA
	V8-455 LeMans & Grand Prix Exc. Calif.	PR
	V8-455 LeMans & Grand Prix, Calif.	PL

GENERAL DESCRIPTION

This transmission, Fig. 1, is a fully automatic unit consisting primarily of a three-element hydraulic torque converter and a compound planetary gear set. Three multiple-disc clutches, two one-way clutches, and two bands provide the friction elements required to obtain the desired functions of the planetary gear set.

The torque converter, the multiple-disc clutches and the one-way clutches couple the engine to the planetary gears through oil pressure, providing three forward speeds and reverse. The torque converter, when required, supplements the gears by multiplying engine torque.

Torque Converter

The torque converter is of welded construction and is serviced as an assembly. The unit is made up of two vaned sections, or halves, that face each other in an oil-filled housing. The pump half of the converter is connected to the engine and the turbine half is connected to the transmission.

When the engine makes the converter pump revolve, it sends oil against the turbine, making it revolve also. The oil then returns in a circular flow back to the converter pump, continuing this flow as long as the engine is running.

Stator

The converter also has a smaller vaned section, called a stator, that funnels the oil back to the converter pump through smaller openings, at increased speed. The speeded up oil directs additional force to the engine-driven converter pump, thereby multiplying engine torque. In other words, without the stator, the unit is nothing more than a fluid coupling.

External Controls

The external control connections to the transmission are:

1. Manual linkage to select the desired operating range.
2. Engine vacuum to operate the vacuum modulator unit.
3. An electrical signal to operate an electric detent solenoid.

Vacuum Modulator

A vacuum modulator is used to sense engine torque input to the transmission automatically. The vacuum modulator

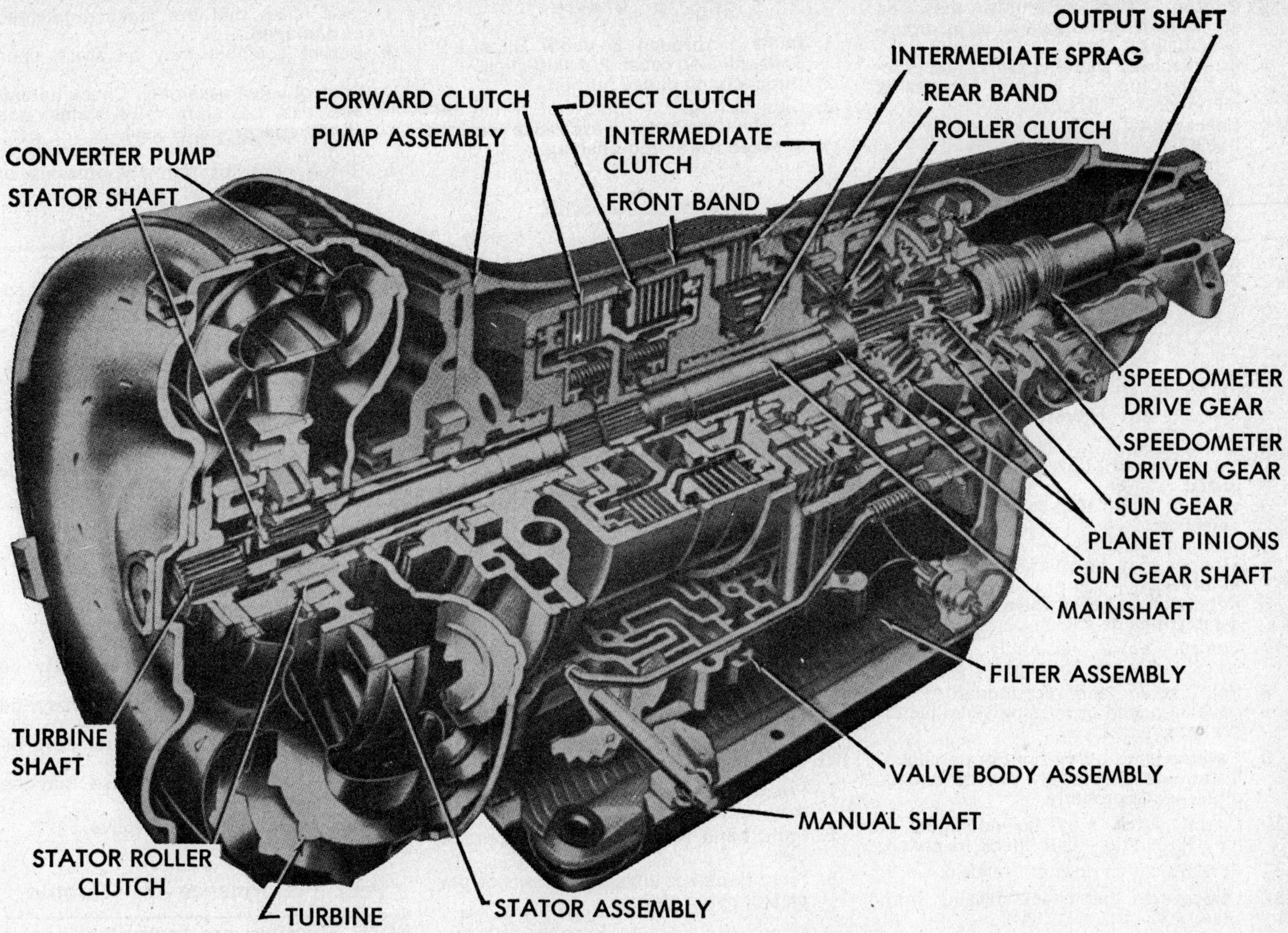

Fig. 1 Cutaway view of transmission assembly

transmits this signal to the pressure regulator, which controls line pressure, so that all torque requirements of the transmission are met and proper shift spacing is obtained at all throttle openings.

Detent Solenoid

The detent solenoid is activated by an electric switch at the carburetor. When the throttle is opened sufficiently to close this switch, the solenoid in the transmission is activated, causing a downshift at speeds below 70 mph. At lower speeds, downshifts will occur at lesser throttle openings without use of the electric switch.

TROUBLE SHOOTING GUIDE

Oil Pressure High or Low

1. Vacuum line or fittings clogged or leaking.
2. Vacuum modulator.
3. Modulator valve.
4. Pressure regulator.
5. Oil pump.
6. Governor.

No Drive In Drive Range

1. Low oil level (check for leaks).
2. Manual control linkage not adjusted properly.
3. Low oil pressure. Check for blocked strainer, defective pressure regulator, pump assembly or pump drive gear. See that tangs have not been damaged by converter.
4. Check control valve assembly to see if manual valve has been disconnected from manual lever pin.
5. Forward clutch may be struck or damaged. Check pump feed circuits to forward clutch including clutch drum ball check.
6. Sprag or roller clutch assembled incorrectly.

1 - 2 Shift At Full Throttle Only

1. Detent switch may be sticking or defective.
2. Detent solenoid may be stuck open, loose or have leaking gasket.
3. Control valve assembly may be leaking, damaged or incorrectly installed.
4. Porous transmission case.

1st Speed Only - No 1 - 2 Shift

1. Governor valve may be sticking.
2. Driven gear in governor assembly loose, worn or damaged.
3. The 1-2 shift valve in control valve assembly stuck closed. Check governor feed channels for blocks, leaks, and position. Also check control valve body gaskets for leaks and damage.
4. Intermediate clutch plug in case may be leaking or blown out.
5. Check for porosity between channels and for blocked governor feed channels in case.
6. Check intermediate clutch for proper operation.

No 2 - 3 Shift - 1st & 2nd Only

1. Detent solenoid may be stuck open.
2. Detent switch may not be properly adjusted.

3. Control valve assembly may be stuck, leaking, damaged, or incorrectly installed.
4. Check direct clutch case center support for broken, leaking or missing oil rings.
5. Check clutch piston seals and piston ball check in clutch assembly.

Moves Forward In Neutral

1. Manual control linkage improperly adjusted.
2. Forward clutch does not release.
3. Oil pump.
4. Internal linkage.

No Drive In Reverse or Slips In Reverse

1. Check oil level.
2. Manual control linkage improperly adjusted.
3. Vacuum modulator assembly may be defective.
4. Vacuum modulator valve sticking.
5. Strainer may be restricted or leaking at intake.
6. Regulator or boost valve in pump assembly may be sticking.
7. Control valve assembly may be stuck, leaking or damaged.
8. Rear servo and accumulator may have damaged or missing servo piston seal ring.
9. Reverse band burned out or damaged. Determine that apply pin or anchor pins engage properly.
10. Direct clutch may be damaged or may have stuck ball check in piston.
11. Forward clutch does not release.
12. Low-reverse ball check missing from case.

Slips In All Ranges & On Starts

1. Check oil level.
2. Vacuum modulator defective.
3. Modulator valve sticking.
4. Strainer assembly plugged or leaking at neck.
5. Pump assembly regulator or boost valve sticking.
6. Leaks from damaged gaskets or cross leaks from porosity of case.
7. Forward and direct clutches burned.

Slips 1 - 2 Shift

1. Incorrect oil level.
2. Vacuum modulator valve sticking.
3. Vacuum modulator defective.
4. Pump pressure regulator valve defective.
5. Porosity between channels in case.
6. Control valve assembly.
7. Pump-to-case gasket may be mispositioned.
8. Intermediate clutch plug in case may be missing or leaking excessively.
9. Intermediate clutch piston seal missing or damaged.
10. Intermediate clutch plates burned.
11. Front or rear accumulator oil ring may be damaged.

Slips 2 - 3 Shift

1. Items 1 through 6 under Slips 1-2 Shift will also cause 2-3 shift slips.
2. Direct clutch plates burned.
3. Oil seal rings on direct clutch may be damaged permitting excessive leaking between tower and bushing.

Rough 1 - 2 Shift

1. Modulator valve sticking.
2. Modulator assembly defective.
3. Pump pressure regulator or boost valve stuck or inoperative.
4. Control valve assembly loosened from case, damaged or mounted with wrong gaskets.
5. Intermediate clutch ball missing or not sealing.
6. Porosity between channels in case.
7. Rear servo accumulator assembly may have oil rings damaged, stuck piston, broken or missing spring or damaged bore.

Rough 2 - 3 Shift

1. Items 1, 2 and 3 under Rough 1-2 Shift will also cause rough 2-3 shift.
2. Front servo accumulator spring broken or missing. Accumulator piston may be sticking.

No Engine Braking in Second Speed

1. Front servo or accumulator oil rings may be leaking.
2. Front band may be broken or burned out.
3. Front bank not engaged on anchor pin and/or servo pin.

No Engine Braking In Low Range

1. Low-reverse check ball may be missing from control valve assembly.
2. Rear servo may have damaged oil seal ring, bore or piston; leaking, apply pressure.
3. Rear band broken, burned out or not engaged on anchor pins or servo pin.

No Part Throttle Downshifts

1. Vacuum modulator assembly.
2. Modulator valve.
3. Regulator valve train.
4. Control valve assembly has stuck 3-2 valve or broken spring.

No Detent Downshifts

1. Detent switch needs fuse, connections tightened or adjustment.
2. Detent solenoid may be inoperative.
3. Detent valve train in control valve assembly malfunctioning.

Low or High Shift Points

1. Oil pressure. Check vacuum modulator assembly, vacuum line connections, modulator valve, and pressure regulator valve train.
2. Governor may have sticking valve or feed holes that are leaking, plugged or damaged.
3. Detent solenoid may be stuck open or loose.
4. Control valve assembly. Check detent, 3-2, and 1-2 shift valve trains, and check spacer plate gaskets for positioning.
5. Check case for porosity, missing or leaking intermediate plug.

Won't Hold In Park

1. Manual control linkage improperly adjusted.
2. Internal linkage defective; check for chamfer on actuator rod sleeve.
3. Parking pawl broken or inoperative.

Excessive Creep At Idle

NOTE: Transmissions have the variable pitch stator.

1. High idle speed.
2. Stator switch inoperative or defective.
3. Stator solenoid defective.
4. Pump may have stator valve train stuck.
5. Pump lead wires disconnected or grounded out.
6. Pump feed circuit to stator may be restricted or blocked.
7. Converter out check valve may be broken or stuck.
8. Turbine shaft may have defective oil seal ring.
9. Stator orifice plug in case may be blocked.
10. Converter assembly defective.

Poor Performance - ¾ Throttle

NOTE: Transmissions having the variable pitch stator.

1. Stator and detent switch inoperative.
2. Items 3 through 10 above will also cause poor performance at ¾ throttle.

Noisy Transmission

1. Pump noises caused by high or low oil level.
2. Cavitation due to plugged strainer, porosity in intake circuit or water in oil.
3. Pump gears may be damaged.
4. Gear noise in low gear of Drive Range-transmission grounded to body.
5. Defective planetary gear set.
6. Clutch noises during application can be worn or burned clutch plates.

Forward Clutch Plates Burned

1. Check ball in clutch housing damaged, stuck or missing.
2. Clutch piston cracked, seals damaged or missing.
3. Low line pressure.
4. Manual valve mispositioned.
5. Restricted oil feed to forward clutch.
6. Pump cover oil seal rings missing, broken or undersize; ring groove oversize.
7. Case valve body face not flat or poro-

sity between channels.
8. Manual valve bent and center land not properly ground.

Intermediate Clutch Plates Burned

1. Constant bleed orifice in center support missing.
2. Rear accumulator piston oil ring damaged or missing.
3. 1-2 accumulator valve stuck in control valve assembly.
4. Intermediate clutch piston seal damaged or missing.
5. Center support bolt loose.
6. Low line pressure.
7. Intermediate clutch plug in case missing.
8. Case valve body face not flat or porosity between channels.
9. Manual valve bent and center land not ground properly.

Direct Clutch Plates Burned

1. Restricted orifice in vacuum line to modulator.
2. Check ball in direct clutch piston damaged, stuck or missing.
3. Defective modulator bellows.
4. Center support bolt loose.
5. Center support oil rings or grooves damaged or missing.
6. Clutch piston seals damaged or missing.
7. Front and rear servo pistons and seals damaged.
8. Manual valve bent and center land not cleaned up.
9. Case valve body face not flat or porosity between channels.
10. Intermediate sprag clutch installed backwards.
11. 3-2 valve, 3-2 spring or 3-2 spacer pin installed in wrong location in 3-2 valve bore.

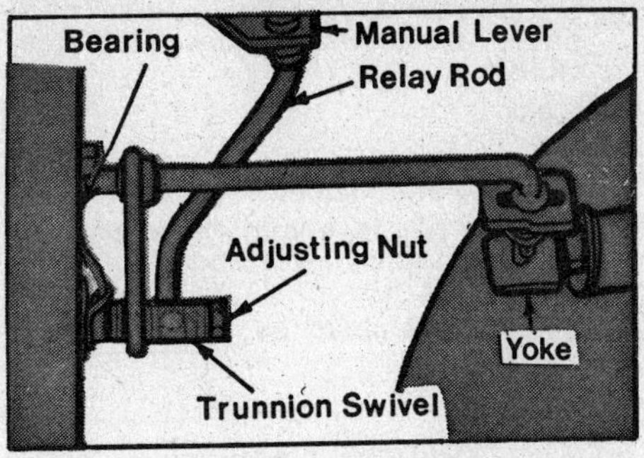

Fig. 2 Manual linkage adjustment. 1969-75 Cadillac (typical)

NOTE: Beginning with 1973 models, Cadillac uses an extended-life Dexron transmission fluid. With this new fluid, strainer replacement and fluid change is now recommended at 100,000 miles under normal operating conditions and 50,000 miles under severe or abnormal service such as trailer towing.

This recommendation applies only to the improved fluid and its availability for service. If the new fluid is not available, the former fluid can be used but then the 24,000 mile maintenance rule will apply.

Beginning with 1973 Chevrolet and late 1973 Buick, Oldsmobile and Pontiac models, a revised type Dexron fluid is used in these transmissions. An early change to a darker color from the usual red color and or a strong odor that is usually associated with overheated fluid is normal, and should not be treated as a positive sign of needed maintenance or unit failure.

The normal maintenance schedule for drain and refill of this type fluid remains unchanged at 24,000 miles under normal service and 12,000 miles under severe operating conditions, such as trailer towing.

CAUTION: *Do not overfill as foaming might occur when the fluid heats up. If fluid level is too low, especially when cold, complete loss of drive may result after quick stops. Extremely low fluid level will result in damage to transmission.*

MAINTENANCE

Checking & Adding Fluid

Fluid level should be checked at every engine oil change. The full ("F") and "ADD" marks on the transmission dipstick are one pint apart and determine the correct fluid level at normal operating temperature (170°F.). *Careful attention to transmission oil temperature is necessary as proper fluid level at low operating temperatures will be below the "ADD" mark on the dipstick. Proper fluid level at higher operating temperatures will rise above the "F" mark.*

Fluid level must always be checked with the car on a level surface, and with the engine running to make certain the converter is full. To determine proper fluid level, proceed as follows:

1. Operate engine at a fast idle for about 1½ minutes with selector lever in park ("P") position.
2. Reduce engine speed to slow idle and check fluid level.
3. With engine running, add Dexron fluid as required to bring it to the proper level.

Draining Bottom Pan Only

1. Disconnect filler tube at bottom pan and allow fluid to drain. Remove and discard filler tube O-ring.
2. Use a new O-ring on filler tube and install tube on pan.
3. Lower car and add three quarts of Dexron transmission fluid through filler tube when replacing intake pipe and strainer assembly. When just draining bottom pan, add only two quarts.
4. Operate engine at a fast idle for about 1½ minutes with selector lever in park ("P") position.
5. Reduce engine speed to slow idle and check fluid level. Then add fluid as required to bring it to the proper level.

Adding Fluid to Fill Dry Transmission and Converter

1. Add seven quarts of fluid through filler tube.
2. Operate engine at a fast idle for about 1½ minutes with selector lever in park ("P") position.
3. Reduce engine speed to slow idle and add three more quarts of fluid.
4. Check fluid level and add as required to bring it to the proper level.

IMPORTANT: Beginning with 1969 General Motors units are equipped with anti-theft systems which lock not only the steering column but the transmission linkage as well. It is essential that this "back drive" linkage also be adjusted when any adjustment to the manual shift linkage is made. This adjustment must be precise as any inaccuracies may result in unit operation without controls in full detent which, in turn, will cause reduced oil pressure and subsequent failure of the transmission.

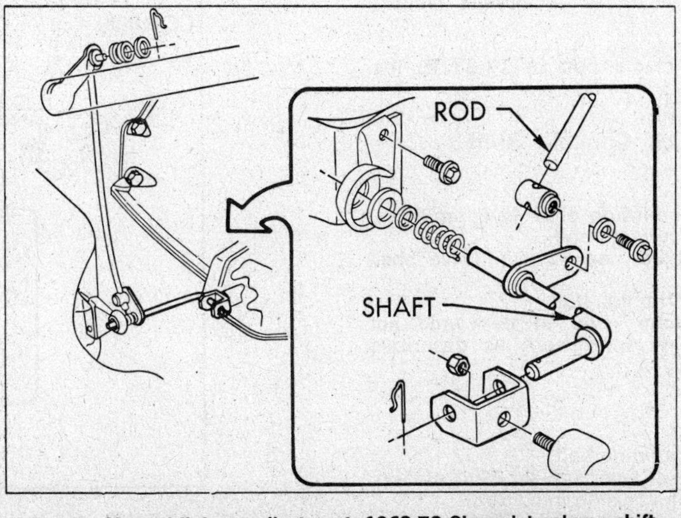

Fig. 3 Manual linkage adjustment. 1969-72 Chevrolet column shift

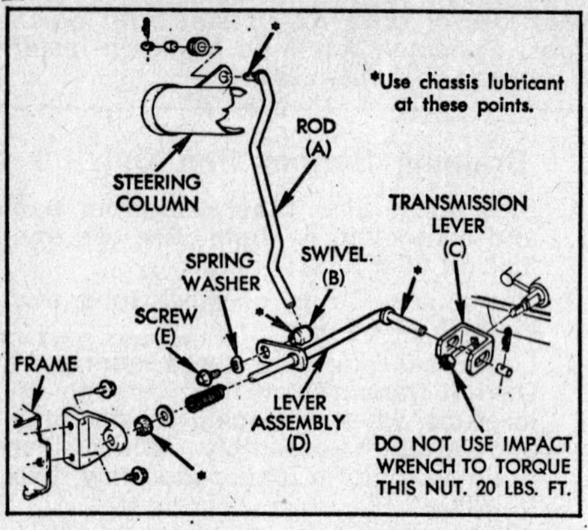

Fig. 4 **Column shift linkage. 1973-75 Camaro, Chevrolet**

BACK DRIVE LINKAGE, ADJUST

Adjust back drive at trunnion so that:

1. Transmission is in full detent in each selector position.
2. With key in Run position and transmission in Reverse, key cannot be removed and steering wheel is not locked.
3. With key in Lock position and transmission in Park, key can be removed and steering wheel is locked.

MANUAL LINKAGE, ADJUST

Buick Column Shift

1969-70

1. Loosen shift rod adjusting clamp.
2. Place selector lever against Drive stop.
3. Place transmission in Drive.
4. Tighten clamp bolt to 17-23 ft-lbs.

1971-75

1. Loosen shift rod adjusting clamp bolts.
2. Place selector lever against Neutral stop.
3. Place transmission in Neutral.
4. Tighten clamp bolt to 17-23 ft. lbs.

Buick Console Shift

1969-70 All

1. Set transmission shift lever in Drive.
2. Loosen trunnion bolt.
3. Set shift bar assembly in Drive position.
4. Tighten trunnion bolt.
5. Set selector lever in park and set back drive adjustment as described previously.

1971-74 All

1. Loosen trunnion bolt.
2. Set selector lever against Neutral stop.
3. Place transmission in Neutral.
4. Tighten trunnion bolt to 6-9 ft. lbs.
5. Set selector lever in Park and set back drive adjustment as described previously .

1975 All

1. Place selector lever and transmission lever in Park position.
2. Position pin to obtain "Free pin" fit in transmission lever and torque nut to 15-25 ft. lbs.
3. Adjust back drive as outlined previously.

1969-75 Cadillac

1. Refer to Fig. 2, loosen nut on steering column manual lever-to-relay rod clamp.
2. Pull relay rod up to position transmission shift valve in PARK; then push rod down to third (neutral) step.
3. Position selector lever in neutral.
4. Tighten clamp nut. Selector lever should enter all positions and indicator pointer should index correctly.

Chevrolet Column Shift 1969-72

1. Place selector lever in Drive as determined by transmission detent and loosen adjusting swivel at cross-shaft, Fig. 3.
2. Move the transmission lever so lever contacts mechanical stop on steering column. Tighten swivel and check adjustment for proper operation.
3. Adjust shift indicator needle, if necessary.

1973-75

1. Place transmission lever in Neutral

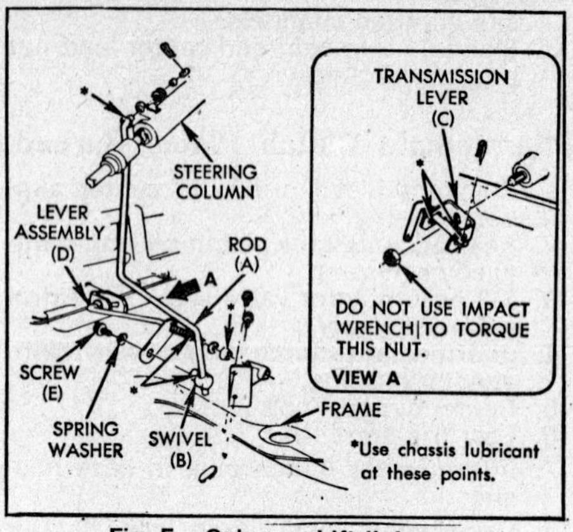

Fig. 5 **Column shift linkage. 1973-75 Chevelle & Monte Carlo**

by moving lever counter-clockwise to L1 detent then clockwise three detent positions to Neutral.

2. Place selector lever in Neutral as determined by mechanical stop on steering column. Do not use indicator as reference.
3. Assemble swivel, spring washer and screw to lever assembly then tighten screw to 20 ft. lbs., Figs. 4 and 5.

Chevy Nova 1969-70 Floor Shift

Refer to Fig. 6, and proceed as follows:

1. Loosely assemble nuts (A) and (B) on lower rod (C).
2. Set transmission lever (D) in Drive.
3. Set control pawl rod (E) in the Neutral or Drive notch of detent (F).
4. Apply load in direction of arrow (Y) on actuating lever (G) until pawl rod

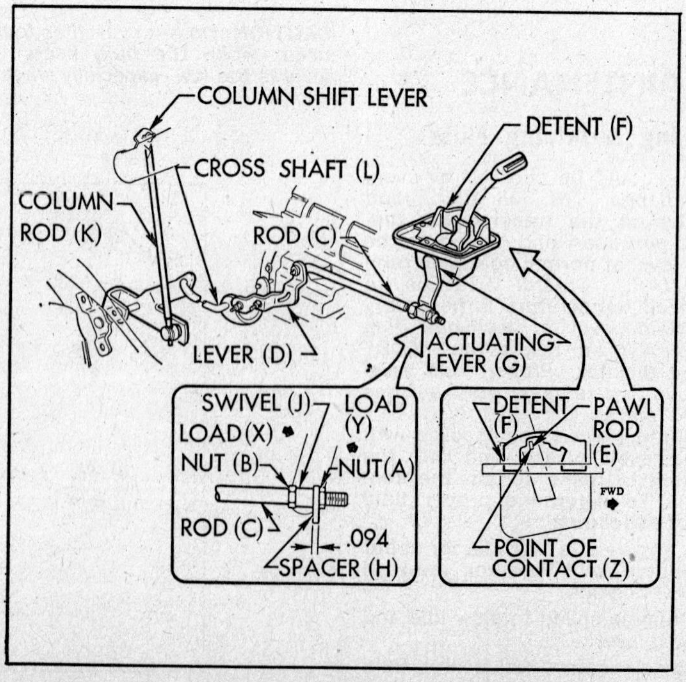

Fig. 6 **Manual linkage adjustment. 1969-70 Nova floor shift**

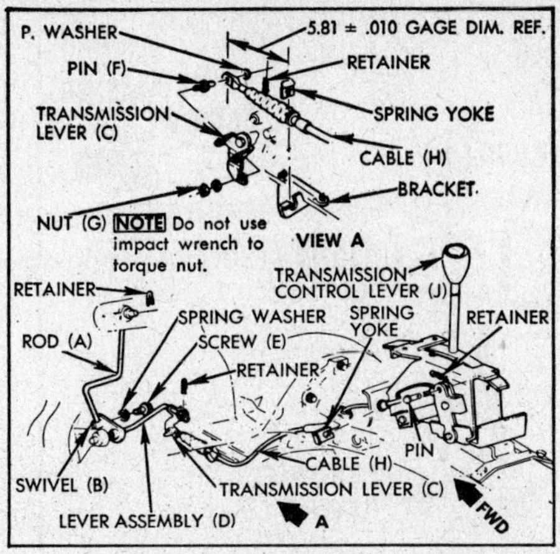

Fig. 7 Console shift adjustment. 1973-75 Camaro, Chevelle & Monte Carlo

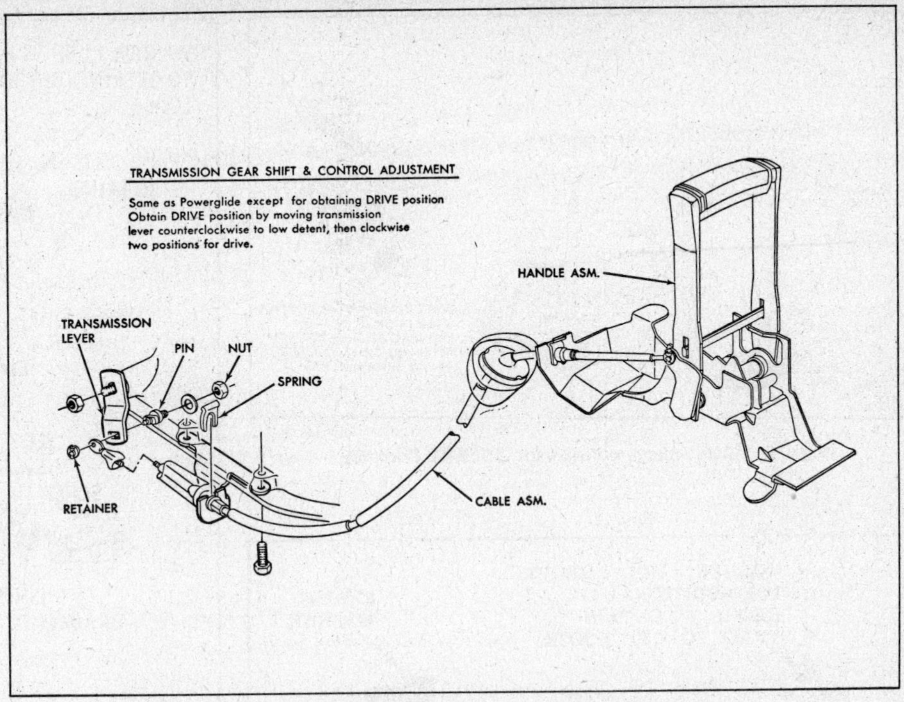

Fig. 8 Console shift linkage. 1970-72 Camaro, Chevelle and Chevrolet

comes in contact with detent at contact point (Z).

5. Place a .094" spacer (H) between nut (A) and swivel (J), run nut (A) until it touches spacer. Tighten nut (B) against swivel and lock swivel between nuts (A) and (B).
6. Place transmission and shift lever in Park.
7. Install column rod (K) to column lever and cross shaft (L).

NOTE: With shift lever in Park the ignition key must move freely to Lock position.

Chevelle, Monte Carlo & Camaro 1973-75 Console Shift

1. Place shift lever (J) in Drive and loosen nut (G) so that pin (F) moves freely in slot of transmission lever (C). Fig. 7.
2. Place transmission lever (C) in Drive by moving lever counter-clockwise to L1 detent then clockwise three detent positions to Drive.
3. Tighten nut (G) to 20 ft. lbs.

Camaro, Chevelle and Chevrolet 1970-72 Floor Shift

1. Place shift lever in Drive position, Fig. 8.
2. Raise vehicle. Disconnect cable from transmission lever. Manually place transmission lever in Drive position.

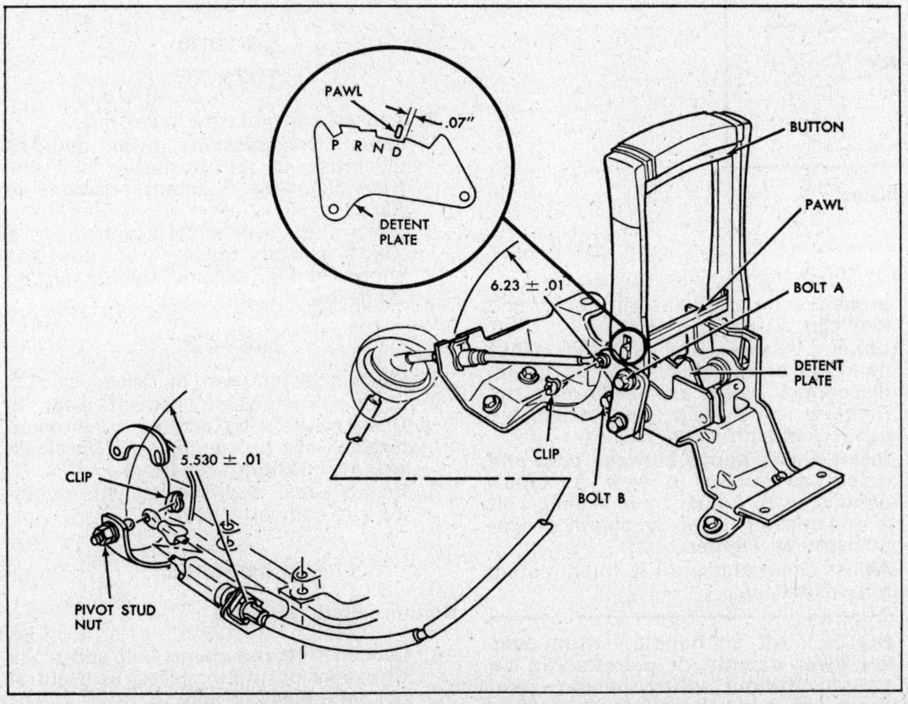

Fig. 9 Console shift linkage adjustment. 1969 Camaro, Chevelle and Chevrolet

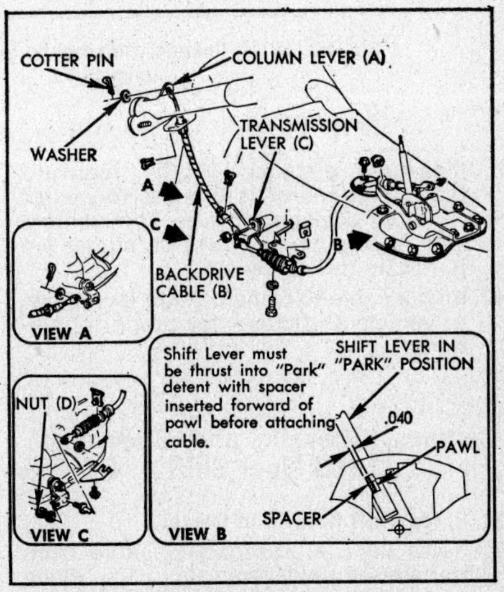

Fig. 10 Floor shift adjustment. 1973-75 Corvette

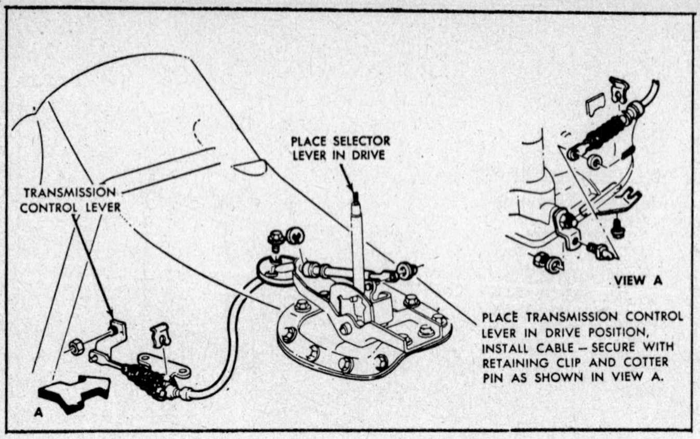

Fig. 11 **Shift linkage adjustment. 1969-72 Corvette**

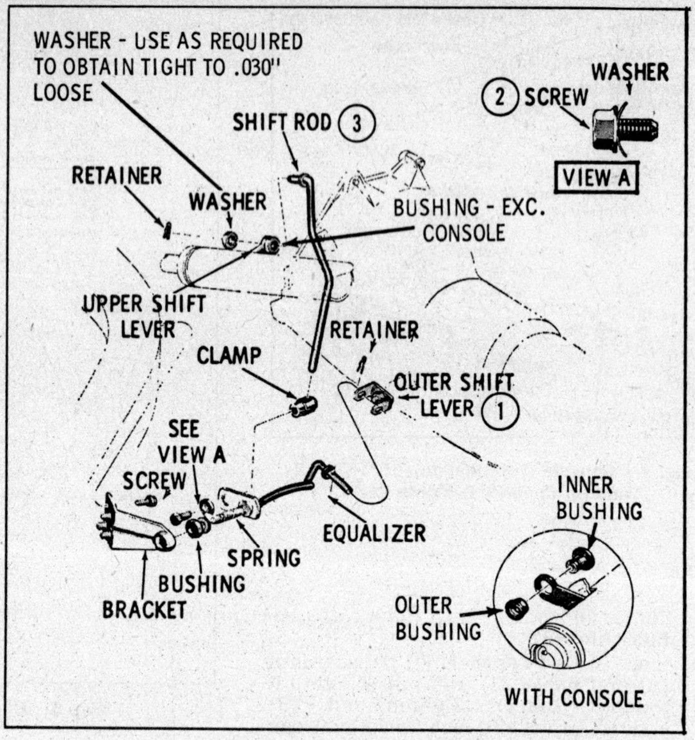

Fig. 12 **Shift linkage adjustment. Oldsmobile full size (typical)**

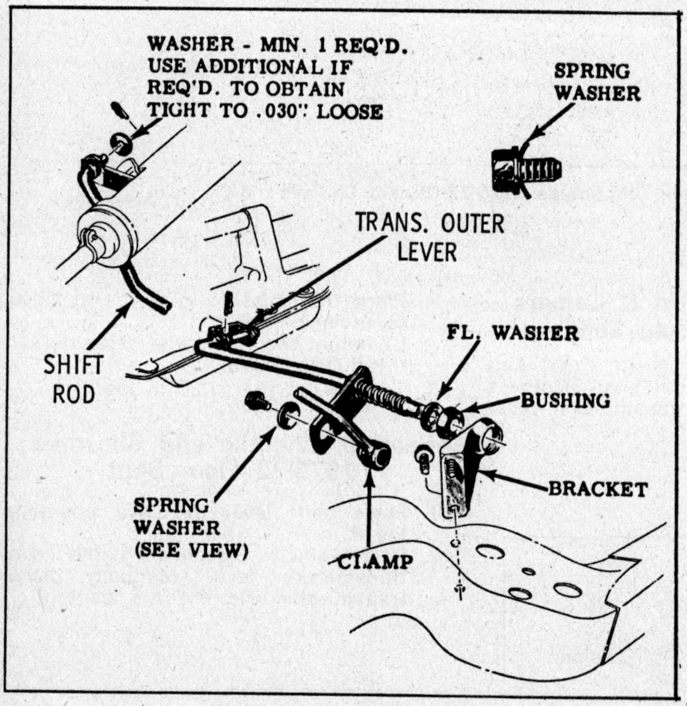

Fig. 13 **Shift linkage. Oldsmobile intermediates (typical)**

3. Measure distance from rearward face of attachment bracket to center of cable attachment pin. This dimension should be $5\frac{1}{2}$". If not, adjust pin to obtain this dimension.
4. Install cable to transmission lever, lower vehicle and check for proper operation.

Camaro, Chevelle and Chevrolet 1969 Floor Shift

1. Place shift handle in Drive.
2. Raise vehicle. Disconnect cable from transmission lever, Fig. 9. Place transmission in Drive range. Check dimension A and adjust by loosening stud nut and moving stud to obtain proper setting. Reinstall cable and lock clip.
3. Lower vehicle and remove shift quadrant cover. Raise quadrant plate, disconnect bulbs and remove plate. Remove cable clip and disengage cable from shift lever.
4. Insert a .07" gauge between pawl and detent as shown in view A. Check dimension B. Adjust by loosening bolt A and moving lever to obtain proper adjustment. Tighten bolt.
5. Adjust cable end until it freely enters pin. Install clip.

NOTE: If lift on handle button does not clear detents or detents can be cleared without lifting handle, pawl engagement can be adjusted by raising or lowering of the detent plate after loosening bolt B. If such an adjustment is made, repeat step 4.

---

6. Remove gauge. Check operation.
7. Reinstall shift plate and cover.

Corvette

1973-75

1. Loosen transmission lever nut.
2. Move transmission lever counterclockwise to its maximum position, then clockwise 5 detent positions to Park.
3. Place shift lever in Park and insert a .040" spacer forward of pawl as shown in Fig. 10 and tighten nut to 20 ft. lbs.

1969-72

1. Place selector lever in Drive, Fig. 12.
2. Place transmission control lever in Drive position by moving lever counterclockwise to low detent, then clockwise two positions for Drive.
3. Install cable and secure with retaining clip and cotter pin.

Oldsmobile

Column Shift, 1971-75

1. Loosen shift rod clamp bolt and place transmission outer lever in Neutral position, Figs. 12 and 13.
2. Push on shift rod until selector lever

is against Neutral position stop in upper steering column.

3. Tighten bolt in clamp on lower end of shift rod to 20 ft. lbs.
4. Check for proper operation.

Column Shift, 1969-70

1. Use enough washers to obtain zero clearance at upper shift lever and shift rod, Figs. 12 and 13.
2. Set outer shift lever in "D".
3. Loosen swivel bolt.
4. Push up on shift rod until selector lever is against the "D" position stop in upper steering column.
5. Tighten swivel bolt to 20 ft-lbs.
6. Check neutral switch adjustment.

Console Shift, 1969-71

1. Place shift lever in Park, Fig. 14.
2. Set transmission outer shift lever in Park position.
3. Set pin to just enter hole in shift cable.
4. Tighten nut and check through all detent positions and recheck adjustment.

Console Shift, 1972-75

1. Loosen shift rod clamp bolt. Fig. 14A.
2. Place selector lever in Park position. Place transmission lever in Park position with ignition key in Lock position.
3. Hold shift rod against Lock stop and tighten clamp screw.
4. Set pin to give free pin fit and tighten nut.
5. Check for proper operation.

Pontiac

Column Shift, 1969-75

1. Loosen screw on adjusting swivel clamp.
2. Set transmission selector lever in Park detent.
3. Set upper gearshift lever in Park position and lock ignition.
4. Push up on gearshift control rod to take up clearance in steering column lock mechanism and tighten screw on adjusting swivel clamp to 20 ft. lbs.

Console Shift, 1969-75

1. Disconnect shift cable from transmission.
2. Adjust back drive linkage as outlined at the front of this section.
3. After adjusting back drive, unlock ignition and set transmission and gear selector in Neutral position.
4. Install cable and tighten nut to 30 ft-lbs.

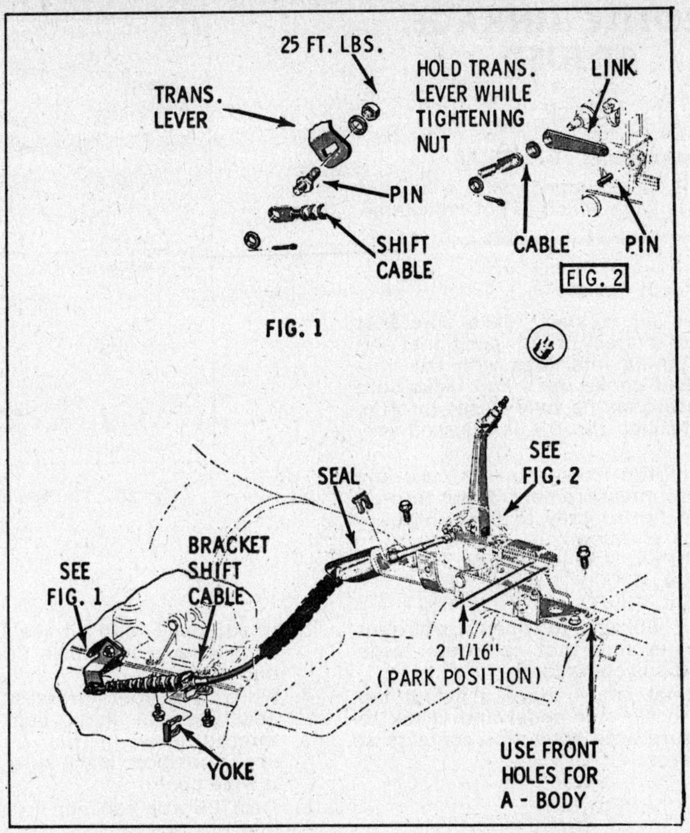

Fig. 14 Console shift adjustment. 1969-71 Oldsmobile

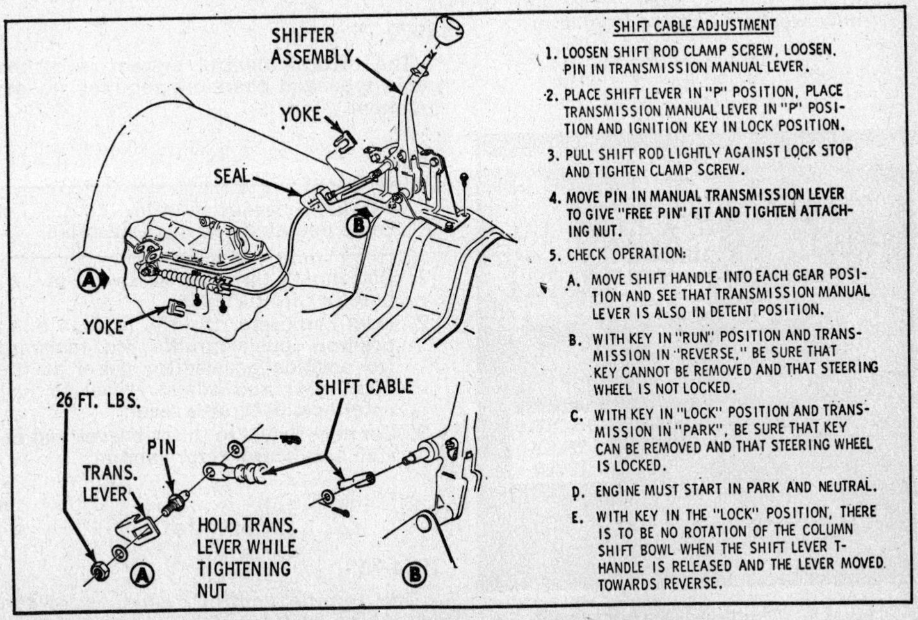

Fig. 14A Console shift adjustment. 1972-75 Oldsmobile

THROTTLE LINKAGE, ADJUST

Buick

1969-70 Special and 1971-75 All

This unit is equipped with a flexible cable type linkage which is not adjustable.

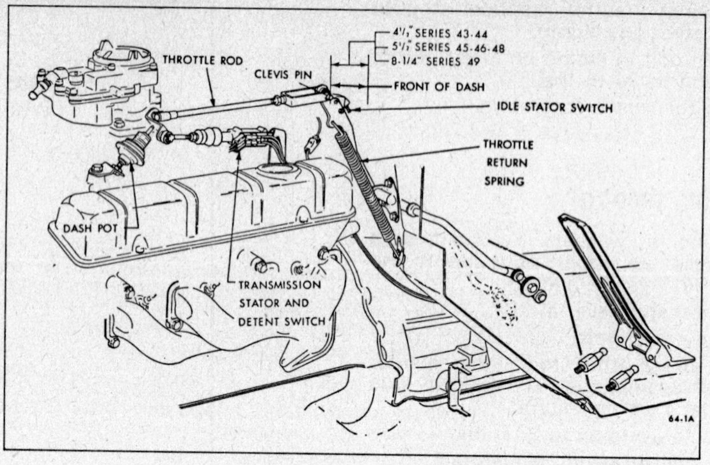

Fig. 15 Throttle linkage adjustment. 1969-70 Buick except Special

1969-70 Except Special

1. Remove air cleaner. Make sure that linkage is free in all positions and that nothing interferes with the linkage. Hold choke open and make sure that return spring fully closes throttle, even through throttle is released very slowly.
2. With throttle linkage in hot curb idle position, measurement from throttle rod pin horizontally to dash, Fig. 15, must be as follows:
   1969-70 Ser. 45, 46, 48000 . 5⅞"
   1969 Ser. 49000 . . . . 8¼"
   1970 Ser. 49000 . . . . 6⅛"
3. Operate linkage to open carburetor and make sure that carburetor wide open stop is contacting.
4. As a final check, have a helper depress accelerator pedal and check to make sure wide open stop contacts at carburetor.

Cadillac

1971-75

Although a cable is used in place of the rod, procedures to adjust remain basically the same as previous models.

1969-70

1. Remove air cleaner. Check linkage for free movement in all positions, and check to see that return spring fully closes the throttle.
2. Remove cotter pin that holds end of throttle rod in relay lever and remove washers and rod from lever.
3. If equipped with Cruise Control, detach linkage at Cruise Control power unit end.
4. While a helper presses accelerator pedal to the floor, hold carburetor throttle lever in full throttle (wide open) position. Make sure choke valve is wide open.
5. Turn throttle rod end in either direction as necessary to allow free entry into bushing on relay lever Fig. 16.
6. With accelerator pedal released, reinstall washer on throttle rod and install rod into bushing in relay lever.
7. Install other washer and then the cotter pin.
8. With accelerator pedal pressed again to floor mat, recheck throttle for wide open position.
9. Install Cruise Control linkage (if equipped) and adjust if necessary.

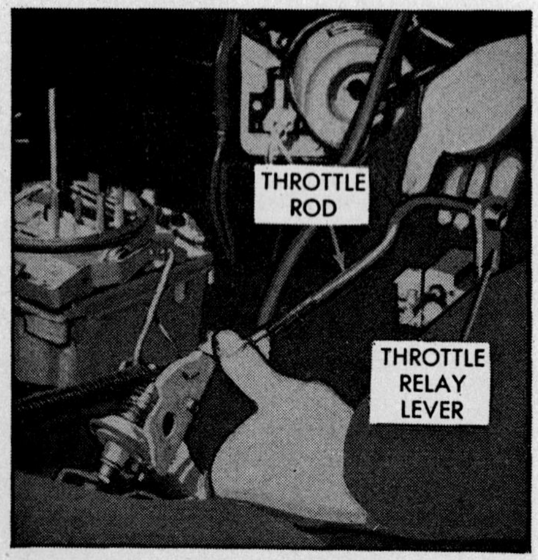

Fig. 16 Throttle linkage adjustment. Cadillac 1969-70

Chevrolet, Chevelle & Camaro

1970-75

The throttle control system is of the cable type and therefore requires no adjustment.

1969

NOTE: Chevelle throttle, control is cable operated and not adjustable.

1. Disconnect throttle rod swivel at carburetor throttle lever.
2. Hold carburetor throttle in wide open position, push throttle rod rearward (to position accelerator pedal at the floor mat) and adjust swivel to just enter hole in throttle lever.
3. Connect swivel to throttle lever and install accelerator return spring.

Corvette

1974-75

The throttle control system is of the cable type and therefore requires no adjustment.

1969-73

1. Loosen cable clamp bolt and hold accelerator pedal to the floor.
2. Move carburetor throttle to wide open position.
3. Tighten cable clamp bolt to 45 in lbs.

Oldsmobile

1970-75

The throttle control system is cable operated and therefore has no adjustment.

Carburetor Rod, 1969

1. With slow idle adjusted and adjusting screw off fast idle cam, disconnect swivel on carburetor rod from auxiliary bellcrank.
2. Push upper lever of auxiliary bellcrank towards cowl until it hits its stop.
3. Pull carburetor rod towards firewall until throttle is wide open.

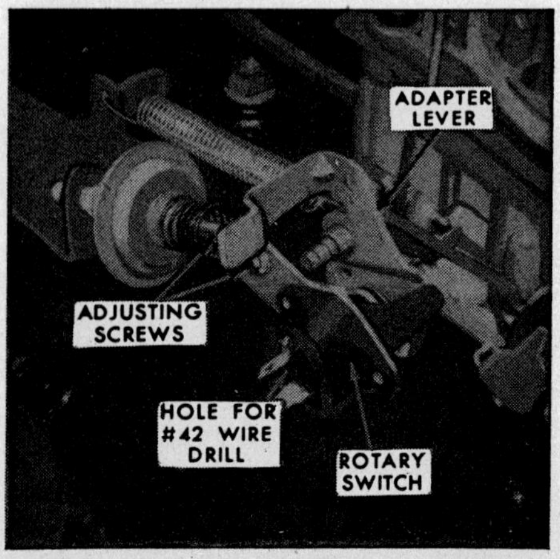

Fig. 17 Detent switch adjustment. 1970-71 Cadillac

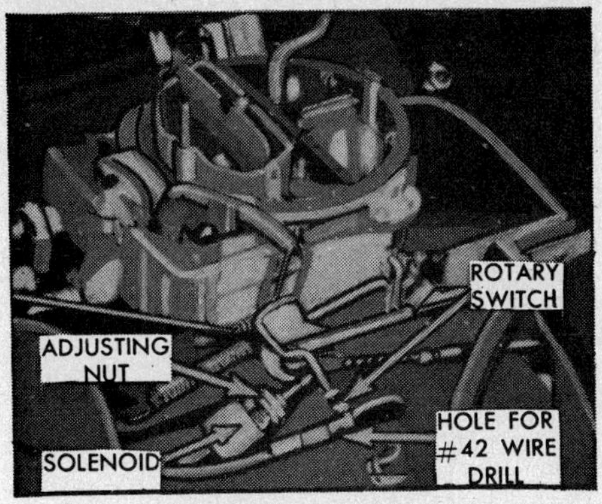

Fig. 18 Detent switch adjustment. 1972-75 Cadillac

4. Adjust swivel until swivel pin enters the "2" notch for 2 bbl. carburetor or the "4" notch for 4 bbl. carburetor.
5. Reconnect carburetor rod to auxiliary bellcrank.

Pontiac

1969 Intermediates: 1970-75 All

The throttle control system is of the cable type and requires no adjustment. On 1969-71 models, a reference dimension of $1\frac{9}{16}$ inch between the accelerator pedal and floor pan is used to check for bent bracket assemblies.

1969 Full Size Models

With accelerator rod and throttle return spring disconnected, insert $\frac{1}{4}''$ diameter gauge pin through hole in mounting bracket and through holes in throttle control lever. Holding carburetor lever extension at full throttle, adjust rod to line up with carburetor extension stud plus three additional turns toward front of car (counterclockwise). Attach rod, remove gauge pin and connect return spring.

DOWNSHIFT SWITCHES

Buick

Detent Switch, 1972-75

Push switch lever all the way towards dash. Final adjustment is made automatically the first time accelerator pedal is depressed to floor.

Detent Switch, 1971

1. Install cable through throttle lever hole and install retainer (nylon) being sure it is seated.
2. Position retainer on upper end of cable.
3. With throttle cable connected to carburetor and throttle lever, fully depress accelerator pedal to adjust retainer.

Detent Switch, 1969-70

With plunger bottomed in switch, fully depress accelerator pedal and tighten lock screw to 16 ft lbs.

Cadillac

1970-75

1. Remove air cleaner.
2. Make certain carburetor is adjusted to specification and that linkage is at low speed idle setting.
3. Loosen two mounting screws and insert a #42 drill through calibrating hole below lower wire terminal extending through to carburetor side of switch, Figs. 17 and 18. Adjust position of switch so that lever just touches the carburetor adapter plate arm.
4. Tighten mounting screws and remove drill.
5. Install air cleaner.

1969

1. Remove carburetor air cleaner.
2. Make sure that low idle speed is properly adjusted.
3. If downshift switch is properly adjusted, a $\frac{1}{8}''$ drill or rod can be inserted through calibrating hole below wire terminal extending through carburetor side of switch, Fig. 19.

NOTE: With this adjustment the stator should break contact $5\frac{1}{2}°$ to $7\frac{1}{2}°$ from closed hot idle throttle and make contact at 40° throttle. The downshift should make contact above 60° throttle.

---

4. If adjustment is necessary, loosen the two $\frac{7}{16}''$ switch mounting screws and position switch for proper alignment as in Step 3.
5. With switch positioned, tighten mounting screws and remove drill or rod from calibrating hole through switch.

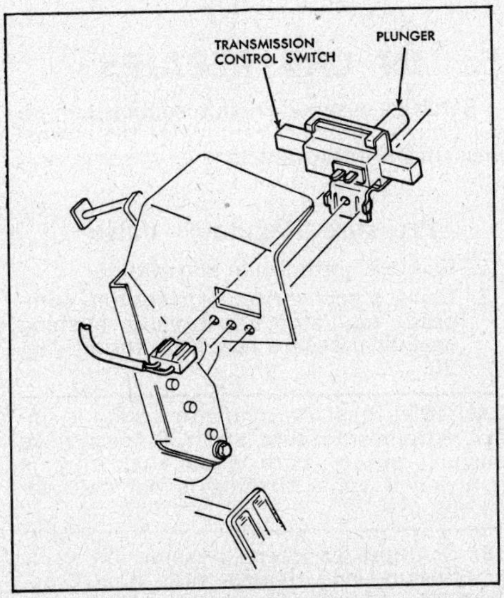

Fig. 20 Detent switch adjustment. 1971-75 Chevrolet Line

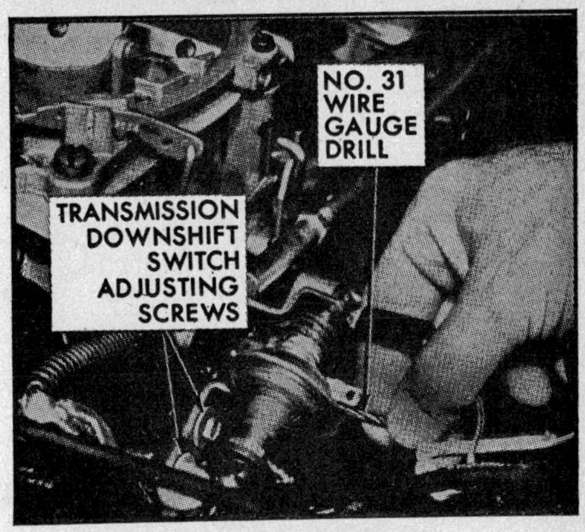

Fig. 19 Downshift adjustment. Cadillac 1969

Chevrolet Line

1971-75

Install switch as shown in Fig. 20. After installing, press switch plunger as far forward as possible. This switch will then adjust itself the first time the accelerator pedal is pushed to floor.

Camaro, Chevrolet and Nova

1970

1. Refer to Fig. 21 and loosen mounting bolts.
2. With choke in open position, move accelerator to wide open throttle position. Depress detent switch plunger till it bottoms in switch. Move switch towards throttle lever paddle until there is only .23″ between lever paddle and plunger.
3. Tighten mounting bolts.

Chevrolet (Except V8-307) & Camaro

1969

1. Loosen mounting bolts.
2. With automatic choke and accelerator linkage in wide open position, depress detent switch plunger until it bottoms in switch. Move switch toward throttle lever paddle to obtain a clearance of .23″ between lever paddle and plunger, Fig. 21.
3. Tighten mounting bolts.

Chevrolet V8-307

1969

1. Loosen adjusting nut, Fig. 22.
2. With automatic choke in wide open position, turn adjusting nut to bring switch forward until plunger is com-

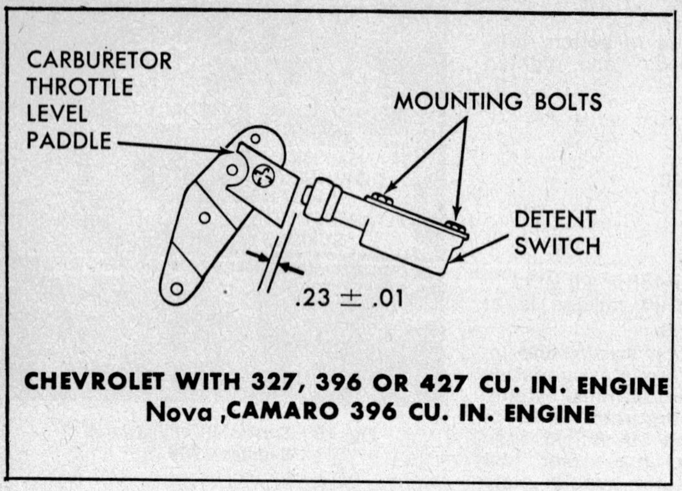

Fig. 21 Detent switch adjustment. Camaro and Chevrolet except V8-307

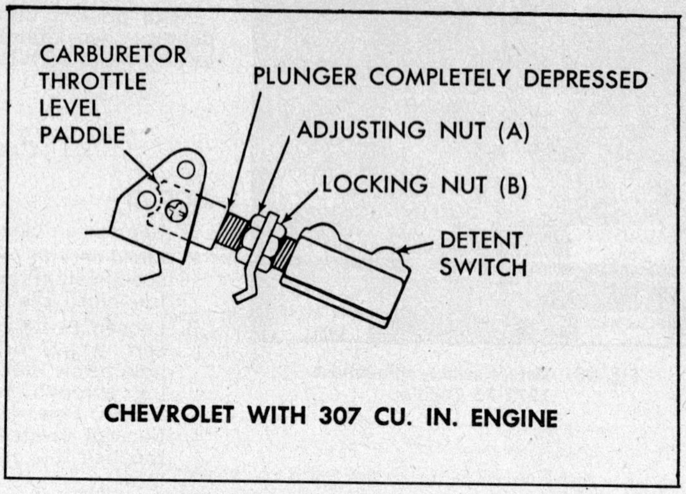

Fig. 22 Detent switch adjustment. Chevrolet V8-307

pletely depressed and threaded barrel of switch contacts throttle lever paddle.

3. Tighten lock nut.

Chevelle and Corvette

1969-70

1. Pull detent switch driver rearward until the hole in the switch body aligns with the hole in the driver. Insert a .092" pin through the aligned holes to a depth of .10" to hold driver in position, Figs. 23 and 24.
2. Loosen mounting bolt.
3. With accelerator pedal in wide open position, move switch forward until driver contacts accelerator lever.
4. Tighten mounting bolt and remove pin.

Oldsmobile

1969-75

1. Push plunger of switch forward until flush with switch housing.
2. Push accelerator pedal to wide open position to set switch.
3. Energizing of switch can be checked with a test light.

Pontiac

1969-75 Exc. 1969 Firebird

After installing switch, fully bottom plunger to insure proper setting then fully depress accelerator pedal, Fig. 25.

1969 Firebird

1. Loosen lock nut. Switch is mounted at carburetor and should not be confused with the idle stop solenoid.
2. With automatic choke in wide open position, adjust switch so that when throttle is in wide open position, detent circuit is closed.

IN CAR REPAIRS

Services outlined in this section can be performed without removing the transmission from the vehicle.

Pressure Regulator Valve

1. Remove bottom pan and strainer.
2. Using a screwdriver or steel rod, compress regulator boost valve bushing against pressure regulator spring, Fig. 26.

CAUTION: Pressure regulator spring is under extreme pressure and will force valve bushing out of bore when snap ring is removed if valve bushing is not held securely.

3. Continue to exert pressure on valve bushing and remove snap ring. Gradually release pressure on valve bushing until spring force is exhausted.
4. Carefully remove regulator boost valve bushing and valve, and pressure regulator spring. Be careful not to drop parts as they will fall out if they are not held.
5. Remove pressure regulator valve and spring retainer. Remove spacers if present.
6. Reverse procedure to install.

NOTE: A solid type pressure regulator valve must be used only in a pump cover with a "Squared Off" (machined) pressure regulator boss. A pressure regulator valve with oil holes and an orifice cup plug may be used with either type pump.

Control Valve Body

1. Remove bottom pan and strainer.
2. Disconnect pressure switch lead wire.
3. Remove control valve body attaching screws and detent roller spring assembly. *Do not remove solenoid attaching screws.*
4. Remove control valve body and governor pipes. If care is used in removing control valve body, the six check

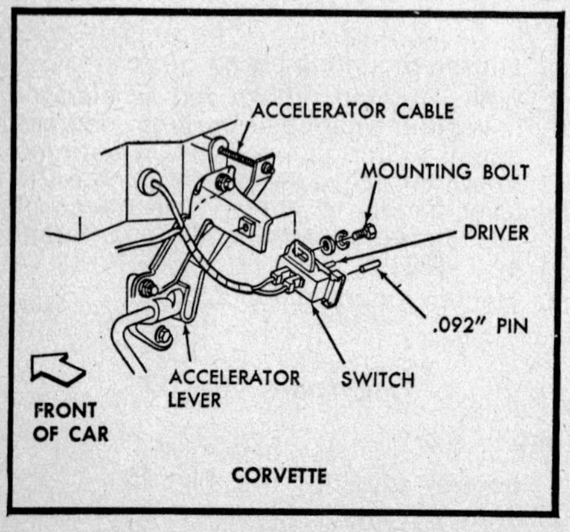

Fig. 23 Detent switch adjustment. 1969-70 Corvette

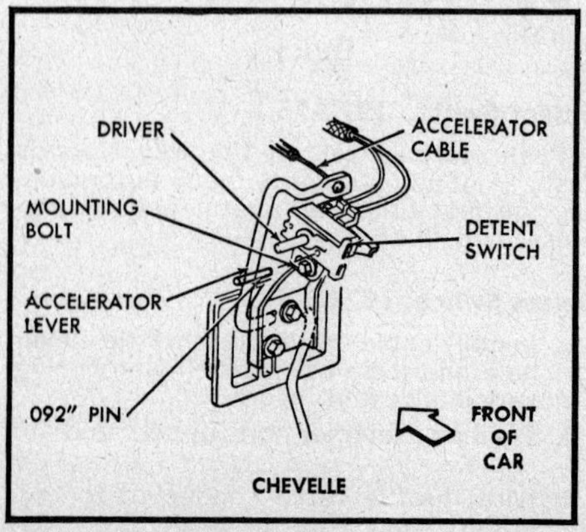

Fig. 24 Detent switch adjustment. 1969-70 Chevelle

balls will stay in place above spacer plate.
5. Remove governor pipes and manual valve from control valve body.
6. Reverse procedure to install.

Governor

1. Remove governor cover and discard gasket.
2. Withdraw governor from case.
3. Reverse procedure to install, using a new gasket.

Modulator & Modulator Valve

1. Remove modulator attaching screw and retainer.
2. Remove modulator assembly from case and discard O-ring seal.
3. Remove modulator valve from case.
4. Reverse procedure to install, using a new O-ring seal.

Parking Linkage

1. Remove bottom pan and oil strainer.
2. Unthread jam nut holding detent lever to manual shaft.
3. Remove manual shaft retaining pin from case.
4. Remove manual shaft and jam nut from case.
5. Remove O-ring seal from manual shaft.
6. Remove parking actuator rod and detent lever assembly.
7. Remove parking pawl bracket, pawl return spring and pawl shaft retainer.
8. Remove parking pawl shaft, O-ring seal and parking pawl.
9. Reverse procedure to install, using new seals and gasket.

Rear Seal

1. Remove propeller shaft.
2. Pry out seal with screwdriver.
3. Install new seal with a suitable seal driver.
4. Install propeller shaft.

TRANSMISSION, REPLACE

Buick, 1969-75

1. Raise and support front and rear of car.
2. Disconnect exhaust crossover pipe, if necessary, and on 1975 models, remove catalytic converter support bracket from transmission.
3. Place suitable jack under transmission.
4. Remove vacuum line from vacuum modulator.
5. Separate cooler lines from transmission.
6. Remove transmission crossmember.
7. Remove detent electrical connector from transmission case.
8. Disconnect speedometer cable.
9. Disconnect shift linkage from transmission.
10. Remove transmission filler pipe.
11. Support engine at oil pan.

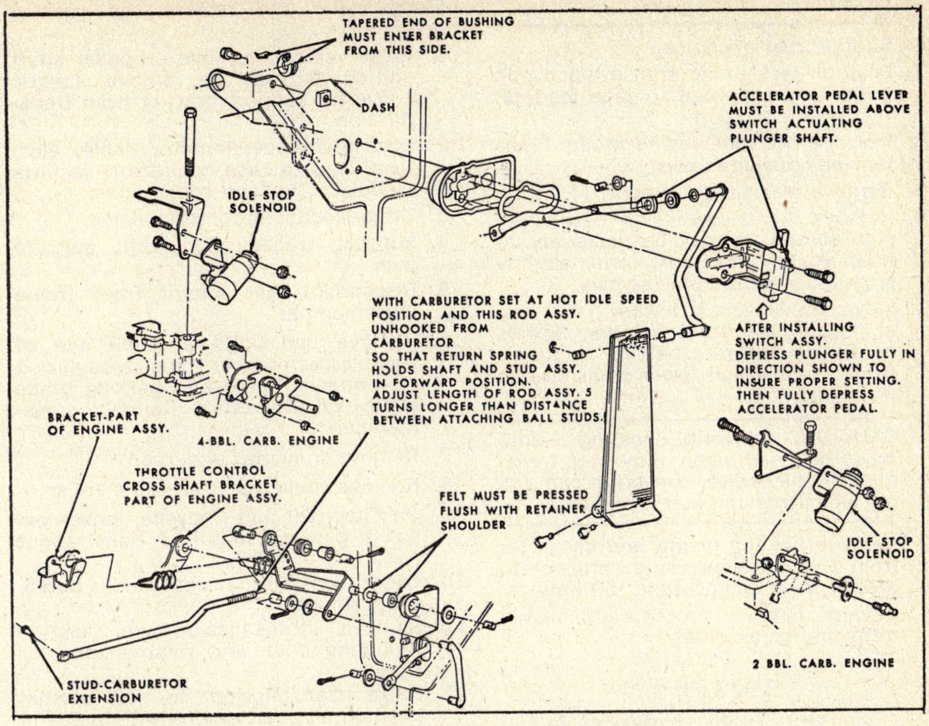

Fig. 25 Detent switch adjustment. 1969-75 Pontiac & 1970-75 Firebird (Typical)

12. Remove flywheel cover pan.
13. Mark flywheel and converter pump for reassembly in same position, then remove three converter pump-to-flywheel bolts.
14. Remove transmission-to-engine bolts.
15. Move transmission rearward to provide clearance between converter pump and crankshaft. Install a suitable holding tool to secure converter. Then lower and remove transmission.
16. Reverse above procedure to install.

Cadillac
1969-72

1. Disconnect negative battery cable.

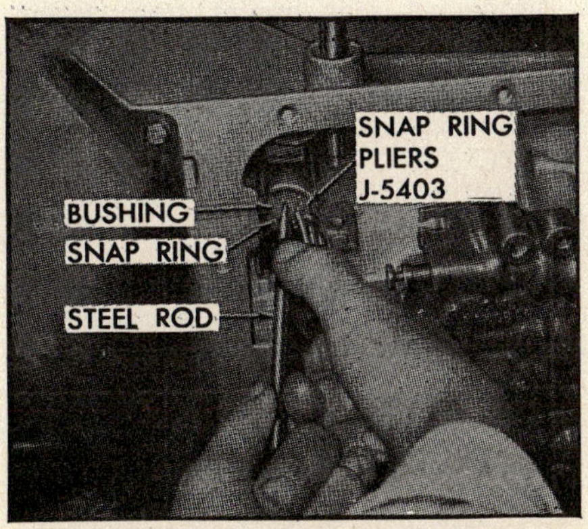

Fig. 26 Removing and installing Pressure regulator valve

2. Raise car on hoist or place on jack stands.
3. Disconnect relay rod from trunnion lever and wire relay rod up out of the way to prevent damage while removing transmission.
4. Remove two screws and bearing from frame side rail.
5. Disconnect trunnion from manual yoke on left side of transmission.
6. Remove speedometer drive cable and disconnect detent wires.
7. Remove transmission filler tube bracket screw from right exhaust manifold.
8. Remove filler tube from transmission case and plug hole in case.
9. Disconnect oil cooler pipes at transmission. Cap pipes and plug connector holes in transmission. Position cooler pipes out of the way.
10. Disconnect vacuum pipe hose from vacuum modulator and position pipe out of the way.
11. Remove resonator support bracket from extension housing.
12. Remove propeller shaft.
13. Unfasten and remove two starter motor brackets and slide starter forward.
14. Remove lower flywheel housing cover and two engine-to-transmission struts.
15. Remove three converter-to-flex plate screws.

NOTE: This is done by inserting a heavy screwdriver in open slot under one of weld nuts on converter, and rotating converter and flex plate until bolts can be reached for removal. Do not pry on flex plate ring gear to rotate converter as flex plate might be damaged.

16. Support rear of engine.
17. Position jack under transmission and raise it just enough to take load off rear engine support.
18. Remove two rear engine mount-to-extension housing screws.
19. Remove rear engine support (4 bolts).
20. Remove six transmission case-to-engine screws. It may be necessary to lower engine and transmission slightly to gain access to upper screws.
21. Move transmission toward rear, disengaging case from locating dowels on engine. Install a holding clamp on front of case and lower transmission from car.

CAUTION: Converter holding clamp must be used when removing transmission otherwise converter can fall out when transmission is removed.

22. Remove holding clamp and converter from transmission, using care not to drop it as it weighs about 50 pounds.
23. Reverse removal procedure to install transmission assembly.

1973-75

1. Disconnect negative battery cable and raise vehicle on hoist.
2. Disconnect transmission linkage and remove speedometer drive cable.
3. Disconnect downshift solenoid connector and Track Master electrical harness, if equipped.
4. Disconnect oil cooler pipes at transmission and cap pipes and plug holes in transmission.
5. Disconnect vacuum modulator hose.
6. Remove starter mount bolts and starter bracket and slide starter forward.
7. Remove propeller shaft.
8. Remove lower flywheel housing cover and three converter to flex plate bolts.

NOTE: This is done by inserting a heavy screwdriver in open slot under one of the welded nuts on converter, and rotating converter and flex plate until bolts can be reached for removal. Do not pry flex plate ring gear to rotate converter as flex plate damage may result.

9. Support rear of engine and remove rear engine mount bolts.
10. Place a jack under transmission and support transmission to remove load from rear engine support.
11. Remove bolts from rear engine support and position out of way.
12. Remove six engine to transmission case bolts. To gain access to upper bolts, lower engine and transmission slightly.
13. Move transmission toward rear of vehicle, disengaging locating dowels and install converter holding tool. Lower transmission from vehicle.
14. Remove converter holding tool and remove converter from shaft.
15. Reverse procedure to install.

Chevrolet Line 1970-75

Before raising car, disconnect negative battery cable and release the parking brake.

1. Raise vehicle, remove propeller shaft and on 1975 models, remove catalytic converter support bracket from transmission.
2. Disconnect speedometer cable, electrical lead to case connector, vacuum modulator line and cooler lines.
3. Disconnect shift control linkage.
4. Support transmission with suitable jack.
5. Disconnect rear mount from frame crossmember.
6. Remove two bolts at each end of frame crossmember (plus through bolt at inside of frame and parking brake pulley on Corvette). Remove crossmember.
7. Remove converter underpan.
8. Remove converter to flywheel bolts.
9. On Chevrolet and Chevelle, loosen exhaust pipe to manifold bolts about ¼".
10. Lower transmission until jack is barely supporting it.
11. Remove transmission to engine mounting bolts and remove oil filler tube.
12. Raise transmission to its normal position, support engine with jack and slide transmission rearward from engine and lower it away from car.

NOTE: Use converter holding tool when lowering transmission or keep rear of transmission lower than front so as not to lose converter.

13. Reverse removal procedure to install transmission.

Chevrolet Line 1969

1. Raise car on hoist.
2. On Camaro models, disconnect parking brake cables, and left exhaust pipe from manifold. Remove underbody reinforcement (convertible models only).
3. On Corvette only, remove both exhaust pipes.
4. Remove propeller shaft.
5. Disconnect speedometer cable, electrical leads, vacuum line and oil cooler pipes.
6. Disconnect shift control linkage.
7. Support transmission with suitable jack.
8. Disconnect rear mount from frame crossmember, remove two bolts at each end of crossmember. Remove crossmember.

NOTE: On Corvette models, through bolt at inside of frame and parking brake pulley must also be removed before removing crossmember.

9. Remove converter under pan.
10. Remove converter to flywheel bolts.
11. On Chevrolet and Chevelle, loosen exhaust pipe to manifold bolts to drop pipes approximately ¼".
12. Lower transmission until jack is barely supporting it, remove engine to transmission bolts, leaving one lower bolt installed. Remove oil filler tube.
13. Raise transmission to its normal position, support engine with jack, remove last attaching bolt and slide transmission rearward and lower it away from vehicle.

CAUTION: Converter holding clamp must be used when removing transmission otherwise converter can fall out when transmission is removed.

14. Remove holding clamp and converter from transmission, using care as it weighs about 50 pounds.
15. Reverse removal procedure to install transmission.

Oldsmobile 1969-75

1. Remove flywheel cover and torque converter attaching bolts.
2. Mark flywheel and converter so they can be installed in same position.
3. Support engine at rear.
4. Disconnect solenoid wires and manual shift linkage at side of transmission.
5. Disconnect oil cooler lines, vacuum modulator line and oil filler pipe.
6. Disconnect parking brake cable.
7. Before removing propeller shaft, scribe marks on drive shaft and companion flange for correct assembly.
8. Disconnect exhaust pipe bracket from rear of crossmember and on 1975 models, remove catalytic converter support bracket from transmission, if necessary.
9. Support transmission, then remove crossmember.
10. Unfasten transmission from engine.
11. Move transmission away from engine, then, before removing transmission, fasten a suitable piece of strap iron to housing to prevent converter from falling out as transmission is removed.

Pontiac 1969-75

1. Disconnect battery and release parking brake. Then raise car.
2. Remove propeller shaft and on 1975 models, remove catalytic converter support bracket from transmission.
3. Disconnect speedometer cable, electrical lead to case connector, vacuum line at modulator, and oil cooler pipes.
4. Disconnect shift control linkage.
5. Support transmission with jack.
6. Disconnect rear mount from transmission and crossmember.
7. Remove crossmember (2 bolts at each end).
8. Remove converter dust shield.
9. Remove converter-to-flex plate bolts.
10. Loosen exhaust pipe to manifold about ¼", and lower transmission until jack is barely supporting it.
11. Remove transmission - to - engine mount bolts.
12. Raise transmission to its normal position, slide it rearward and lower it away from vehicle.

NOTE: When lowering transmission, keep rear of unit lower than front so as not to drop converter.

13. Reverse procedure to install.

GM FRONT WHEEL DRIVE TURBO HYDRA-MATIC

TRANSMISSION IDENTIFICATION

CADILLAC ELDORADO

1969-74 . . . . AJ

OLDSMOBILE TORONADO

1969-70 With High Perf. Eng. . . . . OM

1969-75 Exc. Below . . . . OJ

1974-75 California . . . . OM

DESCRIPTION

This transmission is a fully automatic unit used for front wheel drive applications, Fig. 1. It consists primarily of a three-element hydraulic torque converter, dual sprocket and chain link assembly, compound planetary gear set, three multiple disc clutches, a sprag clutch, a roller clutch, two band assemblies, and a hydraulic control system.

Torque Converter

The torque converter consists of a pump or driving member, a turbine or driven member and a stator or reaction member. The stator is mounted on a one-way roller clutch which allows it to overrun when not used as a reaction member.

The torque converter couples the engine to the planetary gear set through the use of a drive sprocket, a chain link assembly, and a driven sprocket. Clockwise engine torque turns the drive sprocket clockwise. This, in turn, drives the driven sprocket in a clockwise direction. This in effect is a reverse in the direction of engine torque due to the side mounting of the gear unit.

Planetary Gear Set

The gear set provides three forward ratios and reverse. The approximate gear ratios are: First 2½ to 1, second 1½ to 1, third 1.1 to 1, reverse 2.1 to 1. Second and third are also multiplied by a lesser degree.

Converter stall ratio, first gear (2½ x 2) equals 5 to 1. Converter stall ratio, reverse (2.1 x 2) equals 4.2 to 1.

External Controls

External control connections to the transmission are: a) Engine vacuum, b) 12-volt electrical signals, c) manual linkage control.

Engine vacuum is used to operate the vacuum modulator assembly. The vacuum modulator automatically senses any change in torque input to the transmission that the driver induces through a change in accelerator position.

On all models an electrical signal is used to operate an electrical solenoid. The solenoid is activated by a switch at the carburetor. When the throttle is opened sufficiently to close this switch, the solenoid in the transmission is activated, causing a downshift at speeds below approximately 70 mph. At lower speeds, downshifts will occur at lesser speeds without use of the switch.

TROUBLE SHOOTING GUIDE

NOTE: In many of the following diagnosis procedures, it is recommended that air pressure be applied to help in determining if the seal, rings or pistons are stuck, missing or damaged. Therefore, when air is applied, listen carefully for escaping air and piston action as air is applied to a particular area.

No Drive In "D" Range

1. Low oil level. Check for external leaks or vacuum modulator diaphragm leaking.
2. Manual linkage maladjusted. Correct alignment in manual lever shift quadrant.
3. Low oil pressure.
4. Oil strainer O-ring seal missing or damaged, neck weld leaking, strainer blocked.
5. Oil pump pressure regulator stuck or inoperative. Pump drive gear tangs damaged by converter.
6. Case porosity in intake bore.
7. Control valve. Manual valve disconnected from manual lever pin. (Other shift lever positions would also be affected.)
8. Forward clutch does not apply. Piston cracked; seals missing or damaged. These defects can be checked by removing the valve body and applying air pressure to the drive cavity in the case valve body face. Missing, damaged or worn oil rings on driven support housing can also be checked in this manner at the same time because they can also cause the forward clutch not to apply. Clutch plates burned.
9. Roller clutch inoperative. Rollers worn, damaged springs, or damaged races. May be checked by placing selector lever in "L" range.

No Drive In "R" or Slips In Reverse

1. Low oil level.
2. Manual linkage.
3. Oil pressure. Vacuum modulator defective, modulator valve sticking.
4. Restricted strainer, leak at intake pipe or O-ring seal. Pressure regulator or boost valve sticking.
5. Control valve body gaskets leaking or damaged (other malfunctions may also be indicated). Low-reverse check ball missing from case (this will cause no overrun braking in low range). The 2-3 valve train stuck open (this will also cause 1-3 upshifts in drive range). Reverse feed passage not drilled; also check case passages. Apply air to reverse passage in case valve body face.
6. Rear servo and accumulator. Servo piston seal ring broken or missing. Apply air pressure to drilled hole in intermediate clutch passage of case valve body face to check for piston operation and excessive leakage. Band apply pin too short (this may also cause no overrun braking or slip in overrun braking in low range).
7. Rear band burned, loose lining, apply pin or anchor pin not engaged; band broken.
8. Direct clutch outer seal damaged or missing. Clutch plates burned (may be caused by stuck ball check in piston).
9. Forward clutch does not release (will also cause drive in neutral range).

Drive In Neutral

1. Manual linkage maladjusted.
2. Forward clutch does not release (this condition will also cause no reverse).

1st Speed Only—No 1 - 2 Upshift

1. Governor valve sticking; driven gear loose, damaged or worn. If driven gear shows signs of wear or damage, check output flange drive gear for nicks or rough finish.
2. Control valve. The 1-2 shift valve train stuck closed. Dirt, chips or damaged valve in 1-2 shift valve train. Governor feed channels blocked or leaking; pipes out of position. Valve body gaskets leaking or damaged. Case porosity between oil channels. Governor feed passage blocked.
3. Intermediate clutch. Case center support oil rings missing, broken or defective. Clutch piston seals missing, improperly assembled, cut or damaged. Apply air to intermediate clutch passage located in case valve body face to check for these defects.

1 - 2 Shift Obtained Only At Full Throttle

1. Detent switch sticking or defective.
2. Detent solenoid loose, gasket leaking, sticks open, electrical wire pinched between cover and casting.
3. Control valve body gasket leaking or damaged. Detent valve train stuck.

1st & 2nd Speeds Only No 2 - 3 Shift

1. Detent solenoid stuck open (the 2-3 shift would occur at very high speeds) may be diagnosed as no 2-3 shift.
2. Detent switch sticking or defective.
3. Control valve body. The 2-3 valve train stuck with dirt or foreign material. Valve body gaskets leaking or damaged.
4. Direct clutch. Case center support oil rings missing or broken. Clutch piston seals missing, improperly assembled, cut or damaged; piston ball check stuck or missing. Apply air to direct clutch passage in case valve body face to check these conditions.

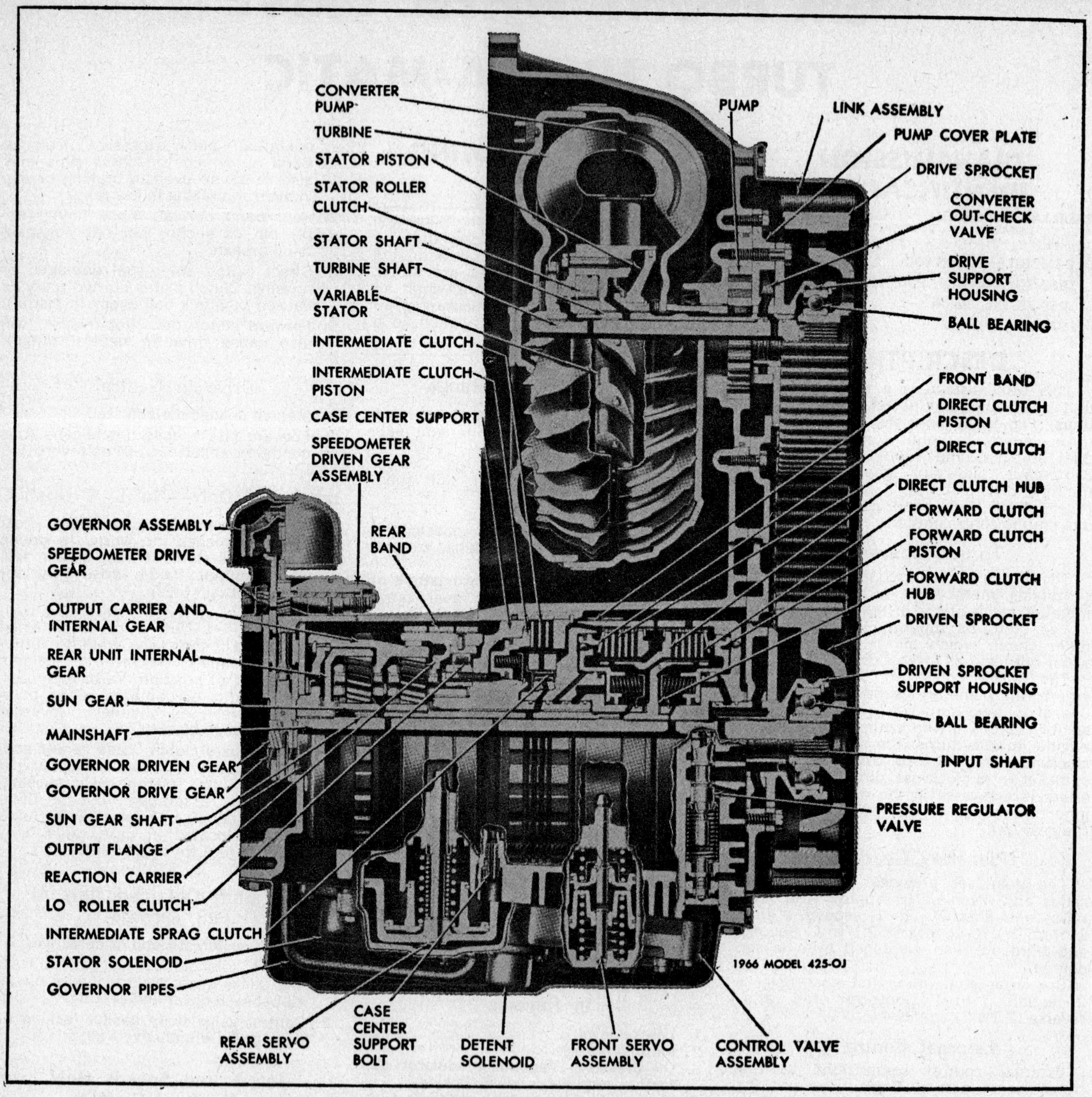

Fig. 1 General Motors Front Wheel Drive Turbo Hydra-Matic

Slips In All Ranges

1. Oil level incorrect.
2. Low oil pressure. Vacuum modulator defective or valve sticking. Oil strainer plugged or leaks at neck; O-ring (case to strainer) missing or damaged. Pressure regulator or boost valve sticking.
3. Case cross channel leaks; porosity.
4. Forward, intermediate and direct clutches slipping. Clutch plates burned. Always look for a primary defect that would cause clutch plates to burn. (Missing feed holes, seals and oil rings, etc., are primary defects).
5. Roller clutch rollers worn; springs or cage damaged, and worn or damaged races (operates normally in low and reverse ranges).

Slips 1 - 2 Shift

1. Oil level incorrect.
2. Low oil pressure. Look for defective vacuum modulator or valve sticking. Pump pressure regulator valve stuck.
3. Front servo accumulator piston cracked or porous, oil ring damaged or missing.
4. Control valve. The 1-2 accumulator valve train (may cause a slip-bump shift). Porous valve body or case valve body face.
5. Rear servo accumulator oil ring missing or damaged; case bore damaged;

piston cracked or damaged.
6. Case porous between oil passages.
7. Intermediate clutch lip seals missing, cut or damaged. Apply air pressure to intermediate clutch passage in case valve body face to check. Clutch plates burned. Case center support leaks in feed circuits (oil rings damaged or grooves damaged) or excessive leak between tower and bushing.

Rough 1 - 2 Shift

1. Oil pressure. Check vacuum modulator for loose fittings, restrictions in line; defective vacuum modulator. Modulator valve stuck. Pressure regulator boost valve stuck.
2. Control valve. 1-2 accumulator valve train; valve body-to-case bolts loose; gaskets inverted, off location, or damaged.
3. Case. Intermediate clutch passage check ball missing or not seating. Case porous between channels.
4. Rear servo accumulator piston stuck. Apply air pressure to 1-2 accumulator passage in case valve body face (you should hear the servo piston move). Broken or missing spring; bore scored or damaged.

Slips 2 - 3 Shift

1. Oil level high or low.
2. Low oil pressure. Modulator defective or valve sticking. Pump pressure regulator valve or boost valve sticking.
3. Control valve. Accumulator piston pin leak at valve body end.
4. Direct clutch piston seals leaking. Case center support oil seal rings damaged or excessive leak between tower and bushing. Apply air to direct clutch passage in case valve body face. If air comes out intermediate passage, center support is defective.

Rough 2 - 3 Shift

1. Oil pressure high. Vacuum modulator defective or valve sticking. Pump pressure regulator valve or boost valve stuck or inoperative.
2. Front servo accumulator spring missing or broken; accumulator piston stuck.

Shifts Occur at too High or too Low Car Speed

1. Oil pressure. Vacuum modulator defective or valve sticking. Leak in vacuum line (engine to transmission). Vacuum modulator line fitting on carburetor blocked. Pump pressure regulator valve or boost valve train stuck.
2. Governor valve stuck or sticking. Feed holes restricted or leaking; pipes damaged or mispositioned.
3. Detent solenoid stuck open or loose on valve body (will cause late shifts).
4. Control valve. Detent valve train sticking; 3-2 valve train sticking; 1-2 shift valve stuck; 1-2 detent valve sticking open (will probably cause early 2-3 shift).

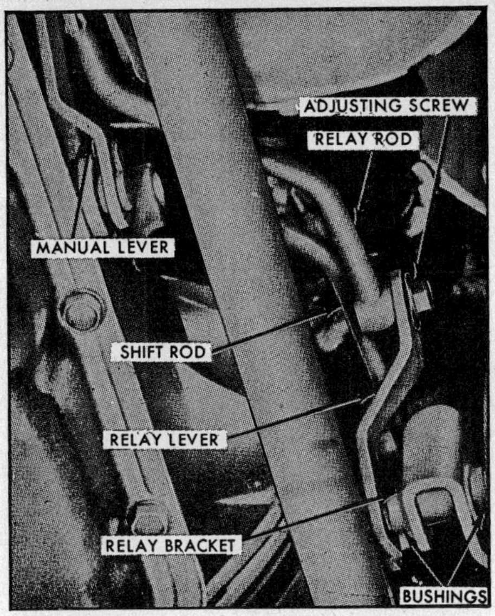

Fig. 2 Manual control linkage adjustment. Cadillac Eldorado (typical)

5. Spacer plate gaskets inverted or mispositioned; orifice holes missing or blocked; check balls missing or mislocated.
6. Case porous in channels or foreign material blocking channels.

No Detent Downshift

1. Detent switch mispositioned or electrical connections loose.
2. Solenoid defective or electrical connections loose.
3. Control valve detent valve train stuck.

No Engine Braking—Super Range 2nd Speed

1. Front servo or accumulator piston rings broken or missing. Case or valve body bores worn oversize, causing excessive leakage.
2. Front band worn or burned (check for cause); band end lugs broken or damaged; band lugs not engaged on anchor pins or servo apply pin (check for cause).

No Engine Braking—Low Range 1st Speed

1. Control valve low-reverse check ball missing from case.
2. Rear servo oil ring damaged or missing; piston damaged or porous, causing a leak in apply pressure.
3. Rear band lining worn or burned (check for cause); band end lugs broken; band ends not engaged on anchor pin or servo apply pin. These items will also cause slip in reverse or no reverse.

Will Not Hold Car In Park Position

1. Manual linkage maladjusted (external).
2. Parking brake lever and actuator rod assembly defective (check for proper actuator spring action). Parking pawl broken or inoperative.

Poor Performance or Rough Idle

1. Stator switch defective or maladjusted.
2. Stator solenoid defective or wire ground to solenoid housing; electrical connection loose; stator valve train stuck (located in valve body); oil feed circuit to stator restricted or blocked (check feed hole in stator shaft); converter-out check valve broken or missing (reed valve located in cover plate under drive support housing).
3. Turbine shaft converter return passage not drilled; oil seal rings broken, worn or missing.
4. Case porous in feed circuit channels or foreign material blocking feed circuit.
5. Converter assembly defective.

Transmission Noise

1. Pump noise. Oil level high or low; water in oil, driving gear assembled upside down; driving or driven gear teeth damaged.
2. Gear noise (1st gear drive range). Check planetary pinions for tooth damage. Check sun gear and front and rear internal gears for tooth finish or damage.
3. Clutch noise during application. Check clutch plates.
4. Sprocket and chain link assembly. Chain link too long (sounds similar to popcorn popping). There will be a rough burr along teeth of drive sprocket if chain link is too long; replace chain link and drive sprocket. Drive or driven sprocket teeth damaged. Engine mounts worn or damaged.

Burned Forward Clutch Plates

1. Check ball in clutch housing damaged, stuck or missing.
2. Clutch piston cracked, seals damaged or missing.
3. Low line pressure.
4. Manual valve mispositioned.
5. Restricted oil feed to forward clutch.
6. Pump cover oil seal rings missing, broken or undersize or ring groove oversize.
7. Case valve body face not flat or porosity between channels.
8. Manual valve bent and center land not ground properly.

Burned Intermediate Clutch Plates

1. Rear accumulator piston oil ring damaged or missing.
2. 1-2 accumulator valve stuck in control valve assembly.
3. Intermediate clutch piston seals damaged or missing.
4. Center support bolt loose.
5. Low line pressure.

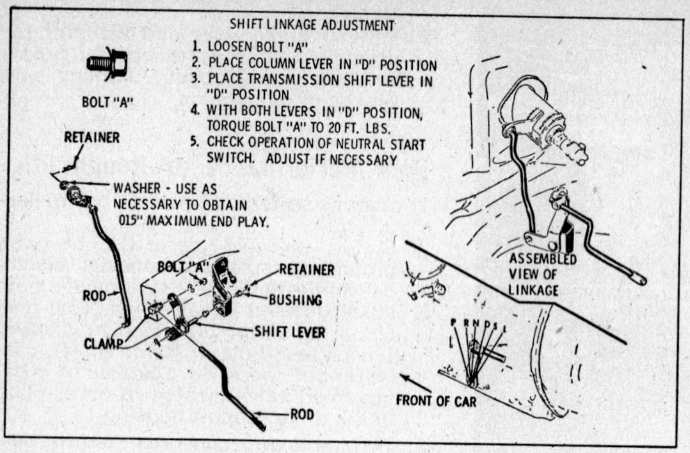

Fig. 3 Shift linkage adjustment. Toronado column shift, 1969-70

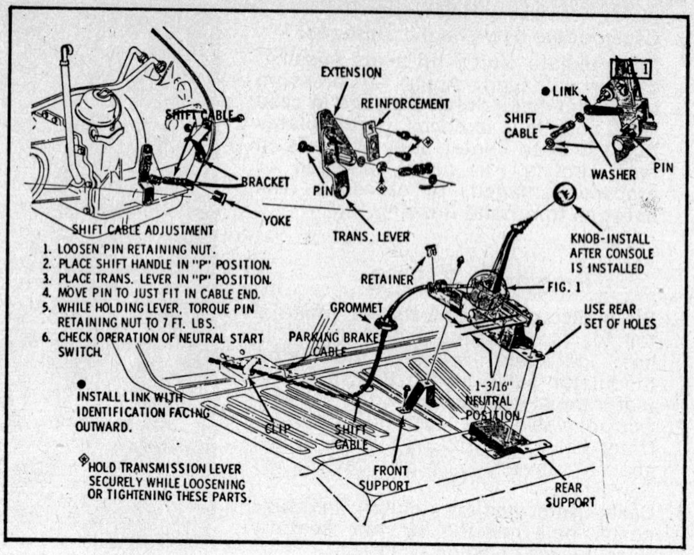

Fig. 4 Shift linkage adjustment. 1969-70 Toronado console shift

6. Intermediate clutch plug in case missing.
7. Case valve body face not flat or porosity between channels.
8. Manual valve bent and center land not ground properly.

Burned Direct Clutch Plates

1. Restricted orifice in vacuum line to modulator.
2. Check ball in direct clutch piston damaged, stuck or missing.
3. Defective modulator bellows.
4. Center support bolt loose.
5. Center support oil rings or grooves damaged or missing.
6. Clutch piston seals damaged or missing.
7. Front and rear servo pistons and seals damaged.
8. Manual valve bent and center land not cleaned up.
9. Case valve body face not flat or porosity between channels.
10. Intermediate sprag clutch installed backwards.

MAINTENANCE

Adding Oil

The fluid level should be checked at every engine oil change interval, and should be changed at 24,000 mile intervals. The fluid level should be checked with the selector lever in PARK position, engine running at idle speed and car on a level surface. The oil indicator and filler tube are located under the hood at the left front corner of the engine. *The filler tube comes out from the final drive housing but it is for the transmission.*

NOTE: If any work is performed on the transmission, it will require the following amounts of oil to bring the oil to the correct level:

1. Pan removed 5½ qts.

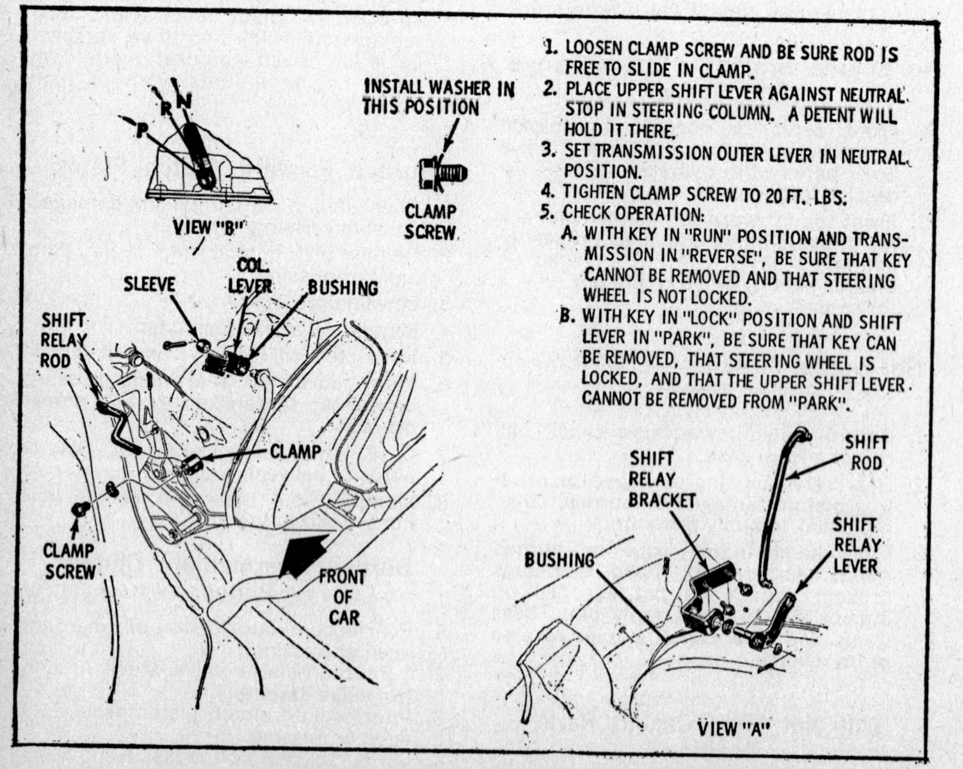

Fig. 4A Shift linkage adjustment. Toronado column shift, 1971-75

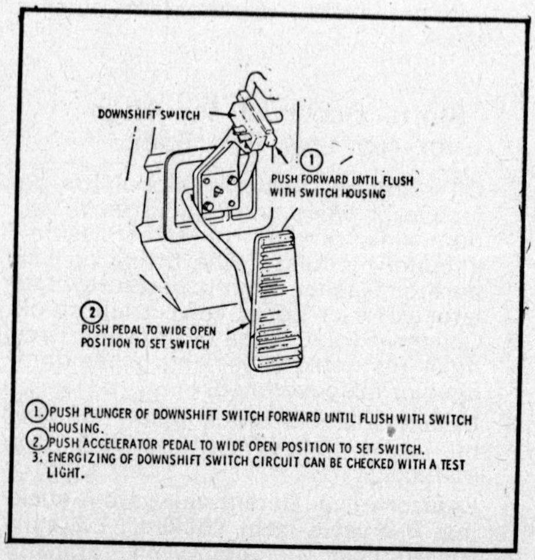

Fig. 4B Downshift switch adjustment. Oldsmobile Toronado

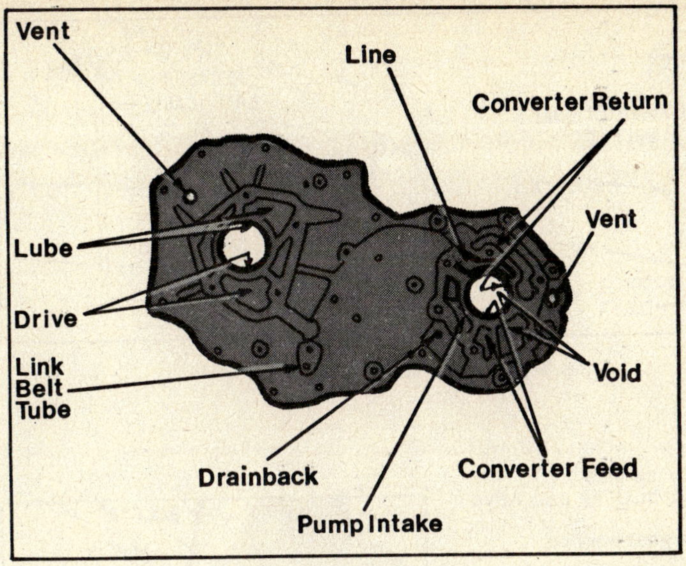

Fig. 5 Location of check balls

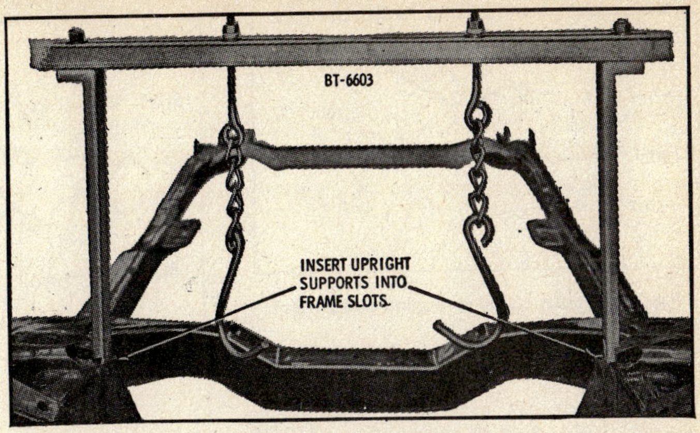

Fig. 6 Installing support bar. Olds Toronado

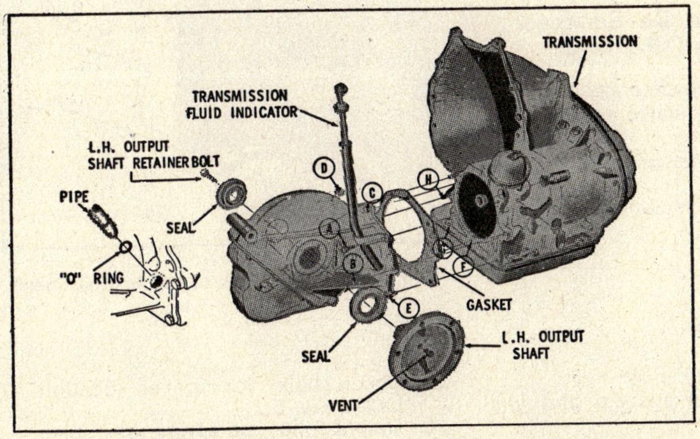

Fig. 7 Transmission attachment. 1969 Oldsmobile Toronado

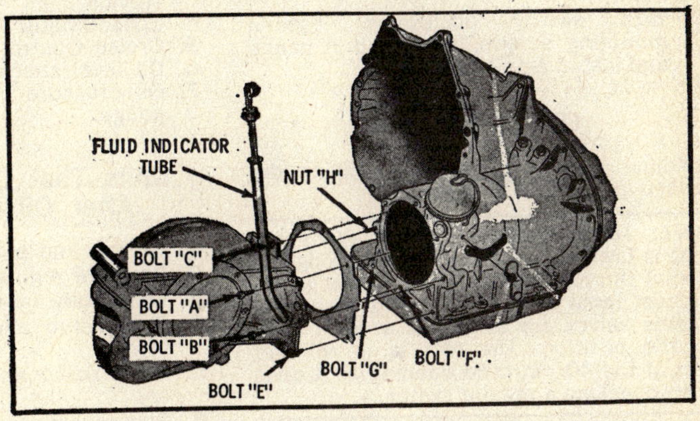

Fig. 7A Transmission attachment. 1970-75 Oldsmobile Toronado

2. Drive cover sprocket housing ½ qt.
3. Converter changed 3½ qts.
4. Total overhaul (total capacity) 13 qts.

Changing Oil

When changing transmission oil, first add 4 quarts, start the engine, and add oil to bring the fluid level to the FULL mark on the dipstick. Use only Dexron automatic transmission oil.

NOTE: Beginning with 1973 models, Cadillac uses an extended-life Dexron transmission fluid. With this new fluid, strainer replacement and fluid change is now recommended at 100,000 miles under normal operating conditions and 50,000 miles under severe or abnormal service such as trailer towing.

This recommendation applies only to the improved fluid and its availability for service. If the new fluid is not available, the former fluid can be used but then the 24,000 mile maintenance rule will apply.

Beginning with late 1973 models, Oldsmobile is using a revised type Dexron fluid. An early change to a darker color from the usual red color and or a strong odor that is usually associated with overheated fluid is normal, and should not be treated as a positive sign of needed maintenance or unit failure.

The normal maintenance schedule for drain and refill of this type fluid remains unchanged at 24,000 miles under normal service and 12,000 miles under severe operating conditions, such as trailer towing.

MANUAL LINKAGE, ADJUST

Cadillac Eldorado

1. Referring to Fig. 2, loosen adjusting screw on relay lever.
2. Pull relay rod up to position transmission shift valve in Park, then push rod down to the third (Neutral) step. Make sure rod is centered in this detent position.
3. Position selector lever in Neutral against quadrant stop in steering column.
4. Tighten relay rod adjusting screw, making sure shift lever is held against Neutral stop while this operation is being performed.

Olds Toronado

Make the adjustment as directed in Figs. 3, 4 and 4A.

DOWNSHIFT SWITCH, ADJUST

Cadillac Eldorado

1. Remove carburetor air cleaner.
2. Make sure that carburetor is properly adjusted and that throttle linkage is at low speed idle setting.
3. On 1969 models, if the downshift switch is properly adjusted, a #31 (wire gauge size) drill can be inserted in the calibrating hole below lower

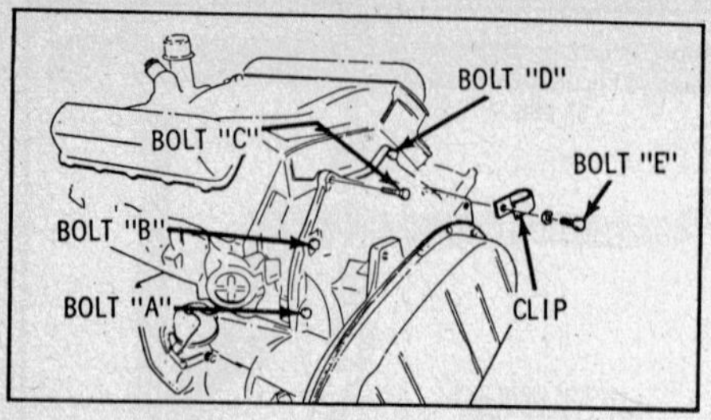

Fig. 8 **Transmission to engine attachment**

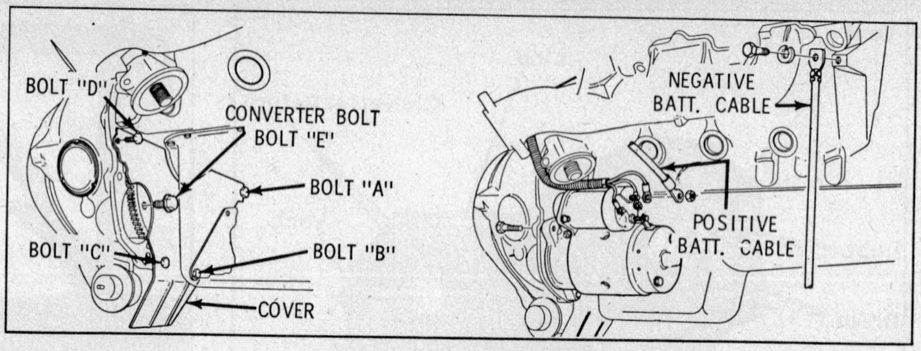

Fig. 9 **Converter attachment**

wire terminal extending through to carburetor side of switch. For 1970-75 models use a #42 (wire gauge size).

4. If adjustment is necessary, loosen the two switch mounting screws and position the switch for proper alignment.
5. With switch positioned, tighten mounting screws, remove drill gauge and install air cleaner.

Olds Toronado

Adjust detent downshift switch as described in Fig. 4B.

NOTE: Trailer Hauling Switch (Y-73 option) is bronze colored. When checking the special throttle switches with a test lamp, the test lamp *should light* only when the throttle valves are at idle or at wide open throttle position. The test lamp *will not light* at the 40° throttle opening as it does when checking a regular switch.

IN CAR REPAIRS

Operations Not Requiring Transmission Removal

1. Oil cooler fitting replacement or adjustment.
2. Governor assembly service.
3. Vacuum modulator, bushing and valve service.
4. Speedometer drive gear service.
5. Cruise Control service.
6. Oil level check.
7. Oil pressure check with oil pressure gauge.

Units That Can Be Serviced After Oil Pan Removal

1. Oil pan and pan-to-case gasket.
2. Pressure regulator valve assembly.
3. Valve body assembly.
4. Rear servo and accumulator assembly.
5. Front servo and accumulator assembly.
6. Governor pipes.
7. Detent solenoid.
8. Stator solenoid.
9. Solenoid connector.
10. Manual linkage.
11. Parking linkage.
12. Valve body-to-case spacers and gaskets.

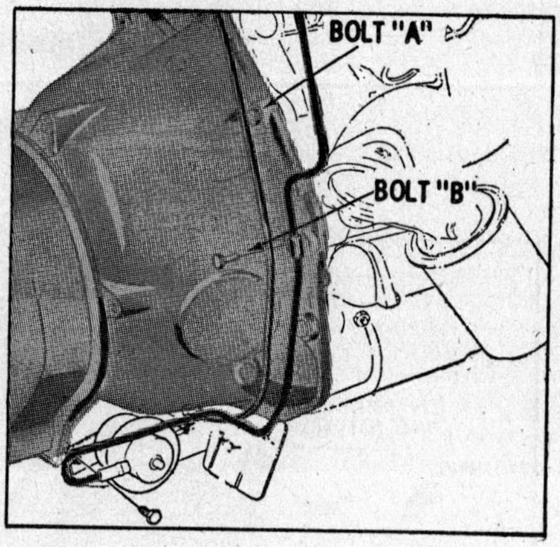

Fig. 10 **Transmission to engine attachment**

13. Check balls for proper location (7 balls), Fig. 5.
14. Detent roller and spring assembly.

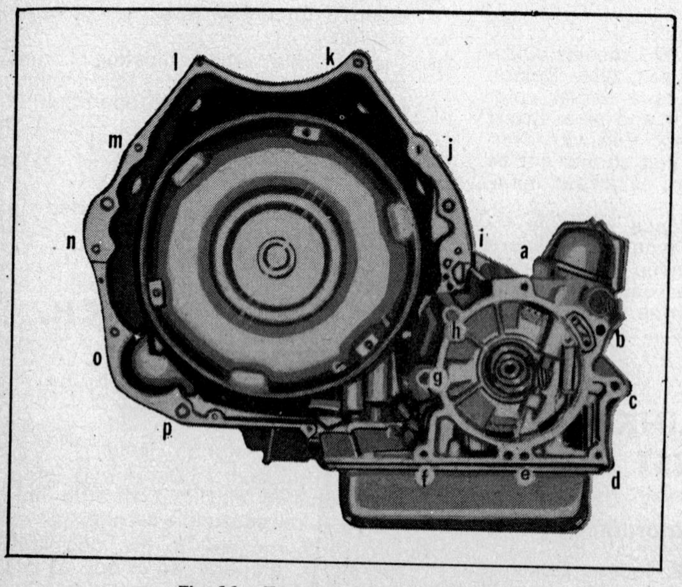

Fig. 11 **Transmission attaching bolt locations. Cadillac Eldorado**

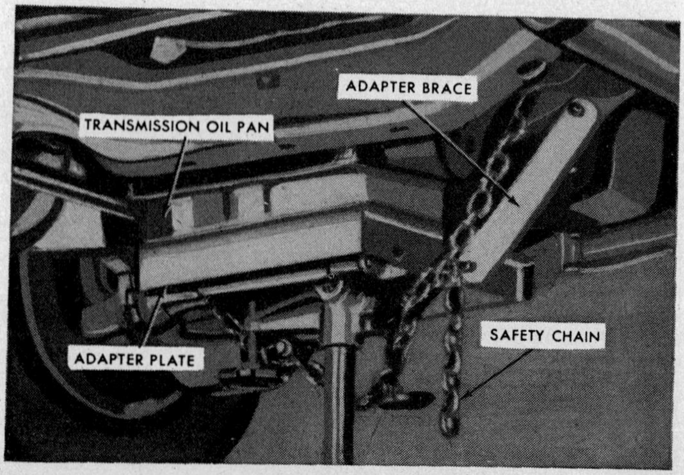

Fig. 12 **Positioning transmission jack to transmission. Cadillac Eldorado**

TRANS., REPLACE

1969-75 Olds Toronado

Removal

1. Disconnect battery.
2. Disconnect oil cooler lines at transmission and speedometer cable at governor. Remove governor and cover opening to prevent entry of dirt.
3. Install a suitable engine support bar, such as shown in Fig. 6.
4. On 1969 models, remove nut D and bolts A, B, and C. Fig. 7. On 1970-75 models, remove three upper final drive to transmission attaching bolts, Fig. 7A.
5. Remove bolts A, B, C, D, Fig. 8.
6. Remove flywheel cover plate bolt, A, Fig. 10.
7. Hoist car and remove starter.
8. Remove bolts B, C and D from flywheel cover plate, Fig. 9.
9. Remove flywheel to converter bolt E, Fig. 9. Rotate flywheel until bolts are removed.
10. Disconnect vacuum modulator line and stator wiring.
11. Install transmission lift.
12. Remove shift linkage.
13. Remove bolts E, F, G and nut H, Fig. 7.
14. Remove bolts A and B, Fig. 10.
15. Remove two upper engine mount bracket-to-transmission bolts.
16. Remove four bracket-to-engine mount bolts.
17. Slide transmission rearward and down.
18. Attach converter holding strap to housing to prevent converter from falling out when transmission is being removed.
19. After transmission is removed from vehicle, the link assembly cover insulator can be removed or installed.

Installation

When installing the transmission the engine mount bracket must be positioned loosely on the link assembly cover until the transmission is in place. Then reverse removal procedure and torque bolts to ft-lb values as follows:

Engine to converter housing	30
Engine bracket to transmission	55
Engine bracket to rubber mount	55
Oil cooler lines to transmission—	
1969-72	30
1973-75	20
Final drive to transmission	25

1969-75 Cadillac Eldorado

Removal—Figs. 11 and 12

1. Disconnect ground cable at battery.
2. Remove transmission dipstick.
3. Remove filler tube.
4. Remove bolts at locations A, B and C, securing final drive case to transmission.
5. Disconnect speedometer cable from governor and detent solenoid connector from transmission case.
6. Disconnect oil cooler pipes from transmission and on 1969 models also from radiator.
7. Cap pipes and plug connector holes in transmission and radiator.
8. Remove bolt securing cooler pipe bracket to final drive bracket and position pipes away from governor.
9. Remove nut at location H, securing final drive case to transmission.
10. Remove bolts at locations I, J, K and L, securing transmission to engine and adapter plate.
11. Remove upper left bolt securing rear engine mount bracket to transmission.
12. Remove ground strap from cowl.
13. Remove upper left nut securing converter cover plate to transmission.

NOTE: Use a 7/16" universal socket and extension and reach underneath left exhaust manifold. Removal of this screw can be facilitated by having a helper under the car, verbally guiding socket onto nut.

14. On 1969-72 models, position cable with looped ends under engine intake manifold and hook looped ends to chain fall. Take up slack in chain fall and cable, putting engine mounts under tension.
15. On all models, position safety chain over top of transmission.
16. Raise vehicle and place on jack stands. Adjust chain fall on 1969-72 models.
17. Disconnect leads from starter motor.
18. Remove bolt at location O, securing starter motor to transmission case and remove ground strap from bolt.
19. Remove bolt at location P and remove starter.
20. Remove three remaining screws securing converter cover plate to transmission and remove cover plate.
21. Position transmission jack.
22. Disconnect electrical connector from transmission connector.
23. Remove pipe from vacuum modulator.
24. Secure transmission to transmission jack adapter plate with safety chain.
25. Remove three flex plate-to-converter attaching bolts.

NOTE: This can be done by installing a 9/16-18 bolt and washer into end of crankshaft at vibration damper, after removing cork plug, and rotating converter and flex plate until bolts are accessible for removal. Do not pry on flex plate ring gear to rotate flex plate and converter as flex plate may be damaged.

26. Remove bolts at locations M and N securing transmission to engine and adapter plate.
27. On left side of transmission, separate relay rod from manual yoke.
28. Remove bolts at locations D, E and F, and nut at location G, securing final drive to transmission.

NOTE: Position drain pan under point where transmission and final drive meet as approximately 1½ quarts of transmission fluid will be lost when transmission and final drive are separated.

29. Remove five bolts and washers securing rear of acromat (cushion) to front cross bar and frame horns and allow acromat to hang free.
30. Through access holes in bottom of front cross bar, remove left bolt and loosen right bolt securing front engine mount to front cross bar. Turn wheels all the way to the left to provide maximum clearance.
31. Have a helper, using a large pry bar, shift engine forward, while you use a small pry bar to help separate transmission from engine and final drive. *Select pry points with care to avoid damaging any components.*
32. After initial separation has been made, allow transmission oil to drain at final drive junction.
33. Remove two bolts on right side securing rear engine mount bracket to transmission.
34. Through access hole in bottom of transmission support bar, remove two bolts, one each side, securing rear mounts to transmission support bar, and position mounts and bracket rearward to underbody.
35. While a helper pries and holds engine forward, move transmission rearward to disengage transmission case from dowels on engine adapter and to disengage final drive from studs on transmission case. Top of transmission should be tilted slightly rearward.
36. Slowly lower transmission, making certain top of transmission case clears flex plate ring gear and splined input shaft of final drive, until converter is approximately half-way exposed from flex plate.
37. Install a suitable clamp to transmission case at location N to avoid possibility of converter becoming disengaged when transmission is removed.
38. Lower transmission from car.

CAUTION: Rear engine mount bracket will follow transmission from car; to avoid damage or injury, remove bracket as soon as there is sufficient clearance.

39. Remove and discard final drive gasket and clean mounting surface of final drive.

Installation

1. Position transmission, on jack, under car.
2. Install new gasket on final drive, after first soaking gasket with transmission fluid.
3. Position rear engine mount bracket on top of transmission support bar against underbody.
4. Raise transmission in place until converter is approximately half-way covered by flex plate, then remove converter holding clamp.
5. While a helper assists in holding engine forward with pry bar, continue raising transmission, making certain top of transmission case clears splined input shaft of final drive, and position to engine.
6. Position transmission to engine and final drive by aligning following points in the order listed, while a helper assists:
   a. Studs on transmission case to mounting holes in final drive.
   b. Guide holes in transmission case to dowels on adapter.
   c. Internal flange on final drive to

transmission.

NOTE: As engagement of splined final drive input shaft to transmission is hidden, extreme care must be taken to avoid damaging transmission and final drive.
To facilitate engagement of final drive splines, rotate one front wheel while a helper holds the other. When alignment is complete and proper, gap between final drive case and transmission should not exceed 1/4".

7. Loosely install bolts (3/8 x 1 1/4) at locations D and F attaching transmission to final drive and bolt (3/8 x 2 1/2) at location N attaching transmission to engine adapter, alternately tightening bolts to avoid cocking transmission. *Do not torque bolts at this time.*
8. Working in engine compartment, loosely install bolt (3/8 x 1 3/8) at location J attaching transmission to adapter. *Do not torque bolt at this time.*
9. Install bolt (3/8 x 1 3/8) at location M attaching transmission to adapter plate. *Do not torque bolt at this time.*
10. Position rear engine mount bracket to transmission and loosely install bolts at locations K and L.
11. Position rear engine mounts and bracket to transmission support bar and loosely install bolts through access holes in bottom of bar, attaching mounts to bar.
12. Reposition engine if necessary and install left bolt securing front engine mount to front cross bar. Tighten both front mount bolts to 30-ft-lbs.
13. Remove transmission jack.
14. Torque rear engine mounts to transmission support bar bolts to 55 ft-lbs. Torque rear engine mounts to transmission bolts (2 on right side) to 55 ft-lbs. Torque transmission to adapter to engine bolts (located N) to 30 ft-lbs. Torque transmission to adapter bolts (location M) to 30 ft-lbs.

NOTE: The procedure for securing the converter to the flex plate outlined in steps 15 through 17 must be strictly followed, otherwise damage to the flex plate and transmission will result from improper installation.

15. Rotate converter until two of the three weld nuts on converter line up with two of the three bolt holes in flex plate. Position converter so that weld nuts are flush with flex plate, making certain converter is not cocked and that pilot in center of converter is properly seated in crankshaft.
16. Loosely install two flex plate to converter bolts through accessible holes in flex plate.
17. Rotate flex plate and converter by rotating bolt previously installed in forward end of crankshaft until third bolt hole is accessible. Install the third flex plate to converter bolt and torque all three bolts to 28-30 foot pounds. Remove bolt from crankshaft and install cork plug.
18. Install hose on vacuum modulator.
19. Install electrical connector to transmission connector.
20. Position converter cover plate to transmission case and install two lower and one upper right bolts securing cover plate to transmission, tightening to 5 ft-lbs.
21. Position starter to transmission case and install bolt at location P.
22. Position ground strap to transmission and install bolt securing ground strap and starter to transmission at location O. Tighten bolts at locations O and P to 25 ft-lbs.
23. Install leads on starter motor.
24. Install bolts at locations C and E and nut at location G, securing transmission to final drive.
25. Torque bolts at locations C through F to 25 foot pounds on 1969-73 models or 35 foot pounds on 1974-75 models.
26. On 1969-72 models, position acromat (cushion) to front cross bar and frame horns and install five bolts and washers.
27. On all models, connect relay rod to manual yoke with a cotter pin.
28. Check operation of manual linkage and adjust.
29. Lower vehicle.
30. Install bolts at locations A and B and the nut at location H. Torque bolts to 25 foot pounds on 1969-73 models and 35 foot pounds on 1974-75 models.
31. Install upper left bolt securing converter cover plate to transmission in the manner described for removing it in Step 13.
32. Install bolt at location I and torque bolts at locations I, J, K, and L to 25 foot pounds.
33. Torque oil cooler pipe connectors at transmission case to 28 foot pounds on 1969-72 models and 20 foot pounds on 1973-75 models. Clean cooler pipe ends with a suitable solvent, connect pipes to transmission and torque pipe fittings to 28 foot pounds on 1969-72 models and 20 foot pounds on 1973-75 models.
34. On 1969 models, connect cooler pipes to radiator and torque fittings to 40 foot pounds. Install cooler pipe clamp.
35. On all models, connect speedometer to governor and detent solenoid connector to transmission case.
36. Install new O-ring seal on transmission oil filler tube through hole in final drive case. Fasten filler tube bracket to exhaust manifold.
37. Install body ground strap to firewall.
38. Connect battery cable, fill transmission with fluid and install hood if previously removed.

CHEVROLET TORQUE DRIVE

GENERAL DESCRIPTION

The "Torque-Drive" transmission is basically a modified Powerglide consisting of a torque converter and a two-speed planetary gear set. As the automatic shifting provisions have been removed the transmission can only be shifted manually.

The selector lever positions are PARK-R-N-Hi-1st. 1st speed should be used for speeds up to 20 mph but never exceeding 55 mph. When in Hi position, do not downshift to 1st at speeds above 55 mph.

LUBRICATION & MAINTENANCE

Lubrication, maintenance and service information as it is covered in the Aluminum Case Powerglide section of this manual will also apply to the "Torque-Drive" with the exception of the following operation.

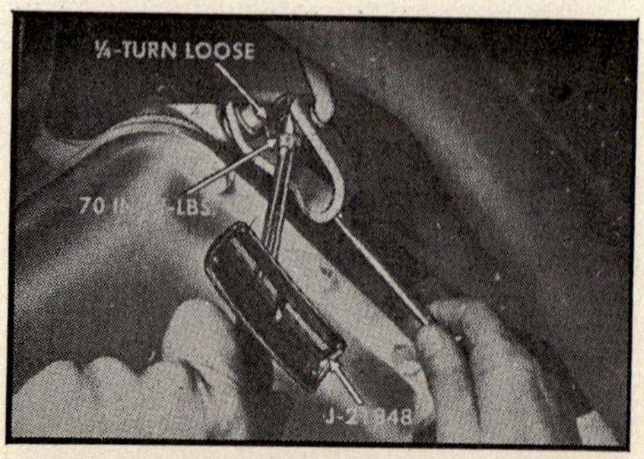

Fig. 1 Low band adjustment (in vehicle)

LOW BAND, ADJUST

Band adjustment should be performed at the first transmission oil change and sooner if slipping is evident.

1. With the selector in Neutral, back off the locknut 1/4 turn. Tighten the adjusting screw to 70 in. lbs., Fig. 1.
2. Next, back off the adjusting screw exactly four turns if the band has more than 6,000 miles on it, or three turns if the band has less than 6,000 miles of use.

CAUTION: Be sure to hold the locknut at 1/4 turn loose during the adjusting procedure. Then tighten locknut. The amount of back off is not an approximate figure; it must be exact.

TRANSMISSION IDENTIFICATION

A production day and shift built number, transmission model and model year are stamped on the 1-2 accumulator cover, which is located on the middle lower right side of the transmission case.

BUICK

Year	Model	CODE
1969	Except Sportwagon	JH
	Sportwagon	JJ
1970	6-250	JE
	8-350 2 bar. carb. Exc. LeSabre	JH
	8-350 4 bar. carb. Exc. LeSabre	JR
	8-350 2 bar. carb. LeSabre	JJ
	8-350 4 bar. carb. LeSabre	JS
1971	6-250	JE
	V8-350 Exc. Below	LA
	V8-350 Gran Sport	MA
	V8-350 LeSabre	KL
1972	Skylark & Custom w/Std. Axle	KA
	V8-350 LeSabre	KL
	V8-350 All Others	KE
1973-74	6-250	JE
	V8-350 Century	KA
	V8-350 LeSabre	KL
1975	V6-231 Skyhawk Exc. Calif.	KX
	V6-231 Skyhawk Calif.	KD
	V6-231 Skylark	KC
	V6-231 Century	KB
	6-250 Apollo	JE
	V8-260 Apollo/Skylark Exc. Calif.	LA
	V8-260 Apollo/Skylark Calif.	NC
	V8-350 Apollo/Skylark	KA
	V8-350 Century	KE
	V8-350 Le Sabre	KL

CHEVROLET

Year	Model	CODE
1969	V8-350 4 bar. carb. w/3.07 or 3.31 rear axle	FD
	With A/C and 3.07, 3.08, 3.31 or 3.36 rear axle	FE
	With 3.07, 3.08 or 3.36 and A/C	FG
	V8-350 4 bar. carb.	FI
	V8-350 2 bar. carb.	FQ
	V8-350 2 bar. carb. Chevelle and Nova	FU
	Taxi W/3.07 rear axle	IF
	V8 Chevy II W/3.36 rear axle	IT
	6 cyl. Chevy II W/3.36 rear axle	IS
	Chevrolet W/3.07 rear axle	TJ
	V8 Chevrolet W/3.08 rear axle and A/C	UX
	V8 Police car	YF
	V8-350 2 bar. carb. Police car	YL
	V8-350 4 bar. carb. Police car	YX
	6 cyl. Police car	YY

CHEVROLET—Cont'd

Year	Model	CODE
1970	6 cyl. Taxi W/3.07 rear axle	GQ
	V8 Taxi W/3.07 rear axle	GT
	Police car	GO, GU
	V8-350 4 bar. carb.	GW
	V8-400 2 bar. carb. Chevrolet	GY
1971	V8-400 2 bar. carb. Station Wagon	HH
	V8-400 2 bar. carb. Chevrolet	HT
	V8-350 4 bar. carb. Camaro, Chevelle and Monte Carlo	HW
	V8-400 2 bar. carb. Monte Carlo	HY
1972	V8-350 2 or 4 bar. carb. Chevelle	JA
	V8-350 2 or 4 bar. carb. Chevelle and Camaro	SA, SB
	V8-350 2 bar. carb. Chevrolet	SD, SH
	V8-400 2 bar. carb. Chevrolet	SG, SJ
1973	6-250 Camaro, Chevelle and Nova	FA
	V8-350 2 or 4 bar. carb. Camaro, Chevelle and Nova	FB
	V8-350 2 bar. carb. Chevrolet	FF, FJ
	V8-307, V8-350 Taxi, Police or Passenger	FD
	V8-400 2 bar. carb. Chevrolet	FH
	6-250 Chevelle	JA
	6-250 Camaro, Chevelle and Nova	TT
	6-250 Camaro and Chevelle	TZ
1974	V8-400 2 Bar. Carb. Chevelle	AW
	V8-350 Camaro Exc. Z28, Chevelle & Nova	FB
	V8-400 4 Bar. Carb. Chevelle	FB
	V8-400 4 Bar. Carb. Chevrolet	FD, FH
	V8-350 Police & Taxi	FH
	V8-400 2 Bar. Carb. Chevrolet	FH
	V8-400 4 Bar. Carb. Chevrolet	FW
	V8-350 Police & Taxi	HA
	6-250 Camaro, Chevelle & Nova	TT
	6-250 Nova	TZ
1975	4-140 Vega (T.H. 250)	DJ
	6-250 All With T.H. 250	TZ
	6-250 Chevelle	DE
	V8-262 Monza 2+2 (T.H. 250)	FU
	V8-262 Nova	DD
	V8-350 4 Bar. carb. Nova	YA
	V8-350 Chevelle	HB
	V8-350 Monte Carlo	JA
	V8-350 Chevrolet	YC
	V8-350 Chevrolet Police	YD
	V8-400 Chevelle	YB

OLDSMOBILE

Year	Model	CODE
1969	F-85 V8-350 2 bar. carb.	JG
	F-85 V8-350 4 bar. carb.	JL
1970	6-250	JE
	8-350 2 bar. carb.	JG
	8-350 4 bar. carb. Exc. below	JL
	8-350 Ram Air	JO
	8-350 Vista Cruiser	JM

OLDSMOBILE—Cont'd

Year	Model	CODE
1971	6-250	JE
	V8-350 Exc. Below	LA
	V8-350 Delta 88	LL
1972		KA, LA
1973	6-250 Omega	JE
	V8-350 Omega & Cutlass 2 bar. carb.	LH
	V8-350 Omega & Cutlass 4 bar. carb.	LA
	V8-350 Vista Cruiser 4 bar. carb.	LE
1974	6-250 Omega	JE
	V8-350 Omega & Cutlass Exc. Calif.	LA
	V8-350 Omega & Cutlass Calif.	LC
	V8-350 Vista Cruiser Exc. Calif.	LE
1975	V6-231 Starfire	KD
	V6-231 Starfire Exc. Calif.	KX
	6-250 Omega	JB
	6-250 Cutlass	JE
	V8-260 Omega Exc. Calif.	LA
	V8-260 Omega Calif.	NC
	V8-260 Cutlass	LH
	V8-350 Omega	KA
	V8-350 Cutlass	LC
	V8-350 Intermediate Wagons Calif.	LC
	V8-350 Intermediate Wagons Exc. Calif.	LF

PONTIAC

Year	Model	CODE
1969	Tempest 6-250 1 bar. carb.	JA
	Firebird 6-250 1 bar. carb.	JB
	Tempest 6-250 4 bar. carb.	JC
	Firebird 6-250 4 bar. carb.	JD
	Tempest & Firebird V8-350 2 bar. carb.	JF
1970	Tempest 6-250	JE
	Tempest 8-350	JF
	Catalina 8-350	JU
1971	6-250	JE
	V8-350 Exc. Below	MA
	V8-350 Catalina	ML
1972	6-250	JE
	V8-307	SB
	V8-350	MA
1973	6-250	FA, JE
	V8-350 2 bar. carb.	MA
	V8-350 W/2.93 rear axle (late)	MC
	V8-350 W/3.08 rear axle (late)	ME
1974	6-250 Firebird & LeMans	JE
	6-250 Ventura	TZ
	V8-350 2 Bar. Carb. w/2.73 or 2.93 rear axle	MA
	V8-350 w/3.08 rear axle	ME
1975	4-140 Astre	DJ
	6-250 Exc. Le Mans	JB
	6-250 Le Mans	JE
	V8-260 Ventura	LA
	V8-350 Ventura	KA
	V8-350 Firebird	MA
	V8-350 Le Mans	ME
	V8-400 Firebird	MF
	V8-400 Firebird	MG

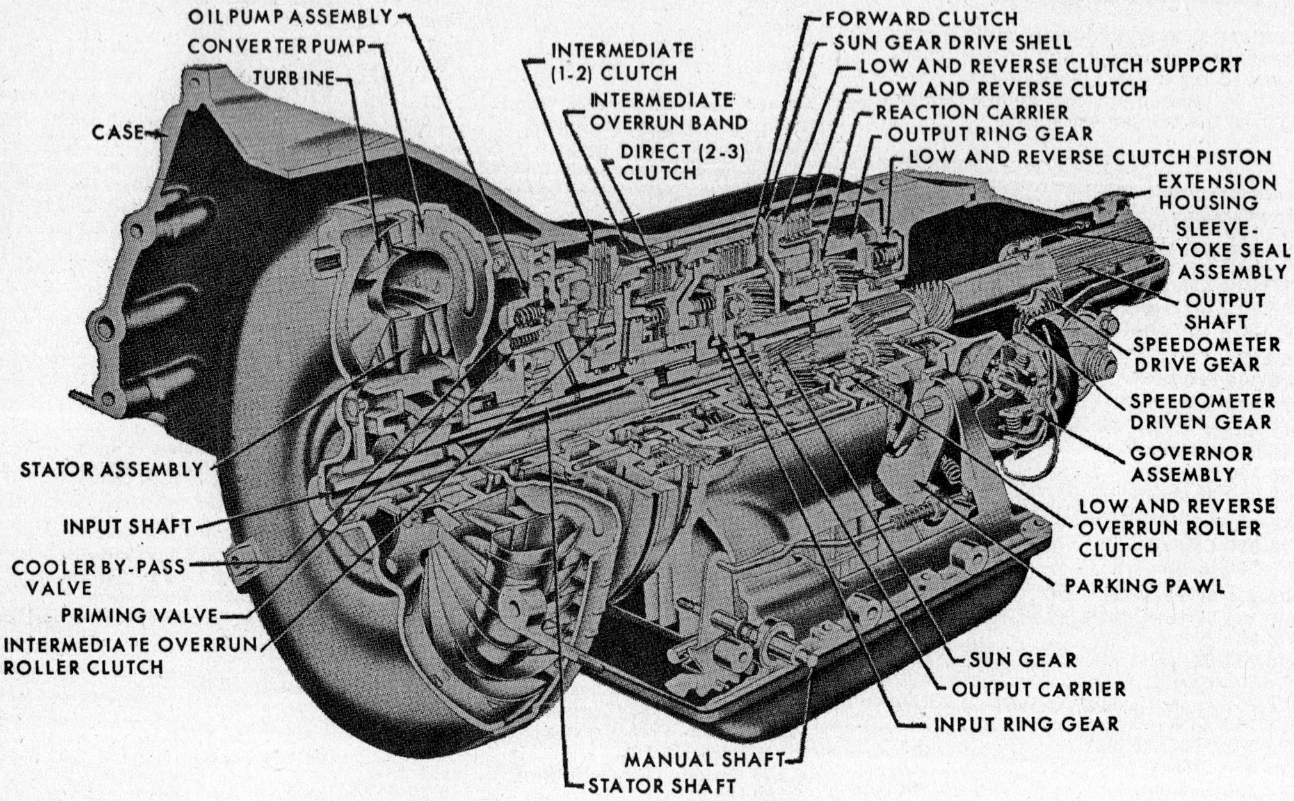

Fig. 1 Cutaway view of Turbo Hydra-Matic 350, 375B transmission

DESCRIPTION

The Turbo Hydra-Matic 250, 350, 375B, Figs. 1 and 2, are fully automatic three speed transmissions consisting of a three element torque converter and a compound planetary gear set. The Turbo Hydra-Matic 350, 375B transmission has four multiple-disc clutches, two roller clutches and a band to provide the required friction elements to obtain the desired function of the planetary gear set. The Turbo Hydra-Matic 250 transmission uses an adjustable intermediate band in place of the intermediate clutch found in the Turbo Hydra-Matic 350. Also, the Turbo Hydra-Matic 250 has three multiple-disc clutches and one roller clutch.

The friction elements couple the engine to the planetary gears through oil pressure, providing three forward speeds and one reverse.

The three element torque converter is of welded construction and is serviced as an assembly. The unit consists of a pump or driving member, a turbine or driven member and a stator assembly. When required, the torque converter supplements the gears by multiplying engine torque.

TROUBLE SHOOTING GUIDE

No Drive In Drive Range

1. Low oil level (check for leaks).
2. Manual control linkage improperly adjusted.
3. Low oil pressure due to blocked strainer, defective pressure regulator, pump assembly or pump drive gear. See that tangs have not been damaged by converter. Check case for porosity in intake bore.
4. Check control valve assembly to be sure manual valve has not been disconnected from inner lever.
5. Forward clutch may be stuck or damaged. Check pump feed circuits to forward clutch, including clutch drum ball check.
6. Roller clutch assembly broken or damaged.

Oil Pressure High or Low

High Pressure:

1. Vacuum line or fittings leaking.
2. Vacuum modulator.
3. Modulator valve.
4. Pressure regulator.
5. Oil pump.

Low Pressure:

1. Vacuum line or fittings obstructed.
2. Vacuum modulator.
3. Modulator valve.
4. Pressure regulator.
5. Governor.
6. Oil pump.

1 - 2 Shift At Full Throttle Only

1. Detent valve may be sticking or linkage may be misadjusted.
2. Vacuum line or fittings leaking.
3. Control valve body gaskets leaking, damaged or incorrectly installed. Detent valve train or 1-2 valve stuck.
4. Check case for porosity.

First Speed Only, No 1 - 2 Shift

T.H.M. 250, 350 & 375B

1. Governor valve may be sticking.

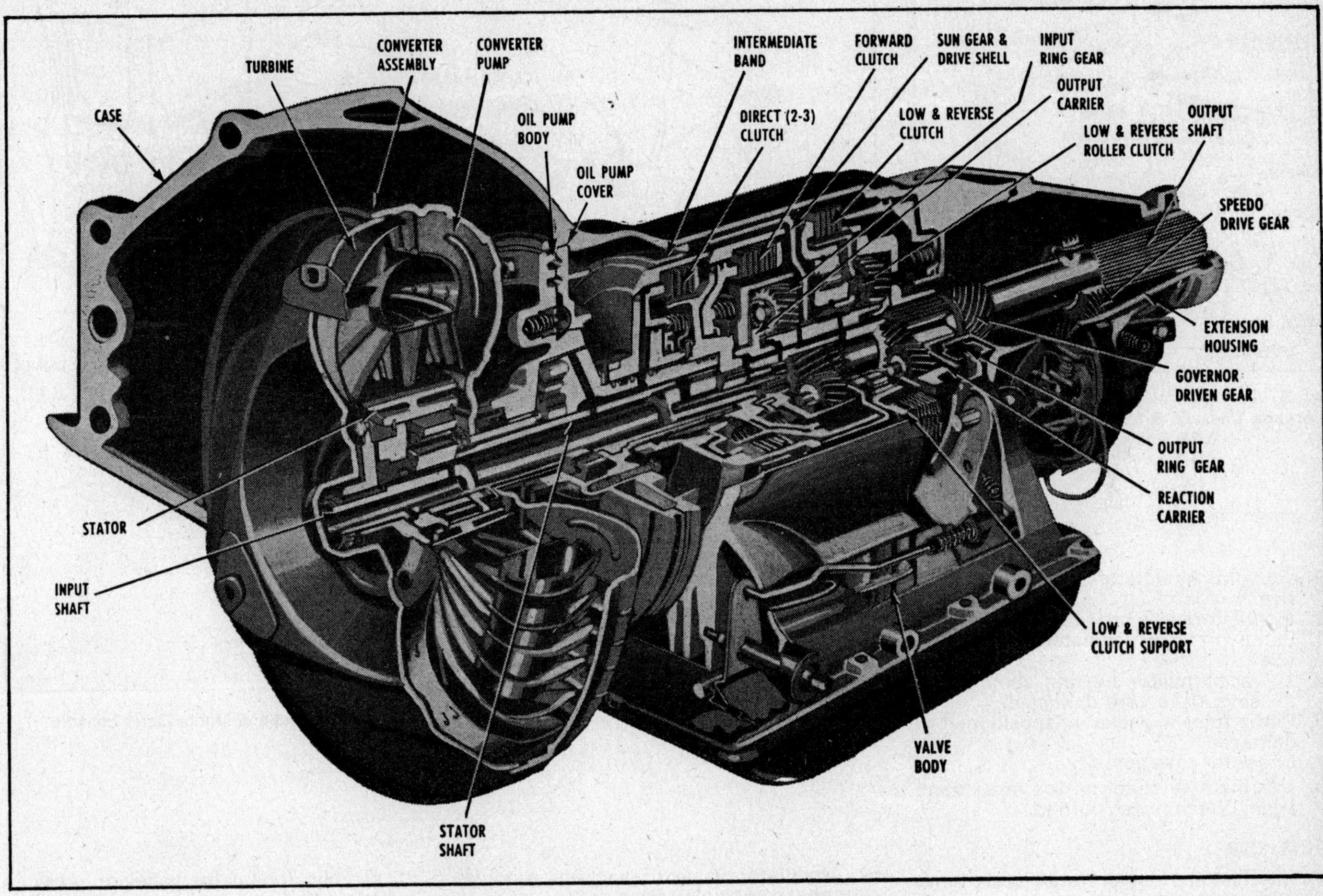

Fig. 2 Cutaway view of Turbo Hydra-Matic 250 transmission

2. Driven gear in governor assembly loose, worn or damaged. If driven gear shows damage, check output shaft drive gear for nicks or rough finish.
3. Control valve governor feed channel blocked or gaskets leaking. 1-2 shift valve train stuck closed.
4. Check case for blocked governor feed channels or for scored governor bore which will allow cross pressure leak. Check case for porosity.
5. Intermediate clutch or seals damaged.
6. Intermediate roller clutch damaged.

T.H.M. 250

1. Intermediate servo piston seals damaged, missing or installed improperly.
2. Intermediate band improperly adjusted.
3. Intermediate servo apply rod broken.

1st & 2nd Only, No 2-3 Shift

1. Control valve 2-3 shift train stuck. Valve body gaskets leaking, damaged or improperly installed.
2. Pump hub-to-direct clutch oil seal rings broken or missing.
3. Direct clutch piston seals damaged. Piston ball check stuck or missing.

No First Speed

T.H.M. 250

1. Intermediate band adjusted too tightly.
2. 1-2 shift valve stuck in upshift position.

T.H.M. 350, 375B

1. Excessive number of clutch plates in intermediate clutch pack.
2. Incorrect intermediate clutch piston.

Moves Forward In Neutral

1. Manual linkage misadjusted.
2. Forward clutch not releasing.

No Drive In Reverse or Slips In Reverse

1. Low oil level.
2. Manual linkage misadjusted.
3. Modulator valve stuck.
4. Modulator and reverse boost valve stuck.
5. Pump hub-to-direct clutch oil seal rings broken or missing.
6. Direct clutch piston seal cut or missing.
7. Low and reverse clutch piston seal cut or missing.
8. Number 1 check ball missing.
9. Control valve body gaskets leaking or damaged.
10. 2-3 valve train stuck in upshifted position.
11. 1-2 valve train stuck in upshifted position.
12. Intermediate servo piston or pin stuck so intermediate band is applied.
13. Low and reverse clutch piston out or seal damaged.
14. Direct clutch plates burned — may be caused by stuck ball check in piston.
15. Forward clutch not releasing.

Slips In All Ranges

1. Low oil level.
2. Vacuum modulator valve defective or sticking.
3. Filter assembly plugged or leaking.
4. Pressure regulator valve stuck.
5. Pump to case gasket damaged.
6. Check case for cross leaks or porosity.
7. Forward clutch slipping.

Slips 1-2 Shift

T.H.M. 250, 350 & 375B

1. Low oil level.

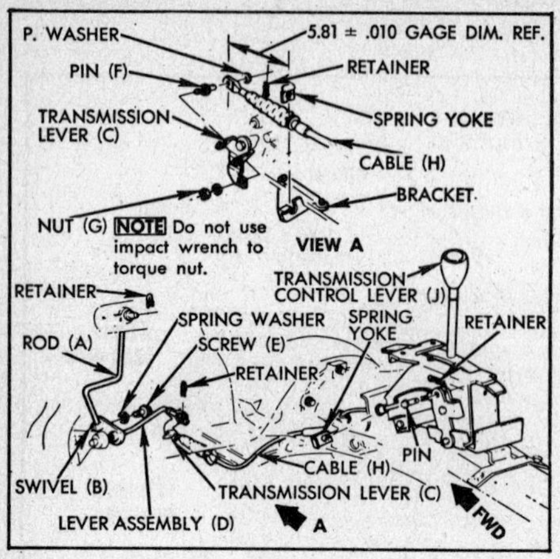

Fig. 3 Manual linkage adjustment. 1973-75 Camaro, Chevelle & Monte Carlo. 1975 Nova & Buick Apollo

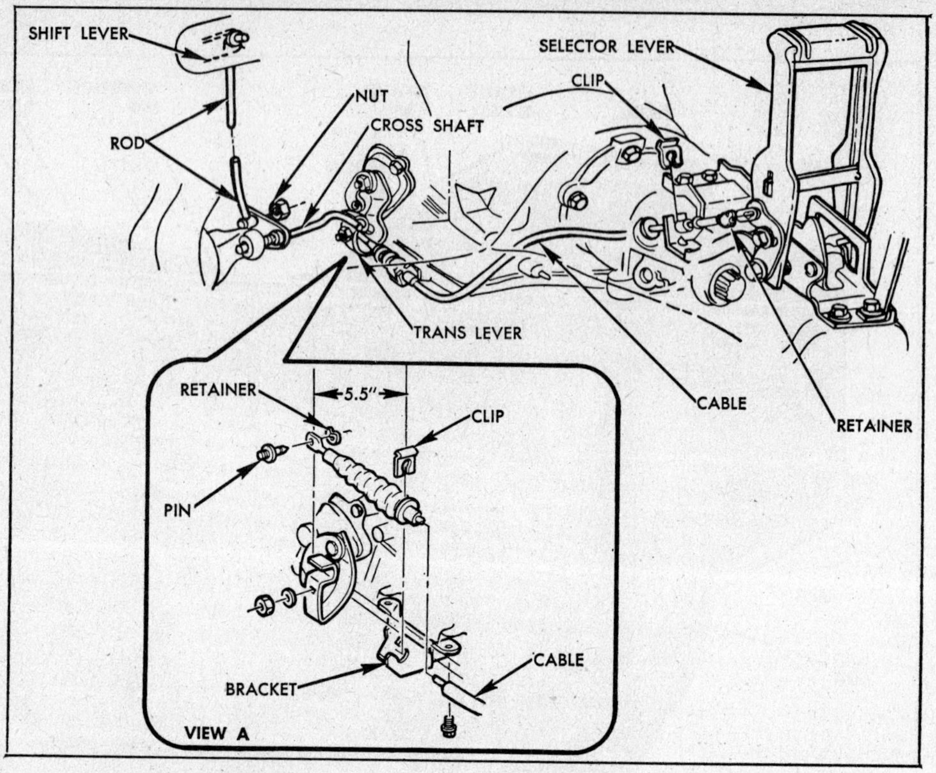

Fig. 4 Manual linkage adjustments. 1969-72 Camaro, Chevelle & Monte Carlo console

2. Vacuum modulator assembly defective.
3. Modulator valve sticking.
4. Pump pressure regulator valve defective.
5. 1-2 accumulator oil ring damaged or missing. Case bore damaged.
6. Pump to case gasket mispositioned or damaged.
7. Check for case porosity.
8. Intermediate clutch piston seals damaged. Clutch plates burned.

T.H.M. 250

1. Intermediate servo piston seals damaged or missing.
2. Burned intermediate band.

T.H.M. 350, 375B

1. 2-3 accumulator oil ring damaged or missing.

Rough 1 - 2 Shift

T.H.M. 250, 350 & 375B

1. Vacuum modulator, check for loose fittings, restrictions in line or defective modulator assembly.
2. Modulator valve stuck.
3. Valve body regulator or boost valve stuck.
4. Pump to case gasket mispositioned or damaged.
5. Check case for porosity.
6. Check 1-2 accumulator assembly for damaged oil rings, stuck piston, broken or missing spring, or damaged case bore.

T.H.M. 250

1. Intermediate band improperly adjusted.
2. Improper or broken servo spring.

T.H.M. 350, 375B

1. Burned intermediate clutch plates.
2. Improper number of intermediate clutch plates.

Slips 2 - 3 Shift

1. Low oil level.
2. Modulator valve or vacuum modulator assembly defective.
3. Pump pressure regulator valve or boost valve; pump to case gasket mispositioned.
4. Check case for porosity.
5. Direct clutch piston seals or ball check leaking.

Rough 2 - 3 Shift

1. High oil pressure. Vacuum leak, modulator valve sticking or pressure regulator or boost valve inoperative.
2. 2-3 accumulator piston stuck, spring broken or missing.

No Engine Braking In Second Speed

1. Intermediate servo or 2-3 accumulator oil rings or bores leaking or accumulator piston stuck.
2. Intermediate band burned or broken.
3. Low oil pressure: Pressure regulator and/or boost valve stuck.

No Engine Braking In 1st Speed

1. Manual low control valve assembly stuck.
2. Low oil pressure: Pressure regulator and/or boost valve stuck.
3. Low and reverse clutch piston inner seal damaged.

No Part Throttle Downshift

1. Oil pressure: Vacuum modulator assembly, modulator valve or pressure regulator valve train malfunctioning.
2. Detent valve and linkage sticking, disconnected or broken.
3. 2-3 shift valve stuck.

No Detent Downshifts

1. 2-3 valve stuck.
2. Detent valve and linkage sticking, disconnected or broken.

Low or High Shift Points

1. Oil pressure: Check engine vacuum at transmission end of modulator pipe.
2. Vacuum modulator assembly vacuum line connections at engine and transmission, modulator valve, pressure regulator valve train.
3. Check governor for sticking valve, restricted or leaking feed holes, damaged pipes or plugged feed line.
4. Detent valve stuck open.
5. 1-2 or 2-3 valve train sticking.
6. Check case for porosity.

Won't Hold In Park

1. Manual linkage misadjusted.
2. Parking brake lever and actuator assembly defective.
3. Parking pawl broken or inoperative.
4. Defective or improperly installed inner lever and actuating rod assembly.
5. Parking lock bracket loose, burred or

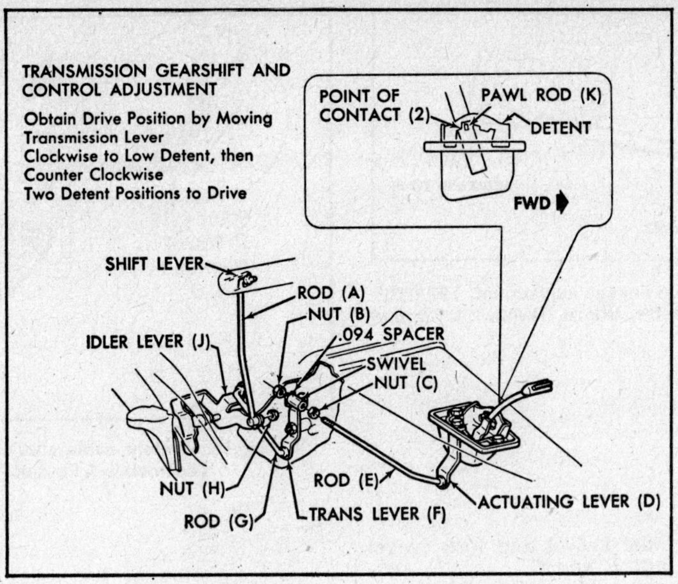

Fig. 5 Manual linkage. Nova console (typical)

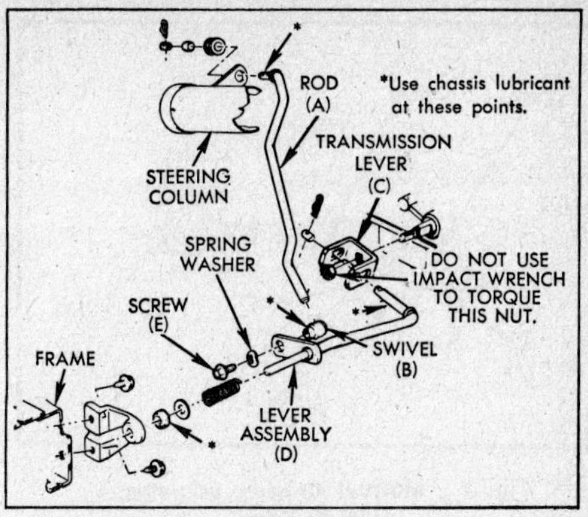

Fig. 6 Column shift linkage. 1973-74 All (Typical)

rough edges, or improperly installed.
6. Parking pawl disengaging spring missing, broken or installed improperly.

Burned Forward Clutch Plates

1. Check ball in clutch drum damaged, stuck or missing.
2. Clutch piston cracked, seals damaged or missing.
3. Low line pressure.
4. Pump cover oil seal rings missing, broken or undersize; ring groove oversize.
5. Transmission case valve body face not flat or porosity between channels.

Burned Intermediate Clutch Plates

T.H.M. 350, 375B

1. Intermediate clutch piston seals damaged or missing.
2. Low line pressure.
3. Transmission case valve body face not flat or porosity between channels.

Burned Intermediate Band

T.H.M. 250

1. Intermediate servo piston seals damaged or missing.
2. Low line pressure.
3. Transmission case valve body face not flat or porosity between channels.

Burned Direct Clutch Plates

1. Restricted orifice in vacuum line to modulator.
2. Check ball in clutch drum damaged, stuck or missing.
3. Defective modulator.
4. Clutch piston cracked, seals damaged or missing.
5. Transmission case valve body face not flat or porosity between channels.

Noisy Transmission

NOTE: Before checking transmission for noise, ensure noise is not coming from water pump, alternator or any belt driven accessory.

Park, Neutral & All Driving Ranges

1. Low fluid level.
2. Plugged or restricted screen.
3. Damaged screen to valve body gasket.
4. Pososity in valve body intake area.
5. Transmission fluid contaminated with water.
6. Pososity at transmission case intake port.
7. Improperly installed case to pump gasket.
8. Pump gears damaged.
9. Driving gear assembled backwards.
10. Crescent interference in pump.
11. Damaged or worn pump oil seals.
12. Loose converter to flywheel bolts.
13. Damaged converter.

1st, 2nd And/Or Reverse Gear

1. Planetary gears or thrust bearings damaged.
2. Damaged input or output ring gear.

Acceleration In Any Gear

1. Transmission case or transmission oil cooler lines contacting underbody.
2. Broken or loose engine mounts.

Squeal At Low Vehicle Speed

1. Speedometer driven gear shaft seal requires lubrication or replacement.

MAINTENANCE

Fluid should be checked every 6,000 miles with engine idling, selector lever in neutral position, parking brake set and transmssion at operating temperature. Use only General Motors Dexron transmission fluid when adding oil. Do not overfill.

Every 24,000 miles, remove drain plug in transmission oil pan and drain transmission oil sump. Add $1\frac{1}{2}$ quarts after replacing plug, check fluid and add enough fluid to bring level to the Full mark.

NOTE: Beginning with 1973 Chevrolet and late 1973 Buick, Oldsmobile and Pontiac models, a revised type Dexron fluid is used in these transmissions. An early change to a darker color from the usual red color and or a strong odor that is usually associated with overheated fluid is normal, and should not be treated as a positive sign of needed maintenance or unit failure.

The normal maintenance schedule for drain and refill of this type fluid remains unchanged at 24,000 miles under normal service and 12,000 miles under severe operating conditions, such as trailer towing.

MANUAL LINKAGE, ADJUST

Buick

Console Shift All Exc., 1975 Apollo

1. Loosen trunnion bolt.
2. Set selector lever against Drive stop.
3. Place transmission in Drive.
4. Tighten trunnion bolt to 6-9 ft-lbs.
5. Set selector lever in Park and set Back Drive Adjustment as outlined below.

Console Shift, 1975 Apollo

Refer to Chevrolet for adjustment procedure.

Column Shift, 1969-70

1. Loosen adjusting clamp bolt.
2. Place selector lever against Drive stop.
3. Place transmission in Drive.
4. Tighten clamp bolt to 17-23 ft. lbs.

Column Shift, 1971-75

1. Loosen adjusting clamp bolt.

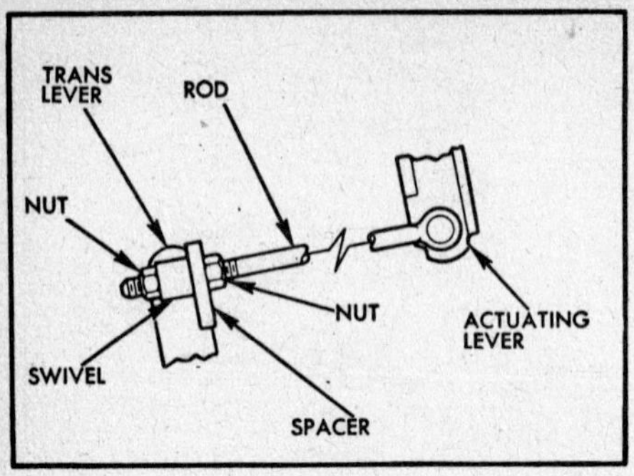

Fig. 7 Manual linkage adjustment. 1972 Vega

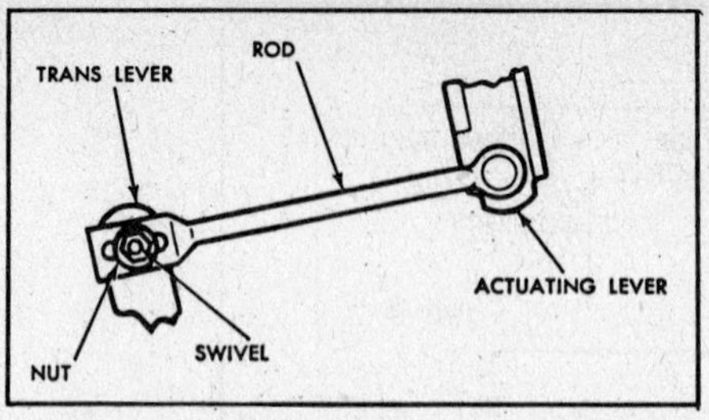

Fig. 8 Manual linkage adjustment. 1973-75 Vega & 1975 Astre, Monza, Skyhawk & Starfire

Fig. 9 Detent cable clip. Buick, Oldsmobile & Pontiac

2. Place selector lever against Neutral stop.
3. Place transmission in Neutral.
4. Tighten clamp bolt to 17-23 ft. lbs.

Chevrolet

Console Shift, Camaro, Chevelle & Monte Carlo 1973-75 & Nova and Buick Apollo 1975

1. Loosen swivel screw so rod is free to move in swivel, Fig. 3.
2. Place transmission control lever in Drive and loosen pin in transmission lever, so it moves in the slot.
3. Move transmission lever counterclockwise to L1 detent and then three detents clockwise to Drive position. Tighten nut on transmission lever to 20 ft. lbs.
4. Place transmission control lever in Park and ignition switch in the Lock position and pull lightly against lock stop, then tighten swivel screw to 20 ft. lbs. Check for proper operation.

Console Shift, Camaro, Chevelle, Chevrolet & Monte Carlo 1969-72

1. Place shift lever in Drive, Fig. 4.
2. Raise vehicle, disconnect cable from transmission lever and manually place transmission lever in Drive. Measure distance from rearward face of attachment bracket to center of cable attachment stud. This dimension should be 5.5″. If not, loosen stud nut and move stud accordingly. Reinstall stud nut and adjust end of cable so that it fits freely on stud. Reinstall clip.

Nova Console Shift, 1969-74

1. Loosen assembly nuts B and C on rod E, Fig. 5.
2. Set transmission lever F in DRIVE.
3. Set pawl rod K in DRIVE notch of detent.
4. Apply load in direction of arrow on actuating lever D until pawl rod K comes in contact with detent at contact point Z.
5. Place a .094″ spacer between nut B and swivel, run nut B until it touches spacer. Remove spacer and tighten nut C against swivel and lock swivel between nuts B and C.
6. Set transmission lever F in PARK and turn ignition switch to LOCK.
7. Install rod G to idler lever J and lever F.
8. Install rod A to shift lever and idler lever J. Loosely attach clamp to idler lever.
9. Remove column lash by rotating shift lever in a downward direction and secure with attaching nut H. The foregoing provides .05″ overtravel gap in notches of the detent.

Column Shift, 1973-75 All

1. Place transmission lever in Neutral by moving lever counter-clockwise to L1 detent then clockwise three detent positions to Neutral.
2. Place selecter lever in Neutral as determined by mechanical stop on steering column. Do not use indicator as reference.
3. Assemble swivel, spring washer and screw to lever assembly then tighten screw to 20 ft. lbs., Fig. 6.

Column Shift, 1969-72 All

1. Place selector lever in Drive as determined by the transmission detent.
2. Loosen adjustment swivel at cross shaft and rotate transmission lever so it contacts the Drive stop in the steering column.
3. Tighten swivel and recheck adjustment.
4. Readjust indicator needle if necessary to agree with the transmission detent positions.
5. Readjust neutral safety switch if necessary.
6. Set selector lever in Park and adjust Bakc Drive as outlined below.

Astre, Monza, Skyhawk, Starfire & Vega

1972

1. Place transmission lever in Neutral by rotating lever clockwise to Park, then two detents counter-clockwise to Neutral.
2. Place shift lever in Neutral.
3. Apply a rearward force on lever until shift lever contacts detent stop. Install a .073 inch spacer between rear nut and swivel and tighten nut finger-tight, Fig. 7.
4. Remove spacer and apply a forward force on lever and tighten rearward nut.

1973-75

1. Loosen nut and perform steps 1 and 2 as outlined in the 1972 Chevrolet Vega adjustment.
2. Insert flats of swivel into slot in rod and tighten nut to 120 pound inches, Fig. 8.

Oldsmobile

Console Shift, Exc. 1973-74 Omega

1. Loosen shift rod clamp screw and pin in transmission manual lever. Place shift handle and transmission manual lever in Park position.
2. With rod held lightly against Park stop, tighten screw in clamp at lower end of shift rod.
3. Move pin to give "free pin" fit in manual lever and tighten.

Console Shift, 1973-74 Omega

1. Transmission manual lever must be free in swivel.
2. Place shift and transmission levers in Neutral.
3. Holding the actuating rod rearward against stop, place a .094″ spacer between rear nut and swivel. Tighten nut to contact spacer, remove spacer, hold rear nut and tighten forward nut to lock swivel in place.
4. Loosen shift rod clamp screw and move shift and transmission manual levers to Park position and ignition switch to Lock position.
5. Pull down lightly on shift rod to contact lock stop and tighten clamp screw.
6. Check operation.

NOTE: With ignition key in the Lock position and transmission selector lever in Park, the key can be removed and the steering wheel should be locked.

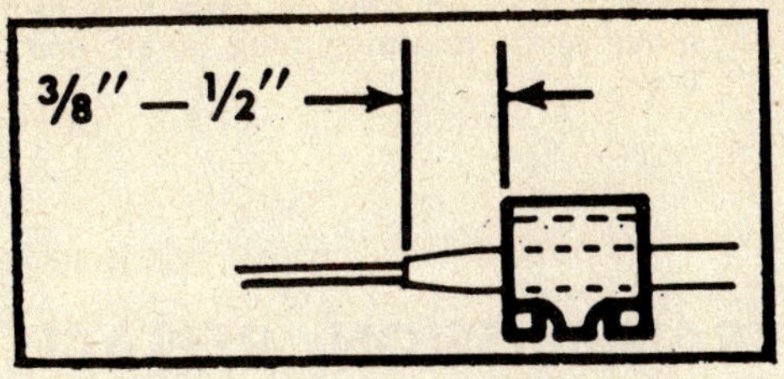

Fig. 10 Rearward downshift cable retainer. 1972 Chevrolet

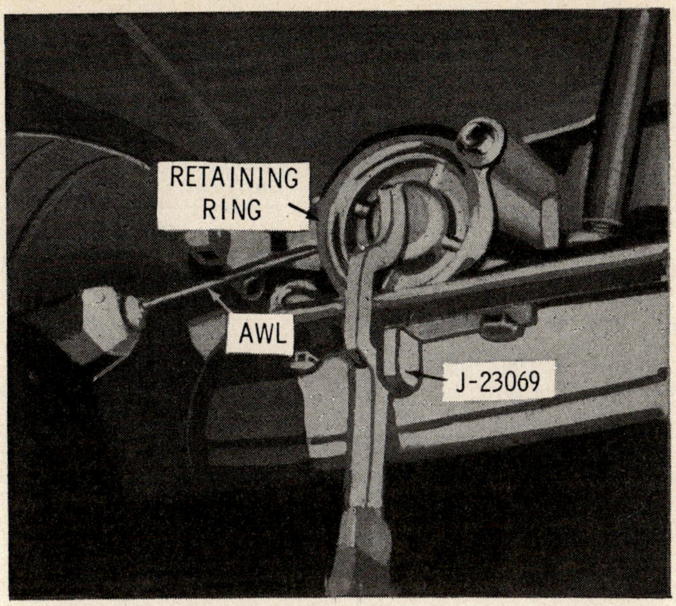

Fig. 11 Intermediate clutch accumulator piston removal

Column Shift, All 1972-75

1. Loosen shift rod clamp screw and place outer lever in the Neutral position. Hold column shift lever in Neutral position and tighten clamp. Check operation.

Column Shift, All 1969-71

1. With shift rod clamp and screw loosely assembled to shift rod, set transmission outer lever in Drive. Check to see that steering column lever is in Drive and tighten clamp on shift rod.

Pontiac

Console Shift, 1969-74 Exc. Ventura

1. Place shift handle in Park and transmission manual lever in Park.
2. With rod held lightly against Park stop, tighten screw in clamp at lower end of rod.
3. Adjust pin in selector lever to freely enter the end of the shift cable and tighten pin nut.

Console Shift, 1975 Exc. Ventura

1. Disconnect shift cable at transmission lever.
2. Adjust back drive, as outlined under "Back Drive, Adjust" procedure.
3. Unlock ignition switch, move transmission lever two detents counterclockwise, then place transmission shift lever against Neutral stop.
4. Assemble cable to transmission lever, torque nut to 20 ft. lbs.

Console Shift, Ventura

Refer to the Oldsmobile Omega "Console Shift" adjustment procedure to adjust the console shift linkage.

Column Shift, All

With shift rod clamp and screw loosely assembled to shift rod, set transmission outer lever in Park position. Check to see that steering column lever is in Park position and tighten clamp on shift rod.

BACK DRIVE, ADJUST

1. Disconnect lower rod at transmission lever.
2. Move transmission lever to Park position.
3. Place transmission selector lever in Park position.
4. Attach lower rod to transmission lever and check for proper operation.

NOTE: Any inaccuracies in the above adjustments may result in premature failure of the transmission due to operation without the controls in full detent. Such operation results in reduced oil pressure and in turn partial engagement of the affected clutches.

THROTTLE LINKAGE, ADJUST

Buick

Buick throttle linkage is of the cable type and no provision is made for adjustment.

Chevrolet

Cable Type Exc. 1969-73 Corvette

No provision is made for adjustment of this type throttle linkage.

Rod Type, 1969

1. Disconnect throttle rod swivel at throttle lever on carburetor.
2. Hold carburetor throttle in wide open position, push throttle rod rearward to place accelerator pedal at the floor mat and adjust swivel to just enter hole in throttle lever.
3. Connect swivel to throttle lever and install accelerator return spring.

Corvette, 1969-73

1. Loosen throttle cable clamp bolt.
2. Hold accelerator pedal to floor against stop.
3. Move carburetor throttle lever to wide open position and tighten throttle cable clamp bolt to 45 in-lbs.

Oldsmobile

Cable Type

Cable type controls require no adjustment.

Pontiac

The throttle control system is of the cable type and requires no adjustment. On 1969-71 models, a reference dimension of 1 9/16 inch between the accelerator pedal and floor pan is used to check for bent bracket assemblies.

DETENT CABLE, ADJUST

Buick, Oldsmobile & Pontiac

The detent cable is adjusted from inside the driver's compartment as follows:

1. With engine off, throttle closed and fast idle off, position retainer clip against the insert on the detent cable, Fig. 9.
2. Grasp throttle pedal lever adjacent to the detent cable and pull the carburetor cable to the wide open oosition. By following this procedure, the detent cable will be adjusted properly.

Chevrolet

On Chevrolet units, the detent cable is adjusted at the carburetor.

1969 All

1. Remove air cleaner.
2. Loosen detent cable screw.
3. With choke off and accelerator linkage properly adjusted, position carburetor lever in the wide open throttle position.
4. Pull detent cable rearward until wide

open throttle stop in transmission is felt.

NOTE: Cable must be pulled through detent position to reach wide open throttle stop in transmission.

5. Tighten detent cable screw and check linkage for proper operation.

1970-71 All, 1972 Exc. Chevrolet, 1973-75 Camaro, Nova & Vega

1. Release snap lock.
2. Place carburetor lever in the wide open throttle position, being sure that lever is against the stop, push snap lock down onto cable until the top is flush with the cable.

NOTE: On models equipped with Quadrajet carburetors, be sure that the secondary lock is disengaged before placing lever in wide open throttle position.

1972 Chevrolet

1. Pass detent cable through lower hole in accelerator lever.
2. Place forward retainer on cable and install in hole in lever.
3. Place rearward retainer on cable as shown in Fig. 10.
4. Accelerator cable must be connected at carburetor and accelerator pedal lever, fully depress accelerator pedal to automatically adjust retainer.

1973-75 Chevrolet, Chevelle & Monte Carlo

1. With the accelerator and detent cables connected, depress accelerator pedal to the wide open throttle position. The ball will then slide into the cable sleeve and permanently adjust the cable length.

INTERMEDIATE BAND, ADJUST

Turbo Hydra-matic 250

Since the Turbo Hydra-matic 250 transmission uses an intermediate band instead of a clutch (used in the Turbo Hydra-matic 350 transmission) to control the operation of the planetary gear sets, it is necessary to adjust the intermediate band as follows:

1. Loosen adjusting screw lock nut, located on case right side, ½ turn.
2. Torque adjusting screw to 30 inch pounds, then back off screw 3 turns.
3. Torque adjusting screw lock nut to 15 foot pounds while holding adjusting screw in position.

IN CAR REPAIRS

Valve Body Assembly

1. Remove oil pan and strainer.
2. Remove retaining pin to disconnect downshift actuating lever bracket, remove valve body attaching bolts and detent roller and spring assembly.
3. Remove valve body assembly while disconnecting manual control valve link from range selector inner lever.

CAUTION: Do not drop valve.

4. Remove manual valve and link from valve body assembly.
5. Reverse procedure to install.

Governor

1. On 1969 Oldsmobile to gain access to the governor, remove shift linkage and crossmember bolts. Raise transmission with a suitable jack, remove crossmember and lower transmission.
2. Remove governor cover retainer and cover.
3. Remove governor.

Intermediate Clutch Accumulator Piston Assembly

1. Remove two oil pan bolts adjacent to accumulator piston cover, install compressor on oil pan lip and retain with these two bolts, Fig. 11.
2. Compress intermediate clutch accumulator piston cover and remove retaining ring piston cover and O ring from case.
3. Remove spring and intermediate clutch accumulator piston.

Vacuum Modulator & Modulator Valve Assembly

1. Disconnect vacuum hose from modulator stem and remove vacuum modulator screw and retainer.
2. Remove modulator and its O ring.
3. Remove modulator valve from case.

Extension Housing Oil Seal

1. Remove propeller shaft.
2. Pry out lip seal with screwdriver or small chisel.

Manual Shaft, Range Selector Inner Lever & Parking Linkage Assemblies

1. Remove oil pan and strainer.
2. Remove manual shaft to case retainer and unthread jam nut holding range selector inner lever to manual shaft.
3. Remove jam nut and remove manual shaft from range selector inner lever and case. *Do not remove manual shaft lip seal unless replacement is required.*
4. Disconnect parking pawl actuating rod from range selector inner lever and remove bolt from case.
5. Remove bolts and parking lock bracket.
6. Remove pawl disengaging spring.
7. If necessary to replace pawl or shaft, clean up bore in case and remove shaft retaining plug, shaft and pawl.

TRANSMISSION, REPLACE BUICK

1969-75

1. Raise car and remove propeller shaft. If necessary, disconnect exhaust crossover pipe. Remove catalytic converter support bracket, if equipped.
2. Place suitable jack under transmission and fasten transmission securely to jack.
3. Remove vacuum line from vacuum modulator.
4. Loosen cooler line nuts and separate cooler lines from transmission.
5. Remove detent cable from accelerator lever assembly. *Do not bend cable.* Remove plastic guide from bracket and slide cable out through slot.
6. Remove detent cable from detent valve link.
7. Remove crossmember.
8. Disconnect speedometer cable, shift linkage and filler pipe. Remove filler pipe.
9. Support engine at oil pan.
10. Remove transmission flywheel cover pan.
11. Mark flywheel and converter for reassembly and remove three flywheel to converter bolts.
12. Be sure transmission is supported by transmission jack and remove transmission case to engine block bolts.
13. Move transmission rearward to provide clearance between converter and crankshaft. Install converter holding tool, lower transmission and remove.

CHEVROLET

1969-75

1. Disconnect negative battery cable and raise car.
2. On 1969 Camaro, disconnect parking brake cables, remove underbody reinforcement plate (convertible). Disconnect left exhaust pipe from manifold.
3. On all models, remove propeller shaft, disconnect speedometer cable, detent cable, modulator vacuum line and oil cooler lines. Disconnect catalytic converter support bracket from transmission.
4. Disconnect shift linkage.
5. Support transmission with suitable jack and remove crossmember.
6. Remove converter under pan.
7. Remove converter to flywheel bolts.
8. On 1969 Chevrolet and Chevelle and all 1970-75 models except Nova, loosen exhaust pipe to manifold bolts approximately ¼ inch. Lower transmission until jack is barely supporting

transmission.

NOTE: On V8 engines, care must be taken not to lower the rear of the transmission too far as the distributor housing may be forced against firewall causing damage to the distributor.

9. Remove transmission to engine mounting bolts and remove oil filler tube at transmission.
10. Raise transmission to its normal position, support engine with jack and slide transmission rearward from engine and lower it away from vehicle.
11. Reverse procedure to install.

OLDSMOBILE

1969-75

1. From inside the car, slide clip to the end of the detent cable.
2. Remove transmission oil level dipstick. Remove catalytic converter support bracket, if equipped.
3. Raise car and remove detent cable from link. Plug hole.
4. Disconnect oil cooler lines at transmission.
5. Remove flywheel cover pan and mark converter and flywheel for reassembly. Remove three flywheel to converter bolts.
6. Disconnect vacuum modulator line. Remove speedometer clip and driven gear. Plug hole.
7. Disconnect shift linkage.
8. Remove propeller shaft.
9. Support transmission with a suitable jack and remove crossmember.
10. Lower transmission slightly and remove transmission to engine bolts.
11. Remove oil level indicator tube and clip holding detent cable to tube.
12. Lower transmission being careful not to damage cooler lines, detent cable, modulator line and shift linkage.
13. Reverse procedure to install.

PONTIAC

1969-75

1. Disconnect battery ground cable and release parking brake.
2. Raise car and remove propeller shaft. Remove catalytic converter support bracket, if equipped.
3. Disconnect speedometer cable, vacuum hose at modulator, detent cable at transmission and shift linkage.

NOTE. When removing detent cable, be careful not to bend it.

4. Support transmission with a suitable jack and remove crossmember.
5. On 1969 Firebird L6, remove driveline damper.
6. Remove converter dust pan, mark flywheel and converter for reassembly and remove flywheel to converter bolts. Make sure converter hub is free of converter.
7. Disconnect transmission filler pipe at engine and remove pipe from transmission.
8. Lower transmission and engine to gain access to cooler line fitting nuts and disconnect cooler lines. On some models it may be necessary to loosen the exhaust system.
9. With transmission in lowered position, remove the transmission to engine bolts.
10. Raise transmission to its normal position, support engine and slide transmission rearward and lower it away from car.

NOTE: When lowering transmission, keep rear of transmission lower than the front so as not to lose the converter.

11. Reverse procedure to install.

BUICK SUPER TURBINE 300
OLDS JETAWAY, 1969
PONTIAC TWO SPEED, 1969

TRANSMISSION IDENTIFICATION

The transmissions can be identified by either the metal tag or the stamping on the low servo cover. The data is stamped on the low servo cover. The identification data includes the year of production, code letters or numbers followed by the transmission serial number. The model application is as follows:

BUICK		CODE
1969	All Sportwagons	MH
	All V8-350 engines except Sportwagons	ME
	All models with 6-250 engine	LC

OLDSMOBILE MODELS		
1969	6-250	LC
	V8-350 Except Delta "88"	MM
	V8-350 Delta "88"	MS
	V8-350 Delta "88" w/Cruise Control	MU

PONTIAC INTERMEDIATE MODELS		
1969	Six w/1 bar. carb.—Tempest	LA
	Six w/1 bar. carb. & air cond.—Tempest	LD
	Six w/1 bar. carb.—Firebird	LF
	Six w/1 bar. carb. & air cond.—Firebird	LH
	V8-350 w/2 bar. carb.	MA
	V8-350 w/2 bar. carb. & air cond.	MC

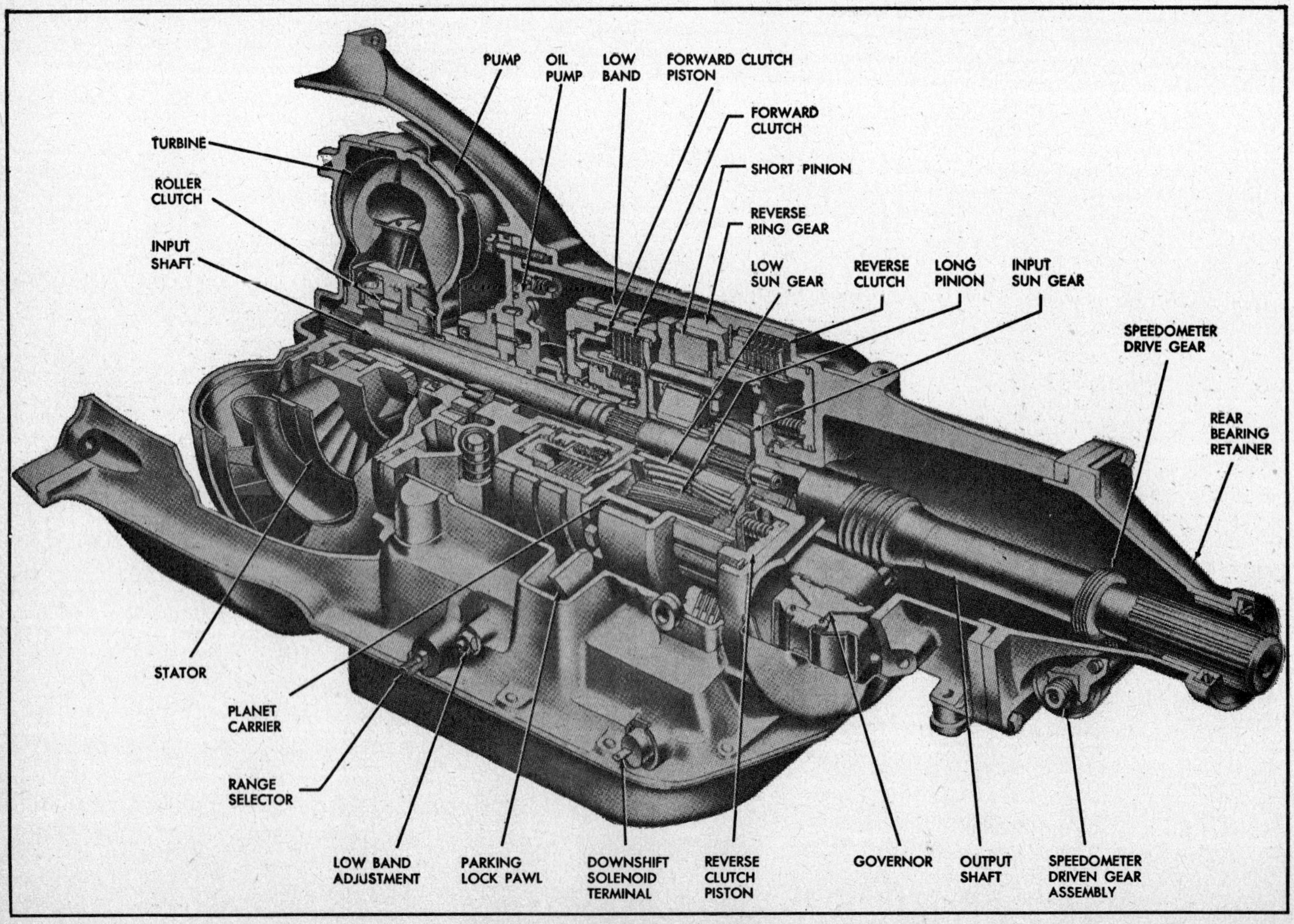

Fig. 1 Cutaway view of transmission

DESCRIPTION

This transmission, Fig. 1, is a combination torque converter and two-speed planetary geared unit. Torque multiplication is obtained hydraulically through the converter and mechanically through the compound planetary gear set. The gear set, in combination with the torque converter, provides a high starting ratio for acceleration from a stop, up steep grades, etc. The torque converter provides torque multiplication for performance and exceptionally smooth operation. It functions as a fluid coupling at normal road load conditions and at higher speeds.

Torque Converter

The torque converter is connected to the engine flywheel and serves as a hydraulic coupling through which engine torque is transmitted to the input shaft. The converter steps up or multiplies engine torque whenever operating conditions demand greater torque than the engine alone can supply.

Converter Pump

The function of the converter pump is to convert engine torque into an energy transmitting a flow of oil to drive the converter turbine into which the oil is projected. The converter pump operates as a centrifugal pump, picking up oil at its center and discharging the oil at its rim. However, the converter is shaped to discharge the oil parallel to its axis in the form of a spinning hollow cylinder.

Fixed Stator

Used on all 1969 units, the performance of the fixed stator is the same as the variable pitch type except that the stator remains at a pre-set angle.

Converter Turbine

The function of the converter turbine is to absorb energy from the oil projected into it by the converter pump and convert the energy into torque and transmit that torque to the input shaft.

External Controls

Vacuum Modulator

The vacuum modulator is used to sense automatically any change in the torque input to the transmission. The vacuum modulator transmits this signal to the pressure regulator, which controls line pressure, so that all torque requirements of the transmission are met and smooth shifts are obtained at all throttle openings.

Detent Solenoid

The detent solenoid is activated by the detent switch in the throttle linkage. When the throttle is fully open, the switch is closed, activating the detent solenoid and causing the transmission to downshift at speeds below approximately 60 mph.

Oil Pump

A positive displacement internal-external gear type oil pump is used to supply oil to fill the converter, for engagement of forward and reverse clutches, for application and release of the low band and to provide oil for lubrication and heat transfer.

Planetary Gear Set

The planetary gear set consists of an input sun gear, low sun gear, long and short pinions, a reverse ring gear and a planet carrier.

The input sun gear is splined to the input shaft. The low sun gear, which is part of the forward clutch assembly, may revolve freely until the low band or high clutch is applied.

The input sun gear is in mesh with three long pinions and the long pinions are in mesh with three short pinions. The short pinions are in mesh with the low sun gear and reverse ring gear.

The input sun gear and short pinions always rotate in the same direction. Application of either the low band or the reverse clutch determines whether the output shaft rotates forward or backward.

Forward Clutch

The forward clutch assembly consists of a drum, piston, cushion ring, springs, piston seals, and a clutch pack. These parts are retained inside the drum by the low sun gear and the flange assembly and retainer ring.

When oil pressure is applied to the piston, the clutch plates are pressed together, connecting the clutch drum to the input shaft through the clutch hub. Engagement of the clutch causes the low sun gear to rotate with the input shaft.

Low Band

The low band is a double-wrap steel band faced with a bonded lining which surrounds the forward clutch drum. The band is hydraulically applied by the low servo piston, and released by spring pressure.

Reverse Clutch

The reverse clutch consists of a piston, cushion ring, inner and outer seal, coil springs, clutch pack, and reaction plate. These parts are retained inside the case by a retaining snap ring.

When oil pressure is applied to the piston, the clutch plates are pressed together, holding the reverse ring gear stationary. This engagement of the clutch causes reverse rotation of the output shaft.

Governor

The governor is located to the rear of the transmission case on the left side and is driven off the output shaft. The purpose of the governor is to generate a speed sensitive modulating oil pressure that increases up to a point with output shaft or car speed.

Valve Body

The valve body assembly is bolted to the bottom of the transmission case and is accessible for service by removing the oil pan. The valve body assembly consists of manual valve, shift valve, modulator limit valve, and high speed downshift timing valve.

TROUBLE SHOOTING GUIDE

Oil Forced Out Of Filler Tube

1. Oil level too high; foaming caused by planet carrier running in oil.
2. Water in oil.
3. Leak in pump suction circuits.

Oil Leaks

1. Check extension oil seal.
2. Check outer shift lever oil seal.
3. Check speedometer driven gear fitting.
4. Check oil cooler pipe connections.
5. Check vacuum modulator assembly and case.

No Drive In Any Position

1. Low oil level.
2. Clogged oil strainer screen or suction pipe loose.
3. Defective pressure regulator valve.
4. Front pump defective.
5. Input shaft broken.

Erratic Operation and Slippage Light to Medium Throttle

1. Low oil level.
2. Clogged oil strainer screen.
3. Servo piston seal leaking.
4. Band facing worn.
5. Low band apply struts disengaged or broken.
6. Vacuum modulator.

Engine Speed Flares On Upshifts

1. Low oil level.
2. Improper band adjustment.
3. Clogged oil strainer screen.
4. Forward clutch not fully engaging.
5. Forward clutch plates worn.
6. Forward clutch piston hanging up.
7. Forward clutch drum relief ball not sealing.
8. Vacuum modulator.

Upshifts Harsh

1. Vacuum modulator line broken or disconnected.
2. Vacuum modulator diaphragm leaks.
3. Vacuum modulator valve stuck.

Closed Throttle (Coast) Downshift Harsh

1. High engine idle speed.
2. Improper low band adjustment.
3. Downshift timing valve malfunction.
4. High main line pressure. Check the following: a) vacuum modulator line broken or disconnected, b) modulator diaphragm ruptured, c) sticking pressure regulator coast valve, pressure regulator valve or vacuum modulator valve.

Car Creeps Excessively In Drive

1. Idle speed too high.

Car Creeps In Neutral

1. Forward clutch not released.
2. Low band not released.

No Drive In Reverse

1. Reverse clutch piston stuck.
2. Reverse clutch plates worn out.
3. Reverse clutch seal leaking excessively.
4. Blocked reverse clutch apply orifice.

MAINTENANCE

Checking Oil Level

The transmission oil level should be checked every 6000 miles. Oil should be added only when the level is near the ADD mark on the dipstick with oil at normal operating temperature.

NOTE: *The difference in oil level between FULL and ADD is one pint.*

To check oil level accurately, the car should be level, the engine should be idled with the transmission oil at normal temperature, and the control lever in Park position.

It is important that the oil level be maintained no higher than the FULL mark. *Do not overfill*, for when the oil level is at the full mark on the dipstick, it is just slightly below the planetary gear unit. If oil is added which brings the level above the full mark, the planetary unit will run in the oil foaming and aerating the oil. This will cause malfunctioning of the transmission assembly due to improper application of the band or clutches and excessive temperature.

If the transmission is found to be consistently low on oil, a thorough inspection should be made to find and correct all external oil leaks. All mating surfaces, such as the oil pan rail, filler tube, governor and modulator should be carefully examined for signs of leakage. The modulator must also be checked to insure that the diaphragm has not ruptured as this would allow transmission oil to be drawn into the intake manifold of the engine. Usually, the exhaust will be excessively smoky if the diaphragm ruptures, due to transmission oil being drawn into the combustion chambers of the engine.

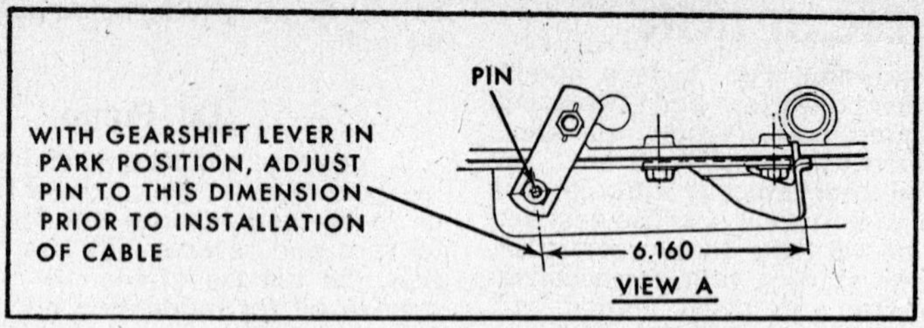

Fig. 2 Console shift adjustment. Pontiac Tempest & Firebird. Dimension for 1969 is 7.260

Draining & Refilling

Draining the transmission oil at 24,000 mile intervals is recommended. Drain the oil by removing the oil pan (no drain plug is provided). Clean oil strainer.

To refill the transmission, replace the oil pan, using a new gasket, and add five (5) pints of transmission fluid, using filler tube and funnel. Start and allow engine to idle in Park position three to five minutes to warm the oil, then check oil level and add as required to bring the level to the Full mark. Assuming that the converter has not been drained (since it is welded) and allowing for normal spillage or drain-down, approximately six pints will be required for refill.

MANUAL LINKAGE, ADJUST

Buick

1969 Column Shift & Back Drive

1. Loosen swivel clamp bolt.
2. Place selector lever against Drive stop.
3. Place transmission lever in Drive detent (2nd from rear).
4. Torque swivel clamp bolt to 17-23 ft-lbs.

1969 Console Shift & Back Drive

1. Loosen shift rod trunnion nut (1968).
2. Place selector lever against Drive stop.
3. Place transmission in Drive detent (2nd from rear).
4. Tighten adjusting clamp bolt to 17-23 ft-llbs or trunnion nut to 6-9 ft-lbs.
5. Adjust back drive: Place transmission in Park push back drive rod up until column lever hits stop, hold lightly and tighten clamp.

Oldsmobile

1969 Console Shift & Back Drive

1. Place shift handle in Park position, transmission in Park position and ignition key in lock.
2. Loosen shift rod clamp and pull shift rod lightly against lock stop. Tighten shift rod clamp.
3. Move shift cable pin to obtain a free pin fit in transmission lever and tighten nut.
4. Check to be sure that with key in Run position and transmission in Reverse, key cannot be removed and steering wheel is not locked. With key in Lock position and transmission in Park, be sure that key can be removed and steering wheel is locked.

1969 Column Shift & Back Drive

1. Set transmission shift lever in Drive.
2. Loosen swivel bolt.
3. Hold upper shift lever against Drive stop in upper steering column. Tighten swivel bolt.
4. Check to be sure that with key in Run position, and transmission in Reverse, key cannot be removed and steering wheel is not locked. With key in Lock position and transmission in Park, be sure that key can be removed and steering wheel is locked.

Pontiac Tempest & Firebird

1969 Column Shift & Back Drive

1. Loosen adjusting swivel clamp.
2. Set transmission in Park position.
3. Set upper gearshift lever in Park position and lock ignition.
4. Push up on gearshift control rod to take up clearance in steering column lock mechanism and tighten swivel clamp to 20 ft-lbs. on Tempest and 30 ft-lbs. on Firebird.

1969 Console Shift & Back Drive

1. Disconnect shift cable from transmission range selector lever pin.
2. Set console gearshift lever in Park and lock ignition.
3. Rotate transmision range selector lever clockwise to Park position and adjust pin on selector to 7.260", Fig. 2.
4. Torque pin nut to 30 ft-lbs. and connect shift cable to pin.

THROTTLE LINKAGE

Buick 1969

Intermediate Models

The flexible cable type linkage is used and is not adjustable.

Le Sabre

1. Remove air cleaner.
2. Make sure linkage is free in all positions.
3. Hold choke open and make sure that return spring fully closes throttle, even though throttle is released very slowly.
4. Adjust engine idle speed and mixture.

5. With throttle linkage at hot idle position, measurement from throttle rod clevis pin to dash must be 5⅞".
6. If measurement is off, shorten or lengthen operating rod as required.
7. Operate linkage to open carburetor and make sure carburetor wide open stop is contacting. If carburetor does not reach wide open position and nothing is interfering with throttle linkage, transmission stator and detent switch must be adjusted as outlined below.

Oldsmobile

1969 Models

These units use cable type accelerator controls and no provision is made for adjustment.

Pontiac Tempest & Firebird

1969

Throttle linkage adjustments cannot be made. A reference dimension of 1⁹⁄₁₆" between the bottom of the accelerator pedal roller and floor pan can be used only as a check for bent bracket assemblies.

DOWNSHIFT SWITCH

Buick, Olds, Pontiac 1969

The downshift switch is in the throttle linkage. To adjust, depress the switch plunger fully to insure proper setting, then depress the accelerator pedal fully.

IN CAR REPAIRS

The following operations can be performed without removing the transmission from the car.

1. Oil pan and strainer.
2. Rear bearing retainer.
3. Vacuum modulator.
4. Valve body.
5. Governor.
6. Low servo.
7. Selector and parking mechanism.
8. Pressure regulator valve.

LOW BAND, ADJUST

The low band adjusting screw is located on the left side of the transmission adjacent to the range selector lever. Adjustment is made as follows:

Remove protective cap, loosen lock nut and tighten adjusting screw 35 to 45 inch pounds with a torque wrench. Then back off *exactly four full turns.* While holding adjusting screw stationary, tighten lock nut securely and replace cap.

TRANSMISSION, REPLACE

Buick

1. Raise and support front and rear of car.
2. Disconnect front exhaust crossover pipe if necessary.
3. Remove propeller shaft.
4. Support transmission with suitable jack.
5. Remove line from vacuum modulator.
6. Separate cooler lines from transmission.
7. Remove cross member.
8. Disconnect speedometer cable.
9. Where necessary, disconnect detent solenoid wire.
10. Disconnect shift linkage from transmission.
11. Remove oil filler pipe.
12. Support engine at oil pan.
13. Remove flywheel cover pan.
14. Mark flywheel and converter for reassembly in same position and remove flywheel to converter bolts.
15. Unfasten transmission from engine.
16. Move transmisson rearward to provide clearance between converter pump and crankshaft.
17. Install holding tool to retain converter, then lower transmission from car.
18. Reverse procedure to install.

Oldsmobile

1. Remove transmission oil filler pipe. Hoist car.
2. Disconnect control wires at transmission, and manual rod from transmission lever.
3. Remove propeller shaft.
4. Remove flywheel dust cover.
5. Support engine with a suitable jack or support bar.
6. Remove transmission cross support bar. *On models with dual exhaust, it may be necessary to disconnect the left hand exhaust pipe at exhaust manifold to provide clearance.*
7. Disconnect and cap oil cooler lines.
8. Disconnect speedometer cable (or speed adapter if so equipped) from speedometer driven gear.
9. Remove three flywheel-to-converter attaching bolts. Mark flywheel and converter so they can be assembled in the same relationship.
10. Support transmission with a suitable lift and remove transmission-to-flywheel housing bolts. *It may be necessary to lower engine slightly to permit removal of the upper bolts.*
11. Carefully move transmission rearward and out of car.
12. Reverse removal procedure to install the unit and adjust the shift linkage as outlined below.

Pontiac Tempest & Firebird

1. Disconnect speedometer cable and remove speedometer driven gear to allow oil to drain during removal procedure.
2. Remove propeller shaft.
3. Disconnect vacuum line and downshift switch lead.
4. Disconnect shift linkage from outer shift lever.
5. Support transmission and remove frame crossmember.
6. Remove flywheel housing bottom cover.
7. After removing flywheel-to-converter bolts, make sure converter hub is free of crankshaft.
8. Lower transmission and engine assembly to gain access to cooler line fittings (V8 only). Disconnect cooler lines, using a crowfoot adapter and a suitable extension. *On some cars it may be necessary to loosen exhaust system.*
9. With transmission in lowered position, remove case-to-engine bolts.
10. Move transmission down and to the rear and install a suitable strap across the converter housing to hold the converter in position until transmission is to be disassembled.
11. Reverse removal procedure to install the unit and adjust shift linkage.

CHEVROLET POWERGLIDE
PONTIAC TWO SPEED, 1970-73

TRANSMISSION IDENTIFICATION

On 1969-73 models the unit number is located on the right rear vertical surface of oil pan. There is no particular code with which a transmission may be identified. Therefore, when ordering parts use the transmission serial number. This unit is also used on 1970-73 Pontiac cars.

DESCRIPTION

This Powerglide is essentially a torque converter coupled to a two-speed transmission, Fig. 1. The gear portion of the transmission is a two-speed compound planetary gear set, permitting a gear reduction of 1.82 to 1 on the light duty version and 1.76 to 1 on the heavy duty version. The shift from low gear to direct drive is automatic, and the vehicle speed at which the shifts occur is determined by the interaction of a governor driven by the output shaft and a throttle valve controlled by the accelerator pedal. Thus, the transmission starts with both the torque converter and the gear box multiplying torque. As vehicle speed increases, the gear box section upshifts to direct drive, leaving only the torque converter for any speed-torque changes required. The torque mutliplication ability of the converter multiplied by the planetary gear reduction allows an overall torque multiplication of approximately 4.55 to 1.

Fig. 1 Sectional view of Powerglide with aluminum case

TROUBLE SHOOTING

Oil Forced Out Of Filler Tube

1. Oil level too high; aeration and foaming caused by planet carrier running in oil.
2. Water in oil.
3. Leak in pump suction circuits.

Oil Leaks

1. Transmission case and extension: extension oil seal, shifter shaft oil seal, speedometer driven gear fitting, pressure taps, oil cooler pipe connections, vacuum modulator and case, transmission oil pan gasket.
2. A very smoky exhaust indicates a ruptured vacuum modulator diaphragm.
3. Converter cover pan; front pump attaching bolts, pump seal ring, pump oil seal, plugged oil drain in front pump, porosity in transmission case.

No Drive In Any Position

1. Low oil level.
2. Clogged oil suction screen.
3. Defective pressure regulator valve.
4. Front pump defective.
5. Input shaft broken.
6. Front pump priming valve stuck.

Erratic Operation and Slippage—Light to Medium Throttle

1. Low oil level.
2. Clogged oil suction screen.
3. Improper band adjustment.
4. Band facing worn.
5. Low band apply linkage disengaged or broken.
6. Servo apply passage blocked.
7. Servo piston ring broken or leaking.
8. Converter stator not holding (rare).

Engine Speed Flares On Upshift

1. Low oil level.
2. Improper band adjustment.
3. Clogged oil suction screen.
4. High clutch partially applied—blocked feed orifice.
5. High clutch plates worn.
6. High clutch seals leak.
7. High clutch piston hung up.
8. High clutch drum relief ball not sealing.
9. Vacuum modulator line plugged.
10. Vacuum modulator defective.

Will Not Upshift

1. Maladjusted manual valve lever.
2. Throttle valve stuck or maladjusted.
3. No rear oil pump output caused by stuck priming valve, sheared drive pin or defective pump.
4. Defective governor.
5. Stuck low-drive valve.

Harsh Upshifts

1. Throttle valve linkage improperly adjusted.
2. Vacuum modulator line broken or disconnected.
3. Vacuum modulator diaphragm leaks.
4. Vacuum modulator valve stuck.
5. Hydraulic modulator valve stuck.
6. Improper low band adjustment.

Harsh Closed Throttle (Coast) Downshifts

1. High engine idle speed.
2. Improper band adjustment.
3. Vacuum modulator line broken or disconnected.
4. Modulator diaphragm ruptured.
5. Sticking hydraulic modulator valve, pressure regulator valve or vacuum modulator valve.
6. Downshift timing valve malfunction.

No Downshift (Direct-to-Low) Accelerator Floored

1. Throttle control linkage improperly adjusted.
2. Sticking shifter valve or throttle and detent valve.

Car Creeps In Neutral

1. Manual control linkage improperly adjusted.
2. High clutch or low band not released.

No Drive In Reverse

1. Manual control linkage improperly adjusted.
2. Reverse clutch piston stuck.
3. Reverse clutch plates worn out.
4. Reverse clutch leaking excessively.
5. Blocked reverse clutch apply orifice.

Improper Shift Points

1. Throttle valve linkage improperly adjusted.

2. Incorrectly adjusted throttle velve.
3. Defective governor.
4. Rear pump priming valve stuck.

Burned Clutch Plates

1. Band adjusting screw backed off more than specified.
2. Improper order of clutch plate assembly.
3. Extended operation with low oil level.
4. Stuck relief ball in clutch drum.
5. Abnormally high speed upshift, probably due to:
   a. Improper governor action.
   b. Transmission operated at high speed in manual "Low".

MAINTENANCE

Oil Level

The transmission oil level should be checked every 1000 miles. Oil should be added only when the level is near the "Add" mark on the dipstick with oil hot or at operating temperature.

In order to check oil level accurately, the engine should be idled with the transmission oil hot and the control lever in neutral "N" position.

It is important that the oil level be maintained no higher than the "Full" mark on the oil level gauge. *Do not overfill for when the oil level is at the full mark on the dipstick, it is just slightly below the planetary gear unit. If additional oil is added, bringing the level above the full mark, the planetary unit will run in the oil, foaming and aerating the oil. This aerated oil carried through the various oil pressure passages may cause malfunction of the transmission assembly, resulting in cavitation noise in the converter and improper band or clutch application.*

NOTE: A revised type Dexron fluid is used in 1973 Chevrolet and late 1973 Pontiac transmissions. An early change to a darker color from the usual red color and or a strong odor usually associated with overheated fluid is normal and should not be treated as a positive sign of needed maintenance or unit failure.

The normal maintenance schedule for drain and refill of this type fluid remains unchanged at 24,000 miles under normal service and 12,000 miles under severe operating conditions, such as trailer towing.

Changing Oil

Periodic draining of the oil pan when equipped with a drain plug is recommended every 12,000 miles under normal operating conditions and more frequently under extreme service usage. It is realized that only a portion of the total transmission fluid can be drained at the oil pan. However, the addition of even this volume of fresh fluid will replenish the additives in the remaining fluid sufficiently to increase transmission durability.

After draining the oil pan, pour two quarts of Dexron transmission fluid into the transmission. Then set the parking brake and operate the transmission through all ranges. With engine idling, transmission selector lever in neutral, and transmission at operating temperature, recheck the fluid level and add fluid as necessary to bring the level to the "Full" mark on the dipstick. Do not overfill as damage to the transmission can result.

Units Without Drain Plug:

When the transmission is to be removed for repairs, drain and refill as follows:

To drain the transmission, carefully loosen the oil pan bolts. Position a receptacle to catch the draining oil. If the transmission is to be removed for repairs, the draining operation may be performed after removal, if desired.

To refill the transmission, remove the dipstick from the filler tube and refill the transmission with Dexron fluid. The engine should then be run at a fast idle speed with the transmission in neutral until the oil warms up. Then add oil as required to raise the fluid level on the dipstick to the "Full" mark. Refill capacity is 1½ quarts.

MANUAL LINKAGE, ADJUST

CAUTION: Shift linkage adjustment must be accurately made. Any inaccuracies may result in premature failure of the transmission due to operation without controls in full detent. Such operation results in reduced oil pressure and in turn partial engagement of the affected clutches. Partial engagement of the clutches with sufficient pressure to cause apparent normal operation of the vehicle will result in failure of the clutches or other internal parts after only a few miles of operation.

1969-73 Chevrolet Column Shift

1. Shift tube and lever assembly must be free in mast jacket.
2. To check for proper adjustment, lift selector lever toward steering wheel. Allow selector lever to be positioned in D by transmission detent.

   NOTE: Do NOT use the indicator pointer as a reference to position the selected lever. When performing linkage adjustment, pointer is adjusted last.

3. Release selector lever. Lever should be inhibited from engaging low range unless the lever is lifted.
4. Lift selector lever towards steering wheel and allow lever to be positioned in N by the transmission detent.
5. Release selector lever. Lever should now be inhibited from engaging reverse range unless lever is lifted.

   NOTE: A properly adjusted linkage will prevent the selector lever from moving from beyond both the neutral detent and the drive detent unless the lever is lifted to pass over the mechanical stop in the steering column.

6. If an adjustment is required, place the selector lever in D as determined by the transmission detent (see Steps 2 and 3).
7. Loosen adjusting swivel at cross shaft and rotate the transmission lever so that it contacts the drive stop in the steering column.
8. Tighten swivel and recheck adjustment.
9. Readjust indicator needle if required.
10. Adjust back drive linkage as described further on.

1969-73 Nova Console Shift

1. Loosen assembly nuts A and B, Fig. 2, on lower rod.
2. Set transmission lever in Drive position.
3. Set pawl rod in the Neutral or Drive notch of detent.
4. Apply load in direction of arrow on actuating lever until pawl rod comes in contact with detent.
5. Place a .094″ spacer between nut and swivel and run nut up until it touches spacer. Tighten nut B against swivel and lock swivel between nuts.
6. Place transmission and shift lever in Park position.
7. Install column rod to column lever, and cross shaft.
8. Set selector lever in Park and adjust Back Drive as described further on.

NOTE: With ignition key in the Lock position and transmission selector lever in Park, the key can be removed and the steering wheel locked.

1973 Vega Console Shift

1. Place transmission lever in Neutral by rotating lever clockwise to Park, then two detents counter-clockwise to Neutral.
2. Place shift lever in Neutral.
3. Insert flats on swivel into slot in rod and tighten nut to 120 pound inches, Fig. 2A.

1971-72 Vega Console Shift

1. Loosen nut and perform steps 1 and 2 as outlined in the 1973 Vega Console Shift adjustment.
2. Apply rearward force on lever until shift lever contacts detent stop. Install a .073 inch spacer between forward nut and swivel and tighten nut against spacer, Fig. 2B.
3. Remove spacer and apply a forward force on lever and tighten rearward nut.

1970-72 Chevelle & Monte Carlo Console Shift

1. Place shift lever in Drive.
2. Raise vehicle. Disconnect cable from transmission lever, Fig. 3.
3. Manually place transmission lever in Drive.
4. Measure distance from rearward face of attachment bracket to center of cable attachment pin. This dimension

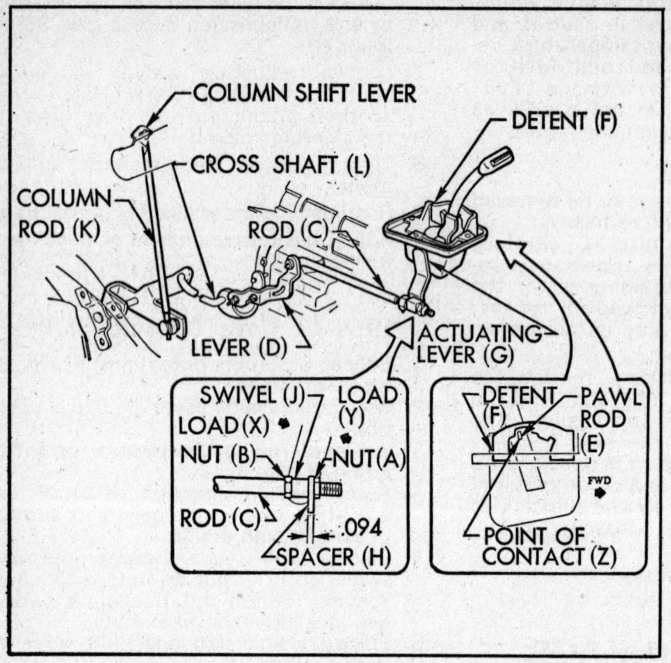

Fig. 2 Manual linkage adjustment. 1969-73 Nova console shift

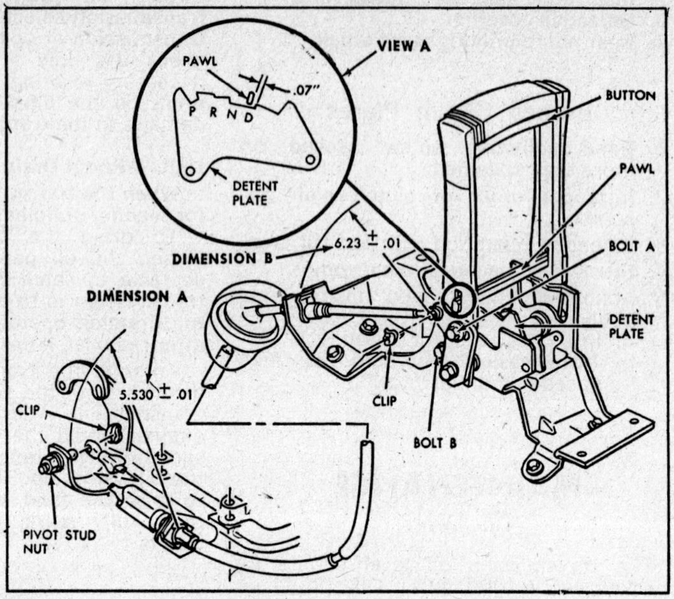

Fig. 3A Floor shift linkage adjustment. 1969 Camaro, Chevelle & Chevrolet

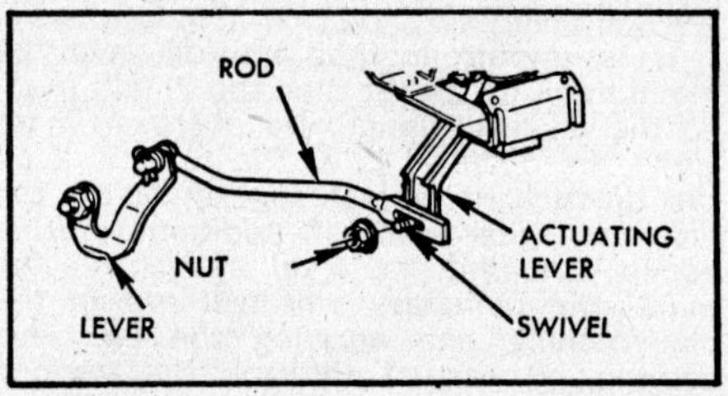

Fig. 2A Manual linkage adjustment. 1973 Vega console shift

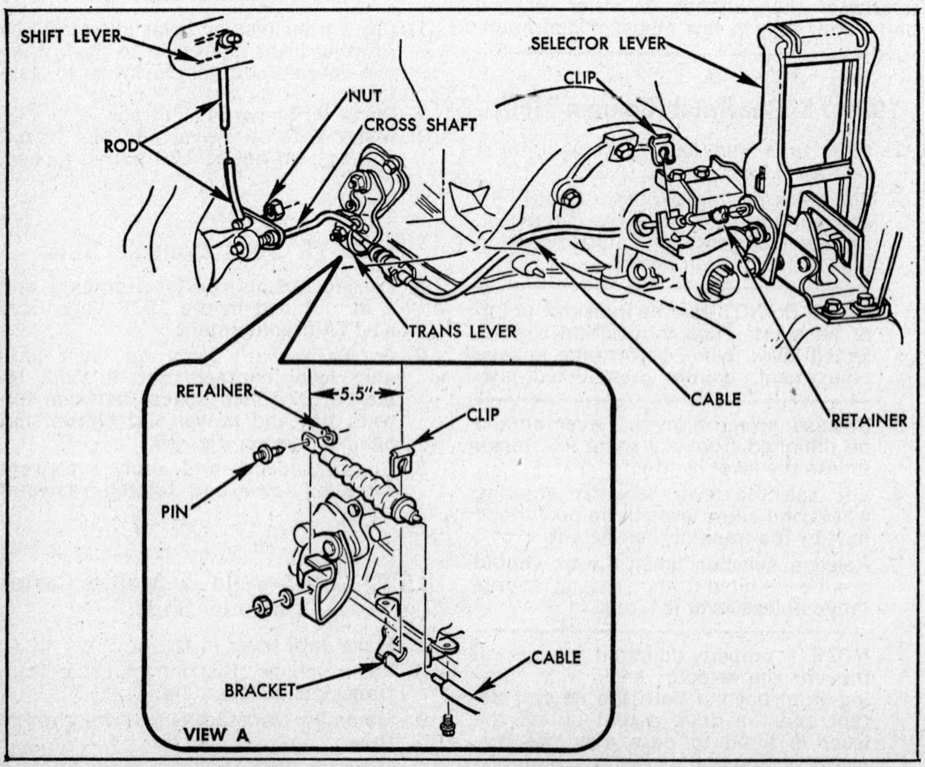

Fig. 3 Floor shift linkage adjustment. 1970-72 Chevelle & Monte Carlo

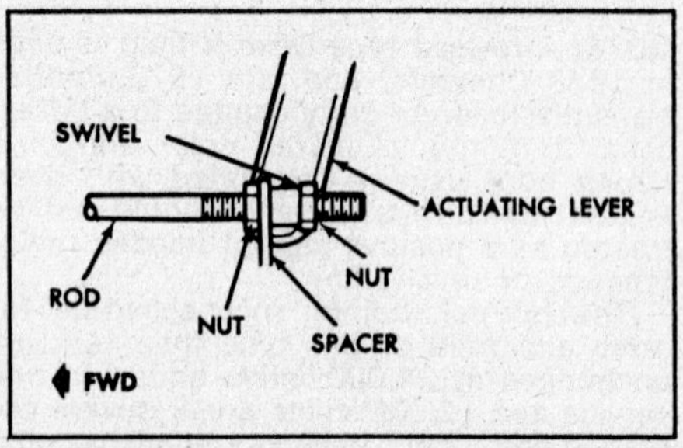

Fig. 2B Manual linkage adjustment. 1971-72 Vega console shift

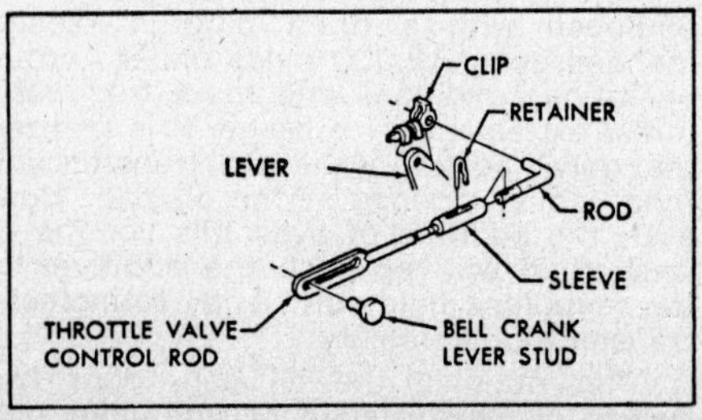

Fig. 3B Throttle linkage adjustment. 1971-73 Vega

should be 5.5". If not, adjust pin to obtain this dimension.

5. Install cable to transmission lever, lower vehicle and check for proper operation.

1969 Floor Shift

Camaro, Chevelle and Chevrolet

1. Place shift handle in Drive.
2. Raise vehicle. Disconnect cable from transmission lever, Fig. 3A. Place transmission in Drive range. Check dimension A and adjust by loosening stud nut and moving stud to obtain proper setting. Reinstall cable and lock clip.
3. Lower vehicle and remove shift quadrant cover. Raise quadrant plate, disconnect bulbs and remove plate. Remove cable clip and disengage cable from shift lever.
4. Insert a .07" gauge between pawl and detent as shown in view A. Check dimension B. Adjust by loosening bolt A and moving lever to obtain proper adjustment. Tighten bolt.
5. Adjust cable end until it freely enters pin. Install clip.

NOTE: If lift on handle button does not clear detents or detents can be cleared without lifting handle, pawl engagement can be adjusted by raising or lowering of the detent plate after loosening bolt B. If such an adjustment is made repeat step 4.

6. Remove gauge. Check operation.
7. Reinstall shift plate and cover.
8. Adjust back drive linkage as outlined below.

1970-73 Pontiac Column Shift

1. Place steering column selector lever in Park position and lock ignition.
2. Loosen screw on adjusting swivel clamp at shaft and lever assembly.
3. Make sure transmission range selector lever is in Park detent.
4. Push up on gearshift control rod to take up clearance in steering lock mechanism and tighten screw on adjusting swivel clamp to 20 ft. lbs.
5. Unlock ignition, readjust indicator needle if necessary to agree with transmission detent positions.

1970-73 Pontiac Console Shift

1. Place console gearshift lever in Park position and lock ignition.
2. Disconnect shift cable from transmission range selector lever by removing nut from pin.
3. Loosen screw on adjusting swivel at the shaft lever assembly.
4. Make sure transmission range selector lever is in Park position.
5. Push up on gearshift control rod to take up clearance in steering column lock mechanism and tighten screw on adjusting swivel to 20 ft. lbs.
6. Unlock ignition and rotate transmission range selector lever counterclockwise by two detent positions.
7. Set the console gearshift lever in Neutral range and move it forward against its stop in Neutral.
8. Assemble the shift cable and pin to the transmission range selector lever, allowing the cable to position its pin in the slot of the selector lever and then install and tighten the nut to 20 ft. lbs.

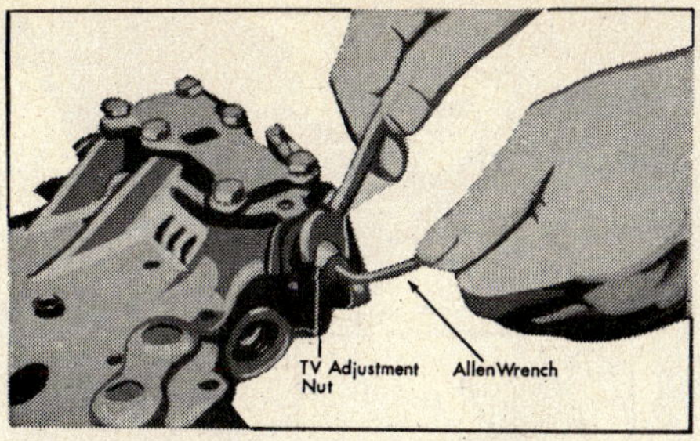

Fig. 4 Adjusting throttle pressure (on bench)

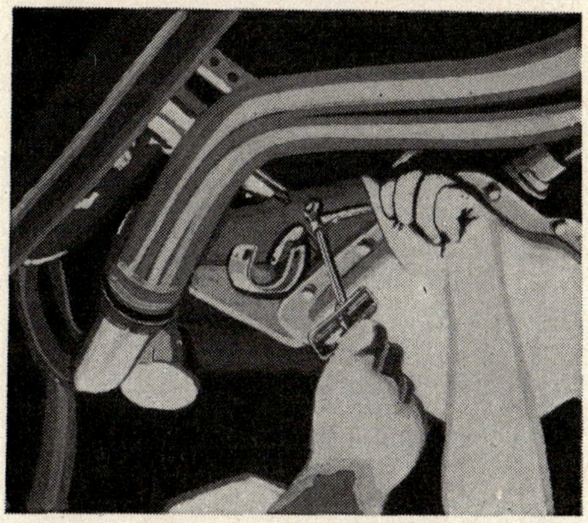

Fig. 4A Adjusting low band (in vehicle)

BACK DRIVE LINKAGE, ADJUST

1. Disconnect lower rod at transmission lever.
2. Move transmission selector lever fully into Park position.
3. Place steering column lever in Park position.
4. Attach lower rod to transmission and check operation.

CAUTION: Any inaccuracies in the above adjustment may result in premature failure of the transmission due to operation without controls in full detent.

THROTTLE LINKAGE, ADJUST

Chevrolet 4 & 6 Cyl. Engines Exc. Vega

With accelerator pedal depressed, bellcrank must be at wide open throttle position.

Dash lever must be $1/64$-$1/16$" off lever stop and transmission lever must be against transmission internal stop.

With accelerator lever in wide open position and transmission lever against stop, adjust swivel to accelerator control lever.

1971-73 Chevrolet Vega

1. Place carburetor bell crank in wide open throttle position and transmission lever against internal stop.
2. Align 90 degree rod with hole in lever, Fig. 3B, and install retainer through sleeve and rod. Attach clip to lever and rod.

Chevrolet & Pontiac V8

1. Remove air cleaner.
2. Disconnect accelerator linkage at carburetor.
3. Disconnect accelerator return and TV rod return springs.
4. With right hand, pull TV upper rod forward until transmission is through detent. With left hand, open carburetor to wide open throttle position. Carburetor must reach wide open throttle position at the same time the ball stud contacts the end of slot in upper TV rod.
5. Adjust swivel on end of upper TV rod to obtain setting described in step 4. Allowable tolerance is about $1/32$".
6. Connect and adjust accelerator linkage.
7. Check for throttle linkage freedom.

Pontiac 6 Cyl. Engine

1. Remove air cleaner.
2. Disconnect TV control rod swivel and clip from carburetor lever and TV return spring from flywheel housing.
3. With right hand, push TV control rod rearward until transmission TV lever is against internal stop of transmission.
4. With left hand, hold carburetor lever in wide open throttle position.
5. Adjust TV control rod swivel so that swivel pin freely enters hole in carburetor lever.
6. Secure swivel and control rod to carburetor lever with clip, connect return spring and check for linkage freeness. Install air cleaner.

THROTTLE VALVE, ADJUST

No provision is made for checking TV pressures. However, if operation of the transmission is such that some adjustment of the throttle valve is indicated, pressures may be raised or lowered by adjusting the position of the jam nut on the throttle valve assembly, Fig. 4.

To raise TV pressure 3 psi, back off the jam nut one full turn. This increases the dimension from the jam nut to the TV valve stop. Conversely, tightening the jam nut one full turn lowers TV pressure 3 psi.

A difference of 3 psi in TV pressure will cause a change of approximately 2 to 3 mph in the wide open throttle upshift point. Smaller pressure adjustments can be made by partial turns of the jam nut. The end of the TV adjusting screw has an allen head so the screw may be held stationary while the jam nut is moved. *Use care when making this adjustment since no pressure tap is provided to check TV pressure.*

LOW BAND, ADJUST

Low band adjustment should be performed at 12,000 mile intervals, or sooner if operating performance indicates low band slippage.

1. Raise vehicle and place selector lever in Neutral.
2. Remove protective cap from transmission adjusting screw.

NOTE: On Chevelle models, to gain clearance between underbody and transmission, it may be necessary to remove rear mount bolts from cross-member and move transmission slightly toward passenger side of vehicle.

Loosen adjusting screw locknut 3/4 turn and hold in this position with a wrench, Fig. 4A.

Using a suitable inch-pound torque wrench as shown, adjust band to 70 inch-lbs and back off four complete turns for a band that has been in operation for 6000 miles or more, or three turns for one in use less than 6000 miles.

CAUTION: Be sure to hold the locknut at 1/4 turn loose during the adjusting procedure. Then tighten locknut. The amount of back off is not an approximate figure; it must be exact.

VACUUM MODULATOR VALVE

1. To remove, disconnect line from vacuum modulator, Fig. 5.
2. Unscrew vacuum modulator from oil pan.
3. Remove vacuum modulator, gasket and valve.
4. Reverse removal procedure to install.

LOW SERVO

1. Remove servo cover and gasket (3 screws).
2. Remove cover oil seal, servo piston and return spring.
3. Reverse removal procedure to install.

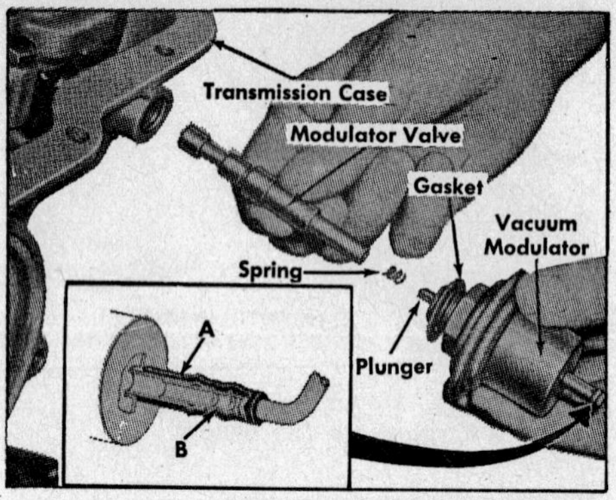

Fig. 5 Vacuum modulator, gasket and valve

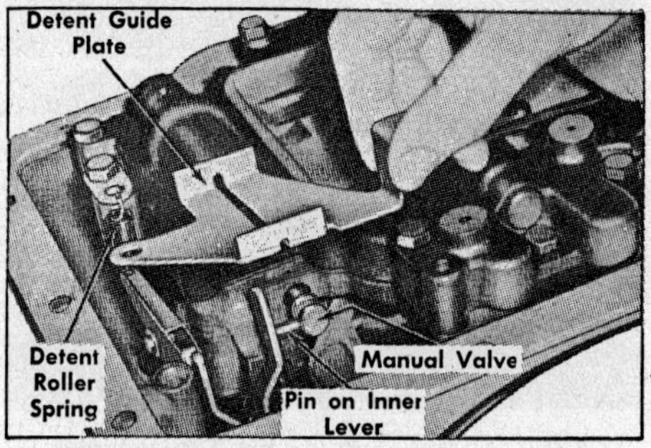

Fig. 6 Detent guide plate installation

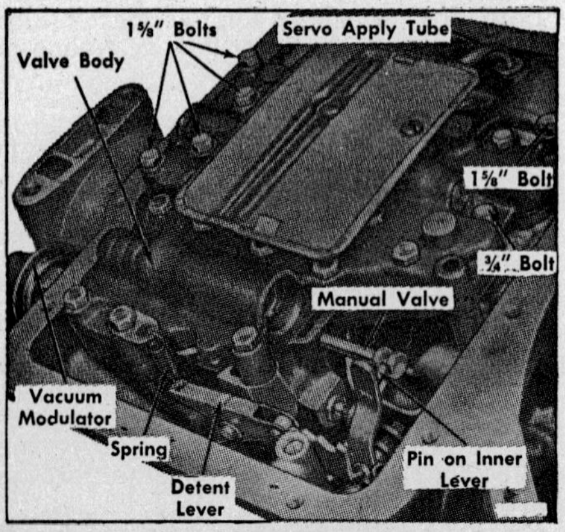

Fig. 7 Control valve assembly installed

Fig. 8 Inner control levers, parking pawl and bracket

CONTROL VALVE BODY

1. To remove control valve, remove vacuum modulator valve as outlined above.
2. Remove oil pan and gasket.
3. Remove two bolts attaching detent guide plate to valve body and transmission case. Remove guide plate and range selector detent roller spring, Fig. 6.
4. Remove remaining control valve bolts, Fig. 7.
5. Carefully remove valve body and gasket, disengaging servo apply tube from transmission case as valve body is removed.
6. Reverse removal procedure to install valve body, being sure range selector detent lever is in position shown in Fig. 7 and that pin on parking lock and range selector inner lever is engaged in slot in manual valve.

PARKING PAWL

After removing the control valve assembly as outlined above, remove the parking pawl, bracket and inner control levers. To make the installation, proceed as follows:

1. Install parking lock pawl and shaft. Install a new "E" ring on shaft.
2. Install parking lock pawl pull-back spring over its boss to rear of pawl. Short leg of spring should locate in hole in parking pawl, Fig. 8.
3. Install parking lock pawl reaction bracket (2 bolts). Fit actuator assembly between parking lock pawl and bracket.
4. Insert outer shift lever into case, making sure to pick up inner shift lever and parking lock assembly. Tighten Allen head nut.
5. Insert outer TV lever and shaft special washer and O-ring seal into case and pick up inner TV lever. Tighten Allen head nut.
6. Install selector lever detent roller.
7. Install valve body as directed above.

EXTENSION HOUSING SEAL

1. Referring to Fig. 9, disconnect propeller shaft from transmission.
2. Use a suitable puller to remove extension rear oil seal.

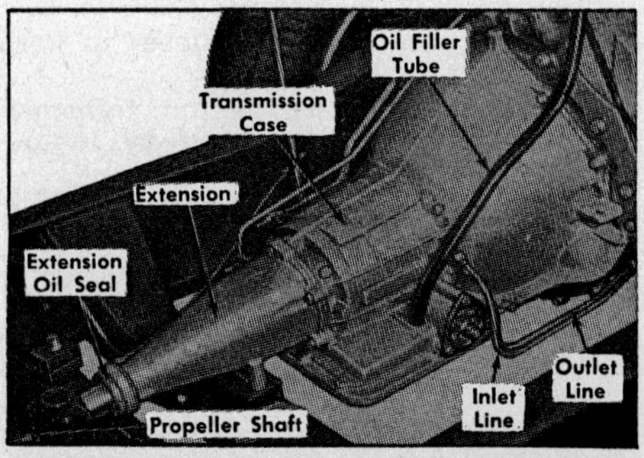

Fig. 9 Exterior of transmission

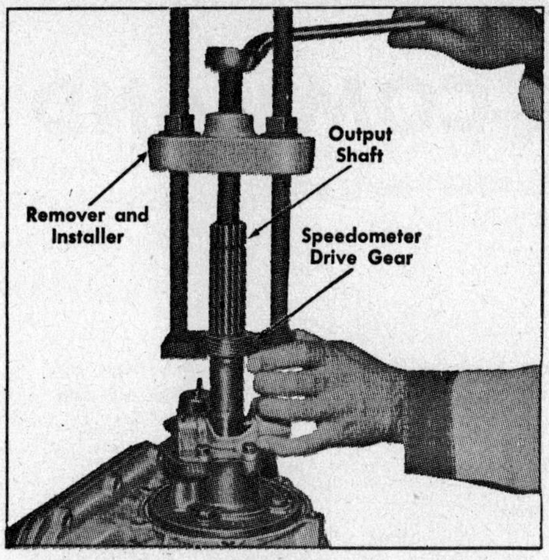

Fig. 10 Removing speedometer drive gear

3. With a suitable installer, drive new seal into bore of extension. Sealing cement should be used at outer diameter of seal to prevent leakage. Wipe off excess cement. Reconnect propeller shaft.

EXTENSION BUSHING

1. Remove extension oil seal as outlined above. Then, using a suitable remover too, pull bushing from rear of extension.
2. Place new bushing in pilot end of a suitable installer and drive bushing into bore of extension. Install rear oil seal.

SPEEDOMETER DRIVEN GEAR

1. Disconnect speedometer drive cable fitting. Remove cap screw and retainer clip holding driven gear in extension and remove gear.
2. Reverse removal procedure to install.

EXTENSION CASE

1. Disconnect propeller shaft from transmission output shaft.
2. Disconnect speedometer cable fitting.
3. Install transmission lift to support transmission and engine.
4. Unfasten extension from support crossmember (2 studs).
5. Unfasten extension from transmission case (5 bolts.)

NOTE: *Remove any shims found between extension and crossmember. Tie shims together as it is vital that exactly the same number of shims be used when the extension is reinstalled as these shims affect the drive line angle.*

6. Reverse removal procedure to install.

SPEEDOMETER DRIVE GEAR

1. Remove extension case as outlined above. Remove speedometer drive gear from output shaft, Fig. 10.
2. Using a suitable installer, install gear on output shaft and replace extension case.

GOVERNOR

1. Remove extension case and speedometer drive gear as outlined above.
2. Remove "C" clip from governor shaft on weight side of governor.
3. Remove shaft and governor valve from opposite side of governor. Remove two Belleville springs, Fig. 11.
4. Loosen governor drive screw and lift governor from output shaft.
5. Reverse removal procedure to install governor. However, be sure concave side of Belleville springs are against transmission output shaft.

TRANSMISSION, REPLACE

1969-73

1. Disconnect oil cooler lines (external cooled models), vacuum modulator line and speedometer drive cable fitting at transmission. Tie lines out of the way.
2. Disconnect manual and TV rods from transmission.
3. Disconnect propeller shaft from transmission.
4. Attach transmission jack on transmission.
5. Disconnect engine rear mount on transmission extension, then disconnect transmission support crossmember and slide rearward. Remove crossmember on 1969 Camaro.
6. Remove converter underpan. Scribe flywheel-converter relationship for reassembly, then remove flywheel-to-converter bolts.

NOTE: The "light" side of the converter is denoted by a "blue" stripe painted across the ends of the converter cover and housing. This marking should be aligned as closely as possible with the "white" stripe painted on the engine side of the flywheel outer rim (heavy side of engine) to maintain balance during assembly.

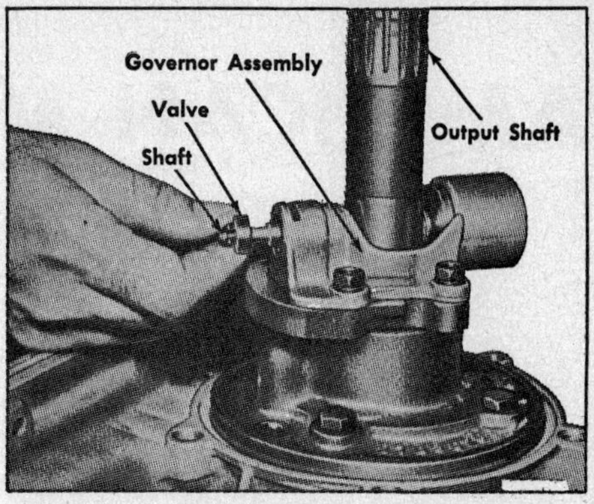

Fig. 11 Removing governor valve and shaft

7. Support engine at oil pan rail with a jack or other suitable brace capable of supporting engine when transmission is removed.
8. Lower rear of transmission slightly so that upper transmission housing-to-engine attaching bolts can be reached, using a universal socket and a long extension. Remove upper bolts.

CAUTION: On V8 engines, care must be taken not to lower rear of transmission to far as the distributor housing may be forced against the dash, causing damage to the distributor. It is best to have an assistant observe clearance of all upper engine components while transmission rear end is being lowered.

9. Remove remainder of transmission-to-engine bolts.
10. Remove transmission by moving it slightly to the rear and downward, then remove from under vehicle.

NOTE: Observe converter when moving transmission rearward. If it does not move with the transmission, pry it free of flywheel before proceeding.

CAUTION: Keep front of transmission upward to prevent converter from falling out. Install a suitable holding strap or length of strong wire across the housing to keep the converter in place.

11. Reverse procedure to install.

CHRYSLER TORQUEFLITE & AMERICAN MOTORS TORQUE-COMMAND

IDENTIFICATION

Transmission identification markings shown in the following application chart are cast in raised letters and numerals on the lower left side of the bell housing. NOTE: There are sufficient variations within each of the main categories listed below to make it necessary to service them by serial number—a stamped 7-digit number appearing on the oil pan side rail.

CHRYSLER CORPORATION

1972-75	6-198, 225 engine	A-904
	6-225 engine, police & taxi	A-727
	V8-318 engine	A-904-LA
	V8-318 engine, police & taxi	A-727
	V8-340, 360, 400, 440 engines	A-727
1969-71	6-170, 6-198, 6-225 engines	A-904-G
	V8-273 engine	A-904-A
	V8-318 engine	A-904-LA

CHRYSLER CORP.—(Continued)

	6-225 police and taxi	A-727-RG
	V8-318, V8-340, V8-360 engines	A-727-A
	V8-383, V8-426, V8-440	A-727-B

AMERICAN MOTORS

1972-75	6-232, 258 engines	904
	6-258, V8-304, 360, 401 Heavy Duty	727
	V8-304 engine	998
	V8-360, 401 engines	727

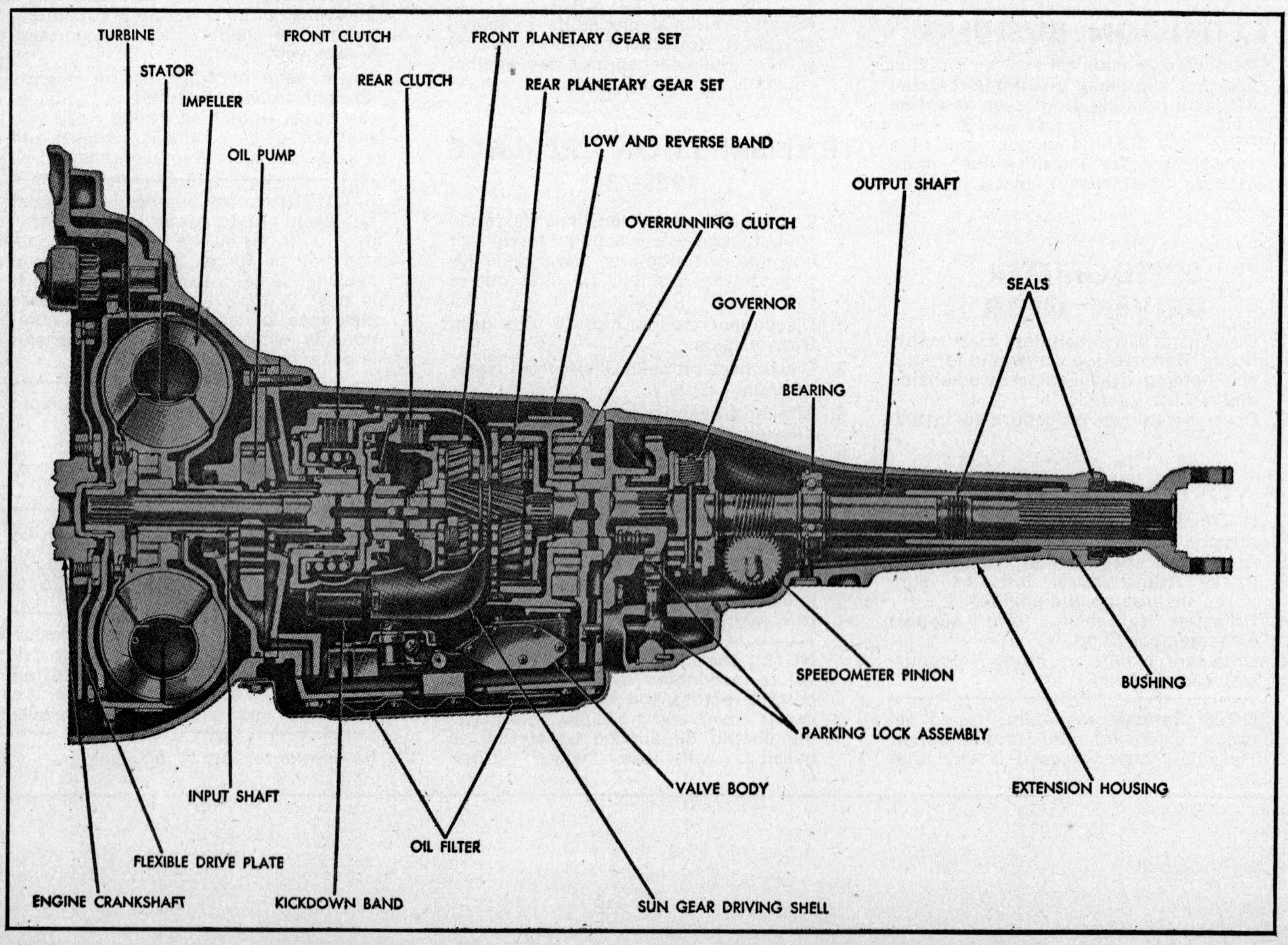

Fig. 1 Series 904 Torqueflite transmission used on 1969-75. 998 model is similar

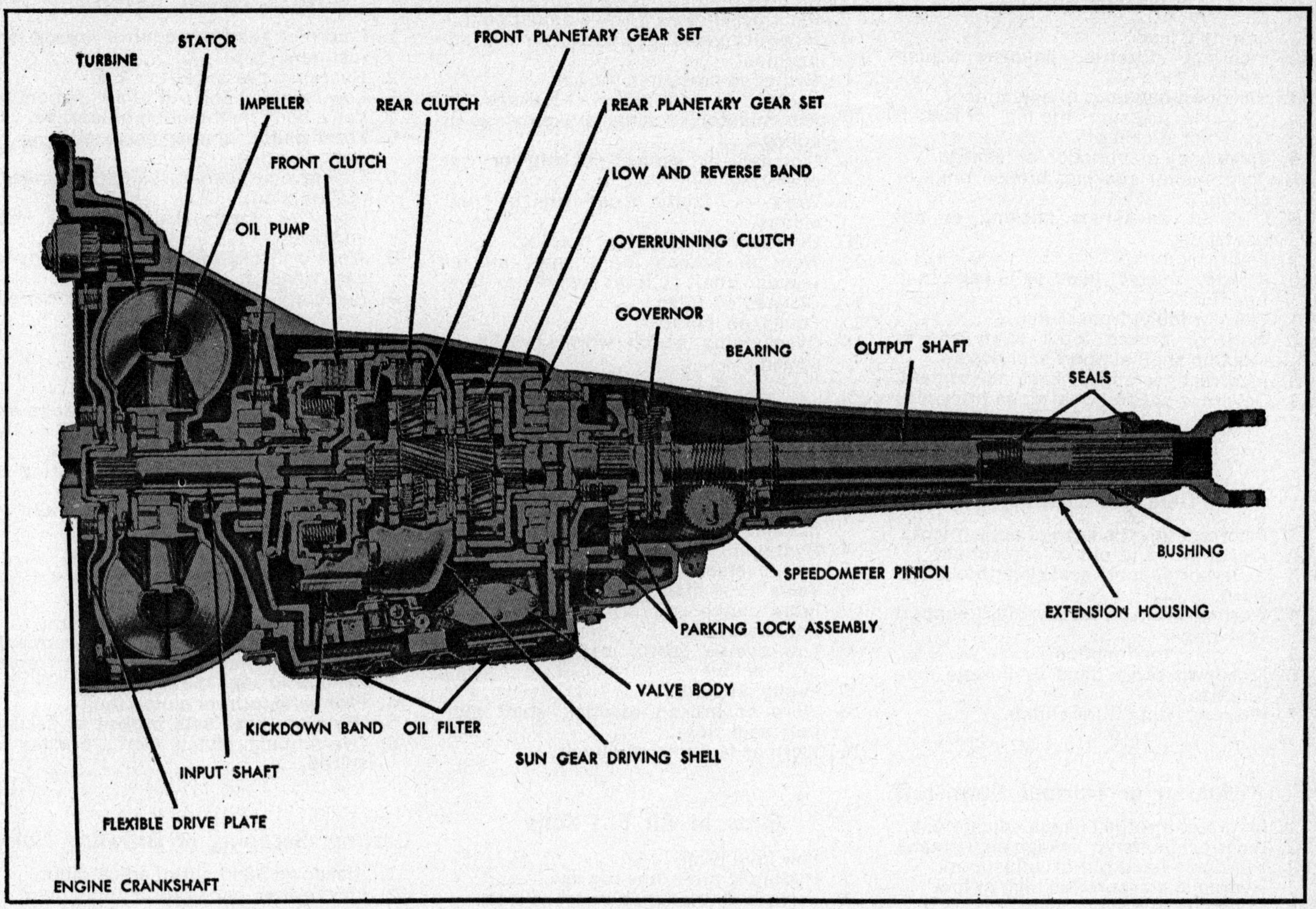

Fig. 2 Series 727 Torqueflite transmission used on 1969-75

DESCRIPTION

These transmissions, Figs. 1 and 2, combine a torque converter with a fully automatic three speed gear system. The converter housing and transmission case are an integral aluminum casting. The transmission consists of two multiple disc clutches, an overrunning (one-way) clutch, two servos and bands and two planetary gear sets to provide three forward speeds and reverse.

The common sun gear of the planetary gear sets is connected to the front clutch by a driving shell that is splined to the sun gear and to the front clutch retainer.

The hydraulic system consists of a single oil pump and a valve body that contains all the valves except the governor valve.

Venting of the transmission is accomplished by a drilled passage through the upper part of the front pump housing.

The torque converter is attached to the engine crankshaft through a flexible driving plate. The converter is cooled by circulating the transmission fluid through an oil-to-water type cooler located in the radiator lower tank. The converter is a sealed assembly that cannot be disassembled.

TROUBLE SHOOTING GUIDE

Harsh Engagement In D - 1 - 2 - R

1. Engine idle speed too high.
2. Hydraulic pressures too high or too low.
3. Low-reverse band out of adjustment.
4. Accumulator sticking, broken rings or spring.
5. Low-reverse servo, band or linkage malfunction.
6. Worn or faulty front and/or rear clutch.
7. Valve body malfunction or leakage.

Delayed Engagement In D - 1 - 2 - R

1. Low fluid level.
2. Incorrect manual linkage adjustment.
3. Oil filter clogged.
4. Hydraulic pressures too high or low.
5. Valve body malfunction or leakage.
6. Accumulator sticking, broken rings or spring.
7. Clutches or servos sticking or not operating.
8. Faulty front oil pump.
9. Worn or faulty front and/or rear clutch.
10. Worn or broken input shaft and/or reaction shaft support seal rings.
11. Aerated fluid.

Runaway or Harsh Upshift and 3 - 2 Kickdown

1. Low fluid level.
2. Incorrect throttle linkage adjustment.
3. Hydraulic pressures too high or low.
4. Kickdown band out of adjustment.
5. Valve body malfunction or leakage.
6. Governor malfunction.
7. Accumulator sticking, broken rings or spring.
8. Clutches or servos sticking or not operating.
9. Kickdown servo, band or linkage malfunction.
10. Worn or faulty front clutch.
11. Worn or broken input shaft and/or reaction shaft support seal rings.
12. Aerated oil.
13. Clogged oil filter.

No Upshift

1. Low fluid level.
2. Incorrect throttle linkage adjustment.
3. Kickdown band out of adjustment.
4. Hydraulic pressures too high or low.
5. Governor sticking.
6. Valve body malfunction or leakage.
7. Accumulator sticking, broken rings or spring.
8. Clutches or servos sticking or not operating.
9. Faulty oil pump.
10. Kickdown servo, band or linkage malfunction.
11. Worn or faulty front clutch.
12. Worn or broken input shaft and/or reaction shaft support seal rings.
13. Incorrect gearshift linkage adjustment.
14. Governor support seal rings broken or worn.

Delayed Upshift

1. Incorrect throttle linkage adjustment.
2. Kickdown band out of adjustment.
3. Governor support seal rings broken or worn.
4. Worn or broken reaction shaft support seal rings.
5. Governor malfunction.
6. Kickdown servo band or linkage malfunction.
7. Worn or faulty front clutch.

No Kickdown or Normal Downshift

1. Incorrect throttle linkage adjustment.
2. Incorrect gearshift linkage adjustment.
3. Kickdown band out of adjustment.
4. Hydraulic pressure too high or low.
5. Governor sticking.
6. Valve body malfunction or leakage.
7. Accumulator sticking, broken rings or spring.
8. Clutches or servos sticking or not operating.
9. Kickdown servo, band or linkage malfunction.
10. Overrunning clutch not holding.

Erratic Shifts

1. Low fluid level.
2. Aerated fluid.
3. Incorrect throttle linkage adjustment.
4. Incorrect gearshift control linkage adjustment.
5. Hydraulic pressures too high or low.
6. Governor sticking.
7. Oil filter clogged.
8. Valve body malfunction or leakage.
9. Clutches or servos sticking or not operating.
10. Faulty oil pump.
11. Worn or broken input shaft and/or reaction shaft support rings.
12. Governor support seal rings broken or worn.
13. Kickdown servo band or linkage malfunction.
14. Worn or faulty front clutch.

Slips In Forward Drive Positions

1. Low oil level.
2. Aerated fluid.
3. Incorrect throttle linkage adjustment.
4. Incorrect gearshift control linkage adjustment.
5. Hydraulic pressures too low.
6. Valve body malfunction or leakage.
7. Accumulator sticking, broken rings or springs.
8. Clutches or servos sticking or not operating.
9. Worn or faulty front and/or rear clutch.
10. Overrunning clutch not holding.
11. Worn or broken input shaft and/or reaction shaft support seal rings.
12. Clogged oil filter.
13. Faulty oil pump.
14. Overrunning clutch worn, broken or seized.

Slips In Reverse Only

1. Low fluid level.
2. Aerated fluid.
3. Incorrect gearshift control linkage adjustment.
4. Hydraulic pressures too high or low.
5. Low-reverse band out of adjustment.
6. Valve body malfunction or leakage.
7. Front clutch or rear servo sticking or not operating.
8. Low-reverse servo, band or linkage malfunction.
9. Faulty oil pump.
10. Worn or broken reaction shaft support seal rings.
11. Worn or faulty front clutch.

Slips In All Positions

1. Low fluid level.
2. Hydraulic pressures too low.
3. Valve body malfunction or leakage.
4. Faulty oil pump.
5. Clutches or servos sticking or not operating.
6. Worn or broken input shaft and/or reaction shaft support seal rings.
7. Oil filter clogged.
8. Aerated oil.

No Drive In Any Position

1. Low fluid level.
2. Hydraulic pressures too low.
3. Oil filter clogged.
4. Valve body malfunction or leakage.
5. Faulty oil pump.
6. Clutches or servos sticking or not operating.
7. Planetary gear sets broken or seized.

No Drive In Forward Drive Positions

1. Hydraulic pressures too low.
2. Valve body malfunction or leakage.
3. Accumulator sticking, broken rings or spring.
4. Clutches or servos, sticking or not operating.
5. Worn or faulty rear clutch.
6. Overrunning clutch not holding.
7. Worn or broken input shaft and/or reaction shaft support seal rings.
8. Low fluid level.
9. Planetary gear sets broken or seized.
10. Overrunning clutch worn, broken or seized.

No Drive In Reverse

1. Incorrect gearshift control linkage adjustment.
2. Hydraulic pressures too low.
3. Low-reverse band out of adjustment.
4. Valve body malfunction or leakage.
5. Front clutch or rear servo sticking or not operating.
6. Low-reverse servo, band or linkage malfunction.
7. Worn or faulty front and/or rear clutch.
8. Worn or broken reaction shaft support seal rings.
9. Planetary gear sets broken or seized.

Drives In Neutral

1. Incorrect gearshift control linkage adjustment.
2. Valve body malfunction or leakage.
3. Rear clutch worn, faulty, dragging or inoperative.
4. Insufficient clutch plate clearance.

Drags or Locks

1. Kickdown band out of adjustment.
2. Low-reverse band out of adjustment.
3. Kickdown and/or low-reverse servo, band or linkage malfunction.
4. Front and/or rear clutch faulty.
5. Planetary gear sets broken or seized.
6. Overrunning clutch worn, broken or seized.

Grating, Scraping or Growling Noise

1. Kickdown band out of adjustment.
2. Low-reverse band out of adjustment.
3. Output shaft bearing and/or bushing damaged.
4. Governor support binding or broken seal rings.
5. Oil pump scored or binding.
6. Front and/or rear clutch faulty.
7. Planetary gear sets broken or seized.
8. Overrunning clutch worn, broken or seized.

Buzzing Noise

1. Low fluid level.
2. Pump sucking air.
3. Valve body malfunction.
4. Overrunning clutch inner race damaged.
5. Aerated oil.

Hard to Fill, Oil Flows Out Filler Tube

1. High fluid level.
2. Breather clogged.
3. Oil filter clogged.
4. Aerated fluid.

Transmission Overheats

1. Low fluid level.
2. Kickdown band adjustment too tight.
3. Low-reverse band adjustment too tight.

4. Faulty cooling system.
5. Cracked or restricted oil cooler line or fitting.
6. Faulty oil pump.
7. Insufficient clutch plate clearance in front and/or rear clutches.
8. Engine idle too low.
9. Hydraulic pressures too low.
10. Incorrect gearshift linkage adjustment.
11. Kickdown band adjustment too tight.

Starter Will Not Energize in Neutral or Park

1. Incorrect gearshift control linkage adjustment.
2. Faulty or incorrectly adjusted neutral starting switch.
3. Broken lead to neutral switch.

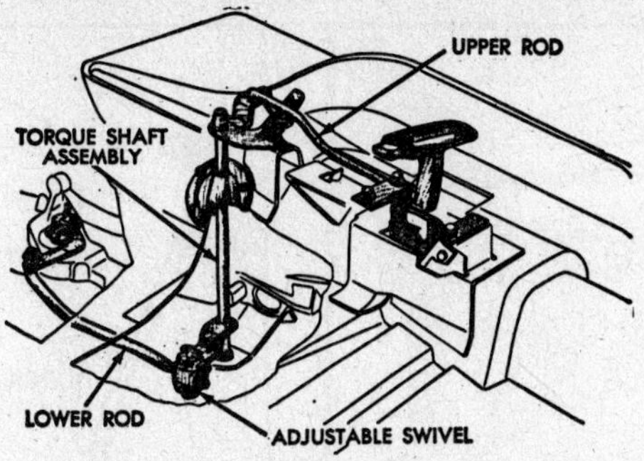

Fig. 3 Console gearshift linkage. 1971-75 Barracuda, Challenger, Charger, Cordoba, Coronet & Satellite

MAINTENANCE

Adding Oil

To check the oil level, apply the parking brake and operate the engine at idle speed with the transmission in Neutral position. Add oil as necessary.

Changing Oil

Fluid and filter changes or band adjustments are not required for average passenger car use. Severe usage such as police, taxi, trailer towing or prolonged operation in city traffic, requires that fluid and filter be changed and bands adjusted every 24,000 miles.

Whenever the factory fill fluid is changed, only fluids of the type labeled DEXRON should be used.

1. Remove drain plug (if equipped) from transmission oil pan and allow oil to drain.

NOTE: *If the oil pan does not have a drain plug, loosen pan bolts and tap pan with a soft mallet to break it loose, permitting fluid to drain.*

2. Remove flywheel access plate, remove torque converter drain plug and allow to drain. Replace drain plug.
3. Remove transmission oil pan, replace filter or clean intake screen and pan, adjust bands and reinstall.
4. Install initial amount of approved automatic transmission fluid through filler tube (five quarts on Series 904 and 998 units; eight quarts on Series 727 units.
5. Start engine and add approximately one quart while engine is idling.
6. Allow engine to idle for about two minutes. Then with parking brake applied, place transmission in neutral.
7. Add oil as necessary to bring to proper level.

BANDS, ADJUST

Kickdown Band

The kickdown band adjusting screw is located on the left side of the transmission case near the throttle lever shaft.

1. Loosen lock nut and back off approximately five turns. Check adjusting screw for free turning in transmission case.
2. Using an inch-pound torque wrench, tighten the band adjusting screw to a reading of 72 inch lbs.
3. Back off adjusting screw 2 turns with the following exceptions:

Chrysler Corp.

6-170 engine	2⅝ turns
V8-426, 1969-71	1½ turns
V8-440 3 carbs. 1971-72	1½ turns
V8-440 dual exhaust, 1971-75	2 turns
All other 1971-75 with A-727 model trans	2½ turns

American Motors

1972

All Exc. V8-360	2
V8-360	2½

1973-75

6-232, 258, V8-304, 360①, 401	2
V8-360	2½
6-232①, V8-304①	2½

①Heavy Duty

Hold adjusting screw in this position and tighten locknut.

Low and Reverse Band

1. Raise vehicle, drain transmission and remove oil pan.
2. Inspect fluid for friction material or metal particles which indicate damaged or worn parts.
3. Loosen adjusting screw lock nut and back off nut approximately five turns. Check adjusting screw for free turning in lever.
4. Using an inch pound torque wrench, tighten band adjusting screw to 72 in. lbs. on all except 1974-75 904 series transmissions. On 1974-75 904 series transmissions, tighten adjusting screw to 41 in. lbs.
5. Back off adjusting screw the amount of turns indicated:

American Motors

1972-73 904 Series . . . 3½ turns

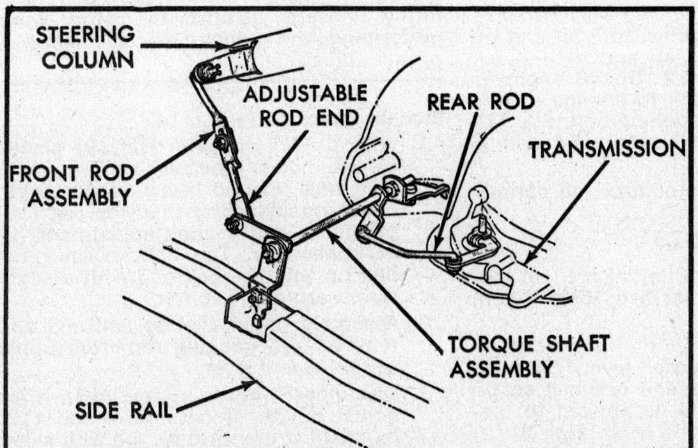

Fig. 4 Column gearshift linkage. 1970-72 Chrysler, Imperial, Fury, Monaco & Polara

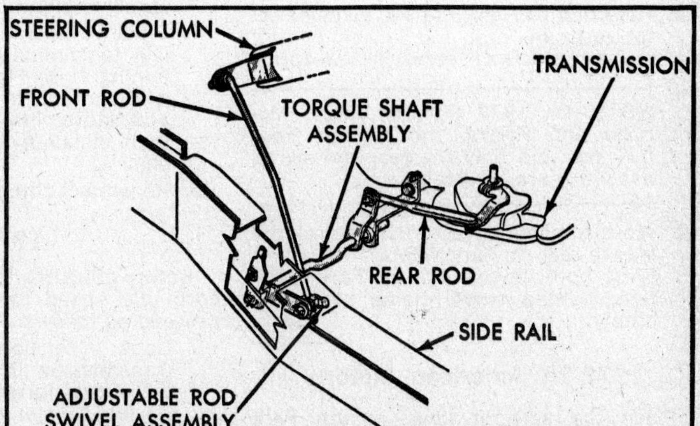

Fig. 5 Column gearshift linkage. 1970-73 Barracuda & Challenger. 1971-73 Coronet, Charger & Satellite. 1973 Chrysler, Imperial, Fury, Polara & Monaco

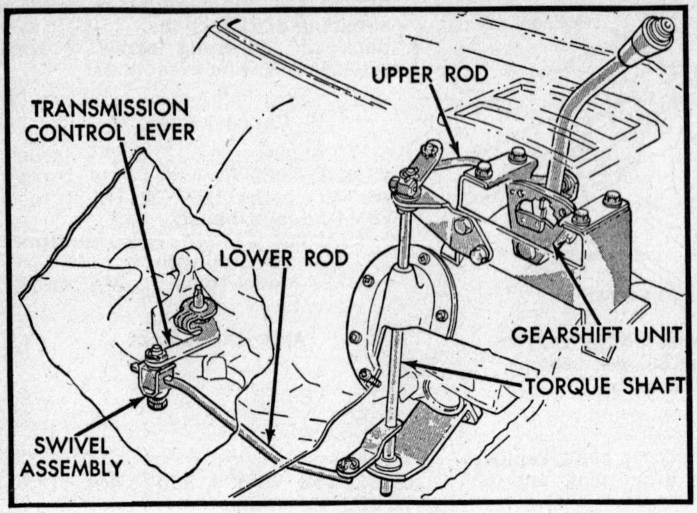

Fig. 6 Console gearshift linkage. 1971-75 Dart & Valiant, 1969 All & 1973 Fury

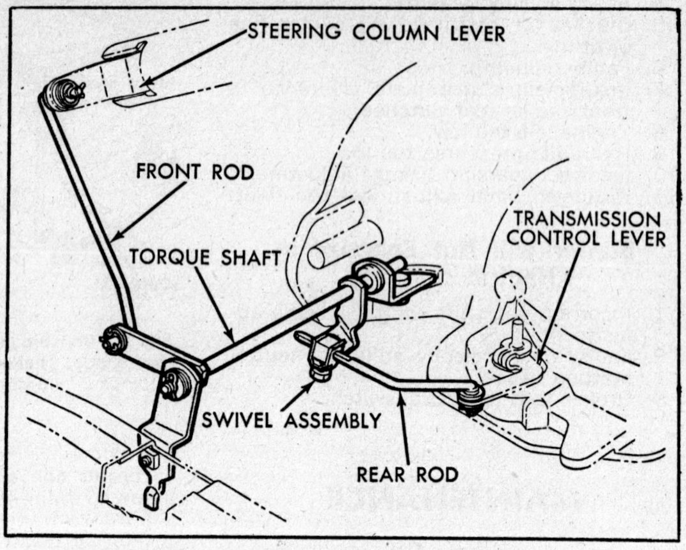

Fig. 7 Column gearshift linkage. 1969 Chrysler, Dodge, Imperial & Plymouth 1970 Belvedere, Charger, Coronet & Satellite; 1970-73 Dart & Valiant

1974-75 904 Series ....7 turns
998 Series ....4 turns
727 Series ....2 turns

Chrysler Corp.

1969-73 904 Series with 6 cyl. ....3¼ turns
1974-75 904 Series with 6 cyl. ....7 turns
1969-75 904 Series with V8-318 ....4 turns
727 Series ....2 turns

6. Hold adjusting screw and tighten to 35 ft. lbs., then install oil pan and refill transmission.

GEARSHIFT CONTROL LINKAGE, ADJUST

1969-75 Chrysler Corp.

1. Referring to Figs. 3 to 9B, place selector lever in Park and loosen control rod swivel clamp screw a few turns.

NOTE: On 1972 Chrysler, Fury, Monaco and Polara, the column front lock rod and only the rear rod swivel assembly are adjustable.

2. Move transmission control lever all the way to rear (in Park detent).
3. With both levers still in Park position, tighten swivel clamp screw securely.

1972-75 American Motors

Place the selector lever in the Park position and place the transmission shift lever in the Park detent. Adjust the shift rod to obtain a "free pin" fit. Check steering column lock for ease of operation. Move selector lever to Neutral position and check safety switch operation.

THROTTLE LINKAGE, ADJUST

American Motors

1974-75

1. Disconnect throttle control spring, then use spring to hold transmission throttle control lever forward against stop, Fig. 10.
2. Block choke open and set throttle off fast idle.

NOTE: On carburetors equipped with a throttle solenoid, energize solenoid and open throttle half way to allow solenoid to lock and return carburetor to idle.

3. Loosen retaining bolt on throttle control adjusting link. On 6 cylinder engines, do not remove spring clip and nylon washer. On V8 engines, remove spring clip and nylon washer from link.
4. On 6 cylinder engines, pull on end of link to eliminate lash and tighten retaining bolt, Fig. 12. On V8 engines, push on end of link to eliminate lash and tighten link retaining bolt, Fig. 11, then install nylon washer and spring clip.
5. Reconnect throttle control rod spring.

1972-73

Before adjusting throttle linkage, set the curb idle speed to specified RPM. Then proceed as follows:

1. Hook a tension spring into hole in transmission throttle lever. Stretch the spring forward and retain it so 8-10 lbs. tension will be applied to the transmission throttle lever, Fig. 10.
2. Remove retainer clip attaching adjustable throttle rod link at the slotted end. Remove tab washer from rod slot and loosen throttle rod slip joint retaining screw.
3. For V8 engines, shorten adjustable rod to remove all slack and tighten slip joint retaining screw, Fig. 11.
4. For six cylinder engines, lengthen adjustable rod to remove all slack and tighten slip joint retaining screw, Fig. 12.
5. Place rod slot over carburetor throttle lever (V8) or bellcrank lever (Six). Install tab washer and retainer clip. The washer tabs must fit into throttle rod slot, Fig. 13.
6. Remove tension spring from transmission throttle lever.

1969-75 Chrysler Corp.

NOTE: Before proceeding with the adjustment, disconnect the choke rod at the carburetor or block the choke valve wide open. Open the throttle slightly to release the fast idle cam, then return carburetor to the hot idle position.

Hold or fasten the transmission lever firmly forward against the stop while performing the adjustment to insure a proper adjustment.

Models Indicated in Fig. 14

1. With a $3/16''$ diameter rod (9) placed in the holes provided in the upper bellcrank (6) and lever, adjust length of intermediate transmission rod (10) by means of threaded adjustment (2) at upper end. The ball socket must line up with the ball end with a slight downward effort on rod.
2. Assemble ball socket to ball end and remove $3/16''$ gauging rod from upper bellcrank and lever.
3. Disconnect return spring (13), then adjust length of carburetor rod (12) by pushing rearward on rod with slight effort and turning threaded adjustment (1). Rear end of slot should contact carburetor lever pin without exerting any forward force when slotted adjuster link (1) is in its normal

operating position against lever pin nut.

4. Assemble slotted adjustment (1) to carburetor lever pin and install washer and retaining pin. Connect transmission linkage return spring (13).
5. Release transmission lever, then check linkage for freedom of operation. Move slotted adjuster link (1) to full rearward position, then allow it to return slowly, making sure it returns to full forward position.
6. Loosen cable clamp nut (4) and adjust position of cable housing ferrule (5) in clamp so that all slack is removed from cable with carburetor at hot idle position. To remove slack from cable, move ferrule in clamp in a direction away from carburetor lever.
7. Back off ferrule 1/4" which provides enough free play between front edge of accelerator shaft lever and dash bracket. Tighten clamp nut to 45 inch-lbs.
8. Connect choke rod or remove blocking fixture.

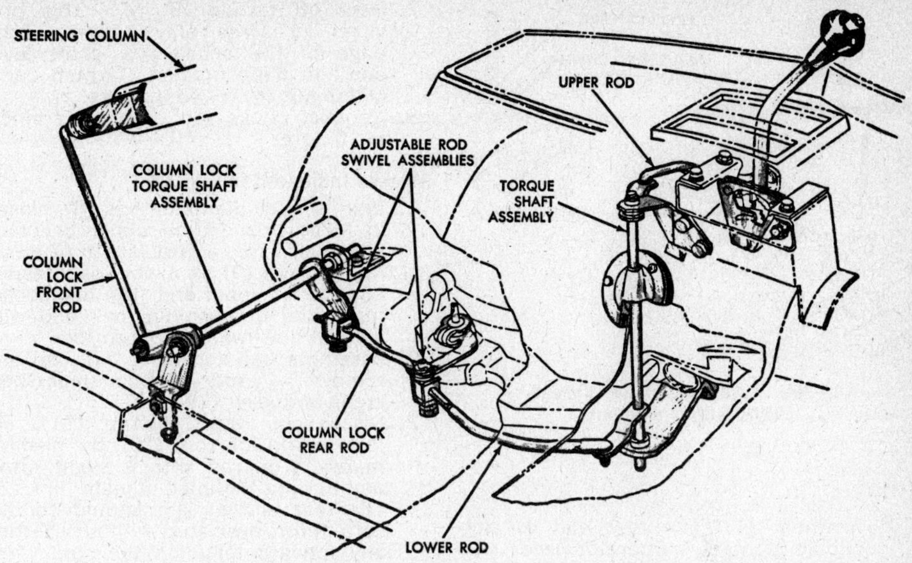

Fig. 8 Console gearshift linkage. 1970 Belvedere, Charger, Coronet, Dart, Satellite & Valiant. 1970-72 Chrysler, Fury, Monaco & Polara

Models Indicated in Fig. 15

1. With a 3/16" diameter rod (9) placed in the holes provided in upper bellcrank and lever, adjust length of intermediate transmission rod (10) by means of threaded adjustment at upper end. The ball socket (2) must line up with the ball end with a slight downward effort on rod.
2. Assemble ball socket (2) to ball end and remove 3/16" rod (9) from upper bellcrank and lever.
3. Disconnect return spring (13), then adjust length of carburetor rod (12) by pushing rearward on rod with a slight effort and turning the threaded adjuster (1). The rear end of slot should contact carburetor lever pin without exerting any forward force on pin when slotted adjuster link (1) is in its normal operating position against lever pin nut.
4. Assemble slotted adjustment (1) to carburetor lever pin and install washer and retaining pin. Assemble transmission linkage return spring (13) in place.
5. Release transmission lever and check linkage for freedom of operation. Move slotted adjuster link (1) to full rearward position and allow it to return slowly, making sure it returns to full forward position.
6. Loosen cable clamp nut (4), adjust position of cable housing ferrule (5) in the clamp so all slack is removed from cable with carburetor at curb idle. To remove slack, move ferrule (5) in the clamp in direction away from carburetor lever.
7. Back off ferrule (5) 1/4". This provides 1/4" free play between front edge of accelerator shaft lever and dash bracket. Tighten cable clamp nut (4) to 45 inch lbs.
8. Connect choke (8) rod or remove blocking fixture.

Models Indicated in Fig. 16

1. Adjust length of transmission rod by loosening slotted link lock nut (12). Pull forward on slotted adjuster link (7) so it contacts carburetor lever pin.
2. Tighten transmission rod adjustment lock nut (12) to 95 inch lbs. Check transmission linkage for freedom of operation by moving slotted adjuster link to full rearward position and allow it to return slowly, making sure it returns to full forward position.
3. When carburetor throttle is opened, the transmission lever (9) should begin its travel at the same time with no vertical movement of lever or vertical movement of rod (10) in the lever.
4. Loosen cable clamp nut (5), adjust position of housing ferrule (6) in the clamp so all slack is removed from cable with carburetor at curb idle. To remove slack, move ferrule (6) in the clamp in direction away from carburetor lever.
5. Back off ferrule (6) 1/4". This provides 1/4" free play between dash mounted accelerator lever and the bracket. Tighten cable clamp nut to 45 inch lbs.
6. Connect choke rod (4) or remove blocking fixture.

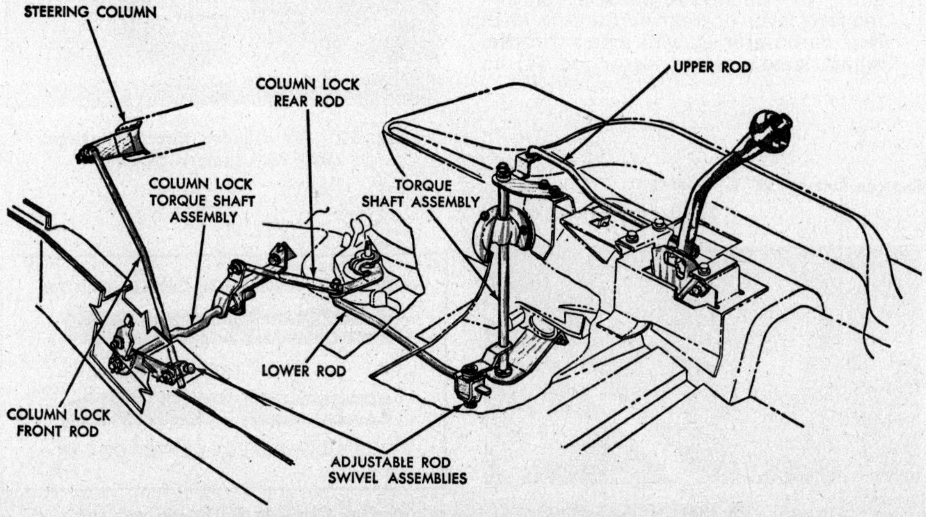

Fig. 9 Console gearshift linkage. 1970 Barracuda & Challenger

Models Indicated in Fig. 17

1. With a 3/4" diameter rod (9) placed in the holes provided in upper bellcrank and lever, adjust length of transmission rod (10) by means of threaded adjustment at upper end. The ball socket (2) must line up with the ball end with a slight downward effort on rod.
2. Assemble ball socket (2) to ball end and remove 3/16" rod (9) from upper bellcrank and lever.
3. Disconnect return spring (13), washer and retainer pin (14), then adjust length of carburetor rod (12) by pushing rearward on rod with a slight effort and turning the threaded ad-

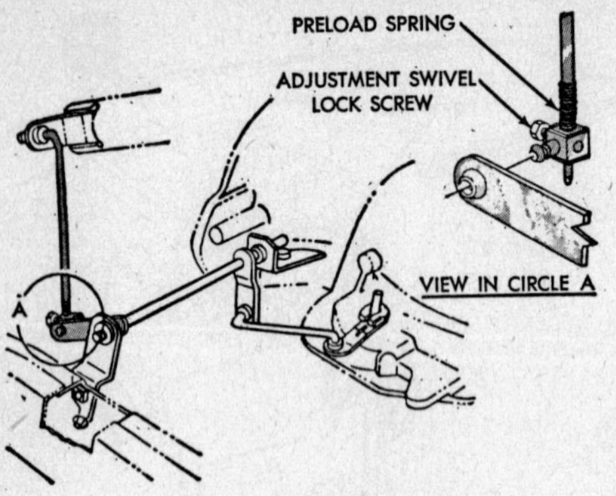

Fig. 9A Column gearshift linkage. 1974-75 Dart & Valiant

justment (1). The rear end of slot should contact carburetor lever pin without exerting any forward force on pin when slotted adjuster link (1) is in its normal operating position against lever pin nut.

4. Assemble slotted adjustment (1) to carburetor lever pin and install washer and retainer pin (14). Assemble transmission linkage return spring (13).
5. Release transmission lever and check linkage for freedom of operation by moving slotted adjuster link (1) to full rearward position, then allow it to return slowly, making sure it returns to the full forward position.
6. Loosen cable clamp nut (4), adjust position of cable housing ferrule (5) in the clamp so that all slack is removed from the cable with carburetor at curb idle. To remove slack, move the ferrule (5) in the clamp in direction away from carburetor lever.

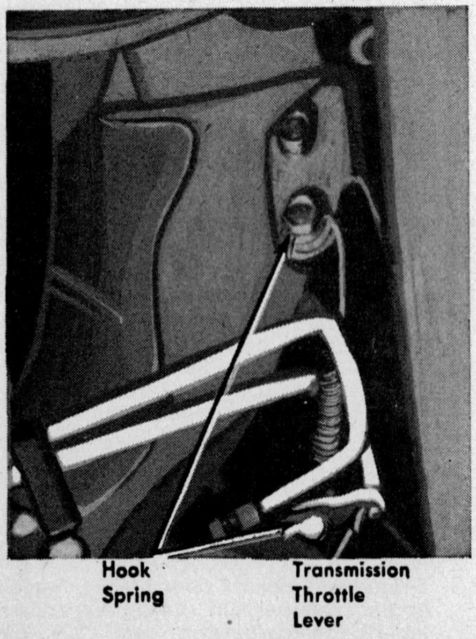

Fig. 10 Install spring on lever. 1972-75 American Motors

7. Back off ferrule (5) $\frac{1}{4}$". This provides $\frac{1}{4}$" free play between front edge of the accelerator shaft lever and the dash bracket. Tighten cable clamp nut (4) to .45 inch lbs.
8. Connect choke rod or remove blocking fixture.

Models Indicated in Fig. 18

1. With a $\frac{3}{16}$" diameter rod (8) placed in holes provided in upper bellcrank and lever (15), adjust length of transmission rod (9) by means of threaded adjuster at upper end. The ball socket must line up with the ball end with a slight downward effort on rod.
2. Assemble ball socket to ball end and remove $\frac{3}{16}$" rod (8) from upper bellcrank and lever (15).
3. Disconnect return spring (11), adjust length of rod (20) by pushing rearward on rod with a slight effort and turning threaded adjuster link (2). The rear end of slot should contact carburetor lever stud without exerting any forward force on the stud when slotted adjuster link is in its normal operating position.
4. Assemble slotted adjuster link (2) to carburetor lever stud and install washer and retainer pin. Assemble linkage return spring (11) in place.
5. Release transmission lever and check linkage for freedom of operation by moving slotted adjuster link (2) to full rearward position, then allow it to return slowly to the full forward position against the stud.
6. Loosen cable clamp nut (12), adjust position of cable housing ferrule. (13) in the clamp (14) so all slack is removed from cable with rear carburetor at curb idle. To remove slack from cable, move ferrule (13) in clamp (14) in direction away from carburetor lever.
7. Back off ferrule (13) $\frac{1}{4}$". This provides $\frac{1}{4}$" free play between front edge of accelerator shaft lever and dash bracket. Tighten clamp (14) to 45 inch lbs.
8. Route cable so it does not interfere with carburetor rod (20) or upper bellcrank (15) throughout full throttle link age travel.
9. Attach carburetor rod assembly (4) between carburetors with slotted rod end (16) attached to outboard side of inboard lever on rear carburetor. With rear carburetor at wide open throttle, adjust length of connector rod (4) so

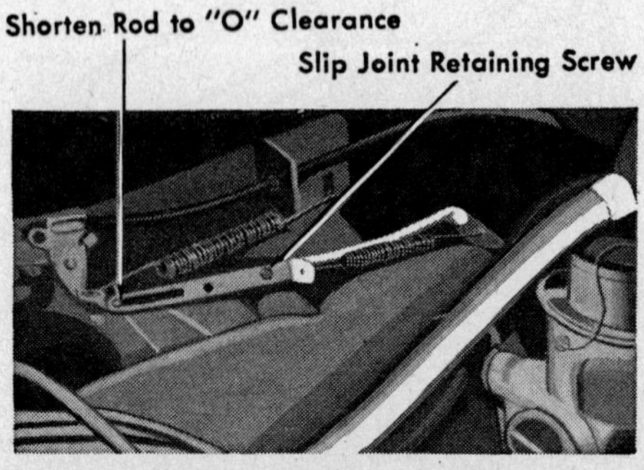

Fig. 11 V8 throttle linkage. 1972-75 American Motors

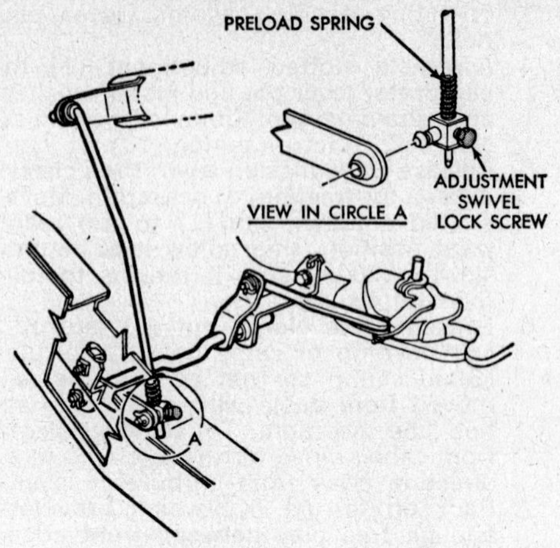

Fig. 9B Column gearshift linkage. 1974-75 Exc. Dart & Valiant

that front carburetor is also at wide open throttle. To lengthen this rod (4) turn adjusting stud (17) clockwise as viewed from front of engine. Tighten lock nut (18).

10. Remove choke valve blocking fixture.

Models Indicated in Fig. 19

1. Disconnect choke at carburetor or block choke valve fully open. Open throttle slightly to release fast idle cam and return carburetor to curb idle position.

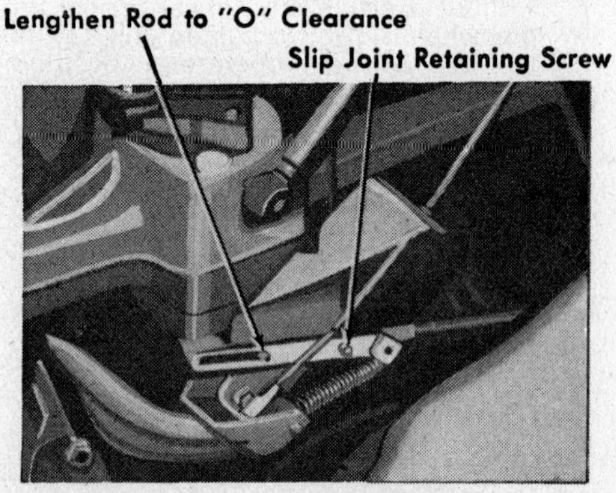

Fig. 12 Six cylinder throttle linkage. 1972-75 American Motors

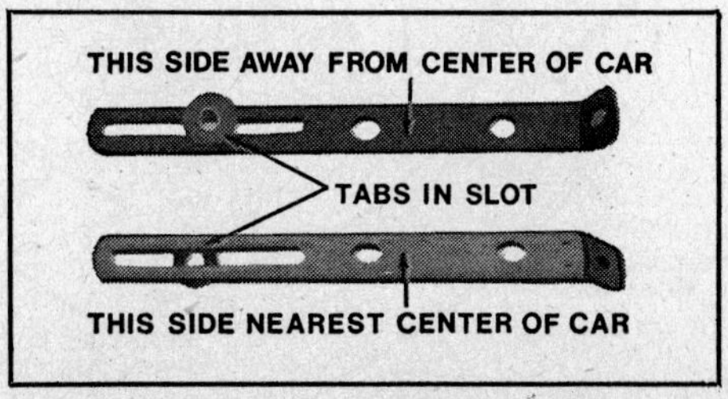

Fig. 13 Tab washer in rod slot. 1972-73 American Motors

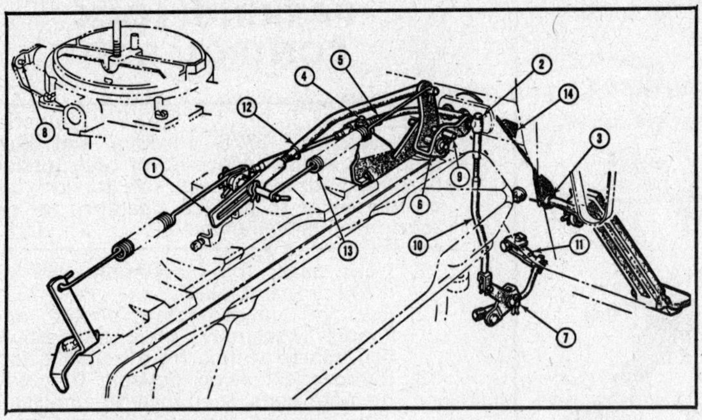

Fig. 14 Throttle linkage adjustment. All 1969 models with 383 and 440 engines. Also 1969 Monaco, Polara & Fury with 318 engine

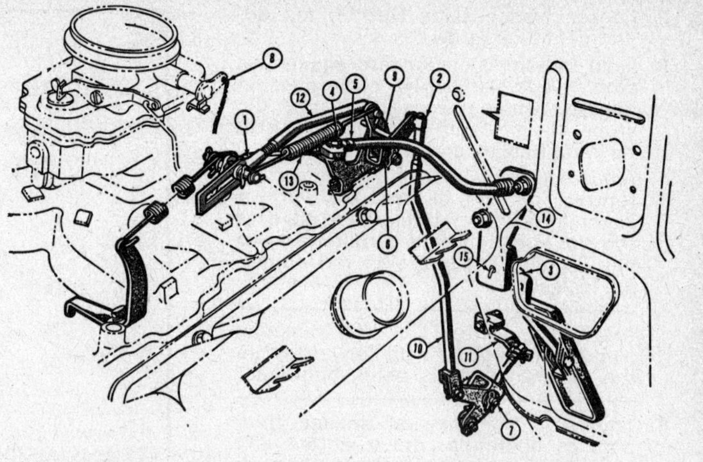

Fig. 15 Throttle linkage adjustment. 1969 Dart, Valiant and Barracuda with 273, 318 or 340 engine. 1970-73 V8 with 3 section rod

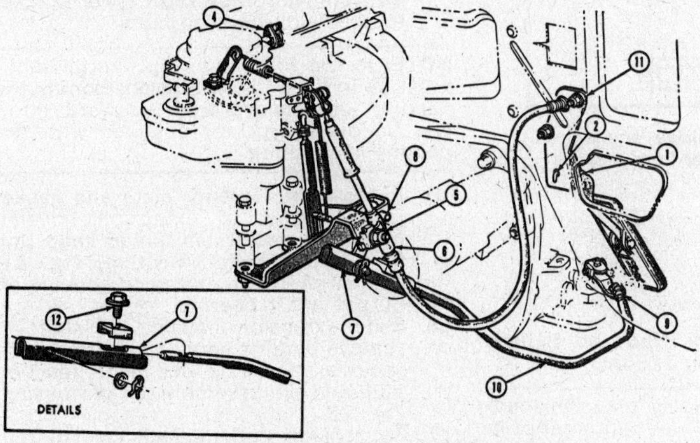

Fig. 16 Throttle linkage adjustment. 1969-73 models with 6-cylinder engine

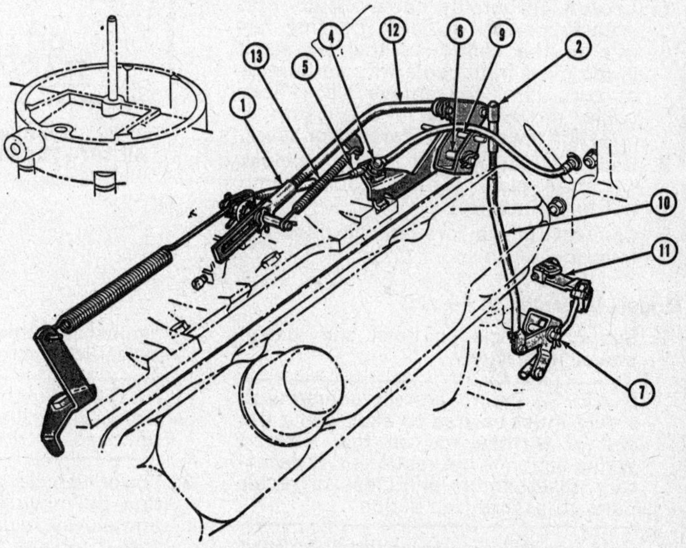

Fig. 17 Throttle linkage adjustment. 1969 Dart and Barracuda with 383 engine

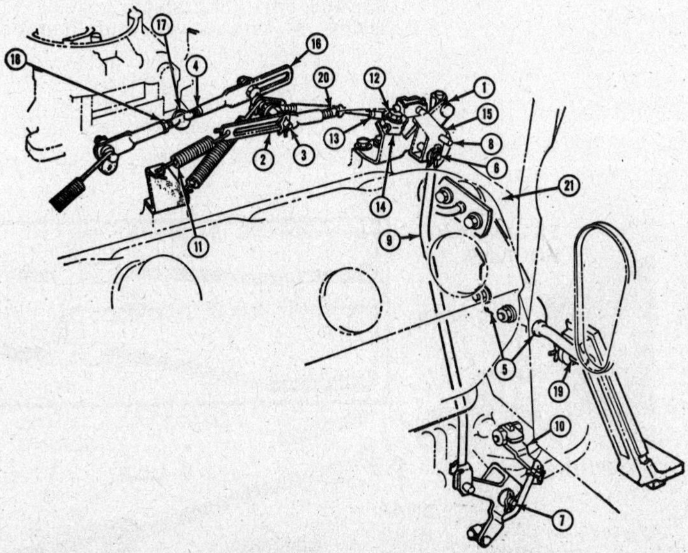

Fig. 18 Throttle linkage adjustment. 1969-71 models with 426 engine

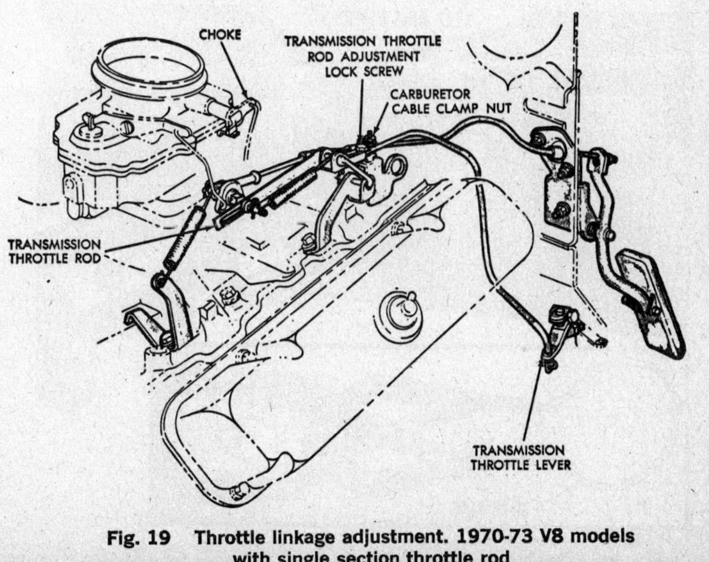

Fig. 19 Throttle linkage adjustment. 1970-73 V8 models with single section throttle rod

2. Loosen transmission throttle rod adjustment lock screw.
3. Hold transmission forward against its stop while adjusting transmission linkage. (On engines with solenoid idle stops, the solenoid plunger must also be in its fully extended position).
4. Adjust the transmission rod by pulling forward on the slotted link with a slight effort so that the rear edge of the slot is against the carburetor lever pin. Tighten transmission rod adjusting lock screw.

NOTE: The slotted link and transmission lever must be held forward while the locking screw is being tightened.

5. To check transmission linkage freedom of operation, move slotted link to the full rearward position, then allow it to return slowly, making sure it returns to the full forward position.
6. Loosen carburetor cable clamp nut. Adjust position of cable housing ferrule in the clamp so that all slack is removed from cable with carburetor at curb idle. To remove slack from cable, move ferrule in the clamp in direction away from carburetor lever.
7. Back off ferrule 1/4". This provides 1/4" free play. Tighten cable clamp nut to 45 inch lbs.
8. Connect choke or remove blocking fixture.

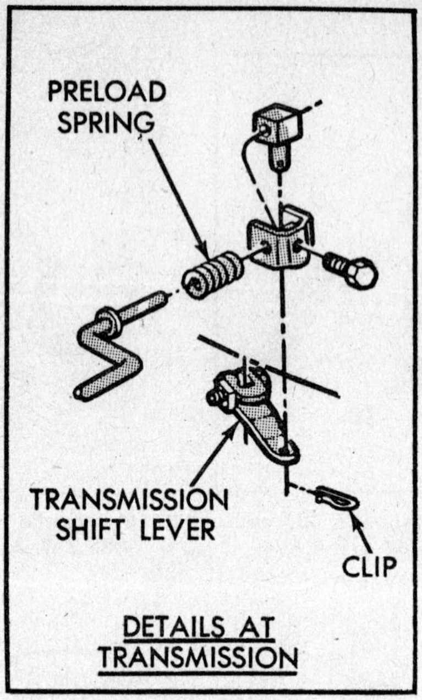

Fig. 20 Throttle linkage adjustment. All 1974-75 Chrysler Corp. models

Models Indicated in Fig. 20

1. Support vehicle on hoist and loosen swivel lock screw.

NOTE: To insure correct adjustment, swivel must be free to slide along flat end of throttle rod so that preload spring action is not restricted. If necessary, disassemble and clean or repair parts to assure free action.

2. Hold transmission lever firmly forward against its internal stop and tighten swivel lock screw to 125 inch lbs.

NOTE: Adjustment is now finished. Linkage backlash was automatically removed by the preload spring.

3. Lower vehicle and test linkage operation by moving throttle rod rearward and slowly releasing it making certain that it returns fully.

EXTENSION HOUSING & PARKING LOCK CONTROL ROD

NOTE: 1973-75 Charger, Coronet, Imperial and Satellite, 1975 Cordoba and Fury models, require unloading of both torsion bars and drop-one side of torsion bar crossmember to provide clearance for extension housing removal.

1. Mark parts for reassembly and remove propeller shaft.
2. Remove speedometer pinion and adapter assembly, then drain about two quarts of fluid from transmission.
3. Remove extension housing to crossmember bolts, then raise transmission with jack and remove crossmember.
4. Remove extension housing to transmission bolts. On console shift models, remove torque shaft lower bracket to extension housing bolts.

NOTE: In the following step, the gearshift must be in "low" therefore positioning the parking lock control rod rearward so it can be disengaged or engaged with the parking lock sprag.

5. Remove two screws, plate and gasket from bottom of extension housing mounting pad, then spread snap ring from output shaft bearing, Fig. 21, and carefully tap extension housing off output shaft bearing.
6. Slide extension housing off shaft to remove parking sprag and spring, then remove snap ring and slide reaction plug and pin assembly out of housing, Fig. 22.
7. To replace parking lock control rod, refer to "Valve Body".

OUTPUT SHAFT OIL SEAL

1. Mark propeller shaft to aid in reassembly and remove propeller shaft being careful not to scratch or nick surface on sliding spline yoke.
2. Using a screwdriver and hammer,

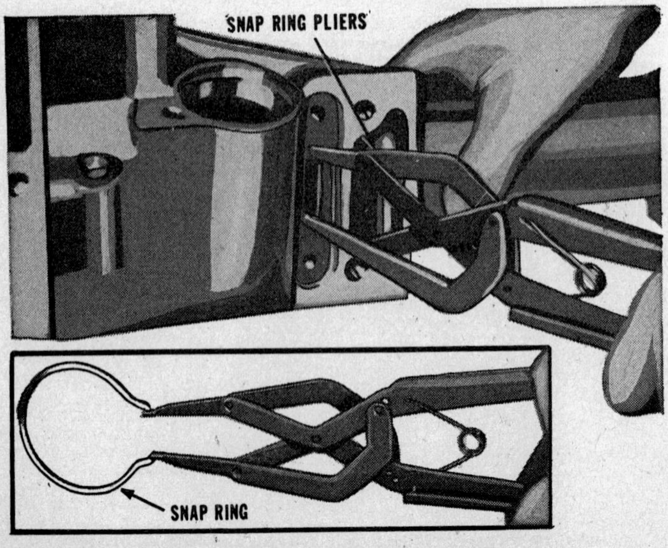

Fig. 21 Removing or installing extension housing snap ring

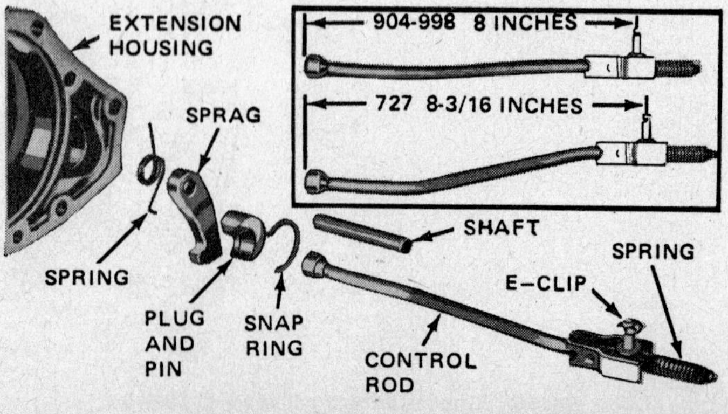

Fig. 22 Parking lock components

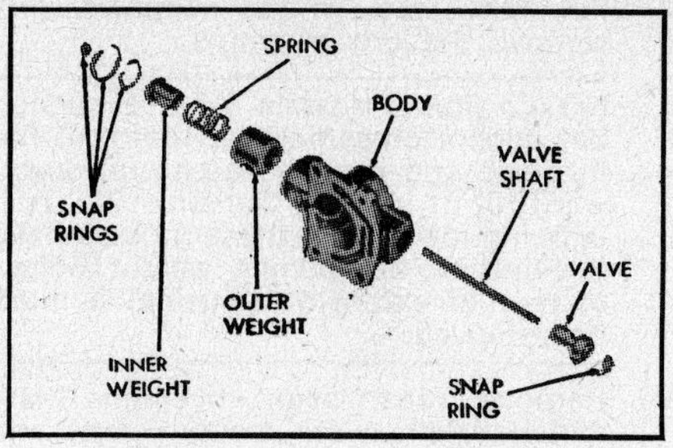

Fig. 23 Governor disassembled

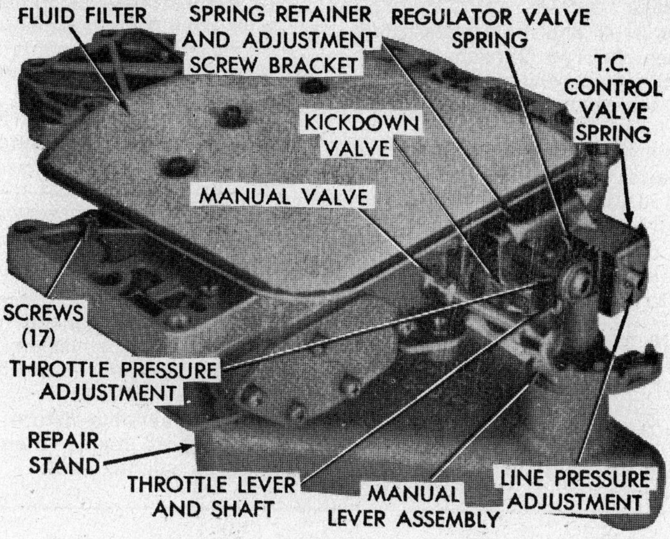

Fig. 24 Valve body external parts

drive between extension housing and seal and remove seal.

3. Position new seal and drive it into extension housing using tool C-3995 or C-3972.
4. Carefully install yoke into housing, then align marks made at removal and install propeller shaft.

GOVERNOR

1. Remove extension housing, then remove output shaft bearing rear snap ring and remove bearing. On 727 Series, remove remaining snap ring from shaft.
2. Remove snap ring, Fig. 23, from weight end of governor valve shaft and remove valve shaft from governor body.
3. Remove large snap ring from weight end of governor housing, and lift out weight assembly.
4. Remove snap ring from inside governor weight and remove inner weight and spring from outer weight.
5. Remove snap ring from behind governor housing, then slide governor housing and parking brake sprag assembly off input shaft. If necessary, separate governor housing from sprag (4 screws).
6. The primary cause of governor operating failure is due to a sticking governor valve or weights. Rough surfaces may be removed with crocus cloth. Thoroughly clean all parts and check for free movement before assembly.
7. Reverse above operations to assemble and install governor.

VALVE BODY

1. Drain transmission and remove oil pan.
2. Loosen clamp bolts and remove throttle and gear selector levers from manual lever, Fig. 24.
3. Remove neutral safety switch and oil filter.
4. Place a drain pan under transmission and remove the ten valve body to transmission bolts. Hold valve body in place while removing bolts.
5. Carefully lower valve body while pulling it forward to disengage parking control rod.

NOTE: It may be necessary to rotate output shaft to permit parking control rod to clear sprag.

6. Remove accumulator piston and spring from transmission case. Inspect piston for nicks, scores and wear. Inspect spring for distortion. Inspect rings for freedom in piston grooves and wear or breakage. Replace parts as necessary.

TRANSMISSION, REPLACE

American Motors

1972-75

CAUTION: The hood must be open to prevent damage to hood and air cleaner when removing rear crossmember.

1. Remove cover from front of bell housing then turn converter until drain plug is at "6 o'clock" position.
2. Drain converter and transmission. Remove filler tube and starter.
3. Mark rear universal joint and yoke for alignment purposes during installation. Then remove propeller shaft.
4. Disconnect speedometer cable, throttle and shift linkages, neutral safety switch and transmission controlled spark switch wires. Remove transmission oil cooler lines.
5. Mark converter and drive plate for alignment purposes during installation. Remove converter to drive plate bolts.
6. Install a suitable jack under transmission and remove crossmember from side sill and rear support cushion.
7. Remove support cushion and adapter from extension housing.
8. Remove transmission to engine attaching bolts and move transmission an adequate distance to the rear to clear crankshaft.
9. Lower transmission, maintaining pressure against converter, until transmission clears engine.
10. Reverse procedure to install.

Chrysler Corp.

1969-75

CAUTION: The transmission and converter must be removed as an assembly, otherwise, the converter drive plate, front pump bushing and oil seal will be damaged. The drive plate will not support the load; therefore, none of the weight of the transmission should be allowed to rest on the plate during removal.

NOTE: Some 400 CID 2-BBL engines (engine number 4T400-1-03 through 4T400-108) were built with a forged crankshaft requiring a different torque converter and damper than the engines using the cast crankshaft.

This forged crankshaft is normally only used in the 400-4-BBL, HP engine with manual transmission.

If replacement of the crankshaft, torque converter, crankshaft damper or short engine is required, it is important that matching parts are used otherwise severe engine vibration will result, (Consult Chrysler Parts Dept.).

The cast crankshaft engine can be easily identified since it has the letter "E" stamped on the engine numbering pad following the built date.

1. Disconnect battery ground cable.

NOTE: Some models will require that the exhaust system be lowered for clearance.

2. Remove engine to transmission struts (if equipped), then disconnect transmission cooler lines and remove starter motor, cooler line bracket and converter access cover.
3. Using a socket wrench on vibration dampener bolt, rotate engine clockwise until convertor drain plug is at bottom. Remove plug and drain converter, than loosen oil pan and drain

transmission.

4. Mark converter and drive plate to aid in reassembly. The crankshaft flange bolt circle, inner and outer circle of holes in the drive plate, and the four tapped holes in front face of converter all have one hole offset so these parts will be installed in the original position. This maintains balance of the engine and converter.
5. Remove converter to drive plate bolts. Rotate engine clockwise using socket wrench to gain access to all bolts.

CAUTION: Do not rotate converter or drive plate by prying with a screwdriver or similar tool as the drive plate might become distorted. Also, the starter should never be engaged if the drive plate is not attached to the converter with at least one bolt or if the transmission case-to-engine bolts have been loosened.

6. Mark drive shaft to aid in reassembly and remove drive shaft.
7. Disconnect neutral and back-up light switch and gearshift and torque shaft assembly from transmission.

NOTE: When disassembling linkage rods from levers which use plastic grommets as retainers, the grommets should be replaced with new ones.

8. Disconnect throttle rod from lever at left side of transmission, then remove linkage bellcrank from transmission, if so equipped.
9. Install a suitable fixture or jack that will support engine, then raise transmission slightly with a jack to relieve the load on the supports, and remove the crossmember.

NOTE: Some models have a torsion bar anchor crossmember that remains in place and requires a careful downward tilt of front of transmission as it is being lowered. If these models have a vibration dampening weight bolted to rear of extension housing, it must be removed.

10. Remove transmission to engine bolts and carefully work transmission and converter assembly rearward off engine block dowels and disengage converter hub from end of crankshaft. Using a small C-clamp on edge of bell housing, hold converter in place during transmission removal.
11. Remove transmission assembly from under vehicle.
12. Reverse procedure to install.

C3 & C4 DUAL RANGE AUTOMATIC

TRANSMISSION IDENTIFICATION

Each transmission may be identified by the tag attached to the low-reverse servo cover bolt. The tag includes the model prefix and suffix, a service identification number and a build date code. The service identification number indicates changes to service details which affect interchangeability when the transmission model is not changed. For interpretation of this number the Ford Master Parts Catalog should be consulted.

YEAR	CAR MODEL	TRANS. MODEL	ENGINE MODEL
	C3 Transmission		
1974	Mustang II	BKB	4-140
1975	Bobcat	75DT-BB	4-140
	Bobcat	75DT-BC	4-140
	Bobcat①	75DT-EA	V6-171
	Bobcat②	75DT-FA	V6-171
	Mustang II	75DT-AA	4-140
	Mustang II	75DT-AD	4-140
	Pinto	75DT-BB	4-140
	Pinto	75DT-BC	4-140
	C4 Transmission		
1969	Fairlane③	PEE-V	V8-302
	Fairlane④	PEE-M	V8-302
	Fairlane③	PEE-AE	6-250
	Fairlane④	PEE-AF	6-250
	Falcon③	PEB-C2	6-170, 200
	Falcon③	PEE-V	V8-302
	Ford③	PEA-A2	6-240
	Ford③	PEA-M1	V8-302
	Ford④	PEA-N1	V8-302
	Montego④	PEE-M	V8-302
	Montego③	PEE-V	V8-302
	Montego③	PEE-AE	6-250
	Montego④	PEE-AF	6-250

YEAR	CAR MODEL	TRANS. MODEL	ENGINE MODEL
	Mustang④	PEB-B2	6-200
	Mustang④	PEE-AC	V8-302
	Mustang④	PEE-AD	6-250
1970	Cougar	PEE-AC2	V8-302
	Fairlane④	PEE-M1	V8-302
	Fairlane③	PEE-V1	V8-302
	Fairlane	PEE-AC1	V8-302
	Fairlane	PEE-AD1	6-250
	Fairlane③	PEE-AE1	6-250
	Fairlane④	PEE-AF1	6-250
	Fairlane③	PEF-D	V8-352
	Fairlane④	PEF-E	V8-351
	Falcon	PEE-M1	V8-301
	Falcon	PEE-V2	V8-302
	Falcon	PEE-AC1	V8-302
	Falcon	PEE-AD1	6-250
	Falcon	PEE-AE1	6-250
	Falcon	PEE-AF1	6-250
	Falcon	PEB	6-200
	Ford	PEA-M2	6 & V8-302
	Ford	PEA-N2	V8-302
	Ford	PEA-A3	6-240
	Ford③	PEF-A	V8-351
	Ford④	PEF-B	V8-351
	Maverick	PEB-D	6-170, 200
	Maverick	PEB-D1	6-170, 200
	Maverick	PEE-AK	6-250
	Montego④	PEE-M1	V8-302
	Montego③	PEE-V1	V8-302
	Montego③	PEE-AE1	6-250
	Montego④	PEE-AF1	6-250
	Montego③	PEF-D	V8-351
	Montego④	PEF-E	V8-351
	Mustang	PEE-AC1	V8-302
	Mustang	PEE-AD1	6-250
	Mustang	PEB-B3	6-200
1971	Comet③	PEB-D2	6-200
	Comet④	PEB-H	6-200
	Comet③	PEE-AK1	6-250

YEAR	CAR MODEL	TRANS. MODEL	ENGINE MODEL
	Comet④	PEE-AM	6-250
	Comet③	PEE-AH1	V8-302
	Comet④	PEE-AL	V8-302
	Fairlane③	PEE-AE2	6-250
	Fairlane④	PEE-AF2	6-250
	Fairlane③	PEE-M2	V8-302
	Fairlane④	PEE-V2	V8-302
	Fairlane③	PEF-D1	V8-351
	Fairlane④	PEF-E1	V8-351
	Ford	PEA-M3	6-240
	Ford③	PEA-M3	V8-302
	Ford④	PEA-N3	V8-302
	Ford③	PEF-A1	V8-351
	Ford④	PEF-B1	V8-351
	Montego③	PEE-AE2	6-250
	Montego④	PEE-AF2	6-250
	Montego④	PEE-M2	V8-302
	Montego③	PEE-V2	V8-302
	Montego③	PEF-D1	V8-351
	Montego④	PEF-E1	V8-351
	Maverick③	PEB-D2	6-200
	Maverick④	PEB-H	6-200
	Maverick③	PEE-AK1	6-250
	Maverick④	PEE-AM	6-250
	Mustang	PEE-AD2	6-250
	Mustang	PEE-AC2	V8-302
	Pinto	PEJ-B	4-122
1972	Comet③	PEB-D3, 4, 5	6-200
	Comet④	PEB-H1,2, 3, 4	6-200
	Comet④	PEE-AK2, 3	6-250
	Comet④	PEE-AM1, 2, 3	6-250
	Comet③	PEE-AH2, 3	V8-302
	Comet④	PEE-AL1, 2, 3	V8-302
	Ford	PEA-M4, 5	6 & V8-302
	Ford	PEF-A2, 3	V8-351
	Maverick③	PEB-D3, 4, 5	6-200
	Maverick④	PEB-H1, 2, 3, 4	6-200

Continued

TRANSMISSION IDENTIFICATION—Continued

YEAR	CAR MODEL	TRANS. MODEL	ENGINE MODEL
1972	Maverick③	PEE-AK2, 3	6-250
	Maverick④	PEE-AM1, 2	6-250
	Maverick③	PEE-AH2, 3	V8-302
	Maverick④	PEE-AL1, 2	V8-302
	Montego③	PEE-AE4, 5	6-250
	Montego④	PEE-AF4, 5	6-250
	Montego③	PEE-V4, 5	V8-302
	Montego④	PEE-M4, 5	V8-302
	Montego③	PEF-D3, 4	V8-351
	Montego④	PEF-E3, 4	V8-351
	Mustang④	PEE-AD3, 4	6-250
	Mustang④	PEE-AC3, 4	V8-302
	Pinto	PEJ-B1, 2	4-122
	Torino③	PEE-AE4, 5	6-250
	Torino④	PEE-AF4, 5	6-250
	Torino③	PEE-V4, 5	V8-302
	Torino④	PEE-M4, 5	V8-302
	Torino③	PEF-D3, 4	V8-351
	Torino④	PEF-E3, 4	V8-351
	Torino④	EDPEF-E4	V8-351
1973	Comet③	PEB-D6	6-200
	Comet④	PEB-H5	6-200
	Comet③	PEE-AK4	6-250
	Comet④	PEE-AM6	6-250
	Comet③	PEE-AH4	V8-302
	Comet④	PEE-AL4	V8-302
	Ford	PEE-M6	V8-302
	Ford	PEF-A4	V8-351
	Maverick③	PEB-D6	6-200
	Maverick④	PEB-H5	6-200
	Maverick③	PEE-AK4	6-250
	Maverick④	PEE-AM4	6-250
	Maverick③	PEE-AH4	V8-302
	Maverick④	PEE-AL4	V8-302
	Montego③	PEE-AE6	6-250
	Montego④	PEE-AF6	6-250
	Montego③	PEE-V6	V8-302
	Montego④	PEE-M6	V8-302
	Montego③	PEF-D5	V8-351
	Montego④	PEF-E5	V8-351
	Mustang④	PEE-AD6	6-250
	Mustang④	PEE-AC6	V8-302
	Pinto	PEJ-B4	4-122
	Torino③	PEE-AE6	6-250
	Torino④	PEE-AF6	6-250
	Torino③	PEE-V6	V8-302
	Torino④	PEE-M6	V8-302
	Torino③	PEF-D5	V8-351
	Torino④	PEF-E5	V8-351
1974	Comet③	PEB-D7	6-200
	Comet④	PEB-H6	6-200
	Comet③	PEE-AK5	6-250
	Comet④	PEE-AM5	6-250
	Comet③	PEE-AH5	V8-302
	Comet④	PEE-AL5	V8-302
	Comet①③	PEE-BH	V8-302
	Comet①④	PEE-BJ	V8-302
	Cougar③	PEF-D6	V8-351
	Cougar④	PEF-E6, 7	V8-351
	Ford	PEF-A5	V8-351
	Maverick③	PEB-D7	6-200
	Maverick④	PEB-H6	6-200
	Maverick③	PEE-AK5	6-250
	Maverick④	PEE-AM5	6-250
	Maverick③	PEE-AH5	V8-302
	Maverick④	PEE-AL5	V8-302
	Maverick①③	PEE-BH	V8-302
	Maverick①④	PEE-BJ	V8-302
	Montego	PEE-V7	V8-302
	Montego	PEF-D6	V8-351
	Mustang II①	PEJ-H	4-140
	Mustang II	PEJ-E	6-171
	Mustang II①	PEJ-J	6-171
	Pinto	PEJ-B6	4-122
	Pinto①	PEJ-G	4-140
	Pinto	PEJ-C	4-140
	Torino③	PEE-V7	V8-302
	Torino④	PEE-M7, 8	V8-302
	Torino③	PEF-D6	V8-351
	Torino④	PEF-E6, 7	V8-351
1975	Bobcat	PEJ-G2	4-140
	Bobcat	PEJ-J1	V6-171
	Comet③	PEB-D8	6-200
	Comet④	PEB-H7	6-200
	Comet③	PEE-AK6	6-250
	Comet④	PEE-AM6	6-250
	Comet③②	PEE-AH6	V8-302
	Comet③②	PEE-AH7	V8-302
	Comet③①	PEE-BH1	V8-302
	Comet④②	PEE-AL6	V8-302
	Comet④②	PEE-AL7	V8-302
	Comet④①	PEE-BJ1	V8-302
	Cougar	PEF-S	V8-351M
	Granada③	PEE-BT	6-250
	Granada④	PEE-AM6	6-250
	Granada③②	PEE-CA	V8-302
	Granada④②	PEE-AL6	V8-302
	Granada④①	PEE-BJ1	V8-302
	Granada③②	PEF-Y	V8-351
	Granada③①	PEF-Q	V8-351
	Granada④②	PEF-Z	V8-351
	Granada④①	PEF-R	V8-351
	Maverick③	PEB-D8	6-200
	Maverick④	PEB-H7	6-200
	Maverick③	PEE-AK6	6-250
	Maverick④	PEE-AM6	6-250
	Maverick③②	PEE-AH6	V8-302
	Maverick③①	PEE-BH1	V8-302
	Maverick④②	PEE-AL6	V8-302
	Maverick④①	PEE-BJ1	V8-302
	Monarch④	PEE-BU	6-250
	Monarch③	PEE-BT	6-250
	Monarch④	PEE-AM6	6-250
	Monarch③②	PEE-CA	V8-302
	Monarch③	PEE-CA1	V8-302
	Monarch④	PEE-AL6	V8-302
	Monarch④	PEE-AL7	V8-302
	Monarch④①	PEE-BJ1	V8-302
	Monarch③②	PEF-Y	V8-351
	Monarch③①	PEF-Q	V8-351
	Monarch④②	PEF-Z	V8-351
	Monarch④①	PEF-R	V8-351
	Montego	PEF-D7	V8-351W
	Montego	PEF-T	V8-351W
	Montego③	PEF-S	V8-351M
	Mustang II	PEJ-J1	V6-171
	Mustang II	PEE-CC	V8-302
	Pinto	PEJ-G2	4-140
	Pinto	PEJ-J1	V6-171
	Torino③	PEF-D7	V8-351W
	Torino③	PEF-T	V8-351W
	Torino④	PEF-E8	V8-351W
	Torino④	PEF-U	V8-351W
	Torino③	PEF-S	V8-351M
	Torino④	PEF-V	V8-351M
	Torino④	PEF-W	V8-351M

①—California.
②—Exc. Calif.
③—Column shift.
④—Floor shift.

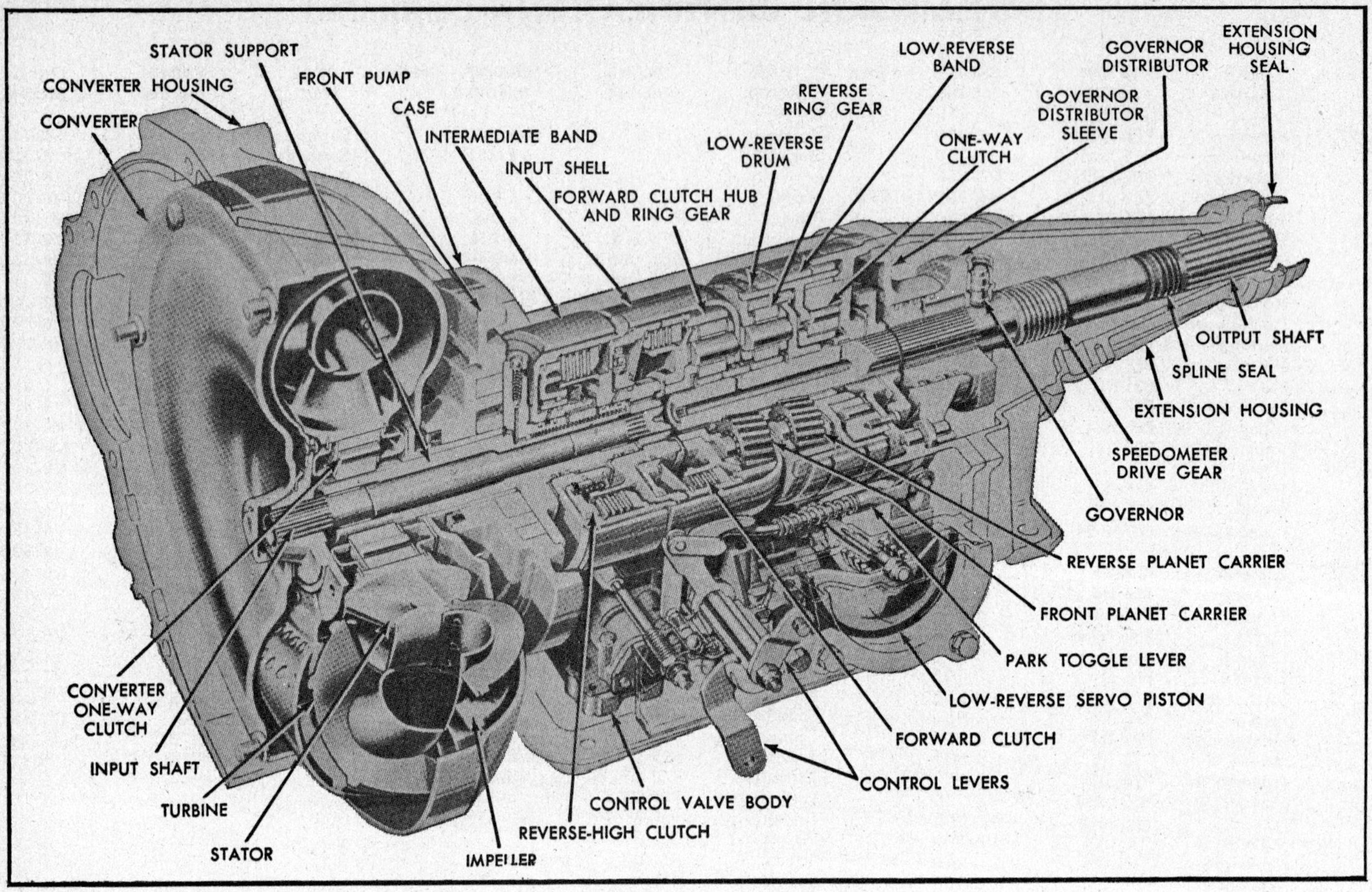

Fig. 1 C4 Dual Range Automatic

DESCRIPTION

The main control assembly incorporates a manually selective first and second gear range. The transmission features a drive range that provides for fully automatic upshifts and downshifts, and manually selected low and second gears.

The transmission consists essentially of a torque converter, a compound planetary gear train, two multiple disc clutches, a one-way clutch and a hydraulic control system, Figs. 1 and 2.

For all normal driving the selector lever is moved to the green dot under "Drive" on the selector quadrant on the steering column or on the floor console. As the throttle is advanced from the idle position, the transmission will upshift automatically to intermediate gear and then to high.

The driver can force downshift the transmission from high to intermediate at speeds up to 65 mph. A detent on the downshift linkage warns the driver when the carburetor is wide open. Accelerator pedal depression through the detent will bring in the downshift.

With the throttle closed the transmission will downshift automatically as the car speed drops to about 10 mph. With the throttle open at any position up to the detent, the downshifts will come in automatically at speeds above 10 mph and in proportion to throttle opening. This prevents engine lugging on steep hill climbing, for example.

When the selector lever is moved to "L" with the transmission in high, the transmission will downshift to intermediate or to low depending on the road speed. At speed above 25 mph, the downshift will be from high to intermediate. At speeds below 25 mph, the downshift will be from high to low. With the selector lever in the "L" position the transmission cannot upshift.

TROUBLE SHOOTING GUIDE

Rough Initial Engagement In D1 or D2

1. Engine idle speed.
2. Vacuum diaphragm unit or tubes restricted, leaking or maladjusted.
3. Check control pressure.
4. Pressure regulator.
5. Valve body.
6. Forward clutch.

1-2 or 2-3 Shift Points Erratic

1. Check fluid level.
2. Vacuum diaphragm unit or tubes restricted, leaking or maladjusted.
3. Intermediate servo.
4. Manual linkage adjustment.
5. Governor.
6. Check control pressure.
7. Valve body.
8. Make air pressure check.

Rough 1-2 Upshifts

1. Vacuum diaphragm unit or tubes restricted, leaking or maladjusted.
2. Intermediate servo.
3. Intermediate band.
4. Check control pressure.
5. Valve body.
6. Pressure regulator.

Rough 2-3 Upshifts

1. Vacuum diaphragm unit or tubes restricted, leaking or maladjusted.
2. Intermediate servo.
3. Check control pressure.
4. Pressure regulator.
5. Intermediate band.
6. Valve body.
7. Make air pressure check.
8. Reverse-high clutch.
9. Reverse-high clutch piston air bleed valve.

Dragged Out 1-2 Shift

1. Check fluid level.

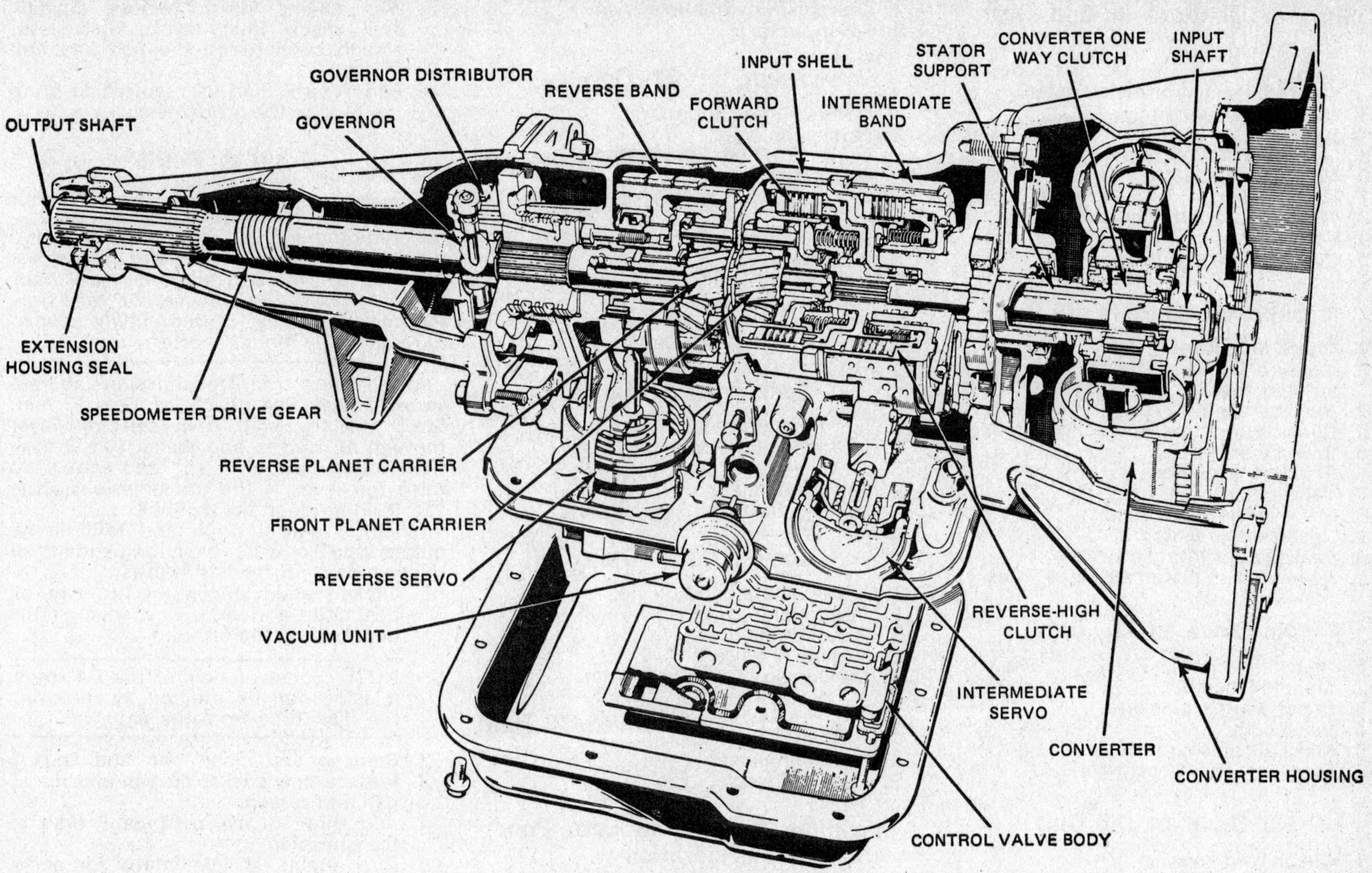

Fig. 2 C3 Dual Range Automatic

2. Vacuum diaphragm unit or tubes restricted, leaking or maladjusted.
3. Intermediate servo.
4. Check control pressure.
5. Intermediate band.
6. Valve body.
7. Pressure regulator.
8. Make air pressure check.
9. Leakage in hydraulic system.

Engine Overspeeds on 2-3 Shift

1. Manual linkage.
2. Check fluid level.
3. Vacuum diaphragm unit or tubes restricted, leaking or maladjusted.
4. Reverse servo.
5. Check control pressure.
6. Valve body.
7. Pressure regulator.
8. Intermediate band.
9. Reverse-high clutch.
10. Reverse-high clutch piston air bleed valve.

No 1-2 or 2-3 Shift

1. Manual linkage.
2. Downshift linkage, including inner lever position.
3. Vacuum diaphragm unit or tubes restricted, leaking or maladjusted.
4. Governor.
5. Check control pressure.
6. Valve body.
7. Intermediate band.
8. Intermediate servo.
9. Reverse-high clutch.
10. Reverse-high clutch piston air bleed valve.

No 3-1 Shift in D1 or 3-2 Shift in D2

1. Governor.
2. Valve body.

No Forced Downshifts

1. Downshift linkage, including inner lever position.
2. Valve body.
3. Vacuum diaphragm unit or tubes restricted, leaking or maladjusted.

Runaway Engine on Forced 3-2 Downshift

1. Check control pressure.
2. Intermediate servo.
3. Intermediate band.
4. Pressure regulator.
5. Valve body.
6. Vacuum diaphragm unit or tubes restricted, leaking or maladjusted.
7. Leakage in hydraulic system.

Rough 3-2 or 3-1 Shift at Closed Throttle

1. Engine idle speed.
2. Vacuum diaphragm unit or tubes restricted, leaking or maladjusted.
3. Intermediate servo.
4. Valve body.
5. Pressure regulator.

Shifts 1-3 in D1 and D2

1. Intermediate band.
2. Intermediate servo.
3. Vacuum diaphragm unit or tubes restricted, leaking or maladjusted.
4. Valve body.
5. Governor.
6. Make air pressure check.

No Engine Braking In 1st Gear —Manual Low

1. Manual linkage.
2. Reverse band.
3. Reverse servo.
4. Valve body.
5. Governor.
6. Make air pressure check.

Slips or Chatters in 1st Gear—D1

1. Check fluid level.
2. Vacuum diaphragm unit or tubes restricted, leaking or maladjusted.
3. Check control pressure.
4. Pressure regulator.
5. Valve body.
6. Forward clutch.
7. Leakage in hydraulic system.
8. Planetary one-way clutch.

Slips or Chatters in 2nd Gear

1. Check fluid level.
2. Vacuum diaphragm unit or tubes restricted, leaking or maladjusted.
3. Intermediate servo.
4. Intermediate band.
5. Check control pressure.
6. Pressure regulator.
7. Valve body.
8. Make air pressure check.
9. Forward clutch.
10. Leakage in hydraulic system.

Slips or Chatters in R

1. Check fluid level.
2. Vacuum diaphragm unit or tubes restricted, leaking or maladjusted.
3. Reverse band.
4. Check control pressure.
5. Reverse servo.
6. Pressure regulator.
7. Valve body.
8. Make air pressure check.
9. Reverse-high clutch.
10. Leakage in hydraulic system.
11. Reverse-high piston air bleed valve.

No Drive in D1 Only

1. Check fluid level.
2. Manual linkage.
3. Check control pressure.
4. Valve body.
5. Make air pressure check.
6. Planetary one-way clutch.

No Drive in D2 Only

1. Check fluid level.
2. Manual linkage.
3. Check control pressure.
4. Intermediate servo.
5. Valve body.
6. Make air pressure check.
7. Leakage in hydraulic system.
8. Planetary one-way clutch.

No Drive in L Only

1. Check fluid level.
2. Manual linkage.
3. Check control pressure.
4. Valve body.
5. Reverse servo.
6. Make air pressure check.
7. Leakage in hydraulic system.
8. Planetary one-way clutch.

No Drive in R Only

1. Check fluid level.
2. Manual linkage.
3. Reverse band.
4. Check control pressure.
5. Reverse servo.
6. Valve body.
7. Make air pressure check.
8. Reverse-high clutch.
9. Leakage in hydraulic system.
10. Reverse-high clutch piston air bleed valve.

No Drive in Any Selector Position

1. Check fluid level.
2. Manual linkage.
3. Check control pressure.
4. Pressure regulator.
5. Valve body.
6. Make air pressure check.
7. Leakage in hydraulic system.
8. Front pump.

Lockup in D1 Only

1. Reverse-high clutch.
2. Parking linkage.
3. Leakage in hydraulic system.

Lockup in D2 Only

1. Reverse band.
2. Reverse servo.
3. Reverse-high clutch.
4. Parking linkage.
5. Leakage in hydraulic system.
6. Planetary one-way clutch.

Lockup in L Only

1. Intermediate band.
2. Intermediate servo.
3. Reverse-high clutch.
4. Parking linkage.
5. Leakage in hydraulic system.

Lockup in R Only

1. Intermediate band.
2. Intermediate servo.
3. Forward clutch
4. Parking linkage.
5. Leakage in hydraulic system.

Parking Lock Binds or Won't Hold

1. Manual linkage.
2. Parking linkage.

Maximum Speed Too Low, Poor Acceleration

1. Engine performance.
2. Brakes bind.
3. Converter one-way clutch.

Noisy in N or P

1. Check fluid level.
2. Pressure regulator.
3. Front pump.
4. Planetary assembly.

Noisy in All Gears

1. Check fluid level.
2. Pressure regulator.
3. Planetary assembly.
4. Forward clutch.
5. Front pump.
6. Planetary one-way clutch.

Car Moves Forward in N

1. Manual linkage.
2. Forward clutch.

MAINTENANCE

Checking Oil Level

1. With transmission at operating temperature, park vehicle on a level surface.
2. Run engine at idle speed with service and parking brakes applied and move selector lever through each range. Return selector lever to Park.
3. With engine idling, remove dipstick and check fluid level. Fluid level should be between the Add and Full marks.
4. Add Type F fluid as required to bring the fluid to the proper level.

Drain & Refill

NOTE: *Normal maintenance and lubrication requirements do not necessitate periodic fluid changes. If a major failure has occurred in the transmission, it will have to be removed for service. At this time the converter must be thoroughly flushed to remove any foreign matter.*

When filling a dry transmission and converter, install five quarts of Type F fluid. Start engine, shift the selector lever through all ranges and place it at P position. Check fluid level and add enough to raise the level in the transmission to the "F" (full) mark on the dipstick.

When a partial drain and refill is required due to front band adjustment or minor repair, proceed as follows:

1. Loosen and remove all but two oil pan bolts and drop one edge of the pan to drain the oil.

NOTE: Some models of the C4 transmission can be drained by removing the filler tube from the pan.

2. Remove and clean pan and screen.
3. Place a new gasket on pan and install pan and screen.
4. Add three quarts of Type F fluid to transmission.
5. Run engine at idle speed for about two minutes.
6. Check oil level and add oil as necessary.
7. Run engine at a fast idle until it reaches normal operating temperature.
8. Shift selector lever through all ranges and then place it in P position.
9. Add fluid as required to bring the level to the full mark.

MANUAL LINKAGE, ADJUST

1969-75

Floor Shift

1. Place transmission selector lever in D position.
2. Raise car and loosen shift rod retaining nut, Fig. 3.
3. Move transmission manual lever to D position (third detent position from rear of transmission).
4. Tighten attaching nut to 10-20 ft. lbs.

NOTE: On some 1970-75 models, after adjusting manual linkage, adjust lock rod as follows:

1. Raise car and loosen rod retaining nut, Fig. 3.
2. Lower car and place shift lever in D position.
3. Align hole in steering column socket casting with column alignment mark and insert a .180" diameter gauge

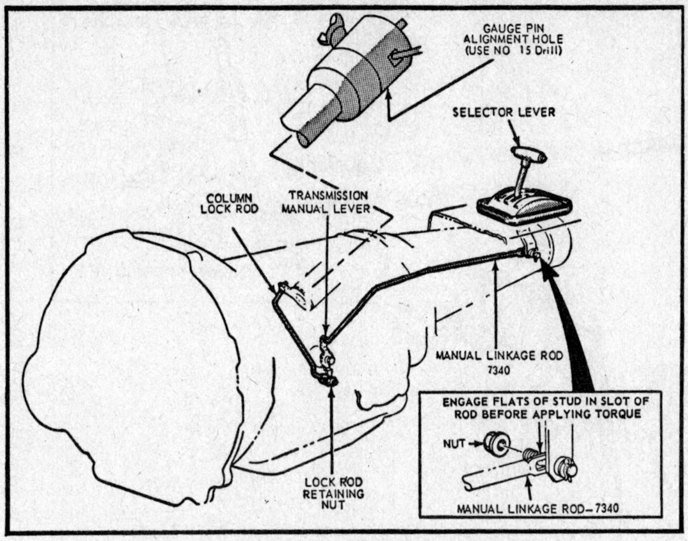

Fig. 3 Manual linkage, floorshift. Typical. Column lock rod is used on some models beginning 1970

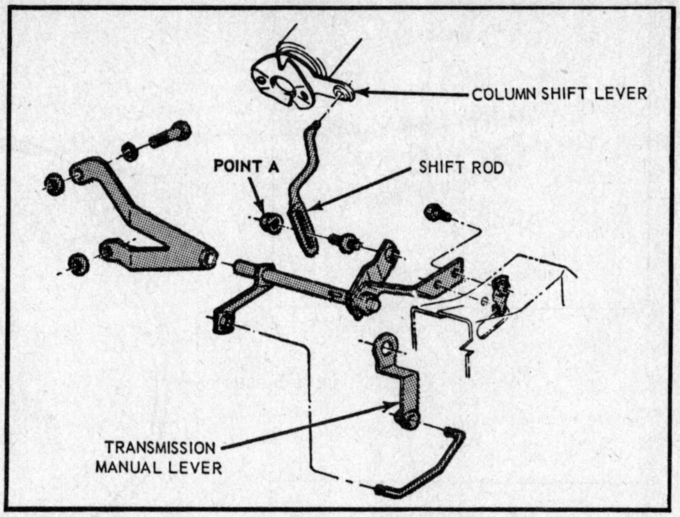

Fig. 4 Manual linkage. 1969-75 Ford column shift. Typical

pin. Column casting must not rotate with pin in place.

4. Raise car and torque lock rod nut to 10-20 ft. lbs.
5. Lower vehicle and remove gauge pin. Check linkage for proper operation.

Column Shift

1. Place selector lever in D position (HI on Maverick with semi-automatic).
2. Loosen shift rod adjusting nut at point A, Figs. 4 through 8.
3. Shift manual lever at transmission to D or HI position. (Third from rear).
4. Tighten adjusting nut to 10-20 ft. lbs.

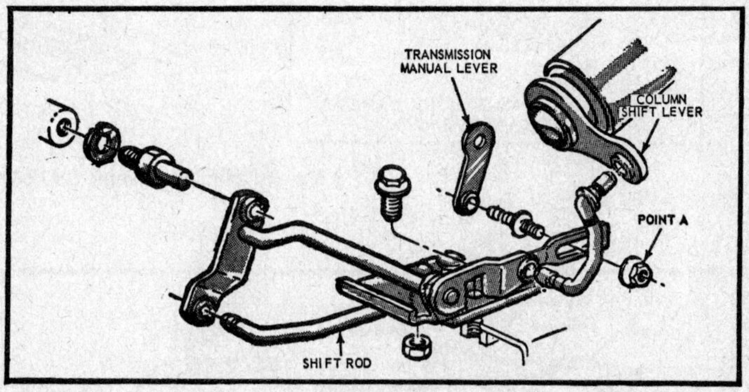

Fig. 5 Manual linkage. 1969-70 Fairlane, Falcon & Montego column shift

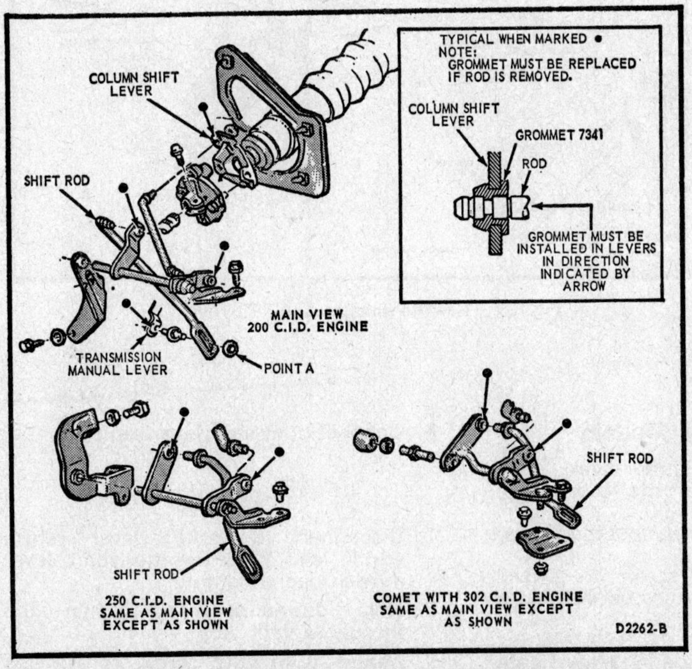

Fig. 6 Manual linkage. 1970-75 Maverick and Comet column shift

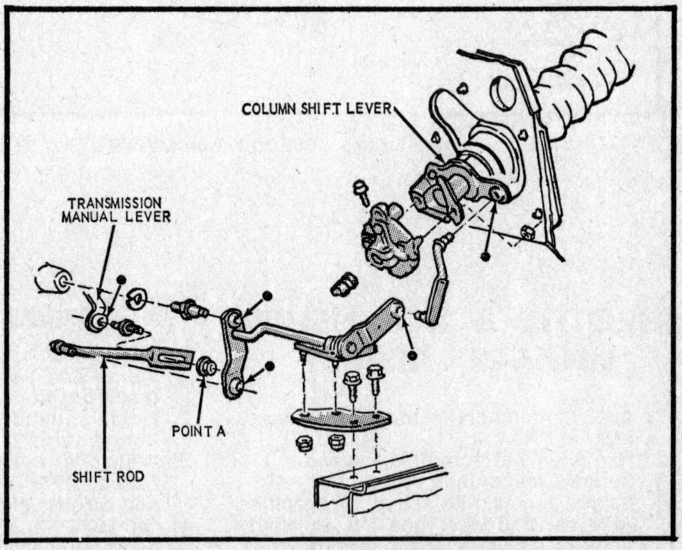

Fig. 7 Manual linkage. 1971 Torino and Montego column shift

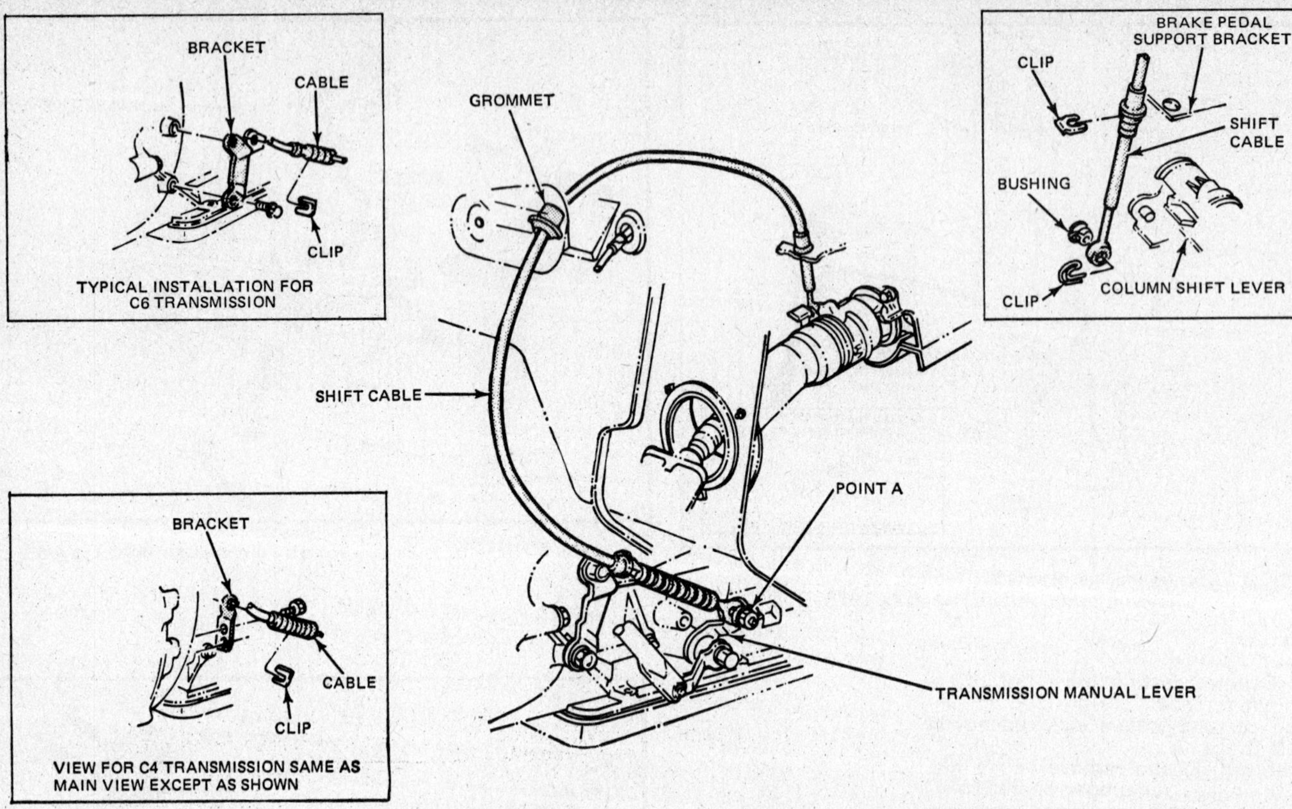

Fig. 8 Manual linkage. 1972-75 Torino & Montego column shift

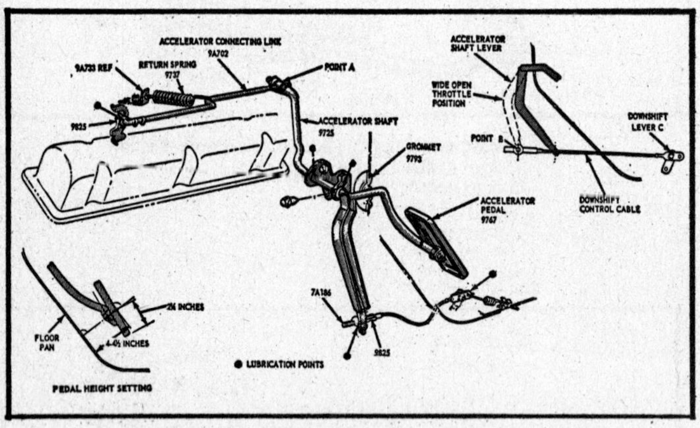

Fig. 9 Throttle linkage. 1969-70 Falcon with V8

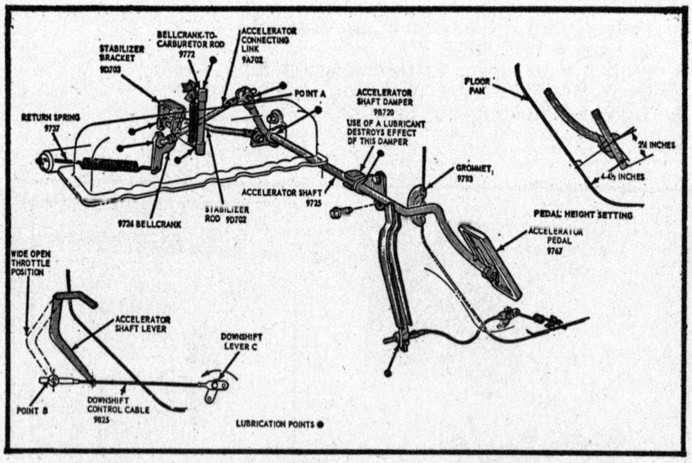

Fig. 10 Throttle linkage. 1969-70 Falcon 6

THROTTLE & DOWNSHIFT LINKAGE, ADJUST

1. Apply parking brake and place selector lever in "N".
2. Run engine at normal idle speed.
3. Connect tachometer to engine.
4. Adjust engine idle speed to specified RPM listed under Tune Up Specifications in individual car chapters.
5. Proceed with the adjustments as outlined below.

1973-75 All Models

1. Disconnect downshift lever return spring and hold throttle lever in wide open position.
2. Hold downshift rod against the through detent stop.
3. Adjust downshift screw to provide 0.01 to 0.08" clearance between screw and throttle arm.
4. On 1973 Mustang, 1973-74 Montego and Torino six cylinder engines, tighten screw lock nut to maintain screw position.
5. Connect downshift lever return spring.

1972 except Pinto

1. Disconnect downshift lever return spring and hold throttle shaft lever in wide open position.
2. Hold downshift rod against the through detent stop.
3. Adjust downshift screw to provide .050-.070" clearance between screw and throttle shaft lever.

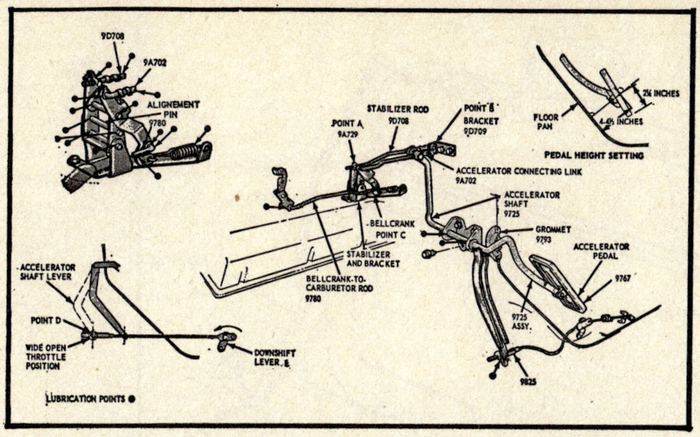

Fig. 11 Throttle linkage. 1969-70 Comet and Fairlane V8

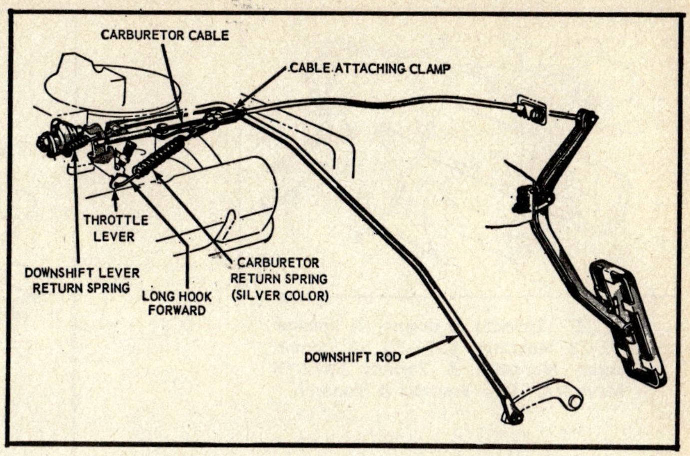

Fig. 12 Throttle linkage. 1969-70 Ford V8

4. On six cylinder engines, tighten locknut to maintain screw position.
5. Connect downshift lever return spring.

1971-72 Pinto

1. Disconnect downshift rod return spring. Hold throttle shaft lever in wide open position and hold downshift rod against the through detent stop.
2. Adjust downshift screw to provide .050-.070″ clearance between screw tip and throttle shaft lever tab.
3. Connect downshift lever spring.

1969-71 Comet, Fairlane, Falcon, Montego & Mustang & 1969-71 Cougar

NOTE: On Fairlane and Montego V8 units, throttle bellcrank must be adjusted first. Disconnect carburetor rod and accelerator connecting link from bellcrank and stabilizer rod from stabilizer. Insert a ¼″ diameter pin through the stabilizer and bracket and adjust the rods so their trunnions freely enter holes. Secure rods and remove pins.

1. On units with rod controlled throttle only, with engine off, check accelerator pedal for a height of 4½″ Figs. 9, 10 and 11. Adjust accelerator connecting link to correct. This step is not required on units with cable controlled throttles.
2. Disconnect downshift control cable from accelerator shaft lever.
3. With carburetor choke in the off position, depress the accelerator and block it in the wide open position.
4. Rotate the downshift lever counterclockwise to place it against internal stop.
5. Except 1969 Mustang V8 and all 1970-71: With the lever held in this position and all slack removed from the cable, adjust the trunnion so that it will slide into the lever. Turn it one additional turn to increase length of cable then secure it to the lever with retaining clip.
6. On 1969 Mustang V8 and all 1970-71, turn adjustment screws on carburetor kick down lever to within .040″ to .080″ gap of contacting pick up surface of carburetor throttle lever.

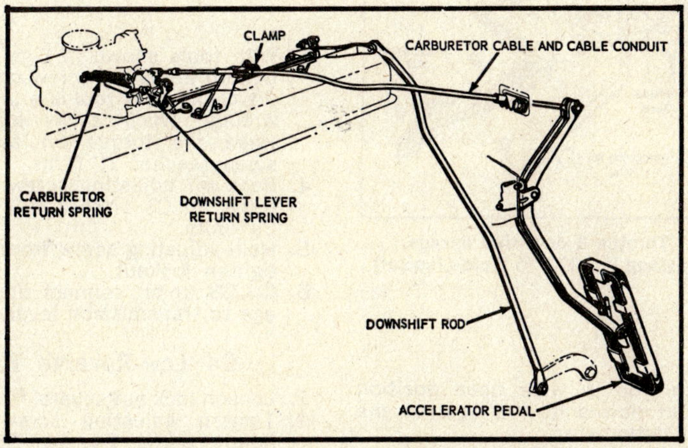

Fig. 13 Throttle linkage. 1969-70 Ford 6

1969-71 Ford

On all engines, the conduit covering the cable at the carburetor end must be evenly nestled between the clamp and the accelerator shaft bracket. Due to the fixed clamping of the cable conduit, accelerator pedal height adjustment is not required.

1. Disconnect the downshift lever return spring.

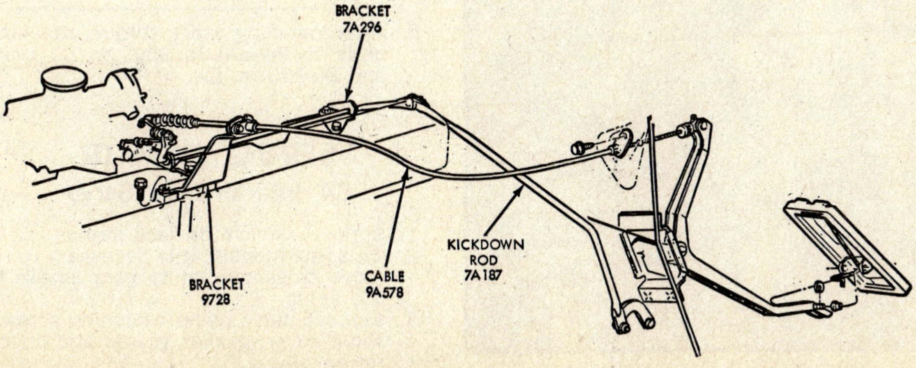

Fig. 14 Throttle & downshift linkage 1971-75 Ford & Mercury (typical)

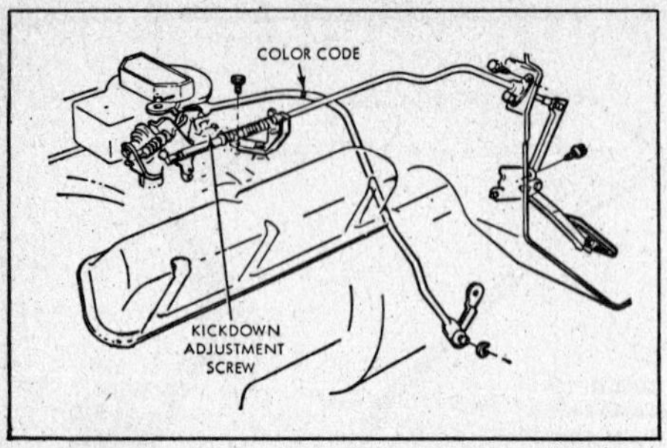

Fig. 15 Throttle & downshift linkage 1971-73 Mustang, 1971-75 V8 Comet, Cougar, Montego, & Torino; 1972-75 Maverick; 1975 Granada & Monarch

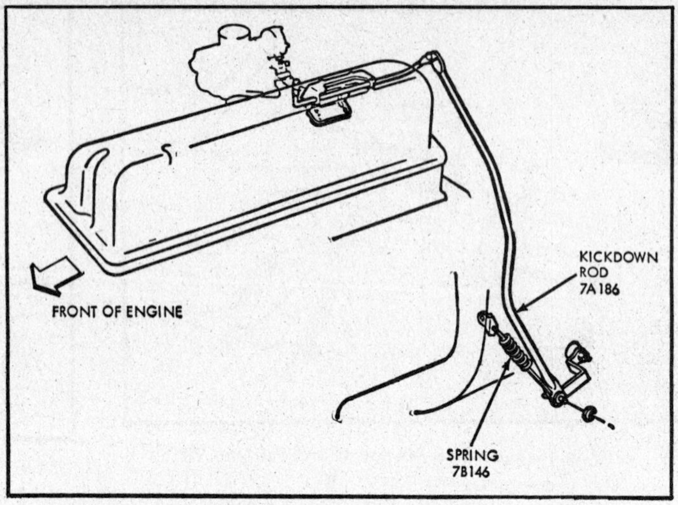

Fig. 16 Downshift linkage 1971-75 6 cyl. Comet, Maverick, Montego, Torino, 1971-73 Mustang, 1975 Granada & Monarch

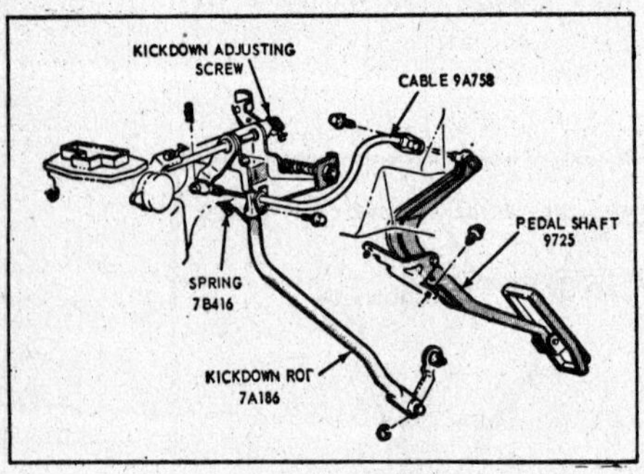

Fig. 17 Throttle & downshift linkage. 1974-75 Mustang II, 1971-75 Pinto (Typical)

2. Hold throttle in wide open position and hold downshift rod against the through detent stop.
3. Adjust the downshift screw to provide .050"-.070" clearance between the screw and the throttle shaft lever. On 6-240, tighten the lock nut to maintain the screw position. End play adjustments is not required.
4. Connect downshift lever return spring.

1969-70 Ford

1. Apply parking brake and place selector lever at N.
2. With engine idling at normal operating temperature, connect a tachometer.

BANDS, ADJUST

NOTE: The intermediate and low-reverse bands adjusting screw locknut must be discarded and a new one installed each time a band is adjusted.

Intermediate Band

C3 & C4

1. On C3 units, disconnect downshift linkage from transmission lever.
2. On all units, discard adjusting screw locknut and install a new locknut.
3. With tools shown in Figs. 18 and 19, tighten adjusting screw until tool handle clicks. *This tool is a pre-set torque wrench which clicks and overruns when the torque on the adjusting screw reaches 10 ft-lbs.*
4. Back off adjusting screw exactly 1½ turns on C3 units and 1¾ turns on C4 units.
5. Hold adjusting screw from turning and tighten locknut.
6. On C3 units, connect downshift linkage to transmission lever.

C4 Low-Reverse Band

1. Loosen lock nut several turns.
2. Tighten adjusting screw until tool handle clicks, Fig. 20. *Tool shown is a pre-set torque wrench which clicks and overruns when the torque on the adjusting screw reaches 10 ft-lbs.*
3. Back off adjusting screw exactly 3 full turns.
4. Hold adjusting screw from turning and tighten lock nut.

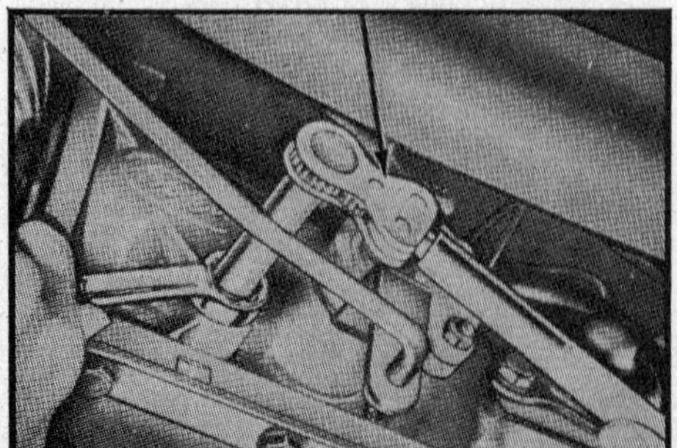

Fig. 18 Intermediate band adjustment. C4

CONTROL VALVE

NOTE: All fasteners used on C3 transmissions are designed to metric specifications.

1. Support vehicle on jack stands.
2. Drain transmission fluid, then remove oil pan, fluid screen, gasket and on early C3 units, remove three spacers.

NOTE: If fluid is to be reused, filter it through a 100 mesh screen.

3. On C3 units:
   a. Remove control valve attaching bolts. Note the different length and location of each bolt.
   b. Carefully remove control valve while unlocking and detaching selector lever connecting rod.
4. On C4 units:
   a. Shift selector lever into PARK and remove the two detent spring to control valve and case bolts.
   b. Remove remaining control valve to case bolts, then while holding manual valve inward, remove control valve.

   NOTE: Failure to hold manual valve inward, while removing control valve, could cause manual valve to become damaged.

5. After installing valve, torque attaching bolts to 84-108 in. lbs. on C3 units and 80-120 in. lbs. on C4 units.

SERVO REPAIR

C4 Intermediate Servo

1. Support vehicle on jack stands.
2. On some models, it is necessary to remove crossmember to gain access to the servo.
3. Remove servo cover attaching screws, servo cover, gasket, piston and piston return spring.
4. Replace piston seals. Lubricate new

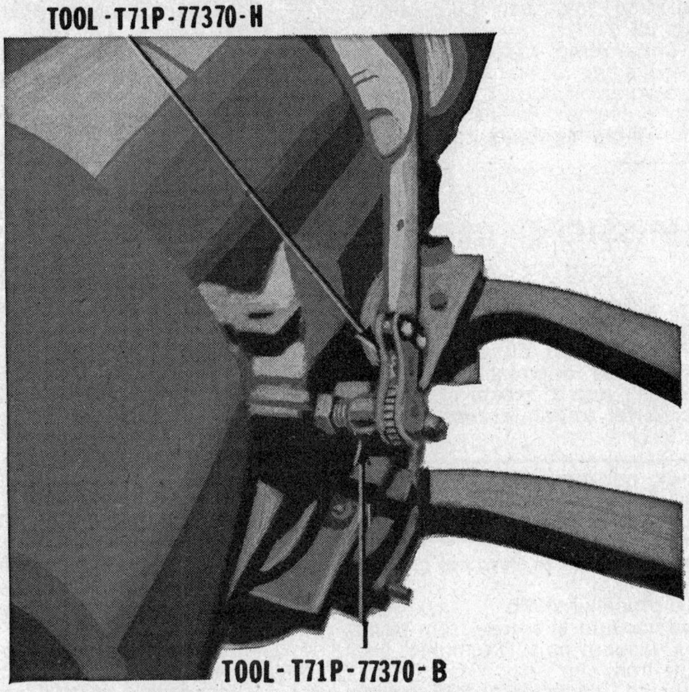

Fig. 19 Intermediate band adjustment. C3

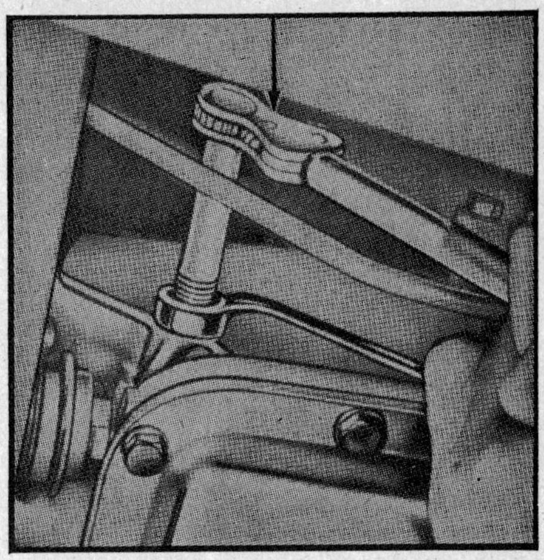

Fig. 20 Low-reverse band adjustment. C4

seals with transmission fluid before installation.

5. Reverse procedure to install.

C4 Low-Reverse Servo

1. Support vehicle on jack stands.
2. Loosen reverse band adjusting screw locknut and torque adjusting screw to 10 foot pounds. With adjusting screw torqued, the band strut is forced against case, preventing the strut from falling out of position when removing servo piston.
3. Remove servo cover retaining bolts, servo cover, seal and servo piston from case.
4. On some models, the piston seal is bonded to the piston, requiring piston replacement. To remove piston from stem, insert a small screwdriver through hole in stem and remove piston retaining nut, piston, accumulator spring and spacer.
5. On all other models, replace piston seals. Lubricate new seals with transmission fluid before installation.
6. Reverse procedure to install.

C3 Rear Servo

1. Support vehicle on jack stands, then drain transmission.
2. Remove oil filter screws, gasket and on early models, three spacers.
3. Remove servo cover retaining screws, servo cover, piston and spring, Fig. 21.
4. Reverse procedure to intsall.

Fig. 21 Rear servo removal. C3

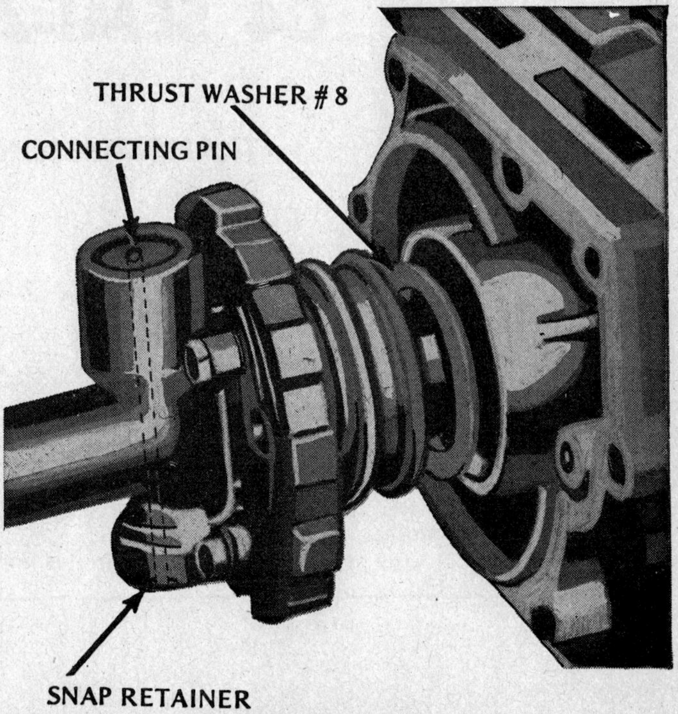

Fig. 22 Governor removal. C3

EXTENSION HOUSING

1. Support vehicle on jack stands and remove driveshaft.

NOTE: Scribe marks on driveshaft yoke and companion flange, to insure proper positioning of driveshaft during assembly.

2. Support transmission with suitable jack and disconnect speedometer cable.

NOTE: On some models, it will be necessary to disconnect the exhaust system from the exhaust manifolds to perform the following step.

3. Remove engine rear support to crossmember attaching bolts or nuts, then raise transmission slightly and remove rear support from extension housing.

NOTE: On some models, it will be necessary to remove crossmember in order to remove rear support from extension housing.

4. Loosen extension housing bolts and allow transmission fluid to drain and remove extension housing.

GOVERNOR

1. Remove extension housing as described previously.
2. On C3 units, remove governor retaining pin snap ring, then the retaining pin, Fig. 22.
3. On all units, remove governor to governor housing retaining bolts and slide governor off output shaft.
4. Reverse procedure to install. Torque governor retaining bolts to 7 to 10 foot pounds.

TRANSMISSION, REPLACE

1969-75

1. Support vehicle on jack stands and remove converter housing lower cover.
2. Drain transmission oil pan and the converter. Use a wrench on crankshaft pulley nut to rotate crankshaft and converter to gain access to drain plug.

NOTE: Do not rotate 2000 cc or 2300 cc engines in opposite direction of normal rotation.

3. Remove converter to flywheel bolts or nuts.
4. Remove propeller shaft.
5. Remove vacuum line hose from transmission vacuum unit. Disconnect vacuum line from clip.
6. If equipped, disconnect TRS switch wire.
7. Remove two extension housing to crossmember bolts.
8. Remove speedometer cable from extension housing.
9. On 1969 models and 1971-75 Comet, Maverick, Mustang and Pinto models, disconnect exhaust pipe from manifold.
10. On 1969 models, disconnect parking brake cable from equalizer lever.
11. On all models, disconnect oil cooler lines from transmission.
12. Remove the manual and kick down linkage rods from transmission shift levers.
13. Where necessary, disconnect the neutral start switch wires.
14. Remove starter.
15. Remove transmission fluid filler tube.
16. On all models except 1971-75 Pinto and 1974-75 Mustang II:
    a. Support transmission with a suitable jack and remove crossmember.
    b. Remove converter housing to engine bolts and lower transmission from vehicle.
17. On 1971-75 Pinto and 1974-75 Mustang II:
    a. Raise transmission slightly with a suitable jack and remove engine rear support to crossmember nut.
    b. Remove crossmember attaching bolts and crossmember.
    c. Lower jack, allowing transmission to hang, and raise front of engine with a suitable jack, thereby gaining access to the two upper converter housing to engine bolts.
    d. Secure transmission to jack, remove the lower converter housing to engine bolts, then the two upper bolts, move transmission rearward and lower from vehicle.
18. Reverse procedure to install.

C4 SEMI-AUTOMATIC

TRANSMISSION IDENTIFICATION

Maverick	Code
1970	PEG-A
1971	PEG-A1

GENERAL DESCRIPTION

This unit is essentially the same as the C4 Cruise-O-Matic except that the automatic shifting provisions have been removed.

The major differences between this unit and the C4 are the control valve body and the manual linkage. The manual linkage adjustment procedures are the same as the C4, however, the shift selector position of "Hi" on this unit is the equivalent of "D" on the C4.

For service procedures on this unit, refer to the C4 Dual Range chapter.

C6 DUAL RANGE AUTOMATIC

TRANSMISSION IDENTIFICATION

An identification tag attached to the servo cover bolt, includes the model prefix and suffix

YEAR	CAR MODEL	TRANS. MODEL	ENGINE MODEL
1969	Cougar②	PGA-AE	V8-390③
	Cougar②	PGB-AF-1	V8-428④
	Fairlane①	PGA-AC	V8-390④
	Fairlane②	PGA-AD	V8-390③
	Fairlane①	PGB-AG	V8-428④
	Fairlane②	PGB-AH	V8-428④
	Ford①	PGA-A3	V8-390③
	Ford②	PGA-J3	V8-390③
	Ford①	PGA-Z	V8-429③
	Ford②	PGA-AA	V8-429③
	Ford①	PGB-F2	V8-428④
	Ford①	PBG-G2	V8-390
	Ford①	PGB-H2	V8-428④
	Ford①	PGB-AD	V8-429④
	Ford②	PGB-AE	V8-429④
	Lincoln	PGC-B	V8-460
	Mark III	PGC-C1	V8-460
	Mercury①	PGA-A3	V8-390③
	Mercury②	PGA-J3	V8-390③
	Mercury①	PGA-Z	V8-429③
	Mercury②	PGA-AA	V8-429③
	Mercury①	PGB-F2	V8-428④
	Mercury①	PGB-G2	V8-390
	Mercury①	PGB-H2	V8-428④
	Mercury①	PGB-AD	V8-429④
	Mercury②	PGB-AE	V8-429④
	Montego①	PGA-AC	V8-390④
	Montego②	PGA-AD	V8-390③
	Montego①	PGB-AG	V8-428④
	Montego②	PGB-AH	V8-428④
	Mustang②	PGA-AE	V8-390③
	Mustang②	PGB-AF-1	V8-428④
	Thunderbird	PGB-J1	V8-429④
1970	Cougar	PGB-AF2	V8-428
	Fairlane①	PJB-A	V8-429
	Fairlane②	PJB-B	V8-429
	Fairlane①	PJB-J	V8-429
	Fairlane①	PJC-A	V8-429
	Fairlane②	PJC-B	V8-429
	Fairlane①	PJC-E	V8-429
	Fairlane②	PJC-F	V8-429
	Ford①	PGA-A4	V8-390
	Ford②	PGA-J4	V8-390
	Ford①	PGB-F3	V8-428
	Ford①	PGB-G3	V8-390
	Ford①③	PJA-A	V8-429
	Ford②③	PJA-B	V8-429
	Ford①④	PJB-C	V8-429
	Ford②④	PJB-D	V8-429
	Ford①④	PJB-F	V8-429
	Lincoln	PJD-CF	V8-460
	Mark III	PJD-BE	V8-460
	Mercury①	PGA-A4	V8-390
	Mercury②	PGA-J4	V8-390
	Mercury①	PGB-F3	V8-428
	Mercury①	PGB-G3	V8-390
	Mercury①③	PJA-A	V8-429
	Mercury②③	PJA-B	V8-429
	Mercury①④	PJB-C	V8-429
	Mercury②④	PJB-D	V8-429
	Mercury①④	PJB-F	V8-429
	Montego①	PJB-A	V8-429
	Montego②	PJB-B	V8-429
	Montego①	PJB-J	V8-429
	Montego①	PJC-A	V8-429
	Montego②	PJC-B	V8-429
	Montego①	PJC-E	V8-429
	Montego②	PJC-F	V8-429
	Mustang	PGB-AF2	V8-428
	Thunderbird	PJB-GH	V8-429
1971	Cougar	PGA-AH	V8-351
	Cougar	PJC-G	V8-429
	Fairlane①	PGA-AF	V8-351
	Fairlane②	PGA-AG	V8-351
	Fairlane①	PJC-A	V8-429
	Fairlane②	PJC-B	V8-429
	Ford①	PGA-A4	V8-390
	Ford②	PGA-J4	V8-390
	Ford①⑤	PGB-G3	V8-390
	Ford①③	PJA-A	V8-429
	Ford②③	PJA-B	V8-429
	Ford①	PJA-C	V8-400
	Ford②	PJA-D	V8-400
	Ford①④	PJB-C	V8-429
	Ford②④	PJB-D	V8-429
	Ford①⑤	PJB-F	V8-429
	Ford①⑤	PJB-K	V8-400
	Ford⑤	PJC-HI	V8-429
	Lincoln	PJD-F	V8-460
	Mark III	PJD-E	V8-460
	Mercury①③	PJA-A	V8-429
	Mercury	PJA-C	V8-400
	Mercury①④	PJB-C	V8-429
	Mercury①⑤	PJB-F	V8-429
	Mercury①⑤	PJB-K	V8-400
	Mercury①⑤	PJC-HI	V8-429
	Montego①	PGA-AF	V8-351
	Montego②	PGA-AG	V8-351
	Montego①	PJC-A	V8-429
	Montego②	PJC-B	V8-429
	Mustang	PGA-AH	V8-351
	Mustang	PJC-G	V8-429
	Thunderbird	PJB-H	V8-429
1972	Cougar③	PGA-AV1	V8-351
	Cougar④	PGA-AU1, 2	V8-351
	Cougar④	PGA-AH1, 2	V8-351
	Ford①	PJA-C2	V8-400
	Ford①	PJA-C2	V8-429
	Ford①	PJA-K	V8-400
	Ford①⑤	PJB-K1	V8-400
	Ford①⑤	PJB-K1	V8-429
	Ford①⑤	PJC-H2	V8-429
	Ford①⑤	PJC-H4	V8-429
	Lincoln	PJD-F1	V8-460
	Mark IV	PJD-E1	V8-460
	Mercury	PJA-C2	V8-400
	Mercury	PJA-C2	V8-429
	Mercury⑤	PJB-K1	V8-400
	Mercury⑤	PJB-K1	V8-429
	Mercury⑤	PJC-H2	V8-429
	Mercury	PJD-G	V8-460
	Montego①	PGA-AF1	V8-351
	Montego②	PGA-AG1	V8-351
	Montego①	PJA-G	V8-400
	Montego①	PJA-G	V8-429
	Montego②	PJA-H	V8-400
	Montego②	PJA-H	V8-429
	Mustang③	PGA-AV1	V8-351
	Mustang④	PGA-AU1	V8-351
	Torino①③	PGA-AW	V8-351
	Torino②③	AGA-AY	V8-351
	Torino①④	PGA-AF1	V8-351
	Torino②④	PGA-AG1	V8-351
	Torino①	PJA-G	V8-400
	Torino①	PJA-G	V8-429
	Torino①③	PJA-L	V8-400
	Torino②③	PJA-M	V8-400
	Torino②	PJA-H	V8-400
	Torino②	PJA-H	V8-429
	Torino①⑤	PJC-J	V8-429
	Thunderbird	PJA-J	V8-400
	Thunderbird	PJA-J	V8-429
	Thunderbird	PJD-E1	V8-460
1973	Cougar②③	PGA-AV2	V8-351
	Cougar②④	PGA-AV3	V8-351
	Ford①	PJA-C3-4	V8-400
	Ford①⑤	PJB-K2-3	V8-400
	Ford①	PJB-C1	V8-429
	Ford①⑤	PJB-F1	V8-429
	Ford①⑤	PJC-H5	V8-460
	Lincoln	PJD-F2	V8-460
	Mark IV	PJD-E2	V8-460
	Mercury①	PJB-K2	V8-400
	Mercury①	PJA-C3-4	V8-400
	Mercury①⑤	PJB-K3	V8-400
	Mercury①	PJB-C1	V8-429
	Mercury①⑤	PJB-F1	V8-429
	Mercury①⑤	PJC-H5	V8-460
	Mercury①	PJD-J1	V8-460
	Montego①	PGA-AW1	V8-351C
	Montego②	PGA-AY1	V8-351C
	Montego①	PGA-AZ1	V8-351W
	Montego②	PGA-BA1	V8-351W
	Montego①④	PGA-AF2	V8-351
	Montego②④	PGA-AG2	V8-351
	Montego①	PJA-G1-2	V8-400
	Montego②	PJA-H1-2	V8-400
	Montego①	PJB-A1-B1	V8-429
	Montego①⑤	PJC-J1	V8-460
	Mustang②	PGA-AV2	V8-351C
	Mustang②	PGA-AV3	V8-351CJ
	Torino①	PGA-AW1	V8-351C
	Torino②	PGA-AY1	V8-351C
	Torino①	PGA-AF2	V8-351CJ
	Torino②	PGA-AG2	V8-351CJ
	Torino①	PGA-AZ1	V8-351W
	Torino②	PGA-BA1	V8-351W
	Torino①	PJA-G1-2	V8-400
	Torino②	PJA-H1-2	V8-400
	Torino①	PJB-A1	V8-429
	Torino②	PJB-B1	V8-429
	Torino①	PJC-J1	V8-460
	Thunderbird	PJA-J1	V8-429
	Thunderbird	PJD-E2	V8-460
1974	Cougar①③	PGA-AW	V8-351
	Cougar①④	PGA-AF3	V8-351
	Cougar②③	PGA-AY2, 3	V8-351
	Cougar②④	PGA-AG3, 4	V8-351
	Cougar①	PJA-G3	V8-400
	Cougar②	PJA-H3, 4	V8-400
	Cougar①	PJD-R	V8-460
	Cougar②	PJD-S, S1	V8-460
	Ford	PJA-C5	V8-400
	Ford⑤	PJC-H6	V8-460
	Ford	PJD-G3	V8-460
	Lincoln	PJD-F4	V8-460
	Mark IV	PJD-E4	V8-460
	Mercury	PJA-C5	V8-400
	Mercury⑤	PJC-H6	V8-460
	Mercury	PJD-G3	V8-460
	Montego③	PGA-AW2	V8-351
	Montego④	PGA-AF3	V8-351
	Montego	PJA-C3	V8-400
	Montego⑤	PJC-J2	V8-460

Continued

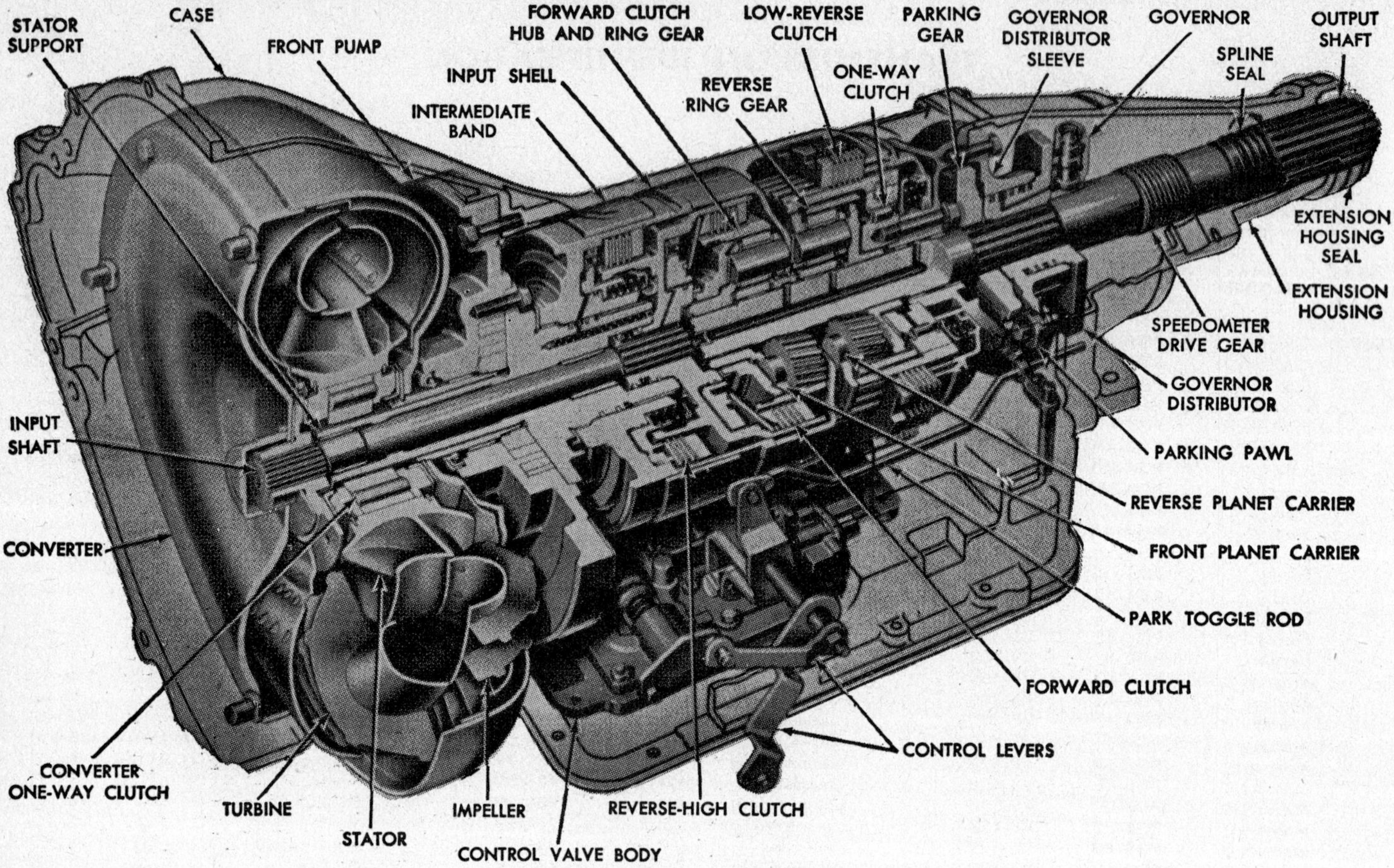

Fig. 1 Sectional view of C6 transmission

TRANSMISSION IDENTIFICATION—Continued

YEAR	CAR MODEL	TRANS. MODEL	ENGINE MODEL
1974	Montego	PJD-R	V8-460
	Torino①③	PGA-AW2	V8-351
	Torino②③	PGA-AY2, 3	V8-351
	Torino①④	PGA-AF3	V8-351
	Torino②④	PGA-AG3, 4	V8-351
	Torino①④	PJA-C3	V8-400
	Torino②④	PJA-H3, 4	V8-400
	Torino①⑤	PJC-J2	V8-460
	Torino①	PJD-R	V8-460
	Torino②	PJD-S, S1	V8-460
	Thunderbird	PJD-E4	V8-460
1975	Cougar①	PJA-V-1	V8-351M
	Cougar②	PJA-W-1	V8-351M
	Cougar①	PJA-G4-5	V8-400
	Cougar①	PJA-R1	V8-400
	Cougar②	PJA-H5-6	V8-400
	Cougar②	PJA-S1	V8-400
	Cougar①⑥	PJD-R1-2	V8-460
	Cougar①⑦	PJD-T1-2	V8-460
	Cougar②⑥	PJD-S2-3	V8-460
	Cougar②⑦	PJD-U1-2	V8-460
	Ford⑤	PJA-C6-7	V8-400
	Ford⑤⑥	PJC-H7-8	V8-460
	Ford⑤⑦	PJC-K-1	V8-460
	Ford⑥	PJD-G4-5	V8-460
	Ford⑦	PJD-P-1-2	V8-460
	Lincoln⑥	PJD-F5-6	V8-460
	Lincoln⑦	PJD-N-1-2	V8-460
	Mark IV⑥	PJD-E5-6	V8-460
	Mark IV⑦	PJD-M-1-2	V8-460
	Mercury⑤	PJA-C6-7	V8-400
	Mercury⑤	PJA-P1	V8-400
	Mercury⑤⑥	PJC-H7-8	V8-460
	Mercury⑤⑦	PJC-K-1	V8-460
	Mercury⑥	PJC-G4-5	V8-460
	Mercury⑦	PJD-P-1-2	V8-460
	Montego①	PJA-V-1	V8-351M
	Montego	PJA-G4-5	V8-400
	Montego	PJA-R1	V8-400
	Montego⑥	PJD R1 2	V8-460
	Montego⑦	PJD-T1-2	V8-460
	Torino①	PJA-V-1	V8-351M
	Torino②	PJA-W-1	V8-351M
	Torino①	PJA-G4-5	V8-400
	Torino②	PJA-H5-6	V8-400
	Torino①⑥	PJD-R1-2	V8-460
	Torino①⑦	PJD-T1-2	V8-460
	Torino②⑥	PJD-S2-3	V8-460
	Torino②⑦	PJD-U1-2	V8-460
	Thunderbird⑥	PJD-E5-6	V8-460
	Thunderbird⑦	PJD-M1-2	V8-460

①—Column shift.
②—Floor shift.
③—Two barrel carburetor.
④—Four barrel carburetor.
⑤—Police or Fleet.
⑥—Exc. Calif.
⑦—Calif.

DESCRIPTION

As shown in Fig. 1, the transmission consists essentially of a torque converter, a compound planetary gear train controlled by one band, three disc clutches and a one-way clutch, and a hydraulic control system.

The transmission is made so that a system of manual and automatic shifting is provided.

Automatic & Manual Shifting 1969-75

This unit has a shift pattern which is indicated on the selector as P-R-N-D-2-1. This refers respectively to Park, Reverse, Neutral, Full Automatic, Second Gear (manual), Low Gear (manual).

In this unit an overriding control is provided which enables the driver to exercise his own judgement with regard to the gear ratios to be selected and an understanding of what is possible greatly enhances the pleasure to be derived from driving the car. No automatic mechanism has the power of anticipation, but the driver can see ahead and has the means for over-riding the automatic mechanism.

Automatic Shift

In "D" position the shift sequence is fully automatic in that the transmission starts in low gear and upshifts through second gear to third or high gear.

Manual Shifting

The shift to 2 or 1 is done manually by shifting the lever from neutral to either position. In "1" position, the transmisson starts in 1st (low gear) and is retained. In "2" position it starts in 2nd gear and remains in 2nd gear, regardless of road speed.

Manual Shift To "1"

Manual shifting from "D" to "1" can also be accomplished any time. Here the transmission immediately shifts to second and remains in second until the predetermined governor control speed allows it to shift down to low gear where it remains. The governor speed control at this point eliminates the possibility of a direct down shift to low gear until the road speed is reduced.

Shift Lever Controls

A shift lever button control is used to shift from neutral to reverse or park, also when shifting from "D" to "2" or "1" position. However, the button control function is not required when shifting from neutral to "D", or to shift forward from "1" to "2" position.

Parking Pawl

The transmission gear train is in neutral in both P and N positions. There is no pressure to any clutch and only the transmission input shaft turns. In park, a pawl engages a parking gear which is splined to the transmission output shaft, Fig. 1, to lock the rear wheels to the transmission main case.

A neutral start switch, mounted on the transmission and operated by the selector linkage, completes the engine cranking circuit in P and N only so that the engine cannot be started in any drive gear.

Forced Downshifts

Forced downshifts (kickdown shifts) from high to second gear are possible at speeds as high as 65 mph in D1 or D2. In D1 it is possible to force a downshift to 1st gear up to 30 mph.

The carburetor is at full throttle before the accelerator is floored. Up to full throttle, a "torque demand" downshift to 2nd is possible up to 40 mph. "Kickdown" shifts require depressing the accelerator to the floor to actuate the downshift valve in the transmission.

TROUBLE SHOOTING GUIDE

No Drive In Forward Speeds

1. Manual linkage adjustment.
2. Check control pressure.
3. Valve body.
4. Make air pressure check.
5. Forward clutch.
6. Leakage in hydraulic system.

Rough Initial Engagement in D, D1, D2 or 2

1. Engine idle speed too high.
2. Vacuum diaphragm unit or tubes restricted, leaking or maladjusted.
3. Check control pressure.
4. Valve body.
5. Forward clutch.

1-2 or 2-3 Shift Points Incorrect or Erratic

1. Check fluid level.
2. Vacuum diaphragm unit or tubes restricted, leaking or maladjusted.
3. Downshift linkage, including inner lever position.
4. Manual linkage adjustment.
5. Governor defective.
6. Check control pressure.
7. Valve body.
8. Make air pressure check.

Rough 1-2 Upshifts

1. Vacuum diaphragm unit or tubes restricted, leaking or maladjusted.
2. Intermediate servo.
3. Intermediate band
4. Check control pressure.
5. Valve body.

Rough 2-3 Shifts

1. Vacuum diaphragm or tubes restricted leaking or maladjusted.
2. Intermediate servo.
3. Check control pressure.
4. Intermediate band.
5. Valve body.
6. Make air pressure check.
7. Reverse-high clutch.
8. Reverse-high clutch piston air bleed valve.

Dragged Out 1-2 Shift

1. Check fluid level.
2. Vacuum diaphragm unit or tubes restricted, leaking or maladjusted.
3. Intermediate servo.
4. Check control pressure.
5. Intermediate band.
6. Valve body.
7. Make air pressure check.
8. Leakage in hydraulic system.

Engine Overspeeds on 2-3 Shift

1. Manual linkage adjustment.
2. Check fluid level.
3. Vacuum diaphragm unit or tubes restricted, leaking or maladjusted.
4. Intermediate servo.
5. Check control pressure.
6. Valve body.
7. Intermediate band.
8. Reverse-high clutch.
9. Reverse-high clutch piston air bleed valve.

No 1-2 or 2-3 Shift

1. Manual linkage adjustment.
2. Downshift linkage including inner lever position.
3. Vacuum diaphragm unit or tubes restricted, leaking or malajusted.
4. Governor.
5. Check control pressure.
6. Valve body.
7. Intermediate band.
8. Intermediate servo.
9. Reverse-high clutch.
10. Leakage in hydraulic system.

No 3-1 Shift In D1, 2 or 3-2 Shift In D2 or D

1. Governor.
2. Valve body.

No Forced Downshifts

1. Downshift linkage, including inner lever position.
2. Check control pressure.
3. Valve body.

Runaway Engine on Forced 3-2 Shift

1. Check control pressure.
2. Intermediate servo.
3. Intermediate band.
4. Valve body.
5. Vacuum diaphragm unit or tubes restricted, leaking or maladjusted.
6. Leakage in hydraulic system.

Rough 3-2 Shift or 3-1 Shift at Closed Throttle

1. Engine idle speed.
2. Vacuum diaphragm unit or tubes restricted, leaking or maladjusted.
3. Intermediate servo.
4. Check control pressure.
5. Valve body.

Shifts 1-3 in D, D1, 2, D2

1. Intermediate band.
2. Intermediate servo.
3. Valve body.
4. Governor.
5. Make air pressure check.

No Engine Braking in 1st Gear—Manual Low Range

1. Manual linkage adjustment.
2. Low-reverse clutch.
3. Valve body.
4. Governor.
5. Make air pressure check.
6. Leakage in hydraulic system.

Creeps Excessively

1. Engine idle speed too high.

Slips or Chatters In 1st Gear, D1

1. Check fluid level.
2. Vacuum diaphragm unit or tubes restricted, leaking or maladjusted.
3. Check control peessure.
4. Valve body.
5. Forward clutch.
6. Leakage in hydraulic system.
7. Planetary one-way clutch.

Slips or Chatters In 2nd Gear

1. Check fluid level.
2. Vacuum diaphragm unit or tubes restricted, leaking or maladjusted.
3. Intermediate servo.
4. Intermediate band.
5. Check control pressure.
6. Valve body.
7. Make air pressure check.
8. Forward clutch.
9. Leakage in hydraulic system.

Slips or Chatters In Reverse

1. Check fluid level.
2. Vacuum diaphragm unit or tubes restricted, leaking or maladjusted.
3. Manual linkage adjustment.
4. Low-reverse clutch.
5. Check control pressure.
6. Valve body.
7. Make air pressure check.
8. Reverse-high clutch.
9. Leakage in hydraulic system.
10. Reverse-high clutch piston air bleed valve.

No Drive In D1 or 2

1. Manual linkage adjustment.
2. Check control pressure.
3. Valve body.
4. Planetary one-way clutch.

No Drive In D, D2

1. Check fluid level.
2. Manual linkage adjustment.
3. Check control pressure.
4. Intermediate servo.
5. Valve body.
6. Make air pressure check.
7. Leakage in hydraulic system.

No Drive In L or 1

1. Check fluid level.
2. Check control pressure.
3. Valve body.
4. Make air pressure check.
5. Leakage in hydraulic system.

No Drive In R Only

1. Check fluid level.
2. Manual linkage adjustment.
3. Low-reverse clutch.
4. Check control pressure.
5. Valve body.
6. Make air pressure check.
7. Reverse-high clutch.
8. Leakage in hydraulic system.
9. Reverse-high clutch piston air bleed valve.

No Drive In Any Selector Position

1. Check fluid level.
2. Manual linkage adjustment.
3. Check control pressure.
4. Valve body.
5. Make air pressure check.
6. Leakage in hydraulic system.
7. Front pump.

Lockup In D1 or 2

1. Valve body.
2. Parking linkage.
3. Leakage in hydraulic system.

Lockup In D2 or D

1. Low-reverse clutch.
2. Valve body.
3. Reverse-high clutch.
4. Parking linkage.
5. Leakage in hydraulic system.
6. Planetary one-way clutch.

Lockup In L or 1

1. Valve body.
2. Parking linkage.
3. Leakage in hydraulic system.

Lockup In R Only

1. Valve body.
2. Forward clutch.
3. Parking linkage.
4. Leakage in hydraulic system.

Parking Lock Binds or Does Not Hold

1. Manual linkage adjustment.
2. Parking linkage.

Transmission Overheats

1. Oil cooler and connections.
2. Valve body.
3. Vacuum diaphragm unit or tubes restricted, leaking or maladjusted.
4. Check control pressure.
5. Converter one-way clutch.
6. Converter pressure check valves.

Maximum Speed Too Low, Poor Acceleration

1. Engine performance.
2. Car brakes.
3. Forward clutch.

Transmission Noisy In N and P

1. Check fluid level.
2. Valve body.
3. Front pump.

Noisy In 1st, 2nd, 3rd or Reverse

1. Check fluid level.
2. Valve body.
3. Planetary assembly.
4. Forward clutch.
5. Reverse-high clutch.
6. Planetary one-way clutch.

Car Moves Forward In N

1. Manual linkage adjustment.
2. Forward clutch.

Fluid Leak

1. Check fluid level.
2. Converter drain plugs.
3. Oil pan gasket, filler tube or seal.
4. Oil cooler and connections.
5. Manual or downshift lever shaft seal.
6. 1/8" pipe plugs in case.
7. Extension housing-to-case gasket.
8. Extension housing rear oil seal.
9. Speedometer driven gear adapter seal.
10. Vacuum diaphragm unit or tubes.
11. Intermediate servo.
12. Engine rear oil seal.

MAINTENANCE

Checking Oil Level

1. Make sure car is on a level floor.
2. Apply parking brake firmly.
3. Run engine at normal idle speed. If transmission fluid is cold, run engine at a fast idle until fluid reaches normal operating temperature. When fluid is warm, slow engine to normal idle speed.
4. Shift selector lever through all positions, then place lever at "P". Do not shut down engine during fluid level checks.
5. Clean all dirt from dipstick cap before removing dipstick from filler tube.
6. Pull dipstick out of tube, wipe it clean and push it all the way back in tube.
7. Pull dipstick out of tube again and check fluid level. If necessary, add enough fluid to raise the level to the "F" mark on dipstick. Do not overfill.

Drain & Refill

NOTE: The Ford Motor Company recommends the use of an automatic transmission fluid with Qualification No. M2C-33F (on container) instead of the conventional Type A fluid. The recommended fluid is said to have a greater coefficient of friction and greater ability to handle maximum engine torques without band or clutch slippage.

Normal maintenance and lubrication requirements do not necessitate periodic fluid changes. If a major failure has occurred in the transmission, it will have to be removed for service. At this time the

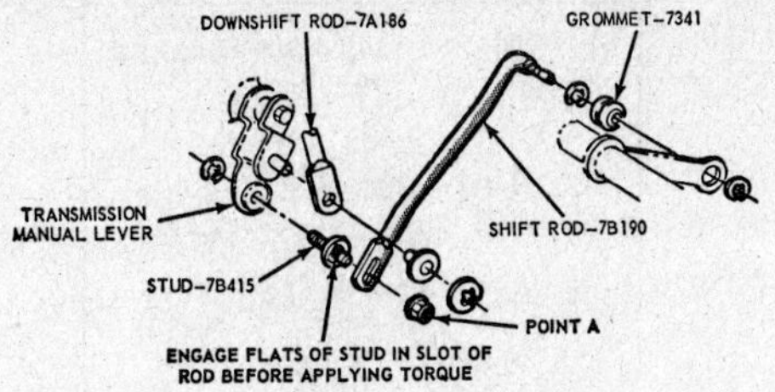

Fig. 2 Column shift linkage. 1969-70 Continental

converter must be thoroughly flushed to remove any foreign matter.

1. To drain the fluid, loosen pan attaching bolts and allow fluid to drain.
2. After fluid has drained to the level of the pan flange, remove pan bolts working from rear and both sides of pan to allow it to drop and drain slowly.
3. When fluid has stopped draining, remove and clean pan and screen. Discard pan gasket.
4. Using a new gasket, install pan.
5. Add 3 quarts of recommended fluid to transmission through filler tube.
6. Run engine at idle speed for 2 minutes, and then run it at a fast idle until it reaches normal operating temperature.
7. Shift selector lever through all positions, place it at "P" and check fluid level.
8. If necessary, add enough fluid to transmission to bring it to the "F" mark on the dipstick.

MANUAL LINKAGE, ADJUST

1969-75

Column Shift

1. Place selector lever in D position.
2. Loosen shift rod adjusting nut, point A in Figs. 2 through 8.
3. Shift transmission manual lever to D, third detent from rear of transmission.
4. Make sure selector lever has not moved from D position then tighten adjusting nut to 10-20 ft. lbs.

Floor Shift

1. Place transmission selector lever in D position.
2. Raise car and loosen shift rod retaining nut, Fig. 9.
3. Move transmission manual lever to D position, third detent from rear of transmission.
4. Tighten retaining nut to 10-20 ft. lbs.

NOTE: On some 1970-73 models, after adjusting manual linkage, adjust lock rod as follows:

1. Raise car and loosen lock rod retaining nut, Fig. 9.
2. Lower car and place shift lever in D position.
3. Align hole steering column socket casting with column alignment mark and insert a .180″ diameter gauge rod. Column casting must not rotate with rod in place.
4. Raise car and torque lock rod nut to 10-20 ft. lbs.

THROTTLE & DOWNSHIFT LINKAGE

Adjusting the throttle linkage is important to be certain the throttle and kickdown systems are properly adjusted. The kickdown system should come in when the accelerator is pressed through detent, and not before detent. See Figs. 8-15.

1973-75 All Models

1. Hold carburetor lever in the wide open throttle position.
2. Hold downshift rod against through detent stop.
3. Adjust downshift screw to provide .01 to .08 inch clearance between screw and throttle arm.
4. Connect downshift lever return spring.

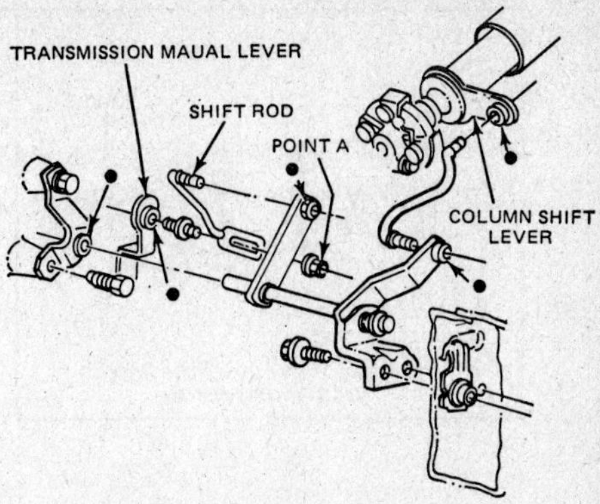

Fig. 3 Column shift linkage. 1971-75 Continental (Typical)

1969-71 Ford & Mercury, 1970-71 Lincoln, 1972 All

On all engines, the conduit covering the cable at the carburetor end must be evenly nestled between the clamp and the accelerator shaft bracket. Due to the fixed clamping of the cable conduit, accelerator pedal height adjustment is not required.

1. Disconnect downshift lever return spring.
2. Hold throttle in wide open position and hold downshift rod against the through detent stop.
3. Adjust the downshift screw to provide .050″-.070″ clearance between the screw and the throttle shaft lever.
4. Connect downshift lever return spring.

1969-71 Mustang & Cougar & 1970-71 Fairlane/Torino & Montego

Because the throttle is cable operated,

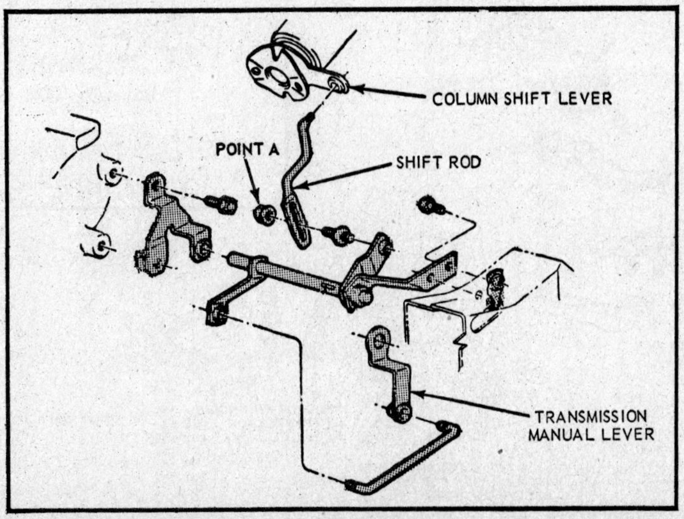

Fig. 4 Column shift linkage. 1969-75 Ford & Mercury (Typical)

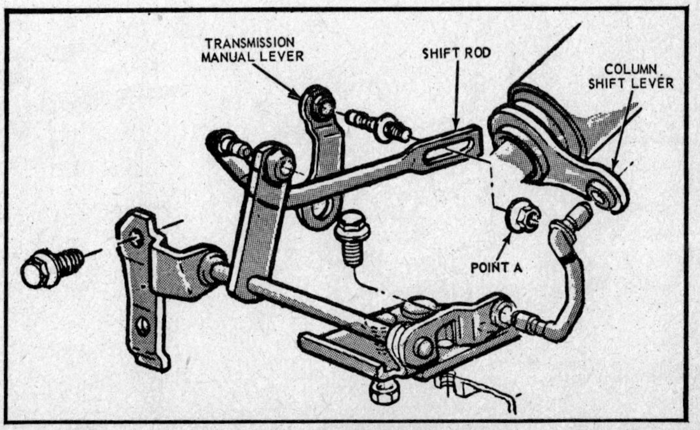

Fig. 5 Column shift linkage. 1969-70 Fairlane & Montego

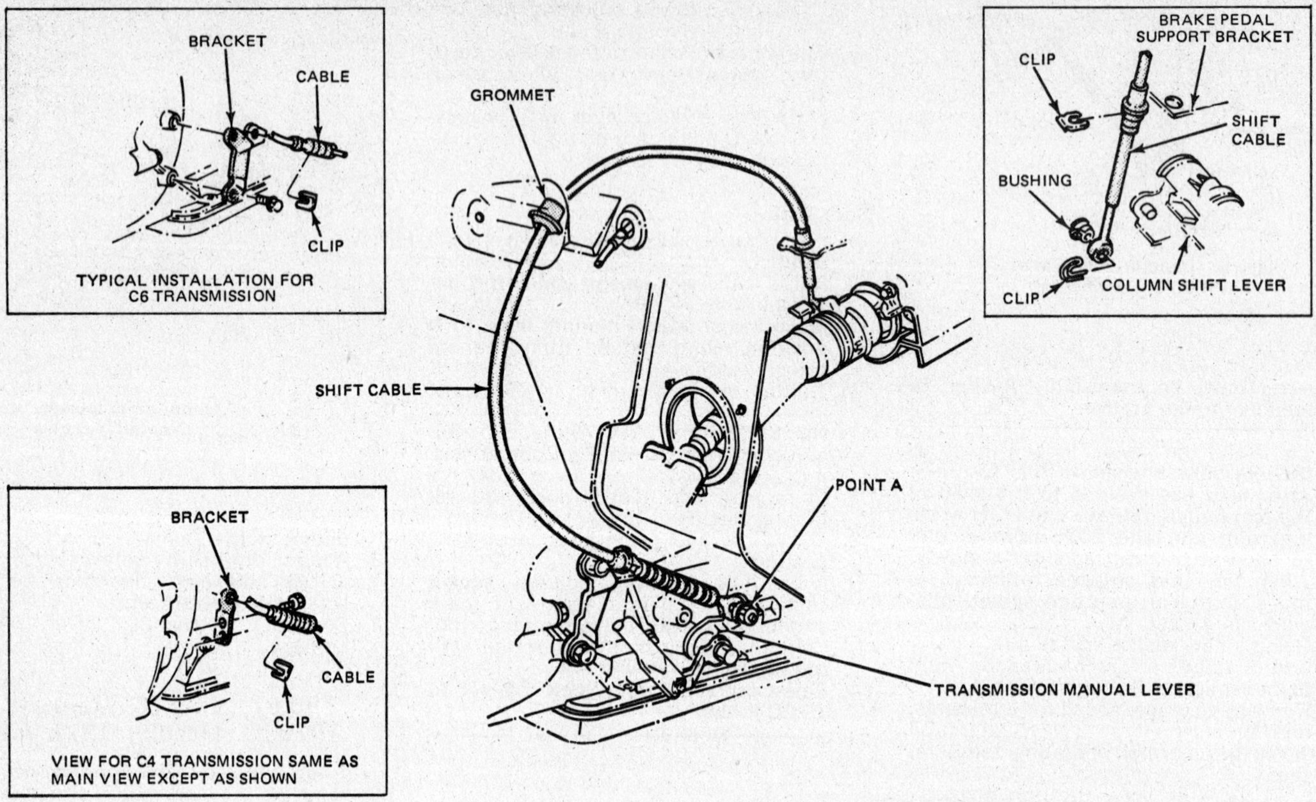

Fig. 6 Column shift linkage. 1971-75 Torino, Montego, Thunderbird & Mark IV

the transmission kickdown is the only adjustment required.

1. Disconnect the throttle and downshift return springs.
2. Hold the carburetor throttle lever in the wide open position against the stop.
3. Hold the transmission in full downshift position against the internal stop.
4. Turn adjustment screw on the carburetor kickdown lever to within .040"-.080" gap of contacting pickup surface of carburetor throttle lever.
5. Release the transmission and carburetor to the normal free position.
6. Install the throttle and downshift return springs.

1969 Fairlane & Montego

1. With engine off, check and adjust accelerator pedal for a height of 4½", Fig. 10. Adjust connecting link at point "A" to obtain correct dimension.
2. Disconnect the downshift control cable or rod at point "B".
3. With the carburetor choke in the off position, depress the accelerator pedal to the floor and block it in the wide open position.

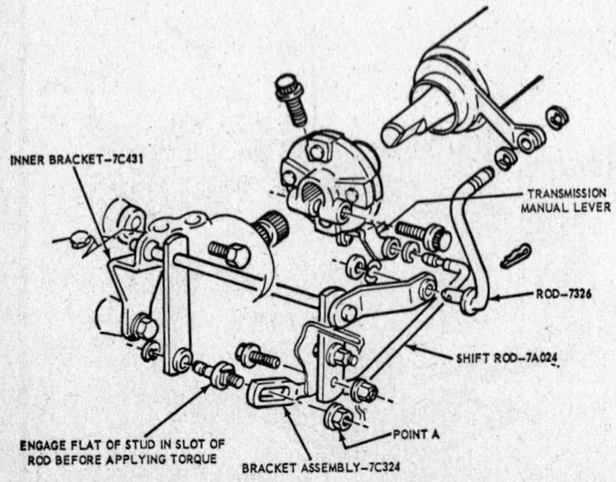

Fig. 7 Column shift linkage. 1969-71 Mark III & Thunderbird

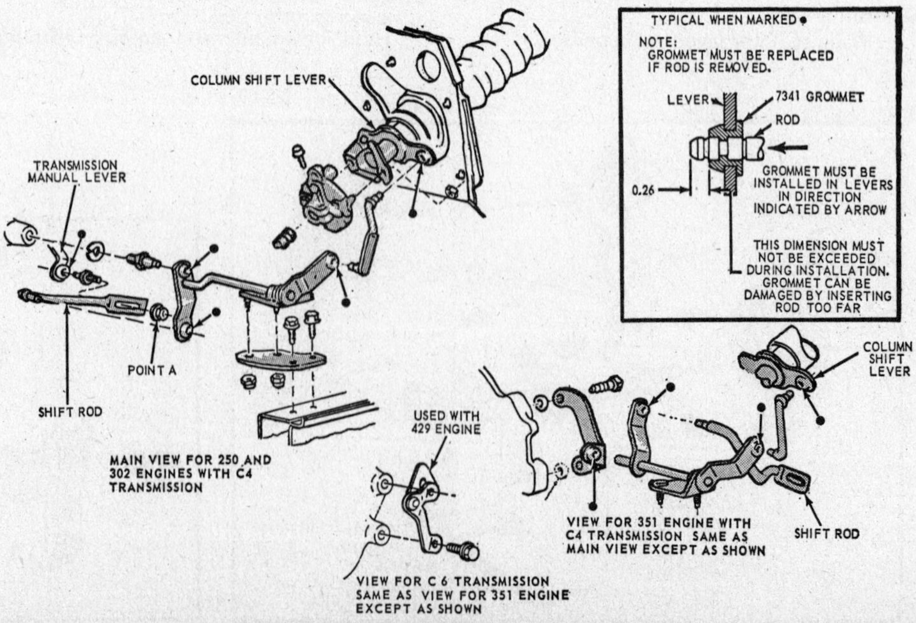

Fig. 8 Column shift linkage. 1971 Montego & Torino

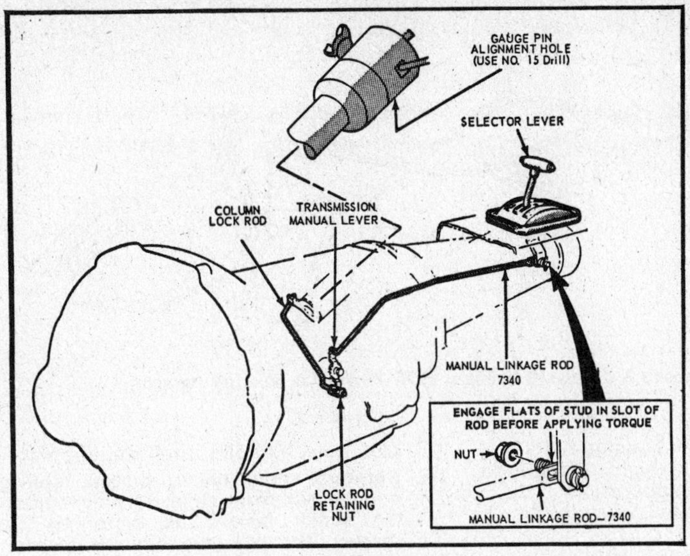

Fig. 9 Manual linkage, floor shift. Typical. Column lock rod is used on some models beginning 1970

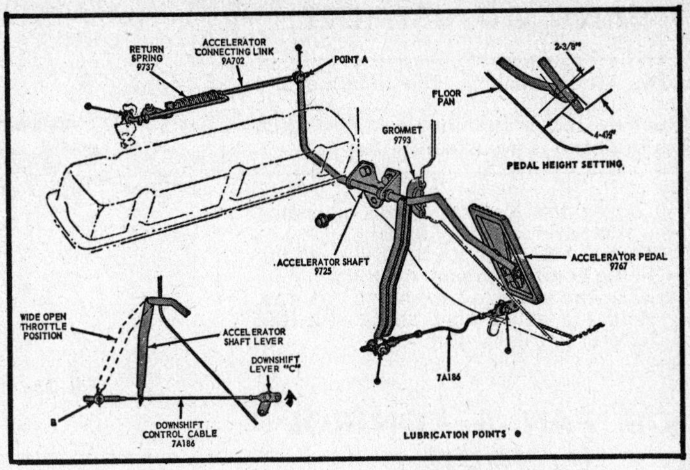

Fig. 10 Throttle linkage. 1969 Fairlane & Montego

4. Rotate the downshift lever counterclockwise to place it against the internal stop.
5. With the lever held in this position and, if cable is used, all slack is removed from the cable, adjust the trunnion so that it will slide into the accelerator shaft lever. Turn it one additional turn clockwise then secure it to the lever with the retaining clip.
6. Remove the clock and release the accelerator linkage.

1969-71 Mark III, IV & Thunderbird

The conduit covering the accelerator cable at the carburetor end must be evenly nestled between the clamp and the accelerator shaft bracket. Accelerator pedal height adjustment is not required.

1. Disconnect the downshift lever return spring.
2. With the throttle in the wide open position, hold the downshift rod against the through detent stop.
3. Adjust the downshift screw to provide .050"-.070" clearance between the screw and the throttle shaft lever. End play adjustment is not required.
4. Connect the downshift return spring.

1969 Lincoln

1. With engine stopped, measure distance from top of accelerator pedal to carpet, Fig. 12. The 3¾" requirement is nominal only. Further adjustment (at accelerator connecting link) after a road test if the kickdown operation is not functioning.
2. Depress accelerator pedal and check for detent feel and kickdown action of the bellcrank.
3. Disconnect downshift rod from bellcrank. Make sure movable outer bracket on bellcrank is up against stop pin on inner mounting bracket on bellcrank.
4. Pull upward and hold downshift rod against transmission internal stop. Adjust length of rod until hole in rod is aligned with ball stud on bellcrank.
5. Lengthen downshift rod one turn and position it on ball stud. Slide spring clip over end of rod. Tighten lock nut.
6. Be sure bellcrank outer bracket remains against stop pin. If it is not against pin, lengthen downshift rod one additional turn. If the downshift rod is adjusted too long the transmission will not upshift because the downshift valve is open to line pressure.

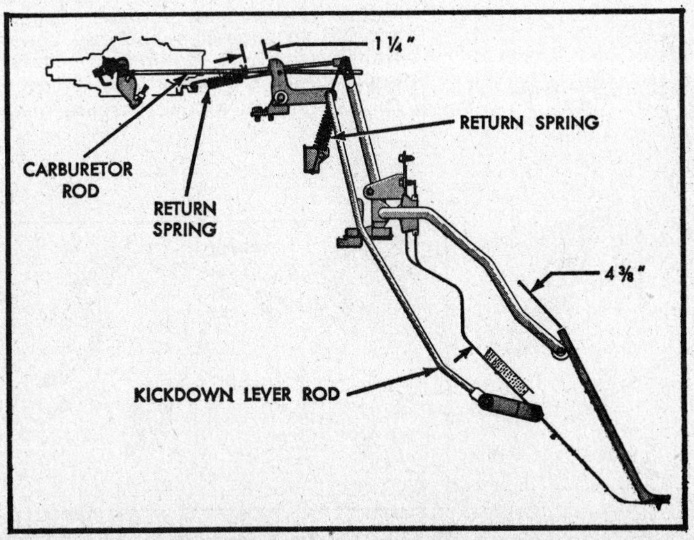

Fig. 11 Thunderbird throttle linkage. 1969 (Typical)

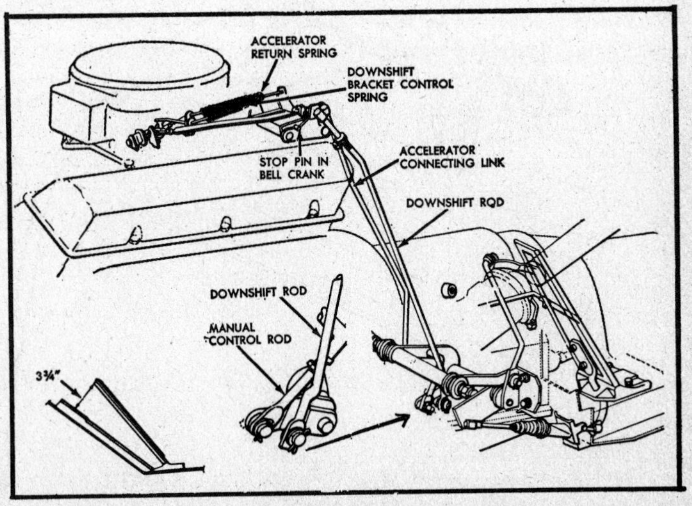

Fig. 12 Lincoln throttle linkage. 1969

BAND ADJUSTMENT

NOTE: When making the intermediate band adjustment, the lock nut must be discarded and a new one installed each time the band is adjusted.

1. Loosen the locknut on the adjusting screw several turns, Fig. 16.
2. Torque the screw to 10 ft-lbs, or until the adjuster wrench overruns.
3. Back the screw off exactly 1½ turns.
4. Hold the adjustment and torque the locknut to the 35-45 ft-lbs.

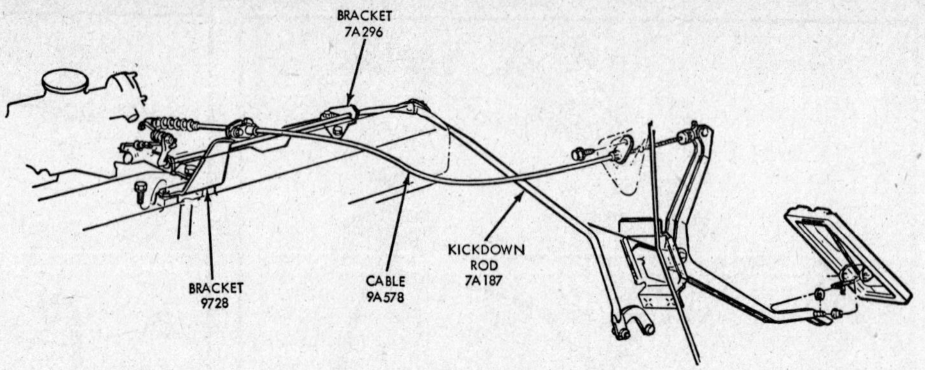

Fig. 13 Throttle & downshift linkage. 1971-75 Ford & Mercury (typical)

OIL PAN & CONTROL VALVE

Removal

1. Raise car on hoist or jack stands.
2. Loosen and remove all but two oil pan bolts from front of case and drop rear edge of pan to drain fluid. Remove and clean pan and screen.
3. Unfasten and remove valve body.

Installation

1. Position valve body to case, making sure that selector and downshift levers are engaged, then install and torque attaching bolts to specifications.
2. Using a new pan gasket, secure pan to case and torque bolts to specifications.
3. Lower car and fill transmission to the correct level with specified fluid.

INTERMEDIATE SERVO

Removal, Exc. Continental

1. Raise car and remove engine rear support-to-extension housing bolts.
2. Raise transmission high enough to relieve weight from support.
3. Remove support (1 bolt).
4. Lower transmission.
5. Place drain pan beneath servo.
6. Remove servo cover-to-case bolts.
7. Loosen band adjusting screw locknut.
8. Remove servo cover, piston, spring and gasket from case, *screwing band adjusting screw inward as piston is removed. This insures that there will be enough tension on the band to keep the struts properly engaged in the band end notches while the piston is removed.*

Removal, Continental

1. Raise vehicle and remove servo cover retaining bolts.
2. Remove manual and downshift control rod splash shield from frame side rail and reinforcement plate from beneath transmission oil pan.
3. Loosen band adjusting screw locknut.
4. Remove engine rear mount to crossmember nuts and with a suitable jack, raise transmission to remove weight from crossmember.
5. Remove engine rear support to extension housing bolts, then the support.
6. Remove servo cover, piston, spring and gasket from case, turning adjusting screw inward as piston is removed. This places enough tension on the band to keep struts properly engaged in band end notches as piston is withdrawn.

Replacing Seal, Fig. 17

1. Apply air pressure to port in servo cover to remove piston and stem.
2. Remove seals from piston.

NOTE: On Continental, Mark III and Mark IV units, replace complete piston and rod assembly if piston or piston sealing lips are damaged.

3. Remove seal from cover.
4. Dip new seals in transmission fluid.
5. Install seals in piston and cover.
6. Dip piston in transmission fluid and install in cover.

Installation, All

1. Position new gasket on servo cover and spring on piston stem.
2. Insert piston stem in case. Secure cover with bolts, taking care to back off band adjusting screw while tightening cover bolts. Make sure that vent tube retaining clip is in place.
3. Raise transmission high enough to install engine rear support. Secure support to extension housing. Lower transmission as required to install support-to-crossmember bolt.
4. On Continental models, secure man-

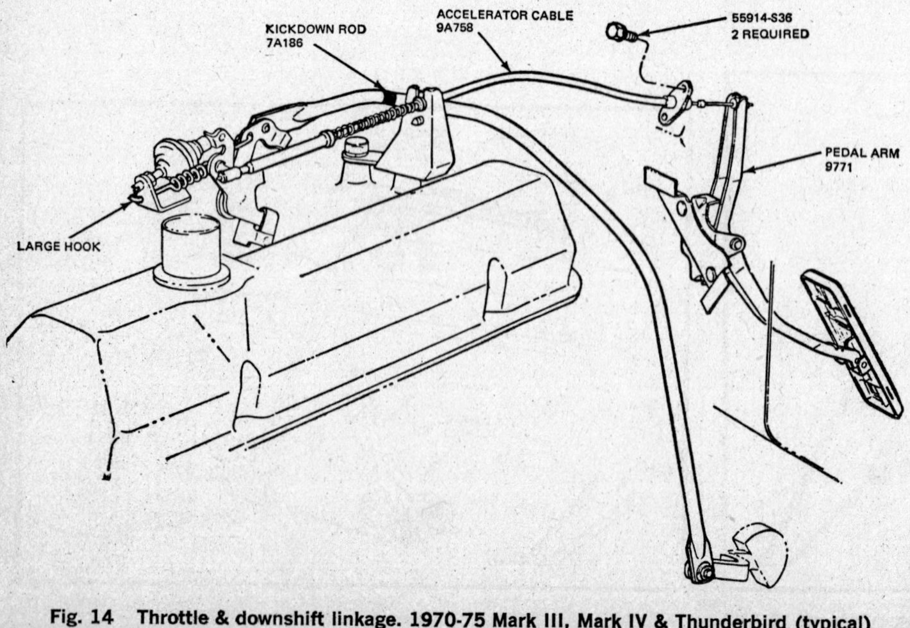

Fig. 14 Throttle & downshift linkage. 1970-75 Mark III, Mark IV & Thunderbird (typical)

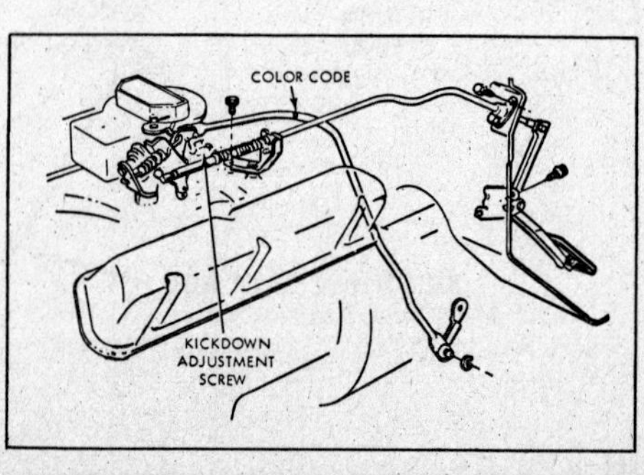

Fig. 15 Throttle & downshift linkage. 1971-75 Cougar, Montego, Mustang, Torino

Fig. 16 Band adjustment

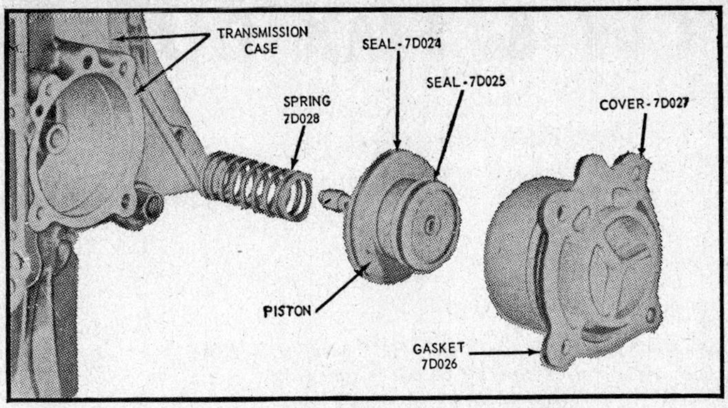

Fig. 17 Intermediate servo disassembled

ual and downshift rod splash shield to frame side rail.
5. Remove jack supporting transmission and adjust the band as outlined previously.
6. Lower car and replenish fluid as required.

EXTENSION HOUSING & GOVERNOR

Removal

1. Raise vehicle and disconnect parking brake cable from equalizer and on Continental models, remove the equalizer.
2. Disconnect drive shaft from rear axle flange and remove from transmission.
3. Disconnect speedometer cable from extension housing.
4. Remove engine rear support to extension housing bolts and on Continental models, remove reinforcement plate from beneath oil pan.
5. Raise transmission slightly with a suitable jack to remove weight from engine rear support.
6. Remove engine rear support to crossmember bolt and the support.
7. Lower transmission to permit access to extension housing bolts. Remove bolts and slide housing off output shaft.
8. Disconnect governor from distributor (4 bolts) and slide governor off output shaft.

Installation, Fig. 18

1. Secure governor to distributor flange.
2. Position new gasket on transmission.
3. Secure extension housing to case.
4. Raise transmission to position engine rear support on crossmember and install the support attaching bolt.
5. Lower transmission and remove jack. Install engine rear support to extension housing bolts and on Continental models, oil pan reinforcement plate.
6. Install speedometer cable.
7. On Continental models, install parking brake cable equalizer and on all models, connect parking brake cable to equalizer. Adjust parking brake.
8. Install drive shaft.
9. Correct transmission fluid level.

TRANSMISSION, REPLACE

NOTE: On models with the neutral safety switch wire harness connected at the dash panel, disconnect the harness before raising the vehicle.

1. Raise vehicle and drain transmission and converter.

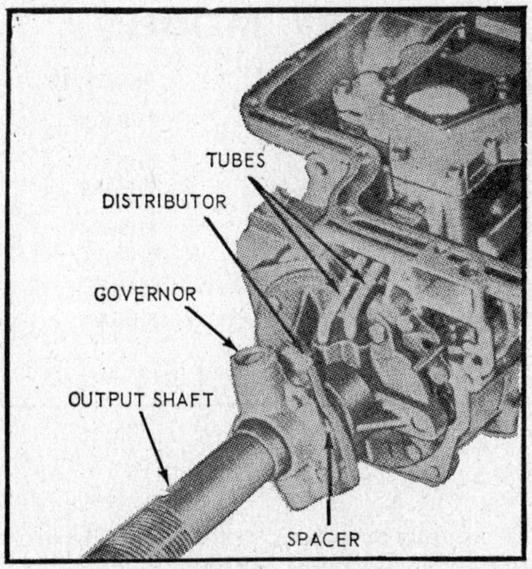

Fig. 18 Governor installed

NOTE: On 1969-75 Continental models, disconnect idler arm from frame side rail.

2. Remove drive shaft and starter.
3. Remove four converter to flywheel attaching bolts.
4. Disconnect parking brake front cable from equalizer.
5. Disconnect speedometer cable and transmission linkage.
6. Disconnect TRS switch wire, if equipped.
7. On 1969-71 Thunderbird, disconnect shift rod bellcrank bracket and allow bracket to hang free.
8. On 1969-75 Continental, remove lower shift rod bellcrank and pry upper shift rod bellcrank from converter housing and allow bellcrank to hang free.
9. Where necessary, disconnect muffler inlet pipes from exhaust manifold.

NOTE: On 1969-73 Mustang and Cougar, disconnect the entire exhaust system and allow it to hang on the rear axle.

10. Support the transmission with a suitable jack, remove parking brake rear cables from the equalizer and remove the crossmember.
11. Lower transmission and remove oil cooler lines, vacuum line and transmission oil filler tube.
12. Secure the transmission to the jack with the chain, remove the converter housing to cylinder block bolts and carefully move the transmission away from the engine, at the same time lowering it to clear the underside of the vehicle.
13. Reverse procedure to install.

CW & FMX DUAL RANGE AUTOMATIC

TRANSMISSION IDENTIFICATION

The identification tag on 1969 FMX units, is attached under the oil pan by a case-to-pan bolt while on 1970-75 FMX units, on the lower right hand extension housing-to-case bolt. On CW units, the identification tag is attached to the transmission case.

CW Transmissions

FORD MODELS

YEAR	ENGINE MODEL	TRANS. MODEL
1973	8-400③⑦	PHE-A
1974	8-400③⑦	PHE-A1

MERCURY MODELS

YEAR	ENGINE MODEL	TRANS. MODEL
1974	8-400③⑦	PHE-A1

FMX Transmissions

FAIRLANE/TORINO & MONTEGO MODELS

YEAR	ENGINE MODEL	TRANS. MODEL
1969	8-351③④	PHB-C
	8-351③①	PHB-D
	8-351②④	PHB-F
	8-351②①	PHB-G
1970	8-351④	PHB-R
	8-351①	PHB-S
1972	8-351	PHB-C
	8-351	PHB-Z
1973	8-351④⑤⑥	PHB-Z4
	8-351④⑤	PHB-C3
1974	V8-302	PHA-J3
	V8-351③	PHB-Z5
1975	8-351⑤	PHB-Z
	8-351⑧	PHB-AJ
	8-351⑧	PHB-AP
	8-400⑨	PHB-AL
	8-400⑩	PHB-AM

FORD MODELS

YEAR	ENGINE MODEL	TRANS. MODEL
1969	6-240④	PHD-A
	8-302④	PHD-B
	8-390④	PHB-A
1970	6-240	PHD-A1
	8-302	PHD-B1
	8-351④	PHB-L1
	8-351①	PHB-V
1971	6-240	PHD-A1, A2
	8-302	PHD-B1, B2
	8-351④	PHB-L1, L2
	8-351①	PHB-V, V1
1972	6-240	PHD-A4,A5
	8-302	PHB-L3, L4
	8-351	PHB-L3, L4
1973	8-351⑤	PHB-L6, 7
	8-351⑥	PHB-AA
1974	V8-351	PHB-AA1
	V8-351	PHB-L8
	V8-400	PHB-AC1
1975	8-351⑧	PHB-AN
	8-400⑨	PHB-AC2
	8-400⑩	PHB-AK

MERCURY MODELS

YEAR	ENGINE MODEL	TRANS. MODEL
1971	8-351	PHB-L1, L2
1972	8-351	PHB-L3
1973	8-351⑤	PHB-L6
	8-351⑥	PHB-AA
1974	V8-351	PHB-AA1
	V8-400	PHB-AC1
1975	8-400⑨	PHB-AC2
	8-400⑩	PHB-AK

MUSTANG (1969-73) & COUGAR MODELS

YEAR	ENGINE MODEL	TRANS. MODEL
1969	8-351③①	PHB-E
	8-351②①	PHB-H
1970	8-351③	PHB-E1
	8-351②	PHB-P
1971	8-351③	PHB-E2, E3
1972	8-302	PHA-H, H1
	8-351	PHB-E4, E5
1973	8-302	PHB-H3
	8-351⑥	PHB-E7
1974	V8-351③	PHB-Z5
1975	8-351⑧	PHB-AP
	8-351⑧	PHB-AJ
	8-400⑨	PHB-AL
	8-400⑩	PHB-AM

①—With floor shift.
②—With 4 barrel carburetor.
③—With 2 barrel carburetor.
④—With column shift.
⑤—Windsor engine.
⑥—Cleveland engine.
⑦—Ford & Mercury sedans with 2.75 rear axle ratio only.
⑧—Modified engine.
⑨—Exc. California.
⑩—California.

DESCRIPTION

Operation

This transmission features a drive range that provides for fully automatic upshifts and downshifts, and manually selected low and second gears. The six selector lever positions provided are P (park), R (reverse), N (neutral), D (automatic drive range), 2 (second gear hold) and 1 (low gear hold).

D is a fully automatic range providing for a first gear start with automatic upshifts to second and high gear occurring at appropriate intervals.

Second gear (2) is a manually selected second gear hold. When the selector lever is moved to 2, the transmission will engage and remain in second gear, regardless of throttle opening or road speed.

Low gear (1) is a manually selected first gear hold. When the selector lever is moved to this position the transmission will remain in first gear. To provide engine braking, moving the lever to this position will cause the transmission to downshift from 2nd when the car speed reaches about 22 to 39 mph depending on axle ratio and tire size.

D—Drive

The normal automatic driving range is indicated by D. In this range the car starts off in first gear and gives the best combination of automatic gear shifts to provide for economy and full power starts. As the accelerator is depressed and the

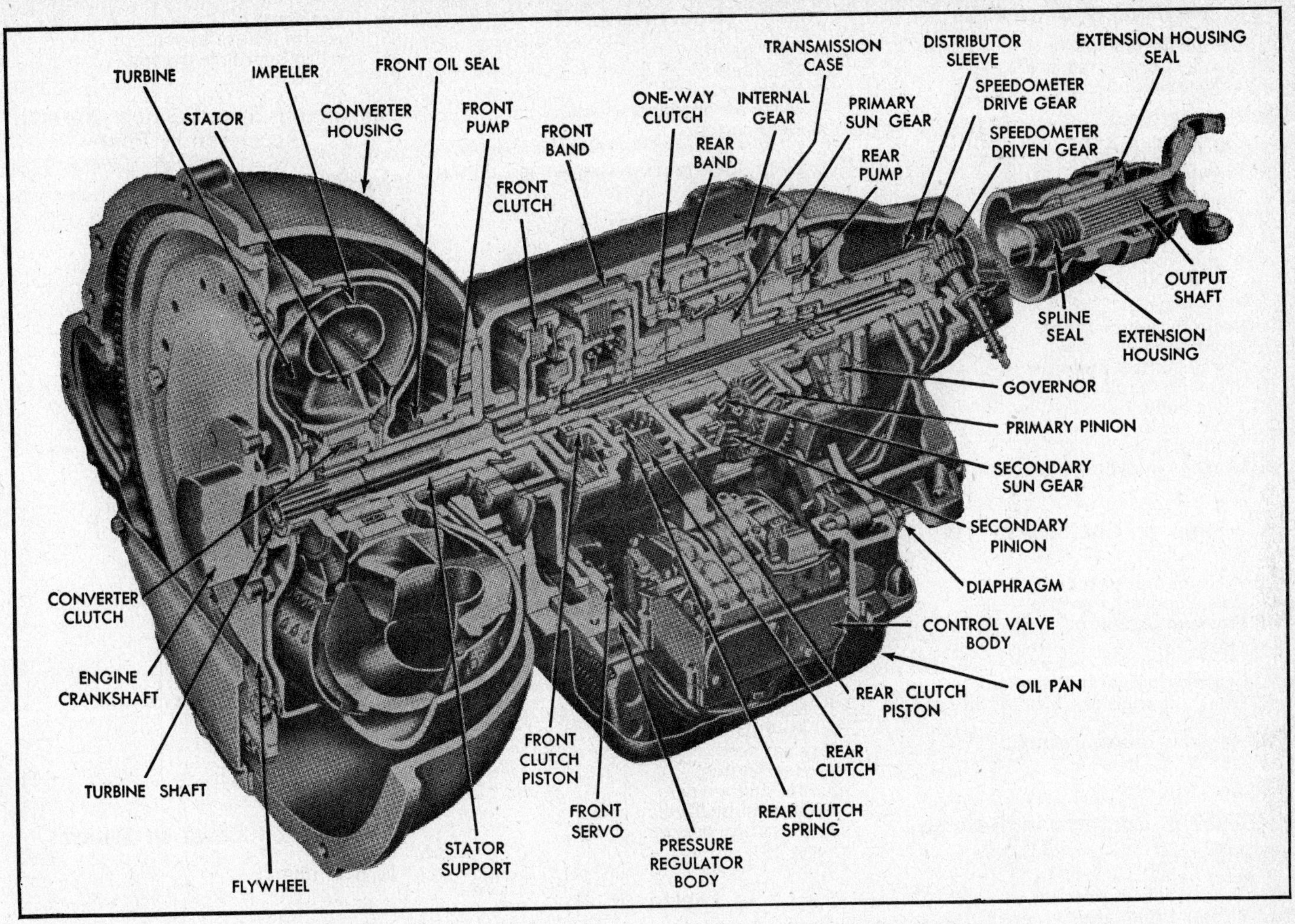

Fig. 1 Cruiseomatic and Mercomatic three speed dual range unit with cast iron case

car picks up speed, automatic shifts to second and high gears will occur. The transmission will automatically downshift as speed decreases. Forced downshifts in D are made by pressing the accelerator pedal all the way to the floor.

2—Second Gear Hold

When the car is started and the shift lever is moved to 2, the car will start off and remain in second gear, regardless of throttle opening or road speed. This range is especially useful for starting the car on icy pavements or other slippery surfaces. Similarly, when engine braking is required and the shift lever is moved from D to 2, the transmission will engage and remain in second gear.

Selector lever position 2 is not a cruising range in the usual sense of the term. While the transmission is capable of limited cruising in second gear, maximum fuel economy and best all-around performance are realized in D range.

1—Low Gear Hold

This range is identical in operation to manual low range except that when the shift lever is moved to 1 to provide engine braking, the automatic shift from second to low gear will occur between 22 and 39 mph (exact shift point will vary with axle ratio and tire size).

TROUBLE SHOOTING GUIDE

Rough Initial Engagement

1. Idle speed.
2. Vacuum unit or tubes.
3. Front band.
4. Check control pressure.
5. Pressure regulator.
6. Valve body.

Shift Points High, Low or Erratic

1. Fluid level.
2. Vacuum unit or tubes.
3. Manual linkage.
4. Governor.
5. Check control pressure.
6. Valve body.
7. Downshift linkage.

Rough 2-3 Shift

1. Manual linkage.
2. Front band.
3. Vacuum unit or tubes.
4. Pressure regulator.
5. Valve body.
6. Front servo.

Engine Overspeeds, 2-3 Shift

1. Vacuum unit or tubes.
2. Front band.
3. Valve body.
4. Pessure regulator.

No 1-2 or 2-3 Shifts

1. Governor.
2. Valve body.
3. Manual linkage.
4. Rear clutch.
5. Front band.
6. Front servo.
7. Leakage in hydraulic system.
8. Pressure regulator.

No Forced Downshifts

1. Downshift linkage.
2. Check control pressure.
3. Valve body.

Rough 3-2 or 3-1 Shifts

1. Engine idle speed.
2. Vacuum unit or tubes.
3. Valve body.

Slips or Chatters in 2nd

1. Fluid level.
2. Vacuum unit or tubes.
3. Front band.
4. Check control pressure.
5. Pressure regulator.
6. Valve body.
7. Front servo.
8. Front clutch.
9. Leakage in hydraulic system.

Slips or Chatters in 1st

1. Fluid level.
2. Vacuum unit or tubes.
3. Check control pressure.
4. Pressure regulator.
5. Valve body.
6. Front clutch.
7. Leakage in hydraulic system.
8. Fluid distributor sleeve in output shaft.
9. Planetary one-way clutch.

Slips or Chatters in Reverse

1. Fluid level.
2. Rear band.
3. Check control pressure.
4. Pressure regulator.
5. Valve body.
6. Rear servo.
7. Rear clutch.
8. Vacuum unit or tubes.
9. Leakage in hydraulic system.
10. Fluid distributor sleeve in output shaft.

No Drive in D or D2

1. Valve body.
2. Make air pressure check.
3. Manual linkage.
4. Front clutch.
5. Leak in hydraulic system.
6. Fluid distributor sleeve in output shaft.

No Drive in D1

1. Manual linkage.
2. Valve body.
3. Planetary one-way clutch.

No Drive in L

1. Manual linkage.
2. Front clutch.
3. Valve body.
4. Make air pressure check.
5. Leak in hydraulic system.
6. Fluid distributor sleeve in output shaft.

No Drive in R

1. Rear band.
2. Rear servo.
3. Valve body.
4. Make air pressure check.
5. Rear clutch.
6. Leak in hydraulic system.
7. Fluid distributor sleeve in output shaft.

No Drive in Any Range

1. Fluid level.
2. Manual linkage.
3. Check control pressure.
4. Pressure regulator.
5. Valve body.
6. Make air pressure check.
7. Leak in hydraulic system.

Lockup in D or D1

1. Manual linkage.
2. Rear servo.
3. Front servo.
4. Rear clutch.
5. Parking linkage.
6. Leak in hydraulic system.

Lockup in D2

1. Manual linkage.
2. Rear band.
3. Rear servo.
4. Rear clutch.
5. Parking linkage.
6. Leak in hydraulic system.
7. Planetary one-way clutch.

Lockup in R

1. Front band.
2. Front servo.
3. Front clutch.
4. Parking lnkage.
5. Leak in hydraulic system.

Lockup in L

1. Front band.
2. Pressure regulator.
3. Valve body.
4. Rear clutch.
5. Parking linkage.
6. Leak in hydraulic system.

Parking Lock Binds or Won't Hold

1. Manual linkage.
2. Parking linkage.

Transmission Overheats

1. Oil cooler and connections.
2. Pressure regulator.
3. Converter one-way clutch.

Engine Runaway on Forced Downshift

1. Front band.
2. Pressure regulator.
3. Valve body.
4. Front servo.
5. Vacuum unit or tubes.
6. Leak in hydraulic system.

Maximum Speed Below Normal, Acceleration Poor

1. Converter one-way clutch.

No 3-1 Downshift

1. Engine idle speed.
2. Vacuum unit or tubes.
3. Valve body.

Noise in Neutral

1. Pressure regulator.
2. Front clutch.
3. Front pump.

Noise in 1-2-3 or R

1. Pressure regulator.
2. Planetary assembly.
3. Front clutch.
4. Rear clutch.
5. Front pump.

Noise in Reverse

1. Pressure regulator.
2. Front pump.

Noise on Coast in Neutral

1. Rear pump.

MAINTENANCE

NOTE: Despite apparent similarities, the CW and FMX transmissions must be serviced as individual units. Although some CW parts are interchangeable with FMX parts, the majority of parts are not interchangeable.

Adding Fluid

The fluid level in the transmission should be checked at 1000-mile intervals Make sure that the car is standing level, and firmly apply the parking brake.

Run the engine at normal idle speed. If the transmission fluid is cold, run the engine at fast idle speed until the fluid reaches normal operating temperature. When the fluid is warm, slow the engine down to normal idle speed, shift the transmission through all ranges and then place the lever or button at P.

Clean all dirt from the transmission fluid dipstick cap before removing the dipstick from the filler tube. Pull the dipstick out of the tube, wipe it clean and push it all the way back into the tube.

Pull the dipstick out again and check the fluid level. If necessary, add enough Type F Automatic Transmission Fluid to the transmission to raise the fluid level to the F (full mark) on the dipstick.

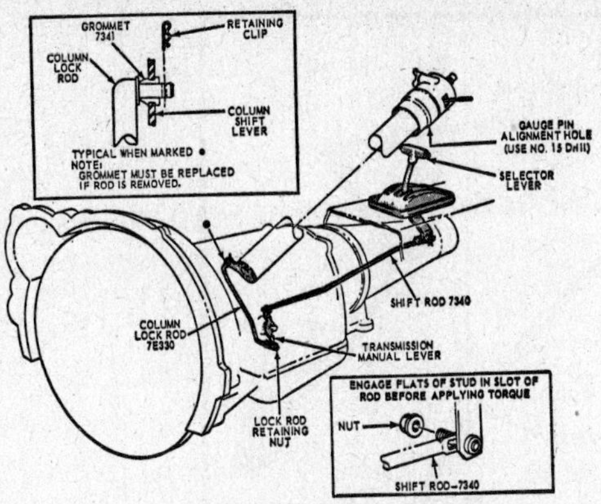

Fig. 2 Manual linkage. 1969-75 floor shift (typical)

Changing Fluid

NOTE: Normal maintenance and lubrication does not require periodic transmission fluid changes. However, a major transmission repair will require that the fluid be drained and be replaced with type F "lifetime" fluid.

1. Raise and support vehicle on hoist, then place a drain pan under transmission.
2. Loosen pan bolts and drain fluid until it has reached level of pan flange, then remove bolts working from rear and both sides of pan allowing to drop and drain slowly.
3. When fluid has completely drained, remove and thoroughly clean pan and filter.
4. Install filter, then using a new gasket install oil pan and add three quarts of fluid.
5. With transmission at operating temperature and parking brake applied, move shift lever through each range, allowing time for transmission to engage, then return shift lever to Park.
6. Check fluid level and add fluid as necessary to bring level between ADD and FULL mark.

MANUAL LINKAGE, ADJUST

Cougar, Fairlane/Torino, Ford, Mercury, Montego & Mustang

Floor Shift 1969-75, Fig. 2

1. Place selector lever in D.
2. Raise vehicle and loosen manual shift rod retaining nut.
3. Move transmission manual lever to D position (third detent position from back of transmission).
4. Torque nut to 10-20 ft-lbs.

Column Shift, 1969-75, Figs. 3, 4 and 5

1. Place selector lever in D.
2. Loosen shift rod adjusting nut at point A. On vehicles with a shift cable, remove nut at point A and remove cable from transmission manual lever stud.
3. Shift transmission manual lever into D.
4. On vehicles equipped with a shift cable, place cable end on transmission lever stud, using care to align flats on stud with flats on cable. Install but do not tighten retaining nut.
5. Making certain that selector lever has not moved from D position, torque retaining nut at point A to 10-20 ft. lbs.

NOTE: After adjusting the manual linkage, the steering column lock rod starting with 1970 models, is adjusted as follows:

1. Raise vehicle and loosen lock rod retaining nut.
2. Lower vehicle and place selector lever in D position tight against the D stop.
3. Align the hole in the steering column socket casting with the column alignment mark and insert a .180" diameter gauge pin. The column casting must not rotate with gauge pin in position.
4. Raise vehicle and torque lock rod retaining nut to 10-20 ft. lbs.
5. Lower vehicle. Remove gauge pin and check linkage for proper operation.

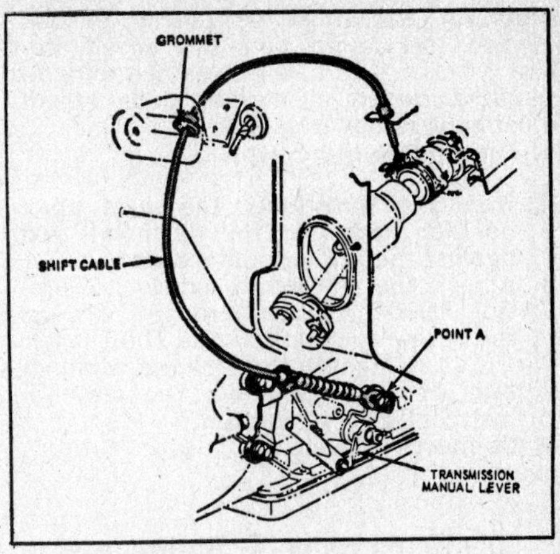

Fig. 3 Manual linkage. 1972-75 Montego & Torino column shift (typical)

THROTTLE LINKAGE, ADJUST

1973-75 All

1. Disconnect downshift lever return spring and hold throttle lever in wide open position.
2. Hold downshift rod against the through detent stop.
3. Adjust downshift screw to provide .01 to .08 inch clearance between screw and throttle arm.
4. Connect downshift lever return spring.

1969-71 Ford & Mercury 1972 All

The conduit covering the cable at the

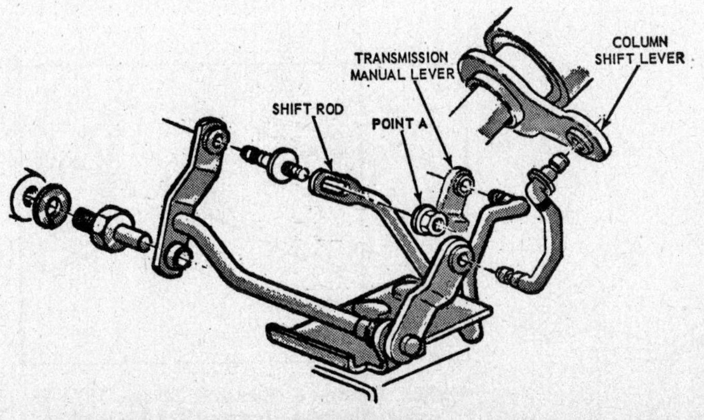

Fig. 4 Manual linkage. 1969-70 Fairlane & Montego, column shift

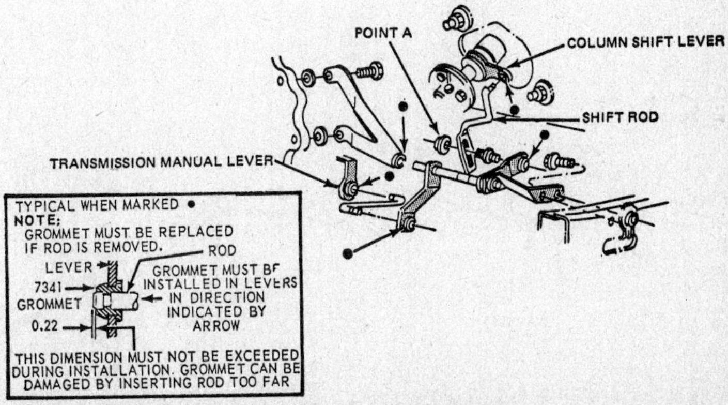

Fig. 5 Manual linkage. 1969-75 Ford & 1971-75 Mercury, column shift

carburetor end must be evenly nestled between the clamp and the accelerator shaft bracket. Due to the fixed clamping of the cable conduit, accelerator pedal height adjustment is not required.

1. Disconnect the downshift lever return spring.
2. Hold the throttle in the wide open position and hold the downshift rod against the through detent stop.
3. Adjust the downshift screw to provide .050"-.070" clearance between the screw and the throttle shaft lever. On Sixes, tighten the lock nut to maintain the screw position. End play adjustment is not required.
4. Connect downshift lever return spring.

1969 Fairlane & Montego

1. With engine off, check accelerator pedal for a height of 4½". Adjust accelerator connecting link to correct.
2. Disconnect downshift control cable from accelerator shaft lever.
3. With carburetor choke in the off position, depress the accelerator and block it in the wide open position.
4. Rotate the downshift lever counter clockwise to place it against the internal stop.
5. With the lever in this position and all slack removed from the cable, adjust the trunnion so that it will slide into the lever. Turn it one additional turn to increase length of cable then secure it to the lever with retaining clip.

1970 Fairlane & Montego 1969-70 Mustang & Cougar

Because the throttle on these models is cable operated, transmission kickdown is the only adjustment required.

1. Disconnect throttle and downshift return springs and disconnect downshift control cable from accelerator shaft lever.
2. With carburetor choke in the off position, depress the accelerator and block it in the wide open position.
3. Rotate the downshift lever counterclockwise to place it against the internal stop.
4. Turn adjustment screws on carburetor kickdown lever to within .040" to .080" gap of contacting pick up surface of carb. throttle lever.

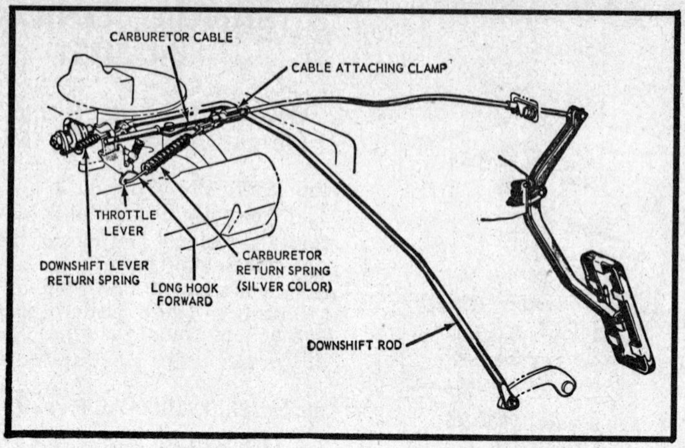

Fig. 6 Throttle linkage. 1969-70 Ford and Mercury

1971 Cougar, Fairlane/Torino, Montego & Mustang

1. Hold transmission in full downshift against stop.
2. Hold carburetor throttle lever in wide open throttle against stop.
3. Turn the adjustment screw on the kickdown lever until a gap of .040-.080 " exists between the carburetor lever and the adjusting screw.
4. Release the transmission and carburetor to the normal free position.
5. Install throttle return spring.

BAND ADJUSTMENTS

NOTE: CW series transmissions incorporate a self-adjusting front band. It is adjusted at time of assembly and requires no further adjustment until unit is overhauled. The rear band should be adjusted at the first 12,000 miles. No further adjustment is required until unit is overhauled.

Front Band

FMX Units

1. Drain fluid from transmission, remove and clean oil pan and screen.
2. Loosen front servo adjusting screw locknut.
3. Pull back on actuating rod and insert a ¼ inch spacer between adjusting screw and servo piston stem, Fig. 9.
4. Tighten adjusting screw to 10 inch-lbs torque. Remove spacer and tighten adjusting screw an additional ¾ turn. Hold adjusting screw stationary and tighten locknut securely.
5. Install oil pan with new gasket and add fluid to transmission.

CW Units

1. Install a .250 inch thick spacer between adjusting screw and servo piston stem, Fig. 10.
2. Using an inch pound torque wrench and an 8 point socket torque adjusting screw to 10 in. lbs. and remove spacer.

NOTE: Adjusting screw has a left hand thread and is self adjusting. After initial adjustment, no further adjustment should be required.

3. Space one way clutch spring .125-.188 inch from lever, otherwise proper automatic adjustments cannot be made.

Rear Band

NOTE: On late 1969 and on all 1970-75 models, there is no access hole in the floor pan to adjust the rear band. With

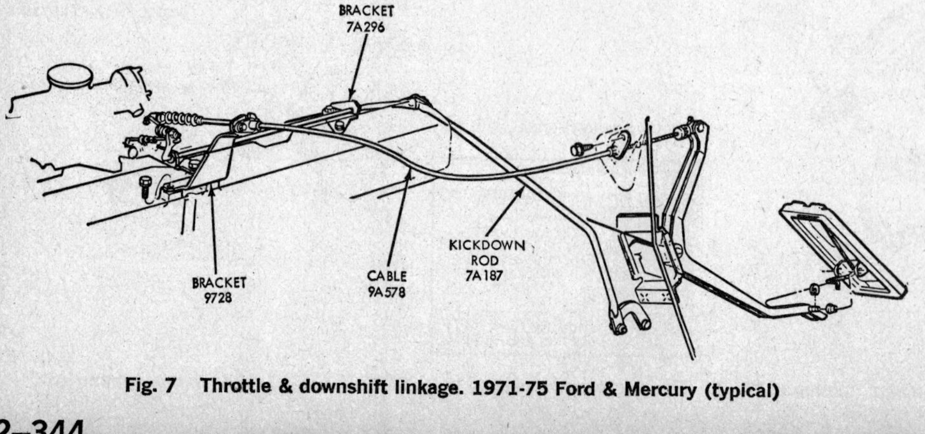

Fig. 7 Throttle & downshift linkage. 1971-75 Ford & Mercury (typical)

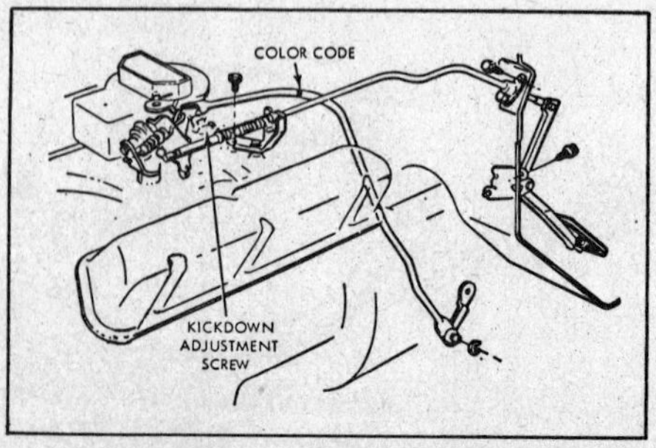

Fig. 8 Throttle & downshift linkage. 1971-75 Cougar, Montego, Torino 1971-73 Mustang (typical)

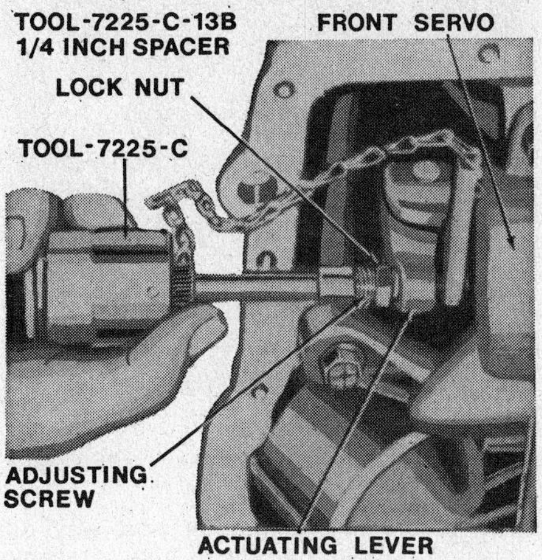

Fig. 9 Front band adjustment. Ford, Mercury and Thunderbird

the use of special tools this band can be adjusted externally as follows, Figs. 11, 12:

1. Loosen rear band adjusting screw locknut. A special tool is required to gain access in limited space.
2. Tighten adjusting screw until special tool clicks. It is preset to overrun when torque reaches 10 ft. lbs.

NOTE: If screw is found to be tighter than 10 ft. lbs., loosen screw and tighten until wrench clicks and breaks.

3. Back off adjusting screw 1½ turns on FMX units and 1¼ turns on CW units.

NOTE: Severe damage may result if adjusting screw is not backed off the exact amount of turns indicated.

4. Hold adjusting screw stationary and tighten locknut securely.

1969 Fairlane, Montego, Mustang & Cougar

1. Drain transmission and remove oil pan and filter.
2. Loosen rear servo adjusting screw lock nut.
3. Pull the adjusting screw end of the actuating lever away from the servo body spacer tool between servo accumulator piston and the adjusting screw, Fig. 13.

NOTE: Be sure that flat surfaces of the tool are positioned squarely between the adjusting screw and the accumulator piston. The tool must not touch the servo piston and the tool handle must not touch the servo piston spring retainer.

4. Using a torque wrench with an Allen head socket, tighten the adjusting screw to 24 in-lbs.
5. Back off the adjusting screw 1½ turns. Hold the adjusting screw and tighten the lock nut securely. Remove spacer tool.
6. Replace oil filter and pan and fill transmission.

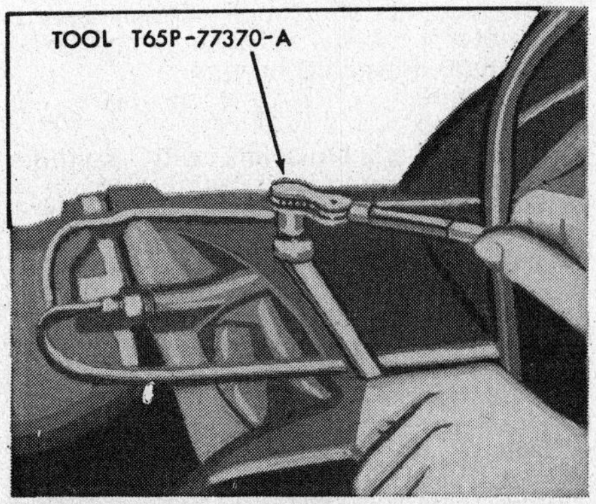

Fig. 11 Rear band adjustment. 1970-73 Cougar & Mustang

1969 Ford

The tool shown in Fig. 14 is especially made for this adjustment. However, a satisfactory adjustment may be made by using a conventional torque wrench. Loosen the lock nut and tighten the adjusting screw to a torque of 10 ft-lbs. Then back off the screw exactly 1½ turns and tighten lock nut. *Severe damage may result to the transmission if the adjusting screw is not backed off exactly 1½ turns.*

CONTROL VALVE BODY

1. Support vehicle on hoist, then drain transmission fluid and remove pan.

NOTE: If fluid is to be reused, filter it through a 100-mesh screen before replacing it in transmission.

2. Disconnect hoses from vacuum diaphragm unit, then using Snap-On tool S8696-A, remove vacuum diaphragm and push rod.
3. Remove fluid screen retaining clip and small compensator pressure tube.
4. Remove main pressure oil tube, by gently prying up end that connects to main control valve, then remove other end of tube from pressure regulator.

NOTE: Tube must be removed in this manner. Failure to do so, could kink or bend tube causing excessive internal transmission leakage.

5. Loosen front servo bolts three turns, then remove the three control valve body screws, and lower valve body while pulling it off front servo tubes, being careful not to damage valve body or tubes.
6. Before installing control valve, check for bent manual valve by rolling it on a flat surface.

NOTE: Before torquing control valve attaching bolts, move valve toward center of case until clearance is less than .050 inch between manual valve and actuating pin.

7. After installing control valve, torque control valve attaching bolts to 8-10 ft. lbs., front servo bolts to 30-35 ft. lbs. and vacuum diaphragm to 15-23 ft. lbs.

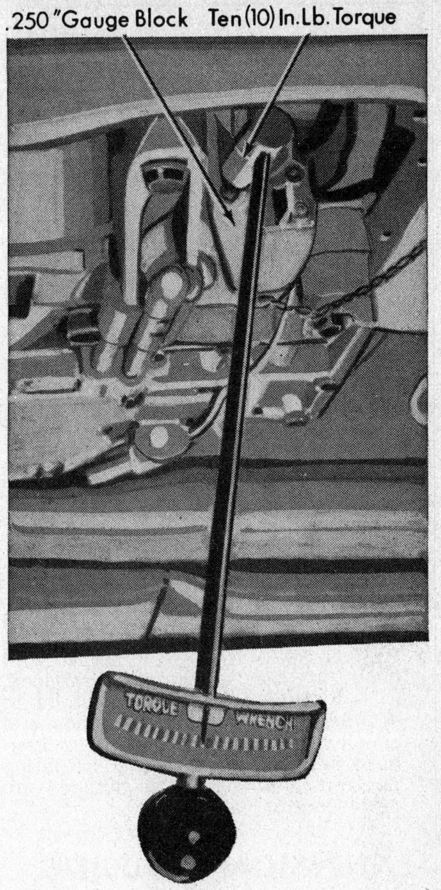

Fig. 10 Adjusting front servo band

8. After completing assembly, adjust front and rear bands. If valve body was replaced, adjust control linkage.

FRONT & REAR SERVOS

1. Drain transmission fluid and remove oil pan and screen.
2. Remove vacuum diaphragm, then loosen control valve body attaching bolts.
3. Remove retaining bolts from servo, then hold actuating strut and remove servo. On 1969 PHB units, remove servo dowel.

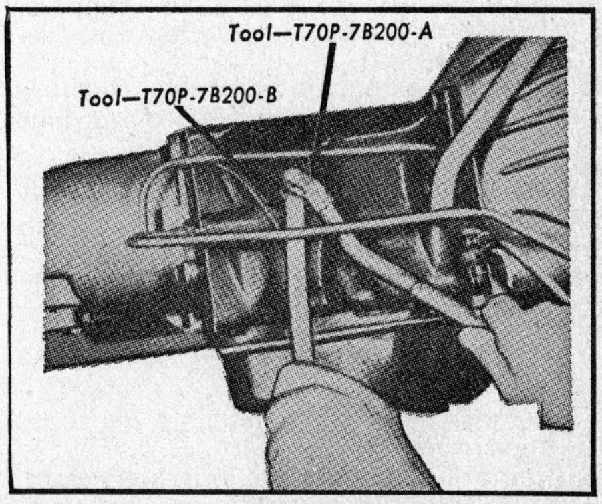

Fig. 12 Rear band adjustment 1970-73 Exc. Cougar & Mustang and all 1974-75

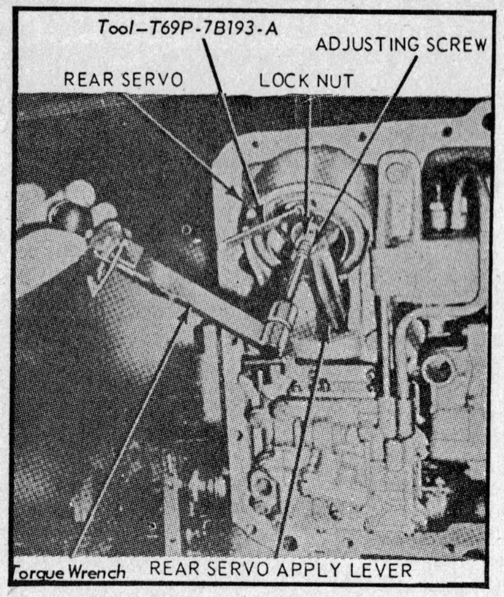

Fig. 13 Adjusting rear band, 1969 Fairlane, Montego, Mustang & Cougar

4. After installation, if front servo was serviced, torque front servo attaching bolts to 30-35 ft. lbs. and adjust front band. If rear servo was serviced, torque rear servo attaching bolts to 40-45 ft. lbs., adjust rear bands and check for less than .050 inch clearance between manual lever actuating pin and manual lever as outlined under "Control Valve."

EXTENSION HOUSING BUSHING & REAR SEAL

Proper removal and installation of extension housing bushing and rear seal will necessitate the use of five specialty tools. When removing bushing and rear seal, the vehicle will have to be raised and driveshaft removed.

EXTENSION HOUSING

1. Drain transmission fluid and remove driveshaft.
2. Disconnect speedometer cable from housing and remove engine rear supports to crossmember nuts.
3. Raise transmission slightly with suitable jack, then remove crossmember to side rail bolts and position the crossmember out of way.
4. Remove the two engine rear support-to-extension housing bolts and remove support.
5. Remove extension housing attaching bolts, then slide housing off output shaft and remove gasket.

NOTE: Hold output shaft and rear support from moving rearward to prevent needle bearing and race from dropping out.

GOVERNOR

1. Remove extension housing as outlined previously.
2. Remove governor to counterweight screws and lift governor from counterweight.

NOTE: When removing governor, hold output shaft and rear support from moving rearward to prevent needle bearing and race from dropping out.

3. When installing governor torque attaching bolts to 50-60 in. lbs.

PRESSURE REGULATOR

1. Drain transmission fluid and remove oil pan and fluid screen.
2. Remove small compensator pressure tube from control valve body and pressure regulator.
3. Remove main pressure oil tube by gently prying up end that connects to main control valve assembly, then remove other end of tube from pressure regulator.

NOTE: Tube must be removed in this manner. Failure to do so, could kink or bend tube, causing excessive internal transmission leakage.

4. Remove pressure regulator spring retainer, springs and spacer.

NOTE: Maintain pressure on retainer to prevent springs from falling out.

5. Remove regulator attaching bolts and remove regulator.
6. After installing regulator, torque attaching bolts to 17-22 ft. lbs.

PARKING PAWL

1. Drain transmission fluid and remove driveshaft.
2. Support rear of transmission and remove crossmember, then remove the two engine rear support-to-extension housing bolts and remove support.
3. Disconnect speedometer cable and remove oil pan and screen.
4. Loosen rear band adjusting screw locknut and tighten adjusting screw to 24 in. lbs.

NOTE: This will hold planetary carrier and clutch assemblies in place during parking pawl removal.

5. Remove small compensator pressure tube from pressure regulator and control valve body.
6. Remove main pressure oil tube by gently prying up end that connects to main valve body assembly, then remove other end of tube from pressure regulator.

NOTE: Tube must be removed in this manner. Failure to do so could kink or bend tube, causing excessive internal transmission leakage.

7. Remove vacuum diaphragm and loosen front servo attaching bolts, then remove valve body attaching bolts and lower valve body while pulling it off servo tubes, being careful not to damage valve body or tubes.
8. Remove rear servo bolts and remove servo and struts.
9. Remove extension housing and output shaft rear support.
10. Using a magnet, remove parking pawl pin from case, then working from inside of case, drive on shoulder of toggle lever pin with a small punch to move retaining plug part way out of case. Using a pair of pliers, remove plug.
11. Slide toggle lever toward front of case. Cock lever to one side to apply pressure on pin, then move toggle to rear of case to move pin outward.
12. Remove pawl and toggle lever as an assembly.
13. During assembly refer to "Control Valve Body" and "Front & Rear Servos" for installation. After assembly, adjust front and rear bands as outlined previously.

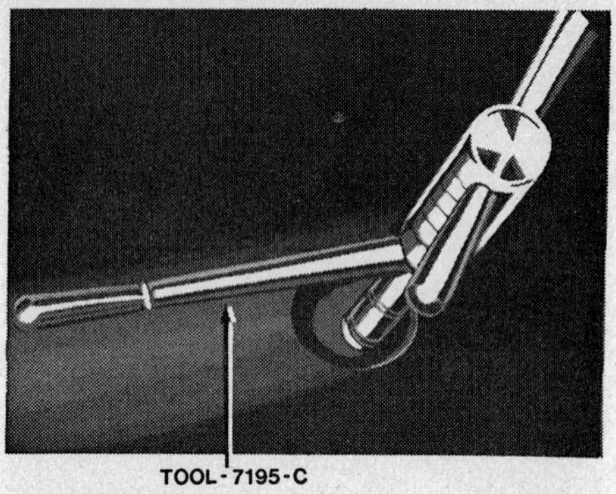

Fig. 14 Rear band adjustment. 1969 Ford

TRANSMISSION, REPLACE

NOTE: On some models, it is necessary to remove the two upper converter housing to engine bolts before raising vehicle.

1. Raise vehicle and place on jack stands, then drain transmission oil pan.
2. Remove converter access cover and remove converter drain plug.

NOTE: On 1969 units, two converter drain plugs are used. Remove one plug, rotate converter 180 degrees and remove the other plug. Do not attempt to rotate converter using a wrench on the converter stud nuts.

3. Remove converter to flywheel attaching nuts, reinstall converter drain plugs and converter housing access cover to hold converter in place when transmission is removed.
4. Remove starter and propeller shaft.
5. Where necessary, disconnect exhaust pipes from manifold.
6. Disconnect oil cooler lines, speedometer cable, vacuum hose and manual downshift linkage.
7. Disconnect TRS switch wire, if equipped.
8. Support transmission with suitable jack and remove crossmember.

NOTE: On some models, it is necessary to remove the engine rear support to transmission bolts before crossmember removal.

9. Lower transmission and remove filler tube and dipstick.
10. Remove converter housing-to-engine bolts. Move transmission and jack rearward and lower away from vehicle.
11. Reverse procedure to install.

AMERICAN MOTORS SHIFT-COMMAND

1969-71

TRANSMISSION IDENTIFICATION

An identification plate is attached to the left side of the transmission case. Included on the plate is the transmission model and the serial number of the unit.

DESCRIPTION

Two basic types of this transmission are used, cast iron and aluminum case. Model application is as follows:

Cast Iron:

Checker Motors: all models.
American Motors: 1969-71 V8 exc. V8-290 w/2 barrel carb and V8-304.

Aluminum Case:

American Motors: 1969-71 Six, V8-290 w/2 barrel carb. and V8-304.

Operation

While valve body and control location vary, gear units and power flow are essentially the same.

These transmissions combine a three-element torque converter and a hydraulically-controlled three speed and reverse planetary gear train. The drive is always through the torque converter and one of the planetary gear ranges.

The torque converter consists of an impeller (pump), a turbine and stator. All these parts operate in a fluid-filled housing which is sealed. The torque converter cannot be serviced and must be replaced as a unit in case of a malfunction.

The planetary gear train in all units transmits power from the torque converter turbine shaft to the transmission output shaft. Hydraulic clutches and servo-operated bands drive or hold certain gears to provide the various output ratios.

When the selector lever is placed in Drive, the transmission starts in 1st, shifts into 2nd, then into Drive. If selector lever is placed in 2nd, the transmission starts and remains in 2nd unless manually shifted into Drive or 1st. Similarly, when selector lever is placed in 1st, transmission starts and remains in 1st unless manually upshifted.

When maximum acceleration is desired in order to pass a slow moving vehicle or to ascend a steep grade, the transmission may be downshifted from 3rd to 2nd by pushing the accelerator to the floor. If pressure is released on the pedal the transmission will automatically upshift to high.

1st is used for going up very steep grades or driving in deep mud, sand or snow. This position is also used for descending steep grades in order to take advantage of engine braking. There is no automatic upshift in the 1 or 2 position regardless of car speed or throttle position. The selector lever may be moved from 1st to drive at any car speed and the transmission will then accomplish all the automatic upshifts.

TROUBLE SHOOTING GUIDE

Harsh Engagement

1. Front clutch seized or plates distorted.
2. Rear clutch seized or plates distorted.

Delayed Forward Engagement

1. Sealing rings missing or broken.
2. Front clutch piston check valve leaks.

Delayed Reverse Engagement

1. Sealing rings missing or broken.

No Engagement

1. Sealing rings missing or broken.
2. Broken input shaft.
3. Front pump drive tangs or converter hub broken.
4. Front pump worn.
5. Defective converter.

No Forward D-1

1. Sealing rings missing or broken.
2. Front clutch slipping, worn plates or faulty parts.
3. One-way (sprag) clutch slipping or incorrectly installed.
4. Front clutch piston check valve leaks.

No Forward D-2

1. Sealing rings missing or broken.
2. Front clutch slipping, worn plates or faulty parts.
3. Front clutch piston check valve leaks.

No Reverse

1. Sealing rings missing or broken.
2. Rear clutch slipping, worn or faulty parts.
3. Rear band worn or broken.

No Neutral

1. Front clutch seized or distorted plates.

No 1-2 Upshift

1. Sealing rings missing or broken.
2. Output shaft plug missing (6 cyl.).

No 2-3 Upshift

1. Sealing rings missing or broken.
2. Rear clutch slipping, worn or faulty parts.
3. Rear clutch piston ball check leaks.
4. Output shaft plug missing (6 cyl.).

Shift Points Too High

1. Sealing rings missing or broken.

Shift Points Too Low

1. Sealing rings missing or broken.

1-2 Delayed Followed Close By 2-3 Shift

1. Sealing rings missing or broken.
2. Front clutch slipping, worn plates or faulty parts.
3. Front band worn or broken.

2-3 Slips

1. Sealing rings missing or broken.
2. Rear clutch slipping, worn or faulty parts.
3. Front band worn or broken.
4. Rear clutch piston ball check leaks.

Harsh 1-2 Shift

1. Front clutch slipping, worn plates or faulty parts.

Harsh 2-3 Shift

1. Rear clutch seized or plates distorted.

1-2 Ties Up

1. Rear clutch seized or plates distorted.
2. One-way (sprag) clutch seized.

No 2-1 in D-1

1. One-way (sprag) clutch slipping or incorrectly installed.
2. Output shaft plug missing (6 cyl.).

No 2-1 in L Range

1. Rear band worn or broken.
2. Output shaft plug missing (6 cyl.).

No 3-2 Shift

1. Front band worn or broken.
2. Output shaft plug missing (6 cyl.).

Shift Points Too High or Too Low

1. Sealing rings missing or broken.

2-1 Slips

1. Front clutch slipping, worn plates or faulty parts.
2. Front pump drive tangs or converter hub broken.
3. Front clutch piston check valve leaks.

3-2 Slips

1. Sealing rings missing or broken.
2. Rear clutch slipping, worn or faulty parts.
3. Front band worn or broken.
4. Rear clutch piston ball check leaks.

3-1 Shift Above 30 MPH

1. Sealing rings missing or broken.
2. Front band worn or broken.

Harsh 2-1 Shift

1. Sealing rings missing or broken.
2. Front clutch slipping, worn plates or faulty parts.
3. One-way (sprag) clutch slipping or incorrectly installed.

Harsh 3-2 Shift

1. Rear clutch slipping, worn or faulty parts.
2. Rear clutch seized or plates distorted.

Slips or Chatters in Reverse

1. Sealing rings missing or broken.
2. Front clutch seized or plates distorted.
3. Rear clutch slipping, worn or faulty parts.
4. Rear band worn or broken.
5. Rear clutch piston ball check leaks.

Reverse Tie Up

1. Sealing rings missing or broken.
2. Front clutch seized or plates distorted.

Low Idle Pressure

1. Sealing rings missing or broken.
2. Front pump worn.

Low Stall Pressure

1. Sealing rings missing or broken.
2. Front pump worn.
3. Output shaft plug missing (6 cyl.).

Stall Speed Too Low

1. Converter.

Stall Speed Too High D-1

1. Broken output shaft.
2. Broken gears.
3. Sealing rings missing or broken.
4. Front clutch slipping, worn plates or faulty parts.
5. One-way (sprag) clutch slipping or incorrectly installed.
6. Broken input shaft.
7. Converter.
8. Front clutch piston check valve leaks.

Reverse Stall Speed Too High

1. Broken output shaft.
2. Broken gears.
3. Rear band worn or broken.
4. Rear clutch slipping, worn or faulty parts.
5. Broken input shaft.
6. Converter.

Poor Acceleration

1. Output shaft plug missing (6 cyl.).
2. Converter.

Noisy in Neutral

1. Rear clutch seized or plates distorted.
2. Front pump.
3. Front clutch hub thrust washer missing (detectable in N, P, R only).
4. Converter.

Noisy in Park

1. Front pump.
2. Front clutch hub thrust washer missing (detectable in N, P, R only).
3. Converter.

Noisy in All Gears

1. Front pump.
2. Planetary assembly.
3. Converter.

Noisy in 1st & 2nd Gears Only

1. Front pump.
2. Planetary assembly.
3. Forward sun gear thrust washer missing.

Park Brake Does Not Hold

1. Parking linkage.

Oil Out Breather

1. Sealing rings missing or broken.
2. Breather baffle missing.

Oil Out Fill Tube

1. Sealing rings missing or broken.
2. Breather baffle missing.

Ties Up in L or D-1, 1st Gear

1. Rear clutch seized or plates distorted.
2. Sealing rings missing or broken.

Ties Up in D-1 or D-2, 2nd & 3rd Gears

1. Rear clutch seized or plates distorted.
2. Sealing rings missing or broken.
3. One-way (sprag) clutch seized.

Chatters - D-1, D-2 or Low

1. Sealing rings missing or broken.
2. Front clutch slipping, worn plates or faulty parts.
3. Front clutch piston check valve leaks.

MAINTENANCE

The fluid level in the transmission should be checked at 4000-mile intervals. Make sure that the car is standing on a level floor, and firmly apply the parking brake.

With transmission fluid warm, run engine at idle speed. Shift transmission through all ranges and return selector lever to Park.

Clean all dirt from the transmission fluid dipstick cap before removing the dipstick from the filler tube. Pull the dipstick out of the tube, wipe it clean and push it all the way back into the tube.

Pull the dipstick out again and check the fluid level. If necessary, add enough automatic transmission fluid to the transmission to raise the level to the "F" or "Full" mark on the dipstick.

Changing Fluid

The transmission fluid should be changed at 24,000-mile intervals. The procedure is as follows:

1. Raise vehicle and place on jack stands.
2. If filler tube is located in the side of oil pan, remove filler tube to drain oil pan. On other units, loosen oil pan bolts and allow fluid to drain to pan flange level. Then, remove pan attaching bolts from one side to drain remaining fluid.
3. When all oil is drained, remove and clean oil pan and screen. Then, using a new oil pan gasket, install oil pan and screen.
4. Connect filler tube, if removed.
5. Install 3 quarts of approved automatic transmission fluid.
6. Run engine at idle speed for about 2 minutes; then add the additional quantity of oil required for the particular transmission being serviced.
7. Run engine at a fast idle speed until it reaches normal operating temperature.
8. Shift transmission through all ranges; then place it in "P" and check fluid level. If necessary, add enough fluid to bring the level up to the "F" or "Full" mark on the dipstick.

MANUAL LINKAGE, ADJUST

American Motors

1969 Column & Console Shift

1. Place selector lever in "N" position.
2. Hold selector lever against neutral stop.
3. Adjust linkage to a free fit in transmission outer lever and connect linkage.

1970-71 Column Shift

1. Place selector lever and transmission lever in Neutral.
2. Adjust shift rod trunnion to a free pin fit.
3. Place selector lever in Park and check column lock.

1970-71 Console Shift

1. Loosen park lock-up rod trunnion lock nuts to allow rod movement in trunnion.
2. Place selector lever and transmission lever in Neutral.
3. Adjust linkage for a free pin fit and connect linkage.
4. Place selector lever in Park position and lock steering column. It may be necessary to move lower column lever upward until it is locked. Tighten first lower then upper trunnion nuts while holding trunnion centered in the column lever.

Checker Motors

1. Place selector lever in Drive as determined by transmission detent and loosen adjusting swivel at cross shaft.
2. Move the transmission lever so lever contacts mechanical stop on steering column. Tighten swivel and check adjustment for proper operation.
3. Adjust shift indicator needle, if necessary.

FRONT BAND

Remove transmission oil pan. Loosen locknut on front servo adjusting screw and insert a .250" metal block between end of adjusting screw and servo piston rod, Fig. 1.

With metal block in place, tighten adjusting screw to a torque of 10 inch-pounds and tighten locknut to 23 ft-lbs torque.

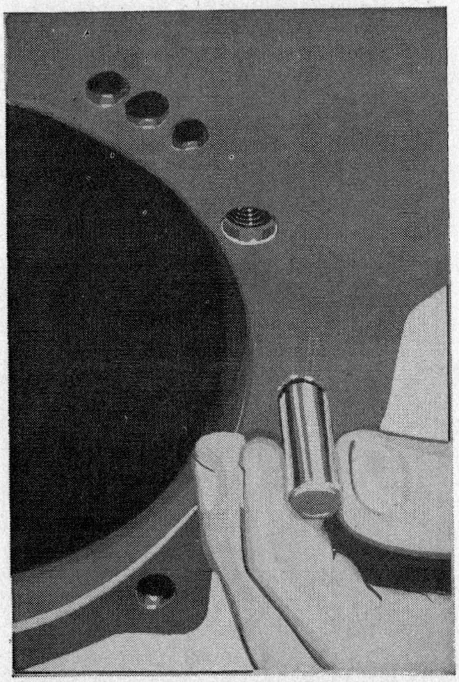

Fig. 2 Removing parking pawl shaft

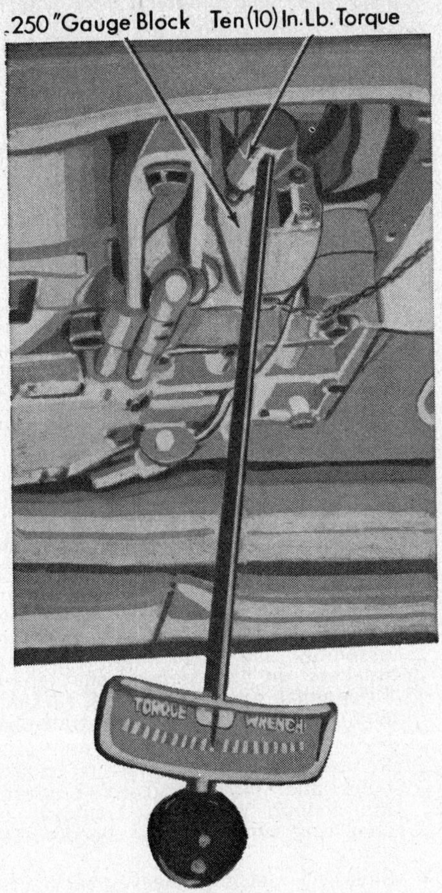

Fig. 1 Adjusting front servo band

NOTE: On cast iron case units, the front servo adjusting screw has a left hand thread.

REAR BAND, ADJUST

NOTE: To gain access to the rear band adjusting screw, it may be necessary to remove the crossmember bolts and lower the transmission on some models.

Aluminum Case

Loosen locknut and tighten adjusting screw to a torque of 10 ft-lbs. Then back off the adjusting screw 3/4 turn and tighten locknut to 28 ft-lbs. torque.

Cast Iron Case

Loosen locknut and tighten adjusting screw to 10 ft-lbs torque. Then back off the adjusting screw 1 1/4 turns and tighten locknut.

NOTE: On 1969 Checker Motors units, back off adjusting screw 1 1/2 turns.

CONTROL VALVE

Aluminum Case

1. Remove filler tube, drain transmission and remove oil pan.
2. Remove vacuum hose from vacuum control unit and remove vacuum unit with push rod attached.
3. Remove rear servo and rear clutch oil tubes.
4. Remove front servo apply and release tubes.
5. Disconnect solenoid wire at case connector.
6. Remove control valve attaching screws and remove control valve from transmission.
7. Reverse procedure to install.

Cast Iron Case

1. Drain transmission and remove oil pan.
2. Remove vacuum control unit and rod.
3. Disconnect solenoid wire at case connector.
4. Loosen front servo adjusting screw until arm contacts servo body.

NOTE: On 1969-71 Units, the front servo adjusting screw has left hand threads. Hold bent tang of adjuster wire in counter-clockwise direction, while adjusting screw is turned clockwise to loosen.

5. Some 1969-71 units have two front servo attaching bolts, loosen these bolts.
6. Remove control valve attaching bolts and disconnect front servo pressure tubes from control valve and remove control valve from transmission.

FRONT & REAR SERVOS

It is recommended that these units not be overhauled or replaced to correct a malfunction without first inspecting bands and drums.

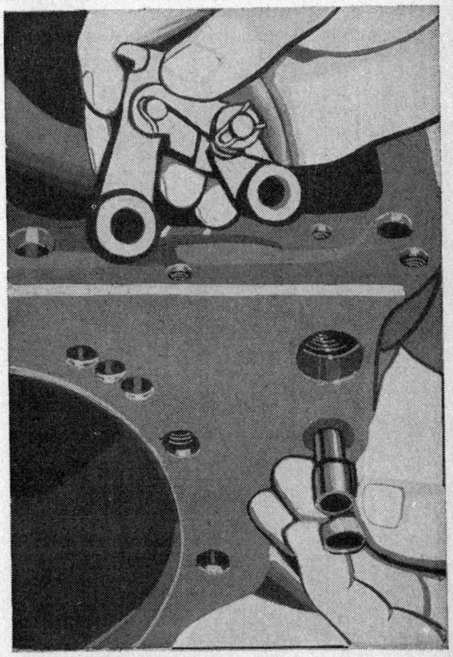

Fig. 3 Removing toggle lever

EXTENSION HOUSING SEAL

Remove propeller shaft or torque tube and propeller shaft from transmission. Pull seal out of extension housing.

Before installing the new seal, inspect the sealing surface of the universal joint yoke for scores. If scores are evident, replace the yoke. Inspect the counterbore in the housing for burrs. Polish all burrs with crocus cloth.

To install the new seal, position it in the bore of the extension housing with the felt side of the seal to the rear. The seal may be driven into the housing with a special tool designed for the purpose.

GOVERNOR

Drain transmission and remove propeller shaft. Support transmission and remove crossmember. Remove speedometer cable from extension housing and on aluminum case units, remove speedometer gear. Remove extension housing using caution not to separate rear case adapter from case as the rear thrust washer may become improperly positioned. Remove two governor body retaining bolts and remove governor from sleeve.

When installing governor into sleeve, be sure side plate faces rearward. Torque retaining bolts to 75 pounds inch.

Remove the governor inspection cover from the extension housing, Rotate the drive shaft to bring the governor body in line with the inspection hole. Remove the two screws which attach the governor body to the counterweight, and remove the body.

Remove the valve from the new governor body. Lubricate the valve with automatic transmission fluid. Install the valve in the body, making sure the valve moves freely in the bore. Install the body in the counterweight. Be sure the fluid passages in the counterweight and body are aligned.

PARKING PAWL, REPLACE

Transmission In Car

Aluminum Case

1. Support engine at rear.
2. Support torque tube with jack.
3. Remove speedometer cable and remove exhaust pipe clamp from lower bracket.
4. Unfasten and remove rear crossmember over exhaust pipe by pulling down on exhaust pipe.
5. Remove oil pan and control valve.
6. Remove parking brake toggle roll pin and remove toggle pin.
7. Unfasten rear extension housing from torque tube adapter and rotate housing clockwise until governor inspection plate is almost level to the bottom. The parking brake anchor pin will then clear extension housing.
8. Remove parking brake anchor pin with a magnet or remove pin from inside of case with needle nose pliers.
9. Remove parking brake toggle link and pawl assembly.
10. Reverse procedure to install.

Cast Iron Case

1. Remove oil pan and screen, pressure regulator and control valve. Disconnect speedometer cable.
2. Completely tighten rear band to prevent movement of planetary assembly and dislocation of thrust washers on the transmission shaft. Disconnect drive shaft or drive shaft and torque tube from the transmission.
3. Remove extension housing-to-case bolts and move housing rearward far enough to permit removal of snap ring which retains speedometer gear.
4. Slide oil delivery sleeve back just far enough so that the oil distributor tubes clear the transmission case.
5. Rotate oil pump housing until parking pawl pin in case is exposed.
6. Disconnect link (parking pawl torsion rod) located between detent lever and torsion lever assembly.
7. Remove hair pin clip retaining torsion lever assembly and remove from shaft.
8. Tap toggle lever pin toward rear of transmission to remove plug and pin, then remove parking pawl pin by working pawl back and forth, Fig. 2.
9. Remove toggle lever and parking pawl assembly from transmission and replace any damaged parts, Fig. 3.
10. Reverse above procedure for reassembly.

TRANSMISSION, REPLACE

1969-71

1. Disconnect battery.
2. Raise car and support with car stands. *Car weight must be on rear springs, therefore, place stands under rear axle tubes.*

NOTE: Before removing the rear crossmember on Javelin and AMX with power steering, it is necessary to open the hood to avoid damage to the hood from the power steering pump wing nut.

3. Disconnect the following:
4. Oil filler tube and drain transmission.
5. Selector linkage at transmission outer manual lever.
6. Speedometer cable at transmission.
7. Vacuum hose and solenoid wire. On 1971 models, disconnect TCS wire, if equipped.
8. Exhaust pipe and remove pipe bracket from converter housing.
9. Position transmission hoist under transmission.
10. Disconnect rear support crossmember from body side sill brackets and transmission.
11. Lower transmission hoist to gain access to upper converter housing to engine bolts and disconnect oil cooler lines.
12. With 199, 232, 258 engines, remove converter housing lower cover.
13. Remove converter access cover (in spacer plate on 290 engine) on V8.
14. Mark converter and drive plate to assure original location upon assembly.
15. Remove converter-to-drive plate capscrews (6-cyl.) stud nuts on V8s.
16. Remove starter mounting bolts and converter housing-to-cylinder block bolts.
17. Push converter housing and converter to rear a sufficient distance to clear crankshaft.

NOTE: Rear of engine tends to raise when transmission weight is removed and may bind the converter in the crankshaft pilot bushing. Blocking the engine up at the front will assist separating converter from crankshaft.

18. Maintain pressure against converter housing and lower assembly until converter housing is clear of engine. Then disconnect propeller shaft and remove transmission from vehicle.
19. Reverse procedure to install.

SEAT BELT INTERLOCK SYSTEMS

INDEX

AMERICAN MOTORS

Fig. 1 shows the components of the seat belt interlock system. In case of a malfunction in the interlock system, a starter relay is located in the engine compartment to permit starting. The relay is located on the right side next to the starter solenoid and is identified with a decal. Fig. 2 is a system wiring diagram.

NOTE: If a "no start" condition occurs, the fault may be in the seat belt interlock system or in the regular starting system. Manually depress and hold the relay button down while an assistant tries to start the engine. If the engine will not crank while the button is held down, the trouble is not in the interlock system.

CAUTION: Anytime an interlock by-pass start or a solenoid jump start is attempted, be sure the parking brake is applied and the automatic transmission is in Park or manual transmission in Neutral.

Trouble Shooting
Exc. Pacer

Interlock By-Pass Check

Place shift lever in neutral for manual transmissions or park for automatics and apply parking brake with front seat position unoccupied. Turn ignition switch to start position. Engine should crank. If not, refer to chart #1 NO START-FRONT SEATS UNOCCUPIED.

No-Buckle Seat Belt Check

Place shift lever in neutral for manual transmissions and park for automatics and apply parking brake. Activate seat sensor switch, but do not buckle seat belt. Turn ignition switch to start position. Engine should not crank, FASTEN BELTS lamp should light and warning buzzer should sound. If engine starts, refer to Improper Start Chart. Turn ignition switch to the off position.

Repeat above step for center and passenger seat positions.

NOTE: Wait five seconds on 1974 vehicles and 10 seconds on 1975 vehicles between each position check.

Warning Buzzer Check

Place shift lever in drive position (release parking brake on manual transmissions). Turn ignition switch to on position and activate seat sensor switch. FASTEN BELTS lamp should come on and buzzer should sound. If not, refer to chart #3 BUZZER AND SAFETY BELT LAMP DIAGNOSIS.

Pre-Buckled, No Start Check

Place shift lever in neutral for manual transmissions and park for automatics and release parking brake. Buckle all front seat belts and while activating each seat sensor switch, turn ignition switch to start position. Engine should not crank. FASTEN BELTS lamp should light and buzzer should sound. If not, refer to chart #4 IMPROPER START.

Proper Start Check

Place shift lever in neutral for manual transmissions or park for automatics with seat belts buckled and apply parking brake. For each front seat position, activate seat sensor switch, buckle and unbuckle seat belt and turn ignition switch to start position. Engine should crank. If

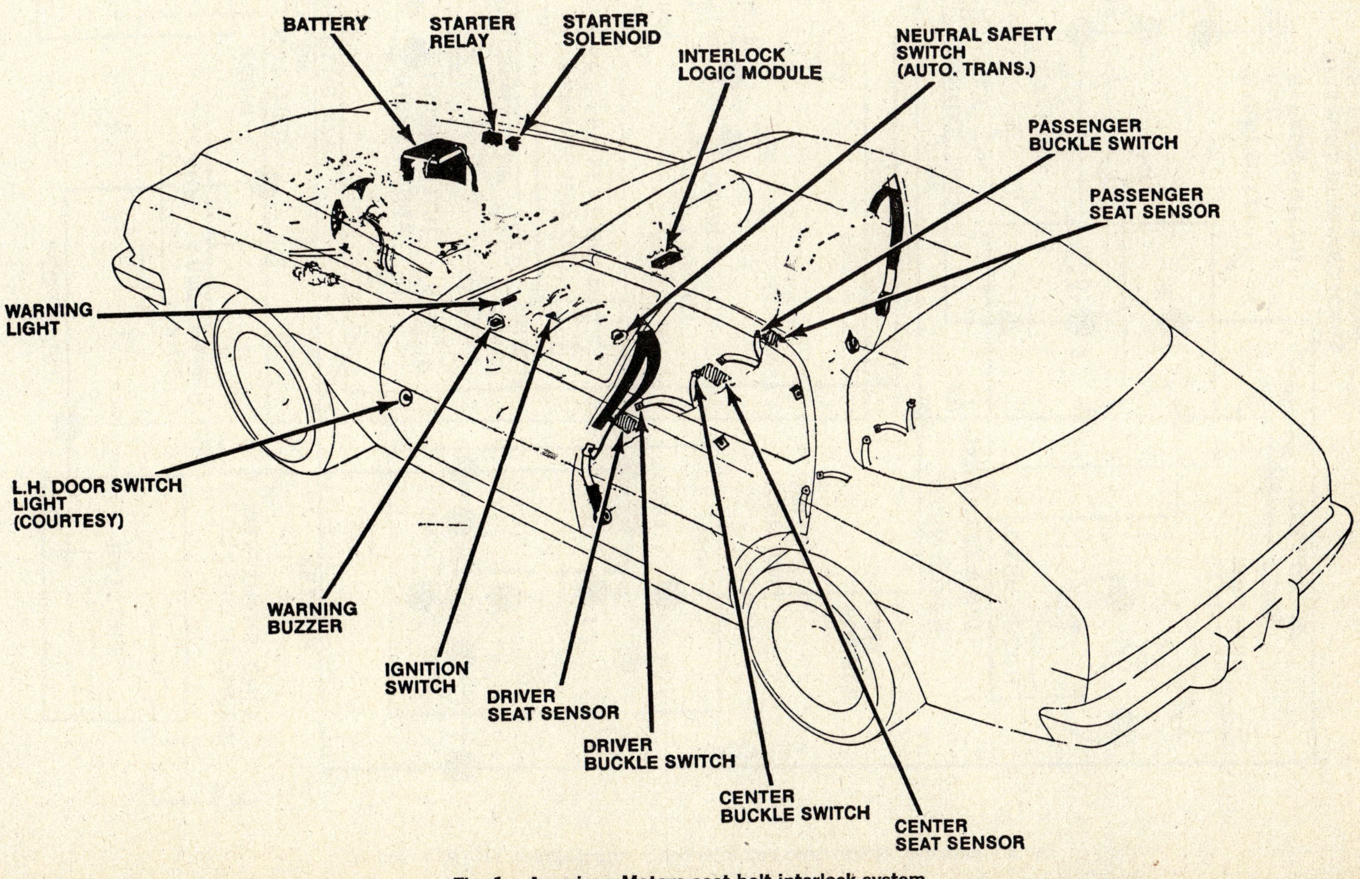

Fig. 1 American Motors seat belt interlock system

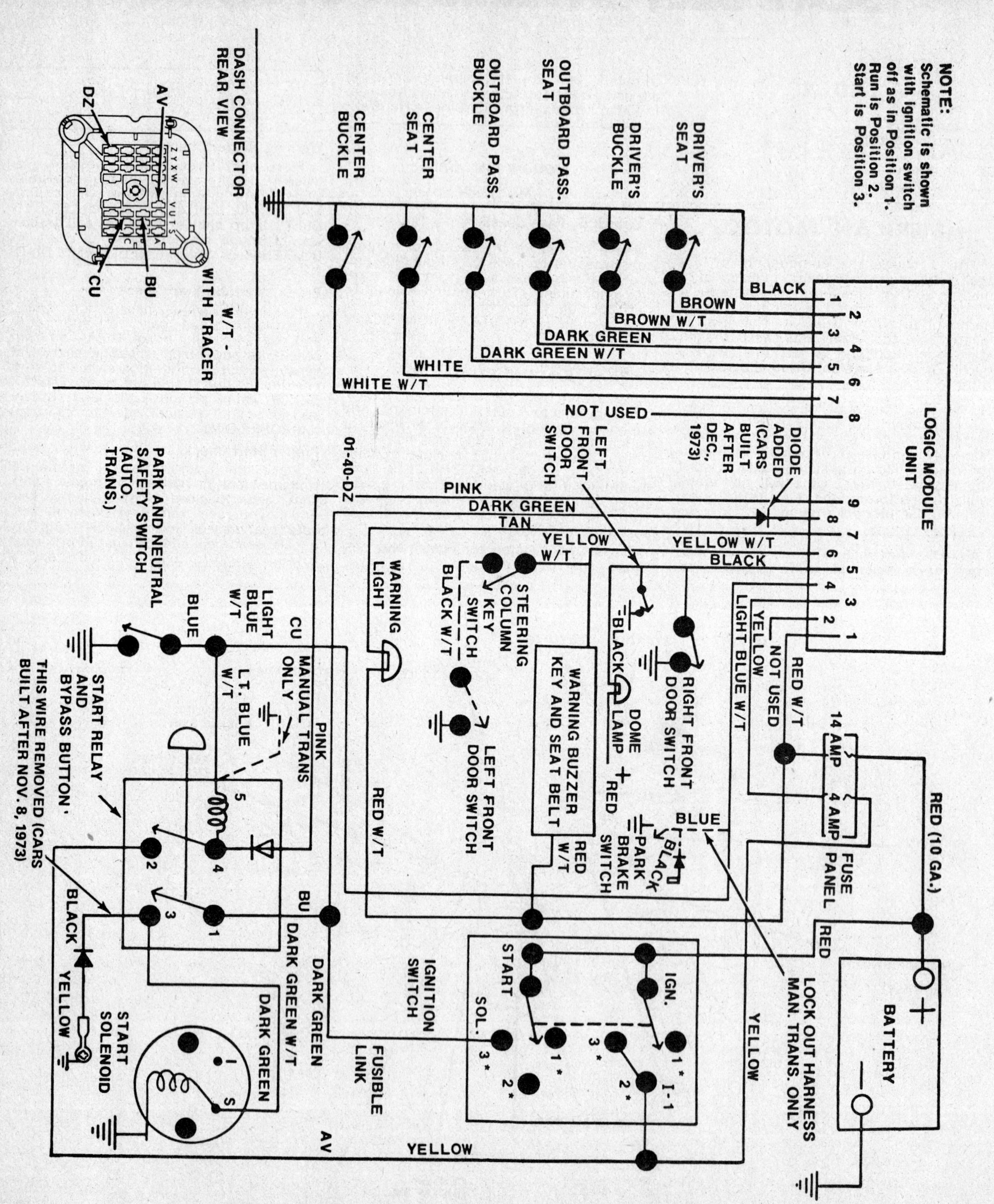

Fig. 2 American Motors seat belt interlock wiring diagram. All 1974 models

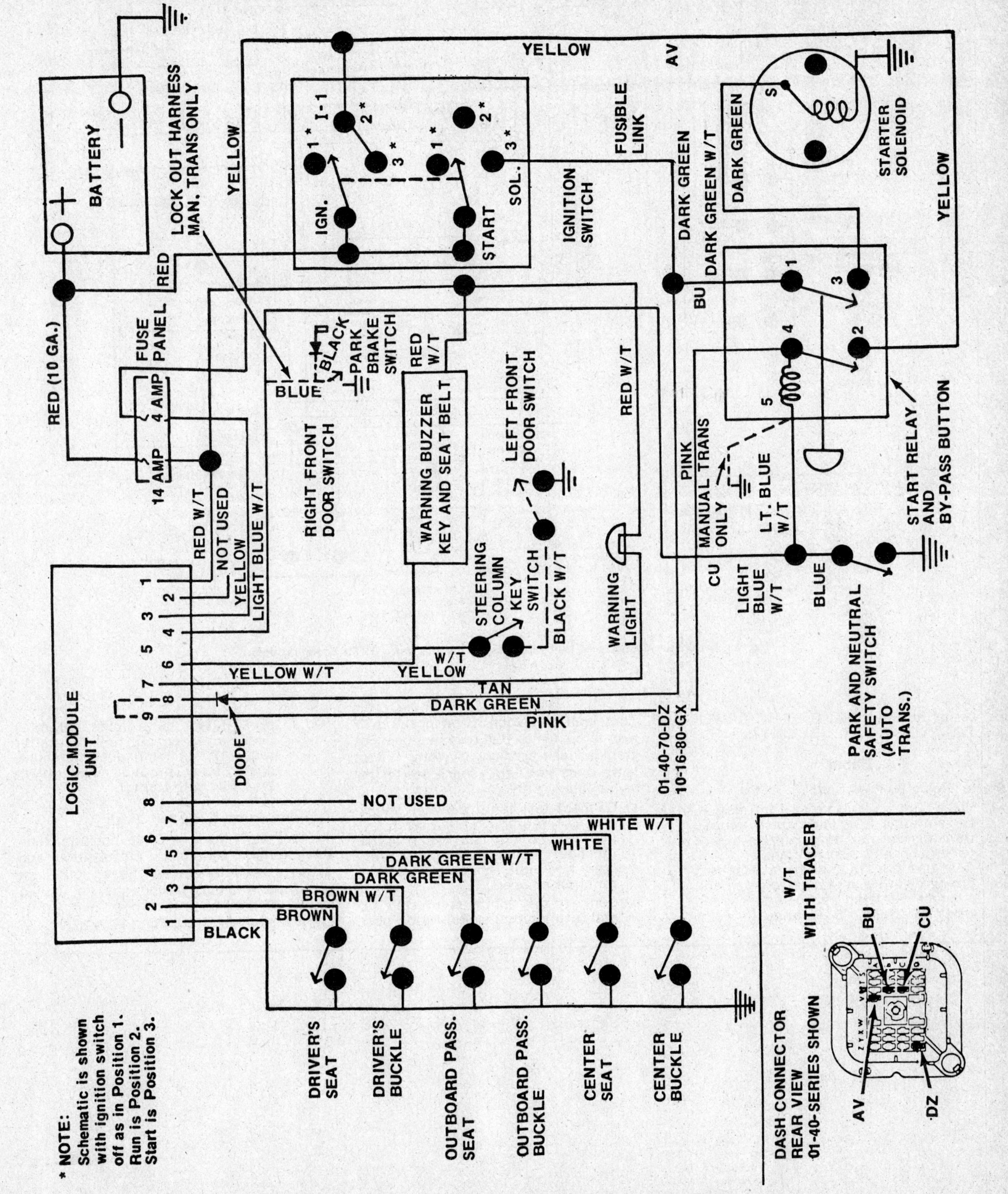

Fig. 3 American Motors seat belt interlock wiring diagram. 1975 models exc. Pacer

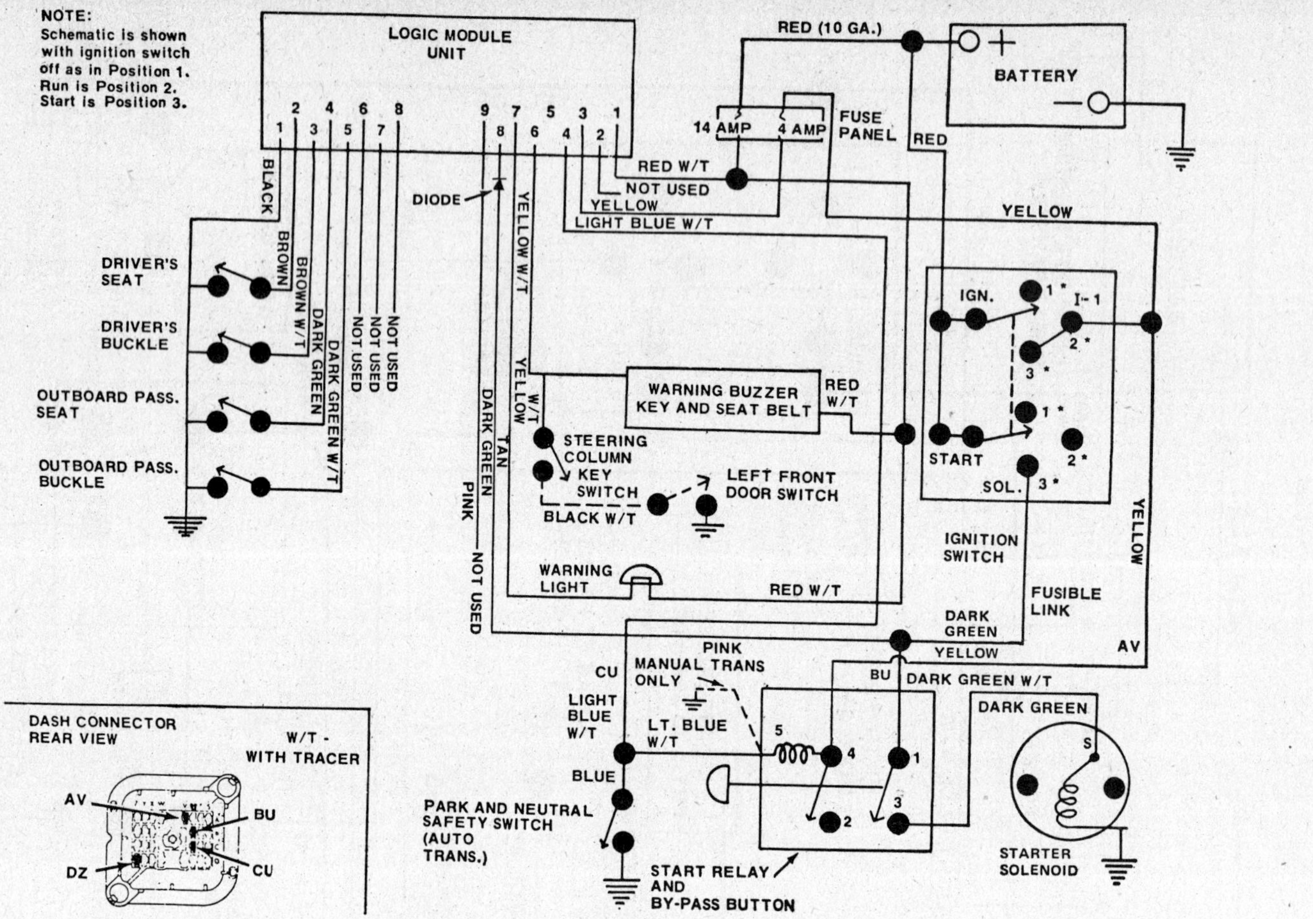

Fig. 4 American Motors seat belt interlock wiring diagram. 1975 Pacer models

not, refer to chart #2 NO START-DRIVER SEATED AND PROPERLY BUCKLED.

Pacer

Starter Relay By-Pass Check

1. Place shift lever in neutral for manual transmission and Park for automatic transmission and turn ignition switch to Start. Engine should crank.
2. If engine does not crank, disconnect dark green wire at starter solenoid and connect test lamp between dark green wire and solenoid S terminal.
3. Turn ignition switch to on position. If test lamp lights but engine does not crank, check cranking system. If test lamp does not light, check starter relay as follows:
   a. Connect test lamp between yellow wire (#4 terminal at starter relay) and ground. If test lamp lights, proceed to next step. If test lamp does not light, check yellow wire for short or open.
   b. Connect test lamp as in step (a) and turn ignition switch to on position. If test lamp lights but relay does not energize, replace starter relay. If test lamp does not light, check relay ground circuit (light blue wire with tracer).

Warning Lamp & Buzzer Check

Place shift lever in Drive position, then turn ignition switch to on position and activate switch sensor. Seat belt light should go on and warning buzzer should sound. If not, refer to Buzzer and Safety Light Diagnosis Chart, Figs. 12 & 13.

CHART 1
NO START—FRONT SEATS UNOCCUPIED

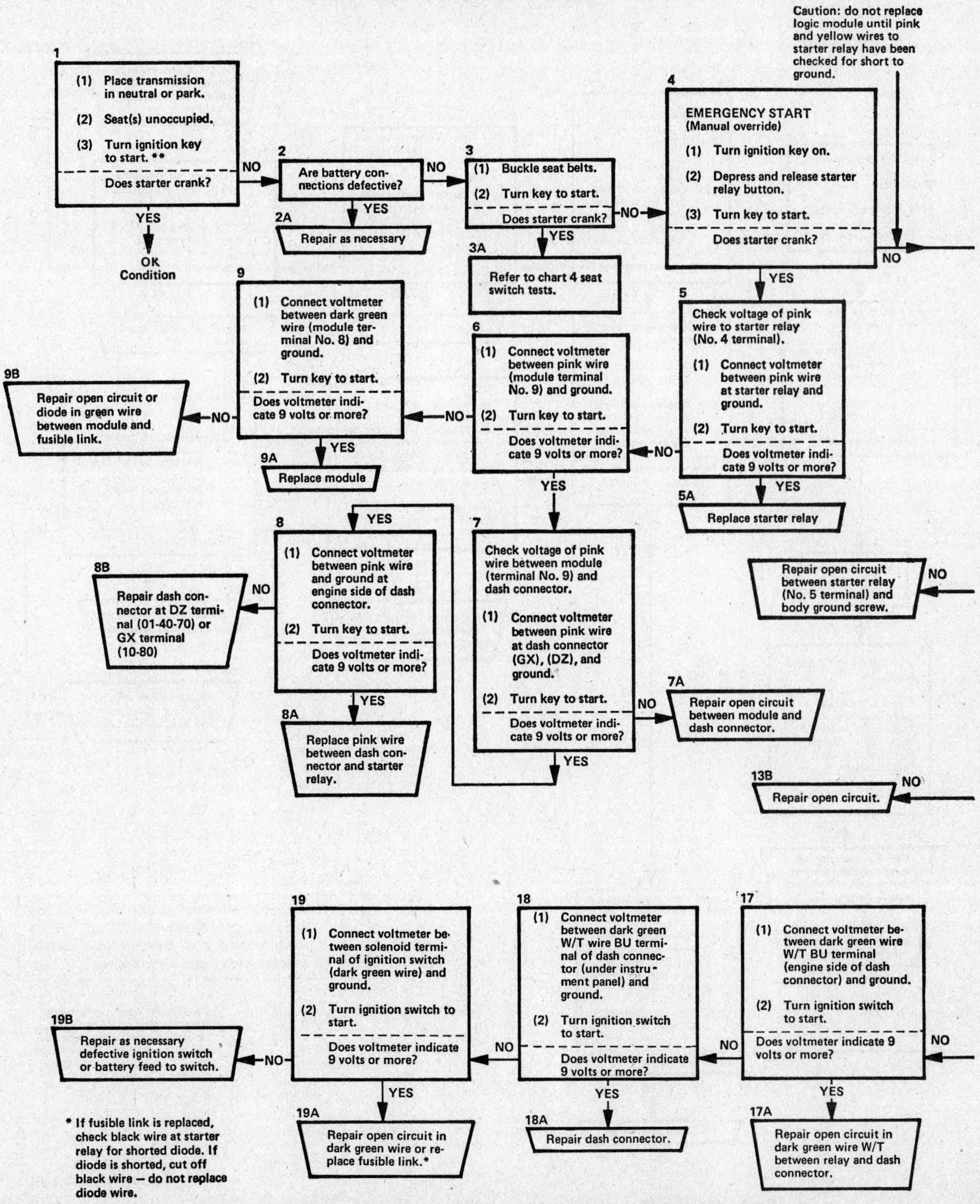

Fig. 5 American Motors seat belt interlock diagnosis chart. All models exc. Pacer

CHART 1

NO START—FRONT SEATS UNOCCUPIED (Continued)

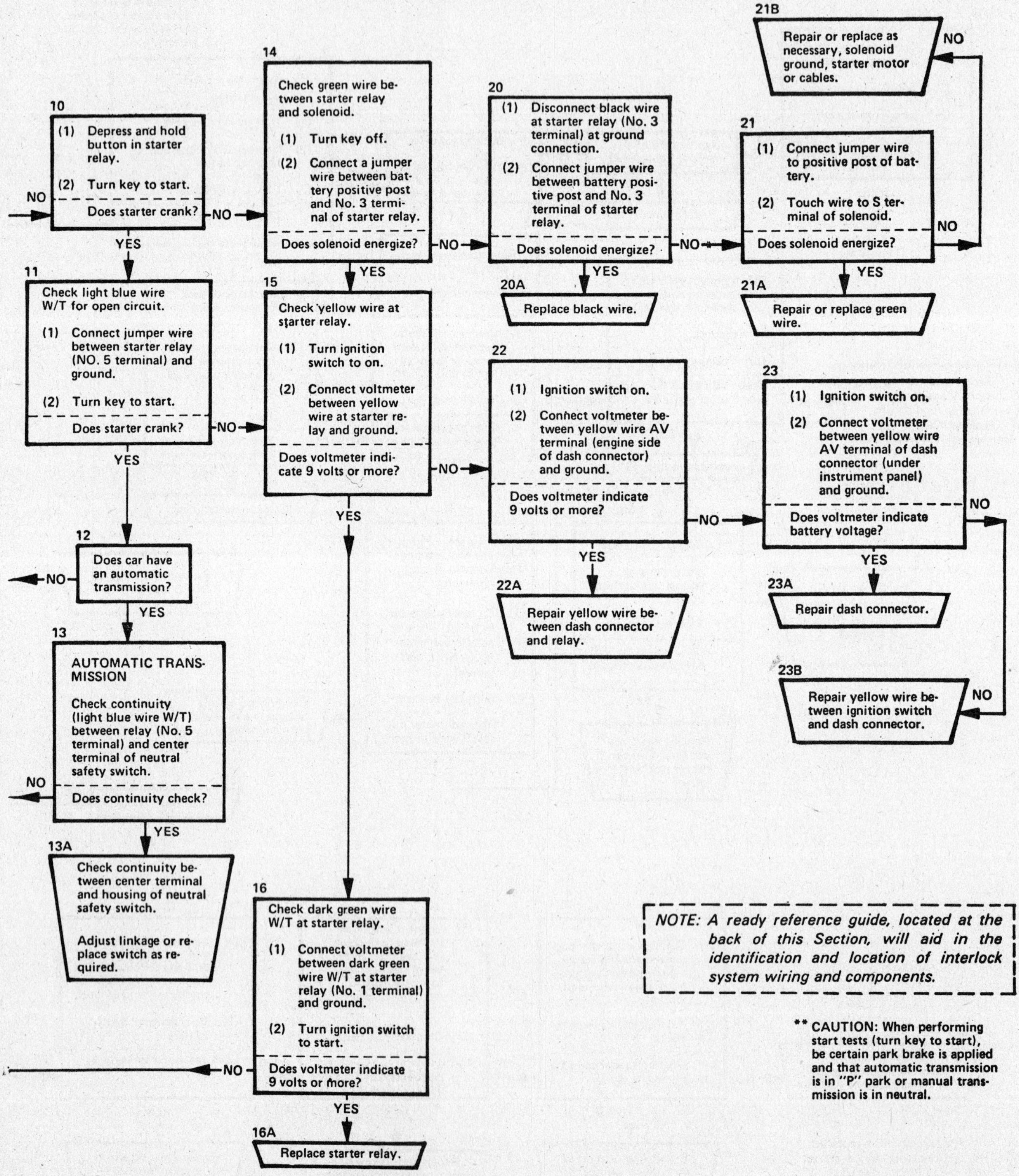

NOTE: A ready reference guide, located at the back of this Section, will aid in the identification and location of interlock system wiring and components.

** CAUTION: When performing start tests (turn key to start), be certain park brake is applied and that automatic transmission is in "P" park or manual transmission is in neutral.

NOTE: W/T = With Tracer

Fig. 6 American Motors seat belt interlock diagnosis chart. All models exc. Pacer.
NOTE: On 1975 models, omit steps 20 and 20A

CHART 2
NO START—DRIVER SEATED AND PROPERLY BUCKLED

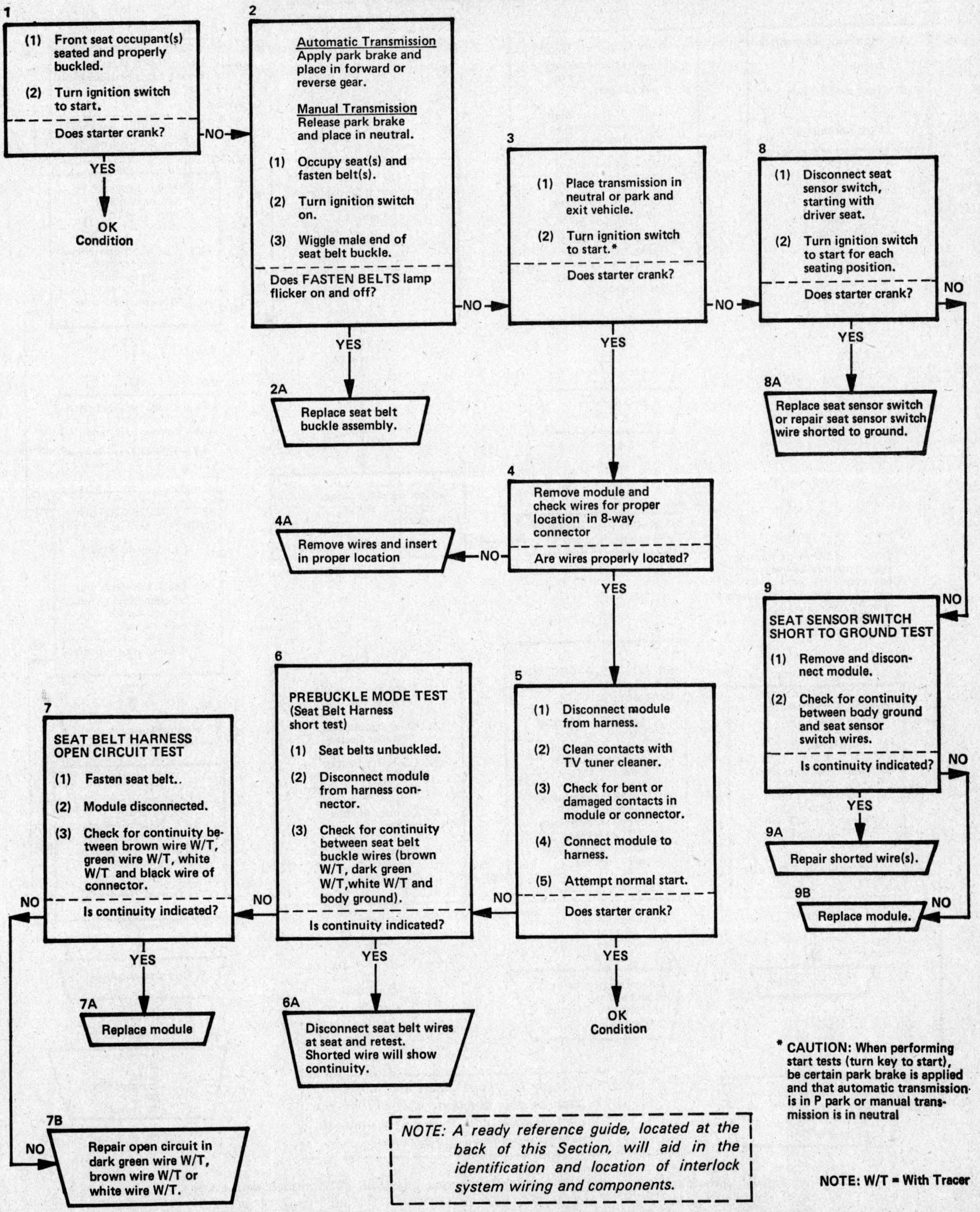

Fig. 7 American Motors seat belt interlock diagnosis chart. All 1974 models

CHART 2

No Start—Driver Seated and Properly Buckled

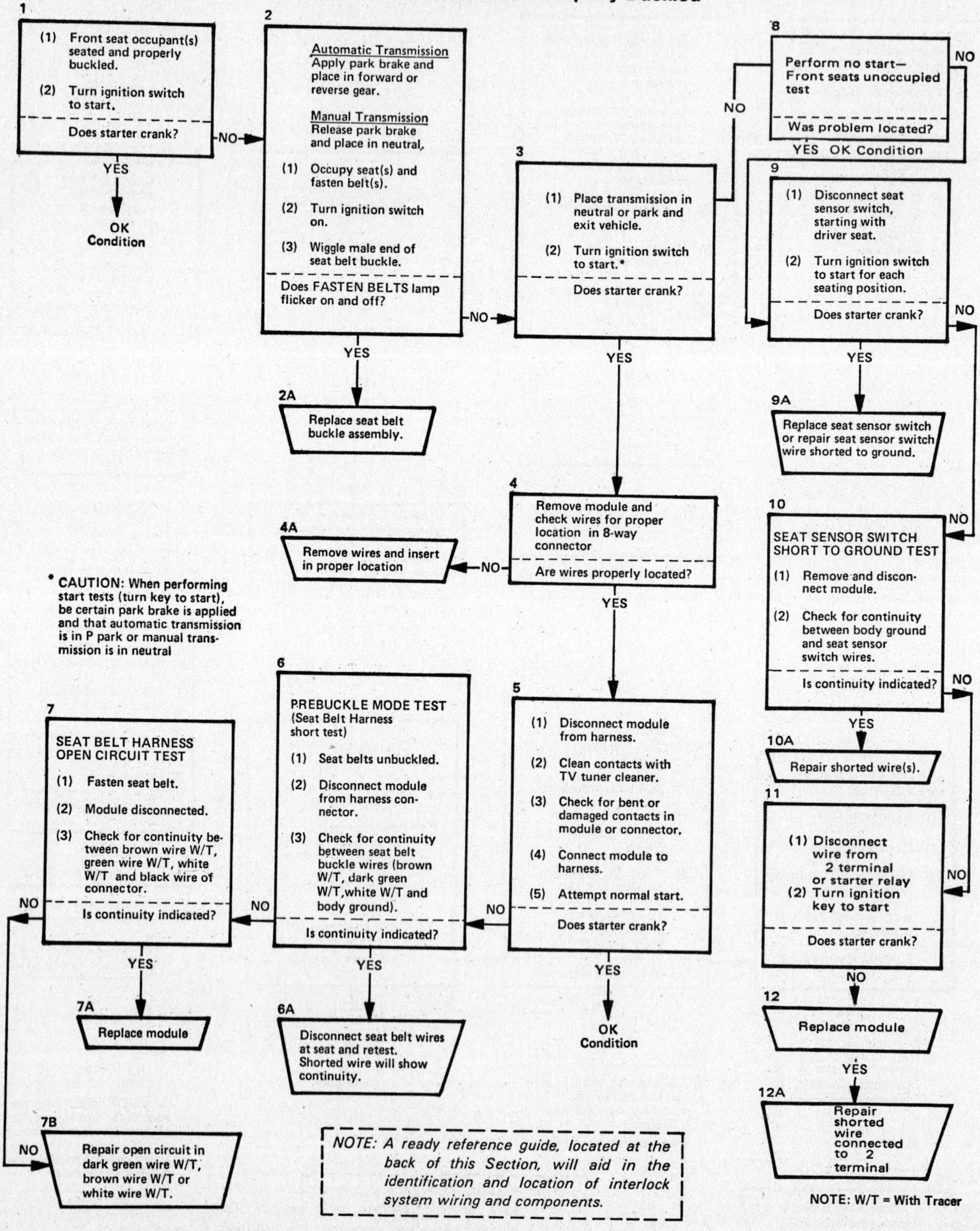

Fig. 8 American Motors seat belt interlock diagnosis chart. All 1975 models exc. Pacer

CHART 3
BUZZER AND SAFETY BELT LAMP DIAGNOSIS

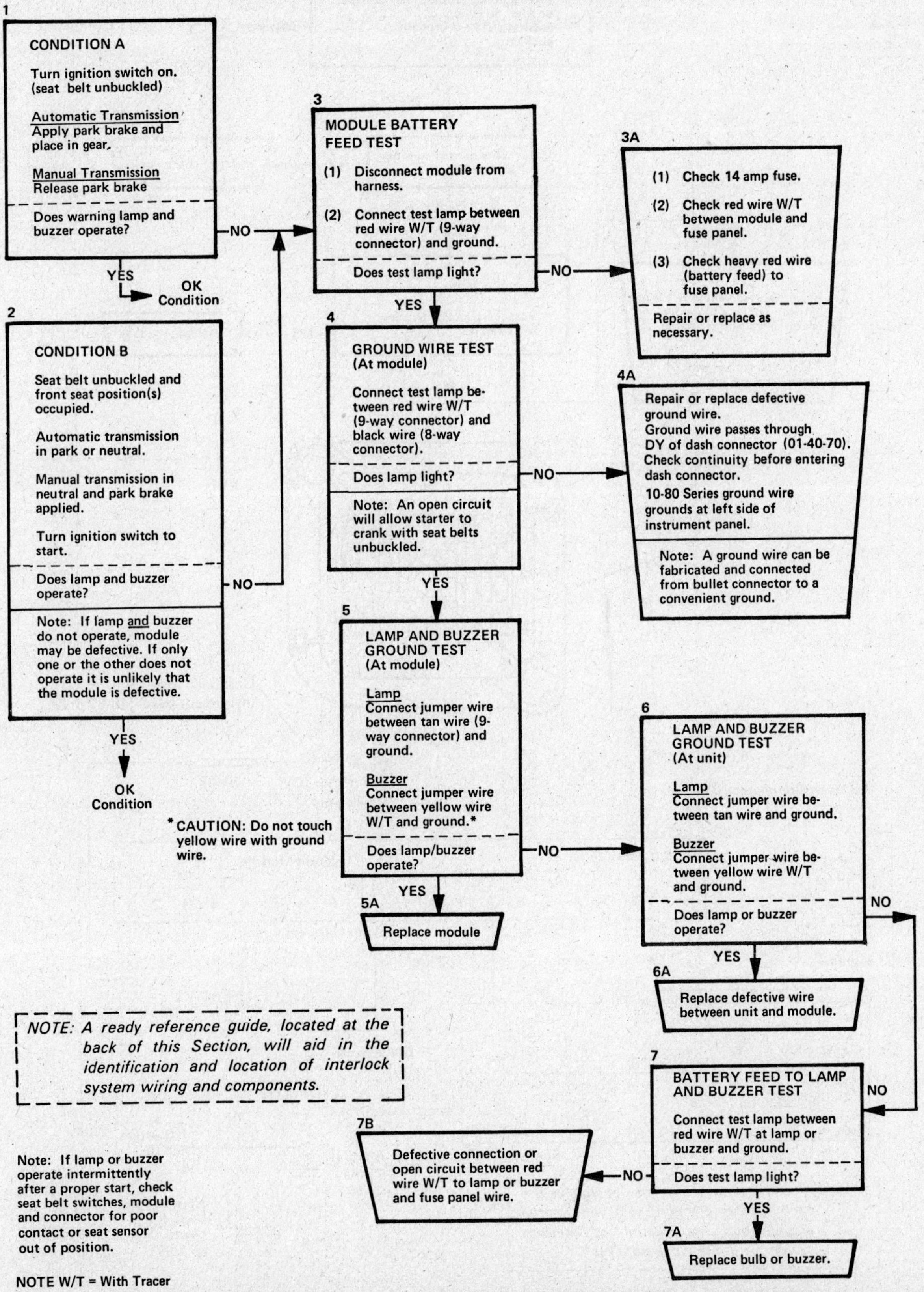

Fig. 9 American Motors seat belt interlock diagnosis chart. All models exc. Pacer

CHART 3
BUZZER AND SAFETY BELT LAMP DIAGNOSIS (Continued)

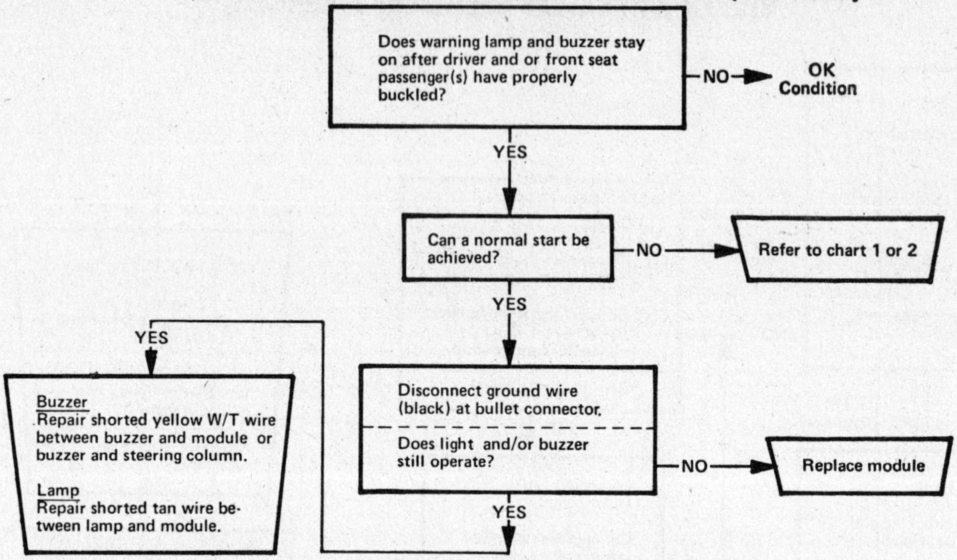

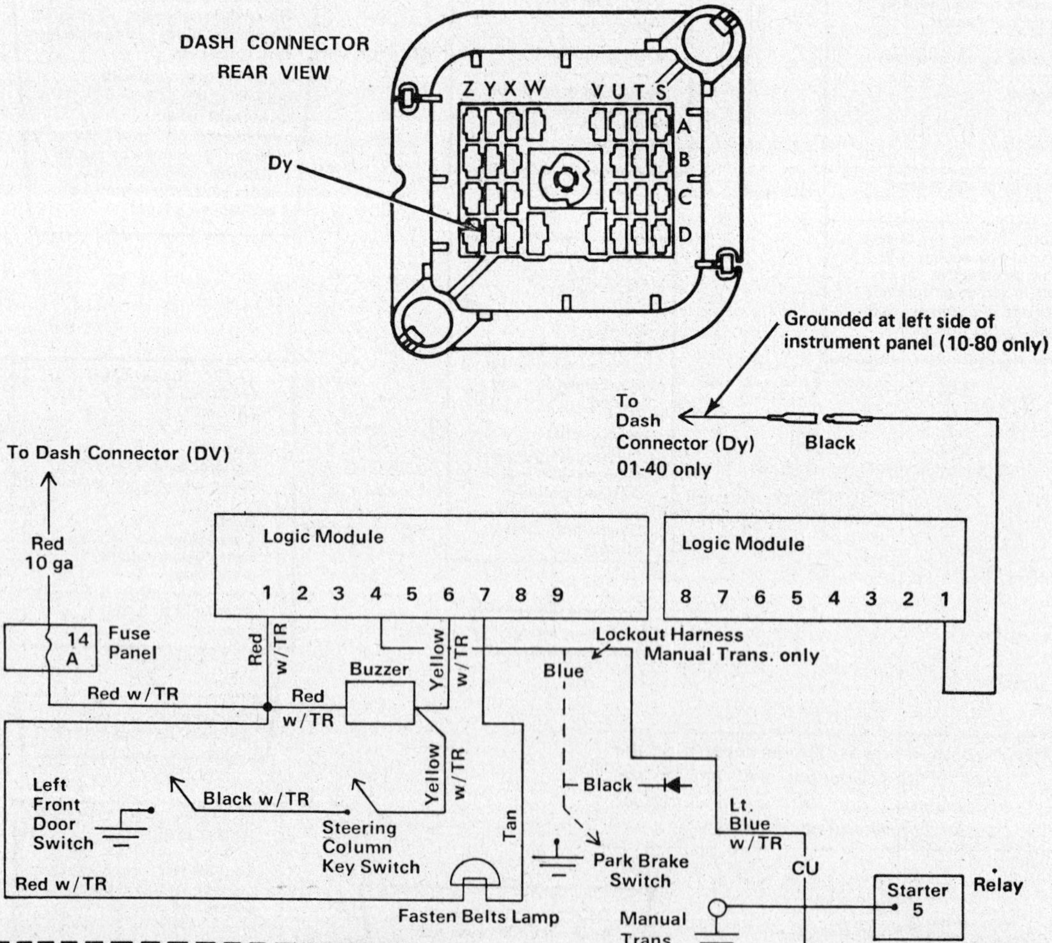

NOTE: A ready reference guide, located at the back of this Section, will aid in the identification and location of interlock system wiring and components.

Fig. 10 American Motors seat belt interlock diagnosis chart. All models exc. Pacer

CHART 4
IMPROPER START—DRIVER SEATED AND NOT BUCKLED

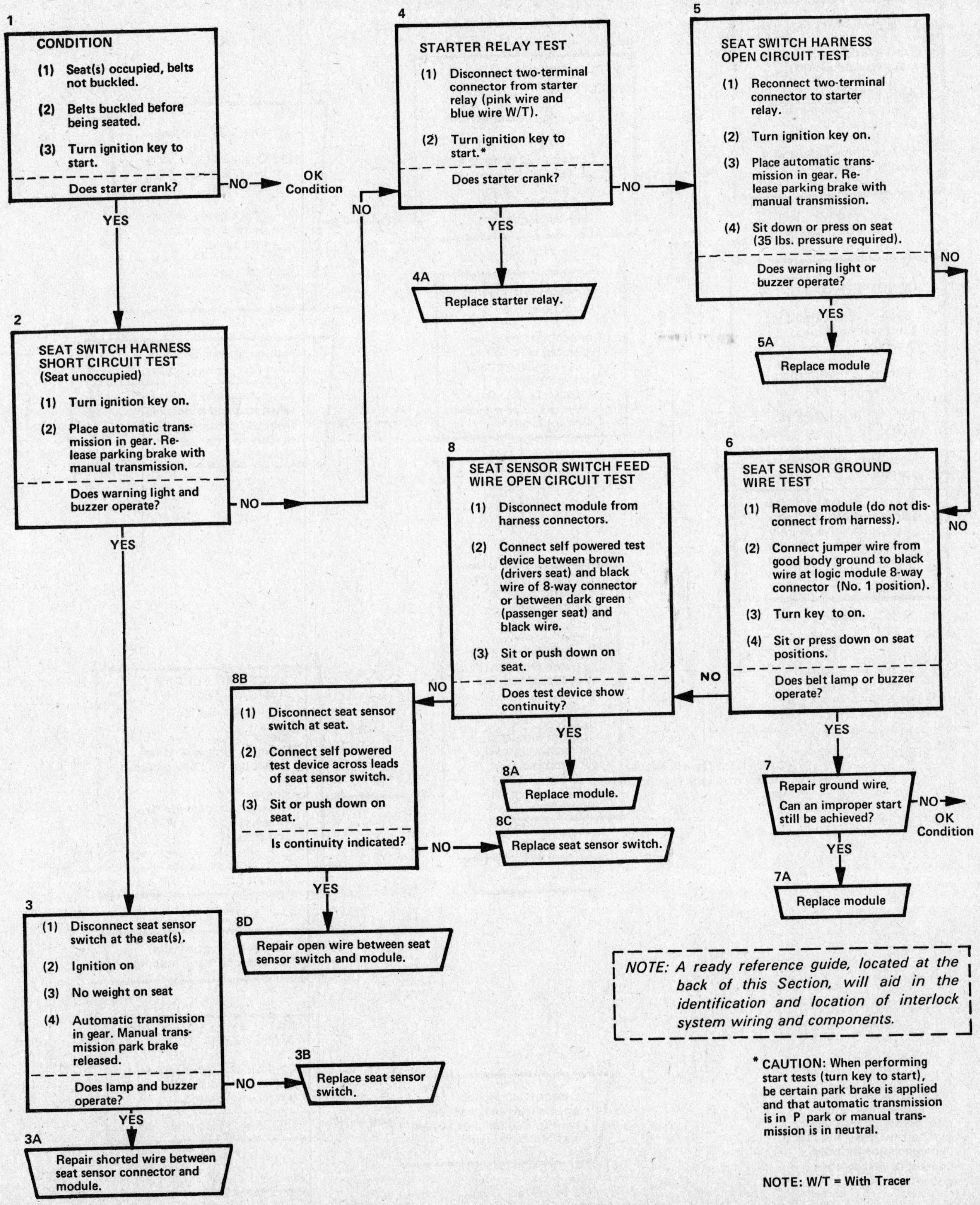

Fig. 11 American Motors seat belt interlock diagnosis chart. All models exc. Pacer

Warning Buzzer and Lamp Diagnosis

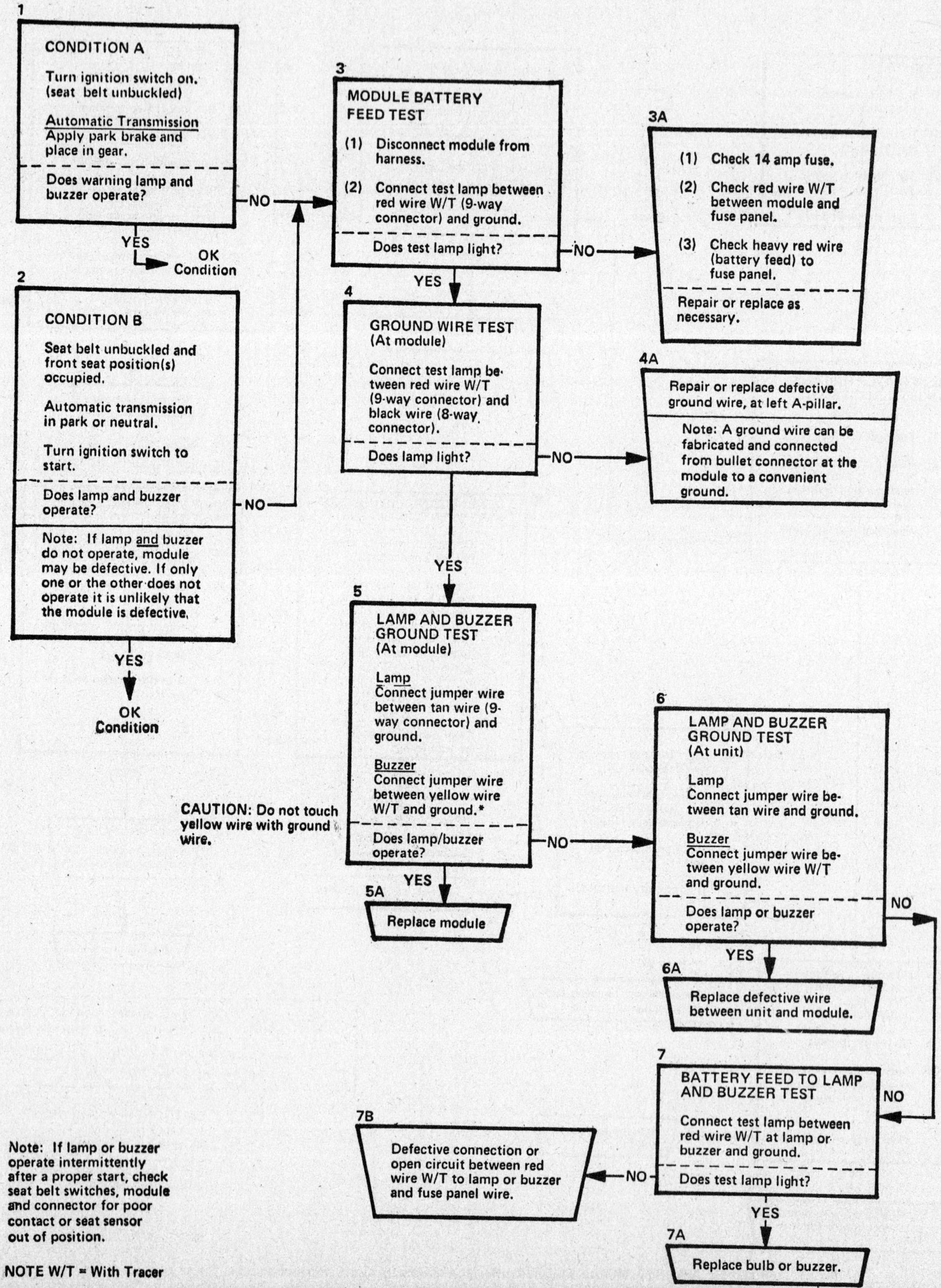

Fig. 12 American Motors seat belt interlock diagnosis chart. 1975 Pacer models

Warning Buzzer and Lamp Diagnosis (Continued)

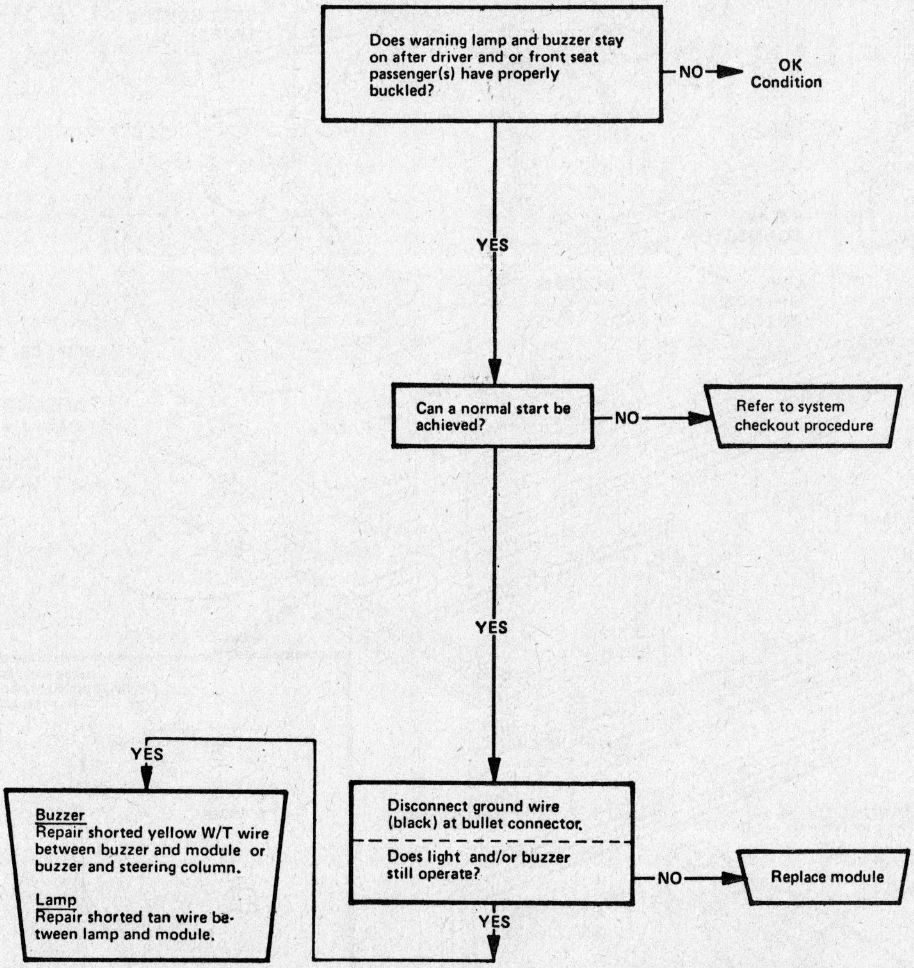

Fig. 13 American Motors seat belt interlock diagnosis chart. 1975 Pacer models

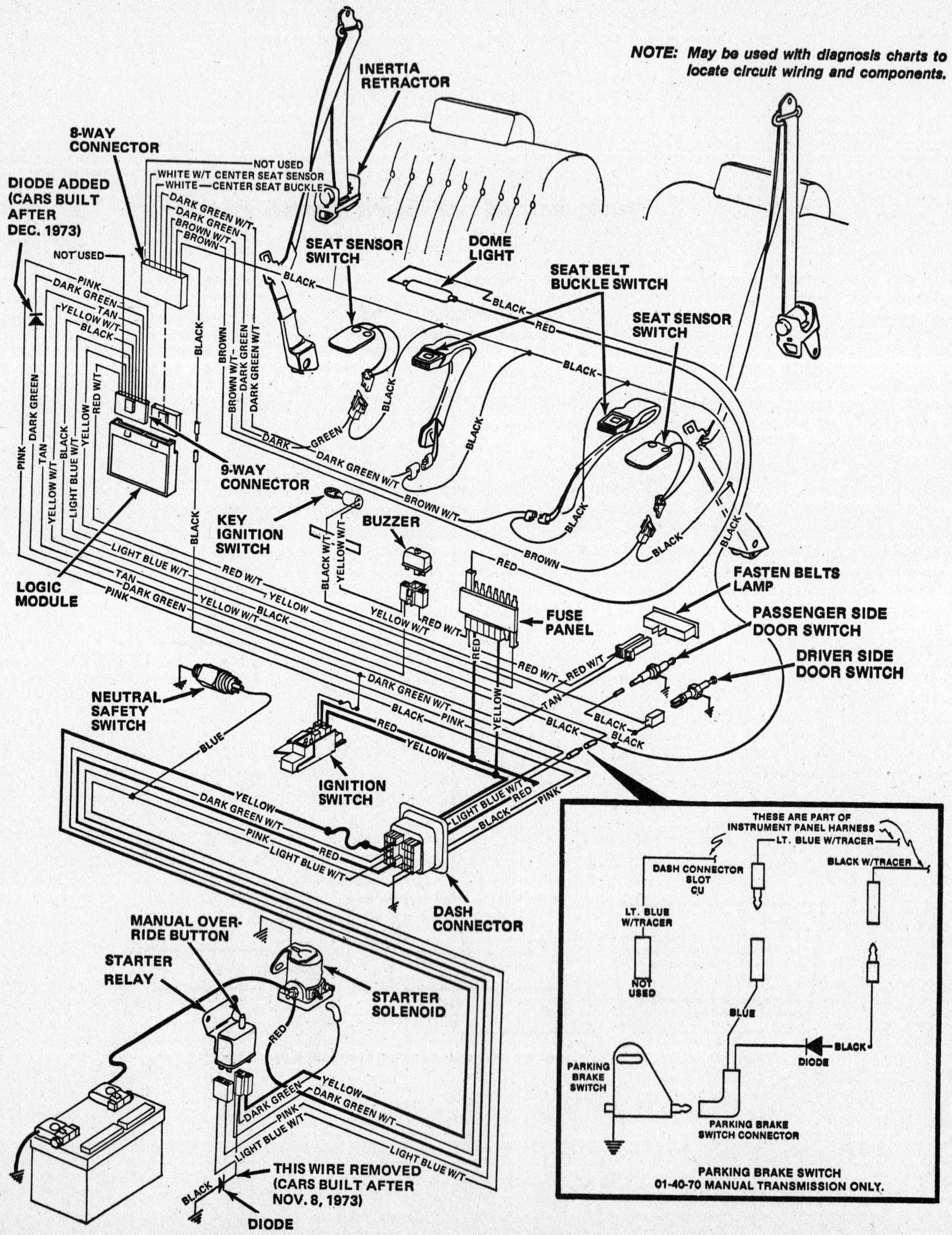

Fig. 14 American Motors seat belt ready reference guide. All 1974 models

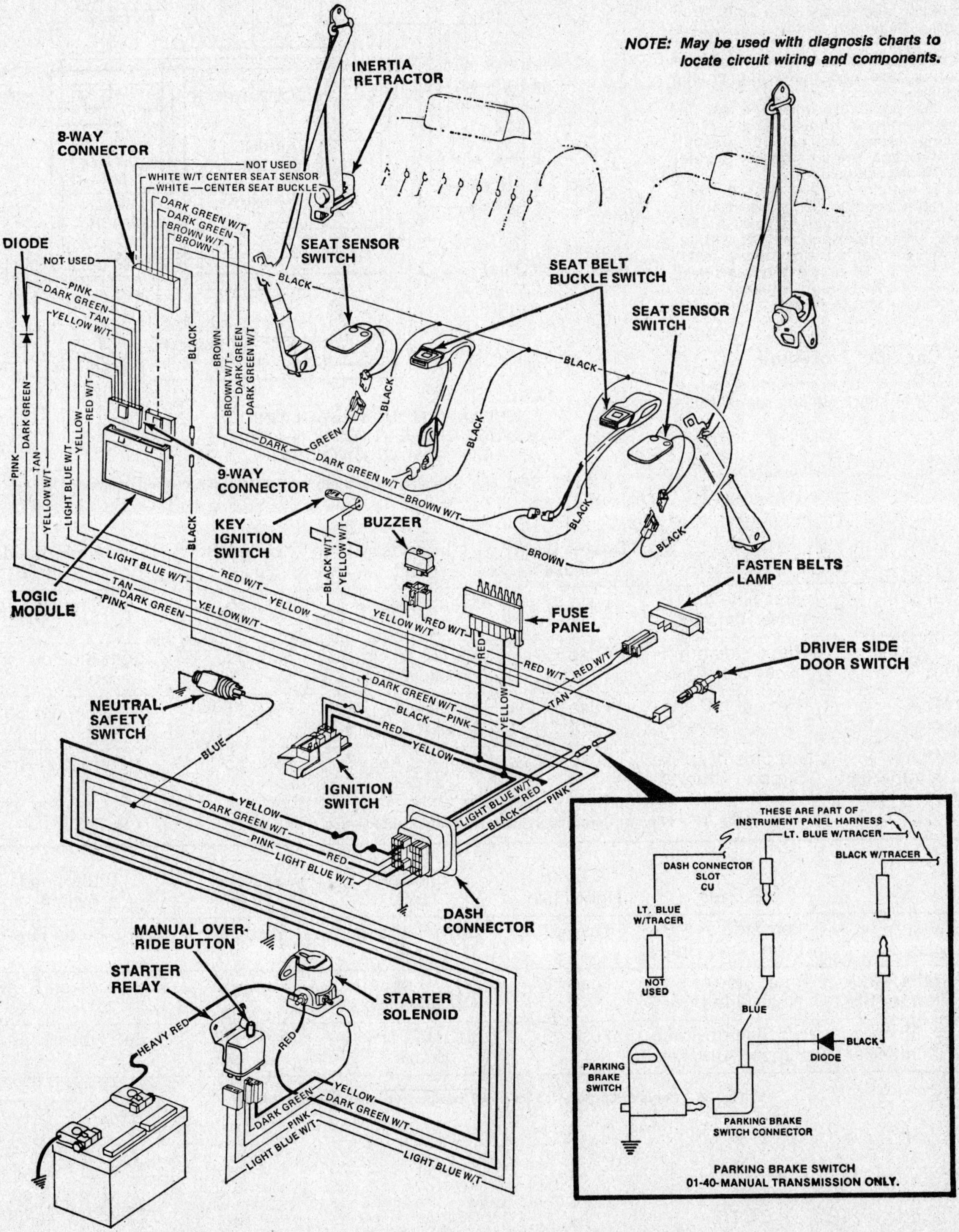

Fig. 15 American Motors seat belt ready reference guide. All 1975 models exc. Pacer

CHRYSLER CORPORATION

To use the Trouble Shooting procedure, perform each and every Checkout Procedure step beginning with step 1 and continuing until the malfunction occurs. Then proceed to the corresponding Trouble Shooting step and follow the procedure outlined.

Begin the procedure with the car in Neutral on automatic transmission cars or the parking brake applied on manual transmission cars. The ignition key should be in the Off or Lock position.

The seat switches in the driver's seat and the right front passenger seat location must be set by having at least a 150 lb. person or equivalent weight sitting momentarily in that position before tests can proceed. Fig. 16 provides a basic wiring diagram of the system. Interlock component location charts, Figs. 17 and 18 are also provided.

Checkout Procedure

Interlock diagnosis charts are provided to assist in the checkout procedure, Figs. 19 and 20.

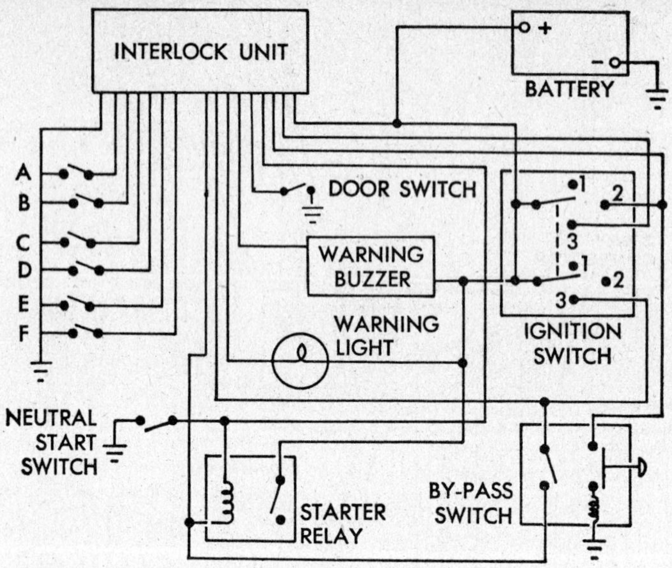

Fig. 16 Chrysler Corporation seat belt interlock wiring diagram

	Buzzer	Buzzer Light Fuse	Interlock Unit	System Fuse	Underhood Switch
Dart & Valiant	Left Side of I/P Above Hand Brake	Fuse #7	Behind Buzzer	Fuse #6	Left Side on Firewall
Charger, Coronet & Satellite	Right Side of I/P Near Side Cowl	Fuse #2	On I/P to Left Of Glove Box	Fuse #1	Right Side on Firewall
Barracuda & Challenger	Rear of I/P and Right of Glove Box	Fuse #4	Left Side of I/P Above A/C Duct	Fuse #1	Left Fender Shield
Chrysler, Fury, Monaco & Imperial	Left Side of Brake Support Bracket	Fuse #3	Mounted Above Buzzer	Fuse #5	Left Hood Hinge

Fig. 17 Chrysler Corporation interlock component location chart. 1974

	Buzzer	Buzzer Light Fuse	Control Unit	System Fuse	Underhood Switch
Dart & Valiant	Left Side of I/P Above Hand Brake	Fuse #7	On I/P of Left Of Glove Box	Fuse #6	Left Side on Firewall
Fury, Coronet, Cordoba & Charger SE	Right Side of I/P Near Side Cowl	Fuse #2	On I/P of Left Of Glove Box	Fuse #5	Right Side on Firewall
Gran Fury, Monaco, Chrysler & Imperial	Left Side of Brake Support Bracket	Fuse #3	Mounted Near Buzzer	Fuse #5	Left Hood Hinge

Fig. 18 Chrysler Corporation interlock component location chart. 1975

1974

Underhood Switch Operational Check

1. Sit in driver's seat with all doors closed but do not buckle belt. Turn ignition key to Start. Engine should not crank. Light and buzzer should go on while key is in Start. Light and buzzer should go off when key is in Run position.
2. Turn ignition key to Off or Lock. Open hood and depress then release the underhood switch. Sit in driver's seat but do not buckle belt. Turn ignition key to Start. Engine should crank. Light and buzzer should go on while key is in Start.
3. Leave key On for 30 seconds, then turn Off and leave ignition Off for 2½ minutes. Get out of driver's seat and reach in car and buckle driver's seat belt. Sit in driver's seat, close all doors again and turn ignition key to Start. Engine should not crank. Light and buzzer should go on while key is in Start.

Interlock and Sensing Switches

4. With driver in seat and pre-buckled as in step 3 above, leave ignition On and for automatic transmission cars move the gear selector to any forward or reverse gear. For manual transmissions release parking brake. Light and buzzer should go On and stay On.
5. Unbuckle driver's belt and light and buzzer should stay On. Kneel on seat and buckle belt across empty seat. In this way weight is applied to driver's seat before belt is buckled. Light and buzzer should go Off.
6. Get out of car and close door. Go to right side of car and open door. Reach in and buckle right passenger seat belt. Sit on seat and close door and light and buzzer should be On.
7. Unbuckle right passenger belt and light and buzzer should remain On.
8. Buckle right passenger belt and light and buzzer should go Off.
9. Reach over and buckle center passenger belt and move to center seating position. The light and buzzer should go On.
10. Unbuckle the center belt and light and buzzer should remain On.
11. Buckle center seat belt and light and buzzer should turn Off within 5 seconds.
12. Unbuckle center seat belt and move to driver's seat, then place automatic transmission shift lever to Neutral or Park. Depress clutch pedal on manual transmission cars. Turn ignition key to Start and engine should not crank. Unbuckle and rebuckle driver's belt. Engine should crank and light should be Off.
13. Turn ignition to Off and unbuckle driver's belt. Next, turn ignition key to Start. Engine should crank. Light and buzzer should operate while key is in Start position.
14. Turn ignition key Off and move to center or right passenger seat for at least 5 seconds. Return to driver's seat and turn ignition key to Start and engine should crank. Light and buzzer should operate while key is in Start.

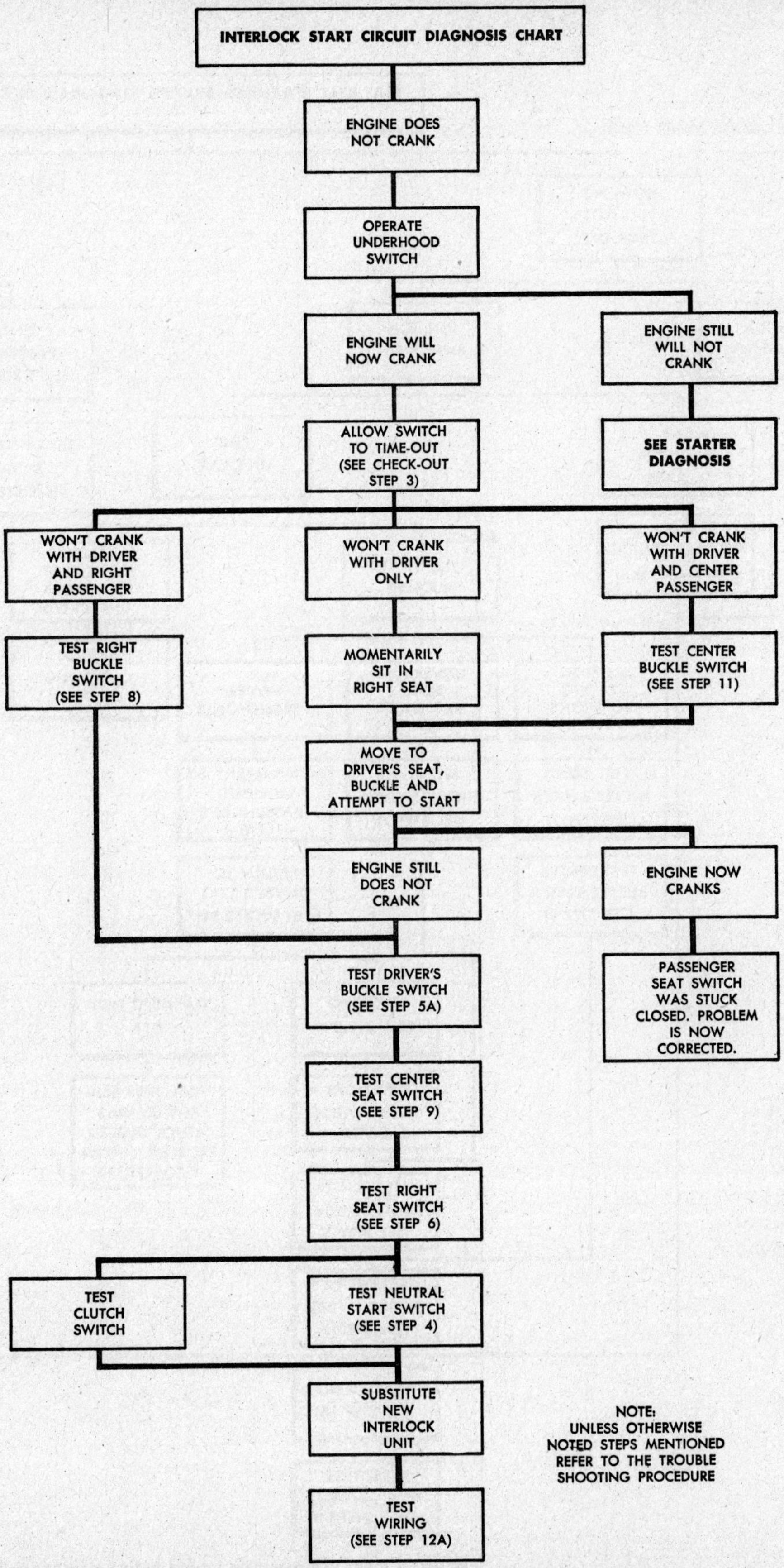

Fig. 19 Chrysler Corporation interlock diagnosis chart. 1974

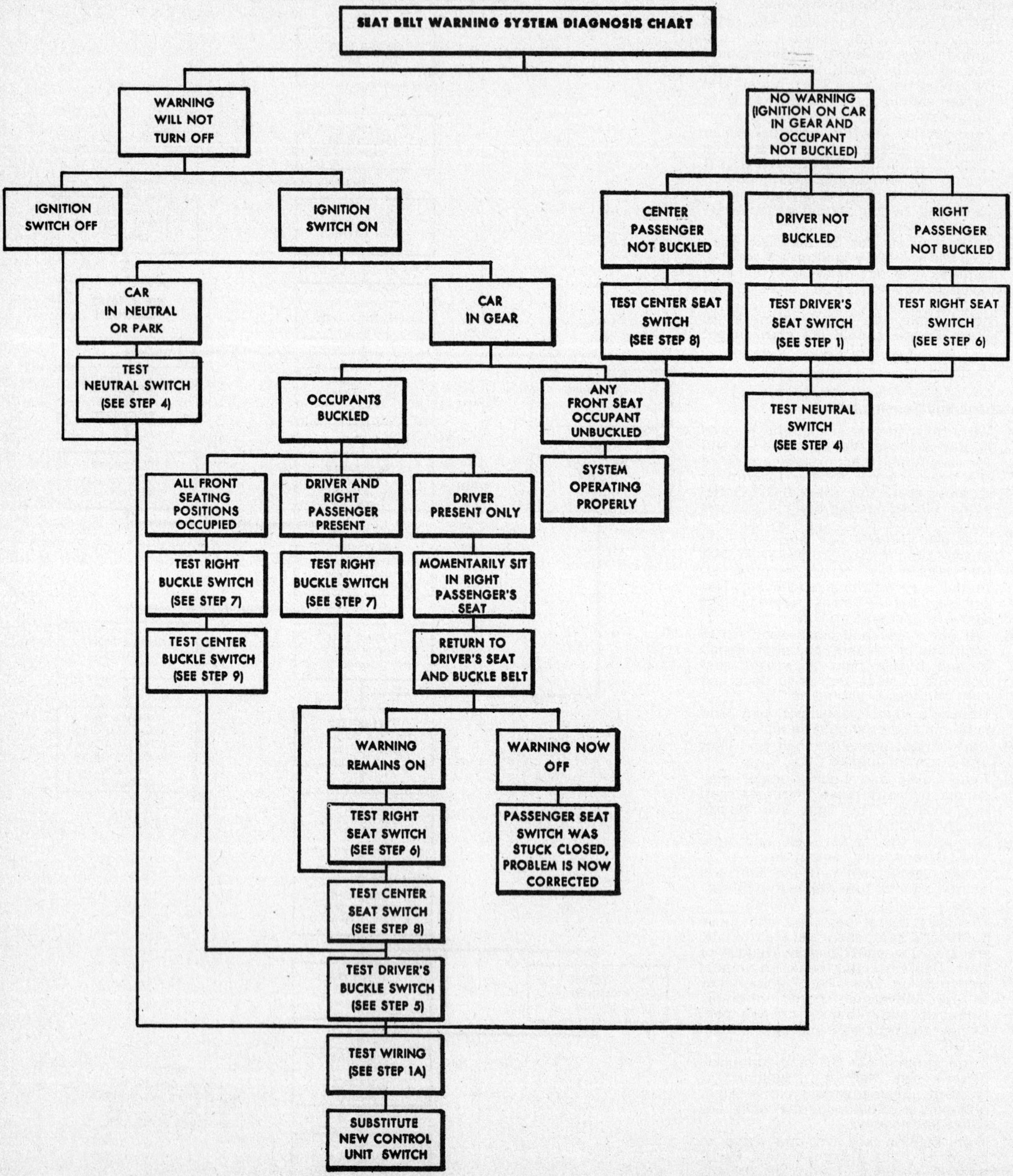

Fig. 20 Chrysler Corporation interlock diagnosis chart. 1975

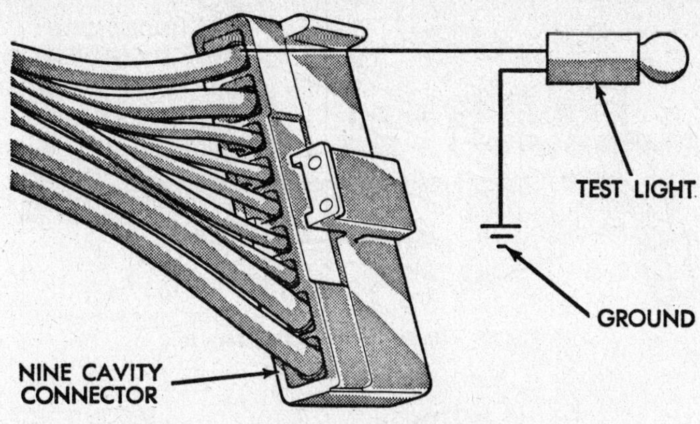

Fig. 21 Battery test

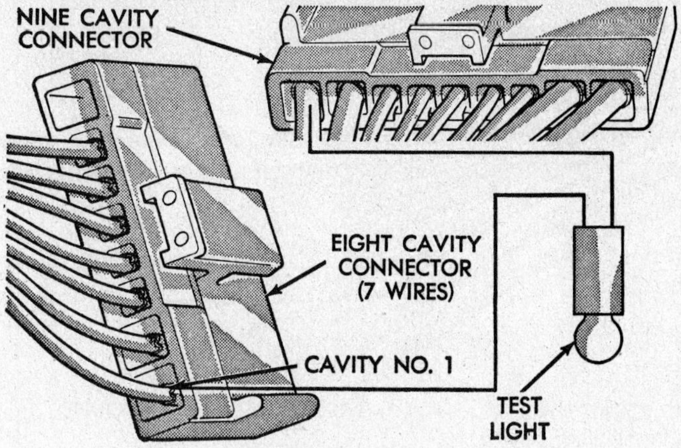

Fig. 22 Ground test

1975

Underhood Switch Operational Check

1. Sit in driver's seat and leave belt unbuckled. Turn ignition key to Start position. Engine should not crank. Light and buzzer should go on while key is in Start position. Light and buzzer should go off when key is in run position.
2. Leave key in Run position, then open hood and depress underhood switch. Sit in driver's seat and leave belt unbuckled. Turn ignition key to Start position. Engine should crank and light and buzzer should go on while key is in Start position.
3. Turn key to Off position and then to Start position. Engine should not crank and light and buzzer should go on while key is in Start position.

Interlock and Sensing Switches

4. With driver in seat, leave ignition On, and for automatic transmission cars, move shift lever to any forward or reverse gear. On manual transmission cars, release parking brake. Light and buzzer should go on and remain on.
5. Buckle belt with key in On position and vehicle in gear as in step 4. Light and buzzer should go off.
6. Sit on right passenger seat. Light and buzzer should be on.
7. Buckle right passenger's belt. Light and buzzer should go off.
8. Sit on center passenger seat. Light and buzzer should be on.
9. Buckle center seat belt. Light and buzzer should go off.
10. Unbuckle center seat belt and buckle driver's belt, then move to driver's seat. Place shift lever in Neutral or Park on automatic transmission vehicles. Depress clutch on manual transmission vehicles.
11. Turn key to Start position. Engine should not crank. Unbuckle and rebuckle driver's belt. Engine should crank and light and buzzer should be off.
12. Turn key to Off position and unbuckle driver's belt. Engine should not crank and light and buzzer should be on while key is in Start position.
13. Turn key to Off position and leave driver's seat for at least 30 seconds. Sit in driver's seat and turn key to Start position. Engine should not crank and light and buzzer should be on while key is in start position.

Diagnosis Step Procedure

The step by step directions can be accomplished with the aid of a voltage test light and a test light jumper wire. When checking circuits at interlock connectors, test only at wire end of connector as shown in various illustrations.

Step 1—Engine Cranks with Driver Not Buckled

Underhood Switch

1. Remove connector from underhood switch, then return to driver's seat and turn ignition key to Start. If engine does not crank, replace underhood switch.

Wiring

1. Remove both connectors from interlock unit, then sit in driver's seat and turn ignition to Start. If engine cranks, a wiring short exists. Trace circuits to correct problem.
2. On 1974 models, if engine does not crank, check for voltage with test light between red wire in 9-way connector and a good ground, Fig. 21.
3. On 1975 models, if engine does not crank, check for voltage with a test light between red wire in 9-way connector and grey wire in 8-way connector, Fig. 22.
4. When connected, test light should turn On. If not, an open circuit exists. Trace circuit and correct.

Driver's Seat Switch

1. Check driver's seat switch by connecting test light to red wire in 9-way connector and to white wire in 8-way connector, Fig. 23.

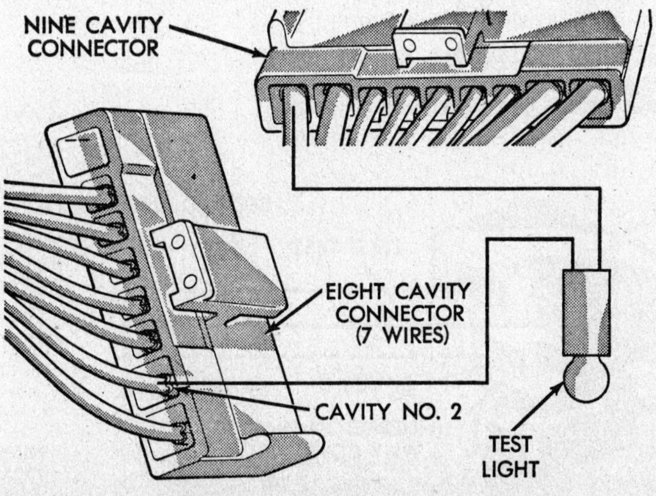

Fig. 23 Driver seat switch test

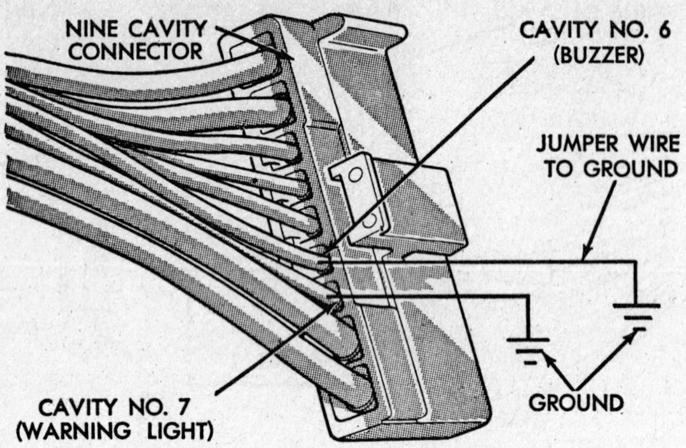

Fig. 24 Buzzer and warning light test

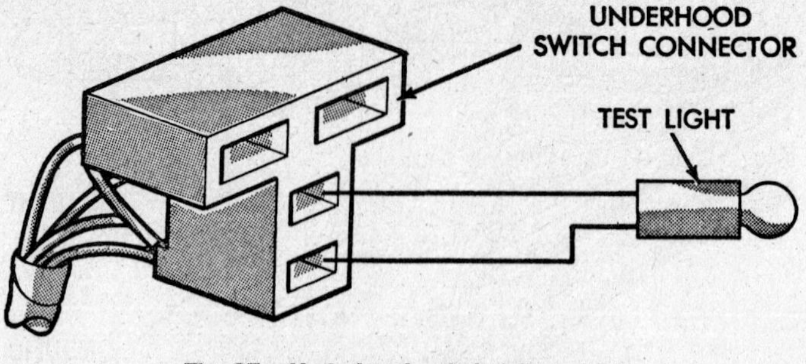

Fig. 25 Underhood switch voltage test

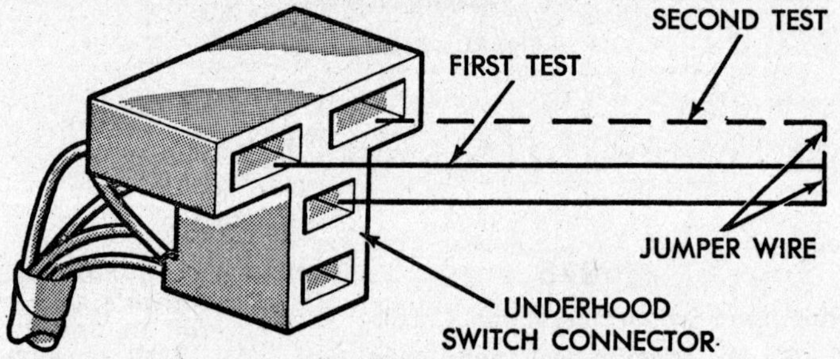

Fig. 26 Underhood switch starter test

2. Test light should be On when driver's seat is occupied and Off when the seat is empty. If switch does not operate properly, use a self powered test light to check switch operation across switch connector terminals under seat. If switch does not operate, replace the switch. If switch is OK, check the white wire between the interlock unit and the switch or grey wire between switch and ground for an open circuit.

Interlock Unit

1. If seat switch, underhood switch and wiring check OK, replace interlock unit.

Step 1A—Warning Light and/or Buzzer do not Turn on when Key is in Start and Off when Key is in Run with Selector in Park or Neutral (Hand Brake Applied for Manual Transmission)

Malfunction of Interlock Unit

A malfunction of the interlock unit will affect both the warning lamp and the buzzer. If only one does not function, the interlock unit is not the cause.

1. If both buzzer and warning lamp do not function, check fuse.
2. Remove both connectors with test light from interlock unit. Check for voltage between red wire in 9-way connector and gray wire in 8-way connector, Fig. 22.
3. Test light should turn On. If not, open circuit exists either in red wire or gray wire. Trace circuits and correct.

Light

1. Connect a jumper wire from dark blue with red tracer in 9-way connector to a good ground, Fig. 24.
2. Warning light should turn On. If not, check bulb and wiring.

Buzzer

1. Connect jumper wire from dark blue wire and ground, Fig. 24. Buzzer should come On. If not, check wiring and buzzer.

Interlock Unit

1. If warning lamp, buzzer and wiring check OK, replace interlock.

Step 2—Engine Does Not Crank Following Actuation of Underhood Switch

Underhood Switch

1. Remove connector from underhood switch. Sit in driver's seat and buckle belt. Turn ignition key to Start. If engine does not crank, check neutral switch (step 4). If neutral switch is OK, check starter system. If engine does crank, then with engine not running, turn ignition switch On; test voltage with light between black wire ning, turn ignition switch On; on 1974 vehicles, check voltage with light between black wire and dark blue wire

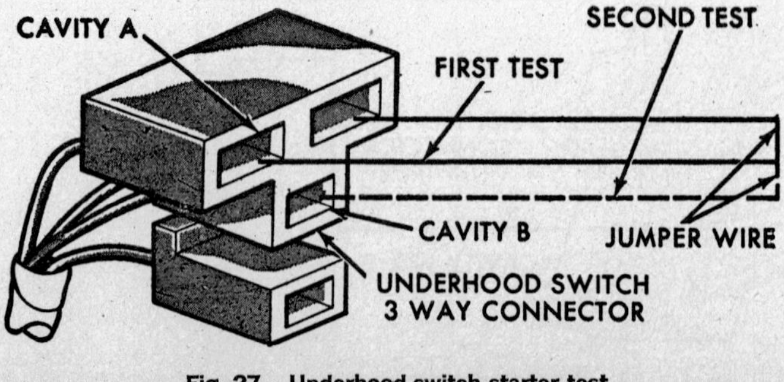

Fig. 27 Underhood switch starter test

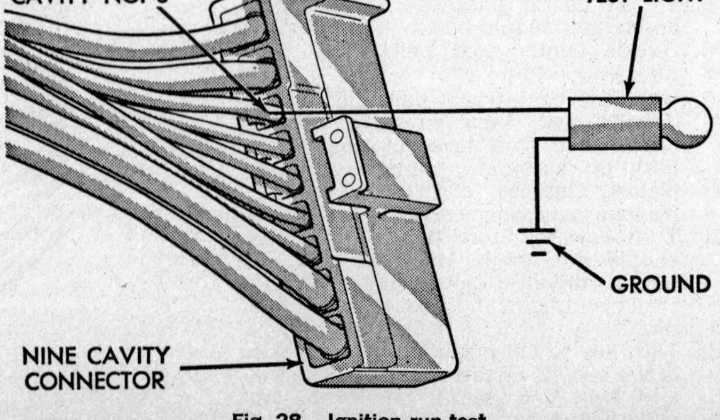

Fig. 28 Ignition run test

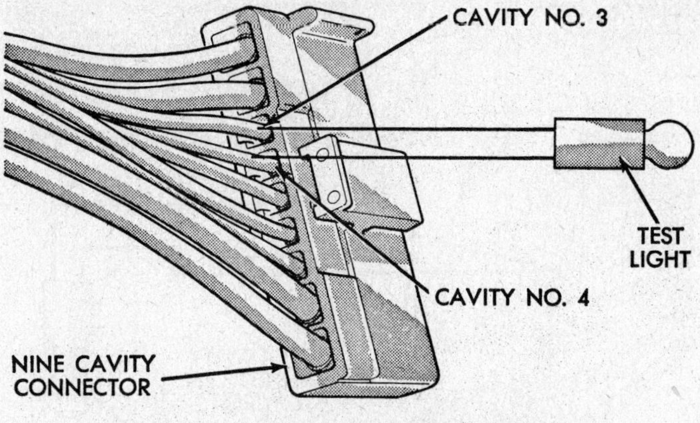

Fig. 29 Neutral start or parking brake test

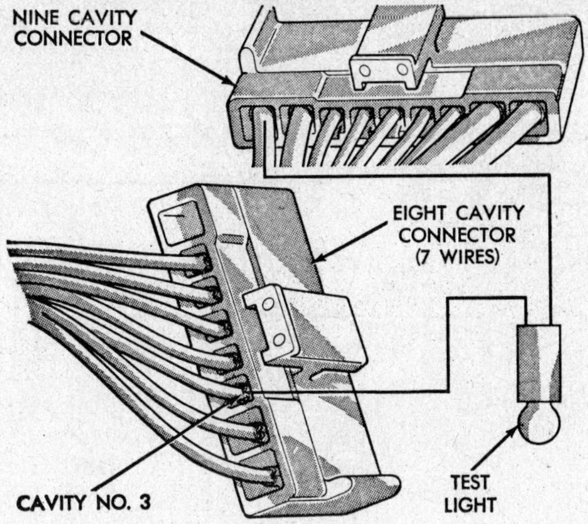

Fig. 30 Driver's buckle switch test

and on 1975 vehicles, check voltage with light between black wire and brown wire, Fig. 25. If test lamp does not light, an open circuit exists. Check wiring.

2. On 1974 vehicles, proceed as follows:
   a. Use test lead and momentarily jump between dark blue wire and first one and then the other yellow wire, Fig. 26. The engine should crank. If not, check wiring.
   b. Connect test light from yellow wire that did not activate starter to ground. Test lamp should light when key is turned to Start. If test lamp does not light the problem is in wiring. If no wiring problem is evident, replace underhood switch.
3. On 1975 vehicles, proceed as follows:
   a. Use test light and momentarily connect jumper between brown wire and cavity A, Fig. 27. Engine should crank. If not, check wiring to starter relay.
   b. Connect test light from cavity B to a good ground, Fig. 27. Test lamp should light when ignition key is turned to start position. If not, problem is in wring. If no wiring problem is evident, replace underhood switch.

Step 2A—Buzzer And Light Not On
See Step 1A.

Step 3—Engine Cranks When it Should Not. Driver Belt Buckled Before Being Seated

Underhood Switch

1. Remove connector at underhood switch. Return to prebuckled driver's seat and turn key to Start. If engine will not now crank, reconnect underhood switch and repeat checkout steps 2 and 3. Make sure car runs for 30 seconds minimum followed by the ignition Off for at least $2\frac{1}{2}$ minutes. If engine still cranks, replace underhood switch.

Wiring

Use step 1.

Driver's Seat Switch

Use Step 1.

Interlock Unit

Use Step 1.

Step 3A—Buzzer And Warning Light Not On
See Step 1A.

Step 4—Light And Buzzer Do Not Go On When Ignition On, Car Is In Gear, and Driver Belt Buckled Before Being Seated

Drivers Seat Switch

Use step 1.

Wiring

Use step 1.

Ignition Run Input

1. Check system fuse. If OK, remove both connectors from interlock unit. Connect test light to dark blue wire with white tracer in 9-way connector to a good ground, Fig. 28. With ignition On, test light should be On. If not, trace wiring for open circuit.

Neutral Start or Parking Brake Switch

1. Connect test light between dark blue wire with white tracer and brown wire with yellow tracer. Ignition On, Fig. 29.
2. For automatic transmission cars, when selector is in Neutral or Park, test light should turn On and when selector is in gear test light should turn Off. For manual transmission cars, when parking brake is applied test light should turn On. When parking brake is released, test light should turn Off. If not, check wiring and switches.

Interlock Unit

1. Ignition run and neutral start or parking brake switches check OK, replace interlock unit.

Step 5—Driver Unbuckled with Ignition On and Transmission in Gear (Parking Brake Released on Manual Transmission) Does Not Activate Warning

1974 Vehicles

Interlock Unit

1. Replace interlock unit.

Step 5A—Warning Does Not Stop After Driver Properly Buckled

Driver's Buckle Switch

1. Remove both connectors from interlock unit. Connect test light to red wire in 9-way connector and to light green wire in 8-way connector, Fig. 30. When belt is unbuckled test light should be Off and when belt is buckled test light should be On. If switch does not operate use a self powered test light to check switch operation at connector attached to switch and located near belt anchor. If switch does not operate properly, replace the buckle webbing and switch assembly. If switch does operate, trouble is in wiring.

Right Passenger Seat Switch

See Step 6.

Center Seat Switch

See Step 9.

Interlock Unit

1. If switch and wiring check OK, replace interlock unit.

Step 6—Right Front Passenger Buckled Before Sitting With Ignition On And Car in Gear Does Not Activate Warning

Right Passenger Seat Switch

1. Remove both connectors from interlock. Connect test light to red wire in 9-way connector and dark green wire in 8-way connector, Fig. 31. Test light should be Off when seat is empty and On when seat is occupied. If switch does not operate use a self powered test light to check switch operation at connector attached to switch and

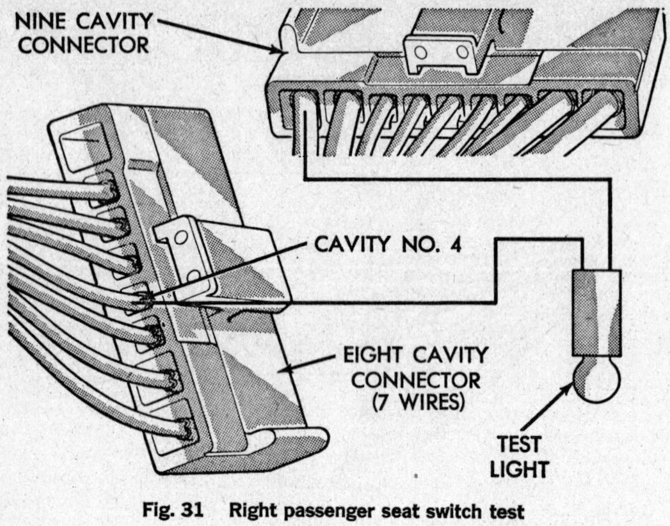

Fig. 31 Right passenger seat switch test

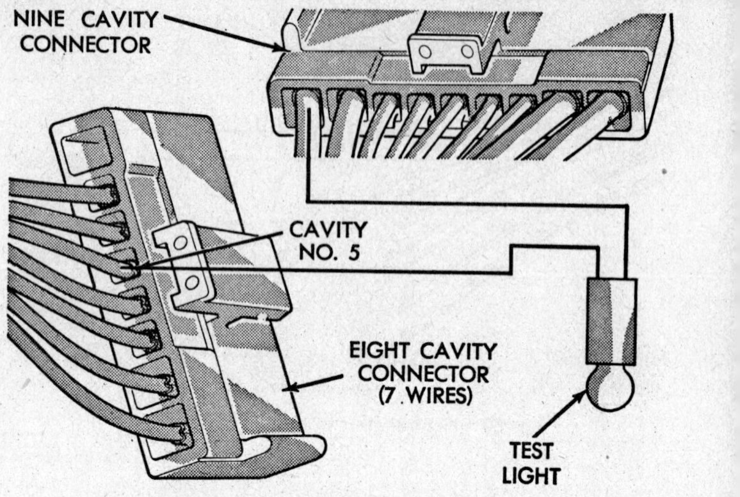

Fig. 32 Right passenger buckle switch test

located under seat. If switch still does not operate, replace switch. If switch now checks OK, problem is in wiring.

Interlock Unit

1. If switch and wiring check OK, replace interlock unit.

Step 7—Right Front Passenger Unbuckled With Ignition On and Car In Gear Does Not Activate Warning

Seat Switch

See Step 6.

Interlock Unit

1. Replace interlock unit.

Step 8—Warning Does Not Stop After Right Seat Passenger Properly Buckles

Right Seat Buckle Switch

1. Remove both connectors from interlock unit. Connect test light to red wire in 9-way connector and red wire with white tracer in 8-way connector, Fig. 32.
2. When belt is unbuckled test light should be Off and when belt is buckled test light should be On. If buckle switch does not operate use a self powered test light to check switch operation at switch connector located at rear of seat near belt anchor. If switch still does not operate, replace belt, webbing and switch assembly. If switch now checks OK, problem is in wiring.

Driver's Seat Switch

See Step 2.

Center Seat Switch

See Step 9.

Interlock Unit

1. If switch and wiring check OK, replace interlock unit.

Step 9—Center Passenger Buckled Before Sitting with Ignition On and Car in Gear Does Not Activate Warning

Center Passenger Seat Switch

1. Remove both connectors from interlock unit. Connect a test light to red wire in 9-way connector and tan wire in 8-way connector, Fig. 33. When seat is empty test light should be Off and when seat is occupied test light should be On. If switch does not operate use a self powered test light to check switch operation at connector attached to switch and located under seat. If switch still does not operate properly, replace switch. If switch now checks OK, problem is in wiring.

Interlock Unit

1. If switch and wiring check OK, replace interlock unit.

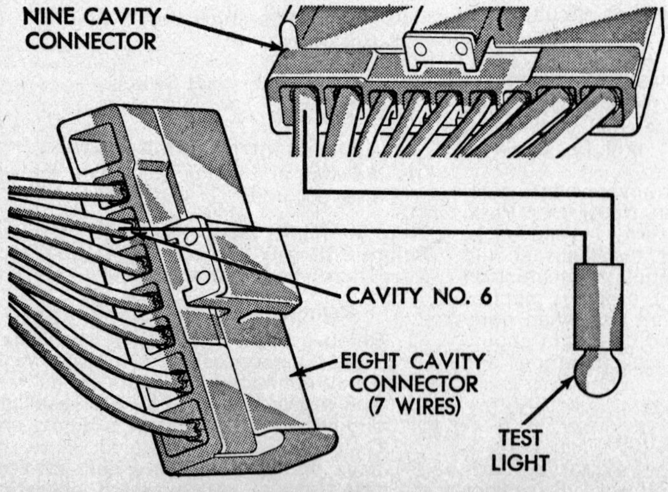

Fig. 33 Center passenger seat switch test

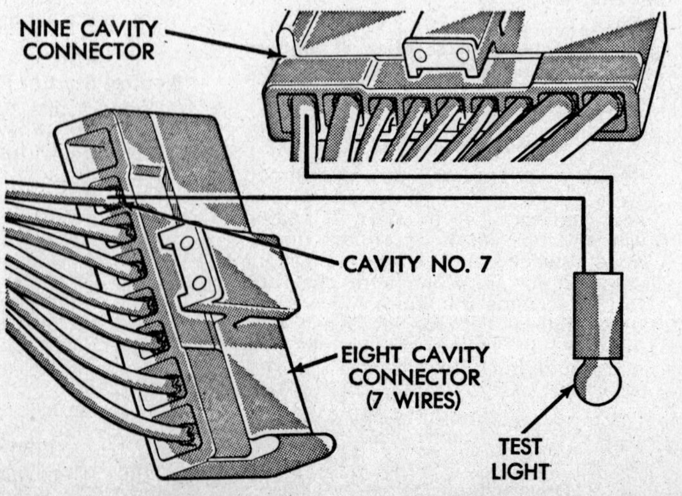

Fig. 34 Center passenger buckle switch test

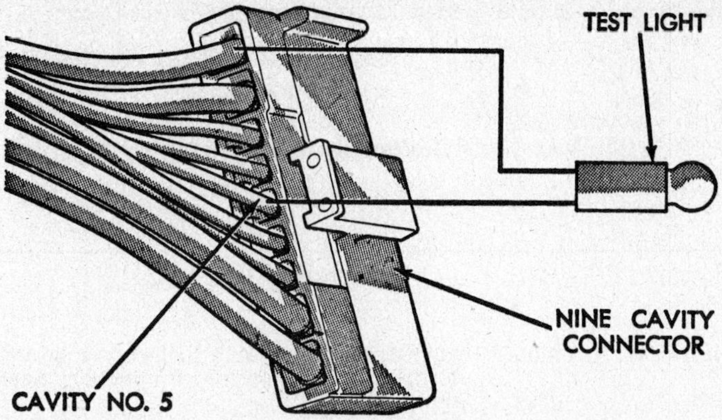

Fig. 35 Door switch test

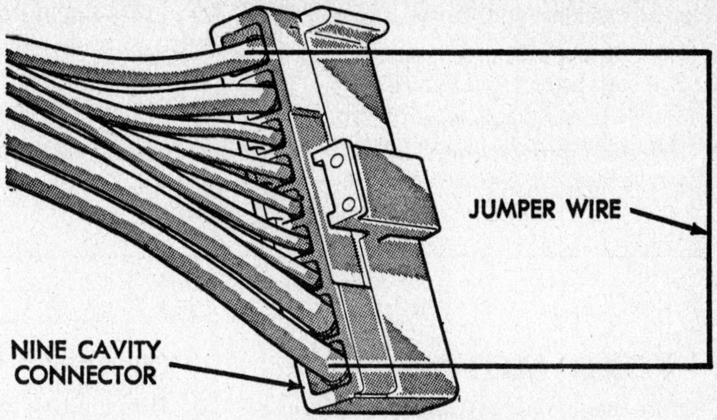

Fig. 36 Interlock starter test

Step 10—Center Seat Passenger Unbuckled with Ignition On and Car in Gear, Warning Not Activated

Interlock Unit

1. Replace interlock unit.

Step 11—Warning Does Not Stop After Center Passenger Properly Buckles

Center Seat Buckle Switch

1. Remove both connectors from interlock unit. Connect test light to red wire in 9-way connector and orange wire in 8-way connector, Fig. 34.
2. Test light should be Off only when belt is unbuckled. If switch does not operate use a self powered test light to check operation of switch at connector attached to switch and located at rear of seat near belt anchor. If switch still does not operate properly, replace buckle, webbing and switch assembly. If switch does operate, problem is in wiring.

Driver's Seat Switch

See Step 1.

Right Passenger Seat Switch

See Step 6.

Interlock Unit

1. If switch and wiring check OK, replace interlock unit.

Step 12—Engine Cranks When it Should Not (Driver Prebuckled)

Door Switch

1. Remove both connectors from interlock unit. Connect test light to red wire and the black wire with white tracer, both in the 9-way connector. Test light should be On when door is open and Off when door is closed, Fig. 35.

See Step 1

1. If wiring and door switch are OK, replace interlock unit.

Step 12A—Engine Does Not Crank When Driver Properly Buckled

See Step 1

See Step 5A.

Wiring

1. Remove both connectors from interlock unit. Using jumper wire, momentarily connect red wire to yellow wire with tracer, Fig. 36, and starter should engage.
2. Engine does not crank. Check for voltage with test light between red wire and ground. If no voltage present, check wiring for open circuit. (Red wire is spliced into Battery circuit). If voltage is present, check yellow wire with tracer for an open circuit (wire goes to starter relay). If wiring checks OK, trouble is in starter system.
3. Engine Cranks. Check for voltage with test light between yellow wire and ground, Fig. 37. Light should be On only when key is turned to Start. If not, check circuit to ignition switch. Check for voltage between brown wire and ground, Fig. 38. Light should be On when key is in Start or Run. If not, check circuit to ignition switch.

Step 12B—Warning Occurs When Driver Properly Buckled

See step 1A for 1975 vehicles.

Interlock Unit

1. Replace interlock unit on 1974 vehicles.

Step 13—Car Does Not Restart When Driver Unbuckles But Does Not Leave Seat

Interlock Unit

1. Replace interlock unit.

Step 13A—Warning Does Not Occur When

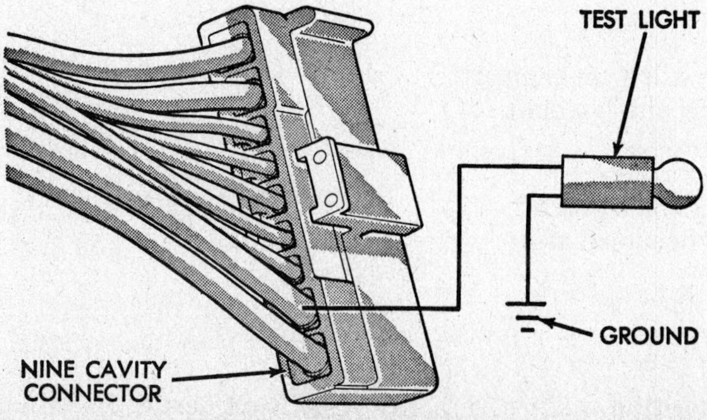

Fig. 37 Starter voltage test

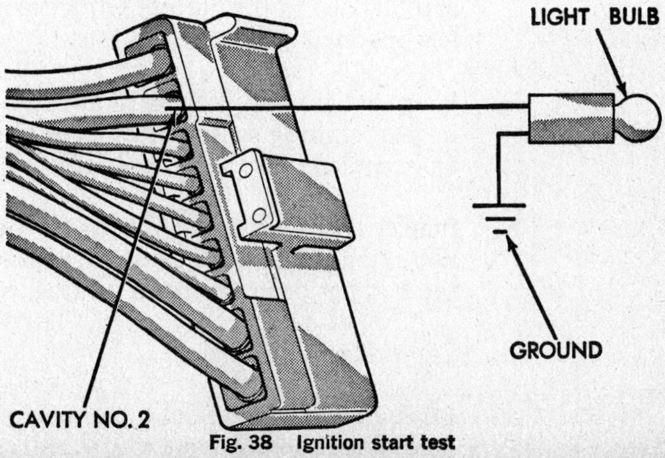

Fig. 38 Ignition start test

Driver Unbuckles and Turns Key to Start

Driver's Buckle Switch

1. See step 5 for 1975 vehicles.

Interlock Unit

1. Replace interlock unit.

Step 14—Engine Cranks When it Should Not

Interlock Unit

1. See step 1 for 1975 vehicles.
2. Replace interlock unit on 1974 vehicles.

Step 14A—Warning Does Not Occur

See step 1A for 1975 vehicles.

Interlock Unit

1. Replace interlock unit.

FORD MOTOR CO.

Before beginning to diagnose and correct problems that may be related to the Seat Belt Interlock System, verify the problem under the conditions which the problem occured.

Some problems could be caused by operator failing to follow operating procedure or sequence requirements that have been designed into the system. The system should operate normally when the following sequence is adhered to.

Starting Vehicle

1. After sitting down, extend and buckle lap belts at all occupied front seats.
2. Place transmission selector in PARK (automatic) or NEUTRAL (manual) while starting.
3. Turn ignition to start.

NOTE: A seat position is occupied whenever a weight of 47 pounds or more is placed on it.

When the vehicle has stalled after initial start and driver has remained in seat vehicle can be restarted.

After following the vehicle operating procedure if the system does not function properly, refer to the diagnosis guide further on in this chapter.

The seat sensor, connector locations, Fig. 39 and the wiring diagrams, Figs. 40 thru 46 will assist in tracing circuits and identifying accessable test points in order to diagnose symptoms of improper operation.

1975 vehicles may be diagnosed in the same manner as 1974 vehicles except for the following:

a. Starter interlock is only activated on the outboard seats. Starter operation is affected only when the outboard seats are occupied and the seat belts have not been buckled sequentially.
b. The center seat sensor will activate only the warning buzzer and light.

Additional diagnosis for 1975 vehicles are provided in charts 7 and 8.

DIAGNOSTIC SEQUENCE CHARTS

A.	Seat belt warning light and buzzer continue to operate when transmission gear selector is placed in PARK (NEUTRAL with manual transmission).	See Chart 1
B.	Starter will not operate when lap belts have been extended and buckled in correct sequence.	See Chart 3
C.	Starter will not operate after the interlock override switch (under hood) has been pressed.	See Chart 4
D.	Warning Light and Buzzer turn on when the car is placed in gear after lap belts have been extended and buckled in the correct sequence and the starter has operated normally.	See Chart 1
E.	Warning light and Buzzer do **not** turn on when car is placed in gear, ignition switch positioned at RUN and lap belts for an occupied seat are not extended or buckled.	See Chart 1
F.	Starter operates when lap belts for an occupied seat are not extended or buckled or when seat is occupied **after** the belt has been extended or buckled.	See Chart 2
G.	Lap belt will not lock.	See Chart 5
H.	Lap belt will not extend to connecting position or will not fully retract.	See Chart 6

CHART 1

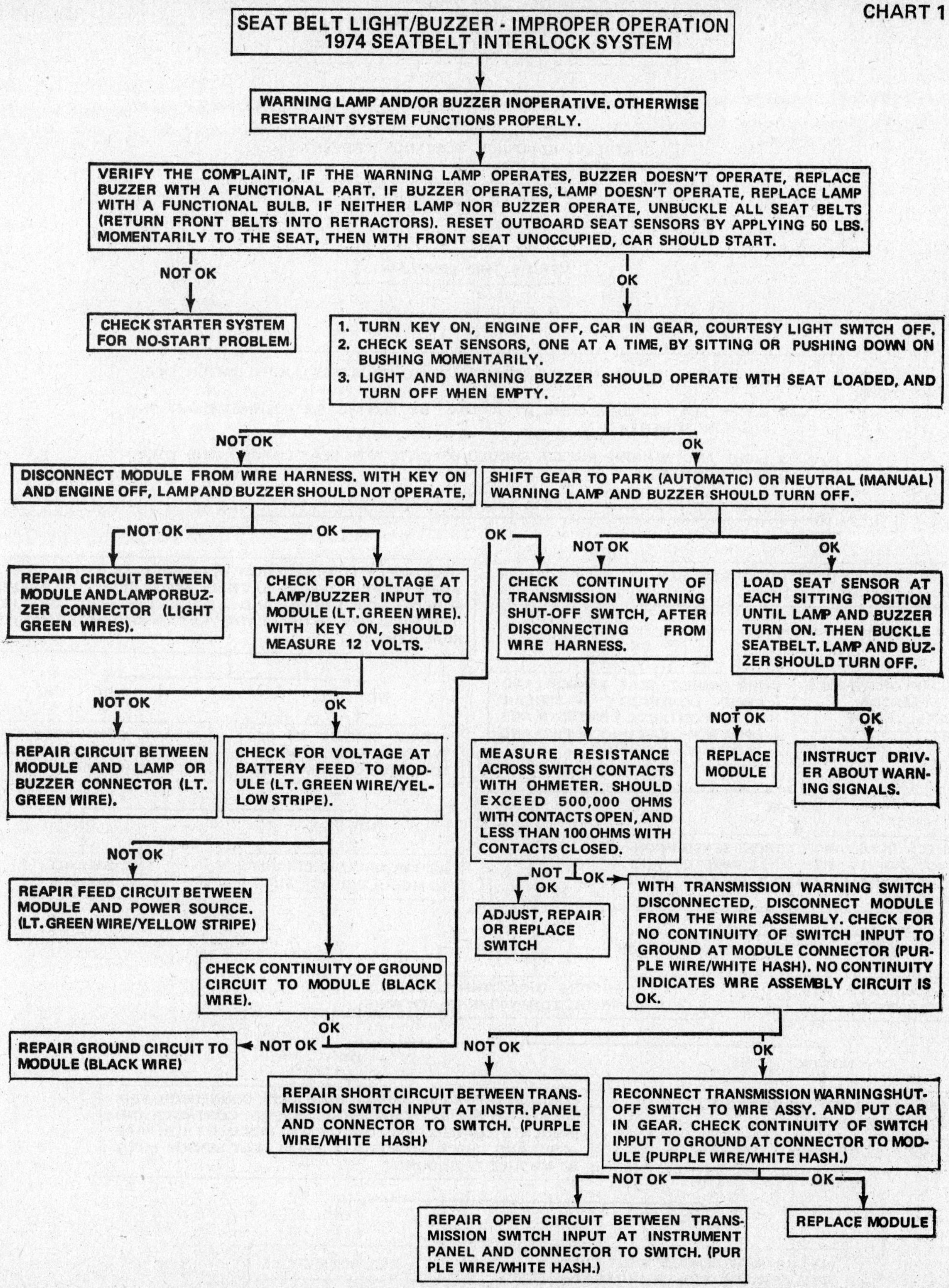
SEAT BELT LIGHT/BUZZER - IMPROPER OPERATION
1974 SEATBELT INTERLOCK SYSTEM
WARNING LAMP AND/OR BUZZER INOPERATIVE. OTHERWISE RESTRAINT SYSTEM FUNCTIONS PROPERLY.
VERIFY THE COMPLAINT, IF THE WARNING LAMP OPERATES, BUZZER DOESN'T OPERATE, REPLACE BUZZER WITH A FUNCTIONAL PART. IF BUZZER OPERATES, LAMP DOESN'T OPERATE, REPLACE LAMP WITH A FUNCTIONAL BULB. IF NEITHER LAMP NOR BUZZER OPERATE, UNBUCKLE ALL SEAT BELTS (RETURN FRONT BELTS INTO RETRACTORS). RESET OUTBOARD SEAT SENSORS BY APPLYING 50 LBS. MOMENTARILY TO THE SEAT, THEN WITH FRONT SEAT UNOCCUPIED, CAR SHOULD START.
NOT OK
CHECK STARTER SYSTEM FOR NO-START PROBLEM
OK
1. TURN KEY ON, ENGINE OFF, CAR IN GEAR, COURTESY LIGHT SWITCH OFF.
2. CHECK SEAT SENSORS, ONE AT A TIME, BY SITTING OR PUSHING DOWN ON BUSHING MOMENTARILY.
3. LIGHT AND WARNING BUZZER SHOULD OPERATE WITH SEAT LOADED, AND TURN OFF WHEN EMPTY.
NOT OK
DISCONNECT MODULE FROM WIRE HARNESS. WITH KEY ON AND ENGINE OFF, LAMP AND BUZZER SHOULD NOT OPERATE.
OK
SHIFT GEAR TO PARK (AUTOMATIC) OR NEUTRAL (MANUAL) WARNING LAMP AND BUZZER SHOULD TURN OFF.
NOT OK
REPAIR CIRCUIT BETWEEN MODULE AND LAMP OR BUZZER CONNECTOR (LIGHT GREEN WIRES).
OK
CHECK FOR VOLTAGE AT LAMP/BUZZER INPUT TO MODULE (LT. GREEN WIRE). WITH KEY ON, SHOULD MEASURE 12 VOLTS.
OK
NOT OK
CHECK CONTINUITY OF TRANSMISSION WARNING SHUT-OFF SWITCH, AFTER DISCONNECTING FROM WIRE HARNESS.
OK
LOAD SEAT SENSOR AT EACH SITTING POSITION UNTIL LAMP AND BUZZER TURN ON. THEN BUCKLE SEATBELT. LAMP AND BUZZER SHOULD TURN OFF.
NOT OK
REPAIR CIRCUIT BETWEEN MODULE AND LAMP OR BUZZER CONNECTOR (LT. GREEN WIRE).
OK
CHECK FOR VOLTAGE AT BATTERY FEED TO MODULE (LT. GREEN WIRE/YELLOW STRIPE).
MEASURE RESISTANCE ACROSS SWITCH CONTACTS WITH OHMETER. SHOULD EXCEED 500,000 OHMS WITH CONTACTS OPEN, AND LESS THAN 100 OHMS WITH CONTACTS CLOSED.
NOT OK
REPLACE MODULE
OK
INSTRUCT DRIVER ABOUT WARNING SIGNALS.
NOT OK
REAPIR FEED CIRCUIT BETWEEN MODULE AND POWER SOURCE. (LT. GREEN WIRE/YELLOW STRIPE)
NOT OK
ADJUST, REPAIR OR REPLACE SWITCH
OK
WITH TRANSMISSION WARNING SWITCH DISCONNECTED, DISCONNECT MODULE FROM THE WIRE ASSEMBLY. CHECK FOR NO CONTINUITY OF SWITCH INPUT TO GROUND AT MODULE CONNECTOR (PURPLE WIRE/WHITE HASH). NO CONTINUITY INDICATES WIRE ASSEMBLY CIRCUIT IS OK.
CHECK CONTINUITY OF GROUND CIRCUIT TO MODULE (BLACK WIRE).
OK
NOT OK
REPAIR GROUND CIRCUIT TO MODULE (BLACK WIRE)
NOT OK
REPAIR SHORT CIRCUIT BETWEEN TRANSMISSION SWITCH INPUT AT INSTR. PANEL AND CONNECTOR TO SWITCH. (PURPLE WIRE/WHITE HASH)
OK
RECONNECT TRANSMISSION WARNING SHUT-OFF SWITCH TO WIRE ASSY. AND PUT CAR IN GEAR. CHECK CONTINUITY OF SWITCH INPUT TO GROUND AT CONNECTOR TO MODULE (PURPLE WIRE/WHITE HASH.)
NOT OK
REPAIR OPEN CIRCUIT BETWEEN TRANSMISSION SWITCH INPUT AT INSTRUMENT PANEL AND CONNECTOR TO SWITCH. (PURPLE WIRE/WHITE HASH.)
OK
REPLACE MODULE

CHART 2

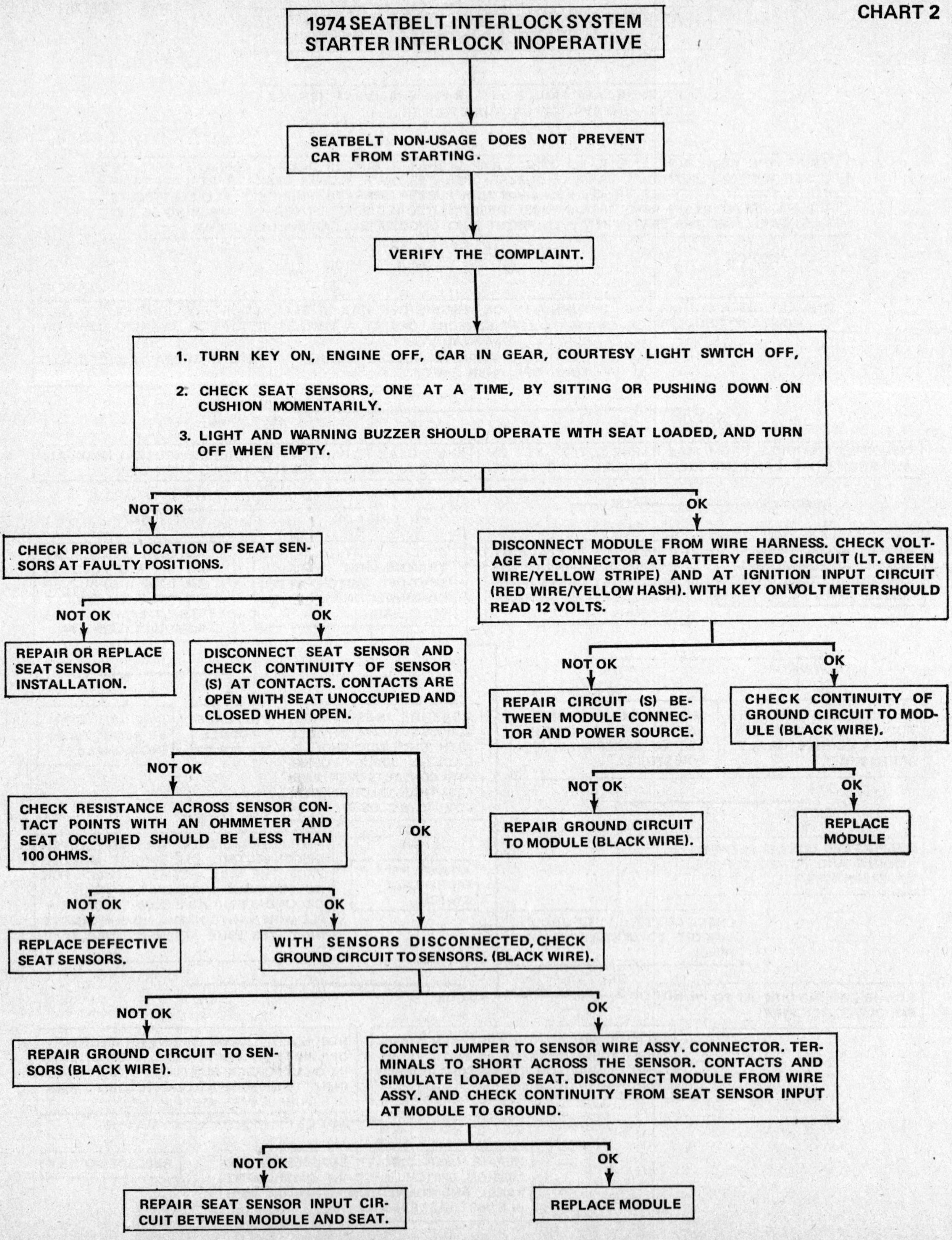
1974 SEATBELT INTERLOCK SYSTEM STARTER INTERLOCK INOPERATIVE
SEATBELT NON-USAGE DOES NOT PREVENT CAR FROM STARTING.
VERIFY THE COMPLAINT.
1. TURN KEY ON, ENGINE OFF, CAR IN GEAR, COURTESY LIGHT SWITCH OFF,
2. CHECK SEAT SENSORS, ONE AT A TIME, BY SITTING OR PUSHING DOWN ON CUSHION MOMENTARILY.
3. LIGHT AND WARNING BUZZER SHOULD OPERATE WITH SEAT LOADED, AND TURN OFF WHEN EMPTY.
NOT OK
CHECK PROPER LOCATION OF SEAT SENSORS AT FAULTY POSITIONS.
NOT OK
REPAIR OR REPLACE SEAT SENSOR INSTALLATION.
OK
DISCONNECT SEAT SENSOR AND CHECK CONTINUITY OF SENSOR (S) AT CONTACTS. CONTACTS ARE OPEN WITH SEAT UNOCCUPIED AND CLOSED WHEN OPEN.
NOT OK
CHECK RESISTANCE ACROSS SENSOR CONTACT POINTS WITH AN OHMMETER AND SEAT OCCUPIED SHOULD BE LESS THAN 100 OHMS.
NOT OK
REPLACE DEFECTIVE SEAT SENSORS.
OK
OK
WITH SENSORS DISCONNECTED, CHECK GROUND CIRCUIT TO SENSORS. (BLACK WIRE).
NOT OK
REPAIR GROUND CIRCUIT TO SENSORS (BLACK WIRE).
OK
CONNECT JUMPER TO SENSOR WIRE ASSY. CONNECTOR. TERMINALS TO SHORT ACROSS THE SENSOR. CONTACTS AND SIMULATE LOADED SEAT. DISCONNECT MODULE FROM WIRE ASSY. AND CHECK CONTINUITY FROM SEAT SENSOR INPUT AT MODULE TO GROUND.
NOT OK
REPAIR SEAT SENSOR INPUT CIRCUIT BETWEEN MODULE AND SEAT.
OK
REPLACE MODULE
OK
DISCONNECT MODULE FROM WIRE HARNESS. CHECK VOLTAGE AT CONNECTOR AT BATTERY FEED CIRCUIT (LT. GREEN WIRE/YELLOW STRIPE) AND AT IGNITION INPUT CIRCUIT (RED WIRE/YELLOW HASH). WITH KEY ON VOLT METER SHOULD READ 12 VOLTS.
NOT OK
REPAIR CIRCUIT (S) BETWEEN MODULE CONNECTOR AND POWER SOURCE.
OK
CHECK CONTINUITY OF GROUND CIRCUIT TO MODULE (BLACK WIRE).
NOT OK
REPAIR GROUND CIRCUIT TO MODULE (BLACK WIRE).
OK
REPLACE MODULE

1974 CHART 3

SEATBELT WARNING AND STARTER INTERLOCK SYSTEM
FAULT DIAGNOSIS

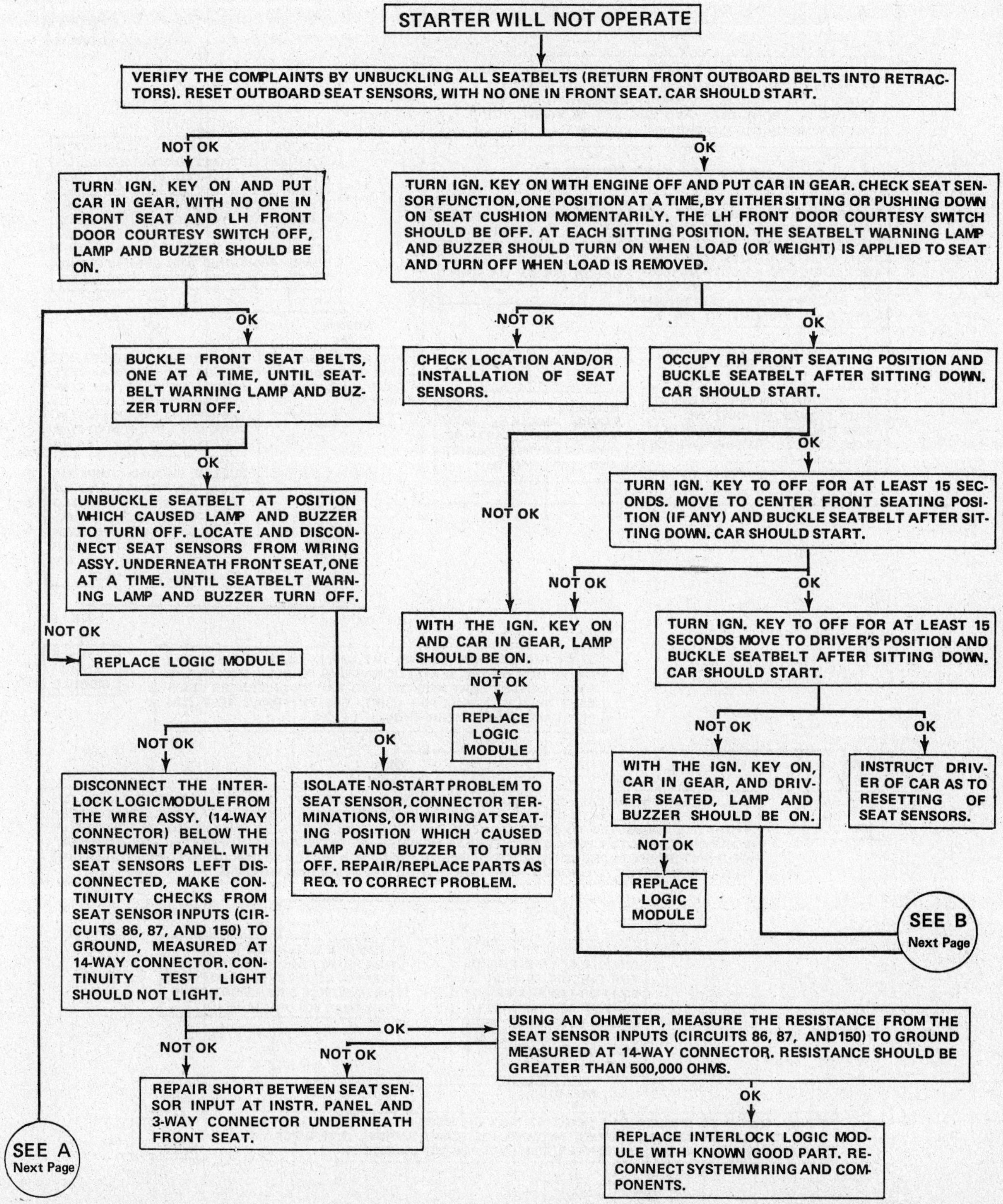

A

OK

CHECK FOR OUTPUT VOLTAGE AT PIN 8 (CIRCUIT NO. 32) OF 14-WAY CONNECTOR THAT MATES WIRE ASSY. INPUTS/ OUTPUTS TO INTERLOCK LOGIC MODULE. WITH NO ONE SITTING IN FRONT SEAT AND IGN. KEY IN START, 10.5 VOLTS MINIMUM SHOULD BE PRESENT AT PIN 8.

NOT OK

CHECK FOR VOLTAGE INPUT AT PIN 9 (CIRCUIT NO. 33) OF 14-WAY CONNECTOR THAT MATES WIRE ASSY. INPUTS/OUTPUTS TO INTERLOCK LOGIC MODULE. WITH NO ONE SITTING IN FRONT SEAT AND IGN. KEY IN START, 10.5 VOLTS MINIMUM SHOULD BE PRESENT AT PIN 9.

OK

REPAIR OPEN WIRE IN CIRCUIT NO. 32 BETWEEN INTERLOCK LOGIC MODULE AND STARTER SYSTEM ACUATOR (RELAY OR SOLENOID).

NOT OK

CHECK FOR OPEN WIRE IN START INPUT SIGNAL (CIRCUIT NO. 33) FROM IGN. SWITCH TO INTERLOCK LOGIC MODULE. REPAIR/REPLACE PARTS AS REQUIRED.

OK

REPLACE INTERLOCK LOGIC MODULE WITH KNOWN GOOD PART, RECONNECT SYSTEM WIRING AND COMPONENTS.

CHART 3 (Cont'd.)

B

REMOVE CONNECTOR TO RH/LH/CENTER SEATBELT RETRACTOR SWITCH (BUCKLE SWITCH FOR CENTER POSITION) AND CHECK OPERATION OF SWITCH USING CONTINUITY CHECKER. SWITCH MAKES DIRECT PATH TO GROUND WITH SEATBELT UNBUCKLED (AND STOWED FULLY INTO THE RETRACTORS FOR RH/LH OUTBOARD POSITIONS) AND OPENS CIRCUIT TO GROUND WITH SEATBELT BUCKLED.

NOT OK

REPLACE BELT AND RETRACTOR ASSY. (OR BUCKLE SWITCH).

OK

TEMPORARILY CONNECT ONE END OF AN EXTERNAL TEST JUMPER WIRE TO THE CONNECTOR WHICH LEADS TO THE RH/LH/CENTER SWITCH. TURN IGN. KEY ON WITH ENGINE OFF AND PUT CAR IN GEAR. SEATBELT LAMP AND BUZZER SHOULD TURN ON WHEN THE OTHER END OF THE TEST WIRE IS SHORTED TO GROUND (SEATING POSITION OCCUPIED) AND TURN OFF WHEN THE OTHER END OF THE TEST WIRE IS REMOVED FROM GROUND.

NOT OK

WITH THE IGN. KEY ON AND THE CAR IN GEAR, MONITOR THE OPERATION OF THE SEATBELT WARNING DEVICES. OCCUPY THE RH/LH/CENTER SEAT AND GROUND THE RH/LH/CENTER SEATBELT SWITCH CONNECTOR USING THE EXTERNAL TEST JUMPER LAMP AND BUZZER SHOULD BE ON.

OK

REPLACE LOGIC MODULE

NOT OK

REPLACE LOGIC MODULE

OK

DISCONNECT THE INTERLOCK LOGIC MODULE FROM THE WIRE ASSY. (14-WAY CONNECTOR) BELOW THE INSTRUMENT PANEL. WITH SEATBELT SWITCH LEFT DISCONNECTED AND THE EXTERNAL TEST JUMPER WIRE STILL CONNECTED TO GROUND THE RH/LH RETRACTOR SWITCH (OR BUCKLE SWITCH FOR CENTER POSITION) CONNECTOR. USE AN OHMMETER TO MEASURE THE RESISTANCE FROM THE RH/LH CENTER SEATBELT SWITCH INPUT (CIRCUIT 42 FOR RH. CIRCUITCIRCUIT 85 FOR LH, AND CIRCUIT 83 FOR CENTER) AT THE 14-WAY CONNECTOR TO GROUND. RESISTANCE SHOULD BE LESS THAN 100 OHMS.

NOT OK

REPAIR OPEN CIRCUIT IN CIRCUIT 42 (FOR RH SIDE FOR CIRCUIT 85 (FOR LH SIDE) OR CIRCUIT 83 (FOR CENTER) BETWEEN INSTR. PANEL INTERLOCK MODULE CONNECTOR AND RH/ LH/CENTER SEATBELT SWITCH CONNECTOR.

OK

REMOVE EXTERNAL TEST JUMPER FROM GROUND, USING OHMMETER CONNECTED AS BEFORE, RESISTANCE SHOULD BE GREATER THAN 500,000 OHMS.

NOT OK

REPAIR SHORT TO GROUND IN CIRCUIT 42 (FOR RH SIDE) OR CIRCUIT 85 (FOR LH SIDE) OR CIRCUIT 83 (FOR CENTER) BETWEEN INSTRUMENT PANEL INTERLOCK MODULE CONNECTOR AND RH/LH/CENTER SEATBELT SWITCH CONNECTOR.

OK

REPLACE LOGIC MODULE

CHART 4

STARTER WILL NOT OPERATE AFTER STARTER INTERLOCK OVER-RIDE SWITCH HAS BEEN PRESSED.

↓

VERIFY COMPLAINT

↓

PERFORM OVER-RIDE SWITCH TEST

- NOT OK → REPLACE SWITCH
- OK → PERFORM CIRCUIT CHECK-OUT AT OVER-RIDE SWITCH CONNECTOR (SEE PAGE14).
  - NOT OK → TRACE CIRCUIT WHICH CHECKS "NOT OK" TO LOCATE OPEN, SHORTED, OR GROUNDED CONDITION.
  - OK → CHECK STARTER SYSTEM FOR NO-START PROBLEM.

CHART 5

CANNOT PULL WEBBING FROM LAP BELT RETRACTOR

↓

VERIFY THE CONDITION

↓

PULL WEBBING FROM THE SHOULDER HARNESS RETRACTOR ALLOWING LAP BELT TO FULLY RETRACT BACK INTO RETRACTOR. PULL WEBBING FROM THE LAP BELT RETRACTOR.

- NOT OK MUSTANG/PINTO/MAVERICK/COMET → VISUALLY CHECK THE WEBBING INSIDE THE SNOUT (LOOK DOWN THE SNOUT OPENING FOR FOREIGN MATERIAL OR TWISTING. REMOVE TWISTING BY UNTWISTING BELT. REMOVE FOREIGN MATERIAL WITH TWEEZERS OR ANY APPROPRIATE TOOL THAT DOES NOT HAVE BURRS OR SHARP EDGES WHICH WOULD DAMAGE WEBBING. ALLOW WEBBING TO RETRACT BACK INTO RETRACTOR. WITHDRAW LAP BELT WEBBING AND BUCKLE UP WHILE IN SEAT. REPEAT FIVE TIMES TO INSURE PROPER OPERATION.
  - OK → NO FURTHER ACTION REQUIRED.
  - NOT OK → TIGHTEN THE WEBBING ON THE SPOOL IN THE RETRACTOR BY TRYING TO PULL THE WEBBING FROM THE RETRACTOR (USE BOTH HANDS TO PULL WEBBING). ALLOW WEBBING TO RETRACT BY PUSHING THE WEBBING INTO THE RETRACTOR. WITHDRAW WEBBING AND INSPECT FOR FOREIGN MATERIAL OR TWISTING. REMOVE FOREIGN MATERIAL OR TWISTING. WITHDRAW LAP BELT AND BUCKLE UP WHILE IN SEAT. REPEAT FIVE TIMES TO INSURE PROPER OPERATION.
    - NOT OK → REPLACE SEAT BELT ASSEMBLY.
    - OK → NO FURTHER ACTION REQUIRED.
- NOT OK ALL OTHER MODELS. → UNSNAP RETRACTOR COVER DOOR AND DISENGAGE LOCKING BAR FROM THE RATCHET GEAR (USE A SMALL COMMON SCREW DRIVER WHICH HAS NO BURRS OR SHARP EDGES). HOLD THE LOCKING BAR DOWN AND PULL THE WEBBING FROM THE RETRACTOR. INSPECT THE WEBBING FOR FOREIGN MATERIAL. ALLOW WEBBING TO RETRACT BACK INTO THE RETRACTOR. WITHDRAW LAP BELT WEBBING AND BUCKLE UP WHILE SITTING IN SEAT. REPEAT FIVE TIMES TO INSURE PROPER OPERATION.
  - NOT OK → REPLACE SEAT BELT ASSEMBLY.
  - OK → NO FURTHER ACTION REQUIRED.
- OK → WITHDRAW LAP BELT WEBBING AND BUCKLE UP WHILE SITTING IN SEAT. REPEAT FIVE TIMES TO INSURE PROPER OPERATION.
  - NOT OK → REPLACE SEAT BELT ASSEMBLY
  - OK → NO FURTHER ACTION REPUIRED.

CHART 6

LAP BELT RETRACTOR DOES NOT LOCK

VERIFY THE CONDITION. WITHDRAW A MINIMUM OF 18 INCHES OF WEBBING FROM LAP BELT RETRACTOR AND CHECK FOR LOCK-UP. (RETRACTOR MUST LOCK-UP AFTER 18 INCHES OF WEBBING WITHDRAWAL).

REPLACE SEAT BELT ASSEMBLY.

UNSNAP RETRACTOR COVER DOOR AND CHECK LOCKING BAR SPRING. PULL WEBBING FROM THE LAP BELT RETRACTOR AND CHECK FOR BROKEN SPRING ON LOCKING BAR BY MOVING LOCKING BAR AWAY FROM RATCHET GEAR. (USE A SMALL COMMON SCREW DRIVER WHICH HAS NO BURRS OR SHARP EDGES). SPRING SHOULD FORCE LOCKING BAR BACK INTO ENGAGEMENT WITH RATCHET GEAR.

- NOT OK: REPLACE SEAT BELT ASSEMBLY.
- OK: CHECK LOCKING BAR FOR FOREIGN MATERIAL. REMOVE FOREIGN MATERIAL. WITHDRAW LAP BELT WEBBING AND BUCKLE UP WHILE SITTING IN SEAT. REPEAT FIVE TIMES TO INSURE PROPER OPERATION.
  - NOT OK: REPLACE SEAT BELT ASSEMBLY.
  - OK: NO FURTHER ACTION REQUIRED.

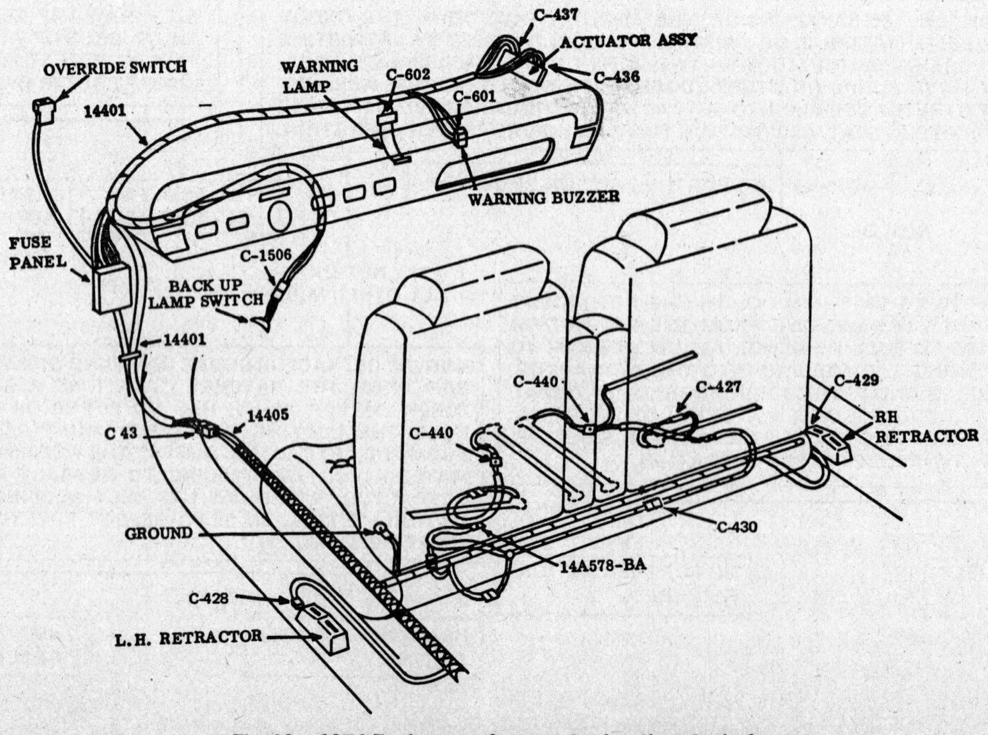

Fig. 39 1974 Ford sensor & connector locations typical

CHECKOUT PROCEDURES AT STARTER INTERLOCK OVERRIDE SWITCH CONNECTOR CHART 7

	IGNITION SWITCH POSITION	TEST OPERATION	TEST RESULT	IF RESULT IS INCORRECT
CIRCUIT 640 VOLTAGE	 ON OR START	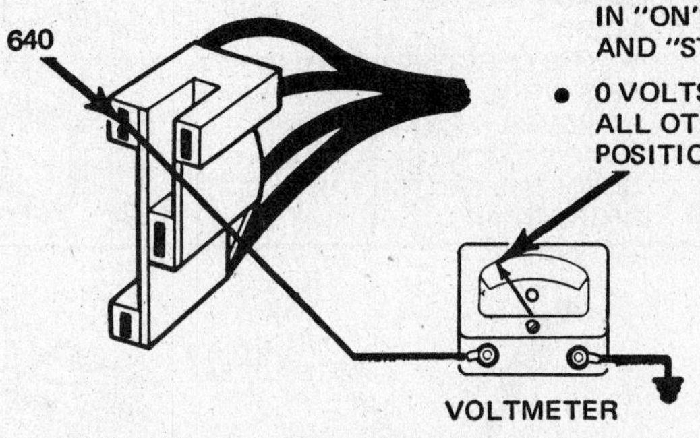	• 12 VOLTS IN "ON" AND "START" • 0 VOLTS IN ALL OTHER POSITIONS	CHECK HARNESS FOR SHORTS AND OPENS
CIRCUIT 57 CONTINUITY	* CONNECTION AT 57 ONLY FOR LINCOLN CONTINENTAL, THUNDERBIRD, AND MARK IV	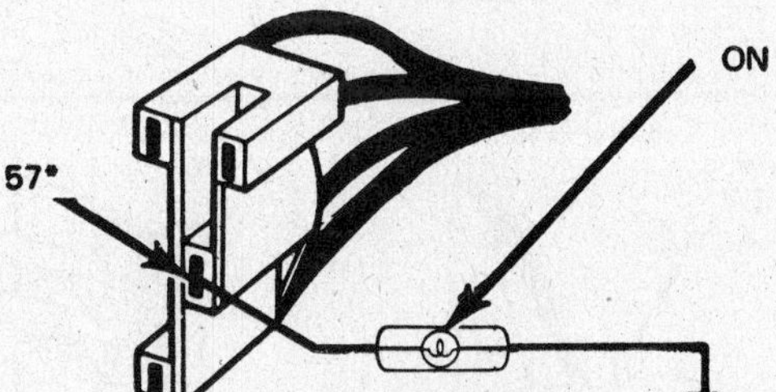		CHECK HARNESS FOR SHORTS AND OPENS
CIRCUIT 33 VOLTAGE	 START	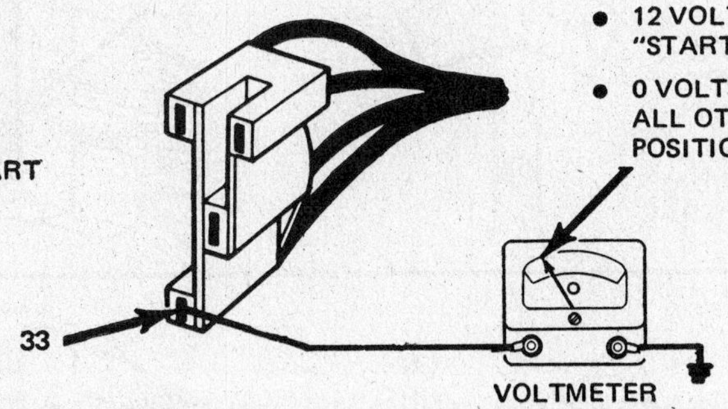	• 12 VOLTS IN "START" • 0 VOLTS IN ALL OTHER POSITIONS	CHECK HARNESS FOR SHORTS AND OPENS
CIRCUIT 32 CONTINUITY		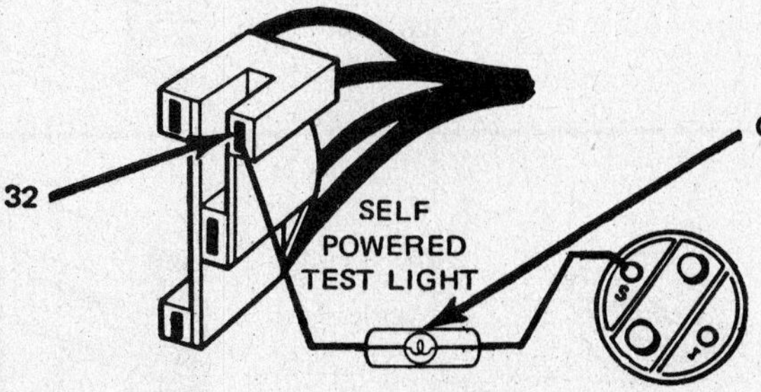		CHECK HARNESS FOR SHORTS AND OPENS

CHECKOUT PROCEDURES AT ACTUATOR MODULE HARNESS CONNECTOR CHART 8

TEST	CIRCUIT NUMBER / COLOR CODE	CHECKOUT CONDITIONS	TEST OPERATION	CORRECT RESULT	IF RESULT IS INCORRECT
(A) BATTERY INPUT (VOLTAGE)	54 — LIGHT GREEN-YELLOW STRIPE	NOTE: • PIN 5 SHOULD BE 'HOT' AT ALL POSITIONS OF THE IGNITION SWITCH-INCLUDING OFF AND LOCK. • THE CURRENT DRAIN IN THIS CIRCUIT IS SO SMALL, HOWEVER, THAT IT WOULD TAKE 2-3 MONTHS TO RUN DOWN THE BATTERY IN A STORED CAR.	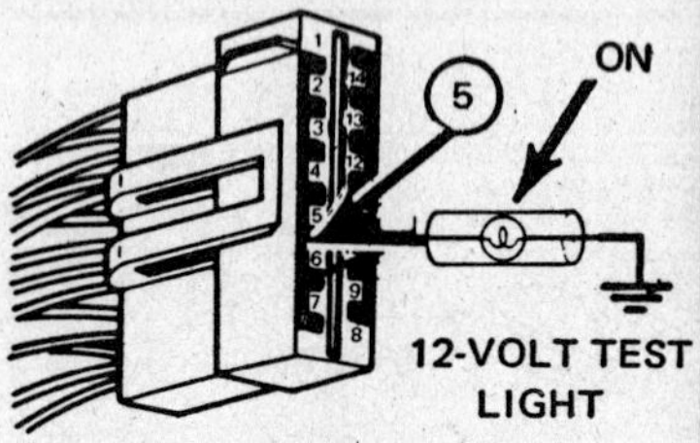 5 / 12-VOLT TEST LIGHT	ON	• TRACE CIRCUIT 54 BACK TO A JUNCTION POINT WITH OTHER CIRCUITS
(B) RIGHT-HAND LAP BELT RETRACTOR SWITCH INPUT (CONTINUITY)	42 — RED-WHITE DOT	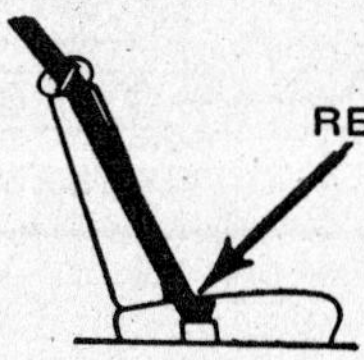 R.H. LAP BELT RETRACTED	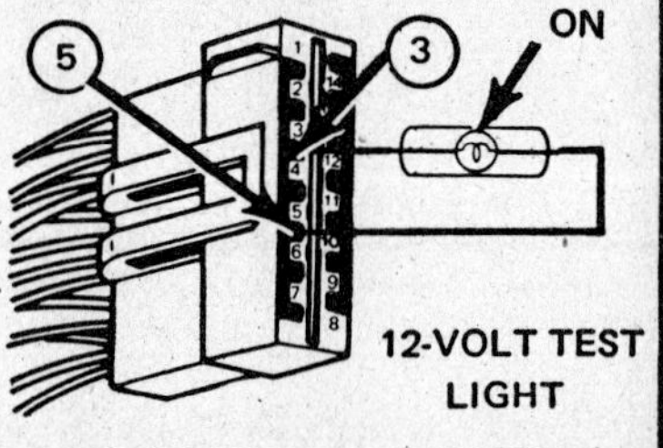 5 / 3 / 12-VOLT TEST LIGHT	ON	• IF LIGHT IS OUT OR IF LIGHT FLICKERS (VARIES IN INTENSITY), TRACE CIRCUIT 42 BACK THROUGH HARNESS CONNECTOR TO RIGHT-HAND SEAT BELT RETRACTOR SWITCH
(C) LEFT-HAND LAP BELT RETRACTOR SWITCH INPUT (CONTINUITY)	85 — BROWN-LIGHT BLUE DOT	L.H. LAP BELT RETRACTED 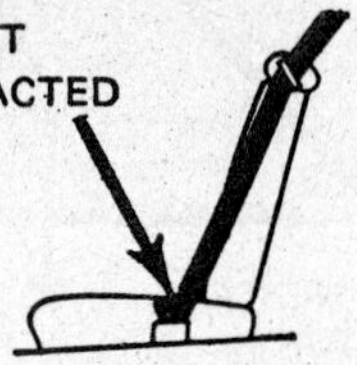	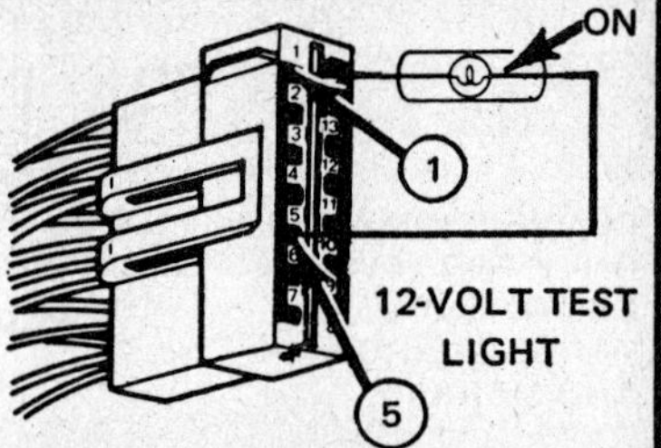 1 / 5 / 12-VOLT TEST LIGHT	ON	• IF LIGHT IS OUT OR IF LIGHT FLICKERS (VARIES IN INTENSITY), TRACE CIRCUIT 85 BACK THROUGH HARNESS CONNECTOR TO LEFT-HAND SEAT BELT RETRACTOR SWITCH
(D) TRANSMISSION SWITCH INPUT (CONTINUITY)	520 — PURPLE-WHITE HASH	AUTO-PARK / MANUAL — NEUTRAL 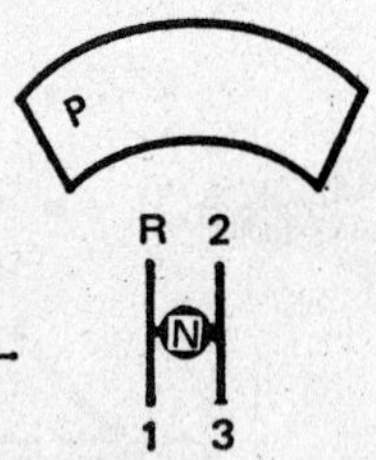 P / R 2 / N / 1 3	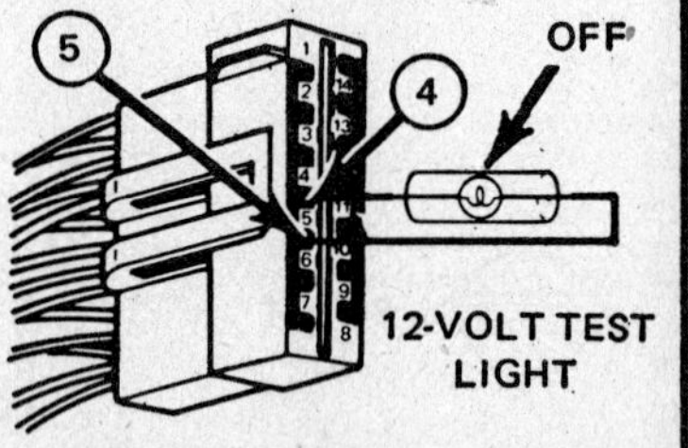 5 / 4 / 12-VOLT TEST LIGHT	OFF	AUTOMATIC — • TRACE CIRCUIT 520 BACK TO BACK-UP LAMP SWITCH MANUAL • TRACE CIRCUIT 520 BACK TO TRANSMISSION SWITCH
(E) IGNITION "ON" INPUT (VOLTAGE)	640 — RED-YELLOW HASH MARK	 ON OR START	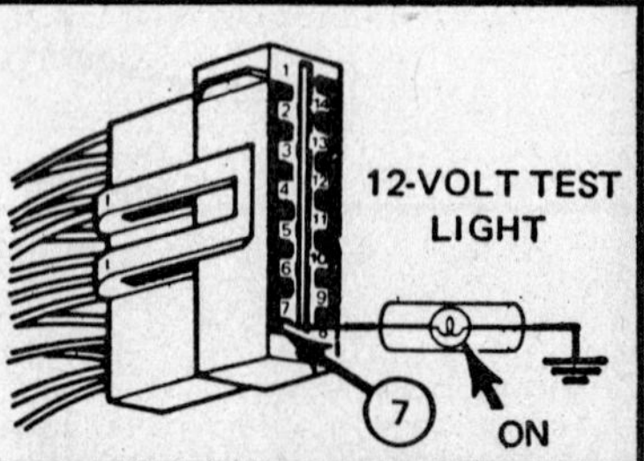 12-VOLT TEST LIGHT / 7	ON	• TRACE CIRCUIT 640 BACK TO A JUNCTION POINT WITH OTHER CIRCUITS
(F) SEAT BELT WARNING LAMP AND BUZZER INPUT (VOLTAGE)	469 — LIGHT GREEN	 ON OR START	 WARNING LIGHT / BUZZER / 10	ON	• TRACE CIRCUIT 469 BACK TO A JUNCTION POINT WITH OTHER CIRCUITS

CHECKOUT PROCEDURES AT ACTUATOR MODULE HARNESS CONNECTOR CHART 8 (cont'd)

TEST	CIRCUIT NUMBER COLOR CODE	CHECKOUT CONDITIONS	TEST OPERATION	CORRECT RESULT	IF RESULT IS INCORRECT
(G) GROUND INPUT (CONTINUITY)	57 — BLACK		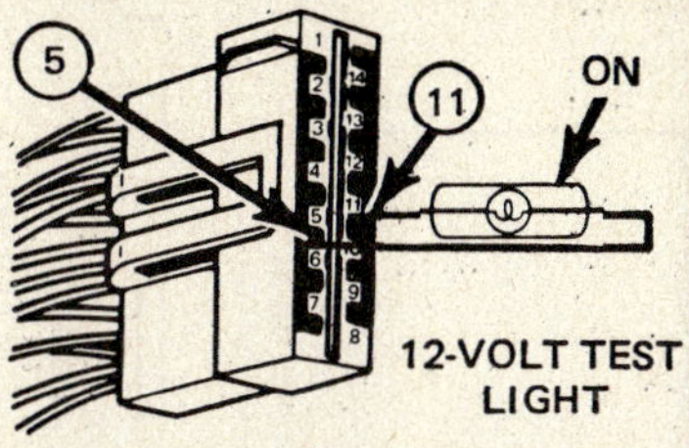 5, 11	ON 12-VOLT TEST LIGHT	• TRACE CIRCUIT 57 BACK TO A JUNCTION POINT WITH OTHER CIRCUITS
(H) RIGHT-HAND SEAT SENSOR INPUT (CONTINUITY)	86 — DARK BLUE-WHITE DOT	 RIGHT SEAT OCCUPIED	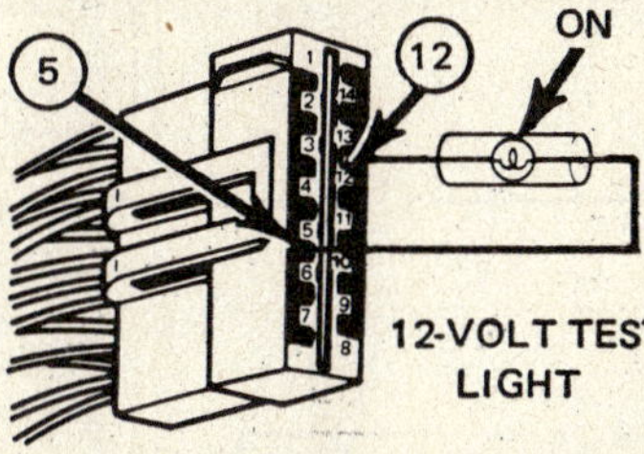 5, 12	ON 12-VOLT TEST LIGHT	• IF LIGHT IS OUT OR IF LIGHT FLICKERS (VARIES IN INTENSITY), TRACE CIRCUIT 86 BACK THROUGH HARNESS CONNECTOR TO RIGHT-HAND SEAT SENSOR
(I) LEFT-HAND SEAT SENSOR INPUT (CONTINUITY)	150 — DARK GREEN-WHITE DOT	NOT APPLICABLE TO CANADIAN VEHICLES 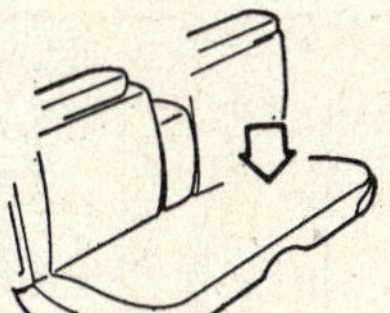 LEFT SEAT OCCUPIED	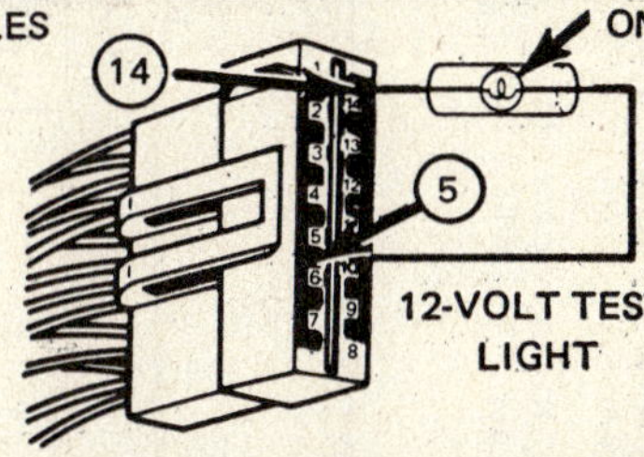 14, 5	ON 12-VOLT TEST LIGHT	• IF LIGHT IS OUT OR IF LIGHT FLICKERS (VARIES IN INTENSITY), TRACE CIRCUIT 150 BACK THROUGH HARNESS CONNECTOR TO LEFT-HAND SEAT SENSOR
(J) CENTER POSITION TEST (CONTINUITY)	83 DARK BLUE-YELLOW DOT 87 TAN-WHITE DOT 107 LIGHT BLUE-PINK DOT	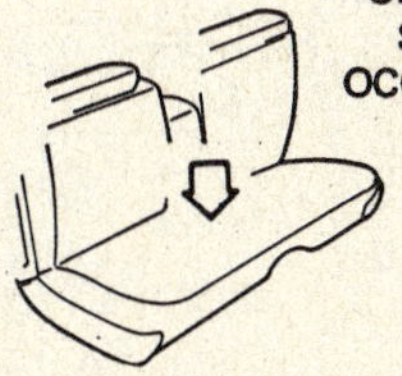 CENTER SEAT OCCUPIED CENTER SEAT BELT UNBUCKLED 	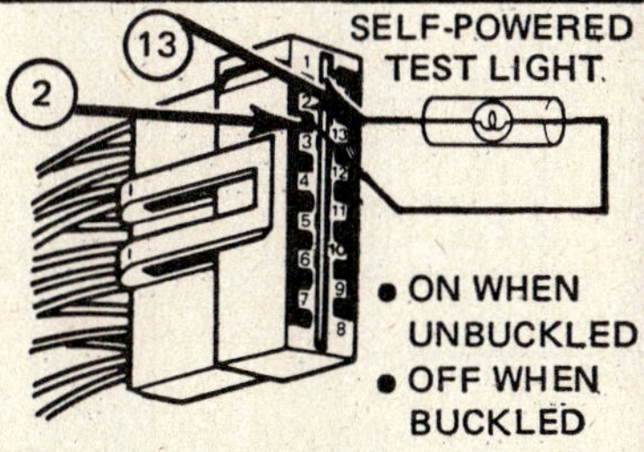 13, 2	SELF-POWERED TEST LIGHT • ON WHEN UNBUCKLED • OFF WHEN BUCKLED	• IF LIGHT IS OUT WHEN BELT IS UNBUCKLED, CHECK —CENTER SEAT SENSOR SWITCH —CIRCUIT 83 —CIRCUIT 87 —CIRCUIT 107
(K) IGNITION "START" INPUT (CONTINUITY)	33 — WHITE-PINK DOT	TRANS. IN PARK (AUTO) OR NEUTRAL (MANUAL) 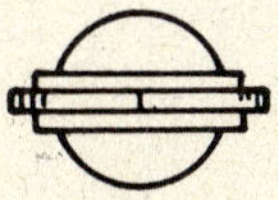 START	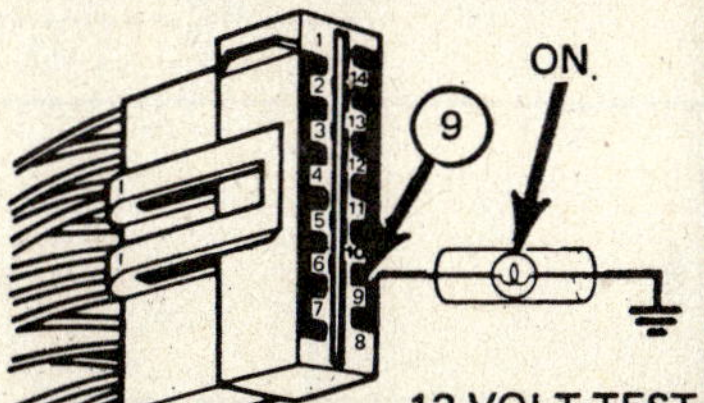 9	ON 12-VOLT TEST LIGHT	• TRACE CIRCUIT 33 BACK TO A JUNCTION POINT WITH OTHER CIRCUITS
(L) STARTER ACTUATOR OUTPUT (RESISTANCE)	32 — RED-LIGHT BLUE STRIPE	NOTE • THIS CIRCUIT CONNECTS DIRECTLY TO THE STARTER RELAY COIL • FAULTY TERMINATIONS OR AN OPEN COIL WILL AFFECT THE READINGS	 8	3 TO 8 OHMS OHMMETER	• TRACE CIRCUIT 32 BACK THROUGH HARNESS CONNECTORS

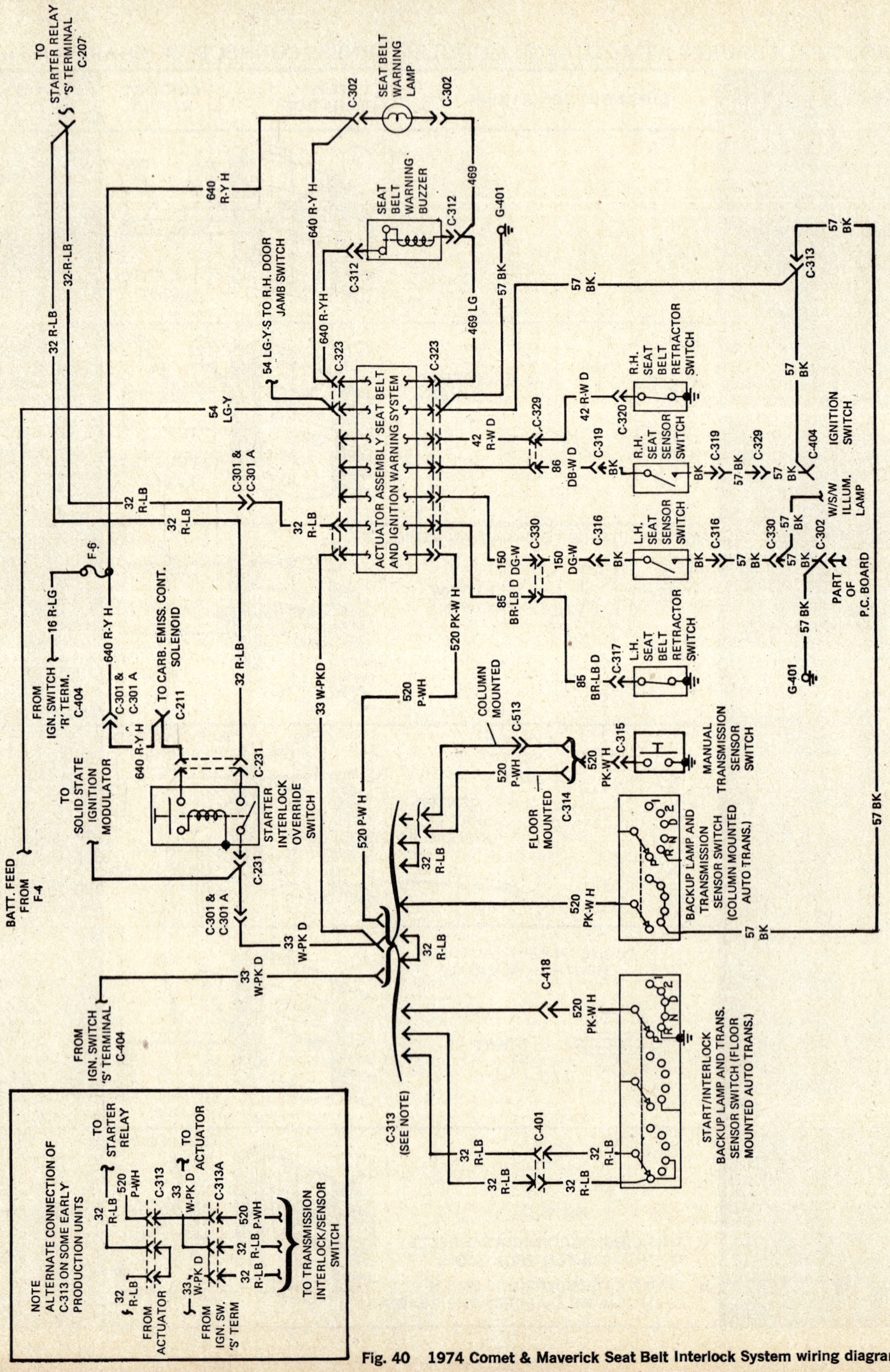

Fig. 40 1974 Comet & Maverick Seat Belt Interlock System wiring diagram

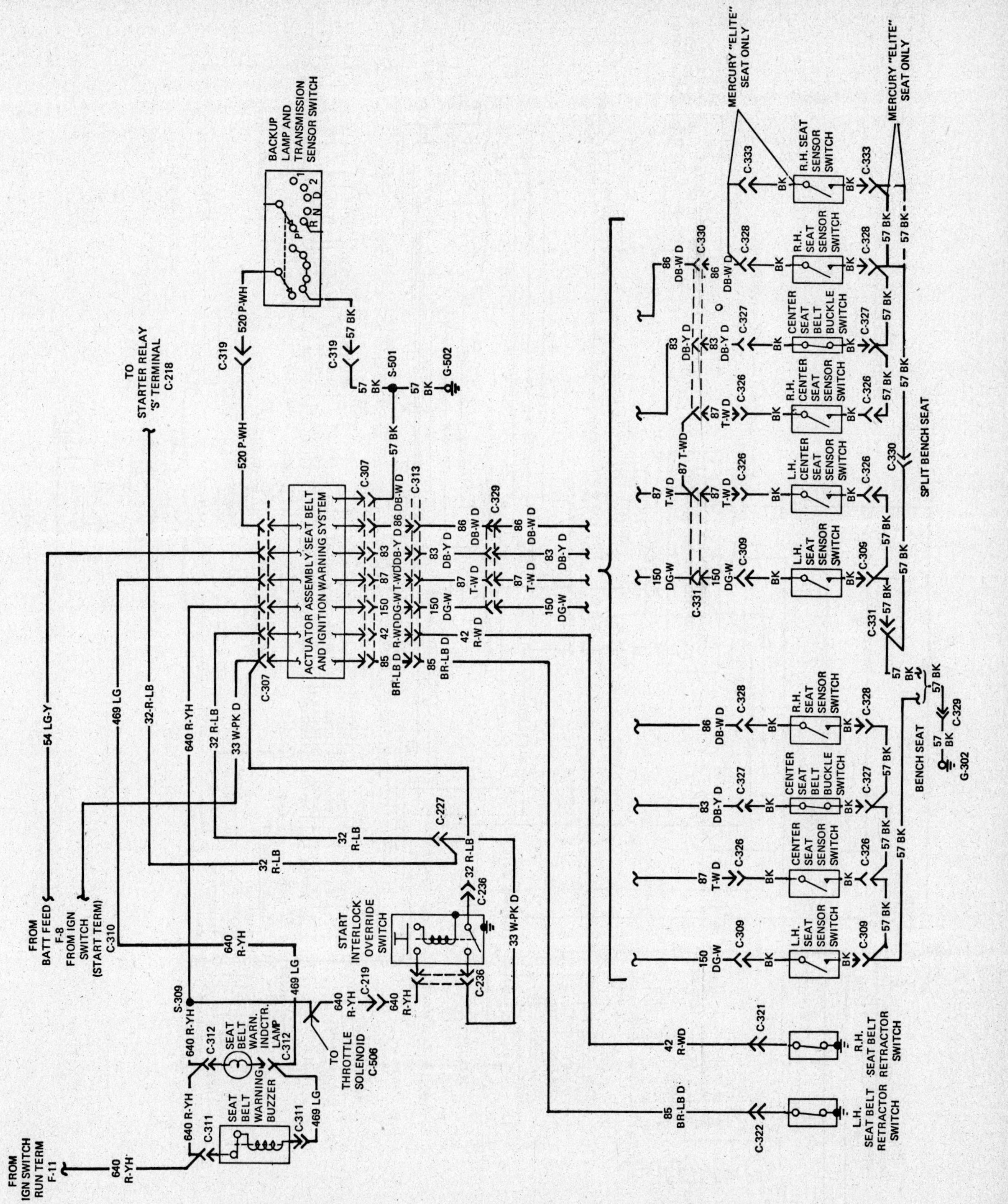

Fig. 41 1974 Ford & Mercury Seat Belt Interlock System wiring diagram

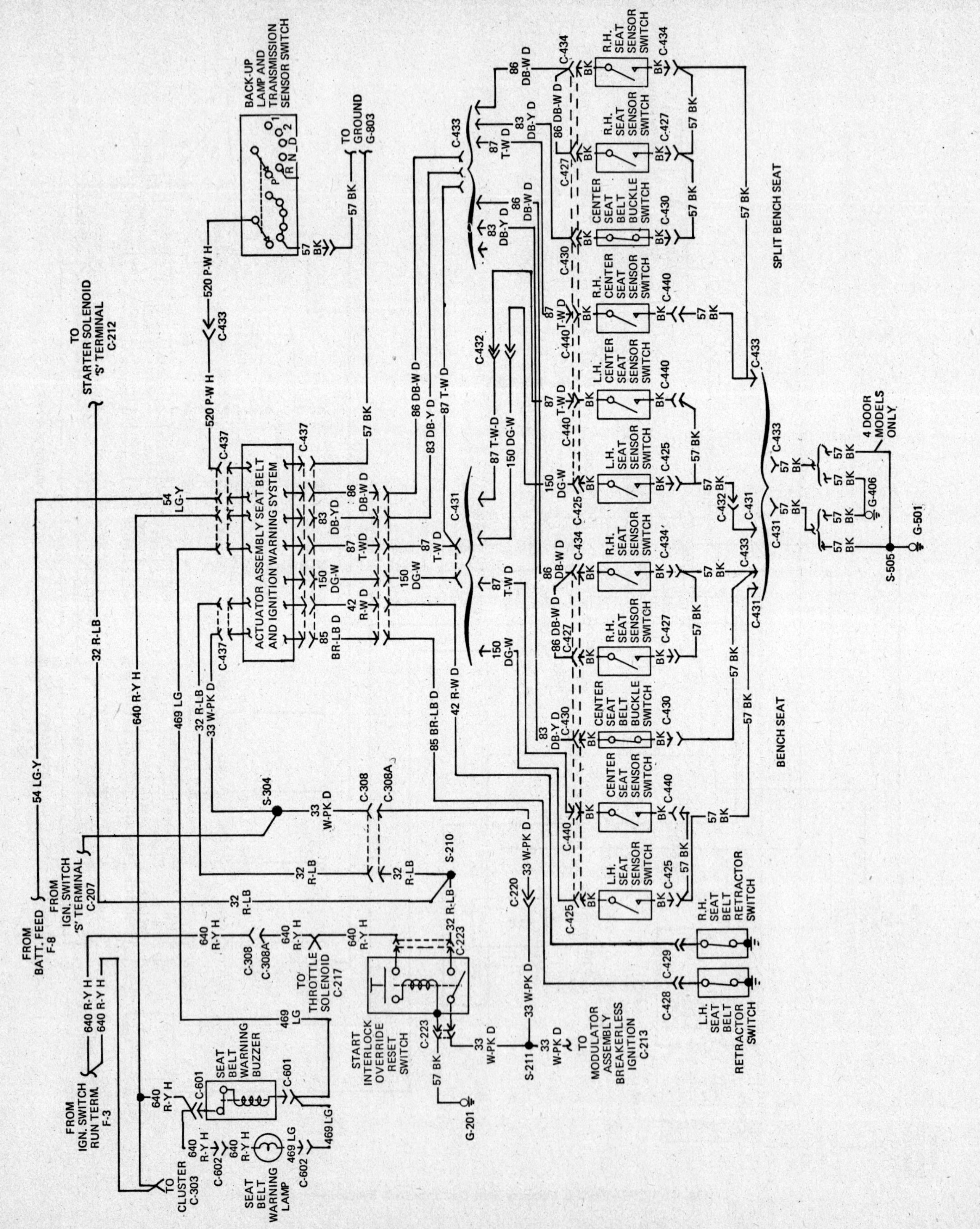

Fig. 42 1974 Lincoln Seat Belt Interlock System wiring diagram

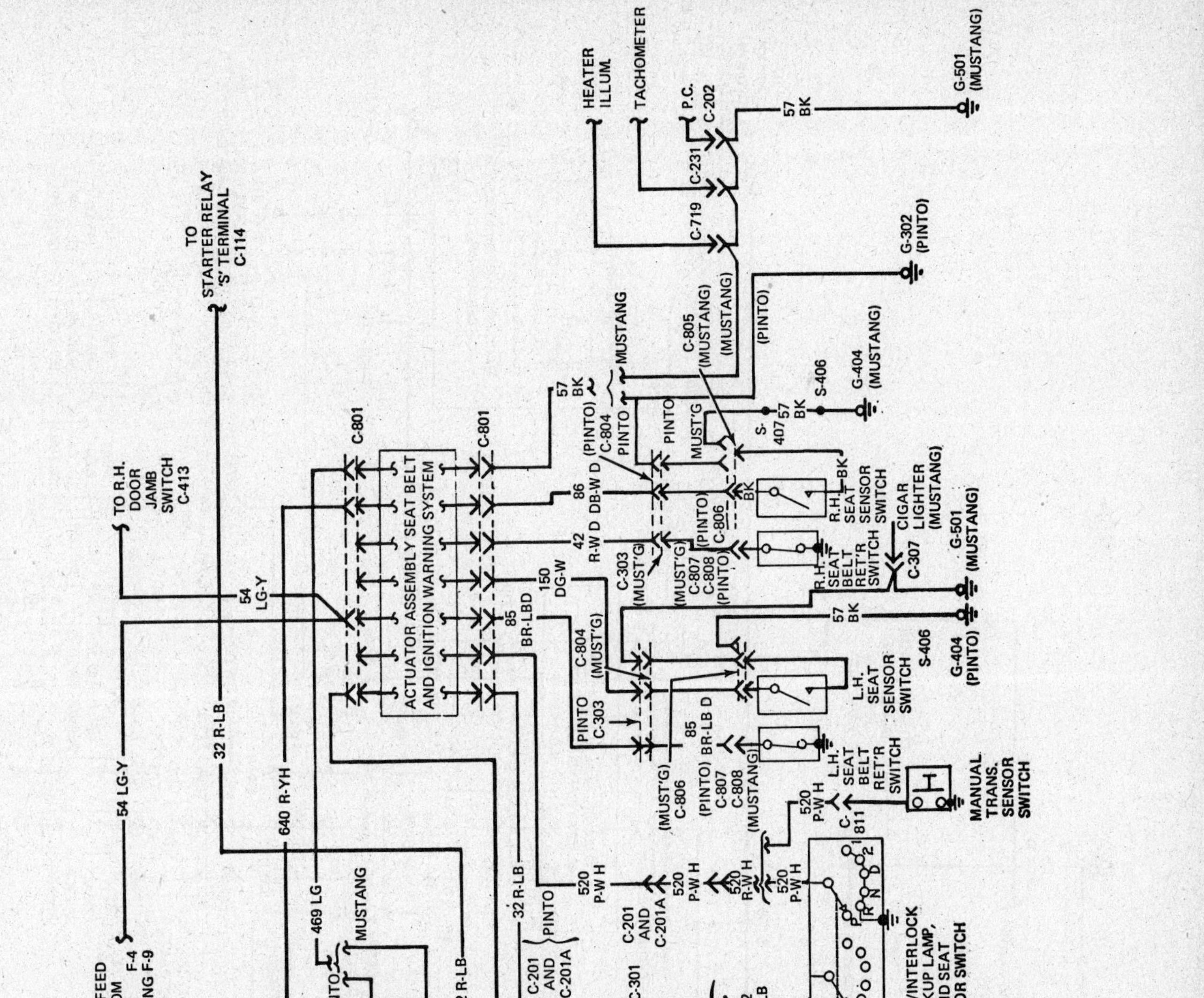

Fig. 43 1974 Mustang II & Pinto Seat Belt Interlock System wiring diagram

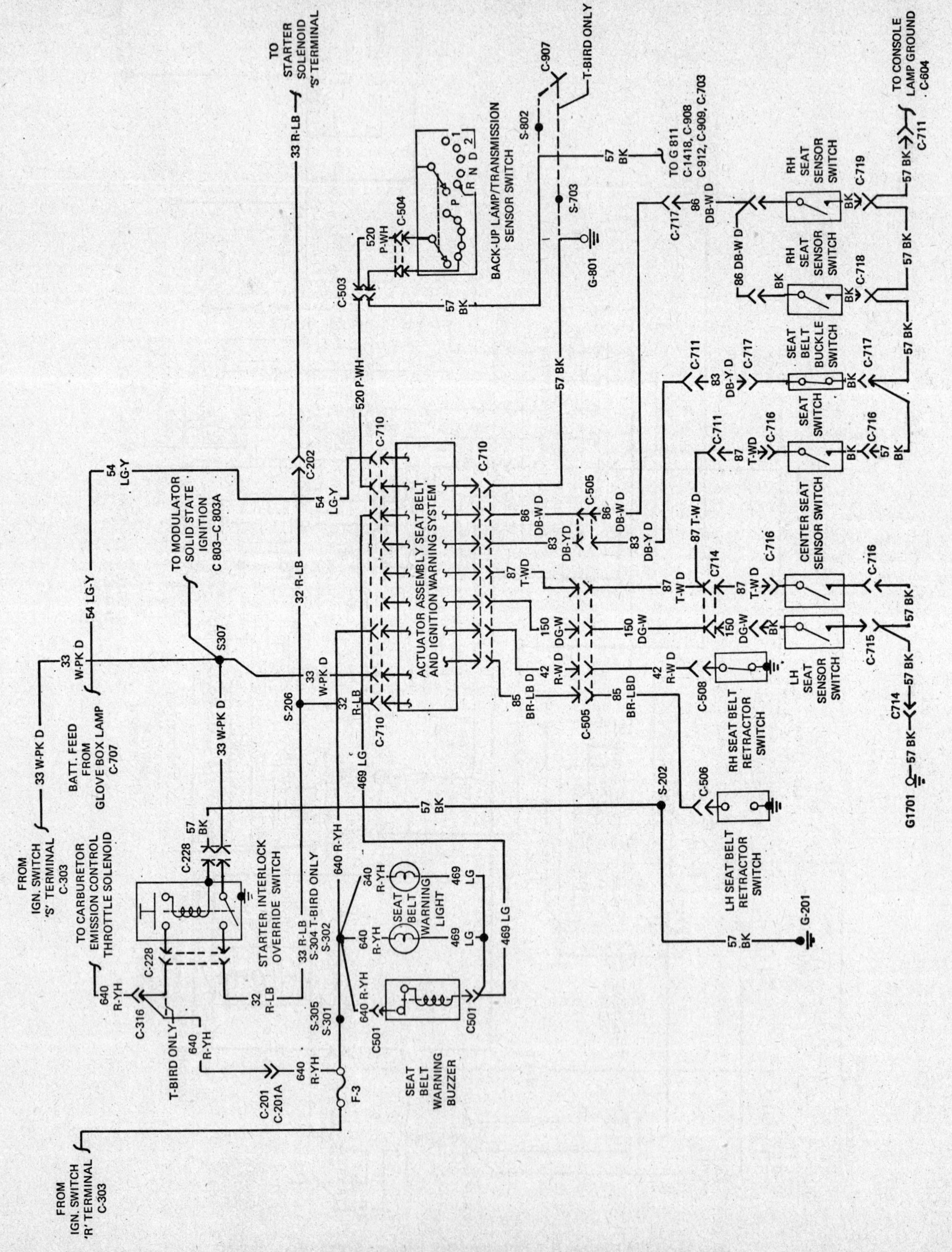

Fig. 44 1974 Mark IV & T-Bird Seat Belt Interlock System wiring diagram

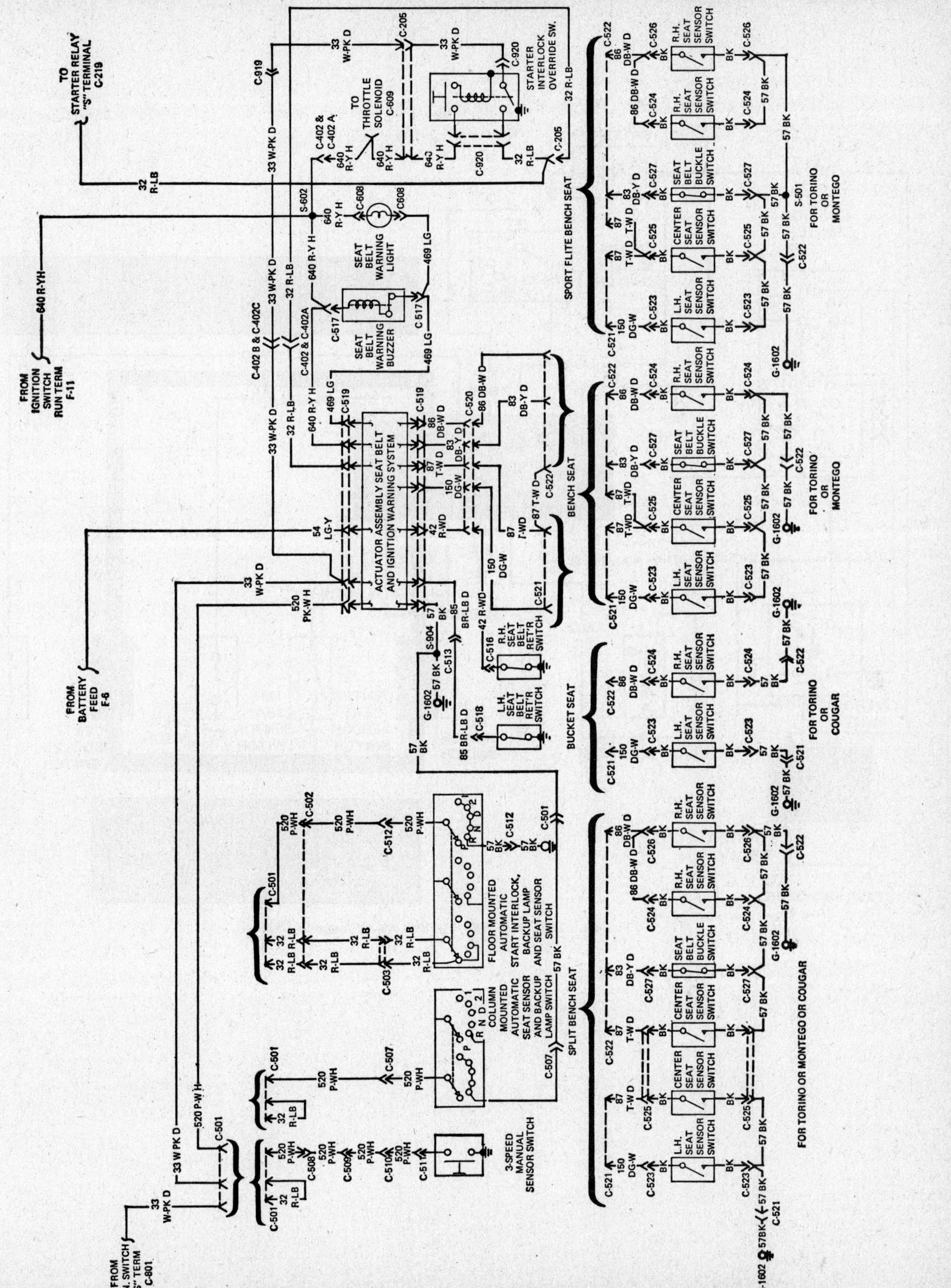

Fig. 45 1974 Torino, Montego & Cougar Seat Belt Interlock System wiring diagram

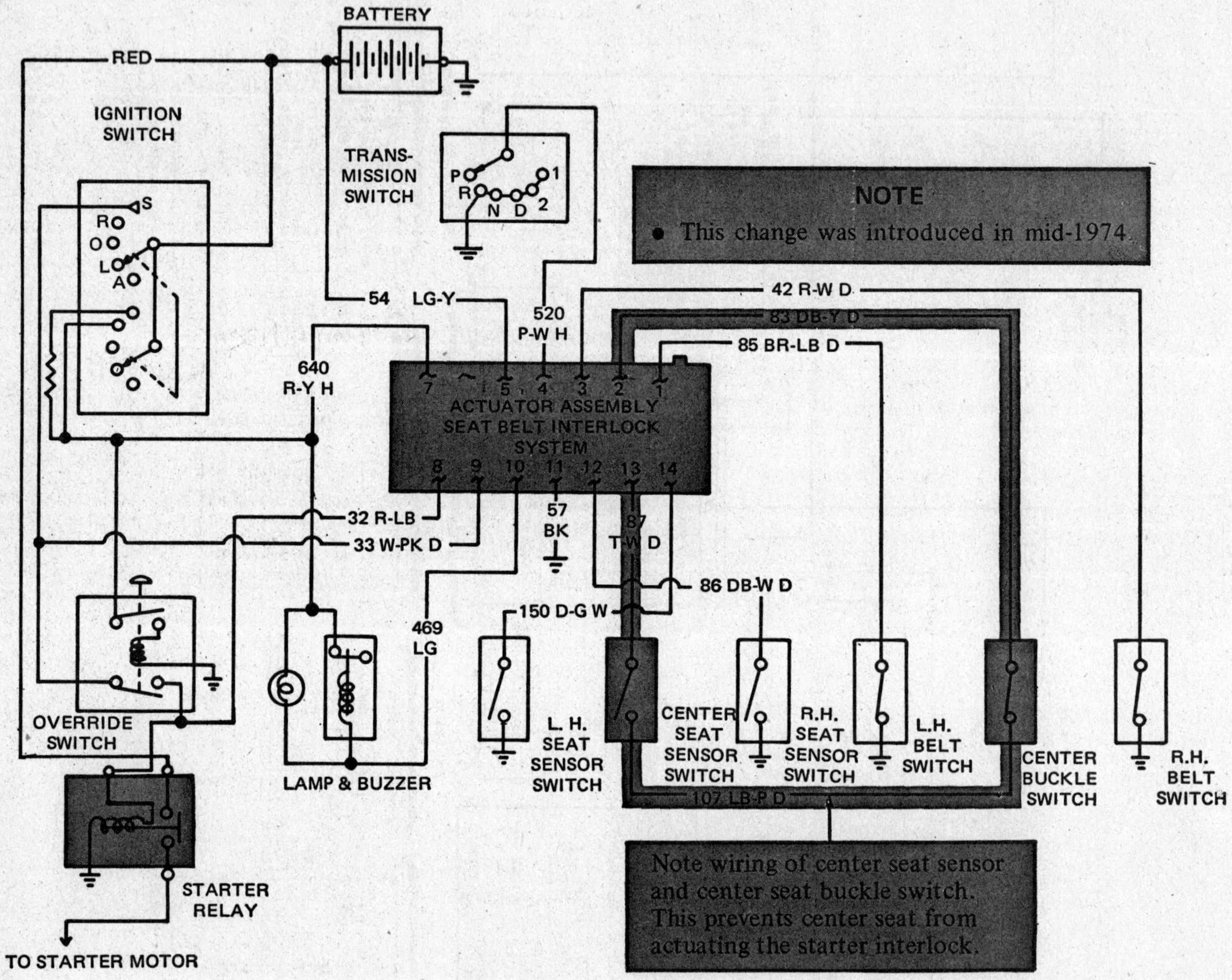

Fig. 46 1975 Seat Belt Interlock wiring diagram (typical)

GENERAL MOTORS CORP.

If the seat belts are not buckled or are buckled before the passengers in the outboard seats are seated, the logic module will activate the interlock relay which will prevent the engine from starting and will activate the warning buzzer and light. The light and buzzer will not come on if the engine is running and a front seat belt is unbuckled if the transmission is in Park or Neutral but will come on if a front seat belt is unbuckled while the engine is running and the transmission in any forward gear.

The logic module incorporates a bounce feature which prevents the interlock system from preventing a start when a buckled in passenger lifts off the seat and deactivates the sensor switch.

If the passenger is off the seat for more than 5-10 seconds the seat belts must be unbuckled and rebuckled before the engine will start.

If the car has been started using the correct starting procedure, it can be restarted with the seat belts unbuckled at front seat occupied positions as long as the driver remains seated. If the driver leaves the car, the interlock relay will activate and the step buckling procedure must be used to start the car.

Mechanic's Start

The vehicle can be started with seat belts in any position if the front seats are unoccupied. To start the engine, reach inside the car and turn the ignition key to "START" without sitting on the seat. The light and buzzer will come on when the front seat is now occupied and the transmission shifted to DRIVE, but will turn off as soon as the seat belts at occupied positions are buckled.

Bypass Relay

A bypass relay on the firewall permits starting the car by bypassing the interlock relay in the event of system failure. To activate the relay, turn the ignition key to "ON." Then, open the hood and press and release the bypass relay button.

NOTE: The override mechanism will be damaged if the button is held in the depressed position.

The engine may then be started by turning the ignition key to "START." The relay will remain engaged until the ignition key is turned to OFF or LOCK.

Electronic Module

Two types of wiring harnesses and electronic modules are used with the seat belt-starter interlock system in 1974. Both wiring harnesses are designed to by-pass the center seat sensor and allow the vehicle to start with an occupant in the center seat position without buckling the seat belt.

The first wiring harness by-passes the center seat sensor through a jumper wire while the other harness by-passes the center seat sensor through the electronic module.

The electronic module with the built-in center seat sensor by-pass can be used with either wiring harness. The module without the built in by-pass can only be used on the wiring harness that incorporates a jumper wire. To determine which wiring harness is being serviced refer to Figs. 47 and 48.

NOTE: All replacement electronic modules will be of the center seat sensor by-pass type.

SERVICE NOTE

A non-sequential 8 second buzzer and light reminder system which does not affect engine start-up was incorporated to replace the Seat Belt Interlock system on mid-year 1975 vehicles.

The buzzer will remain on for 8 seconds after the ignition switch is turned on if the driver's seat belt is unbuckled. If the seat belt is buckled prior to the 8 second turn-off time, the buzzer will go off. The reminder light will remain on at all times for 8 seconds after the ignition switch is turned on regardless if seat belts are buckled.

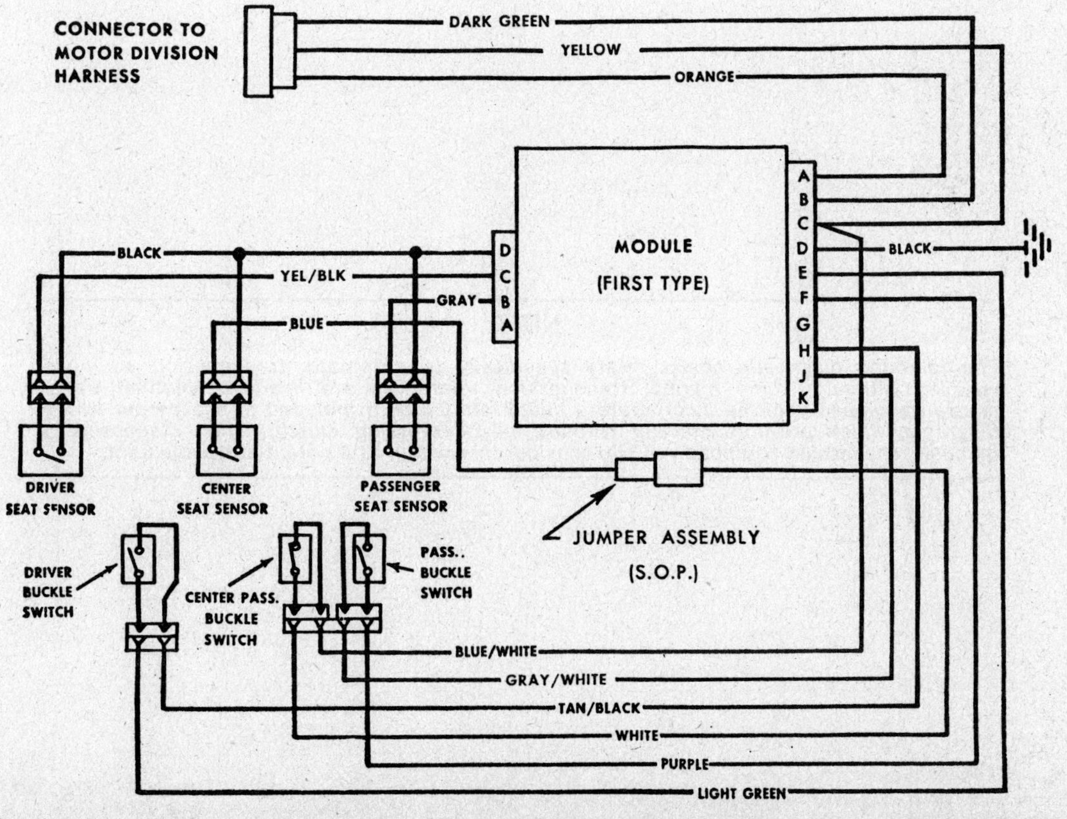

Fig. 47 Electronic module with jumper wire by-pass

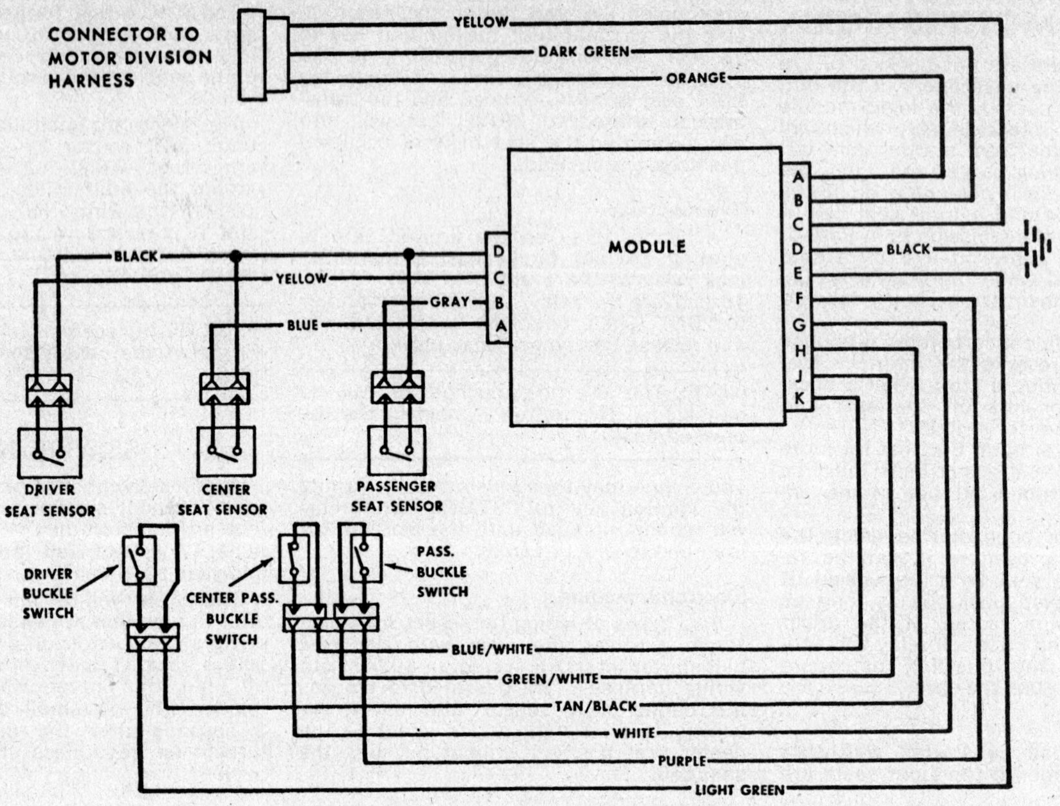

Fig. 48 Electronic module with built in by-pass

NOTE:

The following diagnostic charts relate specifically to automatic transmission equipped vehicles, although some manual transmission references are included. Manual transmission equipped vehicles incorporate a clutch start switch included in the proper wiring diagrams, which prevents starting without fully depressing clutch. When diagnosing a manual transmission equipped vehicle, consider this switch as a potential trouble item.

DIAGNOSIS CHARTS

SEAT BELT / STARTER INTERLOCK SYSTEM

Introduction

This section presents a systematic method of diagnosing and troubleshooting the seat belt/starter interlock system. The charts you will be using are different from the ones you have used before.

They aren't "go — no go" decision trees 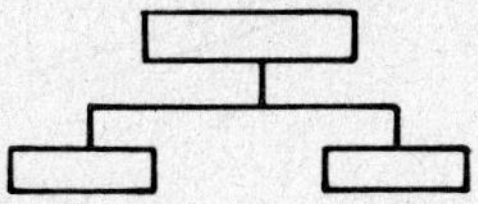 or tables.

Condition	Possible Cause	Correction

Instead the new diagnosis and troubleshooting charts use pictures plus a few words to help you solve a problem,

and symbols have replaced words.

Using the Charts

The charts are divided into three sections: step, sequence, and result.

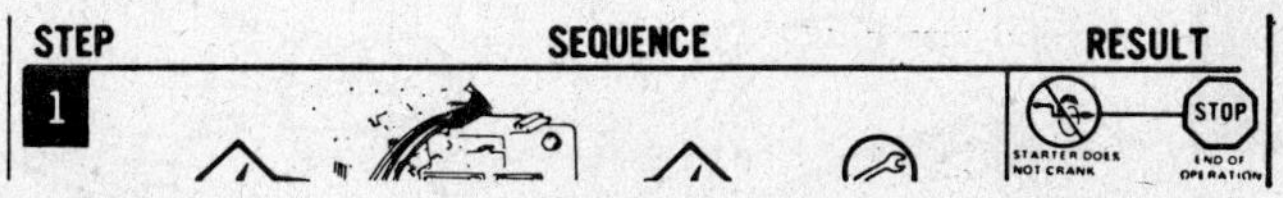

Always start at the first step and go through the complete sequence from left to right.

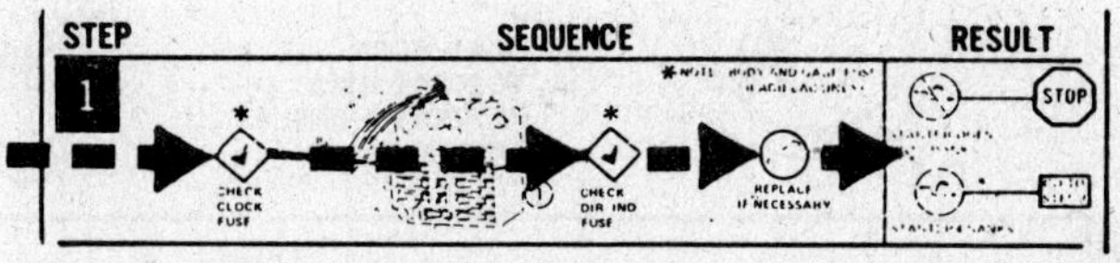

A sequence could be checking the fuses and replacing if necessary. Each sequence ends with a result and tells you the next step to go to.

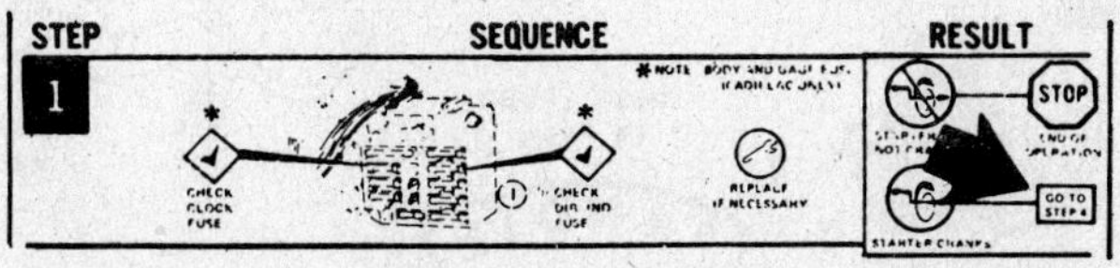

Work through each step of the diagnosis and troubleshooting charts till the system is repaired. 

To find where parts are located in the system just look at the parts locator at the top of each chart.

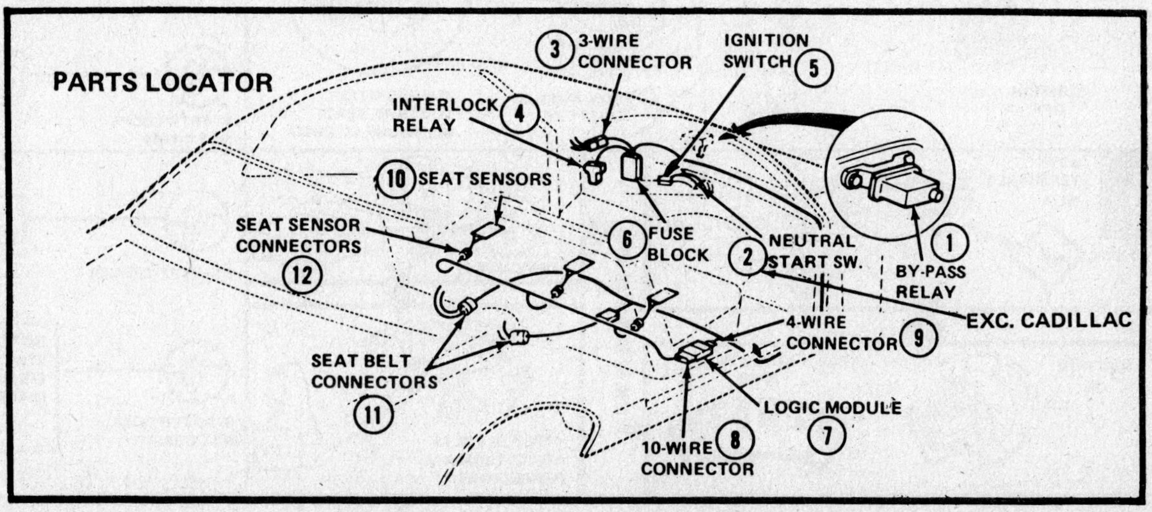

Seat Belt Interlock System. General Motors except Oldsmobile

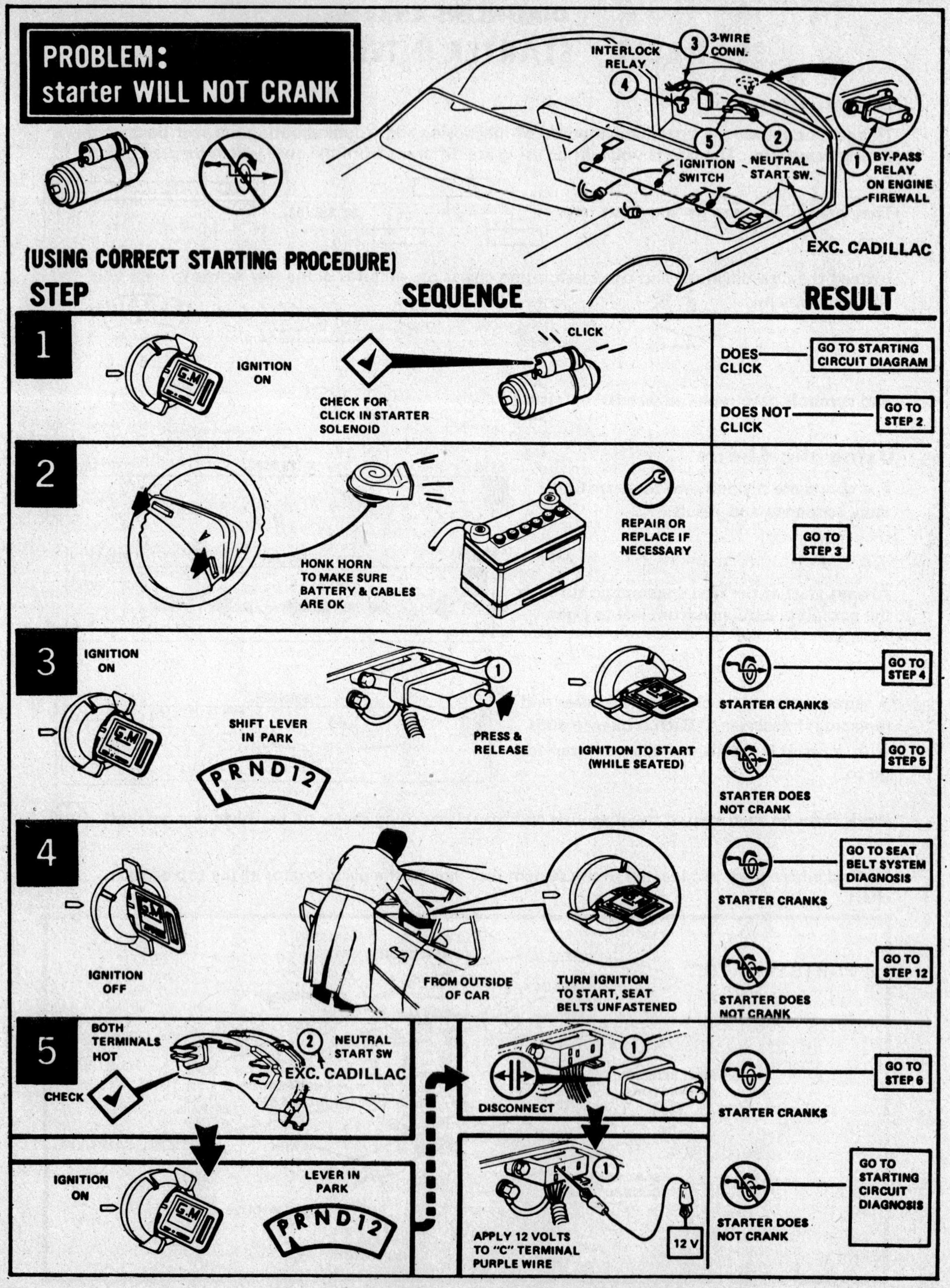

Seat Belt Interlock System. General Motors except Oldsmobile

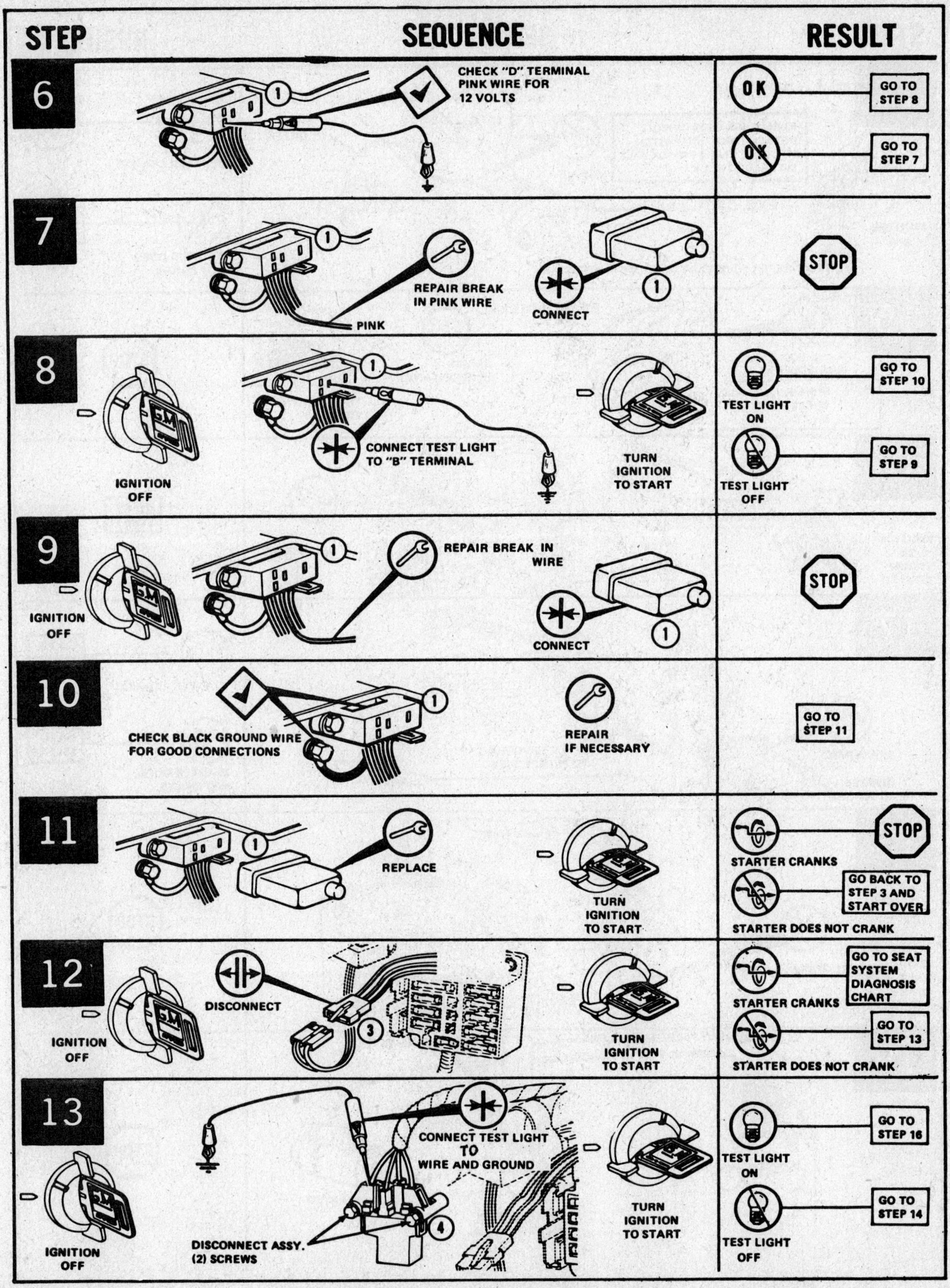

Seat Belt Interlock System. General Motors except Oldsmobile

STEP | **SEQUENCE** | **RESULT**

14

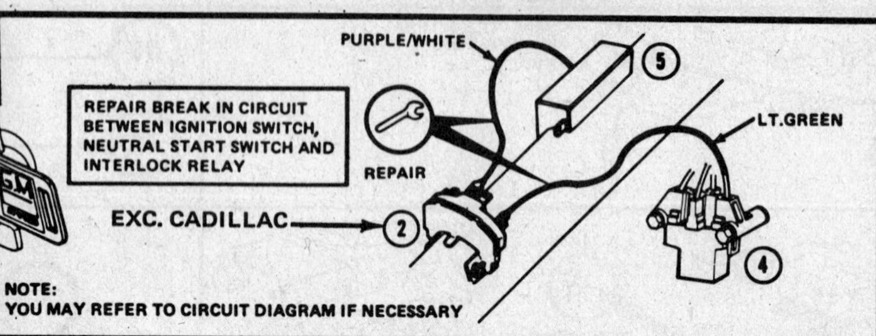

Result:
- STARTER CRANKS — STOP
- STARTER DOES NOT CRANK — GO TO STEP 15

15

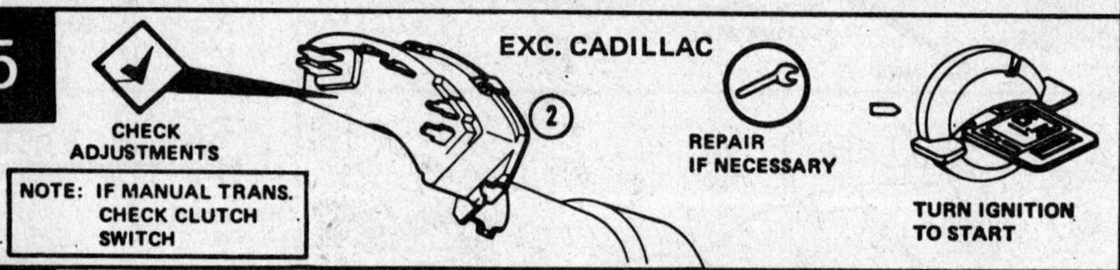

Result: STOP

16

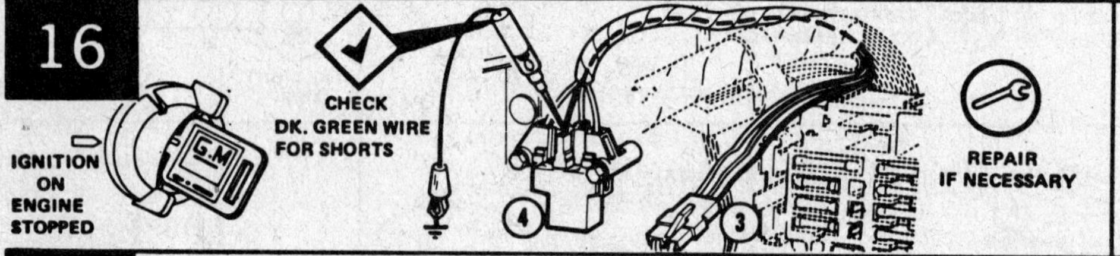

Result: GO TO STEP 17

17

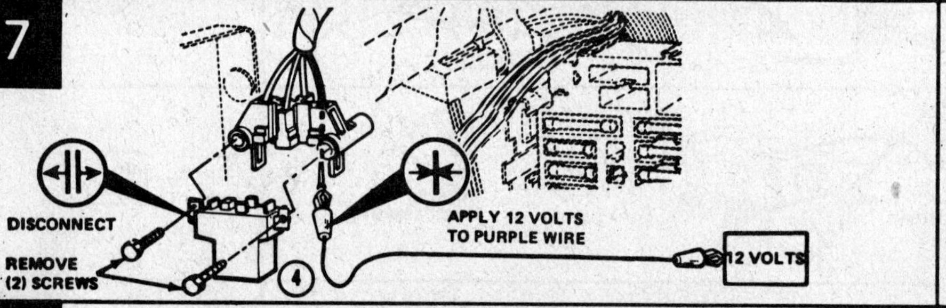

Result:
- STARTER CRANKS — GO TO STEP 18
- STARTER DOES NOT CRANK — GO TO STEP 19

18

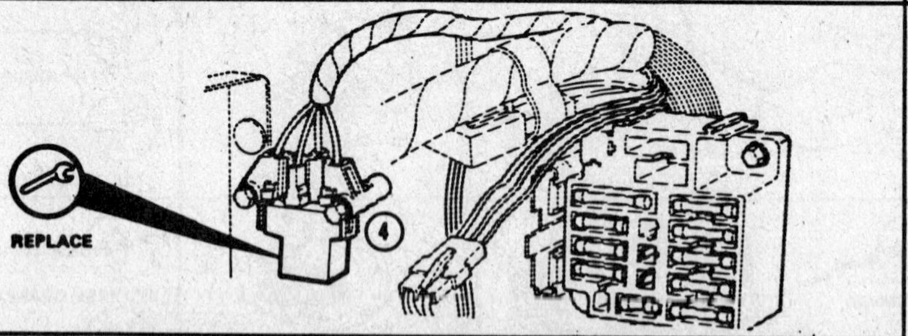

Result: STOP

19

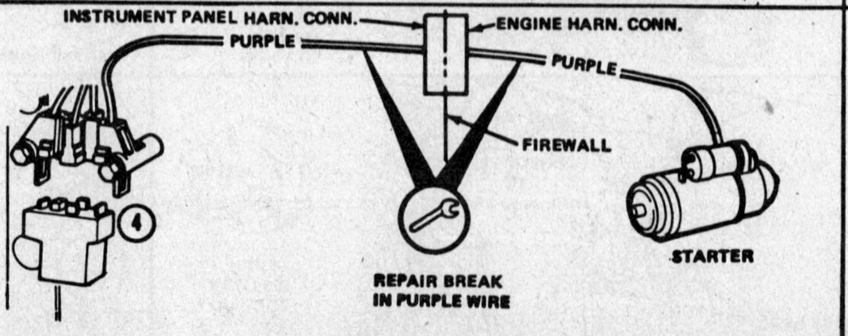

Result: STOP

Seat Belt Interlock System. General Motors except Oldsmobile

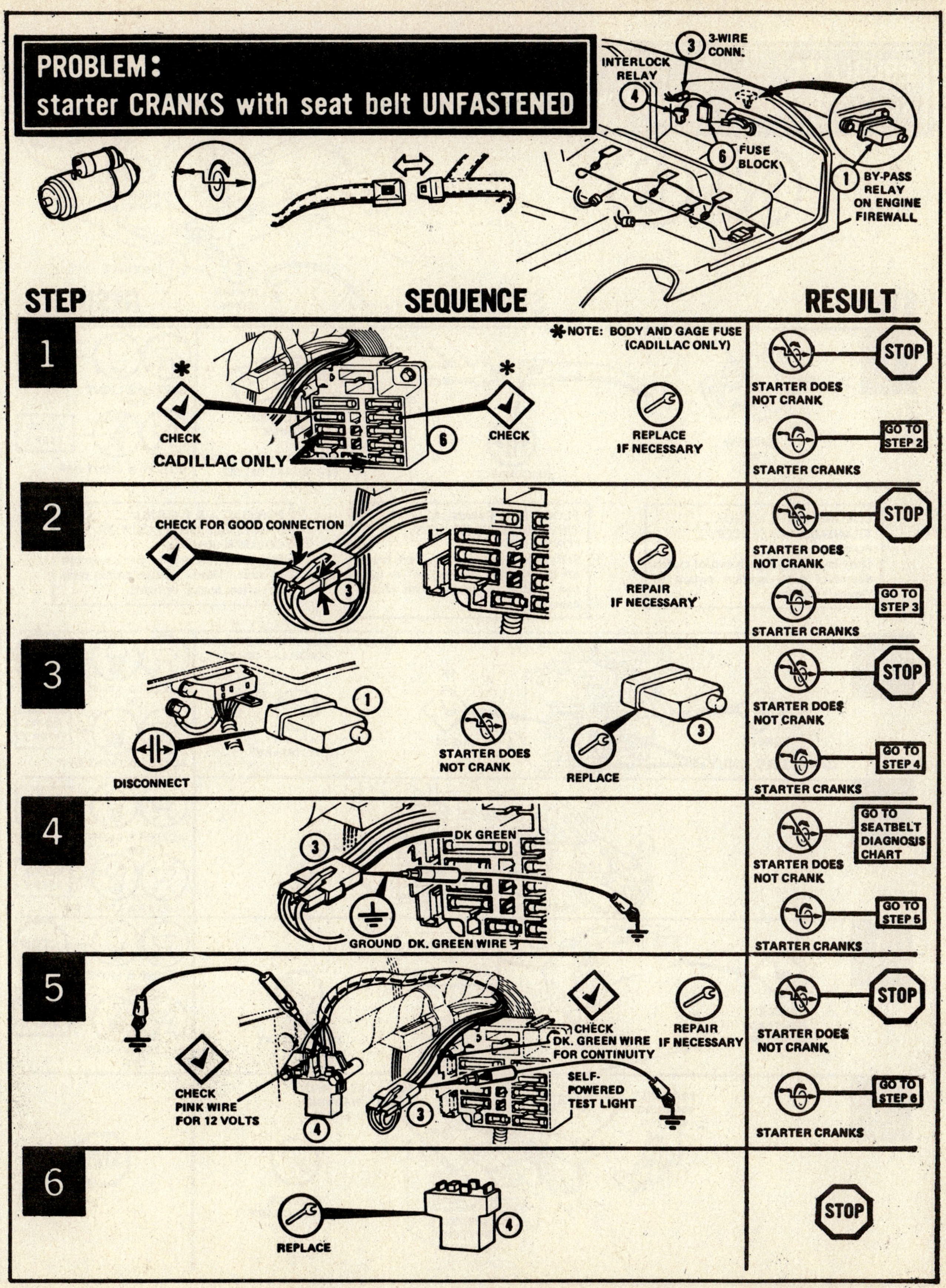

Seat Belt Interlock System. General Motors except Oldsmobile

PROBLEM:
seat belt buzzer & light WILL NOT OPERATE

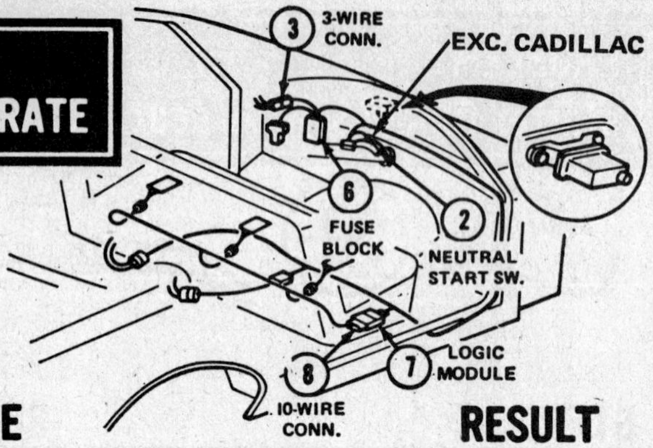

STEP | **SEQUENCE** | **RESULT**

1

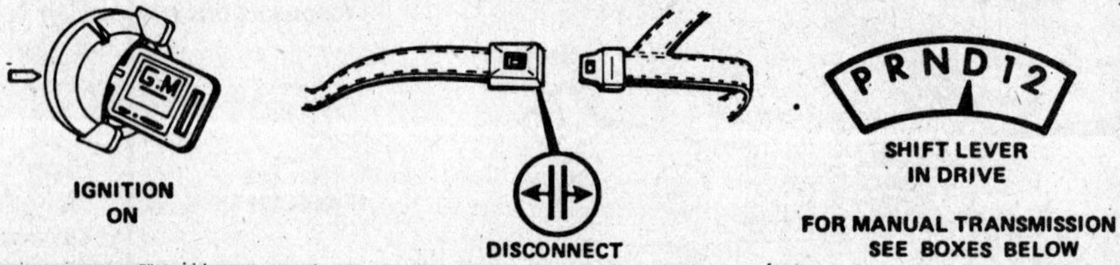

FOR MANUAL TRANSMISSION SEE BOXES BELOW

Result:
- BUZZER & LIGHT ON — STOP
- BUZZER & LIGHT OFF — GO TO STEP 2

COLUMN SHIFT MANUAL TRANSMISSION (EXCEPT PONTIAC A & F): Shift lever out of park and/or clutch depressed with ignition switch to "crank" position.

FLOOR SHIFT MANUAL TRANSMISSION (EXCEPT PONTIAC A & F): Shift lever in any position, Park brake off and ignition switch "on" or ignition switch to "start" with clutch depressed.

PONTIAC A & F MANUAL TRANSMISSION (FLOOR OR COLUMN SHIFT): Park brake released or ignition switch to start. (Note: clutch switch does not affect buzzer or light).

2

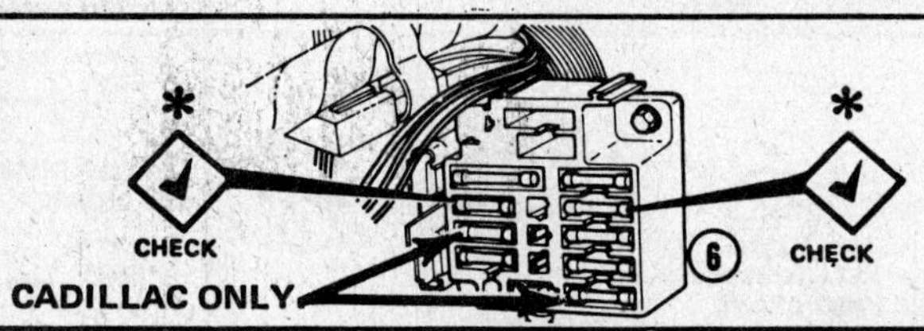

*NOTE: BODY AND GAGE FUSE (CADILLAC ONLY)

REPLACE IF NECESSARY

Result:
- BUZZER & LIGHT ON — STOP
- BUZZER & LIGHT OFF — GO TO STEP 3

3

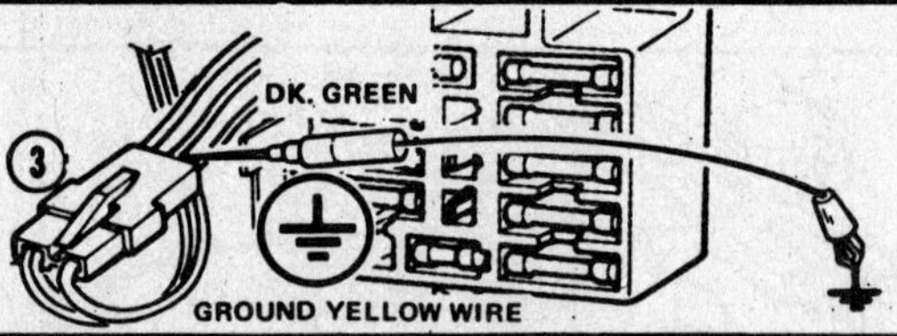

Result:
- BUZZER & LIGHT ON — GO TO STEP 6
- BUZZER & LIGHT OFF — GO TO STEP 4

4

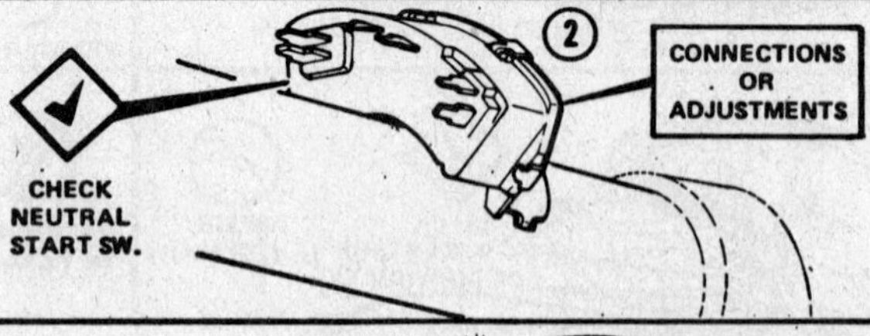

REPAIR IF NECESSARY

Result:
- BUZZER & LIGHT ON — STOP
- BUZZER & LIGHT OFF — GO TO STEP 5

5

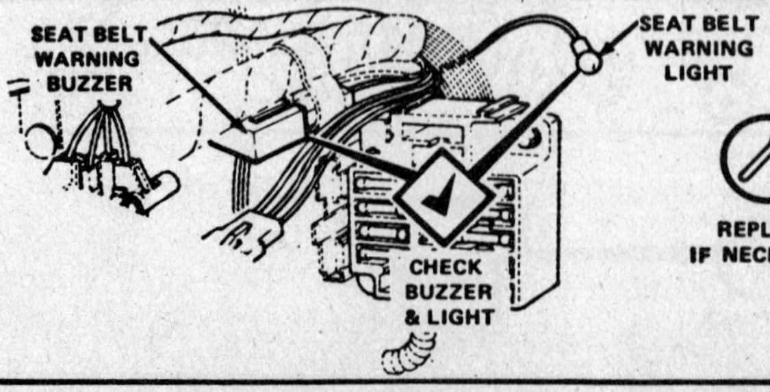

REPLACE IF NECESSARY

Result: STOP

Seat Belt Interlock System. General Motors except Oldsmobile

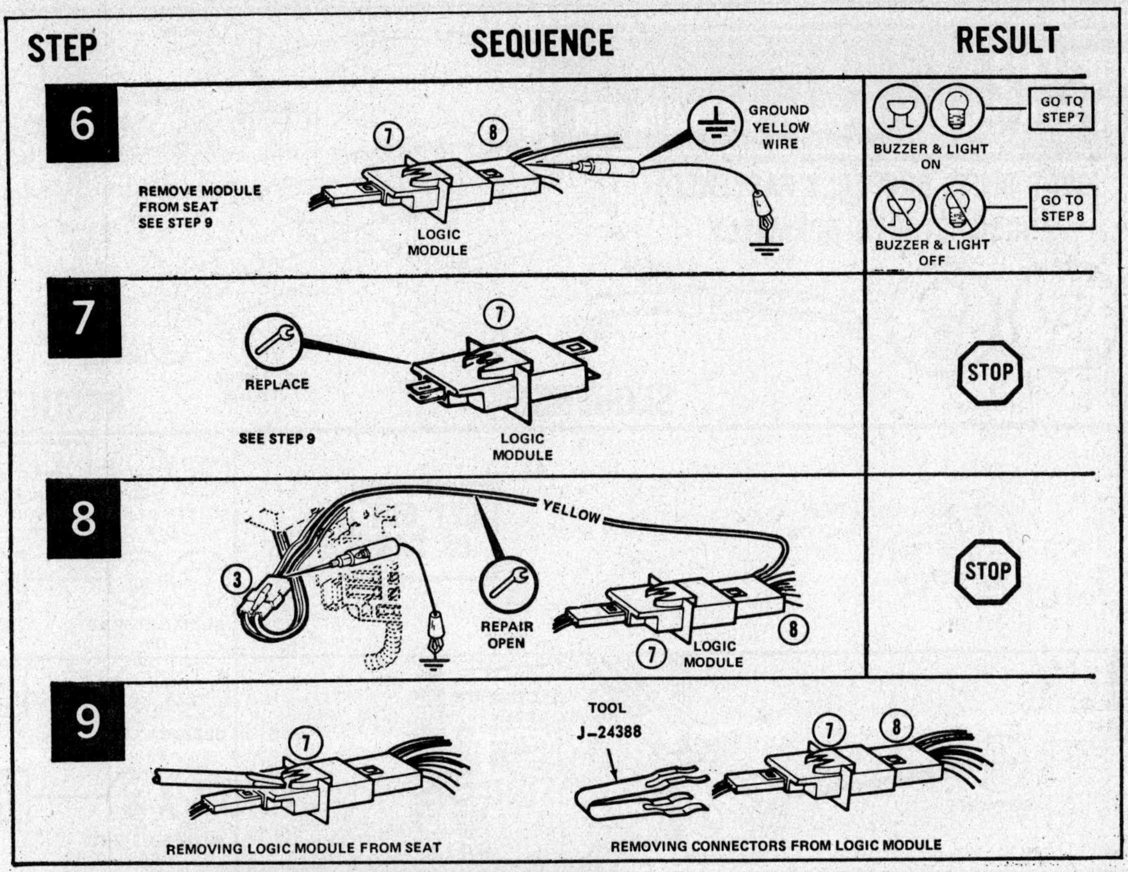

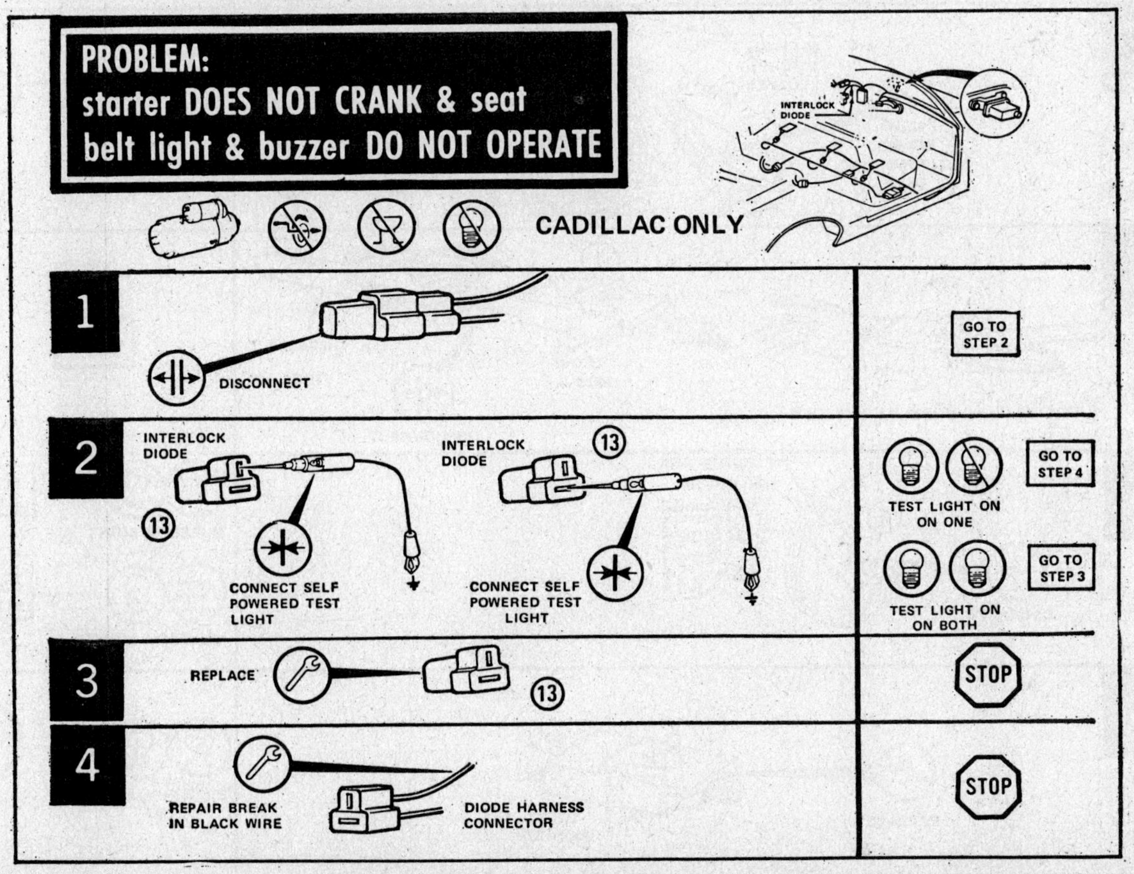

Seat Belt Interlock System. General Motors except Oldsmobile

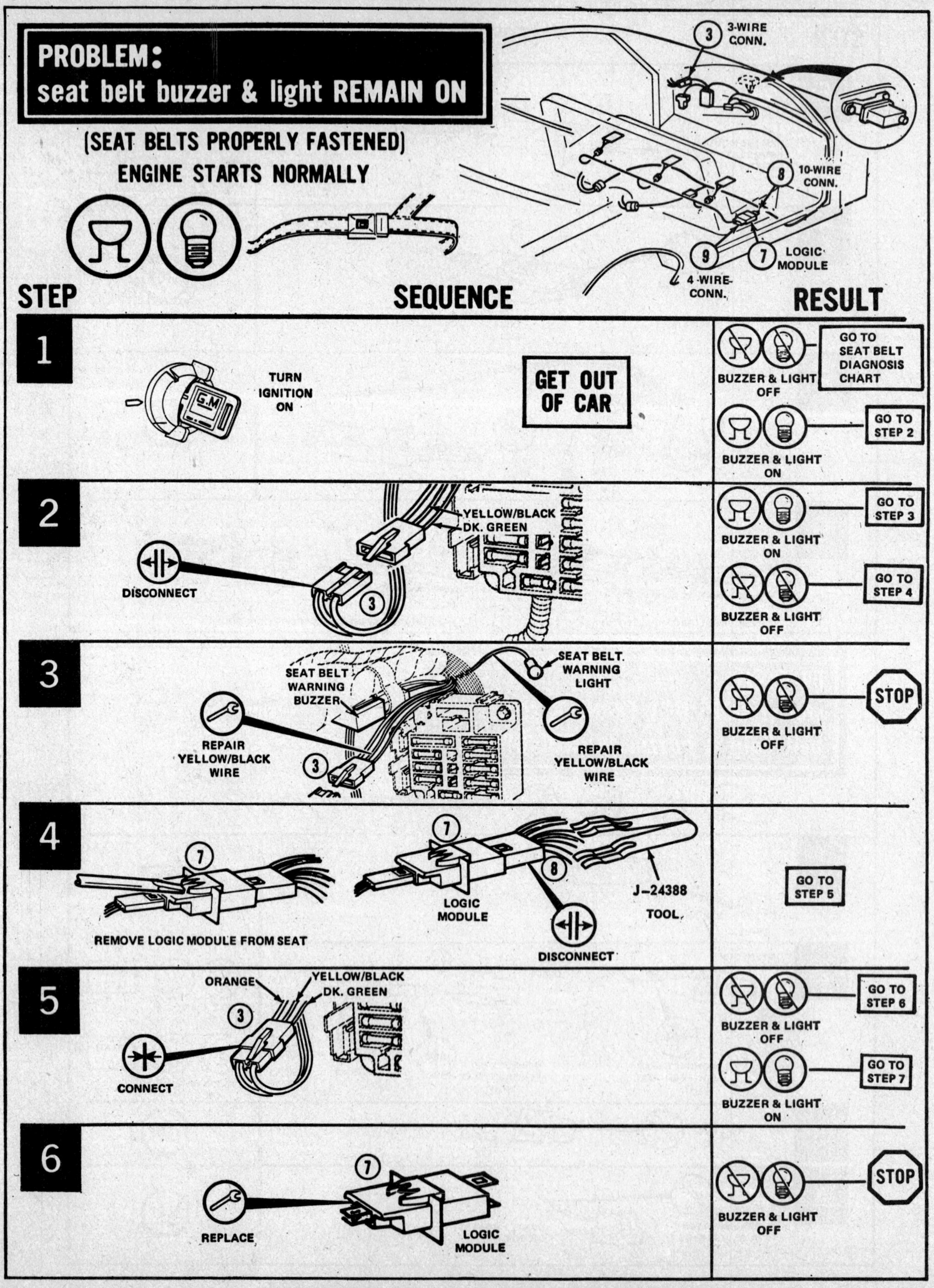

Seat Belt Interlock System. General Motors except Oldsmobile

STEP | **SEQUENCE** | **RESULT**

7

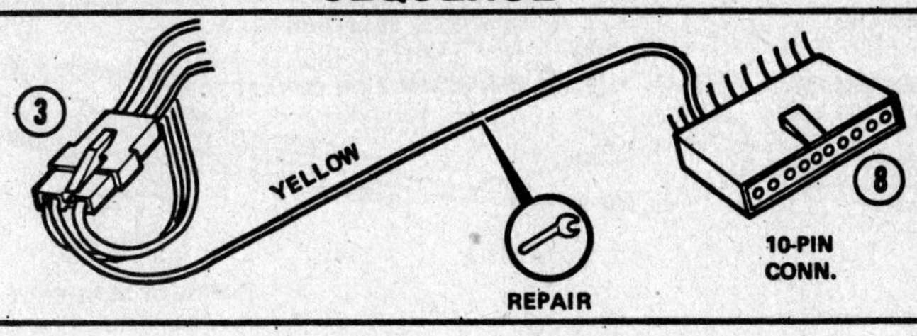

BUZZER & LIGHT OFF

SEAT BELT SYSTEM DIAGNOSIS

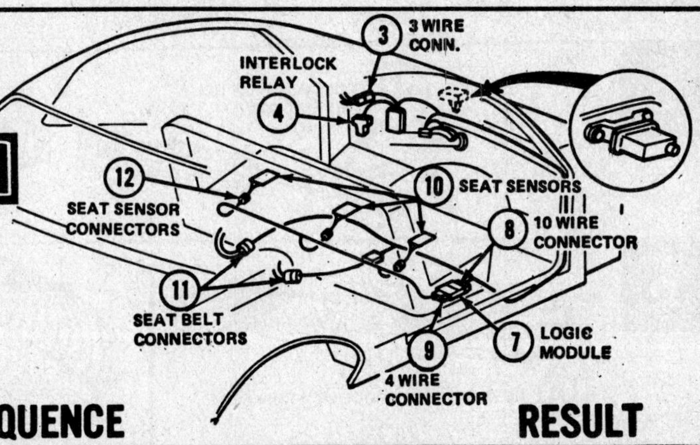

STEP | **SEQUENCE** | **RESULT**

1

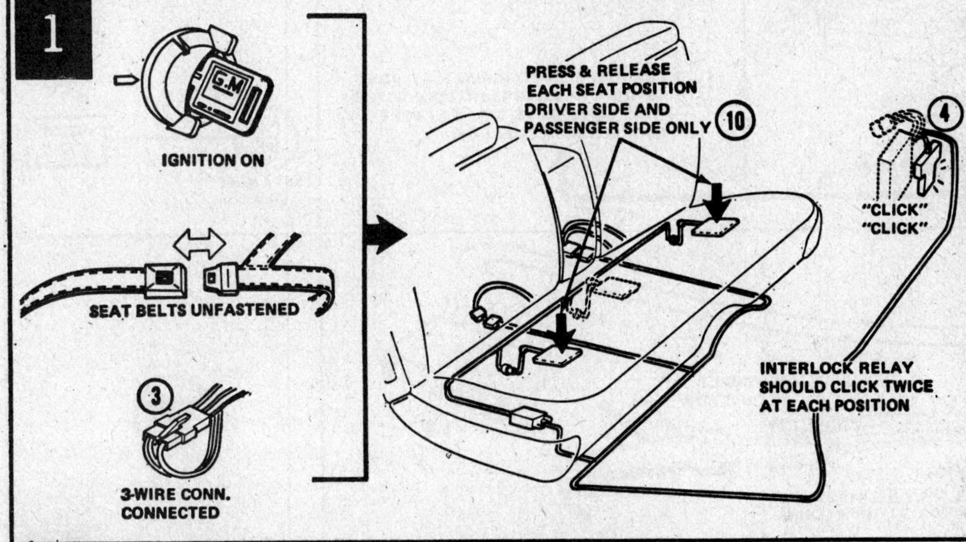

- TWO CLICKS AT BOTH POSITIONS
  GO TO STEP - 2
- TWO CLICKS AT ONE POSITION ONLY
  GO TO STEP - 12
- ONLY ONE CLICK OR NO CLICK AT BOTH POSITIONS
  GO TO STEP - 11

2

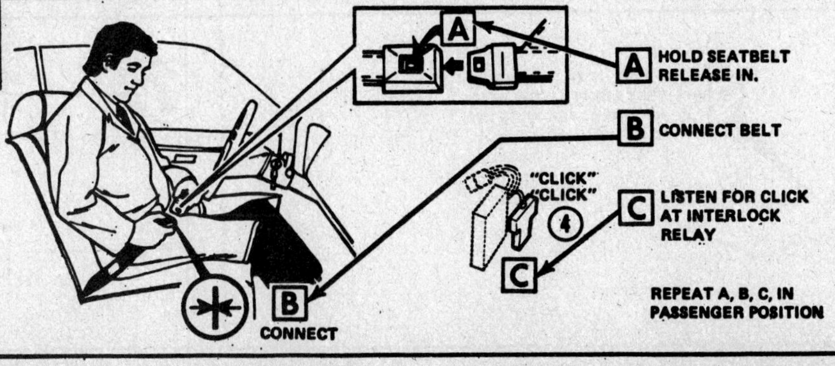

- RELAY DOES NOT CLICK AT EITHER POSITION
  GO TO STEP - 6
- RELAY DOES NOT CLICK AT ONE POSITION
  GO TO STEP - 3

Seat Belt Interlock System. General Motors except Oldsmobile

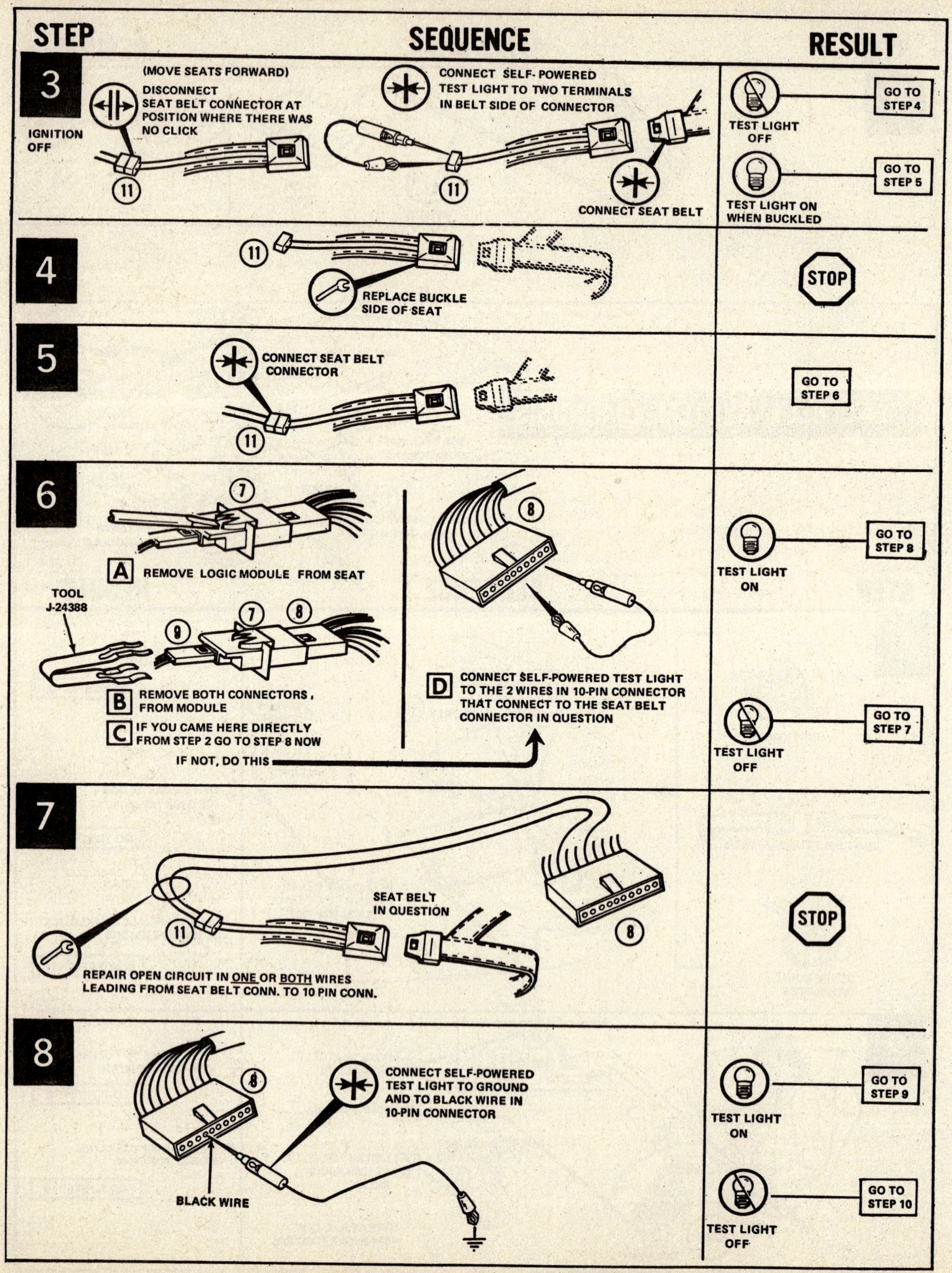

Seat Belt Interlock System. General Motors except Oldsmobile

STEP	SEQUENCE	RESULT
9	 	STOP
10	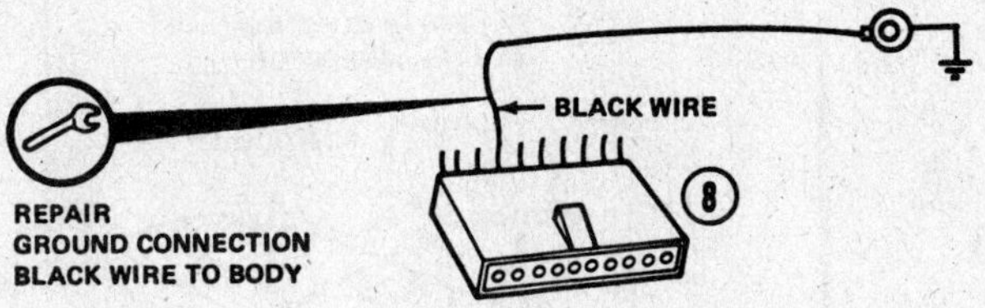 	STOP
11	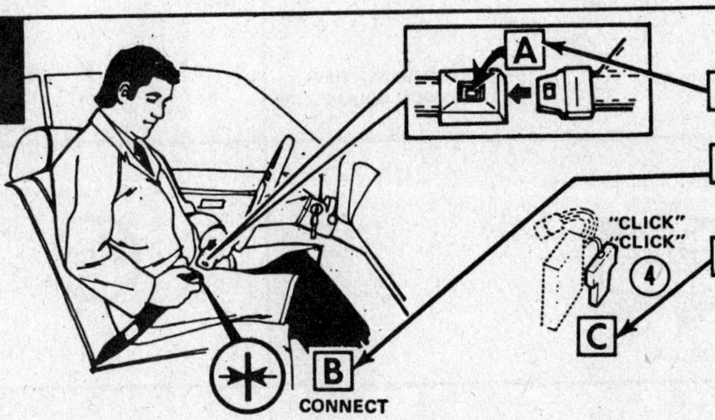 A HOLD SEATBELT RELEASE IN. B CONNECT BELT C LISTEN FOR CLICK AT INTERLOCK RELAY REPEAT A, B, C, IN PASSENGER POSITION	• RELAY CLICKS AT ONE POSITION GO TO STEP - 12 • NEITHER RELAY CLICKS GO TO STEP - 17
12	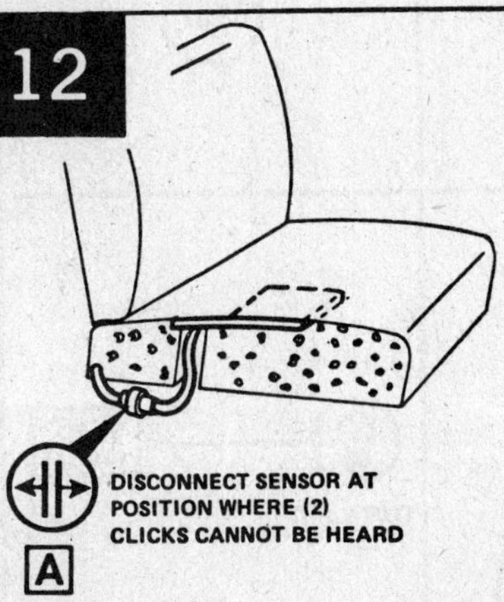 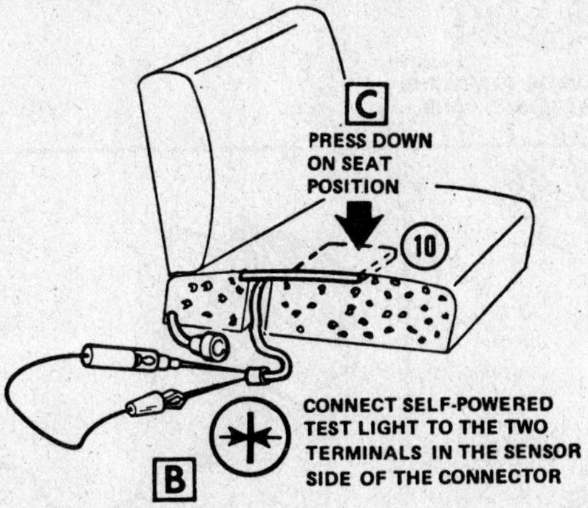 	TEST LIGHT OFF — GO TO STEP 13 TEST LIGHT ON WHEN SEAT DEPRESSED — GO TO STEP 14
13	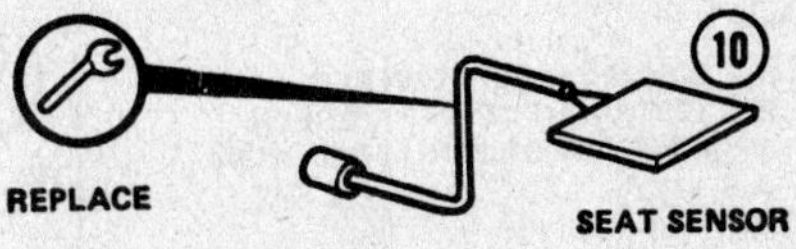 	STOP

Seat Belt Interlock System. General Motors except Oldsmobile

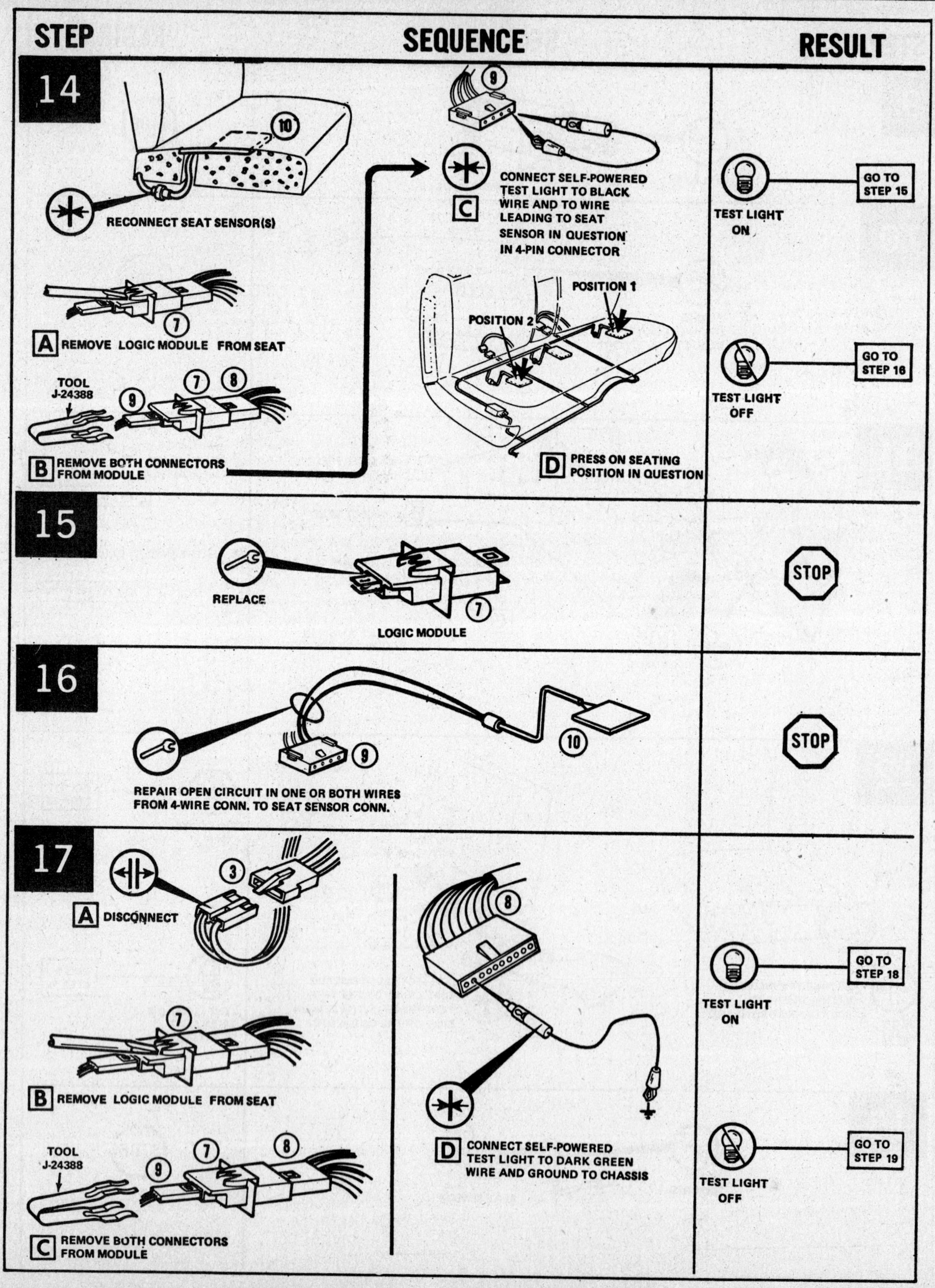

Seat Belt Interlock System. General Motors except Oldsmobile

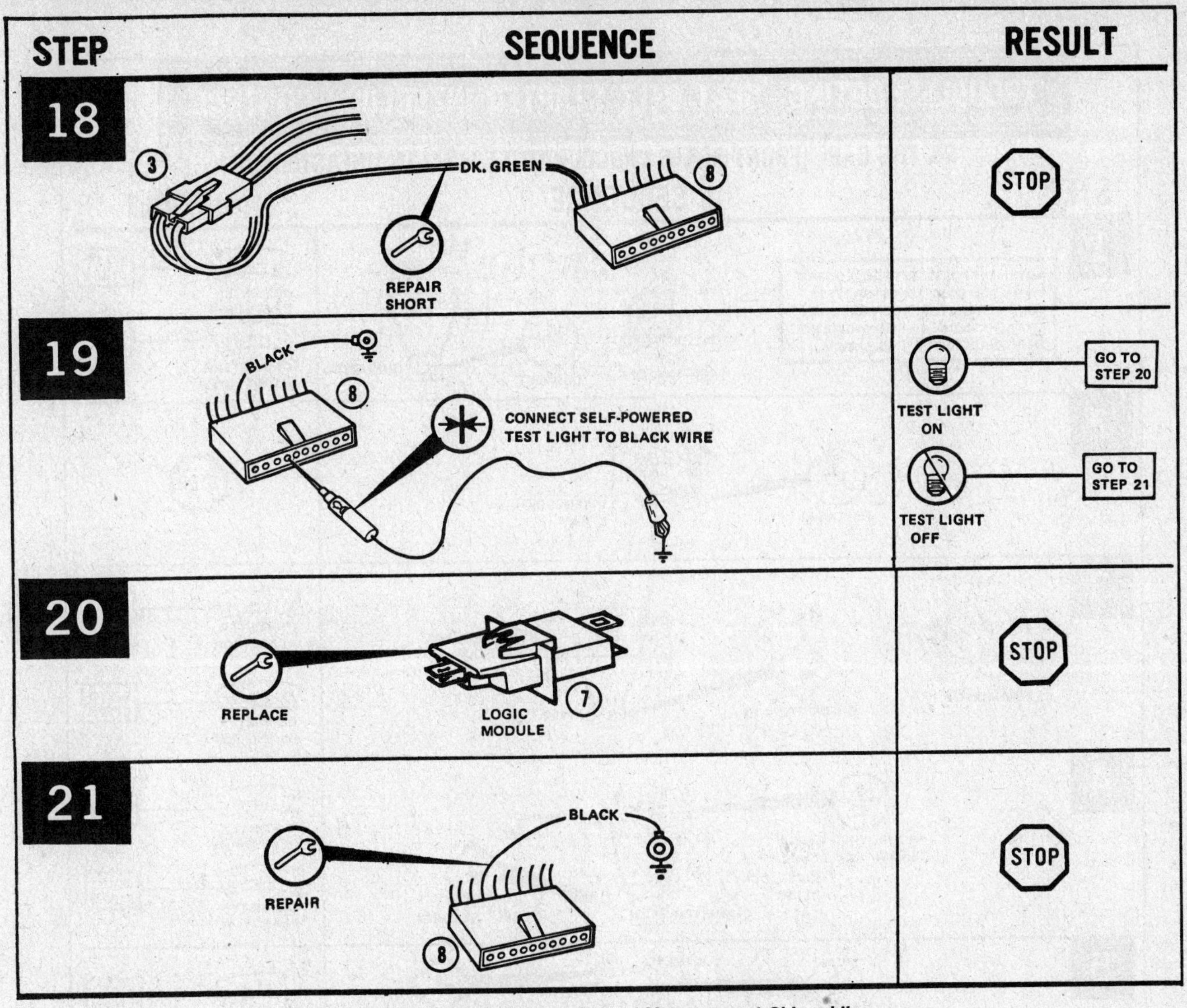

Seat Belt Interlock System. General Motors except Oldsmobile

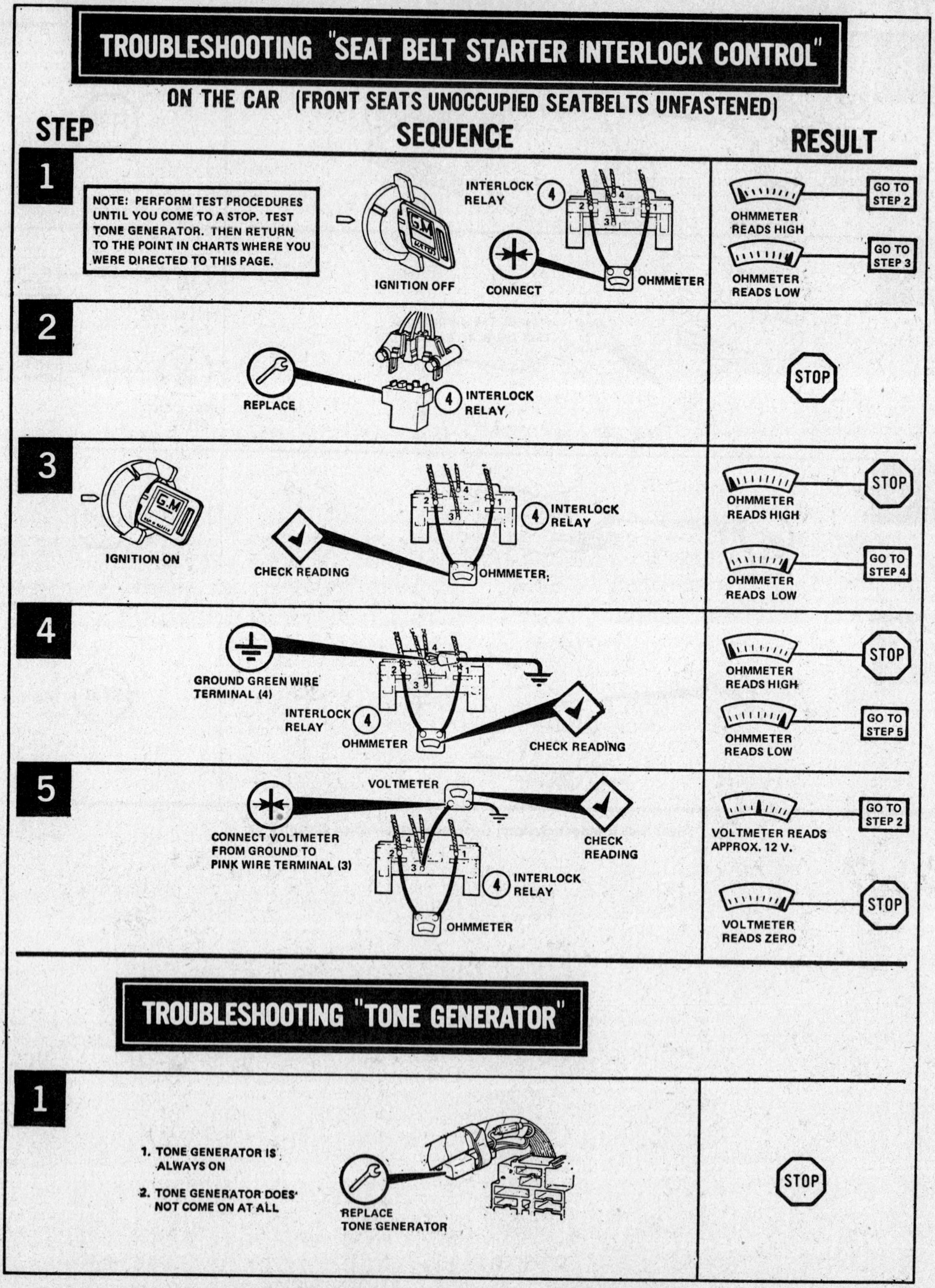

Seat Belt Interlock System. General Motors except Oldsmobile

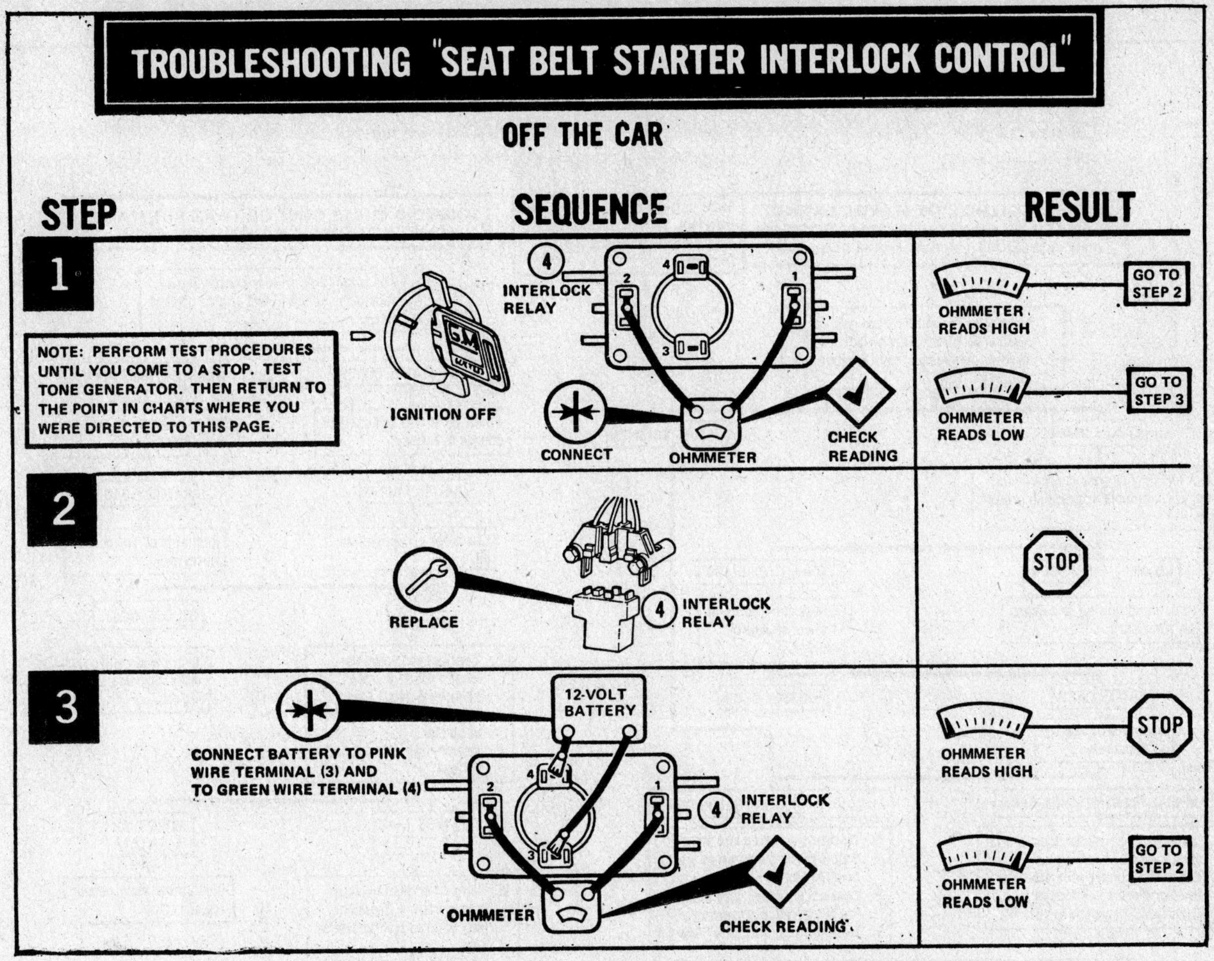

Seat Belt Interlock System. General Motors except Oldsmobile

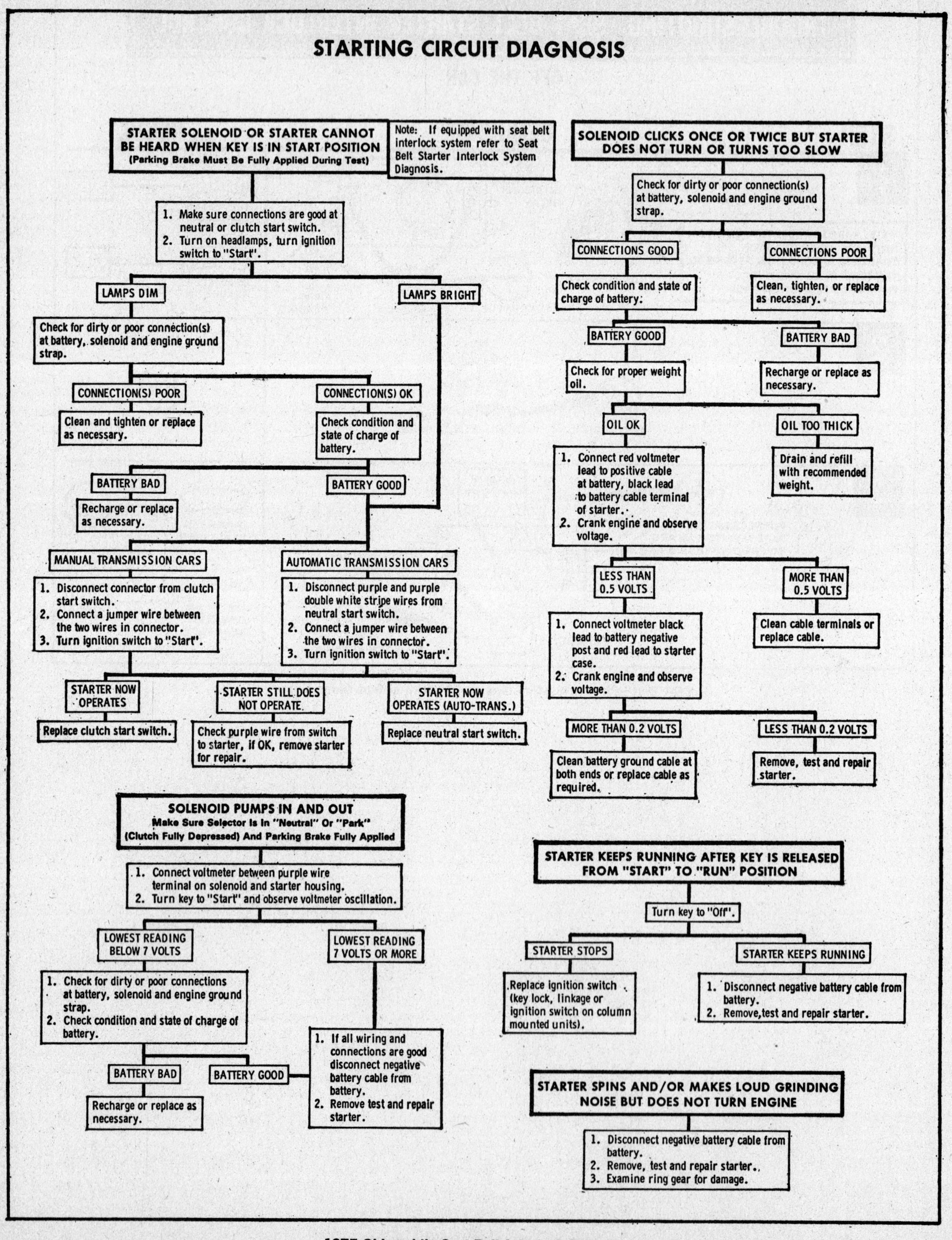

1975 Oldsmobile Seat Belt Interlock System

SEAT BELT STARTER INTERLOCK SYSTEM DIAGNOSIS

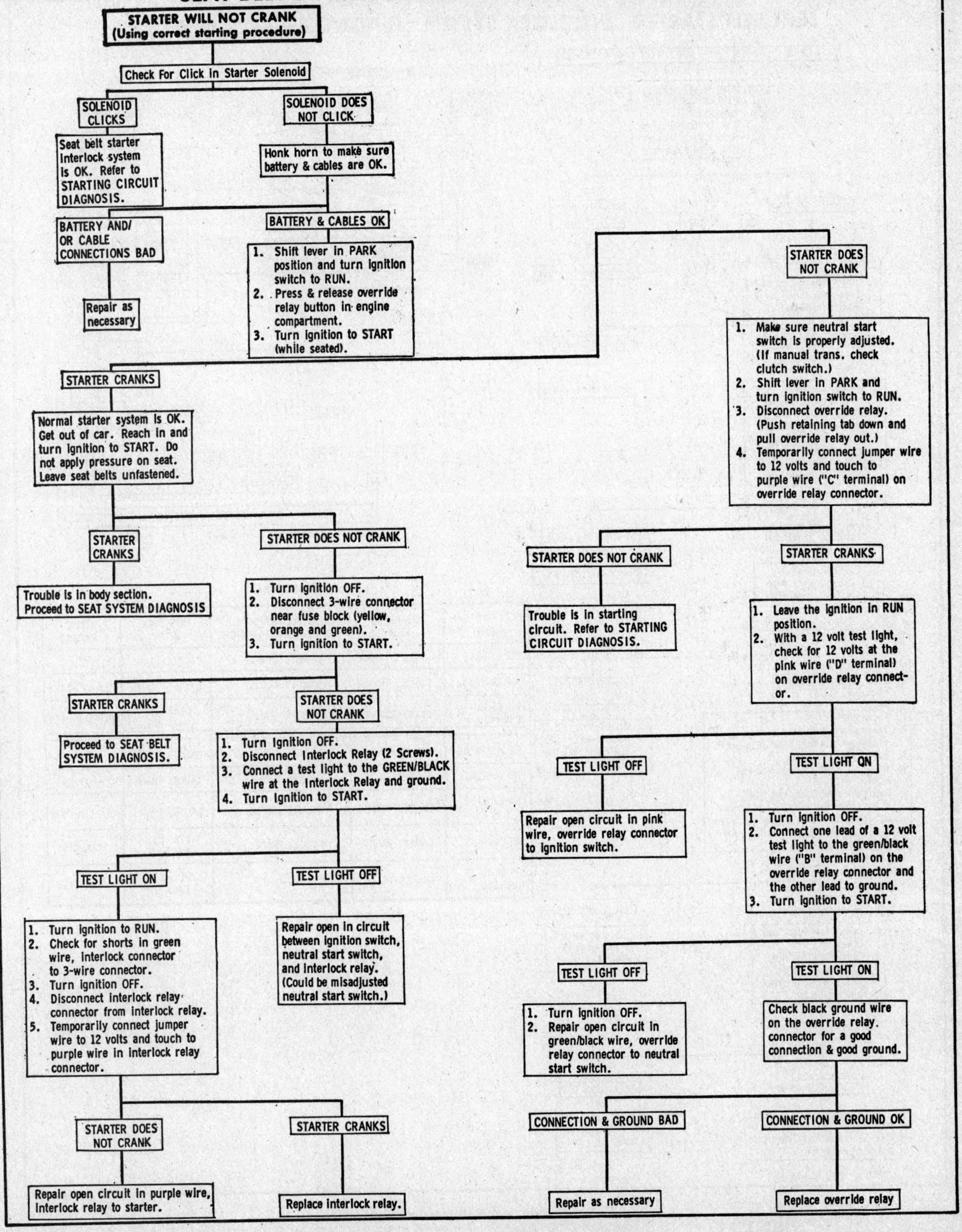

1974-75 Oldsmobile Seat Belt Interlock System

SEAT BELT STARTER INTERLOCK SYSTEM DIAGNOSIS (CONTINUED)

STARTER CRANKS WITH SEAT BELT UNFASTENED

1. Refer to fuse chart and replace fuse if necessary.
2. Turn ignition to Start (while sitting on seat with belt unfastened).

STARTER CRANKS

1. Turn ignition OFF.
2. Check 3-wire connector (near fuse panel) for a good connection.

STARTER DOES NOT CRANK

Trouble was bad fuse

CONNECTION GOOD

1. Remove override relay (in engine compartment) from its connector (push retaining tab down and pull out).
2. Turn ignition to START (while sitting on seat with belt unfastened).

CONNECTION BAD

Repair as necessary

STARTER CRANKS

1. Turn ignition OFF.
2. Ground green wire at 3-wire connector.
3. Turn ignition to START.

STARTER DOES NOT CRANK

Replace override relay

STARTER CRANKS

1. Turn ignition OFF.
2. Remove interlock relay assembly (2) screws (near fuse block).
3. Turn ignition to RUN.
4. Ground one lead of a 12 volt test light.
5. Touch the other test light lead to the pink wire (Starfire and Omega) dk. green dbl. white str. wire (Cutlass 88, 98 & Toronado) at interlock relay connector.

STARTER DOES NOT CRANK

1. Turn ignition OFF.
2. Proceed to SEAT BELT SYSTEM DIAGNOSIS.

TEST LIGHT ON

1. Leave 12 volt test light grounded.
2. Touch the other test light lead to the dk. green wire at interlock relay connector.

TEST LIGHT OFF

Repair open circuit, interlock relay to ignition switch.

TEST LIGHT ON

Repair open circuit, DK. GREEN wire, Interlock Relay connector to 3-wire connector.

TEST LIGHT OFF

Replace interlock relay

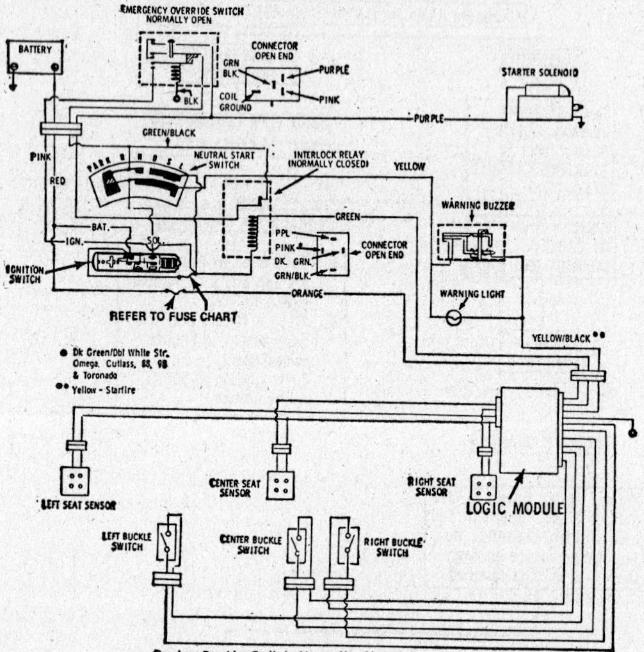

Center Buckle Switch Normally Closed

FUSE CHART

MODEL	OVERRIDE RELAY	SEAT BELT WARNING (NEUTRAL START SWITCH)	INTERLOCK RELAY	LOGIC MODULE
1974 OMEGA	GAUGES-TRANS.	GAUGES-TRANS.	GAUGES-TRANS.	TAIL
1974 CUTLASS	GAUGES-TRANS.	GAUGES-TRANS.	GAUGES-TRANS.	CLK-LTR-CTSY
1974 88-98 TORONADO	GAUGES-TRANS.	GAUGES-TRANS.	DIR. SIG. BACK-UP	CLK-LTR-CTSY
1975 STARFIRE	GAUGES-TRANS.	GAUGES-TRANS.	GAUGES-TRANS.	TAIL
1975 OMEGA	GAUGES-TRANS.	GAUGES-TRANS.	GAUGES-TRANS.	CLK-LTR-CTSY
1975 CUTLASS S. M. T.	GAUGES-TRANS.	GAUGES-TRANS.	GAUGES-TRANS.	CLK-LTR-CTSY
1975 CUTLASS	GAUGES-TRANS.	GAUGES-TRANS.	DIR. SIG. BACK-UP	CLK-CTSY
1975 88-98 TORONADO	GAUGES-TRANS.	GAUGES-TRANS.	DIR. SIG. BACK-UP	CLK-CTSY

1975 Oldsmobile Seat Belt Interlock System

SEAT BELT STARTER INTERLOCK SYSTEM DIAGNOSIS (CONTINUED)

STARTER CRANKS WITH SEAT BELT UNFASTENED

1. Test clock and gauges transmission fuses. Replace if necessary.
2. Turn ignition to START (while sitting on seat with belt unfastened).

STARTER CRANKS

1. Turn ignition OFF.
2. Check 3-wire connector (near fuse block) for a good connection.

CONNECTION GOOD

1. Remove override relay (in engine compartment) from its connector (push retaining tab down and pull out).
2. Turn ignition to START (while sitting on seat with belt unfastened).

STARTER CRANKS

1. Turn ignition OFF.
2. Ground green wire at 3-wire connector.
3. Turn ignition to START.

STARTER CRANKS

1. Turn ignition OFF.
2. Remove interlock relay assembly (2) screws (near fuse block).
3. Turn ignition to RUN.
4. Ground one lead of a 12 volt test light.
5. Touch the other test light lead to the pink wire in interlock relay connector.

TEST LIGHT ON

1. Leave 12 volt test light grounded.
2. Touch the other test light lead to the green wire at interlock relay connector.

TEST LIGHT ON

Repair open circuit, GREEN wire, Interlock Relay connector to 3-wire connector.

TEST LIGHT OFF

Replace interlock relay

TEST LIGHT OFF (pink wire test)

Repair open circuit, interlock relay to ignition switch.

STARTER DOES NOT CRANK (ground green wire test)

1. Turn ignition OFF.
2. Proceed to SEAT BELT SYSTEM DIAGNOSIS.

STARTER DOES NOT CRANK (override relay removed)

Replace override relay

CONNECTION BAD

Repair as necessary

STARTER DOES NOT CRANK (first test)

Trouble was bad fuse

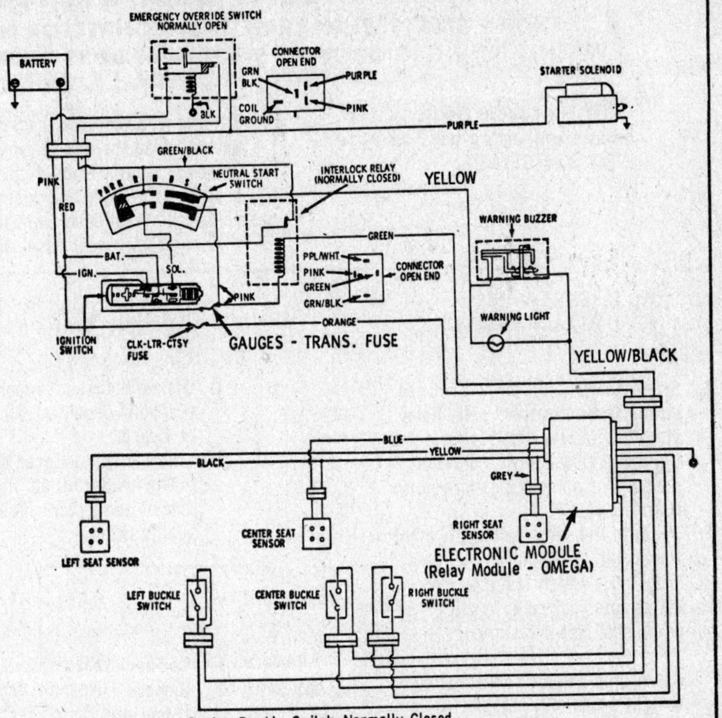

Center Buckle Switch Normally Closed

NOTE: On OMEGA all three buckle switches normally closed.

1974 Oldsmobile Seat Belt Interlock System

SEAT BELT SYSTEM DIAGNOSIS

ENGINE STARTS WHEN THREE WIRE CONNECTOR NEAR FUSE PANEL IS DISCONNECTED OR WHEN TURNING IGNITION TO START, SEAT BELTS UNFASTENED AND NO ONE SITTING ON SEATS.

NOTE: To remove Electronic Module connectors use Tool J-24388 or BT7402A.

Turn ignition to RUN, press and release each seating position (2) and listen for interlock relay. Should click when seat position is pressed and again when it is released. (Seat belts unfastened, three wire connector near fuse panel connected).

TWO CLICKS CAN BE HEARD AT EACH POSITION

1. Sit on seat, hold seat belt release button in and connect belt. Relay should click when belt is fastened. (Holding release button prevents snap lock from making more noise than the relay).
2. Perform the same test at the right hand seating position.
3. Determine which test does not cause interlock relay to click when seat belt is fastened. If relay does not click at any position, skip Steps 4, 5 & 6 and go to TEST LIGHT ON, Step 2.
4. Disconnect seat belt connector determined in previous test.
5. Connect self powered test light to the two terminals in belt side of connector.
6. Connect seat belt. (OMEGA-Disconnected)

TEST LIGHT OFF: Replace buckle side of belt

TEST LIGHT ON:
1. Connect seat belt connector.
2. Remove Electronic Module from seat frame and disconnect 10-wire connector (If Steps 4, 5 & 6 in previous test were not performed, skip Step 3 and go to TEST LIGHT ON.)
3. Connect self-powered test light to the 2 wires in the 10 pin connector or that connects to the seat belt connector.

- TEST LIGHT OFF: Repair open circuit in one or both wires leading to the seat belt connector from the 10 pin connector.
- TEST LIGHT ON: Connect self powered test light to a good ground and to the black wire in the 10 pin connector.
  - TEST LIGHT OFF: Repair ground connection. (Black wire to body).
  - TEST LIGHT ON: Replace Electronic Module

TWO CLICKS CAN BE HEARD BUT NOT AT BOTH POSITIONS

1. Disconnect seat sensor at seat position where two clicks cannot be heard.
2. Connect self powered test light to the two terminals in sensor side of connector. Press on seat position.

TEST LIGHT OFF: Replace seat sensor

TEST LIGHT ON:
1. Connect seat sensor.
2. Remove Electronic Module from seat frame and disconnect 4-wire connector
3. Connect self-powered test light to the black wire in the 4-wire connector and to the other wire leading to the seat sensor determined above.
4. Press on seating position.

- TEST LIGHT ON: Replace Electronic Module
- TEST LIGHT OFF: Repair open circuit in one or both wires from 4 wire connector to seat sensor connector.

TWO CLICKS CANNOT BE HEARD AT ANY POSITION

1. Disconnect three wire connector near fuse panel.
2. Remove Electronic Module from seat frame and disconnect 10-wire connector
3. Connect self-powered test light to the dark green wire in the 10 pin connector and ground.

TEST LIGHT ON: Repair short circuit to ground in dark green wire from 10 pin connector to 3 wire connector near fuse panel.

TEST LIGHT OFF: Connect self powered test light to a good ground and to the black wire in the 10 pin connector.

- TEST LIGHT OFF: Repair ground connection (Black wire to body)
- TEST LIGHT ON: Replace Electronic Module

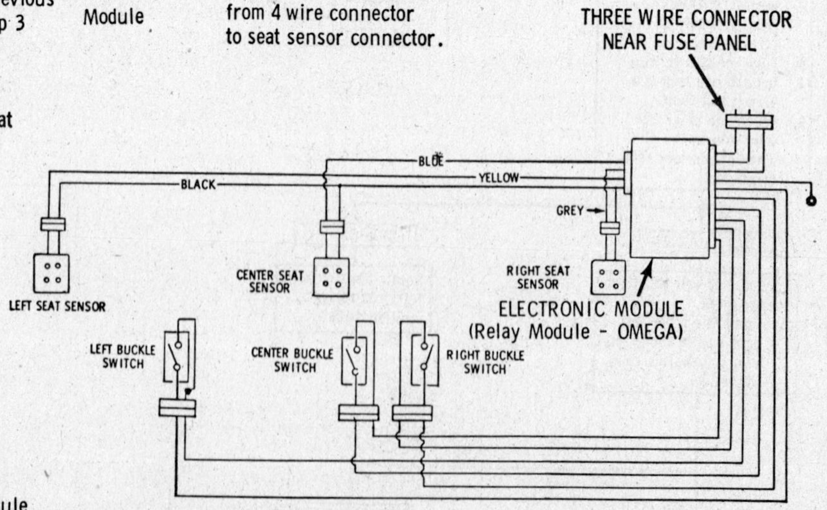

Center Buckle Switch Normally Closed.

NOTE: On OMEGA all three buckle switches normally closed.

1974-75 Oldsmobile Seat Belt Interlock System

SEAT BELT WARNING SYSTEM DIAGNOSIS

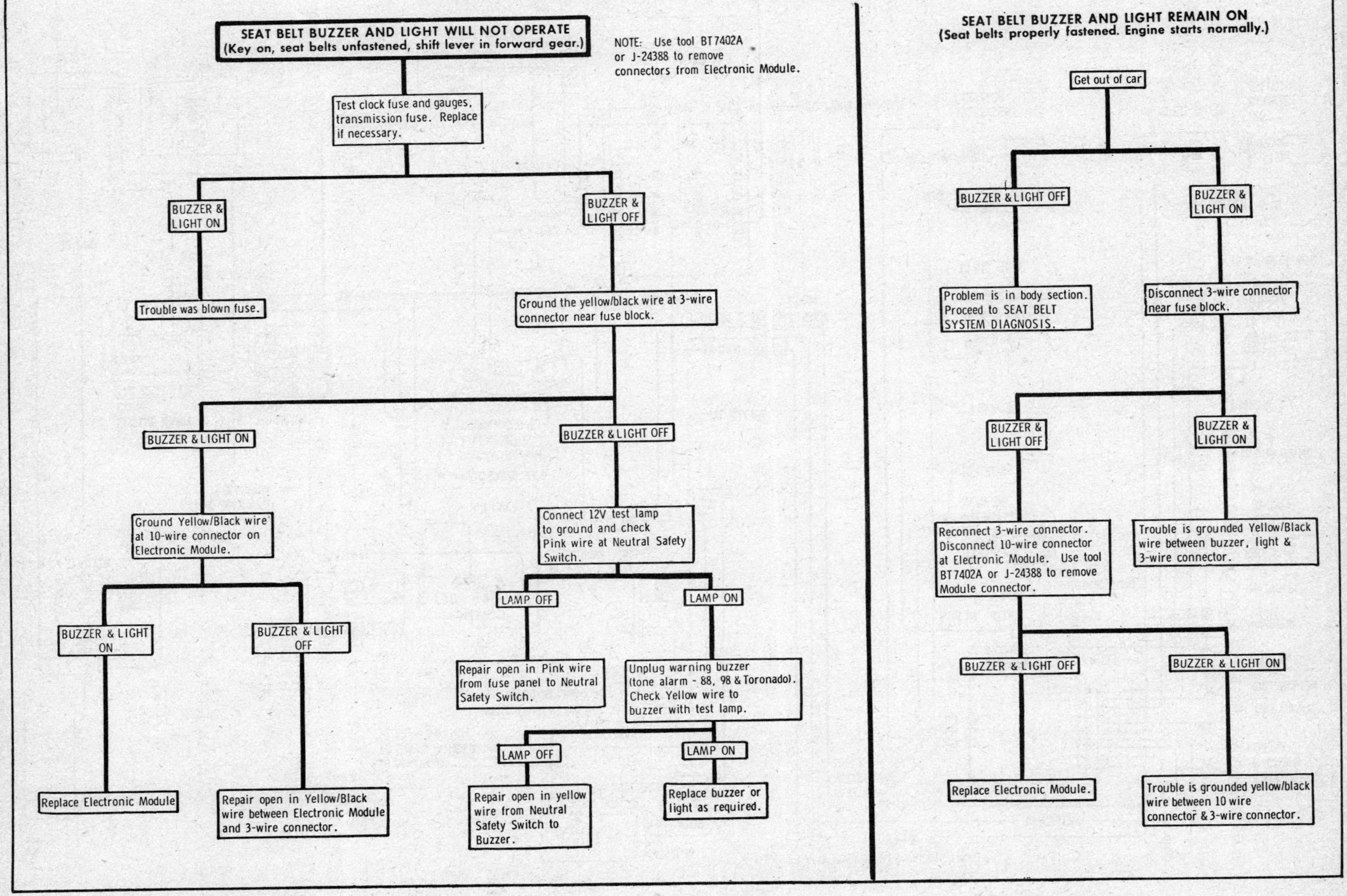

1974-75 Oldsmobile Seat Belt Interlock System

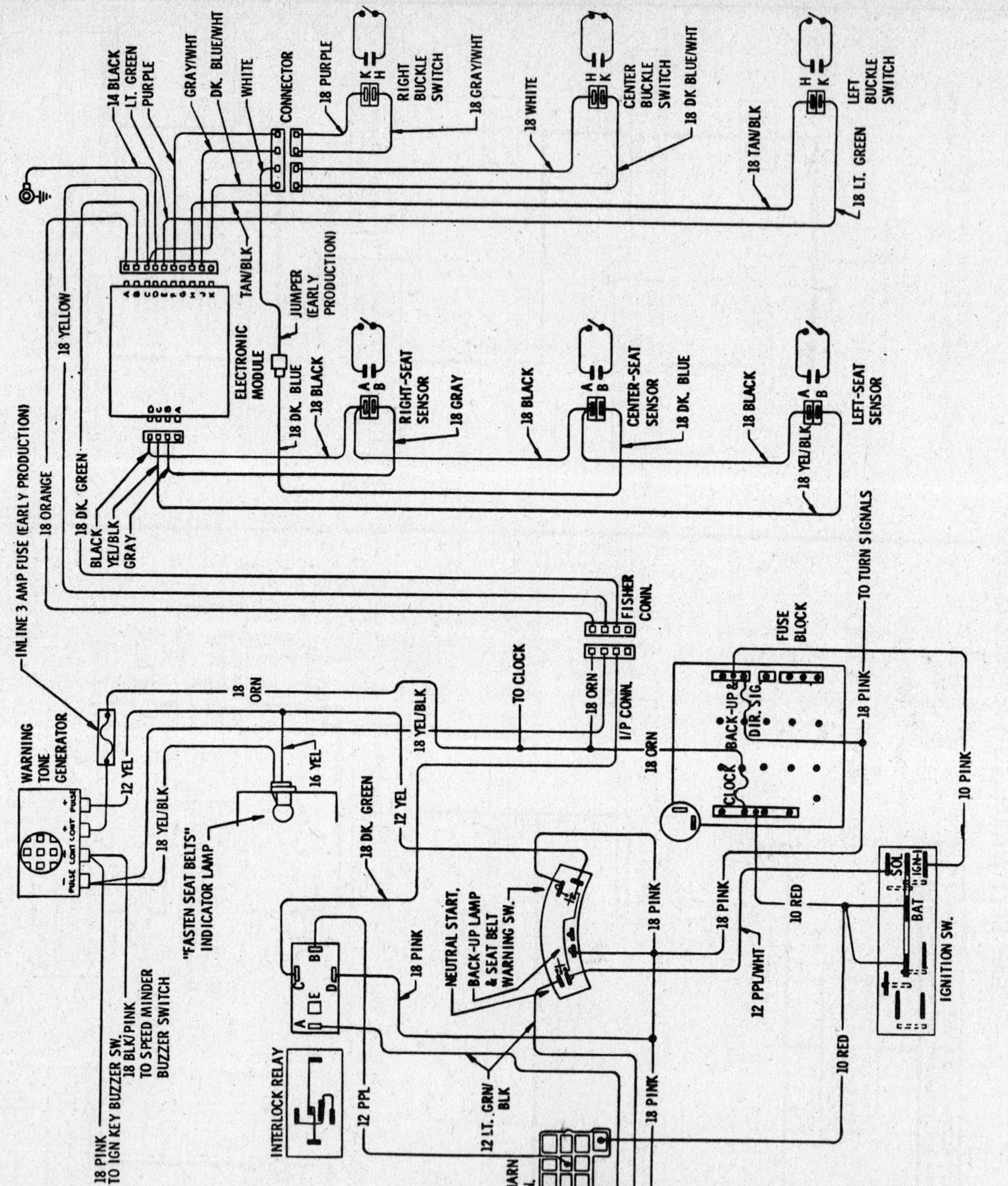

Fig. 49 Interlock system wiring diagram. All 1974 Buick exc. Apollo (Early type)

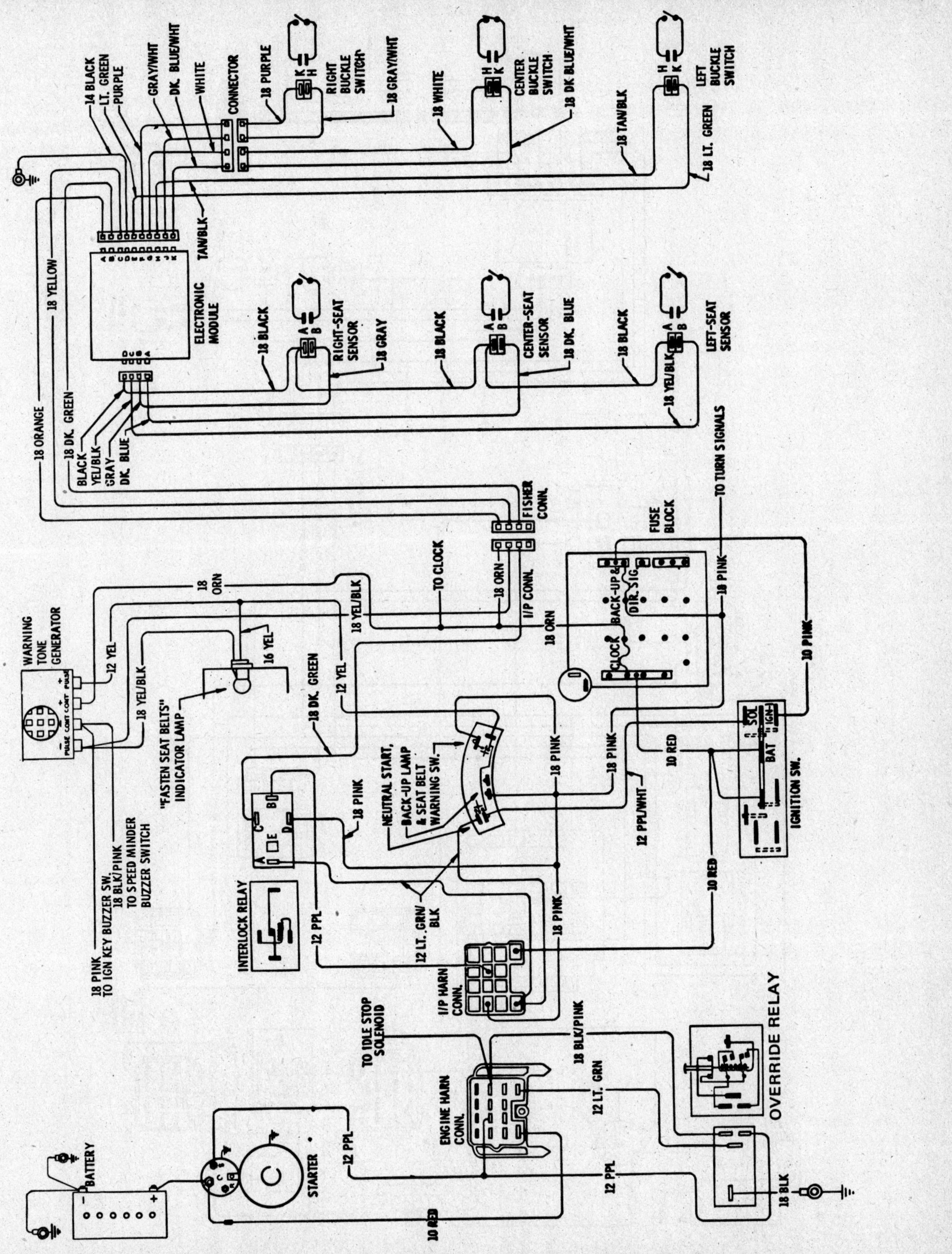

Fig. 50 Interlock system wiring diagram. All 1974 Buick exc. Apollo (Late type)

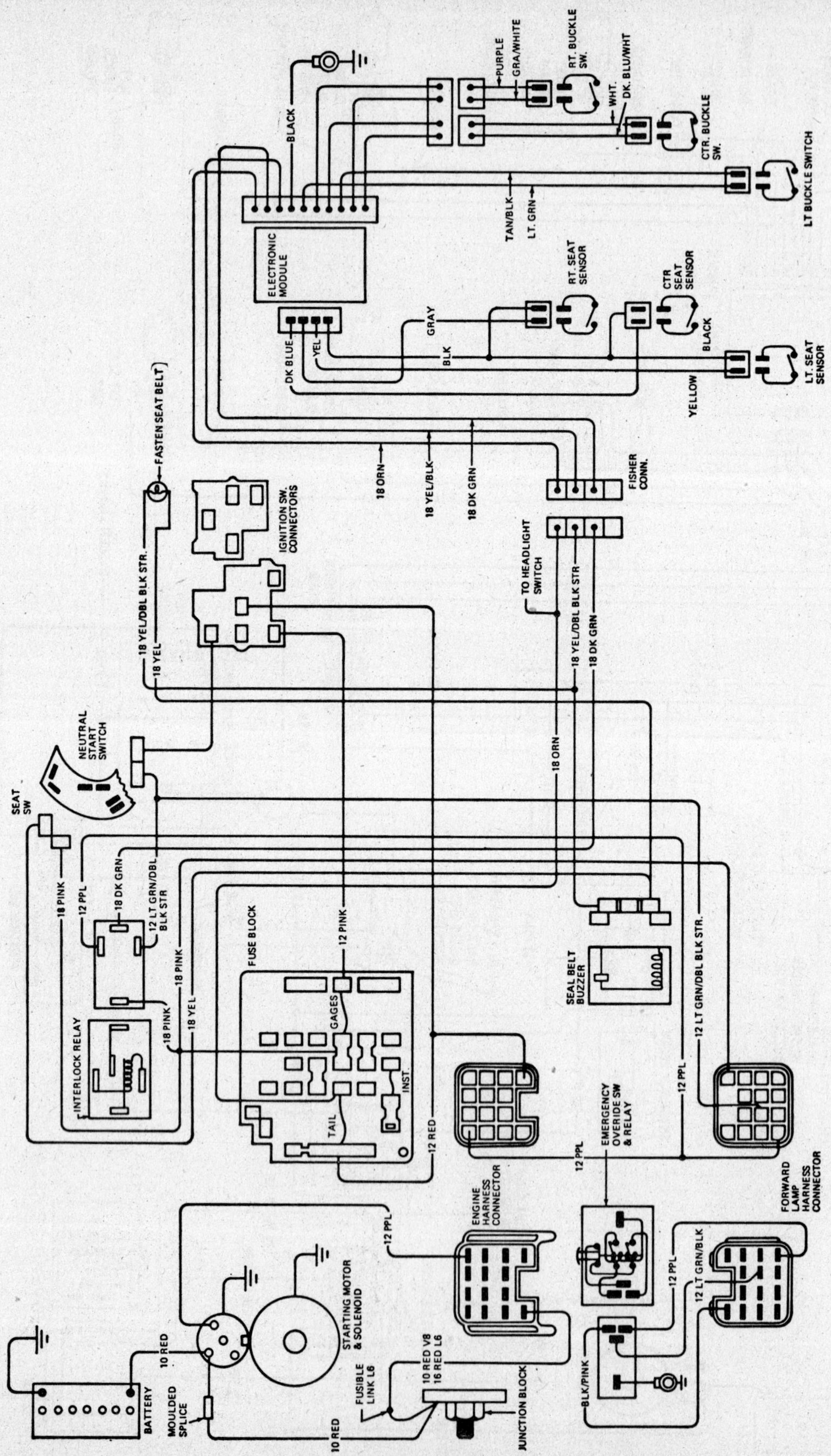

Fig. 51 Interlock system wiring diagram. 1974 Apollo with Auto. Trans.

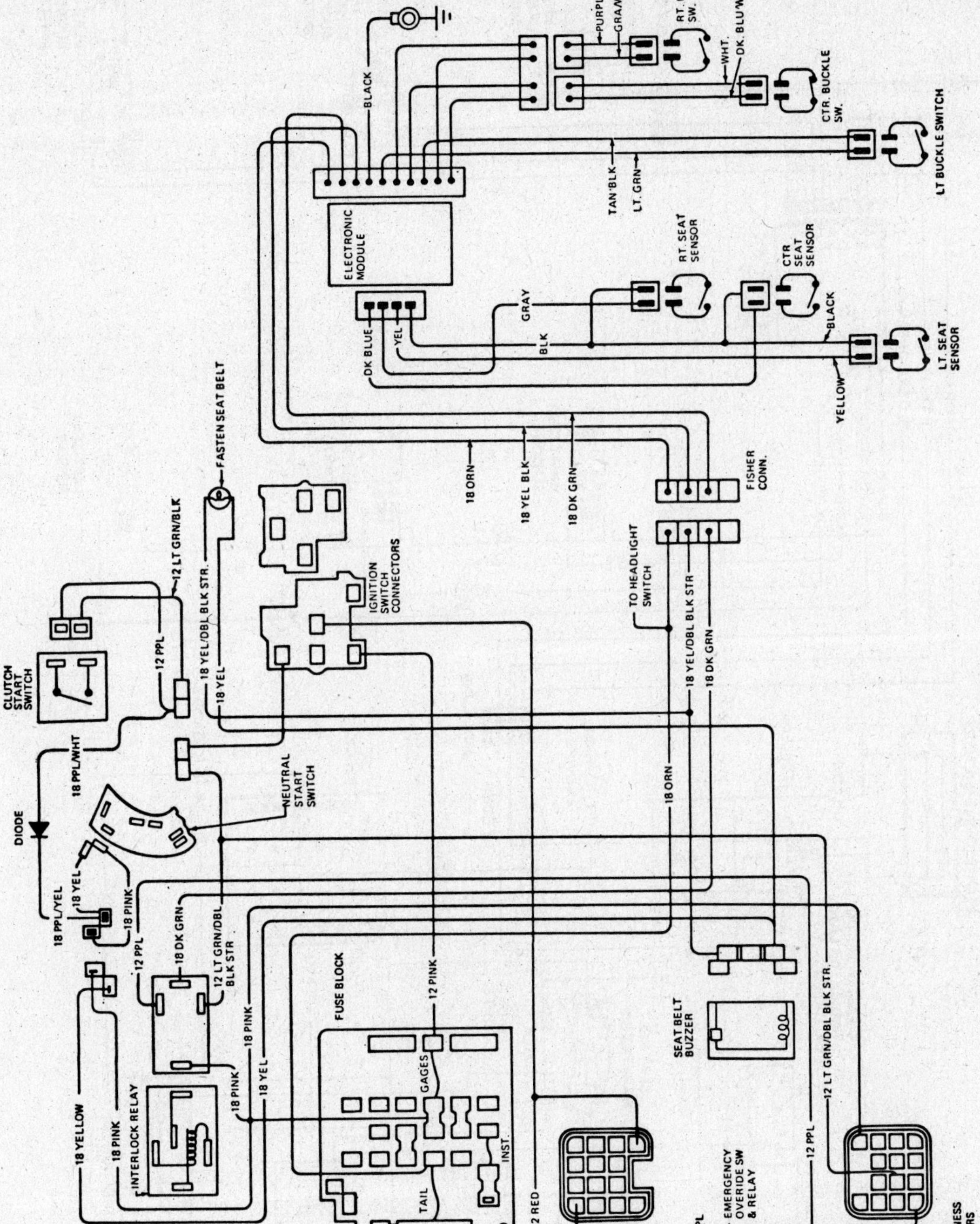

Fig. 52 Interlock system wiring diagram. 1974 Apollo with Manual Trans.

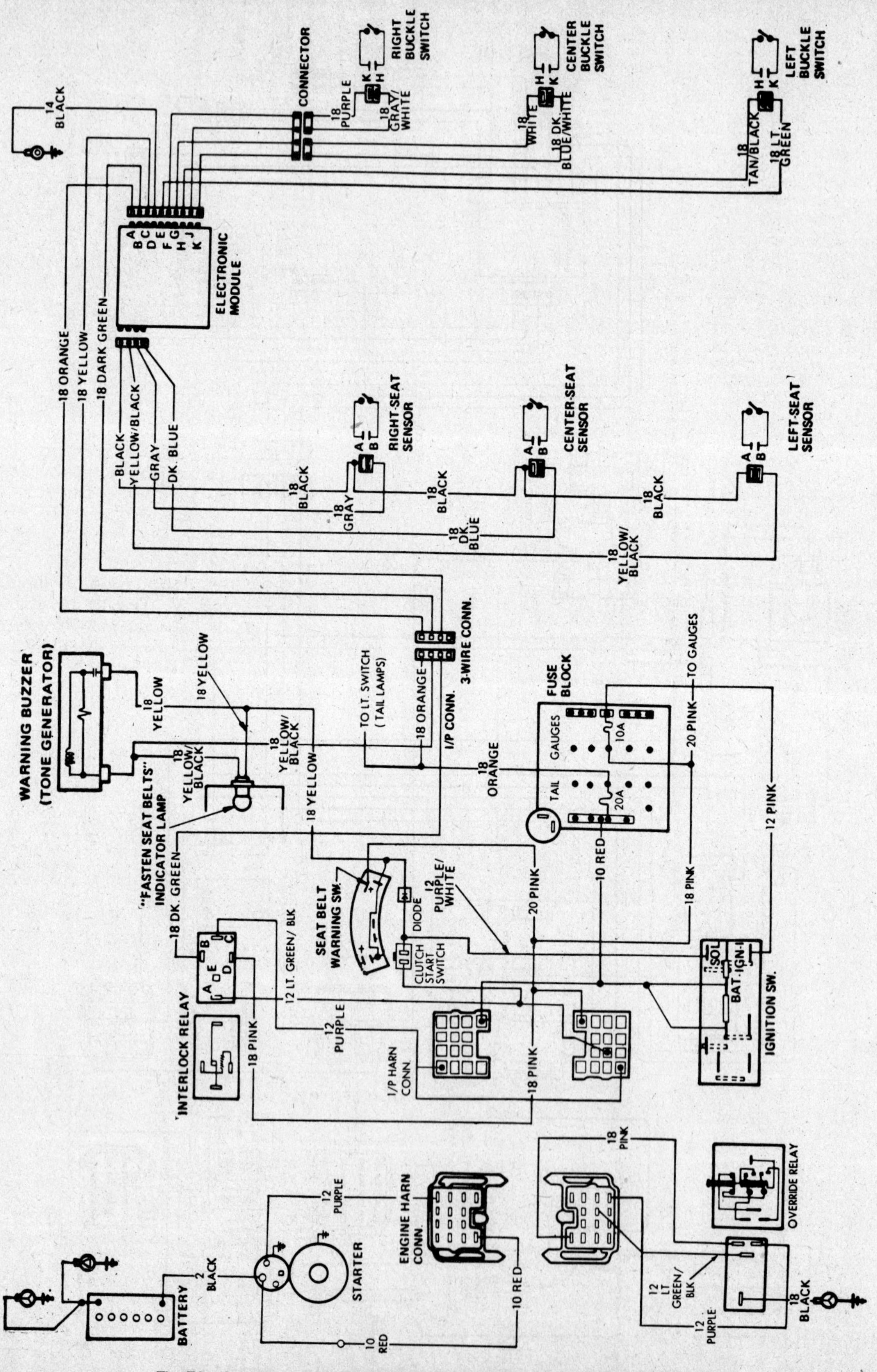

Fig. 53 Interlock system wiring diagram. Chevrolet all, with Manual Trans.

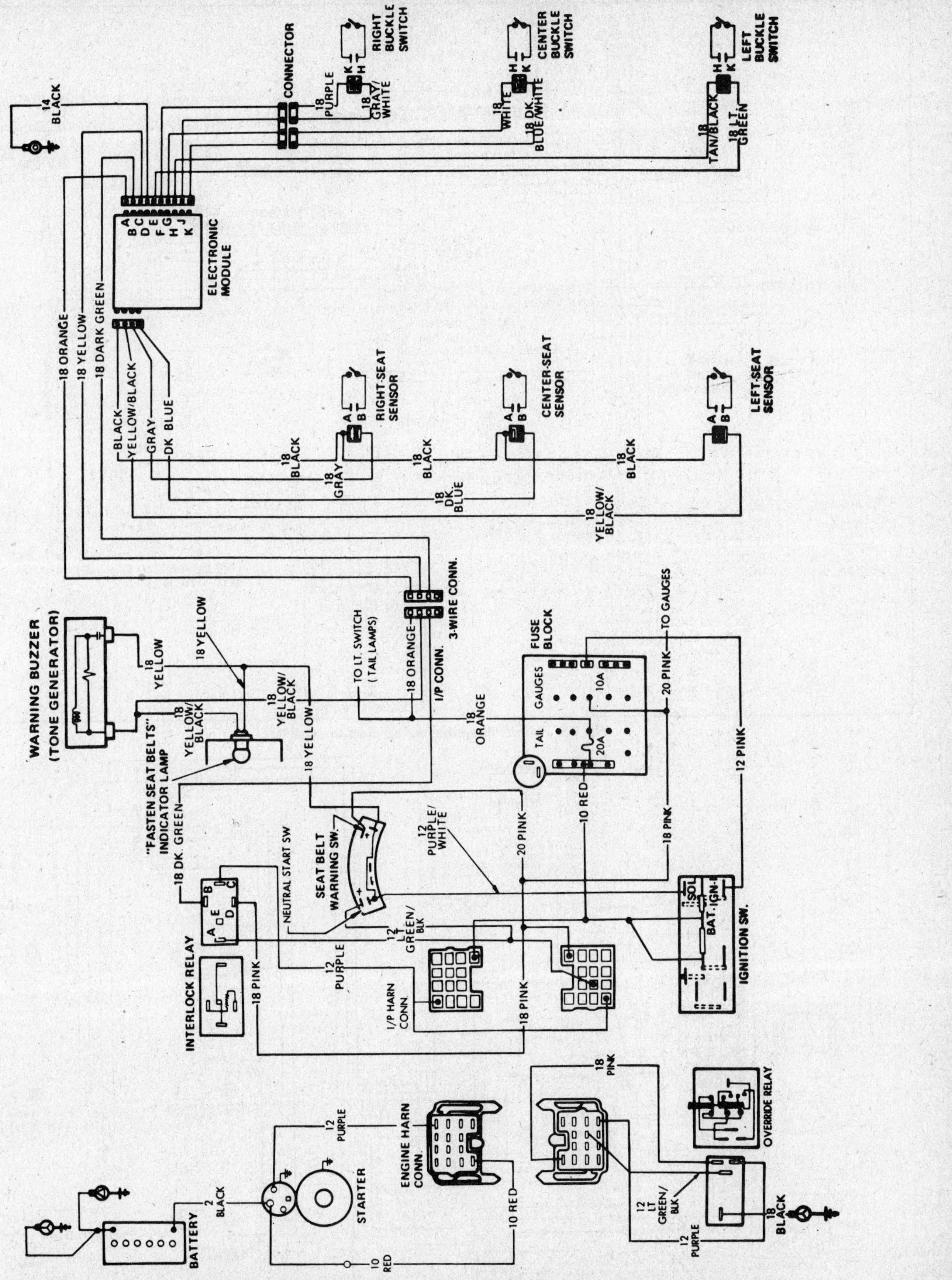

Fig. 54 Interlock system wiring diagram. Chevrolet all & Pontiac Astre with Auto. Trans.

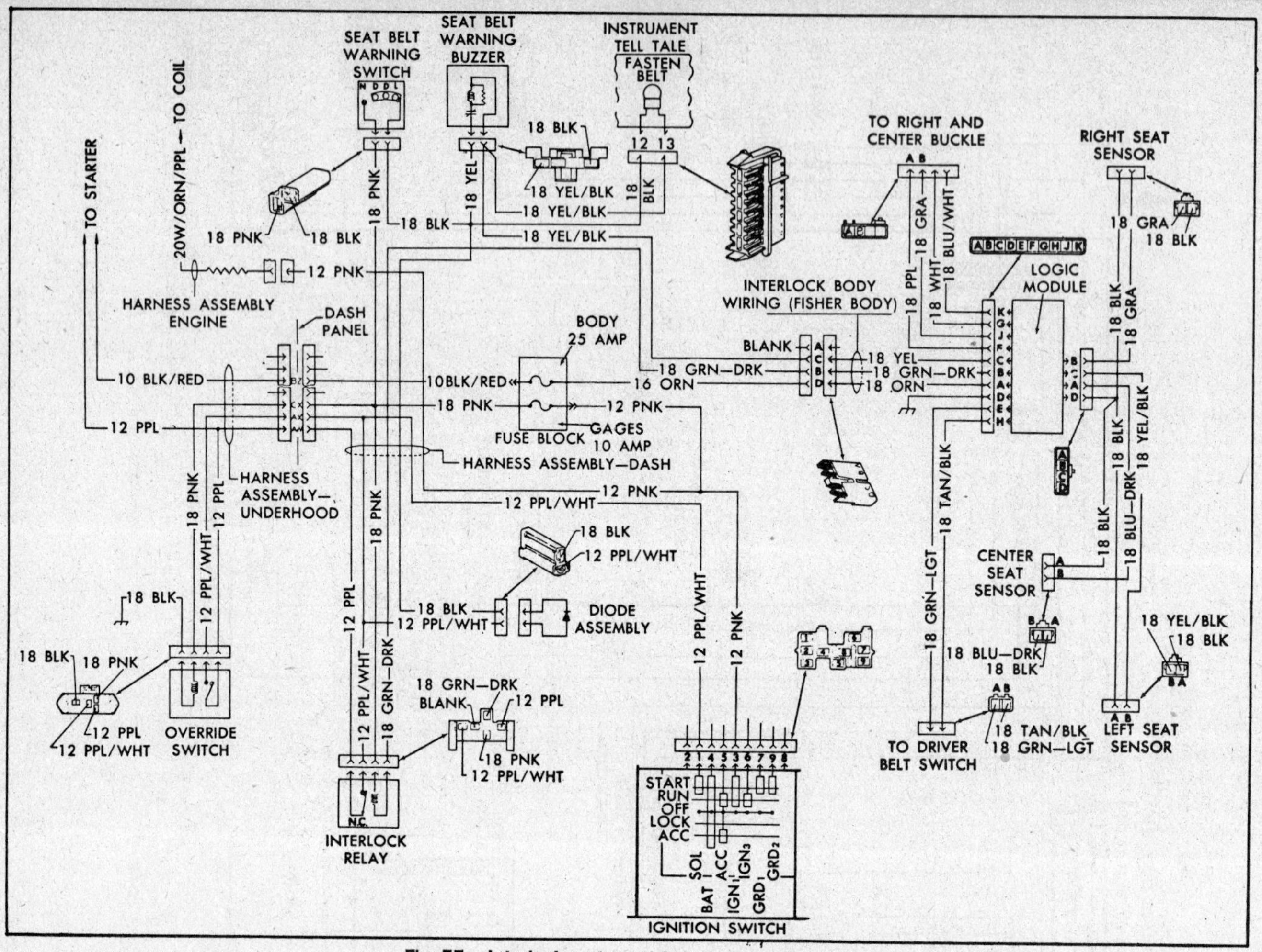

Fig. 55 Interlock system wiring diagram. Cadillac

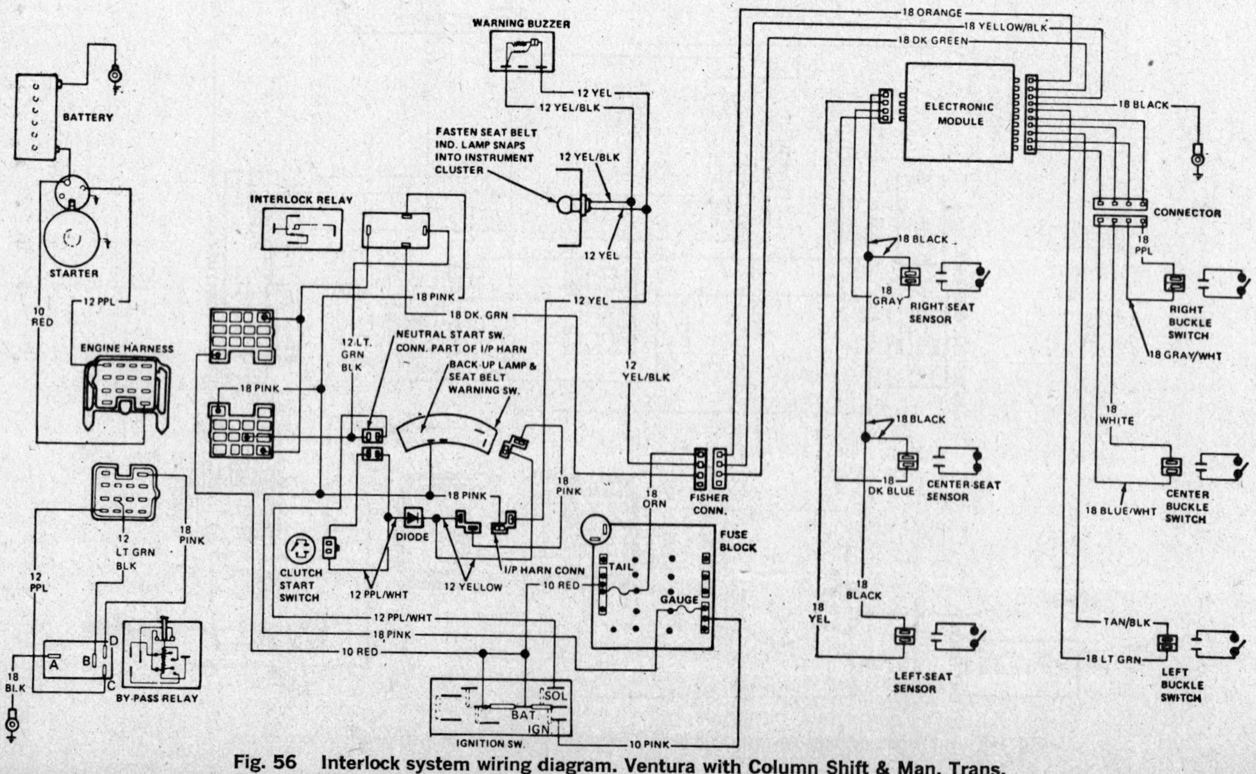

Fig. 56 Interlock system wiring diagram. Ventura with Column Shift & Man. Trans.

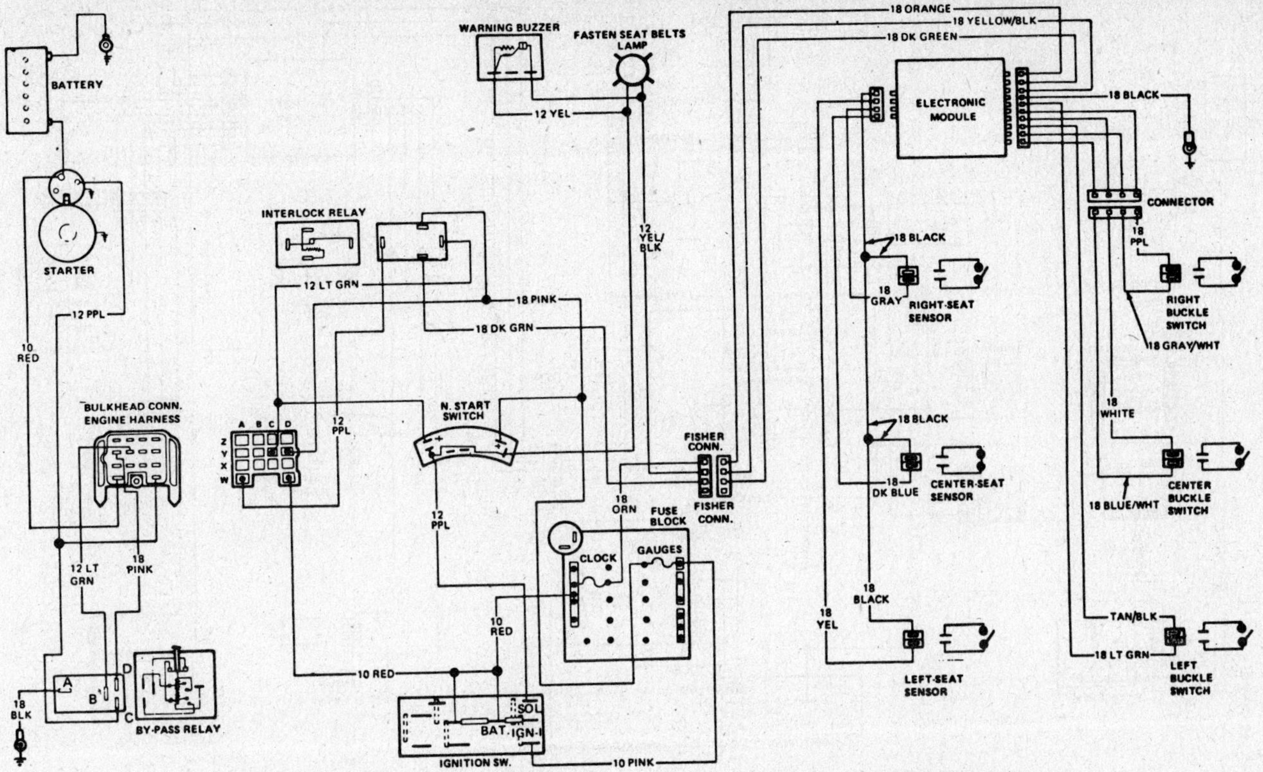

Fig. 57 Interlock system wiring diagram. All Pontiac Auto. Trans. exc. Astre & Ventura

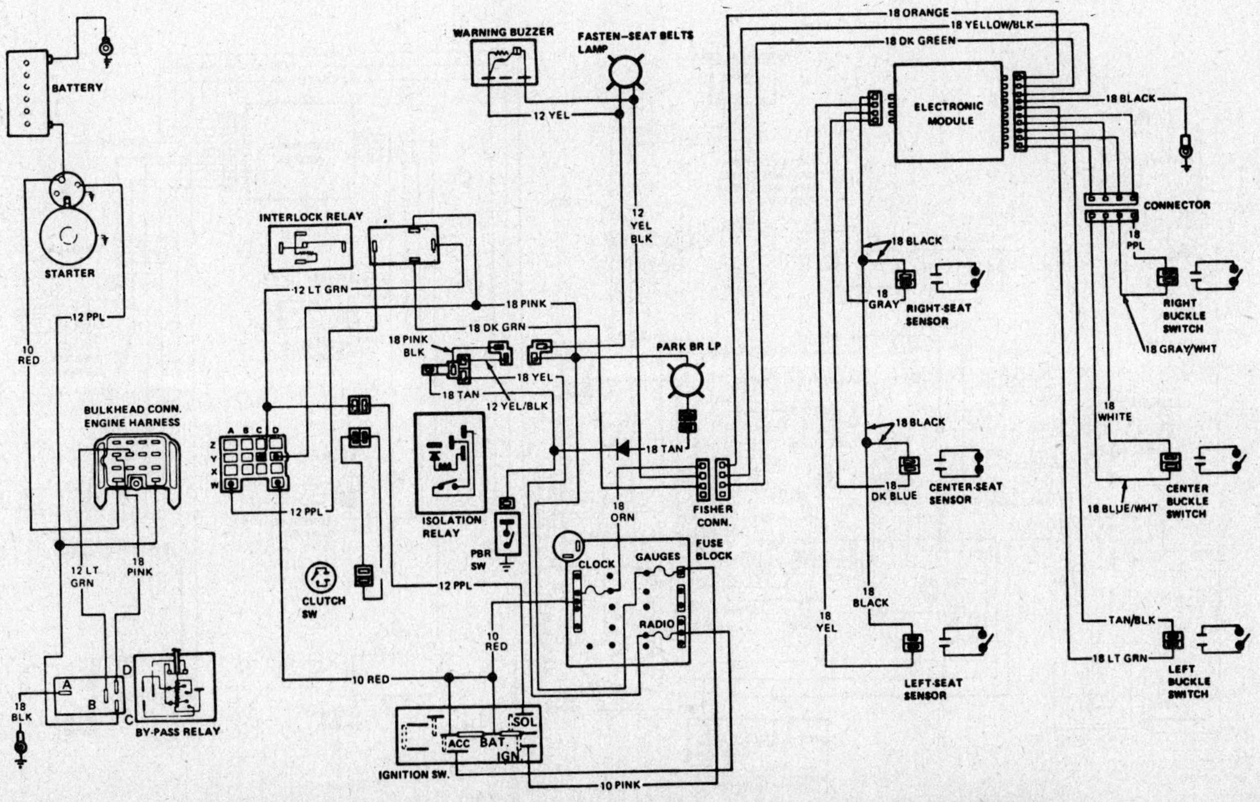

Fig. 58 Interlock system wiring diagram. All Pontiac Manual Trans. exc. Astre & Ventura

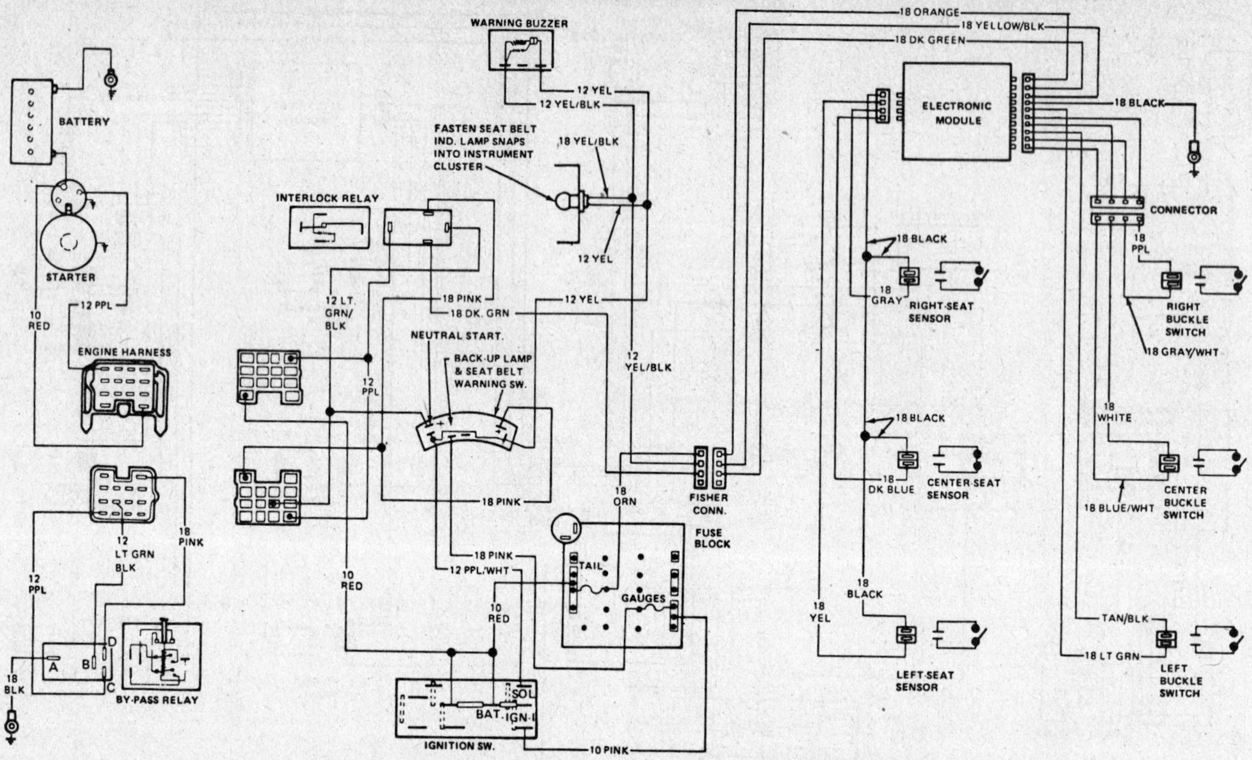

Fig. 59 Interlock system wiring diagram. Ventura with Auto. Trans.

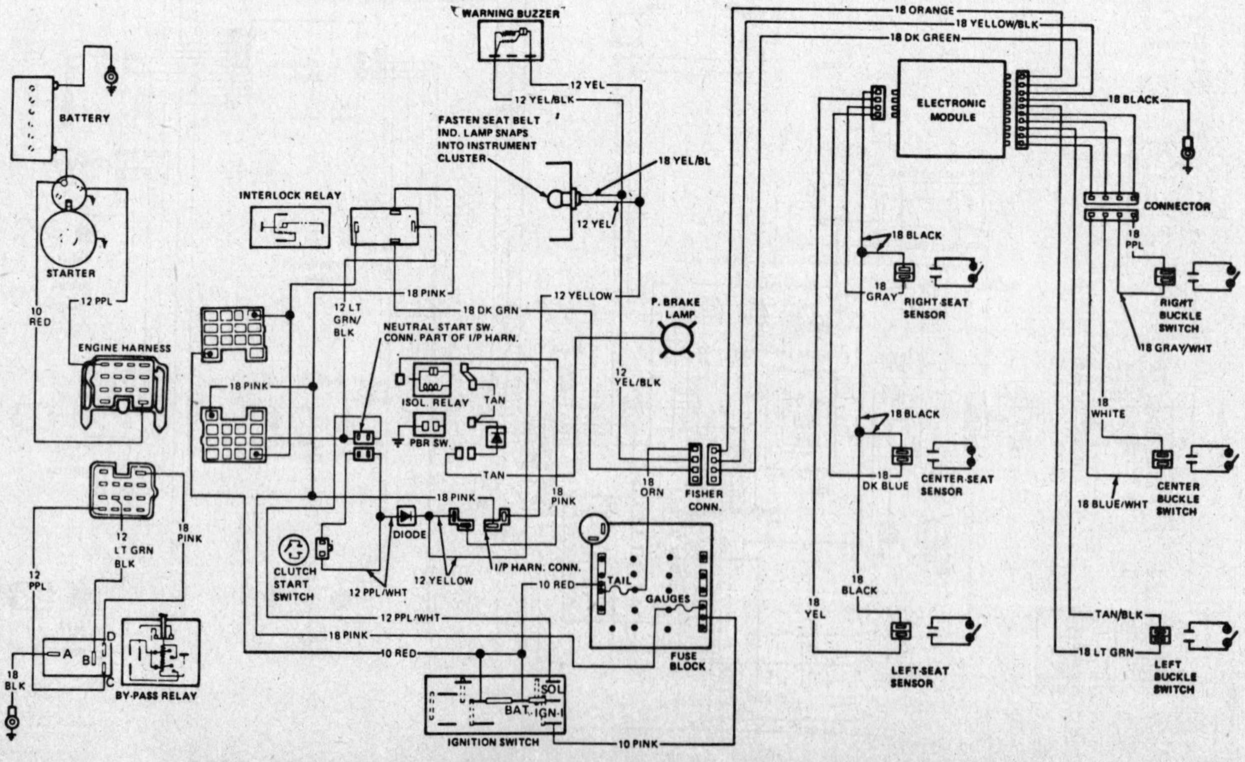

Fig. 60 Interlock system wiring diagram. Ventura & Astre with Floor Shift & Manual Trans.

DISCONNECT PROCEDURES

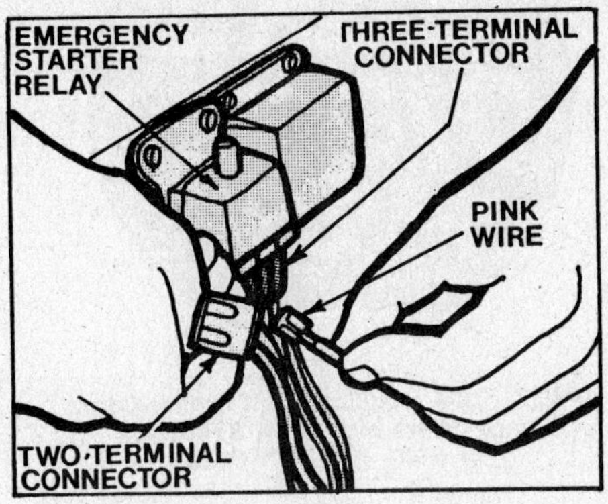

Fig. 1 Removing pink wire from two terminal connector. American Motors

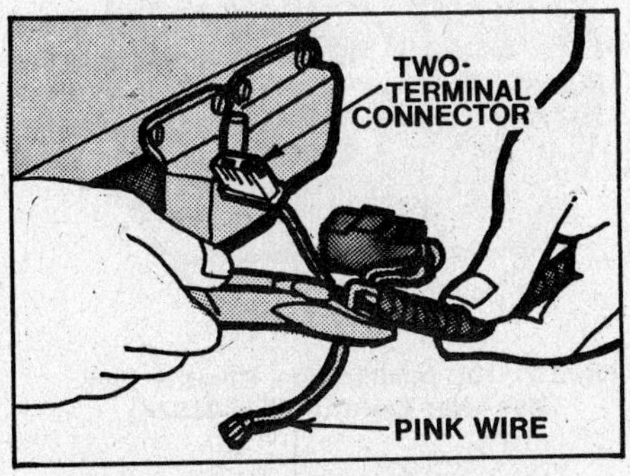

Fig. 2 Cutting pink wire from harness. American Motors

AMERICAN MOTORS

1. Remove pink wire from the two terminal connector at emergency starter relay, Fig. 1, then cut off wire close to the tape junction of wiring harness, Fig. 2.
2. Disconnect yellow wire from the three terminal connector, Fig. 3, then install the wire into the two terminal connector at location of previously removed pink wire, Fig. 4.
3. Remove interlock module bracket. Cut the yellow with-black-tracer wire as close to the module nine terminal connector as possible. Cut off remaining wire close to the taped junction of wiring harness.
4. Check for proper operation.
   A. On 1975 models with the green module, seat belt buckling is non-sequential. Fasten seat belt for occupied seat positions. The reminder light should remain off.
   B. On all other models, seat belt buckling is sequential. Occupants must be seated prior to fastening seat belts or turning ignition switch on. Fasten seat belts for occupied seat positions, turn ignition on and place vehicle in gear. The reminder light should remain off.

NOTE: If an occupant rises off the seat with a front door open, with dome light on or the seat belt unbuckled, it is necessary to rebuckle to turn reminder light off.

CHRYSLER CORP.

NOTE: Remove interlock control unit (ICU) from mounting to gain access to wiring connectors.

Deactivate Warning Buzzer

1. Disconnect nine-cavity wiring connector from ICU.
2. Momentarily ground the wire in cavity No. 6, Fig. 5. Buzzer should activate.
3. Cut wire at cavity No. 6 and tape to harness.

Deactivate Interlock Function

1. Disconnect ICU from wiring harness and remove ICU cover.
2. Refer to Fig. 6 to locate general area of circuit board to be reworked.
3. Using a soldering iron of 50 watts or less, or a soldering gun of 100 watts or less and rosin core solder, bridge the pads on the circuit board, Figs. 7 through 10. If a wire is present at one pad, bend toward the other pad and solder over wire. Do not overheat, however, ensure solder has melted and has a smooth surface. Remove any splashes of solder.
4. Install ICU cover, connect ICU to wiring harness and install into vehicle.
5. Check warning light for proper operation.

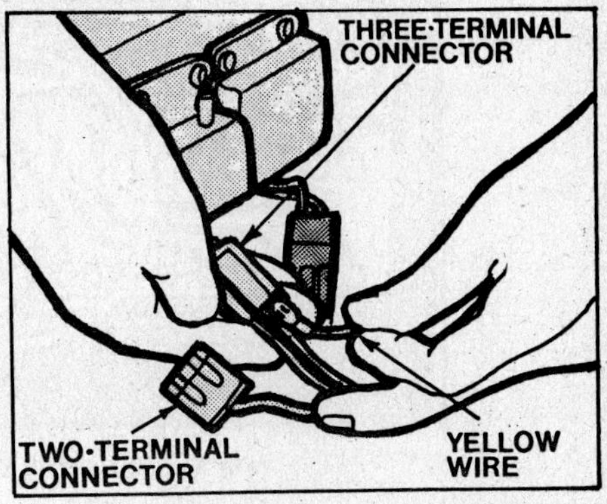

Fig. 3 Removing yellow wire from three terminal connector. American Motors

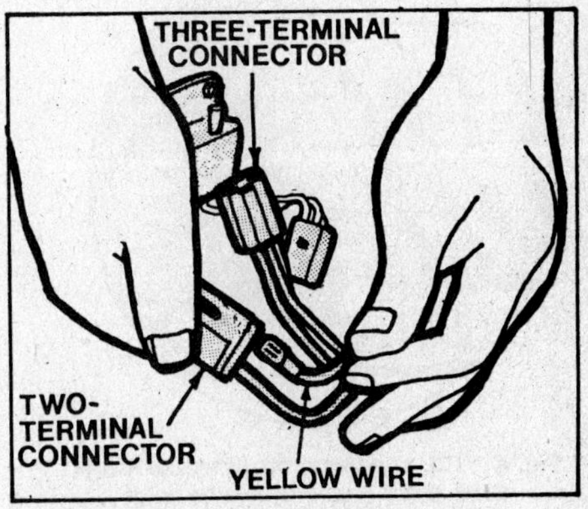

Fig. 4 Installing yellow wire in two terminal connector. American Motors

FORD MOTOR CO.

NOTE: Apply parking brake and remove ignition key prior to performing any rework operations.

1. Referring to Figs. 11 and 12, cut circuit Nos. 32 and 33.
2. On the wiring harness end, splice circuit Nos. 32 and 33 together, using a

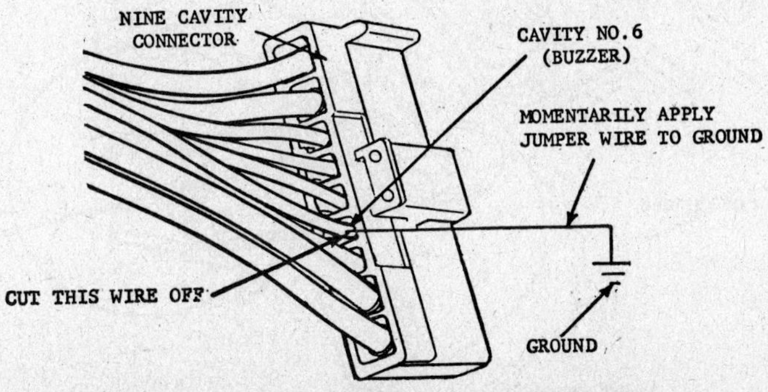

Fig. 5 Deactivating warning buzzer. Chrysler Corp.

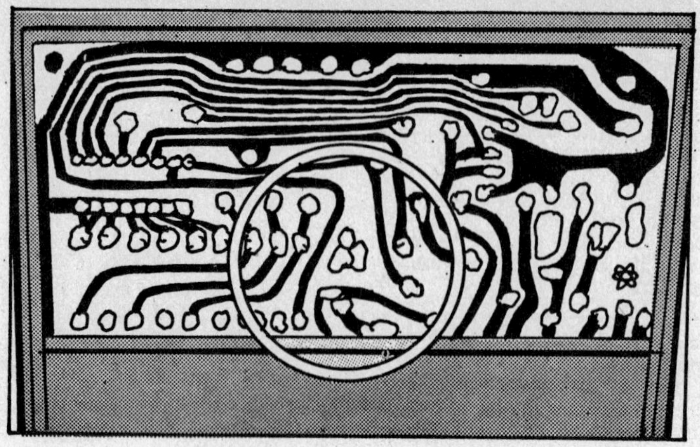

Fig. 6 ICU circuit board. Chrysler Corp.

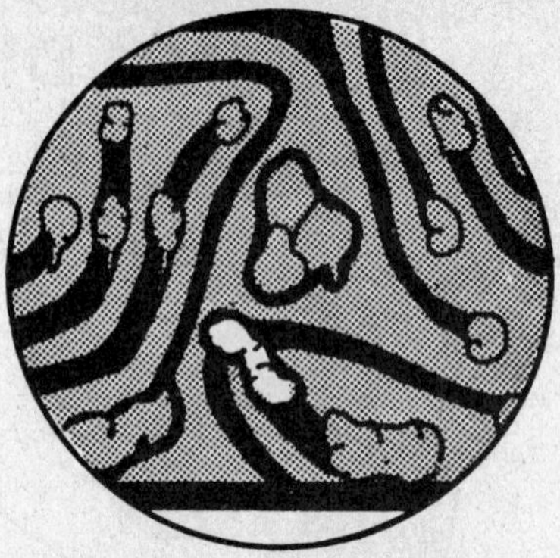

Fig. 7 ICU modification. Chrysler Corp. Case colors blue (P/N 3746653) or white (P/N 3746810)

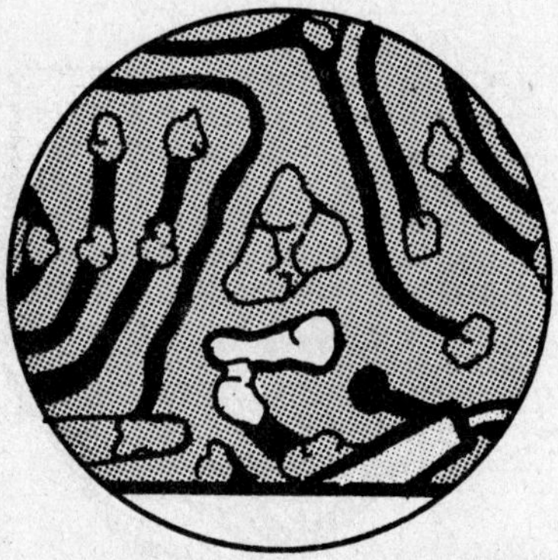

Fig. 8 ICU modification. Chrysler Corp. Case color blue with white paint (P/N 3501372)

Fig. 9 ICU modification. Chrysler Corp. Case color buff (P/N 3746849)

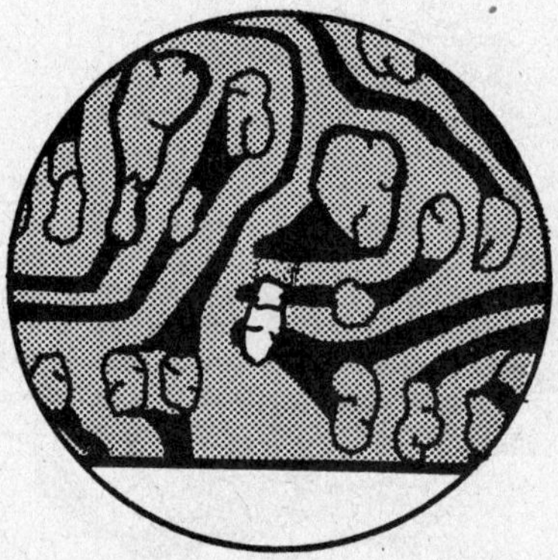

Fig. 10 ICU modification. Chrysler Corp. Case color green (P/N 3501374)

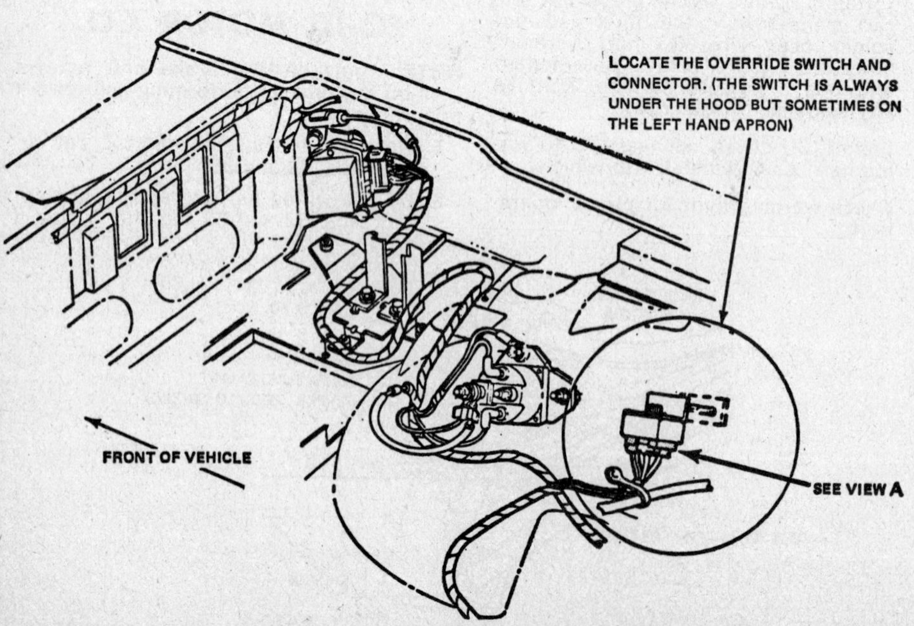

Fig. 11 Override switch location. Ford Motor Co.

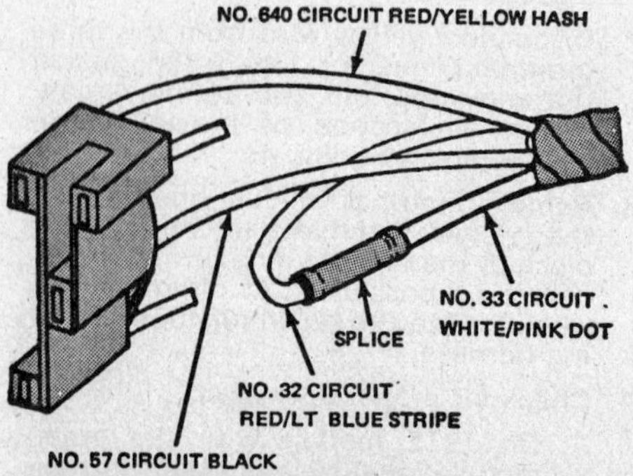

Fig. 12 Deactivating interlock function. Ford Motor Co.

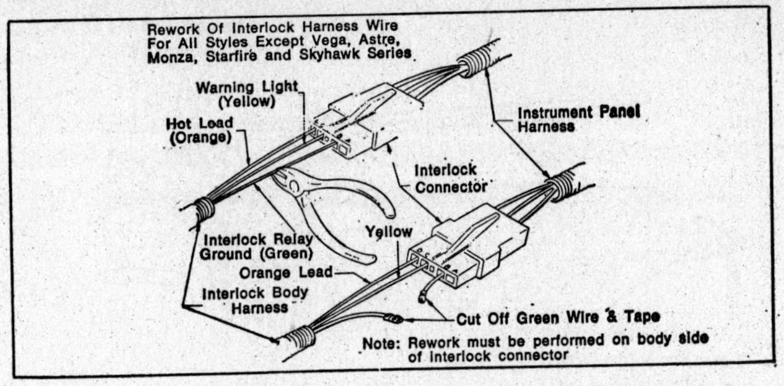

Fig. 13 Deactivating interlock function. General Motors Corp. exc. Astre, Monza, Skyhawk, Starfire & Vega

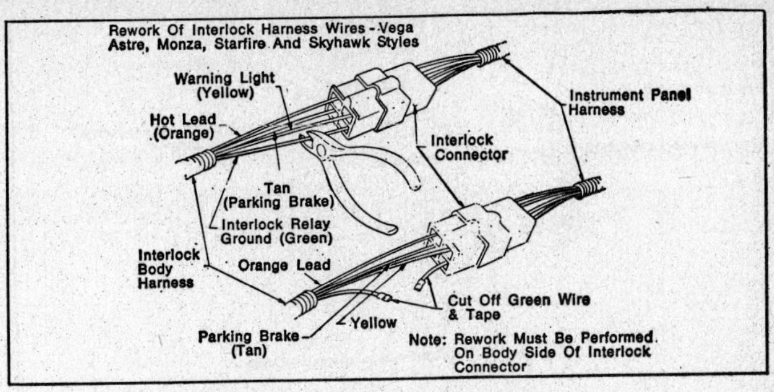

Fig. 14 Deactivating interlock function. Astre, Monza, Skyhawk, Starfire & Vega

B9A-14487-A butt connector on a 16 to 18 gauge wire or a B9A-14487-B connector on a 10 to 14 gauge wire.

NOTE: If the circuit terminals contain two wires, cut and splice all wires from circuits Nos. 32 and 33 into one butt connector.

---

3. Tape complete splice to waterproof, then connect the connector to override switch.
4. Check for proper operation. Buckle seat belt and turn ignition switch to "On" position. If starter cranks with switch in "On" position and in any gear, the wires have been crossed. Check the rework wiring.
5. Remove buzzer from connector and tape connector to harness.

GENERAL MOTORS

Deactivate Interlock Function

1. Disconnect battery ground cable.
2. Locate interlock harness connector under instrument panel left side near or on the fuse block on all models except 1974 Vega. On 1974 Vega models, the interlock connector is located under the parking brake lever cover.
3. Referring to Figs. 13 and 14, cut green wire on body harness side of connector and tape ends.

Deactivate Warning Buzzer

Buick

1. Disconnect buzzer from fuse block or connector, Figs. 15 through 17.
2. On all models except Apollo, Skyhawk and Skylark, remove terminal with the two yellow-with-black-stripe wires from the multiple connector into which the buzzer is connected, tape the terminal and reinstall buzzer, Figs. 13 and 17.

Cadillac

1. Remove steering column lower cover.
2. Disconnect the connect and buzzer from body bracket, disconnect buzzer from harness and reinstall connector on body bracket, Fig. 17.
3. Install steering column lower cover.

Chevrolet

Referring to Figs. 15, 16 and 18, remove buzzer from connector.

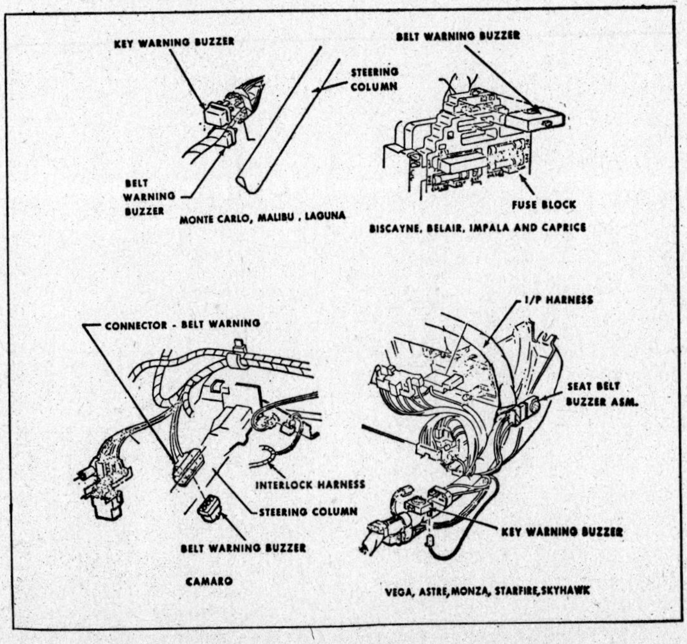

Fig. 15 Deactivating warning buzzer. General Motors Corp.

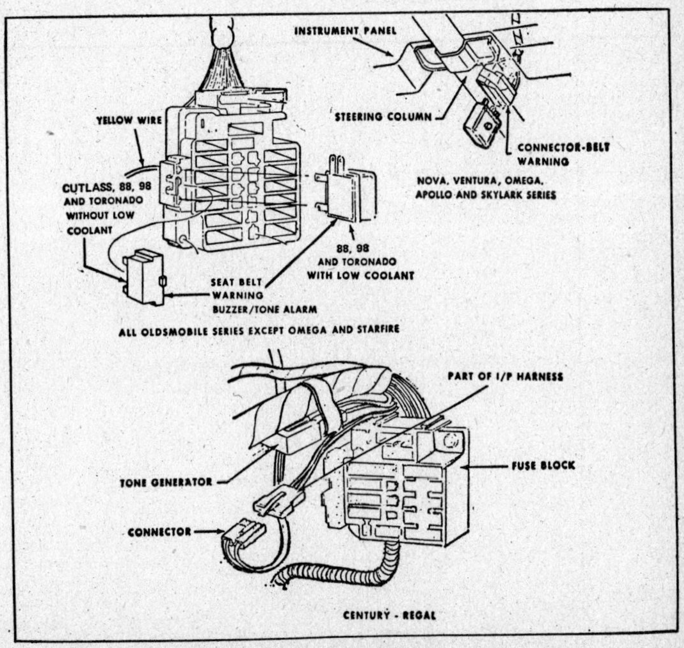

Fig. 16 Deactivating warning buzzer. General Motors Corp.

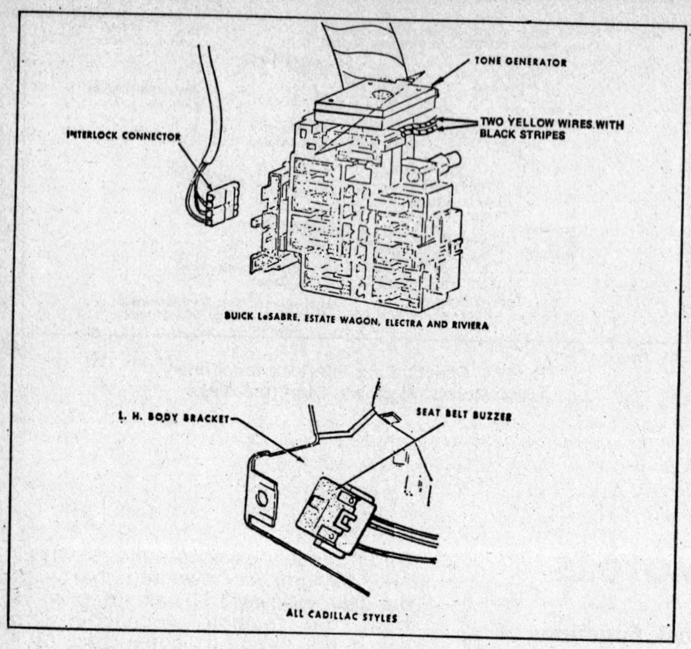

Fig. 17 Deactivating warning buzzer. General Motors Corp.

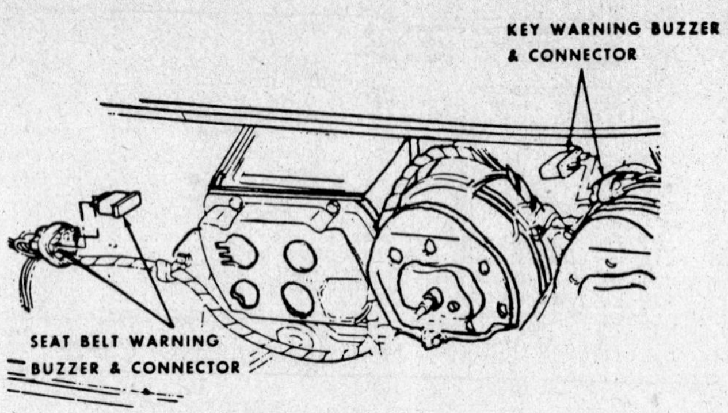

Fig. 18 Deactivating warning buzzer. Corvette.

Oldsmobile

1. On Omega and Starfire models, remove buzzer from connector, Figs. 15 and 16.
2. On models equipped with the low coolant warning and heavy duty cooling systems, cut yellow wire behind buzzer connector and tape ends, Fig. 16.
3. On models except Omega, Starfire and those equipped with low coolant warning and heavy duty cooling systems, remove buzzer from fuse block, Fig. 16.

Pontiac

Remove buzzer from fuse block or connector, Figs. 15, 16 and 19.

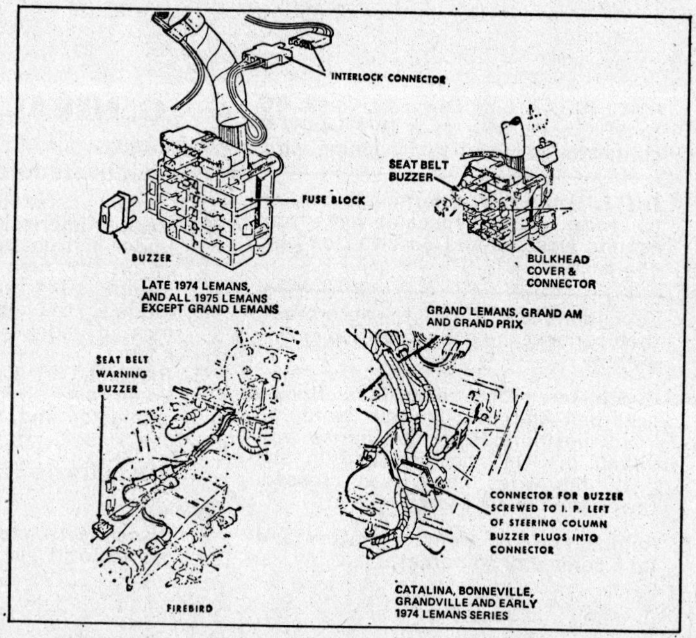

Fig. 19 Deactivating warning buzzer. General Motors Corp.

AIR CUSHION RESTRAINT SYSTEMS

GENERAL MOTORS

The Air Cushion Restraint System is designed to deploy for all front seat occupants when the vehicle is involved in a front end accident situation of sufficient force, up to 30° off the center line of the car. Injury reduction is provided through the use of inflatable air cushion assemblies, which are deployed to restrain the occupants, when an accident is sensed by the crash sensing mechanisms.

The complete system, Fig. 1, consists of the following components: The Driver System, Passenger(s) System, Sensing Mechanisms and the Diagnostic System.

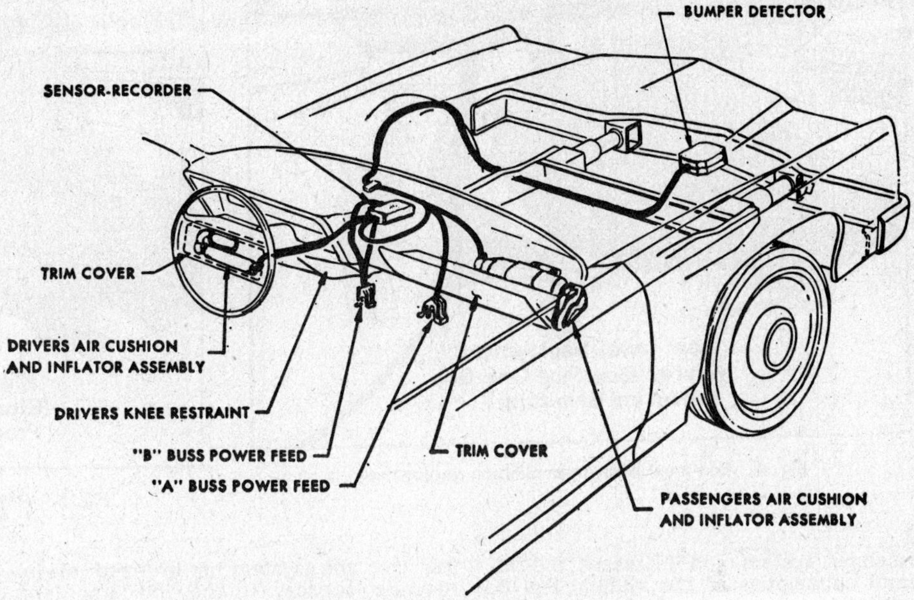

Fig. 1 Air Cushion Restraint System

Operation

Driver System

The driver system consists of an air cushion and inflator assembly which is stored in the steering wheel hub area and covered by a vinyl trim cover, Fig. 2.

When an accident signal is sensed by the sensing mechanisms, the air cushion is deployed by a gas producing generator that inflates the driver cushion, Fig. 2. The trim cover opens at predetermined rupture seams as the system is deployed.

A knee restraint energy absorbing pad and a telescoping steering column complete the driver protection system.

Passenger(s) System

The passenger(s) system consists of an inflator, air cushions and trim cover. The system is positioned low on the instrument panel and to the right of the centerline of the car. This portion of the system provides protection for the right and center front passengers. It is positioned behind a vinyl-skinned, foam trim cover that opens when the system is deployed.

When a signal is received from the crash sensing mechanisms, stored gas and gas producing generators are used to inflate the air cushion assembly, Fig. 3.

Sensing Mechanisms

The system incorporates two devices to control the deployment of the driver and passenger systems, a bumper impulse detector and a passenger compartment crash sensor.

The bumper impulse detector and/or passenger compartment sensor detects "low-level" front end impacts equal to a 12-18 miles per hour barrier crash early in the accident sequence. When a "low-level" crash signal is received, the driver system is deployed, Fig. 2, and the stored gas in the passenger system inflator assembly is released and one of the two gas producing generators is activated to produce "low-level" deployment of the passenger system, Fig. 4.

The passenger compartment sensor also detects "high-level" front end impacts equivalent to an 18 miles per hour, or greater, barrier crash. It activates a second gas producing generator in the

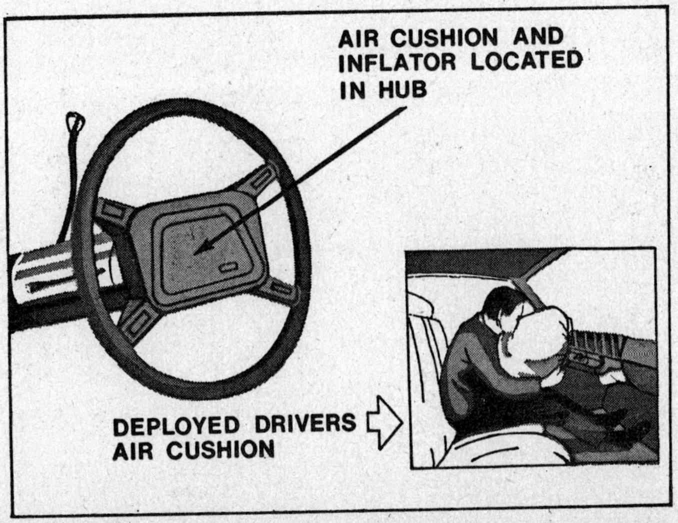

Fig. 2 Driver air cushion deployed

Fig. 3 Passenger air cushion deployed

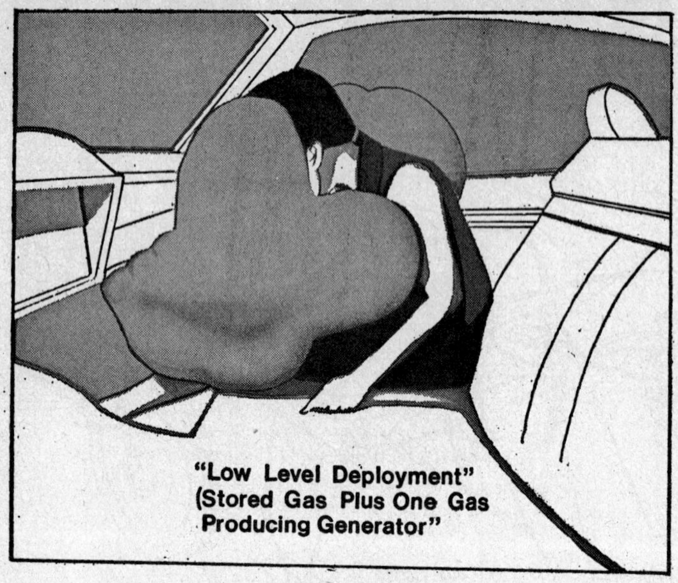

Fig. 4 Low-level passenger cushion deployment

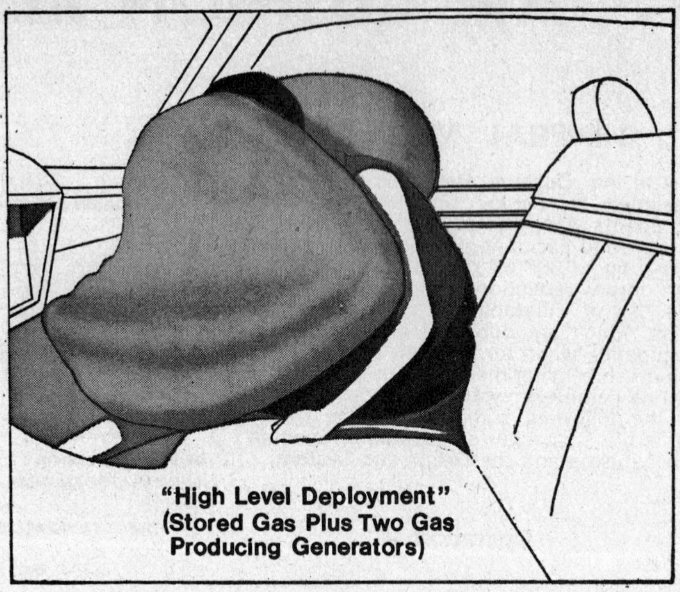

Fig. 5 High-level passenger cushion deployment

passenger system and increases the restraint capabilities of the system, Fig. 5.

Diagnostic System

The complete Air Cushion Restraint System has an integral electronic diagnostic system, which monitors the systems electrical circuits. It indicates to the driver that the system is either operating properly or is in need of immediate service. This is accomplished by an indicator lamp located on the instrument panel that lights when the ignition switch is first turned "ON".

If the indicator lamp does not light when the ignition switch is first turned "ON," remains lighted after 3-9 seconds, or comes on intermittently when driving the car, the system is in need of immediate service. If the indicator lamp comes on when the ignition switch is first turned "ON," and goes out after 3-9 seconds the system is functioning properly.

Service

WARNING: Do not attempt any servicing or disconnecting of the Air Cushion Restraint System instrument panel, instrument panel components or any other electrical components in the adjacent area, until the negative cable has been disconnected from the battery and the terminal end of the cable taped. This procedure must be followed to prevent accidental deployment of the system which could result in personal injury and/or damage to the system components. In addition, care must be exercised to never strike the sensing mechanisms in a manner which could cause inadvertent deployment or improper operation of the system.

A special tester, J-24628, with substitute loads is used to check the Air Cushion Restraint System to determine whether or not a specific part or the entire system is functioning properly. Therefore, any service by other than properly trained personnel, should not be attempted.

FUEL PUMPS

NOTE: Fuel pump pressures are listed in car chapters.

MECHANICAL FUEL PUMP FIG. 1

Operation

During the suction stroke, the rotation of the camshaft eccentric moves the pump rocker arm which in turn pulls the diaphragm up, causing fuel to be drawn in through the one-way inlet valve. This suction also closes the outlet valve. During the return stroke, the diaphragm is forced down by the diaphragm spring. The resulting fuel pressure opens the outlet valve and closes the inlet valve. Fuel is then forced out through the outlet valve to the carburetor.

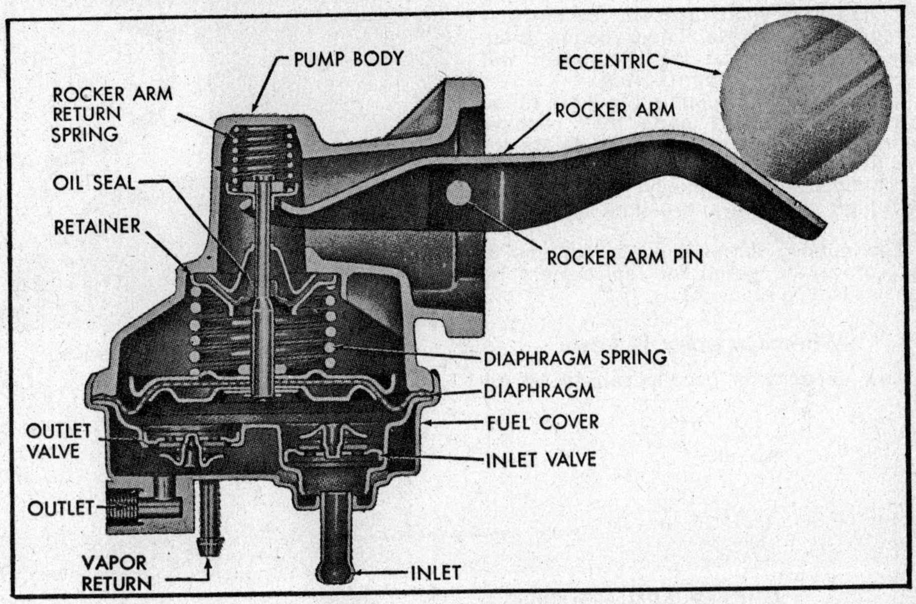

Fig. 1 Single action mechanical fuel pump. Typical

Fuel Pump Performance

It is essential that the fuel pump deliver sufficient fuel to supply the requirements of the engine under all operating conditions and that it maintain sufficient pressure in the line between the pump and carburetor to keep the fuel from boiling and to prevent vapor lock.

Excessive fuel pump pressure holds the carburetor float needle valve off its seat, causing high gasoline level in the float chamber which in turn increases gasoline consumption.

The pump usually delivers a minimum of ten gallons of gasoline per hour at top engine speeds, under an operating pressure of from 2 to 6 psi. The highest operating pressure will be attained at idling speed and the lowest at top speed.

Fuel Pump Tests

The fuel pump can be tested on the car with a pressure gauge, a hose and a pint measuring can. With this equipment, it is possible to check the fuel pump to see if it is delivering the proper amount of gasoline at the correct pressure.

Pressure Test

To make the pressure test, disconnect the fuel pipe at the carburetor inlet and attach the pressure gauge and hose between the carburetor inlet and the disconnected fuel pipe, Fig. 3. Take the pressure reading with the engine running. The pressure should be within the limits listed in the individual car chapters. The pressure should remain constant or return very slowly to zero when the engine is stopped.

Capacity Test

To make this test, connect the hose so the pump will deliver gasoline into the pint measure held at carburetor level. Run the engine at idle speed and note the time it takes to fill the measure. On the average it should take from 20 to 30 seconds, depending on the pump tested.

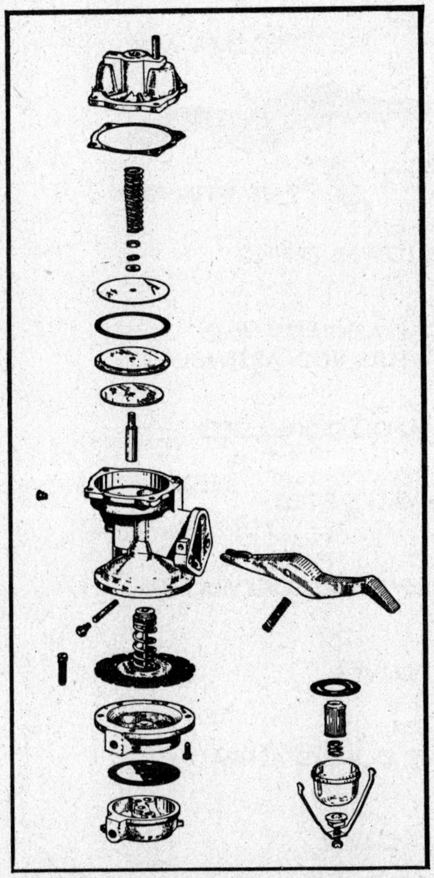

Fig. 2 1969-71 American Motors double action (combination) pump. Exploded

When Pressure Is Low

Low pressure indicates extreme wear on one part, small wear on all parts, ruptured diaphragm, dirty valve or gummy valve seat.

Wear in the pump usually occurs at the rocker arm pivot pin and on the contacting surfaces of the rocker arm and links. Due to the leverage design, wear at these points is multiplied five times in the movement of the diaphragm. It is apparent therefore, that very little wear will materially reduce the stroke of the diaphragm. The worn parts must be replaced for a satisfactory correction.

The diaphragm pull rod has an oil seal around it which prevents the hot oil vapors from the crankcase coming in contact with the diaphragm. If this seal is damaged, the oil vapors have a tendency to shorten the life of the diaphragm.

The first three conditions—extreme wear on one part, small wear on all parts, and ruptured diaphragm—are brought about by usage, while dirty and/or poor fuel is usually the cause of valve trouble.

When Pressure Is High

High pressure is caused by a tight diaphragm, fuel between diaphragm layers, diaphragm spring too strong, pump link frozen to rocker arm.

A tight diaphragm will stretch slightly on the down stroke. As the pump operates, the diaphragm will rebound on the up stroke beyond its normal position, much as a stretched rubber band when it is suddenly released. This rebound will cause a higher than normal pressure in the pump chamber.

A loose diaphragm retainer nut or poor riveting on the diaphragm assembly may allow fuel to seep between the diaphragm layers. This will cause a bulge in the diaphragm and have the same effect as a diaphragm that is too tight.

A diaphragm spring that is too strong also causes a high pressure for the diaphragm will operate longer before pressure of the fuel on the diaphragm will overcome the diaphragm spring.

On a combination pump there are times when the operating parts may become badly corroded and the links freeze to the rocker arm. In this condition the pump operates continually, resulting in a very high pressure and a flooding carburetor.

The remedy for all these conditions is to remove the pump for replacement or repair, using a repair kit.

When Capacity Is Low

Low capacity is usually caused by an air leak in the intake pipe at these points: fuel pipe fitting at pump, bowl flange or diaphragm flange, fuel bowl. (It is assumed that the conditions of too little fuel have already been checked and the pump is the cause of the difficulty.)

Fig. 3 Testing fuel pump pressure

An air leak at fuel pipe fittings indicates either poor installation of pump or a defective fitting. The fitting should be tightened or replaced.

A leak at the diaphragm flange may be caused by a warped cover casting, loose diaphragm cover screws or foreign material between cover casting and diaphragm.

A leak at the bowl flange of the cover casting can usually be corrected by the installation of an extra gasket. A warped top cover indicates that the pump must be replaced.

A chipped glass or bent metal bowl may cause a leak at the bowl flange as may a defective gasket or foreign material between gasket and bowl or cover casting. A chipped glass bowl must be replaced while a dented metal bowl can be straightened.

Fuel Pump Service

NOTE: Most modern fuel pumps are sealed, non-serviceable units. These pumps cannot be repaired and must be replaced as units.

Illustrated are representative fuel pumps among the many serviceable models that have been produced. Before disassembling any pump, scribe a mark across the housings in such a manner that they may be reassembled with inlet and outlet fitting holes in correct location.

When disassembled, clean all parts (except diaphragms) in solvent and blow dry with compressed air. Examine the diaphragm for cracks, torn screws holes or ruptures. If deteriorated, install new diaphragm and pull rod assembly. Check the strainer screen and if found to be corroded or clogged, install a new screen. Check the rocker arm for wear or scoring on that portion that contacts the camshaft eccentric. If arm is scored or worn install a new one.

When reassembling a pump, do not use shellac or other adhesive on a diaphragm.

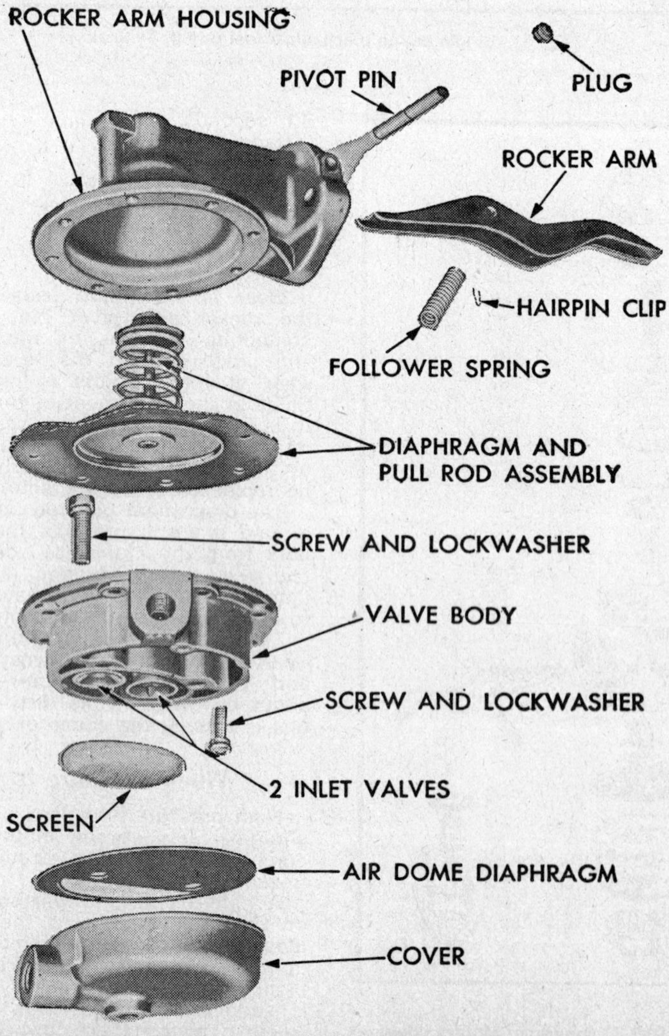

Fig. 4 Carter fuel pump 1969-71 Police Interceptor

ELECTRIC FUEL PUMPS

1969-70 Buick Riviera

These models have a turbine type electric pump located at the lower end of the fuel pick-up pipe in the bottom of the tank, Fig. 5. This pump is controlled by a switch located near the oil filter and hydraulically connected to the engine oil system so engine oil pressure controls the switch.

Operation

During cranking, current is taken from the starter solenoid and supplied to the pump as long as the starter is energized and oil pressure remains below 3 psi.

With the ignition switch in the run position and the engine oil pressure above 3 psi, current is supplied to the pump through another set of contacts in the control switch. If oil pressure drops below 3 psi with the ignition switch in the run position, the control switch opens the current, shutting off the pump.

Fuel Pump Test

1. Turn ignition switch on and be sure oil and generator lights are lit. If not, check fuse.
2. Make sure oil light goes out while cranking engine. If not, remove fuse from in-line holder just above master cylinder and replace with a new 4 amp 5/8" fuse.
3. Disconnect fuel hose from steel pipe and install pressure gauge. With engine idling, minimum pressure is 4½ psi.
4. Insert fuel hose in a suitable container. With engine idling, a pint measure should be filled in 30 seconds.
5. If pressure is low, check voltage at tank connector. If voltage is satisfactory and ground is clean and tight, pump is defective. If only fuel flow is low, check for kinked hoses or lines.
6. If no fuel flow, unplug connector from oil pressure switch and check for current at one of the parallel slots of connector.
7. If current is present at one of the slots, connect a jumper wire between the two slots and again check for flow.
8. If still no fuel flow, raise car and check fuel tank connector for current. If current is present and ground connection is clean and tight, disconnect fuel line at tank and check for fuel flow, check for kinked lines and hoses. If no fuel flow, remove pump.

1971-75 Vega; 1975 Astre, Monza, Skyhawk & Starfire

An electric fuel pump is located in the fuel tank and is an integral part of the tank unit assembly, which includes the fuel gauge metering unit. Since no repairs can be performed on the pump it must be replaced if found faulty.

Operation

The fuel pump is energized when the ignition switch is in the start position. After the engine starts, the pump receives current through the oil pressure switch as long as there is about 2 pounds of oil pressure. If, for any reason, the oil pressure drops below 2 pounds the contact is broken at the pressure switch and the pump is deactivated.

Fuel Pump Test

NOTE: Operating the pump for more than 30 seconds will seriously damage the motor unless submerged in gasoline.

The following checks should be made:

1. Check fuel flow.
2. Check for proper ground and voltage at the pump.
3. If there is no current at the tank, it will be necessary to check back to the source.
4. If there is fuel flow on the cranking cycle but not when the engine runs, jump the oil pressure switch. If this provides fuel flow, replace the switch.

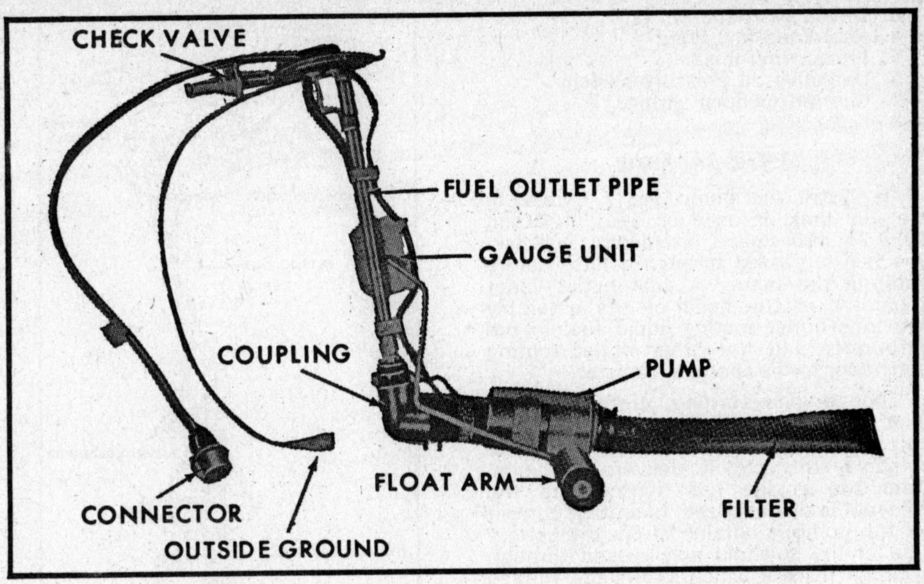

Fig. 5 Electric fuel pump and gauge tank unit assembly. 1969-70 Riviera

1972 Imperial w/Air Injection Pump

The pump used on these models is the Bendix solenoid type, Fig. 6. It is mounted in-line and has a variable output feature controlled by engine demand. To prevent the pump from being powered when it is not needed it receives power through the starting relay when the ignition switch is in start position and through the oil pressure switch when the ignition switch is in run position. Therefore the only times the pump runs is during engine start and when the engine is running and has oil pressure.

The bottom of the pump is apparently removable but there is no reason to remove it since this installation has no filter inside it. There is no service to the pump. If it becomes inoperable it is to be replaced.

The pump is located on the left side of the frame in the kick up area about even with the front of the fuel tank. The pump must be mounted so that the outlet is higher than the inlet.

Pump Test

Should the pump fail to produce fuel or produce insufficient fuel, check the following:

1. Vent in tank restricted (this might also cause tank to collapse).
2. Leaks in fuel line or fittings.

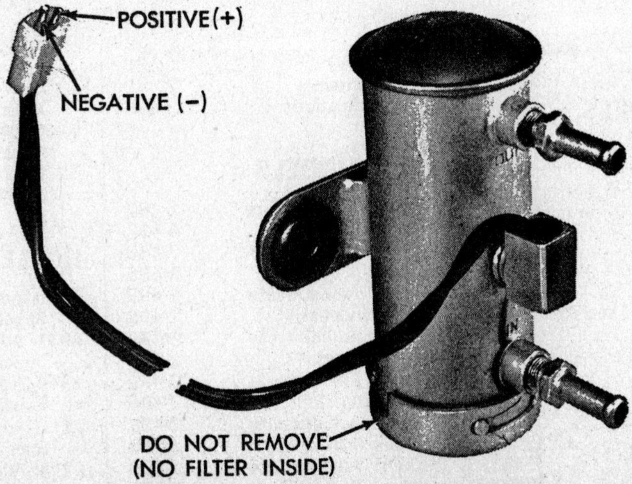

Fig. 6 Electric fuel pump. 1972 Imperial

3. Dirt or restriction in tank.
4. Restricted fuel filter.
5. Frozen fuel lines.
6. Defective oil pressure switch.
7. Shorted or open wiring.

1972-74 Ford

An electric fuel pump, Fig. 7, located in the fuel tank, is used on 1972 429 and 1973-74 460 Police Interceptor vehicles. The fuel is passed through a filter before entering the pump. A dual outlet vapor separator in the head of the assembly with one outlet routing liquid fuel to the carburetor and the other outlet routing fuel vapor to the charcoal canister.

Upon engine starting, current is supplied to the pump through a by-pass circuit from the starter side of the starter solenoid, this circuit delivers 12 volts. After the engine has started and the solenoid is de-energized, operating current for the pump is obtained from the battery side of the solenoid and passed through a circuit resistor which lowers the voltage to 8.5 to 10 volts.

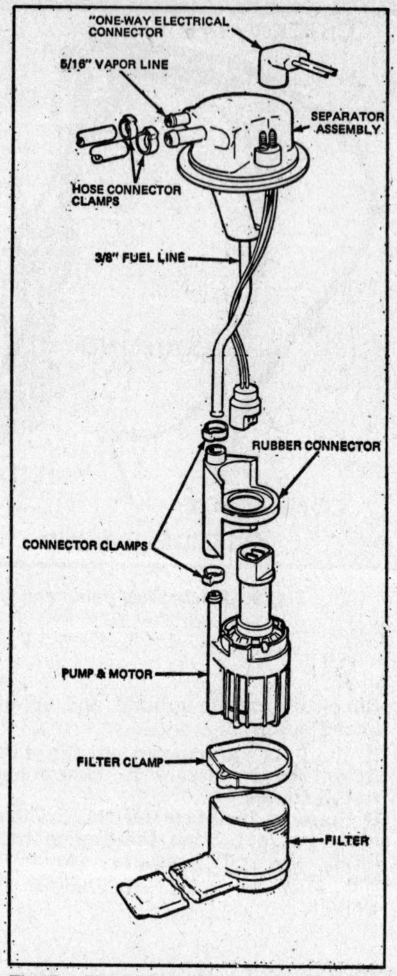

Fig. 7 Electric fuel pump. 1972-74 Ford

NOTE: This pump cannot be overhauled and must be serviced as an assembly.

Current Test

1. Disconnect the connector between oil pressure switch and pump.
2. Using a test lamp, connect one lead to the red/brown strip wire and the other lead to ground. The test lamp should then light, if not, there is an open in the circuit between current source on either side of starter solenoid and fuel pump.

Pressure & Volume Tests

When performing pressure and volume tests on this pump, it is necessary to provide the pump with the full 12 volts. Connect a jumper wire between the battery side of the starter solenoid and the bypass circuit pigtail. Pressure for this pump should be 4 PSI minimum and pump should deliver one pint of fuel in 20 seconds.

CARBURETORS

INDEX

CARBURETION

Since carburetion is dependent in several ways on both compression and ignition, it should always be checked last when tuning an engine. See the car chapter for adjustments for the unit you are interested in.

Before adjusting the carburetor, consider the factors outlined below and which definitely affect engine performance.

Performance Complaints

Flooding, flat spots or other performance complaints are often caused by dirt, or water in the carburetor. To aid in diagnosing the complaint, the carburetor should be carefully removed from the engine without draining the fuel from the bowl. The contents of the fuel bowl can then be examined for contamination as the carburetor is disassembled. A magnet moved through the fuel in the bowl will pick up any iron oxide dust that may have caused needle valve leakage.

Check float setting carefully. Too high a level will cause flooding while too low a level will starve the engine.

Before installing carburetor, fill the bowl with clean fuel and operate the throttle by hand several times to visually check the discharge from pump jets.

Inspect gasketed surfaces between body and air horn. Small nicks or burrs should be smoothed down to eliminate air or fuel leakage. On carburetors having a vacuum piston, be especially particular when inspecting the top surface of the inner wall of the bowl around the vacuum piston passage. A poor seal at this location may contribute to a "cutting out" on turns complaint.

Dirty or Rusty Choke Housing

In cases where it is found that the interior of the choke housing is dirty, gummed or rusty while the carburetor itself is comparatively clean, look for a punctured or eroded manifold heat tube (if one is used).

Manifold Heat Control Valve

An engine equipped with a manifold heat control valve can operate with the valve stuck in either the open or closed position. Because of this, an inoperative valve is frequently overlooked at vehicle lubrication or tune-up.

A valve stuck in the "heat-off" position can result in slow warm up, deposits in combustion chamber, carburetor icing, flat spots during acceleration, low gas mileage and spark plug fouling.

A valve stuck in the "heat-on" position can result in power loss, engine knocking, sticking or burned valves and spark plug burning.

To prevent the possibility of a stuck valve, check and lubricate the valve each time the vehicle is lubricated or tuned-up. Check the operation of the valve manually. To lubricate the valve, place a few drops of penetrating oil on the valve shaft where it passes through the manifold. Then move the valve up and down a few times to work the oil in. *Do not use engine oil to lubricate the valve as it will leave a residue which hampers valve operation.*

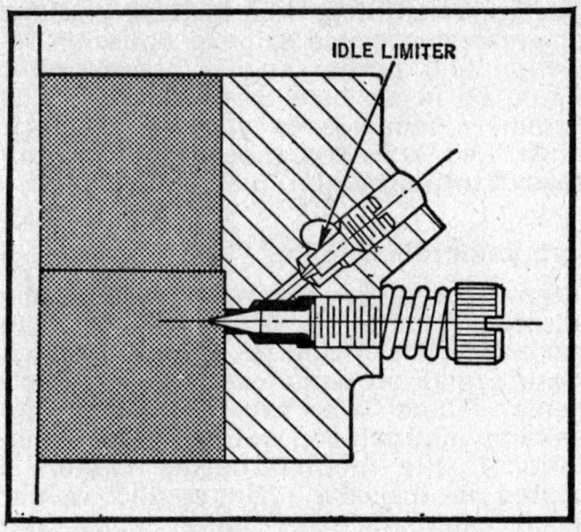

Internal idle mixture limiter

Carburetor Flange

Check the flange for looseness on the manifold. If one of the flange nuts is loose as little as one-half turn, a sufficient amount of air will enter the intake manifold below the throttle plate to destroy engine idle and all engine performance.

If a tight fit cannot be obtained by tightening the nuts, install a new gasket but be sure that all the old gasket material has been removed.

Throttle Linkage

If the throttle linkage is adjusted so that the accelerator pedal will strike the floor board before the throttle plate is wide open, it will result in low top speed.

Fuel Lines

A restriction of the fuel line will result in an apparent vapor lock action or a definite cut-off of gasoline. This can generally be corrected by blowing out the line with compressed air. In some cases, it may be necessary to replace the line.

Fuel Pump

The pump should be tested to make sure that it will draw an adequate supply of fuel from the tank and deliver it to the carburetor under all conditions of operation. If the pump functions inefficiently, proper adjustment and operation of the carburetor is impossible because the fuel will not be maintained at the prescribed level in the idle passages and main discharge jet (or jets) of the carburetor under all operating conditions.

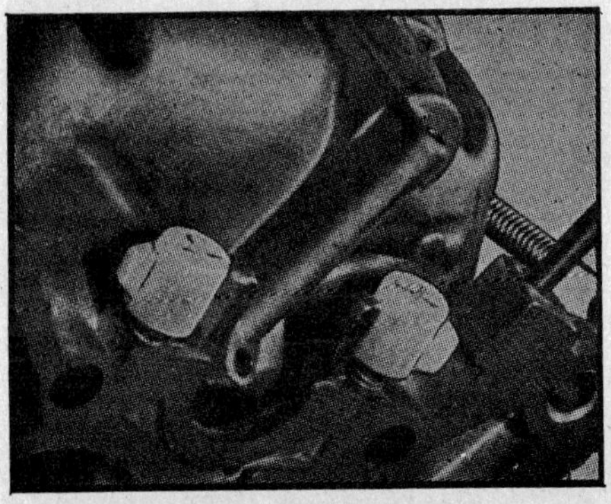

External idle mixture limiter

Fuel Tank

The fuel tank should not be overlooked as a possible source of trouble with carburetion. A shortage of fuel at the fuel pump or carburetor may be caused by pieces of filling station pump hose or other material obstructing the mouth of the feed pipe in the tank, or by a restriction of the air vents in the filler cap and neck.

An unusual amount of dirt, water or gum in the fuel filter indicates that the tank is contaminated with these substances, which should be cleaned out to prevent future failure of the pump or carburetor.

Intake Manifold Leaks

Leakage of air into the intake manifold at any point will affect carburetion and general engine performance. Air may leak into the manifold through the joints at the carburetor or cylinder head, cracks in the manifold, cracks or poor connections in the windshield wiper or windshield washer hose lines, or the connections of any accessories which may be connected to the manifold. All such joints should be tested for leaks.

To test the intake manifold for leaks, apply oil from an oil can along the gasket joints with the engine idling. An air leak is indicated when oil is drawn past the gaskets by the suction of the engine. Tighten the nuts or cap screws holding the manifold to the engine and retest for leaks. If tightening fails to stop the leaks, replace the manifold gaskets. If the new gaskets fail to stop the leaks, carefully inspect the manifold for cracks and test any suspicious area with oil.

Air Cleaner

An air cleaner with a dirty element, will restrict the air flow through the carburetor and cause a rich mixture at high speeds. In such a condition the air cleaner likewise will not properly remove dirt from the air, and the dirt entering the engine will cause rapid formation of carbon, sticking valves, and wear of piston rings and cylinder bores.

Automatic Choke

The choke mechanism must be inspected and cleaned to make sure it is operating freely. Sluggish action or sticking of the choke will cause excessive fuel consumption, poor performance during warm-up, and possibly hard starting.

The choke thermostat should be set in accordance with the average air temperature as well as the volatility of the fuel being used. It is desirable to have the thermostat set as lean as operating conditions permit in order to avoid an over-rich mixture during engine warm up.

Choke Thermostat

If necessary to adjust the choke more than two marks from the specified setting, either rich or lean, it indicates that the thermostat spring may be bent or has lost its tension.

Carter Float Settings

When replacing a solid float needle and seat with the new type resilient seat, the float setting should be reduced 1/32" on AFB, WCFB, WGD and WCD carburetors.

Carburetor Ball Checks

Whenever it becomes necessary to dismantle a carburetor be sure to account for the ball checks that may be found under pump plungers and compensating or power valves.

CARBURETOR IDLE ADJUST

Cars Without Exhaust Emission Controls

There are two basic types of exhaust emission control systems—air injection type and engine modification type. With both types, the slow idle adjustment method, referred to as "Lean Roll" is to be used. This method insures proper idle, ignition timing and mixture settings for greatest possible exhaust emission reduction and proper engine operation. It should be noted here that smooth idle is extremely sensitive to vacuum leaks. If rough idle is noted, check for vacuum leaks at the carburetor, manifold, etc.

Carburetor Idle Limiters

Some carburetors are equipped with idle adjustment limiters which restrict the maximum idle richness of the air/fuel mixture and prevents overly rich adjustments. There are two types of idle limiters: internal and external (see illustrations). The internal needle limiter is located in the idle channel and is not visible externally. This limiter is set and sealed at the factory and, under no circumstances, during normal service or during overhaul, should the seal be removed and adjustments made to this needle.

The other type of idle limiter is an external idle limiter cap installed on the knurled head of the idle mixture adjusting screw. Any adjustment to the idle fuel mixture on carburetors with this type of limiter must be made within the range of the limiter cap.

Under no circumstances may the limiter cap, the stop boss or the power valve cover, which the limiter caps stop against, be mutilated or deformed in any way to render the limiter inoperative. A satisfactory idle is obtainable within the range of the limiter cap.

The addition of idle limiters does not eliminate the need for adjusting idle speed and mixture. All the limiters do is prevent overly rich mixtures, which increase the amount of hydro-carbons emitted into the atmosphere.

1. With engine at operating temperature, set parking brake and block drive wheels.
2. Make sure choke valve is wide open.
3. On C.C.S. equipped vehicles, see that the air cleaner thermostatic valve is open.
4. On carburetors so equipped, hold hot idle compensator hole closed with eraser on pencil.
5. Turn air conditioner off or on according to directions given in *Tune Up Charts* in car chapters.
6. Set idle mixture screw(s) for maximum idle rpm.
7. Adjust speed screw (or idle stop solenoid screw on C.C.S.) to obtain the specified rpm in Drive or Neutral as specified.
8. Set ignition timing according to specifications with vacuum advance line disconnected and hole in manifold plugged.
9. Adjust mixture screw IN to obtain a 20 rpm drop (lean roll).
10. Adjust mixture screw OUT 1/4 turn.
11. Repeat Steps 9 and 10 for second mixture screw (2 and 4 barrel carbs).
12. Readjust speed screw (or solenoid screw) if necessary to obtain specified rpm.
13. On C.C.S. with idle solenoid stop on carburetor, electrically disconnect solenoid and adjust carburetor idle speed screw to obtain 400 rpm in neutral, then reconnect wire to solenoid.

NOTE: Exact instructions for each C.C.S. equipped engine-transmission combination is given for this Lean Roll (low idle) speed method on a decal permanently affixed to the vicinity of the radiator support as well as in the *Tune Up Charts* in the car chapter of this manual.

SERVICE BULLETINS

Carter Thermo-Quad

If a stumble or lag is encountered on acceleration on 1973 cars with V8-340, 400, 440 engines with a Carter Thermo-Quad carburetor, it may be caused by an undersize accelerating pump plunger cup. This can be verified by checking the discharge from the accelerating pump discharge jets.

If the condition is considered to be caused by a deficiency in the pump circuit, the plunger should be replaced. A kit #3780111 is available for this purpose and consists of a plunger and check valve seat. The seat must be replaced because it becomes damaged during removal of the plunger.

Fig. 1 Carter Thermo-quad float lever pins improperly placed

Carter Thermo-Quad

When installing the bowl cover on Carter TQ carburetors, it is important that the float lever pins are correctly positioned (centered in their supports). If the pins are not properly placed, they may be trapped between the gasket surfaces. When the bowl cover screws are tightened, the bowl will crack, Fig. 1.

Rochester Quadrajet

Delco/Rochester advises the possibility exists that the wrong throttle body to float bowl gaskets are being used by servicemen on 1970 and later Quadrajet carburetors. When this wrong substitution is made, vacuum leaks occur and cause rough idle due to air bypassing the primary throttle valves through the canister purge passage in the throttle body because the gasket will not seal this passage.

The difference between the throttle body to bowl gaskets is shown in Fig. 2.

Carter Thermo-Quad

The metering rods are preset to specifications by the manufacturer. No additional adjustment should be attempted since an accurate adjustment cannot be made in the field. An improper adjustment may result in the exhaust emissions exceeding specifications. Also, maladjustment of the metering rods may prevent the carburetor from returning to the idle position from wide open throttle (WOT) due to throttle shaft binding, resulting in poor fuel economy.

Carter BBD

Some 1974 Carter 1¼" BBD carburetors used on V8-318 engine may have been manufactured with the cluster cap gaskets reversed, Fig. 3. This causes fuel leakage into the manifold resulting in flooding, poor hot starting or poor fuel economy.

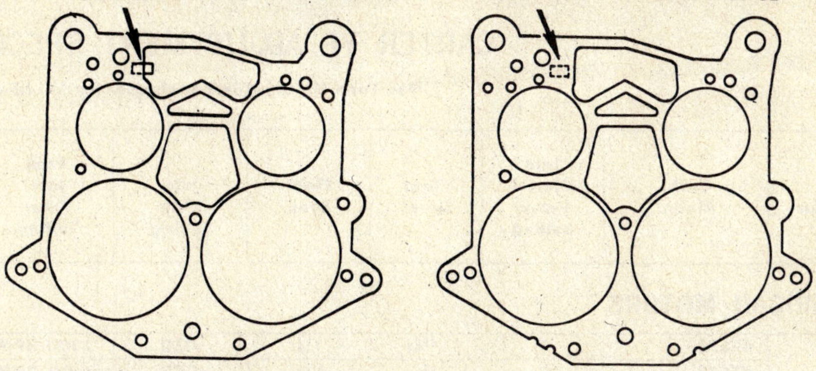

Fig. 2 Delco/Rochester Quadrajet throttle body to float bowl gaskets

Holley 2245

Some difficulty may be encountered when adjusting curb idle speed on these carburetors. This condition may be caused by the side of the carburetor interfering with the throttle lever. To correct this condition, the throttle lever may be filed to remove excess metal causing the interference, Fig. 4.

Motorcraft 2150

On some applications, the clearance between the carburetor choke pulldown vacuum hose and the EGR spacer may be less than ¼ inch. If improper clearance exists, cut ½ inch off hose and recheck clearance.

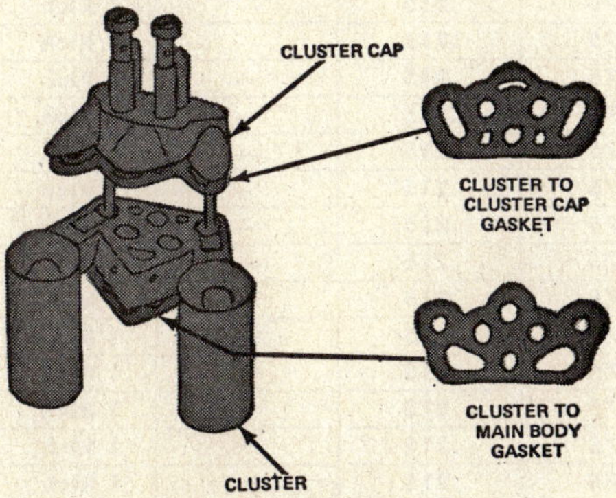

Fig. 3 Cluster cap gasket installation. Carter BBD carburetors

Fig. 4 Removing excess metal from throttle lever. Holley 2245 carburetors

Carter Carburetor Section

CARTER YF ADJUSTMENT SPECIFICATIONS

See Tune Up Chart in car chapter for hot idle speed.

Year	Carb. Model	Idle Mixture Screw Setting	Float Level	Float Drop	Idle Vent Setting	Fast Idle Cam Setting	Dechoke or Unloader Setting	Pulldown Setting	Vacuum Break Setting	Choke Setting
AMERICAN MOTORS										
1970	4767S	①	29/64	1 1/4	.052	2300 RPM	.300	—	—	Index
	4768S	①	29/64	1 1/4	.052	2300 RPM	.325	—	—	Index
	4769S	①	29/64	1 1/4	.055	2300 RPM	.300	—	—	Index
	4770S	①	29/64	1 1/4	.055	2300 RPM	.300	—	—	Index
	4978S	①	29/64	1 1/4	.055	2300 RPM	.300	—	—	1 Rich
1971	6038S	①	29/64	1 1/4	—	2300 RPM	.300	—	—	Index
	6093S	①	29/64	1 1/4	—	2300 RPM	.300	—	—	Index
	6094S	①	29/64	1 1/4	—	2300 RPM	.300	—	—	1 Rich
	6095S	①	29/64	1 1/4	—	2300 RPM	.300	—	—	Index
	6096S	①	29/64	1 1/4	—	2300 RPM	.300	—	—	1 Rich
1972	6199S	①	29/64	1 1/4	—	1600 RPM	.300	.230	—	Index
	6200S	①	29/64	1 1/4	—	1600 RPM	.300	.230	—	Index
	6202S	①	29/64	1 1/4	—	1600 RPM	.300	.230	—	Index
1973	6299S	①	29/64	1 1/4	—	1600 RPM	.275	.215	—	1 Rich
	6300S	①	29/64	1 1/4	—	1600 RPM	.275	.215	—	1 Rich
	6400S	①	29/64	1 1/4	—	1600 RPM	.275	.215	—	1 Rich
	6401S	①	29/64	1 1/4	—	1600 RPM	.275	.215	—	1 Rich
	6421S	①	29/64	1 1/4	—	1600 RPM	.275	.215	—	1 Rich
	6422S	①	29/64	1 1/4	—	1600 RPM	.275	.215	—	1 Rich
	6424	①	15/32	1 3/8	—	1600 RPM	.275	.215	—	1 Rich
	6429	①	15/32	1 3/8	—	1600 RPM	.275	.215	—	1 Rich
	6430	①	15/32	1 3/8	—	1600 RPM	.275	.215	—	1 Rich
	6432	①	15/32	1 3/8	—	1600 RPM	.275	.215	—	1 Rich
1973–74	6423	①	15/32	1 3/8	—	.190	.275	.215	—	1 Rich
	6431	①	15/32	1 3/8	—	.190	.275	.215	—	1 Rich
1974	6510	①	15/32	1 3/8	—	.190	.275	.215	—	1 Rich
	6511	①	15/32	1 3/8	—	.190	.275	.215	—	1 Rich
	7000	①	15/32	1 3/8	—	.190	.275	.215	—	1 Rich
	7001	①	15/32	1 3/8	—	.190	.275	.215	—	1 Rich
	7028	①	15/32	1 3/8	—	.190	.275	.215	—	1 Rich
	7029	①	15/32	1 3/8	—	.190	.275	.215	—	1 Rich
1975	7039	①	15/32	1 3/8	—	.180②	.275	.205	—	1 Rich
	7041	①	15/32	1 3/8	—	.180②	.275	.205	—	1 Rich
	7061	①	15/32	1 3/8	—	.180②	.275	.205	—	1 Rich
	7062	①	15/32	1 3/8	—	.180②	.275	.205	—	1 Rich
	7074	①	15/32	1 3/8	—	.180②	.275	.205	—	1 Rich
1976	7083	①	15/32	1 3/8	—	.185②	.275	.205	—	1 Rich
	7084	①	15/32	1 3/8	—	.185②	.275	.205	—	2 Rich
	7085	①	15/32	1 3/8	—	.185②	.275	.205	—	1 Rich
	7086	①	15/32	1 3/8	—	.185②	.275	.205	—	2 Rich
	7112	①	15/32	1 3/8	—	.185②	.275	.205	—	1 Rich

①—Air/fuel ratio or idle CO% rating is found in Tune Up Specification tables in car chapters.
②—1600 RPM hot on 2nd step of cam with TCS solenoid & EGR disconnected.

CARTER YF ADJUSTMENT SPECIFICATIONS—Continued

See Tune Up Chart in car chapter for hot idle speed.

Year	Carb. Model	Idle Mixture Screw Setting	Float Level	Float Drop	Idle Vent Setting	Fast Idle Cam Setting	Dechoke or Unloader Setting	Pulldown Setting	Vacuum Break Setting	Choke Setting
FORD										
1969	C8AF-BF	②	7/32	1¼	—	.035	.280	.265	—	Index
	C8DF-G	②	7/32	1¼	—	.046	.280	.265	—	Index
	C8DF-H	②	7/32	1¼	—	.040	.280	.265	—	1 Lean
1970	D0AF-A	②	3/8	1¼	—	.029	.250	.225	—	Index
	D0AF-B	②	3/8	1¼	—	.025	.250	.225	—	1 Lean
	D0DF-L	②	3/8	1¼	—	.036	.250	.265	—	Index
	D0DF-M	②	3/8	1¼	—	.031	.250	.265	—	Index
	D0DF-N	②	7/32	1¼	—	.035	.280	.225	—	Index
	D0DF-R	②	7/32	1¼	—	.036	.280	.225	—	1 Rich
	D0DF-S	②	7/32	1¼	—	.035	.280	.225	—	Index
	D0DF-T	②	3/8	1¼	—	.031	.250	.265	—	Index
	D0DF-U	②	7/32	1¼	—	.036	.280	.225	—	1 Rich
	D0DF-V	②	3/8	1¼	—	.036	.250	.265	—	Index
1971	D1DF-EA	②	3/8	1¼	—	.105	.250	.200	—	Index
	D1DF-GA, HA	②	3/8	1¼	—	.170	.250	.230	—	Index
	D1DF-JA, LA	②	3/8	1¼	—	.140	.250	.200	—	Index
	D1DF-KA, MA	②	3/8	1¼	—	.140	.250	.200	—	Index
	D1DF-PA	②	3/8	1¼	—	.190	.250	.200	—	Index
	D1DF-RA	②	3/8	1¼	—	.220	.250	.230	—	Index
1972	D2DF-AA	②	3/8	1¼	—	.105	.280	.170	—	Index
	D2DF-BA	②	3/8	1¼	—	.170	.250	.230	—	Index
	D2DF-CA	②	3/8	1¼	—	.170	.250	.230	—	Index
	D2DF-DA	②	3/8	1¼	—	.140	.250	.200	—	1 Rich
	D2DF-EA	②	3/8	1¼	—	.140	.250	.200	—	1 Rich
	D2AF-JA	②	3/8	1¼	—	.220	.250	.230	—	1 Lean
1973	D3DF-AA	②	3/8	1¼	—	.170	.250	.230	—	Index
	D3DF-CA	②	3/8	1¼	—	.140	.250	.230	—	1 Rich
	D3OF-CA	②	9/16	—	—	.115	.250	.300	—	1 Rich
1974	D4DE-ABA	②	3/8	—	—	.170	.250	.230	—	Index
	D4DE-EA	②	3/8	—	—	.140	.250	.200	—	Index
	D4DE-JA	②	3/8	—	—	.140	.250	.200	—	1 Rich
	D4DE-JB	②	3/8	—	—	.140	.250	.200	—	Index
	D4DE-KA	②	3/8	—	—	.140	.250	.200	—	1 Rich
	D4DE-KB	②	3/8	—	—	.140	.250	.200	—	Index
1975	D5DE-DA	②	3/8	—	—	.140	.250	.290	—	2 Rich
	D5DE-DB	②	23/32	—	—	.140	.250	.290	—	2 Rich
	D5DE-EA	②	3/8	—	—	.140	.250	.290	—	2 Rich
	D5DE-GA	②	3/8	—	—	.140	.250	.290	—	2 Rich
	D5DE-MA	②	3/8	—	—	.140	.250	.290	—	2 Rich
	D5DE-MB	②	23/32	—	—	.140	.250	.290	—	2 Rich
	D5DE-ZA	②	3/8	—	—	.140	.250	.290	—	2 Rich
1976	D6BE-AA	②	25/32	—	—	.140	.250	.260	—	1 Rich
	D6BE-BB	②	25/32	—	—	.140	.250	.260	—	2 Rich
	D6DE-AB	②	25/32	—	—	.140	.250	.290	—	Index
	D6DE-BB	②	25/32	—	—	.140	.250	.230	—	Index

②—Air/fuel ratio or idle CO% rating is found in Tune Up Specification tables in car chapters.

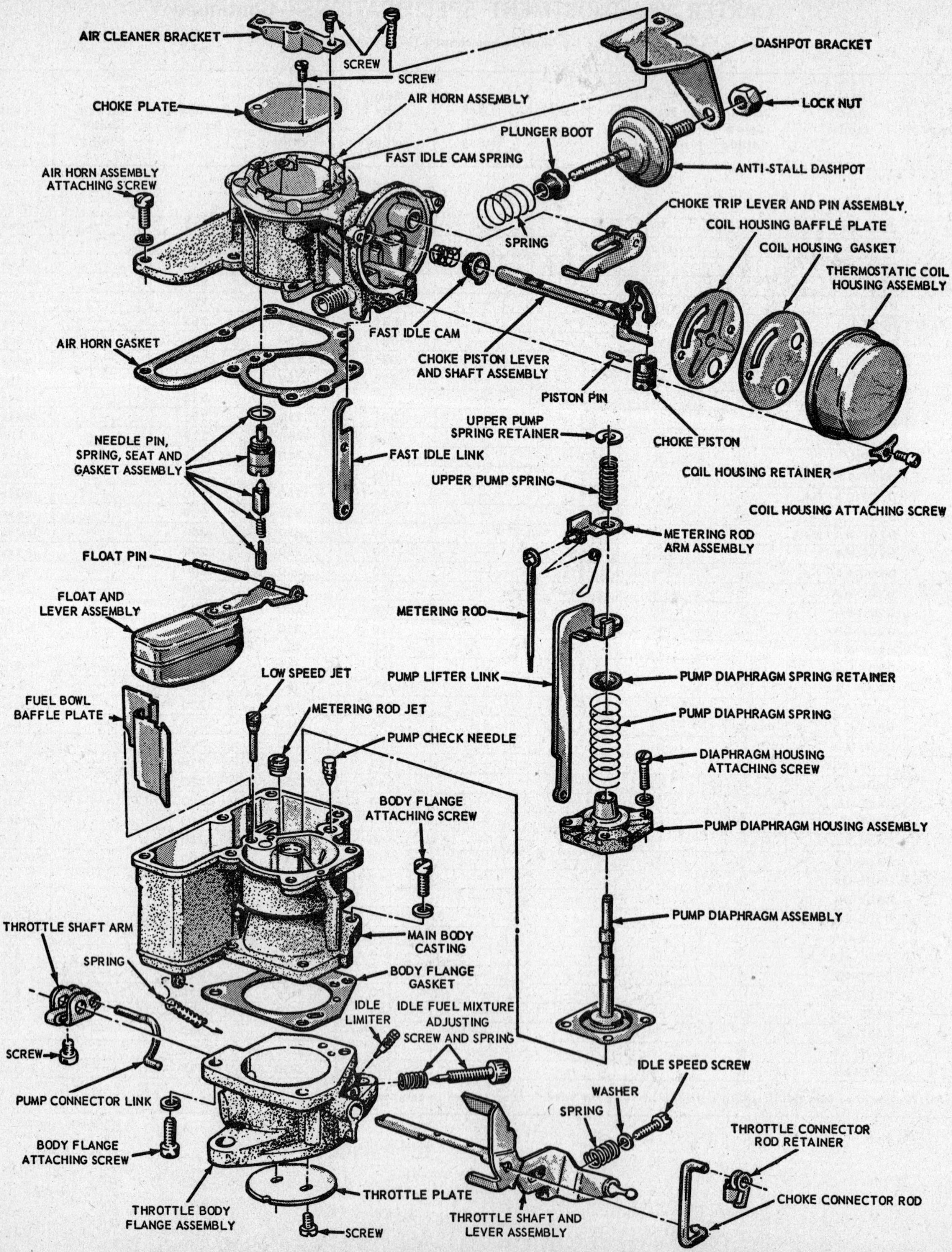

Fig. 1 Carter Model YF carburetor with built-in automatic choke. Note that this unit is provided with an idle limiter screw which is used to prevent an overly rich mixture on cars with exhaust emission control

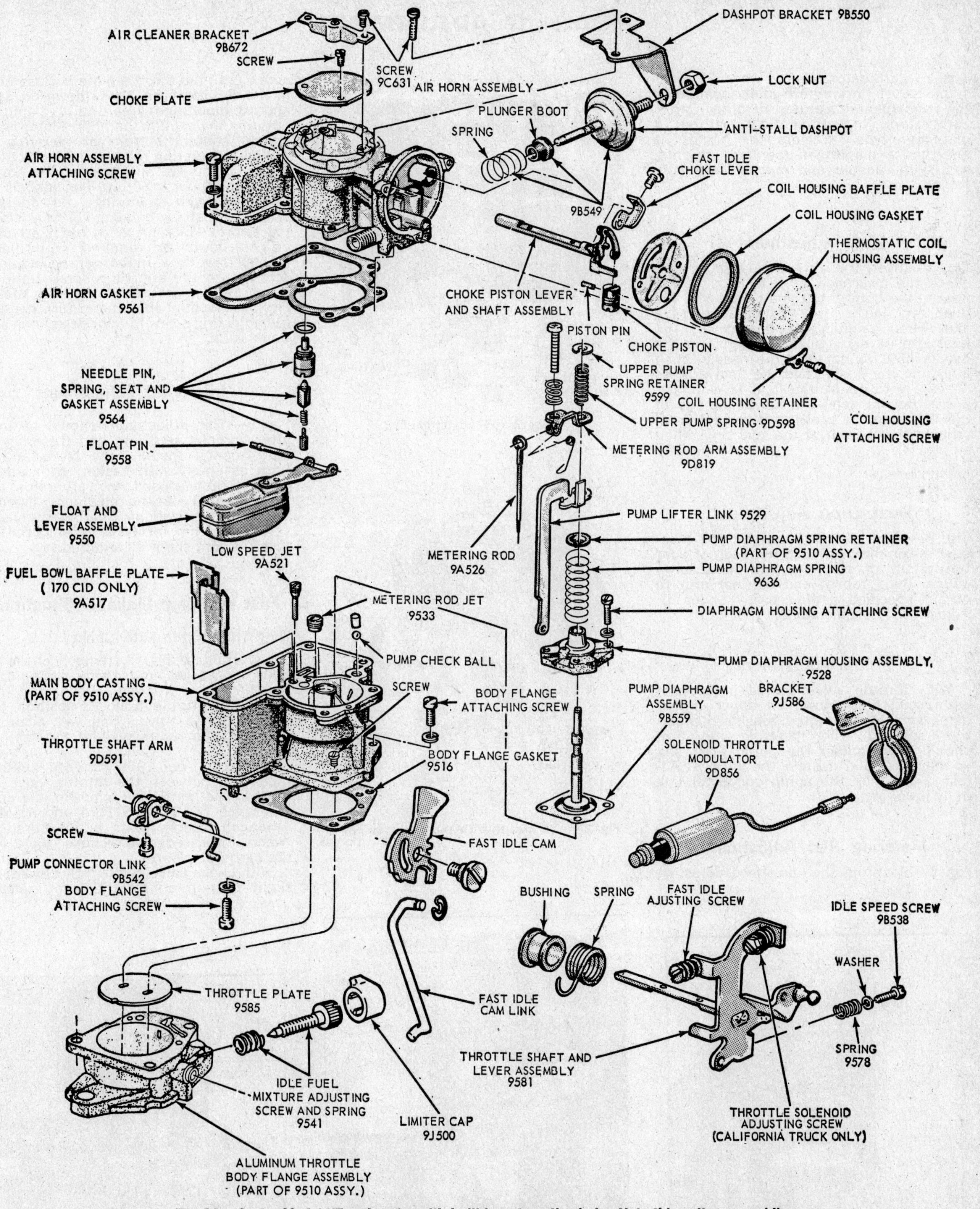

Fig. 1A Carter Model YF carburetor with built-in automatic choke. Note this unit uses an idle limiter cap and throttle positioner solenoid

MODEL YF ADJUSTMENTS

The YF carburetor, Figs. 1 thru 2A, is a single-barrel, downdraft unit combining the fundamental features of other Carter carburetors. In addition, it features a diaphragm-type accelerating pump. It also has a diaphragm-operated metering rod, both vacuum and mechanically controlled.

Float Adjustment

Fig. 3—Invert the air horn assembly, and check the clearance from the top of the float to the bottom of the air horn with the float level gauge. Hold the air horn at eye level when gauging the float level. The float arm (lever) should be resting on the needle pin. Do not load the needle when adjusting the float. Bend the float arm as necessary to adjust the float level (clearance). Do not bend the tab at the end of the float arm. It prevents the float from striking the bottom of the fuel bowl when empty.

Float Drop Adjustment

Fig. 4—Hold air horn upright and measure maximum clearance from top of float to bottom of air horn with float drop gauge. Bend tab at end of float arm to obtain specified setting listed under YF Adjustment Specifications.

Pump Adjustment

With throttle valve seated in bore of carburetor, press down on upper end of diaphragm shaft until it reaches its bottom position. The metering rod arm should now contact the pump lifter link at the outer end nearest the springs. Adjust by bending the pump connector link at its lower angle.

Metering Rod Adjustment

Fig. 5—Back out the idle speed adjusting screw until the throttle plate is closed tight in the throttle bore. Press down on upper end of diaphragm shaft until diaphragm bottoms in vacuum chamber. Metering rod should contact bottom of metering rod well, and metering rod should contact lifter link at the outer end nearest the springs and at supporting lug. For models not equipped with metering rod adjusting screw, adjust by bending lip of metering rod arm to which metering rod is attached, up or down as required. For models equipped with a metering rod adjusting screw, turn the adjusting screw until metering rod just bottoms in the body casting, Fig. 5a. For final adjustment turn metering rod adjusting screw in (clockwise) one additional turn.

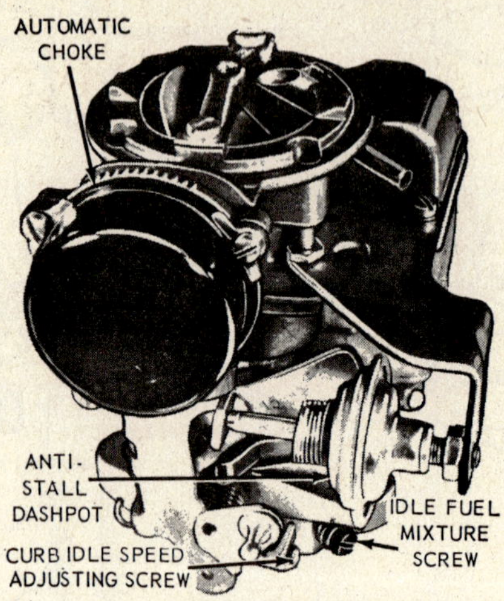

Fig. 2 YF adjustments

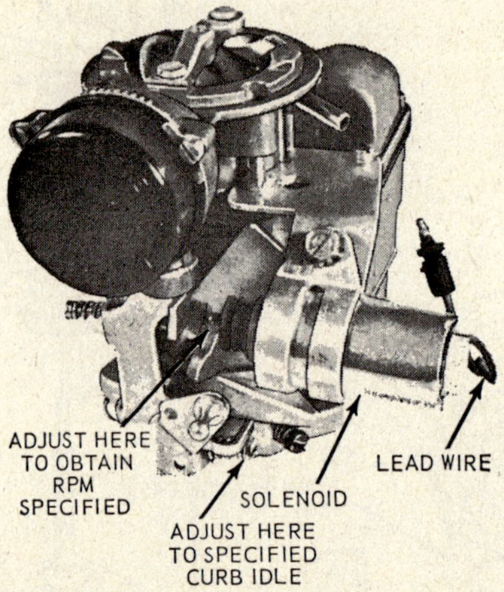

Fig. 2A YF with throttle positioner solenoid

Idle Vent Adjustment

Fig. 6—This adjustment should be made after completing pump and metering rod adjustments. Install bowl cover and air horn assembly with gasket. With throttle valve tightly closed in carburetor bore there should be the clearance listed in the *YF Specifications Chart* between idle vent valve and inside of bowl cover. Adjust idle vent screw as required.

Fast Idle Cam Linkage Adjustment

Unit With Built-In Auto. Choke

Referring to Fig. 7, remove choke coil housing, gasket and baffle plate. Crack throttle valve (barely open) and hold choke valve firmly in closed position, then close throttle valve. This will allow the fast idle cam to revolve to the fast idle position.

Clearance on 1969-70 Ford units is measured between the throttle valve and the throttle bore (side opposite idle port). Clearance on 1971-75 Ford and 1974-75 American Motors units is measured between lower edge of choke plate and the carburetor bore.

With choke valve held tightly closed, and slight tension on throttle lever, refer to

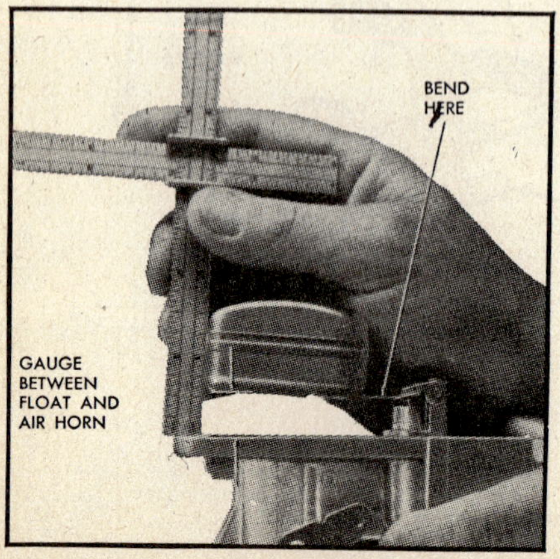

Fig. 3 YF float level adjustment

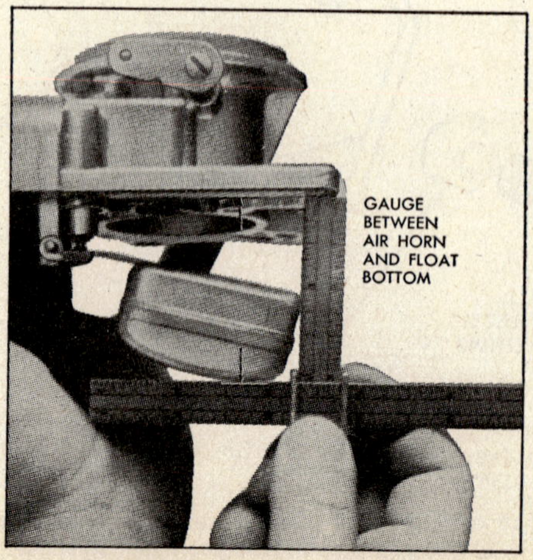

Fig. 4 YF float drop adjustment

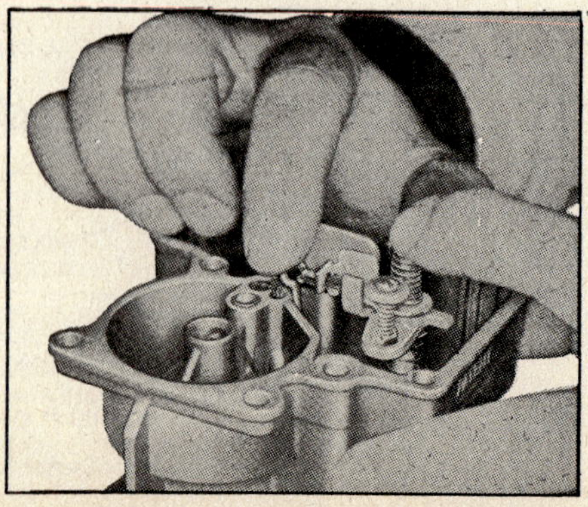

Fig. 5 YF metering rod adjustment

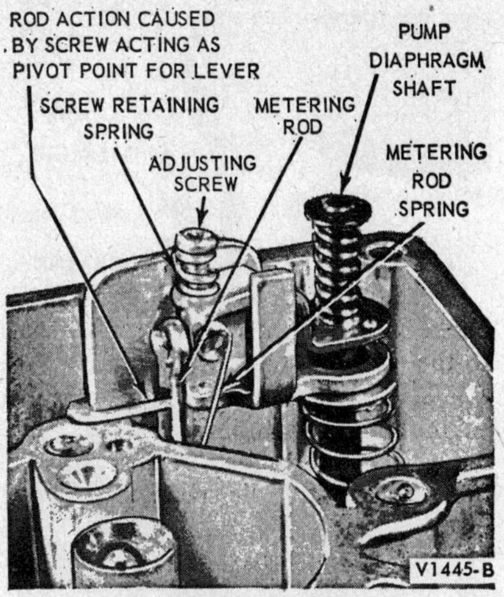

Fig. 5A YF metering rod adjustment

Fig. 6 YF idle vent adjustment

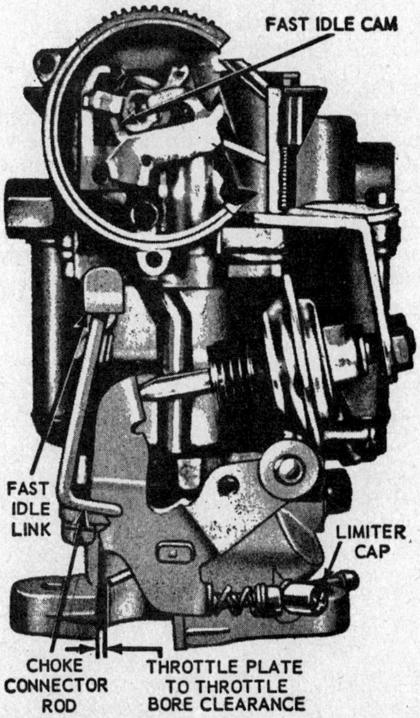

Fig. 7 YF fast idle speed adjustment for Fig. 1 carburetors

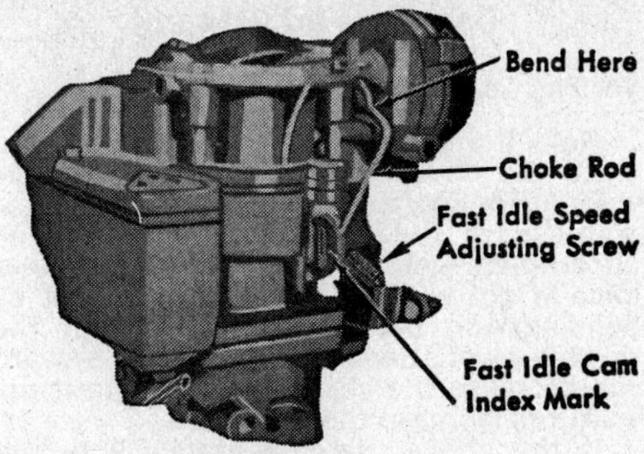

Fig. 7A YF fast idle cam linkage adjustment

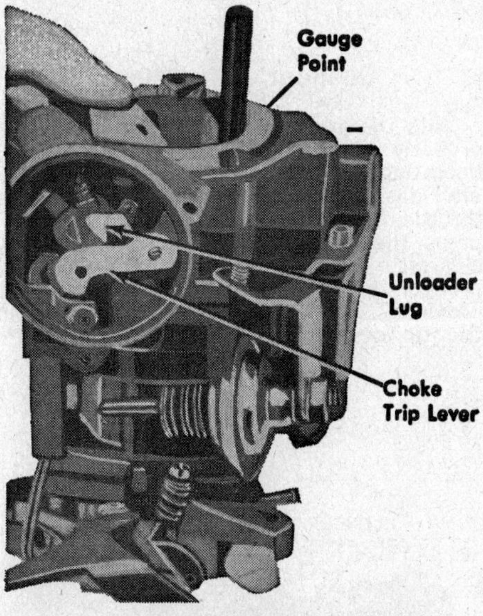

Fig. 8 YF unloader adjustment for Fig. 1 carburetors

YF Specifications when adjusting units with throttle valve to throttle bore clearance settings.

NOTE: Caution: The choke trip lever must not contact the fast idle cam during adjustment. All other units should have the fast idle adjusting screw aligned with the index mark at the back side of the cam. If cam has no index mark, position the fast idle screw on the kickdown step of the fast idle cam against the shoulder of the high step. Adjust linkage by bending the choke rod at elbow, Figs. 7 and 7a.

The 1970-73 American Motors units have engine fast idle RPM settings in place of specified carburetor to choke plate clearance. Refer to *YF Specifications Chart*. Adjustment is same as above.

Choke Unloader Adjustment

With throttle valve held wide open and choke valve held toward closed position with a rubber band, there should be the clearance listed in the *YF Specifications Chart* between lower edge of choke valve and inner air horn wall.

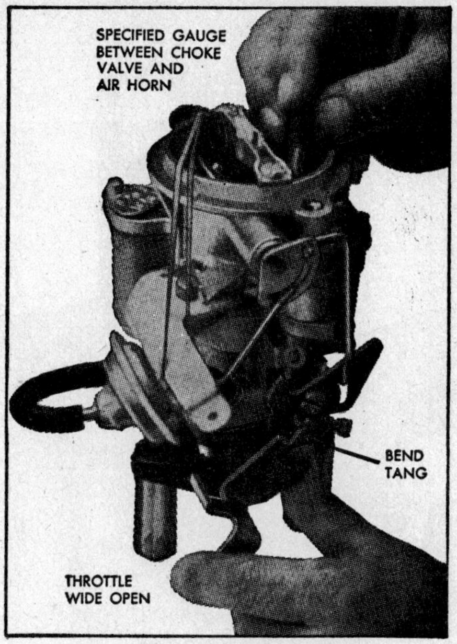

Fig. 9 YF unloader adjustment for Fig. 2 carburetors

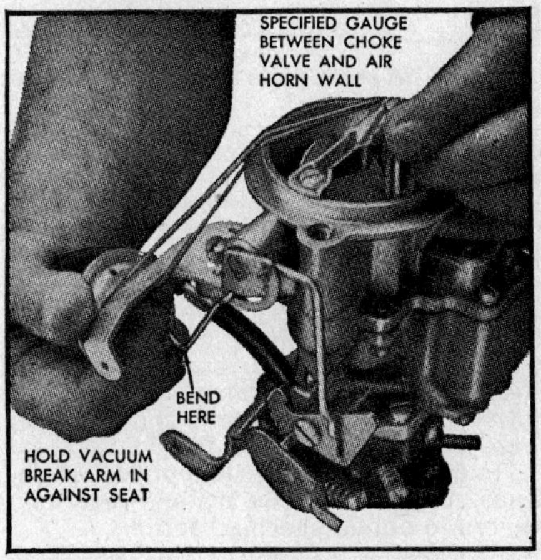

Fig. 10 YF vacuum break adjustment for Fig. 2 carburetors

On Fig. 1 carburetors, adjust by bending unloader lug on choke trip lever, Fig. 8. On Fig. 2 carburetors adjust by bending unloader tang on throttle lever, Fig. 9.

Vacuum Break Adjustment

Fig. 10—With vacuum break arm held against its stop, and choke valve held toward closed position with a rubber band, bend vacuum break link to obtain the clearance listed in the *YF Specifications Chart* between lower edge of choke valve and air horn wall.

Choke Plate Pulldown Adjustment

For Fig. 1a Units

Bend a 0.026 in. diameter wire gauge at

Fig. 11 YF choke plate pulldown adjustment

a 90 degree angle approximately ⅛-inch from one end. Insert the bent end of the gauge between the choke piston slot and the right hand slot in choke housing. Rotate the choke piston lever counterclockwise until gauge is snug in the piston slot. Exert a light pressure on choke piston lever to hold the gauge in place, then use a drill with a diameter equal to the specified pulldown clearance between the lower edge of choke plate and carburetor bore to check clearance, Fig. 11.

To adjust the choke plate pulldown clearance, bend the choke piston lever as required to obtain specified setting.

NOTE: When bending the lever, be careful not to distort the piston link. Install the choke thermostatic spring housing and gasket. Set the housing to specifications.

Dechoke Adjustment

For Fig. 1a Units

Hold the throttle plate fully open and close the choke plate as far as possible without forcing it. Use a drill of specified diameter to check the clearance between choke plate and air horn, Fig. 12. If clearance is not within specification, adjust by bending arm on choke trip lever of the throttle lever. Bending the arm downward will decrease the clearance, bending it upward will increase the clearance.

If the choke plate clearance and fast idle cam linkage adjustment was performed with the carburetor on the engine, adjust the engine idle speed and fuel mixture. Adjust dashpot (if so equipped).

Dashpot Adjustment

With the engine idle speed and mixture properly adjusted, the engine at normal operating temperature, loosen the anti-stall dashpot lock nut, Fig. 13. Hold the throttle in the curb idle position and depress the dashpot plunger. Measure the clearance between the throttle lever and plunger tip. Turn the anti-stall dashpot to provide 7/64″ ± 1/64″ clearance between the tip of the plunger and the throttle lever. Tighten the locknut to secure the adjustment.

Fig. 13 YF dashpot adjustment

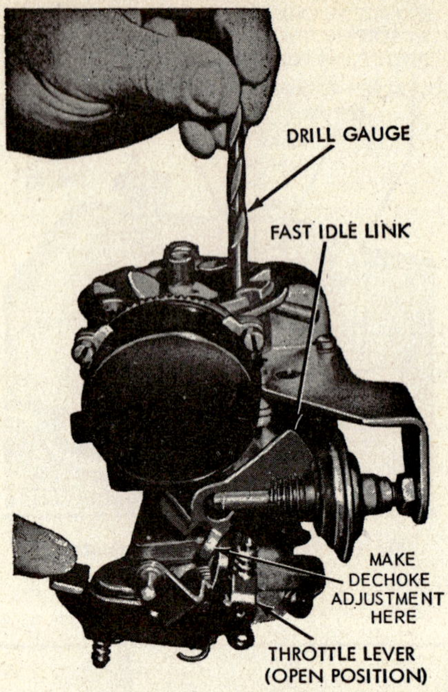

Fig. 12 YF dechoke adjustment

Automatic Choke Adjustment

For Figs. 1, 1a, 2a and 2b Units

Loosen choke cover retaining screws and turn choke cover so that line or Index mark on cover lines up with the specified mark listed in *YF Specifications Chart* on choke housing.

Choke Diaphragm Linkage Adjustment

For Fig. 2 Units

With vacuum diaphragm bottomed, close choke valve as far as possible without forcing. Adjust choke diaphragm connector rod to give the clearance listed in the *YF Specifications Chart* between lower edge of choke valve and inner wall of air horn. Remove connector rod to prevent damage to diaphragm.

CARTER BBS ADJUSTMENT SPECIFICATIONS

See Tune Up Chart in car chapter for hot idle speed.

Year	Carb. Model	Idle Mixture Screw Turns Open	Float Level	Pump Travel Inch	Bowl Vent Drill Size	Choke Unloader Drill Size	Fast Idle Cam Position Drill Size	Choke Vacuum Kick Drill Size	Automatic Choke Setting
DODGE & PLYMOUTH									
1969	4601S	1–2	¼	—	.060	3/16	48 Drill	35 Drill	2 Rich
	4602S	1–2	¼	—	.060	3/16	48 Drill	35 Drill	2 Rich
1970	4715S	①	¼	—	1/32	3/16	48 Drill	35 Drill	2 Rich
	4716S	①	¼	—	1/32	3/16	48 Drill	48 Drill	2 Rich
	4717S	①	¼	5/16	9/32	3/16	48 Drill	35 Drill	2 Rich
	4718S	①	¼	5/16	9/32	3/16	48 Drill	48 Drill	2 Rich
1971	4955S	①	¼	5/16	17/64	3/16	48 Drill	35 Drill	2 Rich
	4956S	①	¼	5/16	17/64	3/16	48 Drill	35 Drill	2 Rich

①—Air/fuel ratio or idle CO% rating is found in Tune Up Specification tables in car chapters.

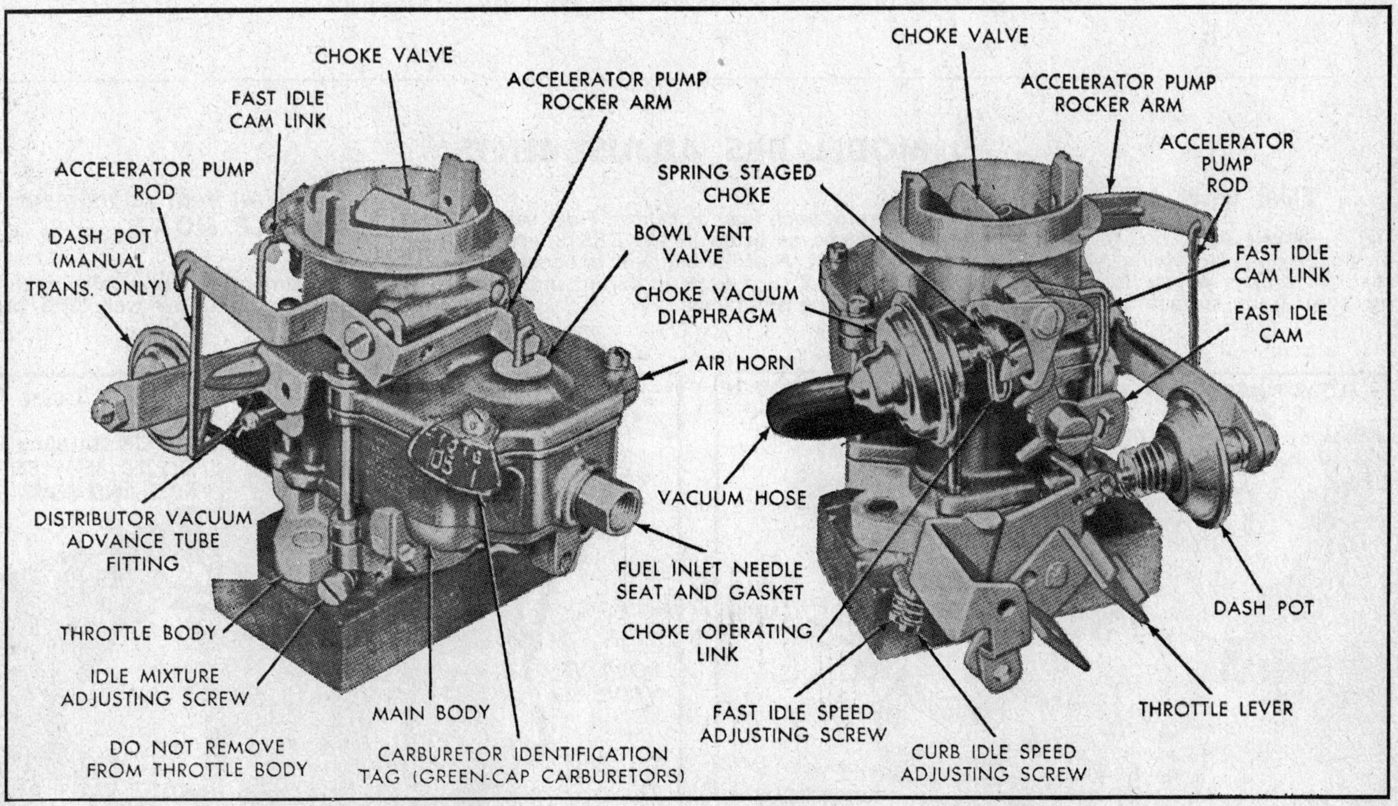

Fig. 1 Exterior of 1969 model BBS carburetor used with C.A.P. (Typical)

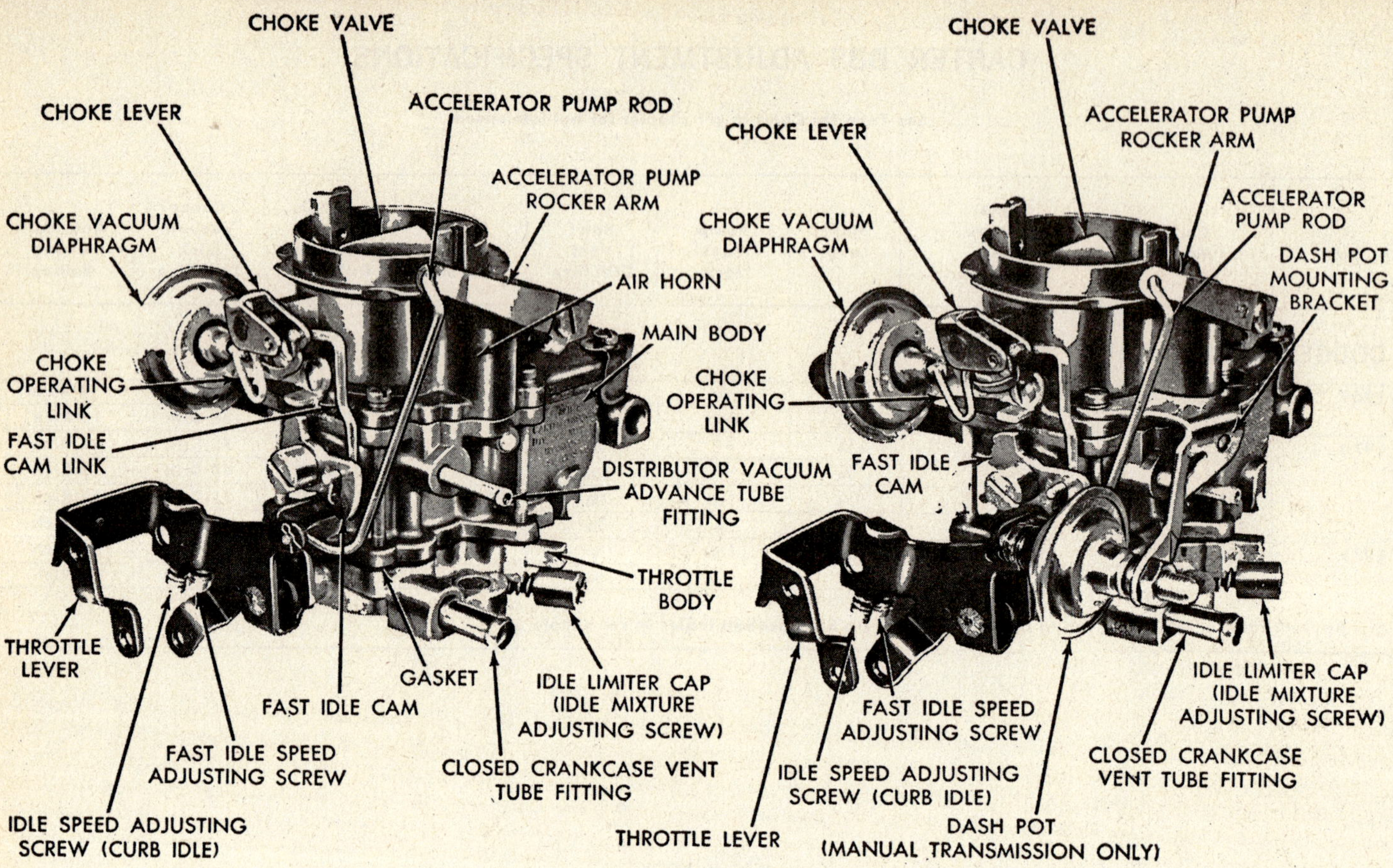

Fig. 2 Exterior of 1970 model BBS carburetor used with Cleaner Air System (CAS)

MODEL BBS ADJUSTMENTS

Float Level Adjustment

Fig. 4—Invert main body so that weight of floats only is forcing needle against seat. If proper gauge is not available, measure from surface of fuel bowl to crown of each float at center. Float setting should be as listed in the *BBS Specifications Chart.* If an adjustment is necessary, bend lip of float lever in or out until correct setting is obtained.

Pump & Bowl Vent Adjustment C.A.P. & C.A.S.

Fig. 5

This adjustment automatically adjusts the accelerator pump as well. The pro-

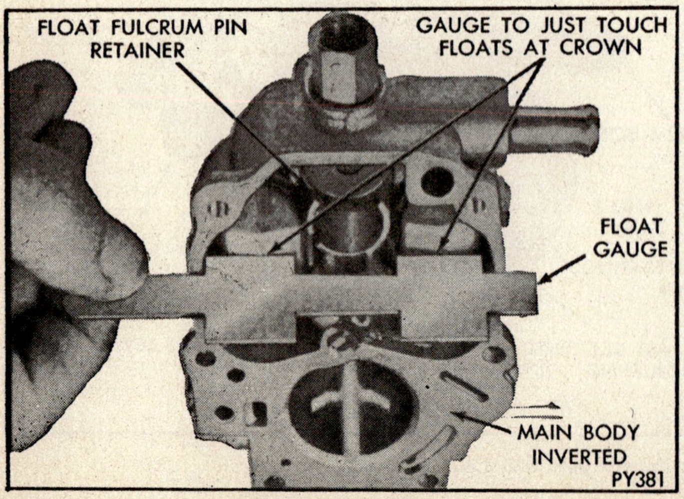

Fig. 4 Checking float level. BBS carburetors

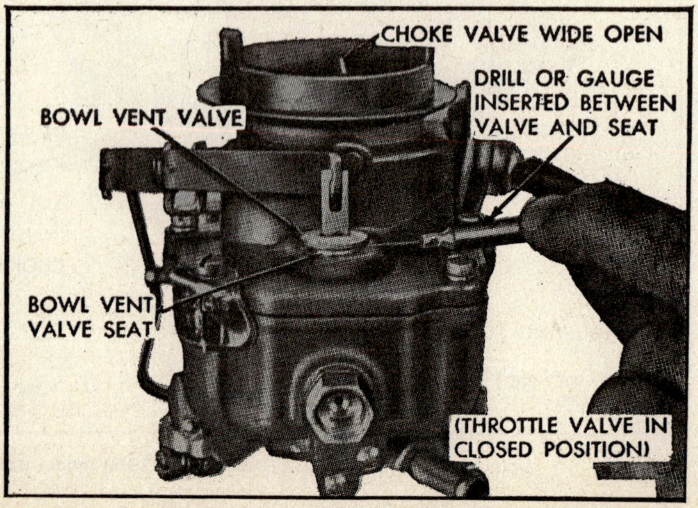

Fig. 5 Bowl vent adjustment. BBS carburetors with C.A.P.

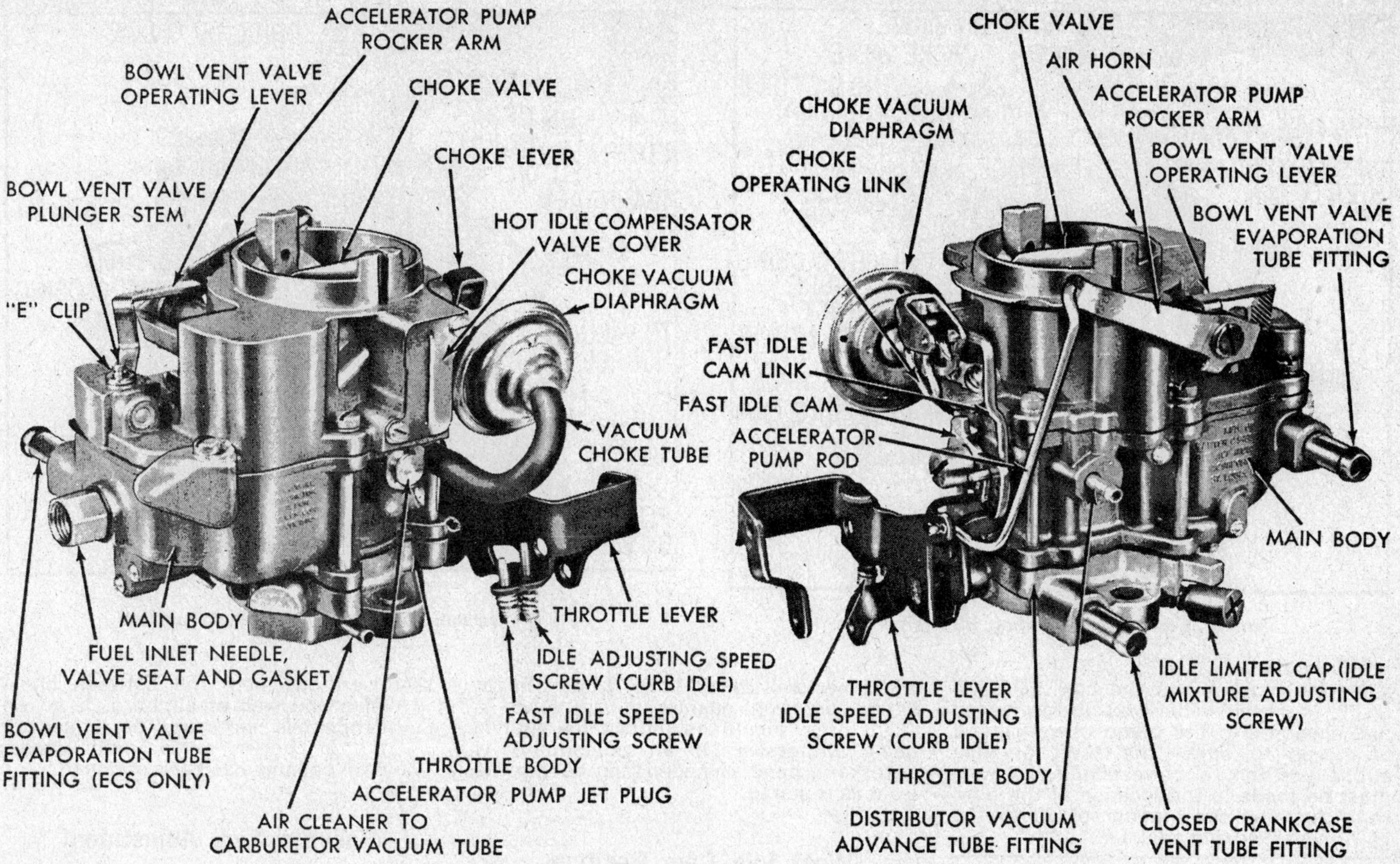

Fig. 3 Exterior of 1970-71 model BBS carburetor used with Evaporation Control System (ECS)

cedure is as follows:

1. Back off idle speed adjusting screw. Open choke valve so that when throttle valve is closed the fast idle adjusting screw will not contact fast idle cam.
2. Be sure pump operating rod is in center hole in throttle lever and that bowl vent clip on pump stem is in center groove.
3. Close throttle valve tightly. It should be possible to insert a drill of the specified size between bowl vent and air horn.
4. If an adjustment is necessary, bend pump operating rod at the lower angle as required to obtain the correct bowl vent opening.

NOTE: This is an important adjustment

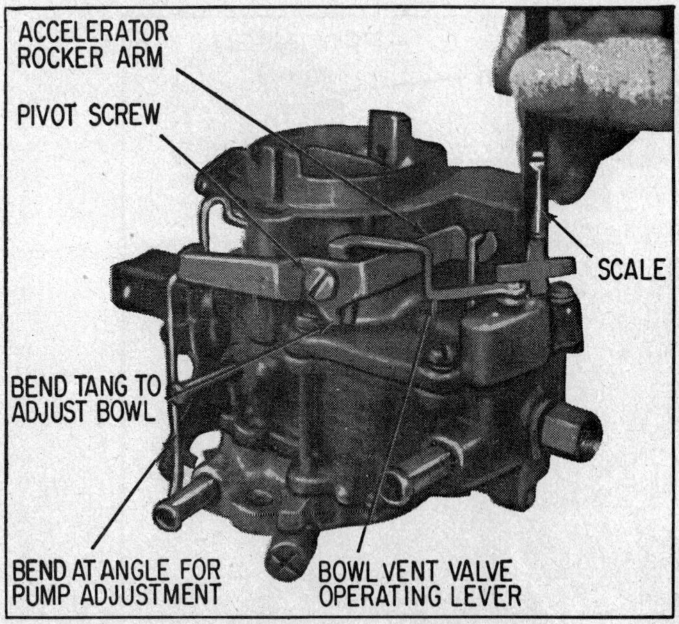

Fig. 6 Bowl vent adjustment. BBS carburetors with E.C.S.

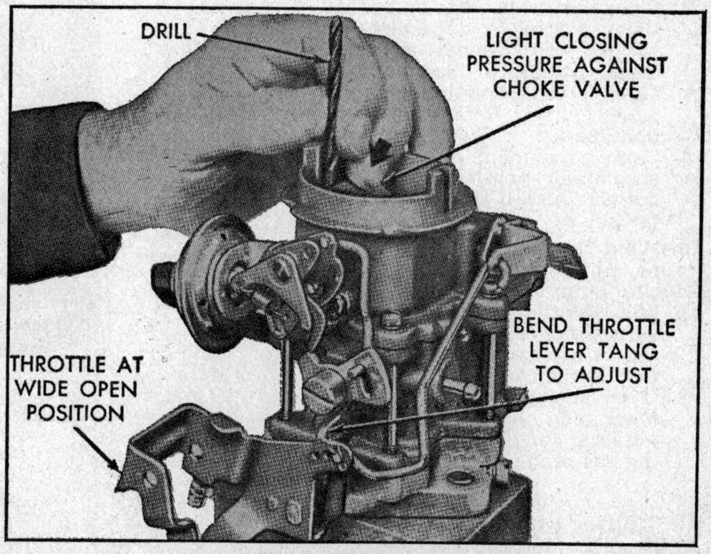

Fig. 7 Checking choke unloader setting (wide open kick). BBS carburetors

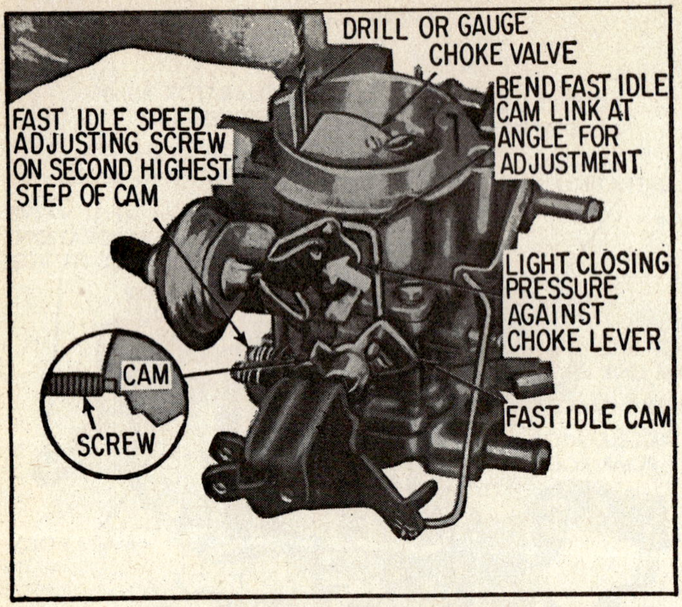

Fig. 8 Fast idle cam position. BBS carburetors

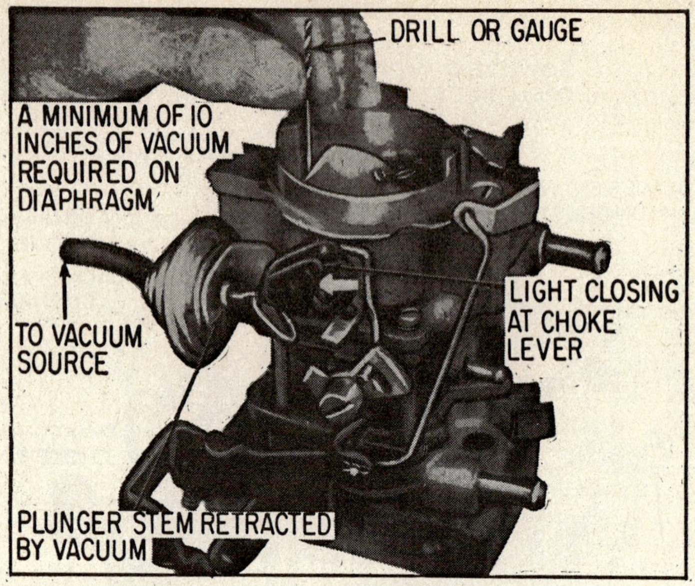

Fig. 9 Vacuum kick adjustment. BBS carburetors

since too much lift at the bowl vent will result in considerable loss in low speed fuel economy. If the pump operating rod is moved to either the short or long stroke position, a corresponding change must be made in the location of the bowl vent clip, and the amount of lift of the bowl rechecked and adjusted.

Pump and Bowl Vent Adjustments E.C.S.

Fig. 6
On the E.C.S. BBS carburetor, it is necessary to check or set the accelerator pump travel before checking the bowl vent valve opening.

Accelerator Pump Travel Adjustment

1. Be sure the accelerator pump rod is in outer hole (long stroke) of throttle lever.
2. Close throttle valve to curb idle.
3. Using a straight edge placed on flat surface (air cleaner mounting surface), measure distance between straight edge and top of accelerator pump plunger. Refer to *BBS Specification Chart.*
4. If an adjustment is necessary, bend accelerator pump rod at angle until correct pump travel has been obtained.

Bowl Vent Valve Adjustment

This adjustment to be made after accelerator pump adjustment.

1. With throttle valve at curb idle, measure distance from top of casting to top of bowl vent valve stem. Refer to *BBS Specifications Chart.*
2. If an adjustment is necessary, bend lower tang on bowl vent valve operating lever (at pivot) until correct opening has been obtained.

Choke Unloader Adjustment

Fig. 7—Hold throttle valve in wide open position. Insert a drill of the specified size between upper edge of choke valve and inner wall of air horn. With a finger lightly pressing against choke valve, a slight drag should be felt as the drill is being withdrawn. If an adjustment is necessary, bend unloader tang on throttle lever as required.

Fast Idle Cam Position

Fig. 8

1. With fast idle speed adjusting screw contacting the second highest step on the fast idle cam, move choke valve toward closed position with light pressure on choke shaft lever.
2. Insert specified drill between choke valve and wall of air horn. If an adjustment is necessary, bend fast idle rod at upper angle until correct valve opening has been obtained.

Vacuum Kick Adjustment

Fig. 9
The choke diaphragm adjustment controls the fuel delivery while the engine is running. It positions the choke valve within the air horn by action of the linkage between the choke shaft and diaphragm. The diaphragm must be ener-

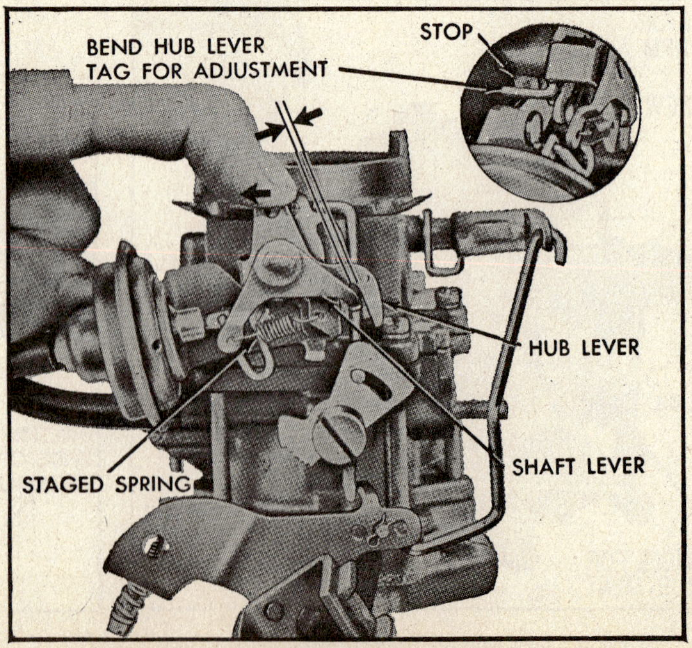

Fig. 10 Spring staged choke adjustment. BBS carburetors

gized to measure the vacuum kick adjustment. Use a vacuum source, or vacuum supplied by another vehicle. Adjust as follows:

1. If adjustment is to be made with the engine running, back off the fast idle speed screw until choke can be closed to the kick position with engine at curb idle.

NOTE: Number of screw turns required so that fast idle can be returned to original adjustment.

If an auxiliary vacuum source is to be used, open throttle valve (engine not running) and move choke to closed position. Release throttle first, then release choke.

2. When using an auxiliary vacuum source, disconnect vacuum hose from carburetor and connect it to hose from vacuum supply with a small length of tube to act as a fitting. Removal of hose from diaphragm may require forces which change the system. Apply a vacuum of 10 inches or more of mercury.
3. Insert specified drill between choke valve and wall of air horn. Apply sufficient closing pressure on lever to which choke rod attaches to provide a minimum stroke valve opening without distortion of diaphragm link. Note that cylindrical stem of diaphragm will extend as internal spring is compressed. This spring must be fully compressed for proper measurement of vacuum kick adjustment.
4. An adjustment will be necessary if a slight drag is not obtained as drill is being removed. Shorten or lengthen diaphragm link to obtain correct choke opening. Length changes should be made carefully by bending (opening or closing) the bend provided in the diaphragm link. *Do not apply twisting or bending force to diaphragm.*
5. Reinstall vacuum hose on correct carburetor fitting. Return fast idle screw to its original location if disturbed as suggested in Step 1.
6. Check the adjustment as follows: With no vacuum applied to diaphragm, *choke valve should move freely between open and closed positions.* If movement is not free, examine linkage for misalignment or interferences caused by bending operation. Repeat adjustment if necessary to provide proper link operation.

Spring Staged Choke Adjustment

Fig. 10

The spring staged choke is a device incorporated in the choke mechanism that limits the choke blade closing torque when cranking the engine at temperatures below zero. Thus the spring staging of the choke is a better match for the engine's starting mixture requirements at low temperatures.

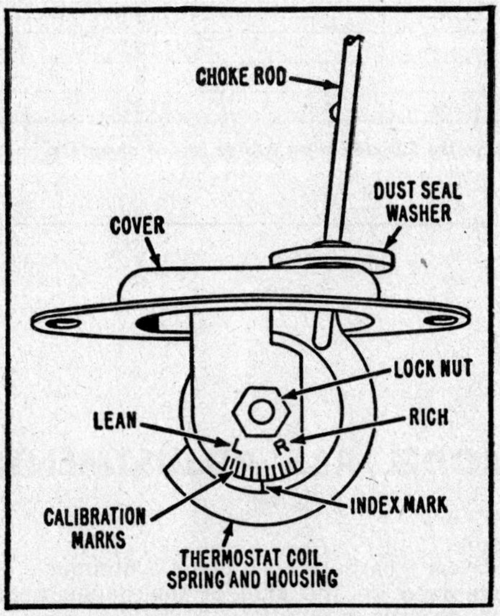

Fig. 11 BBS well-type choke setting

To check the spring staged choke for correct operating clearance, proceed as follows:

1. Push on hub lever with finger at closed choke position. A small opening should exist between shaft and hub levers as indicated.
2. Using a drill or gauge, measure the opening which should be from .010 to .040".
3. If adjustment is necessary, bend hub lever tang until correct opening is obtained.

Choke Adjustment

Fig. 11—Loosen mounting post lock nut and turn mounting post with screwdriver until index mark on disc is positioned as listed in the *BBS Specifications Chart.* Hold in this position with screwdriver and tighten lock nut.

NOTE: Screwdriver may be held in a vise so that one hand may be used to support housing while tightening nut. After adjustment is completed and coil housing and rod and carburetor are installed on engine, lift cover disc and open and close choke valve manually to see if connector rod clears sides of hole in housing cover without binding. If rod does not clear housing cover without binding, replace with a new unit since connecting rod cannot be bent without affecting calibration.

Dashpot Adjustment

Clean Air Carburetors

The dashpot is used only on cars equipped with the Cleaner Air Package and manual transmission, Figs. 1 and 2.

To adjust the dashpot, have the curb idle speed and mixture properly adjusted, and install a tachometer. Position throttle lever so that actuating tab on lever is contacting stem of dashpot but not depressing it. The tachometer should read 2000 rpm if the setting is correct. If not correct, screw dashpot in or out as required, then tighten lock nut on dashpot against the bracket.

CARTER RBS ADJUSTMENT SPECIFICATIONS

See Tune Up Chart in car chapter for hot idle speed.

Year	Carb. Model	Idle Screw (Mixture) Turns Open	Float Level	Accel. Pump	Bowl Vent	Fast Idle Cam Setting RPM	Fast Idle Throttle Plate Setting	Choke Unloader	Pulldown Setting	Choke Setting
AMERICAN MOTORS										
1969	4631S	1	9/16	1/64	5/64	2000	.035	1/8	13/64	Index
	4633S	1	9/16	1/64	5/64	2000	.035	3/16	13/64	Index
	4634S	1	9/16	1/64	5/64	2000	.035	3/16	7/32	2 Rich
	4666S	1	9/16	1/64	5/64	2000	.035	1/8	7/32	2 Rich

Continued

CARTER RBS ADJUSTMENT SPECIFICATIONS—Continued

See Tune Up Chart in car chapter for hot idle speed.

Year	Carb. Model	Idle Mixture Screw Setting	Float Level	Step-Up Rod	Accelerating Pump Stroke	Fast Idle Linkage	Fast Idle Throttle Plate Setting	Dechoke Setting	Pulldown Setting	Dashpot Setting	Choke Setting
FORD ENGINES											
1970	D0ZF-C	①	9/16	—	.400③	See Text	.040	.250	.190	—	Index
	D0ZF-D	①	9/16	—	.400③	See Text	.046	.252	.190	—	1 Rich
	D0ZF-F	①	9/16	—	.400③	See Text	—	.252	.190	7/32	1 Rich
1971	D1ZF-HA, LA	①	9/16	—	.400③	.115②	—	.250	.270	—	Index
	D1ZF-NA, KA	①	9/16	—	.400③	.115②	—	.250	.190	—	1 Rich
1972	D2OF-LA	①	9/16	—	.400③	.115②	—	.250	.300	—	Index
	D2OF-MA	①	9/16	—	.400③	.115②	—	.250	.190	—	1 Rich
	D2OF-SA	①	9/16	—	.400③	.115②	—	.250	.190	—	1 Rich
1973	D3OF-BA	①	9/16	—	.420③	.115②	—	.250	.300	—	Index
	D3OF-CA	①	9/16	—	.400③	.115②	—	.250	.190	—	Index
1974	D4DE-AAA	①	9/16	—	—	.115②	—	.250	.190	—	1 Lean
	D4DE-AB	①	9/16	—	—	.115②	—	.250	.190	—	Index
	D4DE-BB	①	9/16	—	—	.115②	—	.250	.300	—	Index
	D4DE-SB	①	9/16	—	—	.115②	—	.250	.300	—	Index
	D4DE-TA	①	9/16	—	—	.115②	—	.250	.190	—	Index

①—Air/fuel ratio or idle CO% rating is found in Tune Up Specification tables in car chapters.
②—At kickdown.
③—Closed throttle.

MODEL RBS ADJUSTMENTS

This carburetor, Fig. 1 incorporates a single aluminum casting with a pressed steel bowl. Adjustments are readily accessible and most calibration points are located in the single casting.

Fuel pickups are located near the centerline of the carburetor bore to gain the benefits of a concentric bowl carburetor, yet so located that engine heat being radiated through the bore is conducted through the casting but is not readily conducted to the fuel in the bowl. Two internal vapor vents allow rapid fuel vapor dissipation to help provide smooth idle conditions and to minimize hard starting when engine is hot. Also an external vent, mechanically controlled by the throttle, is used on 1974 California units.

A diaphragm-controlled step-up type metering rod controls the fuel supply. The accelerator pump is spring-actuated.

The carburetor is equipped with a vacuum piston automatic choke. On 1974 California units, an electric assist choke is used to open the choke plate within 1-1½ minutes after underhood temperatures reach approximately 60° F. For service procedures, refer to the "Emission Control System" chapter.

The 1973-74 RBS unit incorporates an Exhaust Gas Recirculation port, Fig. 2. The E.G.R. port connects to the primary bore and allows a metered amount of exhaust gas to be fed into the fuel-air mixture.

The carburetor model number is stamped on the side of the flange near the throttle lever.

Float Level Adjustment

Fig. 3—With carburetor inverted, bowl and bowl gasket removed, and only weight of float pressing needle into its seat, measure vertical distance from casting to the small "bump" at outer ends of float. Gauge both ends of float. If the vertical distance is not as listed in the *RBS Specifications Chart*, adjustment can be made by removing float from casting, or by

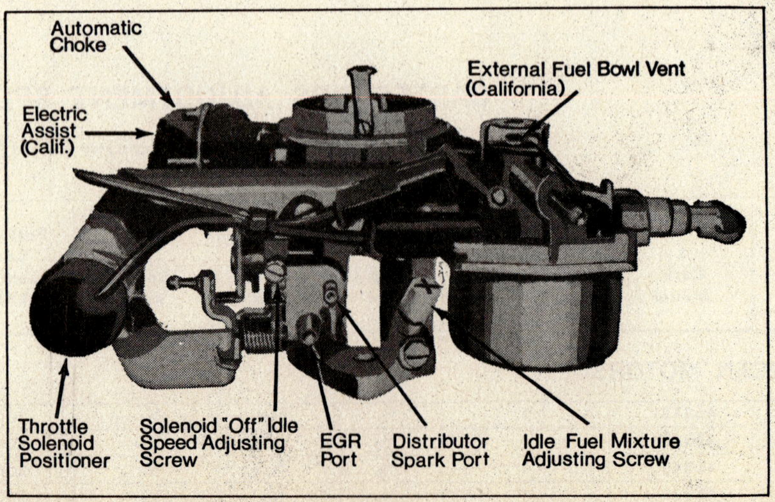

Fig. 2 Carter RBS carburetor side view. 1973-74

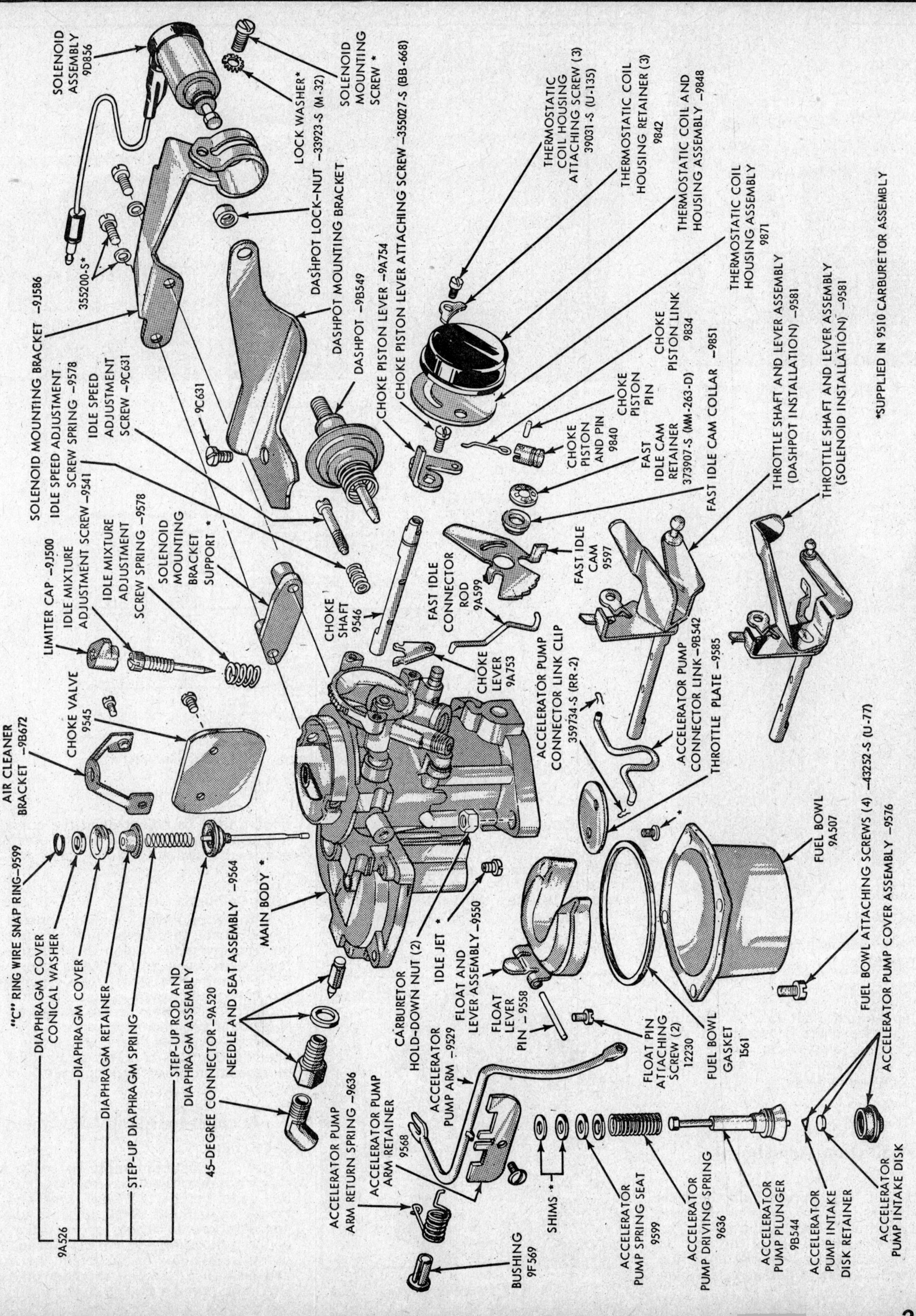

Fig. 1 Carter model RBS carburetor. 1969-74 (typical)

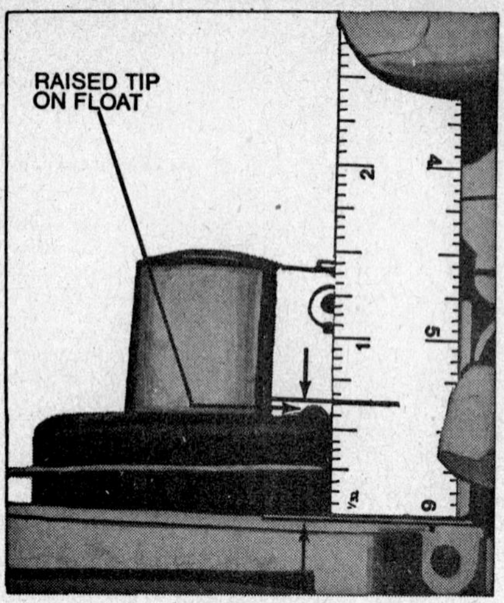

Fig. 3 RBS float level adjustment

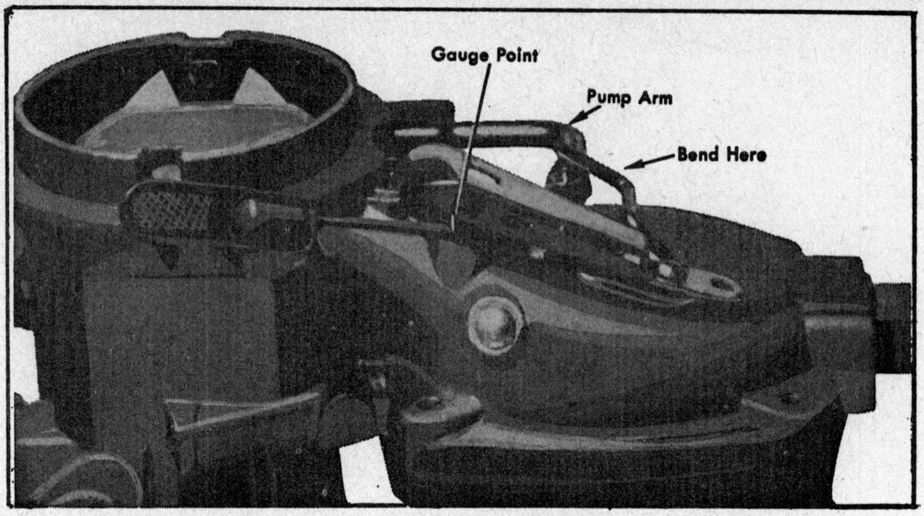

Fig. 5 Bowl vent adjustment

Fig. 4 Adjusting fuel float level

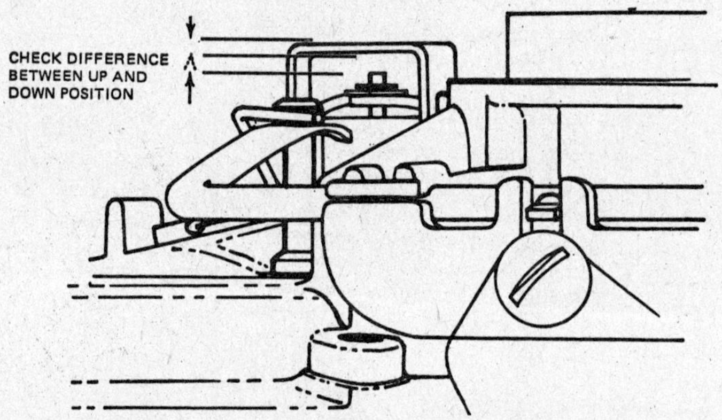

Fig. 5A Bowl vent adjustment. 1974 California units

holding lip end of float bracket securely with needle-nose pliers. However, be sure to hold float lip away from needle when adjusting. To adjust, bend bracket at its narrowest portion, Fig. 4.

Float Drop Adjustment

1970-74

With the air horn upright and the float hanging free measure the vertical distance from the main body casting surface for the fuel bowl to the outer ends of the float on the top side. Adjust by bending tab at the end of the float arm to obtain 1.250" setting.

Bowl Vent Adjustment

1969 Units

With pump stroke adjusted properly and throttle valve fully closed, measure the clearance between the vent valve and the carburetor casting. Refer to *RBS Specifications Chart*. To adjust bend the connector rod at accelerator pump end to obtain the specified clearance, Fig. 5.

1974 California Units

With accelerator pump properly adjusted, place throttle at curb idle position on the extended solenoid. Measure direction between top of vent guard and top of vent valve stem. Then, open throttle until vent valve seats and measure same distance again. The difference between the two measurements is the vent opening, Fig. 5A. Refer to Carter RBS Specification Chart. Adjust by bending link at accelerator pump arm contact point.

Fig. 6 Pump adjustment

Accelerator Pump Adjustment

1969 Units

Pump adjustment must be made each time carburetor is disassembled, and must be made before the bowl vent and unloader adjustment. Back out curb idle adjusting screw and hold choke valve wide open so throttle valve is fully closed. Turn pump adjusting nut to obtain the specified clearance between shoulder on pump shaft and pump arm, Fig. 6.

Fig. 7 Accelerator pump stroke

1970-74 Units

Open the choke plate and back off the idle speed adjustment screw to allow the throttle plate to seat in the bore. Measure the height from the flat surface of the main body casting to the top surface of the pump stem, Fig. 7. Open the throttle wide, then measure the height again. The pump stroke is the difference between the two measurements. Refer to *RBS Specifications Chart.* To adjust the stroke to specifications, open or close the pump connector link at the offset portion, Fig. 8.

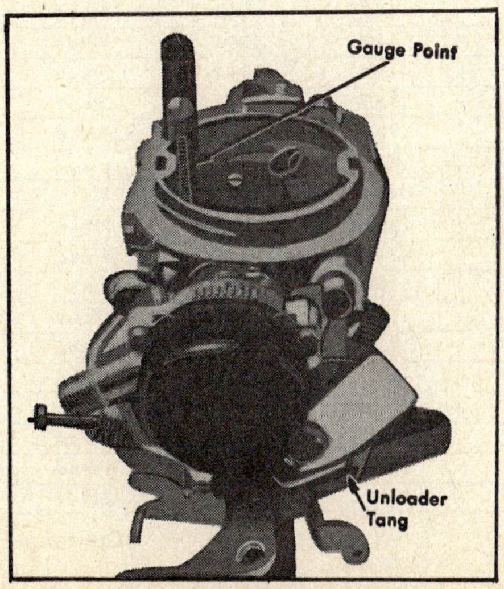

Fig. 10 Unloader adjustment

Fig. 8 Adjusting accelerator pump stroke

Fast Idle Cam Adjustment

1969-70 Units

With choke valve tightly closed and choke connector rod in upper end of slot in cam, align cam index with center of fast idle tang. Adjust by bending choke connector rod at the offset portion.

1971-74 Units

Position the fast idle screw on the kick-down step of the fast idle cam against the shoulder of the high step. Adjust by bending the choke plate connecting rod to obtain the specified clearance between the lower edge of the choke plate and the carburetor bore.

Throttle Plate Clearance

Align the fast idle tang on the throttle lever with the index mark on the fast idle cam. Use a drill between the throttle plate and the throttle bore at the idle port side to check the clearance, Fig. 9. Adjust by bending tang on the throttle lever. Refer to *R.B.S. Specification Chart.*

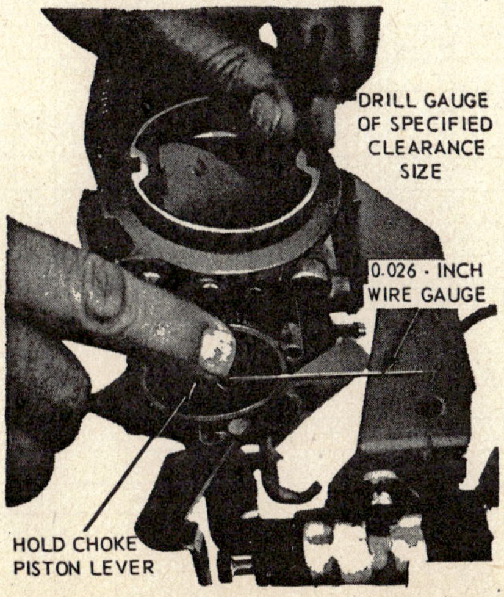

Fig. 11 Adjusting choke plate pulldown

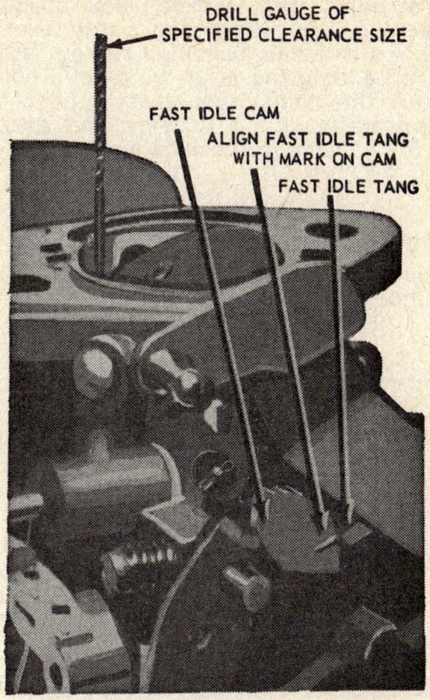

Fig. 9 RBS throttle plate clearance.

Choke Unloader Adjustment

1969 American Motors Units

Fig. 10—With throttle valve wide open, there should be the clearance listed in the *RBS Specifications Chart* between top edge of choke valve and inner wall of air horn. To adjust, bend tang on throttle lever.

Choke Plate Pulldown or Piston Linkage Adjustment

Bend a .026 in. diameter wire gauge at a 90 degree angle approximately ⅛-inch from one end. Insert the bent end of the gauge between the choke piston slot and the right hand slot in choke housing. Rotate the choke piston lever counter clockwise until gauge is snug in the piston slot. Exert a light pressure on choke piston lever to hold the gauge in place, then use a drill with a diameter equal to the specified pulldown clearance between the lower edge of choke plate and carburetor bore to check clearance, Fig. 11.

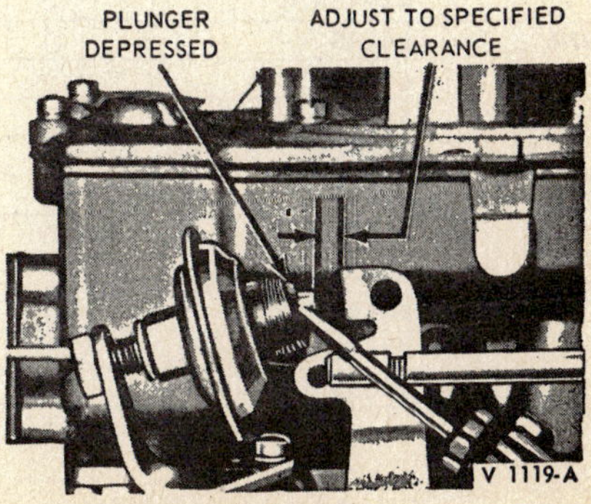

Fig. 12 Typical anti-stall dashpot adjustment

To adjust the choke plate pulldown clearance, bend the choke piston lever as required to obtain specified setting. NOTE: When bending the lever, be careful not to distort the piston link. Install the choke thermostatic spring housing and gasket. Set the housing to specifications.

Dechoke Clearance

1970-74 Ford Units

Hold the throttle lever at the wide open position and close the choke plate as far as possible without forcing it. The dechoke clearance on the 1970 units is measured between the upper edge of the choke plate and the inner wall of the main body. While clearance on 1971-74 units is measured between the lower edge of the choke plate and the inner wall of the main body.

Dashpot Adjustment

Loosen the anti-stall dashpot lock nut, Fig. 12. Hold the throttle in the closed position and depress the plunger. The clearance between the throttle lever and the plunger tip for American Motors is 3/32" and Ford 7/32". Tighten the locknut to secure the adjustment.

Automatic Choke Adjustment

Loosen retainer screws and turn choke cover so that line or index mark on cover lines up with the specified mark on choke housing (see *RBS Specifications Chart*).

CARTER BBD ADJUSTMENT SPECIFICATIONS

See Tune Up Chart in car chapter for hot idle speed.

Year	Carb. Model	Initial Idle Mix. Screws Turns Open	Float Level	Pump Travel Inch	Bowl Vent Drill Size	Choke Unloader Drill Size	Choke Vacuum Kick Drill Size	Fast Idle Cam Position Drill Size	Automatic Choke Setting
CHRYSLER, DODGE & PLYMOUTH									
1969	4605S	2	1/4	—	1/16	1/4	20 Drill	41 Drill	On Index
	4606S	2	1/4	—	1/16	1/4	41 Drill	41 Drill	On Index
	4607S	1	1/4	—	1/16	1/4	20 Drill	41 Drill	On Index
	4608S	1	1/4	—	1/16	1/4	28 Drill	41 Drill	On Index
	4613S	1 1/2	5/16	1	1/16	1/4	20 Drill	30 Drill	2 Rich
	4614S	1 1/2	5/16	1	1/16	1/4	20 Drill	30 Drill	2 Rich
	4474S	1 1/2	5/16	1	1/16	1/4	20 Drill	30 Drill	2 Rich
1970	4721S	①	1/4	—	1/32	1/4	20 Drill	41 Drill	On Index
	4722S	①	1/4	—	1/32	1/4	20 Drill	41 Drill	On Index
	4723S	①	1/4	—	15/64	1/4	20 Drill	41 Drill	On Index
	4724S	①	1/4	—	15/64	1/4	20 Drill	41 Drill	On Index
	4725S	①	11/32	1	1/32	1/4	20 Drill	28 Drill	2 Rich
	4726S	①	11/32	1	1/32	1/4	28 Drill	28 Drill	2 Rich
	4727S	①	11/32	1	1/8	1/4	20 Drill	28 Drill	2 Rich
	4728S	①	11/32	1	3/16	1/4	28 Drill	28 Drill	2 Rich
	4894S	①	11/32	1	1/32	1/4	28 Drill	28 Drill	2 Rich
	4895S	①	1/4	—	1/32	1/4	20 Drill	41 Drill	On Index
1971	4957S	①	1/4	.200②	13/64	1/4	20 Drill	41 Drill	On Index
	4958S	①	1/4	.200②	13/64	1/4	20 Drill	41 Drill	On Index
	4961S	①	11/32	1	3/16	1/4	20 Drill	20 Drill	2 Rich
	4962S	①	11/32	1	3/16	1/4	28 Drill	28 Drill	2 Rich
1972	6149S	①	1/4	.225③	17/64	1/4	.149	.096	Fixed
	6150S	①	1/4	.225③	17/64	1/4	.149	.096	Fixed
	6151S	①	1/4	.225③	17/64	1/4	.149	.096	Fixed
	6152S	①	1/4	.225③	17/64	1/4	.149	.096	Fixed
1973	6316SA	①	1/4	.242③	—	1/4	.150	.095	Fixed
	6317SA	①	1/4	.242③	—	1/4	.130	.095	Fixed
	6343SA④	①	1/4	.242③	—	1/4	.150	.095	Fixed
	6344SA④	①	1/4	.242③	—	1/4	.150	.095	Fixed
1974	6464S	①	1/4	1/2③	—	.325	.150	.095	Fixed
	6465S	①	1/4	1/2③	—	.325	.110	.095	Fixed
	6466S④	①	1/4	1/2③	—	.325	.150	.095	Fixed
	6467S④	①	1/4	1/2③	—	.325	.110	.095	Fixed
1975	8000S	①	1/4	1/2③	—	.280	.130	.070	Fixed
	8001S	①	1/4	1/2③	—	.310	.110	.070	Fixed
	8003S	①	1/4	1/2③	—	.310	.110	.070	Fixed
	8062S	①	1/4	1/2③	—	.310	.110	.070	Fixed
	8064S	①	1/4	1/2③	—	.310	.070	.070	Fixed
	8066S	①	1/4	1/2③	—	.280	.130	.070	Fixed

Continued

CARTER BBD ADJUSTMENT SPECIFICATIONS

See Tune Up Chart in car chapter for hot idle speed.

Year	Carb. Model	Initial Idle Mix. Screws Turns Open	Float Level	Pump Travel Inch	Bowl Vent Drill Size	Choke Unloader Drill Size	Choke Vacuum Kick Drill Size	Fast Idle Cam Position Drill Size	Automatic Choke Setting
CHRYSLER, DODGE & PLYMOUTH									
1976	8071S	①	¼	½③	—	.280	.130	.070	Fixed
	8072S	①	¼	½③	—	.310	.070	.070	Fixed
	8077S	①	¼	½③	—	.280	.110	.070	Fixed
AMERICAN MOTORS									
1976	8067	①	¼	—	—	¼	—	.095	2 Rich
	8073	①	¼	—	—	¼	—	.095	1 Rich

①—Air/fuel ratio or idle CO% rating is found in Tune Up Specification tables in car chapters.
②—Throttle closed.
③—At idle.
④—With California Emission package.

MODEL BBD ADJUSTMENTS

Fig. 1 is an exploded view of the BBD two barrel carburetor. The choke housing containing the thermostatic coil spring is located in a well at the exhaust cross-over passage, Fig. 2.

The carburetor shown in Fig. 3 is a standard model when the vehicle is equipped with either a manual shift or automatic transmission. When used with C.A.P. or C.A.S. equipment, a dashpot (slow closing throttle device) is provided for vehicles with manual transmission only, Fig. 3A.

Note that the 1973-74 E.C.S. equipped unit incorporates a Canister Purge Port, Fig. 3B.

The carburetor show in Fig. 3B is used on the smaller 1970-73 V8 engines when equipped with E.C.S. Note that this carburetor is equipped with a Canister Purge Port.

The carburetor shown in Fig. 3C is used on the larger 1970 V8 engines when equipped with C.A.S.

The carburetors shown in Fig. 4 are used on the larger 1970-71 V8 engines when equipped with E.C.S.

The carburetors shown in Fig. 4A are used on 1974-75 models. On vehicles equipped with manual transmissions, the carburetor is equipped with a dashpot. On 1975 vehicles with automatic transmissions, an idle enrichment system is used to reduce cold engine stalling by the use of an additional metering system which enriches the mixture in the off idle position. This system is controlled by a vacuum diaphragm mounted near the top of the carburetor.

Float Level Adjustment

Fig. 5—With carburetor body inverted so that weight of floats ONLY is forcing needle against its seat, use a T-scale or the tool shown, and check the float level from surface of fuel bowl to crown of each float at center.

If an adjustment is necessary, hold floats on bottom of bowl and bend float lip as required to give the specified dimension.

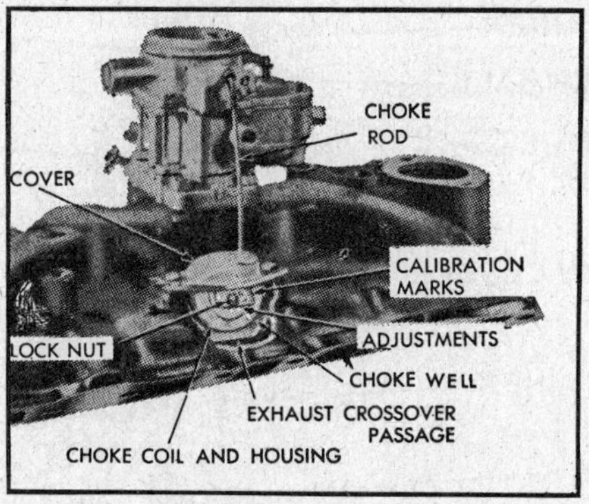

Fig. 2 Well-type automatic choke mounted on intake manifold

CAUTION: When bending the float lip, do not allow the lip to push against the needle as the synthetic rubber tip (if used) can be compressed sufficiently to cause a false setting which will affect correct level of fuel in bowl. After being compressed, the tip is very slow to recover its original shape.

Accelerator Pump

Except 1¼" Bore Units, Fig. 6

1. Back off idle adjusting screw. Open choke valve so that fast idle cam allows throttle valves to fully close. Be sure that pump connector rod is

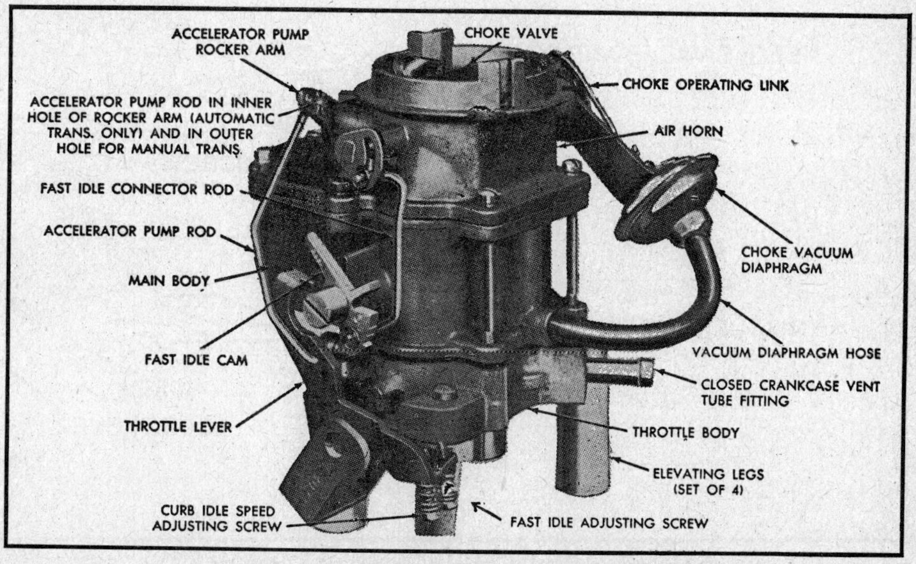

Fig. 3 BBD 1¼" bore carburetor assembly C.A.P. equipped

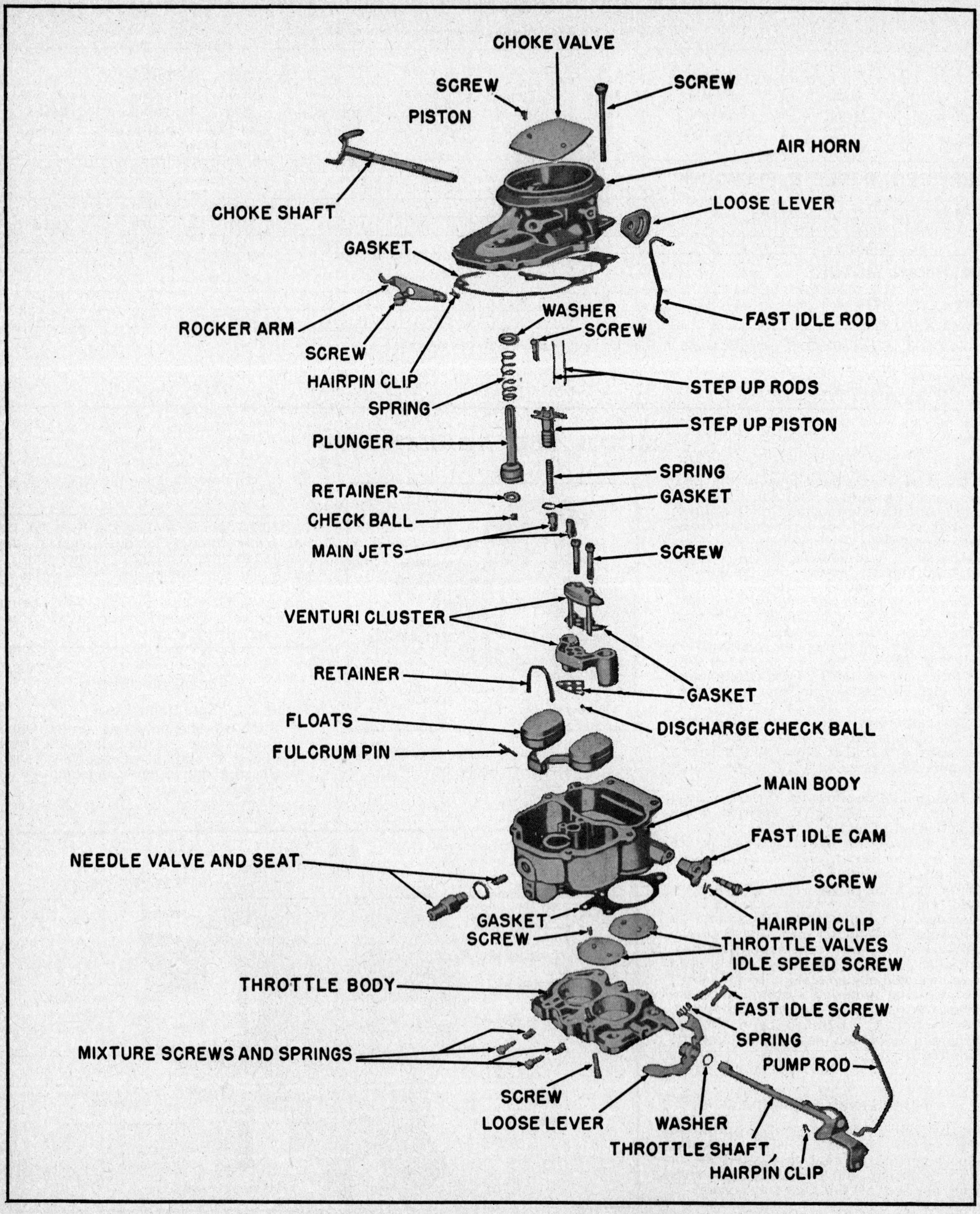

Fig. 1 Carter Model BBD two-barrel carburetor exploded

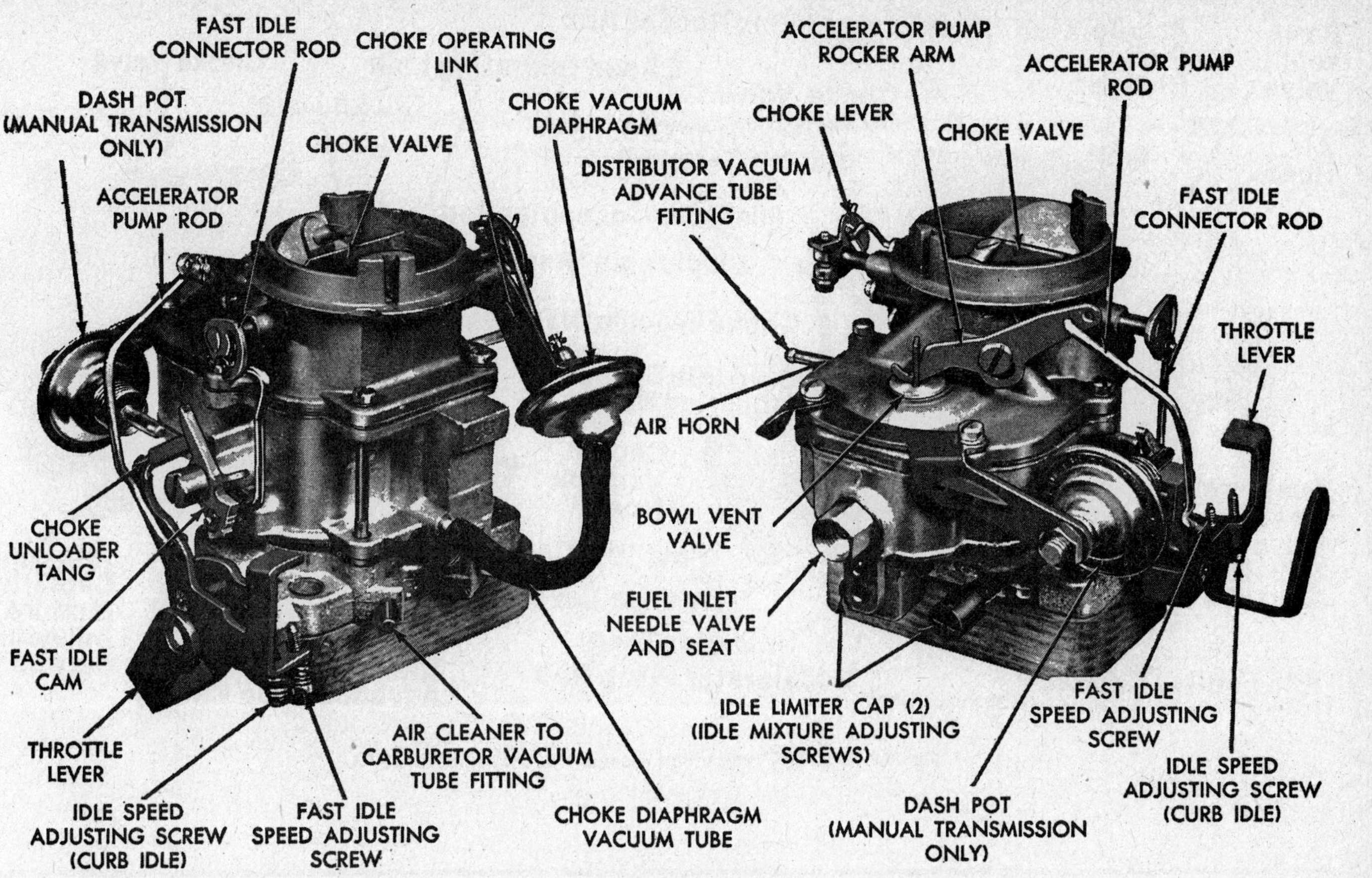

Fig. 3A BBD 1¼" bore carburetor assembly C.A.S. equipped

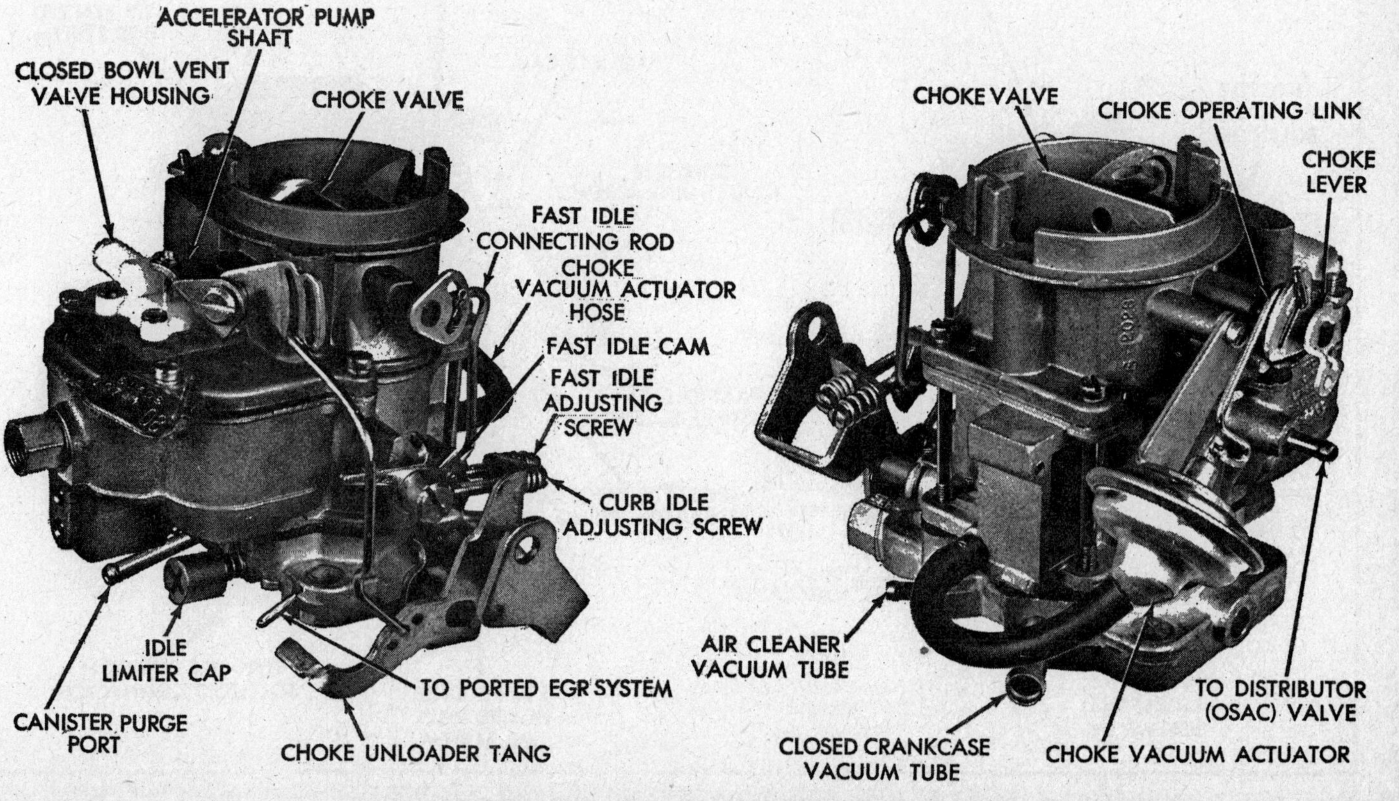

Fig. 3B BBD 1¼" bore carburetor assembly E.C.S. equipped (typical)

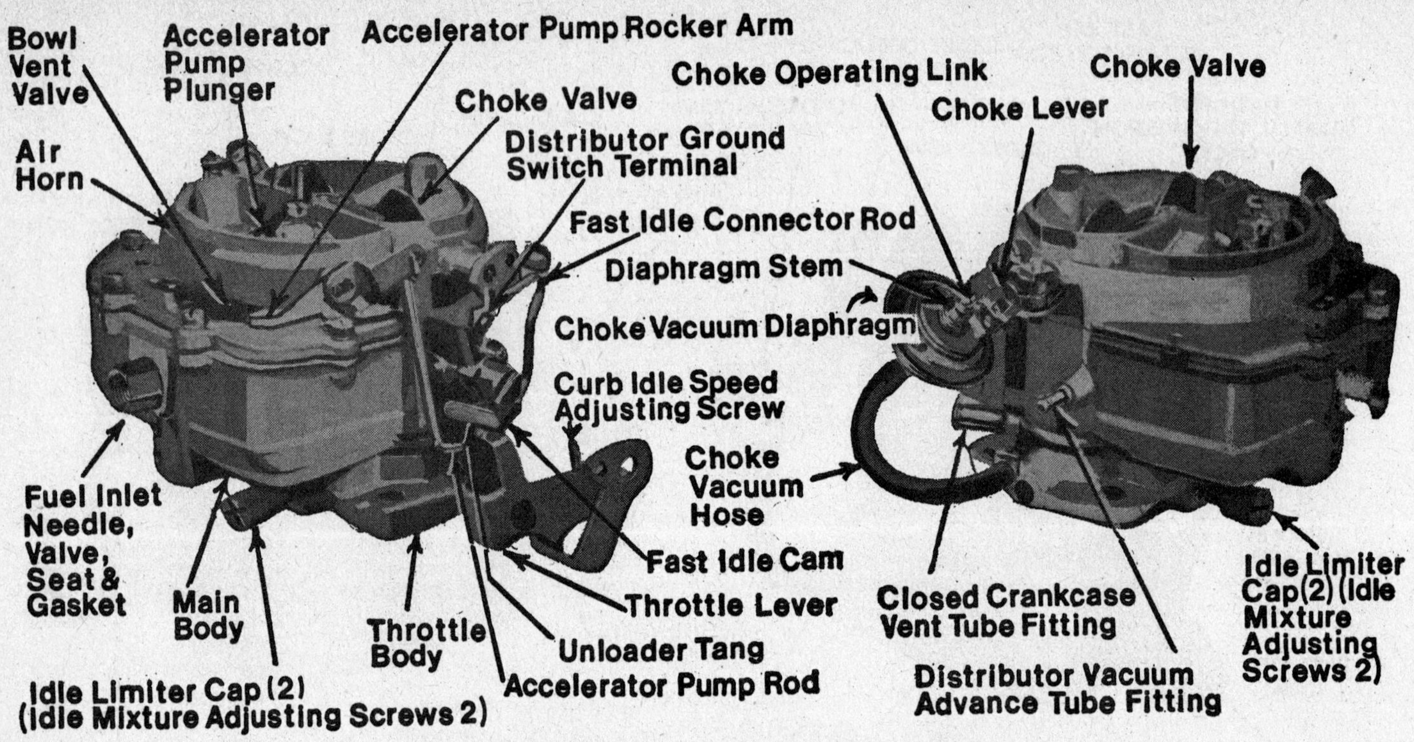

Fig. 3C BBD 1½" carburetor assembly C.A.S. equipped. 1970

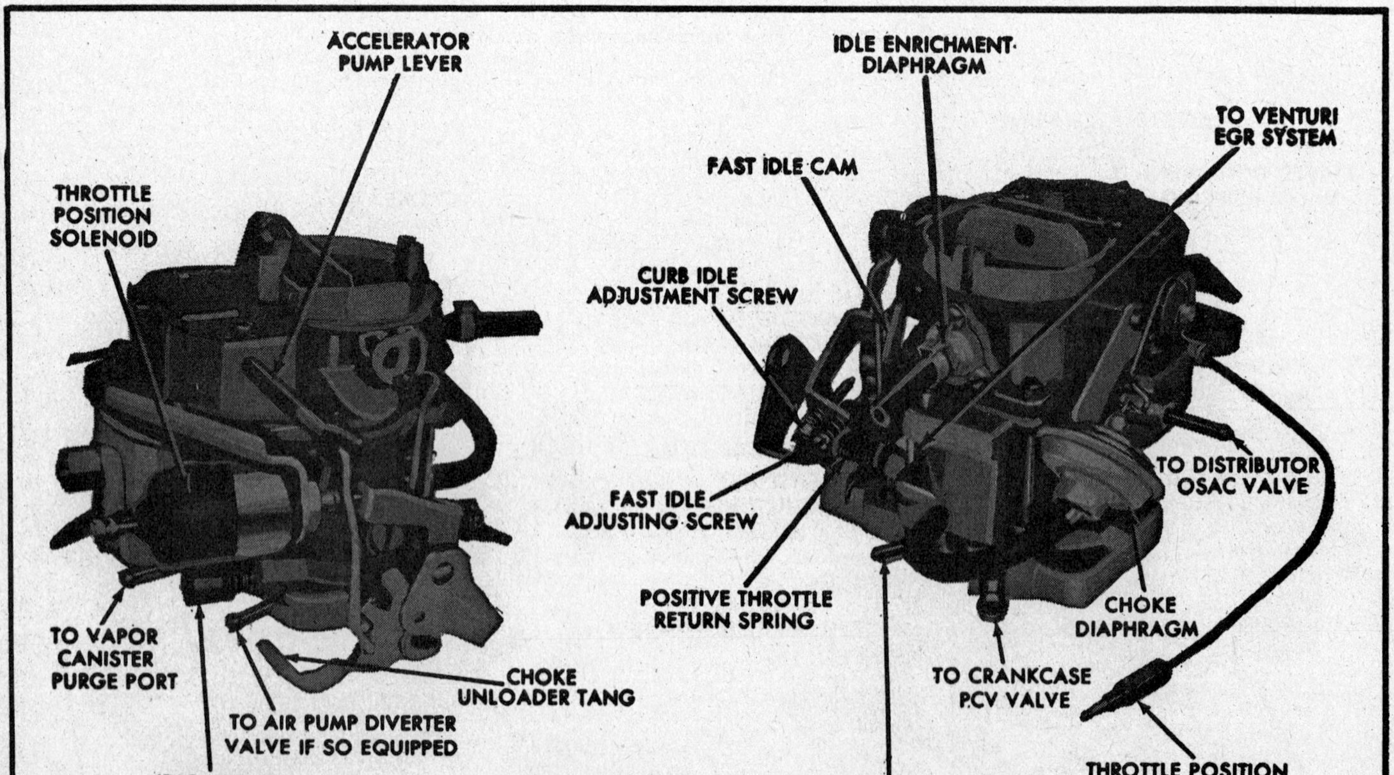

Fig. 4A BBD 1¼" carburetor assembly. 1974-75

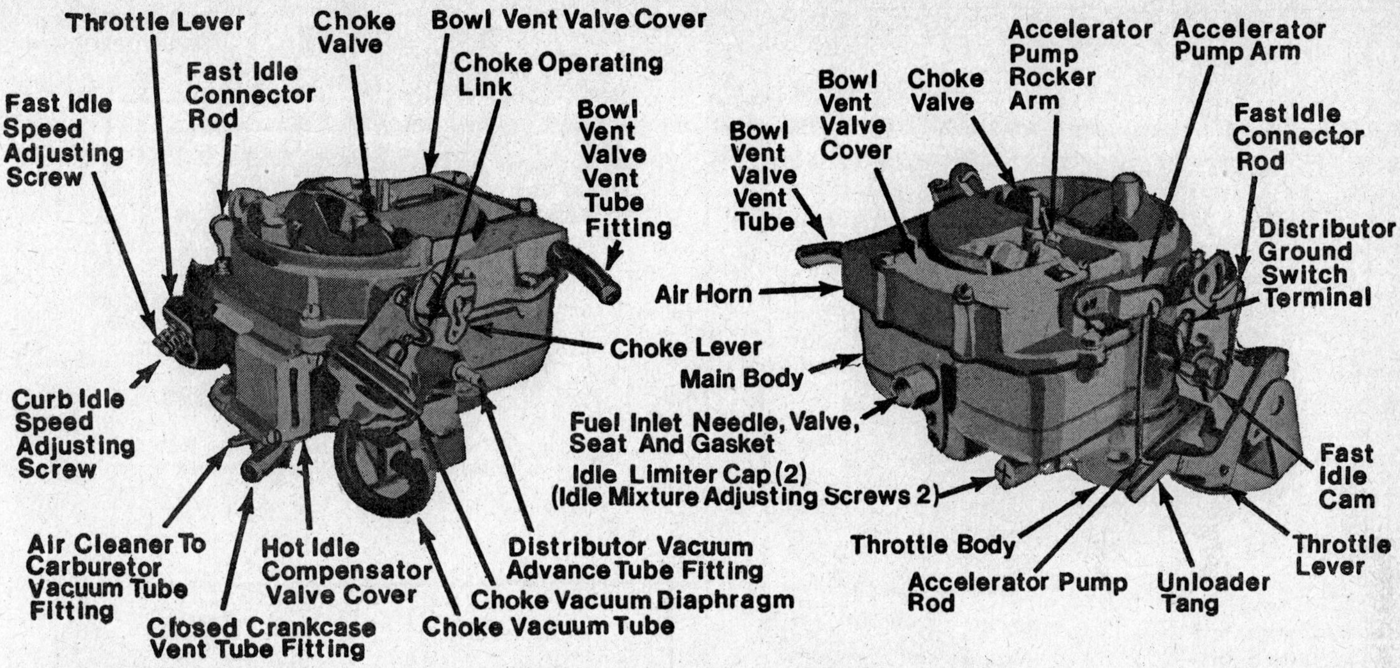

Fig. 4 BBD 1½" bore carburetors E.C.S. equipped (typical)

installed in outer hole of throttle lever.

2. With throttle valves closed tightly, measure distance between top of air horn and end of pump plunger shaft. If the dimension is not as specified, bend pump connector rod at the angle on the rod until correct setting is obtained.

1974 1¼" Bore Units, Fig. 6A

1. Back off curb idle adjusting screw, completely closing throttle valve, then open choke valve, allowing throttle valves to seat in bores. Ensure accelerator pump "S" link is located in outer hole or pump arm.
2. Turn curb idle adjusting screw until screw contacts, stop, then rotate screw two additional turns.
3. Measure distance between air horn surface and top of accelerator pump shaft. Refer to BBD Specifications Chart.
4. Adjust by loosening pump arm adjusting screw and rotating sleeve until proper dimension is obtained. Tighten adjusting screw.

Accelerator Pump and Bowl Vent

C.A.S. Equipped 1¼" Bore Units, Fig. 7

1. Back off idle speed adjusting screw. Open choke valve so that fast idle cam allows throttle valves to close completely.
2. Be sure pump operating rod is in medium stroke hole in throttle lever, and that bowl vent clip on pump stem is on center notch.
3. With throttle valves closed tightly, it should be just possible to insert the specified gauge or drill size between bowl vent and its seat.
4. If an adjustment is necessary, bend pump operating rod at the angle. On CAP carburetors, bend pump operating rod to give the height of pump plunger stem above bowl cover as specified in the table.

E.C.S. Equipped 1¼" Bore Units, Fig. 7A

1. Back off idle speed adjusting screw to

Fig. 5 Checking float level. BBD carburetors

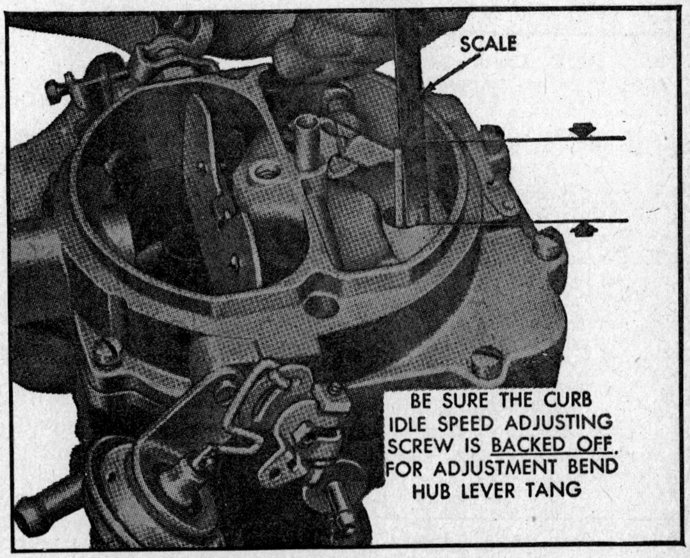

Fig. 6 Checking accelerator pump travel. BBD except 1¼" bore carburetors

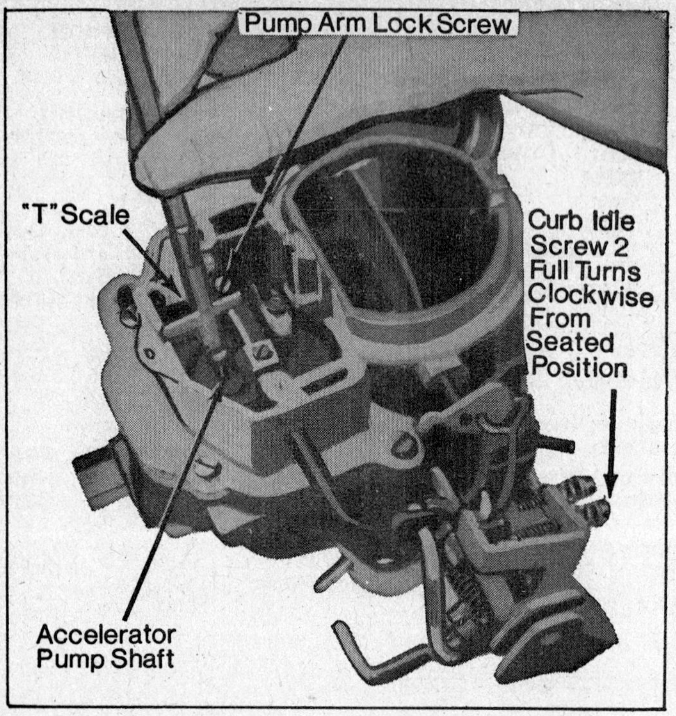

Fig. 6A Accelerator pump setting, 1974-75 BBD 1¼″

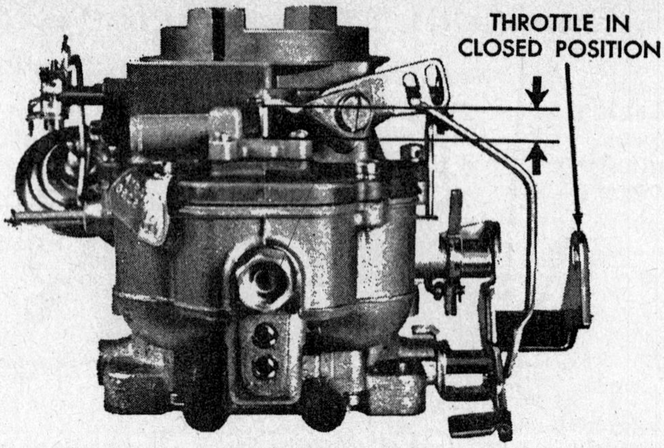

Fig. 7A Checking bowl vent opening BBD 1¼″ bore carburetors E.C.S. equipped

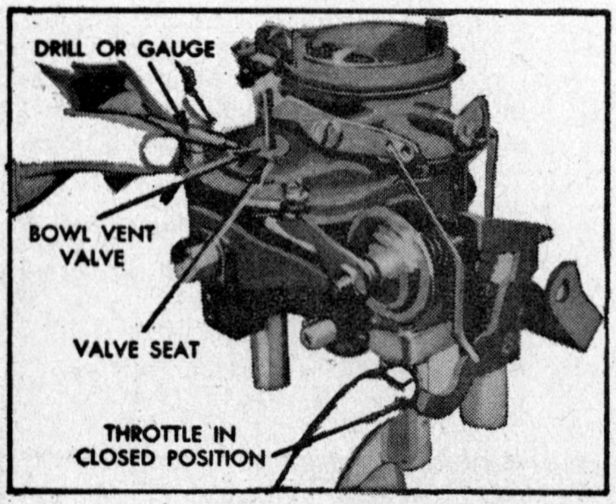

Fig. 7 Checking bowl vent opening. BBD 1¼″ bore carburetors C.A.S. equipped

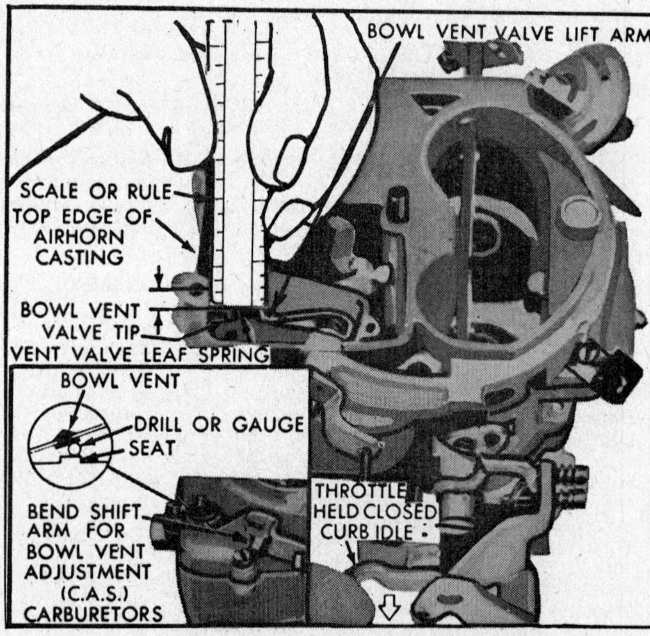

Fig. 7B Measuring bowl vent valve opening BBD 1½″ bore C.A.S. and E.C.S. carburetors

completely close throttle valves. Open choke valve so that fast idle cam allows throttle valves to seat in bores.

2. Be sure accelerator pump operating rod is in medium stroke hole in throttle lever.
3. Close throttle valves tightly. Measure distance between top of vent valve plastic housing and end of pump plunger stem.
4. To adjust pump travel, bend accelerator pump operating rod at lower angle. Refer to BBD Specifications Chart.

NOTE: This is an important adjustment, since too much lift at the bowl vent will result in considerable loss in low speed fuel economy.

Remember that if the pump operating rod is moved to either the short or long stroke position, a corresponding change must be made in the location of the bowl vent clip, and the amount of lift of the bowl vent rechecked.

Bowl Vent Adjustment

Models with Separate Pump Fig. 7B

1½″ Bore Units

1. Open choke valve so that fast idle cam allows valves to close, curb idle.
2. Be sure that pump operating rod is in long stroke hole in throttle lever. Remove bowl vent valve cover if not previously done.
3. On E.C.S. units, close throttle valves tightly. Using a narrow ruler, measure the distance from top of bowl vent valve (rubber tip) to top of air horn casting, Fig. 7B. Refer to Specifications Chart.

If an adjustment is necessary, bend bowl vent lift arm, using a suitable tool, until correct opening has been obtained. On C.A.S. units, with the throttle valves closed, (curb idle). There should be specified clearance between bowl vent valve and seat on air horn. Measure at outermost or largest dimension with a drill shank, Fig. 7B.

If an adjustment is necessary, bend vent valve lifter arm until correct clearance has been obtained.

Note: Important: Do not bend bowl

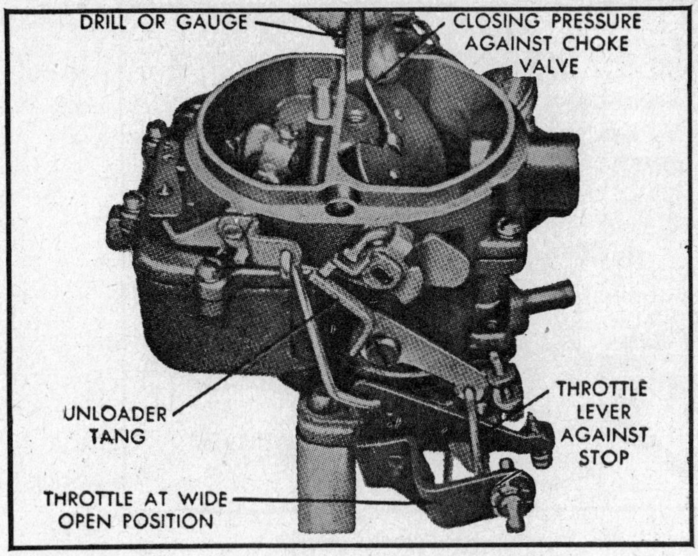

Fig. 8 **Choke unloader setting. BBD carburetors**

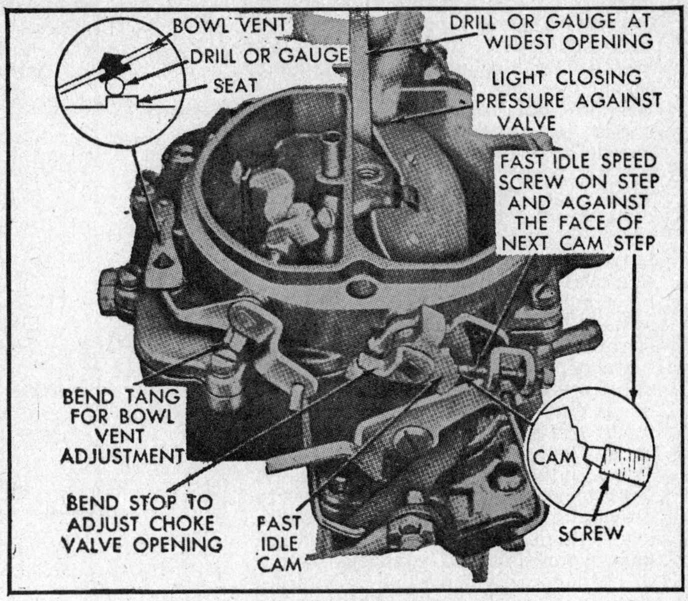

Fig. 10 **Fast idle cam position adjustment. BBD except 1¼" bore carburetors**

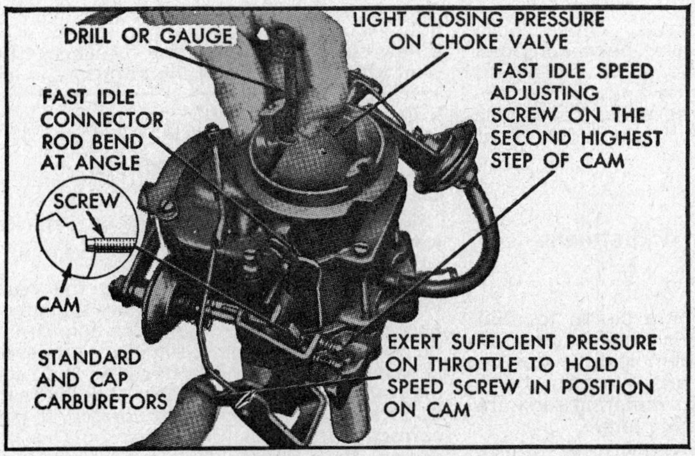

Fig. 9 **Fast idle cam position adjustment. BBD 1¼" bore carburetors**

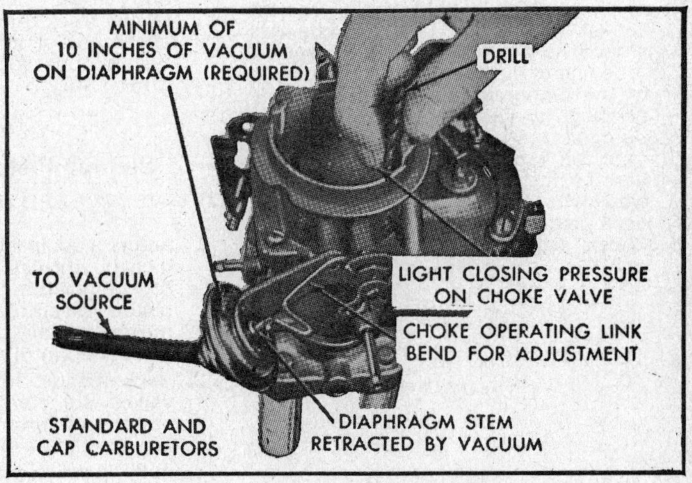

Fig. 11 **Choke vacuum kick setting. BBD 1¼" bore carburetors, 1969-74**

vent valve leaf spring during bending operation or improper vent valve operation will result.

Install bowl vent valve cover and secure with attaching screws.

Choke Unloader Adjustment

Fig. 8—The choke unloader is a mechanical device to partially open the choke valve at wide open throttle. It is used to eliminate choke enrichment during engine cranking. Engines that have been flooded or stalled by excessive choke enrichment can be cleared by the use of the unloader. Adjust as follows:

1. Hold throttle valve in wide open position. Insert the specified drill size between upper edge of choke valve and inner wall of air horn.
2. With a finger lightly pressing against choke valve, a slight drag should be felt as the drill is being withdrawn.
3. If an adjustment is necessary, bend unloader tang on throttle lever until specified opening has been obtained.

Fast Idle Cam Position

Figs. 9 and 10

1. With fast idle adjusting screw contacting second highest step on fast idle cam, move choke valve toward closed position with light pressure on choke shaft lever.
2. Insert the specified size drill between choke valve and air horn wall. An adjustment will be necessary if a slight drag is not obtained as drill is being removed.
3. On all models except 1969 1½ inch bore units, adjust by bending fast idle connector rod at lower angle. On 1969 1½ inch bore units, adjust by bending stop on choke shaft.

Choke Vacuum Kick Adjustment

Figs. 11 and 12

The choke diaphragm adjustment controls the fuel delivery while the engine is running. It positions the choke valve within the air horn by action of the linkage between choke shaft and diaphragm. The diaphragm must be energized to measure the vacuum kick adjustment. Use either a distributor test machine with a vacuum source, or vacuum supplied by another vehicle.

1. If adjustment is to be made with en-

gine running, disconnect fast idle linkage to allow choke to close to kick position with engine at curb idle. If an auxiliary vacuum source is to be used, open throttle valves (engine not running) and move choke to closed position. Release throttle first, then release choke.

2. When using an auxiliary vacuum source, disconnect vacuum hose from carburetor and connect it to hose from vacuum supply with a small length of tube to act as a fitting. Removal of hose from diaphragm may require forces which damage the system. On all 1969-72 units, apply a vacuum of 10 or more inches of mercury and on all 1973-75 units, apply a vacuum of 15 or more inches of mercury.
3. Insert the specified drill size between choke valve and wall of air horn. Apply sufficient closing pressure on lever to which choke rod attaches to provide a minimum choke valve opening without distortion of diaphragm link. Note that the cylindrical stem of diaphragm will extend as internal spring is compressed. This spring must be fully compressed for proper measurement of vacuum kick adjustment.
4. An adjustment will be necessary if a slight drag is not obtained as drill is being removed. Shorten or lengthen diaphragm link to obtain correct choke opening. Length changes should be made carefully by bending (opening or closing) the bend provided in the diaphragm link. *Do not apply twisting or bending force to diaphragm.*
5. Reinstall vacuum hose on correct carburetor fitting. Return fast idle linkage to its original condition if it has been disturbed as in Step 1.
6. Check as follows: With no vacuum applied to diaphragm, choke valve should move freely between open and closed positions. If movement is not free, examine linkage for misalignment or interferences caused by bending operation. Repeat adjustment if necessary.

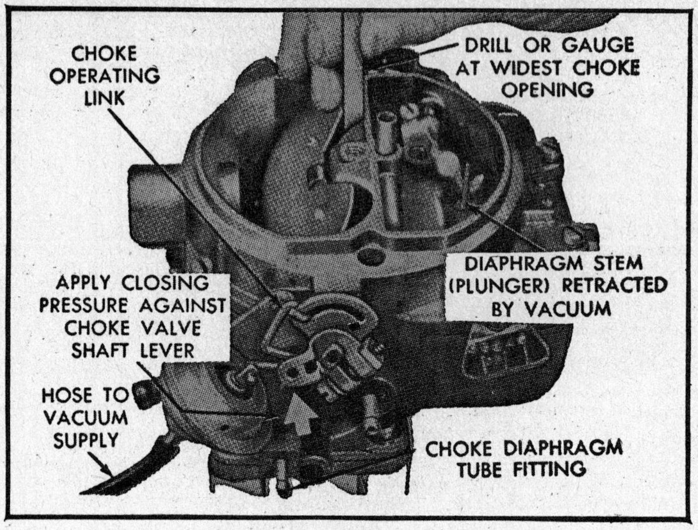

Fig. 12 Choke vacuum kick setting. BBD except 1¼" bore carburetors, 1969-71

Step-up Piston Adjustment

1974-75 1¼" BBD Units, Fig. 13

1. Adjust gap in step-up piston to .035 inch by turning allen head screw on top of piston, then install step-up piston assembly into air horn bore, making certain that metering rods are positioned in metering jets.
2. Back off curb idle screw until throttle valves are completely closed. Count number of turns so that screw can be returned to its original setting.
3. Fully depress piston and while applying moderate pressure on rod lifter, tighten rod lifter screw.
4. Release piston and rod lifter and reset curb idle screw.

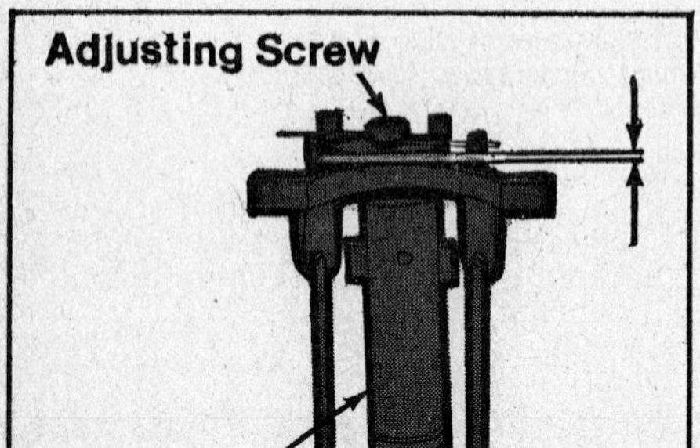

Fig. 13 Step-up piston adjustment

Well-Type Choke Setting

Fig. 13—Loosen mounting post lock nut and turn mounting post with screwdriver until index mark on disc is positioned as listed in the *BBD Specifications Chart*. Hold in this position with screwdriver and lock nut.

NOTE: Screwdriver may be held in vise so that one hand may be used to support housing while tightening nut. After adjustment is completed and coil housing, rod and carburetor are installed on engine, lift cover disc and open and close choke valve manually to see if connector rod clears sides of hole in housing cover without binding. If rod does not clear housing cover without binding, replace with a new unit since connecting rod cannot be bent without affecting calibration.

Dashpot Adjustment

Cleaner Air Carburetors

The dashpot is used only on vehicles with the Cleaner Air Package and manual shift transmission, Fig. 4.

To adjust the dashpot, have the curb idle speed and mixture properly adjusted, and install a tachometer. Position throttle lever so that actuating tab on lever is contacting stem of dashpot but not depressing it. Tachometer should read 2000 rpm if the setting is correct. If not correct, screw dashpot in or out as required, then tighten lock nut on dashpot against the bracket.

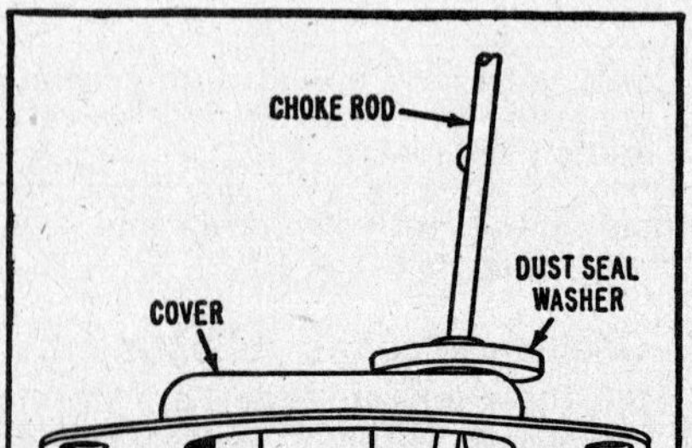

Fig. 14 BBD well-type choke setting

CARTER WCD ADJUSTMENT SPECIFICATIONS

See Tune Up Chart in car chapter for hot idle speed.

Year	Carb. Model	Idle Mixture Screw Setting	Float Level	Pump Setting	Metering Rod	Fast Idle Setting	Choke Unloader	Dashpot Setting	Automatic Choke
AMERICAN MOTORS									
1969	4667S	1½	7/32	See Text	See Text	—	3/16	7/64	On Index
	4668S	1½	7/32	See Text	See Text	—	3/16	7/64	On Index
1970	4816S	①	7/32	See Text	See Text	.021	3/16	3/32	Index
	4817S	①	7/32	See Text	See Text	.024	3/16	3/32	Index
	4950S	①	7/32	See Text	See Text	.024	3/16	3/32	Index

① —Air/fuel ratio or idle CO% rating is found in Tune Up Specification tables in car chapters.

MODEL WCD ADJUSTMENTS

The WCD carburetor, Fig. 1, is a two-barrel unit containing the five basic circuits. The carburetor uses a single needle valve even though two floats are provided. On some WCD units the two floats operate independently of each other so that the highest float always controls the fuel level. This is necessary when the carburetor is mounted with the centerline of the floats parallel to the centerline of the engine.

Float Adjustment

Lateral Adjustment—Referring to Fig. 2 and with bowl cover inverted and gasket removed, place float gauge directly under floats with notched portions of gauge fitted over edges of casting. Sides of float should barely touch vertical uprights of float gauge. Adjustment is made by bending arms of floats.

Vertical Adjustment—With float gauge in same position as shown, floats should just clear horizontal portion of gauge. The vertical distance between top center of float and machined surface of casting must be the dimension given in the *WCD Specifications Chart*. Adjust by bending float arms as required. Remove floats, install bowl cover gasket and reinstall floats.

Pump Adjustment

Fig. 3—Install pump connector link in outer hole (long stroke) of pump arm with ends extending away from counter-shaft arm. Back out throttle lever set screw until throttle valves seat in carburetor bores. Be sure fast idle adjusting screw does not hold throttle open.

Hold straightedge across top of dust cover boss at pump arm. The flat on top of pump arm should be parallel to straightedge. Adjust by bending throttle connector rod to the upper angle.

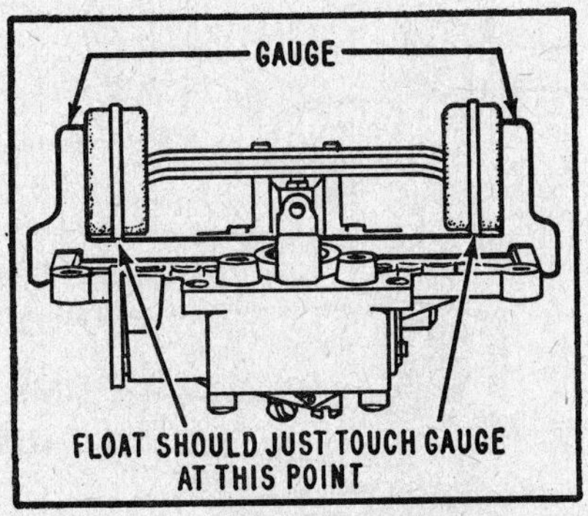

Fig. 2 WCD float level adjustment

Fig. 3 WCD pump adjustment

Metering Rod Adjustment

This adjustment must be made after completing the pump adjustment. No metering rod gauges are necessary. Adjust as follows:

1. Back out throttle lever set screw to allow throttle valves to seat in bores of carburetor and loosen metering rod arm clamp screw.
2. With metering rod in place, press down on vacumeter link until metering rods bottom in carburetor body casting.
3. While holding rods in downward position and throttle valves seated, revolve metering rod arm until finger on arm contacts lip of vacumeter link. Hold in place and carefully tighten clamp screw.

Fast Idle Adjustment

Fig. 4—Loosen choke lever clamp screw on choke shaft. Insert a .010" feeler gauge between lip of fast idle cam and boss of flange casting. Hold choke valve tightly closed and take slack out of linkage by pressing choke lever towards closed position.

With choke valve tightly closed, tighten fast idle adjusting screw until there is the clearance listed in the *WCD Specifications Chart* between throttle valve and carburetor bore (side opposite idle port). Be sure fast idle adjusting screw is on high step of cam or index mark while making this adjustment.

Choke Unloader Adjustment

Fig. 5—With throttle valves wide open, there should be the clearance listed in the *WCD Specifications Chart* between upper edge of choke valve and inner wall of air horn. Adjust by bending unloader lip (ear) on throttle shaft lever.

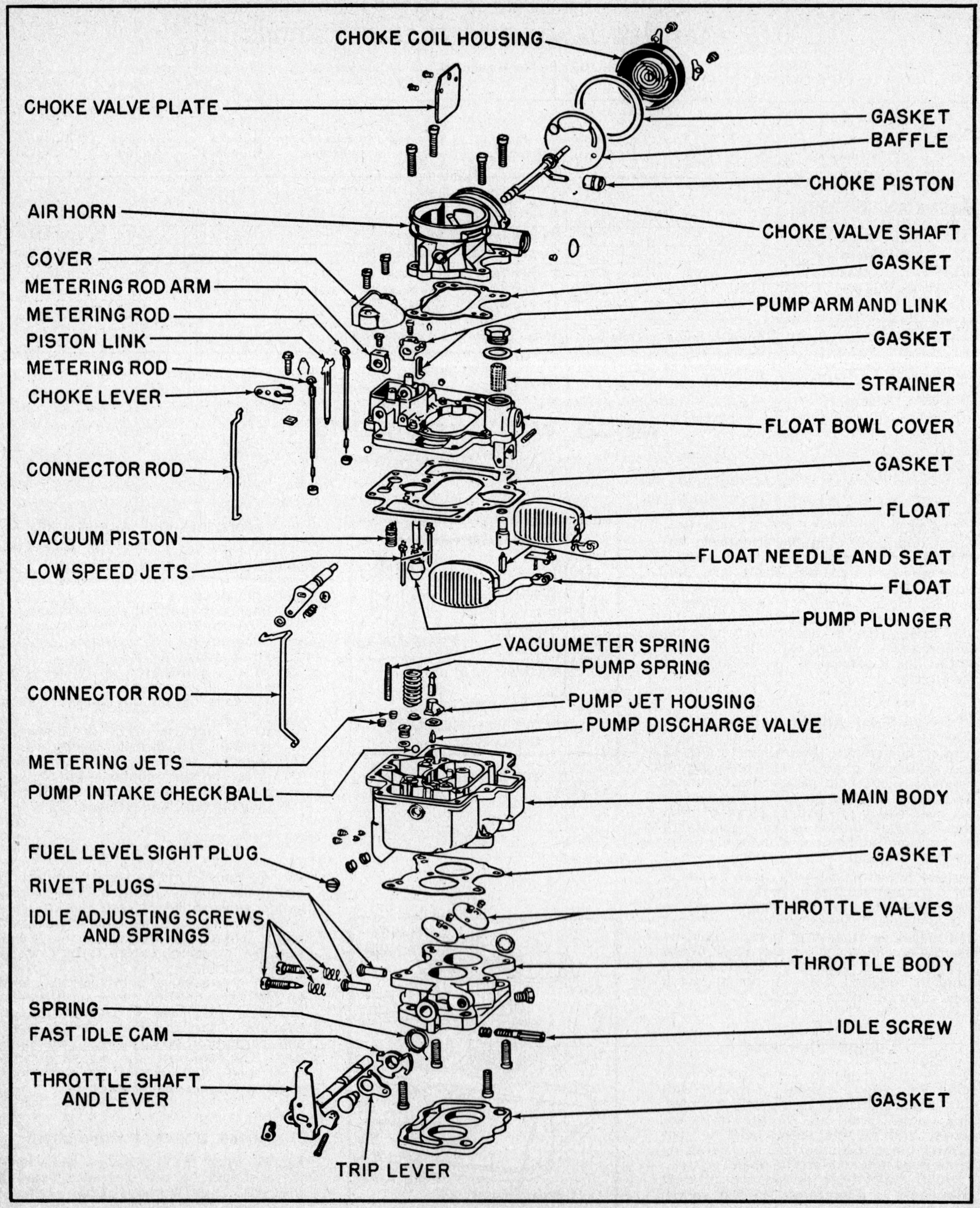

Fig. 1 Carter Model WCD two-barrel carburetor

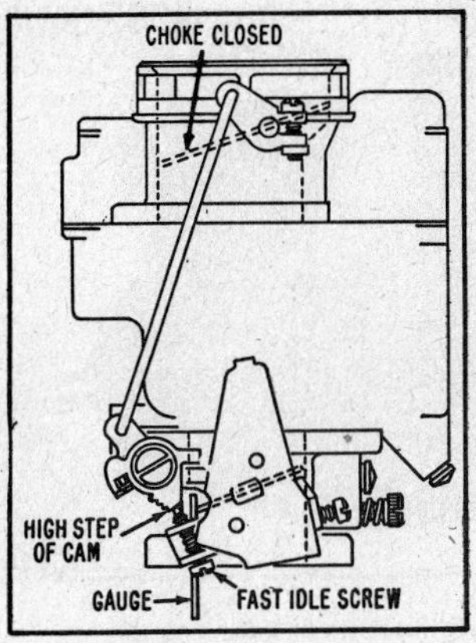

Fig. 4 WCD fast idle adjustment

Dashpot Adjustment

American Mtrs—With throttle valves tightly closed and diaphragm stem fully depressed, adjust dashpot to give the clearance listed in the *WCD Specifications Chart* between dashpot stem and throttle lever.

Automatic Choke Adjustment

Loosen choke cover retaining screws and turn cover so that line or index mark on cover lines up with the specified mark on choke housing (see *WCD Specifications Chart*).

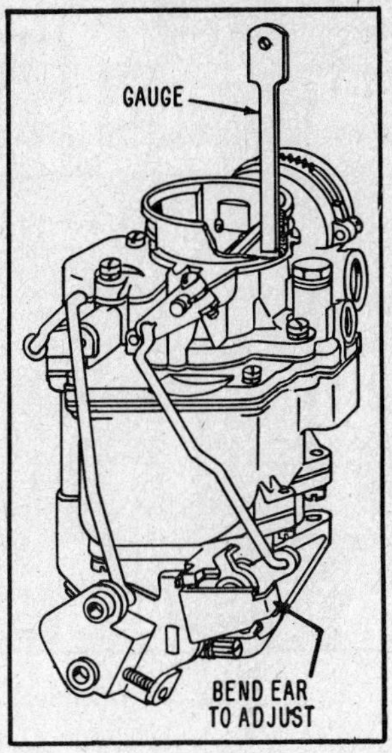

Fig. 5 WCD unloader adjustment

CARTER WGD ADJUSTMENT SPECIFICATIONS

See Tune Up Chart in car chapters for hot idle speed.

Year	Carb. Model	Air/Fuel Ratio	Float Setting	Fast Idle Throttle Valve Setting	Choke Unloader	Choke Diaphragm Linkage	Choke Coil Rod Position
PONTIAC							
1972	6311S	①	5/16	.030	3/16	9/64	Center Notch

①—Air/fuel ratio or idle CO% rating is found in Tune Up Specification tables in car chapters.

MODEL WGD ADJUSTMENTS

MODEL WGD ADJUSTMENTS

The WGD carburetor, Fig. 1, is similar to the WCD model except that the WGD has only one float, a reduced overall height thermostatic coil located on the engine manifold which is connected to the choke valve by a rod and simplified adjustments.

Also a vacuum break diaphragm unit has been incorporated, this is a modified unit in that a tension spring has been added to the choke vacuum break diaphragm plunger. The purpose of the tension or bucking spring is to offset tension of the thermostatic coil in relation to atmospheric temperatures.

Float Adjustment

With air horn inverted, check to see that float is parallel with outer edge of air horn casting, Fig. 2. Adjust by bending float arm. Next, place gauge between air horn

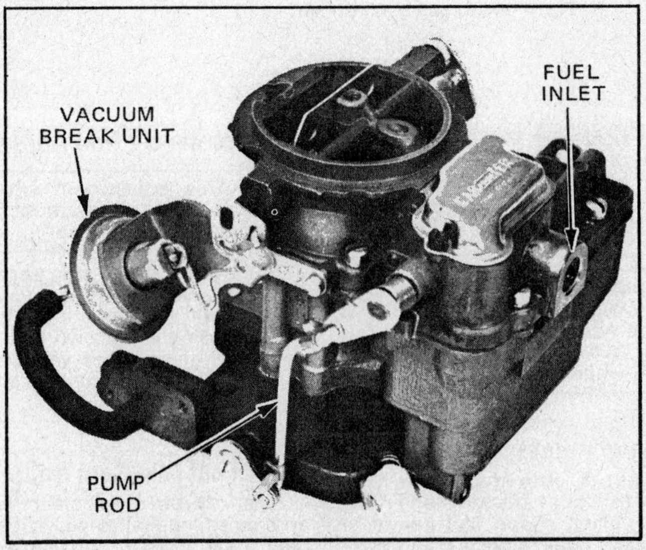

Fig. 1 Carter WGD carburetor assembly

Fig. 2 WGD float level alignment

Fig. 3 WGD checking float level adjustment

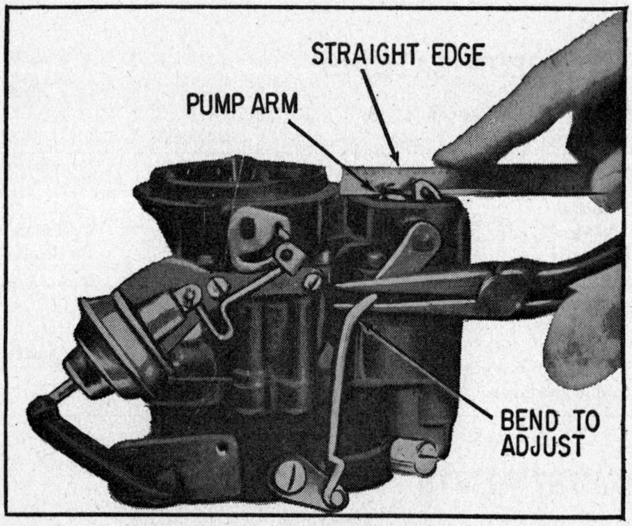

Fig. 4 WGD pump adjustment

Fig. 5 WGD metering rod adjustment

and center of float, Fig. 3. Refer to *WGD Specifications Chart*. Adjust float level by bending float arm until float touches gauge. Float should not have excessive clearance at hinge pin and must operate freely.

NOTE: When adjusting float, care must be exercised to avoid pressing the flared tip needle into the needle seat as a false setting will result. Allow only the float weight to seat needle when gauging.

Pump Adjustment

Fig. 4—Back out throttle stop screw, turn fast idle cam to "hot" position and fully close throttle valves. Place 1/4" gauge or a similar straight edge across dust cover boss. The dust cover boss should be parallel with the top surface of pump arm. Adjust by bending pump rod at offset.

Metering Rod Adjustment

NOTE: This adjustment should be made after the pump adjustment. No metering rod gauges are necessary.

Back out throttle stop screw, fully close throttle valves and press down on vacuum piston link until metering rods bottom. While holding rods down and metering arm tongue against lip of vacuum piston link, carefully tighten metering arm set screw, Fig. 5.

Fast Idle Cam Adjustment

Open throttle to clear fast idle cam and close choke valve. With choke valve held fully closed and stop on fast idle cam against the casting, there should be .005" minimum clearance between inner and outer choke levers, Fig. 6. Adjust by bending outer lever lug, as required, Fig. 7.

NOTE: With choke fully closed, tang on fast idle cam must clear stop on throttle body flange.

Unloader Adjustment

Hold choke closed lightly. Fully open throttle, forcing choke valve open. Check clearance between upper edge of choke valve and wall of air horn, Fig. 8. Refer to *WGD Specifications Chart*. Adjust by bending the unloader arm as required.

Fast Idle Speed Adjustment

With carburetor on engine, rotate fast idle cam until fast idle tang contacts high step of cam. With engine at normal operating temperature adjust fast idle tang to obtain engine speed of 1500 rpm, Fig. 9.

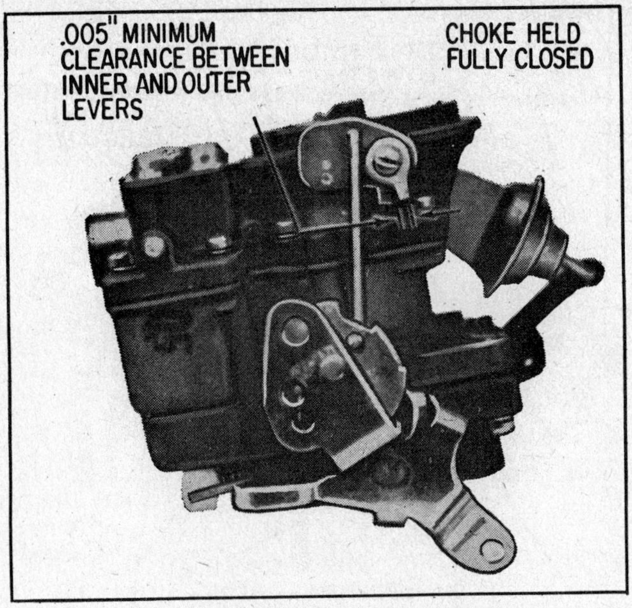

Fig. 6 WGD fast idle cam position

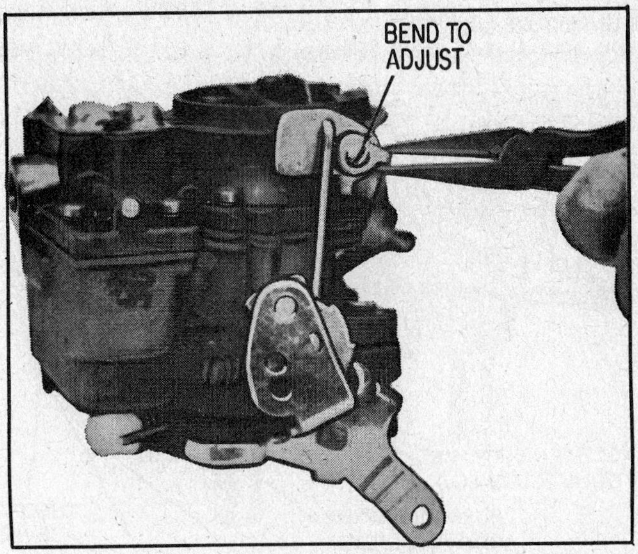

Fig. 7 WGD choke lever adjustment

Fig. 8 WGD choke unloader adjustment

Fig. 9 WGD fast idle speed adjustment unit on engine

CARTER AVS ADJUSTMENT SPECIFICATIONS

See Tune Up Chart in car chapter for hot idle speed.

Year	Carb. Model	Idle Mixture Turns Open	Float Level	Float Drop	Pump Travel	Bowl Vent	Fast Idle Cam Position	Choke Vacuum Break	Secondary Throttle Lever	Secondary Throttle Lockout	Choke Setting
CHRYSLER, DODGE, PLYMOUTH, IMPERIAL											
1969	4611S	1–2	7/32	1/2	7/16	1/8	50 Drill①	35 Drill	21/64	.020	Index
	4612S	1–2	7/32	1/2	7/16	1/8	50 Drill①	50 Drill	21/64	.020	Index
	4615S	1–2	5/16	1/2	7/16	1/8	50 Drill①	35 Drill	21/64	.020	Index
	4616S	1–2	5/16	1/2	7/16	1/8	50 Drill①	50 Drill	21/64	.020	Index
	4617S	1–2	7/32	1/2	7/16	1/8	50 Drill①	25 Drill	11/32	.020	Index
	4618S	1–2	7/32	1/2	7/16	1/8	50 Drill①	35 Drill	3/8	.020	Index
	4638S	1–2	5/16	1/2	7/16	1/8	50 Drill①	50 Drill	21/64	.020	Index
	4639S	1–2	7/32	1/2	7/16	1/8	50 Drill①	50 Drill	21/64	.020	Index
	4640S	1–2	7/32	1/2	7/16	1/8	50 Drill①	35 Drill	1/8	.020	Index
	4682S	—	5/16	1/2	7/16	1/8	50 Drill①	50 Drill	21/64	.020	Index
	4711S	1–2	5/16	1/2	7/16	1/8	50 Drill①	35 Drill	21/64	.020	Index

Continued

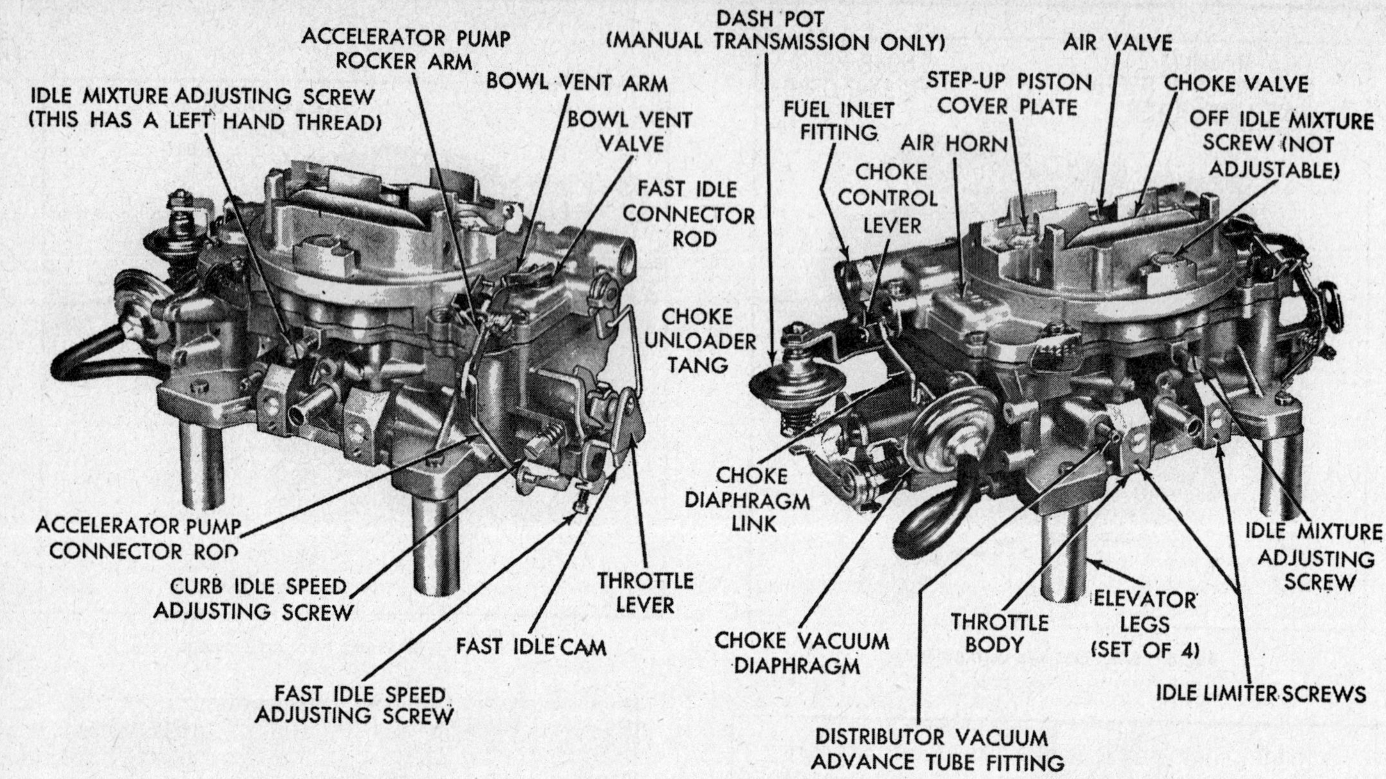

Fig. 1 1969 Carter AVS four-barrel carburetor

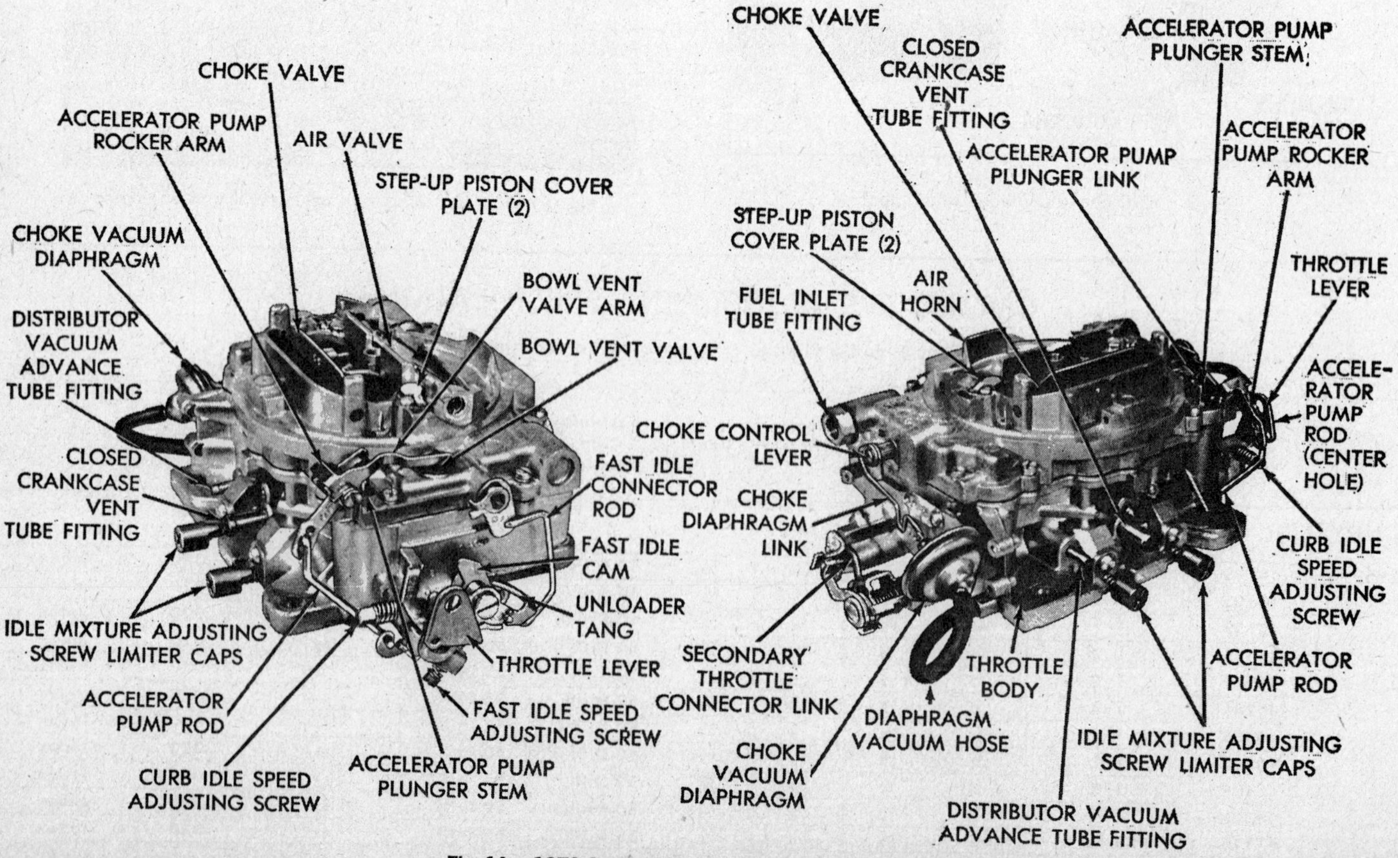

Fig. 1A 1970 Carter AVS four-barrel C.A.S. equipped

CARTER AVS ADJUSTMENT SPECIFICATIONS

See Tune Up Chart in car chapter for hot idle speed.

Year	Carb. Model	Idle Mixture Turns Open	Float Level	Float Drop	Pump Travel	Bowl Vent	Fast Idle Cam Position	Choke Vacuum Break	Secondary Throttle Lever	Secondary Throttle Lockout	Choke Setting
CHRYSLER, DODGE, PLYMOUTH, IMPERIAL											
1970	4732S	②	5/16	1/2	7/16	3/64	50 Drill①	44 Drill	19/64	.020	2 Rich
	4734S	②	5/16	1/2	7/16	3/64	50 Drill①	44 Drill	19/64	.020	2 Rich
	4736S	②	5/16	1/2	7/16	3/64	50 Drill①	44 Drill	19/64	.020	2 Rich
	4737S	②	7/32	1/2	7/16	3/64	50 Drill①	20 Drill	23/64	.020	2 Rich
	4738S	②	7/32	1/2	7/16	3/64	50 Drill①	20 Drill	23/64	.020	Index
	4739S	②	7/32	1/2	7/16	3/64	50 Drill①	20 Drill	23/64	.020	Index
	4740S	②	7/32	1/2	7/16	3/4	50 Drill①	20 Drill	23/64	.020	Index
	4741S	②	7/32	1/2	7/16	3/64	50 Drill①	20 Drill	23/64	.020	Index
	4933S	②	7/32	1/2	7/16	3/64	50 Drill①	35 Drill	19/64	.020	Index
	4934S	②	7/32	1/2	7/16	1/8	50 Drill①	50 Drill	19/64	.020	Index
	4935S	②	7/32	1/2	7/16	1/8	50 Drill①	50 Drill	19/64	.020	Index
	4936S	②	7/32	1/2	7/16	3/4	50 Drill①	35 Drill	19/64	.020	Index
	4937S	②	7/32	1/2	7/16	3/4	50 Drill①	50 Drill	19/64	.020	Index
1971	4966S	②	7/32	1/2	7/16	3/4	50 Drill①	44 Drill	23/64	.020	2 Rich
	4967S	②	7/32	1/2	7/16	3/4	50 Drill①	44 Drill	23/64	.020	2 Rich
	4968S	②	7/32	1/2	7/16	3/4	50 Drill①	44 Drill	23/64	.020	2 Rich
	6125S	②	7/32	1/2	7/16	3/4	50 Drill①	44 Drill	23/64	.020	2 Rich

①—With fast idle adjusting screw contacting second highest speed step on fast idle cam the clearance between choke valve and wall of air horn should be as specified.

②—Air/fuel ratio or idle CO% rating is found in Tune Up Specification tables in car chapters.

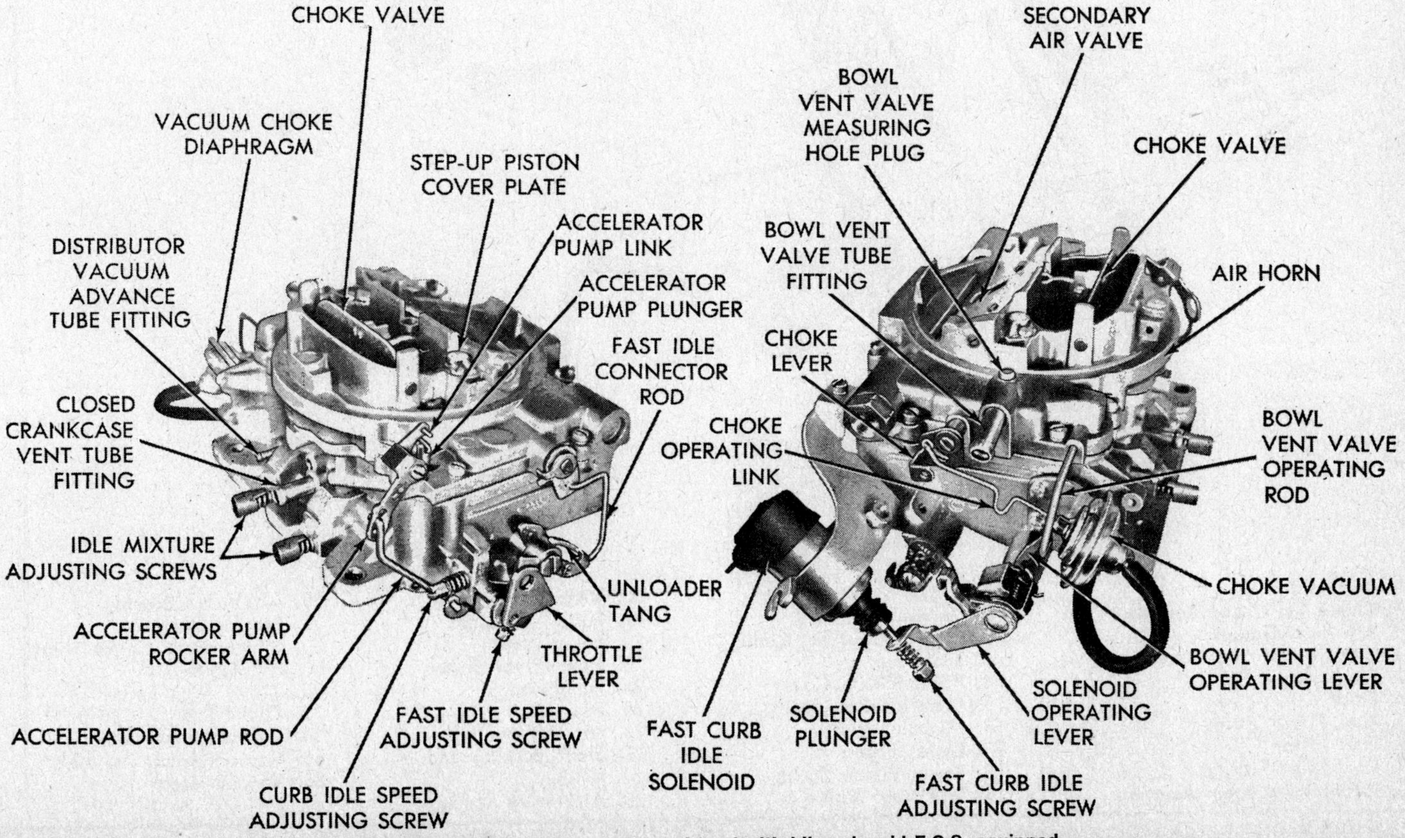

Fig. 1B 1970-71 Carter AVS four-barrel with idle solenoid E.C.S. equipped

MODEL AVS ADJUSTMENTS

The AVS carburetor, Figs. 1 thru 3, is used in conjunction with a temperature sensing choke coil mounted on the intake manifold over the exhaust crossover passage. The AVS is similar to the more familiar AFB which employs a built-in automatic choke.

AVS means "Air Valve Secondary". The spring loaded air valve, located above the secondary fuel nozzles, gives smooth response whenever the secondary throttle valves are actuated.

The primary side of the carburetor uses venturi clusters for fine fuel control in the idle and economy ranges. The use of fuel nozzles pressed into the secondary side of the fuel bowl virtually eliminates secondary bore restriction, thus giving this carburetor high air capacity in the power ranges.

A hot idle compensator, consisting of a bi-metal strip, a valve and a mounting bracket, is located between the secondary bores to supply additional air to the idle mixture during prolonged hot idle periods.

Float Alignment

Fig. 4—Sides of floats should be parallel to edge of casting with minimum clearance between lever and air horn lugs without binding. To adjust, bend float lever.

Float Level Adjustment

Fig. 5—With air horn inverted, air horn gasket in place and float needle seated, slide float gauge between top of the float (at outer end) and air horn gasket. Refer to *AVS Specifications Chart.*

Check other float in same manner. Adjust by bending float arm. After bending arm, recheck the float alignment.

Float Drop Adjustment

Fig. 6—There should be the dimension

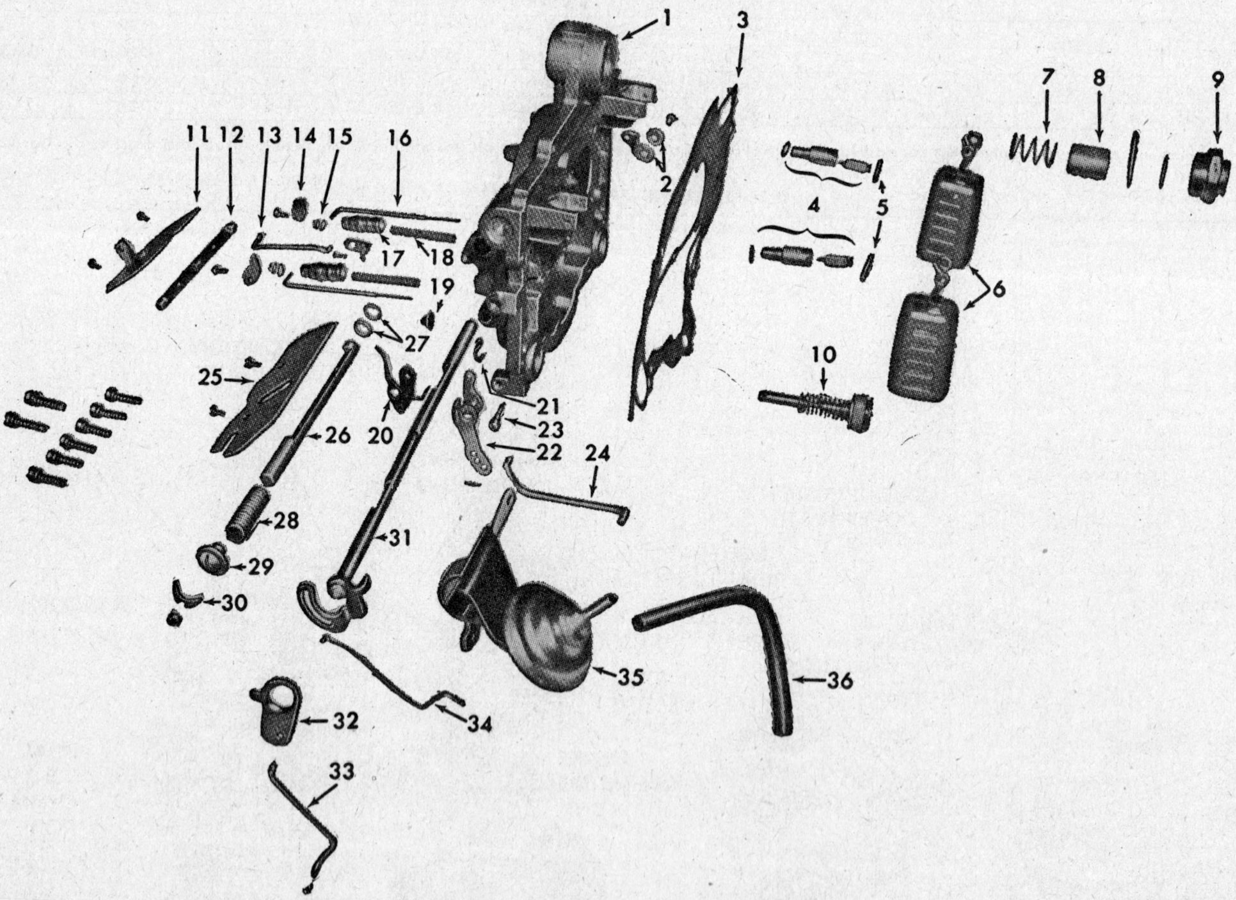

1. Air Horn
2. Choke Lever and Retainer
3. Air Horn Gasket
4. Needle and Seat Assembly
5. Float Hinge Pin
6. Float
7. Fuel Filter Spring
8. Fuel Filter
9. Fuel Inlet Fitting
10. Accelerator Pump Assembly
11. Choke Valve
12. Choke Shaft
13. Choke Actuating Link and Lever
14. Power Piston Cover
15. Metering Rod Spring
16. Metering Rod
17. Power Piston
18. Power Piston Spring
19. Idle Vent Valve
20. Vent Valve Lever
21. Pump "S" Link
22. Pump Lever
23. Pump Pivot Screw
24. Pump Rod
25. Air Valve
26. Air Valve Shaft
27. Shaft Bushing and Washer
28. Air Valve Spring
29. Air Valve Bearing
30. Bearing Retainer
31. Intermediate Choke Shaft and Lever
32. Choke Kick Lever
33. Choke Rod
34. Vacuum Break Link
35. Vacuum Break Assembly
36. Vacuum Break Hose

Fig. 2 AVS air horn parts exploded

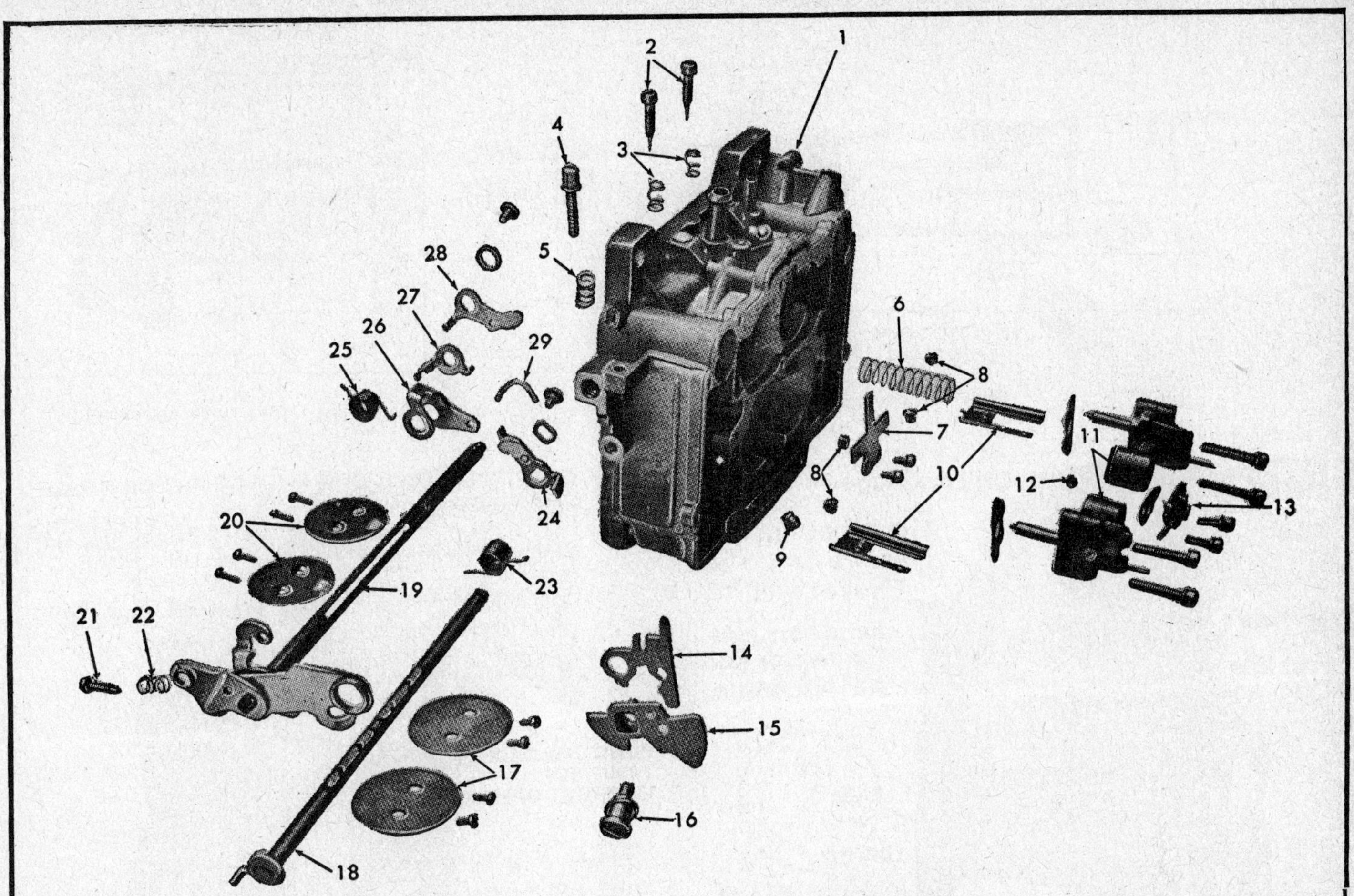

1. Main Body
2. Idle Mixture Needle
3. Idle Mixture Spring
4. Idle Speed Screw
5. Idle Speed Spring
6. Pump Return Spring
7. Hot Idle Compensator
8. Main Metering Jet
9. Pump Inlet Assembly
10. Splash Shield
11. Venturi Cluster
12. Pump Discharge Check Ball
13. Pump Nozzle
14. Lockout Dog
15. Fast Idle Cam
16. Pivot Screw
17. Secondary Throttle Valve
18. Secondary Throttle Shaft
19. Primary Throttle Shaft
20. Primary Throttle Valve
21. Fast Idle Screw
22. Fast Idle Spring
23. Secondary Throttle Shaft Spring
24. Secondary Throttle Shaft Dog Lever and Retainer
25. Primary Throttle Shaft Spring
26. Secondary Throttle Trip Lever
27. Spring Pick Up Lever
28. Primary Throttle Shaft Arm and Retainer
29. Secondary Actuating Link

Fig. 3 AVS main body parts exploded

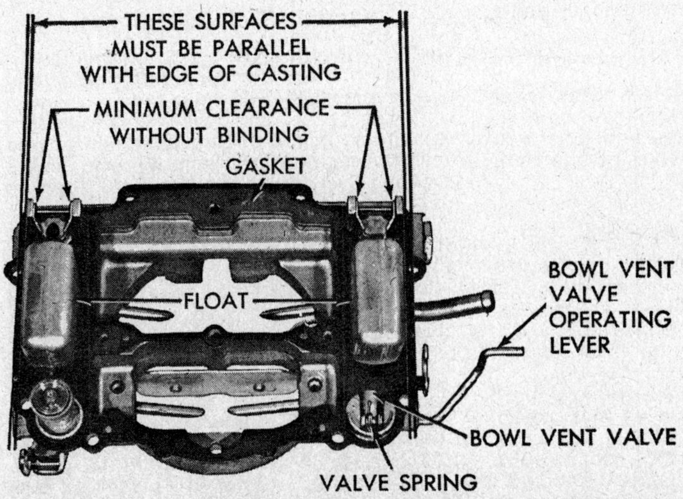

Fig. 4 AVS float alignment

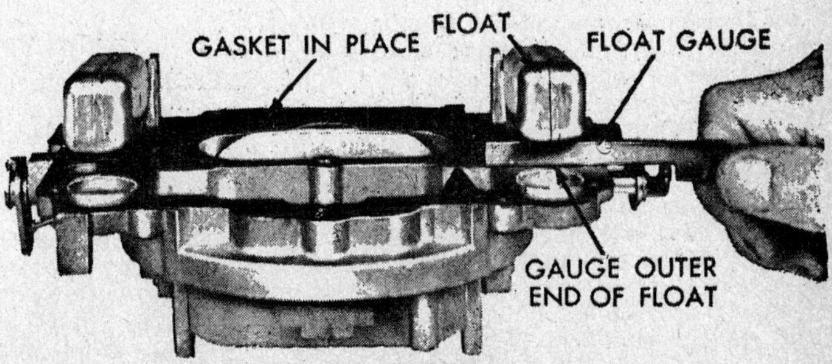

Fig. 5 AVS float level

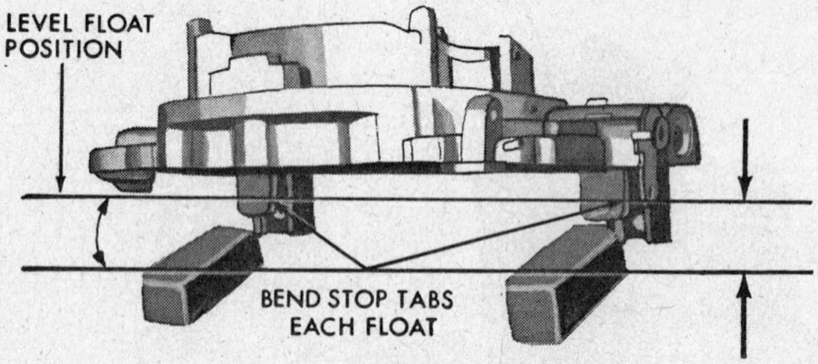

Fig. 6 AVS float drop

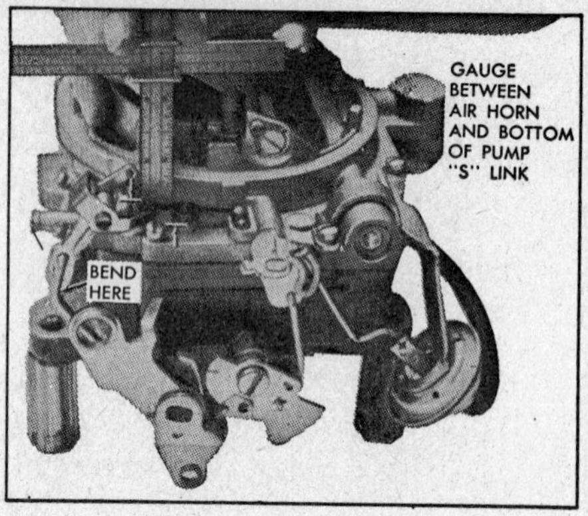

Fig. 7 AVS pump adjustment

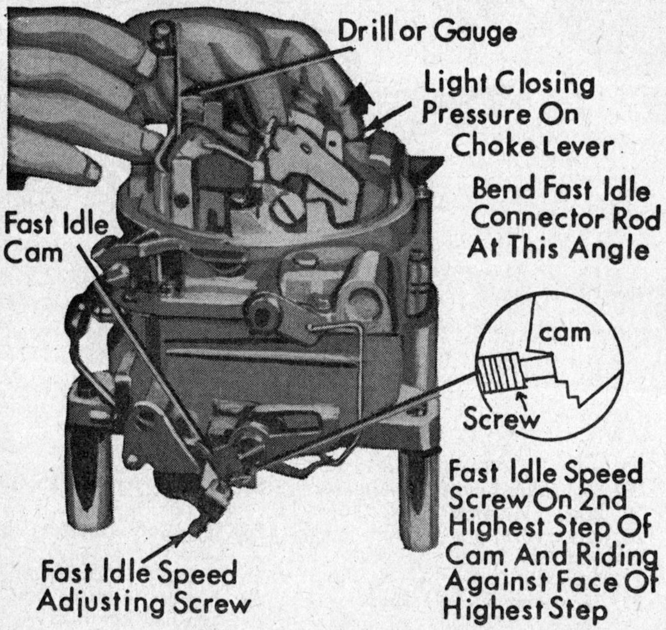

Fig. 8 AVS fast idle cam adjustment

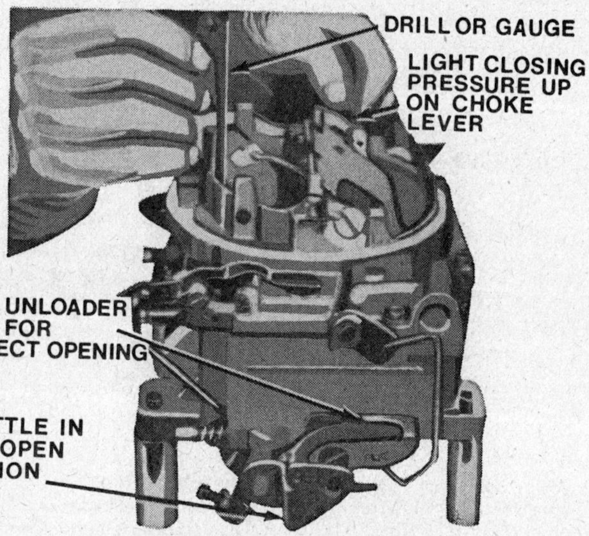

Fig. 9 AVS choke unloader adjustment

listed in the *AVS Specifications Chart* between top of floats (at outer end) and air horn gasket. To adjust, bend stop tabs on float brackets.

Pump Adjustment

Fig. 7—With throttle valves tightly closed there should be the dimension listed in the *AVS Specifications Chart* from top of air horn to top of pump plunger shaft with throttle connector rod in inner hole of pump arm. To adjust, bend throttle connector rod at angle.

Fast Idle Cam Adjustment

Fig. 8—With fast idle speed adjusting screw contacting second highest speed step on fast idle cam, move choke valve toward closed position with light pressure on choke shaft lever.

Insert drill gauge between choke valve and wall of air horn. Refer to *AVS Specifications Chart*. An adjustment will be necessary if a slight drag is not obtained as the drill is being removed. Adjust by bending fast idle connector rod at angle.

Choke Unloader Adjustment

Fig. 9—With throttle valves wide open there should be ¼" clearance between upper edge of choke valve and inner wall of air horn. To adjust, bend unloader tang on fast idle cam.

Bowl Vent Adjustment

C.A.S. Equipped, Fig. 10—With throttle valves tightly closed, insert a drill gauge between air horn and valve at smallest opening. Refer to *AVS Specifications Chart*. Adjust by bending tang on pivot end of lever.

E.C.S. Equipped, Fig. 10A—Remove bowl vent valve checking hole plug in air horn. With throttle valves at closed curb idle position, insert a narrow ruler down through hole. Allow ruler to rest lightly on top of valve. Measure from top of valve to top of air horn casting at opening. Refer to *AVS Specifications Chart*.

Adjust by bending bowl vent valve operating lever. Install new plug and rap lightly to seat, using a hammer.

Vacuum Kick Adjustment

Fig. 11—If adjustment is to be made with engine running, disconnect fast idle linkage to allow choke to close to kick position with engine at curb idle. If an auxiliary vacuum source is to be used, open throttle valves (engine not running) and move choke valve to closed position. Release throttle first, then release choke.

When using an auxiliary vacuum source, disconnect vacuum hose from carburetor and connect it to hose from vacuum supply with a small length of tube to act as a fitting. Removal of hose from diaphragm may require forces which damage the system. Apply a vacuum of 10 or more inches.

Insert drill gauge between choke valve and wall of air horn. Refer to *AVS Specifications Chart*. Apply sufficient closing pressure on lever to which choke rod attaches to provide a minimum choke valve opening without distortion of diaphragm link. Note that on most units, a cylindrical stem extends as an internal

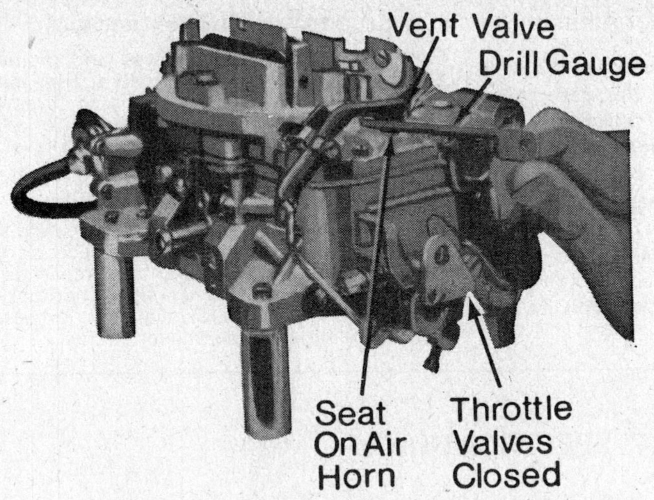

Fig. 10 AVS bowl vent adjustment. C.A.S.

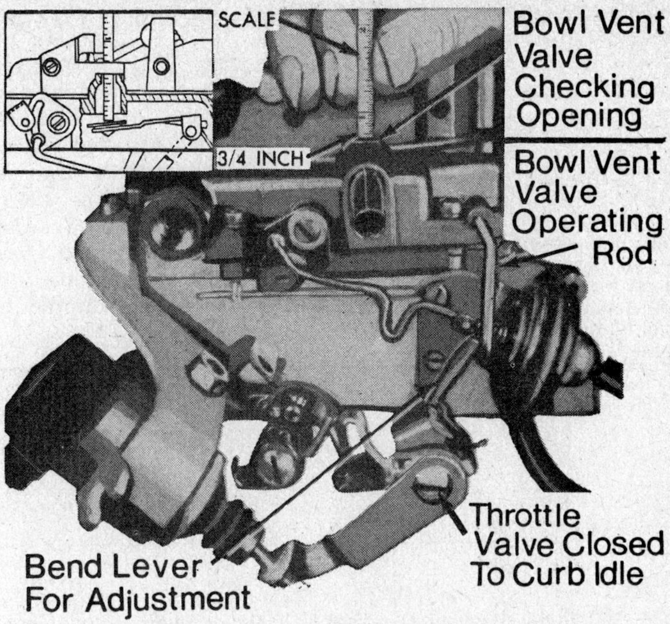

Fig. 10A AVS bowl vent adjustment. E.C.S.

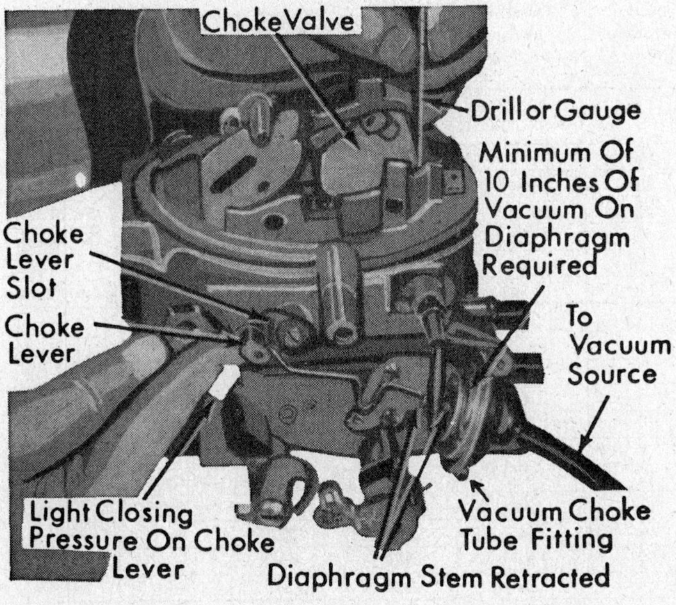

Fig. 11 AVS choke vacuum kick adjustment

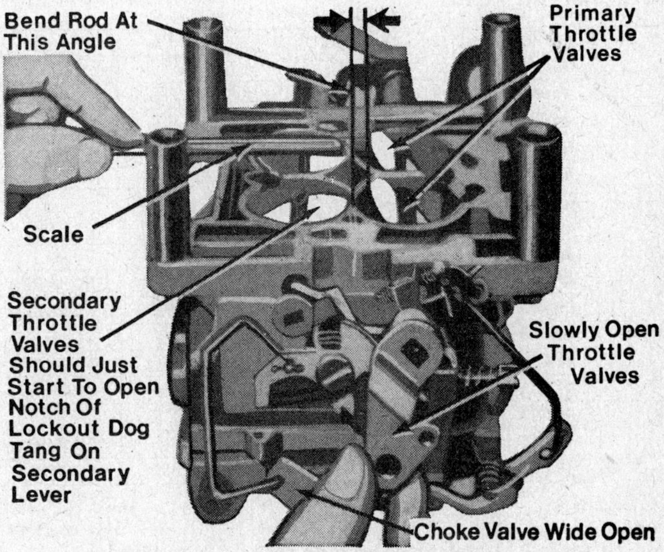

Fig. 13 AVS secondary throttle opening adjustment

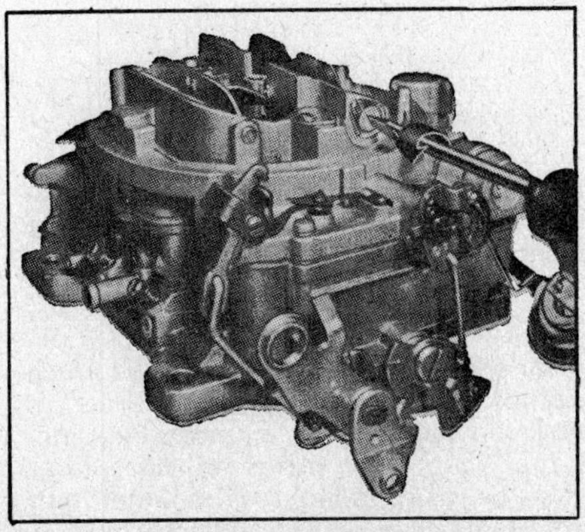

Fig. 12 AVS secondary air valve adjustment

spring is compressed. This spring must be fully compressed for proper measurement of vacuum kick adjustment.

An adjustment will be necessary if a slight drag is not obtained as drill is being removed. Shorten or lengthen diaphragm link to obtain correct choke opening. Adjust by carefully opening or closing the bend provided in diaphragm link. Note: Important: Do not apply twisting or bending force to diaphragm.

Reinstall vacuum hose on correct carburetor fitting. Return fast idle linkage to its original condition if disturbed. Make following check. With no vacuum applied to diaphragm, choke valve should move freely between open and closed positions. If movement is not free, examine linkage for misalignment or interferences caused

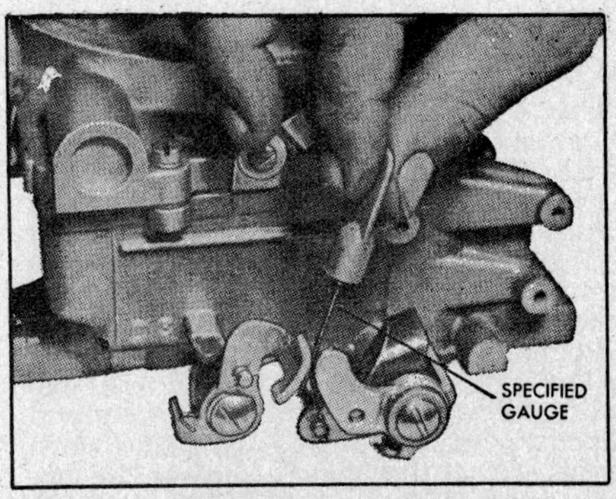

Fig. 14 AVS closing shoe adjustment

by bending operation. Repeat adjustment if necessary to provide proper link operation.

Secondary Air Valve Adjustment

Fig. 12—Loosen lock screw and allow air valve to position itself at wide open position. From wide open position, (spring barely moving valve), turn slotted sleeve two full turns counter clockwise. Hold in this position with finger, then tighten lock screw securely. Check valve for freedom of movement.

Secondary Throttle Lever Adjustment

Fig. 13—Block the choke valve in the wide open position and invert the carburetor. Slowly open the primary throttle valves until specified measurement is obtained between the lower edge of the primary valve and the bore opposite idle port. Refer to AVS *Specifications Chart*. At this measurement, the secondary valves should just start to open. Adjust by bending the secondary throttle operating rod at the angle.

Secondary Throttle Lockout

Fig. 13—Crack throttle valves and manually open and close choke valve. Tang on secondary throttle lever should freely engage in notch of lockout dog. To adjust, bend tang on secondary throttle lever.

Closing Shoe Adjustment

Fig. 14—With primary and secondary throttle valves closed, bend secondary closing shoe to obtain .020″ clearance between positive closing shoes on primary and secondary throttle levers. To adjust, bend shoe on secondary lever.

CARTER AFB ADJUSTMENT SPECIFICATIONS

See Tune Up Chart in car chapters for hot idle speed.

Year	Carb. Model	Idle Mixture Turns Open	Float Level	Float Drop	Pump Travel	Fast Idle Throttle Valve or Cam Position Clearance	Choke Unloader Setting	Secondary Throttle Lever Setting	Secondary Throttle Lockout Setting	Bowl Vent Setting	Choke Piston Linkage Setting	Choke Setting
AMERICAN MOTORS												
1969	4660S	2	11/32	2	21/64	—	5/32	7/16	.015	—	5/64	2 Rich
	4661S	2	11/32	2	21/64	—	5/32	7/16	.015	—	1/8	On Index
	4662S	2	11/32	2	21/64	—	5/32	7/16	.015	—	7/64	On Index
	4663S	2	11/32	2	21/64	—	5/32	7/16	.015	—	1/8	On Index
	4664S	2	11/32	2	21/64	—	5/32	7/16	.015	—	7/64	On Index
	4665S	2	11/32	2	21/64	—	5/32	7/16	.015	—	1/8	On Index
CHRYSLER, IMPERIAL, DODGE, PLYMOUTH												
1969	4619S	—	7/32	3/4	7/16	—	—	17/64	.020	—	—	—
	4620S	1–2	7/32	3/4	7/16	50 Drill②	1/4	17/64	.020	5/32	39 Drill	2 Rich
	4621S	1–2	7/32	3/4	7/16	50 Drill②	1/4	17/64	.020	5/32	39 Drill	2 Rich
1970	4742S	③	7/32	3/4	7/16	—	—	17/64	.020	—	—	—
	4745S	③	7/32	3/4	7/16	50 Drill②	1/4	17/64	.020	3/4①	54 Drill	2 Rich
	4746S	③	7/32	3/4	7/16	50 Drill②	1/4	17/64	.020	3/4①	39 Drill	2 Rich
1971	4969S	③	7/32	3/4	31/64	—	1/4	17/64	.020	3/4①	—	—
	4970S	③	7/32	3/4	31/64	50 Drill②	1/4	17/64	.020	3/4①	—	—
	4971S	③	7/32	3/4	31/64	50 Drill②	—	17/64	.020	—	—	—

①—At curb idle.
②—With fast idle speed adjusting screw contacting second highest speed step on fast idle cam the clearance between choke valve and wall of air horn should be as specified.
③—Air/fuel ratio or idle CO% rating is found in Tune Up Specification tables in car chapters.

MODEL AFB ADJUSTMENTS

The AFB carburetor, Figs. 1 and 2, contains many features, some of which are the locations of the step-up rods and pistons. The step-up rods, pistons and springs are accessible for service without removing the air horn or the carburetor from the engine. The venturi assemblies (primary and secondary) are replaceable and contain many of the calibration points for both the high and low speed systems. One fuel bowl feeds both the primary and secondary nozzles on the right side while the other fuel bowl takes care of the primary and secondary nozzles on the left side. This provides excellent performance in cornering, quick stops and acceleration.

All the major castings of the carburetor are aluminum, with the throttle body integral with the main body. This allows an overall height reduction in the carburetor. The section containing the accelerator pump is termed the primary side of the carburetor; the rear section is the secondary.

Float Alignment

Fig. 3—Sight down side of float to determine if it is parallel to the outer edge of the air horn casting. To adjust, bend float lever by applying just enough pressure to make the adjustment. Apply the pressure on the end of the float with the fingers while supporting the float lever with the thumb.

After aligning the float, remove as much clearance as possible between arms of float lever and lugs on air horn by bending the float lever. Arms of float lever should be as parallel to the inner surfaces of the lugs on the air horn as possible. Floats must operate freely without excess clearance on its hinge pin.

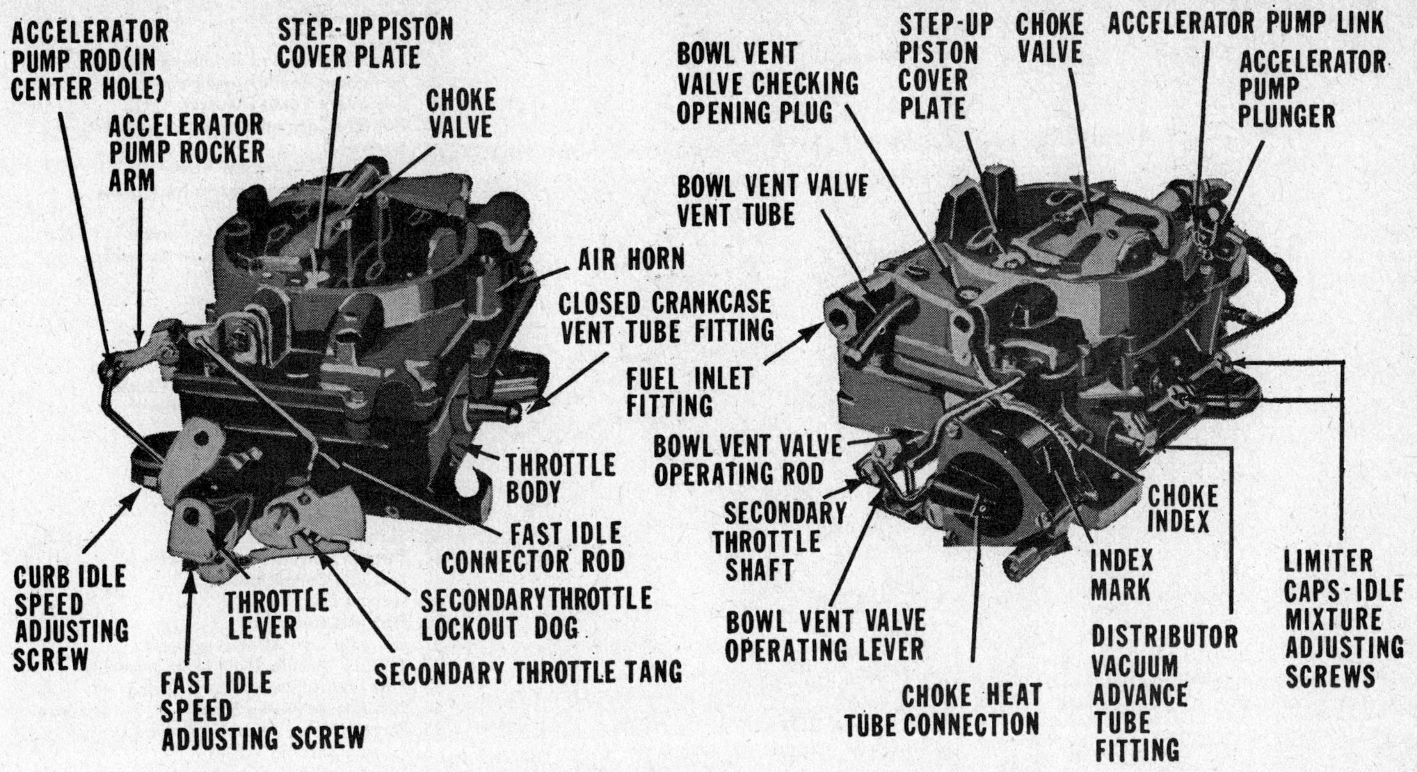

Fig. 1 Typical Carter AFB carburetor assembly

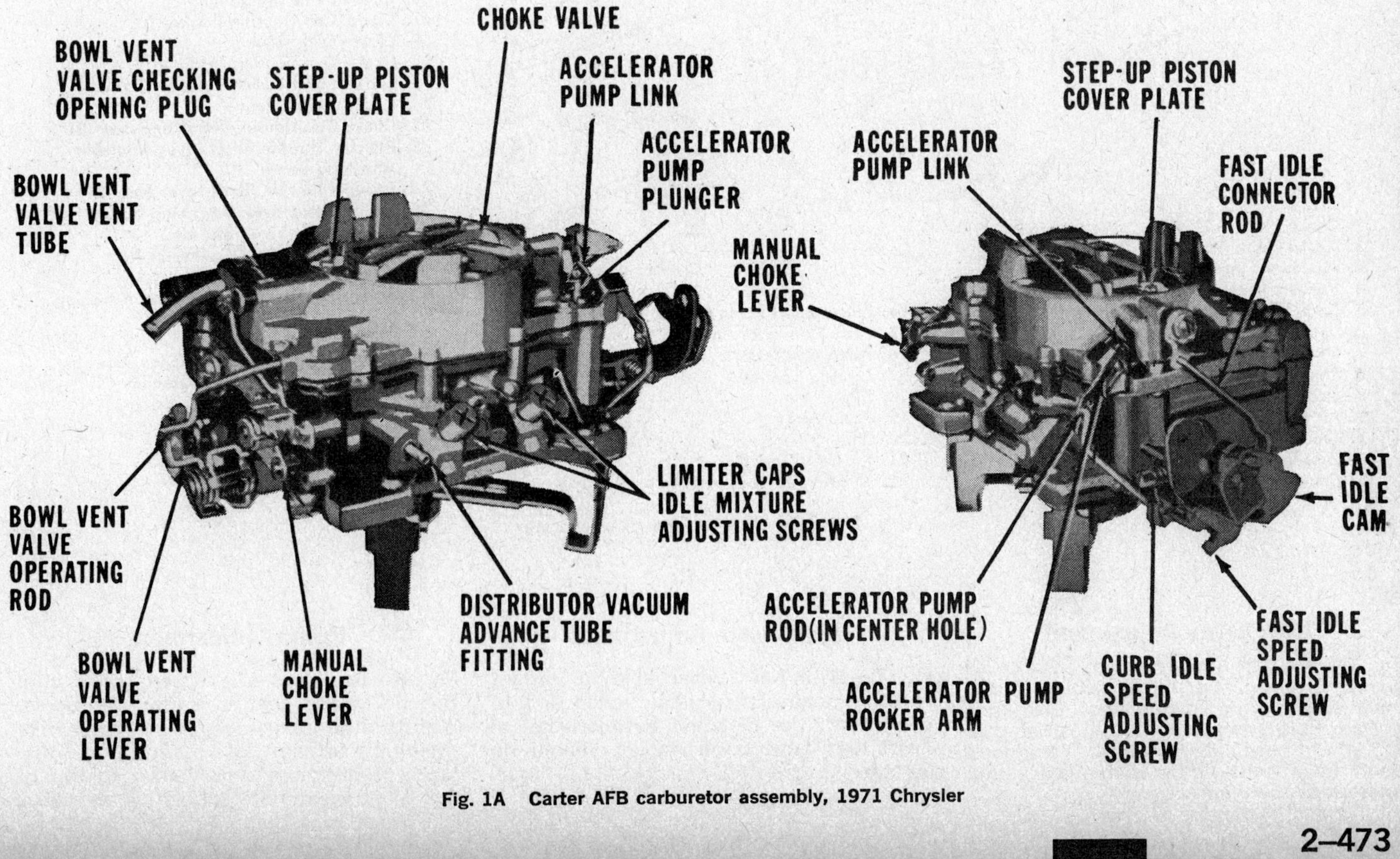

Fig. 1A Carter AFB carburetor assembly, 1971 Chrysler

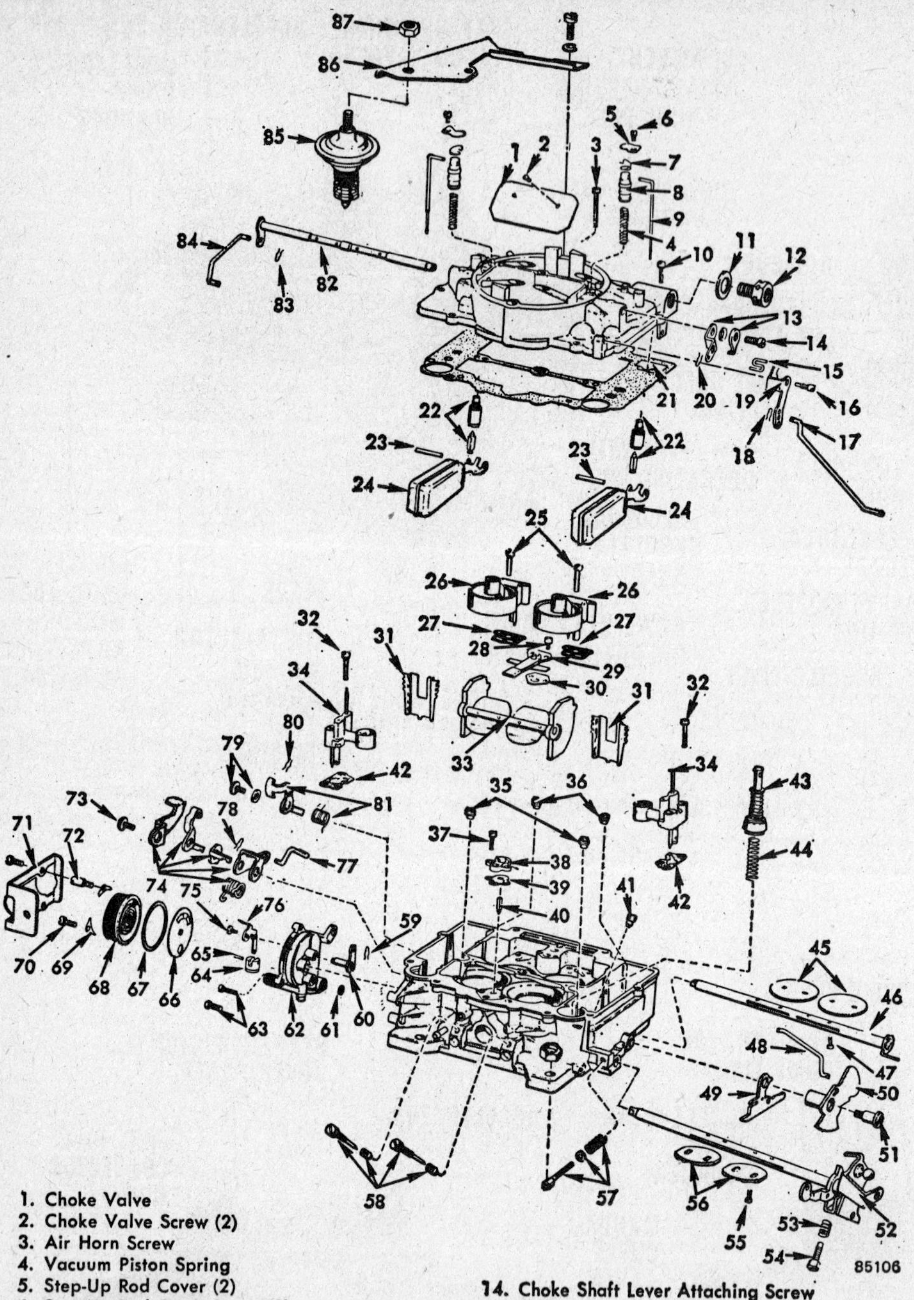

1. Choke Valve
2. Choke Valve Screw (2)
3. Air Horn Screw
4. Vacuum Piston Spring
5. Step-Up Rod Cover (2)
6. Step-Up Rod Cover Screw (2)
7. Step-Up Rod Retainer Spring
8. Step-Up Rod Piston
9. Step-Up Rod
10. Air Horn Attaching Screw (8)
11. Fuel Inlet Fitting Gasket
12. Fuel Inlet Fitting
13. Choke Shaft Lever Assembly
14. Choke Shaft Lever Attaching Screw
15. Pump Connector "S" Link
16. Pump Arm Screw
17. Throttle Connector Rod
18. Clip
19. Pump Arm
20. Clip
21. Air Horn Gasket
22. Needle and Seat Assembly
23. Float Lever Pin
24. Float
25. Secondary Venturi Retaining Screws
26. Secondary Venturi Assembly
27. Secondary Venturi Gasket
28. Hot Idle Compensator Valve Retaining Screws (2)
29. Hot Idle Compensator Valve
30. Hot Idle Compensator Valve Gasket
31. Fuel Bowl Baffle
32. Primary Venturi Attaching Screw
33. Secondary Auxiliary Valve Assembly
34. Primary Venturi Assembly
35. Primary Step-Up Rod Jets
36. Secondary Metering Jets
37. Pump Jet Housing Attaching Screw
38. Pump Jet Housing Assembly
39. Pump Jet Housing Assembly Gasket
40. Pump Discharge Check Needle
41. Pump Intake Ball Check
42. Primary Venturi Gasket
43. Pump Plunger Assembly
44. Lower Pump Spring
45. Secondary Throttle Valves
46. Secondary Throttle Shaft
47. Throttle Valve Attaching Screw (4)
48. Fast Idle Connector Rod
49. Lockout Dog
50. Fast Idle Cam
51. Fast Idle Cam Attaching Screw
52. Primary Throttle Shaft Lever Assembly
53. Fast Idle Adjusting Screw Spring
54. Fast Idle Adjusting Screw
55. Throttle Valve Attaching Screw (4)
56. Primary Throttle Valves
57. Idle Speed Adjusting Screw and Spring
58. Idle Mixture Adjustment Screws and Springs
59. Clip
60. Choke Shaft Lever
61. Choke Housing Gasket
62. Choke Housing
63. Choke Housing Attaching Screws
64. Choke Piston
65. Choke Piston Pin
66. Choke Baffle (Rotating)
67. Choke Coil Housing Gasket
68. Choke Coil Housing
69. Choke Coil Housing Retainer (3)
70. Choke Coil Housing Retainer Screw
71. Choke Heat Shield
72. Choke Coil Housing Retainer Screw (2)
73. Primary Throttle Shaft Lever Assembly Attaching Screw
74. Primary Throttle Shaft Lever Assembly
75. Choke Piston Lever Attaching Screw
76. Choke Piston Lever and Link
77. Primary Secondary Connector Rod
78. Clip
79. Secondary Throttle Lever Shoe Attaching Screw and Washer
80. Clip
81. Secondary Throttle Lever Shoe Assembly
82. Choke Shaft and Lever Assembly
83. Clip
84. Choke Lever Rod
85. Dashpot
86. Dashpot Bracket
87. Dashpot Nut

Fig. 2 Carter AFB four-barrel carburetor

Float Level Adjustment

Fig. 4—With air horn inverted, bowl cover gasket in place and needle seated, clearance between top of float (at outer end) and air horn gasket should be as listed in the *AFB Specifications Chart*. To adjust, bend float arm. Adjust both floats and recheck float alignment.

Float Drop Adjustment

Fig. 5—With bowl cover held in upright position, measure between outer end of each float, the distance between top of floats and bowl cover gasket should be as listed in the *AFB Specifications Chart*. To adjust, bend tabs on float brackets.

Pump Adjustment

Fig. 6—Back out idle speed screw until throttle valves seat in carburetor bores. With throttle connector rod in center hole (medium stroke) of pump arm, distance from top of bowl cover to top of pump plunger shaft should be as listed

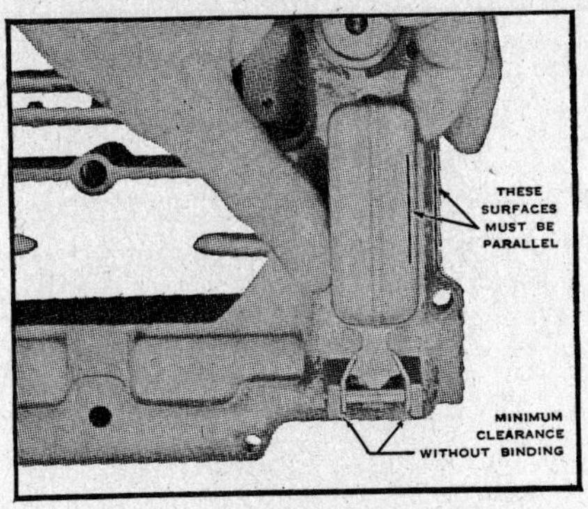

Fig. 3 AFB float alignment

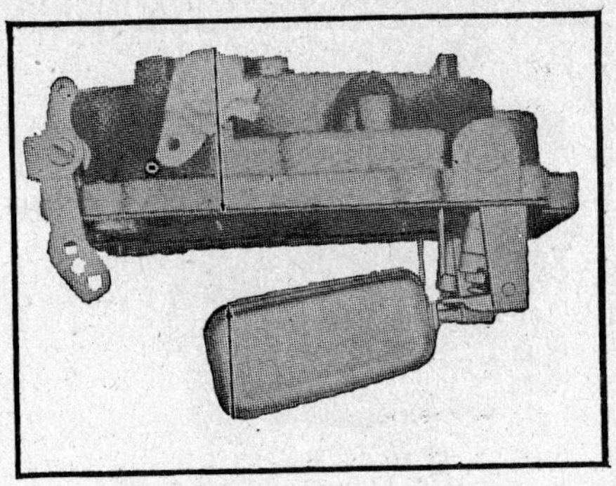

Fig. 5 AFB float drop adjustment

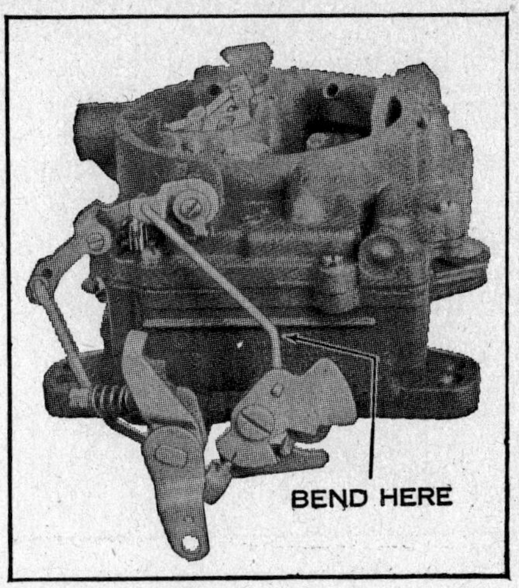

Fig. 7 AFB fast idle linkage adjustment

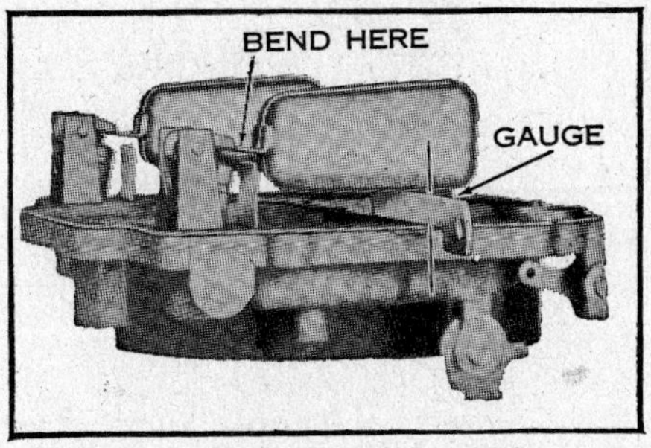

Fig. 4 AFB float level adjustment

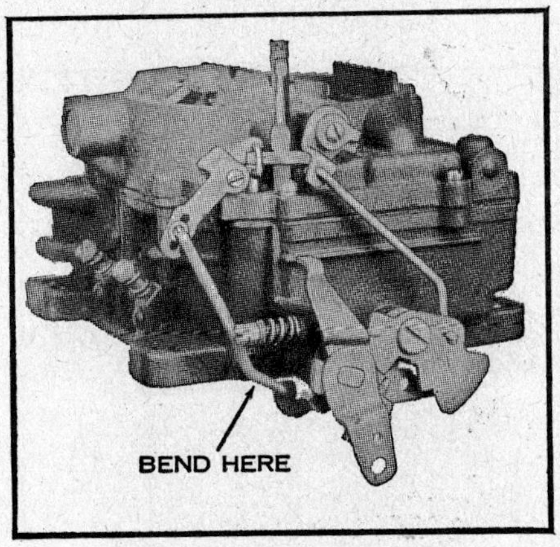

Fig. 6 AFB pump adjustment

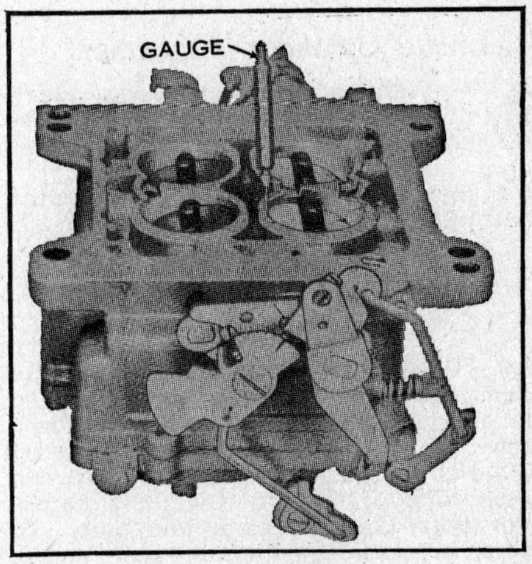

Fig. 8 AFB throttle valve clearance adjustment

in the *AFB Specifications Chart.* Adjust by bending throttle connector rod at its offset angle.

NOTE: Some models require the throttle connector rod to be placed in either the inner hole (long stroke) or outer hole (short stroke). In such cases the chart will indicate the proper hole connection.

Fast Idle Cam Linkage Adjustment

Exc. Chrysler Corp. Fig. 7—With choke valve tightly closed and lug on outer choke shaft lever contacting stop on inner choke shaft lever, align center of fast idle screw with index mark on cam. To adjust, bend fast idle connector rod. On some models it may be necessary to bend stop lug on fast idle cam.

Fast Idle Throttle Valve Clearance

Exc. Chrysler Corp. Fig. 8—With choke valve tightly closed, tighten fast idle adjusting screw on index mark on cam until the clearance between throttle valve and carburetor bore (side opposite idle port) is as listed in the *AFB Specifications Chart.*

Fast Idle Speed Cam Position Adjustment

Fig. 8A—With fast idle speed adjusting screw contacting second highest speed step on fast idle cam, move choke valve toward closed position with light pressure on choke shaft lever.

Insert drill gauge between choke valve and wall of air horn. Refer to *AFB Specifications Chart.* An adjustment is necessary if a slight drag is not obtained as the drill is being removed. Adjust by bending fast idle connector rod at angle.

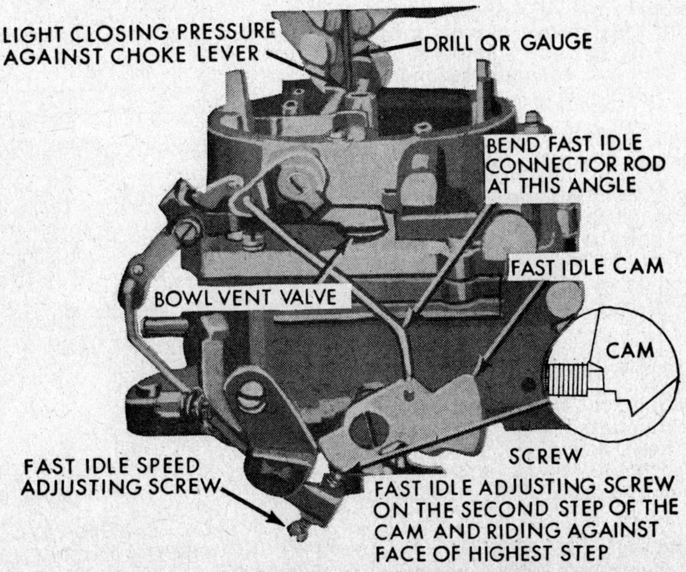

Fig. 8A AFB fast idle cam position adjustment

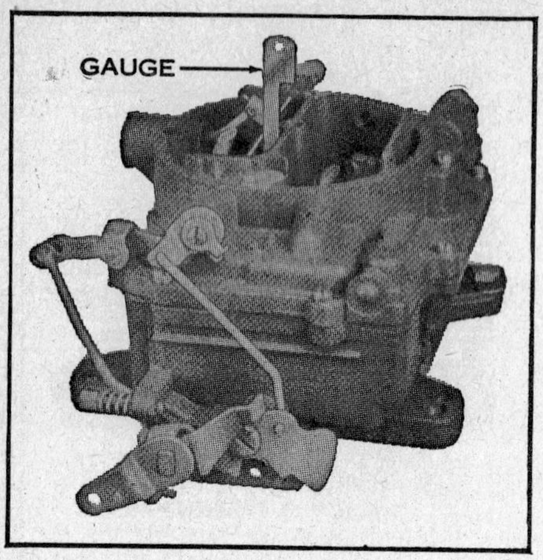

Fig. 9 AFB choke unloader adjustment

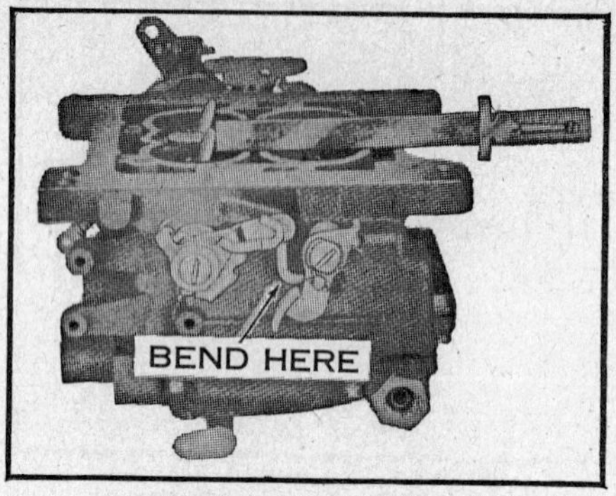

Fig. 10 AFB secondary throttle lever adjustment

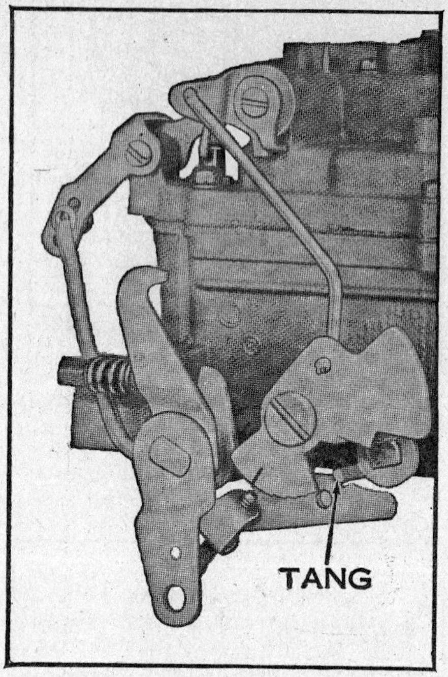

Fig. 12 AFB secondary lockout adjustment

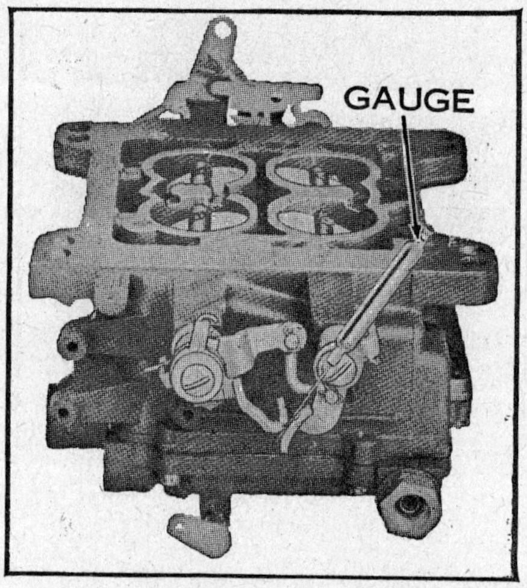

Fig. 11 AFB gauging clearance between positive closing shoes on primary and secondary throttle valves

Choke Unloader Adjustment

Fig. 9—With throttle wide open, clearance between upper edge of choke valve and inner wall of air horn should be as listed in the *AFB Specifications Chart*. To adjust, bend unloader tang on throttle shaft lever.

Secondary Throttle Lever Adjustment

Figs. 10 and 11—Block choke valve wide open. Secondary throttle valves should just start to open when primary throttle valves are opened to the clearance listed in the *AFB Specifications Chart* between lower edge of throttle valve and carburetor bore (side opposite idle port). To adjust, bend throttle operating rod, Fig. 10.

Primary and secondary throttle valves should reach wide open position at the same time.

With primary and secondary throttle valves tightly closed, there should be the clearance listed in the *AFB Specifications Chart* between the positive closing shoes and primary and secondary throttle levers. To adjust, bend shoe on secondary lever.

Secondary Throttle Lockout

Fig. 12—Crack throttle valves and manually open and close choke valve. Tang on secondary throttle lever should freely engage in notch of lockout dog. To adjust, bend tang on secondary throttle lever. Refer to AFB Specifications Chart.

Bowl Vent Valve Adjustment

Except ECS Equipped-Fig. 8A—With throttle valves tightly closed, insert drill gauge between air horn and valve at smallest opening. Refer to *AFB Specifications Chart*. Adjust by bending adjusting tang on pivot end of lever.

ECS Equipped Fig. 13—Remove bowl vent valve checking hole plug in air horn. With throttle valves at closed curb idle position, insert a narrow ruler down through hole. Allow ruler to rest lightly on top of valve. Measure from top of valve to top of air horn casting at opening. Refer to *AFB Specifications Chart*.

Adjust by bending bowl vent valve operating lever. Install new plug and rap lightly to seat, using a hammer.

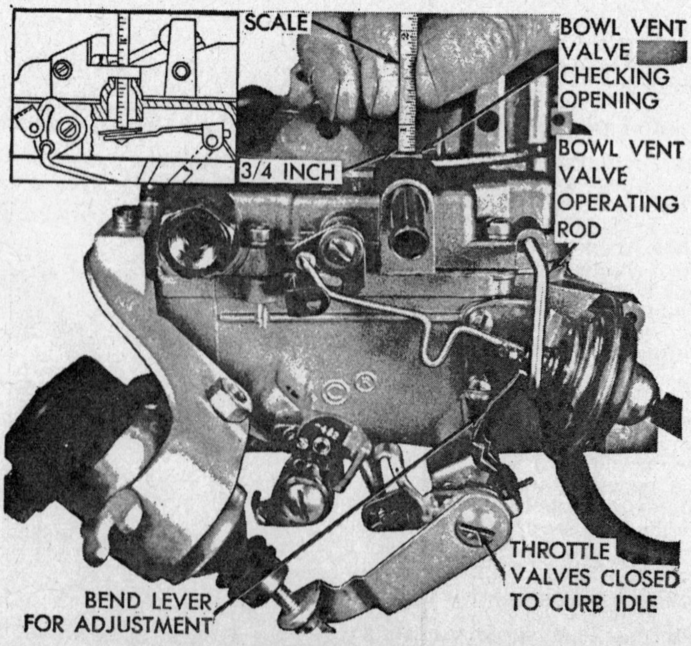

Fig. 13 AFB bowl vent valve adjustment ECS equipped

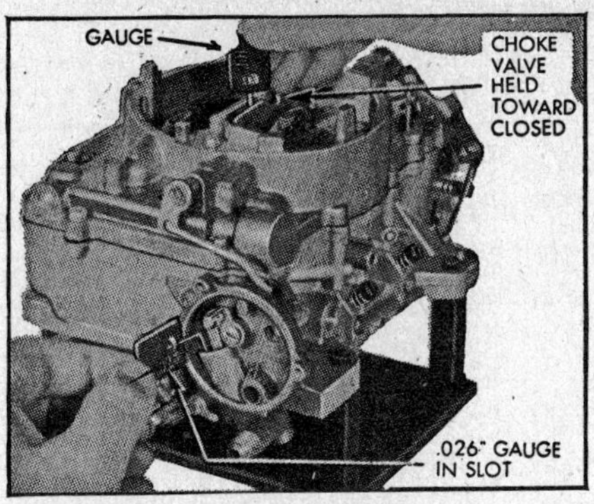

Fig. 14 AFB choke piston clearance

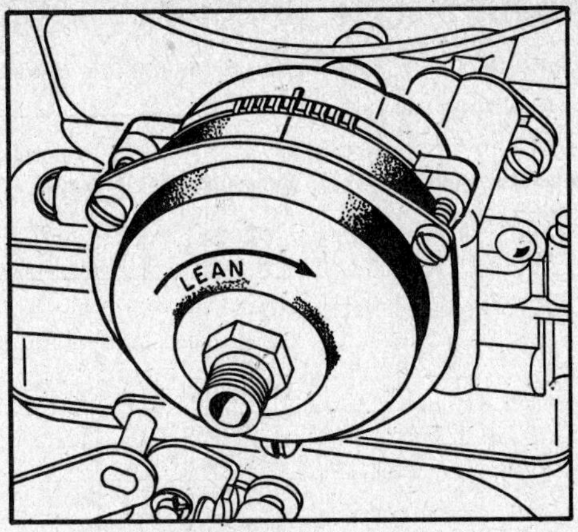

Fig. 15 AFB choke setting (built-in type)

Fig. 16 AFB well-type choke setting (Chrysler line)

Choke Piston Linkage Adjustment

Fig. 14

Bend a .026″ wire gauge at a 90-degree angle approximately 1/8″ from its end. Open choke valve and insert the wire gauge so that bent portion is between top of slot in choke piston cylinder and bottom of slot in piston.

Hold wire gauge in position and close choke valve by pressing on piston lever in choke housing until resistance is felt. There should now be the clearance listed in the *AFB Specifications Chart* between top of choke valve and air horn wall. To adjust, bend choke connector link.

Automatic Choke Setting

Built-In Type, Fig. 15

Loosen retaining screws and turn choke cover so that index mark or line on cover lines up with specified mark on choke housing listed in the *AFB Specifications.*

Well-Type Choke

Loosen mounting post lock nut and turn mounting post with screwdriver until index mark on disc is positioned as listed in the *AFB Specifications Chart.* Hold in this position with a screwdriver and tighten lock nut, Fig. 16

After adjustment is completed and coil housing and carburetor are installed on engine, lift cover disc and open and close choke valve manually to see if connector rod clears sides of hole in housing cover without binding. If binding exists, replace with a new unit since the connector rod cannot be bent without affecting calibration.

Dashpot Adjustment

Chrysler Line with C.A.P.

Use only on cars with manual transmission, make the dashpot adjustment after the fast idle setting. Then with the fast idle screw on highest step of cam, adjust dashpot for a clearance of .052″ between dashpot stem and lever. Tighten lock nut.

CARTER TQ ADJUSTMENT SPECIFICATIONS

See Tune Up Chart in car chapters for hot idle speeds.

Year	Carb. Model	Air/Fuel Ratio	Float Setting	Secondary Throttle Linkage	Secondary Air Valve Opening	Secondary Air Valve Spring	Pump Travel	Choke Control Lever (On Car)	Choke Unloader	Fast Idle R.P.M.	Choke Setting
CHRYSLER, DODGE, IMPERIAL & PLYMOUTH											
1971	4972S	①	1	11/32	31/64	1 1/4 Turn	31/64	5 41/64	11 Drill	1800	2 Rich
	4973S	①	1	11/32	31/64	1 1/4 Turn	31/64	5 41/64	11 Drill	1800	2 Rich
1972	6090S	①	1	②	31/64	1 Turn	31/64	3 3/8③	.190	1900	Fixed
	6138S	①	1	②	29/64	1 Turn	9/16⑥	3 3/8③	.190	1900	Fixed
	6139S	①	1	②	29/64	1 Turn	31/64	3 3/8③	.190	1900	Fixed
	6140S	①	1	②	31/64	1 Turn	9/16⑥	3 3/8③	.190	1900	Fixed
	6165S	①	1	②	31/64	1 Turn	9/16⑥	3 3/8③	.190	2000	Fixed
	6166S	①	1	②	31/64	1 Turn	31/64	3 3/8③	.190	2100	Fixed
1973	6318S	①	1 1/16	②	29/64	1 1/4 Turn	35/64⑦	3 3/8③	.190	1300	Fixed
	6319S	①	1 1/16	②	29/64	1 1/4 Turn	31/64	3 3/8③	.190	1800	Fixed
	6320S	①	1 1/16	②	31/64	1 1/4 Turn	35/64⑦	3 3/8③	.190	1700	Fixed
	6321S	①	1 1/16	②	31/64	1 1/4 Turn	31/64	3 3/8③	.190	1800	Fixed
	6322S	①	1 1/16	②	31/64	1 1/4 Turn	31/64	3 3/8③	.190	1700	Fixed
	6324S	①	1 1/16	②	31/64	1 1/4 Turn	31/64	3 3/8③	.190	1700	Fixed
	6339S	①	1 1/16	②	29/64	1 1/4 Turn	35/64⑦	3 3/8③	.190	1300	Fixed
	6340S④	①	1 1/16	②	29/64	1 1/4 Turn	31/64	3 3/8③	.190	1800	Fixed

Continued

CARTER TQ ADJUSTMENT SPECIFICATIONS—Continued

See Tune Up Chart in car chapters for hot idle speeds.

Year	Carb. Model	Air/Fuel Ratio	Float Setting	Secondary Throttle Linkage	Secondary Air Valve Opening	Secondary Air Valve Spring	Pump Travel	Choke Control Lever (On Car)	Choke Unloader	Fast Idle R.P.M.	Choke Setting
CHRYSLER, DODGE, IMPERIAL & PLYMOUTH—Continued											
1973	6341S④	①	1 1/16	②	29/64	1 1/4 Turn	35/64⑦	3 3/8③	.190	1700	Fixed
	6342S④	①	1 1/16	②	29/64	1 1/4 Turn	31/64	3 3/8③	.190	1800	Fixed
	6410S④	①	1 1/16	②	31/64	1 1/4 Turn	31/64	3 3/8③	.190	1700	Fixed
	6411S④	①	1 1/16	②	31/64	1 1/4 Turn	31/64	3 3/8③	.190	1700	Fixed
1974	6452S	①	⑤	②	1/2	1 1/4 Turn	35/64⑧	3 3/8③	.310	1900	Fixed
	6453S	①	⑤	②	1/2	1 1/4 Turn	31/64	3 3/8③	.310	1900	Fixed
	6454S④	①	⑤	②	1/2	1 1/4 Turn	35/64⑧	3 3/8③	.310	1900	Fixed
	6455S④	①	⑤	②	1/2	1 1/4 Turn	31/64	3 3/8③	.310	1900	Fixed
	6456S	①	⑤	②	1/2	1 1/4 Turn	35/64⑧	3 3/8③	.310	1700	Fixed
	6457S④	①	⑤	②	1/2	1 1/4 Turn	31/64	3 3/8③	.310	1800	Fixed
	6459S④	①	⑤	②	1/2	1 1/4 Turn	31/64	3 3/8③	.310	1800	Fixed
	6460S	①	⑤	②	1/2	1 1/4 Turn	31/64	3 3/8③	.310	1700	Fixed
	6461S④	①	⑤	②	1/2	1 1/4 Turn	31/64	3 3/8③	.310	1700	Fixed
	6462S	①	⑤	②	1/2	1 1/4 Turn	31/64	3 3/8③	.310	1700	Fixed
	6463S④	①	⑤	②	1/2	1 1/4 Turn	31/64	3 3/8③	.310	1700	Fixed
	6487S	①	⑤	②	1/2	1 1/4 Turn	35/64	3 3/8③	.310	2000	Fixed
	6488S④	①	⑤	②	1/2	1 1/4 Turn	35/64	3 3/8③	.310	1800	Fixed
	6489S	①	⑤	②	1/2	1 1/4 Turn	31/64	3 3/8③	.310	2000	Fixed
	6496S	①	⑤	②	1/2	1 1/4 Turn	31/64	3 3/8③	.310	2000	Fixed
	6498S	①	29/32	②	33/64	1 1/4 Turn	31/64	3 3/8③	.310	1900	Fixed
	6499S④	①	29/32	②	33/64	1 1/4 Turn	35/64	3 3/8③	.310	1900	Fixed
	9013S④	①	⑤	②	1/2	1 1/4 Turn	35/64	3 3/8③	.310	1800	Fixed
	9014S	①	⑤	②	1/2	1 1/4 Turn	31/64	3 3/8③	.310	2000	Fixed
	9015S	①	⑤	②	1/2	1 1/4 Turn	31/64	3 3/8③	.310	1700	Fixed
	9016S④	①	⑤	②	1/2	1 1/4 Turn	31/64	3 3/8③	.310	1700	Fixed
	9023S	①	⑤	②	1/2	1 1/4 Turn	31/64	3 3/8③	.310	1700	Fixed
	9024S④	①	⑤	②	1/2	1 1/4 Turn	31/64	3 3/8③	.310	1700	Fixed
1975	9002S	①	29/32	②	1/2	1 1/4 Turn	35/64	3 3/8③	.310	1600	Fixed
	9004S	①	29/32	②	1/2	1 1/4 Turn	35/64	3 3/8③	.310	1600	Fixed
	9008S	①	29/32	②	1/2	1 1/4 Turn	35/64	3 3/8③	.310	1800	Fixed
	9009S	①	29/32	②	1/2	1 1/4 Turn	35/64	3 3/8③	.310	1600	Fixed
	9010S	①	29/32	②	1/2	1 1/4 Turn	35/64	3 3/8③	.310	1600	Fixed
	9011S	①	29/32	②	1/2	1 1/4 Turn	35/64	3 3/8③	.310	1600	Fixed
	9012S	①	29/32	②	1/2	1 1/4 Turn	35/64	3 3/8③	.310	1800	Fixed
	9046S	①	29/32	②	1/2	1 1/4 Turn	35/64	3 3/8③	.310	1800	Fixed
	9053S	①	29/32	②	1/2	1 1/4 Turn	35/64	3 3/8③	.310	1800	Fixed
1976	9002S	①	29/32	②	33/64	1 1/4 Turn	33/64⑨	3 3/8③	.310	1700	Fixed
	9052S	①	29/32	②	33/64	1 1/4 Turn	33/64⑨	3 3/8③	.310	1600	Fixed
	9054S	①	29/32	②	33/64	1 1/4 Turn	33/64⑨	3 3/8③	.310	1800	Fixed
	9055S	①	29/32	②	33/64	1 1/4 Turn	33/64⑨	3 3/8③	.310	1700	Fixed
	9057S	①	29/32	②	33/64	1 1/4 Turn	33/64⑨	3 3/8③	.310	1600	Fixed
	9058S	①	29/32	②	33/64	1 1/4 Turn	31/64	3 3/8③	.310	1600	Fixed
	9059S	①	29/32	②	33/64	1 1/4 Turn	31/64	3 3/8③	.310	1600	Fixed
	9062S	①	29/32	②	33/64	1 1/4 Turn	33/64⑨	3 3/8③	.310	1600	Fixed
	9066S	①	29/32	②	33/64	1 1/4 Turn	33/64⑨	3 3/8③	.310	1600	Fixed

①—For Air/Fuel ratio and idle CO%, see Tune-Up charts in car chapters.
②—Adjust link so primary and secondary stops both contact at same time.
③—Off car.
④—With California Emission package.
⑤—Brass float, 1 inch; cellular plastic float, 29/32 inch.
⑥—Secondary stage pick/up adjustment 13/32 inch.
⑦—Secondary stage pick/up adjustment 23/64 inch.
⑧—Secondary stage pick/up adjustment 21/64 inch.
⑨—Secondary stage pick/up adjustment 5/16 inch.

FORD & MERCURY

1974	D4AE-BB, BC	①	1 1/16	②	.468	—	5/16③	—	.250	1250	Index

①—For Air/Fuel ratio and idle CO%, see Tune-Up charts in car chapters.
②—Adjust link so primary and secondary stops both contact at same time.
③—Measured bottom of "S" link to top of bowl cover.

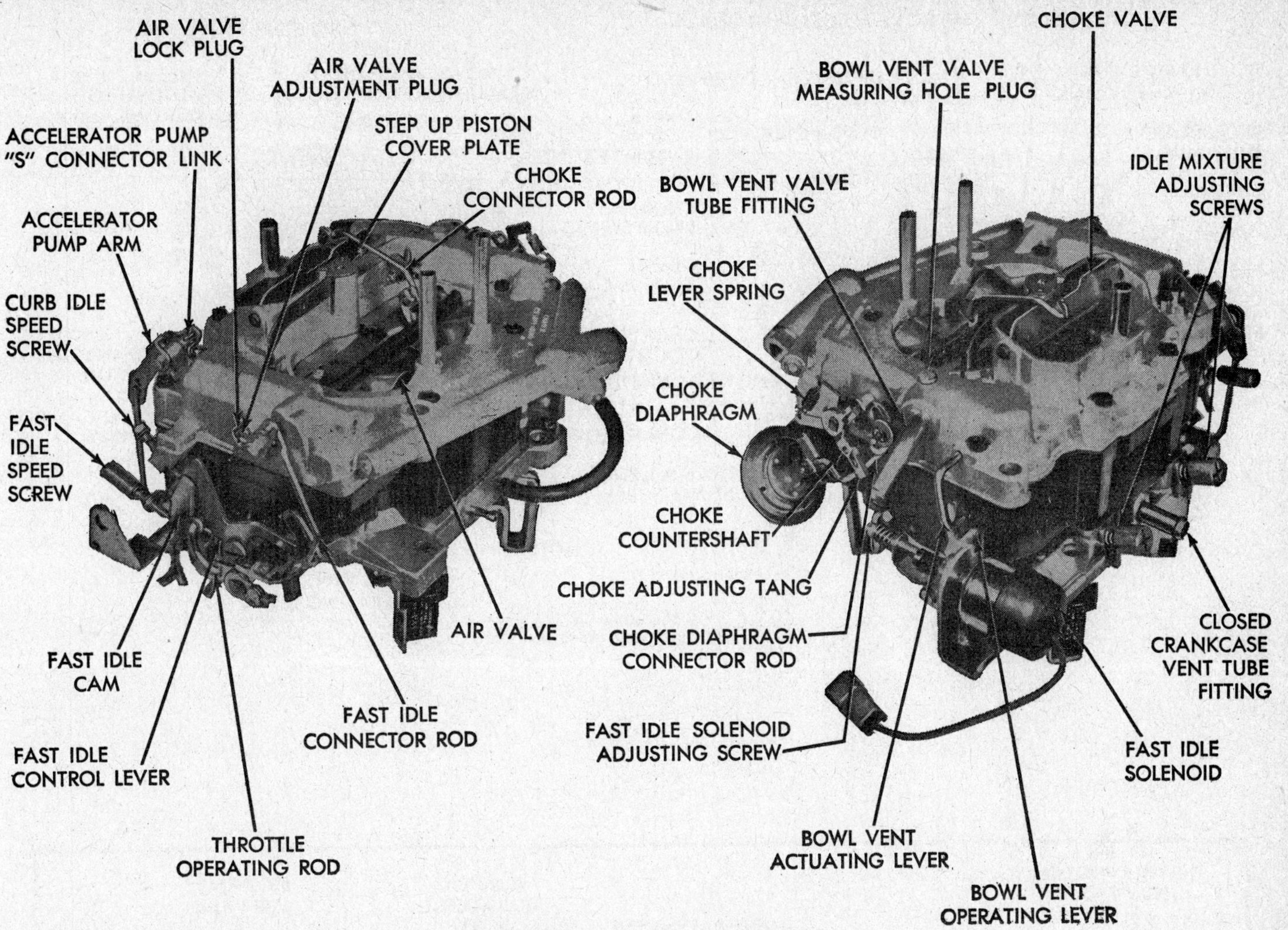

Fig. 1 TQ carburetor assembly. 1972

MODEL TQ ADJUSTMENTS

The TQ (Thermo-Quad) carburetor, Figs. 1 thru 2, is unique in design in that it has a black main body or fuel bowl of molded phenolic resin. This acts as an effective heat insulator. Fuel is kept cooler by about 20 degrees Fahrenheit than in carburetors of all metal design. Another reason for the lower operating temperatures is its suspended design metering system. All calibration points with the exception of the idle adjusting screws, are in the upper aluminum casting or air horn and are in effect suspended in cavities in the plastic main body.

On 1975 vehicles with automatic transmissions, an idle enrichment system is used to reduce cold engine stalling by the use of an additional metering system, which enrichens the mixture in the off idle position. This system is controlled by a vacuum diaphragm mounted near the top of the carburetor.

On most 1975 California units, an altitude compensator is used to maintain the correct air/fuel mixture.

Some 1974 and all 1975 units incorporate cellular plastic floats. Note that the float level settings are different. Refer to TQ Specifications Chart.

Float Setting

Fig. 3—With bowl cover inverted, gasket installed and floats resting on seated needle, the dimension of each float from bowl cover gasket to bottom side of float should be as shown in *TQ Specifications Chart*.

Secondary Throttle Linkage

Fig. 4—Block choke valve in wide open position and invert carburetor. Slowly open the primary throttle valves until it is possible to measure between lower edge of primary valve and its bore. When dimension is as shown in *TQ Specifications Chart*, the secondary valves should just start to open. If necessary to adjust, bend rod until correct dimension is obtained.

Secondary Air Valve Opening

Fig. 5

1. With air valve in closed position, the opening along air valve at its long side must be at its maximum and parallel with air horn gasket surface.
2. With air valve wide open, the opening of the air valve at the short side and air horn must be as shown in *TQ Specifications Chart*. The corner of air valve is notched for adjustment. Bend the corner with a pair of pliers to give proper opening.

Secondary Air Valve Spring Tension

Fig. 6—Loosen air valve lock plug and allow air valve to position itself wide open. With a long screwdriver that will enter center of tool C-4152 positioned on air valve adjustment plug, turn plug counterclockwise until air valve contacts stop lightly, then turn additional turn as specified in *TQ Specifications Chart*. Hold plug with screwdriver and tighten lock plug securely with tool C-4152.

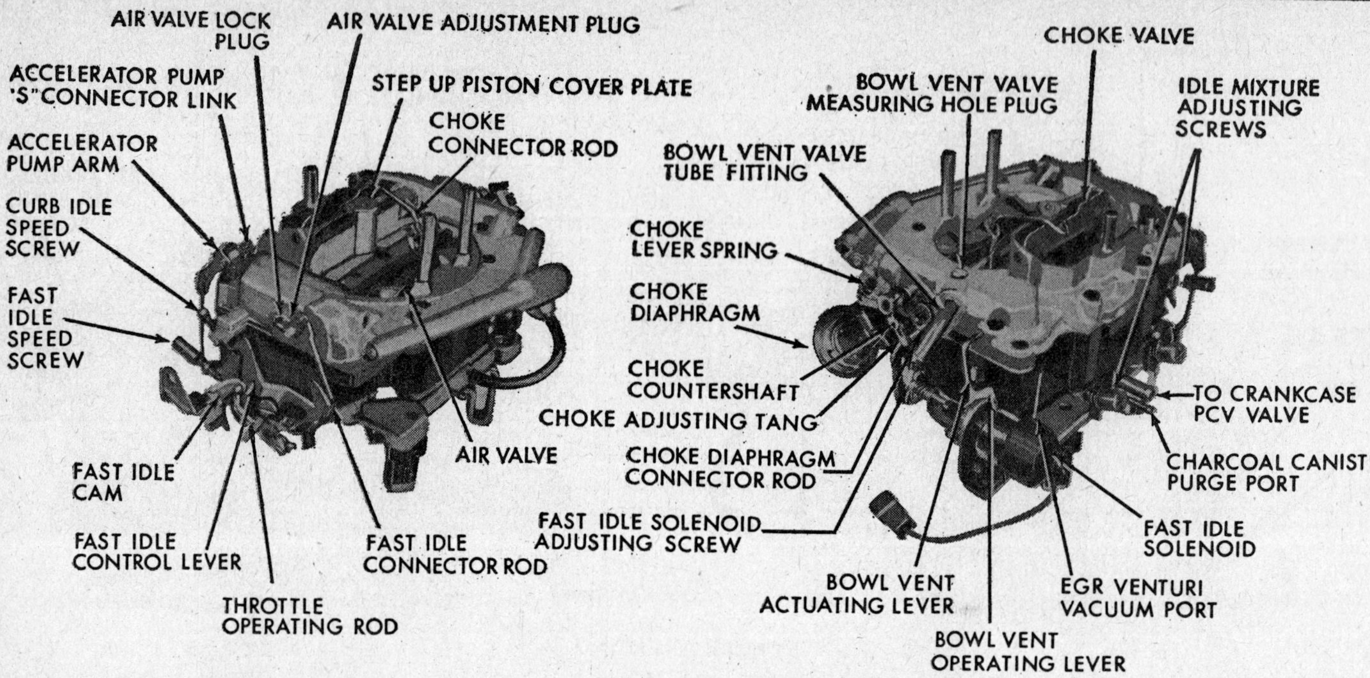

Fig. 1A TQ carburetor assembly. 1973

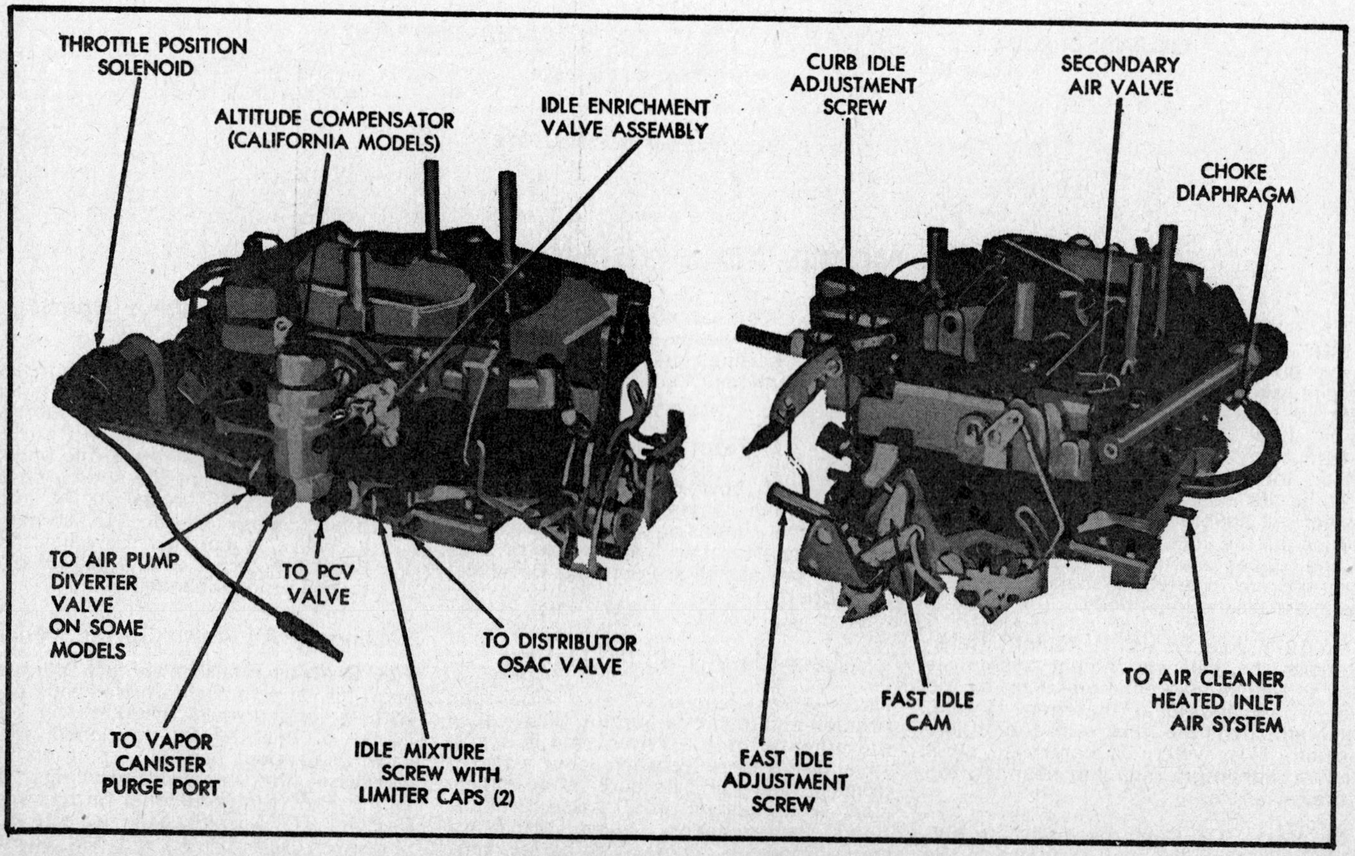

Fig. 1B TQ carburetor assembly. 1974-75 (typical)

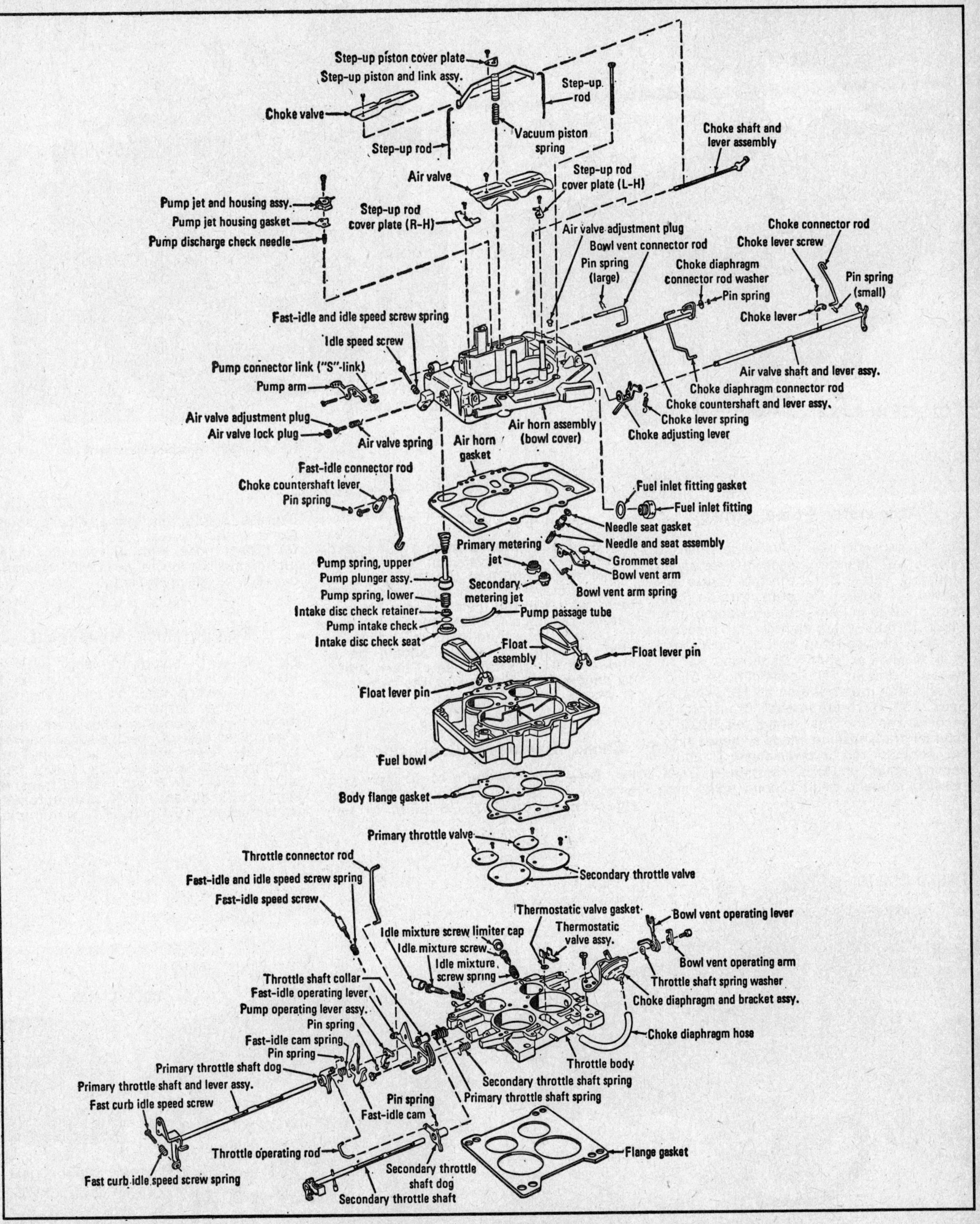

Fig. 2 Carter TQ model exploded view

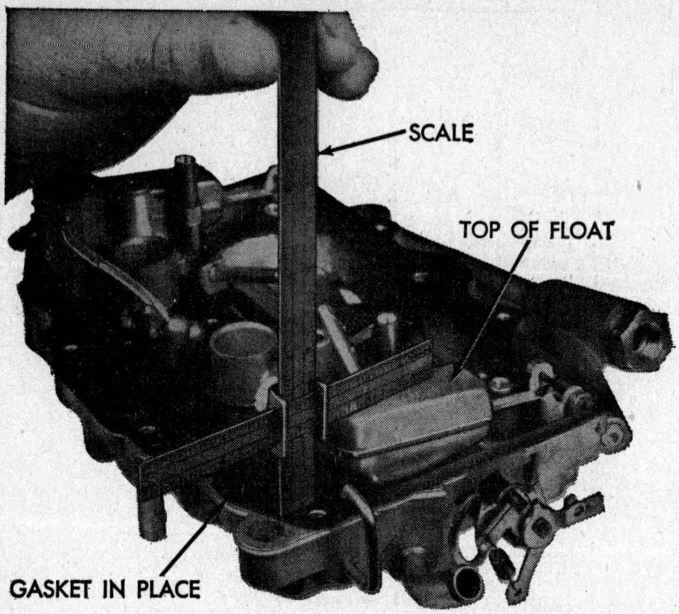

Fig. 3 TQ float setting

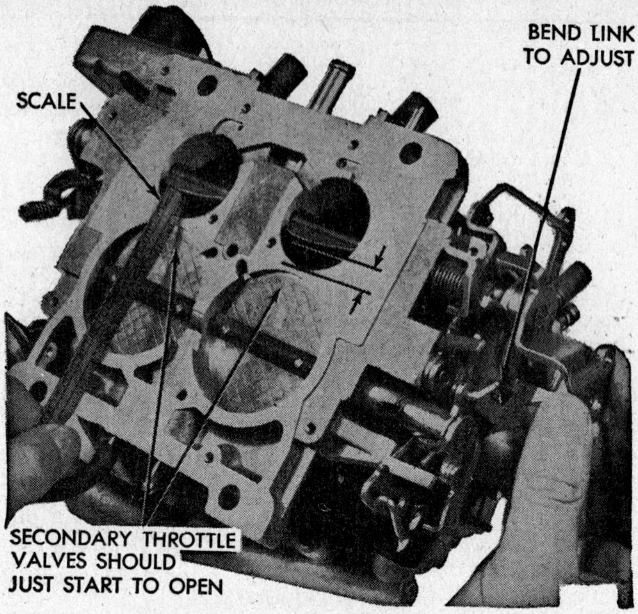

Fig. 4 TQ secondary throttle adjustment

Accelerator Pump Stroke

Fig. 7—Move choke valve wide open to release fast idle cam. Back off idle speed adjusting screw until throttle valves are seated in bores. Be sure throttle connector rod is in center hole of pump arm. Close throttle valve tightly and measure distance between top of bowl cover and end of plunger shaft. Dimension should be as shown in *TQ Specifications Chart.* To adjust pump stroke on all 1971 models and 1972-75 automatic transmission models and the first stage on 1972-75 manual transmission models, bend throttle connector rod at lower angle. To adjust second stage on 1972-75 manual transmission models, bend second stage stop lever.

Choke Control Lever

Fig. 8—Place carburetor on a flat surface and on 1973-75 Chrysler units, disconnect choke diaphragm rod at diaphragm. On all units, close choke by pushing on choke lever with throttle partly open. Measure vertical distance between top of rod hole in control lever and base of carburetor (flat surface). Dimension should be as shown in TQ Specifications Chart. Adjust by bending link connecting the two choke shafts.

Choke Diaphragm Connector Rod

Fig. 9—Apply a vacuum of 10 inches Hg or more on 1970-72 units and 15 inches Hg or more on 1973-75 units to fully depress diaphragm. An auxilliary source like a distributor test machine can be used for this purpose. With air valve closed, adjust connector rod to give .040″ clearance between air valve and stop.

Vacuum Kick Adjustment

Fig. 10—With engine running, back off fast idle speed screw until choke can be closed to kick position at idle. Note number of screw turns so fast idle can be turned back to original adjustment. Insert a gauge or drill of specified size between long side (lower edge) of choke valve and air horn wall as follows: 1971 and 1975 all; .110 inch. 1972 with manual transmission, 1973 all and 1974 with automatic transmission; .160 inch. 1972 with manual

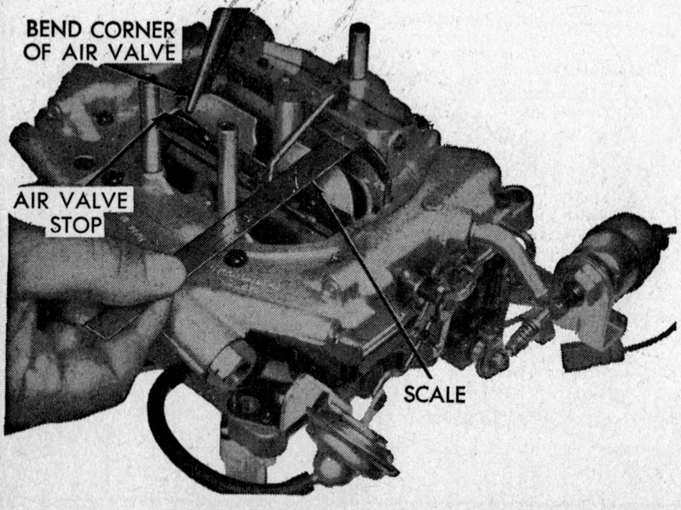

Fig. 5 TQ secondary air valve opening

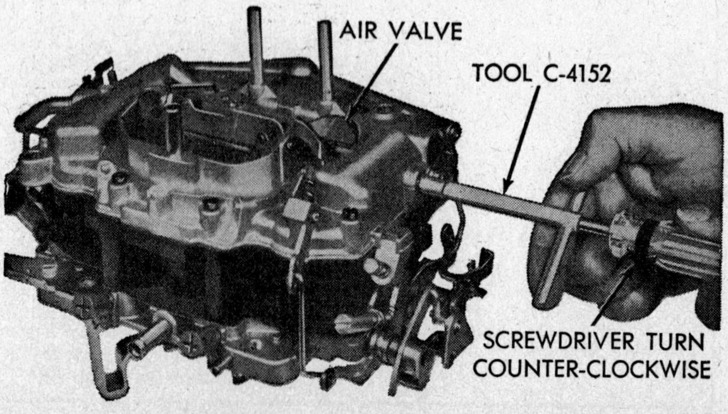

Fig. 6 TQ secondary air valve spring tension

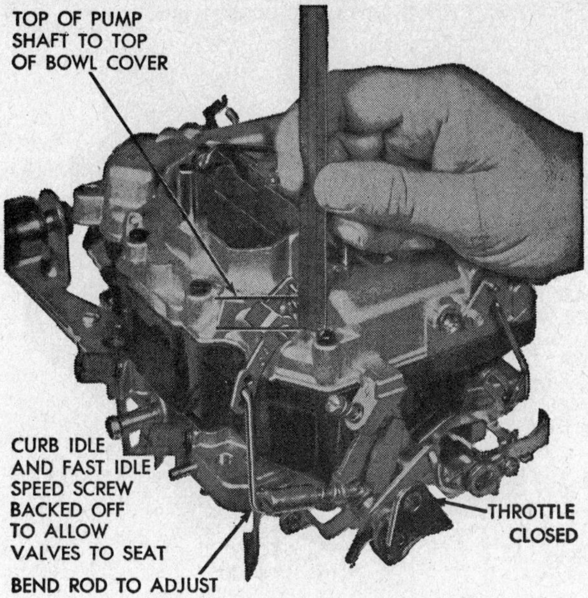

Fig. 7 TQ accelerator pump stroke

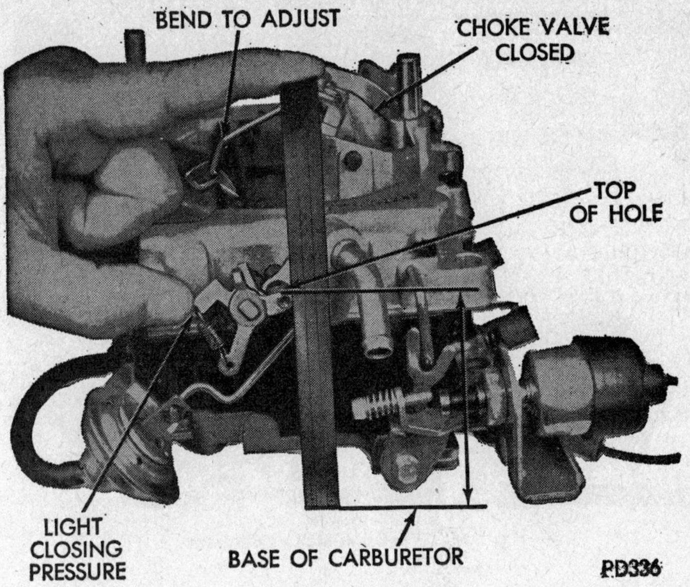

Fig. 8 TQ choke control lever adjustment

transmission; .140 inch. 1974 with automatic transmission; .210 inch. Apply sufficient pressure on choke control lever to provide a minimum choke valve opening. The spring connecting the control lever to the adjustment lever must be fully extended for proper adjustment. Bend tang as shown to change contact with end of diaphragm rod. Do not adjust diaphragm rod. A slight drag should be felt as drill is being removed.

Fig. 9 TQ choke diaphragm connector rod

Fast Idle Cam & Linkage

Fig. 11—With fast idle adjusting screw on second step of fast idle cam, move choke valve toward closed position with light pressure on choke control lever. Clearance between choke valve lower edge and air horn wall should be .110 inch on all 1971-73 units, .100 inch on 1974-75 Chrysler units and .099 inch on 1974 Ford units. Adjust by bending fast idle connector rod at angle.

Choke Unloader Adjustment

Fig. 12—Hold throttle valves in wide open position and insert specified drill between long side (lower edge) of choke valve and inner wall of air horn. With finger lightly pressing against choke valve control lever, a slight drag should be felt as drill is withdrawn. Refer to TQ Specification Chart for proper drill size or dimension. Adjust by bending tang on fast idle control lever.

Secondary Throttle Lockout

Fig. 13—Move choke control lever to open choke position. Measure clearance between lockout lever and stop. Clearance should be .010-.030 inch on all 1971-73 units and .060-.090 inch on 1974-75 units. Adjust by bending tang on fast idle control lever.

Bowl Vent Valve Adjustment

Fig. 14—Remove bowl vent valve checking hole plug in bowl cover. With throttle valves at curb idle, insert a narrow ruler down through hole. Allow ruler to rest lightly on top of valve. Dimension should be .850 inch on all 1971-73 units, .812 inch on 1974-75 Chrysler units and .900 inch on 1974 Ford units. Adjust by bending bowl vent operating lever at notch. Install a new plug.

Fast Idle Speed Cam

Note: On 1974-75 Chrysler units, remove air cleaner and plug vacuum fittings to heated air control and OSAC valves. On 1974 Ford units, remove air cleaner, disconnect vacuum hoses at carburetor spark port and distributor primary diaphragm, then install a jumper hose between the two disconnected hoses. Disconnect and plug EGR vacuum line.

Fig. 15—With engine off and transmission in Park or Neutral, open throttle slightly. Close choke valve until fast idle screw is positioned on second step of cam against shoulder of first step. Start engine and stabilize RPM, then adjust fast idle speed. Refer to TQ Specification Chart.

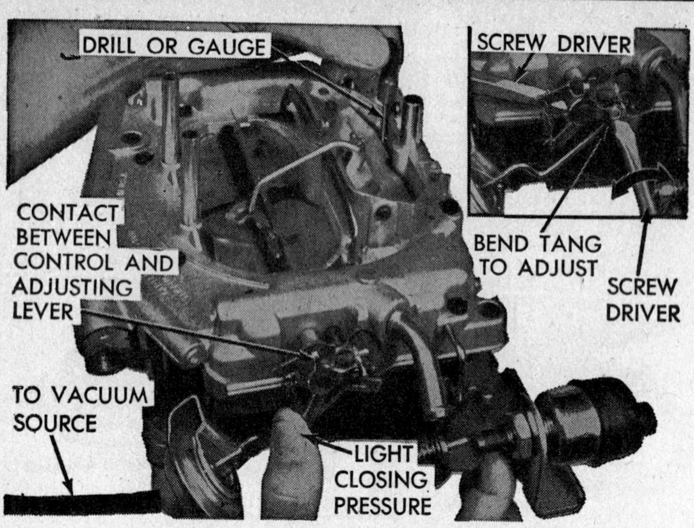

Fig. 10 TQ vacuum kick adjustment

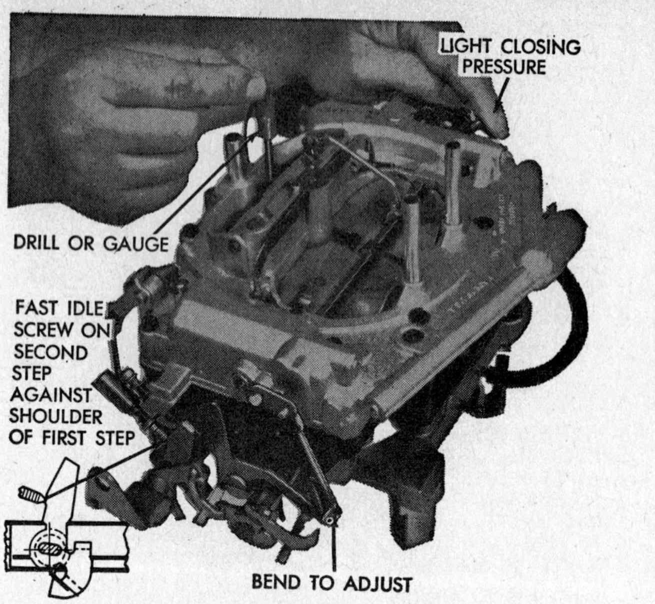

Fig. 11 TQ fast idle cam & linkage adjustment

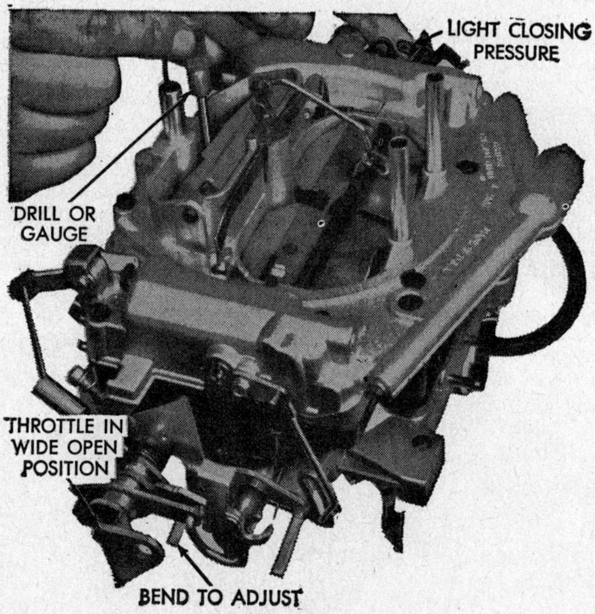

Fig. 12 TQ choke unloader adjustment

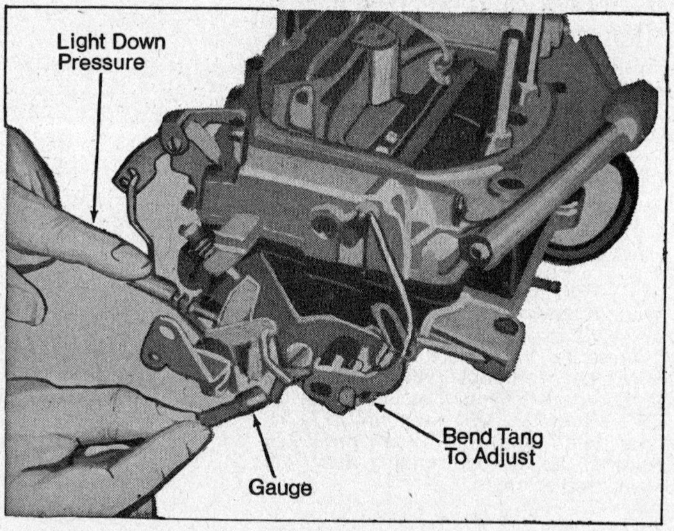

Fig. 13 TQ secondary throttle lockout

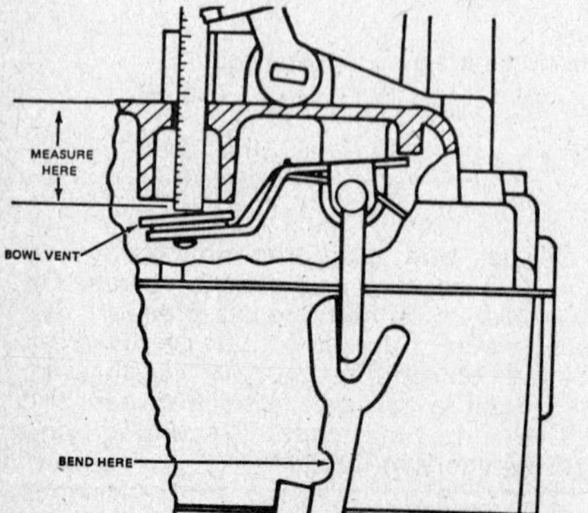

Fig. 14 TQ bowl vent valve adjustment

Fig. 15 TQ fast idle speed cam position

Ford Autolite/Motorcraft Carburetors

MODEL 1250-1V CARB. ADJUSTMENT SPECIFICATIONS

See Tune Up Chart in car chapters for hot idle speeds.

Year	Carb. Model (9510)	Idle Mixture Turns Open	Float Setting	Pump Stroke	Fast Idle Speed	Choke Pulldown	Dechoke Clearance	Choke Link Position	Choke Setting
PINTO									
1971	711-BDA	②	①	.085	1700	.120	.210	Outer Hole	Index
	711-BDB	②	①	.070	1700	.075	.210	Outer Hole	Index
1972	721F-KFA	②	③	.070	1700	.075	.210	Outer Hole	Index
1973	731F-KAA	②	③	.085	—	.075	—	—	Index
	731F-KEA	②	③	.085	1700	.075	.210	Outer Hole	Index

①—With body in vertical position, set at 1.16–1.20.
With body horizontal set at 1.35–1.37 (see Fig. 2).
②—Air/fuel ratio or idle CO% rating is found in Tune Up Specification tables in car chapters.
③—With body in vertical position (dry), set at 1.20.

MODEL 1250-1V ADJUSTMENTS

Fuel & Float Level Adjust

Fig. 2—With carburetor upper body held in vertical position, measure distance between upper body gasket and bottom of float (left view in illustration). Bend tab to adjust.

Turn upper body to horizontal position and measure distance from bottom of float to gasket (right view in illustration). Bend tab to adjust.

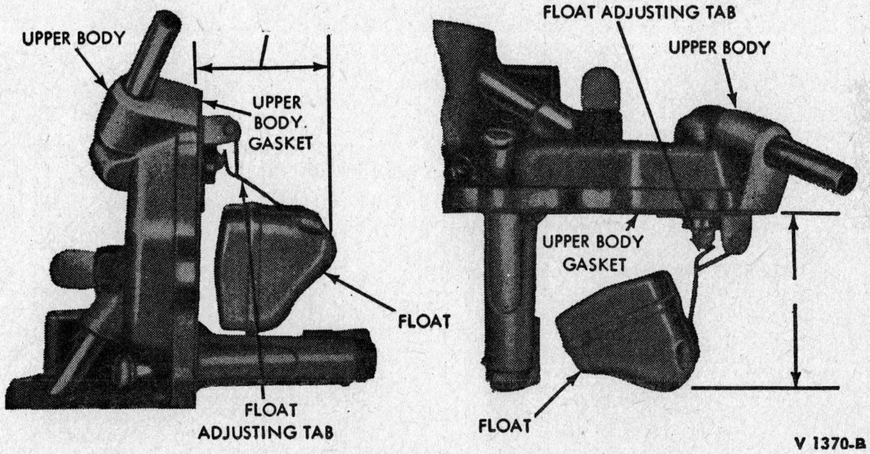

Fig. 2 Fuel and float level setting

Choke Plate Pull-Down

Fig. 3—Remove air cleaner, thermostatic spring and water housing. Depress vacuum piston until vacuum bleed port is revealed and insert a piece of wire .040" thick, suitably bent, into this port. Raise piston to trap wire. With wire and piston held in this position, close the choke plate until its movement is stopped through the linkage. Partially open throttle for fast idle tab to clear cam. Bottom of choke plate should now be as specified from carburetor body. If necessary, bend extension of choke thermostat lever to adjust.

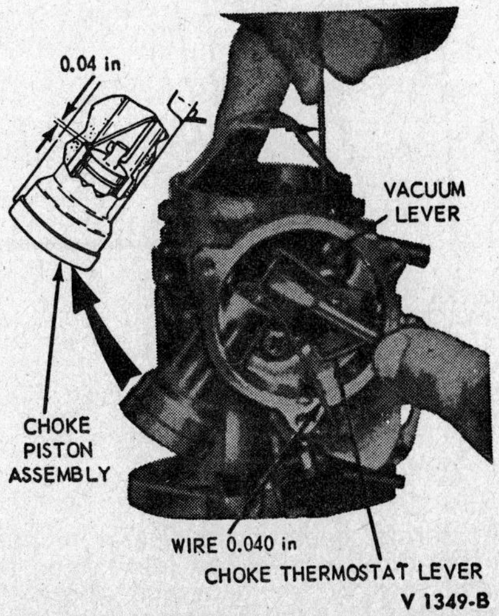

Fig. 3 Choke plate pull-down clearance

Dechoke Adjustment

Fig. 4—Open throttle fully and hold it against stop. Check clearance between bottom of choke plate and carburetor body. Adjust if necessary, by bending the projection on the fast idle cam.

Accelerator Pump Adjustment

Unscrew the throttle stop screw until the throttle plate is fully closed. Depress the accelerator pump diaphragm plunger and check the clearance between the operating lever and the plunger. Refer to *1250-1V Specifications Chart.* Adjust by bending the gooseneck of the pump push rod.

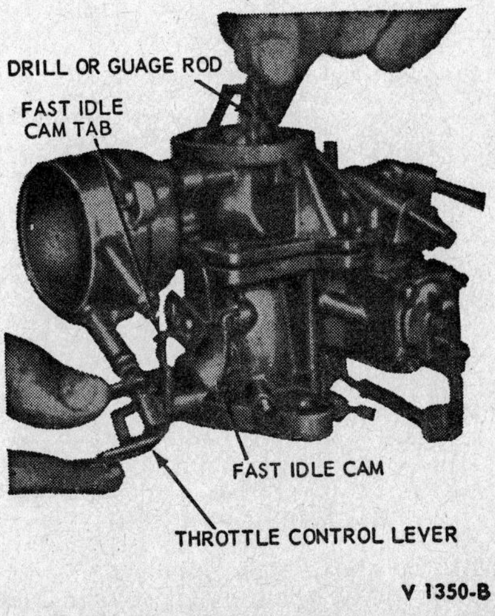

Fig. 4 De-choke adjustment

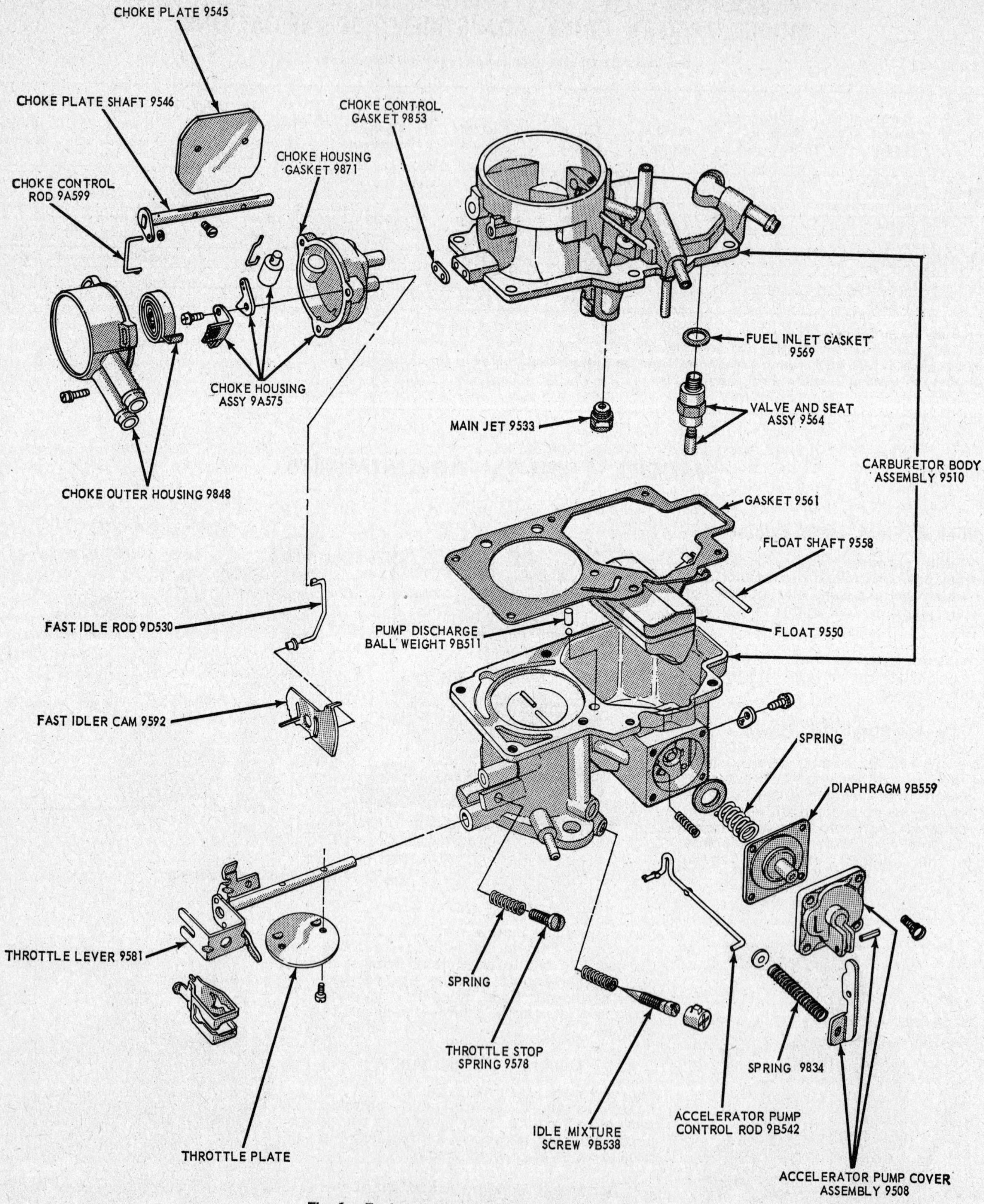

Fig. 1 Exploded view of Model 1250-IV carburetor

Fast Idle Adjustment

With the choke plate pulldown correctly adjusted and held in the pulldown position, check that the throttle lever fast idle tab is on the second step of the fast idle cam at the arrow on the fast idle cam. If necessary, bend the fast idle rod at its existing bend. Install thermostatic spring and water housing.

At normal operating temperature position the throttle lever fast idle tab on the second step of the cam and check the engine speed. Adjust if necessary by bending the tab contacting the fast idle cam.

MODEL 1100-1V CARB. ADJUSTMENT SPECIFICATIONS

See Tune Up Chart in car chapters for hot idle speeds.

Year	Carb. Model (Code 9510)	Idle Screw (Mixture) Turns Open	Float Setting (Dry)	Pump Setting	Dashpot Setting	Fast Idle Cam Linkage Clearance	Fast Idle Speed	Choke Pulldown Clearance	Dechoke Clearance ①	Automatic Choke Setting
FORD & MERCURY										
1969	C8AF-E④	②	1 3/32	.190	.080	—	1600	.280	15/64	3 Lean
	C8OF-A③	1–1½	1 3/32	.190	.080	—	1400	.150	15/64	2 Lean
	C8OF-B④	②	1 3/32	.190	2	—	1500	.130	¼	1 Lean
	C9DF-B	②	1 3/32	.150	3	—	1400	.150	¼	3 Lean
	C9OF-A	②	1 3/32	.190	.080	—	1600	.200	.160	3 Lean
	C9OF-B	②	1 3/32	.190	.080	—	1400	.200	.160	1 Lean
	C9OF-J	②	1 3/32	.190	—	—	1400	.200	.160	1 Lean
	C9OF-K	②	1 3/32	.190	—	—	1600	.200	.160	3 Lean

①—Minimum clearance between choke plate and air horn with throttle plates wide open.
②—Air/fuel ratio or idle CO% rating is found in Tune Up Specifications tables in car chapters.
③—Thermactor system.
④—Imco system.

MODEL 1100-1V ADJUSTMENTS

Figs. 1 and 2 illustrate exterior views of the different types of one-barrel carburetors while Figs. 3 and 4 illustrate the units exploded.

As shown, both types consist of two main assemblies, the upper body (air horn) and the lower (throttle) body. The upper body contains the metering components which include the main and idle fuel systems with power valve, float chamber vent and fuel inlet system. The lower body contains the fuel bowl, accelerating pump, idle mixture adjusting screw (needle).

On 1100 models a built-in hydraulic dashpot is incorporated in the lower body as shown. However, the 1101 model differs in that it has an externally mounted dashpot fastened to a bracket on the upper body.

Float Level Adjustment

Fig. 5—With air horn inverted and gasket removed, measure distance from gasket surface of air horn to top of float. If the dimension is not as listed in the *Ford Specifications Chart*, bend float arm tab as required to obtain the specified dimension.

Vent Valve Adjustment

Fig. 6—With throttle valve fully closed, groove on vent valve rod should be even with open end of vent. To adjust, bend arm on vent valve rod actuating lever (where it contacts pump lever) to align groove with edge of bore.

Pump Adjustment

Position throttle and choke valve link-

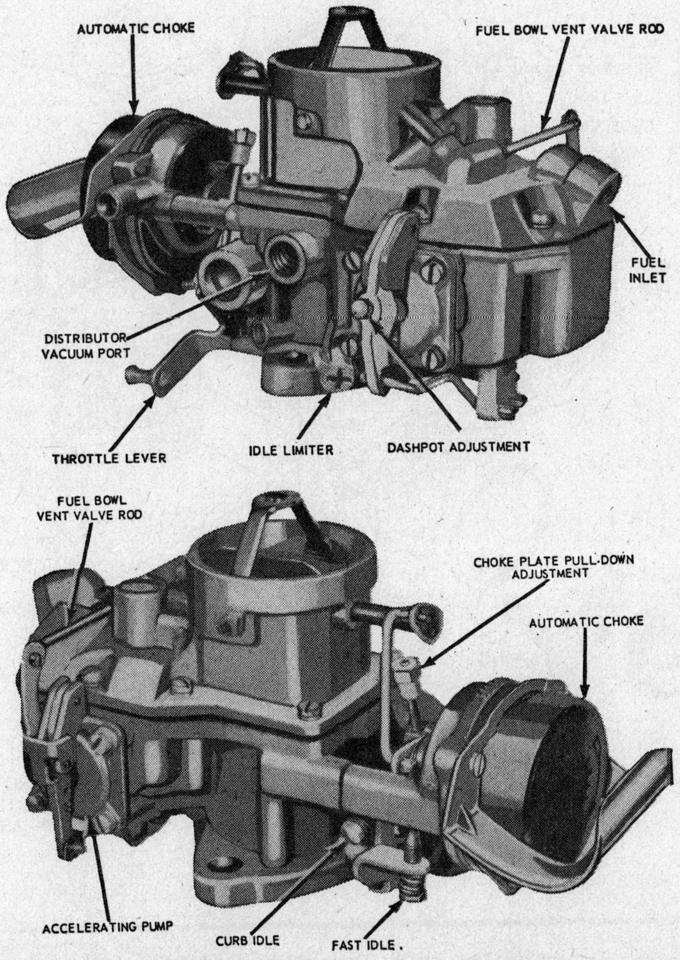

Fig. 1 Autolite 1100 carburetor with automatic choke mounted on throttle body & internal type dashpot. 1969

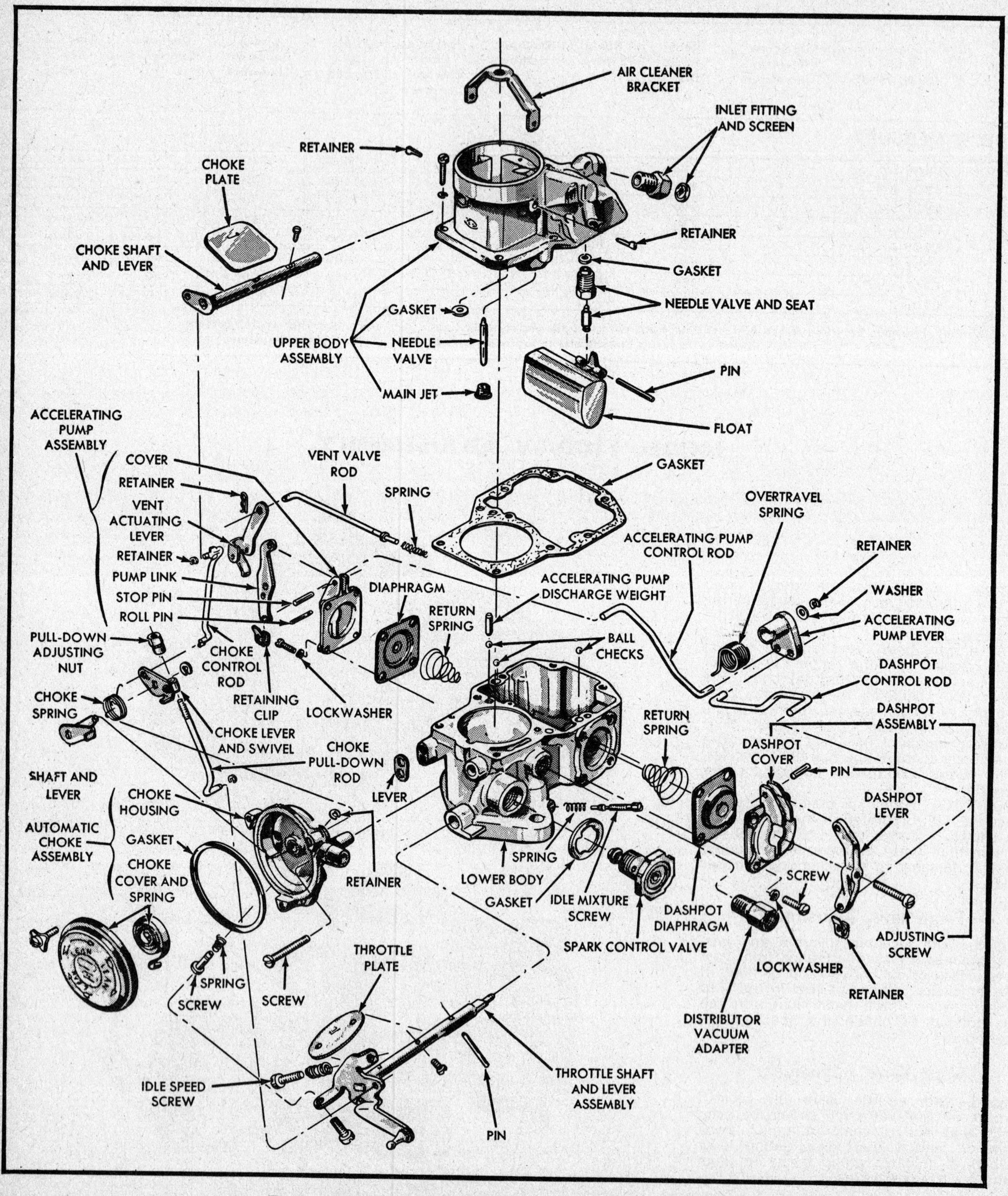

Fig. 3 Exploded view of Autolite 1100 carburetor shown in Fig. 1

age so that throttle valve will be completely closed. Hold throttle valve closed and place a gauge of the thickness listed in the *Ford Specifications Chart* between roll pin and cover surface, Fig. 7. Bend pump actuating rod to obtain specified clearance between cover and roll pin in pump lever.

Acceleration requirements in various climates are satisfied by controlling the amount of fuel discharged from the pump. The pump stroke is controlled by changing the location of the roll pin in the external type lever stop hole, Fig. 8.

For operation in temperatures of 50 degrees and below, place the roll pin in the hole marked "HI". For best performance and fuel economy at normal temperatures and high altitudes (5000 ft.) place roll pin in hole marked "LO".

Dashpot Adjustment

Internal-type, Fig. 9

With throttle position adjusted to the curb idle setting, turn the dashpot adjusting screw outward until it clears the dashpot plunger assembly. Turn the dashpot adjusting screw inward until it initially contacts the dashpot plunger assembly; then, turn the adjusting screw inward (clockwise) the specified number of turns against the dashpot diaphragm plunger assembly. Refer to *1100-1V Specifications Chart.*

External-type

Adjust the throttle position to the curb idle setting, loosen the dashpot locknut.

NOTE: Check the accelerating pump lever and stroke for proper adjustment.

On 1969 units close throttle valve and fully depress diaphragm stem and adjust dashpot.

Refer to *1100-1V Specifications Chart.* Torque the locknut to 7-10 inch lbs.

Automatic Choke Fast Idle Adjustment

Fig. 11—This adjustment refers to Fig. 1 carburetor only. Insert a gauge or drill of the size listed in the *Ford Specifications Chart* between throttle valve and carburetor bore. Close choke valve and turn fast idle adjusting screw inward until it just contacts fast idle cam.

Choke Plate Pulldown Adjustment

Fig. 12—This adjustment refers to Fig. 1 carburetor only. The fast idle adjustment must be made before making this adjustment because the position of the pulldown rod is one of the determining factors affecting throttle-to-choke opening relationship.

Place a drill or gauge of a size listed in the *Ford Specifications Chart* between choke valve and inner wall of air horn. Close choke valve on gauge or drill and hold it securely. Close throttle until fast idle screw touches fast idle cam. Adjust plastic nut to just contact swivel on choke lever.

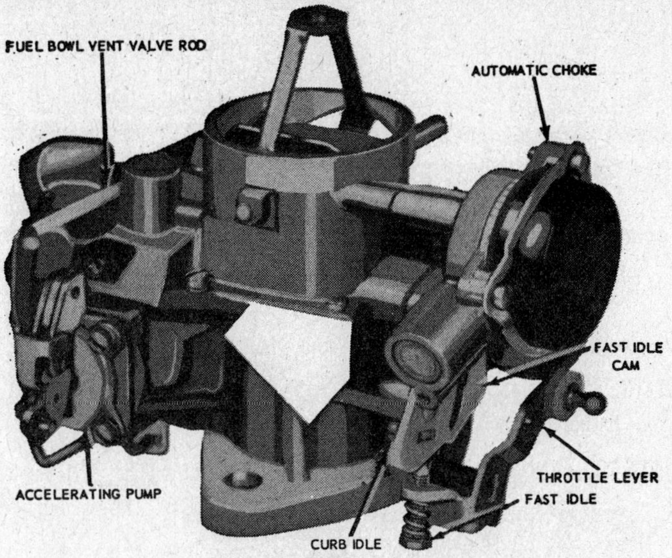

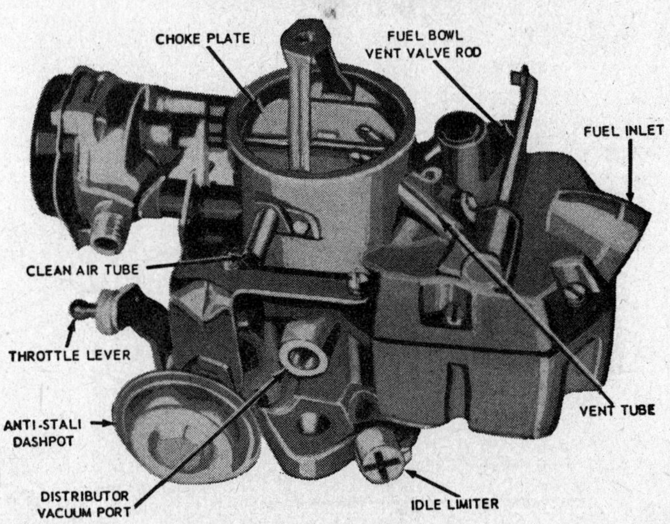

Fig. 2 Autolite 1100 carburetor with automatic choke mounted on air horn & external type dashpot. 1969

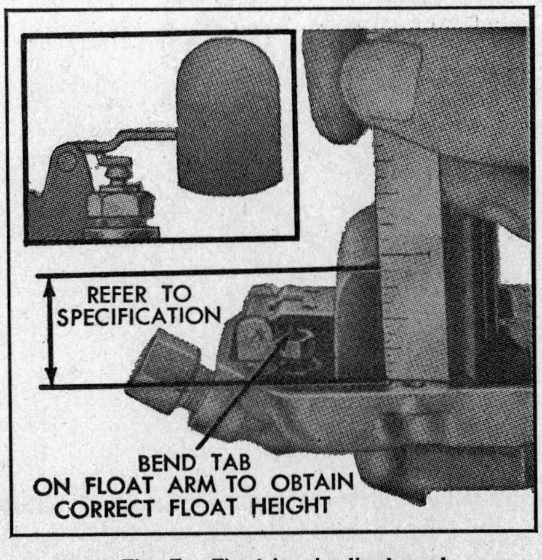

Fig. 5 Float level adjustment

Choke Pulldown & Fast Idle Linkage Adjustment

Figs. 13 and 14—These adjustments refer to Fig. 1 carburetor only.

1. Remove air cleaner and choke thermostatic spring housing from carburetor.
2. Bend a .036" wire gauge as shown in inset, Fig. 13.
3. Block throttle about half open so that fast idle cam does not contact fast idle adjusting screw.
4. Insert bent end of gauge between lower end of piston slot and upper edge of right-hand slot in choke housing, Fig. 13, and pull choke piston lever counterclockwise until gauge is snug in piston slot.
5. Hold gauge in place by exerting light pressure on choke piston lever.
6. Gradually bend rod (link) between choke piston and piston lever until choke plate opens just wide enough

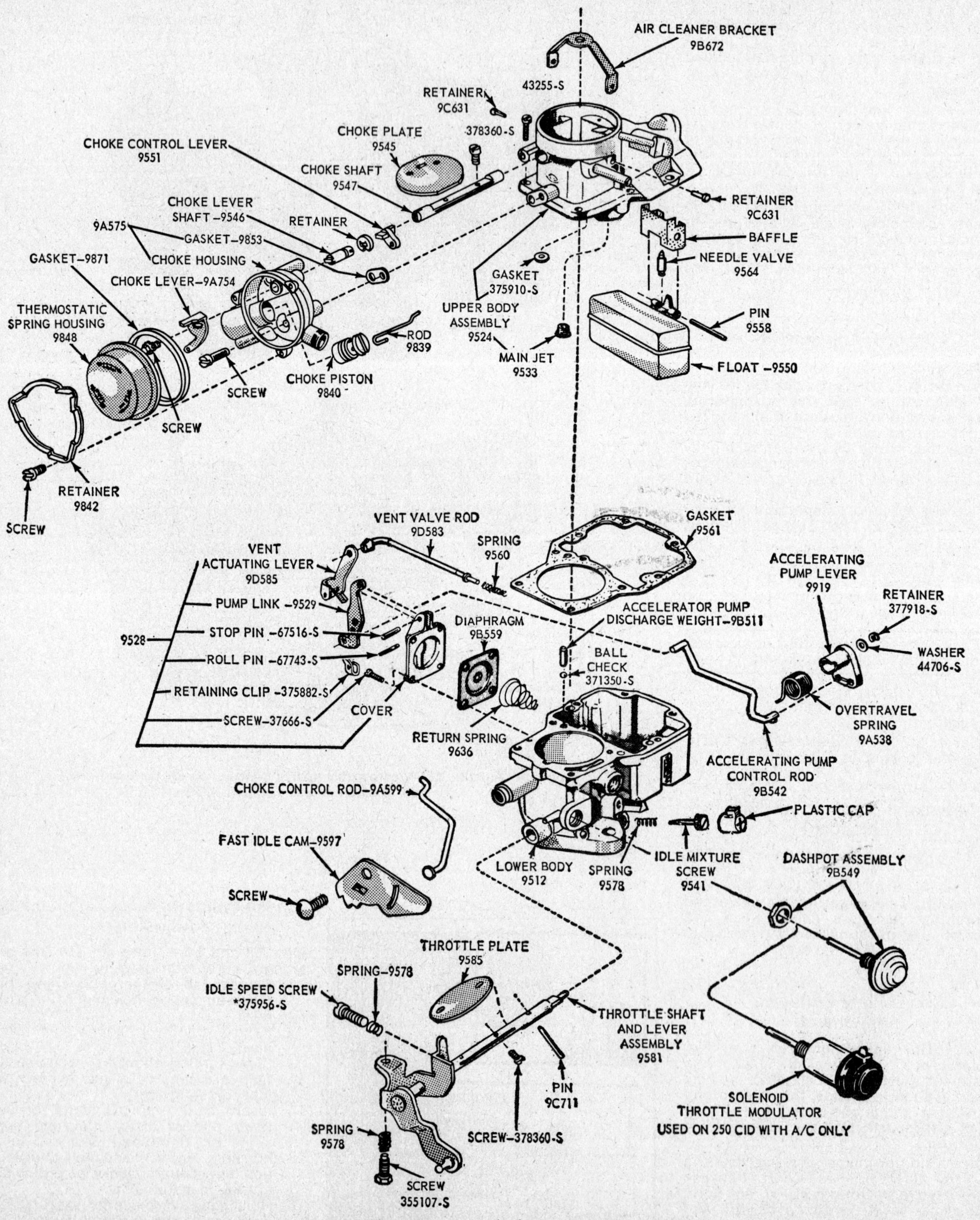

Fig. 4 Exploded view of model 1100 carburetor shown in Fig. 2

Fig. 6 **Vent valve adjustment**

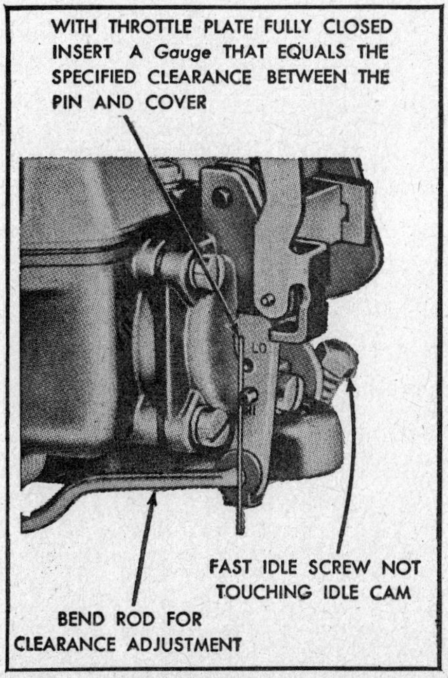

Fig. 7 **Accelerator pump adjustment**

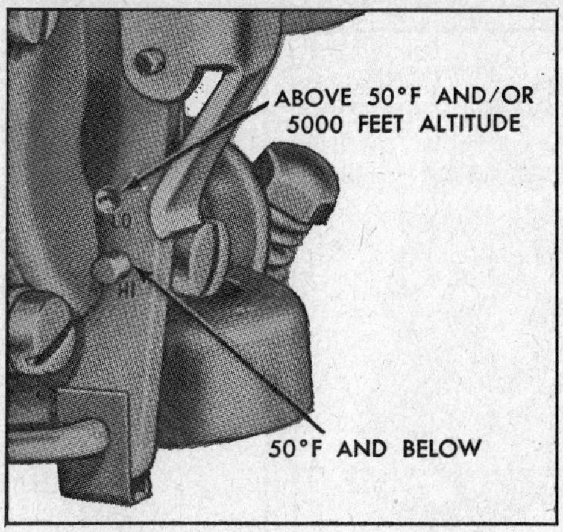

Fig. 8 **Accelerator pump lever adjustment**

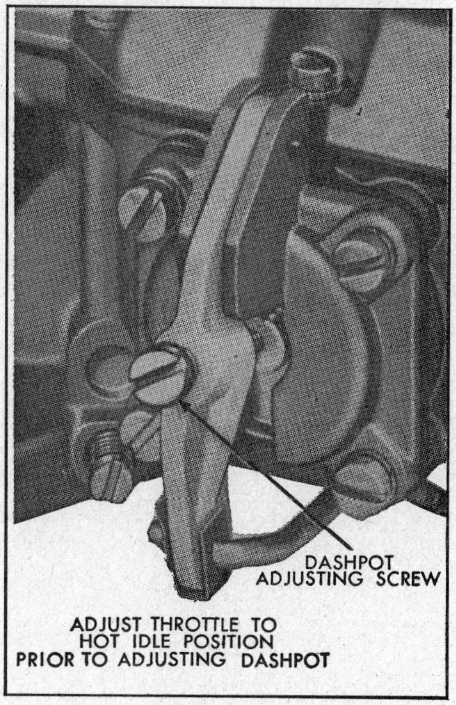

Fig. 9 **Internal type anti-stall dashpot adjustment**

to allow gauge or drill of the specified size between front of choke plate and air horn, Fig. 13 (see *Ford Specifications Chart*).

7. Install thermostatic spring housing and gasket on choke housing and secure with clamp and screws.
8. Rotate spring housing counterclockwise (rich direction) to align center index mark on choke housing with index mark on spring housing; then rotate spring housing 90 degrees counterclockwise.
9. Position fast idle adjusting screw on index mark on fast idle cam, Fig. 14.
10. Adjust fast idle cam linkage to speci-

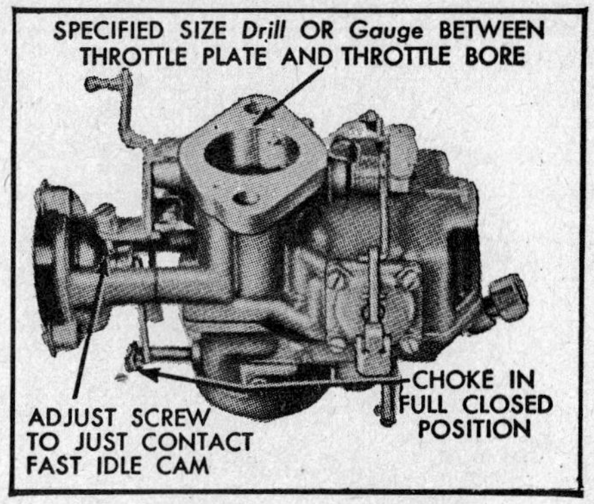

Fig. 11 **Automatic choke fast idle adjustment (Fig. 1)**

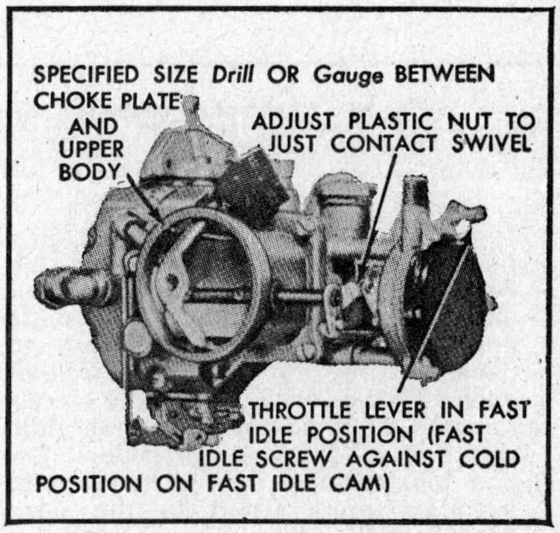

Fig. 12 **Automatic choke linkage adjustment (Fig. 1)**

Fig. 15 **Automatic choke setting**

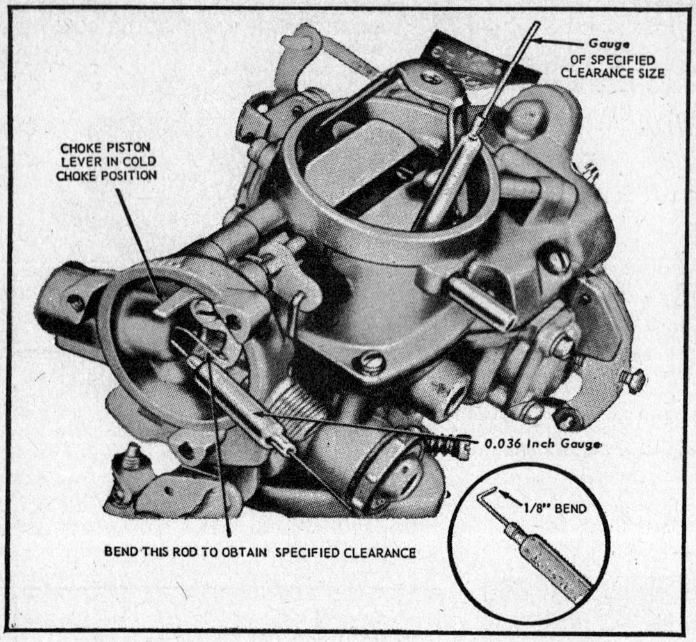

Fig. 13 Choke plate clearance adjustment (Fig. 2)

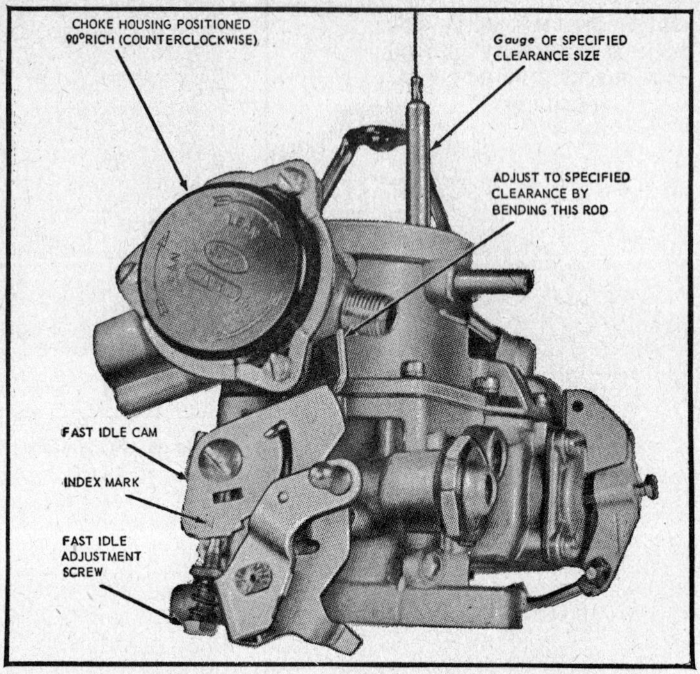

Fig. 14 Fast idle cam linkage adjustment (Fig. 2)

fication by bending choke control rod, Fig. 14, to provide the specified clearance between front of choke plate and air horn. Bend rod inward to decrease (outward to increase) clearance. Make certain fast idle screw remains on index mark of fast idle cam during adjustment procedure.

11. Set thermostatic spring housing to specified mark listed in the *Ford Specifications Chart.*

NOTE: If the foregoing adjustments were made with the carburetor installed on the engine, adjust engine idle speed and mixture, fast idle speed and dashpot as outlined above.

Automatic Choke Setting

Fig. 15—Loosen retaining screws and set thermostatic spring housing to the mark specified in the *Ford Specifications Chart* and tighten retaining screws.

MODEL 1940-1V CARB. SPECIFICATIONS

See Tune Up Chart in car chapters for hot idle speeds.

Carb. Model	Float Setting (Dry)	Pump Setting	Fast Idle Cam	Vacuum Kick	Choke Unloader	Automatic Choke Adjustment	Dashpot Setting	Fast Idle Speed
FORD & MERCURY								
DOPF-A	①	②	5/64	3/16	5/32	Index	—	2100
DOPF-C	①	②	5/64	3/16	5/32	Index	—	2100
DOPF-D	①	②	5/64	3/16	5/32	Index	3/32	2100
DOPF-E	①	②	③	—	④	—	—	—
DOPF-F	①	②	③	—	④	—	—	—
DOPF-G	①	②	③	—	④	—	3/32	—
DOPF-H	①	②	③	—	④	—	3/32	—
DOPF-J	①	②	③	—	④	—	3/32	—
DOPF-K	①	②	1/16	5/32	1/8	Index	3/32	2100
DOPF-L	①	②	1/16	5/32	1/8	Index	3/32	—
DOPF-M	①	②	—	—	—	—	3/32	—
DOPF-N	①	②	—	—	—	—	3/32	—
DOPF-R	①	②	—	—	—	—	3/32	—
DOPF-T	①	②	—	—	—	—	3/32	—

①—Toe in float, flush with gasket surface of main body.
②—27/32″ from centerline of pump link hole in pump rod to edge of 5/16″ boss on main body.
③—No vacuum, second step.
④—At wide open throttle.

MODEL 1940-1V ADJUSTMENTS

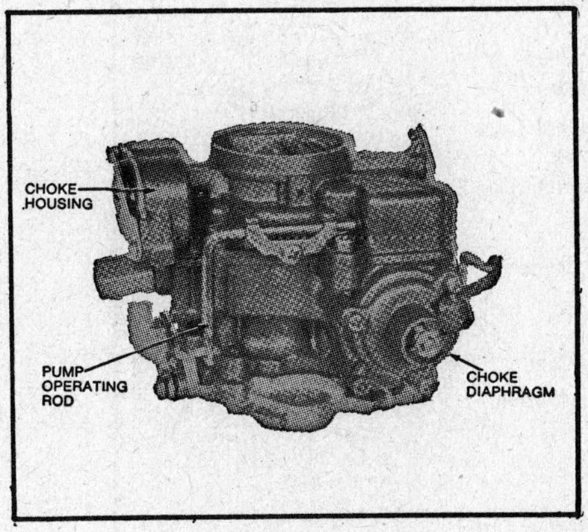

The model 1940-1V, Figs. 1A, 2A, 3A, is a concentric unit, that is, the fuel bowl completely surrounds the venturi, and is available with either manual or automatic choke.

Principal sub-assemblies include a bowl cover, carburetor body and throttle body. A thick gasket between the throttle body and carburetor body retards heat transfer to the fuel to resist percolation.

In addition to the usual circuits, this unit incorporates a spark valve. This is a diaphragm operated, spring loaded valves in the main body connected to the throttle bore port. When the spark valve closes off the port, spark advance vacuum is decreased, retarding the ignition timing.

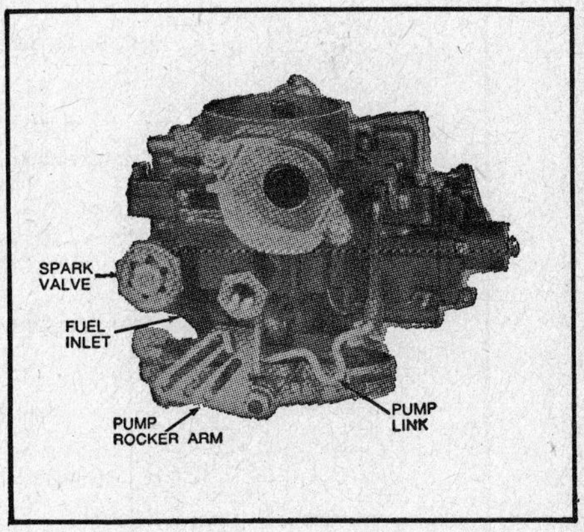

Float Level Adjustment

Fig. 4A—Install float assembly in float shaft cradle. Insert retaining spring and while holding with fingers, invert the bowl. A straight edge placed across the surface of the bowl should just touch the toes of the float. If necessary, bend the float tang to obtain this adjustment.

Pump Adjustment

Fig. 5A—With throttle in the curb idle position, the distance from the vacuum passage casting to the center of the hole in the pump operating rod should be as listed in specifications.

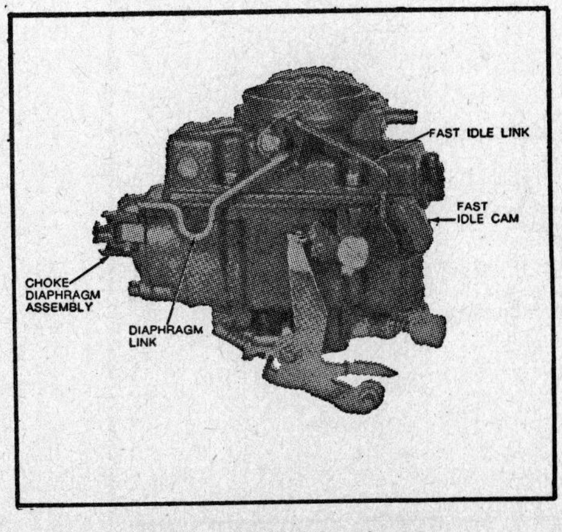

Fig. 1A 1940-1V carburetor with automatic choke

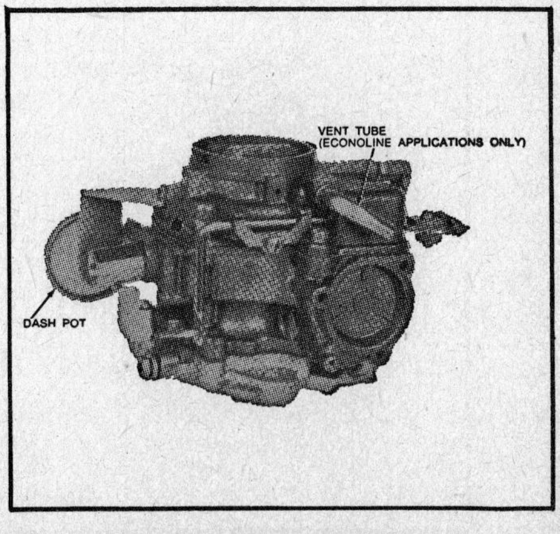

Fig. 2A 1940-1V carburetor with manual choke

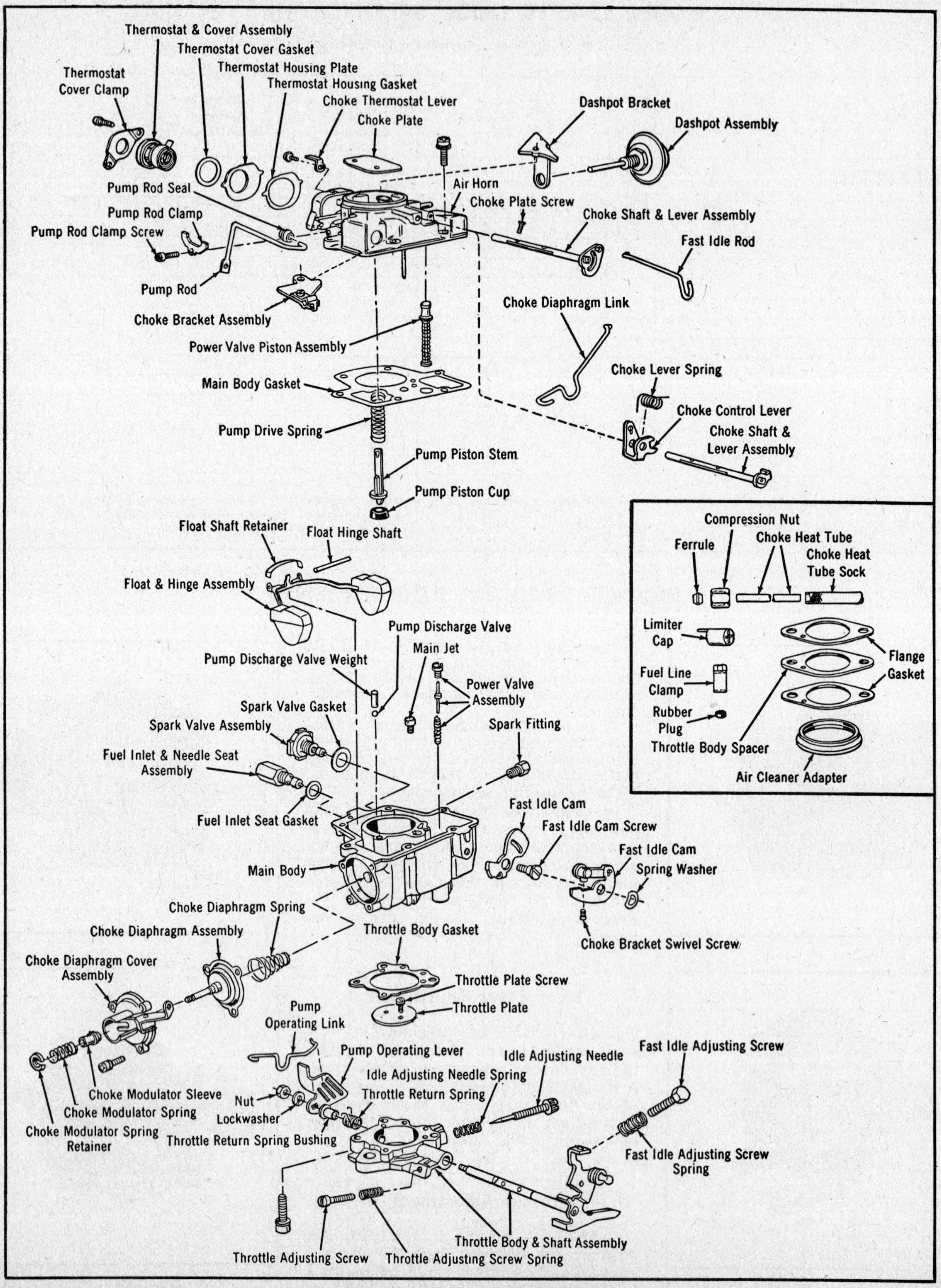

Fig. 3A 1940-1V carburetor. Exploded

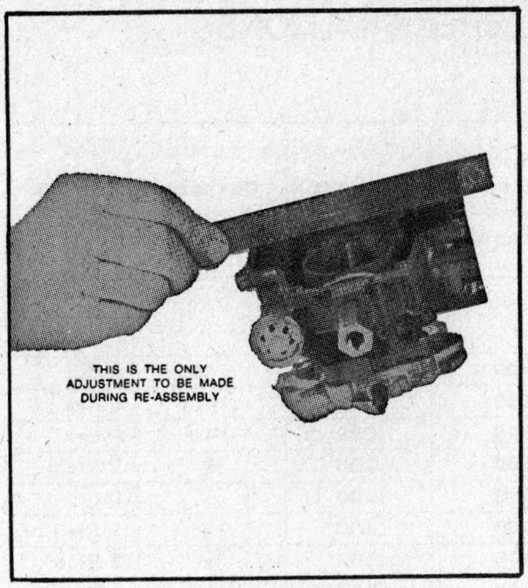

Fig. 4A Checking float setting

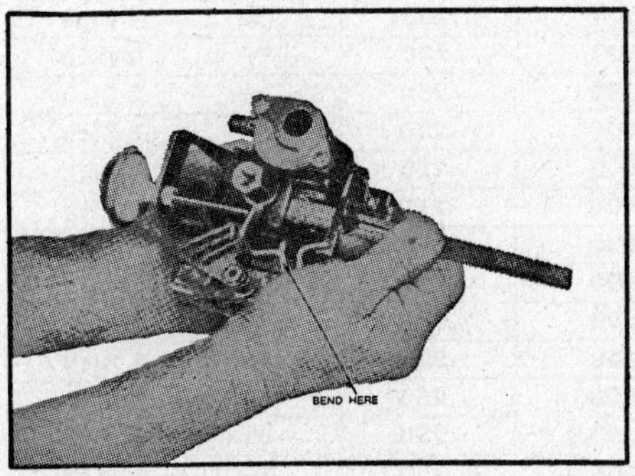

Fig. 5A Pump piston stroke adjustment

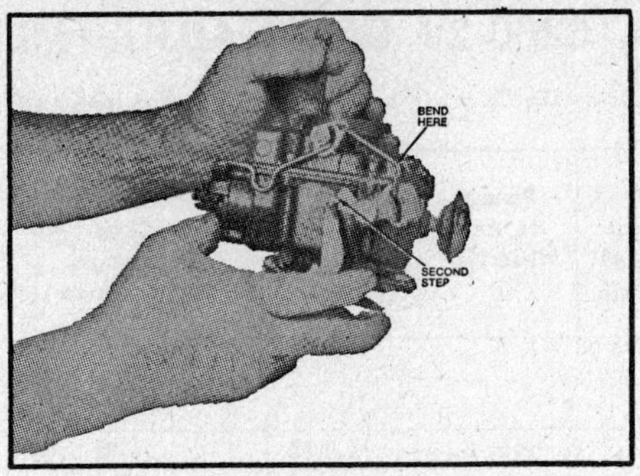

Fig. 6A Fast idle cam adjustment

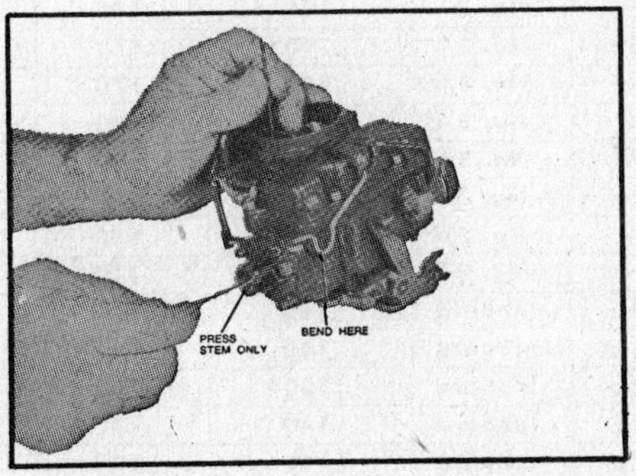

Fig. 7A Vacuum kick adjustment

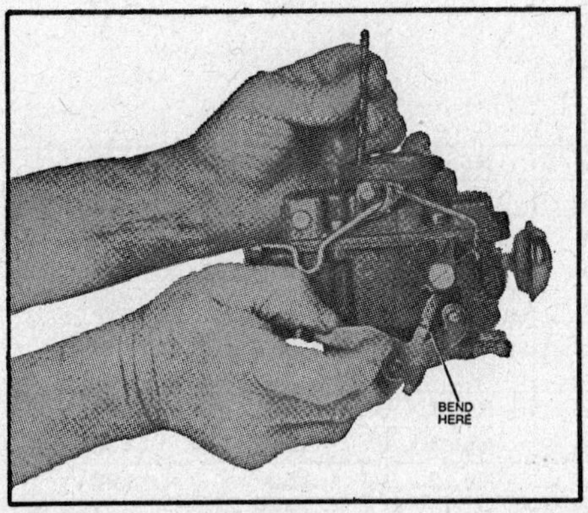

Fig. 8A Unloader adjustment. Type A tang

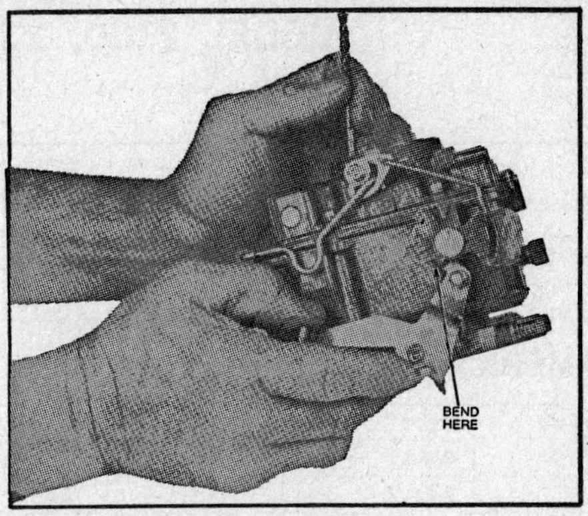

Fig. 9A Unloader adjustment. Type B tang

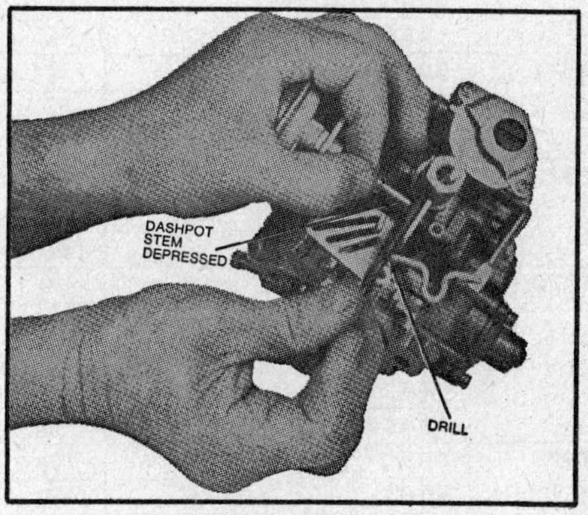

Fig. 10A Dashpot adjustment

Fast Idle Cam

Fig. 6A—With throttle lever contacting the second highest speed step on the fast idle cam, move the choke valve toward closed position with light pressure on choke shaft lever.

Insert specified drill between choke valve and wall of air horn. Adjust by bending fast idle connector rod.

Vacuum Kick

Fig. 7A—Depress diaphragm with a small drift. Insert specified drill between choke valve and wall of air horn while holding pressure on choke lever. Note that the diaphragm extends as spring is compressed. Spring must be fully compressed for proper measurement.

Choke Unloader

Figs. 8A, 9A—With throttle valve in wide open throttle position, insert specified drill between upper edge of choke valve and inner wall of air horn. With finger lightly pressing against shaft lever, a slight drag should be felt as drill is withdrawn. Bend unloader tang on throttle if necessary.

Dashpot

Fig. 10A—With curb idle speed properly adjusted and the engine at idle, depress the dashpot. Insert specified drill between end of stem and tang.

FORD AUTOLITE/MOTORCRAFT CARBURETORS

MODEL 2100, 2150, 6200-2V CARB. ADJUSTMENT SPECIFICATIONS

See Tune Up Chart in car chapters for hot idle speeds.

Year	Carb. Model (Code 9510) ①	Idle Mixture Turns Open	Float Level (Dry)	Fuel Level (Wet)	Pump Setting Hole No. ④	Choke Plate Clearance (Pull down)	Fast Idle Cam Linkage Clearance	Fast Idle Speed (Hot Engine)	Dechoke Clearance ②	Dashpot Setting	Choke Setting
AMERICAN MOTORS											
1969	9HM2	2	1/2	13/16	No. 3	.125	.120	1600	.080	.140	Index
	9HA2	2	1/2	13/16	No. 3	.140	.120	1600	.080	.140	Index
	9ZA2	2	1/2	13/16	No. 3	.140	.120	1600	.080	.140	Index
1970	0DA2	③	3/8	13/16	No. 3	.300	.170	1600	.200	1/8	2 Rich
	0DM2	③	3/8	13/16	No. 3	.260	.170	1600	.200	1/8	Index
	0RA2	③	3/8	13/16	No. 3	.350	.170	1600	.200	1/8	1 Rich
1971	1DA2	③	3/8	13/16	No. 3	.190	.170	1600	.200	1/8	2 Rich
	1DM2	③	3/8	13/16	No. 3	.190	.170	1600	.200	1/8	1 Rich
	1RA2	③	3/8	13/16	No. 3	.190	.170	1600	.200	1/8	2 Rich
1972	2DA2	③	3/8	3/4	No. 3A	.130	.120	1600	.200	9/64	2 Rich
	2DM2	③	3/8	3/4	No. 3A	.140	.130	1600	.200	7/64	1 Rich
	2RA2	③	3/8	3/4	No. 3A	.130	.120	1600	.200	—	2 Rich
1973	3DA2	③	3/8	3/4	No. 3A	.120	.110	1600	.250	—	2 Rich
	3DM2	③	3/8	3/4	No. 3A	.130	.130	1600	.250	9/64	1 Rich
	3RA2	③	3/8	3/4	No. 3A	.120	.110	1600	.250	—	2 Rich
1974	4DA2	③	13/32	25/32	Inboard	.140	.130	1600	.250	.140	1 Rich
	4DA2-E	③	13/32	25/32	Inboard	.140	.130	1600	.250	.140	1 Rich
	4DM2	③	13/32	25/32	Inboard	.130	.130	1600	.250	.140	2 Rich
	4RA2	③	13/32	25/32	Inboard	.140	.130	1600	.250	.140	1 Rich
	4RAC2	③	13/32	25/32	Inboard	.140	.130	1600	.250	.140	1 Rich
1975	5DA2	③	13/32	3/4	—	.140	.130	1600⑦	.250	—	1 Rich
	5DMS	③	13/32	3/4	—	.130	.130	1600⑦	.250	.093	2 Rich
	5RAS	③	13/32	3/4	—	.140	.130	1600⑦	.250	—	1 Rich
1976	6DA2	③	13/32	3/4	—	.140	.130	1600⑦	.250	—	1 Rich
	6DM2	③	13/32	15/16	—	.140	.120	1600⑦	.250	.093	2 Rich
	6RA2	③	13/32	3/4	—	.140	.130	1600⑦	.250	—	1 Rich
FORD AND MERCURY											
1969	C8AF-BD	③	3/8	3/4	No. 3	.130	.110	1400	.060	1/8	2 Rich
	C9AF-A	③	3/8	3/4	No. 2	.120	.110	1600	.060	1/8	Index
	C9AF-B	③	31/64	7/8	No. 3	.210	.170	1300	.060	1/8	1 Rich
	C9AF-C	③	31/64	7/8	No. 3	.130	.100	1500	.060	1/8	2 Rich
	C9AF-J	③	31/64	7/8	No. 3	.130	.100	1500	.060	1/8	2 Rich
	C9OF-C	③	31/64	7/8	No. 3	.120	.100	1600	.060	—	2 Rich
	C9ZF-A	③	9/16	15/16	No. 3	.150	.130	1300	.060	7/64	1 Rich
	C9ZF-B	③	31/64	7/8	No. 3	.120	.100	1600	.060	—	2 Rich
	C9ZF-G	③	3/8	3/4	No. 3	.120	.110	1600	.060	1/8	Index
	C9MF-A	③	31/64	7/8	No. 2	.150	.120	1500	.060	1/8	2 Rich
1970	D0AF-C	③	7/16	13/16	No. 3	.150	.130	1400	.060	—	1 Rich
	D0AF-D	③	7/16	13/16	No. 3	.150	.130	1500	.060	1/8	1 Rich
	D0AF-E	③	7/16	13/16	No. 3	.230	.190	1300	.060	—	2 Lean
	D0AF-F	③	7/16	13/16	No. 3	.200	.170	1600	.060	1/8	2 Lean
	D0AF-G	—	7/16	13/16	No. 3	.210	.170	1400	.060	1/8	2 Lean
	D0AF-J	③	7/16	13/16	No. 3	.200	.160	1400	.060	1/8	1 Rich
	D0AF-T	③	7/16	13/16	No. 3	.200	.160	1400	.060	1/8	—
	D0AF-U	③	7/16	13/16	No. 3	.150	.130	1500	.060	—	—
	D0AF-V	③	7/16	13/16	No. 3	.200	.170	1600	—	—	1 Rich
	D0AF-Y	③	7/16	13/16	No. 3	.210	.170	1400	.060	—	2 Lean
	D0AF-Z	③	7/16	13/16	No. 3	.200	.160	1500	.060	1/8	1 Rich
	D0AF-AA	③	7/16	13/16	No. 3	.200	.160	1500	.060	—	—
	D0AF-AU	—	7/16	13/16	No. 3	.200	.170	1600	.060	—	—
	D0AF-AV	—	7/16	13/16	No. 3	.200	.170	1600	.060	1/8	2 Lean
	D0OF-K	③	7/16	13/16	No. 3	.220	.190	1500	—	—	2 Lean

Continued

MODEL 2100, 2150 6200-2V CARB. ADJUSTMENT SPECIFICATIONS—Continued

See Tune Up Chart in car chapters for hot idle speeds.

Year	Carb. Model (Code 9510) ①	Idle Mixture Turns Open	Float Level (Dry)	Fuel Level (Wet)	Pump Setting Hole No. ④	Choke Plate Clearance (Pull down)	Fast Idle Cam Linkage Clearance	Fast Idle Speed (Hot Engine)	Dechoke Clearance ②	Dashpot Setting	Choke Setting
FORD AND MERCURY—Continued											
1970	D0OF-L	③	7/16	13/16	No. 3	.190	.130	1500	—	1/8	1 Rich
	D0OF-M	③	7/16	13/16	No. 3	.190	.130	1500	—	—	1 Rich
	D0OF-T	—	7/16	13/16	No. 3	.190	.160	1500	.060	1/8	1 Rich
	D0OF-U	—	7/16	13/16	No. 3	.220	.190	1500	.060	—	Index
	D0OF-V	—	7/16	13/16	No. 3	.190	.160	1500	.060	—	1 Rich
1971	D1AF-BA	③	7/16	13/16	No. 2	.150	.130	1500	.060	—	Index
	D1AF-DA	③	7/16	13/16	No. 3	.170	.150	1400	.060	—	1 Rich
	D1AF-FA	③	7/16	13/16	No. 3	.220	.190	1300	.060	—	1 Rich
	D1AF-JA	③	7/16	13/16	No. 3	.190	.130	1600	.060	1/8	Index
	D1AF-KA	③	7/16	13/16	No. 3	.190	.130	1600	.060	—	Index
	D1MF-FA	③	7/16	13/16	No. 3	.200	.160	1400	.060	1/8	1 Rich
	D1MF-JA	③	7/16	13/16	No. 3	.190	.160	1500	.060	1/8	1 Rich
	D1MF-KA	③	7/16	13/16	No. 3	.190	.160	1500	.060	1/8	1 Rich
	D1OF-ABA	③	7/16	13/16	No. 3	.170	.150	1400	.060	—	1 Rich
	D1OF-PA	③	7/16	13/16	No. 3	.230	.190	1500	.060	—	Index
	D1OF-RA	③	7/16	13/16	No. 3	.200	.170	1500	.060	—	1 Rich
	D1OF-YA	③	7/16	13/16	No. 3	.200	.170	1500	.060	—	1 Rich
	D1OF-ZA	③	7/16	13/16	No. 3	.230	.190	1500	.060	—	Index
	D1YF-DA	③	7/16	13/16	No. 3	.200	.160	1500	.060	—	Index
	D1ZF-AA	③	7/16	13/16	No. 2	.150	.130	1500	.060	—	Index
	D1ZF-SA	③	7/16	13/16	No. 3	.200	.170	1500	.060	—	1 Rich
	D1ZF-UA	③	7/16	13/16	No. 3	.200	.170	1500	.060	—	1 Rich
1972	D2AF-FB	③	7/16	13/16	No. 3	.140	.130	1500	.030	1/8	Index
	D2AF-GB	③	7/16	13/16	No. 3	.140	.130	1500	.030	1/8	Index
	D2AF-HA	③	7/16	13/16	No. 2	.150	.130	1400	.060	1/8	1 Rich
	D2AF-UC	③	7/16	13/16	No. 3	.170	.150	1500	.060	—	Index
	D2GF-AA	③	7/16	13/16	No. 2	.150	.130	1400	.060	1/8	1 Rich
	D2GF-BA	③	7/16	13/16	No. 2	.150	.130	1400	.060	—	1 Rich
	D2MF-FB	③	7/16	13/16	No. 4	.180	.150	1500	.060	—	1 Rich
	D2MF-FE	③	7/16	13/16	No. 4	.170	.150	1500	.060	—	Index
	D2MF-FF	③	7/16	13/16	No. 4	.155	.150	1500	.060	—	Index
	D2OF-KA	③	7/16	13/16	No. 2	.140	.130	1400	.060	—	1 Rich
	D2OF-UB	③	7/16	13/16	No. 3	.190	.160	⑤	.030	—	⑥
	D2WF-CA	③	7/16	13/16	No. 3	.190	.160	⑤	.030	—	⑥
	D2ZF-FA	③	7/16	13/16	No. 2	.150	.130	1400	.060	—	1 Rich
	D2ZF-LA	③	7/16	13/16	No. 3	.240	.210	1500	.030	—	1 Rich
1973	D3AF-CE	③	7/16	13/16	No. 2	—	—	1500	—	—	2 Rich
	D3AF-DC	③	7/16	13/16	No. 3	—	—	1500	—	—	3 Rich
	D3GF-AF	③	7/16	13/16	No. 2A	—	—	1400	—	—	3 Rich
	D3GF-BA	③	7/16	13/16	No. 2A	—	—	—	—	—	3 Rich
	D3MF-AE	③	7/16	13/16	No. 3	—	—	1500	—	—	3 Rich
	D3ZF-EA	③	7/16	13/16	No. 2A	—	—	1400	—	—	1 Rich
	D3GF-BB	③	7/16	13/16	No. 2A	—	—	1250	—	—	1 Rich
	D3AF-KA	③	7/16	13/16	No. 3	—	—	1500	—	—	3 Rich
	D3MF-BA	③	7/16	13/16	No. 3	—	—	1500	—	—	3 Rich
1974	D4AE-DA	③	7/16	13/16	No. 2A	—	—	—	—	—	1 Rich
	D4AE-EA	③	7/16	13/16	No. 2A	—	—	—	—	—	3 Rich
	D4AE-FA	③	7/16	13/16	No. 3A	—	—	—	—	—	3 Rich
	D4AE-GA	③	7/16	13/16	No. 3A	—	—	—	—	—	3 Rich
	D4AE-HB	③	7/16	13/16	No. 3A	—	—	—	—	—	3 Rich
	D4AE-KA	③	7/16	13/16	No. 2A	—	—	—	—	—	3 Rich
	D4DE-RB	③	7/16	13/16	No. 2	—	—	—	—	—	3 Rich
	D4DE-LA	③	7/16	13/16	No. 2	—	—	—	—	—	3 Rich
	D4DE-NB	③	7/16	13/16	No. 2	—	—	—	—	—	3 Rich

Continued

MODEL 2100, 2150 6200-2V CARB. ADJUSTMENT SPECIFICATIONS—Continued

See Tune Up Chart in car chapters for hot idle speeds.

Year	Carb. Model (Code 9510) ①	Idle Mixture Turns Open	Float Level (Dry)	Fuel Level (Wet)	Pump Setting Hole No. ④	Choke Plate Clearance (Pull down)	Fast Idle Cam Linkage Clearance	Fast Idle Speed (Hot Engine)	Dechoke Clearance ②	Dashpot Setting	Choke Setting
FORD AND MERCURY—Continued											
1974	D4DE-PA	③	7/16	13/16	No. 2	—	—	—	—	—	3 Rich
	D4DE-VA	③	7/16	13/16	No. 2A	—	—	—	—	—	3 Rich
	D4OE-CA,EA	③	7/16	13/16	No. 2	—	—	—	—	—	3 Rich
	D4OE-FA,PA	③	7/16	13/16	No. 2A	—	—	—	—	—	3 Rich
	D4ME-BA,CA	③	7/16	13/16	No. 3A	—	—	—	—	—	3 Rich
1975	D5AE-AA,EA	③	7/16	13/16	3	.125	—	1500	—	—	3 Rich
	D5DE-AA,BA	③	7/16	13/16	2	.140	—	⑧	—	—	3 Rich
	D5DE-AEA	③	7/16	13/16	2	—	—	⑧	—	—	3 Rich
	D5DE-AFA	③	7/16	13/16	2	—	—	⑧	—	—	3 Rich
	D5DE-HA	③	7/16	13/16	2	.140	—	1400	—	—	3 Rich
	D5DE-JA	③	7/16	13/16	2	.140	—	⑧	—	—	3 Rich
	D5DE-UA	③	7/16	13/16	2	.140	—	1400	—	—	3 Rich
	D5ME-BA,FA	③	7/16	13/16	2	.125	—	1500	—	—	3 Rich
	D5OE-AA	③	7/16	13/16	2	.140	—	1400	—	—	3 Rich
	D5OE-BA,CA	③	7/16	13/16	3	.125	—	1350	—	—	3 Rich
	D5OE-DA	③	7/16	13/16	2	.140	—	1400	—	—	3 Rich
	D5OE-GA	③	7/16	13/16	2	.125	—	1350	—	—	3 Rich
	D5WE-FA	③	7/16	13/16	2	—	—	⑧	—	—	3 Rich
	D5ZE-AC,BC	③	3/8	3/4	2	.145	—	1600	—	—	2 Rich
	D5ZE-BE	③	3/8	3/4	2	.105	.125	1600	—	—	3 Rich
	D5ZE-CC, DC	③	3/8	3/4	2	.145	—	1600	—	—	2 Rich
	D5ZE-JA	③	7/16	13/16	2	.140	—	⑧	—	—	3 Rich
1976	D6AE-HA	③	7/16	13/16	2	.160	.180	1500	—	—	3 Rich
	D6ME-AA	③	7/16	13/16	2	.160	.180	1500	—	—	3 Rich
	D6OE-AA,BA	③	7/16	13/16	3	.160	.180	1400	—	—	3 Rich
	D6OE-CA	③	7/16	13/16	3	.160	.180	1400	—	—	3 Rich
	D6WE-AA	③	7/16	13/16	2	.160	.180	1350	—	—	3 Rich
	D6WE-BA	③	7/16	13/16	2	.160	.180	1350	—	—	2 Rich
	D6ZE-AA,BA	③	3/8	3/4	2	.100	.120	1600	—	—	3 Rich
	D6ZE-CA	③	13/32	3/4	2	.110	.130	1600	—	—	3 Rich
	D6ZE-DA	③	3/8	3/4	3	.110	.130	1600	—	—	3 Rich
	D6ZE-JA	③	7/16	13/16	2	.140	.160	⑧	—	—	3 Rich

①—Stamped on left side of fuel bowl or on tag attached to bowl cover.
②—Minimum clearance between choke plate and air horn wall with throttle plates wide open.
③—Air/fuel ratio or idle CO% rating is found in Tune Up Specification tables in car chapters.
④—With link in inboard hole in pump lever.
⑤—Auto Trans. 1500; Manual Trans. 1400.
⑥—California 1 Rich; all others 2 Rich.
⑦—Hot on 2nd step of cam with TCS solenoid & EGR disconnected.
⑧—Manual trans.; 2000. Auto. trans. on high step of cam; 2100.

MODEL 4100-4V ADJUSTMENT SPECIFICATIONS

See Tune Up Chart in car chapters for hot idle speeds.

Year	Carb. Model (Code 9510) ①	Idle Screws (Mixture) Turns Open	Float Level (Dry)	Fuel Level (Wet)	Pump Setting (Hole No.) ④	Choke Plate Clearance (Pull down)	Fast Idle Cam Linkage Clearance	Fast Idle Speed (Hot Engine)	Secondary Throttle Plate Clearance ②	Dechoke Clearance ③	Dashpot Setting	Choke Setting
1969	C8AF-AE	⑤	⑥	⑦	No. 3	.140	.120	1350	—	.060	7/64	2 Rich

①—Stamped on left side of primary bowl cover or tag attached to blow cover.
②—Additional turns in after screw contacts lever.
③—Minimum clearance between choke plate and air horn with primary throttle plates wide open.
④—With link in inboard hole in pump lever.
⑤—Air/fuel ratio or idle CO% rating is found in Tune Up Specification tables in car chapters.
⑥—Primary floats 17/32″, secondary 1 1/16″.
⑦—Primary floats 29/32″, secondary 1 1/16″.

2100, 2150, 6200-2V & 4100-4V ADJUSTMENTS

Models 2100, 2150, 6200-2V, Figs. 1 thru 4

These carburetors have two main bodies—the air horn and throttle body. The air horn assembly, which serves as a cover for the throttle body, contains the choke plate and vents for the fuel bowl. On all 1969 units and 1972 and 1975 Ford units, an external bowl vent valve is used. On 2150 units installed on V-6 engines, the air horn assembly contains a fuel deceleration system which consists of a metered pickup orifice in the fuel bowl and air/fuel mixing orifices and bleeds.

A choke modulator assembly is incorporated in the 1970-74 units, Fig. 5. While a staged choke system is an added feature used on 1972 Ford installations, Fig. 2. This system, through the use of a bimetal sensor and a series of diaphragms, pulls open the choke plate within 15-60 seconds. The system operates only during times when underhood temperatures are above approximately 60 degrees F. The 1973 units have an additional port for the Exhaust Gas Recirculation (EGR) system, Fig. 3. On 1973-74 Ford units an electric choke system is incorporated which opens choke plate within 1-1½ minutes when underhood temperatures are above approximately 55° to 60°. For service information refer to the "Emission Control System" chapter.

The throttle plate, accelerating pump, power (enrichment) on some valve and fuel bowl are in the throttle body. The choke housing is attached to the throttle body.

The two bodies each contain a main and booster venturi, main fuel discharge, accelerating pump discharge, idle fuel discharge, and a throttle plate. On some 2100 and 6200 units, an antistall dashpot is attached to the carburetor when the

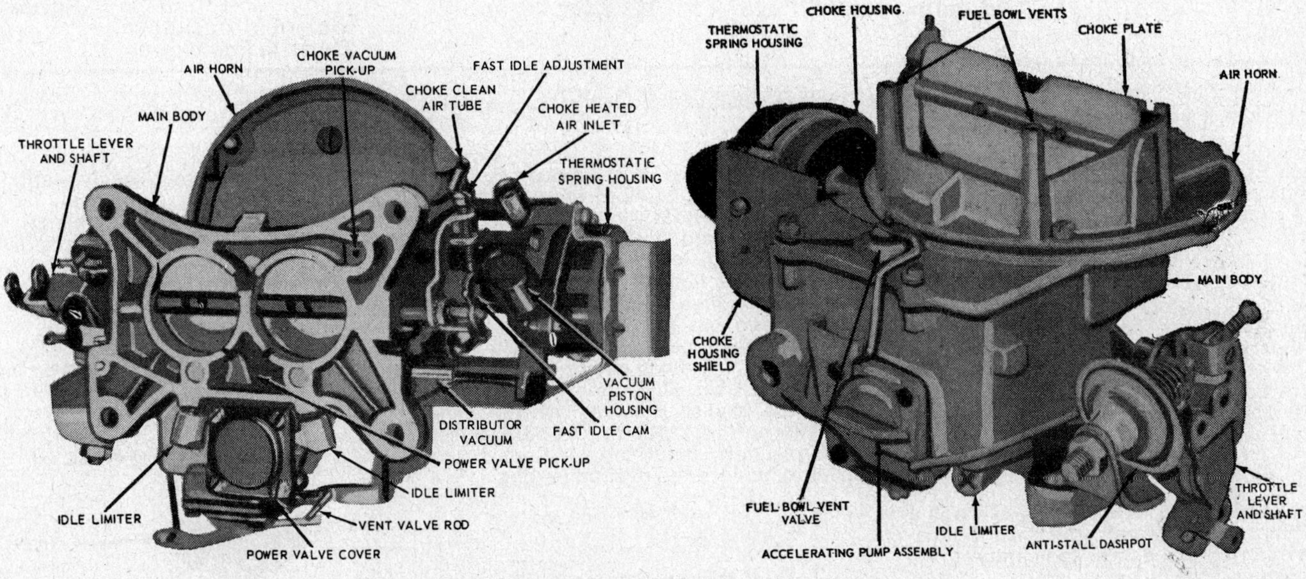

Fig. 1 Typical Autolite/Motorcraft 2100, 6200 carburetor. 1969

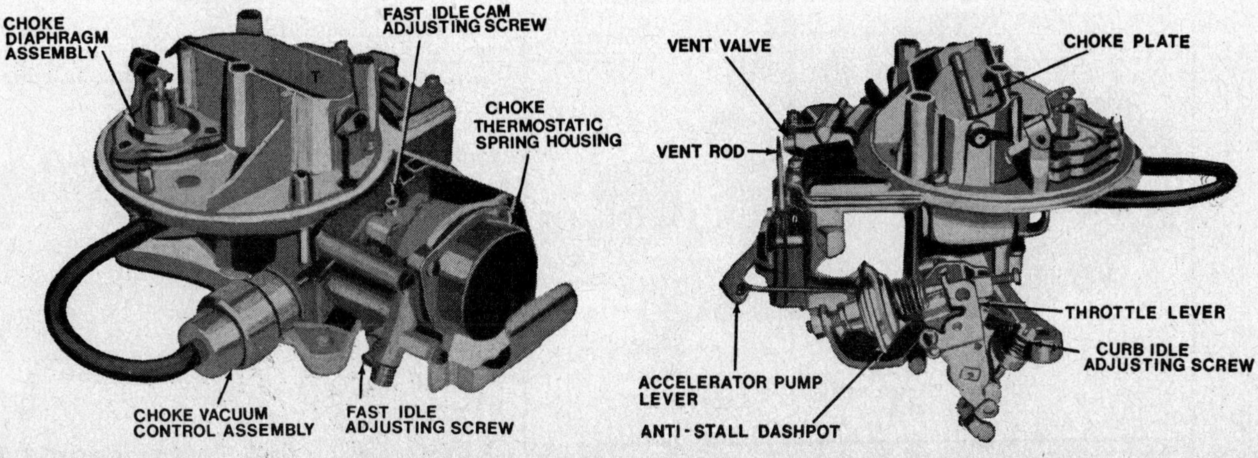

Fig. 2 Typical Autolite/Motorcraft 2100 carburetor. 1970-72

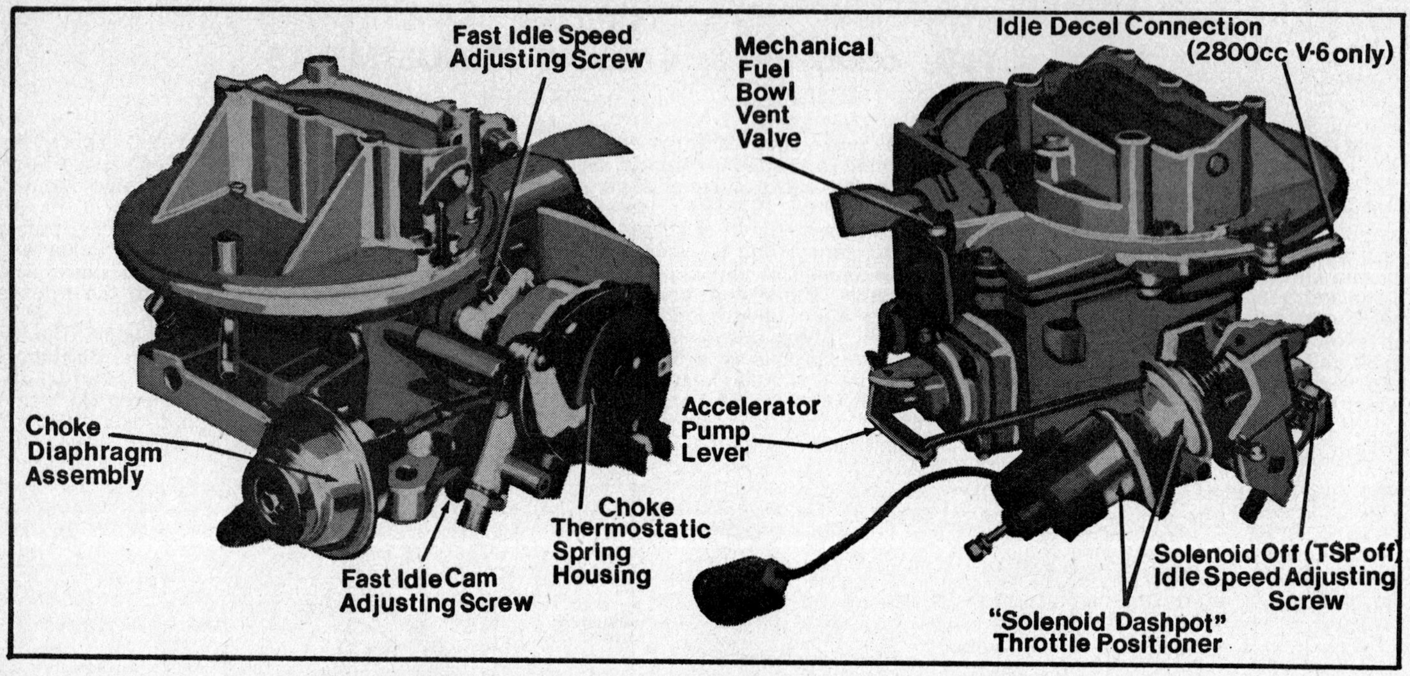

Fig. 2A Typical Autolite/Motorcraft 2150 carburetor. 1975

vehicle is equipped with an automatic transmission.

Autolite 4100-4V

In these carburetors, Figs. 6 and 7, the choke plate magnet has been eliminated but an automatic choke clean air pick-up tube has been added. A rubber hose and steel tube connects the clean air pick-up tube to the automatic choke heat chamber in the right-hand exhaust manifold.

Float Level Adjustment

2100, 2150, 6200, 4100, Fig. 8—This is a preliminary adjustment; the final adjustment must be made after the carburetor is mounted on the engine.

With air horn removed, float raised and fuel inlet needle seated, measure distance between top surface of throttle body and top surface of float. Take measurement near center of float at a point ⅛" from free end of float.

If a cardboard float gauge is used, place the gauge in the corner of the enlarged end section of the fuel bowl as shown. The gauge should touch the float near the end but not on the end radius.

Depress the float tab to seat the fuel inlet needle. The float height is measured from the gasket surface of the throttle body with gasket removed. If the float height is not as listed in the *Ford Specifications Chart*, bend tab on float as required to achieve the desired setting.

Fuel Level Adjustment

Fig. 9—With vehicle on a level surface, operate engine until normal temperature is reached, then stop engine and check fuel level as follows:

1. Remove carburetor air cleaner.
2. Remove air horn retaining screws and carburetor identification tag.
3. Temporarily leave air horn and gasket in position on throttle body and start engine.

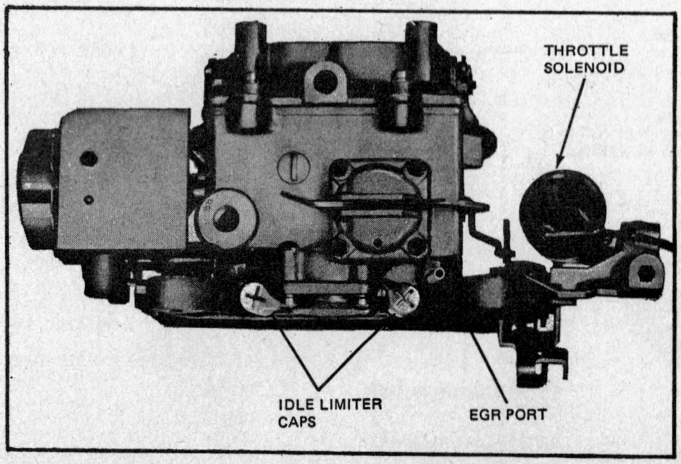

Fig. 3 Typical Autolite/Motorcraft 2100 carburetor. 1973-75

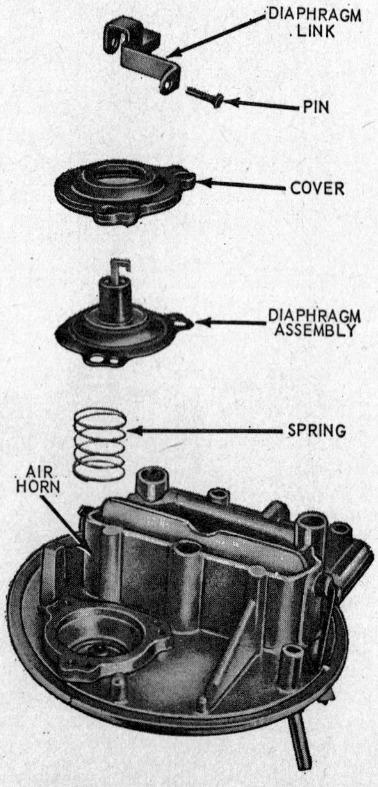

Fig. 5 Autolite/Motorcraft 2100 choke diaphragm assembly

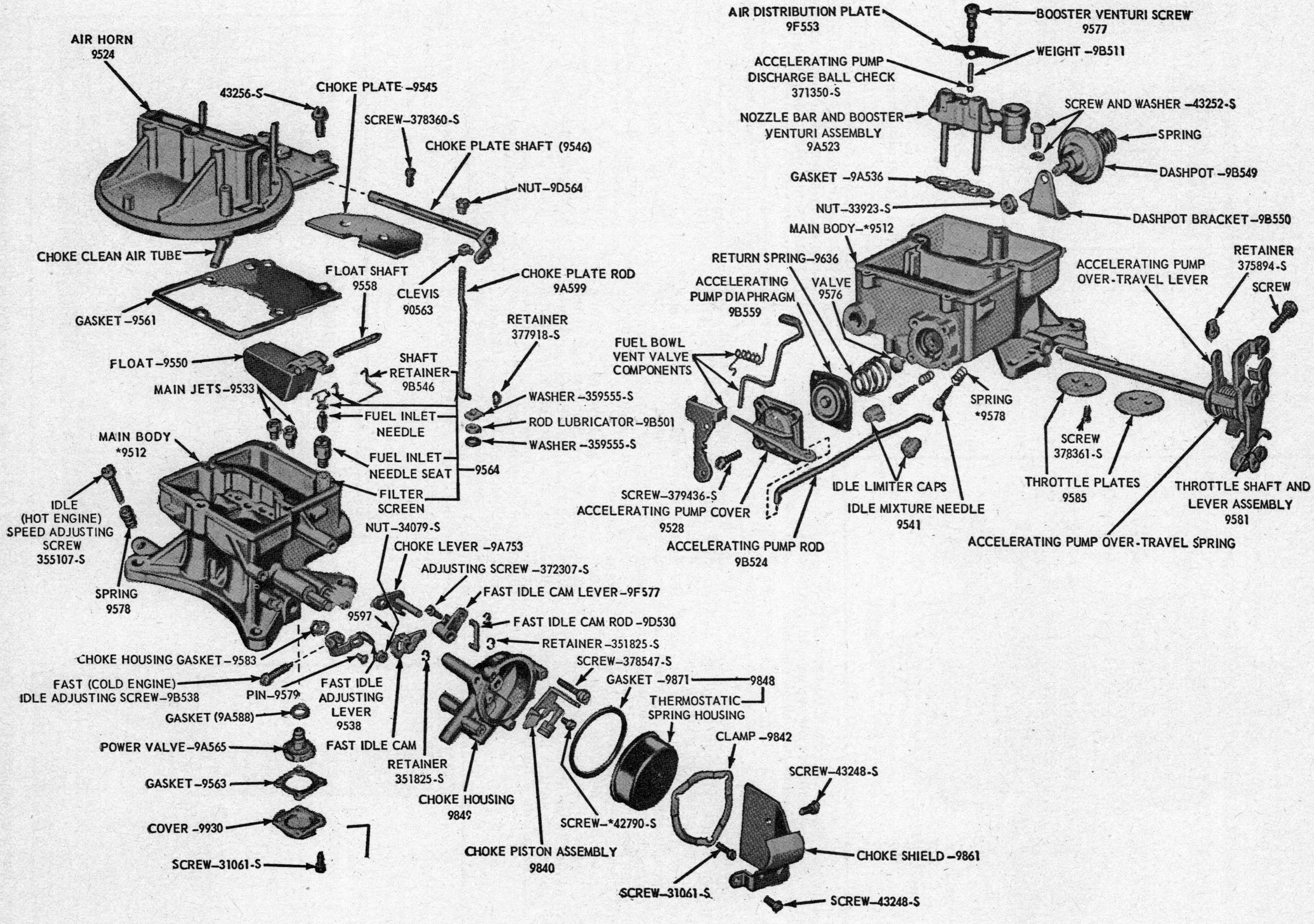

Fig. 4 Exploded view of a typical Autolite 2100, 2150, 6200 two-barrel carburetor. Note that this unit is provided with the idle limiter caps which are used to prevent an overly rich mixture adjustment on cars with exhaust emission control

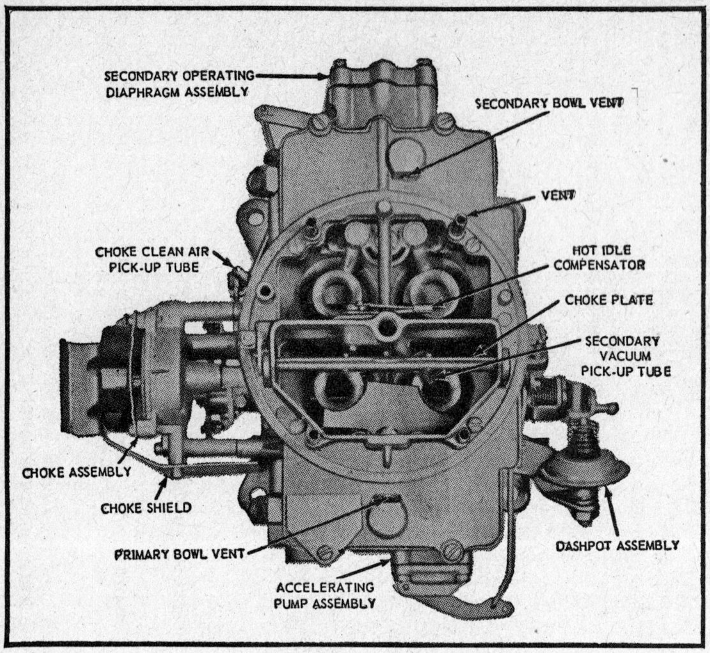

Fig. 6 Autolite 4100 carburetor showing hot idle compensator and choke clean air pick-up tube

4. Allow engine to idle for several minutes, then rotate air horn and remove air horn gasket to gain access to float or floats.
5. While engine is idling, use a standard depth gauge to measure vertical distance from top machined surface of throttle body to level of fuel in bowl. The measurement must be made at least 1/4″ away from any vertical surface to assure an accurate reading.
6. If the fuel level is not as listed in the *Ford Specifications Chart*, stop the engine to avoid any fire hazard due to fuel spray when float setting is disturbed.
7. To adjust fuel level, bend float tab (contacting fuel inlet needle) upward in relation to original position to raise the fuel level, and downward to lower it.
8. Each time an adjustment is made to the float tab to alter the fuel level, the engine must be started and permitted to idle for at least three minutes to stabilize the fuel level. Check fuel level after each adjustment until the specified level is achieved.
9. Assemble carburetor with a new air horn gasket. Then adjust idle speed and mixture, and anti-stall dashpot, if so equipped.

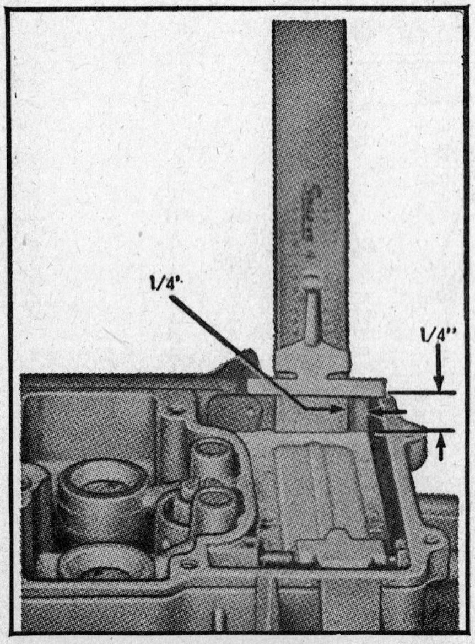

Fig. 9 Fuel level adjustment. 2100, 2150, 6200, 4100

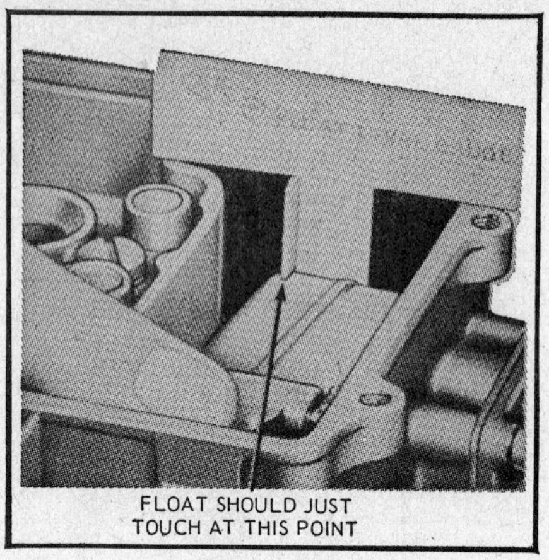

Fig. 8 Float level adjustment. 2100, 2150, 4100

Accelerating Pump Adjustment

Figs. 10 and 11—The primary throttle shaft lever (overtravel lever) has 4 holes and the accelerating pump link has 2 or 4 holes to control the pump stroke for various atmospheric temperatures, operating conditions and specific engine applications.

NOTE: The stroke should not be changed from the specified setting.

1. To release rod from retainer clip, press tab end of clip toward rod. Then, at the same time, press rod away from clip until it is disengaged.
2. Position clip over specified hole in overtravel lever. Press ends of clip together and insert operating rod through clip and lever. Release clip to engage rod.

Secondary Throttle Plate

Fig. 12, 4-Barrel Units

Hold secondary throttle plate closed. Turn secondary throttle shaft lever ad-

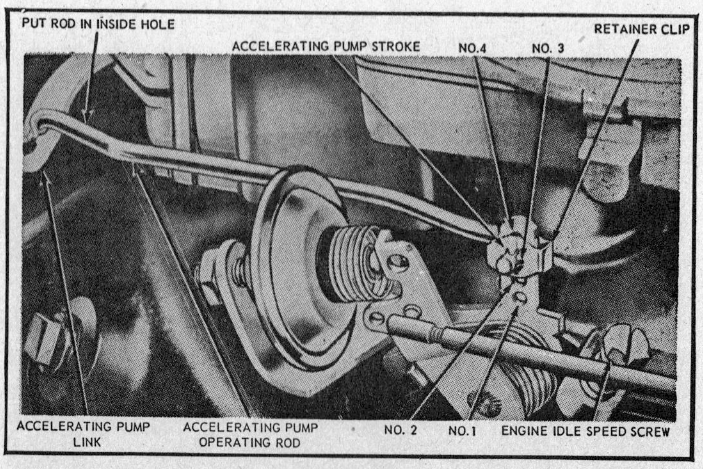

Fig. 10 Pump stroke and idle speed adjusting points. 2100, 6200, 4100

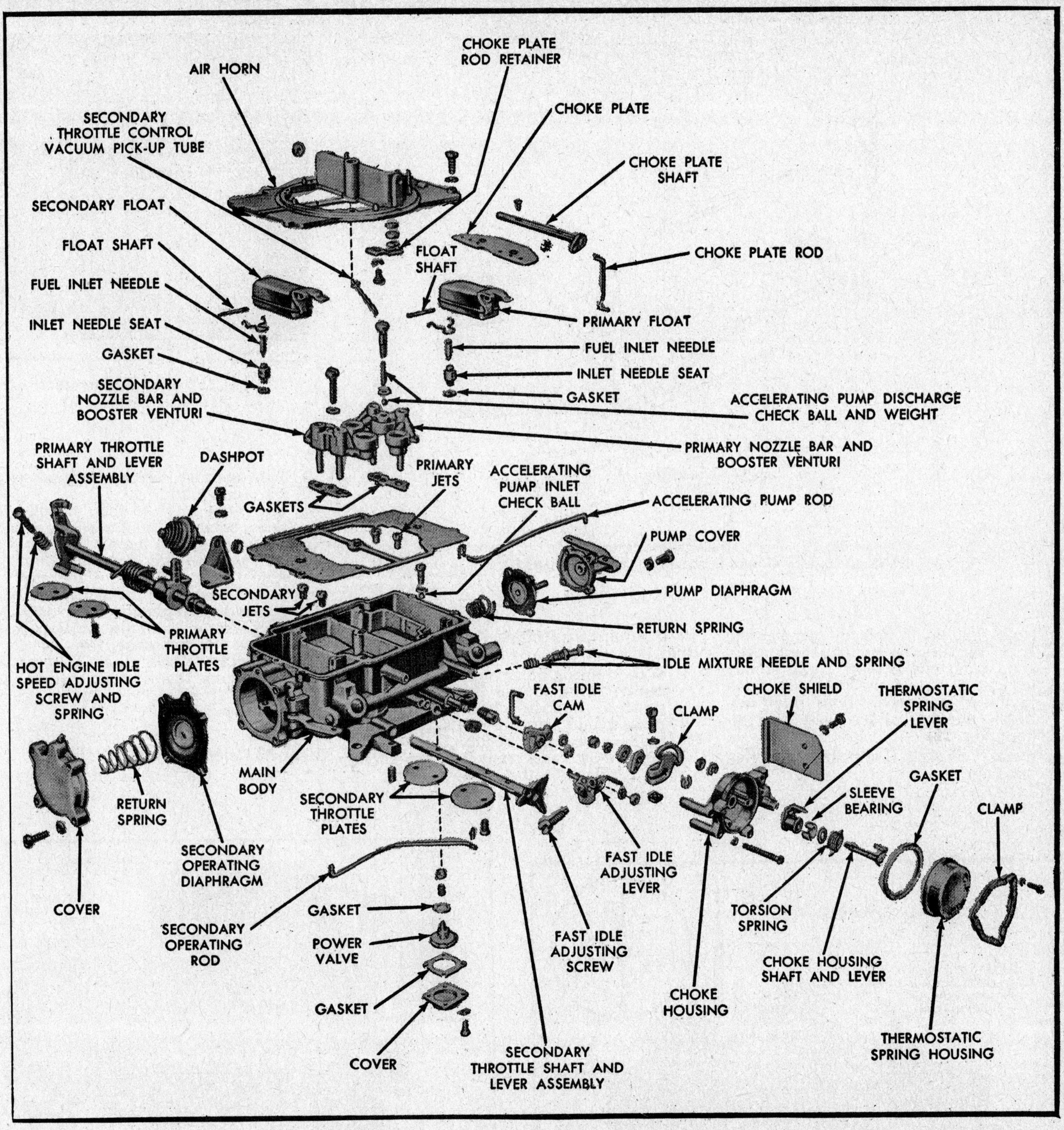

Fig. 7 Exploded view of a typical Autolite 4100 four-barrel carburetor

justing screw out (counterclockwise) until the secondary throttle plates stick in throttle bores. Turn screw in (clockwise) until it just contacts secondary lever. Then turn it the additional number of turns listed in the *Ford Specifications Chart.*

Choke Plate Clearance (Pulldown) Adjustment

1969 Units

1. Bend a .036" pin gauge as shown in inset of Fig. 13.
2. Block throttle about half-way open so that fast idle cam does not contact fast idle adjustment screw.
3. Insert bent edge of gauge between lower edge of piston slot and upper edge of right-hand slot in choke housing as shown.
4. Pull choke countershaft lever counterclockwise until gauge is snug in piston slot.

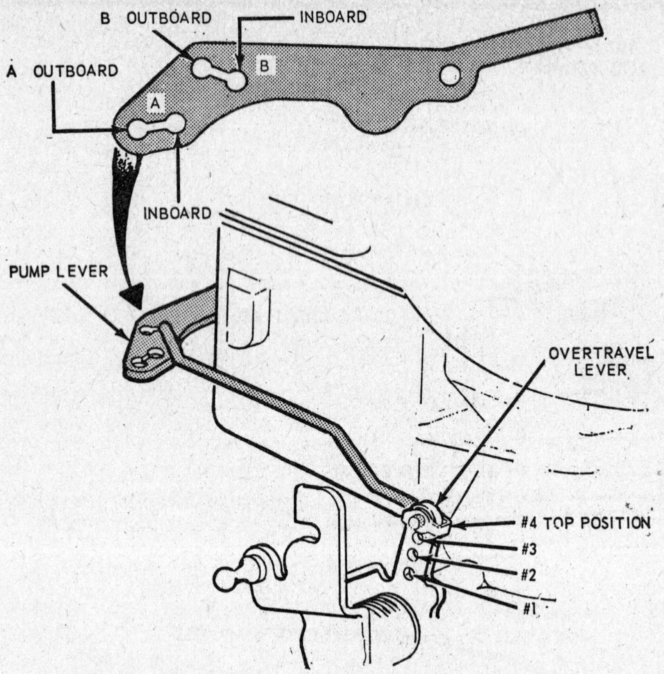

Fig. 11 Pump stroke adjusting points, 2100, 2150, with 4 holes in accelerating pump linkage

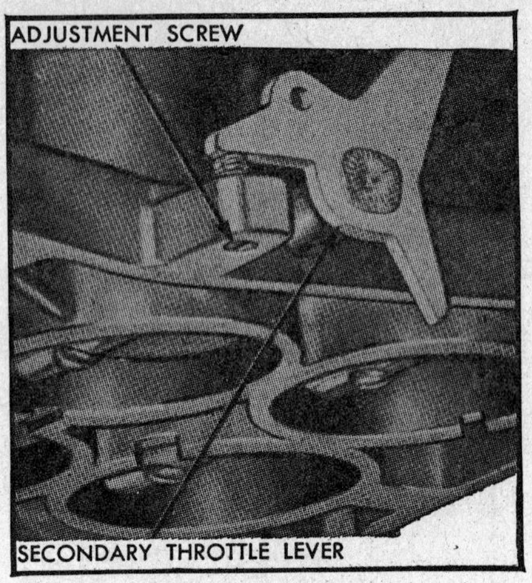

Fig. 12 Secondary throttle plate adjustment. 4100

5. Hold gauge in place by exerting light pressure on countershaft lever, and adjust choke plate clevis adjusting nut to obtain the clearance listed in the *Ford Specifications Chart* between front of choke plate and air horn, Fig. 13.
6. Install choke thermostatic spring housing.

1970 American Motors & 1970-72 Ford Units

1. With the engine at normal operating temperature, loosen the choke thermostatic spring housing retainer screws and set the housing 90 degrees in the rich direction.
2. Disconnect and remove the choke heat tube from the choke housing.
3. Turn the fast idle adjusting screw outward one full turn.
4. Start the engine, then check clearance between the lower edge of the choke plate and the air horn wall, Fig. 14. Refer to Specifications Chart.
5. Adjust by turning the diaphragm stop screw located on the underside of the choke diaphragm housing, clockwise to decrease or counter-clockwise to increase the clearance, Fig. 14.
6. Connect the choke heat tube and set the choke thermostatic spring housing to specifications. Adjust the fast idle speed.

1971-75 American Motors Units

1. Loosen the choke cover retaining

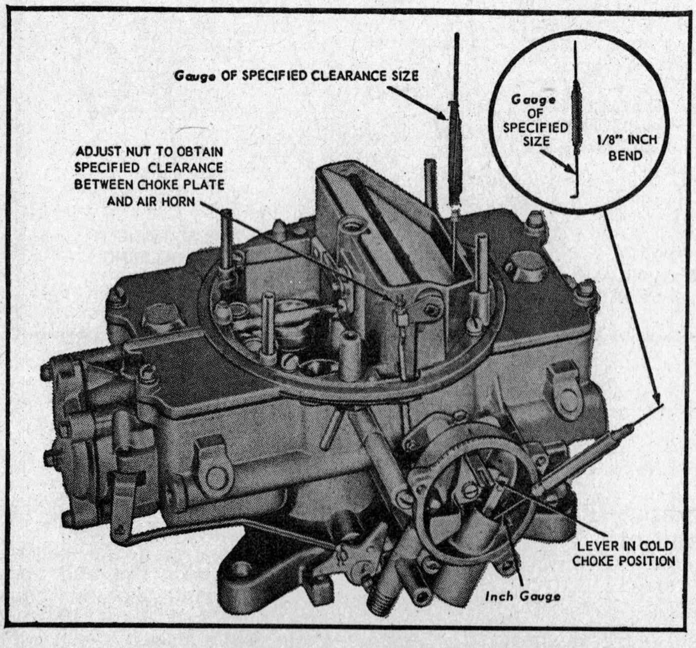

Fig. 13 Vacuum piston type choke plate clearance adjustment, 2100, 6200, 4100; 1969

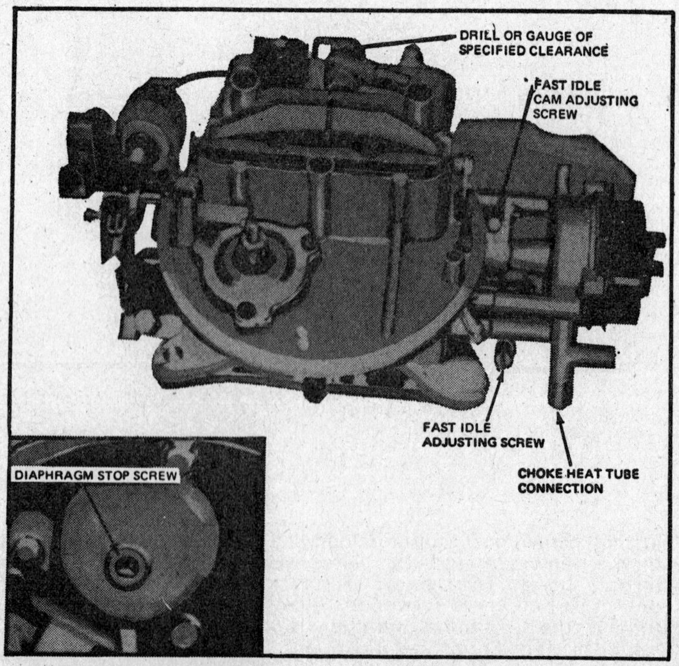

Fig. 14 Choke plate pulldown clearance 2100. 1970 American Motors and 1970-74 Ford

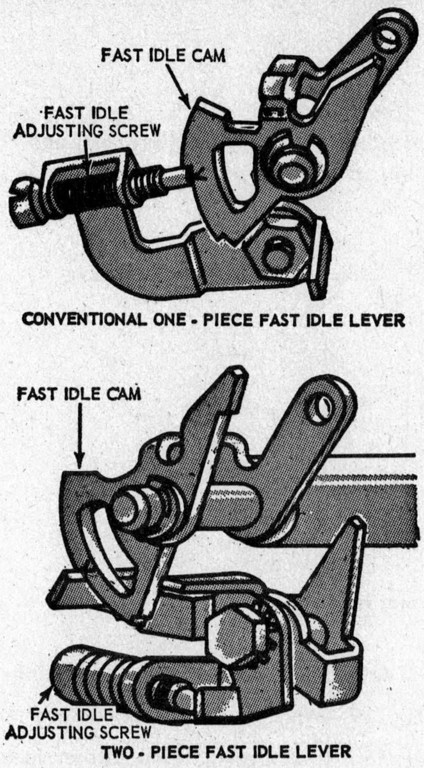

Fig. 15 Autolite/Motorcraft fast idle adjustment 2100, 2150

screws. Rotate the choke cover ¼ turn counterclockwise (rich) from index and tighten the retaining screws. Disconnect the choke heat inlet tube.

2. Align the fast idle speed adjusting screw with the second step of the fast idle cam, Fig. 15.
3. Start the engine without moving the accelerator linkage. Turn the fast idle cam lever adjusting screw out counterclockwise 3 full turns.
4. Measure the clearance between the lower edge of the choke valve and the air horn wall. Refer to *2100 Specifications Chart* for the correct setting. Adjust by grasping the modulator arm securely with a pair of pliers at point "A" and twisting the arm at point "B" with a second pair of pliers. Twist toward the front of the carburetor to increase clearance and toward the rear to decrease clearance, Fig. 16.

CAUTION: Use extreme care while twisting the modulator arm to avoid damaging the nylon piston rod of the modulator assembly.

NOTE: Connect the choke heat tube. Turn the fast idle cam lever adjusting screw in (clockwise) 3 full turns. Do not reset the choke cover until the fast idle cam linkage adjustment has been performed.

1973-74 Ford Units

Since the choke plate pulldown is set in production by means of an air-fuel meter, no specific clearance is indicated for pulldown adjustment. If the vehicle shows indication of leanness during cold starting, decrease the clearance between the choke plate and the air horn wall by 0.020 inch. If the engine shows signs of an overrich condition during cold starting, increase the pulldown clearance by 0.020 inch. If additional adjustment is required always make the adjustments in steps of 0.020 inch. If the original pulldown adjustment is lost, set the clearance between the choke plate and the air horn to 0.160 inch. Then adjust as required in steps of 0.020 inch, Fig. 14.

1975 Ford Units

1. Set throttle on fast idle cam top step, Fig. 15, then loosen choke thermostatic housing retaining screws and set housing 90° in rich direction.
2. Activate pulldown motor by manually forcing pulldown control diaphragm link in direction of applied vacuum or

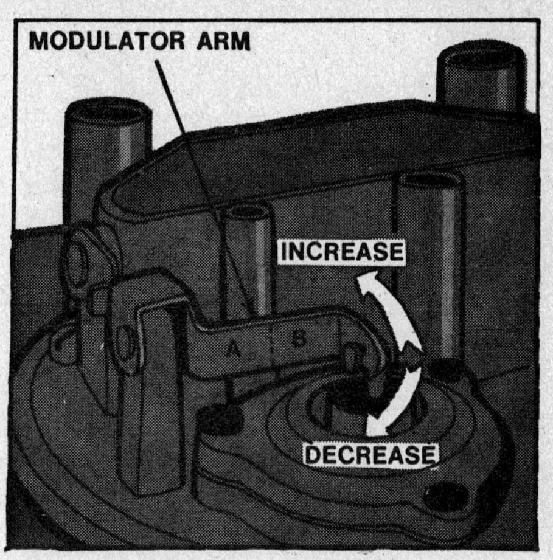

Fig. 16 Choke plate pulldown clearance 2100. 1971-75 American Motors

Fig. 18 Automatic choke adjustment. 2100, 6200, 4100

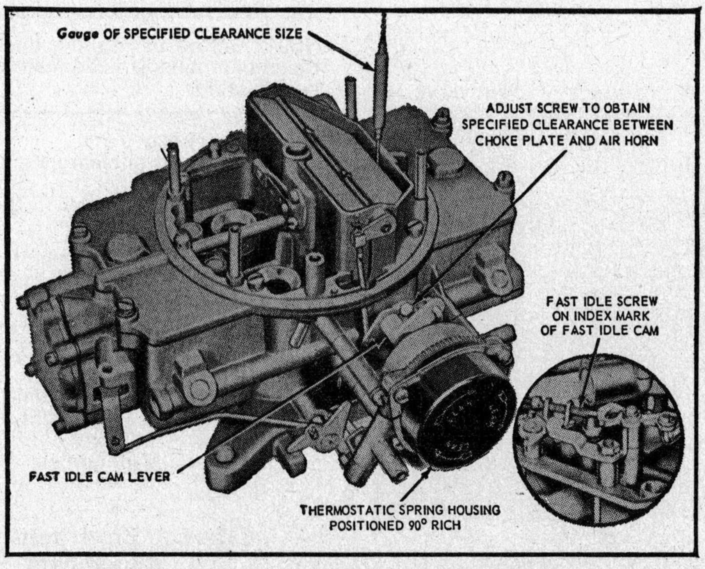

Fig. 17 Vacuum piston type choke fast idle cam linkage adjustment. 2100, 6200, 4100

Fig. 19 Anti-stall dashpot adjustment. 2100, 6200, 4100

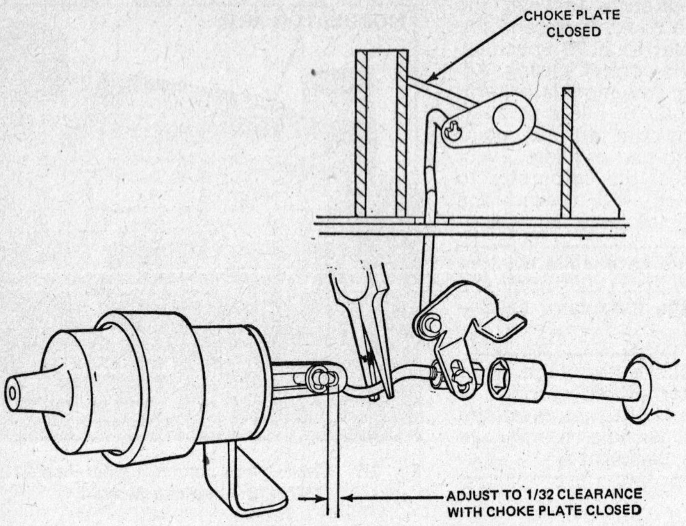

Fig. 20 Autolite/Motorcraft staged choke vacuum control adjustment 2100D. 1972 Ford

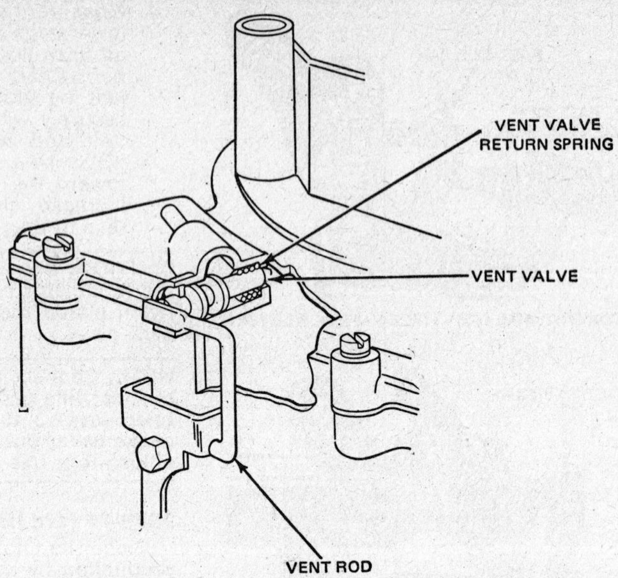

Fig. 22 Fuel bowl vent valve adjustment 2100. 1972 Ford

by applying vacuum to external vacuum tube.

3. Check clearance between choke plate and center of carburetor air horn wall nearest fuel bowl. Refer to 2150 Specifications Chart. If clearance is not as specified, reset by adjusting diaphragm stop on end of choke pulldown diaphragm.

Fast Idle Cam Clearance

1969-75 Units

1. Loosen the choke thermostatic spring housing retainer screws and set the housing 90 degrees in the rich direction.
2. Position the fast idle speed screw at the kickdown step of the fast idle cam. The kickdown step is identified on most units by a V stamped on the cam, Fig. 15. When a two-piece fast idle lever is used, a tang on the top lever will align with the V mark on the cam, Fig. 15.
3. Be sure the cam is at the kickdown position while checking or adjusting the fast idle cam clearance. Check clearance between the lower edge of the choke plate and the air horn wall, Fig. 17. Refer to *2100 Specifications Chart*. Adjust clearance by turning the fast idle cam clearance adjusting screw clockwise to increase and counterclockwise to decrease the clearance.
4. Set the choke thermostatic spring housing to specifications, Fig. 18. Adjust the anti-stall dashpot, idle speed and fuel mixture.

1975 2150 Units

1. Open and close throttle slightly and observe where fast idle screw contacts cam. The fast idle screw should be set to approximate high cam speed and should strike the second highest step on fast idle cam.
2. After adjusting fast idle cam with fast idle screw, allow choke plate to close. Choke plate should close tightly in air horn.
3. If choke plate is loose, an additional adjustment must be made to pulldown diaphragm rod assembly, by loosening lock nut and tightening turnbuckle until choke plate is tight on airhorn.

Automatic Choke Valve Tension

Turn thermostatic spring cover against spring tension until index mark on cover is aligned with mark specified in the *Ford Specifications Chart* on choke housing, Fig. 18.

Fig. 21 Fuel bowl vent valve adjustment 2100, 6200. 1969

Anti-Stall Dashpot Adjustment

Fig. 19—With engine idle speed and mixture properly adjusted and with engine at normal operating temperature, loosen dashpot lock nut. Hold throttle in closed position and depress plunger with screwdriver as shown. Check clearance between throttle lever and plunger tip. If clearance is not as listed in the *Ford Specifications Chart*, turn dashpot in its bracket as required to obtain the desired clearance. Tighten lock nut.

Staged Choke Vacuum Control Adjustment 1972 Ford

NOTE: This adjustment should be made only if the control unit has been replaced, the carburetor has been overhauled or a choke adjustment has been made.

NOTE: Before performing this adjustment, the choke pulldown and fast idle cam must be adjusted.

1. Close choke plate.
2. Measure the clearance between the forward edge of the choke link and the edge of the slot in the choke vacuum lever, Fig. 20.

NOTE: Place C-clip in a position so that gap is visible.

3. Adjust by inserting a 1/4 inch socket and drive on the nylon adjuster, Fig. 20. Hold the choke link with needlenose pliers to prevent flexing, then turn the adjuster until 1/32 inch clearance is obtained.

External Bowl Vent Valve Adjustment

NOTE: The vent valve adjustment should

be performed after the accelerating pump adjustment.

The vent valve, must be closed during normal and wide-open throttle operation, and open to the specified clearance at closed throttle or idle operation.

1969 Units

1. Check clearance between bowl vent valve and air horn casting with throttle positioned at curb idle setting, Fig. 21.
2. Adjust by bending the vent rod at the rod attaching bracket to obtain .065 inch clearance for American Motors installation and .070 inch clearance for Ford installations.

1972 Units

1. Fully depress the vent valve into the valve bore.
2. Measure the clearance between the flat on the vent rod and the fully seated valve, Fig. 22.
3. Adjust by bending the vent rod at the point where it contacts the accelerator pump lever until 3/32 inch clearance is obtained.

MODEL 4300-4V ADJUSTMENT SPECIFICATIONS

See Tune Up Chart in car chapters for hot idle speeds.

Year	Carb. Model (Code 9510) ①	Idle Mixture Turns Open	Float Level (Dry)	Pump Setting (Hole No.)	Choke Plate Clearance (Pulldown)	Fast Idle Cam Linkage Setting	Fast Idle Speed (Hot Engine)	Auxiliary Inlet Valve Setting	Dechoke Clearance	Dashpot Setting	Choke Setting
AMERICAN MOTORS											
1970	OWA4	②	13/16	Center	.170	.190	1600	—	.300	1/8	2 Rich
	OWM4	②	13/16	Center	.190	.200	1600	—	.300	1/16	2 Rich
1971	ITA4	②	13/16	Center	.190	.200	1600	—	.300	1/8	Index
	ITM4	②	13/16	Center	.170	.190	1600	—	.300	1/16	Index
1972	2RA4	②	13/16	Center	.190	.190	1600	—	.300	9/64	1 Rich
	2TA4	②	13/16	Center	.190	.190	1600	—	.300	9/64	1 Rich
	2TM4	②	13/16	Center	.190	.190	1600	—	.300	9/64	1 Rich
1973	3TA4	②	13/16	Center	.190	.160	1600	.060	.275	9/64	2 Rich
	3TA4 Police	②	15/16	Center	.190	.160	1600	.030	.275	9/64	2 Rich
	3TM4	②	13/16	Center	.190	.160	1600	.060	.275	9/64	2 Rich
1974	4TA4	②	13/16	Center	.170	.160	1600	.050	.325	9/64	2 Rich
	4TA4 Police	②	25/32	Center	.170	.160	1600	.030	Preset	9/64	2 Rich
	4TM4	②	13/16	Center	.170	.160	1600	.050	.325	9/64	2 Rich

①—Tag attached to bowl cover.
②—Air/fuel ratio or idle CO% rating is found in Tune Up Specification tables in car chapters.

Year	Carb. Model (Code 9510) ①	Idle Mixture Turns Open	Float Level (Dry)	Pump Setting (Hole No.)	Choke Plate Clearance (Pulldown)	Fast Idle Cam Linkage Setting	Fast Idle Speed (Hot Engine)	Auxiliary Inlet Valve Setting	Dechoke Clearance	Dashpot Setting	Choke Setting
FORD AND MERCURY ENGINES											
1969	C8SF-H	⑤	25/32	#2	.230	.160	1300	1/16	.300	3/32	1 Rich
	C8VF-J	⑤	25/32	#2	.230	.160	1300	1/16	.300	3/32	1 Rich
	C9AF-G	⑤	25/32	#2	.270	.220	1200	1/16	.300	3/32	Index
	C9AF-R	⑤	25/32	#2	.230	.160	1300	1/16	.300	—	1 Rich
	C9OF-D	⑤	13/16	#2	.160	.100	1400	1/16	.300	—	1 Lean
	C9OF-E	⑤	13/16	#3	.250	.230	1400	1/16	.300	—	1 Lean
	C9ZF-C	⑤	13/16	#2	.170	.130	1250	1/16	.300	3/32	2 Lean
	C9ZF-D	⑤	13/16	#2	.160	.100	1400	1/16	.300	—	1 Lean
	C9ZF-E	⑤	13/16	#3	.230	.210	1300	1/16	.300	1/8	Index
	C9ZF-F	⑤	13/16	#3	.250	.230	1400	1/16	.300	—	1 Lean
1970	DOAF-K	⑤	49/64	#2	.220	.170	1300	1/16	.300	.070	Index
	DOAF-L	⑤	25/32	#2	.250	.220	1400	1/16	.300	.070	Index
	DOAF-M	⑤	1.00	#3	.160	.120	1600	1/16	.300	.080	2 Rich
	DOAF-R	⑤	1.00	#3	.160	.120	1600	1/32	.300	—	2 Rich
	DOAF-AD	⑤	1.00	#3	.160	.120	1600	1/16	.300	.080	2 Rich
	DOAF-AJ	⑤	1.00	#3	.160	.120	1600	1/16	.300	.080	2 Rich
	DOAF-AE	⑤	1.00	#3	.160	.120	1600	1/32	.300	—	2 Rich
	DOAF-AK	⑤	1.00	#3	.160	.120	1600	1/32	.300	—	2 Rich
	DOAF-AB	⑤	25/32	#2	.250	.220	1400	1/16	.300	.070	Index
	DOAF-AL	⑤	25/32	#2	.250	.220	1400	1/16	.300	.070	Index
	DOAF-AG	⑤	25/32	#2	.220	.170	1300	1/16	.300	.070	Index
	DOAF-AM	⑤	25/32	#2	.220	.170	1300	1/16	.300	.070	Index
	DOAF-AN	⑤	49/64	#2	.225	.170	1350	1/16	.300	.070	Index
	DOOF-B	⑤	13/16	#2	.180	.160	1250	1/16	.300	—	Index
	DOOF-C	⑤	13/16	#2	.200	.180	1400	1/16	.300	.080	Index
	DOOF-D	⑤	13/16	#2	.180	.160	1250	1/16	.300	—	Index
	DOOF-H	⑤	13/16	#2	.200	.180	1400	1/16	.300	—	Index

Continued

MODEL 4300-4V ADJUSTMENT SPECIFICATIONS—Continued

See Tune Up Chart in car chapters for hot idle speeds.

Year	Carb. Model (Code 9510) ①	Idle Mixture Turns Open	Float Level (Dry)	Pump Setting (Hole No.)	Choke Plate Clearance (Pulldown)	Fast Idle Cam Linkage Setting	Fast Idle Speed (Hot Engine)	Auxiliary Inlet Valve Setting	Dechoke Clearance	Dashpot Setting	Choke Setting
FORD AND MERCURY ENGINES—Continued											
1970	DOOF-Y	⑤	13/16	#2	.180	.160	1250	1/16	.300	—	Index
	DOOF-Z	⑤	13/16	#2	.180	.160	1250	1/16	.300	—	Index
	DOOF-AA	⑤	13/16	#2	.200	.180	1400	1/16	.300	—	Index
	DOOF-AB	⑤	13/16	#2	.180	.160	1250	1/16	.300	—	Index
	DOOF-AC	⑤	13/16	#2	.200	.180	1400	1/16	.300	.080	Index
	DOOF-AD	⑤	13/16	#2	.200	.180	1400	1/16	.300	—	Index
	DOOF-AE	⑤	13/16	#2	.180	.160	1250	1/16	.300	—	Index
	DOSF-A	⑤	25/32	#2	.220	.170	1300	1/16	.300	.070	Index
	DOSF-D	⑤	25/32	#2	.220	.170	1300	1/16	.300	.070	Index
	DOSF-E	⑤	25/32	#2	.220	.170	1300	1/16	.300	.070	Index
	DOVF-A, C	⑤	25/32	#2	.230	.170	1250	1/16	.300	.100	1 Rich
1971	D1AF-MA	⑤	49/64	#2	.220	—	1350	1/16	—	1/16	Index
	D1OF-AAA	⑤	13/16	#2	.200	.180	1400	1/16	—	—	Index
	D1OF-EA	⑤	13/16	#2	.180	.160	1250	1/16	—	—	Index
	D1SF-AA	⑤	49/64	#2	.220	—	1350	1/16	—	1/16	Index
	D1VF-AA	⑤	49/64	#2	.220	.170	1250	1/16	—	.100	1 Rich
	D1ZF-FA	⑤	13/16	#3	.220	.220	1200	1/16	—	—	1 Lean
	D1ZF-GA	⑤	13/16	#3	.200	.200	1200	1/16	—	—	Index
1972	D2AF-AA	⑤	49/64	#1	.220	.200	1350	—	—	—	2 Rich
	D2AF-LA	⑤	49/64	#1	.215	.190	1900	.030	—	—	Index
	D2SF-AA	⑤	49/64	#1	.220	.200	1350	—	—	—	2 Rich
	D2SF-BA	⑤	49/64	#1	.220	.200	1350	—	—	—	2 Rich
	D2VF-AA	⑤	49/64	#1	.230	.200	1250	—	—	—	Index
	D2VF-BA	⑤	49/64	#1	.230	.200	1250	—	—	—	Index
	D2ZF-AA	⑤	13/16	#1	.200	.180	1200	.030	—	—	Index
	D2ZF-BB	⑤	13/16	#1	.200	.200	1200	.030	—	—	Index
	D2ZF-DA	⑤	13/16	#1	.200	.200	1200	.030	—	—	Index
	D2ZF-GA	⑤	13/16	#1	.200	.180	1200	.030	—	—	Index
1973	D3AF-HA	⑤	3/4	#1	.210	.200	1350	1/16	—	—	Index
	D3VF-DA	⑤	3/4	#1	.210	.190	1350	1/16	—	—	Index
	D3ZF-AC	⑤	13/16	#1	.180	.180	1300	1/32	—	—	Index
	D3ZF-BC	⑤	13/16	#1	.170	.170	1300	1/32	—	—	1 Rich
	D3ZF-DC	⑤	13/16	#1	.180	.180	1300	1/32	—	—	Index
	D3AF-EB	⑤	7/8	#1	.200	.200	1900	1/32	—	—	Index
1974	D4AE-AA	⑤	3/4	#1	.230	.200	—	1/16	—	—	Index
	D4AE-NA	⑤	3/4	#1	.220	.200	—	1/16	—	—	Index
	D4OE-AA	⑤	13/16	#1	.180	.180	—	1/32	—	—	Index
	D4OE-BA	⑤	13/16	#1	.170	.170	—	1/32	—	—	Index
	D4TE-ATA	⑤	13/16	#1	.220	.180	—	1/16	—	—	Index
	D4VE-AB	⑤	3/4	#1	.220	.200	—	1/16	—	—	Index

①—Tag attached to bowl cover.

③—8-390 with Imco 2 Rich, with Thermactor 1 rich.

④—8-428 with Imco Index, with Thermactor 1 rich.

⑤—Air/fuel ratio or idle CO% rating is found in Tune Up Specification tables in car chapters.

MODEL 4300-4V ADJUSTMENTS

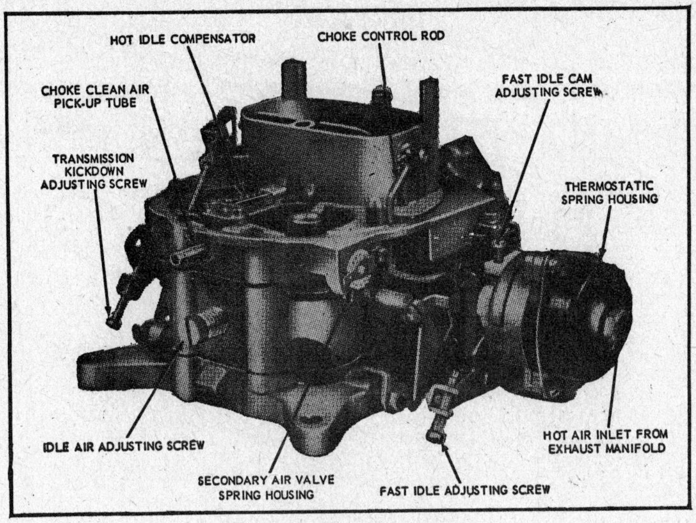

Fig. 1 Right rear view of 4300-4V carburetor

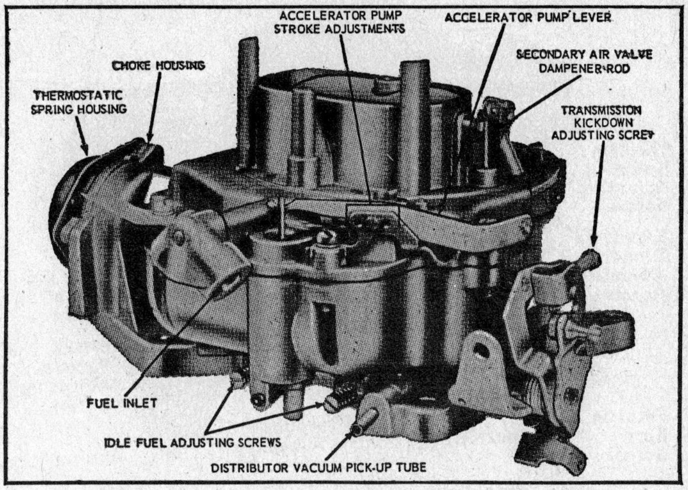

Fig. 2 Left front view of Model 4300-4V carburetor

The model 4300, Figs. 1 thru 6, is a four barrel, three piece, separately cast design consisting of air horn, main body and mounting flange. A cast-in center fuel inlet has provision for a supplementary fuel inlet system. The fuel bowl is vented by an internal balance vent and a mechanical atmospheric vent operates during idle. An idle air by-pass system is designed to provide a consistent idle and a hot idle compensator is used to help stability.

The main (primary) fuel system has booster-type venturi cast integral with the air horn, and the main venturii are cast integral with the main body. The secondary throttle plates are mechanically operated from the primary linkage. On models 4300 and 4300-A, the air valve plates are located above the main venturis, Figs. 1, 2 and 3, an integral hydraulic dashpot dampens sudden movement of the air valve plates to help prevent plate flutter and erratic engine operation. On model 4300-D the air valve plates are located above the secondary bore. Secondary fuel supply is controlled by metering rods which are attached to the air valve plates, Figs. 4 and 6.

The automatic choke system consists of a standard bimetal thermostat. On 1972 Ford units a staged choke release system is used which opens the choke plate within 15-60 seconds when underhood temperatures are above 60°, Figs. 3 and 4. On 1973-74 Ford units an electric choke system is incorporated which opens choke within 1-1½ minutes when underhood temperatures are above approximately 55° to 60°. On 1973-74 American Motors units, an electric choke system is also incorporated which opens choke when underhood temperatures are above approximately 95°. For service information refer to the "Emission Control System" chapter.

A single fuel bowl supplies both the primary and secondary fuel systems. Pontoon-type floats are used to help stabilize fuel level during cornering and hill-climbing. The accelerator pump is of the piston type. It is located in the fuel bowl for more positive displacement and a safeguard against external leaks.

Float Setting

Fig. 7

1. Adjust gauge to specified height.
2. Insert gauge into air horn outboard holes as shown.
3. Check clearance and alignment of float pontoons to gauge. Both pontoons should just touch gauge for proper setting. Align pontoons if necessary by slightly twisting pontoons.
4. If it is necessary to adjust float clearance, bend primary needle tab downward to raise float and upward to lower float.

NOTE: To raise float, insert open end of bending tool to *right* side of float lever tab and between needle and float hinge. Raise float lever off needle and bend tab downward.

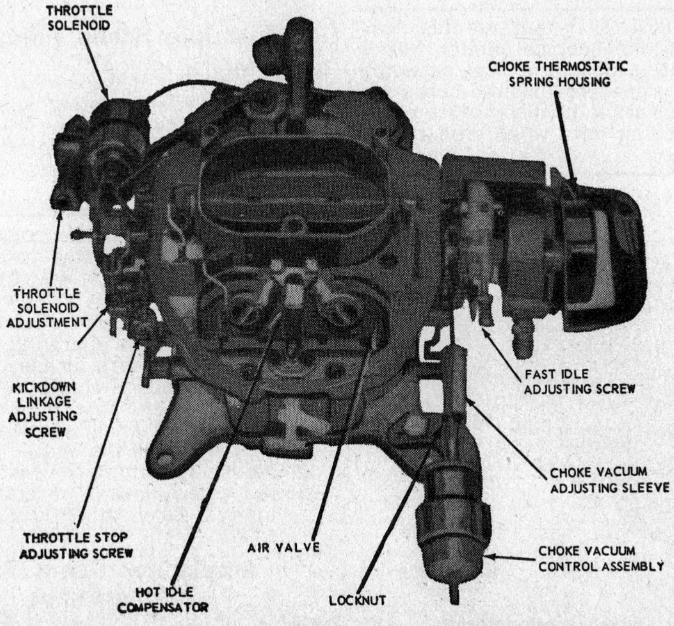

Fig. 3 4300A-4V carburetor with staged choke. 1972

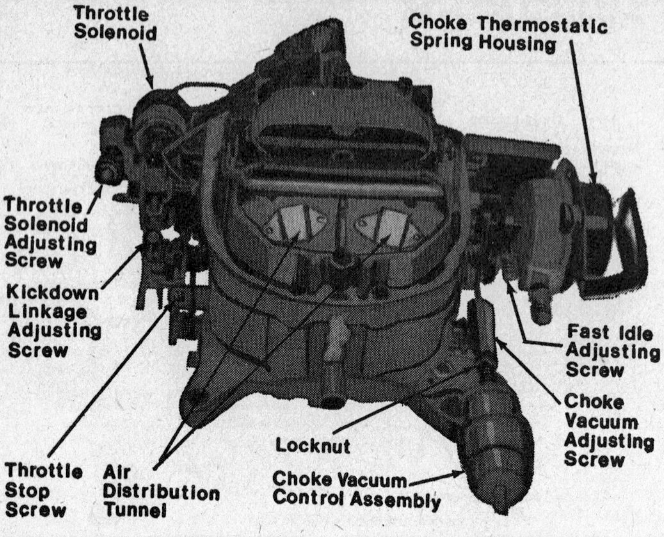

Fig. 4 4300D-4V carburetor with staged choke. 1972

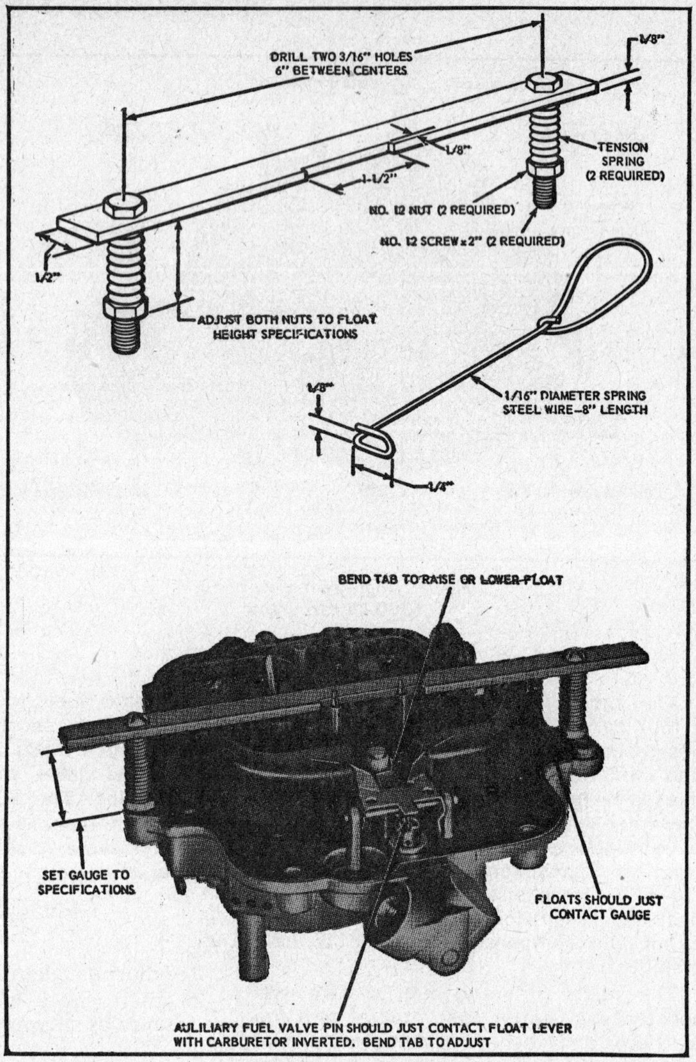

Fig. 7 Float setting

To lower float, insert open end of bending tool to *left* side of float lever tab, between needle and float hinge. Support float lever and bend tab upward.

Auxiliary Valve Setting

Fig. 8, 1969-74 Ford and 1973-74 American Motors units

1. Turn the air horn assembly upright allowing the float to hang freely.
2. Push up on the float until the primary fuel inlet needle lightly contacts its seat.
3. While holding the float in this position, measure the clearance between the float level auxiliary tab and the auxiliary inlet valve plunger.
4. Adjust by bending the tab. Refer to *4300 Specifications Chart.*

NOTE: *IMPORTANT:* To measure this clearance on a police fleet carburetor with a semi-articulating float, the float assembly must be positioned so that the bottom of both pontoons are same distance from gasket surface of air horn when primary fuel inlet is seated.

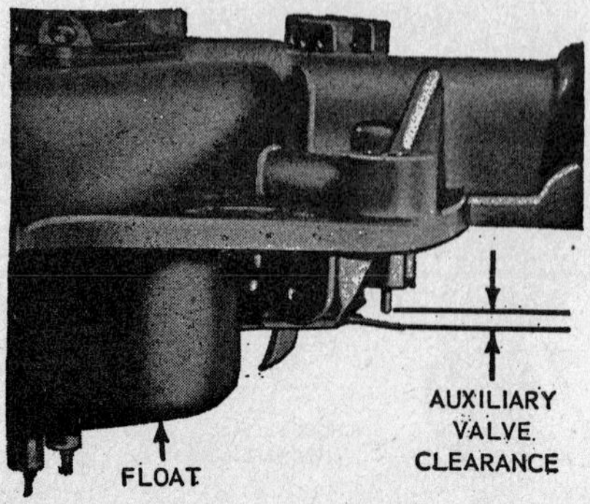

Fig. 8 Auxiliary valve setting. 1969-74 Ford and 1973-74 American Mtrs.

Fast Idle Speed Adjustment

Figs. 1 thru 4

The fast idle adjusting screw contacts one edge of the fast idle cam. Each position on the fast idle cam permits a slower idle rpm as engine temperature rises and choking is reduced.

NOTE: Make certain the curb idle speed and mixture are adjusted to specification before attempting to set the fast idle speed.

1. With the engine operating temperature normalized (hot), air cleaner removed and the tachometer attached, manually rotate the fast idle cam until the fast idle adjusting screw rests on the center step of the cam.
2. Adjust by turning the fast idle adjusting screw inward or outward as required. Refer to *4300 Specifications Chart.*

Accelerator Pump Stroke Adjustment

Fig. 9

The accelerator pump stroke has been

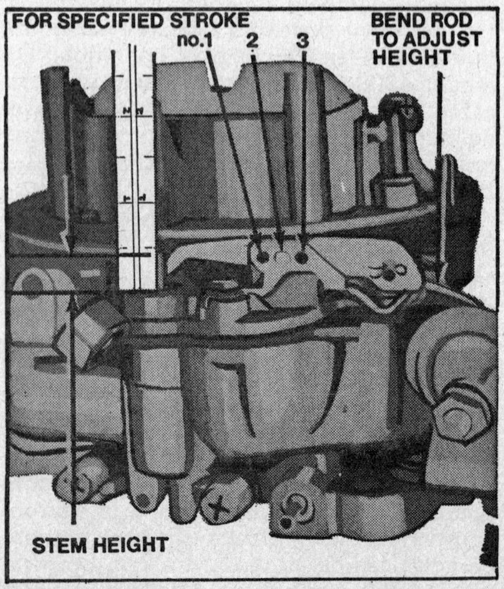

Fig. 9 Accelerator pump stroke adjustment

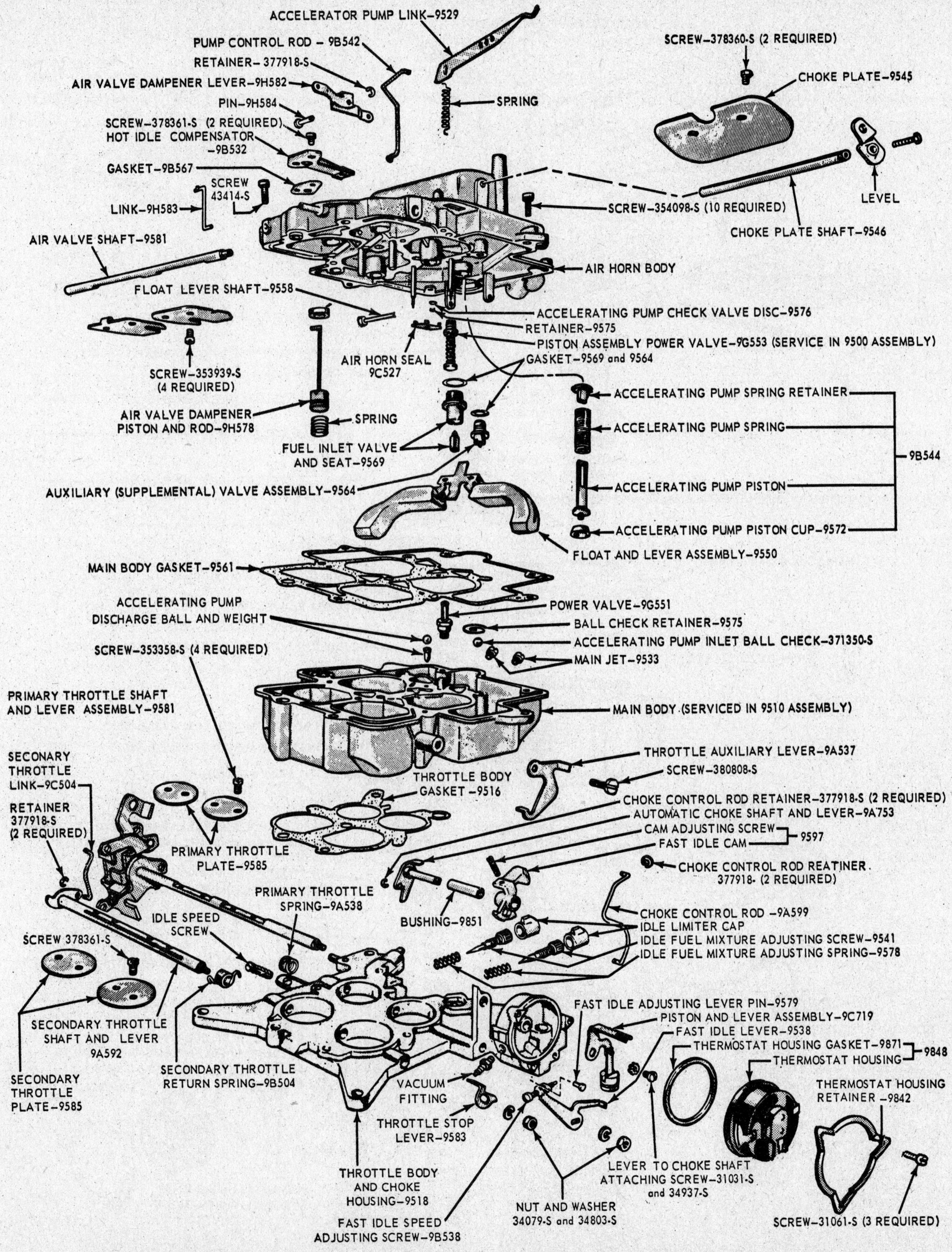

Fig. 5 Exploded view of a typical Autolite 4300A four-barrel carburetor. Note that this unit is provided with idle limiter caps which are used to prevent an overly rich mixture on cars with exhaust emission control

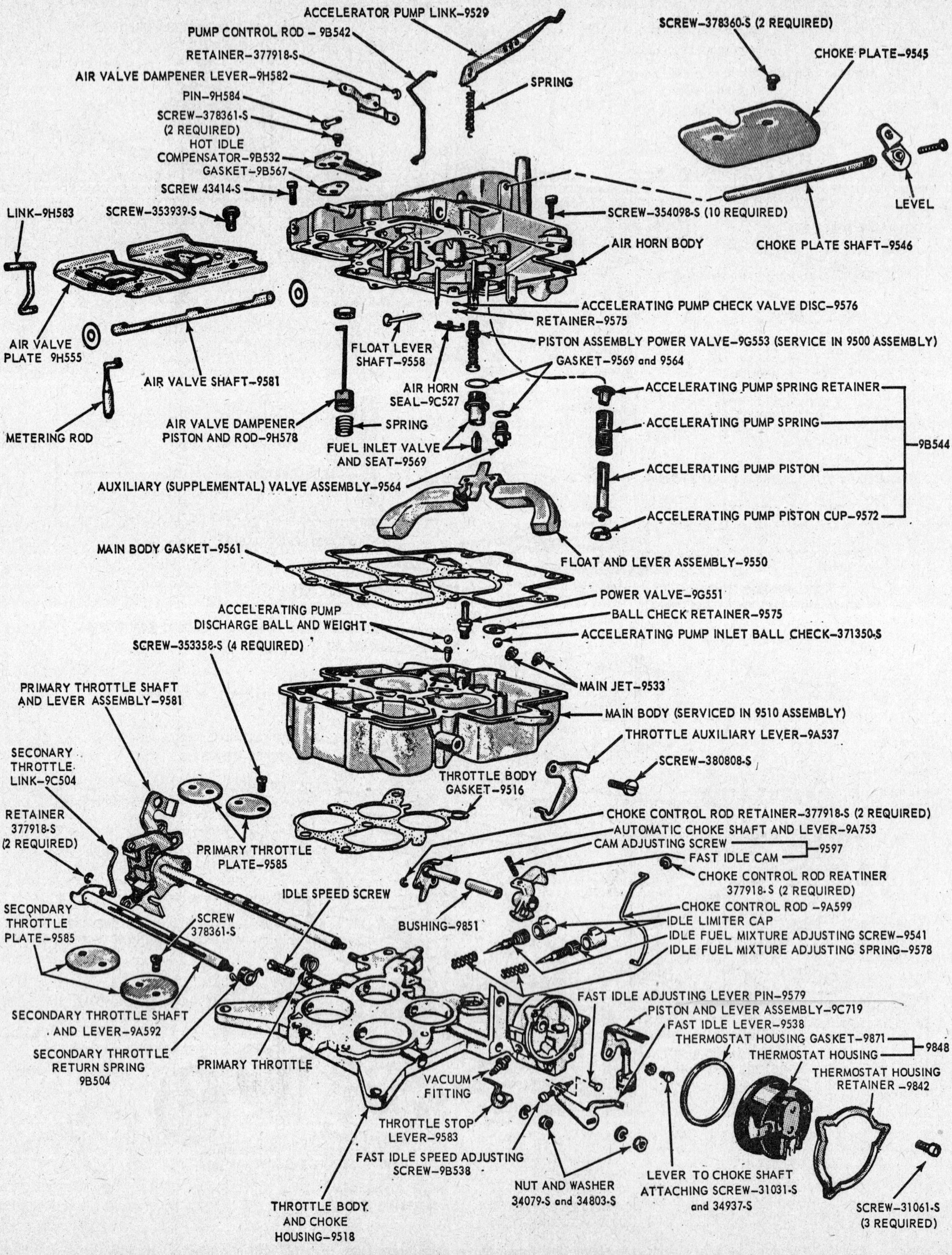

Fig. 6 Exploded view of typical 4300D-4V carburetor

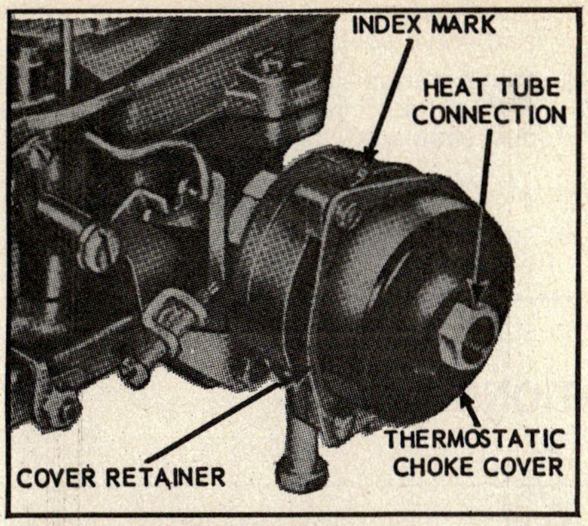

Fig. 10 Automatic choke setting (Typical)

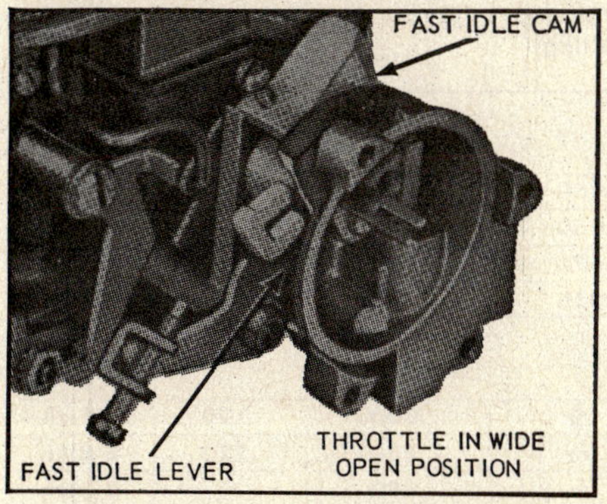

Fig. 11 Dechoke clearance setting

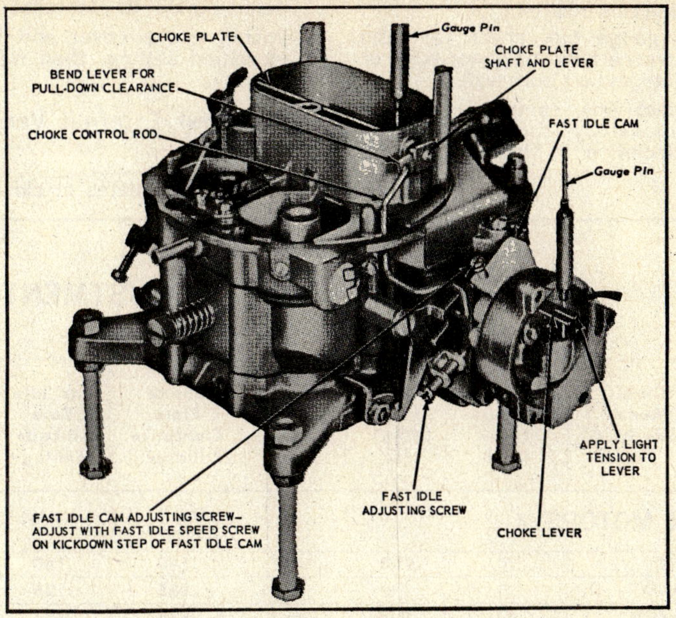

Fig. 12 Choke plate pulldown and fast idle cam adiustment

calibrated to inject a pre-determined quantity of fuel into the air stream and help keep the exhaust emission level of the engine within the specified limits. The additional holes provided for pump stroke adjustment are for adjusting the stroke for specific engine applications.

NOTE: The stroke should not be changed from the specified setting.

If the pump stroke has been changed from the specified setting refer to the following instructions to correct the stroke to specification. Before adjusting the accelerator pump stroke, measure the height of the pump piston. Adjust by bending the pump control rod to correct the piston stem height. Refer to *4300 Specifications Chart.*

Automatic Choke Setting

Fig. 10

1. If carburetor is installed on engine, loosen choke heat tube nut.
2. Loosen choke cover retaining screws.
3. Rotate choke cover clockwise to reduce choking action or counterclockwise to increase choking action.
4. Tighten choke cover screws and choke heat tube nut.

Staged Choke Vacuum Control Adjustment

Figs. 3 and 4

This adjustment should be made only if the control unit has been replaced, carburetor has been overhauled or a choke adjustment has been made.

1. Close choke plate.
2. Back off the adjusting sleeve locknut until it is clear of the adjusting sleeve.
3. Turn the adjusting sleeve inward untii the choke plate just begins to move. Scribe a mark on the upper flat of the sleeve, then back off the sleeve one complete turn.
4. Tighten the locknut firmly against the sleeve to set the adjustment.

NOTE: Do not turn the diaphragm stem.

Fig. 13 Fuel bowl vent valve adjustment

Dechoke Clearance

Fig. 11

1. Open and hold throttle plate to wide open position.
2. Rotate choke plate toward closed position until pawl on fast idle speed lever contacts fast idle cam.
3. Check clearance between lower edge of choke plate and air horn wall.
4. Adjust clearance to specifications by bending pawl on fast idle speed lever forward to increase (backward to decrease) clearance.

Choke Plate Pulldown and Fast Idle Cam

Fig. 12

1. Remove choke cover.
2. Bend a .036" wire gauge at a 90° angle, approximately ⅛" from the end.
3. Insert bent edge of gauge between piston slot and upper edge of right hand slot in choke housing.
4. Rotate automatic choke lever counterclockwise until gauge is snug in piston slot. Exert light pressure on choke lever to hold gauge in place.
5. Using a gauge pin, check pulldown clearance between lower edge of choke plate and air horn wall.
6. Adjust pulldown clearance to specifications by bending adjusting arm on choke shaft lever. Bend downward to increase (upward to decrease) clearance.
7. Remove gauge and install choke cover loosely so it can rotate. Be sure thermostatic spring end is engaged in choke lever slot.
8. Rotate choke cover to 90° rich.
9. Position fast idle adjusting screw end to kickdown step on fast idle cam

and hold in this position.

10. Using a gauge pin, check fast idle cam clearance between lower edge of choke plate and air horn wall.
11. Adjust fast idle cam clearance to specifications by turning adjusting screw clockwise to increase (counterclockwise to decrease) clearance.
12. Install choke cover and rotate it to specified setting, then tighten cover.

Fuel Bowl External Vent Valve

Fig. 13

1. Set throttle plates in closed position.
2. Check clearance between vent valve and valve seat.
3. If clearance is not .070 inch, bend end of vent valve lever downward to decrease and upward to increase.

MODEL 4350-4V ADJUSTMENT SPECIFICATIONS

Year	Carb. Model (Code 9510)	Idle Mixture Turns Open	Float Level (Dry)	Pump Setting (Hole No.)	Choke Plate Clearance (Pulldown)	Fast Idle Cam Linkage Setting	Fast Idle Speed (Hot Engine)	Auxiliary Inlet Valve Setting	Dechoke Clearance	Metering Rod Setting	Choke Setting
AMERICAN MOTORS											
1975	5TA4	①	.900	②	.140	.160	1600③	.050	.325	.120	2 Rich
	5TA4D	①	29/32	—	.135	.135	1600③	.050	.325	—	2 Rich
	5TA4PD	①	15/16	—	.135	.135	1600③	.030	.325	—	2 Rich
1976	6TA4	①	29/32	—	.140	.135	1600③	.050	.325	—	2 Rich
FORD											
1975	D5AE-CA	①	31/32	#1	④	.160	1350	1/32	.300	.120	2 Rich
	D5AE-DA	①	31/32	#1	④	.160	1350	1/32	.300	.120	2 Rich
	D5VE-AD	①	15/16	#1	④	.160	1350	1/16	.300	⑤	2 Rich
	D5VE-BA	①	15/16	#1	④	.160	1350	1/16	.300	⑤	2 Rich
1976	D6AE-CA	①	1.00	#2	⑥	.140	1350	.030	.300	—	2 Rich
	D6AE-DA	①	.960	#2	⑦	.160	1350	.030	.300	—	2 Rich
	D6AE-FA	①	1.00	#2	⑥	.140	1350	.030	.300	—	2 Rich

①—Air/fuel ratio or idle CO% rating is found in Tune Up Specification tables in car chapters.
②—In lower hole of throttle shaft lever assembly.
③—On second step of fast idle cam with TCS solenoid and EGR disconnected.
④—Initial setting; .160". Delayed setting; .190".
⑤—Check setting before disassembly and reset to original setting.
⑥—Initial setting; .140". Delayed setting; .190".
⑦—Initial setting; .160". Delayed setting; .210".

MODEL 4350-4V ADJUSTMENTS

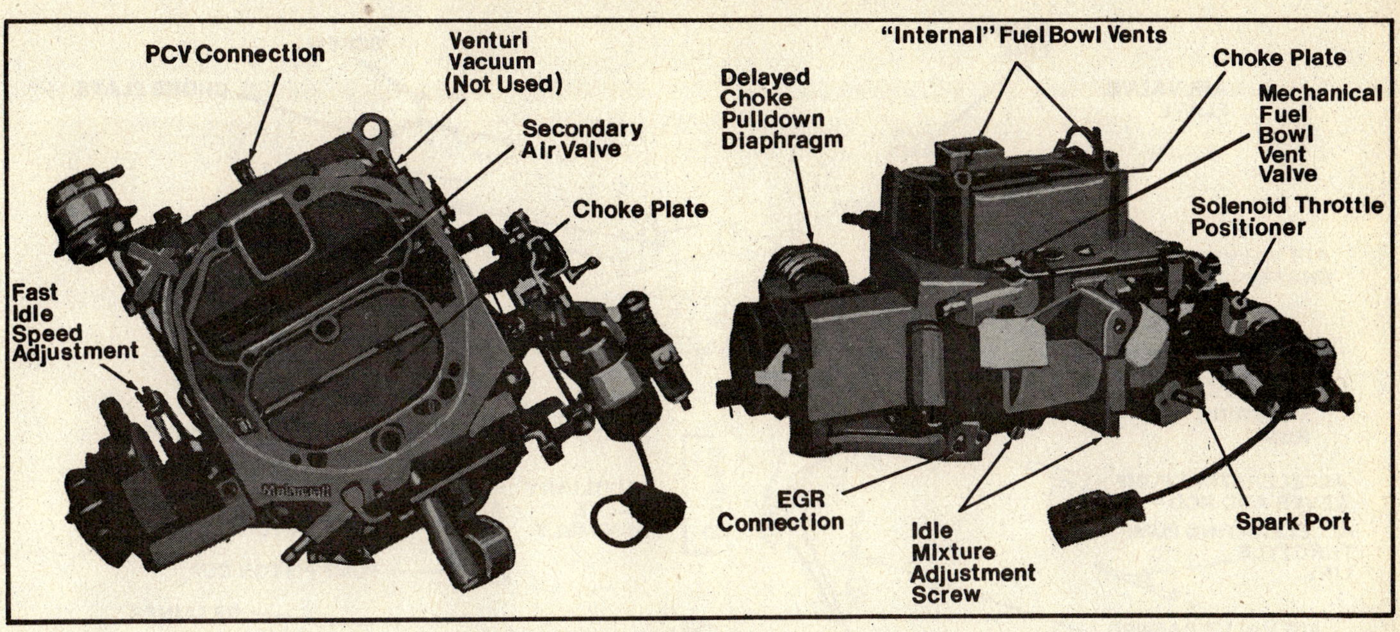

Fig. 1 Motorcraft 4350 4V carburetor assembly

The model 4350, Figs. 1 and 2, is a four barrel, three piece, separately cast design consisting of an upper body, main body and throttle body. The fuel bowl is vented by a mechanical fuel bowl vent valve to the carbon canister which opens when the engine is off and closes when the engine is running.

The upper main body consists of the choke plate, boost venturi assembly, secondary air valve, acceleration pump plunger, mechanical fuel vent valve and fuel inlet and float system.

The main body consists of the fuel bowl, primary fuel metering system vacuum piston and rods.

The throttle body consists of the primary and secondary throttle plates. The primary throttle is actuated by the accelerator linkage and the secondary throttle is actuated by a linkage from the primary throttle lever. A lockout tang coupled to the automatic choke linkage, prevents operation until choke is fully off. The secondary throttle shaft is split and loosely coupled in the center to permit tighter secondary plate seating.

Primary fuel metering is accomplished by tapered metering rods that are raised or lowered according to engine vacuum and are limited by a mechanical system and throttle plate opening.

Fuel inlet is controlled by twin floats on either side of the carburetor. The piston type accelerator pump is suspended into the pump well by a linkage and operating rod which are part of upper body.

The choke system consists of a bimetallic thermostatic choke system which operates when underhood temperatures are below 60°F and an electric assist choke system which operates when underhood temperatures are above 60°F. On Ford models, this carburetor is equipped with a vacuum operated delayed choke system, Fig. 4, which operates in addition to the usual pulldown system. On American Motors models, a thermostatic bypass valve, which is integral with the choke heat tube, helps prevent premature choke valve opening. For service information on the choke system, refer to the "Emission Control System" chapter.

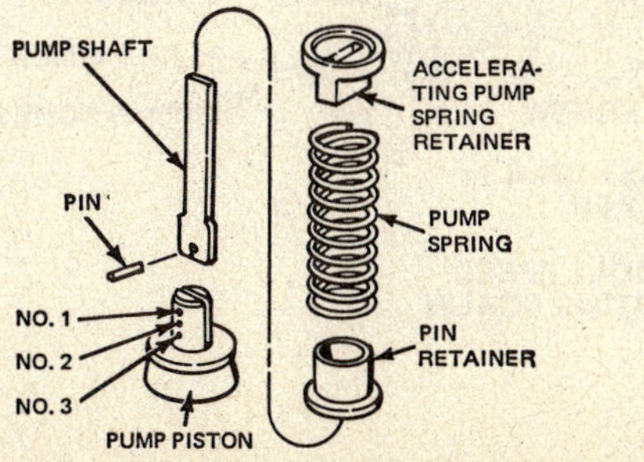

Fig. 3 Accelerator pump stroke adjustment

Accelerator Pump Adjustment

The piston-to-shaft pin position which is the only adjustment, is pre-set to deliver the correct amount of fuel for the engine on which it is installed and should not be changed from the specified setting. Do not attempt to adjust the accelerator pump stroke by turning the vacuum limiter level adjusting nut. This adjustment is pre-set and changing it could affect driveability.

1. Remove air horn assembly, then disconnect accelerator pump from operating lever by depressing spring and sliding arm out of pump shaft slot.
2. Disassemble spring and nylon keeper retaining adjustment pin. Refer to 4350 specifications chart. If pin is not in specified hole, remove it, reposition shaft in specified hole and reinstall pin, Fig. 3.
3. Slide nylon retainer over pin and position spring on shaft, then compress spring on shaft and install pump on arm.

Vent Valve Adjustment

1. With engine at curb idle speed and solenoid throttle positioner (if used)

Fig. 4 Mechanical fuel vent valve adjustment

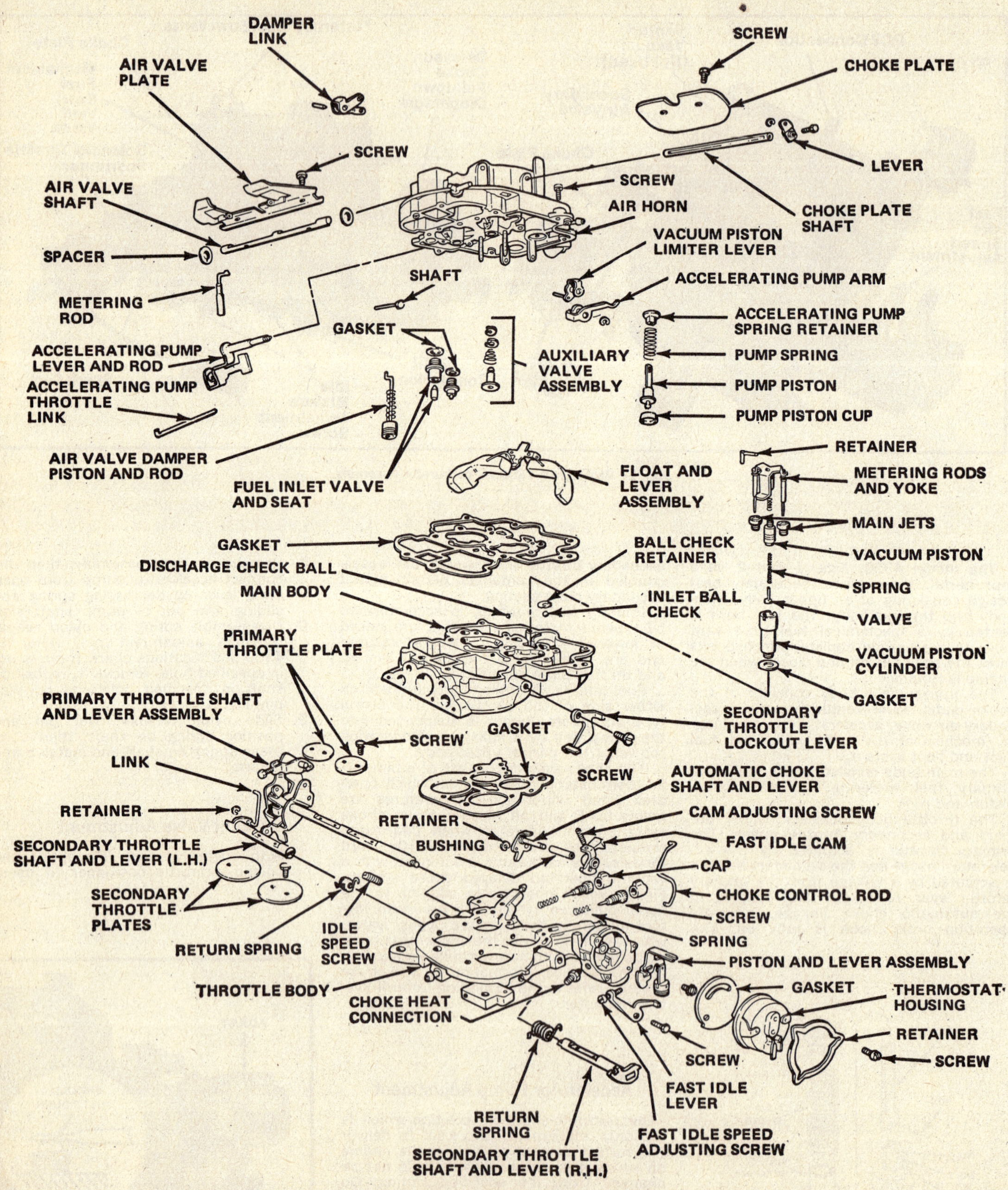

Fig. 2 Exploded view of 4350 4V carburetor

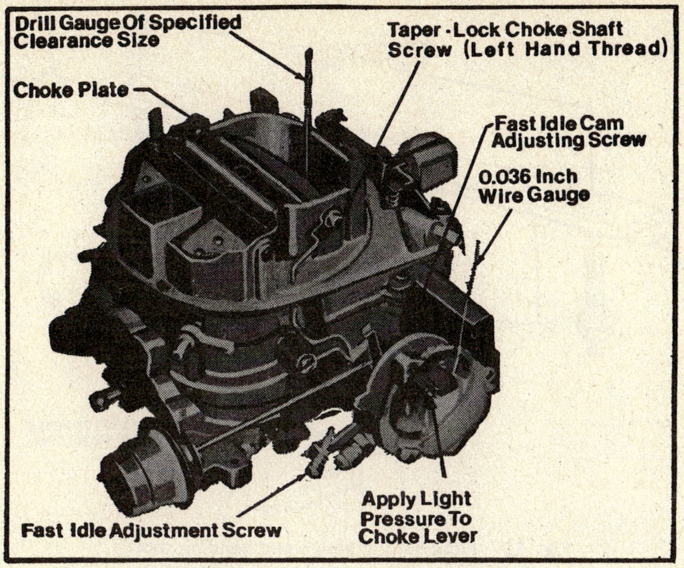

Fig. 5 Initial choke plate pulldown adjustment

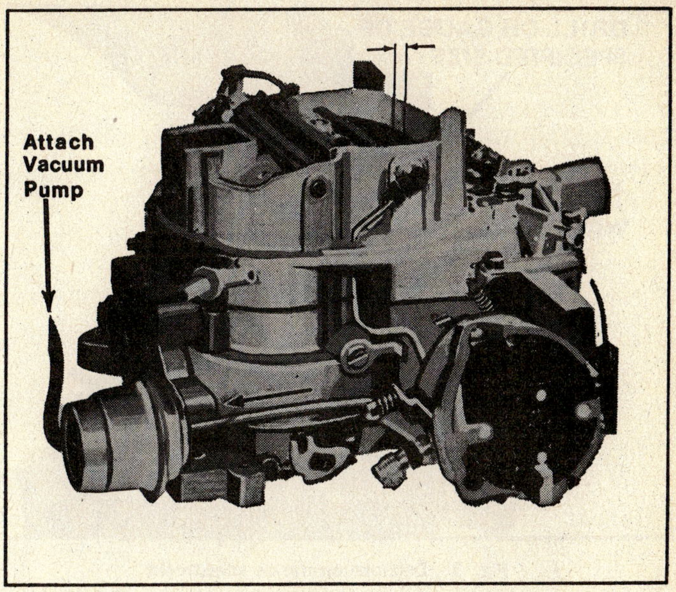

Fig. 6 Choke plate delayed pulldown adjustment

in extended position, adjust vent valve so that arm is just touching valve stem, Fig. 4.

2. With engine off, observe that vent operating linkage pushes down on vent valve stem.

Initial Choke Pulldown

1. Remove choke housing assembly, then open throttle about half-way and position fast idle adjusting screw on high step of fast idle cam.
2. Bend a .036″ wire gauge at 90° angle, about ⅛″ from end, then insert bent end of gauge between choke piston slot and upper edge of right hand slot in choke housing, Fig. 5.
3. Rotate choke lever counterclockwise until gauge is snug in piston slot, then apply light force to lever to hold gauge and move top of choke rod away from carburetor while moving bottom of rod toward carburetor to remove end play from linkage.
4. Using a drill gauge or pin of specified size, check clearance between lower edge of choke plate and wall, Fig. 5. Refer to 4350 specifications chart.
5. Adjust clearance to specifications, by turning lock screw on choke plate shaft 3 full turns in a clockwise direction. Pry choke lever from shaft to break taper lock. Choke lever should rotate freely on choke shaft.
6. With drill gauge and .036″ wire gauge positioned as above, tighten, screw on choke shaft.
7. Install choke housing assembly.

Delayed Choke Pulldown

The delayed choke pulldown opens the choke to a wider setting about 6 to 18

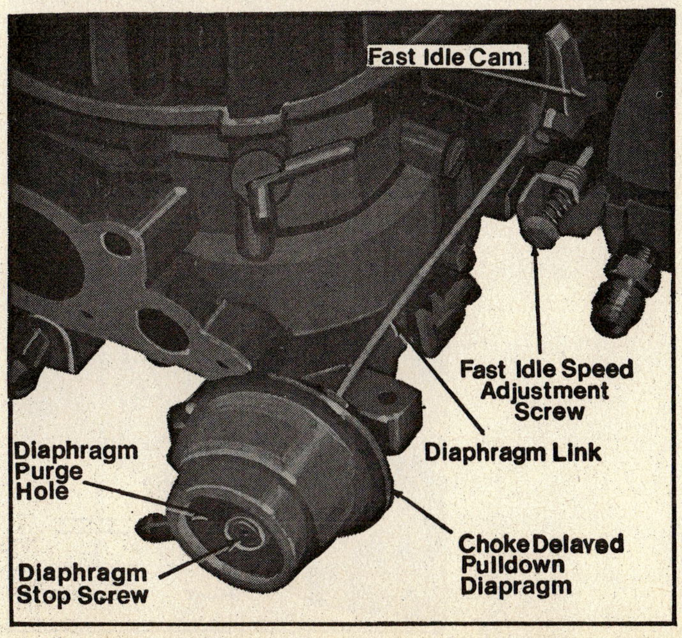

Fig. 7 Choke plate delayed pulldown stop screw adjustment

Fig. 8 Fast idle top step pulloff adjustment

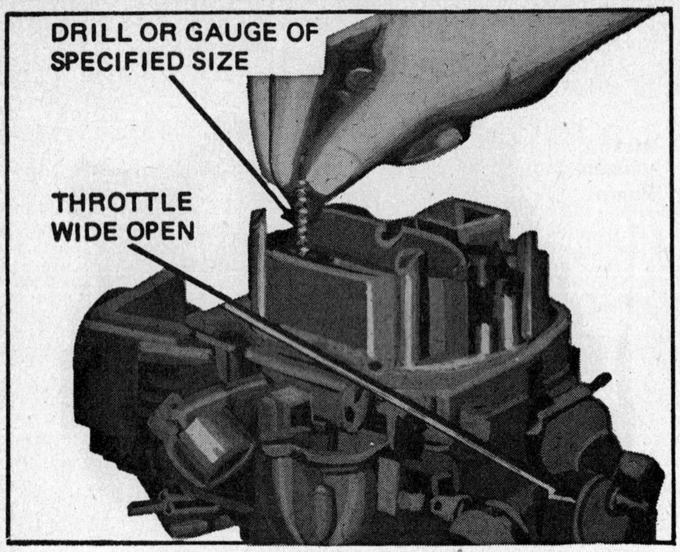

Fig. 9 Dechoke clearance adjustment

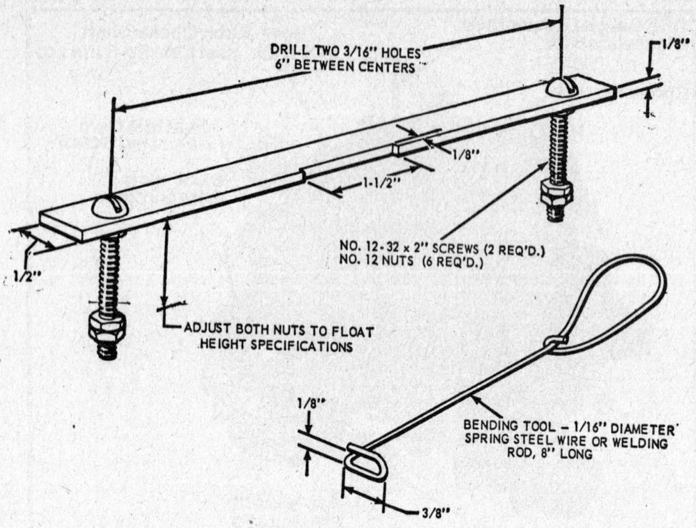

Fig. 10 Float setting gauge and bending tool

seconds after the engine is started.

1. With throttle set on fast idle cam, note position of index marks on choke housing, then loosen retaining screws and rotate cap 90° in closing (rich) direction.
2. Connect a vacuum source of 14-18 inches Hg. to vacuum supply port of delayed choke pulldown diaphragm and check clearance between lower edge of choke plate and wall, Fig. 6. Refer to 4350 specifications chart.
3. Adjust clearance to specifications by turning stop screw on delayed choke pulldown diaphragm, Fig. 7.

Fast Idle Top Step Pulloff

1. Operate delayed pulldown diaphragm manually or by applying 14 to 18 inches Hg.
2. Position fast idle speed screw on top step fast idle cam with choke plate closed. If necessary, rotate choke housing cap to close choke plate.
3. Observe that fast idle speed adjusting screw drops to second step of cam as delayed choke pulldown diaphragm is operated, Fig. 8.

Dechoke (Unloader) Clearance

1. Hold throttle plate fully open and apply pressure on choke valve toward close position.
2. Measure clearance between lower edge of choke valve and air horn wall. Fig. 9. Refer to 4350 Specifications Chart.
3. Adjust clearance to specifications by bending unloader tang which contacts fast idle cam forward to increase or rearward to decrease clearance.

NOTE: Do not bend tang downward from a horizontal plane. After adjustment, make certain that there is at least .070 inch clearance between unloader tang and choke housing with throttle fully open.

Float Setting

To simplify float setting, refer to Fig. 10 for construction of an adjustable float gauge and float tab bending tool.

1. Adjust gauge to specified height and insert gauge into air horn, Fig. 11.
2. Check clearance and alignment of floats to gauge. Refer to 4350 Specifications Chart. Both floats should just touch gauge. If necessary, align floats by twisting floats slightly.
3. To raise float setting, insert open end of bending tool to right side of float lever tab between needle and float hinge. Raise float lever off needle and bend tab downward.
4. To lower float setting, insert open end

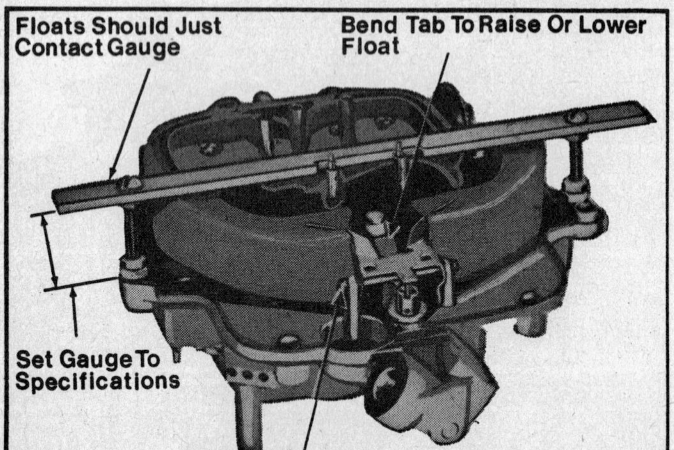

Fig. 11 Float setting

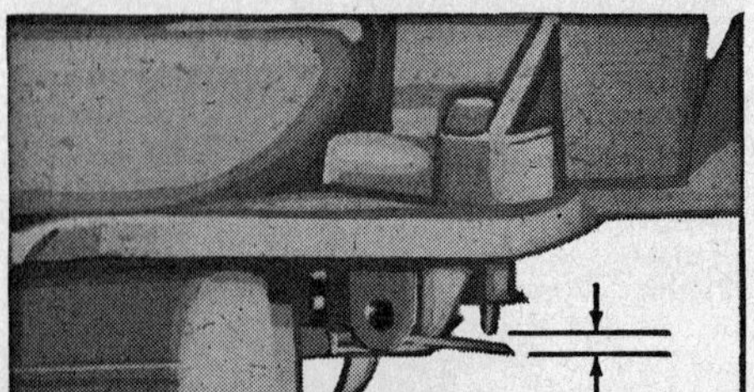

Fig. 12 Auxiliary inlet valve adjustment

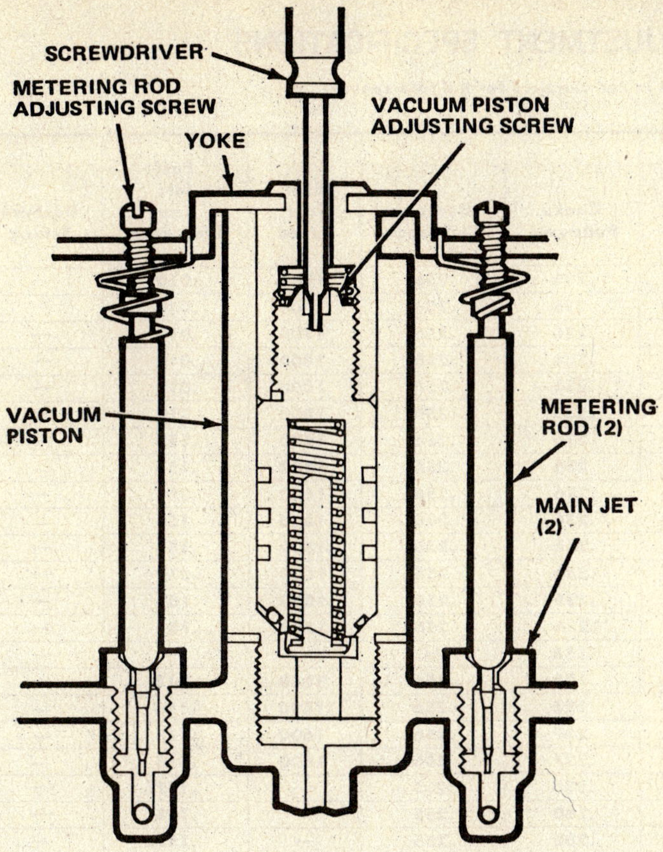

Fig. 13 Metering rod adjustment

of bending tool to left side of float lever tab between needle and float hinge. Support float lever and bend tap upward.

5. If above gauge is not available, invert upper body and measure distance from top of float to gasket surface of upper body. Adjust alignment and height of floats with bending tool as described previously.

Auxiliary Valve Setting

Check auxiliary valve clearance as shown in Fig. 12. If necessary to adjust clearance, use bending tool, Fig. 10.

Vacuum Piston & Metering Rods Adjustment

1. Remove air horn assembly, then depress metering rod hanger and turn vacuum piston adjusting screw and metering rod adjusting screws until metering rod hanger is fully seated against vacuum cylinder top face, Fig. 13.
2. Hold metering rod hanger in full downward position and turn each metering rod adjusting screw clockwise until the metering rod hanger starts to rise. Metering rods are now properly positioned in relation to vacuum piston.
3. Turn vacuum piston adjusting screw clockwise until specified clearance between metering rod hanger and vacuum cylinder top face is obtained.

Fast Idle Cam

1. Loosen choke cover screws and rotate cover to align index marks on cover and housing. Then, rotate cover an additional 90° counterclockwise and tighten attaching screws.
2. Place fast idle speed adjusting screw on kickdown (center) step of fast idle cam. Fully close choke plate and check clearance between air horn and lower edge of choke plate. Adjust by turning fast idle cam adjusting screw, ensuring adjusting screw remains on kickdown step during adjustment.
3. Loosen choke cove screws and rotate cover to specified index mark and tighten cover screws.

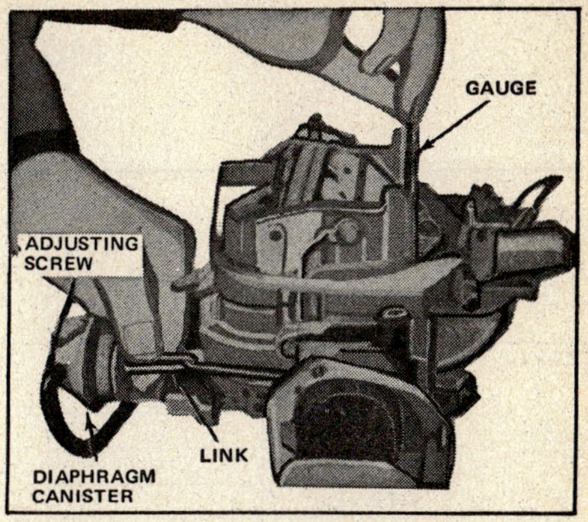

Fig. 14 Initial choke valve clearance adjustment. 1975 American Motors units with vacuum diaphragm

American Motors Units With Vacuum Diaphragm

Initial Choke Valve Clearance

1. Loosen choke cover screws, open throttle and rotate choke cover until choke valve closes. Then, tighten one cover screw to hold cover in place.
2. Close throttle. The fast idle screw should be on top step of cam.
3. Bottom choke diaphragm on setscrew, however, do not press on links, Fig. 14.
4. Measure clearance between air horn and lower edge of choke valve, Fig. 14. Adjust by turning adjusting screw at rear of diaphragm housing.
5. Adjust fast idle cam linkage.
6. Loosen choke cover screw and set choke cover to specifications.

Choke Plate Indexing

1. Loosen choke lever attaching screw and pry lever from choke shaft to permit choke valve rotation.

NOTE: The choke lever attaching screw has a left hand thread.

2. Loosen choke cover screws and rotate cover 90° counterclockwise so the choke shaft lever contacts cam adjusting screw.
3. Position fast idle screw on top step of fast idle cam. Then, back off cam adjusting screw until clearance is obtained between screw and choke shaft lever.
4. Turn cam adjusting screw until it contacts choke shaft lever, then turn 6-7 additional turns.
5. Manually close choke plate and tighten choke shaft lever attaching screw.
6. Adjust initial choke valve clearance and fast idle cam linkage.
7. Set choke cover and tighten cover screws.

MODEL 5200 ADJUSTMENT SPECIFICATIONS

See Tune Up Chart in car chapters for hot idle speeds.

Year	Carb. Model (9510) ①	Idle Mixture Turns Open	Float Level	Pump Setting (Hole)	Choke Pulldown	Dechoke Clearance	Fast Idle Speed	Fast Idle Cam Clearance	Dashpot Setting	Choke Setting
1971	D12F-AA	②	.420	Lower	.236	.256	1800	.010	—	Index
	D12F-BA	②	.420	Lower	.236	.256	1600	.010	—	Index
	D12F-CA	②	.420	Lower	.236	.256	1800	.010	—	Index
	D12F-DA	②	.420	Lower	.236	.256	1800	.010	—	1 Rich
	D12F-EA	②	.420	Lower	.236	.256	1600	.010	—	1 Rich
	D12F-FA	②	.420	Lower	.236	.256	1800	.010	—	1 Rich
1972	D22F-AB	②	.420	#3	.236	.256	1800	.156	—	1 Lean
	D22F-AC	②	.420	#3	.236	.256	1800	.156	—	1 Lean
	D22F-BB	②	.420	#2	.236	.256	1600	.079	—	1 Lean
	D22F-EB	②	.420	#3	.236	.256	1800	.156	—	Index
	D22F-CB	②	.420	#3	.236	.256	1800	.156	—	1 Lean
	D22F-DB	②	.420	#2	.236	.256	1600	.079	—	1 Lean
	D22F-EA	②	.420	#3	.236	.256	1800	.156	—	Index
	D22F-GA	②	.420	#3	.236	.256	1800	.156	—	Index
1973	D32F-AC	②	.420	#2	.158	—	—	—	—	Index
	D32F-BD	②	.420	#2	.158	.256	1600	.118	—	1 Lean
	D32F-CA	②	.420	#2	.158	.256	1800	.158	—	Index
	73TF-AA	②	.420	#2	.237	.256	1600	.118	—	Lean
	73TF-BA	②	.420	#2	.237	.256	1800	.118	—	Index
1974	D42E-AA	②	.460	#2	.280	.255	—	.1575	—	Index
	D42E-AC	②	.460	#2	.280	.255	—	.158	—	Index
	D42E-BA	②	.460	#2	.280	.255	—	.1575	—	1 Rich
	D42E-CB	②	.460	#2	.280	.255	—	.1575	—	Index
	D42E-CD	②	.460	#2	.280	.255	—	.158	—	Index
	D42E-DB	②	.460	#2	.280	.255	—	.1575	—	1 Rich
	D42F-EA	②	.460	#2	.236	.255	—	.1575	—	Index
	D42E-EB	②	.460	#2	.236	.255	—	.158	—	Index
	D42F-FA	②	.460	#2	.236	.255	—	.1575	—	Index
	D42F-GA	②	.460	#2	.236	.255	—	.1575	—	Index
	D42E-KA	②	.460	#2	.280	.255	—	.158	—	1 Rich
	D4ZE-AA	②	.430	#2	.195	.256	—	.195	—	1 Rich
	D4ZE-BA	②	.430	#2	.195	.256	—	.195	—	1 Rich
	D4ZE-BC	②	.435	#2	.195	.255	—	.195	—	1 Rich
	D4ZE-CA	②	.430	#2	.195	.256	—	.195	—	1 Rich
	D4ZE-DA	②	.430	#2	.195	.256	—	.195	—	1 Rich
	D4ZE-DC	②	.435	#2	.195	.255	—	.195	—	1 Rich
1975	D52E-AA	②	.460	#2	.197	.256	1800	.098	—	1 Lean
	D52E-BA	②	.460	#2	.197	.256	1800	.098	—	1 Lean
	D52E-CA	②	.460	#2	.197	.256	1800	.098	—	1 Lean
	D52E-DB	②	.460	#2	.197	.256	1800	.098	—	1 Lean
	D5ZE-EA	②	.460	#2	.197	.256	1800	.098	—	1 Lean
	D5ZE-FA	②	.460	#2	.197	.256	1800	.098	—	1 Lean
	D5ZE-GA	②	.460	#2	.197	.256	1800	.098	—	1 Lean
	D5ZE-HB	②	.460	#2	.197	.256	1800	.098	—	1 Lean
1976	D6EE-BA	②	.460	#2	.236	.256	1800	.118	—	1 Lean
	D6EE-CA	②	.460	#2	.276	.256	1800	.158	—	1 Lean
	D6EE-DA	②	.460	#2	.236	.256	1800	.118	—	1 Lean
	D6ZE-EA	②	.460	#2	.276	.256	1800	.158	—	1 Lean

①—Tag attached to carburetor
②—Air/fuel ratio or idle CO% rating is found in Tune Up Specification tables in car chapters.

MODEL 5200 ADJUSTMENTS

This carburetor is a two stage, two venturi carburetor, Figs. 1 and 5. The primary stage or venturi is smaller than the secondary venturi. The secondary is operated by mechanical linkage.

The primary stage includes a curb idle system, accelerator pump system, idle transfer system, main metering system and power enrichment system.

The secondary stage includes a transfer system, main metering system, and power system. Both the primary and secondary systems draw fuel from a common fuel bowl.

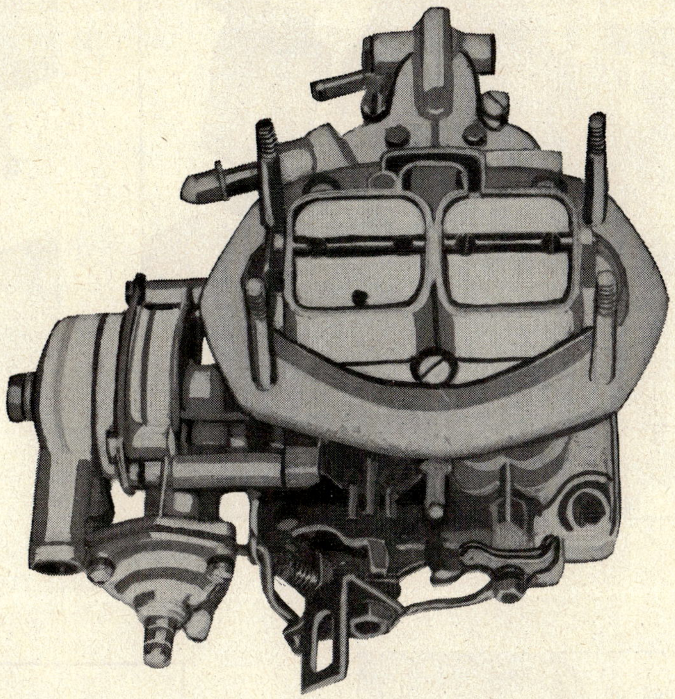

Fig. 1 Model 5200-2V carburetor

Dry Float Setting

Fig. 2—With the bowl cover held in an inverted position and the float tang resting lightly on the spring loaded fuel inlet needle, measure the clearance between the edge of the float and the bowl cover. Adjust clearance by bending the float tang up or down as required, Fig. 3.

NOTE: Do not scratch or damage the tang. Adjust both floats equally.

Dechoke Clearance

Fig. 4—Hold throttle lever in wide open position and take slack out of choke linkage by applying finger pressure to top edge of choke plate. Measure clearance between lower edge of choke plate and air horn wall. Adjust by bending tab on fast idle lever where it touches the fast idle cam.

Choke Plate Vacuum Pull-Down

Fig. 6—Remove three screws and ring retaining choke spring cover and pull the water cover and choke spring cover out of way. Set the fast idle cam on the top step. Push the diaphragm stem back against its stop. Place gauge rod or drill between the lower edge of the choke plate and the air horn wall. Remove the slack from the choke linkage by applying finger pressure to the top edge of the choke plate. Adjust the choke plate-to-air horn clearance by removing the plug from the diaphragm and turning the adjusting screw in or out as required.

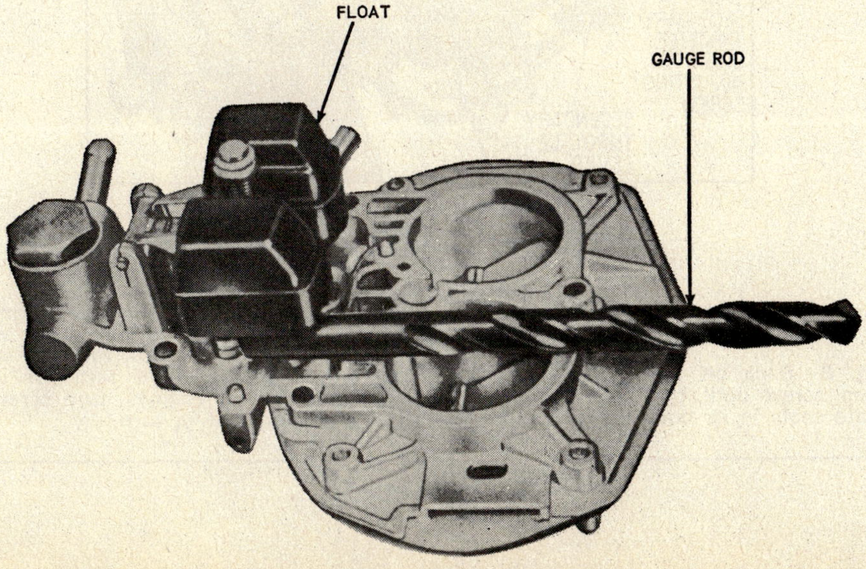

Fig. 2 Dry float setting

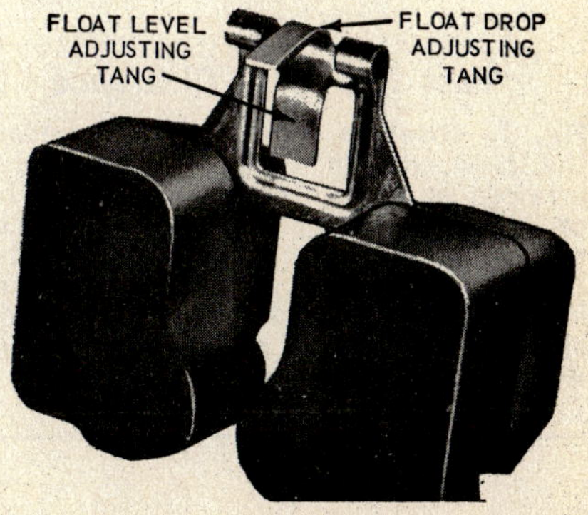

Fig. 3 Float adjusting point

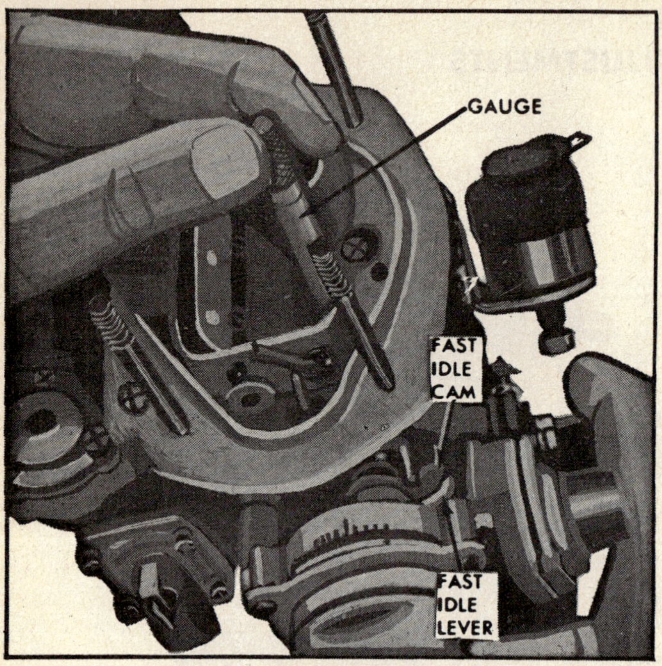

Fig. 4 De-choke adjustment

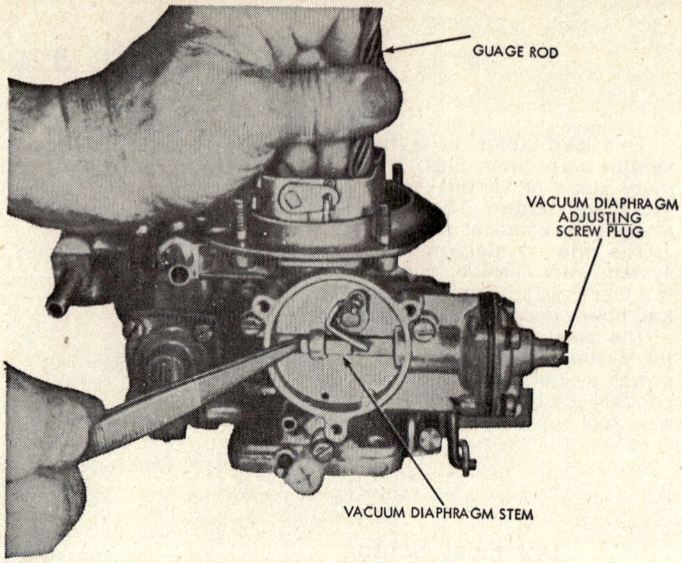

Fig. 6 Choke plate pull-down

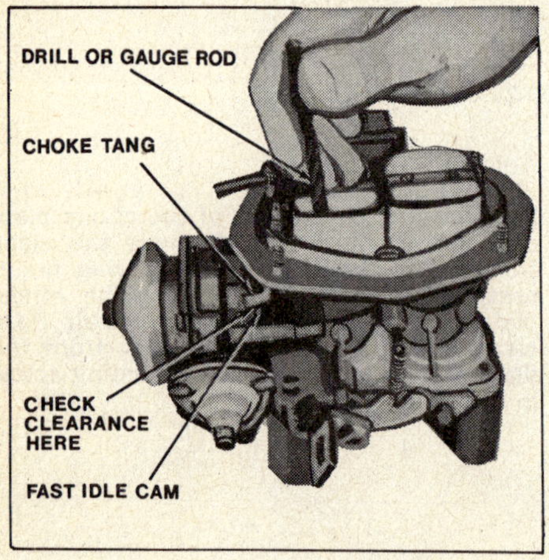

Fig. 7 Fast idle cam clearance

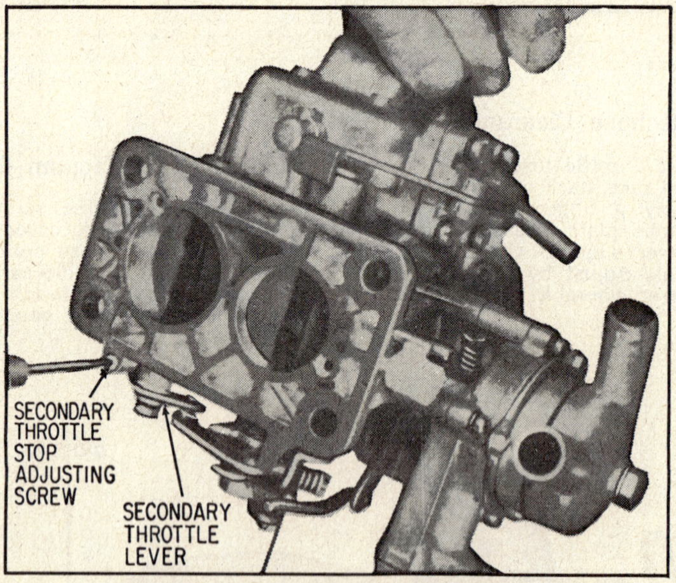

Fig. 8 Secondary throttle stop adjustment

Fast Idle Cam Clearance

Fig. 7—Insert a 5/32 inch drill between the lower edge of the choke plate and the air horn wall. With the fast idle screw held on the second step of the fast idle cam, measure the clearance between the tang of the choke lever and the arm on the fast idle cam. Adjust clearance by bending choke lever tang up or down as required. Refer to *5200 Specifications Chart.*

Secondary Throttle Stop Screw Adjustment

Fig. 8—Back off the secondary throttle stop screw until the secondary throttle plate seats in its bore. Turn the screw in until it touches the tab on the secondary throttle lever. On all units except when used on 2800 cc engines, turn the screw inward an additional $\frac{1}{4}$ turn. On units used on 2800 cc engines, turn screw inward an additional $\frac{3}{4}$ turn.

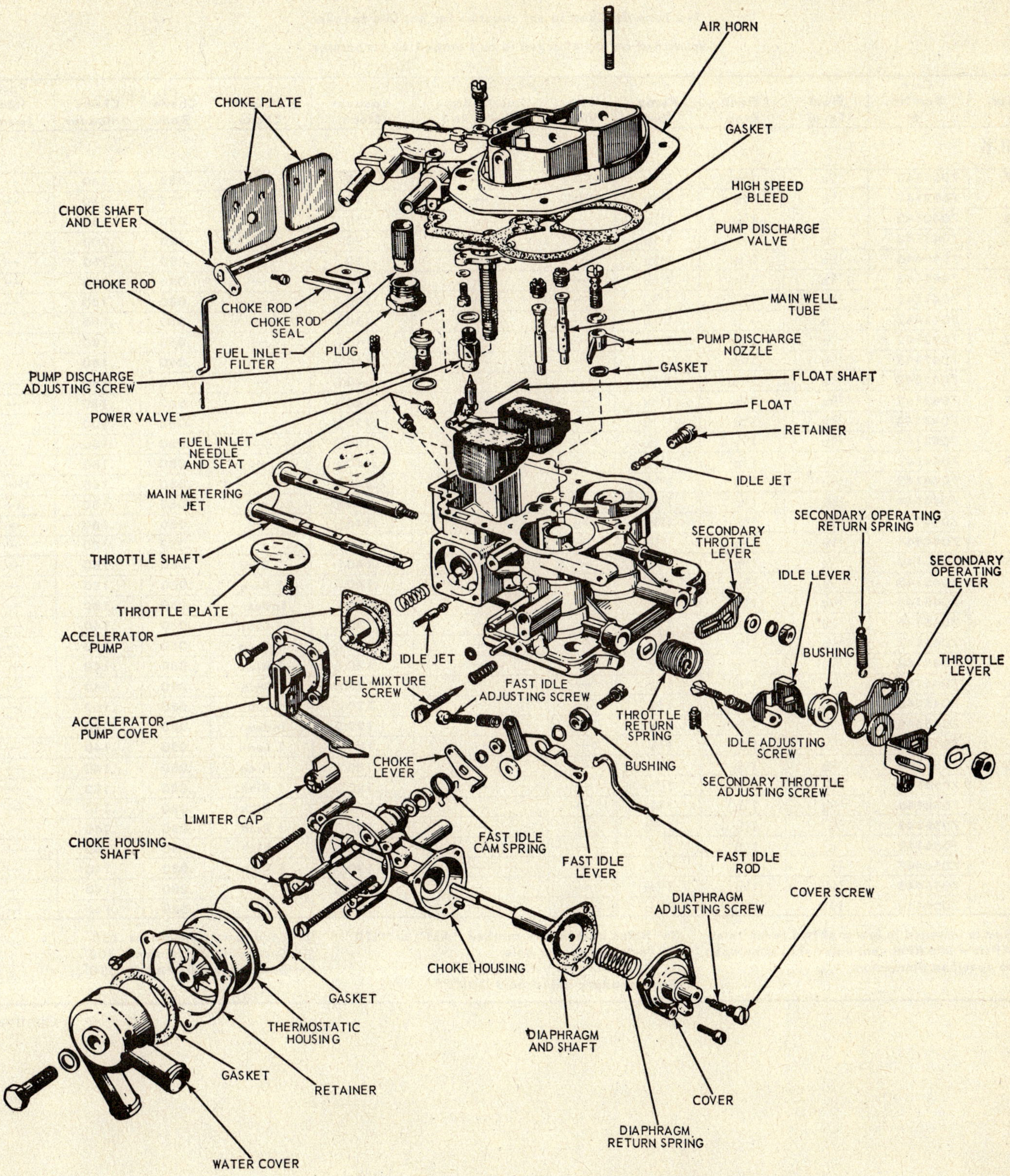

Fig. 5 Exploded view of model 5200 carburetor. All bolts and screws are Metric thread. Early production Metric heads; later production U.S. heads

GM Delco/Rochester Carburetor Section

MODEL 2GC, 2GV CARBURETOR ADJUSTMENT SPECIFICATIONS

See Tune Up Chart in car chapters for hot idle speeds.

★Located on tag attached to or stamped on carburetor.

Year	Carb. Part No. ★	Float Level	Float Drop	Pump Rod	Idle Vent	Intermediate Choke Rod	Vacuum Break	Automatic Choke	Choke Rod	Choke Unloader	Fast Idle Speed
BUICK											
1969	7029140	15/32	1 7/32	1 11/32	.020②	—	.110	③	.055	.140	—
	7029141	15/32	1 7/32	1 11/32	.020②	—	.110	③	.055	.140	—
1970	7040142	15/32	1 7/32	1 13/32	—	—	.150	③	.080	.180	—
	7041143	15/32	1 7/32	1 15/32	—	—	.190	③	.100	.200	—
	7040446	15/32	1 7/32	1 13/32	—	—	.150	③	.080	.180	—
1971	7041143	15/32	1 7/32	1 15/32	—	—	.160④	—	.080	.200	—
	7041142	15/32	1 7/32	1 15/32	—	—	.150④	—	.080	.180	—
	7041442	15/32	1 7/32	1 15/32	—	—	.150④	—	.080	.180	—
1972	7042142	15/32	1 11/32	1 15/32	—	—	.140⑤	—	.080	.180	—
	7042143	15/32	1 11/32	1 15/32	—	—	.160④	—	.080	.180	—
	7042842	15/32	1 11/32	1 15/32	—	—	.140⑤	—	.080	.180	—
1973	7043142	15/32	1 9/32	1 15/32	—	—	.140⑤	—	.080	.180	—
	7043143	15/32	1 9/32	1 15/32	—	—	.150⑤	—	.080	.200	—
	7043144	15/32	1 9/32	1 15/32	—	—	.140⑤	—	.080	.180	—
1974	7044141	15/32	1 9/32	1 15/32	—	—	.160⑥	—	.080	.180	—
	7044142	15/32	1 9/32	1 15/32	—	—	.140⑤	—	.080	.180	—
	7044144	15/32	1 9/32	1 15/32	—	—	.140⑤	—	.080	.180	—
	7044442	15/32	1 9/32	1 15/32	—	—	.140⑤	—	.080	.180	—
	7044444	15/32	1 9/32	1 15/32	—	—	.140⑤	—	.080	.180	—
1975	7045140	15/32	1 9/32	1 3/4	—	—	.140⑤	1 Rich	.080	.180	—
	7045143	15/32	1 9/32	1 3/4	—	—	.140⑤	1 Rich	.080	.180	—
	7045145	15/32	1 9/32	1 3/4	—	—	.120⑤	Index	.080	.140	—
	7045146	15/32	1 9/32	1 3/4	—	—	.120⑤	1 Lean	.080	.140	—
	7045147	15/32	1 9/32	1 3/4	—	—	.120⑤	1 Lean	.080	.140	—
	7045148	15/32	1 9/32	1 3/4	—	—	.120⑤	1 Rich	.080	.140	—
	7045149	15/32	1 9/32	1 3/4	—	—	.120⑤	1 Rich	.080	.140	—
	7045446	15/32	1 9/32	1 3/4	—	—	.120⑤	1 Lean	.080	.140	—
	7045448	15/32	1 9/32	1 3/4	—	—	.120⑤	Index	.080	.140	—
	7045449	15/32	1 9/32	1 3/4	—	—	.120⑤	1 Lean	.080	.140	—
1976	7056140	15/32	1 9/32	1 19/32	—	—	.180④	1 Rich	.080	.180	—
	7056143	15/32	1 9/32	1 19/32	—	—	.180④	1 Rich	.080	.180	—
	7056145	13/32	1 9/32	1 3/4	—	—	.110⑦	1 Rich	.080	.140	—
	7056148	7/16	1 9/32	1 19/32	—	—	.120⑦	1 Rich	.080	.140	—
	7056149	7/16	1 9/32	1 19/32	—	—	.120⑦	1 Rich	.080	.140	—
	7056447	7/16	1 9/32	1 19/32	—	—	.130⑦	1 Rich	.080	.140	—
	7056448	7/16	1 9/32	1 19/32	—	—	.130⑧	—	.080	.140	—
	7056449	7/16	1 9/32	1 19/32	—	—	.130⑧	—	.080	.140	—

①—Rod is installed in lower of two lever holes.
②—At slow idle RPM vent valve should be open to specified dimension.
③—Holes in lever are marked "ALT" or "STD". Install in proper hole.
④—Secondary adjustment .140″.
⑤—Secondary adjustment .120″.
⑥—Secondary adjustment .130″.
⑦—Secondary adjustment, .100″.
⑧—Secondary adjustment, .110″.

Continued

MODEL 2GC, 2GV CARBURETOR ADJUSTMENT SPECIFICATIONS—Continued

See Tune Up Chart in car chapters for hot idle speeds.
★Located on tag attached to or stamped on carburetor.

CHECKER MOTORS

Year	Carb. Part No. ★	Float Level	Float Drop	Pump Rod	Idle Vent	Intermediate Choke Rod	Vacuum Break	Automatic Choke	Choke Rod	Choke Unloader	Fast Idle Speed
1969	7029102	3/4	1 3/8	1 13/32	.020	—	.215	—	.085	.275	2100
	7029104	3/4	1 3/8	1 13/32	.020	—	.215	—	.085	.275	2100
	7029124	3/4	1 3/8	1 7/16	—	—	—	—	—	—	—
	7029127	3/4	1 3/8	1 13/32	.020	—	.215	—	.085	.275	2100
	7029129	3/4	1 3/8	1 13/32	.020	—	.215	—	.085	.275	2100
1970	7040114	23/32	1 3/8	1 17/32	.025	—	.200	—	.085	.325	—
	7040116	23/32	1 3/8	1 17/32	.025	—	.200	—	.085	.325	—
	7040123	23/32	1 3/8	1 17/32	.020	—	—	—	—	—	—
	7040134	25/32	1 11/32	1 17/32	.020	—	.200	—	.085	.325	—
	7040136	25/32	1 11/32	1 17/32	.020	—	.200	—	.085	.325	—
	7040414	23/32	1 3/8	1 17/32	—	—	.200	—	.085	.325	—
	7040416	23/32	1 3/8	1 17/32	—	—	.200	—	.085	.325	—
	7040434	25/32	1 11/32	1 17/32	.020	—	.215	—	.085	.325	—
	7040436	25/32	1 11/32	1 17/32	.020	—	.215	—	.085	.325	—
1971	7041114	25/32	1 5/8	1 17/32	—	—	.170	—	.100	.325	—
	7041123	23/32	1 1/8	1 17/32	.020	—	—	—	—	—	—
1972	7042114	23/32	1 9/32	1 1/2	—	—	.170	—	.100	.325	—
1973	7043114	19/32	1 9/32	1 7/16	—	—	.130	—	.245	.325	—
1974	7044114	19/32	1 9/32	1 3/16	—	—	.130	—	.245	.325	1600
1975	7045114	21/32	31/32	1 5/8	—	.120	.130	—	—	.350	—

CHEVROLET ENGINES

Year	Carb. Part No. ★	Float Level	Float Drop	Pump Rod	Idle Vent	Intermediate Choke Rod	Vacuum Break	Automatic Choke	Choke Rod	Choke Unloader	Fast Idle Speed
1969	7029101	27/32②	1 3/4	1 1/8	.020	—	.100	—	.060	.215	—
	7029102	3/4②	1 3/8	1 13/32	.020	—	.215	—	.085	.275	—
	7029103	27/32②	1 3/4	1 1/8	.020	—	.100	—	.060	.215	—
	7029104	3/4②	1 3/8	1 13/32	.020	—	.215	—	.085	.275	—
	7029110	27/32②	1 3/4	1 1/8	.020	—	.100	—	.060	.215	—
	7029112	27/32②	1 3/4	1 1/8	.020	—	.100	—	.060	.215	—
	7029113	3/4②	1 3/8	1 13/32	.020	—	.200	—	.085	.275	—
	7029114	3/4②	1 3/8	1 13/32	.020	—	.200	—	.085	.275	—
	7029115	3/4②	1 3/8	1 13/32	.020	—	.200	—	.085	.275	—
	7029116	3/4②	1 3/8	1 13/32	.020	—	.200	—	.085	.275	—
	7029117	3/4②	1 3/4	1 13/32	.020	—	.215	—	.085	.275	—
	7029118	3/4②	1 3/4	1 13/32	.020	—	.215	—	.085	.275	—
	7029119	5/8②	1 3/4	1 13/32	.020	—	.215	—	.085	.275	—
	7029120	5/8②	1 3/4	1 13/32	.020	—	.215	—	.085	.275	—
	7029127	3/4②	1 3/8	1 13/32	.020	—	.215	—	.085	.275	—
	7029129	3/4②	1 3/8	1 13/32	.020	—	.215	—	.085	.275	—
1970	7040101	27/32②	1 3/4	1 3/8	.025	—	.130	—	.060	.160	—
	7040102	23/32②	1 3/8	1 17/32	.025	—	.200	—	.085	.325	—
	7040103	27/32②	1 3/4	1 3/8	.025	—	.130	—	.060	.160	—
	7040104	23/32②	1 3/8	1 17/32	.025	—	.200	—	.085	.325	—
	7040110	27/32②	1 3/4	1 3/8	.025	—	.100	—	.060	.215	—
	7040112	27/32②	1 3/4	1 3/8	.025	—	.100	—	.060	.215	—
	7040113	23/32②	1 3/8	1 17/32	.025	—	.215	—	.085	.275	—
	7040114	23/32②	1 3/8	1 17/32	.025	—	.200	—	.085	.325	—
	7040115	23/32②	1 3/8	1 17/32	.025	—	.215	—	.085	.275	—
	7040116	23/32②	1 3/8	1 17/32	.025	—	.200	—	.085	.325	—
	7040117	23/32②	1 3/8	1 17/32	.025	—	.215	—	.085	.325	—
	7040118	23/32②	1 3/8	1 17/32	.025	—	.215	—	.085	.325	—
	7040119	23/32②	1 3/8	1 17/32	.025	—	.215	—	.085	.325	—
	7040120	23/32②	1 3/8	1 17/32	.025	—	.215	—	.085	.325	—

Continued

MODEL 2GC, 2GV CARBURETOR ADJUSTMENT SPECIFICATIONS—Continued

See Tune Up Chart in car chapters for hot idle speeds.

★Located on tag attached to or stamped on carburetor.

Year	Carb. Part No. ★	Float Level	Float Drop	Pump Rod	Idle Vent	Intermediate Choke Rod	Vacuum Break	Automatic Choke	Choke Rod	Choke Unloader	Fast Idle Speed
CHEVROLET ENGINES—Continued											
1970	7040126	25/32②	1 11/32	1 17/32	.020	—	.200	—	.085	.325	—
	7040127	23/32②	1 3/8	1 17/32	.025	—	.215	—	.085	.275	—
	7040128	25/32②	1 11/32	1 17/32	.020	—	.200	—	.085	.325	—
	7040129	23/32②	1 3/8	1 17/32	.025	—	.215	—	.085	.275	—
	7040134	25/32②	1 11/32	1 17/32	.020	—	.200	—	.085	.325	—
	7040136	25/32②	1 11/32	1 17/32	.020	—	.200	—	.085	.325	—
	7040401	27/32②	1 3/4	1 5/16	—	—	.130	—	.060	.160	—
	7040402	23/32②	1 3/8	1 17/32	—	—	.200	—	.085	.325	—
	7040403	27/32②	1 3/4	1 5/16	—	—	.130	—	.060	.160	—
	7040404	23/32②	1 3/8	1 17/32	—	—	.200	—	.085	.325	—
	7040410	27/32②	1 3/4	1 5/16	—	—	.100	—	.060	.215	—
	7040412	27/32②	1 3/4	1 5/16	—	—	.100	—	.060	.215	—
	7040413	23/32②	1 3/8	1 17/32	—	—	.215	—	.085	.275	—
	7040414	23/32②	1 3/8	1 17/32	—	—	.200	—	.085	.325	—
	7040415	23/32②	1 3/8	1 17/32	—	—	.215	—	.085	.275	—
	7040416	23/32②	1 3/8	1 17/32	—	—	.200	—	.085	.325	—
	7040417	23/32②	1 3/8	1 17/32	—	—	.215	—	.085	.325	—
	7040418	23/32②	1 3/8	1 17/32	—	—	.215	—	.085	.325	—
	7040419	23/32②	1 3/8	1 17/32	—	—	.215	—	.085	.325	—
	7040420	23/32②	1 3/8	1 17/32	—	—	.215	—	.085	.325	—
	7040426	25/32②	1 11/32	1 17/32	.020	—	.215	—	.085	.325	—
	7040427	23/32②	1 3/8	1 17/32	—	—	.215	—	.085	.275	—
	7040428	25/32②	1 11/32	1 17/32	.020	—	.215	—	.085	.325	—
	7040429	23/32②	1 3/8	1 17/32	—	—	.215	—	.085	.275	—
	7040434	25/32②	1 11/32	1 17/32	.020	—	.215	—	.085	.325	—
	7040436	25/32②	1 11/32	1 17/32	.020	—	.215	—	.085	.325	—
1971	7041101	27/32②	1 3/4	1 5/16	—	—	.110	—	.075	.215	—
	7041102	25/32②	1 3/8	1 17/32	—	—	.170	—	.100	.325	—
	7041106	9/16②	1 7/8	1 3/16	—	—	.085	—	.080	.200	—
	7041107	9/16②	1 7/8	1 3/16	—	—	.100	—	.080	.200	—
	7041110	27/32②	1 3/4	1 5/16	—	—	.080	—	.040	.215	—
	7041113	23/32②	1 3/8	1 17/32	—	—	.180	—	.100	.325	—
	7041114	25/32②	1 3/8	1 17/32	—	—	.170	—	.100	.325	—
	7041117	23/32②	1 3/8	1 17/32	—	—	.170	—	.100	.325	—
	7041118	23/32②	1 3/8	1 17/32	—	—	.170	—	.100	.325	—
	7041127	23/32②	1 3/8	1 17/32	—	—	.180	—	.100	.325	—
	7041137	23/32②	1 3/8	1 17/32	—	—	.180	—	.100	.325	—
	7041181	21/32②	1 7/8	1 3/8	—	—	.120	—	.080	.180	—
	7041182	21/32②	1 7/8	1 3/8	—	—	.120	—	.080	.180	—
1972	7042100	25/32	1 31/32	1 5/16	—	—	.080	—	.040	.215	—
	7042101	25/32	1 31/32	1 5/16	—	—	.110	—	.075	.215	—
	7042106	19/32	1 7/8	1 1/16	—	—	.085	—	.060	.215	2800
	7042107	19/32	1 7/8	1 1/16	—	—	.100	—	.080	.215	2800
	7042111	23/32	1 9/32	1 1/2	—	—	.180	—	.100	.325	—
	7042112	23/32	1 9/32	1 1/2	—	—	.170	—	.100	.325	—
	7042113	23/32	1 9/32	1 1/2	—	—	.180	—	.100	.325	—
	7042114	23/32	1 9/32	1 1/2	—	—	.170	—	.100	.325	—
	7042118	23/32	1 9/32	1 1/2	—	—	.190	—	.100	.325	—
	7042122	19/32	1 7/8	1 1/16	—	—	.085	—	.060	.215	—
	7042125	19/32	1 7/8	1 1/16	—	—	.100	—	.080	.215	—
	7042820	25/32	1 31/32	1 5/16	—	—	.080	—	.040	.215	—
	7042821	25/32	1 31/32	1 5/16	—	—	.110	—	.075	.215	—

Continued

MODEL 2GC, 2GV CARBURETOR ADJUSTMENT SPECIFICATIONS—Continued

See Tune Up Chart in car chapters for hot idle speeds.
★Located on tag attached to or stamped on carburetor.

Year	Carb. Part No. ★	Float Level	Float Drop	Pump Rod	Idle Vent	Intermediate Choke Rod	Vacuum Break	Automatic Choke	Choke Rod	Choke Unloader	Fast Idle Speed
CHEVROLET ENGINES—Continued											
1972	7042826	19/32	1 7/8	1 1/16	—	—	.085	—	.060	.215	2800
	7042827	19/32	1 7/8	1 1/16	—	—	.100	—	.080	.215	2400
	7042831	23/32	1 9/32	1 1/2	—	—	.180	—	.100	.325	—
	7042832	23/32	1 9/32	1 1/2	—	—	.170	—	.100	.325	—
	7042833	23/32	1 9/32	1 1/2	—	—	.180	—	.100	.325	—
	7042834	23/32	1 9/32	1 1/2	—	—	.170	—	.100	.325	—
	7042836	19/32	1 7/8	1 1/16	—	—	.085	—	.060	.215	—
	7042838	23/32	1 9/32	1 1/2	—	—	.200	—	.100	.325	—
	7042839	19/32	1 7/8	1 1/16	—	—	.100	—	.080	.215	—
1973	7043100	13/16	1 9/32	1 9/32	—	—	.080	—	.150	.215	1600
	7043101	13/16	1 9/32	1 9/32	—	—	.080	—	.150	.215	1600
	7043105	13/16	1 9/32	1 9/32	—	—	.080	—	.150	.215	1600
	7043111	19/32	1 9/32	1 1/2	—	—	.140	—	.200	.250	1600
	7043112	19/32	1 9/32	1 7/16	—	—	.130	—	.245	.325	1600
	7043113	19/32	1 9/32	1 1/2	—	—	.140	—	.200	.250	1600
	7043114	19/32	1 9/32	1 7/16	—	—	.130	—	.245	.325	1600
	7043118	19/32	1 9/32	1 7/16	—	—	.130	—	.245	.325	1600
	7043120	13/16	1 9/32	1 9/32	—	—	.080	—	.150	.215	1600
1974	7044111	19/32	1 9/32	1 21/32	—	—	.140	—	.200	.250	1600
	7044112	19/32	1 9/32	1 9/16	—	—	.130	—	.245	.325	1600
	7044113	19/32	1 9/32	1 21/32	—	—	.140	—	.200	.250	1600
	7044114	19/32	1 9/32	1 9/16	—	—	.130	—	.245	.325	1600
	7044115	19/32	1 9/32	1 21/32	—	—	.140	—	.200	.250	1600
	7044116	19/32	1 9/32	1 9/16	—	—	.130	—	.245	.325	1600
	7044117	19/32	1 9/32	1 9/16	—	—	.140	—	.245	.325	1600
	7044118	19/32	1 9/32	1 9/16	—	—	.130	—	.245	.325	1600
	7044120	19/32	1 9/32	1 21/32	—	—	.130	—	.200	.250	1600
	7044121	19/32	1 9/32	1 21/32	—	—	.140	—	.200	.250	1600
	7044123	19/32	1 9/32	1 21/32	—	—	.140	—	.200	.250	1600
	7044124	19/32	1 9/32	1 9/16	—	—	.130	—	.245	.325	1600
	7044125	19/32	1 9/32	1 21/32	—	—	.140	—	.200	.250	1600
	7044126	19/32	1 9/32	1 21/32	—	—	.140	—	.200	.250	1600
	7044127	19/32	1 9/32	1 21/32	—	—	.140	—	.200	.250	1700
	7044129	19/32	1 9/32	1 21/32	—	—	.140	—	.200	.250	1600
1975	7045101	19/32	1 7/32	1 19/32	—	—	.130	Index	.375	.350	—
	7045102	19/32	1 7/32	1 19/32	—	—	.130	Index	.375	.350	—
	7045105	19/32	1 7/32	1 19/32	—	—	.130	Index	.375	.350	—
	7045106	19/32	1 7/32	1 19/32	—	—	.130	Index	.380	.350	—
	7045111	11/16	31/32	1 5/8	—	—	.130	Index	.400	.350	—
	7045112	11/16	31/32	1 5/8	—	—	.130	Index	.400	.350	—
	7045114	11/16	31/32	1 5/8	—	—	.130	Index	.400	.350	—
	7045115	11/16	31/32	1 5/8	—	—	.130	Index	.400	.350	—
	7045123	11/16	31/32	1 5/8	—	—	.130	Index	.400	.350	—
	7045124	11/16	31/32	1 5/8	—	—	.130	Index	.400	.350	—
	7045401	21/32	1 7/32	1 19/32	—	—	.130	Index	.380	.350	—
	7045402	21/32	1 7/32	1 19/32	—	—	.130	Index	.375	.350	—
	7045405	21/32	1 7/32	1 19/32	—	—	.130	Index	.380	.350	—
	7045406	21/32	1 7/32	1 19/32	—	—	.130	Index	.380	.350	—
1976	17056101	17/32	1 9/32	1 5/8	—	—	.130	1 Lean	.260	.325	—
	17056102	17/32	1 9/32	1 5/8	—	—	.130	Index	.260	.325	—
	17056104	17/32	1 5/32	1 5/8	—	—	.140	Index	.260	.325	—
	17056108	9/16	1 9/32	1 21/32	—	—	.140	Index	.260	.325	1600
	17056110	9/16	1 9/32	1 21/32	—	—	.140	Index	.260	.325	1600
	17056111	9/16	1 9/32	1 21/32	—	—	.140	1 Lean	.260	.325	1600
	17056112	9/16	1 9/32	1 21/32	—	—	.140	Index	.260	.325	1600
	17056113	9/16	1 9/32	1 21/32	—	—	.140	1 Lean	.260	.325	1600
	17056114	21/32	31/32	1 11/16	—	—	.130	1 Rich	.260	.325	1600

Continued

MODEL 2GC, 2GV CARBURETOR ADJUSTMENT SPECIFICATIONS—Continued

See Tune Up Chart in car chapters for hot idle speeds.

★Located on tag attached to or stamped on carburetor.

Year	Carb. Part No. ★	Float Level	Float Drop	Pump Rod	Idle Vent	Intermediate Choke Rod	Vacuum Break	Automatic Choke	Choke Rod	Choke Unloader	Fast Idle Speed
CHEVROLET ENGINES—Continued											
1976	17056404	9/16	1 3/16	1 21/32	—	—	.140	Index	.260	.325	1600
	17056410	9/16	1 9/32	1 11/16	—	—	.140	Index	.260	.325	1600
	17056412	9/16	1 9/32	1 11/16	—	—	.140	Index	.260	.325	1600

②—Gauge from lip at toe of float to air horn gasket.

③—At slow idle RPM vent valve should be opened to specified dimension.

Year	Carb. Part No. ★	Float Level	Float Drop	Pump Rod	Idle Vent	Intermediate Choke Rod	Vacuum Break	Automatic Choke	Choke Rod	Choke Unloader	Fast Idle Speed
DODGE AND PLYMOUTH											
1971	7041180	21/32	1 3/4	1 5/64	—	—	41 Drill	—	—	29 Drill	1800
OLDSMOBILE											
1969	7029155	9/16	1 3/8	1 7/16	.025②	—	.180	1 Lean	.140	.170	900③
	7029156	9/16	1 3/8	1 7/16	.025②	—	.180	On Index	.140	.170	900③
	7029158	9/16	1 3/8	1 7/16	.025②	—	.180	On Index	.140	.170	900③
	7029159	9/16	1 3/8	1 7/16	.025②	—	.180	On Index	.140	.170	900③
1970	7040154	9/16	1 3/8	1 11/32	—	—	.160	On Index	.140	.170	900③
	7040155	9/16	1 3/8	1 11/32	—	—	.160	1 Lean	.140	.170	900③
	7040156	9/16	1 3/8	1 11/32	—	—	.160	On Index	.140	.170	900③
	7040158	9/16	1 3/8	1 11/32	—	—	.160	On Index	.140	.170	900③
	7040159	9/16	1 3/8	1 11/32	—	—	.160	On Index	.140	.170	900③
1971	7041155	17/32	1 3/8	1 11/32	—	—	.200	1 Lean	.140	.170	1000③
	7041156	17/32	1 3/8	1 11/32	—	—	.200	On Index	.140	.170	1000③
	7041159	17/32	1 3/8	1 11/32	—	—	.215	On Index	.140	.170	1000③
1972	7042155	17/32	1 3/8	1 11/32	—	—	.215	1 Lean	.160	.170	1000③
	7042156	17/32	1 3/8	1 11/32	—	—	.200	On Index	.160	.170	1000③
1973	7043152	15/32	1 9/32	1 11/32	—	—	.200	On Index	.160	.250	900③
	7043154	15/32	1 9/32	1 11/32	—	—	.200	On Index	.160	.250	900③
	7043158	15/32	1 9/32	1 11/32	—	—	.200	On Index	.160	.250	900③
1975	7045143	15/32	1 9/32	1 19/32	—	.120	.140①	1 Rich	.080	.180	1800③
	7045147	7/16	1 9/32	1 19/32	—	.120	.120①	1 Lean	.080	.140	1800③
	7045149	7/16	1 9/32	1 19/32	—	.120	.120①	1 Rich	.080	.140	1800③
	7045160	9/16	1 7/32	1 11/32	—	.120	.145④	1 Rich	.085	.180	—
	7045161	9/16	1 7/32	1 11/32	—	.120	.145④	1 Rich	.085	.180	—
	7045449	7/16	1 9/32	1 19/32	—	.120	.120①	1 Lean	.080	.140	1800③
1976	17056143	15/32	1 9/32	1 11/32	—	.120	.140⑤	1 Rich	.080	.180	—
	17056145	7/16	1 9/32	1 19/32	—	.120	.110⑤	1 Rich	.080	.140	—
	17056149	7/16	1 9/32	1 19/32	—	.120	.120⑤	1 Rich	.080	.140	—
	17056447	7/16	1 9/32	1 19/32	—	.120	.130⑥	1 Rich	.080	.140	—
	17056449	7/16	1 9/32	1 19/32	—	.120	.130⑥	1 Rich	.080	.140	—

①—Auxiliary, .120.
②—At slow idle RPM vent valve should open to specified dimension.
③—On low step of cam.
④—Auxiliary, .265.
⑤—Auxiliary, .100.
⑥—Auxiliary, .110.

Continued

MODEL 2GC, 2GV CARBURETOR ADJUSTMENT SPECIFICATIONS—Continued

See Tune Up Chart in car chapters for hot idle speeds.

★Located on tag attached to or stamped on carburetor.

Year	Carb. Part No. ★	Float Level	Float Drop	Pump Rod	Idle Vent	Intermediate Choke Rod	Vacuum Break	Automatic Choke	Choke Rod	Choke Unloader	Fast Idle Speed
PONTIAC											
1969	7028066	9/16	1 3/4	1 11/32	—	—	.170	①	.085	.180	—
	7028071	9/16	1 3/4	1 11/32	—	—	.160	①	.085	.180	—
	7029060	9/16	1 3/4	1 11/32	—	—	.150	①	.085	.180	—
	7029062	9/16	1 3/4	1 11/32	—	—	.150	①	.085	.180	—
1970	7040060	11/16	1 3/8	1 11/32	—	—	.150	①	.085	.180	—
	7040062	9/16	1 3/8	1 11/32	—	—	.150	①	.085	.180	—
	7040064	11/16	1 3/8	1 11/32	—	—	.150	①	.085	.180	—
	7040066	11/16	1 3/8	1 11/32	—	—	.170	①	.085	.180	—
	7040071	9/16	1 3/8	1 11/32	—	—	.160	①	.085	.180	—
	7040072	9/16	1 3/8	1 11/32	—	—	.150	①	.085	.180	—
	7040460	11/16	1 3/8	1 11/32	—	—	.150	①	.085	.180	—
	7040461	11/16	1 3/8	1 11/32	—	—	.150	①	.080	.180	—
	7040462	9/16	1 3/8	1 11/32	—	—	.150	①	.085	.180	—
	7040463	9/16	1 3/8	1 11/32	—	—	.150	①	.085	.180	—
	7040466	11/16	1 3/8	1 11/32	—	—	.170	①	.085	.180	—
	7040471	9/16	1 3/8	1 11/32	—	—	.160	①	.085	.180	—
1971	7041060	11/16	1 3/4	1 11/32	—	—	.125	①	.085	.180	—
	7041061	11/16	1 3/4	1 11/32	—	—	.125	①	.085	.180	—
	7041062	9/16	1 3/4	1 11/32	—	—	.105	①	.085	.180	—
	7041063	9/16	1 3/4	1 11/32	—	—	.105	①	.085	.180	—
	7041064	11/16	1 3/4	1 11/32	—	—	.130	①	.085	.180	—
	7041070	11/16	1 3/4	1 11/32	—	—	.125	①	.085	.180	—
	7041072	9/16	1 3/4	1 11/32	—	—	.105	①	.085	.180	—
	7041074	11/16	1 3/4	1 11/32	—	—	.130	①	.085	.180	—
	7041101	27/32	1 3/4	1 13/32	—	—	.110	①	.075	.200	—
	7041110	27/32	1 3/4	1 13/32	—	—	.080	①	.040	.200	—
	7041171	9/16	1 3/8	1 11/32	—	—	.160	①	.085	.180	—
1972	7042060	11/16	1 9/32	1 11/32	—	—	.150	①	.085	.180	—
	7042061	11/16	1 9/32	1 11/32	—	—	.150	①	.085	.180	—
	7042062	11/16	1 9/32	1 11/32	—	—	.130	①	.085	.180	—
	7042064	11/16	1 9/32	1 11/32	—	—	.150	①	.085	.180	—
	7042065	21/32	1 3/8	1 11/32	—	—	.157	①	.085	.180	—
	7042067	11/16	1 9/32	1 11/32	—	—	.147	①	.085	.180	—
	7042073	21/32	1 3/8	1 11/32	—	—	.130	①	.085	.180	—
	7042076	11/16	1 9/32	1 11/32	—	—	.157	①	.085	.180	—
	7042078	11/16	1 9/32	1 11/32	—	—	.147	①	.085	.180	—
	7042100	25/32	1 31/32	1 5/16	—	—	.080	①	.040	.215	—
	7042101	25/32	1 31/32	1 5/16	—	—	.100	①	.075	.215	—
1973	7043061	21/32	1 9/32	1 11/32	—	—	.180	1 Lean	.085	.180	—
	7043063	21/32	1 9/32	1 5/16	—	—	.170	1 Lean	.085	.180	—
	7043066	21/32	1 9/32	1 11/32	—	—	.180	1 Lean	.085	.180	—
	7043067	21/32	1 9/32	1 11/32	—	—	.180	1 Lean	.085	.180	—
	7043073	21/32	1 9/32	1 5/16	—	—	.180	1 Lean	.085	.180	—
1973-74	7043060	21/32	1 9/32	1 11/32	—	—	.160	1 Lean	.085	.180	—
	7043062	21/32	1 9/32	1 5/16	—	—	.170	1 Lean	.085	.180	—
	7043070	23/32	1 9/32	1 11/32	—	—	.160	1 Lean	.085	.180	—
	7043071	23/32	1 9/32	1 5/16	—	—	.200	1 Lean	.085	.180	—
	7043072	23/32	1 9/32	1 5/16	—	—	.170	1 Lean	.085	.180	—
1974	7044063	21/32	1 9/32	1 5/16	—	—	.160	1 Lean	.085	.180	—
	7044066	21/32	1 9/32	1 11/32	—	—	.180	1 Lean	.085	.180	—
	7044067	21/32	1 9/32	1 11/32	—	—	.180	1 Lean	.085	.180	—

Continued

MODEL 2GC, 2GV CARBURETOR ADJUSTMENT SPECIFICATIONS—Continued

See Tune Up Chart in car chapters for hot idle speeds.

★Located on tag attached to or stamped on carburetor.

Year	Carb. Part No. ★	Float Level	Float Drop	Pump Rod	Idle Vent	Intermediate Choke Rod	Vacuum Break	Automatic Choke	Choke Rod	Choke Unloader	Fast Idle Speed
PONTIAC—Continued											
1975	7045143	15/32	1 7/32	1 13/16	.025	—	.140②	—	.080	.180	—
	7045160	9/16	1 7/32	1 11/32	.025	—	.145③	—	.085	.180	—
	7045162	9/16	1 7/32	1 13/16	.025	—	.145④	—	.085	.180	—
	7045171	9/16	1 7/32	1 13/16	.025	—	.145④	—	.085	.180	—
1976	17056143	15/32	1 9/32	1 11/32	—	—	.140⑤	1 Rich	.080	.180	—
	17056145	7/16	1 9/32	1 19/32	—	—	.110⑤	1 Rich	.080	.140	—
	17056149	7/16	1 9/32	1 19/32	—	—	.120⑤	1 Rich	.080	.140	—
	17056160	9/16	1 9/32	1 11/32	—	—	.165⑥	1 Rich	.085	.180	—
	17056161	9/16	1 9/32	1 11/32	—	—	.165⑥	1 Rich	.085	.180	—
	17056162	9/16	1 9/32	1 11/32	—	—	.165⑥	1 Rich	.085	.180	—
	17056447	7/16	1 9/32	1 19/32	—	—	.130⑦	1 Rich	.080	.140	—
	17056449	7/16	1 9/32	1 19/32	—	—	.130⑦	1 Rich	.080	.140	—

①—With choke valve closed, pull upward on choke rod to the limit of its travel. The end of rod should fit the gauge notch on the choke lever.
②—Rear .120″.
③—Rear .265″.
④—Rear .260″.
⑤—Rear, .100″.
⑥—Rear, .285″.
⑦—Rear, .110″.

MODEL 2GC, 2GV ADJUSTMENTS

Models 2GC and 2GV use an automatic choke, Figs. 1 and 2.

There are two different types of automatic choke systems used on these units; 1) The carburetor mounted thermostatic coil, Fig. 3 and, 2) the well type, Fig. 4.

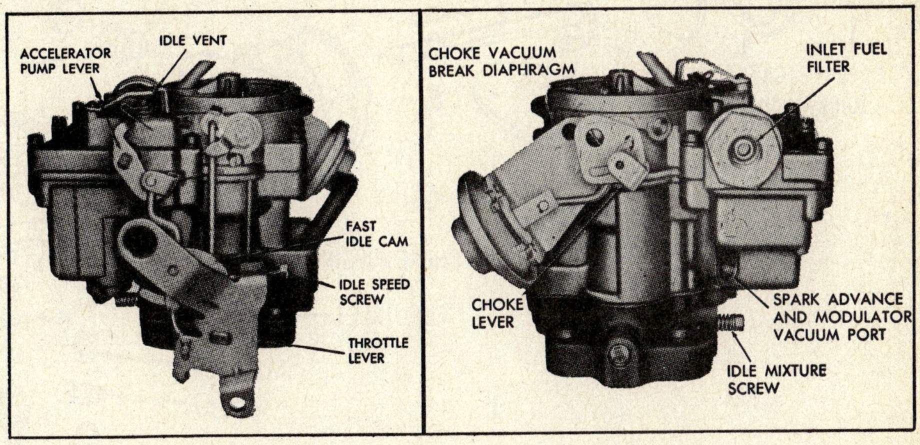

Fig. 1 Rochester Model 2GV two-barrel carburetor (typical)

Float Level Adjustment

Figs. 5 & 6—Adjust float level as directed for the type of float shown in the specification listed in the *Rochester Specifications Chart.*

Float Drop Adjustment

Figs. 7 & 8—Adjust float drop as directed for the type of float shown in the specification listed in the *Rochester Specifications Chart.*

Pump Rod Adjustment

Fig. 9—Back out idle stop screw and completely close throttle valves in bore. Place proper size gauge listed in the *Rochester Specifications Chart* on top of air horn ring. Bend pump rod at lower angle to obtain specified dimension to top of pump rod.

Idle Vent Adjustment

Fig. 10—Open throttle until vent valve just closes. Place proper size gauge on top of air horn ring. Dimension to top of pump rod should be as specified in the *Rochester Specifications Chart.* Adjust by bending tang on pump lever.

Bowl Vent Adjustment

Fig. 11—Set idle speed to specifications, then with idle screw on second step of fast idle cam, vent valve should just be closed. To adjust, turn vent valve screw.

Intermediate Choke Rod Adjustment

1969-70 2GC Models, Fig. 12—Remove thermostat cover and coil assembly, then with fast idle screw on high step of cam and choke valve closed, the edge of choke lever must line up with edge of projection on casting inside choke housing. To adjust, bend choke rod.

1971-75 2GC Models, Fig. 13—Remove thermostat cover and coil assembly, then with fast idle screw on high step of cam and choke valve closed, the edge of choke lever must align with edge of .120" plug gauge as shown. To adjust, bend choke rod at point shown.

Vacuum Break Adjustment

Figs. 14 & 15—Place fast idle screw on high step of cam, then using an outside vacuum source, seat vacuum diaphragm. On purge type vacuum diaphragms, cover bleed hole so that diaphragm will not bleed down. Hold choke towards closed position, then with vacuum diaphragm seated and rod in end of slot in plunger, check clearance between upper edge of choke valve and air horn wall. Refer to *Rochester Specifications Chart.* To adjust, bend vacuum break rod at point shown.

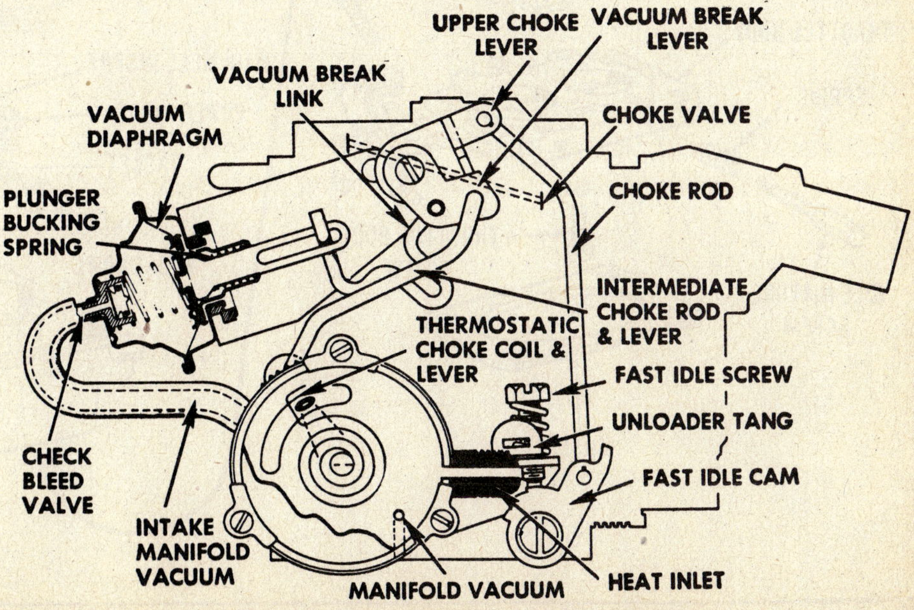

Fig. 3 Carburetor mounted automatic choke

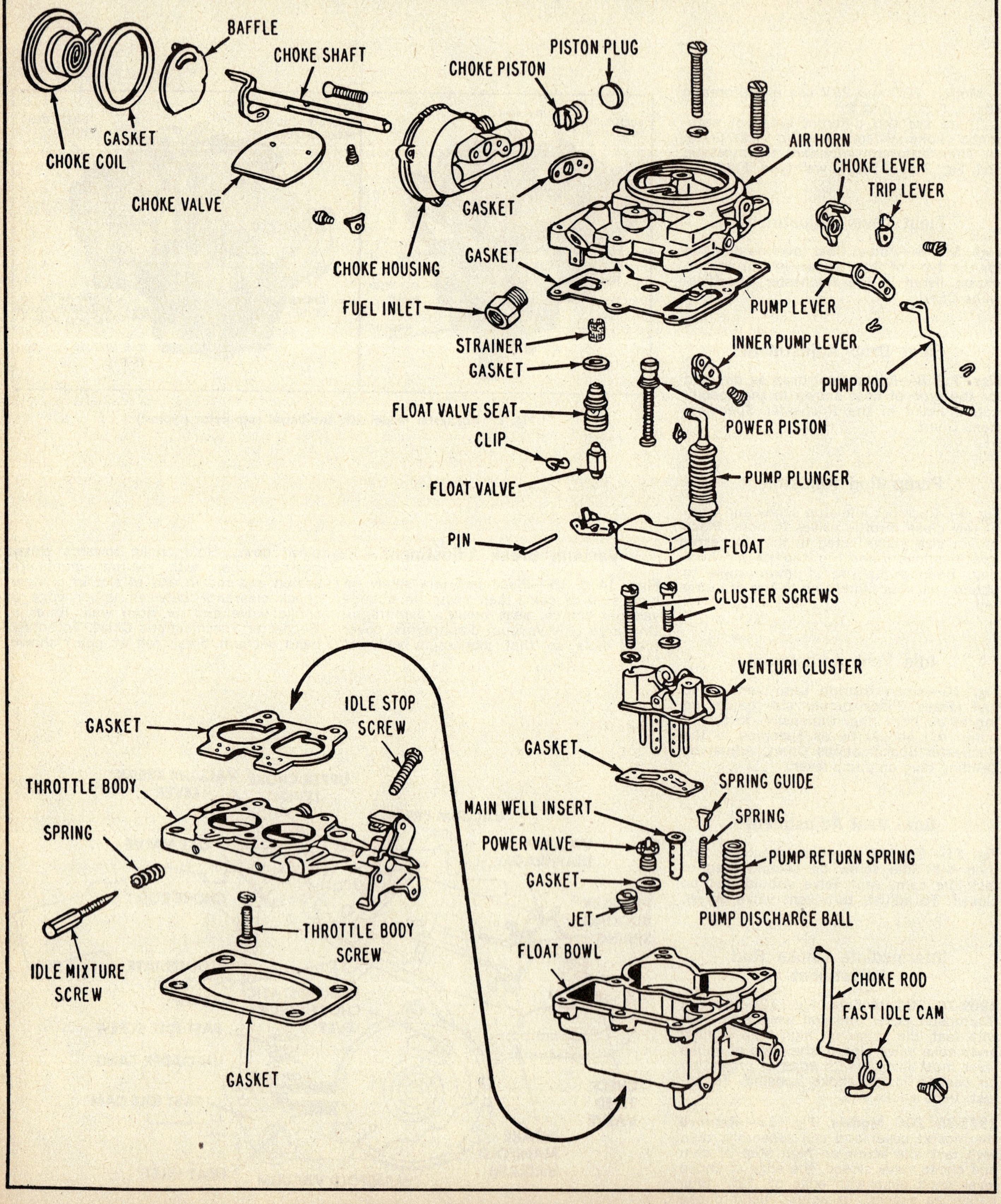

Fig. 2 Exploded view of typical Rochester two-barrel carburetor

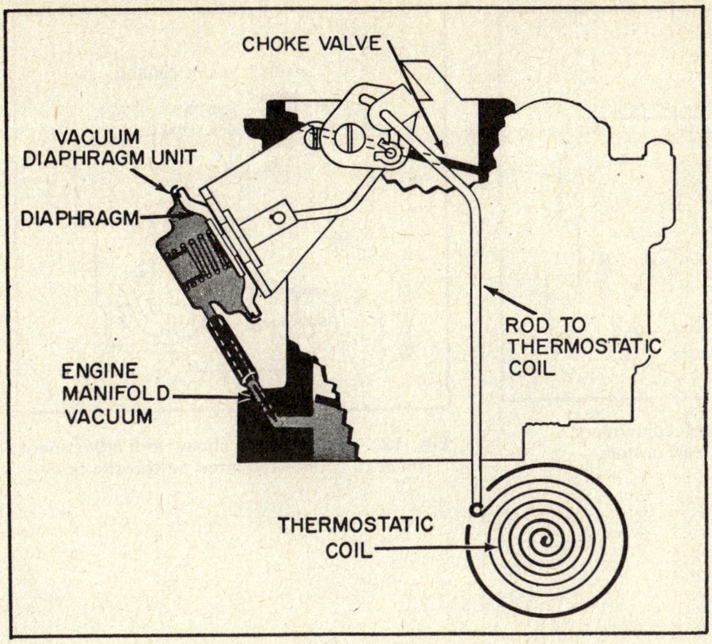

Fig. 4 Well-type choke. 2GV models

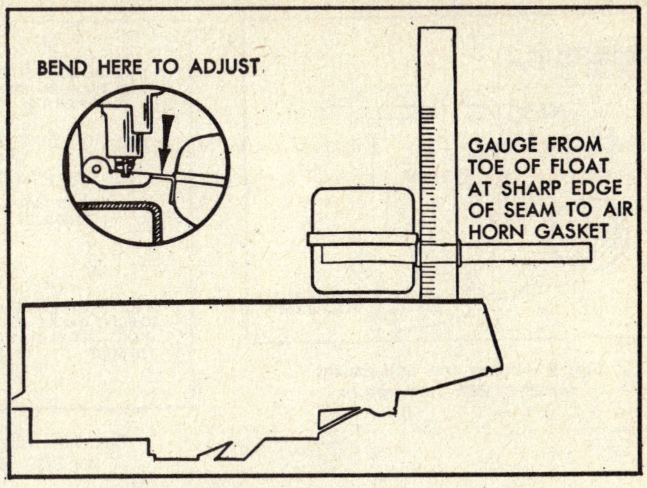

Fig. 5 Float level adjustment (brass float)

Automatic Choke Setting

Carburetor Mounted Choke, Fig. 16—Place idle screw on high step of fast idle cam, then loosen three retaining screws and rotate choke cover against coil tension until index mark is in line with specified point on choke housing (see *Rochester Specifications Chart*).

Choke Rod Adjustment

Fig. 17—It is important to position both slow idle and fast idle screws as follows

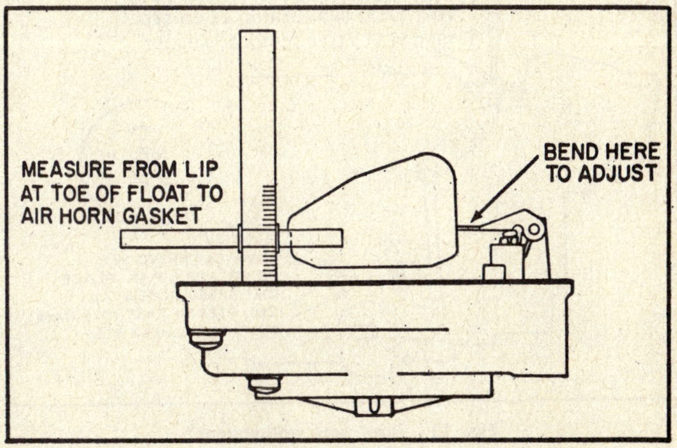

Fig. 6 Float level adjustment (nitrophyl float)

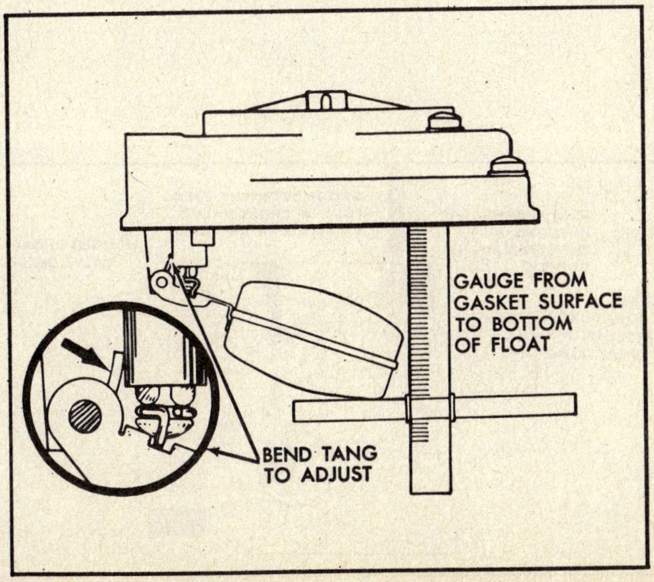

Fig. 7 Float drop adjustment (brass float)

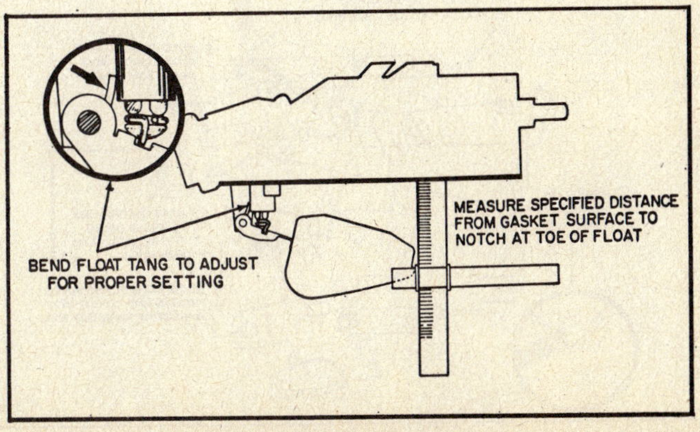

Fig. 8 Float drop adjustment (nitrophyl float)

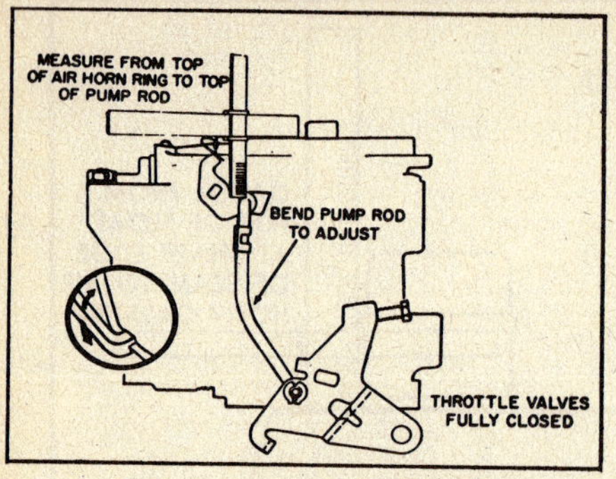

Fig. 9 Pump rod adjustment for 2GC, 2GV carburetors

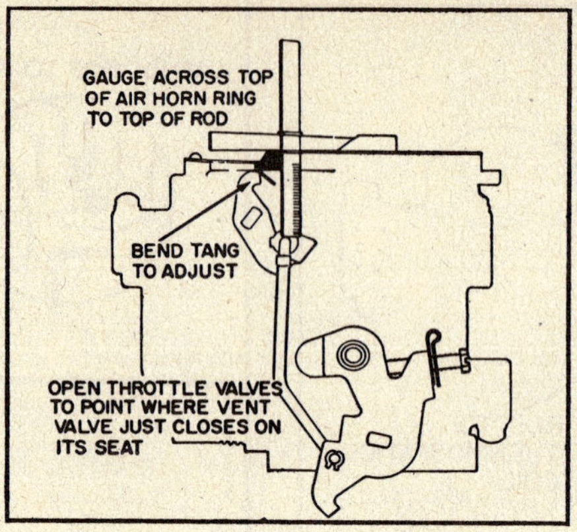

Fig. 10 Idle vent adjustment for 2GC, 2GV carburetors

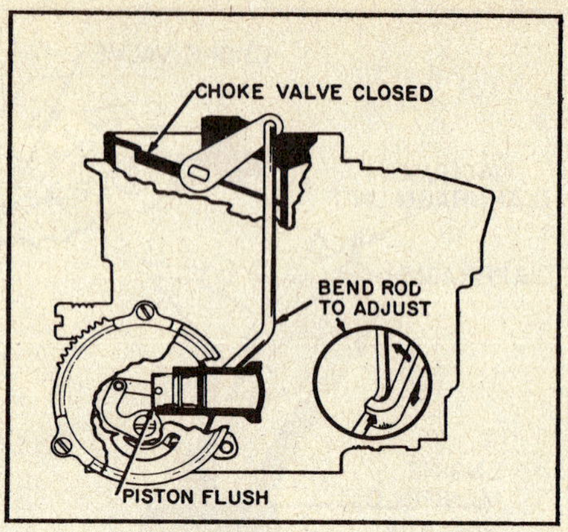

Fig. 12 Intermediate choke rod adjustment for 2GC chokes mounted on throttle body

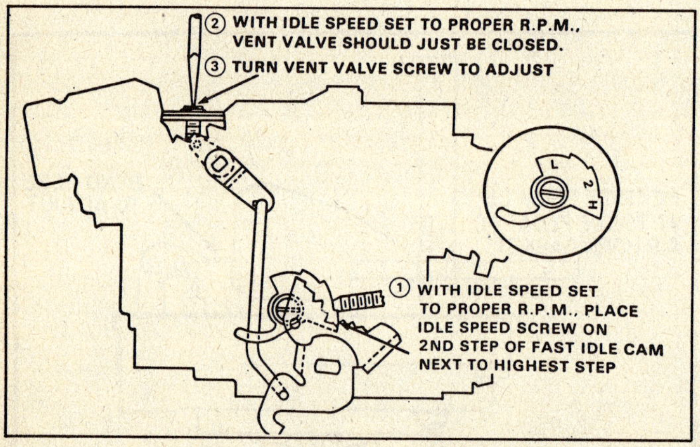

Fig. 11 Bowl vent adjustment

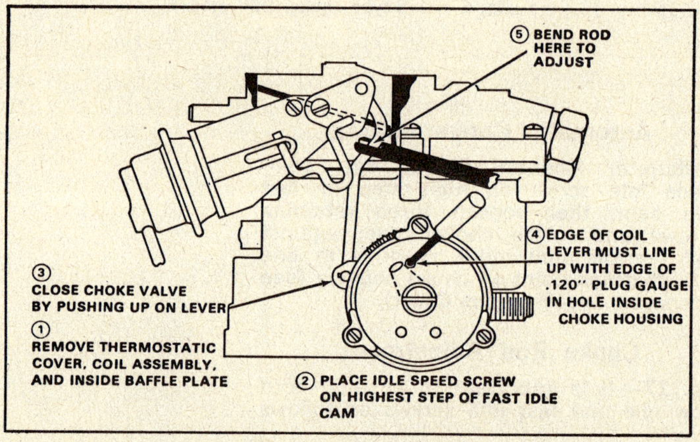

Fig. 13 Intermediate choke rod adjustment

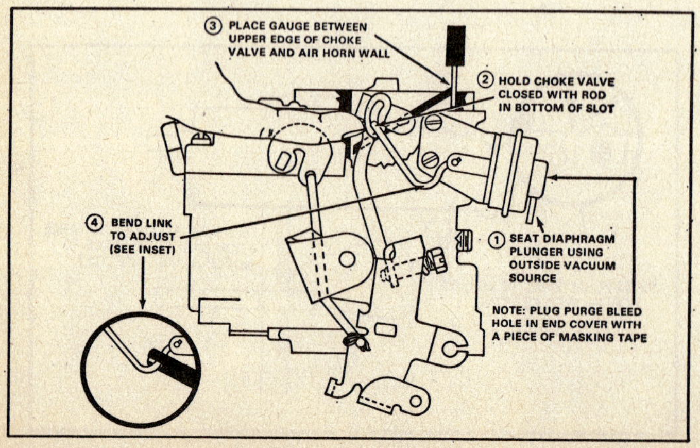

Fig. 14 Primary vacuum break adjustment

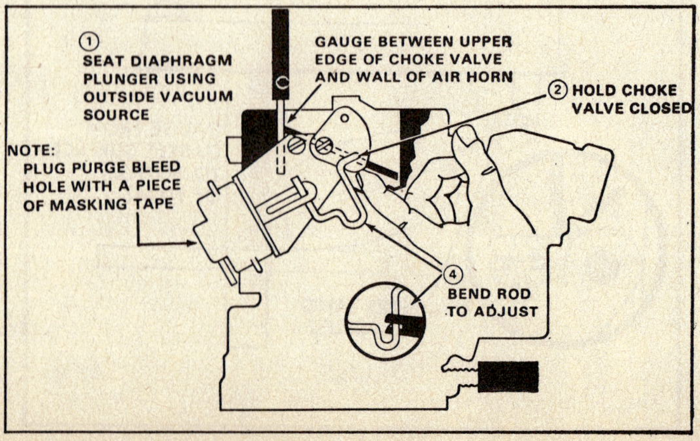

Fig. 15 Auxiliary vacuum break adjustment

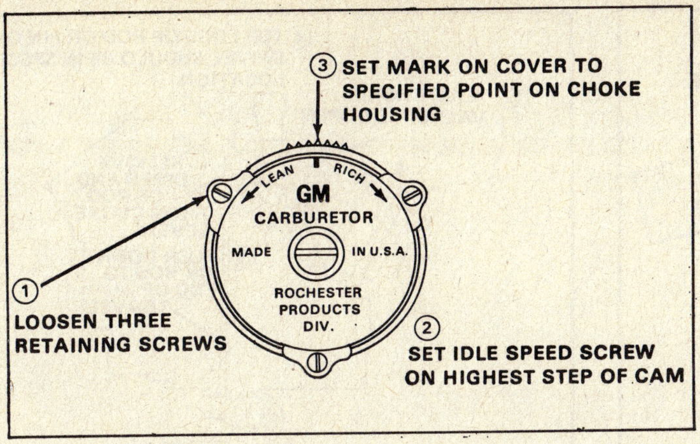

Fig. 16 Automatic choke adjustment for 2GC carburetors

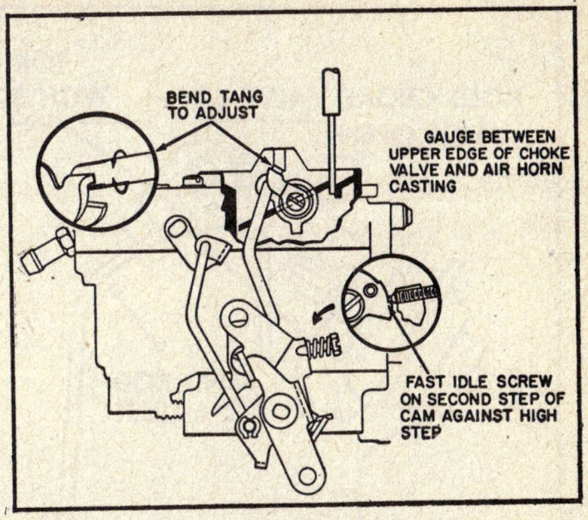

Fig. 17 Choke rod adjustment

before making choke rod adjustment.

1. On models using a single idle stop screw, turn stop screw in until it just contacts bottom step of fast idle cam. Then turn screw in one full turn farther.
2. On models using both a slow idle and a fast idle screw, turn slow idle stop screw in until it just contacts stop. Then turn this screw in one full turn from this point. Next turn the fast idle screw in until it touches bottom step of fast idle cam.
3. On all models, place idle screw on second step of fast idle cam against shoulder of high step. While holding screw in this position, check clearance between upper edge of choke valve and air horn wall as shown. Adjust to specified dimension by bending tang on choke lever and collar assembly (see *Rochester Specifications Chart*).

Choke Unloader Adjustment

Fig. 18—With throttle valves held wide open, the choke valve should be open just enough to admit the specified gauge between upper edge of choke valve and air horn wall (see *Rochester Specifications Chart*). To adjust, bend tang on throttle lever.

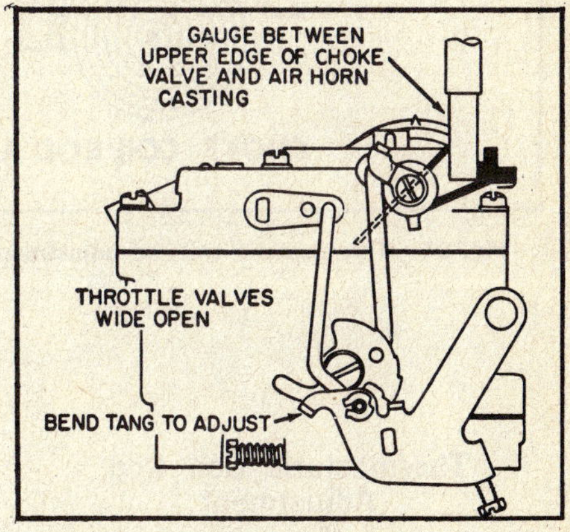

Fig. 18 Choke unloader adjustment for 2GC, 2GV carburetors

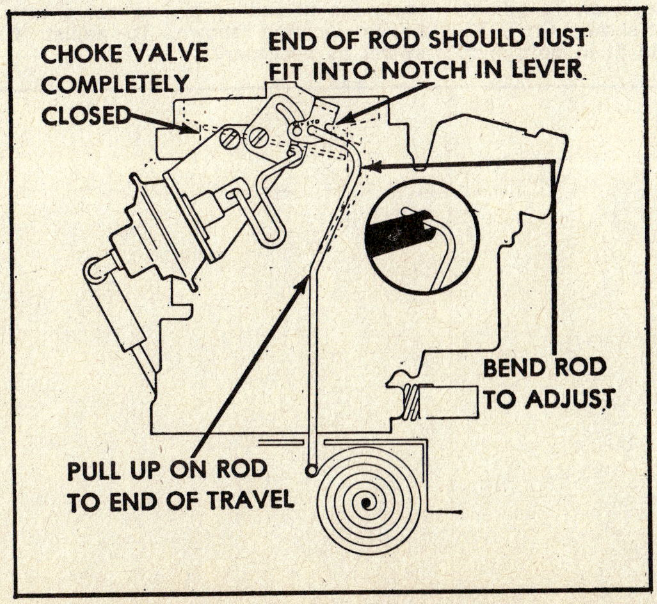

Fig. 19 Thermostatic coil rod adjustment. 1969-74 Buick

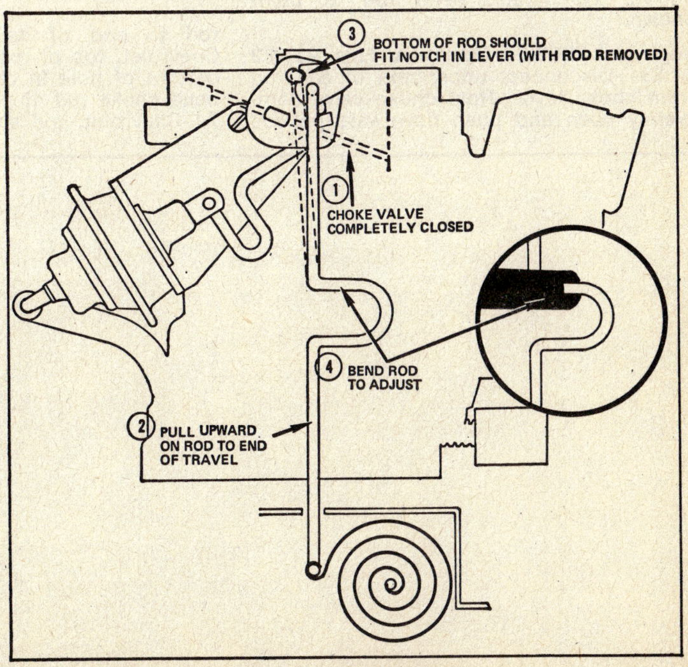

Fig. 20 Thermostatic coil rod adjustment. 1969-72 Pontiac

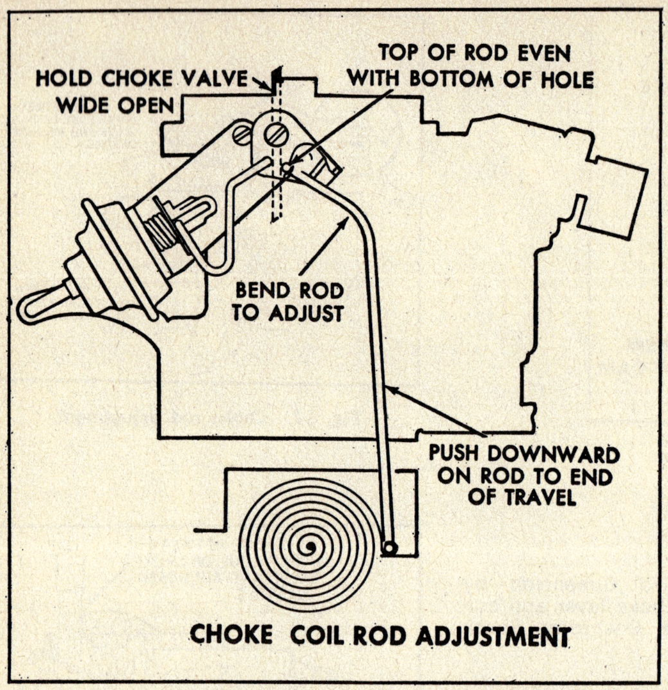

Fig. 21 Thermostatic coil rod adjustment. 1969-71 Chevrolet

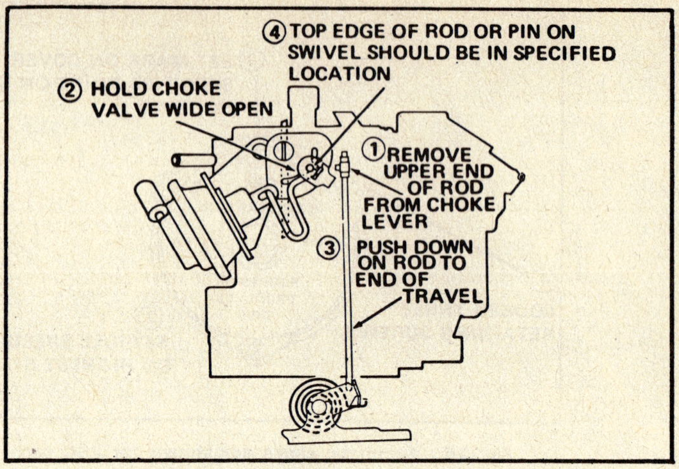

Fig. 22 Thermostatic coil rod adjustment. Chevrolet Vega

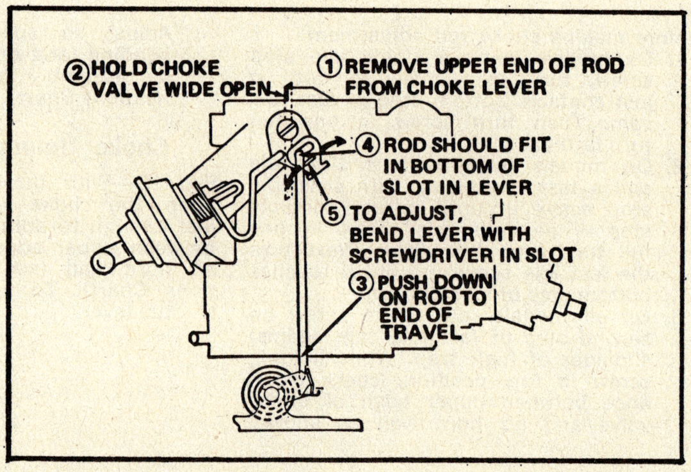

Fig. 23 Thermostatic coil rod adjustment. 1972-74 Chevrolet

Thermostatic Coil Rod Adjustment

1969-74 Buick and 1969-72 Pontiac, Figs. 19 & 20—Disconnect upper end of thermostatic coil rod from choke. Hold choke valve completely closed and pull upward on coil rod to end of its travel. Rod should just fit into notch in lever as shown. To adjust, bend rod at point shown.

Vega and 1969-74 Chevrolet, Figs. 21, 22 & 23—Disconnect upper end of coil rod from choke lever. Hold choke valve completely open and push downward on coil rod to end of its travel. On 1969-71 Chevrolet, top of rod should be even with bottom of hole in choke lever. To adjust, bend choke rod at point shown. On 1972-74 Chevrolet, rod should fit in bottom of slot in lever. To adjust, bend lever with screwdriver in slot as shown. On Vega, top edge of rod or pin on swivel should be in location shown. To adjust, turn swivel as necessary.

QUADRAJET 4MC & 4MV ADJUSTMENT SPECIFICATIONS

See Tune Up Chart in car chapters for hot idle speeds.

Year	Carb. Model	Float Level	Pump Rod Hole	Pump Rod Adj.	Idle Vent	Air Valve	Fast Idle (Bench)	Choke Rod	Vacuum Break	Air Valve Dash-pot	Choke Unloader	Air Valve Lockout	Secondary Metering Rods	Air-Valve Valve Spring Wind-Up
BUICK														
1969	7029240	3/8	Outer	13/32	1/2	—	—	.130	.180	.030	.325	.045	53/64	1/2
	7029241	5/16	Outer	13/32	1/2	—	—	.130	.180	.030	.325	.045	53/64	1/2
	7029242	3/8	Outer	13/32	1/2	—	—	.130	.180	.030	.325	.045	53/64	1/2
	7029243	3/8	Outer	13/32	1/2	—	—	.140	.215	.030	.325	.015	53/64	1/2
	7029244	5/16	Outer	13/32	1/2	—	2 Turns	.130	.190	.030	.325	.045	53/64	1/2
	7029245	5/16	Outer	13/32	1/2	—	2 Turns	.130	.215	.030	.325	.015	53/64	1/2
1970	7040240	3/8	Inner	9/32	—	—	—	.130	.180	.030	.335	—	53/64	1/2
	7040243	3/8	Inner	9/32	—	—	—	.130	.215	.030	.335	—	53/64	1/2
	7040244	5/16	Outer	13/32	—	—	—	.130	.170	.030	.335	.045	53/64	1/2
	7040245	5/16	Outer	13/32	—	—	—	.130	.215	.030	.335	—	53/64	1/2
	7040246	5/16	Inner	9/32	—	—	—	.130	.200	.030	.335	—	53/64	1/2
	7040247	3/8	Inner	9/32	—	—	—	.130	.160	.030	.325	—	53/64	1/2
1971	7041242	3/8	Inner	1/4	—	—	—	.130	.200⑤	.030	.335	—	53/64	1/2
	7041243	13/32	Inner	1/4	—	—	—	.130	.215⑥	.030	.335	—	53/64	1/2
	7041245	15/32	Inner	9/32	—	—	—	.130	.170①	.030	.335	—	53/64	1/2
	7041540	3/8	Inner	1/4	—	—	—	.130	.180⑤	.030	.335	—	53/64	1/2
	7041544	15/32	Inner	9/32	—	—	—	.130	.170①	.030	.335	—	53/64	1/2
1972	7042240	3/8	Inner	1/4	—	—	—	.130	.180⑤	.030	.335	.015	53/64	1/2
	7042242	3/8	Inner	1/4	—	—	—	.130	.200⑤	.030	.335	.015	53/64	1/2
	7042243	13/32	Inner	1/4	—	—	—	.130	.215⑥	.030	.335	.015	53/64	1/2
	7042244	15/32	Inner	9/32	—	—	—	.130	.170①	.030	.335	.015	53/64	1/2
	7042245	15/32	Inner	9/32	—	—	—	.130	.170①	.030	.335	.015	53/64	1/2
	7042940	3/8	Inner	1/4	—	—	—	.130	.180⑤	.030	.335	.015	53/64	1/2
	7042942	3/8	Inner	1/4	—	—	—	.130	.200⑤	.030	.335	.015	53/64	1/2
	7042944	15/32	Inner	9/32	—	—	—	.130	.170①	.030	.335	.015	53/64	1/2
1973	7043240	13/32	Inner	1/4	—	—	2	.130	.215⑤	.030	.325	.015	53/64	7/16
	7043242	13/32	Inner	7/16	—	—	2	.130	.200⑦	.030	.325	.015	53/64	7/16
	7043243	13/32	Inner	7/16	—	—	2	.130	.215⑥	.030	.325	.015	53/64	7/16
	7043244	15/32	Outer	3/8	—	—	2	.130	.170①	.030	.325	.015	53/64	11/16
	7043245	15/32	Inner	9/32	—	—	2	.130	.170①	.030	.325	.015	53/64	11/16
	7043246	15/32	Outer	3/8	—	—	2	.130	.170①	.030	.325	.015	53/64	11/16
1974	7044240	13/32	Inner	1/4	—	—	2	.130	.215⑤	.030	.325	.015	53/64	7/16
	7044241	13/32	Inner	1/4	—	—	2	.130	.215⑤	.030	.325	.015	53/64	7/16
	7044242	13/32	Inner	1/4	—	—	2	.130	.200⑦	.030	.325	.015	53/64	7/16
	7044244	15/32	Outer	3/8	—	—	2	.130	.170①	.030	.325	.015	53/64	11/16
	7044246	15/32	Outer	3/8	—	—	2	.130	.170①	.030	.325	.015	53/64	11/16
	7044540	13/32	Inner	1/4	—	—	2	.130	.215⑤	.030	.325	.015	53/64	7/16
	7044544	15/32	Outer	3/8	—	—	2	.130	.170①	.030	.325	.015	53/64	11/16
	7044546	15/32	Outer	3/8	—	—	2	.130	.170①	.030	.325	.015	53/64	11/16

①—Secondary adjustment .150″.
②—Early 3/8″; late 7/16″.
③—Early, inner; late, outer.
④—Early 9/32″; late 13/32″.
⑤—Secondary adjustment .160″.
⑥—Secondary adjustment .195″.
⑦—Secondary adjustment .180″.
⑧—Rear .120.
⑨—Rear .115.
⑩—Rear .130.

Year	Carb. Model	Float Level	Pump Rod Hole	Pump Rod Adj.	Idle Vent	Air Valve	Fast Idle (Bench)	Choke Rod	Vacuum Break	Air Valve Dash-pot	Choke Unloader	Air Valve Lockout	Secondary Metering Rods	Air-Valve Valve Spring Wind-Up
CADILLAC														
1969	7029232	3/8	Outer	11/32	—	—	1 1/2 Turns	.090	.230	.030	.300	.015	27/32	1/2
	7029233	3/8	Outer	11/32	—	—	1 1/2 Turns	.090	.230	.030	.300	.015	27/32	1/2
1970	7040230	1/4	Outer	11/32	—	—	1 1/2 Turns	.090	.230	.030	.300	.015	.840	7/16
	7040231	1/4	Outer	11/32	—	—	1 1/2 Turns	.090	.230	.030	.300	.015	.840	7/16
	7040232	1/4	Outer	11/32	—	—	1 1/2 Turns	.090	.230	.030	.300	.015	.840	7/16
1971	7041230	1/4	Outer	11/32	—	—	1 1/2 Turns	.090	①	.030	.310	.015	.840	1/2
	7041231	1/4	Outer	11/32	—	—	1 1/2 Turns	.090	①	.030	.310	.015	.840	1/2
	7041232	1/4	Outer	11/32	—	—	1 1/2 Turns	.090	①	.030	.310	.015	.840	1/2

Continued

GM DELCO/ROCHESTER CARBURETORS

QUADRAJET 4MC & 4MV ADJUSTMENT SPECIFICATIONS—Continued

See Tune Up Chart in car chapters for hot idle speeds.

Year	Carb. Model	Float Level	Pump Rod		Idle Vent	Air Valve	Fast Idle (Bench)	Choke Rod	Vacuum Break	Air Valve Dash-pot	Choke Unloader	Air Valve Lockout	Secondary Metering Rods	Air-Valve Valve Spring Wind-Up
			Hole	Adj.										
CADILLAC—Continued														
1972	7042230	1/4	Outer	11/32	—	—	1 1/2 Turns	.090	.190	.030	.310	.030	.840	1/2
	7042231	1/4	Outer	11/32	—	—	1 1/2 Turns	.090	.190	.030	.310	.030	.840	1/2
	7042232	11/32	Outer	11/32	—	—	1 1/2 Turns	.090	.190	.030	.310	.030	.840	1/2
	7042234	1/4	Outer	11/32	—	—	1 1/2 Turns	.090	.110	.030	.310	.030	.840	1/2
	7042235	1/4	Outer	11/32	—	—	1 1/2 Turns	.090	.110	.030	.310	.030	.840	1/2
	7042236	3/8	Outer	11/32	—	—	1 1/2 Turns	.090	.110	.030	.310	.030	.840	1/2
	7042237	3/8	Outer	11/32	—	—	1 1/2 Turns	.090	.110	.030	.310	.030	.840	1/2
	7042238	3/8	Outer	11/32	—	—	1 1/2 Turns	.090	.110	.030	.310	.030	.840	1/2
	7042239	3/8	Outer	11/32	—	—	1 1/2 Turns	.090	.110	.030	.310	.030	.840	1/2
1973	7043230	1/4	Outer	11/32	—	—	1 1/2 Turns	.090	.200	—	.310	.030	.840	5/16
	7043232	3/8	Inner	1/4	—	—	1 1/2 Turns	.090	.200	—	.310	.030	.840	3/8
	7043233	5/16	Outer	11/32	—	—	1 1/2 Turns	.090	.180	—	.310	.030	.840	5/16
	7043234	1/4	Inner	1/4	—	—	1 1/2 Turns	.090	.200	—	.310	.030	.840	5/16
	7043235	3/8	Outer	11/32	—	—	1 1/2 Turns	.090	.215	—	.310	.030	.840	3/8
1974	7044230	1/4	Inner	1/4	—	—	1 1/2 Turns	.110	.185	.025	.300	.030	.840	3/8
	7044232	3/8	Inner	1/4	—	—	1 1/2 Turns	.110	.200	.025	.300	.030	.840	1/2
	7044233	9/32	Outer	11/32	—	—	1 1/2 Turns	.110	.185	.025	.300	.030	.840	3/8
	7044234	1/4	Outer	11/32	—	—	1 1/2 Turns	.090	.200	.025	.300	.030	.840	5/16
	7044235	3/8	Outer	11/32	—	—	1 1/2 Turns	.090	.200	.025	.300	.030	.840	3/8
	7044530	1/4	Inner	1/4	—	—	1 1/2 Turns	.110	.185	.025	.300	.030	.840	3/8
	7044532	3/8	Inner	1/4	—	—	1 1/2 Turns	.110	.200	.025	.300	.030	.840	1/2

①—Early (single stage), .280"; late (two stage), .190".

Year	Carb. Model	Float Level	Pump Rod Hole	Pump Rod Adj.	Idle Vent	Air Valve	Fast Idle (Bench)	Choke Rod	Vacuum Break	Air Valve Dash-pot	Choke Unloader	Air Valve Lockout	Secondary Metering Rods	Air-Valve Valve Spring Wind-Up
CHECKER MOTORS														
1970	7029202	7/32	Inner	—	3/8	—	2 Turns	.100	.180	.015	.450	—	—	—
	7040202	1/4	Inner	—	—	—	2 Turns	.100	.245	.020	.400	—	—	—
	7040502	1/4	Inner	—	—	—	2 Turns	.100	.245	.020	.450	—	—	—
1972	7042202	1/4	Inner	—	—	—	2 Turns	.100	.215	.020	.450	—	—	—
	7042210	3/16	Inner	—	—	—	2 Turns	.100	.215	.020	.450	—	—	—
	7042902	1/4	Inner	—	—	—	2 Turns	.100	.215	.020	.450	—	—	—
	7042910	3/16	Inner	—	—	—	2 Turns	.100	.215	.020	.450	—	—	—
1973	7043202	1/4	Inner	—	—	—	2 Turns	.430	.250	—	.450	—	—	—
1974	7044502	1/4	Inner	—	—	—	—	.430	.230	—	.450	—	—	7/8
CHEVROLET ENGINES														
1969	7029202	7/32	Inner	5/16	3/8	—	2 Turns	.100	.180	.015	.450	.015	—	7/16
	7029203	7/32	Inner	5/16	3/8	—	2 Turns	.100	.245	.015	.450	.015	—	7/16
	7029204	1/4	Inner	5/16	3/8	—	2 Turns	.100	.180	.015	.450	.015	—	13/16
	7029207	3/16	Inner	5/16	3/8	—	2 Turns	.100	.245	.015	.450	.015	—	13/16
	7029215	1/4	Inner	5/16	3/8	—	2 Turns	.100	.245	.015	.450	.015	—	13/16
1970	7040200	1/4	Inner	5/16	—	—	2 Turns	.100	.245	.020	.400	.015	—	13/16
	7040201	1/4	Inner	5/16	—	—	2 Turns	.100	.275	.020	.450	.015	—	13/16
	7040202	1/4	Inner	5/16	—	—	2 Turns	.100	.245	.020	.400	.015	—	7/16
	7040203	1/4	Inner	5/16	—	—	2 Turns	.100	.275	.020	.450	.015	—	7/16
	7040204	1/4	Inner	5/16	—	—	2 Turns	.100	.245	.020	.400	.015	—	13/16
	7040205	1/4	Inner	5/16	—	—	2 Turns	.100	.275	.020	.450	—	—	13/16
	7040207	1/4	Inner	5/16	—	—	2 Turns	.100	.275	.015	.450	—	—	13/16
	7040500	1/4	Inner	5/16	—	—	2 Turns	.100	.245	.020	.450	.015	—	13/16
	7040501	1/4	Inner	5/16	—	—	2 Turns	.100	.275	.020	.450	.015	—	13/16

Continued

QUADRAJET 4MC & 4MV ADJUSTMENT SPECIFICATIONS—Continued

See Tune Up Chart in car chapters for hot idle speeds.

Year	Carb. Model	Float Level	Pump Rod		Idle Vent	Air Valve	Fast Idle (Bench)	Choke Rod	Vacuum Break	Air Valve Dash-pot	Choke Unloader	Air Valve Lockout	Secondary Metering Rods	Air-Valve Valve Spring Wind-Up
			Hole	Adj.										
CHEVROLET ENGINES—Continued														
1970	7040502	1/4	Inner	5/16	—	—	2 Turns	.100	.245	.020	.450	.015	—	7/16
	7040503	1/4	Inner	5/16	—	—	2 Turns	.100	.275	.020	.450	.015	—	7/16
	7040504	1/4	Inner	—	—	—	2 Turns	.100	.245	.020	.400	—	—	13/16
	7040505	1/4	Inner	5/16	—	—	2 Turns	.100	.275	.020	.450	.015	—	13/16
	7040507	3/16	—	5/16	—	—	2 Turns	.100	.275	.020	.450	—	—	13/16
	7040509	1/4	—	5/16	—	—	2 Turns	.100	.275	.020	.450	—	—	13/16
	7040511	1/4	—	5/16	—	—	2 Turns	.100	.245	.020	.450	—	—	7/16
1971	7041200	1/4	—	—	—	—	—	.100	.260	.020	.450	—	—	7/16
	7041201	1/4	Inner	—	—	—	2 Turns	.100	.275	.020	.450	—	—	7/16
	7041202	1/4	Inner	—	—	—	2 Turns	.100	.260	.020	.450	—	—	7/16
	7041203	1/4	Inner	—	—	—	2 Turns	.100	.275	.020	.450	—	—	7/16
	7041204	1/4	Inner	—	—	—	2 Turns	.100	.260	.020	.450	—	—	7/16
	7041205	1/4	Inner	—	—	—	2 Turns	.100	.275	.020	.450	—	—	7/16
	7041212	1/4	Inner	—	—	—	2 Turns	.100	.260	.020	—	—	—	7/16
	7041213	1/4	Inner	—	—	—	2 Turns	.100	.275	.020	.450	—	—	7/16
1972	7042202	1/4	—	3/8	—	—	—	.100	.215	.020	.450	—	—	1/2
	7042203	1/4	—	3/8	—	—	—	.100	.215	.020	.450	—	—	1/2
	7042215	1/4	—	3/8	—	—	—	.100	.250	.020	.450	—	—	7/16
	7042216	1/4	—	3/8	—	—	—	.100	.250	.020	.450	—	—	7/16
	7042217	1/4	—	3/8	—	—	—	.100	.250	.020	.450	—	—	7/16
	7042220	1/4	—	3/8	—	—	—	.100	.250	.020	.450	—	—	7/16
	7042902	1/4	—	3/8	—	—	—	.100	.215	.020	.450	—	—	1/2
	7042903	1/4	—	3/8	—	—	—	.100	.215	.020	.450	—	—	1/2
1973	7043200	1/4	Inner	13/32	—	—	2 Turns	.430	.250	—	.450	.015	—	11/16
	7043201	1/4	Inner	13/32	—	—	2 Turns	.430	.250	—	.450	.015	—	11/16
	7043202	1/4	Inner	13/32	—	—	2 Turns	.430	.250	—	.450	.015	—	1/2
	7043203	1/4	Inner	13/32	—	—	2 Turns	.430	.250	—	.450	.015	—	1/2
	7043212	1/4	Inner	13/32	—	—	2 Turns	.430	.215	—	.450	.015	—	3/4
	7043213	1/4	Inner	13/32	—	—	2 Turns	.430	.215	—	.450	.015	—	3/4
1974	7044201	3/8	Inner	13/32	—	—	2 Turns	.430	.250	.015	.450	—	—	7/16
	7044202	1/4	Inner	13/32	—	—	2 Turns	.430	.230	.015	.450	—	—	7/8
	7044203	1/4	Inner	13/32	—	—	2 Turns	.430	.230	.015	.450	—	—	7/8
	7044206	1/4	Inner	13/32	—	—	2 Turns	.430	.230	.015	.450	—	—	7/8
	7044207	1/4	Inner	13/32	—	—	2 Turns	.430	.230	.015	.450	—	—	7/8
	7044208	1/4	Inner	13/32	—	—	2 Turns	.430	.230	.015	.450	—	—	1 Turn
	7044209	1/4	Inner	13/32	—	—	2 Turns	.430	.230	.015	.450	—	—	1 Turn
	7044210	1/4	Inner	13/32	—	—	2 Turns	.430	.230	.015	.450	—	—	1 Turn
	7044211	1/4	Inner	13/32	—	—	2 Turns	.430	.230	.015	.450	—	—	1 Turn
	7044221	3/8	Inner	13/32	—	—	2 Turns	.430	.250	.015	.450	—	—	7/16
	7044223	3/8	Inner	13/32	—	—	2 Turns	.430	.220	.015	.450	—	—	7/16
	7044225	3/8	Inner	13/32	—	—	2 Turns	.430	.220	.015	.450	—	—	7/16
	7044226	1/4	Inner	13/32	—	—	2 Turns	.430	.230	.015	.450	—	—	3/4
	7044500	3/8	Inner	13/32	—	—	2 Turns	.430	.250	.015	.450	—	—	7/16
	7044502	1/4	Inner	13/32	—	—	2 Turns	.430	.230	.015	.450	—	—	7/8
	7044503	1/4	Inner	13/32	—	—	2 Turns	.430	.230	.015	.450	—	—	7/8
	7044505	3/8	Inner	—	—	—	2 Turns	.430	.250	.015	.450	—	—	7/16
	7044506	1/4	Inner	13/32	—	—	2 Turns	.430	.230	.015	.450	—	—	7/8
	7044507	1/4	Inner	13/32	—	—	2 Turns	.430	.230	.015	.450	—	—	7/8
	7044508	3/8	Inner	13/32	—	—	2 Turns	.430	.250	.015	.450	—	—	7/16
	7044509	1/4	Inner	13/32	—	—	2 Turns	.430	.230	.015	.450	—	—	7/8
	7044526	1/4	Inner	13/32	—	—	2 Turns	.430	.230	.015	.450	—	—	3/4

①—Front .200, Rear .550 ②—Front .180, Rear .170

Continued

QUADRAJET 4MC & 4MV ADJUSTMENT SPECIFICATIONS—Continued

See Tune Up Chart in car chapters for hot idle speeds.

Year	Carb. Model	Float Level	Pump Rod Hole	Pump Rod Adj.	Idle Vent	Air Valve	Fast Idle (Bench)	Choke Rod	Vacuum Break	Air Valve Dash-pot	Choke Unloader	Air Valve Lockout	Secondary Metering Rods	Air-Valve Valve Spring Wind-Up
FORD ENGINES														
1970–71	D0OF-A	11/32	Outer	5/16	—	—	2 Turns	.130	.140	.030	.300	.015	—	—
	D0OF-B	11/32	Outer	5/16	—	—	2 Turns	.166	.190	.030	.300	.015	—	—
	D0OF-E	11/32	Outer	5/16	—	—	2 Turns	.166	.190	.030	.300	.015	—	—
	D0OF-F	11/32	Outer	5/16	—	—	2 Turns	.130	.140	.030	.300	.015	—	—
OLDSMOBILE														
1969	7029250	1/4	Inner	5/16	—	—	2 Turns	.140	.180	.030	.200	.020	—	1/2
	7029251	1/4	Inner	5/16	—	—	2 Turns	.120	.180	.030	.200	.020	—	3/4
	7029252	1/4	Inner	5/16	—	—	2 Turns	.120	.180	.030	.200	.020	—	3/4
	7029253	1/4	Outer	3/8	—	—	—	.120	—	.050	.200	.020	—	3/4
	7029254	1/4	Inner	5/16	—	—	—	.140	—	.030	.200	.020	—	3/4
	7029255	1/4	Inner	5/16	—	—	—	.090	—	.050	.200	.020	—	3/4
1970	7040250	1/4	Inner	3/8	—	—	—	.140	.200	.030	.200	.020	—	1/2
	7040251	1/4	Inner	3/8	—	—	—	.140	.200	.030	.200	.020	—	3/4
	7040252	1/4	Inner	3/8	—	—	—	.170	.200	.030	.200	.020	—	3/4
	7040253	1/4	Inner	3/8	—	—	—	.230	.275	.030	.200	.020	—	3/4
	7040255	1/4	Inner	3/8	—	—	—	.275	.325	.030	.200	.020	—	3/4
	7040256	1/4	Inner	3/8	—	—	—	.275	.325	.030	.200	.020	—	3/4
	7040257	1/4	Inner	3/8	—	—	—	.170	.200	.030	.200	.020	—	3/4
	7040258	1/4	Inner	3/8	—	—	—	.170	.200	.030	.200	.020	—	3/4
1971	7041250	1/4	Inner	3/8	—	—	—	.140	.200	.050	.200	.035	—	1/2
	7041251	1/4	Inner	3/8	—	—	—	.140	.200	.050	.200	.035	—	3/4
	7041252	1/4	Inner	3/8	—	—	—	.170	.200	.050	.200	.035	—	3/4
	7041253	1/4	Inner	3/8	—	—	—	.230	.200	.050	.200	.035	—	3/4
	7041257	1/4	Inner	3/8	—	—	—	.170	.200	.050	.200	.035	—	3/4
1972	7042250	1/4	Inner	3/8	—	—	—	.230	.230	.050	.200	.035	—	1/2
	7042251	1/4	Inner	3/8	—	—	—	.230	.215	.050	.200	.035	—	3/4
	7042252	1/4	Inner	3/8	—	—	—	.230	.215	.050	.200	.035	—	3/4
	7042953	1/4	Inner	3/8	—	—	2 Turns	.230	.275	.050	.200	.035	—	3/4
1973	7043251	1/4	Inner	—	—	—	2 Turns	.230	—	—	.325	—	—	3/4
	7043252	1/4	Inner	—	—	—	2 Turns	.230	—	—	.325	—	—	3/4
	7043253	1/4	Inner	—	—	—	2 Turns	.230	.275	—	.300	.035	—	3/4
	7043259	1/4	Inner	—	—	—	2 Turns	.230	.215	—	.300	.035	—	3/4
1973–74	7043250	1/4	Inner	—	—	—	2 Turns	.230	.200	—	.300	.035	—	1/2
	7043251	1/4	Inner	—	—	—	2 Turns	.230	.200	—	.300	.035	—	3/4
	7043252	1/4	Inner	—	—	—	2 Turns	.230	.200	—	.300	.035	—	3/4
	7043255	1/4	Inner	—	—	—	2 Turns	.230	.200	—	.300	.035	—	1/2
	7043256	1/4	Inner	—	—	—	2 Turns	.230	.200	—	.300	.035	—	1/2
	7043257	1/4	Inner	—	—	—	2 Turns	.230	.200	—	.300	.035	—	1/2
	7043250	1/4	Inner	3/8	—	—	—	.230	.200	.030	.300	.035	.070	1/2
	7043251	1/4	Inner	3/8	—	—	—	.230	.200	.030	.300	.035	.070	3/4
	7043252	1/4	Inner	3/8	—	—	—	.230	.200	.030	.300	.035	.070	3/4
	7043255	1/4	Inner	3/8	—	—	—	.230	.200	.030	.300	.035	.070	1/2
	7043256	1/4	Inner	3/8	—	—	—	.230	.200	.030	.300	.035	.070	1/2
	7043259	1/4	Inner	3/8	—	—	—	.230	.200	.030	.300	.035	.070	—
1974	7044152	—	—	—	—	—	—	—	.200	—	—	—	—	—
	7044557	1/4	Inner	3/8	—	—	2 Turns	.230	.200	.030	.300	.035	.070	3/4
	7044558	1/4	Inner	3/8	—	—	2 Turns	.230	.200	.030	.300	.035	.070	3/4
	7044559	1/4	Inner	3/8	—	—	2 Turns	.230	.275	.030	.275	.035	.070	3/4

①—Rear adjustment .140".
②—Rear adjustment .115".
③—Rear adjustment .180".
④—Rear adjustment .130".
⑤—Rear adjustment .260".
⑥—Secondary lockout.

Year	Carb. Model	Float Level	Pump Rod Hole	Pump Rod Adj.	Idle Vent	Air Valve	Fast Idle (Bench)	Choke Rod	Vacuum Break	Air Valve Dash-pot	Choke Unloader	Air Valve Lockout	Secondary Metering Rods	Air-Valve Valve Spring Wind-Up
PONTIAC														
1969	7028270	1/4	Inner	9/32	3/8	—	3 Turns	.100	.245	.030	.300	.015	53/64	1/2
	7028273	1/4	Inner	9/32	3/8	—	3 Turns	.100	.245	.030	.300	.015	53/64	1/2
	7029260	3/16	Inner	9/32	3/8	—	3 Turns	.100	.150	.030	.300	.015	53/64	1/2
	7029261	3/16	Inner	9/32	3/8	—	3 Turns	.100	.180	.030	.300	.015	53/64	1/2
	7029262	9/32	Inner	9/32	3/8	—	3 Turns	.100	.245	.030	.300	.015	53/64	1/2

Continued

QUADRAJET 4MC & 4MV ADJUSTMENT SPECIFICATIONS—Continued

See Tune Up Chart in car chapters for hot idle speeds.

Year	Carb. Model	Float Level	Pump Rod		Idle Vent	Air Valve	Fast Idle (Bench)	Choke Rod	Vacuum Break	Air Valve Dash-pot	Choke Unloader	Air Valve Lockout	Secondary Metering Rods	Air-Valve Valve Spring Wind-Up
			Hole	Adj.										
PONTIAC—Continued														
1969	7029263	9/32	Inner	9/32	3/8	—	3 Turns	.100	.245	.030	.300	.015	53/64	1/2
	7029268	9/32	Inner	9/32	3/8	—	3 Turns	.100	.245	.030	.300	.015	53/64	1/2
	7029270	9/32	Inner	1/4	3/8	—	3 Turns	.100	.245	.030	.300	.015	53/64	1/2
	7029273	9/32	Inner	1/4	3/8	—	3 Turns	.100	.245	.030	.300	.015	53/64	1/2
1970	7040262	9/32	Inner	—	—	—	2 Turns	.100	.400	.025	.300	.015	—	7/16
	7040263	9/32	Inner	—	—	—	2 Turns	.100	.400	.025	.300	.015	—	7/16
	7040264	9/32	Inner	—	—	—	2 Turns	.100	.400	.025	.300	.015	—	7/16
	7040267	9/32	Inner	—	—	—	2 Turns	.100	.400	.025	.300	.015	—	7/16
	7040268	9/32	Inner	—	—	—	2 Turns	.100	.400	.025	.300	.015	—	7/16
	7040270	9/32	Inner	—	—	—	2 Turns	.100	.245	.025	.300	.015	—	3/4
	7040273	9/32	Inner	—	—	—	2 Turns	.100	.245	.025	.300	.015	—	3/4
	7040274	9/32	Inner	—	—	—	2 Turns	.100	.400	.025	.300	.015	—	7/16
	7040562	9/32	Inner	—	—	—	2 Turns	.100	.400	.025	.300	.015	—	7/16
	7040563	9/32	Inner	—	—	—	2 Turns	.100	.400	.025	.300	.015	—	7/16
	7040564	9/32	Inner	—	—	—	2 Turns	.100	.400	.025	.300	.015	—	7/16
	7040567	9/32	Inner	—	—	—	2 Turns	.100	.400	.025	.300	.015	—	7/16
	7040568	9/32	Inner	—	—	—	2 Turns	.100	.400	.025	.300	.015	—	7/16
	7040570	9/32	Inner	—	—	—	2 Turns	.100	.245	.025	.300	.015	—	3/4
	7040573	9/32	Inner	—	—	—	2 Turns	.100	.245	.025	.300	.015	—	3/4
1971	7041262	9/32	Inner	—	—	—	2 Turns	.100	.240	.025	.300	.015	—	7/16
	7041263	9/32	Inner	—	—	—	2 Turns	.100	.240	.025	.300	.015	—	7/16
	7041267	9/32	Inner	—	—	—	2 Turns	.100	.370	.025	.300	.015	—	1/2
	7041268	9/32	Inner	—	—	—	2 Turns	.100	.430	.025	.300	.015	—	1/2
	7041270	9/32	Inner	—	—	—	2 Turns	.100	.430	.025	.300	.015	—	1/2
	7041271	9/32	Inner	—	—	—	2 Turns	.100	.240	.025	.300	.015	—	7/16
	7041273	9/32	Inner	—	—	—	2 Turns	.100	.370	.025	.300	.015	—	1/2
1972	7042262	13/32	Inner	13/32	—	—	2 Turns	.100	.290	.025	.310	.015	—	7/16
	7042263	13/32	Inner	13/32	—	—	2 Turns	.100	.325	.025	.310	.015	—	11/16
	7042264	13/32	Inner	13/32	—	—	2 Turns	.100	.290	.025	.310	.015	—	5/8
	7042270	13/32	Inner	7/16	—	—	2 Turns	.100	.290	.025	.310	.015	—	7/16
	7042272	13/32	Inner	13/32	—	—	2 Turns	.100	.195	.025	.310	.015	—	—
	7042273	13/32	Inner	7/16	—	—	2 Turns	.100	.325	.025	.310	.015	—	7/16
	7042274	13/32	Inner	13/32	—	—	2 Turns	.100	.195	.025	.310	.015	—	—
	7042276	13/32	Inner	13/32	—	—	2 Turns	.100	.290	.025	.310	.015	—	—
	7042278	13/32	Inner	13/32	—	—	2 Turns	.100	.290	.025	.310	.015	—	—
1973	7043262	13/32	Inner	13/32	—	—	2 Turns	.200	.260	.025	.310	.015	—	3/8
	7043263	13/32	Inner	13/32	—	—	2 Turns	.200	.290	.025	.310	.015	—	5/8
	7043264	13/32	Inner	13/32	—	—	2 Turns	.200	.260	.025	.310	.015	—	1/2
	7043265	13/32	Inner	13/32	—	—	2 Turns	.200	.290	.025	.310	.015	—	9/16
	7043266	13/32		13/32	—	—	—	.205	.260	.025	.310	.015	—	1/2
	7043270	13/32	Inner	—	—	—	2 Turns	.200	.290	—	.310	.015	—	—
	7043272	13/32	Outer	13/32	—	—	2 Turns	.200	.290	.025	.310	.015	—	3/8
	7043273	13/32	Inner	—	—	—	2 Turns	.200	.290	—	.310	.015	—	—
	7043274	13/32	Outer	13/32	—	—	2 Turns	.200	.290	.025	.310	.015	—	9/16
1974	7043263	13/32	Inner	.410	—	—	—	.205	.290	.025	.310	.015	—	5/8
	7044262	13/32	Inner	.410	—	—	2 Turns	.200	.260	.025	.300	.010	—	3/8
	7044266	13/32	Inner	.410	—	—	2 Turns	.200	.260	.025	.300	.010	—	1/2
	7044267	13/32	Inner	.410	—	—	—	.205	.260	.025	.310	.015	—	3/8
	7044268	13/32	Inner	.410	—	—	2 Turns	.200	.260	.025	.300	.010	—	1/2
	7044269	13/32	Inner	.410	—	—	2 Turns	.200	.290	.025	.300	.010	—	1/2
	7044270	13/32	Inner	.410	—	—	2 Turns	.200	.290	.025	.300	.010	—	3/4
	7044272	13/32	Outer	.315	—	—	2 Turns	.200	.290	.025	.300	.010	—	3/8
	7044273	13/32	Inner	.410	—	—	2 Turns	.200	.290	.025	.300	.010	—	3/4
	7044274	13/32	Outer	.315	—	—	2 Turns	.200	.290	.025	.300	.010	—	9/16
	7044560	13/32	Inner	.410	—	—	2 Turns	.200	.260	.025	.300	.010	—	3/8
	7044568	13/32	Inner	.410	—	—	2 Turns	.200	.260	.025	.300	.010	—	1/2

①—Rear .115″. ②—Rear .130″. ③—Rear .265″. ④—Rear .260″.

QUADRAJET 4MV, 4MC ADJUSTMENTS

The Quadrajet unit, Figs. 1, 2, has two stages in operation. The primary (fuel inlet) side has small bores with a triple venturi equipped with plain tube nozzles. The triple venturi feature, plus the smaller primary bores, give a more stable and finer fuel control in the idle and economy ranges of operation. Fuel metering in the primary side is accomplished with tapered metering rods positioned by a manifold vacuum responsive piston.

The secondary side has two very large bores which have greatly increased air capacity to meet all engine demands. The air valve principle is used in the secondary side for metering control and supplements fuel flow from the primary bores. Using the air valve principle, fuel is metered in direct proportion to the air passing through the secondary bores.

The fuel reservoir is centrally located to avoid problems of fuel slosh causing engine turn cut-out and delayed fuel flow to the carburetor bores. The float system uses a single float pontoon for ease of service. The float needle valve is pressure balanced to overcome problems encountered with high fuel pump pressures and to permit use of a small float to control fuel "shut-off' through the large fuel inlet needle seat. It has a synthetic tip which gives added insurance against flooding problems casued by dirt.

The primary side of the carburetor has six systems of operation: float, idle, main metering, power, pump and choke. The secondary side has one metering system which supplements the primary main metering system and receives fuel from a common float chamber.

Model 4MV Choke System

Fig. 3—The choke system consists of a choke valve located in the primary air horn bore, a vacuum diaphragm unit, fast idle cam, connecting linkage, air valve lockout lever and a thermostatic coil. Some applications may use a split choke pick-up spring or a vacuum break modulating spring. The thermostatic coil is located in the engine manifold and is connected to the intermediate choke shaft and lever assembly. Choke operation is controlled by a combination of engine intake manifold vacuum, the offset choke valve, temperature and throttle position.

Model 4MC Choke System

Fig. 4—The choke consists of a choke valve located in the primary air horn bore, a choke housing and vacuum diaphragm assembly, fast idle cam, connecting linkage, air valve lockout lever, and thermostatic coil. Choke operation is controlled by a combination of intake manifold vacuum, the offset choke valve, temperature, and throttle position.

Air Valve Operation

Fig. 5—When the engine reaches a point where the primary bores cannot meet engine air and fuel demands, the primary throttle lever, through connecting linkage to the secondary throttle shaft lever, begins to open the secondary throttle valves. As air flow through the secondary bores creates a low pressure (vacuum) beneath the air valve, atmospheric pressure on top of the air valve forces the air valve open against spring tension. This allows the required air for increased engine speed to flow past the air valve.

Air Valve Dashpot Operation

Fig. 6—The secondary air valve has an attached piston assembly which acts as a damper to prevent oscillation of the valve due to engine pulsations. The damper piston operates in a well that is filled with fuel from the float bowl. The motion of the piston is retarded by fuel which must by-pass the piston when it moves up in the fuel well. The piston is attached loosely to a plunger rod. The rod has a rubber seal which retains the damper piston to the plunger rod and also acts as a valve. The purpose of the valve is to seat on the piston when the air valve opens and the piston rod moves upward. This closes off the area through the center of the piston and slows down the air valve opening to prevent secondary discharge nozzle lag.

ADJUSTMENTS

Float Level Adjustment

Fig. 7—With adjustable T-scale, measure from top of float bowl gasket surface

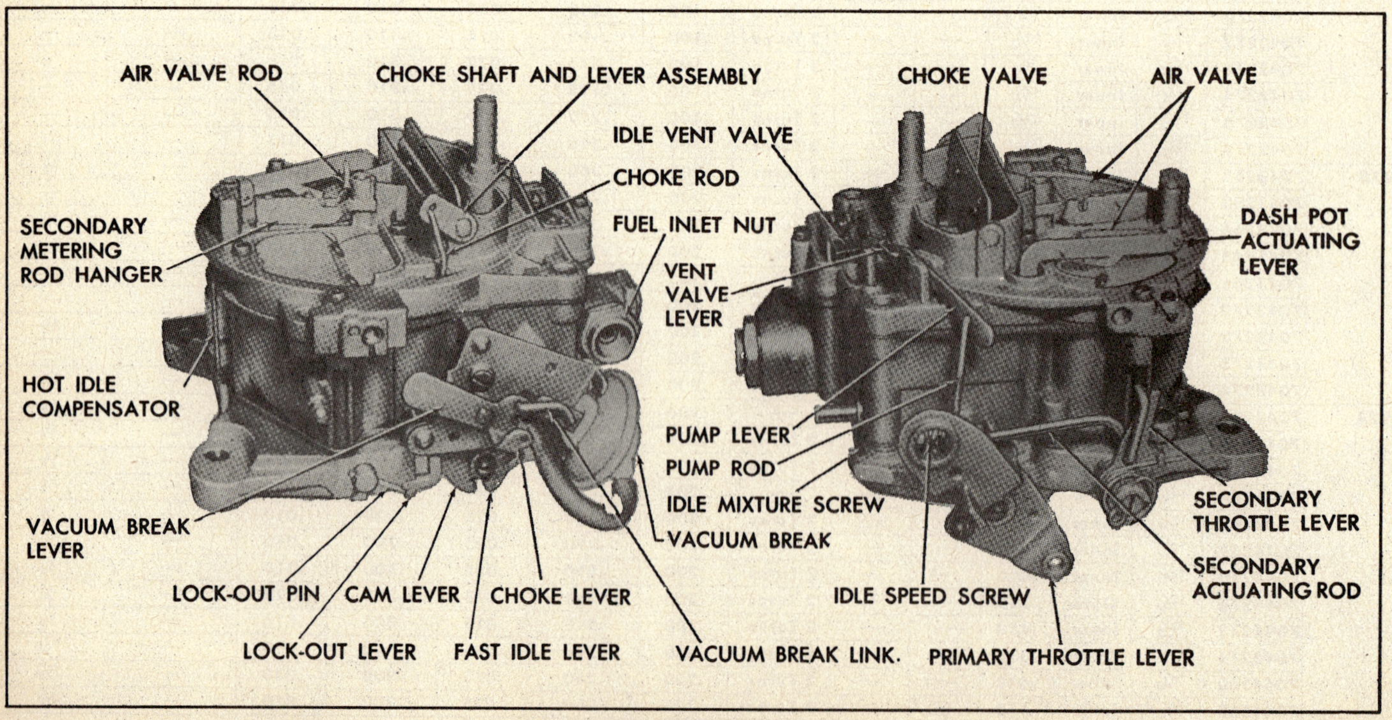

Fig. 1 Quadrajet Model 4MV carburetor

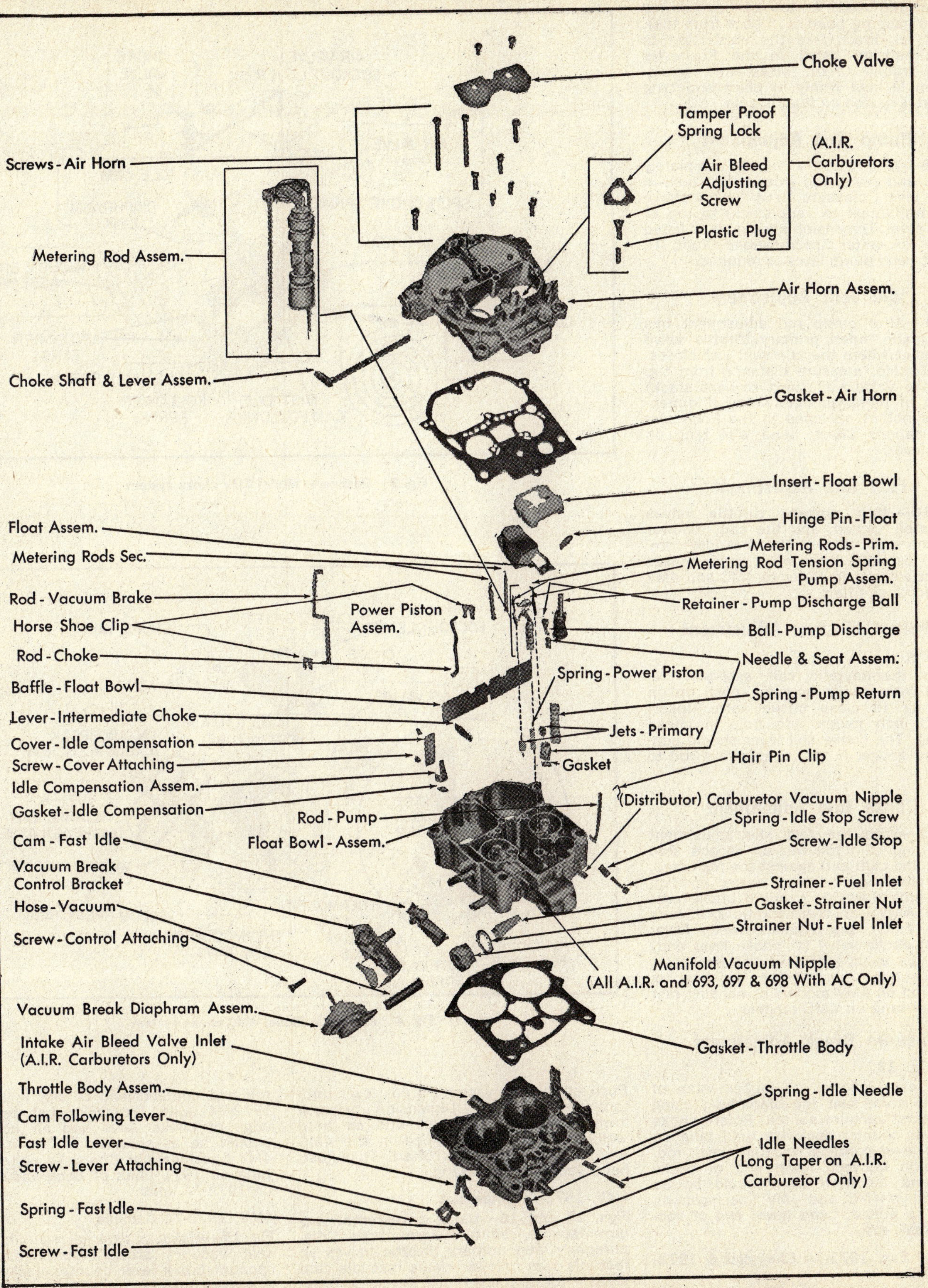

Fig. 2 4MV Quadrajet carburetor. Typical

(gasket removed) to top of float at toe (locate gauging point 1/16" back from toe). Adjust as directed in the illustration to the dimension listed in the Rochester *Specifications Chart*. Make sure retaining pin is held firmly in place and tang of float is seated on float needle.

Pump Rod Adjustment

Fig. 8—With throttle valves completely closed and pump rod in specified hole in pump lever, measure from top of choke valve wall (next to vent stack) to top of pump stem. Dimension should be as listed in the *Rochester Specifications Chart*. To adjust, bend pump lever as required.

Idle Vent Adjustment

Fig. 9—After pump rod adjustment has been made, open primary throttle valve to a point where the idle vent just closes. With T-scale, measure distance from top of choke valve wall (next to vent stack) to top of pump plunger stem. If dimension is not as specified in the *Rochester Specifications Chart*, bend wire tang on pump lever.

Fast Idle Adjustment

Fig. 10—With primary throttle valves completely closed, and the cam follower over the high step of the fast idle cam, adjust fast idle screw after screw makes contact with lever. Refer to *4MC and 4MV Specifications Chart.*

Choke Coil Lever Adjustment

4MC, Fig. 11

With thermostatic coil assembly removed from choke housing, push up on coil tang to close choke valve. Insert a .120 inch gauge into hole in choke housing. The choke coil lever should contact the gauge. If not, bend choke rod to adjust.

Choke Rod Adjustment

Fig. 12—With the fast idle adjustment made, and cam follower on second step of fast idle cam and against the high step, rotate choke valve toward closed position by pushing down on vacuum break lever (Model 4MV) or thermostatic coil tang (Model 4MC). Dimension between lower edge of choke valve (at choke lever end) should be as specified in the *Rochester Specifications Chart*. Adjust by bending choke rod on 4MV models or bending fast idle cam tang on 4MC models.

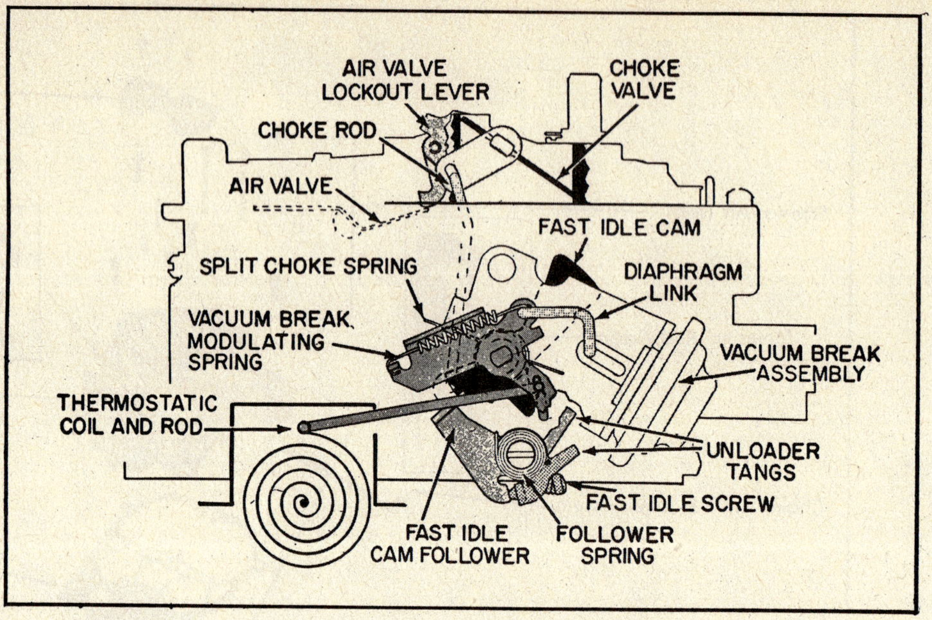

Fig. 3 Rochester Model 4MV choke system

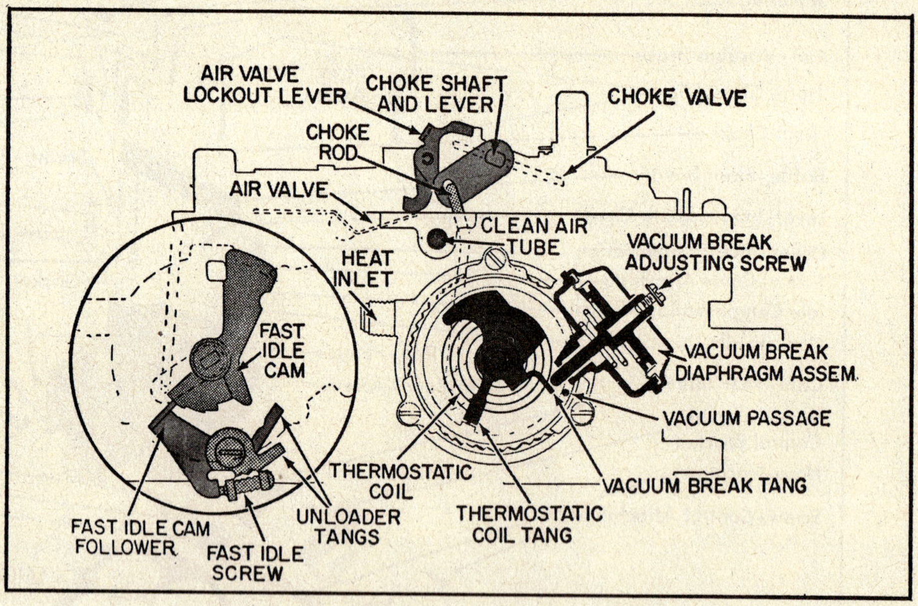

Fig. 4 Rochester Model 4MC choke system

Vacuum Break Adjustment

4MC, Fig. 13

With cam follower on highest step of fast idle cam, seat the diaphragm using an external vacuum source. Rotate choke coil lever counter-clockwise until tang on outside lever contacts vacuum break rod. Dimension between lower end of choke valve and air horn wall should be as specified in *4MC and 4MV Specifications Chart*. To adjust, bend lower end of vacuum break rod.

4MV, All Exc. 1971-74 Chevrolet & 1972-74 Cadillac

Fig. 14—Using an external vacuum source, seat vacuum break diaphragm. Push upward on vacuum break lever until tang contacts rod. Dimension between lower end of choke valve and air horn wall should be as specified in the *4MC and 4MV Specifications Chart*. To adjust, bend vacuum break tang.

4MV, 1971-74 Chevrolet

Figs. 15 and 16—Using an external vacuum source, seat vacuum break diaphragm. Open primary throttle valves so fast idle cam follower clears fast idle cam steps. Rotate choke coil lever counter-clockwise until top edge of cam contacts diaphragm plunger tang on 1971 units or rod end contacts end of slot in lever on 1972-74 units. Dimension between lower edge of choke valve and air horn wall should be as specified in the *4MC and 4MV Specifications Chart*. To adjust, bend tang on 1971 units or bend vacuum link on 1972-74 units.

4MV, 1972-74 Cadillac

Fig. 17—Using an external vacuum source, seat vacuum break diaphragm. Rotate vacuum break lever counter-clockwise until tang contacts rod and the choke vacuum diaphragm plunger spring is compressed. Dimension between lower edge of choke valve and air horn wall should be as specified in the 4MC and 4MV Specifications Chart. To adjust, bend vacuum break tang.

Secondary Vacuum Break, 1970-74 Buick

Fig. 18—Using an external vacuum source, seat secondary vacuum break diaphragm. Rotate choke lever counter-clockwise to pull plunger outward until seated. Dimension between air horn wall and lower edge of choke valve should be as specified in the 4MC and 4MV Specifications Chart. To adjust, bend secondary vacuum break rod.

Air Valve Dashpot Adjustment

Figs. 19 through 21—With vacuum break diaphragm seated, there must be the specified clearance between dashpot rod and end of slot in air valve lever. To adjust, bend rod at air valve end.

Choke Unloader Adjustment

Fig. 22—With choke valve held closed by means of a rubber band on vacuum break lever, open throttle valves fully. With valves in this position, dimension between lower edge of choke valve and air horn wall should be as specified. To adjust, bend tang on fast idle lever.

Air Valve Lockout Adjustment

Fig. 23—Rotate vacuum break lever clockwise until choke valve is wide open. If upper choke lever is slotted, the rod must be in upper end of slot. Open air valve slightly so that edge of air valve is opposite tang on lockout lever, as shown. Measure distance between tang on lockout lever and edge of air valve. Adjust by bending the upper end of lockout lever. Refer to *4MC and 4MV Specifications Chart.*

After adjustment, close choke valve to make sure lower edge of lockout lever clears top edge of air valve for proper locking during choke operation. If the lock-

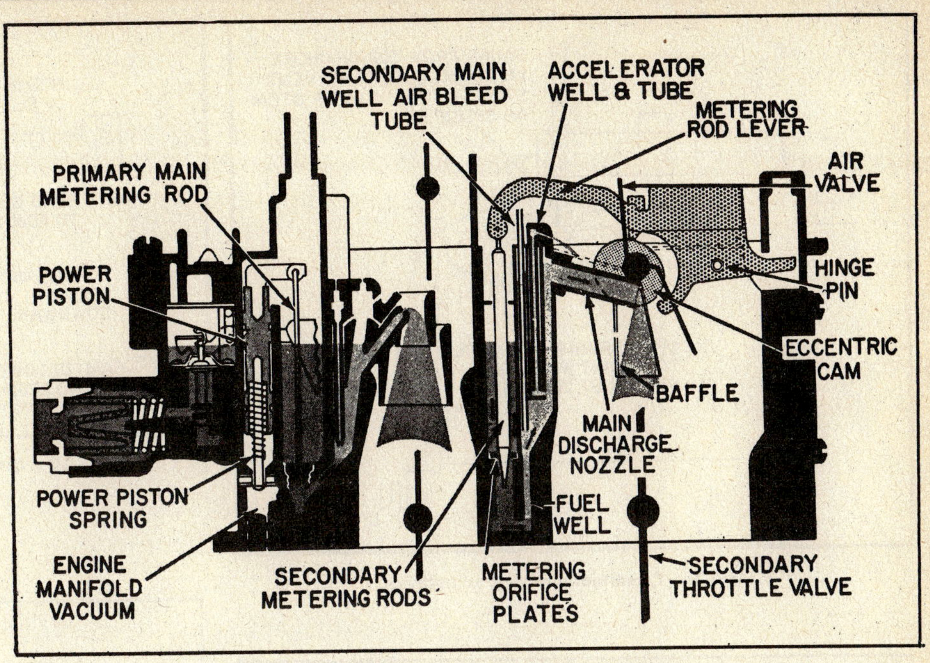

Fig. 5 Diagram of Quadrajet power system

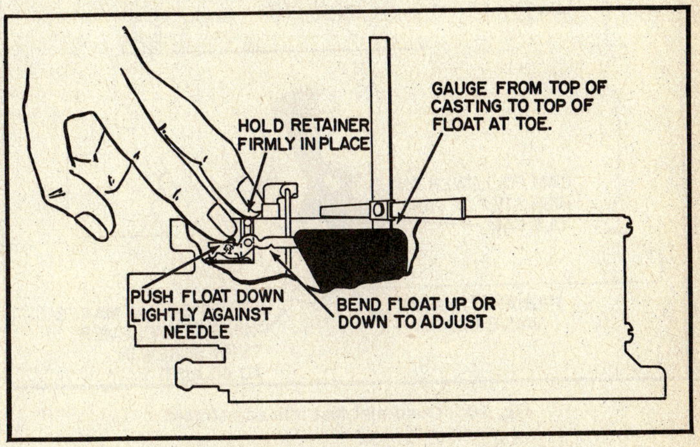

Fig. 7 Quadrajet float level adjustment

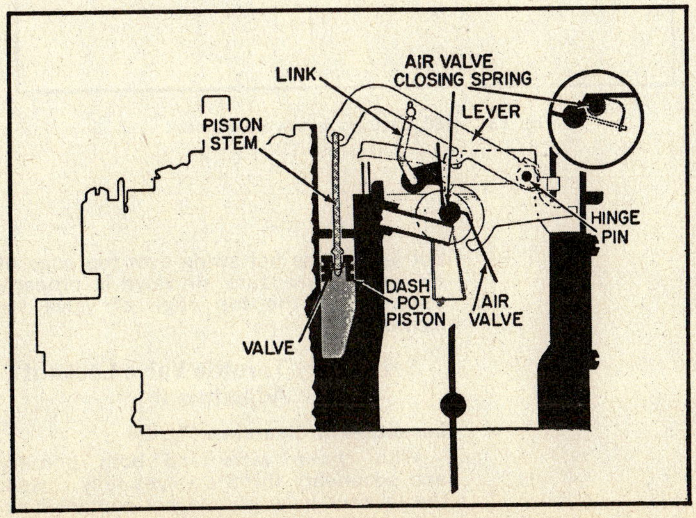

Fig. 6 Quadrajet air valve dashpot

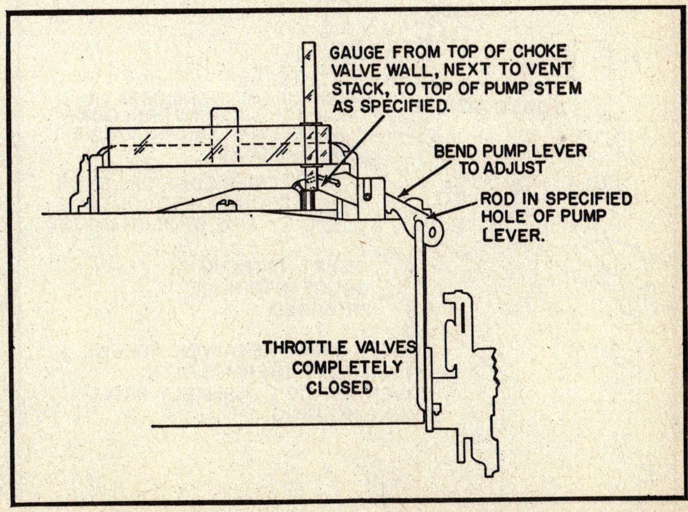

Fig. 8 Quadrajet pump rod adjustment

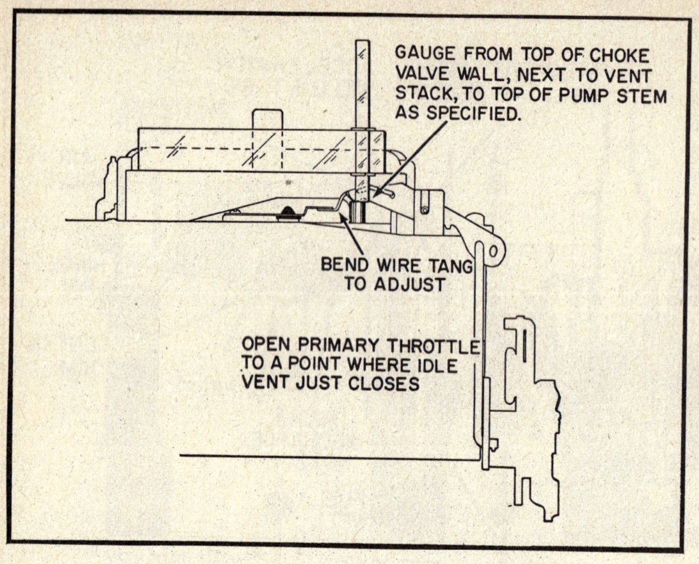

Fig. 9 Quadrajet idle vent adjustment

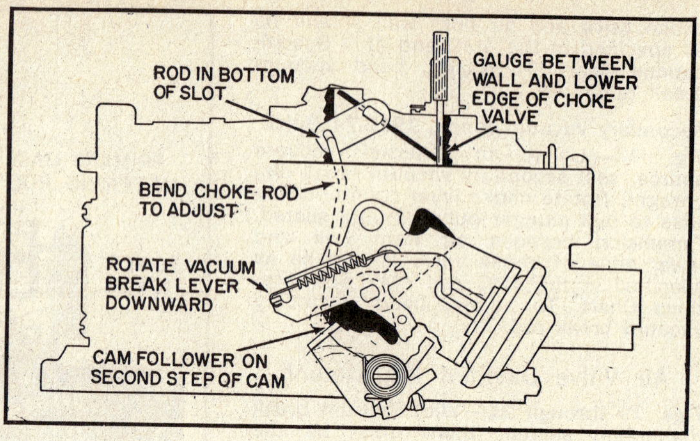

Fig. 12 Quadrajet choke rod adjustment (Typical)

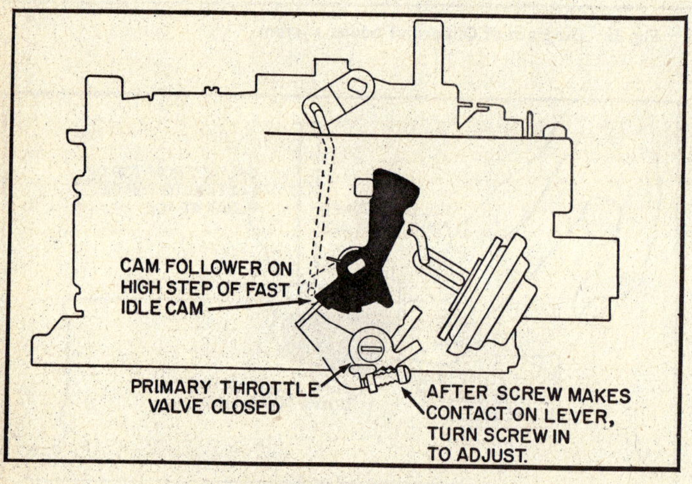

Fig. 10 Quadrajet fast idle adjustment

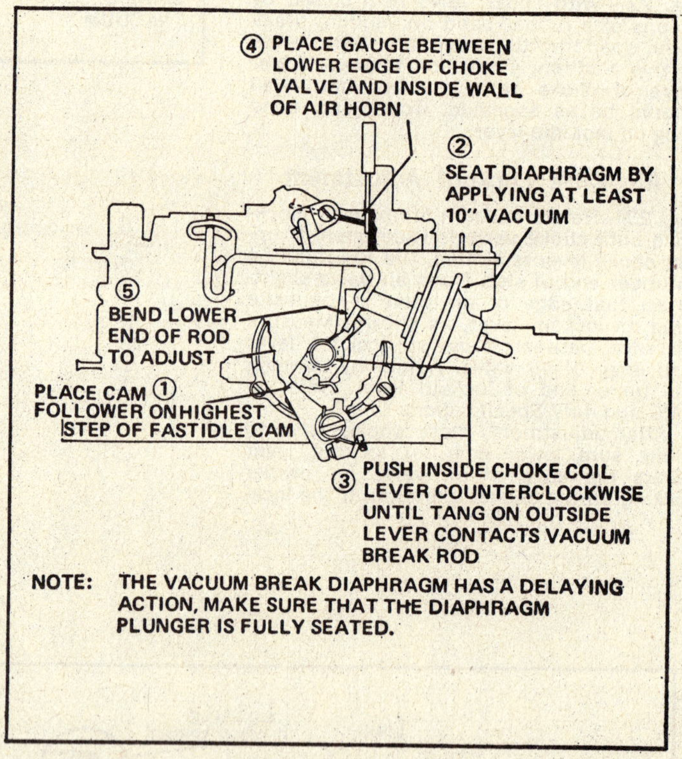

Fig. 13 4MC vacuum break adjustment

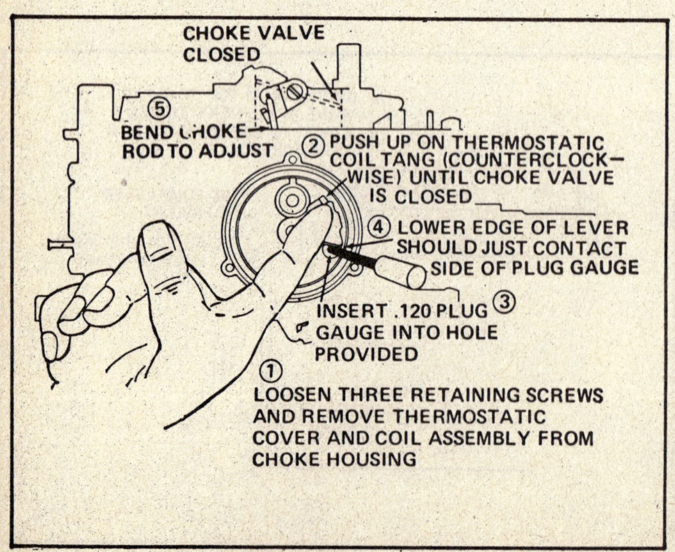

Fig. 11 4MC choke coil lever adjustment

out lever does not swing over top edge of air valve make sure air valve is properly seated. File the top edge of valve for clearance.

Secondary Throttle Valve Lockout Adjustment

Lockout–Pin Clearance Fig. 24

With choke valve and both primary and secondary throttle valves fully closed, the lockout lever should not contact lockout pin. Clearance must not exceed .015". Bend lockout pin to adjust.

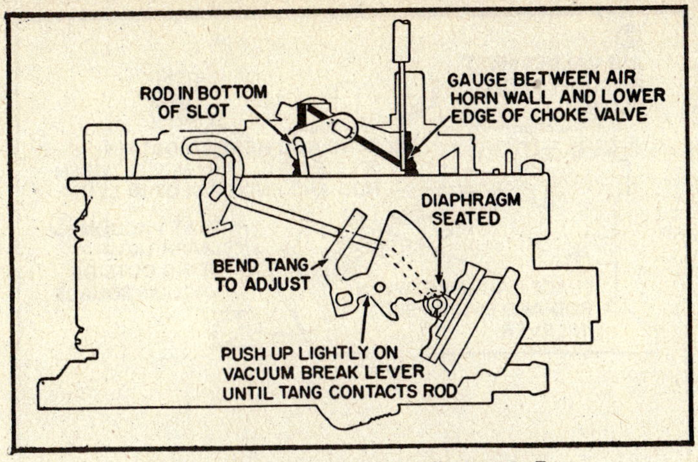

Fig. 14 4MV vacuum break adjustment, Exc. 1971-74 Chevrolet & 1972-74 Cadillac

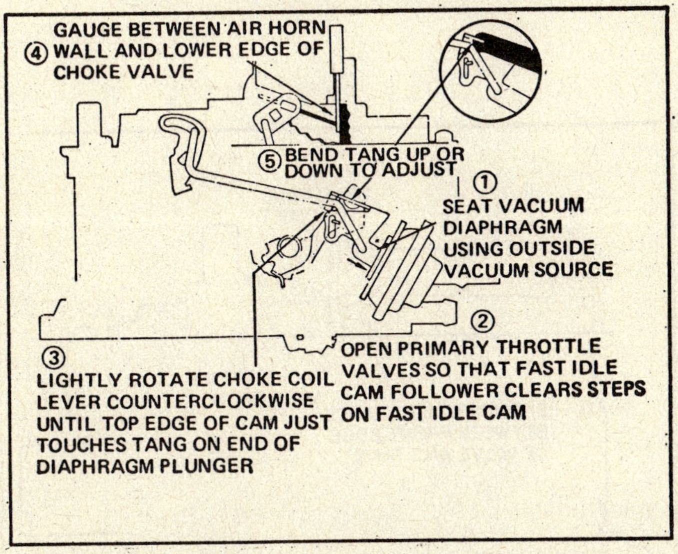

Fig. 15 4MV vacuum break adjustment. 1971 Chevrolet

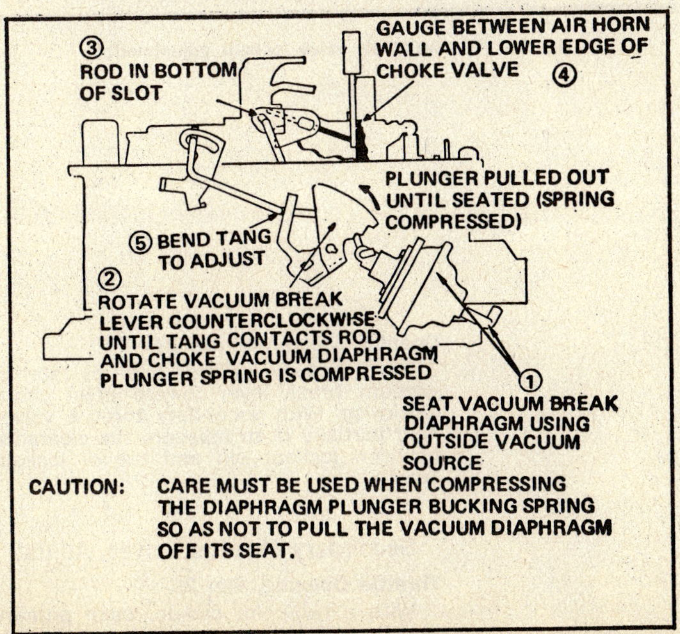

Fig. 17 4MV vacuum break adjustment. 1972-74 Cadillac

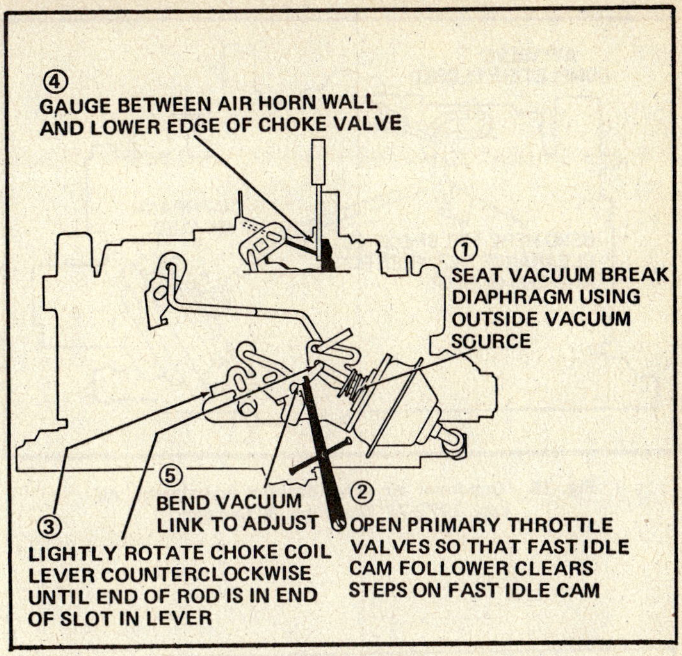

Fig. 16 4MV vacuum break adjustment. 1972-74 Chevrolet

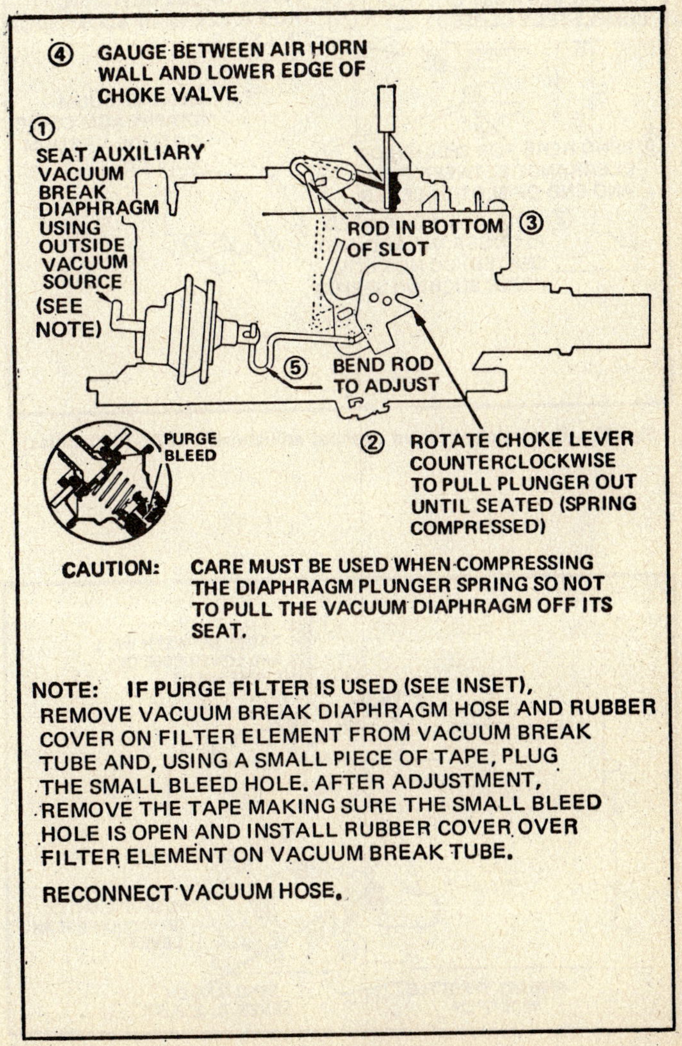

Fig. 18 4MV secondary vacuum break adjustment. 1970-74 Buick

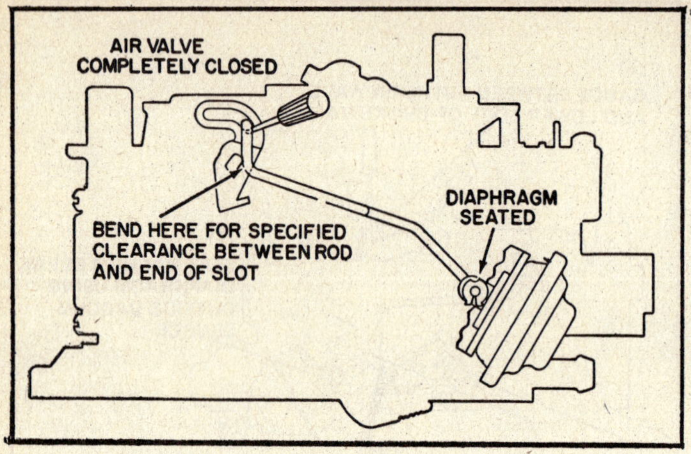

Fig. 19 Quadrajet air valve dashpot adjustment. All exc. 1972-74 Chevrolet & Cadillac

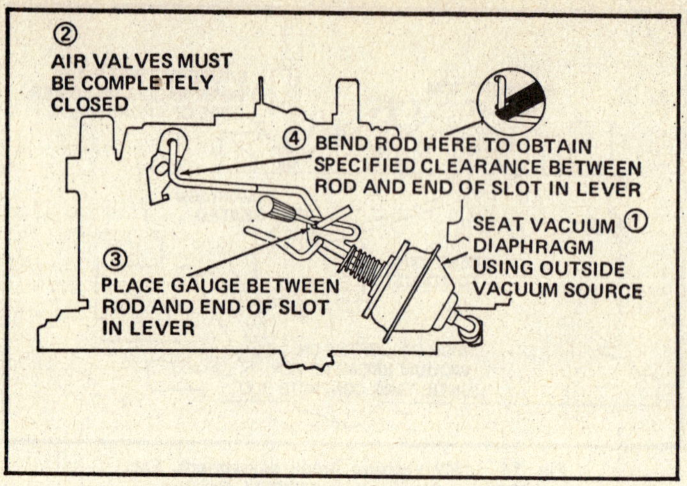

Fig. 20 Quadrajet air valve dashpot adjustment. 1972-74 Chevrolet

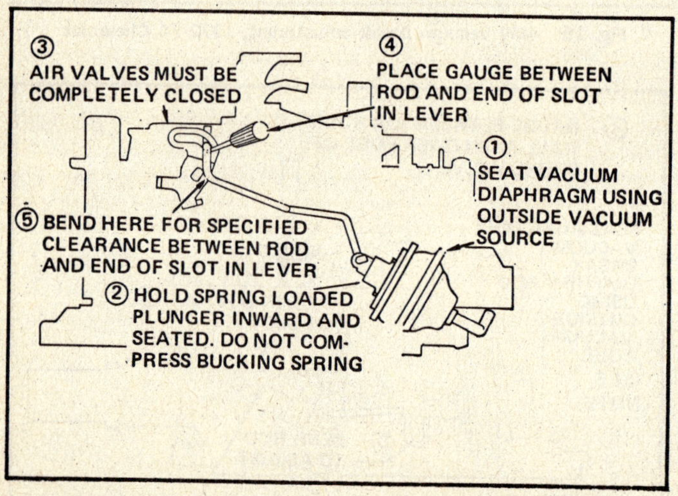

Fig. 21 Quadrajet air valve dashpot adjustment. 1972-74 Cadillac

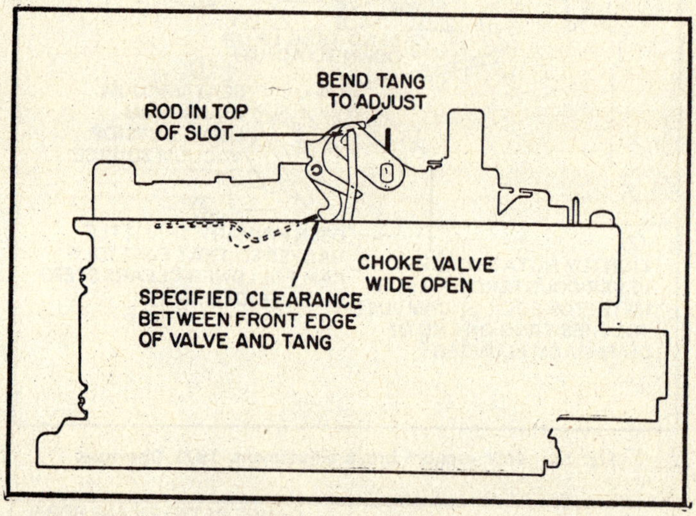

Fig. 23 Quadrajet air valve lockout adjustment

Fig. 22 Quadrajet unloader adjustment (Typical)

Opening Clearance, Fig. 24

Hold choke valve wide open by rotating vacuum break lever toward open choke clockwise. With secondary throttle valves held partially open measure the clearance between lockout pin and toe of lockout lever.

Secondary Throttle Valves, Adjust

Throttle Opening, Fig. 25

With a two-point pickup, open primary throttle valves until actuating link contacts tang on secondary lever. With valves

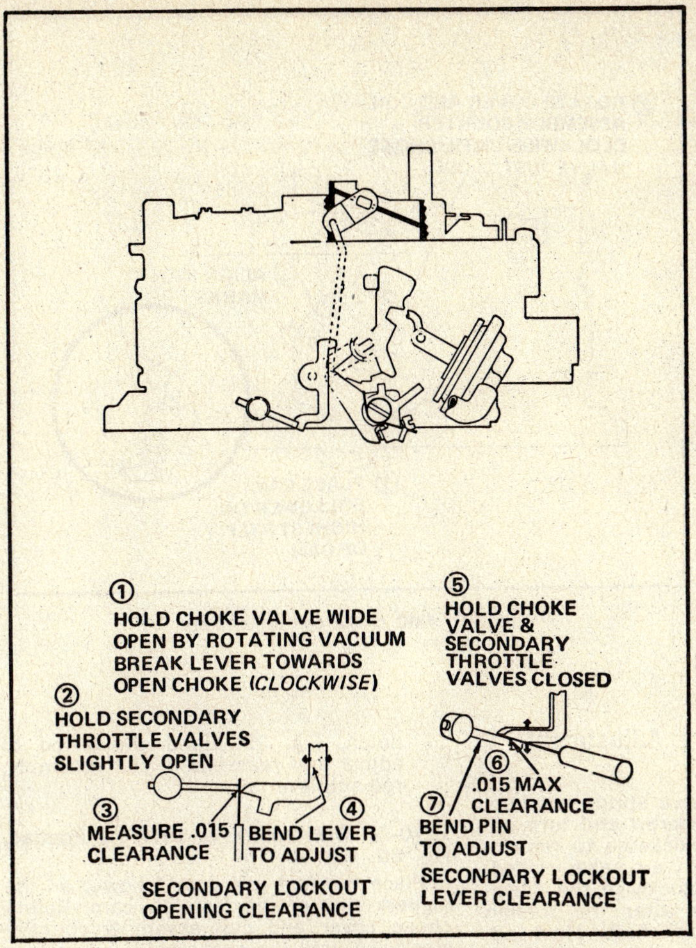

Fig. 24 Quadrajet secondary lockout adjustment

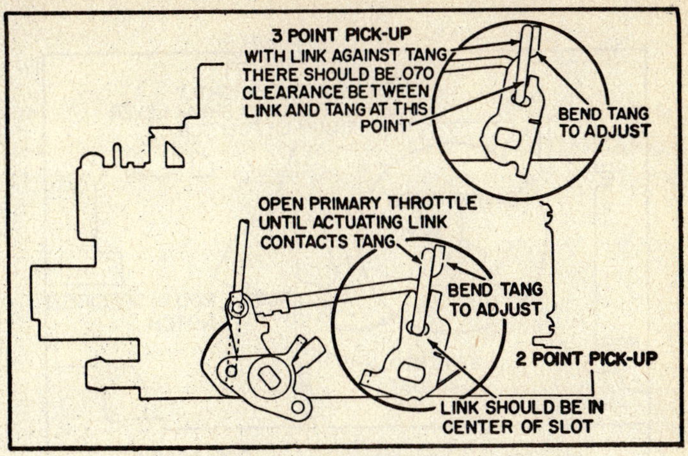

Fig. 25 Quadrajet secondary throttle valves opening adjustment

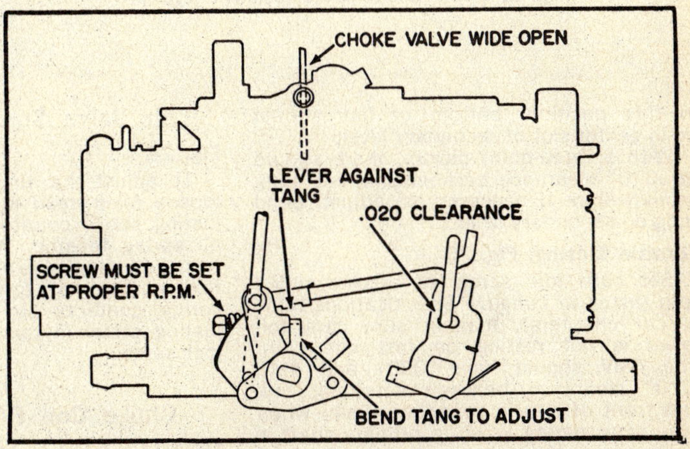

Fig. 26 Quadrajet secondary throttle valves closing adjustment

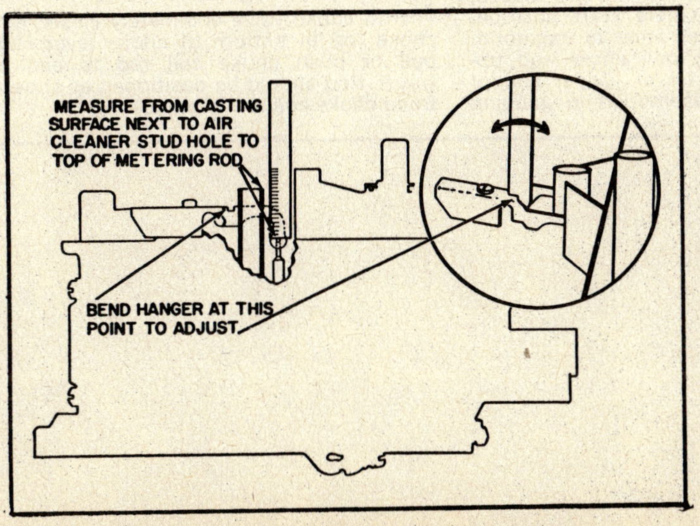

Fig. 27 Quadrajet secondary metering rod adjustment

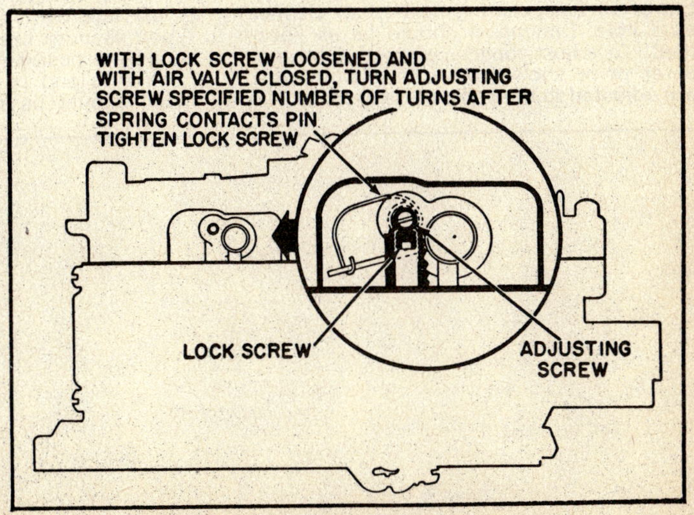

Fig. 28 Quadrajet air valve spring adjustment

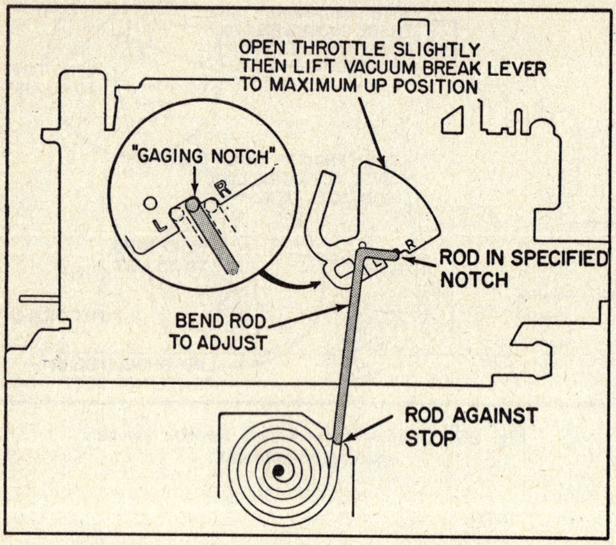

Fig. 29 4MV Choke coil adjustment. Typical

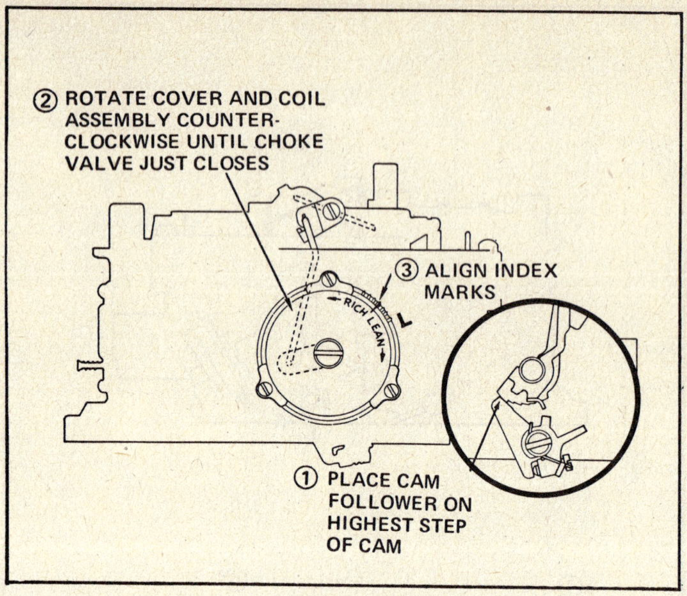

Fig. 30 4MC choke coil adjustment

in this position, bottom of link should be in center slot of secondary lever.

With a three-point pickup, there should be .070" clearance between link and tang as indicated. If necessary to adjust, bend tang on secondary lever.

Throttle Closing, Fig. 26

Set curb idle screw to recommended rpm (listed in *Tune-Up Specifications* table in car chapters), making sure cam follower is not resting on fast idle cam. The 4MV should have .020" and 4MC .070" clearance between actuating link and front of slot in secondary lever when tang of actuating lever on primary shaft is against pin. If necessary to adjust, bend tang on primary actuating lever.

Secondary Metering Rod Adjustment

Fig. 27

Measure from top of metering rod to top of air horn casting next to air cleaner stud hole. Dimension should be as specified. To adjust, bend metering rod hanger at point shown. *Make sure both rods are adjusted to the same dimension.*

Air Valve Spring Adjustment

Fig. 28

To adjust the air valve spring wind-up, loosen Allen head lockscrew and turn adjusting screw counterclockwise to remove all spring tension. With air valve closed, turn adjusting screw clockwise the specified number of turns after the torsion spring contacts pin on shaft. Hold adjusting screw in this position and tighten lock screw.

Choke Coil Rod Adjustment

1973-74 Buick, 1969-74 Cadillac & Chevrolet, 1970-71 Ford, Fig. 29

1. Remove choke coil assmebly to disengage choke rod from vacuum break lever.
2. Reinstall coil assembly but do not install rod into lever.
3. With choke valve completely closed, fast idle cam in cold start position, and vacuum break lever in maximum upward position, pull choke rod upward to end of travel. Upper end of rod should be positioned in gauging notch, Fig. 29. Bend choke rod to adjust and reassemble coil assembly, rod and lever.

1970-74 Oldsmobile & 1973-74 Pontiac, Fig. 30

Place the fast idle cam follower on the highest step of the fast idle cam. Rotate choke cover and coil assembly counterclockwise until the choke valve just closes and the index point on cover aligns with the center index point on the choke housing.

1969 Oldsmobile and 1969-72 Pontiac

With fast idle adjusted and cam follower on second step and against high step of cam, close choke and check dimension between lower edge of choke valve and inner wall of main body. Bend choke rod to adjust.

1969-72 Buick Fig. 29

With choke valve completely closed and choke rod in bottom of choke lever slot, pull or push choke coil rod to end of travel. Rod should be positioned as shown. Bend choke coil rod to adjust.

QUADRAJET M4M SERIES ADJUSTMENT SPECIFICATIONS

See Tune Up Chart in car chapters for hot idle speeds.

Year	Carb. Model	Float Level	Pump Rod		Bowl Vent	Fast Idle (Bench)	Choke Coil Lever	Choke Rod	Vacuum Break		Air-Valve Dash-pot	Choke Setting	Choke Unloader	Air-Valve Valve Spring Wind-Up
			Hole	Adj.					Front	Rear				
BUICK														
1975	7045240	15/32	Inner	3/8	—	3 Turns	.120	.095	.135	.120	.015	1 Rich	.235	7/16
	7045244	3/8	Outer	3/8	—	3 Turns	.120	.095	.130	.115	.015	1 Rich	.235	3/4
	7045246	3/8	Outer	3/8	—	3 Turns	.120	.095	.130	.115	.015	1 Rich	.235	3/4
	7045264	17/32	Inner	9/32	—	3 Turns	.120	.130	.150	.260	.030	Index	.230	1/2
	7045541	3/8	Outer	3/8	—	3 Turns	.120	.095	.145	.130	.015	1 Rich	.235	3/4
	7045544	3/8	Outer	3/8	—	3 Turns	.120	.095	.145	.130	.015	1 Rich	.235	3/4
	7045546	5/16	Outer	15/32	—	3 Turns	.120	.095	.145	.130	.015	1 Rich	.240	3/4
	7045598	7/16	Inner	9/32	—	3 Turns	.120	.095	.135	.120	.015	1 Rich	.240	7/16
1976	17056240	15/32	Inner	3/8	—	—	.120	.095	.135	.120	.015	1 Rich	.250	7/16
	17056244	5/16	Outer	3/8	—	—	.120	.095	.130	.120	.015	1 Rich	.250	3/4
	17056246	5/16	Outer	3/8	.075	—	.120	.095	.130	.120	.015	1 Rich	.250	3/4
	17056540	15/32	Outer	3/8	—	—	.120	.095	.135	.120	.015	1 Rich	.250	7/16
	17056544	5/16	Outer	3/8	—	—	.120	.095	.130	.130	.015	Index	.250	3/4
	17056546	5/16	Outer	3/8	—	—	.120	.095	.130	.130	.015	Index	.250	3/4
CADILLAC														
1975	7045230	15/32	Outer	3/8	.075	2 Turns	.120	.080	.160	.130	.030	2 Rich	.215	7/16
	7045530	15/32	Outer	3/8	.075	2 1/4 Turns	.120	.080	.180	.180	.030	1 Rich	.215	1/2
1976	17056230	13/32	Outer	3/8	—	—	.120	.080	.160	.160	—	2 Rich	.230	3/8
	17056232	13/32	Outer	3/8	—	—	.120	.080	.160	.160	—	2 Rich	.230	3/8
	17056530	7/16	Inner	9/32	—	—	.120	.080	.160	.160	—	2 Rich	.230	3/8
CHEVROLET														
1975	7045200	17/32	Inner	9/32	—	2 Turns	.120	.060	.200	.550	.015	Index	.270	9/16
	7045202	15/32	Inner	9/32	—	3 Turns	.120	.300	.180	.170	.015	Index	.325	7/8
	7045203	15/32	Inner	9/32	—	3 Turns	.120	.300	.180	.170	.015	2 Lean	.325	7/8
	7045206	15/32	Inner	9/32	—	3 Turns	.120	.300	.180	.170	.015	Index	.325	7/8
	7045207	15/32	Inner	9/32	—	3 Turns	.120	.300	.180	.170	.015	2 Lean	.325	7/8
	7045210	15/32	Inner	9/32	—	3 Turns	.120	.300	.180	.170	.015	Index	.325	7/8
	7045211	15/32	Inner	9/32	—	3 Turns	.120	.300	.180	.170	.015	2 Lean	.325	7/8
	7045218	5/16	Inner	9/32	—	3 Turns	.120	.325	.180	.170	.015	Index	.325	3/4
	7045219	5/16	Inner	9/32	—	3 Turns	.120	.325	.180	.170	.015	2 Lean	.325	3/4
	7045220	17/32	Inner	9/32	—	2 Turns	.120	.060	.200	.550	.015	Index	.270	9/16
	7045222	15/32	Inner	9/32	—	3 Turns	.120	.300	.180	.170	.015	Index	.325	7/8
	7045223	15/32	Inner	9/32	—	3 Turns	.120	.300	.180	.170	.015	2 Lean	.325	7/8
	7045224	15/32	Inner	9/32	—	3 Turns	.120	.325	.180	.170	.015	Index	.325	3/4
	7045228	15/32	Inner	9/32	—	3 Turns	.120	.325	.180	.170	.015	Index	.325	3/4
	7045502	15/32	Inner	9/32	—	3 Turns	.120	.300	.190	.170	.015	Index	.325	7/8
	7045504	15/32	Inner	9/32	—	3 Turns	.120	.300	.180	.170	.015	Index	.325	7/8
	7045506	15/32	Inner	9/32	—	3 Turns	.120	.300	.190	.170	.015	Index	.325	7/8
	7045512	17/32	Inner	9/32	—	2 Turns	.120	.060	.180	.550	.015	Index	.270	9/16
	7045517	17/32	Inner	9/32	—	2 Turns	.120	.060	.180	.550	.015	Index	.270	9/16
1976	17056200	13/32	Inner	9/32	—	3 Turns	.120	.190	.240	.160	.015	—	.270	7/8
	17056202	13/32	Inner	9/32	—	3 Turns	.120	.325	.185	—	.015	2 Lean	.325	7/8
	17056203	13/32	Inner	9/32	—	3 Turns	.120	.325	.170	—	.015	3 Lean	.325	7/8
	17056206	13/32	Inner	9/32	—	3 Turns	.120	.325	.185	—	.015	2 Lean	.325	7/8
	17056207	13/32	Inner	9/32	—	3 Turns	.120	.325	.170	—	.015	3 Lean	.325	7/8
	17056210	13/32	Inner	9/32	—	3 Turns	.120	.325	.185	—	.015	2 Lean	.325	1
	17056211	13/32	Inner	9/32	—	3 Turns	.120	.325	.185	—	.015	2 Lean	.325	—
	17056228	13/32	Inner	9/32	—	3 Turns	.120	.325	.185	—	.015	2 Lean	.325	3/4
	17056502	13/32	Inner	9/32	—	3 Turns	.120	.325	.185	—	.015	2 Lean	.325	7/8
	17056506	13/32	Inner	9/32	—	3 Turns	.120	.325	.185	—	.015	2 Lean	.325	7/8
	17056528	13/32	Inner	9/32	—	3 Turns	.120	.325	.185	—	.015	3 Lean	.325	3/4

Continued

QUADRAJET M4M SERIES ADJUSTMENT SPECIFICATIONS—Continued

See Tune Up Chart in car chapters for hot idle speeds.

Year	Carb. Model	Float Level	Pump Rod		Bowl Vent	Fast Idle (Bench)	Choke Coil Lever	Choke Rod	Vacuum Break		Air-Valve Dash-pot	Choke Setting	Choke Unloader	Air-Valve Valve Spring Wind-Up
			Hole	Adj.					Front	Rear				
OLDSMOBILE														
1975	7045183	15/32	Inner	9/32	.075	3 Turns	.120	.135	.190	.140	.030	2 Lean	.230	1/2
	7045184	15/32	Inner	9/32	.075	3 Turns	.120	.135	.190	.140	.030	2 Lean	.230	3/4
	7045185	15/32	Inner	9/32	.075	3 Turns	.120	.135	.190	.140	.030	2 Lean	.230	3/4
	7045246	3/8	Outer	3/8	—	2 Turns	.120	.095	.130	.115	.015	1 Rich	.235	3/4
	7045250	15/32	Inner	9/32	.075	3 Turns	.120	.135	.190	.140	.030	2 Rich	.230	1/4
	7045251	15/32	Inner	9/32	.075	3 Turns	.120	.135	.190	.140	.030	2 Rich	.230	3/4
	7045264	17/32	Inner	9/32	—	3 Turns	.120	.130	.150	.260	.030	Index	.230	1/2
	7045483	15/32	Inner	9/32	.075	3 Turns	.120	.135	.190	.180	.030	2 Rich	.230	1/2
	7045484	15/32	Inner	9/32	.075	3 Turns	.120	.135	.190	.140	.030	2 Rich	.230	3/4
	7045485	15/32	Inner	9/32	.075	3 Turns	.120	.135	.200	.180	.030	2 Rich	.230	3/4
	7045546	5/16	Outer	3/8	—	3 Turns	.120	.095	.145	.130	.015	1 Rich	.240	3/4
	7045550	15/16	Inner	9/32	.075	3 Turns	.120	.135	.190	.140	.030	2 Rich	.230	1/2
	7045551	15/16	Inner	9/32	.075	3 Turns	.120	.135	.190	.140	.030	1 Rich	.230	3/4
1976	17056246	5/16	Outer	3/8	.075	—	.120	.095	.130	.120	.015	1 Rich	.250	3/4
	17056250	13/32	Inner	9/32	—	—	.120	.130	.190	.140	.030	2 Rich	.230	1/2
	17056251	13/32	Inner	9/32	—	—	.120	.130	.190	.140	.030	2 Rich	.230	3/4
	17056252	13/32	Inner	9/32	—	—	.120	.130	.190	.140	.030	2 Rich	.230	3/4
	17056253	13/32	Inner	9/32	—	—	.120	.130	.190	.140	.030	2 Rich	.230	1/2
	17056256	13/32	Inner	9/32	—	—	.120	.130	.190	.140	.030	2 Rich	.230	3/4
	17056546	5/16	Outer	3/8	.075	—	.120	.095	.130	.130	.015	Index	.250	3/4
	17056550	13/32	Inner	9/32	—	—	.120	.130	.190	.140	.030	2 Rich	.230	1/2
	17056551	13/32	Inner	9/32	—	—	.120	.130	.190	.140	.030	1 Rich	.230	3/4
	17056552	13/32	Inner	9/32	—	—	.120	.130	.200	.140	.030	2 Rich	.230	3/4
	17056553	13/32	Inner	9/32	—	—	.120	.130	.190	.140	.030	2 Rich	.230	1/2
	17056556	13/32	Inner	9/32	—	—	.120	.130	.190	.140	.030	1 Rich	.230	3/4
PONTIAC														
1975	7045246	3/8	Outer	3/8	—	2 Turns	.120	.095	.130	.115	.015	1 Rich	.235	3/4
	7045260	1/2	Inner	9/32	—	3 Turns	.120	.130	.150	.260	.030	Index	.230	1/2
	7045262	1/2	Inner	9/32	—	3 Turns	.120	.130	.150	.260	.030	Index	.230	1/2
	7045263	17/32	Inner	9/32	—	3 Turns	.120	.130	.150	.260	.030	Index	.230	1/2
	7045264	17/32	Inner	9/32	—	3 Turns	.120	.130	.150	.260	.030	Index	.230	1/2
	7045266	1/2	Inner	9/32	—	3 Turns	.120	.130	.160	.260	.030	Index	.230	1/2
	7045268	17/32	Inner	9/32	—	3 Turns	.120	.130	.150	.260	.030	Index	.230	7/16
	7045269	17/32	Inner	9/32	—	3 Turns	.120	.130	.160	.265	.030	Index	.230	7/16
	7045274	17/32	Inner	9/32	—	3 Turns	.120	.130	.150	.260	.030	Index	.230	1/2
	7045546	5/16	Outer	15/32	—	3 Turns	.120	.095	.145	.130	.015	1 Rich	.240	3/4
	7045562	17/32	Inner	9/32	—	3 Turns	.120	.130	.150	.260	.030	Index	.230	1/2
	7045564	17/32	Inner	9/32	—	3 Turns	.120	.130	.150	.260	.030	Index	.230	1/2
	7045566	17/32	Inner	9/32	—	3 Turns	.120	.130	.160	.260	.030	Index	.230	1/2
	7045568	17/32	Inner	9/32	—	3 Turns	.120	.130	.150	.260	.030	Index	.230	7/16

Continued

QUADRAJET M4M SERIES ADJUSTMENT SPECIFICATIONS—Continued

See Tune Up Chart in car chapters for hot idle speeds.

Year	Carb. Model	Float Level	Pump Rod		Bowl Vent	Fast Idle (Bench)	Choke Coil Lever	Choke Rod	Vacuum Break		Air-Valve Dash-pot	Choke Setting	Choke Unloader	Air-Valve Valve Spring Wind-Up
			Hole	Adj.					Front	Rear				
PONTIAC—Continued														
1976	17056246	5/16	Outer	3/8	—	—	.120	.095	.130	.120	.015	1 Rich	.250	3/4
	17056261	—	—	—	—	—	—	—	—	—	—	—	—	—
	17056262	17/32	Outer	3/8	—	—	.120	.125	.160	.250	.030	1 Rich	.230	1/2
	17056263	17/32	Outer	3/8	—	—	.120	.125	.170	.250	.030	1 Rich	.230	5/8
	17056264	17/32	Outer	3/8	—	—	.120	.125	.160	.250	.030	1 Rich	.230	1/2
	17056274	17/32	Outer	3/8	—	—	.120	.125	.160	.250	.030	1 Rich	.230	1/2
	17056546	5/16	Outer	3/8	—	—	.120	.095	.130	.130	.015	Index	.250	3/4
	17056562	17/32	Outer	3/8	—	—	.120	.120	.170	.250	.030	1 Rich	.230	1/2
	17056564	17/32	Outer	3/8	—	—	.120	.130	.150	.260	.030	1 Rich	.230	1/2
	17056566	17/32	Outer	3/8	—	—	.120	.120	.170	.250	.030	1 Rich	.230	1/2
	17056568	17/32	Outer	3/8	—	—	.120	.125	.160	.250	.030	1 Rich	.230	1/2

QUADRAJET M4M SERIES ADJUSTMENTS

The M4M Series (M4MC, M4MCA and M4MEA) Quadrajet carburetor, Fig. 1 is similar to the Quadrajet 4MC unit previously described in this chapter. However, some modifications have been made to the basic Quadrajet design.

An Adjustable Part Throttle (A.P.T.) feature incorporates an adjustable metering rod assembly operating in a fixed jet. On some models, a barometric pressure-sensitive aneroid (Bellows) is an integral part of the A.P.T. metering rod assembly. This provides a close tolerance control of fuel flow to the main metering system, thereby controlling air/fuel ratios during part throttle operation.

Some units use a multiple stage power enrichment system with two power pistons. One piston is an auxiliary power piston with a single metering rod operating in a fixed jet. The primary power piston with two metering rods operates in replaceable metering jets. This system provides sensitive control of the air/fuel ratio during light engine power requirements while providing richer mixtures during moderate to heavy engine loads.

All units use a bowl mounted choke housing with a thermostatic coil assembly. Also, a dual vacuum break system is used to improve cold engine warm-up and drive-away performance. Cadillac units use a three stage electric choke with a ceramic resistor for precise timing of the choke valve opening to improve engine warm-up performance. Refer of the "Emission Control System" chapter.

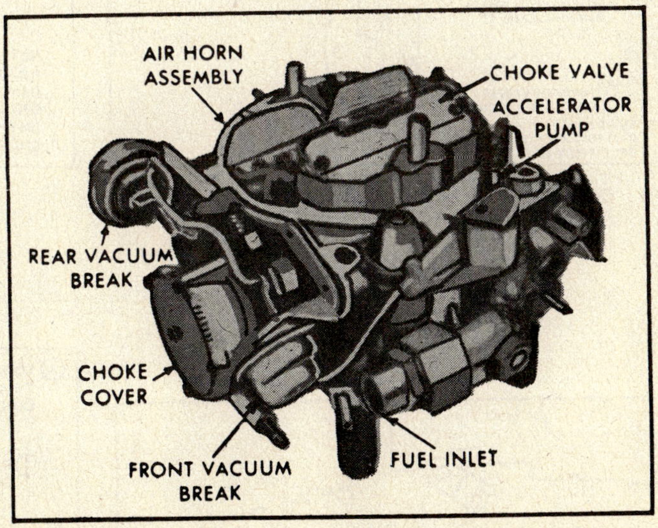

Fig. 1 Quadrajet M4M series carburetor (Typical)

ADJUSTMENTS

Float Level Adjustment

Fig. 2—With an adjustable T-scale, measure distance from top of float bowl gasket surface, with gasket removed, to top of float at a point 1/16 inch back from toe. Dimension should be as specified in the M4M Series Specification Chart. To adjust, bend float arm.

A.P.T. Metering Rod Adjustment

NOTE: This adjustment is performed only when the metering rod, filler spool or aneroid is replaced.

Fig. 3—Mark position of metering rod assembly adjusting screw on cover, then remove cover screws and lift metering rod assembly from float bowl. Rotating the adjusting screw counter-clockwise, count

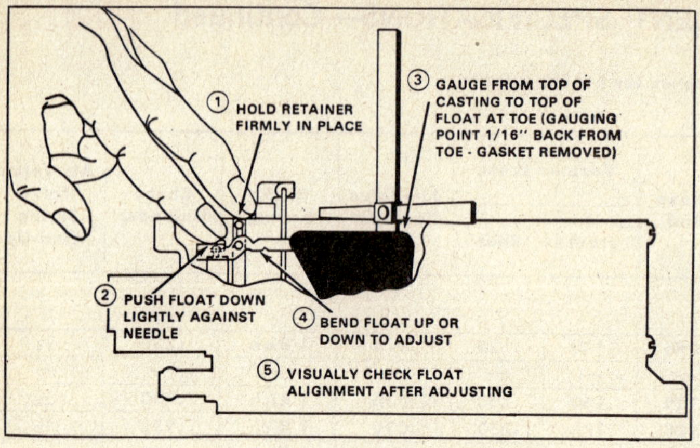

Fig. 2 Quadrajet M4M series float level adjustment

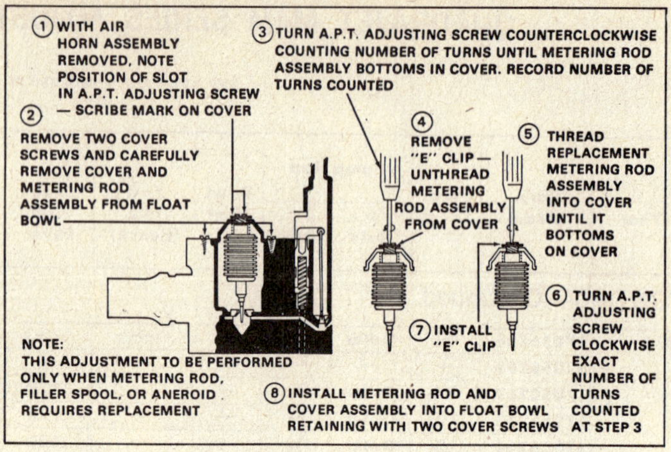

Fig. 3 Quadrajet M4M series A.P.T. metering rod adjustment

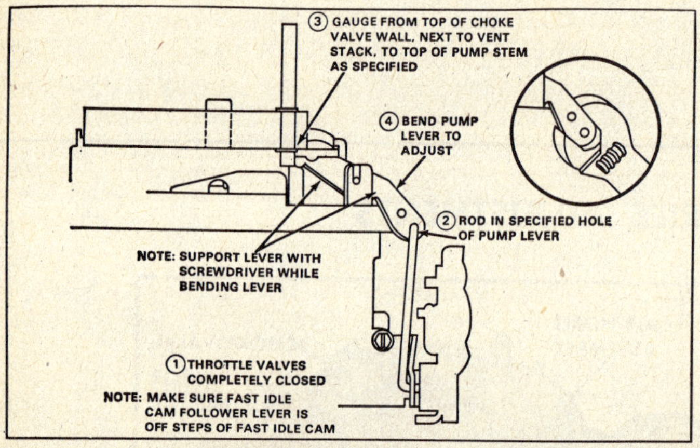

Fig. 4 Quadrajet M4M series pump rod adjustment

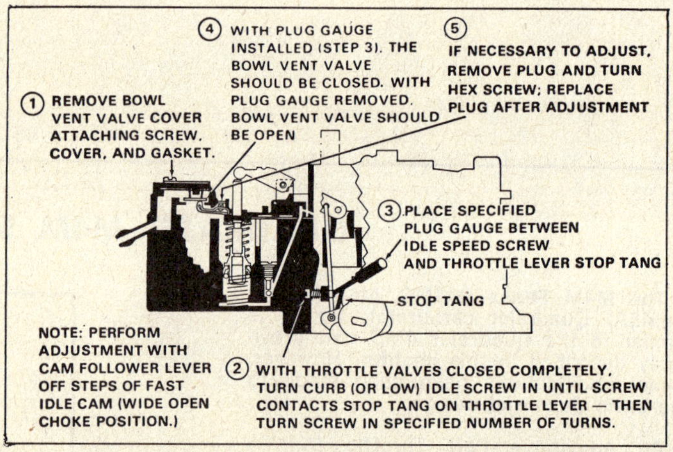

Fig. 5 Quadrajet M4M series bowl vent adjustment

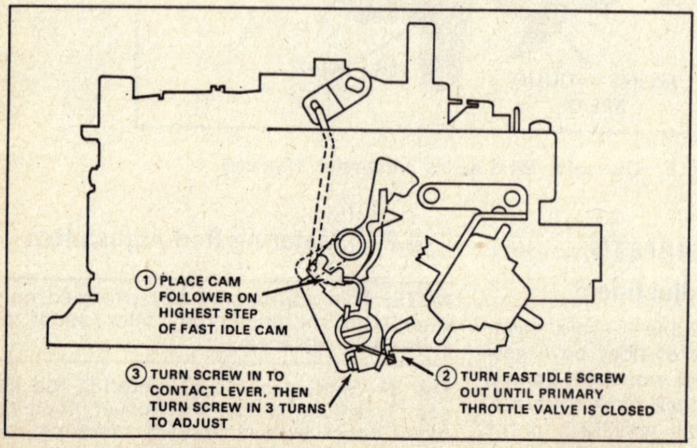

Fig. 6 Quadrajet M4M series fast idle adjustment (Bench)

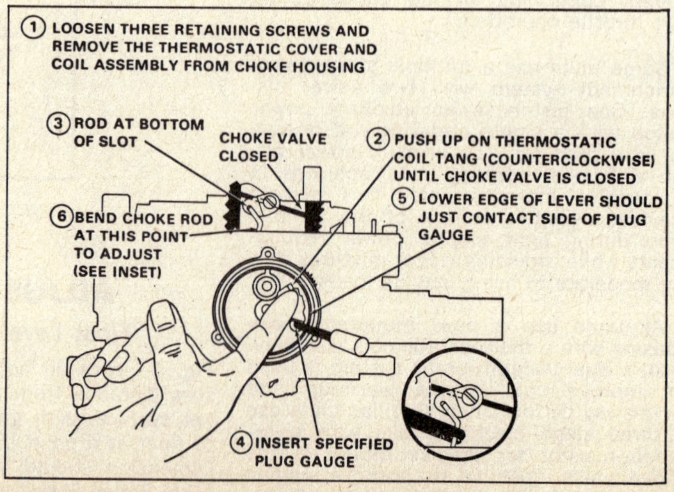

Fig. 7 Quadrajet M4M series choke coil lever adjustment

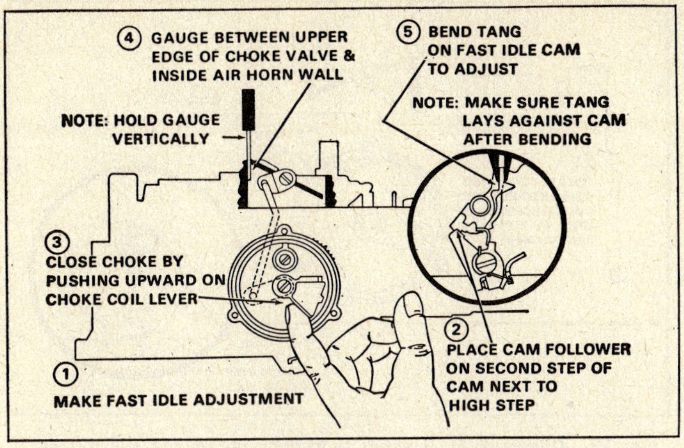

Fig. 8 Quadrajet M4M series choke rod adjustment

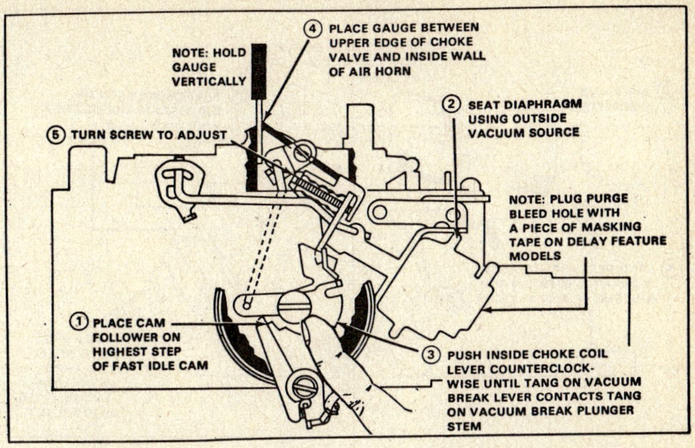

Fig. 9 Quadrajet M4M series front vacuum break adjustment

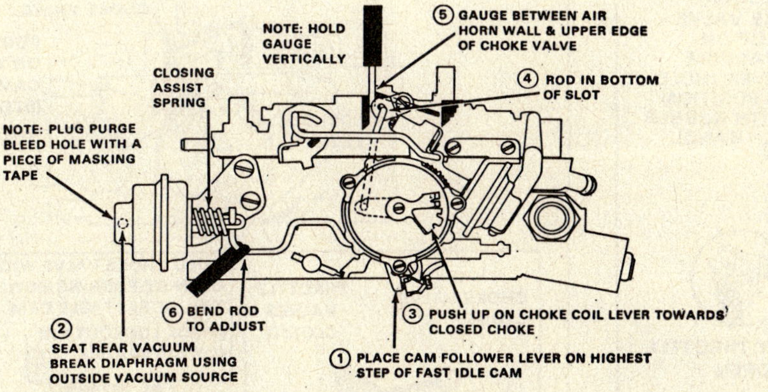

Fig. 10 Quadrajet M4M series rear vacuum break adjustment exc. Chevrolet V8-454

number of turns required to bottom the threaded metering rod assembly in the cover. Note this number for metering rod adjustment.

Remove "E" clip retainer from rod end, then using a small screwdriver, rotate slotted rod clockwise to disengage rod assembly from cover. Use caution when removing the rod assembly since it is spring loaded.

Install tension spring on replacement rod assembly, then thread the assembly into the cover until bottomed. Rotate the adjusting clockwise exactly the same number of turns from adjusting screw alignment mark noted previously. Install "E" clip in rod assembly groove, ensuring clip is securely locked. Place cover and rod assembly in float bowl, aligning cover assembly tab with float bowl tab nearest the fuel inlet nut, then install cover attaching screws.

Pump Rod Adjustment

Fig. 4—With throttle valves completely closed and pump rod in specified hole, measure distance from top of choke valve wall (next to vent stack) to top of pump stem. Dimension should be as specified in the M4M Series Specification Chart. To adjust, bend pump lever.

Bowl Vent Valve Adjustment

Fig. 5—Remove bowl vent valve cover and gasket. With throttle valves completely closed, rotate curb idle screw inward until it contacts throttle lever stop tang, then an additional 1½ turns. Insert the gauge specified in the M4M Series Specification Chart between curb idle screw and throttle lever stop tang. The bowl vent valve should now be closed, then remove the gauge and the valve should open. To adjust, pry out plug located next to pump plunger shaft, then turn bowl vent valve, actuating arm allen head adjusting screw.

Fast Idle Adjustment

Fig. 6—With cam follower on highest step of cam, turn fast idle screw out until primary throttle valve is completely closed. Turn screw in to contact lever, then turn screw in the additional turns listed in the M4M Series Specification Chart.

Choke Coil Lever Adjustment

Fig. 7—With thermostic coil assembly removed from choke housing, push upward on coil tang to close choke valve. Insert gauge specified in the M4M Series Specifi-

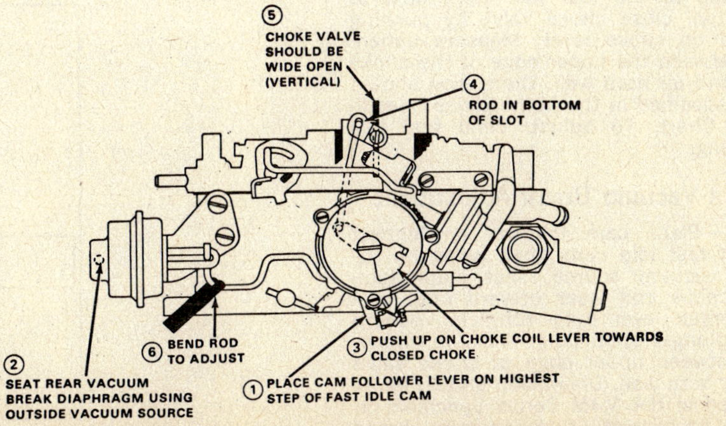

Fig. 10A Quadrajet M4M series rear vacuum break adjustment Chevrolet V8-454

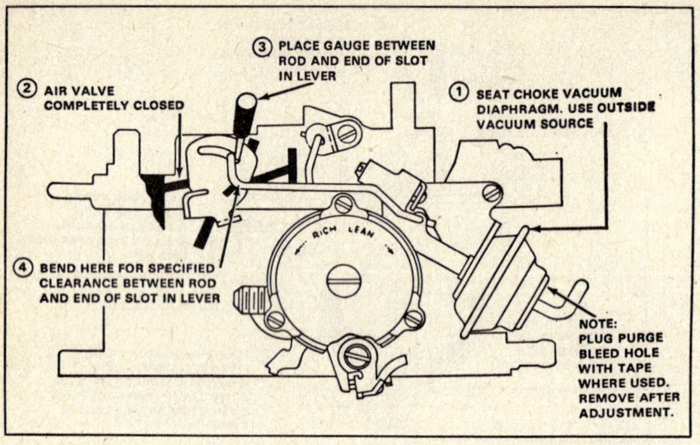

Fig. 11 Quadrajet M4M series air-valve dashpot adjustment

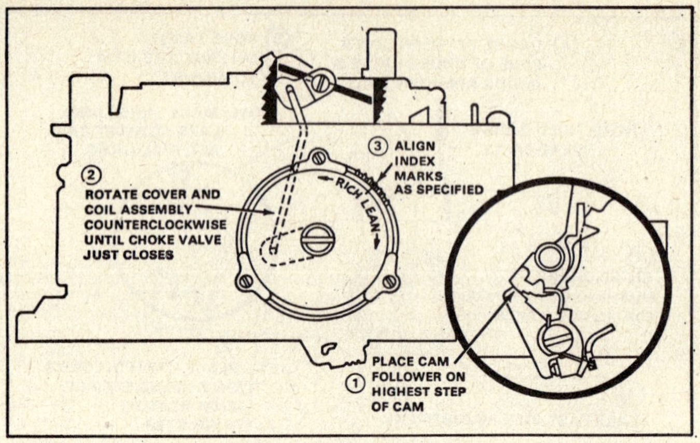

Fig. 12 Quadrajet M4M series automatic choke coil adjustment

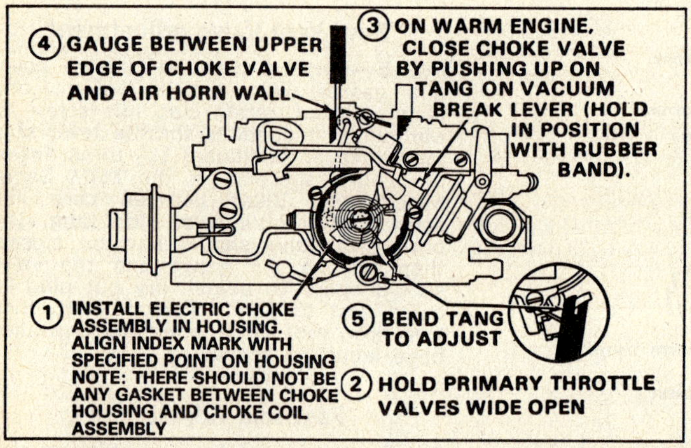

Fig. 13 Quadrajet M4M series choke unloader adjustment

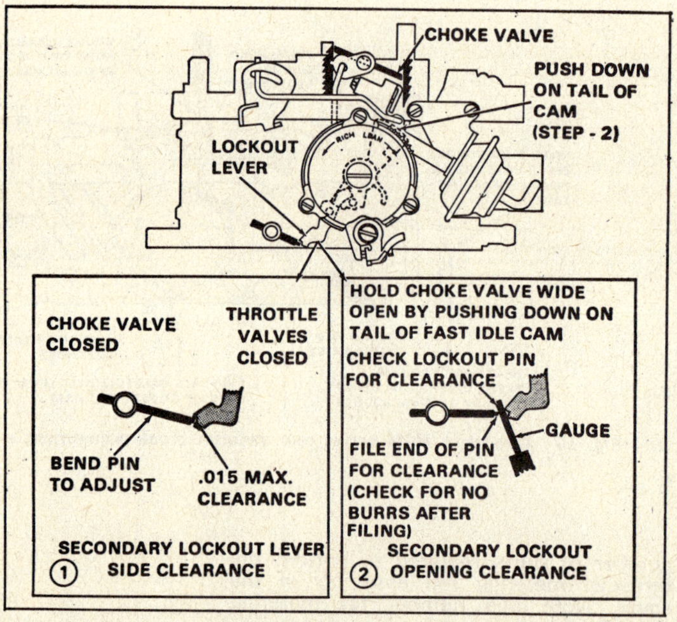

Fig. 14 Quadrajet M4M series secondary throttle valve lockout adjustment

cation Chart into hole in choke housing. The choke coil lever should contact the gauge. If not, bend choke rod to adjust.

Choke Rod Adjustment

Fig. 8—With the fast idle adjustment performed, and the cam follower on the second step of the fast idle cam, next to high step, close choke valve by pushing upward on choke lever. Measure dimension between the upper edge of the choke valve and air horn wall. Dimension should be as specified in the M4M Series Specification Chart. To adjust, bend fast idle cam tang.

Front Vacuum Break Adjustment

Fig. 9—Place cam follower on highest step of fast idle cam, then using an external vacuum source, seat diaphragm. Push choke coil lever upward until vacuum break lever tang contacts vacuum break plunger stem tang. Measure dimension between upper edge of choke valve and air horn wall. Dimension should be as specified in the M4M Series Specification Chart. To adjust, turn vacuum break plunger adjusting screw.

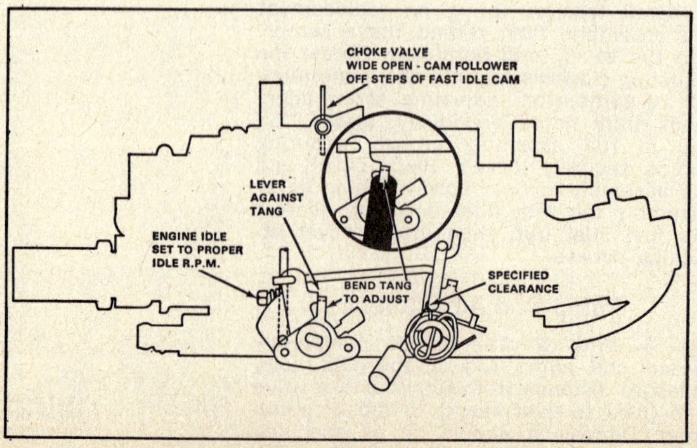

Fig. 15 Quadrajet M4M series secondary throttle valve closing adjustment

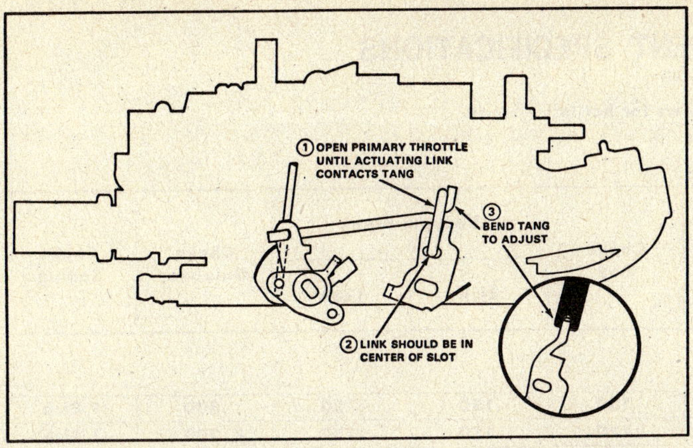

Fig. 16 Quadrajet M4M series secondary throttle valve opening adjustment

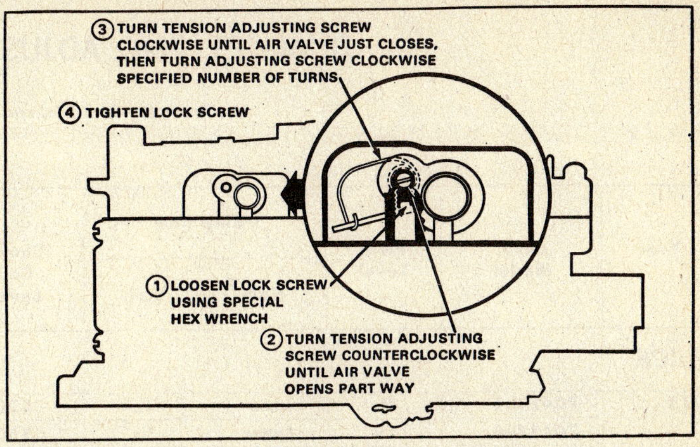

Fig. 17 Quadrajet M4M series air-valve spring wind-up adjustment

Rear Vacuum Break Adjustment

Figs. 10 & 10A—Place cam follower on highest step of fast idle cam, then using an external vacuum source, seat diaphragm. Push upward on choke coil lever toward closed choke position until stem is seated. With choke rod in bottom of slot, insert gauge specified in the M4M Series Specification Chart between upper edge of choke valve and air horn wall on all units except those used on Chevrolet V8-454 engines. On Chevrolet V8-454 units, the choke valve should be wide open (fully vertical position). To adjust, bend vacuum break rod.

Air Valve Dashpot Adjustment

Fig. 11—Using an external vacuum source, seat choke vacuum diaphragm. Then, with the air valve completely closed, place gauge specified in the M4M Series Specification Chart between air valve rod and end of slot in air valve lever. To adjust, bend rod at air valve end.

Choke Coil Adjustment

Fig. 12—Install choke coil and cover assembly. On Cadillac units, do not install a gasket between the choke housing and electric choke assembly. Place fast idle cam follower on highest step of fast idle cam. Rotate coil and cover assembly counter-clockwise until choke just closes, then align index mark on choke cover with the specified index mark on choke housing. Refer to the M4M Series Specification Chart.

Choke Unloader Adjustment

Fig. 13—With the choke coil adjustment performed, close choke, valve and open throttle valve. To close choke valve on a warm engine, push upward on tang of intermediate choke lever that contacts the fast idle cam and hold in place with a rubber band. Measure dimension between upper edge of choke valve and air horn wall. To adjust, bend tang on fast idle lever.

Secondary Throttle Valve Lockout Adjustment

Lockout-Pin Clearance

Fig. 14—With choke and throttle valves fully closed, measure clearance between lockout-pin and lockout lever. Maximum clearance should not exceed .015 inch. To adjust, bend lockout-pin.

Opening Clearance

Fig. 14—Push downward on fast idle cam tail to open choke valve, then push lockout lever counter-clockwise so upper end of lever contacts the round pin in fast idle cam. Measure clearance between lockout-pin and lockout lever toe. Clearance should be .015 inch. To obtain proper clearance, file metal from end of lockout-pin.

Secondary Throttle Valves Adjustment

Throttle Closing

Fig. 15—With curb idle speed set, hold choke valve open with cam follower lever off fast idle cam steps. Measure clearance between forward edge of slot in the secondary throttle valve pick-up lever and the secondary actuating rod. On all units except Pontiac, clearance should be .020 inch. On Pontiac units, the clearance is .025 inch. To adjust, bend secondary closing tang on primary throttle lever.

Throttle Opening

Fig. 16—Open primary throttle lever until link lightly contacts secondary lever tang. The bottom end of the link should be in the center of the slot in secondary lever. To adjust, bend tang on secondary lever.

Air Valve Spring Wind-Up Adjustment

Fig. 17—Remove front vacuum break diaphragm unit and air valve dashpot rod. Loosen lock screw and rotate spring adjusting screw counter-clockwise until air valve is partially open. Hold air valve closed, then rotate adjusting screw clockwise specified number of turns after the spring contacts pin. Refer to the M4M Series Specification Chart. Tighten lock screw and install air valve dashpot rod and the front vacuum break diaphragm unit.

DUAL-JET 2MC ADJUSTMENT SPECIFICATIONS

See Tune Up Chart in car chapters for hot idle speeds.

Year	Carb. Model	Float Level	Pump Rod		Choke Coil Lever	Choke Rod	Vacuum Break		Choke Unloader	Choke Setting
			Hole	Adj.			Rich	Lean		
BUICK										
1975	7045156	5/32	Inner	9/32	.120	.130	.150	.250	.300	1 Rich
	7045298	5/32	Inner	9/32	.120	.130	.150	.250	.300	1 Rich
	7045354	3/16	Outer	5/16	.120	.130	.150	.300	.300	1 Rich
	7045358	3/16	Outer	5/16	.120	.130	.150	.300	.300	1 Rich
1976	17056156	1/8	Inner	9/32	.120	.105	.110	.175	.210	1 Rich
	17056158	1/8	Inner	9/32	.120	.105	.110	.175	.210	1 Rich
	17056454	1/8	Outer	3/16	.120	.105	.110	.175	.210	1 Rich
	17056458	1/8	Outer	3/16	.120	.105	.110	.175	.210	1 Rich
OLDSMOBILE										
1975	7045156	5/32	Inner	9/32	.120	.130	.150	.250	.300	1 Rich
	7045297	3/16	Inner	9/32	.120	.130	.150	.300	.300	1 Rich
	7045298	5/32	Inner	9/32	.120	.130	.150	.250	.300	1 Rich
	7045354	3/16	Outer	5/16	.120	.130	.150	.300	.300	1 Rich
	7045356	5/32	Inner	3/16	.120	.130	.150	.300	.300	Index
	7045358	3/16	Outer	5/16	.120	.130	.150	.300	.300	1 Rich
	7045598	5/32	Inner	3/16	.120	.130	.150	.300	.300	Index
1976	17056156	1/8	Inner	9/32	.120	.105	.110	.175	.210	1 Rich
	17056157	1/8	Outer	3/16	.120	.105	.110	.175	.210	1 Rich
	17056158	1/8	Inner	9/32	.120	.105	.110	.175	.210	1 Rich
	17056454	1/8	Outer	3/16	.120	.105	.110	.210	.210	1 Rich
	17056456	1/8	Outer	3/16	.120	.105	.110	.210	.210	Index
	17056458	1/8	Outer	3/16	.120	.105	.110	.210	.210	1 Rich
	17056459	1/8	Outer	3/16	.120	.105	.110	.210	.210	Index
PONTIAC										
1975	7045156	5/32	Inner	9/32	.120	.130	.150	.250	.300	1 Rich
	7045297	3/16	Inner	9/32	.120	.130	.150	.300	.300	1 Rich
	7045298	5/32	Inner	9/32	.120	.130	.150	.250	.300	1 Rich
	7045356	5/32	Inner	3/16	.120	.130	.150	.300	.300	Index
	7045598	5/32	Inner	3/16	.120	.130	.150	.300	.300	Index
1976	17056156	1/8	Inner	9/32	.120	.105	.110	.175	.210	1 Rich
	17056157	1/8	Outer	3/16	.120	.105	.110	.175	.210	1 Rich
	17056158	1/8	Inner	9/32	.120	.105	.110	.175	.210	1 Rich
	17056454	1/8	Outer	3/16	.120	.105	.110	.210	.210	①
	17056456	1/8	Outer	3/16	.120	.105	.110	.210	.210	①
	17056458	1/8	Outer	3/16	.120	.105	.110	.210	.210	①
	17056459	1/8	Outer	3/16	.120	.105	.110	.210	.210	①

①—LeMans, index; Ventura, 1 rich.

DUAL-JET 2MC ADJUSTMENTS

The Rochester 2MC carburetor (Dual-Jet), Fig. 1, is a two barrel, single stage unit, incorporating the design features of the primary side of the Quadrajet (four barrel) carburetor. The Dual-Jet unit is similar in appearance to the Quadrajet carburetor, however, there are no secondary throttle valves or secondary metering system. The power piston controls fuel metering during all phases of operation.

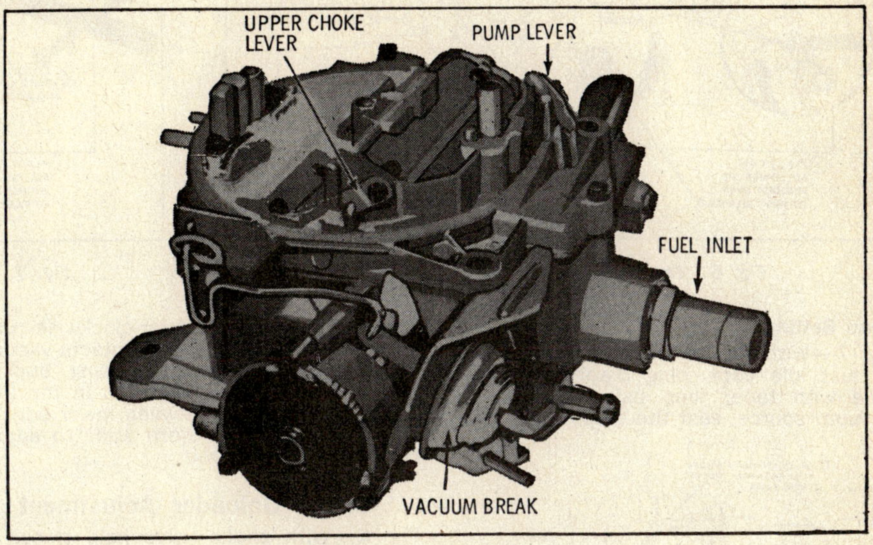

Fig. 1 Rochester Dual-Jet 2MC carburetor

Float Level Adjustment

Fig. 2—With adjustable T-scale, measure from top of float bowl gasket surface (gasket removed) to top of float at toe (locate gauging point $^{3}/_{16}$" back from toe). Adjust as directed in the illustration to the dimension listed in the *2MC Specifications Chart.* Make sure retaining pin is held firmly in place and tang of float is seated on float needle.

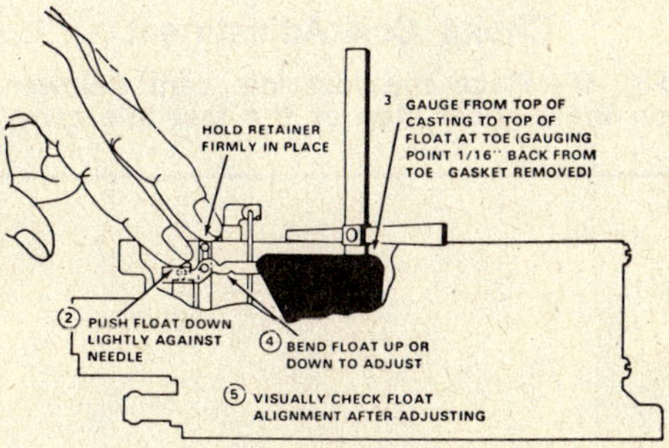

Fig. 2 Float level adjustment

Pump Rod Adjustment

Fig. 3—With throttle valves completely closed and pump rod in specified hole in pump lever, measure from top of choke valve wall (next to vent stack) to top of pump stem. Dimension should be as listed in the *2MC Specifications Chart.* To adjust, bend pump lever as required.

Choke Coil Lever Adjustment

Fig. 4—With thermostatic coil assembly removed, push upward on coil tang until choke valve closes. Insert gauge specified in the 2MC Specifications Chart into choke housing hole. The lower edge of choke coil lever should just contact gauge. To adjust, bend choke rod as required.

Choke Rod Adjustment

Fig. 5—With fast idle adjustment made, and cam follower on second step of fast idle cam and against the high step, push upward on choke coil lever until choke valve closes. Dimension between upper edge of choke valve and air horn wall should be as specified in the 2MC Specifications Chart. Adjust by bending tang on intermediate choke lever.

Vacuum Break Adjustment

Rich Setting

Fig. 6—With cam follower on highest step of fast idle cam, plug diaphragm bleed hole with tape, then using an external vacuum source, seat the diaphragm. Push choke coil lever counter-clockwise until outside lever tang contacts vacuum break rod and the bucking spring is compressed. Insert gauge specified in the 2MC Specification Chart between the lower edge of choke valve and air horn wall. To adjust, bend lower end of rod.

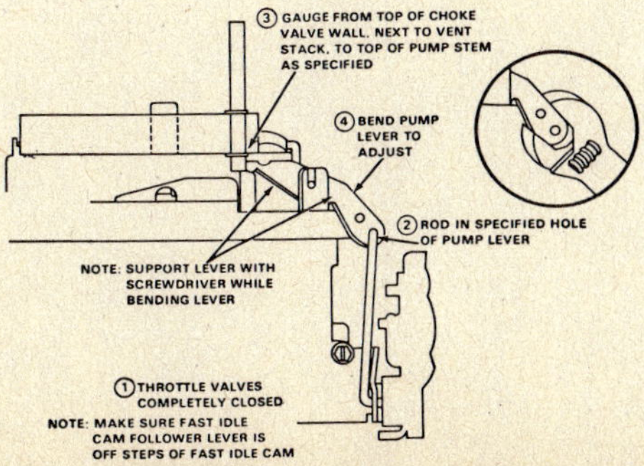

Fig. 3 Pump rod adjustment

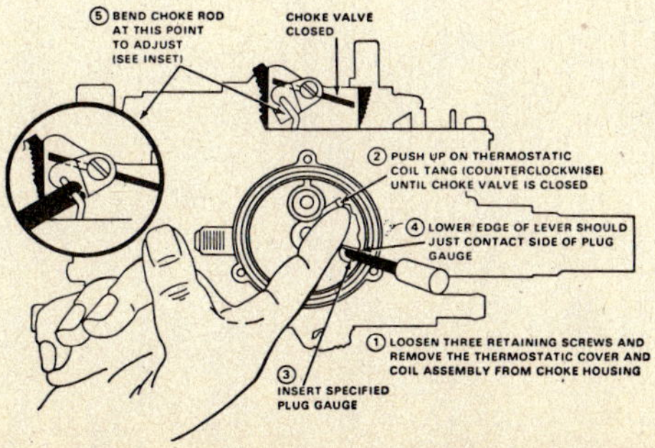

Fig. 4 Choke coil lever adjustment

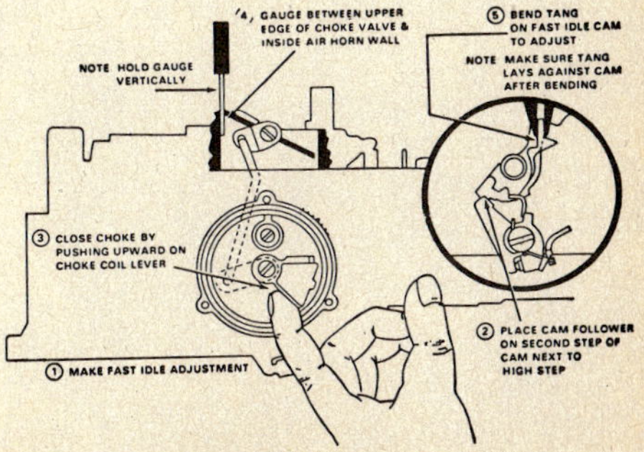

Fig. 5 Choke rod adjustment

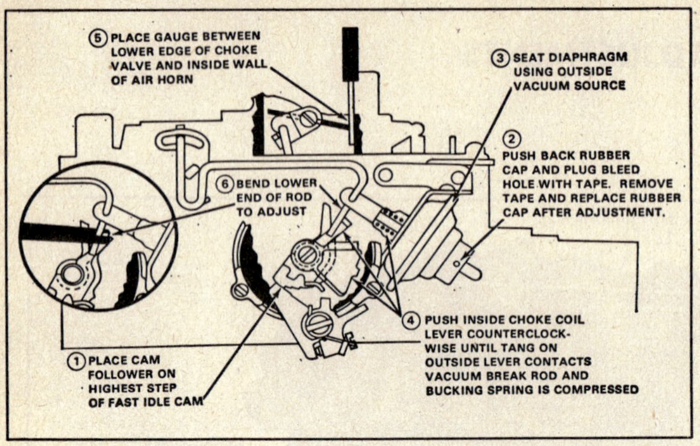

Fig. 6 Vacuum break adjustment, rich setting

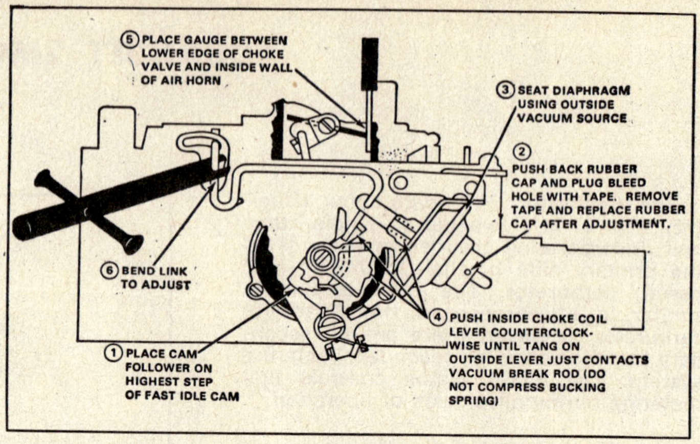

Fig. 7 Vacuum break adjustment, lean setting

Lean Setting

Fig. 7—With cam follower on highest step of fast idle cam, plug diaphragm bleed hole with tape, then using an external vacuum source, seat the diaphragm. Push choke coil lever counter-clockwise until outside lever tang lightly contacts vacuum break rod without compressing bucking spring. Insert gauge specified in the 2MC Specification Chart between lower edge of choke valve and air horn wall. To adjust, bend upper end of rod.

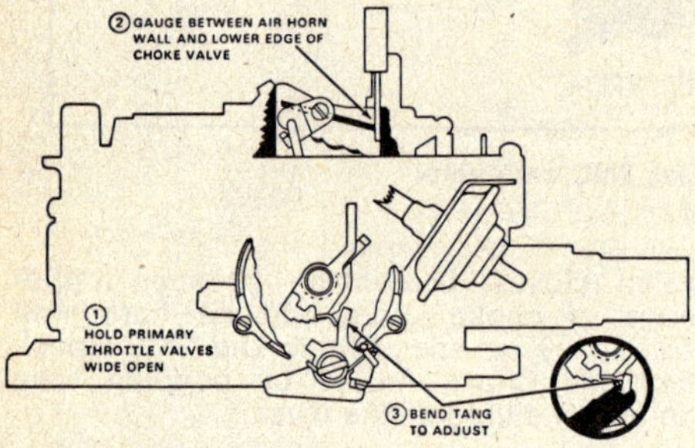

Fig. 8 Choke unloader adjustment

Choke Unloader Adjustment

Fig. 8—With choke valve held closed by means of a rubber band on vacuum break lever, open throttle valves fully. With valves in this position, dimension between lower edge of choke valve and air horn wall should be as specified. To adjust, bend tang on fast idle lever.

Choke Coil Adjustment

Fig. 9—Place the fast idle cam follower on the highest step of the fast idle cam. Rotate choke cover and coil assembly counter-clockwise until the choke valve just closes and the index point on cover aligns with the specified index point on the choke housing.

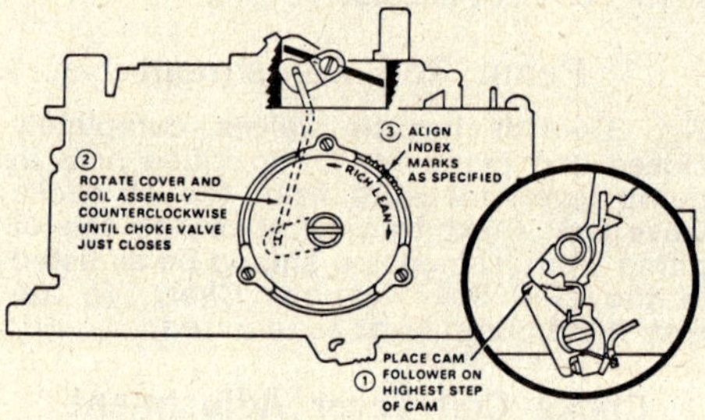

Fig. 9 Choke coil adjustment

MONOJET M, MV ADJUSTMENT SPECIFICATIONS

See Tune Up Chart in car chapters for hot idle speeds.

Year	Carb. Part No. ①	Initial Idle Mix Screw Turns Open	Float Level	Metering Rod	Idle Vent	Fast Idle Off Car	Choke Rod	Vacuum Break	Unloader	Fast Idle R.P.M.
BUICK										
1969	7029014	2	1/4	.070	.050	—	.170	.245	.350	620④
	7029047	2	9/32	.140	.020	—	.190	.275	.350	720④
1970	7040014	2	1/4	.070	.050	—	.170	.245	.350	650④
	7040015	2	1/4	.140	.050	—	.200	.275	.350	900④
	7040017	—	1/4	.070	—	.100	.190	.230	.350	900④
1971	7041014	—	1/4	.080	.050	—	.160	.225	.500	500④
	7041017	—	1/4	.080	.050	—	.180	.225	.350	550④
1973	7043014	—	1/4	.080	—	—	.245	.300	.500	1800
	7043017	—	11/32	.080	—	—	.275	.350	.500	1800
1974	7044014	—	11/32	.080	—	—	.230	.275	.500	1800
	7044017	—	11/32	.080	—	—	.275	.350	.500	1800
1975	7045012	—	11/32	.080	—	—	.160	.200⑦	.275	—
	7045013	—	11/32	.080	—	—	.275	.350⑧	.275	—
	7045314	—	11/32	.080	—	—	.230	.275⑧	.275	—
CHECKER MOTORS										
1969	7029014	1½–2	1/4	.070	.050	.100	.170	.245	.350	2400
	7029015	1½–2	1/4	.090	.050	.100	.200	.245	.350	2400
	7029017	1½–2	1/4	.090	.050	.100	.200	.245	.350	2400
1970	7040014	—	1/4	.070	—	.100	.170	.200	.350	2400
1971	7041014	—	1/4	.080	—	.100	.160	.200	.350	900④
1972	7042014	—	1/4	.080	—	—	.125	.190	.500	2400
	7042984	—	1/4	.080	—	—	.125	.190	.500	2400
1973	7043014	—	1/4	.080	—	—	.245	.300	.500	1800
1974	7044014	—	19/64	.079	—	—	.230	.275	.500	1800
	7044314	—	19/64	.073	—	—	.245	.300	.500	1800
1975	7045012	—	11/32	.080	—	—	⑩	.200⑦	.215	—
	7045314	—	11/32	.080	—	—	⑩	.275⑧	.275	—
CHEVROLET ENGINES										
1969	7029008	3	1/4	.080	.050	.100	.150	—	—	2400⑤
	7029014	3	1/4	.070	.050	.100	.170	.245	.350	2400⑤
	7029015	3	1/4	.090	.050	.100	.200	.275	.350	2400⑤
	7029017	3	1/4	.090	.050	.100	.200	.275	.350	2400⑤
1970	7040008	—	1/4	.080	—	.100	.200	—	—	2400⑤
	7040014	—	1/4	.070	—	.110	.170	.200	.350	2400⑤
	7040017	—	1/4	.090	—	.100	.190	.160	.350	2400⑤
1971	7041014	—	1/4	.080	—	.100	.160	.200	.350	—
	7041017	—	1/4	.080	—	.100	.180	.230	.350	—
	7041023	—	1/16	—	—	.110	.120	.200	.350	—
	7041024	—	1/16	—	—	.110	.080	.140	.350	—
1972	7042014	—	1/4	.080	—	—	.125	.190	.500	2400
	7042017	—	1/4	.078	—	—	.150	.225	.500	2400
	7042023	—	1/8	—	—	.110	.130	.200	.350	2400
	7042024	—	1/16	—	—	.110	.070	.120	.350	2400
	7042984	—	1/4	.078	—	—	.125	.190	.500	2400
	7042987	—	1/4	.076	—	—	.150	.225	.500	2400
	7042993	—	1/8	—	—	.110	.130	.200	.350	2400
	7042994	—	1/16	—	—	.110	.070	.120	.350	2400

Continued

MONOJET M, MV ADJUSTMENT SPECIFICATIONS—Continued

See Tune Up Chart in car chapters for hot idle speeds.

Year	Carb. Part No. ①	Initial Idle Mix Screw Turns Open	Float Level	Metering Rod	Idle Vent	Fast Idle Off Car	Choke Rod	Vacuum Break	Unloader	Fast Idle R.P.M.
CHEVROLET—Continued										
1973	7043014	—	1/4	.080	—	—	.245	.300	.500	1800
	7043017	—	11/32	.080	—	—	.275	.350	.500	1800
	7043023	—	1/8	—	—	—	.090	.140	.375	2000
	7043024	—	1/8	—	—	—	.070	.120	.375	2200
	7043033	—	11/32	—	—	—	.090	.140	.375	2000
	7043034	—	1/8	—	—	—	.070	.120	.375	2200
	7043323	—	11/32	—	—	—	.090	.140	.375	2000
	7043324	—	1/8	—	—	—	.070	.120	.375	2200
	7043333	—	11/32	—	—	—	.090	.140	.375	2000
	7043334	—	1/8	—	—	—	.070	.120	.375	2200
1974	7044014	—	11/32	.070	—	—	.230	.275	.500	1800
	7044017	—	11/32	.070	—	—	.275	.350	.500	1800
	7044023	—	1/8	—	—	—	.080	.130	.375	2000
	7044024	—	1/8	—	—	—	.080	.130	.375	2200
	7044033	—	1/8	—	—	—	.080	.130	.375	2000
	7044034	—	1/8	—	—	—	.080	.130	.375	2200
	7044314	—	11/32	.080	—	—	.245	.300	.500	1800
	7044323	—	1/8	—	—	—	.080	.130	.375	2000
	7044324	—	1/8	—	—	—	.080	.130	.375	2200
	7044333	—	1/8	—	—	—	.080	.130	.375	2000
	7044334	—	1/8	—	—	—	.080	.130	.375	2200
	7044336	—	1/8	—	—	—	.245	.300	.500	1800
	7044337	—	1/8	.080	—	—	.080	.130	.375	2000
	7044339	—	11/32	—	—	—	.245	.300	.500	1800
	7044340	—	1/8	.080	—	—	.080	.130	.375	2200
1975	7045012	—	11/32	.080	—	—	.160	.200⑦	.215	—
	7045013	—	11/32	.080	—	—	.275	.350⑧	.275	—
	7045018	—	1/8	.080	—	—	.230	.275⑧	.275	—
	7045024	—	1/8	—	—	—	.080	.100⑨	.375	—
	7045025	—	1/8	—	—	—	.080	.100⑨	.375	—
	7045027	—	1/8	—	—	—	.075	.350⑧	.375	—
	7045028	—	1/8	—	—	—	.080	.100⑨	.375	—
	7045029	—	1/8	—	—	—	.080	.100⑨	.375	—
	7045038	—	11/32	.084	—	—	.080	.100⑧	.375	—
	7045314	—	11/32	.080	—	—	.230	.275⑧	.275	—
1976	17056012	—	11/32	.084	—	—	.100	.140⑪	.260	2100
	17056013	—	11/32	.082	—	—	.140	.140⑭	.260	2100
	17056016	—	11/32	.080	—	—	.115	.140	.260	2200
	17056018	—	11/32	.084	—	—	.100	.140⑪	.260	2100
	17056022	—	1/8	—	—	—	.045	.060⑨	.215	1500
	17056023	—	1/8	—	—	—	.045	.060⑨	.215	2200
	17056026	—	1/8	—	—	—	.045	.060⑨	.215	2200
	17056027	—	1/8	—	—	—	.045	.060⑨	.215	1500
	17056314	—	11/32	.083	—	—	.135	.150⑭	.260	1700
OLDSMOBILE										
1969	7029014	⑥	1/4	.070	.050	—	.170	.245	.350	750④
	7029057	⑥	5/16	.120	.030	—	.180	.260	.350	750④
1970	7040014	⑥	1/4	.070	—	—	.170	.200	.350	900④
	7040017	⑥	1/4	.070	—	—	.190	.225	.350	750④
1971	7043014	⑥	1/4	.070	—	—	.160	.200	.350	900④
	7043017	⑥	1/4	.070	—	—	.180	.225	.350	750④

Continued

MONOJET M, MV ADJUSTMENT SPECIFICATIONS—Continued

See Tune Up Chart in car chapters for hot idle speeds.

Year	Carb. Part No. ①	Initial Idle Mix Screw Turns Open	Float Level	Metering Rod	Idle Vent	Fast Idle Off Car	Choke Rod	Vacuum Break	Unloader	Fast Idle R.P.M.
OLDSMOBILE—Continued										
1973	7043014	⑥	¼	.080	—	—	.245	.300	.500	1800
	7043017	⑥	11/32	.080	—	—	.275	.350	.500	1800
1974	7044014	⑥	11/32	.080	—	—	.230	.275	.500	1800
	7044017	⑥	11/32	.080	—	—	.275	.350	.500	1800
	7044314	⑥	11/32	.080	—	—	.245	.300	.500	1800
1975	7045012	⑥	11/32	.080	—	—	.160	.200⑦	.275	—
	7045013	⑥	11/32	.080	—	—	.275	.350⑧	.275	—
	7045314	⑥	11/32	.080	—	—	.230	.275⑧	.275	—
1976	17056012	⑥	—	—	—	—	.100	.140⑪	.265	—
	17056013	⑥	—	—	—	—	.140	.165⑫	.265	—
	17056014	⑥	—	—	—	—	.100	.140⑪	.265	—
	17056018	⑥	—	—	—	—	.100	.140⑬	.265	—
	17056314	⑥	—	—	—	—	.135	.150⑬	.265	—
PONTIAC										
1969	7029165	5	9/32	.085	.040	.120	.200	.275	.450	2400⑤
	7029166	5	9/32	.085	.040	.130	.180	.260	.450	2800⑤
	7029167	5	9/32	.085	.040	.120	.200	.275	.450	2600⑤
	7029168	5	9/32	.085	.040	.130	.180	.260	.450	2800⑤
1970	7040014	⑥	¼	.100	—	—	.170	.200	.350	—
	7040017	⑥	¼	.100	—	—	.190	.230	.350	—
1971	7041014	⑥	¼	.080	—	—	.160	.200	.350	—
	7041017	⑥	¼	.078	—	—	.180	.225	.350	—
1972	7042014	⑥	¼	.080	—	—	.160	.200	.500	2400
	7042017	⑥	¼	.080	—	—	.180	.230	.500	2400
	7042984	⑥	¼	.080	—	—	.160	.200	.500	2400
	7042987	⑥	¼	.080	—	—	.180	.230	.500	2400
1973	7043014	⑥	11/32	.080	—	—	.245	.300	.500	1800
	7043017	⑥	11/32	.080	—	—	.275	.350	.500	1800
1974	7044017	⑥	11/32	.080	—	—	.275	.350	.500	1800
	7044041	⑥	11/32	.079	—	—	.230	.275	.500	1800
	7044314	⑥	11/32	.080	—	—	.245	.300	.500	1800
1975	7045012	⑥	11/32	.080	—	—	.160	.200⑥	.275	—
	7045013	⑥	11/32	.080	—	—	.275	.350⑦	.275	—
	7045314	⑥	11/32	.080	—	—	.230	.275⑦	.275	—
	7045024	⑥	⅛	—	—	—	.080	.100⑨	.215	2200
	7045025	⑥	⅛	—	—	—	.080	.100⑨	.215	2000
	7045028	⑥	⅛	—	—	—	.080	.100⑨	.215	2200
	7045029	⑥	⅛	—	—	—	.080	.100⑨	.215	2000
1976	17056012	⑥	11/32	—	—	—	.100	.140⑪	.265	—
	17056013	⑥	11/32	—	—	—	.140	.165⑫	.265	—
	17056014	⑥	11/32	—	—	—	.100	.140⑪	.265	—
	17056018	⑥	11/32	—	—	—	.100	.140⑪	.265	—
	17056022	⑥	⅛	—	—	—	.045	.060⑨	.215	—
	17056023	⑥	⅛	—	—	—	.045	.055⑨	.215	—
	17056026	⑥	⅛	—	—	—	.045	.060⑨	.215	—
	17056027	⑥	⅛	—	—	—	.045	.060⑨	.215	—
	17056314	⑥	11/32	—	—	—	.135	.150⑬	.265	—

①—On tag attached to carburetor.
②—20 RPM above slow idle speed.
③—Turns in from slow idle position.
④—On low step of cam.
⑤—On high step of cam.
⑥—Air/fuel ratio or idle CO% rating is found in Tune Up Specification tables in car chapters.
⑦—Rear .215.
⑧—Rear .312.
⑨—Rear .450.
⑩—Bottom of rod even with top of hole.
⑪—Auxiliary, .265.
⑫—Auxiliary, .320.
⑬—Auxiliary, .260.
⑭—Auxiliary, .325.

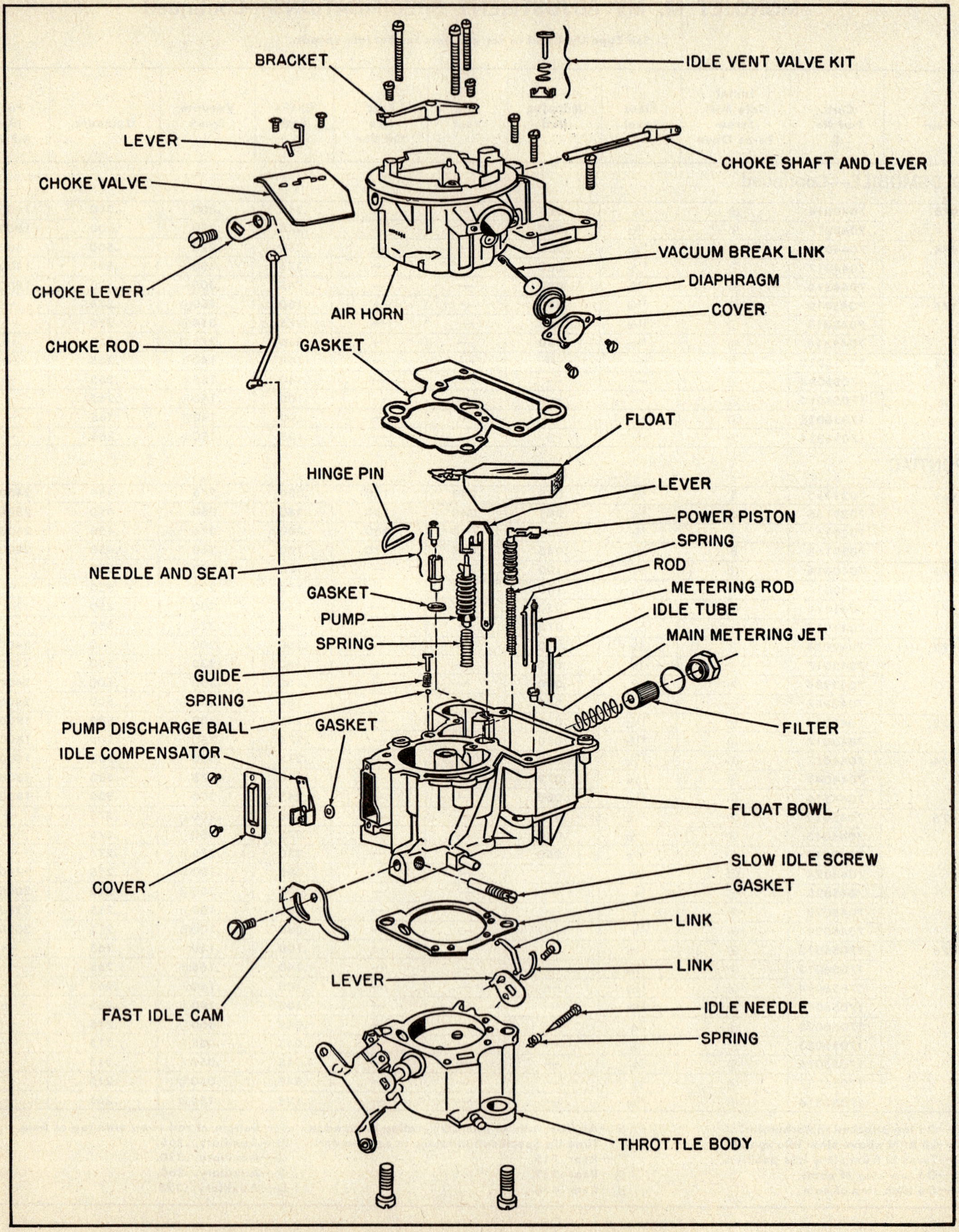

Fig. 1 Monojet Model MV carburetor. Model M has manual choke

MONOJET M & MV ADJUSTMENTS

The Monojet carburetor is a single-bore downdraft unit with a triple venturi coupled with a refined metering system which results in a unit having superior fuel mixture control and performance.

A plain tube nozzle is used in conjuction with the multiple venturi. Fuel flow through the main metering system is controlled by a mechanically and vacuum operated variable orifice jet. This consists of a specially tapered rod which operates in the fixed orifice main metering jet and is connected directly by linkage to the main throttle shaft. A vacuum-operated enrichment system is used in conjunction with the main metering system to provide good performance during moderate to heavy accelerations.

A separate and adjustable idle system is used in conjunction with the main metering system to meet fuel mixture requirements during engine idle and low speed operation. The off-idle discharge port is of a vertical slot design which gives good transition between curb idle and main metering system operation.

The idle system incorporates a hot idle compensator on some models where necessary to maintain smooth engine idle during periods of extreme hot engine operation.

The main metering system has an adjustable flow feature which enables production to control the fuel mixture more accurately then attained heretofore.

The Monojet carburetor is designed so that a manual or automatic choke system can be used. The conventional choke valve is located in the air horn bore. On automatic choke models, the vacuum diaphragm unit is an integral part of the air horn. The automatic choke coil is manifold mounted and connects to the choke valve shaft by connecting linkage.

The choke system has a new feature to give added enrichment during cold start. This feature greatly reduces starting time and yet allows the use of low torque thermostatic coils for increased economy.

The carburetor has internally balanced venting through a vent hole in the air horn. An external idle vent valve is used on some models where necessary for improved hot engine idle and starting.

Float Level Adjustment

Fig. 2

1. Hold float retaining pin firmly in place and float arm against top of float needle by pushing downward on float arm at point between needle seat and hinge pin as shown.
2. With adjustable T-scale, measure distance from top of float at toe to float bowl gasket surface (gasket removed). Measurement should be made at a point $1/16''$ in from end of flat surface at float toe (not on radius).
3. Bend float pontoon up or down at float arm junction to adjust.

Metering Rod Adjustment

Fig. 3

1. Remove metering rod by holding throttle valve wide open. Push down-

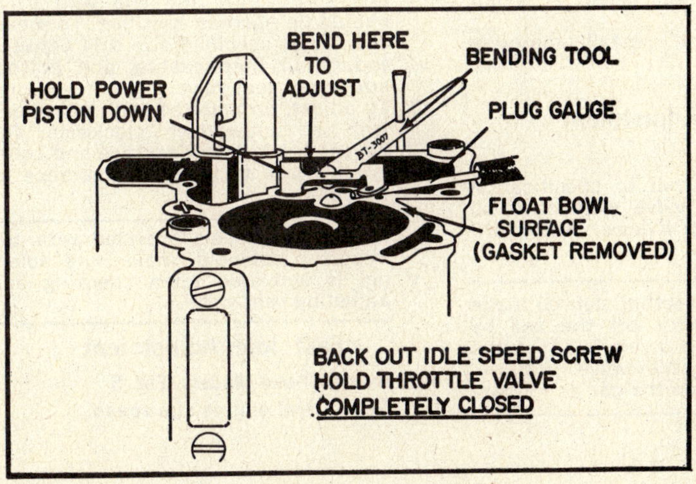

Fig. 3 Monojet metering rod adjustment

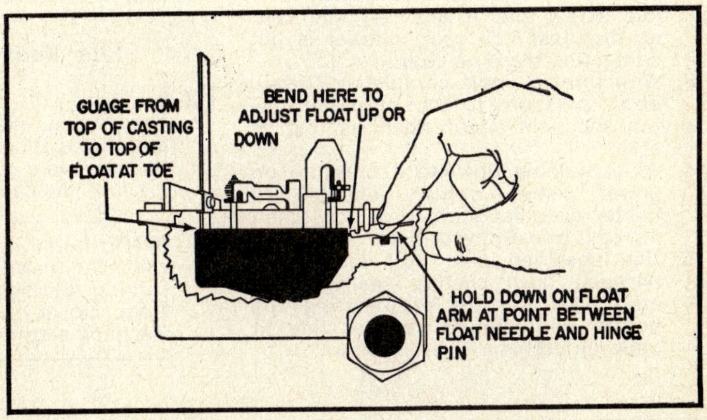

Fig. 2 Monojet float level adjustment

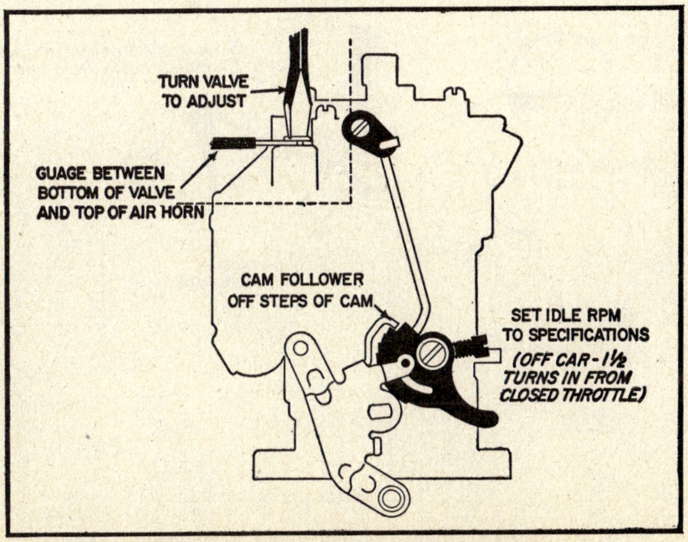

Fig. 4 Monojet idle vent adjustment

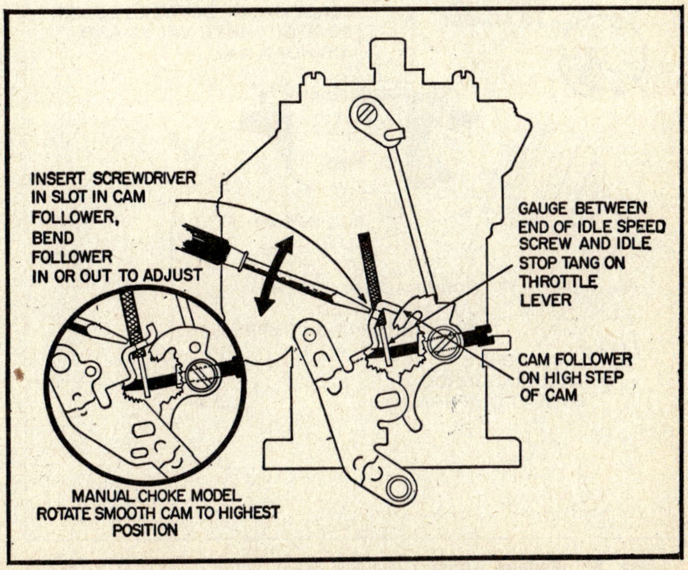

Fig. 5 Monojet fast idle adjustment (off car)

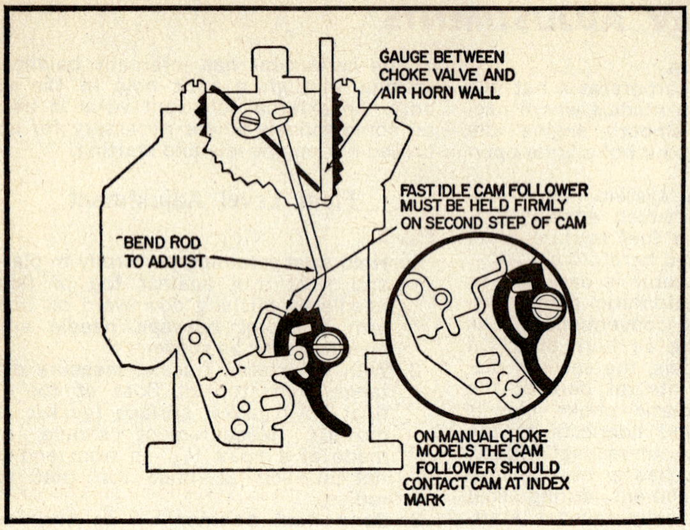

Fig. 6 Monojet choke rod adjustment

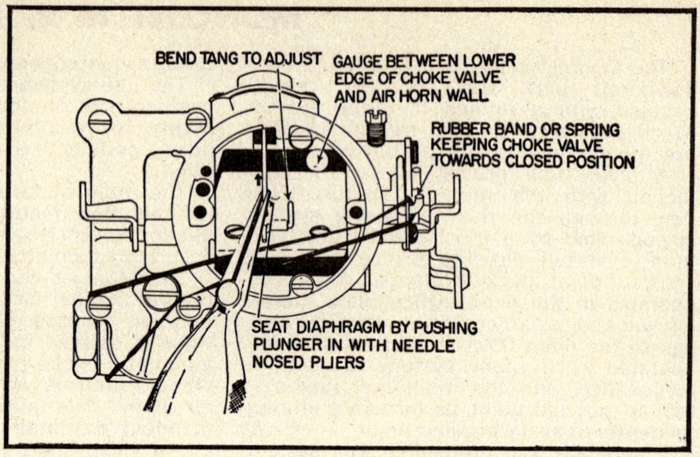

Fig. 7 Monojet vacuum break adjustment. 1969-71

ward on metering rod against spring tension, then slide metering rod out of slot in holder and remove from main metering jet.

2. To check adjustment, back out slow idle screw and rotate fast idle cam so that fast idle cam follower is not contacting steps on cam.
3. With throttle valve completely closed, apply pressure to top of power piston and hold piston down against its stop.
4. While holding downward pressure on power piston, swing metering rod holder over flat surface of bowl casting next to carburetor bore.
5. Use specified size drill and insert between bowl casting sealing bead and lower surface of metering rod holder. Drill should have a slide fit between both surfaces as shown.
6. To adjust, carefully bend metering rod holder up or down at point shown.
7. After adjustment, install metering rod.

Idle Vent Adjustment

Fig. 4

1. Set engine idle rpm to specification and hold choke valve wide open so that fast idle cam follower is not hitting fast idle cam.

NOTE: Initial idle setting can be made with the carburetor off the car by turning idle speed screw in $1\frac{1}{2}$ turns from closed throttle valve position. Recheck setting on the car as follows:

2. With throttle stop screw held against idle stop screw, the idle vent valve should be open as specified. To measure, insert specified size drill between top of air horn casting and bottom surface of vent valve.
3. To adjust, turn slotted vent valve head with a screwdriver clockwise (inward) to decrease clearance and counterclockwise to increase clearance as required.

NOTE: On models provided with the idle stop solenoid, make sure solenoid is activated when checking and adjusting vent valve.

Fast Idle Adjustment

Automatic Choke Models, Fig. 5

1. Set normal engine idle speed.

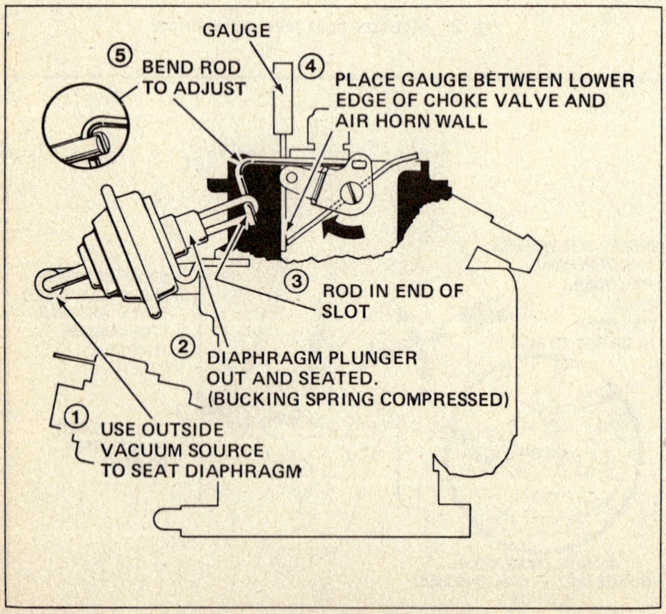

Fig. 8 Monojet primary vacuum break adjustment. 1972-75 Chevrolet Vega

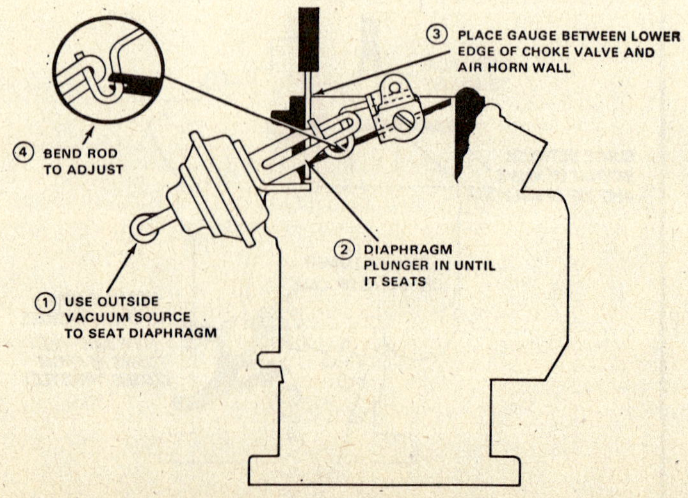

Fig. 9 Monojet primary vacuum break adjustment. 1972-75 All exc. Chevrolet Vega

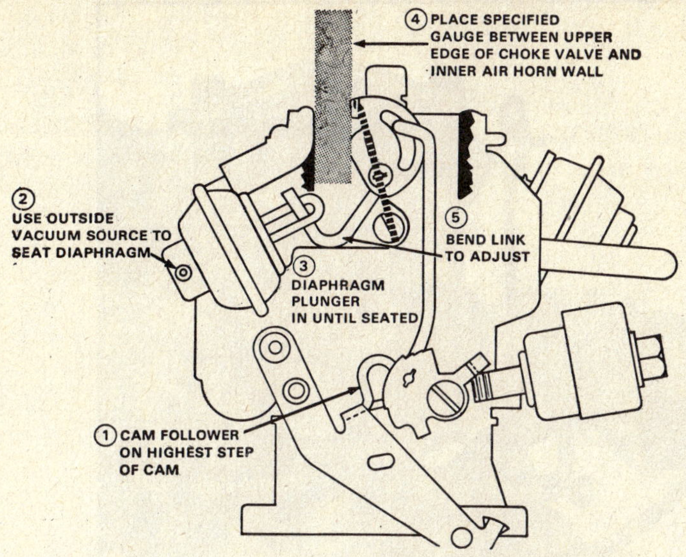

Fig. 10 Monojet auxiliary vacuum break adjustment. 1975

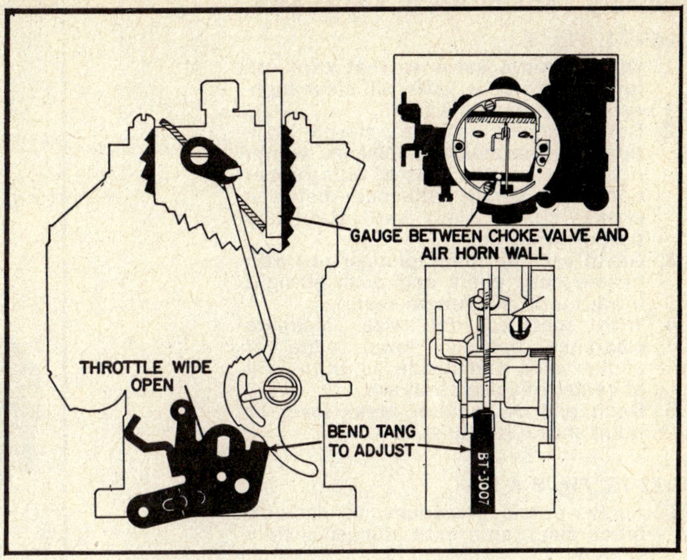

Fig. 11 Monojet unloader adjustment

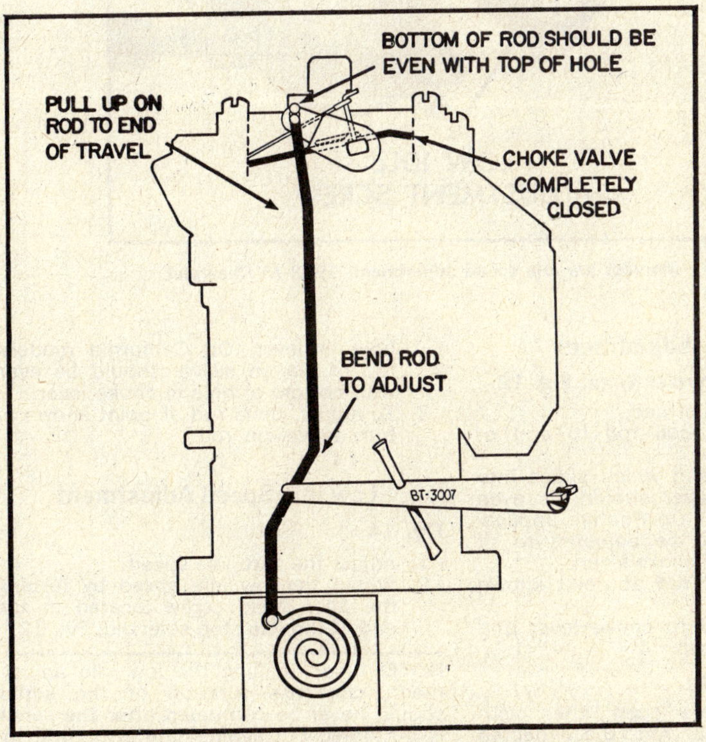

Fig. 12 Monojet choke coil adjustment. Typical

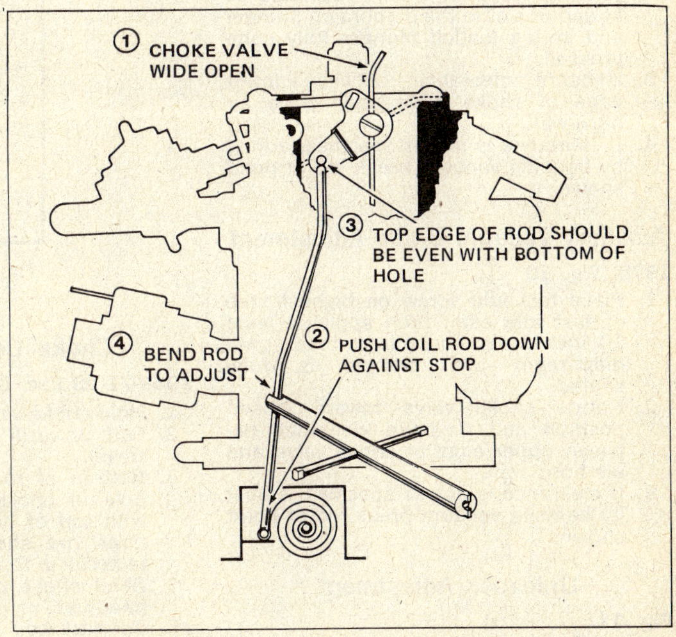

Fig. 13 Monojet choke coil adjustment. Typical

2. Place fast idle cam follower tang on highest step of cam.
3. With tang held against cam, check clearance between end of slow idle speed screw and idle stop tang on throttle lever. It should be as specified.
4. To adjust, insert screwdriver in slot provided in fast idle cam follower tang and bend inwards (towards cam) or outward to obtain specified dimension.

Manual Choke Models

Use same procedure as above except in Step 2 rotate fast idle cam clockwise to its farthest up position.

Choke Rod Adjustment

Automatic Choke Models, Fig. 6

1. With fast idle adjustment made, place fast idle cam follower on second step of fast idle cam and hold firmly against the rise to the high step.
2. Rotate choke towards direction of closed choke by applying force to choke coil lever.
3. Bend choke rod at point shown to give specified opening between lower edge of choke valve (at center of valve) and inside air horn wall.

Manual Choke Models

Use same procedure as above except in Step 1. As there are no steps on the manual choke cam, the index line on side of cam should be lined up with contact point of fast idle cam follower tang.

Primary Vacuum Break Adjustment

1969-71, Fig. 7

1. Open throttle valve so that cam follower on throttle lever will clear highest step on fast idle cam.
2. Rotate choke valve to closed position. If thermostatic coil is warm, hold choke valve closed with rubber band or spring attached between choke shaft lever and stationary part of carburetor.
3. Grasp vacuum break plunger rod with needle nose pliers and push straight inward until diaphragm seats.
4. With specified drill size, measure clearance between lower edge of choke valve and inside air horn wall at center of valve as shown.
5. Bend end of vacuum break lever at point shown to adjust.

1972-75, Fig. 8 & 9

1. Apply outside vacuum to vacuum break diaphragm until plunger is fully seated.
2. With diaphragm in seated position, push choke valve to the closed choke position. Vacuum break rod should be at end of slot in the diaphragm plunger and spring loaded plunger fully compressed.
3. Measure clearance between lower edge of choke valve and inside air horn wall.
4. If clearance is not as specified, adjust by bending vacuum break rod at point shown.

Auxiliary Vacuum Brake Adjustment

1975, Fig. 10

1. Place fast idle screw on highest step of fast idle cam, then apply at least 10 inches Hg. to vacuum break until diaphragm and plunger is fully seated.
2. Rotate choke valve toward closed position and measure clearance between upper edge of choke valve and air horn wall.
3. If clearance is not as specified, adjust by bending vacuum break rod at point shown.

Unloader Adjustment

Fig. 11

1. Hold choke valve in closed position by applying a light force to choke coil lever.
2. Rotate throttle lever to wide open throttle valve position.
3. Bend unloader tang on throttle lever to obtain specified dimension between lower edge of choke valve (at center) and air horn wall.

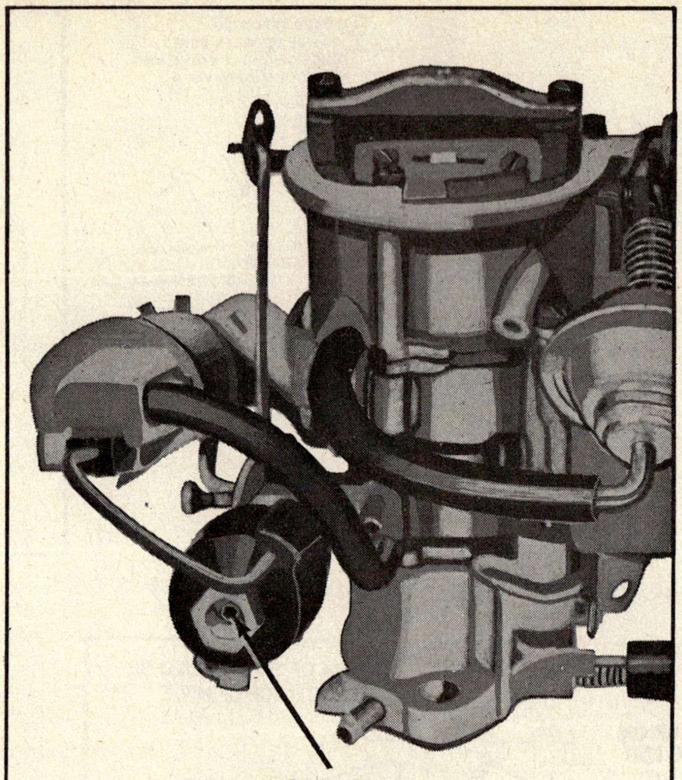

Fig. 14 Monojet lcw idle speed adjustment, 1972-73 Chevrolet

Choke Coil Adjustment

1969-71 All Exc. Chevrolet Vega, Fig. 12

1. Hold choke valve closed.
2. Pull upward on coil rod to end of travel.
3. Bottom of rod end which slides into hole in choke lever should be even with top of hole. *On Pontiac applications rod should be adjusted to fit in notch in top of choke lever.*
4. Bend choke coil rod at point shown to adjust.
5. Connect coil rod to choke lever and install retaining clip.

Chevrolet Vega, Fig. 13

1. Hold choke valve open, then with thermostatic coil rod disconnected from choke lever, push downward on rod to end of its travel.
2. On all except California models, top of rod should be even with bottom of hole in lever. On California models, top of pin on swivel should be even with bottom of hole in choke lever.
3. To adjust, bend rod at point shown or turn swivel on rod.

Low Idle Speed Adjustment

Fig. 14

1. Adjust the curb idle speed.
2. Adjust the low idle speed by turning the allen head screw located in the end of the idle stop solenoid, Fig. 12.

NOTE: When making the low idle adjustment, clockwise rotation of the screw should never be continued after the screw is bottomed out against the armature. Increased rotation will result in raising the solenoid cover up through the staked housing.

Holley Carburetor Section

ONE & TWO BARREL CARB. ADJUSTMENT SPECIFICATIONS

See Tune Up Chart in car chapter for hot idle speeds.

Year	Carb. Part No.①	Carb. Model	Idle Mixture Turns Open	Float Level (Dry)	Fuel Level (Wet)	Pump Setting	Bowl Vent Clearance	Fast Idle		Choke Unloader Clearance	Anti-Stall Dashpot Clearance	Choke Setting
								Bench	On Car			
AMERICAN MOTORS												
1969	4294A	1931	1	5/16	—	②	1/16	—	1600⑦	15/64	—	1 Rich
CHEVROLET ENGINES												
1969–70	R4055-A	2300	1½	.350	③	.015	.085	.025	2200	.250	—	⑥
	R4056-A	2300	1½	.350	③	.015	.085	.025	2200	.250	—	⑥
	R3659-A	2300	—	.350	③	—	—	—	—	—	—	—

Year	Carb. Part No.①	Carb. Model	Idle Mixture Turns Open	Float Level (Dry)	Fuel Level (Wet)	Pump Setting	Bowl Vent Clearance	Fast Idle		Choke Unloader Clearance	Vacuum Kick Drill Size	Cam Position Drill Size	Choke Setting
								Bench	On Car				
CHRYSLER ENGINES													
1969	R-4161-A	1920	⑱	See Text	27/32	⑥	3/32	—	1600⑪	9/32⑮	#39	#52	2 Rich
	R-4162-A	1920	⑱	See Text	27/32	⑥	3/32	—	1800⑪	9/32⑮	#50	#52	2 Rich
	R-4163-A	1920	⑱	See Text	27/32	⑥	3/32	—	1600⑪	9/32⑮	#39	#52	2 Rich
	R-4164-A	1920	⑱	See Text	27/32	⑥	3/32	—	1800⑪	9/32⑮	#50	#52	2 Rich
	R-4165-A	1920	⑱	See Text	27/32	⑥	3/32	—	1700⑪	9/32⑮	#39	#52	—
	R-4391-A	2300	⑱	9/16	③	.015	.080–.125	.059	2200	5/32	—	—	See Text
	R-4392A	2300	⑱	9/16	③	.015	.080–.125	.059	1800	5/32	—	—	See Text
	R-4393A	2300	⑱	3/4	③	—	—	—	—	—	—	—	—
	R-4394A	2300	⑱	3/4	③	—	—	—	—	—	—	—	—
1970	R-4351A	1920	⑱	See Text	27/32	⑥	3/32	—	1600⑪	④	#39	#52	2 Rich
	R-4352A	1920	⑱	See Text	27/32	⑥	3/32	—	1800⑪	④	#50	#52	2 Rich
	R-4353A	1920	⑱	See Text	27/32	⑥	3/32	—	1600⑪	④	#39	#52	2 Rich
	R-4354A	1920	⑱	See Text	27/32	⑥	3/32	—	1800⑪	④	#50	#52	2 Rich
	R-4355A	1920	⑱	See Text	27/32	⑥	3/32	—	1700⑪	④	#39	#52	2 Rich
	R-4363A	1920	⑱	See Text	27/32	⑥	3/32	—	1700⑪	④	#39	#52	2 Rich
	R-4371A	2210	⑱	.200	—	—	5/64	—	1700⑪	11/64	#28	#35	2 Rich
	R-4175AF	2300	⑱	⑰	③	—	—	—	—	—	—	—	—
	R-4144A	2300	⑱	⑰	③	.015	.101	.059	1800	5/32	#50	#53	2 Rich
	R-4365AR	2300	⑱	⑰	③	—	—	—	—	—	—	—	—
	R-4374A	2300	⑱	⑰	③	.015	.101	.059	2200	5/32	#28	#53	2 Rich
	R-4375A	2300	⑱	⑰	③	.015	.101	.059	2200	5/32	#28	#53	2 Rich
	R-4376A	2300	⑱	⑰	③	.015	.101	.059	1800	5/32	#50	#53	2 Rich
	R-4382AF	2300	⑱	⑰	③	—	—	—	—	—	—	—	—
	R-4383AR	2300	⑱	⑰	③	—	—	—	—	—	—	—	—
1971	R-4655A	1920	⑱	See Text	27/32	—	1/32	.063	1600	9/32	#39	#52	2 Rich
	R-4656A	1920	⑱	See Text	27/32	—	1/32	.063	1900	5/32	#39	#52	2 Rich
	R-4659A	1920	⑱	See Text	27/32	—	1/32	.063	1800	9/32	#39	#52	2 Rich
	R-6363A	1920	⑱	See Text	27/32	—	1/64	.063	2000	9/32	—	—	2 Rich
	R-6364A	1920	⑱	See Text	27/32	—	1/64	.063	1900	9/32	—	—	2 Rich
	R-4373A	2210	⑱	.200	—	9/16	5/64	.110	1700	11/64	—	—	2 Rich
	R-4665A	2210	⑱	.200	—	9/16	.015	.110	1800	1/4	#28	#35	2 Rich
	R-4666A	2210	⑱	.200	—	9/16	.015	.110	1800	1/4	#30	#35	2 Rich
	R-4669A	2300	⑱	⑰	③	.015	.101	.059	1800	5/32	—	—	2 Rich
	R-4670A	2300	⑱	⑰	③	.015	.101	.059	1800	5/32	—	—	2 Rich
	R-4671A	2300	⑱	⑰	③	—	—	—	—	—	—	—	—
	R-4672A	2300	⑱	⑰	③	—	—	—	—	—	—	—	—
	R-4789A	2300	⑱	⑰	③	—	—	—	—	—	—	—	—

Continued

ONE & TWO BARREL CARB. ADJUSTMENT SPECIFICATIONS—Continued

See Tune Up Chart in car chapter for hot idle speeds.

Year	Carb. Part No.①	Carb. Model	Idle Mixture Turns Open	Float Level (Dry)	Fuel Level (Wet)	Pump Setting	Bowl Vent Clearance	Fast Idle		Choke Unloader Clearance	Vacuum Kick Drill Size	Cam Position Drill Size	Choke Setting
								Bench	On Car				
CHRYSLER ENGINES—Continued													
1971	R-4790A	2300	⑱	⑰	③	—	—	—	—	—	—	—	—
	R-4791A	2300	⑱	⑰	③	.015	.101	.059	2600	5/32	#28	#53	On Index
	R-4792A	2300	⑱	⑰	③	.015	.101	.059	2800	5/32	#39	#53	On Index
1972	R-6153A	1920	⑱	See Text	27/32	—	.015	.063	2000	④	.100	.064	Fixed
	R-6154A	1920	⑱	See Text	27/32	—	.015	.063	2000	④	.100	.064	Fixed
	R-6155A	1920	⑱	See Text	27/32	—	.015	.063	2000	④	.100	.064	Fixed
	R-6156A	1920	⑱	See Text	27/32	—	.015	.063	2000	④	.100	.064	Fixed
	R-6159A	1920	⑱	See Text	27/32	—	.015	.063	1900	④	.100	.064	Fixed
	R-6363A	1920	⑱	See Text	27/32	—	.015	.063	2000	④	.100	.064	Fixed
	R-6364A	1920	⑱	See Text	27/32	—	.015	.063	1900	④	.100	.064	Fixed
	R-6365A	1920	⑱	See Text	27/32	—	.015	.063	2000	④	.100	.064	Fixed
	R-6366A	1920	⑱	See Text	27/32	—	.015	.063	2000	④	.100	.064	Fixed
	R-6162A	2210	⑱	.180	—	.285	.015	.110	1900	.170	.100	.064	Fixed
	R-6164A	2210	⑱	.180	—	.250	.015	.110	2000	.170	.100	.064	Fixed
	R-6368A	2210	⑱	.180	—	.285	.015	.110	1900	.170	.100	.064	Fixed
	R-6370A	2210	⑱	.180	—	.285	.015	.110	2000	.170	.100	.064	Fixed
	R-6404A	2300	⑱	⑰	③	.015	.015	—	1800	.150	.130	.110	Fixed
	R-6405A	2300	⑱	—	—	—	—	—	—	—	.130	.110	—
	R-6406A	2300	⑱	—	—	—	—	—	—	—	.130	.110	—
1973	R-6447A	1920	⑱	.260	—	—	.015	.065	2000	④	.100	.065	Fixed
	R-6448A	1920	⑱	.260	—	—	.015	.045	1700	④	.080	.045	Fixed
	R-6593A⑨	1920	⑱	.260	—	—	.015	.065	2000	④	.100	.065	Fixed
	R-6594A	1920	⑱	.260	—	—	.015	.065	1700	④	.100	.065	Fixed
	R-6595A	1920	⑱	.260	—	—	.015	.065	2000	④	.100	.065	Fixed
	R-6596A⑨	1920	⑱	.260	—	—	.015	.065	1700	④	.100	.065	Fixed
	R-6452A	2210	⑱	.180	—	.250	.015	.110	1900	.170	.150	.110	Fixed
	R-6454A	2210	⑱	.180	—	.250	.015	.110	1800	.170	.150	.110	Fixed
	R-6472A⑨	2210	⑱	.180	—	.250	.015	.110	1800	.170	.150	.110	Fixed
	R-6575A⑨	2210	⑱	.180	—	.250	.015	.110	1900	.170	.150	.110	Fixed
1974	R-6721A	1945	⑱	3/64	—	11/16	—	.080	1600	.250	.140	.080	Fixed
	R-6722A	1945	⑱	3/64	—	13/16	—	.080	1800	.250	.090	.080	Fixed
	R-6723A	1945	⑱	3/64	—	11/16	—	.080	1600	.250	.140	.080	Fixed
	R-6724A	1945	⑱	3/64	—	3/4	—	.080	1800	.250	.080	.080	Fixed
	R-6725A⑨	1945	⑱	3/64	—	11/16	—	.080	1600	.250	.140	.080	Fixed
	R-6726A⑨	1945	⑱	3/64	—	3/4	—	.080	1800	.250	.090	.080	Fixed
	R-6731A	2245	⑱	.180	—	.255	.015	.110	1800	.170	.150	.110	Fixed
	R-6737A	2245	⑱	.180	—	.255	.015	.110	1600	.170	.150	.110	Fixed
1975	R-7017A	1945	⑱	3/64	—	2 7/32	—	.080	1600	.250	.130	.080	Fixed
	R-7018A	1945	⑱	3/64	—	2 21/64	—	.080	1700	.250	.090	.080	Fixed
	R-7019A	1945	⑱	3/64	—	2 7/32	—	.080	1600	.250	.130	.080	Fixed
	R-7020A	1945	⑱	3/64	—	2 21/64	—	.080	1700	.250	.090	.080	Fixed
	R-7027A	2245	⑱	3/16	—	1/4	.015	.110	1600	.170	.150	.110	Fixed
	R-7226A	2245	⑱	3/16	—	1/4	.015	.110	1600	.170	.150	.110	Fixed
	R-7329A	1945	⑱	3/64	—	2 7/32	—	.080	1700	.250	.130	.080	Fixed

Continued

ONE & TWO BARREL CARB. ADJUSTMENT SPECIFICATIONS—Continued

See Tune Up Chart in car chapter for hot idle speeds.

Year	Carb. Part No.(1)	Carb. Model	Idle Mixture Turns Open	Float Level (Dry)	Fuel Level (Wet)	Pump Setting	Bowl Vent Clearance	Fast Idle		Choke Unloader Clearance	Vacuum Kick Drill Size	Cam Position Drill Size	Choke Setting
								Bench	On Car				
CHRYSLER ENGINES—Continued													
1976	R-7356A	1945	(18)	(19)	—	2 7/32	1/16	.080	1600	.250	.110	.080	Fixed
	R-7357A	1945	(18)	(19)	—	2 21/32	1/16	.080	1700	.250	.100	.080	Fixed
	R-7360A	1945	(18)	(19)	—	2 7/32	—	.080	1600	.250	.110	.080	Fixed
	R-7361A	1945	(18)	3/64	—	2 21/32	—	.080	1700	.250	.100	.080	Fixed
	R-7364A	2245	(18)	3/16	—	5/16	.025	.110	1600	.170	.150	.110	Fixed
	R-7366A	2245	(18)	3/16	—	5/16	.025	.110	1600	.170	.150	.110	Fixed

(1)—Located on tag attached to carburetor or on casting.
(2)—Seasonal setting holes in lever; long stroke for cold weather, short stroke for warm weather.
(3)—At lower edge of sight plug opening.
(4)—Unloader automatically set when fast idle cam is adjusted.
(6)—Seasonal setting; use center hole for moderate weather, long stroke hole for cold weather, and short stroke hole for warm weather.
(7)—Engine hot and screw on second step of cam.
(8)—See "Fast idle index adjustment" in text.
(9)—With California Emission package.
(10)—With headlights on A-C operating (if equipped); engine hot and screw on lowest step of cam.
(11)—Engine hot and screw on highest step of cam.
(12)—Engine hot and screw on second highest step of cam.
(13)—Lever touching screw; then tighten screw 1/4 turn.
(14)—Between throttle valve and carburetor bore.
(15)—With fast idle speed adjusting screw contacting second highest step on fast idle cam there should be 1/16" clearance between choke valve and wall of air horn. When this adjustment is correct the choke unloader clearance should be as specified in the chart.
(16)—Top of rod should be even with bottom of hole.
(17)—Center float in bowl with bowl inverted.
(18)—Air/fuel ratio or idle CO% rating is found in Tune Up chart in car chapters.
(19)—Flush with top of bowl cover gasket with bowl inverted.

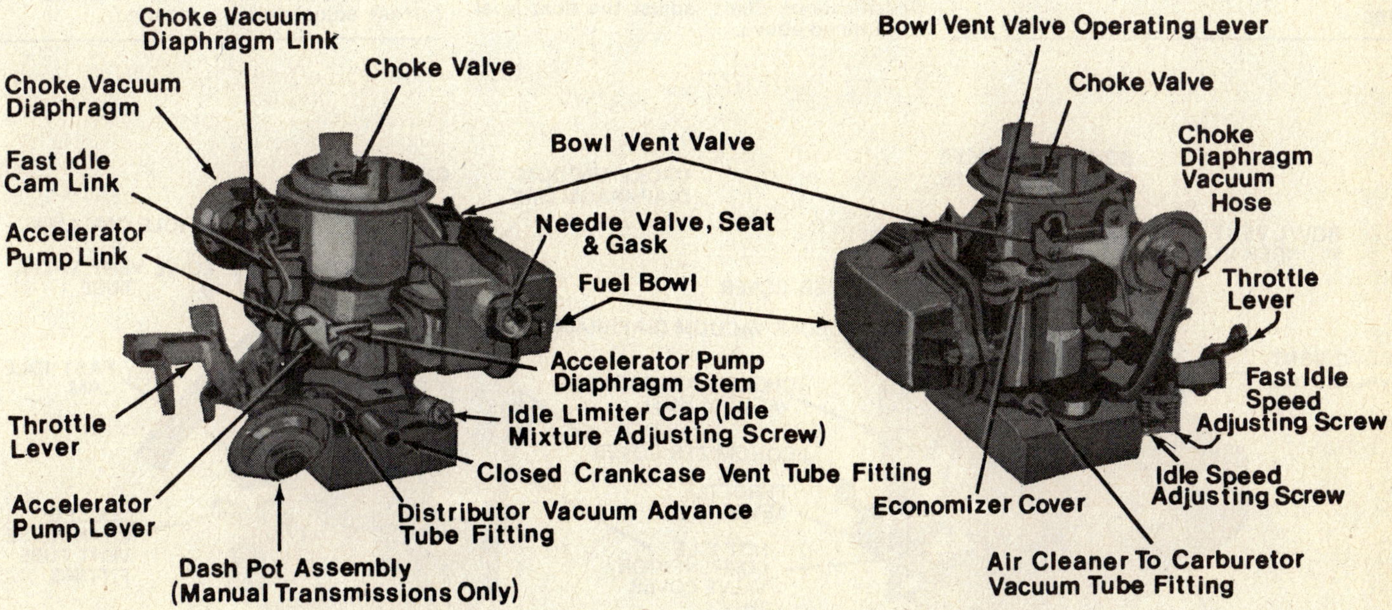

Fig. 1 Holley model 1920 single-barrel carburetor. C.A.S. equipped 1970-72 typical

ONE & TWO BARREL CARBURETOR ADJUSTMENTS

MODEL 1920

On these units, Figs. 1 thru 4, the choke valve located in the carburetor bore is connected to a well type choke.

Some models are equipped with a spring-staged choke, shown in Fig. 12, which is a device incorporated in the choke mechanism that limits the choke valve closing torque when cranking the engine at temperatures below zero. The spring-staging of the choke is suited for starting mixture requirements at both low and moderate temperatures. On 1973 Chrysler units, an electric choke system is incorporated to open choke at approximately 63°. For service information refer to the "Emission Control System" chapter.

The accelerator pump is a diaphragm, spring driven type operated by a lever connected to the throttle shaft.

A two stage power valve mounted in the metering body and actuated by manifold vacuum delivers additional fuel for full power and high speed operation.

Float Level Setting

NOTE: Do not allow float tab to contact float needle head during the adjustment procedure as the rubber tip of the needle can be compressed, giving a false reading.

1969-72 Units

Fig. 5—With carburetor inverted slide float gauge into position and test setting on "touch" leg of gauge. Float should just touch gauge. Reverse gauge and test "no touch" leg. Float should just clear gauge. To adjust, bend float tab which touches head of fuel inlet needle, using needle nose pliers.

1973 Units

Fig. 6—With the carburetor inverted, measure from the top of the float to the upper wall of the main body with the gauge against the cast rib approximately two inches from the float hinge pin. Be sure the gauge is parallel with the top of the float. Refer to *Holley Specifications Chart* for proper dry float setting. Adjust by bending the float tab which touches the head of the fuel inlet needle using needle nosed pliers.

Checking Fuel Level

Fig. 7—With engine running and vehicle on a level floor, measure fuel level through the economizer diaphragm opening. Use a 6" scale with a depth gauge, measure distance from machined surface of the opening to the exact fuel surface. If the level is not as listed in the *Holley Specifications Chart*, adjust the float level as outlined above.

Float Bowl Vent Valve Setting

Fig. 8—With throttle valve closed, bowl vent should be adjusted so that the shank of a drill of the size listed in the *Holley Specifications Chart* can be inserted between valve and surface of carburetor body. Adjust by bending bowl vent operating lever up or down as required. Be sure vent rod does not bind in the guide after adjusting.

Fast Idle Cam Position and Choke Unloader Adjustment

Fig. 9—With fast idle speed adjusting screw contacting second highest step on fast idle cam, move choke valve toward closed position with light pressure on choke shaft lever. Insert specified gauge between top of choke valve and wall of air horn. Refer to *Holley Specifications Chart.* Adjust by bending fast idle link at angle, until correct valve opening has been obtained.

NOTE: IMPORTANT: When the correct fast idle cam position adjustment has been made, the choke unloader (wide open kick) adjustment has also been obtained. No further adjustment is required.

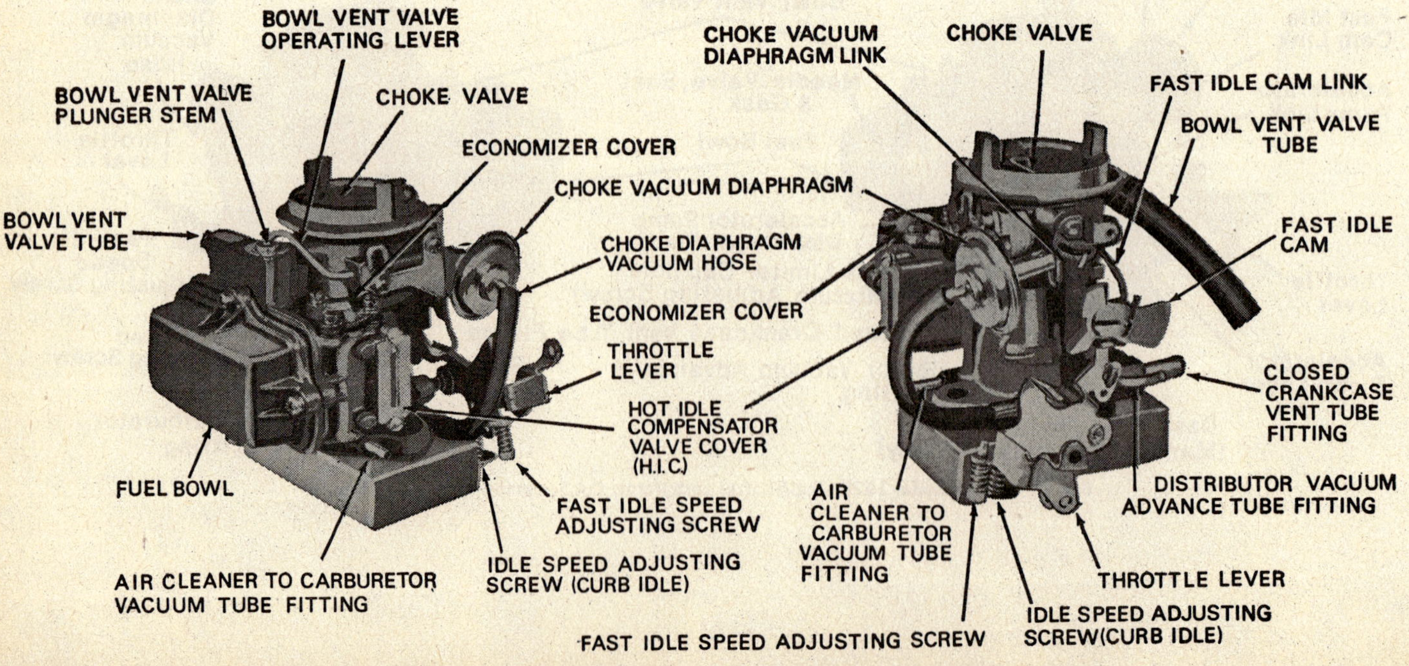

Fig. 2 Holley model 1920 single-barrel carburetor. E.C.S. equipped 1970-72 typical

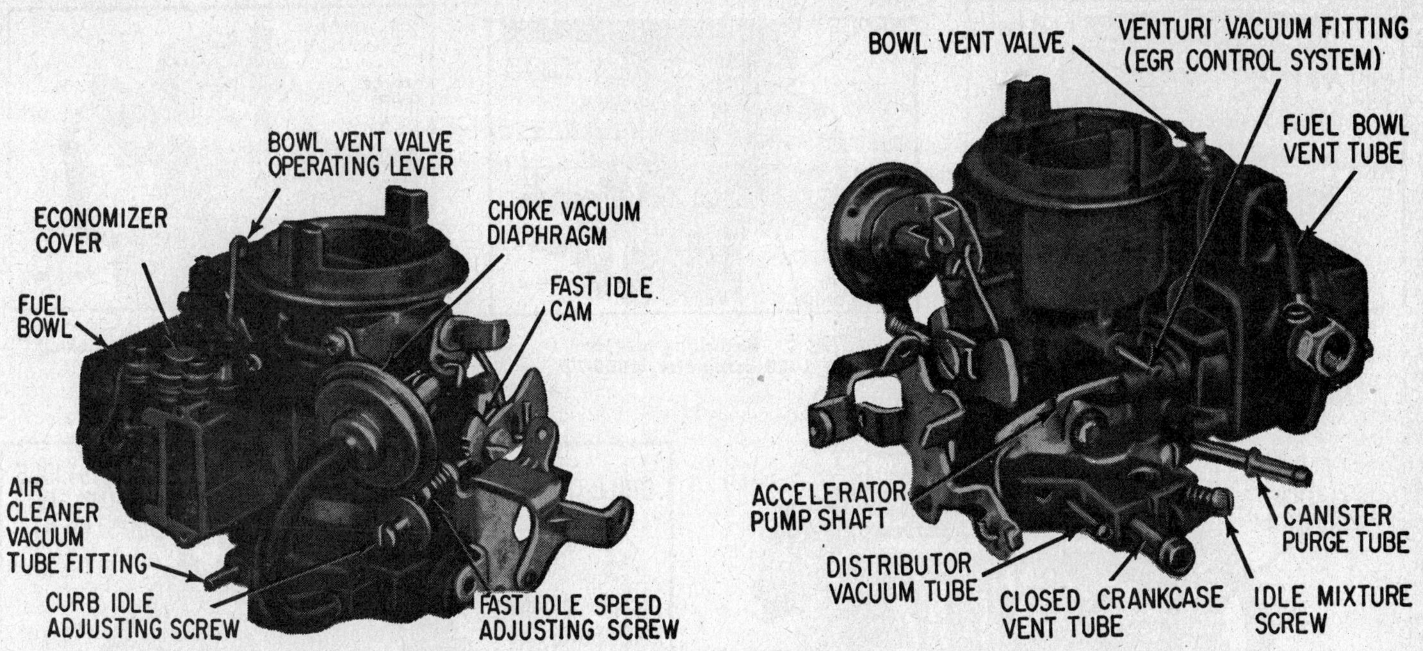

Fig. 3 Holley model 1920 single-barrel carburetor. 1973 typical

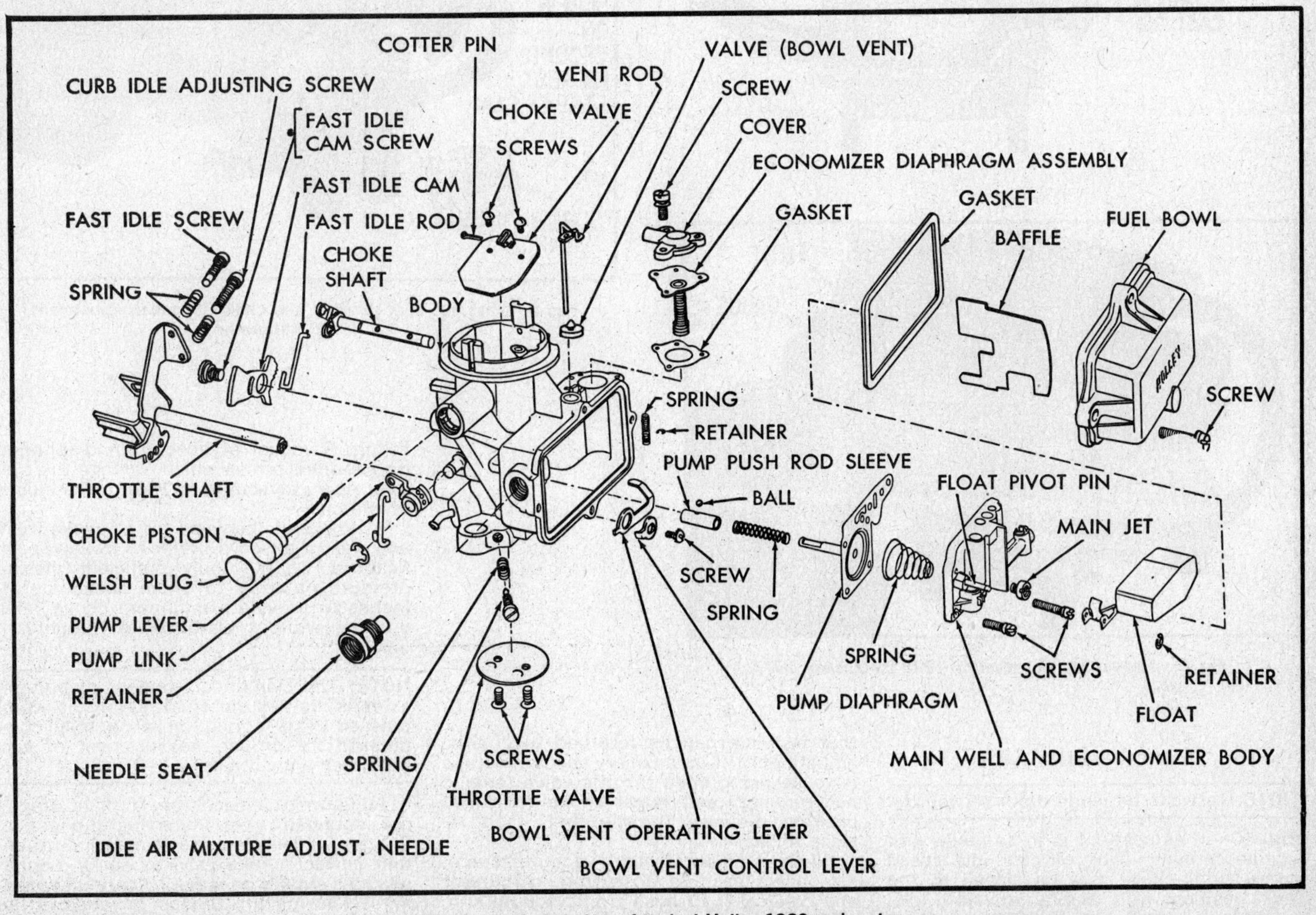

Fig. 4 Exploded view of typical Holley 1920 carburetor

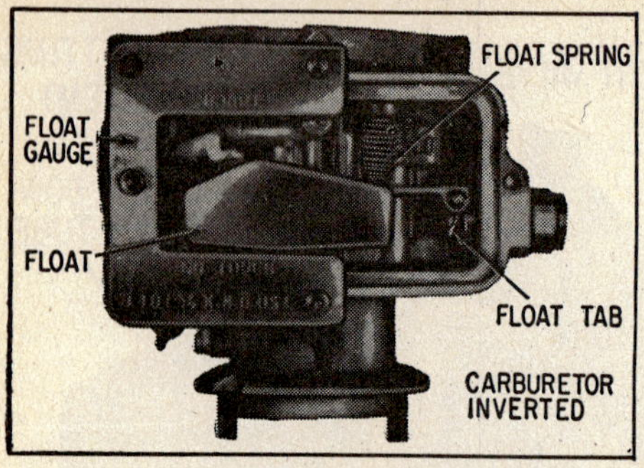

Fig. 5 Measuring float level on 1920 carburetor, 1969-72

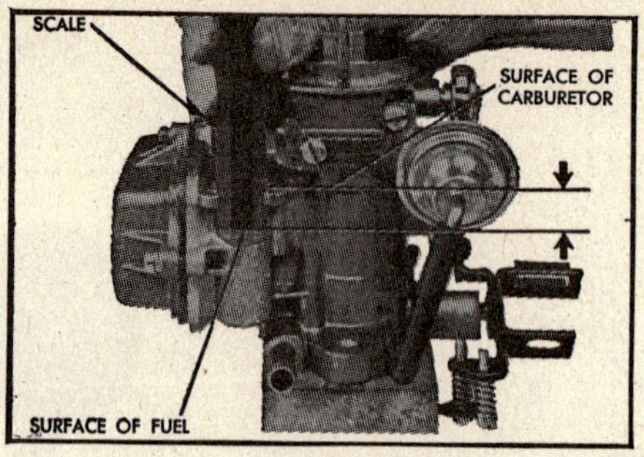

Fig. 7 Measuring fuel level on 1920 carburetor, 1969-72

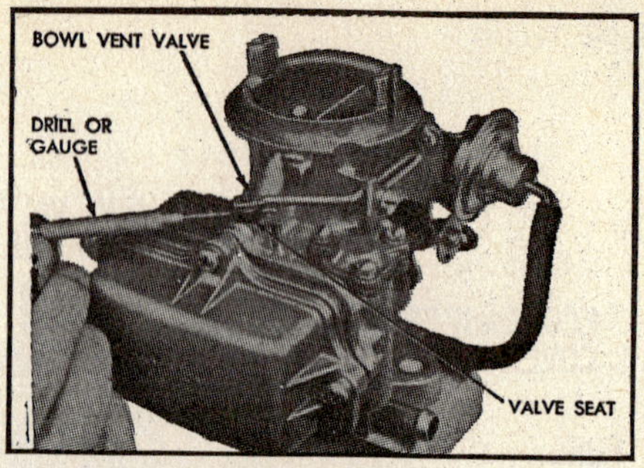

Fig. 8 Measuring bowl vent opening on 1920 carburetor

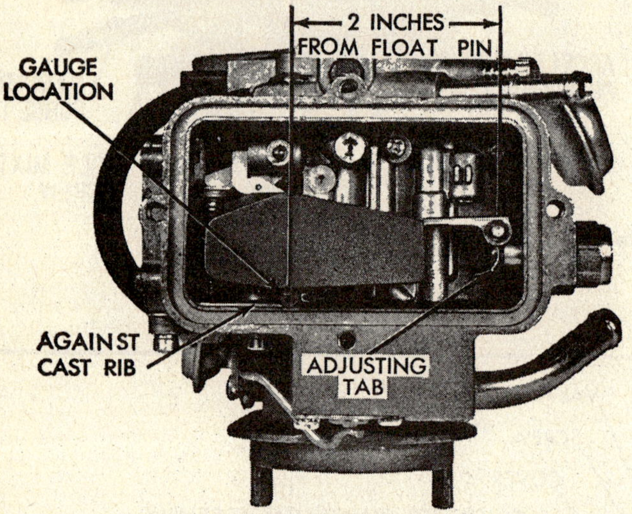

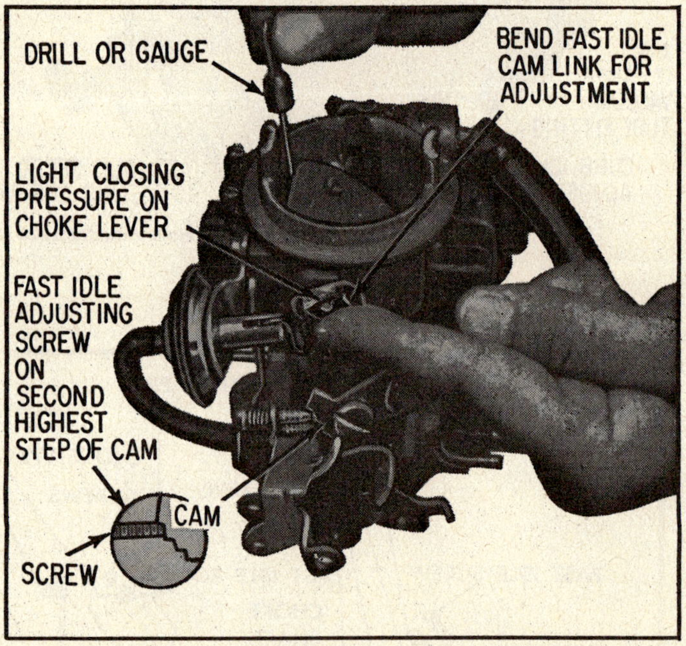

Fig. 9 Fast idle cam position and choke unloader adjustment on 1920 carburetor

Fig. 6 Measuring float level on 1920 carburetor. 1973

Choke Vacuum Kick

NOTE: Test can be made on or off vehicle.

Fig. 10—If adjustment is to be made with engine running, back off fast idle speed screw until choke can be closed to the kick position with engine at curb idle. (Note number of screw turns required so that fast idle can be returned to original adjustment). If an auxiliary vacuum source is to be used, open throttle valve (engine not running) and move choke to closed position. Release throttle first, then release choke.

When using an auxiliary vacuum source, disconnect vacuum hose from carburetor and connect it to hose from vacuum supply with a small length of tube to act as a fitting. Removal of hose from diaphragm may require forces which damage the system. Apply a vacuum of 15 or more inches of mercury.

Insert gauge between top of choke valve and wall of air horn. Refer to *Holley Specifications Chart*. Apply sufficient closing pressure on lever to which choke rod attaches to provide a minimum choke valve opening without distortion of diaphragm link.

NOTE: The cylindrical stem of diaphragm extends as the internal spring is compressed. This spring must be fully compressed for proper measurement of vacuum kick adjustment.

Adjustment is necessary if slight drag is not obtained when removing the gauge. Shorten or lengthen diaphragm link to obtain correct choke valve opening. Length changes should be made by carefully opening or closing the U-bend provided in the link. Improper bending causes contact be-

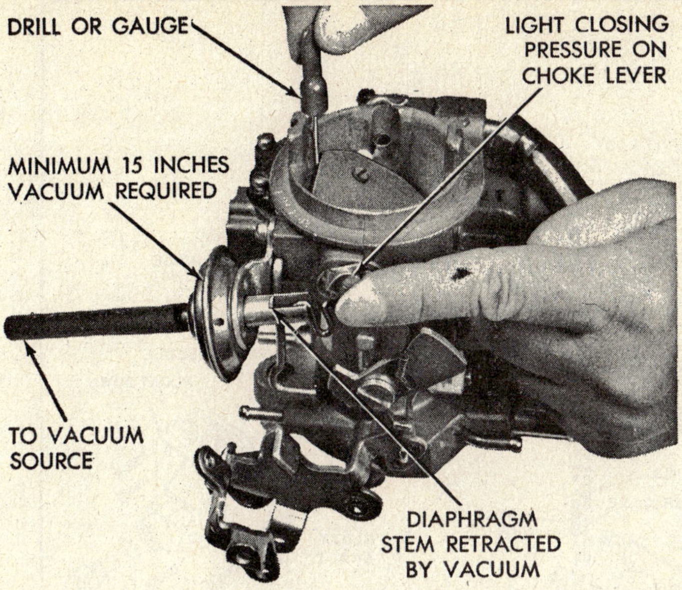

Fig. 10 Choke vacuum kick adjustment on 1920 carburetor

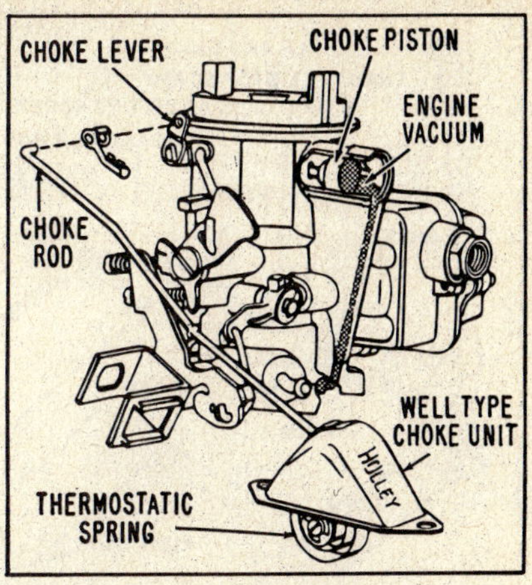

Fig. 11 Well-type automatic choke used with 1920 carburetor

tween the U-section and the diaphragm assembly.

NOTE: Do not apply twisting or bending force to diaphragm.

After completion of adjustment, reinstall vacuum hose on correct carburetor fitting. Return fast idle screw to its original location if disturbed. Make following check. With no vacuum applied to diaphragm, the choke valve should move freely between open and closed positions. If movement is not free, examine linkage for misalignment or interferences caused by bending operation.

Well-Type Automatic Choke

Fig. 11—To function properly, it is important that all parts be clean and move freely. Other than an occasional cleaning, the choke requires no attention. However, it is important that the choke control unit work freely in the well and at the choke shaft. Move the choke rod up and down to check for free movement on the pivot. If the unit binds, a new choke unit should be installed.

This type choke is serviced only as a unit. Do not attempt to repair or change the setting.

When installing the choke unit, be certain that the coil housing does not contact the sides of the well in the exhaust manifold. Any contact at this point will affect choke operation. Do not lubricate any parts of the choke or the control unit. This causes an accumulation of dirt which will result in binding of the mechanism.

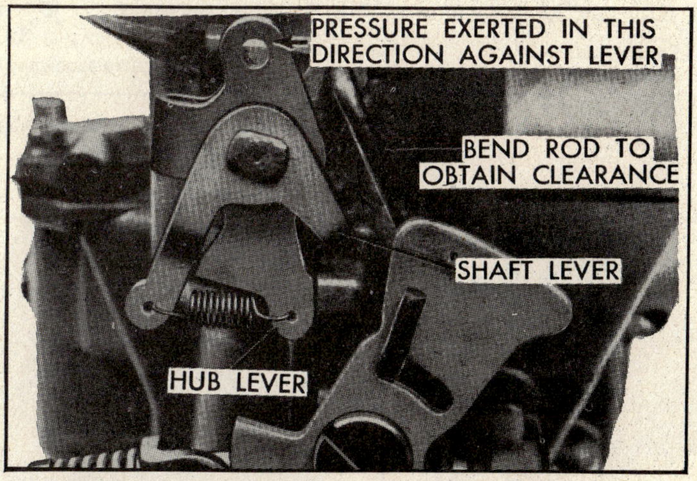

Fig. 12 Spring-staged choke adjustment on 1920 carburetor

Spring Staged Choke Adjustment

To test the adjustment on carburetors equipped with the feature, press against the choke lever firmly, Fig. 12. Measure clearance between hub lever and shaft lever. If the clearance is not within .010″ and .025″, bend fast idle rod slightly until normal clearance is obtained. The cam position and unloader setting resulting from such bending of the choke link are satisfactory.

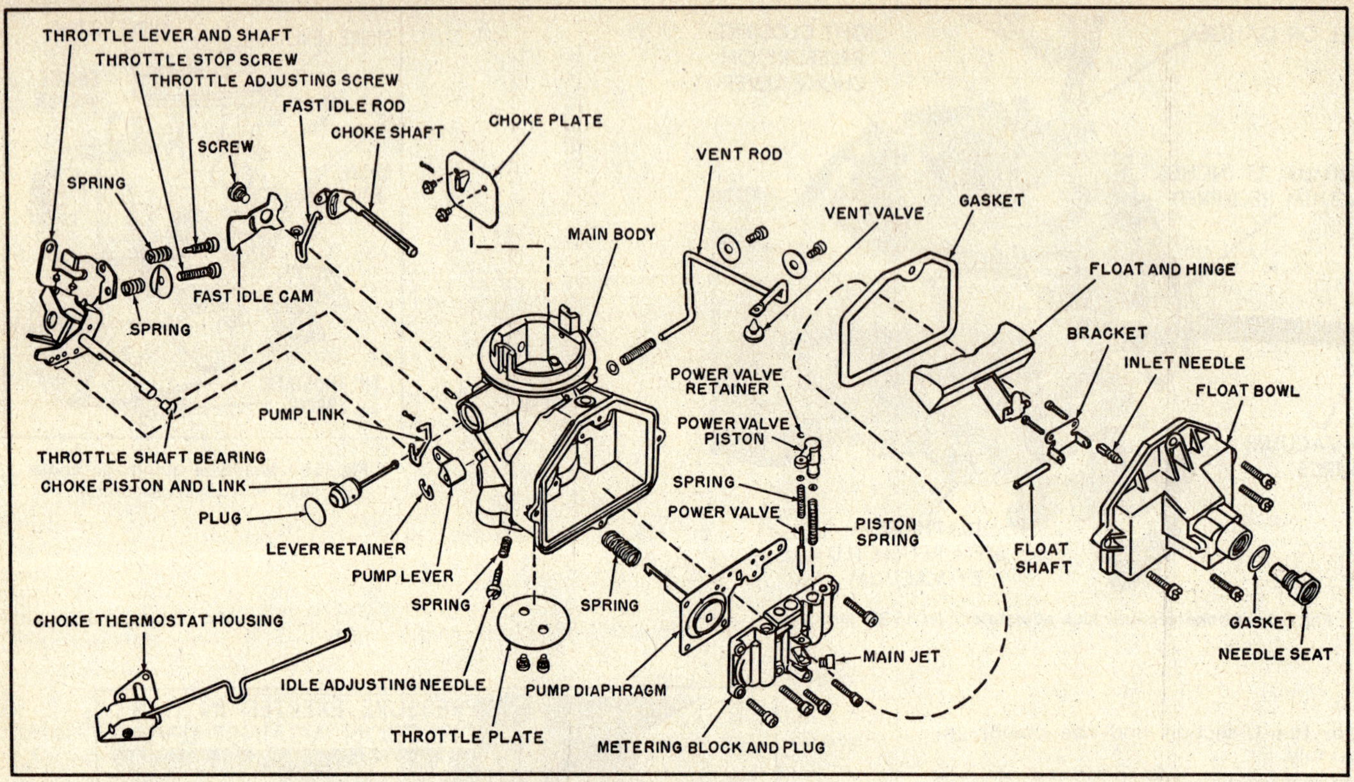

Fig. 2 Exploded view of Holley 1931 carburetor

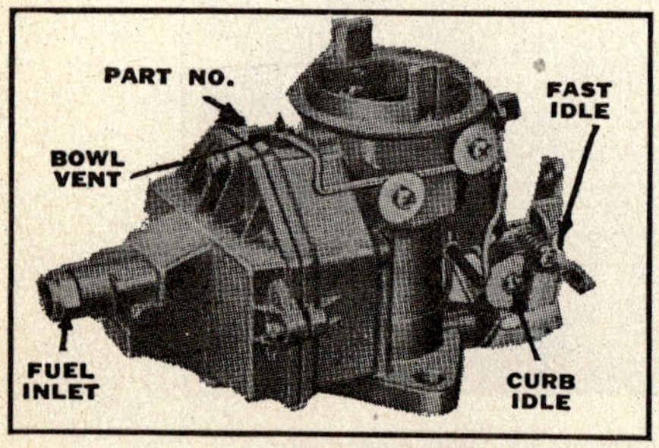

Fig. 1 Holley Model 1931 single-barrel carburetor

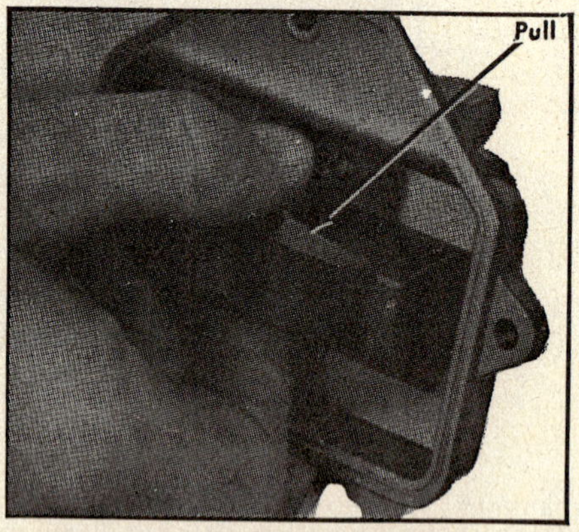

Fig. 4 Lowering float by pulling it out on 1931 carburetor

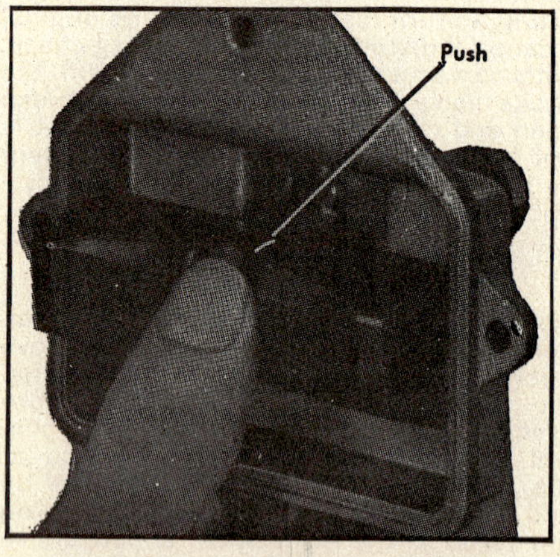

Fig. 5 Raising float by pushing it in on 1931 carburetor

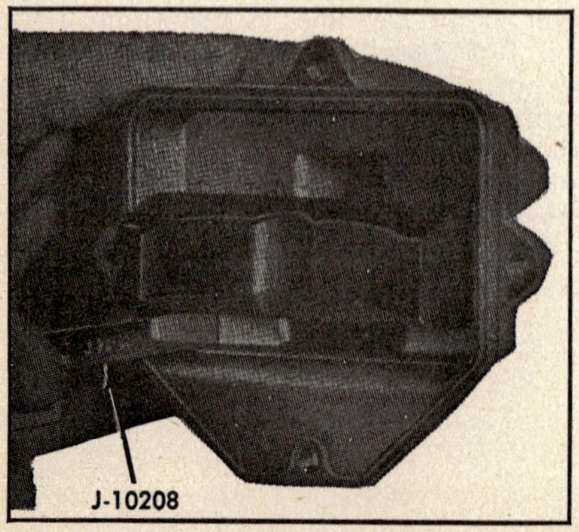

Fig. 3 Checking float level on 1931 carburetor

MODEL 1931

This carburetor, Figs. 1 and 2, features a one-piece main body and throttle body casting together with a large capacity fuel bowl. The fuel inlet and float assembly, located in the center of the fuel bowl cover, maintains a stable fuel level for best performance on turns. The large capacity fuel bowl is designed to efficiently handle vapor loaded fuels and fuel vapors.

The automatic choke is mounted in a heat sink on the exhaust manifold and is connected to the carburetor by a choke rod.

Float Adjustment

Figs. 3, 4, 5—Invert carburetor fuel bowl cover and check float setting at both ends, Fig. 3. If float adjustment is necessary the float may be lowered by pulling on the center of the float, Fig. 4. The float may be raised by pushing on the center of the float, Fig. 5.

CAUTION: During float adjustment do not allow the float tab to contact the fuel inlet needle as the resilient tip of the needle can be damaged or compressed, resulting in an improper float setting and a leaky needle and seat.

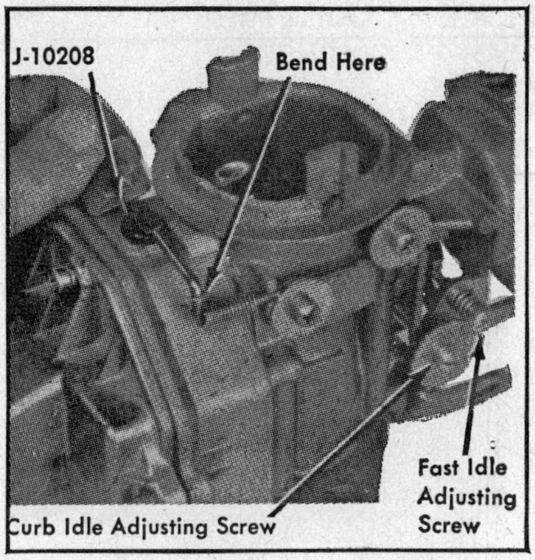

Fig. 6 Bowl vent adjustment on 1931 carburetor

If the proper gauge is not available, measure the distance between the roof of the float cover and top of float. If the setting is not as listed in the *Holley Specifications Chart,* adjust as required.

Bowl Vent Setting

Fig. 6—With throttle set at curb idle, clearance between vent valve and seat should be as listed in the *Holley Specifications Chart*. If an adjustment is necessary, bend vent rod at the horizontal portion above the fuel bowl. Check operation of vent rod for binding.

Accelerator Pump Setting

The accelerator link is set in the middle hole in the throttle lever for normal driving conditions. In the event a richer pump discharge is required, place the pump link in the outer hole of the throttle lever. For a leaner pump discharge, place the pump link in the inner hole of the throttle lever.

Fig. 7 Choke unloader adjustment on 1931 carburetor

Fast Idle Adjustment

With the fast idle screw resting on second step of fast idle cam, and engine at normal operating temperature, the fast idle rpm should be as listed in the *Holley Specifications Chart*. Turning the fast idle screw in clockwise increases (counterclockwise decreases) speed.

Choke Unloader Setting

Fig. 7—With throttle valve held in wide open position and choke plate rotated toward closed position, the distance between top edge of choke plate and flat portion of air horn should be as listed in the *Holley Specifications Chart*. Adjust by bending tab on throttle lever.

Fig. 8 Choke piston stop adjustment on 1931 carburetor

Choke Piston Stop Adjustment

Fig. 8—Hold choke piston against stop screw with a wire inserted in slot above choke piston link. Rotate choke plate toward closed position until link is firm. The distance between top edge of choke plate and flat portion of air horn should be $^3/_{16}$". Adjust by turning piston stop screw in or out as required.

Automatic Choke Setting

The adjustment is made by loosening choke cover screws and rotating cover in the desired direction as indicated by an arrow on the cover. The choke should be set to the mark specified in the *Holley Specifications Chart* for all normal driving. Never set the choke more than two graduations in either direction of the specified setting.

MODEL 1945

This single barrel carburetor, Fig. 1, utilizes dual cellular plastic floats to control the fuel level, thus permitting high angularity operation during the most severe operating conditions. Also, the float construction eliminates the possibility of a malfunction due to a punctured float.

On 1974-75 Chrysler units, an electric choke system is incorporated to open the choke at approximately 60 degrees F. For service information, refer to the "Emission Control System" chapter.

The accelerator pump is of the piston type and is operated by a rod and a link connected to the throttle lever.

The power enrichment system on all units, consists of a power valve installed near the center of the carburetor body and a vacuum piston located in the bowl cover. On 1974 California and all 1975 units, in addition to the vacuum operated enrichment system, a spring loaded mechanical modulator rod opens the power valve at 80 degrees of throttle opening regardless of engine vacuum.

On 1975 vehicles with automatic transmissions, an idle enrichment system is used to reduce cold engine stalling by the use of an additional metering system which enrichens the mixture in the off idle position. This system is controlled by a vacuum diaphragm mounted near the top of the carburetor.

Dry Float Setting

Fig. 2—Hold float fulcrum retaining pin in position and invert carburetor bowl. Place a straight edge across surface of bowl, contracting float toes. Remove straight edge and measure distance float dropped from surface of fuel bowl. Refer to Holley Specifications Chart. Adjust by bending float tang to obtain proper dimension.

Fast Idle Cam Position Adjustment

Fig. 3—With fast idle speed adjusting screw contacting second highest step on fast idle cam, move choke valve toward closed position with light pressure on choke shaft lever. Insert specified gauge between top of choke valve and wall of air horn. Refer to *Holley Specifications Chart*. An adjustment will be necessary if a slight drag is not obtained as drill shank is being removed. Adjust by bending fast idle link at lower angle, until correct valve opening has been obtained.

Choke Vacuum Kick Adjustment

NOTE: Test can be made on or off vehicle.

Fig. 4—If adjustment is to be made with engine running, back off fast idle speed screw until choke can be closed to the kick position with engine at curb idle. (Note number of screw turns required so that fast idle can be returned to original adjustment). If an auxiliary vacuum source is to be used, open throttle valve (engine not running) and move choke to closed position. Release throttle first, then release choke.

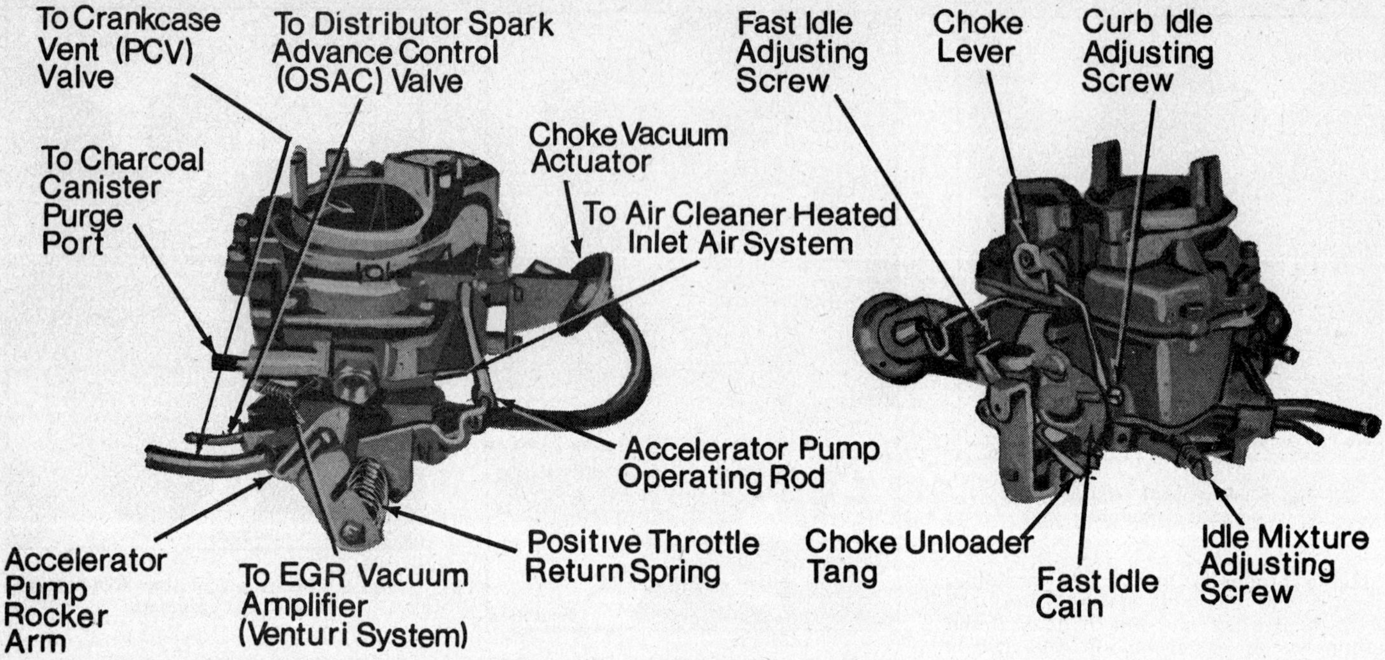

Fig. 1 Holley model 1945 single barrel carburetor

When using an auxiliary vacuum source, disconnect vacuum hose from carburetor and connect it to hose from vacuum supply with a small length of tube to act as a fitting. Removal of hose from diaphragm may require forces which damage the system. Apply a vacuum of 15 or more inches of mercury.

Insert gauge between top of choke valve and wall of air horn. Refer to *Holley Specifications Chart.* Apply sufficient closing pressure on lever to which choke rod attaches to provide a minimum choke valve opening without distortion of diaphragm link.

NOTE: The cylindrical stem of diaphragm extends as the internal spring is compressed. This spring must be fully compressed for proper measurement of vacuum kick adjustment.

Adjustment is necessary if slight drag is not obtained when removing the gauge. Shorten or lengthen diaphragm link to obtain correct choke valve opening. Length changes should be made by carefully opening or closing the U-bend provided in the link. Improper bending causes contact between the U-section and the diaphragm assembly.

NOTE: Do not apply twisting or bending force to diaphragm.

After completion of adjustment, reinstall vacuum hose on correct carburetor fitting.

Return fast idle screw to its original location if disturbed. Make following check. With no vacuum applied to diaphragm, the choke valve should move freely between open and closed positions. If movement is not free, examine linkage for misalignment or interferences caused by bending operation.

Choke Unloader (Wide Open Kick) Adjustment

Fig. 5—With throttle valves in wide open position, insert drill gauge between upper edge of choke valve and inner wall of air

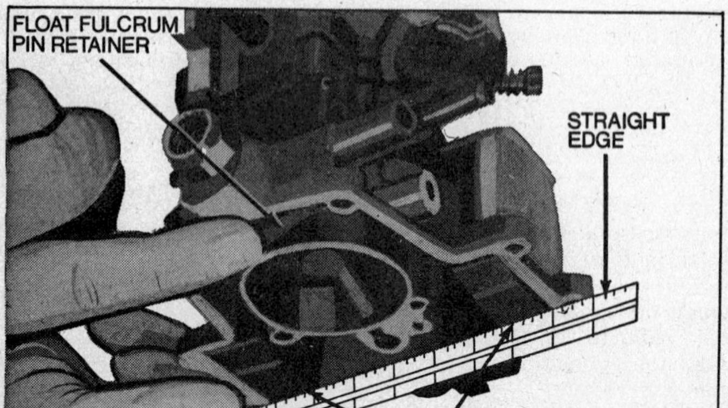

Fig. 2 Measuring float level. 1945 carburetor

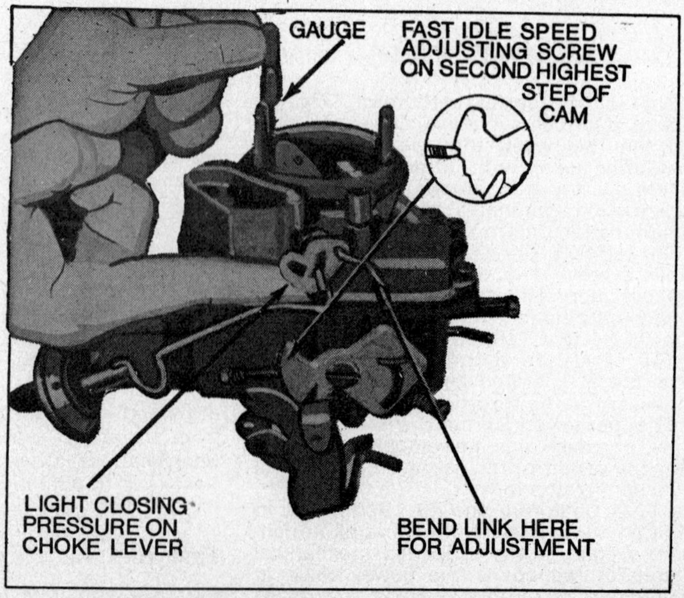

Fig. 3 Fast idle cam position adjustment. 1945 carburetor

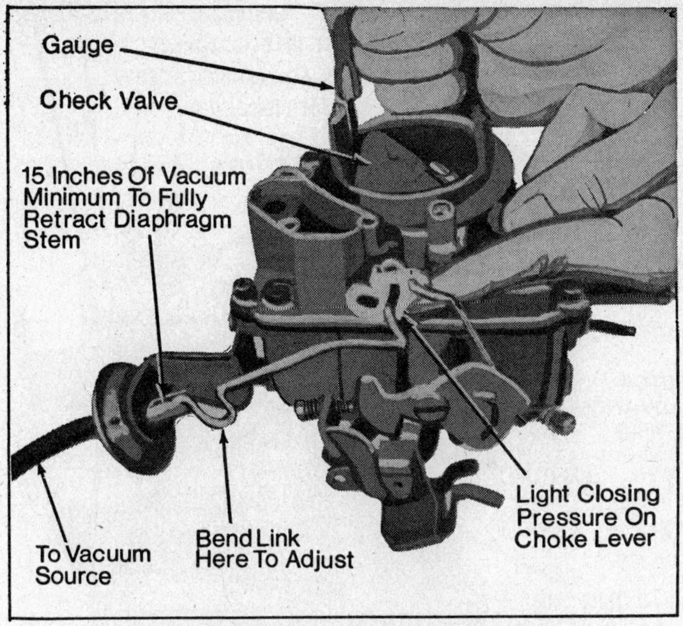

Fig. 4 Choke vacuum kick adjustment. 1945 carburetor

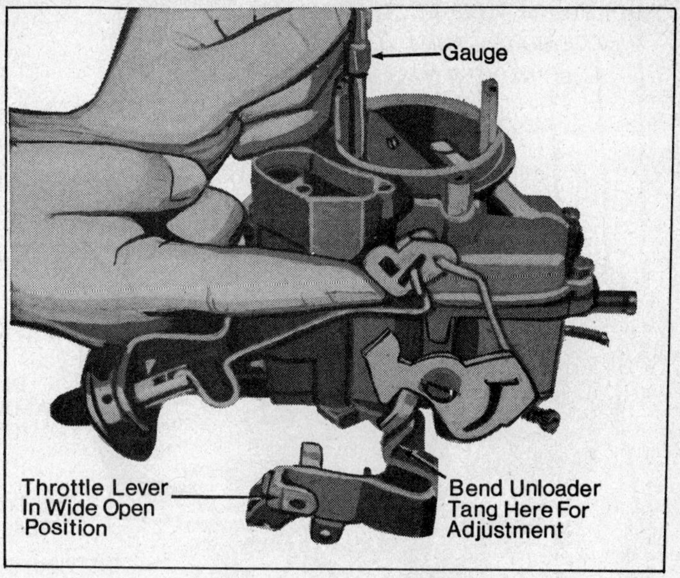

Fig. 5 Choke unloader adjustment. 1945 carburetor

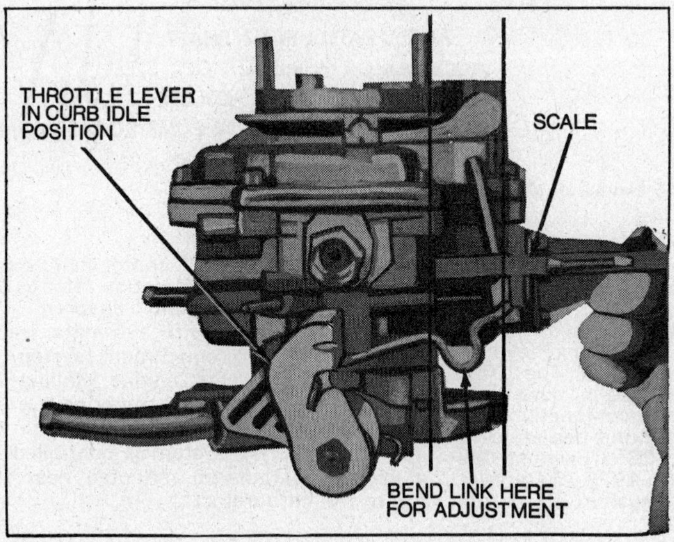

Fig. 6 Accelerator pump adjustment. 1945 carburetor

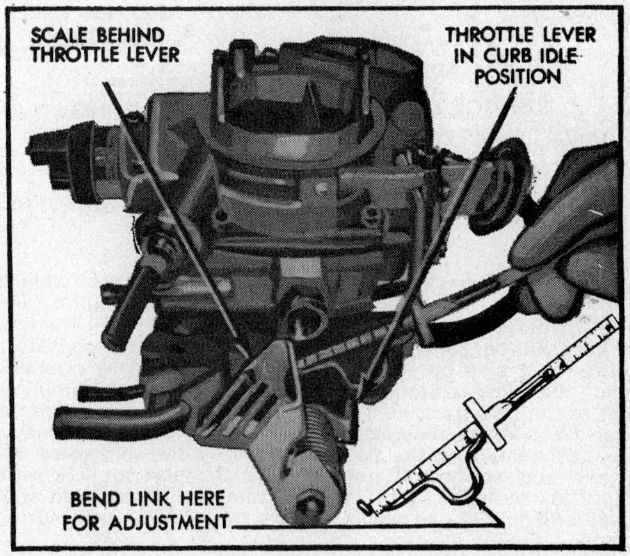

Fig. 7 Acclerator pump adjustment, 1975. 1945 carburetor

horn. Refer to *Holley Specifications Chart.* With a finger lightly pressing against shaft lever, a slight drag should be felt as drill is being withdrawn. Adjust by bending unloader tang on throttle lever until correct opening has been obtained.

Accelerator Pump Setting

Figs. 6 & 7—With throttle at curb idle position, on 1974 units, measure distance from vacuum passage casting to outer edge of hole in pump operating rod and on 1975 units, measure length of pump operating link. Refer to Holley Specifications Chart. Adjust by bending link between throttle lever and pump operating rod.

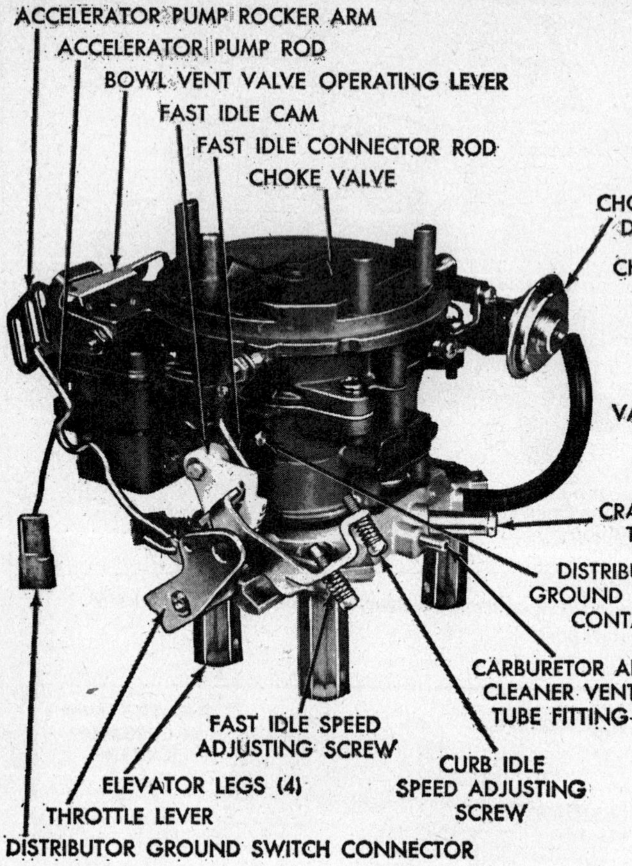

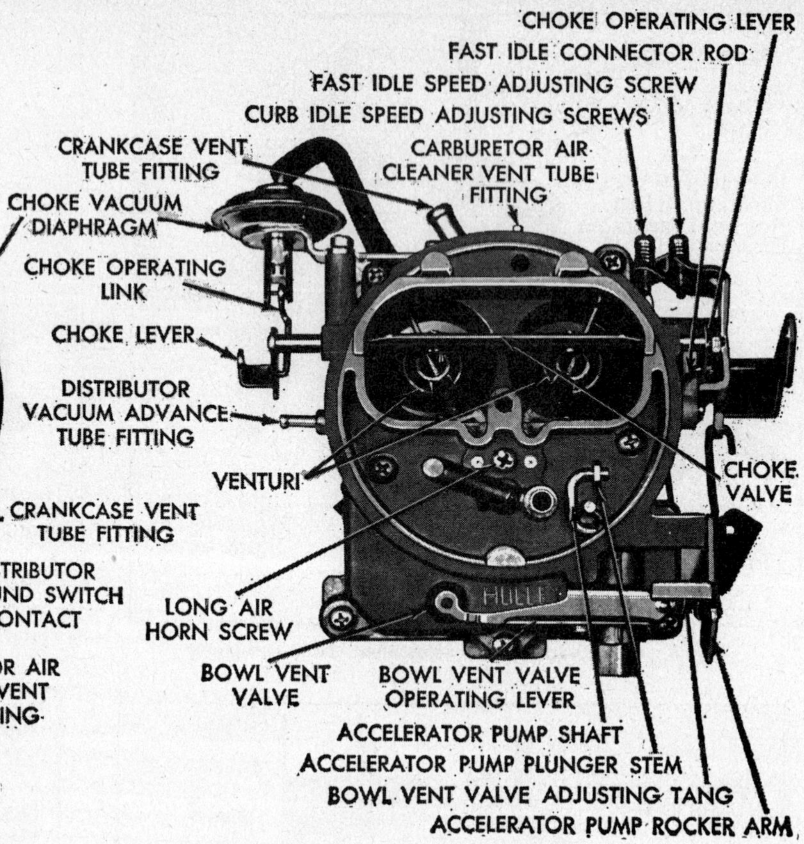

Fig. 1 Holley models 2210 & 2245 two-barrel typical carburetors

MODELS 2210 & 2245

This carburetor, Fig. 1, is a two-barrel unit but can be considered as two carburetors built side by side into one unit, utilizing the same fuel and air inlets. Each throat of the carburetor has its own throttle valve and main metering systems and are supplemented by the float, accelerating, idle and power systems. The 1970 version is equipped with a distributor ground switch, Fig. 1, which retards the distributor when the carburetor is at curb idle, resulting in better emission control. The 1971 version is equipped with a hot idle compensator valve which is a thermostatically operated air bleed to relieve an overrich condition at idle, Fig. 2. The 1971-75 units have incorporated a bowl vent valve tube which works in conjunction with the vent valve. The 1973 unit has an extra port for use with the (EGR) Exhaust Gas Recirculation system. On 1973-75 Chrysler units, an electric choke system is incorporated to open choke at approximately 60° to 63°. For service information refer to the "Emission Control System" chapter.

On 1975 vehicles with automatic transmissions, an idle enrichment system is used to reduce cold engine stalling by the use of an additional metering system which enrichens the mixture in the off idle position. This system is controlled by a vacuum diaphragm mounted near the top of the carburetor.

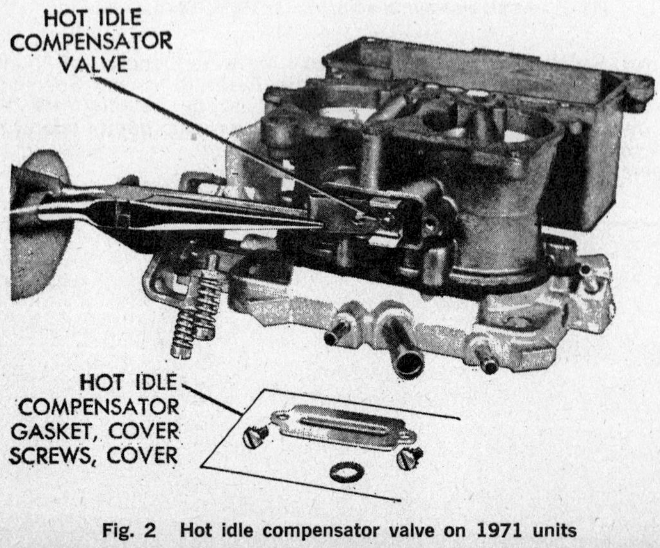

Fig. 2 Hot idle compensator valve on 1971 units

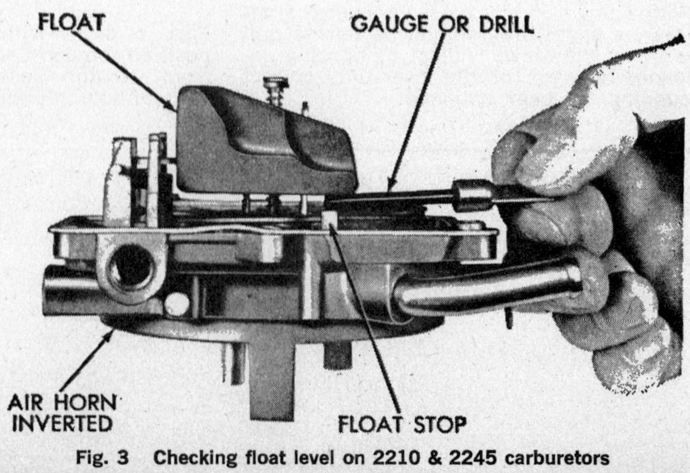

Fig. 3 Checking float level on 2210 & 2245 carburetors

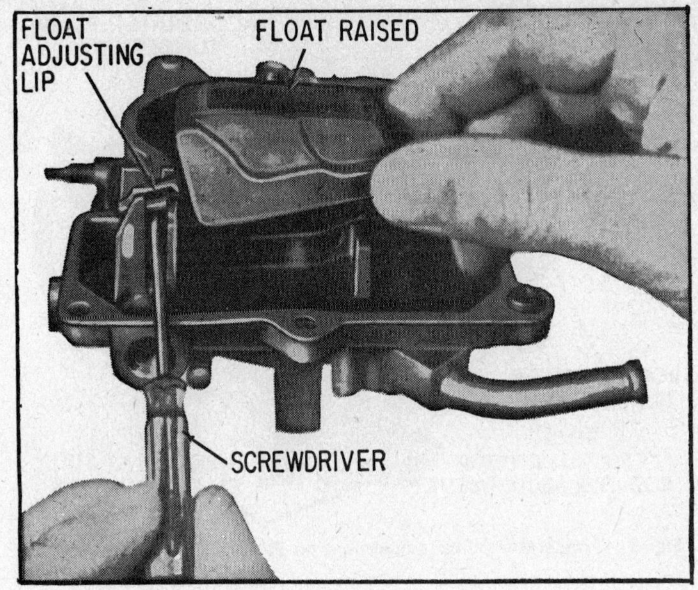

Fig. 4 Adjusting float on 2210 & 2245 carburetors

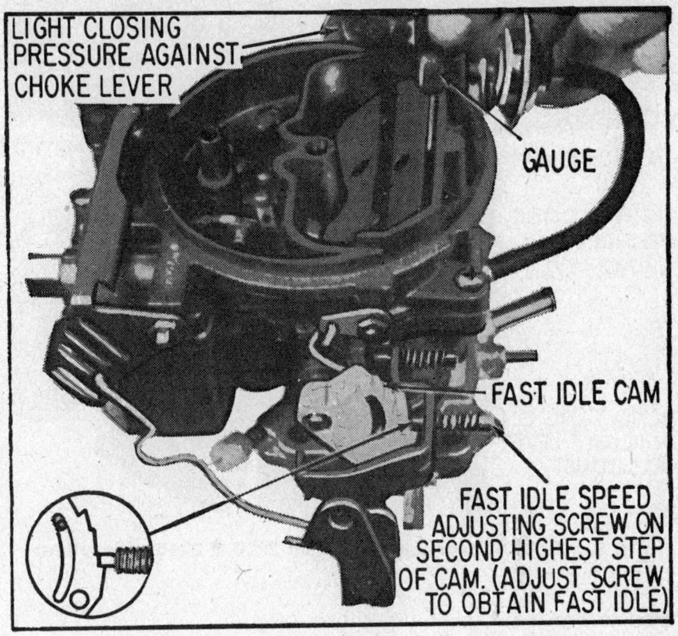

Fig. 6 Fast idle cam position adjustment on 2210 & 2245 carburetors

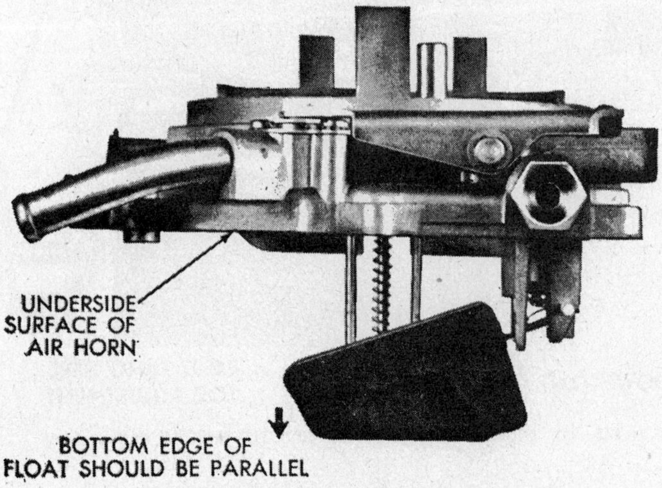

Fig. 5 Checking float drop on 2210 & 2245 carburetors

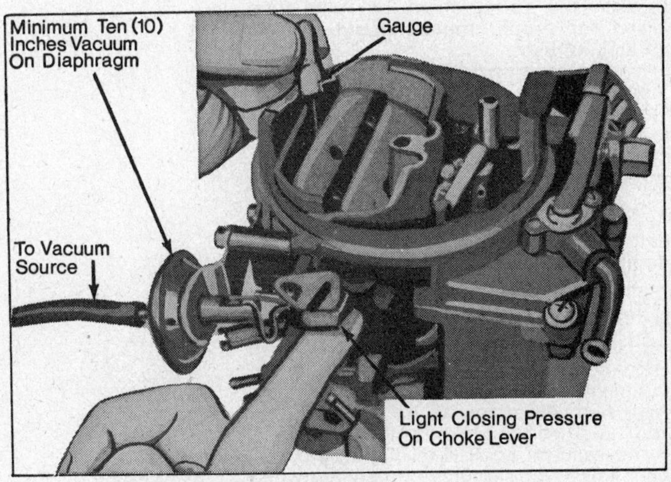

Fig. 7 Vacuum kick adjustment on 2210 & 2245 carburetors

Float Adjustment

Invert air horn so that weight of float only is forcing needle against seat. Measure the clearance between top of float and float stop, Fig. 3. Be sure drill gauge is perfectly level when measuring. Refer to *Holley Specifications Chart.* Adjust by bending float lip toward or away from needle, using a narrow blade screwdriver, Fig. 4, until correct clearance of setting has been obtained.

Float Drop Adjustment

Check float drop, by holding air horn in an upright position. The bottom edge of float should be parallel to underside surface of air horn, Fig. 5. Adjust by bending tang on float arm until parallel surfaces have been obtained.

Fast Idle Cam Position Adjustment

Fig. 6—With fast idle speed adjusting screw contacting second highest step on fast idle cam, move choke valve toward closed position with light pressure on choke shaft lever. Insert specified gauge between top of choke valve and wall of air horn. Refer to *Holley Specifications Chart.* An adjustment will be necessary if a slight drag is not obtained as drill shank is being removed. Adjust by bending fast idle link at angle, until correct valve opening has been obtained.

Choke Vacuum Kick Adjustment

NOTE: Test can be made on or off vehicle.

Fig. 7—If adjustment is to be made with engine running, back off fast idle speed screw until choke can be closed to the kick position with engine at curb idle. (Note number of screw turns required so that fast idle can be returned to original adjustment). If an auxiliary vacuum source is to be used, open throttle valve (engine not running) and move choke to closed position. Release throttle first, then release choke.

When using an auxiliary vacuum source, disconnect vacuum hose from carburetor and connect it to hose from vacuum supply with a small length of tube to act as a fitting. Removal of hose from diaphragm may require forces which damage the system. Apply a vacuum of 10 or more inches of mercury.

Insert gauge between top of choke valve and wall of air horn. Refer to *Holley Specifications Chart.* Apply sufficient closing pressure on lever to which choke rod at-

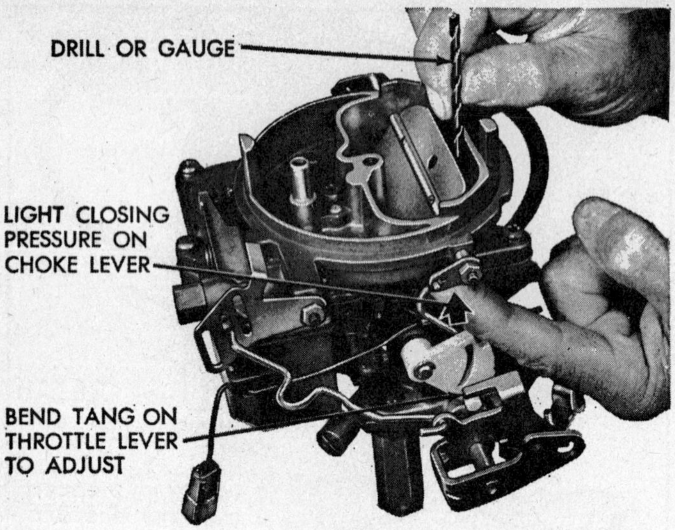

Fig. 8 Choke unloader adjustment on 2210 & 2245 carburetors

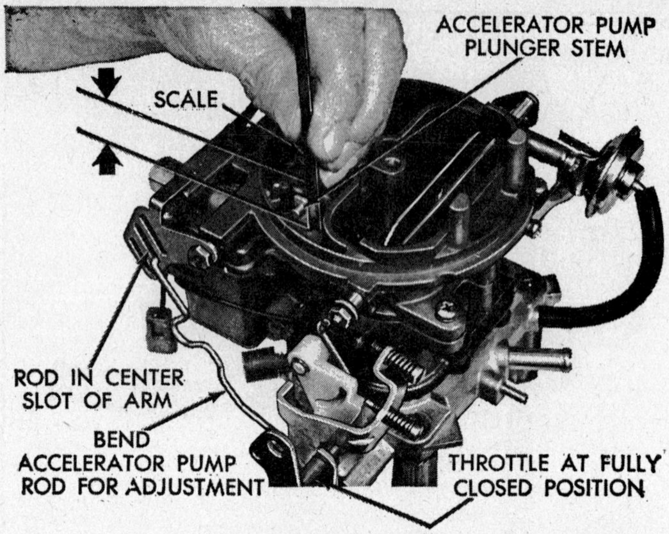

Fig. 9 Accelerator pump adjustment on 2210 & 2245 carburetors

taches to provide a minimum choke valve opening without distortion of diaphragm link.

NOTE: The cylindrical stem of diaphragm extends as the internal spring is compressed. This spring must be fully compressed for proper measurement of vacuum kick adjustment.

Adjustment is necessary if slight drag is not obtained when removing the gauge. Shorten or lengthen diaphragm link to obtain correct choke valve opening. Length changes should be made by carefully opening or closing the U-bend provided in the link. Improper bending causes contact between the U-section and the diaphragm assembly.

NOTE: Do not apply twisting or bending force to diaphragm.

After completion of adjustment, reinstall vacuum hose on correct carburetor fitting. Return fast idle screw to its original location if disturbed. Make following check. With no vacuum applied to diaphragm, the choke valve should move freely between open and closed positions. If movement is not free, examine linkage for misalignment or interferences caused by bending operation.

Choke Unloader (Wide Open Kick) Adjustment

Fig. 8—With throttle valves in wide open position, insert drill gauge between upper edge of choke valve and inner wall of air horn. Refer to *Holley Specifications Chart*. With a finger lightly pressing against shaft lever, a slight drag should be felt as drill is being withdrawn. Adjust by bending unloader tang on throttle lever until correct opening has been obtained.

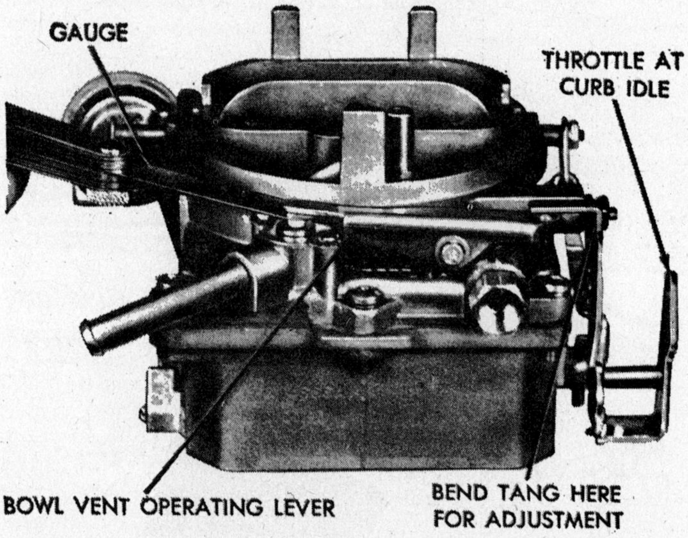

Fig. 10 Bowl vent adjustment on 2210 & 2245 carburetors

Accelerator Pump Adjustment

Fig. 9—Back off curb idle speed adjusting screw. Open choke valve so that fast idle cam allows throttle valves to be completely seated in bores. Be sure that pump connector rod is installed in correct slot of accelerator pump rocker arm. First slot for manual transmissions next to attaching screw.

Close throttle valves tightly. Measure the distance between top of air horn and end of plunger shaft. Refer to *Holley Specifications Chart*. Adjust pump travel by bending pump operating rod, at loop of rod, until correct setting has been obtained.

Bowl Vent Valve Clearance Adjustment

Fig. 10—With the throttle valves at curb idle, it should be possible to insert the specified gauge between the bowl vent valve plunger stem and operating rod. Refer to *Holley Specifications Chart*. Adjust by bending the tang on pump lever to change arc of contact with throttle lever, until correct clearance has been obtained.

MODEL 2300

The 2300 carburetor, Fig. 1 and 2, is used only as a compound installation. This system utilizes two types of Holley two-barrels, primary, Fig. 1, mounted in the center and the secondarys, Fig. 2, mounted fore and aft. The Fig. 2, units contain all the regulatory systems with the exception of chokes, power enrichment valve, accelerating pump, idle system and spark advance. The throttle operation of the primary carburetor is conventional whereas the secondary units are equipped with throttle control vacuum diaphragms, Fig. 2, for the purpose of opening the secondary throttles which closes mechanically. The choke used only on the primary unit is controlled by a temperature sensing choke coil mounted on the intake manifold, over the exhaust crossover passage, Fig. 3.

The only adjustments required on the secondary units are the float level and the wet fuel level. All other adjustments are made on the primary unit.

Float Adjustment

Make a preliminary float adjustment by inverting fuel bowl and turn adjustable needle and seat until top of float is specified distance from top of fuel bowl, Fig. 4. Refer to *Holley Specifications Chart.* Do not fully tighten lock screw. Snug screw to temporarily retain adjustment.

NOTE: Final adjustment of the float is made on the vehicle.

Wet Fuel Level

Fig. 5—With car on a level floor and engine idling, remove sight plug from fuel bowl. Fuel level should be in line with threads at bottom of sight plug hole. To adjust, loosen lock screw and turn adjusting nut as required to reise or lower fuel level.

Automatic Choke Control Lever Setting

Adjustment of the choke control lever is necessary to provide correct relationship between choke valve, thermostatic coil spring and the fast idle cam. It should be checked and adjusted (if necessary), as preparation of the choke system linkage before making the Vacuum Kick, Cam Position or Unloader adjustment. These three adjustments must be made after qualification of the choke control lever.

NOTE: Improper bending of choke rod will result in binding.

Chevrolet

Close the choke rod by applying slight pressure on choke control lever, the thermostatic choke rod should be even with top of choke rod hole. Adjust by bending choke rod at upper angle.

Chrysler

Open the throttle to mid-position. Close the choke valve by applying slight pressure on choke control lever. The top of choke rod hole in control lever should be 3-49/64" above choke pad with carburetor on engine, or 1-23/32" above carburetor base with carburetor on bench, Fig. 6. Adjust by bending choke shaft rod at point indicated.

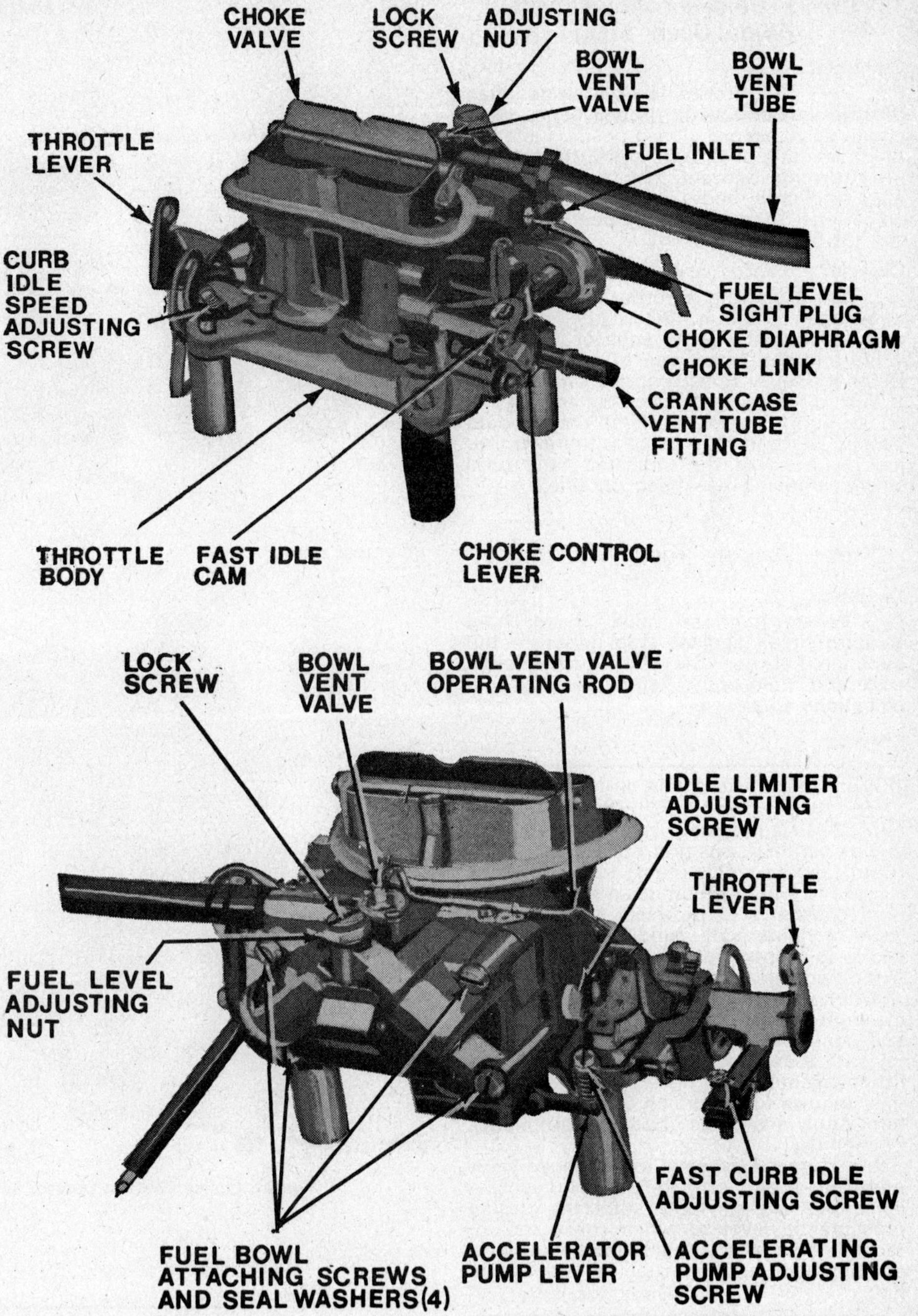

Fig. 1 Holley 2300 carburetor assembly primary center unit

Fast Idle Cam Position Adjustment

Chevrolet

Fig. 7—With throttle slightly open, close choke plate positioning fast idle lever against top step of fast idle cam. Adjust the fast idle screw to obtain the clearance listed in the *Holley Specifications Chart* between throttle valve and bore on idle transfer slot side of carburetor. Adjust by bending the idle lever.

Chrysler

Fig. 8—With fast idle speed adjusting screw contacting second highest step on fast idle cam, move choke valve toward closed position with light pressure on choke shaft lever. Insert specified gauge between top of choke valve and wall of air horn. Refer to *Holley Specifications Chart.* An adjustment will be necessary if a slight drag is not obtained as drill shank is being removed. Adjust by bending cam position adjusting tang.

Choke Unloader Adjustment (Wide Open Kick)

Chevrolet

Fig. 9—Hold throttle lever in wide open throttle position with rubber band. Hold choke valve toward closed position against unloader tang of throttle shaft, then measure opening between choke valve lower edge and main body. Refer to *Holley Specifications Chart*. Adjust by bending choke rod (at off-set bend).

Chrysler

Fig. 10—Hold the throttle valves in the wide open position. Insert the specified drill between the upper edge of the choke valve and the inner wall of the air horn. Refer to *Holley Specifications Chart*.

With a finger lightly pressing against the choke control lever, a slight drag should be felt as the drill is being withdrawn. Adjust by bending the indicated tang until correct opening has been obtained.

Choke Vacuum Kick Adjustment

Chevrolet

Fig. 11—With choke valve closed, hold vacuum break against stop. Measure the distance between the choke valve lower edge and main body. Refer to *Holley Specifications Chart*.

Chrysler

NOTE: Test can be made on or off vehicle.

Fig. 12—If adjustment is to be made with engine running, position the fast idle tang (Cam position adjustment) to allow choke closure to kick position. If an auxiliary vacuum source is to be used, open throttle valve (engine not running) and move choke to closed position. Release throttle first, then release choke.

When using an auxiliary vacuum source, disconnect vacuum hose from carburetor and connect it to hose from vacuum supply with a small length of tube to act as a fitting. Removal of hose from diaphragm may require forces which damage the system. Apply a vacuum of 10 or more inches of mercury.

Insert gauge between top of choke valve and wall of air horn. Refer to *Holley Specifications Chart*. Apply sufficient closing pressure on lever to which choke rod attaches to provide a minimum choke valve opening without distortion of diaphragm link.

NOTE: The cylindrical stem of diaphragm extends as the internal spring is compressed. This spring must be fully compressed for proper measurement of vacuum kick adjustment.

Adjustment is necessary if slight drag is not obtained when removing the gauge. Shorten or lengthen diaphragm link to obtain correct choke valve opening. Length changes should be made by carefully opening or closing the U-bend provided in the link. Improper bending causes contact between the U-section and the diaphragm assembly.

NOTE: Do not apply twisting or bending force to diaphragm.

After completion of adjustment, reinstall

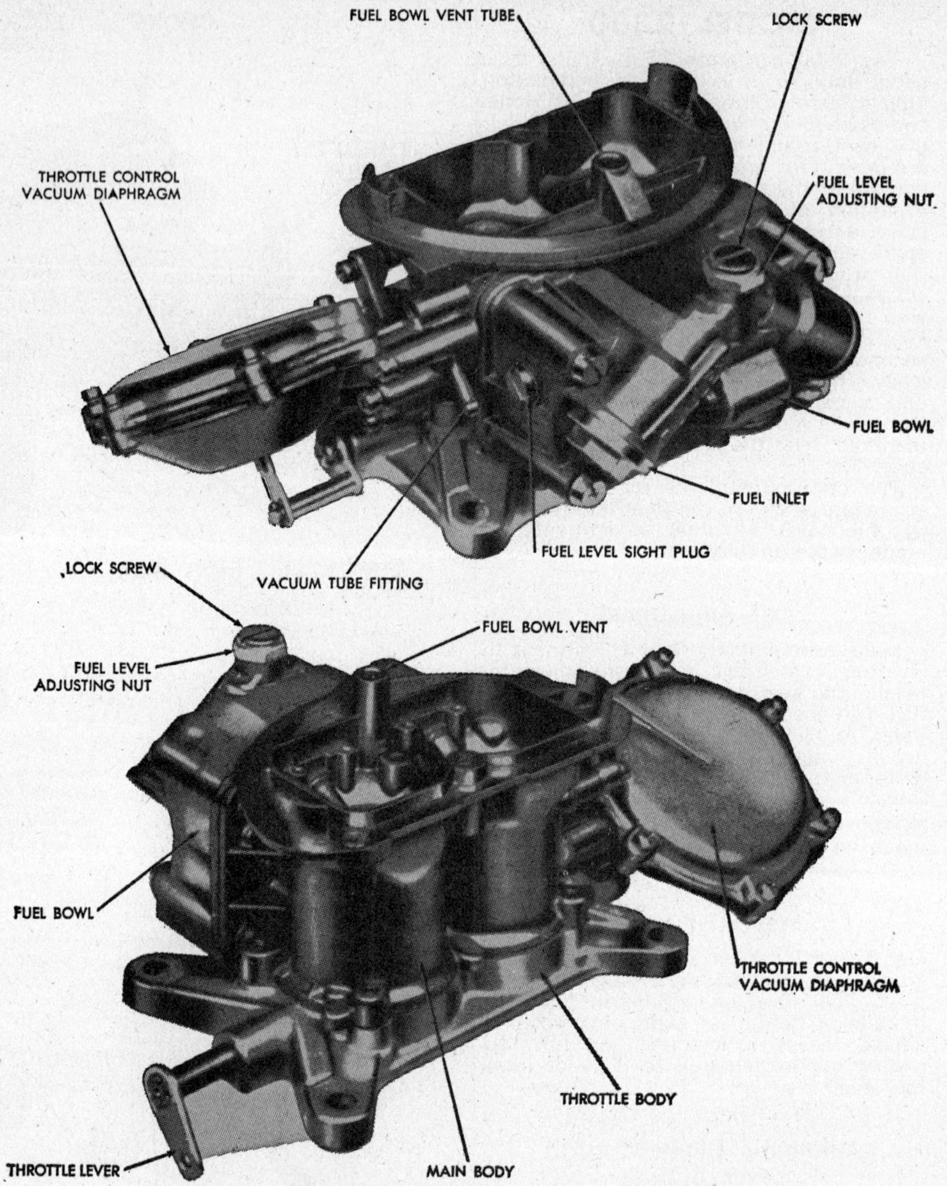

Fig. 2 Holley 2300 carburetor assembly secondary outboard front or rear

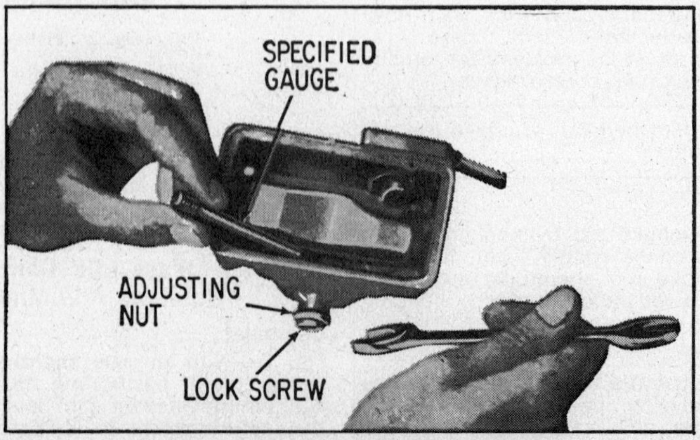

Fig. 4 Float adjustment on 2300 carburetor

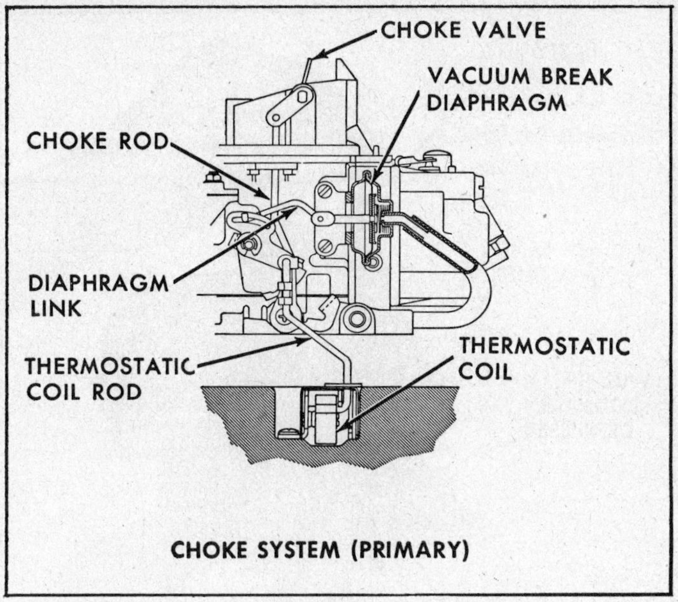

Fig. 3 Well type choke. 2300 carburetor

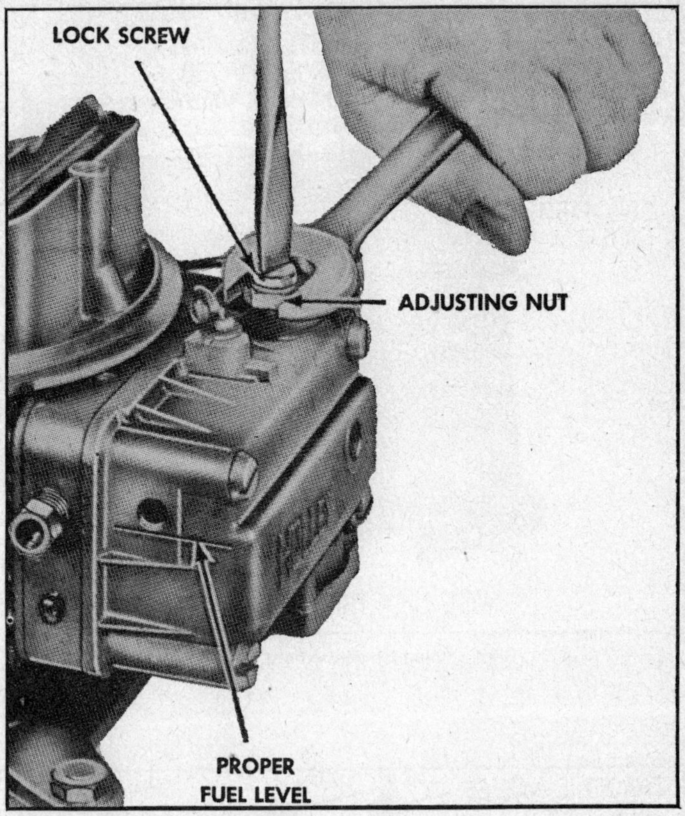

Fig. 5 Fuel level adjustment on 2300 carburetor

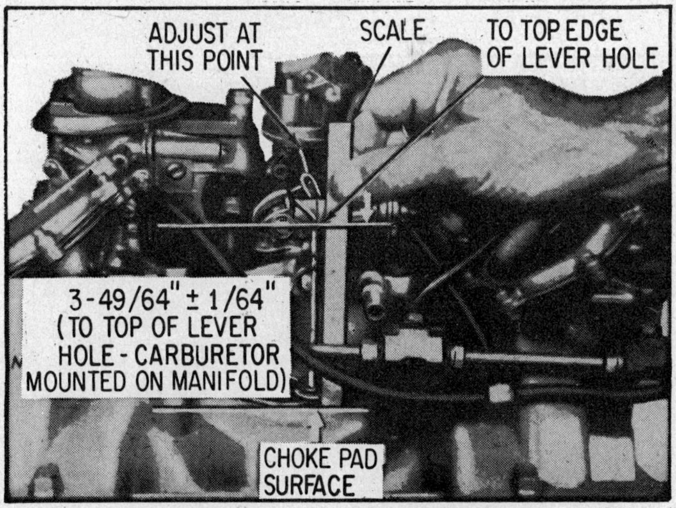

Fig. 6 Automatic choke control lever adjustment, Chrysler 2300

vacuum hose on correct carburetor fitting. Return fast idle screw to its original location if disturbed. Make following check. With no vacuum applied to diaphragm, the choke valve should move freely between open and closed positions. If movement is not free, examine linkage for misalignment or interferences caused by bending operation.

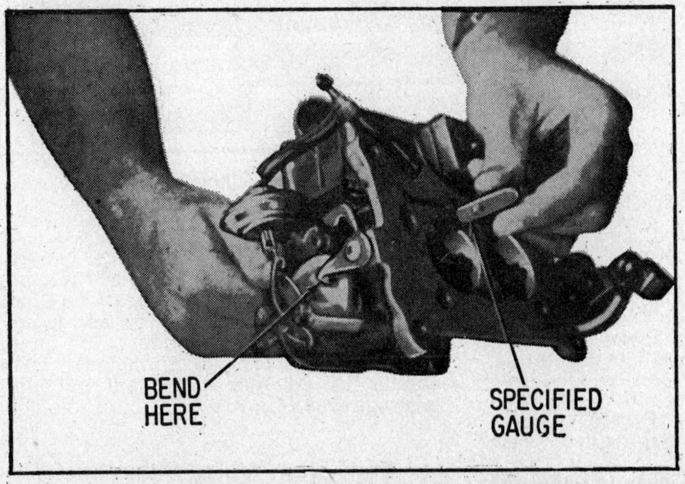

Fig. 7 Fast idle cam adjustment, Chevrolet 2300

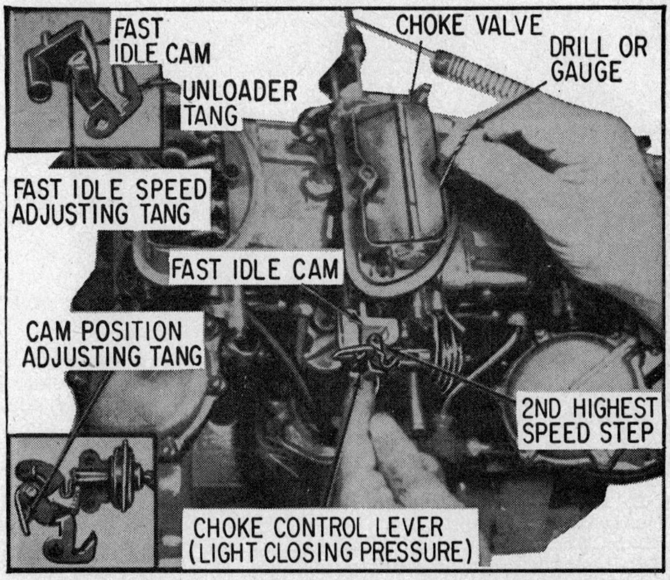

Fig. 8 Fast idle cam and speed adjustment, Chrysler 2300

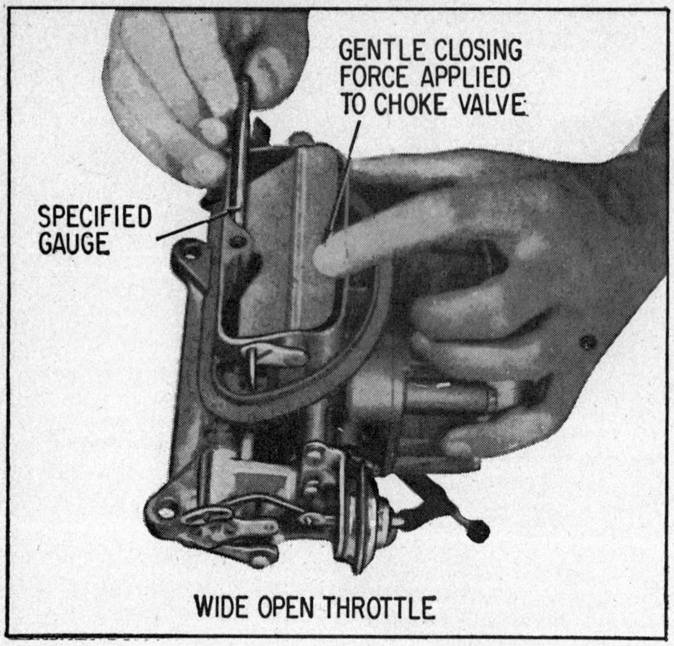

Fig. 9 Choke unloader adjustment, Chevrolet 2300

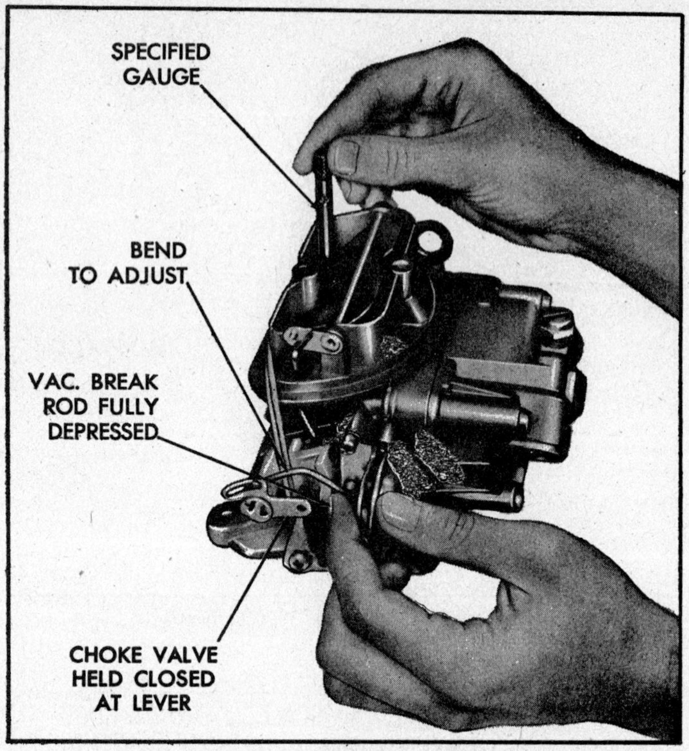

Fig. 11 Vacuum kick adjustment, Chevrolet 2300

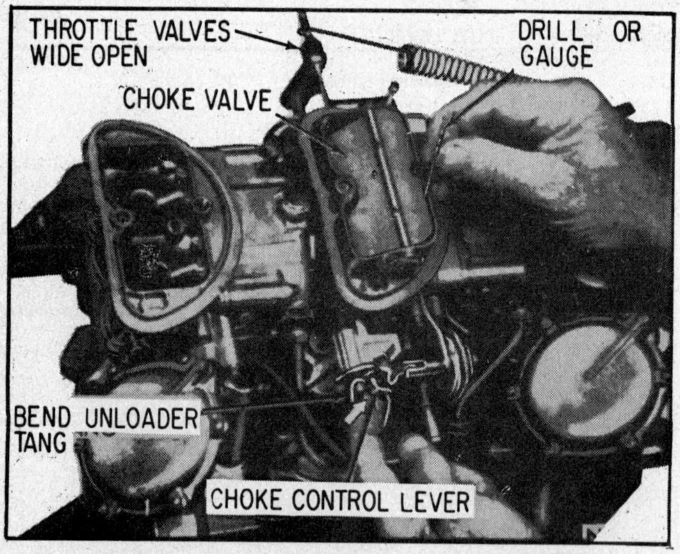

Fig. 10 Choke unloader adjustment, Chrysler 2300

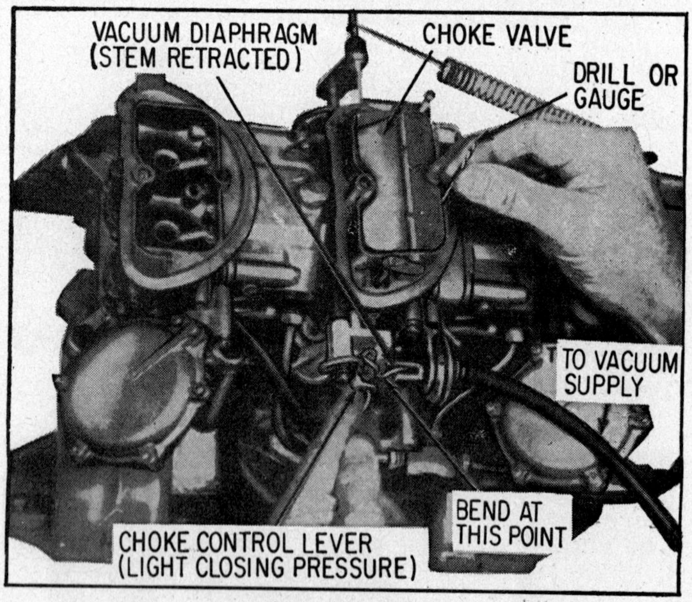

Fig. 12 Vacuum kick adjustment, Chrysler 2300

Fast Idle Adjustment (On Vehicle)

Chrysler

Fig. 8—Close choke valve until fast idle screw tang can be positioned on the second highest-speed step of the fast idle cam. Start the engine and determine the stabilized speed. Bend the fast idle tang by use of a screwdriver placed in the tang slot to secure the specified speed. Refer to *Holley Specifications Chart.*

NOTE: Bend only in a direction perpendicular to the contact surface of the cam. Movement in any other direction changes the cam position adjustment. Bend only when tang is clear of cam. Stopping the engine between adjustments is not necessary. However, reposition the fast idle tang on the cam after each speed adjustment to provide correct throttle closing torque.

Accelerator Pump Lever Adjustment

Fig. 13—With throttle lever in wide open position and pump lever fully compressed (down). Measure the clearance between spring adjusting nut and arm of the pump lever. Refer to *Holley Specifications Chart.* Adjust by turning nut or screw as required while holding opposite end. (The pump operating lever is not threaded). There should be no free movement of pump leverage when throttle is at curb idle.

Bowl Vent Valve Adjustment

Fig. 14—With throttle valves at fast curb

Fig. 13 Adjusting accelerator pump lever clearance

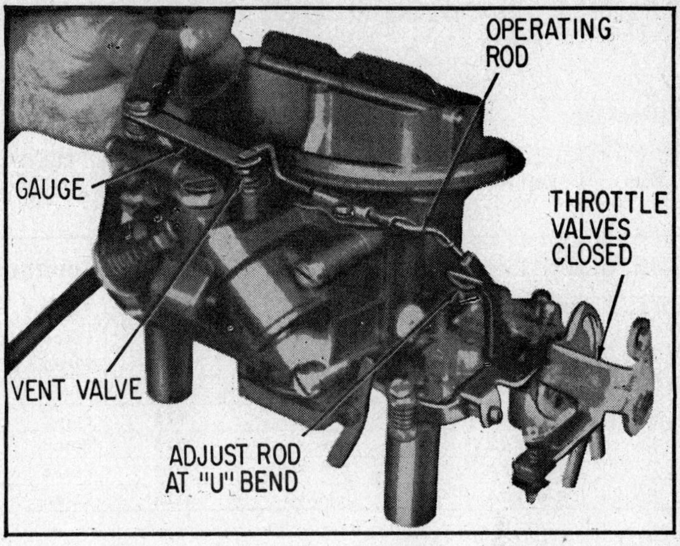

Fig. 14 Adjusting bowl vent valve clearance

idle, insert drill or gauge between the bowl vent valve and the bowl vent rod with the fast curb idle speed properly set. Refer to *Holley Specifications Chart*. Adjust by bending rod to change arc of contact with throttle lever, until correct clearance has been obtained.

4150 & 4160 CARB. ADJUSTMENT SPECIFICATIONS

See Tune Up Chart in car chapters for hot idle speeds.

Year	Carb. Part No. (1)	Float Level (Dry)	Fuel Level (Wet)	Pump Lever Clearance	Choke Setting	Choke Unloader Clearance	Bowl Vent Clearance	Fast Idle Bench	Fast Idle On Car	Choke Vacuum Break	Dashpot Setting
CHEVROLET ENGINES											
1969–70	R4053-A	(10)	(4)	.015	(13)	.350	—	.025	2200	.300	—
	R4296-A	(10)	(4)	.015	(13)	.350	—	.025	2200	.350	—
	R4346	(10)	(4)	.015	(13)	.350	—	.025	2200	.300	—
	R4492-A	(10)	(4)	.015	(13)	.350	—	.025	2200	.350	—
	R4557-A	(10)	(4)	.015	(13)	.350	—	.025	2200	.350	—
1971	R4800-A	(2)	(4)	.015	1.320(5)	.350	—	.025	2200	.350	—
	R4801-A	(2)	(4)	.015	1.320(5)	.350	—	.025	2200	.350	—
	R4802-A	(2)	(4)	.015	1.320(5)	.350	—	.025	2200	.350	—
	R4803-A	(2)	(4)	.015	1.320(5)	.350	—	.025	2200	.350	—
1972	R6238-A	(2)	(4)	.015	1.320(5)	.350	—	.025	2350	.350	—
	R6239-A	(2)	(4)	.015	1.320(5)	.350	—	.025	2350	.350	—
CHRYSLER, IMPERIAL, DODGE, PLYMOUTH											
1969	R-4166-A	(11)	(9)	.015	See Text	5/32	5/64	46 Drill	1400(8)	—	—
	R-4440-A	(11)	(9)	.015	See Text	5/32	5/64	46 Drill	1500(14)	—	—
1970–71	R4360-A	(11)	(9)	.015	2 Rich	25 Drill	72 Drill	53 Drill	1600	46 Drill	—
	R4366-A	(11)	(9)	.015	2 Rich	25 Drill	5/64	53 Drill	1600	46 Drill	—
1971	R4668-A	(11)	(9)	.015	2 Rich	25 Drill	.015	53 Drill	1700	46 Drill	—
	R4735-A	(11)	(9)	.015	2 Rich	25 Drill	.015	53 Drill	1700	46 Drill	—
	R6191-A	(11)	(9)	.015	2 Rich	25 Drill	.015	53 Drill	1800	18 Drill	—
	R6193-A	(11)	(9)	.015	2 Rich	25 Drill	.015	53 Drill	1800	18 Drill	—

Continued

4150 & 4160 CARB. ADJUSTMENT SPECIFICATIONS—Continued

See Tune Up Chart in car chapters for hot idle speeds.

CHRYSLER, IMPERIAL, DODGE, PLYMOUTH—Continued

Year	Carb. Part No. ①	Float Level (Dry)	Fuel Level (Wet)	Pump Lever Clearance	Choke Setting	Choke Unloader Clearance	Bowl Vent Clearance	Fast Idle		Choke Vacuum Break	Dashpot Setting
								Bench	On Car		
1972	R-6160A	⑦	⑨	.015	Fixed	.150	.015	.060	1600	.080	—
	R-6252A	⑦	⑨	.015	Fixed	.150	.015	.060	1800	.140	—
	R-6253A	⑦	⑨	.015	Fixed	.150	.015	.060	1600	.080	—
	R-6254A	⑦	⑨	.015	Fixed	.150	.015	.060	1800	.140	—
	R-6255A	⑦	⑨	.015	Fixed	.150	.015	.060	1600	.080	—
	R-6256A	⑦	⑨	.015	Fixed	.150	.015	.060	2000	.140	—
	R-6257A	⑦	⑨	.015	Fixed	.150	.015	.060	1800	.080	—
	R-6290A	⑦	⑨	.015	Fixed	.150	.015	.060	1500	.080	—

FORD ENGINES

Year	Carb. Part No. ①	Float Level (Dry)	Fuel Level (Wet)	Pump Lever Clearance	Choke Setting	Choke Unloader Clearance	Bowl Vent Clearance	Fast Idle Bench	Fast Idle On Car	Choke Vacuum Break	Dashpot Setting
1969	C9AF-M	See Text	④	No. 2⑮	2 Rich	—	.060–.090	.060	1350⑫	—	.100
	C9AF-N	See Text	④	No. 2⑮	1 Rich	—	.060–.090	.080	1550⑫	—	.100
	C9OF-H	See Text	④	No. 2⑮	1 Rich	—	.060–.090	.080	1550⑫	—	.100
1970	D0OF-N	See Text	④	—	2 Rich	—	—	—	2200	—	—
	D0OF-R	See Text	④	—	2 Rich	—	—	—	2400	—	—
	D0OF-S	See Text	④	—	—	—	—	—	—	—	—
	D0ZF-AA	See Text	④	.015	—	—	—	—	1900	—	.140
	D0ZF-AB	See Text	④	.015	—	—	—	—	2100	—	.200
	D0ZF-AC	See Text	④	.015	—	—	—	—	1900	—	—
	D0ZF-AD	See Text	④	.015	—	—	—	—	2100	—	—
	D0ZF-Z	See Text	④	—	—	—	—	—	—	—	—
1971	D1ZF-VA	See Text	④	.015	—	—	—	—	2100	—	1/8
	D1ZF-XA	See Text	④	.015	2 Rich	19/64	—	—	2400	—	—
	D1ZF-YA	See Text	④	.015	2 Rich	19/64	—	—	2200	—	7/64

①—Located on tag attached to carburetor, on casting or on choke plate flange.
②—Float centered in bowl.
③—Primary .350″, secondary .450″.
④—Use sight plug hole in fuel bowl as outlined in text.
⑤—Bottom of throttle body to center of hole in operating lever.
⑥—With lever touching screw, then tighten 1/4 turn more.
⑦—Primary .110″, secondary .204″.
⑧—No. 5 Step on cam.
⑨—Primary 9/16″, secondary 13/16″.
⑩—Primary .350″, secondary .500″.
⑪—Primary 15/64″, secondary 17/64″.
⑫—On top step of cam.
⑬—Top of rod even with bottom of hole.
⑭—No. 2 Step on cam.
⑮—Hole setting.

MODELS 4150, 4160 ADJUSTMENTS

These carburetors, Figs. 1, 2 and 3 are four barrel units whose primary sides contain all the basic systems that make up a complete carburetor. The secondary sides of these units contain a fuel transfer and by-pass system which operates when a greater quantity of fuel-air mixture is required.

As shown in Fig. 1, some late model 4150 units have a central fuel inlet whereas other 4150 and 4160 units have a side inlet.

As shown in Figs. 1 and 4, fuel inlet systems vary slightly between models. In some units, the floats are hinged in the center while in others they are hinged at the float end.

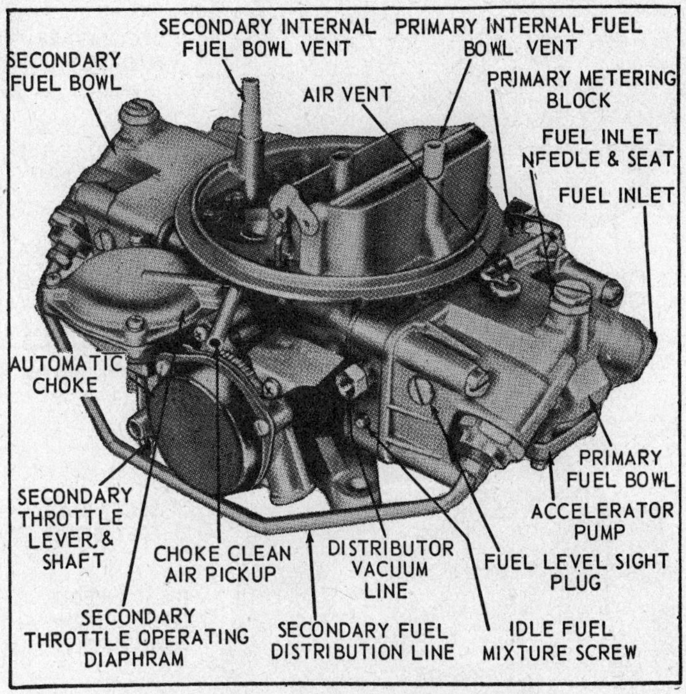

Fig. 4 Holley 4150 carburetor featuring an external fuel distributor tube connecting primary and secondary fuel inlets. These units also feature a choke clean air pick-up tube

Dry Float Setting

Except Fig. 4 Units—Invert fuel bowl. Loosen lock screw enough to allow adjusting nut to rotate. Turn adjusting nut to proper setting:

On Ford units, adjust float parallel to bowl floor, Fig. 5.

On Chevrolet & Chrysler units, whether the float is hinged at the center or the end, set level to obtain specified dimension between float and bowl with bowl inverted, Fig. 5.

For Fig. 4 Units Only—Referring to Fig. 6, adjust float so that its center is an equal distance from top and bottom of fuel bowl with fuel bowl inverted. After carburetor is installed on engine, check and adjust fuel level.

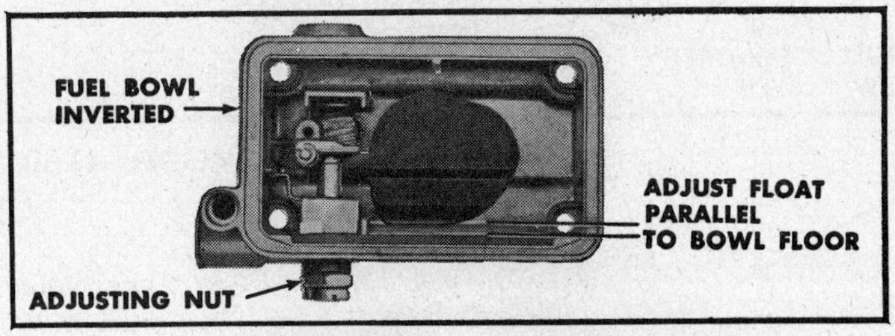

Fig. 5 Float adjustments except Fig. 4 type Ford units. On Chevrolet and Chrysler, adjust to specified dimension

Fuel Level Adjustment

Fig. 7—With car on a level floor and engine idling, fuel level should be at the level with the threads on the bottom of the sight plug hole (plus or minus 1/16"). To adjust, loosen lock screw and turn adjusting nut as required.

Secondary Throttle Plate Adjustment

Late Model Units—Referring to Fig. 8, back the secondary throttle stop screw out until secondary throttle plates are closed in the bores. Turn screw in (clockwise) until it touches stop on secondary throttle lever; then turn it in an additional ½ turn.

Accelerating Pump Adjustment

Late Model Units—Using a feeler gauge and with primary throttle plates in wide open position, there should be the clearance listed in the *Holley Specifications Chart* between pump operating lever adjusting screw and pump arm when pump arm is fully depressed manually, Fig. 9. Turn adjusting screw in to increase (out to decrease) clearance. One-half turn of adjusting screw is equal to .015".

To satisfy acceleration requirements in various climates, the pump discharge can be adjusted, Fig. 9. The bottom hole (No. 2) in the cam provides a maximum pump discharge for extreme cold weather and the top hole (No. 1) provides the minimum pump discharge for warm weather operation.

Late Model Units—Referring to Fig. 10, the pump cam screw should be in the No. 1 position on the throttle lever for all normal operating conditions. For extreme cold weather the No. 2 position can be used to provide maximum pump discharge.

With throttle plates held in wide open position, there should be the clearance listed in the *Holley Specifications Chart* between the pump diaphragm actuating lever and the lower portion of the pump override spring screw.

This adjustment *must* be rechecked with the throttle plates closed to make certain that there is no lag between throttle linkage and pump lever. The slightest movement of the throttle lever must correspondingly actuate the pump lever. Should there be any lag, a stumble or flat spot will result. To eliminate the lag, lengthen the adjusting screw.

Bowl Vent Valve Adjustment

Fig. 11—The fuel bowl vent valve clearance must be adjusted whenever the accelerator pump lever and/or pump stroke

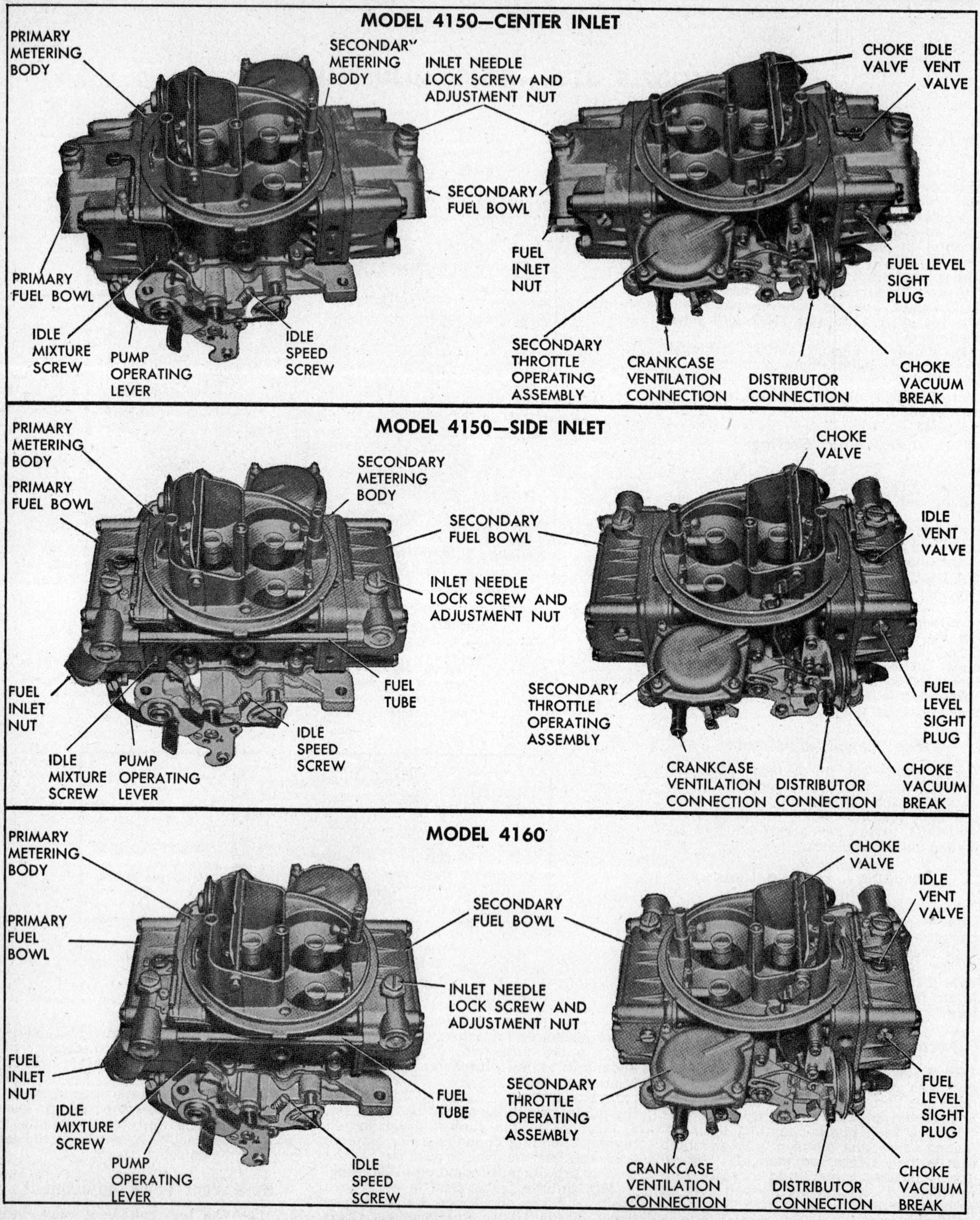

Fig. 1 Holley Models 4150 and 4160 late model four-barrel carburetors

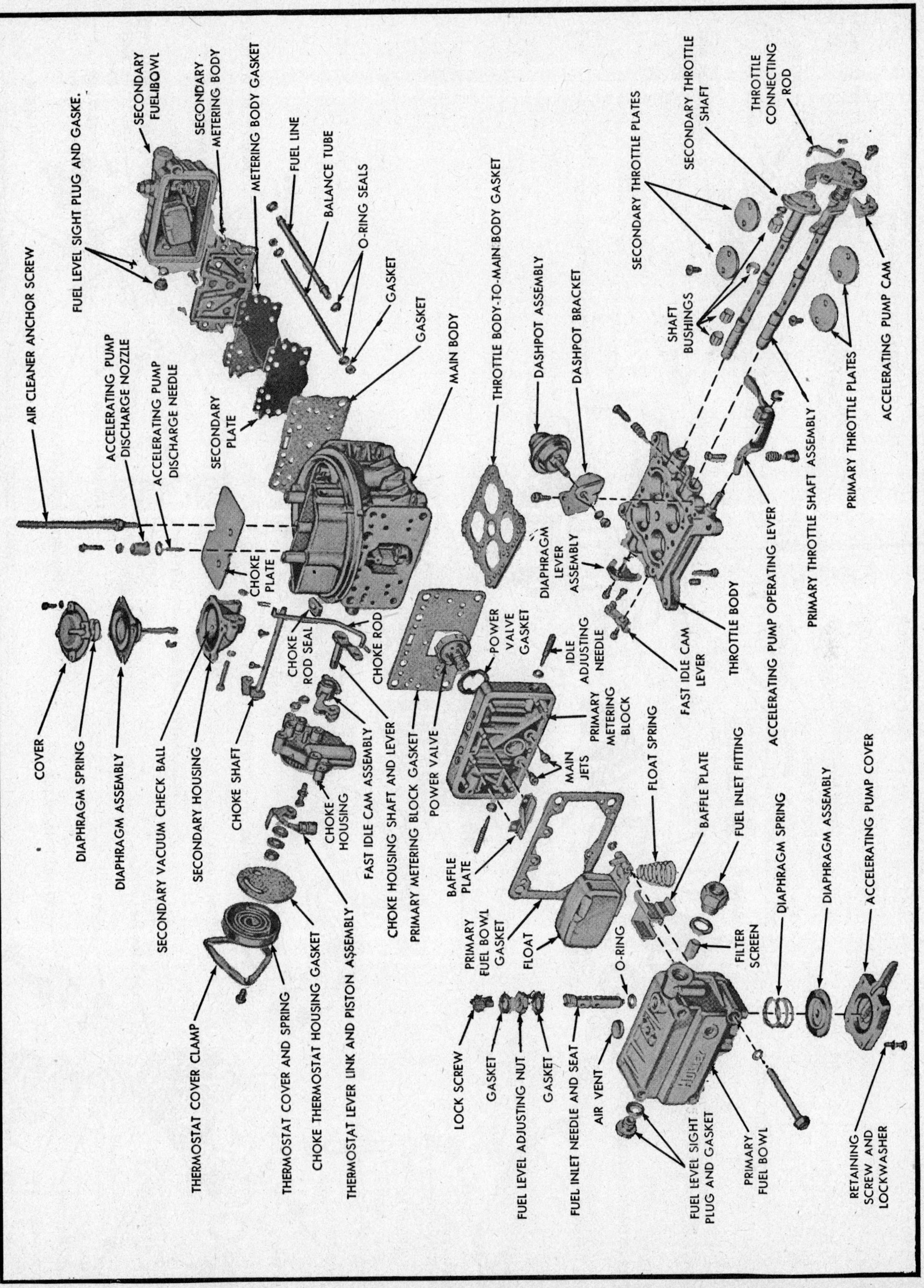

Fig. 2 Exploded view of a typical side fuel inlet 4150 and 4160 carburetor

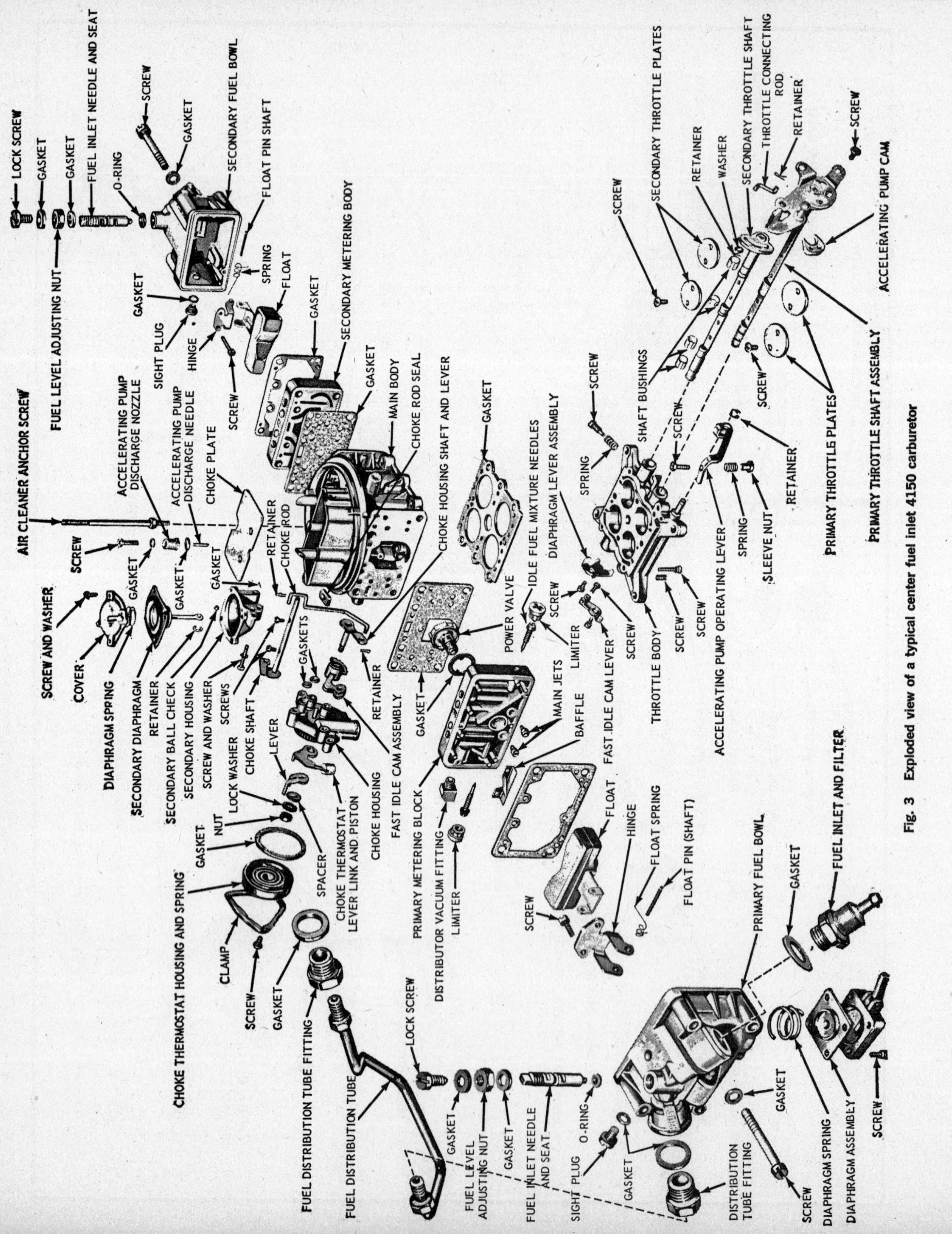

Fig. 3 Exploded view of a typical center fuel inlet 4150 carburetor

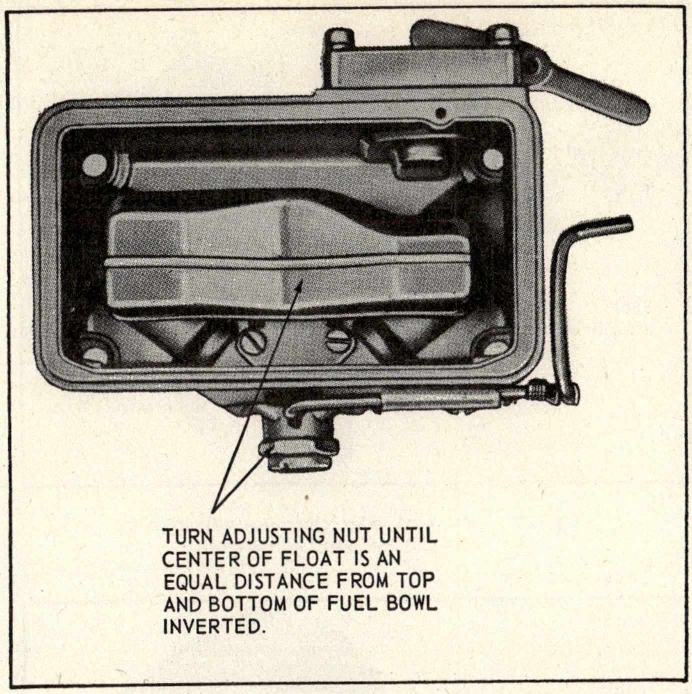

Fig. 6 Float adjustment on Fig. 4 type Ford units

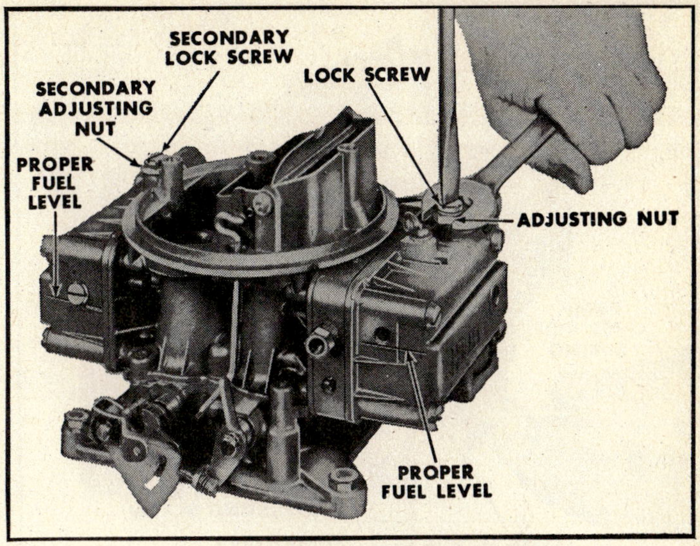

Fig. 7 Float level adjustment on all 4150 and 4160 carburetors

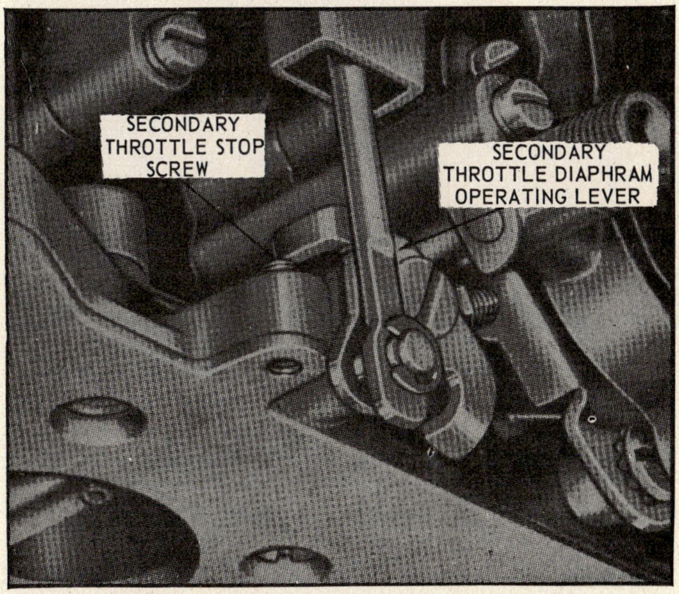

Fig. 8 Secondary throttle plate adjustment on Fig. 11 type carburetors

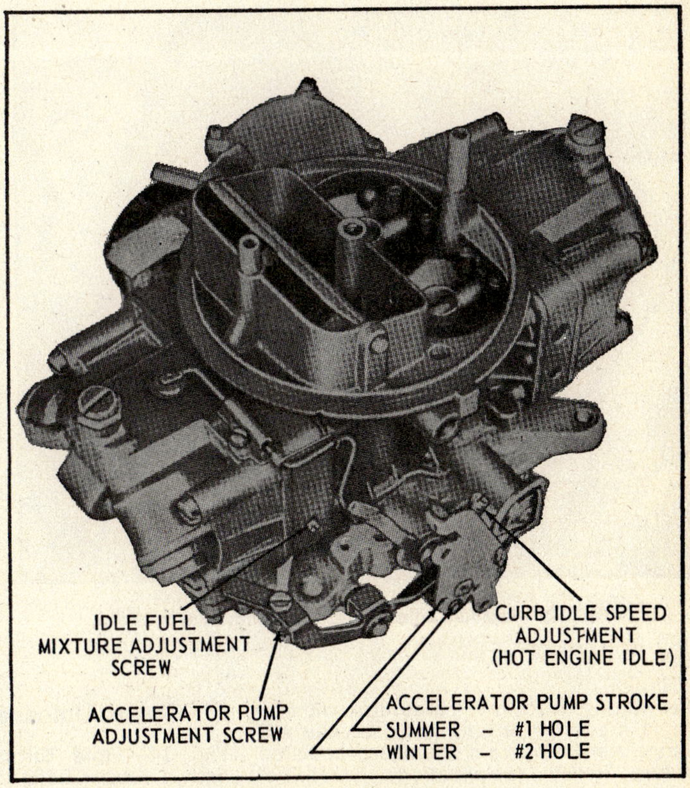

Fig. 9 Carburetor adjustment points on late type carburetors

adjustments have been changed. A change in the pump adjustments will affect the fuel bowl vent valve clearance.

With engine temperature stabilized and engine operating at curb idle speed, check clearance between bottom of vent valve and top of fuel bowl at vent opening. If clearance is not as listed in the *Holley Specifications Chart*, bend vent rod as required.

Fast Idle Speed

Units With Built-in Choke—With hot idle speed properly adjusted, set fast idle speed with transmission in neutral, engine at normal operating temperature, tachometer attached, headlamps turned on, and air conditioner operating (if so equipped). Referring to Fig. 12, align kickdown step of fast idle cam with the adjusting screw, and turn adjusting screw to obtain the specified rpm listed in the *Holley Specifications Chart*.

Units With Well Type Choke—With fast idle lever on high step of cam and choke valve wide open (engine warm) set fast idle to give the rpm listed in *Holley Specifications Chart*. Adjust fast idle screw on 4150 units or bend fast idle lever on 4160 units.

Fast Idle Cam Setting

Units With Well Type Choke—The fast

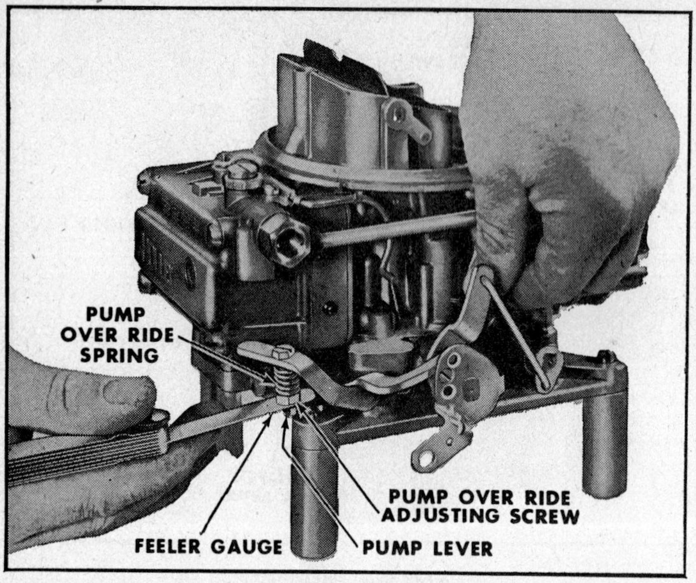

Fig. 10 Pump adjustment on Fig. 9 type carburetors

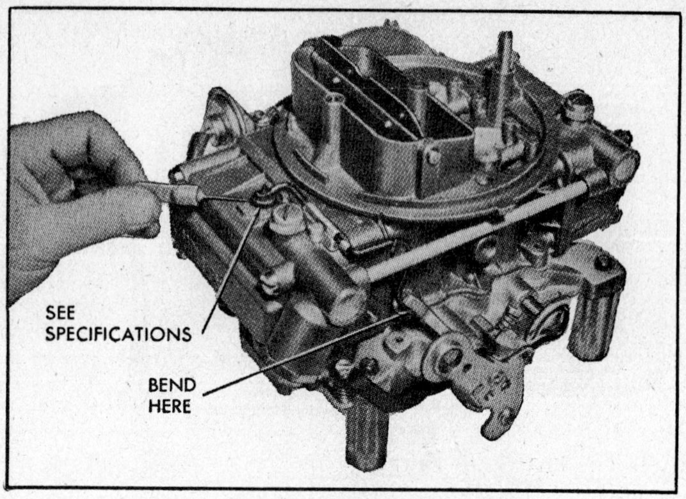

Fig. 11 Bowl vent valve clearance adjustment on 4150 and 4160 carburetors

Fig. 12 Fast idle adjustment. Units with built-in choke

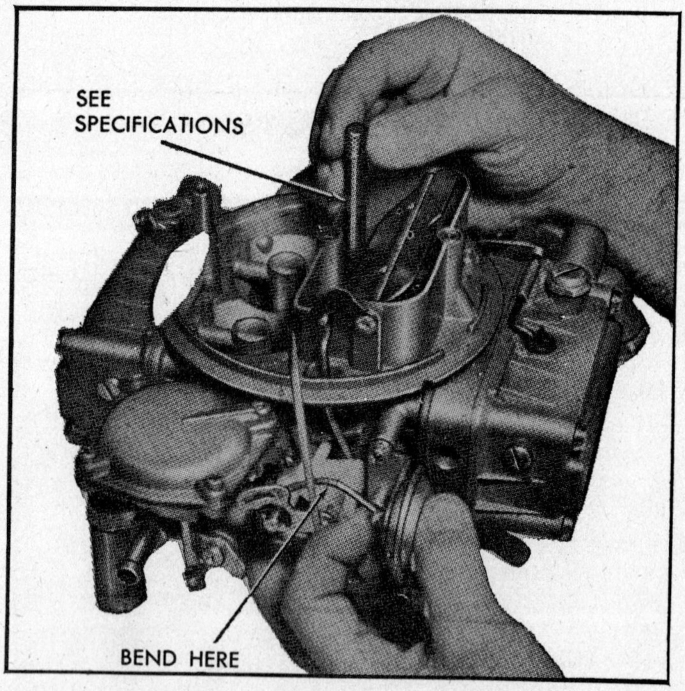

Fig. 13 Vacuum break adjustment. Well type choke

idle setting with the carburetor off the car is made by turning the fast idle adjusting screw to obtain the clearance listed in the *Holley Specifications Chart* between throttle valve and carburetor bore (idle port side) with fast idle screw on high step of cam.

Vacuum Break Adjustment

Fig. 13—On models so equipped, hold choke valve closed with a rubber band. Hold vacuum break in against its stop. Measure distance between lower edge of choke valve and wall of air horn. If the clearance is not as listed in the *Holley Specifications Chart*, bend vacuum break link to adjust.

Choke Unloader Adjustment

Fig. 14—Hold throttle lever wide open with a rubber band. Hold choke valve toward closed position toward unloader tang of throttle shaft. Then measure opening between lower edge of choke valve and air horn wall. If the opening is not as listed in the *Holley Specifications Chart*, bend choke rod at offset angle.

Idle Mixture Adjustment

1968-69 Chrysler Corp.

The screw located next to the primary bowl vent on top of choke air horn is the only means to be used to adjust carburetor idle mixture. Unless an engine exhaust analyzer is available. Rough idle and low speed surge on vehicles may be the result of improper idle setting balance between the right and left carburetor bores. To correct this condition:

1. Remove the cup plugs in the sides of the primary metering block. A small sharp punch is recommended for removing cup plugs.
2. With engine thoroughly warmed up, install an exhaust gas analyzer and adjust carburetor idle speed and mixture.
3. Turn the single idle adjusting screw (located on top of air horn) counterclockwise (rich direction) until it is seated, then move it ¾ turn clockwise (lean direction). Do not disturb this screw during the other steps of this procedure.

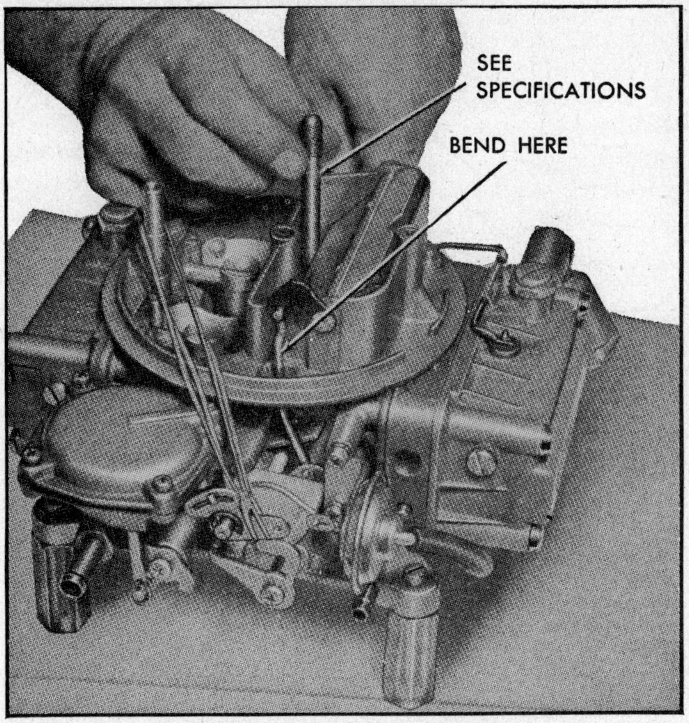

Fig. 14 Choke unloader adjustment. Well type choke

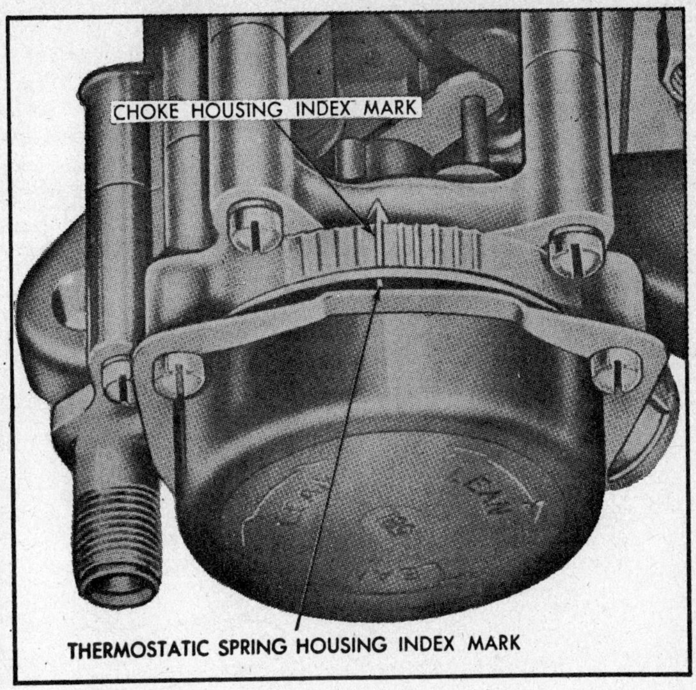

Fig. 15 Built-in automatic choke adjustment

4. With a narrow screw driver, turn the two idle limiter screws clockwise until they are both seated. These screws have prevailing torque features which may require rather high effort to turn screws. It is very important that this prevailing torque is overcome and the screws completely seat against idle discharge ports.
5. Turn both idle limiter screws 1½ to 2 turns counterclockwise as a starting point (experience will dictate more or less turns as a rough setting but both screws should be turned equally).
6. Start engine and set specified idle speed for engines with 300 or more miles. Set 75 rpm below specifications if under 50 miles or 50 rpm below specifications if 50 to 300 miles are on engine.
7. Observe air/fuel ratio reading on exhaust gas analyser. Turn each screw $^1/_{16}$ inch turn richer (counterclockwise) and note change in air/fuel meter reading.

IMPORTANT: When adjusting mixture screws to obtain air/fuel ratio specified, do not turn the mixture screw more than $^1/_{16}$ turn at a time of 10 second increments. The combustion analyzer is so sensitive that the ratio must be changed in very small increments if accurate readings are to be obtained. The meters read in air/fuel ratio so that a higher reading indicates a leaner mixture and visa versa.

8. It is very important that both idle limiter screws be turned the same amount on each adjustment so that as finally set both screws will be the same number of turns from the seated position.
9. Install cup plugs over limiter screws.

Built-In Choke Setting

Fig. 15—Loosen retaining screws and turn choke cover so that index mark or line on choke cover lines up with specified mark on choke housing. See *Holley Specifications Chart.*

Well Type Choke Adjustment

With throttle half open and choke valve closed, bend the choke rod to obtain specified clearance for Chevrolet units. On Chrysler units, follow same procedure so that the top of the choke rod hole in the control lever is $1^{23}/_{32}''$ above the bottom of the carburetor base.

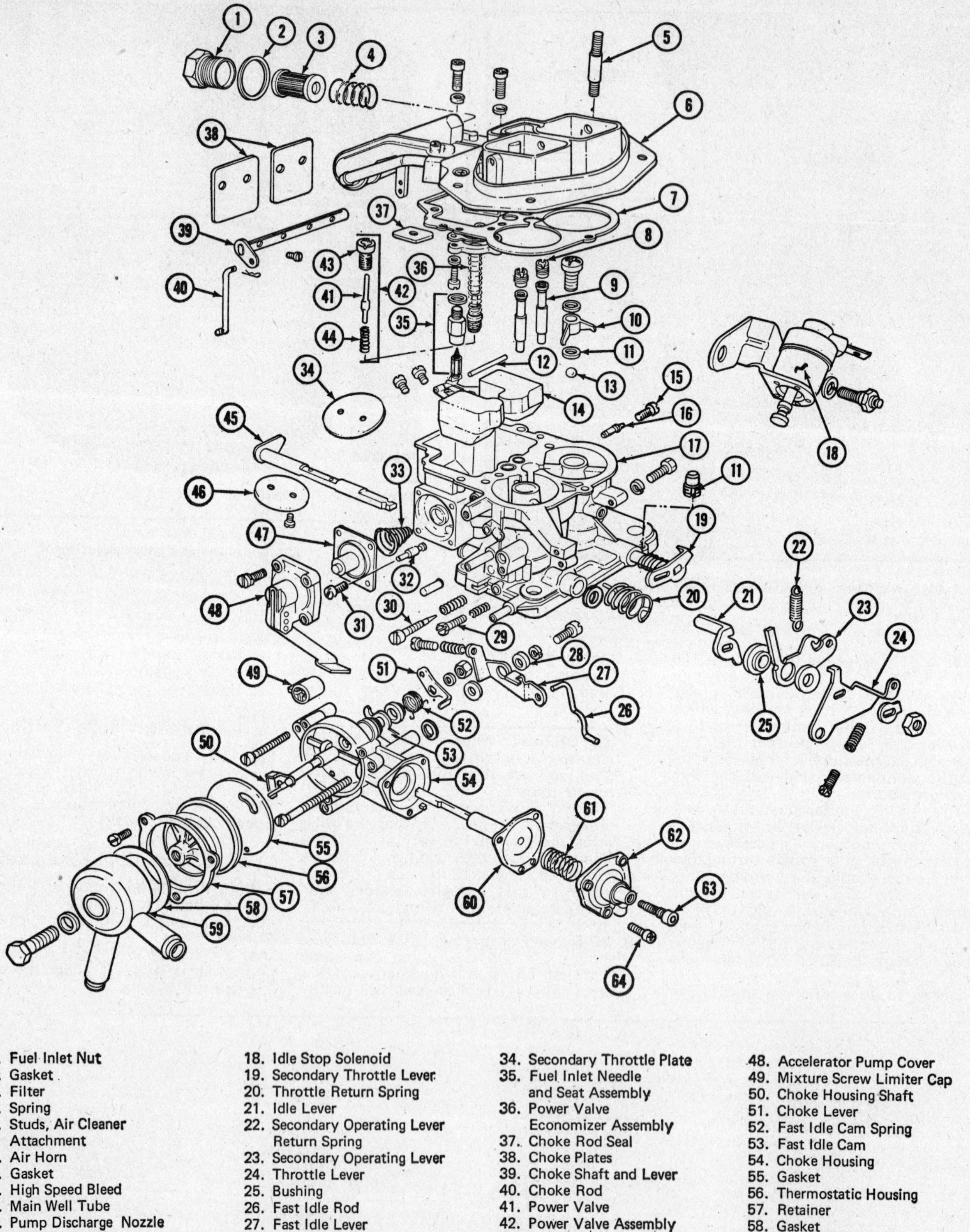

1. Fuel Inlet Nut
2. Gasket
3. Filter
4. Spring
5. Studs, Air Cleaner Attachment
6. Air Horn
7. Gasket
8. High Speed Bleed
9. Main Well Tube
10. Pump Discharge Nozzle
11. Gasket
12. Float Shaft
13. Discharge Check Ball
14. Float
15. Retainer
16. Secondary Idle Jet
17. Carburetor Body Assembly
18. Idle Stop Solenoid
19. Secondary Throttle Lever
20. Throttle Return Spring
21. Idle Lever
22. Secondary Operating Lever Return Spring
23. Secondary Operating Lever
24. Throttle Lever
25. Bushing
26. Fast Idle Rod
27. Fast Idle Lever
28. Bushing
29. Low Idle Screw
30. Fuel Mixture Screw
31. Retainer
32. Primary Idle Jet
33. Return Spring
34. Secondary Throttle Plate
35. Fuel Inlet Needle and Seat Assembly
36. Power Valve Economizer Assembly
37. Choke Rod Seal
38. Choke Plates
39. Choke Shaft and Lever
40. Choke Rod
41. Power Valve
42. Power Valve Assembly
43. Seat
44. Spring
45. Primary Throttle Shaft and Lever Assembly
46. Primary Throttle Plate
47. Accelerator Pump
48. Accelerator Pump Cover
49. Mixture Screw Limiter Cap
50. Choke Housing Shaft
51. Choke Lever
52. Fast Idle Cam Spring
53. Fast Idle Cam
54. Choke Housing
55. Gasket
56. Thermostatic Housing
57. Retainer
58. Gasket
59. Water Cover
60. Diaphragm and Shaft
61. Return Spring
62. Choke Diaphragm Cover
63. Hex Head Screw
64. Cover Screw

Fig. 2 Exploded view of a typical Holley 5210 carburetor

5210 CARB. ADJUSTMENT SPECIFICATIONS

See Tune Up Chart in car chapter for hot idle speeds, air/fuel ratio and idle CO% rating.

Year	Carb. Part No.	Carb. Model	Float Level (Dry)	Float Drop	Pump Position	Fast Idle Cam Index	Vacuum Plate Pulldown	Fast Idle Setting	Choke Setting
CHEVROLET MONZA & VEGA, PONTIAC ASTRE & SUNBIRD									
1973	R-6477A	5210	.420	1″	#3	.140	.300	2000	1 Rich
	R-6478A	5210	.420	1″	#2	.140	.300	2200	2 Rich
	R-6580A	5210	.420	1″	#2	.140	.300	2200	2 Rich
	R-6581A	5210	.420	1″	#3	.140	.300	2000	1 Rich
1974	338168	5210-C	.420	1″	#2	.140	.400	2200	3½ Rich
	338170	5210-C	.420	1″	#2	.140	.400	2200	3½ Rich
	338179	5210-C	.420	1″	#3	.140	.300	2000	2½ Rich
	338181	5210-C	.420	1″	#3	.140	.300	2000	2½ Rich
1975	348659	5210-C	.420	1″	—	—	.325	1600	3 Rich
	348660	5210-C	.420	1″	#2	.110	.300	1600	4 Rich
	348661	5210-C	.420	1″	—	—	.275	1600	3 Rich
	348662	5210-C	.420	1″	#2	.110	.275	1600	4 Rich
	348663	5210-C	.420	1″	—	—	.325	1600	3 Rich
	348664	5210-C	.420	1″	#2	.110	.300	1600	4 Rich
	348665	5210-C	.420	1″	—	—	.275	1600	3 Rich
	348666	5210-C	.420	1″	#2	.110	.275	1600	4 Rich
1976	366829	5210-C	.420	1″	#3	.320	.313	2200	2 Rich
	366830	5210-C	.420	1″	#2	.320	.288	2200	3 Rich
	366831	5210-C	.420	1″	#3	.320	.313	2200	2 Rich
	366832	5210-C	.420	1″	#2	.320	.288	2200	3 Rich
	366833	5210-C	.420	1″	#3	.320	.268	2200	2 Rich
	366834	5210-C	.420	1″	#2	.320	.268	2200	3 Rich
	366840	5210-C	.420	1″	#2	.320	.268	2200	3 Rich
	366841	5210-C	.420	1″	#3	.320	.268	2200	2 Rich

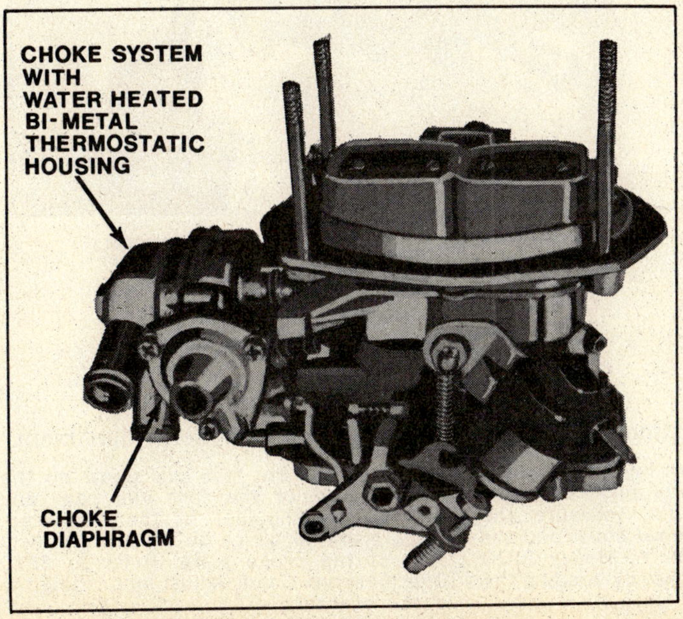

Fig. 1 Holley 5210 2 barrel carburetor

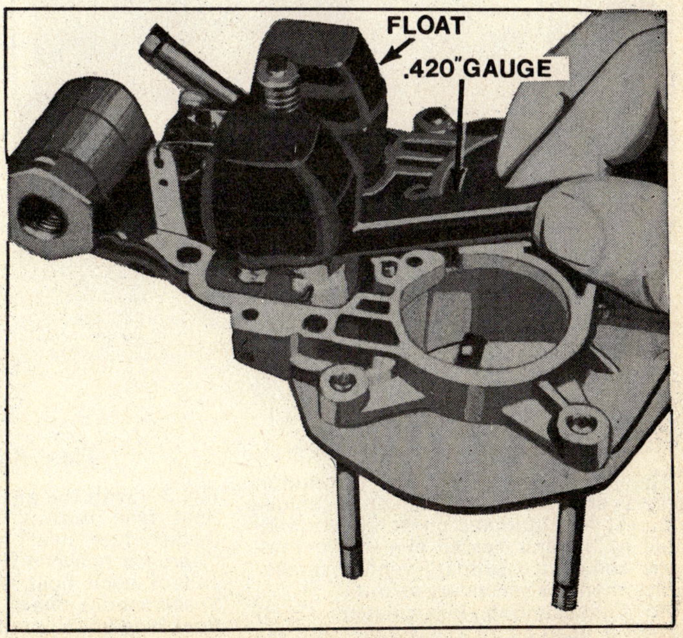

Fig. 3 Checking float level on 5210 carburetor

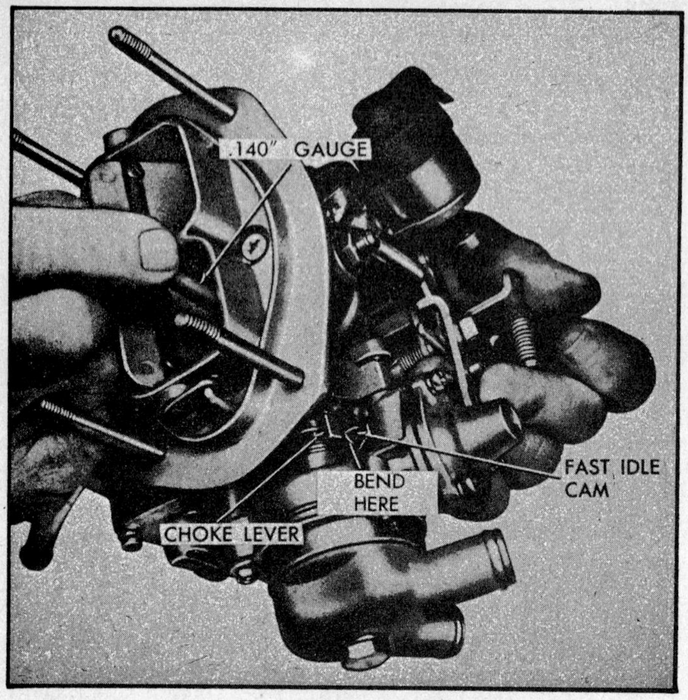

Fig. 4 Fast idle cam index adjustment 5210 Holley carburetor

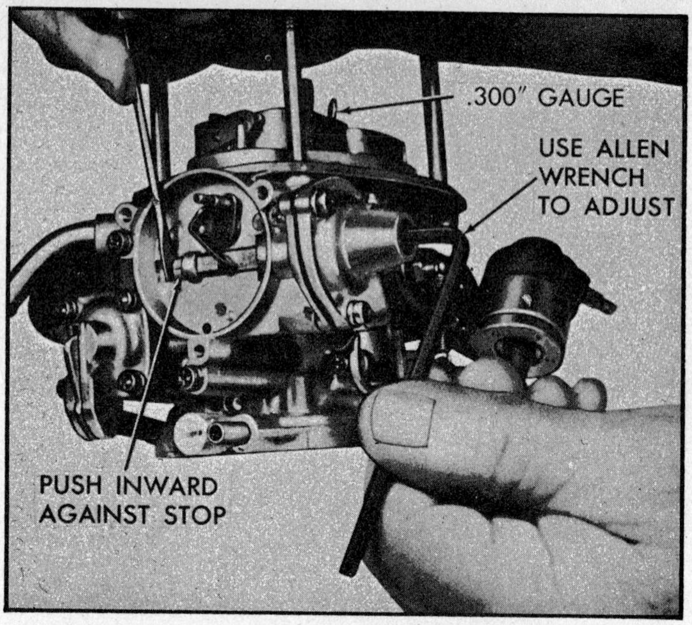

Fig. 5 Vacuum pull down adjustment 5210 Holley carburetor

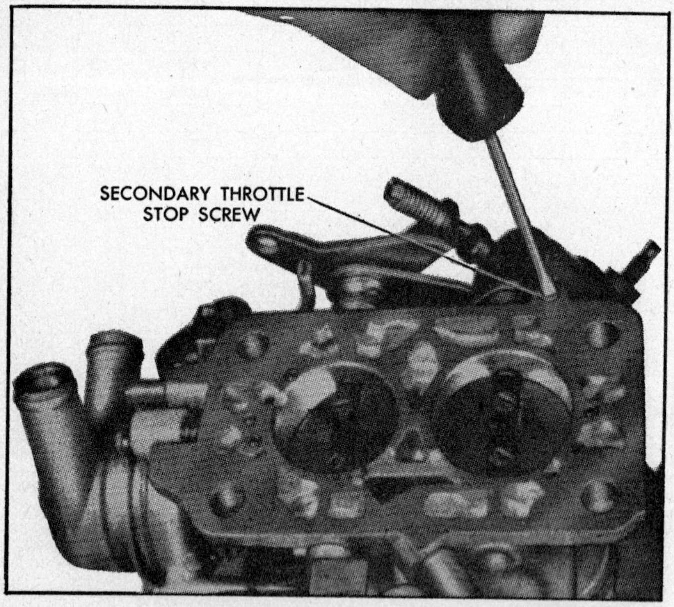

Fig. 6 Secondary throttle plate adjustment

Fig. 7 Fast idle adjustment 5210 Holley carburetor

MODEL 5210

The Holley 5210 two-barrel carburetor, Figs. 1 and 2, has a number of unique features. An automatic choke system activated by a water heated bi-metal thermostatic coil and a primary venturi smaller in size than the secondary venturi.

An Exhaust Gas Recirculation (EGR) system is used on all applications with the EGR valve located in the intake manifold.

Float Adjustment

Fig. 3—With the air horn inverted and the float tang resting lightly on the spring loaded fuel inlet needle, measure the clearance between the bowl cover and the end of each float. Refer to *Holley 5210 Specifications Chart*. Adjust by bending the float tang as required.

Fast Idle Cam Index Adjustment

Fig. 4—Place the fast idle screw on the second step of the fast idle cam and against the shoulder of the first step. Place a drill or gauge on the down stream side of the choke plate. Refer to *5210 Specifications Chart*. Adjust by bending the choke lever tang.

NOTE: The dechoke is automatically set when the fast idle cam index is adjusted.

Vacuum Plate Pulldown Adjustment

Fig. 5—Remove the three hex headed screws and ring retaining the choke bimetal cover. Do not remove the choke water housing screw if adjusting on the car. Pull the choke water housing and bimetal cover assembly out of the way. With a screwdriver or suitable tool, push the diaphragm stem back against the stop. Place drill or gauge on the down stream side of the primary choke plate. Take all the slack out of the linkage. Refer to *5210 Specifications Chart*. Adjust by turning adjusting screw in or out with an 5/32 inch Allen wrench.

Secondary Throttle Plate Adjustment

Fig. 6—Back out the secondary throttle stop screw, until the secondary throttle plate seats in the carburetor bore. Turn screw in until it makes contact with tab on the secondary throttle lever, then turn screw in an additional ¼ turn.

Fast Idle Adjustment

Fig. 7—With engine at operating temperature, air cleaner off, choke off, A/C off and EGR valve free to operate, position fast idle screw on second step of fast idle cam and adjust by turning fast idle screw in or out. Refer to 5210 Specifications Chart.

Unloader (Dechoke) Adjustment

Fig. 8—On 1973 units, this adjustment is automatically set when Fast Idle Cam Index Adjustment is performed. Position throttle to wide open position and insert specified drill or gauge on downstream side of choke plate. If adjustment is necessary, bend tang on fast idle lever. Refer to 5210 Specifications Chart.

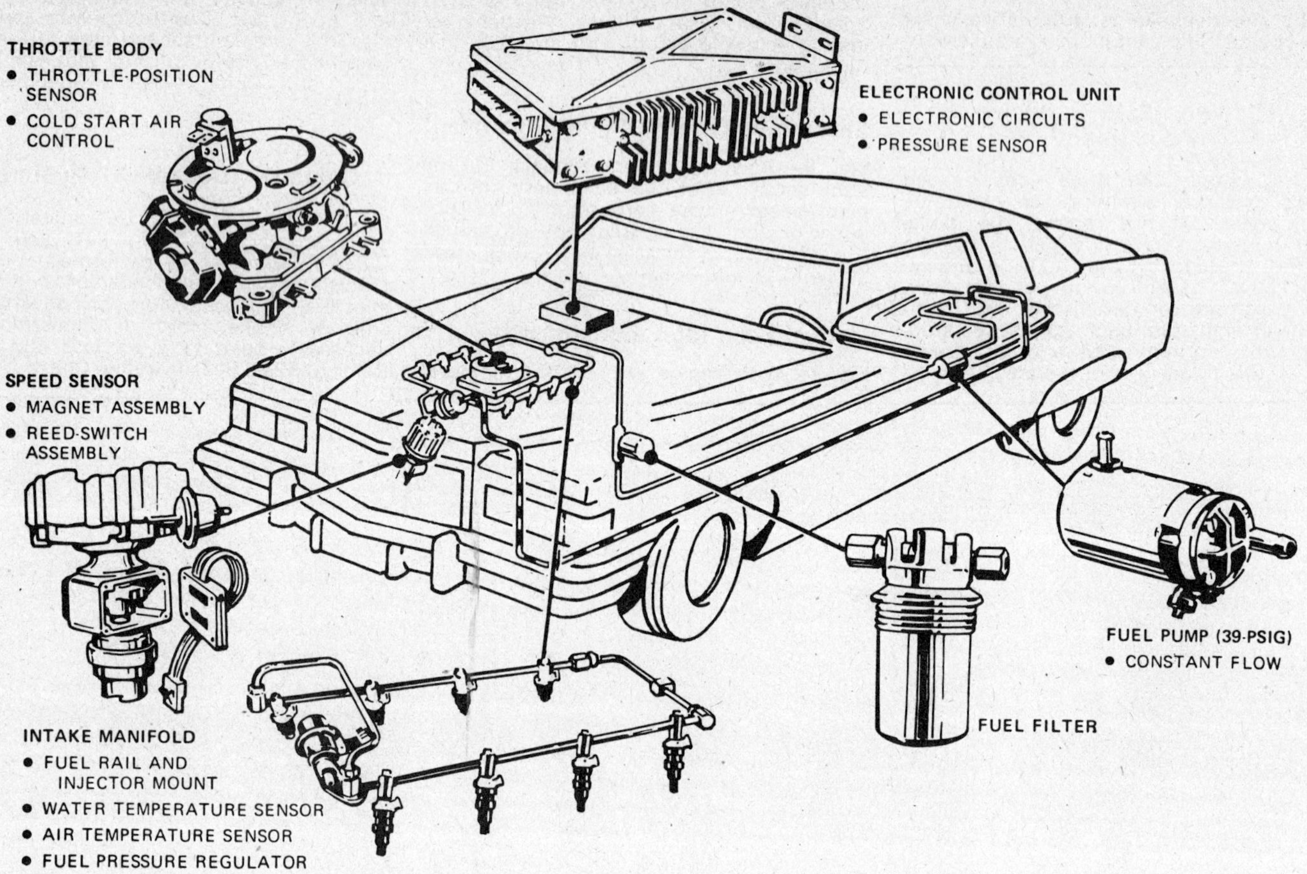

Fig. 1 Cadillac electronic fuel injection (EFI) system. (Typical)

ELECTRONIC FUEL INJECTION SYSTEM

The Bendix Electronic Fuel Injection System (EFI), Figs. 1 and 2, provides a means for controlling exhaust emissions within the required limits by precisely controlling the air/fuel ration. This is accomplished by monitoring selected engine operating conditions and electronically metering the fuel requirements to meet those conditions.

EFI is basically a system in which electrically actuated fuel metering valves, when actuated, injects a quantity of metered fuel into the engine. The injectors are mounted in the intake manifold with the metering tip pointed toward the intake valve head. This arrangement is known as port injection. The injector opening is timed so the fuel charge is injected prior to the cylinder intake stroke.

Fuel is supplied to the injector through the fuel rail at high pressure to obtain proper fuel atomization and also prevents vapor formation in the fuel system during high temperature operating conditions. When the solenoid operated injector valves are energized, the injector metering valve moves to the full open position. Since the pressure differential across the valve is constant, the fuel quantity is changed by varying the time the injector is held open.

This system is a two group system in which the injectors are divided into two groups of two or four each. On Cadillac applications, cylinder numbers 1, 2, 7, and 8 form one group and cylinders 3, 4, 5, and 6 comprise the second group. On Cosworth Vega applications, cylinders 1 and 2 form one group and cylinders 3 and 4 form the second. All injectors in a group are actuated simultaneously. Each of the groups are actuated once for every revolution of the camshaft or two revolutions of the crankshaft.

NOTE: On the Cosworth Vega, under loaded, wide open throttle operations, the injectors are actuated simultaneously every crankshaft revolution.

The amount of air entering the engine is measured by monitoring intake manifold absolute pressure (MAP), inlet air temperature and engine speed. This information allows the ECU to compute the air flow rate and consequently, the fuel flow rate required to achieve the proper air/fuel ratio.

The prevailing engine operating conditions are monitored by sensors which provides the ECU with the information. The ECU then converts this multi-variable information into an injector pulse width which opens the injector for the proper duration and time with respect to engine firing order.

SUB-SYSTEMS DESCRIPTION

The EFI system is comprised of the following major sub-systems: Fuel Delivery, Air Induction, Sensors and ECU. The functions of these separate sub-systems are integrated into the EFI system operation.

Fuel Delivery

The fuel delivery sub-system incorporates an in-tank boost pump, chassis mounted fuel pump, fuel filter, fuel rails, injector valves, fuel pressure regulator and fuel supply and return lines. The fuel pumps are activated by the ECU when the ignition is turned "On" and the engine is cranking or operating. If the engine stalls or the starter is not engaged, the fuel pumps will deactivate in approximately one second. Fuel is pumped from the fuel tank through the supply line and filter to the fuel rails. The injectors supply fuel to the cylinders when energized by the ECU. Excess fuel is returned to the fuel tank.

Fuel Tank

The fuel tank used with Cadillac applications incorporates a reservoir below the sending unit, in-tank boost pump assembly. This "Bath tub" shaped reservoir ensures a constant supply of fuel for the in-tank pump at low fuel levels or under severe maneuvering conditions. The fuel returned to the tank is directed into the reservoir as an additional means of keeping the in-tank pump intake below the fuel level.

Fuel Pumps

Two electric fuel pumps are used with the EFI system. The in-tank boost pump, located inside the fuel tank and is an integral part of the fuel tank gauge unit, is used to supply fuel to the chassis-mounted pump and also, prevent vapor lock on the suction side of the system, Fig. 3. The chassis-mounted fuel pump is a constant-displacement, roller-vane pump driven by a 12 volt electric motor. This pump incorporates a check valve to prevent back-flow and to maintain fuel pressure when the pump is "Off". An internal relief valve opens at an excessive pressure (55-95 PSI). The pump is mounted under the vehicle, forward of the left rear wheel on all models except Cadillac Eldorado and Cosworth Vega. On Cadillac Eldorado and Cosworth Vega, the pump is located forward of the right rear wheel.

The in-tank and chassis-mounted fuel pumps are electrically connected in parallel to the ECU and are protected by a 10 amp in-line fuse located in the harness near the ECU.

Fuel Filter

The fuel filter consists of a casing with an internal, throw-away paper filter element, Fig. 4. The filter element should be replaced at the first 7500 miles, then at 22,500 miles and every 22,500 miles after. On Cadillac models except Seville, the filter is mounted on a bracket on the lower left front of the engine. On Cadillac Seville models, the filter is mounted on a bracket near the left rear wheel. On Cosworth Vega models, the filter is mounted on the frame in the engine compartment left side.

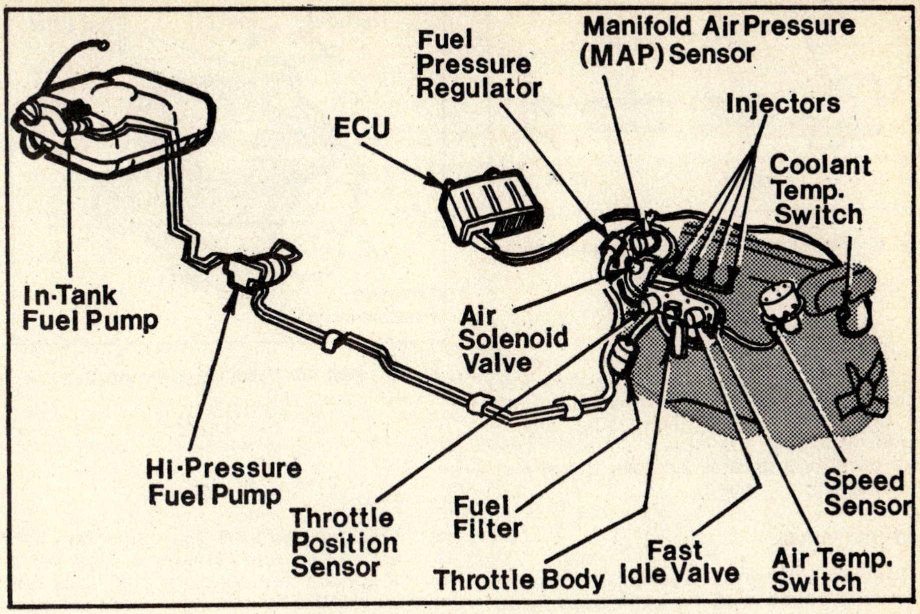

Fig. 2 Cosworth Vega electronic fuel injection (EFI) system

Fuel Pressure Regulator

The fuel pressure regulator contains an air chamber and fuel chamber separated by a spring-loaded diaphragm. The air chamber is connected through a hose to the throttle body. The air pressure in this chamber is intake manifold pressure (MAP). The changing MAP and the spring, controls the diaphragm valve action, thereby opening or closing an orifice in the fuel chamber. The regulator, connected to the fuel rail and intake manifold, maintains a constant 39 PSI differential across the injectors. Excess fuel is returned to the fuel tank.

Injector Valves

The injector valves are solenoid operated pintle valves which meters fuel to each cylinder. Based upon a pulsed signal from the ECU, the valve opens for the proper time interval (Pulse-width), delivering the exact amount of fuel required. When energized, the valve sprays fuel in fine droplets and when de-energized, prevents further fuel flow to the engine. On Cadillac models, the injector valves are located in the intake manifold above the intake valve for each cylinder. On Cosworth Vega, the injector valves are mounted in the fuel rail and are positioned to direct the fuel to the top side of the intake valves. Use caution when handling the injector valves to avoid damage to the metering tips.

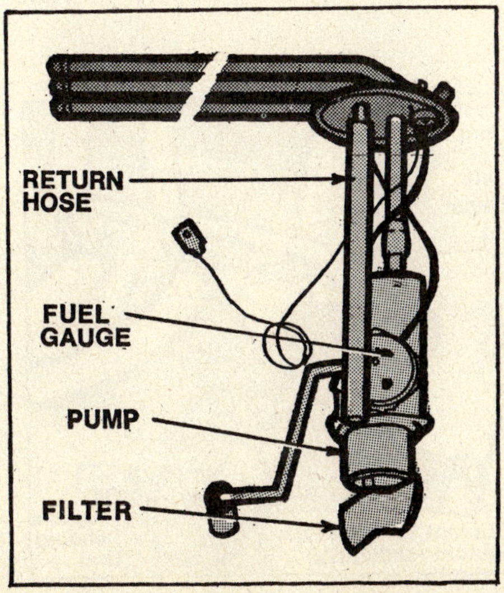

Fig. 3 In-tank fuel pump & gauge assembly. (Typical)

Air Induction System

The air induction sub-system consists of the throttle body assembly, fast idle valve assembly, air solenoid valve (Cosworth Vega) and intake manifold. Combustion air enters the throttle body where the primary air flow is controlled by the throttle valves. An adjustable idle by-pass air passage is incorporated within the throttle body, allowing a regulated amount of air to by-pass the throttle valves. The throttle valves are preset slightly open when the throttle lever is resting against the idle stop.

Additional air for cold starts and warm-up is provided through the fast idle valve. On Cadillac applications, the fast idle valve is mounted in the top of the throttle body. On Cosworth Vega, the fast idle valve is mounted in the right side of the cylinder head under the intake manifold.

Intake Manifold (Cadillac)

The dual plane intake manifold is basically the same as for carbureted engines. The EFI intake manifold, however, distributes only air to the cylinder intake ports. Also, the EFI manifold incorporates ports for injection valve installation and for an air temperature sensor. No exhaust heat passage is used in this intake manifold. The exhaust passages from the right cylinder head are used for EGR only.

Throttle Body

The throttle body consists of a housing with two bores and two shaft mounted throttle valves connected to the accelerater pedal by mechanical linkage. Fittings are incorporated on the body to accommodate vacuum connections. The end of the throttle shaft controls the throttle position switch. An adjustable set screw on the throttle body front permits adjustment of

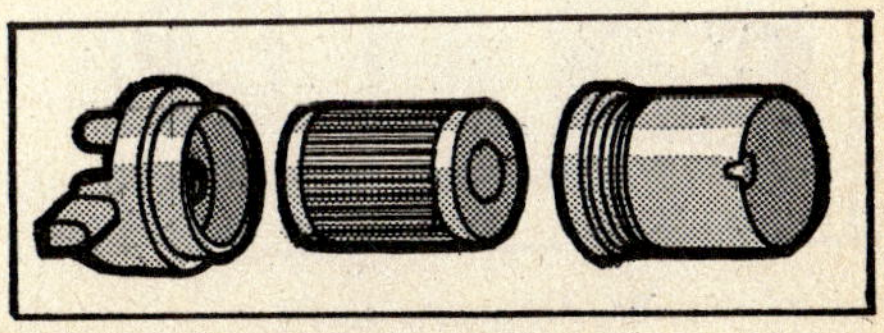

Fig. 4 Fuel filter

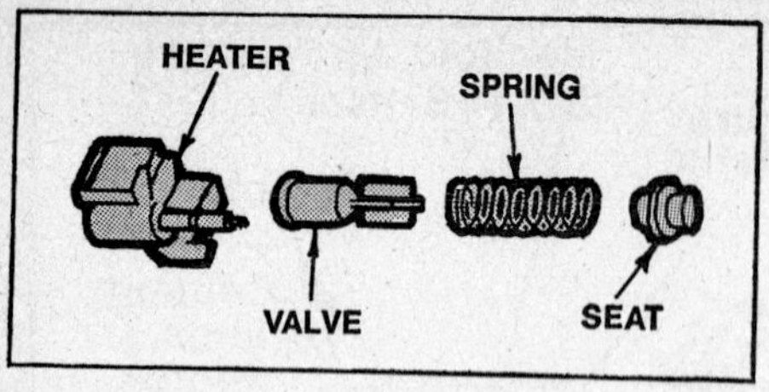

Fig. 5 Fast idle valve, disassembled. Cadillac

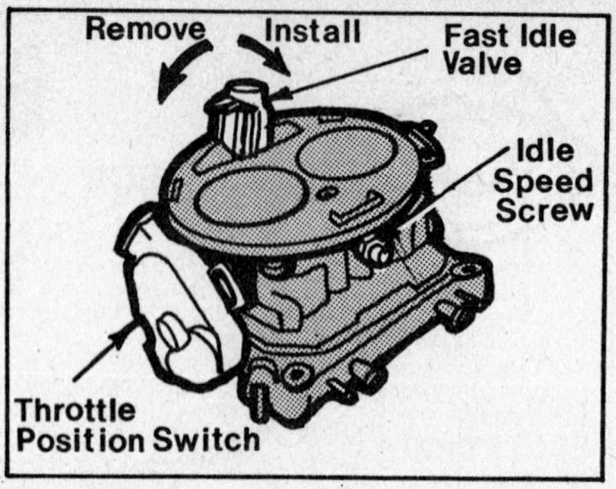

Fig. 6 Fast idle valve replacement. Cadillac

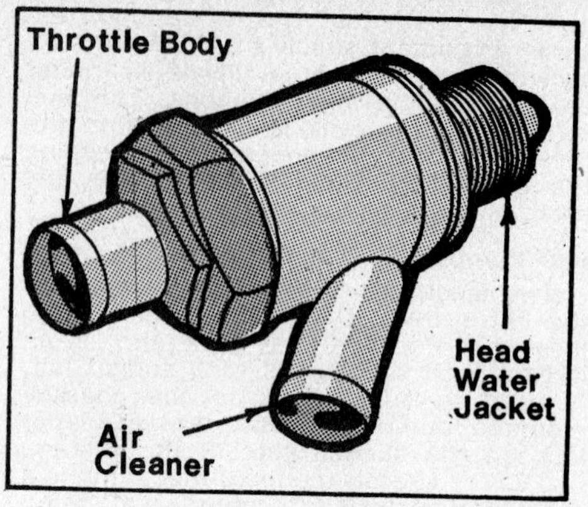

Fig. 7 Fast idle valve. Cosworth Vega

warm engine idle speed. Turning the screw clockwise restricts air flow through the idle air channel, in turn decreasing engine speed. Rotating the screw counter-clockwise permits a greater air flow, thereby increasing engine speed.

Fast Idle Valve

The fast idle valve used on Cadillac applications consists of a plastic body housing an electric heater, a spring and plunger and a thermal sensitive element, Fig. 5. This device, installed in the top of the throttle body, Fig. 6, is electrically connected to the fuel pump circuit through the ECU. When a cold engine is started, the valve opens, allowing air to by-pass the throttle. The heater warms the thermal element which then forces the spring and plunger toward an orifice, reducing the by-pass air flow, in turn decreasing engine speed. The fast idle valve has no further effect after the thermal element temperature reaches 140° F. The valve closure rate is a function of time and temperature. The warmer the ambient temperature, the faster the valve will close. At 68° F, the valve willl close in approximately 90 seconds or at −20° F, the valve requires approximately five minutes to close.

The fast idle valve, Fig. 7, used on the Cosworth Vega, is mounted in the right side of the cylinder head under the intake manifold and performs the same function as the Cadillac unit. However, this valve senses coolant temperature and does not have an electric heater. As coolant temperature increases, the thermal element in the valve expands, moving the internal metering valve toward the closed position, thereby decreasing air flow to the engine.

Air Solenoid Valve (Cosworth Vega)

The air solenoid valve is used to supplement the air supply from the fast idle valve to the throttle during engine warm-up, Fig. 8. This ECU controlled pintle valve is mounted on the sensor cluster bracket at the cowl. The signal from the ECU is directly related to engine coolant temperature. The valve is closed when the choke cycle is complete and curb idle is obtained.

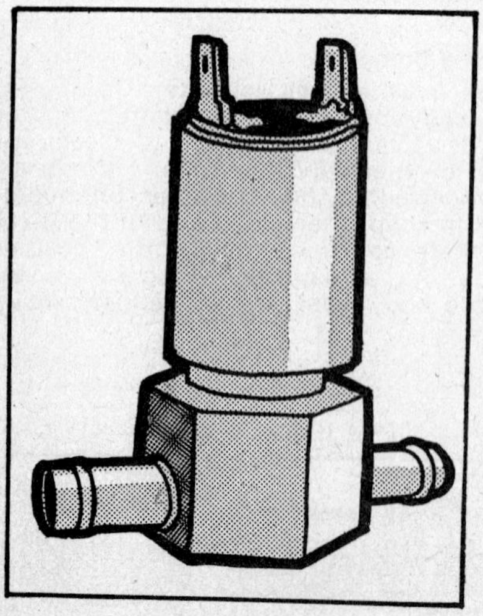

Fig. 8 Air solenoid valve. Cosworth Vega

Engine Sensors

The engine sensors are electrically connected to the ECU and operate independently of each other. Each sensor transmits a signal to the ECU, relating to a specific engine operating condition. The ECU analyzes these signals and transmits the appropriate commands.

Manifold Absolute Pressure Sensor

The manifold absolute pressure (MAP) sensor monitors changes in intake manifold pressure resulting from engine load, speed, barometric pressure and altitude variations. These pressure changes are transmitted to the ECU. As intake manifold pressure increases, additional fuel is required. The MAP sensor transmits this information to the ECU which then increases the injector pulse-width. Conversely, as manifold pressure decreases, the injector pulse-width will be decreased. On Cadillac applications, the MAP sensor is mounted within the ECU and is connected to the throttle body through a manifold pressure line, routed with the engine harness. On Cosworth Vega applications, the MAP sensor is mounted on the sensor bracket at the cowl.

Throttle Position Switch

The throttle position switch, mounted on the throttle body and is connected to the throttle shaft, senses throttle shaft movement and position. The sensor transmits this information to the ECU which then determines fuel requirements for the particular situation.

Temperature Sensors

The temperature sensors, coolant and air, are comprised of a coil of high temperature nickel wire sealed into an epoxy case, and molded into a brass housing. The resistance of the wire changes as a function of temperature. Low temperature provides low resistance and as temperature rises, resistance also rises. The voltage drop across the sensors is monitored by the ECU.

On all Cadillac applications, the air temperature sensor is located at the rear of the intake manifold. On Cadillac models except Seville, the coolant temperature sensor is located on the heater hose fitting at the rear of the right cylinder head. On Cadillac Seville, the coolant temperature sensor is located at the left front of the intake manifold.

On Cosworth Vega, the air temperature

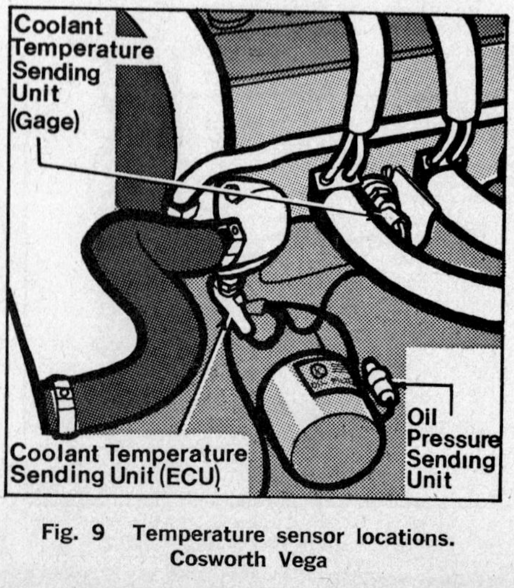

Fig. 9 Temperature sensor locations. Cosworth Vega

sensor is mounted in the bottom of the intake manifold, in the No. 3 intake runner. The coolant temperature sensor is located in the thermostat housing, Fig. 9.

NOTE: Both temperature sensors are identical and are interchangeable.

Speed Sensor

The speed sensor is incorporated within the HEI distributor assembly and consists of two components.

a. Two reed switches mounted to a plastic housing, attached to the distributor shaft housing.
b. Rotor with two magnets, attached to and rotating with the distributor shaft, Fig. 10.

The rotating rotor activates the reed switches, providing for the following:

a. synchronization of the ECU and proper injector group with intake valve timing.
b. engine speed for fuel scheduling.

Electronic Control Unit

The ECU is a preprogrammed analog computer consisting of custom electronic circuits housed in a steel case. The units is installed above the glove box and is electrically connected to the vehicle power supply and other EFI components by a harness routed through the firewall. When the ignition switch is placed in the "ON" or "CRANK" position, the ECU receives power from the vehicle battery. The ECU then receives signals from the EFI sensors and activates the electric fuel pumps, fast idle valve, injection valves and EGR solenoid.

The desired air/fuel ratios for various operating conditions are designed into the ECU. As the signals are received from the sensors, the ECU processes those signals and computes fuel requirements. The ECU then issues commands to the injectors to open for a specific time duration. The duration of the command pulses vary as operating conditions change. All injector valves in each group open simultaneously.

TROUBLE-SHOOTING

NOTE: Complete diagnosis and trouble-shooting of the EFI system requires the use of special testing equipment. Tester No. J-25400 should be used for Cadillac systems and tester No. J-24706 used for Cosworth Vega systems. Instructions for the use of these testers are supplied with the units.

The following trouble shooting charts relate to the Cadillac EFI system not requiring the use of special testing equipment.

Engine Cranks But Will Not Start

1. Blown 10 amp in-line fuel pump fuse, located below instrument panel near ECU jumper harness connectors. To check, refer to Note 1 at end of "Trouble-Shooting" section.
2. Open circuit in 12 purple wire between starter solenoid and ECU.
3. Open circuit in 18 dark green wire between alternator "BAT" terminal and ECU (fusible link). To check, refer to Note 1 at end of "Trouble-Shooting" section.
4. Faulty connection at ECU jumper harness or at ECU.
5. Faulty connection at fuel pump jumper harness 14 dark green wire, located below instrument panel near ECU jumper harness. To check, refer to Note 1 at end of "Trouble-Shooting" section.
6. Faulty connection at engine coolant sensor or open circuit in sensor or wiring. This condition will affect EFI operation on cold engines only. To check, refer to Note 2 at end of "Trouble-Shooting" section.
7. Faulty connection at speed sensor or speed sensor stuck closed.
8. Faulty chassis-mounted fuel pump.
9. Defective throttle position switch. To check, disconnect switch and engine should start.
10. Restricted fuel flow.

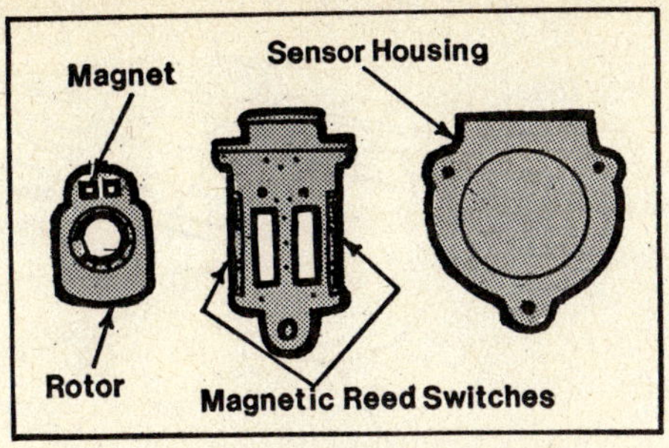

Fig. 10 Speed sensor, disassembled. Cosworth Vega

Hard Starting

1. Open engine coolant sensor, cold or partially warm engine. To check, refer to Note 2 at end of "Trouble-Shooting" section.
2. Defective throttle position switch. To check, disconnect switch and engine should start.
3. Faulty chassis-mounted fuel pump.
4. Malfunctioning fuel pressure regulator.

Poor Fuel Economy

1. Leaking or disconnected MAP sensor hose.
2. Disconnected vacuum hose at fuel pressure regulator or throttle body.
3. Defective air temperature or coolant sensor. To check, refer to Note 3 at end of "Trouble-Shooting" section.

Engine Stalls After Starting

1. Open circuit in 12 black/yellow ignition signal wire between fuse block and ECU or faulty connection at the connector, located below instrument panel near ECU jumper harness.
2. Faulty connection at engine coolant sensor or open circuit in sensor or wiring, cold or warm engines only. To check, refer to Note 2 at end of "Trouble-Shooting" section.

Rough Idle

1. Disconnected, leaking or restricted MAP sensor hose. If plastic harness line requires replacement, replace entire EFI engine harness.
2. Faulty connection at air temperature or coolant sensor or open circuit in sensor or wiring, cold engine conditions only. To check, refer to Note 2 at end of "Trouble-Shooting" section.
3. Faulty electrical connections at injector valves.
4. Shorted engine coolant sensor. To check, refer to Note 3 at end of "Trouble-Shooting" section.

Prolonged Fast Idle

1. Faulty electrical connection at fast idle valve or open circuit in heating element.
2. Throttle position switch not adjusted properly.
3. Vacuum leak.

Engine Hesitates Or Stumbles On Acceleration

1. Disconnected, leaking or restricted MAP sensor hose. If plastic harness line requires replacement, replace entire EFI engine harness.
2. Throttle position switch not adjusted properly or switch defective.
3. Intermittent malfunction of speed sensor.
4. Faulty connection at 6 pin connector of ECU jumper harness or at ECU.
5. Faulty electrical connection at EGR solenoid or solenoid open, cold engine conditions only.

Lack Of High Speed Performance

1. Throttle position switch not adjusted properly, wide open throttle position only.
2. Malfunctioning throttle position switch.
3. Defective chassis-mounted fuel pump.
4. Intermittent malfunction of speed sensor.
5. Blocked or restricted fuel filter.
6. Open circuit in 12 purple wire between starter solenoid and ECU.

Trouble-Shooting Notes

1. Turn ignition switch to "ON" position and listen for chassis-mounted fuel pump actuation (Whine). Pump should actuate for one second.
2. Connect an ohmmeter to EFI temperature sensor terminals. If sensor resistance is greater than 1600 ohms, sensor replacement is required.
3. Connect an ohmmeter to EFI temperature sensor terminals. If sensor resistance is less than 700 ohms, sensor replacement is required.

EFI SYSTEM SERVICE

CAUTION: Fuel in this system may be under pressure and could spray out when loosening fittings. To relieve fuel pressure on early production models without a pressure fitting in the rear fuel rail, cover the fitting being loosened with a shop cloth so the fuel will be absorbed. On later production models with a "Schrader" type fitting, remove fitting protective cap, loosely install tool J-5420 on fitting, place a shop cloth or suitable container to absorb or contain fuel and tighten tool J-5420.

NOTE: A conical-shaped metal gasket (P/N 1608786) should be installed when servicing fuel rail fitting connections on Cadillac vehicles. This gasket is not used in production, however, should be inserted in the inverted flare fitting when a fuel rail connection is separated for EFI system servicing. This gasket minimizes the possibility of fuel leakage under pressure. The gasket is not reusable and must be replaced each time a fitting is disassembled.

Electronic Control Unit, Replace

Cadillac Exc. Seville

1. Disconnect battery ground cable.
2. Remove right, left and right center A/C outlet grilles, then working through the openings, remove bolt and two nuts securing instrument panel pad to panel support.
3. Remove instrument panel pad to horizontal support screws, pull pad outward, disconnect electrical connector from W/S switch and remove instrument panel pad.

NOTE: To permit removal or installation of instrument panel pad, place selector lever in "LO" range and on vehicles equipped with tilt steering wheel, place wheel in lowest position.

4. Remove MAP sensor hose from ECU, Fig. 11.
5. Remove the three ECU mounting screws, then the ECU and disconnect electrical connectors.
6. Reverse procedure to install.

Cadillac Seville

1. Disconnect battery ground cable.
2. Loosen ECU cover to ECU right hand forward screw, then remove the three remaining ECU cover to ECU screws, Fig. 12.
3. Remove ECU to instrument panel support screws.
4. Lower ECU, disconnect electrical connectors and MAP hose and remove ECU from vehicle.
5. Reverse procedure to install.

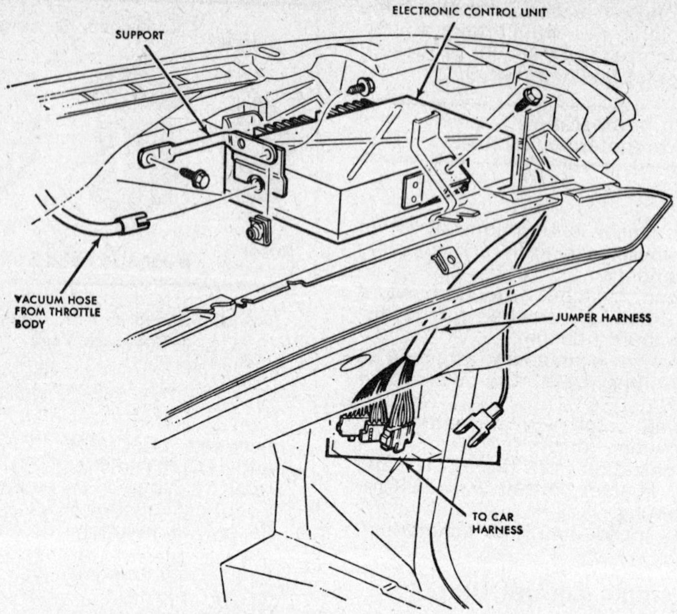

Fig. 11 ECU installation. Cadillac exc. Seville

Cosworth Vega

1. Disconnect battery ground cable.
2. Remove glove box, then the ECU left mounting bracket to instrument panel reinforcement screw, Fig. 13.
3. Support ECU and remove the right mounting bracket to support screw. Tilt ECU to one side, push toward firewall and lower from panel. It may be necessary to remove support for clearance.
4. Disconnect electrical connector from ECU and remove from vehicle.
5. Reverse procedure to install.

Fast Idle Valve, Replace

Cadillac

1. Remove air cleaner, disconnect fast idle valve heater and remove air cleaner stud.
2. Push downward on heater and rotate 90° counter-clockwise to remove, Fig. 6.
3. Remove fast idle valve, spring and seat from throttle body, Fig. 5.
4. Reverse procedure to install. Ensure tabs on fast idle valve heater are aligned with throttle body cut-out when compressing spring.

Cosworth Vega

1. Disconnect hoses from fast idle valve, Fig. 4.
2. Rotate fast idle valve counter-clockwise to remove. Note position of filtered air inlet tube.
3. Reverse procedure to install. Rotate valve clockwise to install.

Throttle Position Switch, Replace

Cadillac

1. Remove throttle body as outlined under "Throttle Body, Replace" procedure.
2. Remove switch mounting screws, then the switch from throttle body.

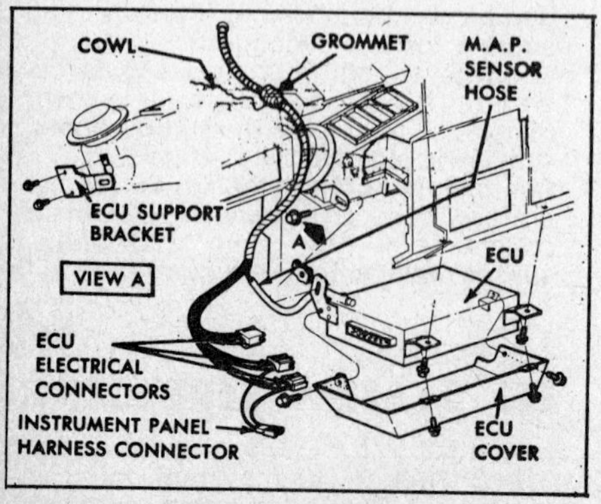

Fig. 12 ECU installation. Cadillac Seville

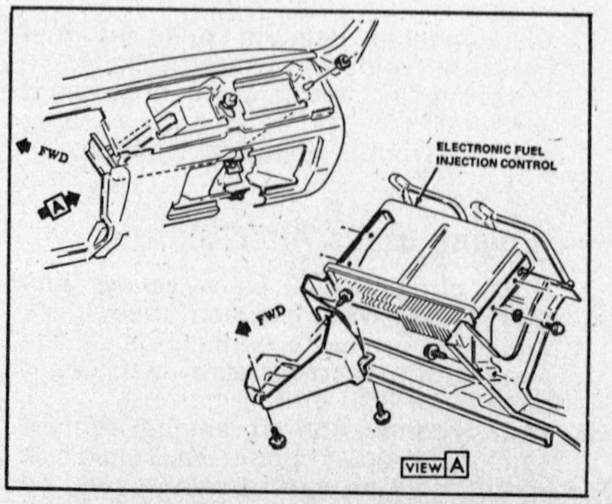

Fig. 13 ECU installation. Cosworth Vega

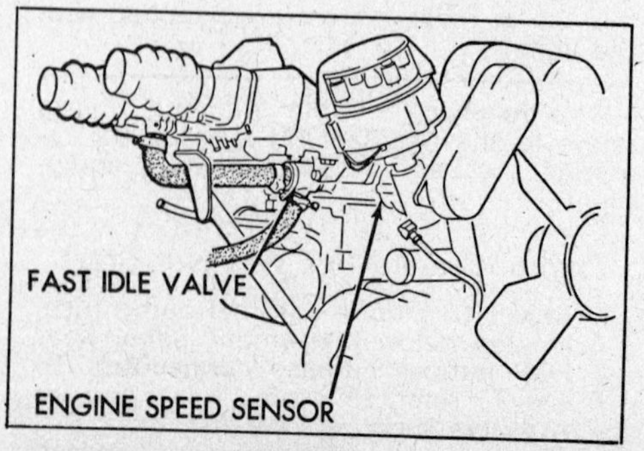

Fig. 14 Fast idle valve installation. Cosworth Vega

3. Install switch with switch tab engaging flat on throttle shaft and loosely install mounting screws.
4. Adjust switch as outlined under "Adjustments".
5. Install throttle body.

Cosworth Vega

1. Disconnect battery ground cable.
2. Disconnect electrical connector from throttle position switch.
3. Remove switch retaining screws and the switch.
4. Reverse procedure to install.

Throttle Body, Replace

Cadillac

1. Remove air cleaner and disconnect throttle return springs from throttle lever, Fig. 15.
2. Remove retainer and cruise control chain from throttle lever, if equipped.
3. Remove clip, then throttle cable from throttle lever.
4. Remove left rear throttle body mounting screw and one screw securing throttle bracket to intake manifold.
5. Disengage downshift switch from throttle lever and place bracket, switch and linkage aside.
6. Disconnect throttle position switch and fast idle valve electrical connectors, then slide fast idle valve wiring from throttle body notch.
7. Remove throttle body vacuum hoses.

NOTE: When removing power brake vacuum line, use a backup wrench to prevent damage to fitting.

---

8. Remove remaining throttle body mounting screws, then the throttle body.
9. The following components are not included in a new throttle body assembly and should be removed as necessary: throttle position switch; fast idle valve seat, spring, valve and heater assembly; power brake vacuum fitting.
10. Reverse procedure to install. Torque mounting screws to 15 ft. lbs.

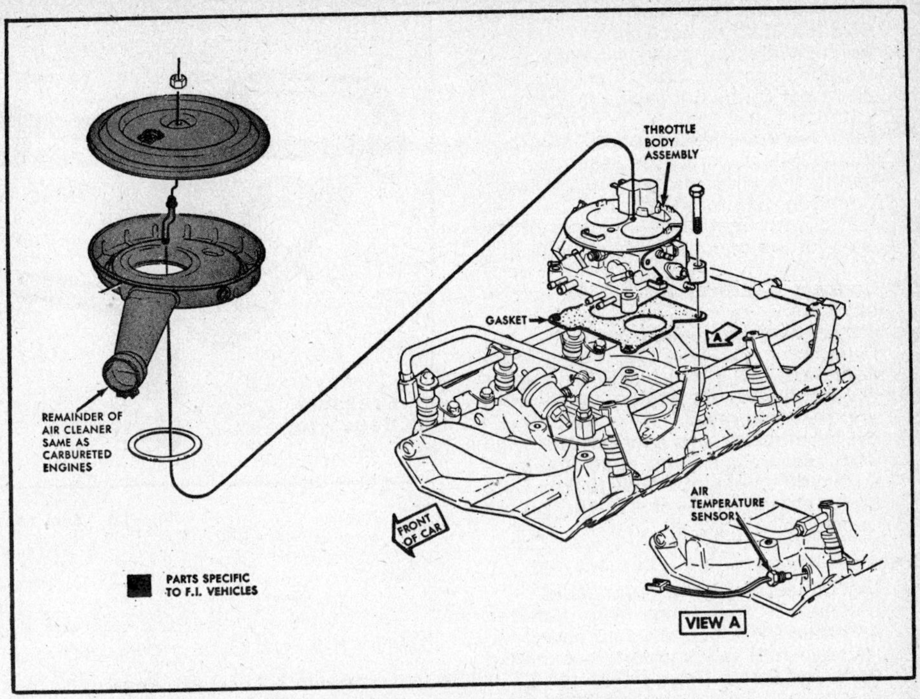

Fig. 15 Throttle body installation. Cadillac

Cosworth Vega

1. Disconnect battery ground cable.
2. Disconnect throttle cable and return springs, Fig. 16.
3. Disconnect PCV hose and the air cleaner boot.
4. Disconnect all lines, hoses and the electrical connector from throttle body, Fig. 17.
5. Remove throttle body retaining nuts and the throttle body.
6. Remove throttle position sensor.
7. Reverse procedure to install.

Injector Valve, Replace

Cadillac

1. Disconnect battery ground cable and on, models except Seville, remove front and rear fuel rails as outlined under "Fuel Rail, Replace" procedure.
2. Remove electrical conduit from injector brackets.
3. Remove injector brackets and grommets, then disconnect electrical leads from injectors on fuel rail being removed.
4. Remove injectors and fuel rail as an assembly, Fig. 18. Some injectors may remain in the intake manifold.

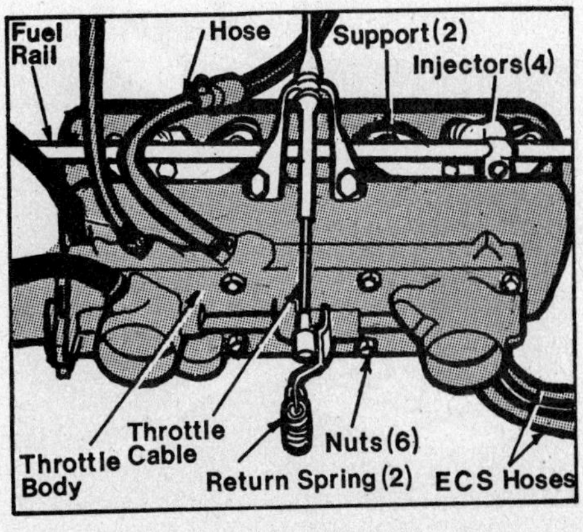

Fig. 16 Throttle body installation. Cosworth Vega

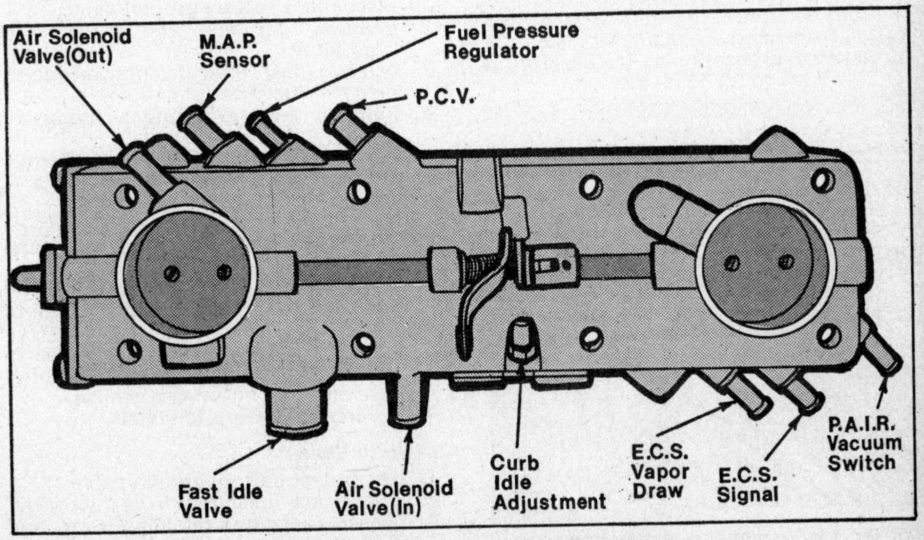

Fig. 17 Throttle body. Cosworth Vega

5. Remove injectors from fuel rail or intake manifold as necessary.
6. Replace injector valve "O" rings.
7. Install injectors into fuel rail with electrical connector facing inboard.
8. Lubricate and install "O" ring in intake manifold injector port.
9. Install fuel rail-injector assembly, ensuring the injectors are properly positioned in manifold "O" ring.
10. Install grommets and injectors brackets. Torque injector bracket bolts to 5 ft. lbs.
11. Connect electrical conduit to injector brackets.
12. Connect injector leads as follows: Cylinders 1, 2, 7, 8—red/black wires Cylinders 3, 4, 5, 6—black/white wires
13. On models except Seville, install front and rear fuel rails.
14. On all models, turn ignition "On" and "Off" several times to build fuel pressure and check for fuel leaks. Then start engine and check for proper operation.

Cosworth Vega

1. Disconnect battery ground cable.
2. Disconnect fuel lines from fuel rail and the PCV hose from cam cover.
3. Remove fuel rail support to manifold bolts and lift rail from injectors.
4. Disconnect electrical connectors from injectors and pull injectors from manifold.
5. Replace injector "O" rings.
6. Reverse procedure to install.

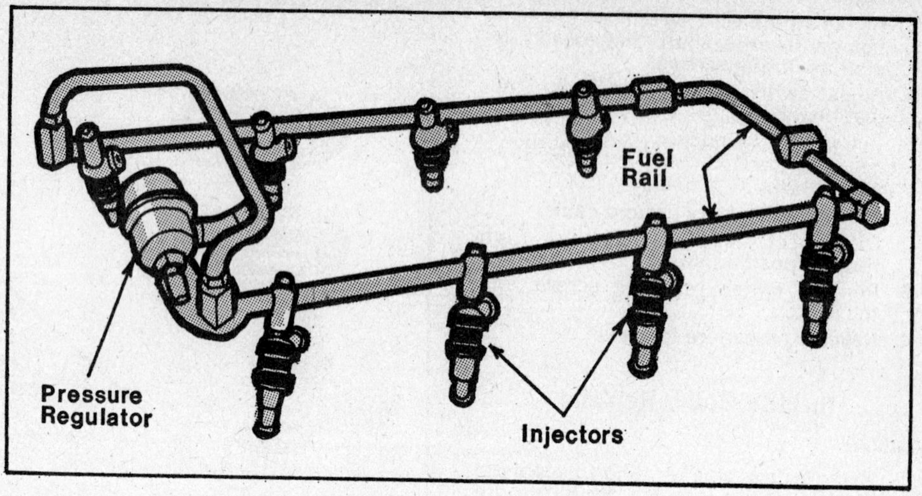

Fig. 18 Fuel rails & injectors. Cadillac

Fuel Rail, Replace

Cadillac Exc. Seville

NOTE: Right and left side fuel rail removal is described under the "Injector Valve, Replace" procedure.

1. If removing front fuel rail, remove flexible hose clamp.
2. If removing rear fuel rail, remove fuel inlet line.
3. Remove flare nut on each side rail fitting.

NOTE: Use a back-up wrench to prevent damage to fitting.

4. Remove fuel rail from vehicle. On front fuel rail removal, disengage fuel rail from pressure regulator hose.
5. Reverse procedure to install. Torque flare nuts to 25-35 ft. lbs.

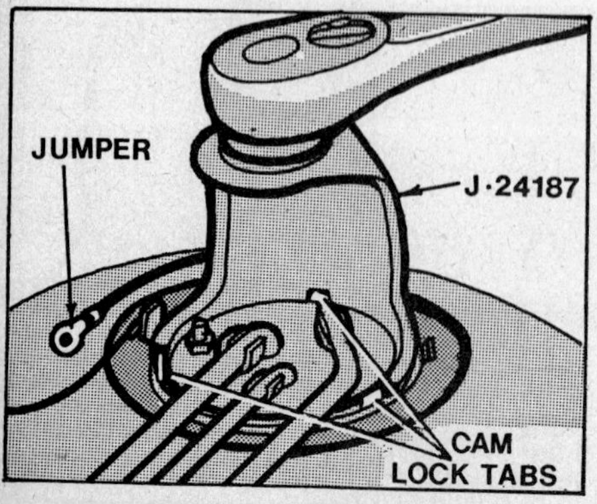

Fig. 19 In-tank fuel pump assembly lock ring removal. Cadillac

Cadillac Seville & Cosworth Vega

Refer to the "Injector Valve, Replace" procedures for fuel rail service on these vehicles.

Fuel Pressure Regulator, Replace

1. Disconnect battery ground cable.
2. Remove vacuum hose from regulator, then the fuel hose clamps.
3. Remove nut securing regulator to bracket, then work regulator off fuel lines and out of bracket. On Cosworth Vega units, note position of fuel inlet line.
4. Reverse procedure to install.

Fuel Filter, Replace

NOTE: The fuel filter element may be replaced by unscrewing the bottom cover and replacing the element, Fig. 4.

Cadillac Exc. Seville

1. Disconnect battery ground cable.
2. Remove fuel inlet and outlet lines from filter.
3. Remove fuel filter to bracket bolts, then the fuel filter.
4. Remove inlet and outlet fittings if needed for a new assembly.
5. Reverse procedure to install. Torque filter mounting screws to 12 ft. lbs.

Cadillac Seville

1. Disconnect battery ground cable and raise vehicle.
2. Disconnect fuel filter inlet hose and remove two screws securing filter to bracket.
3. Pull filter downward to gain access to outlet hose, then remove the outlet hose and fuel filter from vehicle.
4. Reverse procedure to install.

Cosworth Vega

1. Disconnect battery ground cable.
2. Disconnect fuel lines from fuel filter.
3. Loosen fuel filter retaining strap and remove filter assembly.
4. Reverse procedure to install.

Chassis-Mounted Fuel Pump, Replace

Cadillac

NOTE: The chassis-mounted fuel pump is located under the vehicle, forward of the left rear wheel on all models except Eldorado. On Eldorado models, the pump is mounted forward of the right rear wheel.

1. Remove fuel inlet and outlet hoses.
2. Pull back rubber boot and remove nuts from electrical terminals, then the electrical leads.
3. Remove fuel pump to mounting bracket screws, then the fuel pump assembly.
4. Reverse procedure to install. Torque fuel pump mounting screws to 25 inch lbs. Connect 14 dark green wire to the positive terminal and the 14 black wire to the negative terminal.

Fig. 20 In-tank fuel pump assembly lock ring removal. Cosworth Vega

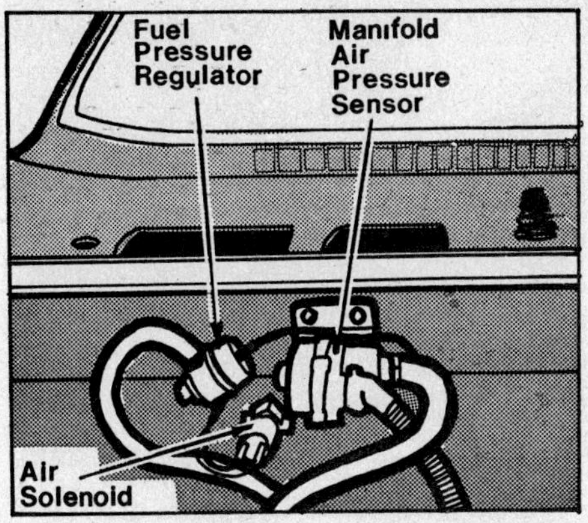

Fig. 21 Sensor cluster. Cosworth Vega

Cosworth Vega

1. Disconnect battery ground cable and raise vehicle.
2. Disconnect fuel lines from pump.
3. Remove two shield screws and the cable strap securing electrical harness to pump assembly.
4. Disconnect electrical connectors from pump and remove pump.
5. Reverse procedure to install.

In-Tank Fuel Pump, Replace

Cadillac

1. Remove fuel tank as outlined under "Fuel Tank, Replace" procedure.
2. Remove fuel gauge tank unit locknuts and the fuel pump feed wires to tank unit.
3. Using tool J-24187, disengage lock ring from fuel tank, Fig. 19. Remove tool and lift gauge-pump unit from tank.
4. Reverse procedure to install.

Cosworth Vega

1. Remove fuel tank as outlined under "Fuel Tank, Replace" procedure.
2. Unscrew retaining ring using spanner wrench J-22554, Fig. 20, and remove pump-tank unit assembly.
3. Remove flat wire conductor from plastic clip on fuel tube.
4. Squeeze clamp and pull pump straight back about ½ inch.
5. Remove two nuts and lockwashers and conductor wires from pump terminals.
6. Squeeze clamp and pull pump straight back to remove it from tank unit. Take care to prevent bending of circular support bracket.
7. Slide replacement pump through circular support bracket until it rests against rubber coupling. Make sure pump has rubber isolator and saran strainer attached.
8. Attach two conductor wires to pump terminals using lockwashers and nuts being certain flat conductor is attached to terminal located on side away from float arm.
9. Squeeze clamp and push pump into rubber coupling.
10. Replace flat wire conductor in plastic clip on fuel pick up tube.
11. Install unit into tank.

Fuel Tank, Replace

Cadillac

1. Disconnect battery ground cable, open fuel filler door and disconnect sending unit feed wire.
2. Siphon fuel from tank.
3. Raise vehicle.
4. Disconnect ground wire from crossmember, fuel supply, ECS and fuel return lines.
5. Support tank and remove screws securing tank support straps, then lower tank and disconnect fuel pump lead.
6. Reverse procedure to install. Torque tank support strap screws to 25 ft. lbs.

Cosworth Vega

1. Disconnect battery ground cable.
2. Disconnect meter and pump wires at rear wiring harness connector.
3. Raise vehicle on hoist and drain fuel tank.
4. Disconnect fuel line hose at gauge unit pick up line.
5. Disconnect tank vent lines to vapor separator.
6. Remove gauge ground wire screw at underbody floorpan.
7. Remove tank straps bolts and lower tank carefully.
8. Reverse procedure to install.

MAP Sensor, Replace

Cadillac

The MAP sensor is located within the ECU on these vehicles and cannot be serviced separately.

Cosworth Vega

1. Disconnect battery ground cable.
2. Disconnect vacuum line and the electrical harness from sensor, Fig. 21.
3. Remove screws at top and bottom of sensor, then the sensor.
4. Reverse procedure to install.

Temperature Sensors, Replace

Cadillac

1. Disconnect battery ground cable and the sensor lead.
2. Remove air temperature sensor from right rear of intake manifold.
3. Remove coolant temperature sensor on V8-350 from intake manifold left front and on V8-500, remove sensor from heater hose outlet on rear of right hand cylinder head, Fig. 22.
4. Reverse procedure to install. Torque sensors to 15 ft. lbs.

Cosworth Vega

1. Disconnect battery ground cable and the sensor lead.
2. Remove air temperature sensor from the bottom of the intake manifold in the No. 3 intake runner, Fig. 9.
3. Remove coolant temperature sensor from thermostat housing.
4. Reverse procedure to install. Torque air temperature sensor to 20 ft. lbs. and the coolant temperature sensor to 28 ft. lbs.

Speed Sensor, Replace

Cadillac

Speed sensor replacement on these vehicles require the disassembly of the HEI distributor. Refer to the Tune Up chapter, "Electronic Ignition System" for service procedures.

Fig. 22 Coolant temperature sensor. Cadillac exc. Seville

Cosworth Vega

1. Disconnect battery ground cable and the speed sensor electrical.
2. Remove speed sensor retainer screws and the retainer, Fig. 23.
3. Note location speed sensor electrical receptacle, then remove speed sensor.
4. Reverse procedure to install. Ensure electrical receptacle is positioned properly.

ADJUSTMENTS

Idle Speed, Adjust

Cadillac

NOTE: Before performing this adjustment, disconnect and plug the distributor vacuum advance hose to eliminate the possibility of advancing the ignition timing, thereby increasing engine RPM.

1. Disconnect and plug vacuum hose at parking brake vacuum release cylinder. Apply parking brake and chock wheels. Disconnect and plug air leveling compressor hose at air cleaner.
2. Connect a tachometer and with engine at operating temperature, place selector lever in "Drive" and turn A/C off.
3. Loosen idle air by-pass adjusting screw locknut and rotate adjusting screw to obtain 600 RPM, Fig. 24. Tighten locknut.

Cosworth Vega

1. Connect tachometer to engine and warm-up engine.
2. Loosen idle stop adjusting screw lock nut on throttle body and rotate idle stop adjusting screw to obtain 1600 RPM, then tighten lock nut.

Throttle Position Switch, Adjust

Cadillac

1. Loosen switch mounting screws.
2. Hold throttle valves in the idle posi-

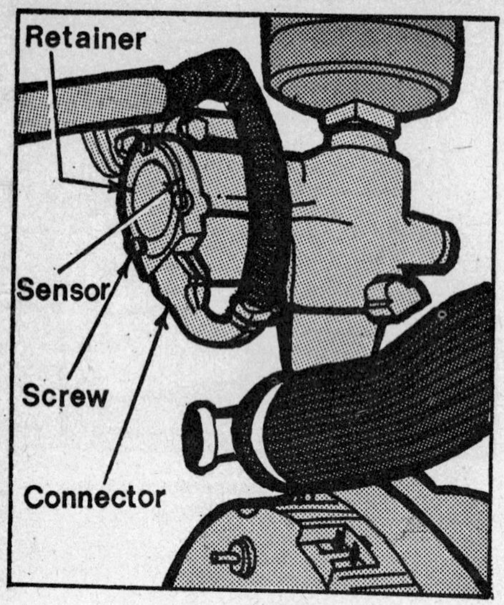

Fig. 23 Speed sensor installation. Cosworth Vega

tion, turn switch counter-clockwise until end-stop is reached and torque switch mounting screws to 11 ft. lbs.

3. Ensure throttle valves return to the throttle stop. If not, repeat adjustment procedure.

Cosworth Vega

Adjustment of the throttle position switch on these vehicles requires the use of a special tester.

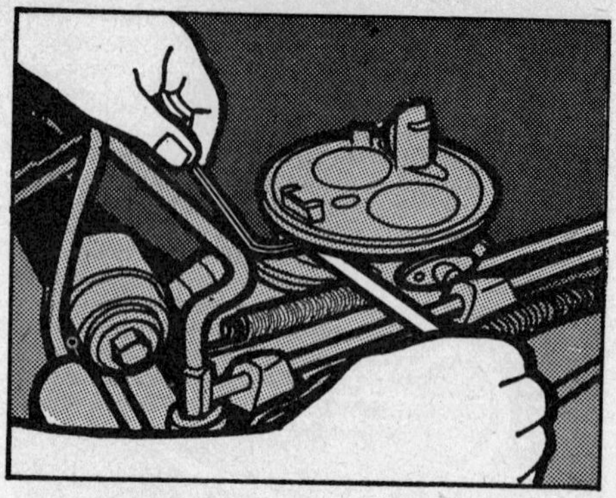

Fig. 24 Idle speed adjustment. Cadillac

TUNE UP SERVICE

CONTENTS

Tune up service has become increasingly important to the modern automotive engine with its vastly improved power and performance. Improved fuel and electrical systems and especially the exhaust emission controls with their inherently critical settings, engines have become more sensitive to usage and operating conditions, which have a decided effect on power and performance. It is important, therefore, that this service be performed on the engine every spring and fall or more often if conditions warrant.

In addition to the servicing of spark plugs, ignition points and condenser, a proper tune up includes a number of tests to check the condition of the engine and its related systems and uncover sources of future problems.

TUNE UP PROCEDURE

Since a quality tune up is dependent upon the proper operation of a number of systems, we have listed here in a logical sequence, the steps to be followed.

1. Diagnosis. This to consist of a compression test, cylinder balance test, oscilloscope check, manifold vacuum test, charging circuit and cranking voltage test.
2. Service spark plugs.
3. Service ignition system, secondary wiring, distributor and coil.
4. Check and service battery and charging system.
5. Service manifold heat valve, if used.
6. Service carburetor, linkage, fuel and air filters.
7. Check operation of various emission control devices.

Once these mechanical checks have been performed the tune up can be finalized. The final steps are:

1. Setting the dwell.
2. Setting slow idle, idle fuel mixture, choke and ignition timing.
3. Adjusting the fast idle.
4. Checking ignition output and secondary resistance.
5. Road test.

DIAGNOSIS & TESTING

Before a satisfactory tune up can be performed, the existing condition of the engine and its related systems must be determined. A tune up should not be attempted if tests indicate internal engine problems such as burnt valves, worn rings, blown head gasket, etc., until such conditions have been corrected.

Oscilloscope Test

Although oscilloscopes differ in many ways, they all display a light or "trace" on a screen which measures the voltage present at a given point and time. As the ignition system operates, its voltage creates a pattern on the screen. This pattern, when read in accordance with the manufacturers instructions for the particular unit, indicates the condition of the entire ignition system.

Compression Test

An engine cannot be tuned to develop maximum power and smooth performance unless the proper compression is obtained in each cylinder.

CAUTION: When cranking the engine for a compression test or any other reason, the coil high tension cable should be removed from the distributor cap and grounded to the engine block or the distributor primary grounded with jumper wire.

---

1. Remove any foreign matter from around spark plugs by blowing out plug area with compressed air. Then remove plugs.
2. Remove air cleaner and block throttle and choke in wide open position.
3. Insert compression gauge firmly in spark plug opening and crank engine through at least four compression strokes to obtain highest possible reading.
4. Test and record compression of each cylinder. Compression should read within the limits given in the *Tune Up Charts* in the car chapters. Refer to chart, Fig. 1, for minimum compression pressures as indicated by percentages.
5. If one or more cylinders read low, inject about a tablespoon of engine oil on top of pistons in the low reading cylinders. Crank engine several times and recheck compression.
6. If compression is now higher, it indicates worn piston rings. If compression does not improve, valves are sticking or seating poorly. If two adjacent cylinders show low compression and injecting oil does not improve the condition, the cause

MAXIMUM PSI	MINIMUM PSI 75%	MINIMUM PSI 80%	MAXIMUM PSI	MINIMUM PSI 75%	MINIMUM PSI 80%	MAXIMUM PSI	MINIMUM PSI 75%	MINIMUM PSI 80%
134	101	107	174	131	139	214	160	171
136	102	109	176	132	141	216	162	173
138	104	110	178	133	142	218	163	174
140	105	112	180	135	144	220	165	176
142	107	114	182	136	146	222	166	178
144	108	115	184	138	147	224	168	179
146	110	117	186	140	149	226	169	181
148	111	118	188	141	150	228	171	182
150	113	120	190	142	152	230	172	184
152	114	122	192	144	154	232	174	186
154	115	123	194	145	155	234	175	187
156	117	125	196	147	157	236	177	189
158	118	126	198	148	158	238	178	190
160	120	128	200	150	160	240	180	192
162	121	130	202	151	162	242	181	194
164	123	131	204	153	163	244	183	195
166	124	133	206	154	165	246	184	197
168	126	134	208	156	166	248	186	198
170	127	136	210	157	168	250	187	200
172	129	138	212	158	170			

Fig. 1 Compression pressure limit chart

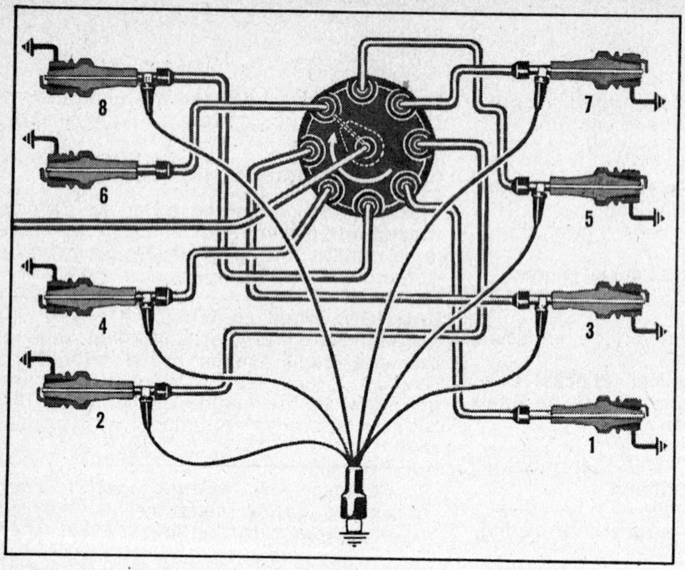

Fig. 1A Cylinder balance test connections. The firing order in this example is 1-8-4-3-6-5-7-2. Therefore, the cylinders to be tested together are 1-6, 8-5, 4-7, 3-2, using the grounding leads as shown

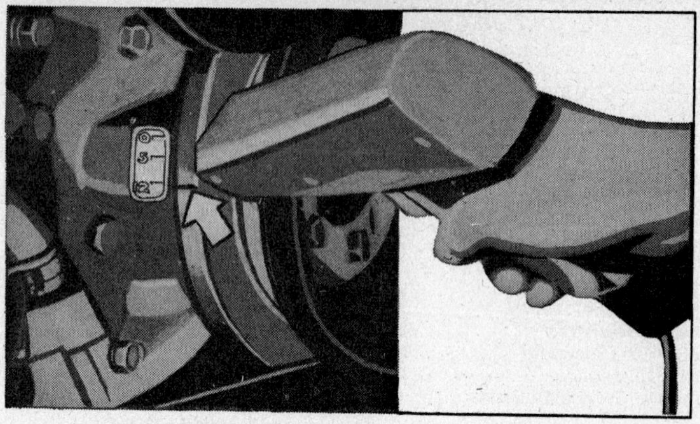

Fig. 2 Checking ignition timing with timing lights

may be a head gasket leak between cylinders.

Cylinder Balance

CAUTION: When performing engine diagnosis on 1975 vehicles equipped with catalytic converters, a "Cylinder Balance Test", Fig. 1A, is *not* to be performed as damage to the catalytic converter will result. The alternate method of shorting out one cylinder at a time and noting the rpm drop of each individual cylinder can be used, but test should be performed as rapidly as possible.

It is sometimes difficult to locate a weak cylinder especially in an eight cylinder engine. A compression test, for example, will not locate a leaky intake manifold, a valve not opening properly due to a worn camshaft, or a defective spark plug.

With the cylinder balance test, the power output of one cylinder may be checked against another, using a set of grounding leads, Fig. 1A. When the power of each cylinder is not equal, the engine will lose power and run roughly. The cylinder balance test is as follows:

1. Connect a tachometer and vacuum gauge.
2. Start engine and run it at a fast idle.
3. Ground large clip of grounding leads and connect individual leads to all spark plugs *except the pair being tested, Fig. 1A.*
4. Divide the firing order in half and write down the first half over the second half. The cylinders to be tested together appear one over the other.

Firing Order	Pairs Tested
1-8-4-3-6-5-7-2	1-6, 8-5, 4-7, 3-2
1-2-7-8-4-5-6-3	1-4, 2-5, 7-6, 8-3
1-5-4-8-6-3-7-2	1-6, 5-3, 4-7, 8-2
1-5-4-2-6-3-7-8	1-6, 5-3, 4-7, 2-8
1-8-7-3-6-5-4-2	1-6, 8-5, 7-4, 3-2
1-3-7-2-6-5-4-8	1-6, 3-5, 7-4, 2-8
1-5-6-3-4-2-7-8	1-4, 5-2, 6-7, 3-8
1-6-5-4-3-2	1-4, 6-3, 5-2
1-5-3-6-2-4	1-6, 5-2, 3-4
1-4-5-2-3-6	1-2, 4-3, 5-6
1-3-4-2	1-4, 3-2
1-4-3-2	1-3, 4-2

5. Operate engine on each pair of cylinders in turn and note engine rpm and manifold vacuum for each pair. A variation of more than one inch of vacuum or 40 rpm between pairs of cylinders being tested indicates that the cylinders are off balance.
6. To isolate one weak cylinder, short out one bank of cylinders at a time. The bank giving the lower readings will include the weak cylinder.

Manifold Vacuum Test

NOTE: On some high performance engines, the camshaft provides such a great degree of valve overlap that vacuum at idle speed will be too low for an accurate test. For this reason, a vacuum test made on such units is of little value.

Manifold vacuum is affected by carburetor adjustment, valve timing, ignition timing, valve condition, cylinder compression, condition of positive crankcase ventilation system and leakage of manifold, carburetor, carburetor spacer or cylinder head gaskets.

Because abnormal gauge readings may indicate that more than one of the above factors are at fault, use care in analyzing an abnormal reading. For example, if the vacuum is low, the correction of one item may increase the vacuum enough to indicate that the trouble has been corrected. It is important, therefore, that each cause of an abnormal reading be investigated and further tests conducted, where necessary, to arrive at the correct diagnosis of the trouble. To check manifold vacuum, proceed as follows:

1. Bring engine to operating temperature.
2. Connect an accurate vacuum gauge to the intake manifold.
3. Operate engine at recommended idle speed.
4. Check vacuum reading on gauge.

Test Conclusions

NORMAL READING: 18 inches or more. Allowance should be made for the effect of altitude on gauge reading.

LOW & STEADY: Loss of power in all cylinders possibly caused by late ignition or valve timing, or loss of compression.

VERY LOW: Intake manifold, carburetor spacer or head gasket leak.

NEEDLE FLUCTUATES STEADILY AS SPEED INCREASES: Partial or complete loss of power in one or more cylinders caused by a leaky head or manifold gasket, burnt valve, weak valve spring or a defect in the ignition system.

GRADUAL DROP IN READING AT IDLE SPEED: Excessive back pressure in exhaust system.

INTERMITTENT FLUCTUATION: Defect in ignition system or sticking valve.

SLOW FLUCTUATION OR DRIFTING OF NEEDLE: Improper idle mixture, carburetor, carburetor spacer, intake manifold gasket leak or restricted crankcase ventilation system.

Cranking Voltage Test

The condition of the starting circuit can be checked by connecting a voltmeter across the battery posts, grounding the coil so the engine will not fire and cranking the engine. If, during cranking, the voltage reading drops below 9.6 volts, there is high resistance in the circuit.

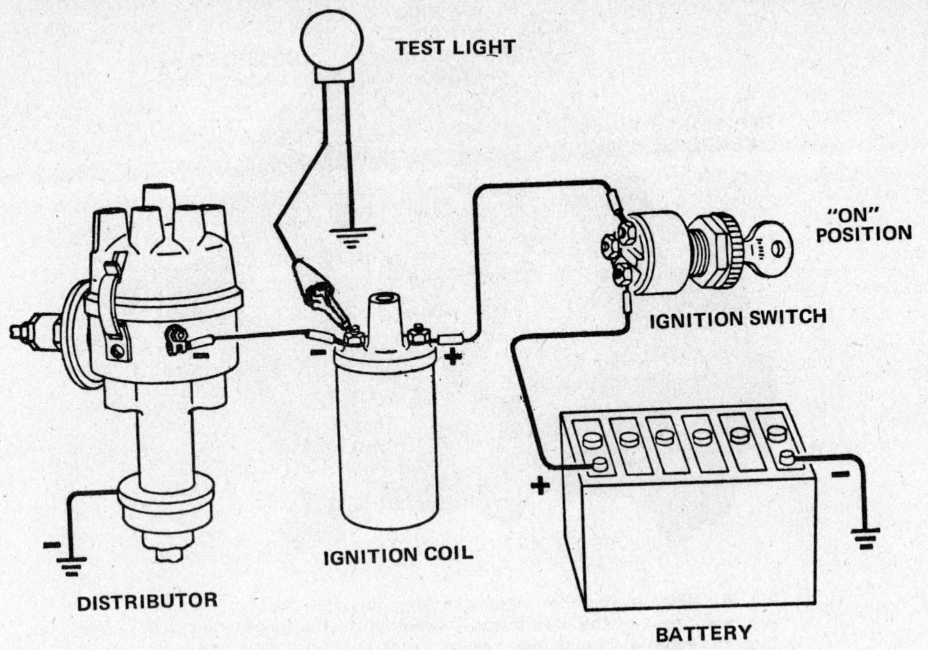

Fig. 3 Jumper light circuit for static ignition timing

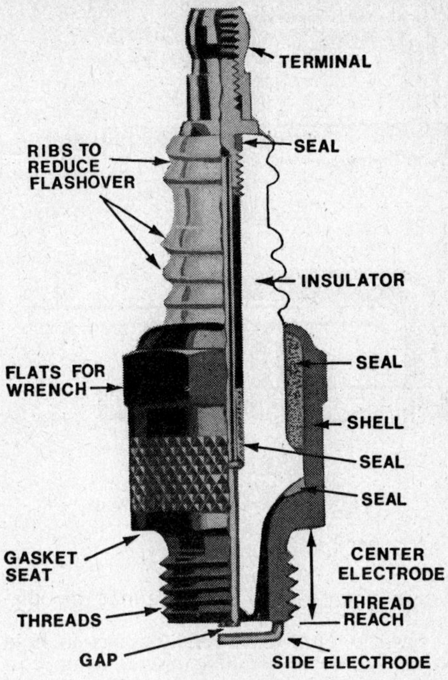

Fig. 4 Spark plug details

Charging Circuit Test

The performance of the charging circuit should be checked during any tune-up. See specific unit section of this manual for test procedures.

Ignition Timing

The use of a timing light, Fig. 2, is recommended for checking and setting ignition timing. This setting is critical especially on units equipped with emission control systems.

Some engines have a provision for monolithic method of ignition timing utilizing a timing receptacle designed to accept an electronic probe. The receptacle is usually mounted at the front of the engine and the electronic probe is connected to electronic equipment which reads out the engine timing. These engines can also be timed using a timing light.

NOTE: When setting timing, be sure idle speed is set lower than speed at which centrifugal advance begins. See *Distributor Specifications Chart* in car chapters.

NOTE: The timing light should be connected to the proper spark plug lead by the use of an adapter. The boots around the connections should not be pierced to connect the light as this can cause spark arcing and misfiring.

Lacking a timing light, the timing can be set with the engine stopped by using a jumper light. Be sure to use a light bulb that corresponds with the vehicle voltage, Fig. 3.

1. Rotate engine until No. 1 cylinder is positioned at the specified timing mark.
2. Connect jumper light between distributor ignition terminal and ground.
3. Turn on ignition switch.
4. Loosen distributor and turn it in the direction of normal rotation until the points just close (light out). Then slowly turn the distributor in the opposite direction just to the exact point that the light goes on. Tighten distributor in this position.

Combustion Efficiency Test

This test checks the carburetor air/fuel mixture by measuring the amount of various chemicals present in the engine exhaust under different conditions. By following the manufacturer's instructions for the specific unit, the carburetor idle, intermediate, high speed and accelerator pump circuits can be checked for proper operation.

This test is especially important when working on engines equipped with exhaust emission controls because of the more critical mixture adjustments on such units.

SERVICE

Spark Plugs

1. Examine firing ends of plugs for evidence of oil fouling gas fouling, burned or overheated condition. *Oil fouling is usually identified by wet, sludgy deposits caused by excessive oil consumption. Gas fouling is identified by dry, black, fluffy deposits caused by incomplete combustion. Burned or overheated spark plugs are identified by white, burned or blistered insulator nose and badly burned electrodes. Improper fuel, insufficient cooling or improper ignition timing normally are the cause. Normal conditions are usually identified by white powdery deposits or rusty-brown to grayish-tan powdery deposits.*
2. Clean plugs with a suitable sand blast cleaner following the manufacturers instructions.
3. Remove carbon and other deposits from threads with a stiff wire brush.

NOTE: Do not use brush to clean electrodes as small pieces of wire can stick inside plug and later cause misfiring.

4. Dress electrodes with a small file to secure flat, parallel surfaces on both center and side electrodes, Fig. 4.
5. Use a round wire gauge to check the gap, Fig. 5, and adjust by bending the side (never center) electrode to the proper specifications as shown in the *Tune Up Charts* in the car chapters.
6. If gaskets are used, place new ones on plugs and torque plugs to specification.

IMPORTANT

Improper installation of spark plugs is one of the greatest single causes of unsatisfactory spark plug performance. Improper installation is the result of one or more of the following practices: 1) Installation of plugs with insufficient torque to fully seat the gasket; 2) excessive torque which changes gap settings; 3) installation of plugs on dirty gasket seal; 4) installation of plugs to corroded spark plug hole threads.

Failure to install plugs properly will cause them to operate at excessively high temperatures and result in reduced operating life under mild operation or complete destruction under severe operation

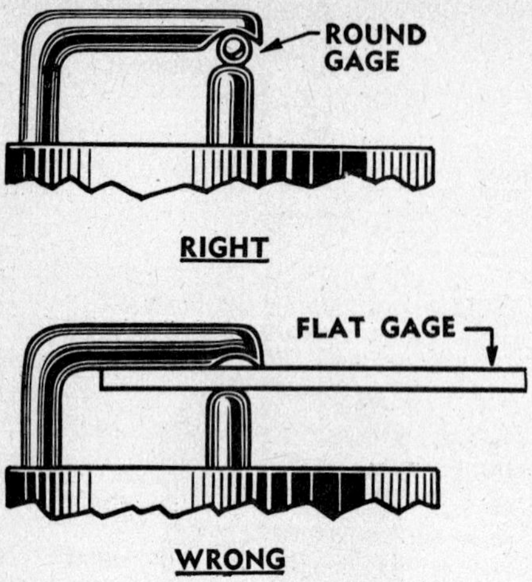

Fig. 5 Correct and incorrect spark plug gauges

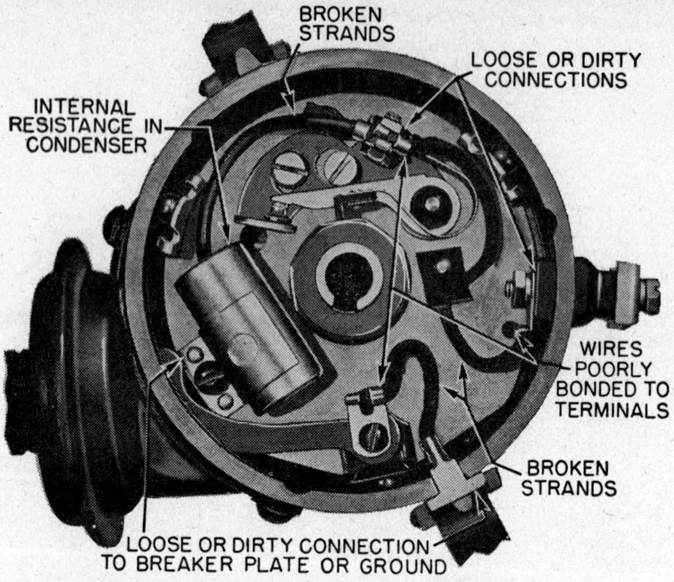

Fig. 6 What to look for when checking for high resistance in the primary circuit of the distributor. In addition to the points indicated, look for external circuit high resistance at ignition switch terminals, ammeter terminals, coil terminals and broken or poorly insulated wires in this circuit

where the intense heat cannot be dissipated rapidly enough.

Always remove carbon deposits in hole threads before installing plugs. When corrosion is present in threads, normal torque is not sufficient to compress the plug gasket (if used) and early failure from overheating will result.

Always use a new gasket (if required) and wipe seats in head clean. The gasket must be fully compressed on clean seats to complete heat transfer and provide a gas tight seal in the cylinder. For this reason as well as the necessity of maintaining correct plug gap, the use of correct torque is extremely important during installation.

Ignition System

1. Check to be sure all connections are clean and tight. Repair or replace any wires that are frayed, loose or damaged, Fig. 6. Replace brittle or damaged spark plug wires.
2. Remove distributor cap, clean and inspect for cracks, carbon tracks and burned or corroded terminals. Replace cap if necessary.
3. Clean rotor and inspect for damage or deterioration. Replace rotor if necessary.
4. Check distributor centrifugal advance mechanism (if used) by turning distributor rotor in direction of running rotation as far as possible, then release rotor to see if springs return it to its retarded position. If rotor does not return readily, the distributor must be disassembled and cause of trouble corrected.
5. Check to see that the vacuum spark control operates freely by turning the movable breaker plate (if used) or distributor housing in a direction opposite to that of running rotation to see if the spring returns it to the retarded position. Any stiffness in the operation of the spark control will affect ignition timing. Correct any interference or binding condition noted.
6. Examine distributor points and clean or replace if necessary. Points with an overall gray color and only slight roughness or pitting need not be replaced.
7. Dirty points should be cleaned with a clean point file. Use only a few strokes of the file. The file should not be used on other metals and should not be allowed to become dirty or greasy. *Never use emery cloth or sandpaper to clean points since particles will embed and cause arcing and rapid burning of points.* Do not attempt to remove all roughness nor dress the point surfaces down smooth. Merely remove scale or dirt.
8. Replace points that are badly burned or pitted. Where burned or badly pitted points are encountered, the ignition system and engine should be checked to determine the cause of the trouble so it can be eliminated. Unless the condition causing point burning is corrected, new points will provide no better service than the old points. See *distributor* chapter for an analysis of point burning or pitting, and for proper installation of points & condenser.

Battery & Cables

Inspect for signs of corrosion on battery, cables and surrounding area, loose or broken carriers, cracked or bulged cases, dirt and acid, electrolyte leakage and low electrolyte level. Fill cells to proper level with distilled water or water passed through a "demineralizer."

The top of the battery should be clean and the battery hold-down bolts properly tightened.

For best results when cleaning batteries, wash first with a dilute ammonia or soda solution to neutralize any acid present and then flush off with clean water. Care must be taken to keep vent plugs tight so that the neutralizing solution does not enter the battery.

To insure good contact, the battery cables should be tight on the battery posts. Oil battery terminal felt washer. If the battery posts or cables terminals are corroded, the cables should be cleaned separately with a soda solution and a wire brush.

If the battery has remained undercharged, check for a loose generator belt, defective generator, high resistance in charging circuit, oxidized voltage regulator contact points, or a low voltage setting.

If the battery has been using too much water, the voltage regulator setting is too high.

Fuel System

All fuel filters and the air cleaner element should be serviced during a tune up, including the sintered bronze "stone" used in some units.

Since carburetion is dependent in several ways on both compression and ignition, it should always be checked last when tuning an engine. Refer to the Carburetor Chapter for pertinent data on specific units.

Service Bulletins

Beginning with 1972 some General Motors V-8 cars may be equipped with the new Uni-Set breaker points (one piece points and condenser), Fig. 7. Due to the limited availability of the Uni-Set for service, you should be aware of the alternatives to circumvent this problem.

1. A conversion kit is available (Part No. 1876065) which contains a breaker plate, contact set, radio shields, condenser and wick.
2. If the existing breaker plate has tapped holes for the condenser clamp and radio shields, it need not be changed and the conversion kit is not necessary. You can install the previous two piece points and condenser but you must obtain and install the radio shields.

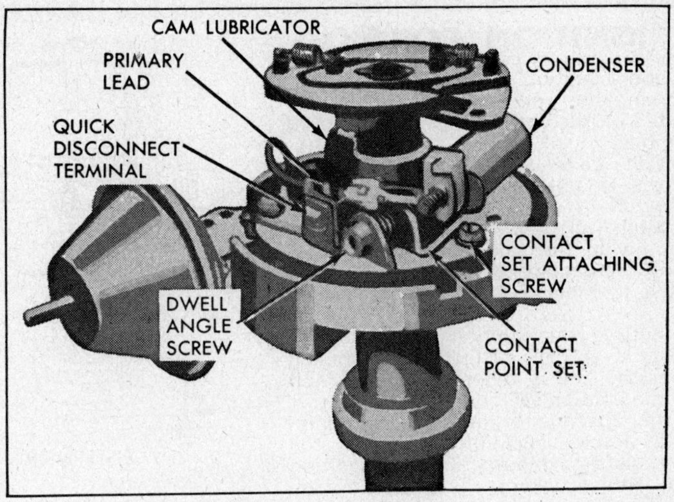

Fig. 7 Delco-Remy one-piece breaker points & condenser (Uni-Set)

Ignition Coils & Resistors

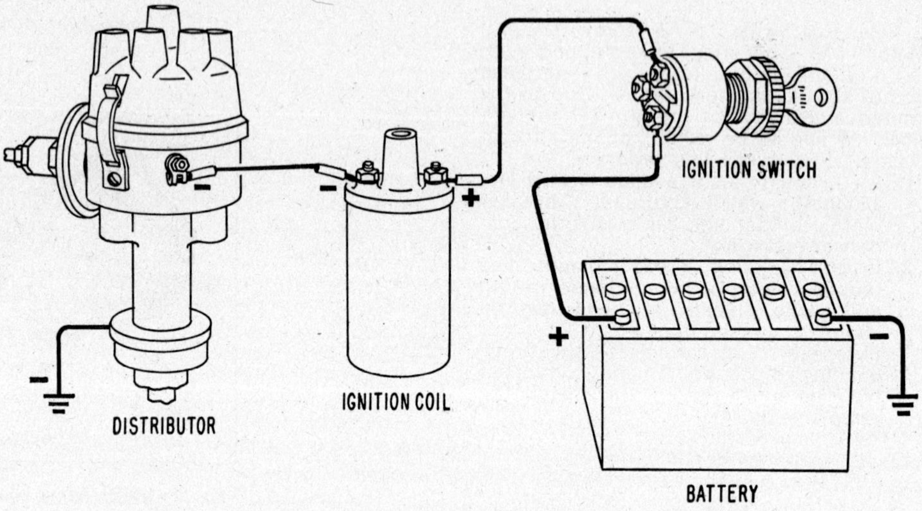

Fig. 1 Wiring connections for coil with negative grounded system

IGNITION COILS

If poor ignition performance is obtained and the coil is suspected, it may be tested on the car or it may be removed for the test.

Ignition coils are often condemned when the trouble is actually in the ignition switch. A completely defective ignition switch will produce an open primary circuit, giving the same indications as if the coil were completely dead. A partly defective ignition switch will cause a weak spark.

By cutting the ignition switch out of the circuit, it can easily be determined whether the coil is defective or the fault lies in the ignition switch.

In the absence of any testing equipment a simple check of an ignition coil can be made as follows: Turn on ignition switch with breaker points closed. Remove the high tension cable from the center socket of the distributor cap and hold it 1/4" to 3/8" away from a clean spot on the engine. If the coil and other units connected to it are in good condition a spark should jump from the wire to the engine as the points are opened. If not, use a jumper wire from the distributor terminal to the engine; if the primary is in good condition a spark will occur.

All ignition coils with metal containers can be tested for gounded windings by placing one test clip on a clean part of the metal container and touching the other clip to the primary and high tension terminals. If the lamp lights or tiny sparks appear at the points of contact, the windings are grounded and the coil should be replaced.

Coil Polarity

Most coils are marked positive and negative at the primary terminals. When installing or connecting a coil be sure to make the connections as shown in Fig. 1. A reversal of this polarity may affect the performance of the engine (or the radio).

If the coil is not marked as to its polarity, it can be checked by holding any high tension wire about 1/4" away from its spark plug terminal with the engine running. Insert the point of a wooden lead pencil between the spark plug and the wire, Fig. 2. If the spark flares and has a slight orange tinge on the spark plug side of the pencil, polarity is correct. If the spark flares on the cable side, coil connections should be reversed.

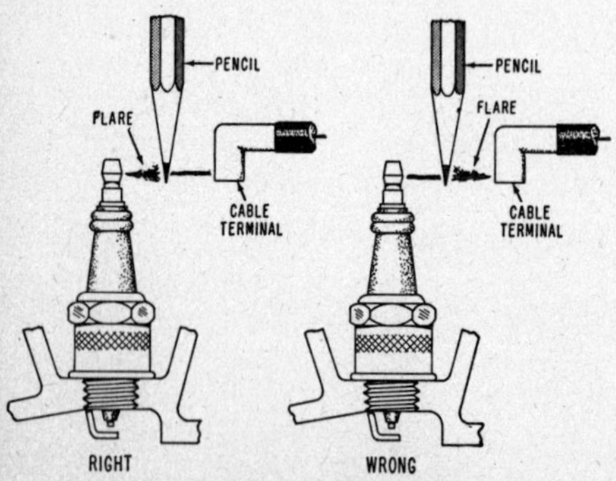

Fig. 2 Checking coil polarity

IGNITION RESISTORS

The ignition coils used with 12-volt systems are specially designed *6-volt coils* which operate with a resistor connected in series with the primary ignition circuit. The purpose of the resistor is to prolong the service life of the distributor breaker points.

Block Type Ballast Resistors

This type resistor came into existance with the introduction of the 12-volt battery. Its basic purpose is to allow full battery voltage to the ignition coil during engine starting, and to reduce battery voltage to the coil when the engine is running. The higher voltage during starts means easier starts. But sustained high voltage to the breaker points can cause point failure. The reduced voltage during engine operation increases break-

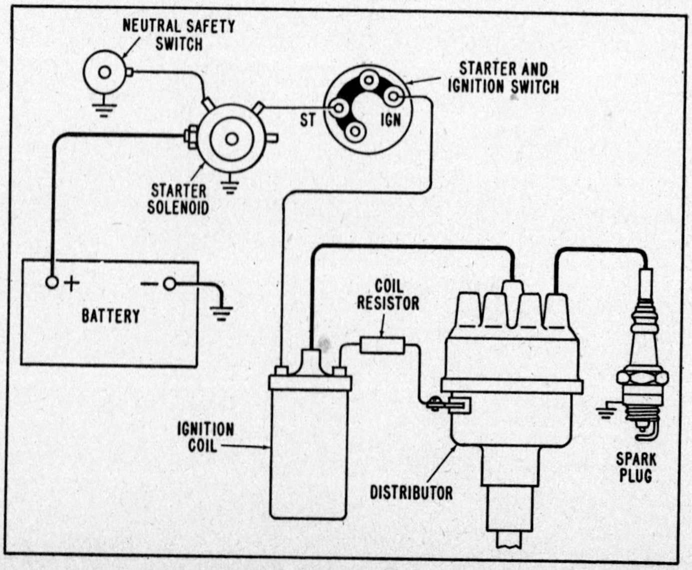

Fig. 3 Chrysler ignition circuit diagram with a temperature sensitive block-type resistor

er point life.

These resistors are normally very dependable. But if one fails, it can be one of the most difficult of all ignition malfunctions to diagnose. An open resistor means that no current reaches the coil and the engine cannot operate. It is possible for the resistance wire to warp or bend enough to touch the side of the case. When this happens, the engine may continue to run but the overall performance will be poor.

Resistors can change value. Any creeping change in resistance value of a ballast resistor is invariably an increase in resistance. This means that coil output to the spark plugs is reduced proportionately. If a ballast resistor is slowly increasing in value the engine could gradually deliver less and less horsepower, particularly under high load conditions. An unsuspecting mechanic could unsuccessfully try to get the engine back to where it will deliver acceptable power output by changing spark plugs, adjusting timing, etc. Replacement of the faulty resistor is the only cure in this case.

To check a ballast resistor, replace it with one of known good quality. Then road-test the vehicle for improved performance.

It is important to remember that new spark plugs can temporarily mask the need for resistor replacement because new plugs require less voltage to fire.

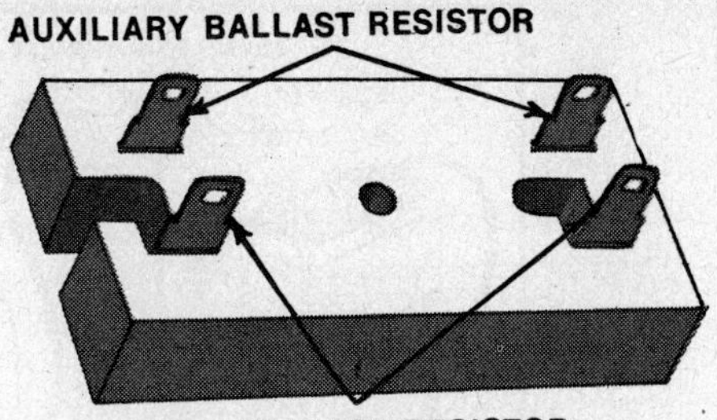

Fig. 4 Dual ballast resistor, Chrysler Corp. electronic ignition system

This is why a new-plug tune-up may prove satisfactory for a time. But if the ballast resistor is faulty, eventually the engine will misfire under load.

On Chrysler Corp. standard ignition systems, the resistor has a lower resistance value when cold than when hot. When the ignition is first turned "On", more current will flow through the primary windings of the ignition coil for a very short time until the resistor heats up.

SERVICE NOTE: If the engine fires when the ignition switch is turned on but quits when the switch is released to its running position, it indicates that the resistor is defective and must be replaced.

A dual ballast resistor, Fig. 4, is used on Chrysler Corp. vehicles equipped with electronic ignition. The normal side of the resistor is a compensating resistance in the primary circuit. At low engine speeds, current is maintained for a longer period of time in this side of the resistor, causing the unit to heat up, in turn increasing resistance. This increased resistance reduces the primary circuit voltage, protecting the ignition coil from high voltage at low engine speeds.

As engine speed increases, the period of time in which current is maintained in the normal side of the resistor is shorter, causing the unit to cool, in turn decreasing resistance. This decreased resistance permits the primary circuit voltage to increase for high speed operation.

During engine start, the normal side of the ballast resistor is by-passed, allowing full battery voltage to be applied to the primary circuit.

The auxiliary side of the ballast resistor limits voltage to the control unit, thereby protecting the unit.

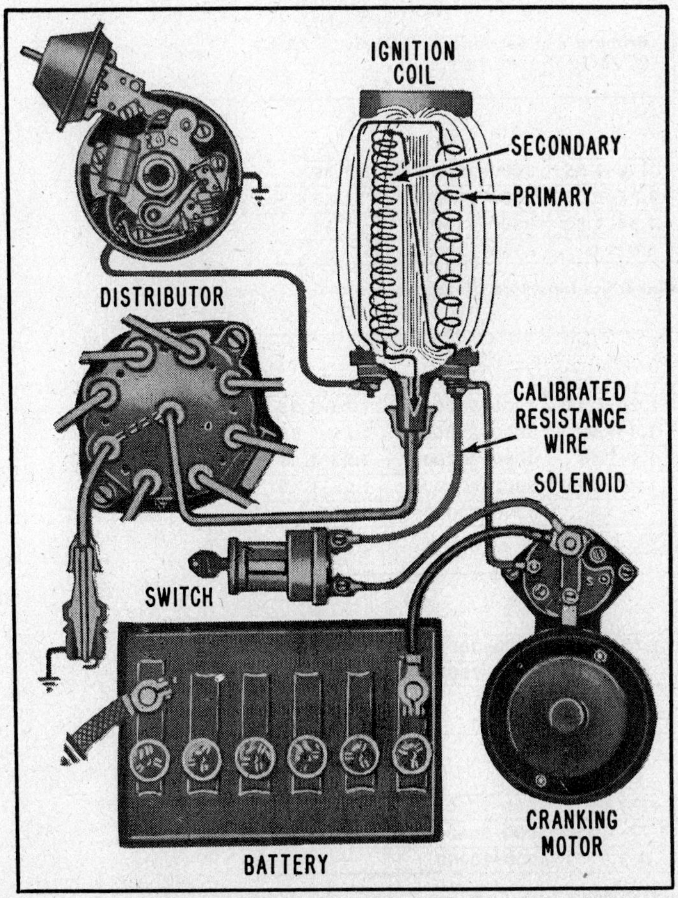

Fig. 5 Delco-Remy ignition circuit diagram with a resistance wire connected to a two-terminal ignition switch

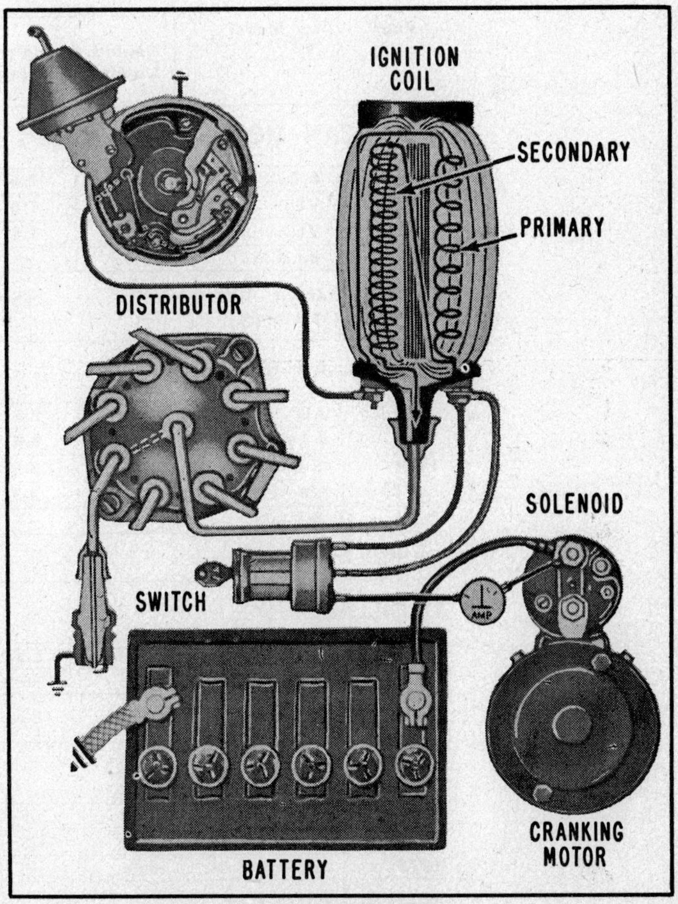

Fig. 6 Delco-Remy ignition circuit diagram with a resistance wire connected to a three-terminal ignition switch

Wire Type Resistors

The special resistance wires used with 12-volt systems are five to six feet long and contained in the regular wiring harness. The wire is made of stainless steel or special alloy, plastic-coated and covered with a glass braid. There is a relatively small temperature rise and the resistance wire is switched out of the circuit for starting and back in again for running.

On Delco-Remy systems Figs. 5 and 6, the resistor is by-passed by means of a "finger" inside the solenoid switch housing which is attached to the additional switch terminal.

On Ford systems, Fig. 7, the resistor is by-passed through a terminal on the starter relay which is connected directly to the positive terminal of the coil.

SERVICE NOTE: If the engine fires when the ignition switch is turned on but quits when the switch is released to the "run" position, it indicates that the resistance wire has lost its continuity or there is a bad connection at the resistor terminals. If the wire is defective, it must be replaced, either by a ballast type resistor or a new length of resistance wire.

IMPORTANT: Do not attempt to operate the engine for an extended length of time with the resistor shorted out by means of a jumper wire as the breaker points will burn up in short order.

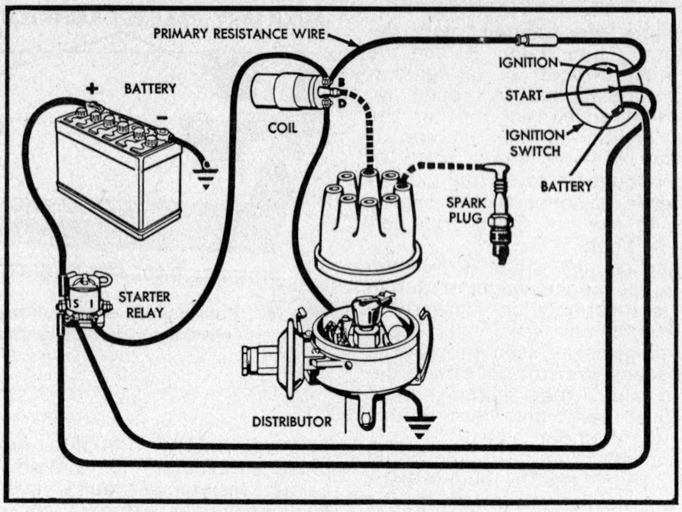

Fig. 7 Ford ignition circuit diagram with a resistance wire connected to a three-terminal ignition switch

IGNITION COIL & RESISTOR SPECIFICATIONS

Year	Model	Coil Draw, Amps.		Coil Resistance, Ohms		Ignition Resistor Ohms @ 75°F.
		Engine Stopped	Engine Idling	Primary @ 75°F.	Secondary @ 75°F.	
AMERICAN MOTORS—All Models						
1969–74	6 Cyl.	3.5	1.6	1.40–1.65	3000–20000	1.80
	V8①	3.5	1.6	1.77–2.05	3000–20000	1.35
	V8②	3.5	1.6	1.64–1.80	9300–11800	1.35
1975	B.I.D.③	—	—	1.0–2.0	8000–12000	—

①—Delco-Remy.
②—American Motors
③—Breakerless Inductive Discharge.

Year	Model	Engine Stopped	Engine Idling	Primary @ 75°F.	Secondary @ 75°F.	Ignition Resistor Ohms @ 75°F.
BUICK—All Models						
1969–73	All	3.8	2.3	1.28–1.42	7200–9500	1.75–1.85
1973	6 Cyl.	4.0	3.9	1.4–1.65	3000–20000	1.75–1.85
1974	6 Cyl.	4.0	3.9	1.4–1.65	3000–20000	1.25–1.45
	V8	3.8	2.3	1.77–2.05	3000–20000	1.25–1.45
1975	H.E.I.①	—	—	0–1.0	6000–30000	—

①—High Energy Ignition.

Year	Model	Engine Stopped	Engine Idling	Primary @ 75°F.	Secondary @ 75°F.	Ignition Resistor Ohms @ 75°F.
CADILLAC						
1969–74	Exc. Below	2.4	1.25	1.77–2.01	3000–20000	1.30–1.35
1974–75	H.E.I.①	—	—	.4–.5	8000–9500	—

①—High Energy Ignition.

Year	Model	Engine Stopped	Engine Idling	Primary @ 75°F.	Secondary @ 75°F.	Ignition Resistor Ohms @ 75°F.
CHECKER MOTORS						
1969–74	6 Cyl.	4.0	1.8	1.41–1.65①	3000–20000	1.80
	V8	4.0	1.8	1.77–2.05	3000–20000	1.35
1975	H.E.I.②	—	—	0–1.0	6000–30000	—

①—1969-70, 1.41-1.63.
②—High Energy Ignition.

Continued

IGNITION COIL & RESISTOR SPECIFICATIONS—Continued

Year	Model	Coil Draw, Amps.		Coil Resistance, Ohms		Ignition Resistor Ohms @ 75°F.
		Engine Stopped	Engine Idling	Primary @ 75°F.	Secondary @ 75°F.	

CHEVROLET—(ALL) MODELS

Year	Model	Engine Stopped	Engine Idling	Primary @ 75°F.	Secondary @ 75°F.	Ignition Resistor Ohms @ 75°F.
1969–74	4 & 6 Cyl.	4.0	1.8	1.41–1.65①	3000–20000	1.80
	V8 Std. Ign.	4.0	1.8	1.77–2.05	3000–20000	1.35
	Trans. Ign.	4.0	1.8	.41– .51	3000–20000	.43– .68
1974-75	H.E.I.②	—	—	0–1.0	6000–30000	—

①—1969-70, 1.41-1.63. ②—High Energy Ignition.

CHRYSLER, DODGE, PLYMOUTH, IMPERIAL—All Models

Year	Model	Engine Stopped	Engine Idling	Primary @ 75°F.	Secondary @ 75°F.	Ignition Resistor Ohms @ 75°F.
1969–72	①	3.0	1.9	1.65–1.79	9400–11700	.50– .60
	②	3.0	1.9	1.41–1.55	9200–10600	.50– .60
1973–74	①	3.0	1.9	1.60–1.79	9400–11700	.50– .60③
	②	3.0	1.9	1.41–1.55	8000–10200	.50– .60③
1975	①	3.0	1.9	1.60–1.79	9400–117000	.50–.60③
	②	3.0	1.9	1.41–1.62	8000–112000	.50–.60③

①Prestolite coils. ②Essex coils.
③Auxiliary (control unit side) 4.75—5.75 ohms.

FORD, MERCURY, LINCOLN, THUNDERBIRD—All Models

Year	Model	Engine Stopped	Engine Idling	Primary @ 75°F.	Secondary @ 75°F.	Ignition Resistor Ohms @ 75°F.
1969–74	Std. Ign.	4.5	2.5	1.40–1.54	7600–8800	1.30–1.40
1974–75	Breakerless	—	—	1.0–2.0	7000–13000	1.30–1.40

OLDSMOBILE—All Models

Year	Model	Engine Stopped	Engine Idling	Primary @ 75°F.	Secondary @ 75°F.	Ignition Resistor Ohms @ 75°F.
1969–70	Toronado	4.0	2.2	1.77–2.05	6500–9500	1.35
	Delmont	6.0	1.35	1.77–2.05	6500–9500	1.35
	6 Cyl.	4.0	1.8	1.45–1.63	6500–9500	1.85
	All Others	4.0	2.0	1.77–2.05	6500–9500	1.35
1971–72	6 Cyl.	4.0	1.8	1.45–1.63	6500–9500	1.85
	V8	4.0	2.2	1.77–2.05	6500–9500	1.35
1973-74	Toronado	4.0	2.2	1.77–2.05	6500–20000	1.35
	6 Cyl.	4.0	1.8	1.40–1.65	3000–20000	1.80
	V8	4.0	2.0	1.77–2.05	6500–20000	1.35
1974-75	H.E.I.①	—	—	0–1.0	6000–30000	—

①—High Energy Ignition.

PONTIAC—All Models

Year	Model	Engine Stopped	Engine Idling	Primary @ 75°F.	Secondary @ 75°F.	Ignition Resistor Ohms @ 75°F.
1969	Pontiac	3.2–3.6	1.9–2.3	1.7–2.0	3000–20000	...
	6 Cyl.	3.5	2.8	1.4–1.7	3000–20000	...
	Temp., F-Bird	3.4	2.1	1.7–2.0	3000–20000	...
1970	Six	3.4	2.1	1.4–1.7①	3000–20000	...
	V8	3.4	2.1	1.7–2.0	3000–20000	...
1971–72	Six	3.5②	2.8③	1.4–1.65	3000–20000	...
	V8	3.4	2.1	1.7–2.0	3000–20000	...
1973-74	Six	...	...	1.4–1.65	3000–20000	...
	V8	...	...	1.8–2.2	3000–20000	...
	Unitized	...	...	0–.5	6000–9000	...
1975	H.E.I.④	—	—	0–1.0	6000–30000	—

①—Firebird 1.4–1.65 ②—Ventura II 4.0 ③—Ventura II 1.8 ④—High Energy Ignition.

Standard Ignition Distributors

CONTENTS

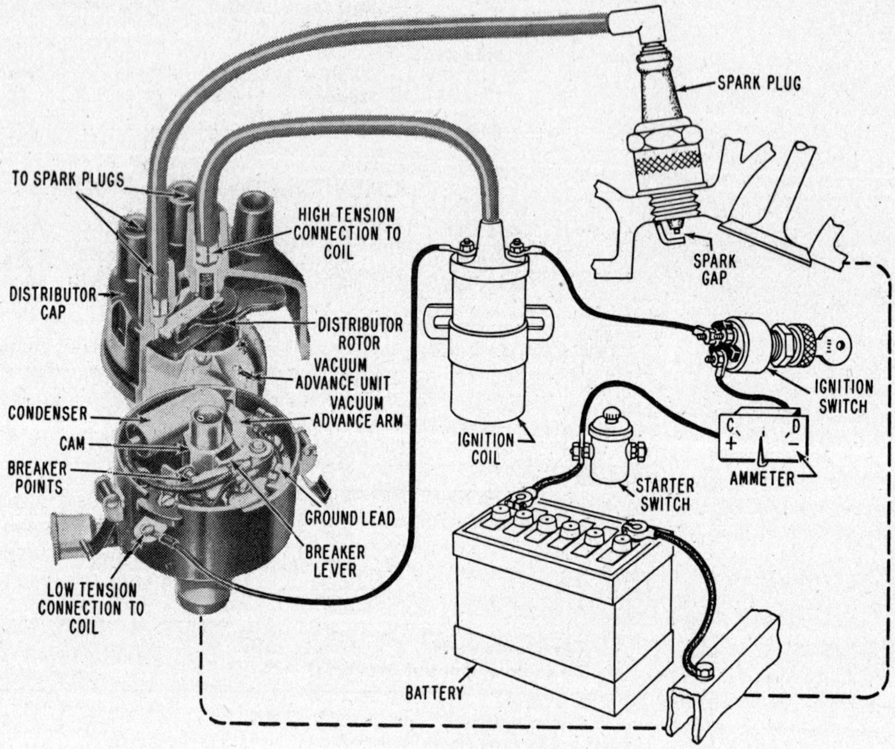

Fig. 1 Typical ignition system

BREAKER CONTACT POINTS

Contact Analysis

The normal color of points should be a light gray. If the contact surfaces are black it is usually caused by oil vapor or grease from the cam. If they are blue, the cause is usually excessive heating due to improper alignment, high resistance or open condenser circuit.

If the contacts develop a crater or depression on one point and a high spot of metal on the other, the cause is an electrolytic action transferring metal from one contact to the other, Fig. 2, due to an unbalanced ignition system, which can sometimes be improved by a slight change in condenser capacity. If the mound is on the positive point, Fig. 3, install a condenser of greater capacity; if on the negative point, Fig. 4, use a condenser of lesser capacity.

One of the most common causes of point failure is the presence of oil or grease on the contact surfaces, usually from over-lubrication of the wick at the top of the cam or too much grease on the rubbing block of the breaker arm.

Breaker Point Gap

If points are set too close, arcing and burning will occur, causing hard starting and poor low speed performance. If points are set too wide, the cam angle or dwell will be too small to allow saturation of the coil at high engine speeds, resulting in weak spark.

Contact point opening has a direct bearing on cam angle or dwell which is the number of degrees that the breaker cam rotates from the time the points close until they open again, Fig. 5. The cam angle or dwell increases as point opening is decreased and vice versa. If point gap is set with a feeler gauge, the cam angle or dwell should be checked either with a portable dwell meter or by installing the distributor in a distributor tester.

Breaker Arm Spring Tension

Breaker arm spring tension is important. If the tension is too great the arm will bounce, causing an interruption of the current in the coil and misfiring. If the spring tension is too little, the rubbing block will not follow the cam, causing a variation in cam dwell. The spring tension should always be set at the high limit as given in the *Distributor Specifications* chart in the car chapter, as it will be reduced as the rubbing block wears.

Hook a spring scale on the breaker arm and pull in a straight line as shown in Fig. 6. Take a reading as the points start to separate under the slow and steady pull of the scale. If the tension is not within specifications, loosen the screw that holds the end of the point spring and slide the end of the spring in or out as necessary. Tighten the screw and recheck the spring tension.

Breaker Point Alignment

Check alignment of points with points closed, Fig. 7. Align new points where necessary but do not attempt to align used points. Instead, replace used points where serious misalignment is observed. After aligning points, adjust point gap.

Adjusting Breaker Gap

Specifications for breaker gap, *as measured with a feeler gauge,* are listed in the *Tune Up Specifications* in the car chapters. However, if at all possible, this should be set on a distributor tester, with a dial indicator, Fig. 8, or by hooking up a portable dwell meter with the distributor cap and rotor removed and, while cranking the engine, setting the dwell.

NOTE: When setting the dwell with the engine cranking, be sure to ground the coil secondary lead and do not operate the starter for sustained periods at a time.

This eliminates the possibility of an incorrect gap because of rough points, Fig. 9.

The advantage of a distributor testing machine is that it not only measures cam angle or dwell but it also uncovers irregularities between cam lobes, point bounce, alignment of rubbing block with cam, alignment of contacts and breaker arm spring tension.

Setting Dual Points

The distributor used on some Chrysler Corp. and Ford Co. vehicles contain two sets of points which permit additional current build-up in the primary winding of the coil, Fig. 10. Thus, maximum voltage is induced in the secondary winding.

The two sets of points are connected

Fig. 2 Showing how metal from one contact transfers to the other

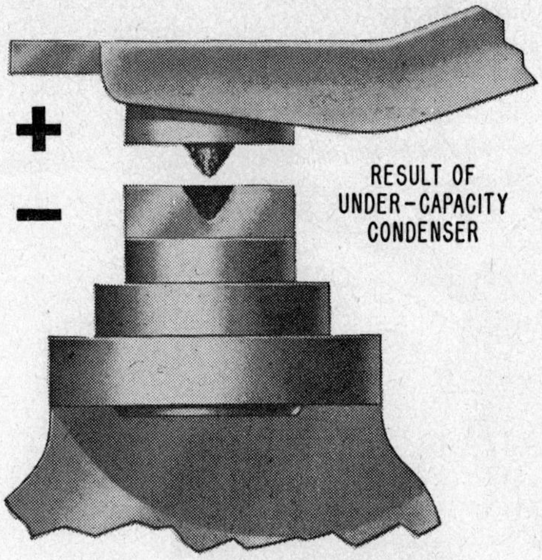

Fig. 3 Mound on positive point

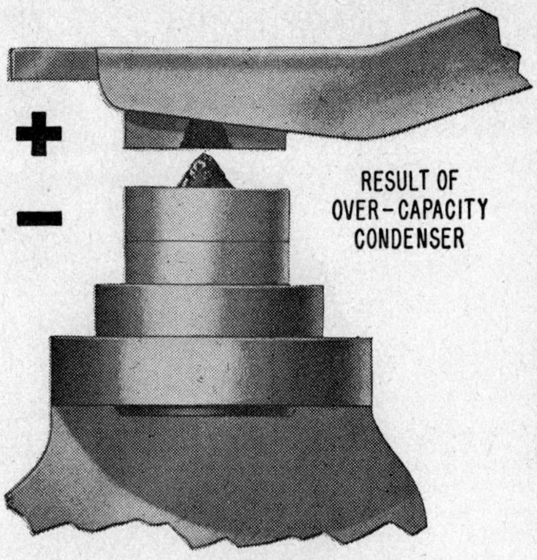

Fig. 4 Mound on negative point

in parallel and are positioned in relation to the 8-lobe cam so as to provide approximately a 7 to 8 degree overlap of points opening and closing. One set of points (circuit *maker* points) closes the primary circuit in the coil and the second set of points (circuit *breaker* points) opens the circuit, causing a spark at the plug. Immediately after the spark occurs, the circuit *maker* points are closed ahead of the circuit *breaker* points, thus providing a circuit to build-up the primary winding. As the cam rotates further, the secondary points close and just before the secondary points open, the primary points open 7 degrees ahead.

Since the "make" and "break" points are timed to close and open at the exact instant necessary for efficient engine operation, adjustment of the points is an important factor in correct distributor operation.

Feeler Gauge or Dial Indicator Method—Rotate the distributor shaft until the breaker arm rubbing block of one set of points is on the high spot of the cam. Then, with a screwdriver blade in the triangular opening, close or open the points to the proper clearance by turning the screwdriver blade against the stationary point plate. Check the clearance with a clean wire gauge or dial indicator. Then turn the distributor shaft until the rubbing block of the second set of points is on the high spot of the cam and adjust the second set of points in the same manner.

Dwell Meter Method—If this method is used *block one set of points open* with a piece of wrapping paper or cardboard. Then adjust the other set of points to the correct dwell angle. Now block open the first set of points and adjust the

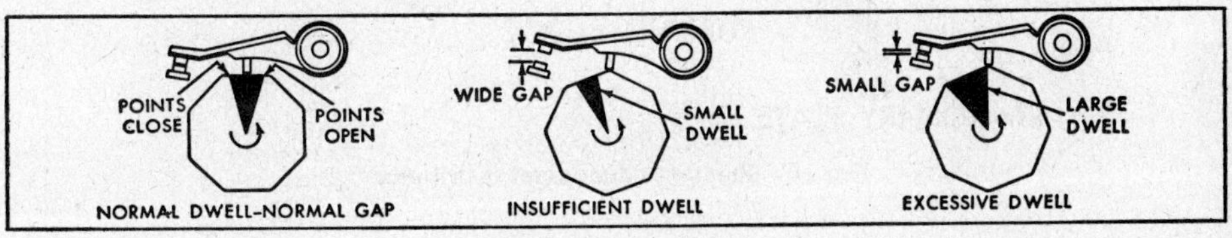

Fig. 5 Cam angle or dwell

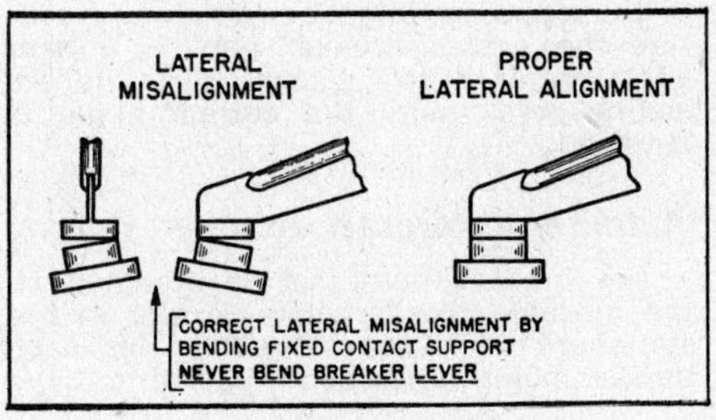

Fig. 7 Breaker point alignment

Fig. 6 Measuring breaker spring tension

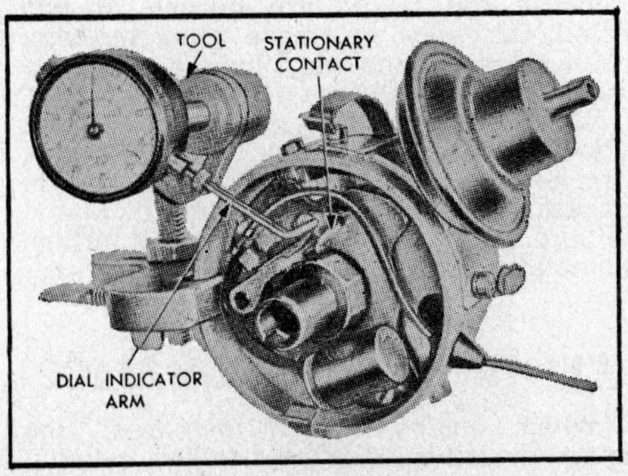

Fig. 8 Dial indicator for measuring breaker gap

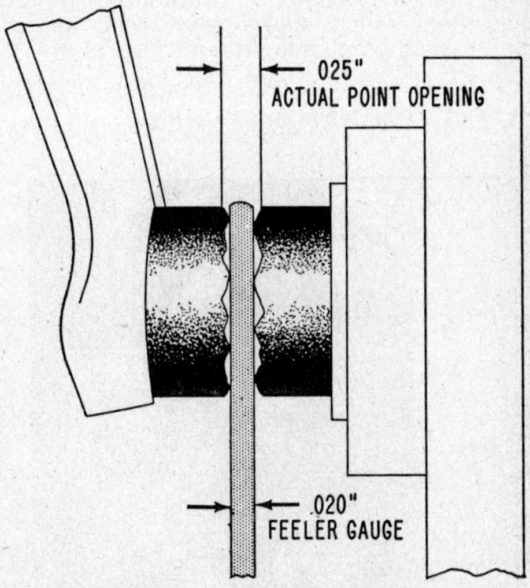

Fig. 9 Why flat feeler gauge will not provide accurate point spacing if points are rough

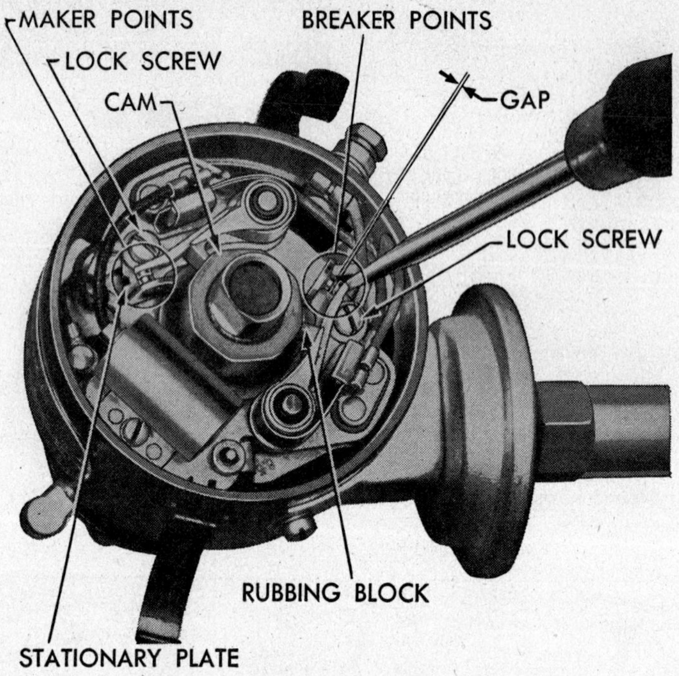

Fig. 10 Prestolite dual point distributor

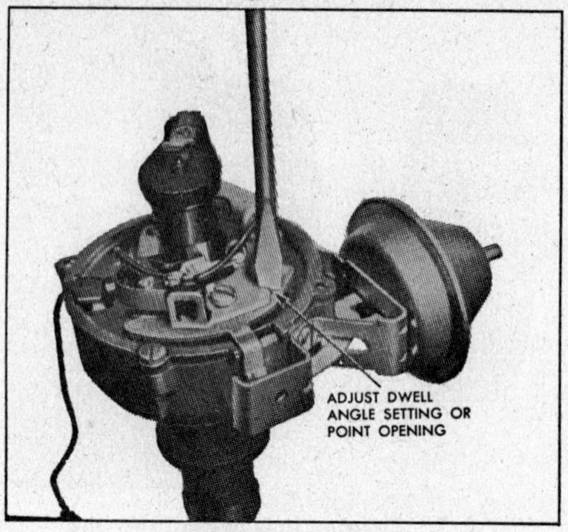

Fig. 11 Breaker point setting. Delco-Remy internally adjusted points

second set. After both sets of points have been adjusted, allow them to operate together while checking that their total dwell angle measures up to manufacturer's specifications.

Prestolite & Chrysler Single Point Set Adjustment

Adjustment of breaker gap is accomplished by loosening the lock screw in the stationary point and moving the adjusting screw as required to obtain the correct breaker gap.

Delco-Remy Internal Adjustment Breaker Points

A slot is provided in the contact point assembly which allows easy dwell angle or breaker point adjustment, Fig. 11.

Delco-Remy External Adjustment Breaker Points

With engine running at idle speed, the breaker gap is adjusted by first raising the window provided in the cap and inserting a "hex" wrench into the adjusting screw, Fig. 12. Turn the adjusting screw clockwise until the engine begins to misfire. Then give the wrench one-half turn in the opposite direction which will provide the proper breaker gap. If a cam angle meter is to be used, turn the adjusting screw until the correct angle is obtained.

Autolite/Motorcraft Breaker Points

The breaker point set is attached to the movable breaker plate. A slot in the stationary point bracket allows for easy breaker point adjustment, Fig. 13.

CONDENSER

A condenser should not be condemned because the points are burned or oxidized. Oil vapor, or grease from the cam, or high resistance may be the cause of such a condition.

Condensers should be tested with a good condenser tester for leakage, break-down, capacity, and resistance in series in the condenser circuit. Manufacturers of condenser testers furnish complete instructions as to their use.

CENTRIFUGAL ADVANCE

When engine speed increases, the spark must be introduced in the cylinder earlier in the cycle in order that the fuel charge can be ignited and will have time to burn and deliver its power to the piston. To provide this spark advance based on engine speed, the centrifugal governor mechanism is used.

This mechanism, Fig. 14, consists of centrifugal advance weights which throw out against spring tension as the engine speed increases. This movement imparts, through a toggle arrangement, rotational motion to the breaker cam or plate, depending on the model, causing it to rotate a number of degrees with respect to the distributor drive shaft. This causes the points to be opened and closed earlier in the cycle so the spark is delivered to the cylinder earlier.

In servicing the distributor, all weights should be removed from the hinge pins, cleaned and checked for excessive wear, either in the weights or pins, or the plate which is slotted for the movement of the pins on top of the governor weights. Replacement should be made if there is any appreciable wear in the slots, as any wear at this point would change the characteristic of the spark advance.

If these parts are in good condition, the hinge pins should be lubricated before being reassembled, by greasing the hinge pins and filling the pockets in the governor weights with grease. Do not use vaseline for this purpose as its melting point is comparatively low.

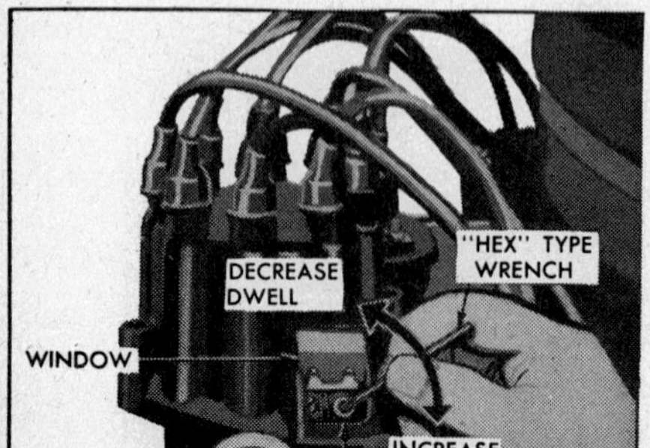

Fig. 12 Adjust breaker point gap through window in distributor cap

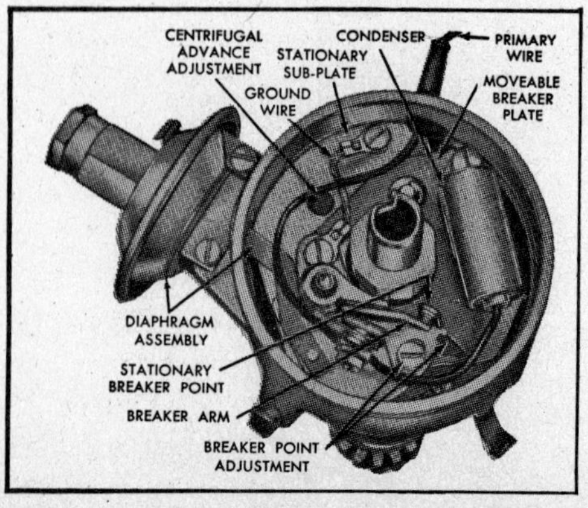

Fig. 13 Breaker plate installation. Autolite/Motorcraft dual advance distributor (Typical)

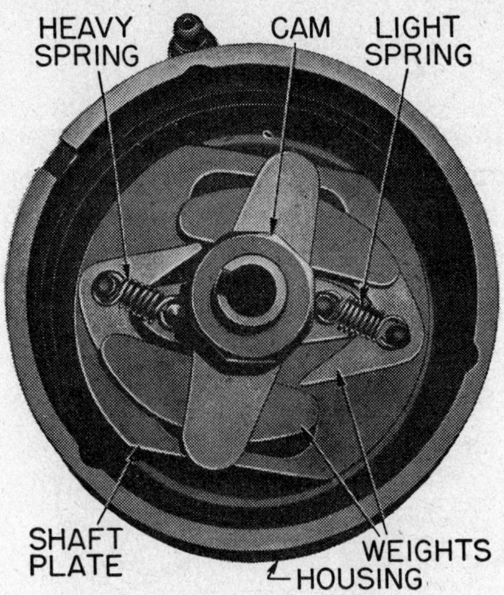

Fig. 14 Top view of Delco-Remy distributor with breaker plate removed to show centrifugal governor mechanism

When installing new centrifugal governor assemblies, it is important that the spacer washers between the housing and shaft be installed correctly. If incorrectly installed, the governor assembly will be too high, causing it to rub against the bottom of the breaker plate.

On some distributors, both springs are alike, while on others there is one heavy and one light spring, as in Fig. 14. Another combination that may be found is an additional flat spring on the outside of the outer spring posts, Fig. 15. As the governor speed is increased, the flat springs are first pulled against the posts by the eyes of the coil springs to provide a rapid spark advance of a few degrees before the coil springs pull against the spring posts.

VACUUM ADVANCE

Conventional Type

The vacuum advance unit consists of a spring loaded diaphragm, which is connected through linkage to the distributor breaker plate. The spring loaded side of the diaphragm is connected through a vacuum line to the carburetor or intake manifold. As vacuum increases, the diaphragm is drawn toward the source of vacuum, the diaphragm linkage is pulled with it and the breaker plate, attached to the linkage, is turned to advance the timing.

Exhaust Emission Control Types

Chrysler C.A.P.

The Chrysler vacuum advance units incorporate a vacuum advance control valve in the advance vacuum circuit to provide the necessary retard during closed throttle operation.

Chrysler C.A.S.

Some Chrysler C.A.S. engines have a solenoid incorporated in the distributor vacuum advance mechanism to retard the ignition timing when the throttle is closed. At closed throttle, electrical contacts on the carburetor throttle stop, with idle adjusting screw in the closed position, cause the distributor solenoid to energize. This retards the ignition timing to provide reduced exhaust emissions under hot idle conditions. Cold or part throttle starting is not penalized because the distributor solenoid is not energized unless the hot idle adjusting screw is against the throttle stop contact. Timing must be set at closed throttle to give accurate setting.

Ford Dual-Diaphragm

This unit consists of two independent diaphragms. The outer diaphragm uses carburetor vacuum to advance timing. The inner diaphragm uses intake manifold vacuum to provide additional retard during closed throttle operation.

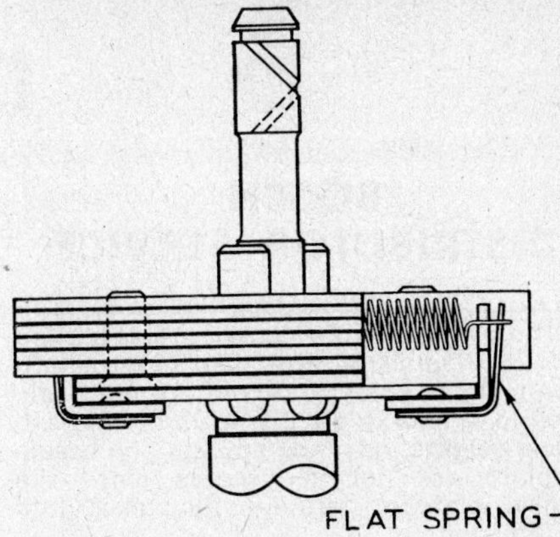

Fig. 15 Flat spring used on some governors to provide a rapid spark advance

Ford Single Diaphragm

This unit operates in the same manner as conventional units.

General Motors CCS

In this system, the advance is the ported type, that is the vacuum take-off is located above the throttle plate(s) so that during periods of closed throttle operation there is little or no vacuum reaching the advance unit and timing is retarded. As soon as the throttle is cracked, vacuum reaches the advance unit and timing is advanced.

NOTE: Some models of the Chevrolet V8-307 engine use the conventional type advance unit.

Distributor Service

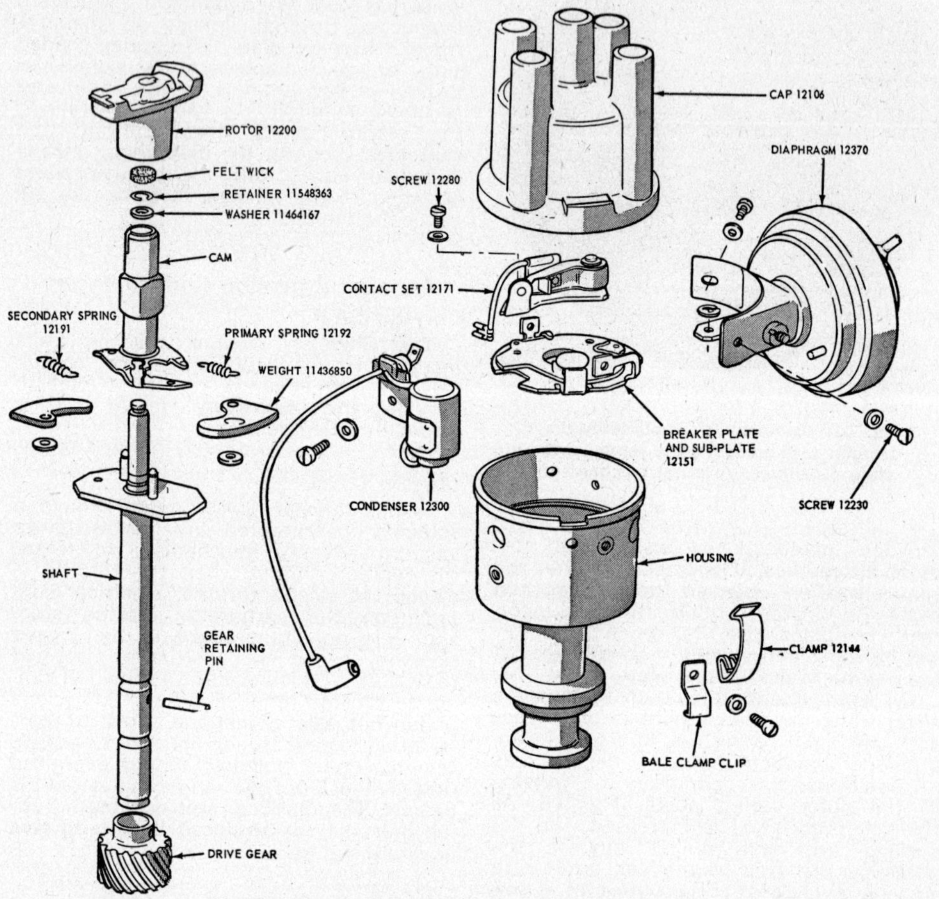

Fig. 16 Exploded view of Bosch distributor

BOSCH DISTRIBUTOR SERVICE

NOTE: On 1971-74 units, replacement parts for servicing the distributor shaft, drive gear, bushing and cam are not available. In addition on 1974 units, the advance weights, advance springs and breaker plate are not serviceable since the breaker plate is permanently staked into the housing.

If distributor has been disassembled, refer to Fig. 16 at reassembly.

1. Lubricate pivot pins and install advance weights. Lubricate shaft and install cam assembly.
2. Install advance springs and advance plate, securing plate with clip and screws.
3. Insert grommet and condenser wire through hole in housing and install condenser and contact points.
4. Install vacuum advance unit. Hook vacuum advance rod over pin in advance plate and install snap ring.

PRESTOLITE & CHRYSLER DISTRIBUTOR SERVICE

Shaft & Bushing Wear Test

1. Remove distributor from vehicle and clamp distributor in a vise. Use extreme caution not to damage distributor.
2. Install a dial indicator on housing so plunger rests against moveable contact arm when rubbing block is on the highest point of a cam.
3. Place a wire loop around distributor shaft and hook a spring scale on the other end of the loop. Apply a one pound pull in line with indicator plunger and read movement on indicator. Movement must not exceed .006 inch. If movement exceeds limit, replace either housing or shaft on Chrysler distributors or bushing or shaft on Prestolite distributors to bring movement back within tolerance.

Distributor Reassembly

If the distributor has been disassembled, reassemble as follows, referring to Figs. 17 and 18 for guidance.

1. Check operation of centrifugal weights and weight springs for distortion. Lubricate governor weights.
2. Inspect all bearing surfaces and pivot pins for roughness, binding, or excessive looseness.
3. Install cam spacer (chamfered end down) on distributor shaft.
4. Slide cam and yoke on distributor shaft. Engage weight lugs with slots in yoke as shown in Fig. 19. Install cam retaining clip, being sure it is properly seated in distributor shaft groove.
5. Lubricate and install two concave washers for Prestolite distributors, or a single flat thrust washer for Chrysler distributors. Position washers on shaft and slide shaft into distributor body. On Chrysler six cylinder distributors, if drive gear is worn or damaged, replace as follows:
   a. Install thrust washer and old gear on shaft and install pin. Scribe a line centered between two gear teeth from center to edge of shaft and in line with center of rotor electrode.
   b. Remove pin and gear. Clean burrs from around pin hole and install new gear.
   c. Place a .007 inch feeler gauge between gear and thrust washer and drill a .124-.129 inch hole in gear and shaft about 90 degrees from old hole in shaft and with scribe line centered between gear teeth and in line with center of rotor electrode as shown in Fig. 18A.

NOTE: If new pin hole location appears to interfere with shaft oil groove, rotate gear to centerline of next pair of gear teeth and align with scribe line on shaft.

   d. Install pin.
6. Position lower thrust washer and drive collar on lower end of shaft and install retainer pin.
7. Install oiler wick and oiler.
8. Install breaker plate assembly, align condenser lead, breaker point spring, primary lead and install attaching screw.
9. Install felt wick in top of cam.
10. Attach vacuum advance unit arm to breaker plate and install retainer. Install vacuum unit attaching screws and washers.
11. Test breaker arm spring tension and adjust breaker gap.
12. Lubricate felt pad in top of distributor cam with 3 to 5 drops of light engine oil and install rotor.

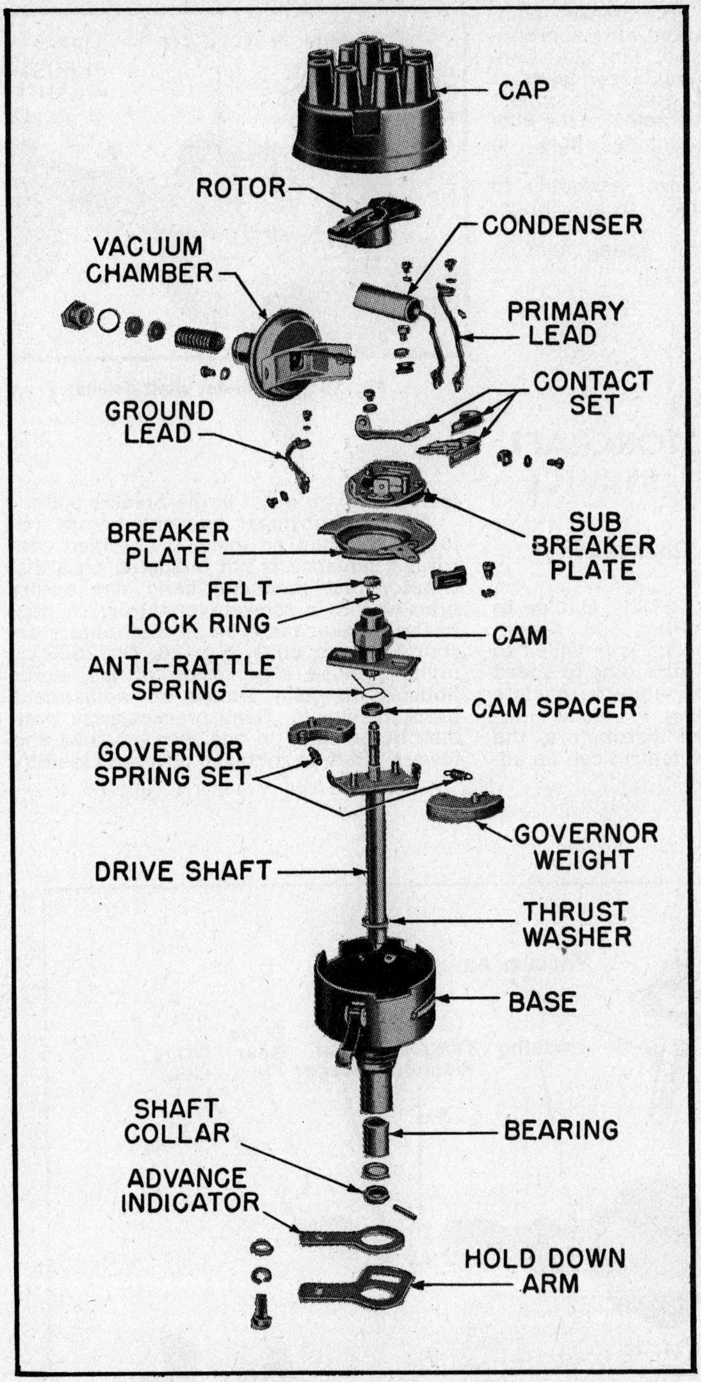

Fig. 17 Exploded view of typical Prestolite distributor

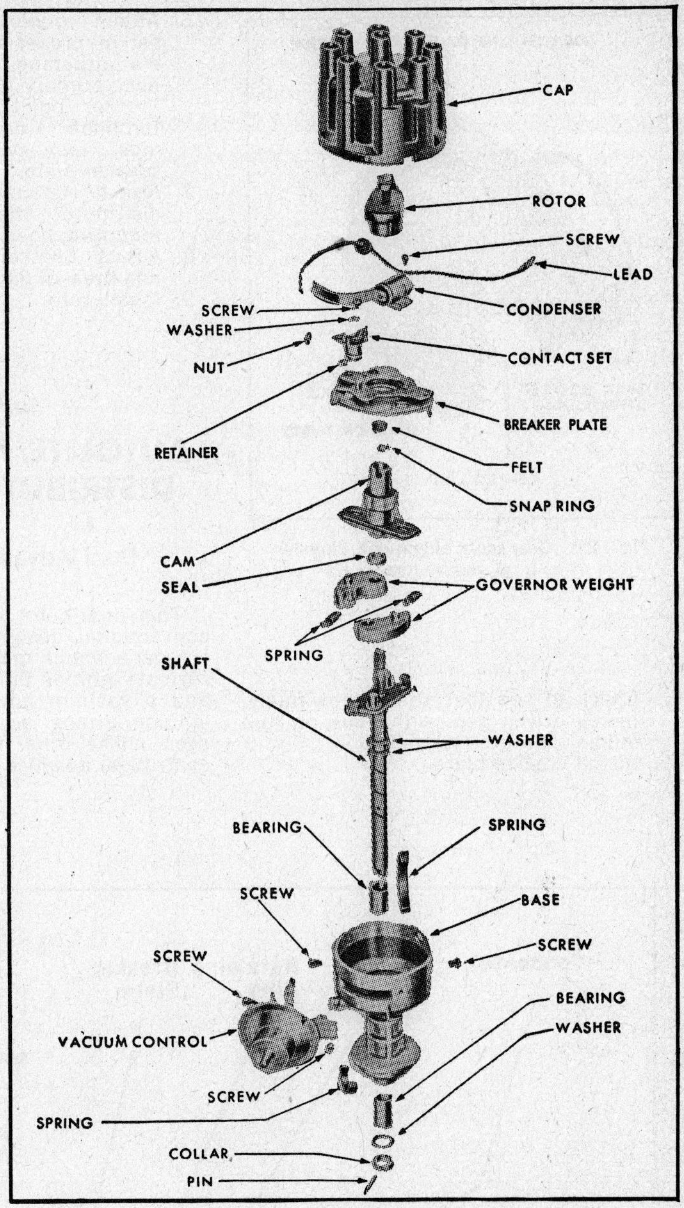

Fig. 18 Exploded view of typical Chrysler V-8 distributor

DELCO-REMY DISTRIBUTOR SERVICE

External Adjustment Type

If the distributor has been disassembled, refer to Fig. 20 for guidance when reassembling.

1. Fill housing lubricating cavity with proper lubricant and install new plastic seal and felt washer.
2. Install vacuum advance unit, breaker plate and spring retainer on upper bushing.
3. Slide weight cam over mainshaft and install weights and springs.
4. Insert mainshaft assembly into housing, indexing assembly with drive gear and washers.
5. On Chevrolet Corvette units, install tachometer drive gear.
6. On all units, slide gear shims and drive gear over mainshaft and install retaining pin. Ensure shaft rotates freely.
7. Install breaker points and condenser.

Internal Adjustment Type

If the distributor has been disassembled, refer to Fig. 21 when reassembling. Fig. 22 shows the details of the breaker plate and attaching parts.

1. Replace cam assembly to shaft. Lubricate top end of shaft with light engine oil prior to replacing.
2. Install weights on their pivot pins. Install springs, weight cover and stop plate.
3. Lubricate shaft and install in housing.
4. Install thrust washers and driven gear to shaft and secure with roll pins. On Vega distributors, install damper cup on shaft and install pin.

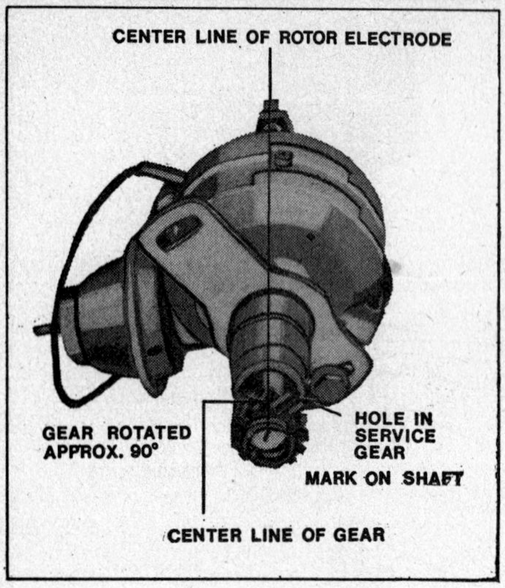

Fig. 18A Gear tooth alignment, Chrysler 6 cyl. distributor

Check to see that shaft turns freely. Install driven gear with mark on hub in line with rotor segment.

5. Install breaker plate.
6. Attach condenser and breaker point set in proper location with appropriate attaching screws, Fib. 22. Connect primary and condenser leads to breaker point set quick disconnect terminal. *Contact point set pilot must engage matching hole in breaker plate.*
7. Attach vacuum control assembly to distributor housing, using upper mounting holes.
8. Adjust breaker arm spring tension and breaker gap.
9. Install rotor.

FORD AUTOLITE/MOTORCRAFT DISTRIBUTOR SERVICE

Dual Advance Distributor

This distributor, Fig. 23, is similar to conventional design in that both a centrifugal advance mechanism is provided to regulate ignition timing according to speed and a vacuum advance unit to regulate ignition timing according to load. However, unlike other make distributors, the centrifugal advance mechanism can be adjusted through a slot in the breaker plate.

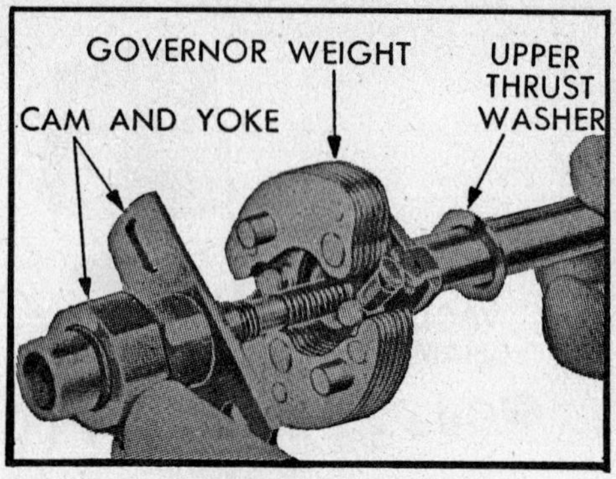

Fig. 19 Distributor shaft details

Adjust centrifugal advance before adjusting vacuum advance. If specified centrifugal advance is not indicated on a distributor test machine, bend one spring bracket with a screwdriver through a hole in the breaker plate on all distributors except 2800 cc units, Fig. 24. On 2800 cc units, remove plug inside of distributor housing to gain access to adjustment bracket, Fig. 25. Bend bracket away from distributor shaft to decrease advance and toward shaft to increase advance. Identify

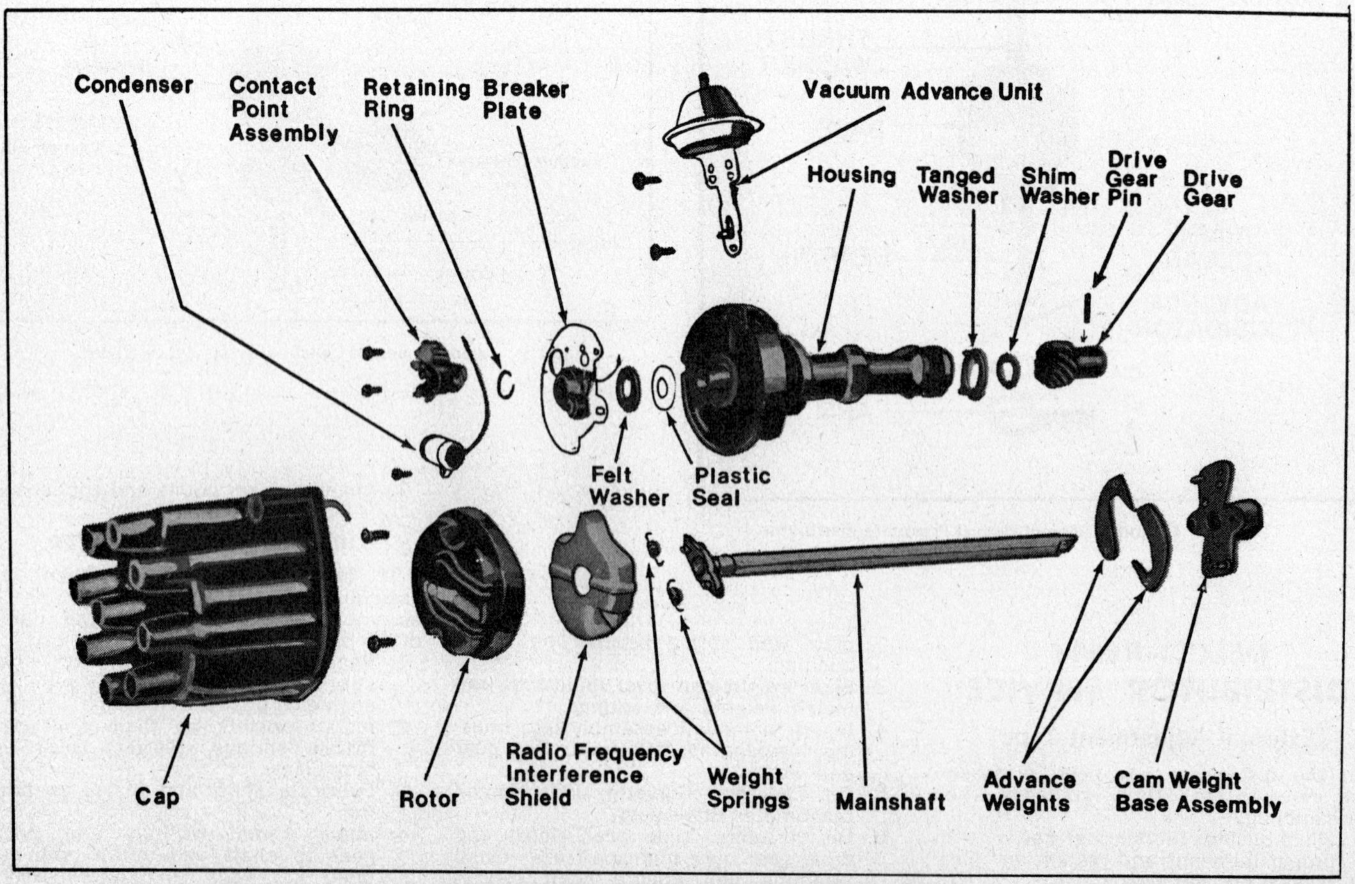

Fig. 20 Delco-Remy external adjustment distributor

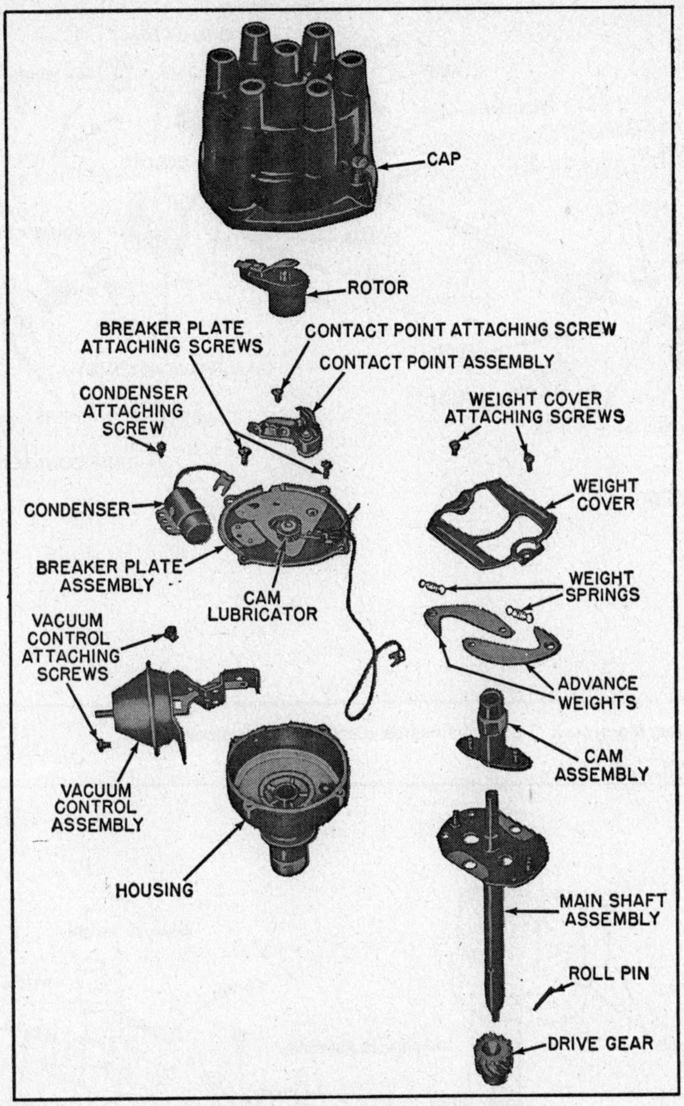

Fig. 21 Exploded view of typical Delco-Remy internal adjustment distributor

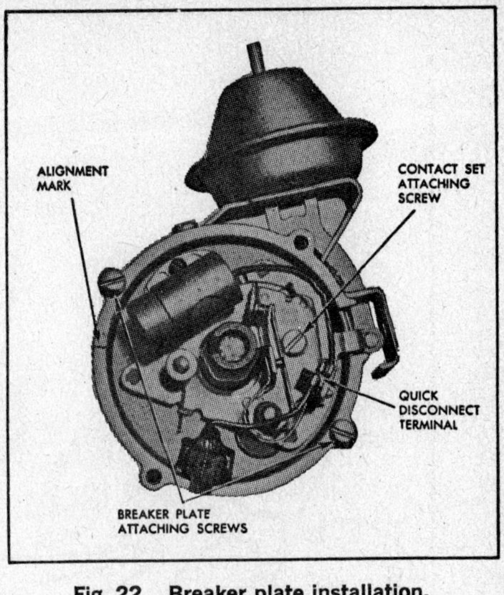

Fig. 22 Breaker plate installation. Delco-Remy internal adjustment distributor

bracket after adjustment is made. After an adjustment has been made to one spring, check the minimum advance point again. Then operate distributor at the specified rpm to give an advance just below maximum. If this advance is not up to specifications, bend the other spring bracket to give the correct advance.

Vacuum advance on 1969-71 six and 8 cylinder and 1971-72 four cylinder units can be adjusted by changing the calibrated washers between the vacuum chamber spring and nut, Fig. 26. The addition of one washer will decrease advance and the removal of a washer will increase advance.

Vacuum advance on all 1972 distributors except four cylinder units and all 1973-74 distributors except 2300 cc and 2800 cc units can be adjusted by turning an allen head screw inside vacuum unit, Fig. 27. Turning the screw clockwise increases advance and turning it counterclockwise decreases advance. Vacuum advance on 2300 cc and 2800 cc units is pre-set and cannot be adjusted.

Distributor Service

Six & V8 Units

If the distributor has been disassembled, refer to Fig. 24 for guidance upon reassembly.

1. Oil shaft and slide it into distributor body.
2. Place collar in position on shaft and align holes in collar and shaft, then install a new pin.
3. Install distributor cap clamps.
4. Check shaft end play with feeler gauge placed between collar and base of distributor. If shaft end play is not within .024-.035", replace shaft and gear.
5. Fill grooves in weight pivot pin with ball bearing grease.
6. Position weights in distributor.
7. Install weight springs, being sure proper weight, spring and adjustment bracket are assembled together.
8. Install upper thrust washer.
9. Fill grooves in upper portion of distributor shaft with ball bearing grease.
10. Install cam assembly, being sure that slots in cam engage pins in weights.
11. Install cam retainer. Apply a light film of cam lubricant to cam lobes. Saturate wick with 10W engine oil.

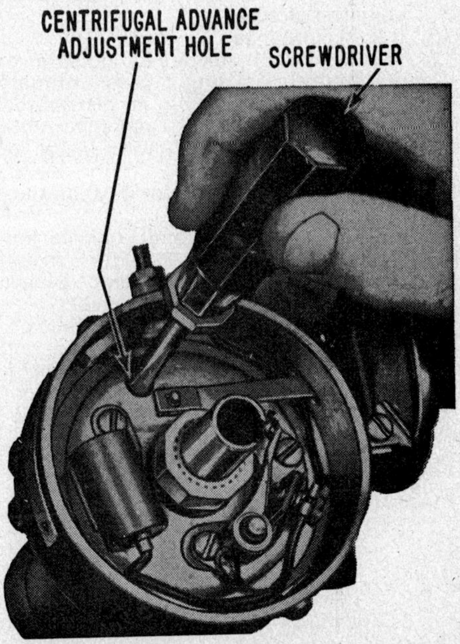

Fig. 24 Centrifugal advance adjustment all exc. 2800 cc units

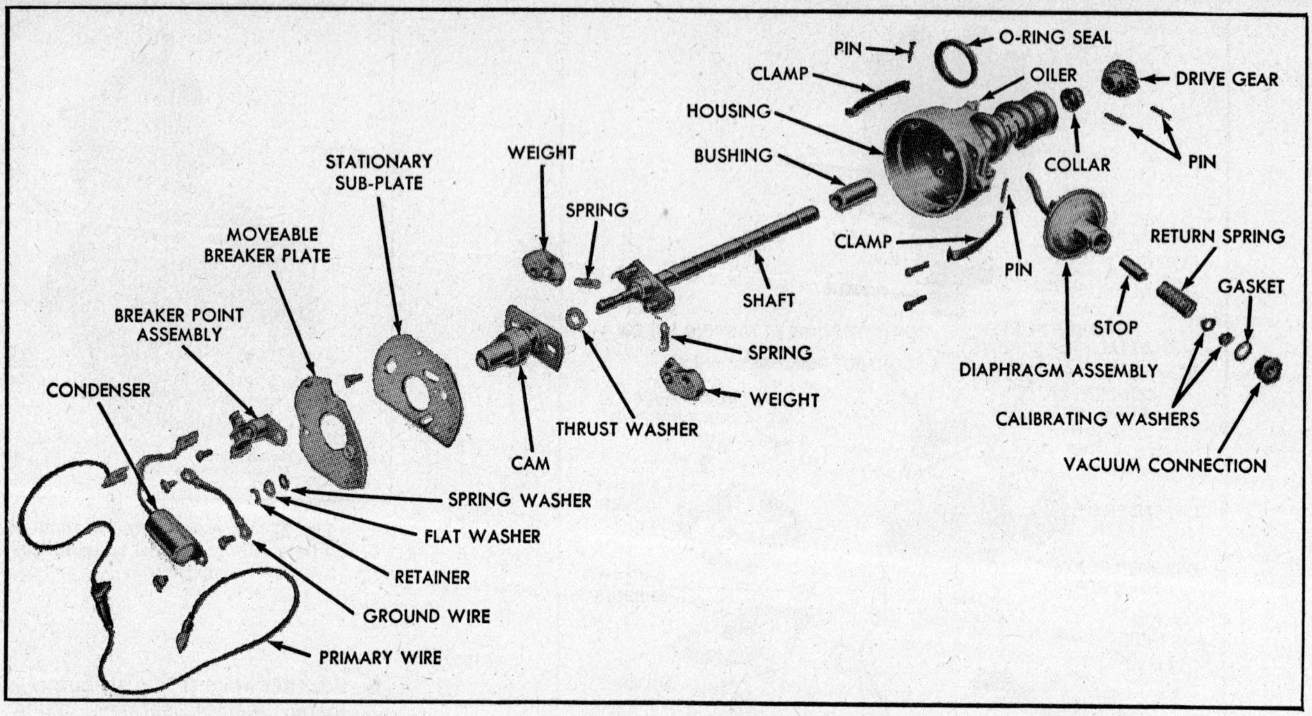

Fig. 23 Exploded view of Autolite/Motorcraft V-8 dual-advance distributor. (Typical)

Install wick in cam.

12. Position stationary sub plate in distributor. Install one end of ground wire under plate retaining screw closest to diaphragm mounting flange.
13. Position movable breaker plate in distributor. Install spring washer on pivot pin. Place flat washer on spring washer. Be sure protruding edges of spring washer are facing upward. Install retainer.
14. Install new breaker point assembly. Install ground wire on breaker point attaching screw furthest from point adjustment slot.
15. Install condenser.
16. Working from inside to outside of distributor housing, pass primary wire through opening in distributor. Pull wire through opening until locating stop is flush with inside of distributor.
17. Connect condenser wire and primary wire to breaker points.
18. Position diaphragm and hook its link over pin on breaker plate. Install diaphragm attaching screws. Secure link with retainer. Install oil seal.
19. Adjust breaker arm spring tension, align and adjust breaker points and check and adjust cam dwell, centrifugal and vacuum advance.

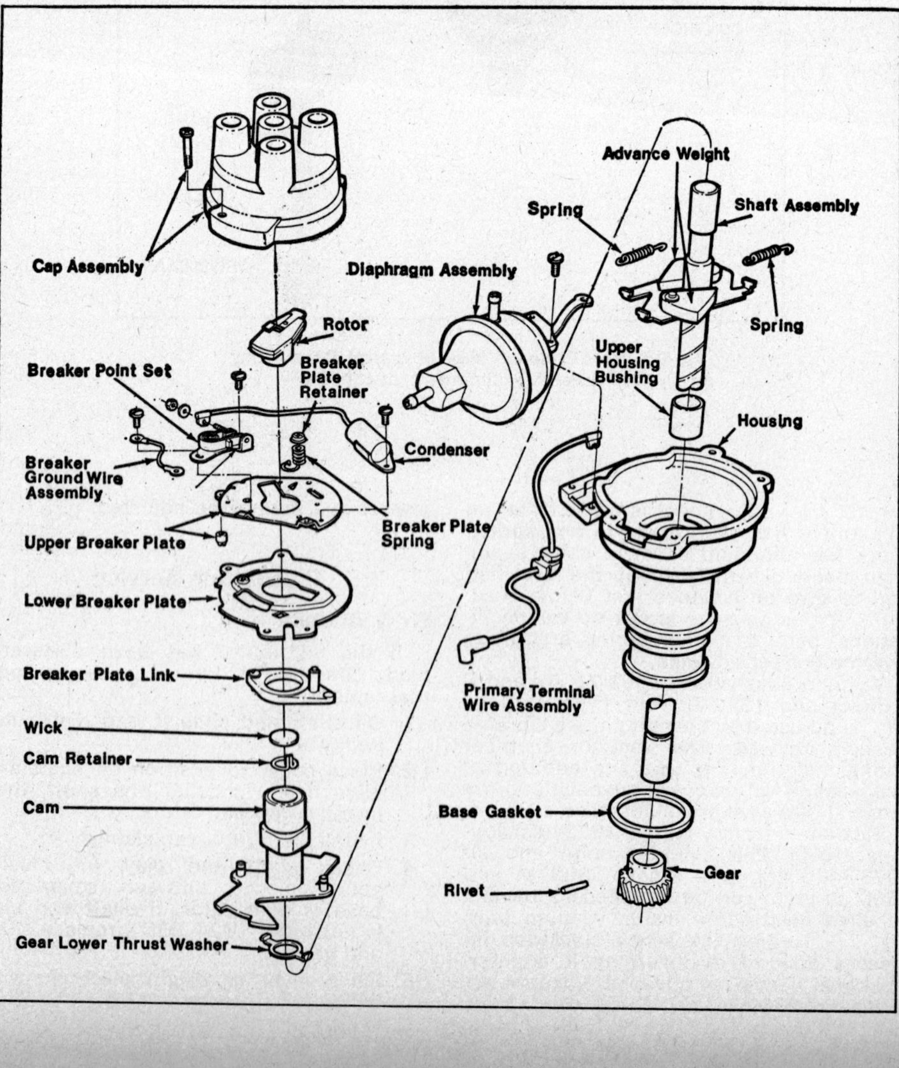

Fig. 28 Exploded view of Motorcraft 2300 cc dual advance distributor

Fig. 25 Centrifugal advance adjustment. 2800 cc distributor

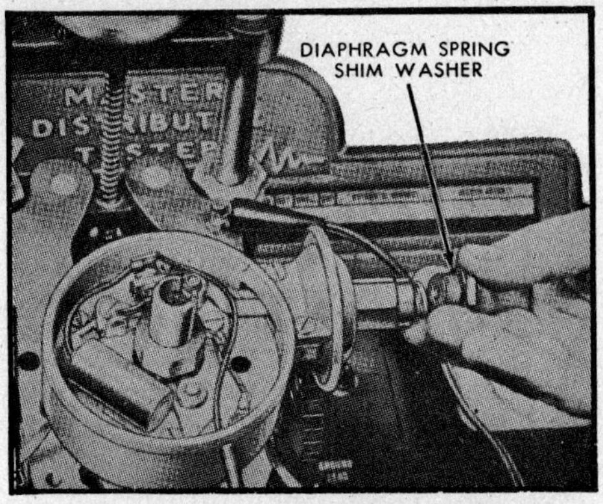

Fig. 26 Vacuum advance adjustment Autolite/Motorcraft dual advance distributor. 1969-71 6 & 8 cyl. & 1971-72 4 cyl.

Fig. 27 Vacuum advance adjustment. Autolite/Motorcraft 1972 6 & 8 cyl. & all 1973-74 distributors exc. 2300 cc & 2800 cc units

2300 cc & 2800 cc Units

If distributor has been disassembled, refer to Figs. 28 and 29 at reassembly.

1. Lubricate top of distributor shaft and install cam assembly, then retaining ring and felt wick.
2. Lubricate pivot pins and install advance weights and on 2800 cc units, "C" clips. On all units, install advance springs.
3. Place breaker plate into housing, ensuring mounting holes align. It may be necessary to lightly tap breaker plate to seat plate on swedged perches in housing.
4. Install vacuum advance unit, breaker points and cam wiper.
5. On 2300 cc units, route primary wire and grommet through side of housing. Install condenser and connect leads to terminal.
6. On 2800 cc units, place terminal and grommet into slot on side of housing and install condenser. Connect breaker point lead to terminal.
7. On all units, install rotor.

1600 cc Units

If distributor has been disassembled, refer to Fig. 30 at reassembly.

1. Install thrust washers onto shaft, slide shaft into housing and install thrust washer, wave washer and drive gear. Install tension pin.
2. Install cam assembly making sure advance stop is in correct slot and install snap ring. Replace felt wick.
3. Install vacuum advance unit.
4. Install advance springs to posts from which they were removed, lubricate governor weight pivots and install governor weights with flat edge adjacent to cam spindle and install spring clips.
5. Install grommet in lower plate leaving sufficient wire to reach contact point connection. Connect ground spring to pivot post and place upper plate on lower plate, engaging the hold down spindle in keyhole slot and install the

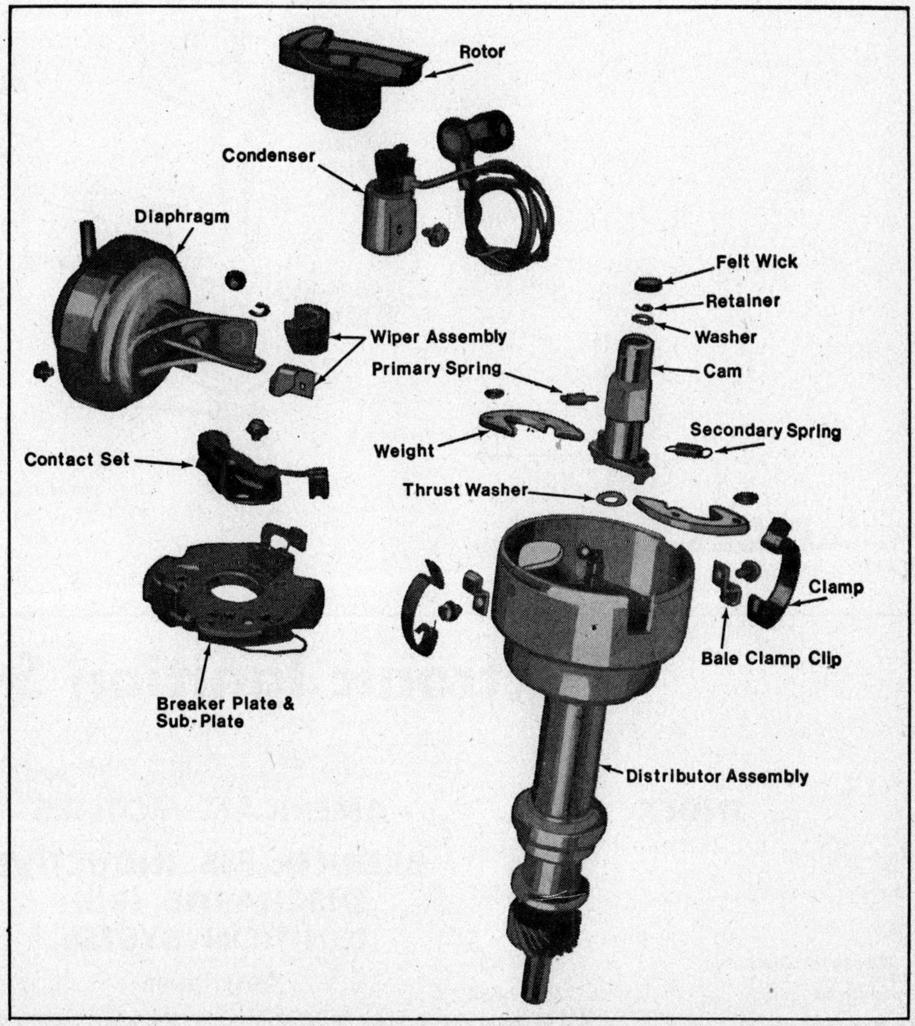

Fig. 29 Exploded view of Motorcraft 2800 cc dual advance distributor

two spring washers, flat washer and large snap ring.

6. Check clearance between upper and lower plates, beneath nylon bearing nearest to the hold down pin. Maximum clearance is .010 inch. If clearance is excessive, thread nut further onto hold down screw.
7. Install contact points and condenser onto breaker plate assembly and install the assembly and secure with two screws.

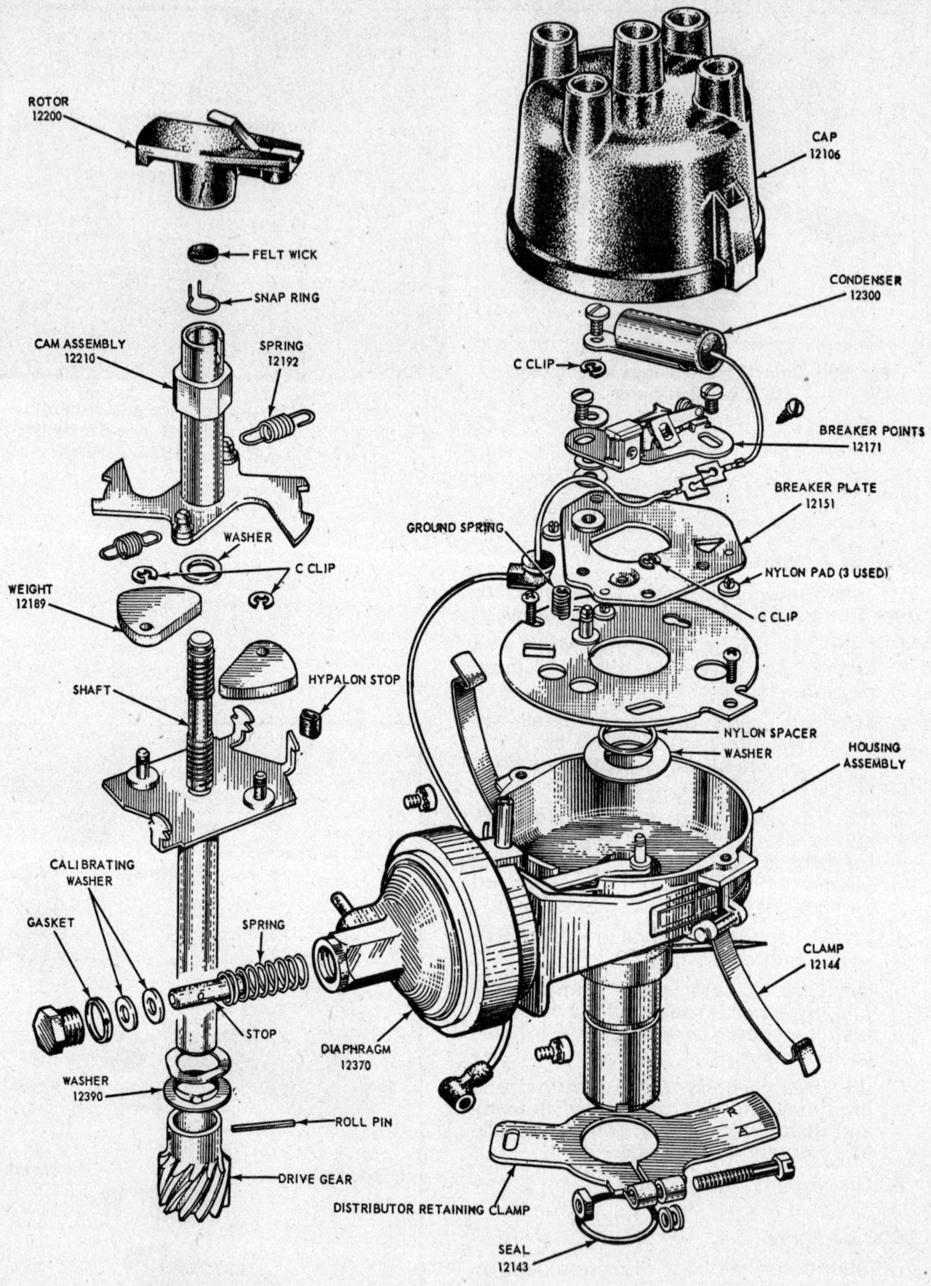

Fig. 30 Exploded view of Autolite/ Motorcraft 1600 cc dual advance distributor

Electronic Ignition Systems

INDEX

AMERICAN MOTORS

BREAKERLESS INDUCTIVE DISCHARGE (BID) IGNITION SYSTEM

Description

The BID ignition system incorporates four major units; an electronic control unit, ignition coil, distributor and high tension wires, Fig. 1. The electronic control unit is a solid-state, moisture resistant module with the components sealed in a potting compound to resist vibration and environmental conditions. Since the control unit has an internal current regulator, a resistance wire or ballast resistor is not necessary in the primary circuit. Battery voltage is applied to the ignition coil positive terminal when the ignition switch is in the "On" or "Start" position, therefore, an ignition system bypass is not required in this system. The primary coil circuit is electronically regulated by this unit.

The ignition coil is of standard construction and requires no special service. The function of the ignition coil in the BID ignition system is the same as for conventional ignition systems.

The distributor is conventional except the contact points, condenser and cam are replaced by a sensor and trigger wheel, and since no wearing occurs between the trigger wheel and sensor, dwell angle remains constant and requires no adjustment. The sensor is a small coil of fine wire and receives an alternating current signal from the electronic control unit. The senor develops an electromagnetic field used to detect the presence of metal which are the leading edges of the trigger wheel teeth.

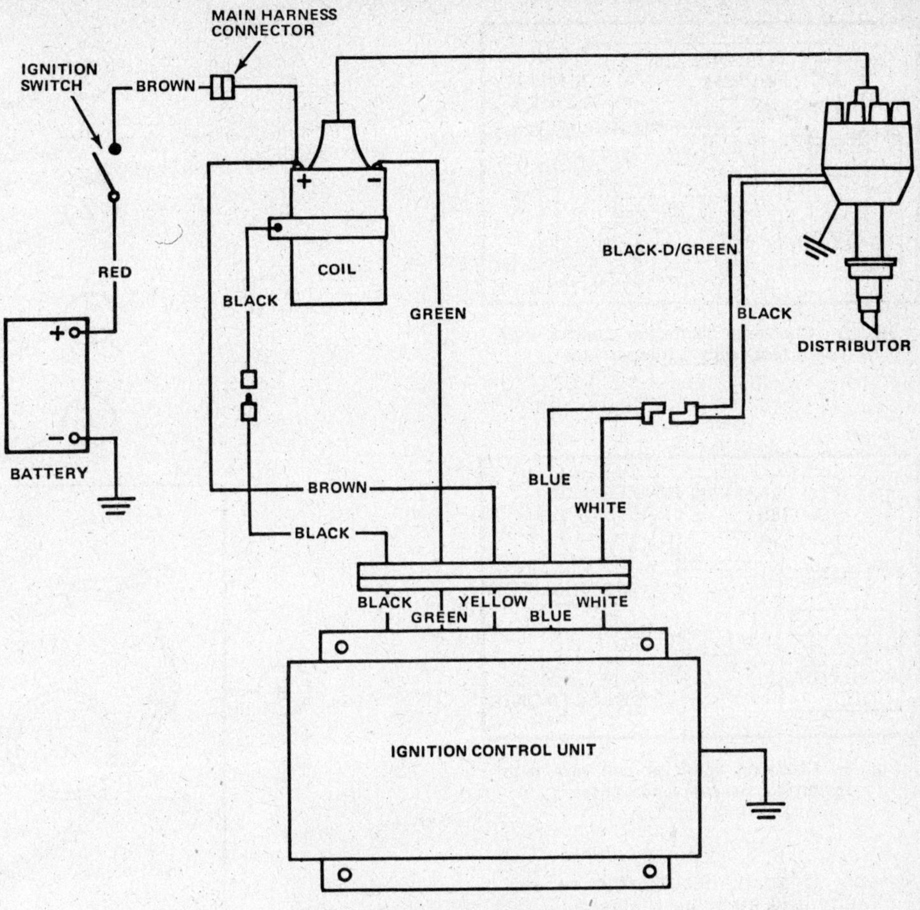

Fig. 1 BID ignition system wiring

Operation

When the ignition switch is placed in the "Start" or "Run" position, the control unit is activated. An oscillator within the control unit excites the sensor coil, in turn developing the electromagnetic field. When a leading edge of a trigger wheel tooth enters the electromagnetic field, the tooth reduces the sensor oscillation strength to a predetermined level, in turn activating the demodulator circuit. The demodulator circuit controls a power transistor located in series with the coil primary circuit. The power transistor switches the coil primary circuit off, thereby inducing a high voltage in the coil secondary winding. The high voltage is then delivered to the spark plugs through the distributor rotor, cap and high tension wires.

Troubleshooting

Ensure all electrical connections are correct before proceeding with the following checks.

1. Disconnect coil wire from distributor, hold wire approximately ½ inch from a suitable ground and crank engine, Fig. 2. If spark jumps between the coil wire and the ground, system is satisfactory.
2. If no spark occurs in step No. 1, connect a No. 57 test lamp between coil positive terminal and the ground, Fig. 3. Turn ignition switch to "On" and "Start" positions. If test lamp does not light in both positions, check ignition switch and wiring between battery and ignition coil.
3. If test lamp lights in both ignition switch positions, connect test lamp across ignition coil terminals, disconnect distributor connector and turn ignition switch to "On" position, Fig. 4. If test lamp does not light, check electronic control unit for proper ground.
4. If test lamp lights in step No. 3, connect a jumper wire across electronic control lead terminals, Fig. 5. If test lamp remains lit, replace electronic control unit.
5. If test lamp goes out in step No. 4, remove test lamp and hold coil wire approximately ½ inch from a suitable ground. Intermittantly short the electronic control lead terminals and check for spark at coil wire gap, Fig. 6. If no spark occurs, replace ignition

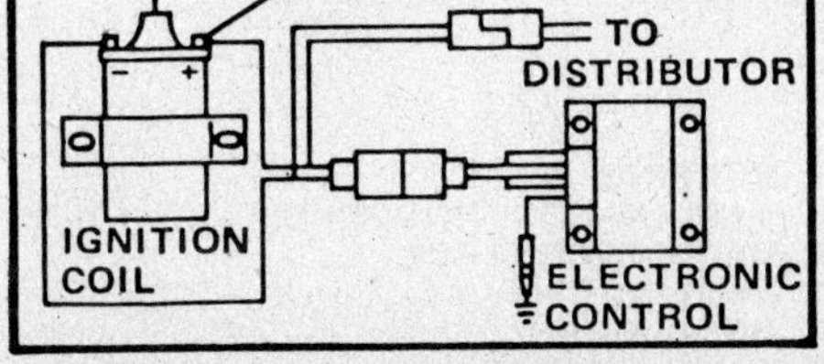

Fig. 2 Checking spark at coil wire

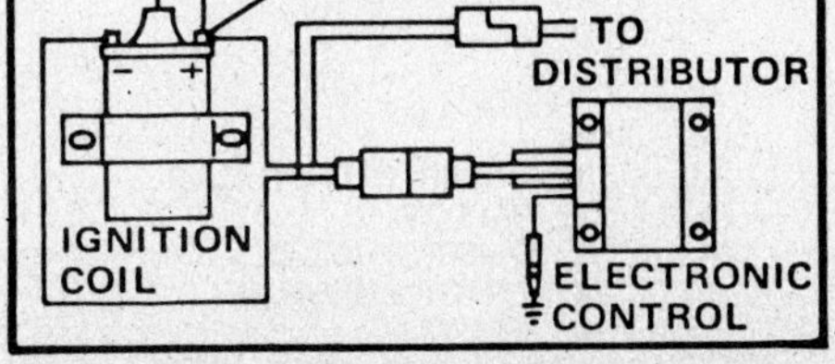

Fig. 3 Checking battery to ignition coil wiring

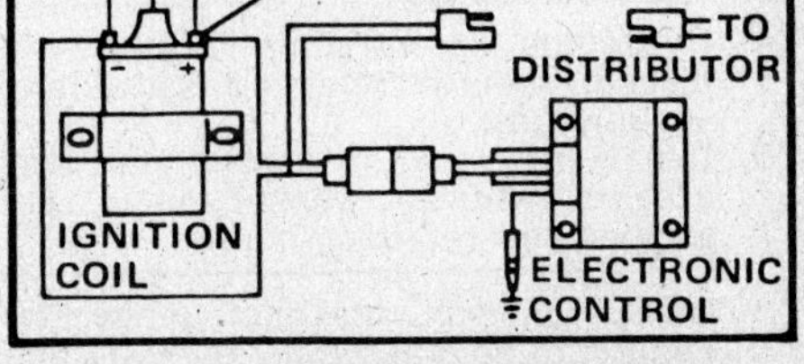

Fig. 4 Checking electronic control unit with test lamp

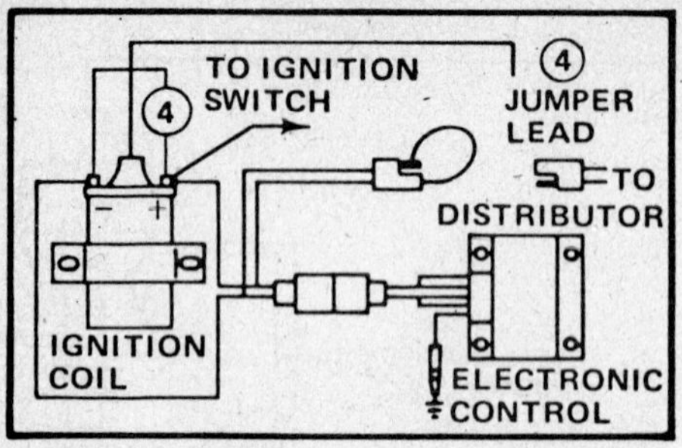

Fig. 5 Checking electronic control unit with test lamp & jumper wire

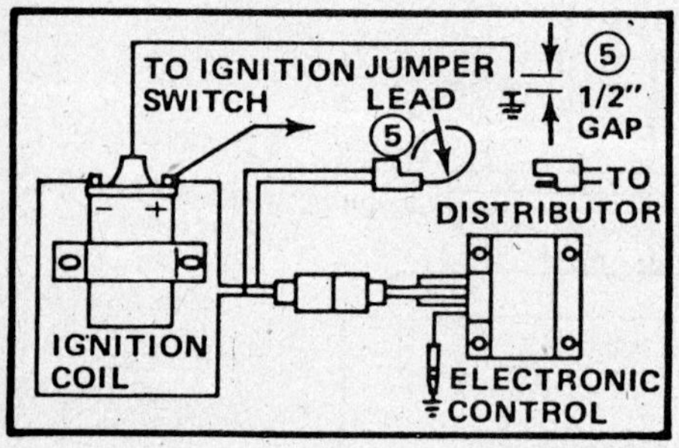

Fig. 6 Checking spark at coil wire with electronic control leads shorted

coil. If spark occurs, the sensor is faulty and must be replaced.

Component Replacement

1. Place distributor in a suitable holding fixture and remove cap, rotor and dust shield, Fig. 7.
2. Using a small gear puller, remove trigger wheel. Ensure puller jaws are gripping trigger wheel inner shoulder to prevent trigger wheel damage. Also, use a thick flat washer or nut as a spacer and do not press against small center shaft.
3. Loosen sensor locking screw approximately three turns, lift sensor lead grommet from distributor bowl and pull sensor leads from slot around sensor spring pivot pin. Release sensor spring, ensure spring clears sensor leads and slide sensor from bracket.

NOTE: The sensor locking screw utilizes a tamper proof head design and requires tool J-25097 for removal. However, if special tool is not available, use a small needlenose plier to remove screw. The service (replacement) sensor has a standard slotted head screw.

4. If vacuum control unit is to be replaced, remove retaining screw and vacuum unit.
5. Install new vacuum control unit and assemble sensor, sensor guide, flat washer and retaining screw.

NOTE: Install retaining screw far enough to hold assembly together and ensure it does not protrude past bottom of sensor.

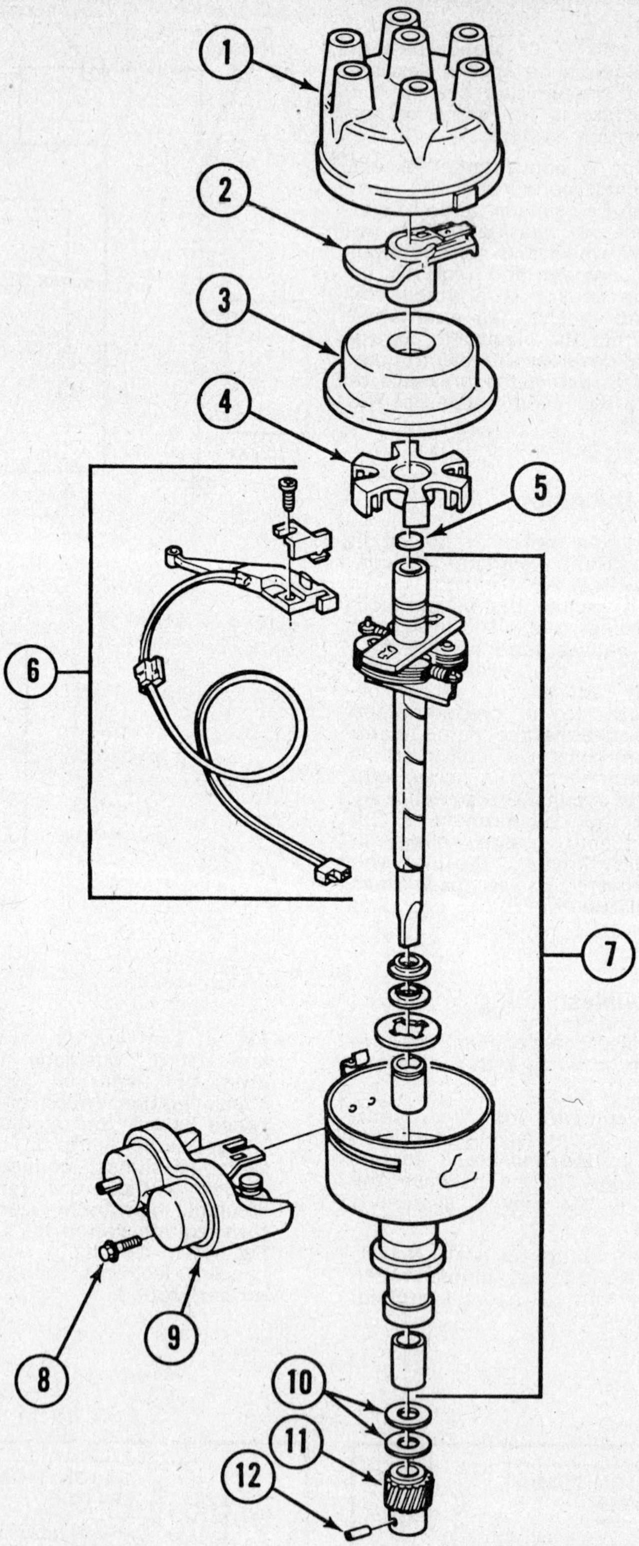

1. DISTRIBUTOR CAP
2. ROTOR
3. DUST SHIELD
4. TRIGGER WHEEL
5. FELT
6. SENSOR ASSEMBLY
7. HOUSING
8. VACUUM CONTROL SCREW
9. VACUUM CONTROL
10. SHIM
11. DRIVE GEAR
12. PIN

Fig. 7 BID distributor, exploded view

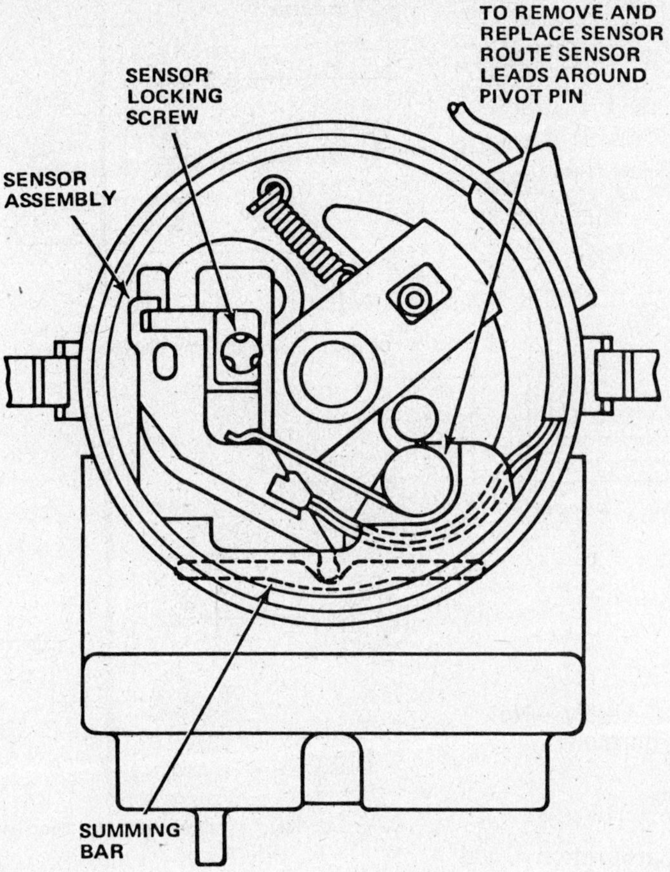

Fig. 8 Sensor installation

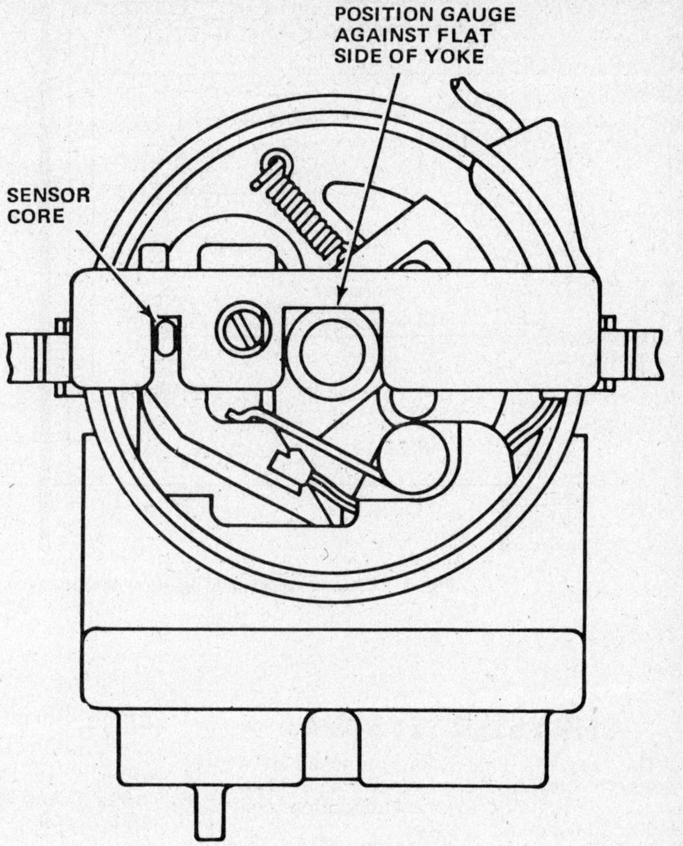

Fig. 9 Positioning sensor

6. If vacuum control has been replaced and original sensor is being used, replace special head screw with standard slotted head screw.
7. Install sensor assembly on vacuum chamber bracket, ensuring tip of sensor located properly in summing bar. Place sensor spring on sensor and route sensor leads around spring pivot pin, Fig. 8. Install sensor lead grommet and position leads away from trigger wheel.
8. Install sensor positioning gauge over yoke, ensure gauge is against flat of shaft, and move sensor sideways until gauge can be positioned. Snug retaining screw and check sensor position by removing and installing gauge, Fig. 9. When gauge can be removed and replaced without sensor side movement, sensor is positioned properly. Tighten retaining screw and check sensor position.
9. Place trigger wheel on yoke and check if sensor core is positioned approximately in center of trigger wheel legs. Bend a .050 inch gauge wire to dimension specified in Fig. 10, and place between trigger wheel legs and sensor base. Press trigger wheel onto yoke until legs contact gauge wire.
10. Apply 3 to 5 drops of light engine oil to felt wick in top of yoke, then install dust shield, rotor and cap.

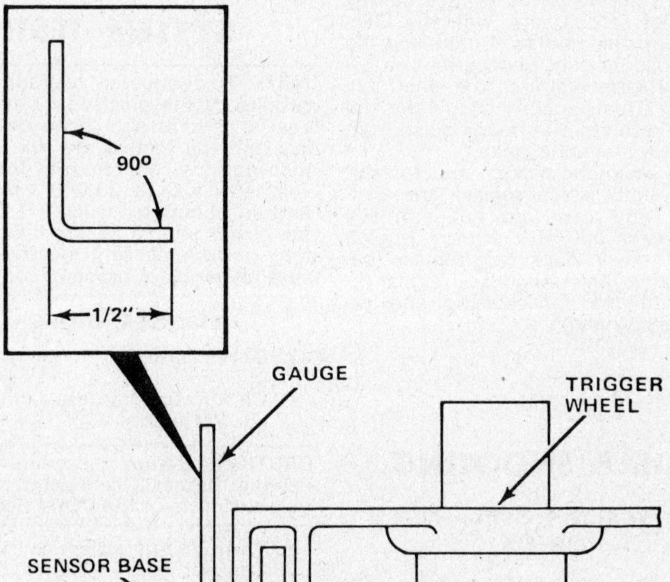

Fig. 10 Trigger wheel installation

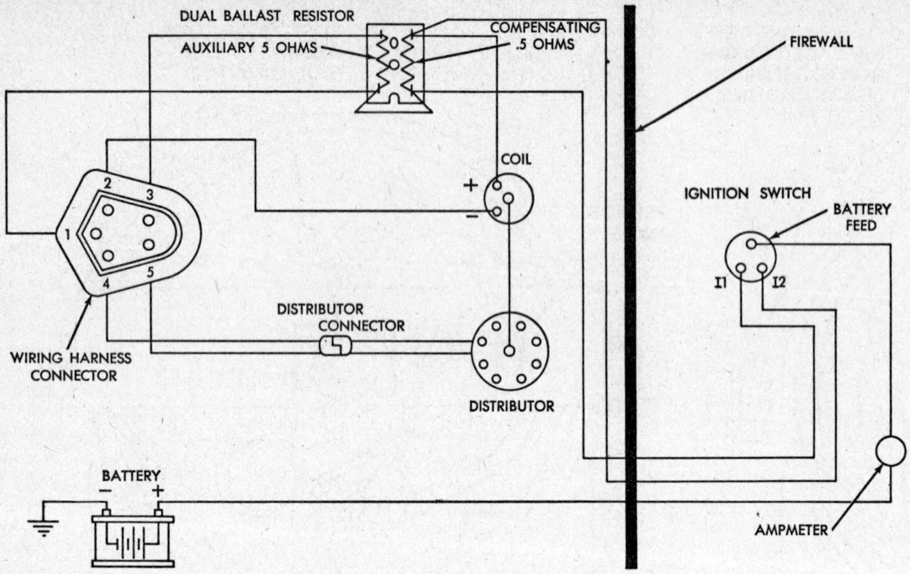

Fig. 1 Chrysler electronic ignition system wiring

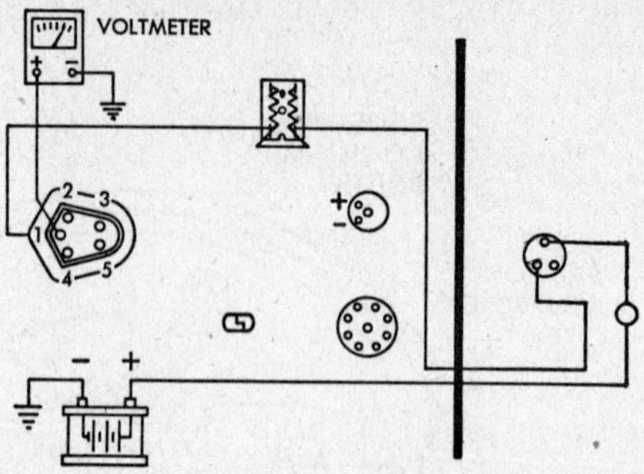

Fig. 2 Harness wiring test, No. 1 cavity

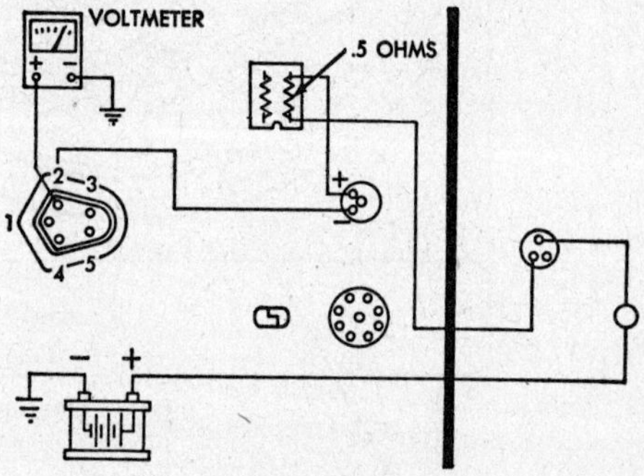

Fig. 2A Harness wiring test, No. 2 cavity

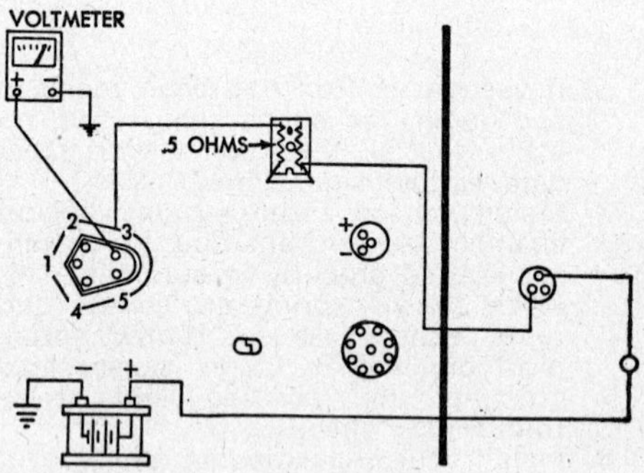

Fig. 2B Harness wiring test, No. 3 cavity

CHRYSLER SYSTEM

This system, Fig. 1, is composed of a magnetic distributor, an electronic control unit, a wiring harness, a production coil and a dual ballast resistor.

The distributor is essentially the same as the conventional type except the contacts have been replaced by a pickup coil and the cam by a reluctor. With a conventional contact type system, the voltage necessary to fire the spark plugs is developed by interrupting the current flowing through the primary of the ignition coil by opening a set of contacts. With the Electronic System, the voltage is produced the same way except that the current is interrupted by a transistor in the electronic control unit. This happens each time the control unit receives a "timing" pulse from the distributor magnetic pickup.

Since the magnetic pickup, reluctor and the control unit, which replace the contact points and cam, do not normally change or wear out with service, engine timing and dwell does not require periodic adjusting. This minimizes regular ignition maintenance of cleaning and replacing the spark plugs.

TROUBLE SHOOTING

Engine Will Not Start—Fuel System OK

1. Wiring harness electrical terminals covered with grease.
2. Dual ballast.
3. Faulty ignition coil.
4. Faulty pickup or improper pickup air gap.
5. Faulty wiring.
6. Faulty control unit.

Engine Surges Severely—Not Lean Carburetor

1. Wiring.
2. Faulty pickup leads.
3. Ignition coil.

Engine Misses—Carburetion Good

1. Spark plugs.
2. Secondary cables.
3. Ignition coil.
4. Wiring.
5. Control unit.

SYSTEM TESTING

NOTE: To completely test components and circuits of the electronic ignition system, special testers should be used. However, in event the testers are not available, the following procedures may be utilized. A voltmeter with a 20,000 volt/ohm rating and an ohmmeter using a 1.5 volt battery for power should be used for testing. Before performing any electrical tests, ensure all wiring is properly connected.

Harness Wiring Test

1. Check battery voltage and note reading.
2. Disconnect harness connector from control unit.

CAUTION: Before disconnecting or connecting harness connector, ensure ignition switch is in the "Off" position.

3. Turn ignition switch to "On" position.
4. Connect the voltmeter between harness connector cavity No. 1 and the ground. Voltage reading should be within 1 volt of battery voltage earlier noted. If not, check circuit between cavity No. 1 and the battery, Fig. 2.
5. Test harness connector cavities numbers 2 and 3 in same manner. If voltage is not within specifications, check circuits between cavities numbers 2 and 3 and the battery, Figs. 2A and 2B.
6. Turn ignition switch to "Off" position.

Distributor Pick-up Coil Test

1. Conect an ohmmeter between harness connector cavities numbers 4 and 5. Resistance reading should be 150 to 900 ohms.
2. If reading is not as specified in above step, disconnect distributor dual lead connector and connect ohmmeter between the two leads on distributor

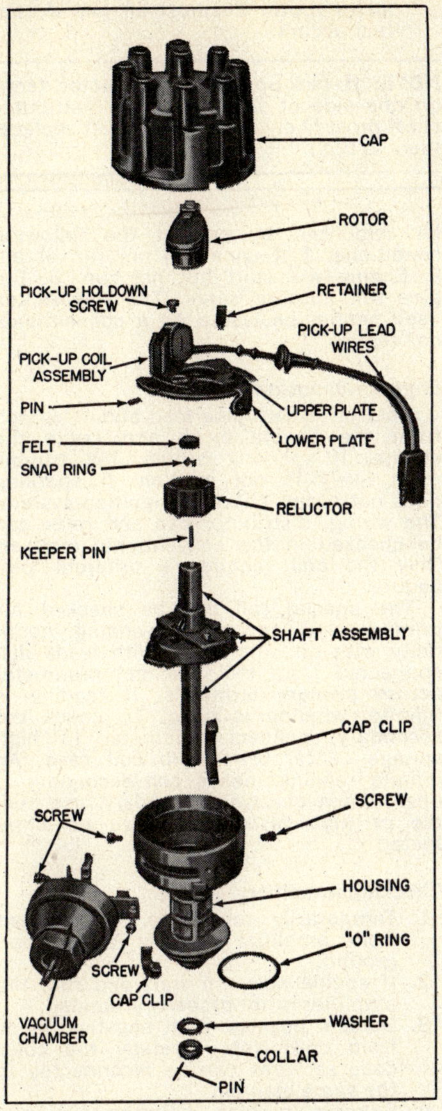

Fig. 3 Exploded view of typical Chrysler electronic distributor

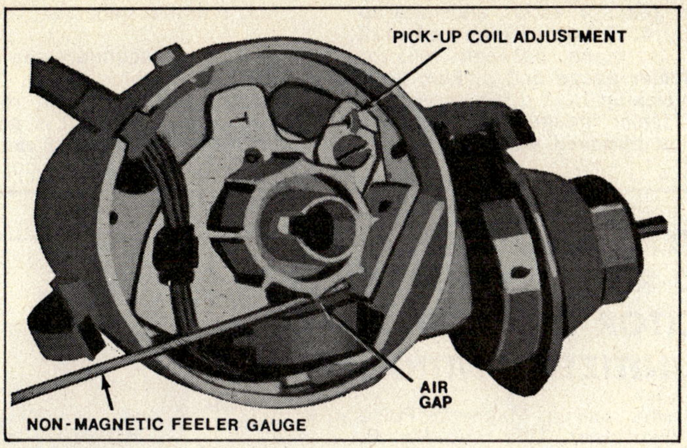

Fig. 4 Air gap adjustment

side of connector. If resistance is not between 150 and 900 ohms, replace pick-up coil.

3. Connect one ohmmeter lead to a good ground and the other lead to either connector of the distributor. If ohmmeter shows a reading, the pickup coil must be replaced.

Control Unit Ground Circuit Test

Connect an ohmmeter between control unit connector pin No. 5 and the ground. If ohmmeter indicates infinite resistance, tighten bolts securing control unit to firewall and recheck resistance. If reading is still infinite, replace control unit.

Distributor Shaft & Bushing Wear Test

1. Remove distributor from vehicle and clamp distributor in a vise. Use extreme caution not to damage distributor.
2. Attach a dial indicator to housing so plunger rests against reluctor sleeve.
3. Place a wire loop around reluctor sleeve and hook a spring scale on the other end of the loop. Apply a one pound pull in line with indicator plunger and read movement on indicator. Movement must not exceed .006 inch. If movement exceeds limit, replace either housing or shaft to bring movement back within tolerance.

DISTRIBUTOR SERVICE

Distributor Disassemble

1. Remove rotor and vacuum advance unit.
2. Remove reluctor. If reluctor cannot be pulled off easily, use two screwdrivers with a 7/16 inch blade to pry up from bottom of reluctor. Use extreme caution not to damage teeth on reluctor.
3. Remove two screws from lower plate and lift out lower and upper plates and pick-up coil as an assembly.
4. On six cylinder distributors, remove drive gear retaining pin and slide gear from shaft. Support gear so that gear teeth will not be damaged when pin is driven from shaft. On V8 distributors, remove shaft retaining pin and slide retainer from shaft.
5. Remove lower thrust washer by cleaning burrs from around pin hole.
6. Push up on shaft and remove through top of housing.

Distributor Assemble

1. Lubricate and test operation of governor weights. Inspect weight springs for distortion and bearing surfaces and pins for damage.
2. Lubricate upper thrust washer and install onto shaft. Install shaft into housing.
3. On six cylinder distributors, if drive gear is damaged, replace gear as outlined in the Standard Distributor section in this chapter. On V8 distributors, install shaft retainer and pin.
4. Install lower and upper plates and pick-up coil as an assembly.
5. Connect vacuum advance unit arm to pick-up plate and install vacuum unit mounting screws.
6. Place reluctor keeper pin into position on reluctor sleeve and slide reluctor downward on sleeve and press into place firmly. On V-8 distributors, the reluctor is installed with the two arrows on top.

NOTE: Distributors on V-8 engines may rotate either clockwise or counter-clockwise. The arrow at the keeper pin must point in the direction of distributor rotation. If not, remove reluctor and turn it 180 degrees and reinstall it. When removing reluctor, use care not to lose keeper pin.

7. Place 1 drop of light engine oil on felt pad on top of reluctor and install rotor.

Pick-up Replacement & Air Gap Adjustment

1. With distributor removed from vehicle, perform Steps 1 to 3 as outlined in Distributor Disassemble.
2. Remove pick-up coil and upper plate by depressing retainer clip and moving it away from mounting stud. Pick-up coil cannot be removed from upper plate.
3. Lightly lubricate upper plate pivot pin and lower plate support pins with distributor lubricant. Install upper plate pivot pin through smallest hole in lower plate and install retainer clip.

NOTE: The upper plate must ride on the support pins on the lower plate.

4. Install lower and upper plates and pick-up coil as an assembly and install distributor into vehicle.
5. To set air gap, line up one reluctor tooth with pick-up pole and install a non-magnetic .008 inch feeler gauge

between reluctor tooth and pick-up pole, Fig. 4. Rotate pick-up coil until contact is made between reluctor tooth, feeler gauge and pick-up pole. Tighten pick-up hold down screw and remove feeler gauge. Feeler gauge should be removed without force. If not, readjust gap.

6. Perform a secondary gap check with a .010 inch feeler gauge. Do not force feeler gauge between reluctor tooth and pick-up since it is possible to do so. Apply vacuum to vacuum control unit. Pick-up should not contact reluctor tooth. Readjust air gap if contact occurs.

NOTE: If pick-up contacts reluctor teeth on one side of shaft only, the distributor shaft most likely is bent and shaft replacement is required.

GENERAL MOTORS DELCO-REMY SYSTEMS CAPACITOR DISCHARGE (CD) & UNITIZED IGNITION

Both systems use a Magnetic Pulse Breakerless distributor. This unit, Fig. 5, resembles a conventional distributor. However, in the Magnetic Pulse unit, an iron timer core replaces the conventional breaker cam, Fig. 6. The timer core, which has equally spaced projections, one for each cylinder, rotates inside a magnetic pick-up assembly, which replaces the conventional breaker plate assembly.

The magnetic pick-up assembly consists of ceramic permanent magnet, a pole piece and a pick-up coil. The pole piece is metal plate having equally spaced internal teeth, one for each cylinder. The magnetic pick-up assembly is mounted over the main bearing on the distributor housing and is actuated by the vacuum advance unit. A conventional centrifugal advance is also used.

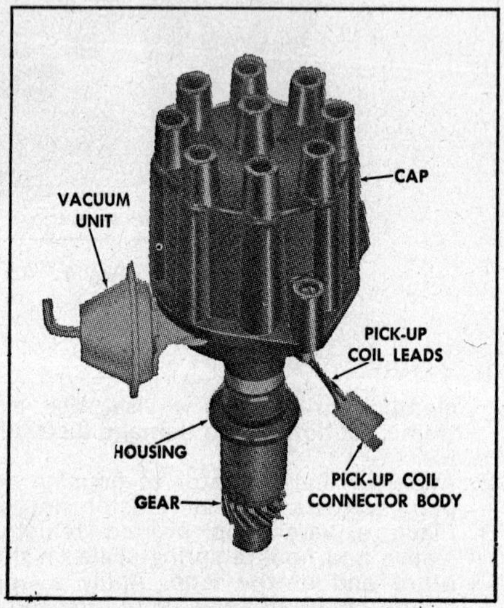

Fig. 5 Delco-Remy Magnetic Pulse distributor

Ignition Pulse Amplifier

Capacitor Discharge Amplifier

This unit, Fig. 7, consists of transistors, diodes, resistors, a thyristor and a transformer. These are mounted on a printed circuit board to make the amplifier a solid state unit with no moving parts, with a capacity of delivering 30,000 volts.

The ignition coil primary is connected across a high voltage capacitor (condenser), which is charged to about 300 volts during the time the spark plugs are not firing. On impulse signal voltage from the distributor, the capacitor discharges this high voltage into the coil primary.

Due to the transformer action in the coil, the high voltage primary is increased many times to produce the high voltage secondary. Fig. 8 shows a typical circuit diagram.

Voltage is supplied to the transformer which operates through a rectifying bridge circuit of four diodes to keep the capacitor charged. A zener diode limits this charge to 300 volts. This capacitor voltage is maintained at its maximum value during cranking even though battery voltage may be well below its normal 12 volts.

As the engine turns, the vanes on the rotating timer core in the distributor line up with the internal teeth on the pole piece. This establishes a magnetic path through the center of the pick-up coil. This voltage is amplified then applied at the gate of the thyristor, causing it to turn on. The charged capacitor then discharges through the thyristor and primary winding of the coil, inducing high voltage in the secondary winding to fire the spark plugs. This special ignition coil acts as a step up transformer to fire the spark plugs when the primary current *increases.*
This contrasts with the conventional ignition system in which the secondary voltage is induced when the distributor contacts open and the primary current *decreases.*

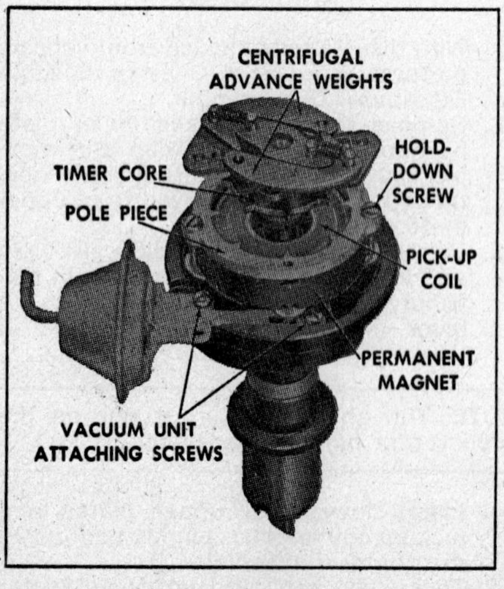

Fig. 6 Magnetic pulse distributor components

Trouble Shooting Procedure

Faulty engine performance usually will be evidenced by one of the following conditions: 1. Engine will not run at all. 2. Engine will start but not run. 3. Engine will miss or surge. *The special coil used cannot be tested on a conventional coil tester.*

Engine will not run at all

Hold one spark plug lead about ¼ inch from the engine block and crank the engine. If sparking occurs, the trouble most likely is not ignition. If sparking does not occur, check the ignition system. The wiring, distributor cap and rotor can be checked in the conventional manner. Only the coil requires a different procedure.

The special coil can be checked for primary and secondary winding continuity with an ohmmeter: With leads disconnected from coil, connect ohmmeter across primary terminals. If reading is infinite, winding is open. To check the secondary, connect ohmmeter to high voltage center tower and coil case. An infinite reading means coil secondary is open. *When checking secondary, use middle or high resistance range on ohmmeter.*

Checking Amplifier:

1. Temporarily connect a jumper lead from amplifier housing to a good ground.
2. If engine now will start and run, the amplifier is not properly grounded.
3. Detach positive and negative leads from coil. *Note carefully the color code so wires can be reconnected in the same manner.*
4. Connect a bulb between the two leads, Fig. 9.
5. Crank engine.

Fig. 7 Capacitor Discharge (CD) amplifier unit

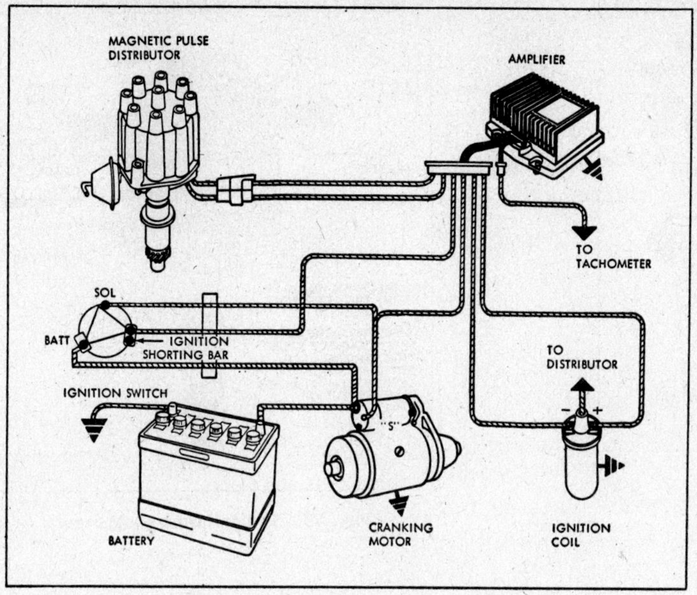

Fig. 8 CD ignition circuit (typical)

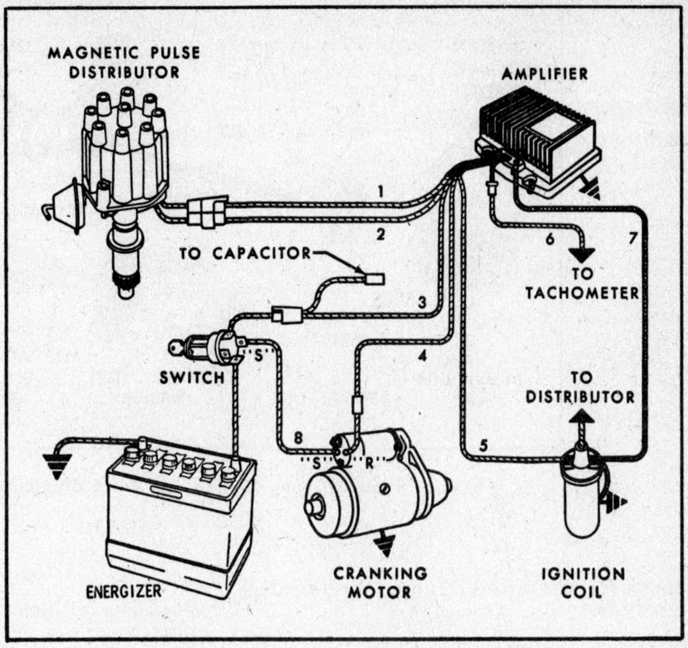

Fig. 10 Pictorial diagram of CD system

6. If bulb flickers on and off, amplifier is operating properly. In this case, recheck secondary system for the cause of "no run" condition.
7. If bulb does not flicker on and off, check distributor.

Distributor checks:

1. On CD system, be sure that the two distributor leads are connected to distributor connector body, Fig. 10.
2. With distributor connector disconnected from harness connector, connect an ohmmeter (1), Fig. 11, to the two terminals on distributor connector.
3. Connect a test stand vacuum source to the distributor and observe ohmmeter reading throughout vacuum range. (Distributor need not be removed from engine.)
4. Any reading outside the 550-750 ohm range indicates a defective pick-up coil in distributor.
5. Remove one ohmmeter (2) lead, Fig. 11, from connector body and connect to ground.
6. Observe ohmmeter reading throughout vacuum range.
7. Any reading less than infinite indicates a defective pick-up coil.
8. Reconnect harness connector to distributor connector.

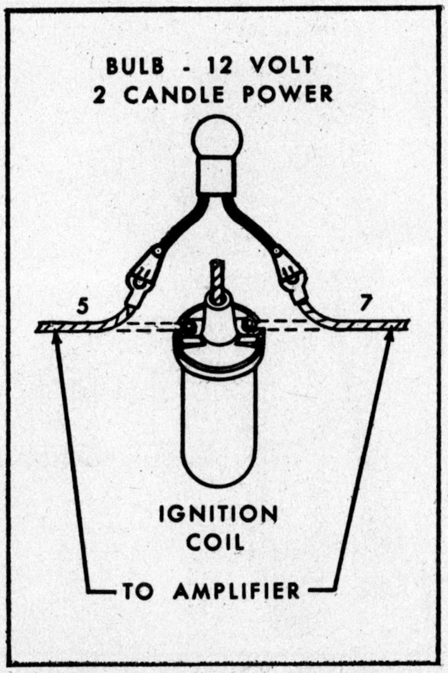

Fig. 9 Amplifier output test

Continuity checks-CD system

Carefully inspect all wiring connections to be sure that they are clean and tight. If satisfactory, disconnect amplifier No. 3 and No. 4 leads, Fig. 10 from the two connectors, then proceed as follows:

1. Connect voltmeter from ground to No. 4 connector lead.
2. Turn switch to "Start" position.
3. If reading is zero, circuit is open between connector body and battery.
4. If reading is obtained, connect voltmeter from ground to No. 3 connector lead.
5. Turn switch to the run position.
6. If reading is zero, circuit is open between connector body and switch.
7. If reading is obtained, replace amplifier.

Engine will start but not run-CD System

If engine starts but then stops when switch is returned to the run position, proceed as follows:

1. Be sure that leads are properly connected to No. 3 lead connector body.
2. If satisfactory, connect a voltmeter from ground to the terminal connector inside the connector.
3. Turn switch to run position.
4. If reading is zero, lead between connector and ignition switch is open.
5. If reading is obtained, replace amplifier.

Engine miss or surge:

The vehicle fuel system should be checked in the usual manner. If satisfactory, check the ignition system in the usual manner except the special coil which should be checked with a ohmmeter as described above.

A poorly ground amplifier can cause an engine miss or surge. If it is properly grounded and the plugs, wiring, cap and coil are satisfactory, the most likely

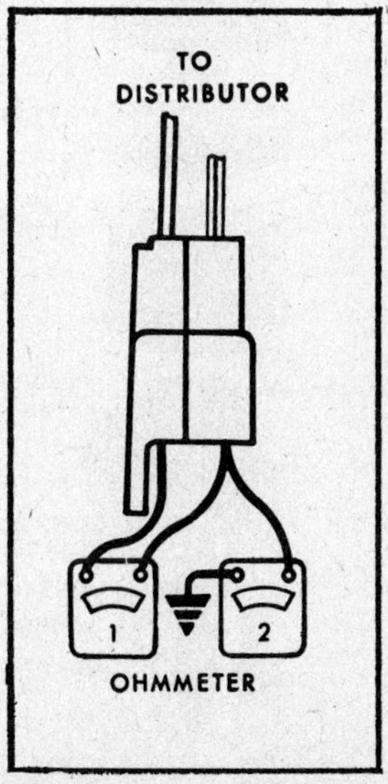

Fig. 11 Distributor test

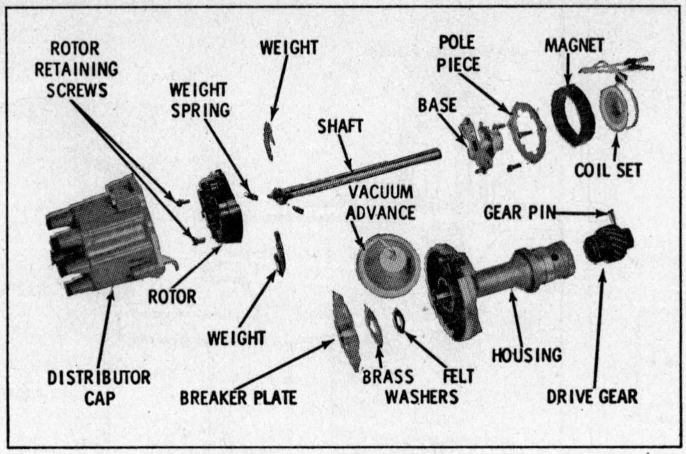

Fig. 12 Exploded view of Magnetic Pulse distributor

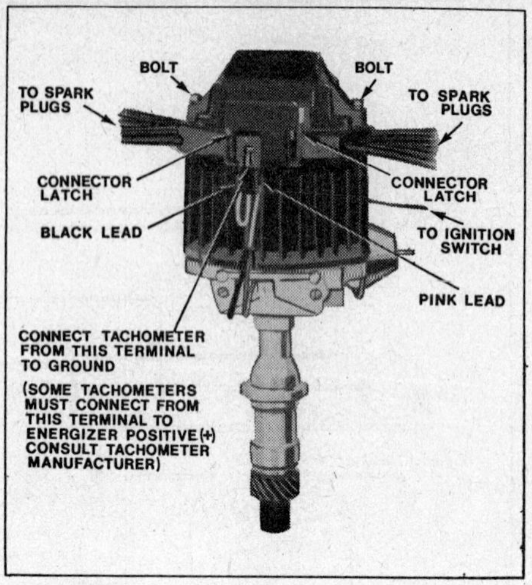

Fig. 13 Typical Unit Distributor

cause for the miss or surge is a defective amplifier.

Distributor Service

Remove distributor in the usual manner, being sure to note position of rotor, then pull distributor up until rotor just stops turning and again note position of rotor. To insure correct timing of the distributor, it must be installed with the rotor correctly positioned as noted above.

If necessary to remove secondary wires from cap, mark position on cap tower for lead to No. 1 cylinder. This will aid in reinstallation of leads.

If the engine has been turned after the distributor was removed, it will be necessary to install a jumper wire and crank engine until the timing mark on vibration damper indexes with the proper mark on the engine front cover. If both valves of No. 1 cylinder are closed, the piston will be on top dead center of the firing stroke.

Fig. 12 shows an exploded view of the distributor.

No adjustments can be made on either system and no periodic maintenance is required.

Unit Ignition System

The "Unit Distributor" utilizes an all-electronic module, pickup coil and timer core in place of the conventional ignition points and condenser. Point pitting and rubbing block wear resulting in retarded ignition timing, is eliminated. Since the coil is part of the Unit Distributor there is no need for distributor-to-coil primary (breaker points to coil negative lead) or secondary lead (high voltage lead).

The main features making the Unit Ignition System unique are shown in Figs. 13 and 14.

A magnetic pickup assembly located over the shaft contains a permanent magnet, a pole piece with internal teeth, and a pickup coil. When the teeth of the timer

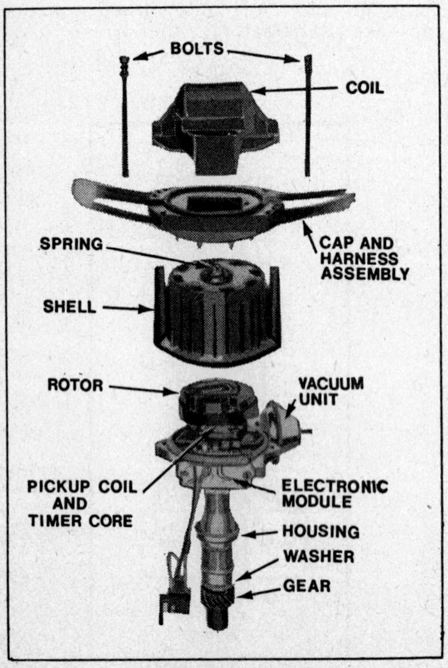

Fig. 14 Exploded view of Unit Distributor

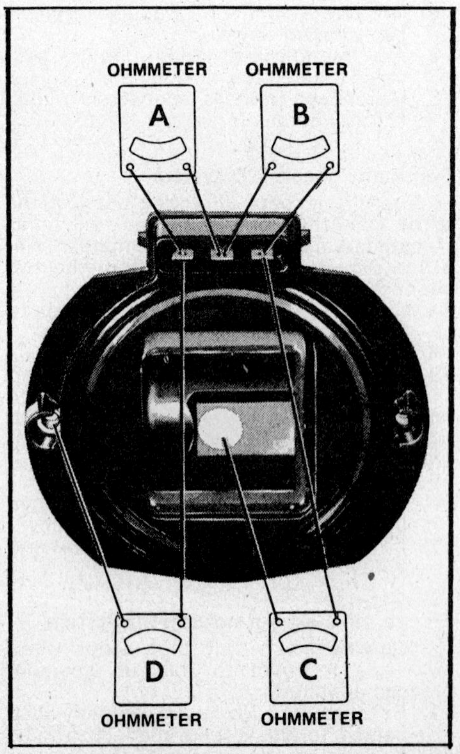

Fig. 15 Ignition Coil test. Unit Distributor

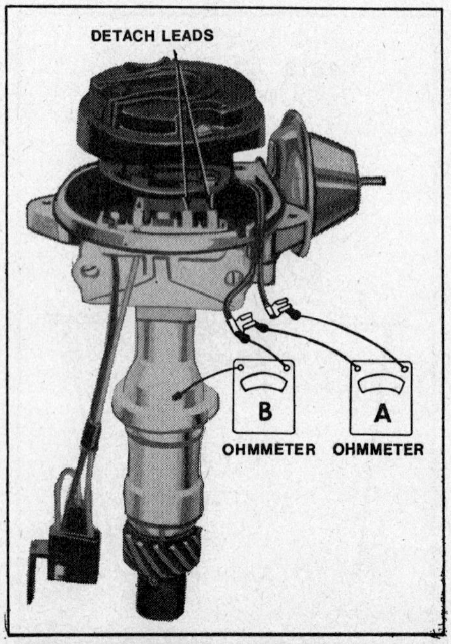

Fig. 16 Pickup coil test. Unit Distributor

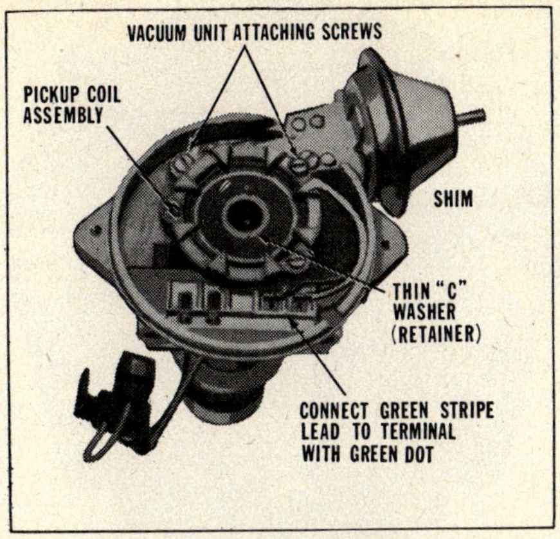

Fig. 17 Unit Distributor with shaft assembly removed

core rotating inside the pole piece line up with the teeth of the pole piece, the induced voltage in the pickup coil signals the all-electronic module to open the ignition coil primary circuit. When this occurs, the primary current decreases and a high voltage is induced in the ignition coil secondary winding. This high voltage is directed through the rotor and high voltage leads to fire the spark plugs.

The magnetic pickup assembly is mounted over the main bearing on the distributor housing, and is rotated by the vacuum control unit, thus providing vacuum advance. The timer core is rotated about the shaft by conventional advance weights, thus providing centrifugal advance.

NOTE: *IMPORTANT*: When making compression checks, disconnect ignition switch connector from Unit Ignition System.

No periodic lubrication is required. Engine oil lubricates the lower bushing, and an oil-filled reservoir provides lubrication for upper bushing.

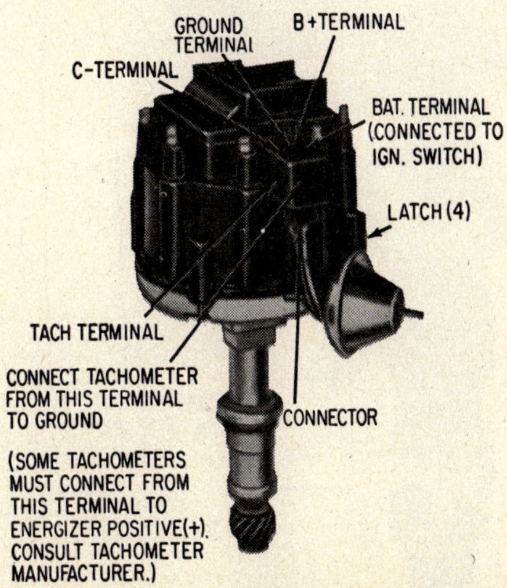

Fig. 18 H.E.I. distributor external components. V6 & V8 engines

Trouble Diagnosis

Insure that the black and pink leads are connected, Fig. 13. Tighten both bolts, Fig. 13. Loose bolts may cause poor performance and radio interference.

NOTE: Before performing extensive diagnostic operations determine if fuel system is operating properly.

On Vehicle

Engine Will Not Start

1. Disconnected or loose ignition switch connector, Fig. 13.
   a. Connect voltmeter from ignition switch connector to ground.
   b. Turn on ignition switch.
   c. If reading is zero, circuit is open between connector and ignition switch.
   d. If reading is Energizer voltage, hold one spark lead with insulated pliers about 1/4" from dry area of engine block while cranking engine. If a strong spark is present, ignition would appear to be trouble free. If no spark, follow "On Test Bench" procedure.

Engine Will Start, But Will Not Run. Rough Engine Operation, Miss or Surge

1. Spark plug lead arc-over, voltage leaking to ground or faulty spark plugs.
   a. If no defects are found, follow "On Test Bench" procedure.

On Test Bench

NOTE: Unit disassembled, Fig. 14

Engine Will Not Start. Engine Will Start, But Will Not Run. Rough Engine Operation, Miss or Surge.

1. Inspect coil, eight inserts, shell and rotor for arc-over or leakage.
   a. Connect ohmmeter, Fig. 15.
   b. Tests "A" and "B" should have practically a zero result. If infinite on either reading, replace coil.
   c. Test "C" should result in 6000-9000 ohms. If outside range, replace coil.
   d. Test "D" should result in infinity. If not, replace coil.
   e. Connect vacuum source to vacuum unit.
   f. Connect ohmmeter, Fig. 16 and observe ohmmeter throughout vacuum range.
   g. If test "A" reads less than 650 ohms, or more than 850 ohms at any time, replace pickup coil.
   h. If test "B" does not read infinity at any time, replace pickup coil.
2. If no defects have been found, replace electronic module.

Pickup Coil Or Vacuum Unit Replacement, Fig. 17

1. Remove unit from engine.
2. Drive retaining pin out through gear.
3. Remove rotor and shaft assembly from housing.

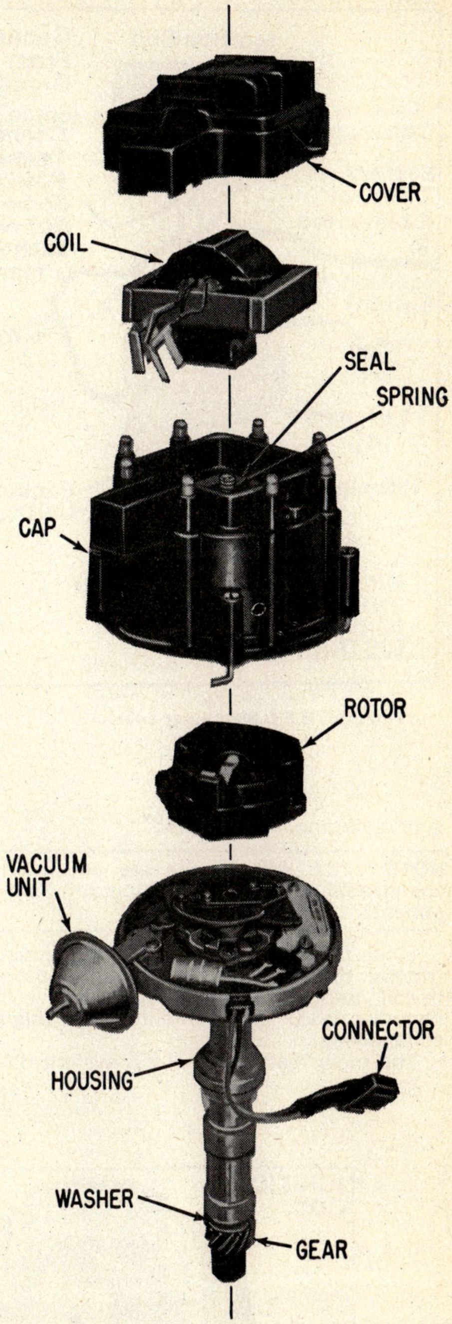

Fig. 19 H.E.I. distributor internal components. V6 & V8 engines

4. Remove shim and "C" washer.
5. Remove pickup coil or vacuum unit.

HIGH ENERGY IGNITION SYSTEM (H.E.I.)

The H.E.I. system utilizes an all-electronic module, pickup coil and timer core in place of the conventional ignition points and condenser (the condenser is used for noise suppression only). Point pitting and rubbing block wear resulting in retarded

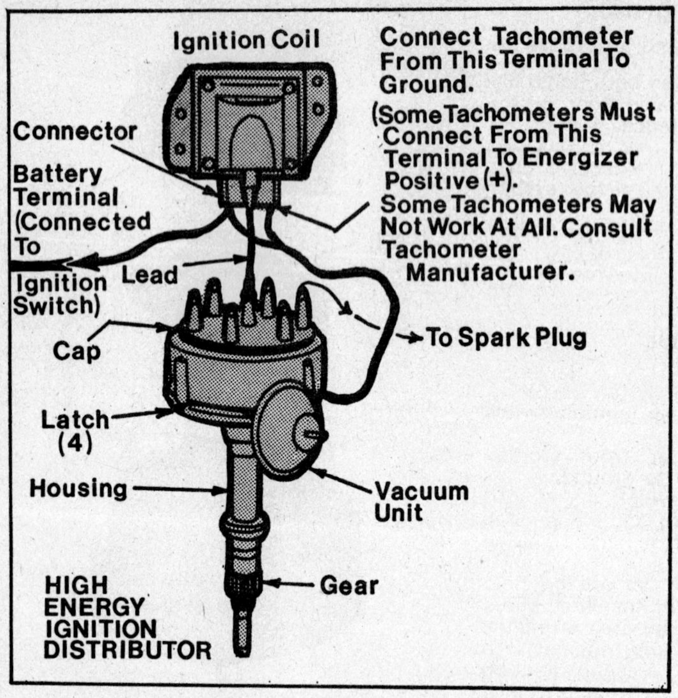

Fig. 20 H.E.I. distributor components. Inline 4 & 6 cyl. engines

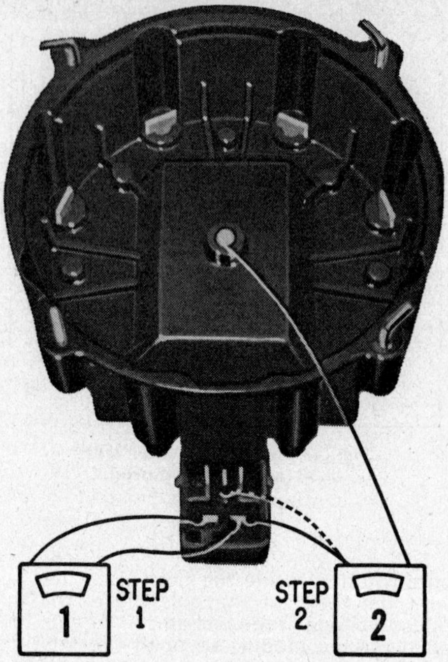

Fig. 21 H.E.I. distributor ignition coil ohmmeter test. V6 & V8 engines

ignition timing, is eliminated.

NOTE: H.E.I. components are not interchangeable with "Unit Distributor" Components.

Since the coil is part of the H.E.I. distributor there is no need for distributor-to-coil primary (breaker points to coil negative lead) or secondary lead (high voltage lead).

The main features of H.E.I. system dif-

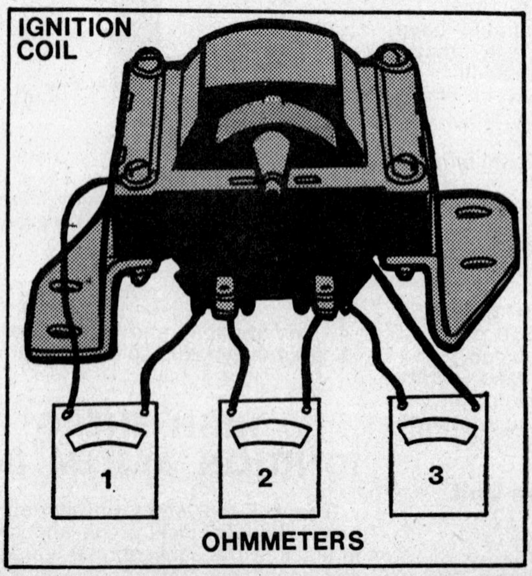

Fig. 22 H.E.I. distributor ignition coil ohmmeter test. Inline 4 & 6 cyl. engines

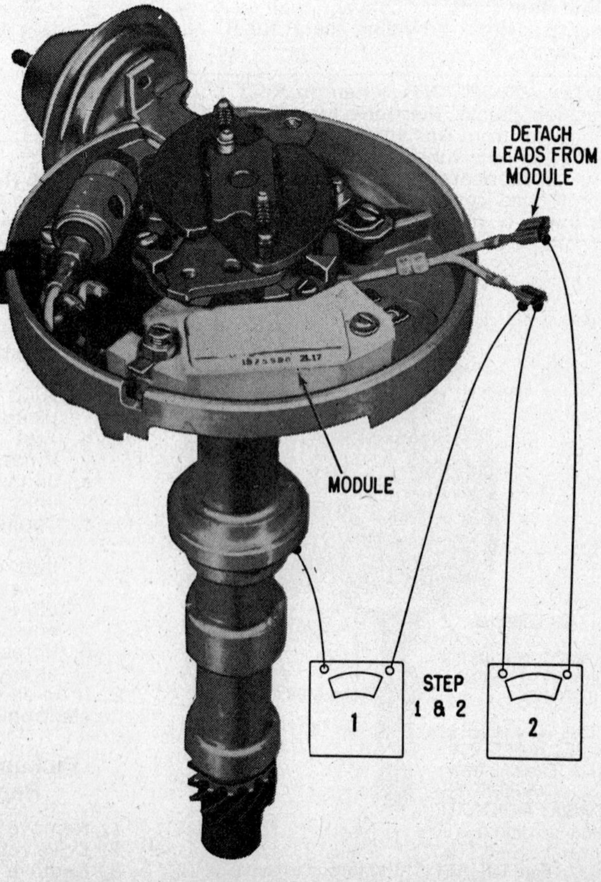

Fig. 23 Distributor pickup coil ohmmeter test

ferentiating this system from the "Unit Ignition" system are shown in Figs. 18, 19 and 20.

The magnetic pickup consists of a rotating timer core attached to the distributor shaft, a stationary pole piece, permanent magnet and pickup coil.

When the distributor shaft rotates, the teeth of the timer core line up and pass the teeth of the pole piece inducing voltage in the pickup coil which signals the all-electronic module to open the ignition coil primary circuit. Maximum inductance occurs at the moment the timer core teeth are lined up with the teeth on the pole piece. At the instant the timer core teeth start to pass the pole teeth, the primary current decreases and a high voltage is induced in the ignition coil secondary winding and is directed through the rotor and high voltage leads to fire the spark plugs.

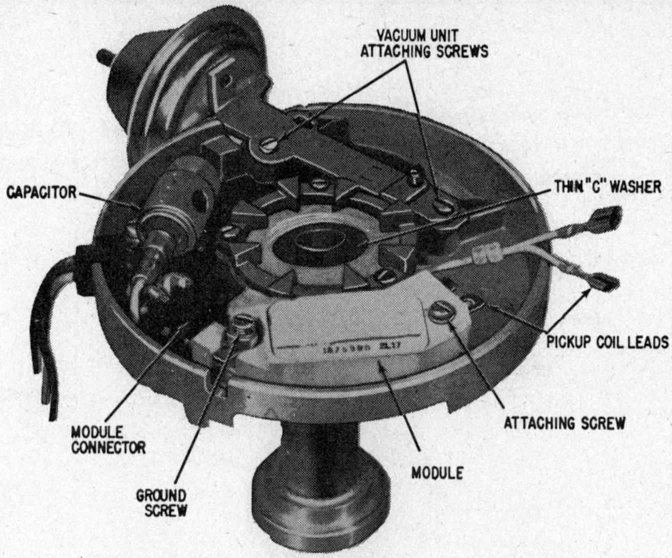

Fig. 24 H.E.I. distributor component replacement

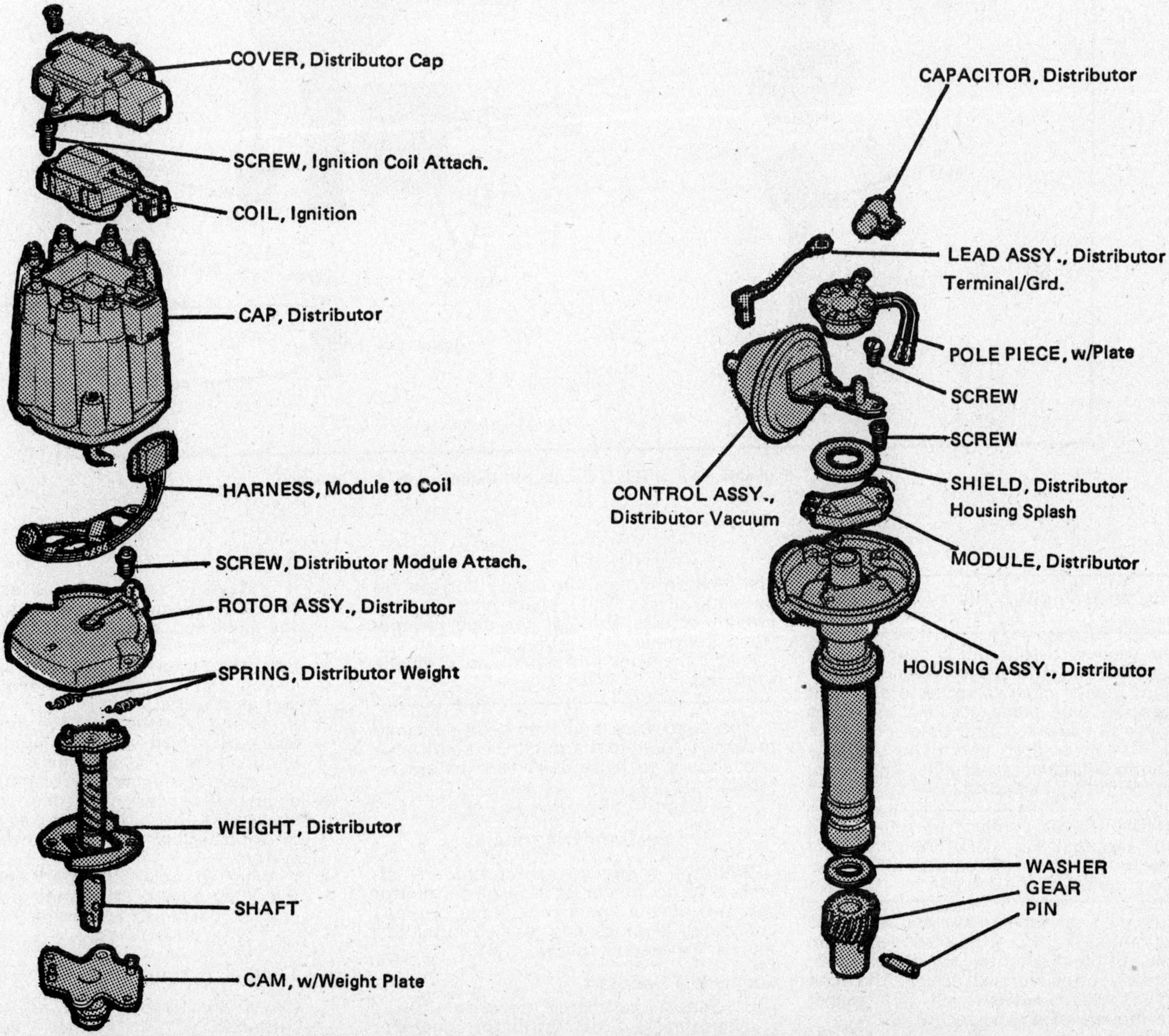

Fig. 25 Exploded view of H.E.I. distributor (Typical). All except Cadillac with fuel injection

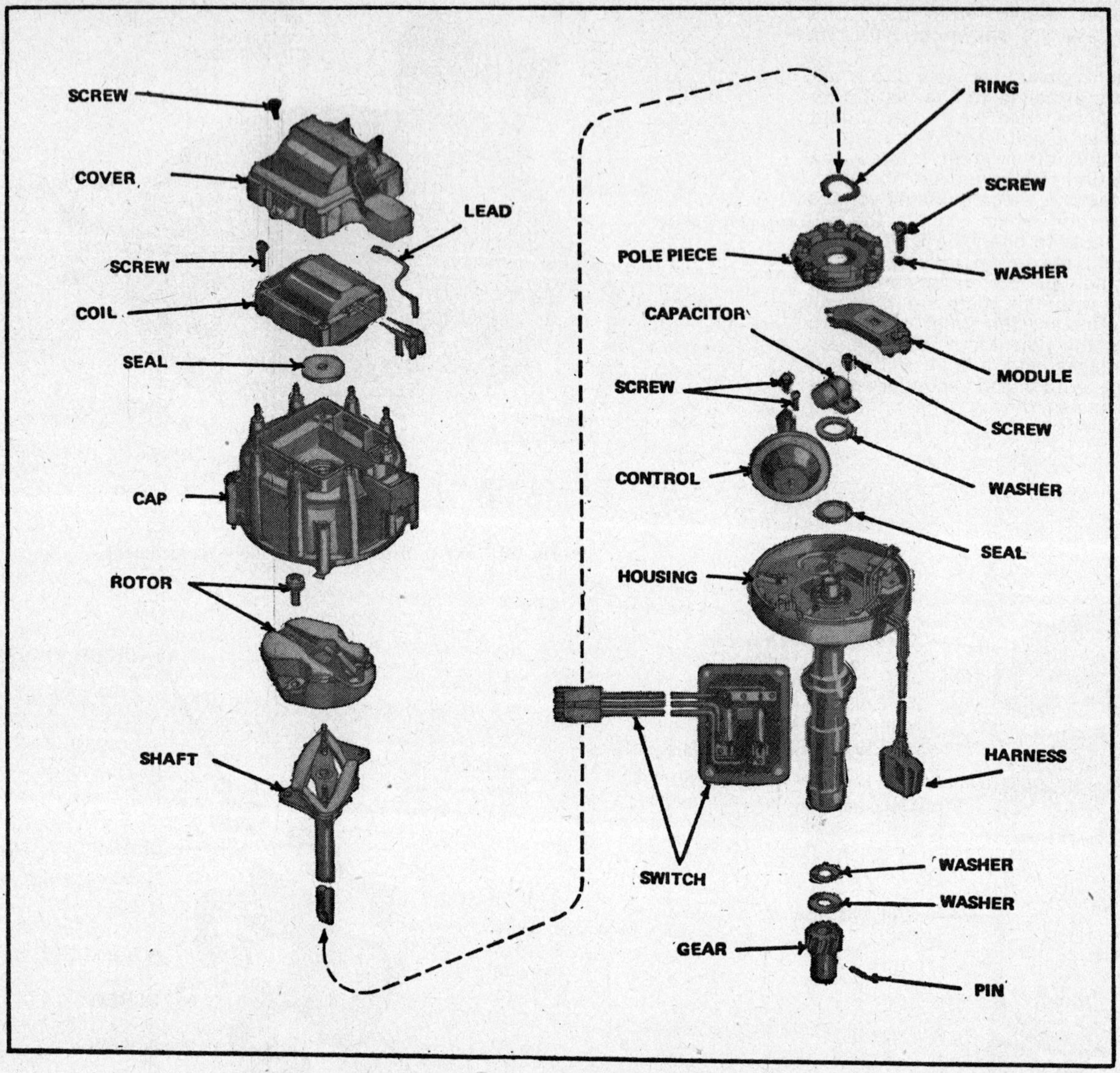

Fig. 26 Exploded view of H.E.I. distributor. Cadillac with fuel injection

NOTE: Since this is a full 12 volt system it does not require a resistance wire.

The vacuum diaphragm is connected by linkage to the pole piece. When the diaphragm moves against spring pressure it rotates the pole piece allowing the poles to advance relative to the timer core. The timer core is rotated about the shaft by conventional advance weights, thus providing centrifugal advance.

CAUTION: Never connect to ground the "tach" terminal, Fig. 18, of the distributor connector as this will damage the electronic circuitry of the module.

A convenient tachometer connection is incorporated in the wiring connector on the side of the distributor, Fig. 18. However due to its transistorized design, the high energy ignition system will not trigger some models of engine tachometers.

NOTE: When using a timing light to adjust ignition timing, the connection should be made at the No. 1 spark plug. Forcing foreign objects through the boot at the No. 1 terminal of the distributor cap will damage the boot and could cause engine misfiring.

The spark plug boot has been designed to form a tight seal around the spark plug and should be twisted ½ turn before removal.

System Diagnosis

With the wiring connector properly attached to connector at side of distributor cap and all the spark plug leads properly connected at plugs and at distributor terminals. Proceed as follows:

Engine Will Not Start

1. Connect voltmeter between "BAT" terminal lead on distributor connector and ground and turn on ignition switch.
2. If voltage is zero, there is an open circuit between the distributor and the bulkhead connector; or between the bulkhead connector and the ignition switch; or between the ignition switch and the starter solenoid. Repair as required.
3. If reading is battery voltage, hold one spark plug lead with insulated pliers approximately ¼ inch away from a dry area of engine block and crank engine. If a spark is visible, the distributor has been eliminated as source of trouble. Check spark plugs and fuel system.
4. If there is no visible spark, perform the "Component Checkout" and proceed as described further on.

Engine Starts But Runs Rough

1. Check for proper fuel delivery to carburetor.
2. Check all vacuum hoses for leakage.
3. Visually inspect and listen for sparks

jumping to ground.
4. Check ignition timing.
5. Check centrifugal advance mechanism for proper operation.
6. Remove spark plugs and check for unusual defects, such as very wide gap, abnormal fouling, cracked insulators (inside and out), etc.
7. If no defects are found, perform the "Component Checkout" procedure as described below.

Component Checkout

1. Remove cap and coil assembly.
2. Inspect cap, coil and rotor for spark arc-over.
3. On V6 and V8 engines:
   a. Connect ohmmeter, Fig. 21, step 1. If ohmmeter reading is other than zero or very near to zero, the ignition coil must be replaced.
   b. If no ohmmeter reading was observed in step 1, reconnect ohmmeter both ways, Fig. 21, step 2. If both ohmmeter readings are infinite on high scale, replace ignition coil.
   c. Connect an external vacuum source to the vacuum advance unit. Replace vacuum unit is inoperative.
4. On inline 4 and 6 cylinder engines:
   a. Connect ohmmeter, Fig. 22, step 1. If reading is not infinite, replace coil.
   b. Connect ohmmeter, Fig. 22, step 2. If reading is not zero or near zero, replace coil.
   c. Connect ohmmeter, Fig. 22, step 3. If reading is infinite, replace coil.
5. If vacuum unit is operating properly, connect ohmmeter, Fig. 23, step 1. If ohmmeter reading on middle scale is not infinite at all times, pick-up coil must be replaced.
6. With ohmmeter connected, Fig. 23, step 2, reading should be within 650 to 850 ohms for all except 1975 Buick, Chevrolet and Pontiac and within 500 to 1500 ohms for 1975 Buick, Chevrolet and Pontiac.

COMPONENTS REPLACE

Ignition Coil Replacement, Fig. 19

V6 & V8 Engines

1. Remove screws holding distributor cover to distributor cap and remove distributor cover.
2. Remove four screws holding coil to cap.
3. Remove harness connector and battery wire from side of distributor cap.
4. Push coil leads out of position in cap and remove coil.
5. Reverse procedure to install.

Inline 4 & 6 Cyl. Engines

1. Disconnect ignition switch to coil lead from coil.
2. Disconnect coil to distributor leads from coil.
3. Remove coil to engine retaining screws and remove coil.
4. Reverse procedure to install.

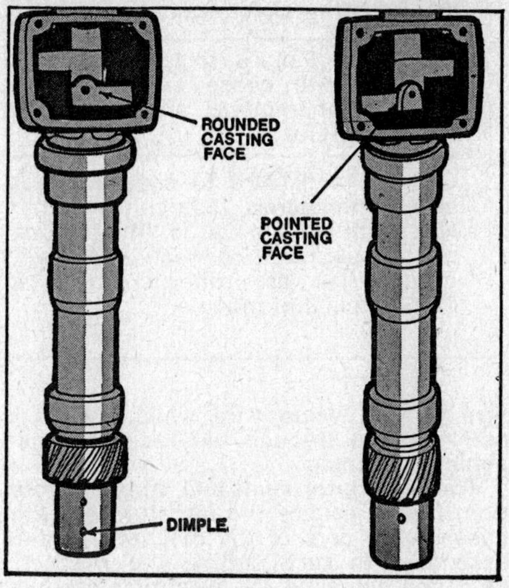

Fig. 27 Speed sensor location

Module Replacement, Fig. 24

1. Disconnect wiring harness connector at side of distributor cap and remove distributor.
2. Remove rotor and disconnect wires from module terminals.
3. Remove two mounting screws and remove module.

CAUTION: At installation, coat bottom of new module with dielectric lubricant (furnished with new module) to aid in heat transfer into distributor housing. Failure to apply lubricant will cause excessive heat at module and premature module failure.

4. Reverse procedure to install.

Pole Piece, Magnet or Pick-Up Coil Replacement, Fig. 24

Removal

1. With distributor removed, disconnect wires at module terminals.
2. Remove roll pin from drive gear by driving out with 1/8 inch diameter drift punch.
3. Remove gear, shim and the tanged washer from distributor shaft. Remove any burrs that may have been caused by removal of pin.
4. Remove distributor shaft from housing.
5. Remove washer from upper end of distributor housing.

NOTE: Bushings in the housing are not serviceable.

6. Remove three screws securing pole piece to housing and remove pole piece, magnet and pick-up coil.

Installation

1. Install pick-up coil, magnet and pole piece and loosely install three screws holding pole piece.
2. With washer installed at top of housing, install distributor shaft and rotate to check for proper clearance between pole piece teeth and timer core teeth.
3. If necessary, realign pole piece to provide adequate clearance and secure properly.
4. Install tanged washer, shim and drive gear (teeth up) to bottom of shaft. Align drive gear and install new roll pin.

DISTRIBUTOR SERVICE
Exc. Cadillac w/Fuel Injection

NOTE: If distributor has been disassembled, refer to Fig. 25 during reassembly.

1. Liberally apply silicone to bottom of module and install module and retaining screws.

CAUTION: Failure to apply silicone lubricant will cause excessive heat build-up at module and premature module failure.

2. Install capacitor and harness making certain that ground lead is under capacitor and retaining screw.
3. Install vacuum advance unit and retaining screws.
4. Place felt wick under pick-up coil plate and lubricate with engine oil, then install pick-up coil and retain with waved C-washer.
5. Install distributor shaft into housing, then install drive gear onto shaft and install new roll pin.
6. On V8 engine distributors, install arc seal, coil and coil cover.

Cadillac w/Fuel Injection

NOTE: Close adherence of the following procedure is important, as several components may be installed in more than one way and could affect engine performance. Refer to Fig. 26 during assembly.

1. Install vacuum advance unit, then place felt washer over reservoir and position shim over felt washer.
2. Place pick-up coil retainer on housing with vacuum advance arm over actuating pin of vacuum advance mechanism and secure with lock ring.
3. Install pick-up coil magnet and pole piece and loosely install the three pole piece retaining screws.
4. Install washer onto top of housing and install timer core on distributor shaft.
5. Install distributor shaft and rotate to check for even clearance between pole piece and timer core. Align pole piece as necessary to provide even clearance and install the three retaining screws.
6. Pull out distributor shaft until speed sensor rotating magnet can be installed into distributor housing and push distributor shaft into position.
7. Install tanged washer (teeth up), shim and drive gear onto bottom of shaft. Note that distributor gear has a dimple on one side next to roll pin hole, Fig. 27. Align drive gear so that dimple is on same side of shaft as rotor pointer, then temporarily install rotor to assure correct alignment and install new roll pin.

8. Align speed sensor rotating magnet, Fig. 27, and install new roll pin.
9. Install plug in distributor housing behind speed sensor, then using a new gasket, position speed sensor onto distributor housing with harness coming out of top of sensor and install retaining screws.
10. Position capacitor onto housing and loosely install retaining screw, then install connector to "B" and "C" terminals of module. Liberally apply silicone lubricant to bottom of module and install module retaining screws.

CAUTION: Failure to apply silicone lubricant will cause excessive heat build-up at module and premature module failure.

11. Connect pink wire to capacitor stud and tighten screw, then connect white wire from pick-up coil to "W" terminal of module and connect white with green stripe wire from pick-up coil to "G" terminal of module.
12. Install advance cam over pins on shaft with stamped numbers facing down and install retaining rings.
13. Install advance weights with letters or numbers facing down and install springs with large loop in-board and offset up. Check weights for proper operation.
14. Install rotor and retaining screws.

NOTE: Assure that notch on side of rotor engages properly.

FORD SYSTEM

1973-75 Breakerless (B/L) Solid State Ignition System

The (B/L) ignition system does not use ignition points and is controlled by an electronic module, also a new oil filled coil is incorporated. Total diagnosis of system requires only a volt-ohmmeter tester.

The electronic module, Fig. 1, is the brain of this system and is well protected from outside elements such as heat and shock. The heat sink containing all the electronic devices is sealed in a mixture of epoxy and sand. This module can not be disassembled and must be replaced if malfunctioning.

The conventional ignition coils are not to be used with this system. The proper coil is easily identified as it is all blue and terminals are labeled differently from conventional ignition coils "BAT" (battery) and "DEC" (Distributor Electronic Control), Fig. 2.

The ignition switch energizes the module through the white wire while engine is cranking and through the red wire when engine is running.

The distributor shaft and armature rotation, Fig. 3, causes the armature poles to pass by the core of the magnetic pick-up assembly. In turn cutting the magnetic field and signaling the electronic module, Figs. 4 and 5, through the orange and purple wires to break the primary ignition current, thus inducing secondary voltage in the coil to fire the spark plugs. The coil is then energized again by the primary circuit and ready for the next spark cycle. This primary circuit is controlled by a timing circuit in the module.

The B/L ignition system is protected against electrical current produced during normal vehicle operation and against reverse polarity or high voltage accidentally applied if vehicle is jump started.

NOTE: IMPORTANT: The ignition system will be damaged if other than volt-ohm test procedures are used to check alternator output. This alternator test procedure is outlined in the "Ford Autolite/Motorcraft Alternator" section, under "Voltmeter Test."

CAUTION: Do not use the volt-amp test procedure or any other test that utilizes a knife switch on the battery terminal.

Since the interval between the time that the module activates the primary ignition circuit and the time the distributor signal turns it off varies with engine speed. Consequently, a dwell measurement is insignificant.

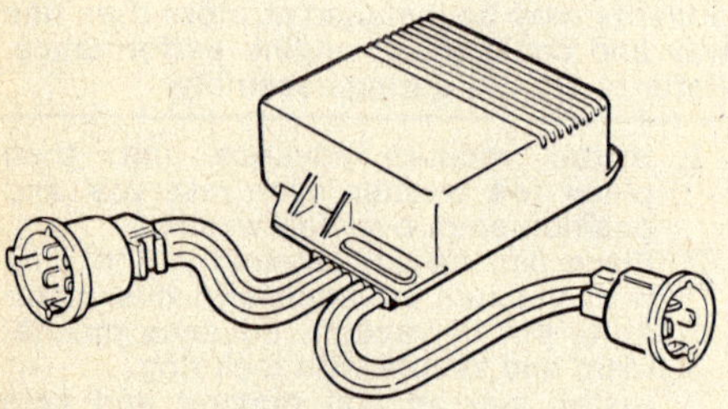

Fig. 1 Ford electronic module

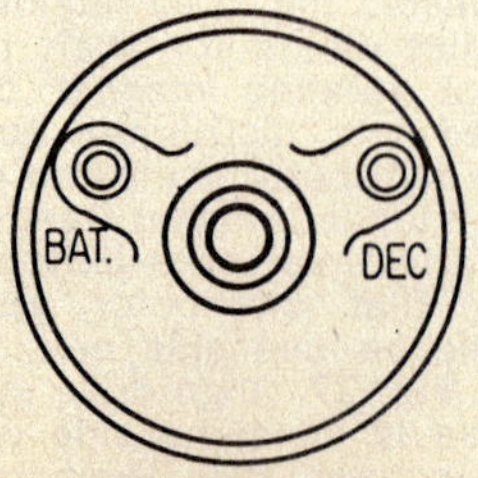

Fig. 2 Ford electronic ignition coil identification

System Diagnosis

If the ignition system is suspected of a malfunction inspect for loose connections and perform the "Secondary Ignition Checkout" procedure in the Trouble Shooting Chapter elsewhere in this manual. Since the secondary circuit is identical to that of a conventional ignition system.

If no spark is observed during the above test, check the ignition coil high tension wire, replace if damaged. If no damage is observed at the coil wire, disconnect the three-way and four-way connectors at the electronic module and make tests at the harness connectors.

NOTE: Do not make tests at the module terminals.

Voltage Tests At Harness Connectors, Figs. 6 & 7

NOTE: If all the following tests comply with specifications replace the module.

Key On

1. Check for battery voltage between pin #3 for 1973-74 or #4 for 1975 and engine ground. If voltage is less than specified, the voltage feed wire to the module is damaged and must be repaired.
2. Check for battery voltage between pin #5 for 1973-74 or #1 for 1975 and

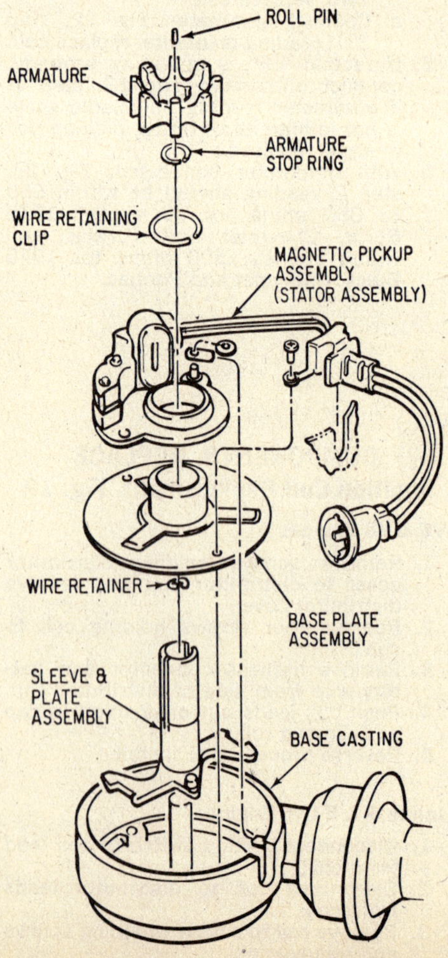

Fig. 3 Ford breakerless (B/L) distributor

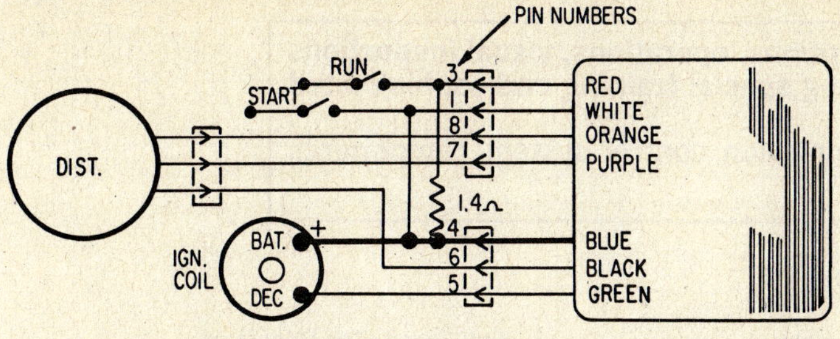

Fig. 4 **1973-74 Ford breakerless ignition primary circuit**

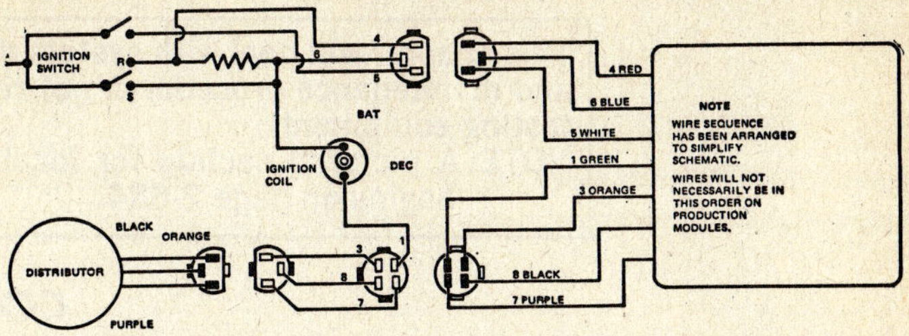

Fig. 5 **1975 Ford breakerless ignition primary circuit**

engine ground. If voltage is less than specified proceed as follows:

a. Without disconnecting the coil, connect voltmeter between coil "BAT" terminal and engine ground.
b. Connect a jumper wire between the coil "DEC" terminal and engine ground.
c. With all lights and accessories off, turn on the ignition switch.
d. A satisfactory primary circuit between the battery and the coil will register 4.9 to 7.9 volts.
e. If less than 4.9 volts register on voltmeter, check for worn primary circuit insulation, broken wire strands or loose-corroded terminals.
f. If a greater than a 7.9 voltage is registered on voltmeter, check and replace if necessary the resistance wire.

Cranking Engine

1. Check for 8 to 12 volts between pin #1 for 1973-74 or pin #5 for 1975 and engine ground. If voltage is not within specifications, the voltage feed wire to the module is damaged.
2. Check for 8 to 12 volts between pin #5 and ground. If voltage is not within specifications, the ignition bypass circuit is open or grounded between starter solenoid or ignition switch and pin #5. Also check primary connections at the coil.
3. Check for ½ volt oscillation (using the 2.5 volt scale) between pin #7 and pin #8. If the voltmeter does not register this oscillation proceed as follows:
   a. Perform same test at the three wire pigtail (with the distributor disconnected), in turn eliminating the orange and purple wires as a cause of malfunction. If no oscillation is present at pigtail, visually inspect distributor components. Make sure that the toothed armature is not damaged, is tight on sleeve and secured properly with the alignment pin, Fig. 3. If armature is not damaged and is rotating properly when cranking the engine and voltmeter is not oscillating, replace the magnetic pickup (stator assembly).

Resistance Test At Harness Connectors, Fig. 6

Key Off

1. Check for resistances of: 400 to 800 ohms between pin #7 and #8 for 1973-74 or pin #3 for 1975, 0 ohms between pin #6 for 1973-74 or pin #8 for 1975 and engine ground and for 70,000 ohms or more between ground and either pin #7 or pin #8. If any of these values do not comply with specifications proceed as follows:
   a. Perform same test at the three wire pigtail (with the distributor disconnected), in turn eliminating the orange and purple wires as a cause of malfunction.
   b. If any of the above checks do not comply with specifications, the magnetic pick-up assembly (stator assembly) is not functioning and must be replaced.
2. Check for a resistance of 7000 to 13,000 ohms between pin #3 for 1973-74 or pin #4 for 1975 and coil tower. Also check for a resistance of 1.0 to 2.0 ohms between pin #5 for 1973-74 or pin #1 for 1975 and pin #4 for 1973-74 or pin #6 for 1975. If either check is not within specifications, diagnose coil separately from rest of system. Coil primary resistance must be within 1.0 to 2.0 ohms and coil secondary resistance must be within 7,000 to 13,000 ohms. If not within specifications follow procedures for testing standard ignition coils as outlined in the "TUNE UP SERVICE" chapter under "Ignition Coils & Resistors."
3. Check for a resistance of more than 4.0 ohms between pin #5 for 1973-74 or pin #1 for 1975 and engine ground. If resistance is less than specified, locate the short to ground either at the coil "DEC" terminal or in the green wire, Fig. 4.
4. If a resistance of 1.0 to 2.0 ohms is not obtained between pins #3 for 1973-74 or pin #6 for 1975 and #4, replace the primary resistance wire.

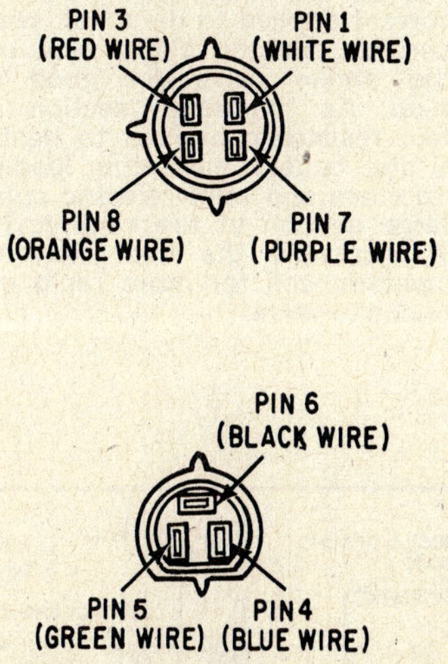

Fig. 6 **1973-74 Ford breakerless ignition female harness connectors system test points**

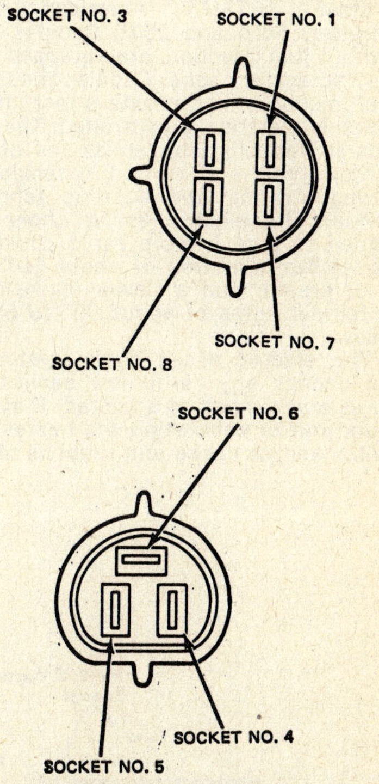

Fig. 7 **1975 Ford breakerless ignition female harness connectors (system test points)**

EMISSION CONTROL SYSTEMS

This chapter will deal with system descriptions, operations, visual inspections and maintenance procedures not requiring special training and sophisticated testing equipment.
NOTE: A pictorial section for locating emission control system components begins on page 2-684.

CONTENTS

ELECTRIC ASSIST CHOKE

Most 1973-75 American Motors, Chrysler, Ford and 1975 Cadillac models without fuel injection, are equipped with an electric assist choke. Cadillac models with fuel injection, incorporate a fast idle valve which is electrically controlled. This device aids in reducing the emissions of hydrocarbon (HC) and carbon monoxide (CO) during starting and warmup (choke on) period. The electric assist choke is designed to give a more rapid choke opening at temperatures of about 60° to 65° F. or greater and a slower choke opening at temperatures of about 60° to 65° F. or below.

The electric assist choke system does not change any carburetor service procedures and cannot be adjusted. If system is found out of calibration the heater control switch and/or choke unit must be replaced.

American Motors & Ford

Fig. 1—The electric choke system consists of a choke cap, thermostatic spring, a bimetal temperature sensing disc (switch), and a ceramic positive temperature coefficient (PTC) heater. The choke is powered from terminal or tap of the alternator. Current is constantly supplied to the ambient temperature switch. The system is grounded through a ground strap connected to the carburetor body. At temperatures below approximately 60 degrees, the switch opens and no current is supplied to the ceramic heater located within the thermostatic spring. Normal thermostatic spring choking action then occurs. At temperatures above approximately 60-65 degrees, the temperature sensing switch closes and current is supplied to the ceramic heater. As the heater warms, it causes the thermostatic spring to pull the choke plates open within 1-1½ minutes.

Cadillac

Fig. 1A—The main components of the electric assist choke system consist of the thermostatic coil, ceramic resistor, cover, bi-metal snap disc and contact spring. The ceramic resistor is divided into a small center section for gradual heating and a large outer section for rapid heating of the thermostatic coil. The electric actuated ceramic resistor heats the thermostatic coil, gradually relaxing coil tension and allowing the choke valve to open.

At air temperatures below 50° F., electric current applied to the small section of the ceramic resistor, allows slow opening of the choke valve for good engine warm-up. As the small section of the ceramic resistor continues to heat, a bi-metal disc causes the spring loaded contact to close and apply electric current to the large section of the ceramic resistor which increases the heat flow to the thermostatic coil for more rapid opening of the choke valve.

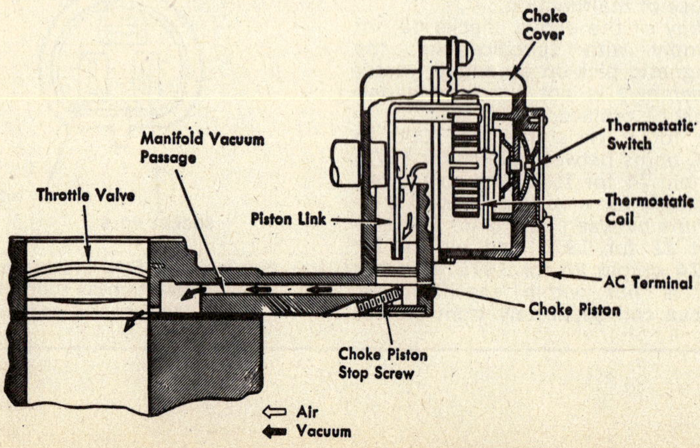

Fig. 1 American Motors and Ford electric assist choke

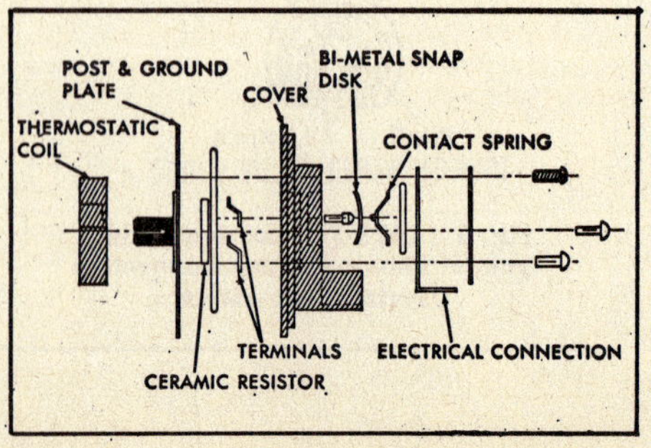

Fig. 1A Electric choke assembly. 1975 Cadillac

At air temperatures between 50-70° F., electric current applied to the small section, or both the small and large sections of the ceramic resistor, will produce the amount of heat required to control the choke valve position for good engine operation in these temperature ranges.

At air temperatures above 70° F., electric current applied to the small section of the ceramic resistor and through the spring contact to the large section of the ceramic resistor, provides rapid heating of the thermostatic coil for quicker choke valve opening when leaner air/fuel mixtures are required at warmer temperatures.

For operation and servicing of the fast idle valve, refer to the "Fuel Injection Section".

Chrysler

NOTE: The wattage of the choke heater is part of the choke calibration and may change from year to year.

Fig. 2—The control switch is connected to the ignition switch from which electrical power is obtained and transfered through electrical connection to the control switch. The 1973 switch serves two purposes:

1. Above 63° F. the control switch will energize the choke heater.
2. After a period of time, the control switch will de-energize the choke heater. The shut-down will result after the control switch warms to about 110° F. by engine heat and a small electrical heater within the switch. Since the heater control switch is mounted to the engine and near the carburetor, some winter operation may energize the choke heater. This could happen after the choke has opened without benefit of electric heat. If this happens it will have no adverse effect on engine operation, and will soon be turned off.

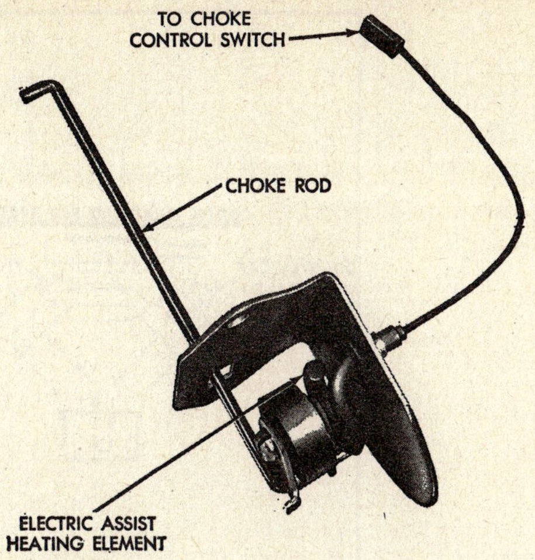

Fig. 2 Chrysler electric assist choke

The 1974 switch serves three purposes:

1. Below 58° F. the control switch will partially energize the choke heater.
2. Above 58° F. the control switch will fully energize the choke heater.
3. The control switch will de-energize the choke heater at approximately 110° F. During winter operations, engines will experience three stages of choke heat; partial heat during engine warm-up, full heat after engine warm-up and no heat well after engine warm-up. Engine starts during summer temperatures will not experience the partial heat stage.

The 1975 single stage control switch serves two purposes:

1. Above 68° F., the control switch will energize the choke heater.
2. At about 130° F., the control switch will de-energize the choke heater.

Since the control switch is engine mounted, some cold engine operation may energize the choke heater, especially if the choke valve has been opened without elec-

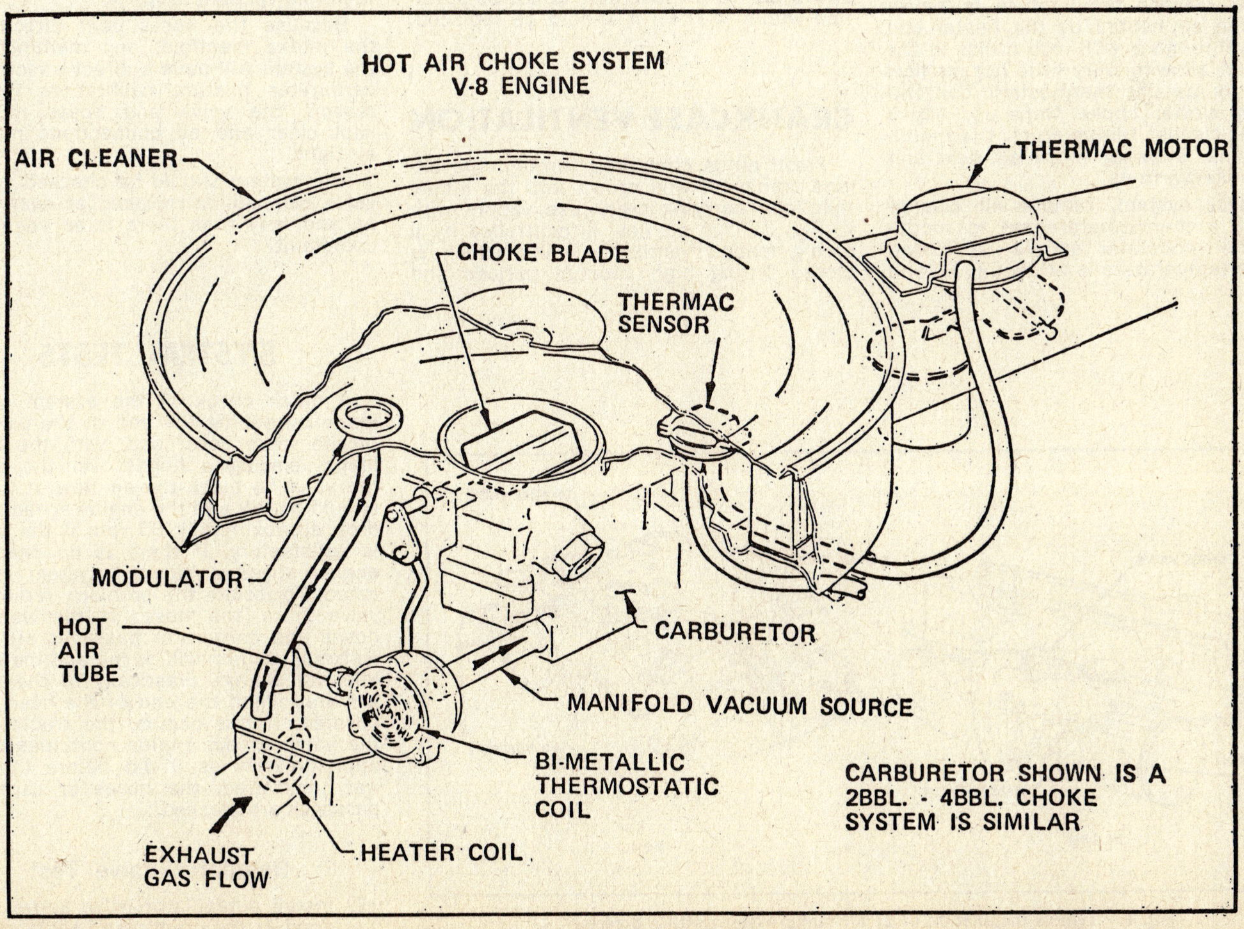

Fig. 2 A Choke Hot Air Modulator System

tric heat. This condition which has no adverse effect on engine operation, will return to normal.

The 1975 dual stage control switch serves four purposes:

1. Below 58° F., the control switch will partially energize the choke heater.
2. Between 58-68° F., the control switch will either partially energize or fully energize the choke heater.
3. Above 68° F., the control switch will fully energize the choke heater.
4. At about 130° F., the control switch will de-energize the choke heater.

NOTE: The heating element should not be exposed to or immersed in any fluid for any purpose. An electric short in the wiring to the heater or within the heater will be a short of the ignition system.

CHOKE HOT AIR MODULATOR

General Motors

All 1975 Buick V6-231, V8-350 and V8-455 and Oldsmobile Omega and Starfire V6-231 and V8-350, incorporate a choke hot air modulator located in the bottom of the air cleaner housing to provide heated filtered air to the choke housing, Fig. 2A. When air cleaner temperature is below 68° F., the air heated by the heater coil will pass through a .005 inch orifice in the modulator, allowing very little hot air flow over the bi-metallic thermostatic coil and providing slower choke warm-up. Above 68° F. air cleaner temperature, the modulator opens, allowing more air flow and faster choke warm-up.

To check system, remove air cleaner and tape a thermometer next to modulator. The modulator should be closed when the temperature is below 68° F., and should be open when the temperature is above 68° F. If modulator is not as specified within ±7° F., it should be replaced.

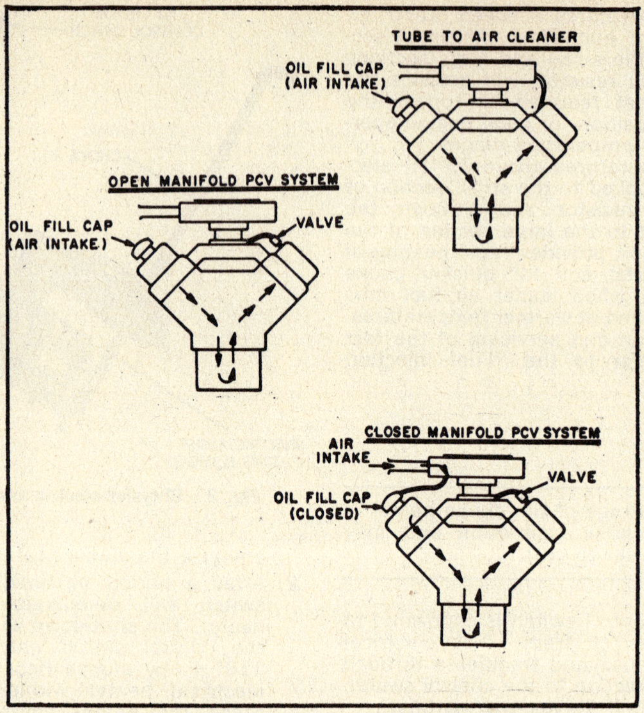

Fig. 3 Various PCV systems

CRANKCASE VENTILATION

Fresh air is circulated through the engine crankcase and drawn into the intake manifold, carrying crankcase vapors with it, Fig. 3. This air flow is controlled by a spring loaded ventilation valve which is closed during high vacuum periods and opens to provide maximum ventilation as manifold vacuum drops off.

Because the hoses vent directly into the intake manifold, any malfunction in the system will have a direct effect on the carburetor mixture calibration. For this reason, the valve and hoses must be kept clear and all connections must be air tight.

The system should be checked and the valve cleaned or replaced at every tune-up and, any time there is an engine idle complaint.

SYSTEM TESTS

A quick check of the system can be made by pulling the end of the valve out of the valve cover and, with the engine idling, placing a finger over the end of the valve to block the air flow. A vacuum should be felt and the engine speed should drop approximately 50 rpm if the system is satisfactory. If there is no change in engine speed a clogged system is indicated. To isolate the problem, remove the valve from the hose. If the ventilator hoses and carburetor passages are clear, a strong vacuum will be felt and the engine idle will change drastically or the engine will stall when the end of the hose is uncovered. If this occurs, the trouble is in the valve. If the engine continues to idle approximately as it did before the hose was uncovered, the hoses or carburetor passages are blocked.

Regulator Valve Test

1. Install a regulator valve known to be good in the crankcase ventilation system.

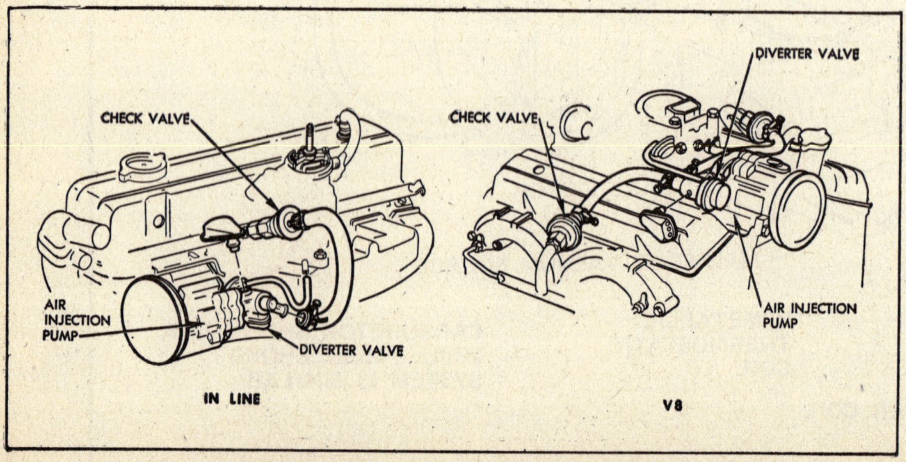

Fig. 4 Typical installation of an air pump system with a diverter valve, otherwise known as an air by-pass valve. 1969-75

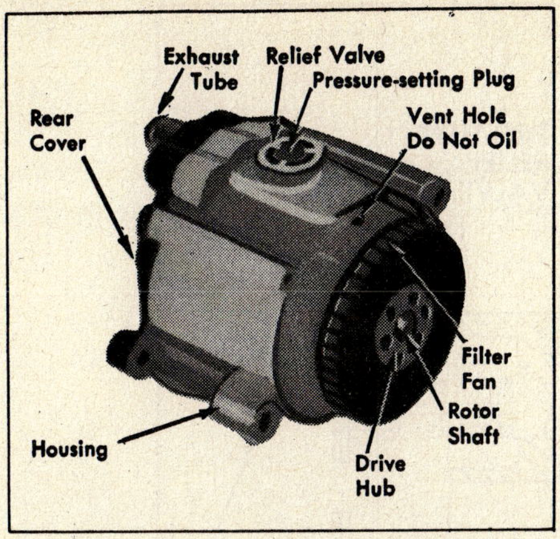

Fig. 6 Air injection pump with integral centrifugal air filter and pressure relief valve

2. Start engine and compare engine idle condition to the prior idle condition.
3. If the loping or rough idle condition remains when the good regulator valve is installed, the crankcase ventilation system is not at fault. Further engine component diagnosis will have to be made to find the cause of the malfunction.
4. If the idle condition proves satisfactory, replace the regulator valve and clean hoses, fittings, etc.

EXHAUST EMISSION CONTROLS AIR PUMP SYSTEMS

American Motors Air Guard, Chrysler Air Injection, Ford Thermactor & GM Air Injection Reactor (A.I.R.)

All air pump systems, Fig. 4 consist of an air pump, Figs. 6 and 7, injection tubes (one for each cylinder), a diverter or air by-pass valve, check valves (one for In Line engines, two for V8s), air manifolds, tubes and hoses necessary to connect the various components.

Carburetors, distributors and diverter or air by-pass valves are designed for specific engine applications and should not be interchanged with other units.

In this system, fresh air is pumped into the exhaust system in the area of the exhaust valves, to ignite and burn the unburned portion of exhaust gases in the exhaust system, thus minimizing exhaust contaminations.

The mixture control or backfire by-pass valve, Fig. 8, when triggered by a sharp increase in manifold vacuum, supplies the intake manifold with fresh air to lean out the fuel/air mixture and prevent backfire.

The diverter or air by-pass valve, Figs. 9 and 10, when triggered by a sharp increase in manifold vacuum, shuts off the injected air to the exhaust and prevents backfiring during this richer period. On engine overrun the total air supply is dumped through the muffler on the diverter or air by-pass valve. At high engine speeds the excess air is dumped through the pressure relief valve when the relief valve is part of the air pump, Fig. 6, and through the diverter or air by-pass valve when the pressure relief valve is part of the diverter or air by-pass valve, Fig. 10.

The check valve(s) prevents air pump damage due to exhaust back flow, which can occur even under normal operating conditions.

Although a properly operating system will effectively reduce emissions, if any system component or engine component operating in conjunction with the system should malfunction, emissions may be increased.

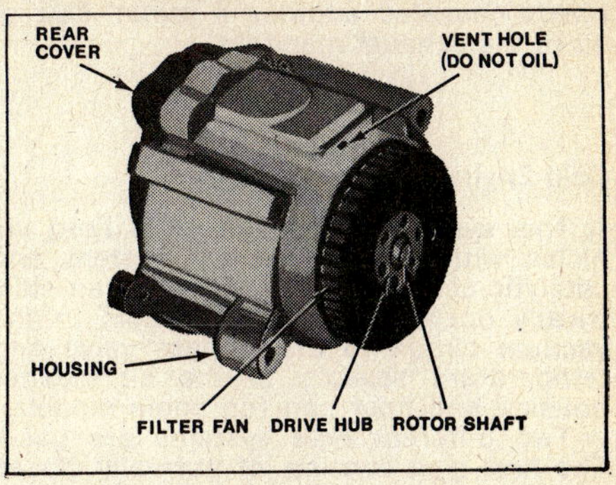

Fig. 7 Air injection pump with integral centrifugal air filter without pressure valve

Vacuum Differential Valve (VDV)

On 1975 Ford vehicles with the Thermactor system and catalytic converters, a (VDV) Fig. 10A, is used to control the operation of the air bypass valve. Under normal operation, vacuum applied through the VDV holds the valve upward, blocking the vent port and allowing Thermactor air flow. During acceleration or deceleration or in case of system failure, the VDV momentarily cuts off vacuum flow to the bypass valve, diverting the Thermactor air flow to atmosphere. In case of excessive pressure or system restriction, the excess pressure will unseat the valve in the lower part of the bypass valve, allowing a partial flow of air to atmosphere. At the same time, the valve in the upper part of the valve re-

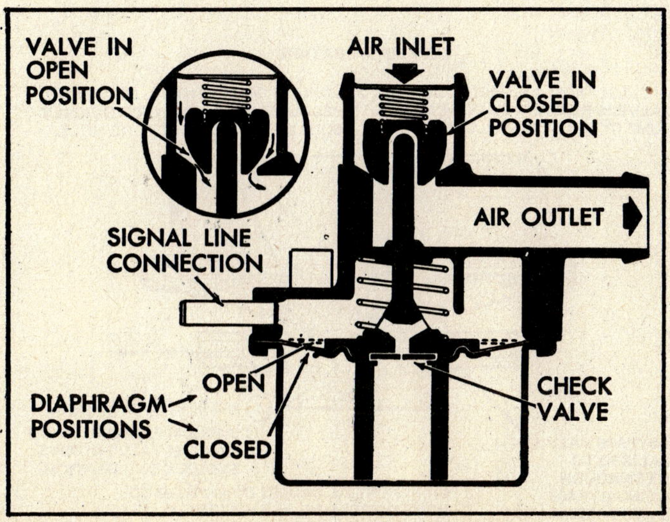

Fig. 8 Typical mixture control or backfire by-pass valve

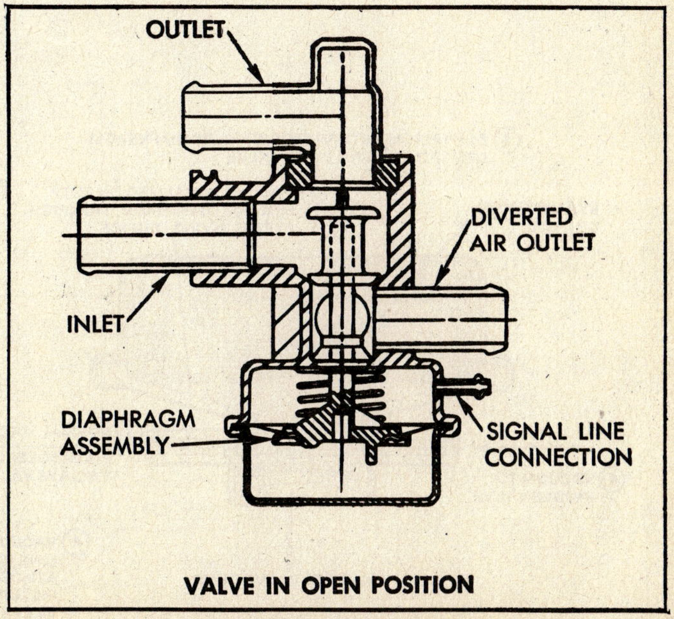

Fig. 9 Typical diverter or air by-pass valve without pressure relief valve

mains unseated allowing a partial flow of air to the exhaust manifold.

Cold Engine Lockout System

This system is used on 1975 Ford vehicles with the Thermactor system and catalytic converters. It consists of an electrically operated vacuum solenoid in the vacuum circuit to the by-pass valve and temperature sensors in the air cleaner housing and floor pan (on some models).

Two different type systems are used. The first type consists of normally closed switches that provide current to the solenoid only when the engine is warm. The second type consists of normally open switches that provide current to the solenoid only when the engine is cold. Operation of both systems is as follows:

Normally Closed Switches, Type 1, Figs. 10B & 10C

With engine at normal operating temperature and all components operating properly, electrical path is provided through the solenoid vacuum valve, air cleaner sensor, floor pan temperature sensor (on some models) and to ground. This provides vacuum flow through the vacuum solenoid valve, VDV and to the bypass valve, allowing Thermactor air flow.

If any of the switches should operate due to low intake air temperature or overheated floor pan (on some models), the vacuum solenoid will stop vacuum flow, causing the bypass valve to divert Thermactor air to atmosphere.

Normally Open Switches, Type 2, Figs. 10D & 10E

With engine at normal operating temperature and all components operating properly, the solenoid vacuum valve is inoperative, thereby providing vacuum flow through the solenoid vacuum valve, VDV and to the bypass valve, allowing Thermactor air flow. With the engine cold, the air cleaner sensor operates, providing current flow through the solenoid vacuum to ground. If the floor pan sensor operates (on some models), it will provide current flow through the solenoid to ground.

If the solenoid vacuum valve operates, it stops the vacuum flow to the bypass valve, causing the Thermactor air flow to be diverted to atmosphere.

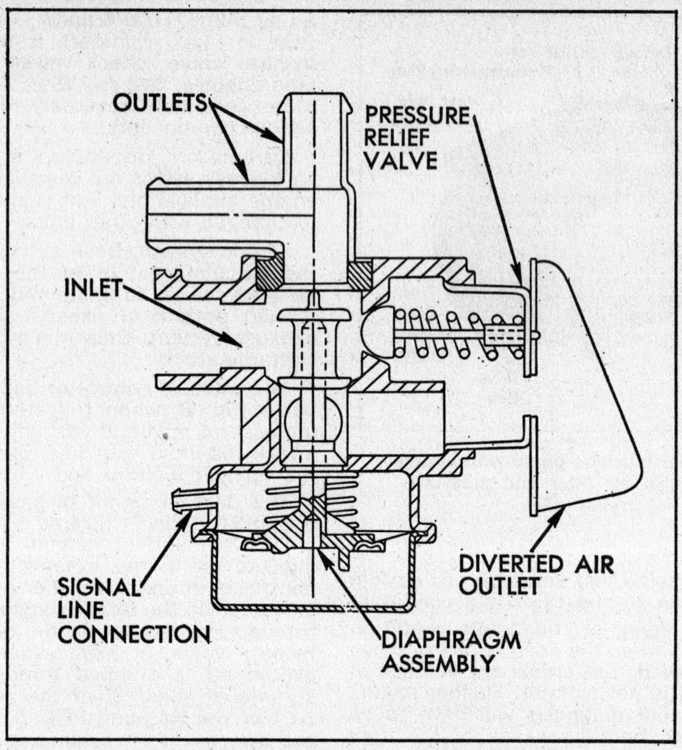

Fig. 10 Typical diverter or air by-pass valve with integral pressure relief valve

Maintenance

Engine tune-up should be checked whenever the air pump system seems to be malfunctioning, especially items affecting air/fuel ratio.

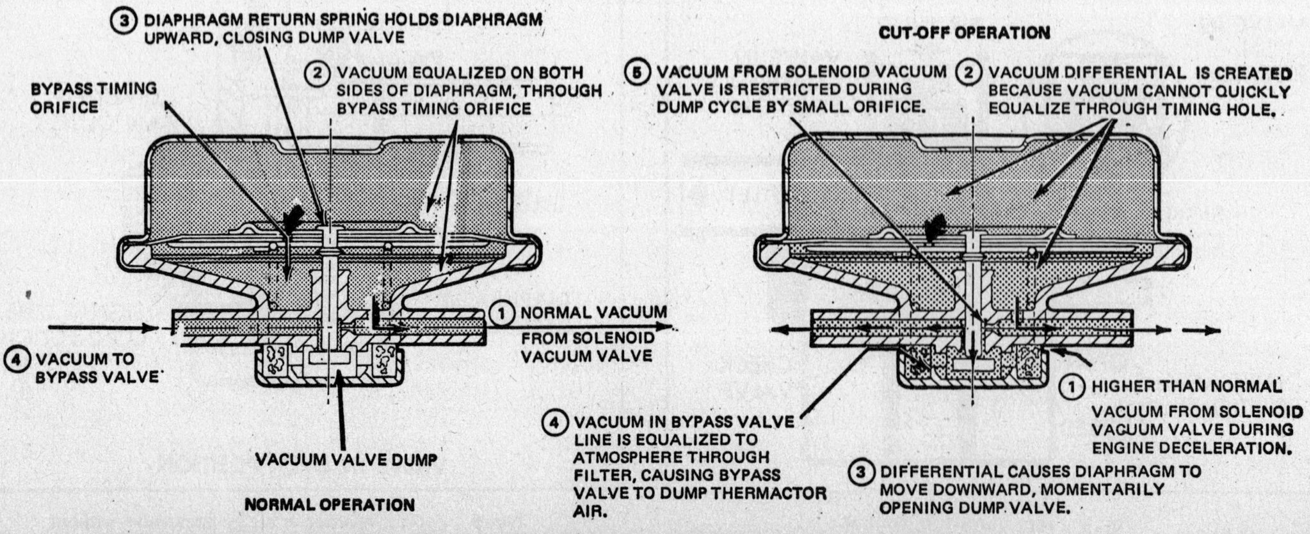

Fig. 10A Vacuum differential valve. 1975 Ford with catalytic converter

Because of the similarity of many parts, typical illustrations and procedures are given in the following text.

Air Manifold, Hose and Tube; Fig. 11

1. Inspect all hoses and tubes for deterioration or cracks.
2. Check hose and tube routing as interference may cause wear.
3. Check all hose and tube connections.
4. Check the pressure side of the system for leaks with a soapy water solution. With the pump running, bubbles will form if a leak exists.
5. When replacing any hose or tube, note routing before removal.

CAUTION: The hoses used with this system are made of special material to withstand high temperature. No other type should be used.

Check Valves:

1. Check valves should be inspected whenever the hose is disconnected from the valve or check valve failure is suspected.

NOTE: Any indication of exhaust gases in the air pump indicates check valve failure.

2. Orally blow through the check valve (toward air manifold) then attempt to suck back. Flow should be toward air manifold only.
3. When replacing a check valve, be careful not to bend or twist the air manifold.

Diverter or Air By-Pass Valve:

1. Check condition and routing of all lines, especially the signal line. All lines must be secure without crimps or leaks.
2. Disconnect signal line at valve. A vacuum signal must be available with engine running.
3. With engine warmed up to operating temperature and carburetor at curb idle, no air should be escaping through the valve's muffler. Manually open and quickly close the throttle; a momentary blast of air should discharge through the valve's muffler for at least one second. Defective valve should be replaced.

CAUTION: Although sometimes similar in appearance, these valves are designed to meet particular requirements of various engines; therefore, be sure to install the correct valve for the engine being serviced.

4. To replace a valve, disconnect vacuum signal line and valve exhaust hose(s).
5. Remove diverter or air by-pass valve from pump; also muffler from valve assembly, noting angle of attachment.
6. Install muffler to new valve at angle previously noted.
7. Install diverter or air by-pass valve to pump or bracket with new gasket.
8. Install outlet and vacuum lines and check system for leaks.

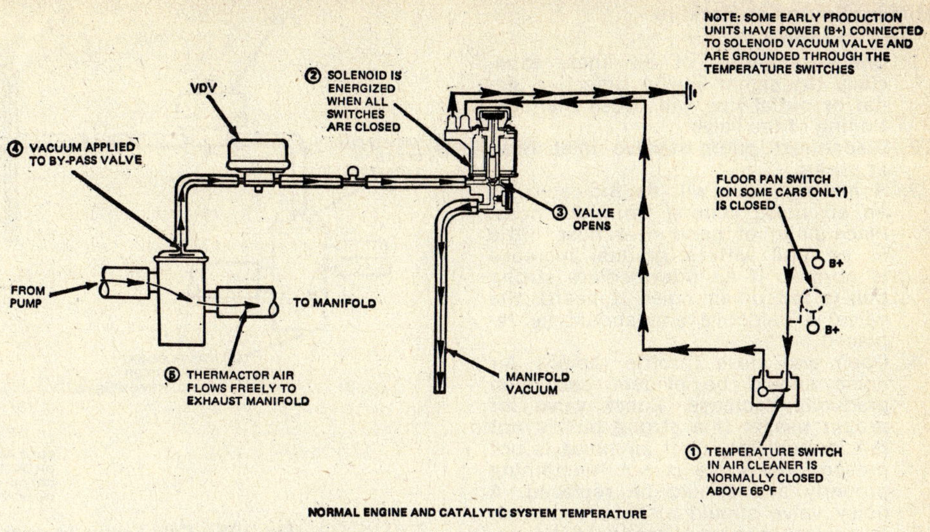

Fig. 10B Cold Engine Lockout System. Normally closed switches

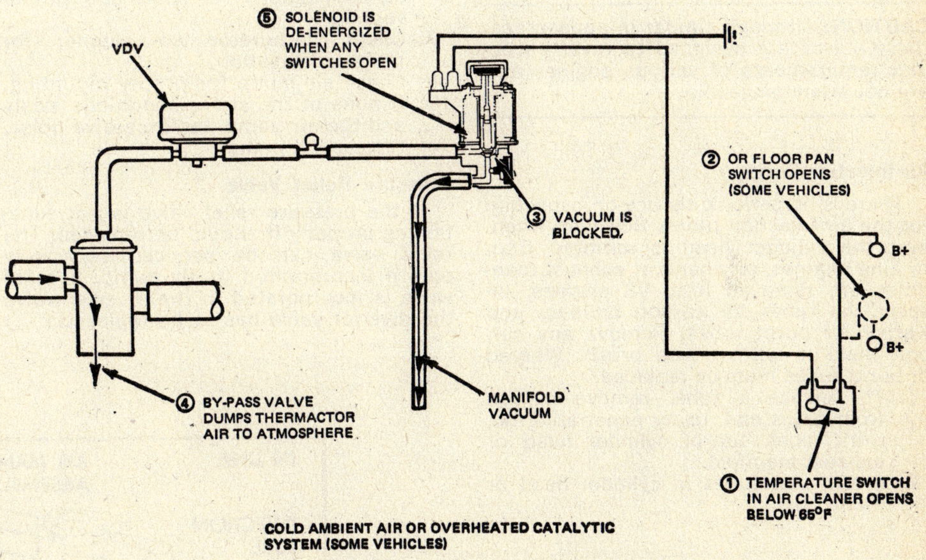

Fig. 10C Cold Engine Lockout System. Normally closed switches

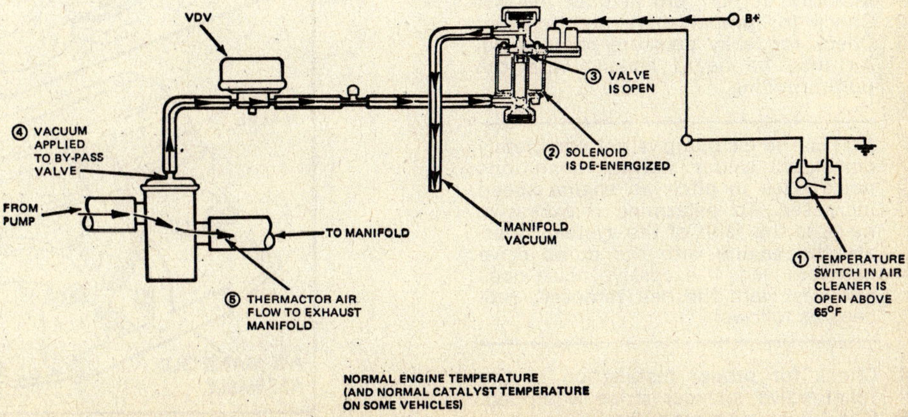

Fig. 10D Cold Engine Lockout System. Normally open switches

Mixture Control or Backfire By-Pass Valve:

1. Check condition of all lines, especially the signal line. A defective signal or outlet line will cause malfunctioning of the valve.
2. Disconnect pump-to-valve inlet hose at pump.
3. A leaking valve will be indicated by an air noise coming from the hose. Place palm of hand over hose; little or no pull with a gradual increase is normal. If an intermediate strong pull is felt or air noise is heard, the valve is defective and should be replaced.
4. Open and close throttle rapidly. Air noise should be evident and then gradually decrease. Check valve for proper usage. If a strong pull is not felt immediately or if air noise is not present, the valve is not functioning properly and should be replaced. A noisy valve should also be replaced.
5. To replace a valve, disconnect the signal line, air inlet and outlet hoses then remove valve.
6. Install new valve and connect air outlet, inlet and signal line hoses.

CAUTION: Although similar in appearance, these valves are designed to meet particular requirements of various engines and are not interchangeable.

Air Injection Tube:

There is no periodic service or inspection for the air injection tubes. However, whenever the cylinder head is removed from In Line engines, or whenever exhaust manifolds are removed from V8 engines, inspect the tubes for carbon build-up and warped or burnt tubes. Remove any carbon build-up with a wire brush. Warped or burnt tubes must be replaced.

1. To replace a tube, remove carbon from tubes and, using penetrating oil, work tubes out of cylinder head or exhaust manifold.
2. Install new tubes in cylinder head or manifold.

Air Injection Pump:

1. Accelerate engine to about 1500 rpm and observe air flow from hose(s). If air flow increases as engine is accelerated, the pump is operating satisfactorily. If not, proceed as follows:
2. Check for proper drive belt tension.
3. Check for leaky pressure relief valve. Air may be heard leaking with the pump running.

NOTE: The air pump is not completely noiseless. Under normal conditions noise rises in pitch as engine speed increases. To determine if excessive noise is the fault of the system, operate the engine with the pump drive belt removed. If excessive noise does not exist with the belt removed, proceed as follows:

---

4. Check for proper installation of the relief valve silencer if so equipped.
5. Check for seized air pump.
6. Check hoses, tubes, air manifolds and all connections for leaks and proper routing.
7. Check carburetor air cleaner for proper installation.
8. Check air pump for proper mounting.
9. If none of the above conditions exists and the air pump has excessive noise, replace the pump.

Pressure Relief Valve

If the pressure relief valve is not functioning properly it should be replaced. The relief valve can be serviced separately only if incorporated in air pump, if relief valve is incorporated in the diverter valve the diverter valve has to be replaced.

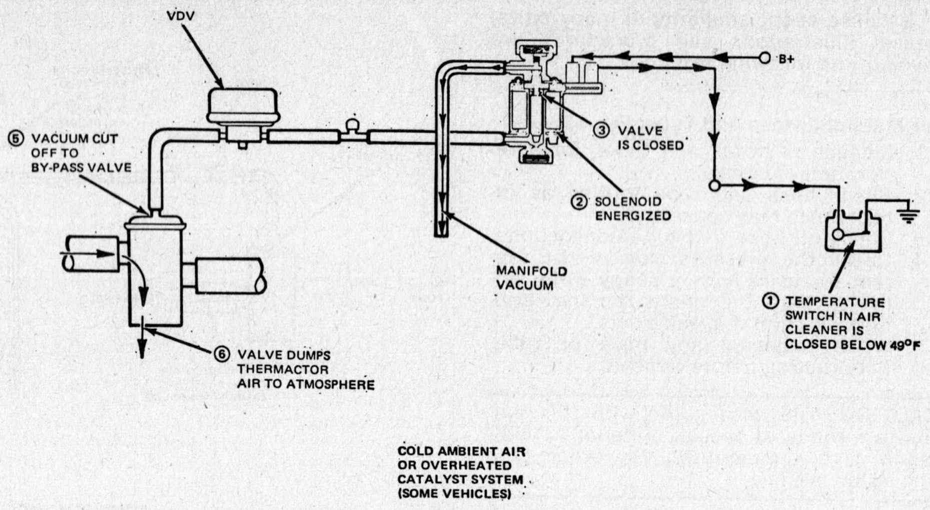

Fig. 10E Cold Engine Lockout System. Normally open switches

PULSE AIR INJECTION REACTOR (PAIR)

Chevrolet Cosworth Vega

This system, Fig. 11A, utilizes exhaust pressure pulsations to draw fresh air from the air cleaner into the exhaust system.

This system consists of one air "check valve", Fig. 11B, located on each of the four exhaust manifold legs adjacent to the cylinder head, and a pulse air shut off valve through which filtered air flows to the check valves.

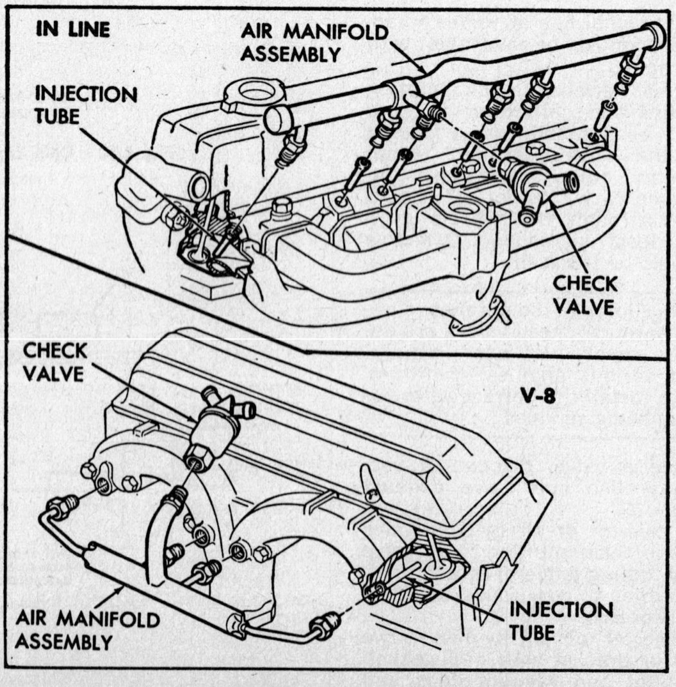

Fig. 11 Typical air manifold installations

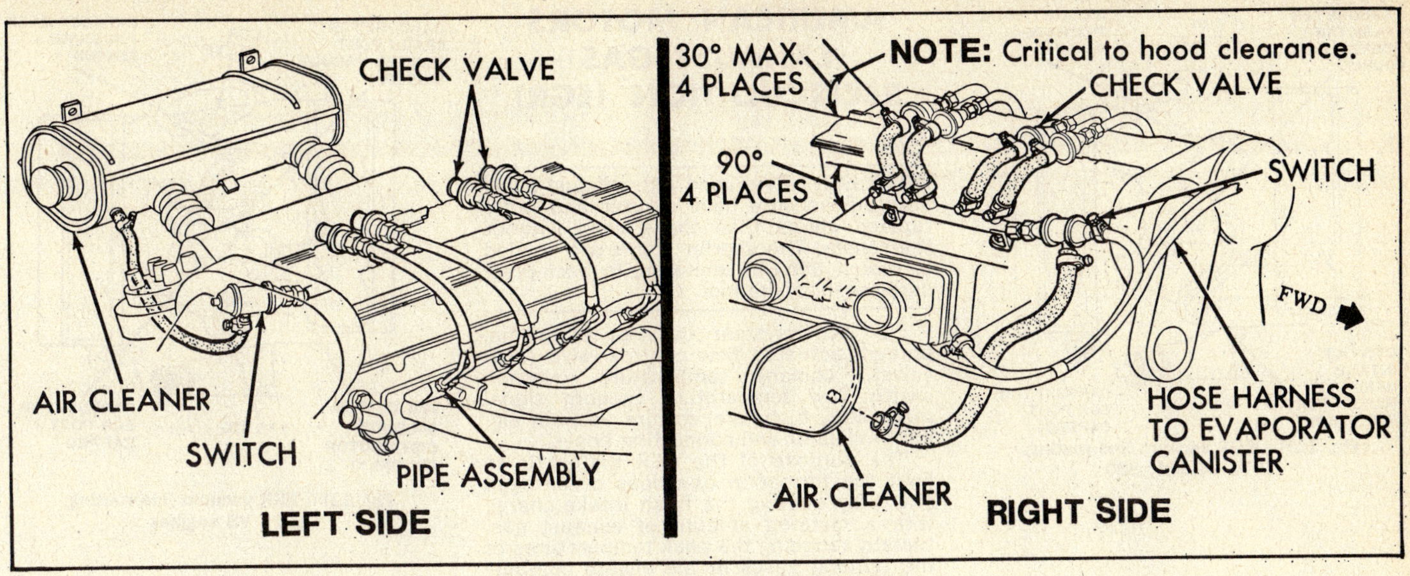

Fig. 11A Pulse Air Injection Reactor System

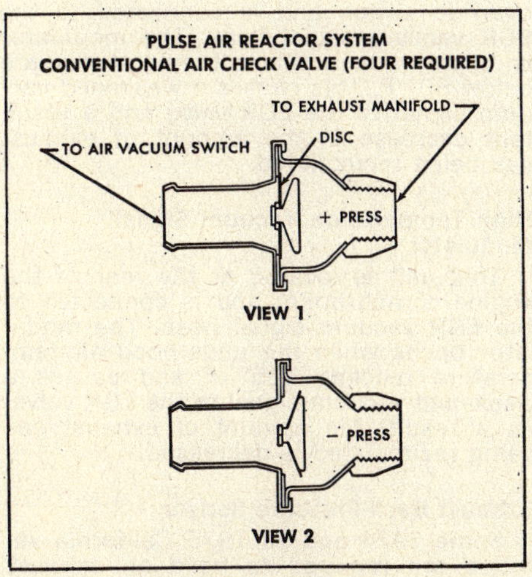

Fig. 11B PAIR System check valve

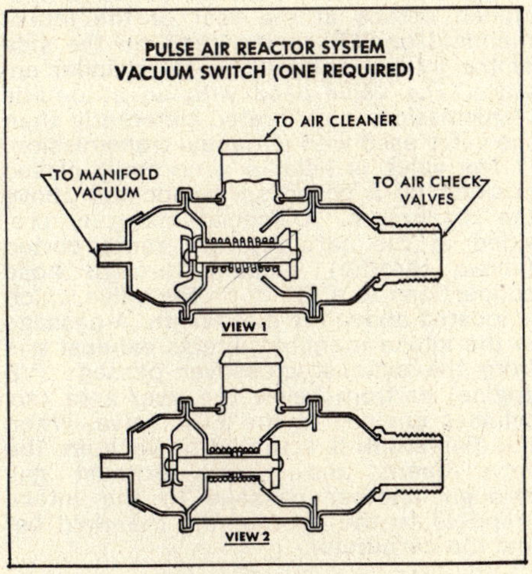

Fig. 11C PAIR System air shut off valve

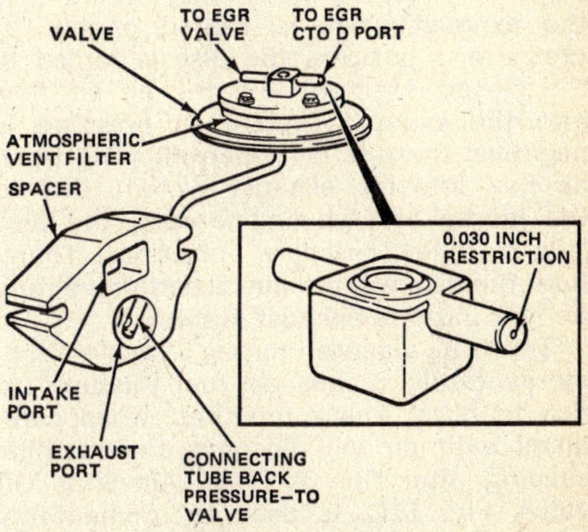

Fig. 11D Exhaust back pressure transducer sensor. (Typical)

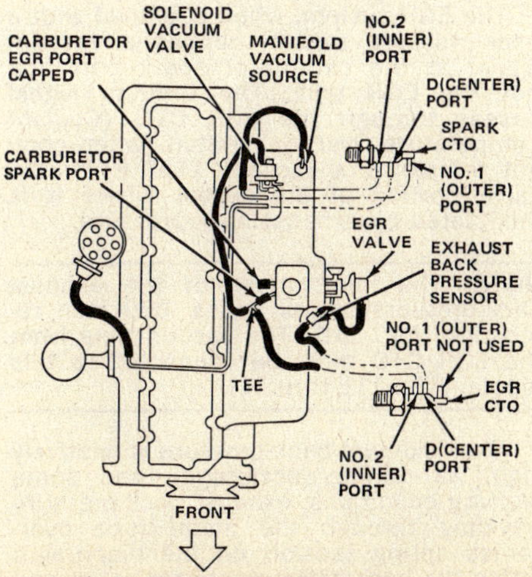

Fig. 11E EGR vacuum line routing. 1974 6 cyl. less Air Guard

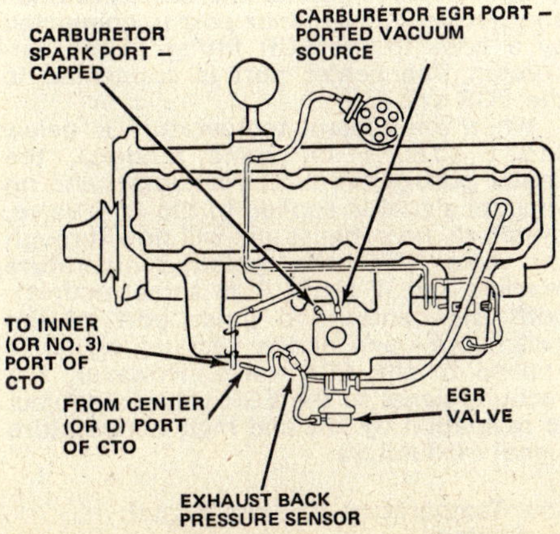

Fig. 11F EGR vacuum line routing. 1974 6 cyl. with Air Guard

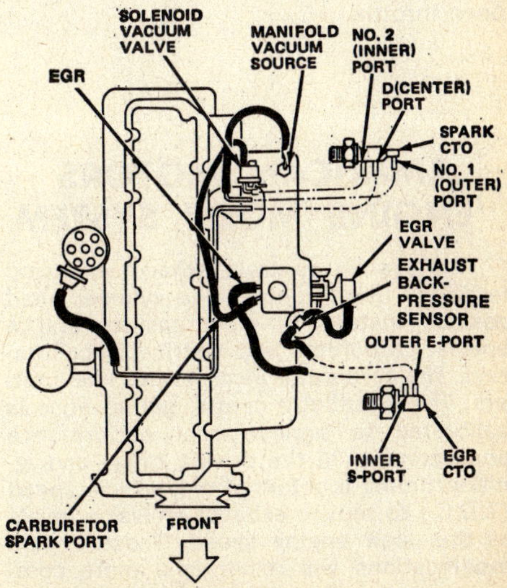

Fig. 11G EGR vacuum line routing. 1975 6 cylinder

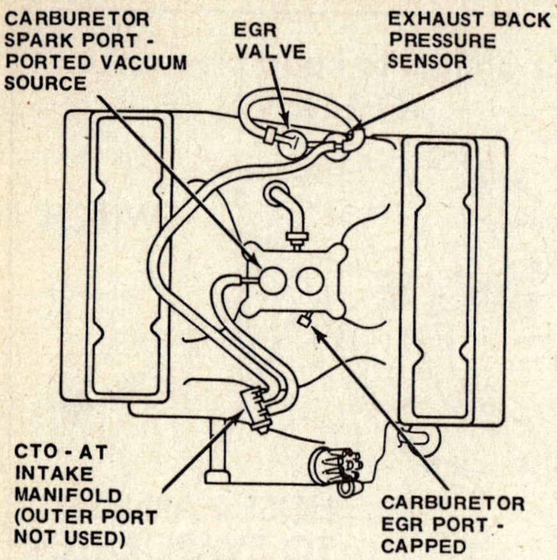

Fig. 11H EGR vacuum line routing. 1974 V8-304, 360

Engine operation causes a pulsating flow of exhaust gasses which are of positive or negative pressure, depending on whether the exhaust valve is seated or not. If pressure is positive, the disc is forced to the closed position, allowing no air flow into the exhaust system. If pressure is negative, the disc will open allowing fresh air flow into the exhaust system. Due to the inertia at high engine rpm, the disc fails to follow pressure pulsations. Therefore, the disc will remain closed preventing air flow into the exhaust system.

Throttle closure during deceleration, momentarilly causes air fuel mixtures to rich to burn. These mixtures, when combined with air will become combustible, causing after fire. A Pulse Air Shut Off Valve, Fig. 11C, is used to momentarily divert air from the exhaust system to prevent after fire. This valve is calibrated to close only at vacuums exceeding idle vacuum which occur only during engine overrun. A spring is used to hold the valve open during idle, part throttle and wide open throttle.

AMERICAN MOTORS "ENGINE MOD" SYSTEM

This system controls exhaust emission levels by using composition cylinder head gaskets instead of steel gaskets and a special carburetor and distributor calibration. The carburetor incorporates idle limiters. The distributor centrifugal advance is calibrated to provide best performance and economy in the driving range and ignition timing is retarded only at idle speed (T.D.C.) to reduce exhaust emission levels at this slow engine speed. These engine modifications will result in a more complete combustion. Thermostatically controlled air cleaners are also used on most units to speed up engine warm-up.

AMERICAN MOTORS EXHAUST GAS RECIRCULATION (EGR)

NOTE: EGR valves used with exhaust back-pressure sensors have shorter and more tapered stems than those used without the sensor. The shorter stems are needed to match the EGR metering to back-pressure sensor operation.

The EGR system consists of a diaphragm actuated flow control valve (EGR valve), coolant temperature override switch, low temperature vacuum signal modulator, high temperature vacuum signal modulator and connecting hoses.

The purpose of the EGR system is to limit the formation of oxides of nitrogen (NOx) by diluting the fresh intake charge with a metered amount of exhaust gas, thereby reducing the peak temperatures of the burning gases in the engine combustion chambers.

EGR Valve

The EGR valve is mounted on a machined surace at the rear of the intake manifold on V-8 engines and on the side of the intake manifold on six cylinder engines. The valve used with an automatic transmission is calibrated differently than the valve used with a manual transmission.

The valve is held in a normally closed position by a coiled spring located above the diaphragm. A special fitting is provided at the carburetor to route ported (above throttle) vacuum through hose connections to a fitting on the valve which is located above the diaphragm. A passage in the intake manifold directs exhaust gas from the exhaust crossover passage (V8 engine) or from below the riser area (six cylinder engine) to the EGR valve. When the diaphragm is actuated by vacuum, the valve opens and meters exhaust gas through another passage in the intake manifold to the floor of the manifold below the carburetor.

Coolant Temperature Override Switch

This switch is located at the coolant passage of the intake manifold (adjacent to oil filler tube) on a V8 engine or at the left side of the engine block (formerly the drain plug location) on a six cylinder engine. The outer part of the switch is open and not used. The inner port is connected by a hose to the EGR fitting at the carburetor. The center port is connected to the EGR valve.

When the coolant temperature is below 115° F (160° F on some models), the center port of the switch is closed and no vacuum signal is applied to the EGR valve, therefore, no exhaust gas will flow through the valve. When the coolant temperature reaches 115° F (160° F on some models), both the center and inner port of the switch are open and a vacuum signal is applied to the EGR valve. However, the vacuum signal to the EGR valve is subject to regulation by low and high temperature signal modulators.

Low Temperature Vacuum Signal Modulator

This unit is located at the left side of the front upper crossmember, just ahead of the radiator, on the same mounting bracket as the TCS ambient temperature override switch and is connected to the EGR vacuum signal hose. The modulator is open when ambient temperatures are below 60° F. This causes a weakened vacuum signal to the EGR valve and a resultant decrease in the amount of exhaust gas being recirculated.

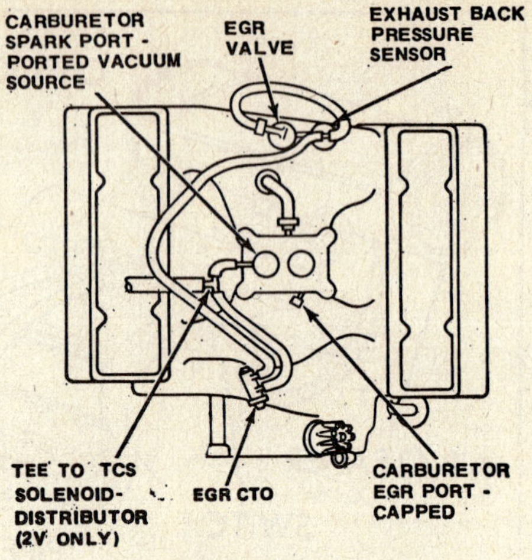

Fig. 11I EGR vacuum line routing. 1975 V8 engines

High Temperature Vacuum Signal Modulator

This unit is located at the rear of the engine compartment and is connected to the EGR vacuum signal hose. The modulator opens when the underhood air temperature reaches 115° F and causes a weakened vacuum signal to the EGR valve. As a result, the amount of exhaust gas being recirculated is decreased.

Exhaust Back-Pressure Sensor

Some 1974 and all 1975 California vehicles are required to have an exhaust back pressure transducer sensor, Fig. 11D. This device consists of a diaphragm valve, a spacer and a metal tube. The EGR valve is mounted to the sensor spacer and is modulated by the sensor.

The EGR system, when equipped with a back-pressure sensor, obtains a vacuum signal at the carburetor spark port and not the EGR port. The vacuum signal passes through the EGR CTO (Coolant Temperature Override) switch (when coolant temperature exceeds 115° F) to the valve portion of the sensor where it is modulated by exhaust back-pressure.

NOTE: The inlet nipple of the exhaust back-pressure sensor has a .030 inch restriction, Fig. 11D. The vacuum line from the EGR CTO must be connected to this nipple, Figs. 11E thru 11I.

When exhaust back-pressure is relatively high, as during acceleration and some cruising conditions, exhaust back-pressure traveling through the metal tube overcomes spring tension on the diaphragm within the back-pressure sensor valve, and closes the valve atmospheric vent.

With the back-pressure sensor valve no

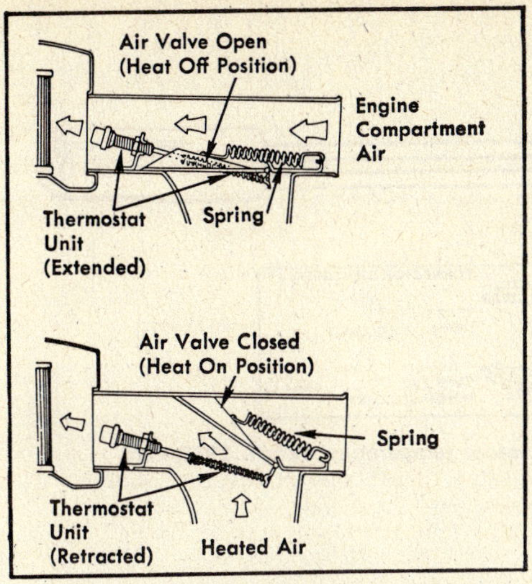

Fig. 12 Temperature operated duct and valve assembly

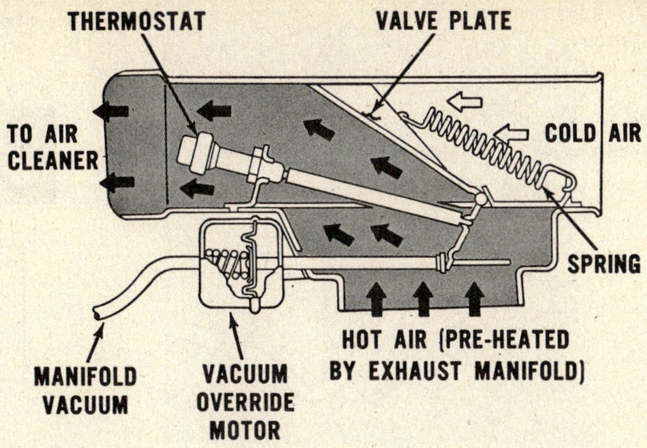

Fig. 13 Duct and valve assembly with vacuum override motor

longer vented to atmosphere, the vacuum signal now passes through the back-pressure sensor valve, and the EGR valve. When vacuum signals the EGR valve, exhaust gas recirculation commences.

When exhaust back-pressure is too low to overcome diaphragm spring tension, the vacuum signal is vented to atmosphere and does not pass through to the EGR valve. With no vacuum signal applied to the EGR valve, exhaust gas does not recirculate.

All six cylinder and some V8-304, 360 engines incorporate a steel restrictor plate under the exhaust back-pressure sensor. The restrictor plate limits the rate of EGR flow, thereby improving driveability. Note that gaskets are used on both sides of the plate.

The back-pressure sensor is not serviceable and must be replaced if defective.

THERMOSTATIC CONTROLLED AIR CLEANER, TAC & AUTO-THERM AIR CLEANER SYSTEMS

American Motors Six & All Ford

Temperature Controlled

Carburetor air temperature is thermostatically controlled by the air duct and valve assembly. Air from the engine compartment, or heated air from the shrouded exhaust manifold is supplied to the engine Fig. 12.

During the engine warm-up period when the air temperature entering the air duct is less than 105° F, the thermostat is in the retracted position and the air valve is held in the closed position by the air valve spring, thus shutting off the air from the engine compartment. Air is then drawn from the shroud at the exhaust manifold.

As the temperature of the air passing the thermostat unit rises, the thermostat starts to open and pulls the air valve down. This allows cooler air from the engine compartment to enter the air cleaner. When the temperature of the air reaches 130° F, the air valve is in the open position so that only engine compartment air is allowed to enter the air cleaner.

Vacuum Controlled

Some Ford systems incorporate a vacuum override motor, Fig. 13. This motor during cold acceleration periods provides additional air to carburetor. The decrease in intake manifold vacuum during acceleration causes the vacuum override motor to override the thermostat control, opening the system to both engine compartment air and heated air from the exhaust manifold shroud.

Yet another Ford application of same system uses a vacuum motor, Fig. 14, installed on the perimeter of the air cleaner to take the place of the vacuum override motor. When the manifold vacuum is low, during heavy engine loading or high speed operation, a spring in the vacuum motor opens the motor valve plate into the air cleaner. This provides the maximum air supply for greater volumetric efficiency.

Thermostatically Controlled Vacuum Operated

A vacuum operated duct valve with a thermostatic bi-metal control, Fig. 15, is used on some installations. The valve in the duct assembly is in an open position when the engine is not operating. When the engine is operating at below normal operating temperature, manifold vacuum is routed through the bi-metal switch to the vacuum motor to close the duct valve allowing only heated air to enter the air cleaner. When the engine reaches normal operating temperature the bi-metal switch opens an air bleed which eliminates the vacuum and the duct valve opens allowing only cold air to enter the air cleaner. During periods of acceleration the duct valve

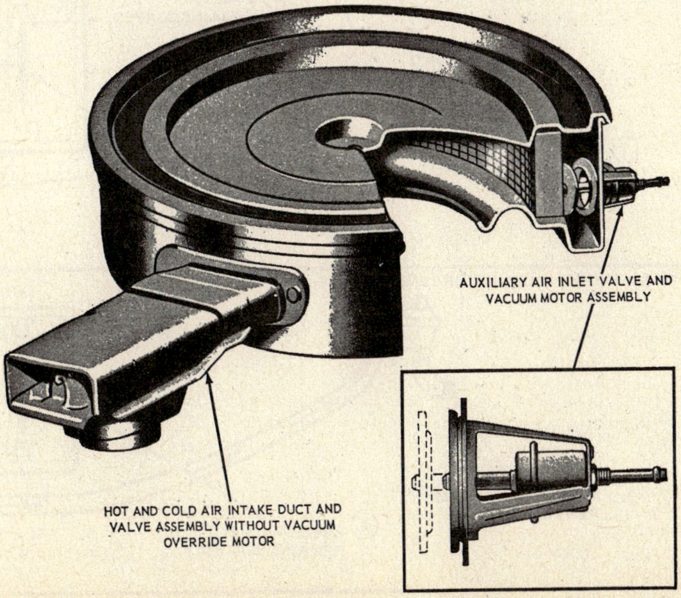

Fig. 14 Air cleaner with auxiliary air inlet valve and vacuum motor

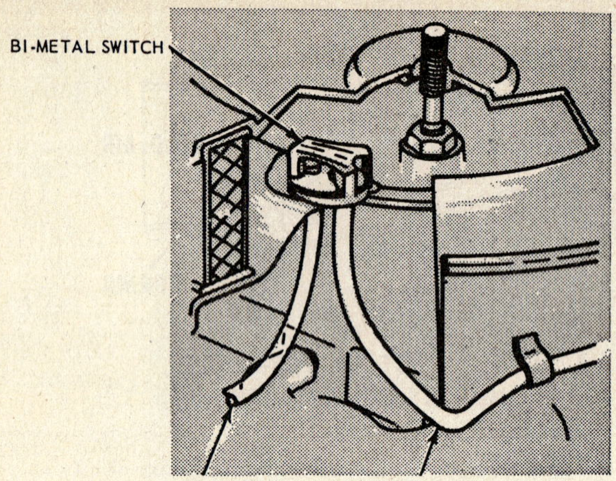

Fig. 15 Vacuum operated duct and valve assembly

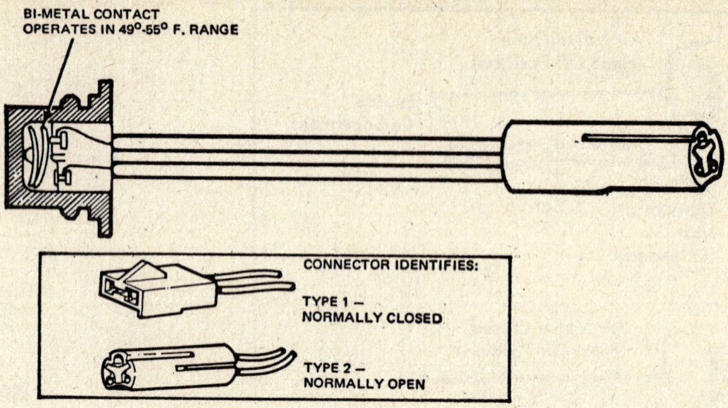

Fig. 15B Air cleaner temperature switch. 1975 Ford

will open regardless of temperature due to the loss of manifold vacuum.

The Ford V8-428 Police engine incorporates a thermostatic choke bleed valve in the air cleaner. The bleed valve allows cold air to be blended with the hot air from the choke stove at underhood temperatures below 50 degrees F.

Cold Weather Modulator

This modulator, Fig. 15A, located in the air cleaner housing of some 1975 Ford vehicles, prevents the air cleaner door from opening to non-heated outside air at

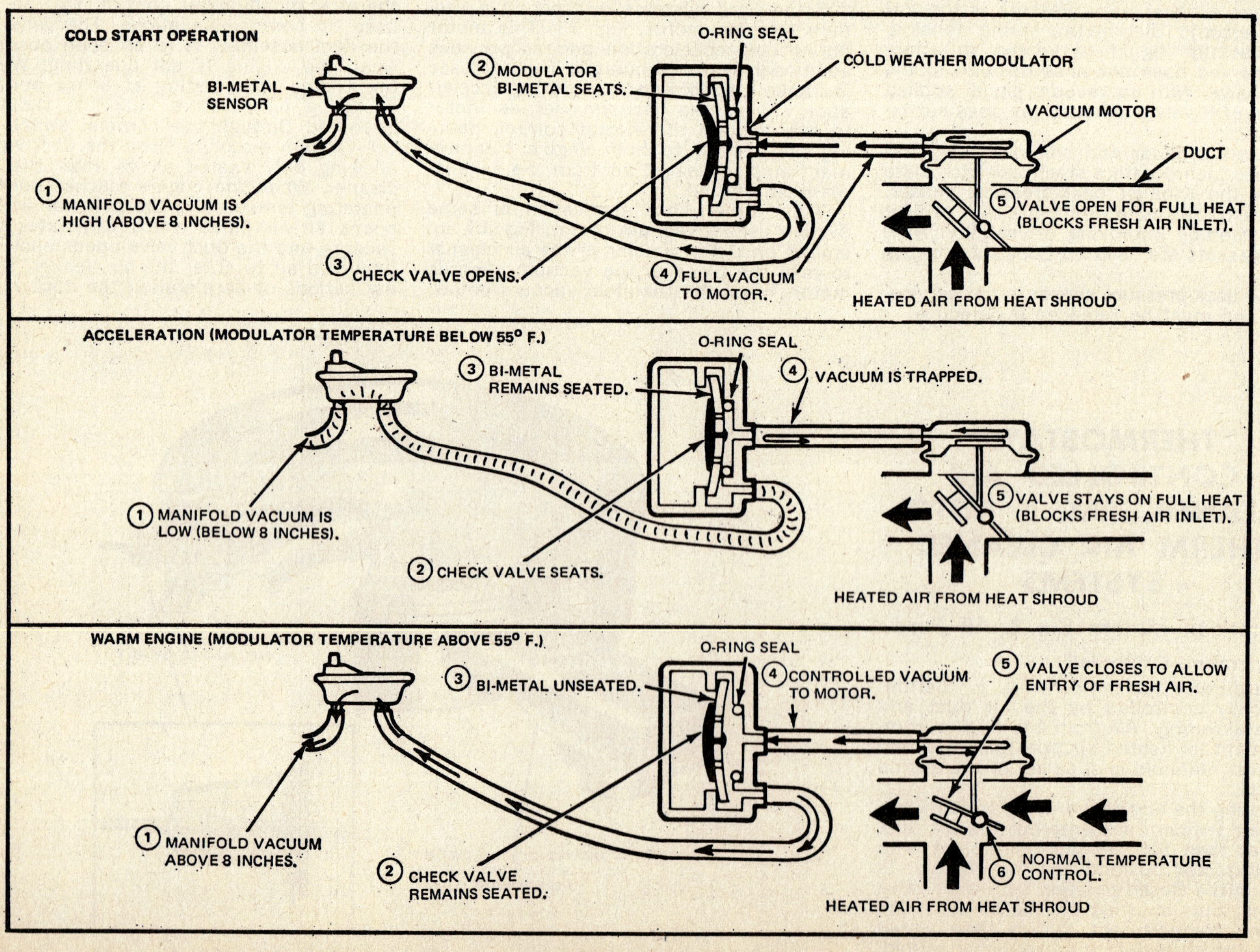

Fig. 15A Cold Weather Modulator. 1975 Ford

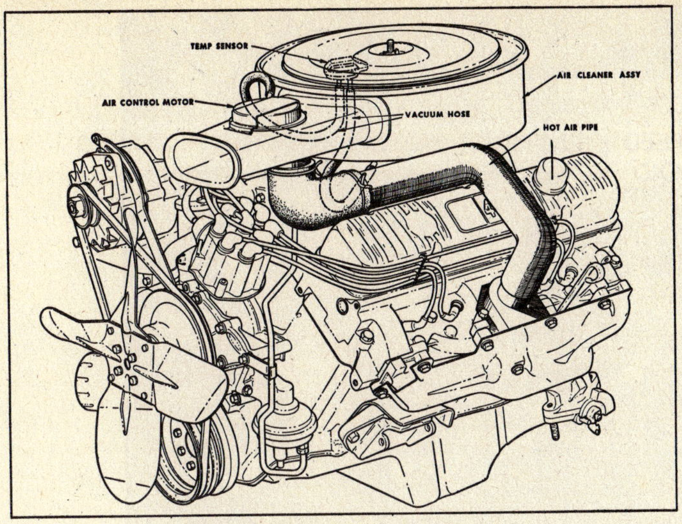

Fig. 16 Controlled Combustion System installed on a V8 engine

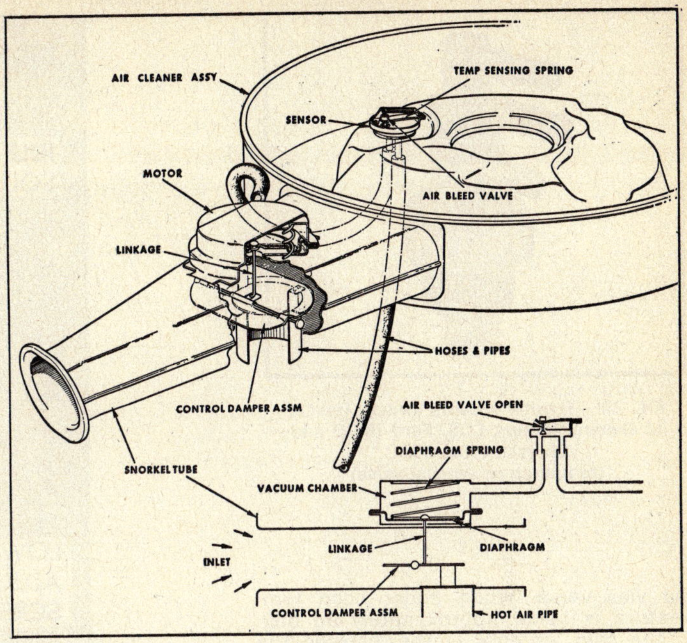

Fig. 17 Cold air door open

ambient temperatures below 55° F. At ambient temperatures above 55° F., the modulator is inoperative. During acceleration at ambient temperatures below 55° F., the modulator located in-line between the bi-metal sensor and vacuum duct motor, will close off vacuum to the motor and hold the duct open.

Air Cleaner Temperature Switch

1975 vehicles with CTAV systems and those with Thermactor systems used with catalytic converters, are equipped with a bi-metallic temperature switch mounted in the air cleaner housing. The two type switches used may be identified by the connector, Fig. 15B. The first type has normally closed contacts, while the second type has normally open contacts. Refer to "Cold Engine Lockout" under "Air Pump Systems", for operation of these switches.

American Motors V8 & All General Motors

Figs. 16 thru 19—Carburetor air temperature is controlled by a pair of doors, located in the air cleaner snorkel, which channel either pre-heated or under hood air to the carburetor.

Pre-heated air is obtained by passing under hood air through ducts surrounding the exhaust manifold, causing it to pick up heat from the manifold surface. The heated air is then drawn up through a pipe to the air cleaner snorkel.

Under hood air is picked up at the air cleaner snorkel in the conventional manner.

The two air mixing doors work together so that as one opens, the other closes

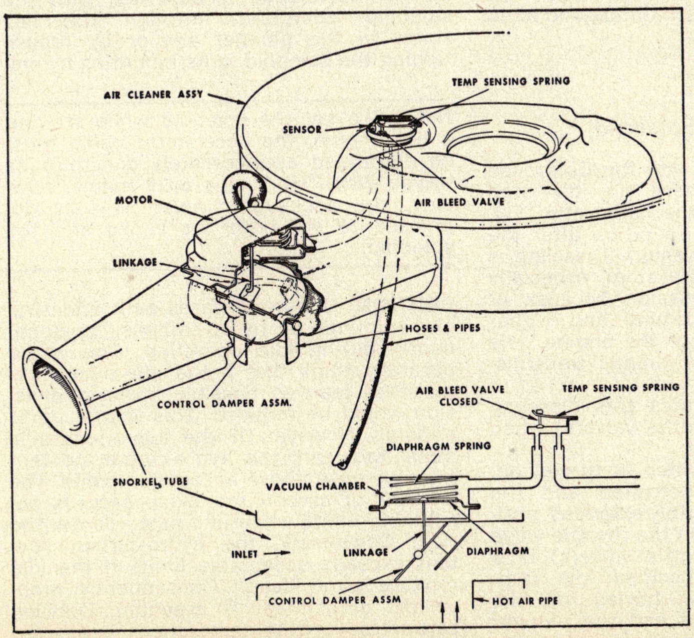

Fig. 18 Hot air door open

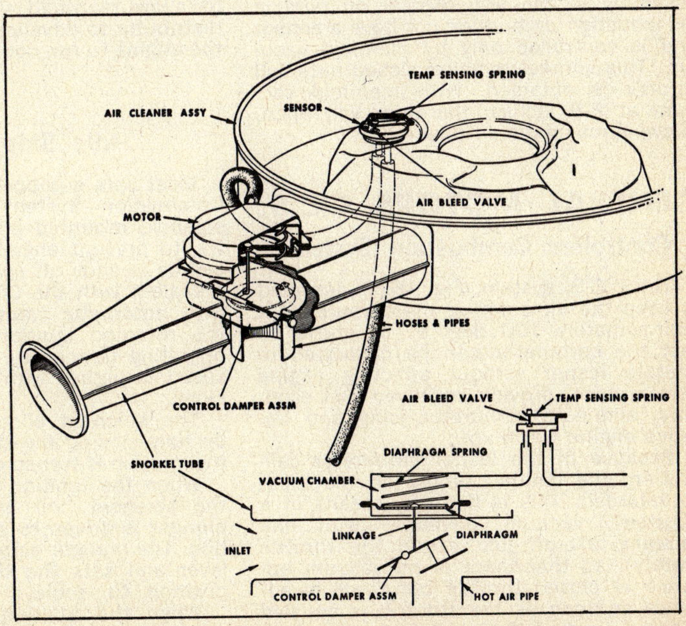

Fig. 19 Cold and hot air doors both partially open

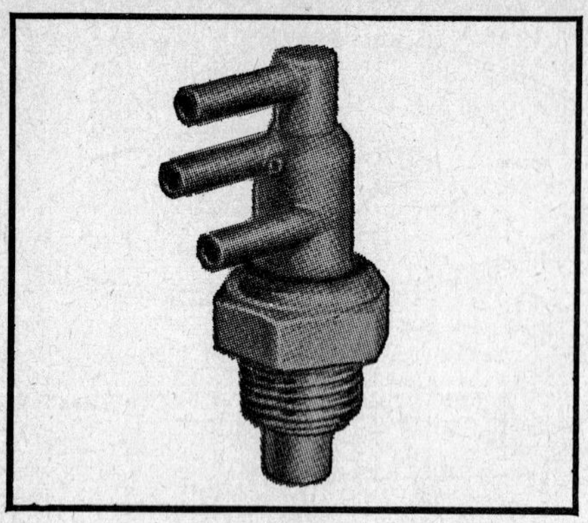

Fig. 20 Thermo vacuum switch. Typical of General Motors CCS, Ford IMCO and Chrysler NOx, OSAC (temperature sensing valve)

Fig. 21 Idle stop solenoid

and vice versa. When under hood temperature is below approximately 90 deg. F., the cold air door closes, causing the hot air door to open. Hot air from the exhaust manifold stove is then drawn into the carburetor. As the under hood temperature increases, the cold air door begins to open until the temperature reaches approximately 115° to 130 deg. F, at which time the cold air door is fully open and the hot air door is fully closed.

The doors are controlled by a vacuum motor mounted on the air cleaner snorkel. This motor, in turn, is controlled by a sensor inside the air cleaner which regulates the amount of vacuum present in the vacuum motor according to air cleaner temperature. Whenever manifold vacuum drops below 5-8 inches, depending on the unit, the diaphragm spring in the motor will open the cold air door wide in order to provide maximum air flow.

The vacuum motor and control door assembly in the left snorkel on outside air induction units does not have a sensor and is controlled only by manifold vacuum. This snorkel remains closed until full throttle is obtained. With manifold vacuum at 6-8 inches, the door will open, allowing maximum air flow.

GENERAL MOTORS C.C.S.

Controlled Combustion System

The C.C.S. system, Fig. 16, is designed to keep the air entering the carburetor at approximately 100 deg. F. or above so that the carburetor can be calibrated to operate leaner without affecting engine performance, provide improved fuel economy, eliminate carburetor icing and improve engine warm-up.

Because of the leaner carburetor calibration, the ignition timing at idle must be retarded. This is done by means of a "ported" vacuum advance, with the vacuum take-off just above the throttle plate(s), so that there is no vacuum advance at closed throttle but there is advance as soon as the throttle is cracked slightly. Ignition timing is set at or near TDC and centrifugal advance does not start until approximately 1000 rpm.

Thermo Vacuum Switch

Because of the increased possibility of engine overheating at idle with the C.C.S. calibration, the thermo vacuum switch Fig. 20 is added to the system on some engines. This switch senses engine coolant temperature and, if temperature reaches 220 deg. F., the switch valve moves to allow manifold vacuum to reach the distributor to advance the timing and allow the engine to run cooler.

Idle Stop Solenoid

Most cars equipped with the Controlled Combustion System have an idle stop solenoid mounted on the carburetor, Fig. 21, to prevent engine operation after the engine is shut off (dieseling). Dieseling is prevalent with the CCS system, especially with automatic transmission because of the retarded ignition timing and higher operating temperature of the engine. Idle speed is higher with automatic transmissions.

The solenoid eliminates this dieseling by fully closing the throttle valve(s) when the ignition is turned off.

When the ignition switch is turned on, the solenoid coil is activated and the plunger is driven to its full extended position. The plunger acts on the throttle valve lever and sets the throttle valve(s) in a position to achieve specified idle rpm.

When the ignition is turned off, the solenoid is de-energized and the plunger retracts into the solenoid, causing the throttle valve(s) to close to a position controlled by the low idle adjusting screw. At this point the throttle valve(s) is open only enough to allow the engine to run at a much lower rpm. This lower setting keeps the throttle valve(s) from completely closing and scuffing the throttle bore(s).

The lower idle speed setting is made with a set screw when the solenoid is electrically disconnected. The setting for normal idle speed is adjusted, with the solenoid energized, through the hex screw in the plunger and/or by repositioning the solenoid in its mounting clamp.

NOTE: To set the solenoid while starting a hot engine, the accelerator pedal must be depressed approximately one third its travel. When starting a cold engine, however, the accelerator must be fully depressed to set both the choke and the solenoid.

CAUTION: Unburned gases emit undesirable amounts of hydro-carbons, predominantly during idle operation. One would naturally think that if the idle mixture is made as lean as possible, exhaust emission would be reduced. This is true up to a point. However, if the idle mixture is made too lean, the hydro-carbon content may increase above acceptable limits. The amount of carbon monoxide exceeds acceptable limits when idle mixtures are too rich. Conversely, the hydro-carbon content exceeds acceptable limits if the idle mixture is too "lean". Consequently, proper idle adjustment is essential. Detailed carburetor setting instructions are given in the carburetor chapter.

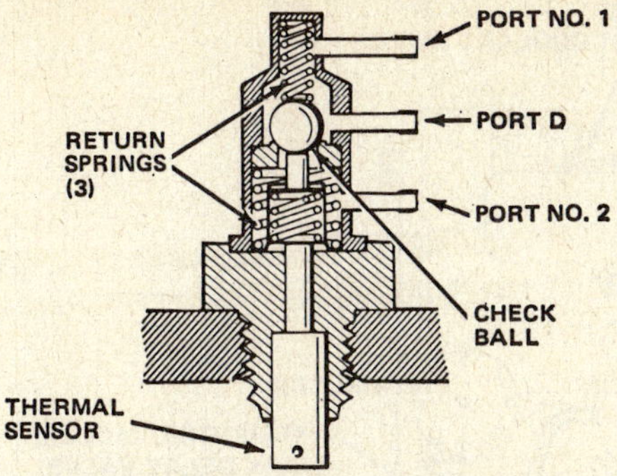

Fig. 21A Coolant temperature override switch. American Motors

TRANSMISSION CONTROLLED SPARK (TCS)

American Motors

This system is designed to provide vacuum spark advance during high gear operation and certain engine conditions. The resultant is a lower peak combustion pressure and temperature during the power stroke, significantly reducing exhaust emissions.

Basically the system incorporates an ambient temperature override switch, solenoid vacuum valve, solenoid control switch and in some systems a coolant temperature override switch.

The ambient temperature override switch, used on 1971-73 vehicles, senses ambient temperatures and completes the electrical circuit from the battery to the solenoid vacuum valve when ambient temperatures are above 63° F.

The solenoid vacuum valve is attached to the intake manifold. When the valve is energized, carburetor ported vacuum is blocked and the distributor vacuum line is vented to atmosphere through a port in the valve, resulting in no vacuum advance. When the valve is de-energized, ported vacuum is applied to the distributor resulting in normal vacuum advance.

The solenoid control switch, located at the transmission, opens or closes in relation to car speed automatic transmission equipped cars or gear range manual transmission equipped. At speeds above 34 MPH on automatic transmission equipped cars, high gear on manual transmission equipped cars, the switch opens and breaks the ground circuit to the solenoid vacuum valve. The control switch closes and completes the ground circuit to the solenoid vacuum valve on automatic transmission equipped vehicles at speeds under 25 MPH on 1971-73 models or 35 MPH on 1974-75 models. On manual transmission equipped vehicles, the switch closes when the transmission is in the lower gear ranges.

On automatic transmission equipped vehicles, the switch is automatically operated by the speedometer gear speed on 1971-73 models or transmission governor

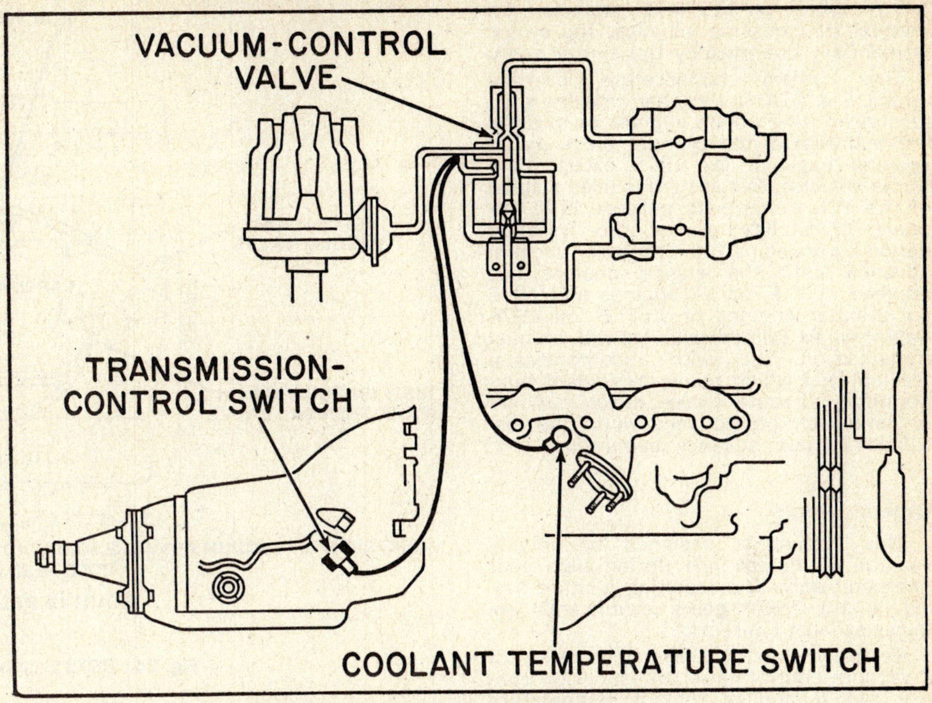

Fig. 22 Typical T.C.S. emission control

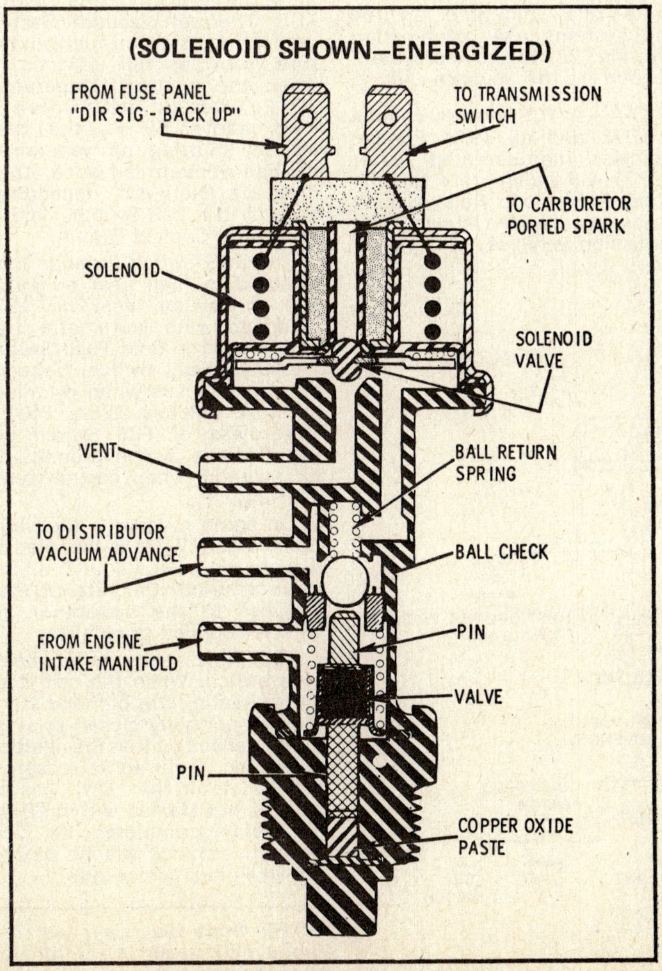

Fig. 23 TVS-TCS combination valve

oil pressure on 1974-75 models. On manual transmission vehicles, the switch is manually operated by the shifter shaft.

The coolant temperature override switch, Fig. 21A, is threaded into the thermostat housing on V8 engines or into the left rear side of the cylinder block on six cylinder engines. On 1975 except California vehicles, this system is used instead of the TCS system. Its purpose is to improve driveability during the warm-up period by providing full distributor vacuum advance until the engine coolant has reached 160° F, on V8 engines and 1975 six cylinder engines or 115° F. on 1974 California six cylinder engines with manual transmission. The switch incorporates a thermal unit which reacts to coolant temperatures to route either intake manifold or carburetor ported vacuum to the distributor vacuum advance diaphragm.

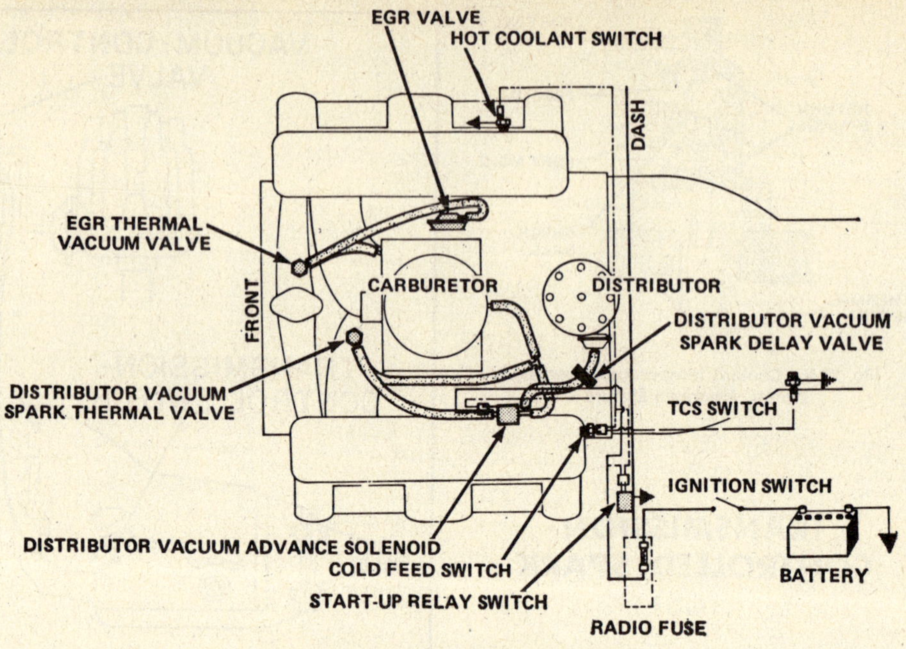

Fig. 24 1973 Pontiac mid-year TCS system

General Motors

This system is designed to provide vacuum spark advance during high gear operation only. The resulting ignition timing in the lower gears significantly reduces exhaust emissions.

Basically, the T.C.S. system consists of a vacuum control valve, transmission control switch and a coolant temperature switch. The valve controls the vacuum signal to the distributor vacuum advance unit in response to a signal from either switch, Fig. 22.

The vacuum control valve is of solenoid design and is installed in the vacuum line between the carburetor and the distributor. When energized by the transmission control switch, the valve blocks the vacuum source and vents the advance unit through the carburetor air horn.

When the engine is cold, and in some units when the engine temperature exceeds 210°F, the coolant temperature switch overrides the system and provides full manifold vacuum to the advance unit, advancing timing in turn lowering engine operating temperature. This switch operates through a relay mounted on the firewall. Some systems use combination TVS-TCS valve, Fig. 23, to provide system over-riding vacuum to the advance unit.

Figs. 24 and 24A—A TCS system change on mid-year 1973 and all 1974 Pontiac models (vehicles manufactured after March 15, 1973) will incorporate the following revisions: A Start-Up Relay Switch supplies full advance (ported advance on manual transmission vehicles) in any gear for 20 seconds after engine start.

A Distributor Vacuum Spark Thermal Valve on 1973 models or Distributor Spark-EGR Thermal Vacuum Switch on 1974 models, sensing air-fuel mixture temperature, provides full advance in any gear when the mixture temperature is below 62° F. When air-fuel mixture temperature rises above 62° F., the thermal valve closes, shutting off vacuum, in turn to provide vacuum advance, the distributor must be energized, depending on condition, by the TCS Switch, Cold Feed Switch or the Hot Coolant Switch.

The TCS Switch grounds the distributor solenoid in high gear on all applications and in reverse gear on 1974 vehicles equipped with automatic transmissions only when the Cold Feed Switch is closed. The Cold Feed Switch, depending on application, closes when cylinder head temperature reaches 125°, 140° or 155° F. Regardless of TCS Switch position, the Hot Coolant Switch grounds the distributor solenoid when coolant temperature is over 240° F.

On some models, a Distributor Vacuum Spark Delay Valve is installed between the distributor solenoid and distributor. This valve restricts the rate of initial vacuum supplied to the distributor, full vacuum will be supplied gradually.

The Time Relay is an electrical on-off type switch. When the coil is energized, it starts heating the bi-metal strip and opening the normally closed relay points within 20 seconds after the ignition switch is turned on. If for some reason the ignition key is left in the "ON" position and vehicle is not started within 20 seconds and the relay completes its "countdown", vacuum advance will be blocked off until the relay has cooled off.

NOTE: Once the relay has run one cycle (about 20 seconds), after the key has been turned on, the relay must cool off before it will reactivate, even if the key is

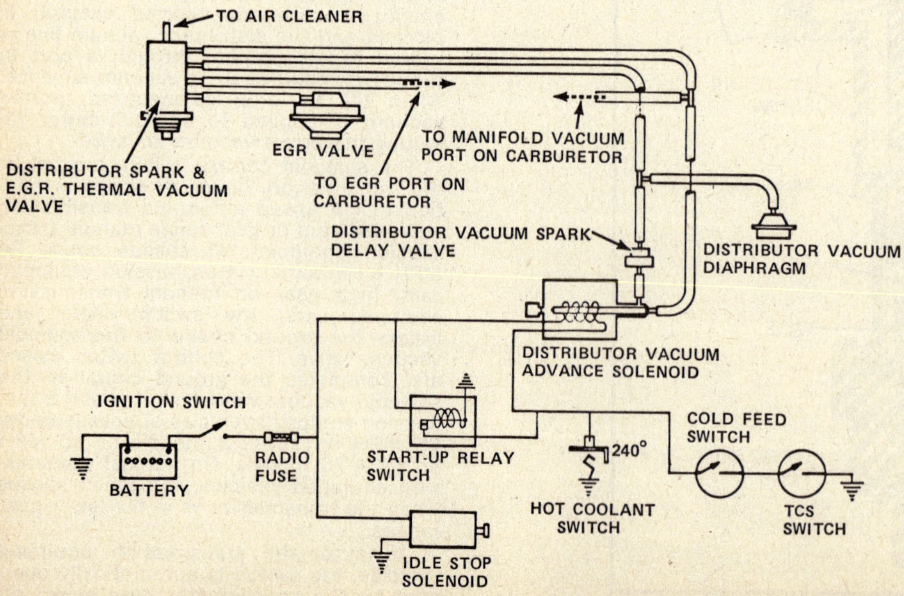

Fig. 24A 1974 Pontiac TCS-EGR system. V8 (Typical)

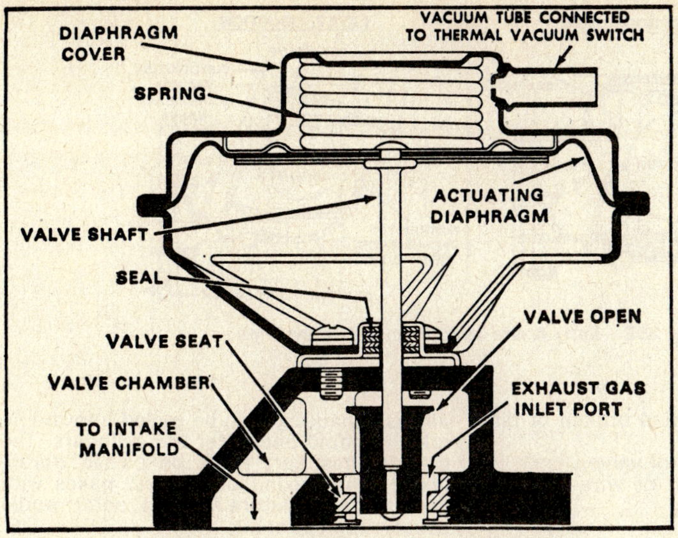

Fig. 25 Single diaphragm EGR valve cross section

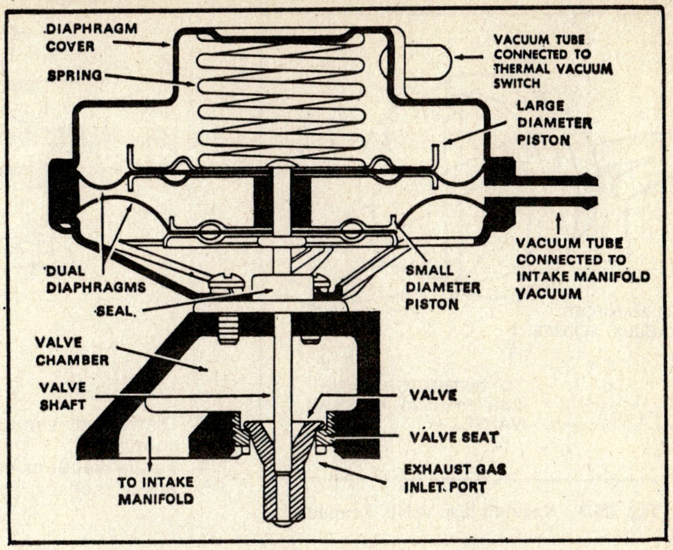

Fig. 25A Dual diaphragm EGR valve cross section

switched "OFF" and then turned "ON" again.

GENERAL MOTORS C.E.C.

Combined Emission Control System

This system is designed to provide vacuum spark advance during high gear operation only, as does the T.C.S. system, but the C.E.C. solenoid vacuum switch also regulates curb idle and high gear deceleration throttle positions, further reducing emissions.

The C.E.C. system consists of a vacuum control solenoid valve, transmission control switch, coolant temperature switch and a time-delay relay.

When the solenoid is in the non-energized position, vacuum to the distributor vacuum advance unit is shut off and the distributor is vented to the atmosphere through a filter at the opposite end of the solenoid. When the solenoid is energized by one of the switches or the relay, the vacuum port is uncovered and the plunger is seated at the opposite end, shutting off the clean air vent. This routes vacuum to the distributor. The solenoid is energized in high gear by the transmission switch. The coolant temperature switch overrides the transmission switch to energize the solenoid and provide vacuum advance below 82° F., for 1969-72 vehicles and 93° for 1973-74 vehicles. The time-delay is 93° is incorporated into the circuit to energize the solenoid for approximately 15 seconds on 1969-72 vehicles and 20 seconds on 1973-74 vehicles, after the ignition key is turned on.

When the solenoid plunger is in the non-energized position, it allows the throttle to close to the curb idle setting. When the solenoid plunger is in the energized position, it keeps the throttle open to the high gear deceleration position which controls hydrocarbon emission.

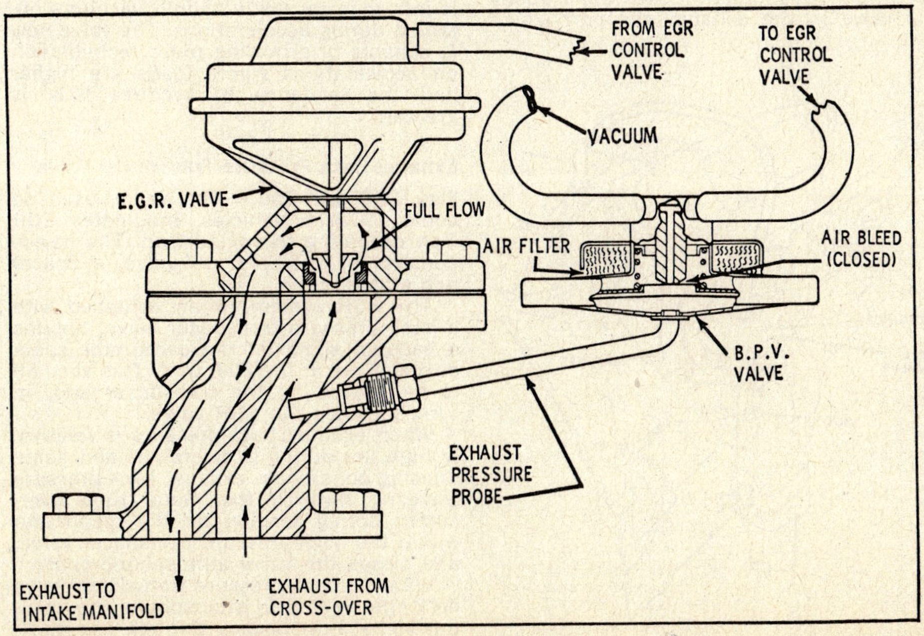

Fig. 25B Exhaust back pressure transducer. Oldsmobile

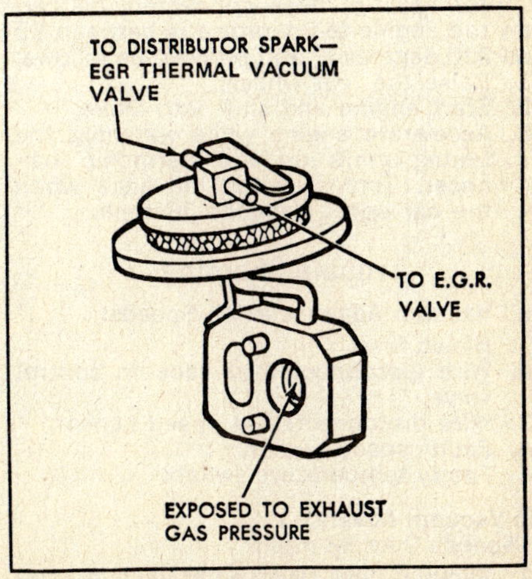

Fig. 25C Exhaust back pressure transducer. Pontiac

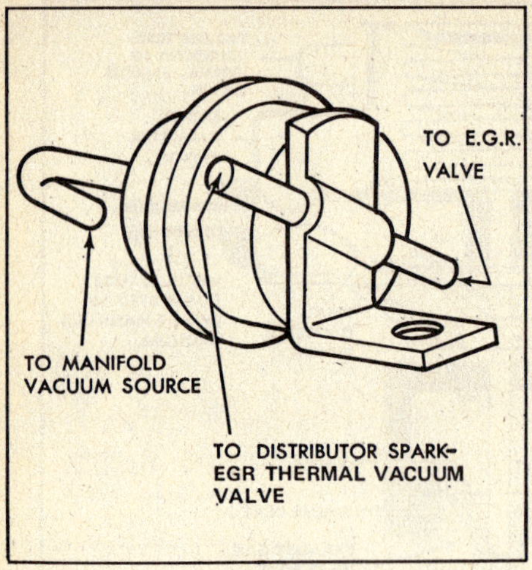

Fig. 25D Vacuum bias valve. Pontiac

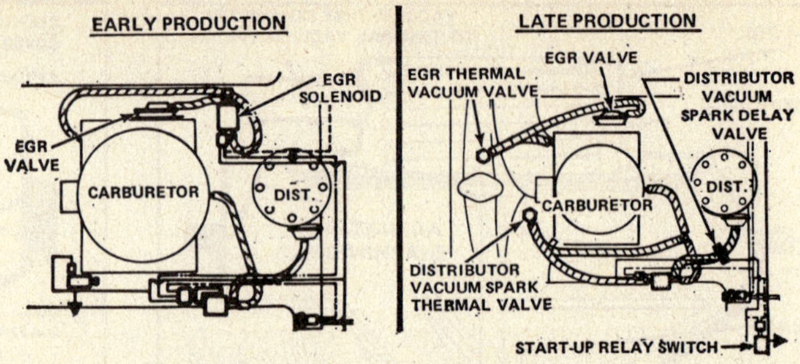

Fig. 25E Early & late 1973 Pontiac EGR systems

GENERAL MOTORS S.C.S.

Speed Controlled Spark

This system is essentially the same as the T.C.S. system except that here the vacuum spark advance is controlled by the vehicles speed rather than the gear position of the transmission.

The S.C.S. system consists of a vacuum advance solenoid valve, a speed sensing switch and a temperature switch. The vacuum control valve is installed in the vacuum line between the carburetor and the distributor. When energized, this valve cuts off vacuum to the distributor. The speed sensing switch de-energizes the vacuum control valve at speeds above 38 mph allowing normal vacuum spark advance. Abnormal operating temperatures (below 85 degrees F and above 220 degrees F) will cause the temperature switch to de-energize the vacuum control valve, over-riding the speed sensing switch.

Testing

To test the S.C.S. system, leave the distributor vacuum hose connected, be certain the engine temperature is between 95 and 230 degrees F, and proceed as follows:

1. Raise the rear wheels.
2. Start engine and shift into Drive.
3. Accelerate engine while watching the timing mark on the harmonic balancer. Timing should advance when the car speed exceeds 38 mph.

Trouble Shooting

Full Vacuum Advance at all Speeds:

1. Blown fuse.
2. Wire disconnected at vacuum control valve.
3. Wire disconnected at speed sensor.
4. Faulty speed sensor.
5. Faulty temperature switch.

No Vacuum Advance at Speeds Over 38 mph:

1. Vacuum lines reversed at vacuum control valve.
2. Foreign matter in vacuum control valve.
3. Distributor vacuum line broken or disconnected.
4. Faulty vacuum control valve.
5. Speed sensor switch or wire shorted.

GENERAL MOTORS E.G.R.

Exhaust Gas Recirculation

Figs. 25 and 25A—This system is used to reduce oxides of nitrogen emissions at the engine's exhaust. This is accomplished by introducing exhaust gases into the intake manifold at throttle positions other than idle. It consists of an E.G.R. valve mounted on a special intake manifold. The exhaust gas intake port of the E.G.R. valve is connected to the intake manifold exhaust crossover channels where it can pick up exhaust gases.

As the throttle valves are opened and the engine speeds up, vacuum is applied to a vacuum diaphragm in the E.G.R. valve through a connecting tube. When the vacuum reaches approximately 3″ hg., the diaphragm moves upward against spring tension and is in the full-up position at approximately 7″ to 8″ hg. of vacuum. This diaphragm is connected by a shaft to a valve which closes off the exhaust gas port. As the diaphragm moves up, it opens the valve in the exhaust gas port which allows exhaust gas to be pulled into the intake manifold and enter the cylinders. The exhaust gas port must be closed during idle as the mixing of exhaust gases with the fuel air mixture at this point would cause rough running.

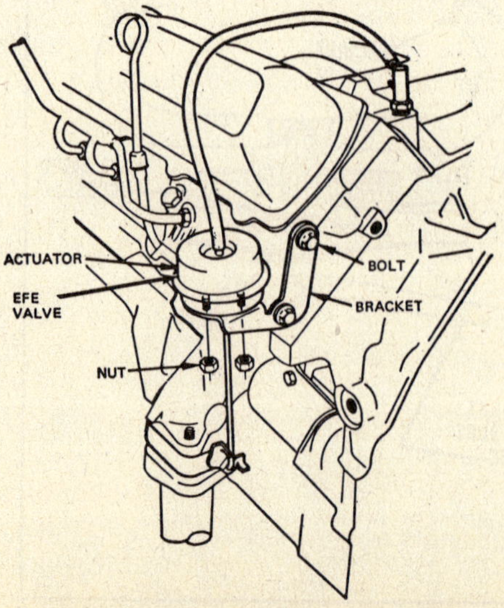

Fig. 25F Typical EFE valve installation. General Motors

The dual diaphragm EGR valve, Fig. 25A, is designed to provide increased exhaust gas recirculation rates when engine loads increase.

NOTE: Manifold vacuum is used as the signal to indicate the engine load.

The valve is similar to the single diaphragm valve except that a second diaphragm has been added to the valve and is connected to the upper diaphragm, with a spacer thus both diaphragms move together. A manifold vacuum signal is applied to the volume between the two diaphragms. The upper diaphragm has a larger diameter piston than the lower diaphragm, therefore the load caused by the manifold vacuum between the two diaphragms aids the spring load. Thus as the engine load increases and manifold vacuum decreases the combined load of the spring and the vacuum chamber are reduced allowing the valve to open further for a given EGR vacuum signal.

Therefore, for high intake manifold vacuums (such as cruising), the opening is less than for low manifold vacuums obtained during accelerations. The valve now is capable of providing more recirculation on accelerations where loads are higher and the tendency to produce NOx is greater.

Exhaust Back Pressure Transducer Valve

Figs. 25B and 25C.—This valve, used on some 1974-75 vehicles, modulates EGR flow according to engine load. The device consists of a diaphragm valve, a spacer and a metal tube.

The EGR system, when equipped with a back pressure transducer valve, obtains a vacuum signal at the carburetor spark port and not at the EGR port. This vacuum is modulated by the transducer and, in turn activates the EGR valve.

When exhaust back-pressure is relatively high, as during acceleration and some cruising conditions, exhaust back-pressure traveling through the metal tube overcomes spring tension on the diaphragm within the back-pressure transducer valve, and closes the valve atmospheric vent.

With the back-pressure transducer valve no longer vented to atmosphere, the vacuum signal now passes through the back-pressure transducer valve, and the EGR valve. When vacuum signals the EGR valve, exhaust gas recirculation commences.

When exhaust back-pressure is too low to overcome diaphragm spring tension, the vacuum signal is vented to atmosphere and does not pass through to the EGR valve. With no vacuum signal applied to the EGR valve, exhaust gas does not recirculate.

Vacuum Bias Valve

Fig. 25D—This valve, used on some 1974-75 Pontiac vehicles, California models, is located between the EGR valve and the distributor spark-EGR thermal vacuum valve. At high manifold vacuum conditions (such as cruising), the VBV decreases EGR flow and in turn, acts to reduce surge. When NOx formation is high and manifold vacuum is low as during acceleration, the VBV does not reduce EGR flow.

Fig. 25E—An EGR system change on all mid-year 1973 V8 and 1974-75 six cylinder Pontiac models (vehicles manufactured after March 15, 1973) will incorporate a new EGR Thermal Vacuum Valve which senses engine coolant temperature. This thermal valve, located between the carburetor and EGR valve, controls vacuum to the EGR valve. When coolant temperature is below 95° F. on 1973 models or 100° F. on 1974-75 models, the thermal valve isolates the EGR valve from the vacuum source, therefore eliminating exhaust gas recirculation. When coolant temperature is above 95° F. on 1973 models or 100° F. on 1974-75 models, the thermal valve opens and allows vacuum to actuate the EGR valve, allowing ported recirculation.

Fig. 24A—On 1974-75 Pontiac vehicles equipped with V8 engines, the EGR Thermal Vacuum Valve and the Distributor Spark Thermal vacuum Valve is incorporated into one assembly. This Distributor Spark-EGR Thermal Vacuum Valve senses air-fuel mixture temperature. This thermal valve is located between the EGR Valve and its vacuum source. The valve prevents actuation of the EGR valve when air-fuel mixture temperature is below 62° F. When air-fuel mixture temperature exceeds 62° F., the valve opens and allows vacuum to activate the EGR valve.

The E.G.R. valve cannot be disassembled and no actual service is required on it. However, it can be checked for proper operation as follows:

All Except 1973 Mid-Year & 1974-75 Pontiac EGR System

1. Check the exhaust gas valve shaft for movement by opening the throttle to 1200-1500 rpm. The shaft should move upward and return to its original position when the engine speed is allowed to drop to idle.
2. An outside vacuum source can be connected to the vacuum supply port in the top of the E.G.R. valve. The valve shaft should reach the top of its travel at 7″ to 10″ hg. of vacuum and the vacuum should not leak down.
3. If the E.G.R. valve does not operate correctly, it must be replaced.

1973 Mid-Year & 1974 Pontiac EGR System

1. With engine at operating temperature, remove air cleaner. Open throttle part way, release throttle and observe EGR valve diaphragm movement through cut away portion on inboard side. The EGR valve should open with throttle opening.
2. If no diaphragm movement is observed, insure that vacuum hoses are properly connected. If so, using a vacuum gauge, check for vacuum at hose connected to the EGR valve. If vacuum is present, the EGR valve is faulty and must be replaced.
3. If no vacuum was present at EGR hose, check for vacuum at thermal valve side of hose running from carburetor. If vacuum is present, the interconnecting hose or the EGR valve may be plugged, if there is no obstruction.
4. If no vacuum is present at thermal valve side of hose running from carburetor, the hose between the carburetor and thermal valve may be obstructed or the carburetor is faulty.

1975 Pontiac Exc. Ventura EGR System

NOTE: It is recommended that choke secondary vacuum break TVV be disconnected from air cleaner body instead of disconnecting hoses. Also, if system has a BPT, a separate check must be made as described further on.

1. Remove air cleaner and plug vacuum hose, then warm engine to operating temperature and connect a tachometer.
2. While observing EGR valve movement, open throttle and release. The EGR valve should open when the throttle is opened and should close when the throttle is released.
3. Disconnect hose from EGR valve and plug hose, then place cam follower on second highest step of cam.
4. Connect air cleaner vacuum hose to EGR valve. Engine speed should drop at least 200 rpm with automatic transmission and at least 150 rpm with manual transmission.

Pontiac Ventura V8-260 EGR System

1. Remove air cleaner and plug vacuum fitting, then disconnect and plug vacuum hose from distributor and EGR valve.
2. With A/C off and engine idling at operating temperature, place cam follower on second step of fast idle cam and note engine speed using a tachometer.
3. Connect air cleaner vacuum hose to EGR valve. Engine speed should drop at least 250 rpm with automatic transmission and at least 100 rpm with manual transmission.
4. If engine speed does not drop as specified above, remove EGR valve and clean EGR ports in intake manifold and clean EGR valve assembly.
5. Install EGR valve and recheck as outlined previously. If engine speed does not drop as specified, replace EGR valve.

Pontiac Ventura V8-350 EGR System

1. With engine temperature above 120° F., place cam follower on fast idle step of cam to hold engine speed at 1400-1500 rpm.
2. Place fingers beneath EGR valve to feel movement of diaphragm, then disconnect EGR vacuum hose. Diaphragm should move down and engine speed should increase.
3. Reconnect EGR vacuum hose. Diaphragm should move upward and engine speed should decrease.
4. If diaphragm does not move, check for vacuum at EGR hose. If no vacuum is present, check for plugged or leaking hoses or fittings or defective EFE/EGR switch.
5. If EGR diaphragm moves with no change in engine speed, check for blockage in EGR manifold passages.

G. M. CURB IDLE SETTING PROCEDURES

C.C.S., T.C.S., T.V.S., & C.E.C. Systems

NOTE: This adjustment must be made with the engine at normal operating temperature. This is particularly important on engines equipped with a thermo vacuum switch.

1. Disconnect fuel tank line from vapor canister.
2. Disconnect distributor vacuum hose at distributor and plug hose.
3. Set dwell and timing (in that order) at specified RPM.
4. Adjust carburetor speed screw to specified RPM in tune-up charts in car chapters. If equipped with C.E.C. *DO NOT ADJUST SOLENOID SCREW*, see Note. C.E.C. Valve adjustment is to be made only after replacement of the solenoid, major carburetor overhaul, or removal of the throttle body. In any of these cases, adjustment should be made as follows:
   a. With slow idle speed set to specified RPM, manually extend C.E.C. Valve plunger to contact throttle lever.
   b. Adjust plunger length to obtain the proper engine RPM as listed in *Tune-Up Specifications Chart.*
5. With auto. trans. in Park and manual trans. in Neutral, adjust fast idle speed. *Refer to Carburetor Specifications Charts and procedures.*

GENERAL MOTORS EFE

Early Fuel Evaporation

This system is used on 1975 vehicles to provide a source of rapid heat for quick induction system warm-up during cold engine operation. Rapid heat is more desireable because it provides for better fuel evaporation and a more uniform mixture.

The vacuum operated EFE heat control valve mounted between the exhaust manifold and pipe, Fig. 25F, directs a portion of the exhaust gasses through the intake manifold and to the base of the carburetor during engine warm-up. A thermal vacuum switch (TVS) is used to regulate the EFE valve at or below a predetermined temperature. Refer to chart.

Buick	120° F.
Cadillac	150° F.
Chevrolet	
Six Cylinder	150° F.
4-140, V8-350, 400	180° F.
V8-454	100° F.
Oldsmobile	120° F.
Pontiac	150° F.

When engine coolant is at or below the specified temperature, the TVS will cause vacuum to flow to the EFE control valve, diverting the exhaust gasses through the intake manifold to the base of the carburetor. At temperatures above those specified, the TVS closes off vacuum to the EFE control valve, stopping the flow of exhaust gasses through the intake manifold.

CHRYSLER LEAN BURN ENGINE ELECTRONIC SPARK ADVANCE

This system, Fig. 26, used on some 1976 V8-400 and 440 engines, allows the engine to run at an air/fuel ratio of 18 to 1. Since the engine runs lean, NOx, HC and CO emissions are reduced without the use of EGR, OSAC, air pumps, catalytic converters, vacuum advance units, some special heat shields and all interconnecting hoses. However, for California applications, the use of catalytic converters and heat shields is retained.

Six sensors are mounted on the engine which monitor the following operating conditions:

1. Engine condition when started (Hot or cold start).
2. Engine RPM and piston position in relation to TDC.
3. Intake manifold vacuum.
4. Throttle position, Fig. 26A.
5. Rate of change in throttle position.
6. Carburetor air temperature.
7. Coolant temperature.

The information from the sensors is sent to the spark control computer which calculates the precise instant the fuel charge should be fired in each cylinder to provide the best combination of driveability, fuel economy and emission control.

The computer, Fig. 26B, which is the brain of the electronic spark advance system, sends signals for each cylinder to the electronic distributor. The distributor, in turn, controls the firing of the spark plugs. The distributor incorporates two magnetic pick-ups, Fig. 26C. One pick-up is used during engine start and operation for sixty seconds after which the second pick-up is activated and the "Start" pick-up is deactivated, regardless of operating conditions. Also, the system incorporates a "Limp-In" mode whereby, if ignition failure occurs vehicle can still be driven into a service station.

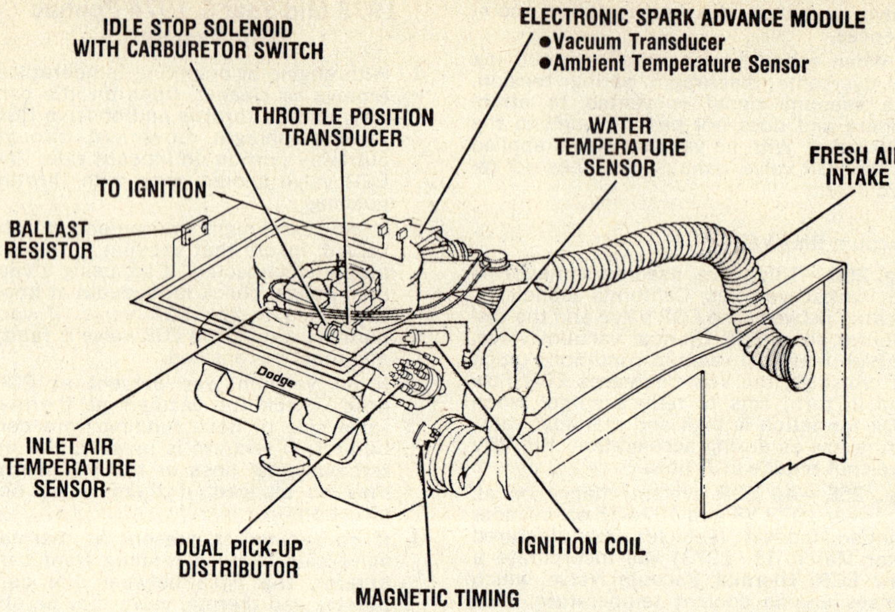

Fig. 26 Chrysler Electronic Spark Advance System

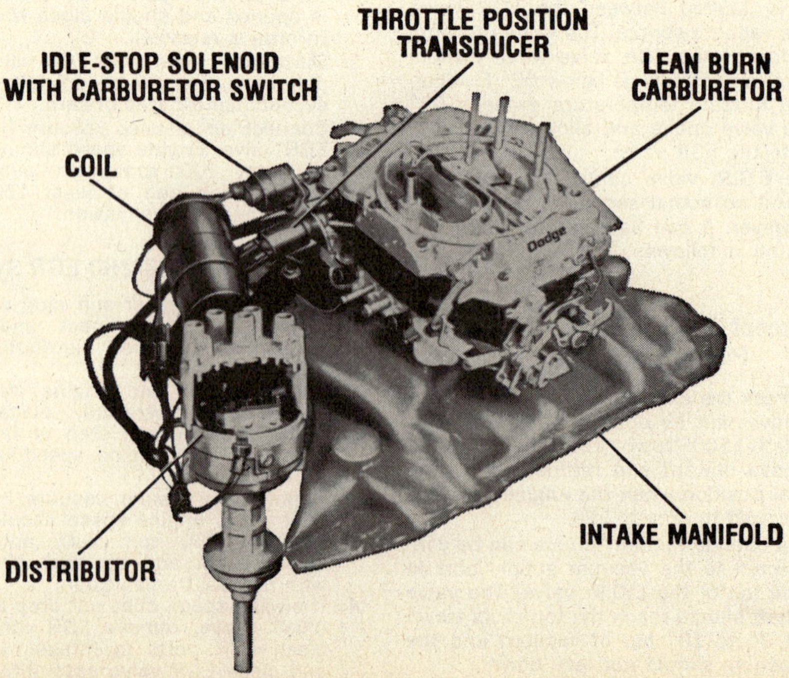

Fig. 26A Chrysler Electronic Spark Advance System solenoid locations

CHRYSLER C.A.P.

Cleaner Air Package

The Cleaner Air Package is in addition to the Positive Crankcase Ventilation (PCV). The PCV device is designed to control the emission of hydrocarbon vapors from the crankcase, whereas the Cleaner Air Package controls the emission of hydrocarbon vapors (unburned gasoline) and carbon monoxide in the vehicle exhaust.

The Cleaner Air Package is engineered to continuously control carburetion and ignition at the best settings for performance and combustion during all driving conditions. These adjustments keep unburned gasoline and carbon monoxide in the exhaust at a minimum concentration.

Only three special components are involved in the CAP installation, Fig. 27. The carburetor and distributor have been re-designed and a new component—the Vacuum Advance Control Valve—has been added. In addition, some cooling system components have been changed to handle the increased heat rejection at idle and low speed.

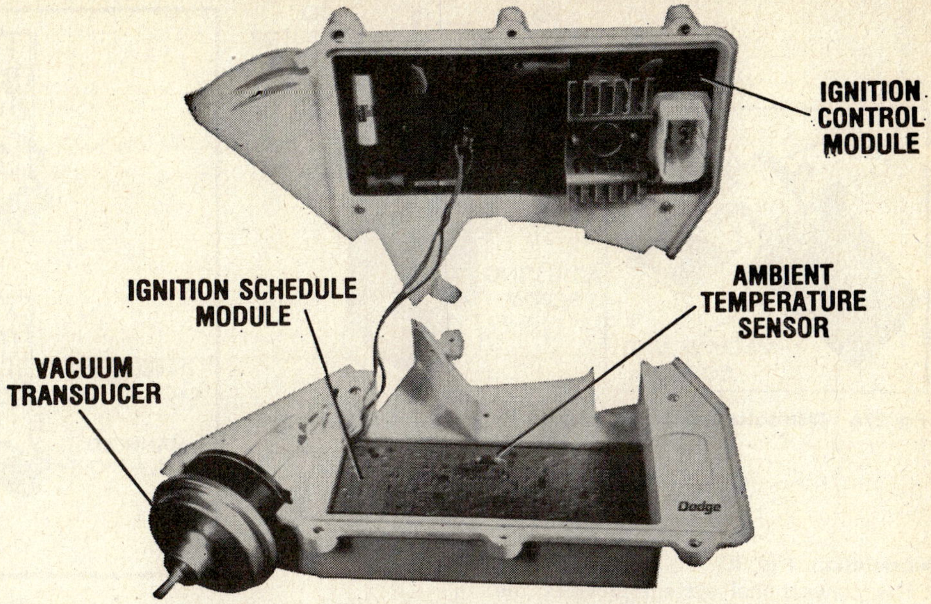

Fig. 26B Chrysler Electronic Spark Advance System spark advance module

Special Carburetor and Ignition

The carburetor is specially calibrated to provide leaner mixtures at idle and low speed operation. The distributor is designed to give retarded timing at idle. The vacuum advance control valve, in conjunction with the distributor, provides advanced timing during deceleration.

CAP idle timing for all engines is retarded instead of advanced. Exhaust emission is reduced at idle by using leaner air/fuel mixtures, increased engine speed, and retarded ignition timing. The higher air flow at this idle condition approximates the desirable conditions of cruise. CAP, therefore, is designed to operate with late timing during idle, and with conventional spark advance during acceleration and cruise.

Vacuum Advance Control Valve, 1969 6-170 with Std. Trans. & 1969 V8-426 Hemi

Early ignition of the air/fuel mixture is needed during deceleration to provide the most efficient combustion and reduced exhaust emissions. The vacuum advance control valve provides that additional spark advance during deceleration. The vacuum advance control valve is connected by vacuum hoses to the carburetor, to the intake manifold, and to the distributor vacuum chamber, Fig. 27A.

Carburetor vacuum and manifold vacuum act on the vacuum advance control valve. From these two signals, the vacuum advance control valve senses engine speed and load conditions, and relays a vacuum signal to the distributor to vary spark timing when necessary to reduce emissions to an acceptable level.

Engine Idle, Fig. 28

The initial idle timing is retarded as much as 15 degrees from conventional timing. The vacuum advance control valve does not affect timing at idle because the distributor vacuum chamber receives the same vacuum signal as in the conventional system, namely, carburetor vacuum. At idle it is not strong enough to overcome the distributor vacuum diaphragm spring.

Manifold vacuum acts on the vacuum control valve diaphragm, but is not strong enough to overcome the vacuum control valve spring. So the spring holds the vacuum control valve closed to manifold vacuum and open the low carburetor vacuum.

Acceleration, Fig. 29

During acceleration and cruise, manifold vacuum is not strong enough to actuate the CAP control valve. Thus the CAP system operates in the same manner as the conventional system. The throttle blade opens enough to permit the distributor vacuum advance to function, and spark timing is advanced according to the amount of vacuum created by the pumping action of the pistons.

Fig. 26C Chrysler Electronic Spark Advance System dual pick-up distributor

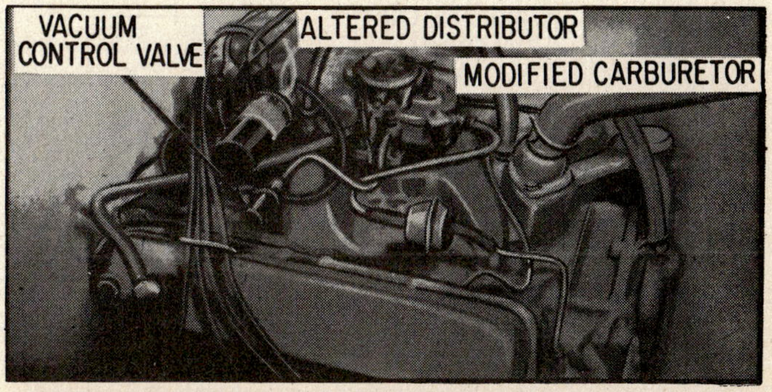

Fig. 27 Chrysler's Cleaner Air Package (CAP)

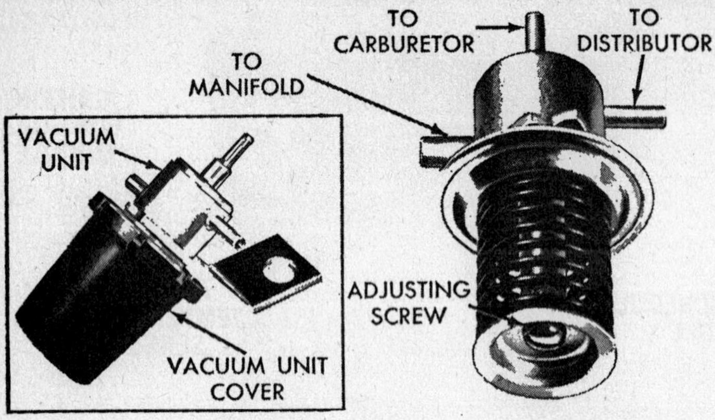

Fig. 27A Distributor vacuum advanced control valve, deceleration valve

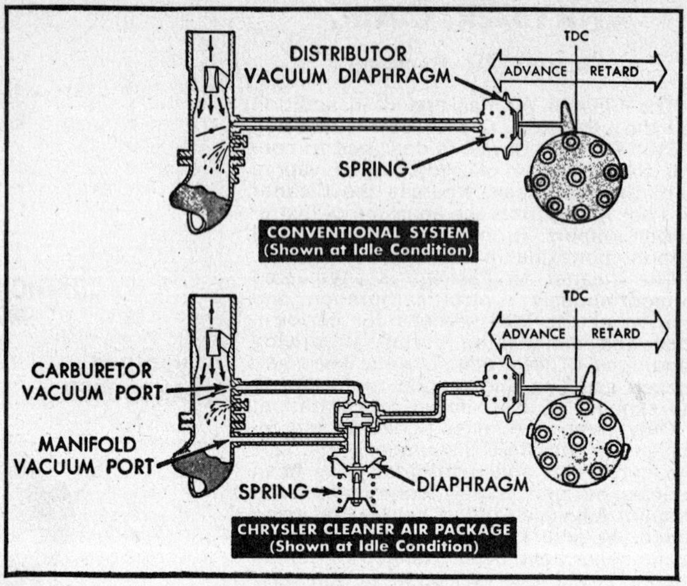

Fig. 28 Engine idle condition

Deceleration, Fig. 30

The conventional system provides the highest emissions under deceleration conditions. Carburetor vacuum is too weak to overcome the distributor advance diaphragm spring.

Manifold vacuum is at its strongest under deceleration conditions. Therefore, the CAP system uses manifold vacuum rather than carburetor vacuum to control spark timing.

Manifold vacuum is strong enough to overcome the vacuum advance control valve spring and the distributor vacuum diaphragm spring, moving spark timing to the maximum advance condition.

CHRYSLER C.A.S.

Cleaner Air System

This system employs higher inlet air temperature, higher idle speeds, retarded ignition timing, leaner carburetor mixtures and lower compression to reduce exhaust emissions, Fig. 31.

Heated Air System

Used on most 1970-75 Chrysler engines, this system uses a thermostatically controlled air cleaner to maintain an inlet air temperature of 95 to 105 deg. F. Temperature is controlled by intake manifold vacuum, a temperature sensor and a vacuum diaphragm which operates the heat

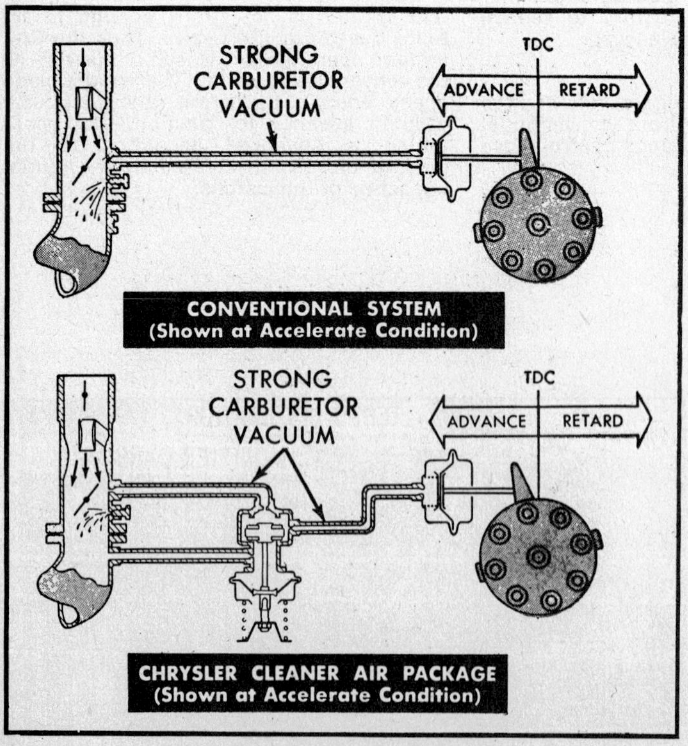

Fig. 29 During acceleration

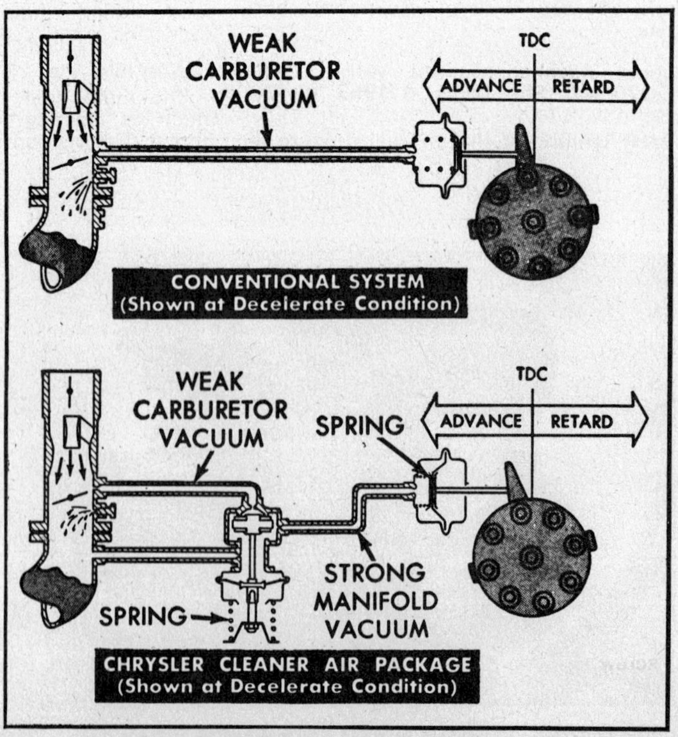

Fig. 30 During deceleration

control door in the air cleaner snorkel.

A vacuum hose connects the base of the carburetor to one side of the sensor. Another hose connects the other side of the sensor to the vacuum diaphragm.

The sensor contains a bimetallic strip which controls a small air valve. This valve controls the amount of vacuum to be applied to the diaphragm.

When the temperature at the sensor is below 95 deg. F, the valve is closed and manifold vacuum is applied to the diaphragm, lifting the heat control door. This allows heated air from the exhaust manifold stove to enter the air cleaner.

When the temperature at the sensor rises above 105 deg. F, the valve in the sensor opens and decreases the vacuum at the diaphragm and the spring in the diaphragm housing pushes the heat control down, closing off the air flow from the stove and opening the snorkel to outside air.

Since the diaphragm is opposed by a spring, it requires not less than 5"Hg. to lift the heat control door off the floor of the snorkel and not greater than 9"Hg. to raise the door to the top of the snorkel.

Under hard acceleration when intake manifold vacuum drops sharply, the spring forces the heat control door to the bottom of the snorkel, closing off the hot air and admitting cooler under hood air to eliminate any undue resistance to free breathing of the engine.

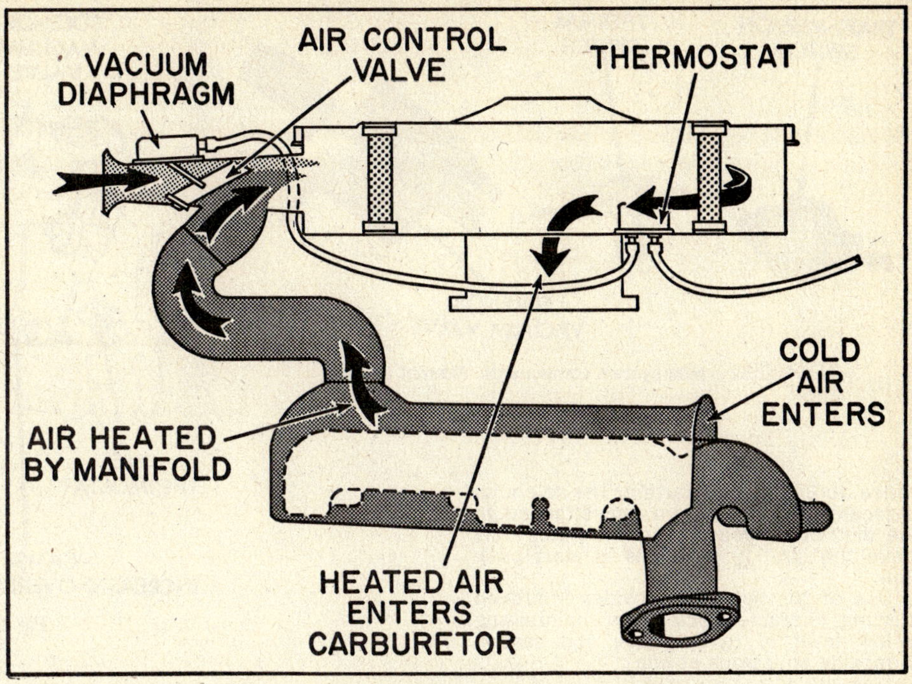

Fig. 31 Chrysler Corp. Heated Air Inlet system

Dual Snorkel

The dual snorkel air cleaner performs at low temperature and above 105 deg. F basically like a single snorkel air cleaner except that on deep throttle acceleration, both snorkels are open (when manifold vacuum drops between 5"Hg).

The "non-heat" air snorkel is connected to manifold vacuum through a tee in the vacuum hose between the carburetor and the sensor.

Check the second snorkel vacuum diaphragm as one with the heat connector.

Idle Speed Solenoid

Because of the high idle speeds used on some high performance engines, these engines have an electrical solenoid throttle stop which holds the throttle at the correct idle position when energized but de-energizes when the ignition is turned off, allowing the throttle blades to close more completely, thereby eliminating the possibility of "afterrun" or "dieseling".

Distributor Solenoid

Some engines have a solenoid incorporated in the distributor vacuum advance mechanism to retard the ignition timing when the throttle is closed. At closed throttle, and with the idle adjusting screw in the closed position, electrical contacts on the carburetor throttle stop cause the distributor solenoid to energize. This retards the ignition timing to reduce emissions during hot idle conditions. Cold or part throttle starting is not penalized because the distributor solenoid is not energized unless the hot idle adjusting screw is against the throttle stop contact.

NOTE: Ignition timing must be set at closed throttle to give accurate setting.

CHRYSLER NOx

Oxides of Nitrogen

The NOx system controls nitrogen oxides emissions by allowing vacuum spark advance only in high gear (manual transmission), or above 30 mph (automatic transmission), and with the use of an increased overlap camshaft and a 185 degree F coolant thermostat. Vacuum to the distributor is controlled by a solenoid vacuum valve mounted in the line between the carburetor vacuum port and the distributor. When the solenoid is energized the plunger shuts off vacuum to the distributor and vents it to the atmosphere. When it is de-energized the plunger opens allowing normal vacuum spark advance. There are two separate systems employed to control the solenoid vacuum valve, one for vehicles equipped with manual transmissions and the other for those having automatic transmissions.

Manual Transmission

The NOx system for manual transmissions consists of a solenoid vacuum valve, transmission switch and in the 1971 system a thermal switch, Fig. 32. The solenoid vacuum valve is mounted and operates as explained above. The transmission switch is mounted on the transmission housing and is used to sense the transmission gear position. It remains closed ("on") in any gear below high which energizes the solenoid thereby preventing vacuum spark advance. It opens when the top gear is selected permitting normal vacuum spark advance. The thermal switch is mounted on the firewall and senses ambient air temperature. If the temperature is below 70 degrees F, this switch will be open. This breaks the circuit between the transmission switch and the solenoid valve leaving the NOx system inoperative and allowing normal vacuum spark advance in all gears.

Automatic Transmission

The NOx system for automatic transmission equipped vehicles consists of a solenoid vacuum valve, a speed switch and a control unit assembly, Fig. 33. The solenoid vacuum valve is mounted and operates as explained above. The speed switch senses the vehicle speed and is mounted in line with the speedometer cable. The control unit assembly mounts on the fire wall. It contains three parts, the control module, the thermal switch and the vacuum switch. It senses ambient temperature and manifold vacuum. These components work together for one purpose, to prevent vacuum spark advance under following conditions:

1. Temperature above 70 degrees F.
2. Speeds below 30 mph.
3. Acceleration necessary on 1971 vehicles only.

Whenever all conditions are present, the solenoid vacuum valve will be energized shutting off vacuum to the distributor.

CHRYSLER IGNITION

Electronic Ignition

A better control of exhaust emissions is achieved through the use of the "Electronic Ignition". By eliminating the breaker points, engine misfiring and increased emissions caused by worn or misadjusted breaker points is eliminated.

Distributor Solenoid, 1972-74

A start only solenoid is used on some distributors to provide additional spark ad-

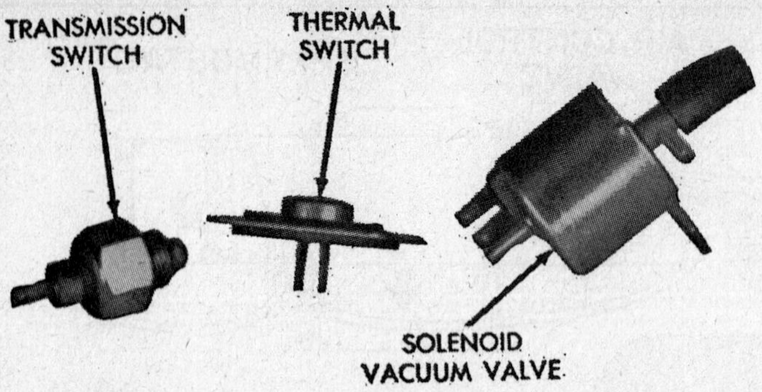

Fig. 32 NOx system components. Manual trans.

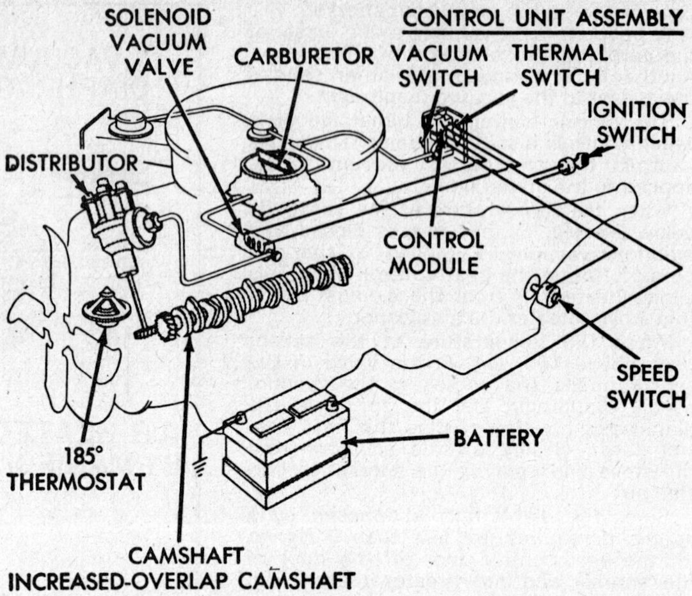

Fig. 33 NOx system. Automatic trans.

vance during engine starting. The solenoid, is located in the vacuum unit attached to the distributor housing and operates only while the ignition switch is in start position.

Use of the solenoid provides improved starting characteristics while maintaining a low level of hydrocarbon and carbon monoxide emissions at idle.

CHRYSLER ORIFICE SPARK ADVANCE CONTROL (OSAC)

Fig. 34—The OSAC system is used on all 1973-75 engines, to aid in the control of NOx (Oxides of Nitrogen). The system controls the vacuum to the vacuum advance actuator of the distributor.

A tiny orifice is incorporated in the OSAC valve which delays the change in ported vacuum to the distributor by about 17 seconds (27 seconds on some 1974-75 applications) when going from idle to part throttle. When going from part throttle to idle, the change in ported vacuum to the distributor will be instantaneous. On 1973 models, the valve will only delay the ported vacuum signal when the ambient temperature is about 60° F. or above. Vacuum is obtained by a vacuum tap just above the throttle plates of the carburetor. This type of tap provides no vacuum at idle, but provides manifold vacuum as soon as the throttle plates are opened slightly. Proper operation of this valve depends on air tight fittings and hoses and on freedom from sticking or plugging due to deposits.

CHRYSLER TEMPERATURE OPERATED VACUUM BY-PASS VALVE

This vacuum by-pass or Thermal Ignition Control (TIC) valve, Fig. 20, is used on some engines to reduce the possibility of engine overheating under extremely high temperature operating conditions. When engine coolant temperature at idle reaches 225° F., the valve opens automatically and applies manifold vacuum directly to the distributor for normal vacuum spark advance. This will by-pass the NOx or OSAC system. This increases en-

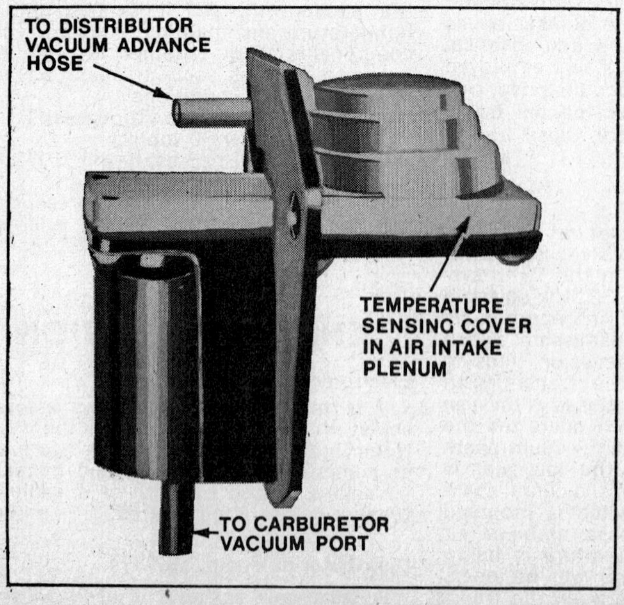

Fig. 34 Chrysler OSAC valve

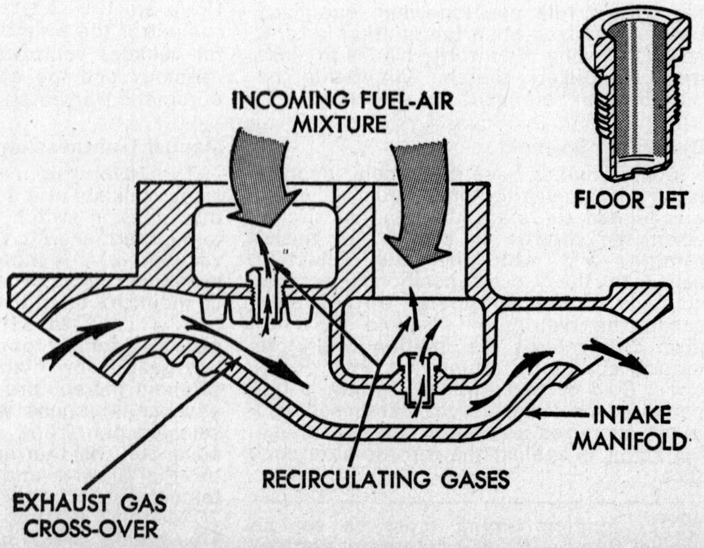

Fig. 35 EGR floor jets V-8 configuration

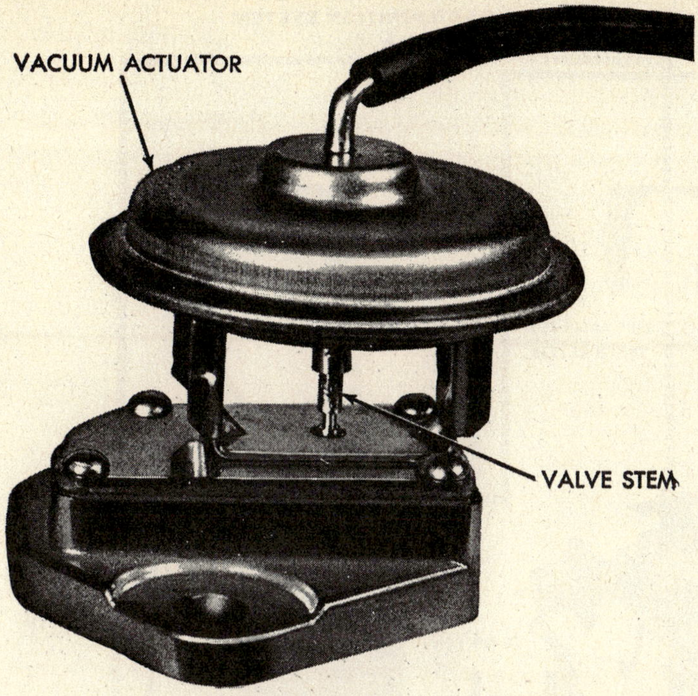

Fig. 36 Chrysler EGR control valve

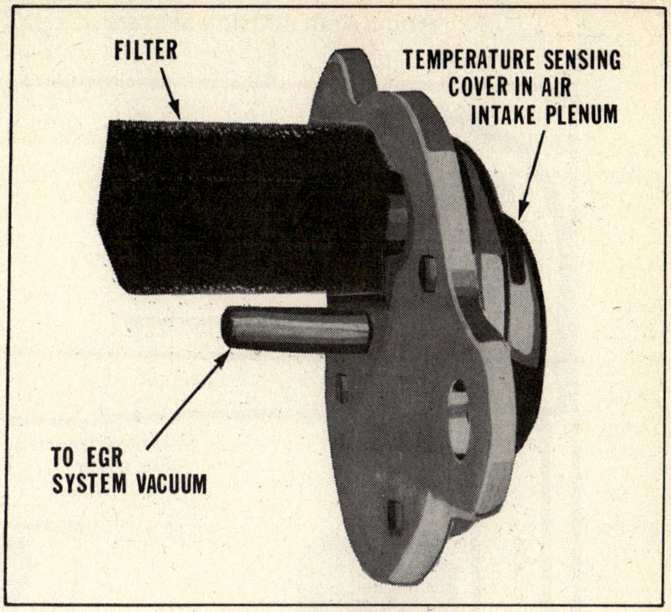

Fig. 37 Chrysler EGR temperature control valve

gine idle speed and provides additional engine cooling. When the engine has cooled to normal operating temperature, the NOx or OSAC system is restored to normal operation.

CHRYSLER EXHAUST GAS RECIRCULATION (EGR)

In this system, exhaust gases are circulated to dilute the incoming fuel air mixture. Dilution of the incoming mixture lowers peak flame temperatures during combustion and thus limits the formation of NOx.

Floor Jet Exhaust Gas System 1972-73

In this system the exhaust gases are introduced into the intake manifold through jets in the floor below the carburetor, Fig. 35. An orifice in each jet allows a controlled amount of exhaust gas to be drawn through by engine vacuum to dilute incoming fuel and air. In eight cylinder engines, exhaust gases are taken from the intake manifold exhaust crossover passage. While in six cylinder engines, gases are taken from the exhaust manifold "plenum" chamber located at the "hot spot" below the carburetor riser.

NOTE: In 1973-75 two additional systems are used to control the rate of exhaust gas recirculation, depending on engine model. These systems are: Ported Vacuum Control System and Venturi Vacuum Control System.

Both systems use the same type exhaust gas recirculation (EGR) control valve, Fig. 36, only the method of controlling the valve is different. The valve is a vacuum actuated, poppet type unit used to modulate exhaust gas flow from the exhaust gas crossover into the incoming air fuel mixture.

Venturi Vacuum Control System

The venturi vacuum control system utilizes a vacuum tap at the throat of the carburetor venturi to provide a control signal. This vacuum signal is amplified to the level required to operate the EGR control valve. Elimination of recycle at wide open throttle is accomplished by a dump diaphragm which compares venturi and manifold vacuum to determine when wide open throttle is achieved. At wide open throttle, the internal reservoir is "dumped", limiting output to the EGR valve to manifold vacuum. The valve opening point is set above the manifold vacuums available at wide open throttle.

NOTE: This system is dependent primarily on engine intake airflow as indicated by the venturi signal, and is also affected by intake vacuum and exhaust gas back pressure.

Ported Vacuum Control System

The ported vacuum control system utilizes a slot type port in the carburetor throttle body which is exposed to an increasing ratio of manifold vacuum as the throttle blade opens. This throttle bore port is connected through an external nipple directly to the EGR valve. The flow rate is dependent on three variables, 1) manifold vacuum, 2) throttle position, and 3) exhaust gas back pressure. Recycle at wide open throttle is eliminated by calibrating the valve opening point above manifold vacuums available at wide open throttle as port vacuum cannot exceed manifold vacuum. Elimination of wide open throttle recycle provides maximum performance.

Temperature Control Valve

The plenum mounted temperature control valve, Fig. 37, is utilized on the ported vacuum control system and the venturi vacuum control system. The valve reduces the recycle rate at low ambient temperature for improved driveability. The unit contains a temperature sensitive bimetal disc which senses plenum air temperature. The snap action of the disc unplugs a calibrated orifice to provide the bleed air. Calibration is protected by an air filter unit.

Coolant Control Exhaust Gas Recirculation (CCEGR)

1974-75 engines using EGR are equipped with a CCEGR valve mounted in the radiator top tank. When coolant temperature in the top tank reaches 65° F, the valve opens so that vacuum is applied to open the EGR valve. On some 1974 engines, a similar CCEGR valve set for 90° F. is mounted in the thermostat housing and on 1975 V8-360 a valve set for 125° F. is mounted in the engine block.

EGR Delay System

Some 1974-75 vehicles are equipped with an EGR Delay System, which has an

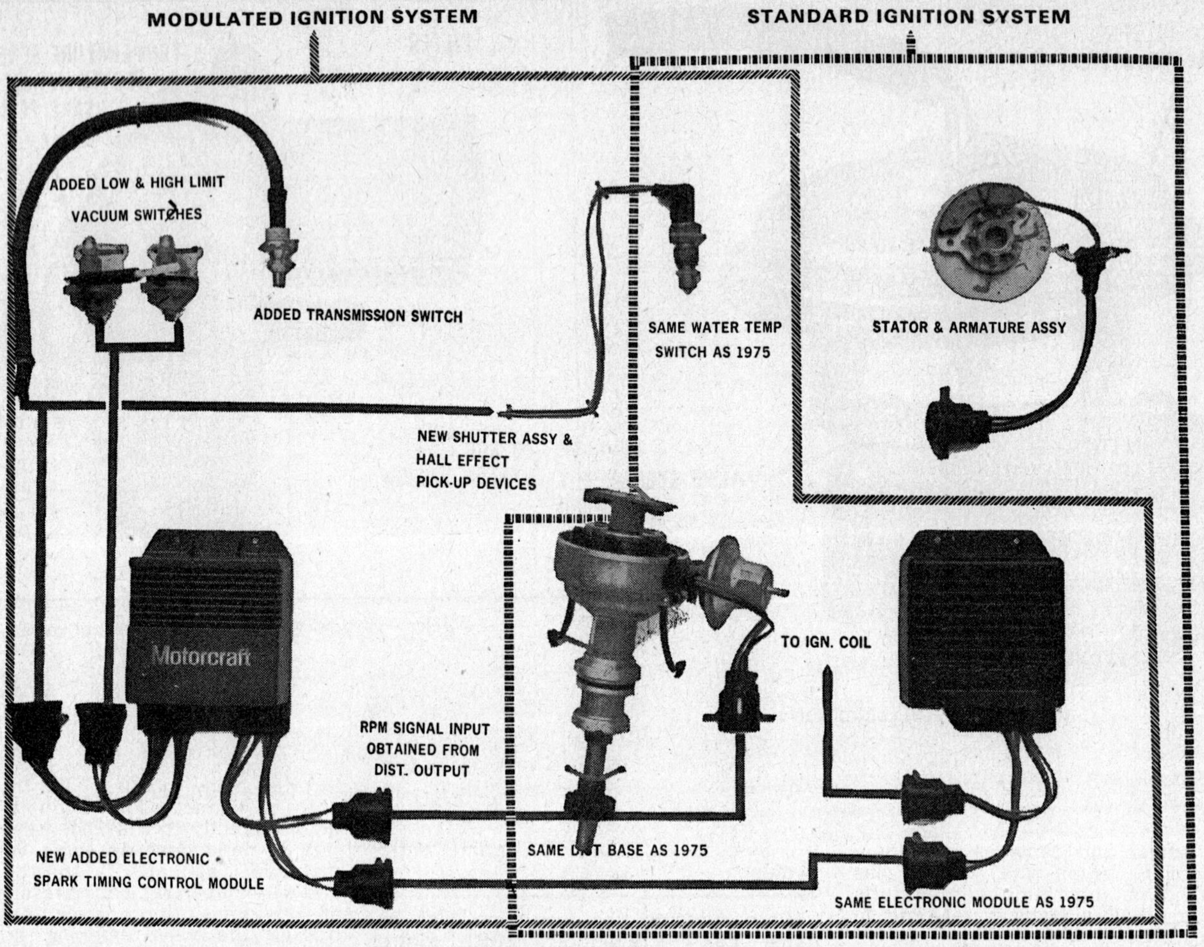

Fig. 38 Ford Computer Controlled Timing System

electrical timer mounted on the dash panel in the engine compartment controlling an engine mounted solenoid. This solenoid which is connected with vacuum hoses to the carburetor venturi and vacuum amplifier, prevents EGR operation for about 35 seconds after engine start up.

EGR Maintenance Reminder System

This reminder system on 1975 vehicles uses a warning light in the instrument cluster, to alert the driver of the vehicle at 15,000 mile intervals to have the EGR system checked for the following:

1. Vacuum hose leakage
2. EGR valve operation
3. CCEGR valve operation
4. Timer and solenoid operation
5. Intake manifold and EGR valve flow passages

Once the EGR system has been properly serviced, the reminder system must be reset for the next 15,000 mile interval and to turn off the warning light. If the EGR switch has to be replaced before a 15,000 mile interval, the EGR system maintenance check must be performed and the new switch installed.

FORD COMPUTER CONTROLLED TIMING

This system, Fig. 38, used on 1976 Mark IV models, provides ignition timing-control compensation to supplement the normal vacuum and centrifugal spark-advance mechanisms under all engine operating conditions. The system incorporates a distributor, electronic spark control module and engine sensors.

The distributor has two separate timing pick-ups, off-set from each other by the desired advance increment. One pick-up is used for initial timing and the other for advance. The electronic spark control module evaluates the engine operating conditions and determines when to advance the spark. The module also can generate a time delay in the signal to retard the spark under certain operating conditions which is used for engine "Knock" control.

The sensors detect engine RPM, coolant temperature, manifold vacuum and transmission gear.

FORD IMCO SYSTEM

Improved Exhaust Emission Control System

This system combines a thermostatically controlled air cleaner and higher engine operating temperature with leaner carburetor calibration and later ignition timing under closed throttle operating conditions.

Dual Diaphragm Distributor

Fig. 38A—In addition to the conventional centrifugal and vacuum advance control units, this unit uses a separate diaphragm to retard the spark timing under closed throttle conditions. The advance diaphragm is connected to the carburetor above the throttle plate(s) so that when the throttle is opened, the timing is advanced. The retard diaphragm is connected to the intake manifold so that during closed throttle operation, when manifold vacuum is high, the timing is retarded to provide more complete combustion.

Distributor Vacuum Control Valve

This valve, Fig. 39, is exposed to cooling system temperature and when coolant temperature exceeds normal limits during long idle periods, the valve opens a vacuum passage to the advance diaphragm of the distributor which speeds up the engine idle lowering the temperature.

To test the valve, proceed as follows:

1. With engine at operating temperature, connect a tachometer to the engine.
2. Note engine idle rpm with transmission in neutral.
3. Disconnect the vacuum hose from the intake manifold at the temperature sensing valve and plug the hose.
4. If idle speed does not change when hose is disconnected, the valve is acceptable to this point. If idle drops 100 rpm or move, valve must be replaced.
5. Reinstall vacuum line then cover radiator sufficiently to raise engine coolant temperature above thermostat setting.

NOTE: Be sure all-season coolant is up to specification and exercise caution so that the engine does not become unduly overheated.

6. Continue to run the engine until the high temperature lamp comes on or the temperature gauge reaches the high end of the band. At this point, the engine idle speed should increase by approximately 100 rpm. If it has not, the valve is defective and must be replaced.

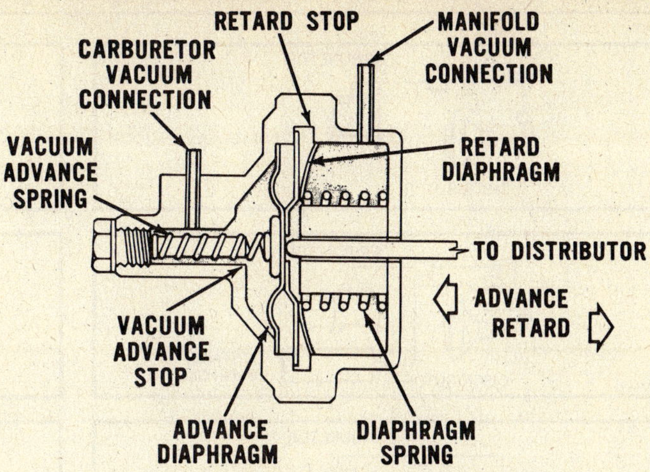

Fig. 38A Dual diaphragm vacuum advance valve

Distributor Vacuum Advance Control Valve

This valve, Figs. 27 and 39, provides the necessary ignition advance during acceleration periods to provide the most efficient combustion and reduce emissions. To check the valve, proceed as follows:

1. With engine at operating temperature and idle speed correctly set, connect a tachometer to the engine.
2. Remove cover from valve and slowly turn the adjusting screw counterclockwise without exerting any inward pressure. After five and no more than six turns, the idle speed should suddenly increase to approximately 1000 rpm.

NOTE: Any more than six turns out will release the compressed spring and washer.

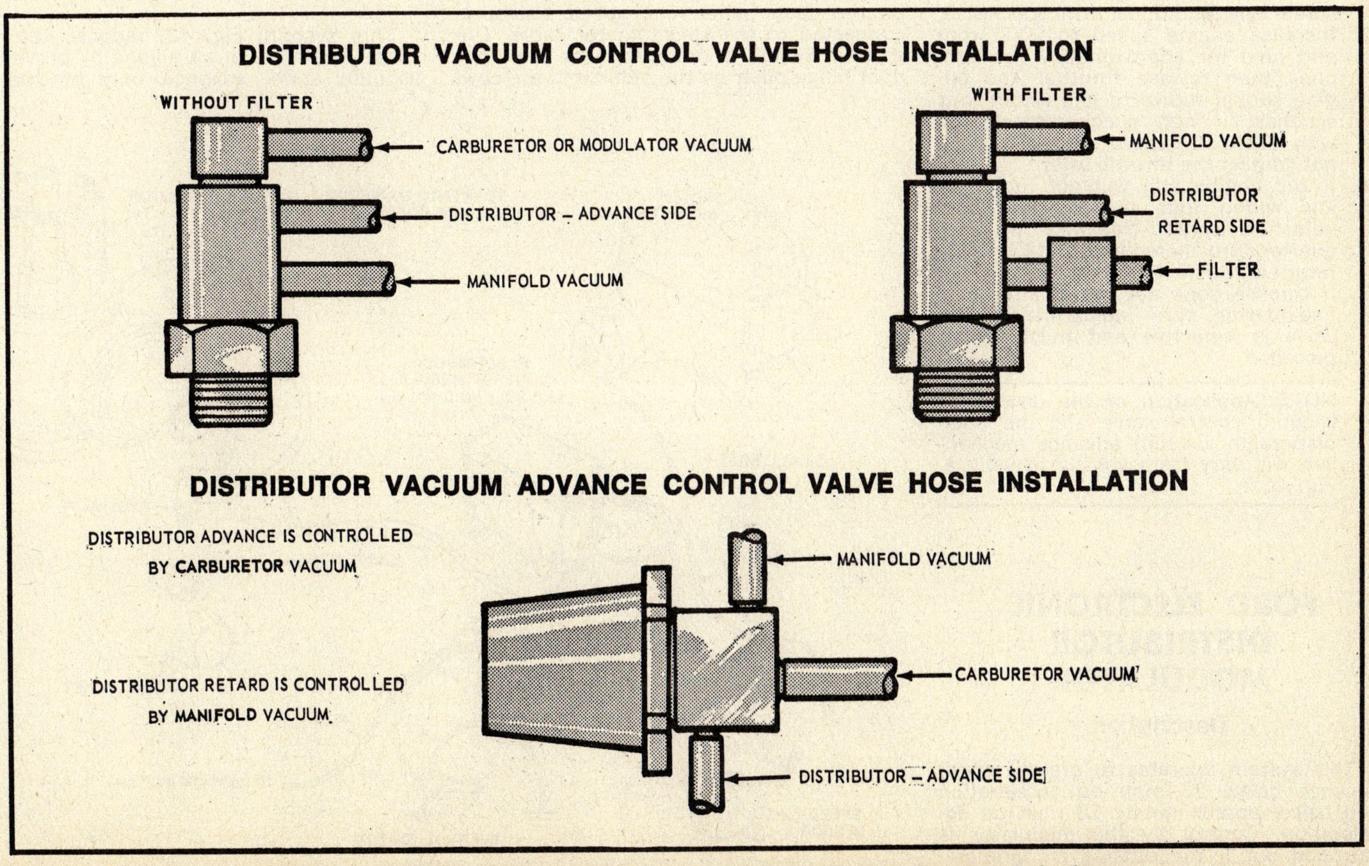

Fig. 39 Distributor vacuum control valve and distributor vacuum advance control valve

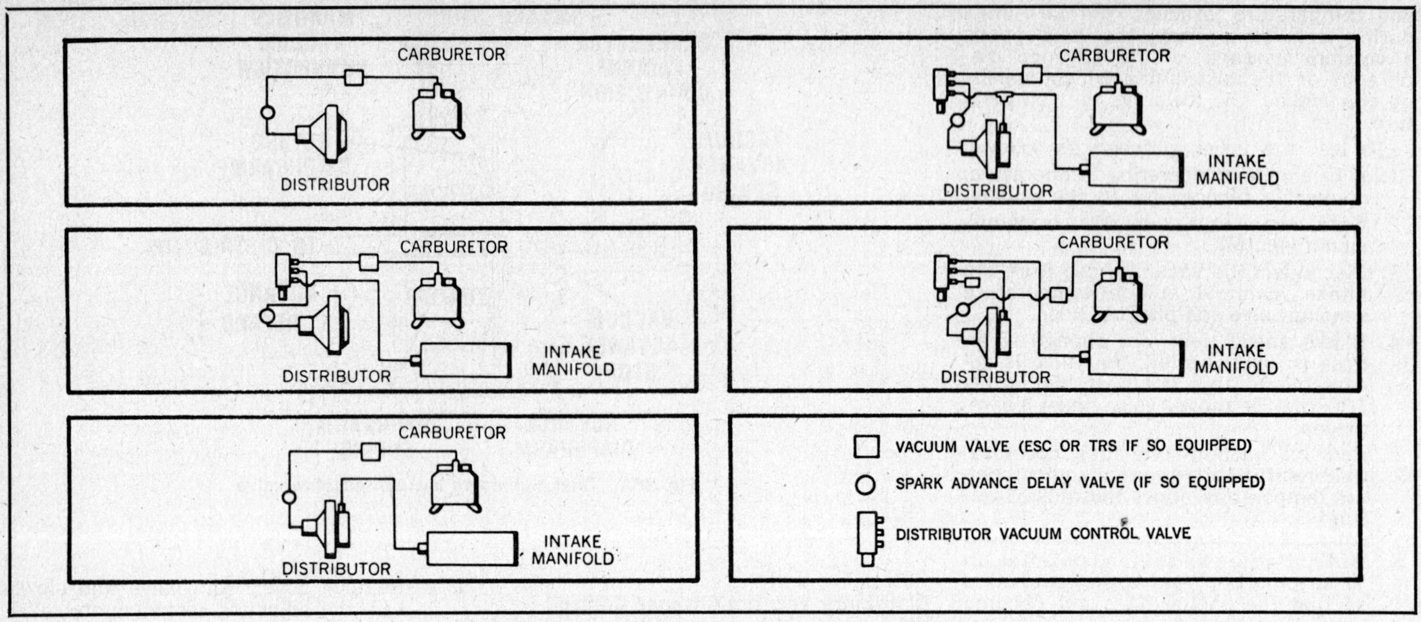

Fig. 40 Ford distributor vacuum system schematics

3. If idle speed does not increase after the sixth turn, push inward on the end of the spring and release. Idle speed will increase.
4. After valve has been triggered to the higher rpm, slowly turn the adjusting screw clockwise until idle speed drops back down to proper level. Turn screw one additional turn clockwise.
5. Increase engine speed to 2000 rpm and hold for approximately five seconds then release throttle. The engine should return to idle within four seconds. If not, check return time with dashpot backed off so it does not contact the throttle lever.
6. If the engine still will not return to idle within four seconds, turn the adjusting screw clockwise in one-quarter turn increments until throttle return is satisfactory.
7. If throttle does not return after four one-quarter turn adjustments, the valve is defective and must be replaced.

NOTE: Application of the distributor vacuum control valve and the dual-diaphragm vacuum advance mechanism will vary from vehicle to vehicle, Fig. 40.

FORD ELECTRONIC DISTRIBUTOR MODULATOR

Description

This system operates to prevent spark advance below 23 mph on acceleration and below approximately 18 mph on deceleration. Control by the modulator is canceled out if the outside air temperature is below 58 deg. F, allowing the distributor to operate through the standard vacuum control system, Fig. 41.

The modulator system consists of four components: speed sensor, thermal switch, and electrical control amplifier-solenoid valve. The control amplifier and solenoid valve are combined in one assembly and mounted in the passenger compartment on the dash panel. The speed sensor is connected to the speedometer cable. The thermal switch is mounted near the front door hinge pillar on the outside of the cowl panel. It may be mounted on either the right or left side.

FORD E.S.C. SYSTEM

Electronic Spark Control System

This system, Fig. 42, reduces the exhaust emissions of an engine by providing vacuum spark advance only at speeds

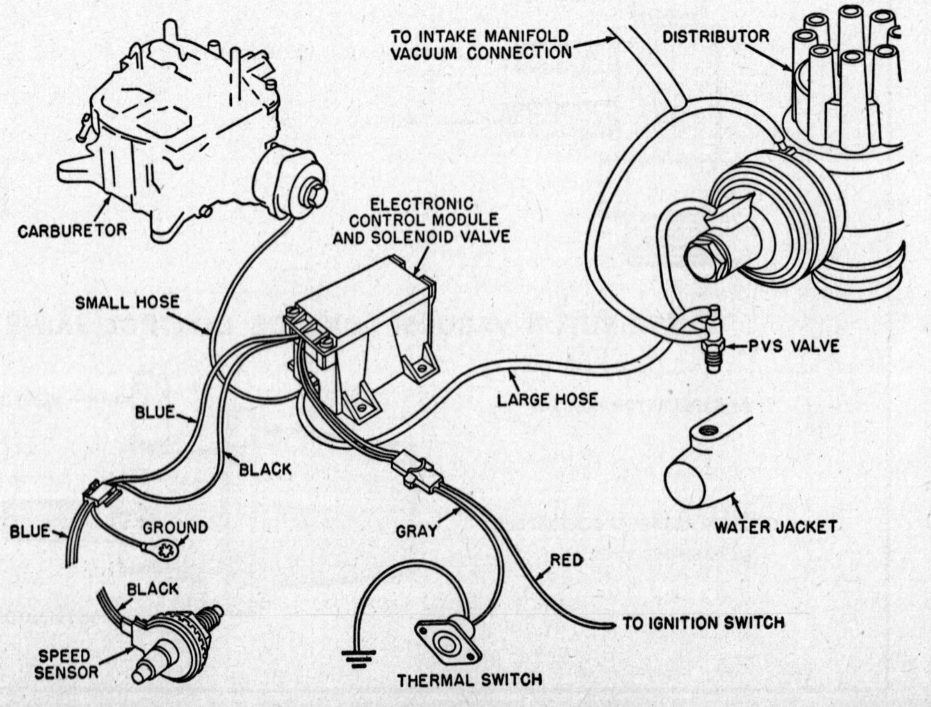

Fig. 41 Electronic Distributor Modulator

above 24 to 33 mph (depending on the engine application). It consists of a speed sensor, an electronic amplifier, an outside air temperature switch and a vacuum control valve. The vacuum control valve is inserted between the carburetor vacuum advance port and the distributor primary advance connection. This valve is normally open, but when energized electrically by the electronic amplifier it closes to cut off vacuum to the primary vacuum advance unit on the distributor thus preventing vacuum spark advance. The temperature switch, which is mounted in either the right or left A-pillar, senses outside air temperature. A temperature below 49 degrees F will cause the switch contacts to open, thereby de-energizing the vacuum valve and allowing normal vacuum advance at all speeds. A temperature of 60 degrees plus or minus 5 degrees F causes the contacts to close, thereby cutting off vacuum to the advance side of the distributor at speeds below 24 to 33 mph. On deceleration the vacuum advance cut-out speed is approximately 18 mph.

On some applications the vacuum hose connections between the carburetor and distributor may route through a PVS valve. This valve serves as a by-pass or safety override switch. When the coolant temperature reaches 230 degrees F, manifold vacuum is applied directly to the primary (advance) side of the distributor advancing the timing and thereby lowering operating temperature.

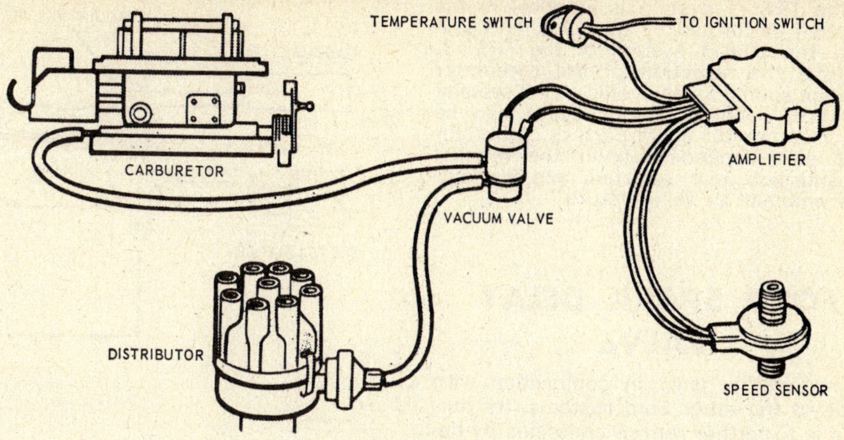

Fig. 42 Ford Electronic Spark Control system

FORD HIGH SPEED EGR MODULATOR SUB-SYSTEM

The high speed EGR modulator sub-system used on some V-8 engines, Fig. 42A is basically the same in operation as the ESC system described previously. This system cuts off exhaust gas recirculation flow by stopping vacuum flow from the EGR port to the EGR valve at speeds above 64 mph, inturn improving driveability.

The vacuum solenoid valve installed in the vacuum line is normally open (not energized), allowing vacuum flow from the EGR port to the EGR valve. The EGR system remains functional when the valve is not energized.

The speed sensor driven by the speedometer cable, produces an electric signal directly proportional to vehicle road speed, signalling the amplifier to energize the vacuum solenoid valve at which time the electronic module receives the signal from the speed sensor and amplifies it to provide a usable signal to the vacuum solenoid valve.

When the vehicle speed exceeds approximately 64 mph (trigger speed of the amplifier) the circuit to the ignition switch is completed and the normally open vacuum solenoid valve is energized. The plunger moves upwards and shuts off the EGR port vacuum and the vent at the bottom of the vacuum valve is opened, bleeding vacuum from the EGR valve and hose. Spring force closes the EGR valve which remains non functional until the vacuum solenoid valve is de-energized, at speeds below approximately 64 mph.

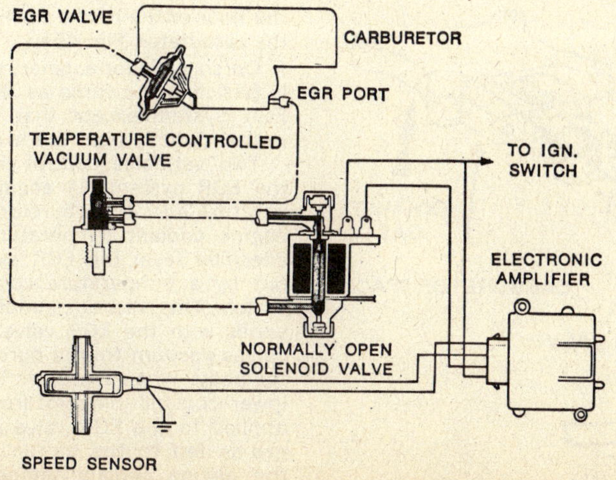

Fig. 42A Ford high speed EGR modulator sub-system components

NOTE: There is a continuous internal vacuum bleed provided by the vent at the top of EGR valve. Whether the valve is in a closed or open position, this vent purges the vacuum supply hose from carburetor of any gasoline vapor.

FORD T.R.S. SYSTEM

Transmission Regulated Spark Control System

This system, Fig. 43, reduces the exhaust emissions of an engine by providing vacuum spark advance only in high gear. It consists of a vacuum control valve, an outside air temperature switch, and a transmission switch. The vacuum control valve is inserted between the carburetor vacuum advance port and the distributor primary advance connection. This valve is normally open, but when energized electrically by the transmission switch it closes to cut off vacuum to the primary vacuum advance unit on the distributor thus preventing vacuum spark advance. The temperature switch, which is mounted in either the right or left A-pillar, senses outside air temperature. A temperature below 49 degrees F will cause the switch contacts to open, thereby de-energizing the vacuum valve and allowing normal vacuum advance in all gears. A temperature of 60 degrees plus or minus 5 degrees F causes the contacts to close, thereby cutting off vacuum to the advance side of the distributor in all but high gear.

FORD T.R.S. + 1

Transmission Regulated Spark +1

Fig. 44—The TRS+1 system consists of two separate vacuum control systems, that are electrically controlled by input information from a manual transmission gear selector switch, and an outside ambient air temperature switch. The TRS function of the TRS+1 system is identical to the function performed by the 1972 TRS system. The plus 1 system of the TRS+1, controls the selection of the carburetor vacuum source for the vehicle EGR system. The EGR vacuum supply source can be either carburetor spark port, or carburetor EGR port depending upon the manual transmission gear selected, and the outside ambient air temperature.

FORD SPARK DELAY VALVE

This unit is used in conjunction with some of the other Ford systems. Its purpose is to further reduce emissions by delaying the spark advance during rapid acceleration and by cutting off advance immediately upon deceleration.

This plastic disc-shaped valve is installed in the carburetor vacuum line at the distributor advance diaphragm. It is a one way valve and will not operate if installed backwards. The black side of the valve must be toward the carburetor. This valve cannot be repaired. On 1975 vehicles, this valve can be checked for proper operation.

NOTE: On all systems which employ the dual diaphragm distributor the line which has high vacuum at idle (normal operating temperature) is connected to the secondary (retard) side of the distributor vacuum advance unit. This is the connection closest to the distributor cap.

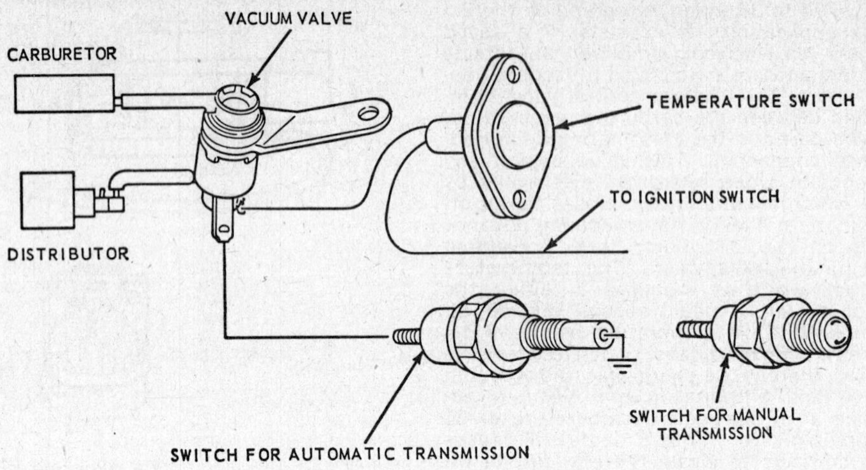

Fig. 43 Ford TRS system

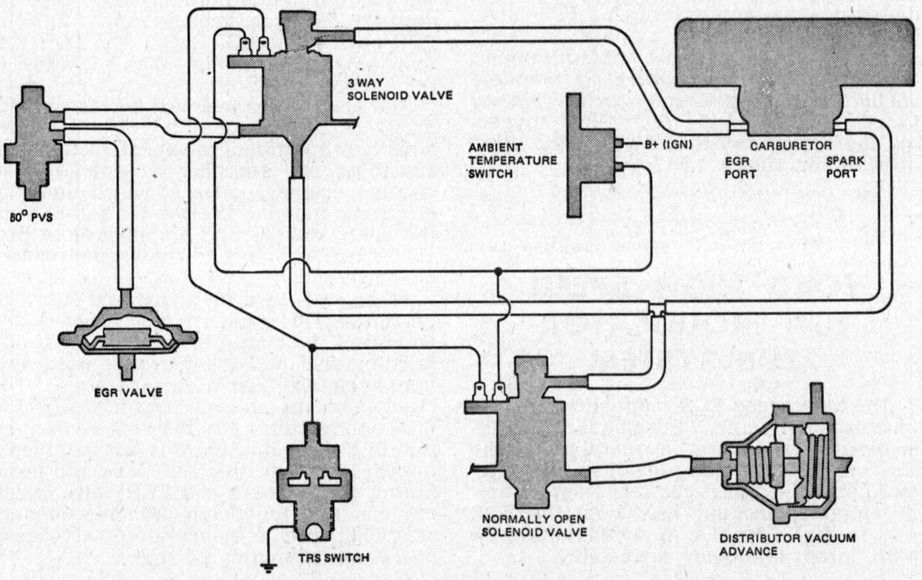

Fig. 44 Ford transmission regulated spark one system—TRS+1

FORD EGR SYSTEM

Exhaust Gas Recirculation

In this system the exhaust gases are metered through the EGR valve to a passage in the carburetor spacer, or on some 1974-75 eight cylinder engines, through two drilled passages in the floor of the intake manifold riser under the carburetor therefore diluting the air fuel mixture entering the combustion chambers. Dilution of the incoming mixture lowers peak flame temperatures during combustion and thus limits the formation of nitrogen oxides (NOx).

Most eight cylinder engines use the "Spacer Entry" EGR System which has the EGR valve mounted on the rear of the carburetor spacer, Fig. 45. The exhaust gases are taken from a drilled passage in the exhaust crossover of the intake manifold. The exhaust gas is then routed through a metered EGR valve to a passage in the carburetor spacer and fed into the primary bore. Some 1974 eight cylinder engines use the "Floor Entry" EGR system, which has the EGR valve mounted on the rear of the intake manifold. The EGR valve controls the exhaust gases that enter specially cast passages in the manifold from the exhaust crossover passage. When the valve opens, the exhaust crossover is then opened to the two drilled passages in the floor of the intake manifold riser under the carburetor, Fig. 45A.

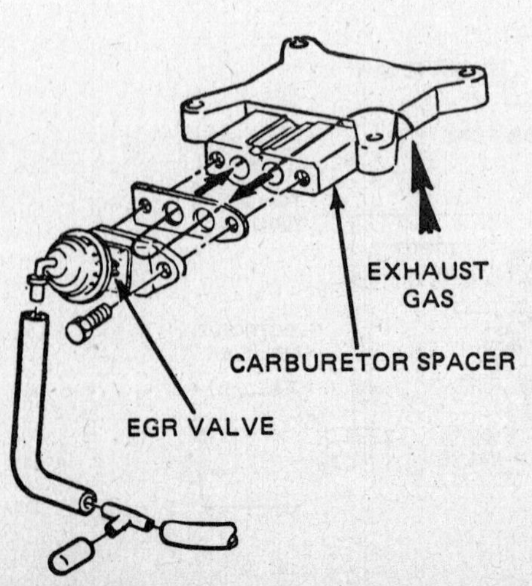

Fig. 45 Ford V8 EGR valve hook up

On six cylinder engines, the EGR system is basically the same as the Spacer Entry EGR System except that exhaust gas is routed directly from the exhaust manifold.

Two variables control the operation of the EGR system, 1) engine coolant temperature and 2) carburetor vacuum. When engine coolant temperature is below the specified level the EGR system is locked out by a temperature controlled vacuum switch. This vacuum switch is installed in series with the EGR valve. This valve receives vacuum from a port in the carburetor body. When the valve is closed due to lower coolant temperature, no vacuum is applied to the EGR valve and no exhaust gas is fed to the air-fuel mixture. When the engine coolant temperature reaches the specified level, the valve opens allow-

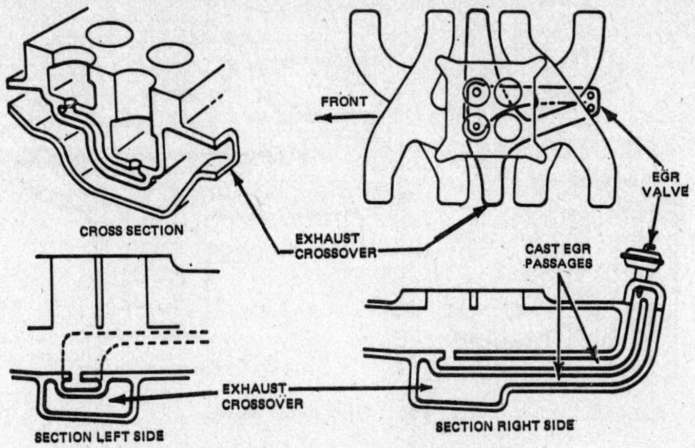

Fig. 45A Ford Floor Entry EGR System

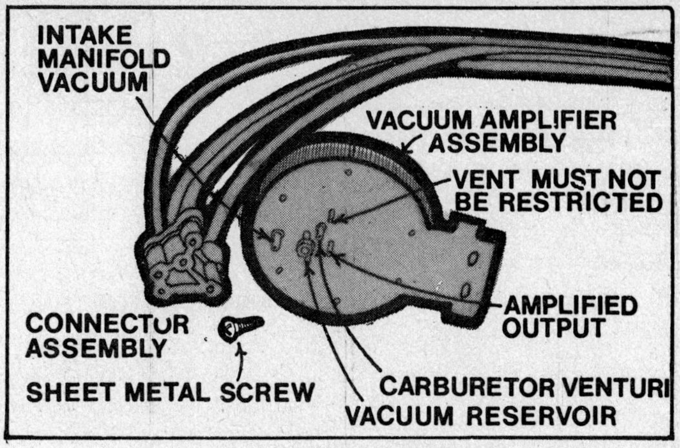

Fig. 45B Venturi Vacuum Amplifier

ing vacuum to be applied to the EGR valve. Exhaust gas is then fed to the air-fuel mixture.

The second factor controlling EGR operation is carburetor vacuum. The location of the EGR port in the carburetor determines at what point vacuum is sent to the EGR valve. Vacuum should be fed to the EGR vacuum control valve when the primary throttle plate reaches a position corresponding to a road speed of approximately 20 mph under light acceleration.

A Venturi Vacuum Amplifier, Fig. 45B, used in 1974-75, uses a weak venturi-vacuum signal to produce a strong intake manifold vacuum to operate the EGR valve, thereby achieving an accurate, repeatable and almost exact proportion between venturi airflow and EGR flow. This assists in controlling oxides of nitrogen with minimal sacrifice in driveability.

There are three types of EGR valves, the poppet type, modulating type and the tapered stem type.

NOTE: If the tapered stem valve is plugged or causes rough idle due to leakage, it should be replaced.

The poppet type valve, Fig. 46, consists of springloaded diaphragm, and a valve stem and valve operating in an enclosed valve body. At approximately 3 inch Hg of vacuum, the valve begins to open. The valve stem is pulled forward unseating the valve and allowing exhaust gas to flow into the valve chamber. Venturi vacuum will then pull the gas from the chamber into the air-fuel flow and then into the combustion chambers. Once the valve has been unseated the only means of limiting exhaust gas flow is the size of the flow restrictor placed in the inlet port of the valve body. The size of the restrictor will vary according to engine application.

On the modulating type valve, Fig. 47, an additional disc has been added to the valve stem below the main valve. The modulating valve operates exactly like the poppet valve when vacuum is between approximately 3 in Hg and 10.5 in Hg. When vacuum reaches approximately 10.5 inches, the lower disc (high vacuum flow restrictor) approaches the shoulders of the valve seat and restricts the flow of exhaust gas. The purpose of the modulation of gas flow is to improve driveability on certain engine models.

NOTE: The EGR valve and vacuum control valve cannot be repaired and must be replaced if damaged.

Cold Start Cycle (CSC)

The EGR/CSC System regulates both distributor advance and EGR valve operation according to coolant temperature by sequentially switching vacuum signals. The major system components are, a 95° F EGR-PVS (Ported Vacuum Switch) valve, a SDV (Spark Delay Valve) and a vacuum check valve, Fig. 45C.

When engine coolant temperature is below 82° F, the EGR-PVS valve admits carburetor EGR port vacuum (at about 2500 RPM) directly to the distributor advance diaphragm, through the one way check valve. At the same time, the EGR-PVS valve shuts off carburetor EGR vacuum to the EGR valve and transmission diaphragm.

When engine coolant temperature is 95°F and above, the EGR-PVS valve is actuated and directs carburetor EGR vacuum to the EGR valve and transmission diaphragm instead of the distributor. At temperatures between 82° and 95° F, the EGR-PVS valve may be open, closed or in mid position.

The Spark Delay Valve (SDV) delays carburetor vacuum to the distributor advance by restricting the vacuum signal

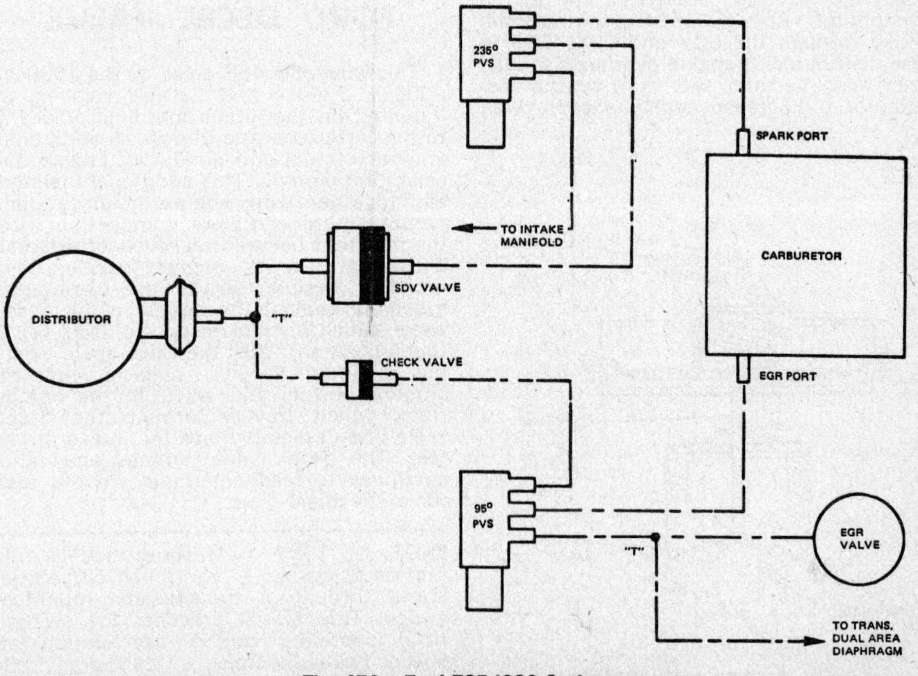

Fig. 45C Ford EGR/CSC System

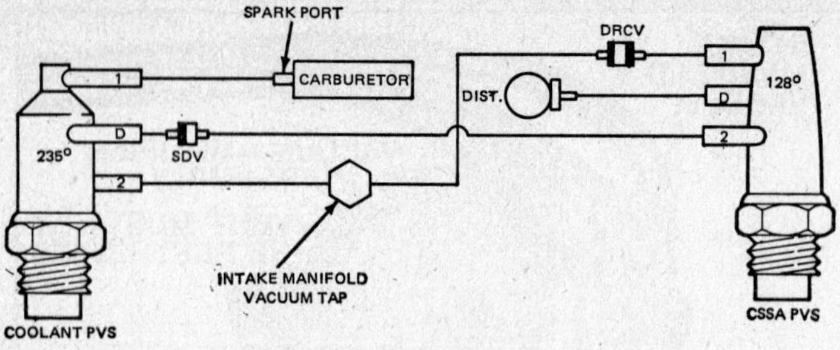

Fg. 45D Cold Start Spark Advance System. 1975 Ford

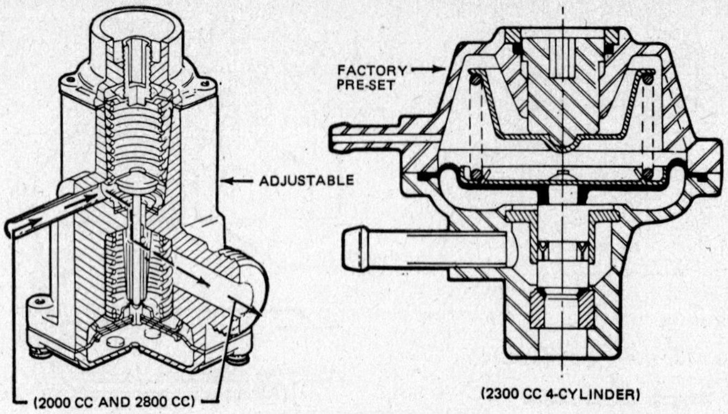

Fig. 45E Ford decel valve

through the SDV for a predetermined time. During normal acceleration, little or no vacuum is admitted to the distributor advance diaphragm until acceleration is completed and engine coolant temperature is 95° F or higher.

The check valve blocks off vacuum signal from the SDV to the EGR-PVS valve so that carburetor spark vacuum will not be dissipated when the EGR-PVS valve is actuated above 95° F.

The 235° F PVS valve which is not part of the EGR-PVS system is connected to the distributor vacuum advance to prevent engine overheating as on previous models.

Cold Start Spark Advance (CSSA)

Used on some 1975 V8-460 engines, this system, Fig. 45D, is added to the distributor control to provide manifold vacuum when engine coolant is below 125° F. by providing vacuum from the intake manifold, through the Distributor Retard Control Valve (DRCV), the CSSA PVS and to the distributor. At 125° F. and above, the vacuum flows from the carburetor spark port, through the cooling PVS, the SDV, the CSSA PVS and to the distributor. Above 235° F., the vacuum flows through the SDV and CSSA PVS to the distributor. If engine overheats at idle, increased vacuum will flow to the distributor to increase engine speed. When engine coolant temperature decreases, spark advance will be controlled by the carburetor spark port.

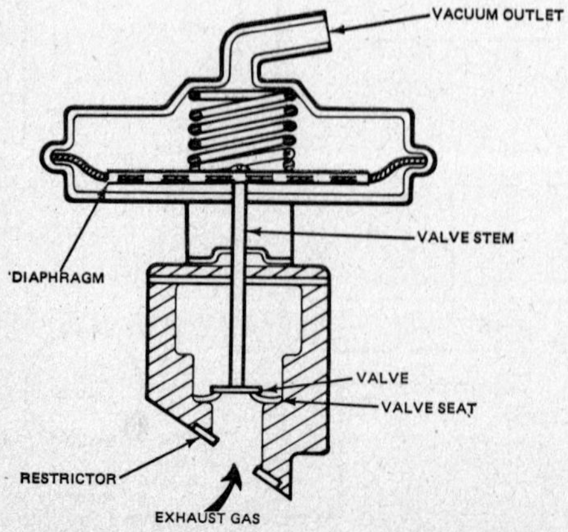

Fig. 46 Ford poppet-type EGR valve

DUAL-AREA DIAPHRAGM

On 1973-75 vehicles, new dual-area diaphragms are used, Fig. 48. These diaphragms offset effects of engines using the EGR system and equipped with automatic transmissions. The new diaphragms permit vehicles to function with satisfactory shift spacing and shift feel.

To test, remove the vacuum diaphragm and test unit using an outside vacuum source. Set regulator on tester to 18 in. Hg with end of vacuum hose blocked off then connect vacuum hose to vacuum diaphragm unit. If unit does not hold 18 in. Hg. reading, the diaphragm is leaking and must be replaced.

FORD DECEL VALVE

This valve, Fig 45E, used on the 1600cc, 2000cc, 2300cc and 2800cc engines, is mounted on the intake manifold adjacent to the carburetor and meters an additional amount of fuel and air during engine deceleration periods. This additional fuel and air, together with engine modifications, permits more complete combustion with the resultant being lower levels of exhaust emissions. During engine deceleration, manifold vacuum forces the diaphragm assembly against the spring in the decel valve, which in turn raises the decel valve (open position). With the valve open, existing manifold vacuum pulls a metered amount of fuel and air from the carburetor, which travels through the decel valve body assembly into the intake manifold. The decel valve remains open and continues to feed additional air and fuel for a specifield time.

NOTE: On 1975 V6 Mustang models, the fuel decel system is controlled by vehicle speed through a transmission mounted switch. This switch prevents the system from operating during deceleration at speeds below 11 mph.

FORD DVB SYSTEM

Delay Vacuum By-Pass System

This system is designed to delay distributor vacuum advance under varying ambient temperatures and vehicle speeds. The system consists of a solenoid vacuum valve (SV), spark delay valve (SDV), and a check valve (one-way valve), Fig. 49.

The spark delay valve SDV is a two-way flow device used for controlling distributor vacuum advance. It delays the flow of vacuum in one direction through a sintered metal restrictor, yet permits an instantaneous flow in the other direction through a one-way check valve.

When the vehicle is accelerating, the direction of air flow is from the distributor toward the carburetor through the restrictor.

The check valve is held closed by the pressure differential between the SDV inlet and outlet connections. After a specified time period, (depending on restrictor porosity) the vacuum will bleed through the restrictor and equalize the pressure at both connections.

When the vehicle decelerates, the pressure differential is reversed, the direction of air flow is reversed, and the check valve opens, to instantly equalize air pressure across the valves.

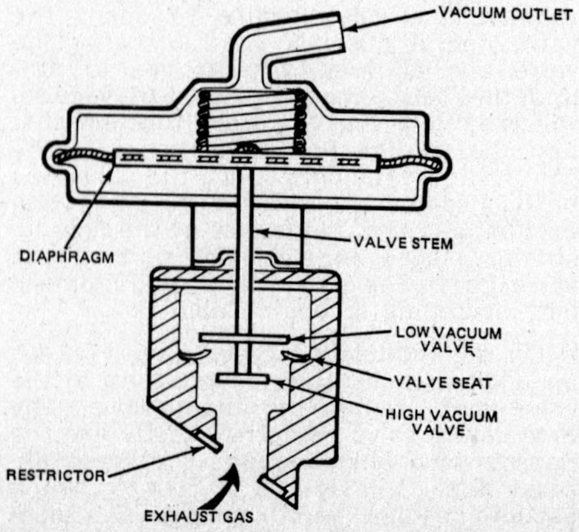

Fig. 47 Ford modulating type EGR valve

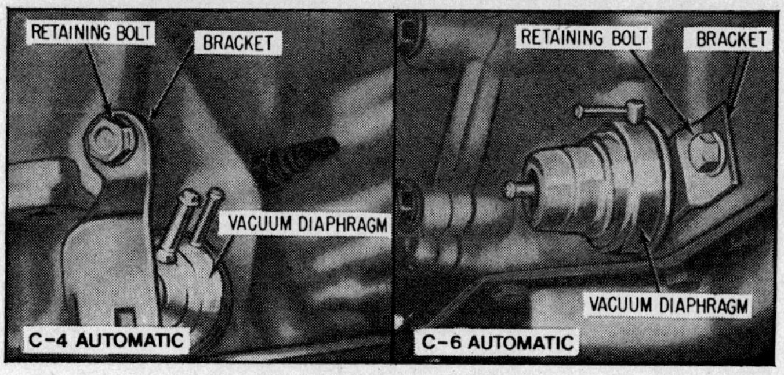

Fig. 48 1973-75 Dual diaphragm vacuum modulator

Since different vehicle engine and transmission combinations require different time delay periods to effectively meet emission requirements. Five different type SDV valves are used in the various vehicle engine applications. Each type is color coded.

Since the SDV is not affected by ambient air temperature changes, it would act to retard the distributor spark advance during cold weather when exhaust emissions can be met normally without the retarding action. For this reason, the DVB system is also connected between the vacuum source at the carburetor spark port and the distributor primary diaphragm in parallel with the SDV system.

At temperatures of 60 degrees F and above. the normally open solenoid vacuum valve receives battery voltage from the ambient temperature switch whenever the ignition switch is on closing the solenoid valve and shutting off vacuum to the distributor. The temperature switch is open at temperatures of 49 degrees F and below and no signal is sent to the solenoid valve. Thus, the solenoid valve will remain open admitting carburetor spark port vacuum to the distributor primary vacuum diaphrgam directly through the open solenoid valve.

NOTE: The DVB system is energized only when the weather is warm and allows no direct flow of vacuum through the solenoid valve.

The DVB system check valve prevents the outlet of the SDV valve from being vented to the atmosphere to cause a loss of vacuum to the distributor resulting in no distributor advance when the DVB is energized. The check valve must be installed in the vacuum line as shown or the SDV valve will not permit the check valve to close.

FORD T.A.V. SYSTEM

Temperature Activated Vacuum

This system selects either the carburetor spark port vacuum, or the carburetor EGR port vacuum as a function of outside ambient air temperature.

The EGR system can be used in addition to the TAV system, although systems work independently of each other.

The TAV system, Fig. 50, consists of an ambient temperature switch, a three-way vacuum valve, and an external inline vacuum bleed. The three-way vacuum valve is used to select the carburetor vacuum source that is supplied to the distributor vacuum advance mechanism. The ambient temperature switch provides the switching circuit to determine which vacuum source will be selected as a function of outside air temperature. The inline vacuum bleed function is to purge the vacuum line in the TAV system of any excessive gasoline vapors.

The basic difference between a TAV system and the standard IMCO system is the selective control feature provided by TAV system for distributor vacuum advance as a function of outside air temperature.

When the ambient air temperature is above 60 degrees F the three-way vacuum valve is energized, therefore the EGR vacuum is controlling the distributor advance. When the ambient air temperature is below 49 degrees F the three-way vacuum valve is de-energized, therefore the spark port vacuum is controlling the distributor advance.

NOTE: The TAV system controlls spark advance below 49 degrees F. while the EGR system controls spark advance above 60 degrees F.

FORD CTAV SYSTEM

Cold Temperature Activated Vacuum

This system operates basically the same as the TAV system previously discussed

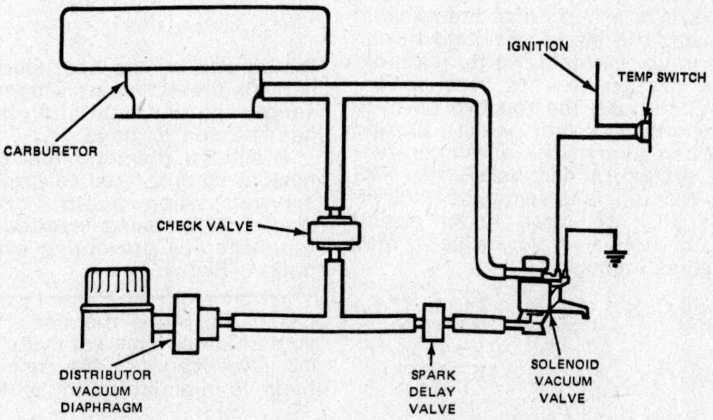

Fig. 49 Ford delay vacuum by-pass system—DVB

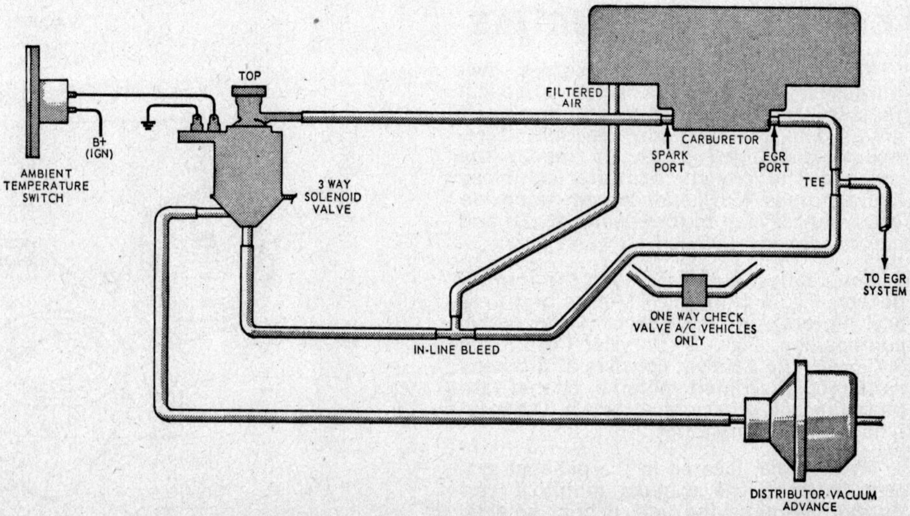

Fig. 50 Ford temperature activated vacuum system—TAV

except that a latching relay, Fig. 51, has been added. The latching relay, activated by temperature switch closing remains energized regardless of temperature switch position which prevents system cycling due to minor ambient temperature changes.

The temperature switch energizes the three-way vacuum valve and latching relay when ambient temperature is above 65° F. When ambient temperature is below 49° F, the system is inoperative and the distributor diaphragm and EGR valve receives vacuum directly from its respective carburetor ports.

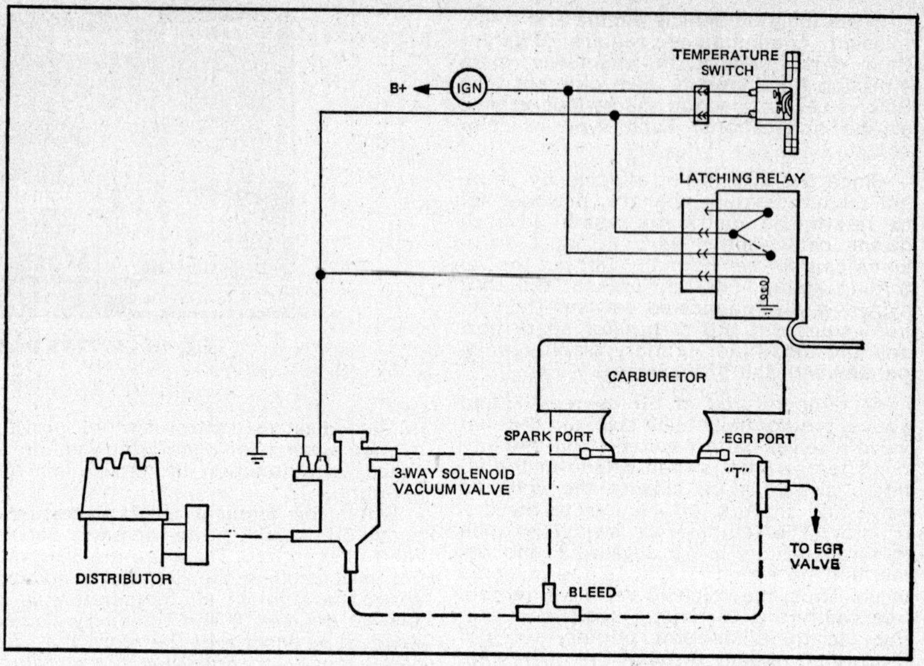

Fig. 51 Ford cold temperature activated vacuum system—CTAV

FORD HCV SYSTEM

Vacuum Operated Exhaust Heat Control Valve

Used on all 1975 vehicles except those equipped with the 2300cc and V8-460 engines, this system, Fig. 52, provides quick induction system warm-up for better cold engine fuel vaporization.

A vacuum operated heat control valve mounted between the exhaust manifold and pipe, directs a portion of the exhaust gases through the intake manifold during engine warm-up. On cold starts, manifold vacuum is directed to the heat control valve (HCV) through the top two ports in HCV PVS (ported vacuum switch), closing the HCV. When engine coolant temperature reaches a predetermined value, the PVS closes off vacuum and vents the PVS allowing the HCV to close under spring tension. The three PVS valves used may be identified as follows:

PVS Body Color	Opening Temp. (° F.)
Black	92-98
Blue	125-131
Purple	157-163

CATALYTIC CONVERTERS

The catalytic converter serves two purposes: it permits a faster chemical reaction to take place and although it enters into the chemical reaction, it remains unchanged, ready to repeat the process. The catalytic converter combines hydrocarbons (HC) and carbon monoxide (CO) with oxygen to form water (H_2O) and carbon dioxide (CO_2).

The catalyst is structured in the form of pellets, Fig. 1 (American Motors and General Motors), or a honeycomb monolithic composition, Figs. 2 (Chrysler Corp.) and 3 (Ford). The catalyst consists of a porous substrate of an inert material, coated with platinum and other noble metals, the catalytically active materials.

This device, located in the exhaust system between the exhaust manifold and muffler, requires the use of heat shields, in some cases, due to its high operating temperatures. The heat shields are necessary to protect chassis components, passenger compartment and other areas from heat related damage.

A smaller diameter fuel tank filler tube neck is incorporated to prevent the larger service station pump nozzle, used for leaded fuels, being inserted into the filler tube, thereby preventing system contamination, Fig. 4.

CAUTION: Since the use of leaded fuels contaminates the catalysts, deteriorating its effectiveness, the use of unleaded fuels is mandatory in vehicles equipped with catalytic converters. The catalytic converter can tolerate very small amounts of leaded fuels without permanently reducing the catalyst effectiveness.

Most Chrysler Corp. vehicles equipped with catalytic converters, have a Catalyst Protection System, Fig. 5, which prevents overheating of the converter during high speed deceleration. This overheating condition occurs when the catalyst tries to oxidize the increased amount of unburned hydrocarbons during deceleration.

This system consists of an electronic

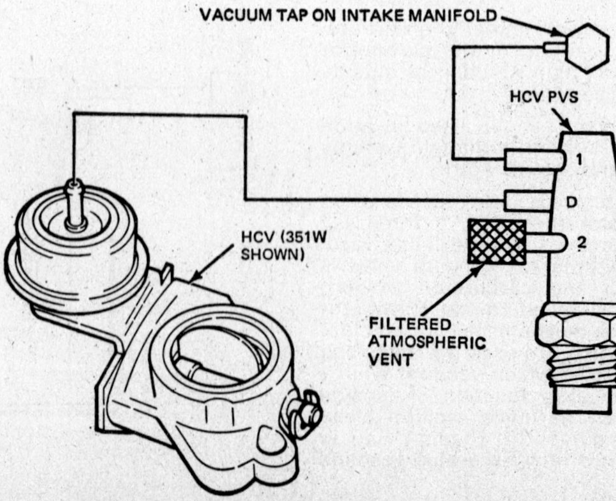

Fig. 52 Heat Control Valve System. 1975 Ford

engine speed switch that receives ignition pulses from the electronic control unit and a carburetor mounted throttle position solenoid. At engine speeds above 2000 rpm, the solenoid will energize to hold the throttle slightly open. During deceleration, the throttle will be held in this position until the engine speed drops below 2000 rpm at which time the throttle will be returned to normal curb idle speed.

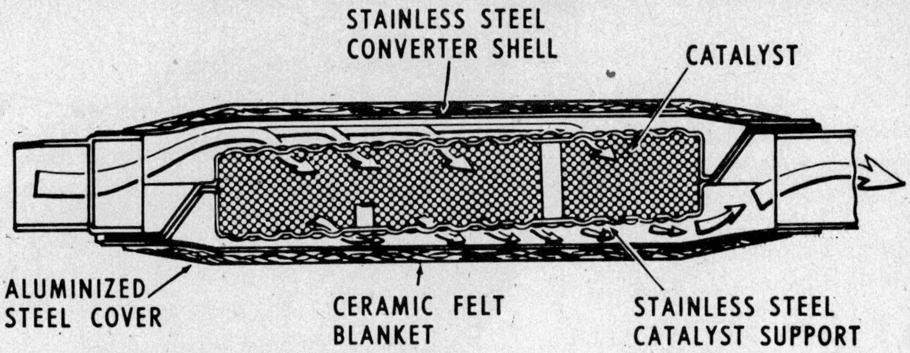

Fig. 1 American Motors & General Motors catalytic converter

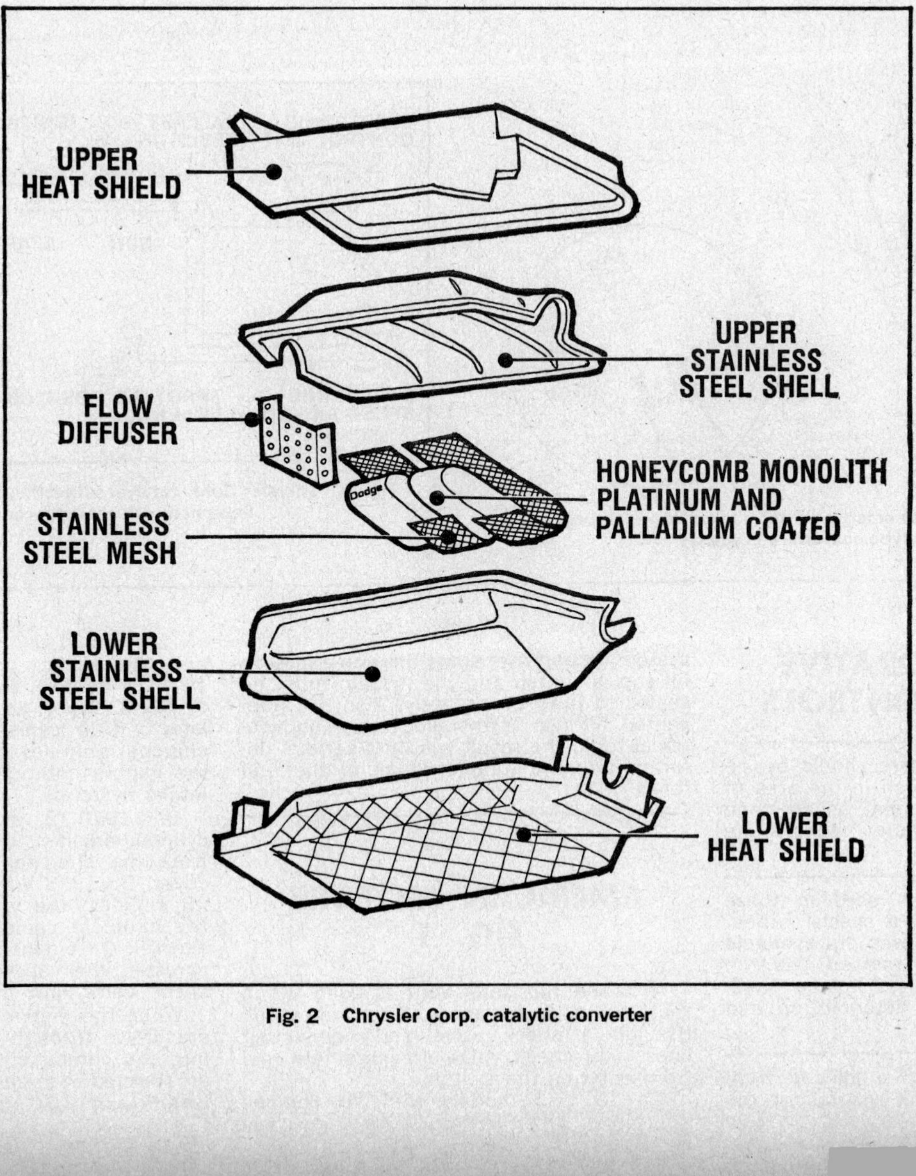

Fig. 2 Chrysler Corp. catalytic converter

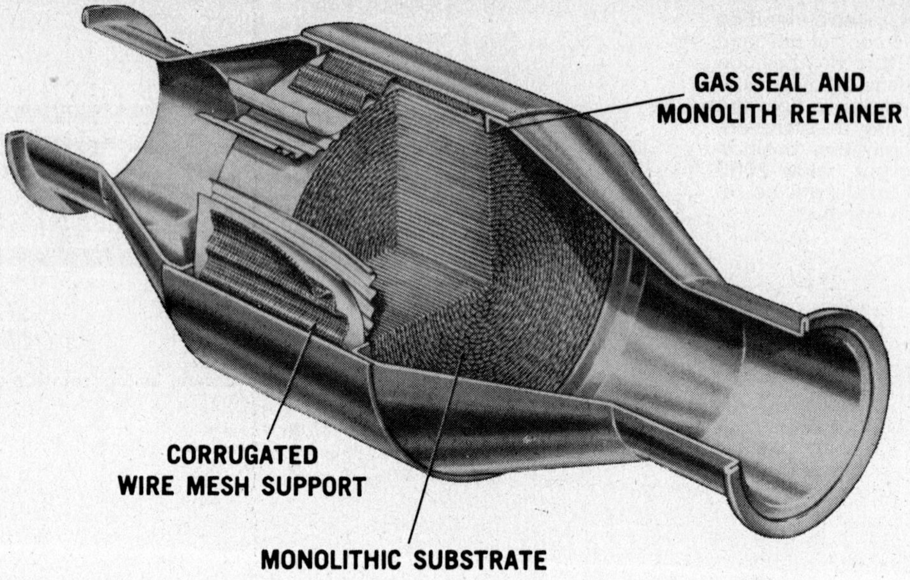

Fig. 3 Ford catalytic converter

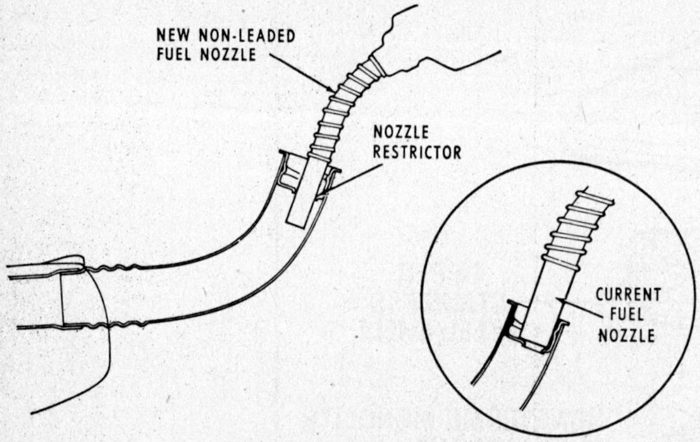

Fig. 4 Fuel tank filler safety neck for all vehicles equipped with catalytic converters (Typical)

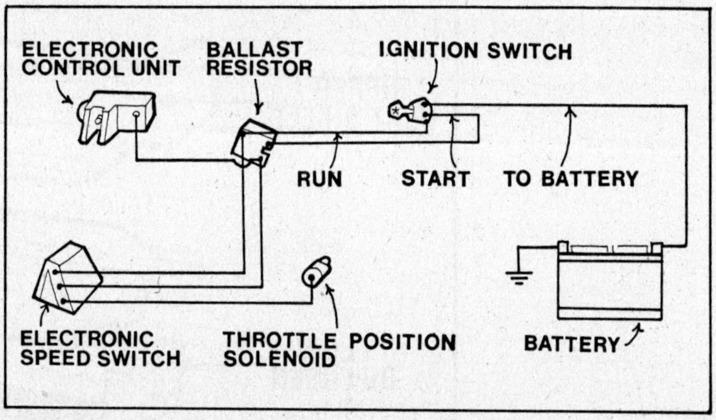

Fig. 5 Chrysler Corp. catalyst protection system for all vehicles equipped with catalytic converters

FUEL EVAPORATIVE EMISSION CONTROLS

CAUTION: The utmost care should be exercised when using a torch in the area of the fuel evaporation system as an open flame near these hoses may cause a fire and ultimate explosion.

NOTE: Vapor line hoses used in these systems are made from a special rubber material. Bulk service hoses are available for service and will be marked "EVAP". Ordinary fuel hoses should not be used, as they are subject to deterioration and may clog system.

CAUTION: Installation of a fill cap from a non-emission fuel tank will render the system inoperative, since the non-emission fill cap is vented and the system must be sealed to function properly. Also if a non-vented fill cap is installed on a conventional tank, the result will be a serious deformation or a total collapse of the fuel tank.

AMERICAN MOTORS FIG. 1

A closed fuel tank vent is used which routes the fuel vapors from the tank through a check valve and connecting lines. The check valve prevents raw fuel from entering the system.

On 1971-72 models with V8 engines and automatic transmissions and all 1973-75 models, the vapors are stored in a "Fuel Vapor Storage (Charcoal) Canister". This canister contains activated charcoal granules which absorb and store the vapors until they are drawn into the intake system.

On 1970-72 engines and 1973 six cylinder engines, these vapors are routed from the charcoal canister to the valve cover. On 1973 V-8 engines and all 1974-75 engines, the vapors are routed from the charcoal canister to the air cleaner snorkel. On engines, not using a charcoal canister, the vapors are routed from the check valve into the valve cover.

When the engine is started, the vapors are drawn through the intake system and into the combustion chamber where they are burned with the normal air/fuel mixture.

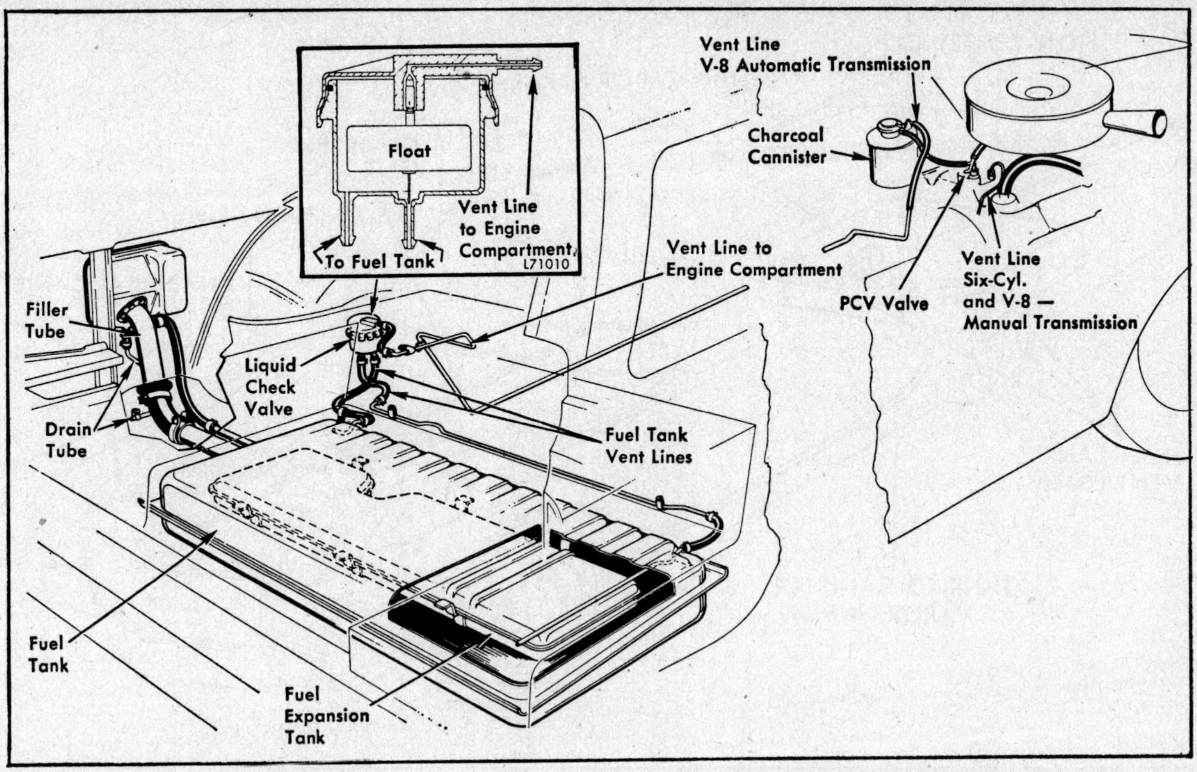

Fig. 1 Evaporative emission control system with charcoal canister. Typical American Motors

Most fuel tanks incorporate an integral fuel expansion tank to provide an air displacement area for normal fuel expansion, unless the tank itself is designed to provide an adequate air displacement area for fuel expansion.

The filler cap includes a two-way relief valve which is closed to atmosphere under normal operating conditions and opens only when an abnormal pressure or vacuum develops within the tank. It is normal to occasionally encounter an air pressure release when removing the filler cap.

GENERAL MOTORS FIG. 2

Fuel vapor is drawn from the tank, through a separator and lines, to a carbon-filled canister located in the engine compartment. Constant purging of vapors from the canister is accomplished through a calibrated orifice in the canister center connection to the PCV hose and/or the air cleaner snorkel.

General Motors fuel tanks include a fill limiting baffle which allows space for normal fuel expansion. To prevent damage to the tank from excessive internal or external pressure resulting from the closed system, a filler cap is used, which includes a two-way relief valve which opens only when an abnormal pressure or vacuum develops within the tank.

A liquid vapor separator is provided to prevent liquid fuel from entering the system. Liquid fuel entering the separator is spilled back into the tank while raw fuel vapor is passed into the lines.

The carbon canister provides a storage place for the raw fuel vapor when the engine is not running. When the engine is running, vapors are drawn from the canister into the engine.

Maintenance:

The only service required is that the filter mounted at the bottom of the canister be replaced at recommended intervals.

NOTE: The purge valve (used on some applications), can be repaired without replacing complete vapor canister, using repair kit, part No. 7041344. The service is NOT a routine maintenance item and should be performed if damage or parts are missing.

CHRYSLER CORP. FIG. 3

1969-71 System

This is a closed system venting fuel vapors through lines to the engine crankcase by way of the crankcase inlet air cleaner. With the engine running, the vapors are purged from the crankcase with the normal crankcase vapor.

The fuel tank incorporates an expansion tank to provide space for air displacement during normal fuel expansion. The filler cap is a special unit which incorporates a relief valve.

A liquid vapor separator is included between the tank and the vapor line to insure against liquid fuel entering the system. Liquid fuel is spilled back into the tank while raw fuel vapor is admitted into the vapor line.

FIGS. 4, 4A & 4B

1972-75 System

In this system when the fuel tank is filled to the base of the filler tube, vapors can no longer escape, they become trapped above the fuel. Vapor flow through the vent line is blocked by the limiting valve; and the filler tube is blocked by fuel preventing more fuel to enter the tank. At any time pressures in the tank rise above operating pressures of the limiting valve, about ½ psi the valve opens and allows vapors to flow forward to the charcoal canister. Due to the configuration of the fuel tank on some models and all station wagons, vapor separator tanks are not required. The charcoal canister is a feature on all models for the storage of fuel vapors from the fuel tank and carburetor bowl. A vacuum port located in the base of the carburetor governs vapor flow to the engine. On some models, each corner of the fuel tank is vented and each of the hoses from these vents is connected to a vapor separator. A tube from the separator leads to the charcoal canister. Evaporated fuel vapor from the fuel

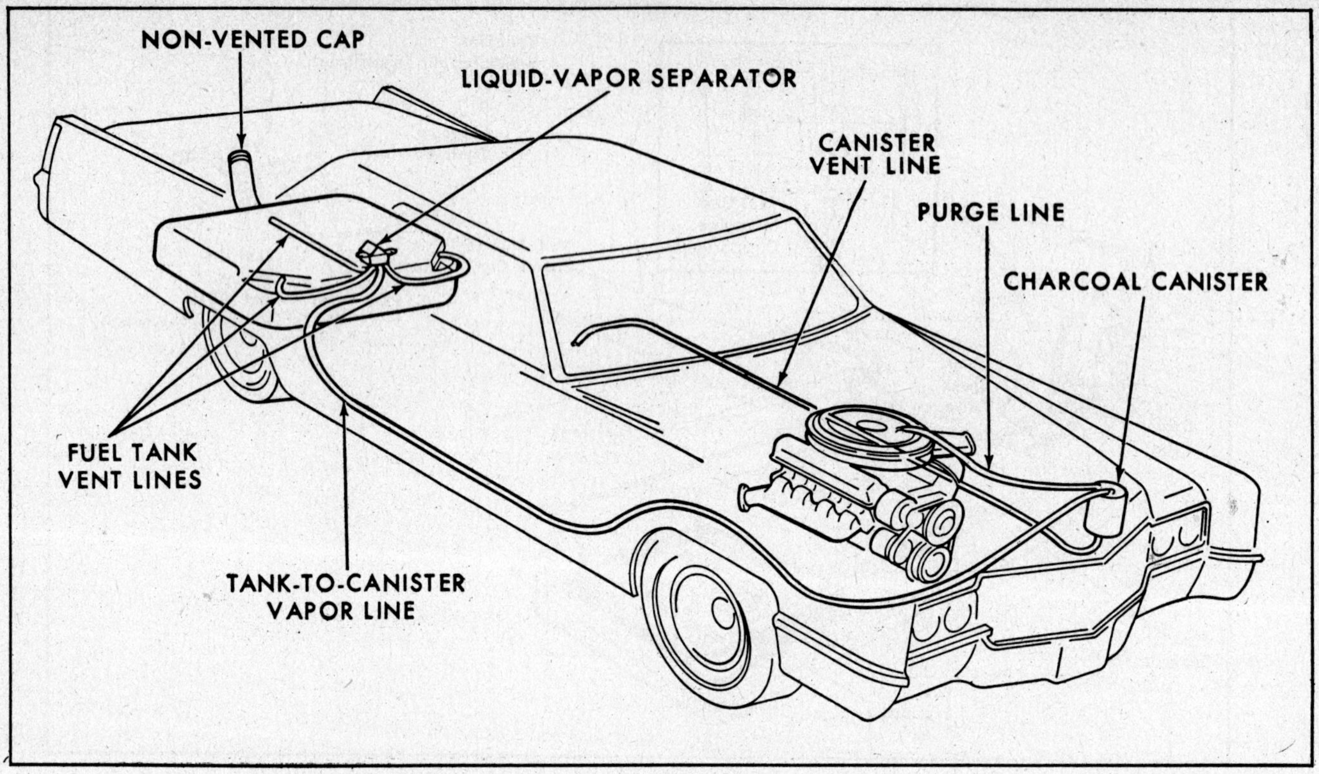

Fig. 2 Evaporative emission control system. General Motors (typical)

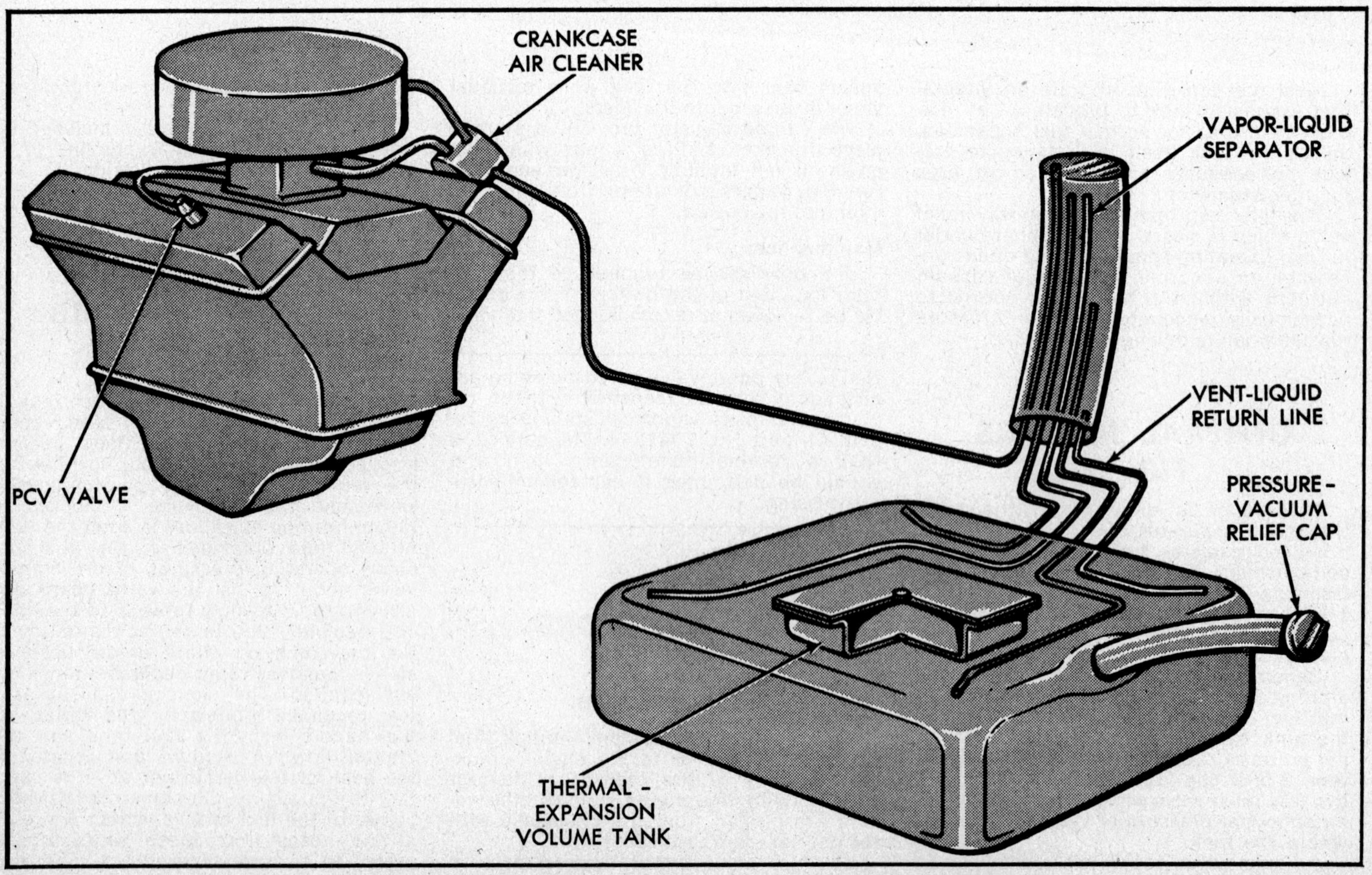

Fig. 3 Evaporative emission control system. Chrysler Corp., 1969-71

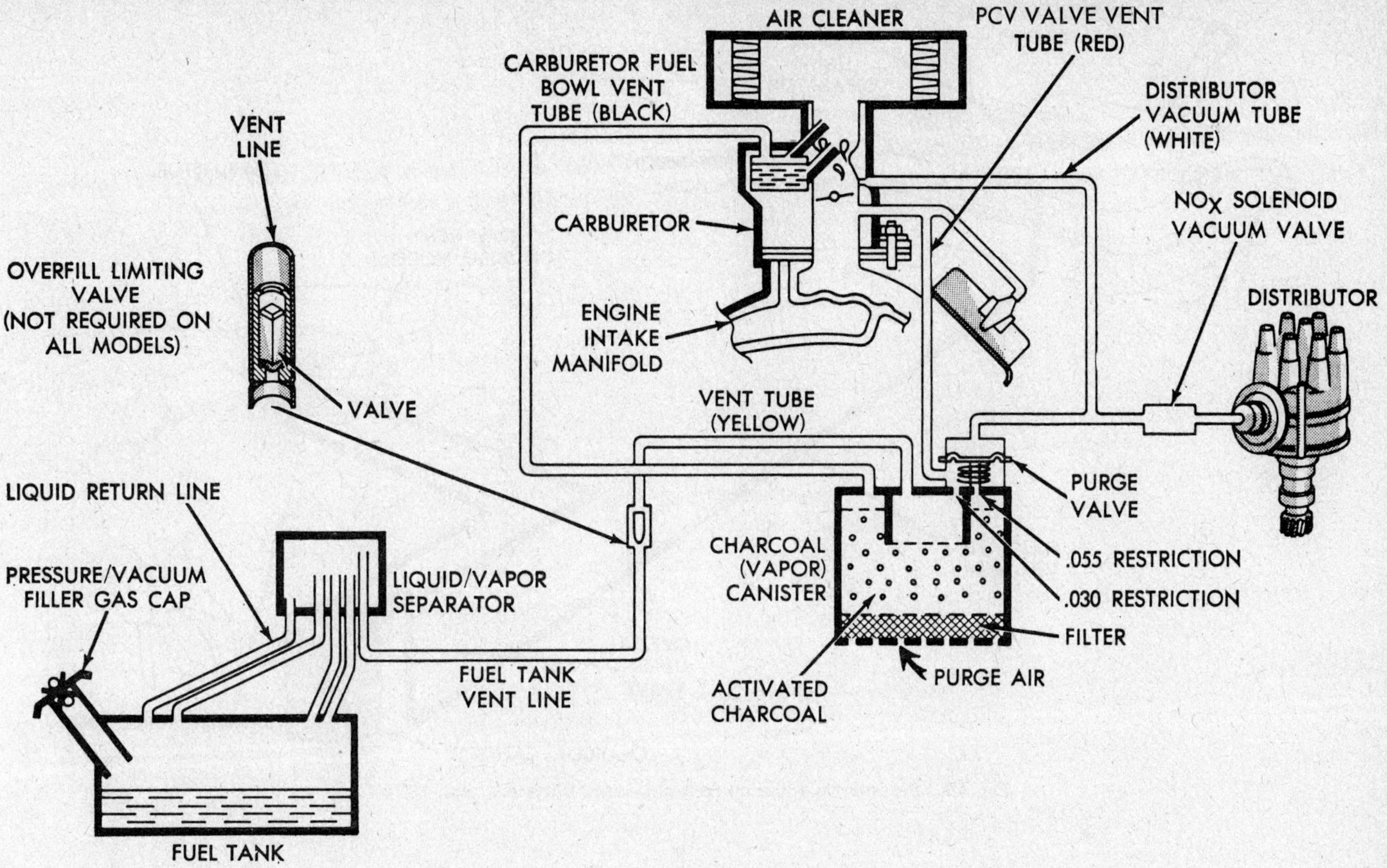

Fig. 4 Evaporative emission control system. Chrysler Corp., 1972

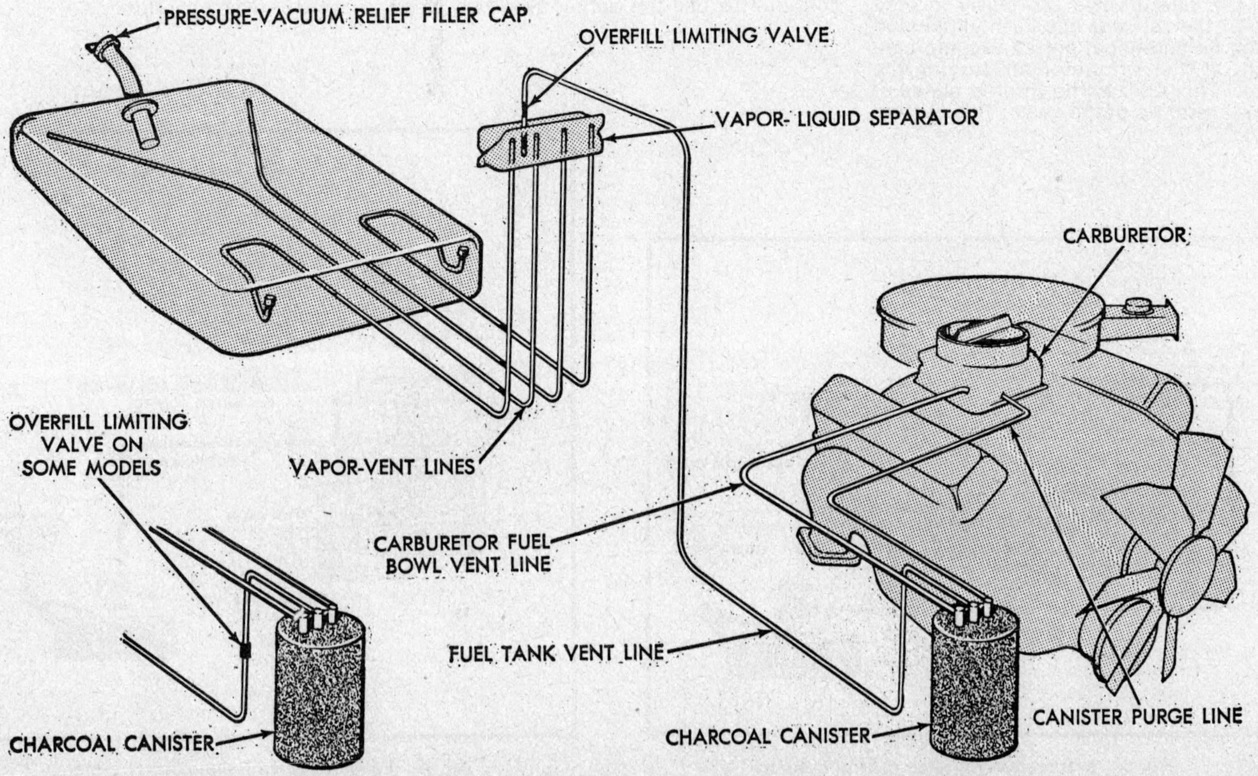

Fig. 4A Evaporative emission control system. Chrysler Corp., 1973

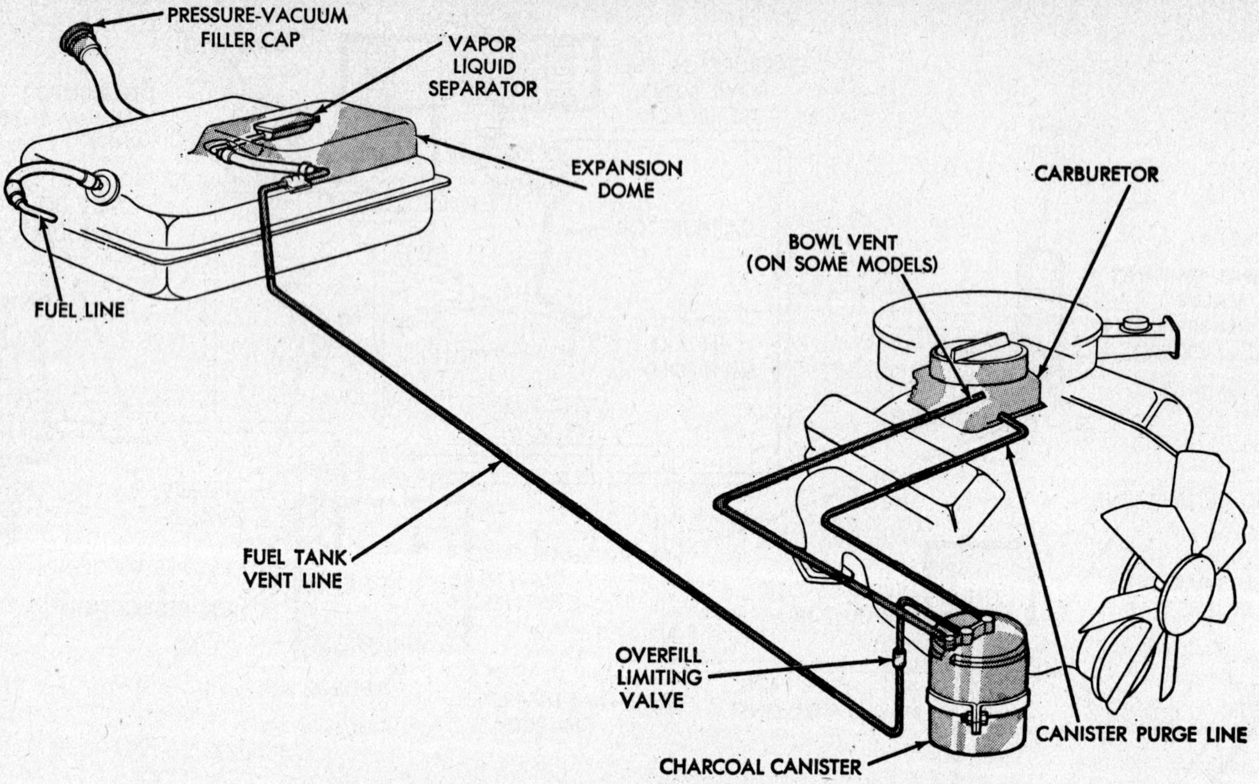

Fig. 4B Evaporative emission control system. Chrysler Corp., 1974-75

tank, flows through the separator to the canister. The canister used in 1973-75 vehicles will have three hoses and no purge valve. The purge valve previously located on top of the canister has been eliminated by using an additional ported vacuum connection on the carburetor for purging the canister. This utilizes the throttle plates of the carburetor as purge valve. This system will improve hot idle quality by eliminating canister purging during idle. Some limited production, High Performance vehicles will continue to use the earlier type two stage canister which utilizes an integral purge valve. This canister can be identified by four hose connections while the new type canister uses only three.

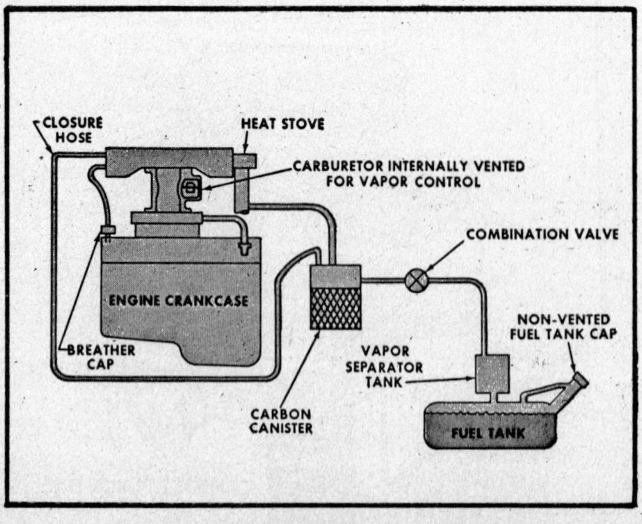

Fig. 5 Evaporative emission control system. Ford carbon storage type. 1970

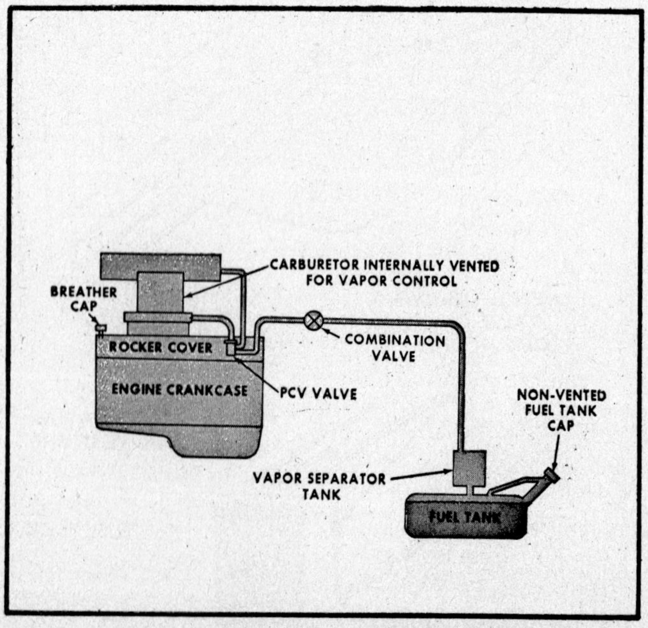

Fig. 6 Evaporative emission control system. Ford crankcase storage type. 1970

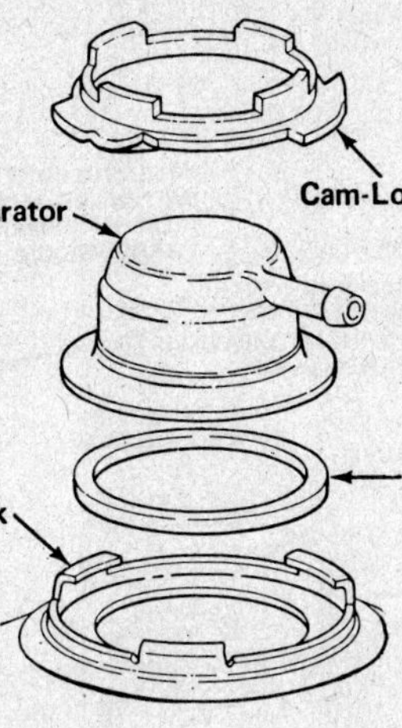

Fig. 7 Tank mounted vapor separator

Maintenance

The only service normally required for the system, is to replace the filter mounted in the bottom of the canister at 12,000 mile intervals.

Overfill Limiting Valve

The overfill limiting valve, located in the engine compartment on some models is not serviceable, in the event that replacement is required, cut the old valve out using a tubing cutter. Flare the end of the existing tube to insure a good vapor seal and install the replacement valve in the same position as the old valve. If overfill limiting valve is part of the vapor separator, the entire vapor separator must be replaced if the overfill limiting valve requires replacement.

NOTE: It is important that all overfill limiting valves be installed as vertical as possible, in order to function properly.

Pressure-Vacuum Filler Cap

The fuel tank is sealed with a special engineered pressure vacuum relief filler cap. The relief valves in the cap are a safety feature, and operate only to prevent excessive pressure or vacuum in the tank caused by malfunction in the system or damage to the vent lines.

FORD MOTOR CO. FIGS. 5 & 6

1970

In this system, fuel vapor is transported either to a carbon canister or the engine crankcase, depending on the engine application. It is then drawn into the engine, from the carbon canister to the air cleaner, or from the crankcase through the PCV system.

The fuel tank is designed to limit the fill capacity to provide space for normal fuel expansion. A separate tank is located above the fuel tank to separate vapor from liquid and prevent any liquid from entering the system.

A combination valve on the forward side of the fuel tank isolates the fuel tank from engine pressures and allows vapor to escape from the separator to the carbon canister. It also relieves excessive fuel tank pressure and allows fresh air to be drawn into the tank as fuel is used.

1971-75

The operation of the system is similar to the system used in 1970, but it has been simplified by the following modifications:

Fill Control Vent System

The fill control vent system which provides positive control of fuel height during fill operations is made possible by the design of the filler pipe and by vent lines within the filler neck or fuel tank. This system is designed so that about 10% of tank capacity will remain empty when the tank is filled. This space allows for thermal fuel expansion and temporary storage of fuel vapors.

Pressure and Vacuum Relief Valve

The pressure and vacuum system operates through the use of a sealed fill cap with a built-in pressure and vacuum relief valve. Under normal operating conditions, the valve opens to relieve pressure when it exceeds $\frac{3}{4}$ to $1\frac{1}{4}$ psi. When fuel tank vacuum reaches $\frac{1}{2}$ inch mercury maximum, the valve opens allowing air to enter the system.

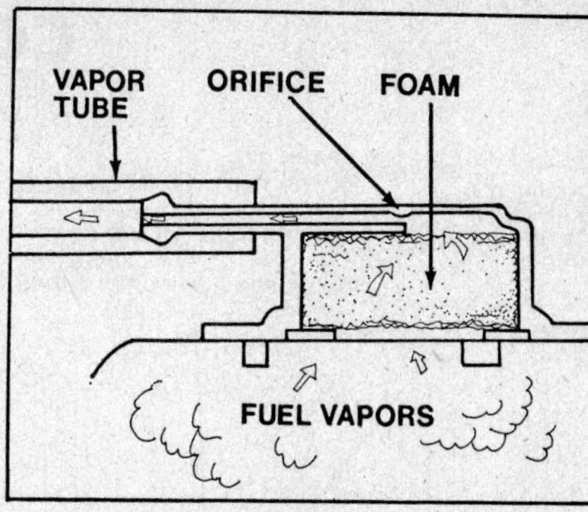

Fig. 8 Vapor separator crossection

Vapor Vent and Storage System

This system, on vertically mounted fuel tanks consists of a vapor separator, Fig. 7, mounted on the uppermost surface of the tank. The empty space at the top of the tank provides adequate breathing space for the vapor separator. Horizontally mounted fuel tanks use a raised mounting section for the vapor separator. This raised section provides additional breathing space for the vapor separator since the space allowed for thermal expansion of fuel is not as deep as it is on vertically mounted tanks.

The vapor separator which acts as a baffle to prevent fuel from entering the charcoal canister, Fig. 8, consists of a small hole in the outlet connected to the vapor tube plus open cell foam to separate liquid fuel and fuel vapors. The fuel vapors in the tank go through the opening in the vapor separator and into the vapor tube.

Fuel Vapor Return System

A fuel vapor return system is used on some engines to reduce the amount of fuel vapor entering the carburetor. It consists of a fuel vapor separator installed in the fuel supply line between the pump and carburetor and a one piece vapor return line from the separator to the fuel tank. Fuel vapors are collected in the separator and routed to the fuel tank where they recondense or are contained by the evaporative emission control system.

COMPONENTS LOCATION

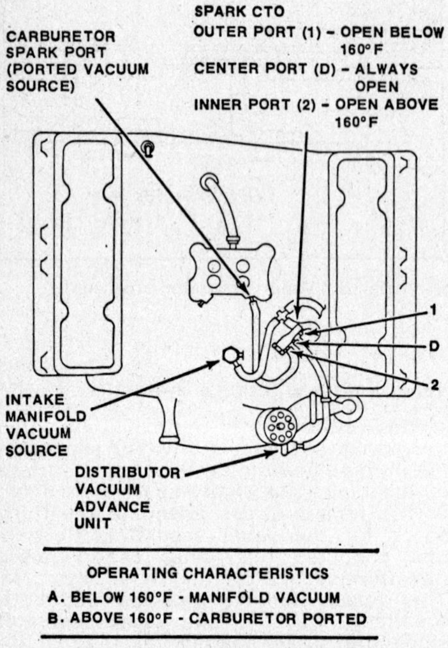

Fig. CL1 Spark CTO system vacuum hose routing. 1974 V8-401 and all 1975 V8 exc. Calif. American Motors

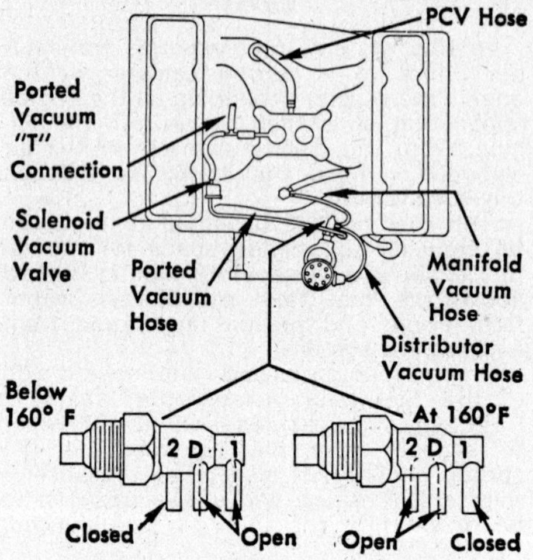

Fig. CL4 CTO system vacuum hose routing. 1972 American Motors

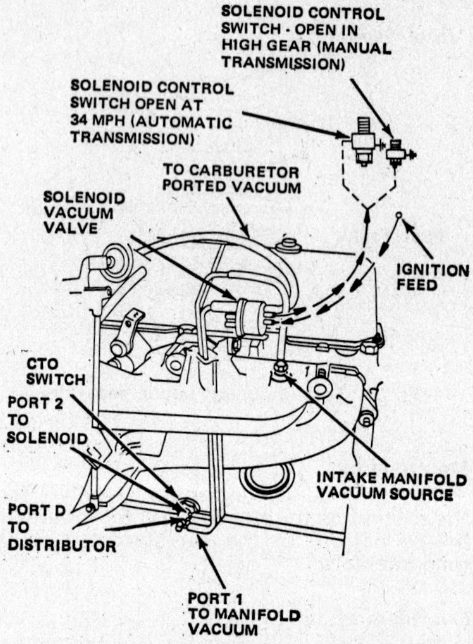

Fig. CL5 TCS system vacuum and wiring. 1972-74 American Motors 6-232

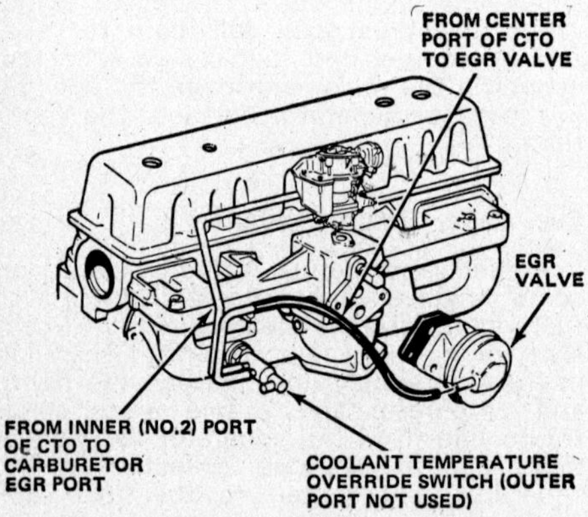

Fig. CL2 EGR system vacuum hose routing. 1974-75 American Motors six cylinder

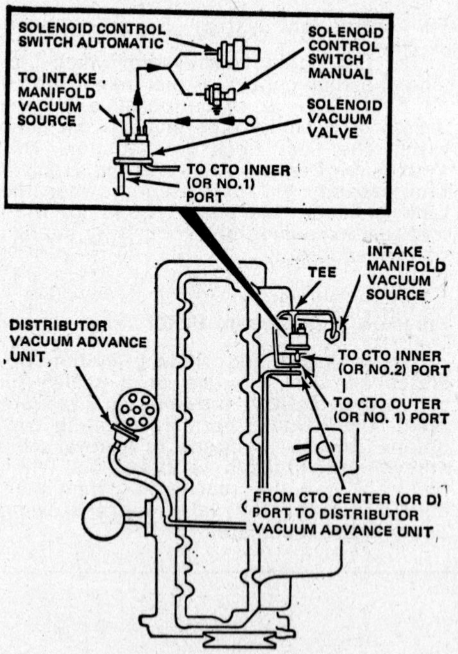

Fig. CL6 TCS system vacuum and wiring. 1972-74 American Motors Matador with 6-258

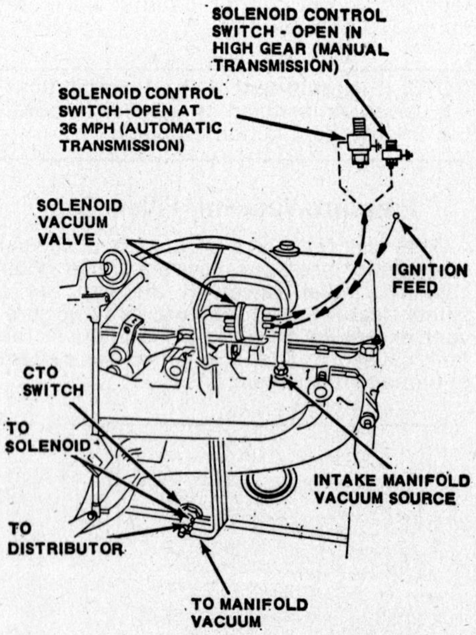

Fig. CL7 TCS vacuum and wiring. 1975 American Motors 6 cylinder

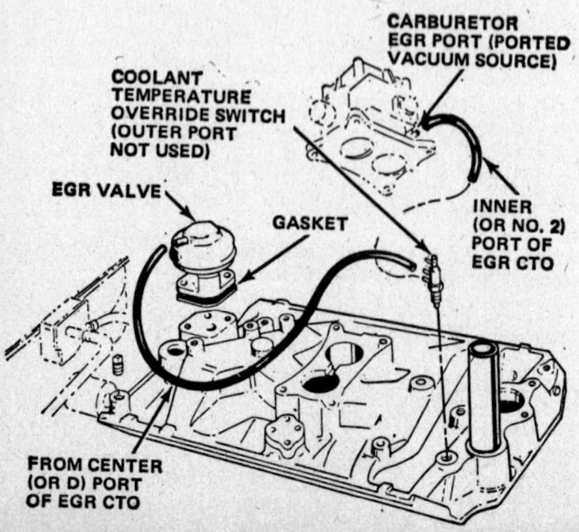

Fig. CL3 EGR system vacuum hose routing. 1974-75 American Motors V8

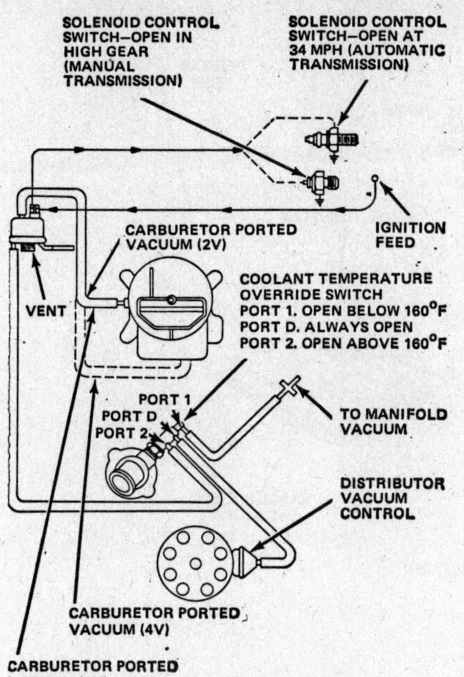

Fig. CL8 TCS system vacuum and wiring. American Motors V8-390 & 401

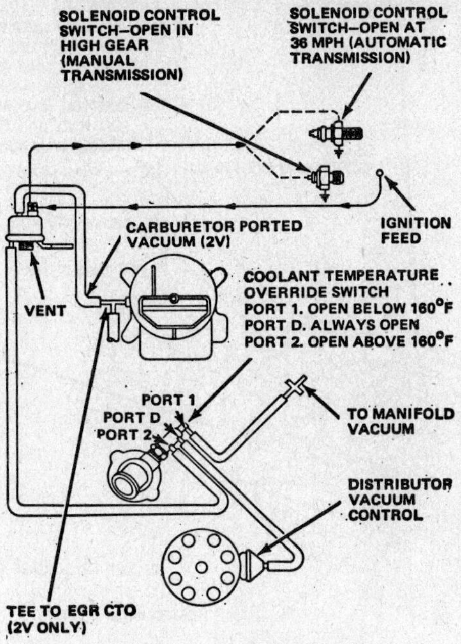

Fig. CL9 TCS vacuum and wiring. 1975 American Motors V8

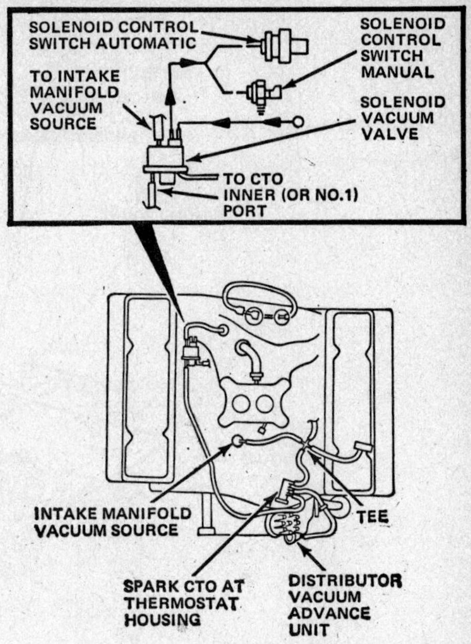

Fig. CL10 TCS system vacuum and wiring. American Motors V8-304 & 360

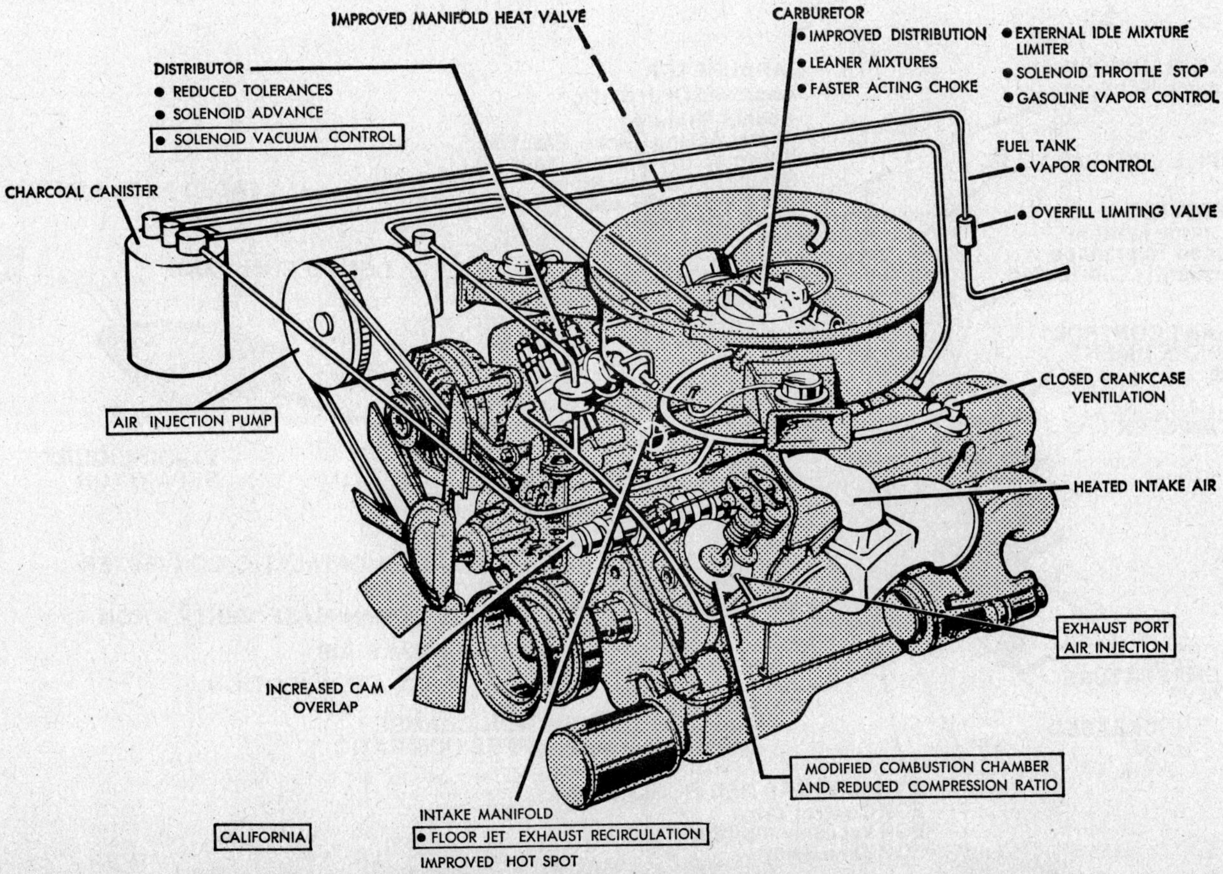

Fig. CL11 Chrysler Emission Control System. 1972

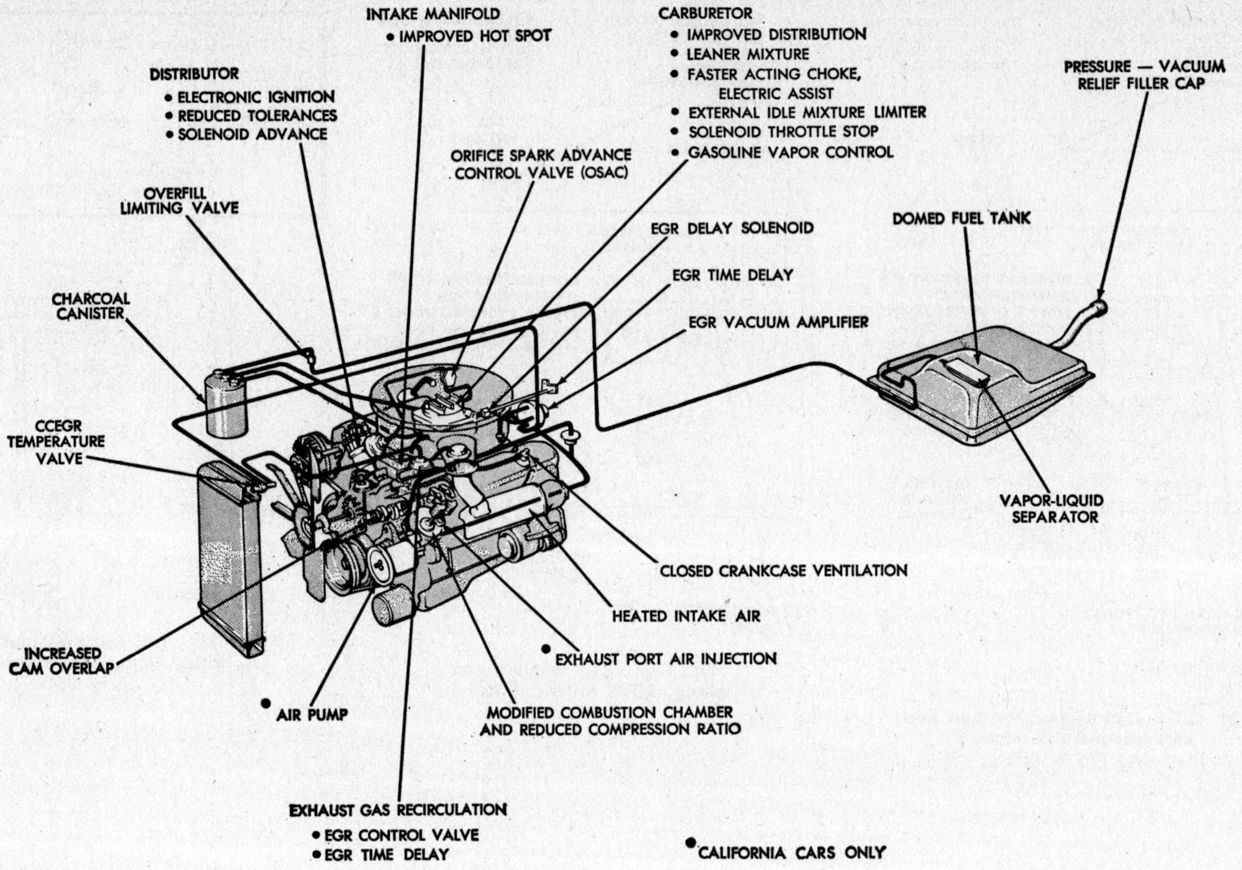

Fig. CL12 Chrysler Emission Control System. 1973-74

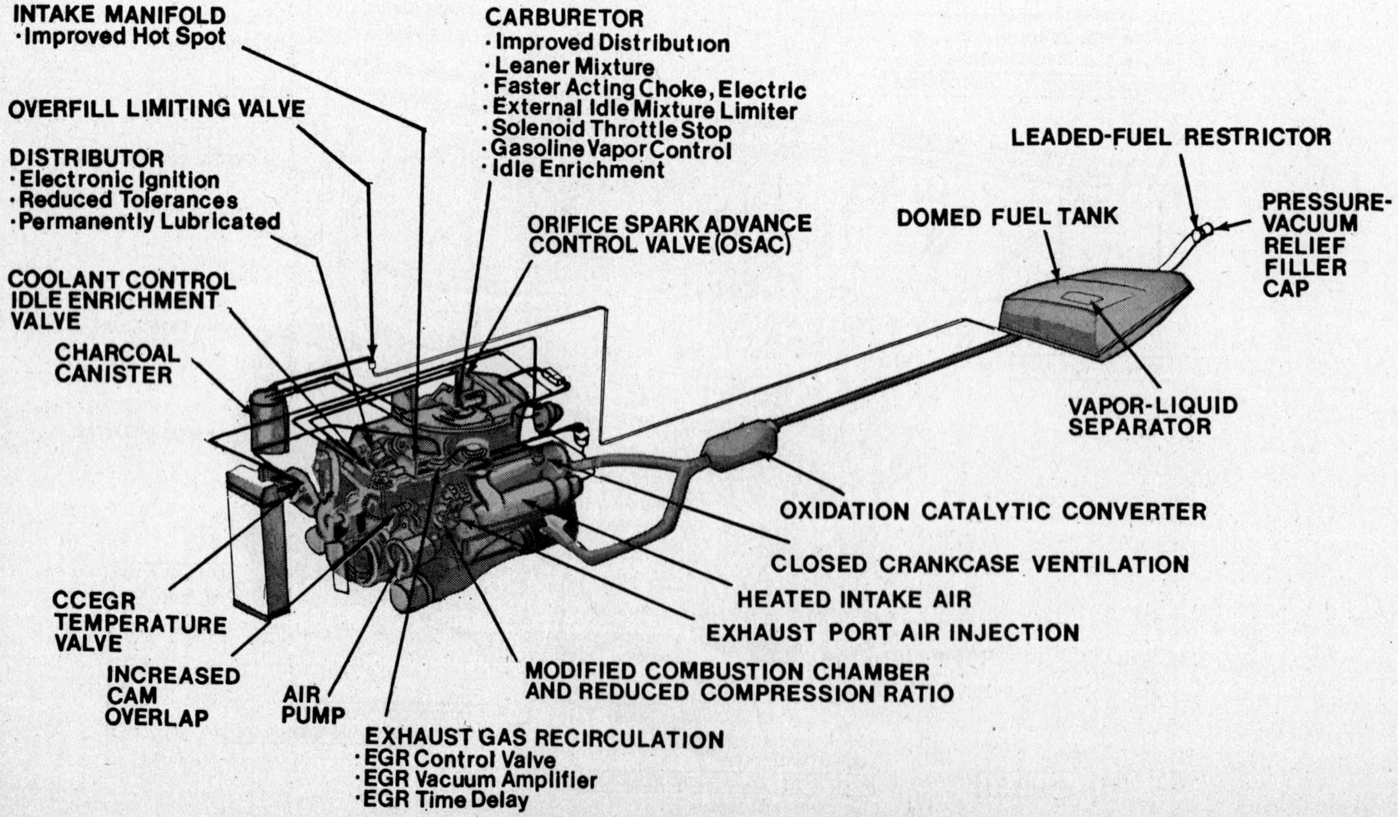

Fig. CL13 Chrysler Emission Control System. 1975

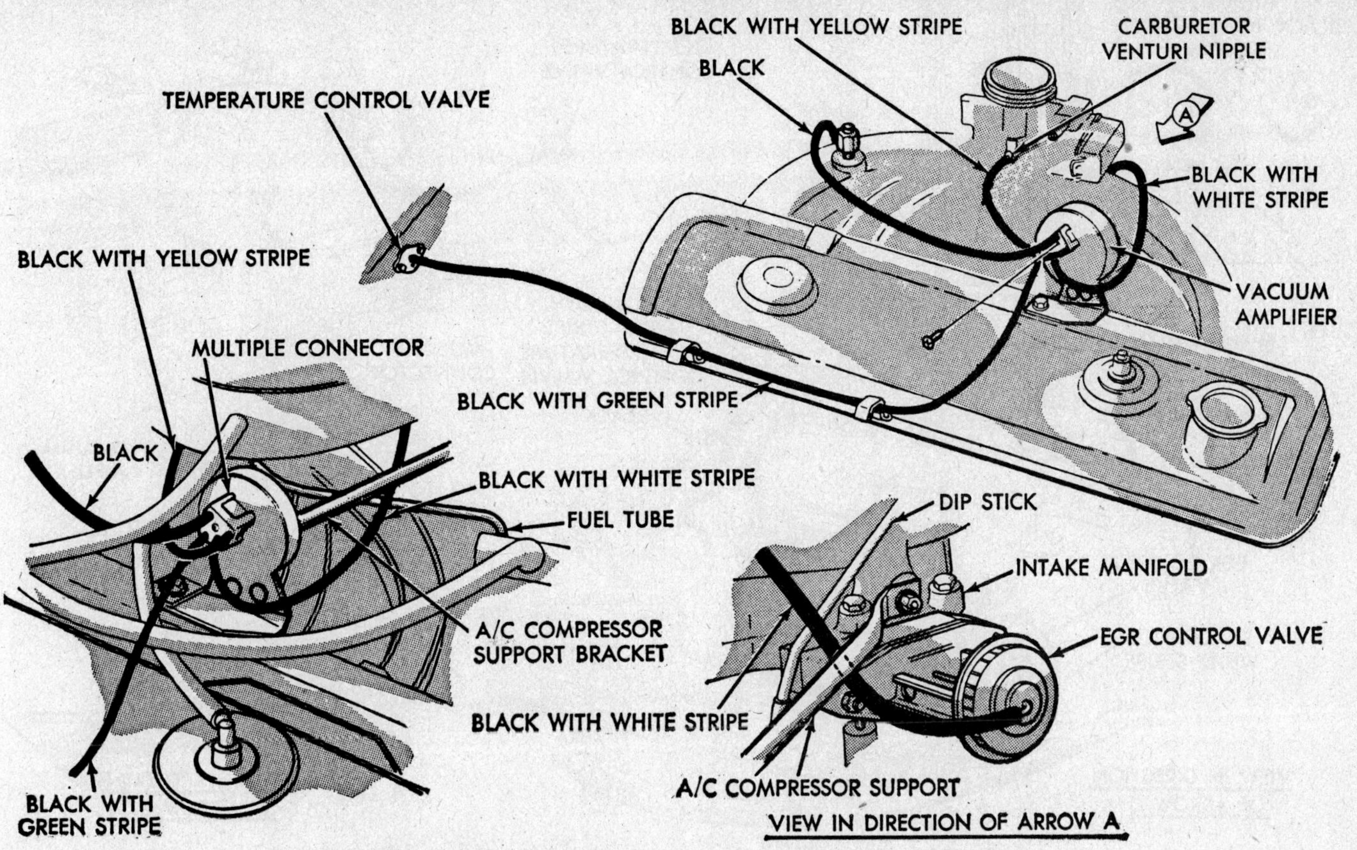

Fig. CL14 EGR system vacuum hose routing. 1973 Chrysler 6-198, 225

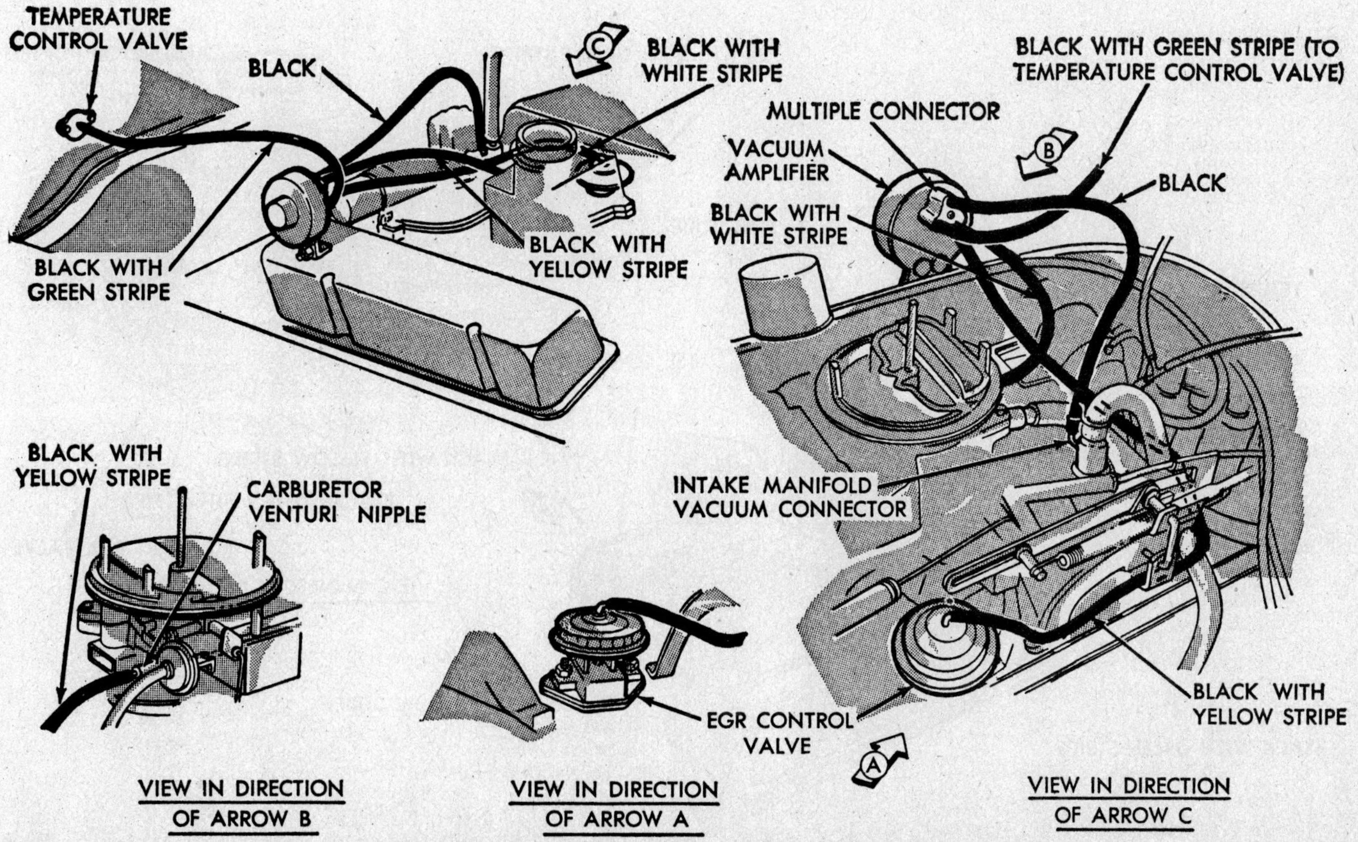

Fig. CL15 EGR system vacuum hose routing. 1973 Chrysler V8-360

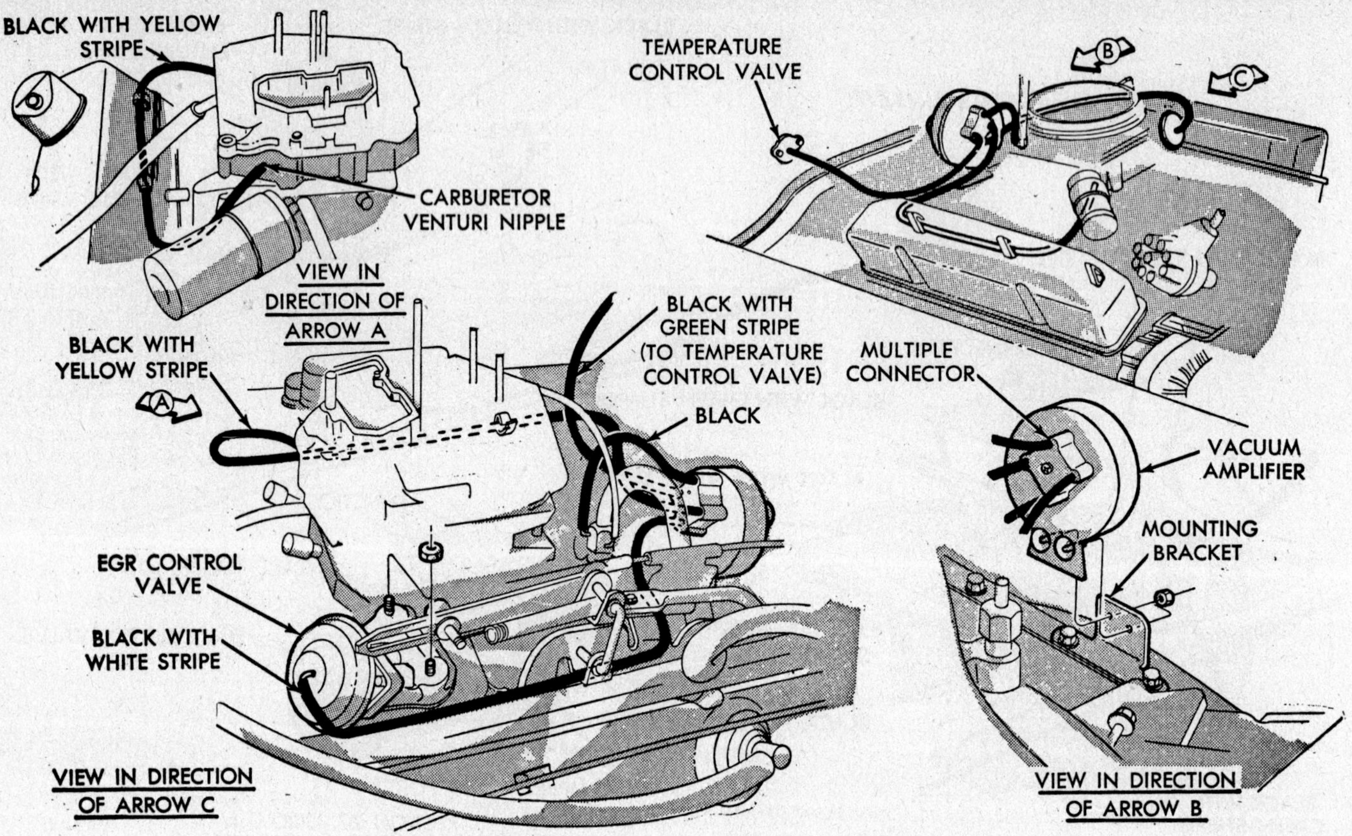

Fig. CL16 EGR system vacuum hose routing. 1973 Chrysler V8-440

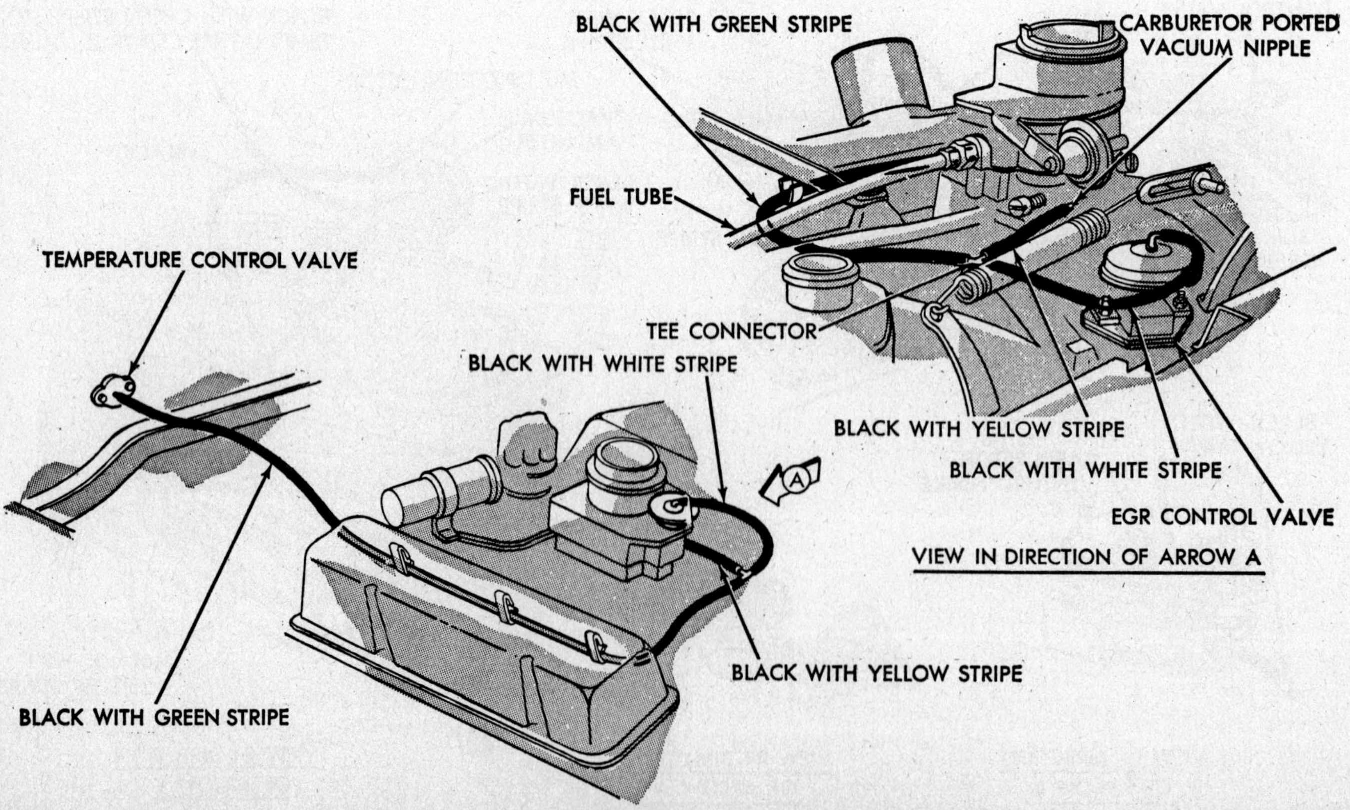

Fig. CL17 EGR system vacuum hose routing. 1973 Chrysler V8-318

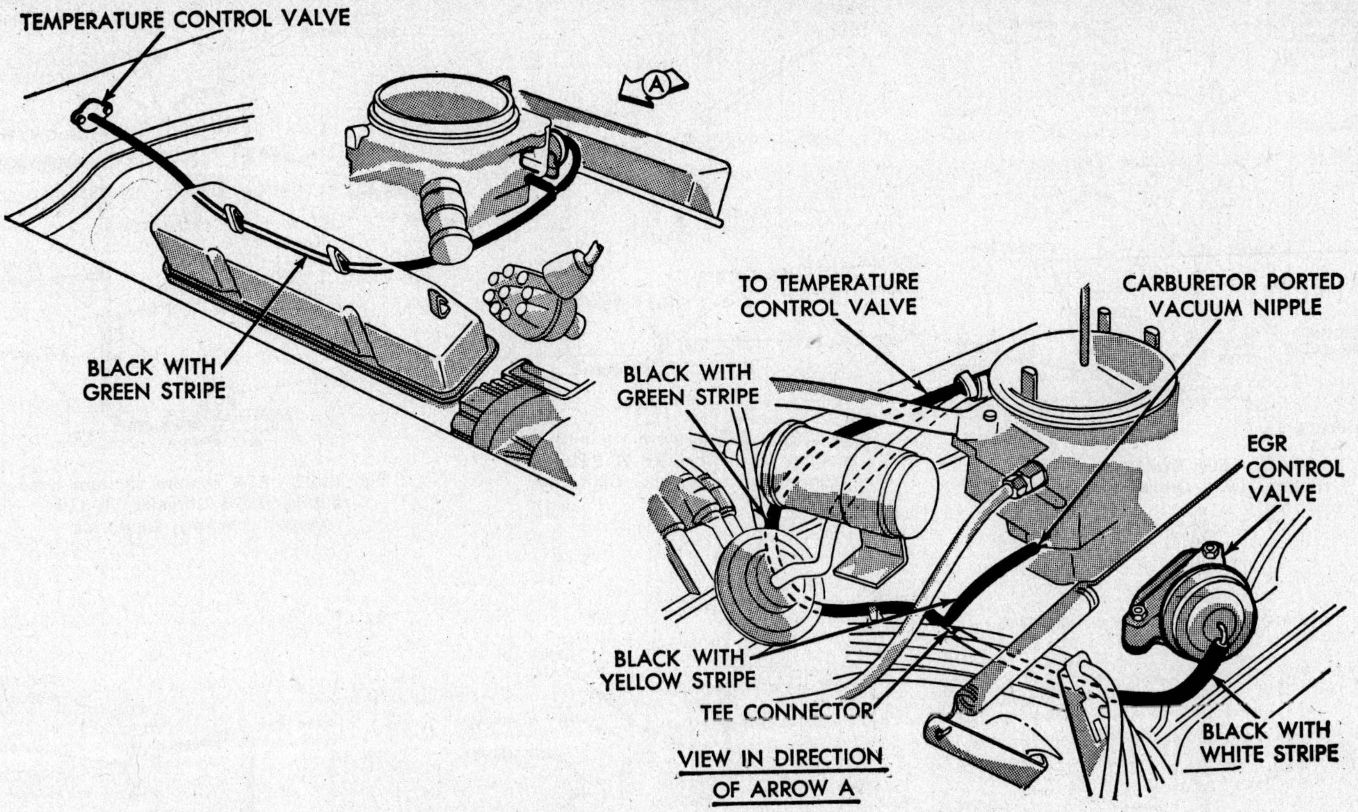

Fig. CL18 EGR system vacuum hose routing. 1973 Chrysler V8-400

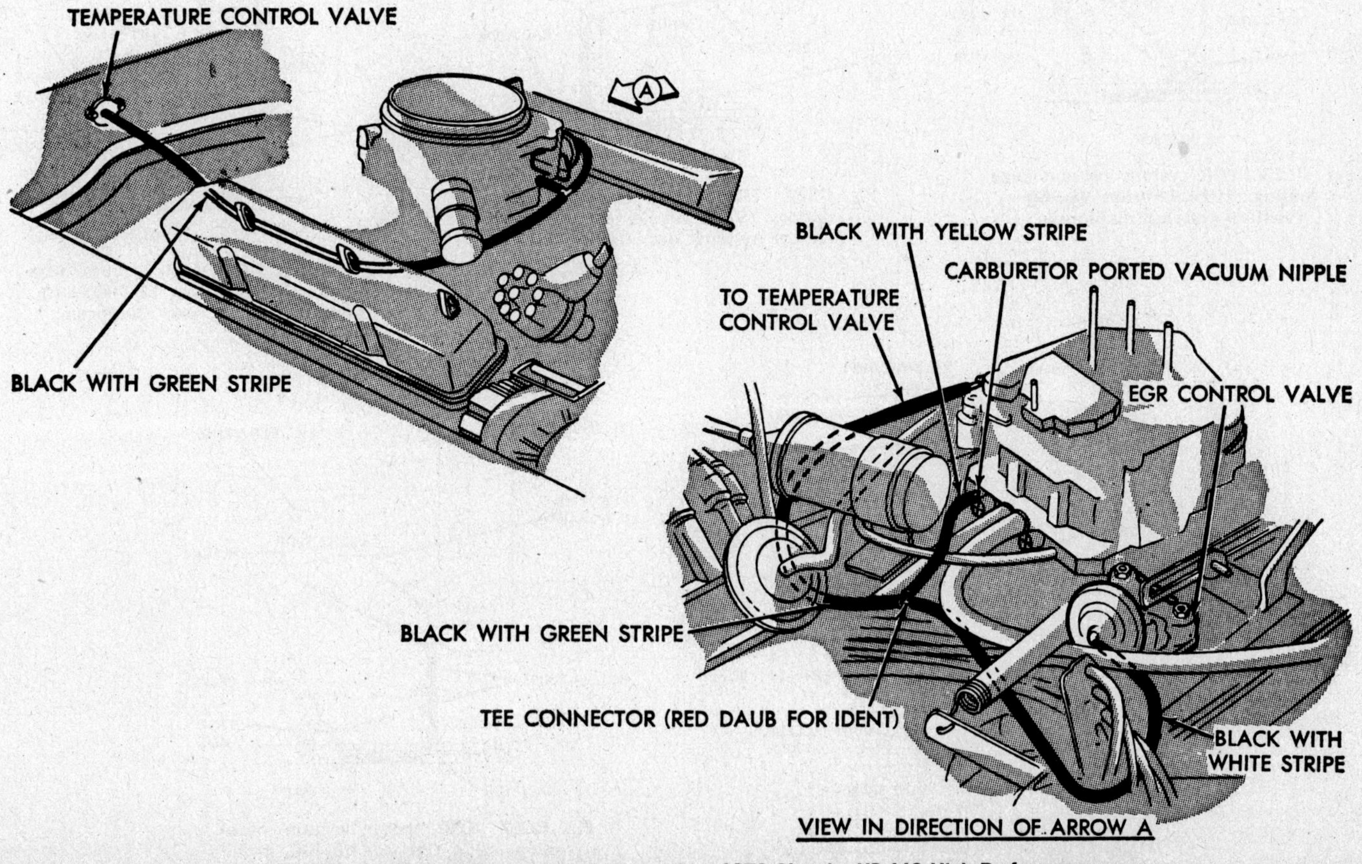

Fig. CL19 EGR system vacuum hose routing. 1973 Chrysler V8-440 High Performance

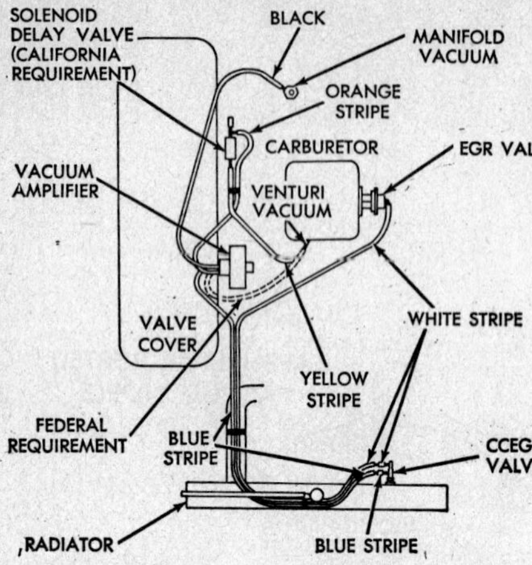

Fig. CL20 EGR system vacuum hose routing. 1974 Chrysler 6-198, 225

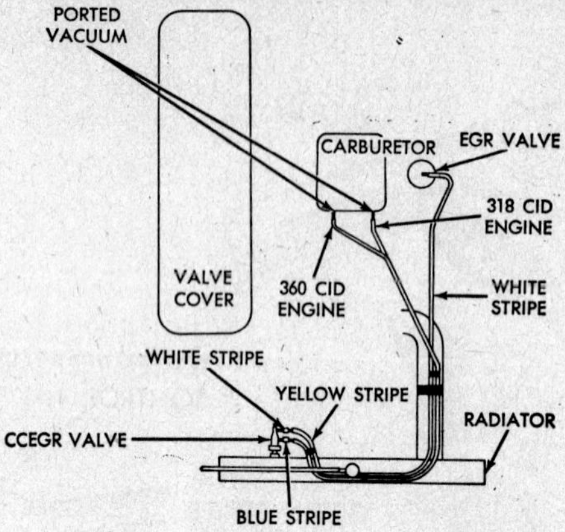

Fig. CL21 EGR system vacuum hose routing. 1974 Chrysler V8-318, 360 (ported vacuum) exc. California

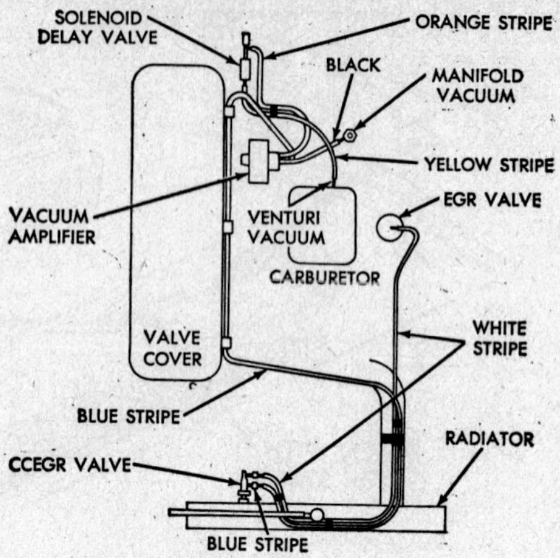

Fig. CL22 EGR system vacuum hose routing. 1974 Chrysler V8-318 (venturi system), California

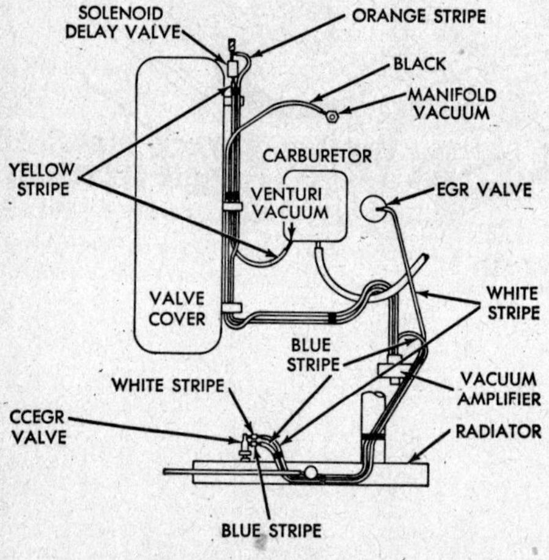

Fig. CL23 EGR system vacuum hose routing. 1974 Chrysler V8-360 (venturi system), California

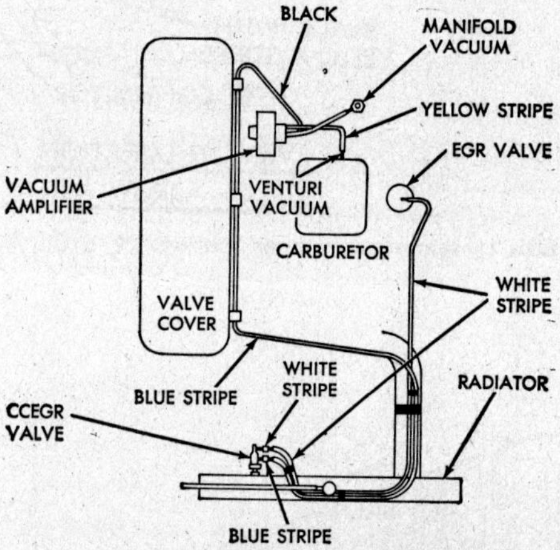

Fig. CL24 EGR system vacuum hose routing. 1974 Chrysler V8-360 (venturi system) exc. California

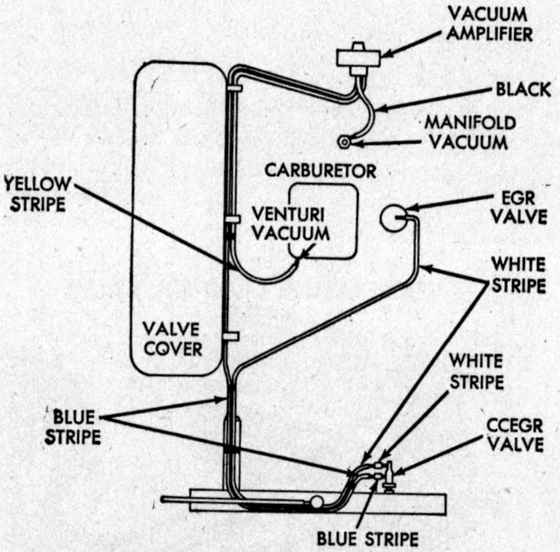

Fig. CL25 EGR system vacuum hose routing. 1974 Chrysler V8-400, 440 (venturi system) exc. California

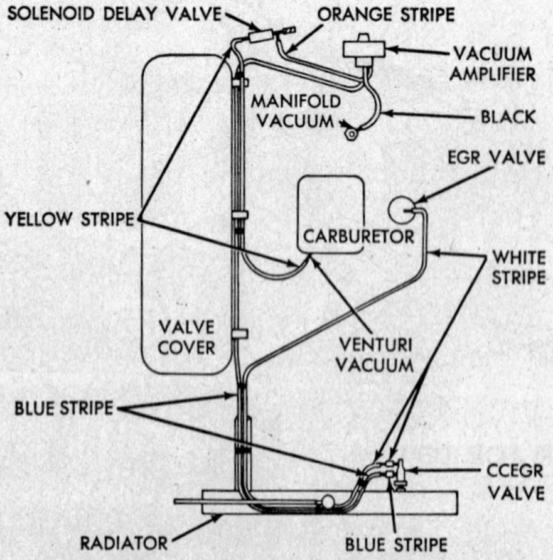

Fig. CL26 EGR system vacuum hose routing. 1974 Chrysler V8-400, 440 (venturi system), California

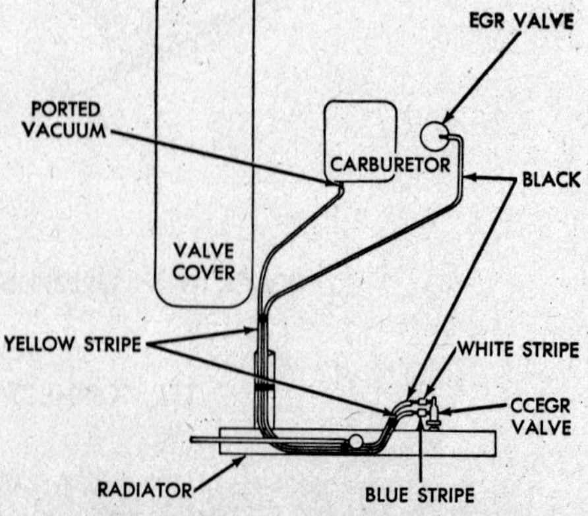

Fig. CL27 EGR system vacuum hose routing. 1974 Chrysler V8-400, 440 (ported system), exc. California

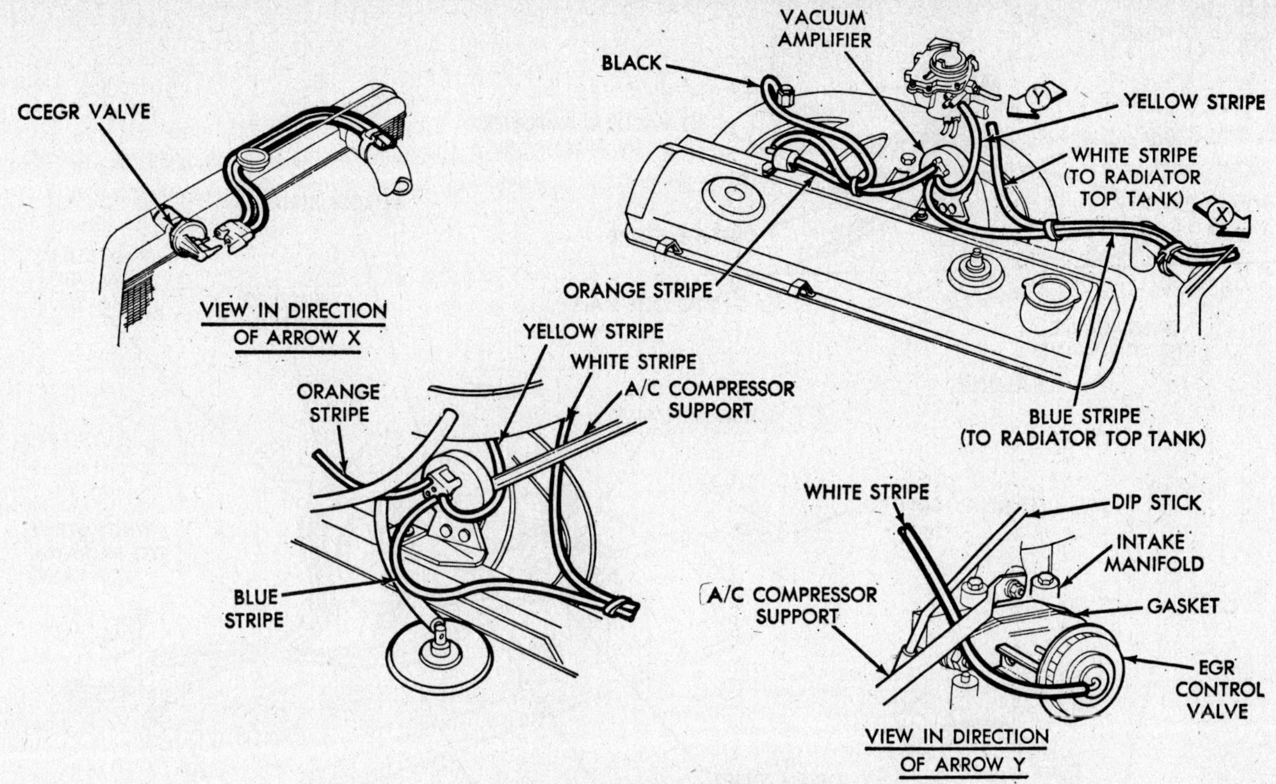

Fig. CL28 ERG system vacuum hose routing. 1975 Chrysler Corp. 6 cylinder with manual transmission

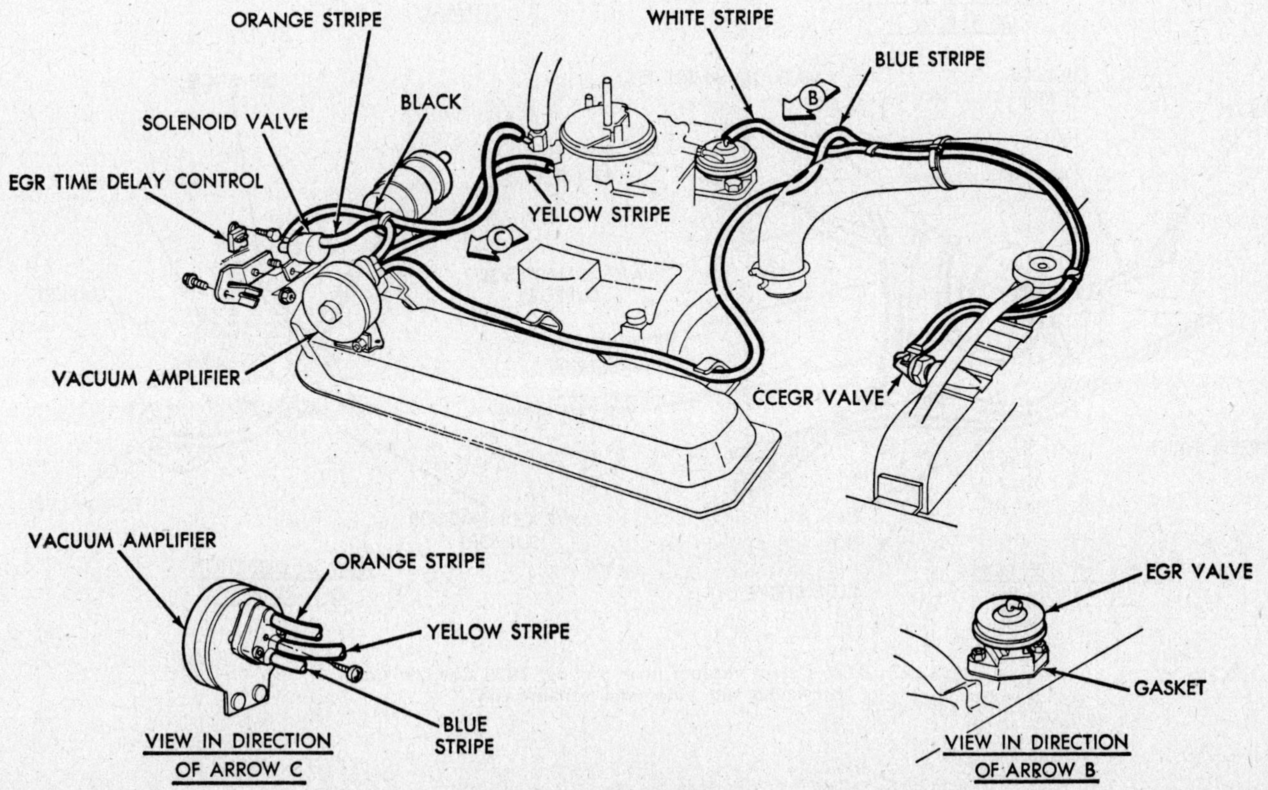

Fig. CL30 EGR system vacuum hose routing. 1975 Chrysler Corp. exc. Calif. V8-318 with manual transmission

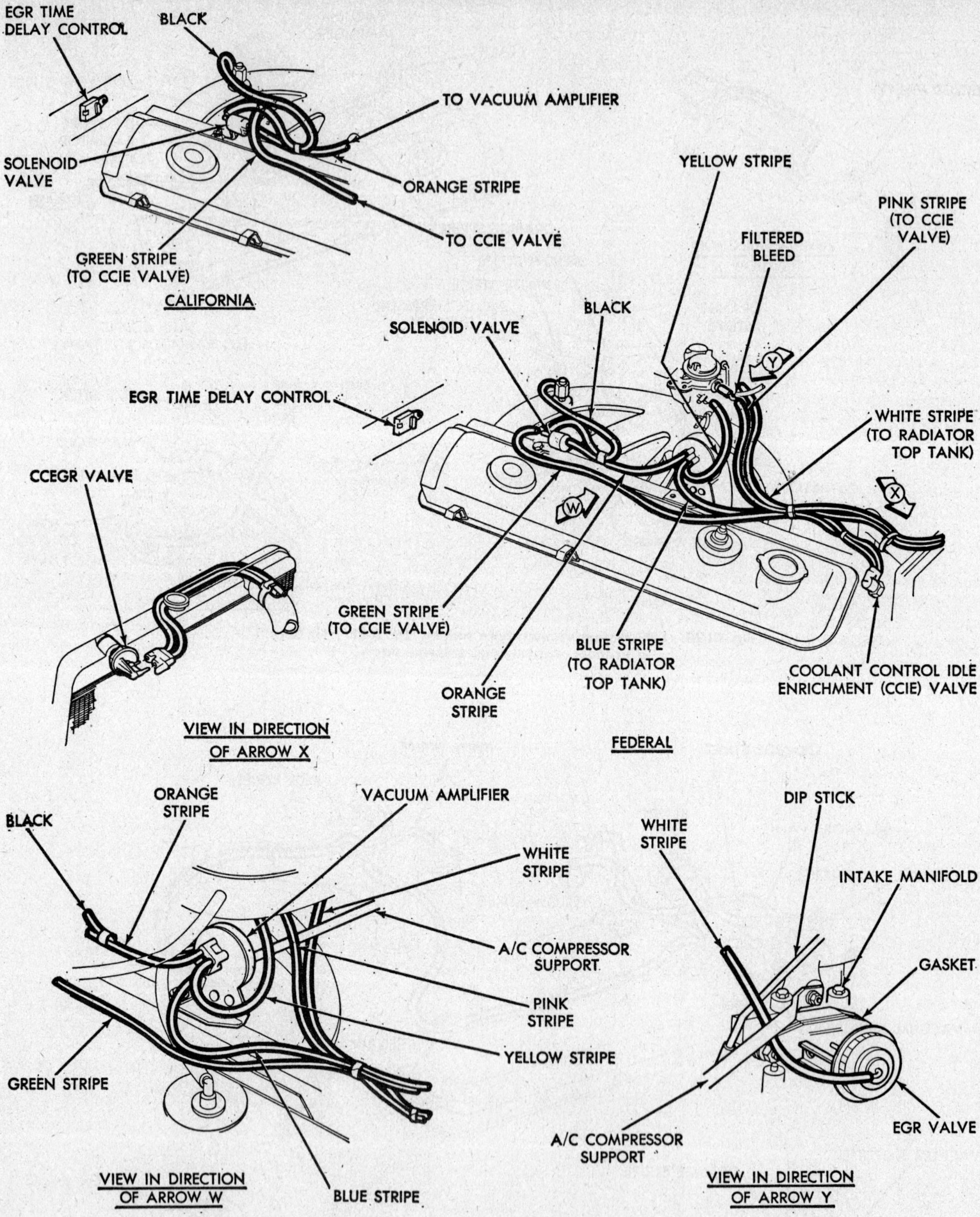

Fig. CL29 EGR system vacuum hose routing. 1975 Chrysler Corp. 6 cylinder with automatic transmission

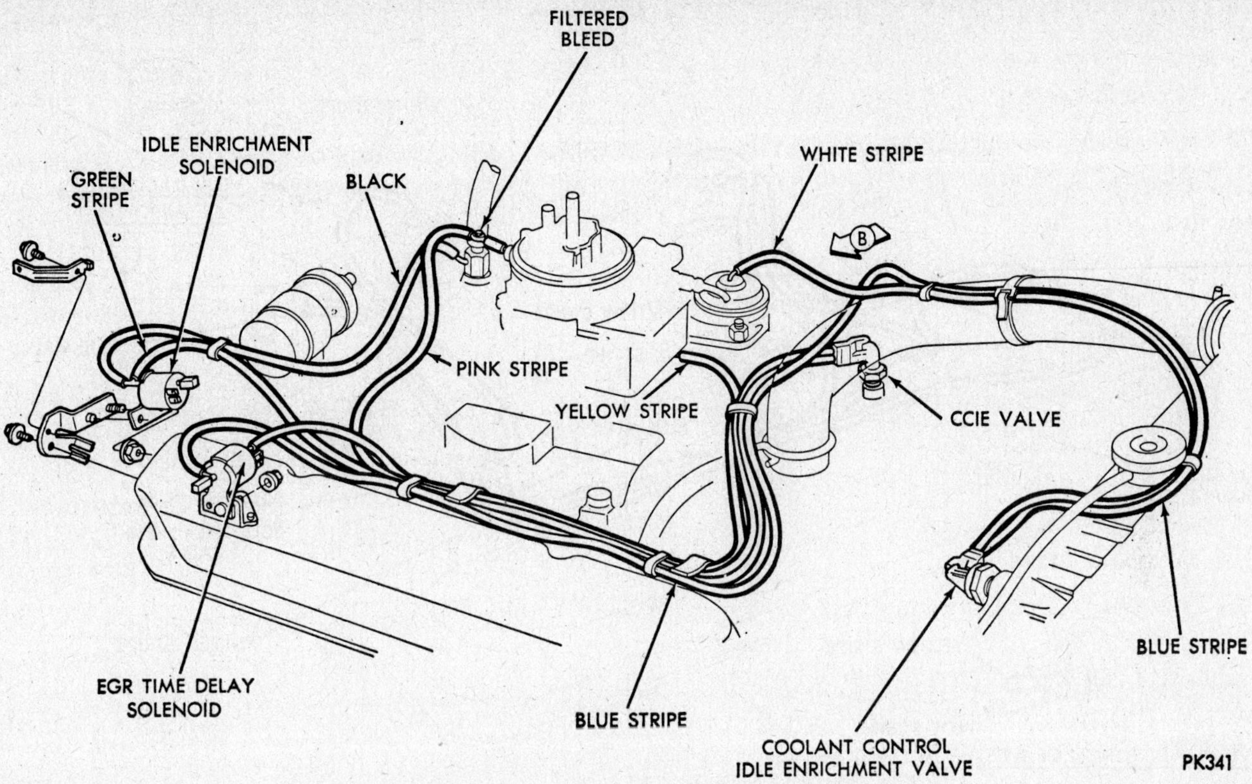

Fig. CL31 EGR system vacuum hose routing. 1975 Chrysler Corp. exc. Calif. V8-318 with automatic transmission

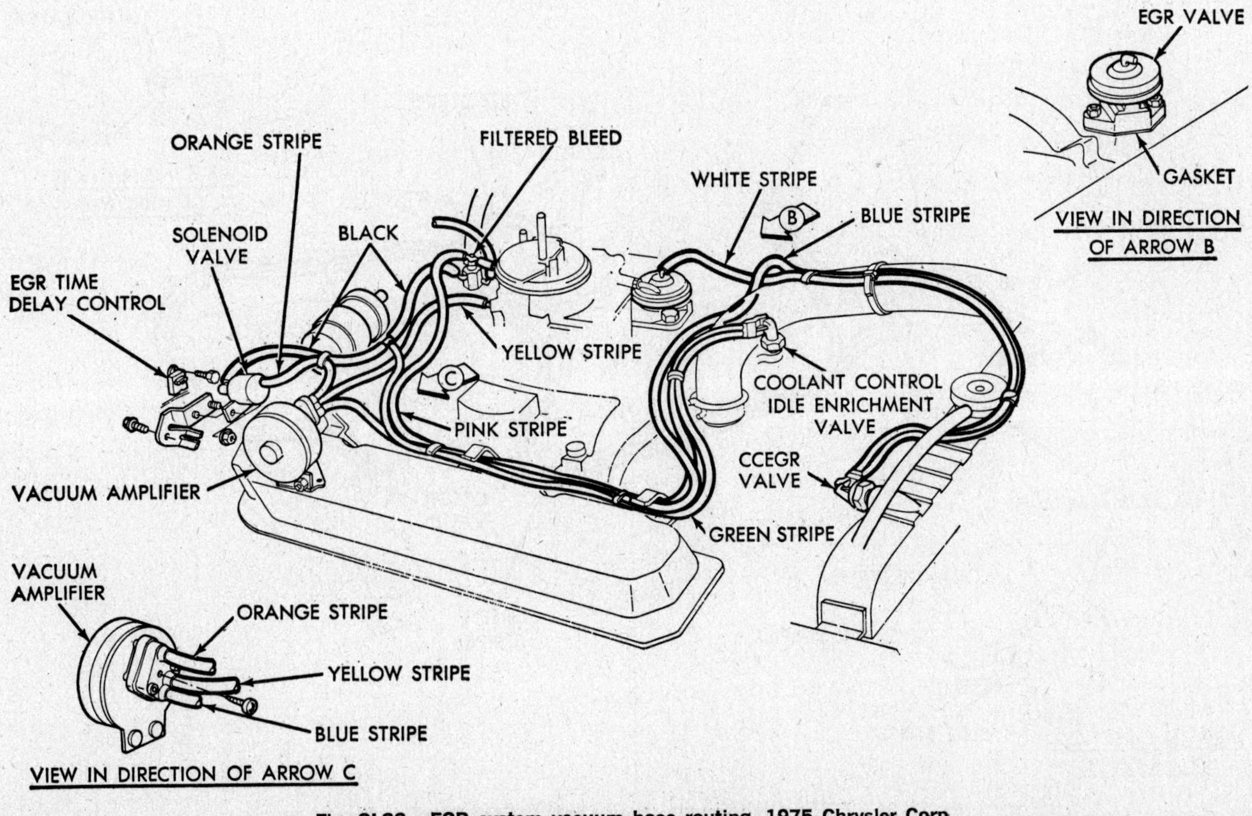

Fig. CL32 EGR system vacuum hose routing. 1975 Chrysler Corp. Calif. V8-318 with automatic transmission

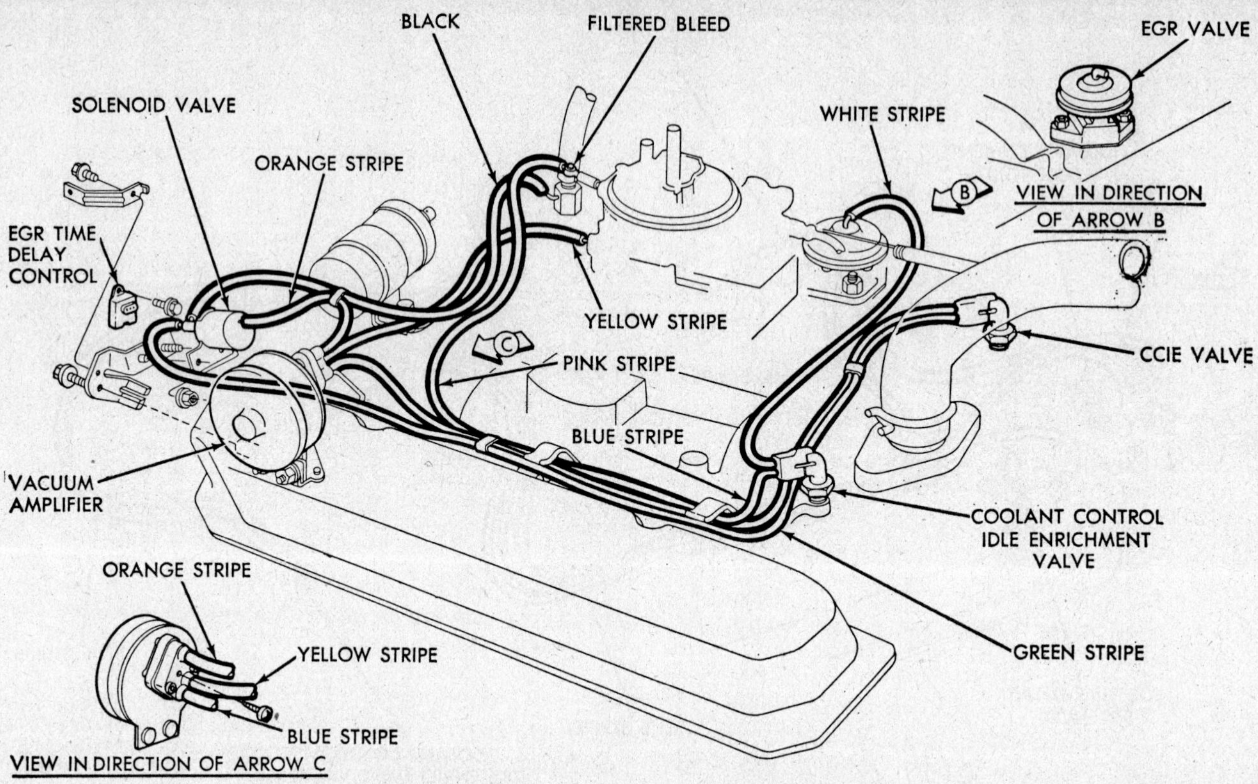

Fig. CL33 EGR system vacuum hose routing. 1975 Chrysler Corp. exc. Calif. V8-360 2 bar. carb. with automatic transmission

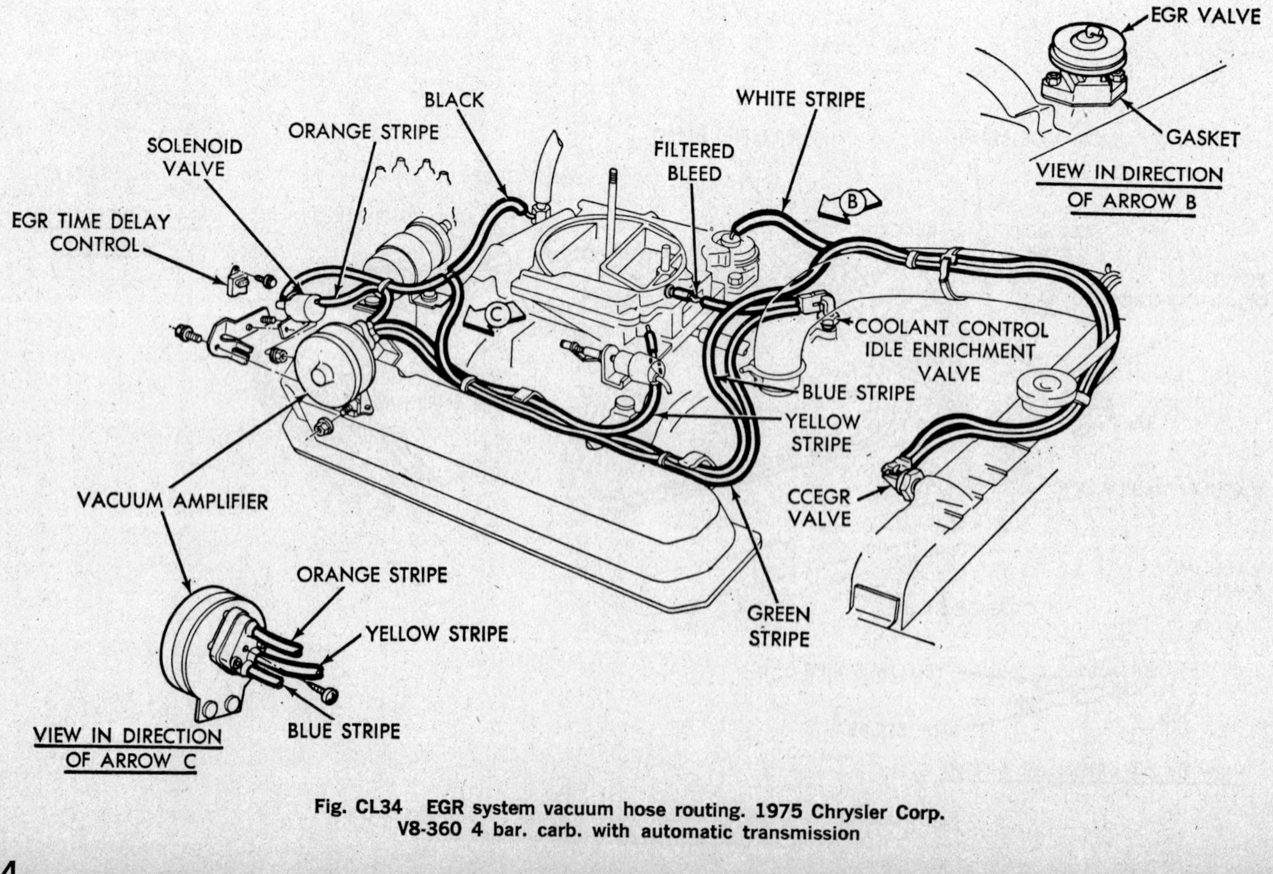

Fig. CL34 EGR system vacuum hose routing. 1975 Chrysler Corp. V8-360 4 bar. carb. with automatic transmission

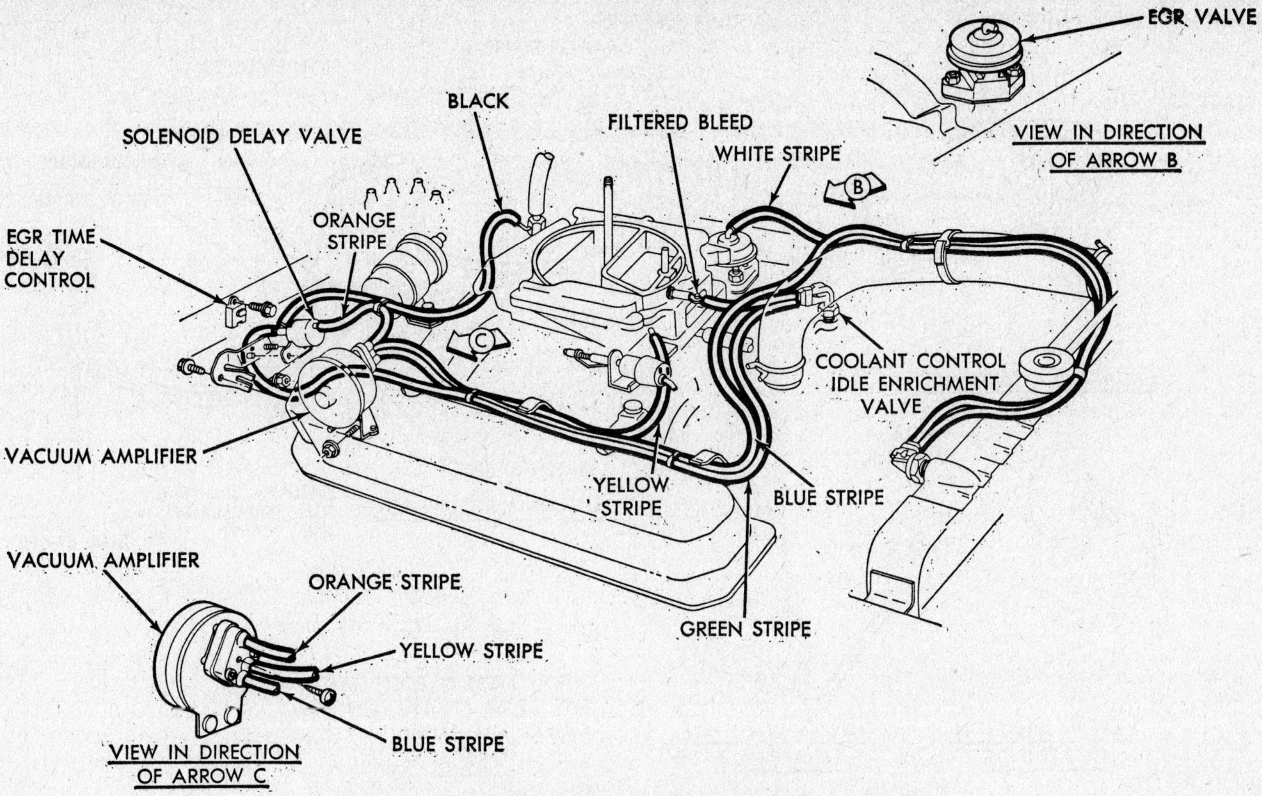

Fig. CL35 EGR system vacuum hose routing. 1975 Chrysler Corp. Calif. V8-360 Hi Performance

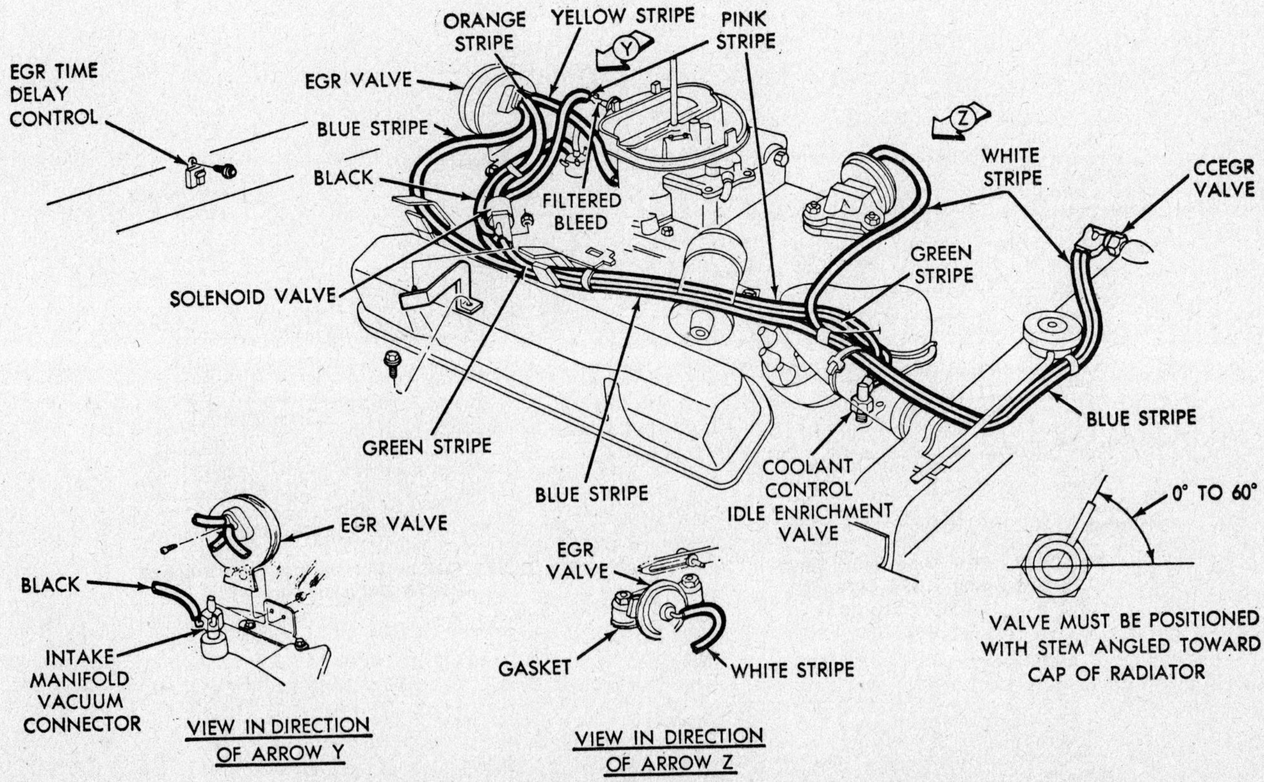

Fig. CL36 EGR system vacuum hose routing. 1975 Chrysler Corp. exc. Calif. V8-400 2 bar. carb.

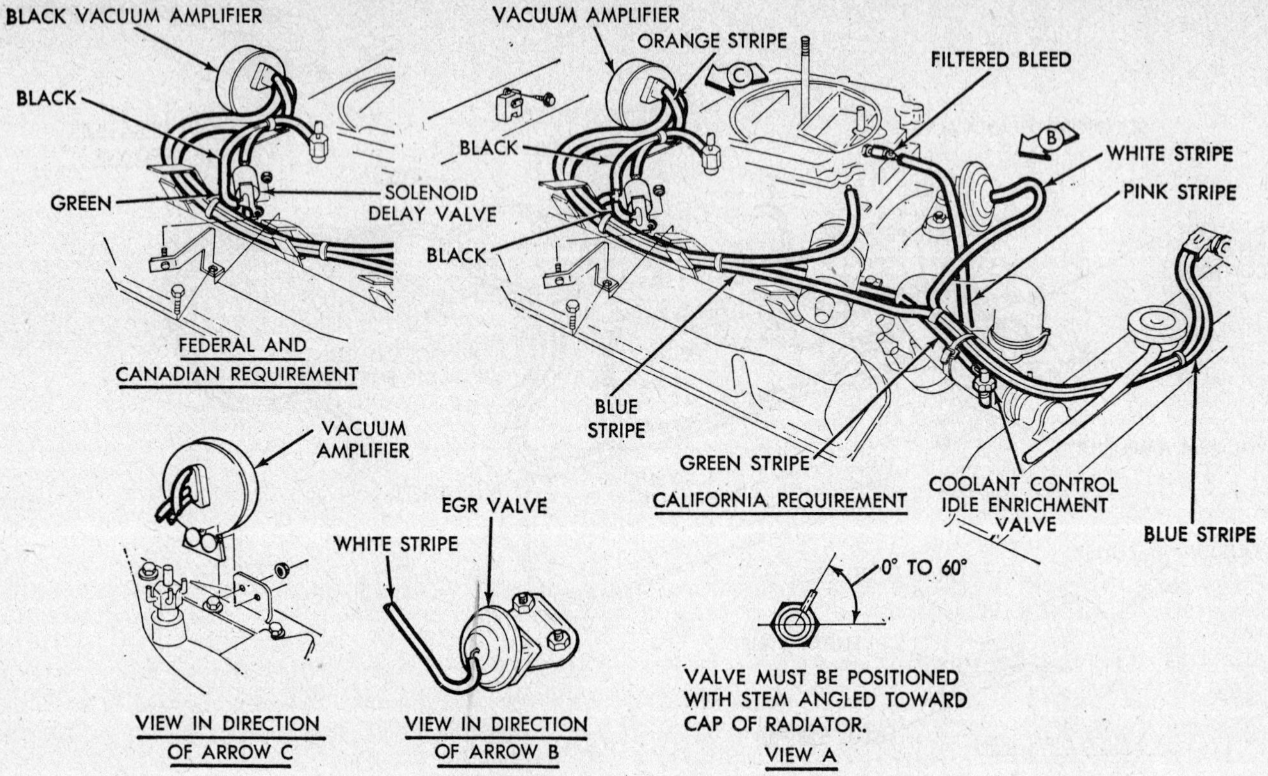

Fig. CL37 EGR system vacuum hose routing. 1975 Chrysler Corp. V8-400, 440 4 bar. carb.

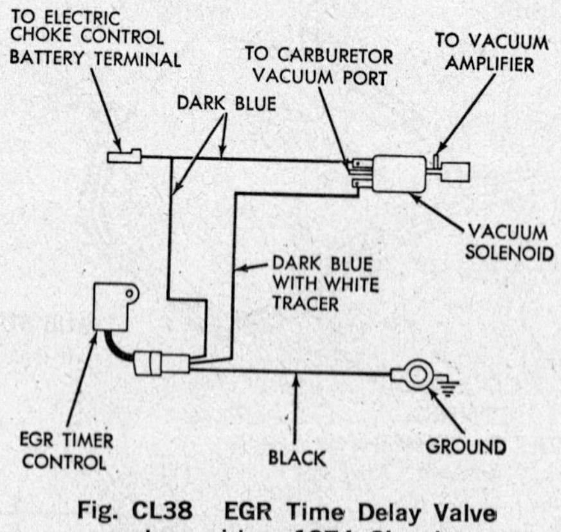

Fig. CL38 EGR Time Delay Valve system wiring. 1974 Chrysler

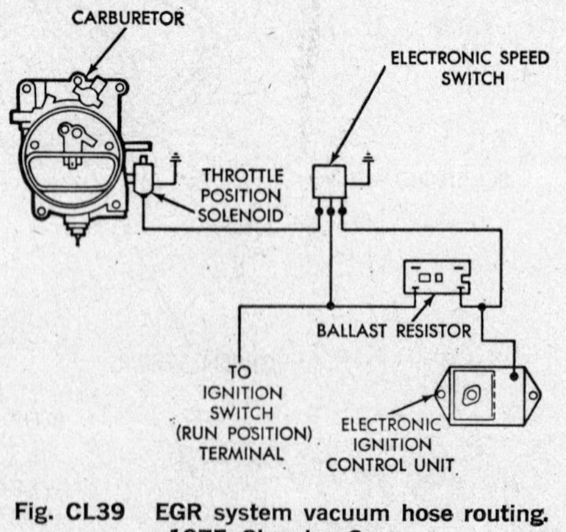

Fig. CL39 EGR system vacuum hose routing. 1975 Chrysler Corp.

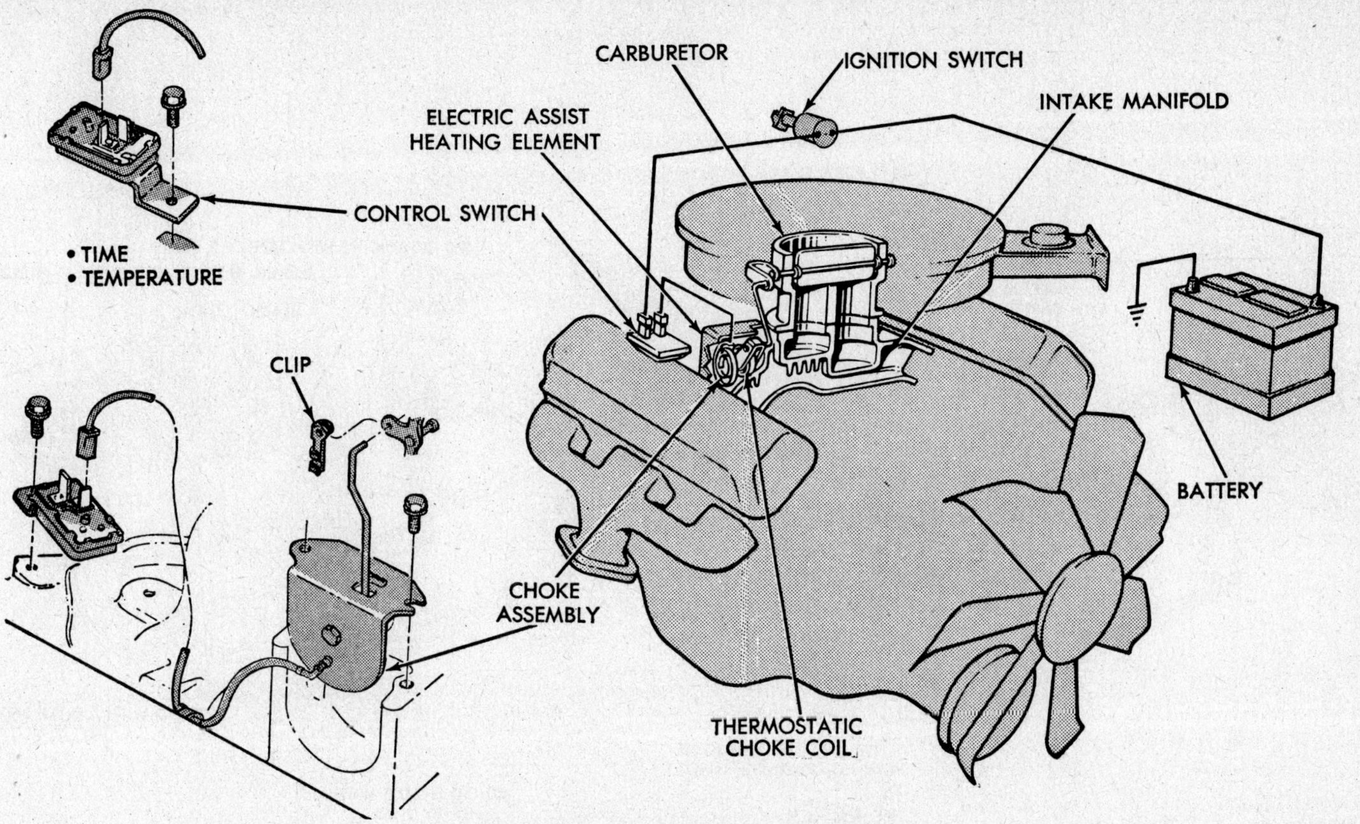

Fig. CL40 Electric assist choke. 1973-75 Chrysler

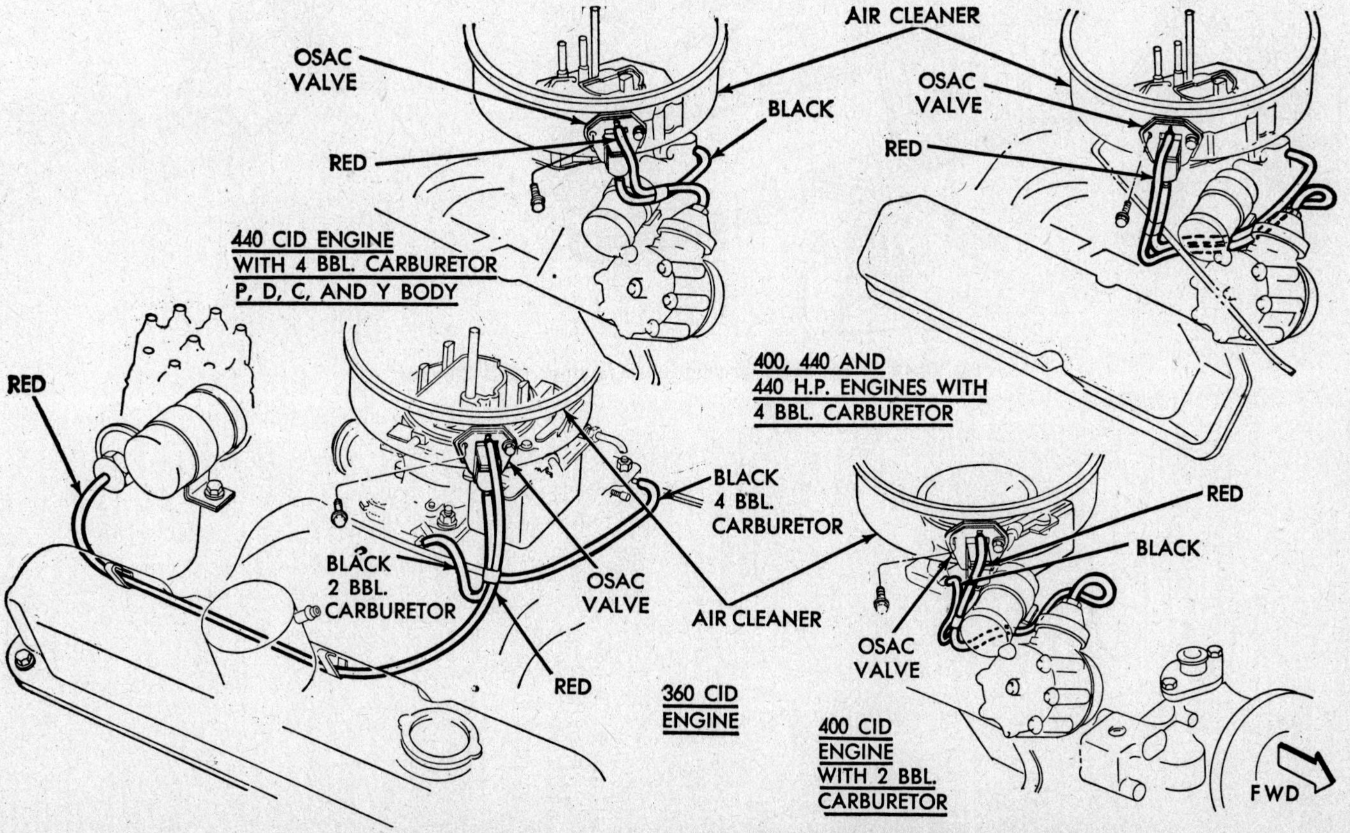

Fig. CL41 OSAC valve vacuum hose routing. 1974-75 Chrysler V8-360, 400, 440

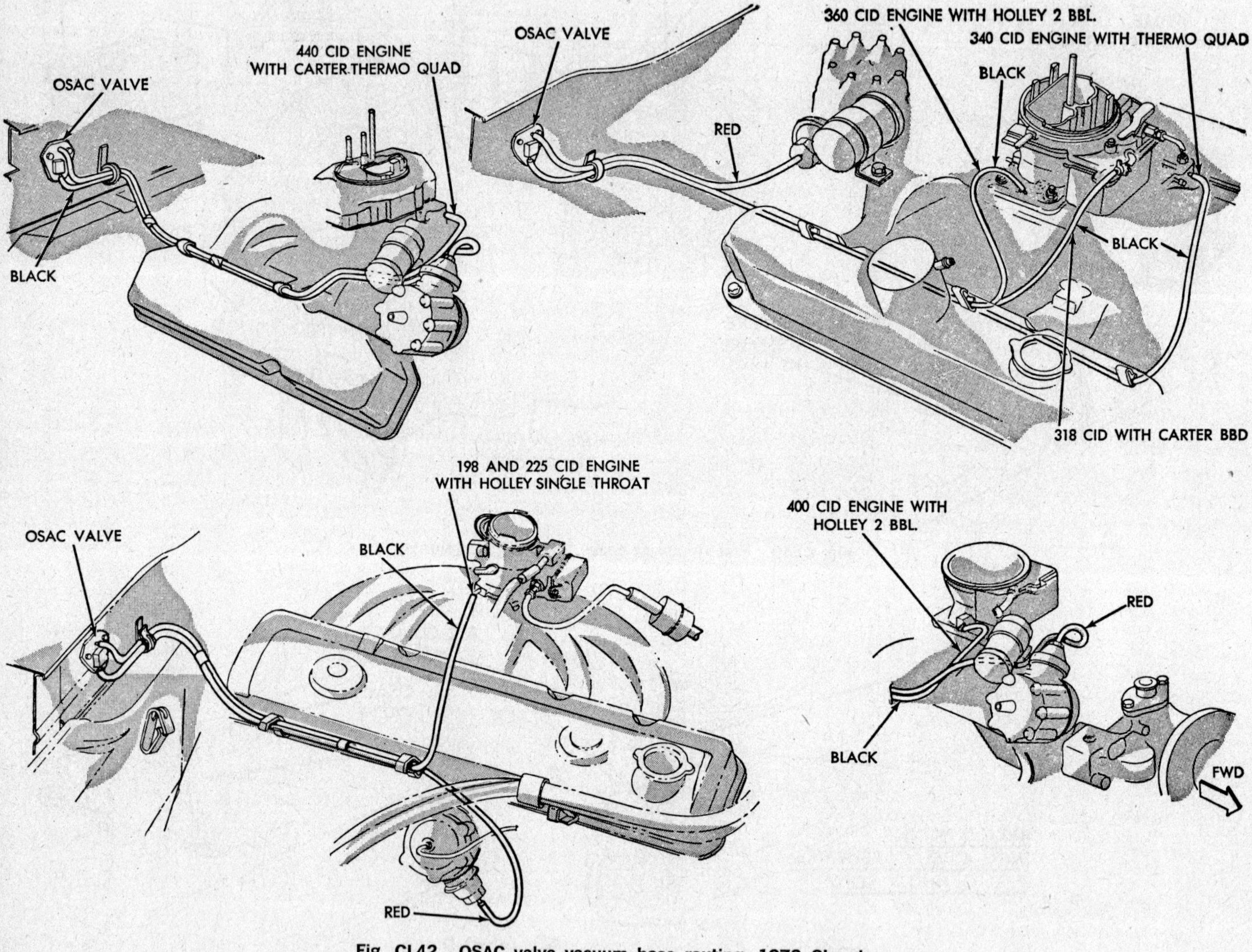

Fig. CL42 OSAC valve vacuum hose routing. 1973 Chrysler

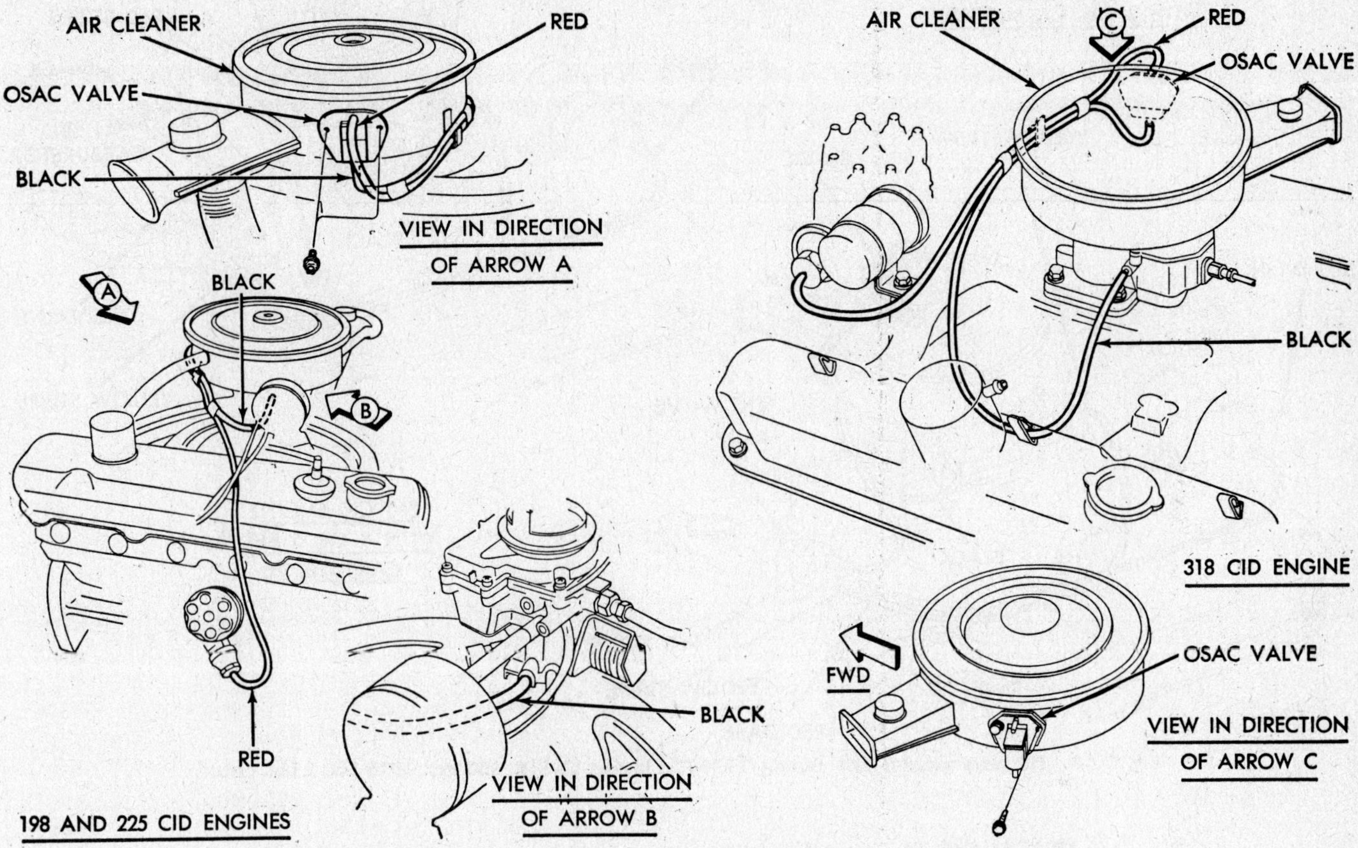

Fig. CL43 OSAC valve vacuum hose routing. 1974-75 Chrysler 6-198, 225 and V8-318

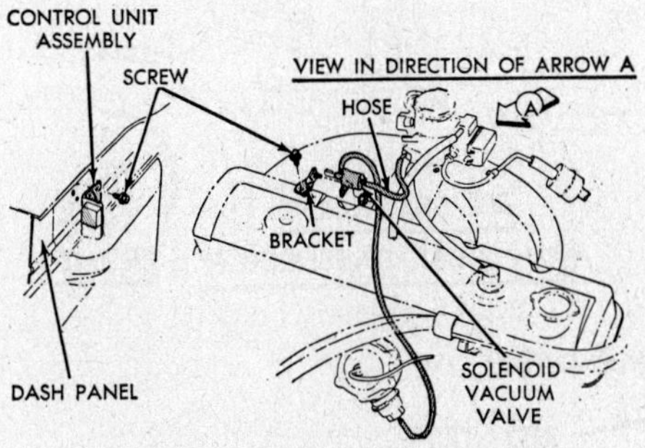

Fig. CL44 Solenoid vacuum hose routing. 1971-72 Chrysler six cylinder

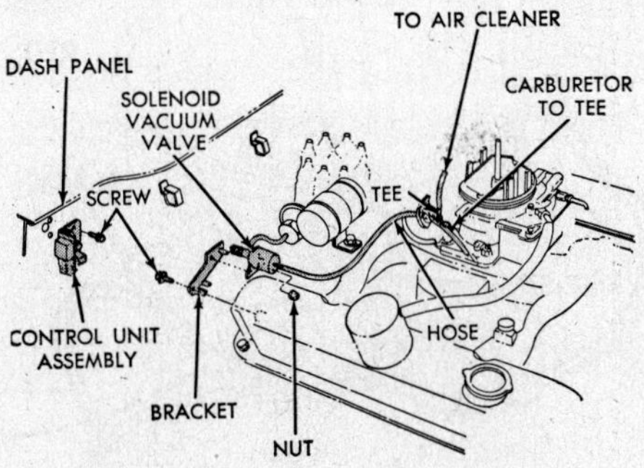

Fig. CL45 Solenoid vacuum hose routing. 1971-72 Chrysler V8-318, 340, 360

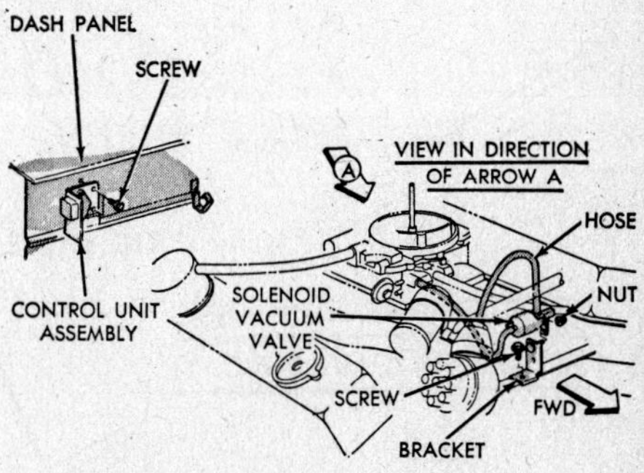

Fig. CL46 Solenoid vacuum hose routing. 1971-72 Chrysler V8-383, 400, 440

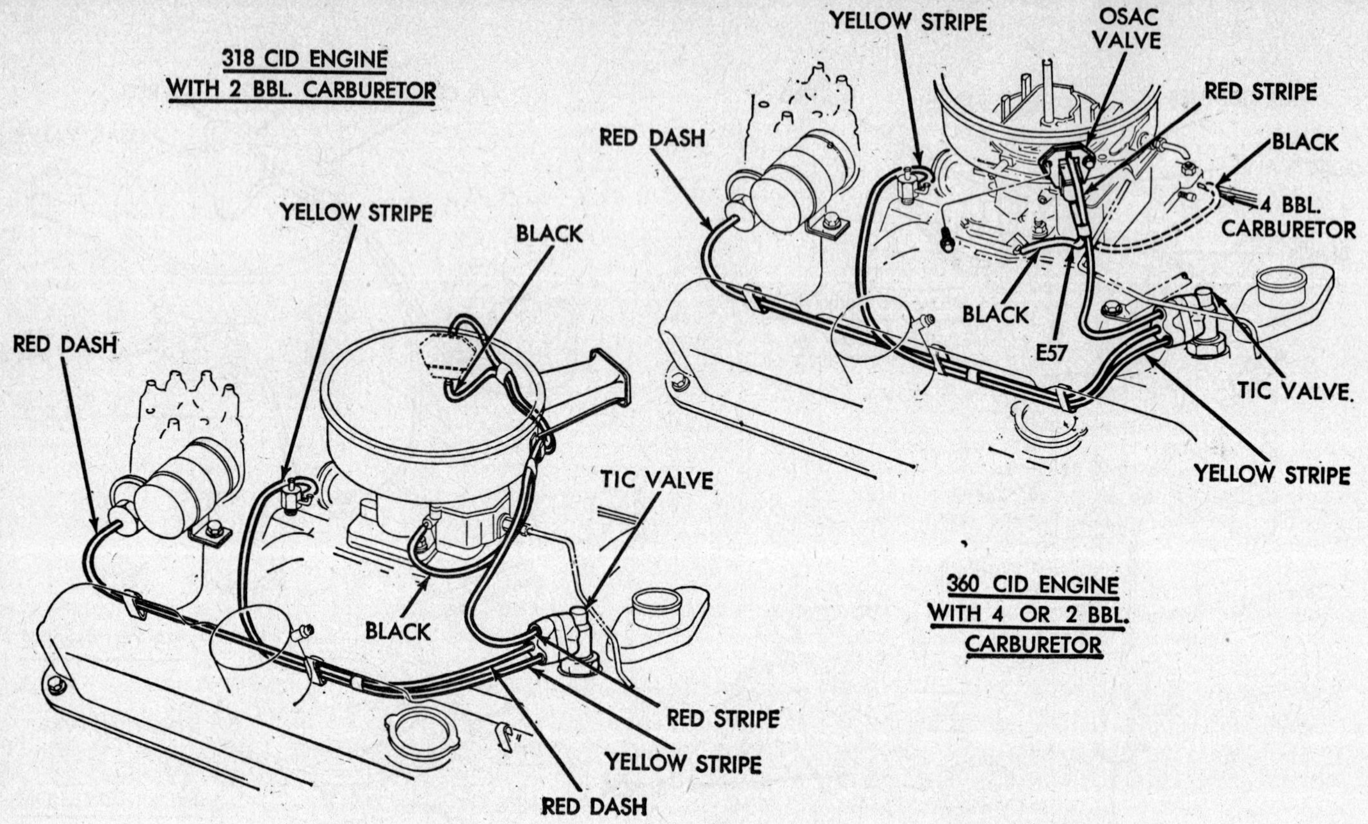

Fig. CL47 TIC valve vacuum hose routing. 1974-75 Chrysler V8-318, 360 exc. 1975 360 2 bbl. carb.

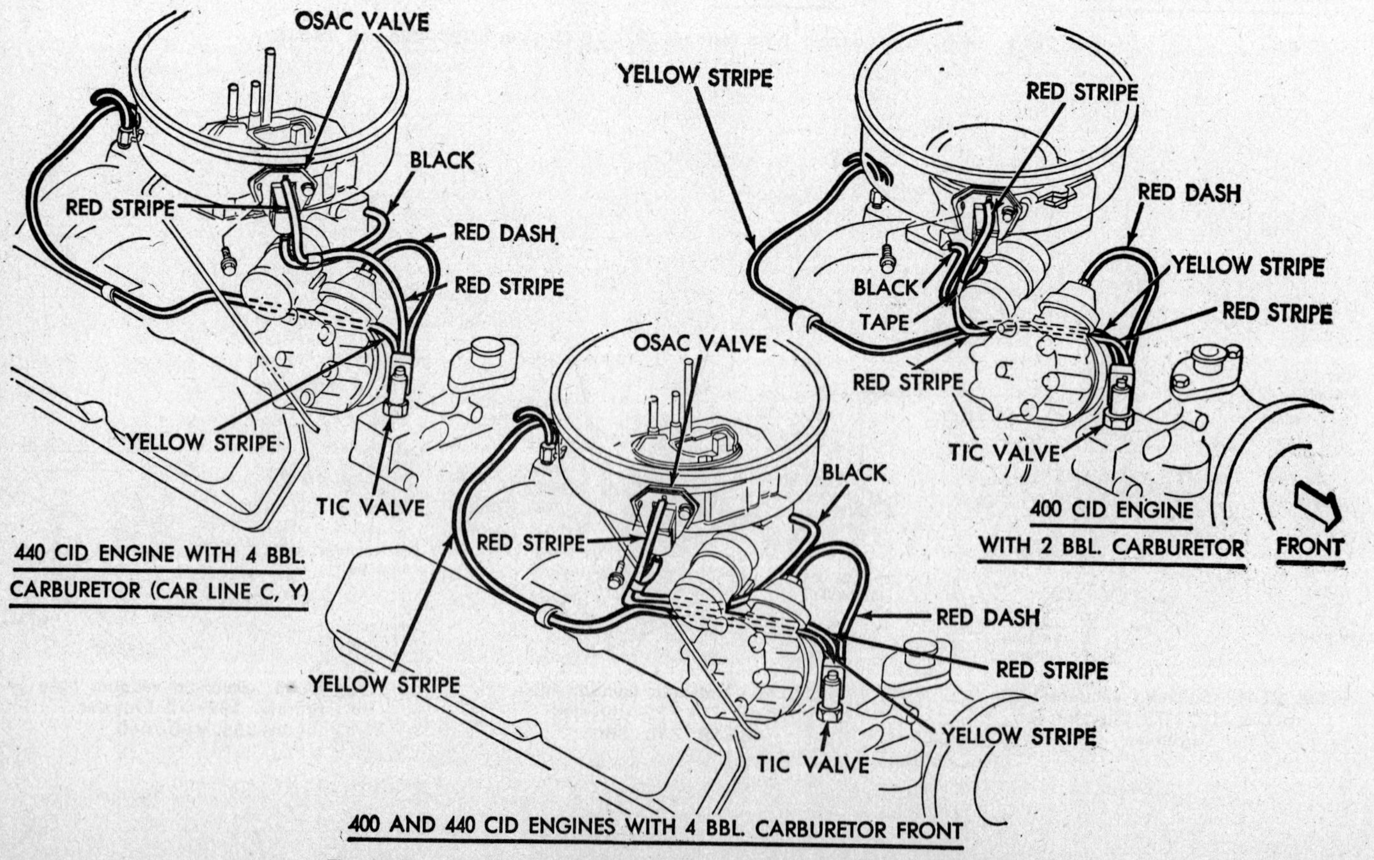

Fig. CL48 TIC valve vacuum hose routing. 1974-75 Chrysler V8-400, 440

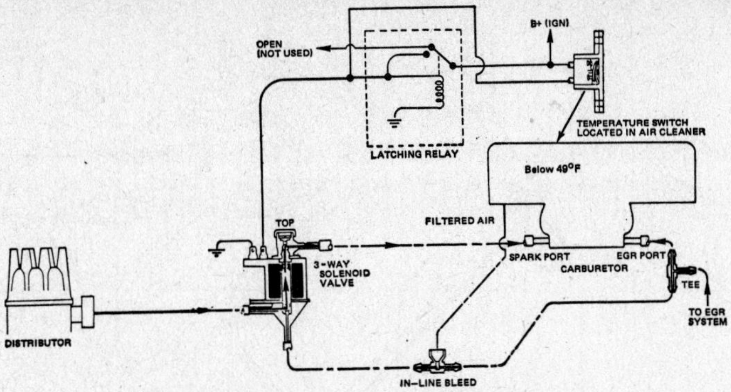

Fig. CL49 CTAV system wiring. 1973-75 Ford

Fig. CL50 Decel valve. Ford 2000cc & 2800cc engines shown

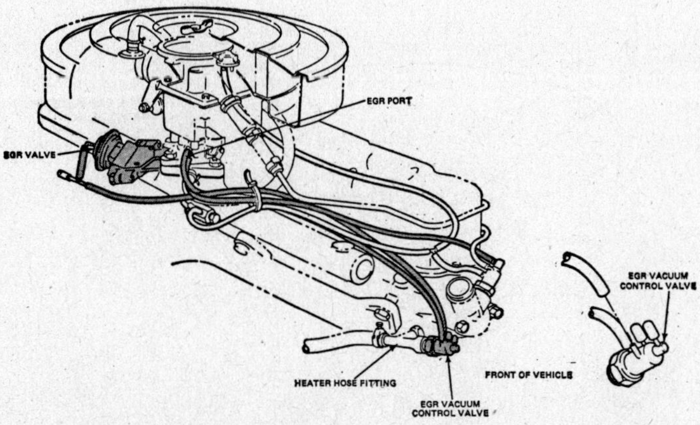

Fig. CL51 EGR system vacuum hose routing. 1973 Ford 6-200

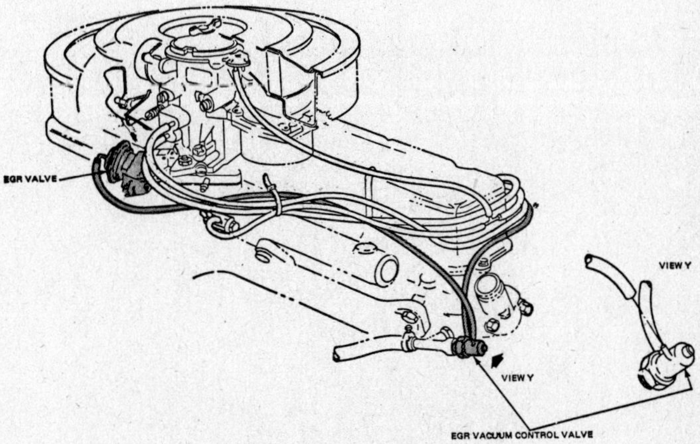

Fig. CL52 EGR system vacuum hose routing. 1973 Ford 6-250 with manual transmission

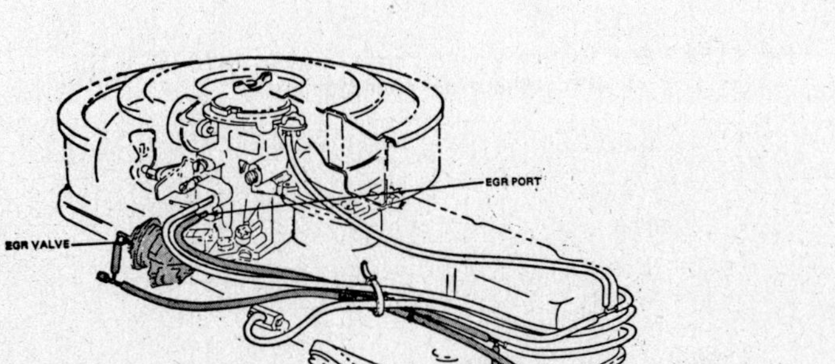

Fig. CL53 EGR system vacuum hose routing. 1973 Ford 6-250 with automatic transmission

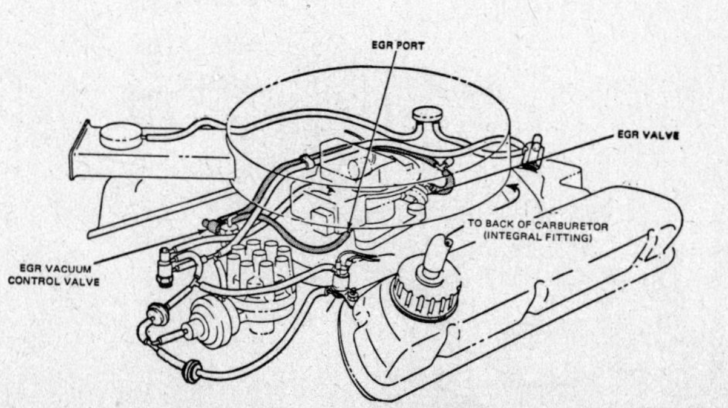

Fig. CL54 EGR system vacuum hose routing. 1973 Ford V8-302

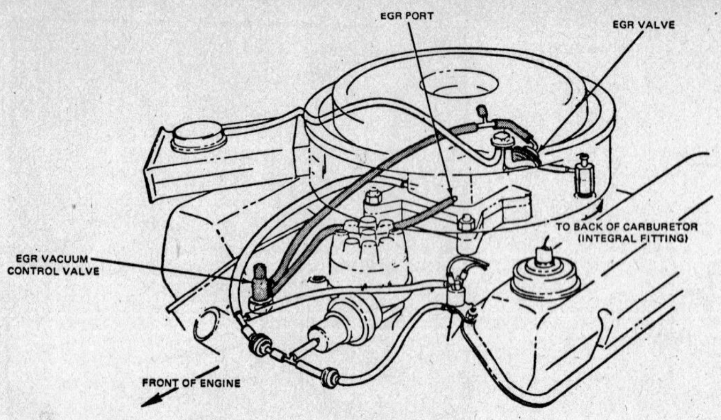

Fig. CL55 EGR system vacuum hose routing. 1973 Ford V8-351W

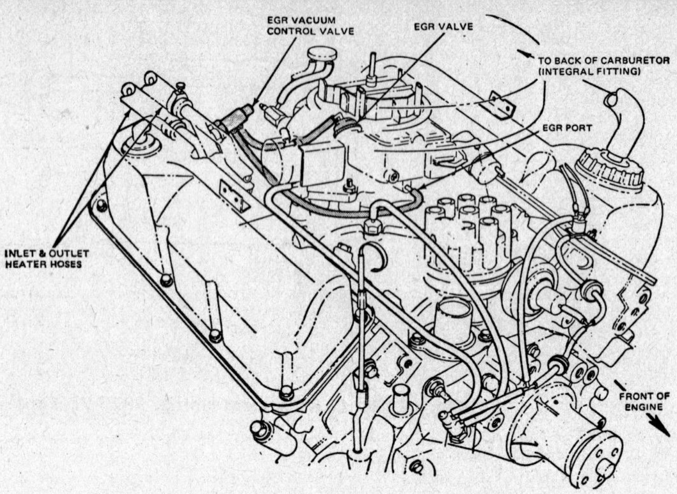

Fig. CL56 EGR system vacuum hose routing. 1973 Ford V8-351C with air conditioning

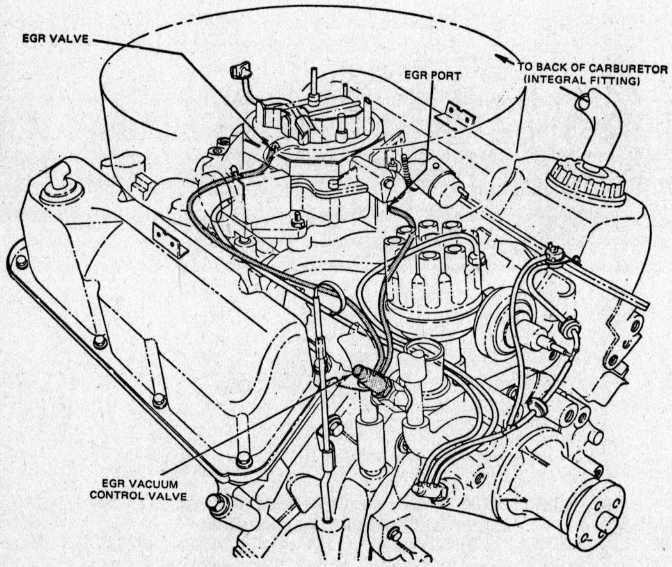

Fig. CL57 EGR system vacuum hose routing. 1973 Ford V8-351C, 400 and V8-351 CJ with air conditioning

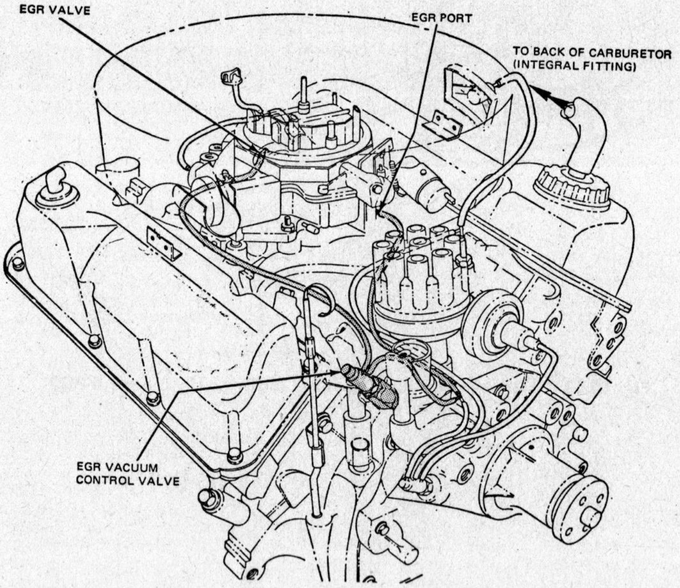

Fig. CL58 EGR system vacuum hose routing. 1973 V8-351C, 400 without air conditioning

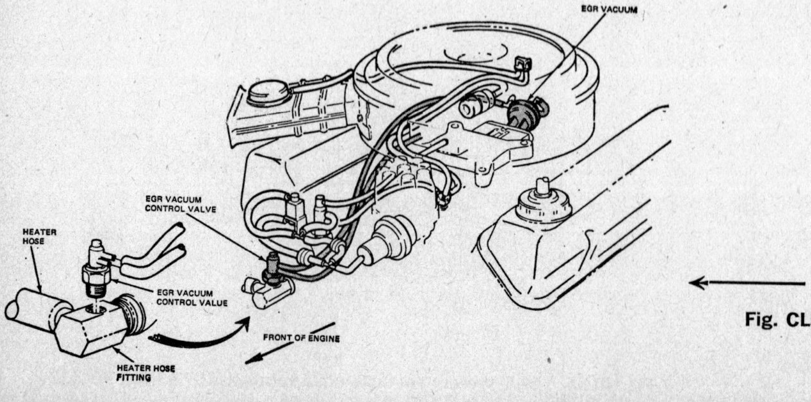

Fig. CL59 EGR system vacuum hose routing. 1973 V8-429, 460

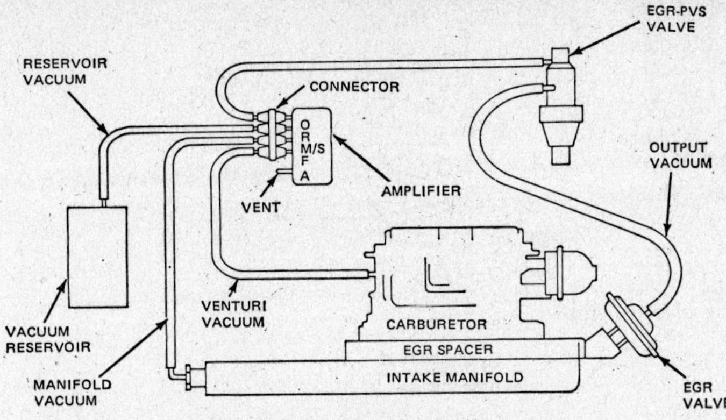

WITH SINGLE CONNECTOR VVA (1974 TYPE)

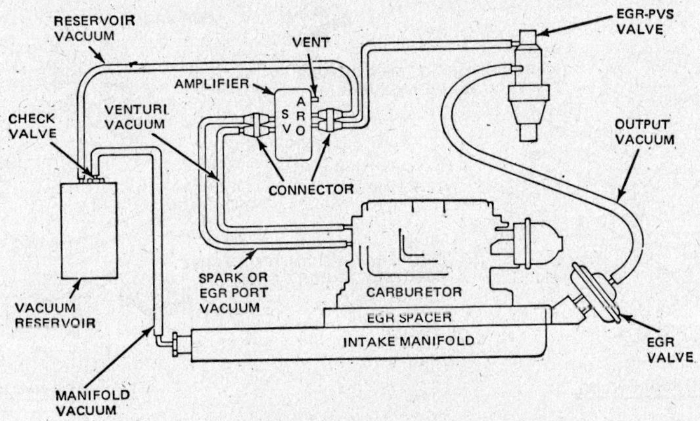

Fig. CL60 EGR system vacuum hose routing. 1974-75 Ford (typical)

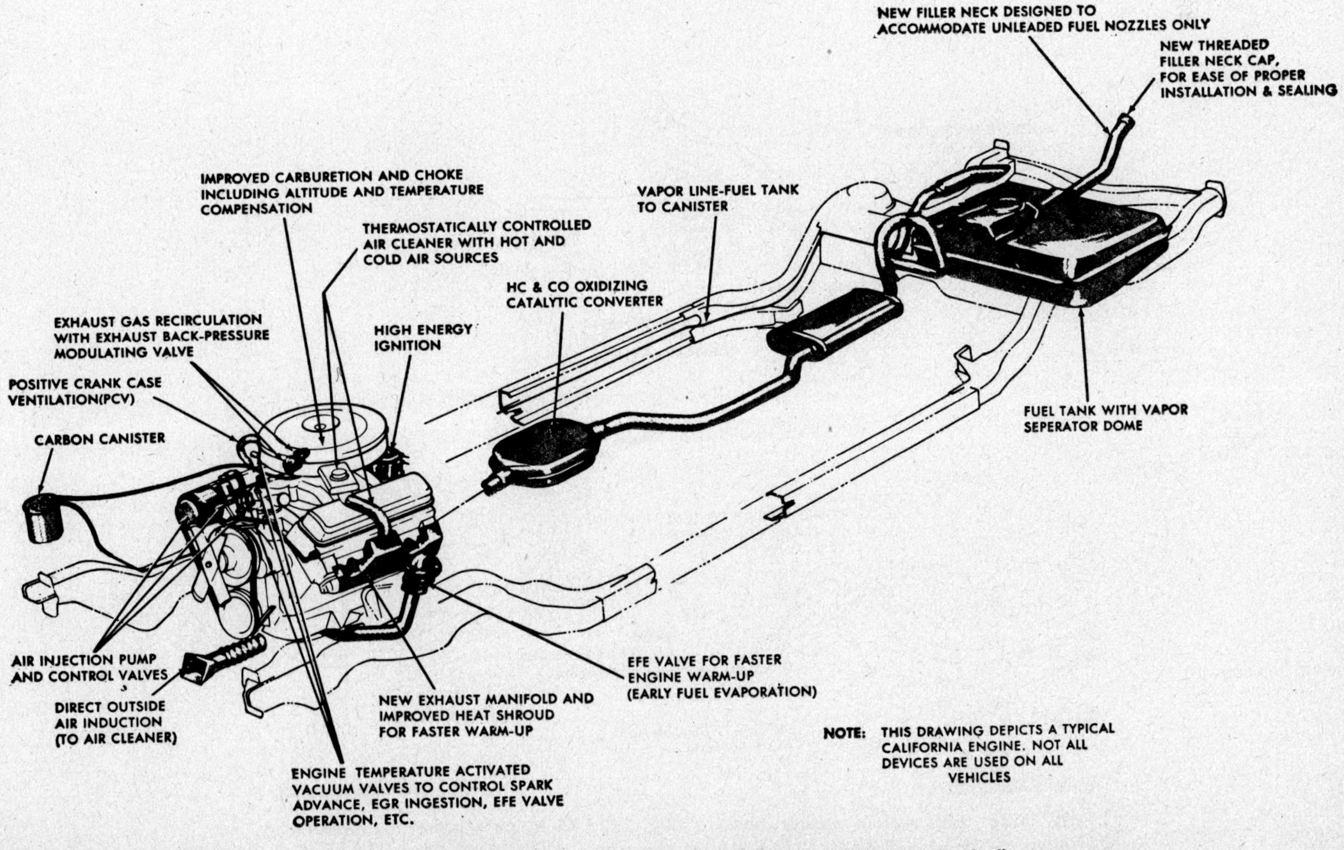

Fig. CL61 Emission Control System. 1975 General Motors (typical)

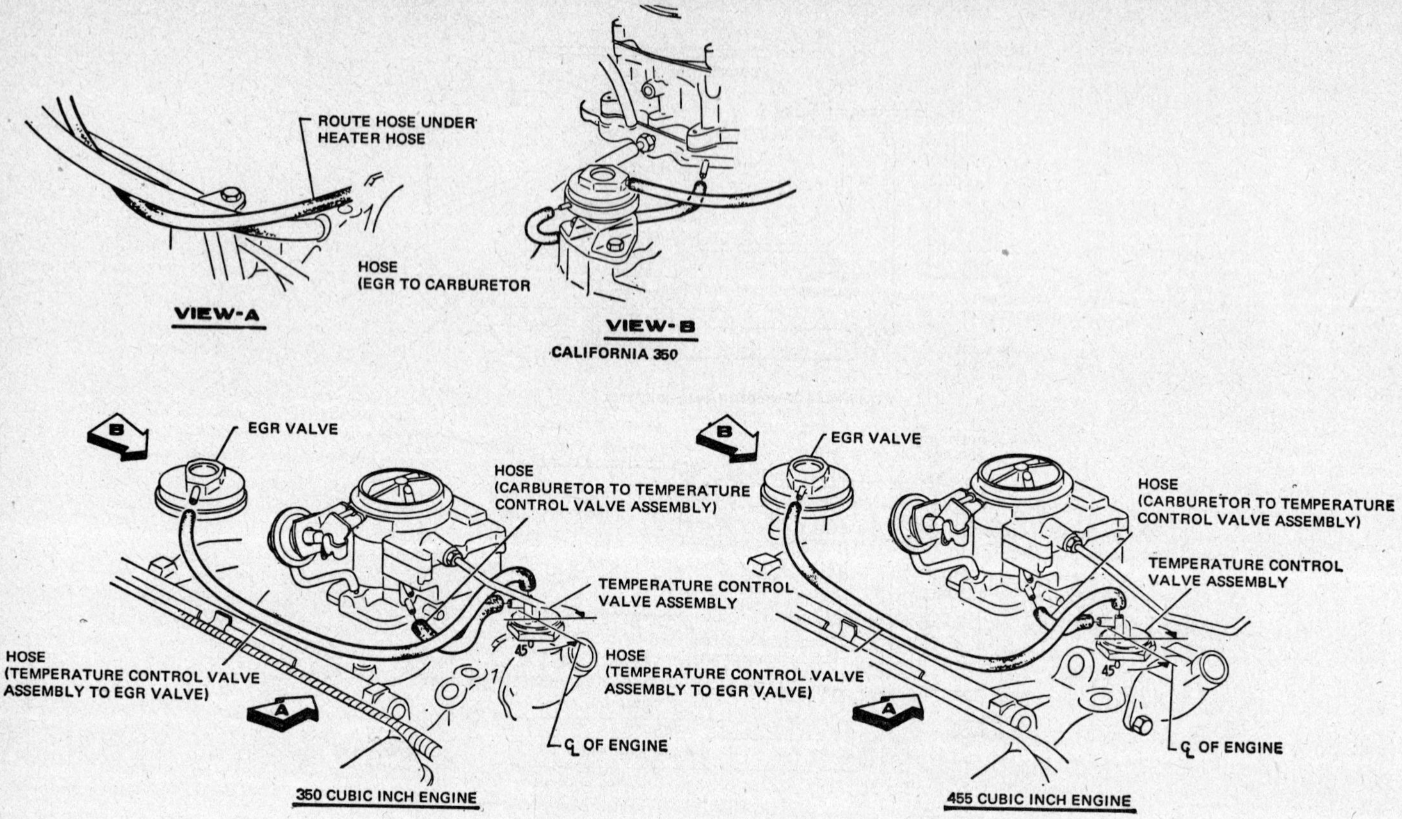

Fig. CL62 EGR system vacuum hose routing. 1973-74 Buick 2 barrel carburetors

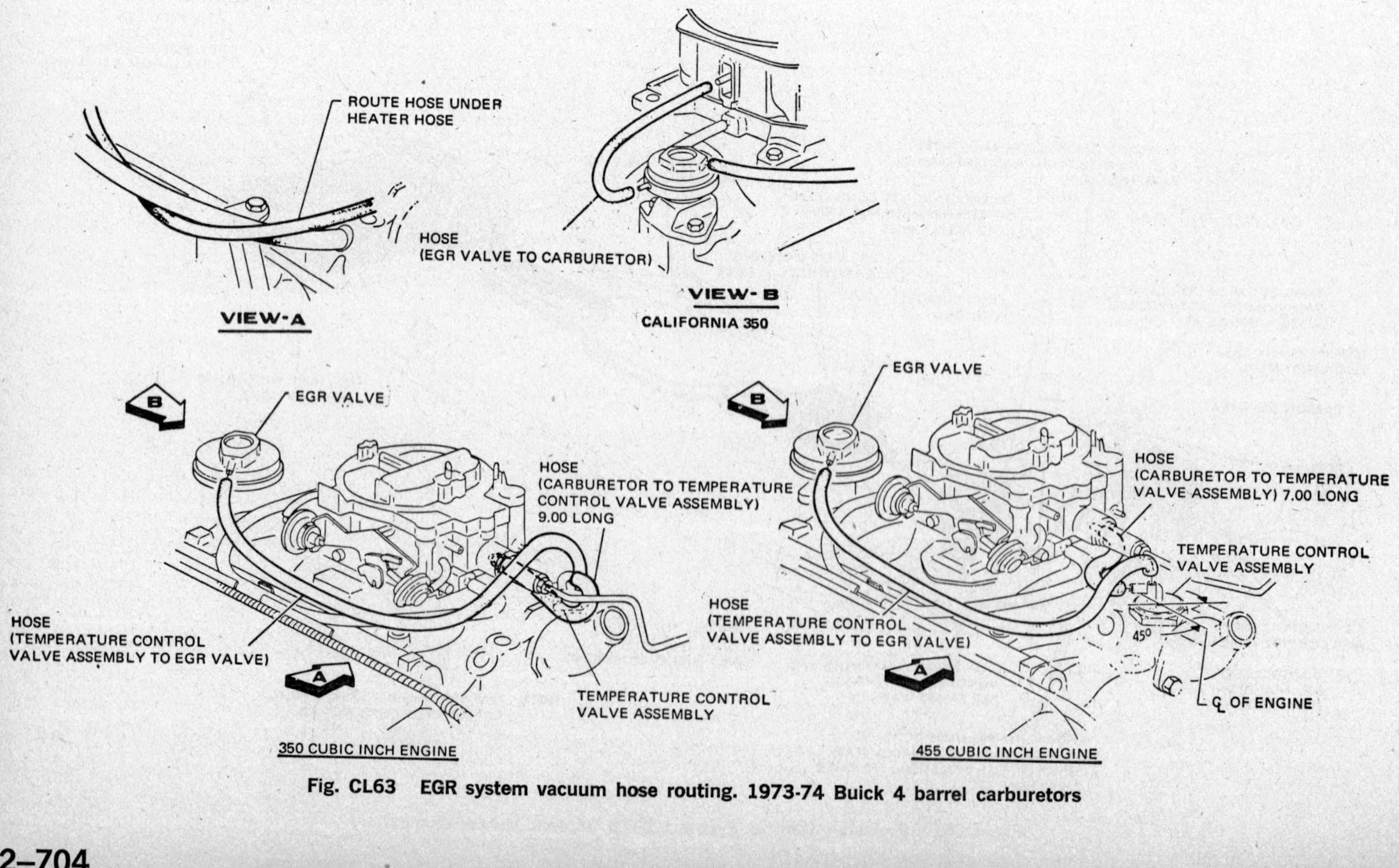

Fig. CL63 EGR system vacuum hose routing. 1973-74 Buick 4 barrel carburetors

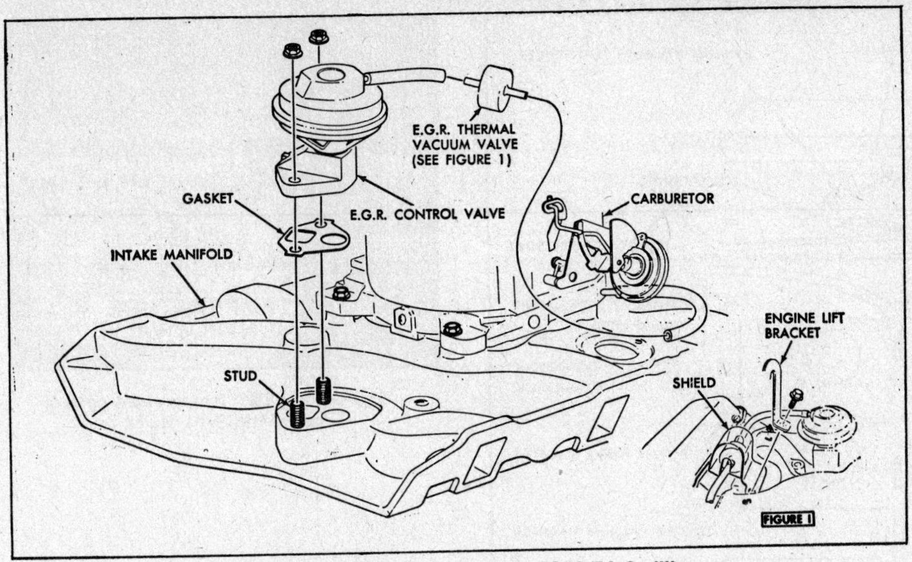

Fig. CL64 EGR valve mounting. 1973-74 Cadillac

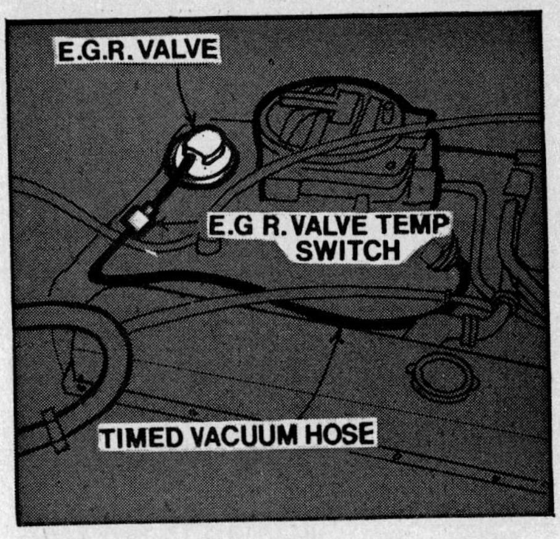

Fig. CL66 EGR temperature control switch. 1973-75 Cadillac

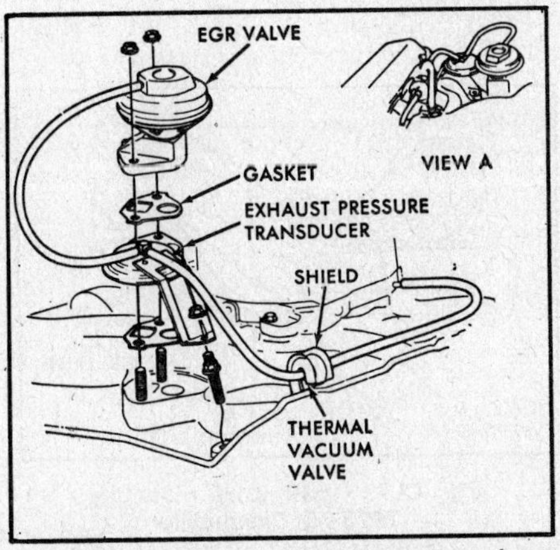

Fig. CL67 Exhaust back pressure transducer. 1975 Cadillac without fuel injection

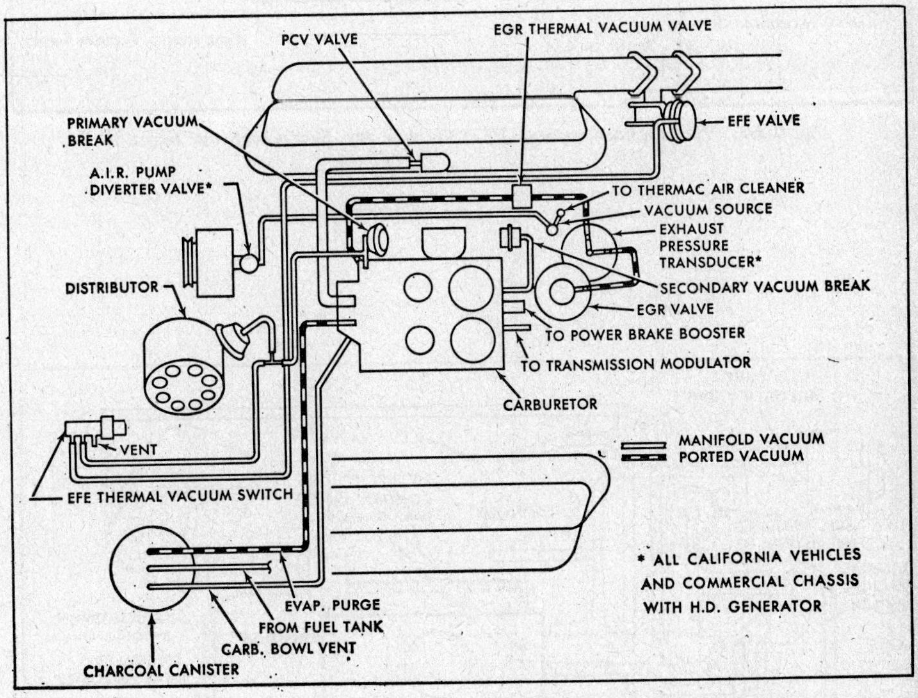

Fig. CL65 Vacuum hose routing. 1975 Cadillac without fuel injection

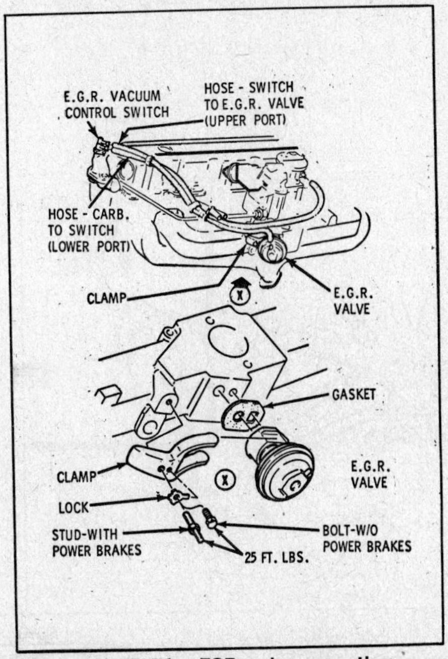

Fig. CL71 EGR valve mounting. General Motors 6 cylinder (typical)

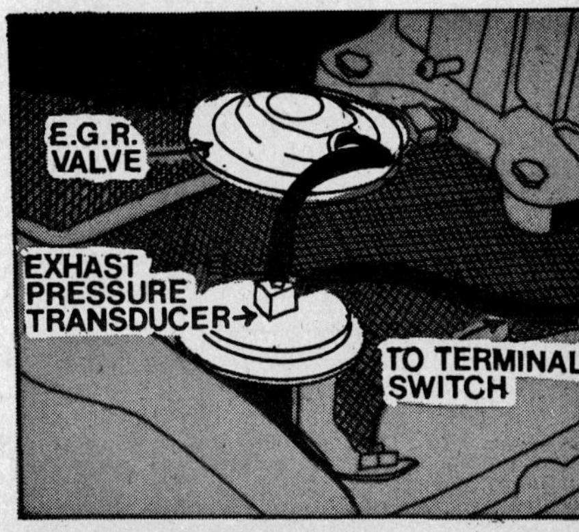

Fig. CL68 Exhaust back pressure transducer valve. 1974 Cadillac

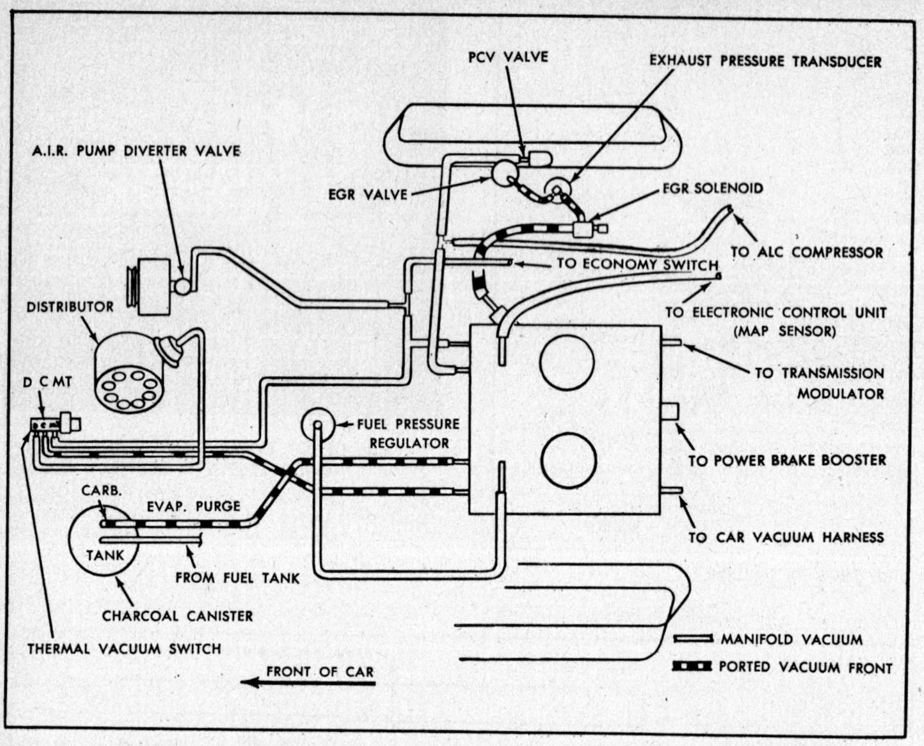

Fig. CL69 Vacuum hose routing. 1975 Cadillac exc. Seville with fuel injection

Fig. CL72 EGR temperature control valve. 1973-74 Buick

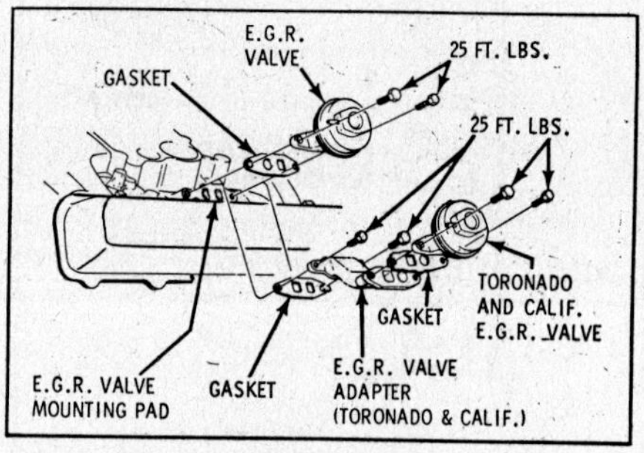

Fig. CL73 EGR valve mounting. 1973-75 Oldsmobile

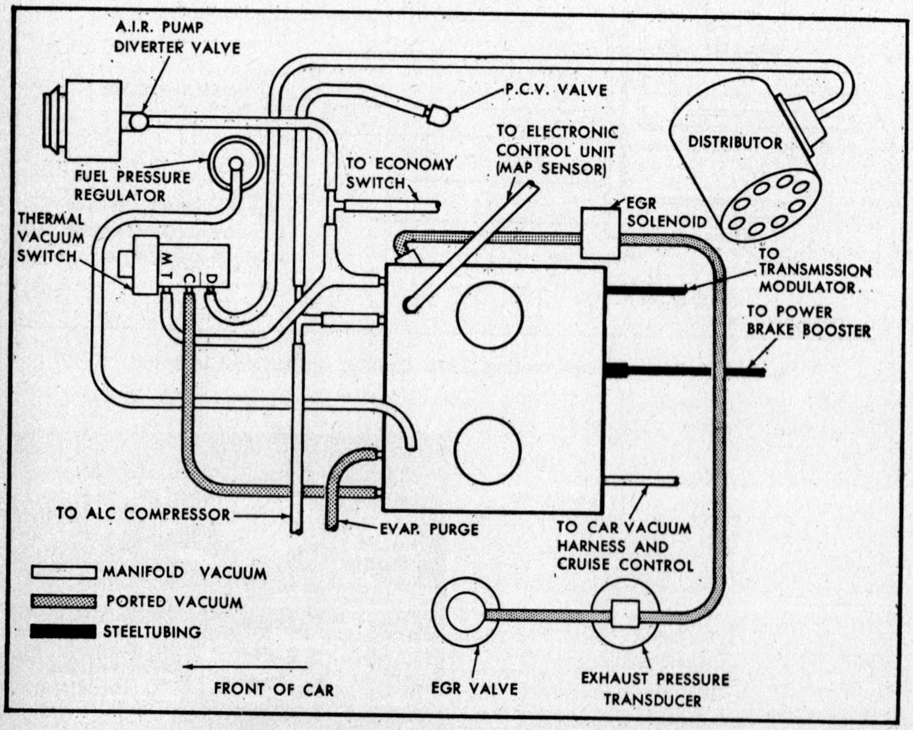

Fig. CL70 Vacuum hose routing. 1976 Cadillac Seville

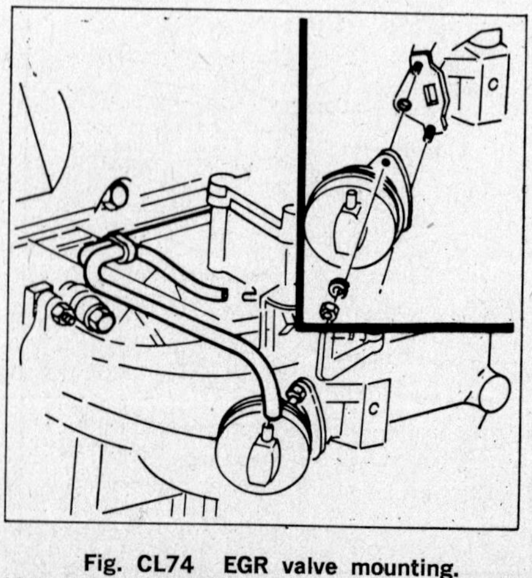

Fig. CL74 EGR valve mounting. 1973-75 Chevrolet Vega & Pontiac Astre 4-140

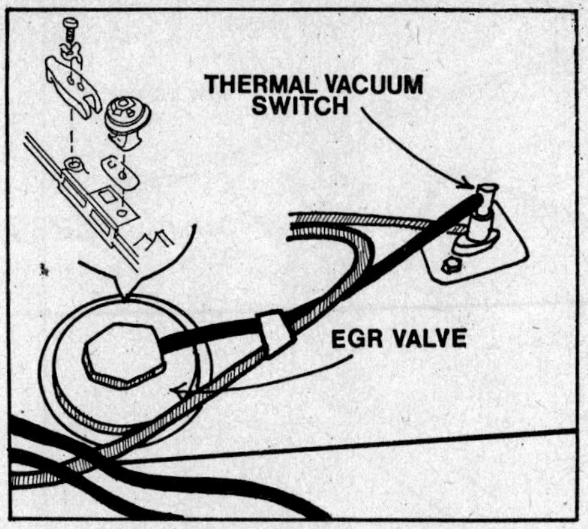

Fig. CL75 EGR & TVS location. 1973-75 Chevrolet small V8 engines

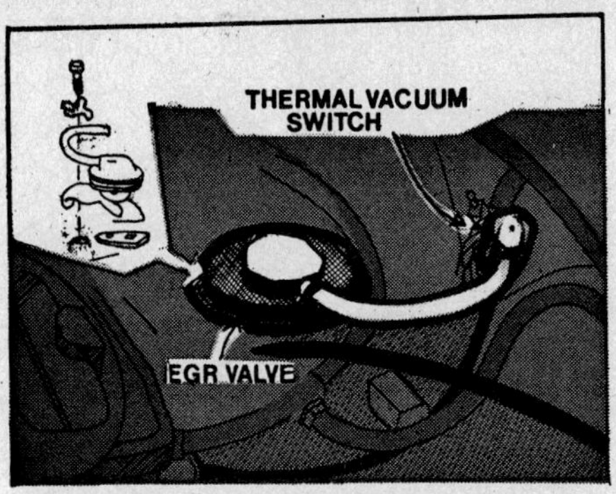

Fig. CL76 EGR & TVS location. 1973-75 Chevrolet large V8 engines

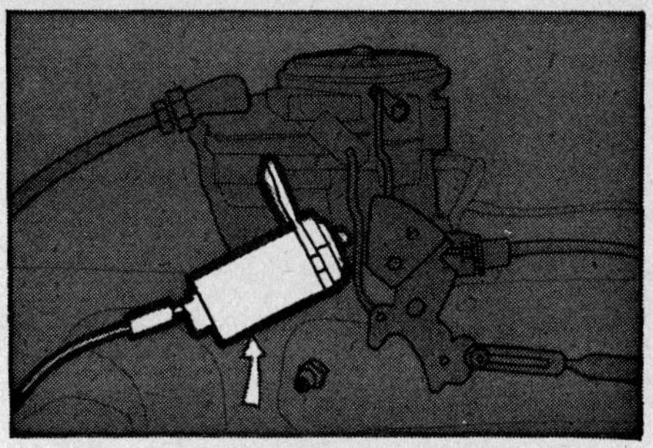

Fig. CL77 Idle stop solenoid. General Motors (typical)

Fig. CL78 Vacuum advance solenoid. 1972-75 Chevrolet small V8

Fig. CL79 Vacuum advance solenoid. 1972-74 Chevrolet large V8

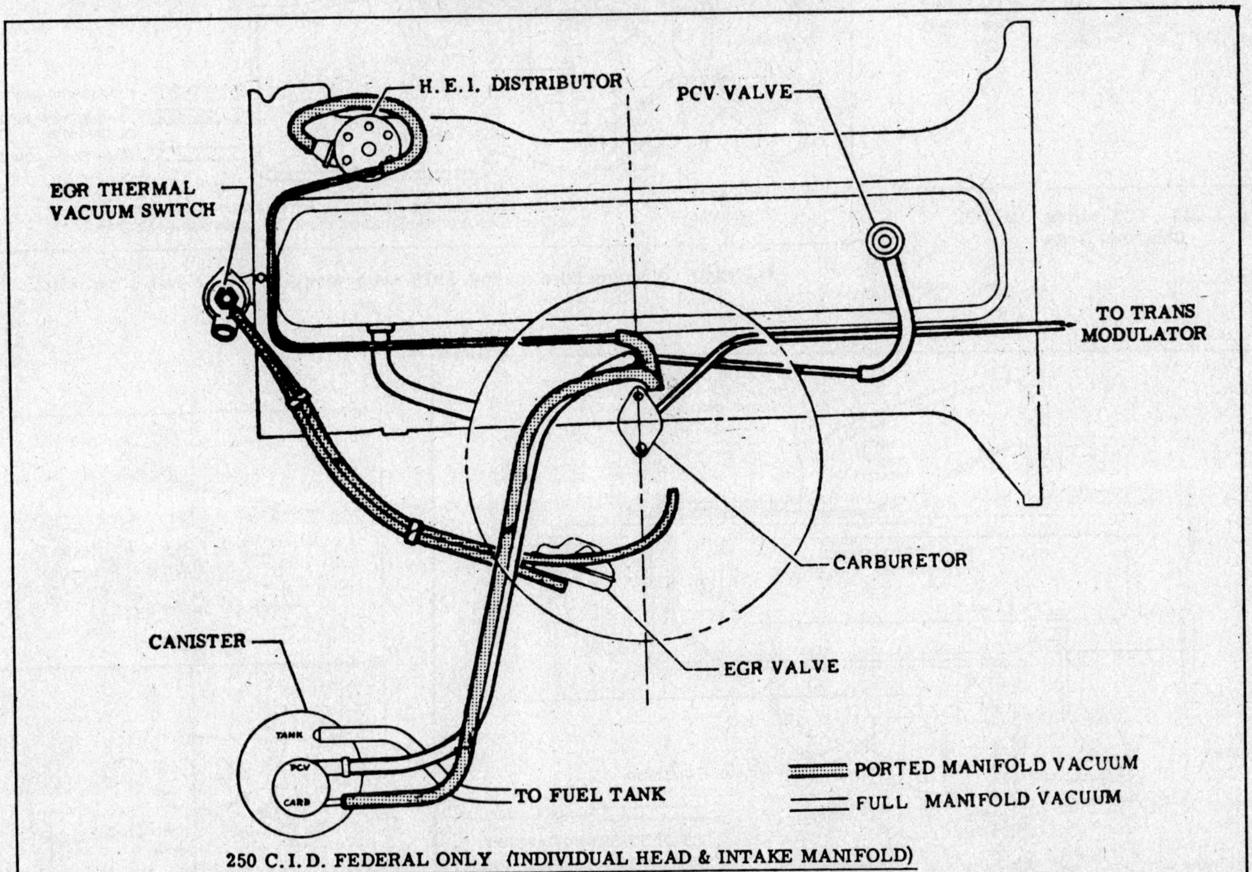

Fig. CL81 Vacuum hose routing. 1975 General Motors 6 cylinder with individual head and intake manifold, exc. Calif.

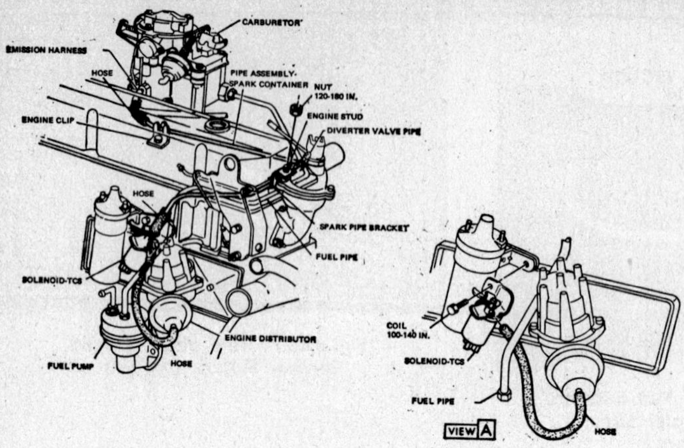

Fig. CL80 TCS vacuum hose routing. General Motors 6 cylinder (typical)

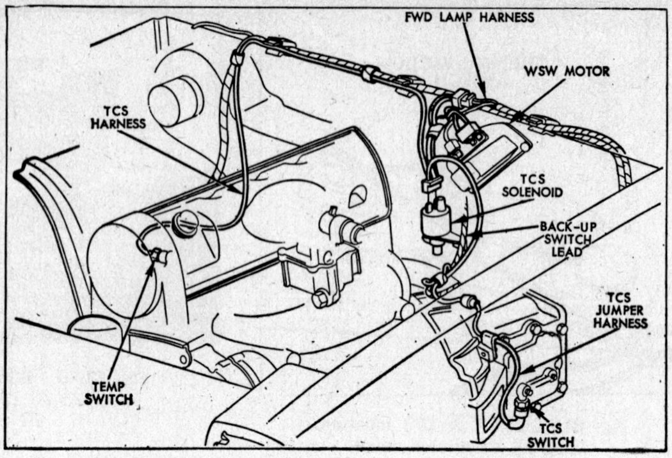

Fig. CL85 TCS wiring. 1973-74 Chevrolet Vega

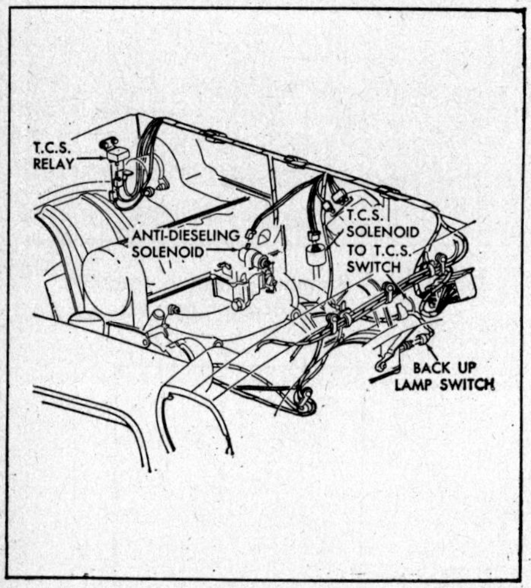

Fig. CL84 TCS wiring. 1971-72 Chevrolet Vega

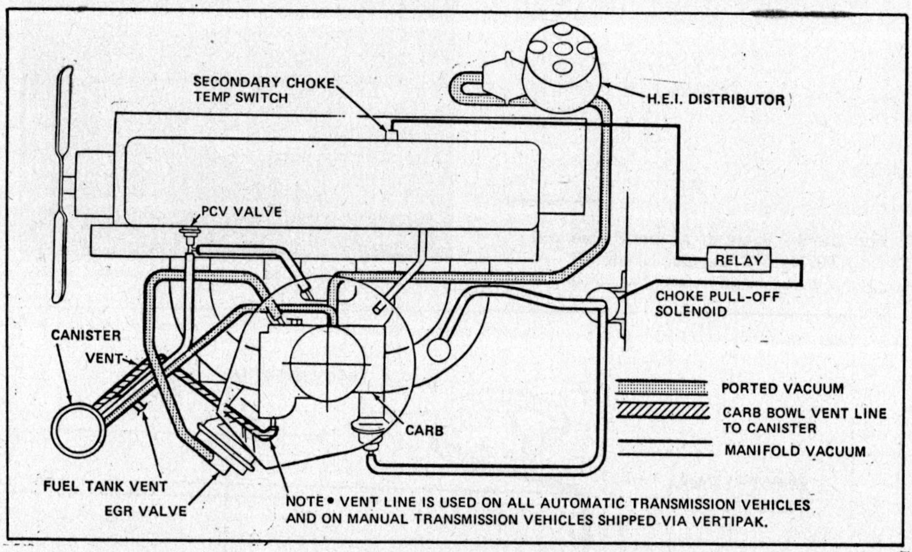

Fig. CL86 Vacuum hose routing. 1975 Astre, Monza & Vega 4-140, 1 bar. carb.

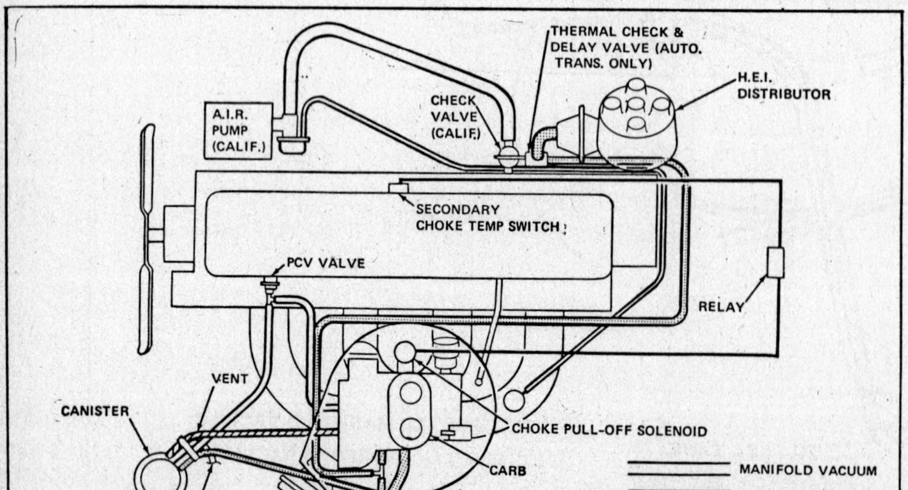

Fig. CL87 Vacuum hose routing. 1975 Astre, Monza & Vega 4-140, 2 bar. carb.

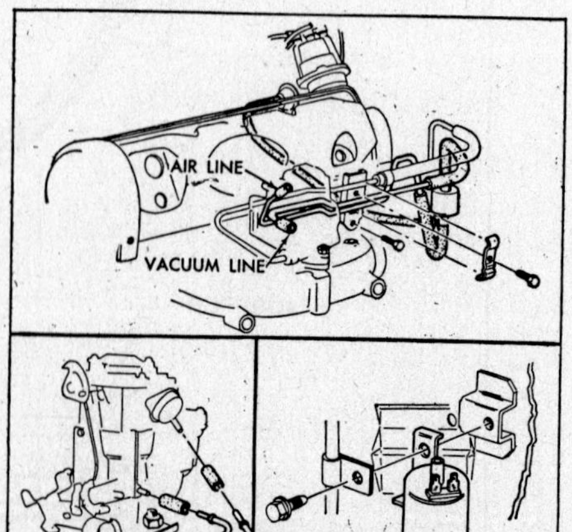

Fig. CL88 TCS vacuum hose routing. 1971-72 Chevrolet Vega

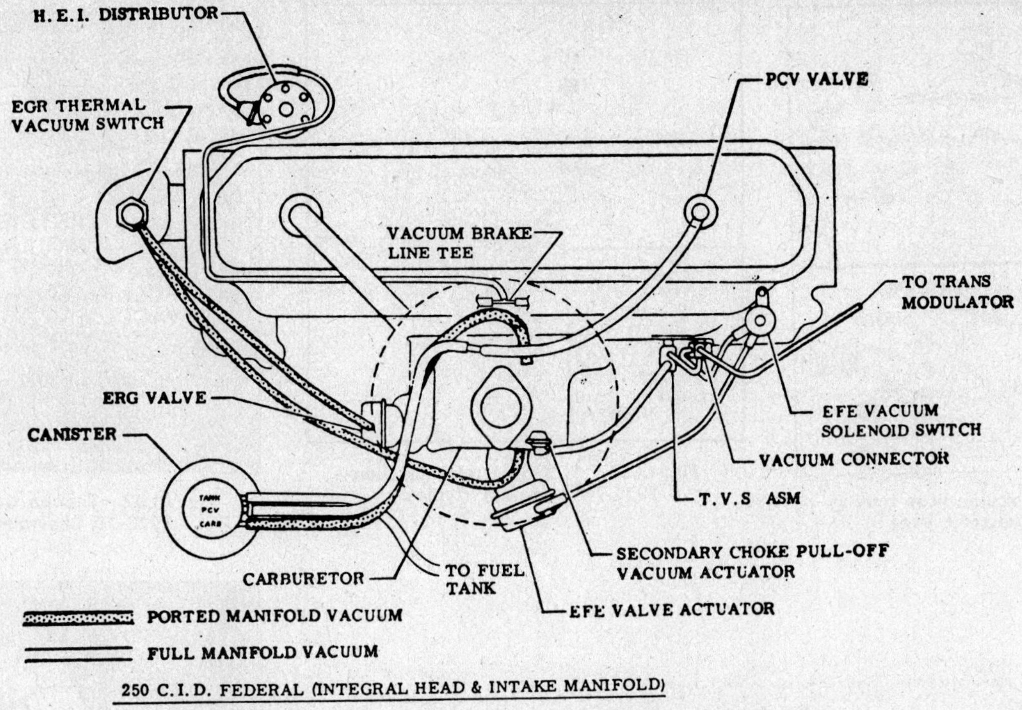

Fig. CL82 Vacuum hose routing. 1975 General Motors 6 cylinder with integral head and intake manifold, exc. Calif.

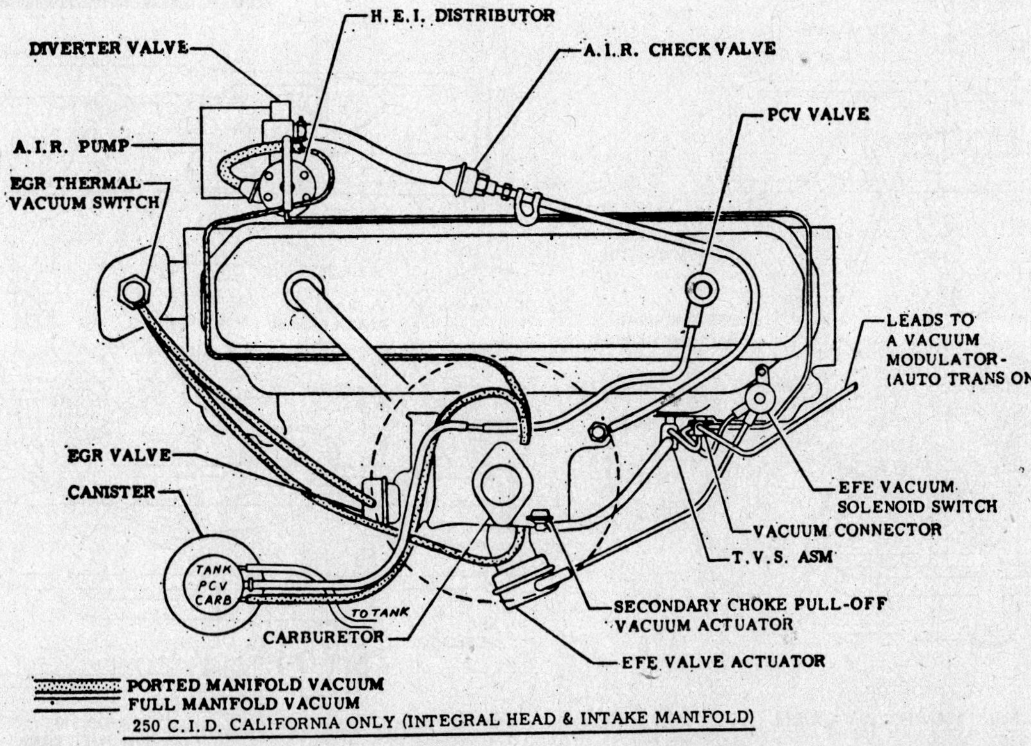

Fig. CL83 Vacuum hose routing. 1975 General Motors Calif. 6 cylinder with integral head and intake manifold

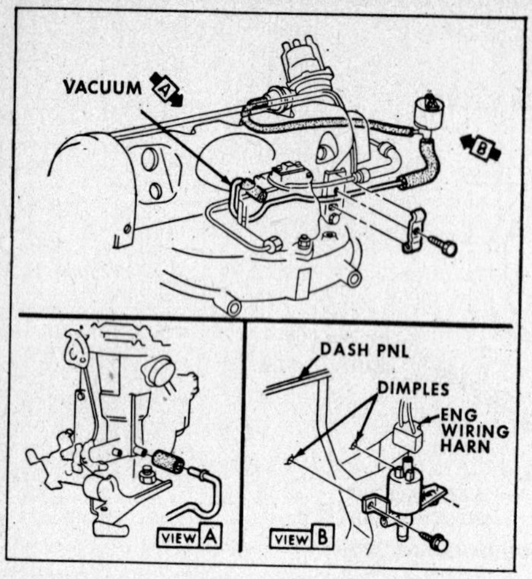

Fig. CL89 TCS vacuum hose routing. 1973-74 Chevrolet Vega

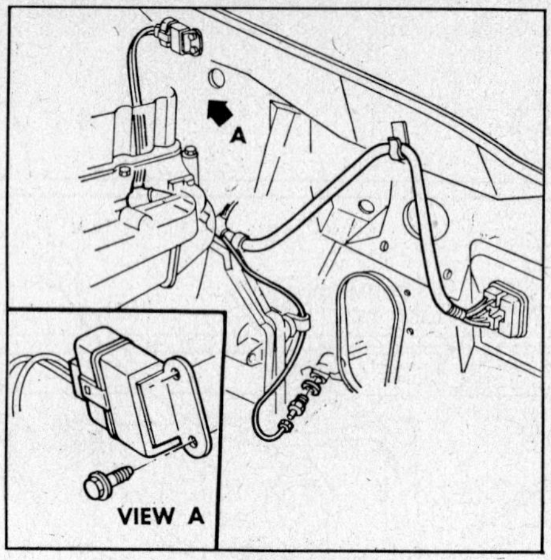

Fig. CL91 TCS time relay location 1971-74 Chevrolet (typical)

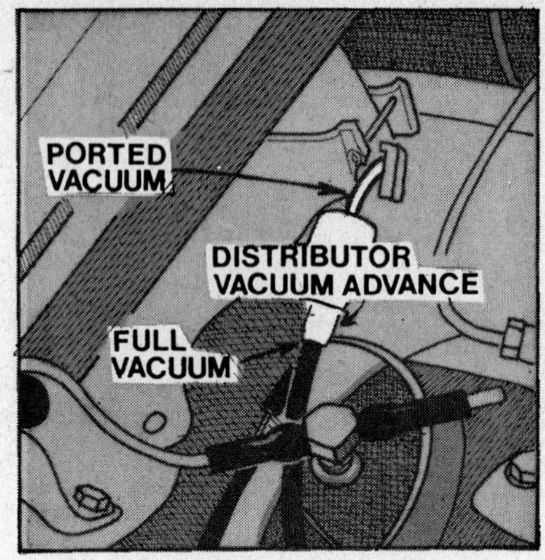

Fig. CL92 Thermo Override Switch. 1973-75 Chevrolet Corvette

Fig. CL90 TCS location. 1971-74 Chevrolet Vega →

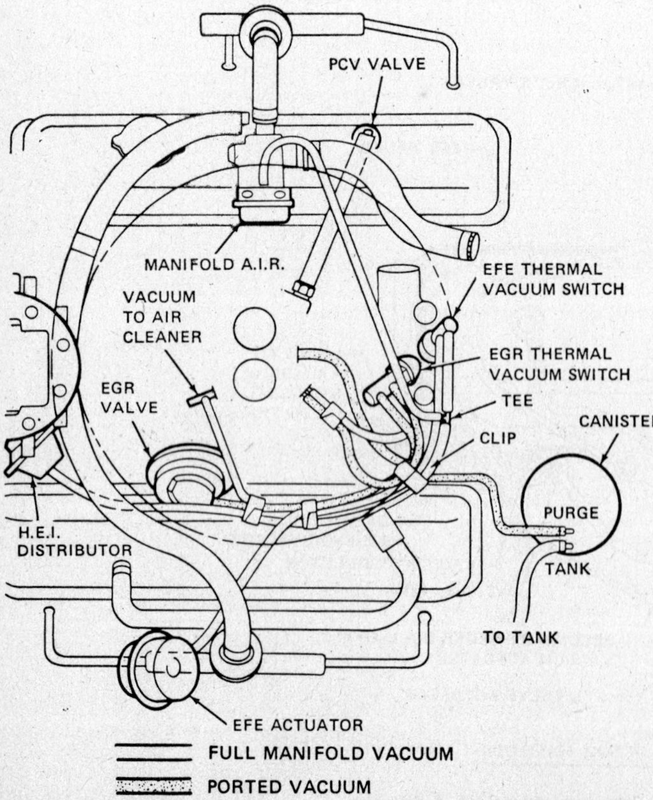

Fig. CL93 Vacuum hose routing. 1975 Monza V8-262

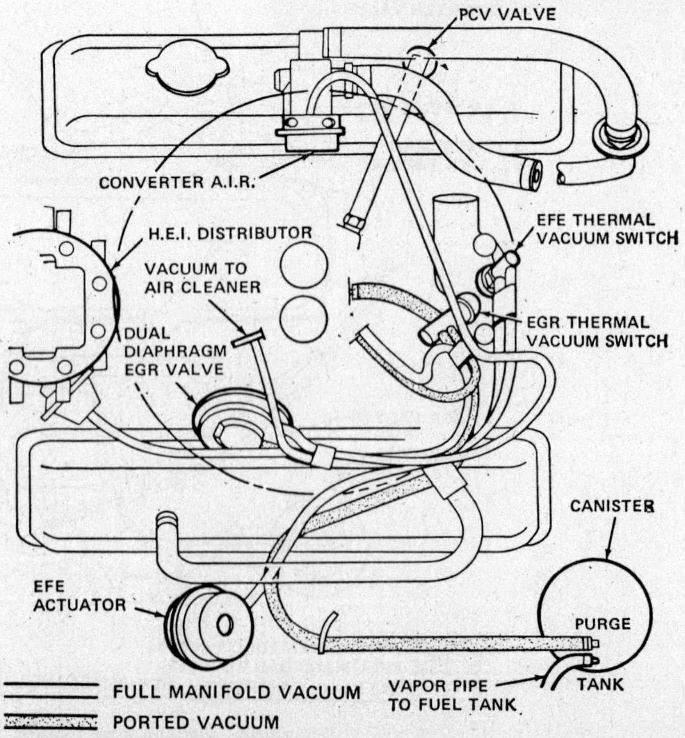

Fig. CL94 Vacuum hose routing. 1975 Monza V8-350

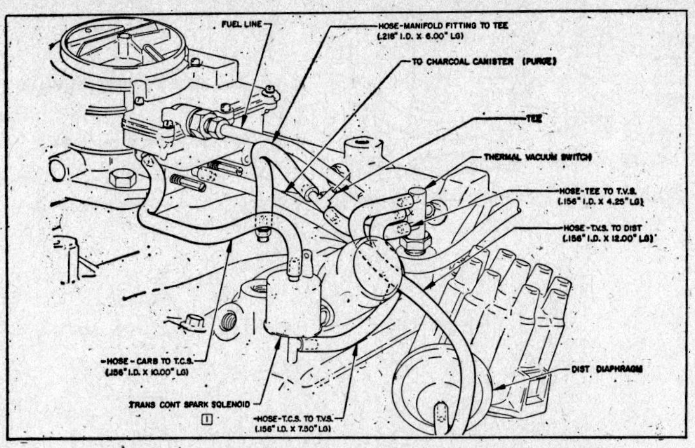

Fig. CL95 Vacuum hose routing. 1971 Buick Le Sabre V8-350 2 bar. carb. with A/C or HD cooling system

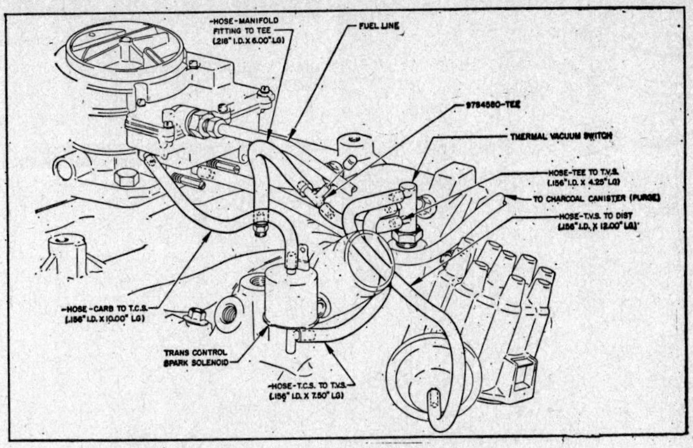

Fig. CL96 Vacuum hose routing. 1971 Buick Skylark V8-350 2 bar. carb. with HD cooling system

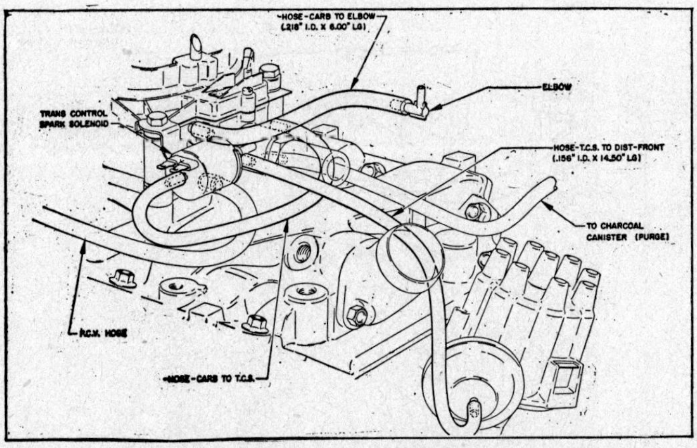

Fig. CL97 Vacuum hose routing. 1971 Buick Skylark V8-350 4 bar. car. except HD cooling system or A/C

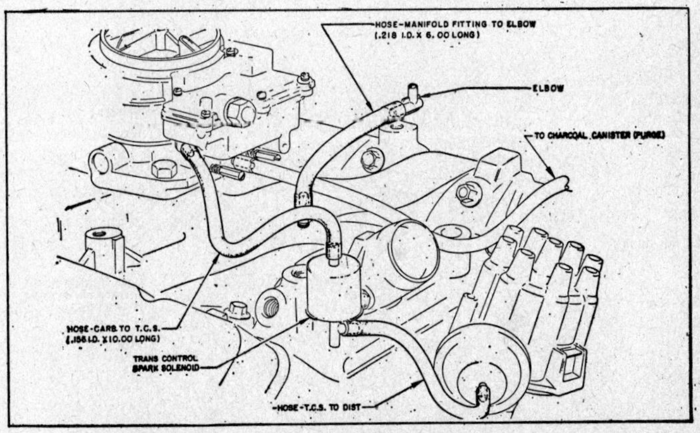

Fig. CL98 Vacuum hose routing. 1971 Buick Le Sabre V8-350 2 bar. carb. except A/C or HD cooling system and Skylark V8-350 2 bar. carb. except HD cooling system

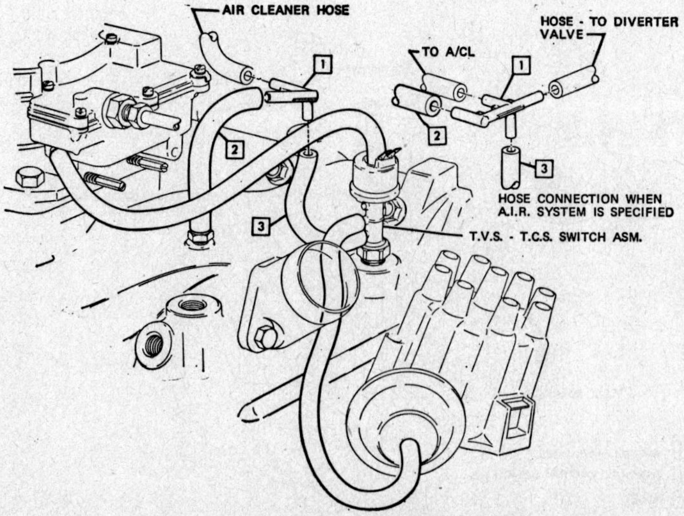

Fig. CL99 Vacuum hose routing. 1972 Buick V8-350 2 bar. carb. with A/C

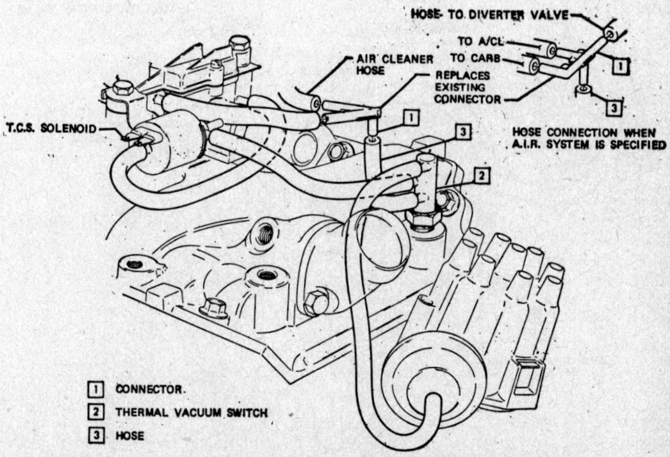

Fig. CL100 Vacuum hose routing. 1972 Buick V8-350 4 bar. carb. with HD cooling system except A/C

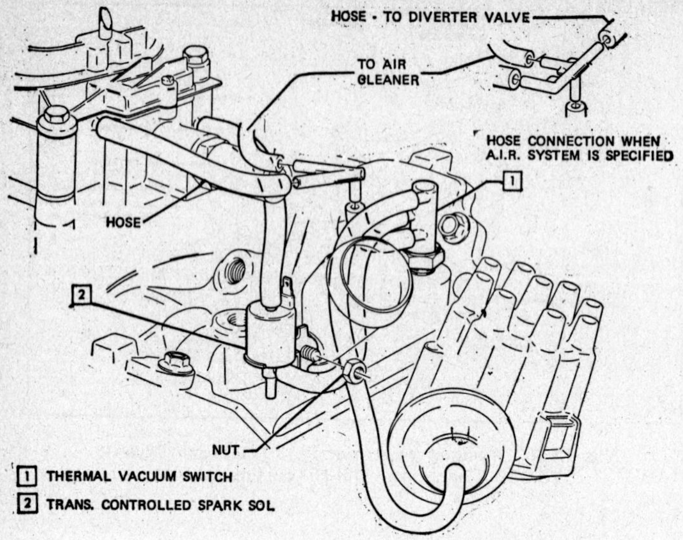

Fig. CL101 Vacuum hose routing. 1972 Buick V8-350 4 bar. carb. with A/C

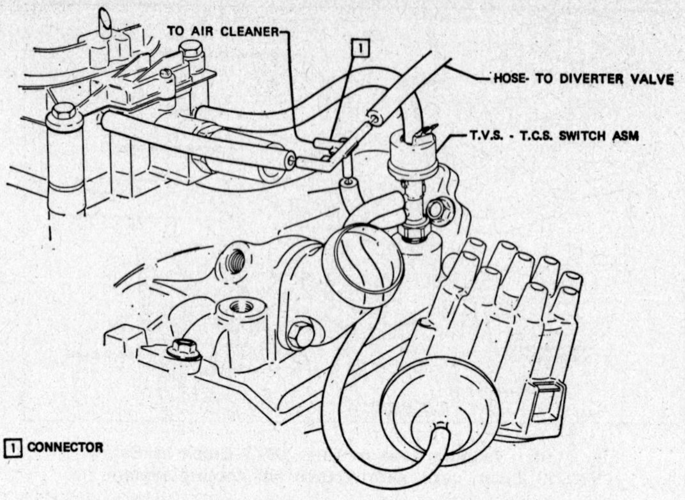

Fig. CL102 Vacuum hose routing. 1972 Buick Centurion, Electra, Estate Wagon, Le Sabre and Riviera V8-455 with A/C

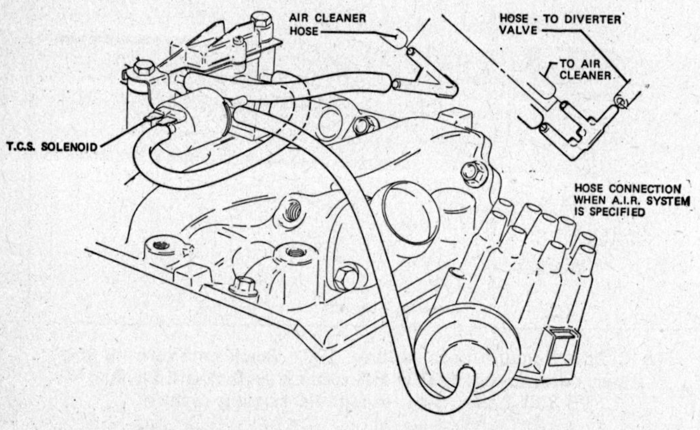

Fig. CL103 Vacuum hose routing. 1972 Buick Le Sabre and Skylark except A/C and HD cooling system

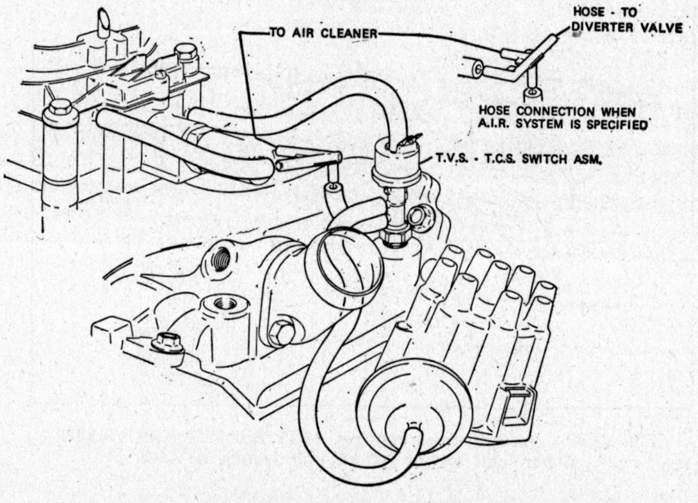

Fig. CL104 Vacuum hose routing. 1972 Buick Le Sabre & Skylark V8-350 4 bar. carb. with A/C

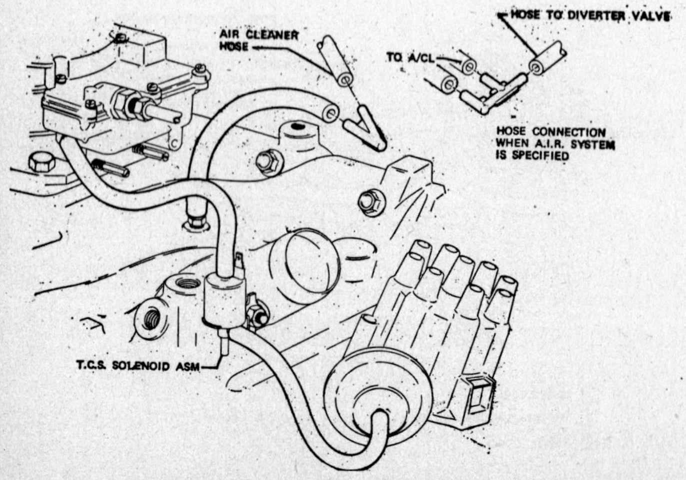

Fig. CL105 Vacuum hose routing. 1972 Buick V8-350 2 bar. carb. except A/C or HD cooling system

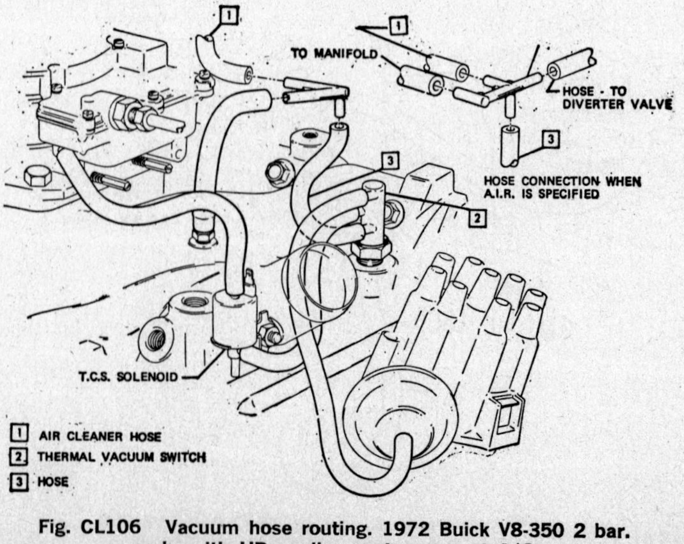

Fig. CL106 Vacuum hose routing. 1972 Buick V8-350 2 bar. carb. with HD cooling system except A/C

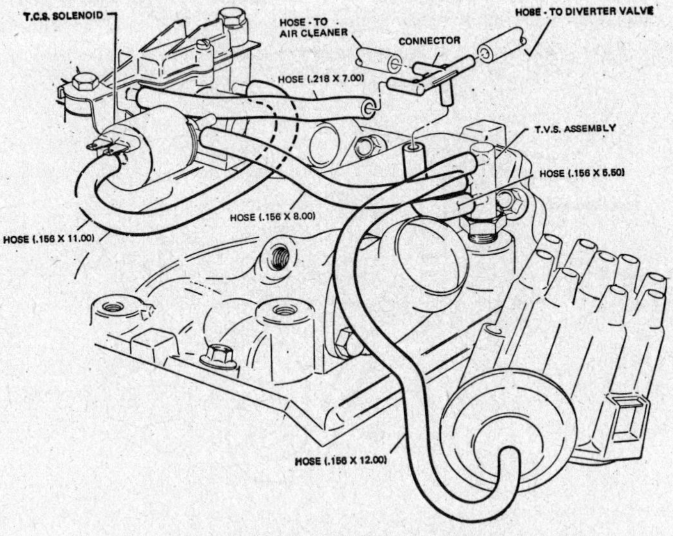

Fig. CL107 Vacuum hose routing. 1973 Buick V8-455 Stage 1 with automatic transmission

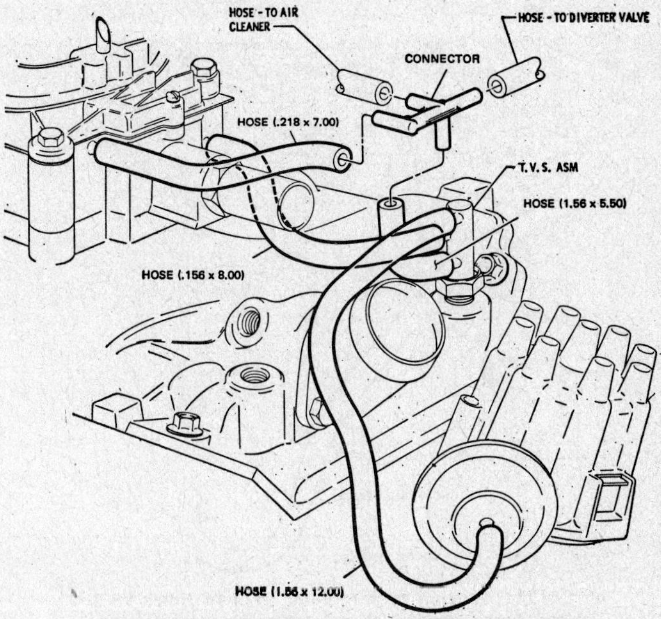

Fig. CL108 Vacuum hose routing. 1973 Buick V8-350 4 bar. carb. with A/C and automatic transmission and all V8-455 with automatic transmission except Stage 1

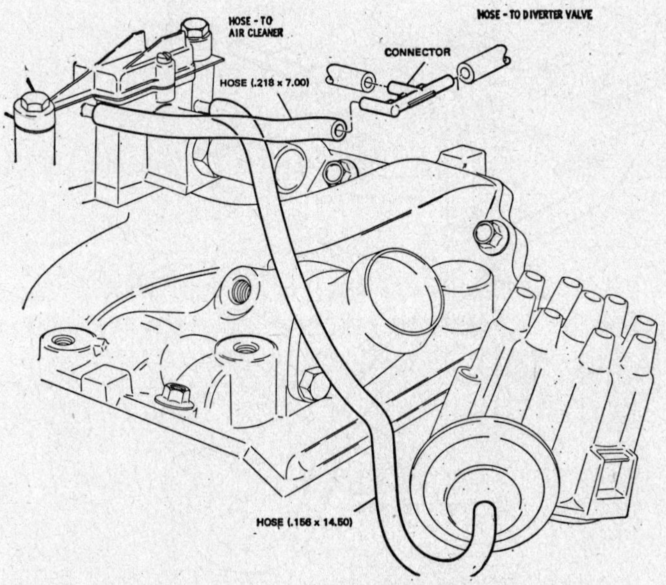

Fig. CL109 Vacuum hose routing. 1973 Buick V8-350 4 bar. carb. except A/C or automatic transmission

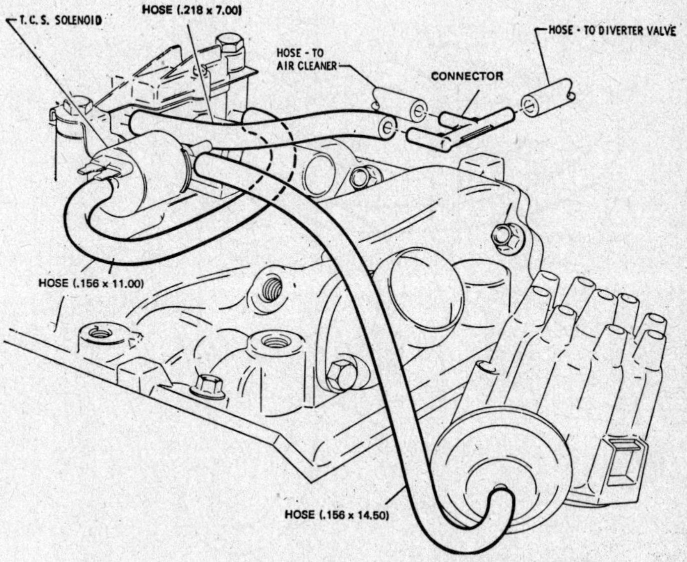

Fig. CL110 Vacuum hose routing. 1973 Buick V8-350 4 bar. with manual transmission

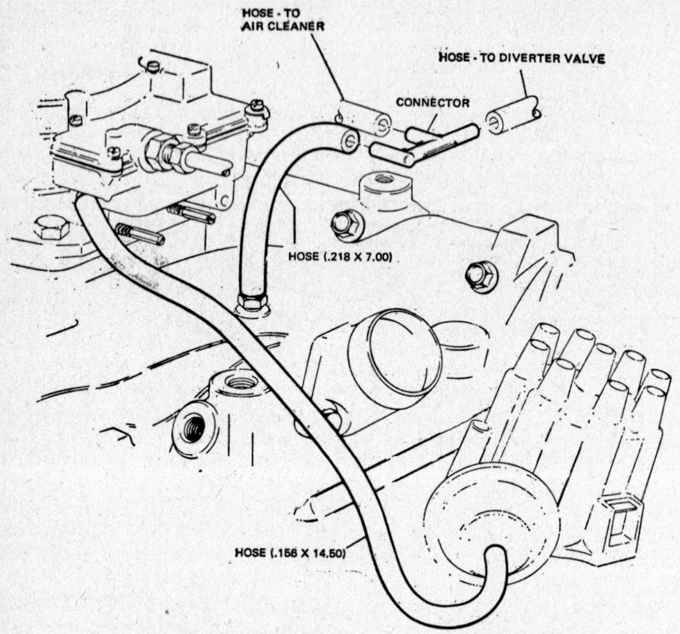

Fig. CL111 Vacuum hose routing. 1973-74 Buick V8-350 2 bar. carb. except A/C and automatic transmission

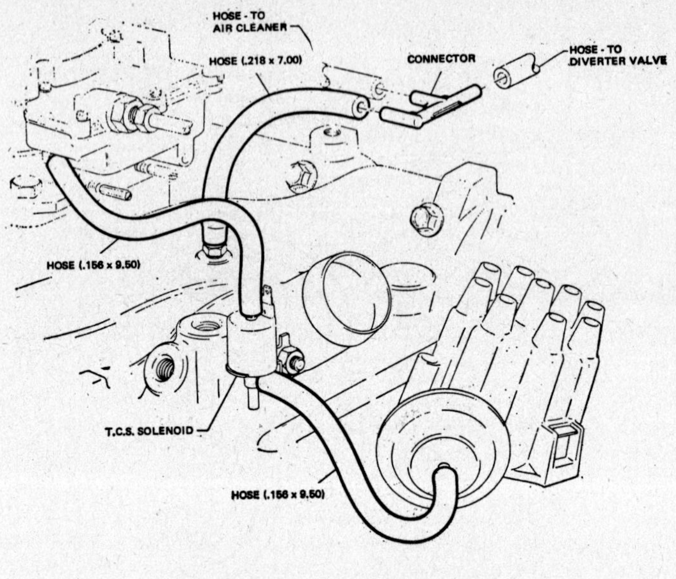

Fig. CL112 Vacuum hose routing. 1972-73 Buick V8-350 2 bar. carb. with manual transmission

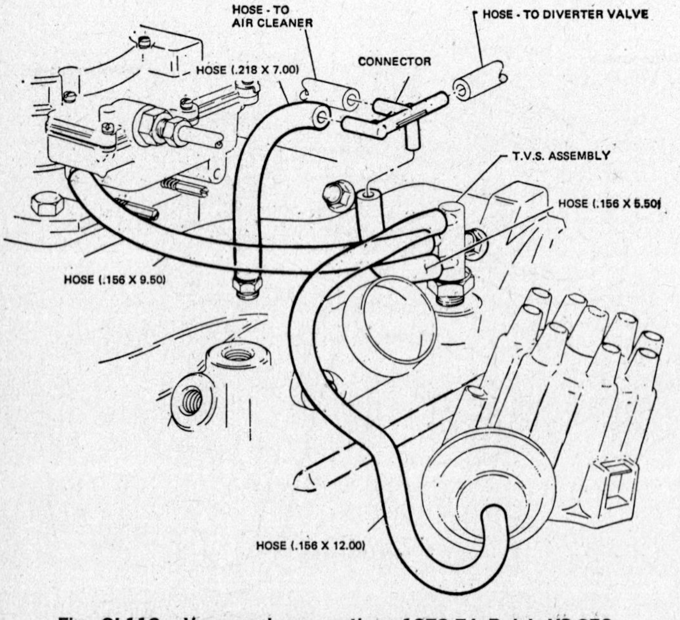

Fig. CL113 Vacuum hose routing. 1973-74 Buick V8-350 2 bar. carb. with A/C and HD cooling system

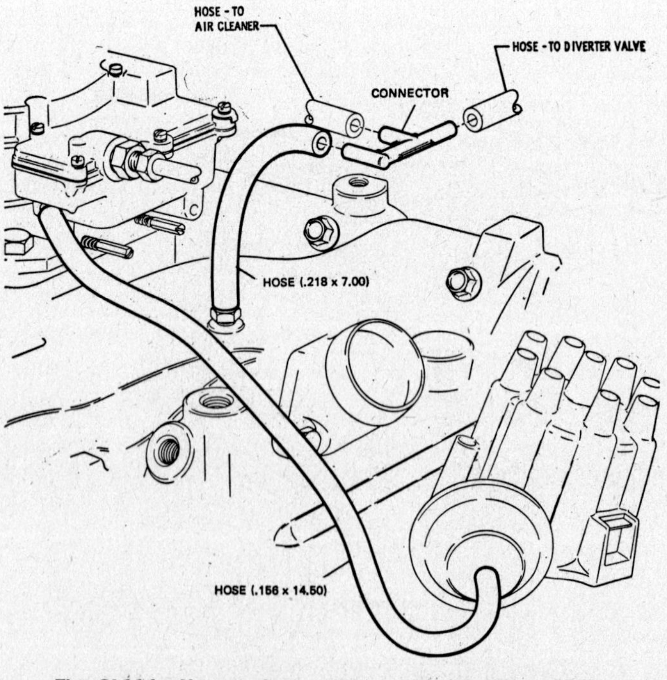

Fig. CL114 Vacuum hose routing. 1973-74 Buick V8-350 2 bar. carb. except A/C or HD cooling system

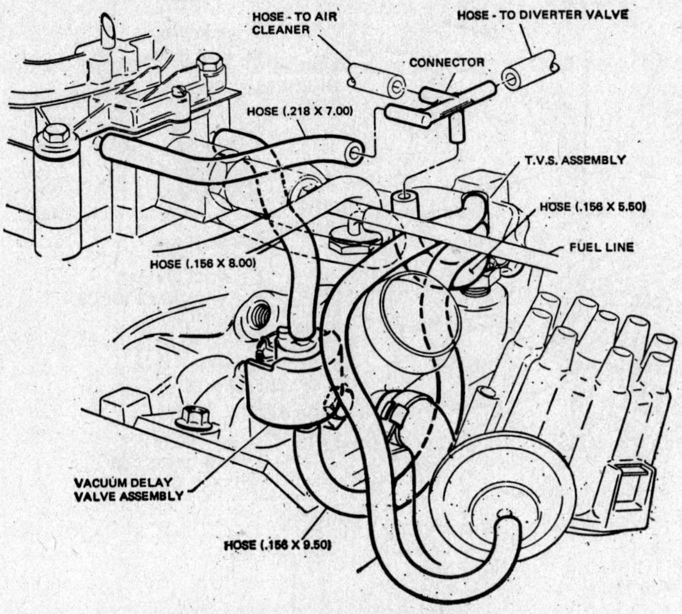

Fig. CL115 Vacuum hose routing. 1973-74 Buick V8-350 4 bar. carb.

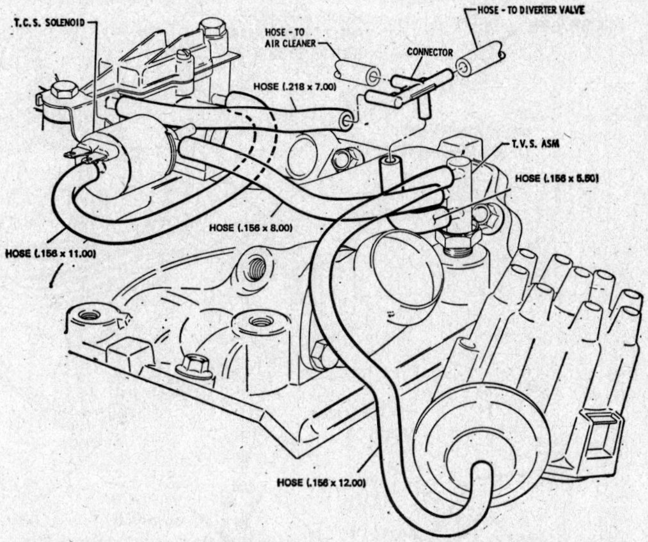

Fig. CL116 Vacuum hose routing. 1972-74 Buick intermediate models with Stage 1 engine

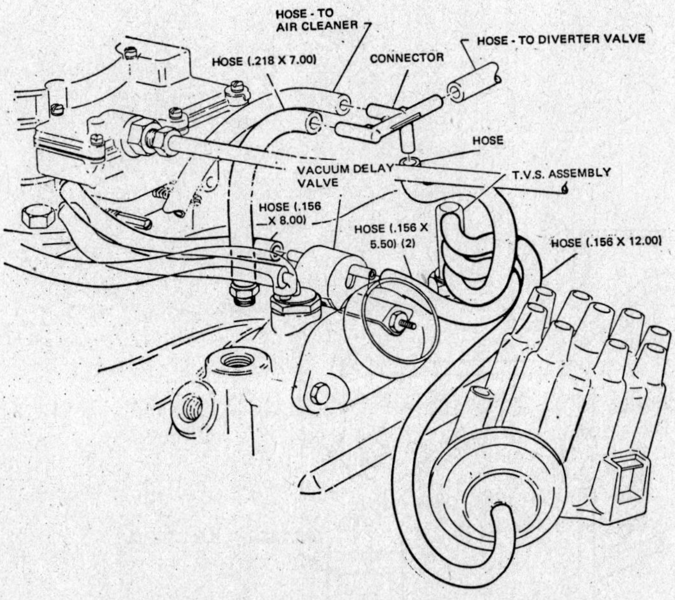

Fig. CL117 Vacuum hose routing. 1974 Buick intermediate models and Le Sabre with V8-455 2 bar. carb.

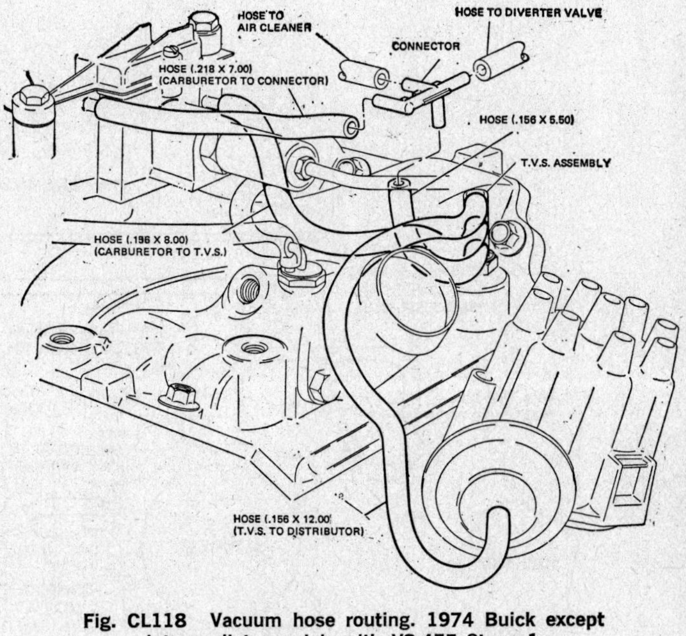

Fig. CL118 Vacuum hose routing. 1974 Buick except intermediate models with V8-455 Stage 1

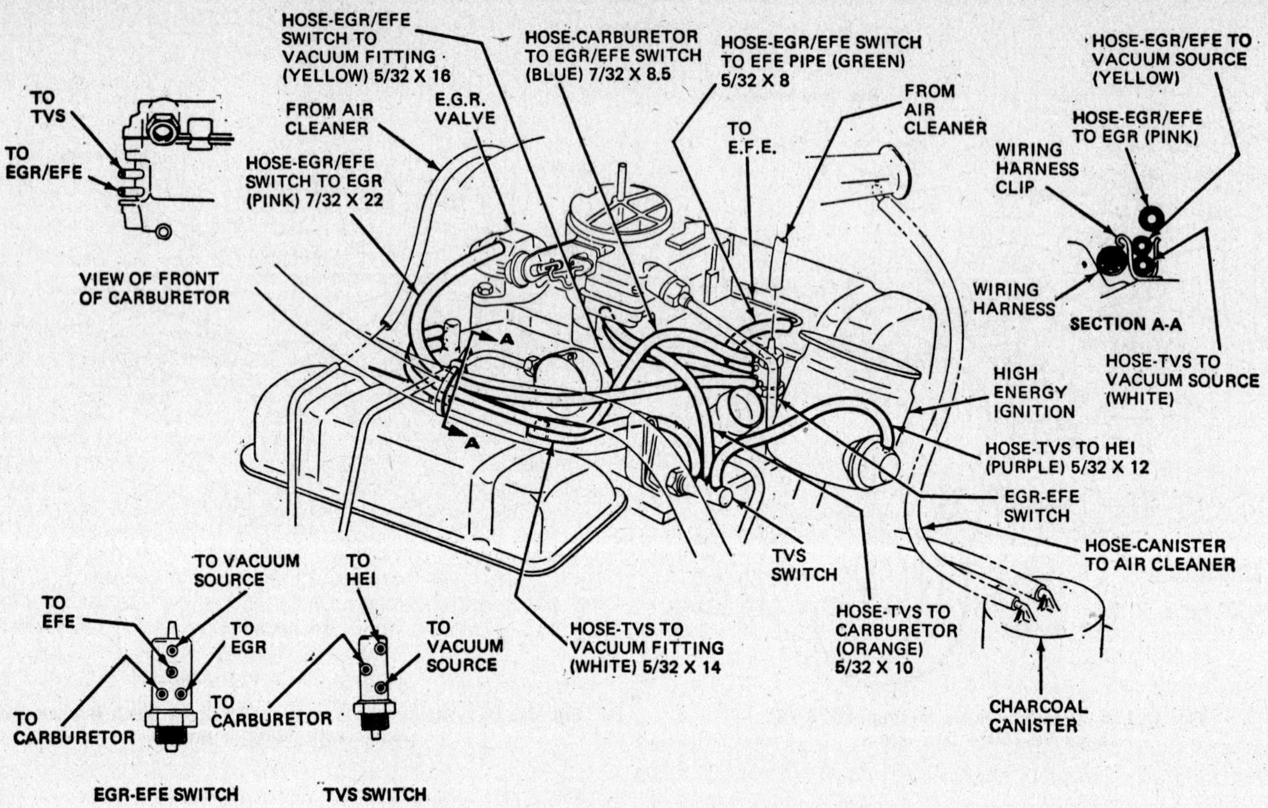

Fig. CL119 Vacuum hose routing. 1975 Buick V6-231

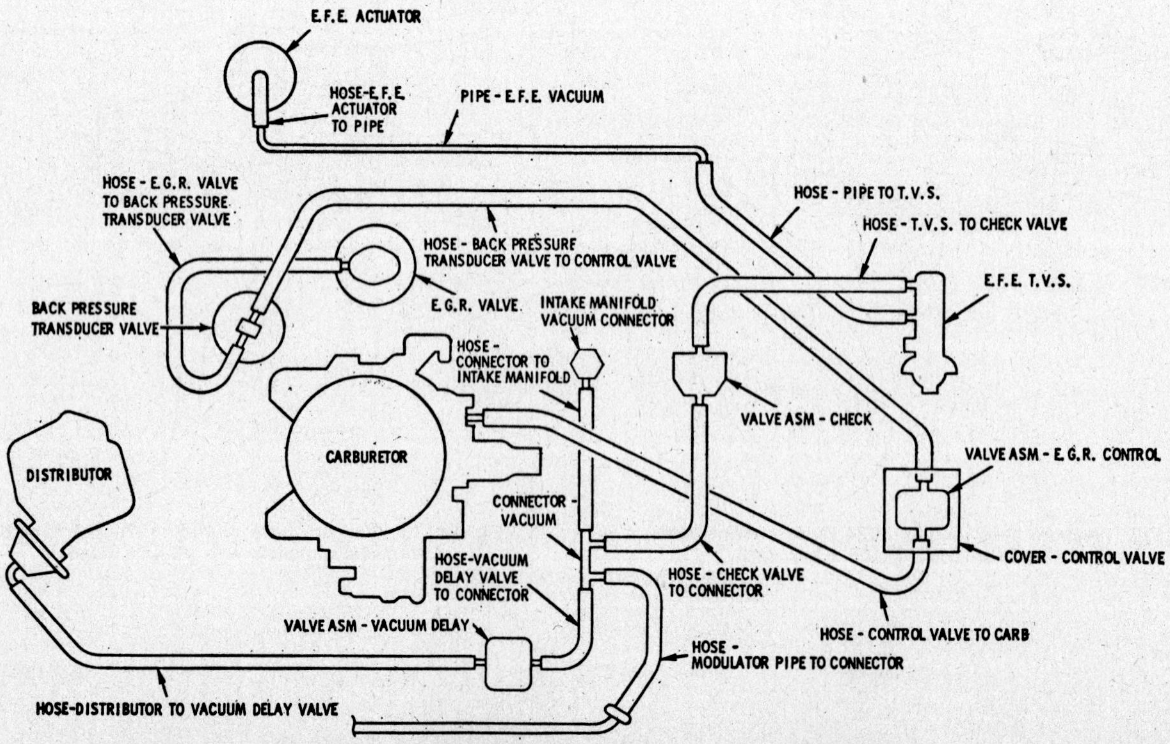

Fig. CL120 Vacuum hose routing. 1975 Buick V8-260 exc. Calif.

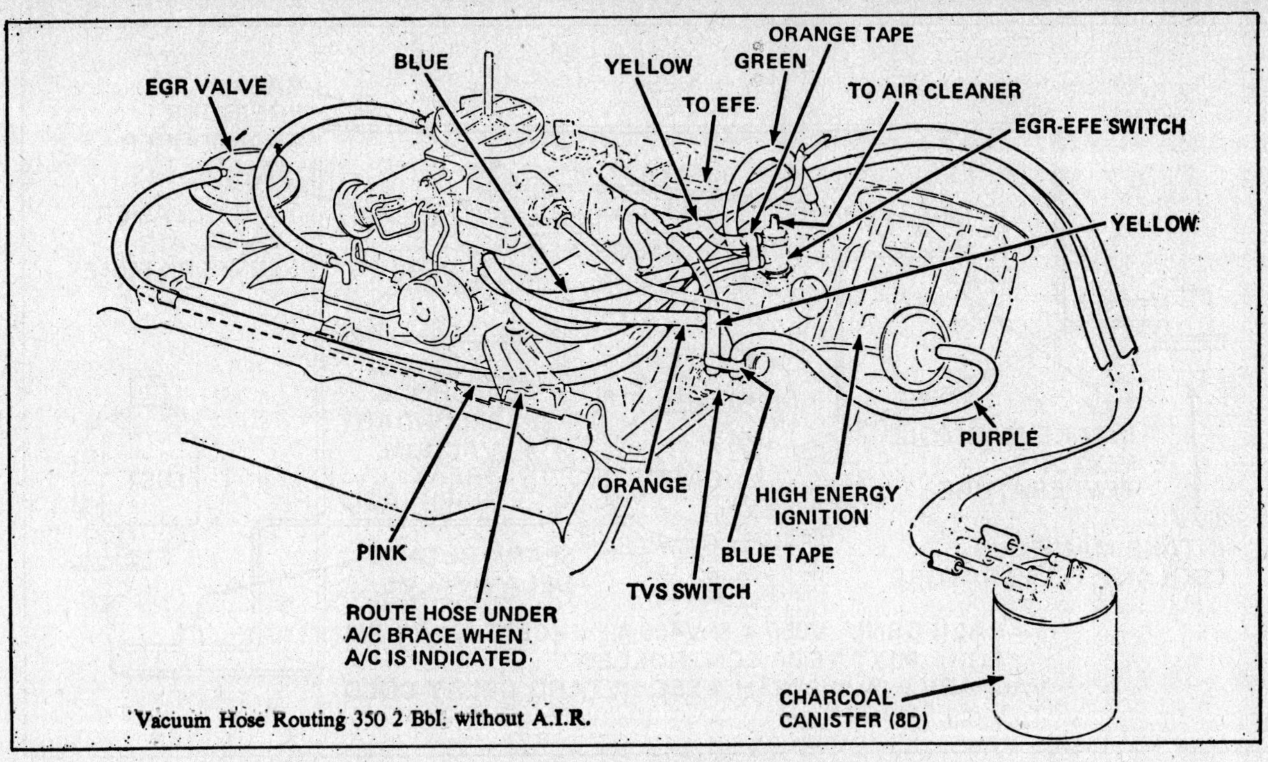

Fig. CL121 Vacuum hose routing. 1975 Buick V8-350 2 bar. carb. and Pontiac V8-350 without AIR

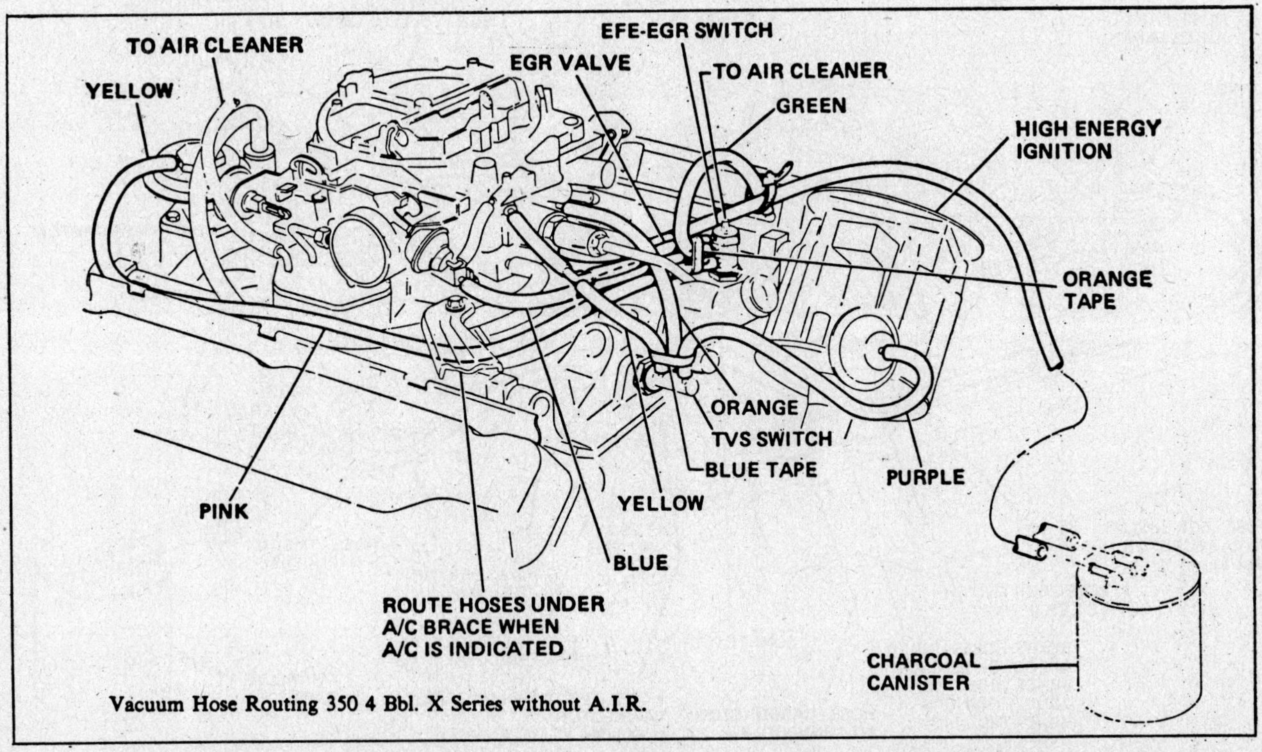

Fig. CL122 Vacuum hose routing. 1975 Buick and Pontiac V8-350 4 bar. carb. without AIR

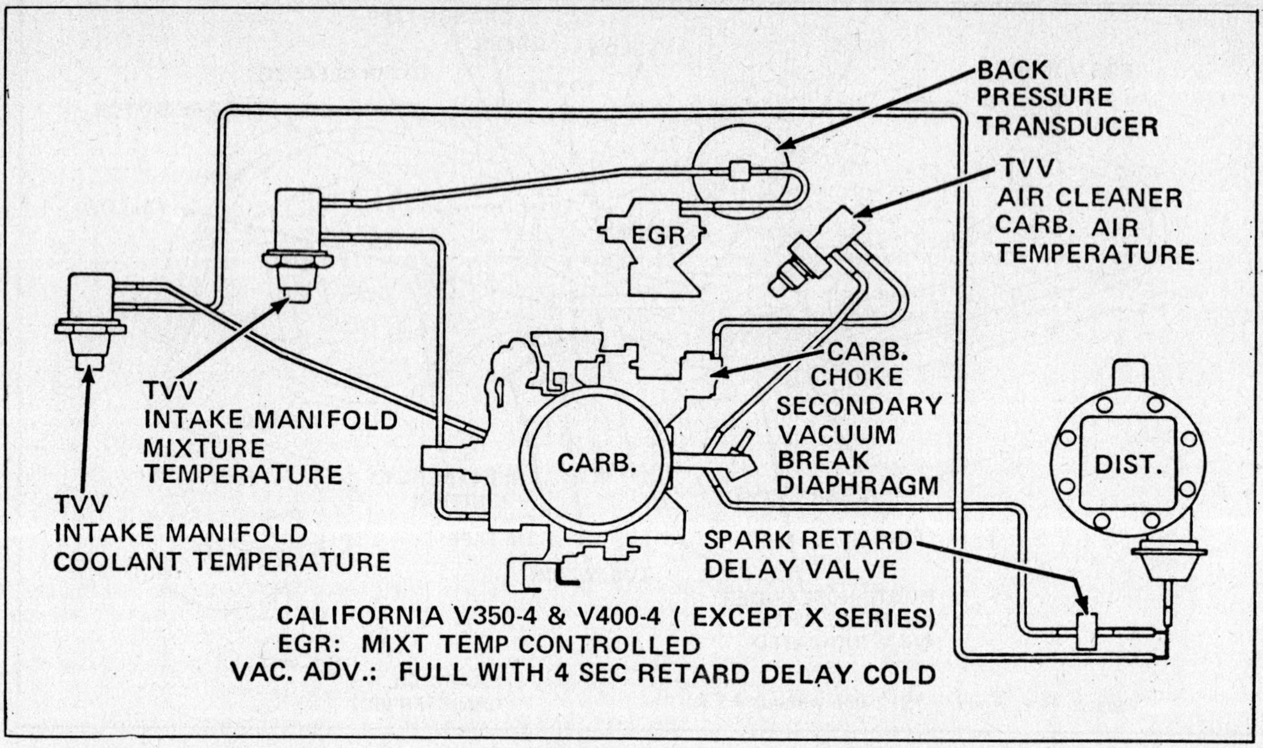

Fig. CL122A Vacuum hose routing. 1975 Pontiac exc. Ventura V8-350, 400 Calif.

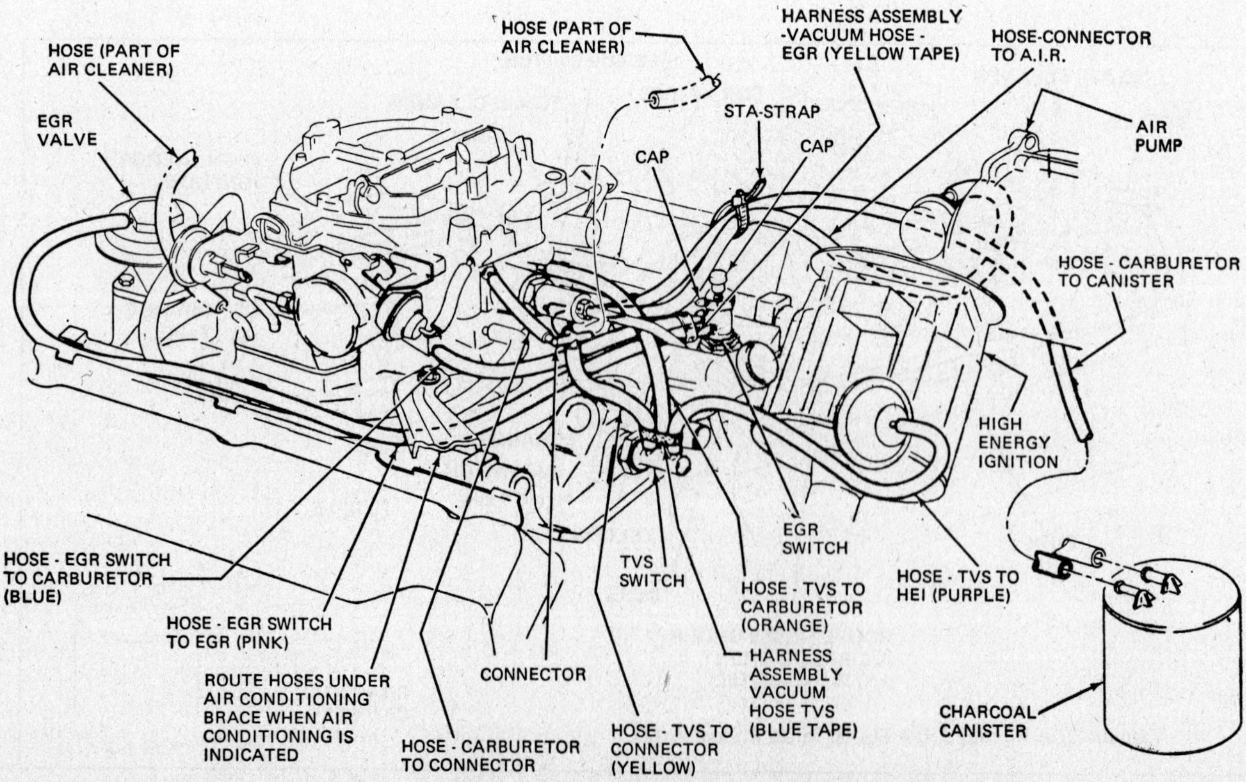

Fig. CL123 Vacuum hose routing. 1975 Buick V8-350 4 bar. carb. with AIR and EFE system

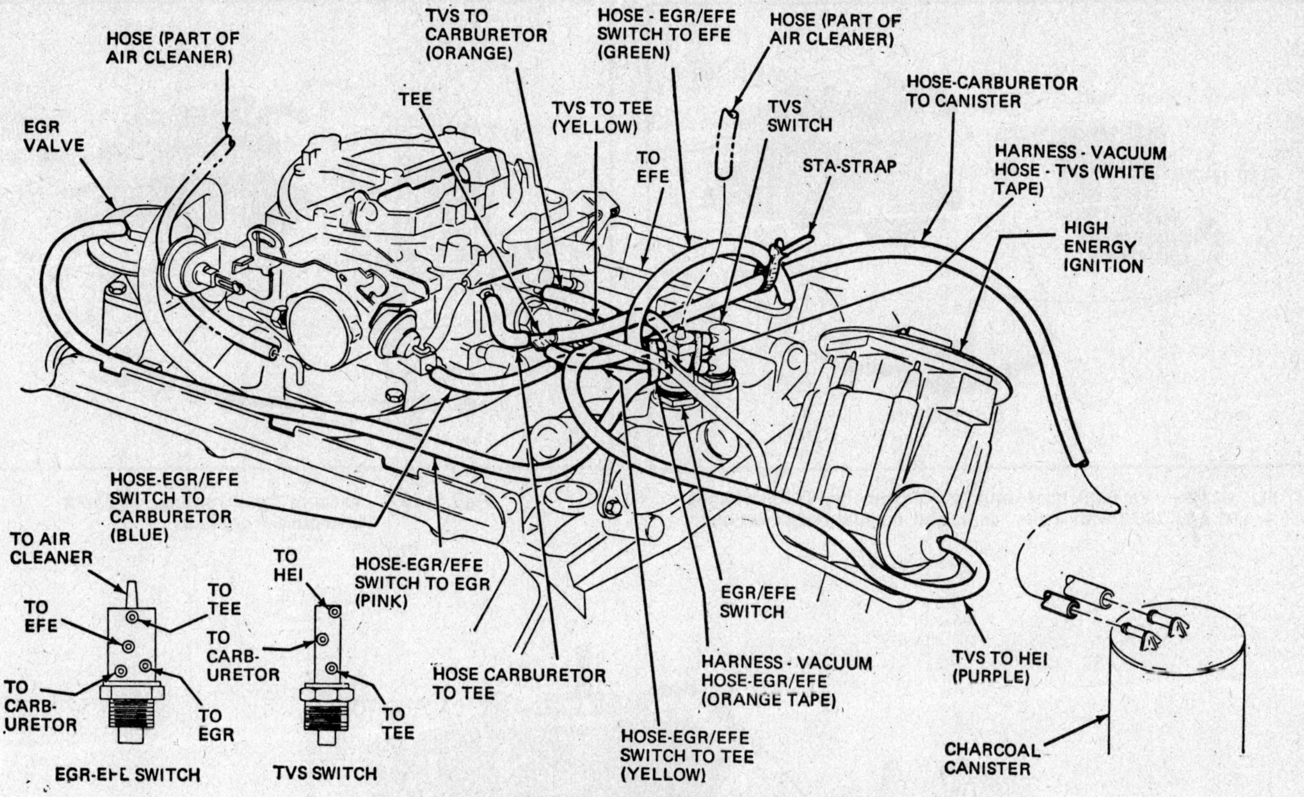

Fig. CL124 Vacuum hose routing. 1975 Buick V8-455

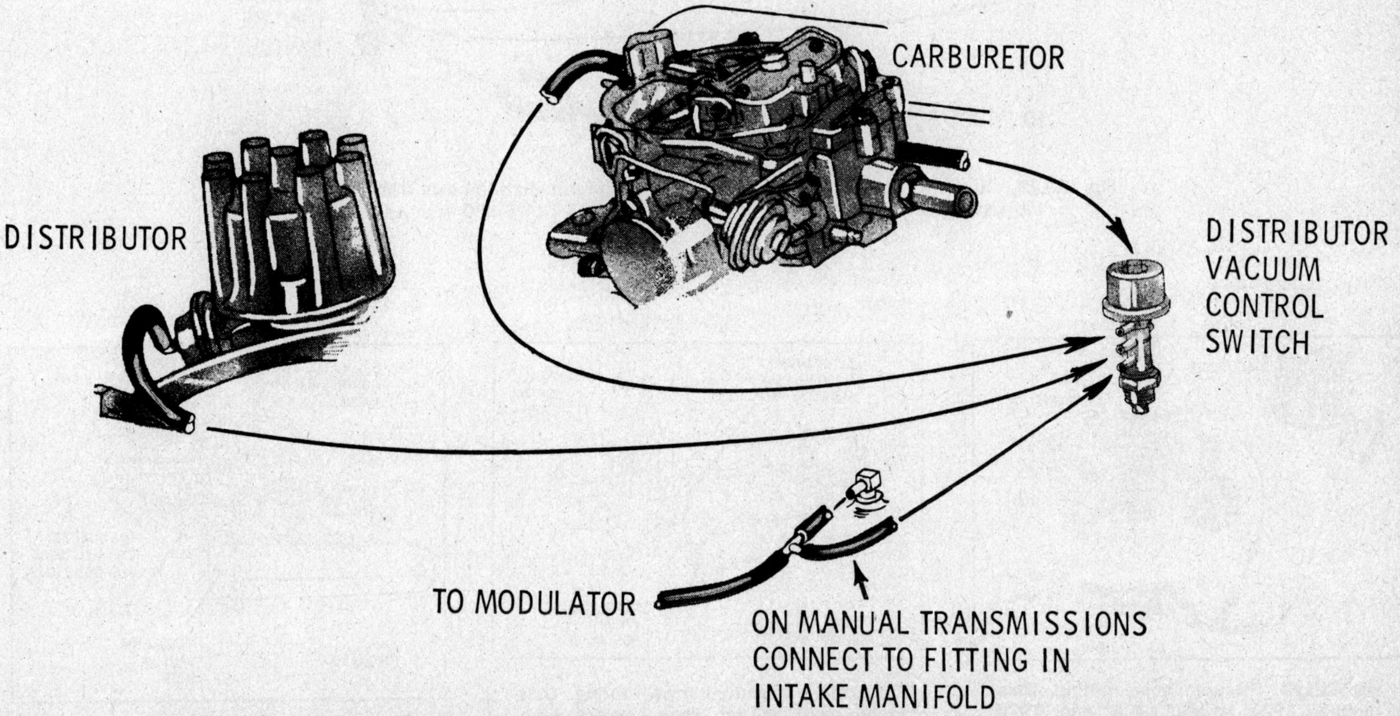

Fig. CL125 Vacuum hose routing. 1971-72 Oldsmobile except Toronado and V8-350 with manual transmission

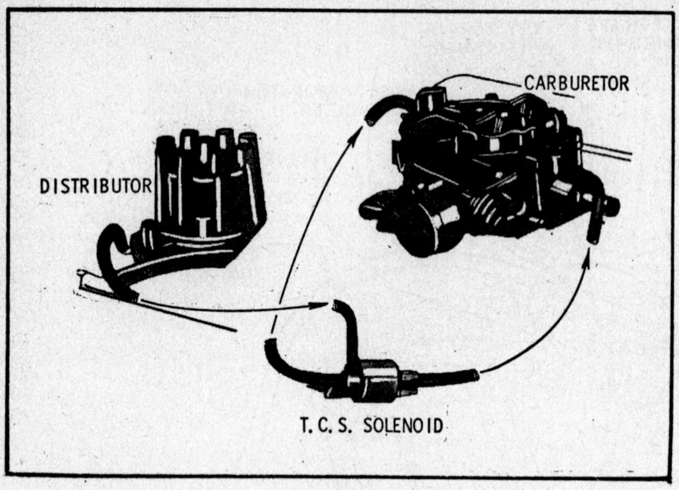

Fig. CL126 Vacuum hose routing. Oldsmobile, 1971 without A/C and 1972 with 4 bar. carb. and manual transmission

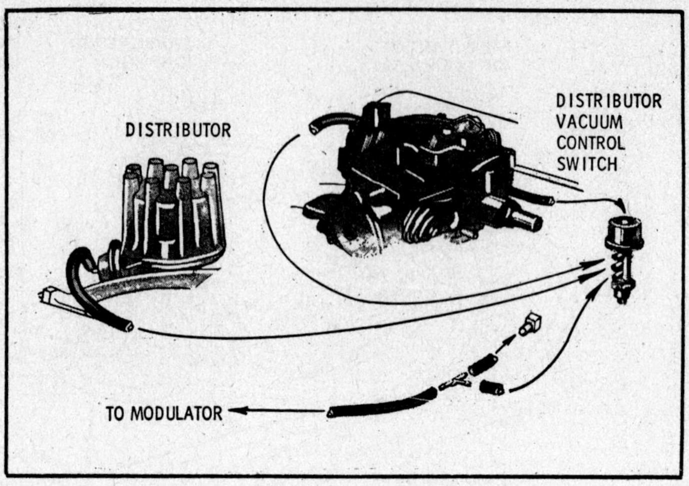

Fig. CL127 Vacuum hose routing. 1971-72 Oldsmobile Toronado

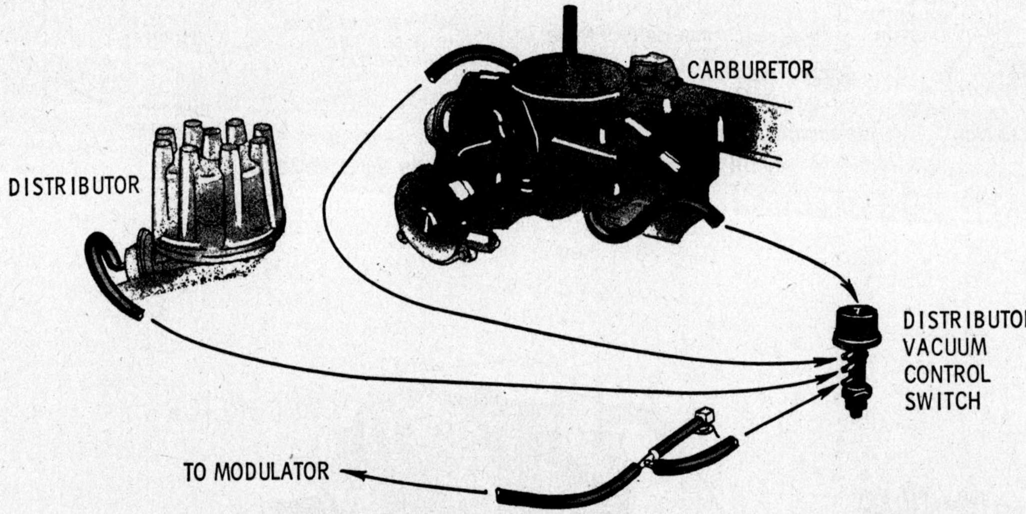

Fig. CL128 Vacuum hose routing. Oldsmobile, 1971 intermediate models with V8-455, full size models with V8-455 and A/C and all 1971-72 V8-350 with A/C

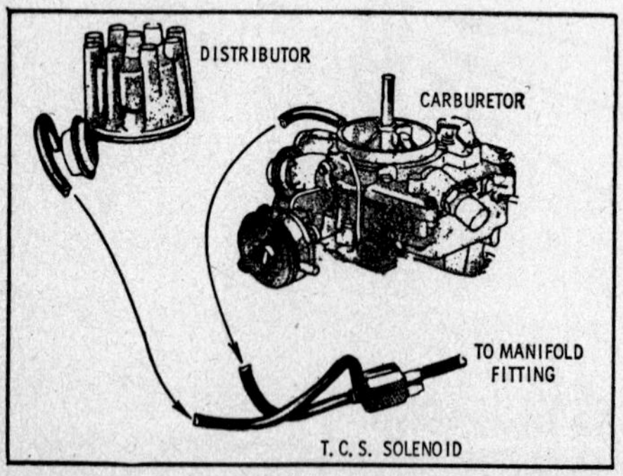

Fig. CL129 Vacuum hose routing. Oldsmobile, 1971 without A/C and 1972 V8-350 with manual transmission and no A/C

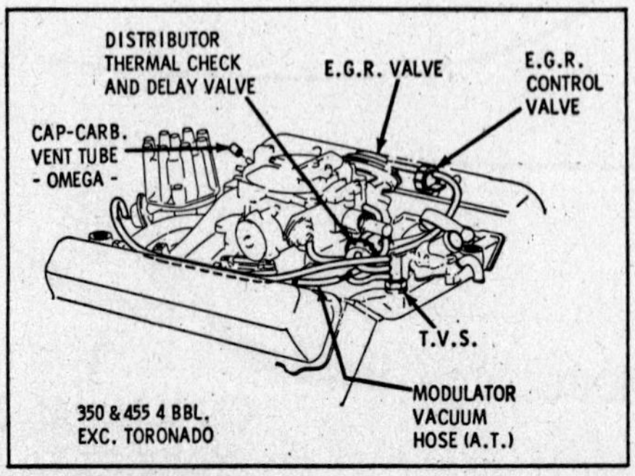

Fig. CL130 Vacuum hose routing. Oldsmobile with 4 bar. carb. except Toronado and V8-350 with manual transmission

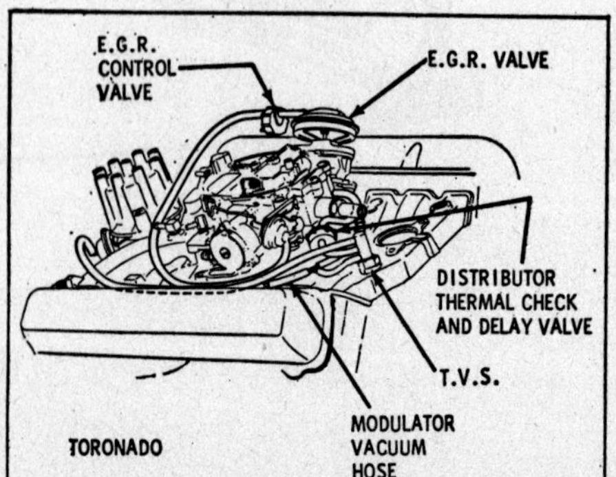

Fig. CL131 Vacuum hose routing. 1973 Oldsmobile Toronado

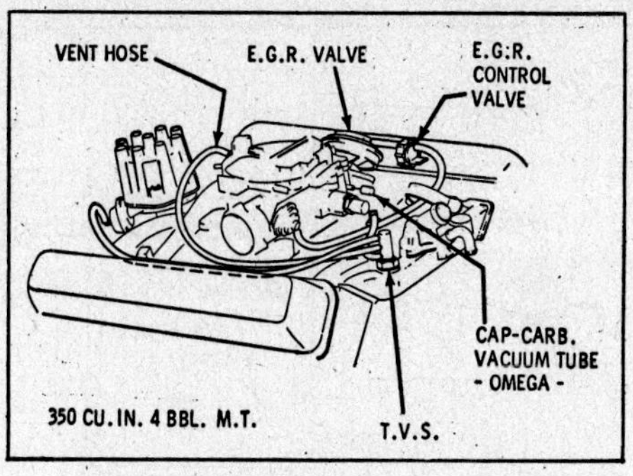

Fig. CL132 Vacuum hose routing. 1973 Oldsmobile with 2 bar. carb.

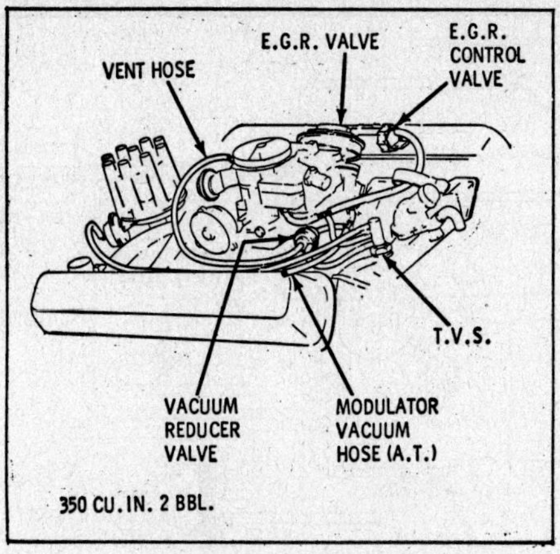

Fig. CL133 Vacuum hose routing. 1973 Oldsmobile V8-350 4 bar. carb.

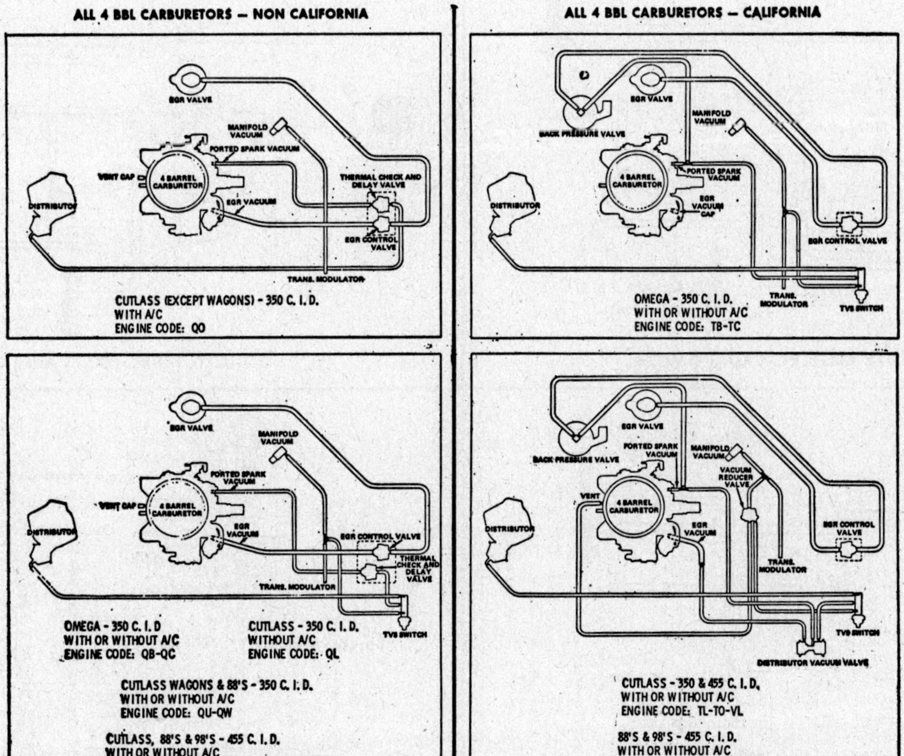

Fig. CL134 Vacuum hose routing. 1974 Oldsmobile

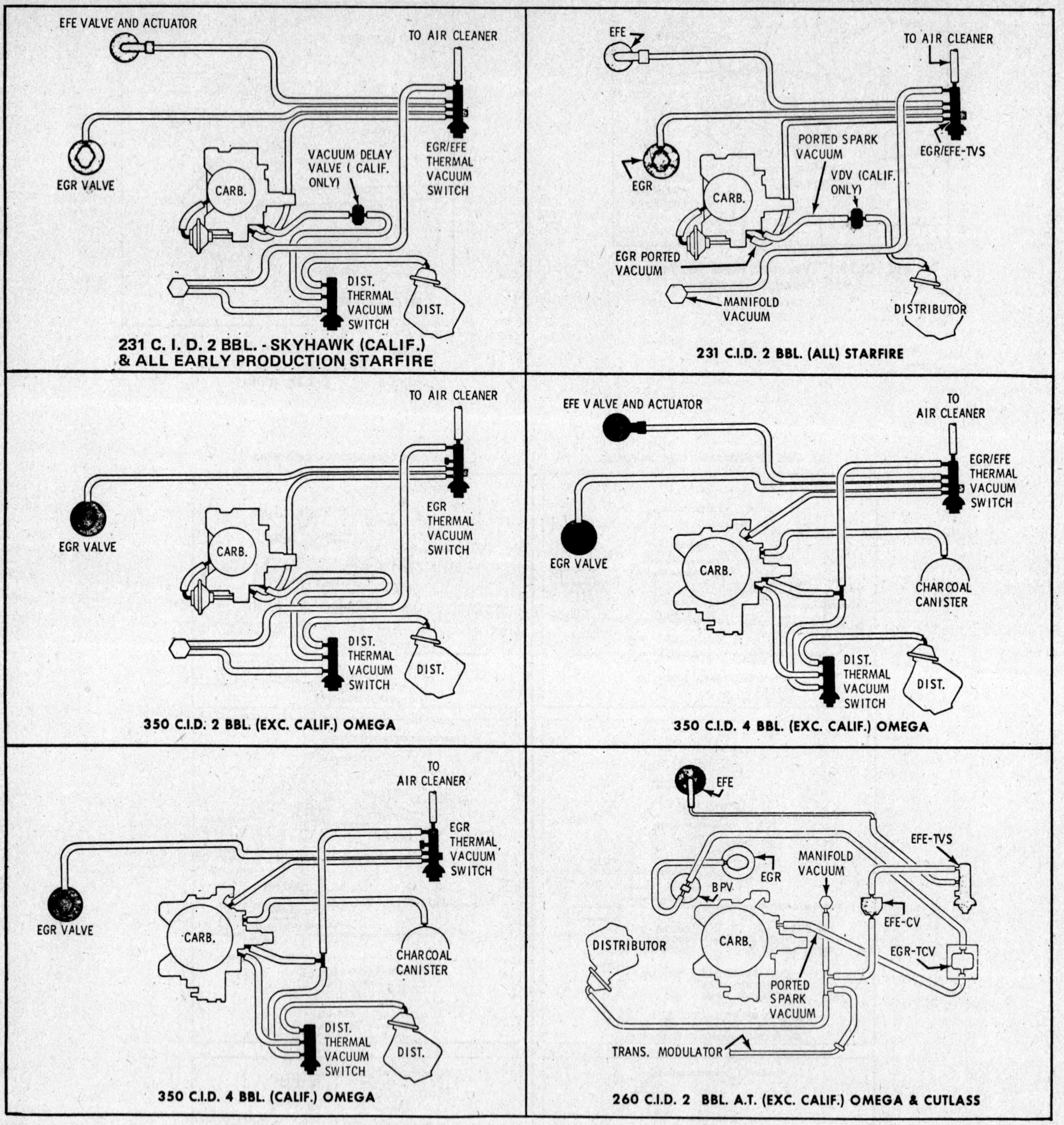

Fig. CL135 Vacuum hose routing. 1975 Buick and Oldsmobile

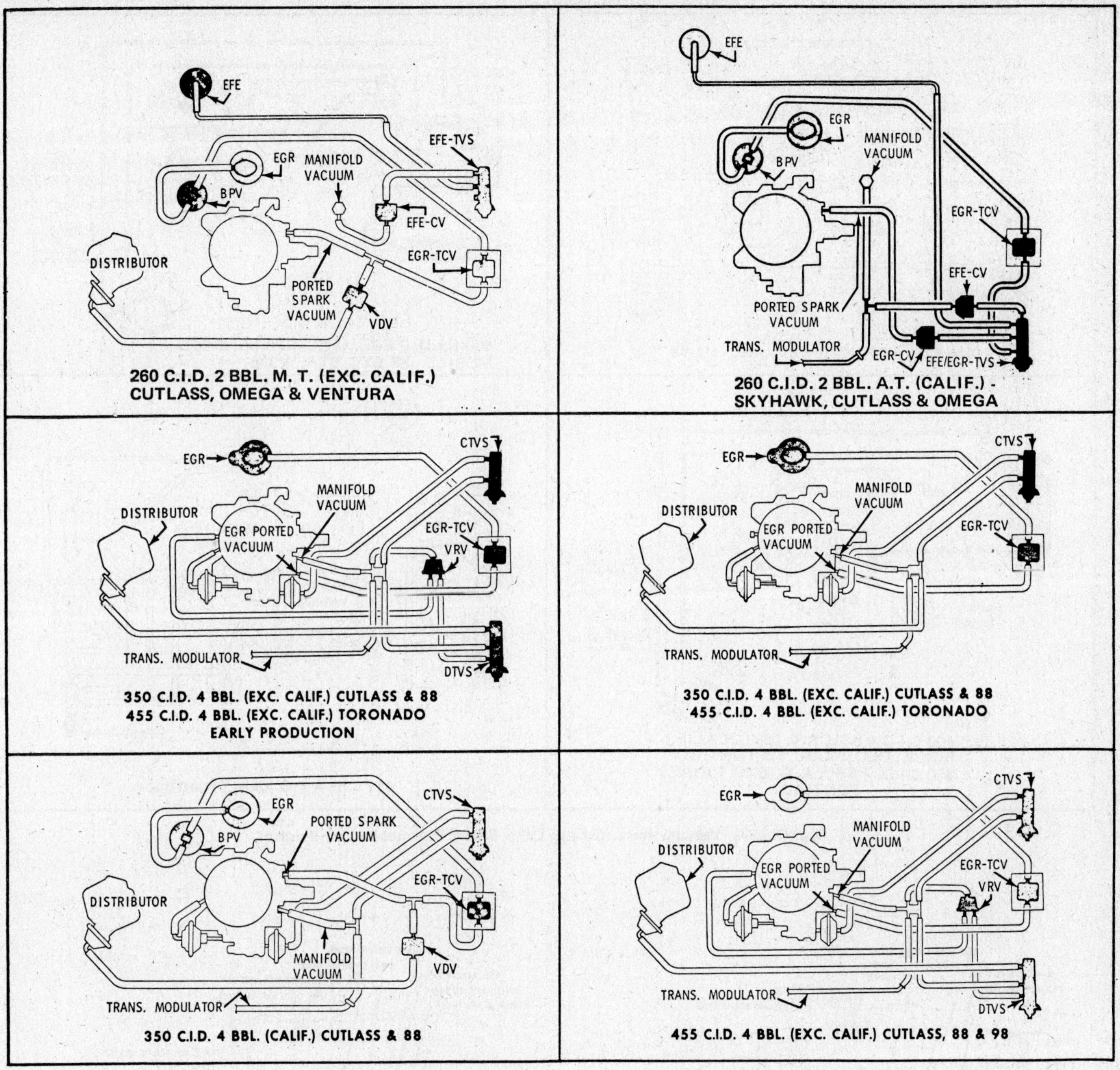

Fig. CL136 Vacuum hose routing. 1975 Buick, Oldsmobile, and Pontiac

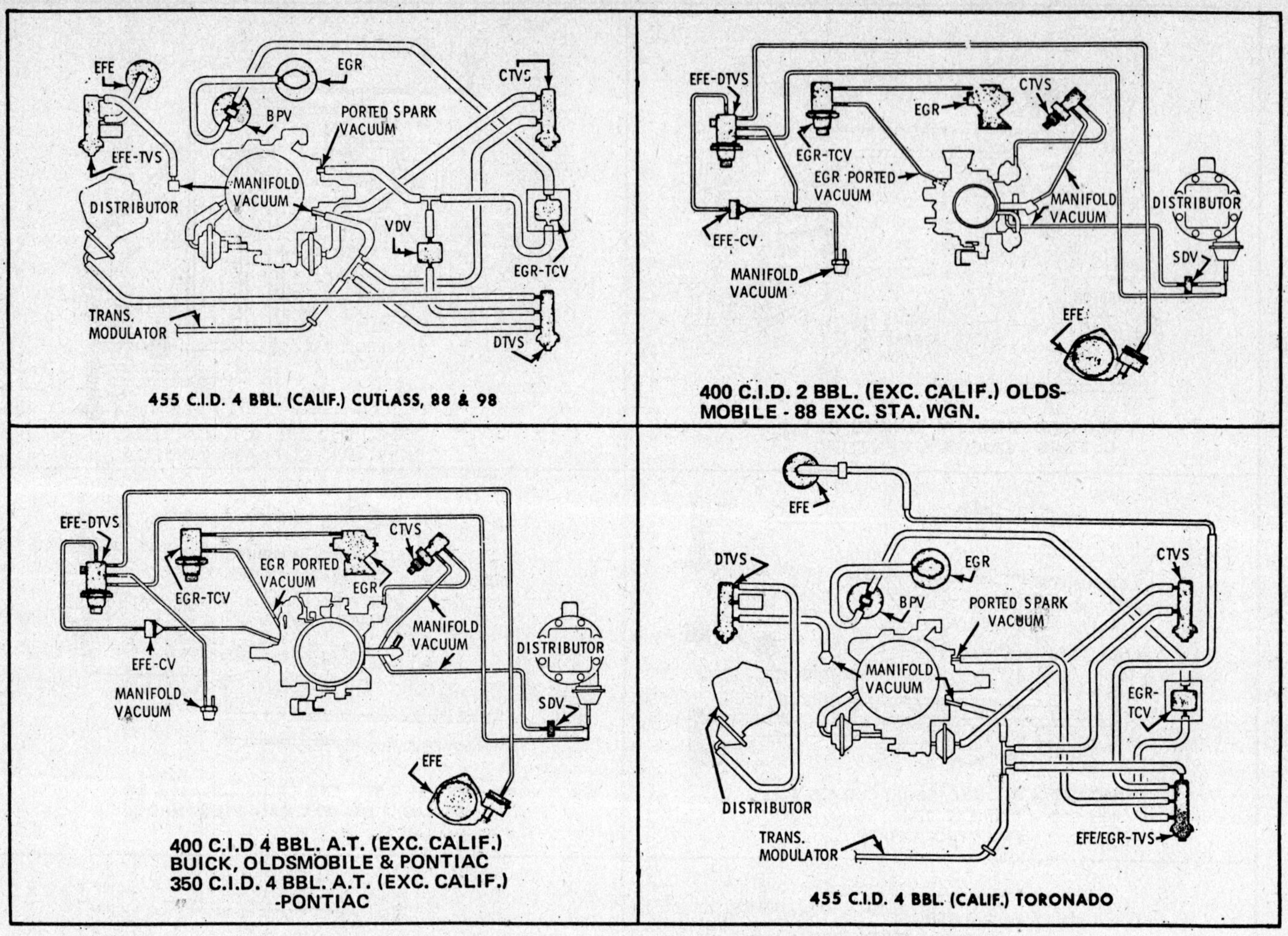

Fig. CL137 Vacuum hose routing. 1975 Buick, Oldsmobile and Pontiac

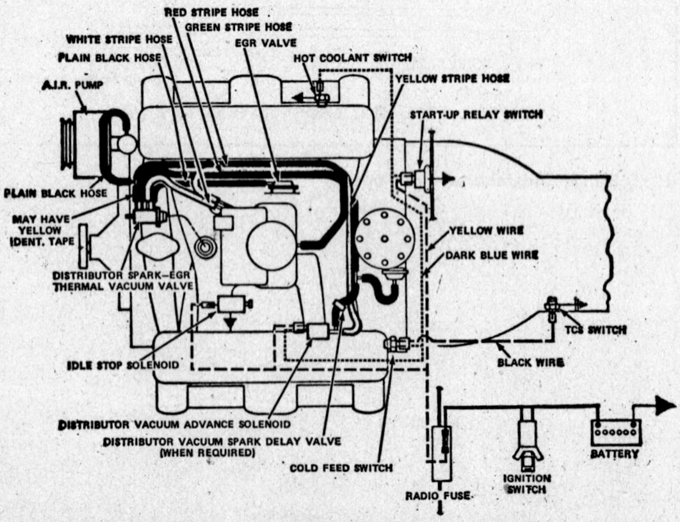

Fig. CL138 Vacuum and wiring. 1974 Pontiac with 2 bar. carb. and Air Injector Reactor

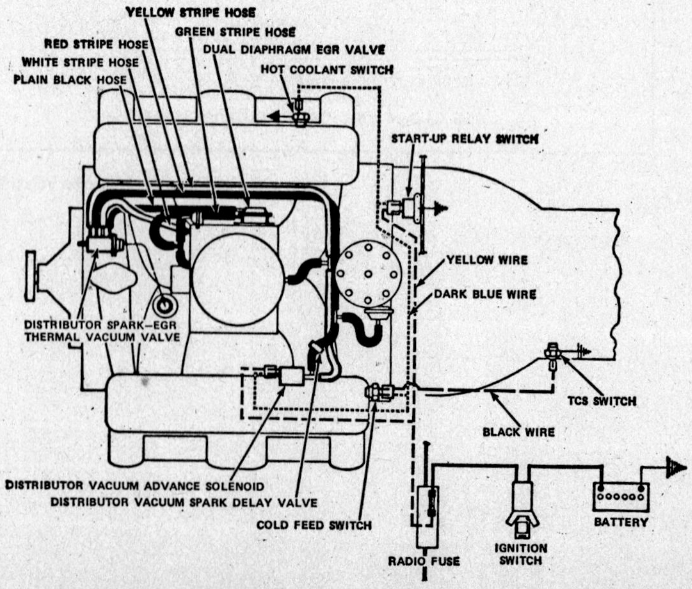

Fig. CL139 Vacuum and wiring. 1974 Pontiac with 2 bar. carb., automatic transmission and vacuum bias valve

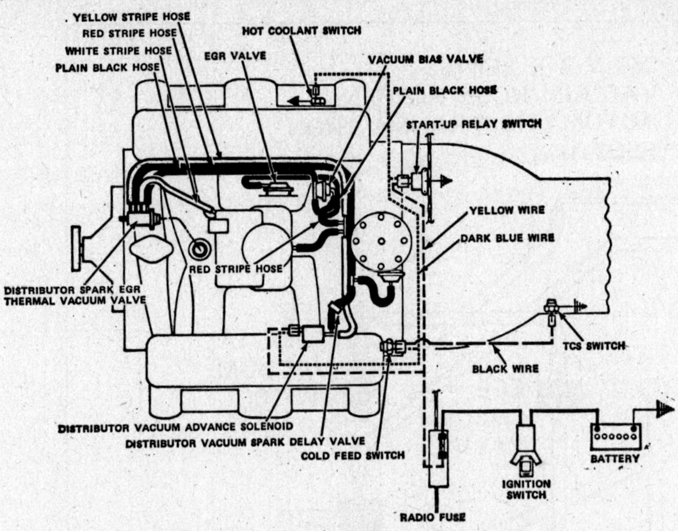

Fig. CL140 Vacuum and wiring. 1974 Pontiac with 4 bar. carb. and automatic transmission with dual diaphragm

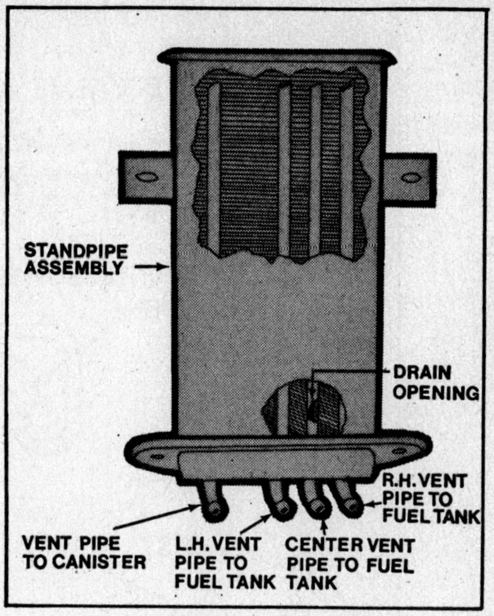

Fig. CL141 Vapor separator. General Motors (typical)

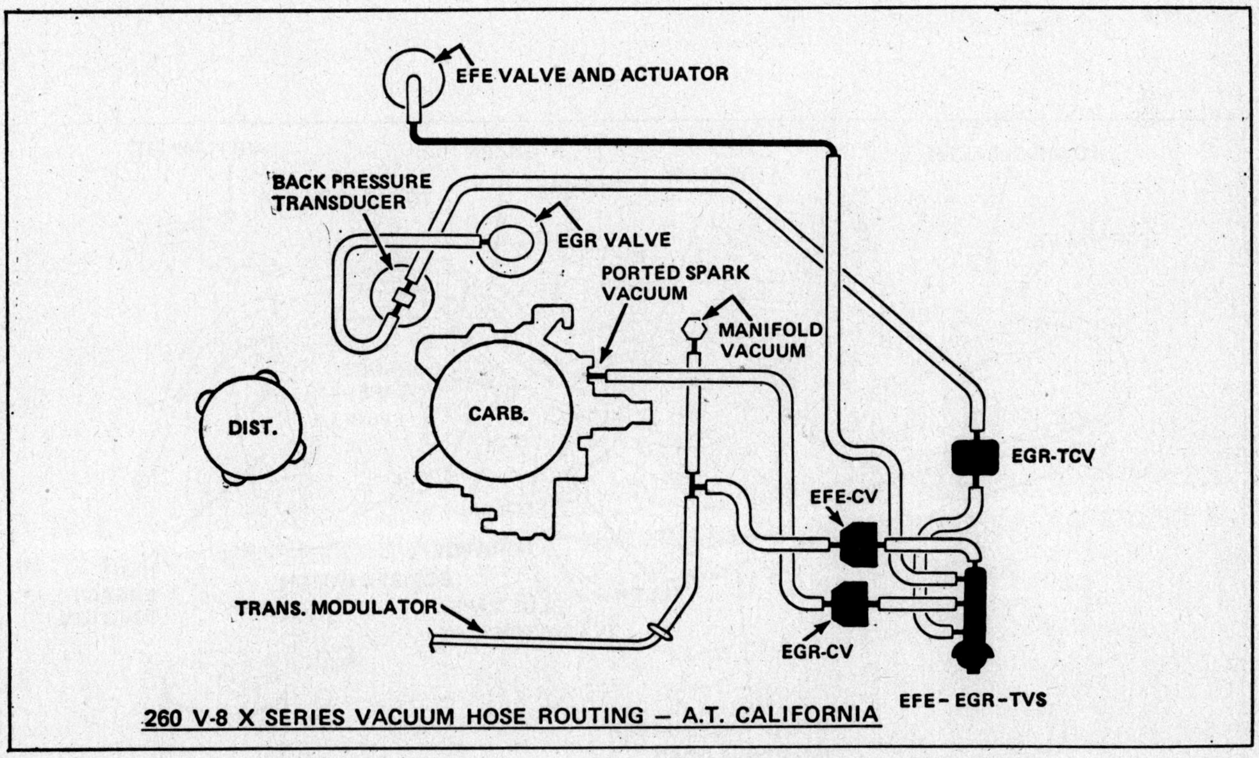

Fig. CL142 Vacuum hose routing. 1975 Ventura V8-260 Calif. with automatic transmission

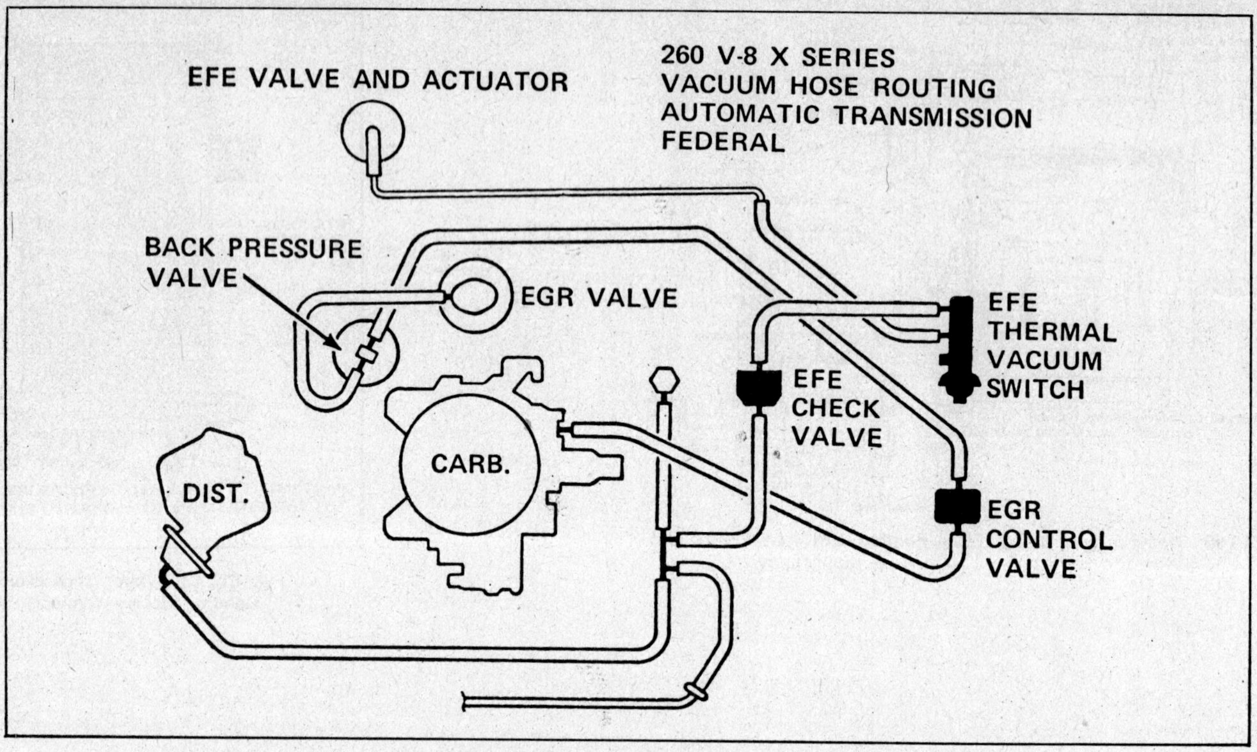

Fig. CL143 Vacuum hose routing. 1975 Ventura V8-260 exc. Calif. with automatic transmission

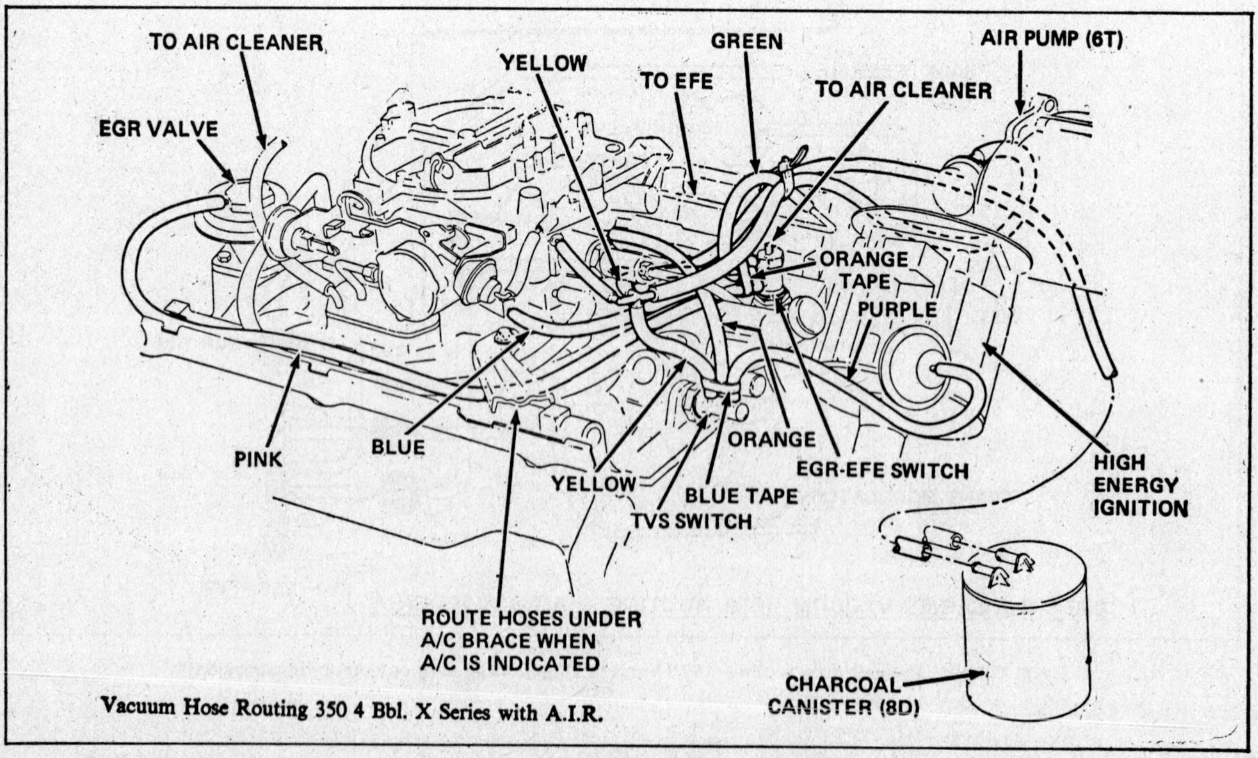

Fig. CL144 Vacuum hose routing. 1975 Ventura V8-350 4 bar. carb. with AIR

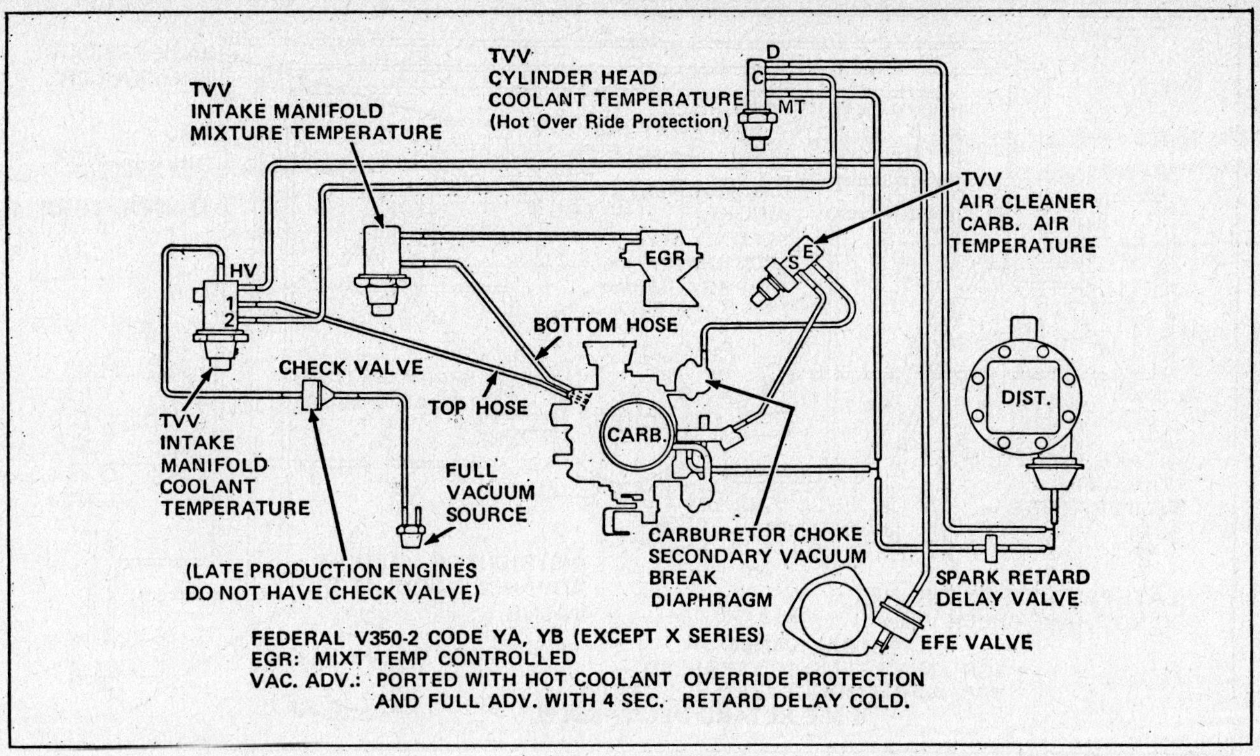

Fig. CL145 Vacuum hose routing. 1975 Pontiac V8-350 2 bar. carb. exc. Calif. exc. Ventura

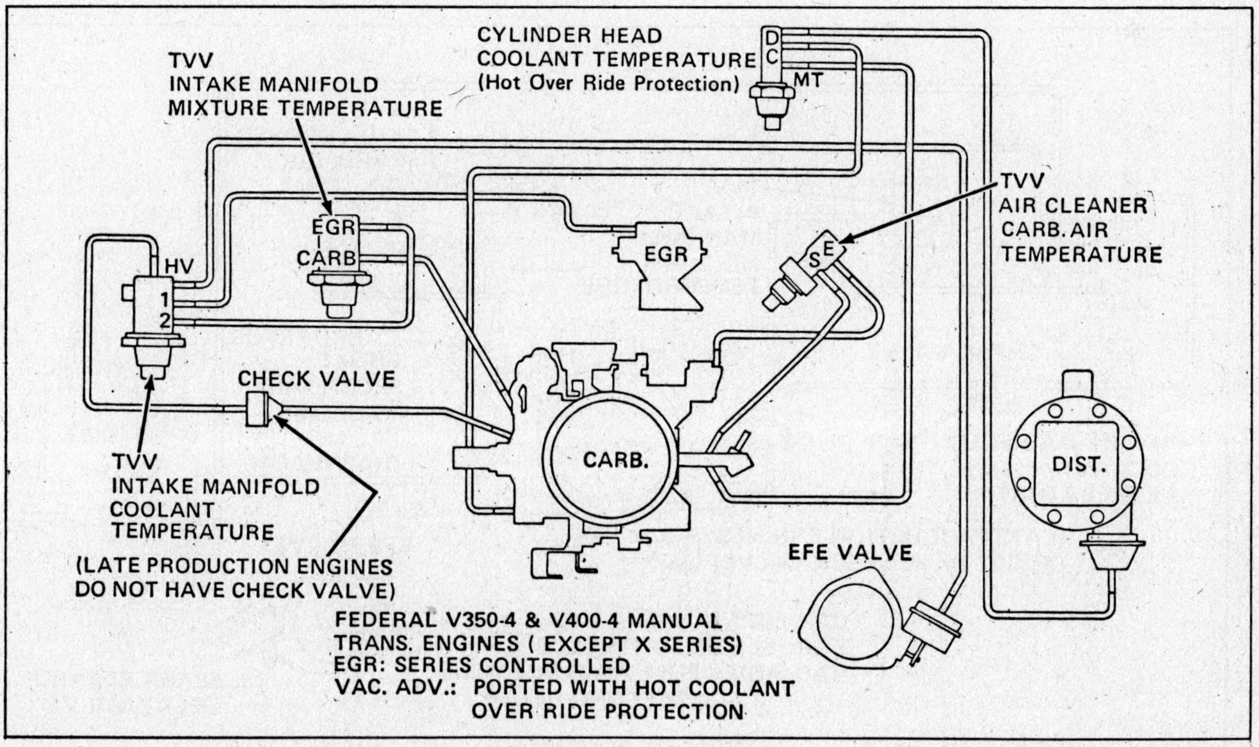

Fig. CL146 Vacuum hose routing. 1975 Pontiac V8-350, 400 4 bar. carb. with manual transmission exc. Calif. & Ventura

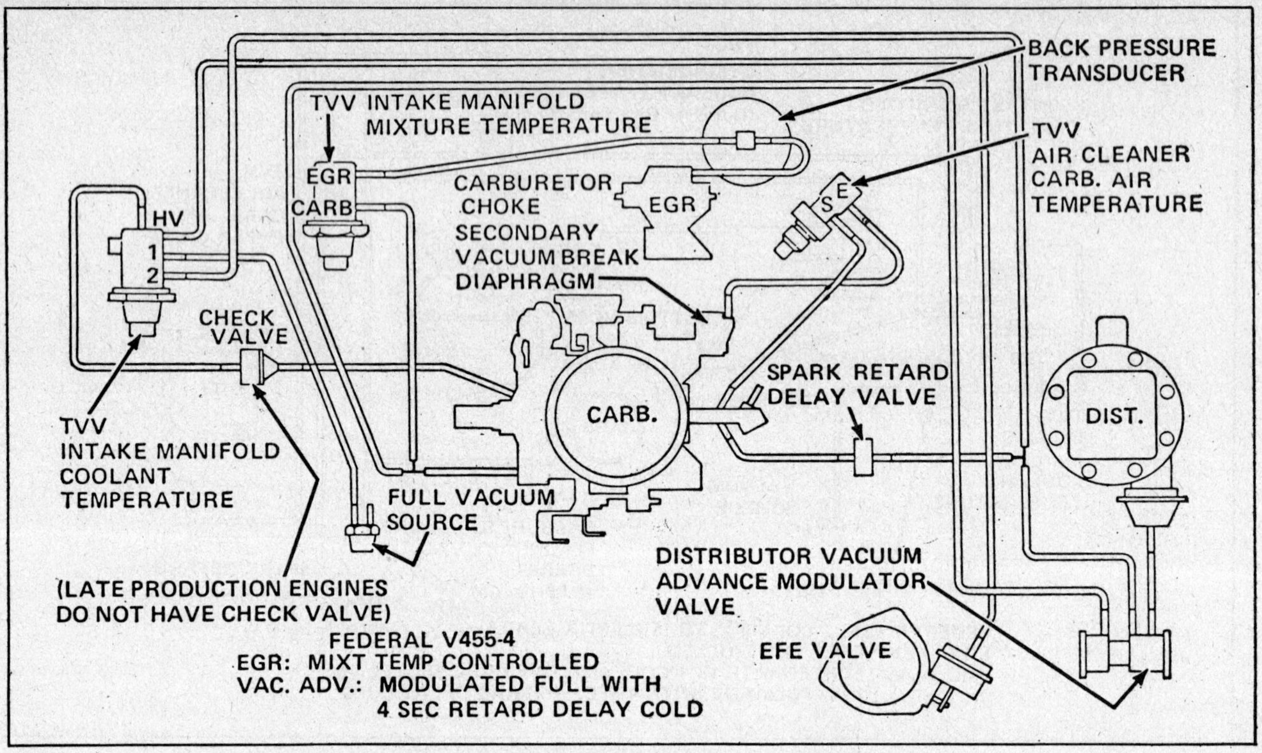

Fig. CL147 Vacuum hose routing. 1975 Pontiac V8-455 exc. Calif.

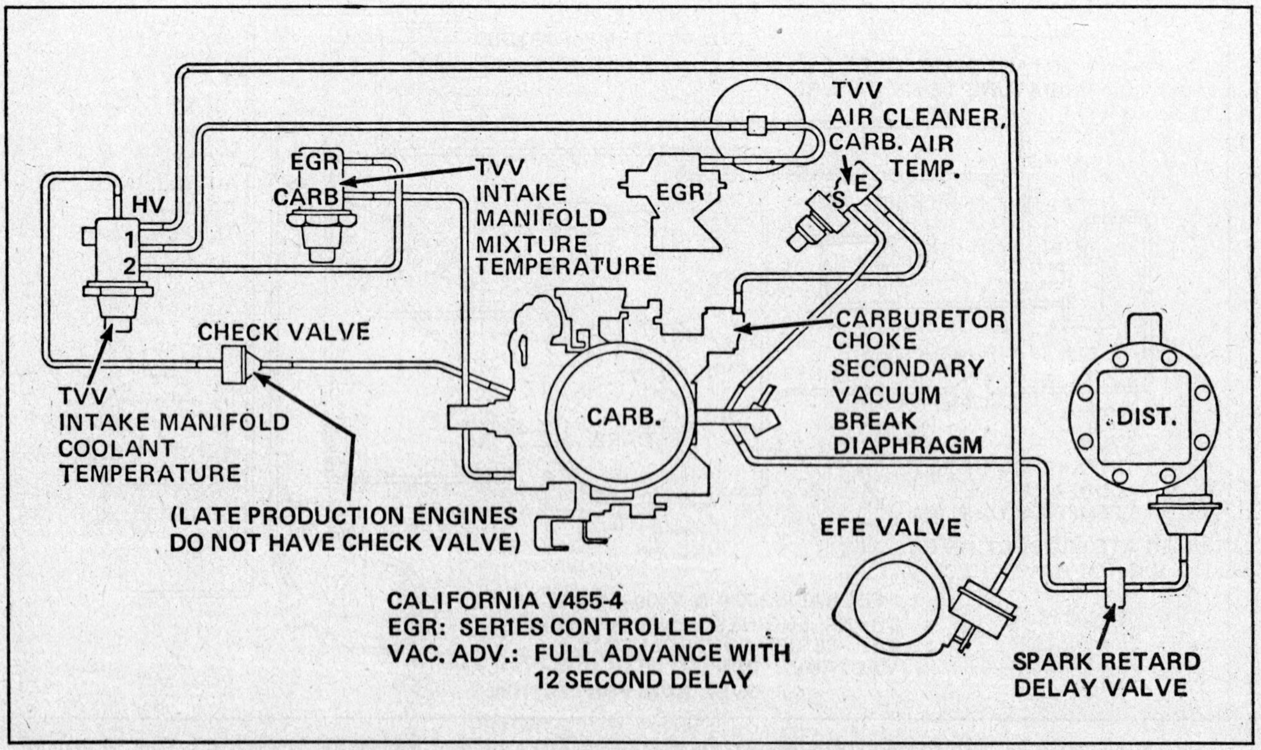

Fig. CL148 Vacuum hose routing. 1975 Pontiac V8-455 Calif.

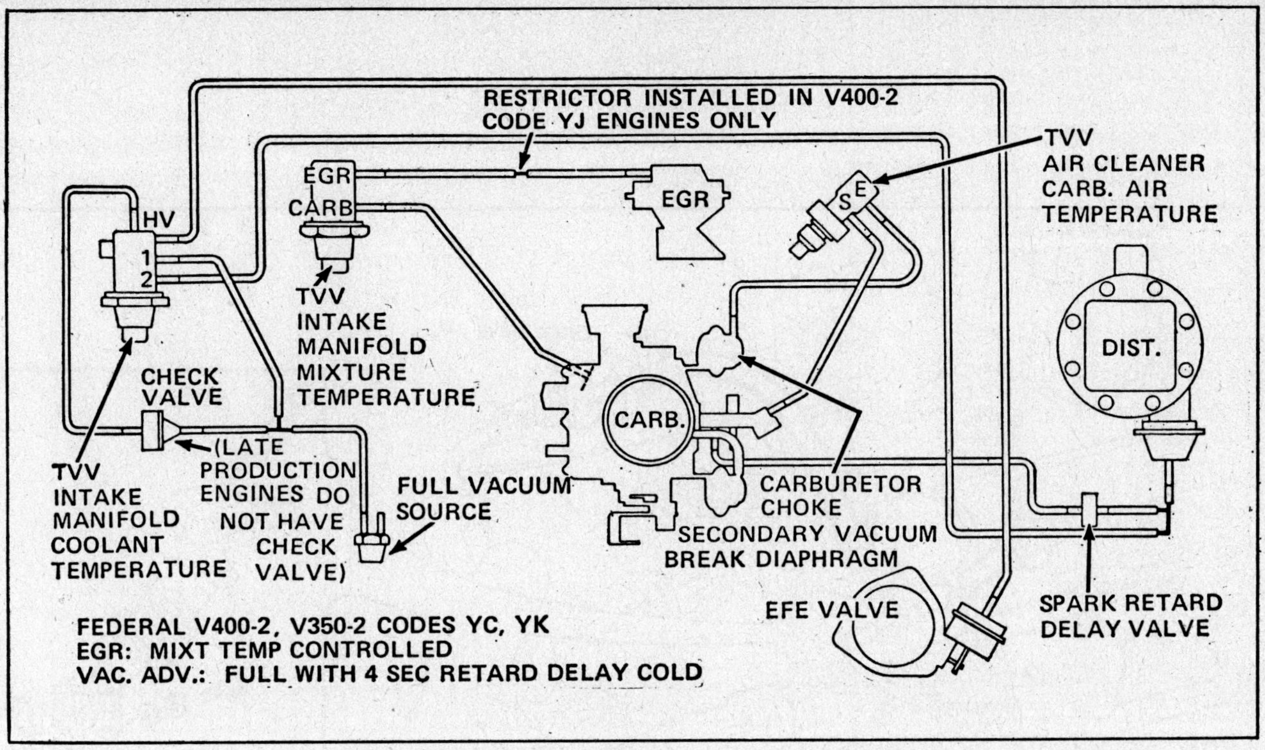

Fig. CL149 Vacuum hose routing. 1975 Pontiac V8-350 (Codes YC, YK) & V8-400 2 bar. carb. exc. Calif.

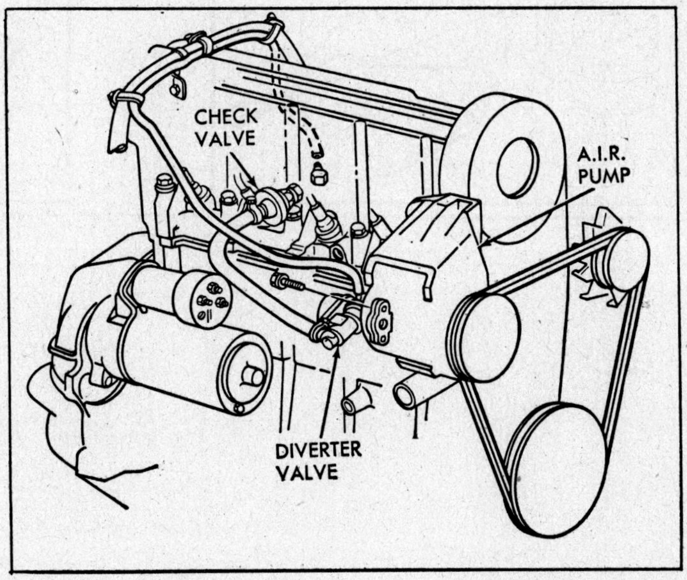

Fig. CL150 Air Injector Reactor components. Chevrolet Vega & Pontiac Astre 4-140

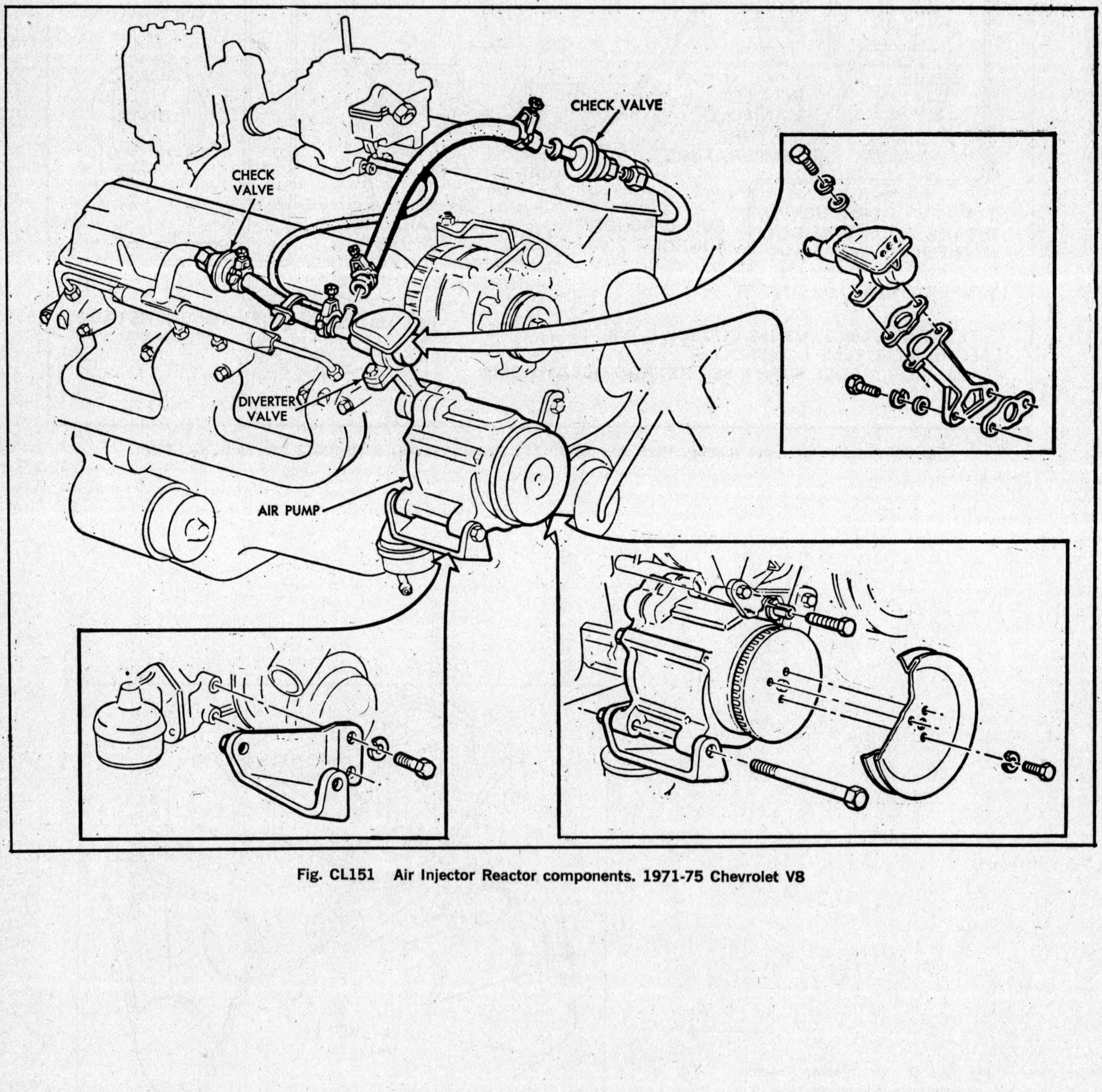

Fig. CL151 Air Injector Reactor components. 1971-75 Chevrolet V8

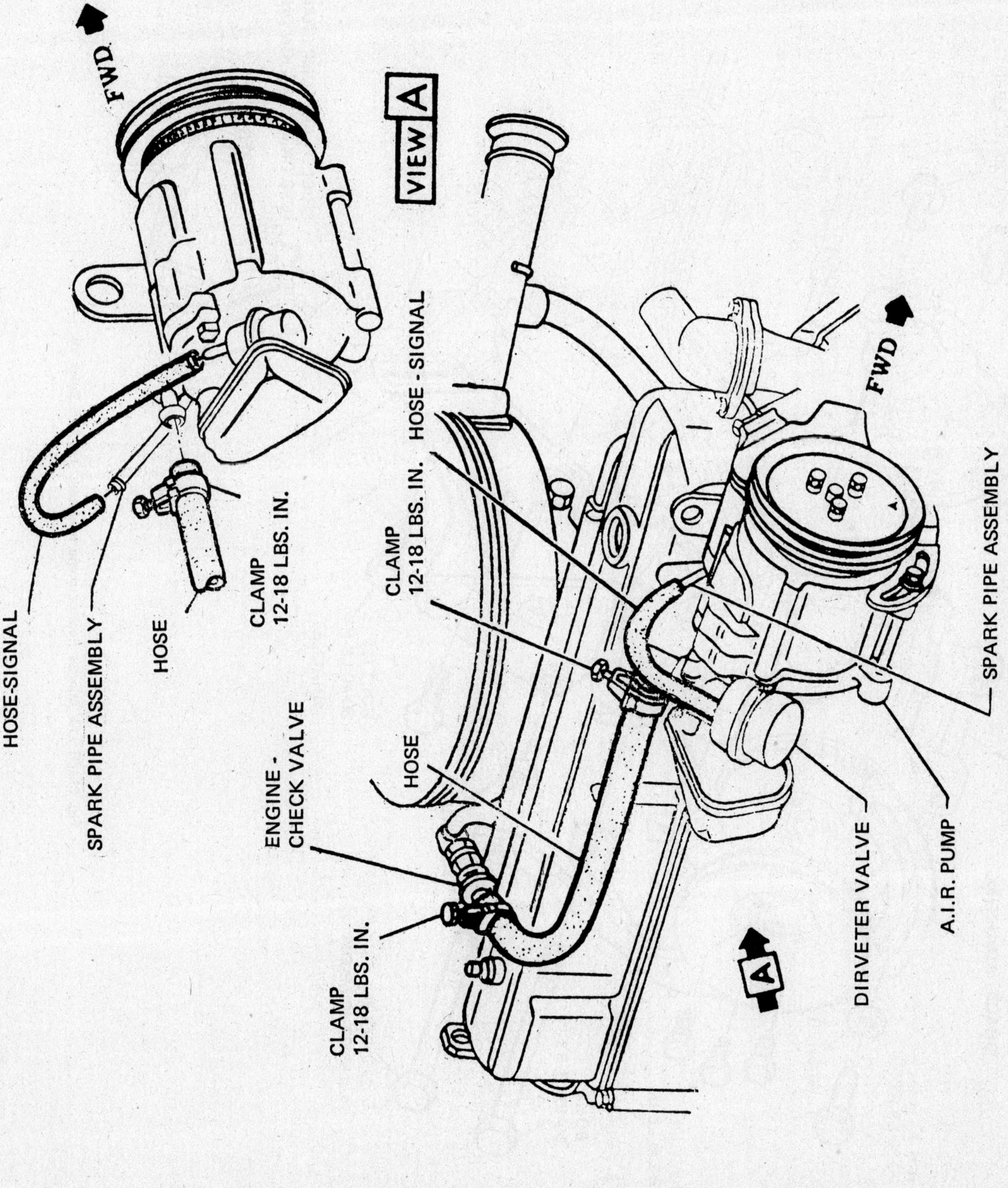

Fig. CL152 Air Injector Reactor components. General Motors 6 cylinder (typical)

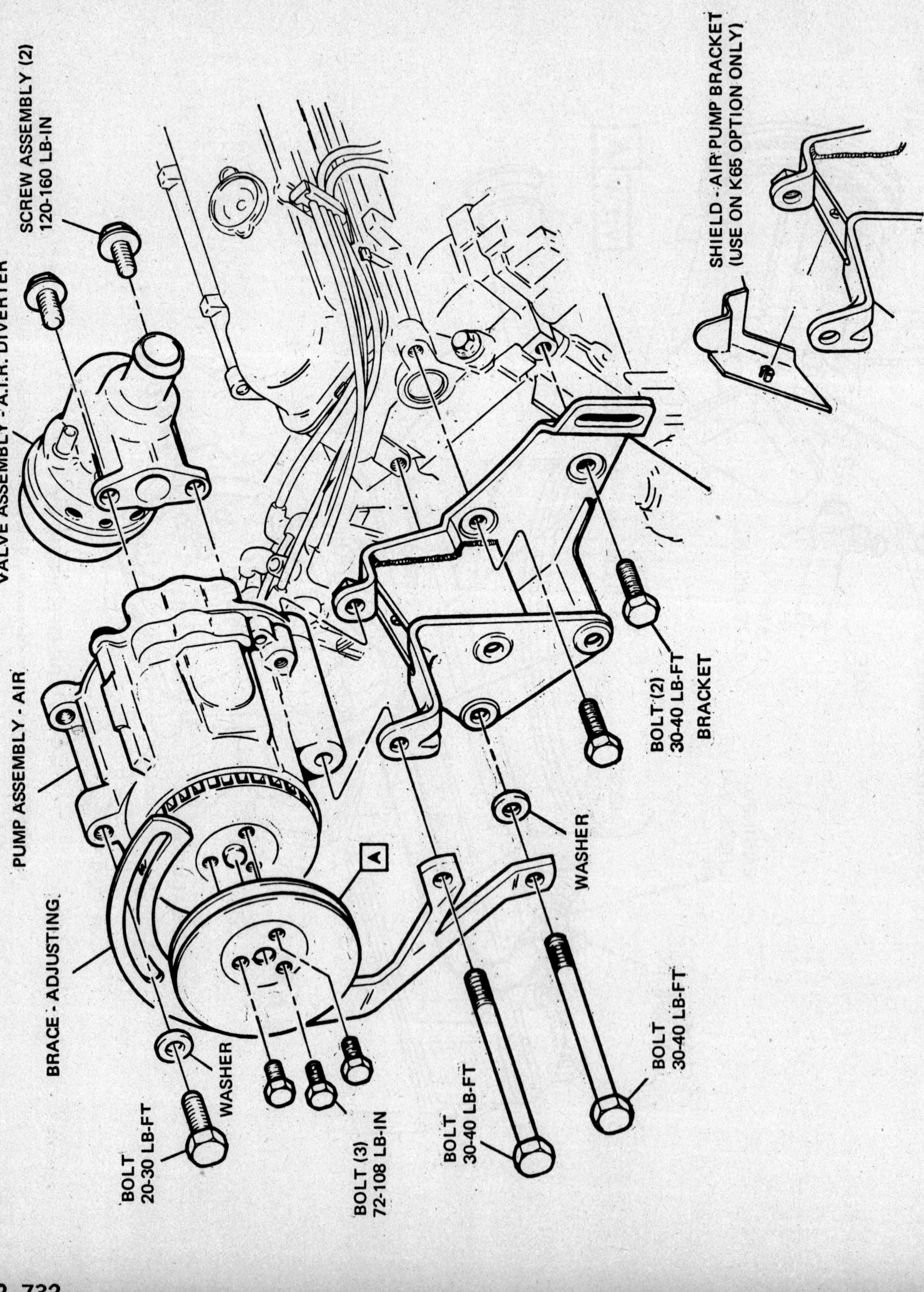

Fig. CL153 Air Injector Reactor pump mounting. 1972-75 Buick V8

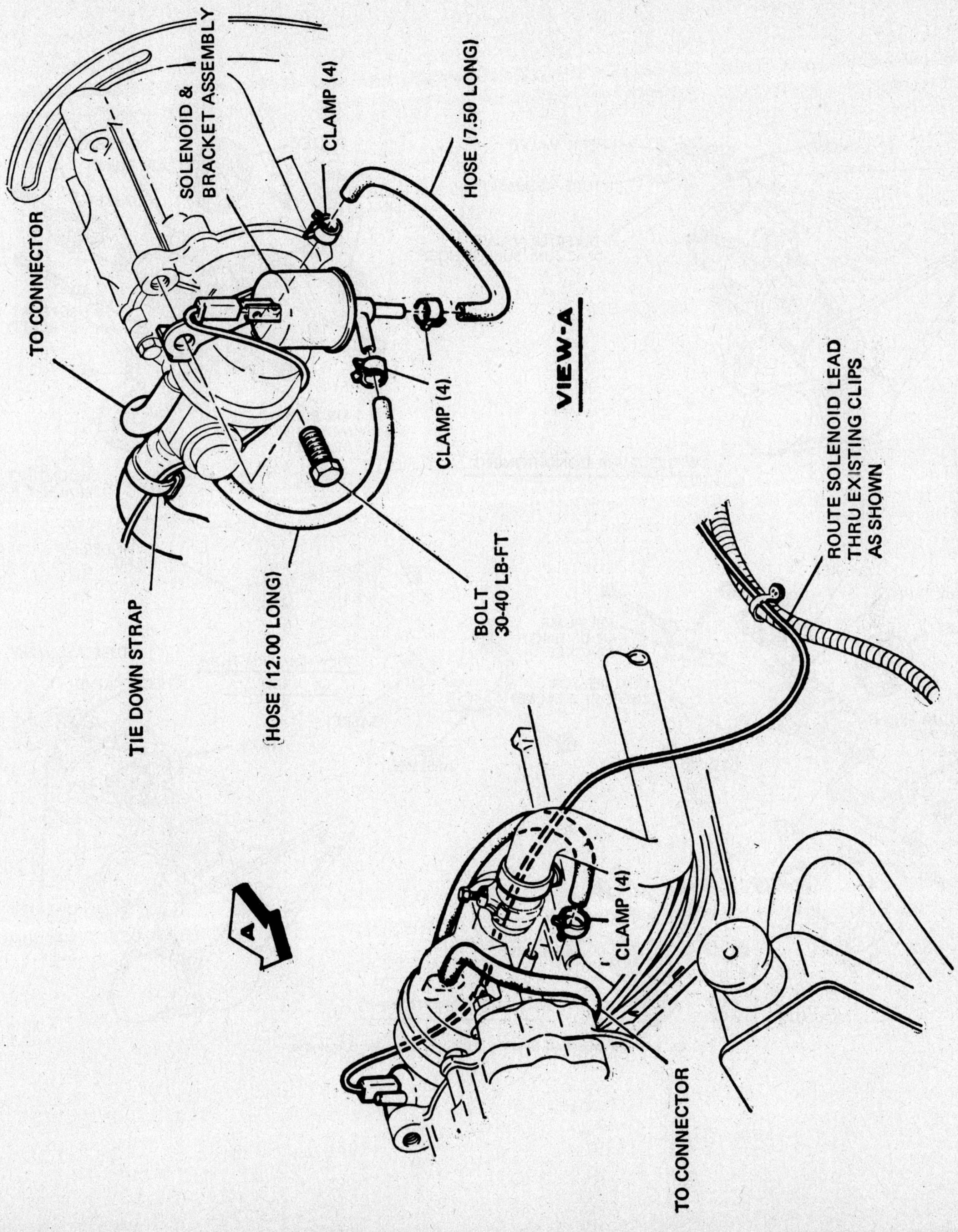

Fig. CL154 Air Injector Reactor components, 1973-74 Buick V8 with Max-Trac

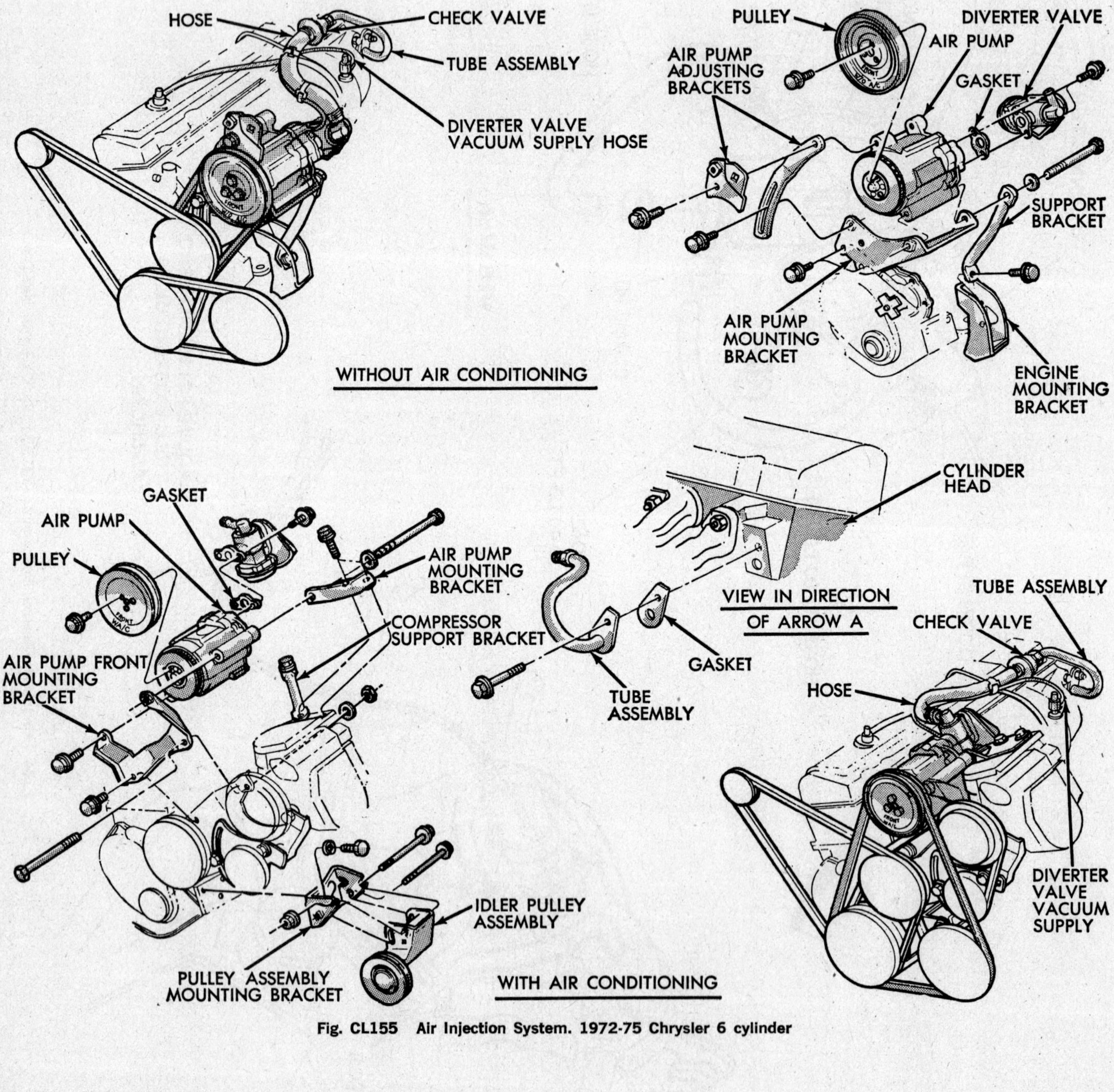

Fig. CL155 Air Injection System. 1972-75 Chrysler 6 cylinder

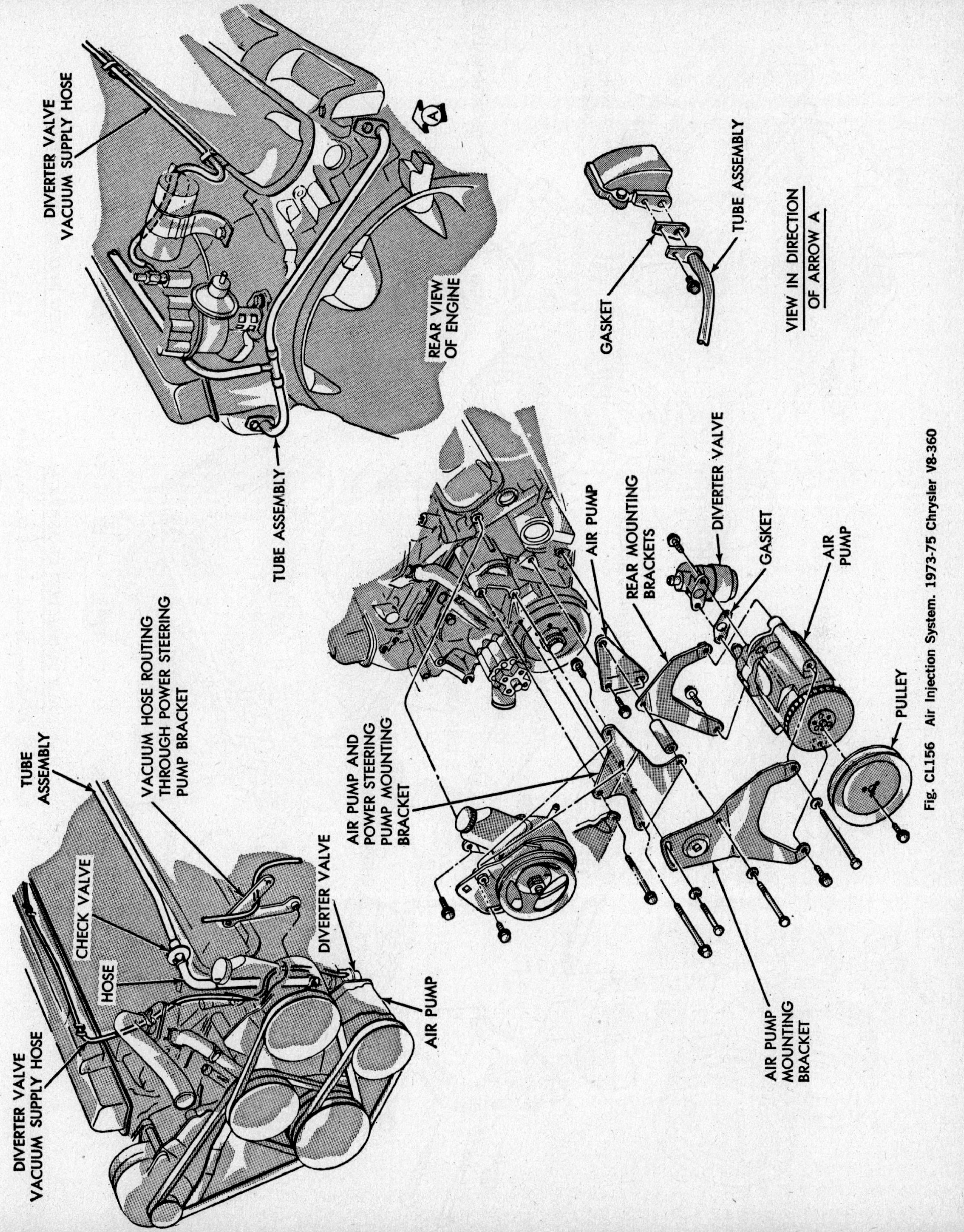

Fig. CL156 Air Injection System. 1973-75 Chrysler V8-360

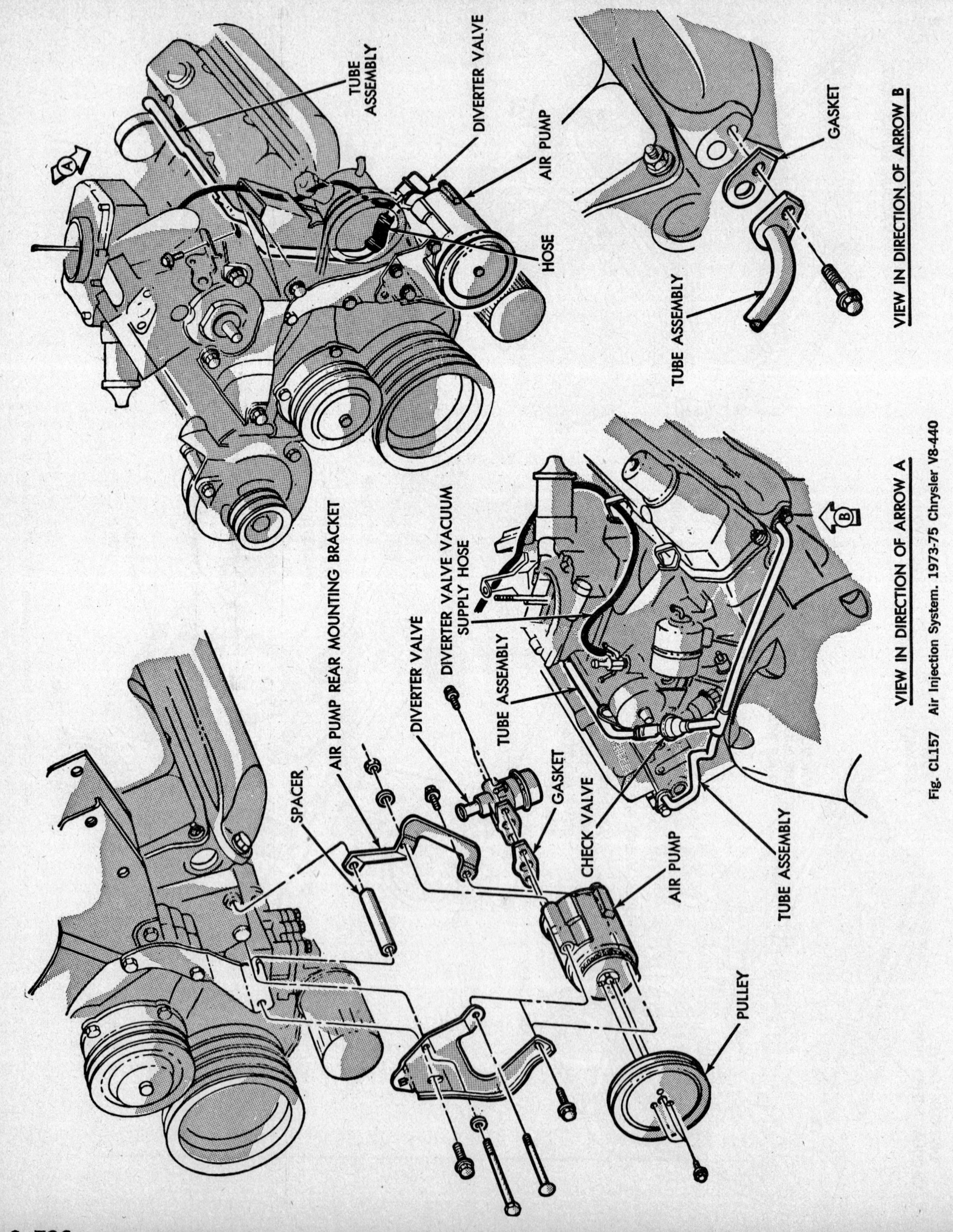

Fig. CL157 Air Injection System. 1973-75 Chrysler V8-440

SYSTEM SERVICING

CONTENTS

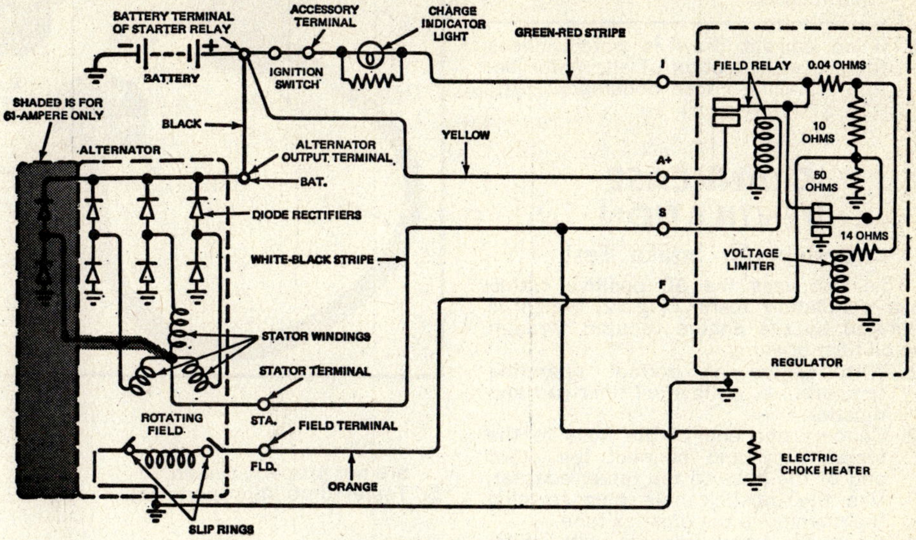

Fig. 1 Ford electric choke wiring schematic

ELECTRIC ASSIST CHOKE

Testing

1973 American Motors & Ford

If the system, Fig. 1, operates at stabilized temperatures below 55 degrees either the temperature sensing disc is broken or the switch contact points have become welded at the heater.

If the system does not operate at stabilized temperatures above 110 degrees the problem is either dirt in the switch area, improper plating on the heater contact points or a broken temperature sensing disc. If the choke does not release within the specified time, check to be sure the bi-metal spring is connected to the choke lever tank. Also check to be sure the ceramic heater is bonded in the choke cap.

NOTE: If damage occurs to any part of the electric choke cap, the entire cap must be replaced as an assembly.

1974-75 American Motors

1. Disconnect electrical lead from choke housing, then connect a test lamp between choke housing terminal and electrical lead.
2. Disconnect choke heat tube from choke housing and start engine.
3. Using a small thermometer, position thermometer bulb end inside heat tube passage in choke housing, then read thermometer and observe test lamp.
4. Lamp should be out between 60-80° F and should be on at temperatures below or above these temperatures.

NOTE: If test lamp did not light, check 7 volt source from alternator. If test lamp did light within 60-80° F, replace choke cover.

Chrysler

Choke Control Switch Test

1. Before starting the engine, remove both electrical connectors from the control switch.
2. Connect a test light to the load (small) terminal of the control switch and to ground.
3. Start engine and allow it to reach normal operating temperature.
4. Then apply 12 volts to the "BAT" terminal of the control switch. The test light should light, it may light for only a few seconds or it may remain lit for a longer duration but must not remain lit for more than 5 minutes. If the above does not occur, the control switch is defective and must be replaced.

Choke Heating Element Test

1. Disconnect electric heating element at the control switch.
2. Connect an ohmmeter lead to the crimped juction of this wire at the choke end.

NOTE: Do not connect to the steel heater casing.

3. Hold the other ohmmeter lead against the choke housing or engine manifold.
4. Electrical resistance of 4-6 ohms for 1973 and 4-12 ohms for 1974-75 units is normal. Meter readings indicating an open or a short, replace choke assembly.

NOTE: The electric assist choke system does not change any carburetor service procedures and cannot be adjusted. However, the choke linkage and shaft must move freely hot or cold.

If the system is found out of calibration, a new heater control switch or choke unit must be installed.

1974-75 Ford

1. Remove air cleaner and make certain that it was not interfering with the choke plate, then check linkages and fast idle cam for freedom of operation.

NOTE: If delay valve is used, make certain it is operating properly.

2. On all except Mustang II and Pinto, disconnect hot air supply tube from choke housing, then using Choke Tester LRE-34618 or equivalent, check for correct operation of choke.
3. Disconnect stator lead and connect a 0-3 amp ammeter between choke lead connector and stator lead, then start engine and observe reading.
4. On all except Mustang II and Pinto, cool choke until temperature is less than 55° F. If current flow is noted, cap is defective.
5. At about 5 minutes of engine operation, a current reading of .3-1 amps should be noted for all except Mustang II and Pinto and .3-.75 amps for Mustang II and Pinto.

NOTE: On Mustang II and Pinto, make certain that coolant is flowing through choke housing. If choke housing is not hot, check hoses for obstructions.

6. If no current draw is noted, check for proper operation of alternator before replacing choke housing.

CRANKCASE VENTILATION

System Air Intake Test

This test uses the AC positive crankcase ventilation tester, Fig. 2, which is operated by the engine vacuum through the oil filler opening.

1. With engine at normal operating temperature, remove oil filler cap and dipstick.
2. Connect one end of the hose to the tester body and connect the other end of the hose to the tester adapter.
3. Use the dipstick hole plug to plug the opening in the dipstick tube.
4. Insert the tester adapter in the filler cap opening and turn the selector knob to No. 2 (Fig. 2).
5. If the vehicle has a system with the tube from the air cleaner going into the oil filler cap, disconnect the tube at the filler cap and plug the tube.
6. Start engine and let it idle.
7. With plugs secure and tube free of kinks, hold tester body upright and note color in the tester windows. Following lists the various colors and probable cause or related condition of the system.

GREEN: System operating properly.

GREEN & YELLOW

1. Regulator valve or system partially plugged.
2. Slight kink in tester hose.
3. Slight engine blow-by.
4. Plugs from kit or engine vacuum lines are not properly sealed.
5. Tester knob improperly set.

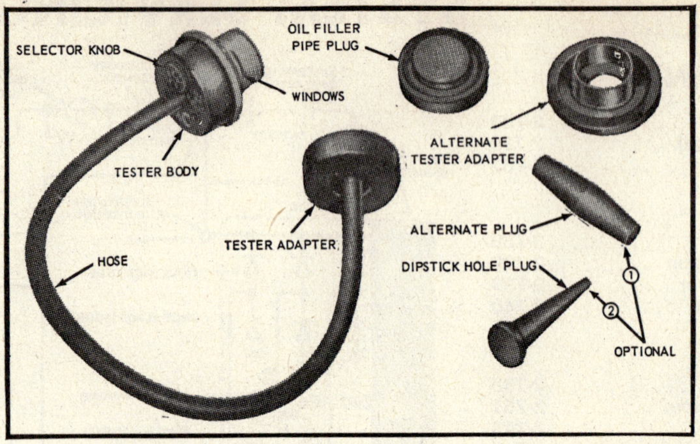

Fig. 2 AC positive crankcase ventilation system tester

YELLOW

1. Regulator valve or system partially plugged.
2. Tester hose kinked or blocked.
3. Blow-by a maximum capacity of regulator valve.
4. Plugs from kit or engine vacuum lines are not properly sealed.
5. Tester knob improperly set.

YELLOW & RED

1. Regulator valve or system partially or fully plugged.
2. More engine blow-by than regulator valve can handle.
3. Vent hose plugged or collapsed.

RED

1. Regulator valve or system fully plugged or stuck.
2. Vent hose plugged or collapsed.
3. Extreme blow-by.

EXHAUST EMISSION CONTROLS AIR PUMP SYSTEMS

American Motors Air Guard, Chrysler Air Injection, Ford Thermactor & GM Air Injection Reactor (A.I.R.)

Pressure Relief Valve:

When the pressure relief valve is incorporated in the diverter or air by-pass valve, the complete valve must be replaced. If the relief valve is in the pump, proceed as follows:

1. Pull relief valve from pump, Fig. 3.
2. Using a 15/16" socket, Fig. 4, tap relief valve into the housing until valve shoulders on housing.

NOTE: Various length pressure setting plugs designed for the particular requirements of the vehicle being serviced determine the pressure required to open the relief valve. Usually, the pressure setting plugs are color coded. To remove the

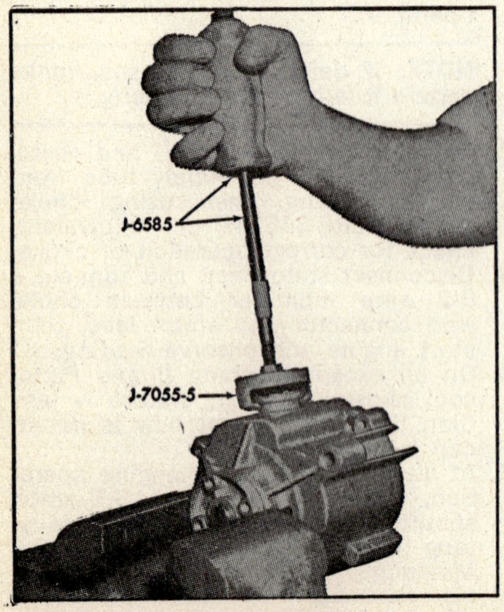

Fig. 3 Removing pressure relief valve with slide hammer tool shown

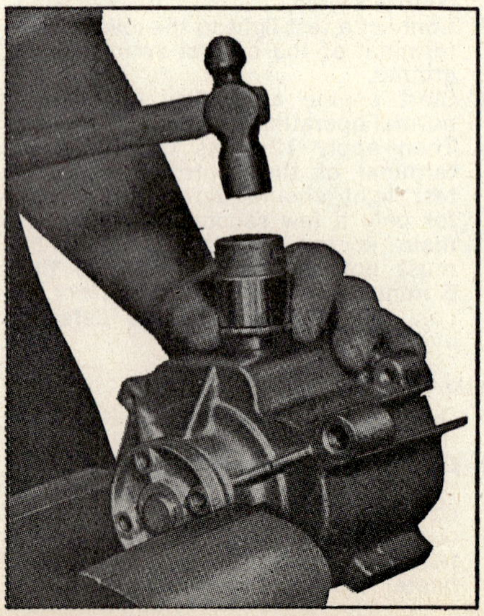

Fig. 4 Installing pressure relief valve

Fig. 5 Removing centrifugal type pump air filter

pressure setting plugs, carefully unlock the legs from the inside surface of the relief valve with a small screwdriver. To install the plug, carefully push it into the relief valve until the legs lock. If a pressure relief valve is to be reused, be sure the leg angles are sufficient to lock plug in place.

Centrifugal Pump Filter:

1. To remove filter, Fig. 5, remove drive belt and pump pulley. Pry loose outer disc of filter fan, being careful to prevent fragments from entering the air intake hole.
2. Install the new filter by drawing it on with the pulley and pulley bolts, by alternately tightening the bolts. Make certain that the outer edge of the filter slips into the housing. The slight amount of interference with the housing bore is normal. Do not attempt to install a filter by hammering or pressing it on.

Conventional Type Filter Fig. 6

Remove and replace the air filter element as suggested by the illustration. The filter element is not cleanable; it must be replaced. Position the assembled air horn and filter element in the air cleaner body, being sure the tang is fitted into the slot.

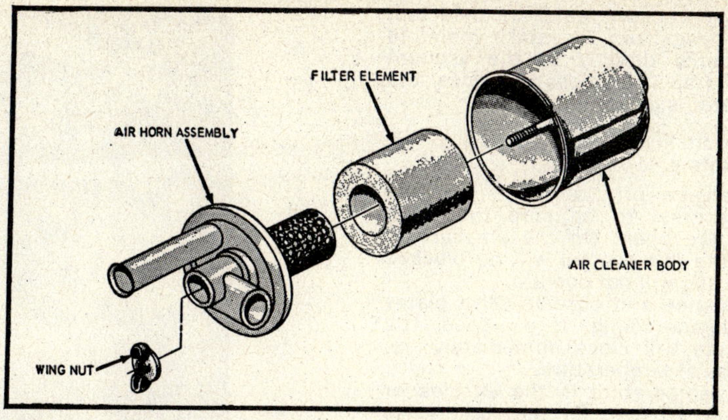

Fig. 6 Conventional type pump air filter

AMERICAN MOTORS EXHAUST GAS RECIRCULATION (EGR)

The EGR valve on the intake manifold must be removed, cleaned and inspected at the mileage intervals as specified. The exhaust gas passages in the intake manifold also should be inspected and cleaned as required.

After cleaning, open the EGR valve manually by pressing down on the pintle, then release.

NOTE: If the valve does not return to the fully closed position, it must be replaced.

Inspect the exhaust gas inlet and discharge passages of the intake manifold for any restriction caused by carbon or lead deposits.

On six cylinder engines, lead or carbon deposits build up rapidly in the exhaust gas discharge passage upper hole. If the deposits cannot be removed a 9/16 inch drill bit may be used.

Testing

With engine idling at operating temperature, manually depress EGR diaphragm, lifting the pintle off its seat. A sudden drop of about 200 RPM should be noticed, indicating that the EGR valve is closing off exhaust from the intake passages. If there is no change in engine RPM and the engine idles properly, the EGR passage to the intake manifold is blocked. If engine idles poorly and lifting the pintle off its seat does not improve idle RPM, the EGR valve is sticking open, defective or there is a flaw in the intake manifold.

Connect a tee in the EGR vacuum signal line near the EGR valve, then connect a vacuum gauge to the tee. Place fingers against EGR valve diaphragm, then slowly accelerate engine and notice amount of vacuum indicated when EGR valve diaphragm begins to move. Refer to the following table for correct values as indicated by part number.

Part Number	Vacuum Required	
	Start Open	Max. Open
7030881	2.8-3.2	6.9-7.3
7040176	2.8-3.2	5.3-5.7
7043589	1.8-2.2	2.8-3.4
7046390	2.8-3.2	4.5-6.6
17050470	1.8-2.2	3.8-4.2
17050471	1.8-2.2	3.8-4.2
17050472	1.8-2.2	3.8-4.2

Continue to accelerate engine until vacuum required to achieve maximum recirculation is obtained. Diaphragm should be deeply depressed with no leakage indicated.

American Motors Exhaust Back Pressure Sensor

Testing

1. Inspect all EGR vacuum lines for leaks and correct routing.

NOTE: Make certain vacuum line from EGR CTO is connected to nipple with .030 inch restriction.

2. Using a tee fitting, connect a vacuum gauge in line between EGR valve and exhaust back pressure sensor.
3. With engine idling there should be no vacuum indicated on gauge.

NOTE: If vacuum is indicated at idle speed, check for correct vacuum line connections. Make certain manifold vacuum is not being used as a vacuum source. If carburetor is providing vacuum, check for partially open throttle plate which could be providing premature ported vacuum to back pressure sensor.

4. Accelerate engine to 2000 rpm and observe vacuum gauge for no reading when coolant temperature is below 115° F and vacuum reading when coolant temperature is above 115° F.
5. If no vacuum was indicated during test, make certain vacuum is being applied to inlet side of back pressure sensor. If vacuum connections are correct, remove back pressure sensor and inspect spacer port and tube for restrictions. Deposits caused by carbon or lead deposits can be removed using a spiral wire brush. If no vacuum is obtained, replace back pressure sensor.

TAC, THERMOSTATIC CONTROLLED AIR CLEANER & AUTO-THERM AIR CLEANER SYSTEM

Tests

American Motors Six & All Ford

Immerse the air cleaner snorkel in a container of cold water. Make certain the thermostat unit is covered by the water. Place a thermometer in the water and observe the temperature while heating the water slowly. With the water temperature at 105 degrees F. or less, the air valve must be in the closed (heat on) position.

With water temperature at 130 degrees F., the air valve must be fully open (heat off) position. If the air valve does not open and close at the temperatures specified, check the valve and spring mechanism for a binding condition. If the valve mechanism is in satisfactory condition, the thermostat unit is defective and the complete assembly must be replaced.

American Motors V8

With air cleaner assembly at room temperature, the air valve should be fully open to outside air. Install the air cleaner assembly on engine and connect hot tube and manifold vacuum hose. Start engine and the air valve should be fully closed to outside air.

Move the engine to warm up to operating temperature and then the air valve should be fully open to outside air. If the air valve does not close at room temperature with vacuum applied, check for a mechanical bind in the snorkel, vacuum motor linkage disconnected or vaccum leaks in hoses or connections at the vacuum motor thermal sensor and intake manifold.

If the air valve mechanism is operating

freely and no vacuum leaks are detected, connect a hose from an intake manifold vacuum source directly to the vacuum motor. If the air valve does not close, the vacuum motor is defective.

General Motors Functional Checks
Vacuum Motors

1. With engine off, look into left and right snorkel to be sure the valve plates are open. *On the smaller engines the left snorkel will be blocked closed and will not operate.*
2. Start engine and observe valve plates with engine idling. If equipped, the left plate will close immediately, regardless of temperature.
3. If the temperature in the air cleaner is below 90 deg. F., the right valve plate will also close to outside air. With temperatures above approximately 100 deg. F., the valve will remain open.
4. If vacuum motors fail to operate with vacuum directly applied, replace motor.

Temperature Sensor:

1. With engine temperature below 85 deg. F., tape a thermometer in air cleaner, Fig. 7.
2. Install air cleaner cover and start engine.
3. When cold air door starts to open, remove air cleaner cover and check temperature. If temperature is below approximately 85 deg. F., sensor must be replaced.

NOTE: If hoses have been removed from connectors for any reason, it is very important that they be correctly positioned for proper operation.

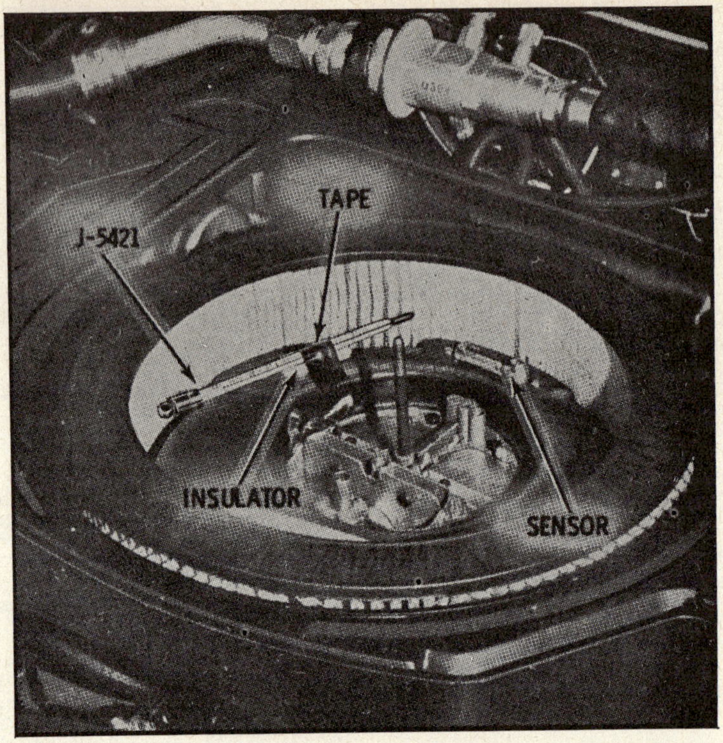

Fig. 7 Checking temperature sensor with thermometer. CCS System

TRANSMISSION CONTROLLED SPARK SYSTEM (TCS)

1972-73 American Motors

A vacuum gauge and a probe type current tester can be used to check for proper operation of the TCS system and the coolant temperature override switch for test connections. With ambient temperature above 63 degrees F. Check all vacuum hose connections proper route. Begin tests with engine coolant temperature below 160 degrees F.

Test

1. Turn ignition switch on. Disconnect the TCS harness connector. Connect the tester wire lead to a good ground and touch the probe to the disconnected feed wire, Fig. 8, "Test Point A." The tester bulb should light, if not, check the ignition feed circuit through the main harness connector fuse.
2. Use a jumper wire temporarily connect the ignition feed wire (female terminal) to the male terminal of the ambient temperature override switch wire. Ground tester wire lead and touch the probe to the open female terminal of the ambient temperature override switch wire, Fig. 8, "Test Point A." The tester bulb should light, if not, there is a defect in the ambient temperature switch or switch wires.

   To check switch operation below 63 degrees F., place a wet cloth over the bracket end of the switch. The light should go out within a short time. If not, the switch is defective.

NOTE: Freon gas may also be used to cool the switch.

3. Disconnect the wire connector from the solenoid vacuum valve. Ground the test lead and touch the probe first to one terminal of the wire connector and then to the other, Fig. 8, "Test Point C." The tester bulb should light at one of the terminals, if not, the ignition feed wire connected to the TCS connector is defective.

CAUTION: Do not touch the tester probe to the terminals of the solenoid vacuum valve with the wires connected because the solenoid control switch located at the transmission will be damaged.

4. With lead connected to the positive battery cable touch the probe to the solenoid vacuum valve ground wire, Fig. 8, "Test Point D." The tester bulb should light, if not, the solenoid control switch at the transmission or the switch wire is defective.
5. V8 engines only: Disconnect coolant temperature override switch port "D." Connect a vacuum gauge to this port, Fig. 8, "Test Point E." Start the engine, it should indicate manifold vacuum. When the engine coolant temperature reaches 160 degrees F. The vacuum reading should drop approximately 4-6 inches. When running a cold engine the gauge does not indicate manifold vacuum or if the vacuum reading does not drop when the coolant temperature reaches 160 degrees F., the coolant temperature override switch is defective.
6. Disconnect the vacuum hose from the front port, Fig. 8, "Test Point F" and connect vacuum gauge. Raise the vehicle rear wheels off the floor. On manual transmission equipped vehicles, place the gear selector in high gear and run the engine above idle speed. Carburetor ported vacuum should be indicated on the vacuum gauge. On automatic transmission equipped vehicles, place the gear selector in "Drive" and slowly increase engine speed while observing the vacuum gauge. When the speedometer needle reaches approximately 34 mph, the needle of the vacuum gauge indicates carburetor ported vacuum. If no vacuum is indicated, the solenoid vacuum valve is defective or no carburetor ported vacuum is being applied to the valve.

1974-75 Test and Adjustment

1. Disconnect wire from solenoid control switch and connect a 12 volt test lamp in series between switch wire and terminal.

2. Raise and support vehicle so that rear wheels are free to rotate.
3. Observe speedometer and test lamp. At speeds between 33-37 mph, the lamp should go off. If test lamp remains on above this range, the solenoid control switch can be adjusted by turning the 1/16 inch allen head screw clockwise to increase opening speed and counterclockwise to decrease opening speed, Fig. 8A. Adjust switch opening speed to 35 mph.

General Motors All Except 1973 Mid-Year & 1974 Pontiac System Functional Checks

T.C.S. can be checked for proper operation by connecting a vacuum gauge into the vacuum line between the distributor and the vacuum control valve.

There should be full vacuum only when the transmission is in high gear with the engine at operating temperature. The vacuum check should be made during a road test.

Trouble Shooting

Full Vacuum in all Gears:

1. Blown fuse.
2. Wire disconnected at vacuum control valve.
3. Wire disconnected at transmission control valve.
4. Faulty transmission control switch.
5. Energized coolant temperature switch.

No Vacuum in High Gear:

1. Vacuum lines reversed at vacuum control valve.
2. Foreign matter in vacuum control valve.
3. Faulty vacuum control valve.
4. Distributor vacuum line broken or disconnected.
5. Transmission control switch or wire shorted.

Solenoid & Valve Test

1. Disconnect hoses and electrical connector.
2. Connect a hose to the carburetor port on top of the solenoid and blow through hose. Air should come out distributor center port.
3. Plugging distributor center port should shut off air flow through solenoid if valve is functioning properly.
4. Connect one electrical terminal to a 12 volt source and ground other terminal with a jumper.
5. Remove hose from carburetor on top of the solenoid and connect to distributor center port.
6. Blow through hose. Air should come out the top vent port. If not, valve is defective.
7. Plug the top vent port of valve and blow into hose. Air should be blocked off. If air comes out the bottom engine intake manifold port or the carburetor port on top of the solenoid, the valve is defective.

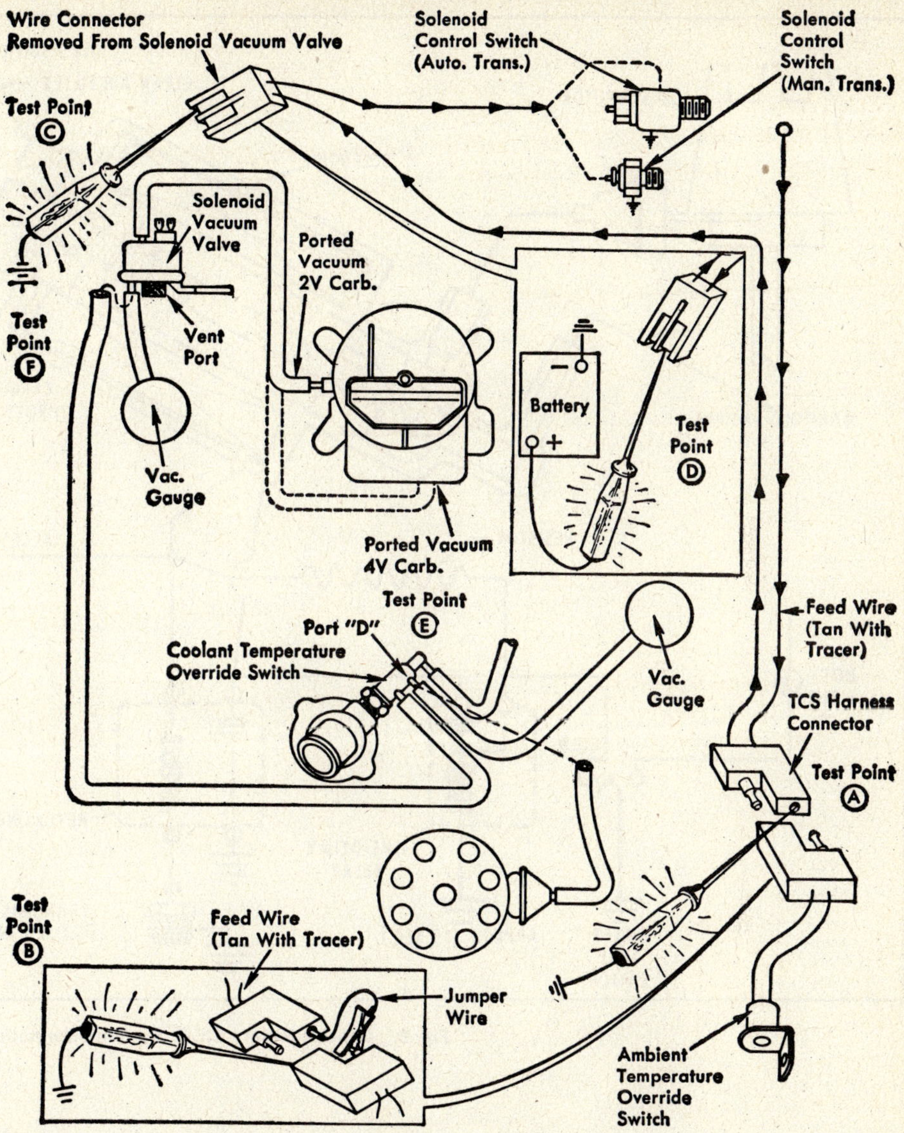

Fig. 8 American Motors TCS system test connections

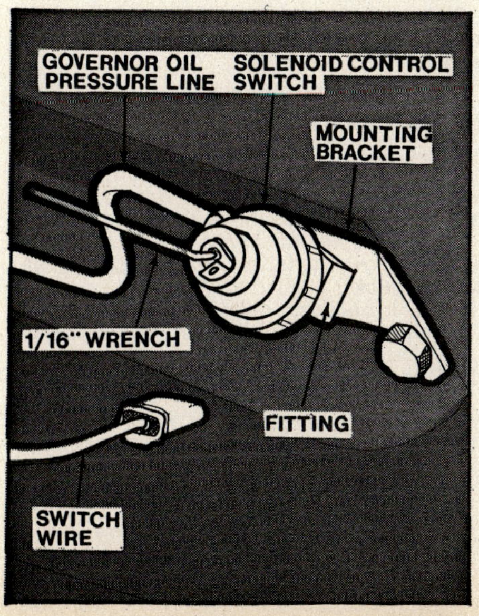

Fig. 8A TCS solenoid adjustment. 1974-75

Temperature Vacuum Switch Test

The temperature vacuum switch section should be tested with the engine running at fast idle, at normal temperature, valve closed. Connect a vacuum gauge to the distributor vacuum port. A zero reading for a duration of one-minute or longer would indicate a properly functioning valve.

NOTE: Vacuum control hoses must be in good condition, correctly installed and fit tightly.

Transmission Control Switch Test

This switch should be tested by removing hoses and electrical connections and

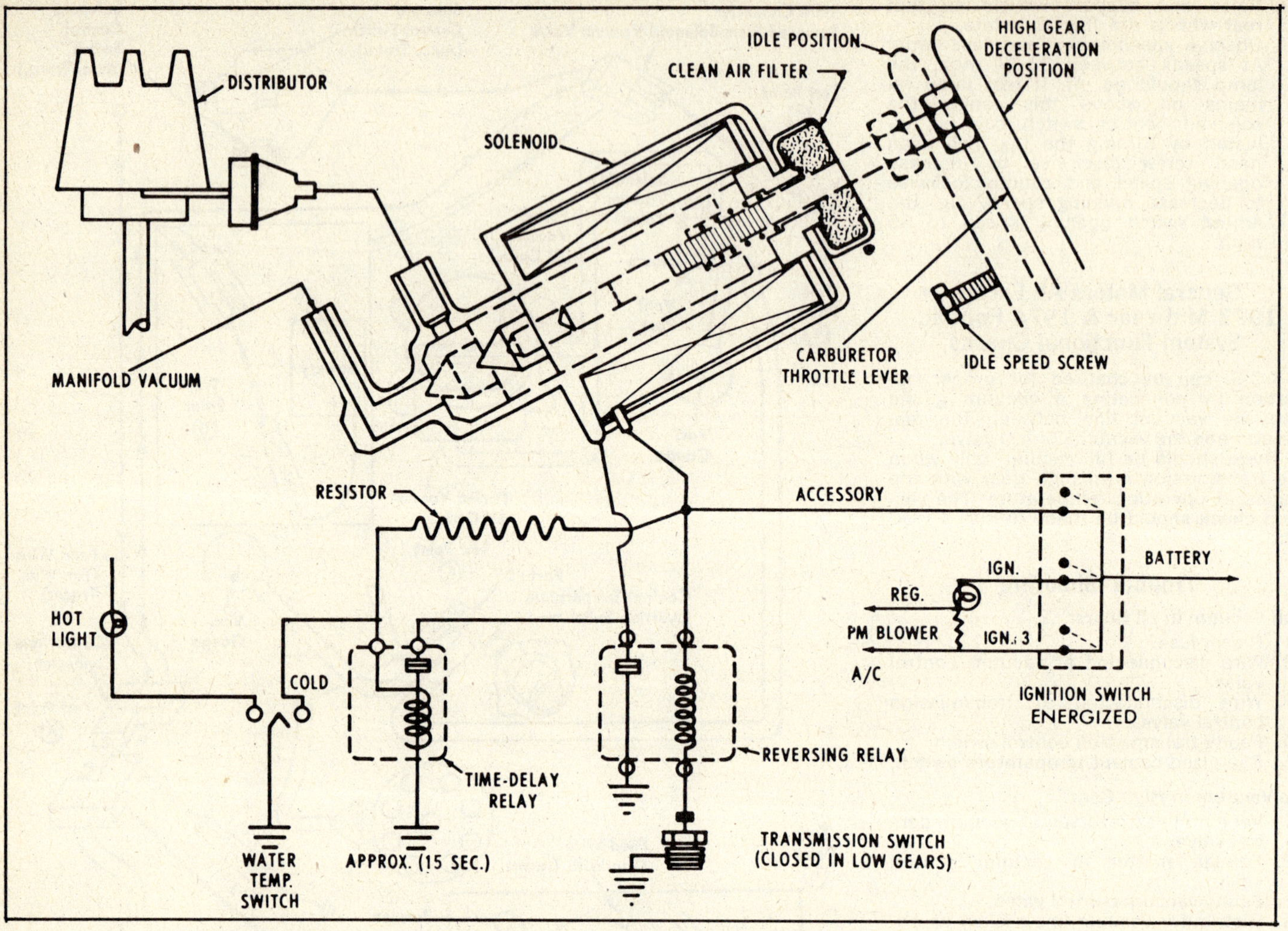

Fig. 9 General Motors C.E.C. system schematic

installing a test hose at the distributor port. Air blown into the test hose should come out of the vacuum port. A similar test made by energizing the unit solenoid with a jumper wire, should divert air to the vent port. If either test fails, valve is defective.

1973 Mid-Year & 1974 Pontiac System Diagnosis

Prior to performing any test, the engine must be at operating temperature and coolant temperature must not exceed 240° F. Insure vacuum hose and electrical connections are correct. On models with Distributor Spark Delay Valve, connect a vacuum gauge to the "Dist" fitting on the delay valve. On models without delay valve, connect vacuum gauge to the vertical fitting at the rear of the distributor vacuum advance solenoid.

NOTE: On models using a spark delay valve, vacuum will slowly rise to full vacuum in 10 to 15 seconds. Also when performing the following tests on vehicles with a manual transmission, open throttle slightly to uncover carburetor vacuum port.

Start-Up Vacuum Test

Start engine and observe vacuum gauge. Vacuum should be present for 20 seconds.

High Gear Vacuum Test

On manual transmission vehicles, shift into high gear and on automatic transmission vehicles, shift into Reverse. Vacuum should be present in these gears.

If vacuum was observed at all times during the above tests, proceed with the following checks:

1. If no vacuum delay was present on vehicles equipped with spark delay valve, replace the delay valve.
2. Check for a stuck open Distributor Vacuum Spark Thermal Valve by removing the vacuum hose running from the front fitting of the distributor solenoid and checking for vacuum at hose. If vacuum is present, replace the thermal valve.
3. If no vacuum is noted at front fitting of distributor solenoid, the distributor solenoid plunger may be stuck in energized position. To check, remove electrical connector from solenoid and if vacuum is still present, replace distributor solenoid.
4. If vacuum is lost after disconnecting solenoid electrical connector, the Start-Up Relay may be energizing the solenoid. To check, remove electrical connector from relay and if vacuum is lost again, replace Start-Up Relay.
5. If vacuum is still present after disconnecting Start-Up Relay electrical connector, the Hot Coolant Switch may be energized or stuck in closed position. To check, remove electrical connector from coolant switch and if vacuum is lost, replace Hot Coolant Switch if it has been determined that coolant temperature is over 240° F.
6. If vacuum is present after disconnecting Hot Coolant Switch electrical connector, the TCS Switch may be stuck in closed position. To check, remove electrical connector from TCS Switch and if vacuum is lost again, replace TCS Switch.
7. If vacuum is still present after disconnecting TCS Switch electrical connector, check system for grounded leads.

If no vacuum was observed during

"Start-Up" vacuum test proceed with following checks:

1. Check for an open circuit in Start-Up Relay using a jumper wire connected between blue wire at Start-Up Relay and ground, energizing relay. If relay does energize, a vacuum reading will be observed, indicating a defective Start-Up Relay which must be replaced. If Start-Up relay does not energize, no vacuum reading will be observed, indicating failure of relay is not due to lack of ground.
2. Connect a test lamp between yellow wire of Start-Up Relay and ground to check if voltage is available to Start-Up Relay. If lamp does not light, check radio fuse and wiring from Start-Up Relay to battery as relay is not receiving voltage.
3. If lamp lights, in above check, connect test lamp between yellow wire at distributor solenoid and ground, to check if voltage is available to solenoid. If lamp does not light, check radio fuse and wiring from distributor solenoid to the battery.
4. If lamp lights in above check, connect a jumper wire between blue wire at distributor solenoid and ground with engine running to energize solenoid. If no vacuum reading was observed, solenoid was not energized and may be internally plugged or has an open circuit, inturn solenoid must be replaced.
5. If a vacuum reading was observed, solenoid was energized producing an audible click, the system is operating normally.

If no vacuum reading was observed during "High Gear" vacuum test, proceed with following checks:

1. Remove TCS Switch connector and ground the black wire with engine running as Cold Feed Switch may be closed. If a vacuum reading is observed, the distributor solenoid is energized, indicating that TCS Switch is defective and must be replaced.
2. If no vacuum reading was observed in above check, inspect black wire between Cold Feed Switch and TCS Switch for an open circuit.

GENERAL MOTORS C.E.C.

Combined Emission Control System Functional Checks

The C.E.C. system, Fig. 9 is tested in the same manner as the T.C.S. system.

GENERAL MOTORS BACK PRESSURE TRANSDUCER VALVE

Testing

1974-75 Oldsmobile & Pontiac

1. Remove air cleaner and plug manifold vacuum fitting, then with A/C off, and engine idling at operating temperature, place cam follower on fast idle step of cam.
2. Check vacuum on back pressure transducer valve and record reading, then using a tee fitting, connect a vacuum gauge to EGR control valve side of back pressure valve. If vacuum reading is not within 1.7-2.7 inches of mercury, disconnect and plug hose from EGR valve. Vacuum reading should be the same or within 2 inches of mercury of that recorded previously. If not, replace back pressure transducer valve.

CHRYSLER C.A.P.

Cleaner Air Package

Vacuum Advance Control Valve Adjustment

1. With engine idle and ignition timing set correctly and engine at operating temperature, connect a tachometer to the engine.
2. Using a tee connector, connect a vacuum gauge into the distributor vacuum line. If the engine is equipped with a dashpot, adjust it so it does not contact the throttle lever.
3. Speed engine up to 2000 rpm in neutral and hold speed for approximately 5 seconds. Release throttle and observe vacuum gauge. When throttle is released, distributor vacuum should increase to 16"Hg. and remain there for a minimum of 1 second. Distributor vacuum must fall to 6"Hg. within 3 seconds after throttle is released.
4. If necessary, remove cover and turn spring end adjusting screw counterclockwise to increase time vacuum remains above 6"Hg. One turn will change setting by approximately ½" Hg. If valve cannot be adjusted, it must be replaced.
5. If necessary, readjust dashpot.

Service Procedure

Periodic servicing of the system is required to maintain good engine performance and prevent malfunction of the system because of the combustion products deposited in the valve, hose and carburetor passages. Every six months clean and service the system, and replace the valve every year. To service the systems, proceed as follows:

1. In cases of severe service, such as experienced in police, taxicab or other operation involving short trips with prolonged idling, it is recommended that the system be checked for operation with every oil change and serviced as required.
2. With engine idling, remove ventilator valve assembly from rocker arm cover. If valve is working freely, a hissing noise will be heard as air passes through the valve, and a strong vacuum should be felt when a finger is placed over the valve inlet.
3. If valve is working properly, reinstall ventilator valve assembly and remove oil filler breather cap. With engine idling, loosely hold a piece of stiff paper over the oil filler pipe. After allowing about a minute for crankcase pressure to reduce, the paper should be drawn against the filler pipe with noticeable force. If this occurs a final check should be made to be certain the valve shuttle is free.
4. Shut off engine, remove valve and shake it vigorously. A clicking noise should be heard if valve shuttle is free. If noise is heard, valve is satisfactory and no further service is necessary.
5. If valve does not click when shaken, or the paper is not drawn against the filler pipe with noticeable force, replace valve and recheck the system.

NOTE: Do not attempt to clean the valve. Replace it with a new one that has a black end washer.

6. Remove ventilator valve hose from cap and carburetor. Inspect for deposits and clean if necessary.
7. Remove breather cap and wash it thoroughly in kerosene or other suitable solvent to remove old oil and dirt. Reoil filter element.
8. Service air cleaner element.
9. Remove carburetor. Hand turn a suitable drill through the passages to dislodge foreign particles.

CAUTION: Use a drill that will clean the passages without removing any metal. Blow passages clean. It is not necessary to disassemble the carburetor for this service.

CHRYSLER C.A.S.

Cleaner Air System

Service Procedure

To determine whether the system is functioning properly, proceed as follows:

1. Make sure all vacuum hoses and the stove to air cleaner flexible connector are properly attached and in good condition.
2. With engine cold, and ambient temperature in the engine compartment below 100 deg. F, the heat control door in the snorkel should be in the up (heat on) position.
3. With the engine warmed up and running, check the air temperature entering the snorkel or at the sensor. When temperature is 105 deg. F or higher, the door should be in the down (heat off) position.
4. Remove air cleaner from the engine and allow it to cool down to 90 to 95 deg., F. With 20"Hg. vacuum applied to the sensor, the door should be in the up (heat on) position. If not, check the vacuum diaphragm.
5. To check the vacuum diaphragm, apply vacuum directly to the diaphragm with a vacuum gauge in the line and a bleed valve to control the vacuum inserted in the line between the gauge and the vacuum source. Apply 20" Hg. to the diaphragm and close off the line. The diaphragm should hold this vacuum for five minutes. Release the vacuum then with the use of the bleed valve, build the vacuum slowly and observe the operation of the door. The door should lift off the bottom of the snorkel at 5"Hg. and reach the full open position with no more than 9"Hg. If it does not, the diaphragm must be replaced.

6. If the diaphragm performs properly but proper temperature is not maintained, replace the sensor and repeat the temperature checks.

CHRYSLER NOx

Oxides of Nitrogen

Manual Transmission

To test the NOx system on a manual transmission equipped vehicle, be certain that the plenum chamber temperature is well above 70 degrees F and proceed as follows:

1. Place the transmission in Neutral.
2. Turn on the ignition switch.
3. Disconnect the wire from the B+ connector of the ballast resistor while holding the solenoid vacuum valve. You should feel it de-energize.
4. Select high gear of the transmission. You should feel no energizing of the solenoid valve whether the wire is connected or disconnected. These steps will indicate whether the system is operating correctly electrically. If it does not operate normally, proceed to test the components separately.
5. To test the solenoid vacuum valve, first remove the connector to expose the valve blade terminals.
6. Jump the piggy back connector on the ballast resistor to one of the solenoid vacuum valve blade terminals. Jump the other blade terminal on the valve to ground. With the ignition switch on the solenoid should energize. If it does not energize it should be replaced.
7. To test the thermal switch, disconnect the dual plug from it and use a short wire to jump the plug and make a circuit.
8. Place the transmission in Neutral and turn the ignition switch on. The solenoid vacuum valve should energize. If the plenum chamber temperature is well above 70 degrees F and the solenoid vacuum valve does not energize when the thermal switch is connected but does energize when the plug is jumped, the thermal switch is defective and should be replaced.
9. To test the transmission switch, place the transmission in Neutral and turn the ignition switch on.
10. Remove the connector from the switch and jump it to a good ground. If the solenoid vacuum valve functions when this is done, the transmission switch is probably defective. Before replacing it, however, check the switch for proper torque in the transmission housing. This should be 180 in. lbs. to insure proper grounding. If proper torque does not relieve the problem, replace the switch.

Automatic Transmission

To test the NOx system on an automatic transmission equipped vehicle, be certain that the plenum chamber temperature is well above 70 degrees F and proceed as follows:

1. Raise the rear wheels.
2. Use a tee fitting, place a vacuum guage between the distributor and the solenoid vacuum valve.
3. Disconnect and plug the vacuum line at the vacuum switch on the control unit assembly.
4. Start the engine and run it at a fast idle (above 850 rpm). The vacuum gauge should read zero.
5. Remove the electrical lead from the control unit "T" connector and the vacuum gauge should register vacuum. Reconnect the wire and the gauge should drop to zero.
6. Reconnect the vacuum line to the vacuum switch and disconnect the single wire lead from the control unit to the speed switch. The gauge should register vacuum.
7. Place the transmission in Drive. Sharp acceleration will cause the vacuum gauge to drop to zero suddenly.
8. Limit the speed to 40 mph and as the rpm stabilize the gauge should suddenly register vacuum.
9. Reconnect speed switch lead and disconnect and plug the vacuum line at the vacuum switch. The gauge should register zero.
10. Accelerate the engine to above 30 mph. The gauge should register normal vacuum.
11. If the solenoid vacuum switch itself is found to be in proper operating condition and yet no solenoid action occurred in the foregoing steps, replace the control unit assembly.

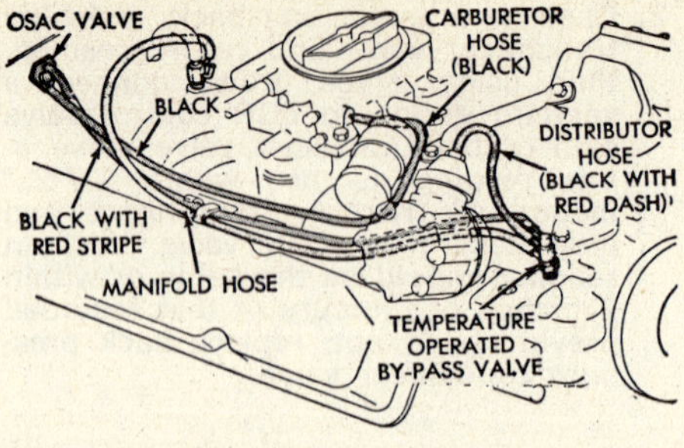

Fig. 10 Chrysler by-pass valve hose routing

CHRYSLER IGNITION

Refer to Tune Up Service chapter for system trouble shooting.

CHRYSLER ORIFICE SPARK ADVANCE CONTROL (OSAC)

Test

NOTE: Ambient temperature in area of vehicle must be above 68 degrees F. for this test.

Inspect hose connections. Replace any hoses that may allow air leakage. Check operation of valve. With engine running at 2000 rpm in neutral, disconnect hose at the OSAC valve leading to distributor and attach a vacuum gauge to this port. If very gradual increase in vacuum is observed, approximately 15 seconds to a stabilized level, the valve is operating properly (some variations between engines). If vacuum immediately goes up to manifold vacuum or there is no vacuum, the OSAC valve is not operating properly and must be replaced.

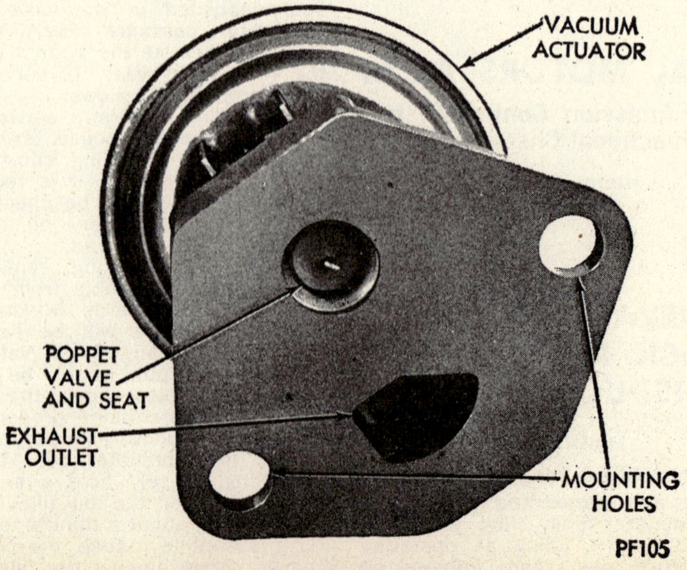

Fig. 11 Chrysler EGR control valve

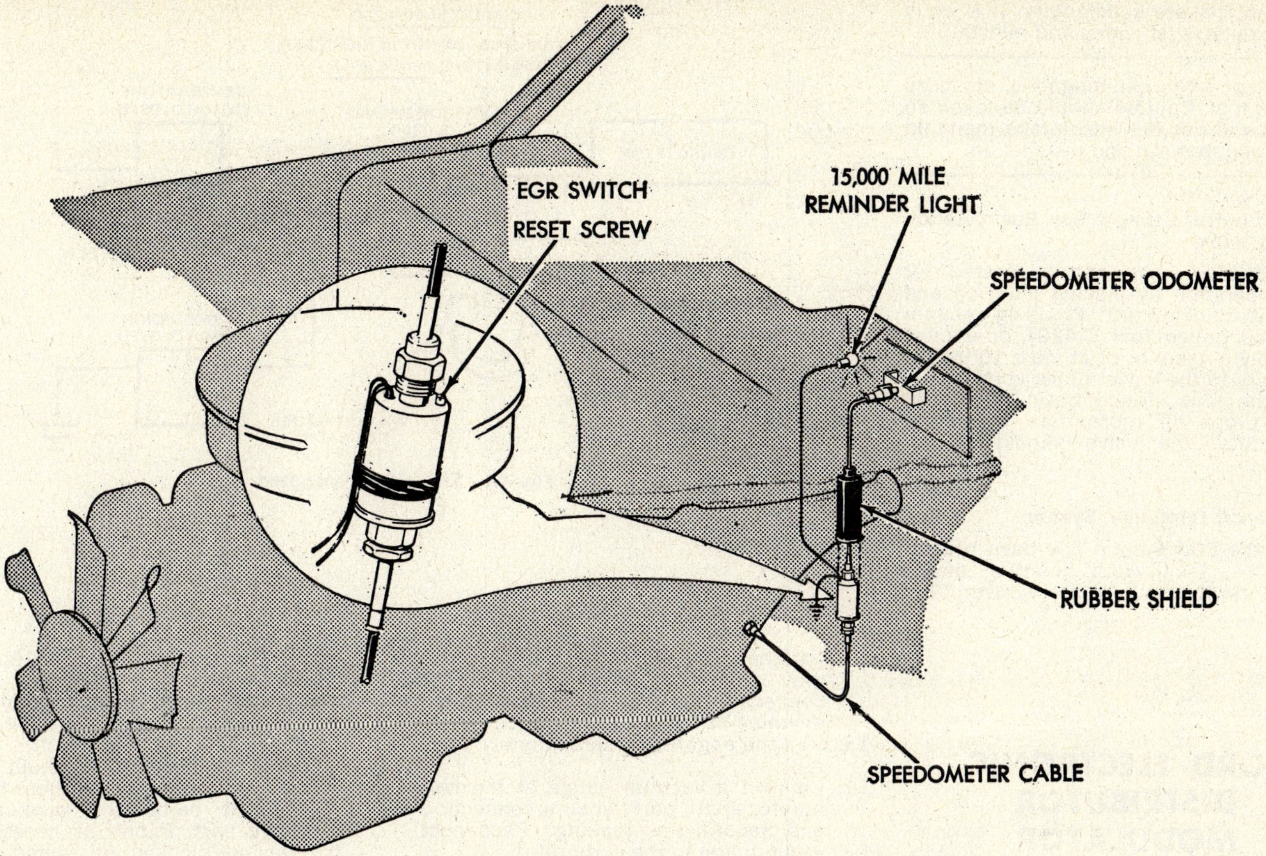

Fig. 11A EGR Maintenance Reminder System. 1975 Chrysler Corp.

CHRYSLER TEMPERATURE OPERATED VACUUM BY-PASS VALVE

Tests

1. Check vacuum hose routing and installation, Fig. 10. Check coolant level.
2. Attach a tachometer to engine.
3. Bring engine to normal operating temperature with choke wide open.
4. Adjust idle to 600 rpm. (For testing purposes only).
5. Disconnect the hose from manifold vacuum port and plug hose.
6. Check engine idle speed, no change indicates valve is holding, if idle speed drops 100 rpm or more the valve is defective.
7. Re-install the manifold vacuum port hose.
8. Cover radiator to increase engine coolant temperature.

NOTE: Do not overheat engine.

---

9. Idle engine until temperature is approximately 225 degrees F. if engine idle speed has not increased 100 rpm or more before gauge indicator reaches top of normal bar, which is 235 degrees F. or 225 degrees F. on numerical gauge. The valve is defective.

CHRYSLER EXHAUST GAS RECIRCULATION

Service

Every 12,000 miles or 12 months, the complete EGR system should be inspected and tested. To assure proper operation of this system all passages and moving parts must operate properly free of deposits, also hoses and connections must be free from leaks.

NOTE: Ambient temperature in the area of the vehicle must be above 68 degrees F. for this test.

Warm up engine and allow the engine to idle, in neutral, with the carburetor throttle closed; then abruptly accelerate the engine to approximately 2000 rpm, but not over 3000 rpm. Visible movement of the EGR valve stem should occur during this operation, this can be determined by change in the relative position of the groove on the EGR valve stem. This operation should be repeated several times to confirm movement. Movement of the stem indicates that the control system is functioning correctly.

If control system is functioning properly, the valve and passages may be checked for exhaust gas flow by applying a vacuum of at least 10 inches Hg directly to the EGR control valve with the engine warm and idling in neutral. Idle speed should drop 150 rpm or more when vacuum is applied. This reduction in idle speed confirms that exhaust gas recirculation is taking place. If the speed change does not occur, or is less than the specified, exhaust deposits in the EGR valve or intake manifold EGR passages are indicated.

EGR Control Valve Service

The EGR valve should be inspected for deposits with particular attention to the poppet and seat area, Fig. 11. If deposits exceed a thin film condition, the EGR valve should be cleaned. If wear of the stem or other moving components is noted, the valve should be replaced.

NOTE: Extreme care should be exercised during the cleaning operation to prevent spilling of solvent on the valve diaphragm, as this will cause diaphragm failure. Do not push on diaphragm to operate valve, use vacuum source only.

Floor Jet Exhaust Gas System

Every 12,000 miles of operation inspect the floor jets for deposit buildup. With engine off and carburetor air cleaner removed, hold choke and throttle valves open. Using a suitable light, inspect the floor jets visually through the carburetor. If there is an open path through the orifice the jets are satisfactory. If a jet is plugged, remove jet, clean and reinstall.

NOTE: Floor jet is non-magnetic, stainless steel material. Caution should be taken so that jets will not fall into intake manifold ports during removal and installation.

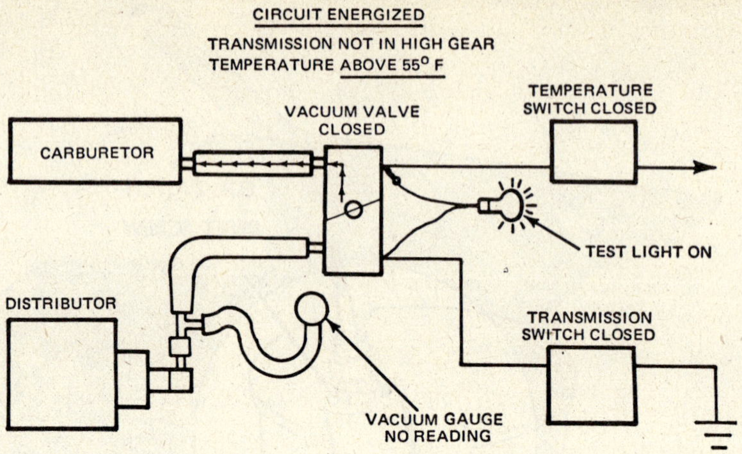

Fig. 12 TRS circuit energized

Coolant Control Exhaust Gas Recirculation (CCEGR) Valve

The CCEGR valve can be tested for proper operation by placing it in ice and cooling it to below 40° F. Using vacuum pump and gauge tool C-4207 or equivalent, apply a vacuum of at least 10 inches of mercury to the valve nipple corresponding to the blue striped hose. If vacuum reading drops off more than 1 inch in one minute, the valve should be replaced.

Maintenance Reminder System

Once the EGR system has been properly serviced, the switch counting device may be reset using a screwdriver, Fig. 11A.

FORD ELECTRONIC DISTRIBUTOR MODULATOR

Testing

1. Connect a vacuum gauge to the large hose connection of the electronic module.
2. Elevate the rear wheels of the vehicle.
3. Start the engine. With the transmission in neutral, the vacuum gauge should read zero.
4. With transmission in gear, slowly accelerate to approximately 35 mph.
5. Vacuum should cut in between 21 to 31 mph. Vacuum reading should be at least 6"Hg. at 25 mph.
6. Allow vehicle to coast down from 25 to 15 mph. At some speed in this range, the vacuum should drop to zero and remain there.
7. With the transmission in neutral and engine running, chill the thermal switch and place the engine on fast idle, approximately 1500 rpm. There should be a vacuum reading. If unit checks out at this point, it is operating properly. If not, proceed as follows.
8. Check supply voltage at the red wire of the quick disconnect (female end) with ignition switch on the On position. Meter should read battery voltage.
9. If meter reads zero, check the fuse and wiring from the ignition switch.
10. Check the thermal switch by disconnecting the yellow quick disconnect and inserting an ohmeter from the gray wire (female end) to ground.
11. Place one hand on the thermal switch and allow switch to warm up. Switch should now be open. Chill the switch and it should close.
12. Leave plug disconnected and, using a jumper, connect the two red wires together. Connect a jumper from the gray wire of the male plug. Do not connect the other end of this wire to anything at this time.
13. Run the engine at approximately 1500 rpm.
14. Connect a vacuum gauge at the carburetor spark port vacuum connection and record the reading. Reconnect vacuum line to the carburetor.
15. Connect vacuum gauge at the large hose connection of the solenoid valve.
16. There should be no vacuum reading.
17. Ground gray jumper wire. Vacuum reading should now match that of Step 14. If not, replace the module.
18. Remove jumper wire from thermal switch. Leave switch disconnected.
19. With vehicle on hoist and running at 30 mph, vacuum gauge should show a reading.
20. Check speed sensor continuity. Resistance of the sensor is 40-60 Ohms at room temperature. Resistance of the sensor to ground should be infinite. If not, replace the sensor.

FORD E.S.C. SYSTEM

Electronic Spark Control System

Testing

To test the E.S.C. system, proceed as follows:

1. Raise the rear wheels.
2. Disconnect the vacuum hose at the distributor primary advance diaphragm and connect it to a vacuum gauge.
3. Ensure that the temperature switch is above 55 degrees F by warming it with the hand or a hot sponge.
4. Start the engine and put the transmission in drive.
5. The vacuum gauge should register zero until the appropriate 23, 28 or 33 mph speed is attained, at which point it should register at least 6" Hg.
6. If there is no vacuum in any case at over 33 mph, there are four possible causes:
   a. No vacuum at the carburetor vacuum advance port. With the transmission in Park or Neutral, block open the throttle to about 1500 rpm and determine if there is vacuum at carburetor vacuum advance port. Repair as necessary.
   b. Pinched, blocked, misrouted or disconnected hoses. Follow vacuum hose from carburetor to PVS valve and through the PVS valve to the ESC vacuum control valve.
   c. Inoperative ESC control system. Disconnect one or both electrical leads at ESC vacuum control valve and follow vacuum to the gauge at the distributor.
   d. Throttle down the engine.
   e. If there is no vacuum in step c, reconnect the ESC vacuum control valve.
7. If there is vacuum below the cut-in speeds, trouble exists in the ESC system which prevents the valve from being energized. Therefore, it is necessary to test the individual system componets:
   a. Temperature Switch Test: This test can be made in the vehicle or on the bench. Disconnect the multiple plug and connect an ohmmeter to both wire terminals from the switch. Place the switch in the palm of the hand and allow sufficient time to warm it. If the ohmmeter registers a reading with the switch at 60 degrees F or higher, the switch is good.

      To test the switch at low temperature, chill it with ice or water which is below 49 degrees F. At this temperature the switch should be open and no reading should register on the ohmmeter. If the switch is good, check the power supply.
   b. Power Supply Test: Ground the lead on a test lamp and check for voltage at the temperature switch connector of the instrument panel wiring (circuit No. 640-red-yellow-

hash). With ignition on, the lamp should light. If no voltage is present at the terminal, replace or repair the wiring to the ignition switch or replace the ignition switch. If the previous steps have not located the problem, replace the electronic amplifier.

c. Speed Sensor Test: Disconnect the speed sensor at the multiple plug and check the sensor for continuity using an ohmmeter. Resistance of the speed sensor at room temperature is 40-60 ohms.

Also, check the resistance of the speed sensor to ground. There should be no continuity between the black wire and the case. Replace the speed sensor if the resistance readings are incorrect.

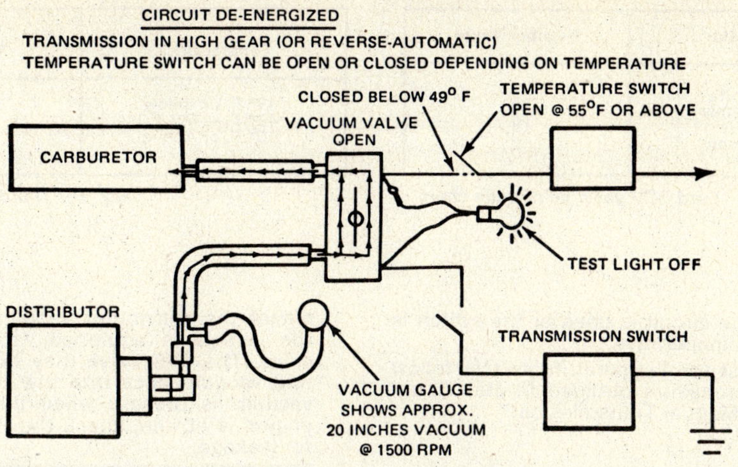

Fig. 13 TRS circuit de-energized

FORD HIGH SPEED EGR MODULATOR SUB-SYSTEM

Testing

1. Disconnect vacuum hose from EGR valve and install a "tee" fitting and connect a vacuum gauge with a long hose into the vacuum line.
2. Raise front and rear wheels off ground, with vacuum gauge visible from the driver's seat.
3. Start engine and allow to run at fast idle for about 3 to 4 minutes. At curb idle speed the vacuum gauge should read zero.
4. Place transmission in third gear for manual transmission and in drive for automatic transmission. Observe the speedometer and vacuum gauge. Increase engine speed. The vacuum reading should increase. At about 67 mph the vacuum reading should drop to zero, indicating that the EGR High-Speed Modulator sub-system is functioning properly.
5. If system is not functioning properly check the vacuum solenoid valve. Disconnect the electrical leads from the valve to isolate it from the electronic amplifier.

CAUTION: Never connect any test jumper or test light to the valve unless it is isolated from the amplifier, as severe damage to the amplifier will result.

6. Connect jumper wires from the valve terminals to the battery and to a ground, operate engine in neutral at about 1500 rpm. The valve should close when power is applied. If there is a vacuum reading replace the valve.
7. If there is no vacuum reading with the valve energized, check for power at the amplifier connector using a self powered test light. If there is no power, check and repair the ignition switch circuit. If there is power check ground connection. If grounded properly, check speed sensor for continuity with an ohmmeter placed across sensor. The resistance should be 40 to 60 ohms. If the speed sensor is functioning properly, replace the amplifier.

FORD T.R.S. SYSTEM

Transmission Regulated Spark Control System

Testing

Manual Transmission

With a vacuum gauge connected to the T.R.S. system, Fig. 12, between the vacuum control valve and the distributor advance unit, and the temperature switch known to be above 55 degrees F, proceed as follows:

1. Start the engine in Neutral. No vacuum should be indicated on the gauge (circuit energized).
2. Disengage the clutch and increase engine speed to 1000 to 1500 rpm. The vacuum indication should still remain at zero.
3. Disengage the clutch then place the transmission in high gear. With the engine running between 1000-1500 rpm and the clutch disengaged, a vacuum of at least 6 inches of mercury should be indicated on the gauge. Make sure the engine is stopped before engaging the clutch.
4. To test transmission switch, proceed as follows: Disconnect both wires from the vacuum control valve. Connect the transmission switch lead from the valve blade terminal in series with a test light to the positive terminal of the battery. With the engine and ignition off, move the gear selector through all positions. The light should stay on in all but high gear. If the light stays on in high gear, the circuit is grounded or the switch is inoperative. If the light does not go on in any gear, the circuit is open or again the switch is inoperative.
5. To test the temperature switch, proceed as follows: Disconnect both wires from the vacuum control valve. Connect the temperature switch lead from the blade terminal of the vacuum control valve in series with a test light to ground. Turn on the ignition switch and warm the temperature switch with the palm of the hand or a hot sponge. The light should go on when the temperature of the switch exceeds 65 degrees F. Cool the switch with an aerosol spray (such as starting fluid) or ice until its temperature is below 49 degrees F and the light should go out. If the light does not go on when warmed, the circuit is open or grounded or the switch is inoperative. If it does not go out when cooled, either the cooling process was not sufficient or the switch is inoperative.

Automatic Transmission

With a vacuum gauge connected to the T.R.S. system, Fig. 12, between the vacuum control valve and the distributor advance unit, and the temperature switch known to be above 65 degrees F, proceed as follows:

1. Start the engine (Park or Neutral). No vacuum should be indicated on the gauge (circuit energized).
2. With foot brake applied firmly, shift into reverse.
3. Disconnect one wire from the vacuum control valve to de-energize the circuit, Fig. 13. Shift into neutral and increase engine speed from 1000 to 1500 rpm. Vacuum should be indicated on the gauge. Reconnect the wire to the valve after the test is made.

NOTE: Malfunction of the electrical circuit affects the vacuum, but the vacuum portion has no effect on the electrical. Therefore, in the event of no vacuum, disconnect both hoses at the valve and connect both hoses together with a nipple, and then repeat step 3.

4. To test transmission switch, proceed as follows: Disconnect both wires from the vacuum control valve. Connect the transmission switch lead from the valve blade terminal in series with a test light to the positive terminal of the battery. Start the engine, apply foot brake and move shift lever through all positions. The light should go out in the reverse position. If the light stays on in all positions, the circuit is grounded or the switch is faulty. If the light does not go on at

Color	Number Value
White	5
Yellow	10
Blue	15
Green	20
Red	30

Fig. 14 Ford SDV valve color code chart

Temperature	Temperature Switch	Spark Advance Path
49° F. or Below	Open	DVB – Direct
49° F. —60° F.	Open or Closed	DVB or SDV
60° F. or Above	Closed	SDV – Delayed

Fig. 15 Ford DVB system operating temperatures

all, the circuit is open or the switch is again inoperative.

5. To test the temperature switch, follow the procedure outlined in step 5 under "Manual Transmission."

FORD DVB SYSTEM

Delay Vacuum By-Pass

Tests

1. Disconnect the ignition feed wire to the DVB vacuum valve.
2. Connect a vacuum gauge to the distributor primary vacuum hose.
3. Start the engine, vacuum should be present as the engine speed increases. The vacuum should fluctuate in direct response to the throttle setting.
4. Connect the ignition feed wire to the DVB vacuum valve.
5. Repeat step 3. The vacuum gauge should show a much slower response to throttle change. The vacuum supply is now being controlled by the SDV valve restrictor.
6. Refer to, Fig. 15, for proper DVB temperature switch activation and spark advance path.

NOTE: When servicing the SDV valve, replace with proper color coded valve, Fig. 14.

CAUTION: The vacuum hose and wiring connections for a DVB system must be connected properly, or the system will not function.

FORD TAV SYSTEM

Temperature Activated Vacuum

System Checks

The TAV system vacuum hoses and wiring connection must be connected properly, or the system will not function.

1. The EGR and spark port vacuum must be connected to the correct inlet of the three-way vacuum valve.
2. The TAV system three-way vacuum valve must be adequately grounded to the cylinder block.
3. The TAV system three-way vacuum valve must be energized when the ignition switch is in the run position, and the ambient temperature is above 65 degrees F. Check power to the valve with a test light connected between the b+ terminal of the valve and chassis ground. The light should be lit, if the above conditions are present.
4. A false vacuum reading may be present at the distributor vacuum inlet if the vehicle is equipped with a PVS valve. The PVS valve may leak manifold vacuum back into the system. If vacuum is present when the vehicle engine is at idle, check the PVS valve for leakage.

NOTE: A check valve is installed in the EGR vacuum supply line on all vehicles equipped with air conditioning to prevent vacuum feed back into the EGR system. If there is feed back into EGR system replace check valve.

FORD CTAV SYSTEM

Cold Temperature Activated Vacuum

NOTE: Prior to performing any test on this system, ensure that temperature switch is above 65° F.

Vacuum System Test

1. With a tachometer connected, disconnect three-way solenoid vacuum valve ground lead. Connect a vacuum gauge to distributor end of hose running from solenoid vacuum valve and if no vacuum reading is observed at 1500 RPM, check vacuum source back to carburetor spark port.
2. If a vacuum reading of approximately 15 inches Hg was observed in above check, reconnect solenoid vacuum valve ground lead. If a vacuum reading is still noted, check solenoid vacuum valve ground and electrical leads back to ignition switch.
3. If after reconnecting solenoid vacuum valve ground lead, vacuum is low or non-existant, run engine at 3000 RPM and if no vacuum is observed, check vacuum source back to carburetor EGR port.
4. If a vacuum reading of approximately 9 inches Hg was observed at 3000 RPM, system is functioning normally.

Electrical System Test

1. Disconnect temperature switch connector and connect a test lamp between the solenoid vacuum valve ground terminal and the ground. Turn ignition "On". If lamp lights, the latching relay is defective, requiring replacement.
2. If lamp does not light in above check, reconnect temperature switch connector. If lamp still does not light, check temperature switch and wiring back to ignition switch.
3. If lamp lights in above check, again disconnect temperature switch connector. If lamp goes out, replace latching relay. If lamp remains lit, connect the test lamp to temperature switch connector and cool temperature switch to below 49° F. If lamp lights, the temperature switch is faulty, requiring replacement.
4. If lamp does not light when temperature switch is cooled to below 49° F, the system is operating normally.

FORD CSSA SYSTEM

Cold Start Spark Advance System

NOTE: If engine is pinging, perform steps 1 thru 4

1. Connect tachometer and start engine.

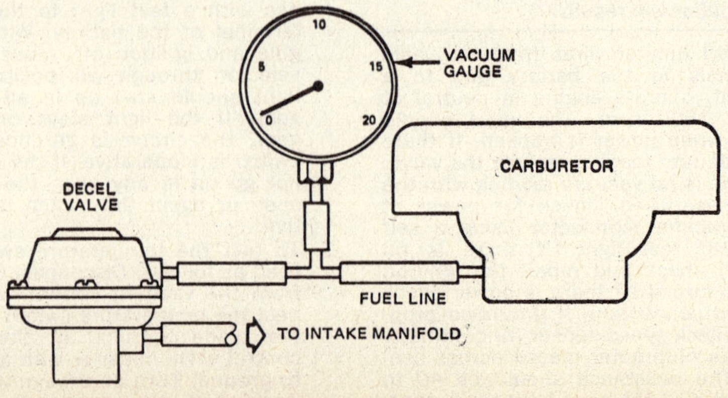

Fig. 16 Decel valve test connections. 1974 2800cc engine

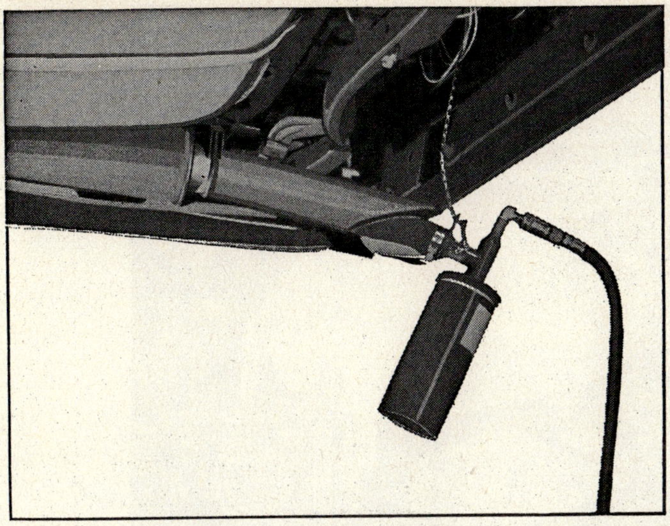

Fig. 17 Aspirator installation

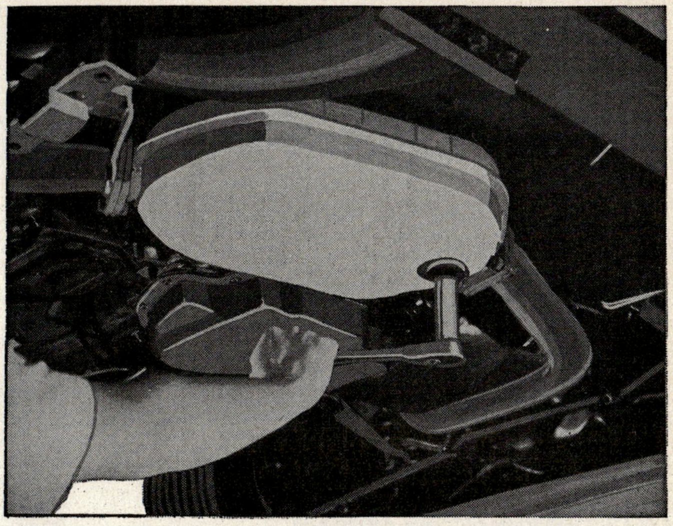

Fig. 18 Catalytic converter fill plug removal & installation

CAUTION: Do not leave transmission in drive for more than 15 seconds while performing steps 2 and 4.

2. With parking brake applied, brake pedal depressed and transmission in drive, accelerate engine and observe rpm at which spark knock or pinging occurs.
3. Place transmission in neutral and accelerate engine to 1500 rpm to allow transmission to cool. Repeat this step three times.
4. Disconnect and plug vacuum advance line from distributor. Repeat step 2 to recorded rpm.
5. If spark knock or pinging reoccurred, stop engine and proceed to next step. If spark knock or pinging did not reoccur, the CSSA PVS and or retard delay valve (if used) is faulty. Repair or replace as necessary.
6. Remove spark delay valve (SDV), or vacuum check valve, or spark delay valve and check as described further on.
7. Using a tee fitting, connect a vacuum gauge to distributor advance hose and disconnect and plug vacuum hose from bottom port of Cold Start PVS.
8. With engine idling at normal operating temperature, there should be no vacuum reading. If vacuum is obtained, the Cold Start PVS should be replaced.
9. Reconnect hose to bottom port of Cold Start PVS and disconnect and plug hose from top port of Cold Start PVS.
10. On vehicles with a cooling PVS, check vacuum at distributor with engine at normal idle. If reading is 2 inches Hg. or less, the cooling PVS is operating properly. If vacuum reading is greater than specified, disconnect and plug the top vacuum line from the carburetor to the cooling PVS. If reading is still greater than specified, the cooling PVS should be replaced.
11. With engine idling in neutral, momentarilly open the throtte half way.
12. Observe vacuum gauge for a quick rise and fall as throttle is opened and closed. If vacuum is obtained, the spark advance system is operating properly. If no vacuum is obtained, check vacuum line for leakage or obstruction and correct as necessary.

FORD (SDV)

Spark Delay Valve

NOTE: To perform the following procedure, an external vacuum source capable of maintaining a minimum constant 10 inches Hg. is required.

Mono Delay Valve

1. Set external vacuum source to 10 inches Hg. and connect black side of delay valve to vacuum source.
2. Connect a vacuum gauge with a 24 inch hose to colored side of delay valve.
3. Apply 10 inches Hg. vacuum and observe time in seconds for gauge to read from 0-8 inches Hg. The minimum and maximum time for gauge to read 8 inches Hg. should be as follows:

Color	I.D. No.	Time in Seconds	
		Min.	Max.
Black and Gray	1	1	4
Black and Brown	2	2	5
Black and White	5	4	12
Black and Yellow	10	5.8	14
Black and Blue	15	7	16
Black and Green	20	9	20
Black and Orange	30	13	24
Black and Red	40	15	28

Dual Delay Valve

1. Set external vacuum source to 10 inches Hg. and connect vacuum gauge with a 24 inch hose to DIST nipple of delay valve.
2. Connect black side of delay valve and CARB nipple of delay valve to vacuum source.

NOTE: Avoid applying vacuum to CARB nipple while applying vacuum to black side of valve.

3. Apply 10 inches Hg. of vacuum and observe time in seconds for gauge to read from 0-8 inches Hg. The minimum and maximum time for gauge to read 8 inches Hg. should be as follows:

Color	I.D. No.	Time in Seconds	
		Min.	Max.
Black and White	5	4	12
Black and Yellow	10	5.8	14
Black and Blue	15	7	16
Black and Green	20	9	20
White and Brown	2	2	2
White and Green	20	9	20

Retard Delay Valve

1. Set external vacuum source to 10 inches Hg. and connect colored side of delay valve to vacuum source.
2. Connect a vacuum gauge with a 24 inch hose to white side of delay valve.
3. Apply 10 inches Hg. of vacuum and observe time in seconds for gauge to read from 0-8 inches Hg. The minimum and maximum time for

gauge to read 8 inches Hg. should be as follows:

Color	I.D. No.	Time in Seconds Min.	Max.
White/Brown	2	2	5
White/Green	20	9	20
Black/White	5	4	12
Black/Yellow	10	5.8	14
Black/Blue	15	7	16
Black/Green	20	9	20

FORD DECEL VALVE

Adjustments

1971-74 1600cc, 2000cc, 2800cc & 1974-75 2300cc Engines

Note: On vehicles equipped with a PVS interlocked fuel decel system, assure that vacuum is available at decel valve with engine idling and at normal operating temperature. If low or no vacuum is obtained, check PVS ports, fittings and vacuum lines for leaks and obstructions and correct as necessary.

1. Connect tachometer to engine and bring engine up to operating temperature.
2. Disconnect rubber hose between decel valve and carburetor at decel valve and cap nipple on valve.

NOTE: Make certain that timing, CO or idle mixture and idle speed are set to specifications.

3. Raise engine speed to 3000 rpm and hold for about 5 seconds, then release throttle. If engine does not return to idle, check for binding linkage and correct as necessary.
4. Remove cap from decel valve and connect a vacuum gauge using a tee fitting so that valve remains operational.

Note: On vehicles so equipped, check operation of fuel decel valve (FDV) override units by connecting a vacuum gauge using a tee fitting between the manifold-to-valve vacuum line. With override units disconnected and engine idling at normal operating temperature, manifold vacuum reading should be obtained. Disconnect gauge and leave override unit disconnected until test is completed.

5. On 1974 2300cc engines, connect a vacuum hose between small nipple on valve and manifold vacuum source, Fig. 16.
6. Raise engine speed to 3000 rpm and hold for about 5 seconds, then release throttle and observe time required for vacuum to reach zero which should be 2-5 seconds.
7. On some 1974 2300cc engines, if valve is not within specifications, it must be replaced and retested. On 1600cc, 2000cc, some 1975 2300cc and 2800cc engines the decel valve may be adjusted by turning the adjusting screw inward to reduce the time the valve is opened and outward to increase it. If decel valve cannot be adjusted within the limits of the adjusting screw, it should be replaced.

1975 2800cc Engine

1. Using a tee fitting, connect a vacuum gauge between solenoid valve and decel valve, then with rear wheels raised, start engine and place transmission in drive.
2. At below the 11 mph cut-in speed, a vacuum reading should be obtained.
3. If no vacuum reading is obtained below the cut-in speed, connect jumper wires between solenoid valve and battery and ground. With engine at 1500 rpm and in neutral, there should be no vacuum reading. If vacuum reading is obtained, replace valve.
4. With ignition on, check for 12 volts at solenoid. If proper voltage is not obtained, check for proper ground or open circuit.
5. Using an ohmmeter, check speed sensor for 40 ohms resistance.

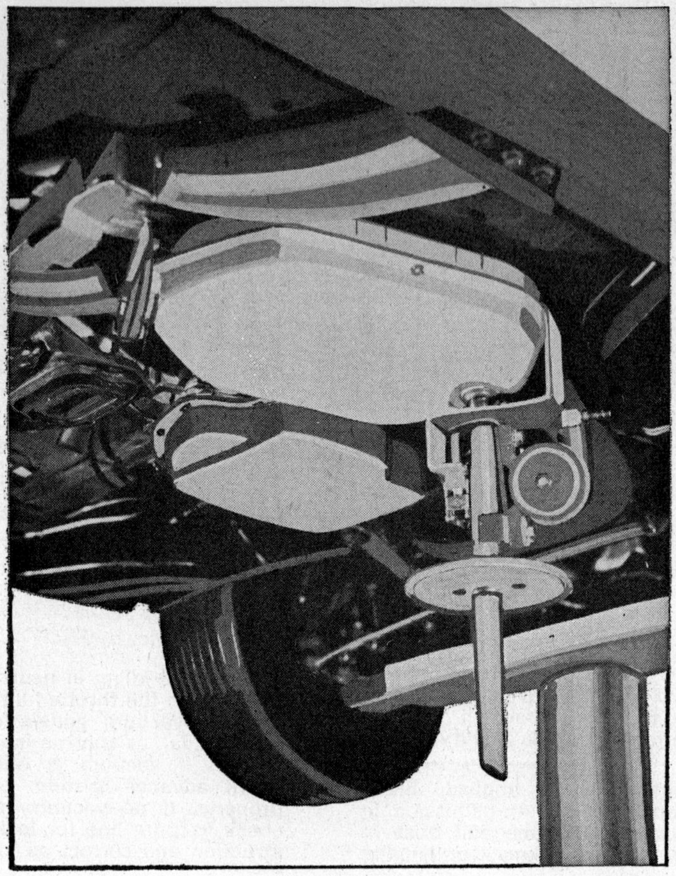

Fig. 19 Vibrator installation

CATALYTIC CONVERTER SYSTEM

American Motors & General Motors

Catalyst Removal

1. Install aspirator, Fig. 17, and connect an air line to aspirator, creating a vacuum in the converter, thereby holding pellets in place when fill plug is removed.
2. Using a ¾ inch Allen wrench or tool J-25077-3, Fig. 18, remove converter fill plug.
3. Install vibrator, Fig. 19, and empty catalyst container to converter.
4. Disconnect air line from aspirator and reconnect line to vibrator. Catalyst will now drain from converter, Fig. 20.
5. When all catalyst has been drained from converter, remove container and discard the used catalyst.

Catalyst Installation

1. Fill container with recommended replacement catalyst and install fill tube extension to fixture.
2. Connect air line to aspirator and vibrator, then secure catalyst container to the fixture, Fig. 20.

3. When catalyst stops flowing, disconnect air line from vibrator and ensure catalyst has filled converter flush with fill plug hole. Add catalyst, if necessary.
4. Apply a suitable anti-seize compound to fill plug threads, then install and torque fill plug to 50 ft. lbs.
5. Disconnect air line from aspirator, then remove aspirator.

Chrysler Corp. & Ford

The catalytic converters used in these vehicles are not serviceable. After determining that the catalyst has lost its effectiveness, the catalytic converter assembly must be replaced.

Fig. 20 Draining & refilling catalytic converter

CATALYST PROTECTION SYSTEM

1975-76 Chrysler

1. Disconnect wire from throttle position solenoid and hold throttle wide open. Apply battery voltage to solenoid. If solenoid does not extend and remain extended, replace unit.
2. Connect a tachometer and start engine, then apply battery voltage to solenoid. Open throttle sufficiently to assure that solenoid fully extends when current is applied. With solenoid energized, engine speed should be 1500 rpm ±50 rpm. Adjust to specifications as required. Allow time for OSAC to provide vacuum and engine speed to stabilize.
3. Disconnect solenoid wire from battery voltage and reconnect it to electronic speed switch.
4. Slowly accelerate engine from idle. As engine speed passes approximately 2000 rpm, the solenoid should extend and remain extended. Slowly decelerate engine. At or before engine speed reaches 1800 rpm, the solenoid should deactivate. If solenoid does not respond properly, the speed switch must be replaced.

OSCILLOSCOPES

As will be seen in this scope description, oscilloscope type test equipment can show the technician a great deal of information. Looking back to the mechanical steps in a tune-up, many of them are aided considerably by the use of a scope. The compression test, can be by-passed in most cases if the scope shows a normal spark pattern for all cylinders. Spark plug servicing is made more certain through the use of a scope because it can tell how the plugs were working before they were removed, and time can be saved by avoiding the work of cleaning plugs that are firing poorly.

Ignition system servicing is probably the part of a tune-up that leads into more trouble than any other step. It is nearly impossible to see bad secondary wiring on a car. Reversed coil polarity escapes the eyes of most mechanics. Poor or borderline condensers do not show up unless tested separately. Weak coils with shorted turns are not visible unless separately tested. Yet all of these items and many more are checked quickly and accurately just by looking at the pattern on an oscilloscope. The dwell adjustment can and should be done with meter equipment. However, is the dwell the same for all cylinders? Is the distributor firing each cylinder at the same crankshaft position as the number one cylinder? These tricky problems can be shown quickly on the face of a scope.

It should be re-emphasized here that we have just covered some of the major scope patterns and applications. The scope can be utilized for many other checks such as coil output, primary circuit output, etc. Read the equipment manufacturer's instructions to determine the other areas in which the scope can be utilized.

INTRODUCTION

A scope is an instrument which gives a live picture of the ignition of an engine. It also reflects other engine performance factors.

The "Scope," shows what is going on every instant. The average technician, with but little training, can use these units to quickly diagnose troubles. The purpose of the following discussion is to demonstrate patterns, which readily indicate the common faults. Actual use of the equipment will point out the real simplicity and convenience of this type of instrument.

Most engine ailments are due to ignition. As a result, oscilloscopes are most valuable as ignition testers. By means of accessories, some oscilloscopes may be used to check such things as valve performance, detonation, preignition and vibration.

The picture obtained on the oscilloscope is a simple one, having just four events or signals to remember. Study Fig. 1 and remember the four basic sections. Some scopes present a pattern that is upside down to this one, but the principles are the same. This is due to variations in scope and pick-up design. Most scope checks are performed at 1000 to 1500 engine rpm. Below 1000 rpm, most scopes tend to be unstable and therefore might produce misleading information. At speeds above 1500 rpm, the coil and condenser section of the pattern begins to disappear and thus cannot be readily interpreted.

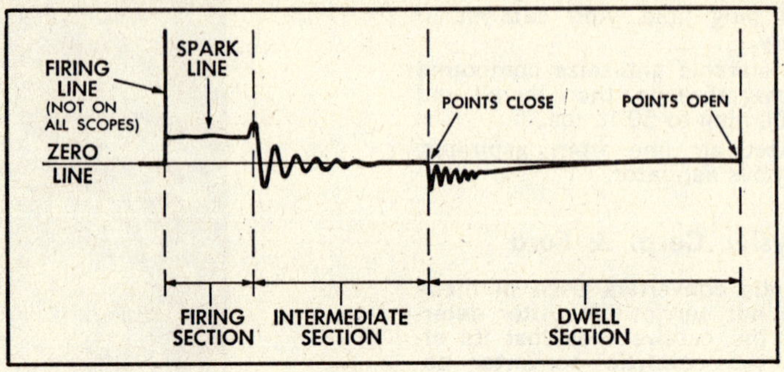

Fig. 1 Typical oscilloscope pattern

The spark plug begins firing at the left edge of the line and continues firing until it goes out at the end of the spark line. A normal spark line is 4° to 7° (read on horizontal scale of scope) at 1000 rpm. If the spark line is not normal, then the spark plug is not firing correctly. This is probably due to plugs, secondary wiring, rotor, cap or combustion chamber problems. The vertical line at the beginning of the spark line shows the voltage required to break down the gaps in the secondary system. These gaps are; rotor to cap, spark plug, and any poor secondary wire connections. This line is not visible on all scopes. The height of the firing line indicates the firing voltage of the cylinder. It is normal for this height to vary under normal conditions. Faulty operation is indicated by a line of different height than the rest. Just how much variation in line height is acceptable and how much will cause trouble will vary between brands of scopes. This can be caused by incorrect rotor gap, broken, burned or shorted high tension wires, cocked distributor cap, improperly connected wire terminals, wide or worn plug gap or lean fuel mixture.

Following the spark line are the coil and condenser wiggles. There must be at least 4 wiggles in this section (at about 1000 rpm). It is common to find 5 or 6. If there are less than the required number then either the coil or condenser is defective. This will be further pinpointed later.

The next signal is the points close signal. Notice that it begins with a large wiggle and then tapers off like a ski jump. It must be this shape. If the points close signal is misshaped, then the points are not closing correctly.

The line ends when the points open. Since it is desirable to open the points with a clean break the line should end with a clean abrupt stop. If the line ends with a deformation in the form of a hook, the points are arcing as they open.

The duration the points are closed, is commonly termed degrees of dwell. This can be measured directly on the pattern through the use of a scale which is placed under the pattern.

The above information is all that needs to be remembered in order to intelligently use a scope.

IMPORTANT CHECK-POINTS

Firing Voltage and Available Voltage

"Firing Voltage" or "Required Voltage" as the terms imply, is the amount of voltage required to establish a spark across the electrodes of a spark plug. In a running engine, the actual amount of voltage required to fire a particular plug at a particular instant depends on the net results of many factors. Some of the factors which can influence this firing voltage are; rotor gap, breaks in the secondary wires, spark plug gap, spark plug electrode shape, improperly connected wire terminals, temperature, compression pressure, air-fuel mixture, engine speed and load.

Actual "Firing Voltages" can easily be measured at any reasonable engine speed by observing the height of the firing line on each cylinder's secondary pattern as it is displayed on the scope screen. At any given engine speed, the firing voltage of all of the cylinders of an engine should generally be fairly uniform and within a normal range for that particular engine.

"Available Voltage" is a term or phrase used to indicate the maximum voltage an ignition system is able to produce under a given operating condition. Due to its operating characteristics, an ignition coil will produce its maximum secondary voltage whenever it attempts to fire an impossible gap, such as when a spark plug wire is removed from a spark plug and held at a distance from any ground. Ignition system design is such that available voltage is always greater than the ignition requirements generally encountered under most operating conditions. The difference in available voltage and that actually required to fire the spark plugs is commonly referred to as "Ignition Reserve."

The maximum secondary voltage available from any given ignition system depends on the combined effect of coil design, coil condition, applied primary voltage, primary circuit resistance, distributor contact condition, dwell angle and engine speed. When in good working order, most ignition systems are capable of producing well over 20,000 volts; in fact, some systems may even produce as high as 30,000 volts or more. However, should any of the factors involved in the operation of the ignition system deteriorate from their normal condition or adjustment, it will usually result in a change in available voltage values. Therefore, measuring the available voltage on "coil reserve" provides a quick means of determining the overall efficiency of any particular system.

Firing Line or Spark Line?

Since the firing line is not shown on all types of scopes, how can the above two checkpoints be covered on a unit that does not show the firing line? This question is frequently asked, and the following is presented in explanation. As may be readily seen, the same end results are obtained on both types of scopes with respect to firing voltage and coil reserve.

The firing of the spark plug involves two separate events. The first of these is the creation of the peak, firing or ionization voltage. This is followed by the second event which is the arc maintaining voltage or spark line.

The peak voltage is the voltage required to ionize the gap or cause the arc to form in the spark plug gap. The peak voltage level is also affected however, by temperature, compression and air fuel mixture. Consequently, some momentary limited peak voltage variations are considered quite normal.

Since current does not flow in the secondary until the arc is formed across the plug, the peak voltage is created with no current flowing in the secondary. In order to detect resistance in a circuit, current must be passed through the circuit. Therefore, peak voltage readings do not indicate resistance in the secondary.

Once the arc has been formed across the spark plug gap, the voltage reduces to a value needed to maintain the arc across the plug. As long as the coil can supply the lower voltage, the arc will remain. The duration of this spark forms the spark line. Since current is passing through the secondary while the arc is formed, *the spark line will indicate all gaps and resistance as variations in the length of the spark line.*

As mentioned before, the spark will burn as long as the coil can supply the proper energy. The spark line ends only because the coil has run out of energy. This then proves that the *spark line length is a measure of the reserve of the coil.* If the coil has very little reserve after ionizing the gap, all spark lines will be short. If the reserve is high, all spark lines will be long. This will not indicate the maximum reserve of the coil due to its built in capacity, but does indicate the voltage left in the coil after the plug gap is fired. This "reserve" or remaining voltage is basically controlled by degrees of dwell and resistance in the secondary circuit.

Coil Polarity

Proper coil polarity is of utmost importance to efficient ignition system operation. Leading manufacturers of ignition system components agree that the polarity of the ignition system secondary circuit should always be "positive" ground, regardless of the ground polarity of the vehicle's electrical system. Should secondary polarity be reversed, due to improper installation of the coil or battery, or installation of a coil not designed for a particular system (not wound in the proper direction), the firing voltage requirements of the vehicle may increase from 20 to 40% under certain operating conditions. Coil polarity, whether proper or reversed, is indicated by the position the secondary waveform assumes on the screen of the Scope.

Secondary Insulation

When an ignition coil produces a surge of high voltage as a result of the collapse of its magnetic field, this voltage is transmitted from the coil tower through the coil wire, distributor cap, rotor and a secondary wire and is finally delivered to a spark plug to create the spark across its electrodes. It must be realized that the insulation of all parts of the secondary circuit must be of high quality and in good condition to insure reliable ignition system function under all operating conditions. Should the insulation of one or more of these secondary components be defective or have deteriorated to some degree, it may have little or no affect on ignition system operation when the firing voltage requirements of the vehicle are moderate. However, when firing voltage requirements are high, this defective insulation may leak or breakdown, allowing much of the voltage surge to be lost to ground resulting in misfire of one or more plugs.

When a spark plug wire is removed from the plug to measure the available voltage of a system, the insulation of the secondary components transmitting the coils high voltage surge are also being stressed to the maximum and any leakage of this voltage, whether intermittent or otherwise, will readily be indicated by the shape of the pattern observed on the Scope's screen.

Secondary Resistance

Resistance is a factor present to some degree in every portion of all electrical circuits. It is the factor which has a hindering effect on the flow of electrical current. In ignition system design, some manufacturers specifically build a reasonable amount of resistance into the secondary circuit for the purpose of controlling secondary current flow in an effort to extend spark plug electrode life, and for the purpose of minimizing any possible interference with radio and television reception. Resistance of this nature is generally referred to as "suppression" resistance, and may be located in any one of a number of places such as in the rotor, the distributor cap, the secondary wires or in the spark plug. Suppression resistances, provided they do not exceed the values specified by the ignition system or vehicle manufacturer, have no adverse effects on ignition system performance.

Occasionally, suppressors will deteriorate or additional resistance will develop in the secondary circuit due to improperly connected wires and terminals, corrosion in the coil or distributor cap towers, or burned and corroded distributor cap inserts or rotor tip. Usually the adverse effects of this abnormal secondary resistance are most noticeable on engine performance during operating conditions that require high firing voltages. However, as the abnormal secondary resistance increases, its effect on engine performance will increase proportionally. The effects most noticeable will be shorter contact point and spark plug life. Even with the added unwanted resistance the engine will not "miss" until the voltage to fire the plugs exceeds the maximum coil reserve, however, the additional voltage will exceed the capacity that the condenser is designed to operate in, thereby causing the points to arc.

Other forms of resistance which can vary the duration of the spark plug firing or vary the voltage level required to maintain the spark exist in the form of engine compression and spark plug gap.

The height, length and shape of the spark line as observed on the scope may vary, depending on the combination of the above mentioned factors present at the time the spark plug is fired. The height of the spark line indicates the voltage required to maintain the spark across the plug electrodes; while the length of the line indicates duration of the spark. The shape and appearance of the spark line; whether it is horizontal or slanted, smooth or irregular depends on how much resistance is encountered and how much the resistance value changes during the firing.

Coil and Condenser Condition

The Intermediate Section, usually observed as a series of diminishing wiggles representing the dissipation of the energy remaining in the coil after the spark plug has ceased firing, can very easily be examined when each cylinder's pattern is displayed individually.

The number of oscillations that can be observed depends on a number of factors such as, dwell angle, engine speed, duration of spark, degree of coil saturation, and coil and condenser condition. Note the rate at which these wiggles diminish. Normally these wiggles will diminish gradually; but should the system contain a shorted coil or leaky condenser, these wiggles will diminish to zero rapidly. In such a case, probably only one or two wiggles will be seen at an engine speed of 1500 rpm providing the dwell angle and firing section are normal.

It should be particularly noted that under operating conditions where the firing line is quite long or the dwell angle is greater than specified, or when observing the patterns of a system utilizing dual points; the intermediate section may appear to be shortened by the closing of the breaker points before all of the coil energy has been dissipated. Under these conditions, fewer than normal wiggles having a fairly high amplitude at the instant of point closing may be displayed. This indicates that the energy level in the coil is still quite high at the time of point close and does not necessarily mean that the coil or condenser are defective.

Breaker Point Action

The Dwell Section begins when the distributor contacts close and extends to the instant at which the contacts open to fire the next cylinder. In analyzing this section of the pattern, the point close and the point open portions should be carefully observed. Normally, when the points close, this action is seen as a short vertical line followed by a series of diminishing wiggles. Note particularly that the *first line should be higher* than any of the oscillations following. Should it be observed that the first line is not as long as one or more of the following, it indicates poor breaker point contact. On the other hand, should the wiggles start to die out and then start again, it indicates a point bounce condition.

When the breaker points open, the end of the dwell section should appear as a clean right angle formed by the horizontal dwell line and the vertical firing line of the next pattern. Point arcing upon opening of the breaker points will be seen at the right end of the dwell section, just prior to the firing line of the next pattern. This will appear as a false start to firing, followed a short interval later by the actual firing line, or by a hook at the end of the points close signal.

Breaker Point Dwell

Proper distributor point dwell is important to the overall operation and efficiency of the ignition system in that it should be set to a specific value to assure adequate coil saturation to meet the firing requirements at all engine speeds.

Dwell Angle can accurately be measured on most scopes.

Distributor Cam Lobe Accuracy

The accuracy of the distributor cam determines the ignition timing relationship of all cylinders. Should one or more lobes of the distributor cam become worn or should the distributor shaft be bent, uneven timing of the various cylinders would result.

Cam lobe accuracy can be checked by superimposing all of the cylinders on parade type pattern scopes. On a multiline raster scope, the vertical alignment of all of the points—open signals shows immediately any condition that affects timing from one cylinder to the next.

Spark Plug Load

Spark plug voltage requirements can vary considerably depending on numerous factors as was previously outlined under "Firing Voltage". In particular, the effect of certain spark plug conditions such as rounded plug electrodes, extremely wide gaps, fouling and flash-over due to dirty plug insulators or deteriorated boots is generally more evident when an engine is operating under a load.

One simple method of loading an engine for test purposes is by "snap acceleration". This is accomplished by quickly opening the throttle and allowing the engine to accelerate from 1000 rpm to approximately 2000 rpm before closing the throttle. The rapid opening of the throttle momentarily places a load on the engine. It is during the momentary period when the engine is operating under a load that the firing and spark section of the patterns should be observed for unusual characteristics. Normally, the requirements of all cylinders will rise, but the firing and spark lines of all patterns should remain uniform in appearance.

Naturally, the requirement of all plugs will progressively increase as a result of electrode wear and will be indicated by higher than normal firing lines or shorter spark lines while the engine is under load. To be particularly noted are the individual voltages that are considerably different than the other cylinders of the engine.

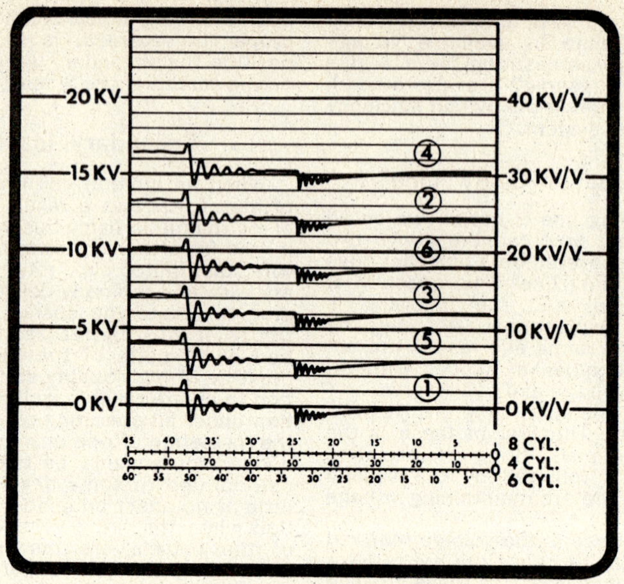

Fig. 2 Standard multiline raster pattern of an in-line 6 cylinder engine

TYPES OF PATTERNS

There are many brands of scopes on the market. Each one has variations in design that produce slightly different patterns.

Since all but a very few show the pattern or waveform that is developed on the secondary or spark plug part of the ignition circuit, secondary patterns are shown here.

Some people find the primary pattern effective in detecting breaker, condenser and coil malfunctioning, used in conjunction with the secondary pattern. Most scopes that show a secondary pattern can be readily adapted to read a primary pattern if desired. (Reference the manufacturer's instruction guide for primary wave pattern instructions and interpretations.)

Many scopes have a parade type of pattern, where one cylinder follows another in a line across the screen. On most of these scopes, the pattern may be expanded to view each pattern in greater detail.

Another type is the multi-line type of pattern. By use of the comparison system using any type pattern, the trouble may be quickly isolated. If the pattern is different on one line only, the trouble is unique to that particular cylinder. If all of the lines differ from the standard pattern, the trouble is common to all. By determining what portion of the pattern is involved, the trouble may be further isolated.

Since each scope has its own features and methods of operation, we leave the fine details of operation up to the manufacturers of the equipment.

Before preceding it should be emphasized here that the particular manufacturer's instructions should be thoroughly read and understood before the equipment is hooked-up or any scope pattern interpretations are made. The remainder of the presentation will discuss *basic* scope patterns. Variations from these may be noted depending on the specific equipment being used.

There are many things that can happen to an ignition system. Variations in compression, valving, carburetion and manifolding can cause pattern changes. To show every possible pattern change would be difficult to do and impossible to use. There are a few common pattern variations that do cover most problems. These patterns are shown and described below to serve as a basic scope guide.

As mentioned there are two methods of showing the pattern of the ignition system on most scopes. These are Multiline type, sometimes referred to as "Raster" pattern or Parade type. Both these methods show the same basic pattern shown in Fig. 1 but display this pattern for each cylinder in a different arrangement. Both these type methods of displaying the basic scope pattern will be discussed.

Multi-Line Display

The multi-line display, Fig. 2, shows each cylinder on a separate line. When the Scope is hooked-up correctly, the first line (bottom line on most scopes) represents #1 cylinder. The remaining patterns appearing above the first line represent the remaining cylinders in their firing order. Fig. 2 represents a multiline display of a 6 cylinder engine that is operating correctly. The encircled numbers appearing above the dwell section of each pattern identify which cylinder that pattern represents. (Note: The firing order of a 6 cylinder engine is 1, 5, 3, 6, 2, 4).

NOTE—Some scopes display a pattern that is inverted to the patterns shown in this text and display #1 cylinder on the top and the remainder underneath in their respective firing orders.

It should be remembered that voltage is measured in a vertical direction while crank angle is measured in the horizontal.

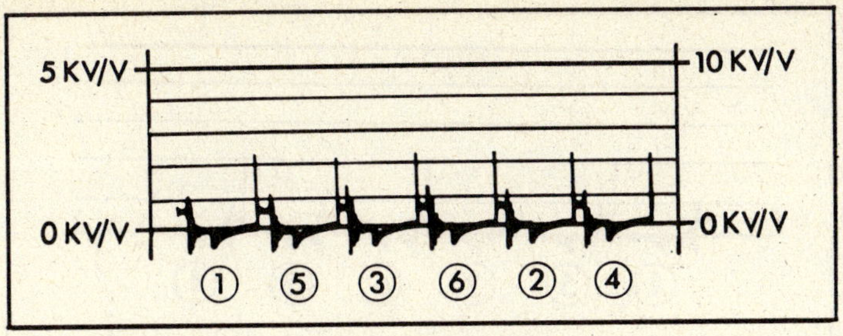

Fig. 3 Standard parade pattern of an in-line 6 cylinder engine

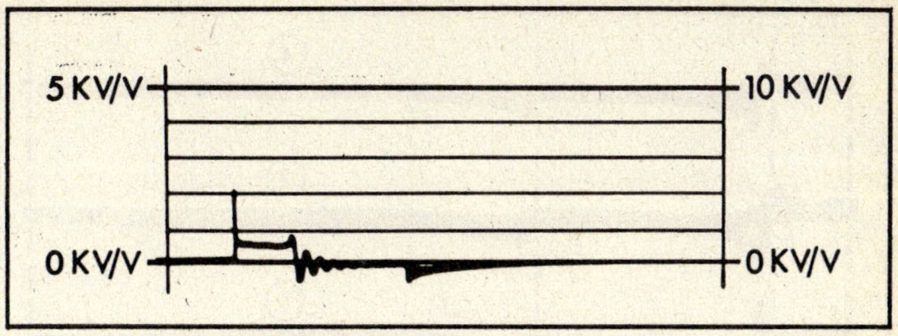

Fig. 4 Standard parade pattern expanded

In checking this pattern it is easy to note that all spark lines are of satisfactory and equal length. The coil and condenser wiggles contain more than the minimum number. The points close signal is shaped like a ski jump and the points open with a clean break. The dwell when checked with the scale agrees with specifications.

It is also significant that all the points close signals and point open signals stack up one beneath the next indicating the same dwell for each cylinder and the same respective piston position for each points opening and therefore each spark plug firing.

Parade Type Display

The parade display presents the basic scope pattern for each cylinder in one continuous line. Fig. 3 shows a parade type display for a 6 cylinder engine that is operating correctly. Each cylinder is represented by a complete wave form. As with the multi-line display the cylinders are represented in their firing order going from left to right across the entire display. The encircled numbers appearing below each pattern identifies which cylinder that pattern represents.

NOTE: The firing order of a 6 cylinder engine is 1, 5, 3, 6, 2, 4).

It should be noted that the "firing line" of #1 cylinder will appear at the extreme right of the trace on most scopes because most scopes use the signal of #1 firing to trigger the scope to start a new trace.

Almost all scopes that have the ability to display a parade pattern also have the ability to expand the wave form of any one cylinder. Fig. 4 shows the expanded pattern of one of the cylinders shown in Fig. 3.

ANALYZING SCOPE PATTERNS

Now that the two basic methods of displaying scope patterns have been covered, we will discuss some of the frequently occurring patterns encountered.

NOTE: *Patterns gained on a 6 cylinder engine will be discussed in the majority of the remaining text and shown in the illustrations for the sake of simplicity. However, the same displays would be gained on a four, six or V8 engine. The only difference would lie in the number of patterns formed. Interpretation of variances between patterns would be the same as brought out here.*

In Fig. 5, there is no problem in picking out the fact that there is difficulty in one cylinder. Counting in firing order, for a 6 cylinder engine, 1-5-3, the trouble is in #3 cylinder. It will be noted that there is no spark line, and the logical conclusion is that there is no spark. This is absolutely correct. Fig. 6 shows this condition on a parade-type pattern.

The energy that is not consumed in the spark goes into the coil and condenser wiggles. In the case of this particular cylinder there is no energy consumed in the spark, so all of it goes into the coil and condenser wiggles. This relationship between the spark line and the coil and condenser wiggles always holds.

The opposite condition is shown Fig. 7. Here cylinder #3 is shorted, giving a very long spark line. It will be noted that the spark line in the shorted cylinder is about 2 to 2½ times as long as normal. This condition always holds. It will also be noted that the amplitude of coil and condenser wiggles is very small. However, the minimum number still exists. Nearly all of the energy is consumed in the spark and very little of it remains to go into the coil and condenser wiggles.

You may ask why is there any signal at all since the spark plug is shorted. Remember that in creating the spark there are two gaps; one in the spark plug and the other in the distributor, between the rotor and the cap contacts. The signal which you see is a result of

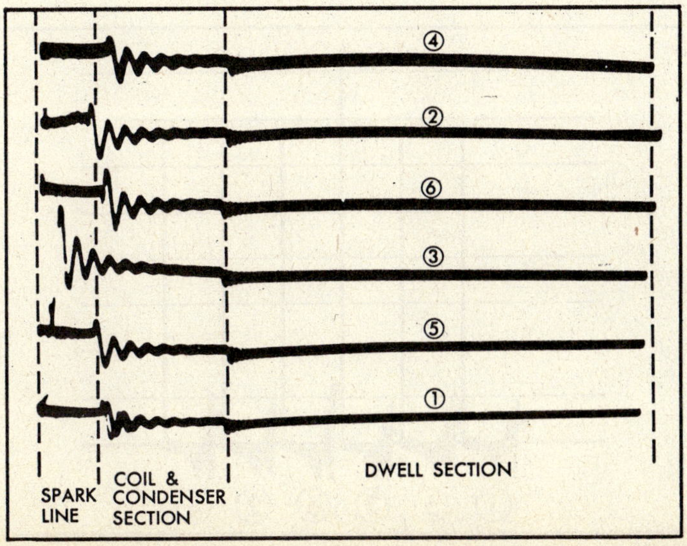

Fig. 5 No spark at #3 cylinder

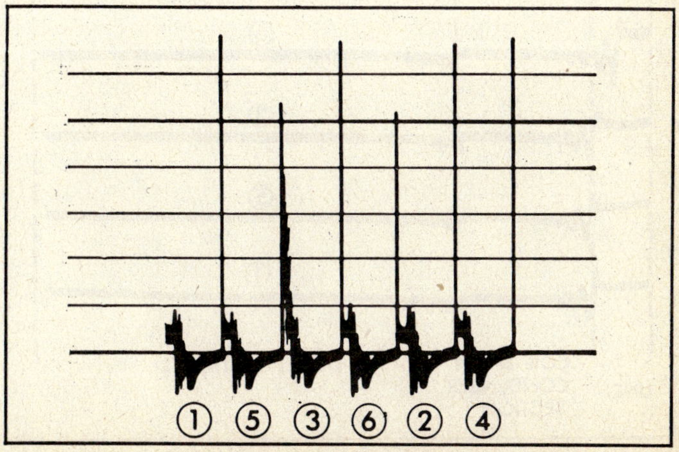

Fig. 6 No spark at #3 cylinder

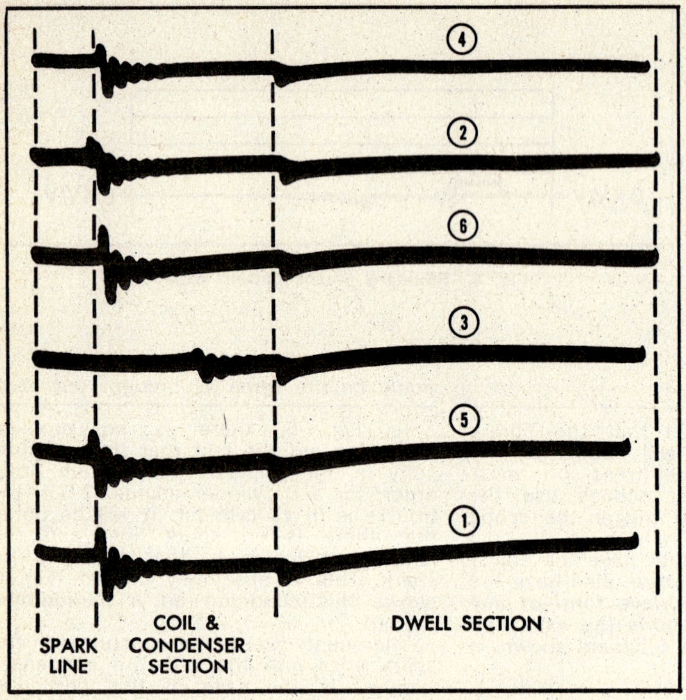

Fig. 7 Shorted spark plug or wire at #3 cylinder

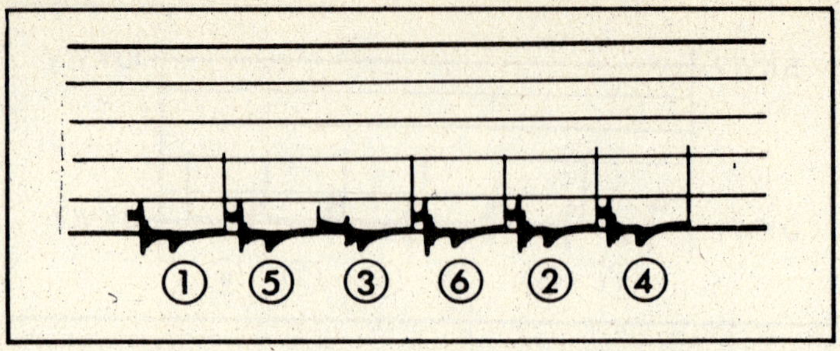

Fig. 8 Shorted spark plug or wire at #3 cylinder

the spark jumping the distributor gap. Actually, then the spark line is a function of the ability to create a spark. Hence, the spark line is a function of the length of the spark plug gap, the distributor gap, or any gap due to a loose connection or a defect, the condition of the electrodes, the compression in the cylinder, and the air-fuel mixture. Techniques are available for pin pointing which area is at fault, such as a camshaft, manifolding, carburetion, spark plug or wiring.

Fig. 8 shows this same condition, shorted plug or grounded plug wire, on a parade pattern.

The two extremes have been shown. Any degree of an in-between condition is also possible.

Fig. 9 shows a cylinder (#6) with a short spark line. Note that the amplitude of the coil and condenser wiggles is larger than normal, as would be expected. Thinking back to the previous pictures, when there is an open plug there is no spark line. This picture, then, is a case where the gap is greater than normal.

Fig. 10 shows this condition on a parade type of pattern.

Fig. 11 shows a partial short. If a vertical line is drawn through the beginning of the coil and condenser wiggles, it is readily apparent that the spark line of the line representing #2 cylinder is longer than normal. These figures show only one cycle of the engine. The engine is running continuously and there are variations from moment to moment due to all of the variables which were discussed. As a result, there are constant changes in the spark line. Minor changes can be ignored because if trouble exists it will show up rather definitely. Fig. 11 shows about as small a variation as would cause concern. This isn't creating any particular trouble at the moment but it is probably worth while checking to locate a potential trouble spot.

In Fig. 12, the spark lines are not horizontal as they have been previously. Instead, the spark lines have a curving slope downward to the right. This is the effect of resistance. In this particular case, the resistance appears in all of the spark lines indicating that it is something common to

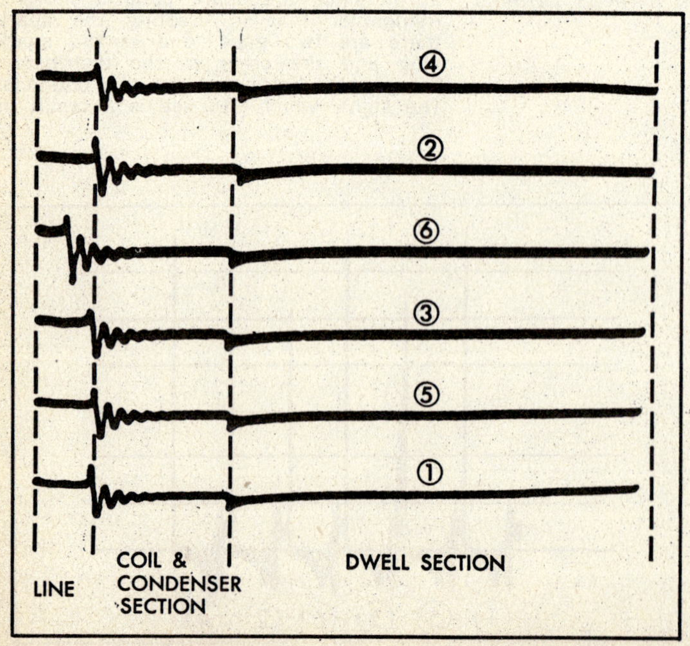

Fig. 9 Wide plug gap or poor connection at #6 cylinder

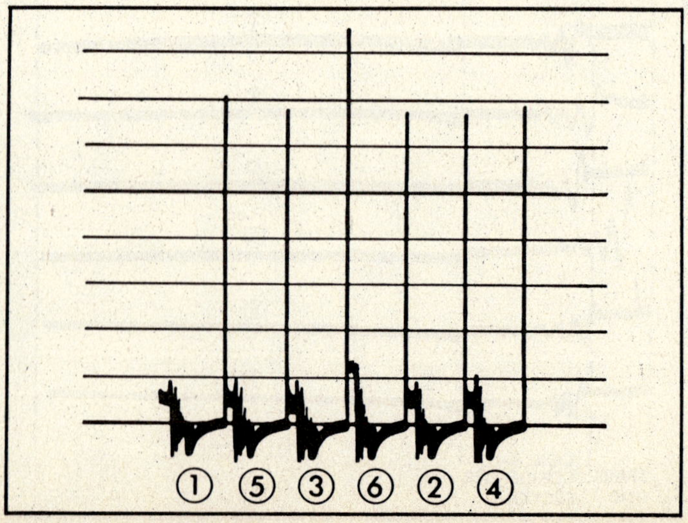

Fig. 10 Wide plug gap or poor connection at #6 cylinder

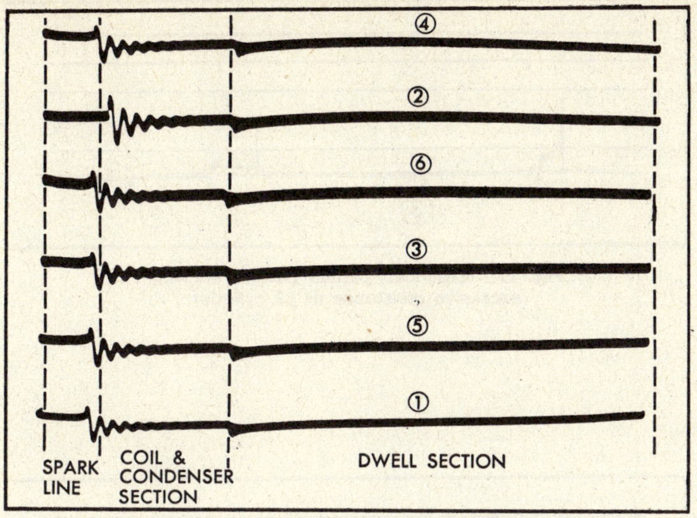

Fig. 11 Partial short at #2 cylinder

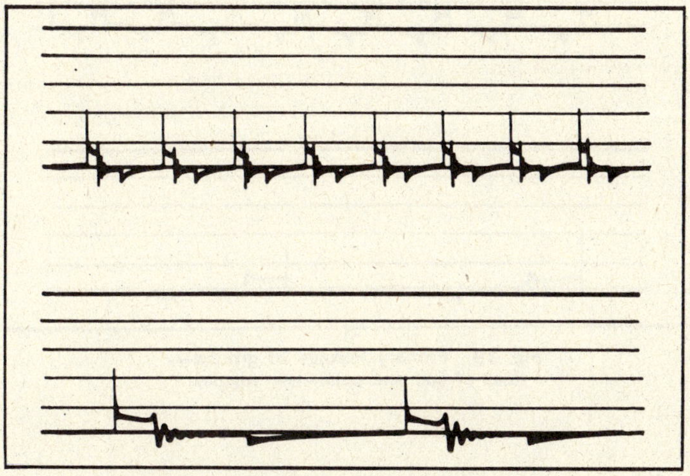

Fig. 13 Resistance affecting all spark lines

all of them. This is the type of picture which will be obtained when there is a suppressor in the circuit or secondary resistance wiring is used. Fig. 13 shows this type of pattern for a parade type pattern for a V-8 engine. The lower portion of Fig. 13 is an expanded display of two successive cylinders displayed above.

Secondary resistance in a parade pattern is indicated when the spark line begins at a comparatively high point on the firing line and slants downward excessively toward the Intermediate Section. Spark duration is reduced considerably and excessive oscillations or "hash" is evident on the spark line. High resistance causing this type of scope pattern may be present in the coil tower, coil wire, rotor, distributor cap towers, plug wire or in the plugs; and may be in the form of deteriorated or damaged suppressors or corroded secondary wire connections.

Fig. 14 represents a scope display of a V-8 engine which has developed *excessive* secondary resistance in a portion of the circuit which is common to all cylinders.

Figs. 15 and 16 show a condition where the resistance effect appears on

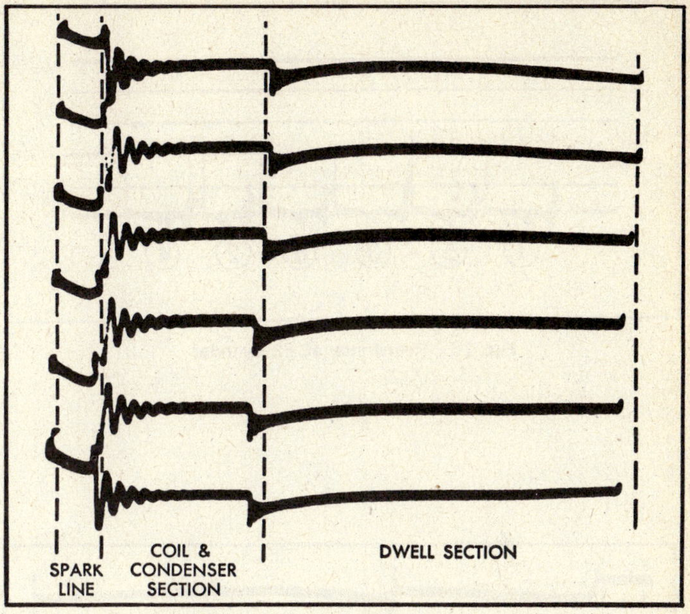

Fig. 12 Resistance affecting all spark lines

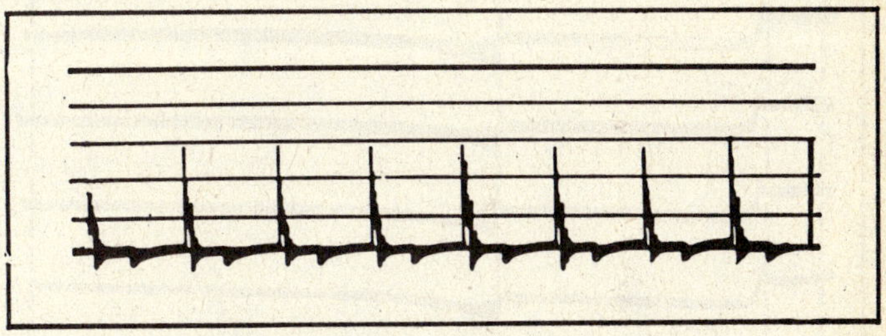

Fig. 14 Excessive resistance affecting all spark lines

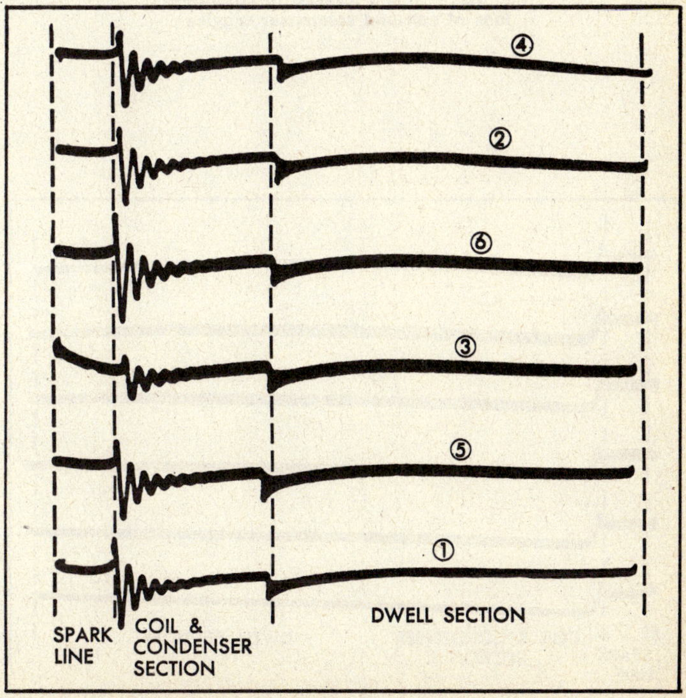

Fig. 15 Resistance at #3 cylinder

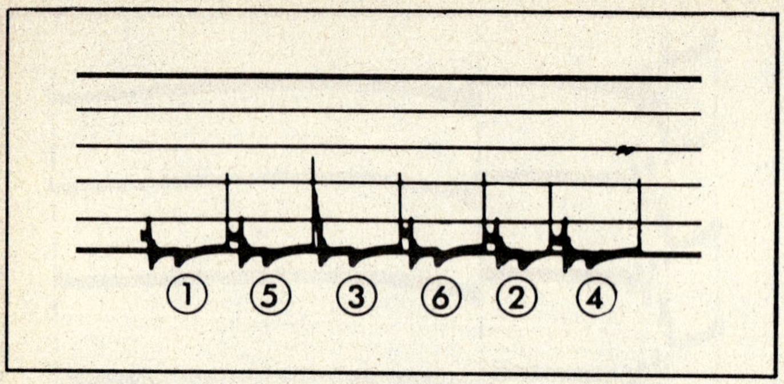

Fig. 16 Resistance at #3 cylinder

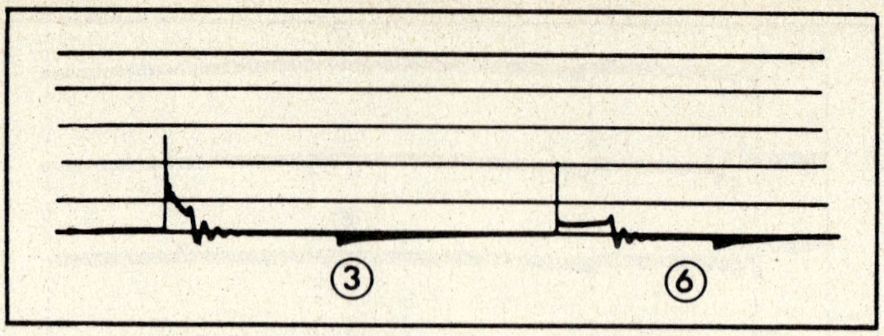

Fig. 17 Expanded parade pattern showing excessive resistance at #3 cylinder

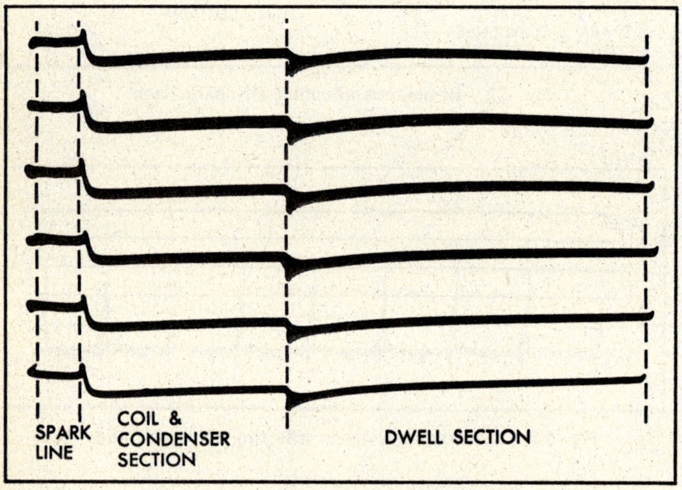

Fig. 18 Primary leakage to ground—loss of coil and condenser wiggles

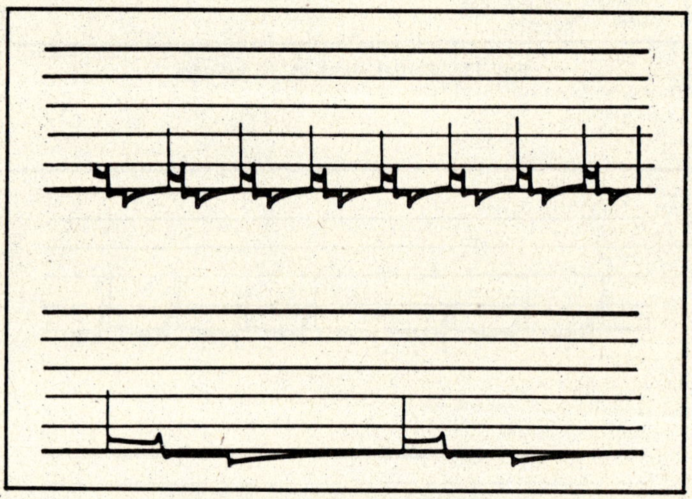

Fig. 19 Primary leakage to ground—loss of coil and condenser wiggles

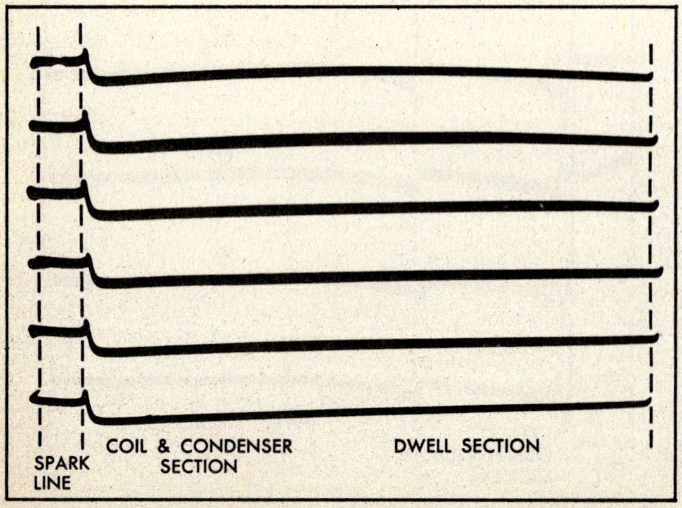

Fig. 20 Defective coil or coil tower

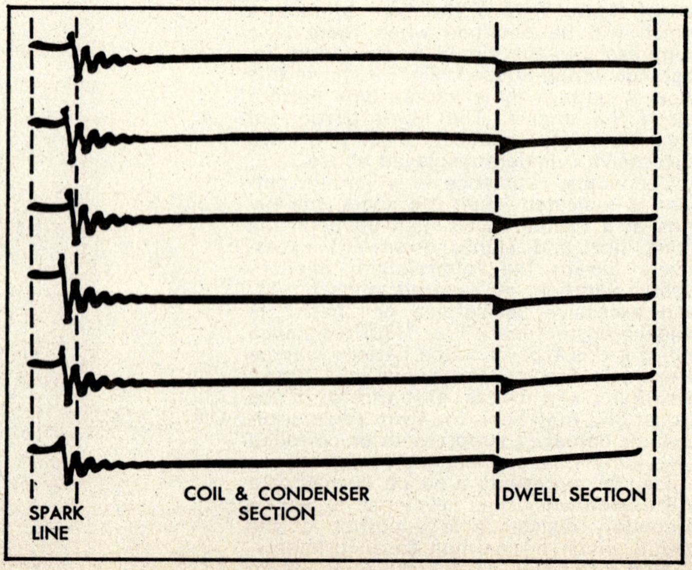

Fig. 21 Short dwell

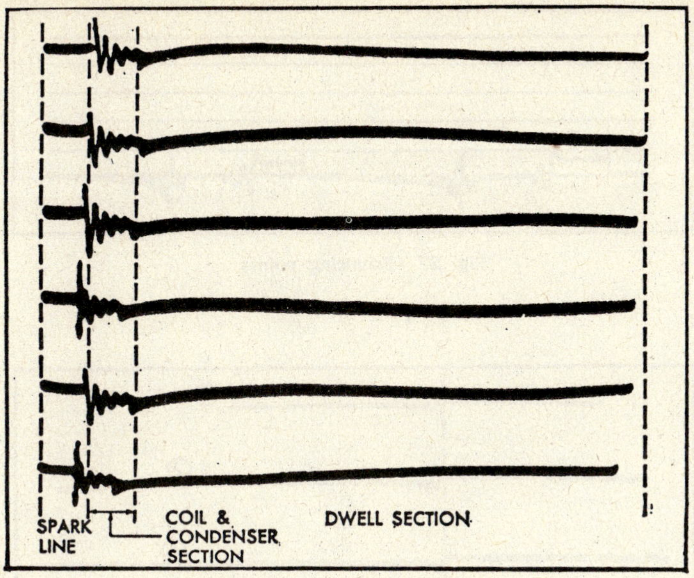

Fig. 22 Long dwell

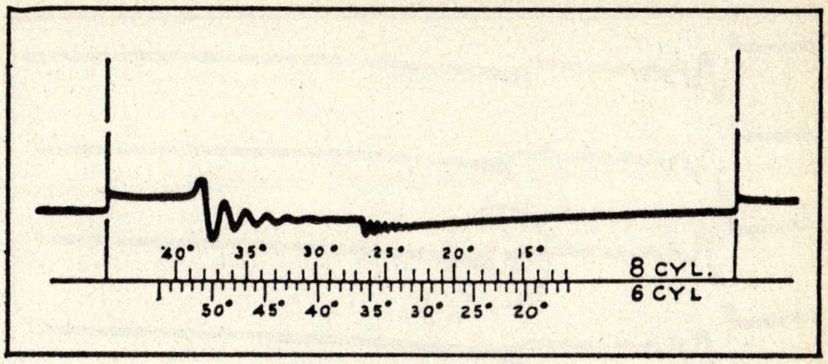

Fig. 23 Correct dwell

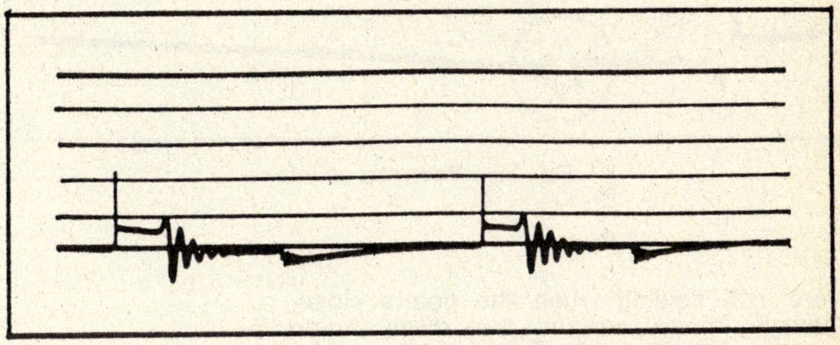

Fig. 25 Incorrectly aligned points

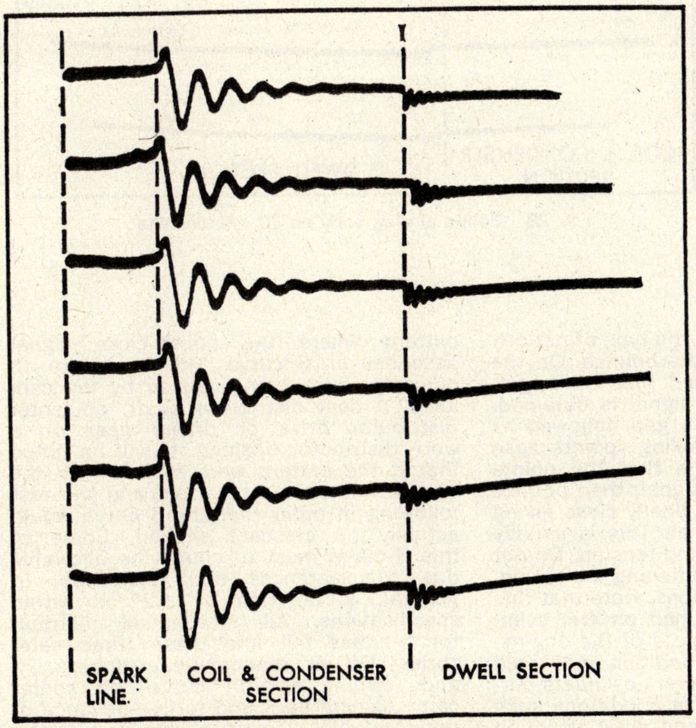

Fig. 24 Improperly closing points

only one line (cylinder #3). The resistance is thus associated with that particular cylinder. This may be obtained under two different conditions. If this pattern is obtained at low speeds with no load on the engine, it is due to a resistance in the wiring. Either the spark plug wire of the connections are at fault.

Fig. 17 shows an expanded pattern of cylinders #3 and #6. Note the spark line section of #3 compared to #6.

If a normal pattern is obtained at low speeds, but this picture is obtained at high speeds and under load, the trouble is due to a fouling spark plug. Fouling is essentially a resistance condition. Deposits are left on the spark plug and as the engine heats up, the resistance of the deposits decreases. If the deposits are sufficiently severe and get warm enough, the resistance drops to the point where the arc follows the deposits rather than firing across the spark plug gap.

The effect of resistance is additive. If resistance appears in one of the spark plug leads of a car that has a radio suppressor, that line will appear to be as much different from the others as it appears to be in this particular figure.

Fig. 18 shows the coil and condenser wiggles have been lost. As you may gather from the previous figures, somewhere between four to six wiggles will normally be present. At about 1000 rpm, there is but one. You will also note that the remainder of the pattern is normal. This type of pattern indicates primary leakage to ground, which is most likely through a defective condenser. Fig. 19 shows the same condition in a parade type pattern of an 8 cylinder engine. The lower portion of Fig. 19 is an expanded form of two consecutive cylinders.

If both coil and condenser wiggles and points close signal are missing or are very small, Fig. 20, the coil or coil tower is probably defective. The spark may still be satisfactory. This is due to the fact that the scope pickup is on the secondary side of the ignition system, and on the lead between the coil and distributor center tower. Thus, the points close signal which normally shows on the scope by being reflected through the coil secondary windings, cannot show because of a defective coil or coil tower connection. This indicates that as with most things, it is possible to see the trouble long before it becomes annoying.

It is easy to see in Fig. 21, that the dwell is very short. This is likely to give high speed miss because the points are closed for such a short period of time that the coil cannot become properly energized to give ignition.

Fig. 22 shows the opposite condition. This is nice in that the coil gets plenty of energy stored in it during the dwell period. However, note that when the points close in this particular figure the coil and condenser wiggles have not yet died out. This indicates that there is high primary cur-

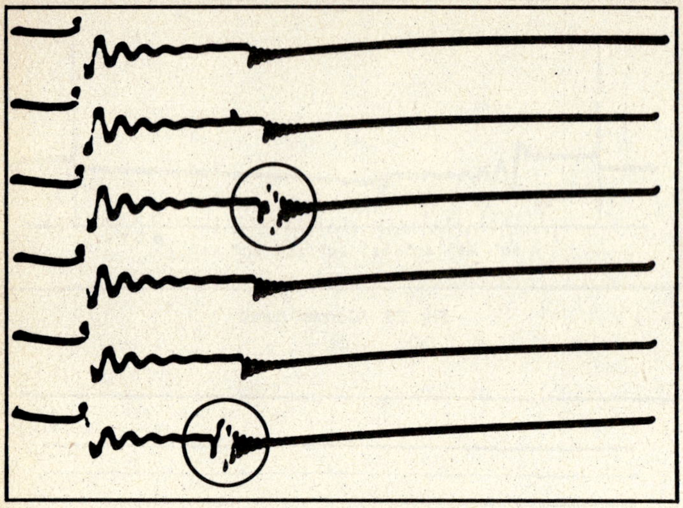

Fig. 26 Bouncing points

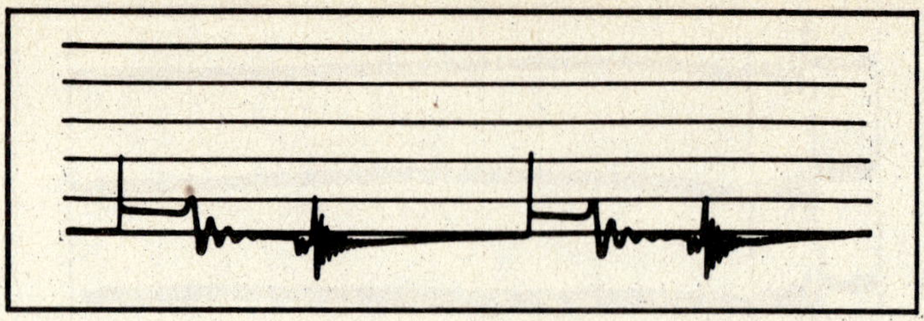

Fig. 27 Bouncing points

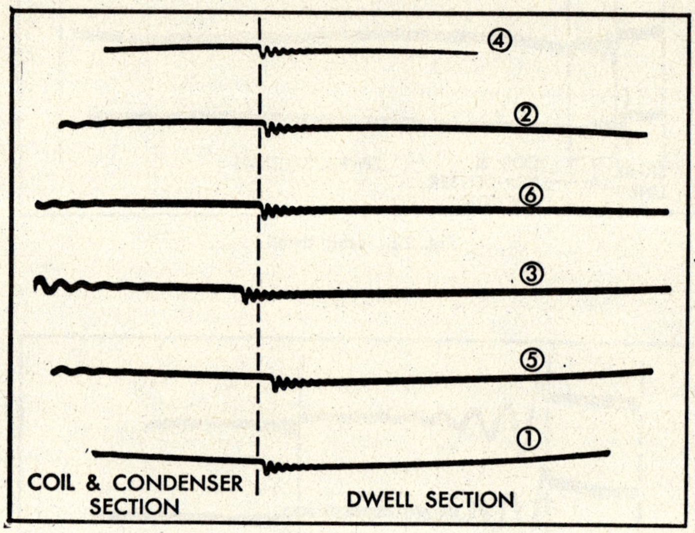

Fig. 28 Points closing early on #3 cylinder line

rent still flowing when the points close. Always make certain the dwell angle agrees with the manufacturer's specifications.

Fig. 23 represents one cylinder's pattern of an eight cylinder engine indicating a dwell period of 26½ degrees, on a parade type scope with the pattern extended.

One thing should be pointed out at this time. The pattern is drawn out from left to right in respect to time. When the speed of the engine is doubled, anything that relates to time will be twice as long. Dwell is a function of crank angle. The percent of line length for dwell remains constant. However, the coil and condenser wiggles are a function of time rather than crank angle, so they will expand as the speed of the engine is increased. As a result if you look at a pattern at low speed and find that there is a space between the end of the coil and condenser wiggles and the points close signal, you will still want to speed up the engine to see if there is difficulty at high speeds. It is possible that the car owner travels a lot and uses the engine at high speed. By adjusting dwell to the low end of the specifications, this problem of primary current flowing when the points close at high speed can be minimized. Always follow specifications; do not go out of the given range.

Fig. 24 shows a 6 cylinder engine pattern that indicates improperly closing points. In Fig. 24 we have expanded the picture in width by means of a control on the instrument so that the right hand edge of the pattern is off the face of the scope. This is done for emphasis so you can better see the points-close signal. Whenever there are wiggles, the first peak is normally the largest. With the points-close signal in this particular case that is not true. Note that in most instances the first peak is not as large. This tells us that the points are not closing properly. This is resistance, usually caused by misaligned points. It may show on all cylinders, or it may only on some, due to distributor cam variations coupled with rough point surfaces.

Fig. 25 shows an expanded parade pattern of misaligned points.

Figs. 26 and 27 show the type of pattern obtained when the points bounce. On the first line it will be noted that the beginning of a points close signal is obtained, but then there is a little gap, followed by a relatively normal-looking points-close signal. What happens is that the points start to close giving a signal, then bounce open giving a void, and finally close giving a relatively normal signal. This is usually caused by low point spring tension. Do not correct this by simply stiffening the spring. Always follow specifications. Note that this point bounce was obtained on two cylinders in this particular cycle of the engine. This is an intermittent condition and would probably happen on different cylinders during the cycle. Intermittent conditions such as this, sort of flicker and draw attention like a flashing beacon.

Fig. 28 shows variation in the points-close signal from one cylinder to the next with respect to time. This is an expanded multi-line pattern showing only the points-close signal. The points close on cylinder #3 slightly before they close on the other cylinders. This is due to a difference in the cam lobe. The tolerance on this is plus or minus 2 distributor degrees. Actually the difficulty is not as important here as it is in the points-open signal because it only affects dwell time, hence the primary problem is that of getting enough energy into the coil.

Fig. 29 shows an expanded multi-line pattern where the points-close signal describes an S curve. This is due to an off-center cam, being caused by the cam itself, a bent distributor shaft, off-center distributor drive or driven gear or a worn distributor bushing. It will be noted that if the pattern were started with the second line being at bottom and the rest following in order that the S curve would actually be crescent shaped. Some of this S curve must of course be allowable due to clearances and tolerances. In general, a variation of ±2° is within specifications. All mechanical distributor troubles fall into these three categories—Off at random—cam lobes; Off and forming smooth curve—off center cam; Jiggling back and forth—advance or drive troubles. These checks for erratic distributor troubles should be conducted at 1000 rpm or higher. Checks at lower engine speeds may show aggravated variations due to erratic engine speed.

On parade-type patterns, superimposing the scope patterns of all cylinders, and expanding the display horizontally to completely fill the space between the two vertical lines on the screen permits any inaccuracy of the distributor cam to be indicated and measured on the dwell scale. The display illustrated in Fig. 30 shows the superimposed patterns of an eight cylinder engine and indicates a total variation of ½ degree, at the point close portion of the pattern.

Fig. 31 is an expanded multiline pat-

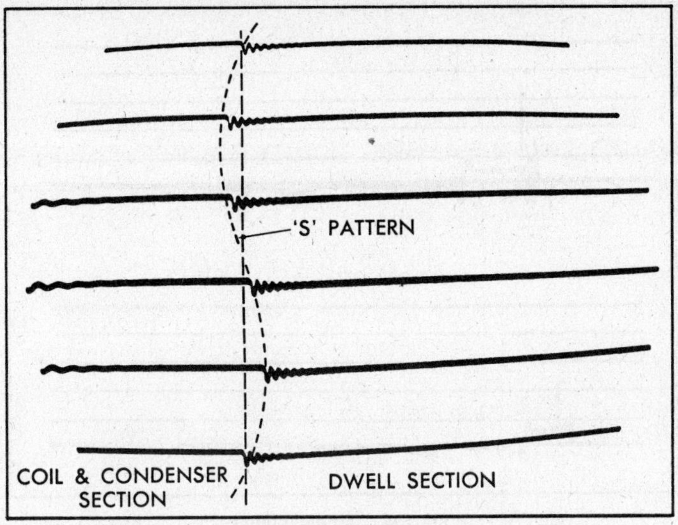

Fig. 29 "S" curve formed by points close signals

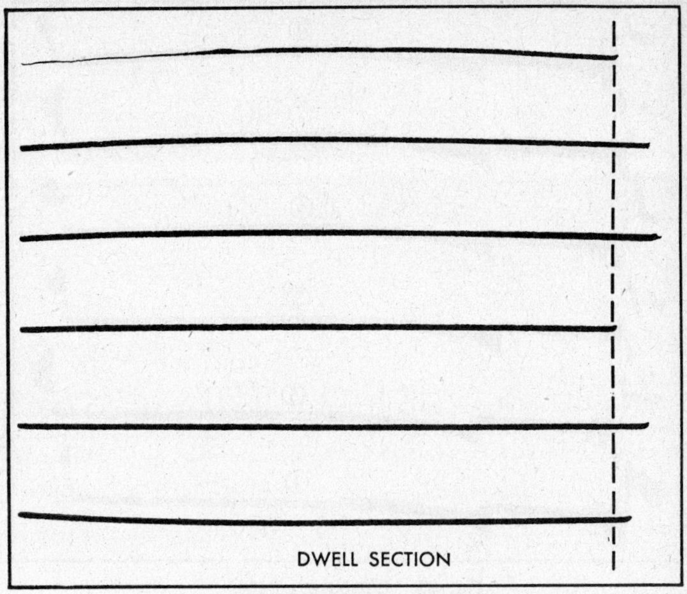

Fig. 31 Variation in points open signal (difference in cylinder-to-cylinder XXX timing

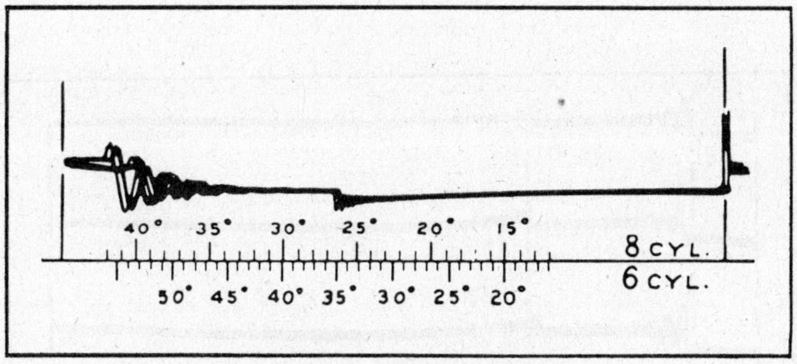

Fig. 30 ½ degree inaccuracy

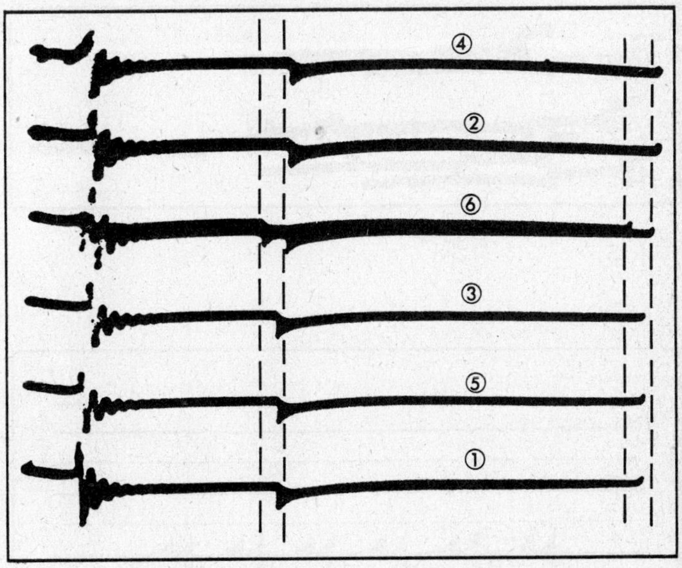

Fig. 32 Double exposure on line representing #6 cylinder shows variation in time of points close and points open

tern showing only the right hand end of the lines. It will be recalled that in a standard pattern, the end of the line or points-open signals line up one beneath the next. Variations, such as are shown in this figure, are due to mechanical tolerances such as differences in the cam lobes which causes the points to open too soon or too late. Various manufacturers have different tolerances, but in no case is the tolerance greater than plus and minus 2 distributor degrees. Anything worse than this should be corrected by repair. The differences in the time of points-open are essentially differences in cylinder-to-cylinder timing. The variation can be measured directly with the dwell scale in distributor degrees. In order to determine the cylinder to cylinder timing variation in crankshaft degrees simply double the scale reading. This information is of particular value since the standard timing light provides a timing check on one cylinder only. It is conceivable that the only cam lobe misshaped or worn could be the one causing cylinder #1 to fire. Then by timing #1 correctly the remaining cylinders would be mistimed. A quick check with the oscilloscope will prevent this.

Fig. 32 shows two complete engine cycles. This shows the pattern for two successive firings of each cylinder. It will be noted that there is a variation of about 2° in the time at which the points close and open for two successive firings of the cylinder #6. This would be apparent when viewing the oscilloscope pattern in the "sideward" jiggling of the points close and points open signals on all cylinders. The reason for this is either a loose distributor drive train or erratic advance mechanism. If the distributor drive is loose the pattern defect should disappear when the engine speed is slowly increased. Increasing the engine speed takes up all slack in the crankshaft to distributor drive linkage.

In Fig. 33 the points-open signal is not normal. Instead of the line ending clean, fish hooks are quite prominent. The points are arcing as they open. This generally indicates a partially open condenser and/or burned points.

It is interesting to note the effect of bad points on ignition. In looking at all of the tails at the right hand side of the pattern, notice that about the cleanest looking is the first one (bottom line). The spark line associated with this is the spark line of the second cylinder in firing order (#5). This looks pretty decent. The tail at the end of the second line is rather bad. Compare this with the third spark line. Actually, in this particular cycle only about two spark plugs fire well, and the remainder are rather poor. Considering the fact that the same set of points produced all these signals, normal ignition cannot occur.

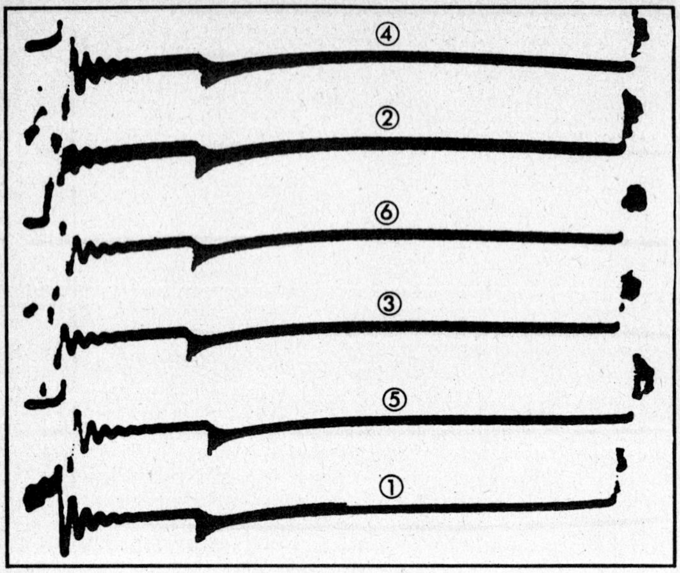

Fig. 33 Defective points

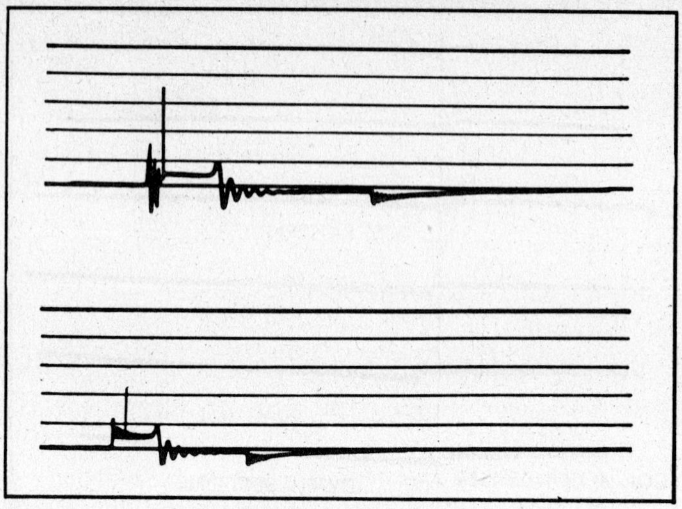

Fig. 34 Arcing points

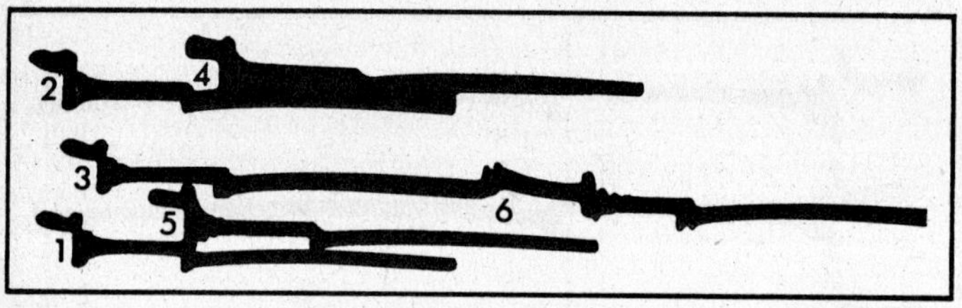

Fig. 35 Cross-fire

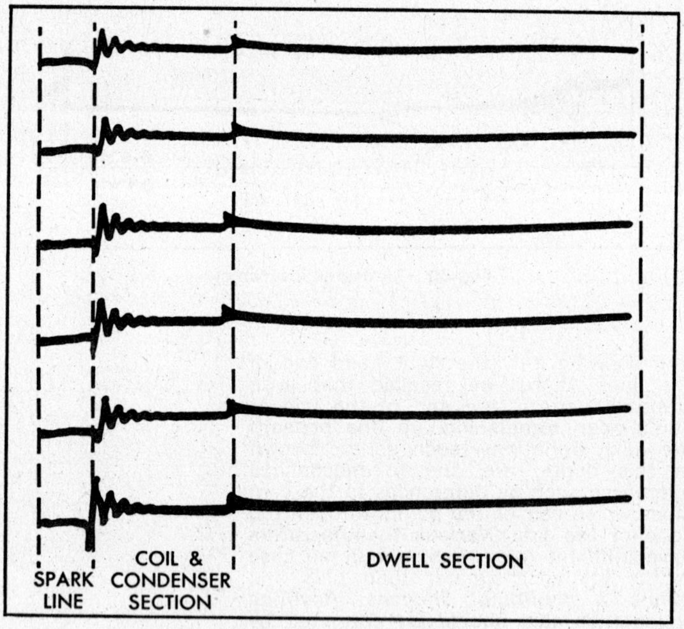

Fig. 36 Reversed polarity

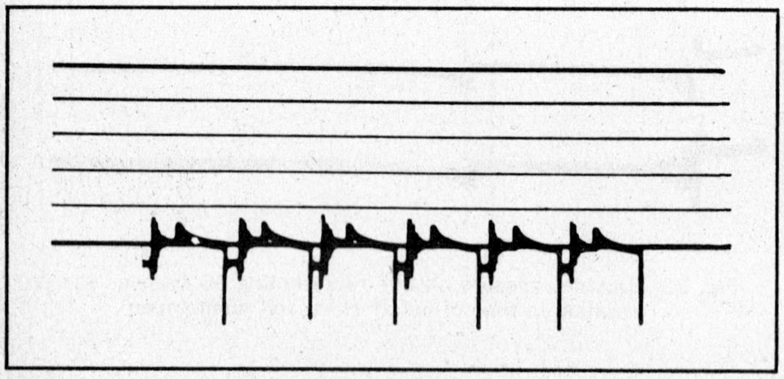

Fig. 37 Reversed polarity

Some intermittent arcing is normal and actually is necessary for good point life.

Fig. 34 shows expanded parade scope patterns indicating arcing breaker points.

A multi-line pattern is designed so that one line starts beneath the next at the left hand edge of the pattern. In spite of this, Fig. 35 shows that the patterns are obtained at random. This is the type of picture obtained when cross fire is occurring. The cylinders are firing at random and show up in exactly the same manner on the screen. This figure shows a particularly bad condition. Cross-fire is usually caused by poor secondary wiring routing or insulation, or a defective distributor cap.

In Fig. 36 the pattern is up-side down. This means that the plug is firing with positive voltage rather than the normal negative. It requires up to 30% more voltage to fire the spark plug when the polarity is reversed. The reason for this is that a hot electrode emits electrons easier than a cooler electrode. Thus, with electrons flowing from negative to positive, the center or hottest electrode should be negative. This is caused by reversed coil or battery leads, or the wrong type of coil installed. Fig. 37 shows reversed polarity on a parade-type pattern.

If unique trouble is encountered that wasn't shown, the equipment manufacturer's guide books should be referred to and the trouble identified by comparison to those illustrated. It is apparent then, that the engine diagnosis with an oscilloscope is quick, simple, and assures diagnosis with a minimum of effort.

W/S WIPER MOTOR SERVICE

CONTENTS

AMERICAN MOTORS

The 12 volt electric operated wiper motor rotates clockwise at the drive end, while the crank rotates counter-clockwise when viewed from front. Motor bearings are pre-lubricated and require no oiling unless motor has been disassembled, if so at assembly, add 10 drops of medium engine oil to felts at armature shaft bearings and drain excess oil, fill gear chamber ¾ full with approved gear grease. The maximum end play is .005", it can be adjusted by turning screw at drive end of armature shaft.

An intermittent governor which provides a pause between wipe cycles is available on some 1973-75 models.

NOTE: Replace screw if nylon lock is worn and will not hold setting. Field coil draw at 70°F. Shunt Winding is 1.52 to 1.67 amperes at 12.0 volts. No load draw at 70°F is 3.0 maximum at 14.0 volts using full field.

NOTE: The wiper arms and blades must not be moved manually from side to side or damage will result.

TROUBLE DIAGNOSIS

1969-72

Wiper Motor Does Not Operate

Disconnect brown supply line from the wiper switch, Fig. 1, and check it for battery voltage, with the ignition switch in the accessory position. If no voltage, correct condition of supply line through ignition switch.

Wiper motor circuit is protected by a 6 amp. circuit breaker in wiper control switch.

No High Speed Operation

If current is present at the brown wire supply line, disconnect the three wire plug and the blue wire plug from the wiper switch, Fig. 2. Connect a jumper from the supply line to the black wire in the harness. Motor should operate. If not, check black wire and connections. If no positive results, replace motor.

No Low Speed Operation

Connect both red and black wires with jumper to supply line and ground yellow wire, Fig. 3. Motor should run at low speed. If not, check red and yellow wires for continuity. If wires and connections are intact, motor is at fault.

Wiper Will Not Park

Connect black and yellow wires together, connect battery supply to blue wire and ground the red wire, Fig. 4. Motor should reverse to park position. If motor does not run, the park switch may be open. With the above connections intact, place a jumper wire from blue to green at the motor momentarily to run motor.

Wiper switch case must be grounded for park or variable speeds.

If motor still does not run, check blue wire at motor contact switch plate for battery voltage. If there is no voltage present, repair the blue wire.

If motor continues to run instead of parking, replace the park switch plate or cam drive plate.

If all above check prove satisfactory, replace wiper control switch on instrument panel.

1971-72
Wiper Blade Park Position Not Correct

If the sweep of the wiper blades is correct but the park position is not, check the assembly and adjustment, Figs. 5, 6 and 7. These items do not affect the sweep area of the wiper blades.

As illustrated in Fig. 5, correct park position of the wiper motor crank is 13° up from a line extending from the center of the crank drive and parallel with the worm drive housing. The actuator cam must also be installed as shown with the stamped letter "R" facing away from the motor.

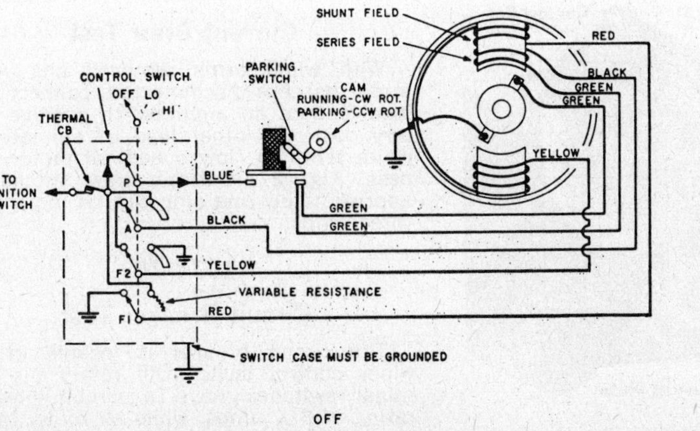

Fig. 1 Internal circuitry

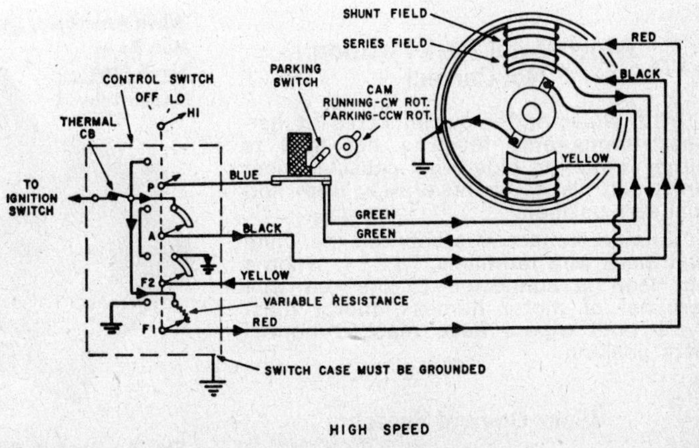

Fig. 2 Current flow-high speed position

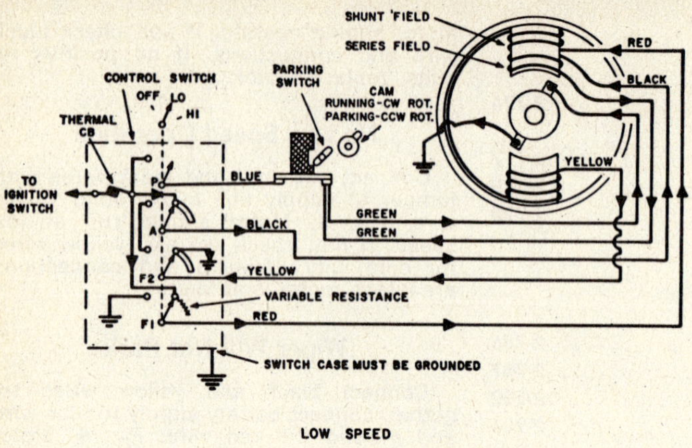

Fig. 3 Current flow-low speed position

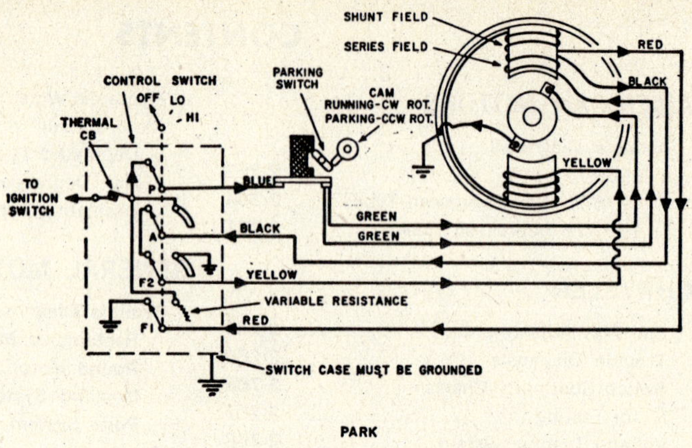

Fig. 4 Current flow-park position

Fig. 5 Correct park position

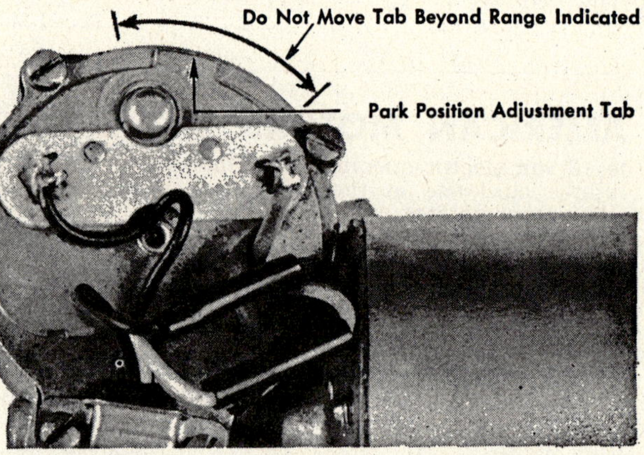

Fig. 6 Tab adjustment-wiper park position

Minor adjustments to the park position can be made by moving the adjustment tab. Do not move the tab clockwise or counterclockwise beyond the points indicated in Fig. 6.

Incorrect assembly by the contact plate and the stud will cause incorrect parking of the wiper blades. Correct positioning of the components, Fig. 7.

1973-75

Wiper Motor Park Position Not Correct

With wiper motor disconnected at harness, temporarily touch a hot lead to either white or blue wire causing wiper arms and blades to move away from normal park position.

Connect jumper wire between the white and black wire terminals, Fig. 34. When a hot lead is connected to the red wire terminal of motor harness, motor must cycle until wipers have reached normal park position.

Wiper Control Switch

Use test lamp or ohmmeter to check switch continuity at various switch positions, Fig. 8. If the intermittent wipe cycle does not operate at low and high speed, check variable resistance between #4 and #5 terminals. This resistance controls governor operation for intermittent wipe.

With knob rotated to full counterclockwise position, ohmmeter should indicate approximately 7000 ohms. As control knob is rotated clockwise, resistance should decrease to a minimum of approximately 600 ohms. If continuity and resistance do not exist as specified, switch must be replaced.

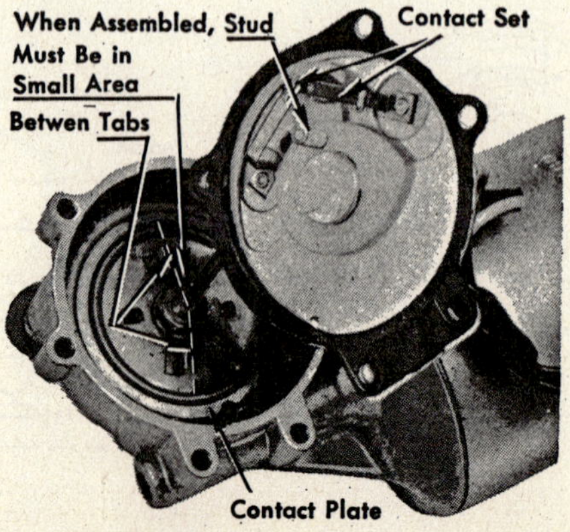

Fig. 7 Correct alignment of contact plate in park position

Intermittent Governor

If intermittent wipe cycle is not functioning properly, check related components such as motor, control switch and connecting wires. If all components are functioning properly, governor must be replaced, Fig. 9.

Current Draw Test

With wiper arms removed and wiper motor harness disconnected, connect positive lead of an ammeter to positive battery post and other lead of ammeter to white terminal (low speed) of motor harness, Fig. 27. Current draw should be approximately one amp but no more than three amps.

1974-75

Circuit Breaker Test

The circuit breaker is located in the wiper control switch. All rotary pin terminal switches have a circuit breaker rating of 8¼ amps, while all rotary blade terminals have a rating of 7 amps.

Two separate tests are required to

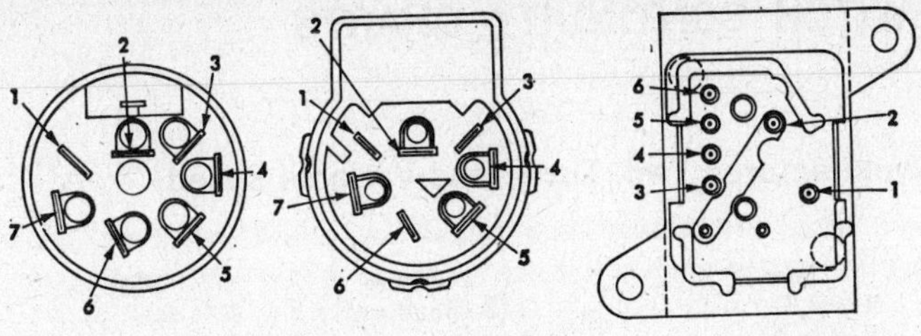

	STANDARD 01-10-40-80 Series	STANDARD 70 Series	INTERMITTENT 01-10-40-80 Series
Off or Park	1-2 3-4	1-2 3-4	1-2 4-5
Low Speed	1-2-3	1-2 4-5	1-2 5 to Case 4-5
High Speed	1-2-5	1-2 5-6	1-2-3 5 to Case 4-5
Intermittent			1-2 5 to Case 4-5 Variable Resist. 1-9K to 7K
Wash	1-2 6-7		1-2 6-7

Fig. 8 Wiper switch continuity test. 1973-75

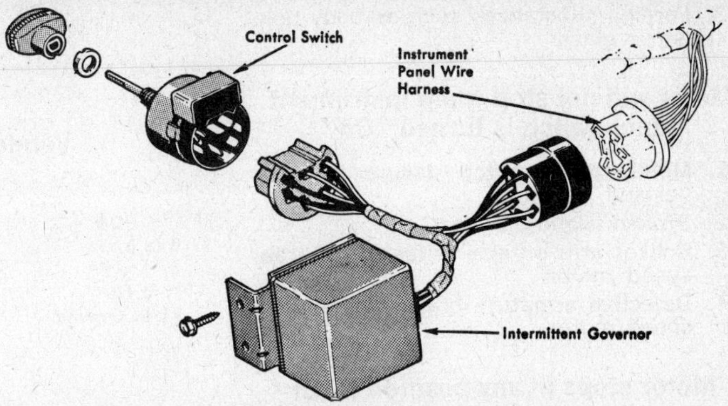

Fig. 9 Intermittent governor. 1973-75

Fig. 10 End play adjustment

check for correct circuit breaker operation.

Test 1

Short tester leads together and adjust current draw until it equals circuit breaker rating, then connect switch to tester as shown in Fig. 28, and leave connected for ten minutes. Current reading should remain at rated current. If circuit breaker opens during the ten minutes, replace wiper switch assembly.

Test 2

Short tester leads together and adjust current draw to twice the rated current, then connect switch to tester as shown in Fig. 28. Current reading should drop to zero within 30 seconds. If it takes longer than 30 seconds for circuit breaker to open, replace wiper switch assembly.

CHRYSLER

Three wiper systems are offered, a three speed, two speed and a variable speed.

The three speed motor is controlled by two resistors in the field circuit of the wiper switch. The two speed motor is controlled by a resistor in the field circuit of the wiper switch. The variable speed motor is controlled by a rheostat built into the wiper switch.

The variable and three speed switches are designed to provide a circuit to the motor, to reverse the current to field winding which reverses the direction of the armature. A circuit breaker, integral with the wiper switch protects the circuitry.

A depressed park feature of the three speed and the variable speed systems is accomplished by reversing the rotation of the wiper motor and the use of an eccentric motor shaft. When the wiper switch is turned "OFF", the motor is reversed, the motor inner shaft stops and the outer shaft rotates 180 degrees lengthening the linkage slightly to park the blades in the depressed position. The linkage shortens again when the motor is turned "ON" by reversing the action. The two speed wiper system on non-concealed wipers will complete the wipe cycle and return to the park position when the switch is turned off. The blades park in the lowest portion of the wiper pattern.

The two speed wiper system used on concealed wiper systems accomplishes its depressed parking action by reversing the rotation of the motor and the use of a parking cam on the motor crank pin. When the wiper switch is turned "OFF", the motor direction is reversed and at the same time, the parking cam rotates 180 degrees, lengthening the drive link slightly to park the blades in a depressed position. Motor operation in the wipe direction returns the cam to "run" position to restore the normal link and wipe pattern.

End Play Adjustment

Most windshield wiper motors have a provision to adjust the armature shaft end play, by turning the adjustment screw in until it bottoms and backoff 1/8 turn (1969-71 models), 1/16 turn (1972-75 models), Fig. 10. This adjustment can be made without removing the wiper motor from the vehicle.

TROUBLE DIAGNOSIS

1969-75 ALL MODELS

Wiper fails to operate

1. Binding linkage.
2. Faulty instrument panel switch.
3. Linkage disconnected.
4. Faulty motor.
5. Open or grounded wiring.

Wiper blades not parking properly

1. Arm set at incorrect position.
2. Motor park switch timing incorrect.

Blades slap against windshield mouldings

1. Improperly adjusted wiper arm.
2. Looseness of the motor crank or other drive parts.

Blades chatter

1. Twisted arm holds blade at wrong angle to glass.
2. Bent or damaged blades.

3. Foreign substances such as body polish on glass.

Motor will not stop when instrument panel switch is turned "Off"

1. Motor park switch failure in the "closed" position.
2. Broken island on gear.
3. Rocker arm outside outer cam, three speed motor.
4. Defective armature brake switch, two speed motor.

Motor stops in any position when instrument panel switch is turned "Off"

1. Motor park switch failure in the open position.
2. Open parking circuit.
3. Open field circuit.

No speed control

1. Open circuit in red or green wiring (variable or 3 speed only); brown or red wiring (2 speed only).
2. Defective control switch.

Motor draws excessive current

1. Jammed gearbox mechanism.
2. Shorted park switch.
3. Shorted motor.
4. Gearbox unlatched (variable and three speed only).

Motor overheated

1. Binding linkage.

Motor stops intermittently in any position when instrument switch is turned "Off" (variable and three speed only)

1. Gearbox unlatched.
2. Intermittent park switch continuity.

ADDITIONAL TROUBLE DIAGNOSIS FOR ALL 1969-75 MODELS

Motor runs but output crank does not turn

1. Stripped intermediate gear (two speed).
2. Stripped output gear.
3. Output gear slips on output shaft (two speed).
4. Crank not fastened properly to output gear shaft.
5. Broken latch slips on output shaft.

Motor will not run at any speed when testing as described in "motor testing" section

1. Common brush stuck.
2. Open circuit in armature at commutator.

SWITCH CONTINUITY CHARTS

Vendor Manufactured Two, Three and Variable Speed

Off	Low	Medium (3 speed only)	High
B to B/U B/U to P A to F2 F1 to Ground	B to B/U B/U to A A to F1 F2 to Ground P-open①	B to B/U B/U to A F1 to R1 (A to R1③) F2 to Ground P-open	B to B/U B/U to A A through the resistor to F1③ F2 to Ground P-open F1 to R2④

①—Chrysler, Imperial and Dodge 1969-71; Plymouth 1969-71.

②—Replaces F1 to R1 on following models, 1969 Charger, Coronet, Dart; 1970 Charger, Coronet.

③—As the switch knob of the variable speed switch is rotated the resistance shown on the ohmmeter should vary from high to a low reading in a smooth rate of change.

④—1969-70 All Models.

Fig. 11 1969-73 Chrysler, Imperial; 1970-73 Dodge

Three Speed

Off	Low	Medium	High
B to P A to F2 F1 to Ground	B to A A to F1 F2 to Ground P-open	B to A A through the resistor to F1 F2 to Ground P-open	B to A A through the resistor to F1 F2 to Ground P-open

Fig. 11A 1974-75 All models

Chrysler Manufactured Two Speed

Off	Low	High
B to B/U B to P A to F2 F1-open	B to B/U B to P B to A F2-open F1-open	B to B/U B to P B to F1 F2-open A-open

Fig. 12 1969-70 Belvedere, Satellite, Charger, Coronet

Chrysler Manufactured Two Speed (Non Concealed Wipers)

Off	Low	High
B to B/U B to P1 A to P2 H-open	B to B/U B to P1 B to A P2-open H-open	B to B/U B to P1 B to H P2-open A-open

Fig. 13 1969 Dart, Barracuda, Valiant; 1970 Dart; 1971-75 All Models

Chrysler Manufactured Two Speed (Concealed Wipers)

Off	Low	High
B to B/U B to P A to Ground F2-open H-open	B to B/U B to A F2 to Ground P-open H-open	B to B/U B to H F2 to Ground P-open① A-open

①—Applies only to 1969 Polara, Monaco; 1970 Fury, Barracuda and all 1971-73 concealed wipers.

Fig. 14 1969 Chrysler, Imperial, Polara, Monaco; 1970 Polara, Monaco, Fury, Barracuda; 1971-73 All Models

Two Speed (With Concealed Wipers)

Off	Low	High
B to P A to Ground F2-open H-open	B to A F2 to Ground P-open H-open	B to H F2 to Ground P-open A-open

Fig. 14A 1974-75 All Models

Chrysler Manufactured Two Speed

Off	Low	High
B to B/U	B to B/U	B to B/U
B to P	B to A	B to F1
A to Ground	F2 to Ground	F2 to Ground
F2-open	F1-open	P-open
F1-open	H-open	A-open

Fig. 15 1970 Challenger

Chrysler Manufactured Two Speed

Off	Low	High
B to B/U	B to B/U	B to B/U
B to P	B to A	B to F1
A to Ground	P-open	F2 to Ground
F2-open	F1-open	A-open
F1-open	F2-Ground	

Fig. 16 1970 Chrysler and Imperial

3. Spring off common brush (two speed).
4. Open in brown, lead circuit (three speed).
5. Open green lead circuit (two speed concealed).
6. Open solder terminal on park switch (three speed).

No speed control (motor runs at one speed only)

1. Open circuit in red or green wiring (three speed); red or brown wiring (two speed).
2. Defective control switch.
3. Brush sticking in holder (two speed).
4. No brush spring (two speed).
5. Broken brush holder (two speed).
6. Low speed torque limiting resistor open (two speed).

Motor runs slower at "high" instrument panel switch position than at "medium"

1. Faulty instrument panel switch.
2. Medium speed field circuit open.

Motor will not stop when instrument panel switch is turned "off"

1. Broken island on gear.
2. Lead not soldered properly at park switch (three speed).
3. Rocker arm outside outer cam (three speed).
4. Motor housing not grounded (two speed with concealed wipers).
5. Defective armature brake switch (two speed).
6. Broken brush holder (two speed).

Motor draws excessive current

1. Shorted brush lead or field winding (three speed).
2. Shorted or burned armature.
3. Seized armature bearing (two speed).
4. Bearings loose in housing (two speed).
5. Loose or chipped magnets and loose clips or chips interfering with armature (two speed).
6. Broken brush holder (two speed).
7. Binding linkage.

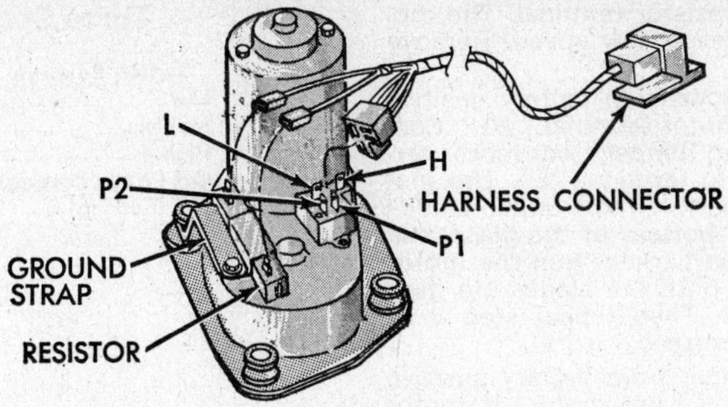

Fig. 17 Two speed motor-Chrysler manufactured

Switch Tests

To test the switch, disconnect the wiring to the switch and remove the switch from the instrument panel. For removal and installation of the wiper switch, see car chapters.

Using a continuity tester or an ohmmeter, test for continuity (no resistance) between the contact terminals of the switch, see Figs. 11 thru 16. For test purposes the "Park" position is the switch "Off" position. The "Low" speed is the position immediately past the "Off" detent and the "High" position is at the extreme position of the switch travel. Bench testing of the two speed switches does not require the use of a 12 volt battery.

MOTOR TESTING

CAUTION: The testing procedure for two speed motors used with non-concealed wipers differs from that used when testing two speed motors used with concealed wipers. Use of the improper procedure will result in a false indication of motor conditions.

Chrysler Manufactured Two Speed Motor (with non-concealed wipers)

1. Disconnect motor leads at motor, Fig. 17. Connect jumper from battery positive terminal to motor terminal "H". Motor should run at high speed. (The ground circuit is completed to the car body through a brass ground strap. Be certain that this is a good connection). Remove jumper.
2. Connect jumper from battery positive terminal to resistor terminal. Connect second jumper from terminal "L" to

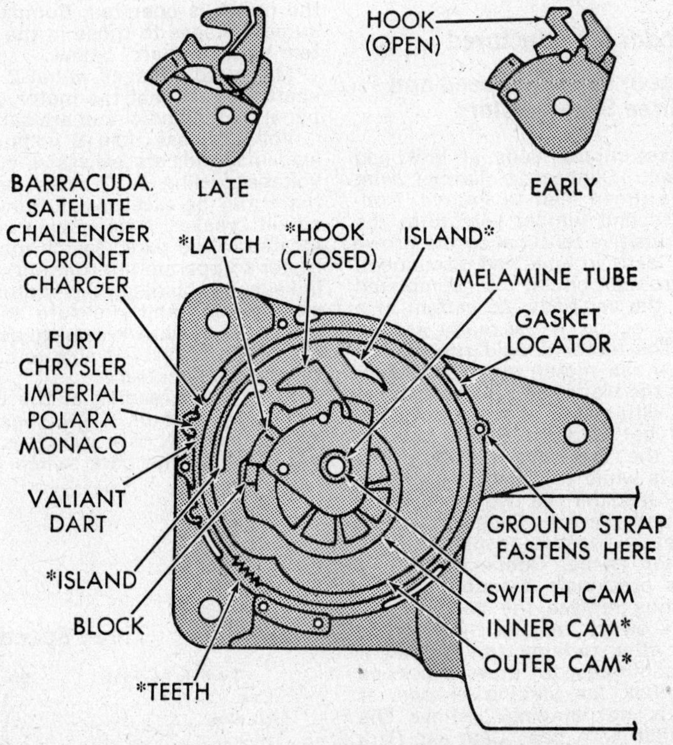

Fig. 18 Gear and park switch replacement. 1971

the second resistor terminal. The motor should run at low speed. Remove jumpers.

3. Connect jumper from battery positive terminal to motor terminal. "P1". Connect a second jumper from motor terminal "P2" to terminal "L". The motor should park. (If the wiper blades are near the bottom of the glass, the motor may be parked. Run the motor as in step 2 until the blades are high on the glass. Then repeat step 3 to observe parking).
4. Connect jumper from battery positive to terminal "L" on motor. If motor runs now but did not run in step 2, replace resistor assembly, and repeat above steps.

Two Speed Reversing Motor (with concealed wipers)

1. Disconnect motor leads at motor, Fig. 17. Connect jumper from battery positive terminal to motor terminal "H". Connect second jumper from terminal "P2" to ground. Motor should run at high speed. Remove jumpers.
2. Connect jumper from battery positive terminal to resistor terminal. Connect second jumper from terminal "L" to the second resistor terminal. Connect a third jumper from terminal "P2" to ground. The motor should run forward at low speed. Remove jumpers.
3. Connect jumper from battery positive terminal to motor terminal "P1". Connect a second jumper from motor terminal "L" to ground. Connect a third jumper from the motor ground strap to ground. The motor should run in reverse rotation for at least a half revolution and park.

Vendor Manufactured

Two Speed, Variable Speed and Three Speed Motors

1. Disconnect motor leads at bulkhead disconnect. Connect a jumper wire from the green lead to ground. Connect a second jumper wire from the battery positive terminal to the brown and red leads in bulkhead disconnect. (The ground circuit is completed through the car body, be certain of a good connection at the motor ground strap). The motor should run continuously. If the motor runs disconnect leads as the blades sweep downwards. Proceed with step 2. If the motor does not run but appears to be shorted analyze the excessive current draw as follows while the motor is still on the car. Unsolder the two green leads at the park switch. Fasten the two green leads together, isolated from the motor frame. Connect the test leads as previously described. If the motor runs replace the park switch. The park switch may or may not be shorted after running the motor. If motor continues to draw excessive current look for binding linkage. If linkage is not binding, remove the park switch plate, see "Gear and Park Switch Replacement". If switch plate is jammed and difficult to remove or motor runs when jumpers are connected again replace the latch assembly as it is unlatched. If motor continues to draw excessive current, remove and replace motor assembly. Do not proceed with step 2 until the green wires are resoldered to the park switch. Disconnect leads as the wiper blades sweep downward.
2. Connect jumper wire from green lead to brown lead. Connect red lead to ground. Connect third jumper wire from battery positive terminal to blue lead. The wiper should run to the park position.

CAUTION: Motor can be damaged if not wired correctly.

VOLTAGE TESTING WIPER SYSTEMS

Three Speed

Remove the blade and arm assemblies to simulate wet glass load. Note that a faulty circuit may damage the wiper system if allowed to run continuously on dry glass.

Measure the voltage at the four motor terminals at the bulkhead disconnect while the motor is operated. Compare the measured voltages to those in the "Wiper System Voltage Chart" below.

Measured voltages within 2 volts of the chart indicate that the motor, park switch, panel switch and circuit are satisfactory.

Voltages less than 8 volts at the blue lead may indicate excessive current draw, voltage cycling to 0 volts indicates that there may be excessive current cycling the circuit breaker. A defective circuit breaker located in the panel switch may cause the motor to operate intermittently. Check the linkage for binding, the motor for excessive current and measure the armature current in the brown lead to ascertain that the circuit breaker is able to carry its continuous rated current.

A "O" volt reading at the blue lead in LOW, MEDIUM or HIGH speeds may indicate that the park switch is open, that the leads on the park switch are not soldered, or that the brown lead circuit to the series field is open or shorted. If the motor runs only when it is turned OFF and does not run when turned to LOW, MEDIUM or HIGH the brown lead is probably opened at the series field splice. Remove the splice insulation material, some brown vinyl insulation, clean the nylon film or solder to expose the solid bare copper wire, twist the splice mechanically, solder and tape to repair. Check the leads in the motor pigtail for damage to be repaired. Inspect the park switch to see if the blade is shorted to the cover plate. See "Motor Testing", "Switch Testing", and "Trouble Diagnosis" to isolate defective components.

A 12 volt measurement at the red terminal at MEDIUM and HIGH speeds indicates that there is an open in the shunt field circuit. A 12 volt reading at the green lead indicates that the field circuit is open between the disconnect and the panel switch, the panel switch is defective or that the panel switch is not grounded, and the motor is usually undamaged. However, a "O" volt measurement on the green lead indicates that the field is open in the motor pigtail, either the red or green wire splice or inside the motor. Note: An open inside the motor requires replacement.

A "O" volt measurement while running at the red terminal at LOW, MEDIUM or HIGH indicates that there is an open circuit from the panel switch to the disconnect, that the panel switch is defective or that a MEDIUM speed resistor circuit is open.

When both the brown lead and blue lead measure 12 volts after the motor has been turned OFF and parked, the park switch may have been damaged so contacts do not open or the panel switch may be defective. To isolate the defect, unsolder the blue lead from the park switch. If both the blue lead and brown lead are still measuring 12 volts see "Gear and Park Switch Replacement" to repair.

Medium Speed Resistor Service

The medium speed resistor is mounted beneath the steering column on some models. Six volts at the red lead at the bulkhead disconnect indicates that the me-

Three Speed Wiper System Voltage Chart, 1969-74

Switch Position	Brown	Red	Green	Blue	Battery
Low.............	12	12	0	8 to 10	12
Medium..........	12	6	0	8 to 10	12
High.............	12	3	0	8 to 10	12
Off (While Parking)	12	0	12	12	12
Off (Parking)......	0	0	0	12	12

Three Speed Wiper System Voltage Chart, 1975

Switch Position	Brown	Red	Green	Blue	Battery
Low.................	12	12	0	10 to 12	12
Medium.............	12	5	0	10 to 12	12
High................	12	2.5	0	10 to 12	12
Off (While Parking)...	10 to 12	0	10 to 12	12	12
Off (Parked).........	0	0	0	12	12

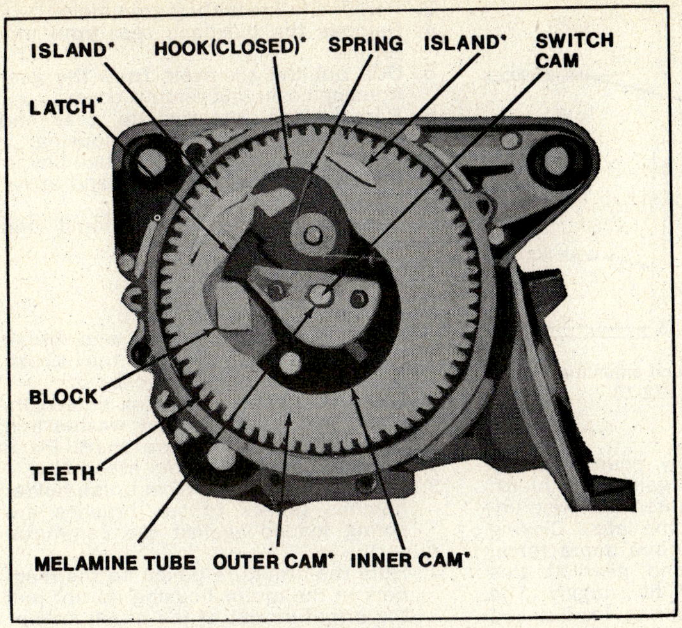

Fig. 19 Gear and park switch-hook closed. 1972-75

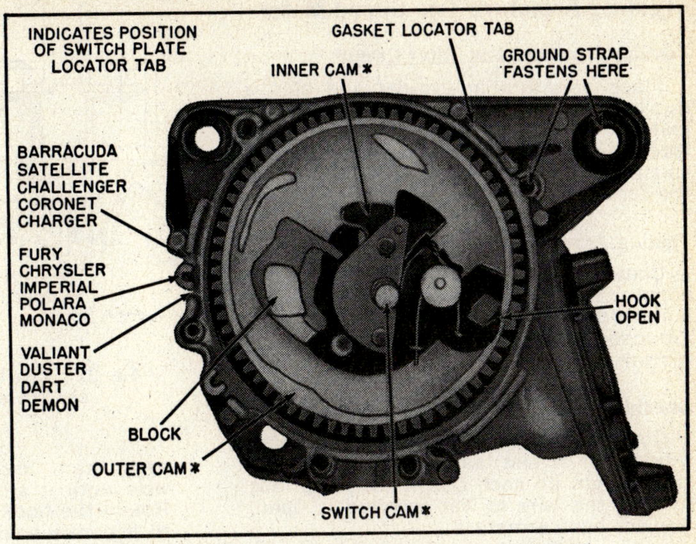

Fig. 20 Gear and park switch-hook open. 1972-75

dium speed circuit is satisfactory. To test the circuit at the resistor remove the steering column cover, and the resistor retaining screw to allow access to the resistor terminals with meter probes. Note that the resistor may be hot to the touch after operation. The resistor should have 12 volts on the source terminal and 6 volts on the other terminal (referenced to ground) while operating at medium speed. Next test the source at the "A" terminal on the panel switch to the resistor and the medium speed resistor to the "R1" terminal of the panel switch for continuity. Repair and replace as needed.

Two Speed

Remove the arm and blade assemblies to simulate wet glass load. Note that a faulty circuit may damage the wiper system if allowed to run continuously on dry glass.

Measure the voltage at four motor terminals while the motor is operated. Compare the measured voltages to those in the "Wiper System Voltage Chart" below. Measured voltages within 2 volts of the chart usually indicate that the motor, park switch, panel switch, and circuit are satisfactory. However, it may be necessary to run the motor (with the ignition switch in the accessory position) for a half hour to warm the motor to observe intermittent conditions or to cause the motor to run faster and possibly coast through park. Switching the motor from low to off several (twenty) times will detect open commutator bars on the armature. The motor must be allowed adequate time to park during the open commutator test.

A motor that fails to run may have the following symptoms: L-12 volts and H-12 volts at both low and high switch positions. Possible causes are: a stuck common brush (see Motor Repair-Two Speed), a spring off the common brush, an open ground strap to the body, or an open commutator bar under the common brush (see Testing Armature).

L-12 volts and H-"0" volts at low speed and motor doesn't run on low speed. Investigate for stuck low speed brush or the spring off the low speed brush.

L-"0" volts and H-12 volts at high speed switch position and the motor does not run on high speed. Investigate for a stuck high speed brush or the spring off the high speed brush.

L-8 volts and H-5 volts at low speed and L-9 volts and H-12 volts at high speed and the motor does not run. The circuit breaker should be cycling. Therefore, much of the time there is "0" volts at the motor terminals. The motor is stalled due to broken magnets, loose motor bearing and retainer, stripped gears, or binding linkage. Remove the output crank nut and crank to isolate linkage. Repeat the voltage test to isolate whether the cause is in the linkage or motor.

If "0" volts is observed at P1 when the panel switch is turned off the circuit to the motor may be open in the wiring, the panel switch circuit breaker or a short in the park switch. Remove the connector from the motor assembly. Measure the voltage at the P1 and H female conductors in the connector at off and high control switch positions. Allow a minute for an open circuit breaker to close. If 12 volts recurs, remove the P1 lead from the connector. Retest the motor with the P1 lead isolated at low and high speed for L and H voltage measurements. If the motor now operates correctly, this test identifies the failure as a shorted park switch.

If the motor does not function properly test the circuit breaker at its rated 6 amps load which it should be able to carry indefinitely. If the circuit breaker cycles at a 6 amp load, replace the control switch. After replacing a panel switch, repeat the above voltage measurements since the motor may have overloaded the circuit breaker causing switch failure. Be cautious that the faulty circuit is rectified so that the failure does not recur.

Wiper System Voltage Chart

Two Speed Non-Concealed

Instrument Panel Switch Position	L	H	P2	P1	POS Battery
Low	11	8	12 or 0①	12	12
High	18	12	12 or 0①	12	12
Off (While Parking)	11	8	12	12	12
Off (Parked)	0	0	0	12	12

①—The P2 Terminal will have a 12 volt potential except for a momentary drop to "0" volts when the park switch opens during each wipe cycle near the Park position.

Two Speed Concealed

Instrument Panel Switch Position	L	H	P2	P1	POS Battery
Low	11	8	0	0	12
High	18	12	0	0	12
Off (While Parking)	1	4	12	12	12
Off (Parked)	0	0	0	12	12

NOTE: Operate the motor for a half hour after repairs are made (with the arms and blades removed) to ascertain that proper service has been effected and to allow the brushes to reseat minimizing noise.

If P2 measures "0" volts and P1 measures 12 volts, the park switch circuit is open in the motor. The switch may be improperly adjusted or contaminated with greasy materials. Adjust the switch or replace the gear housing, brush holder and park switch assembly to correct.

Testing Armature Two Speed Motor

Testing Armature For Short Circuit

Place armature in growler and hold a thin steel blade parallel to the core and just above it, while slowly rotating armature in growler. A shorted armature will cause blade to vibrate and be attracted to the core. Replace armature if shorted.

Testing Armature For Ground

Contact armature shaft and each of the commutator riser bars with a pair of test lamp probes. If lamp lights, it indicates a grounded armature. Replace grounded armature.

Checking For Open Commutator Bars

Contact adjacent commutator riser bars with a pair of test lamp probes. The lamp should light on each bar. If the lamp fails to light the wire to bar junction is open. Replace open armature.

Gear and Park Switch Replacement

1971-75 Variable and Three Speed Motors

1. Run motor out of park to close hook and to prevent damage to park switch. If motor does not run check for open or excessive current. See "Motor Testing".
2. Remove the five switch plate screws and ground strap.
3. Carefully remove park switch plate and gasket. Hook should be closed, Figs. 18 and 19. If open motor is unlatched, replace latch assembly.
4. Remove motor from car, if necessary to replace park switch. See "Motor Removal" in car chapters.
5. Remove gear and latch assembly from gear box.
6. Inspect all components for damage or wear paying special attention to the following, Figs. 18 and 19.
   a. Bent parking switch blade.
   b. Melted switch cam.
   c. Chipped, scraped or broken islands.
   d. Chipped or scraped inner or outer cam.
   e. Bent hook where it latches to inner cam.
   f. Loose hook (early).
   g. High reinforcement pin in inner cam.
   h. Clearance between the block and latch assembly exceeding .025 may cause noise. Measure with a feeler gauge.

NOTE: When replacing components, lubricate all areas marked (*), Figs. 18, 19, 20, and metal surfaces with lubricant contained in package.

7. To replace switch plate unsolder switch plate leads. Solder leads to new switch plate, making sure the two green leads are together in their original location.
8. Insert latch assembly into gear. The early 1971 latch assembly is to be inserted open, Fig. 18. Rotate the early latch assembly counter clockwise until it stops against the block. Rotate the hook counter clockwise until it closes against the latch. Closing the latch should require some force. Re-examine latch and gear at this time if they do not fit snuggly. The latch assembly is spring loaded and inserted closed with the latch against the block, Fig. 18. (The late 1971 latch assembly may be used with either a steel inner cam (preferred) or the plastic inner cam). Insert gear and latch assembly, with washer between gear and gearbox, with the gear approximately in the position, Figs. 18 and 19.

NOTE: The gear and latch assembly must be in the position, Figs. 18 and 19, with the latch closed when the switch plate is installed to prevent damage to switch plate.

9. Install the melamine tube on the switch cam, Figs. 18 and 20. Place a small amount of grease on the tip of the switch cam taking care to keep grease off the melamine tube.

NOTE: The contacts and tube are not used on some models. The tube must be used on all models when the service park switch package with contacts is installed.

10. Install the gasket with the slot over the gasket locator, Figs. 18 and 20.
11. Rotate the rocker arm lever, which is riveted to the switch plate, counter clockwise to its stop. Install the switch plate so that the tab aligns with the correct notch, Figs. 18 and 20. Replace the five screws, making sure the ground strap is under the screw next to the gasket locator.
12. Bench test motor according to "Motor Testing". If test's indicate motor operates correctly, install motor.

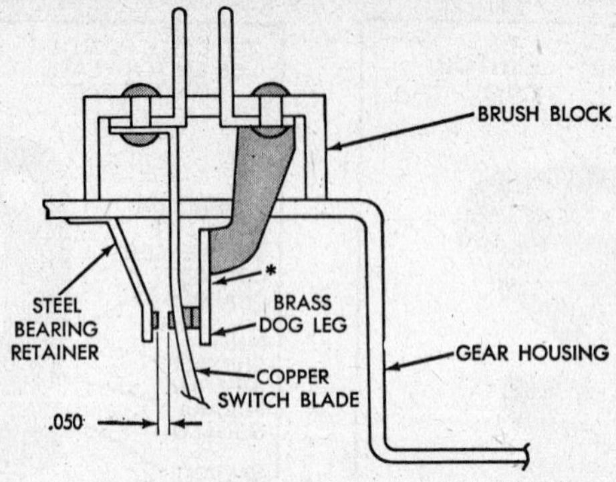

Fig. 21 Checking and adjusting contact gap. 1972-73

MOTOR REPAIR 1972-73 TWO SPEED

Replacement of Gear Housing & Brush Holder Assembly

1. Mount wiper motor in a vise.
2. Remove the mounting grommet spacers.
3. Remove the mounting grommets.
4. Remove the bulkhead seal from motor.
5. Drill out the six rivets from the gear housing cover and remove cover.
6. Remove the intermediate gear and wave washer from the gear housing.
7. Remove motor housing through bolts.
8. Remove the motor housing and armature assembly.
9. Remove the flat washers and one spring washer.

Assembly

1. Using new gear housing and brush holder assembly, hook the brush leads in brush holder notches.
2. Clean commutator with ink eraser, install a flat washer, spring washer and flat washer on armature shaft. Place armature in brush holder assembly.
3. Release brush leads from brush holder notches (check to see brushes are spring loaded against the commutator).
4. Hold the armature pinion so the magnets in the motor housing do not pull the armature out of the brush holder. Line motor housing assembly with brush holder and place over the armature. Install motor housing.
5. Install motor housing through bolts.
6. Gently tap motor with mallet to align bearings. Prior to beginning assembly of the gear housing the contact gap and contact alignment must be checked and adjusted as necessary.
7. Check and verify that the alignment of the contacts on the brass dogleg, copper switch blade, and steel bearing retainer are aligned within 90% of total point contact. Adjust by bending the brass dogleg or the steel bearing retainer.

CAUTION: Do not bend or make any adjustment to the copper switch blade.

8. Adjust the point gap between the brass dogleg and the copper switch blade to obtain a clearance of .050 (fifty thousandths) with the copper switch blade in contact with the steel bearing retainer. This adjustment is accomplished by slightly bending the brass dogleg in the area indicated by the asterisk (*) only, Fig. 21.
9. Lubricate intermediate gear shaft, install the wave washer (convex side of washer must face gear) and the intermediate gear on the shaft.
10. Install the park cam, (the park cam Vee notch must face output gear) flat washer and coil spring (small end first) on the output gear shaft. (With concealed wipers).
11. Rotate park cam so that it does not contact park switch or intermediate gear while assembling the cover assembly. (With concealed wipers).
12. Rotate the output gear so the cam does not contact the park switch while assembling. (With non-concealed wipers).
13. Install cover using screws, lock washers and nuts furnished. Be sure the two cover locating pins are aligned with their hole in the gear housing.
14. Install the mounting grommets.
15. Remove the resistor and ground strap

from old gear housing and install on new gear housing.

16. Bench test motor for proper operation.
17. Install the mounting grommet spacers. (Be sure ground strap is under one spacer).
18. Install the bulkhead seal on motor.

Replacement of Armature or Motor Housing

1. Mount the motor assembly in a vise as shown above. Remove only the through bolts, armature and motor housing, and the flat and spring washers.
2. Clean the brush assembly.
3. Reassemble by performing steps 10 through 12 above on the gearbox assembly. Since the gearbox is assembled it is necessary to rotate the armature counter clockwise against the intermediate gear so the magnets in the motor housing do not pull the armature out of the brush holder.
4. Align the motor housing assembly with the brush holder and place over the armature. Install motor housing. Install motor housing through bolts. Gently tap the motor housing with a small mallet to align bearings.
5. Bench test the motor for proper operation.

Replacement of Intermediate Gear or Output Gear

Mount motor assembly in a vise. Remove gear housing cover, as detailed in "Replacement of Gear Housing and Brush Holder Assembly".

Intermediate Gear

1. Remove the damaged intermediate gear and wave washer from the intermediate gear shaft.
2. Clean the grease and plastic chips from housing and armature shaft.
3. Lubricate the new intermediate gear teeth and bore with grease furnished in package (keep grease off park switch contacts).
4. Install the new wave washer (convex side of washer must face gear) and the intermediate gear on intermediate shaft.

Output Gear

1. Remove the park cam, flat washer and coil spring from output gear shaft, (models equipped with concealed wipers).
2. Remove the output shaft seal from the gear housing cover, (if so equipped).
3. Remove the snap ring from the output shaft. Remove the three washers.
4. Remove the output gear and flat washer from cover assembly (remove burrs from output gear shaft if trouble is encountered removing from cover).
5. Lubricate the output gear teeth and inner and outer cams (keep grease off the park switch contacts).

Assembly

1. Install flat washer on output gear shaft and slide output gear shaft into cover assembly.
2. Place small flat washer, spring washer and large flat washer on output gear shaft. Install snap ring.
3. Clean seal (pack with grease) and install over output gear shaft (if so equipped).
4. Rotate the output gear so the outer cam does not contact the park switch while assembling the cover assembly, (non-concealed wipers).
5. Install the park cam, (the park cam Vee notch must face the output gear) flat washer and coil spring (small end first) on the output gear shaft, (with concealed wipers).
6. Rotate park cam so that it does not contact park switch or intermediate gear while assembling the cover assembly, (with concealed wipers).
7. Install cover using screws, lock washers and nuts furnished. Be sure the two cover locating pins are aligned with their hole in the gear housing.
8. Bench test motor for proper operation.
9. Install the mounting grommets.
10. Install the mounting grommet spacers. (Be sure ground strap is under one spacer).
11. Install the bulkhead seal on motor.

FORD (ALL) & 1974-75 AMERICAN MOTORS

Depressed Park Type

Four wiper systems are offered thru the years. A intermittent wiper governor can be used in conjunction with the motor to provide intermittent wiper action. All switches used with these systems operate same way electrically with some deviation in terminal arrangement which inturn varies switch continuity testing procedures.

The one and two speed non-depressed wiper motors are actuated by a permanent magnet, rotary type electric motor. The two wiper arms and blades are mounted on a pivot shaft, one at each side of the windshield. The pivot shafts are connected to the motor by linkage arms and attaching clips.

The hydraulically operated system obtains the hydraulic power for the motor from the power steering unit. The hydraulic fluid flows from the pump through the steering gear to the wiper motor, and then to the fluid reservoir. During wiper operation, a part of the fluid is bypassed through the motor by the pressure regulator valve on the motor.

The new design depressed-park wiper motor is a two speed permanent magnet type motor, with the brush end plate at one end of the housing and a gear housing at the other end. The park switch is located in the gear cover and the park mechanism is located in the output arm.

The two speed oscillating type electric motor is mounted on the cowl panel. The drive link and pivot plate assembly has two integral links and is bolted to the right side of the cowl panel. One of the links connects to the motor crank pin; the other connects to the right pivot shaft assembly. A third link, integral with the left pivot shaft assembly, also connects to the right pivot shaft.

TROUBLE DIAGNOSIS

Windshield Wiper Inoperative (Hydraulic)

1. Control cable not properly adjusted.
2. Control cable broken.

Inoperative or Slow Wiper

1. Binding linkage.
2. Defective switch.
3. Defective wiper motor(s).
4. Defective wiring or circuit breaker.

Continuous Wiper Action with Selector At Intermittent Position

1. Loose, broken, or plugged vacuum ernor.
hose from engine to control to gov-
2. Ruptured governor or governor switch diaphragm.
3. Defective control selector switch.

Excessive Dwell Time During Intermittent Operation

1. Pinched hose from lower governor fitting to control switch rear fitting.
2. Plugged orifice in control selector dwell regulator.

Wipers Will Not Stop or Will Not Stop in Park Position

1. Open or broken wire.
2. Defective wiper motor.
3. Defective park switch.
4. Defective wiper switch.

Noisy Wiper System

1. Loose or misaligned wiper motor mounting.
2. Improper wiper arms or blades.
3. Loose or worn linkage.

Electric Wiper Motor Testing Wiper Motor Current Draw Tests

To perform the electrical tests on 1969-70 models & 1975 Pacer, it will be necessary to remove the motor from the vehicle for access to the terminals, while the 1971-75 motor tests can be performed with the wiper motor installed in the vehicle with linkage disconnected.

Motor Test

Depressed Park Type

1969-75 Full Size Models & 1975 AMC Pacer

The motor terminals are too small to make the necessary test connections without using connector sleeves and wires between the motor terminals and the test equipment, Figs. 22 and 23.

Connect the positive (red) lead from the test equipment to the center terminal on the motor end plate, and connect the green lead from the tester to the battery positive post, Fig. 23. Connect a jumper wire from the battery negative post to the low speed terminal on the motor end plate and read the current draw. Move the jumper wire from the low speed terminal to the high speed terminal, and read the high speed current draw. In either case the current draw should not exceed 3.5 amperes. If the current draw does exceed 3.5 amperes, check the output arm and windlatch mechanism for binding or damage before replacing the motor.

Non-Depressed Park Type

1969-75 Intermediate Models

Disconnect the linkage from the motor and disconnect the electrical plug to test the motor on the vehicle. Connect the green lead from the test equipment to the battery positive post, Fig. 24. Connect the positive (red) lead from the tester first to the low speed connection and then to the high speed connections at the connector plug as shown. In either case, the current draw should not exceed 3 amperes.

Circuit Breaker Test 1969-71

Place the switch in off position and connect the variable resistance unit of a Rotunda Volt-Amp Tester, Fig. 25. The Lincoln Continental switch is shown, Figs. 26 thru 32. Show terminal identification on other vehicles.

NOTE: Set the master control to maximum resistance position initially to prevent damage to the circuitry and ammeter.

Low Current Pass Test

Adjust the variable resistance until the ammeter indicates amperage in chart below. The ammeter should indicate the specified current for ten minutes. If the circuit will not indicate as specified for ten minutes (current drops to zero), replace the switch.

LOW CURRENT PASS TEST CHART

Year	Model	Low Current in Amperes
1969-71	FORD	
	Full Size, Exc. 1970 Thunderbird	8¼
	Intermediate	7
	MERCURY	
	Full Size, Exc. 1970 Continental Mark III	8¼
	Intermediate	7
1970	Thunderbird & Continental Mark III	7½
1972-75	All with rotary pin terminal switches	8¼
	All with rotary blade terminal switches	7①

①—Exc. Mustang II; Mustang II 6 amps.

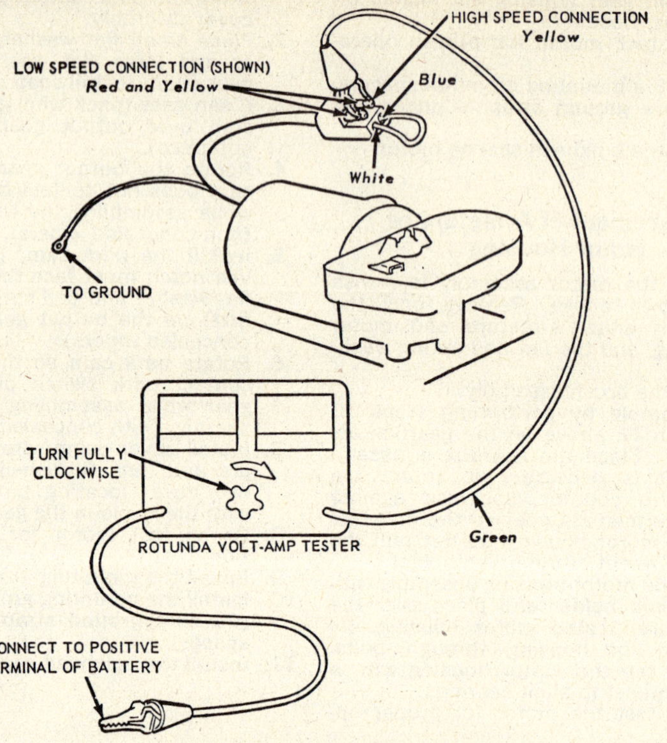

Fig. 22 Wiper motor current draw test connections. 1970 Thunderbird and Mark III

High Current Pass Test

Adjust the master control until the ammeter indicates amperage in chart below. The circuit should open within 30 seconds (current drops to zero), otherwise the switch must be replaced.

HIGH CURRENT PASS TEST CHART

Year	Model	High Current in Amperes
1969-71	FORD	
	Full Size, Exc. 1970 Thunderbird	16½
	Intermediate	14
	MERCURY	
	Full Size, Exc. 1970 Continental Mark III	16½
	Intermediate	14
1970	Thunderbird & Continental Mark III	15
1972-75	All with rotary pin terminal switches	16½
	All with rotary blade terminal switches	14①

①—Exc. Mustang II; Mustang II 12 amps.

Windshield Wiper Control Switch Continuity Test

Check the continuity between terminals of the switch, Figs. 26 thru 32. For a system without intermittent control either a self powered test light or an ohmmeter can be used.

For a system with intermittent control an ohmmeter must be used. A zero ohms reading indicates continuity between all terminals. As the switch knob is rotated to the left, resistance should begin to show on the ohmmeter and it should increase to 6000 to 7000 ohms on 1969 Intermediate and 5600 to 8400 ohms on all 1971-75 as the knob reaches the extreme left position.

If the continuity and/or resistance does not exist between the terminals, Figs. 25 and 27 thru 32, replace the switch. If continuity does exist as specified, check the wiring and repair or replace as necessary.

Parking Switch Test

Depressed Park Type Exc. 1970 Mark III & Thunderbird

After the Current Draw Test, the leads should be disconnected from the motor so that the output arm stops approximately 90 degrees past the park position for this test.

Using the appropriate connector sleeves and wires, make electrical connections to the motor from a 12 volt battery, Fig. 23. With this hookup, the motor output arm should move in the following cycle:

Rotate in the normal direction.

Reverse direction of rotation for approximately 10 degrees to 15 degrees.

Stop rotating while the crank pin is moved radially outward in a semicircular motion.

Stop in the final park position with the current draw going to zero.

If the output arm rotates in the reverse direction more than 15 degrees or if the motor stalls or jams while the out-

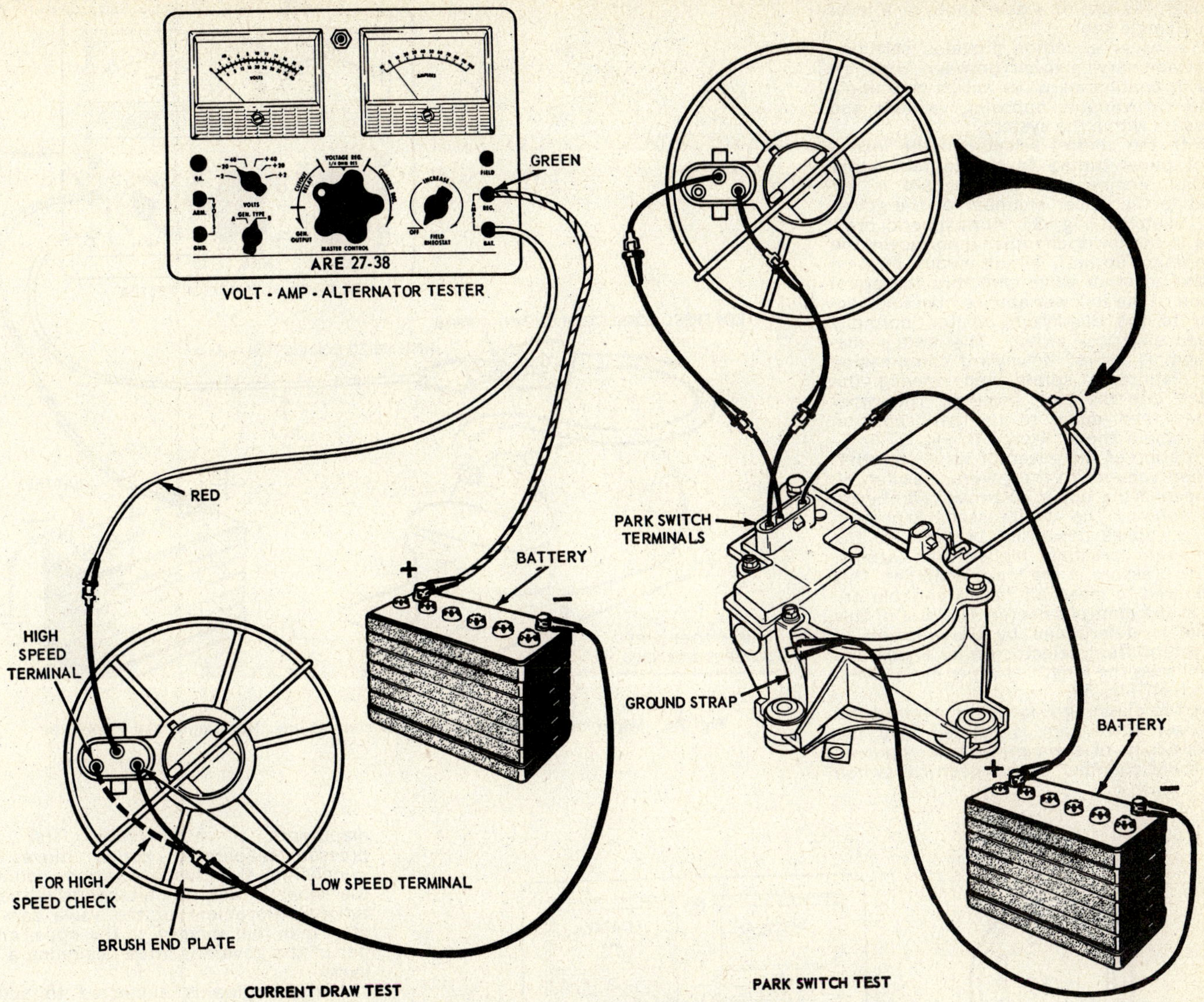

Fig. 23 Windshield wiper motor test. 1970-75 Depressed Park Type

put arm is rotating in the reverse direction replace the arm and windlatch assembly with the appropriate kit.

For any other deviation from the above cycle, replace the gear cover and switch assembly with the appropriate kit.

Non-Depressed Park Type & 1970 Mark III & Thunderbird

Stop the wiper system with the ignition switch so that the wiper blades are not in park position. Connect jumper wires, Figs. 34 and 35. The wipers should not run more than one full cycle and then park. If the motor will not park or will not run to park position, replace the park switch. If the motor stops correctly check the windshield wiper manual control switch and the wiring for continuity.

Noisy Wiper System

Noisy system operation could be caused by loose or misaligned motor mounting, improper wiper arms or blades, or loose or worn linkage.

Disconnect the linkage from the motor and run the motor to determine whether the noise is caused by the motor. Check the motor mounting for misalignment or grounding against cowl sheet metal.

Visually inspect the linkage for loose attaching bolts, loose pivots in the linkage, and loose wiper arms or blades.

Wiper Motor Park Switch Adjustment

With jumper wires connected, Figs. 34 and 35, the motor should move to the park position. If it does, the motor is all right and the fault lies in the panel switch or wiring. If the motor does not park, proceed with the following park switch adjustment, Fig. 36.

Remove the motor thru bolts and remove the motor cup and armature.

Rotate the output shaft until the park switch lower contacts are firmly closed as shown.

Rotate the adjusting screw clockwise until the park switch lower contacts just open.

Rotate the adjusting screw counterclockwise one full turn.

Check the bridge to assure that the legs are contacting the brush plate when the lower contacts are closed.

Install the armature, motor cup and thru bolts.

INTERMITTENT GOVERNOR

Electronic Control

If intermittent operation is unsatisfactory, first check the motor current draw and the control switch and all connecting wires for continuity. If the motor, switch and connecting wires are OK, replace the electronic governor, Fig. 37.

The intermittent wiper governor is an electro-pneumatic switch that controls the motor to provide intermittent wiping action.

The intermittent wiper is operated

through the use of a dual knob, or a lever and a single knob.

The wiper governor provides selection of either varying dwell between wipe cycles or continuous wiper action thru application of variable opposing vacuum and pressure within the system.

With the control selector in the instrument panel turned to intermittent wiper control, engine manifold vacuum is applied to the upper chamber of the governor. Sketch I, Fig. 37. Atmospheric pressure in the lower chamber then moves the diaphragm upward. Simultaneous rotation of the governor valve cam thru the spiral action of the follower applies manifold vacuum to the diaphragm of the normally closed electrical switch. The switch diaphragm is moved downward compressing the diaphragm spring and moving the switch into the park position. The wiper blades then move to the park position and remain there. Sketch II, Fig. 37.

Rotation of the governor valve cam that applied vacuum to the governor switch also opened the upper governor chamber to atmosphere. The compressed diaphragm spring applies downward pressure to the diaphragm creating higher than atmospheric pressure in the lower chamber. This is allowed to bleed off thru a variable orifice in the control selector. The size of this orifice is determined by the position of the intermittent selector (inner knob) and establishes the length of wiper cycle dwell. Sketch II, Fig. 37.

As the diaphragm moves downward the governor valve cam is rotated thru the spiral action of the cam follower opening the lower chamber of the electrical switch

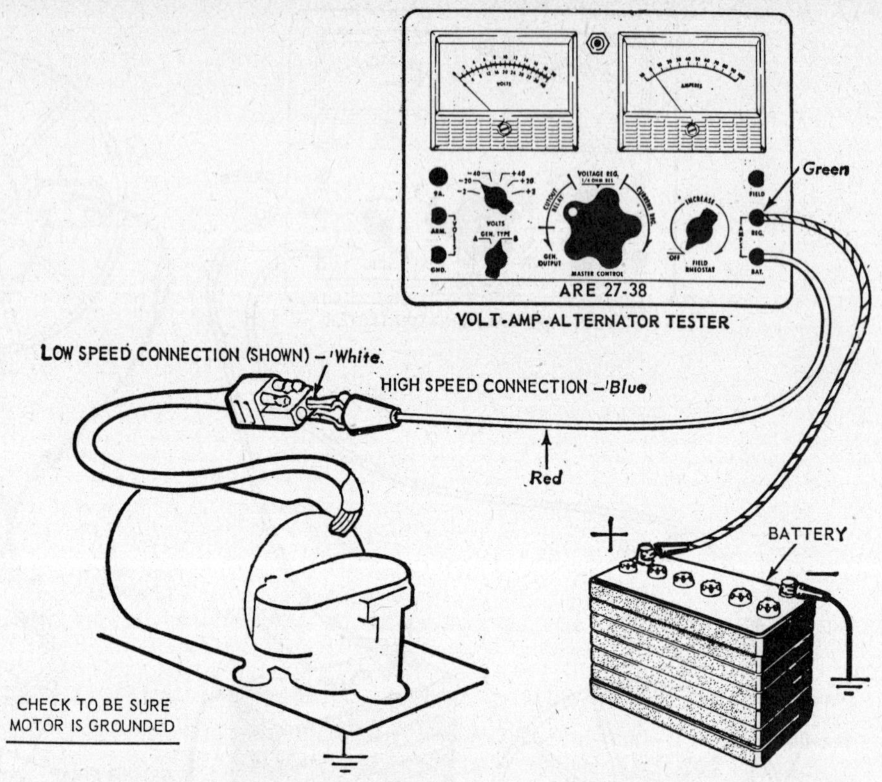

Fig. 24 Wiper motor current draw test connections. Non-Depressed Park Type

diaphragm to atmosphere. The compressed diaphragm spring moves the switch into the closed position, actuating the wiper motor. Simultaneously, the rotational movement of the valve cam applies manifold vacuum to the upper chamber of the governor, thus beginning a new cycle.

If the engine is subjected to sudden acceleration as in passing, the available manifold vacuum will be greatly reduced. When the vacuum loss becomes sufficient to allow the electrical switch diaphragm spring to overcome the vacuum, the switch will be moved to the closed position providing continuous wiper operation. This will continue until the engine has returned to an operating condition that will provide normal manifold vacuum. The wipers will then return to whatever wiper cycle dwell had been previously selected.

In the same manner as above, under engine acceleration, if any loss of vacuum or pressure in the intermittent control system occurs due to a mechanical failure or malfunction, the wipers will automatically shift to continuous operation, thereby rendering the system fail safe. Sketch III, Fig. 37.

HYDRAULIC WIPER SYSTEM

The windshield wiper motor, Fig. 38 is hydraulically operated. The hydraulic power for the motor is obtained from the power steering unit. The hydraulic fluid flows from the pump through the steering gear to the wiper motor, and then to the fluid reservoir. During wiper operation, a part of the fluid is bypassed through the motor by the pressure regulator valve on the

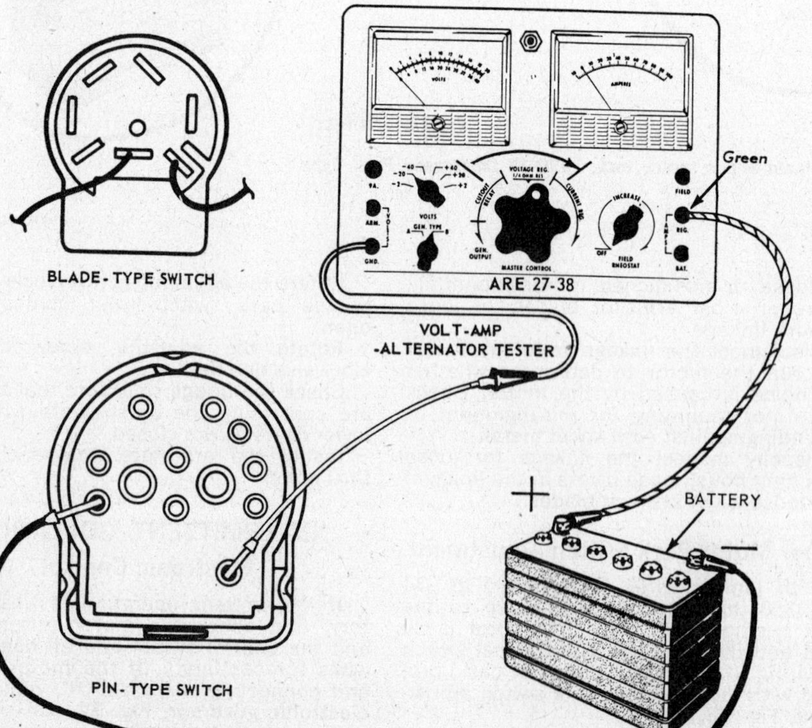

Fig. 25 Typical wiper switch circuit breaker test connections

motor. The speed of the wiper is controlled by adjusting the pressure regulator valve on the wiper motor with a bowden wire control operated from the knob in the center of the pod to the right of the main instrument cluster.

The only adjustment required on the hydraulic motor is the control cable adjustment.

1. Remove the carburetor air cleaner.
2. Remove the screws retaining the hydraulic line seal to the dash and pull the seal back over the lines.
3. Loosen the control cable retaining screw. Insert the control cable into the mounting slot and position the wire end in the motor drive lever slot, Fig. 38.
4. Push the motor drive lever to its extreme position, maintain this position of the motor drive lever, and force the cable housing in the opposite direction until a positive stop is felt.
5. Release the control cable housing, hold the motor control lever in its extreme position and tighten the cable mounting screws.
6. Position the hydraulic seal to the dash and install the retaining screws.
7. Install the carburetor air cleaner.

MOTOR REPAIR

Motor Parts Replacement

The wiper motor can be serviced as a complete assembly, but all parts can not be serviced. Removal and Installation procedures are given here for those parts that are serviced. To replace any of the

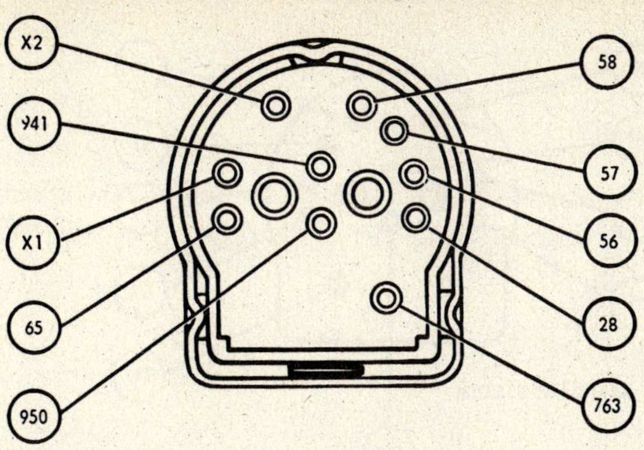

Using a Rotunda ohmmeter, check the continuity through the switch terminals. Continuity should exist only as indicated in the following chart. Replace the switch if it does not meet the specified requirements.

SWITCH POSITION	CONTINUITY BETWEEN TERMINALS
OFF (PARK)	65-763 28-56 X1-X2
INTERMITTENT	65-763 56-57 56-X2 X1-X2 – (100-900 ohms @ 1st detent and up to 7000 ohms through 10 detents.)
LOW	65-763 56-57 56-X1-X2
HIGH	65-763 57-58 58-X1-X2
WASHER ON	941-950

Fig. 26 Wiper switch continuity test connections. 1969-70 Lincoln Continental

STANDARD SYSTEM

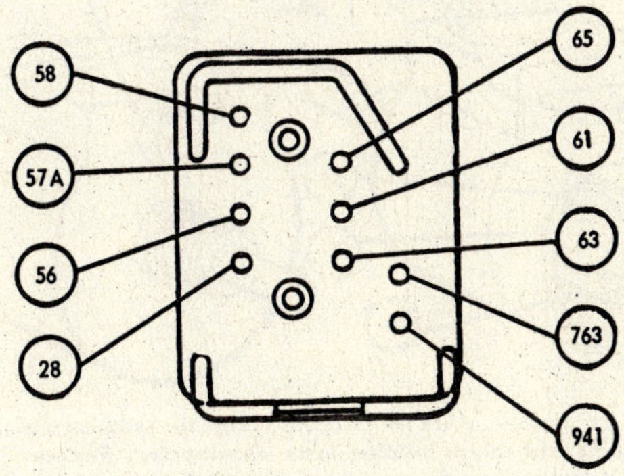

INTERMITTENT SYSTEM

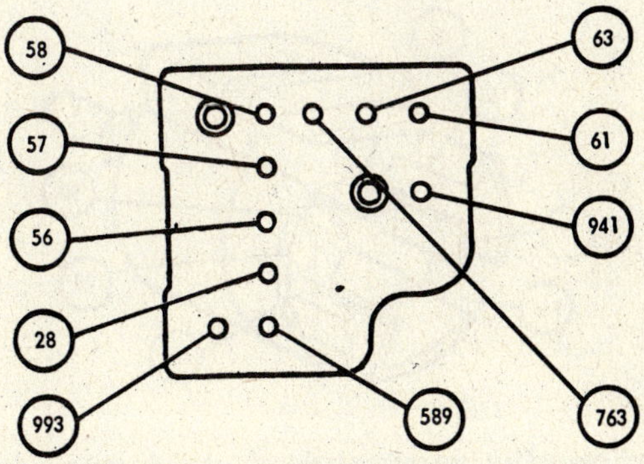

Using a self-powered test lamp or Rotunda ohmmeter, check the continuity through the switch terminals. Continuity should exist only as indicated in the following charts. Replace the switch if it does not meet the specified requirements.

SWITCH POSITION	CONTINUITY BETWEEN TERMINALS	COLOR CODE OF WIRING TO WIPER SWITCH
OFF (PARK)	65 - 763 61 - 63 28 - 56	28 Black 56 Blue 57-A Black
WIPER LOW/WASHER ON	61 - 65 - 763 - 941 56 - 57 - A	58 White 61 Yellow 63 Red
HIGH	61 - 65 - 763 57 - A - 58	65 Green 763 Orange-White 941 Black-White

SWITCH POSITION	CONTINUITY BETWEEN TERMINALS	COLOR CODE OF WIRING TO WIPER SWITCH
OFF (PARK)	28 - 56 61 - 63	28 Black 56 Blue
INTERMITTENT	763 - 61 589 - 28 993 - 57	57 Black 58 White 61 Yellow
LOW	763 - 61 56 - 57	63 Red 589 Orange
HIGH	763 - 61 57 - 58	763 Orange-White 993 Brown-White
WASHER ON	763 - 941	

Fig. 27 Wiper switch continuity test connections. 1969

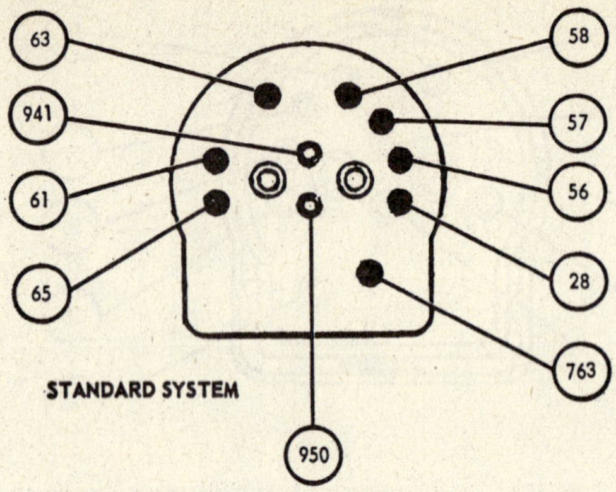

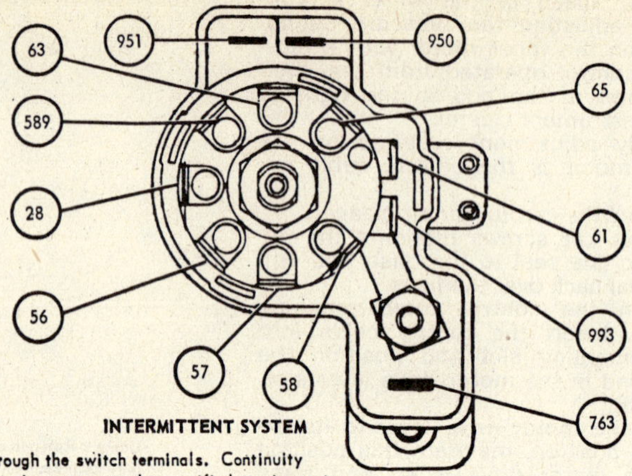

Using a self-powered test lamp or Rotunda ohmmeter, check the continuity through the switch terminals. Continuity should exist only as indicated in the following charts. Replace the switch if it does not meet the specified requirements.

SWITCH POSITION	CONTINUITY BETWEEN TERMINALS	COLOR CODE OF WIRING TO WIPER SWITCH
OFF (PARK)	763 • 65	28 Black
	61 • 63	56 Blue
	28 • 56	57 Black
LOW	763 • 61 • 65	58 White
	56 • 57	61 Yellow
HIGH	763 • 65	63 Red
	57 • 58	65 Green
WASHER ON	763 • 65	763 Orange-White
	941 • 950	941 Black
		950 White-Black

SWITCH POSITION	CONTINUITY BETWEEN TERMINALS	COLOR CODE OF WIRING TO WIPER SWITCH
OFF (PARK)	763 • 65	28 Black
	28 • 56	56 Blue
	61 • 63	57 Black
INTERMITTENT	763 • 61 • 65	58 White
	28 • 589	61 Yellow
	57 • 993	63 Red
LOW	763 • 61 • 65	65 Green
	56 • 57	589 Orange
HIGH	763 • 61 • 65	763 Orange-White
	57 • 58	950 Green-Black
WASHER ON	763 • 65	951 Green
	950 • 951	993 Brown-White

Fig. 28 Wiper switch continuity test connections. 1969-70 Mercury Full Size

STANDARD SYSTEM

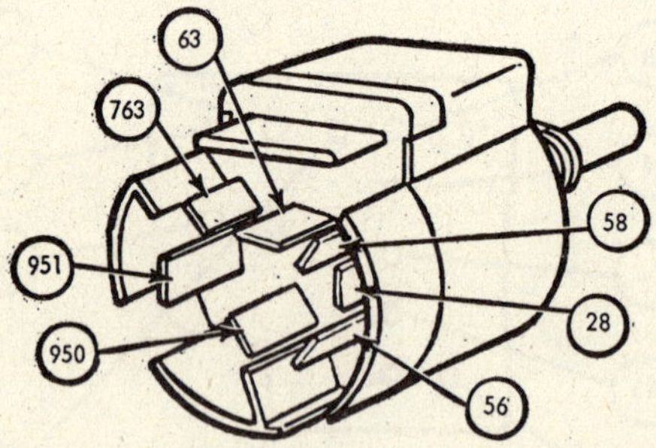

INTERMITTENT SYSTEM

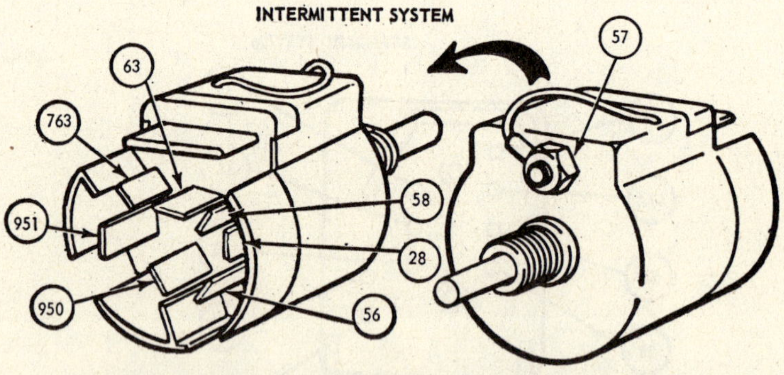

Using a self-powered test lamp or Rotunda ohmmeter, check the continuity through the switch terminals. Continuity should exist only as indicated in the following chart. Replace the switch if it does not meet the specified requirements.

SWITCH POSITION	CONTINUITY BETWEEN TERMINALS
OFF (PARK)	763-63
	28-58
LOW	763-63-58
HIGH	763-63-56
WASHER ON	951-950

COLOR CODE OF WIRING TO WIPER SWITCH
28 Black
56 Blue
58 White
63 Red
763 Orange-White
950 White-Black
951 Green-Black

Using a Rotunda ohmmeter, check the continuity through the switch terminals. Continuity should exist only as indicated in the following chart. Replace the switch if it does not meet the specified requirements.

SWITCH POSITION	CONTINUITY BETWEEN TERMINALS
OFF (PARK)	56-28
	63-763
INTERMITTENT	56-57
	56-28 —(6000 to 7000 ohms
	63-763 Resistance)
LOW	56-28-57
	63-763
HIGH	56-28-57
	58-63-763

Fig. 29 Wiper switch continuity test connections. 1969-70 All Intermediate

STANDARD SYSTEM

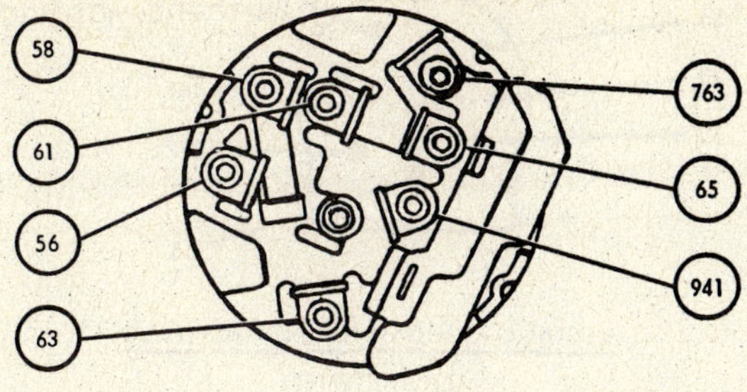

INTERMITTENT SYSTEM

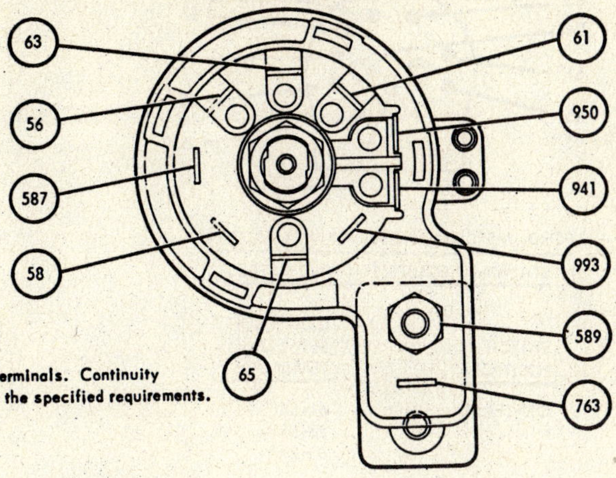

Using a self-powered test lamp or Rotunda ohmmeter, check the continuity through the switch terminals. Continuity should exist only as indicated in the following charts. Replace the switch if it does not meet the specified requirements.

SWITCH POSITION	CONTINUITY BETWEEN TERMINALS
OFF (PARK)	65-763 61-58 63-56
LOW	763-61-63 56-58
HIGH	763-61 56-58
WASHER ON	763-941

COLOR CODE OF WIRING TO WIPER SWITCHES

56	Blue	589	Orange
58	White	763	Orange-White Stripe
61	Yellow	941	Green (Standard System)
63	Red	941	Black-White Stripe (Intermittent System)
65	Green	950	White- Black Stripe

SWITCH POSITION	CONTINUITY BETWEEN TERMINALS
OFF (PARK)	763-65-589 61-58 63-56
INTERMITTENT	763-61-63-589 56-993 58-587
LOW	763-61-63-589 56-58
HIGH	763-61-589 56-58
WASHER ON	941-950

Fig. 30 Wiper switch continuity test connections. 1970 Thunderbird and Continental Mark III

following parts the motor has to be removed from the vehicle as outlined in the foregoing procedure.

1969-75 Depressed Type

Cover and Switch Assembly

Remove the four cover retaining screws and switch, inspect and replace parts if necessary. Be sure to tighten the cover screw that connects the park switch ground strap to the gear box.

Brush End Plate

Carefully observe the original position of the bale clip and pry it off with a screwdriver. Remove the end plate with plug and seal.

When installing the end plate, use a fine wire probe through the hub opening to position the brushes on the commutator. Rotate end plate to position the key in the notch and assemble the plug. Reinstall the bale retainer carefully with a screw driver to avoid overbending.

Output Arm, Windlatch and Cover Plate

Removal

1. With a sharp pointed tool, puncture and pry off the soft plastic plug from the upper surface of the arm, Fig. 39.
2. Remove the retaining bolt from the shaft, being certain not to rotate the shaft.
3. Remove the arm from the shaft with a suitable puller, Fig. 40. *Do not attempt to drift the arm from the shaft.*
4. Note the correct positioning of the windlatch in its guide for proper reassembly, Fig. 41. Lift the plastic windlatch out of the integral cover plate and guide.
5. Unclip the windlatch spring from the guide and remove it, Fig. 42.

Installation

1. Snap the windlatch spring in place, Fig. 42.
2. If the windlatch is not lubricated, apply Lubricate to both sides of the windlatch. (Replacement units are supplied lubricated). Position the windlatch in the integral cover plate and guide, Fig. 41.
3. Position the arm and cam assembly on the shaft so that the locating marks are aligned, Figs. 43 and 44. Note that the service replacement arm, Fig. 44, comes with the cam pin engaged in the detent. Therefore, when installed with the correct cam-to-shaft relationship (marks aligned), the arm will be 180 degrees out of parked position. It will be returned to park position when you perform the final step of this procedure.

STANDARD SWITCH

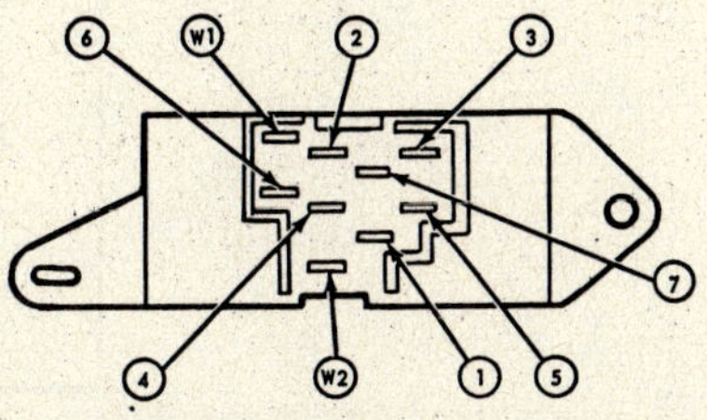

SWITCH POSITION	CONTINUITY BETWEEN TERMINALS
OFF	1-5, 3-7
LO	1-4, 2-7
HI	1-4, 2-6
WASH	W1-W2

* RESISTANCE BETWEEN TERMINALS F AND C VARIES FROM 500Ω ± 400 AT MIN. DWELL TO 10KΩ ± 20% AT MAX. DWELL

INTERMITTENT SWITCH

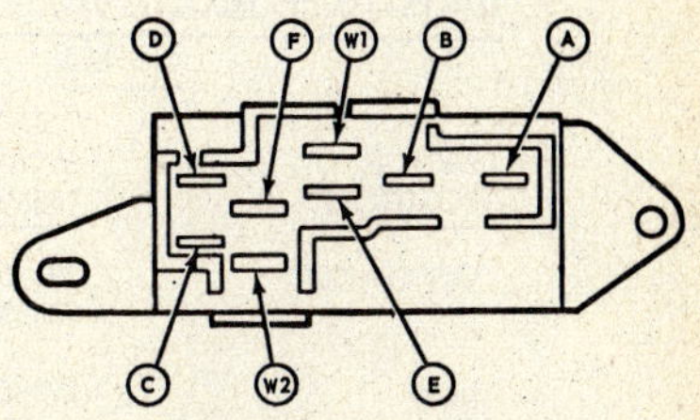

SWITCH POSITION	CONTINUITY BETWEEN TERMINALS
OFF	A-E
INT.	B-E-F, *
LO	B-E-F-C
HI	D-B-F-C
WASH	W1-W2

Fig. 32 Wiper switch continuity test connections. 1973-75 Ford & Mercury & 1975 American Motors Pacer

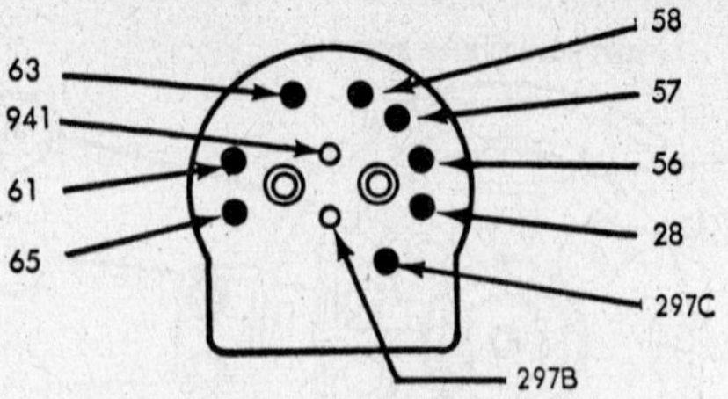

FORD, MERCURY, METEOR AND LINCOLN

STANDARD EQUIPMENT AND 2 SPEED

SWITCH POSITION	CONTINUITY BETWEEN TERMINALS
OFF (PARK)	61-63 28-56 65-297
LOW	61-65-297 56-57
HIGH	57-58 65-297-61
WASH	941-298-LINCOLN 941-296-FORD, MERCURY, METEOR

INTERMITTENT

SWITCH POSITION	CONTINUITY BETWEEN TERMINALS
OFF (PARK)	28-56 65-297 61-63
LOW	56-61-63-57 65-297
HIGH	65-297 57-58-61-63
INTERMITTENT	56-57-63 65-297 VARIABLE RESISTANCE BETWEEN 61 AND 63-MIN. 100-900 ohms MAX. 5600-8400 ohms
WASH	298-941 LINCOLN 941-296 FORD, MERCURY, METEOR

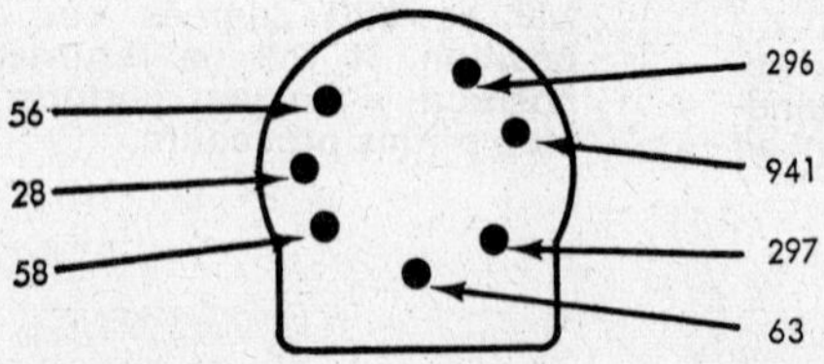

MUSTANG-COUGAR-MAVERICK TORINO-COMET

2-SPEED

SWITCH POSITION	CONTINUITY BETWEEN TERMINALS
OFF (PARK)	297-63 28-58
LOW	297-58-63
HIGH	297-63-56
WASH	296-941

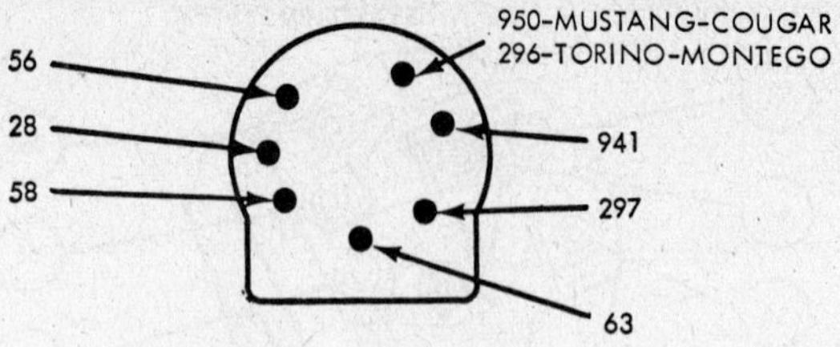

MUSTANG-COUGAR-TORINO-MONTEGO

INTERMITTENT

SWITCH POSITION	CONTINUITY BETWEEN TERMINALS
OFF (PARK)	28-56 63-297
LOW	63-297 56-57-28
HIGH	56-57-28 58-63-297
INTERMITTENT	63-297 56-57 VARIABLE RESISTANCE BETWEEN 28-57-MIN. 100-900 ohms MAX. 5600-8400 ohms
WASH	296-941

WIRING COLOR CODES

28-BLACK-PINK DOT
56-DARK BLUE-ORANGE STRIPE
57-BLACK
58-WHITE
61-YELLOW-RED STRIPE
63-RED
65-DARK GREEN
296-WHITE-PURPLE STRIPE
297-BLACK-LIGHT GREEN HASH
298-PURPLE-ORANGE HASH
941-BLACK-WHITE STRIPE
950-WHITE-BLACK STRIPE

Fig. 31 Wiper switch continuity test connections. 1971-72

4. Draw the arm and cam assembly onto the output shaft with the retaining screw, applying 8-10 ft-lbs of torque.

NOTE: Before tightening the screw, however, slide the windlatch in its guide to make sure that the upstanding tab clears the underside of the arm as it is being drawn onto the shaft.

5. Install a new soft plastic plug, Fig. 39.
6. Make sure that the output arm is in park position, Fig. 43.

1969-75 Two-Speed Electric Wiper Motor—Non Depressed Type

Disassembly

1. Remove the gear cover retaining screws, ground terminal and cover, Fig. 45.
2. Remove the gear and pinion retainer.
3. Remove the idler gear and pinion and output gear and shaft.
4. Remove the motor through bolts, motor housing, switch terminal insulator sleeve, and armature. *Do not pound the motor housing magnet assembly as the ceramic magnets may be damaged.*
5. Mark the position of the output arm with respect to the output shaft, for assembly. Remove the output arm retaining nut, output arm, spring washer, flat washer, output gearshaft assembly, thrust washer, and parking switch lever and parking switch lever washer.
6. Remove the brushes and brush springs.
7. Remove the brush plate and switch assembly, and remove the switch contact to parking lever pin from the gear housing.

Cleaning and Inspection

1. Clean the gear housing of all old grease. *Do not allow any cleaning fluid to contact the armature shaft and output shaft bearings.*
2. Wipe all other parts with a clean cloth.
3. Inspect the gear housing for cracks or distortion. Replace a cracked or distorted housing.
4. Check all shafts, bushings, and gears for scored surfaces. Replace defective parts. It is essential that the idler gear shaft with the push-nut be reset to the original, proper depth in the gear housing.

Assembly

1. Install the parking switch lever washer.
2. Install the parking switch lever on the gear and pinion shaft with the cam rider pointing toward the gear housing output shaft hole. Make certain that the lever bottoms against the casting.
3. Apply a film of Sun Prestige grease to the output gear teeth and shaft bearing surface. Insert the shaft in the bearing. Make certain that the parking switch lever is clear of the cam and gear assembly.
4. Place the spacer washer and spring washer on the shaft, position the output arm on the shaft in the marked position from which it was removed, and install the mounting nut.
5. Position the brush springs and brushes in the holders and wrap wire around them to hold them in the fully retracted position. Push the insulated brush connector onto the switch terminal.
6. Place the switch-contact to parking-lever pin in the gear housing. Position the brush plate assembly to the housing and install the mounting screws.
7. Apply Sun Prestige grease to the ball bearing in the end of the armature shaft. Position the armature shaft in the gear housing and remove the brush retracting wires.
8. Holding the armature in position, install the terminal insulating sleeve, motor housing and magnet assembly, and through bolts. Seal the area where the terminal insulator sleeve seats against the motor and gear housing.
9. Apply Sun Prestige grease to the worm gear and idler gear, and install the idler gear, thrust washer and retainer.
10. Apply a generous amount of Sun Prestige grease to the area around the end of the armature shaft. Install the gear housing cover (and ground terminal if used).

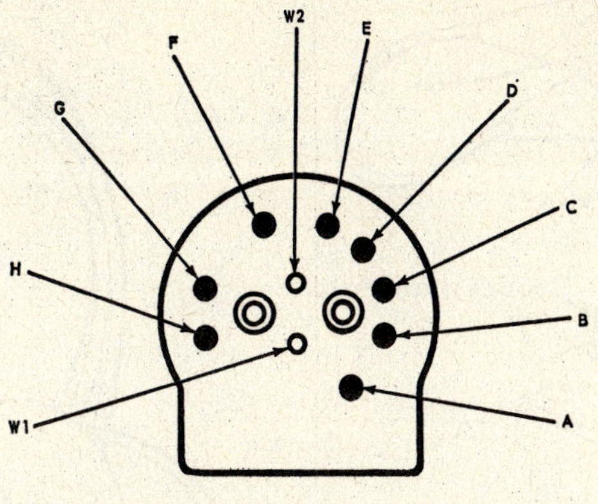

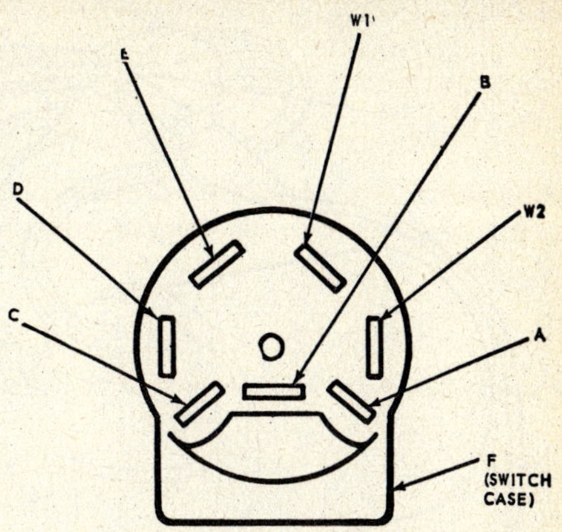

PIN-TYPE SWITCH
STANDARD EQUIPMENT (2 SPEED)

SWITCH POSITION	CONTINUITY BETWEEN TERMINALS
OFF (PARK)	B-C A-H F-G
LOW	C-D A-G-H
HIGH	D-E A-G-H
WASH	W,-W2

INTERMITTENT

SWITCH POSITION	CONTINUITY BETWEEN TERMINALS
OFF (PARK)	B-C A-H F-G
LOW	C-D-F-G A-H
HIGH	D-E-F-G A-H
INTERMITTEINT	A-H C-D-F VARIABLE RESISTANCE BETWEEN ;F AND G—MIN. 100-900 ohms MAX. 5600-8400 ohms
WASH	W1-W2

BLADE-TYPE SWITCH (2 SPEED)

SWITCH POSITION	CONTINUITY BETWEEN TERMINALS
OFF (PARK)	C-D A-B
LOW	A-B-C
HIGH	A-B-E
WASH	W1-W2

INTERMITTENT

SWITCH POSITION	CONTINUITY BETWEEN TERMINALS
OFF (PARK)	A-B D-E
LOW	A-B D-E-F
HIGH	D-E-F A-B-C
INTERMITTENT	E-F A-B VARIABLE RESISTANCE BETWEEN D-E-MIN. 100-900 ohms MAX. 5600-8400 ohms
WASH	296-941

Fig. 33 Wiper switch continuity test connections. 1973-75 exc. Ford and Mercury & 1974-75 American Motors Matador 2 dr. Coupe

Oscillating Type Two—Speed Electric Wiper Motor

Disassembly

The two-speed electric motor may be disassembled for service of the drive mechanism parts.

1. Remove the gear housing cover plate and gasket, Fig. 46.
2. Remove the output shaft retainer and spacer washer.
3. Remove the crankpin bearing retainer

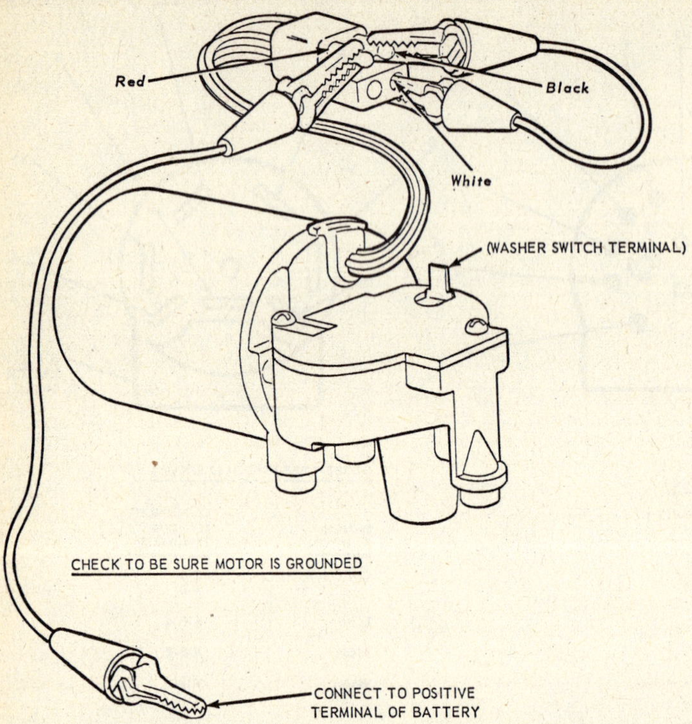

Fig. 34 Wiper motor park switch test connections. Non-Depressed Type

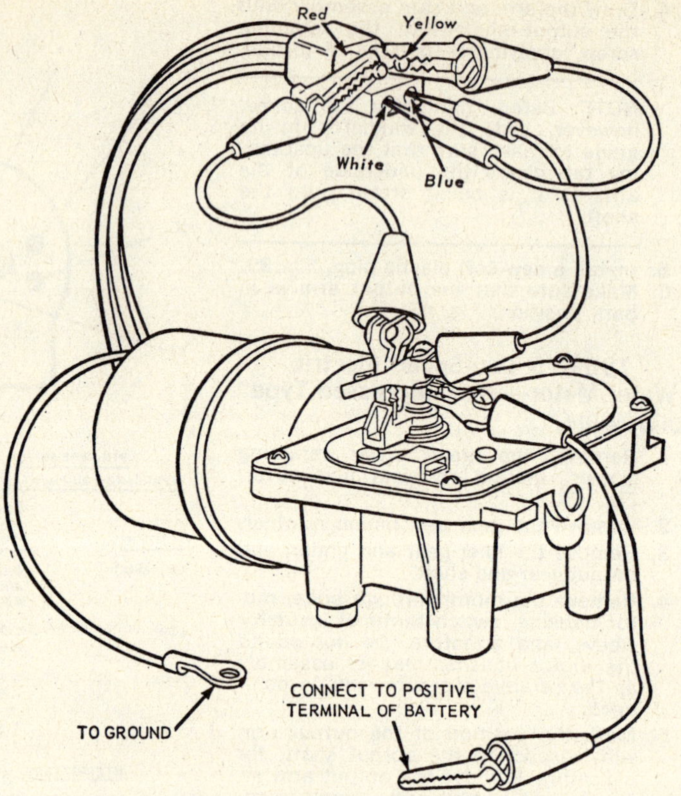

Fig. 35 Wiper motor park switch test connections. 1970 T-Bird and Mark III

and remove the spacer washer and cam return spring assembly.

4. Remove the arm and link assembly.
5. Remove the crankpin bearing cam.
6. Remove the input gear retainer and outer spacer shim, and remove the input gear and inner spacer shim.
7. Remove the wiper arm lever nut and lock washer.
8. Remove the wiper arm lever and spacer, and remove the output shaft and gear assembly from the housing.
9. The output gear may be removed from its shaft by tapping with a fiber hammer. Be careful not to damage the end of the shaft.

NOTE: The worm drive gear and armature assembly is not serviced.

Cleaning and Inspection

1. Clean the gear housing of all old grease. *Do not allow any cleaning fluid to contact the armature shaft and output shaft bearings.*
2. Wipe all other parts with a clean cloth.
3. Inspect the gear housing for cracks or distortion. Replace a cracked or distorted housing.
4. Check all shafts, bushings, and gears for scored surfaces. Replace defective parts.

Assembly

1. Tighten the motor cover. Adjust the motor shaft end play to 0.000-0.005

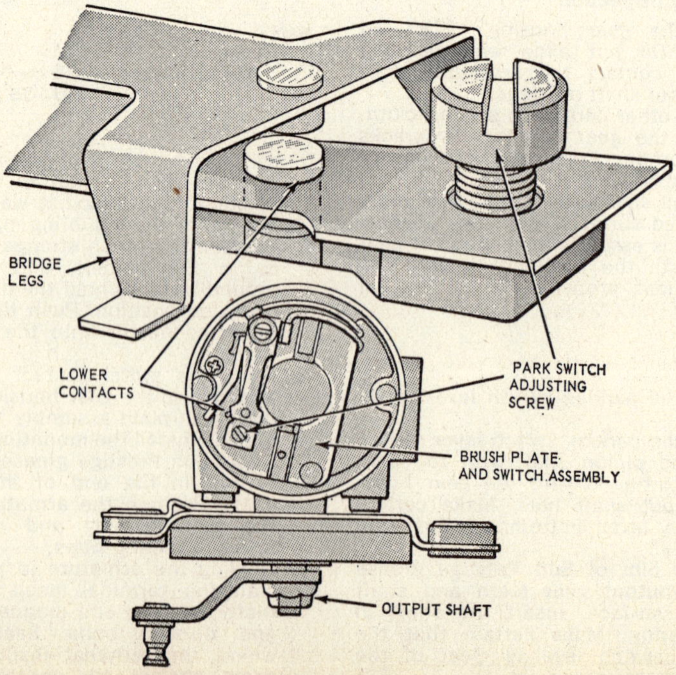

Fig. 36 Windshield wiper motor park switch adjustment

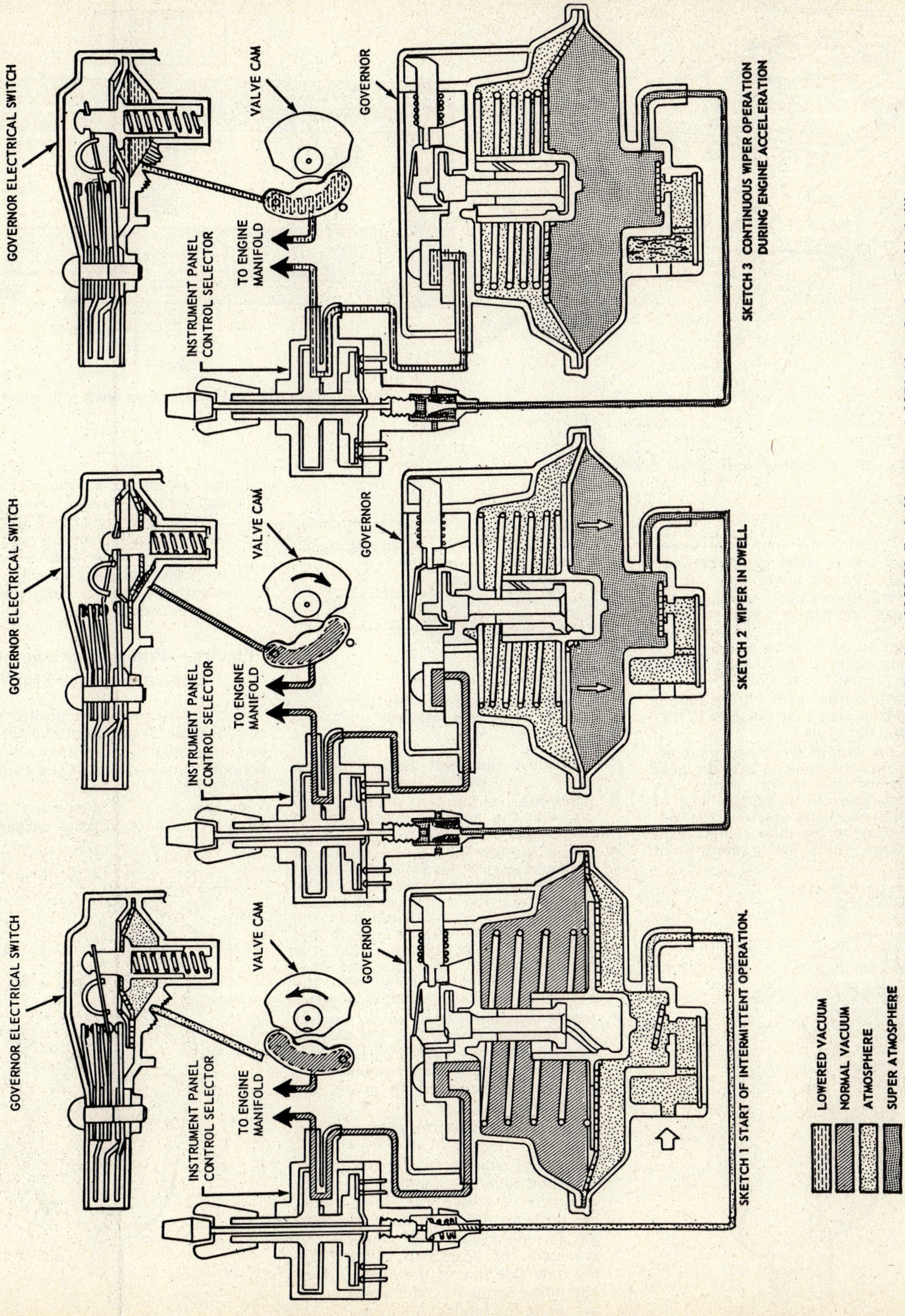

Fig. 37 Electro pneumatic governor intermittent windshield wiper. 1969-70 Ford, Mercury and 1970 Linc. Cont. and Mark III

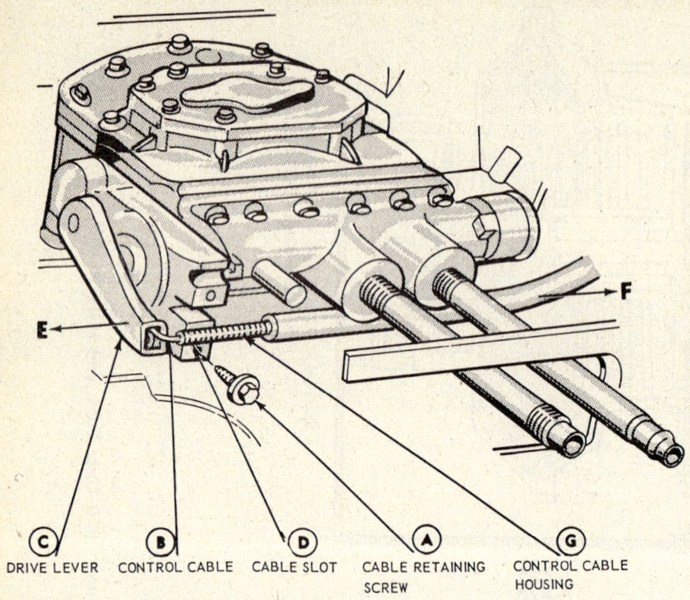

Fig. 38 Windshield wiper control cable adjustment

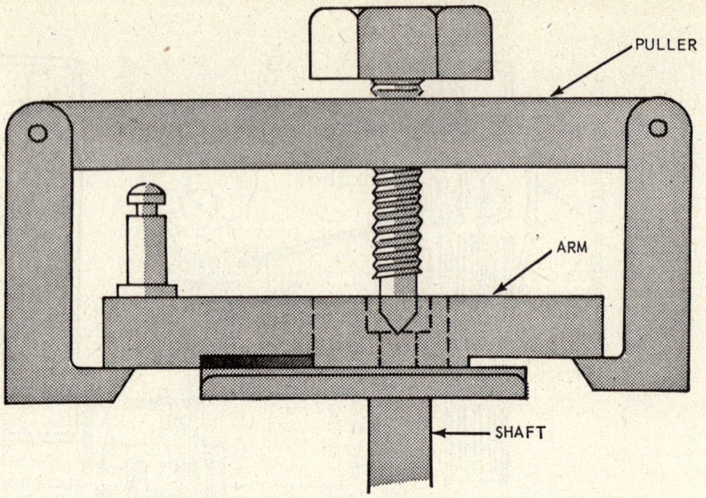

Fig. 40 Removing output arm from shaft with puller

inch by turning the shaft stop screw. Measure with a feeler gauge between the stop screw and the motor shaft.

2. Install the input gear shim on the input gear shaft and install the gear in the housing. Adjust the end play to 0.005 to 0.010 inch by adding or removing shims under the input gear retainer. Install the retainer.
3. Install the output gear on the output shaft. Make sure that the gear is bottomed on the shaft.
4. Install the output shaft and gear assembly into the housing with the gear teeth facing the motor. Install one spacer washer to the outside end of the output shaft and assemble the wiper arm lever to the output shaft, with the linkage studs facing away from

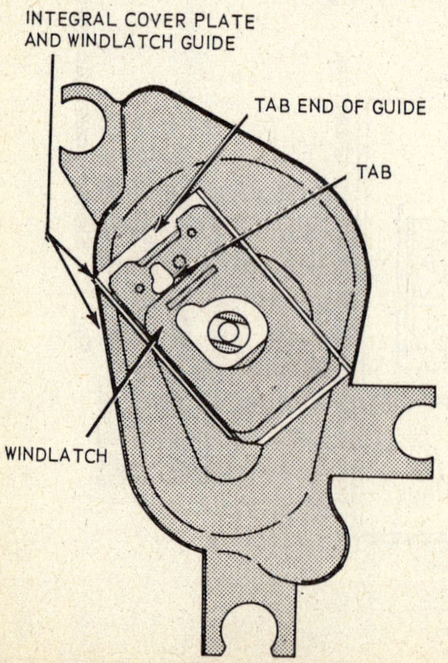

Fig. 41 Windlatch installation

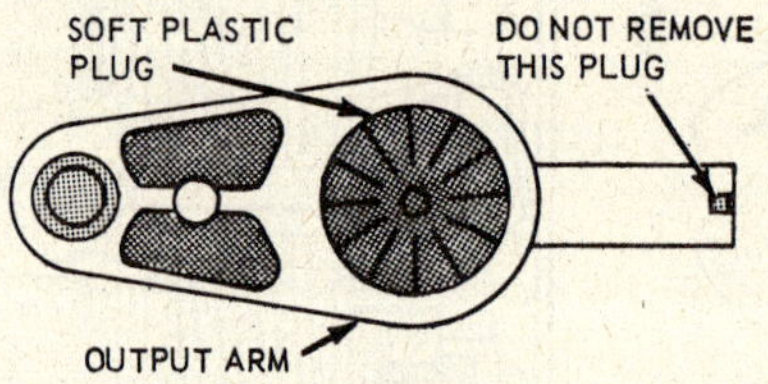

Fig. 39 Output arm with soft plastic plug installed

and above the shaft. Secure the lever with a lock washer and nut.

5. Place the bearing cam on the crankpin with the small diameter portion of the cam facing outward.
6. Install the arm and link assembly to the bearing cam. As the arm is placed on the shaft, the gears must be meshed and the link which is riveted to the arm must be installed to the output shaft at the same time. Proper gear indexing is obtained when the bottom tooth of the arm and gear segment will be in mesh with the bottom valley of the output shaft gear.
7. Install the output shaft spacer washer and retainer. Check the end play of the output shaft (0.005-0.010 inch). Remove or install spacer washers under the shaft retainer to adjust the end play.
8. Install the cam return spring assembly.
9. Install the bearing spacer and retainer. If the retainer cannot be installed, one or more coils of the spring clutch are probably out of place. If the bearing has excessive end play on the crankpin, the projection of the bearing may ride out of the semi-circular slot in the end plate. Add spacer washers under the retainer if necessary.
10. Apply generous amounts of Sun Prestige grease to all moving parts. Install the gear housing cover plate.

When operating the unit on the bench, do not place hands or fingers between the wiper lever and the case, or inside the gear housing, as considerable power is developed by the gear reduction.

Electro—Pneumatic Governor and Control Selector Switch

The electro-pneumatic governor and the control selector switch are serviced as assemblies only. Should any internal component become defective the complete assembly must be replaced.

In the event of suspected partial or total blockage of internal passages, light

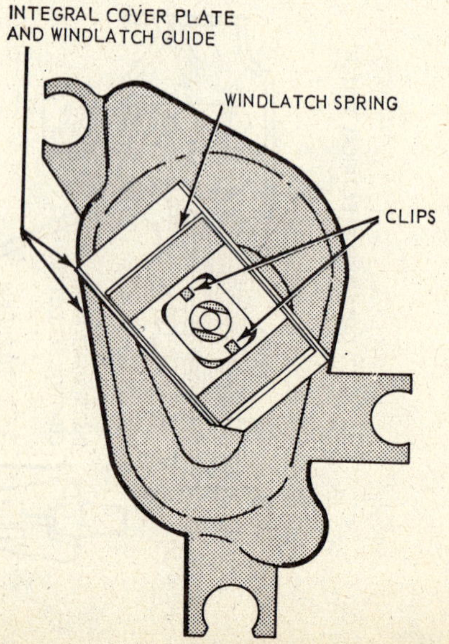

Fig. 42 Windlatch spring installation

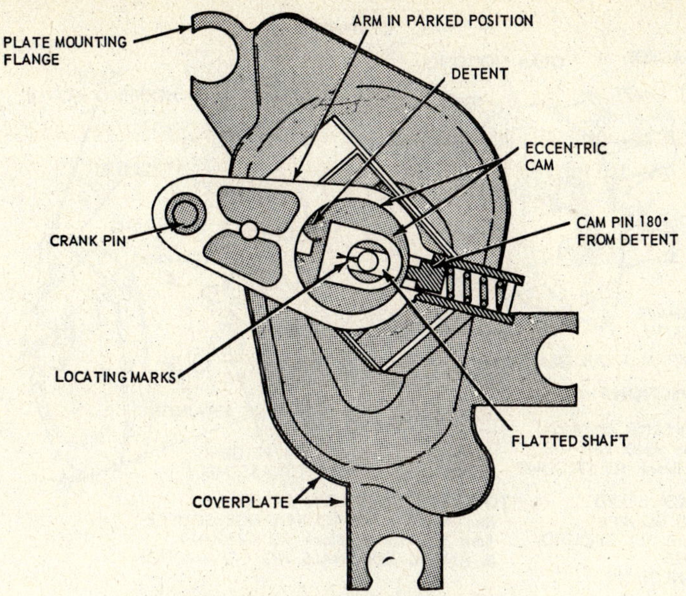

Fig. 43 Windshield wiper motor output arm in parked position

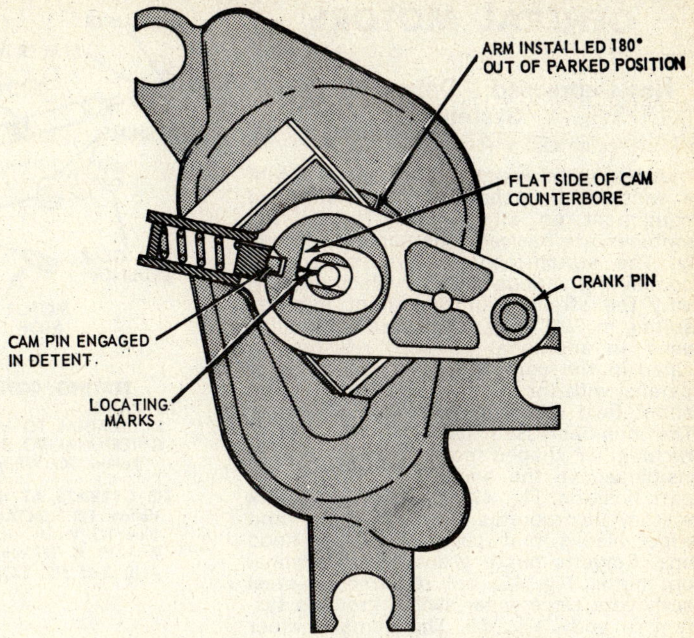

Fig. 44 Service replacement arm installation

air pressure may be applied in an attempt to remove the cause of blockage.

Although the vacuum hoses are an integral part of the wiring harness assembly, individual hoses may be replaced if required.

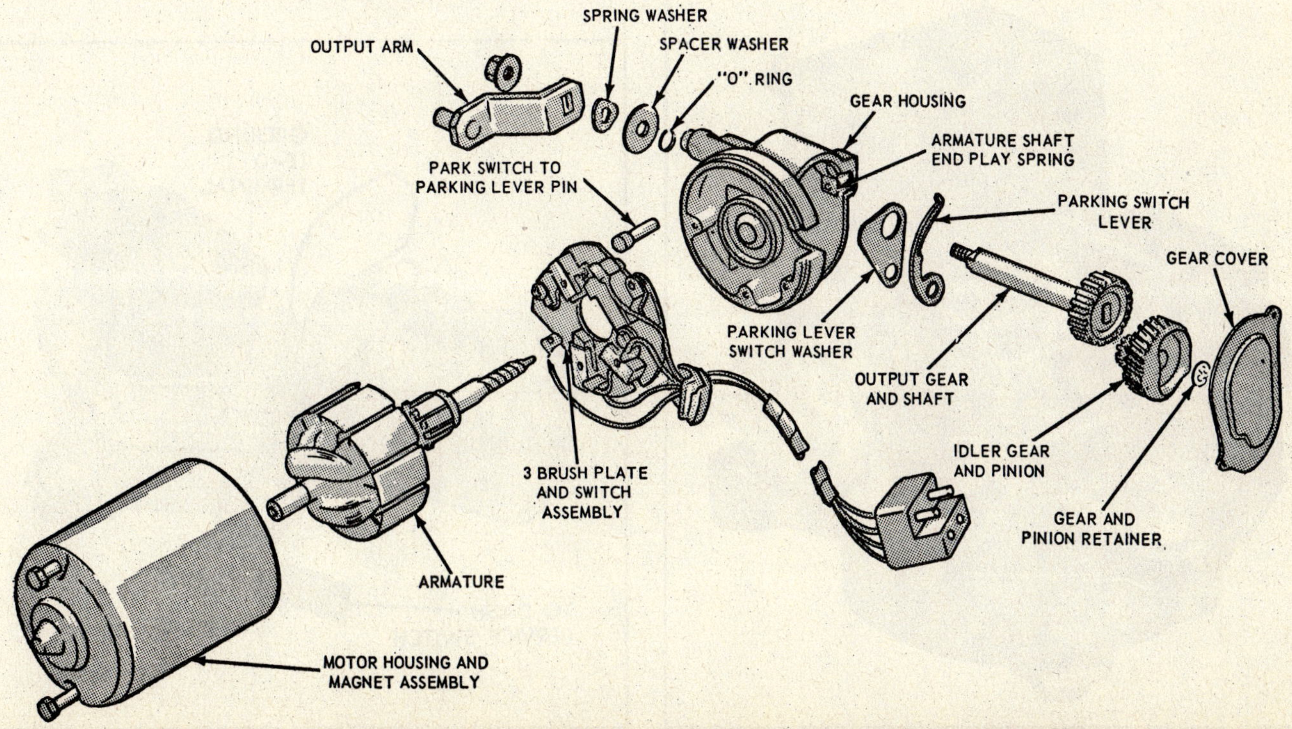

Fig. 45 Disassembled wiper motor, non-depressed type. 1969-75 Ford & 1973-74 American Motors

GENERAL MOTORS

Non-Depressed & Depressed Park Systems

A two-speed wiper motor equipped with a windshield washer system is standard equipment on all models. The control switches are basically the same electrically. The switch operates the system by completing various circuits which in turn vary the strength of the shunt field resulting in "LO" and "HI" speeds. Cadillac uses an additional 13 ohm resistor, located in the control switch connected in parallel with the 20 ohm resistor from the shunt field resulting in a "MED" speed. The non-depressed park system uses a rectangular shaped motor with the blades visible above the hood line when in the park position, Fig. 47. The depressed park system incorporates a round motor and wiper blades that park below the hood line. Corvette has a ground lead terminal on motor, Fig. 48, not required on steel body cars since wiper switch grounds system to body, Fig. 49. The demand wiper system also uses same round motor. However, this system requires the use of an extra relay (demand relay), a dual set of holding switches, Fig. 50, and an extra switch located in the end of the shift lever. The pulse wiper motor except for the washer pump and internal wiring, is similar to the round wiper motor. The pulse wiper motor has three leads coming out of the motor grommet while the depressed park motor has only two. The third lead (black cotton braided) is used to provide an external ground for the armature circuit. The green lead which is attached to the gearbox relay, provides a feed circuit to the pulse relay coil which is mounted on the washer pump frame.

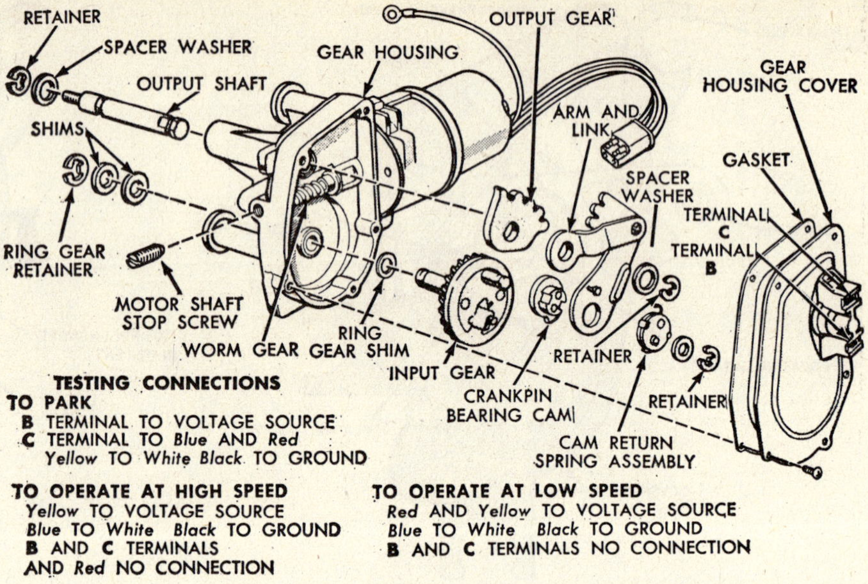

Fig. 46 Two-speed wiper motor—Oscillating Type

NOTE: A modified pulse wiper system is used on all 1975 intermediate and late 1975 full size models. The modified type pulse wiper motor can be identified from the original type by the light grey washer pump cover and two leads coming from the motor grommet. The original type wiper motor has a dark grey washer pump cover and three leads coming from the motor grommet.

Both systems use a tandem wipe pattern, however the depressed park system incorporates an articulated arm on the drivers side, which results in increased

Fig. 47 Non-Depressed park wiper

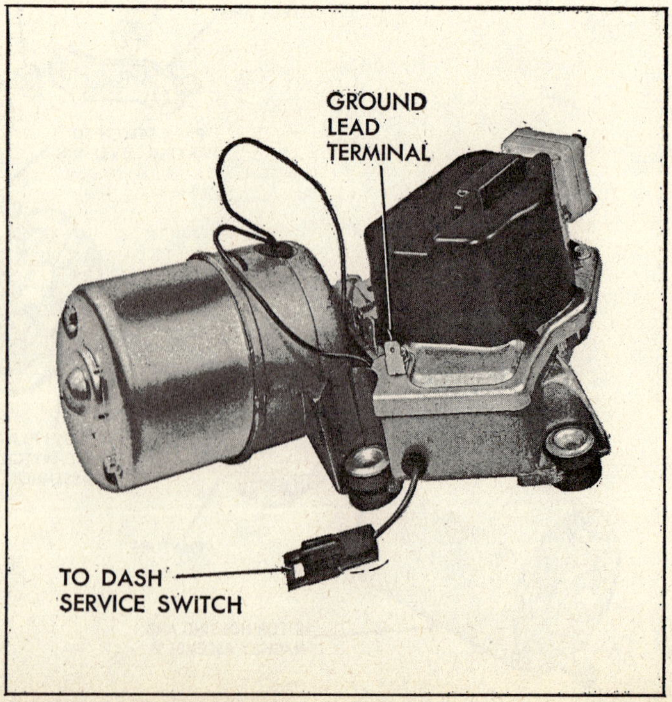

Fig. 48 Depressed park wiper. Corvette

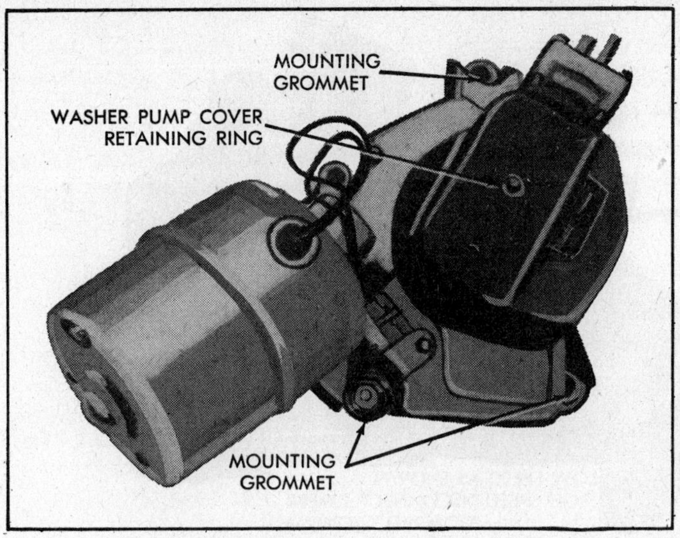

Fig. 49 Depressed park 2-speed wiper

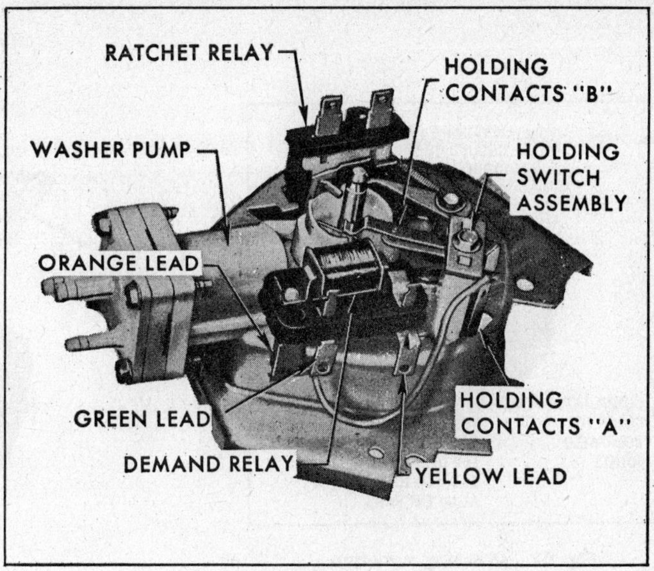

Fig. 50 Demand wiper assembly

wiper coverage for maximum forward visibility.

TROUBLE DIAGNOSIS

TWO SPEED-RECTANGULAR MOTOR
Wiper On Car Fig. 51

NOTES: *Important:* Ignition switch must be "ON" for all electrical tests. Wiper operated correctly independently of car wiring and switch.

Wiper inoperative or intermittent

1. Blown fuse.
2. Open circuit in feed wire (No. 2 terminal on wiper motor).
3. Loose mounting of wiper switch.
4. Open circuit in wire to wiper switch (No. 1 terminal on wiper motor).
5. Defective wiper switch.

Wiper will not shut off

Wiper has both "LO" and "HI" speeds

1. Grounded wire (No. 1 terminal on wiper motor) to wiper switch.

Wiper has "LO" speed only

1. Defective wiper switch.
2. Grounded wire (No. 3 terminal on wiper motor to wiper switch).

Wiper has "HI" speed only

1. Defective wiper switch.
2. Open circuit in wire (No. 3 terminal on wiper motor) to wiper switch.

Wiper has "HI" speed only

1. Open circuit in wire (No. 3 terminal

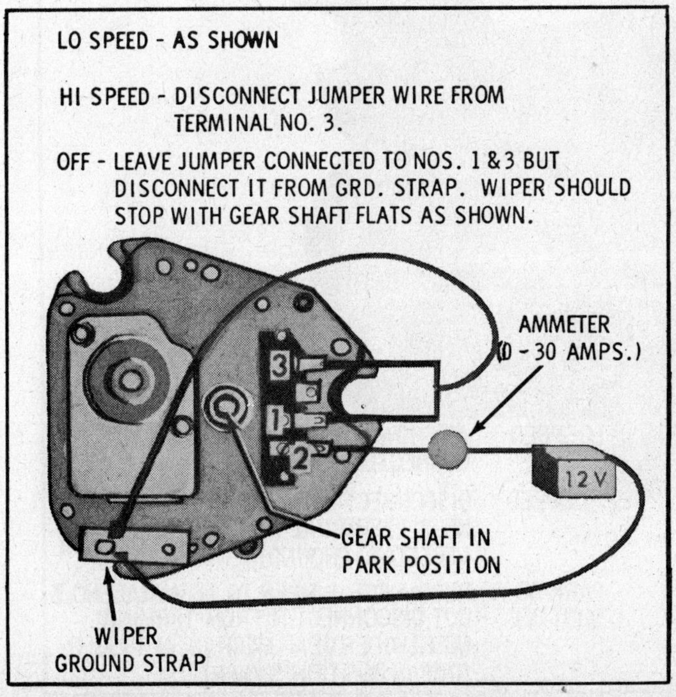

Fig. 51 Jumper wire connections

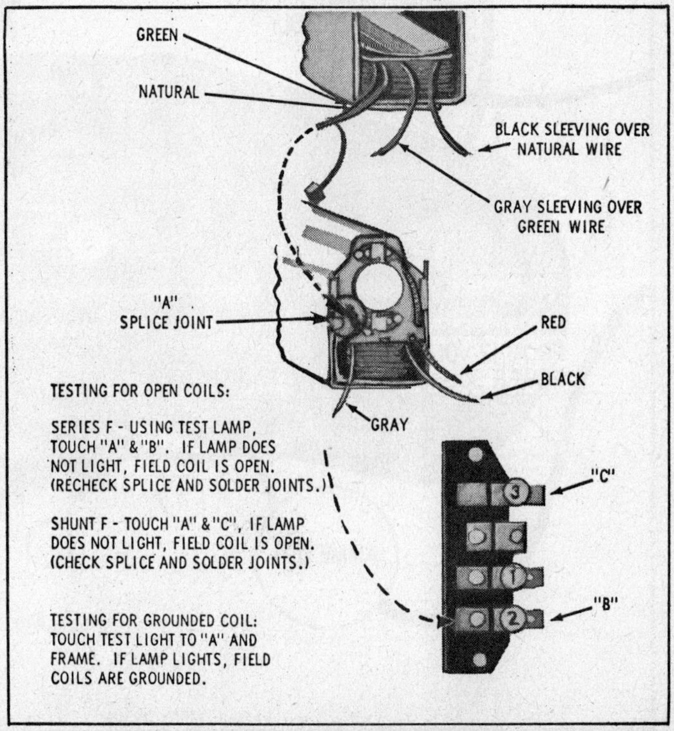

Fig. 53 Testing field coils

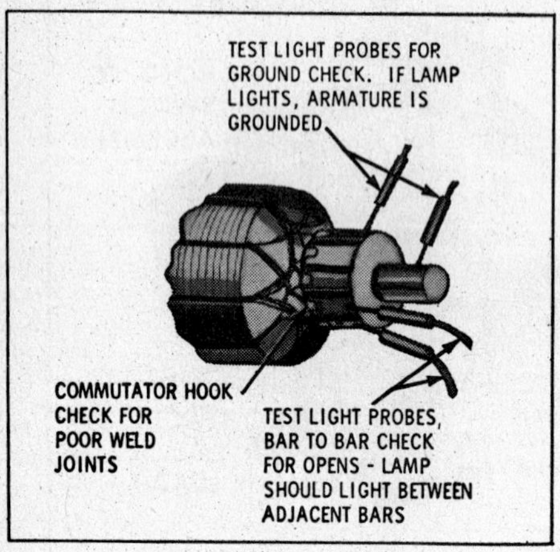

Fig. 52 Checking armature

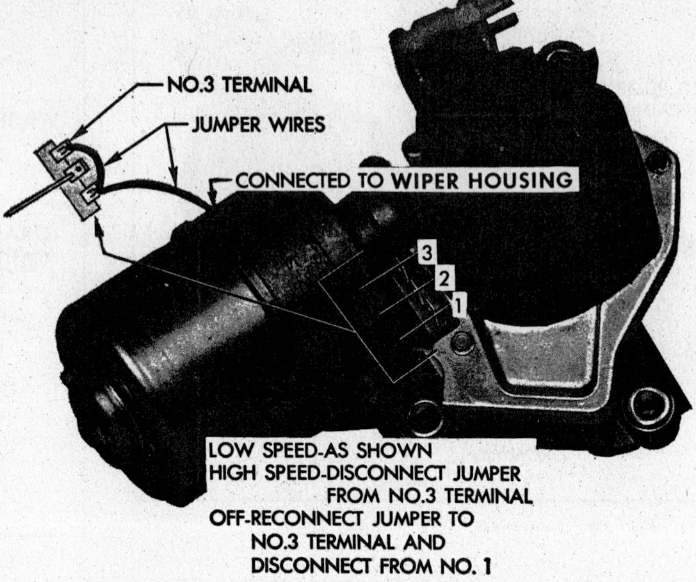

Fig. 54 Connections to operate wiper independent of instrument panel switch

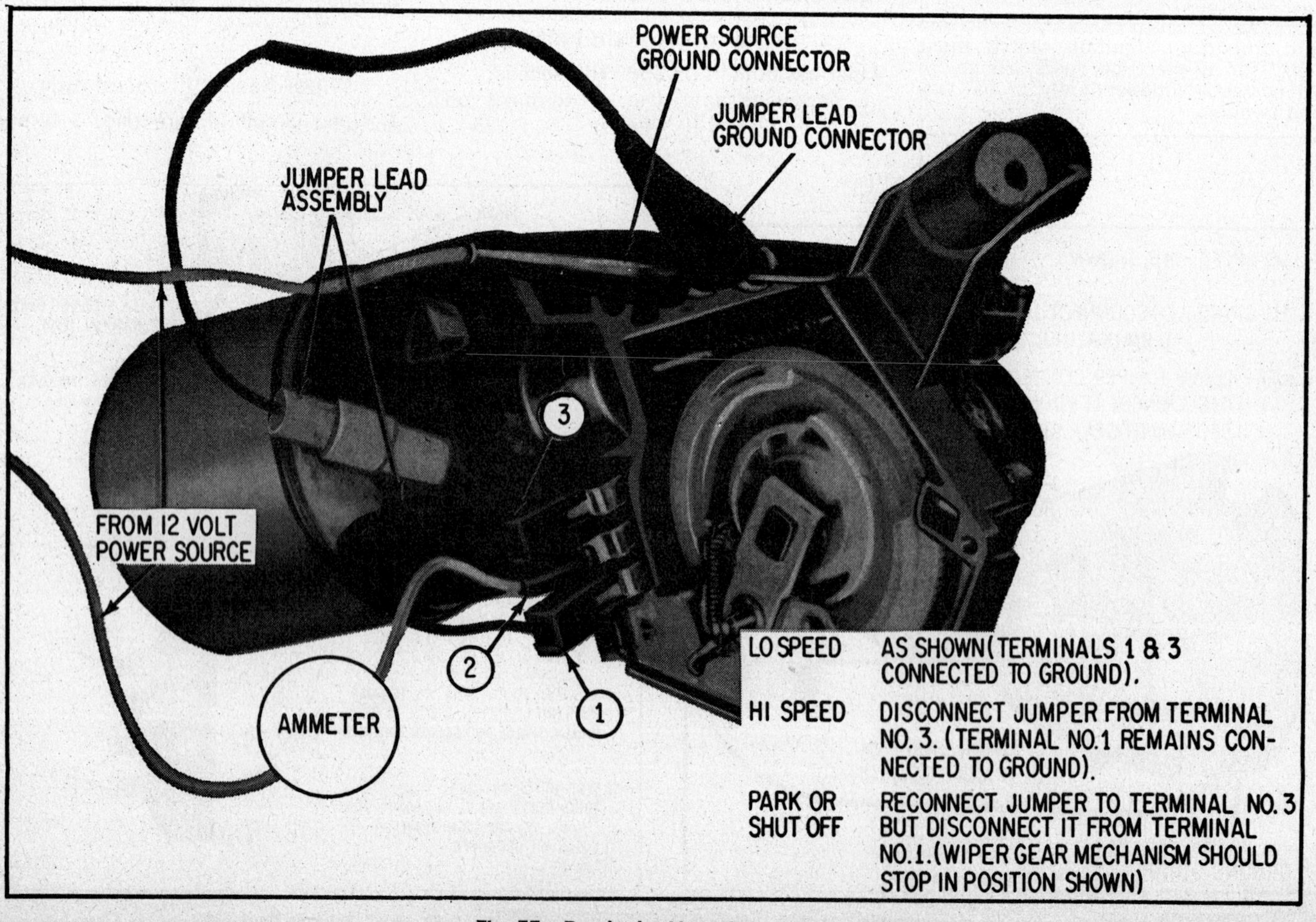

Fig. 55 Bench checking wiper motor

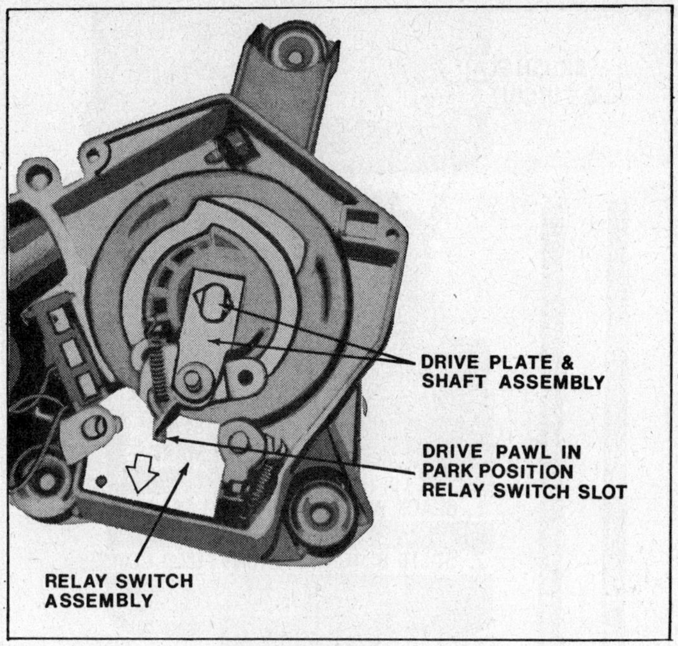

Fig. 56 Testing relay switch

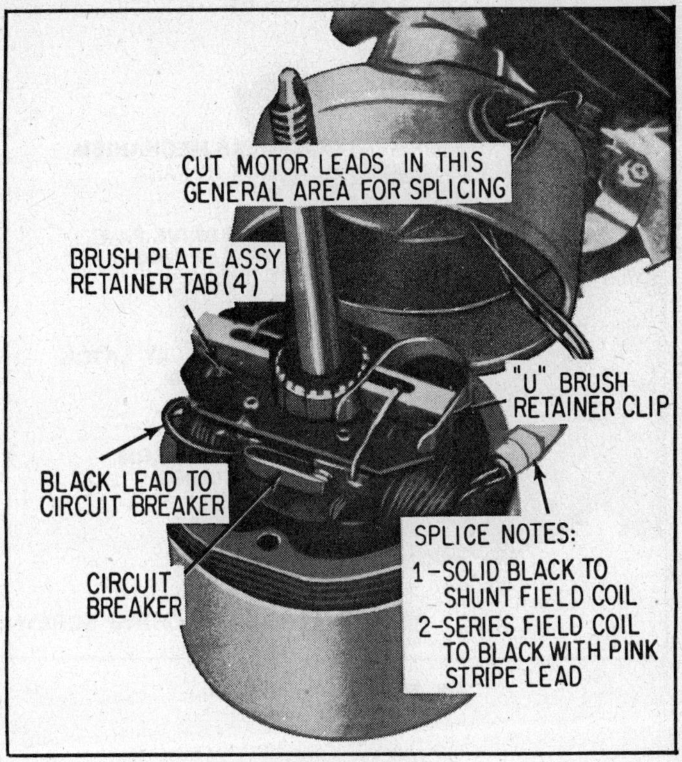

Fig. 57 Brush plate assembly

on wiper motor to wiper switch).
2. Defective wiper switch.

Wiper has "LO" speed only

1. Grounded wire (No. 3 terminal on wiper motor to wiper switch).
2. Defective wiper switch.

Blades do not return to full park position

1. Loose wiper motor ground strap connection.

Wiper Off Car (Refer to Figs. 51, 52, 53)

Wiper inoperative or intermittent

1. Broken or damaged gear train (only if inoperative).
2. Poor solder connections at terminal board.
3. Loose splice joints at brush plate.
4. Brushes binding in brush holder.
5. Open circuit in armature.

Wiper will not shut off (Wiper has normal "HI" and "LO" speed)

1. Defective park switch.
2. Grounded red lead wire.

Wiper will not shut off (Wiper has "LO" speed only)

1. Grounded shunt field coil, Fig. 53.
2. Grounded black wire.

Wiper will not shut off (Wiper has "HI" speed only)

1. Open circuit in shunt field coil.
2. Open circuit in black wire.

Wiper has "LO" speed only

1. Grounded wire (No. 3 terminal on wiper motor) to wiper switch.
2. Defective wiper switch.

Blades do not return to full park position

1 Loose wiper ground strap connection.
2. Park switch defective or contacts dirty.

High speed too fast

1. Resistor defective.

TWO SPEED-ROUND MOTOR (EXC. DEMAND SYSTEM)

Wiper on Car, Fig. 54

NOTES: *Important:* Ignition switch must be "on" for all electrical tests.
Wiper operated correctly independently of car wiring and switch.

Wiper inoperative or intermittent

1. Open lead wire from wiper terminal No. 1 to wiper switch.
2. Wiper switch not securely mounted.
3. Wiper switch defective.

Wiper will not shut off (Blades make full wipe stroke)

1. Grounded lead wire from wiper terminal No. 1 to wiper switch.
2. Corroded wiper terminals.
3. Defective wiper switch.

Wiper will not shut off (Blades move up and down about 15 degrees from park position)

1. Open in lead wire from wiper terminal No. 3 to wiper switch.
2. Wiper switch mounting loose.
3. Wiper switch defective.

"Hi" speed only

1. Open lead wire from wiper terminal No. 3 to wiper switch.
2. Wiper switch defective.

"Lo" speed only

1. Grounded lead from wiper terminal No. 3 to wiper switch.
2. Defective wiper switch.

"Hi" speed in "Med" position (Cadillac only)

1. Open medium speed resistor.

NOTE: The following checking procedures should be used in conjunction with "Wiper Off Car" (round motor) trouble diagnosis.

Wiper inoperative

1. Remove washer pump to gain access to relay switch assembly.
2. Connect 12-volt source to wiper, feed side to center terminal, ground side to gear housing, Fig. 55. Do not connect jumper to terminal 1 and 3.
3. To determine if relay coil is open, connect test lamp to wiper terminal No. 1. Test lamp should light.
4. Test relay switch as follows:
   a. If gear mechanism is in full park position insert a small screwdriver into the switch slot (between the drive pawl and the relay latch arm) and push relay latch arm downward and toward the relay coil in direction of the arrow, Fig. 56. Next, remove a small amount of insulation from black lead with pink tracer and touch test lamp to exposed wire.
   b. If test lamp lights but motor doesn't run, proceed to step 5.
   c. If test lamp doesn't light, relay switch defective.

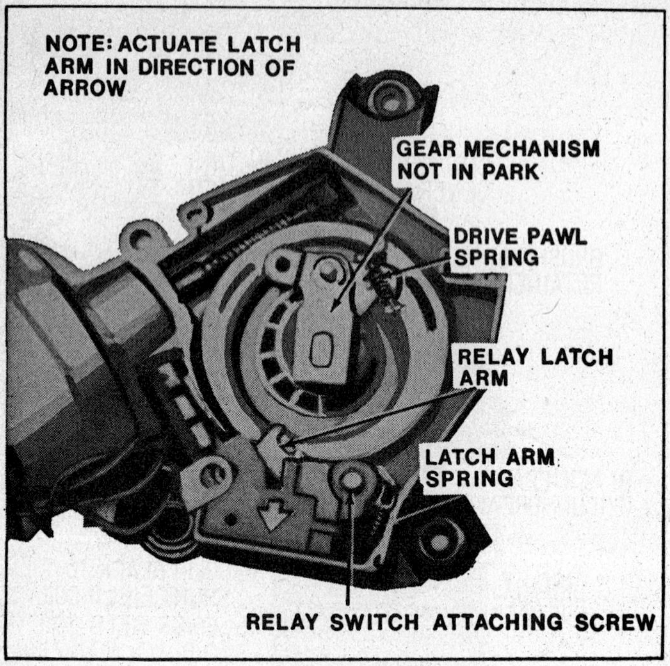

Fig. 58 Latch mechanism

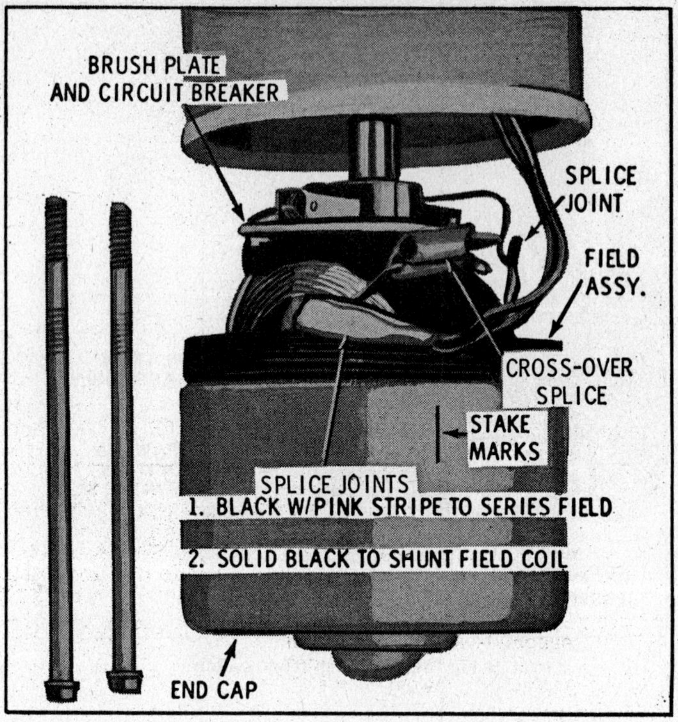

Fig. 59 Field and end cap assembly

NOTE: Cover exposed wire with tape after the test.

5. Disassemble motor section and check the following:
   a. Hung brush.
   b. Solder connections at brush holders.
   c. Splice joints at field coil connections to leads.
   d. Open armature.
   e. Circuit breaker ground connection on field lamination.
   f. To check the circuit breaker, disassemble motor section of wiper and visually inspect the circuit breaker for dirty or burned contacts or solder connections to circuit breaker terminals, Fig. 57.

Wiper will not shut off crankarm rotates 360 degrees

1. Observe if relay latch arm spring is connected properly, Fig. 58.
2. Manually operate latch arm to check it for possible binding condition.
3. If items in 1 and 2 check out, connect power source to wiper and connect jumper wire from terminal No. 3 to wiper housing. DO NOT make any connections from terminal No. 1. Manually actuate latch arm in direction of arrow, Fig. 58, and observe if it remains in energized position (inside plastic switch housing out of path of gear drive pawl). If it remains in energized position, check for grounded red lead from coil to terminal No. 1. If red lead is not grounded, coil is probably grounded internally and relay switch should be replaced.

Wiper will not shut off recycles

NOTE: Crankarm oscillates in a somewhat horizontal plane and is accompanied by a loud "knock" with each revolution of the gear.

1. Check that drive pawl and relay latch arm springs are properly connected, Fig. 58.
2. Check wiper for "LO" speed operation, Fig. 55. If wiper has "HI" speed only, check the following items:
   a. Solder joint at No. 3 wiper terminal.
   b. Splice joint-field coil cross-over splice, Fig. 59.
   c. Splice joint-black lead to field coil.
3. Check relay switch as follows:
   a. Remove small amount of insulation from black lead with pink stripe and connect test light between exposed wire and wiper housing.
   b. Connect positive side of power source to terminal No. 2 and negative side to motor case. Install jumper wire from terminal No. 1 to motor case. Observe if test light goes out once for each revolution of gear or if light glows steadily. If light glows steadily, relay switch contacts are not opening and switch is defective. If light goes out each time drive pawl moves into relay switch slot, relay switch is functioning correctly.

Wiper has one speed, "LO"

1. Check for grounded condition in the internal black lead that connects to wiper terminal No. 3, Fig. 55.
2. Disassemble motor section of wiper and check for grounded shunt field coil, Fig. 60.

Wiper has excessive speed in "HI" but "LO" speed is normal

1. Check for open 20 ohm resistor and the resistor ground connection, Fig. 61.

Intermittent operation

1. Check solder connections at wiper terminal board.
2. Connect wiper to operate in "LO" speed, Fig. 55. Connect ammeter (range 0-30 amps) in feed wire circuit to wiper and observe current draw. Allow motor to run until it becomes hot see Specification Chart, Fig. 72.
   a. If current draw is normal and wiper cycles on and off, a weak circuit breaker is indicated. Replace brush plate assembly.
   b. If current draw exceeds specification, proceed to Steps 3, 4 and 5.
3. Adjust armature end play as required and recheck current draw.
4. Adjust gear assembly end play as required and recheck current draw.
5. If adjustments in Step 3 and 4 fail to correct excessive current draw condition, disassemble motor section of wiper and check armature on growler for shorted or grounded condition.

Wiper off Car

Wiper inoperative (Motor doesn't run)

1. Open relay coil.

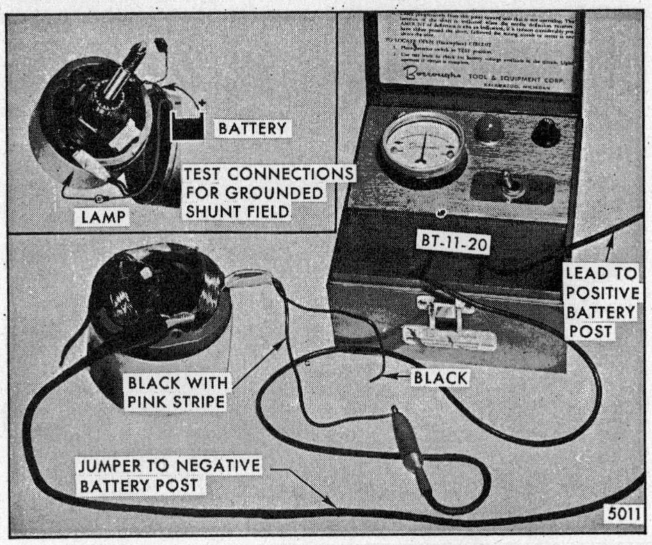

Fig. 60 Testing field coils

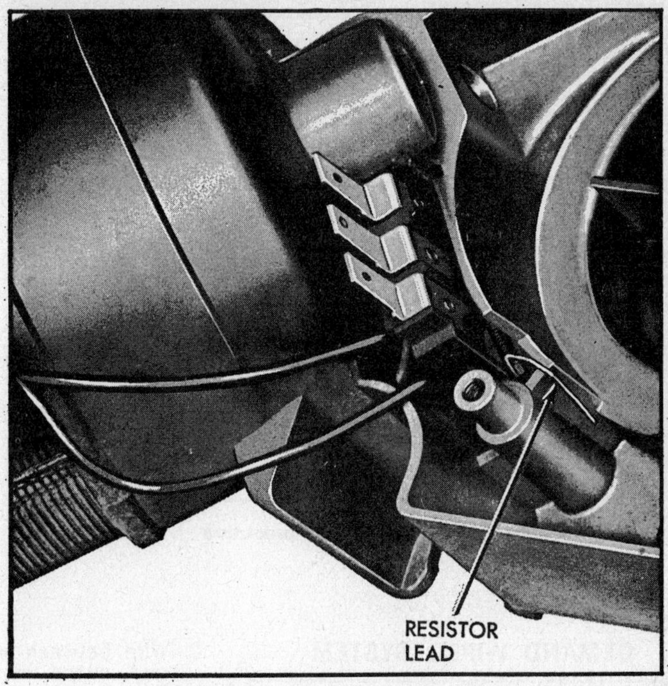

Fig. 61 Terminal board resistor

2. Circuit breaker open.
3. Open armature.
4. Motor series field open.
5. Brushes sticking.
6. Defective solder joints-relay switch.
7. Binding condition-relay latch arm.

Wiper will not shut off (Crankarm rotates through 360 degrees)

1. Relay coil grounded.
2. Relay latch spring disconnected or broken.
3. Latch arm binding.

Wiper will not shut off (Crankarm moves back and forth in a horizontal plane accompanied by a loud "knock")

1. Relay switch contacts shorting together.
2. Drive pawl spring disconnected.
3. Wiper has one speed, "Hi", caused by open shunt field.

Wiper has one speed, "Hi"

1. Shunt field open.
2. Defective soldering at terminal No. 3 on wiper terminal board.

Wiper has one speed, "Lo"

1. Shunt field internally grounded.
2. Shunt field lead to terminal board (black) grounded.
3. Shorted armature.

Wiper has excessive speed in "Hi"; "Lo" speed normal

1. Open speed resistor.
2. Poor resistor ground connection.

Wiper stops at random (Crankarm stops rotating immediately and does not return to full park position

1. Relay switch contacts dirty or broken.

Intermittent operation

1. Defective circuit breaker (weak).
2. Circuit breaker tripping because of shorted armature and/or fields causing motor to draw excessive current.

No apparent trouble on bench test but fails occasionally on car

1. Armature end play tight.
2. Gear assembly end play tight.
3. Loose solder or weld joints.

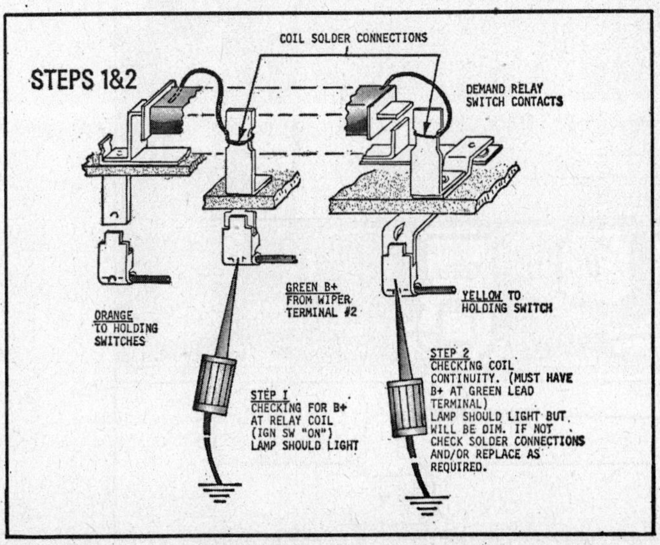

Fig. 62 Checking procedure

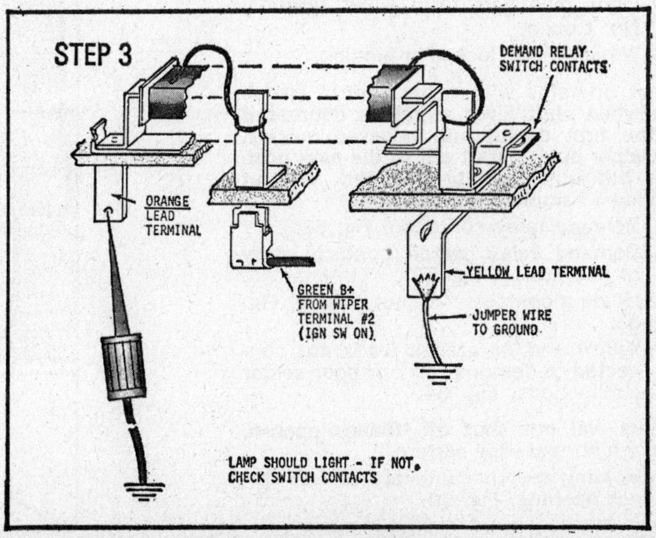

Fig. 63 Checking procedure

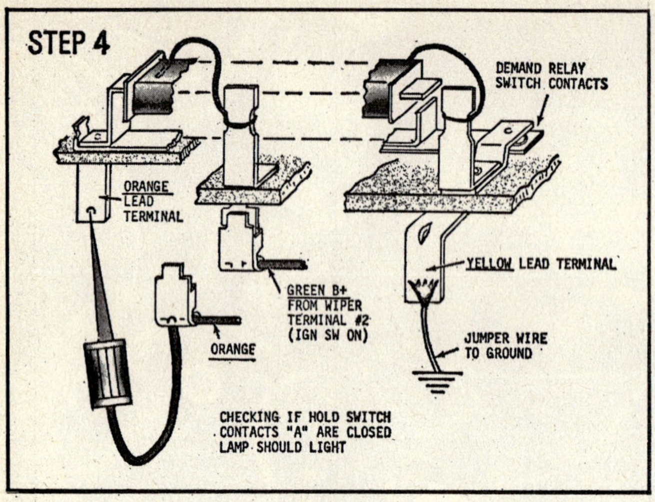

Fig. 64 Checking procedure

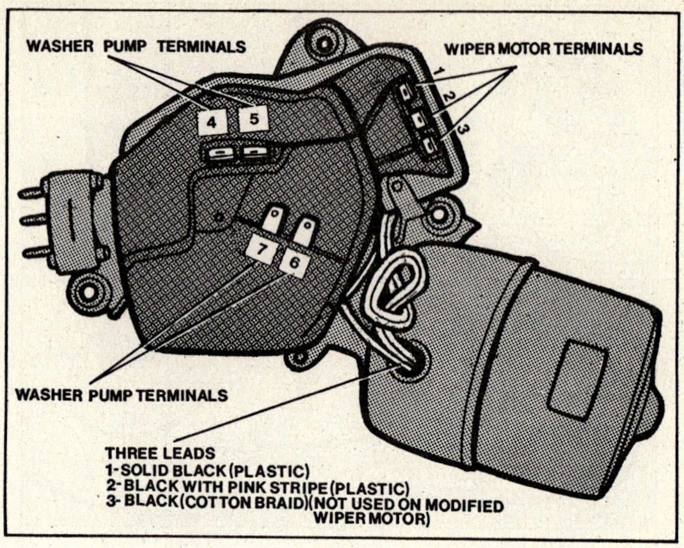

Fig. 65 Pulse wiper leads and terminals

DEMAND WIPER SYSTEM

NOTE: Unless otherwise stated, the following procedures can be performed with the wiper motor installed on car.

Wiper motor operates correctly with wiper control switch but will not start when shift lever switch is used in either detent position

1. Shift lever switch ground connection loose.
2. Light blue wire to shift lever switch open.
3. Shift lever switch defective.

Wiper inoperative when either the wiper control or shift lever switches are used, Fig. 62

NOTE: If wiper inoperative after following checks refer to "Wiper Off Car" (round motor) diagnosis.

1. Blown fuse.
2. Light blue wire from wiper terminal No. 1 open.
3. Wiper linkage locked or binding.

Wiper operates with wiper control switch but when shift lever switch is depressed to the first detent and released quickly, the wiper blades start out of the park position but quickly return to park without making a complete wiping stroke

1. Demand relay coil open, Fig. 62.
2. Demand relay switch contacts dirty or not closing, Fig. 62.
3. Holding contacts "A" not closing, Fig. 63.
4. Yellow and/or orange leads not connected to demand relay or poor solder connections, Fig. 64.

Wipers will not shut off. Blades operate through normal wipe pattern

1. Holding switch contacts "A" or "B" not opening, Fig. 50.
2. Wire between wiper terminal No. 1 and wiper control switch or shift lever switch grounded.
3. Defective wiper control or shift lever switch.
4. Wiper gear box relay coil grounded.

Wipers won't shut off. Blades move up and down approximately 15 degrees in and out of park position and wipers have "HI" speed only

1. Wiper motor defective.
2. Shunt field open.
3. Wire between terminal No. 3 and wiper control switch open.
4. Wiper control switch defective.

Wipers have "Lo" speed only and shut off correctly

1. Defective wiper control switch.
2. Black wire from wiper terminal No. 3 to wiper control switch is grounded.
3. Wiper motor defective. Remove motor and refer to "Wiper Off Car" (round motor) diagnosis.

Intermittent operation with wiper control switch "On"

1. Loose wiper ground strap.
2. Wiper control switch loose.
3. Loose wiring connection.
4. Defective wiper motor. Remove motor and refer to "Wiper Off Car' (round motor) diagnosis.

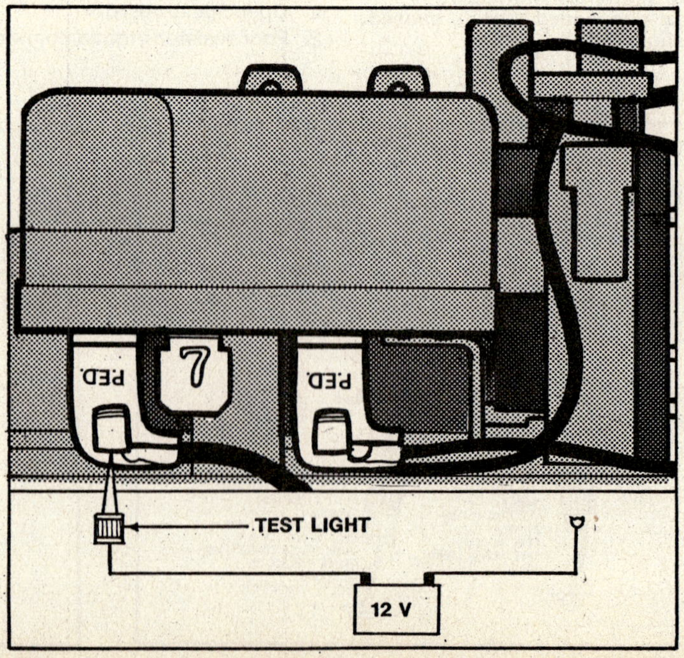

Fig. 66 Test lamp connection, wiper will not pulse (runs continuously)

PULSE WIPER SYSTEM EXC. MODIFIED PULSE TYPE

Wiper Inoperative In All Modes

Test No. 1

1. With ignition "On," check for voltage at wiper terminal No. 2, Fig. 65. If voltage is not present, check for a blown fuse or an open circuit in the fuse block to wiper wiring.
2. If voltage is present in Step 1, connect a jumper wire between wiper terminal No. 1 and the ground and a second jumper wire between wiper terminal No. 6 and the ground, Fig. 65. Turn ignition "On" and place wiper switch in "Lo." If wiper fails to operate, proceed to Test No. 2.
3. If wiper operates in Step 2, disconnect jumper wire from wiper terminal No. 1. If wiper then shuts "Off," check for: an open circuit in wiring between wiper terminal No. 1 and dash switch, an improper dash switch ground wire connection or a defective dash switch.
4. If wiper continues to run in Step 3, disconnect jumper wire from wiper terminal No. 6. If wiper then shuts "Off," check for: an open circuit in wiring between wiper terminal No. 6 and dash switch or a defective dash switch.
5. If wiper continues to run in Step 4, check for: improper motor or dash switch ground connection or any other improper wiring connections that could cause an intermittent condition.

Test No. 2

1. Connect a test lamp between the black wire with pink strip and the ground, Fig. 65. Turn ignition "On" and wiper switch to "Lo." If test lamp does not light, the gear box relay is defective, requiring replacement.
2. If test lamps lights in Step 1, remove washer pump cover, then re-install wiring on wiper-washer unit. Disconnect black wire with cotton braid insulation, Fig. 65, and connect a jumper between the disconnected black wire and the ground. Turn ignition "On" and place wiper switch in "Lo." If wiper motor runs, check all connections at pulse relay. If connections are satisfactory, pulse relay is defective.

Wiper Inoperative In Delay Mode

NOTE: On Buick and Oldsmobile units, wiper also inoperative in "Lo" mode.

1. Disconnect wiring connector from wiper terminals Nos. 6 and 7, then connect a 12 volt test lamp to an external power source and also to connector terminal No. 7. If test lamp lights, check wiring between wiper terminal and dash switch for grounded circuit.
2. If test lamp does not light in Step 1, disconnect wiring connectors from wiper terminals Nos. 1, 2, 3 and 6, 7, then connect an ohmmeter between connector terminals Nos. 2 and 7. Turn ignition "On" and place dash switch in all speeds. Observe ohmmeter readings. For "Minimum Delay," use the 1K scale and for "Maximum Delay," use 100 K scale. Ohmmeter should never indicate less than 400 ohms in any switch position. The "Maximum Delay" reading should be 1 megaohm (1,000,000 ohms). Also, the ohmmeter should indicate a continuous increase in resistance when dash switch is moved from "Minimum" to "Maximum" positions.
3. If correct reading was obtained in Step 2, the timing device in wiper is defective, requiring replacement. If ohmmeter readings were less than 400 ohms, replace dash switch. If problem still exists, replace timing device in wiper.

Wiper Operates Only In Delay Mode

Check for an open circuit between wiper terminal No. 6 and dash switch or a defective dash switch.

Wiper Will Not Pulse (Runs Continuously In Delay)

1. Disconnect wiring connector from wiper terminals Nos. 6 and 7, then remove lead from connector that connects to No. 6 wiper terminal. Reinstall the disconnected connector. Place dash switch in "Delay" position and turn ignition "On." If wiper operates in "Pulse" mode, check wiring between wiper terminal No. 6 and dash switch for an open circuit or determine if dash switch is defective.
2. If wiper runs steadily (will not "Pulse"), remove the washer pump cover and reconnect wiring to wiper unit terminals. Disconnect black wire with cotton braid insulation from over-ride switch terminal. Place dash switch in "Delay" and turn ignition "On." If wiper still operates, check for an improperly grounded brush lead or that black wire with cotton braid insulation is not internally.
3. If wiper does not run in Step 2, remove wiper unit from vehicle, then the washer pump unit from wiper motor gear box. Connect a 12 volt test lamp to an external power source and also to pulse relay, Fig. 66. If test lamp lights, replace pulse relay.
4. If test lamp does not light in Step 3, using a screwdriver, momentarily bridge across "Hold" switch to discharge timing device capacitor. Connect a 12 volt test lamp to an external power source and a jumper wire to pulse relay, Fig. 67. Then, manually rotate washer pump drive cam counter-clockwise and observe the "Hold" switch contacts and test lamp.
5. If contacts open and close and the test lamp remains lit, replace timing device. If contacts do not open and close properly, adjust "Hold" switch or replace switch, if defective. If test lamp does not light when contacts are closed, "Hold" switch contacts are dirty and require cleaning.

Wiper Will Not Shut Off

1. Disconnect wiring connector from wiper terminals Nos. 1, 2, 3 and remove lead from connector that connects to wiper terminal No. 1, then

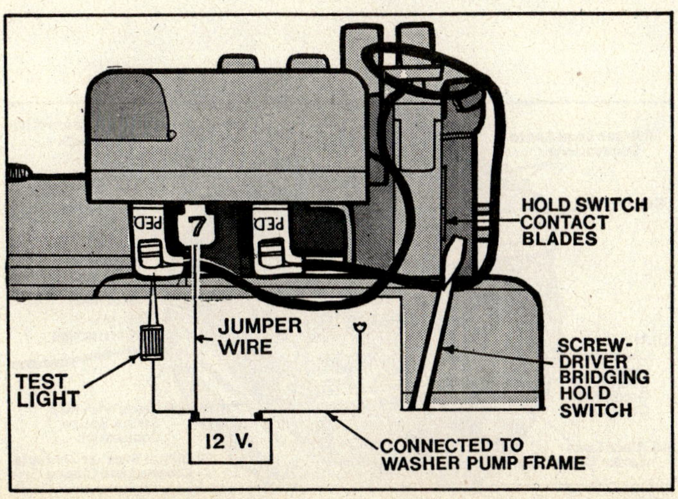

Fig. 67 Test lamp connection, wiper will not pulse (runs continuously)

Fig. 68 Disconnecting yellow lead from wiper terminal No. 1A

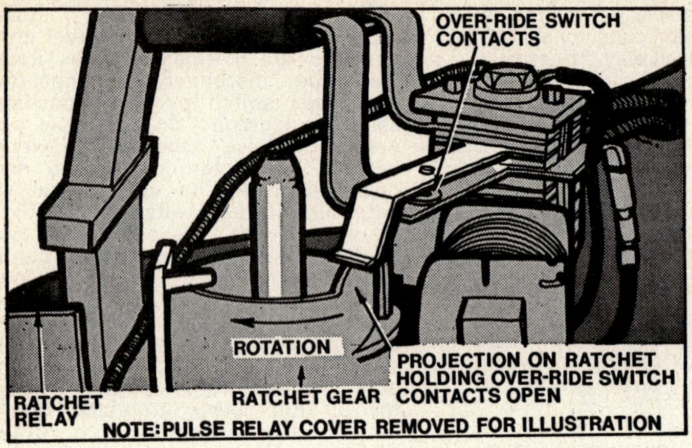

Fig. 69 Checking washer pump ratchet gear position

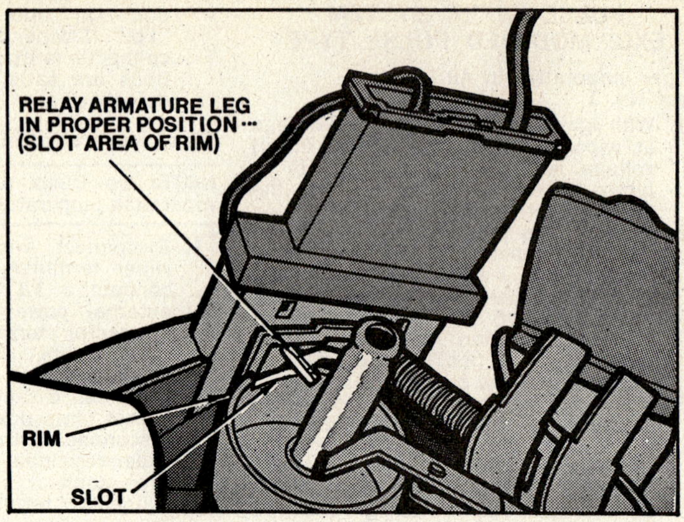

Fig. 70 Checking ratchet relay armature position

re-install connector. Place dash switch in "Off" and turn ignition "On." If wiper shuts "Off," check wiring between wiper terminal No. 1 and dash switch for a grounded circuit or a defective.

2. If wiper does not shut "Off" in Step 1, remove washer pump cover. Disconnect yellow lead from wiper terminal No. 1A, Fig. 68, then reconnect all other wiring. Place dash switch in "Off" and turn ignition "On". If wiper continues to operate, replace gearbox relay.
3. If wiper shuts "Off" in Step 2, check position of washer pump ratchet gear, Fig. 69. If ratchet gear is properly positioned and the over-ride switch contacts are closed, adjust contact gap to .025 inch by bending the lower contact. If ratchet gear is positioned properly and the over-ride switch contacts are open, ensure that yellow lead solder connection to insulator stack is not grounded.
4. If ratchet gear is not properly positioned in Step 3, check ratchet relay armature position, Fig. 70 and 71. If armature is positioned properly, perform testing procedure outlined under "Wiper Will Not Pulse".
5. If ratchet relay armature is improperly positioned in Step 4, reconnect all disconnected wiring and operate wiper unit through a "Wash" cycle. If wiper will not shut "Off" and ratchet has not rotated to open over-ride switch contacts, perform testing procedure outlined under "Wiper Will Not Pulse".
6. If wiper completes "Wash" cycle and shuts "Off" in Step 5, and the override switch contacts are actuated by ratchet gear, the armature is re-programmed to proper position. Operate system through 2 or 3 additional "Wash" cycles to ensure that no other condition caused the relay armature to become mispositioned.

Wiper Does Not Shut Off— Blades Move In & Out of Park

1. Connect a jumper wire between wiper terminal No. 3 and the ground. Place dash switch in "Off" and turn ignition "On". If wiper has "Lo" speed and shuts "Off", check wiring between wiper terminal and dash switch for an open circuit or determine if dash switch is faulty.
2. If wiper has "Hi" speed only in Step 1, remove wiper unit from vehicle and visually inspect motor windings and gearbox.
3. If wiper has both "Lo" and "Hi" speeds, replace gearbox relay.

Wiper Has Lo Speed Only

1. Disconnect wiring connector from wiper terminals Nos. 1, 2, 3 and remove lead from connector that connects to wiper terminal No. 3, then reinstall connector. Place dash switch in "Hi" and turn ignition "On". If wiper has "Hi" speed only, check wiring between wiper terminal and dash switch for a grounded circuit or determine if dash switch is defective.
2. If wiper has "Lo" speed only in Step 1, remove wiper unit from vehicle

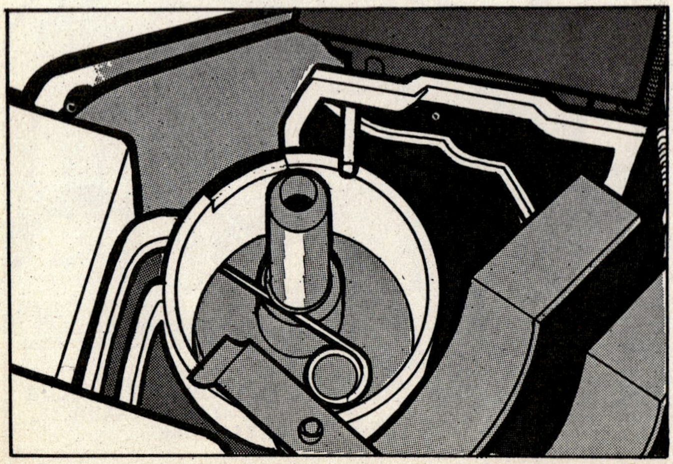

Fig. 71 Checking ratchet relay armature position

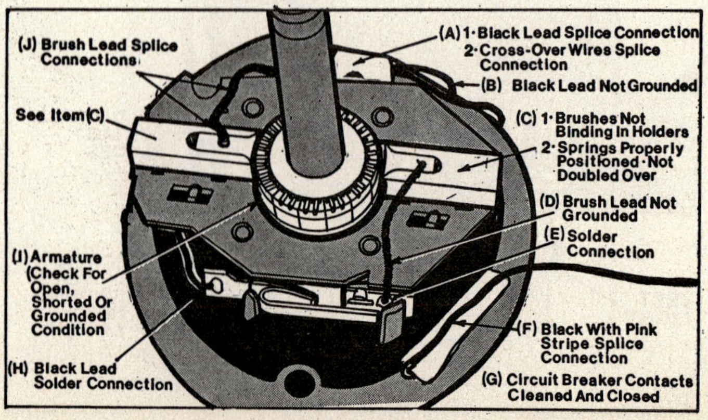

Fig. 72 Inspecting motor windings & gearbox

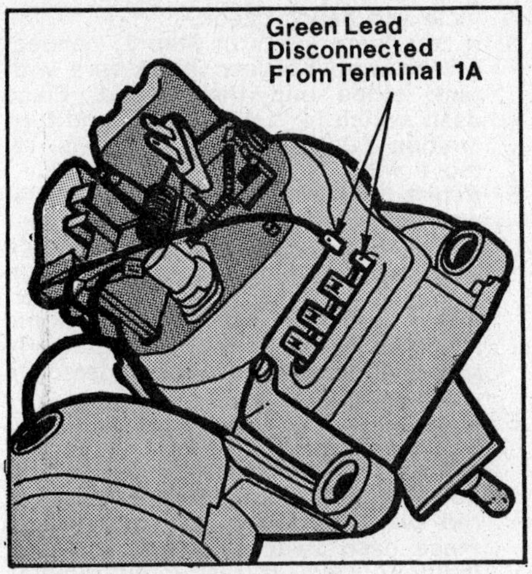

Fig. 73 Disconnecting green lead from wiper terminal No. 1A

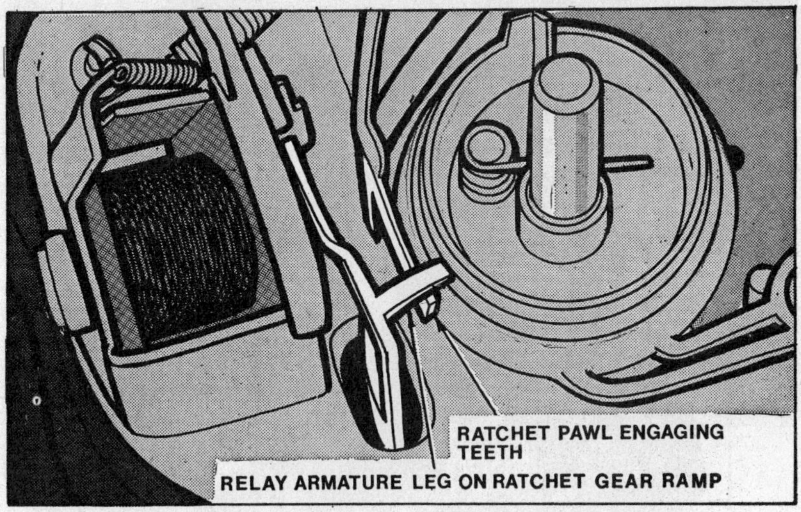

Fig. 74 Checking washer relay armature leg position No. 1

and visually inspect motor windings and gearbox, Fig. 72.

Wiper Operates Intermittently

1. Remove wiper system fuse from fuse block and connect an armature across the fuse block terminals. On Buick and Oldsmobile units, disconnect wiring connector from wiper terminals Nos. 6 and 7, then connect a jumper wire between wiper terminal No. 6 and the ground. Turn ignition "On" and place dash switch in "Lo". Operate washer and, after two minutes of wiper operation, observe ammeter and note lowest current reading.
2. If lowest current reading was 5-6.5 amps in Step 1, a weak wiper motor circuit breaker is indicated. Replace circuit breaker and recheck wiper.
3. If lowest current reading exceeded 6.5 amps in Step 1, replace wiper blade elements and repeat test. If lowest current reading now obtained is below 6.5 amps, the condition is corrected.
4. If current reading still exceeds 6.5 amps in Step 3, disconnect wiper linkage from crank arm and repeat test. If lowest current reading obtained is below 6.5 amps, check for binding in the wiper linkage.
5. If lowest current reading still exceeds 6.5 amps in Step 4, remove wiper unit from vehicle and check: armature end play, for binding gears or armature, or a shorted armature.

MODIFIED PULSE WIPER SYSTEM

Wiper Inoperative

1. Connect a test lamp to wiper terminal No. 2, Fig. 65, and turn ignition "On". If test lamp does not light, check for blown wiper fuse. If fuse is satisfactory, check for an open circuit between wiper terminal No. 2 and fuse block.
2. If fuse is blown in Step 1, disconnect wiring connector from wiper terminals 1, 2, 3, replace fuse and turn ignition "On". If fuse blows, check for a grounded circuit between wiper terminal No. 2 and fuse block.
3. If fuse does not blow in Step 2, reconnect wiring connector to wiper terminals and, with ignition "On", place dash switch in "Lo" position. If fuse blows, remove wiper from ve-

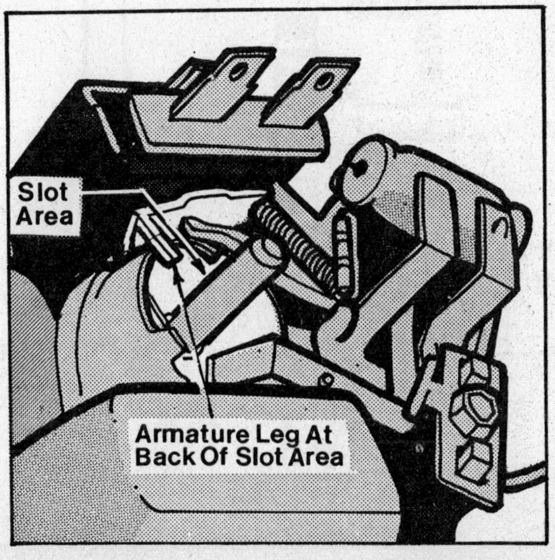

Fig. 75 Checking washer relay armature leg position No. 2

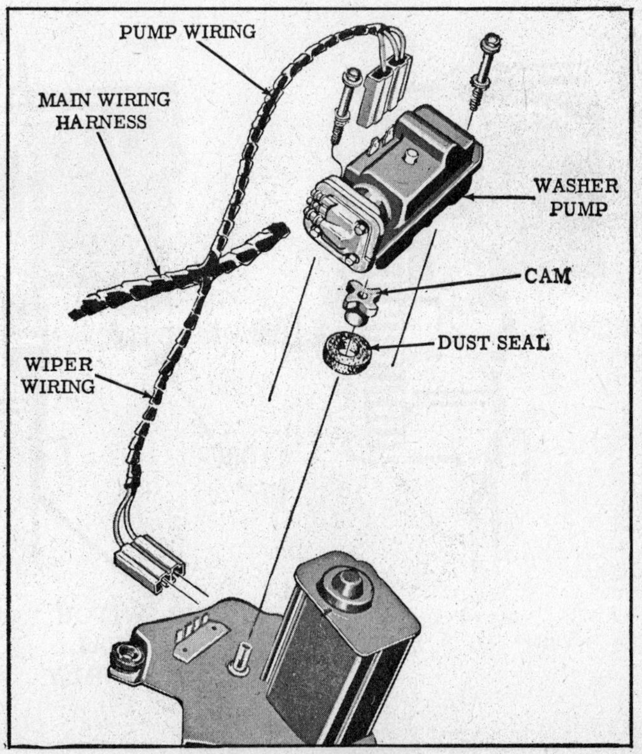

Fig. 76 Washer mounting

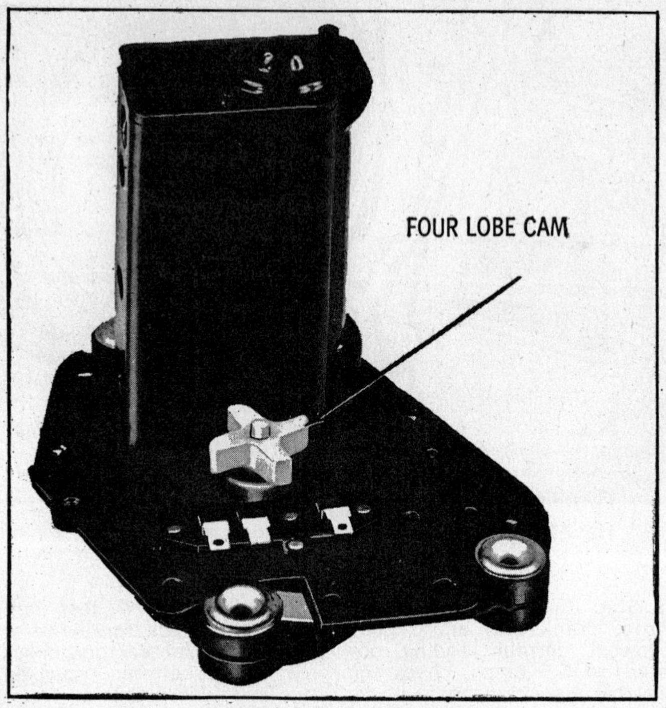

Fig. 77 Cam positioning

hicle and bench check.

4. If test lamp lights in Step 1, connect a test lamp between black wire with pink stripe and the ground. Place dash switch in "Lo" position and turn ignition "On". If test lamp lights, remove wiper unit and bench check.
5. If test lamp remains "Off" in Step 4, connect a jumper wire between wiper terminal No. 1 and the ground. Place dash switch in "Lo" and turn ignition "On". If wiper operates, check for: open circuit between wiper terminal No. 1 and dash switch, an improperly grounded dash switch or a defective dash switch.
6. If wiper remains inoperative in Step 5, remove washer pump cover and reconnect wiring to wiper unit. Connect a test lamp between pulse relay red wire terminal and the ground. Place dash switch in "Lo" and turn ignition "On". If test lamp remains "Off", replace gearbox relay.
7. If test lamp lights in Step 6, connect a jumper wire between pulse relay yellow wire terminal and the ground. Place dash switch in "Lo" and turn ignition "On". If wiper remains inoperative, replace pulse relay. If wiper operates, replace timing device assembly.

Wiper Inoperative In Delay

1. Disconnect wiring connector from wiper terminals Nos. 6 and 7, then connect a voltmeter between No. 7 connector terminal No. 7 and the ground. Place dash switch in "Mid" delay and turn ignition "On". If voltmeter indicates 12 volts, replace tim-

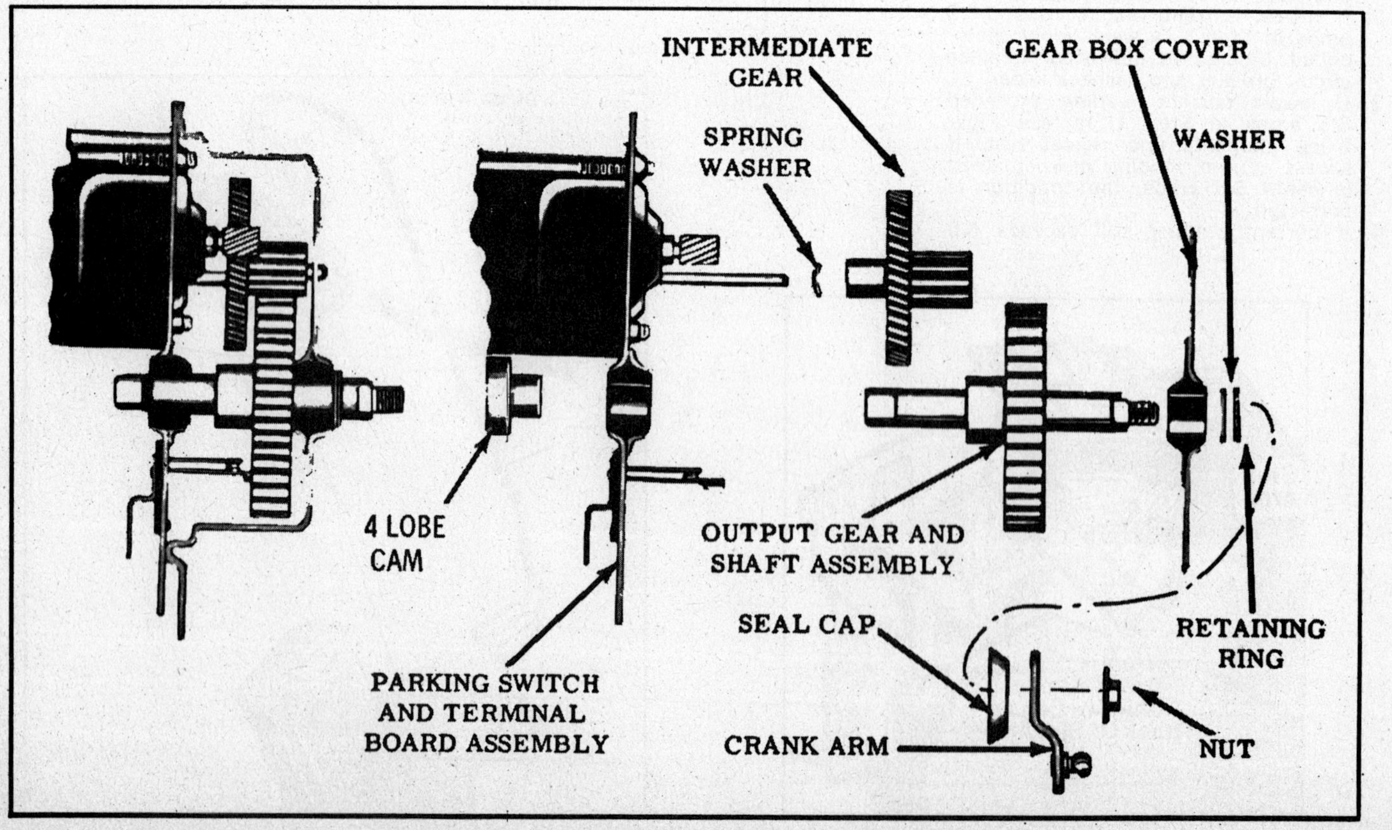

Fig. 78 Gear box

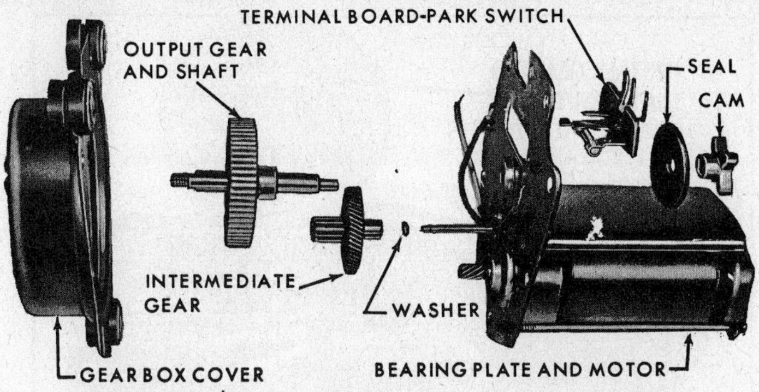

Fig. 80 Gear train and terminal board exploded view

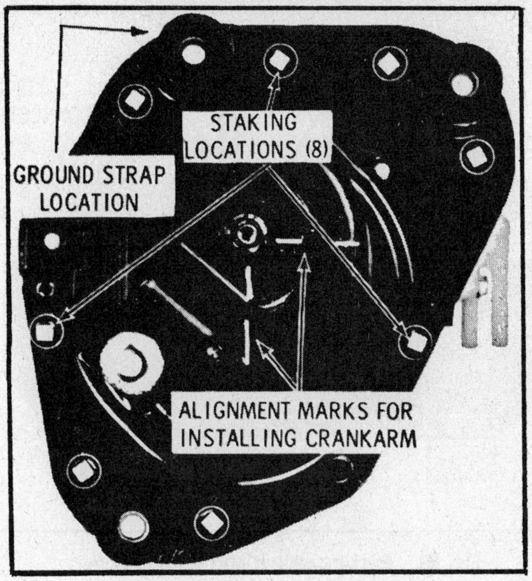

Fig. 79 Staking locations

ing device assembly.

2. If voltmeter reading in Step 1 is zero volts, check B+ circuit to dash switch, wiring between dash switch and wiper and to ensure dash switch is satisfactory.

Wiper Will Not Pulse (Runs Continuously In Delay)

1. Disconnect wiring connector from wiper terminals Nos. 6 and 7, remove lead from connector that connects to wiper terminal No. 6, then re-install connector. Place dash switch in "Mid" delay and turn ignition "On". If wiper operates properly, check for: grounded circuit between wiper terminal No. 6 and dash switch or a defective dash switch.
2. If wiper runs continuously in Step 1, remove washer pump cover, reconnect wiring and disconnect yellow lead from pulse relay. Place dash switch in "Mid" delay and turn ignition "On". If wiper is then inoperative, replace timing device. If wiper does not pulse, replace pulse relay.

Wiper Will Not Shut Off

1. Disconnect wiring connector from wiper terminals Nos. 1, 2, 3, then remove lead that connects to wiper terminal No. 1, then re-install connector. Place dash switch in "Off" and turn ignition "On". If wiper shuts "Off", check for a grounded circuit between wiper terminal No. 1 and dash switch.
2. If wiper continues to run in Step 1, remove washer pump cover and reconnect all disconnected wiring. Disconnect green lead from wiper terminal 1A, Fig. 73, place dash switch in "Off" and turn ignition "On". If wiper continues to run, replace gearbox relay.
3. If wiper shuts "Off" in Step 2, check if ratchet gear is opening over-ride switch contacts. If gear is correctly positioned and the switch contacts are open, check the disconnected green leads for grounds.
4. If ratchet gear is properly positioned and the switch contacts are closed, torque mounting screw to 14 inch lbs., and adjust contact gap to .025 inch.

Wiper Will Not Shut Off & Washer Runs Continuously

1. Remove washer pump cover and reconnect wiring. Place dash switch "Off" and turn ignition "On". Check position of washer relay armature leg with unit operating. If relay armature is in No. 1 position, Fig. 74, check that ratchet gear dog spring engages gear teeth. Tighten mounting screw or replace as required.
2. If relay armature leg is in No. 2 position, Fig. 75, push wash button and check if relay armature legdrops on armature leg ramp and pump completes wash cycle and wiper shuts "Off" in position 1. If relay armature leg drops back into slot, check for loose coil on pole piece or a missing relay armature spring.
3. If relay armature leg drops on ramp, pump completes cycle and wiper shuts "Off" in Step 2, check for loose coil on pole piece, a missing relay armature or if a burr is present on end of armature leg, file off burr and recheck unit.

Wiper Will Not Shut Off—Blades Move In & Out Of Park

1. With ignition "On", check wiper operation in "Lo" and "Hi" speeds. If wiper operates in both speeds, replace gearbox relay.

NOTE: Wiper will not delay if operates only in "Hi" speed.

2. If wiper operates in "Hi" speed only in Step 1, connect a jumper wire between wiper terminal No. 3 and the ground. Place dash switch in "Lo" and turn ignition "On". If wiper still has "Hi" speed only, remove wiper from vehicle and check for an open in the shunt field.
3. If wiper has "Lo" speed in Step 2, check for an open circuit between wiper terminal No. 3 and dash switch or a defective dash switch.

Wiper Has Hi Speed Only

1. Connect a jumper wire between wiper terminal and the ground. Place dash switch in "Lo" and turn ignition "On". If wiper still has "Hi" speed only, remove wiper from vehicle and check for open shunt field.
2. If wiper has "Lo" speed in Step 1, check for open circuit between wiper terminal No. 3 and dash switch or a defective dash switch.

Wiper Has Lo Speed Only

1. Disconnect wiring connector from terminals 1, 2, 3 and remove lead that connects to wiper terminal No. 3, then re-install connector. Place dash switch in "Hi" and turn ignition "On". If wiper still has "Lo" speed only, remove wiper from vehicle and check for a grounded shunt field.
2. If wiper has "Hi" speed in Step 1, check for grounded circuit between wiper terminal No. 3 and dash switch or a defective dash switch.

Wiper Operates Intermittently

1. Remove wiper system fuse from fuse block and connect an ammeter across fuse block terminals. Place dash switch in "Lo" and turn ignition "On". Operate washer and after two minutes of operation, observe ammeter for lowest current reading. If lowest current reading was 5-6.5 amps, a weak circuit breaker is indicated. Replace brush plate and circuit breaker assembly in wiper.
2. If lowest current reading exceeds 6.5 amps, replace wiper blade elements and repeat Step 1. If current draw is less than 6.5 amps, condition is corrected.
3. If lowest current reading in Step 2 still exceeds 6.5 amps, disconnect wiper linkage from crank arm and repeat check. If current draw is below 6.5 amps, check for binding wiper linkage. If current draw still exceeds 6.5 amps, check: armature end play, binding gear or armature, or a shorted armature.

DISASSEMBLY & ASSEMBLY PROCEDURES

The disassembly-reassembly procedures for the wiper are broken down into two major areas: the gear box section and motor section. Round motor sections may be serviced independently.

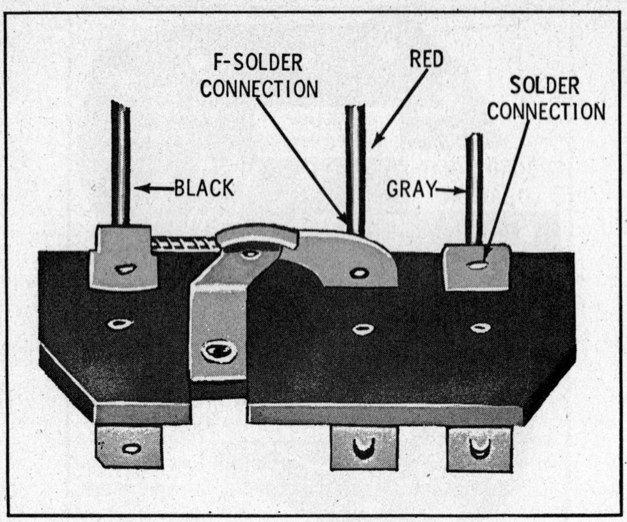

Fig. 81 Parking switch and terminal board

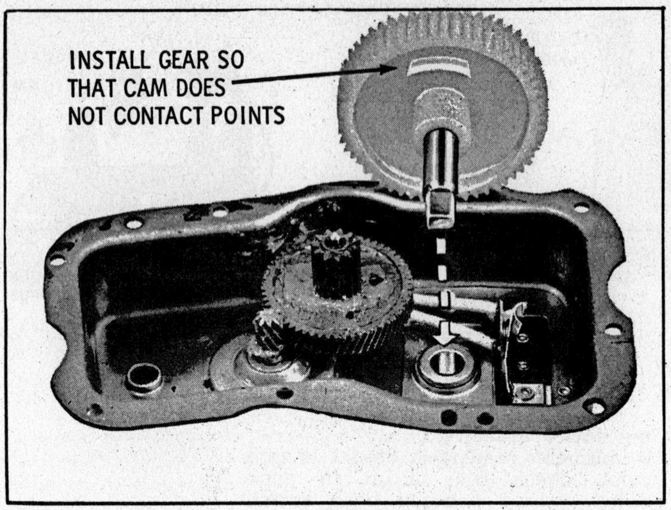

Fig. 82 Gear Installation

Two Speed-Rectangular Motor

Gear Box Disassembly

1. Remove washer pump as follows:
   a. Remove the two washer pump mounting screws, Fig. 76, and carefully lift the washer pump off the wiper motor.
   b. Remove the washer pump four lobe drive cam, Fig. 77. The cam is a press fit on the wiper gearshaft and it may be necessary to use a puller.
   c. Remove the felt washer from the wiper gearshaft.
2. Clamp crankarm in a vise and remove crankarm retaining nut.

NOTE: Failure to clamp crankarm may result in stripping of wiper gears.

3. Remove crankarm, seal cap, Tru-Arc retaining ring and end play washers, Fig. 78.

NOTE: Seal cap should be cleaned and repacked with a high melting point, waterproof type medium grade grease before reassembly.

4. Drill or punch out rivets that secure gear box cover, Fig. 79. Use a 9/32″ drill.

NOTE: Mark ground strap location and save ground strap for reassembly.

5. Remove output gear and shaft assembly: then, slide intermediate gear and pinion assembly off shaft.
6. If required, remove terminal board and park switch assembly as follows:
   a. Note position of motor leads on terminals, then unsolder.
   b. Drill out rivets that secure terminal board and park switch ground strap to plate. Use 7/64″ drill.

NOTE: Screws, nuts and washers for attaching a replacement terminal board park switch assembly are included with the replacement assembly.

Gear Box Reassembly

NOTE: Lubricate all gear teeth with Multifak No. 2 or Prestige No. 42 or equivalent.

1. If park switch and terminal board assembly were removed, Fig. 80, reinstall replacement assembly using the attaching screws and nuts included in the service package. Resolder leads to terminals, Fig. 81.
2. Install wave washer and intermediate gear on intermediate gearshaft.
3. Install output gear and shaft assembly with cam at least 90° away from the park switch position, Fig. 82.
4. Assemble gear box cover to wiper. Be careful to locate cover over locating dowels and intermediate gearshaft.
5. Secure cover to gear housing or plate over dowels. Be sure to reinstall ground strap.

NOTE: Screws, nuts and lockwashers for reassembling cover to wiper are contained in the service repair package.

6. Reassemble end play washers and retaining ring over output gearshaft, Fig. 78. Use end play washers as required to obtain .005″ maximum end play.
7. Install seal cap.
8. To reassemble crankarm in proper po-

Current Draw (AMPS)

Car	Year	No Load Low	No Load High	On Car Dry W/S① Low	On Car Dry W/S① High	Stall
Buick	1969-70	3.6 Max.		4.5 Max.		12 Max.
	1971-75	4.0 Max.	3.5 Max.			12 Max.
Chevrolet	1969	3.6 Max.		4.5 Max.		12 Max.
	1970	4.0 Max.	3.5 Max.			14 Max.
	1971-75	4.0 Max.	3.5 Max.			12 Max.
Oldsmobile	1969-70	3.6 Max.		4.5 Max.		12 Max.
	1971-75	4.0 Max.	3.5 Max.			12 Max.
Pontiac	1969	4.5 Max.	3.5 Max.	5.0 Max.	4.0 Max.	28 Max.
	1970	5.0 Max.	4.0 Max.	5.5 Max.②	4.5 Max.②	18 Max.
	1971-75	4.0 Max.	3.5 Max.			12 Max.

①—Current draw specifications listed for on car check are subject to fluctuations depending upon the condition of blades, windshield, battery, engine running or turned off. The figures shown are only a guide and reflect the average current draw throughout a complete wiping stroke.
②—Wet windshield.

Fig. 83 Specification Chart Rectangular Motor

sition, operate wiper to park or OFF position, Fig. 51, and install crankarm so that index marks on crankarm line up with those on the gear box cover, Fig. 79.

NOTE: Clamp crankarm in vise before securing the retaining nut.

9. Operate wiper, Fig. 51, and check performance per data in specification table, Fig. 83.
10. Reassemble washer pump to wiper, reversing disassembly steps 1a through 1c and observe precautions listed below.
    a. Support crankarm end of wiper output shaft (threaded end) and using a suitable mallet, drive the four lobe cam on the wiper output shaft until it bottoms against the shoulders of the shaft flats, Fig. 84.
    b. Position four lobe cam, Fig. 77. It may be necessary to manually rotate crankarm.

Fig. 84 Install washer pump drive cam

Motor Disassembly

1. Disassemble gear box as required to gain access to internal solder connections at wiper terminal board and unsolder motor leads from terminals.

NOTE: Step 1 necessary for frame and field replacement only.

2. Remove motor tie bolts, Fig. 85.
3. Hold end cap against frame and field and disengage complete motor section from gear box.
4. Turn motor section as required to gain access to brush plate assembly and release brush spring pressure against brushes, Fig. 86.
5. Move brushes away from armature commutator and remove armature and end cap from frame and field assembly.
6. Remove end cap from end of armature shaft.

NOTE: Be careful not to lose the plastic thrust plug in end of armature.

7. Remove end play washers from commutator end of armature shaft. When reassembling armature in wiper, install washers, Fig. 87.
8. To replace brushes, cut brush pigtail approximately ¼" from splicing clip. Splice the new brush pigtail to the ¼" of pigtail left from the original brush.

NOTE: Splicing clips are provided in the replacement brush packages.

Motor Reassembly

Reverse disassembly Steps 1 through 7 and reassemble gear box. Lubricate the motor assembly with Multifak No. 2 or Prestige No. 42 or equivalent.

NOTE: Insure brush plate mounting brackets are properly seated into housing.

Two Speed-Round Motor

Motor Disassembly

1. Before removal of brush plate and circuit breaker, scribe a reference line along the side of the casting and end cap to insure proper reassembly.
2. Remove the two motor tie bolts.
3. Feed exposed excess length of motor leads through the casting grommet and carefully back the case and field assembly plus the armature away from the casting, Fig. 59.

NOTE: It may be necessary to remove the armature end play adjusting screw and insert a rod through the opening in order to apply pressure against the end of the armature.

4. Unsolder the black lead from circuit breaker.
5. Straighten out the 4 tabs that secure the brush plate to the field coil retainers, Fig. 57.

NOTE: Be careful not to break any of the retainer tabs.

6. Install "U" shaped brush retainer clip

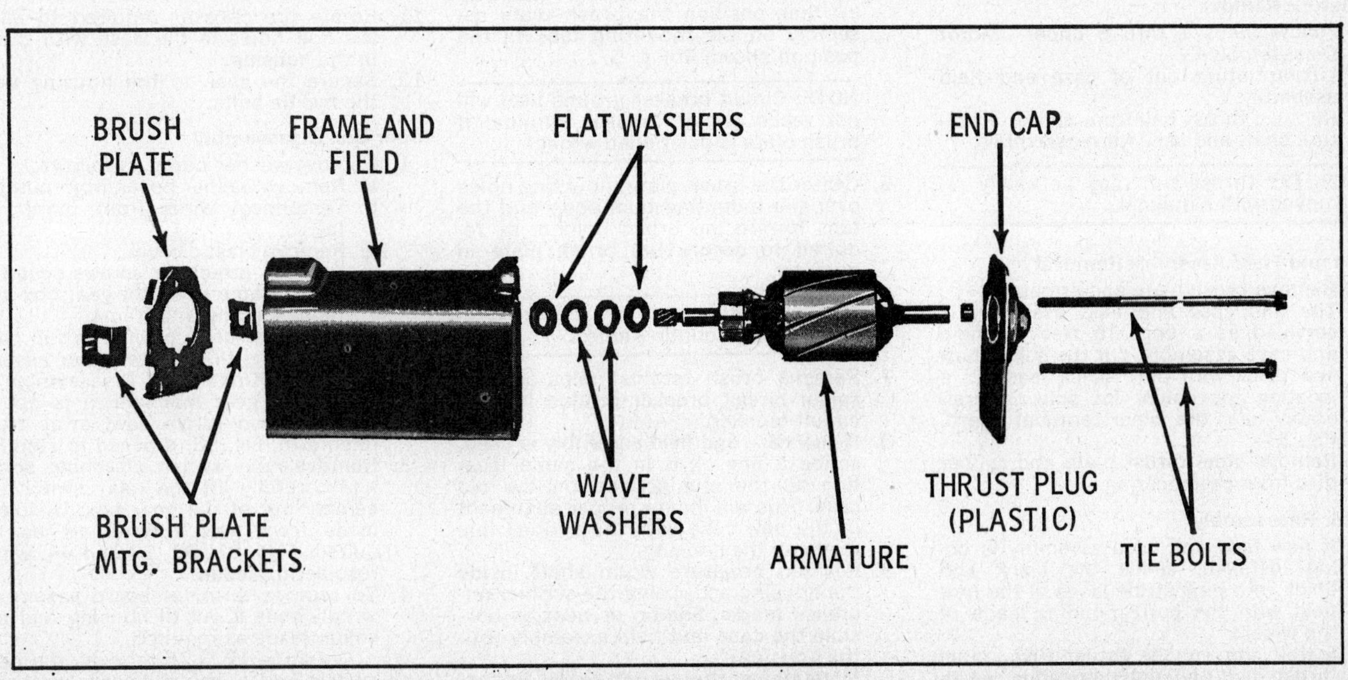

Fig. 85 Wiper motor exploded

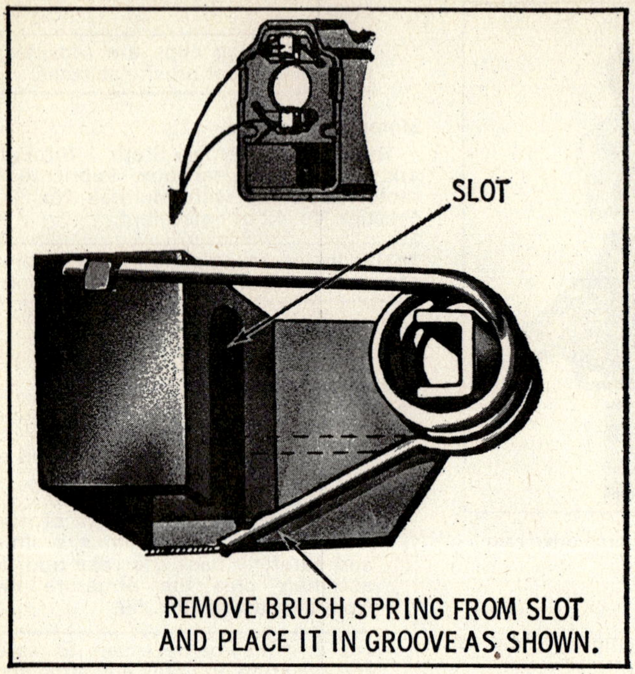

Fig. 86 Removing brush spring

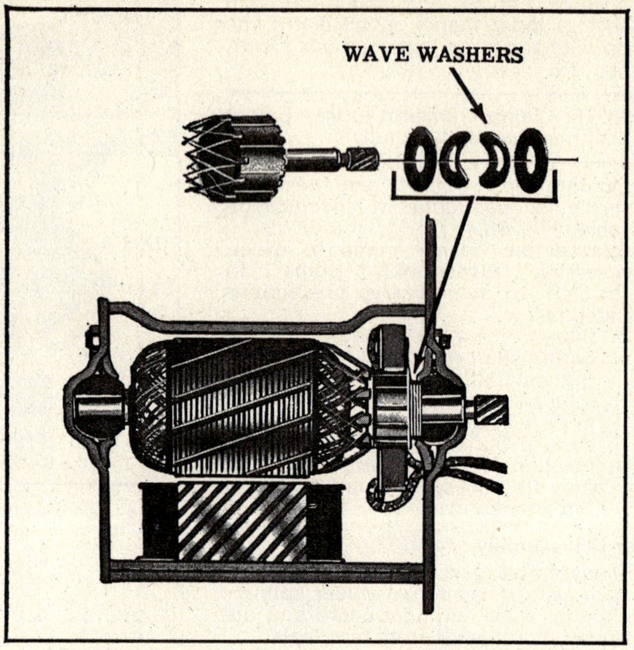

Fig. 87 Armature end play washers

over brush holder that has brush lead attached to circuit breaker, Fig. 57.

7. Holding the opposite brush from that retained in Step 6, carefully lift the brush holder off the mounting tabs far enough to clear the armature commutator, Fig. 88.
8. Allow the brush held in Step 7 to move out of its holder. Remove the brush spring and lift the brush holder off the armature shaft.

Armature Removal

1. Follow Steps 1 thru 8 under "Motor Disassembly."
2. Lift armature out of case and field assembly.
3. Remove thrust ball from end of armature shaft and save for reassembly.

NOTE: Thrust ball may be easily removed with a magnet.

Case and Field Assembly Removal

1. Remove brush plate and armature.
2. The end case and field assembly is serviced as a unit. To free the field and case assembly, cut the solid black and black with pink stripe leads in a location convenient for splicing preferably near the wiper terminal board, Fig. 57.
3. Remove steel thrust plate and rubber disc from case bearing.

Motor Reassembly

1. If new field and case assembly is being installed, splice the black and black with pink stripe leads of the new field with the corresponding leads of the wiper.
2. Install the rubber thrust disc, steel thrust disc and felt lubricating washer in the case assembly bearing in the order indicated.
3. Lubricate end of armature shaft that fits in case bearing with recommended type grease Multifak No. 2 or Prestige No. 42 or equivalent. Next, install thrust ball in end of shaft.
4. Assemble armature in the case and field assembly.
5. Position the partially assembled brush plate, Fig. 89, over the armature shaft far enough to allow reassembly of the remaining brush in its brush holder; then position the brush plate assembly on the mounting tabs in the position shown in Fig. 57.

NOTE: Circuit breaker ground lead will not reach circuit breaker terminal if brush plate is positioned wrong.

6. Center the brush plate mounting holes over the mounting tabs and bend the tabs toward the brush holders as required to secure the brush plate in position.

NOTE: Be sure tabs are centered in brush plate mounting holes.

7. Remove brush retainer clips and resolder circuit breaker ground lead to circuit breaker.
8. If new case and field assembly is used, scribe a line on it in the same location as the one scribed on the old case. This will insure proper alignment of the new case with the scribed line made on the housing.
9. Position armature worm shaft inside the housing and, using the scribed reference marks, line up as near as possible the case and field assembly with the housing.
10. Maintaining the armature in its assembled position in the case, start the armature worm shaft through the field and housing bearing until it starts to mesh with the drive gear. At the same time carefully pull the excess black and black with pink stripe leads through the housing grommet.

NOTE: It may be necessary at this point to rotate armature slightly before the armature worm will engage with drive gear teeth.

11. Rotate the case as required to align the bolt holes in the case with those in the housing.
12. Secure the case to the housing with the two tie bolts.

Gear Box Disassembly

1. Remove washer pump as follows:
   a. Remove washer hoses from pump.
   b. Disconnect wires from pump relay.
   c. Remove plastic cover.
   d. Remove attaching screws securing pump frame to motor gear box and remove pump and frame.
2. If wiper gear drive pawl is in full park position, Fig. 90, remove gear assembly. (See "Drive Gear Disassembly").
   If wiper gear mechanism is not in park position (drive pawl away from latch arm, Fig. 58, proceed to step 3).
3. Remove relay switch attaching screw and carefully lift the relay switch assembly out of the gear box. Unsolder leads from switch terminals as required, Figs. 91, 92, 93 and 94, when resoldering leads.
4. To remove terminal board assembly, simply slide it out of housing and unsolder leads as required.
   On some 1973-75 models, the park switch and terminal board are combined and serviced as one unit.

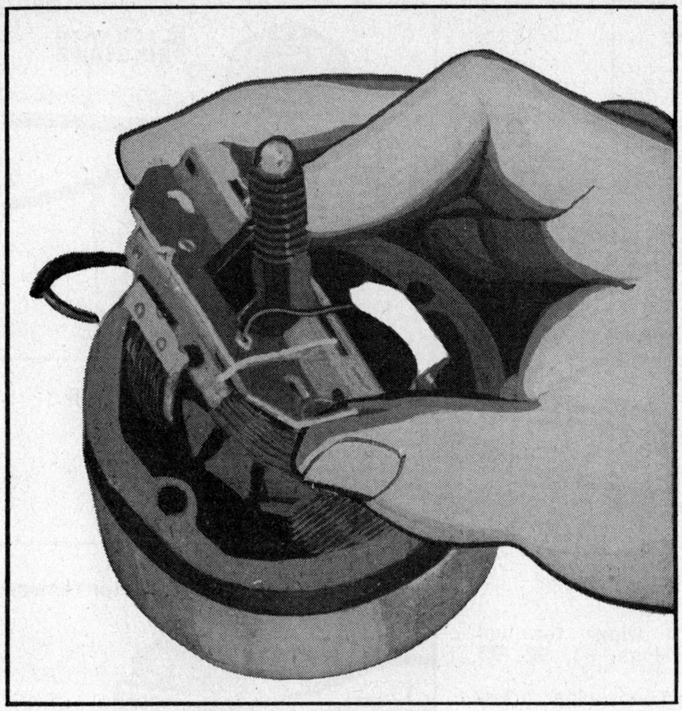

Fig. 88 Removing brush holder

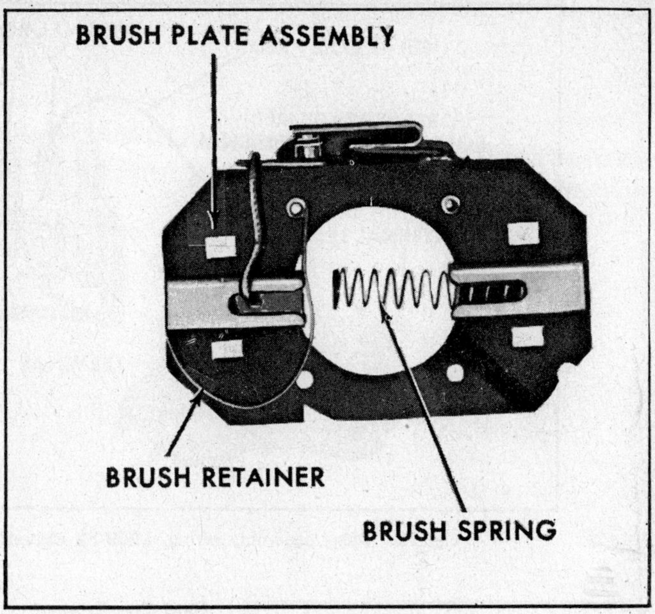

Fig. 89 Brush plate assembly removed

Fig. 90 Wiper shut "off" in park position

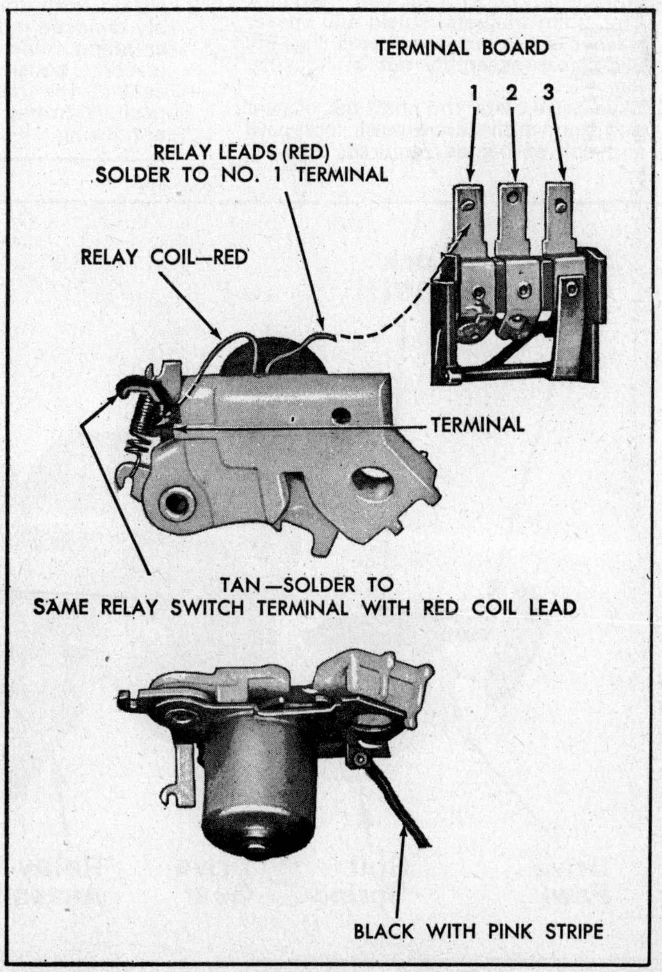

Fig. 91 Relay assembly wiring. 1969-73 except 1969-72 Corvette

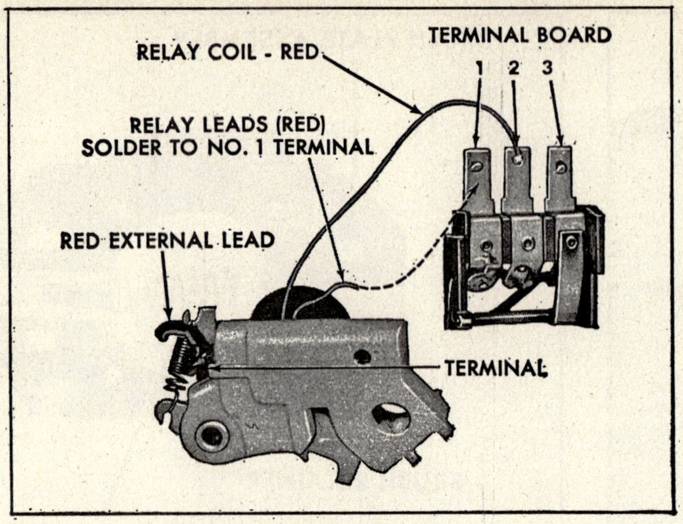

Fig. 92 Relay assembly wiring. 1969-72 Corvette

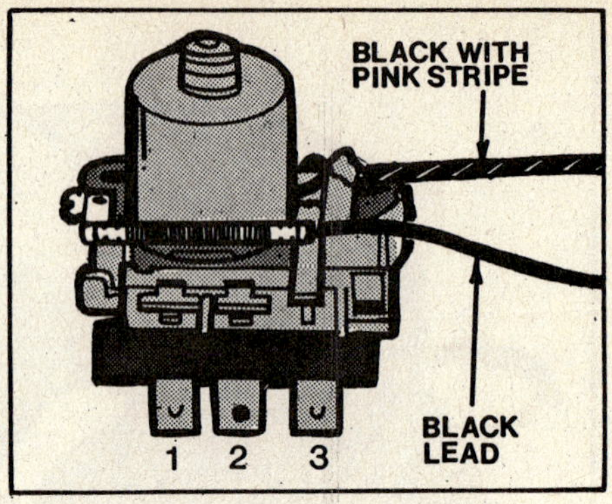

Fig. 93 Relay assembly wiring. 1974-75 exc. Pulse type

NOTE: Tan wire to No. 2 terminal black wire to No. 3 terminal.

Drive Gear Disassembly

1. Remove crankarm retaining nut, crankarm, rubber seal cap, retaining ring, shim washers, shield and spacer washer in the order indicated, Fig. 95.
2. Slide gear assembly out of housing, Fig. 96.
3. Slide drive plate and shaft out of gear and remove the drive pawl, lock pawl and coil spring as required, Fig. 97.

Gear Box Reassembly

1. Resolder leads to wiper terminal board as required, Figs. 91, 92, 93 and 94.
2. Slide terminal board into wiper housing being careful to position the terminal board resistor lead, Fig. 61.

NOTE: With the relay switch assembly replaced in the housing and washer pump reinstalled, the relay switch plastic housing applies pressure against the resistor lead to form a positive ground connection to the wiper housing.

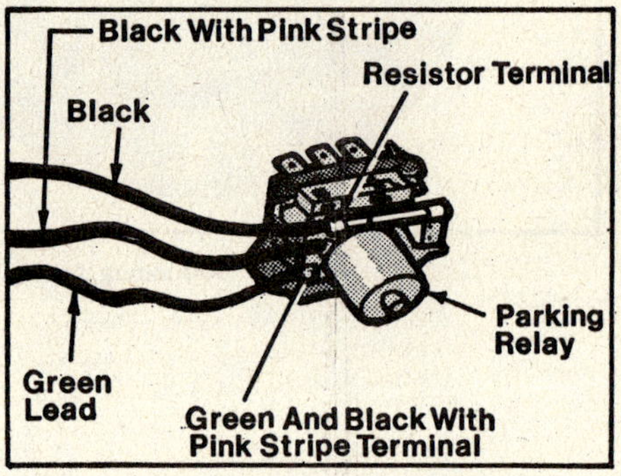

Fig. 94 Relay assembly wiring. 1974-75 Pulse type

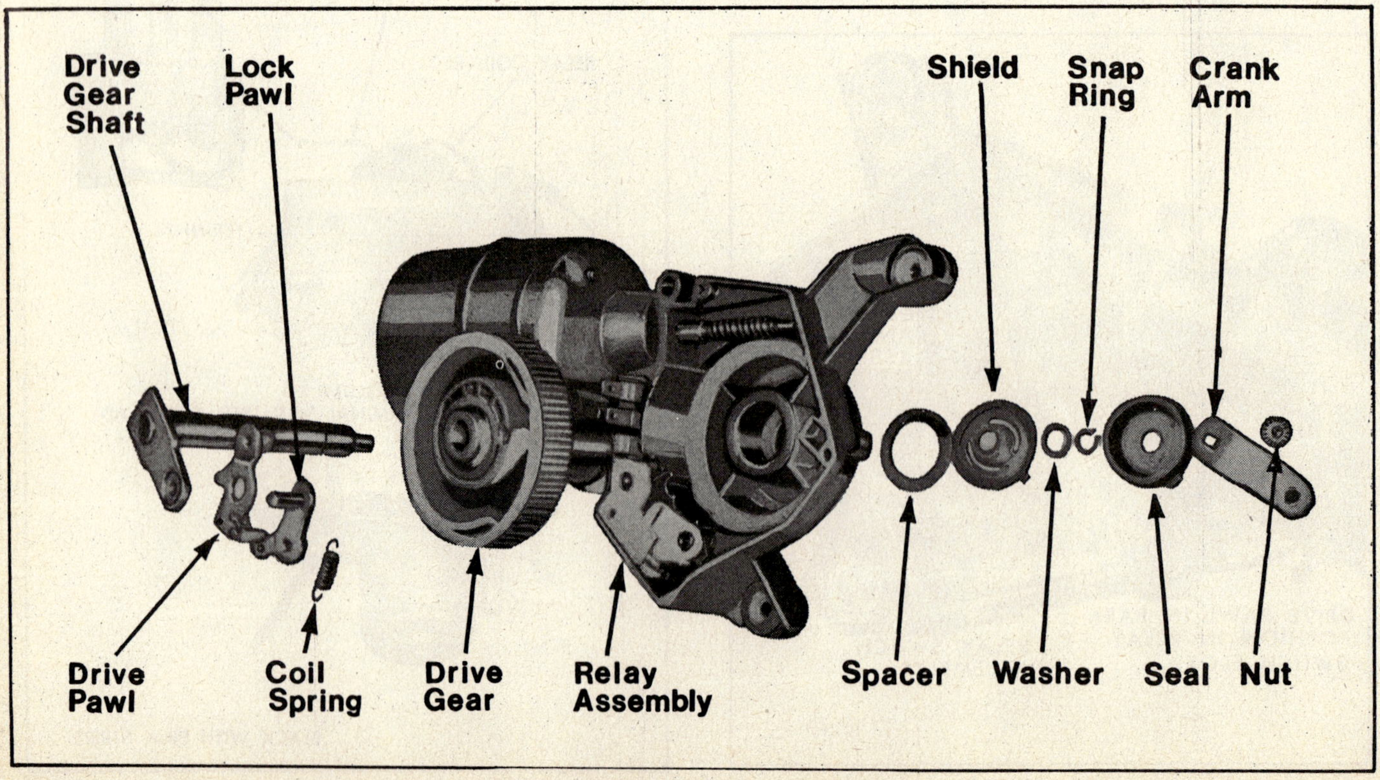

Fig. 95 Drive gear mechanism assembly

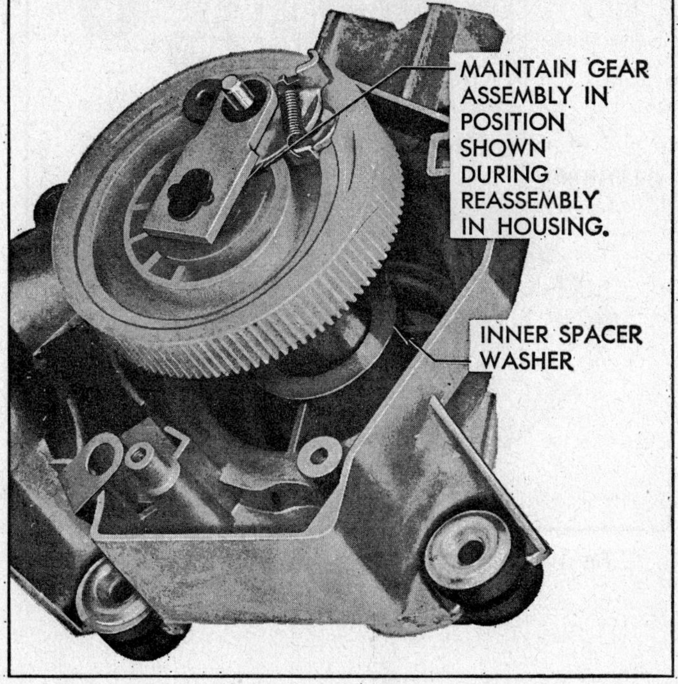

Fig. 96 Removing gear

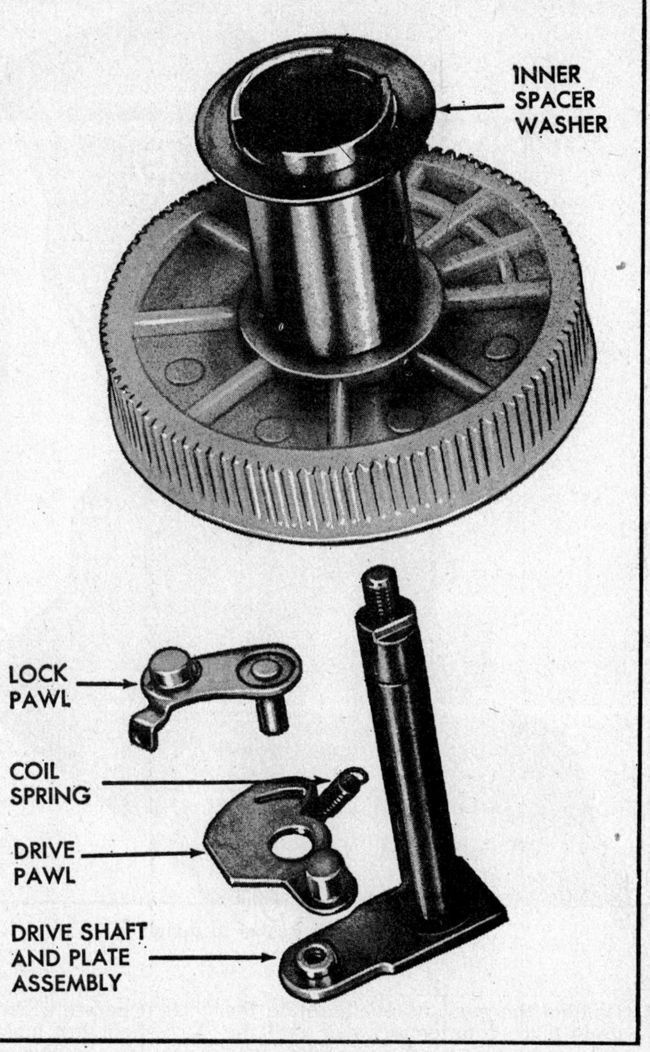

Fig. 97 Drive gear and shaft assembly

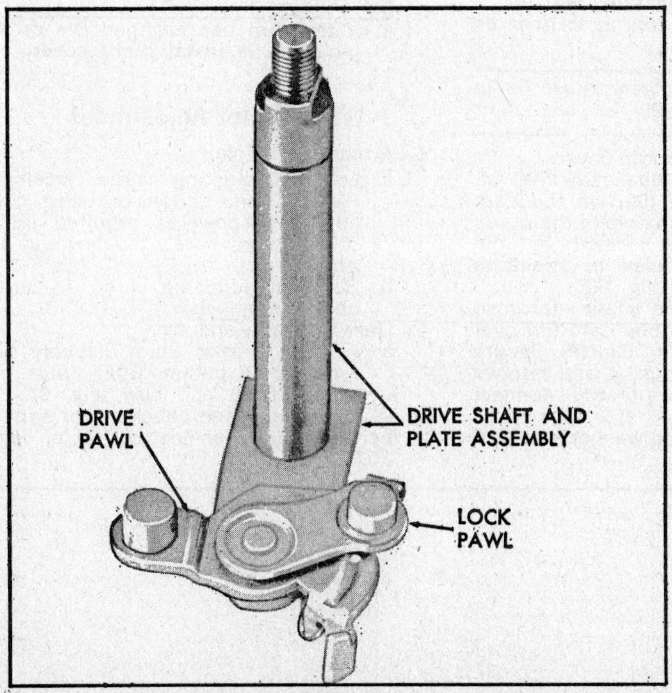

Fig. 98 Lock and drive pawl assembly

3. Resolder leads to relay switch assembly as required.
4. Position relay switch assembly in housing.

NOTE: Be very careful to route leads in such a manner as to avoid having them pinched between relay and wiper housing.

5. Install relay switch attaching screw.
6. Reassemble washer pump to gear, Step 9 under "Drive Gear Reassembly" if drive gear has not been disassembled.

Drive Gear Reassembly

1. Position drive pawl on drive plate, Fig. 97.
2. Assemble lock pawl over drive pawl, Fig. 98.
3. Slide gear and tube over the drive shaft. (Move drive and lock pawls as required to allow their respective pins to fit in the gear guide channel).

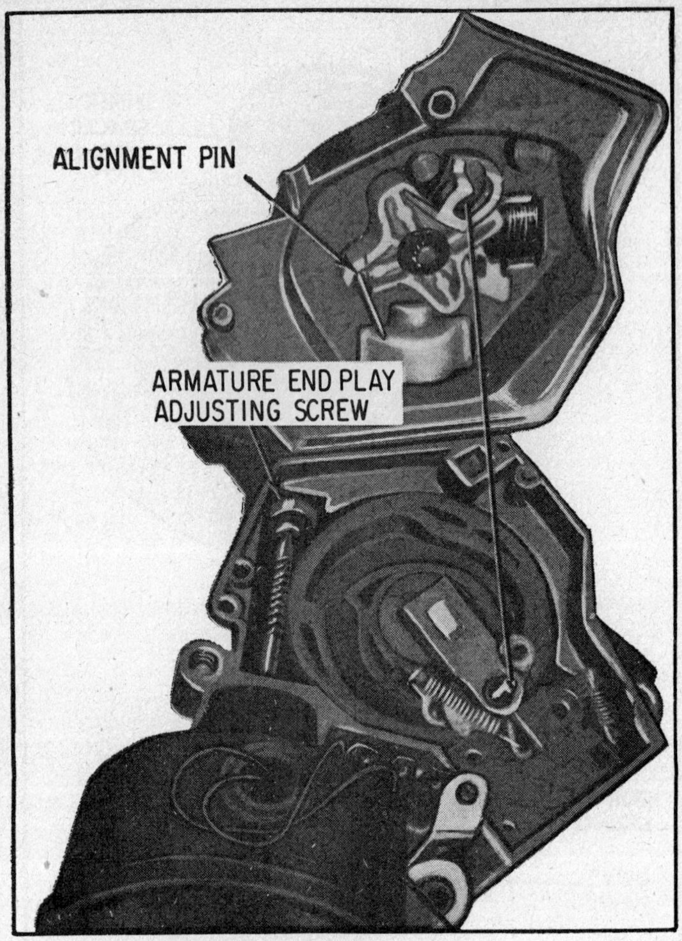

Fig. 99 Installing washer to pump

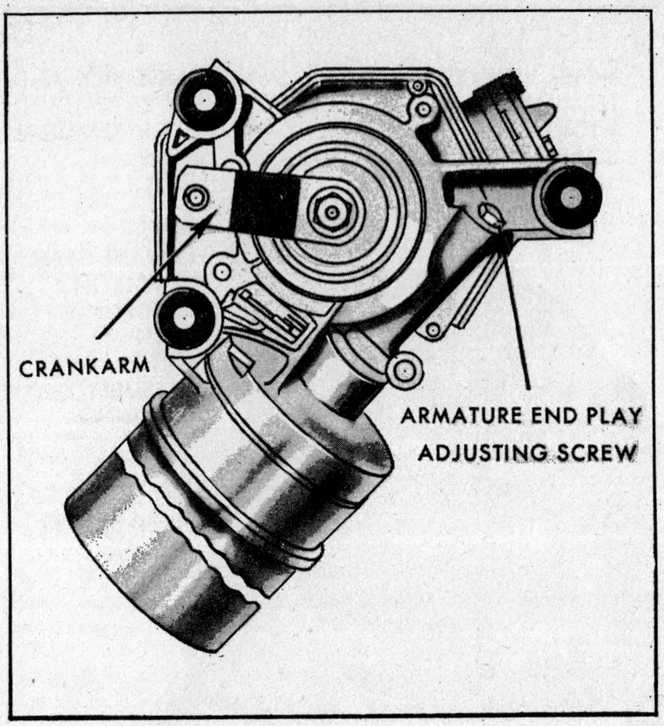

Fig. 100 Wiper motor and crank arm in park position

4. Holding the gear, manually rotate the drive plate counterclockwise until the drive and lock pawl guide pins snap into their respective pockets in the gear, Fig. 90.
5. Reinstall coil spring between lock and drive pawls.

NOTE: Be very careful to maintain lock and drive pawl guide pins in their respective pockets during step 6.

6. Assemble inner spacer washer over gearshaft and assemble gear mechanism in housing so that it is positioned with respect to the housing in the approximate location, Fig. 96.
7. Reassemble the outer spacer washer, shield, shim washers, as required to obtain .004" (plus or minus .002") end play, snap ring and rubber seal cap in the order indicated, Fig. 95.
8. Operate wiper to "park" or "off" position and install crankarm, Fig. 90.
9. Reassemble washer pump to gear as follows:

NOTE: Wiper motor gear must be in park position, Fig. 99.

a. Remove plastic pump cover.
b. Rotate the four-lobe cam until index hole (.125" dia.) in the cam is aligned with the hole in the pump mounting plate. Insert a pin through both holes to maintain cam in position, Fig. 99.
c. Position pump on wiper motor so that slot in four-lobe cam fits over the gear drive pin, Fig. 99. Secure pump to gear housing and remove locator pin, temporarily connect wiring connector.
d. Turn on wiper and washer pump to check operation.

NOTE: A loud knocking noise would indicate that the pump cam has not engaged the drive pin properly.

e. If the cam has engaged the drive pin properly, install pump cover.

Wiper Motor Adjustment

1. Armature end play:
   a. Loosen adjusting screw locknut, Fig. 100, and tighten or loosen the adjusting screw as required until end of screw barely touches end of armature.
   b. Back off adjusting screw 1/4 turn and tighten locknut.
2. Gear assembly end play:
   a. Add or remove shim washers as required to obtain .004" (plus or minus .002) end play, Fig. 95.
3. Operate wiper and check performance per data in specification table, Fig. 101.

Current Draw (Amps)

Car	Year	No Load Low	No Load High	On Car Dry WyS① Low	On Car Dry WyS① High	Stall	Shunt Field Resistance
Buick							
Full Size	1969	5.0 Max.	4.0 Max.	5.5 Max.	4.5 Max.	12 Max.	—
	1970	5.0 Max.	—	—	—	26 Max.	—
	1971-75	6.0 Max.	4.5 Max.	—	—	29 Max.	—
	1974-75③	6.0 Max.	4.5 Max.	—	—	29 Max.	—
Intermediate	1969	5.0 Max.	4.0 Max.	5.5 Max.	4.5 Max.	12 Max.	—
	1970	5.0 Max.	—	—	—	19 Max.	—
	1971-74	5.0 Max.	4.0 Max.	—	—	18 Max.	—
	1974-75③	6.0 Max.	4.5 Max.	—	—	29 Max.	—
	1975	6.0 Max.	4.5 Max.	—	—	29 Max.	—
Cadillac	1971-75	6.0 Max.	4.5 Max.	—	—	29 Max.	—
	1974-75③	6.0 Max.	4.5 Max.	—	—	29 Max.	—
Chevrolet							
Full Size	1969	4.5 Max.	—	5.0 Max.	—	15 Max.	5 to 6 Ohms
	1970-75	6.0 Max.	4.5 Max.	—	—	29 Max.	5 to 6 Ohms
	1974-75③	6.0 Max.	4.5 Max.	—	—	29 Max.	—
Intermediate	1969	4.5 Max.	—	5.0 Max.	—	15 Max.	5 to 6 Ohms
	1970-74	5.0 Max.	4.0 Max.	—	—	18 Max.	5 to 6 Ohms
	1974-75③	6.0 Max.	4.5 Max.	—	—	29 Max.	—
	1975④	6.0 Max.	4.5 Max.	—	—	29 Max.	—
	1975⑤	5.0 Max.	4.0 Max.	—	—	18 Max.	—
Corvette	1969	4.5 Max.	—	5.0 Max.	—	15 Max.	5 to 6 Ohms
	1970-75	6.0 Max.	4.5 Max.	—	—	29 Max.	5 to 6 Ohms
Oldsmobile							
Full Size	1969-70	6.0 Max.	4.5 Max.	6.5 Max.②	5.0 Max.②	29 Max.	—
	1971-75	6.0 Max.	4.5 Max.	—	—	29 Max.	—
	1974-75③	6.0 Max.	4.5 Max.	—	—	29 Max.	—
Intermediate	1969	5.0 Max.	4.0 Max.	4.5 Max.②	—	—	—
	1970	5.0 Max.	4.0 Max.	5.5 Max.②	4.5 Max.②	18 Max.	—
	1971-74	5.0 Max.	4.0 Max.	—	—	18 Max.	—
	1974-75③	6.0 Max.	4.5 Max.	—	—	29 Max.	—
	1975	6.0 Max.	4.5 Max.	—	—	29 Max.	—
Pontiac							
Full Size	1969	6.0 Max.	4.5 Max.	6.5 Max.	5.0 Max.	28 Max.	—
	1970	6.0 Max.	4.5 Max.	6.5 Max.②	5.0 Max.②	29 Max.	—
	1971-75	6.0 Max.	4.5 Max.	—	—	29 Max.	—
	1974-75③	6.0 Max.	4.5 Max.	—	—	29 Max.	—
Intermediate and Grand Prix	1969	6.0 Max.	4.5 Max.	5.5 Max.	4.5 Max.	18 Max.	—
	1970	6.0 Max.	4.5 Max.	6.5 Max.②	5.0 Max.②	29 Max.	—
	1971-74	5.0 Max.	4.0 Max.	—	—	18 Max.	—
	1974-75③	6.0 Max.	4.5 Max.	—	—	29 Max.	—
	1975⑥	6.0 Max.	4.5 Max.	—	—	29 Max.	—
	1975⑦	5.0 Max.	4.0 Max.	—	—	18 Max.	—

①—Current draw specifications listed for on car check are subject to fluctuations depending upon the condition of blades, windshield, battery, engine running or turned off The figures shown are only a guide and reflect the average current draw throughout a complete wiping stroke

②—Wet windshield
③—Pulse Wiper
④—Exc. Camaro.
⑤—Camaro.
⑥—Exc. Firebird.
⑦—Firebird.

Fig. 101 Specification Table Two Speed Round Motor

GENERAL ENGINE SERVICE

INDEX

This chapter is designed to remind the experienced mechanic and to acquaint the novice of what is required to do a first rate job on a modern engine so that chances of neglecting some vital phase of the repair work will be minimized.

We will deal here with operations that are common to almost all modern engines. Unique engines, such as the Pontiac overhead camshaft engine design is covered in the respective chapters. The procedures given are the general requirements for removing, inspecting, repairing and replacing various parts. Refer to the car chapters for specific details that may have to be taken into consideration on a particular engine and vehicle. We suggest that the car chapter be consulted first for specifications and any other special remarks relative to the work being done.

Particular attention should be given to the installation of gaskets. For this reason the following information is supplied.

Gasket Tips

Whenever it becomes necessary to remove and replace a component that requires a gasket, there are some basic instructions which will help to insure a leak-proof joint and thereby avoid having to repeat a part of the job. They are as follows:

1. After removal of component, carefully remove all traces of old gasket material from metal surfaces.
2. Be sure metal surfaces are not scored or burred.
3. If an adhesive is required it is usually applied to the metal surface and then only in thin dabs. The exception to this rule is cylinder head gaskets which should have the adhesive applied directly to them to avoid blocking cooling passages in the head or block.
4. Allow the adhesive to set before installing the gasket. This prevents the gasket from sliding.
5. Upon reinstallation of the part, do not overtighten the screws or bolts. Overtightening will distort the gasket and contribute to premature failure of the gasketed joint.

A word of caution; always select the proper gasket required for the job. Anything less will probably result in early failure.

Cylinder Head & Valves

CYLINDER HEAD, REPLACE

A cylinder head is secured to the cylinder block by studs and nuts or by cap screws. Do not remove a cylinder head when the engine is hot as head warpage may result. Before removing a cylinder head it is necessary to remove any parts that are in the way and to detach any connections between the head and the rest of the vehicle.

1. Disconnect wiring to temperature sender, oil pressure sender and any other switches or solenoids.
2. Remove high tension cables from spark plugs.
3. Remove carburetor air cleaner after disconnecting any hoses or wires.
4. If water pump is attached to the head remove the fan belt.
5. Remove or detach any other item that may interfere with removal of head.
6. Detach vacuum advance tube at carburetor.
7. If ignition coil is attached to head, detach the high tension cable running from coil to distributor at distributor.
8. Detach fuel line at carburetor.
9. Detach throttle linkage at carburetor.
10. Remove carburetor and intake manifold as a unit.
11. Detach exhaust manifold from head or disconnect exhaust pipe from manifold, whichever is easier.
12. Remove rocker arm cover.
13. Remove rocker arms.
14. Remove push rod cover, if used.
15. Pull push rods up through head as they are likely to be damaged if the head is removed first. Keep push rods in proper sequence so they may be reinstalled in their original locations.
16. Remove cylinder head nuts or cap screws. If cap screws are used, be sure to note their length as they are removed. If there are some long and some short ones be sure to note which holes the long ones come from so that when the head is replaced the long ones will be installed in the proper holes. If a long screw is installed in a hole where a short one belongs and turned down tight the cylinder block may be damaged beyond repair. However, if you fail to note the location of the long screws, measure the depth of all holes with any handy rod or pencil and be guided accordingly.

NOTES

Once the cylinder head nuts or cap

screws have been removed, the head is ready to be lifted off. If the head is large or very heavy, two men may be required. If it does not come off easily, jolt it crosswise and lengthwise with a rawhide or plastic mallet. If it still sticks, crank the engine with the throttle open and ignition off to loosen the head by the force of compression. Before doing this, however, install spark plugs in the end cylinders to hold the compression and install a pair of nuts or cap screws. Draw them down finger tight and then back off about two turns. Crank the engine with the starter to see if the head will break loose.

If this fails, install two eye bolts in the head and apply a chain hoist or a crane just as though you were going to lift the engine out of the car.

Installation

Handle the head carefully. Do not allow it to strike any open valves as this may bend them. Before replacing the head be sure the gasket surfaces of the head and block are perfectly clean, otherwise the gasket may leak when the head is installed. Also be sure that there is no carbon or dirt in the cylinder head bolt holes, or any foreign matter in the bolt holes in the block. If there is, blow them out with compressed air. Dirt in the holes may prevent drawing the bolts up tight.

On some engines, the bolt holes in the block extend into the water space, in which case the end of the bolt should be coated with white lead or sealing compound to prevent leakage.

If there are any burrs around the edges of the bolt holes in the block it is advisable to remove them with a medium file as these burrs may prevent the gasket from fitting tightly. In fact, all burrs on both the head and block should be removed.

It should be noted that the cylinder head gasket may have bolt or stud holes which are slightly enlarged on the under face to take care of burrs. If the holes are slightly larger on one side, this side should contact the block. Some gaskets are marked "Top" or have lettering on the top side. Always install a new gasket as the old one may leak even though it appears to be perfect.

Place the gasket on the cylinder block

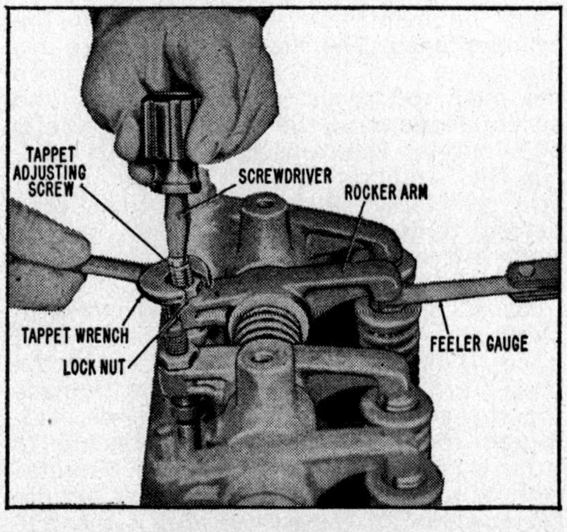

Fig. 2 Adjusting valve lash on overhead valve engine

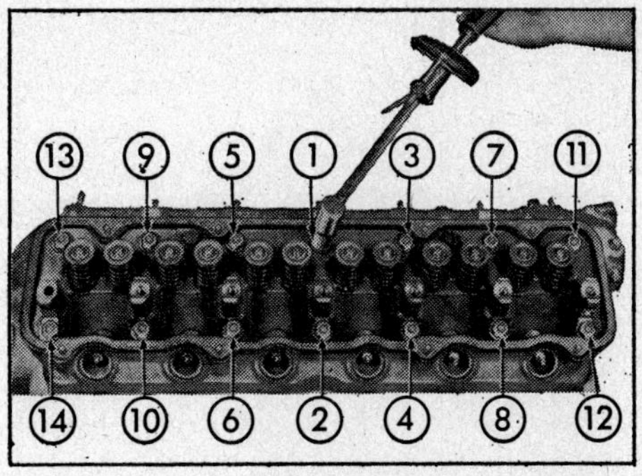

Fig. 1 Using a torque wrench to tighten a typical cylinder head. The limber torque wrench handle bends when it is pulled by hand but the pointer remains stationary. Thus the pointer indicates the torque in foot-pounds

to be certain that it fits. If all bolt holes and water passage holes do not line up, turn the gasket upside down or end for end and inspect it again. Some gaskets are marked "Front" at the front end; if so, be sure to install it this way.

On vehicles that use cylinder head cap screws instead of studs it is desirable to install two "pilot" studs in the block to guide the head into place. These studs are made by sawing the heads off of two spare cylinder head cap screws. Screw them lightly into bolt holes at opposite ends of the block. After these studs are installed, put the gasket in place and then install the head.

Lubricate the threads on the cap screws and the bosses on the cylinder head. Install the cap screws and turn them until they are nearly tight. Then remove the two pilot studs and install cap screws in these holes. Use a torque wrench Fig. 1, to draw the bolts up tight. It is important to tighten the bolts in proper sequence and to tighten them just the right amount in order to prevent distortion of the cylinder block or cylinder head, as well as to insure a tight gasket.

Usually the scale on the torque wrench reads in pounds-feet of torque. But if the wrench reads according to some arbitrary scale, it is customary for the wrench manufacturer to supply instructions with the tool so that the wrench scale readings can be translated into pounds-feet. The term "pounds-feet" simply means the force in pounds applied to the wrench handle at a radius of one foot. Note that the correct term is pounds-feet and not foot-pounds, although foot-pounds is almost universally used.

All vehicle manufacturers specify the pounds-feet torque which should be used in tightening their cylinder heads and also the sequence in which final tightening should be done.

Be careful to tighten the cylinder head cap screws or stud nuts exactly to the vehicle maker's specifications. Tightening them a little more or less than the specified amount is objectionable because "more" may cause distortion and "less" may permit gasket leakage.

If the cylinder block is equipped with studs, put the gasket on the block, install the cylinder head and lubricate the threads on the studs. Put on the nuts and draw them up until they are nearly tight. Then finish the job with the torque wrench as previously described.

After the head is installed, attach all other parts which have been removed from the engine except the valve covers or the rocker arm cover, whichever is used.

NOTE: *Some manufacturers specify that once the head is torqued to specifications it should not be disturbed. Others recommend that a final tightening should be made after the engine is warmed up to operating temperature.*

On overhead valve engines having mechanical valve lifters or hydraulic lifters provided with an adjustment, always check and adjust the valve operating clearance (or valve lash) after the final head tightening is made.

VALVE LASH ADJUSTMENT

Valves must be adjusted after they have been ground or refaced. Also note that after valves have been in use for some thousands of miles they may go out of adjustment. Thus, if the clearance becomes too small, the valve eventually may leak because it is unable to seat fully. If valves are too noisy the clearance should be checked and the valves adjusted to the proper specifications.

On some cars the clearance specifications for intake and exhaust valves are the same while on others more clearance is specified for the exhaust valves than for intake valves.

If the car is often driven at high speed for long periods it may be desirable to increase the exhaust valve clearance by .003" for overhead valve engines.

If intake and exhaust valves have different clearance specifications, it is necessary to determine the location of intake and exhaust valves in each cylinder. Usually this can be done by inspecting the branch arms of the intake and exhaust manifolds where they are

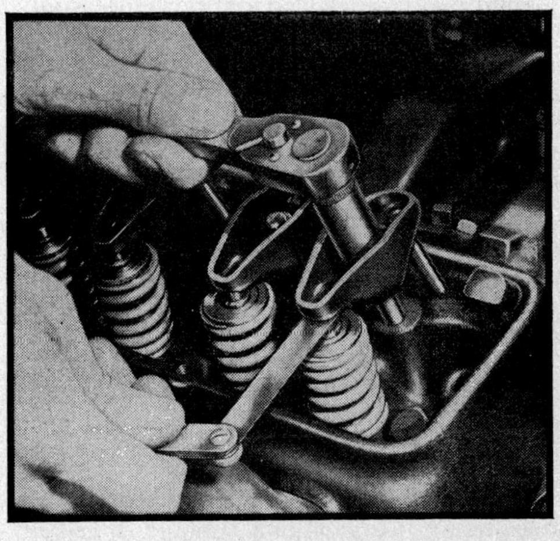

Fig. 3 Valve adjustment on engine with each rocker arm mounted on its individual stud

bolted to the engine. Make a chalk mark to indicate the location of each exhaust valve. The valve arrangement for all engines is given in the car chapters.

Shaft Mounted Rocker Arm Valve Adjustment

Valves in these engines are adjusted by turning the screw in the end of the rocker arm after loosening the lock nut just enough to free the screw, Fig. 2. Insert a feeler gauge of the correct thickness between the valve stem and the tip of the rocker.

During the instant that the valve is closed, if a very slight drag is felt on the feeler the adjustment is correct. To "feel" the drag, pull gently on the feeler while turning the screw. When the adjustment is correct, hold the screw against turning and tighten the lock nut. Then check the clearance again in order to be sure that tightening the lock nut did not change the adjustment.

Adjustment With Hydraulic Valve Lifters

On some overhead valve engines with hydraulic valve lifters an initial adjustment is specified after a valve grind job is performed or whenever the rocker mechanism has been disturbed for any reason. Adjustment is made with the engine stopped. Have valve closed when making adjustment. The procedure is to turn the adjusting screw clockwise while moving the push rod up and down until all slack is removed. Be careful not to force the hydraulic plunger down into the lifter body. After removing all slack, turn the adjusting screw the additional number of turns specified for the particular engine, which will place the hydraulic lifter plunger in the center of its travel. When the adjustment is correct, tighten the lock nut.

Stud Mounted Rocker Arm Valve Adjustment

Instead of the conventional rocker arms and shafts, a set of rocker arms is used, each one mounted on its individual stud, which is pressed or threaded into the cylinder head.

On engines with mechanical valve lifters, the adjustment is made with engine idling. Turn the self-locking rocker arm stud nuts as required to obtain the specified clearance, Fig. 3.

On engines with hydraulic valve lifters, the adjustment is made with the engine stopped. The valves being adjusted, however, must be closed (TDC on firing stroke). The procedure is to back off the rocker arm stud nut until there is play in the valve push rod. Then tighten the nut just enough to remove all play. This may be determined by rocking the push rod as the nut is tightened, Fig. 4. When the rod does not readily move in relation to the rocker arm, the clearance has been eliminated. The adjusting nut should be tightened the additional amount specified for the particular engine being serviced to place the hydraulic plunger in the center of its travel.

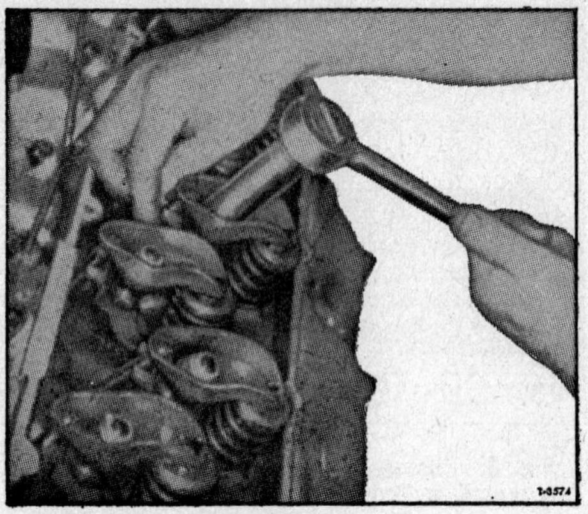

Fig. 4 Valve adjustment on stud mounted rocker arms with hydraulic valve lifters

Cold Valve Adjustment

On some engines it is recommended that the valves be adjusted cold although many car makers specify a warm adjustment.

In most cases the need for adjusting valves is because they have just been ground. The adjustment must be made with valves closed and this is readily done by bringing the piston up on top dead center at the end of the compression stroke.

Warm Valve Adjustment

Before adjusting valve clearance, it is extremely important that the engine be thoroughly warmed up for about 30 minutes to normalize the expansion of all parts. This is very important because during the warm-up period the valve clearances will change considerably. To adjust the valves during or before this warm-up period will produce clearances which will be far from correct after the engine reaches normal operating temperature.

Covering the radiator will not materially hasten this normalizing process because even with the water temperature quickly raised, it does not change the rate at which the oil temperature increases and becomes stabilized, or the engine parts become normalized.

The actual temperature of the oil is not as important as stabilizing the oil temperature. The expansion or contraction of the valve mechanism, cylinder head and block are relative to this oil temperature. These parts stop expanding and valve clearance changes cease to take place only after the oil temperature is stabilized. Then the valves are adjusted with the engine running at normal idling speed.

Some mechanics prefer to adjust the exhaust valves first and then the intake. This is sound practice if the exhaust valves require a different clearance than the intake valves. But if the clearance is the same for both intake and exhaust valves, it is better to start at the front and work toward the rear of the engine.

If the exhaust valves require a greater clearance than the intake valves, set either the intakes first and then all the exhaust valves. In this way only one thickness feeler gauge is in use at a time and the chances of using the wrong feeler gauge are largely eliminated.

ROCKER ARMS

On some engines, rocker arms oscillate on stationary tubular shafts mounted on suitable supports fastened to the cylinder head. On other engines, the rocker arms are each mounted on a stud pressed or threaded into the head and fastened by a nut which is also used for adjusting the valve clearance as already explained. This design, eliminates the conventional shaft.

With conventional rocker arms and shafts, the upper end of the push rod usually terminates in a cup which fits around a ball on the rocker arm. The lower end of the push rod rests on the valve lifter tappet. The engine oil pump supplies oil under pressure to the hollow rocker arm shafts. Drilled openings in the shaft deliver oil to the rocker arm bearings. From here oil flows through a channel in the rocker arm which supplies lubricant to the cup and ball at the top of the long push rods.

Rocker Arm Service

After removing the rocker arm cover, if the rocker arms are found to be heavily sludged, stuck or damaged, the assembly should be removed for cleaning and opening up of oil passages. If none of these conditions is apparent, run the engine with the rocker arm cover off. While the engine is running, check the rocker arms and shafts to see that oil is being distributed to all the shaft bearings and to the outer end of the arms. Correct lubrication of these parts is indicated when oil drips from the bearing edge of each rocker arm and spreads over the push rod ends.

Oil failing to show at these points is usually caused by dirt in the oil feed pipe, in which case it should be removed and cleaned, preferably by blowing out with compressed air.

Too much oil flowing is a definite indication of worn rocker arms and shafts, which calls for replacement of these parts.

Before removing or disassembling a rocker arm assembly, note carefully the arrangement of the parts and look for any identifying marks which indicate their location (see car chapters).

The rocker arm assembly should always be removed before removing the cylinder head. The first step is to remove the rocker arm cover plate. Then remove the push rod cover (if used). Detach the oil connections at the rocker arms shafts or supports. Then unscrew the bolts holding the supports and lift off the rocker arm assembly. Remove the push rods before removing the head and replace them after installing the head to avoid accidentally damaging them.

Dress up minor defects on rocker arm shaft and in the rocker arm bore with a hone. If the pad on the valve end of the rocker arm has a grooved radius, replace the rocker arm. On some engines with forged rocker arms it is permissible to true this surface by grinding. Engines with stamped type rocker arms require replacement of the rocker arm if the pad is worn.

In disassembling the rocker arms from

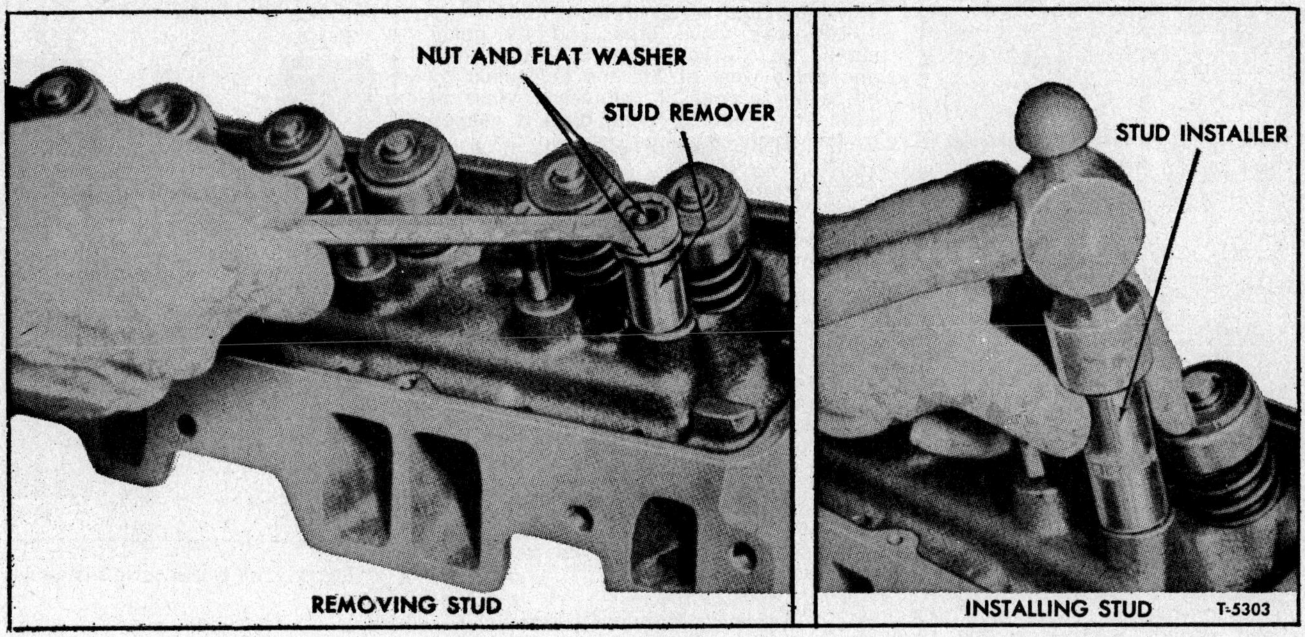

Fig. 5 Removing and installing rocker arm studs

their shafts, it is most important to place the parts along the back of the work bench in exactly the same relationship they assume when on the engine, otherwise there may be difficulty in assembling them properly. This is particularly important because rocker arms are not all alike; some are at right angles to the rocker shafts and others may be a few degrees more or less than a right angle (see car chapters).

When assembling rocker arms to shafts, squirt oil on the shafts to provide initial lubrication when the engine is started.

ROCKER ARM STUDS

If it is necessary to replace a pressed in rocker arm stud, rocker arm tool kits are available which contain a stud remover, two oversize reamers, and a stud installer.

Rocker arm studs that are broken or have damaged threads may be replaced with standard size studs. Loose studs in the head may be replaced with two different oversize studs (one larger than the other) which are available for service. Details are given in the car chapters.

Fig. 6 Reaming valve guide

When going from a standard size stud to the larger of the two oversizes available, always use the smaller of the two reamers in the kit before finish reaming with the larger reamer.

Assemble stud remover tool as shown, Fig. 5. Thread the nut onto stud with wrench to extract stud from head.

If the stud is broken off flush with the stud boss, use on "Easy-Out" to remove the broken stud, following the instructions of the tool manufacturer.

If a loose stud is being replaced, ream the stud bore with the proper reamer (or reamers in sequence) for the selected oversize stud, Fig. 6. *Make sure the metal particles do not enter the valve area.*

Screw the new stud into the driver of the kit installer and coat the end of the stud with Lubriplate. Align the stud and installer with the stud bore, then tap the driver until it bottoms, Fig. 5. When the installer contacts the stud boss, the stud is installed to its correct height.

PUSH RODS

Clean push rods in a suitable solvent and blow out the oil passage in the push rod with compressed air if rods have oil holes. Check the ends of the push rods for nicks, grooves, roughness or excessive wear.

Push rods can be visually checked for straightness while they are installed in the engine by rotating them with the valve closed.

VALVES, REPLACE

If a numbered rack is not available to receive the valves as they are removed, one can be made by drilling enough holes in a board to accommodate them. This is not essential when valve faces and seats are to be reconditioned and valve guides replaced. But when valve guides are still serviceable and not removed, each valve should be returned to its own guide because the extent of wear between one guide and another may vary enough to cause trouble later on if valves are switched. Then, too, exhaust valves are often made of greater heat resisting material than are intake valves. In most cases, however, intake valve heads are larger in diameter than exhaust valve heads, in which case switching them would readily be noticed since they would not seat in the head or block properly.

Overhead Valves, Remove

There are various types of tools for compressing the valve springs but the type most generally used is the C-shaped design, Fig. 7. The forked jaw is applied to the valve spring seat and then the screw plunger is turned clockwise until it touches the head of the valve. Pushing

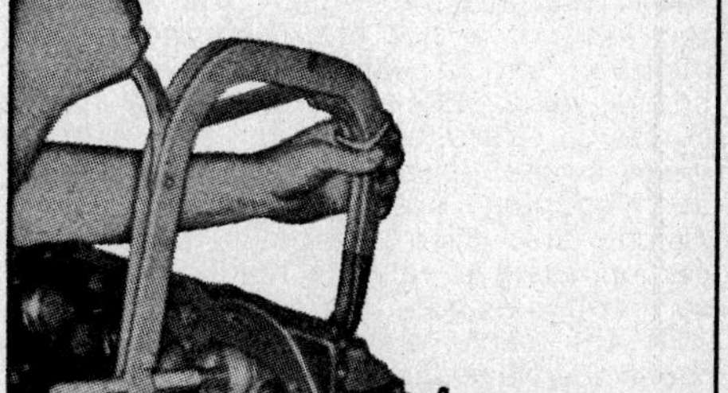

Fig. 7 Removing valve spring

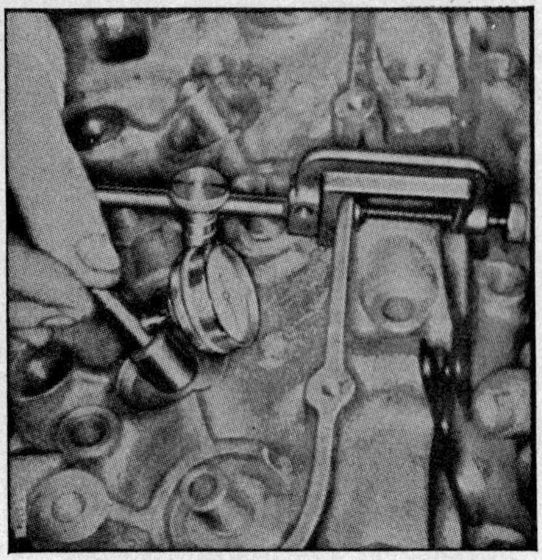

Fig. 8 Measuring valve stem-to-guide clearance using a dial indicator

the operating handle compresses the spring and permits removal of the retainer lock.

Overhead Valves, Install

On most modern overhead valve engines, oil seals are used on the valve stems to prevent the seeping of oil into the combustion chambers.

When new seals are installed, they should be checked with a small suction cup placed on top of the valve spring retainer. If there is no leakage the cup should stick in place.

VALVE INSPECTION

After the valves have been cleaned they should be inspected carefully for defects. Excessive heat may cause the valve to bend or warp. Such a valve should be discarded.

Examine both the upper and lower sides of the head for cracks. If even a small crack is found, scrap the valve.

Inspect the face of each exhaust valve for burning. Even if only a small burned spot is found the valve should be scrapped because the metal adjacent to the spot has probably been damaged and further trouble can be expected if such a valve is used again. The appearance of a burned valve is readily recognized. It looks like a poker which has remained too long in a hot fire. A burned valve, however, should not be confused with a pitted valve. Pits are caused by slight corrosion resulting from hot exhaust gases sweeping past the face of the valve and gouging out shallow depressions which are then partially filled with carbon. Such a valve is readily refinished so that it is as good as new.

If the valve stem is noticeably corroded the valve should be discarded.

If the tip of the valve stem is scratched it should be smoothed up with a valve refacing tool. If the top face of the tappet adjusting screw (if used) has a depression worn in it (caused by loose valve stem guide, loose tappet guide or both) a new adjusting screw should be installed or the old screw should be smoothed up in a valve refacing tool.

Both the valve stem and its guide should be checked for wear. Valve stem and guide wear at the top is caused by off-center seating of the valve. Wear of stem and guide at the bottom is caused by too much clearance between the valve tappet and its guide.

VALVE GUIDES

The exact stem-to-guide clearance varies considerably in different makes of engines (see car chapters). Usually, greater clearance is specified for the exhaust valve than for the intake by the engine manufacturer because the exhaust runs much hotter and therefore expands more.

If the clearance is more than .002" greater than specified, new valves and guides should be installed. On engines without removable valve guides, valves with oversize stems should be installed and the valve guides reamed to provide the specified clearance.

Reaming Valves Guides

After installing new guides, or if valves with oversize stems are to be installed, the new guides or guide holes should be reamed. Lubricate the reamer blades with light oil. Insert the reamer in the valve guide and turn it slowly clockwise while pressing down gently.

When reaming is completed, remove the reamer by continuing to turn it clockwise while lifting it out of the hole. Never turn a reamer in the opposite direction as it will damage the cutting blades.

Non-adjustable reamers are made for some popular makes of engines, the reamer being the correct diameter for a given standard size or oversize valve stem.

Measuring Valve Stem-to-Guide Clearance

The stem-to-guide clearance may be measured in four ways: With a micrometer; with a dial gauge; with 1/16" wide shim stock; with Go and No-Go gauges. The surfaces should be perfectly clean before any measurements are made.

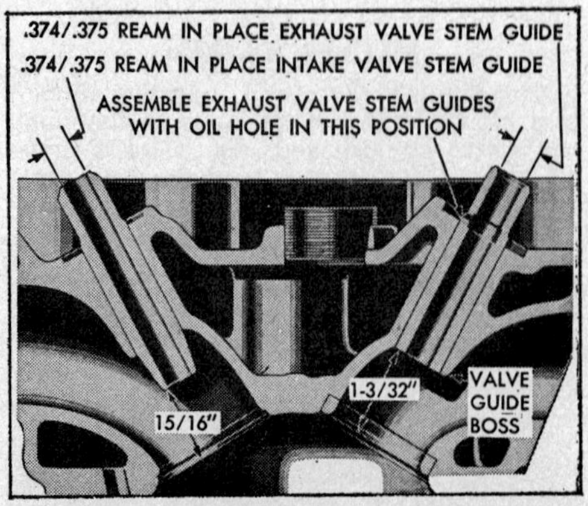

Fig. 10 Make note of distance guide extends into port

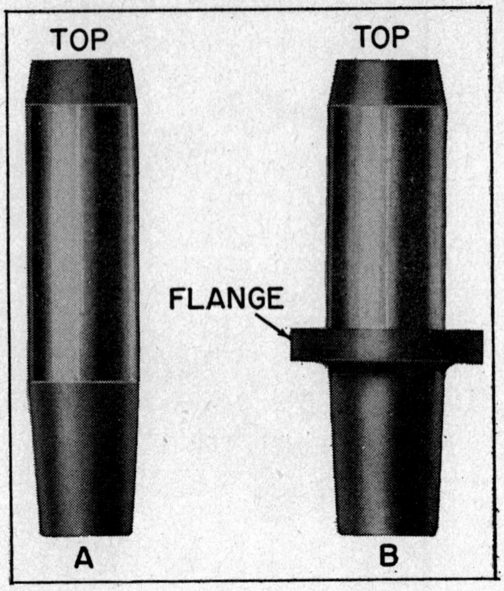

Fig. 9 Two typical valve stem guides

After measuring the clearance and it is found to be excessive, install a new valve. If this does not reduce the clearance appreciably, install a new guide. If the engine is one which has no removable valve guides, ream the guide hole to accommodate the next oversize valve.

When using a dial gauge to measure valve-to-guide clearance, attach gauge to head so that when valve is inserted into guide, the indicator stem can be positioned at right angle to valve stem, Fig. 8. Move the valve to and from the gauge. If the clearance is more than it should be install a new valve and guide.

The Go and No-Go type plug gauges are quite popular because of the facility with which guides can be checked. However, there are two undesirable guide conditions which are difficult to check with gauges of this type: elliptical or egg-shaped bore wear, and bell-mouthing at the port end of the guide. Careful guide inspection will detect egg-shaped wear, and the careful use of the No-Go gauge will tend to show the degree of bell-mouthing.

Measure the diameter of the valve stem with a micrometer and write down the reading. Then measure the diameter of the valve guide with an inside caliper and note the reading. Subtract the former reading from the latter in order to determine the clearance. If the clearance is more than it should be a new valve and/or guide should be installed.

Stem-to-guide clearance may be checked with a 1/16" wide shim stock which is available in various thicknesses. Insert a strip of the shim stock in the guide and then insert the valve stem. By trial and error, select a thickness of shim stock which provides a very slight drag when the valve is moved up and down in the guide. If, for example, a piece of shim stock .003" thick provides a slight drag, then the stem-to-guide clearance is .003".

Valve Guide, Replace

The method to use in removing and replacing a valve guide depends on its construction. The most common type is

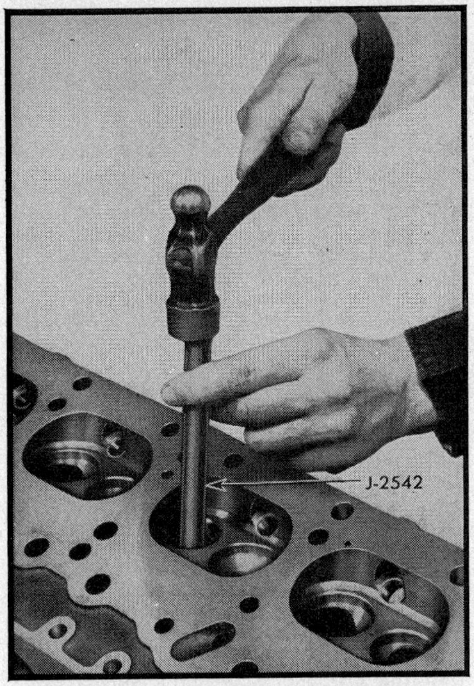

Fig. 11 Removing a valve guide with hammer and drift

shown at left in Fig. 9. It is pressed into the guide hole and held in position by friction. The guide shown in the right view, is used on a few of the older engines. Its flange is intended to fit snugly on top of the cylinder head on overhead valve engines.

On overhead valve engines the straight valve guide can be removed from either the top or bottom of the head, whichever is more convenient.

Barrel type lifters are readily lifted out after the valves and springs are removed but mushroom type lifters cannot be taken out without removing the camshaft unless the lifter guides operate in removable guide brackets which is the case in some older engines.

Before removing a guide, measure the distance it sticks out into the valve port with a steel scale and record this figure. Note that the distance may be different for intake and exhaust guides, Fig. 10.

If the guide has a flange, Fig. 9, it is not necessary to make this measurement because the flange automatically positions the guide.

Fig. 11 shows how a valve guide is driven out with a drift and hammer. The drift has a pilot on its lower end which should be the same diameter as the valve stem.

To install a valve guide, first lubricate its outer surface with engine oil. It may be installed with a hammer and drift, Fig. 12. The pilot on the drift should fit the valve guide hole in order to align the guide while it is being hammered into place. Center the guide in the hole and tap the drift gently until the guide is firmly entered in the hole. The force of the hammer blows can then be increased somewhat. Be careful not to drive the guide in too far. On overhead valve engines, the guide can be driven back by turning the head over and using a drift.

To insure that the guide is driven in just the right distance, special tools are made for given engines. Fig. 12 shows such a tool with a large shoulder which butts against the head when the valve guide is driven in far enough.

Check the clearance between the new guides and valves, and ream the guides if the clearance is too little. Light reaming may also be required if the guides have been slightly distorted during installation.

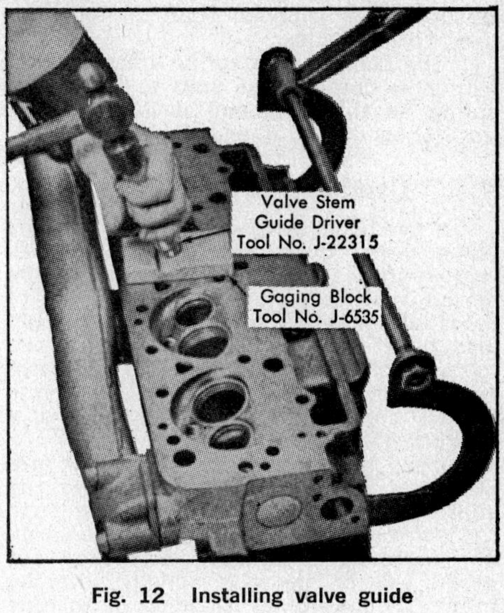

Fig. 12 Installing valve guide

VALVE SPRINGS

Valve Spring Testing

Valves will not operate properly if their springs are weak. Full strength springs are essential if the valves are to follow their cams while closing. If the springs are weak, the valves close later than they should, the result being loss of power, overheated exhaust valves, and increased valve noise. Even slightly weak springs may cause trouble at maximum engine speed.

It is not good practice to do a valve job without testing the springs, because weak springs may be the real cause of the necessity for the valve job, in which case, another valve job will be required long before it would otherwise be necessary.

For example, if a given spring calls for a pressure of 125 pounds when compressed to 1½" and is less than 112 pounds, a new spring should be installed.

There are a number of tools on the market for measuring spring pressure, Figs. 13 and 14.

With the tool shown in Fig. 13, the valve spring is installed between the two plates. Then the handle is pressed down until the spring is compressed to the specified length. The pounds pressure exerted by the spring is registered on the dial.

Fig. 14 shows a fixture which permits using a torque wrench to test the valve spring strength. The hand wheel is adjusted so that the distance from it to the top of the screw equals the specified compressed length of the valve spring. Then the valve spring is put in place and the handle pulled down until its arm contacts the top of the screw. Then the number of pounds pull on the torque wrench is read.

When testing springs, be careful not to compress them beyond the specified length as further compression may permanently weaken them.

Valve Spring, Replace

A new spring can be installed without

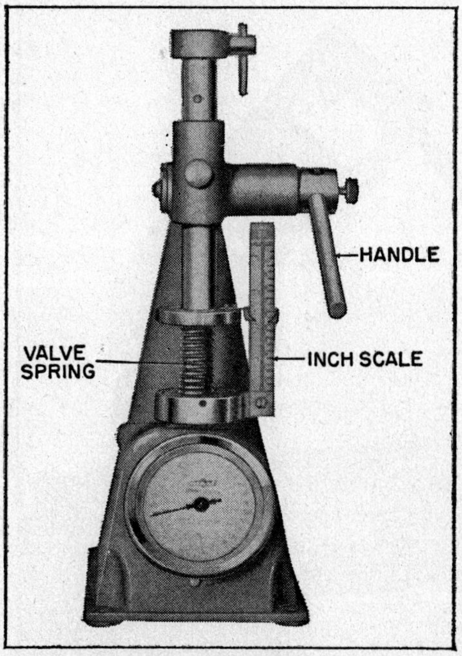

Fig. 13 Valve spring tester with dial gauge to indicate spring pressure

Fig. 14 Valve spring tester used with torque wrench to indicate spring pressure

removing the cylinder head on all overhead valve engines.

If the coils on the spring are closer together at one end, be sure to install the spring so that this end is in the same position as on the other springs.

Overhead Valve Springs

On overhead valve engines remove the rocker arm cover. Before attempting to remove the spring the valve should be closed so that minimum compression of the spring is necessary. The piston should also be at or near the top of its stroke to prevent the possibility of the valve falling down into the bore while the broken spring is being removed and a new one installed.

Loosen the locknut on the rocker arm adjusting stud (if used), and screw the stud upward so that the top of the push rod can be moved out of the rocker arm. The rocker arm can now be moved out of the way of the valve spring, either by tilting it upward or pushing it to one side.

Using a suitable valve spring compressor, remove the valve spring lock and spring. Then install the new spring. To hold the valve up while removing or replacing the spring, insert a bent rod through the spark plug hole, pressing its inner end against the head of the valve.

Valve Spring Installed Height

When valves and seats are reground the position of the valve in the head is changed so as to lessen the valve spring tension. Without proper valve spring tension the valve does not seat long enough or it may not seat completely. Since the valve is cooled by transferring heat from the valve head to the seat and thence to the coolant, improper valve spring tension will cause worn, pitted and distorted valves which result in loss of compression and power as well as poor gasoline mileage.

When valves, springs, retainers and locks are installed, measure the assembled height of the valve springs from the surface of the cylinder head spring pad to the underside of the spring retainer as shown in Fig. 15. If the assembled height is greater than the dimension given in the *Valve Specification Chart* in the car chapter, install a spacer or shim of proper thickness between cylinder head spring pad and spring to bring the assembled height to specifications.

Do not install spacers unless necessary. Excessive use of spacers will result in overstressing valve springs and overloading camshaft lobes which could lead to spring breakage and worn comshaft lobes.

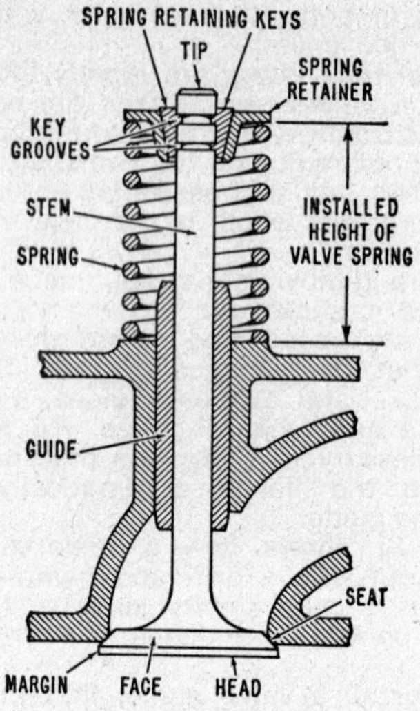

Fig. 15 Checking installed height of valve spring

GRINDING VALVE STEMS

For Proper Hydraulic Lifter Clearance

If a valve grind job is needed, and if shims are found under the valve springs, very likely the valves, if reground, would not leave sufficient margin on the valve head for satisfactory operation (see Fig. 15). Since the maximum thickness of shimming that can be used satisfactorily is 1/16", and since the acceptable valve head margin is approximately 3/64" to 5/64", one can readily see that there would be insufficient margin left on the valve head if it was reground. Therefore, it is recommended that new valves be installed.

After grinding the valves, if shims are required to restore the correct valve spring installed height, the valve stem should be ground off the same amount as the thickness of shims required. This will automatically provide the proper static clearance between the hydraulic lifter and the push rod.

VALVE GRINDING

It seems almost unnecessary to explain that the purpose of grinding valves is to cure valve leakage. Valve grinding smooths up the face of the valve seat and valve face so that the two refinished surfaces mate so perfectly that leakage is eliminated.

It is customary to use two specially designed electrically-driven grinding tools, one for the valve and the other for the seat.

Electric Valve Face Grinder

There are two kinds of tools. One re-

Fig. 16 Electric valve face grinder

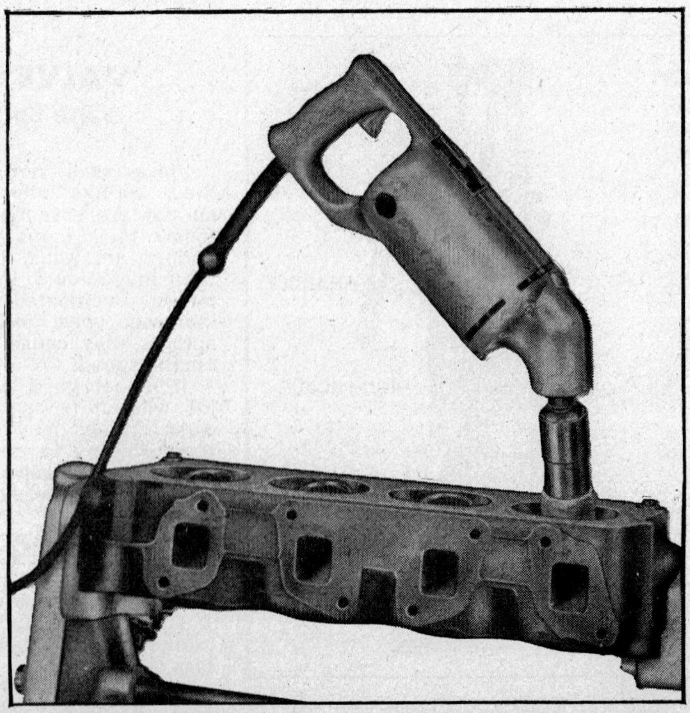

Fig. 17 Electric valve seat refacer

faces the valve, Fig. 16, and the other refinishes the valve seat, Fig. 17. The valve seat refinishing tool is equipped with a grinding wheel that has the same angle as the valve seat.

In the valve refinishing tool, Fig. 16, the face of the valve is parallel to the face of the grinding wheel. This means that the stem of a 45-degree angle valve makes a 45-degree angle with the face of the wheel.

Both the grinding wheel and the valve rotate. In the tool shown, the grinding wheel and its carriage can be moved right or left by swinging the lever. In this way the rim of the wheel is aligned with the face of the valve.

The workhead is mounted on a cross-feed carriage which can be moved toward the grinding wheel or away from it by turning the hand wheel. To grind a valve, the valve face is moved into light contact with the grinding wheel.

Electric Valve Seat Grinder

A grinding wheel operated by an electric drill is used to reface valve seats. A steel cutter may be used except in the case of valve seat inserts which are too hard for the cutter to do a satisfactory job.

One such grinder has a concentric grinding wheel which contacts the seat all around its rim. This design is often made so that the wheel vibrates or bounces off of the seat about once every revolution. The grinding wheel is lined up on the seat by means of a pilot which is inserted in the valve guide. Only a few seconds are required to reface a cast iron seat, a little more time is necessary for refacing the much harder valve seat inserts. This type tool is always used in conjunction with a pilot inserted through the valve guide so that a concentric refinishing job will be obtained.

Hand Grinding

To hand-grind a valve into its seat, a thin coating of valve grinding compound is spread on the face of the valve. A common valve grinding tool has a rubber suction (vacuum) cup which is pressed against the valve and then it is rotated between the two hands while pressing down just enough to force the valve against its seat. A light spring, just strong enough to lift the valve, should be placed under the head of the valve. After every few oscillations the spring is allowed to lift the valve and the valve is rotated through a part of a revolution. Then the valve is ground for a few more oscillations, and so on. About every minute the valve is removed and both the valve face and the seat are inspected. When the seat has been ground down so that bright metal shows all around its circumference, and when a similar band of smooth surface appears on the valve face, the valve may be considered finished.

Hand grinding is a slow, time-consuming job and the accuracy obtained is not likely to be as good as with suitable power tools, especially if the valve stem is somewhat loose in its guide.

However, slight grinding or lapping of valves and seats that have been refinished by power tools is considered good practice. The valve is hand lapped in its seat with very light pressure and just long enough to be sure the valve is perfectly tight on its seat.

After each valve is ground, both the valve and its seat should be carefully cleaned of abrasive with a rag moistened with gasoline or kerosene.

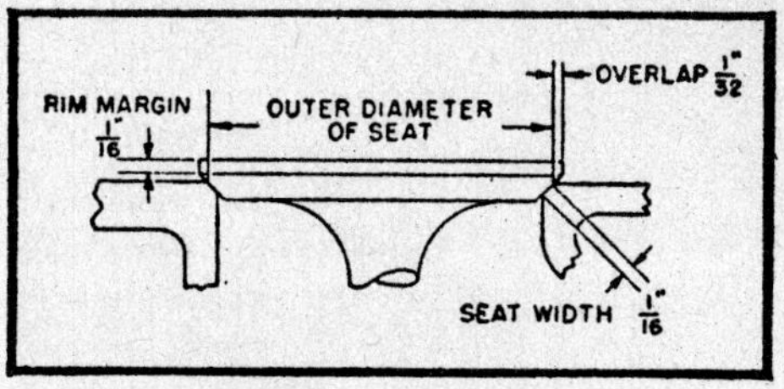

Fig. 18 Important valve and seat dimensions

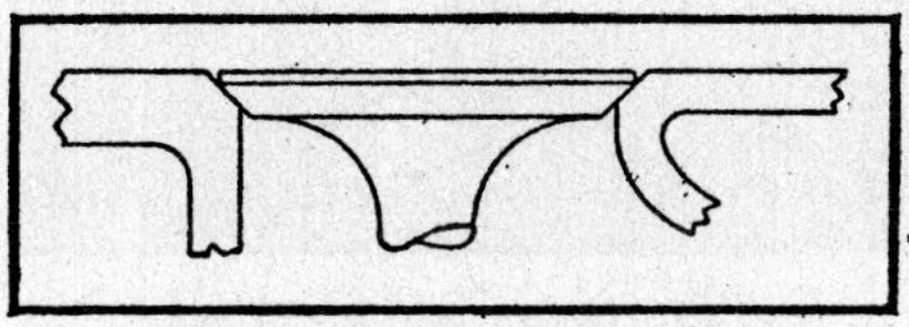

Fig. 19 Grinding valve seat alters dimensions shown in Fig. 18

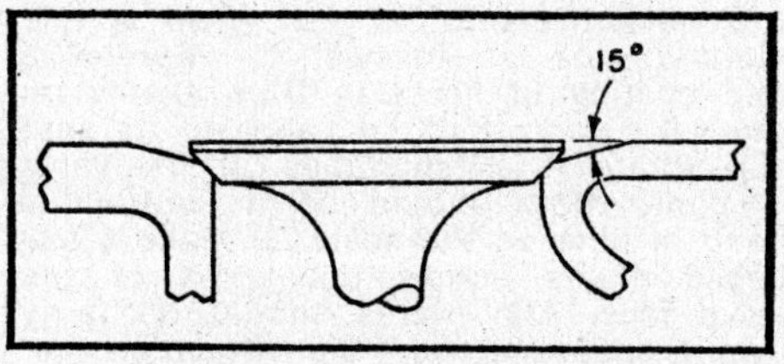

Fig. 20 Overlap of valve is restored by a 15-degree bevel

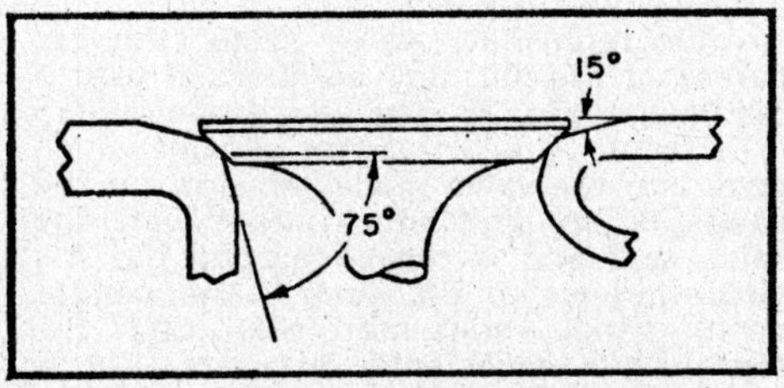

Fig. 21 If seat is too wide after beveling top (Fig. 20) the throat is beveled with a 75-degree stone as indicated

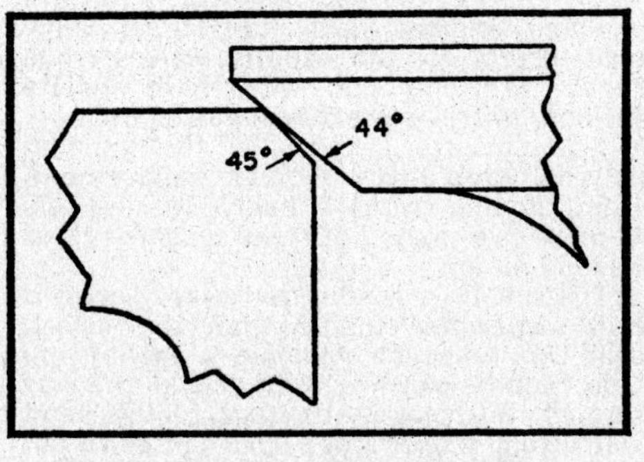

Fig. 22 Diagram showing a 44-degree face resting on a 45-degree seat

Valve Seat Widths

When grinding valve seats it is important to have the seat the correct width. If the seat is too narrow, it may wear unduly because it is not wide enough properly to resist the continual pounding against it. Also, an exhaust valve seat that is too narrow may overheat the valve because of inadequate contact area between valve and seat.

On the other hand, if the valve seat is too wide, the valve may not seat tightly. Other things being equal the wider the seat the more likely it is to leak.

The reason the valve with the wider seat is more likely to leak is that it has less pressure per square inch to hold it down. If the valve spring pull is 100 pounds and the valve face area for the $1/16$" width were .293 square inch, then the pressure on the seat would be 341 pounds per square inch. However, if this seat were enlarged to a width of $3/16$", the pressure would be reduced to 114 pounds per square inch because the area would be three times as great for the same pounds pressure of the spring.

Therefore, for best valve service, the width of the seat should be neither too much nor too little. The recommendations of the various car makers should be carefully followed, but if in doubt as to what width is best, you can't go wrong if you make the seat width $1/16$", Fig. 18. Note that this picture represents a typical valve in a *new* engine.

The face of the valve should extend about $1/32$" beyond the face of the seat, Fig. 18. This overlapping is desirable so that the valve face will have full contact with the seat. On the other hand, too much overlap invites overheating of the rim of the exhaust valve.

After a valve is ground, if its rim margin is less than $1/32$" thick, the valve should be discarded because a thinner rim may become so hot that it will cause pre-ignition.

When a valve seat is ground, both the width of the seat and the outer diameter of the seat are increased. Fig. 19 shows the outer diameter of the seat has increased, the width of the seat is greater, the rim of the valve no longer overlaps the outer circumference of the seat, the valve has moved closer to its tappet.

Usually it is only necessary to remove a few thousandths of an inch from the seat to smooth it up. In fact so little may be removed that seat width etc., may not be appreciably affected. On the other hand, if the seat is badly burned, it is necessary to grind off all the burned metal, in which case, the dimensions mentioned may be greatly increased as shown.

The outer diameter of the seat can be reduced by means of a 15-degree grinding wheel, Fig. 20. Enough material is ground off the seat so that its outer diameter is reduced to what it was in a new engine and overlap is now correct. This operation also reduces the width of the seat.

But if the seat is still too wide, its width should be reduced to the correct dimension with a 75-degree grinding wheel, Fig. 21.

Removing material from the top of the seat is called topping while removing it from the throat of the valve port is called throating. The wheel angles for topping usually are either 15 or 20 degrees and for throating 60 to 75 degrees.

When considerable material is removed from the valve seat, the inner rim of the seat may extend down into the rough casting which forms the valve port. If this rim is uneven, the throating wheel should be used to make the rim perfectly circular but in doing so be careful not to make the seat too narrow.

In some cases where exhaust valve burning is difficult to overcome, it has been found very satisfactory to make the angle of the valve face ¼ to 1 degree flatter than the seat face, Fig. 22. In other words, with a 45 degree seat, the angle of the valve face is made somewhere between 44¾ and 44 degrees. (The angle between the valve face and the seat is called the interference angle.) Thus the valve seats tightly only at the outer edge of the seat as shown. The narrow contact between valve and seat means that the unit pressure is very high between the narrow contacting surfaces, hence the tight seating. Also because of the high contact pressure any particles lodging on face or seat are likely to be quickly pulverized and blown away.

The use of an interference angle naturally reduces the contact area between valve face and seat and therefore it might be assumed that the flow of heat from the hot exhaust valve would be throttled and exhaust temperature would be dangerously increased. Nevertheless, it is apparent that with a narrow contact the pressure with which the valve face pushes against the seat is so high and the contact is so intimate that heat flows from valve to seat very readily whereas with valves and seats having exactly the same angle the full width contact means that the intensity of pressure between them is relatively low and therefore the contact is less intimate. Hence, heat transfer from valve to seat is not as good as might otherwise be expected.

It should also be noted that exhaust valves usually cause not trouble until they begin to leak. Hence, since the narrow contact suppresses leakage the valve job is likely to last longer. And with leakage almost entirely absent, the actual exhaust valve temperature may be less with a narrow contact than with a wider one which permits some leakage right from the start. This leakage may not be enough to be noticed in the operation of the engine but it may be sufficient to elevate the valve temperature to a point which hastens the destruction of the valve job.

The interference angle, as it is called, is obtained by adjusting the workhead on the valve grinding tool to somewhere between 44 and 44¾ degrees (in the case of a 45-degree seat). For a 30-degree seat, the valve face angle should be 29 to 29¾ degrees.

After refinishing valve faces and seats with power tools, it is a good plan to grind them lightly by hand. Spread just a little fine valve grinding compound on the face of the valve. Then grind the valve for just a few strokes but do not exert any pressure. The weight of the valve and the hand tool is sufficient.

After the valve grinding job is finished,

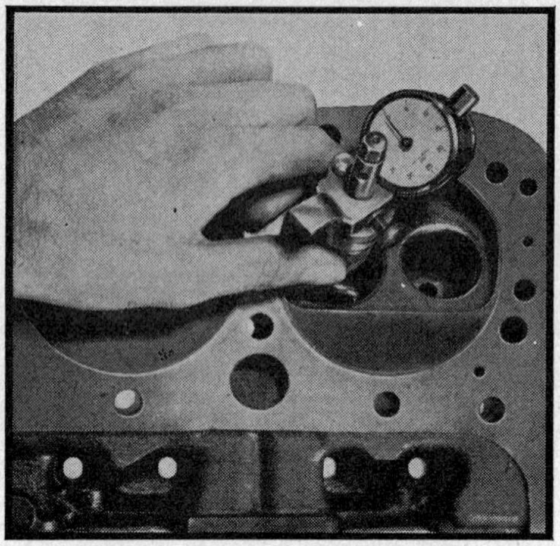

Fig. 23 Testing accuracy of valve seat with a dial gauge

it is desirable to determine how tightly the valve fits into its seat. Two simple methods are as follows: (1) Spread a thin coating of Prussian Blue evenly on the valve face. Put the valve on its seat and rotate it a few times. If the valve is tight, there should be a continuous band of blue on the seat. (2) Make a few pencil marks across the width of the valve face. The marks should be fairly evenly spaced around the circumference. Put the valve on its seat and rotate it a few times. Remove the valve and inspect the pencil marks. If all of them show signs of being rubbed by the seat, the job is satisfactory.

After refinishing valve seats, the accuracy of the job may be checked with a dial gauge, Fig. 23. The gauge is mounted in a fixture which includes a pilot which goes into the valve guide. An arm on the gauge is brought into contact with the valve seat and is moved around the full circumference of the seat. If the pointer on the gauge varies more than .002″ the seat is not sufficiently accurate. For a really good job the variation should be .001″ or less.

VALVE SEAT INSERTS

Some engines have hard steel valve seat inserts for the exhaust valves while others are equipped with these inserts for both intake and exhaust valves.

Seat inserts, especially for exhaust valves, occasionally require replacement. Hard driving with a badly leaking exhaust valve may burn the exhaust seat beyond repair.

The insert is readily removed. Using a drill somewhat smaller than the insert, drill two holes at opposite sides of the insert but be careful not to drill all the way through the seat as damage to the head will result. Use a sharp cold chisel to cut through the holes and thus split the seat in two halves.

After the seat has been removed, clean the counterbore in the block so that the new seat, when installed, will sit squarely in the recess. Note also that dirt in the recess will prevent proper conduction of heat from the seat to the block.

The seat is slightly larger in diameter than the recess. Therefore, to install a new seat, pack it in "dry ice" for at least 15 minutes in order to contract it. Line the cold seat up squarely with the recess and use a hammer and block of wood to force it in place.

If a standard size insert is too loose, the recess should be enlarged by counterboring to fit an oversize seat. Check the diameter of the oversize seat before counterboring. For best results, the insert should be .003″ larger than the diameter of the recess when both are at the same temperature (room temperature) although an insert which is anywhere from .002″ to .004″ larger is considered satisfactory.

Valve seat inserts are so hard that when they need refacing a high-speed grinding stone should be used. They cannot be ground satisfactorily by lapping with valve grinding compound.

HYDRAULIC VALVE LIFTERS

The purpose of the hydraulic valve lifter (or tappet) is to provide perfect adjustment at all times between the rocker arm and the valve stem on overhead valve engines. The oil-operated hydraulic lifter provides this perfect adjustment automatically.

An ideal hydraulic lifter would be one that keeps in constant contact with the push rod regardless of how the valve stem or push rod and cylinder block expand and contract under variations in temperature. With constant contact, the clearance is always zero when the valve is closed.

With this ideal of zero clearance, valve noise is reduced to a minimum. During the valve opening process, the lifter is already in contact with the pushrod and therefore the noise of contact is eliminated. During the closing process, the valve comes gradually to rest on its seat instead of hitting it with a sharp noisy blow.

This zero clearance obviously cannot be obtained with screw-adjusted valve lifters but is obtained with automatically adjusted hydraulic lifters.

The principle of the hydraulic lifter is easy to understand. At the base of the plunger is a small chamber filled with oil. Since oil (and all other liquids) is practically incompressible, the oil trapped in this chamber provides a positive drive from cam to valve stem. When the valve is being opened by the cam, the drive is just as positive as it would be if a steel spacer were substituted for the oil. It is a mistake to assume that the oil gives a cushioning effect. On the contrary, the oil provides a perfectly rigid drive.

Although the plunger diameter has a clearance of a mere fraction of a thousandth of an inch, it is obvious that there must be some slight leakage between the plunger and its bore whenever the valve is open because of the high pressure of the valve spring which pushes down on the plunger. To compensate for this leakage, oil is supplied under pressure to the oil chamber beneath the plunger.

The leakage rate is determined by the amount of clearance between the plunger and the plunger bore. The desired clearance is a fraction of a thousandth of an inch and this clearance must be closely approximated on all lifters made for a

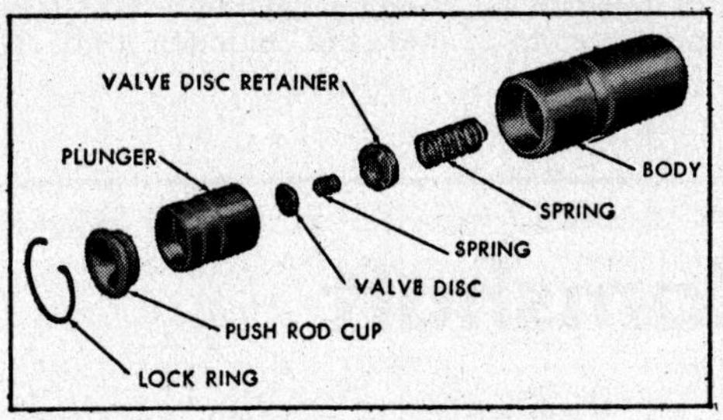

Fig. 24 Typical hydraulic valve lifter. See car chapters for variations of this unit

given model of engine. Therefore after the plungers and bores have been manufactured as accurately as possible they are classified into a series of sizes from a trifle small to a trifle large. Small plungers are assembled to small bores, large to large, and likewise for the sizes in between. In this way satisfactory uniform clearance is obtained in all plungers and bores even though individual plungers and bores may vary in size by a small amount. This "selective," fitting of plungers and bores explains why it is important during repair work that a given plunger always be returned to its own bore.

If the plunger clearance is too great, either because of wear or because of mismating of plunger and bore, the oil may leak away faster than it is supplied to the oil chamber. The chamber will never be full and therefore the lifter will be noisy.

If the plunger is too tight in its bore because of mismating, the plunger action may be sluggish and noisy lifters may result. Or more likely, the leakage rate will be so slow that the valve is held off its seat when it should be closed.

Hydraulic Lifter Troubles

The most frequent causes of hydraulic valve lifter troubles are dirty oil or failure to change the oil filter at suggested intervals. When these causes are eliminated, the trouble will generally correct itself without costly disassembly.

The easiest method for locating a noisy valve lifter is by the use of a piece of garden hose about 4 feet long. Place one end of the hose near each valve in progression and listen through the other end. In this manner the sound is localized, making it easy to determine which lifter is at fault.

Another method is to place a finger on the valve spring retainer. If the lifter is not functioning properly, a distinct shock will be felt when the valve returns to its seat.

In most cases where noise exists in one or more lifter, all lifter units should be removed and cleaned. If dirt, varnish or carbon is found to exist in one unit, it more than likely exists in all the units.

The oil level in the oil pan should never be above the "Full" mark on the dipstick nor below the "Add Oil" mark. Either condition could be responsible for noisy valve lifters.

If the oil level is too high it is possible that the connecting rods can dip into the oil when the engine is running and create foaming. This foam (air mixed with oil) is fed to the hydraulic lifter by the oil pump, and the air gets into the lifters, causing them to go flat and allow the valves to seat noisily.

Low oil level may allow the oil pump to take in air which, when fed to the lifters, causes them to lose length and allows the valves to seat noisily. Any leaks on the intake side of the oil pump through which air can be drawn will create the same lifter action. When the lifter noise is due to aeration, it may be intermittent or constant, and it will always cause more than one lifter to be noisy. When the oil level and leaks have been corrected, the engine should be run at a fast idle speed for sufficient time to allow all of the air inside of the lifter to be worked out.

It should be remembered that worn valve guides or cocked valve springs are sometimes mistaken for noisy lifters. If such is the case, the noise in all probability will be dampened by applying side thrust with your thumb on the valve spring. If the noise is not appreciably reduced, it can then be assumed that the noise is in the hydraulic lifter.

Servicing Hydraulic Lifters

Inasmuch as the removable plunger type lifters have not been in use for quite a number of years, the following discussion is concerned only with the unit type lifter.

With this type lifter, Fig. 24, the units may be lifted out with the fingers. However, if the lifter is stuck in its guide due to varnish or gum substances, use pliers with taped jaws or pull it out with a hooked wire or a magnet, Fig. 25.

If there is a ring of carbon on the guide just above the top of the lifter it will be necessary to remove the carbon before the lifter can be taken out. Soften up the carbon with a solvent and remove with a stiff bristle brush.

As shown in the illustration, these units can be completely disassembled. The first step is to remove the lock wire by prying it out of its groove with a small screwdriver or awl. Hold the thumb over the opening to prevent the lock wire from flying away when it comes loose.

If there is a rim of carbon at the upper end of the plunger bore, soak the unit in solvent (lacquer thinner is recommended) and remove the carbon with a stiff bristle brush. The carbon will interfere with the removal of the plunger.

Next turn the unit upside down and strike it against the palm of the hand to force the assembly out. Do this carefully so that the parts, when they do come out, do not scatter. If pounding the lifter against the palm fails to break the plunger loose, pound it on a block of wood.

Be careful not to mix up plungers and bodies. To avoid this possibility, disassemble and assemble them one at a time. Wash the parts in solvent and then examine them for defects such as scratches, pits and dents. If any are found, discard the complete unit.

Insert the bare plunger in the body. It should move smoothly when pushed gently with a finger. In some cases it will even fall of its own weight if the unit is held vertical. If there is extra resistance, discard the assembly.

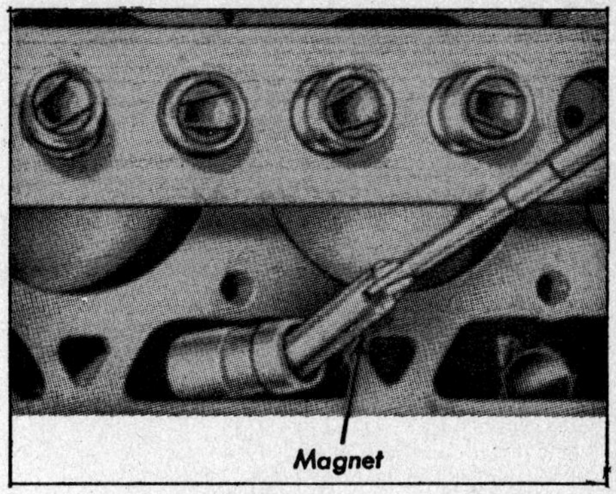

Fig. 25 Removing valve lifter with a magnet rod

To check for leakage, make an air compression test. To do this, turn the plunger upside down and install the check valve and the retainer but do not install the spring. While holding the plunger assembly with one hand, slide the body down over the plunger. Still holding the unit upside down, pump the plunger with the little finger or with a screwdriver. If the plunger bounces, the unit is all right. On the other hand if it does not bounce, examine the check valve for damage. If a ball valve is used, examine it for a flat spot and, if one is found, install a new ball of the same size. However, if the ball or flat valve is in good condition and the compression is poor, the plunger leaks and therefore the complete unit should be discarded.

Assembling Lifters

To assemble, turn the plunger upside down and install the check valve, its retainer and plunger spring. Slip the lifter body on the plunger assembly. Turn the assembly over and install the lock wire, being sure it snaps into its groove in the lifter body. A discarded plunger or a special tool for the purpose may be used to press the plunger down in order to insert the lock wire.

If neither a special tool nor a discarded plunger is available, the following method may be used. Assemble the parts as before. Then slide the body down over the plunger. Take a small piece of wire (a paper clip will do) or a small punch, and insert it in the oil hole in the body. Press the plunger down slowly almost as far as it will go with one finger of the other hand until the wire or punch slips into the oil hole in the plunger. This locks the plunger. Install the plunger cap. Insert the open end of the lock ring in the lifter body. Then force the opposite side of the lock ring into the groove with a small, blunt tool such as a screwdriver.

It is recommended that lifters of this type be filled with engine oil before replacing them in the engine. However, many mechanics rarely take the time to do this. In other words, they are installed dry and then filled by the oil pressure supplied by the idling engine. Apparently no harm is done by neglecting to fill the lifters. However, as there is some risk of scuffing, at least dip the plungers in engine oil before installing them in the engine.

A better method, which is quick and easy, is to put the dry lifter in a pan of engine oil and pump the plunger until the lifter is reasonably full. The oil level in the pan, of course, should be high enough to cover the oil hole in the side of the lifter body. The plunger should be pumped just a few strokes until the plunger travel becomes a minimum. It is possible to pump the plunger until it is rigid.

Timing Gears, Chain & Camshaft

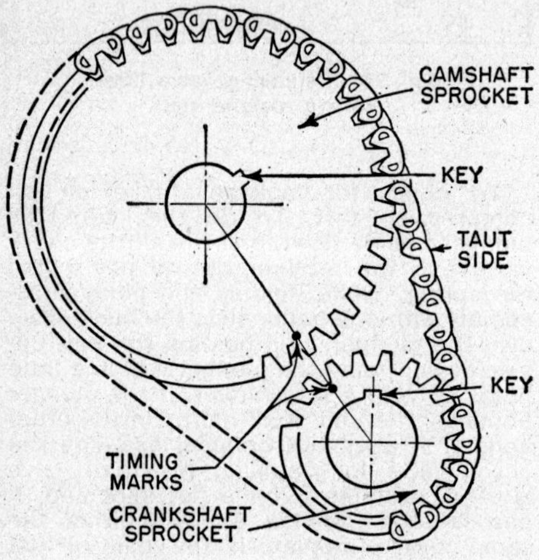

Fig. 26 Example of sprocket marks lined up for correct valve timing. See car chapters for specific requirements

VALVE TIMING

When the camshaft is driven by gears it is customary to mark the gears with either two or three timing marks. When three marks are used the timing will be correct when the marks are meshed. When two marks are used, one mark will most likely be on a crankshaft gear tooth and the other in the valley between two camshaft gear teeth and these two marks should be in a direct line facing each other.

In the case of timing chain and sprockets there are also two common methods of marking. In Fig. 26 it will be seen that the marks on the sprockets line up with a line drawn between the centers of the crankshaft and camshaft. In some cases the correct camshaft setting is determined by the number of chain links or pins between the two marks on the sprockets (see car chapters).

Chain slippage is rare on modern cars because the chains are relatively short (the shorter the chain the less it stretches as it wears). However, it does happen on occasion. Usually the engine will start but it will run with very little power. *The easiest way to determine if the chain has slipped in an otherwise good running engine is to remove the carburetor air cleaner and crank the engine with the starter. If very little suction is felt on the palm of the hand when it is used to cover the carburetor air horn it is a sure indication that the valve timing is out.*

CAMSHAFT END THRUST

It should be realized that the camshaft must be held in a fore and aft position so that camshaft bearings and bearing journals line up and so that the cams are in line with their valve lifters. There are the following common ways of doing this.

A thrust plate or washer is bolted to the front of the cylinder block. The hole in the plate is smaller than the outer diameter of the hub of the sprocket and also smaller than the front camshaft journal. Therefore the plate prevents the camshaft from moving either to front or rear of its correct position.

On all engines equipped with thrust plates it is desirable to install a new thrust plate whenever a new camshaft gear is installed. There should be a slight amount of clearance at the thrust plate, say .002″.

However, on most engines with thrust plates the parts are so designed that when the camshaft gear is properly installed the correct clearance is obtained. For example, in one instance there is a narrow ring spacer washer between the camshaft and the hole in the thrust plate. The washer insures correct clearance because it is a trifle thicker than the thrust plate and thus prevents drawing up the gear so tight that it might clamp itself to the thrust plate.

A different construction has a shoulder on the camshaft sprocket hub which butts against the camshaft and thus provides the required thrust plate clearance.

In another case there is a thrust washer that prevents the cam sprocket from moving rearward while a spring-backed plunger prevents the camshaft from moving forward.

The thrust plate or washer should not be confused with the engine front plate. Fig. 27 makes this point clear. The thrust washer, which is between 3 and 4 inches in diameter, lies within a hole in the front plate. A thrust spacer is located between the camshaft and a hole in the thrust plate. This spacer insures correct lengthwise clearance between cam gear hub and thrust washer. The latter, it should be noted, is bolted to the front of the block.

TIMING GEARS, REPLACE

The camshaft gear on automotive engines is usually made of a highly compressed fibrous composition material and therefore is called a fibre or composition gear. Some engines, however, have aluminum gears. The crankshaft gear is always steel.

The fibre gear provides a quieter drive than a metal gear. The fibre gear wears somewhat faster than the steel crankshaft gear. Therefore, it the gears are noisy it

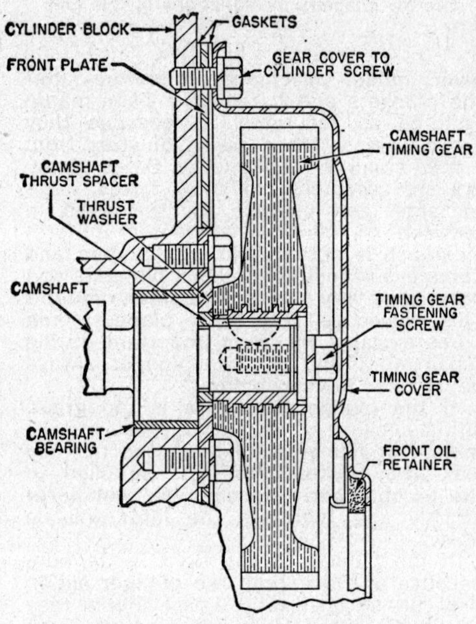

Fig. 27 Drawing shows relationship between thrust washer and front plate

may be sufficient to install only a camshaft gear. In some cases camshaft gears with teeth of oversize thickness are available. Thus, this so-called oversize gear compensates for the wear of the teeth on the camshaft gear, assuming it is not replaced.

Clearance between teeth of a new pair of gears when installed, Fig. 28, should be somewhere between .002″ and .004″. The same clearance limits are desirable if an oversize camshaft gear is mated with a worn crankshaft gear.

However, it should not be forgotten that a gear with a slight wobble may be noisy. Therefore, if tooth clearance is satisfactory, check the gears for runout before removing them, Fig. 29. If runout is more than .004″, check for burrs on the gear hub and camshaft where the gear is mounted. If the runout cannot be brought below .004″, it is desirable to install a new gear. The runout of the crankshaft gear should not be more than .003″.

Removal

If the camshaft gear is not retained by a nut or bolt in the end of the shaft, it is a safe assumption that the gear is a press fit and that the power of an arbor press will be necessary to remove the gear. This means that the camshaft must be removed from the car.

Fig. 28 Checking clearance between timing gears

Fig. 29 Using a dial gauge to check camshaft gear for runout. Screwdriver is used as a pry bar to rotate gear

Fig. 30 Puller removing camshaft gear. Same tool is used to remove crank gear

If the camshaft gear is bolted on it is readily removed by taking off the bolts or screws holding it.

If the camshaft gear is retained by a nut on the end of the shaft, either a pry bar may be used to remove it, if it is not on too tight. Otherwise use a puller of which there are various types.

In Fig. 30 there are two threaded holes in the hub of the gear. The puller screws are inserted in these holes to the full depth of the hub, otherwise the screw threads may be damaged or stripped. Then turn the center screw of the puller to remove the gear. The same tool is used to remove the crankshaft gear.

Installation

If a pusher tool is not available the new crankshaft gear can be installed with a soft hammer or hammer and hollow drift, such as a piece of pipe.

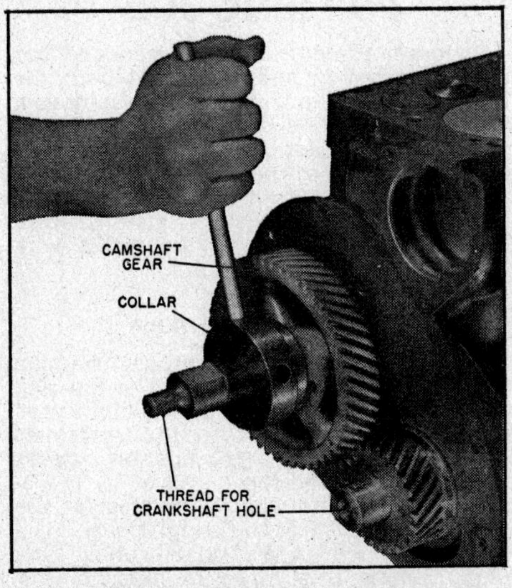

Fig. 31 Pusher tool installing camshaft gear. Same tool is used to install crank gear

After the camshaft and crankshaft gears have been reinstalled on the engine, it is a good plan to check the gears for runout with a dial indicator, Fig. 29.

Camshaft gear run-out of more than .004″ is not desirable. The figure is .003″ for the crankshaft gear. To check either gear for run-out place the tip of the dial gauge in contact with the rim of the gear are shown and rotate the gear through one revolution. High and low readings for the camshaft gear should not vary by more than .004″ for the cam gear and .003″ for the crank gear.

Fig. 31 shows a tool installing the camshaft gear. The same tool installs the crank gear.

TIMING CHAIN AND SPROCKETS, REPLACE

When both the chain and the sprockets are new there should be about ½″ sag or slack as shown by the broken white line, Fig. 32, if the chain is pushed as indicated by the white arrow. Note that this test is made on the slack side of the chain and that the other side of the chain is under tension because the rotation of the crankshaft sprocket pulls on the chain to rotate the camshaft sprocket. The motion of the sprockets and chain is shown by the white arrows. A chain that is too tight will be noisy.

It is desirable to put on a new chain when wear allows the chain to sag one inch when pressed inward on the slack side.

The camshaft and crankshaft sprockets should also be examined for wear and replaced if visibly worn or scored. Badly worn sprockets will shorten the life of a new chain and may cause the drive to become objectionably noisy.

To remove the chain it is always necessary to remove the camshaft sprocket and sometimes it is necessary to remove both sprockets and the chain as a unit. See car chapters for details.

CAMSHAFT BEARINGS, REPLACE

The installation of new camshaft bearings is not a common operation. They have comparatively light loads to carry as compared with crankshaft bearings. Therefore if the camshaft bearings are properly lubricated with clean oil they will usually last for the life of the engine.

If, for some reason, the camshaft bearings do not receive sufficient oil they may burn out, that is, the bearing surface may melt and thus be ruined, in which case new ones will be required.

Also, if examination with a narrow feeler gauge shows that the clearance is .006″ or more, new bearings are called for if satisfactory valve action and valve life are to be obtained.

Fig. 32 Slack in new chain may be ½ inch

If the engine has been removed from the car for an overhaul, it is desirable to install new camshaft bearings if the clearance exceeds .0035" or .004".

Various tools are made for removing and replacing camshaft bearings, but usually they consist of piloted drifts to pull out the bearings and to push them in place.

The rear camshaft bearing rests in a hole in the cylinder block which must be sealed at the rear to prevent oil leakage. A bolted-on plate is sometimes used although on most engines a disc of metal with a slight crown is driven into the hole and thus expanded. The disc is called a welch plug, freeze plug or expansion plug. Once the camshaft is removed the plug can be driven out so that the rear bearing may be removed.

If an oil leak should develop at the plug it can be pried out after drilling or punching a hole in its center and then a new one can be driven into place.

There are several of these plugs along the side of the block in addition to the one at the rear. They are used merely to plug up the holes which are necessary to allow the sand to be removed after the block has been cast. The sand is used in the casting process to form water passages (water jackets) for the cooling system.

Pistons, Rings & Rods

EXTERNAL OIL LOSS

Since an abnormal amount of oil may be leaving the engine in any one or any combination of ways, it is good practice to make a thorough examination of the engine before it is disassembled for repairs.

Dense, blue smoke coming from the exhaust while the engine is running at a fast idle speed indicates that too much oil is being used.

Leaky gaskets, leaking metal-to-metal joints, leaky oil seals, or any other sources of oil leaving the crankcase should be located as definitely as possible, because after disassembly is once started, it is more difficult to locate such sources of trouble.

Fresh oil on the clutch housing, oil pan, fuel pump, edges of valve covers, external oil lines, distributor shaft housing, base of crankcase filler tube, or on the bottom of the timing case cover is a fairly good indication of an external leak close to that point.

The hole in a cylinder block at the rear end of the camshaft is sealed with either a welch plug or a cover plate. Quite often oil leaking downward from this opening is found on the rear end of the oil pan or on the bottom of the clutch housing and is mistaken for a rear main bearing leak.

Any area underneath an engine that is washed clean with oil is an indication that there may be an external leak from the crankcase somewhere in front of the washed area.

The crankcase ventilation (PCV) system should be examined to see if it is partially or completely clogged, thus causing high crankcase pressure to force oil out through joints that normally would not leak.

INTERNAL OIL LOSS

If the symptoms point to leaky rings, remove the spark plugs. If they are oily, the combustion chambers are receiving too much oil but do not forget that this may be due to: (1) Stuck rings which perhaps can be freed with a solvent. (2) A leaking vacuum pump diaphragm, if used. (3) Leaky intake and exhaust valve guides.

If blue smoke is seen at the exhaust, especially when the throttle is abruptly closed after decelerating from speed the rings are either worn or stuck.

If spark plugs are wet after engine has been idling for ten minutes but are dry after a few miles of driving, the cause is probably leaky intake valve guides.

If spark plugs are wet after road driving, either the vacuum pump leaks or the rings leak due to wear or sticking. Be sure the screws on the vacuum diaphragm are tight before making this test.

Worn main, rod or camshaft bearings cause increased oil consumption. The oil which escapes from these bearings is thrown in all directions by the whirling crankshaft and therefore an excess amount is tossed onto the cylinder walls. Some of this excess oil makes its way to the combustion chamber where it is burned.

Leakage of oil from the bearings causes a drop in oil pressure. Therefore, when deciding whether or not a ring job is necessary it is important to know whether worn bearings are contributing to the high oil consumption. So be sure to check the oil pressure—with the engine warm. If the oil pressure is 5 to 10 pounds below normal there is a good chance that bearings are worn. But do not forget that reduced oil pressure may be caused by various other troubles including worn oil pump parts.

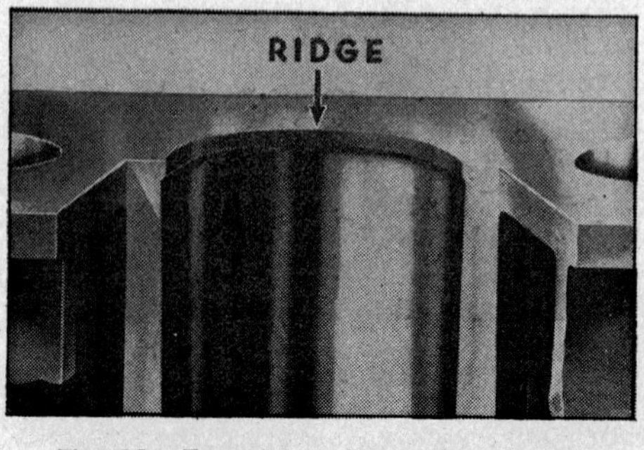

Fig. 33 Top piston ring undercuts cylinder bore leaving a ridge which must be removed before piston is taken out

CYLINDER BLOCK INSPECTION

Before doing any work on rings, pistons, etc., it is advisable to inspect the block for defects.

Examine cylinder bores, top of block, and water jackets for cracks. Do not overlook the inside of the valve lifter chamber. If cracks are not large enough to be seen they may be revealed by wet or rusted areas.

A minute crack not visible to the naked eye may be detected by coating the suspected area with a mixture of 25% kerosene and 75% light engine oil. Mix up some zinc oxide in wood alcohol to form a thin, white "paint". Apply a coat to the surface. If there is a crack the zinc oxide will be discolored. Note that there are several liquids on the market for detecting cracks. Pressure testing the block will usually indicate the presence of a crack. An auto parts jobber/machine shop is usually equipped to make this test.

DISASSEMBLE ENGINE FOR RING JOB

In order to remove the piston and rod assemblies, it is necessary that both the cylinder head and oil pan be removed. The head must be taken off because the piston assemblies must be removed through the top of the block because there is not enough room in the crankshaft area to permit the removal of pistons from below. (The procedure for doing this work is covered in the car chapters.)

Ring Ridge, Remove

Before removing the pistons, examine the bores for ridges, Fig. 33. The top ring on the piston is usually poorly lubricated because it is exposed to the extremely hot, high-pressure combustion gases which blow or boil the oil off of it. Therefore cylinder bore wear is greatest at the upper end of the top ring travel. The worn bore leaves a "ridge" above the ring. Unless this ridge is negligible, it should be cut away before the piston can be pushed out. If a ridge can be felt by rubbing the finger nail over the bore, it

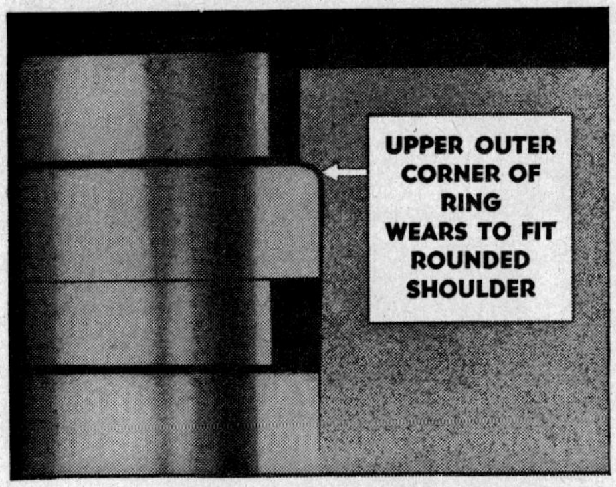

Fig. 34 The old ring has a rounded shoulder

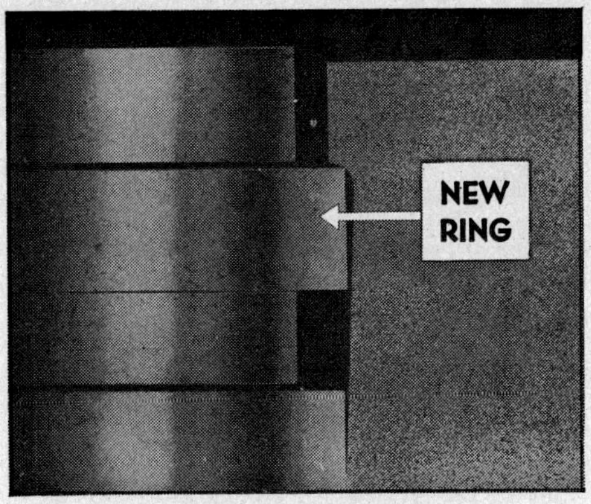

Fig. 35 The new ring has a sharp edge or shoulder

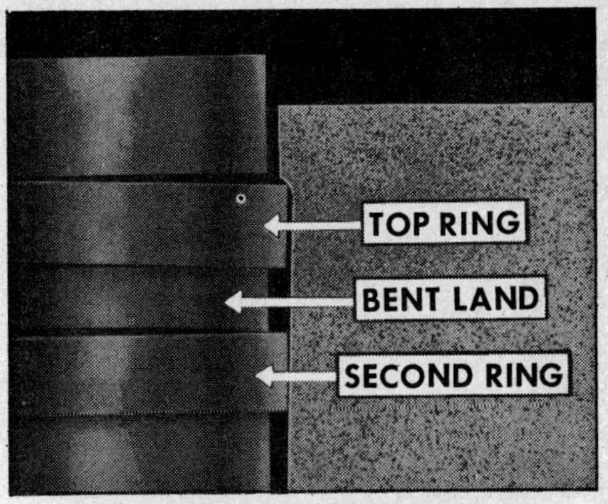

Fig. 36 The new ring has bent the land below it. The land clamps against the second ring and prevents it from functioning properly

must be removed to avoid damage to new rings.

An attempt to force the piston up out of the bore may cause the ridge to bend or break the ring land just below the top ring.

Note also that even if the piston is coaxed past the ridge without suffering damage, it is still necessary to cut away the ridge if new rings are to be installed. In explanation, note that the old top ring has a rounded upper edge, Fig. 34, which has worn itself into the ridge. When a new ring is installed with a sharp edge, Fig. 35, the edge will strike the ridge and may bend the land, Fig. 36, so that the second ring is clamped tight in its groove and therefore cannot function. A clicking sould will be heard when the ring strikes the ridge. The ridge is cut away by means of a special tool called a ridge reamer.

Use of Ridge Reamer

There are a number of ridge reamers commercially available and the manufacturer's instructions should be carefully followed. But regardless of what make is used, it must be correctly installed in the bore and the cutting edge must be adjusted to the surface of the ridge. Care must be exercised not to allow the reamer to cut down into the bore below ridge more than 1/32". Cutting too far down into the bore will prevent the top ring from functioning properly and will greatly shorten its life.

Fig. 37 shows a typical ridge reamer installed in a bore, ready to remove the ridge. The edge of the cutter on all ridge reamers stands at a small angle to the bore, usually about two degrees. Therefore it cuts a slightly tapered hole which blends smoothly with the bore.

To remove the ridge, proceed as follows:

1. Clean the top of the block.
2. Rotate the crankshaft to move the piston down to lower dead center. Wipe the bore with a clean rag wet with oil. Place a cloth over the piston to catch the cuttings from the ridge reamer. (After the ridge has been removed, turn the crankshaft half a revolution to bring the piston up to the top of the block, and remove the cloth with the cuttings.)
3. Lubricate the cutter with a few drops of oil. The maker recommends using a special oil.
4. Install the reamer in the bore as shown.
5. Expand the guide shoes by turning the hex head screw at the top until it is finger tight.
6. Then turn the hex head ½ turn to adjust the tension of the cutter.
7. Use a socket wrench to turn the cutter—clockwise only. If cylinder head studs are in the way use a deep socket. Press down firmly on the wrench with the left hand while turning it with the right. While removing the ridge inspect the progress frequently. This can be done without removing the reamer from the bore.
8. Turn the hex head screw counterclockwise to remove the pressure on the shoes and lift the reamer out of the bore.

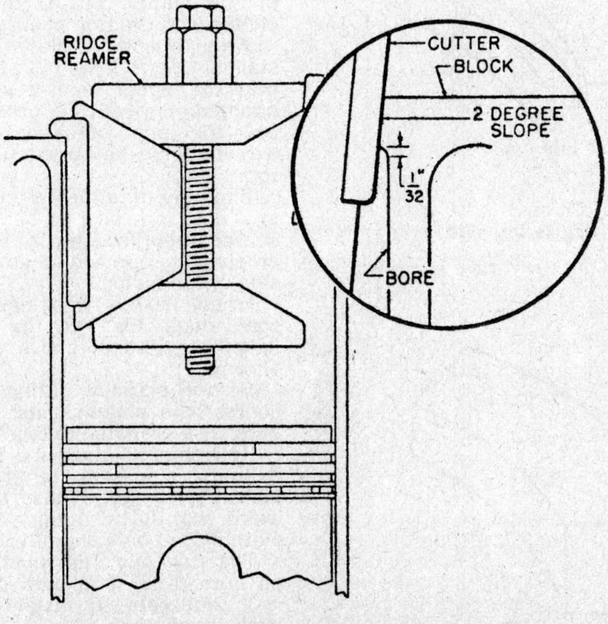

Fig. 37 The cutter of the ridge reamer has a two-degree slope. The slope shown is purposely exaggerated

Remove Pistons and Rods

Before removing the connecting rod caps, note whether the rods and caps are numbered to indicate the cylinder to which they are fitted, Fig. 38. Also note on which side of the engine the numbers are placed. On In-Line engines, the numbers are usually on the camshaft side of the engine. On V8 engines there is no set rule as to which side the numbers are placed so be sure to note the location of the numbers before taking out the pistons and rods.

If the rods are not numbered, identify them with punch marks. Used pistons and rod bearings work best in their original cylinders because pistons conform to the wear pattern of the bores and rod bearings do likewise on the crankshaft.

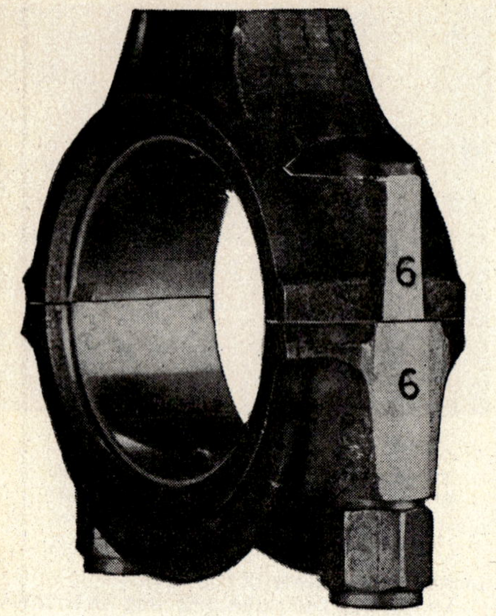

Fig. 38 Connecting rod numbering

Never number a rod with file marks. A file mark may start a crack which will eventually cause the rod to break.

In all engines with four or more cylinders, the rods move up and down in pairs. Therefore it is convenient to remove them in pairs. Rotate the crankshaft until a pair of rod caps are in the most convenient position for removal of rod bolt nuts. Remove the nuts and take off the caps.

Push the two pistons up through and out of the bores. Be careful the rods do not scratch or nick the crankshaft journals. To avoid this, plastic sleeves may be used over the rod bolts to protect the crankshaft as the rod is pushed up into the cylinder.

Rotate the crankshaft to bring two more rods in position. Remove them, and continue in this manner until all pistons and rods are removed.

What Rings To Use

Each piston ring maker offers a variety of ring sets for both new and worn bores. Each set is designed to best serve a particular situation as regards the condition of the bores.

However, ring sets may be divided roughly into three kinds, as follows:

1. For perfectly reconditioned bores.
2. For bores not perfectly reconditioned or bores with moderate taper—say up to .007" or thereabouts. These sets are also preferred by some repairmen for perfect bores on the theory that they do a better job when bores have become worn.
3. Badly worn bores with taper up to .015" or even .020".

Rings for these three kinds of bores for a three-ring piston might be:

1. Two regular compression rings and one regular oil control ring—same as used in a new engine.
2. One regular compression ring, one more flexible compression ring, and a more flexible oil ring.
3. Same compression rings as No. 2 but a very flexible oil ring.

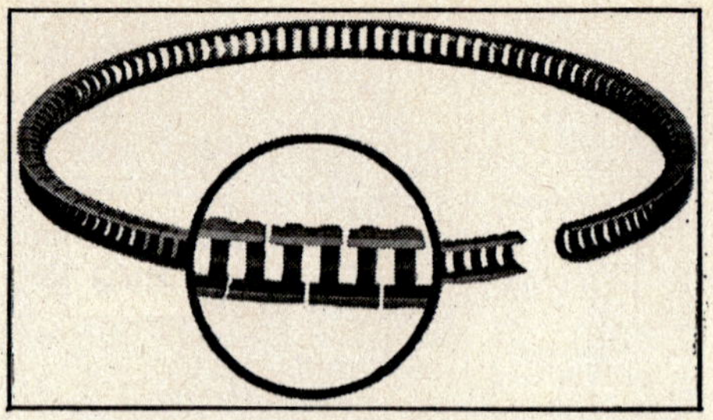

Fig. 39 The U-flex oil ring should be removed first and replaced last

The foregoing is offered to give the reader some idea as to the variety of sets available to meet the different conditions of bore wear.

The situation can be briefly summed up by stating that the greater the taper the more flexible the rings (except the top one) must be in order to follow the bore. The rings are contracted when at the bottom of the bore and must have the ability to expand on the upstroke as the bore grows larger. And if the bore is out-of-round, the rings must do their best to shape themselves to the contour. This is especially important in connection with the bottom oil control ring.

Removing Piston Rings

Hold the piston upright by clamping the rod in a vise. Remove the top ring, then second, third, and fourth if there is one. All piston rings should be removed upward. Never try to slide a ring down over the piston skirt as there is danger of scratching the skirt. When replacing rings, install the bottom one first, and the top ring last. There is one exception: A U-Flex oil ring, Fig. 39, should be removed first and replaced last.

A piston ring removing and replacing tool is almost a necessity. There are numerous designs on the market. One type is shown in Fig. 40. The rings are likely to be distorted if removed and replaced by hand.

Fig. 40 Tool for removing and installing piston rings. Tool expands ends of ring when handles are compressed

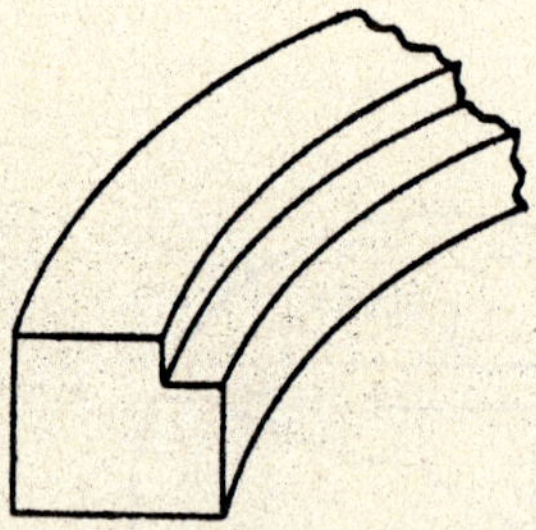

Fig. 41 Grooved inner edge of ring is top side

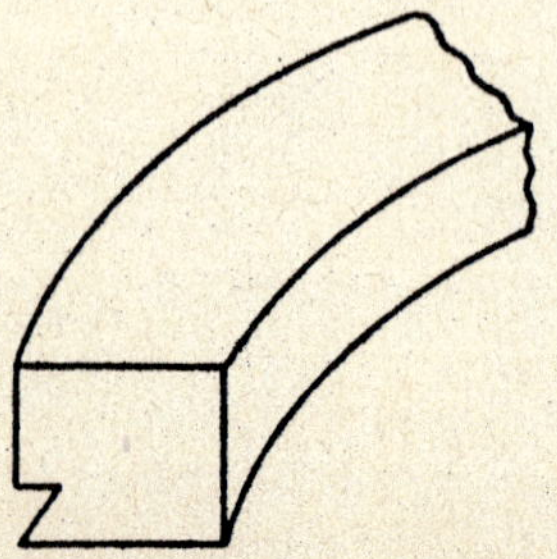

Fig. 42 Grooved lower edge of ring is bottom side

Fitting Piston Rings

Some of the rings in a set may have a top and bottom. That is, one side of the ring must be installed toward the head of the piston. Follow the directions included with the ring package.

When a compression ring must be installed "right side up" it will be found that the upper inner edge is beveled or notched, Fig. 41. In other words, whenever the inner corner is modified in any way it should be on the upside of the piston.

Similarly if either a compression ring or a cast iron oil ring has its lower, outer corner modified, as in Fig. 42, for example, this corner should be on the down side of the piston.

When flexible oil rings have two flat steel rings, Fig. 43, the steel rings are usually installed above and below, as shown.

Before actually fitting the rings, of course, the pistons must be cleaned and otherwise prepared to receive them.

Roll each ring around in its groove. If it binds, check the groove for dirt and burrs. If it still binds, reduce the ring width slightly by rubbing with fine emery cloth placed on a smooth surface.

The ring end gap must be checked to be sure it is sufficient. Usually it is but it is not safe to install a ring without checking the gap.

If the bores have been fully reconditioned either by a rigid hone or reboring,

Fig. 43 Steel rings are placed above and below this flexible oil ring

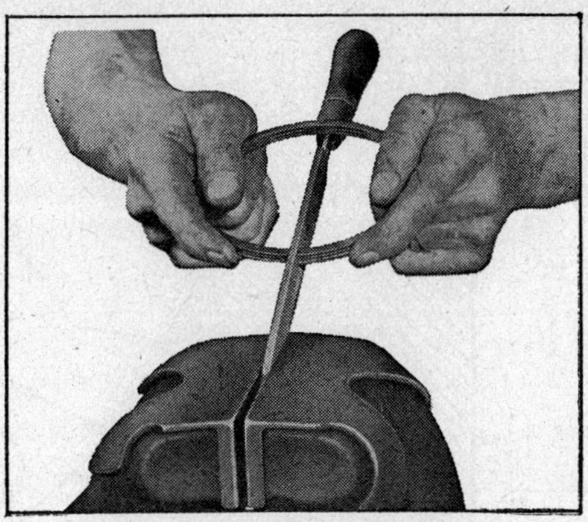

Fig. 44 Filing ring gap by means of file mounted in a vise

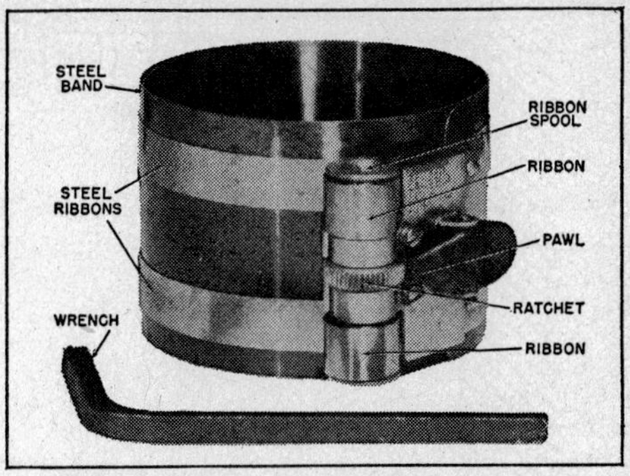

Fig. 45 Ratchet compressor can be adjusted to any size of piston. Spool is turned by inserting wrench in square hole at top. Circular ratchet is attached to spool

place the ring in the bore near the top. Be sure that the ring is straight across the bore by turning a piston upside down, inserting it in the bore and pushing on the ring.

If the rings are being installed in worn bores, the gap should be checked with the ring about half way down in the bore, that is, at the lower end of the ring travel where wear is least. Check the gap with a feeler gauge (see specifications in car chapter). The purpose of the gap is to allow for ring expansion when rings are hot. The gap must be large enough so that there is still some gap clearance when the rings are at their hottest. The gap must never be less than the specified minimum.

But if any rings are found with less than the minimum gap the ends should be filed. Filing should be done carefully because gaps that are much too large may permit too much leakage of oil and gas pressure through the gap. Clamp a file in a vise as shown in Fig. 44. Press the ends of the ring against the file while moving the ring slowly back and forth. Try to keep the ends of the ring flat against the file. Check the ring in the bore occasionally until gap is correct.

A U-Flex oil ring, Fig. 39, does not require a gap. The ends of the ring are designed to be in contact at all times. This ring is so constructed that it is resilient around its circumference and therefore its expansion when heated is automatically absorbed. Before installing this ring be sure it is the correct size for the bore.

Install Rings On Pistons

Mount the rod in a vise with the piston upright to support the piston while the rings are being put in their grooves.

A piston ring replacing tool, Fig. 40, should be used if ring distortion and breakage is to be avoided. All rings above the piston pin should be installed from bottom to top, that is, the lowest ring first and the top ring last. If there is a ring below the piston pin, it may be installed either before or after the others.

After the rings are in place be sure they are free and do not bind. Check them by moving them laterally and also by rotating them in the grooves. If binding exists, correct the condition before proceeding with the work.

Before installing pistons and rods in the block be sure to stagger the ring gaps. Rotate until adjacent rings are a half circumference apart. Or if preferred, space the gaps one-third revolution apart on three-ring pistons and one-fourth revolution apart on four-ring pistons. If the piston pin is pressure lubricated by a passage drilled in the rod, be careful that the oil ring gaps do not line up with the piston pin. Lubricate the rings, ring grooves, piston skirts and bores with engine oil.

The last thing to do before installing a piston is to check the marks on piston and rod, so that it is correctly installed. Be careful, also, to install the piston in its correct cylinder bore (see car chapter for piston and rod assembly details).

Fig. 45 shows a ring compressor that can be adjusted to the piston diameter. It consists of an overlapping spring steel band placed within two steel ribbons whose ends are attached to the clamp spool. The device is adjusted to the approximate diameter of the piston. Then it is placed over the upper end of the piston with its top flush with the top of the piston. Tighten the band by inserting the wrench in the square hole in the top of the clamp shaft. When rings are fully compressed, back the ratchet off one notch. Insert the piston in the bore and tap the piston into place gently with the handle of a hammer.

CLEAN AND INSPECT PISTONS

When cleaning aluminum pistons, do not use a caustic solvent such as lye, washing soda, etc., as they dissolve aluminum. A caustic solvent is readily recognized because it makes the fingers feel soapy—and if strong enough will smart and burn.

Use a stiff brush. If a wire brush is used be careful not to scratch the piston skirt or the ring grooves.

Hard carbon is best removed from the ring grooves by a special tool made for the purpose. There are a number on the market. Lacking such a tool, a satisfactory ring groove cleaner can be made by breaking a piston ring in two. Square off the broken end by filing or grinding. Be sure the edge is sharp and not rounded. Insert the end of the ring in the groove and push the ring forward to clean out the groove.

On some pistons, there is a narrow groove above the top ring which acts as a heat dam to prevent too much heat flowing from head to skirt. Carbon in this groove reduces the effectiveness of the dam, so be sure to clean this groove also.

Clean out the oil drain holes in the lower grooves by running a drill through

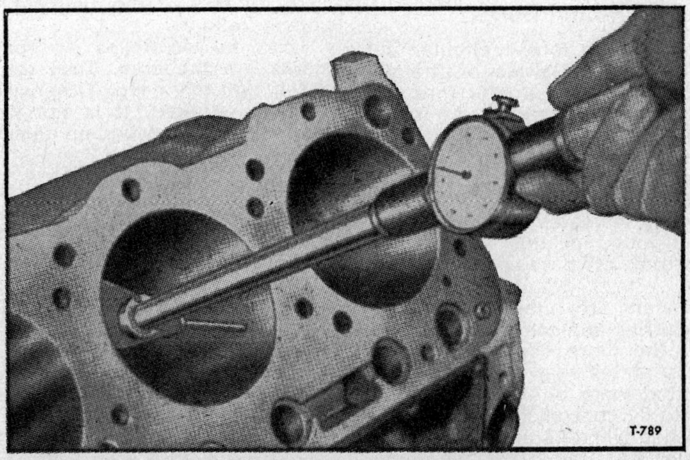

Fig. 46 Measuring cylinder bore

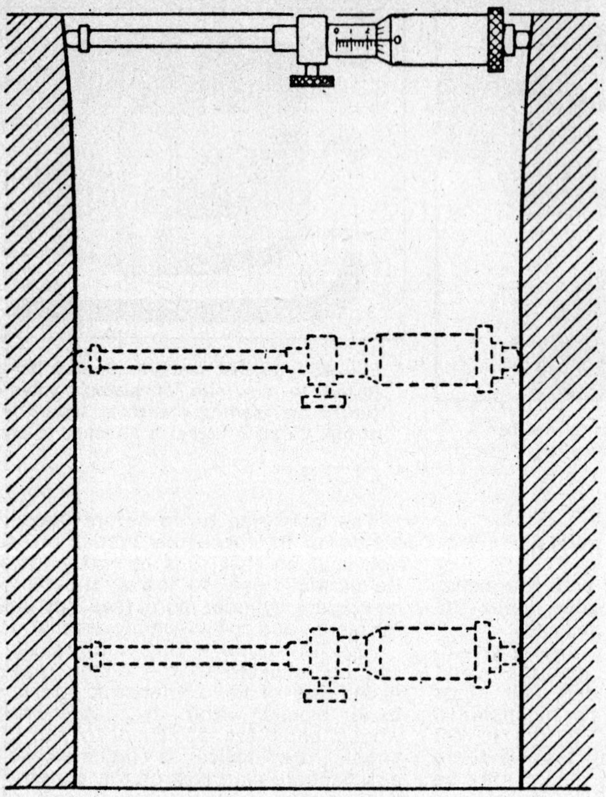

Fig. 47 Cylinder bore wear can be measured with an inside micrometer

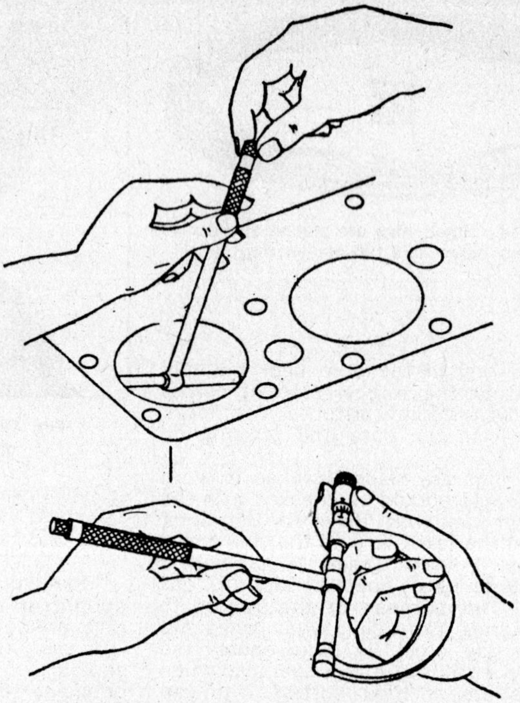

Fig. 48 Diameter of bore is measured with a telescope gauge (top), then the gauge is measured with an outside micrometer (bottom)

them by hand. The drill should be the same size as the holes. It is best to mount the drill in a small hand reamer chuck to use as a handle.

After the pistons have been thoroughly cleaned they should be carefully examined. Look for cracks in the piston head, the ring lands and the skirts. Look for a small hole burned in the top of the piston, caused by excessive pinging. Needless to say, a piston with cracks or holes in it must be discarded. Pistons with badly scored skirts should be scrapped.

CHECK CYLINDER BORES

Before cleaning or inspecting the pistons it is advisable to measure the cylinder bores. If bore taper is more than .015″, the cylinders should be rebored and new pistons and rings installed. Hence it is a waste of time to inspect the pistons if new pistons must be installed.

The bores in a new engine may not be standard size. A car maker may produce a small percentage of engines with oversize bores—say .010″ oversize. Therefore, if the specified bore is 3.250″, the bore of some engines may be 3.260″. Consequently if a re-ring job is done on the engine, .010″ oversize rings should be used.

Cylinders may also be oversize because they have been rebored one or more times in the past. If so, if new rings are to be installed in worn bores, select the ring oversize for the job.

There are three sets of measurements to determine *taper*, *out-of-round* and *wavy* bore. Usually it is sufficient to measure the bore for taper and forget the other two. If taper is more than .015″, the cylinders should be rebored. If less then .015″ install new rings.

Cylinder Bore Taper

This measurement is taken crosswise of the engine because greatest wear can be expected to be crosswise. The bore is measured: (1) At the top of the ring travel just below the point where the ridge was removed. (2) At the bottom of the bore. Subtract measurement 2 from 1 to get the taper. Example: Diameter at top, 3.019″; at bottom, 3.002″; taper is .017″.

Out-Of-Round Bores

Worn bores are no longer circular but oval. The diameter crosswise of the engine is larger than the diameter fore and aft. This measurement should be taken at top and bottom of bore—and at intermediate points if deemed necessary. The difference between the crosswise diameter and fore-and-aft diameter at the upper end of the ring travel indicates the amount the bore is out-of-round at this point. Similarly for bottom of bore and intermediate points.

If the cylinders are out-of-round more than .005″ reboring is recommended. However, one of the more conservative ring manufacturers states that rings installed in cylinders that were out-of-round .010″ have satisfactorily controlled oil consumption, adding that cylinders are more nearly round when warm. This company points out that if the rings are able to follow the taper, the out-of-roundness of cold cylinders can usually be ignored.

Wavy Bore

Measure bore at several points from top to bottom crosswise of engine to determine whether the surface has waves or high and low spots. Also measure bore waviness length-wise of engine.

Look for the possibility of a ridge at the lower end of the bore or the lower end of the ring travel. The ridge can be honed out.

If the cylinder has waves, holes or pockets more than .001″ deep, hone them out completely or at least until the depth is .001″ or less. If too deep for economical honing, rebore.

Look for scratches or gouges—score marks—in the bore. They can be honed out if not too deep. Otherwise, reboring may be required. It is not necessary to rebore the other cylinders unless they need it.

CYLINDER BORE MEASURING INSTRUMENTS

Cylinder bores can be measured by (1) a dial gauge, (2) an inside micrometer, (3) a telescope gauge and outside micrometer. Note that micrometers may go out of adjustment, therefore, always check the zero point before taking a measurement, and adjust if necessary. Cylinder bore tap-

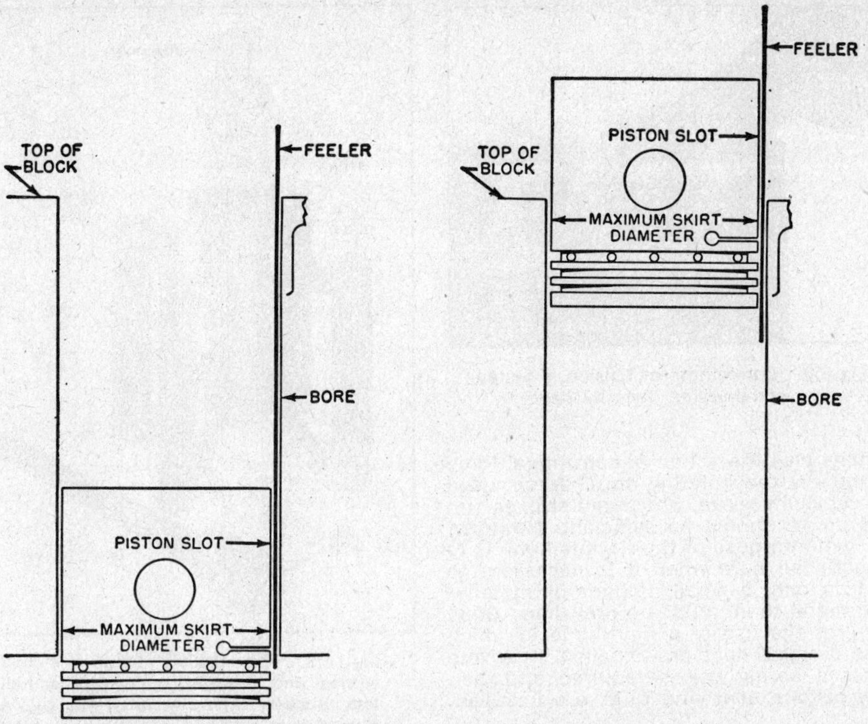

Fig. 49 Measuring cylinder bore taper with feeler gauge. For upper measurement maximum skirt diameter should be taken ½″ down from top of block. For lower measurement maximum skirt diameter should be close to bottom of bore

er can also be measured with a set of long feeler gauges.

Dial Gauge Method

Adjust the thrust pin of the dial gauge, Fig. 46, so that it must be pushed in about ¼″ when it is inserted in the bore. The best way to use the dial gauge is to adjust an outside micrometer to the exact size of the original bore. Then fit the dial gauge between the contacts of the micrometer and adjust the dial carefully to read zero or dispense with the outside micrometer and set the dial gauge to read zero when close to the bottom of the bore where wear is slight.

Check for Taper, Out-of-Round and Wavy bore as already explained. For example, the reading of the dial gauge at the top of the bore crosswise, minus the reading at the bottom of the bore, gives the amount of taper in thousandths of an inch.

Inside Micrometer Method

With the inside micrometer, Fig. 47, measure the bore at the various points mentioned above and note the readings.

Telescope Gauge and Outside Micrometer Method

The telescope gauge, Fig. 48, consists of a rod slidably mounted in a tube. Insert the gauge in the bore and adjust it until it just fits the bore at the place selected. Then remove the gauge and measure the diameter with an outside micrometer.

Fig. 50 Strips of abrasive cloth are clamped to hub of this glaze buster

Feeler Gauge Method

The taper can also be measured with a set of long feeler gauges or by various thicknesses of shim stock. This method is not as accurate as the micrometer or dial gauge methods. Be sure the surfaces involved are free of oil.

One of the old pistons may be used for this job. Feelers of various thickness are used to measure the approximate clearance at top and bottom of bore. The difference in feeler thicknesses gives the taper.

The feeler clearance is measured crosswise of the engine because this is the direction of greatest wear. Fig. 49 shows how the feeler is used. This piston should be upside down with the piston pin fore and aft of the engine. Insert the feeler and the piston together, with the lower end of the feeler extending down beyond the piston skirt. The piston skirt is largest near the top of the skirt. Therefore this part of the skirt should be opposite the top and bottom of the bore when the feeler measurements are taken. At top and bottom, try feelers of various thicknesses until one is found that can be moved with a very gentle pull—while the piston is held stationary in the bore. Consider this the clearance at top or bottom as the case may be. The difference in feeler thickness required at top and bottom indicates the amount of taper.

Feeler gauges come in various thicknesses from .001″ up by thousandths or half thousandths. In checking piston clearances usually a ½″ wide gauge is used but occasionally the one inch wide gauge is specified. The length should be about 12 inches for most cylinder bores. For accurate work the gauge should not be kinked, bent or ragged.

Glaze Buster Or Breaker

NOTE: On Chevrolet Monza or Vega and Pontiac Astre aluminum block engines, do not deglaze cylinder bores found to be in satisfactory condition.

A glaze buster is designed solely to remove the glaze on the cylinder bore. The glaze buster shown in Fig. 50 consists of numerous pieces of special abrasive cloth which are clamped in slots in the hub. The pieces may be reversed or replaced by loosening the nut.

The buster is operated by an electric drill. Dip the abrasive in kerosene. Install the drill and move up and down in bore for 30 seconds. Then give it another coating of kerosene and operate it for another 30 seconds. Wipe bore clean.

The glaze breaker, Figs. 51 and 52, has three "pads" (made of laminations of felt and rubber) covered with abrasive sheets furnished in both medium and fine grit sizes. The three "tension" plates at the base of the pads can be adjusted to give the correct pad stiffness for all cylinders from 3″ to 4¼″ or even larger. The plates are correctly adjusted when they are readily compressed by hand, Fig. 52, so that they can be inserted in the cylinder. For cylinders smaller than 3 inches (down to 2.6″ bore) the tension plates are removed.

The glaze breaker is mounted in an electric drill and 8 or 10 complete strokes

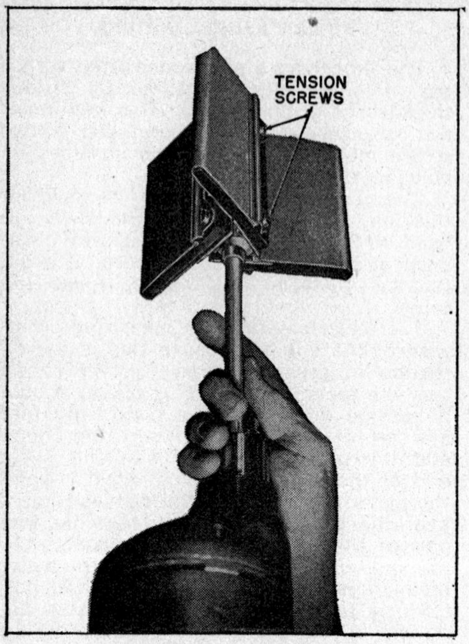

Fig. 51 Three felt and rubber pads are covered with abrasive cloth

(up and down) are usually sufficient to remove the glaze. To reduce flying abrasive to a minimum the pads should be well soaked in oil.

Flexible Hone

A flexible hone can be used for removing a small amount of metal from the bore—up to .003″.

A typical flexible hone is shown in Figs. 53 and 54. It has two stones and two guides made of a felt composition. The hone is rotated by an electric drill. The stones and guides are pressed against the bore by both spring tension and centrifugal force. The hone shown is suitable for bore diameters from 2⁹⁄₁₆″ to 4¾″.

Insert hone in cylinder and start motor. Move hone up and down at the rate of about 30 times per minute. Allow hone to protrude slightly at top and bottom of strokes. Examine bore after 20 or 30 strokes and repeat if necessary—until required amount of metal has been removed.

Coarse and fine stones are available. Use coarse stone for fast cutting and fine stones for finishing. If only a slight amount of metal is to be removed, use fine stones.

Carefully observe the manufacturer's instructions as to care and use of hone. If felt guides are impregnated, the hone may be used dry. Dry honing gives faster cutting. Wet honing (with kerosene) gives smoother finish. Kerosene is conveniently squirted on stones by means of ordinary oil can.

Stones must be kept clean. If dirty or gummed, clean with wire brush and kerosene.

For glaze removal, use 150 grit stones with rapid up and down motion of hone—fast enough so that the scratches the hone makes on the bore surface are at an angle of about 45 degrees.

The honing stones are expanded by

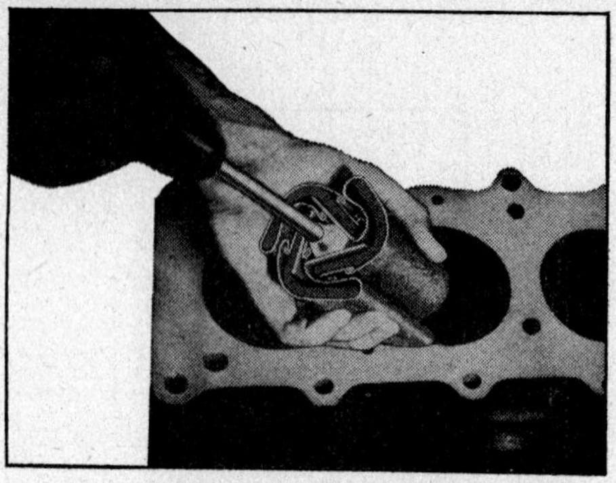

Fig. 52 Correctness of tension is tested by compressing pads by hand

springs plus the action of centrifugal force acting on the rotating tool. When used for removing glaze, about ten strokes (up and down) should be sufficient. However, the main purpose of the flexible hone is to smooth the bore when it is necessary to remove only a small amount of metal—not more than .003″. More than .003″ requires the use of a rigid hone in which case oversize rings are installed. In severe cases a boring bar is required and then new pistons, pins and rings are installed.

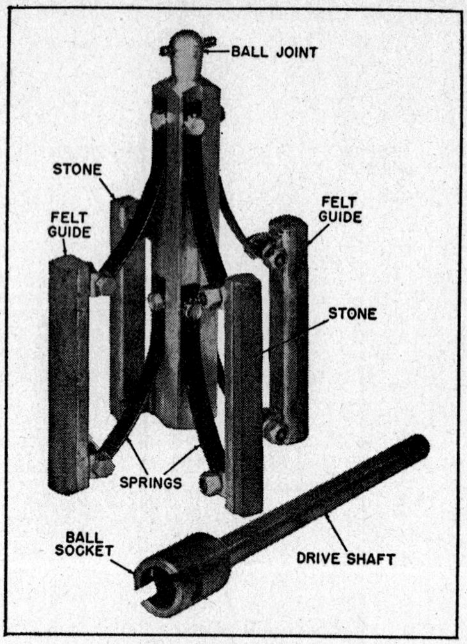

Fig. 53 This flexible hone has two stones and two felt guides. The ball and socket drive permits honing a bore that is recessed into the body cowl

SERVICE BULLETIN

Aluminum Block Procedures

Since the Chevrolet Monza, Vega and Pontiac Astre models utilize an aluminum block engine, special procedures must be observed when reconditioning the cylinder bores. The cylinder bores may be honed using conventional equipment, however, the finishing procedure requires that a silicon lap be performed to obtain a satisfactory finished bore surface. This silicon lap is necessary to remove a sufficient amount of aluminum from the bore, leaving pure silicon particles prominent to form the bore wear surface.

Fig. 54 Hone installed in cylinder bore

Cylinder Block Preparation

1. Wash cylinder block with a noncaustic cleaning solvent.
2. Clean and inspect all gasket surfaces.
3. Remove oil galley plugs and clean oil passages. The plugs may be removed with a punch or drilled and pried out.
4. Remove any ridge or deposits from the cylinder bore upper end.
5. Wash bores with hot water and detergent. After washing, swab bores with light engine oil and clean cloth, then wipe with a clean dry cloth.

NOTE: Do not clean cylinder bores with gasoline or kerosene.

Honing

The cylinder bores may be honed in the conventional manner, however, use a fatty base honing oil. Also, fabricate a 3½ x ⅛ inch aluminum disc to use as a stop for the refinishing stones at the bottom of the bores. Adjust the downward travel stop on the honing stand until the stone holder comes within ¼ inch of the aluminum disc. Hone the bores with equipment which operates at approximately 350 RPM.

NOTE: Failure to use aluminum disc at bottom of bores may result in refinishing stone damage.

Roughing

Using a 280 grit roughing stone set, hone bores to within .003 inch of the finished size and surface finish of 25-30 microinches. The stock removal rate will be approximately .0035 inch per minute. During this operation, flood the bore with a fatty base honing oil using a continual spray pump or apply continuously with an oil squirt can.

Finishing

Using a 400 grit stone set, remove approximately .002 inch of stock and surface finish will be approximately 15-20 microinches. The stock removal rate will be approximately .002 inch per minute. Flood cylinder with honing oil as described in the above procedure.

Polishing

Before performing this operation, ensure all traces of the previous finishing abrasives are removed. Using a 600 grit stone set and keeping stone tension against cylinder wall fairly tight, remove approximately .001 inch of stock. The stock removal rate is approximately .0007 inch per minute with a maximum surface finish of 8 microinches. Flood cylinder with honing oil as previously described.

NOTE: Before proceeding to the next finishing operation, the silicon lap, fit the pistons to the bores.

Silicon Lap

Saturate the felt lapping pads with honing oil and insert pads into holder. Coat cylinder walls and pads with lapping compound (Sunnen M33 or equivalent). Tighten pinion adjustment nut snugly and lap for $1\frac{1}{2}$-2 minutes, stroking the same as for honing. Do not use honing oil since it will wash away the silicon compound. Periodically tighten pinion adjustment nut. The cylinder wall will have a dark-gray finish, approximately 12-18 microinches, and no significant stock removal. Keep overstroke to a minimum during this operation, thereby avoiding premature wear of the felt pads. Replace the felt pads as required to avoid metal contact with the cylinder walls.

After finishing all operations, clean cylinder bores with hot water and detergent. Scrub the cylinder walls with a non-metallic stiff bristle brush and rinse with hot water. It is important to perform a good cleaning operation since if any of the abrasive material remains in the cylinder bores, rapid wear of the new piston rings and cylinder walls will occur. Also, the bearings lubricated with the contaminated oil will be affected.

Swab cylinder bores with a light engine oil and a clean cloth, then wipe with a clean dry cloth. Do not clean bores with gasoline or kerosene. Clean remainder of cylinder block to remove excess material spread during the honing operation.

Upon engine reassembly, some additional items of consideration are as follows:

1. Any stripped bolt holes may be repaired with a Heli-Coil insert.
2. On models prior to 1973, it is mandatory to install a new water pump.
3. Torque engine oil pressure switch to 50 inch pounds.
4. Use only the proper head gasket, identified by three additional cooling holes located next to the Siamese bore section, PN 333997 or 334000.
5. All units must be equipped with a coolant recovery system.
6. If engine was severely overheated, replace the thermostat.
7. Maintain antifreeze to provide minus 20° F or lower protection.

Check For Loose Piston Pins

Before installing new rings, check the piston pins for looseness. If wear is considerable it can be felt. Hold the piston in one hand and grasp the rod with the other hand. Push and pull on the rod and also rock it back and forth in the plane of the piston pin. If there is noticeable play, the pin is too loose.

A more delicate test for wear in pin holes is to squirt some light oil on the end of the pin (after removing lock rings), wipe off the surplus and then note whether there is any motion in the oil film between pin and hole when the rod is rocked.

If the piston pins are pressure lubricated through a rifle-drilled hole in the connecting rod, excessive oil consumption may result if the pins are loose in their piston holes, since the extra clearance between pins and piston permits extra oil to flow onto the cylinder bores.

Piston Pins Too Tight

Sometimes when pistons are removed, the pins will be found to be too tight, as indicated by the fact that some effort is required to rock the piston on the rod. Usually this is caused by a gum or varnish deposit on the wearing surfaces. Remove the pins and clean the surfaces with a suitable solvent. If difficulty is experienced in removing the pins, heat the pistons in boiling water to soften the deposit.

PISTONS AND PINS, REPLACE

When new pistons are required the job should be farmed out to an automotive parts jobber who has the rigid hone, boring bar and the equipment to align rods and fit piston pins.

Crankshaft & Bearings

Bearing Oil Leak Detector

Leakage is assumed to be excessive if oil pressure is below normal or if bearing clearances are somewhat greater than they should be.

Oil leakage increases very rapidly with increase in bearing clearance. For example, in a certain engine where specified bearing clearance is .0015″ the leakage is at a satisfactory minimum. If we say this leakage rate is equal to 1, a clearance of .003″ gives 5 times the leakage while .006″ clearance permits 25 times the leakage.

The only sure way to determine the amount of leakage is to use a bearing leak detector. With it the amount of leakage at each bearing can be seen. Guesswork is eliminated.

Without the aid of a leak detector, objectionably leaky bearings may be overlooked and on the other hand bearings in satisfactory condition may be discarded because the repairman does not want to risk the possibility of doing a ring or piston job only to find that oil consumption is still too high because of leaky bearings.

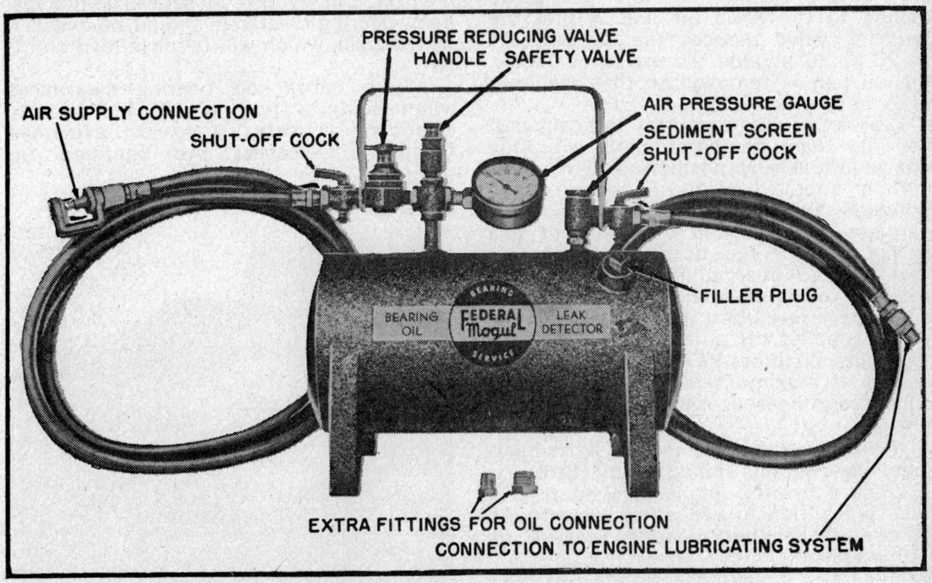

Fig. 55 Engine bearing oil leak detector is used to apply oil pressure to bearings when oil pan is off. Amount of leakage indicates whether or not bearings should be replaced

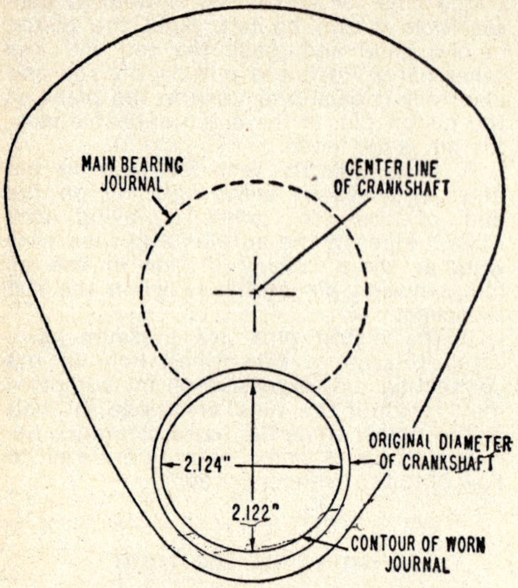

Fig. 56 This rod journal is .002" out-of-round (2.124" minus 2.122")

The leak detector can also be used to determine whether newly installed bearings are too tight or too loose.

If the leakage at a given bearing is 15 to 20 drops per minute (at 25 pounds pressure) the bearing is in good condition. The bearing may still be considered serviceable if the leakage is from 30 to 120 drops per minute. If leakage is less than 15 drops, either the oil passage to this bearing is clogged or the bearing is too tight. Leakage of more than 120 drops per minute indicates a worn bearing, which should be replaced with a new one.

A typical oil leak detector is shown in Fig. 55. It is a compact, easily handled unit with an oil tank a little more than a foot long. It is filled with SAE 20 oil to the level of the filler plug. The hose at the right is connected to the engine oil pressure system. The left hose is attached to the shop air line. A pressure reducing valve reduces the air pressure to 20 to 25 pounds. To use the detector, the oil pan is removed so that the leakage can be observed.

Leakage of main bearings and camshaft bearings can be determined with the crankshaft in any position. Full oil pressure is supplied to these bearings continuously throughout each revolution regardless of crankshaft position. But this is not true of rod bearings. Remember that the rod bearing is fed by a drilled passage in the crankshaft which registers once every revolution with the oil channel which supplies the adjacent main bearing. Therefore oil pressure to the rod bearing is only at maximum when the crankshaft oil passage lines up with its supply channel.

If the leak detector indicates loose bearings, new bearings should be installed.

When a bearing job is finished, it is a wise precaution to use a leak detector to determine whether bearing leakage is within reason.

Bearing Troubles

If one rod bearing knocks, it may be a sign of failure of additional bearings because of faulty lubrication. On the other hand, the lack of lubrication may be due to clogged oil passages leading to this bearing, in which case all or most of the other bearings will be undamaged. Other causes of failure of one rod bearing are: (1) Binding between bearing and journal due to too tight a fit or to a misaligned rod. (2) A slug of abrasive particles which by chance are delivered only to this bearing—and score the bearing. (3) Defect in manufacture of this bearing. (4) Out-of-round rod bore. (5) Dirt between rod bore and bearing shell prevents proper seating of bearing. (6) Bearing improperly installed. (7) Slightly loose rod cap. (8) New bearing installed on rough crankpin, crankpin out-of-round or tapered too much.

Similar considerations apply to main bearings.

Bearing Knocks

If rod bearings knock, the upper and lower halves should be replaced. Sometimes just one rod will fail—often because of a clogged oil passage leading to this bearing. If so, the other rod bearings may be okay—but be sure to inspect them and check for wear. Clean out the oil passages in the crankshaft as described further on.

If rod bearings are worn, the lower halves of the main bearings should also be inspected. If the lower halves are okay the upper halves can be assumed to be in good condition. However, if the lower halves must be replaced the upper halves must also.

If one main bearing develops a knock it may be caused by a clogged oil passage to this bearing, in which case all other bearings may be okay. However, to be on the safe side all main and rod bearings should be inspected and their clearances checked.

If rod bearing wear is suspected because of low oil pressure and high oil consumption—and if an oil leak detector is not available—looseness of rod bearings can often be determined by shaking the rod. But even if looseness is not evident, bearing clearance should be checked by methods which will be described shortly.

Always check rod bearing clearances when doing a ring job. If the bearing clearance is satisfactory—as explained further on—inspect the bearings for faults. The bearing surfaces should be smooth and free of blemishes such as deep scratches, pits, melted or broken metal. If any of these faults are noted the bearings should be discarded.

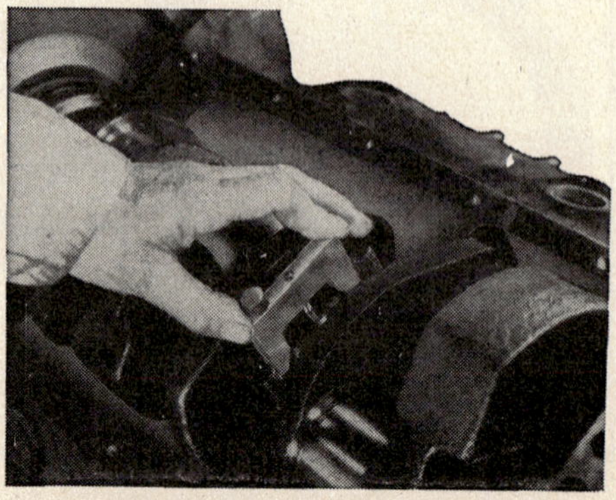

Fig. 58 V-block gauge is pressed against main bearing journal

Fig. 57 V-block crankshaft gauge

Checking Crankshaft For Wear

Before checking bearings for wear, the crankshaft journals should be inspected for surface smoothness, nicks, and for out-of-round and taper. If this checking indicates that the journals must be reground, there is no need to check bearing clearances for wear since the reground shaft requires new undersize bearings.

Rod journals can be reground without removing the crankshaft by portable equipment, whereas regrinding of the main bearing journals requires removal of the crankshaft because the work must be done in a special crankshaft grinding machine.

The rod and main bearings journals should also be inspected for smoothness. If the journal surface is scored or pitted the journal should be reground unless the rod journals can be smoothed up with a lapping tool if the indentations are not too deep.

A good test for satisfactory journal smoothness is to rub the tips of the fingers over the surface. If the roughness can be felt, reduced bearing life will result and lapping or grinding of the crankshaft is called for.

After a crankshaft has been reground, check the smoothness by rubbing a penny lengthwise of the journal. If a copper line shows on the journal, it should be lapped smooth.

In the case of a rod bearing journal, note that the loads imposed on the journal are greatest in the region of upper and lower dead center. Therefore with the rod at dead center, journal wear will be greatest in the up-and-down direction and least in the crosswise direction, Fig. 56.

To measure the journal, bring it to the lower dead center position and mike it up-and-down and crosswise at the middle of the journal. The difference between the two measurements is the amount of out-of-round. For example, if the crosswise di-

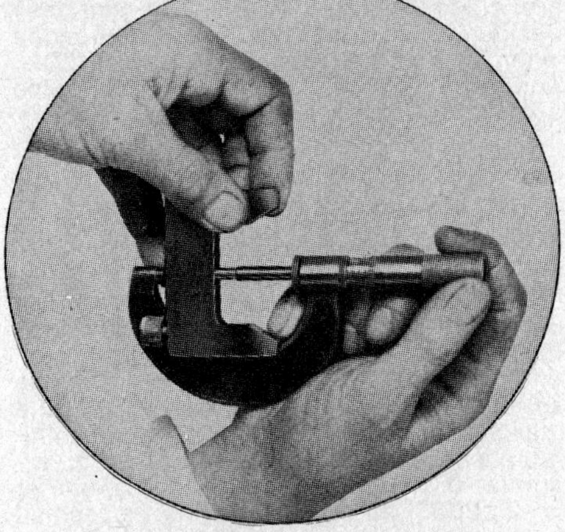

Fig. 59 Distance between upper and lower buttons equals the radius of the main bearing journal

ameter is 2.124″ and the vertical diameter is 2.122″, the out-of-round is .002″, Fig. 56.

The journal should also be miked at both ends on the vertical diameter to determine taper.

For maximum bearing life, obviously out-of-round and taper should be negligible. Also the greater the out-of-round and taper, the more the bearing life is reduced.

For excellent bearing life, out-of-round should not exceed .001″ and taper should be less than .001″—preferably much less. This means that the crankshaft should be reground if taper exceeds .001″.

Practical experience proves that very good bearing life is obtained when the out-of-round does not exceed .0015″ and taper is not more than .001″.

"Miking" Rod Journals

The first thing to do is use a micrometer to measure the amount of out-of-round and taper of the rod journals (also called crankpin journals).

Taper is likely to be greater at the vertical diameter, Fig. 56, therefore measure the vertical diameter at both ends of the journal. On some given engine with rod journals a nominal size of 2$\frac{1}{8}$″, the vertical diameter of one end of the journal might be 2.1235″ and the other end 2.1230″. The taper is the difference between these two dimensions, or .0005″.

A rod journal wears more rapidly on its vertical diameter. Therefore the horizontal diameter of a worn journal is always greater than the vertical diameter. Measure the horizontal and vertical diameters near the middle of the journal. The horizontal diameter might be 2.124″ and the vertical diameter 2.123″. The difference between the two diameters is the out-of-round .001″ in this case. Therefore the journals do not need regrinding provided their surfaces are smooth because .0015″ is usually considered the out-of-round limit.

However, assume that the horizontal diameter is 2.1237″ and the vertical diameter is 2.1216″. The difference (out-of-round) is .0021″. The journal is .0006″ over the usual out-of-round limit of .0015″. The crankshaft should be reground.

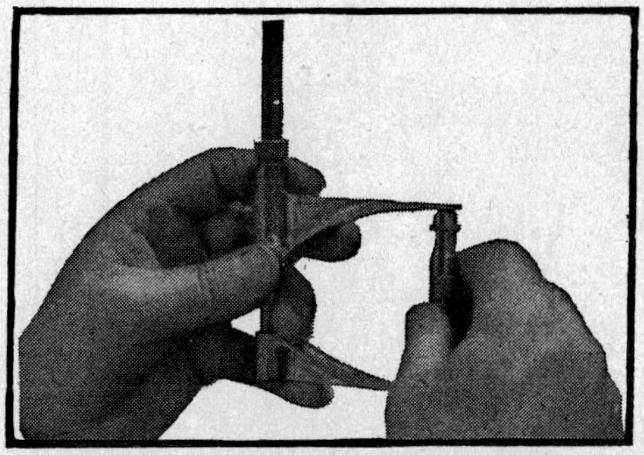

Fig. 61 Measuring distance between jaws of crankshaft caliper to determine diameter of main bearing journal

"Miking" Main Bearing Journals

If the crankshaft has been removed from the engine, an outside micrometer can be used.

However, it is often desirable to mike the main journals with the crankshaft in place. The V-block gauge, Figs. 57, 58, 59, permits miking all crankshafts whether in place or removed from engine.

The crankshaft caliper, Figs. 60 and 61 can be used it the upper halves of the main bearings can be removed.

Whenever the crankshaft is miked in place by either of these methods, be sure that it is properly supported by its bearings. Either remove and replace the caps one by one, or remove all the intermediate caps and mike their journals. Then put these caps back and remove the end caps.

The V-block gauge consists of two pads which make a 30-degree angle with its center line. The upper bottom, as it is called, is really a spring backed plunger locked by tightening the knurled nut.

To use the gauge, loosen the knurled nut, press the plunger all the way down and lock it by tightening the knurled nut. Place the V pads in firm contact with the journal, Fig. 58. Hold it there with one hand and loosen the nut with the other until the plunger clicks against the journal. While still holding the gauge tight, lock the plunger by tightening the nut. Measure the distance between plunger and button with a micrometer, Fig. 59. This distance equals the radius of the crankshaft. Multiply the reading by two to obtain the diameter.

Be sure the micrometer adjustment is accurate because any error in measurement is doubled. All surfaces must be perfectly clean before using the gauge to avoid error in measurement.

What Size Bearings To Use

Replaceable main and rod bearings are generally available in the standard size and under sizes of .001″, .002″, .010″, .020″, and up in some cases.

Automotive machine shops have semi-finished bearings which can be machined to any desired undersize up to .060″.

Fig. 60 The main bearing journal diameter is determined with a crankshaft caliper and then the distance between the jaws on the caliper is measured with an inside micrometer (Fig. 61). The jaws can be accurately adjusted to the journal diameter

If a crankshaft is reground it is customary to remove .010″ of metal and fit .010″ undersize bearings unless a greater undersize is necessary to smooth up badly damaged journals. In some cases where too great a cut is required to clean up a journal, it is advisable to build up the journal with metal and regrind it to the standard size, a job that is often done by operators of large fleets of vehicles.

The standard size bearing and the .001″ and .002″ undersize bearings are intended for installation on crankshafts which do not require regrinding.

Methods Of Checking Bearing Clearance

There are three methods of checking rod and main bearing clearances: (1) Shim method, (2) Plastigage method, (3) Micrometer method.

Fig. 62 Checking bearing clearance with piece of shim stock

If main bearings are checked by the Plastigage method with the engine in the chassis, the crankshaft must be supported in order to take up clearance between the upper bearing insert and crankshaft journal. This can be done by tightening bearing caps or adjacent bearings with .005" to .015" cardboard (such as a calling card) between lower bearing shell and journal. Use extreme care when this is done to avoid unnecessary strain on the crankshaft and bearings or a false measurement may be obtained. *Be sure to remove cardboard.*

Checking Clearance By Shim Method

We know that a standard size bearing will fit but if the journal is worn it may give more clearance than necessary. So the preference is for a .002" undersize bearing if it will fit, or a .001" undersize bearing, assuming that the .002" undersize is too tight and the standard size allows more clearance than needed.

A piece of brass or copper shim stock is used, Fig. 62. Its width should be $\frac{1}{2}$" and its length should be slightly less than the length of the bearing. Most shops use a .002" shim thickness. The edges of the shim should be smooth and free from burrs. Install a new .002" undersize bearing in the rod and cap and coat it with clean engine oil. Lubricate the shim with engine oil and place it lengthwise in the bottom of the rod bearing cap. Install the rod on the crankshaft and draw the nuts up to specified torque wrench tightness. Turn the crankshaft by hand through one revolution. If there is a slight drag, the bearing fit is okay. If there is no drag, the bearing clearance may be less than ideal but still fairly satisfactory.

The "loose" bearing, however, should be reasonably satisfactory provided the clearance is not more than .004". Therefore, try a piece of .004" shim for drag. Both car and bearing makers are reluctant to state positively what the maximum permissible bearing clearance is. But one of the best of them gives a maximum worn clearance of .0045" for rods and mains. Maximum clearance for new bearings on this make of car is .002". Also note that many car makers specify a maximum clearance for new bearings of .003". Therefore the .004" figure does not seem excessive.

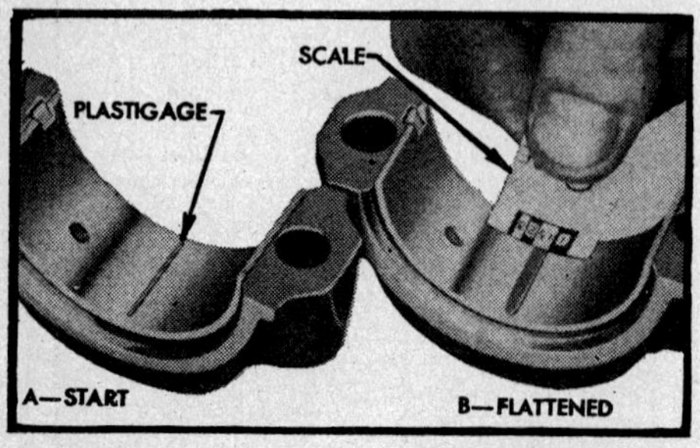

Fig. 64 A round piece of Plastigage is placed on bearing (left). Then the width of the flattened strip is measured with the printed scale on the paper gauge which shows bearing clearance is slightly less than .0015"

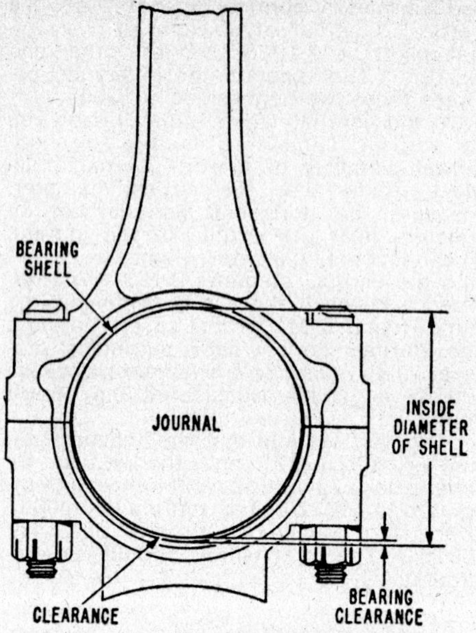

Fig. 63 The journal diameter plus bearing clearance equals inside diameter of bearing

Instead of turning the crankshaft to determine rod bearing drag, the rod bearing may be checked for fit by tapping it lightly with a hammer to move it lengthwise of the rod journal. If only a light tap is required the fit is not too tight.

When the fit of the .002" undersize bearing is checked with a .002" shim, if the drag is heavy instead of light, the bearing is too tight.

Therefore, try a .001" undersize bearing. Use a .002" shim as before. If drag is heavy instead of light a standard size bearing must be used.

Regardless of the size of bearing used it is a good plan to be sure that the new bearing is perfectly free on its journal when the shim is removed and the rod installed. Move it back and forth lengthwise by hand. If it is not perfectly free, remove the rod bearing and find out why. It may be binding at the edges or it may be the wrong size.

If brass or copper shim stock is not handy, paper may be used instead. Paper also has the advantage that there is no danger of marring the bearing Mike up various thicknesses of paper which are handy. Use a strip of paper $\frac{1}{2}$" wide and slightly shorter than the length of the bearing. With it with oil before using.

The .002" shim stock which is generally used is conservative. It insures plenty of clearance. There isn't much danger of fitting the bearings too tight. It conforms to the machinist's rule of thumb that bearing clearance ought to be .001" for each inch of bearing diameter.

However, a .0015" shim is likely to do a more accurate job than a .002" shim. A rod or main bearing clearance of .0015" is within the clearance range of most car engines.

The virtue in using a .0015" shim is simply that a .002" shim may tell you that a .002" undersize bearing is too tight whereas the .0015" shim may indicate the .002" undersize is okay. Therefore if you are guided by the .002" shim you may decide to use a standard size bearing when you would be better off with .002" undersize. To a lesser degree this also applies to .001" undersize bearings.

Checking Clearance By Micrometer Method

Before explaining the micrometer method of deciding what size bearing to use, it is necessary to discuss bearing diameters, journals diameters and specified bearing clearances, Fig. 63.

Standard size bearing shells are made so that their inside diameter equals the maximum journal diameter plus the minimum specified bearing clearance. For example, if the maximum journal diameter is 2.1250" and the minimum specified bearing clearance is .0005", then the inside diameter of the bearing is 2.1255".

If a standard size bearing shell has an inside diameter of 2.1255" as mentioned above, the undersize bearings are so many thousandths less than this figure. Therefore the .002" undersize would be 2.1235" and the .010" undersize would be 2.1155", and so on.

Likewise if the minimum specified bearing clearance is .001" and the maximum diameter of a new journal is 2.125", the inside diameter of the bearing is the sum of the two figures or 2.126". The .002" undersize would be 2.124" and so on.

It is impossible to make any part of any machine—a bearing journal for example—to perfect accuracy such as 2.125"—no more and no less. Therefore most car makers usually allow a tolerance of .001" although a few use a tolerance of .0005". Thus the journal diameter might vary from 2.125" to 2.124".

The dimensions just given for a $2\frac{1}{8}$" rod journal, namely 2.125" to 2.124" mean that in a new engine of this particular make the journals may have a di-

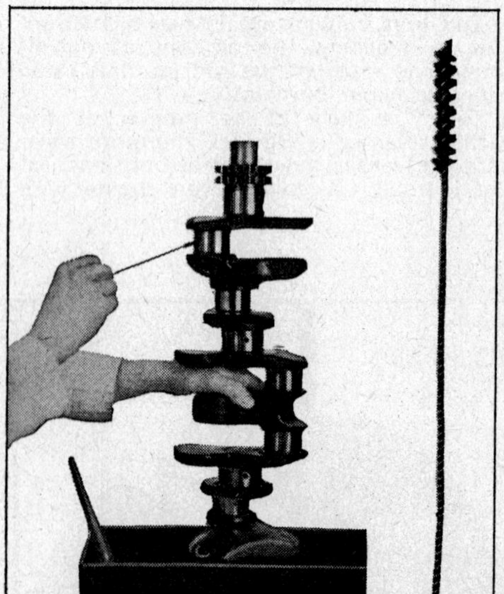

Fig. 65 Crankshaft oil passages are cleaned with a rifle brush

Fig. 66 How to expand thick shell bearing

ameter of 2.125″ or 2.124″ or any where in between these limits. In other words the journal tolerance is said to be .001″.

Likewise bearing clearance has a tolerance range, for example, .0005″ to .0015″. This means that a clearance of less than .0005″ is considered too little and a clearance of more than .002″ is too much.

Whether to install a standard size rod bearing or a .001″ to .002″ undersized bearing can be readily determined by a micrometer—without the use of shim stock or Plastigage except perhaps as a final check when the new bearings are installed. The check is insurance against a bearing whose size is not what it is supposed to be. The check also shows up any errors in reading the micrometer. This method has the advantage that the correct bearing size can be selected with the certainty that it will fit. No time is wasted trying first one bearing size and then another.

The first step is to measure the rod journal for out-of-round and taper, as already described, to decide whether or not the journal should be reground.

If journals do not need regrinding, the next step is to decide whether standard size bearings or .001″ or .002″ undersize will provide the best fit.

Which of the three sizes of bearings to install depends on the maximum diameter of the worn rod journal. If the rod journal when new had a nominal diameter of 2.125″, use a standard size bearing if the journal diameter measures between 2.125″ and 2.124″.

If the rod journal diameter is between 2.124″ and 2.123″, use a .001″ undersize bearing, if available.

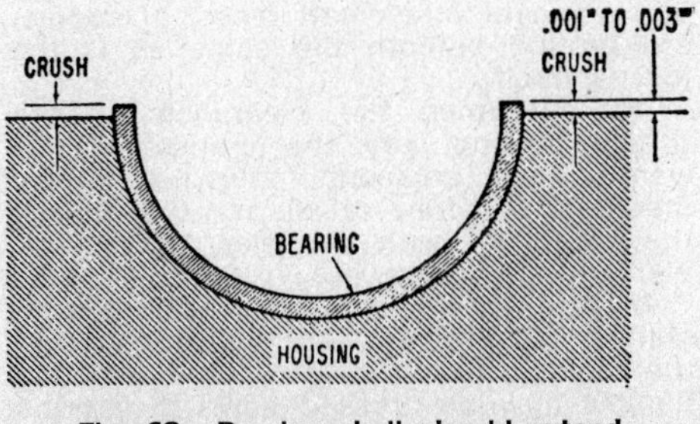

Fig. 68 Bearing shell should extend .001″ to .003″ beyond face of cap

If the diameter is between 2.123″ and 2.122″, use a .002″ undersize bearing.

Fit a standard size bearing if the diameter of the worn journal is not more than .001″ under par.

Checking Clearance By Plastigage Method

Plastigage is placed on the lower half of the bearing, Fig. 64, and the cap is installed and tightened to the specified torque. This flattens the Plastigage to whatever the bearing clearance is. The cap and bearing are removed.

The greater the width of the strip the less the bearing clearance and vice versa. The paper scale measures the clearance in thousandths of an inch. Actually the scale can be read to ½ thousandth or less.

Plastigage is intended to be used for checking worn bearings and also for checking newly installed bearings to make sure that the clearance is within the required limits.

The Plastigage comes in a sealed envelope. It is made for three ranges of bearing clearances each with a different color of wax, as follows: Green, .001″ to .003″; red, .002″ to .006″; blue, .004″ to .009″. The blue is intended for very large crankshafts. For automobiles, the red range is usually used for checking worn bearings. But if new undersize bearings are installed, it may be necessary to use the green range to determine whether or not the new bearing is too tight. Bear in mind that the green range measures as little as .001″ clearance while the minimum for the red range is .002″.

Before using Plastigage, clean the journals and the bearings so that all oil is removed. The Plastigage is soluble in oil and therefore oil will affect the accuracy of the reading.

Do not rotate the crankshaft when using Plastigage.

As explained previously, a rod journal wears more vertically than horizontally. Consequently if Plastigage is placed at the center of the bearing with the rod at lower dead center, maximum clearance will be registered. Likewise, clearance is a minimum at 90 degrees from this point. Best results are obtained if bearing clearance is measured at the halfway point. The rod journal is placed 30 degrees from lower dead center and the Plastigage is placed ¼″ away from the vertical axis.

To decide whether to install a standard size bearing or .001″ or .002″ undersize compare the clearance indicated by the flattened Plastigage with the specified clearance. For example, let us assume that this clearance is .0005″ to .0025″.

If the Plastigage reading is about .003″ or less, try a standard size bearing. Install the bearing and check with Plastigage. If the clearance is less than the specified .0025″, the clearance fit is okay.

If the Plastigage reading is .0035″ more or less, try a .002″ undersize bearing which reduces the clearance to .0015″.

If the clearance is .0025″ to .0030″, try a .001″ undersize bearing if available.

If an undersize bearing is installed be sure that it is a free fit on its journal. If not, the bearing is too small.

For main bearings, the procedure is the same as for rod bearings except the crankshaft can be checked in any position.

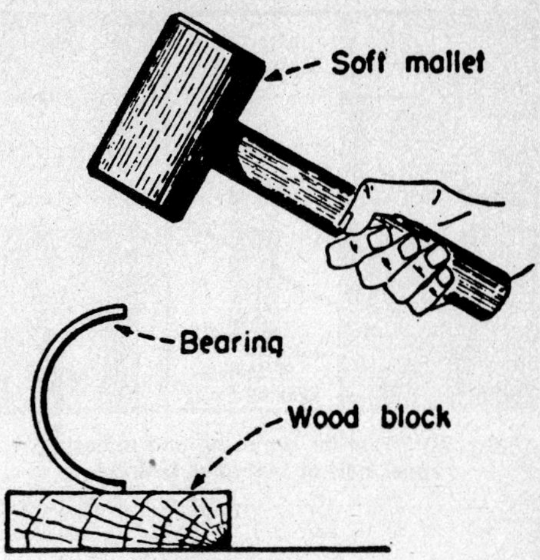

Fig. 67 How to reduce spread of thick shell bearing

When checking main bearings with Plastigage, all caps should be tight.

If the bearings are badly worn, a paper shim should be installed in each main bearing except the one being checked. Otherwise the weight of the crankshaft will press down on the Plastigage and therefore the Plastigage reading will be too small.

Clean Engine Oil Passage

Before installing new bearings it is desirable to clean out the engine oil passages in both cylinder block and crankshaft. Also clean the inside of the crankcase and the oil pan.

If the engine is in the car, use kerosene or a special solvent in conjunction with compressed air to blow out the passages. Automotive jobbers can supply you with a suitable nozzle—and a special solvent.

If the engine is out of the car, the cylinder block and crankshaft should be soaked in a special cleaning tank before any work is done on either the block, the crankshaft or any other engine parts. When dirt has been loosened up, blow out all oil passages in the block and crankcase with compressed air. If this cleaning equipment is not available, use the method described in the preceding paragraph.

Whenever there are oil passages to be cleaned, a rifle brush, Fig. 65, can be used in conjunction with any of the methods

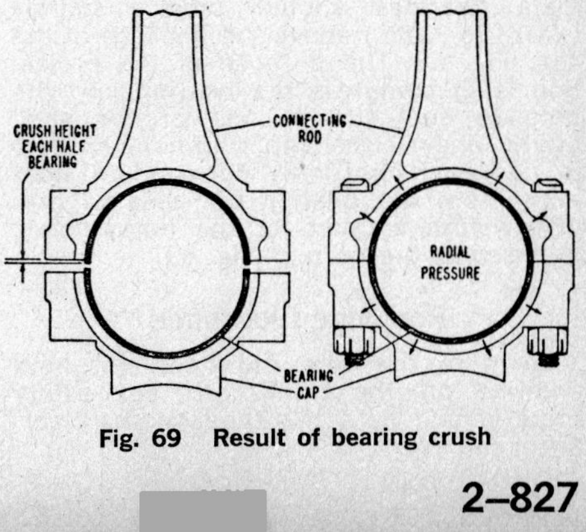

Fig. 69 Result of bearing crush

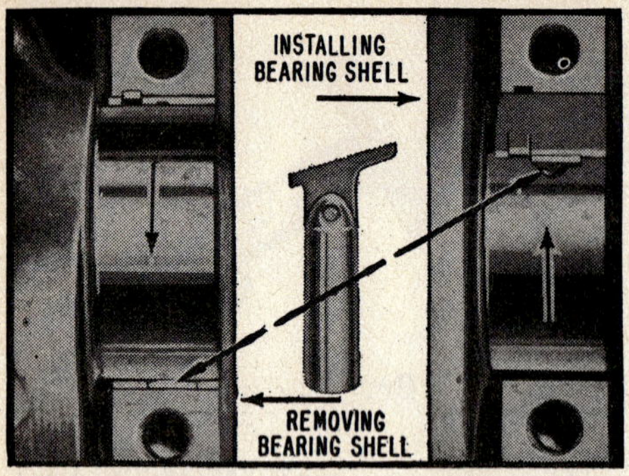

Fig. 70 Tool for removing and replacing upper half of thin shell bearing

described above. Or, if necessary, the rifle brush can be used alone. Wet the brush with a suitable solvent or, if possible, submerge the shaft in solvent and allow it to soak for a few minutes. Then use the rifle brush.

Rod Bearings, Install

Upper and lower bearings may be identical or the upper and lower bearings may have different oil grooving. If the rod has an oil spurt hole and a drilled passage to lubricate the piston pin there will be holes in the upper bearing half to allow the delivery of oil to the spurt hole and the drilled passage. Therefore, when assembling the upper half of the bearing to the rod be sure that the holes are properly aligned.

When installing each bearing half, be sure that the lip on one edge of the bearing is accurately seated in its notch in the rod or cap.

Bearing Spread

Both main and rod bearings are made with a slight spread so that they will stay in place in the bearing bore or bearing cap when inserted. This makes for ease of installation. If a thin shell bearing does not have enough spread to stay in place, expand it by pulling slightly on the ends with the fingers.

If the bearing is a thick-shell type it can be expanded by placing on a flat surface, Fig. 66, and tapping it with a soft mallet. If thick shell bearing has too much spread, reduce spread by tapping it as shown in Fig. 67.

Bearing Crush Fit

The lower halves of thin shell rod or main bearings should project slightly (.001" to .003") above the flat face of the cap, Fig. 68. The purpose of the projection is to compress the bearing into the bore to seat the outside of the shell tightly against the cap. Intimate contact is necessary not only to conduct heat away from the bearing but also to provide a firm support for the heavy loads imposed on the bearing, Fig. 69.

Rod Side Clearance

Be sure that the rod does not bind endwise on the crankshaft. For safety there should be a few thousandths clearance, say .004".

MAIN BEARINGS, REPLACE

To install main bearings shells, remove and replace the bearing caps one at a time. It may facilitate the removal and replacement of the upper shells if all main bearing cap nuts are first loosened ½ turn.

The upper shell is pushed out of its seat bore by rotating the crankshaft after inserting a flat-headed tool (providing bearing shells are not doweled), Fig. 70, in the oil hole in the crankshaft. The thickness of the head must be less than the thickness of the bearing shell. Lacking this tool, the head of a common nail may be used instead. The oil hole is at an angle and the end of the nail head or tool should be parallel to the journal. One end of the shell has a lip on it which fits into a notch in the bearing bore, Fig. 71. It should be obvious that the shell should be removed by pushing on the opposite end. But when installing a new shell, push on the end which has the lip. When the bearing is in place be sure that the lip is accurately seated in the notch. To install the shell, first dip it in engine oil. Then insert the unnotched end in the notched side of the bearing bore. Push it in as far as it will go by hand and then finish the job by rotating the crankshaft. If a heavy push is required, find out why. Probably the shell was not started squarely.

Check Bearing Installation

When installing new bearings, as described above, check each rod and main bearing after it is installed to be sure it is not too tight. If a bearing binds, it is often due to sharp edges on the sides of the bearing shell in which case, dull the edges very slightly with a scraping tool. There is, of course, also the possibility that the bearings are the wrong size.

If the crankshaft is reground and undersize bearings are installed, it is to be expected that bearings and journals will fit together perfectly—and usually they do. But be sure that the crankshaft can be rotated freely when all main bearings have been installed and the caps drawn up to the correct torque wrench readings. Likewise when each rod is installed, see that it can be moved by hand lengthwise on the journal.

If any bearing binds, remove it to determine cause. Always tighten bearing cap nuts to torque wrench specifications. Then check bearings with shim or Plastigage to be sure that the clearance is not too much or too little.

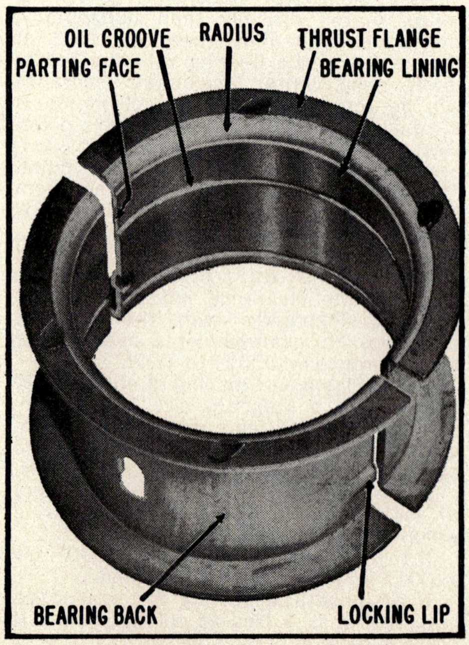

Fig. 72 Typical crankshaft thrust bearing. Note that ends of bearing halves have flanges to take thrust of crankshaft

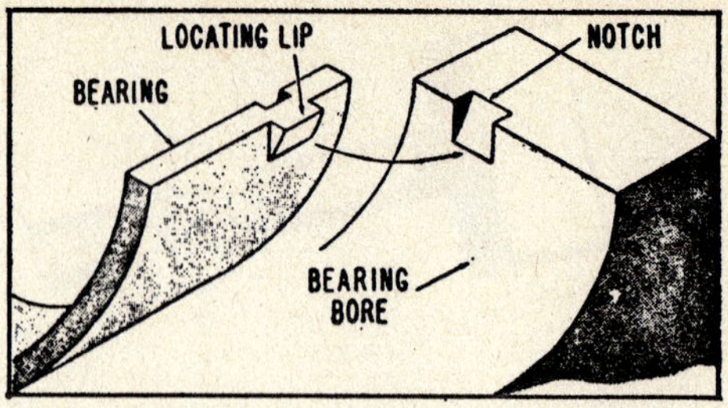

Fig. 71 Lip on bearing should fit into notch shown

CRANKSHAFT END PLAY

There is a certain amount of endwise thrust on the crankshaft and this thrust is controlled to a specified limit by various means, usually by flanges on one of the main bearings. On cars with manual shift transmission, when the clutch pedal is depressed the crankshaft and flywheel assembly is pushed forward. On automatic transmissions with a torque converter, the thrust is exerted rearward.

However, even if there never was any thrust on a crankshaft a thrust bearing would be desirable in order to confine the movement of the crankshaft lengthwise. Any main bearing may be selected to function as a thrust bearing by the engine designer.

The endwise motion or clearance at the thrust bearing must be less than the endwise clearance at the other main bearings (and at the rod bearings) because the crankshaft must be allowed to rub against the edges of these bearings or bearing trouble will ensue.

Fig. 72 illustrates a typical thrust bearing. The ends of the bearing inserts are flanged and coated with bearing metal. The adjacent crankshaft faces, of course, are ground smooth the same as is the journal itself.

To determine the clearance at the thrust bearing, pry the crankshaft forward at an adjacent main bearing by means of a screw driver and then check the clearance with a feeler gauge, Fig. 73, or with a dial gauge, Fig. 74.

If new thrust bearing shells are installed, check the clearance. If less than the specified minimum, the bearing flanges may be dressed down to get the desired clearance or a new bearing should be tried. Measure the distance between

Fig. 73 Checking thrust bearing clearance with feeler gauge

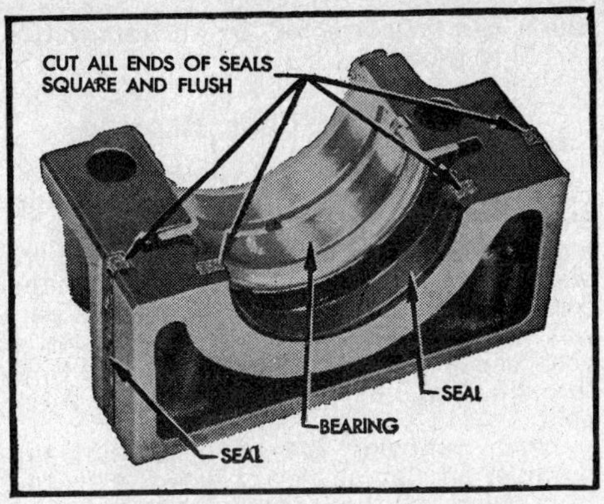

Fig. 75 Showing rear main bearing cap with fabric oil seal and also cork seals at sides of cap

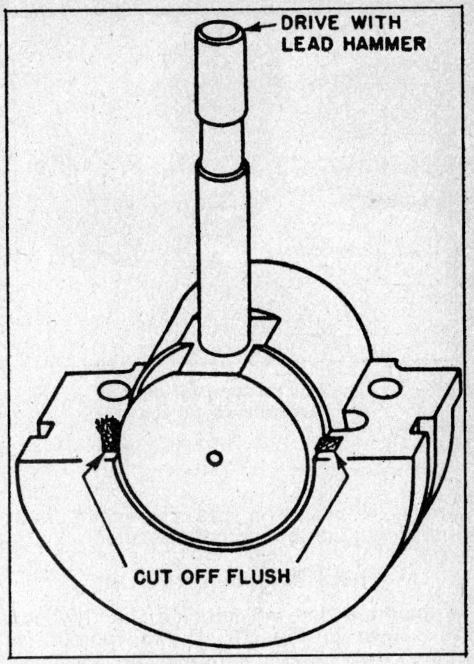

Fig. 76 Tool shown is being used to hold oil seal in place while ends are trimmed off. The tool is also used to install seal firmly in groove. A razor blade or sharp knife is used for trimming excess material

Fig. 74 Checking thrust bearing clearance with dial gauge

the thrust faces on the crankshaft with a micrometer. Then subtract the minimum thrust clearance from this length. The difference is the length that the bearing should be reduced. If the clearance is too great, a new thrust bearing should be installed and clearance checked.

When a crankshaft is reground, it is often necessary to remove a little metal from the thrust faces on the crankshaft. This may require the installation of a thrust bearing with flanges thicker than standard. The automotive machine shop which grinds the crankshaft should supply a thrust bearing to give the correct clearance.

REAR MAIN BEARING OIL SEAL

It is important that oil be prevented from leaking out of the rear main bearing, not only for the sake of oil economy but also because it may ruin the clutch facings if the car has a clutch.

Some engines have a single piece rubber seal which is fit to the groove in the flange at the rear of the block.

However, the most widely used type of rear main bearing oil seal is shown in Fig. 75. It consists of two lengths of wicking or fabric, the upper and lower halves being carried in suitable grooves in the bearing structure. The illustration shows the lower seal in the bearing cap.

To renew any type of rear main bearing oil seal, it is obviously necessary to remove the oil pan. Renewing the lower oil seal only will very often stop the leak. However, whenever the crankshaft is removed for other work, it is good practice to install new seals, both upper and lower, regardless of how effective they appear to be.

In servicing the most common type of oil seal, Fig. 75, bear in mind that the materials used for side seals of this type include synthetic rubber or cork, fibre composition or wood blocks manufactured to the correct size. In replacing these side seals, use the same material as removed unless it has been substituted by the car maker by another material.

To replace the seal, remove the rear main bearing cap and bearing insert. Take out the old seal, clean the groove and install a length of the new seal with the ends sticking up above the face of the cap. Force the seal firmly in place by means of the tool shown in Fig. 76. Lacking this tool or its equivalent, at least use a round rod to force the seal in the groove, Fig. 77. With the side seals installed, the cap assembly should be as shown in Fig. 75.

Although the usual practice is to remove the crankshaft when the upper half of the seal is to be replaced it is possible to do a satisfactory job without removing the crankshaft as follows:

To remove the seal, use needle-nosed pliers to grasp the end of the seal most accessible. Pull the seal downward while rotating the crankshaft slowly in the direction that the seal is being removed.

To install the new seal, fasten a length of wire or strong string such as fishing line securely to one end of the new seal. See that the point of fastening is not bulky and that it is not over 3/8" from the end of the seal. Coat the seal with Lubriplate. Pass the free end of the wire or string up over the crankshaft at the point where the seal is to be installed. Then exert a firm, steady pull on the wire or string and at the same time rotate the crankshaft slowly in the direction of the pull. This will help to move the seal in position. When the installation is completed, trim the ends of the seal flush with the engine block.

CRANKSHAFT REMOVAL

Although it is possible to remove the crankshaft on many cars without removing the engine from the chassis it is not practical because the usual reasons for taking out the crankshaft are for major service operations that can be done best with the engine out of the car.

With the engine out of the car, mount it on an engine stand or on a low workbench. Remove cylinder head or heads, manifolds, etc. with engine right side up. Then turn engine upside down and remove oil pan, vibration damper, fan pulley, flywheel, timing case cover, etc. Take out the pistons and rods. Be sure the main bearing caps are suitably marked as to their location before removing them. Then lift out the crankshaft.

FLYWHEEL SERVICE

If the surface of the flywheel is scored, cracked or burned, it should be refinished in a lathe. If more than a slight amount of metal must be removed, a new flywheel should be installed. It is important that the rear surface of the flywheel be smooth to avoid undue wear of the clutch facing.

Also examine the flywheel bolt holes. If these are worn (due to running with a loose flywheel) the flywheel should be replaced. In other words, the bolts should fit into the bolt holes rather snugly.

Before installing the flywheel on the crankshaft, check the clutch shaft pilot bearing in the crankshaft for wear, and be sure that the mating surfaces are clean and free from burrs. Put the fly-

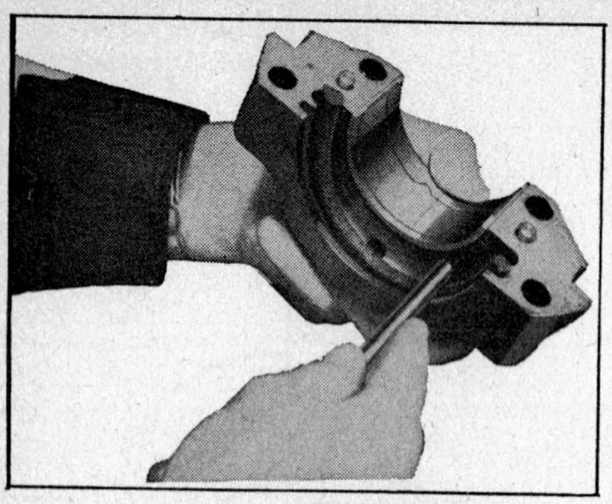

Fig. 77 Pressing oil seal in place with round rod

wheel in place on the crankshaft and install and tighten the bolts securely.

Check Flywheel Run-Out

Mount a dial indicator on the flywheel as shown in Fig. 78. If the run-out is more than .008", remove the flywheel from the crankshaft and examine all contacting surfaces for burrs and dirt. If any are found, remove them. Then install the flywheel on the crankshaft and check the run-out again. If the run-out is caused by an untrue crankshaft or flywheel, the condition can be corrected by an automotive machine shop.

Flywheel Ring Gear, Replace

If the starter ring gear has some broken teeth or if the teeth are badly worn or chipped, the gear should be replaced with a new one. If a new gear is to be installed, remove the flywheel from the crankshaft. Then cut the ring gear part way through with a hacksaw and use a cold chisel to cut the gear all the way through. The gear usually falls off when split.

While removing the gear, support the flywheel on flat blocks of wood, with the front face of the flywheel upward. The blocks should be located within the circumference of the gear so as not to interfere with its removal.

Before installing the new gear, be sure to remove any dirt or rough spots from the flywheel surface that contacts the ring gear and do likewise for the inner circumference of the gear.

Note that the chamfered teeth on the gear should be on the "same side" as the starter pinion. In other words, if the pinion is forward of the flywheel the chamfered teeth should be forward when installed on the flywheel—and should point rearward if the starter pinion is to the rear of the flywheel. Remember that the teeth are chamfered to make it easier for the pinion to engage the ring gear.

To install the ring gear on the flywheel, the gear is expanded by heating and then the flywheel is slipped into it.

Fig. 78 Measuring flywheel run-out with dial gauge

To do this, place the ring gear on a flat plate on the floor. Take a piece of soldering wire the length of the ring gear circumference and lay it carefully on the upper face of the gear. Press it down against the gear so that it makes good contact at all points. Warm the gear slowly with a welding torch, running the flame around and around the gear. Do not direct the flame at the solder. When the gear is warm enough to melt the solder all around, brush off the solder and insert the flywheel into the gear. Be sure that the gear is tight against the flange on the flywheel. When the gear has cooled, it should be perfectly tight on the flywheel.

THREAD REPAIRS

The Heli-Coil® Thread Repair System or similar product, is the most effective method of thread repair which easily and quickly restores worn or damaged threads. This repair consists of drilling, tapping and installing the insert.

Drilling, Fig. 1

Using specified size drill, drill damaged threads to appropriate depth.

Tapping, Fig. 2

Using the specified tap supplied with kit, tap the drilled hole. The tap will cut the proper thread size to accept the insert.

Installation, Fig. 3

Select proper insert and installing tool. Screw insert into tool until driving tang is fully seated in slot, then install insert into tapped hole. Top of insert must be 1/4 to 1/2 turn below top surface of hole.

Tang Removal

Select a punch or rod with a square end which fits snugly into insert. Place punch or rod into insert until it rests on tang, then while holding tool squarely, strike with hammer.

Spark Plug Thread Repair

Ream & Tap

The reamer/tap supplied with kit, uses a threaded pilot and reamer. Check size designation on shank to assure correct size.

1. Pack flutes with heavy grease to contain chips.
2. Thread pilot into hole and rotate until new threads are cut.
3. Blow remaining chips out through ports.

Installation

1. Set stop collar to contact tool body when mandrel projects through prewinder to approximate distance of plug reach.
2. Retract mandrel from body and place insert on threaded portion.
3. Place tool body over mandrel, then rotate mandrel clockwise until insert is engaged in 1 or 2 threads of body.
4. Place tool squarely over tapped hole and rotate mandrel while holding body. Wind insert into hole until collar contacts body. Top of insert must be 1/4 to 1/2 turn below top surface.

Tang Removal

Using long nose pliers, bend tang up and down to snap it off at notch.

Fig. 79 Drilling hole to specified dimension

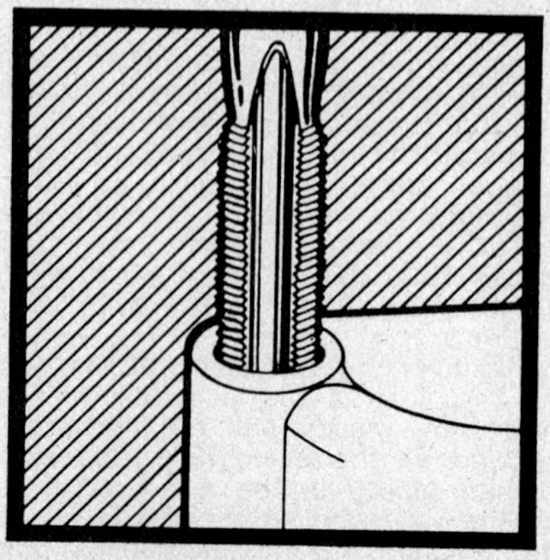

Fig. 80 Tapping drilled hole

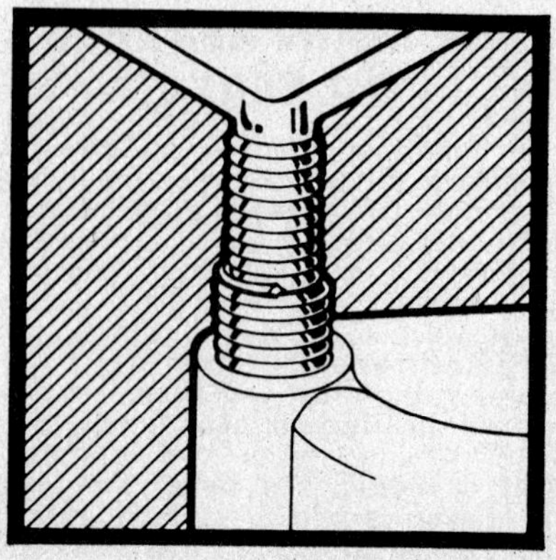

Fig. 81 Installing insert

POWER BRAKE UNITS

A Picture Guide to Power Unit Overhaul

INDEX

APPLICATION

	Fig. No.
AMERICAN MOTORS	
1969-74	BM-1
1969-74	BM-2
1975	see note 1
BUICK	
1969-71	BM-1
1970-75	DM-2
1972-75	BM-2
1974-75	DM-1
BUICK SPECIAL, APOLLO, SKYLARK & CENTURY	
1969-75	DM-1
1974-75	DM-2
CADILLAC	
1969-71	DM-1
1969-75	DM-2
CAMARO	
1969-75	DM-1
1969-71	BM-1
1969-71	BM-2
CHEVROLET	
1969-75	BM-1
1969-75	DM-1
1969-75	BM-2
1972	DM-2
CHEVY II & NOVA	
1969-75	DM-1
1969-71	BM-2
1969-71	BM-1
CHEVELLE & MONTE CARLO	
1969-73	BM-1
1969-71	BM-2
1969-75	DM-1
1973-75	DM-2
CHEVROLET VEGA & PONTIAC ASTRE	
1975	DM-1
CHECKER MOTORS	
1969-74	DM-1
CHRYSLER	
1969-75	BM-1
1969-70	MR-2
1969-75	BM-2
1975	MR-1
COMET	
1969-70	MR-2
1969-70	BM-2
1975	see note 1
CORVETTE	
1969-75	BM-2
COUGAR	
1969-70	MR-2
1969-70	BM-2
1971-73	BM-1
1974-75	see note 1
DART	
1969-74	MR-2
1971-75	MR-1
DODGE	
1969-75	BM-1
1969-74	MR-2
1969-75	BM-2
1970-75	MR-1
FAIRLANE & TORINO	
1969-70	MR-2
1969-71	BM-2
1972-73	BM-1
1974-75	see note 1
FALCON	
1969-70	MR-2
1969-70	BM-2
FORD	
1969-70	MR-2
1969-73	BM-1
1974-75	see note 1
FORD PINTO & MERCURY BOBCAT	
1975	see note 1
GRANADA & MONARCH	
1975	see note 1
IMPERIAL	
1969-75	BM-2
1975	BM-1
LINCOLN	
1969-73	BM-2
1974-75	see note 1
1974-75	HB-1
MAVERICK	
1975	see note 1
MERCURY	
1969-70	MR-2
1969-73	BM-1
1974-75	see note 1
1975	HB-1
MONTEGO	
1969-70	MR-2
1969-71	BM-2
1972-73	BM-1
1974-75	see note 1
MONZA, SKYHAWK & STARFIRE	
1975	DM-1
MUSTANG	
1969-70	MR-2
1969-70	BM-2
1971-73	BM-1
MUSTANG II	
1974-75	see note 1
OLDSMOBILE	
1969	BM-1
1969-74	DM-1
1969-75	BM-2
1970-75	DM-2
OLDS F-85, CUTLASS & OMEGA	
1969-75	DM-1
1975	DM-2
1975	BM-2
PLYMOUTH	
1969-75	BM-1
1969-74	MR-2
1969-75	BM-2
1970-75	MR-1
PONTIAC	
1969-72	BM-1
1969-75	DM-1
1969-75	BM-2
1969-75	DM-2
THUNDERBIRD	
1969-73	BM-2
1974-75	see note 1
1975	see note 1
VALIANT	
1969-74	MR-2
1971-75	MR-1

1 The only services to be performed on the booster unit are, replacement of the grommet, check valve and adjustment of the push rod. If any other malfunctions are apparent replace unit.

The secret of power brake service and repair work is a good understanding of the operation of the different brakes available on today's cars.

Mechanical assist units employ a power unit that is connected only to the brake pedal or linkage and helps push the pedal down. Hydraulic assist types, on the other hand, are inserted in the hydraulic system and directly increase hydraulic fluid pressure. In the hydraulic assist unit known as the integral type, the power unit is attached to the master cylinder; in the other, it is attached to an auxiliary slave cylinder.

Power units are similar in operation and get their energy by opposing engine vacuum to atmospheric pressure. A piston and cylinder, flexible diaphragm or bellows utilize this energy to provide brake assistance. The fundamental difference between the different types lies simply in how the power unit is suspended when the brakes are not in use.

It is obviously important to know whether a power unit is air suspended or vacuum suspended. Air-suspended units are under atmospheric pressure until the brakes are applied. Then engine vacuum is admitted, causing the piston or diaphragm to move or the bellows to collapse. Vacuum-suspended types are balanced with engine vacuum until the brake pedal is depressed, allowing atmospheric pressure to unbalance the unit and apply force to the brake system.

General Service

Regardless of whether the brakes are air or vacuum suspended, or have integral or auxiliary hydraulic cylinders, certain general service procedures apply. Only top quality, clean brake fluid should be used in power brakes. More seals and valves are used with power brake systems than with ordinary brakes, so an inferior brake fluid will do much more damage. For the same reason, be sure all dirt is kept out of the system.

The fact that brakes will operate even if the power unit fails gives us a clue to successful power brake service. This means the conventional brake system is left intact and a power unit is simply added to the existing system. Troubleshooting is then exactly the same until we get to the power unit. As with conventional hydraulic brakes, a spongy pedal with power brakes still means air in the system and grease on the linings will still make the brakes grab. Keep in mind, however, that power brakes give a higher line pressure, thus making leaks more critical.

Power units do not require adjustment. Either they work or they don't. If they don't, the various valves and connections are simply replaced. The only exception is that the power units themselves sometimes have an adjustable connection to the brake pedal or linkage.

Another thing that is helpful in power brake overhaul is the manner in which the units wear. Repairs are infrequent until the unit has been in service for a long time. Then, when a malfunction does occur, it means the whole unit should be overhauled. Complete overhaul kits are available and all the parts in the kit should be used in the overhaul.

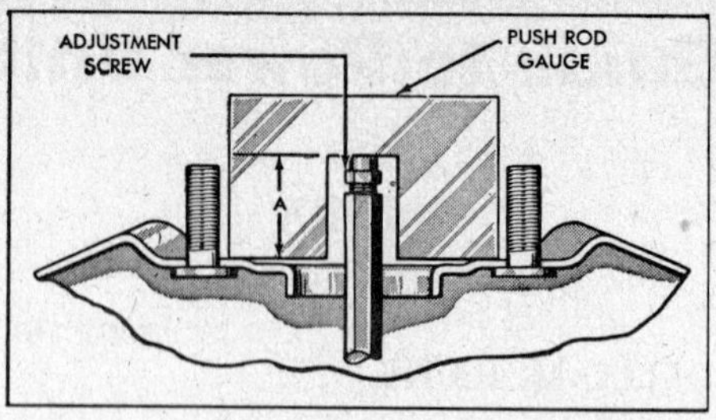

Fig. 1 Typical push rod screw adjustment

PRECAUTIONS

After disassembling a power brake unit, soak all metal parts in a suitable solvent. Use only alcohol on rubber parts or parts containing rubber. After the parts have been thoroughly cleaned and rinsed in solvent, the metal parts that come in contact with hydraulic brake fluid or rubber parts should be rewashed in clean alcohol before assembly. Use an air hose to blow dirt and cleaning fluid from the recesses and internal passages. *Always use all the parts furnished in the repair kit. Discard all old rubber parts.*

Push Rod Adjustment

Most power units of the integral type (unit combined with master cylinder) are provided with a means of adjusting the position of the push rod. This adjustment is necessary to be sure the master cylinder compensating port is open when the brakes are released.

There are two methods that can be used when making this adjustment: gauge method and air method. Usually, if the power unit push rod requires an adjustment the Power Unit Repair Kit for the unit being serviced includes a gauge. The gauge measures from the end of the push rod to the power unit shell, Fig. 1.

Air Method

1. Be sure master cylinder attaching nuts are tight.
2. Remove master cylinder filler cap.
3. With brake released, force compressed air into the hydraulic outlet of the master cylinder.
4. If air passes through the compensating port, which is the smaller of the two holes in the bottom of the master cylinder reservoir, the adjustment is satisfactory.
5. If air does not flow through the compensating port, adjust the push rod as required, either by means of the adjustment screw (if provided) or by adding shims between the master cylinder and power unit shell unitl the air flows freely.

Checking Complaints

Complaints about power brake operation should be handled as if two separate systems exist. Check for faults in the regular brake system first. If it is O.K., start looking over the power brake circuit. For a quick check of proper power unit operation, press the brake pedal firmly and then start the engine. The pedal should fall away slightly and less pressure should be needed to maintain the pedal in any position. On vacuum-suspended power units, air will rush into the air intake when the brakes are applied.

Another check begins with removal of the stoplight switch and installation of a pressure gauge. Take a reading with the engine off and the power unit not operating. Maintaining the same pedal height, start the engine and take another reading. There should be a substantial pressure increase in the second reading.

Pedal free travel and total travel are critical on cars equipped with power brakes. As a general rule, brakes should be adjusted or relined if the pedal is closer than $1\frac{1}{2}$ in. from the floor with the brakes applied. Free travel should be kept strictly to specifications.

Take a manifold vacuum reading if the power unit isn't giving enough assistance. Remember, though, that some of the new V-8's have less than 15 in. of vacuum at idle. If manifold vacuum is abnormally low, tune the engine and then try the power brakes again. Naturally, loose vacuum lines and clogged air intake filters will cut down power brake efficiency. Most units have a check valve that retains some vacuum in the system when the engine is off. A vacuum gauge check of this valve will tell you when it is restricted or stuck open or closed.

Failure of the brakes to release in nine out of 10 cases is caused by a tight or misaligned connection between the power unit and the brake linkage. If this connection is free, look for a broken piston, diaphragm or bellows return spring. The power unit will have to be disassembled for this check and, if the spring is in good condition, continue going through the power unit, checking the fluid, air and vacuum valves.

A simple check of the hydraulic system should be made before proceeding. Loosen the connection between the power unit and the wheel cylinder lines. If the brakes release, the trouble is in the power unit hydraulic circuit. If the brakes still will not release, look for a restricted brake line or similar difficulties in the regular hydraulic circuit.

A residual pressure check valve is usually included immediately under the brake line connection on hydraulic assist power brakes. This valve maintains a slight hydraulic pressure on the brake lines and wheel cylinders to give better pedal response. If it is sticking, the brakes may not release.

Power brakes that have a hard pedal are usually suffering from a milder form of the same ills that cause complete power unit failure. Collapsed or leaking vacuum lines or insufficient manifold vacuum, as well as punctured diaphragms or bellows and leaky piston seals, all lead to weak power unit operation. A steady hiss when the brake is held down means a vacuum leak that will cause poor power unit operation.

Do not immediately condemn the power unit if the brakes grab. First look for all the usual causes, such as greasy linings or scored drums. Then investigate the power unit. When the trouble has been traced to the power unit, check for a damaged reaction control. The reaction control is usually made up of a diaphragm, spring and valving that tends to resist pedal action. It is put in the system to give the pedal "feel."

BENDIX DIAPHRAGM TYPES

Figs. BM-1 & BM-2

These units are of the vacuum suspended type. Some units are of the single diaphragm type while others are of the tandem diaphragm type. Both single piston and double piston or split system type master cylinders are used.

The vacuum suspended diaphragm type units utilize engine manifold vacuum and atmospheric pressure for its power. It consists of three basic elements combined into a single power unit. The three basic elements of the single diaphragm type are:

1. A vacuum power section which includes a front and rear shell, a power diaphragm, a return spring and a push rod.
2. A control valve, built integral with the power diaphragm and connected through a valve rod to the brake pedal, controls the degree of brake application or release in accordance with the pressure applied to the brake pedal.
3. A hydraulic master cylinder, attached to the vacuum power section which contains all of the elements of the conventional brake master cylinder except for the push rod, supplies fluid under pressure to the wheel brakes in proportion to the pressure applied to the brake pedal.

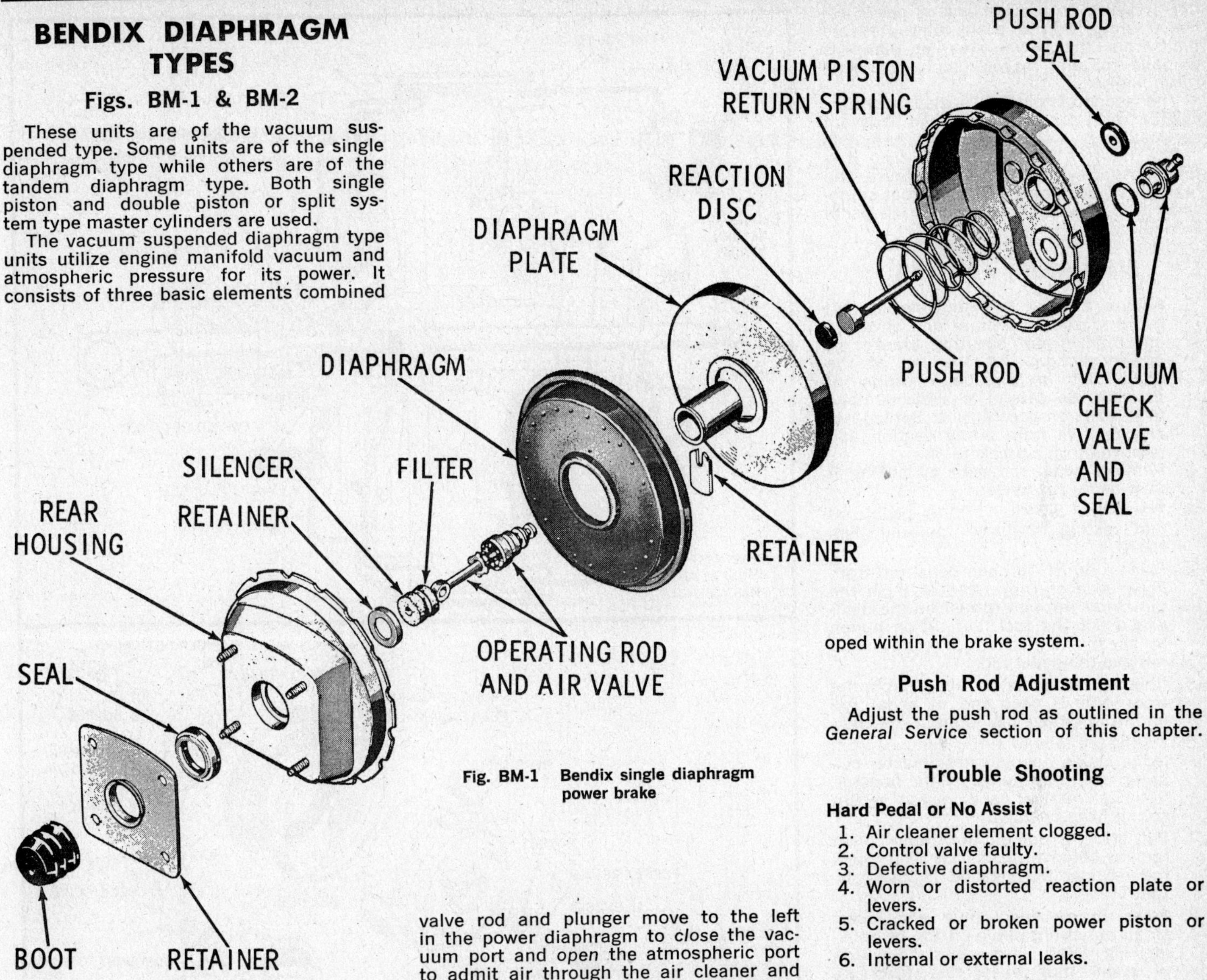

Fig. BM-1 Bendix single diaphragm power brake

Operation

Upon application of the brakes, the valve rod and plunger move to the left in the power diaphragm to *close* the vacuum port and *open* the atmospheric port to admit air through the air cleaner and valve to the rear diaphragm chamber. With vacuum present in the front chamber and atmospheric pressure in the rear chamber, a force is developed to move the power diaphragm, hydraulic push-rod and hydraulic piston or pistons to close the compensating port or ports and force fluid under pressure through the residual check valve or valves and lines into the front and rear wheel cylinders to actuate the brakes.

As pressure is developed within the master cylinder a counter force acting through the hydraulic push-rod and reaction disc against the vacuum power diaphragm and valve plunger sets up a reaction force opposing the force applied to the valve rod and plunger. This reaction force tends to *close* the atmospheric port and *reopen* the vacuum port. Since this force is in opposition to the force applied to the brake pedal by the driver it gives the driver a "feel" of the amount of brake applied. The proportion of reactive force applied to the valve plunger thru the reaction disc is designed into the Master-Vac to assure maximum power consistent with maintaining pedal feel. The reaction force is in direct proportion to the hydraulic pressure developed within the brake system.

Push Rod Adjustment

Adjust the push rod as outlined in the *General Service* section of this chapter.

Trouble Shooting

Hard Pedal or No Assist

1. Air cleaner element clogged.
2. Control valve faulty.
3. Defective diaphragm.
4. Worn or distorted reaction plate or levers.
5. Cracked or broken power piston or levers.
6. Internal or external leaks.

Brakes Grab

1. Control valve defective or sticking.
2. Bind in linkage.
3. Reaction diaphragm leaking.
4. Worn or distorted levers or plate.

No or Slow Release

1. Push rod adjustment incorrect.
2. Linkage binding.
3. Return spring defective.

HYDRO-BOOST

Description

The Hydro-Boost system, Fig. HB-1 provides an additional cylinder in the brake system. This cylinder contains no brake fluid. The Hydro-Boost cylinder is hydraulically operated, by pressurized fluid from the power steering pump, providing power assist to operate a dual master cylinder brake system.

The booster is composed of two sections; the linkage section and the power section, Fig. HB-2.

A spring accumulator incorporated in the Hydro-Boost stores sufficient fluid under pressure, to provide several power assisted stops in the event that pressure fluid from the power steering pump is stopped for any reason (such as a ruptured line).

The accumulator is a sealed hydraulic cylinder with a port at each end. It is not serviceable.

The booster is designed so that if a total absence of power assist occurs, the brakes can be applied manually, but somewhat greater pressure on the brake pedal is required.

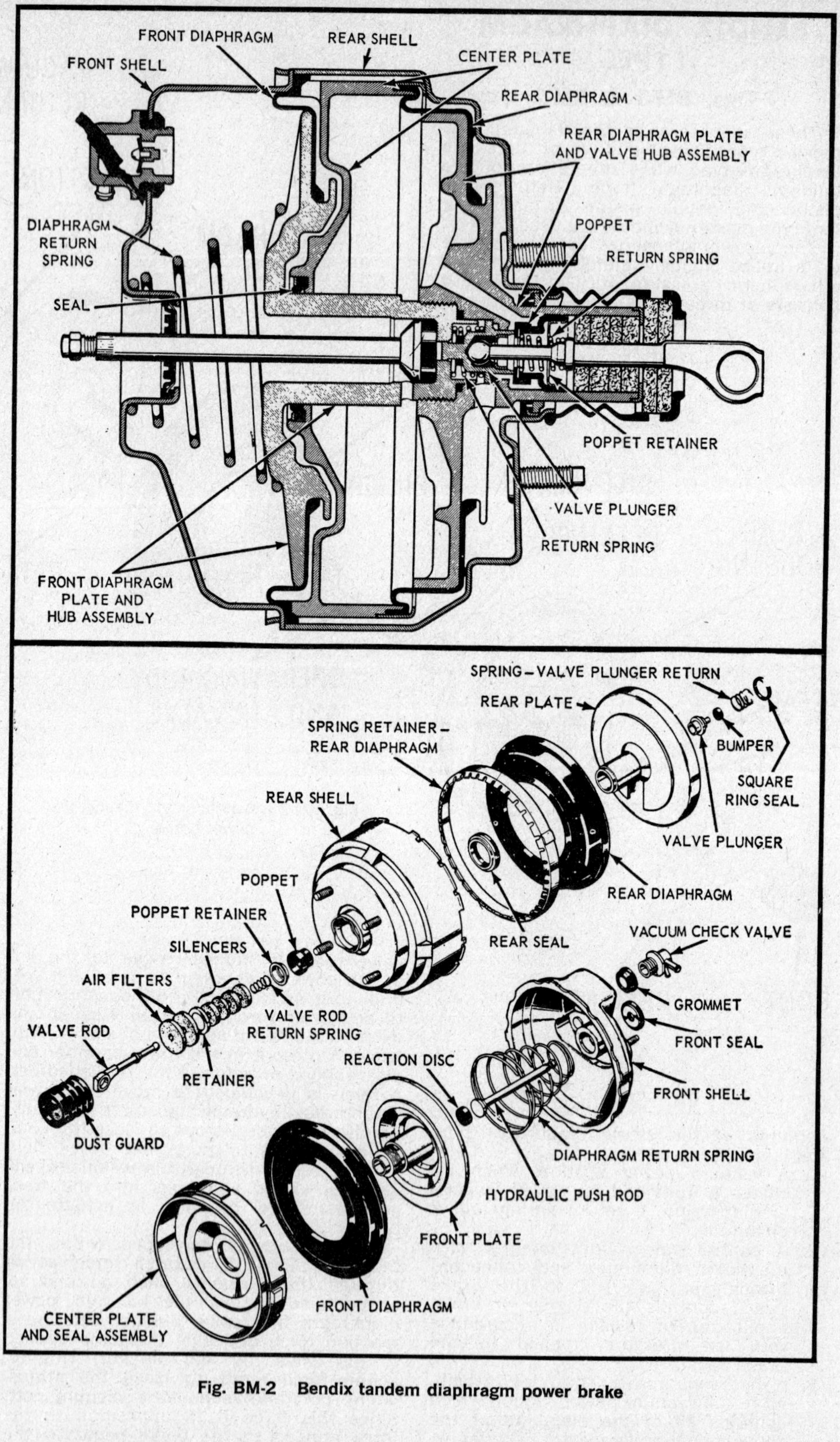

Fig. BM-2 Bendix tandem diaphragm power brake

Disassembly, Fig. HB-1

1. Secure booster in vise (bracket end up) and use a hammer and chisel to cut bracket nut securing bracket to power section. Cut the nut at the open slot in threaded portion of housing. Be careful to avoid damage to threads on booster hub. Spread nut and remove from power section and remove linkage bracket.
2. Remove pedal rod boot by pulling it over pedal rod eyelet.
3. Place Tool J-24569 around pedal rod and resting on input rod end, Fig. HB-4.
4. Place a punch through pedal rod from lower side of Tool J-24569. Push the punch on through to rest on the higher side of the tool. Lift up on punch to shear the pedal rod retainer and remove the pedal rod.
5. Remove remnants of rubber grommet from groove near end of pedal rod and from groove inside input rod end.
6. Disengage tabs of spring retainer from ledge inside opening near master cylinder mounting flange of the booster. Remove retainer and piston return spring from opening.
7. Pull straight out on the output push rod to remove push rod and push rod retainer from inside booster piston.
8. Press in on spool plug and insert small punch into hole on top of housing, Fig. HB-5. This unseats one side of spool plug snap ring from its groove in bore. Remove snap ring.
9. Use pliers to remove spool ring from the bore. Remove "O" ring seal from plug and discard "O" ring. Remove spool spring from bore.
10. Place booster cover in vise equipped with soft jawed devices and remove the five screws securing booster housing to cover.
11. Remove booster from vise and while holding unit over a pan, separate the cover from housing. Remove the large seal ring from groove in cover and discard the seal.
12. Press in on end of spool assembly and use Tool J-24549 to remove the spiral type snap ring from groove near forward end of spool, Fig. HB-6.
13. Remove input rod and piston assembly and the spool assembly from booster housing.
14. Remove input rod seals from input rod end and the piston seal from the piston bore in the housing. Discard seals.
15. Remove the plunger, seat, spacer and ball from accumulator valve bore in flange of booster housing. Remove "O" ring from the seat and discard the "O" rings.
16. Thread a screw extractor into the opening in the check valve in the bottom of the accumulator valve bore and remove the check valve from the bottom of the bore. Discard check valve and "O" ring.

NOTE: The screw extractor damages the seat in the check valve. A new check valve, "O" ring and ball must be installed whenever the check valve is removed from accumulator valve

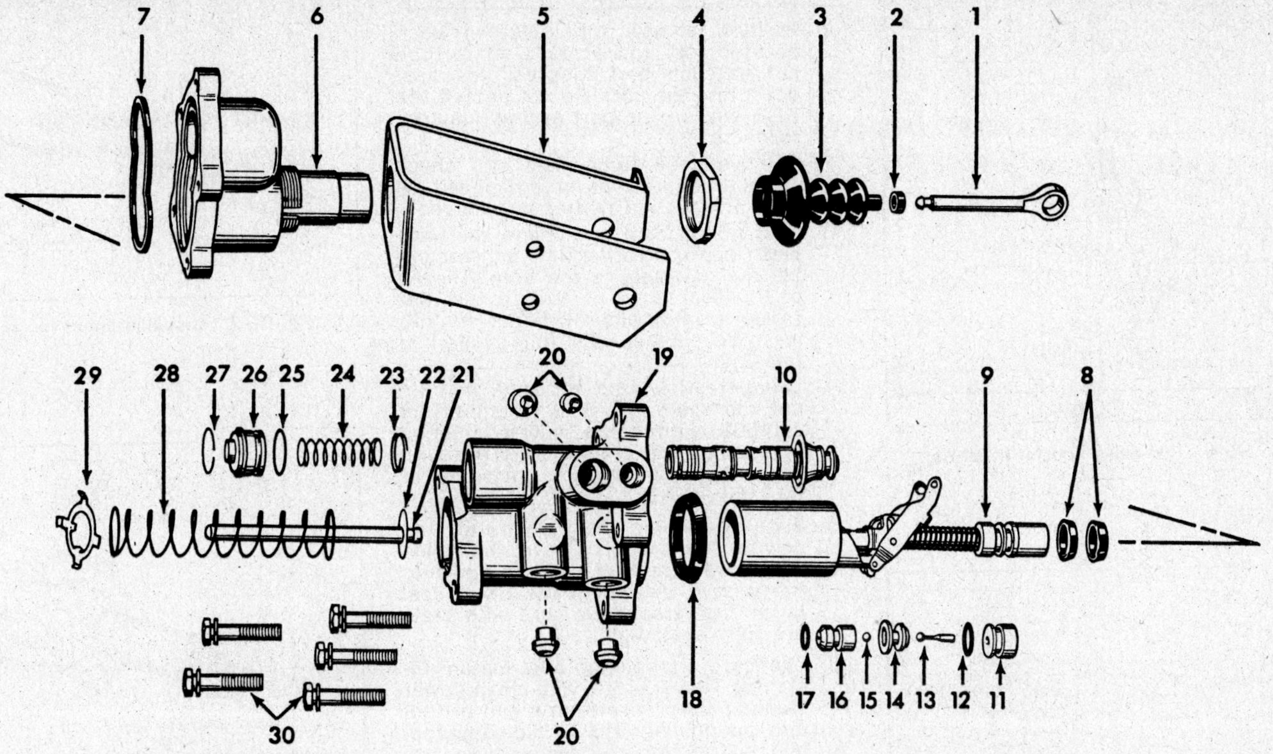

1. Pedal Push Rod
2. Pedal Rod Retainer
3. Boot
4. Bracket Nut
5. Linkage Bracket
6. Booster Cover
7. Cover to Housing Seal
8. Input Rod Seals
9. Input Rod and Piston Assy.
10. Spool Assembly
11. Plunger Seat
12. "O" Ring Seal
13. Plunger
14. Spacer
15. Check Valve Ball
16. Accumulator Check Valve
17. "O" Ring Seal
18. Piston Seal
19. Booster Housing
20. Tube Seat Inserts
21. Output Push Rod
22. Push Rod Retainer
23. Spiral Snap Ring
24. Spool Spring
25. Plug "O" Ring
26. Spool Plug
27. Snap Ring
28. Piston Return Spring
29. Spring Retainer
30. Housing to Cover Bolts.

Fig. HB-1 Exploded view of hydraulic booster (Hydro-Boost)

bore.

17. Use a ¼" or a 5/16" spiral flute type screw extractor to remove tube seats from booster ports.

Cleaning and Inspection

1. Clean all parts in a suitable solvent being careful to avoid losing small parts.

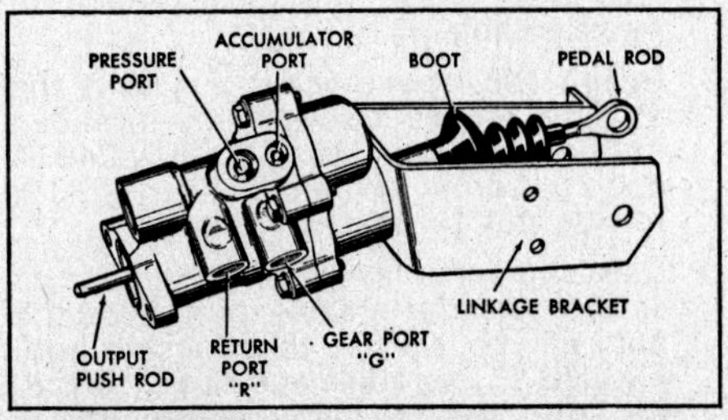

Fig. HB-2 Hydro-Boost cylinder assembly

2. Inspect valve spool and valve spool bore in booster housing for corrosion, nicks, scoring or other damage. Discoloration of the spool or bore, particularly in the grooves is not harmful.
3. If the valve spool or the spool bore has nicks or scoring that can be felt with a fingernail, particularly on the lands, the spool and housing should be replaced as an assembly.

NOTE: The clearance between the valve spool and the spool bore of the housing is important. Because of this, the spool and housing make as selective assembly and therefore can only be replaced as an assembly.

4. Inspect the input rod and piston assembly for corrosion, nicks, scoring or excessive wear. If the piston is damaged, the input rod and piston assembly should be replaced.
5. Inspect piston bore in booster housing for corrosion, nicks, scoring or other damage. If the bore is damaged, the valve spool and housing should be replaced as an assembly.

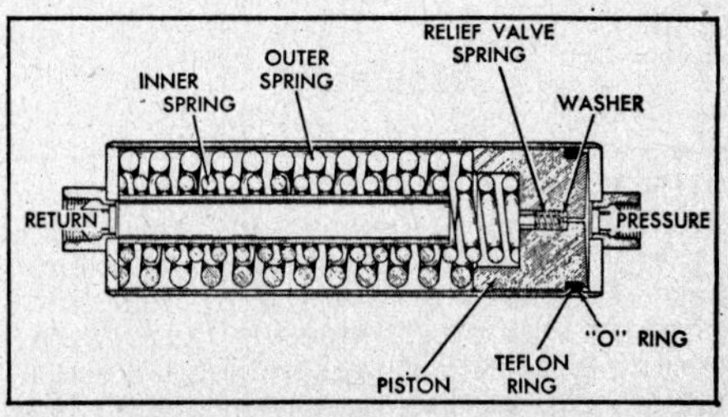

Fig. HB-3 Hydro-Boost accumulator

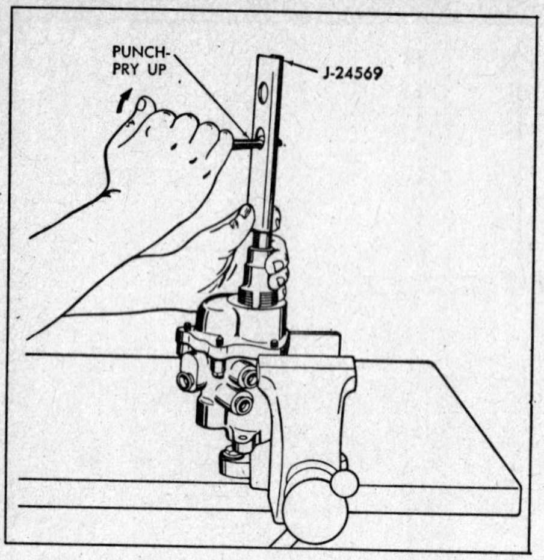

Fig. HB-4 Removing booster pedal rod

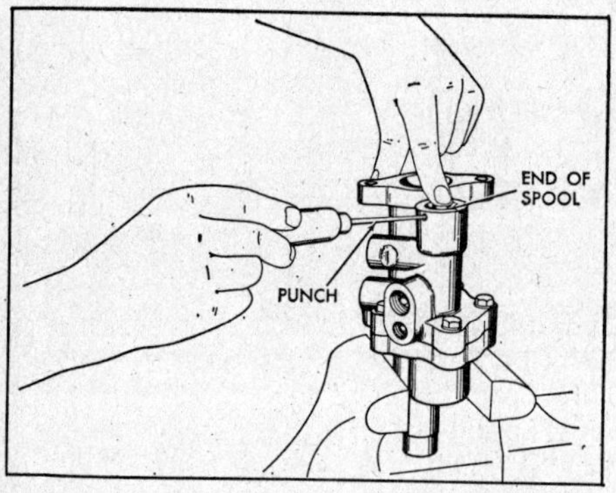

Fig. HB-5 Removing spool plug from booster

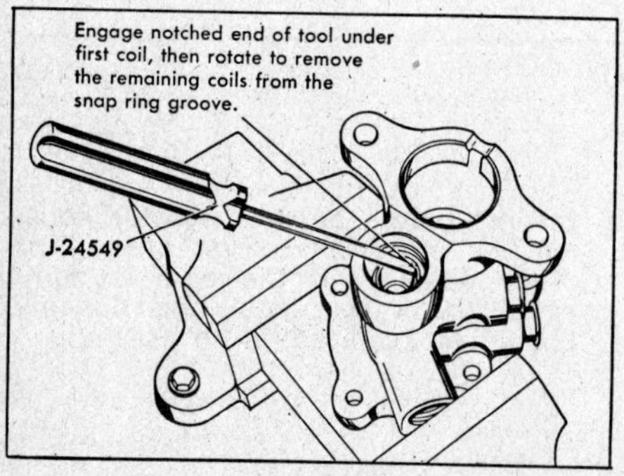

Fig. HB-6 Removing spiral snap ring

Assembly

CAUTION: Be sure to keep all parts clean until reassembly. Re-wash at reassembly if there is any occasion to doubt cleanliness. Lubricate all seals and metal friction parts with power steering fluid. Whenever the booster is disassembled, all seals, tube inserts, spiral snap rings, check valve and ball should be replaced. All of these parts are included in a seal kit.

1. Position a new tube seat in each booster port and screw a spare tube nut in each port to press the seat down into the port. Do not tighten the tube nuts in the port as this may deface the seats.
2. Remove spare tube nuts and check for aluminum chips in the ports. Be sure any foreign matter is removed.
3. Coat piston bore and piston seal with clean power steering fluid and assemble the new seal in the bore. The lip of the seal must be toward the rear (away from master cylinder mounting). Be sure seal is fully seated in housing.
4. Lubricate the input rod end, new input rod seals and Seal Installer Tool J-24553 with clean power steering fluid. Slide the seals on the rod with lip of the cups toward open end of tool, Fig. HB-7.
5. Slide tool over input rod end and down to the second groove; then slide the forward seal off the tool and into the groove. Assemble the other seal in the first groove. Be sure both seals are fully seated.
6. Lubricate the piston and Piston Installing Tool J-24551 with clean power steering fluid. Insert large end of tool into piston, Fig. HB-8. Slide the tool and piston into the piston bore and through the piston seal.
7. Assemble the new "O" ring onto the new accumulator check valve and dip the assembly in clean power steering fluid. Insert the check valve into the accumulator valve recess in the housing flange.
8. Assemble the new ball and spacer in the same recess.
9. Assemble the new "O" ring onto the charging valve plunger seat and insert plunger into the seat. Dip the assembly in clean power steering fluid and insert it into the charging valve recess.
10. Dip spool assembly in clean fluid and insert assembly into spool bore in housing. Be sure the pivot pins on the upper end of the input rod lever assembly are engaged in the groove in the sleeve. Remove Tool J-24551 from piston assembly.
11. Separate the two parts of Snap Ring Installing Tool J-24550 and place the new spiral type spool retaining ring on the tool as shown in Fig. HB-9. Insert rounded end of tool into spool, then while pressing on rear of spool, slide retaining ring off tool and into groove near forward end of spool by pressing in on tool sleeve. Check to be sure retaining ring is fully seated.
12. Position a new housing seal in the groove in housing cover. Then join booster housing and cover and secure with five screws. Torque screws to 18-26 ft. lbs.
13. Assemble a new "O" ring seal on the spool plug. Insert spool spring and spool plug in forward end of spool bore. Press in on plug and assemble the plug snap ring in the groove in the bore.

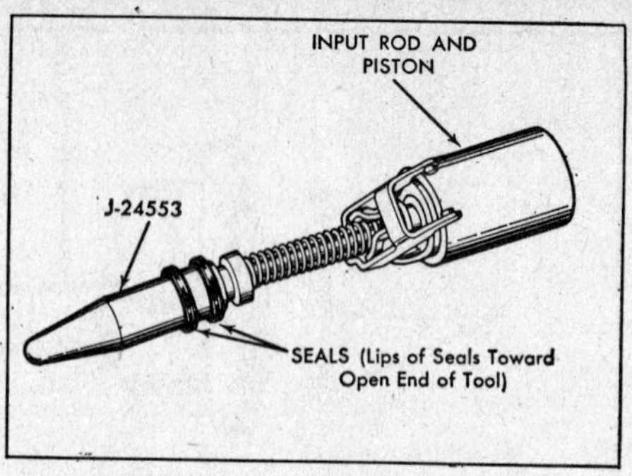

Fig. HB-7 Installing input rod seals

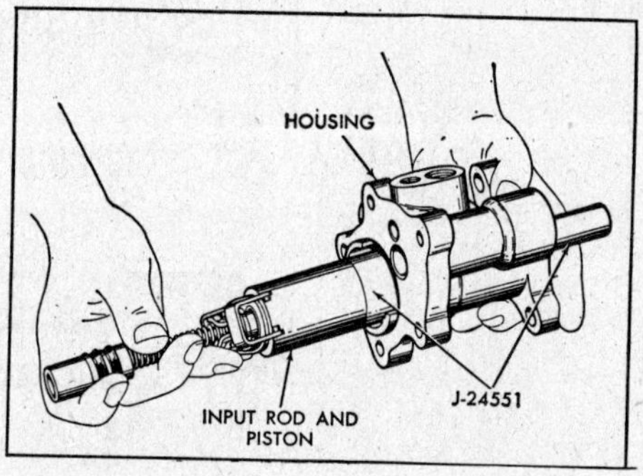

Fig. HB-8 Installing input rod and piston assembly into booster

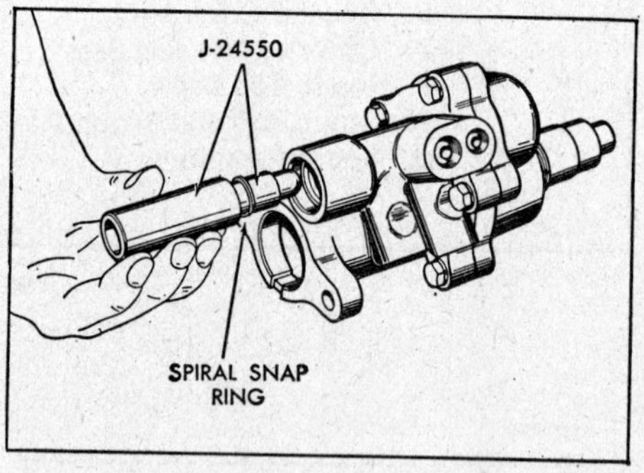

Fig. HB-9 Installing spiral snap ring

14. Position linkage bracket on booster. The tab on inside diameter of the large hole in the bracket should fit into slot in the threaded portion of booster hub.
15. Install the new bracket nut with the staking groove outward on threaded hub of booster. Using Tool J-24554 and a torque wrench, Fig. HB-10, torque nut to 95-120 ft. lbs.
16. Use a hammer and a small punch inserted into the staking groove of the nut, at the slot in the booster hub, Fig. HB-11, to stake the nut in place. Be sure that the outer thread of the nut is upset.

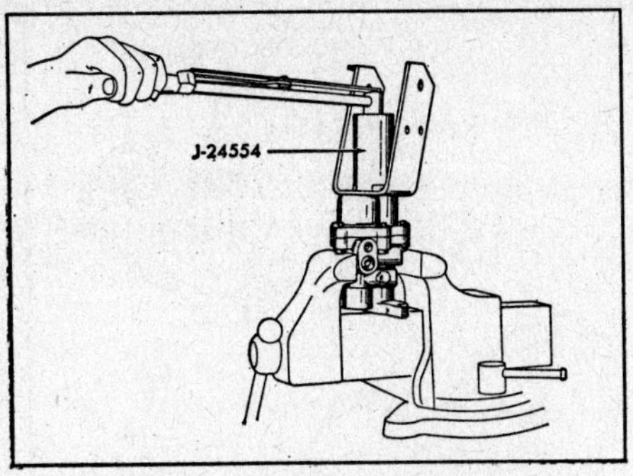

Fig. HB-10 Installing linkage bracket on booster

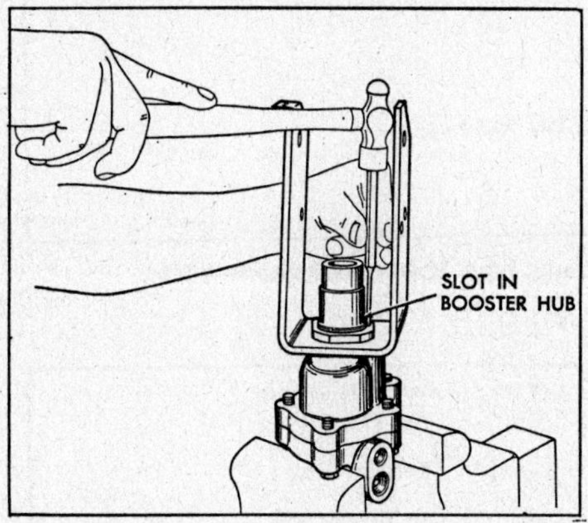

Fig. HB-11 Staking linkage bracket nut

17. Assemble a new boot on the pedal rod. Then assemble a new grommet in the groove near the end of pedal rod.
18. Moisten the grommet with water to ease assembly and insert the grommet end of the pedal rod into the input rod end of the booster housing. Push on the end of the pedal rod to seat the grommet in the groove inside the housing.

NOTE: When the grommet is fully seated the pedal rod will rotate freely with no binding.

19. Slide boot on pedal rod and assemble the large end of the boot on the hub of the power section.

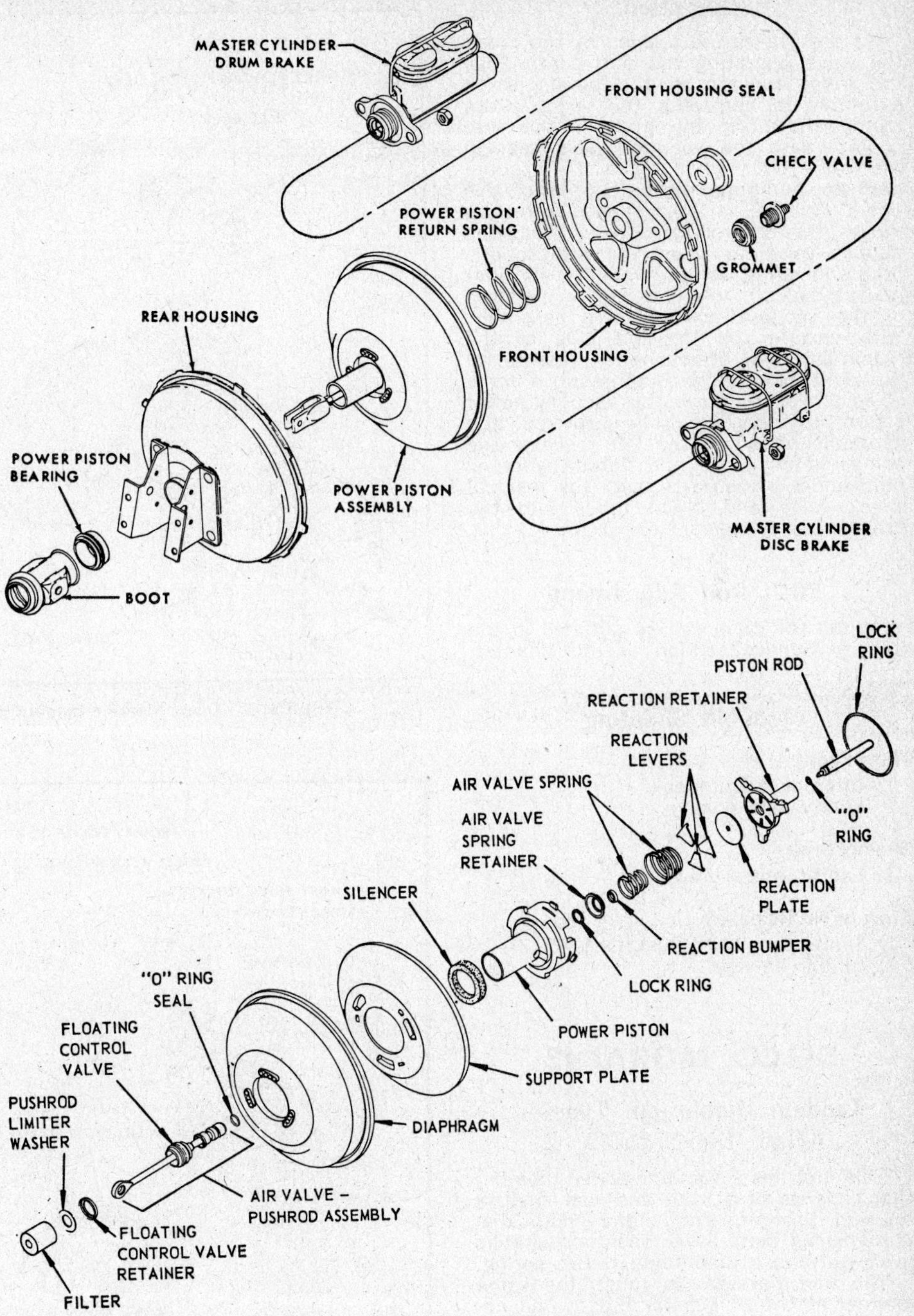

Fig. DM-1 Delco Moraine single diaphragm type

DELCO MORAINE

Single Diaphragm Type—Fig. DM-1

This power brake unit is a combination vacuum-hydraulic brake booster of the vacuum suspended type which utilizes engine intake manifold vacuum and atmospheric pressure for its power.

The unit consists of a vacuum power section and a hydraulic master cylinder section. The vacuum power section contains a power piston with rolling diaphragm and houses the control valve, the reaction mechanism and power piston return spring.

The control valve is made up of an air valve and floating vacuum control valve assembly. The reaction mechanism consists of a hydraulic piston, reaction plate and a series of levers. The valve operating rod, which operates the air valve, projects from the power section and is connected to the brake pedal linkage.

The hydraulic push rod operates against the master cylinder piston. A split system (tandem piston) type master cylinder is incorporated into some units. The front half of the master cylinder in the split system operates the rear brakes while the rear half of the master cylinder operates the front brakes (see HYDRAULIC BRAKE SYSTEM chapter).

A vacuum check valve, attached to the front vacuum chamber and connected to the intake manifold, traps vacuum in the power unit at the highest level of vacuum.

Operation

As the brakes are applied by the driver the valve operating rod and control piston move forward in the power piston assembly to compress the valve return spring and bring the poppet valve into contact with the vacuum valve seat in the valve housing to *close* the vacuum port. Any additional movement of the valve operating rod in the applied direction moves the control valve away from the poppet valve to *open* the atmospheric port and admit air through the air filter and passages to the chamber at the right of the vacuum power piston assembly. With vacuum on the left side of the diaphragm and atmospheric pressure on the right side of the diaphragm, a force is developed to move the vacuum power piston assembly, hydraulic push rod, and hydraulic piston to the left to close the compensating port and force hydraulic fluid under pressure through the residual check valve and brake tubes into the brake wheel cylinders.

Push Rod Adjustment

Adjust the push rod as outlined in the *General Service* section of this chapter.

Trouble Shooting

Hard Pedal

1. Internal vacuum leak.
2. Faulty control valve.

Brakes Grab

1. Faulty control valve.

Slow or No Release

1. Faulty push rod adjustment.
2. Bind in linkage.

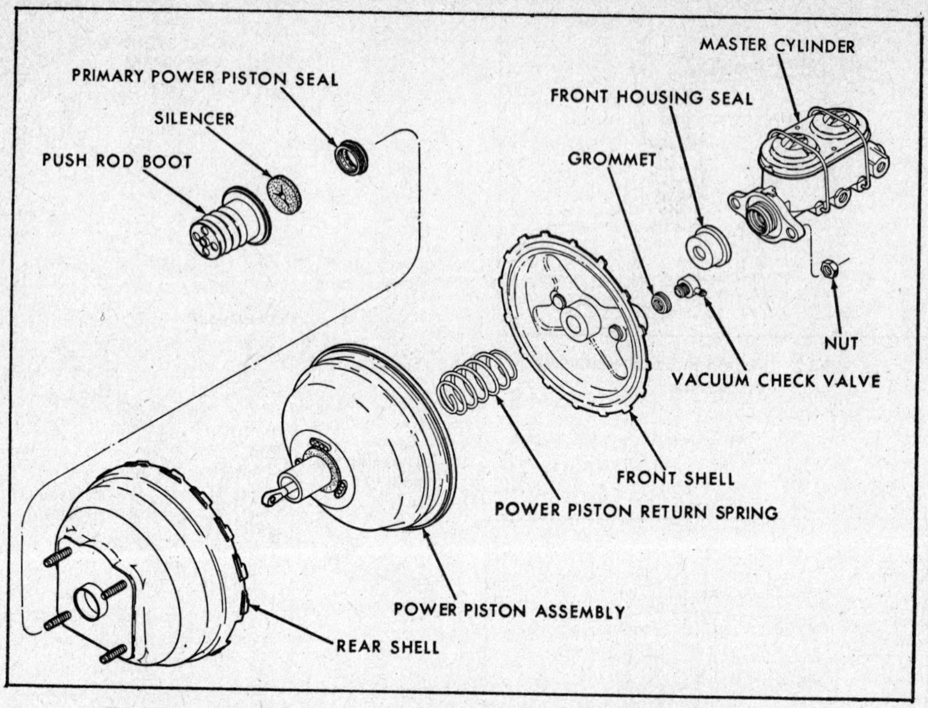

Fig. DM-2 Delco Moraine tandem diaphragm type. Power head disassembled

DELCO MORAINE

Tandem Diaphragm Type—Figs. DM-2, DM-3

This unit has a vacuum power chamber that consists of a front and rear shell, a housing divider, front and rear diaphragm and plate assemblies, a hydraulic push rod and a diaphragm return spring.

The unit operates in much the same manner as the Single Diaphragm unit described above. The diaphragm and plate assemblies utilize the pressure differential created by engine intake manifold vacuum and atmospheric pressure to assist the hydraulic push rod.

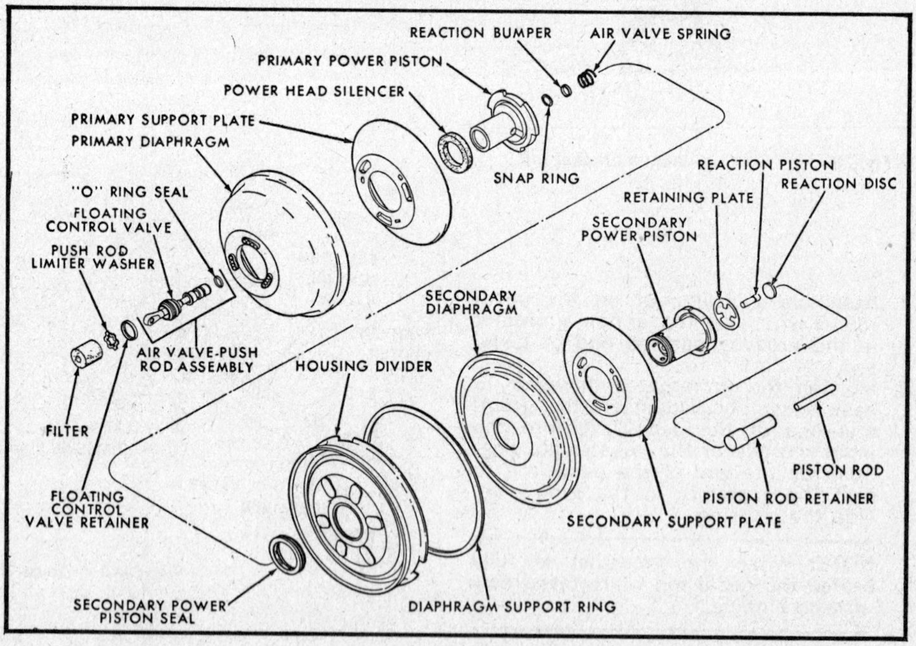

Fig. DM-3 Delco Moraine tandem diaphragm type. Power piston disassembled

Trouble Shooting

Hard Pedal

1. Internal Vacuum Leak.
2. Faulty Control Valve.

Brakes Grab

1. Faulty Control Valve.

Slow or No Release

1. Faulty push rod adjustment.
2. Bind in linkage.

MIDLAND-ROSS

Diaphragm Type—Figs. MR-1 & MR-2

This power unit is installed on the engine side of dash and connected to the brake pedal through a lever assembly and push rod link.

The power unit consists of a vacuum chamber, air valve, control valve plunger assembly, diaphragm and an air chamber.

Atmospheric pressure is present at all times in the air chamber at the front side of the air valve. The air intake to the air chamber is protected by a filter. The air chamber is separated from the vacuum chamber by the bellows assembly

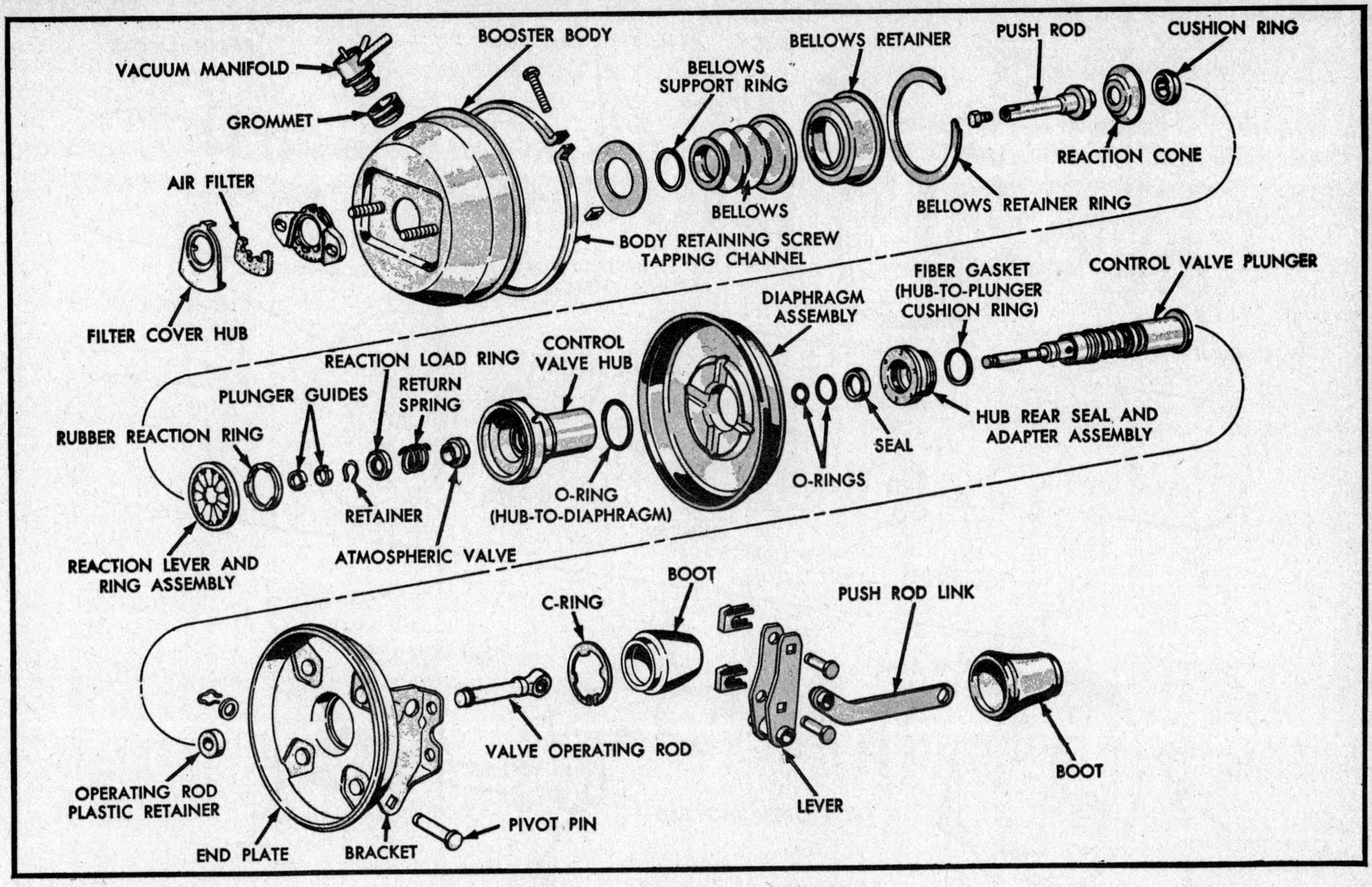

Fig. MR-1 Midland-Ross Diaphragm type

within the vacuum chamber.

Vacuum is present at all times in that area of the vacuum chamber forward of the diaphragm. Vacuum is supplied through a hose from the engine intake manifold to the vacuum manifold and check valve on the power unit body. With this integral check valve and vacuum chamber, it is possible to obtain several assisted brake applications with the engine shut off. This arrangement makes a vacuum reservoir unnecessary.

Either vacuum from the forward side of the diaphragm or air from the bellows (air chamber) can be connected to the rear side of the diaphragm through porting in the control valve hub and plunger assembly.

Certain applications use a tandem diaphragm unit. These units are essentially the same as shown in Fig. MR-2 with the exception that an additional diaphragm is added to the control hub, Fig. MR-2A.

Operation

As the brakes are applied by the driver, the valve operating rod and control valve plunger move forward to compress the valve plunger return spring and close the vacuum port. Further movement of the actuating valve opens the atmospheric port to admit atmospheric pressure to the rear side of the power diaphragm. With vacuum on the front side of the diaphragm and atmospheric pressure on the rear side of the diaphragm, the diaphragm, push rod and master cylinder piston move forward to close the compensating port and force hydraulic fluid under pressure through the residual check valve and brake lines to the wheel brakes. As hydraulic pressure is developed within the hydraulic system, an opposing force (counter force) through the reaction levers and ring assembly transmits the reaction force back through the actuating control valve assembly and valve operating rod to the brake pedal. This reaction force tends to reduce the power application and is the means of providing the proper "pedal feel" to the driver.

Push Rod Adjustment

Adjust the push rod as outlined in the *General Service* section of this chapter.

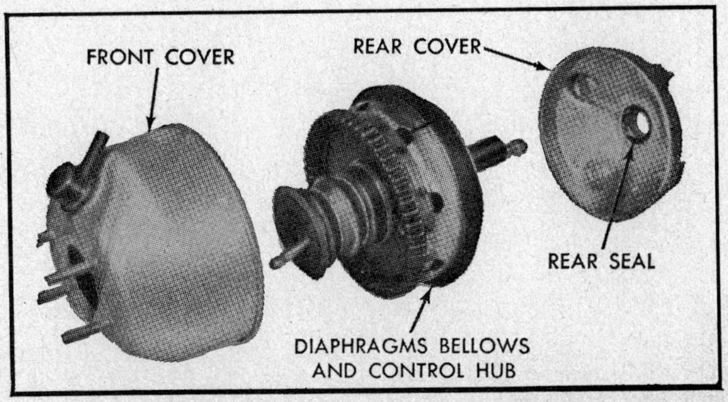

Fig. MR-2A Midland-Ross tandem diaphragm unit

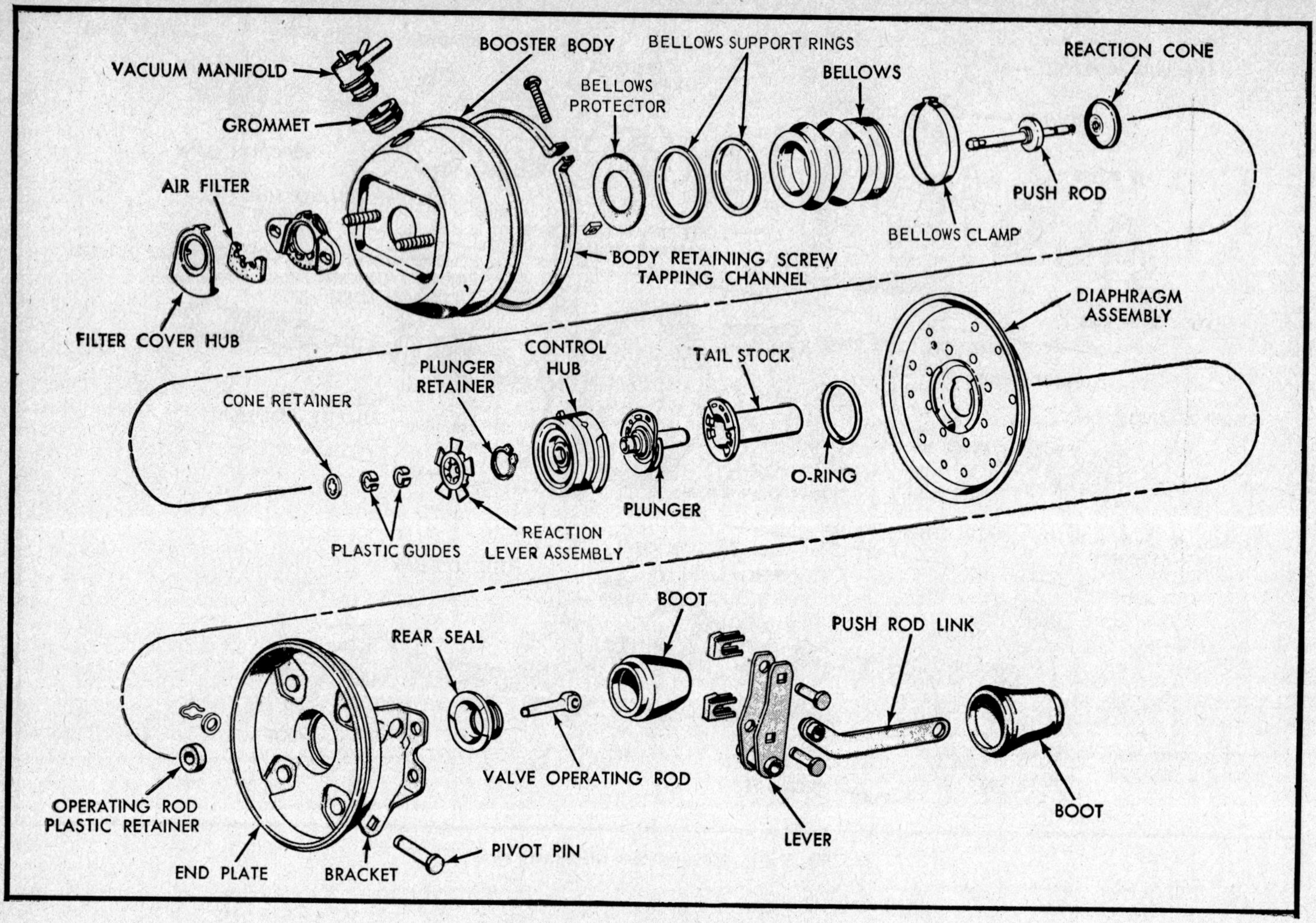

Fig. MR-2 Midland-Ross Diaphragm type

Bleeding The System

The bleeding operation is accomplished in the same manner as outlined in the *Hydraulic Brake System* chapter for conventional brakes. However, some units have a bleeder valve on the power brake master cylinder. In bleeding the brakes, first bleed the master cylinder, then proceed with the wheel cylinders.

Trouble Shooting

Hard Pedal

1. Leak in bellows.
2. Diaphragm out of place in housing.

Brakes Grab

1. Sticking actuating valve.

Brakes Self-Apply when Engine is Started

1. Leak in rear housing.
2. Diaphragm out of position in housing allowing air into rear chamber.
3. Sticking or unseated air valve.

POWER STEERING

> **NOTE:** Automatic transmission fluid should not be used in power steering systems except in small quantities and then only to bring the fluid level up to the fill mark. Power steering fluid must withstand very high pressures (about 1400 psi) whereas transmission fluid does not have the need to operate at these pressures. If a considerable amount of transmission fluid has been added to a power steering system as in an emergency situation, the system should be drained and flushed and refilled with power steering fluid as soon as possible to avoid damage.

[1]—Service procedures given in car chapter.
[2] 1975 models with V8-351 less A/C.
[3] 1974 models with V8-302; 1975 models with V8-351 less A/C.

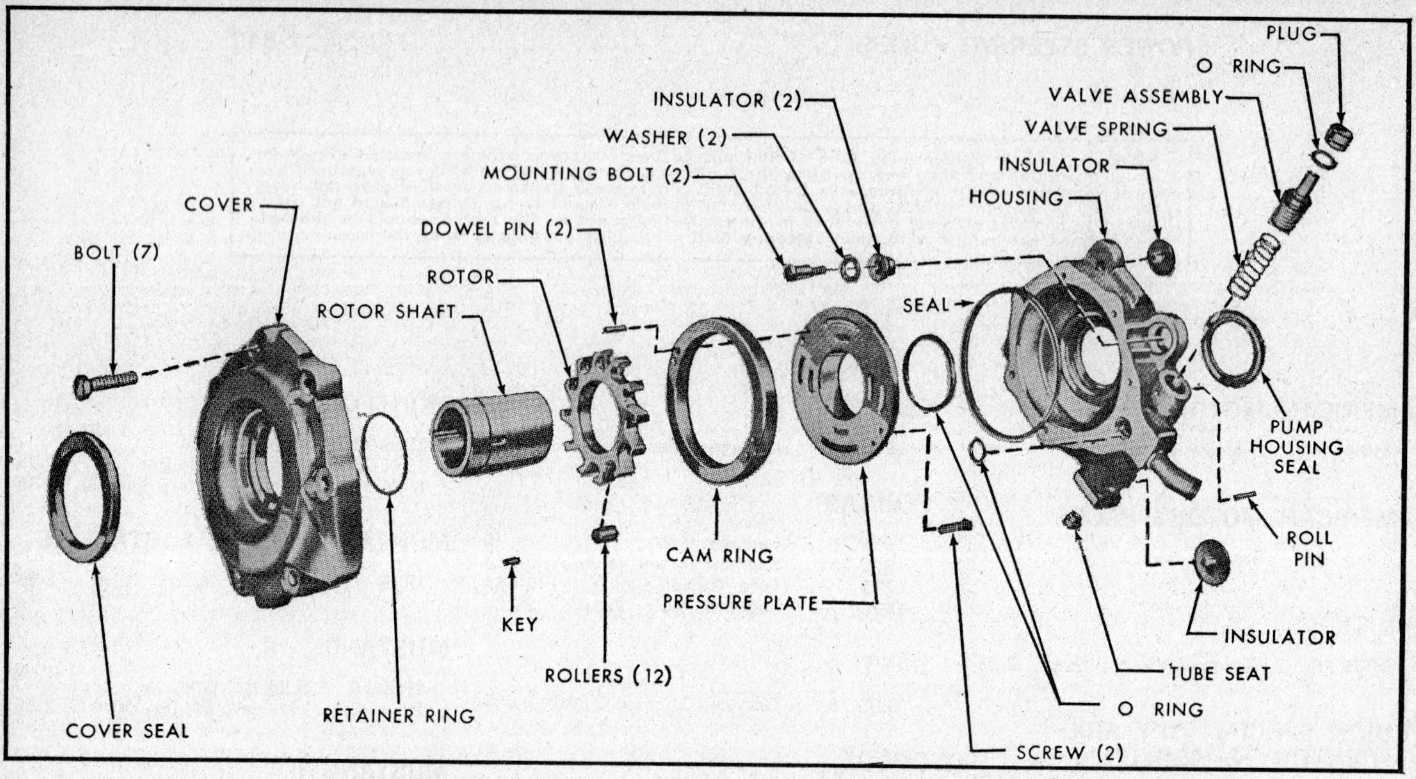

Fig. 1 Lincoln & Mark III crankshaft driven roller type pump, exploded. Typical

ROLLER TYPE PUMP

This type pump, Fig. 1, consists of the reservoir, pressure relief valve, filter element, and the rotating pump cam.

Thoroughly clean exterior of pump and reservoir, using care so that dirt does not enter pump or reservoir upon disassembly. During all assembly operations, extreme caution must be used to prevent any dirt from entering pump. All parts should be lightly oiled during reassembly to provide initial lubrication.

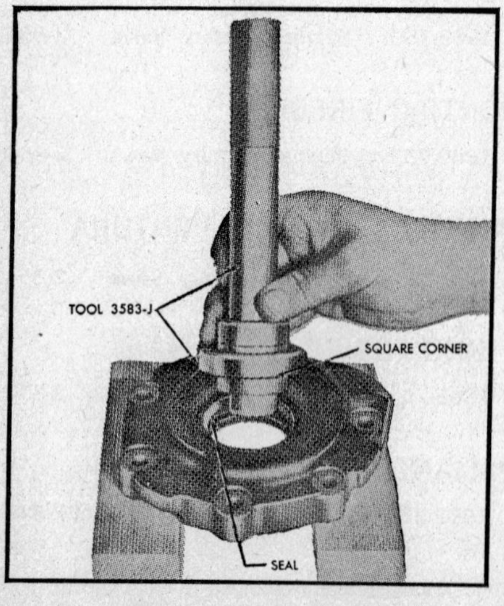

Fig. 2 Installing rotor shaft seals. Lincoln & Mark III

Lincoln & Mark III Units

Disassembly

1. Remove pump cover-to-housing bolts, place pump cover side down on bench and lift housing from cover. Remove O rings, Fig. 1.
2. Remove pressure plate screws and remove plate. Lift cam ring off dowel pins.
3. Remove rollers, lift rotor from shaft and remove drive key from shaft. Remove dowel pins from cover.
4. Remove rotor shaft from cover. *Do not remove snap ring from shaft unless ring or shaft is damaged.*
5. Pry rotor shaft seals out of cover and housing.
6. *Do not disturb the control valve unless valve has been determined to be at fault.* To remove valve, drive out roll pin. Remove plug and O ring, control valve and spring. If plug or valve is stuck, tap housing with a soft face hammer. Remove O ring from plug.

Reassembly

1. Using new O ring, install control valve spring and valve in housing with small hole in valve toward spring. Install plug and roll pin.
2. Install new seals in housing and cover, Fig. 2, then install rotor shaft in cover, Fig. 3.
3. Install dowel pins in cover, install drive key on shaft and install rotor on shaft with the counterbored side of the rotor toward cover.
4. Install cam ring on dowel pins with chamfered edge upward. Place rollers in rotor.
5. Install pressure plate and torque screws to 20 in-lbs.
6. Install new O rings and install housing on cover. Torque bolts to 15-20 ft-lbs.

Chrysler Corp. Unit

Disassembly

1. Using a suitable puller, remove pulley.
2. Remove mounting bracket and reservoir and place pump in vise with shaft down. *Use vise with soft protective jaws.*
3. Using a punch, tap cover and retaining ring around until one end of ring lines up with hole in body. Insert punch through hole far enough to disengage ring from groove. Remove ring, Fig. 4.

Fig. 3 Installing rotor shaft. Lincoln & Mark III

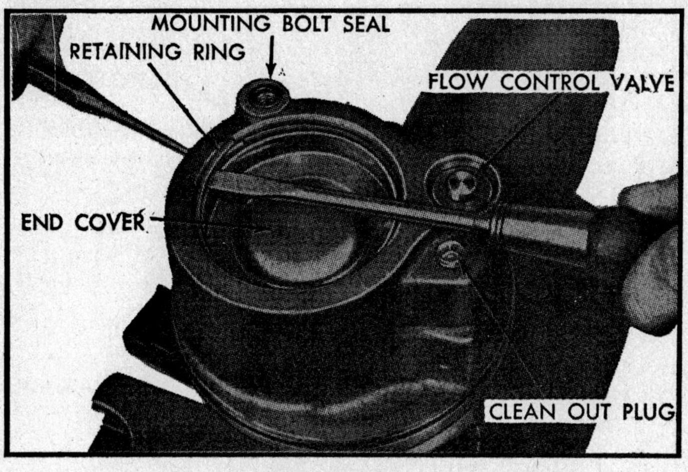

Fig. 4 **Removing end cover retaining ring Chrysler Corp. units**

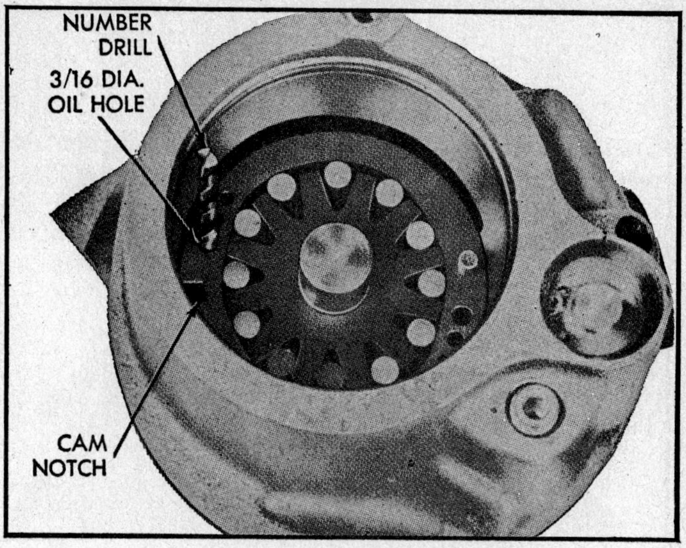

Fig. 6 **Checking dowel pin alignment. Chrysler Corp. units**

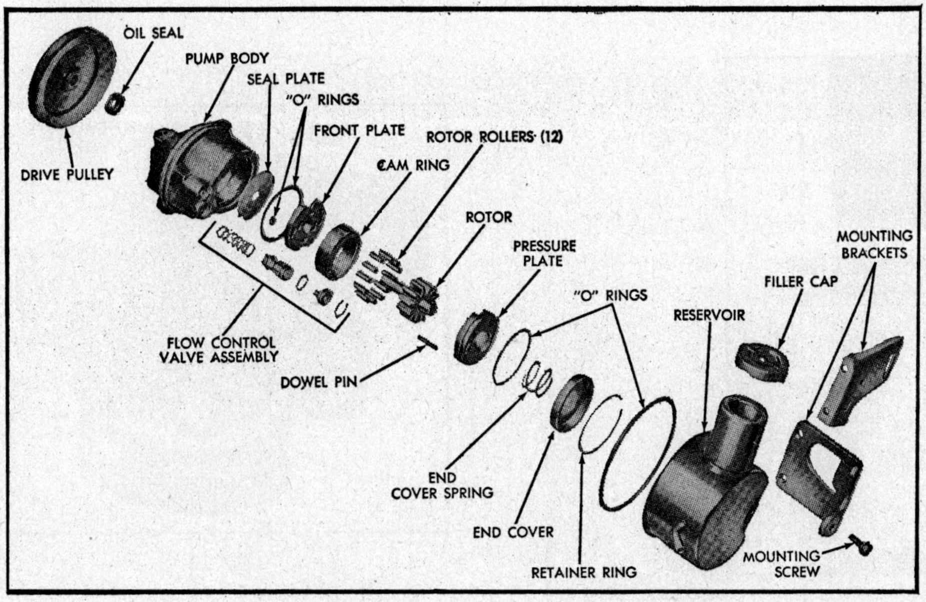

Fig. 5 **Chrysler Corp. roller type pump. Exploded**

4. Tap end cover to jar it loose. Spring under cover should push it up.
5. Remove pump from vise, place it in inverted position on a clean, flat surface and tap end of shaft to loosen rotating group. Remove pump body.
6. Remove cam ring, thrust plate, snap ring, bore plug, flow control valve and spring, Fig. 5.

Reassembly

1. Install new shaft seal, "O" rings, brass seal plate and fiber gasket in bottom of housing bore. Align index notch with dowel pin hole in housing.

NOTE: Use extreme care in lining up dowel pin. Pump can be completely assembled with dowel pin improperly positioned in end plates and not indexed in housing.

2. Place dowel pin in cam ring and position cam ring inside pump bore. Notch on cam ring must be away from pulley end of pump. If cam ring has two notches, install with machined notch (sharp corners) away from pulley. If end of dowel pin in cam ring is more than $3/16''$ above surface of installed cam ring, it is not seated in index hole in housing.
3. Install rotor and shaft in cam ring and place rollers in cavities of rotor. Lubricate all parts with power steering fluid.
4. Rotate shaft by hand to be sure rollers are correctly seated. To check dowel pin alignment, insert a #13-16 drill into large oil hole in cam ring next to cam notch, Fig. 6.
5. Install new O ring on pressure plate and align index notch in plate with dowel pin, Fig. 7. Seat plate on cam ring, Fig. 8, and remove drill. Be sure plate is squarely seated on cam ring.
6. Install coil spring, end cover and retaining ring, flow control valve, reservoir, and mounting bracket.
7. Install pulley, Fig. 9.

American Motors Type

Disassembly

1. Clean exterior of pump.
2. Remove reservoir cover.
3. Remove $5/16''$ stud with puller or two nuts, Fig. 10.
4. Remove reinforcing plate and reservoir.
5. Remove O-ring seals between reservoir and pump body and discard seals.
6. Remove pulley and woodruff key from shaft.
7. Remove seal from body, using a suitable seal puller, Fig. 11.
8. Remove cover-to-body mounting screws and studs. Separate cover from body, lifting cover vertically from body to prevent internal parts from falling out.
9. Before removing carrier, mark a line across the carrier and cam, Fig. 12. This will permit correct reassembly as the carrier has a drive side and must be installed as shown. Do not scratch surfaces when making marks.
10. Remove rollers, carrier and drive pin from shaft. Then lift out cam and cam retaining pin, Fig. 13.
11. To remove combination flow control and relief valve in pump cover, Fig. 14, remove spring retainer cap and O-ring seal. Discard seal.
12. Lift out flow control valve spring.
13. Tap cover on wooden block to remove combination flow control and relief valve, Fig. 15. The relief valve is an internal part of the flow control valve and is not to be disassembled.

Reassembly, Fig. 16

1. Install shaft into body.

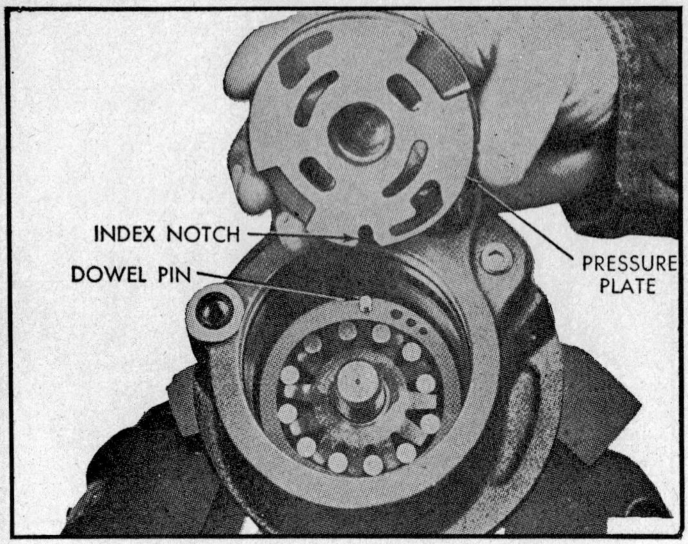

Fig. 7 Installing pressure plate. Chrysler Corp. units

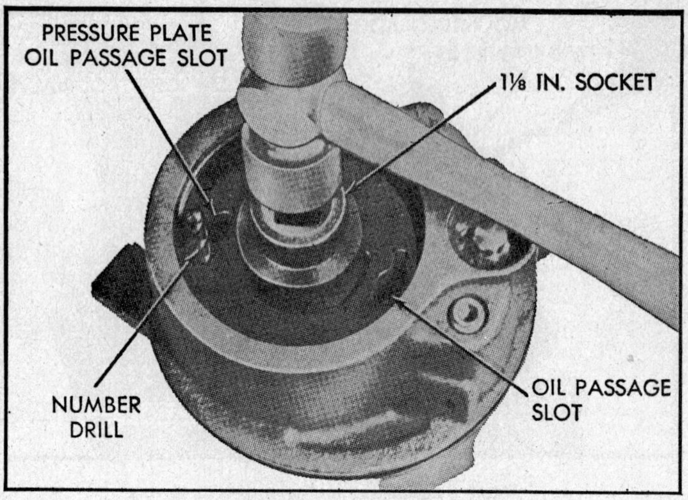

Fig. 8 Seating pressure plate. Chrysler Corp. units

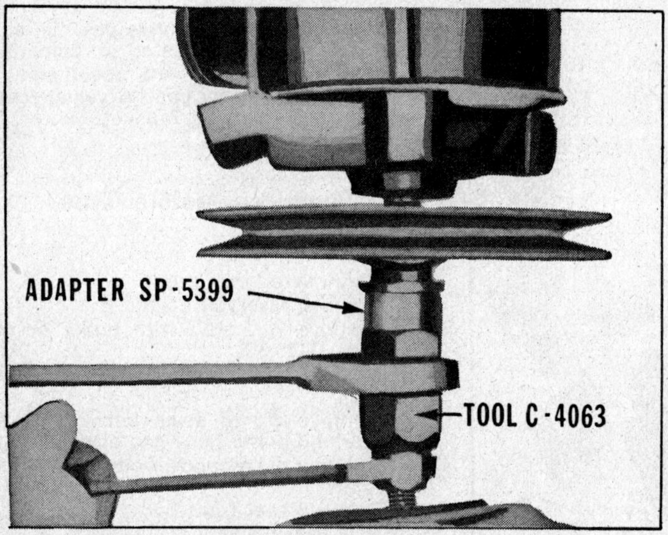

Fig. 9 Installing drive pulley. Chrysler Corp. units

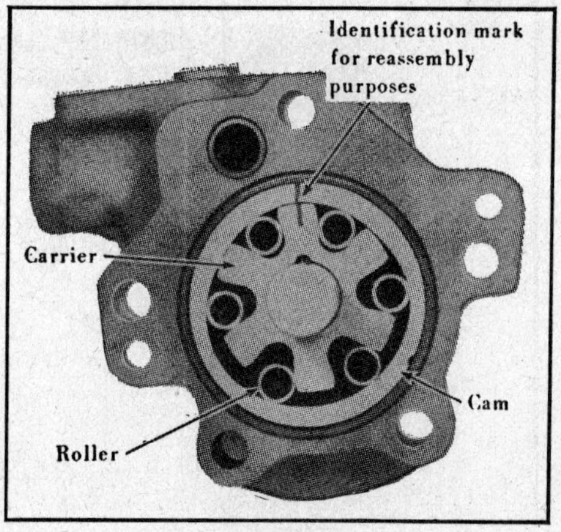

Fig. 12

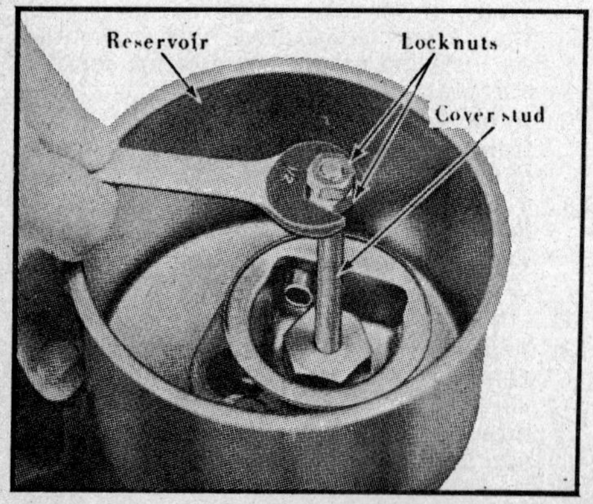

Fig. 10

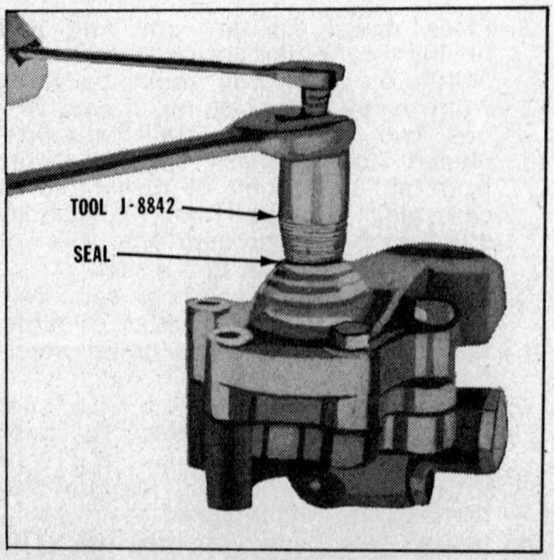

Fig. 11

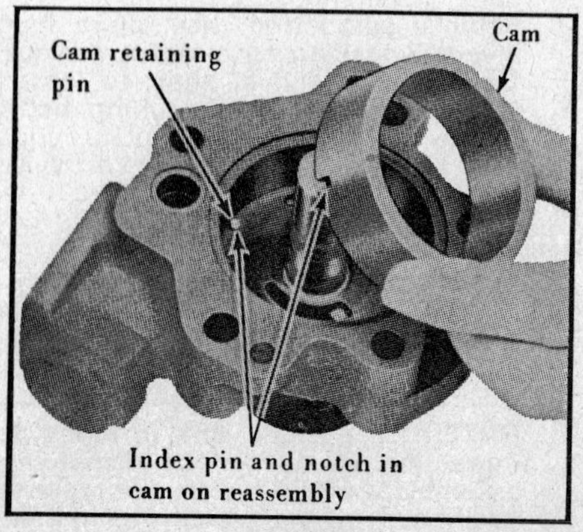

Fig. 13

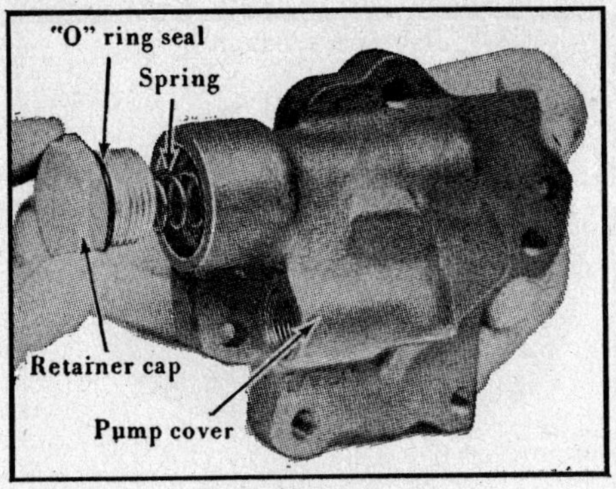

Fig. 14

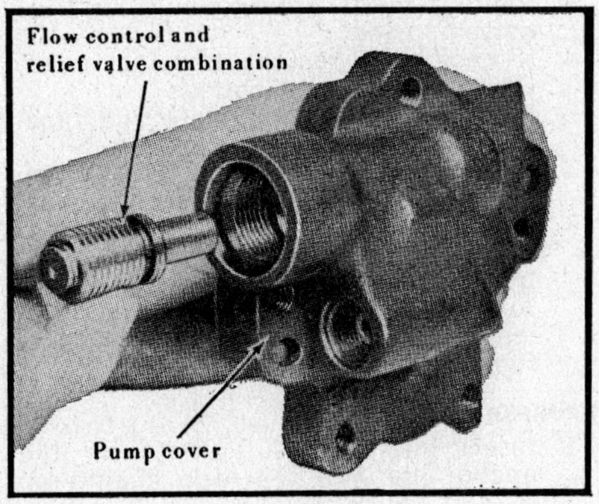

Fig. 15

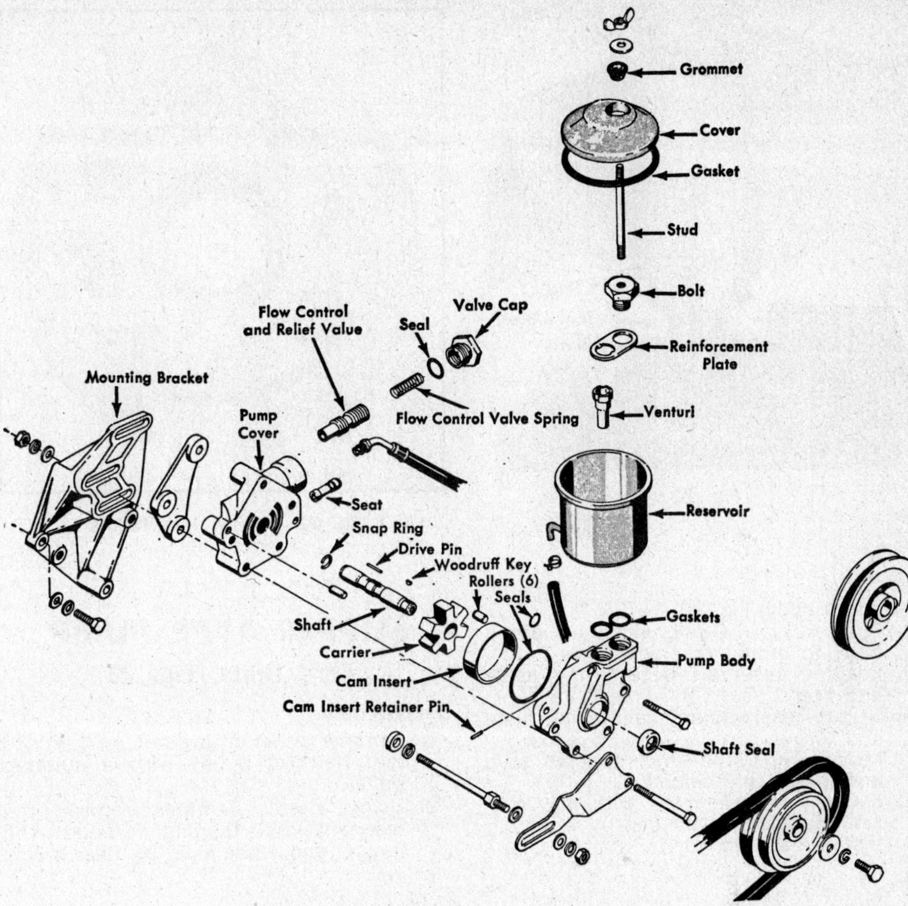

Fig. 16 American Motors roller type pump

2. Install cam retaining pin into body.
3. Install cam into body, making sure notch in cam indexes with retaining pin and cam is seated firmly in pocket.
4. Install carrier drive pin into groove in pump shaft.
5. Install carrier into pump body, Fig. 17, making sure line previously marked on carrier indexes with line marked on cam.
6. Install rollers into pockets of carrier.
7. After the cam, carrier and rollers have been placed in pump body, the end clearance of carrier and rollers in body should be checked, Fig. 18 and 19. If end clearance exceeds .0015", replace carrier or rollers.
8. Install new O-rings seals in body.
9. Position cover on body and torque screws to 25 ft-lbs. Rotate pump shaft and check for free rotation.
10. Install flow control valve into pump cover, making sure exposed ball enters bore first.
11. Install flow control valve spring.
12. Install new O-ring seal on flow control valve retainer cap and install into pump cover. Torque retainer cap to 30-35 ft-lbs.
13. Install new seal with lip facing toward

Fig. 17

Fig. 18

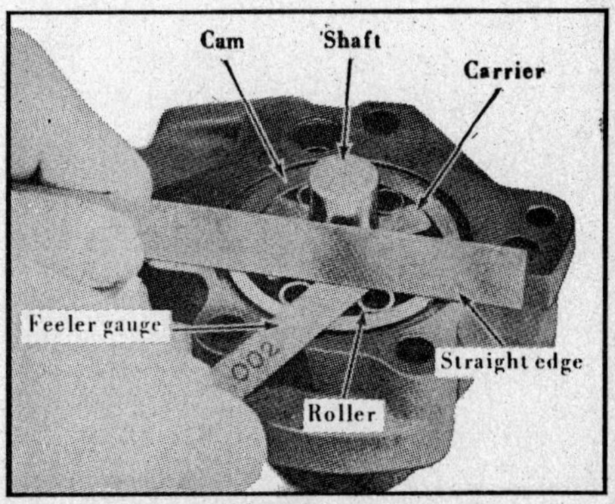

Fig. 19

Fig. 20

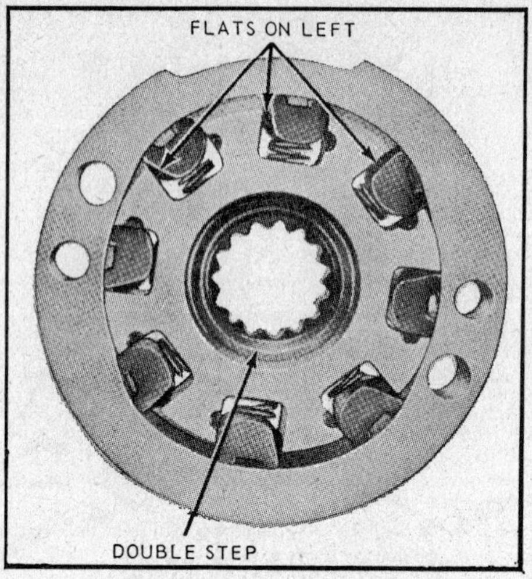

Fig. 22 Slipper installation. Ford units

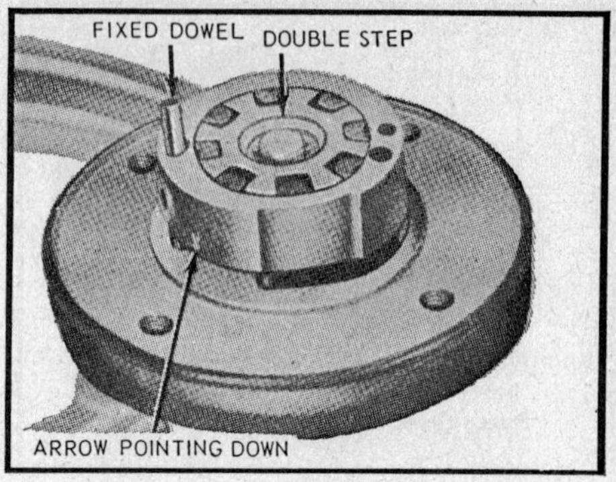

Fig. 23 Cam and rotor installation. Ford units

pump body, Fig. 20.

14. Position new O-ring seals on pump body to mount reservoir.
15. Position reservoir over O-ring seals and body.
16. Secure reservoir and reinforcing plate and torque screws to 30-35 ft-lbs.
17. Install $\frac{5}{16}$" stud into support stud and torque to 60 inch-lbs.
18. Install reservoir cover with new gasket. Install reservoir cap.
19. Install woodruff key in shaft.

SLIPPER TYPE PUMP

Ford Units, Fig. 21

Disassembly

1. Remove outlet fitting nut and service tag. Remove pulley with a suitable puller.
2. Using a soft hammer, remove the reservoir by tapping around the flange and underside of filler neck.
3. Remove pump housing bolts and remove housing.
4. If necessary, remove housing cover, O ring and pressure springs.
5. Remove retainer end plate and pressure plate.
6. Remove the loose fitting dowel pin. Fixed dowel pin remains in housing.
7. Remove rotor assembly, being careful to keep the springs and slippers from falling out. Do not disassemble further unless the lower plate, housing plate, rotor shaft and/or seal is to be replaced.

Reassembly

1. Insert lower pressure plate on the anchor pin with the wide chamfered

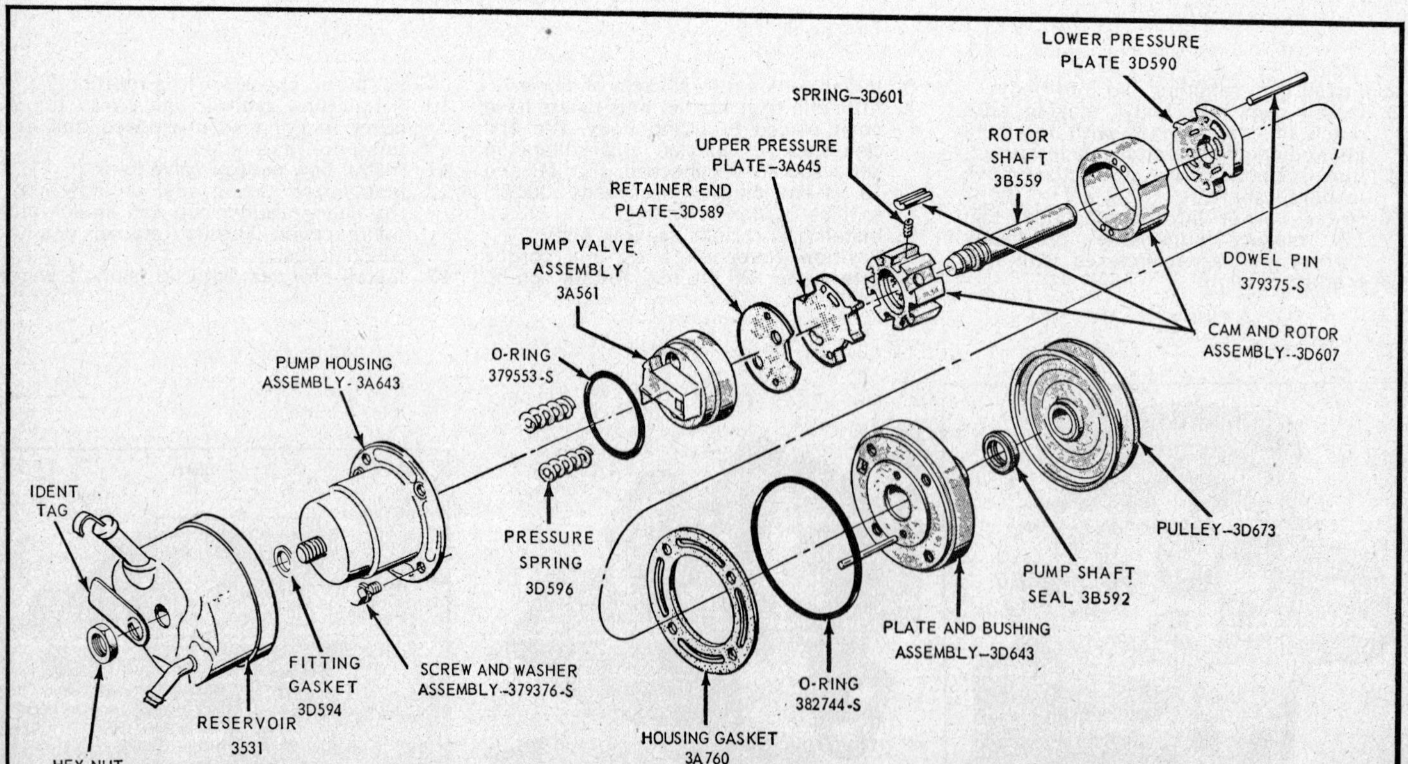

Fig. 21 Ford slipper type pump. Exploded

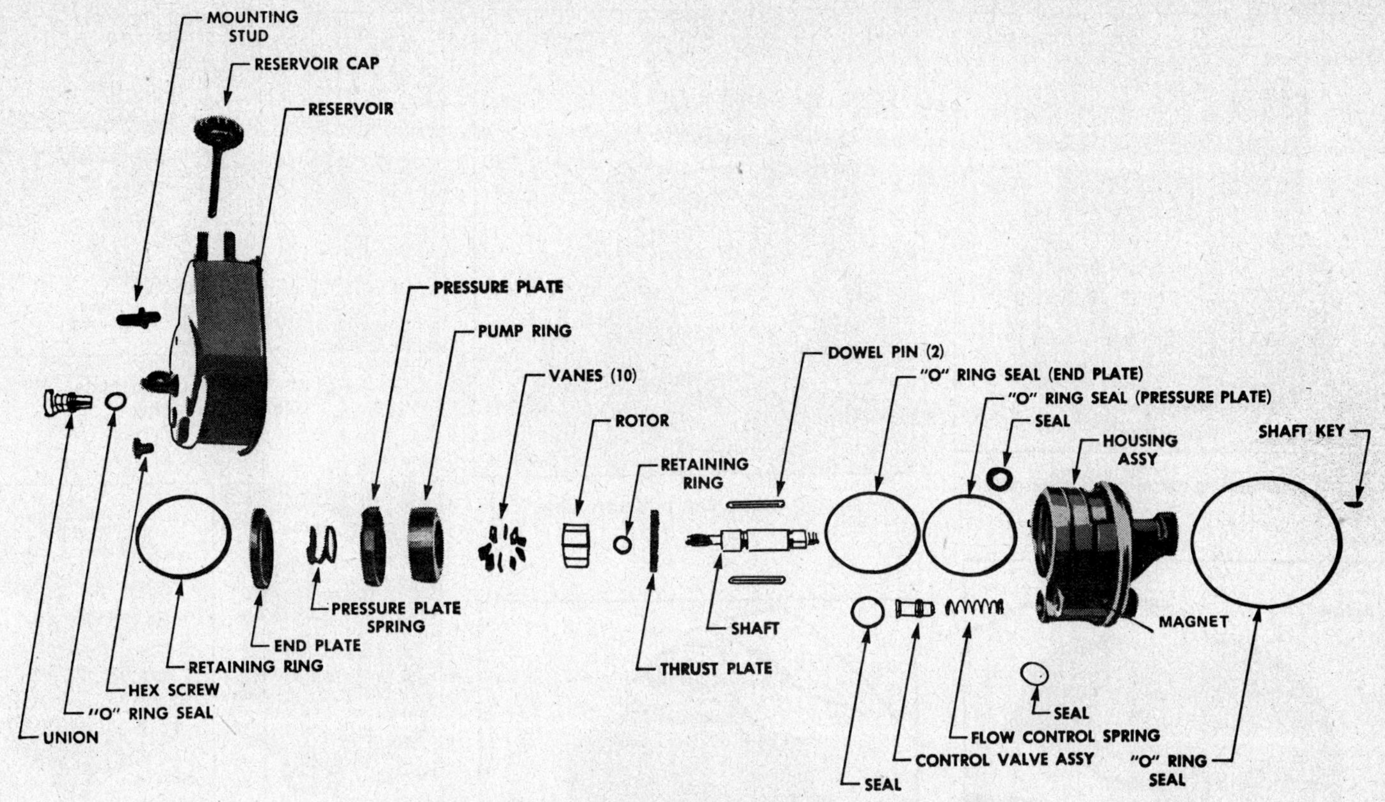

Fig. 24 Vane type power steering pump disassembled (typical)

slots at the center hole facing up.

2. Lubricate rotor shaft and insert it into lower pressure plate and housing plate.
3. If rotor assembly is disassembled, hold cam insert with notch on outside diameter of cam at top and arrow on outside diameter of cam pointing downward.
4. Insert rotor into cam with double step on rotor facing upward.
5. With rotor extended upward halfway out of cam, insert a spring into a rotor spring pocket.
6. Using a slipper, compress the spring and install the slipper with groove facing upward (toward cam notch). Flats on side of slipper should be on the left, Fig. 22. Repeat this step until all slippers and springs have been installed.

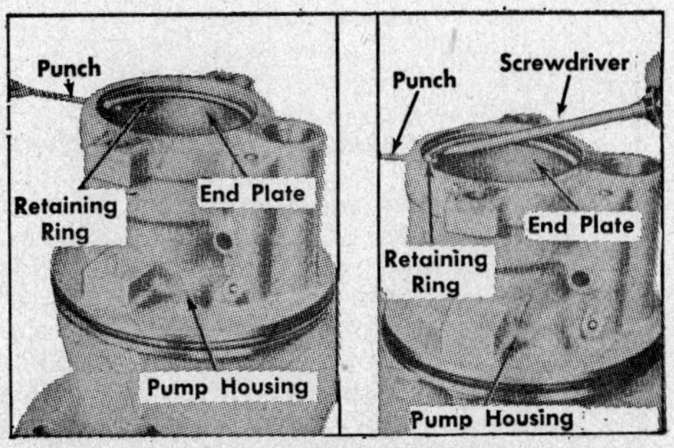

Fig. 25 Removing end plate retaining ring

7. Install cam and rotor assembly into pump housing plate, Fig. 23. If necessary, turn rotor shaft slightly until spline teeth mesh, allowing cam and rotor to drop down into position.
8. Insert loose fitting dowel pin through cam insert and lower plate into hole in housing. When properly installed, height of both dowels will be equal. Lubricate cam and rotor.
9. Install upper pressure plate and retainer end plate.
10. Place pump valve on top of retainer end plate with large exhaust slot in line with notches of previously assembled parts. If valve has been installed correctly, valve stem will be in line with lube return hole in housing plate.
11. Install new gasket on pump housing plate and insert pressure plate springs into pockets in valve assembly.
12. Using fabricated studs, to align cover and housing, install housing, using even, downward pressure. The pressure plate springs must not be moved out of position. Remove guide studs.
13. Install housing retaining bolts and torque evenly to 30 ft-lbs.
14. Install a ⅜ x 16 bolt finger tight into the end of the rotor shaft and, using a torque wrench, check input torque of the shaft. Torque should not exceed 15 in-lbs. If it does, loosen retaining bolts slightly, rotate shaft and re-torque bolts evenly. Recheck torque. *The pump must not be used if shaft torque exceeds 15 in-lbs.*
15. Install "O" ring on housing, then tap housing onto reservoir with a mallet.
16. Install and torque valve outlet fitting nut to 45 foot lbs.
17. If removed, install pulley with a suitable tool.

VANE TYPE PUMP

Disassembly

1. Referring to Fig. 24, carefully mount pump in vise with soft jaws and remove reservoir cap.

NOTE: *When clamping pump in vise, do not exert excessive force on front hub as this may distort drive shaft*

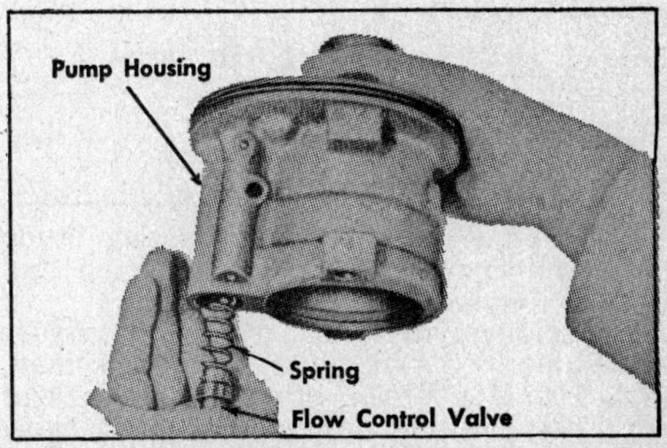

Fig. 26 Removing flow control valve and spring

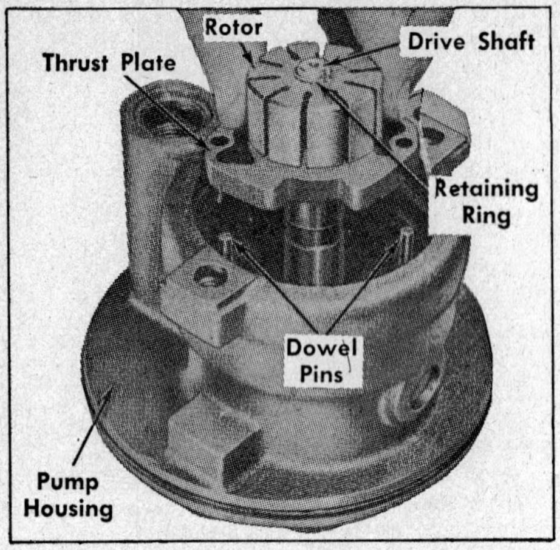

Fig. 27 Installing drive shaft assembly

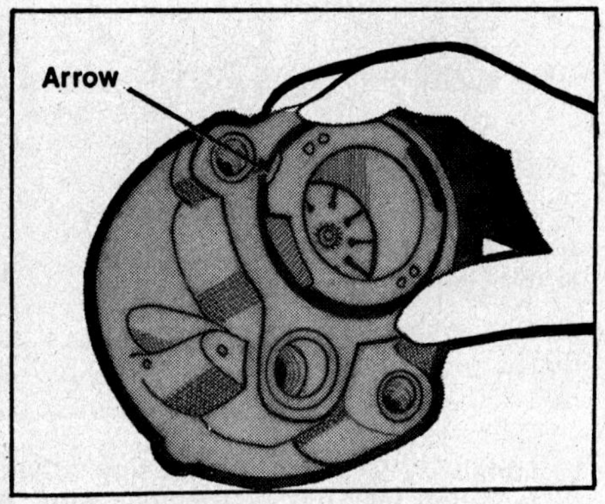

Fig. 28 Installing pump ring

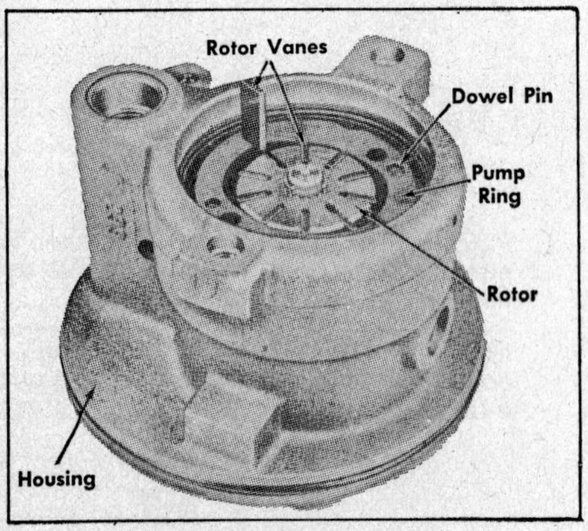

Fig. 29 Installing rotor vanes

bushing in housing.

2. Remove rear pump mounting studs and O-ring seal, pressure union and O-ring seal.
3. Remove reservoir, second pressure union "O" ring and magnet, if used.
4. Using a small punch, depress end plate retaining ring enough to allow removal from groove. Use the ⅛" diameter hole in housing, Fig. 25. Then remove retaining ring with

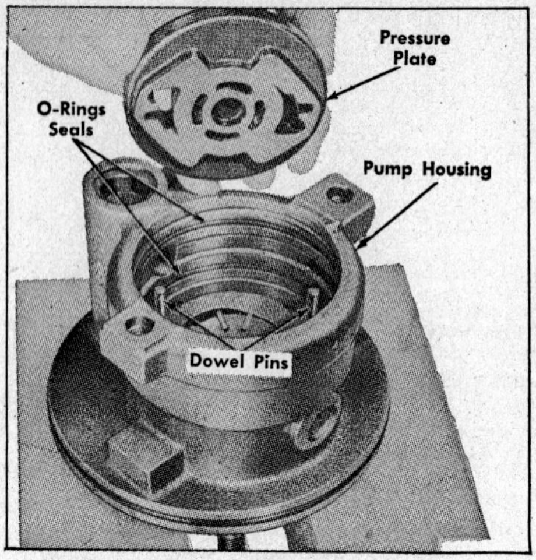

Fig. 30 Installing pressure plate

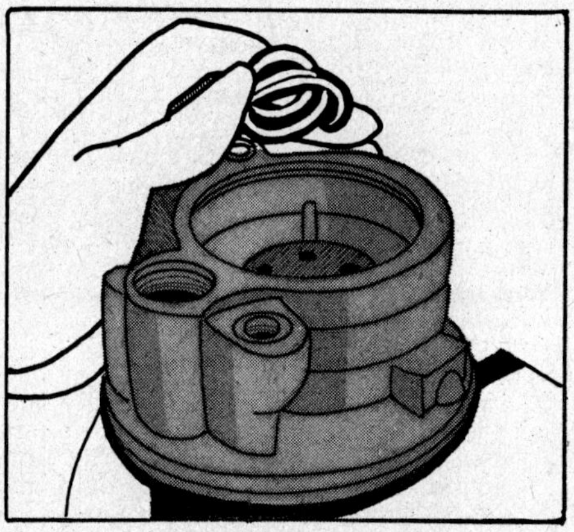

Fig. 31 Installing pressure plate spring

screwdriver.

5. Remove end plate which is spring-loaded and will usually sit above housing level. If end plate sticks, a slight rocking action should free it.
6. Remove two pressure springs from pump housing.
7. Remove flow control valve and spring by inverting housing. Do not disassemble flow control valve as it is serviced as a unit and pre-set at the factory, Fig. 26.
8. Remove drive shaft key from slot in shaft. Then with end of shaft pointed downward, press down until shaft is free.
9. Turn assembly over and remove drive shaft and rotary group.
10. Remove rotor retaining ring from groove in drive shaft. Remove rotor and thrust plate from shaft, Fig. 24.
11. Remove and discard O-ring seals from pump housing. Remove drive shaft oil seal only if inspection shows necessity for replacement.

Inspection

1. Clean all parts except the drive shaft oil seal in cleaning fluid. The seal

Fig. 32 Installing end plate and retaining ring

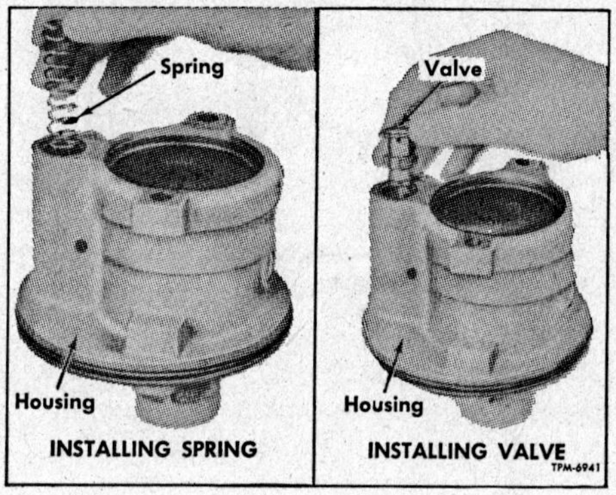

Fig. 33 Installing flow control valve and spring

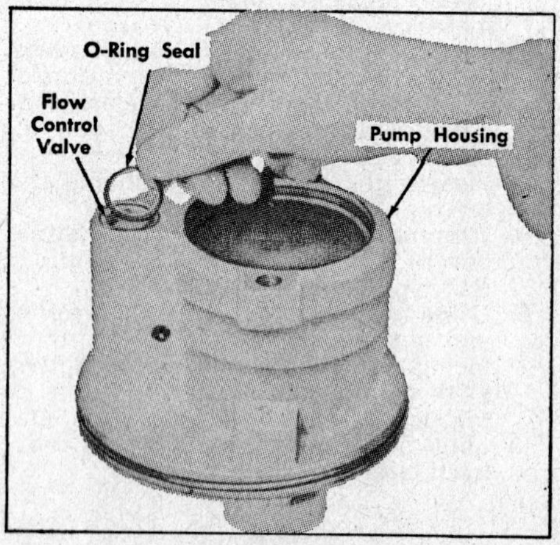

Fig. 34 Installing pressure union O-ring seal

will be damaged if immersed in cleaning fluid.

2. Check fit of vanes in slots of rotor for tightness or excessive looseness. Vanes must fit snugly but slide freely in slots in rotor. Tight fit of vanes in rotor can usually be corrected by thorough cleaning. Replace rotor if excessive looseness exists between rotor and vanes. Replace vanes if worn or scored.
3. Examine machined surface of pump ring for roughness or wear. Replace

ring if roughness cannot be corrected with crocus cloth.

4. Inspect thrust plate, pressure plate and end plate for wear, scores or other damage.
5. Inspect pump housing for cracks or damage. Check housing for evidence of wear or scoring.
6. Check all springs for free length, distortion or collapsed coils.
7. Inspect locating dowel pins for distortion.
8. Examine outer diameter of flow control valve for scoring or roughness. Slight damage may be cleaned up with crocus cloth. Check valve assembly for freedom of movement in bore of pump housing.
9. Check all oil passages in pump parts for obstruction. Use a piece of tag wire to clean out holes.
10. Check bushing in pump housing for wear or damage.

Reassembly

1. Lubricate new O-ring seals with fluid and place in pump housing.
2. If drive shaft oil seal was removed during disassembly, lubricate new seal with fluid. Then install seal in pump body front hub, using a suitable seal driver.
3. Position thrust plate and rotor on drive shaft. Then install rotor retaining ring in groove of drive shaft.
4. Place pump housing on work bench as shown in Fig. 27 and install two pump ring dowel pins in bore of housing.
5. Install drive shaft assembly in housing, making sure shaft seats properly.
6. Install pump ring in housing over two dowel pins with rotation arrow on ring to rear of pump housing, Fig. 28.

NOTE: *Arrow on outer edge of pump ring points in direction of pump rotation (counterclockwise when viewed from rear of pump).*

7. Install rotor vanes in slots of pump rotor, with radius edge of vanes toward outside of rotor, Fig. 29.
8. Position pump housing on two wood blocks and push parts down into place as shown in Fig. 30.
9. Lubricate outside diameter and chamfer of pressure plate with fluid to insure against damage to O-ring seals in housing. Then install pressure plate in housing over dowel pins.

NOTE: *Ported face of pressure plate goes toward pump ring, Fig. 30.*

10. Using a $2\frac{3}{4}$" diameter sleeve to apply pressure to outer edge of pressure plate only, press pressure plate in to seat.

NOTE: *Pressure plate will travel about $\frac{1}{16}$" to seat. Do not press or hammer on center of pressure plate as to do so will cause permanent distortion resulting in pump failure.*

11. Install pressure plate spring in pressure plate center groove, Fig. 31.
12. Lubricate outside diameter and chamfer of end plate with fluid to insure against damage to O-ring seal in housing. Then press end plate into housing. While holding end plate down in housing, install end plate retaining ring. Be sure end plate is completely and solidly seated in groove of housing, Fig. 32.
13. Remove assembly from press and install flow control spring and valve into housing bore with hex head screw entering housing bore first, Fig. 33.
14. Install magnet into housing, if used.
15. Position pressure union O-ring seal in bore of housing over flow control valve, Fig. 34.
16. Mount pump housing in vise with soft jaws and install pump reservoir.
17. Install pressure union and new O-ring seal in reservoir. Tighten union firmly.
18. Install mounting stud and new O-ring seal in reservoir. Tighten stud securely.
19. Install drive shaft key in slot in shaft. Support shaft on opposite side while installing key. Install reservoir cap.

Chrysler Constant Control Type Gear

The Constant Control Full Time Power Steering consists of a hydraulic pressure pump, a power steering gear and connecting hoses. The power steering gear consists of a gear housing, containing a gear shaft and sector shaft; a power piston with gear teeth milled into the side of the piston which is in constant mesh with the gear shaft sector teeth; and a worm shaft which connects the steering wheel to the power unit piston through a coupling.

Pump Pressure

1. Ensure fluid level and drive belt tension are correct, then connect a pressure gauge and shut-off valve between pump and steering gear with shut-off valve located between pressure gauge and steering gear.
2. Open shut-off valve, start engine and allow fluid to reach operating temperature.
3. Turn wheels from stop to stop and note pressure reading. If pressure is less than specified, close shut-off valve and note pressure. If pressure is still less specified, then pump is faulty. If pressure is within specifications, the steering gear is defective. Refer to specifications chart below for proper pump pressure.

CHRYSLER	PRESSURE
1969	1000-1100
1970 (.94 Pump)	1075-1200
1970-72 (1.06 Pump)	1100-1300
1971-72 (.94 Pump)	1100-1200
1973-75 (1.06 Pump)	1150-1300
1975 (.94 Pump)	1200-1300

IMPERIAL	
1969-73	1200-1300
1974-75 (1.06 Pump)	1150-1300

DODGE & DART	
1969 Dart	800-850
1969 Coronet	950-1000
1969 Polara, 880, Monaco	1000-1100
1970-72 Dart	750-850
1970-72 V8	
(Exc. Monaco & Polara)	950-1075
1970-72 Six	
(Exc. Monaco & Polara)	750-850
1970 .94 Monaco & Polara	1075-1200
1970 1.06 Monaco & Polara V8	1100-1300
1970 1.06 Monaco &	
Polara Six	1000-1300
1971-72 .94 Monaco & Polara	1100-1200
1971-72 1.06 Monaco & Polara	
6-225	1000-1300
V8-318	950-1050
V8-383, 400, 440	1100-1300
1973-75 Dart & Challenger—	
.94 Pump	900-1000
1.06 Pump	950-1050
1973-75 Coronet & Charger—	
.94 Pump	1200-1300
1.06 Pump (Six)	950-1300
1.06 Pump (V8)	1150-1300
1973 Monaco & Polara Six	950-1300
1973 Monaco & Polara V8—	
.94 Pump	1200-1300
1.06 Pump	1150-1300
1974-75 Monaco & Polara—	
.94 Pump	1200-1300
1.06 Pump	1150-1300

PLYMOUTH	
1969 Belvedere Six	800-850
1969 Belvedere V8	950-1000
1969 Fury Six	950-1000
1969 Fury V8	1000-1100
1970 .94 Exc. Fury	950-1075
1970 .94 Fury	1075-1200
1970-72 1.06 Fury Six	1000-1300
1970-72 1.06 Fury V8	
(Exc. V8-318)	1100-1300
1970-72 1.06 Fury V8-318	950-1050
1970-72 1.06 Satellite Six	750-850
1970-71 1.06 Satellite V8	950-1050
1971-72 .94 Fury	1100-1200
1971-72 .94 Exc. Fury	975-1075
1973-74 Satellite—	
.94 Pump	1200-1300
1.06 Pump (Six)	950-1300
1.06 Pump (V8)	1150-1300

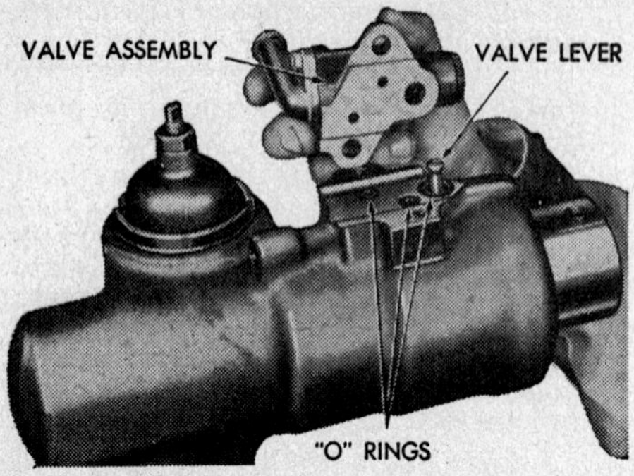

Fig. CC1 Removing or installing valve body assembly

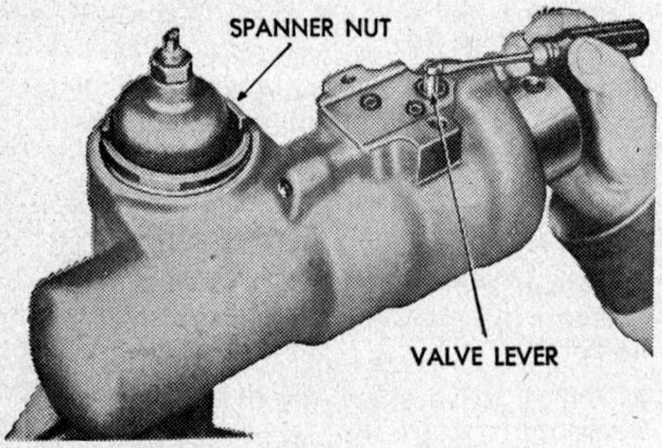

Fig. CC2 Removing valve lever

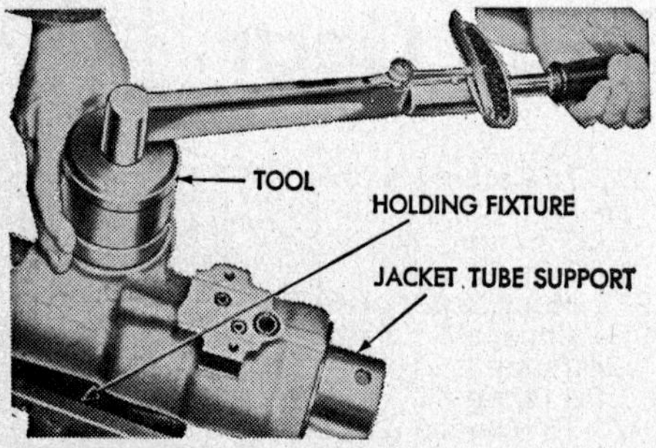

Fig. CC3 Removing or installing gear shaft cover nut

1973 Fury Six	950-1300
1973-74 Fury V8—	
.94 Pump	1200-1300
1.06 Pump	1150-1300
1975 Fury—	
.94 Pump	1200-1300
1.06 Pump Six	950-1300
1.06 Pump V8	1150-1300
Gran Fury—	
.94 Pump	1200-1300
1.06	1150-1300

VALIANT & BARRACUDA

1969	800-850
1970-72 Valiant	750-850
1970 Barracuda	950-1075
1971-72 Barracuda	
6-198, 225	750-850
V8 (.94 Pump)	975-1075
V8 (1.06 Pump)	950-1050
1973-75 .94 Pump	900-1000
1973-75 1.06 Pump	950-1050

Steering Gear Repairs

Prior to disassembly, clean the gear assembly thoroughly in a suitable solvent. Crocus cloth may be used to remove small nicks and burrs provided it is used carefully. When used on the steering gear valve, use extreme care not to round off the sharp edge portions of the two lands located between the valve drilled holes. Remove and discard all "O" ring seals, using new ones lubricated with fluid when reassembling.

Disassemble

1. Drain steering gear by rotating wormshaft from stop to stop.
2. Remove valve body and three "O" rings, Fig. CC1.
3. Remove valve lever by prying under spherical head, Fig. CC2. *Use care not to collapse slotted end of valve lever as this will destroy bearing tolerances of head.*
4. Remove gear shaft cover nut, Fig. CC3.
5. Rotate wormshaft to position sector shaft teeth at center of piston travel, then loosen power train nut.
6. Install tool C-3786 on sector shaft threaded end and slide tool into housing until tool and shaft engages bearings.
7. Rotate wormshaft to left stop, compressing power train components, then remove power train nut and housing head tang washer. Using a screwdriver, pry on piston teeth to remove complete power train, Fig. CC6.

NOTE: The cylinder head, center race and spacer assembly and the housing head must be kept in close contact since the reaction rings may become disengaged from grooves in cylinder head and housing. Also, the center spacer may separate from center race and jam in the housing, in turn damaging the spacer and/or housing when removing the power train.

---

8. Remove worm shaft upper oil seal, Fig. CC7, and disassemble column jacket support. Reaction seal may be removed from groove in face of jacket support by blowing air pressure into ferrule chamber. Make sure passage from ferrule chamber to upper reaction chamber is unobstructed.

Column Jacket Support, Assemble

Install worm shaft upper oil seal with sealing lip toward bearing, Fig. CC8. Tool C-3650 should be used to drive seal until tool bottoms on casting to obtain proper compression of rubber seal. Lubricate reaction seal and install in groove in face of column jacket support with flat side of seal out.

Cylinder Head

Disassemble—Remove two "O" rings in outer grooves in head. Remove "O" ring in groove in face of cylinder head with air pressure into oil hole located in groove

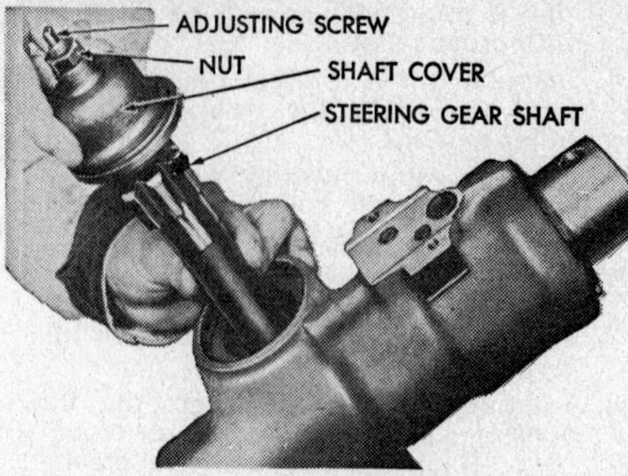

Fig. CC4 Removing or installing gear shaft and cover

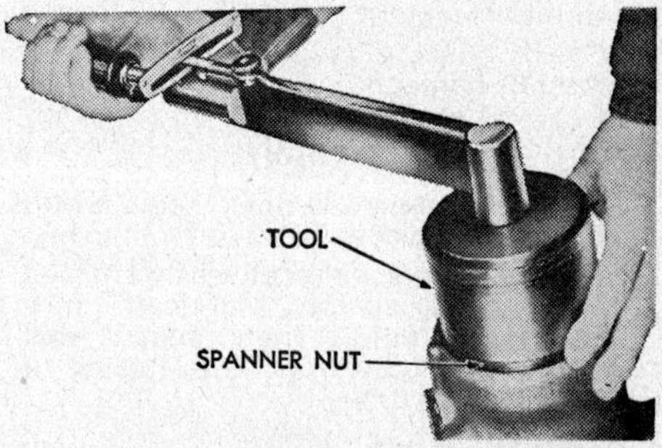

Fig. CC5 Removing steering column support nut

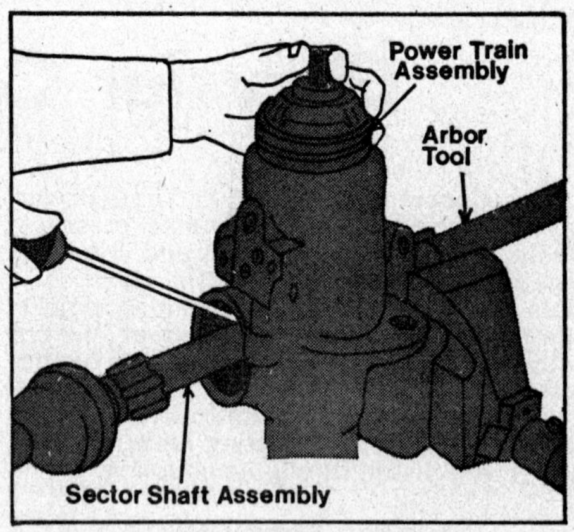

Fig. CC6 Removing or installing power train

between two "O" ring grooves, Fig. CC9. Replace cylinder head seal if necessary, Fig. CC10. Check oil passage in ferrule for obstruction. Check lands of cylinder head for burrs.

Assemble—Lubricate and install two large "O" rings in grooves on head. Install lower reaction seal in groove in face of head. The small "O" ring for ferrule groove should be installed after worm shaft bearing preload has been established, otherwise "O" ring will be damaged by reaction springs.

Steering Valve

Disassemble—Remove outlet fitting,

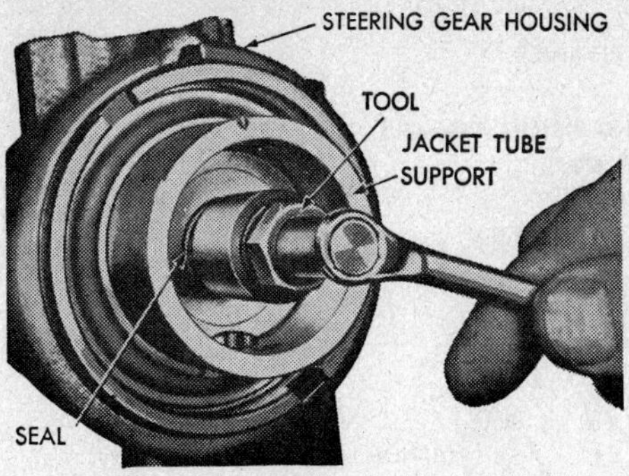

Fig. CC7 Removing worm shaft upper oil seal

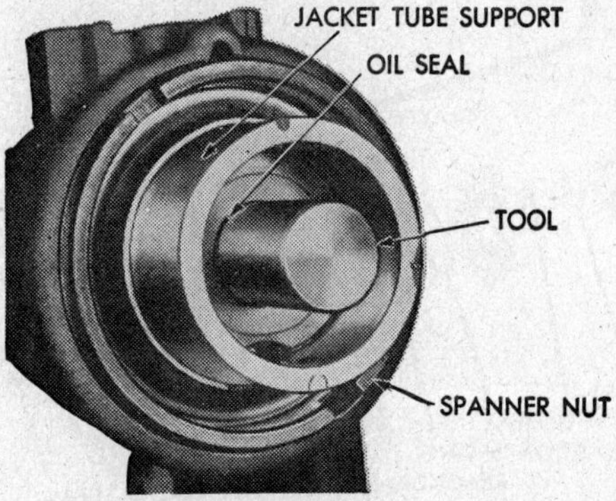

Fig. CC8 Installing worm shaft upper oil seal

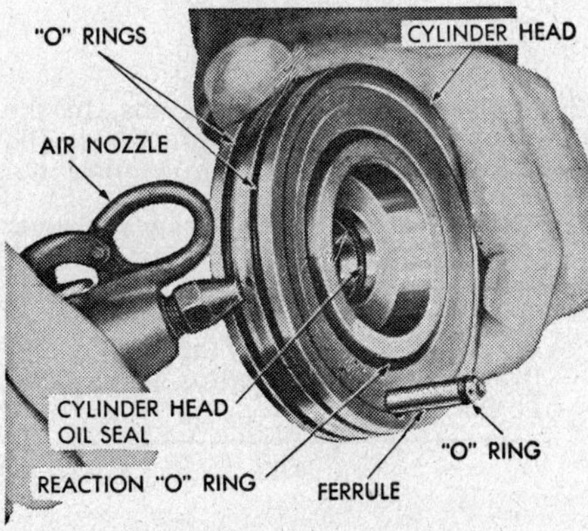

Fig. CC9 Removing reaction ring from cylinder head

spring, valve piston and cushion spring. Remove valve body (2 screws) from steering valve. Shake out valve piston. If spool valve or valve housing is damaged, replace complete valve assembly. Do not remove valve end plug unless inspection indicates a leak at seal.

Assemble — Fig. CC11. Install steering valve in housing. Valve lever hole should align with steering gear valve lever opening in bottom of valve housing. Valve must fit smoothly in housing without striking or binding. If valve end plug was removed install new seal and tighten plug to 25 lb. ft. torque. Install cushion spring, then lubricate and install valve piston, nose end first, into valve body. Install spring on top of piston and outlet plug and torque to 20 ft. lbs. Assemble control valve body to steering valve body and torque attaching screws to 95 inch lbs. Install new copper sealing gasket and fitting in threaded hole on top of valve body and torque to 30 lb. ft.

Gear Shaft

Disassemble — Remove adjusting screw lock nut and unscrew cover from adjusting screw. Remove screw and washer from, "T" slot in end of gear shaft. Remove date tag gasket from top of cover and large "O" ring from base of cover.

Assemble—Place adjusting screw washer over screw and slide both into "T" slot of shaft. Screw cover into adjusting screw until gear shaft bottoms in cover. Install date tag gasket over adjusting screw into position at top of cover. Install lock nut on adjusting screw but do not tighten. Install large "O" ring in groove on lower face of gear cover.

Steering Gear Housing

Disassemble — The equipment shown in Fig. CC12 should be used for this operation. After removing the oil seal snap ring and back-up ring, remove the oil seal by sliding the threaded portion of the tool's adapter over end of gear shaft and install nut section of tool (C-3350) on shaft. Maintain pressure on adapter with nut of tool while turning adapter, forcing it into seal until it has bottomed in seal. Apply two half rings and retainer over both portions of tool. As hex nut is removed from shaft seal will be pulled from housing.

Assemble—Install gear shaft oil seal in housing (lip of seal toward needle bearing) using Tool C-3350. Place adapter against seal and tool nut on threaded end of gear shaft. Tighten tool nut until adapter shoulder contacts housing, Fig. CC13. Install oil seal back-up ring, oil seal and snap ring.

Power Train, Assemble

Fig. CC14. Position piston assembly on bench with worm shaft up. Slide cylinder head (ferrule up) on worm shaft and against piston flange, making sure gap on worm shaft ring is closed to avoid breaking ring. Lubricate and install in the following order:

1. Lower thrust bearing race (thick).
2. Lower thrust bearing.
3. Lower reaction spring over ferrule.
4. Lower reaction ring with flange up so that ring protrudes through reaction spring.
5. Center bearing race, indexing control hole with hole in center bearing race.
6. Install upper thrust bearing, upper thrust bearing race (thin) and new worm shaft thrust bearing nut.
7. Tighten nut as follows: Turn worm shaft clockwise ½ turn and hold shaft in this position while tightening nut to 50 lb. ft. torque.

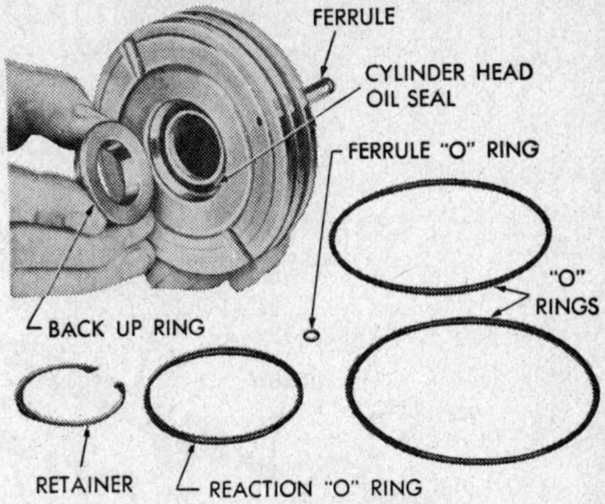

Fig. CC10 Removing cylinder head seal

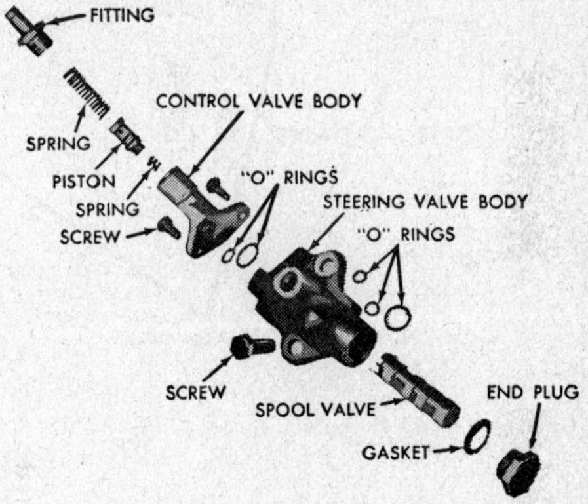

Fig. CC11 Control valve disassembled

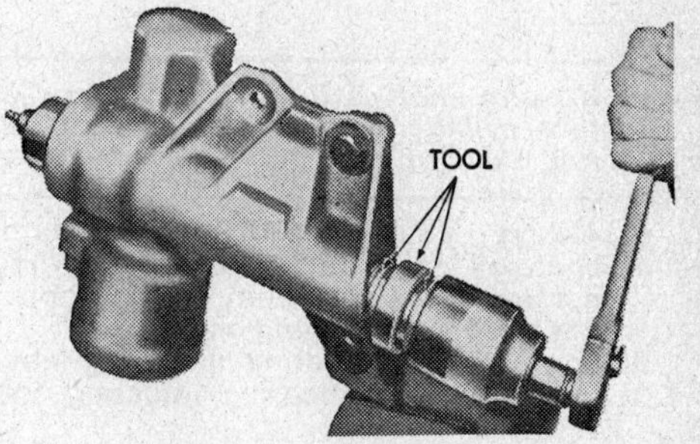

Fig. CC12 Removing gear shaft oil seal

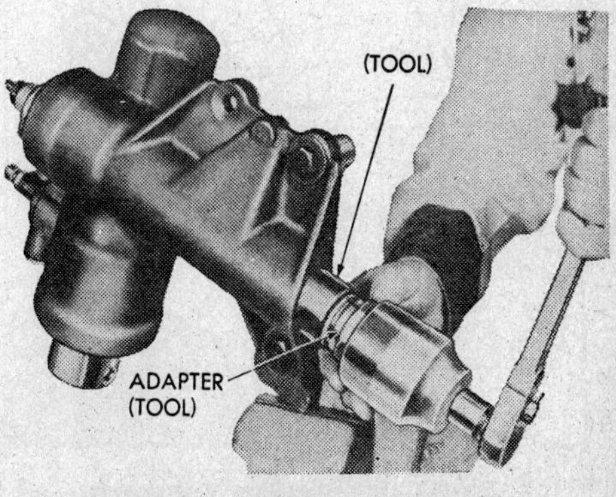

Fig. CC13 Installing gear shaft oil seal

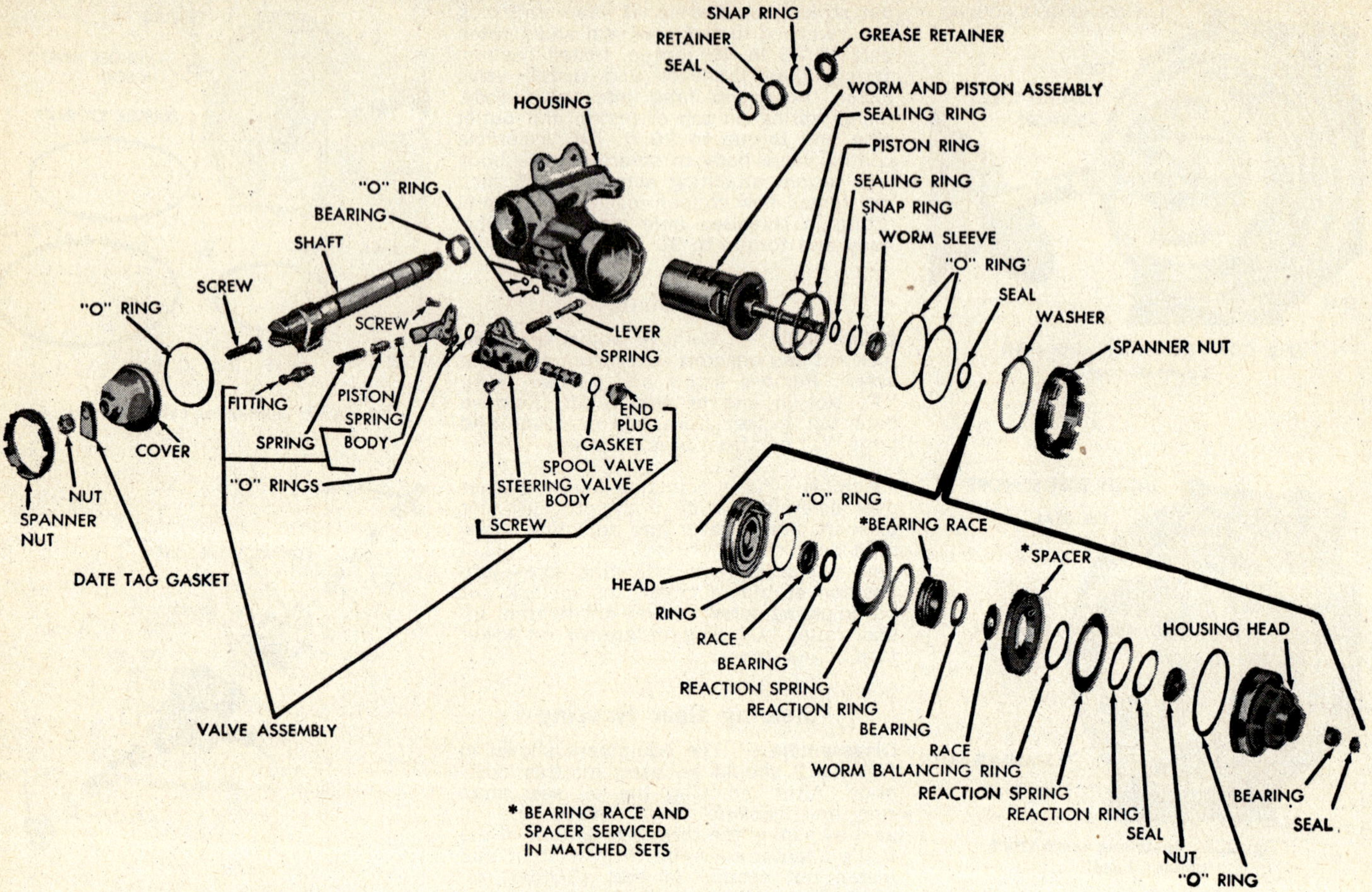

Fig. CC14 Steering gear disassembled

If worm shaft is turned more than ½ turn cylinder head seal will clear oil ring on worm shaft.

8. Always check position of worm shaft oil ring before bottoming cylinder head against worm piston shoulder to avoid damaging oil ring.
9. Loosen worm bearing adjusting nut.
10. Retighten worm bearing adjusting nut to give a bearing torque of 16 to 24 oz. Check by placing rounds of cord around center bearing race. Make a loop in one end and hook the loop of a spring scale in the cord loop. Pulling on cord will cause bearing race to rotate, Fig. CC15.
11. If adjusting nut is tightened properly, reading on scale will be 16 to 24 oz. Stake flange of adjusting nut into depression in worm shaft to lock securely.
12. Stake upper part of adjusting nut into wormshaft knurled area in four places, 90° apart. Check bearing preload after first staking operation and adjust, if necessary, by tapping the adjusting nut with a mallet.
13. Apply a torque of 20 ft. lbs. to the adjusting nut in both directions. The nut should not mcve, ensuring nut was properly staked.
14. Install center bearing spacer over bearing race to engage dowel pin with slot in center bearing race.

Make sure valve lever hole in center bearing race and center spacer are properly aligned.

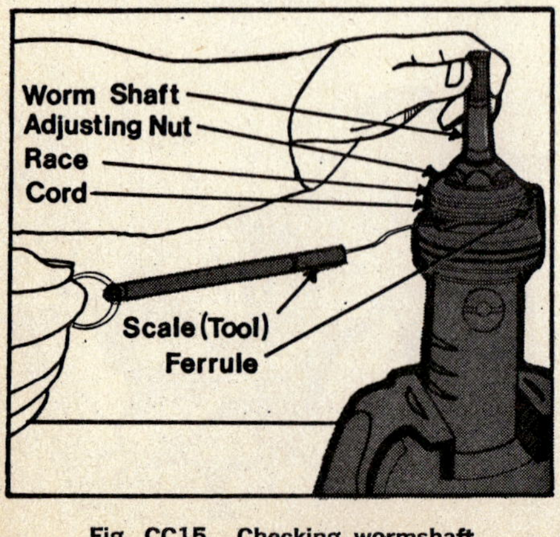

Fig. CC15 Checking wormshaft bearing preload

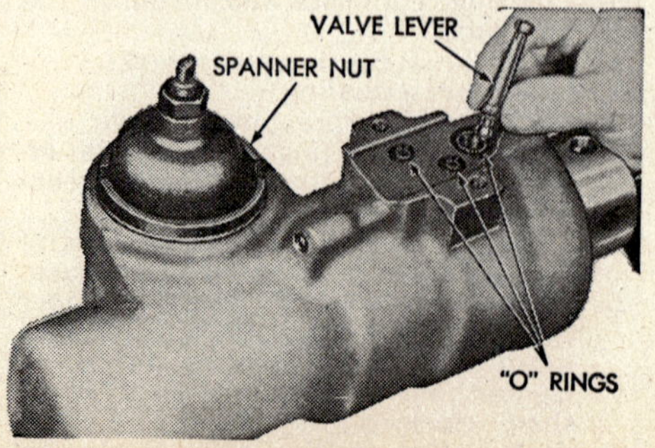

Fig. CC16 Installing valve lever

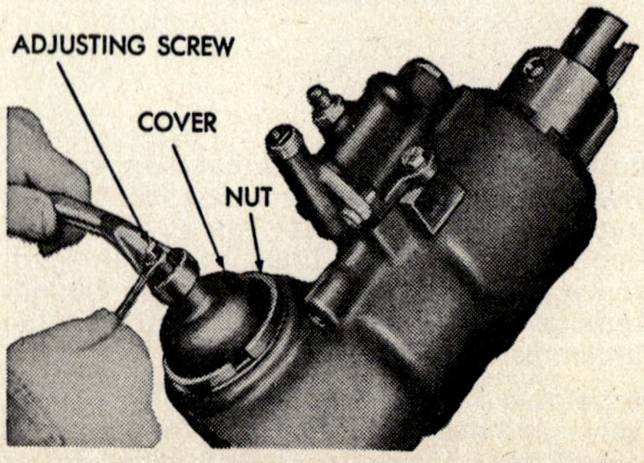

Fig. CC17 Adjusting gear shaft adjusting screw

Steering Gear, Assemble

1. With steering gear housing fastened in holding fixture, lubricate power train housing bore with fluid and install power train assembly, Fig. CC5. Face piston teeth to the right and valve lever hole in center race and spacer in "Up" position.

NOTE: Ensure cylinder head is bottomed against housing shoulder.

2. With valve lever hole in center bearing race and spacer aligned with valve lever hole in gear housing, install valve lever (double bearing end first) into center bearing race and spacer, ensuring valve lever slots are parallel to wormshaft. It may be necessary to gently tap end of valve lever to seat lower pivot point in center race.
3. Install housing head tang washer, indexing with groove in housing, and steering column support nut and torque nut to 150-200 ft. lbs. Ensure valve lever remains centered in housing hole by rotating wormshaft until piston bottoms in both directions and observe valve lever action. The valve lever must be in center of hole and return to the center position when wormshaft torque is relieved.
4. Install valve lever spring (small end first), position power piston at center of travel and install sector shaft and cover assembly, indexing sector teeth with piston rack teeth. Ensure sector shaft cover "O" ring is installed properly.
5. Install and torque cover nut to 110-200 ft. lbs.
6. Install valve body onto housing with valve lever entering hole in valve spool. Ensure "O" rings are installed properly, then install and torque valve retaining screws to 7 ft. lbs.

Tests and Adjustments

1. Fill reservoir to level mark.
2. Connect test hoses to hydraulic pump on car with pressure gauge installed between pump and steering gear to register pressures.
3. Start engine and operate at idle to bring steering gear to normal operating temperature.
4. Center the valve until unit is not self-steering. Tap valve body retaining screw heads to move valve body up on steering housing or end plug to move valve body down.
5. Rotate wormshaft through range of travel to bleed air from system, then refill reservoir.
6. With steering gear on center, tighten sector shaft adjusting screw until backlash is removed from steering arm. If the power train was removed, tighten adjusting screw an additional 1¼ turns, then tighten locknut.
7. Operate unit through several cycles, aligning piston rack and sector teeth.
8. With gear on center, readjust sector shaft backlash. Loosen adjusting screw until backlash is present, then tighten adjusting screw until backlash is removed. Tighten adjusting screw an additional ⅜ to ½ turn and torque locknut to 28 ft. lbs.
9. Recenter valve body.

TROUBLE SHOOTING

Hard Steering

1. Tires not properly inflated.
2. Low oil level in reservoir.
3. Loose pump belt.
4. Oil on pump belts.
5. Steering linkage needs lubrication.
6. Power steering pump output low.
7. Cross shaft adjustment too tight.
8. Pressure control valve stuck in closed position.
9. External oil leaks.
10. Defective or damaged valve lever.
11. Dirt or chips in steering gear.
12. Damaged column support worm shaft bearings.
13. Damaged thrust bearings or excessive preload adjustment.
14. Rough or hard to turn worm and piston assembly.
15. Excessive internal leakage.

Poor Recovery From Turns

1. Tires not properly inflated.
2. Steering linkage binding.
3. Improper wheel alignment.
4. Damaged or defective steering tube bearings.
5. Steering column jacket and steering gear not properly aligned.
6. Improper cross shaft mesh adjustment.
7. Pressure control valve piston stuck open.
8. Column support spanner nut loose.
9. Defective or damaged valve lever.
10. Improper worm thrust bearing adjustment.
11. Burrs or nicks in reaction ring grooves in cylinder head or column support.
12. Defective or damaged cylinder head worm shaft seal ring.
13. Dirt or chips in steering gear unit.
14. Rough or catchy worm and piston assembly.

Self-Steering or Leads to Either Side

1. Tires not properly inflated.
2. Improper wheel alignment.
3. Steering wheel off center when car is traveling straight ahead.
4. Valve body out of adjustment.
5. Valve lever damaged.
6. Column support spanner nut loose.

Temporary Increase In Effort When Turning Steering Wheel

1. Low oil level.
2. Loose pump belt.
3. Oil on pump belts.
4. Binding steering linkage.
5. Engine idle too slow.
6. Defective power steering pump.
7. Air in system.
8. External adjustment.
9. Improper cross shaft adjustment.
10. Excessive internal leakage.

Excessive Steering Wheel Free Play

1. Improper cross shaft adjustment.
2. Column support spanner nut loose.
3. Improper worm thrust bearing adjustment.
4. Improper front wheel bearing adjustment.

Lack of Assistance in One Direction

1. Oil leaking past worm shaft cast iron seal ring or ferrule "O" ring.

Lack of Assistance in Both Directions

1. Broken "O" ring on worm piston.
2. Piston end plug loose.
3. Pump belt slipping.
4. Pump output low.

Noises

1. Buzzing noise in neutral only is caused by sticking pressure control valve.
2. Noisy power pump.
3. Damaged hydraulic lines.
4. Pressure control valve sticking.
5. Improper sector shaft adjustment.
6. Air in system.

Lack of Effort, Left Turn

1. Left turn reaction seal "O" ring missing, damaged or worn.
2. Left turn reaction oil passage not drilled in housing or cylinder head.
3. Left turn reaction ring sticking in cylinder head.

Lack of Effort, Right Turn

1. Right turn U-shaped reaction seal missing, damaged or worn.
2. Right turn reaction oil passage not drilled in housing head or ferrule pin.
3. Right turn reaction ring sticking in housing head.

Lack of Effort in Both Directions

1. Improper sector shaft adjustment.
2. Pressure plates installed on wrong side of reaction rings.

Ford Torsion Bar Power Steering

Description

The power steering unit, Fig. F1, is a torsion bar type of hydraulic-assisted system. This system furnishes power to reduce the amount of turning effort required at the steering wheel. It also reduces road shock and vibrations.

The unit includes a worm and one piece rack-piston which is meshed to the gear teeth on the steering sector shaft. The unit also includes a hydraulic valve, valve actuator, input shaft and torsion bar assembly which are mounted on the end of the worm shaft and operated by a twisting action of the torsion bar.

The gear unit is designed with the one piece rack-piston, worm and sector shaft in the one housing and the valve spool in an attaching housing. This makes possible internal fluid passages between valve and cylinder, thus eliminating all external lines and hoses except the pressure and return hoses between pump and gear.

The power cylinder is an integral part of the gear housing. The piston is double acting in that fluid pressure may be applied to either side of the piston.

A selective metal shim, located in the valve housing of the gear is for the purpose of tailoring the steering gear efforts. If efforts are not satisfactory they can be changed by increasing or decreasing shim thickness as follows:

Efforts heavy to the left—increase shim thickness. Efforts light to the left—decrease shim thickness. A change of one shim size will increase or decrease steering efforts approximately 1½ inch lbs. on 1969-71 units and 2 inch lbs. on 1972-74 units. Shims are available in the following selective thicknesses:

.0057 to .0063″
.0077 to .0083″
.0097 to .0103″
.0117 to .0123″
.0137 to .0143″

Operation

The operation of the hydraulic control valve spool is governed by the twisting of a torsion bar. All effort applied to the steering wheel is transmitted directly through the input shaft and torsion bar to the worm and piston. Any resistance to the turning of the front wheels results in twisting of the bar. The twisting of the bar increases as the front wheel turning effort increases. The control valve spool, actuated by the twisting of the torsion bar, directs fluid to the side of the piston where hydraulic assistance is required.

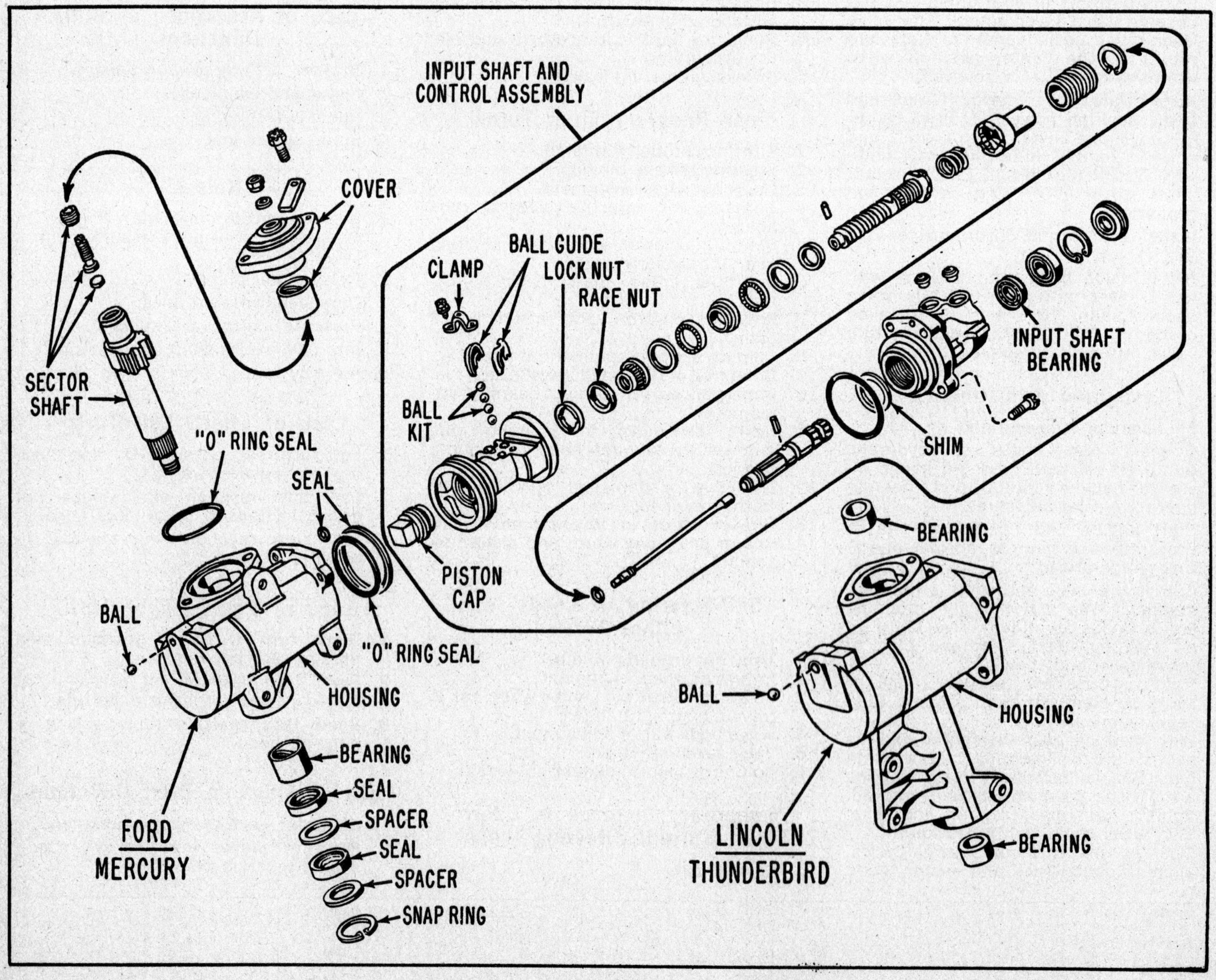

Fig. F1 Ford power steering gear disassembled

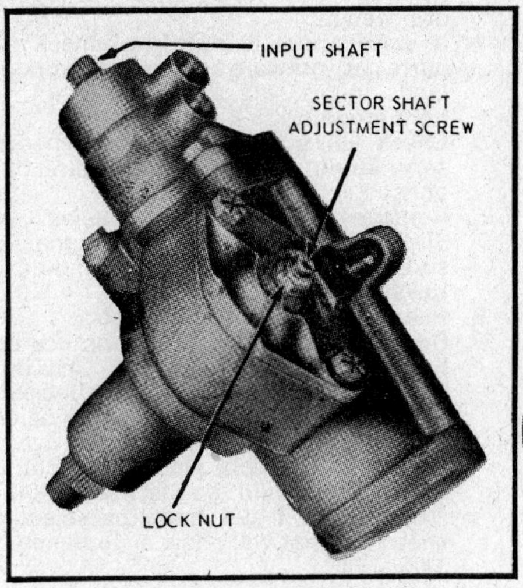

Fig. F2 Adjusting mesh load

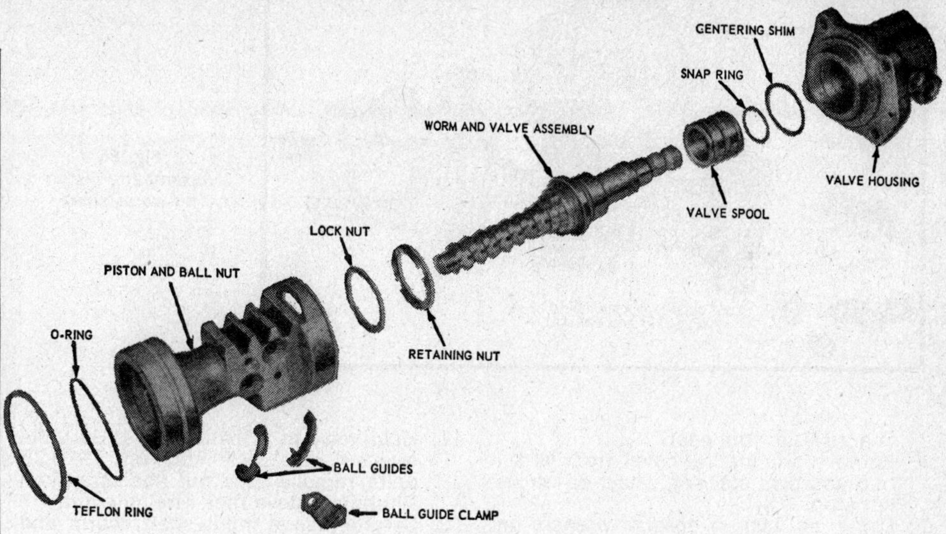

Fig. F3 Ball nut and valve housing disassembled

As the torsion bar twists, its radial motion is transferred into axial motion by three helical threads. Thus, the valve is moved off center, and fluid is directed to one side of the piston or the other.

Power Steering Pump

The power steering pump is a belt-driven slipper type pump which is integral with the reservoir. It is constructed so that the reservoir is attached to the rear side of the pump housing front plate and the pump body is encased within the reservoir.

Pump pressure is 1000 to 1150 psi. If pump is defective, replace it with a new pump as individual parts are not furnished.

IN-CAR ADJUSTMENTS & REPAIRS

Valve Spool Centering Check

NOTE: The "out-of-car" procedure for valve centering check is the same as for the "in-car" except the torque and simultaneous pressure reading must be made at the right and left stops instead of either side of center.

1. Install a 2000 psi pressure gauge in pressure line between pump outlet port and steering gear inlet port. *Make sure that valve on gauge in is fully open position.*
2. Check fluid level in reservoir and replenish as required.
3. Start engine and cycle steering wheel from stop-to-stop to bring steering lubricant up to normal operating temperature. Stop engine and recheck reservoir. Add fluid as necessary.
4. With engine running at a fast idle speed (1000 rpm) and steering wheel centered, attach an inch-pound torque wrench to steering wheel retaining nut. Apply sufficient torque to wrench in each direction (either side of center) to get a gauge reading of 250 psi.
5. The torque reading should be the same in both directions. If the difference between readings exceed 4 inch lbs., remove steering gear from vehicle and change thickness of valve centering shim.

NOTE: On 1975 models if the difference between readings is more than 4 inch lbs., the shaft and control assemblies must be replaced.

On 1969-71 units, use only one shim and on 1972-74 units, use as many shims as necessary, not exceeding .030 inch total thickness. If steering effort is heavy to the left, increase shim thickness and if steering effort is light to left, decrease shim thickness.

Steering Gear Adjustments

Preload (thrust bearing adjustment) and worm-to-rack preload cannot be changed in service. The only adjustment that can be performed is the total over-center position load to eliminate excessive lash between sector and rack teeth.

1. Disconnect pitman arm from sector shaft.
2. Disconnect fluid return line at reservoir and cap reservoir return line pipe.
3. Place end of return line in a clean container and cycle steering wheel in both directions as required to dis-

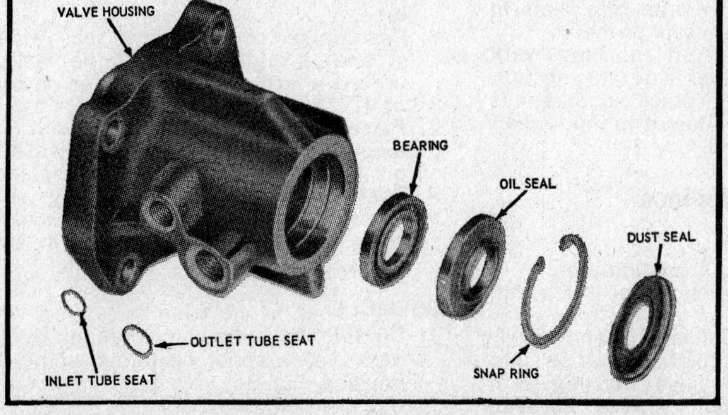

Fig. F4 Valve housing disassembled

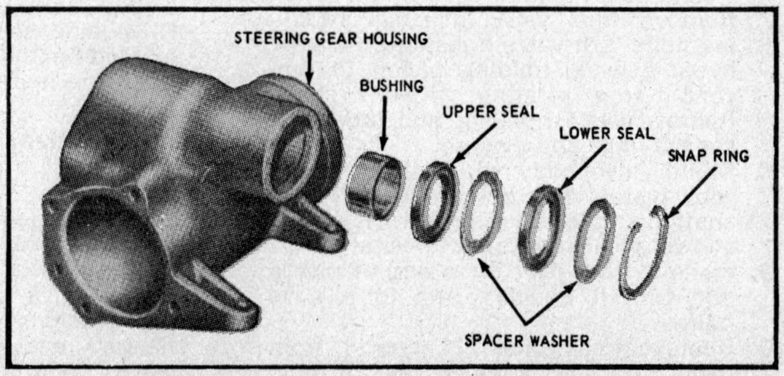

Fig. F5 Steering gear housing disassembled

Fig. F6 Assembling piston on worm shaft

charge fluid from gear.
4. Remove ornamental cover from wheel hub and turn steering wheel 45° from left stop.
5. Using an inch-lb torque wrench on steering wheel nut, determine torque required to rotate shaft slowly through an approximately ⅛ turn from the 45° position.
6. Turn steering gear back to center, then determine torque required to rotate shaft back and forth across center position.
7. Loosen adjuster nut and turn adjusting screw until reading is greater than torque measured 45° from stop by 8-9 inch lbs. on 1969-70 units and 11-12 inch lbs. on 1972-75 units.
8. Recheck readings and replace pitman arm and steering wheel.
9. Connect fluid return line and replenish reservoir.

STEERING GEAR REPAIRS

Disassembly

1. Hold steering gear over drain pan in an inverted position and cycle input shaft six times to drain remaining fluid from gear.
2. Remove lock nut and brass thrust washer from adjusting screw.
3. Turn input shaft to either stop, then turn it back approximately 1¾ turns to center the gear.
4. Remove two sector shaft cover screws and identification tag.
5. Tap lower end of sector shaft with a soft-faced hammer to loosen it, then lift cover and shaft from housing as a unit. Discard O-ring.
6. Turn sector shaft cover counterclockwise off adjuster screw.
7. Remove four valve housing attaching bolts. Lift valve housing from gear housing while holding piston to prevent it from rotating off worm shaft. Remove valve housing and lube passage O-rings and discard.
8. Stand valve body and piston on end with piston end down. Rotate input shaft counterclockwise out of piston, allowing balls to drop into piston.
9. Place a cloth over open end of piston and turn it upside down to remove balls.
10. Remove ball guide (2 screws) from ball nut and remove clamp and guides, Fig. F3.
11. Hold valve in a fixture, on 1969 units remove race lock nut, on 1970-75 units remove race nut set screw, on all units remove race retaining nut.
12. Carefully slide input shaft, worm and valve out of valve housing.

CAUTION: Do to the close diametrical clearance between valve spool and housing, the slightest cocking of the spool may cause it to jam in housing.

13. Remove shim from valve housing.

Valve Housing, Replace

1. Referring to Fig. F4, remove dust seal and snap ring.
2. Invert housing and, from opposite seal end, gently tap bearing and seal out of housing.

CAUTION: Exercise care when inserting and removing tool to prevent damage to valve bore in housing.

3. Remove oil inlet and outlet tube seats with "EZ-out" if damaged.
4. Coat tube seats with vaseline and position them in housing. Install and tighten tube nuts to press seats to proper location.
5. Coat bearing and seal surface in housing with a film of vaseline.
6. Position bearing in valve housing. Seat bearing in housing, making sure that bearing is free to rotate.
7. Dip new oil seal in gear lube. Then place it in housing with metal side of seal facing outward. Drive seal into housing until outer edge of seal does not quite clear snap ring.
8. Place snap ring in housing. Then drive on ring until snap ring seats in its groove to locate seal properly.
9. Place dust seal in housing with dished side (rubber side) facing out. Drive dust seal in place so that it is located behind undercut in input shaft when it is installed.

Worm & Valve, Replace

1969-74 Units

1. Remove snap ring from actuator.
2. Slide control valve spool off worm shaft, Fig. F3.
3. Install valve spool evenly and slowly with a slight oscillating motion into flanged end of valve housing with valve identification groove between spool lands outward. Check for freedom of valve movement within housing working area. Spool should enter housing bore freely and fall by its own weight.
4. If valve spool is not free, check for burrs at outward edges of working lands in housing and remove with a hard stone.
5. Check valve for burrs and, if necessary, stone valve in a radial direction only.
6. Remove valve spool from housing.
7. Slide spool onto worm shaft, making sure that groove in spool annulus is toward worm.
8. Install snap ring to retain spool.
9. On 1969-71 units, check clearance between spool and snap ring. On 1969 units clearance should be between .002-.005 inch, if not install a selective snap ring that will allow a clearance of .003 inch. On 1970-71 units clearance should be between .0005-.0035 inch, if not install a selective snap ring that will allow a clearance of .002.

Worm & Valve Sleeves, Replace

1975 Units

1. Cut valve sleeve rings from valve sleeve, then position worm end of assembly in soft jawed vice.
2. Using tool kit No. T75L-3517-A, install four valve sleeve rings. After installing rings ensure they turn freely in grooves.

Piston & Ball Nut, Replace

1. Remove teflon ring and O-ring from piston and ball nut, Fig. F3.
2. Dip a new O-ring in gear lube and install on piston and ball nut.
3. Install new teflon ring on piston and ball nut, being careful not to stretch it any more than necessary.

Gear Housing

DISASSEMBLE:

1. Remove snap ring and spacer washer, Fig. F5, from lower end of gear housing.
2. Remove lower seal from housing and lift spacer washer from housing.
3. Remove upper seal in same manner.
4. Press bushing from housing if worn or damaged (2 bushings on Lincoln and T-Bird). Press new bushing into housing.
5. Dip both sector shaft seals in gear lube.
6. Position sector shaft inner seal (widest one) into housing with lip facing inward. Place a spacer washer on top of it.
7. Place outer seal in housing with lip facing inward. Place spacer washer on top of it and press into place.
8. Position snap ring in housing and press it in place to properly locate seals and engage snap ring in the groove.

ASSEMBLE:

1. On 1969-74 units place required thickness valve spool centering shim in housing, Fig. 3.
Use one shim only.
2. Install worm and valve in housing.

3. Install retaining nut in housing and torque to 55-65 ft-lbs.
4. On 1969 units install locknut and torque to 20-30 ft. lbs. On 1970-75 units install race nut set screw, on 1970-73 units torque to 20-25 in. lbs., on 1974-75 units torque to 15-25 in. lbs.
5. Place piston on bench with ball guide holes facing up. Insert worm shaft into piston so that first groove is in alignment with hole nearest to center of piston, Fig. F6.
6. Place ball guide into piston. Place 27 balls in guide, turning worm clockwise (viewed from input end of shaft). If all balls have not been fed into guide upon reaching right stop, rotate input shaft in one direction and then in the other while installing balls. After balls have been installed, do not rotate input shaft or piston more than 3½ turns off the right stop to prevent balls from falling out of circuit.
7. Securing guides to ball nut with clamp.
8. Position new lube passage O-ring in counterbore of gear housing.
9. Apply vaseline to teflon seal.
10. Place a new O-ring on valve housing.
11. Slide piston and valve into gear housing, being careful not to damage teflon seal.
12. Align lube passage in valve housing with one in gear housing, and install but do not tighten attaching bolts at this time.
13. Rotate ball nut so that teeth are in same plane as sector teeth. Tighten four valve housing attaching bolts.
14. Position sector shaft cover O-ring in gear housing. Turn input shaft as required to center piston.
15. Apply vaseline to sector shaft journal, then position sector shaft and cover into gear housing. Install identification tag and air conditioner line mounting bracket (if so equipped) and two sector shaft cover bolts.
16. Attach an inch-lb torque wrench to input shaft and adjust mesh load as outlined previously.

TROUBLE SHOOTING

Hard Steering

1. Low or uneven tire pressure.
2. Improper gear adjustment.
3. Improper wheel alignment.
4. Low fluid level.
5. Twisted or bent suspension parts, frame and linkage components.
6. Tight wheel bearings.
7. Steering spindle bent.
8. Pump belt out of adjustment.
9. Pump output low.
10. Air in system.
11. Valve spool out of adjustment.
12. Valve spool sticking.
13. Steering linkage binding.

Hard Steering Straight Ahead

1. Steering adjustment too tight.
2. Steering gear shaft binding.

Hard Steering While Turning or Parking

1. Oil level low.
2. Pump pressure low.
3. Pressure loss in steering gear due to leakage past "O" rings.
4. Pressure loss between valve spool and sleeve.
5. Pressure loss past piston ring or scored housing bore.

Loose Steering

1. Loose wheel bearings.
2. Loose tie rod ends or linkage.
3. Worn ball joints.
4. Worn suspension parts.
5. Insufficient mesh load.
6. Insufficient worm bearing preload.
7. Valve spool out of adjustment.

Erratic Steering

1. Oil or brake fluid on brake lining.
2. Out of round brake drums.
3. Improperly adjusted brakes.
4. Under-inflated tires.
5. Broken spring or other details in suspension system.
6. Improper caster adjustment.
7. Fluid level low.

Binding or Poor Recovery

1. Steering gear shaft binding.
2. Steering gear out of adjustment.
3. Steering linkage binding.
4. Valve spool binding due to dirt or burred edges.
5. Valve spool out of adjustment.
6. Interference at sector shaft and ball stud.

Loss of Power Assist

1. Pump inoperative.
2. Hydraulic lines damaged.
3. Power cylinder damaged.
4. Valve spool out of adjustment.

Loss of Power Assist In One Direction

1. Valve spool out of adjustment.

Noisy Pump

1. Air being drawn into pump.
2. Lines touching other parts of car.
3. Oil level low.
4. Excessive back pressure caused by obstructions in lines.
5. Excessive wear of internal parts.

Poor Return of Steering Gear to Center

1. Valve spool sticking.
2. Valve spool out of adjustment.
3. All items given under "Binding or Poor Recovery".

Steering Wheel Surge While Turning

1. Valve spool sticking.
2. Excessive internal leakage.
3. Belt slippage.

Ford Rack & Pinion Power Steering

Description

This power rack and pinion steering gear, Fig. FR 1, is a hydraulic-mechanical unit, using an integral piston and rack to provide power assisted steering control. Internal valving controls pump flow and pressure as required during operation. The unit consists of a rotory hydraulic control valve connected to the input shaft and a boost cylinder integral with the rack.

Operation

The rotory control valve utilizes the relative rotational motion of the input shaft and valve sleeve to control fluid flow. As the steering wheel is turned, the resistance of the wheels and weight of the vehicle cause a torsion bar to deflect, Fig. FR 1. This torsion bar deflection changes position of rotory valve and sleeve ports, thereby directing fluid under pressure to the proper end of the power cylinder. The pressure differential acting on the piston attached to the rack, provides the power assist.

The control valve is forced back to a centered position by the torsion bar when steering effort is removed. Pressure is then equalized on each side of the piston and the front wheels tend to return to a straight ahead position.

Adjustments

Rack yoke bearing preload is the only service adjustment required. This adjustment is performed with the steering gear removed from the vehicle. Refer to the individual car chapters for steering gear removal.

1. With exterior of steering gear thoroughly cleaned, remove external pressure line assemblies and drain fluid.
2. Secure steering gear in holding fixture, #T57L-500-B, or a suitable vise.
3. Place tool #T74P-3504-R onto an inch pound torque wrench and position assembly on input shaft splines.
4. Loosen yoke plug locknut with tool #T74P-3504-U and insert tool #T74P-3504-W with an inch pound torque wrench into the drilled holes in yoke plug, Fig. FR2.
5. With rack at the center of travel,

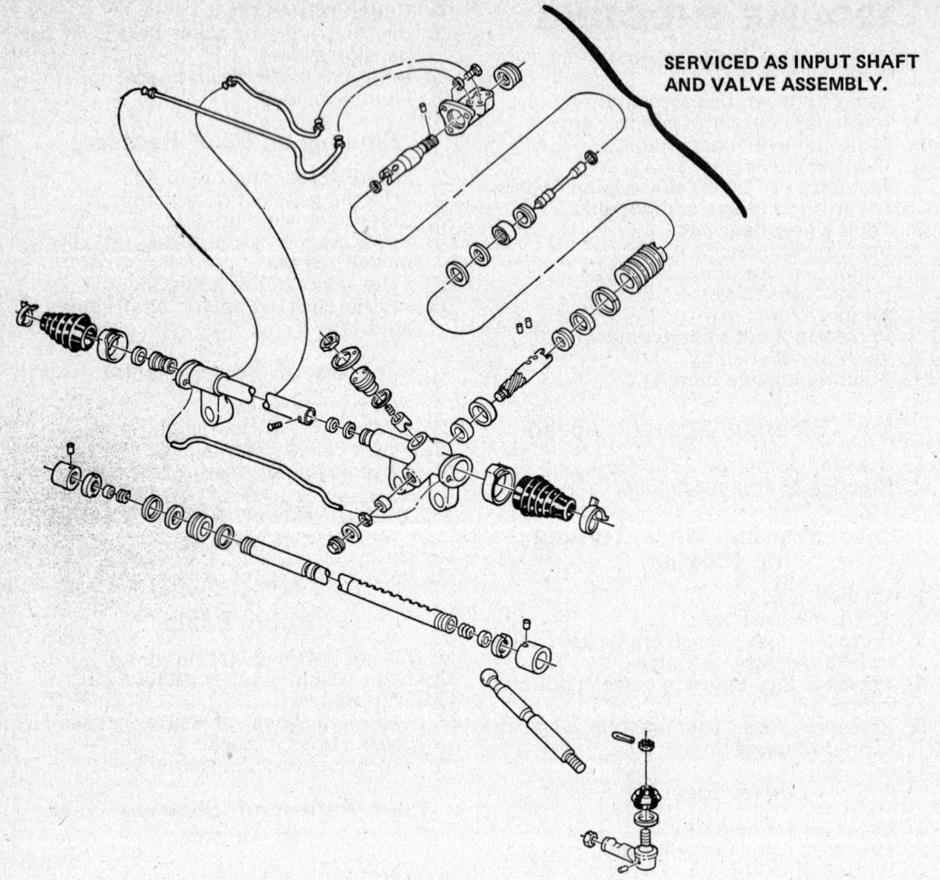

Fig. FR1 Steering gear disassembled

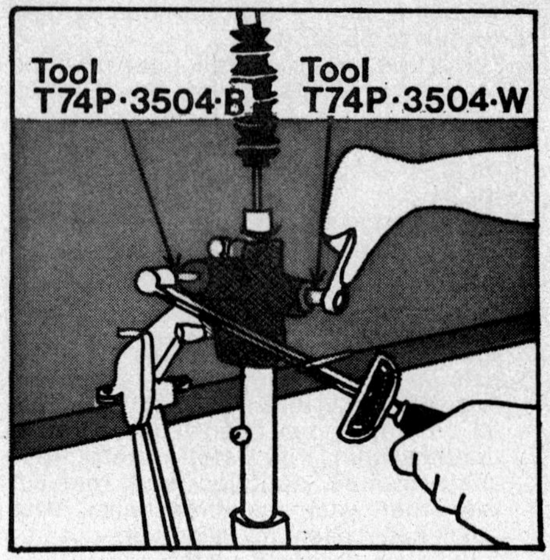

Fig. FR2 Adjusting yoke bearing preload

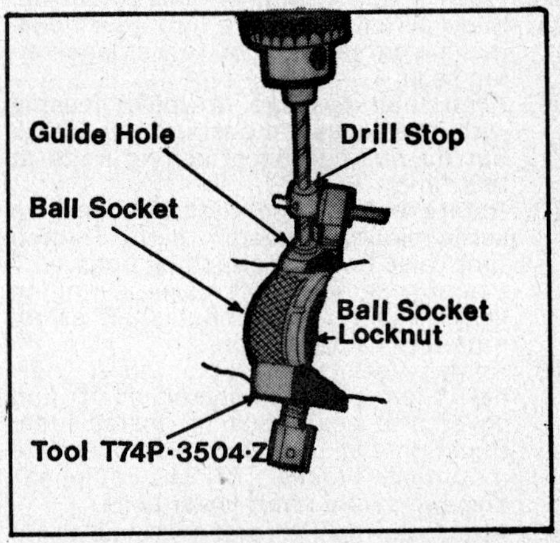

Fig. FR3 Ball socket retaining pin removal

torque yoke plug to 45-50 inch lbs., then back off yoke plug until the torque required to rotate the input shaft is 7 to 15 inch lbs.

6. Install tool #T74P-3504-U on yoke plug locknut and torque the locknut to 44-60 foot lbs. while holding yoke plug with tool #T74P-3504-W to prevent rotation.

NOTE: Ensure that yoke plug does not rotate when yoke plug locknut is torqued since preload can be changed. Recheck input shaft torque.

7. Remove tools and install external pressure lines with new copper flare gaskets.

STEERING GEAR REPAIRS

Tie Rod Ends, Bellows & Ball Joint Sockets

Disassembly

1. With steering gear assembly secured in holding fixture, tool #T57L-500-B, or a suitable vise, loosen tie rod jam nuts on outer ends adjacent to the sockets and remove sockets and jam nuts.
2. Remove bellows clamps, drain lubricant and remove bellows and breather tube.
3. A drill fixture, tool #T74P-3504-Z, is used when drilling out the ball socket retaining pin. Insert a 5/32 inch twist drill into the tool depth gauge, butt the stop against tool and secure with set screw.
4. Place protective cloths on each side of ball socket and install tool #T74P-3504-Z on ball socket with guide hole aligned with socket retaining pin. Place cone points of alignment and locking screws between locknut and ball socket and tighten securely, Fig. FR3.
5. Insert twist drill in guide hole and drill out retaining pin, then remove tie rod and socket with tool #T74P-3504-S.
6. Remove locknut, inner thrust bearing and rack spring from recess in rack end.

Assembly

1. Install new rack spring and assemble ball socket to tie rod, then install new inner thrust bearing in ball socket and lubricate assembly with lubricant, P/N D2AZ-19580-B.
2. Install new ball joint locknut on rack end and thread ball socket onto rack until tie rod movement becomes stiff and ball socket is tight.
3. Install ball joint and jam nut on tie rod end and using a spring scale, measure force required to move the tie rod in the socket which should be 4-6 pounds.
4. Hold ball housing with tool #T74P-3504-S and torque locknut against ball socket with a suitable tool to 25-35 foot lbs.
5. Remove tools and install drill fixture, too #T74P-3504-Z, on ball socket. Using procedure found in steps 3 through 5 under "Disassembly", drill new retaining pin hole on line of contact of locknut and ball socket. The new hole must not enter notches of locknut or old ball socket hole.
6. Insert new retaining pin into hole and tap until flush with ball housing and locknut surfaces. Stake metal of ball housing and locknut over pin to secure tie rod adjustment. Apply a reverse torque of 35 ft. lbs. on locknut to test proper installation of retaining pin. The joint should not loosen under this load.
7. Install bellows, breather tube and new bellows to housing clamps.
8. Install 2½ ounces of lubricant, P/N D2AZ-19580-B, into bellows and secure bellows to tie rod with new clamps.

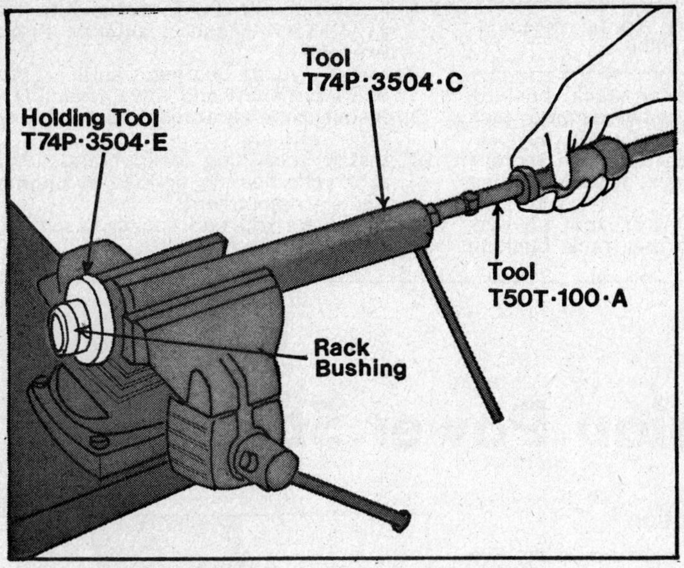

Fig. FR4 Rack bushing oil seal removal

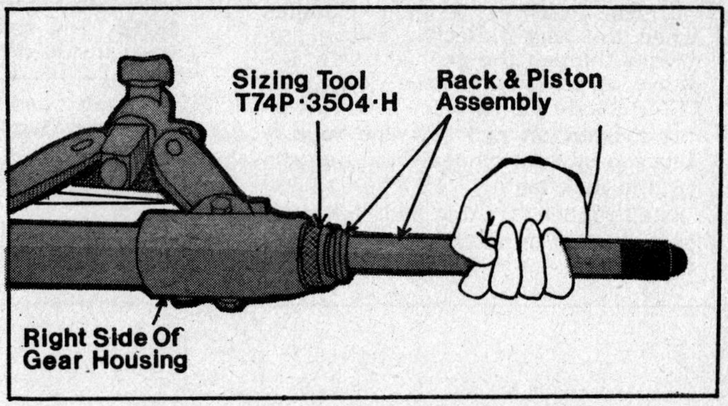

Fig. FR5 Rack & piston assembly installation

Input Shaft & Valve Assembly

Disassembly

1. Mount steering gear in holding fixture or suitable vise and remove external pressure lines and gaskets.
2. Loosen yoke plug locknut as outlined under "Adjustments".
3. Remove pinion bearing plug with 15/16 inch end of tool #T74P-3504-U.
4. Remove pinion bearing locknut while holding input shaft stationary with tool #T74P-3504-R.
5. Remove valve housing retaining bolts, move rack to left stop and mark position of flat tooth on input shaft splines to the valve housing face to aid assembly. Remove input shaft and valve assembly from gear housing.

NOTE: The valve housing, dust shield, oil seal, support bearing, thrust washers and thrust needle bearing may be replaced without removing the input shaft and valve from gear housing.

6. Remove pinion bearing, "O" ring, valve housing to gear housing "O" ring and input shaft oil seal.
7. Slide valve housing splined end of input shaft and remove input shaft needle bearing and two thrust washers from housing.
8. Remove teflon "O" rings from valve using care not to damage the lands and grooves.
9. Remove input shaft support bearing and oil seal from valve housing bore.
10. Pry input shaft dust shield from valve housing with a suitable tool using care not to damage housing surfaces.
11. If damaged, remove pressure and return tube seats with a suitable slide hammer and tool #T74P-3504-L.

Assembly

1. If removed, install new pressure and return tube seats.
2. Install a new input shaft oil seal in valve housing with lip facing toward housing and bottomed in the bore.
3. Fill input shaft dust shield bore with lubricant, ESA-M1C45-A, then install dust shield.
4. Lubricate and install input shaft support bearing with tool #T74P-3504-P. This tool will control the installation depth of the bearing, thereby avoiding contact with the oil seal.
5. Install teflon "O" rings on valve assembly.
6. Lubricate and install thrust needle bearing with a thrust washer on each side, over input shaft and against valve.
7. Lubricate valve assembly and insert, splined end first, into large opening in tool #T74P-3504-H. Place assembly over valve bore and push through until assembly bottoms and the full spline passes through dust shield, then remove tool.
8. Install valve housing to gear housing "O" ring on protruding flange on gear housing.
9. Install a new input shaft oil seal in gear housing with lip facing toward valve assembly.
10. Install pinion bearing through plug bore in the lower gear housing, then place an "O" ring around bearing adjacent to gear housing.
11. Move rack to left stop and install valve assembly in gear housing bore, aligning flat tooth on input shaft splines with alignment mark made during disassembly. Install and torque valve housing retaining bolts to 12-15 foot lbs.
12. Install pinion bearing locknut on pinion shaft.
13. Modify a pinion bearing plug by drilling a hole in the cap large enough to pass a 9/16 inch socket through. Thread modified plug into bore housing and tighten to hold bearing securely. Torque pinion bearing locknut to 44-66 foot lbs. while holding input shaft stationary with tool #T74P-3504-R. Remove modified pinion bearing plug.
14. Install and torque pinion bearing plug to 60-100 foot lbs. with a suitable torque wrench and tool #T74P-3504-U. Stake the plug in the slots in gear housing, ensuring previous stake marks are not aligned with slots.
15. Install flare nuts in pressure line fittings on gear and valve housings, then install external pressure lines.
16. Install tube bracket with tabs engaged with slots in gear housing.
17. Install yoke plug locknut and adjust rack yoke bearing preload as outlined under "Adjustments".

Gear Housing & Rack Assembly

Disassembly

1. Remove tie rods, ball joint sockets and input shaft and valve assembly as previously described.
2. Remove yoke plug locknut, tube bracket, yoke plug, yoke spring and yoke bearing from gear housing.
3. From gear right side, push rack inward until it bottoms and remove rack bushing locknut with tool #T74P-3504-B.
4. Gently pull rack from housing right side until piston contacts aluminum rack bushing, then pull rack and bushing from housing.
5. Remove internal high pressure oil seal with tool #T74P-3504-C and a suitable slide hammer.
6. Remove teflon "O" ring and rubber "O" ring from rack piston.
7. Remove rack bushing oil seal, Fig. FR4, then "O" ring from bushing.

Assembly

1. Install high pressure oil seal into gear housing.
2. Install rubber "O" ring in rack piston groove, then teflon "O" ring over rubber "O" ring.
3. Place protective sleeves, tools, #T74P-3504-J, K, over rack and piston assembly.
4. Insert smaller end of tool #T74P-3504-H into gear housing right side. Lubricate rack and piston assembly with fluid, then install rack with gear tooth end first into the tool, Fig. FR 5. Gently push rack into housing until leading edge engages internal

oil seal. Rack is properly installed when the long protective sleeve protrudes through the gear left side. Remove protective sleeve and tool #T74P-3504-H, then place a ball housing locknut on rack left end to protect the internal oil seal from damage by the rack teeth.
5. Install rubber "O" ring and high pressure oil seal on aluminum rack bushing.

NOTE: The seal lip spring faces toward inside of bushing.

6. Lubricate and install rack bushing with seal facing outward onto rack and into housing bore until the locknut threads are visable, then remove the protective sleeve. Install bushing locknut.
7. Place a ball housing locknut on rack right side and torque rack bushing locknut to 80-120 foot lbs. with tool #T74P-3504-B and a suitable torque wrench.
8. Install yoke bearing, spring, plug and input shaft and valve assembly.
9. Install tube bracket with tabs engaged with slots in gear housing.
10. Install yoke plug locknut and adjust rack yoke bearing preload as outlined under "Adjustments".
11. Install tie rod and socket assemblies as previously outlined.

Saginaw Rack & Pinion Power Steering

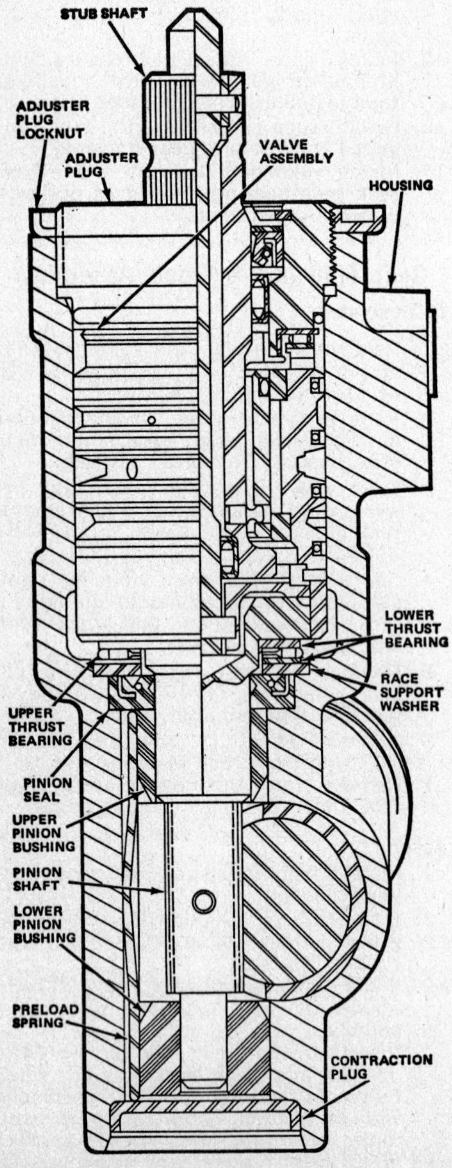

Fig. SRP2 Pinion shaft & valve body assembly

Description

This power steering gear assembly, Fig. SRP1, incorporates an integral tube and housing, containing a pinion shaft and steering rack. The tube and housing are joined by a plastic injection-bonding process. The pinion shaft is supported in the housing by thrust bearings and bushings. A bushing and bulkhead assembly supports the steering in the tube.

Operation

A rotary-type valve body, Fig. SRP2, is used to control the hydraulic steering assist. Fluid under pressure is directed to the gear housing and into the valve body. The valve body then directs fluid to the power cylinder.

A spool valve, connected to the stub shaft by a locating pin, rotates within the valve body. Fluid directional passages, machined into the spool valve, are aligned with fluid passages in the valve body as the spool valve rotates. Fluid is directed through these passages, into either side of the power cylinder through the externally mounted oil lines.

DISASSEMBLY

1. Clamp steering gear housing in a soft-jawed vise.
2. Cut boot clamp from steering gear housing end, then slide boot to expose rack teeth.
3. Rotate flexible coupling, moving rack to housing end of gear assembly, to further expose rack teeth.
4. Inspect rack teeth. If teeth are damaged, replace steering gear assembly.

NOTE: If rack teeth have machining marks or appear excessively bright and shiny, steering gear replacement is not necessary since these are normal conditions.

5. Remove flexible coupling pinch bolt, the separate flexible coupling from stub shaft.
6. Remove adjuster plug lock nut, adjuster plug and valve body, then lift pinion shaft, rotate counter-clockwise and remove from housing.
7. Inspect pinion shaft teeth. If teeth are damaged, replace steering gear assembly.

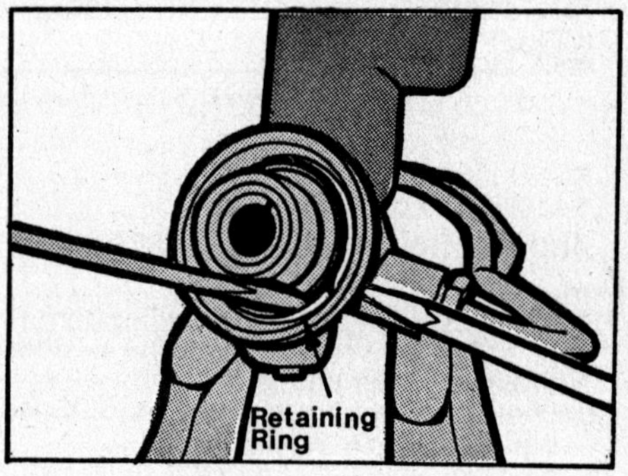

Fig. SRP3 Removing bulkhead retaining ring

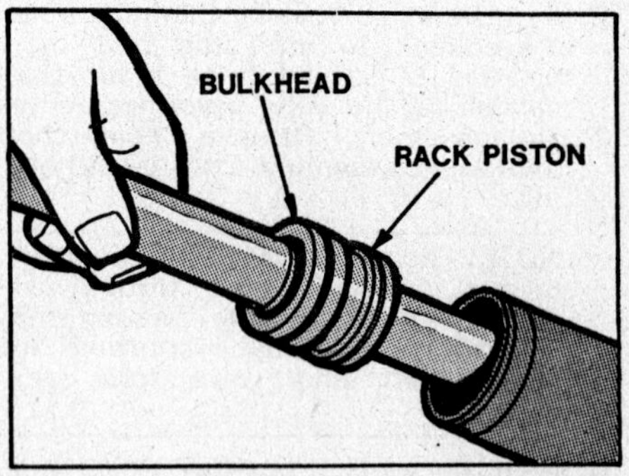

Fig. SRP4 Removing steering rack & bulkhead

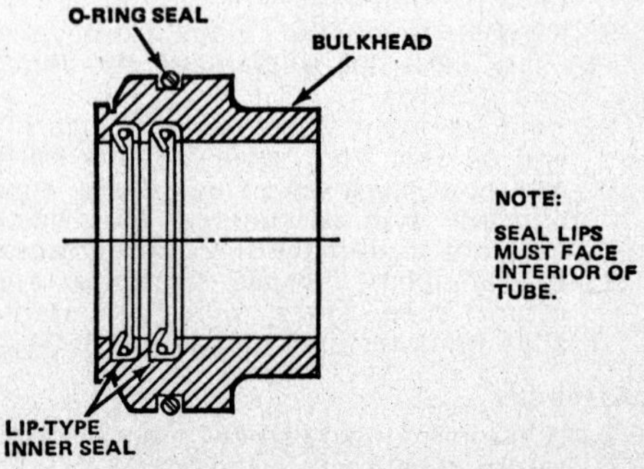

Fig. SRP5 Bulkhead seals installation

8. Remove contraction plug, lower pinion bushing and preload spring from housing.
9. Center the steering rack in tube and install pinion shaft, valve body and adjuster plug into housing. Ensure valve body is seated properly.
10. Mark position of adjuster tubes on tie rods, loosen adjuster tube clamps and remove adjuster tubes and tie rod ends.
11. Cut all boot clamps from boots, then remove boots.
12. Remove shock dampener rings from jam nuts, clamp inner tie rod housing in a vise and loosen jam nuts.
13. Loosen tie rod housing set screws, then remove tie rod housings, ball seats, ball seat springs, jam nuts and shock dampener rings from steering rack. Remove valve body, adjuster plug and pinion shaft from housing.
14. Remove pinion thrust bearing, conical thrust bearing races and support washer from housing.
15. Mark position of the breather tube on steering gear assembly since the breather tube must be installed in original position to ensure proper sealing. Remove breather tube from mounting grommet hole, then the breather tube from gear assembly.
16. Insert a suitable punch through access hole in housing tube forcing bulkhead retaining ring from groove, then pry ring from housing tube, Fig. SRP3.
17. Pull steering rack and bulkhead from tube. Then, remove bulkhead from rack.

NOTE: The rack piston will force bulkhead from tube, Fig. SRP4.

18. Pry rack bushing from housing.
19. Insert tool J-25507 into housing tube bore, then using a brass drift, 12 to 14 inches long, drive inner rack seal from tube.
20. Remove upper pinion bushing and shaft seal from housing.

SUB-ASSEMBLY SERVICE

Bulkhead

Replace the "O" ring and the two inner lip-type seals. Note position of the inner seals for proper installation, Fig. SRP5.

Rack Piston

Replace rack piston seal. Use caution not to score or scratch piston when removing.

Adjuster Plug Assembly

1. Remove thrust bearing retainer, Fig. SRP6. Do not damage needle bearing bore.
2. Remove thrust bearing spacer, thrust bearing and the bearing races.
3. Remove adjuster plug "O" ring.
4. Remove stub shaft seal retaining snap ring, then pry dust seal and oil seal from adjuster plug.
5. Remove needle bearing.
6. Reverse procedure to assemble.

Valve Body & Stub Shaft

NOTE: The valve body and stub shaft as-

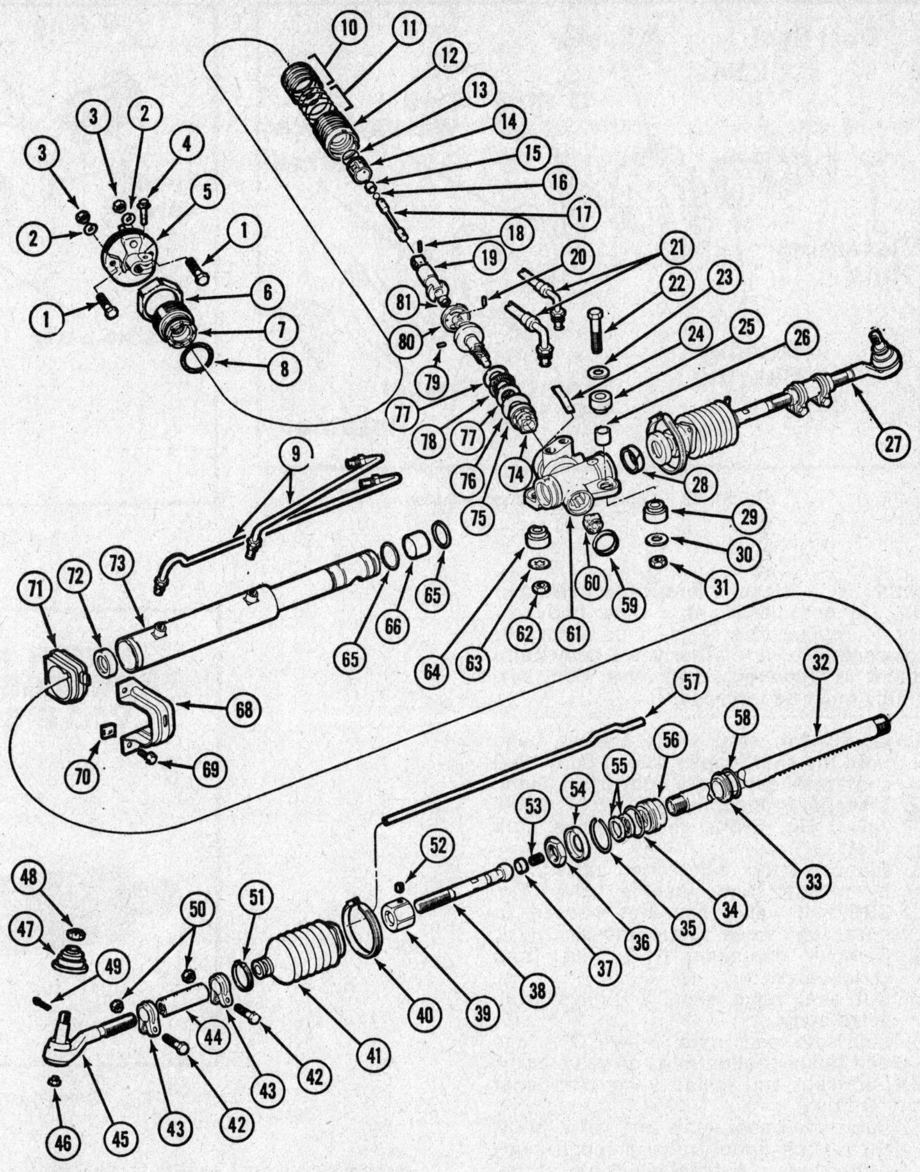

1. FLEXIBLE COUPLING - TO - INTERMEDIATE SHAFT ATTACHING BOLT
2. LOCKWASHER
3. NUT
4. PINCH BOLT
5. FLEXIBLE COUPLING
6. ADJUSTER PLUG LOCKNUT
7. ADJUSTER PLUG ASSEMBLY
8. ADJUSTER PLUG O-RING
9. OIL LINES
10. VALVE BODY SEAL RINGS
11. VALVE BODY O-RINGS
12. VALVE BODY
13. SPOOL VALVE DAMPER O-RING
14. SPOOL VALVE
15. TORSION BAR BUSHING (INCLUDED IN STUB SHAFT)
16. TORSION BAR SEAL RING (INCLUDED IN STUB SHAFT)
17. TORSION BAR (INCLUDED IN STUB SHAFT)
18. DRIVE PIN (INCLUDED IN STUB SHAFT)
19. STUB SHAFT
20. DRIVE PIN (INCLUDED IN STUB SHAFT)
21. POWER STEERING HOSES
22. MOUNTING BOLT
23. WASHER
24. PRELOAD SPRING
25. GROMMET
26. BUSHING
27. STEERING LINKAGE (ASSEMBLED)
28. RACK BUSHING
29. GROMMET
30. WASHER
31. NUT
32. STEERING RACK
33. RACK PISTON
34. BULKHEAD O-RING
35. BULKHEAD RETAINING RING
36. JAM NUT
37. BALL SEAT
38. INNER TIE ROD
39. INNER TIE ROD HOUSING
40. BOOT CLAMP
41. BOOT
42. ADJUSTER TUBE CLAMP BOLT
43. ADJUSTER TUBE CLAMP
44. ADJUSTER TUBE
45. TIE ROD END
46. LUBE PLUG
47. TIE ROD END SEAL
48. TIE ROD END-NUT
49. COTTER PIN
50. ADJUSTER TUBE CLAMP NUT
51. BOOT CLAMP
52. TIE ROD HOUSING SET SCREW
53. BALL SEAT SPRING
54. SHOCK DAMPENER RING
55. BULKHEAD SEALS (LIP TYPE)
56. BULKHEAD
57. BREATHER TUBE
58. RACK PISTON SEAL RING
59. CONTRACTION PLUG
60. LOWER PINION BUSHING
61. HOUSING (INCLUDED IN TUBE AND HOUSING ASSEMBLY)
62. NUT
63. WASHER
64. GROMMET
65. PLASTIC INJECTION RING
66. LOCATING BUSHING
67. PLASTIC INJECTION RING
68. MOUNTING CLAMP
69. BOLT
70. MOUNTING CLIP
71. MOUNTING GROMMET
72. INNER RACK SEAL
73. TUBE AND POWER CYLINDER (INCLUDED IN TUBE AND HOUSING ASSEMBLY)
74. UPPER PINION BUSHING
75. PINION SHAFT SEAL (LIP TYPE)
76. SUPPORT WASHER
77. CONICAL THRUST BEARING RACE
78. THRUST BEARING
79. DRIVE PIN (INCLUDED IN STUB SHAFT)
80. SHAFT CAP (INCLUDED IN STUB SHAFT)
81. TORSION BAR BUSHING (INCLUDED IN STUB SHAFT)

Fig. SRP1 Saginaw rack & pinion power steering gear, disassembled

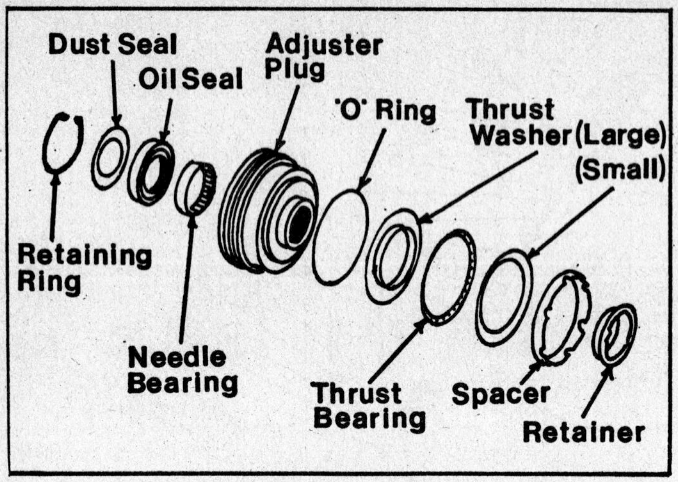

Fig. SRP6 Adjuster plug, disassembled

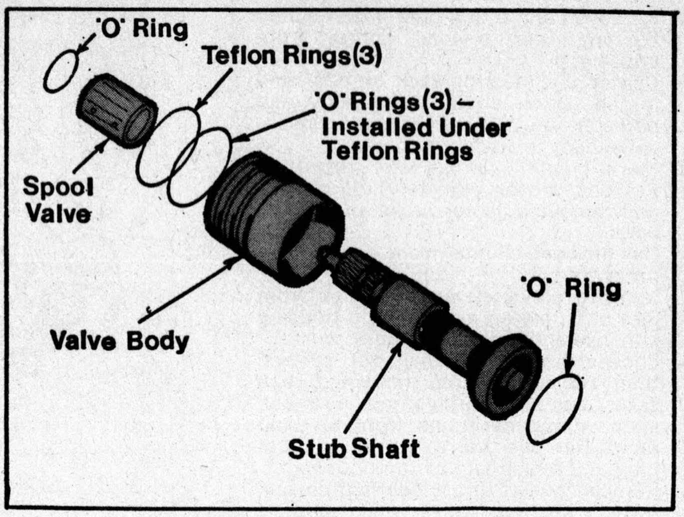

Fig. SRP7 Valve body, disassembled

sembly is a close tolerance constructed unit. Do not disassemble valve body except to replace the seals, if necessary. If replacement of any other valve body component is required, the valve body assembly must be replaced.

1. Lightly tap stub shaft against bend until the shaft cap is free of body. Pull shaft until cap clears body by $\frac{1}{4}$ inch.
2. Disengage locating pin from spool valve, Fig. SRP2, then remove stub shaft.
3. Remove spool valve from valve body by rotating and pushing valve, Fig. SRP7. If valve becomes cocked in bore, align valve, then remove.
4. Remove dampener "O" ring from spool valve.
5. Cut seal rings and "O" rings from valve body.
6. Lubricate and install new "O" rings and teflon sealing rings on valve body.
7. Lubricate and install valve dampener "O" ring.
8. Lubricate spool valve and valve body, then push spool valve through body until shaft pin hole is visible at opposite end.
9. Lubricate stub shaft and insert into spool valve until shaft pin can be installed into spool valve.
10. Align shaft cap notch with valve body pin and press spool valve and shaft assembly into valve body, Fig. SRP8.

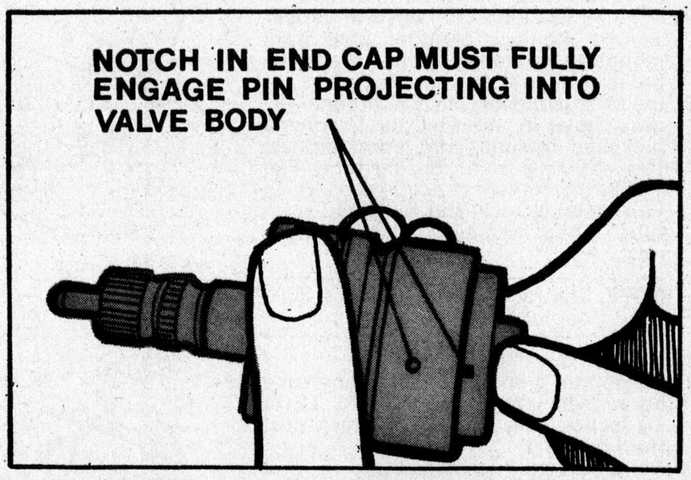

Fig. SRP8 Installing stub shaft assembly into valve body

ASSEMBLY

1. Install rack bushing into housing.
2. Install seal protector tool J-25509 over rack threads and cover remaining rack teeth with cardboard. Lubricate inner rack seal and slide onto rack with sealing lip facing toward rack piston (metal surface of seal facing away from piston).
3. Lubricate inner rack seal, bulkhead retaining ring groove and rack teeth with a suitable chassis lubricant.
4. Push seal protector J-25508 into tube and insert steering rack through protector, into tube. Push rack to seat inner rack seal in end of tube.
5. Install upper pinion bushing with chamfered side facing downward and start pinion shaft seal with sealing lip facing toward bore into seat in housing, Fig. SRP9. Place support washer on top of pinion shaft seal and, using a suitable tool, press on support washer until pinion shaft seal and support washer are fully seated in housing.
6. Lubricate pinion shaft thrust bearing and the races with vaseline. Install a race, bearing and a race on pinion shaft.
7. Center steering rack. Distance from end of rack to machined inner face of housing should be 4 inches, Fig. SRP10.
8. Install pinion shaft into housing with drive pin located between 3 and 4 o'clock positions until bottomed, Fig. SRP11.
9. Reset distance between rack end machined inner housing face to 4 inches. The pinion shaft drive pin should now be located at the 12 o'clock position, Fig. SRP11. If drive pin is mispositioned, remove pinion shaft and repeat steps 7 through 9.
10. Align valve body notch with drive pin in pinion shaft, then install valve body into housing. Ensure pinion shaft drive lugs engage slots in stub shaft cap. If valve body is properly installed, the fluid return hole will be exposed, Fig. SRP12. If not, pinion shaft is not seated, spool valve locating pins are misaligned or the valve body stub shaft locating pins are misaligned.
11. Install and tighten adjuster plug until bottomed. Place an alignment mark between adjuster plug spanner hole and the housing. Measure $\frac{3}{16}$ to $\frac{1}{4}$ inch counter-clockwise and remark housing, Fig. SRP13. Back off adjuster plug to second mark. Install and torque adjuster plug lock nut to 80 ft. lbs.
12. Install seal protector tool J-25509 over rack threads and slide bulkhead over tool and into tube. Then, using a suitable brass drift, drive bulkhead until bottomed against counterbore in tube. Install bulkhead retaining ring with gap in ring $\frac{1}{4}$ inch from access hole.
13. Fill housing with a suitable lithium based grease. Do not overfill.
14. Install preload spring with the center hump bearing against the housing, allowing the spring to extend approxi-

Fig. SRP9 Installing pinion shaft seal

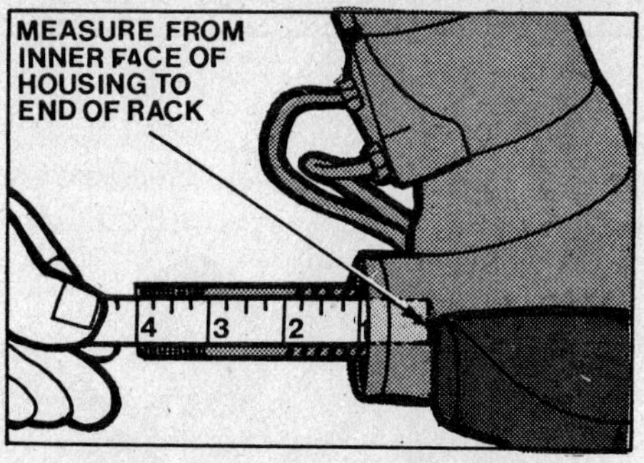

Fig. SRP10 Centering steering rack

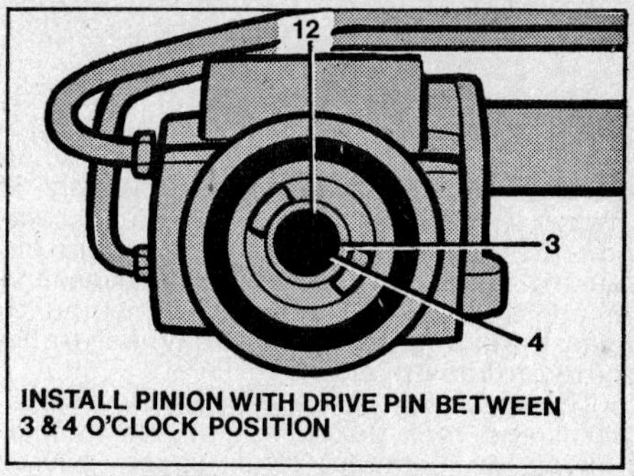

Fig. SRP11 Centering pinion shaft

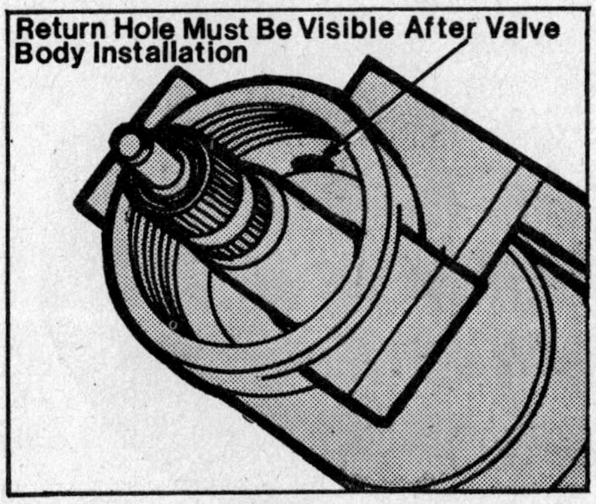

Fig. SRP12 Valve body installation

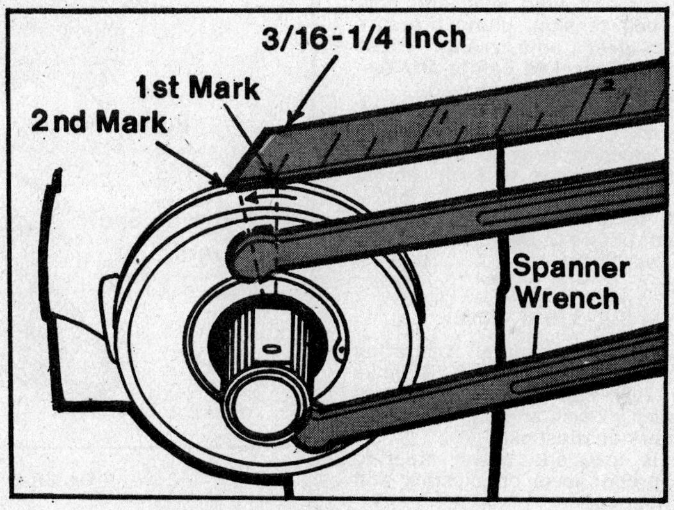

Fig. SRP13 Adjuster plug adjustment

Fig. SRP14 Lower pinion bushing installation

mately ¼ inch from housing end. Spring end must enter upper pinion bushing. Using needlenose pliers, hold preload spring against housing and install bushing into housing with chamfered end facing downward, Fig. SRP14.

15. Seat contraction plug in housing.
16. Install grommet and mounting clamp on tube, if removed.
17. Install shock dampener rings on ends of steering rack with open ends facing outward, then install the jam nuts.
18. Lubricate all inner tie rod wear surfaces with a suitable lithium based grease. Also, pack tie rod housings with same type grease.
19. Assemble and install inner tie rod assemblies onto rack. Torque tie rod housings to 25 ft. lbs. while rocking the inner tie rod, thereby relieving grease lock. Then, loosen housings ⅛ turn. Tie rod should rotate freely.
20. Torque housing set screws to 60 inch lbs.
21. Clamp tie rod housings in a vise and torque jam nuts to 100 ft. lbs.
22. Slide shock dampener rings over jam nuts.
23. Install boot on mounting bracket side of housing with hole in boot aligning with hole in mounting grommet. Slide short end of breather tube through the grommet and boot holes. The long end of tube should lie against housing.
24. Install opposite side boot with hole in boot aligned with breather tube.
25. Slide outer collars of boots over inner tie rod undercuts. Install small boot clamps and compress with tool J-22610.
26. Install adjuster tubes and tie rod ends using alignment marks made during disassembly. A minimum of three threads must be visible on each end of adjuster tubes. Also, the number of threads per side should not differ by more than three threads.
27. Install flexible coupling on stub shaft and torque pinch bolt to 30 ft. lbs.
28. Install large boot clamps and compress with tool J-22610. Position clamp ear approximately ¾ inch from breather tube before compressing clamp.

Saginaw Rotary Valve Type Gear

DESCRIPTION

This type power steering gear, Fig. SR1, operates entirely by displacing oil to provide hydraulic pressure assists only when turning. As the entire gear assembly is always filled with oil, all internal components of the gear are immersed in oil, making periodic lubrication unnecessary. In addition this oil acts as a cushion to absorb road shocks that may be transmitted to the driver.

The steering shaft, hydraulic valve, worm and rack piston nut are all in line, making a compact and space saving steering gear. All oil passages are internal except the pressure and return hoses between gear and pump.

MAINTENANCE

Four factors affect power operation of the steering system: fluid level and condition, drive belt tension, pump pressure and steering gear adjustment. These should always be checked before any major service operations are performed.

Conditions such as, hard or loose steering, road shock or vibrations are not always due to steering gear or pump, but are often related instead to such factors as low tire pressure and front end alignment. These factors should be checked and corrected before any adjustment of the steering gear is made.

Checking Fluid Level

1. Run engine to normal operating temperature, then shut it off. Remove reservoir filler cap and check oil level on dipstick. Level should be up to "Full" mark on dipstick.
2. If level is low, add power steering fluid to proper level on dipstick and replace filler cap.

NOTE: *When adding less than one pint of fluid, it is permissible to use automatic transmission fluid. When adding larger quantities or when making a complete fluid change, it is recommended that special power steering fluid be used.*

---

3. When checking fluid level after system has been serviced, air must be bled from the system as follows:

Bleeding System

1. With wheels turned all the way to the left, add the recommended fluid to the "Add" mark on dipstick.
2. Start engine. While running at fast idle, recheck fluid level and, if necessary, add fluid to the "Add" mark on dipstick.
3. Bleed system by turning wheels from side to side without hitting stops. Maintain fluid level just above internal pump casting. Fluid with air in it will have a light tan appearance. This air must be eliminated from fluid before normal steering action can be obtained.
4. Return wheels to center position and continue to run engine for two or three minutes, then shut it off.
5. Road test vehicle to make sure steering functions normally and is free from noise.
6. Recheck fluid level, making sure level is at the "Full" mark.

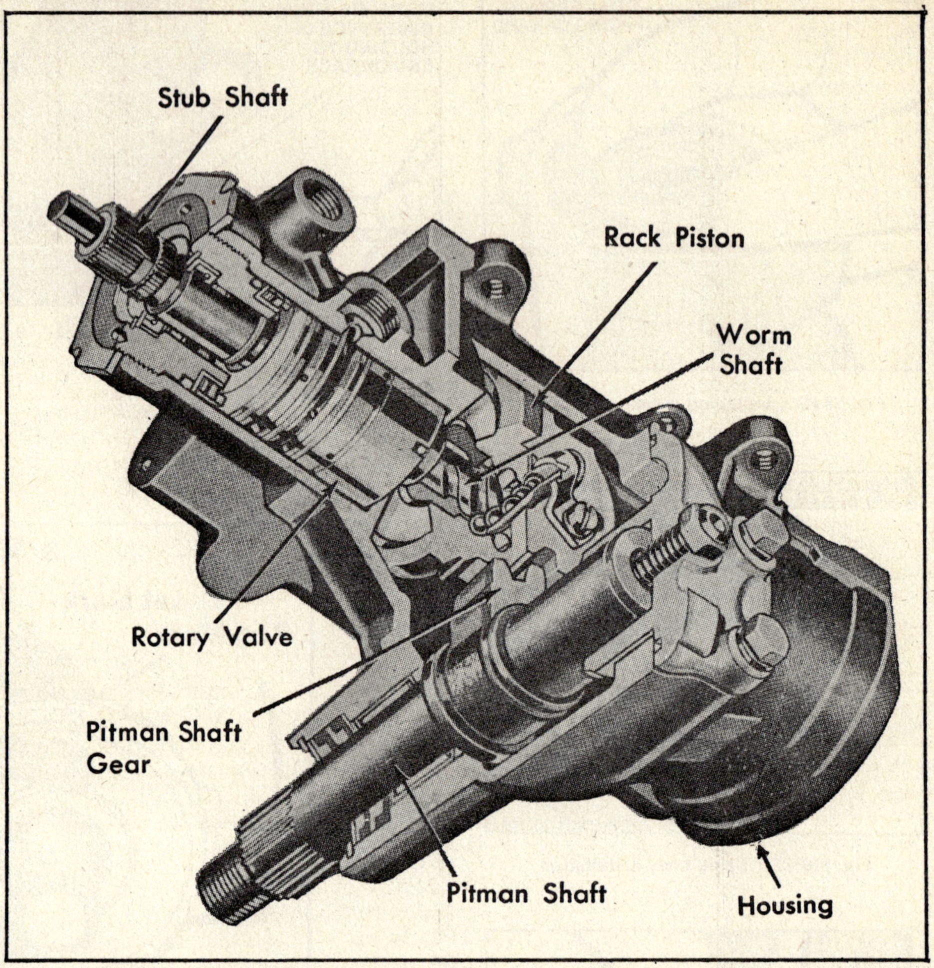

Fig. SR-1 Saginaw rotary valve power steering gear

Checking Pump Pressure

With pump reservoir filled with oil to the proper level, turn steering wheel several times from one extreme to the other until all air is bled from the system. Air in the system will be indicated by bubbles in the reservoir.

Connect a pressure gauge between pump and steering gear. Then, with engine idling and steering gear at normal operating temperature, turn steering wheel to either extreme position, hold it momentarily against the stop and note the pressure indicated on the gauge. The pressures that should obtain are as follows:

AMERICAN MOTORS	PRESSURE
1969-73 V8	1100
1969-73 Six	665-735
1974 V8	1100-1200
1974 Six	1000-1100
1975 All	1100-1200
BUICK	
1969-70 Exc. Below	1100-1200
1969-71 Variable Ratio	1250-1350
1972-75	Minimum 1000
BUICK SPECIAL, SKYLARK, CENTURY & APOLLO	
1969-70 Exc. Below	1100-1200
Variable Ratio	1250-1350
1971 Six	1350-1450
1971 V8	1250-1350
1972-75	Minimum 1000
CADILLAC	
1969-73	1350-1450
1974-75	1425-1450
CAMARO	
1969-71 V8	1350-1450
1969-71 Six	1200-1300
1972-73	1350-1450
1974	1250-1350
1975	1350-1450
CHEVROLET	
1969 Exc. Below	1000-1100

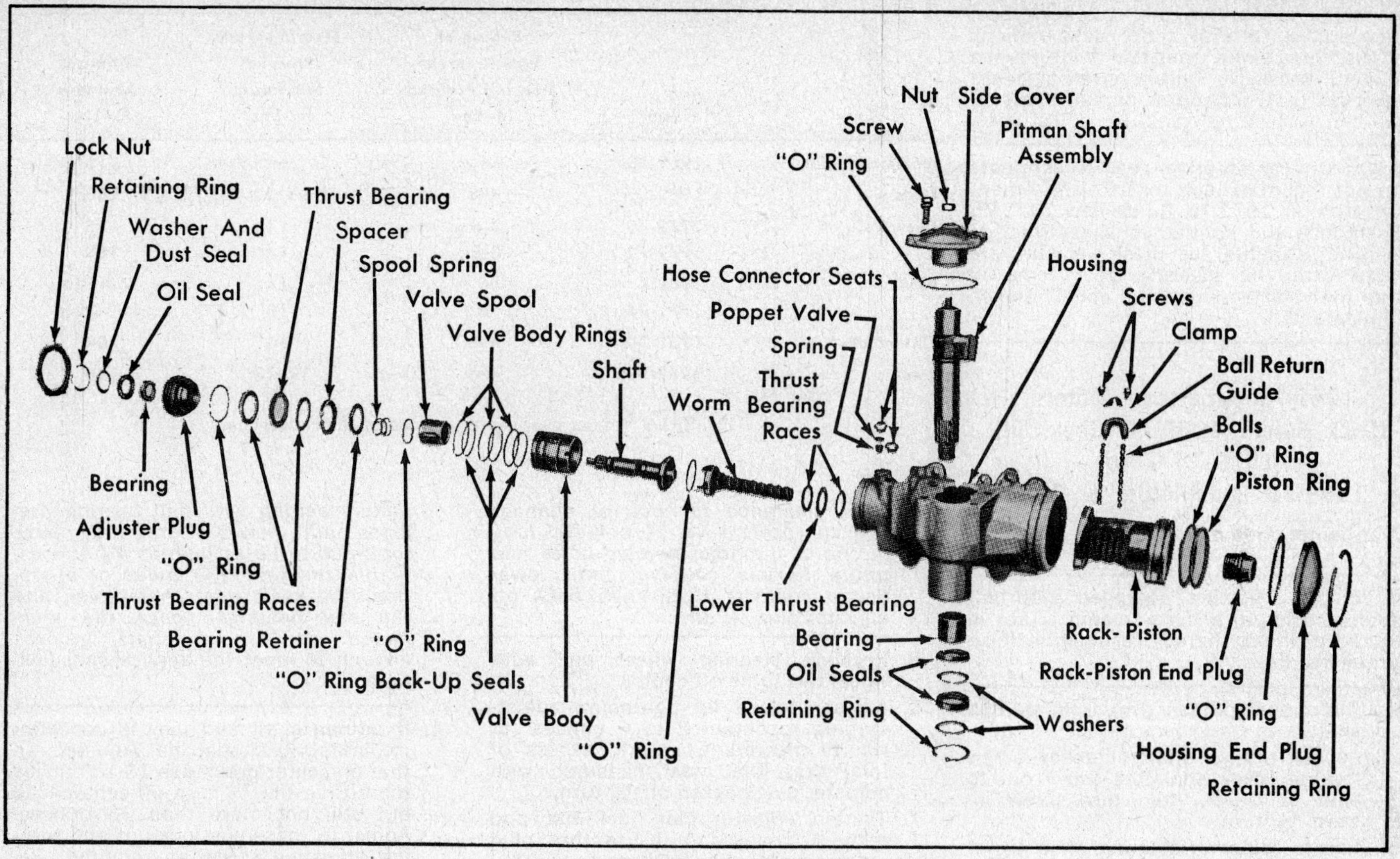

Fig. SR2 Power steering gear disassembled

1969 V8-396, 427 (Exc. Variable Ratio)	1100-1200
1969 Variable Ratio	1350-1450
1970	1350-1450
1971-73 Exc. Below	1350-1450
1971-72 Wagons	900-1000
1974	1250-1350
1975	1350-1450

CHEVELLE & MONTE CARLO

1969-71 Exc. Below	900-1000
1970-71 Variable Ratio (Six)	1200-1300
1970-71 Variable Ratio (V8)	1350-1450
1972-73 Exc. Below	1350-1450
1972 Wagons	900-1000
1974	1250-1350
1975	1350-1450

CHEVY NOVA

1969	900-1000
1970-71 V8	1350-1450
1970-71 Six	1200-1300
1972-74	1350-1450
1972-73	1350-1450
1974	1250-1350
1975	1350-1450

CHEVY VEGA & PONTIAC ASTRE

1971	900-1000
1972-75	750-850

CHECKER MOTORS

1969-70	1000-1100
1971-73	1350-1450
1974	1250-1350
1975	1350-1450

FORD

1969	1000-1150

OLDSMOBILE

1969-71	1250 Min.
1972	1250-1350
1973-74	1350-1450

OLDS F85, CUTLASS & OMEGA

1969-71 Six	850 Min.
1969-71 V8	1100 Min.
1972	1250-1350
1973-74	1350-1450

OLDSMOBILE TORONADO

1969-71	1350 Min.
1972	1250-1350
1973-74	1350-1450

PONTIAC

1969 Catalina W/AC	900-1000
1969 Catalina	1000-1100
1969 All Others	1100-1200
1970	1250-1350
1971	1350-1450
1972-74	Minimum 1300
1975	Minimum 1350

PONTIAC FIREBIRD

1969 W/AC	900-1000
1969 L/AC	1000-1100
1970 Six	1200-1300
1970 V8	1250-1350
1971 Six	1200-1300
1971 V8	1350-1450
1972-74	Minimum 1300
1975	Minimum 1350

PONTIAC TEMPEST & VENTURA II

1969	1000-1100
1970 Six	1200-1300
1970 V8	1250-1350
1971 Six	1200-1300
1971 V8	1350-1450
1972-74	Minimum 1300
1975	Minimum 1350

GEAR ADJUSTMENTS ON CAR

When the steering gear is thought to be out of adjustment, a quick check can be made by moving the steering wheel back and forth with short slow motions at the "on-center" position with engine shut off. Excessive looseness felt or heard indicates that either the pitman shaft or the thrust bearing requires adjustment.

NOTE: On car steering gear adjustments are not recommended for 1974-75 American Motors, 1970-75 Buick and 1971-75 Oldsmobile and Pontiac vehicles. Steering gear adjustments for these vehicles are made with the steering gear removed from the vehicle as outlined under "Bench Adjustments".

Car	Year	Excess of Ball & Thrust Bearing Preload In.-Lbs.	Steering Gear Preload Maximum In.-Lbs.	Pitman Arm Nut Ft.-Lbs.
American Motors	1969-73	6	—	115
Buick	1969	4–8	15–20	140
Chevrolet	1969-70	3-6	14	140
	1971-72	3-6	14	180
Ford	1969	6	14	150-225
	1971-74	6	18	150-225
Oldsmobile	1969-70	4–8	16	200
Pontiac	1969-70	4–8	14–18	140

Fig. SR1A In-car over-center preload specification chart

1969-73 American Motors, 1969 Buick, 1969-72 Chevrolet, 1969-75 Ford & 1969-70 Oldsmobile & Pontiac

Over-Center Adjustment

NOTE: On vehicles equipped with tilt steering column, ensure column is not in a tilted position, thereby resulting in false torque readings.

1. Disconnect pitman arm from pitman shaft.
2. Loosen pitman shaft adjusting screw locknut, back adjusting screw out to limit of travel, then turn screw inward ½ turn.
3. Locate center of steering wheel travel and using an inch pound torque wrench connected to the steering shaft nut, rotate wheel through center of travel and note torque reading (combined ball and thrust bearing preload).
4. Referring to Fig. SR 1A, tighten adjusting screw until torque reading obtained is within the specified excess of the ball and thrust bearing preload noted in the above step, however, total preload must not exceed specifications. Tighten locknut and recheck preload.
5. Connect pitman arm to pitman shaft and torque attaching nut to specifications, Fig. SR 1A.

1969-75 Cadillac

THRUST BEARING PRELOAD

1. Remove pitman arm from drag link with a suitable puller.
2. Check for any distortion or binding in flexible coupling, and correct as necessary.
3. Turn steering wheel two turns "off center" (just away from stops) and use a spring tension scale with a piece of tag wire or string to measure pull on steering wheel through an arc not exceeding one inch. Total pull should be between 4 and 12 ounces.
4. If "off-center" pull is greater or less than specified, loosen adjuster plug lock nut and back off adjuster ⅛ turn, using a suitable spanner wrench.

NOTE—*If position of holes in adjuster plug is such that there is insufficient clearance for use of spanner wrench, insert a No. 10 bolt ¾" long in one of the adjuster plug holes and rotate flexible coupling until lower flange contacts bolt. Then back off adjuster plug ⅛ turn.*

5. Recheck steering wheel pull with wheel two turns off center.
6. Adjust preload by tightening adjuster plug to obtain 2 to 6 ounces (at rim of steering wheel) in excess of total drag that was measured with adjuster plug backed off ⅛ turn.
7. Tighten adjuster plug lock nut and recheck preload to insure that plug did not move when tightened.

WORM & BALL PRELOAD

1. Locate center of steering wheel travel and turn steering wheel ½ turn "off-center".
2. With wheel in this position, measure and record total pull through an arc not exceeding one inch. Due to worm and ball preload, the total pull should be 2 to 8 ounces in excess of two turns "off-center" pull previously recorded.
3. If total pull is less than one ounce or more than nine ounces in excess of the two turns "off-center" pull, steering gear must be removed and disassembled for replacement of worm shaft balls.
4. Check for roughness in worm and rack-piston by turning wheel between ¼ turn and 1 turn "off-center" on each side. Noticeable roughness also requires worm replacement.

WORM SHAFT END PLAY

1. Find exact center of steering wheel travel. Tight spot due to pitman shaft should extend for ¼ turn to either side.
2. With wheel "on-center", measure and record total pull through an arc of three inches or less.
3. Reading should not be more than 36 ounces, but should be at least eight ounces (never more than 20) more than a reading taken ½ turn off-center.

NOTE—*On new steering gears, the factory setting with ball bearing preloads and a new seal drag may cause total pull to be as high as 40 ounces. Within the first 100 miles of operation, the seals seat themselves and the ball bearings polish the rack-piston and worm shaft grooves enough to meet the service specifications.*

4. If pitman shaft end play is not within limits, it should be adjusted so that on-center preload is 16-18 ounces more than the ½ turn off-center load but still not more than 36 ounces. Adjust by loosening lock nut and turning adjusting screw as required. Recheck pull after tightening lock nut.
5. Check for rough spots in pitman shaft sector and power piston rack by turning wheel between ¼ turn to the right and ¼ turn to the left. A visual inspection of gear teeth will reveal necessary correction if rough spots are found. Remember, however, teeth are not absolutely smooth but are machined to hold lubricant.

STEERING GEAR

Disassemble

1. Referring to Fig. SR2, position gear in vise with gear housing end plug facing up.
2. Rotate gear housing end plug retaining ring so that one end of ring is over hole in housing. Spring one end of ring with a punch to allow screwdriver to be inserted to lift ring out, Fig. SR3.
3. Rotate coupling flange counterclockwise until rack-piston just forces end plug out of housing. Remove plug from housing.

CAUTION—*Do not rotate any further than necessary or balls will fall out of their circuit and pitman shaft teeth and rack-piston will become disengaged.*

4. Remove rack-piston end plug, using a ½" square drive.

NOTE—*To make removal easier, tap rack-piston end plug with a plastic* *mallet to unseat threads. This is important as end plug is tightened to 75-80 ft-lbs during assembly and could break during removal if not handled carefully.*

5. Remove lock nut from pitman shaft adjuster screw and discard.
6. Remove four side cover retaining screws and washers from cover.
7. Rotate pitman shaft adjuster screw with an Allen wrench until side cover is lifted free of housing.
8. Separate side cover from pitman shaft; discard side cover O-ring seal.
9. Turn coupling flange until pitman shaft teeth are centered in housing.
10. Tap end of pitman shaft with a soft mallet and slide pitman shaft out of housing.
11. Remove housing end plug O-ring seal and discard.
12. Insert Rack-Piston Arbor into rack-piston against end of worm, Fig. SR4. Turn coupling flange counterclockwise, while holding tool tightly against worm, to force rack-piston onto arbor, and remove rack-piston from gear housing.
13. Remove stub shaft-to-coupling flange retaining screw and remove flange.
14. Remove adjuster plug lock nut by breaking it loose with hammer and punch, and remove lock nut from housing.
15. Loosen adjuster plug and remove from housing, Fig. SR5.
16. Remove valve assembly by grasping stub shaft and pulling out.
17. Remove worm, lower thrust bearing and races from upper end of housing.

Replacing Valve Spool Dampener "O" Ring (Only If Gear Squawks)

The rotary valve assembly includes the valve body, valve spool and stub shaft assembly. All these parts are precision units and are hydraulically balanced at the factory. Under no circumstances are parts in this unit to be replaced or interchanged with other parts or units. If unit parts are scored or damaged the entire rotary valve assembly is to be replaced.

1. To replace the valve spool dampener "O" ring, work spool spring into bearing diameter of stub shaft and remove spool spring.
2. Tap end of stub shaft gently against workbench to remove valve spool. *The diametrical clearance between the valve body and spool may be as low as .0004". The slightest cocking of the spool may jam it in the valve body.*
3. Remove valve spool dampener "O" ring.
4. Install new "O" ring in valve spool groove, then lubricate seal in automatic transmission fluid. Do not allow seal to twist in groove.
5. With notch end of spool towards valve body, install spool, aligning spool notch with pin in stub shaft, Fig. SR6.

6. *Because of the small clearance between spool and valve body, extreme care must be taken when assembling these parts. Push spool evenly and slowly with a slight oscillating motion until spool reaches drive pin. Before pushing spool completely in, make sure dampener "O" ring seal is evenly distributed in spool groove. Slowly push spool completely in, with extreme care taken not to cut or pinch "O" ring seal.*

7. Slide spool spring over stub shaft and work spring into position.

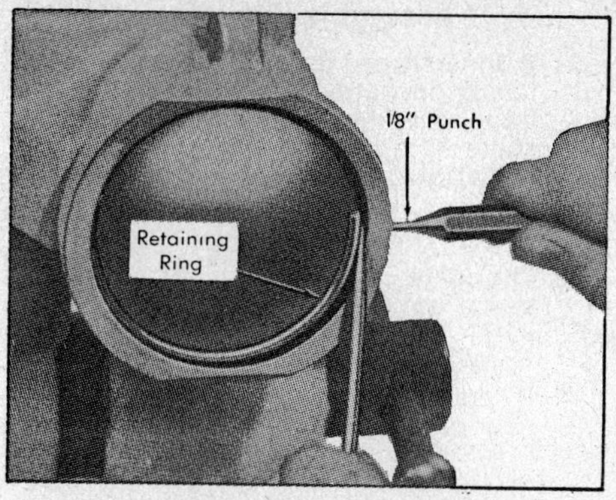

Fig. SR3 Removing end plug retaining ring

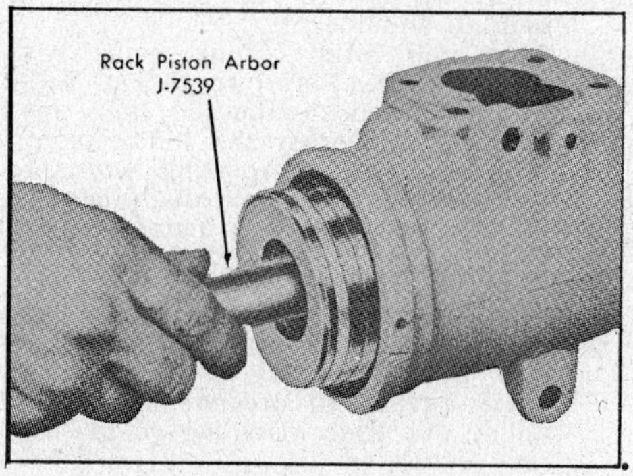

Fig. SR4 Installing rack-piston arbor tool in end of worm

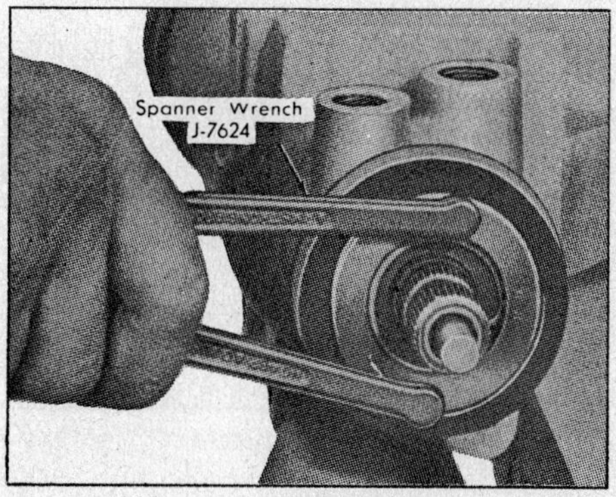

Fig. SR5 Removing adjuster plug

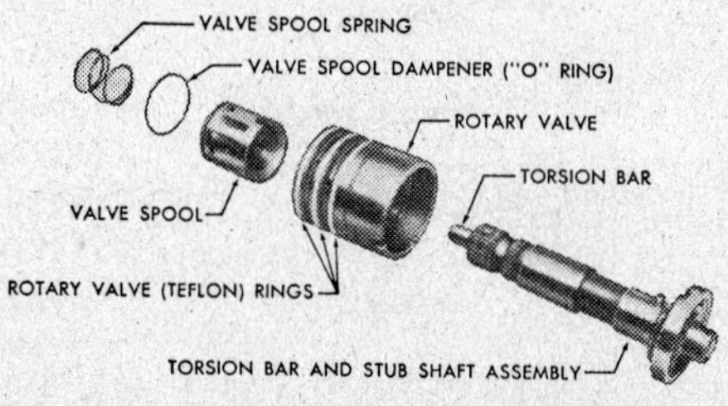

Fig. SR6 Rotary valve parts

Disassemble Rotary Valve

1. If used, work spool spring onto bearing diameter of stub shaft and remove spool spring.
2. Tap end of stub shaft gently against workbench to remove valve spool.

Because of the slight clearance between valve body and spool, the slightest cocking of spool may jam it in valve body. If slight cocking occurs, make a gentle attempt to reverse removal procedure. If this does not free spool, it has become cocked in valve body bore and may be removed later.

3. Remove and discard valve spool dampener "O" ring.
4. Remove stub shaft, torsion bar (small diameter bar extending through stub shaft) and valve cap by tapping end of torsion bar lightly with a plastic hammer. This will dislodge cap from valve body cap pin, Fig. SR6. Do not disassemble stub shaft as these parts are pinned together and serviced only as an assembly.
5. If valve spool has become cocked as mentioned above, first inspect parts to determine in which direction the spool is cocked. A few very light taps with a plastic hammer should align and free the spool in the bore. Remove and discard "O" ring dampener seal from spool.
6. Carefully remove valve body Teflon rings and ring back-up "O" ring seals.

Disassemble Housing

1. Remove pitman shaft outer dust seal retaining ring, Fig. SR8.
2. Remove outer dust seal.
3. Remove seal (double lip) by inserting offset screwdriver between seal and back-up washer and prying out of housing.
4. Remove back-up washer.
5. Remove seal (single lip) by cutting and collapsing seal.
6. Remove pitman shaft needle bearings (if necessary) with a suitable driver.
7. If connectors are to be removed, tap threads in holes of connectors using 5/16-18 NF tap. Remove connectors by using threaded bolt into tapped holes with washer and nut as extractor.

ASSEMBLE SUB-ASSEMBLIES

Refer to Fig. SR2 and lubricate all parts as they are assembled.

1. Screw lash adjuster through side

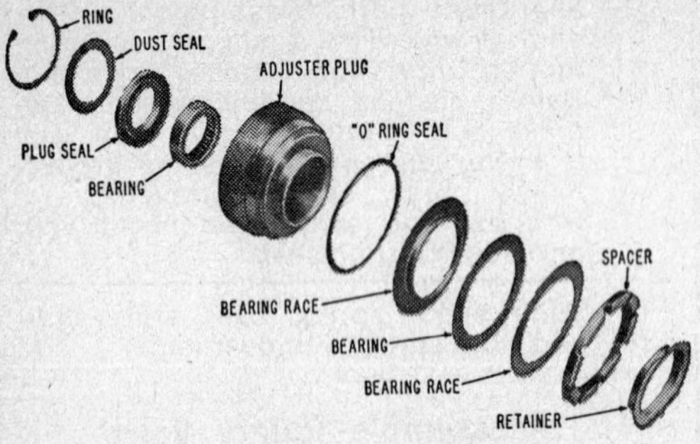

Fig. SR7 Adjuster plug parts

cover until cover bottoms on pitman shaft gear. Install lash adjuster lock nut while holding lash adjuster with 7/32" allen wrench.

2. Mount adjuster plug in vise with soft jaws.
3. If it has been removed, assemble needle bearing by pressing towards thrust bearing end of adjuster plug against identification end of bearing. End of bearing to be flush with bottom surface of stub shaft seal bore.
4. Install stub shaft far enough to provide clearance for dust seal and retaining ring. Install new dust seal with rubber surface outward. Install new retaining ring.
5. Assemble large thrust bearing race, thrust bearing, small race and spacer (with grooves up) on adjuster plug, Fig. SR7, and secure with retainer.

Assembly Rotary Valve

1. Assemble one valve body Teflon ring back-up "O" ring seal in each groove in valve body, being sure seals do not become twisted.
2. Assemble valve Teflon rings in ring grooves over "O" ring seals by carefully slipping rings over valve body. The rings may appear loose or twisted in the grooves but the heat of the oil during subsequent operation will cause them to straighten.
3. Install valve spool dampener "O" ring seal in valve spool groove, being sure it is not twisted.
4. Insert stub shaft through valve spool, engaging spool locking pin.
5. Pull valve spool and stub shaft assembly into valve body, aligning stub shaft notch and valve body pin.
6. If used, slide spool spring over stub shaft and place into position.

---

7. Lubricate cap-to-worm "O" ring and install in valve body. *During assembly of the valve, if the stub shaft and cap is allowed to slip out of engagement with the valve body pin, the spool will be permitted to enter the valve body too far. The dampener "O" ring seal may expand into valve body oil grooves, preventing removal of spool.* If this happens, remove spool spring and disassemble rotary valve. Press on spool until "O" ring seal is cut and can be removed. Install new "O" ring and reassemble.

Assemble Housing

1. With stamped end of needle bearing facing outward, drive bearing into bore from outside of housing until flush. Make sure bearings rotate freely.
2. Lubricate cavity between lips of pitman shaft (double lip) seal with power steering fluid.
3. Lubricate and install pitman shaft seals as shown in Fig. SR8. Make sure seal lips are properly positioned, retaining ring is seated.
4. If connectors were removed, install new ones by driving them into place.

Assemble Rack-Piston and Worm

1. Lubricate and install new ring back-up seal and Teflon piston ring on rack-piston nut, being careful ring and seal do not twist during installation.
2. Insert worm into rack-piston nut to bearing shoulder.
3. Align ball return guide holes with worm groove. Referring to Fig. SR9, load specified number of balls into guide hole nearest the teflon piston ring while slowly rotating worm to left, feeding balls through circuit, Fig. SR10. If balls are installed properly, the worm should turn out of the rack-piston nut.
4. Fill one-half of ball return guide with remaining balls. Place other guide over balls and plug each end with heavy grease to prevent balls from falling out when installing guides into rack-piston nut.
5. Insert guides into guide holes of rack-piston nut. Make sure black ball in guide is installed next to white ball in rack piston. Guides should fit loosely.

Check Worm Preload

The worm groove is ground with a high point in the center. When the rack-piston nut passes over this high point, a preload of a 1 to 4 inch pounds torque should be obtained.

1. With worm pointing up, clamp rack-piston nut in a vise with soft jaws (do not hold rack-piston nut in area of Teflon ring).
2. Place valve on worm, engaging worm drive pin.
3. Rotate worm until it extends $1\frac{1}{4}$" from rack-piston nut to thrust bearing face, Fig. SR11. This is the center position.
4. Attach an inch pound torque wrench with socket on stub shaft, Fig. SR11. Oscillate wrench through a total arc of approximately 60° in both directions several times and take a reading. The highest reading obtained with worm rotating should be between 1 and 4 inch pounds. Take a torque reading pulling the torque wrench to the right and a reading pulling the wrench to the left. Total both readings and take one-half of this total as the average torque. *Do not use a torque wrench having maximum torque reading of more than 100 inch pounds.*

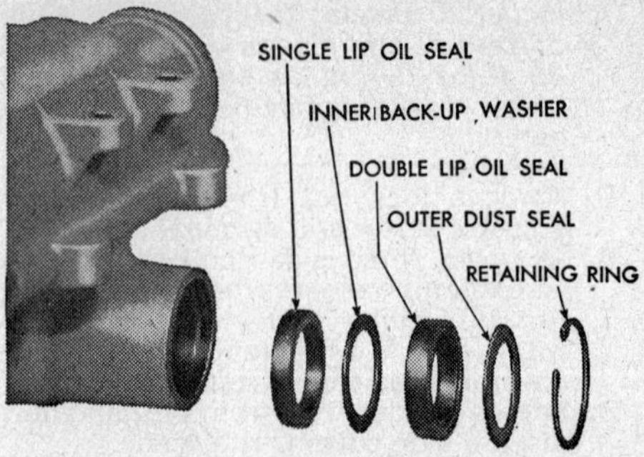

Fig. SR8 Pitman shaft seals and washers

Refitting Rack-Piston Balls

Do not refit balls unless the steering is loose. If such is the case, a thrust adjustment and over-center adjustment should correct the problem if it lies in

Model	Year	Rack-Piston Ball Load
American Motors	1969-73	16
	1974-75	18
Buick	1969-70 Constant Ratio	16
	1969-73 Variable Ratio	18
	1974-75	17
Cadillac	1969-75	17
Chevrolet	1969-70 Constant Ratio	15
	1969-75 Variable Ratio	17
Ford & Mercury	1969-73	16
	1974-75	17
Oldsmobile	1969-71	16
	1972-75	17
Pontiac	1969-71 Constant Ratio	15
	1969-75 Variable Ratio	17

Fig. SR9 Rack-Piston ball specification chart

Fig. SR10 Loading rack-piston nut

the steering gear. If balls are pitted or rough, then select the proper ball size for proper adjustment.

If the torque reading obtained above is too high or too low (on new balls only), disassemble and reassemble, using the next size smaller (or larger) balls and recheck worm preload.

A rack-piston nut with a ball size of 7 does not have a number stamped on the flat surface. For ball sizes other than 7, the ball size is stamped on the flat surface of the rack-piston nut. Ball sizes are numbered from 6 to 11, with 6 being the smallest and 11 the largest. Ball sizes are graduated in increments of .00008″ from .28117″ to .28157″.

STEERING GEAR, ASSEMBLE

Install Rack-Piston Worm and Valve

1. Install valve and worm in housing as a unit, Fig. SR12.
2. Install new "O" rings on adjuster plug.
3. Install adjuster plug on stub shaft in gear housing finger tight, Fig. SR13.
4. Holding a suitable Teflon ring compressor sleeve tightly against shoulder of gear housing, insert rack-piston nut and arbor into housing, holding arbor until it contacts worm end, Fig. SR14.
5. Holding arbor tight against worm, turn stub shaft flange and worm to draw ball nut onto worm and into housing until arbor is free. Be certain that no balls drop out.
6. Remove arbor and sleeve.

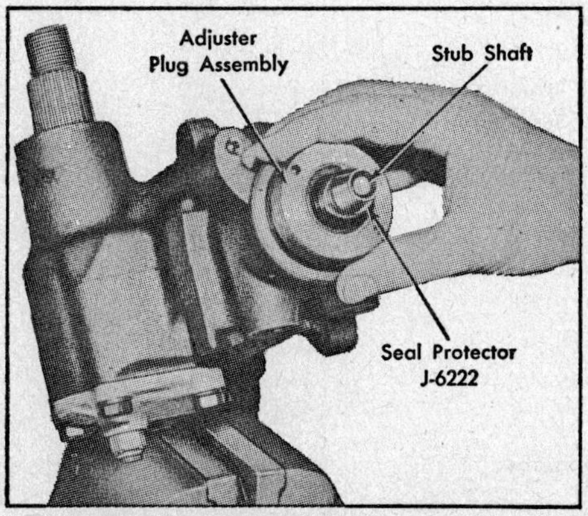

Fig. SR13 Installing adjuster plug

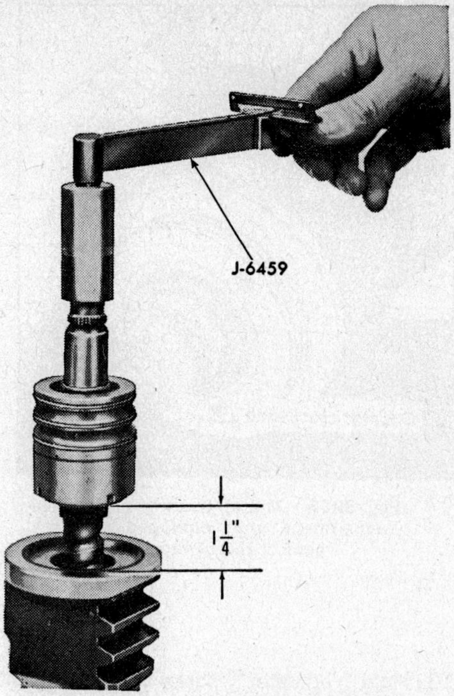

Fig. SR11 Checking worm preload

Install Rack-Piston Nut End Plug

1. Install new "O" ring seal on end plug, being careful not to allow seal to twist in groove.
2. Install end plug in rack-piston nut by pressing into place.
3. Install end plug retaining ring, being sure ring is bottomed in its groove.

Install Housing Lower End Plug

1. Install new housing end plug "O" ring seal.
2. Insert end plug into gear housing and seat against "O" ring. Slight pressure may be necessary to seat end plug properly.
3. Install end plug retainer ring so end of ring extends over and at least ½″ beyond ring removal assist hole.

Install Pitman Shaft Gear and Side Cover

1. Install stub shaft flange and turn steering worm until center groove of rack-piston is aligned with center of pitman shaft needle bearings.
2. Install new side cover "O" ring.
3. Install pitman shaft gear so that center tooth of gear meshes with center groove of rack-piston. Make sure side cover "O" ring is in place before pushing cover against housing.
4. Install and torque side cover screws to 25-30 ft. lbs. torque.

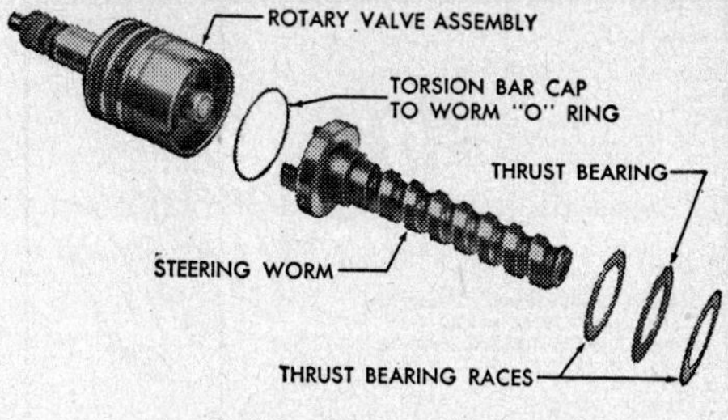

Fig. SR12 Worm shaft and rotary valve parts

Install Stub Shaft Flange

Rotate stub shaft slightly over two complete turns from either extreme right or left turn. This will place gear in center position, which will place torsion bar pin through stub shaft in a straight up and down position.

Rotate flange so the bolt with the large head is in top position. Visually align the block tooth in the serrations of the flange with the torsion bar pin and install flange on stub shaft. Tighten flange clamping bolt to a torque of 25-30 ft. lbs.

SUB-ASSEMBLIES, INSTALL

1. Position gear housing in vise with adjuster plug end facing up.
2. Lubricate worm shaft, lower thrust bearing and races with power steering fluid, then position thrust bearing and races on worm.

NOTE: Some 1974-75 steering gear assemblies may incorporate conical type thrust races, Fig. SR15. When servicing a steering gear equipped with the flat type thrust races, it is recommended that the flat thrust

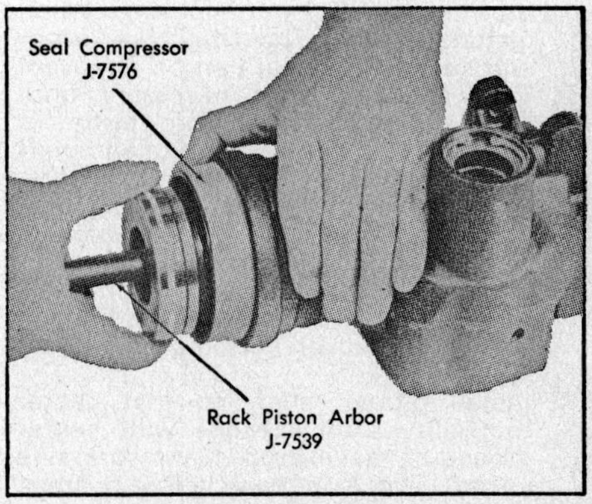

Fig. SR14 Installing rack-piston seal compressor

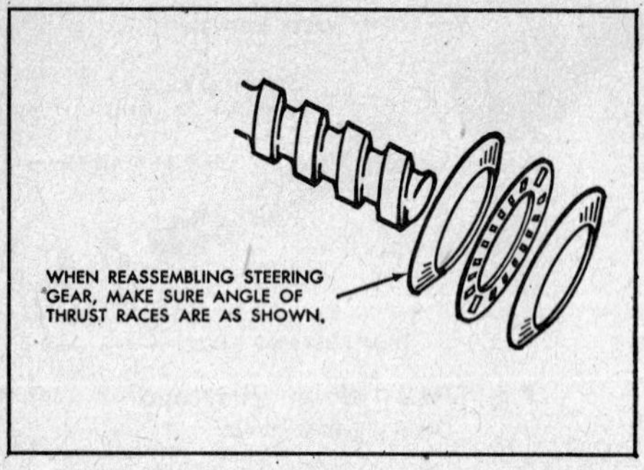

Fig. SR15 Conical type thrust races

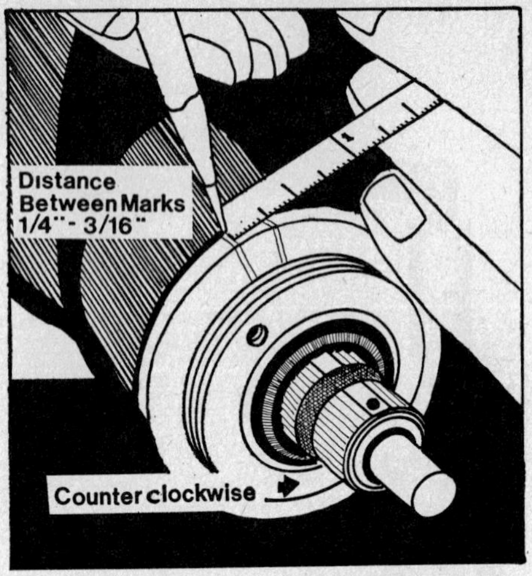

Fig. SR16 Thrust bearing preload adjustment, units equipped with conical thrust races

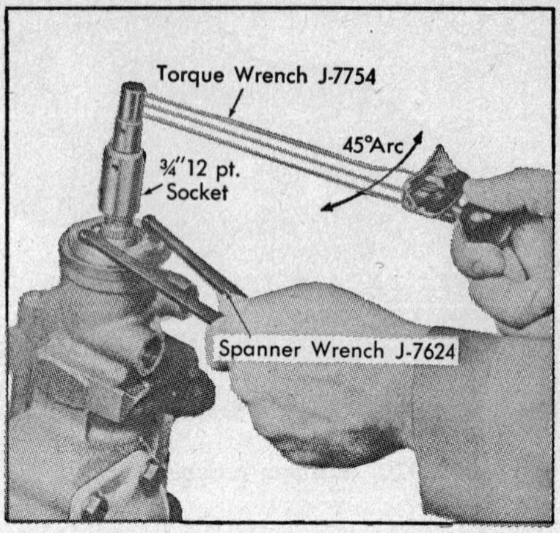

Fig. SR17 Adjusting thrust bearing preload

races are replaced with the new conical type. The thrust bearing for these units remains unchanged.

3. Align valve body drive pin on worm with narrow pin slot on valve body. Be sure O-ring seal between valve body and worm head is installed.
4. Position valve body and worm shaft in housing as a unit.

CAUTION—*Do not push against stub shaft as this might cause stub shaft and cap to pull out of valve body, allowing spool seal to slip into valve body oil grooves. Valve assembly can be installed by pushing on outer diameter of valve body housing with the fingers of both hands. Make certain that Teflon rings are not binding on inside of housing. Valve is properly seated when oil return hole in gear housing is fully visible.*

5. Place a suitable seal protector over end of stub shaft, Fig. SR13.
6. Lubricate new adjuster plug O-ring seal with power steering fluid and install in groove on adjuster plug.
7. Install adjuster plug over end of stub shaft and tighten just enough to make certain that all parts are properly seated in gear housing. Remove seal protector.
8. Install adjuster plug lock nut loosely on adjuster plug.
9. Install seal compressor in gear housing, holding it tightly against shoulder in housing, Fig. SR14.
10. Insert rack-piston into housing until arbor engages worm. Turn stub shaft clockwise, using a 3/4" wrench, to draw rack-piston into housing. When piston ring is in housing bore, arbor and seal compressor can be removed.
11. Turn stub shaft as necessary until middle rack groove in rack-piston is aligned with center of pitman shaft needle bearing.
12. Lubricate new side cover O-ring seal and install in groove in face of side cover.
13. Assemble side cover on pitman shaft by screwing cover onto adjusting screw until cover bottoms on pitman shaft.
14. Install pitman shaft so that center tooth in sector meshes with center groove of rack-piston. Make sure side cover O-ring is in place before pushing side cover down on gear housing.
15. Install side cover screws and lockwashers and tighten to 30 ft-lbs.
16. Hold adjuster screw with Allen wrench and install new adjuster lock nut halfway on adjuster screw.
17. Install rack-piston end plug in rack-piston and tighten plug to 75-80 ft-lbs.
18. Lubricate new housing end plug O-ring seal with power steering fluid and install in gear housing.
19. Insert end plug into gear housing and seat against O-ring seal. Slight tapping with a mallet may be necessary to seat end plug properly.
20. Snap end plug retainer ring into place with fingers. Slight tapping may be required to bottom ring in housing securely. Install coupling flange.

BENCH ADJUSTMENTS

Thrust Bearing Preload

NOTE: If thrust race usage is unknown (flat or conical races), perform adjustment outlined under the "Units with Conical Thrust Races" procedure. If preload measured is less than 4 inch pounds or greater than 10 inch pounds, use the adjustment procedure outlined for the particular model steering gear.

Units with Conical Thrust Races

1. Tighten adjuster plug until plug and thrust bearing bottoms.
2. Draw a line on gear housing aligning

Model	Year	Excess of Preload In.-Lbs.
American Motors	1969-73	1–3
	1974-75	3–4
Buick	1969	2
	1970-71	1–3
	1972-75	3–4
Cadillac	1969-72	1–3
	1973-75	3–4
Chevrolet	1969-70	½–2
	1971-72	2
	1973-75	3–4
Ford	1969-73	1–3
	1974-75	3–4
Oldsmobile	1969-70	1–3
	1971	2
	1972-75	3–4
Pontiac	1969-70①	½–2
	1969-70②	1–3
	1971-75	3–4

①—Catalina with A/C & all intermediates.
②—Full size exc. Catalina with A/C.

Fig. SR18 On-bench thrust bearing specification chart

with one adjuster plug hole. Draw a second line on gear housing 1/4 to 3/16 inch from first line in the counter-clockwise direction, Fig. SR16.
3. Turn adjuster plug counter-clockwise until hole aligns with second line, then tighten locknut while holding adjuster plug securely to prevent plug rotation.
4. Rotate stub shaft to right stop, then back 1/4 turn. Using an inch pound torque wrench to rotate stub shaft, measure preload. Reading should be between 4 and 10 inch pounds.

1974 American Motors, Buick, Cadillac, Oldsmobile & Pontiac

1. Tighten adjuster plug snugly, then back off 1 turn.
2. Rotate stub shaft to right stop, then back 1/4 turn on American Motors, Cadillac and Oldsmobile or 1/2 turn on Buick and Pontiac units.
3. Measure drag torque using an inch pound torque wrench to rotate stub shaft, Fig. SR17.
4. Tighten adjuster plug snugly, then while rotating stub shaft with the torque wrench, loosen adjuster plug until torque reading obtained is 3 to 4 inch pounds greater than drag torque.
5. Tighten locknut and recheck preload.

1972-73 Oldsmobile; 1973 Cadillac & Chevrolet

1. Tighten adjuster plug snugly, then back off 1/4 turn.
2. Measure drag torque using an inch pound torque wrench to rotate stub shaft, Fig. SR17.
3. While rotating stub shaft with the torque wrench, tighten adjuster plug until torque reading obtained is 3 to 4 inch pounds greater than drag torque.
4. Tighten locknut and recheck preload.

1969-71 Oldsmobile; 1969-72 Cadillac & Chevrolet; 1969-73 American Motors, Buick & Pontiac; 1969-74 Ford

1. Tighten adjuster plug snugly, then on all models except 1971-73 Pontiac, back off 1/8 turn. On 1971-73 Pontiac, back plug off 1/2 turn.
2. On all models, rotate stub shaft to right stop, then back 1/2 turn.
3. Measure drag torque using an inch pound torque wrench to rotate stub shaft, Fig. SR17.
4. Referring to Fig. SR18, tighten adjuster plug while rotating stub shaft with torque wrench until torque reading obtained is specified inch pounds greater than drag torque.
5. Tighten locknut and recheck preload.

Over-Center Preload

1. Rotate stub shaft from stop to stop, counting number of turns, then turn back exactly 1/2 number of turns to locate center of travel.
2. Measure combined ball and thrust bearing preload using an inch pound torque wrench to rotate stub shaft through the over center position.
3. Referring to Fig. SR19, tighten adjusting screw while rotating stub shaft with torque wrench until torque reading obtained is specified inch pounds greater than combined ball and thrust bearing preload, Fig. SR20. Note that total steering gear preload must not exceed specifications.

Model	Year	Excess of Ball & Thrust Bearing Preload In.-Lbs.	Steering Gear Preload Maximum In.-Lbs.
American Motors	1969-73	4–8	—
	1974-75	①	②
Buick	1969	8	18
	1970-72	8–10	18
	1973-75	①	②
Cadillac	1969-73	8–10	20
	1974-75	①	②
Chevrolet	1969-74	3–6	14
	1975	4–5	14
Ford	1969	4–8	14
	1971-74	4–8	18
	1975	①	②
Oldsmobile	1969-70	4–8	16
	1971	3–6	16
	1972-75	①	②
Pontiac	1969-70	4–8	18
	1971-75	①	②

①—New gear, 4–8 in. lbs.; used gear 4–5 in. lbs.
②—New gear, 18 in. lbs.; used gear, 14 in. lbs.

Fig. SR19 On-bench over-center preload specification chart

TROUBLE SHOOTING

Hard Steering

1. Frozen steering shaft bearings.
2. Lower coupling flange rubbing against adjuster.
3. Steering adjustment tight.
4. Low oil level in pump.
5. Loose pump belt.
6. Insufficient oil pressure.
7. Improper tire inflation.
8. Flow control valve sticking.
9. Pump leaking internally.
10. Steering gear leaking internally.

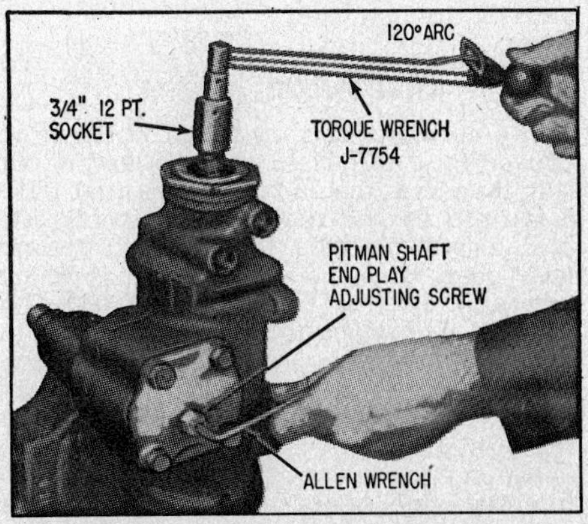

Fig. SR20 Adjusting over-center preload

Poor Return of Steering

1. Frozen steering shaft bearings.
2. Lower coupling flange rubbing against adjuster.
3. Tires not inflated properly.
4. Incorrect caster and toe-in.
5. Tight steering linkage.
6. Steering gear misalignment.
7. Tight suspension ball joints.
8. Steering adjustment tight.
9. Thrust bearing adjustment tight.
10. Tight sector-to-rack piston adjustment.
11. Rack piston nut and worm preload too tight.
12. Sticky valve spool.
13. Lack of lubricant in suspension joints and steering linkage.
14. Steering wheel rubbing against directional signal housing.
15. Rubber spacer in shift tube is rubbing steering shaft.

Car Leads to One Side

1. Front end misalignment.
2. Unbalanced or badly worn valve.
3. Improper tire inflation.

Momentary Increase In Effort When Turning Wheel Fast

1. Low oil level in pump.
2. Pump belt slipping.
3. High internal leakage.

External Oil Leaks

1. Loose hose connections.

2. Damaged hose.
3. Side cover O-ring seal.
4. Pitman shaft seals.
5. Housing end plug seal.
6. Adjuster plug seals.
7. Torsion bar seal.
8. Defective housing.

Steering Gear Noise

1. A rattle or chuckle noise caused by loose over-center adjustment.
2. A hissing sound caused by gear being loose on frame.

Excessive Wheel Kickback or Loose Steering

1. Lash in steering linkage.
2. Air in system.
3. Excessive lash between pitman shaft sector and rack piston.
4. Loose thrust bearing adjustment.
5. Ball nut and worm preload.
6. Steering gear loose on frame.
7. Steering linkage worn.
8. Improper front wheel bearing adjustment.
9. Poppet valve worn.

Wheel Surges or Jerks

1. Loose pump belt.
2. Fluid level low.
3. Steering linkage hitting engine oil pan at full turn.
4. Sticky flow control valve.

Hard Steering When Parking

1. Loose pump belt.
2. Low oil level in reservoir.
3. Lack of lubrication in linkage or front suspension.
4. Tires not properly inflated.
5. Insufficient oil pressure.
6. Low oil pressure due to restriction in hoses.
7. Low oil pressure due to worn piston ring or scored housing bore.
8. Pressure loss due to leakage at valve rings, valve body-to-worm seal or rack piston end plug seal.
9. Pressure loss due to loose fit of spool in valve body or leaky valve body.

Valve Squawk

1. Cut or worn dampener ring on valve spool.
2. Loose or worn rotary valve parts.
3. Loose belt.
4. Glazed belt.
5. Flow control valve sticking or binding.

No Effort Required to Turn

1. Broken torsion bar.

REAR AXLE SERVICE

TROUBLE DIAGNOSIS

External Noises

When a rear axle is suspected of being noisy, it is advisable to make a thorough test to determine whether the noise originates in the tires, road surface, front wheel bearings, engine, transmission, or rear axle assembly. Noise that originates in other places cannot be corrected by adjustment or replacement of parts in the rear axle assembly.

Road Noise

Some road surfaces, such as brick or rough surfaced concrete, cause noise that may be mistaken for tire or rear axle noise. Driving on a different type of road, such as smooth asphalt or dirt, will quickly show whether the road surface is the cause of noise. Road noise usually is the same on drive or coast.

Tire Noise

Tire noise may easily be mistaken for rear axle noise even though the noisy tires may be located on the front wheels. Tires worn unevenly or having surfaces of non-skid divisions worn in saw-tooth fashion are usually noisy, and may produce vibrations that seem to originate elsewhere in the vehicle. This is particularly true with low tire pressure. Some designs of non-skid threads may be more noisy than others, even when tires are new.

Test for Tire Noise

Tire noise changes with different road surfaces, but rear axle noise does not. Temporarily inflating all tires to approximately 50 pounds pressure, *for test purposes only*, will materially alter noise caused by tires, but will not affect noise caused by rear axle. Rear axle noise usually ceases when coasting at speeds under 30 mph; however, tire noise continues but with lower tone as car speed is reduced. Rear axle noise usually changes when comparing "pull" and "coast" but tire noise remains about the same.

Front Wheel Bearing Noise

Loose or rough front wheel bearings will cause noise that may be confused with rear axle noises; however, front wheel bearing noise does not change when comparing "pull" and "coast". Light application of brake while holding car speed steadily will often cause wheel bearing noise to diminish as this takes some weight off the bearings. Front wheel bearings may be easily checked for noise by jacking up the wheels and spinning them, also by shaking them to determine if bearings are loose.

Engine & Transmission Noises

Sometimes a noise that seems to originate in the rear axle is caused by the engine or transmission. To determine which unit is actually causing the noise, observe approximate car speeds and conditions under which the noise is most pronounced; then stop the car in a quiet place to avoid interfering noises. With transmission in neutral, run engine slowly up and down through engine speeds corresponding to car speed at which the noise was most pronounced. If a similar noise is produced with car standing, it is caused by the engine or transmission, and not the rear axle.

Rear Axle Noises

If a careful test shows that noise is not caused by external items described above, it is then reasonable to assume that noise is caused by the rear axle assembly. The rear axle should be tested on a smooth level road to avoid road noise. It is not advisable to test rear axle for noise by running with rear wheels jacked up.

Noises in the rear axle assembly may be caused by a faulty propeller shaft, faulty rear wheel bearings, faulty differential or pinion shaft bearings, misalignment between two universal joints, or worn differential side gears and pinions. Noises may also be caused by mismatched, improperly adjusted or scored ring and pinion gear sets.

Rear Wheel Bearing Noise

A rough rear wheel bearing produces a vibration or growl that continues with car coasting and transmission in neutral. A brinelled rear wheel bearing causes a knock or click approximately every two revolutions of the rear wheel, since the bearing rollers do not travel at the same speed as the rear axle or wheel. With rear wheels jacked up, spin rear wheels *by hand* while listening at the hubs for evidence of rough or brinelled wheel bearing.

Differential Side Gear & Pinion Noise

Differential side gears and pinions seldom cause noise since their movement is relatively slight on straight ahead driving. Noise produced by these gears will be most pronounced on turns.

Pinion Bearing Noise

Rough or brinelled pinion bearings produce a continuous low pitch whirring or scraping noise starting at relatively low speed.

Side Bearing Noise

Defective side bearings produce a continuous rough noise of a lower pitch as compared to pinion bearing noise. The noise may fluctuate during rear wheel bearing test.

Ring & Pinion Gear Noise

Noise produced by the ring and pinion gear set generally shows up as drive noise, coast noise, or float noise.

1. Drive noise is produced during vehicle acceleration.
2. Coast noise is noise produced while allowing car to coast with throttle closed.
3. Float noise is noise occurring while just maintaining constant car speed at light throttle on a level road.

Drive, coast and float noises will vary in tone with speed, and will be very rough and irregular if the differential or pinion shaft bearings are rough, worn or loose.

Fig. 1 Marking differential adjusting nuts and bearing caps before dismantling the assembly

Body Boom Noise or Vibration

Such objectionable noises at 55-65 mph can be caused by an unbalanced propeller shaft. Excessive looseness at the spline can contribute to this unbalance. Other items that may also contribute to noise problem are:

1. Undercoating or mud on the propeller shaft causing unbalance.
2. Shaft balance weights missing (if equipped).
3. Shaft damage such as bending, dents or nicks.
4. Tire-type roughness. Switch tires from a known good car to determine tire fault.

Check for Propeller Shaft Vibration

Objectionable vibrations at high speed (65 mph or higher) may be caused by a propeller shaft that is out of balance. Out of balance may be due to a bent shaft.

To determine whether propeller shaft is causing vibration, drive car through speed range and note speed at which vibration is most pronounced. Shift transmission into a lower gear and drive car at same speed as when vibration was most pronounced in direct drive. Note effect on vibration.

If the vibration is still present at the same engine speed whether in direct drive or in a lower gear, since the propeller shaft speed varies, this cannot be the fault. If the vibration decreases or is eliminated in the lower gear, then the propeller shaft is out of balance and should be rebalanced.

Oil Leaks

1. Oil coming from the drain hole under axle housing at the brake backing plate is caused by a leaking axle shaft seal or a leaking wheel bearing inner gasket.
2. Oil coming from between rear pinion flange slinger and carrier is caused by a leaking pinion seal.
3. Even after a point of leakage has been determined, it is difficult to tell whether the oil is leaking past the lip of the seal or past the outside diameter of the seal. Therefore, it is a good idea to make sure the leak is stopped by using a non-hardening sealing compound around the outside diameter of the seal.
4. Oil coming from the junction of the axle tubes to carrier, or at puddle weld holes, are difficult to repair. Welding should never be done in attempts to stop leaks in these areas. If leak is severe, complete housing should be replaced.

General Instructions

In rear axle service work, there are minor and major operations. Minor operations, such as the removal of axle shafts, bearings and oil seals, can be performed in most cases without removing the differential carrier from the axle housing. (For detailed instructions covering the removal, installation and adjustment of axle shafts, bearings and oil seals, see the car chapters.)

Major operations, such as the replacement of ring gear and pinion, must be performed with the differential carrier removed from the axle housing. The preliminary work required to remove the differential carrier depends upon the construction employed. In some cases, a removable cover is attached to the back of the axle housing, while in others it is welded to the housing. (Procedure for removing differential carriers and axle assemblies is given in the car chapters.)

Disassembly

Before dismantling the carrier, bear in mind that the factory adjustment of ring gear and pinion sets should not be disturbed unless absolutely necessary. For example, should it become necessary to tear down a differential to replace some part other than a ring gear and pinion, do not disturb the drive pinion adjustment in either disassembling or assembling the

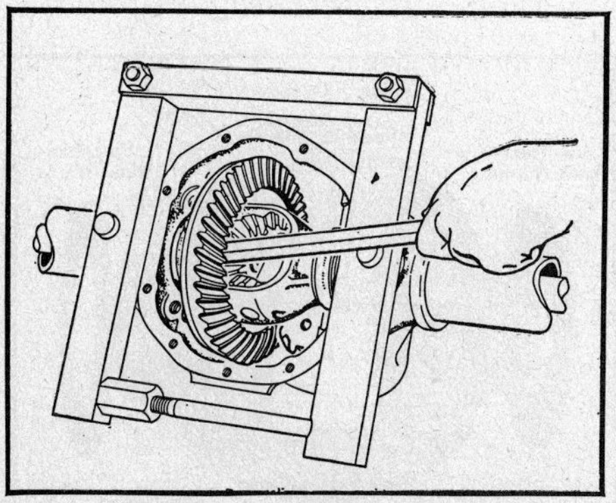

Fig. 3 Prying differential from carrier on type having shim adjusted bearings. If spreader tool shown is not available, pry out differential as shown in Fig. 4

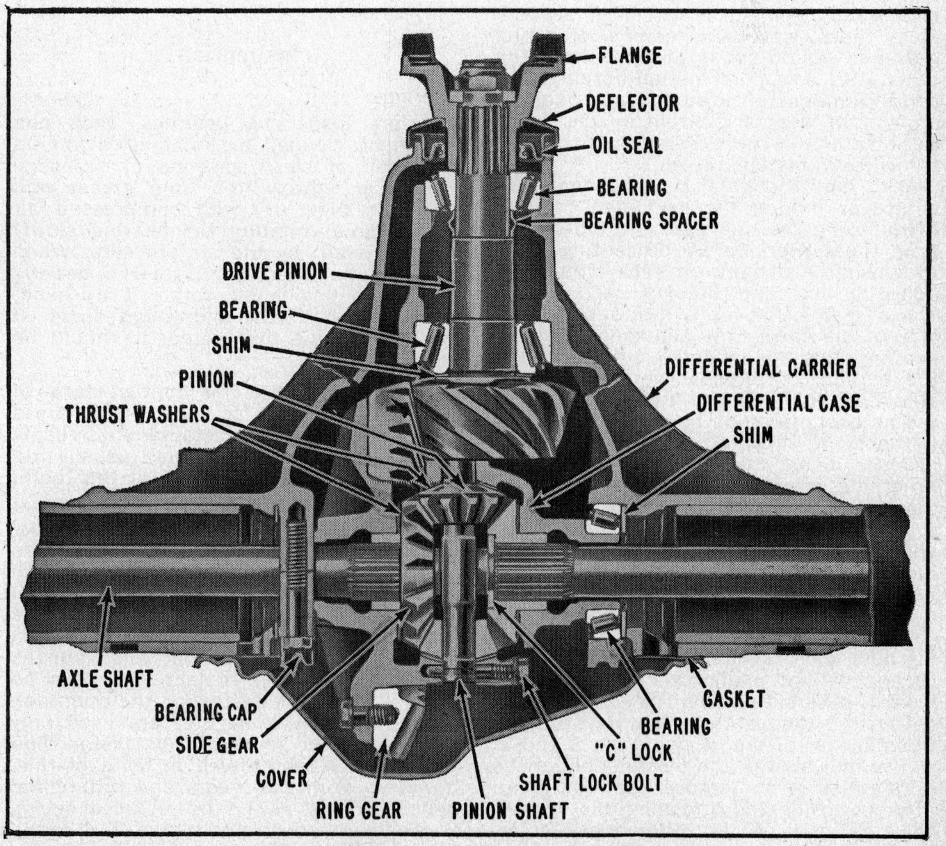

Fig. 2 Differential carrier of the type having shims for adjusting differential bearings

Fig. 4 Using two pry bars to remove differential from carrier on type having shim-adjusted bearings

unit—assuming of course that it was correct beforehand. Instead, only adjust the ring gear when assembling the unit—which will increase the chances of securing a setting precisely the same as the original setting. In addition, backlash between the ring gear and pinion should be checked and noted before dismantling the unit in order that the gear may be adjusted to the original backlash when making the final adjustments.

It should be remembered that ring gears and pinions are matched at the factory and should remain matched in service. Many times, when one member of a ring gear and pinion set is divorced from the other, the new combination results in a certain amount of noise and an abnormal amount of wear and strain on the teeth. Because of this, service replacements are furnished in matched sets.

With the differential carrier mounted in a vise or fixture, the next step consists of removing the ring gear and differential from the carrier. Before proceeding, however, scratch a mark on each differential adjusting nut and bearing cap, Fig. 1. These marks not only locate on which side of the carrier the adjusting nuts and bearing caps are mounted but they show the number of threads exposed on each adjusting nut, and the exact position of the nuts with relation to the bearing caps. If the same ring gear and pinion are to be used again, these marks will serve to bring the original adjustment into being with a minimum of time and effort. (If the differential bearings are adjusted by shims, Fig. 2, simply tie each set of shims together and carefully lay them aside until needed).

After removing the differential bearing cap screws, take off the caps and adjusting nuts and lift the differential out of the carrier. Where shims are used for adjusting differential bearings, remove the caps from the bearings and pry the differential out of the carrier. Figs. 3 and 4.

The first step in the disassembly of the differential is to remove the ring gear. After removing the attaching cap screws, drive the gear off the differential case by tapping lightly with a soft-faced hammer or mallet. Drive out the differential pinion shaft lock pin, Fig. 5, and push out the shaft (sometimes a lock screw is used). The side gears, pinions, thrust washers, thrust block (if used) and spacers will then be loose and may be picked out of the differential case, Fig. 6. The differential bearings may then be removed from the case with a suitable puller.

The next step is to disassemble the drive pinion. However, it should be removed and disassembled only when it is necessary to replace a part, or to examine closely all parts in connection with a major inspection and overhaul operation.

In this connection, however, if the differential carrier was satisfactory from the standpoint of noise before being taken apart, or if it is disassembled for the replacement of a part other than a ring gear and pinion, the location and quantity of shims should be noted. The shims then should be cleaned, tagged and tied together and carefully hung on a hook or set to one side so that, upon reassembly, the original setting may be obtained without delay.

When removing and replacing pinion oil seals, care should be taken to avoid cutting the seal on the sharp edges of the pinion shaft threads, keyway or splines. If the seal is cut and is not replaced, serious leakage of lubricant is almost certain to result.

The best way to guard against cutting the seal is to wrap a piece of shim stock around the pinion shaft so that the oil seal will not come in contact with any sharp edges when the seal is slid off and on. Be sure to blunt or dull the edges of the shim stock so it will not cut the seal.

Inspection

Bearings

Before inspecting bearings, each one should be cleaned by swishing it around in a pail of clean gasoline or kerosene until it is entirely free from grease and oil. Then blow dry with compressed air (if available) rotating the bearing slowly by hand—not by the air pressure. When cleaned and dried, lubricate the bearing with clean engine oil, rotate it by hand, and check for wear and rough spots. If any defects are discovered, it should be discarded.

Score marks on the contact face of gear teeth show up as light colored areas which are caused by instantaneous fusing of the mating gear teeth, and usually run from the bottom to the top of the tooth.

A scoring condition may be the result of insufficient lubrication, the improper lubricant for the load required for the gears, dirty lubricant, or improper tooth contact. The gears in the differential which usually cause noisy operation from these conditions are the ring gear and pinion. The side gears and differential pinions rarely give trouble. In fact, they can be badly worn without affecting the operation of the differential, as they are used only when one rear wheel travels faster than the opposite rear wheel. In spite of this, however, worn side gears and differential pinions should always be replaced during an overhaul operation. If not, the condition will become worse and eventually, through failure of these parts, another tear-down will be required.

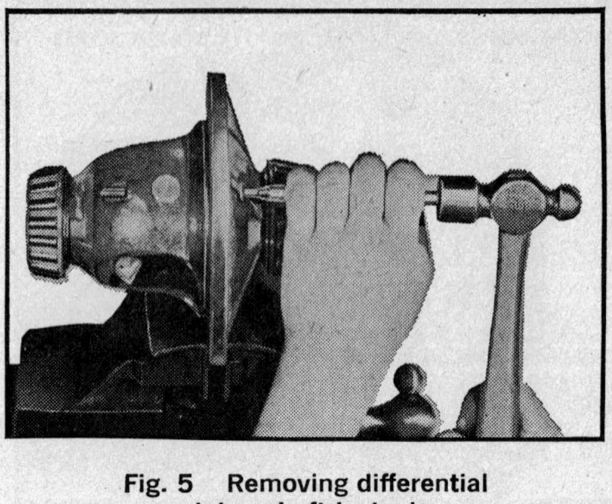

Fig. 5 Removing differential pinion shaft lock pin

Assembly

The sequence of operations in assembling any piece of mechanism is usually performed in the reverse order from which it was disassembled. Differential carrier units are no exception although the sequence given below may be changed to suit shop conditions. For example, instead of starting with the differential case and ring gear, the drive pinion may be set up and adjusted in the carrier first.

Checking Diff. Case Runout

Mount the differential case on V blocks, Fig. 7, and check the run-out of the ring gear pilot and case flange. To make this check, mount the dial indicator as shown, with the indicator needle set to zero and the contacting pin just touching the surface to be checked.

Rotate the differential case in the V blocks for one complete revolution and note the amount of deflection on the needle. If the deflection is in excess of the maximum allowable run-out (usually .002 inch) replace the differential case or true it up on a lathe.

Differential Parts

All parts of the differential, including gears, differential case, carrier, and the axle housing, should be washed thoroughly in clean gasoline or kerosene. Particular attention should be given to remote corners and recesses in the carrier and axle housing to be sure that all muck, grit, gear chips and old grease are removed.

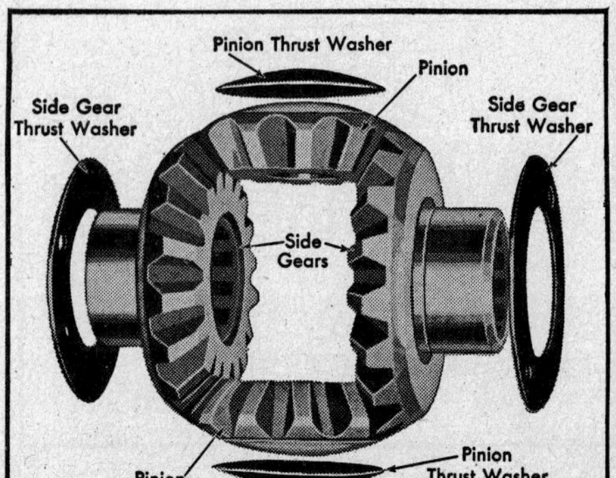

Fig. 6 Layout of differential parts (typical)

Fig. 7 Checking ring gear mounting flange for run-out

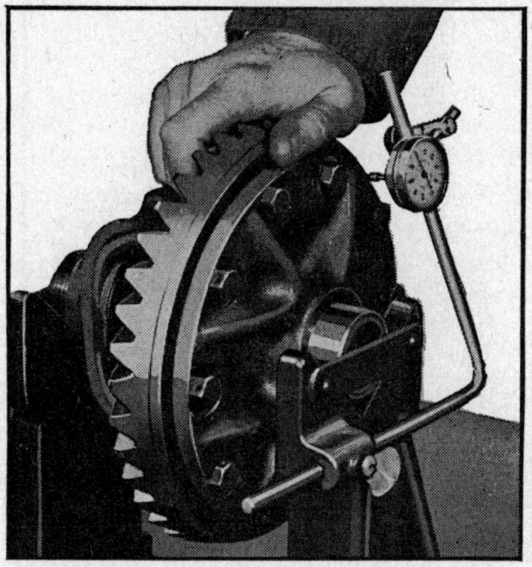

Fig. 8 Checking back face of ring gear for run-out

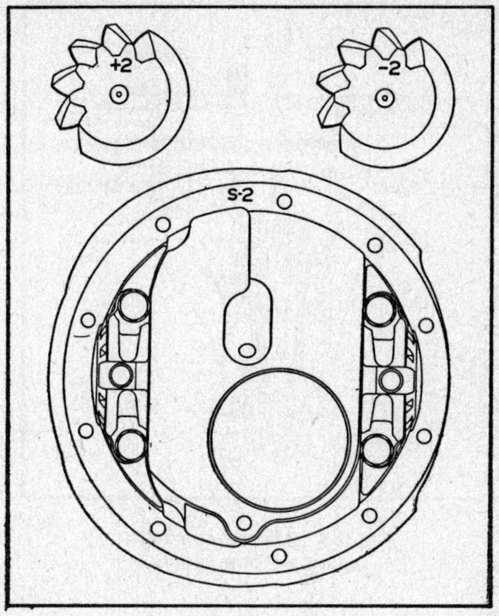

Fig. 10 Example of pinion depth markings

Examine the ring gear and pinion for cracked, chipped or scored teeth. Inspect the bearing surfaces, also the side gears and differential pinion surfaces of the differential case for evidence of roughness or scoring. Inspect the thrust faces of the differential pinions and check their fit on the pinion shaft. Inspect the thrust surfaces of the side gears and check their fit on the axle shafts. Any gears showing any of the above conditions should be scrapped. If the case is equipped with side gear and pinion gear thrust washers, inspect the washers and replace if worn.

Ring Gear, Install

If the ring gear is the bolt-on type, cut the heads from five or six $1\frac{1}{2}$ inch cap screws having the same thread as the ring gear cap screws. (These are to be used as guide pins and can be kept for future jobs.) Cut screwdriver slots in the end of these guide pins and screw them in every other hole in the ring gear. Be sure the back face of the ring gear and differential flange face are clean and free from nicks and burrs, and any rust preventive film that may be present.

Slip the ring gear with the guide pins on the differential case and tap it in place with a mallet. Install the regular cap screws in the open holes and draw them up evenly and tightly. Remove the guide pins and install the other cap screws. New lock washers should be used and if the cap screw heads are drilled, use new wire to lock them in place.

Mount the assembly in V blocks, Fig. 8, and check the run-out of the back face of the ring gear in the same manner as described above for the differential case. If the deflection of the needle is in excess of the maximum allowable run-out (usually .006 inch) the ring gear may not be bolted evenly against the flange face. Sometimes this check is made with the side bearings mounted on the differential case hubs. If so, and the run-out is excessive, the bearings may be mounted incorrectly or they may be cocked on the hubs.

Assembly Of Differential

When installing the differential gears and thrust washers (if used), coat these parts with differential lubricant to facilitate holding them in place while replacing the differential pinion shaft.

Slip the side gear thrust washers and side gears in the differential case. Roll the differential pinions in place, rotating the side gears until the hole through the center of the pinions is in line with the hole for the pinion shaft in the case. Install the axle shaft thrust block (if used) and push the pinion shaft in place so that the hole in the shaft for the lock pin or screw registers with the hole in the case. Install the lock pin. If a lock screw is used, install a new lock washer and set the screw up tight.

Differential Bearings, Install

Install the differential side bearings, with the thick side of the inner races toward the case. Be sure they are not cocked and that the inner races seat squarely against the hubs of the differential case. This completes the assembly of the differential unit.

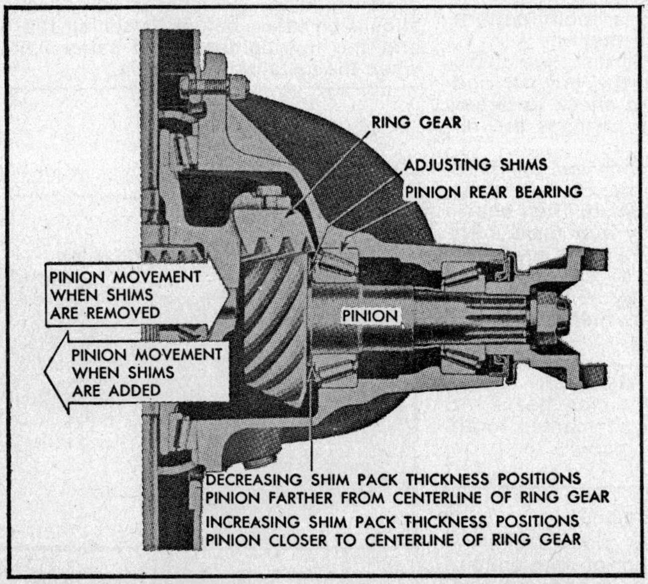

Fig. 11 Effect of shim pack thickness on pinion position

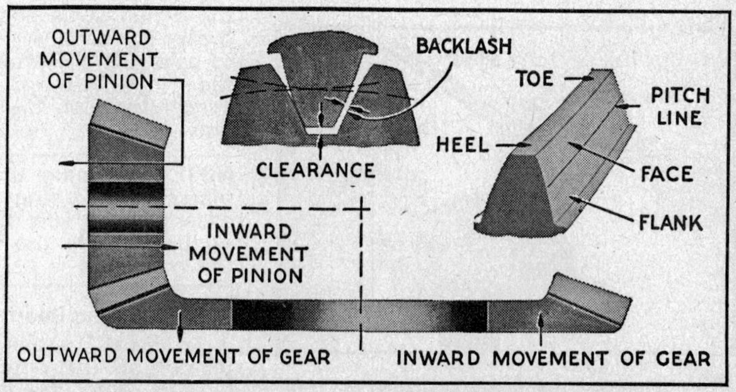

Fig. 12 Showing gear movement and tooth terminology

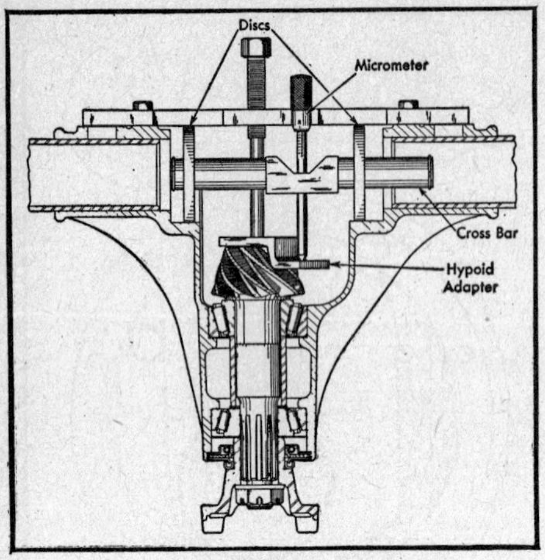

Fig. 13 Micrometer type pinion setting gauge

Fig. 14 Dial indicator type of pinion setting gauge

Fig. 15 Checking end play of drive pinion

Drive Pinion, Install

Due to the fact that all gears are not made on the same machine, a definite setting for pinion depth in the carrier to bring the ring gear and pinion exactly on the same cone centers is determined in manufacture by testing each ring gear and pinion set. These setting marks are usually etched on the toe end of the pinion and read "O", or plus (+) or minus (−) a nominal amount in thousandths of an inch. Fig. 10 is an example showing a pinion marked plus .002″ (+2) and another marked minus .002″ (−2).

A plus mark indicates that the pinion must be moved outward (shims added); a minus mark indicates that the pinion must be moved inward (shims removed), Fig. 11. Fig. 12 explains gear movements and tooth terminology.

In order to make these depth settings in accordance with the marks on the pinion, a pinion setting gauge is helpful. There are several of these gauges commercially available, each differing in detail but not in principle. Measurements are taken either by a micrometer or by a dial gauge.

As shown in Figs. 13 and 14, these gauges are provided with a measuring rod for setting the pinion, and are made adjustable to accommodate all width differential carriers. Manufacturers of these gauges usually provide complete instructions for their use, and data giving the nominal dimension for the various model pinions on which the gauge can be used.

NOTE: If a pinion setting gauge is not available, a satisfactory job can be done by referring to the *Rear Axle Specifications* tables in the car chapters and then proceed as follows:

1. Measure the thickness of the original pinion depth shim removed from the rear axle.
2. Use the same thickness shim upon installation of the new pinion. If any further shim change is necessary, it will be indicated in a tooth pattern check as described further on.
3. If the original shim is lost, substitute a nominal shim for the original and use the tooth pattern check to determine if further shim changes are required.

Install the pinion in the carrier, using the shims from the old pinion. Then check the pinion shaft end play and make necessary corrections. Two methods are used to establish the preload: by shims behind the front bearing, or by a collapsible spacer between the front and rear bearings.

NOTE: On some units, there are timing marks on the ring and pinion gears. At assembly, be sure to mesh marked tooth on pinion with the two marked teeth on the ring gear.

Shim Adjusted Bearing Preload

To check the end play, place a dial indicator on the outer end of the pinion, Fig. 15. Work the pinion in and out and note the amount of end play indicated on the gauge. If the job calls for a preload of .002″, for example, and the indicated end play is .004″, .006″ inch of shimming should be removed to give the required preload of .004″.

Collapsible Spacer Preload Adjustment

Fig. 16 shows the method of setting the preload when the specification calls for the adjustment in inch-pounds; Fig. 17 shows the spring scale method when the specification calls for pounds pull adjustment.

To adjust the preload, draw up the pinion flange nut until the spacer starts to buckle. Check the pull with the torque wrench or the spring scale.

This adjustment must be made every time the flange nut is removed or loosened. If the adjustment is made with the differential unit in the carrier, the rear wheels must be jacked off the floor.

NOTE: A new spacer is required when a new ring gear and pinion set is installed, when one or both pinion bearings is changed, when a new carrier casting is used, or when the pinion depth adjusting shim thickness is increased.

When the same pinion flange is removed and reinstalled, as when an oil seal is replaced, checking the pinion bearing preload is not necessary if care is taken to tighten the flange nut exactly to its previous position. Should a new pinion flange be required, a torque wrench or spring scale reading (whichever is specified) should be taken before loosening the nut, and the nut tightened the same amount when the installation is made.

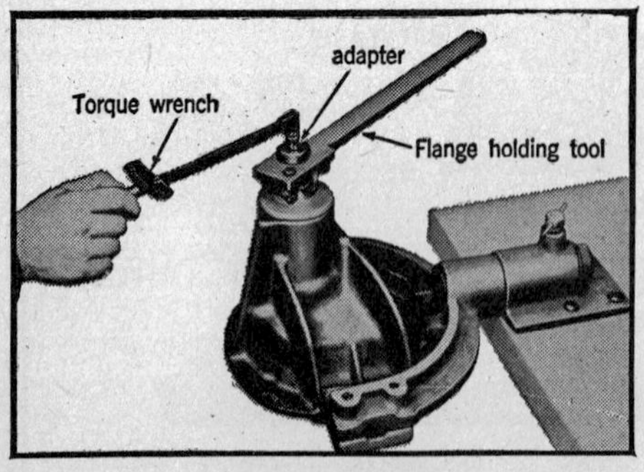

Fig. 16 Using torque wrench to adjust pinion bearing preload

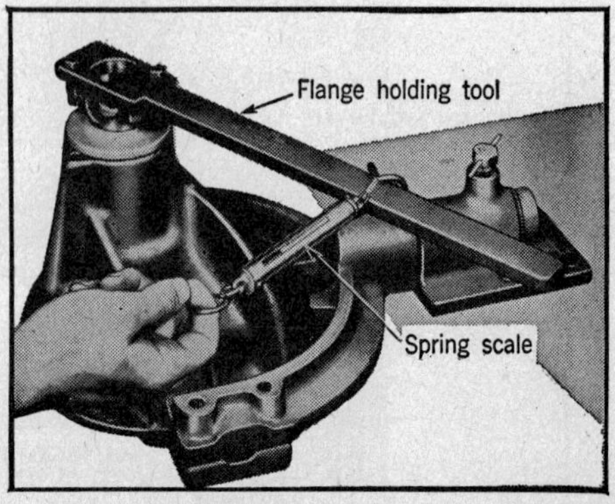

Fig. 17 Using spring scale to adjust pinion bearing preload

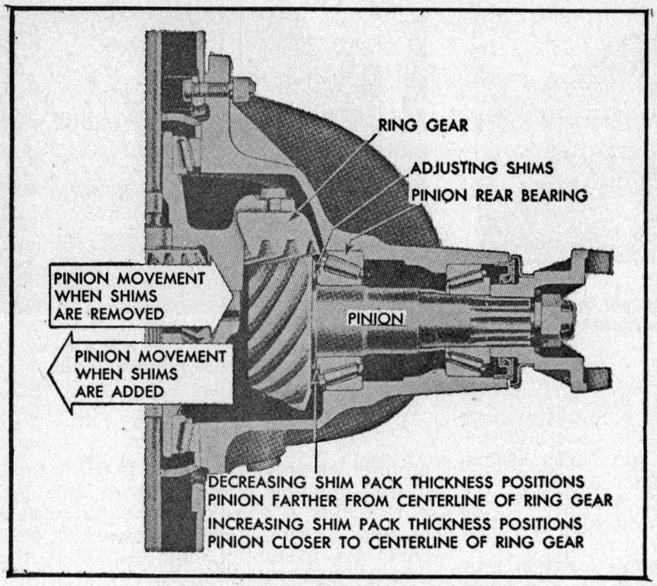

Fig. 18 Effects of shim pack thickness on pinion position

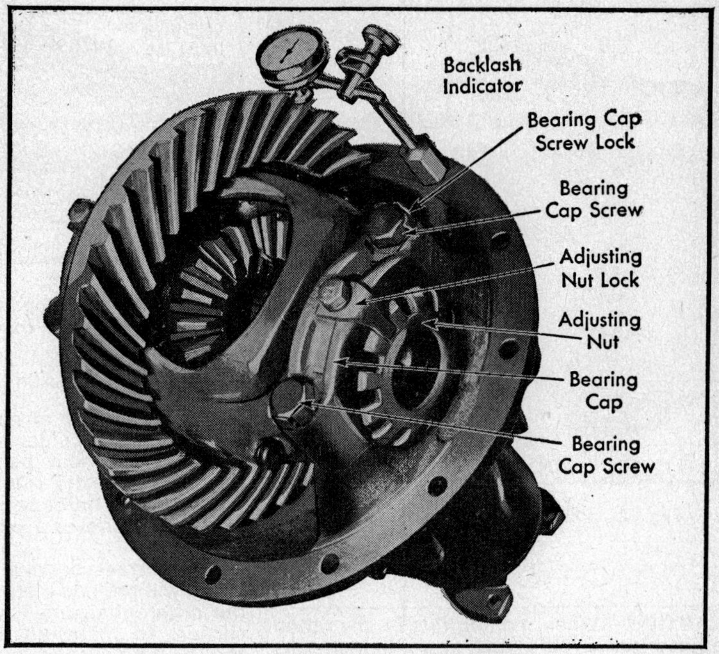

Fig. 19 Checking backlash between ring gear and pinion

Rear Axle Adjustments

Four adjustments can be made which will affect tooth contact patterns: side bearing preload, drive pinion bearing preload, ring gear and pinion backlash, and position of drive pinion in carrier. The effect of bearing preloads are not readily apparent on (hand loaded) red lead tests; however, these adjustments should be within specifications before proceeding with backlash and drive pinion adjustments.

Backlash is adjusted by means of side bearing adjusting nuts or shims which move the entire case and ring gear assembly closer to or farther away from the drive pinion. The adjusting nuts or shims are also used to set the side bearing preload.

The position of the drive pinion is adjusted by increasing or decreasing the shim pack between the pinion head and the inner race of the rear bearing. The shim pack is used in the differential to compensate for manufacturing tolerances. Increasing shim pack thickness will move the pinion closer to the centerline of the ring gear, Fig. 18. Decreasing shim pack thickness will move pinion farther away from center line of ring gear.

Fig. 20 Measuring clearance between bearing cup and differential case on shim-adjusted differential bearings

Differential Bearing Preload

Upon completion of the pinion setting operation, assemble the differential unit in the carrier and proceed as follows:

If the same ring gear and pinion are being used again, and if the adjustment was correct before the unit was dismantled, adjust the bearings according to the marks, Fig. 1, that were scratched on the bearing caps and adjusting nuts. If the differential bearings are adjusted by shims, assemble the unit with the same quantity of shims used originally.

If a new ring gear and pinion are being used, or if an adjustment of the old gears was necessary, proceed as follows, being sure to leave off the drive pinion bearing oil seal to eliminate any drag from this source when making adjustments.

To make the adjustment on units having adjusting nuts, place the differential in the carrier and install the adjusting nuts, taking care to slide these nuts alongside the bearings so that the threads on the nuts fit properly into the threads of the carrier. Install the bearing caps, being sure the marks on the caps line up with the marks on the carrier.

Start the cap screws and drop the bearing caps on the adjusting nuts, using the screws as a pilot so as not to cross the threads. Tighten the cap screws and then loosen them from 1 to 1½ turns.

Loosen the right adjusting nut until the ring gear and differential case are loose in the bearings—usually four to five notches. Tighten the left-hand nut against the bearing race, removing all backlash between the ring gear and pinion, then back off three or four notches.

Tighten the right-hand nut until the bearing race starts to turn—indicating tension on the bearing. Continue to tighten one to two notches more, then back off the nut until the bearing race stops turning. Retighten this same nut until the bearing race again starts to turn, continuing for two notches more and whatever further tightening is necessary to bring the slot in the nut in line with the bearing cap lock, Fig. 19. Tighten the bearing cap screws and check the ring gear and pinion backlash.

Backlash, Adjust

To check the backlash, mount a dial indicator on the carrier, Fig. 19. To assure an accurate reading, make sure the dial indicator is square with the face of the gear tooth it contacts, and that the pointer is set to zero.

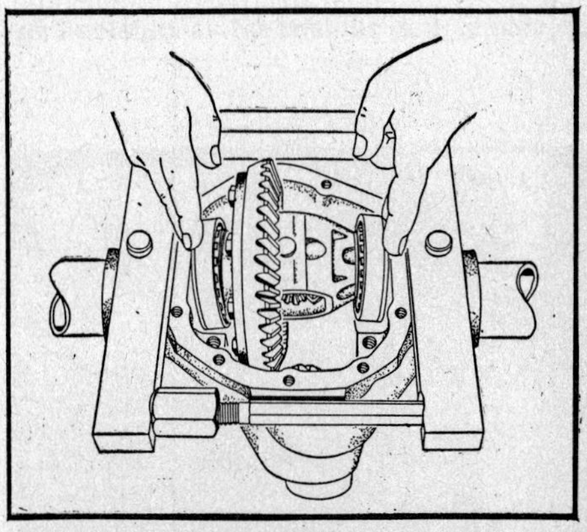

Fig. 21 If spreader tool shown is not available, tap bearing cups lightly with a mallet to install differential unit

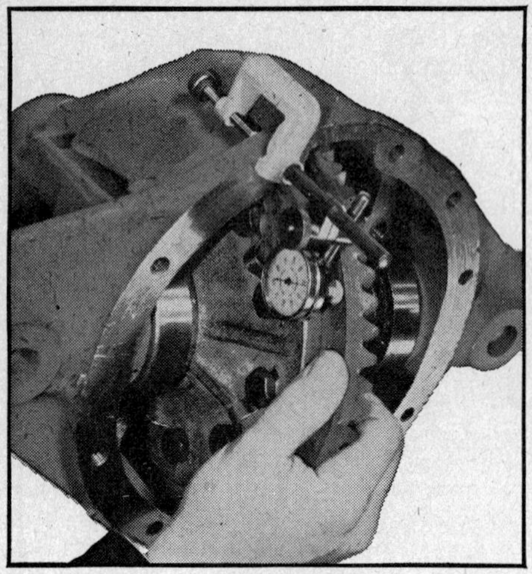

Fig. 22 Checking ring gear for run-out

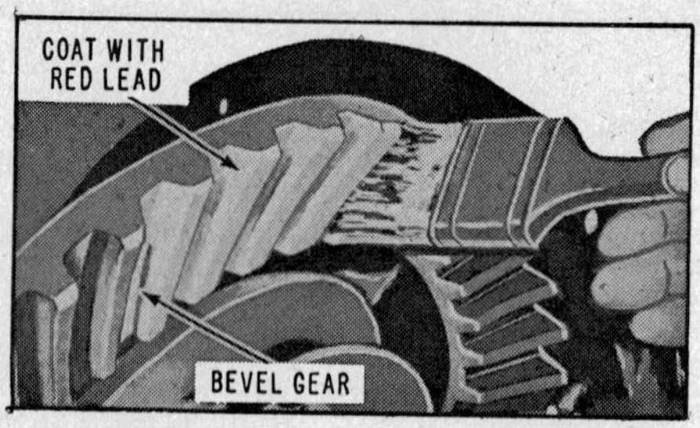

Fig. 23 Painting ring gear teeth to check tooth contact pattern

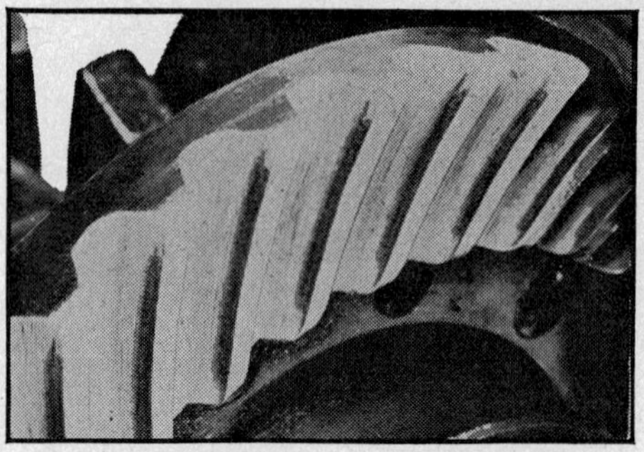

Fig. 24 Contact too high and narrow. Move pinion toward center of axle

CAUTION—When making this or any other adjustment of the ring gear, tighten the cap screws of the differential bearing caps after each adjustment has been made. This will avoid "creeping" of the ring gear in any direction during the process. However, do not lock the cap screws or adjusting nuts until all adjustments have been completed.

Check the backlash by rocking the ring gear back and forth and noting the amount indicated on the gauge. If the backlash is not within the limits specified by the car manufacturer, an adjustment is required. If the backlash is greater than the allowable maximum, loosen the right-hand nut one notch and tighten the left-hand nut one notch. If the lash is less than the allowable minimum, loosen the left-hand nut one notch and tighten the right-hand nut one notch.

When the adjustment is correct, tighten the bearing caps securely, then recheck the backlash. Install the adjustment nut locks. If the cap screw heads are drilled, lock them with new wire. If nut locks are employed, Fig. 19, be sure to set the ears against the sides of the cap screw heads. Use new lock washers if this locking method is used.

If the differential bearings are adjusted by shims, the adjustment is obtained by the shims located at each side of the case between the shoulder of the case and the inner race of the bearing. These shims also establish the ring gear position with the pinion. Therefore, backlash must be checked whenever a bearing adjustment is made.

The correct bearing adjustment is one which will provide a specified preload when the differential unit is assembled into the carrier. To make an adjustment, remove the bearing cones and shims, using a suitable puller. Reinstall the bearing cones without the shims and place the assembly in the housing with the bearing cups. Force the unit to one side and check the clearance between the bearing cup and differential case with feeler gauges, Fig. 20. When the clearance is determined, add sufficient number of shims to obtain the specified bearing preload.

Remove the differential bearings and install equal thicknesses of shims on each side and replace the bearings. Reinstall the differential unit in the housing. This operation is made easier by cocking the differential bearing cups slightly when the differential is placed in the housing and then tapping them lightly with a mallet, Fig. 21. During this operation, be sure the ring gear teeth mesh with the pinion teeth before tapping the bearings in place. After the cups are in position, install the bearing caps. The caps and the gasket surface of the housing are sometimes marked on one side with a horizontal numeral and on the other side with a vertical numeral. The position of the numerals should coincide when installing the bearing caps.

Tighten the bearing caps securely and if there is any backlash between the ring gear and pinion, check the back face of the ring gear for run-out, Fig. 22, in a manner already described. If the total indicator reading is in excess of .003 inch, it indicates a sprung differential case or an improperly installed ring gear. In either case, the assembly must be taken apart and rechecked thoroughly.

Mount the backlash gauge indicator on the carrier, and start checking the backlash between the ring gear and pinion. In the event the lash is not within the proper limits it will be necessary to change the arrangement of the shims back of the bearings—which are supplied in thicknesses of .003, .005, .010 and .030 inch, any combination of which may be used to obtain the desired result. Make corrections in backlash, bearing in mind that shims removed from one side must be installed at the opposite side so that the total shim thickness of the right and left side will remain unchanged, and the bearing preload adjustment undisturbed.

NOTE—If a pinion setting gauge was not used to locate the pinion, tooth contact between the ring gear and pinion should be checked at this time, and in the following manner:

Gear Tooth Contact

Paint about 12 teeth of the ring gear, Fig. 23, with red lead of suitable consistency and use it sparingly. When the pinion is rotated, the paint is squeezed away by contact of the teeth, leaving bare areas showing the exact size, shape and location of the contacts. Sharper contact can be obtained by applying a small amount of resistance to the ring gear when rotating the pinion. This is done by

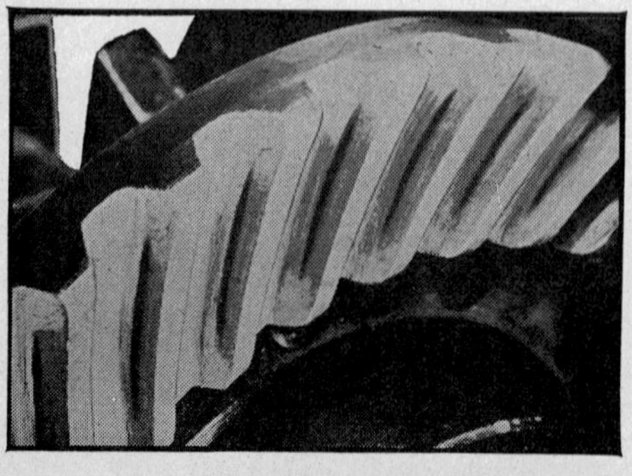

Fig. 25 Contact too low and narrow. Move pinion away from center of axle

Fig. 26 Short toe contact. Move ring gear away from pinion

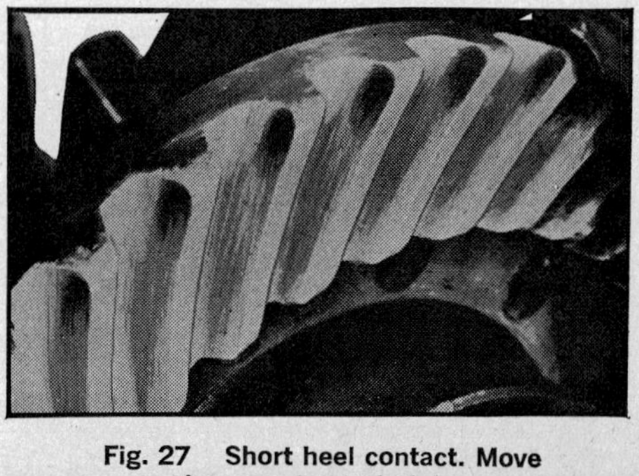

Fig. 27 Short heel contact. Move ring gear toward pinion

Fig. 28 Correct tooth contact

holding a block of wood against the back of the ring gear, and using a wrench to rotate the pinion. By examining the contact marks, it can be determined whether contact is too high or too low, or too near the heel or toe.

If the contact is high and narrow, Fig. 24, remove shims so as to move the drive pinion in toward the toe of the gear teeth (toward the center of axle) to lower the contact area. This adjustment will decrease the backlash between the ring gear and pinion, which can be corrected by moving the gear away from the pinion. Several adjustments of both ring gear and pinion may be necessary before correct contact and backlash are secured, whether it is this one or those described below.

A low, narrow contact, Fig. 25, requires the addition of shims to move the pinion out from the gear (away from the center of the axle) a sufficient amount to move the contact area to the proper location. To obtain correct backlash, move the ring gear in toward the pinion.

Fig. 26 shows a short toe contact which is corrected by moving the ring gear away from the pinion. After establishing the correct contact, the proper backlash is obtained by moving the pinion in toward the toe of the ring gear teeth (toward center of axle).

Fig. 27 shows a short heel contact which is corrected by moving the ring gear in toward the pinion. Correct backlash is then obtained by moving the pinion out toward the heel of the ring gear teeth (away from center of axle).

When adjustments have been properly made, the contact shown in Fig. 28 will be secured. This adjustment provides a quiet operating gear set which, because the load is distributed over the teeth within the proper area, will provide normal service.

Tighten the bearing caps securely, then recheck the backlash. If correct, install the adjusting nut locks, lock the adjustment and install the differential carrier unit in the axle housing, being sure the gasket is in good condition.

Some manufacturers provide thrust blocks of several thicknesses to maintain correct axle shaft end play. Where this construction is used, install the axle shafts and check the clearance as shown in Fig. 29. If the clearance does not come within the allowable limits, a new thrust block should be installed.

Fig. 29 Measuring end play between axle shaft and thrust block

Lubrication

The differential carrier should be lubricated with the exact quantity, grade and type lubricant recommended by the manufacturer of the particular unit being serviced. The housing should never be filled above the level of the bottom of the filler plug hole.

Over-filling causes blown grease seals at the wheels, resulting in leakage and an accumulation of dirt and grease around the wheels, tires and brakes. On the other hand, if the lubricant is not kept up to the proper level, gears and bearings may run dry, resulting in excessive wear and noise.

SERVICE NOTE—All the foregoing is general information for servicing differential carrier units used on passenger cars included in this manual. To complete the picture, however, specific data such as specifications, drive pinion set-ups, special differential designs, propeller shaft installations on torque tube drive cars pertaining to each make of car are given in the car chapters.

LOCKING TYPE DIFFERENTIALS

The main purpose of these differentials is that they supply a greater percentage of torque to the wheel with better traction and still furnishes to the wheel with poorer traction as much torque as it is able to absorb under the circumstances.

In the conventional differential the wheel easiest to turn receives the power. Therefore, when traction conditions under the rear wheels are not the same, the driving force is limited by the wheel with the poorer traction (easiest to turn) even through one wheel is on good traction surface.

These locking differentials do not permit shock loads or full engine torque to be transmitted to one axle shaft. They provide power to both rear wheels and maintain the differential action that is necessary when the vehicle is turning a corner to permit the outer wheel to turn faster than the inner wheel.

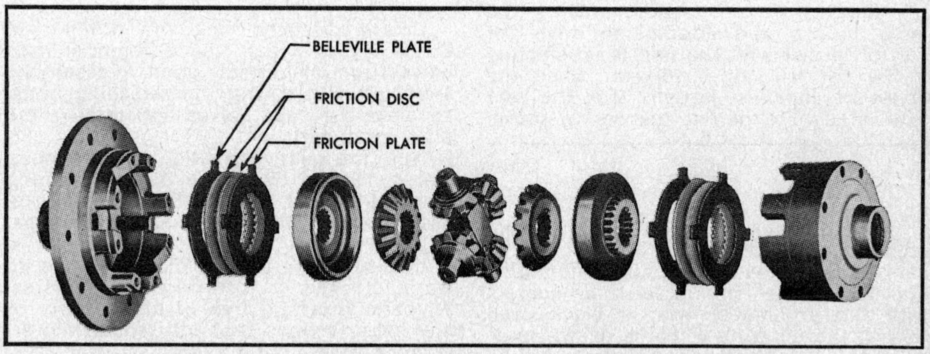

Fig. 1 Plate clutch type of POWR-LOK differential (4 stack unit) with Belleville spring plates at outside of stacks. Install Belleville plate with convex side toward case

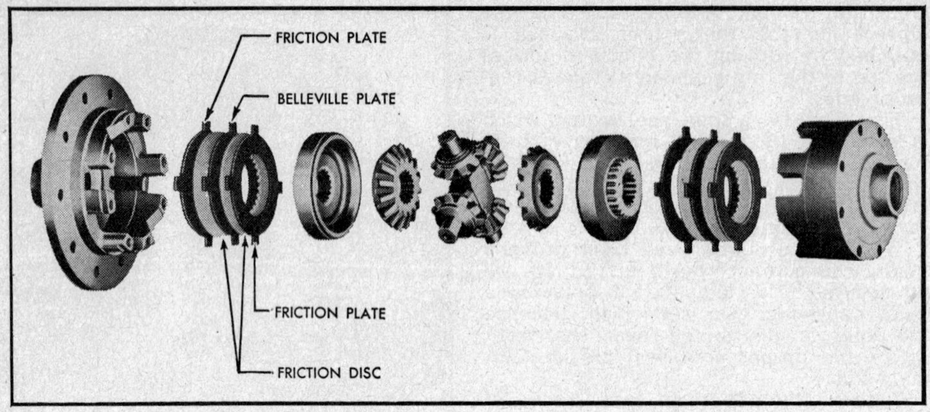

Fig. 2 Plate clutch type of POWR-LOK differential (5 stack unit) with Belleville spring plates in center of stacks. Install Belleville plate with convex side toward case.

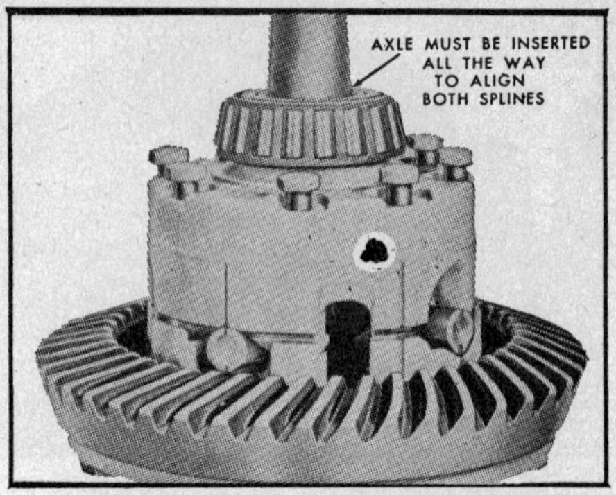

Fig. 4 Aligning splines of side gear and side gear ring before tightening attaching bolts

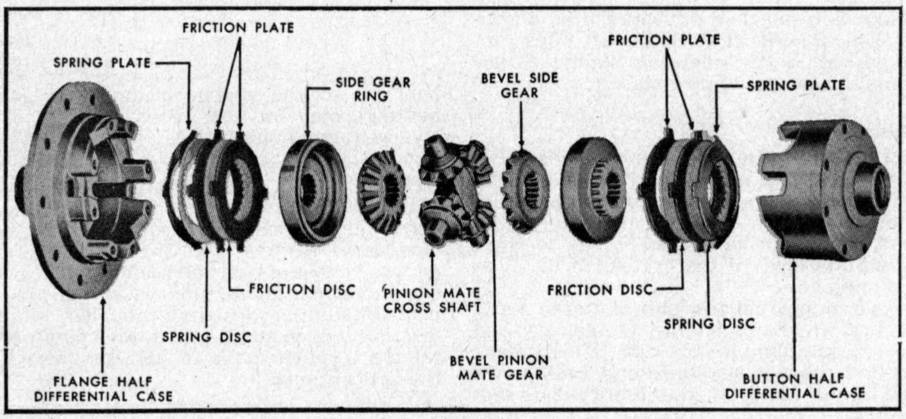

Fig. 3 Plate clutch type of POWR-LOK differential (5 stack unit) with Belleville spring plates at outside of stacks

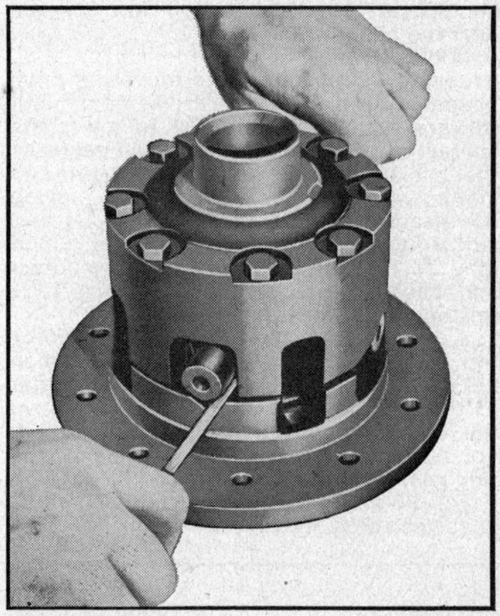

Fig. 5 Checking clearance between pinion mate shaft and V in case. Feeler gauges must be inserted on both ends of same shaft and on opposite sides of V

Operational Test

To determine if the differential is functioning properly, place a roller-mounted floor jack far outboard under the rear axle housing and slightly raise one wheel off the floor. With one person guiding the jack, another can attempt to drive the car forward slowly. The unit is satisfactory if the car will move forward since the cones or clutches partially lock the rear axle enough to permit the car to move.

NOTE

"On car" type wheel balancers are not recommended for use on rear wheels of cars equipped with a locking differential. The rear wheel will drive if in contact with the ground or with a block even though the opposite wheel is raised. *However, this type of wheel balancer may be used by removing the wheel opposite of the one being balanced with the complete assembly raised. See additional "Service Note" at the end of this chapter.*

DANA PLATE TYPE

Except Trac Loc

Differential Repairs

Before disassembling the differential, scribe a line across the differential case halves for alignment upon reassembly. Then, after removing the attaching bolts, separate the case halves and remove the differential parts.

Note the relative position of the friction plates, friction discs and Belleville spring plates (if used) so that they may be reassembled in the proper sequence, Figs. 1, 2 and 3.

Inspect all parts. See that there are no worn, cracked or distorted clutch plates. All parts must be free of nicks, burrs or any imperfections that will reduce the efficient operation of the unit.

1. Assemble differential as suggested in Figs. 1, 2 and 3.
2. Make sure the side gear retainer will rotate with a slight drag when in the case. Repeat for opposite side.
3. Install side gear on ring gear flange half.
4. With ring gear flange half in a vertical position, install one shaft and gears, making certain that notch in shaft is up.
5. Install side gear in other half of case.
6. Hold the remaining case half through the bearing trunnion and install it on the ring gear flange half, being sure scribed mark made previously is in alignment.
7. Insert axle shafts to align side gear ring with bevel side gear, Fig. 4. If this is not done there may be some difficulty when installing axle shafts during final assembly. Tighten attaching bolts evenly to 35-45 ft. lbs.
8. Check clearance between pinion mate shaft and the V of the case, Fig. 5. Do this by placing shim stock or feeler gauges on both ends of the same shaft and on opposite sides of the V. The clearance should be from zero to .020" maximum.

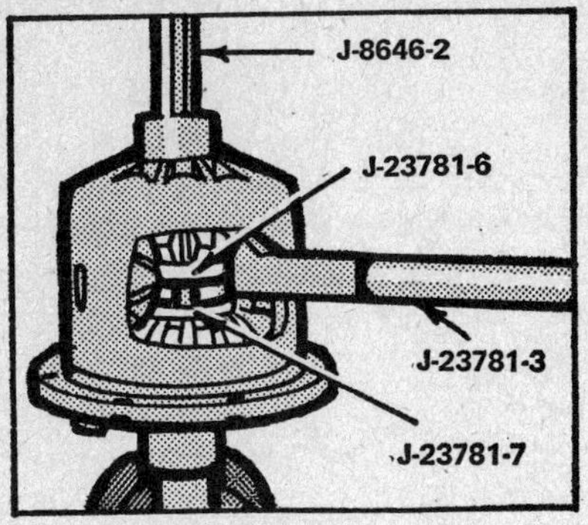

Fig. 7 Installing special tools

Installation Note

Make sure the spline end of the axle shaft does not interfere with the pinion mate shafts. This can be checked with a steel tape by measuring from the bottom of the axle shaft bearing bore to the pinion mate shafts. Then measure the axle shafts from the corresponding point of the bearing to the end of the spline. The minimum clearance required is 1/8". Grind off the spline end of the axle shaft if it is too long. Check the other axle shaft in the same manner.

CAUTION: Use rear axle lubricant of the type specially recommended for locking differentials. Failure to use the proper lubricant will result in chattering or other serious malfunction.

AMERICAN MOTORS TRAC-LOK

Disassembly

1. Remove ring gear bolts, then using a brass drift tap gear from case, Fig. 6.
2. Using two screw drivers, remove cross shaft snap rings and with brass drift, tap cross shaft from case then remove shaft thrust block.
3. Using tools in Fig. 7, position differential gears away from pinions.
4. Using a .030 inch feeler gauge, remove conical washers.
5. Remove pinion and differential gears from case.
6. Remove retaining clips from plate tangs then separate plates and discs.

NOTE: Note position of plates and discs before disassembling.

Assembly

NOTE: When lubricating parts during assembly always use special locking differential lubricant.

1. Lubricate then install Belleville spring, discs and plates on differential gear splines, Fig. 8.
2. Install retaining clips on plate tangs, ensure clips are completely seated on tangs.
3. Install differential gear and clutch packs in case, ensure retaining clips are properly seated in case recesses.
4. Install plates and discs on other differential gear and install in same manner.
5. Install tool J-23781-3 in top differential gear and tools J-2378-6 and -7 between differential gears.
6. Position pinion gears in case, Fig. 9, then with tool rotate differential gear until pinion gear holes align with case bores.
7. Lubricate and install conical washers, using a screw driver.

NOTE: Ensure pinion gear and conical washer holes are aligned with case bores.

8. Remove tools, then install shaft thrust block.
9. Lubricate and install cross shaft, ensure snap ring grooves are exposed, install snap rings.
10. Position ring gear on case, install and torque bolts to 105 ft. lbs. on American Motors models and 45-50 ft. lbs. on Checker models.

NOTE: Tighten bolts evenly so that ring gear will be properly seated on case.

FORD TRACTION-LOK

Disassembly

1. Referring to Fig. 10 remove ring gear-to-differential case bolts. Remove ring gear.
2. With case in press or using two bolts and nuts through ring gear mounting holes to compress case halves and overcome preload spring tension, loosen two Allen or Phillips head screws until only a few threads of the screws are still engaged then remove case from press and tap cover to spring it loose. Remove screws.
3. With cover facing down, remove case, preload spring plate and springs.
4. From cover, remove side gear, clutch guides, hub and clutch discs.
5. Drive pinion shaft lock pins from case.
6. With brass drift, drive long pinion shaft from case, driving from end opposite lock pin hole.

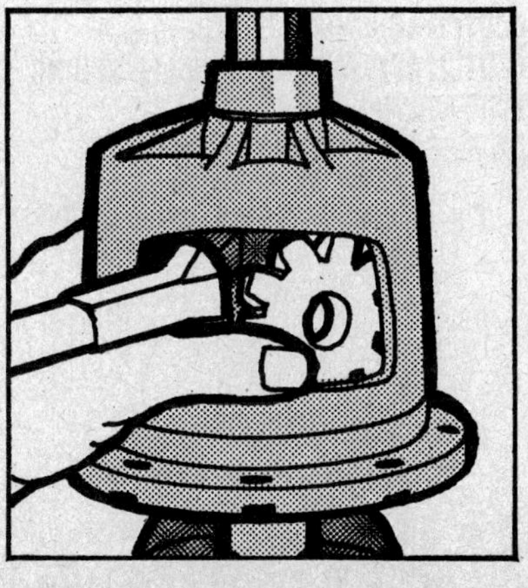

Fig. 9 Installing pinion gears

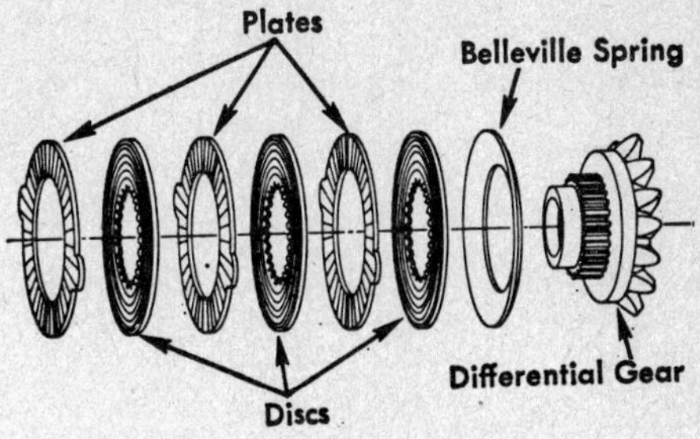

Fig. 8 Installing Belleville spring, discs and plates on differential gear

7. Remove two short pinion shafts by driving each shaft from center outward. Lift out center block, pinion gears, thrust washers and side gear.
8. Remove differential bearings.

Reassembly

1. With case mounted in a soft jawed vise, place a side gear in case.
2. Install pinion gears, center block and pinion shafts. Position center block so long shaft is driven through rough side and short shaft through machined side. Install lock pins.
3. Install pre-load springs and position pre-load plate over springs, straddling center block over narrower machined width.
4. Install selective shim in cover.
5. Starting with a composite plate, install clutch plates alternately. New plates should be soaked in gear lubricant before installation.
6. Install four guides and side gear.
7. Place two halves in press, compress springs and install two Allen or Phillips head screws and tighten evenly.
8. Install ring gear and tighten evenly to 65-80 ft-lbs.

FORD (DAP) DETROIT AUTOMOTIVE TYPE

Disassembly

1. Install differential case with cover down onto a suitable press. Apply enough pressure to hold preload springs between case and cover and remove case half screws. Remove assembly from press and remove case cover.
2. Referring to Fig. 11, remove side gear, spring and retainer, hold out ring and driven clutch assembly from either side. Remove center drive member.

Reassembly

1. Place tool T66L-4204-A2 in a vise and place right hand case over tool.
2. Referring to Fig. 11, install differential parts by first sliding right hand side gear onto tool. When installing center drive member, make sure key or slot in center cam of member is placed between opening in hold out ring.

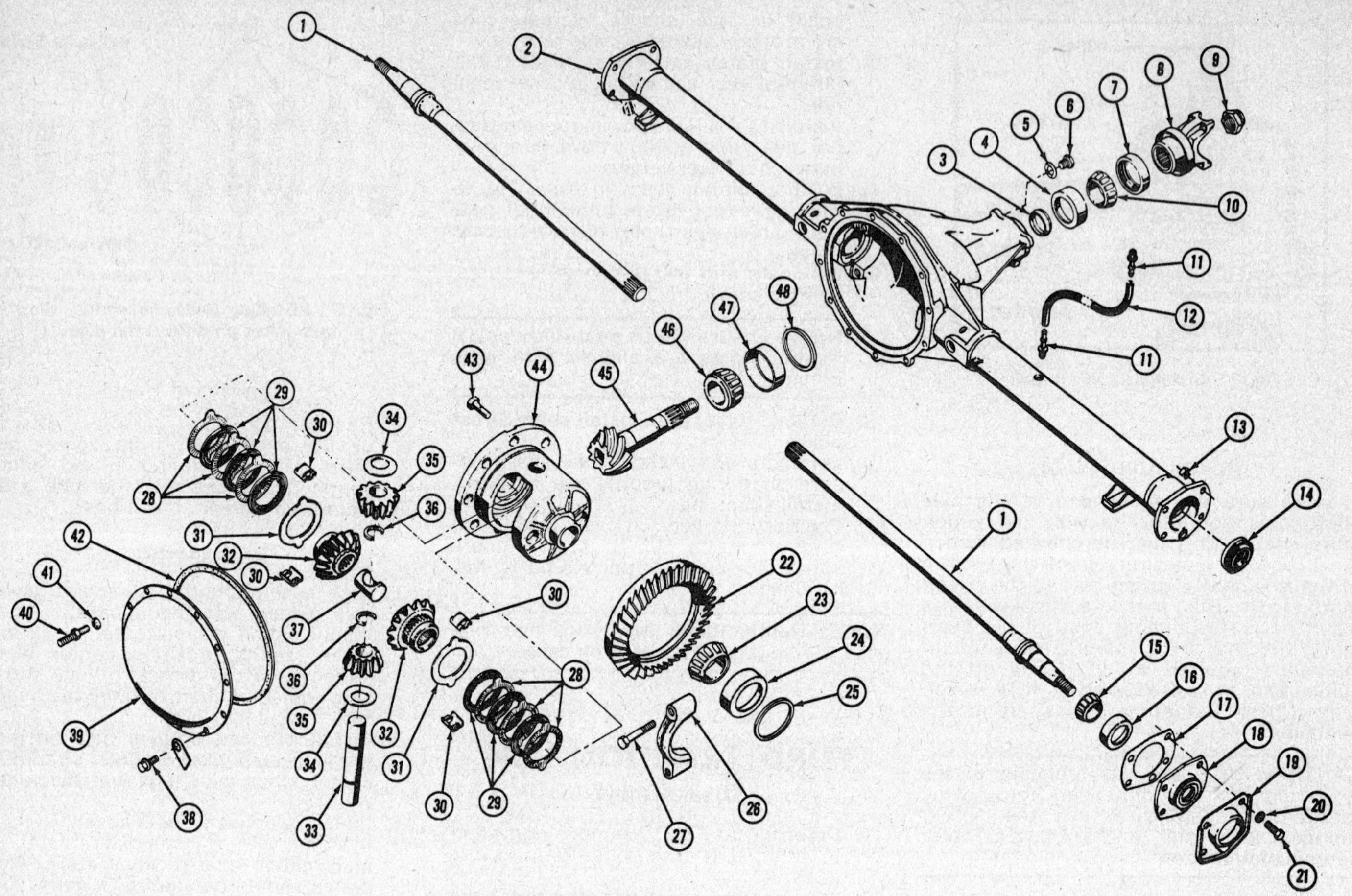

1. AXLE SHAFT
2. REAR AXLE HOUSING
3. FRONT COLLAPSIBLE SPACER
5. FILLER PLUG GASKET
6. FILLER PLUG
7. PINION OIL SEAL
8. UNIVERSAL JOINT YOKE
9. PINION NUT
10. FRONT PINION BEARING
11. BREATHER (2)
12. BREATHER HOSE
13. NUT
14. AXLE SHAFT INNER OIL SEAL
15. AXLE SHAFT BEARING
16. AXLE SHAFT BEARING CUP
17. AXLE SHAFT BEARING SHIM
18. AXLE SHAFT OIL SEAL
19. AXLE SHAFT OIL SEAL RETAINER
20. WASHER
21. BOLT
22. DRIVE GEAR
23. DIFFERENTIAL BEARING
24. DIFFERENTIAL BEARING CUP
25. DIFFERENTIAL BEARING SHIM
26. DIFFERENTIAL BEARING CAP
27. BOLT
28. CLUTCH ASSEMBLY PLATES
29. CLUTCH ASSEMBLY DISCS
30. CLUTCH ASSEMBLY RETAINER CLIP
31. CLUTCH ASSEMBLY BELLEVILLE SPRING
32. DIFFERENTIAL GEAR
33. DIFFERENTIAL PIONION SHAFT
34. DIFFERENTIAL PINION THRUST WASHER
35. DIFFERENTIAL PINION GEAR
36. DIFFERENTIAL PINION SHAFT SNAP RING
37. DIFFERENTIAL PINION SHAFT THRUST BLOCK
38. BOLT
39. HOUSING COVER
40. STUD
41. WASHER
42. HOUSING COVER GASKET
43. BOLT
44. DIFFERENTIAL CASE
45. DRIVE PINION
46. REAR PINION BEARING
47. REAR PINION BEARING CUP
48. PINION DEPTH ADJUSTING SHIM

Fig. 6 Trac Lok rear axle assembly

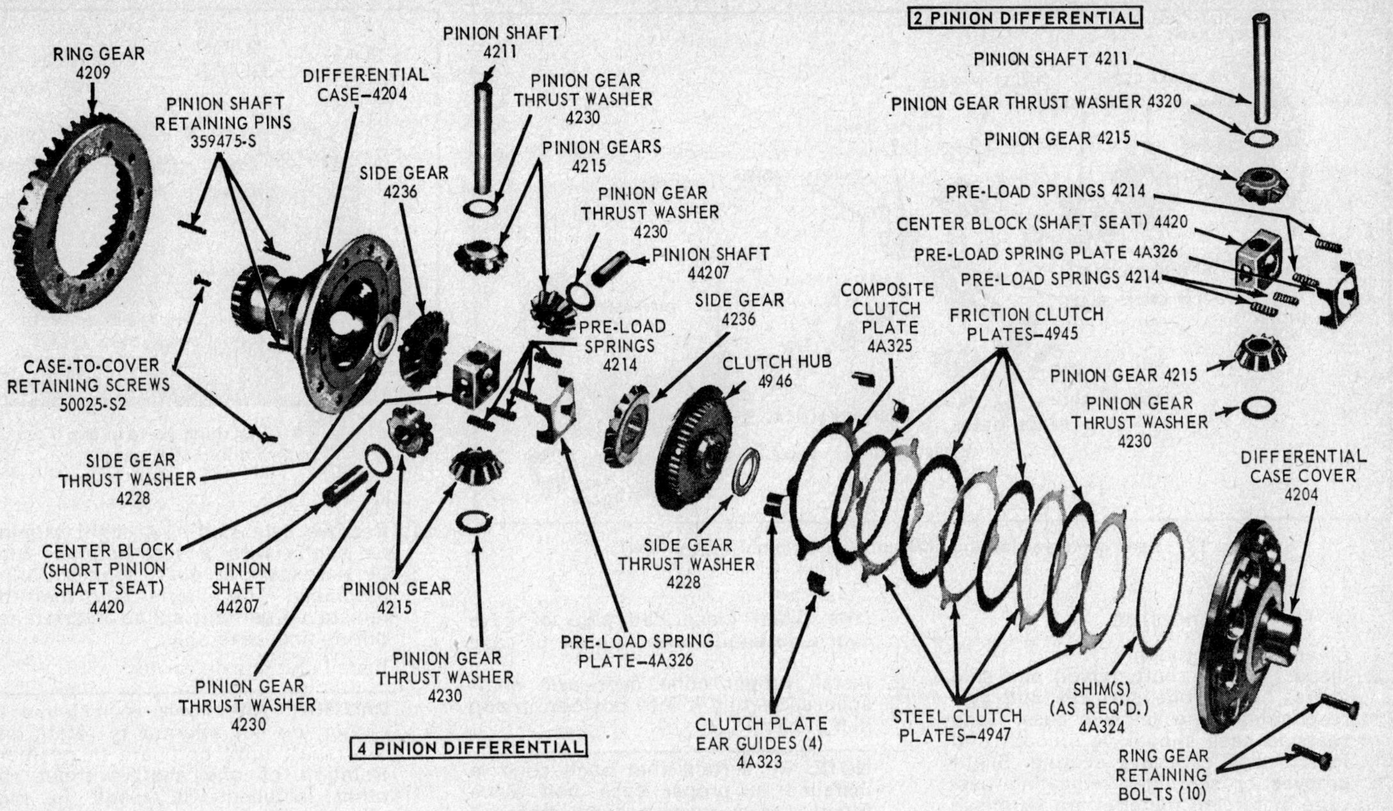

Fig. 10 Ford Traction-Lok. Exploded

3. Place cover on case assembly, aligning case to cover retaining screw holes, and with hand pressure, rotate cover right and left to index splines and mate cover and case halves. Remove assembly from tool T66L-4204-A2 and rest on cover side and install case half retaining screws.

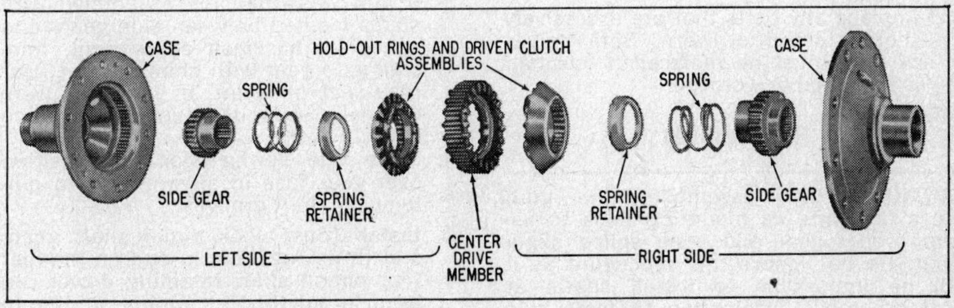

Fig. 11 Ford (DAP) locking differential

DANA CONE TYPE

This unit, Figs. 12, 13, has spiral-threaded cone brakes installed behind the differential side gears. These brakes are statically spring preloaded to provide an internal resistance to differential action within the case itself. This preload assures an adequate amount of pull under extremely low tractive conditions.

Disassembly

1. Clamp case in vise by ring gear or by case flange.
2. Mark flange half of case and cover half with a center punch or paint to provide alignment for reassembly.
3. Loosen bolts holding cover case halves. Remove assembly from vise and remove bolts.
4. Remove parts from each case half, keeping all parts, including side gear shims if present, together with their respective half.

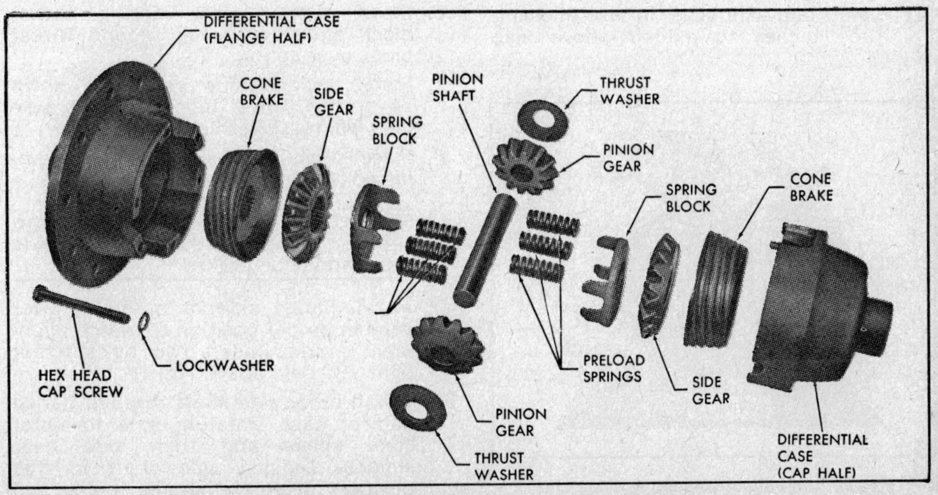

Fig. 12 Dana cone type locking differential disassesmbled (2 pinion)

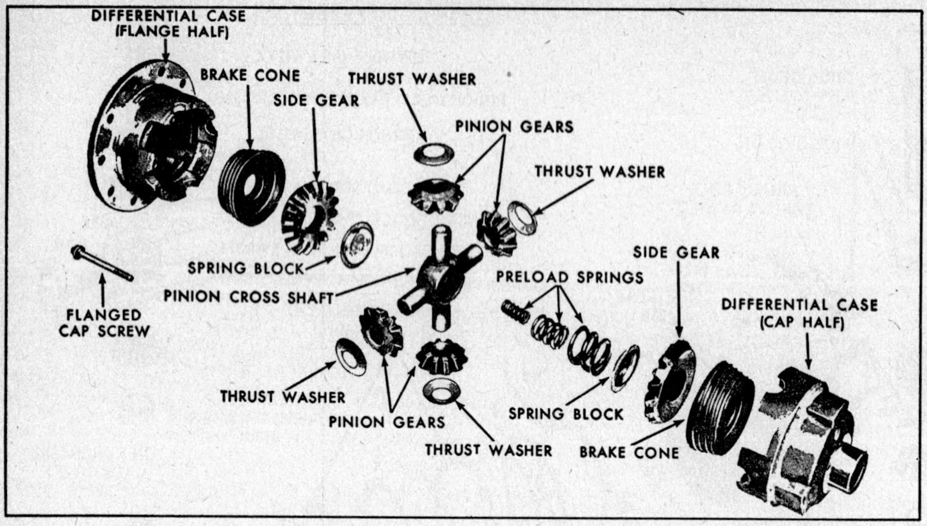

Fig. 13 Dana cone type locking differential disassembled (4 pinion)

Inspection

1. Clean parts thoroughly.
2. Inspect pinion shaft, pinion and side gears, brake cone surfaces and corresponding cone seats in case. Cone seats in case should be smooth and free of any excessive scoring. Slight grooves or scratches indicating passage of foreign material are permissible and normal. The land surfaces on the heavy spirals of the male cones will duplicate the case surface condition.
3. Replace any parts that are excessively scored, pitted or worn. Both halves of case must be replaced if one half is damaged or worn.

Reassembly

CAUTION: When assembling the unit, use axle shafts as mounting tools to assure proper gear and cone spline alignment. Do not ignore this procedure as it will be impossible to install shafts at final assembly. Attempting to force axle shafts into position may result in damage to spring thrust blocks.

---

1. Clamp one axle shaft in vise allowing three inches to extend above vise jaws. Then place cap side of case over axle shafts with interior of case up.
2. Install proper cone over axle shaft splines, seating it into position in cap half of case.

NOTE: Be certain that each cone is installed in proper case half since tapers and surfaces become matched and their positions should not be changed.

---

3. If unit was originally assembled with shims located between side gears and cones for backlash adjustment, reinstall side gear with shim so that gear may seat on shim. If no shims were originally used, reassemble the same way.
4. Place one spring block in position over gear face in alignment with pinion gear shaft grooves.
5. Install thrust block, pinion shaft, gears and thrust washers in such a manner that pinion shaft retaining dowel can be inserted through pinion shaft into differential case. This prevents pinion shaft from sliding out and causing damage to carrier assembly, Fig. 14.
6. Insert springs into spring thrust block and then place second thrust block over springs, Fig. 15.
7. Install second side gear face down on spring thrust block so that side gear will mesh with pinion gears.
8. Place shim, if provided, and remaining cone over side gear.

NOTE: Vega units incorporate two shims on each side between side gears and cone brakes.

---

9. Install flange side of assembly over cone in proper position to match alignment marks. Insert two bolts finger tight 180 deg. apart, Fig. 16.
10. Install other axle shaft through flange half of case, rotating axle to enter cone splies and then side gear splines. Leaving axle shaft in this position insert remaining bolts and torque to 15-18 ft-lbs.

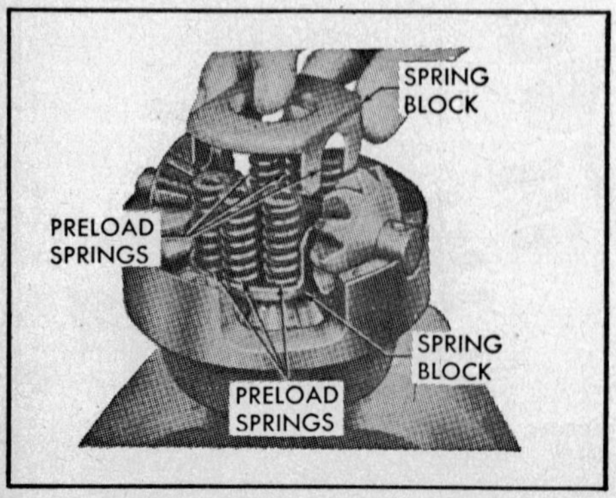

Fig. 15 Installing preload springs and thrust block in differential case

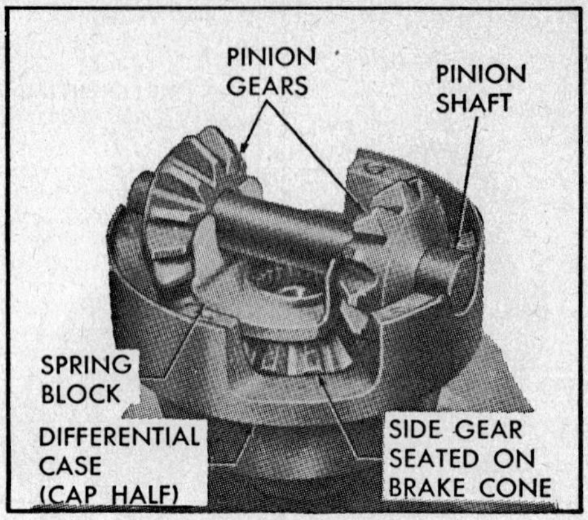

Fig. 14 Installing parts in cap half of differential case

11. Remove axle shafts. A slight tapping on shafts with a soft hammer may be necessary to align splines during assembly. The shafts can then be reinstalled without spline interference during final assembly.
12. Install unit in axle carrier.

CAUTION: After unit is installed in carrier, do not attempt to rotate one axle shaft until both are in position. Rotation of one shaft without the other installed will result in misalignment of cone and side gear splines and may prevent entry of second axle shaft.

---

13. Test differential operation as outlined at the beginning of this chapter.

EATON TYPE

Disassembly, Fig. 17

1. Remove pinion shaft lock screw and

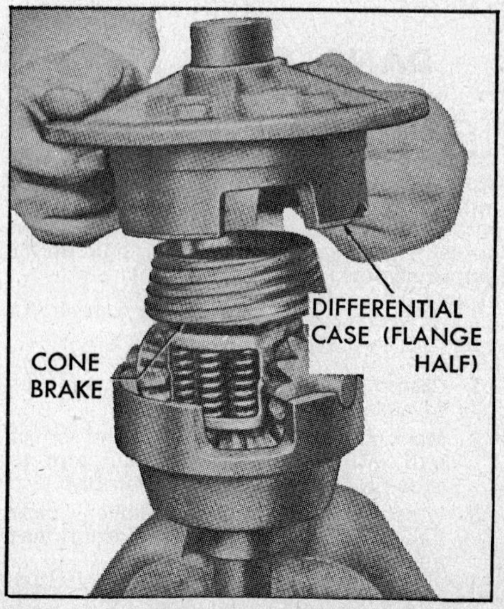

Fig. 16 Installing flange half of differential case

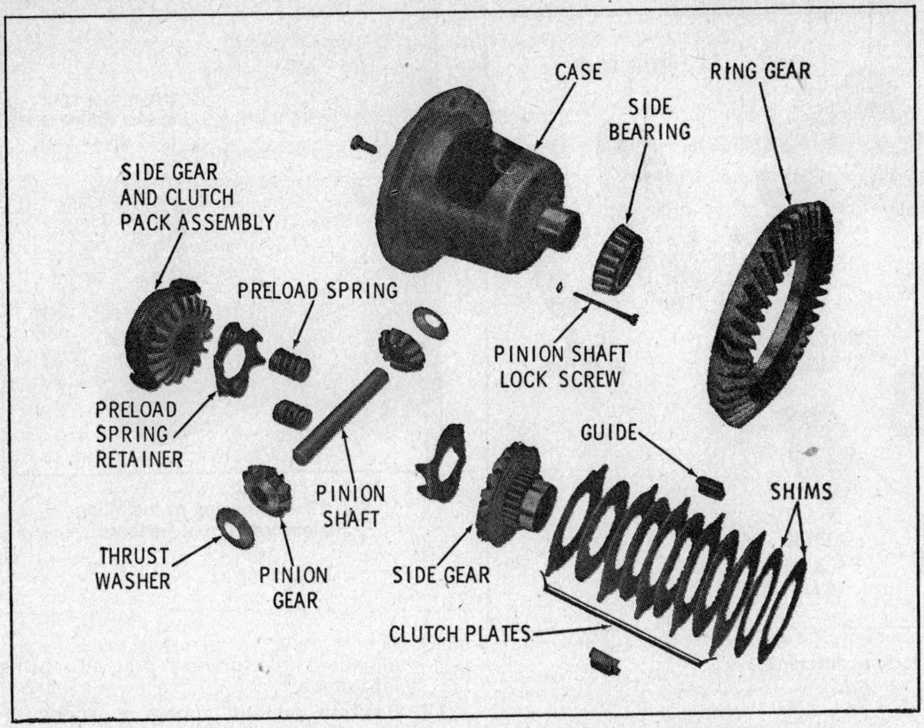

Fig. 17 Eaton plate-type locking differential. Two preload springs (shown) are used on intermediate-size cars; four preload springs are employed with full-size car applications

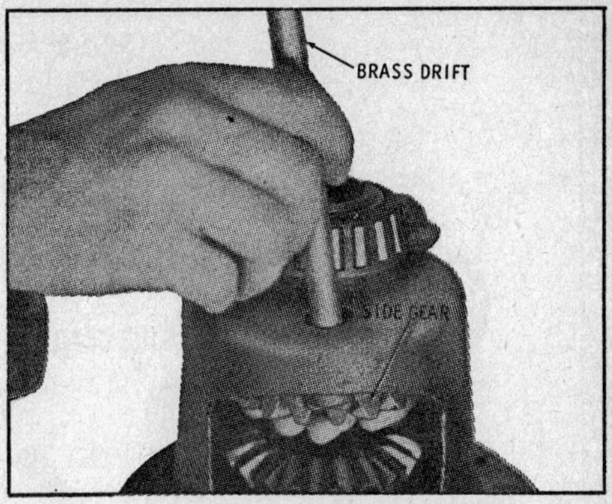

Fig. 18 Removing side gear

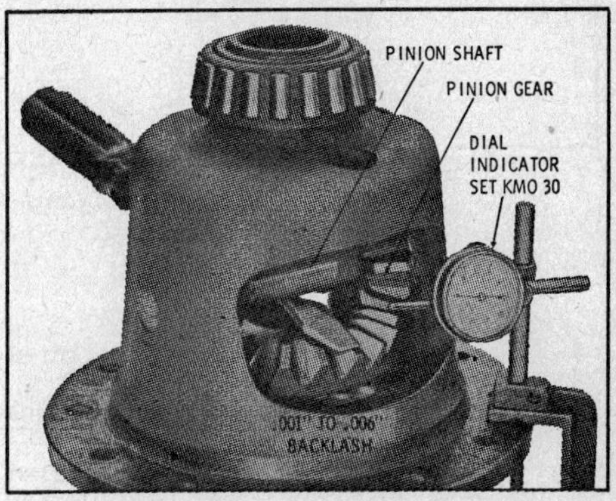

Fig. 19 Checking side gear-to-pinion backlash

pull pinion shaft from case.

2. Remove preload spring retainer and springs.
3. Rotate side gears until pinions are in open area of case and pick out pinions and thrust washers.
4. Remove a side gear, clutch pack and shims, noting shim location in case to aid in reassembly. Remove side gear clutch pack and shims from opposite side. If side gear or clutch pack cannot be removed readily, drive it out with a brass drift, Fig. 18.
5. Remove clutch plate guides and separate shims and clutch plates from side gears.

NOTE: Keep clutch plates in their original location in clutch pack.

Inspection

1. If pinion shaft, pinions or side gears are excessively scored, pitted or worn, replace parts affected.
2. Inspect clutch plates for scored, worn, cracked or distorted condition. If any of these conditions exist, new clutch plates must be installed.

Reassembly

1. Alternately position nine clutch plates on side gear, starting and ending with a plate with external lugs.
2. Install two clutch guides over clutch plate lugs.
3. Install same shims which were removed or an equal amount on clutch plate.
4. Repeat above steps on other clutch pack.
5. Install one side gear with clutch pack and shims in case.
6. Position two pinion gears and thrust washers on side gear and install pinion shaft.
7. Compress clutch pack by inserting a screwdriver or wedge between side gear and pinion shaft.
8. Install a suitable dial indicator with contact button against pinion gear, Fig. 19.
9. Rotate pinion gear. Clearance should be .001" to .006".
10. If clearance is more than .006", add shims between clutch pack and case. If clearance is less than .001", remove shims. A .002" shim will change clearance about .001". Recheck clearance after adding or subtracting shims.
11. Remove side gear and repeat procedure with remaining clutch pack on opposite side of case.
12. Remove pinion shaft, pinions and thrust washers.
13. Install remaining side gear and clutch pack with correct shims in case.
14. Place pinion gears on side gears and rotate into correct position.
15. Compress preload springs and drive preload retainer and springs between side gears.
16. Insert thrust washers behind pinion gears.
17. Install pinion shaft and retain with lock bolt, tightening lock bolt to 15-25 ft-lbs.
18. Check side gear splined hole to be certain it is in line with hole in preload spring retainer. Spring retainer can be moved slightly to correct alignment.
19. Test operation of differential as outlined at the beginning of this chapter.

GM CORP. TYPE

Disassembly, Fig. 20

1. Remove pinion shaft lock screw, then remove pinion shaft from case.
2. Using a brass drift drive preload spring from case.
3. Remove pinion gears and washers, followed by the side gears.
4. Remove clutch plate guides and separate shims and clutch plates from side gears. Keep clutch plates in their original location in clutch pack.
5. Clean and inspect all parts and replace as required.

Reassembly

1. Apply recommended lubricant to clutch plates & discs.
2. Alternately position clutch plates on side gear, starting and ending with a clutch plate having external lugs.
3. Place spacer against plate having external lugs, then install shims. *Make sure to install the same spacer and shims or an equal amount on the*

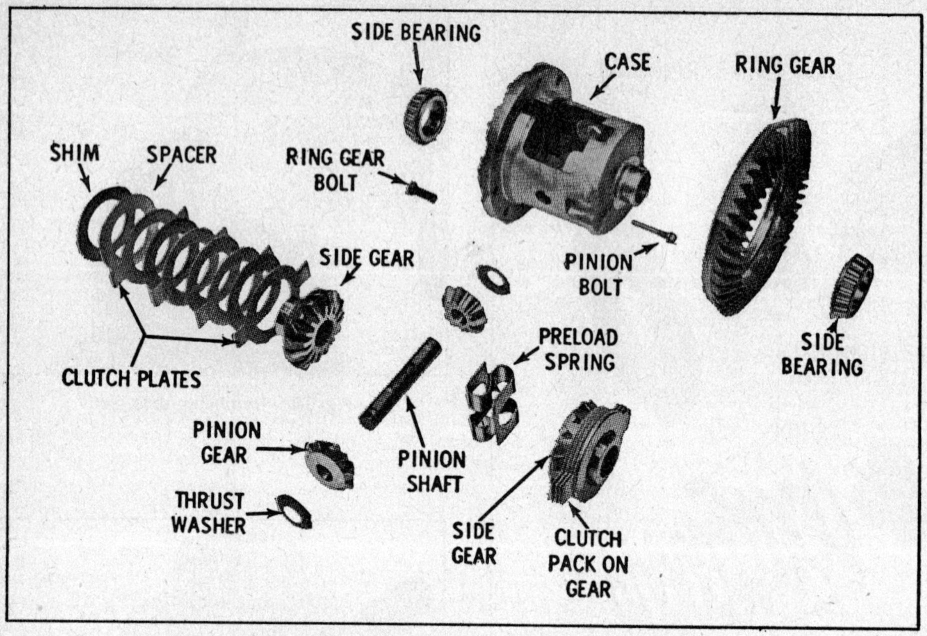

Fig. 20 GM Corp. Type locking differential

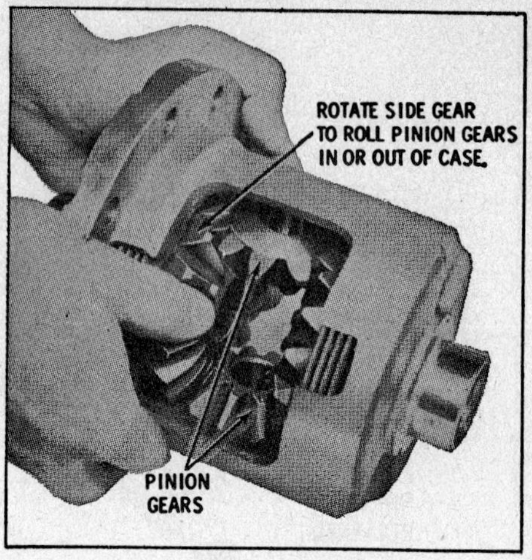

Fig. 22 Removing or installing pinions and thrust washers

clutch pack for a starting point.

4. Repeat foregoing procedure on other clutch pack.
5. Install one side gear with clutch pack and shims in case.
6. Position two pinion gears and thrust washers on side gear and install pinion shaft.
7. Compress clutch pack by inserting a screwdriver between side gear and pinion shaft.
8. Install dial indicator with contact button against pinion gear, Fig. 21.
9. Rotate pinion gear, clearance should be .001-.006 inch on all models except Seville. On Seville clearance should be .005-.008 inch.
10. On Seville models, if clearance is not within limits the unit must be replaced. On all models except Seville, if clearance is more than .006 inch, add shims between clutch pack and case. If clearance is less .001 inch, remove shims. A .002 inch shim will change clearance about .001 inch. Recheck clearance after adding or subtracting shims.
11. Remove side gear and repeat procedure with opposite clutch on opposite side of case.
12. Remove pinion shaft, pinions and thrust washers.
13. Install remaining side gear and clutch pack with correct shims in case.
14. Place pinion gears on side gears and rotate into correct position, Fig. 22.
15. Install thrust washers behind pinion gears and align.
16. Insert pinion shaft into case, through thrust washer and part way into pinion gear. This will keep pinion gears aligned while driving preload spring into place.
17. Position preload spring as shown in Fig. 23, next to side gears and drive into place.
18. Push pinion shaft into position and align lock screw hole in shaft with hole in case. Install lock screw and torque to 15 ft. lbs. on all models except Seville and to 25 ft. lbs. on Seville models.

SERVICE NOTES

Differential Chatter

Locking differential chatter is usually more likely to be caused by a lubrication problem than a mechanical problem. The condition can often be corrected by simply driving the vehicle in 5 fairly tight circles

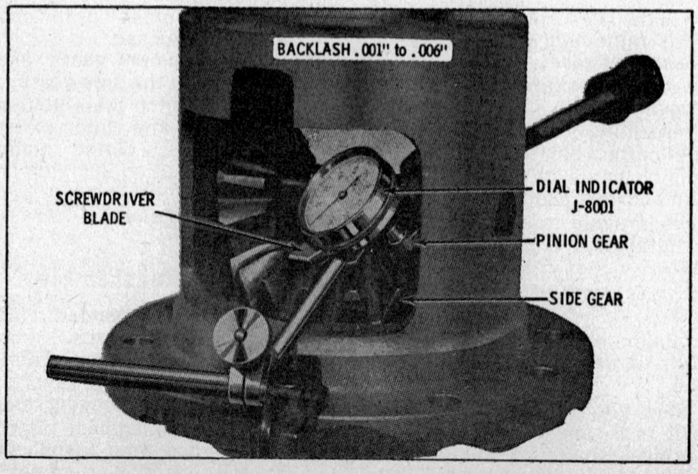

Fig. 21 Checking side gear-to-pinion backlash

Fig. 23 Installing preload spring

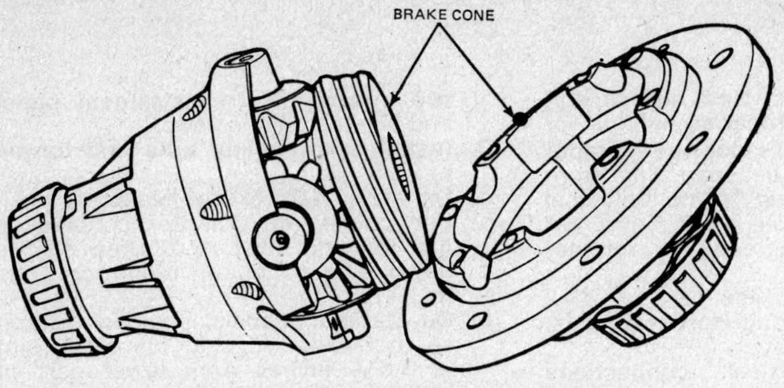

Fig. 24 Borg Warner locking differential

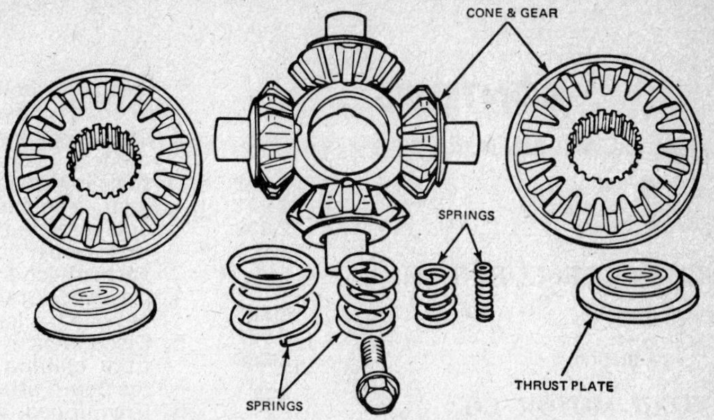

Fig. 25 Borg Warner locking differential disassembled

clockwise and then in 5 circles counter-clockwise. This allows the lubricant to work in between the clutch plates which should provide smooth operation. However, if chatter still exists, proceed as follows:

1. Siphon the lubricant from the axle and refill with a lubricant specifically designed for these axles. The vehicle should then be driven 25-50 miles at which time the chatter should disappear. If the chatter still exists proceed with step 2.
2. Disassemble and clean the differential carrier. Install new friction plates which have been soaked for ½ hour in the special lubricant. Reassemble the differential and install in vehicle. Refill the axle with special lubricant and drive vehicle in tight circles in both clockwise and counter-clockwise directions. If chatter still exists, the vehicle must be driven 25-50 miles to allow the lubricant to work in between the plates. The chattering should then disappear.

BORG WARNER

Disassembly

1. With differential case clamped in a vise, loosen case half retaining screws, leaving 3 to 4 threads engaged. Then loosen vise slightly and tap flange half with a mallet to separate halves, Fig. 24.
2. Remove case from vise, then remove case half retaining screws and flange half of case.
3. Remove side gear and brake cone assemblies, pre-load springs, pinion cross shaft and pinion gears, Fig. 25. Note location of brake cones for proper reassembly.

Assembly

1. Clamp an axle shaft in a vise with approximately 3 inches extending above the vise and place cap half of case over shaft with inside of case facing upward.
2. Place proper brake cone over axle and seat cone into case half, then install in proper order the remaining differential parts.
3. Place flange half of case over second brake cone and install two case half retaining screws finger tight.
4. Install second axle shaft through flange half of case and rotate shaft to align splines. With axle shafts in position, install and tighten case half retaining screws, then remove axle shafts.

STEERING COLUMNS

CONTENTS

AMERICAN MOTORS

CHRYSLER CORPORATION

FORD MOTOR CO.

GENERAL MOTORS CORP.

SERVICE NOTE

When servicing collapsible steering columns, care should be exercised since they are extremely susceptible to damage. Dropping of or leaning on column or striking sharp blows on end of steering shaft or shift levers could loosen or shear plastic fasteners which maintain column rigidity.

It is important that only the specified screws, bolts and nuts be used during the mandatory reassembly sequence and torqued to specifications to insure proper breakaway action of column under impact. Care should also be exercised in using excessively long bolts as they may prevent a portion of the steering column from collapsing under impact.

When removing or installing, steering wheel, ignition switch or lock, stop light switch or turn signal switch, adjusting transmission linkage, or installing and adjusting neutral-start or back-up light switch, refer to appropriate car chapter.

If a shift tube shows a sheared plastic injection, a new shift tube must be installed. If a steering shaft shows a sheared plastic, but it is not bent, it can be repaired by using a Service Steering Shaft Repair Kit part number 3514996 for Chrysler Corporation and 7810077 for General Motors. The kit contains instructions and dimensions for all steering columns. American Motors and Ford Motor Company do not furnish a repair kit, therefore, if the steering column has been collapsed a complete new column must be installed and on some models the attaching brackets will shear under impact and must be replaced.

AMERICAN MOTORS

Figs. 1 Thru 3

Removal, 1969-75

1. With battery ground cable disconnected, remove two flexible coupling to flange bolt nuts. Note size of bolt shanks to insure proper assembly.
2. On 1969-74 Ambassador and Matador models, remove the snap ring from the lower steering shaft universal joint, remove lower shaft from car and remove bearing caps and retainer from upper shaft.
3. Disconnect shift linkage.
4. On 1975 Gremlin and Hornet models, remove package tray.
5. Disconnect electrical connections from column and remove column to toe board attachments.
6. If equipped, disconnect shift indicator cable from shift bowl.
7. Remove lower instrument panel crash pad trim plate and disconnect bracket capsule from instrument panel.
8. Remove column support bracket and remove column.

NOTE: Set bracket aside to protect breakaway capsules. Bracket capsules are slotted to permit column adjustment.

NOTE: On Pacer models, if intermediate shaft is removed from column, mark shaft and column for reference during reassembly.

Installation, 1969-74

1. Attach mounting bracket to column and torque bolts to 15 ft. lbs.

CAUTION: Do not use substitute bolts.

2. Position column into place and loosely attach to instrument panel at the rear mounting bracket attaching studs.

NOTE: Make certain that instrument panel is always supported when either dash mounting or gear mounting is connected.

3. Attach column at coupling. On Ambassador and Matador models, install bearing caps and retainer on upper shaft and install U joint and lower shaft as an assembly.
4. Pull steering column upward and position flexible coupling in a flat position (undistorted). Maintain upward pressure and torque mounting bracket attaching nuts to 10 ft. lbs.

NOTE: Correct torque on bolts and nuts is essential to insure the breakaway action of the bracket in the event of a collision.

5. Connect electrical components, shift linkages and all other panels. Check for proper operation.

Installation, 1975

1. Attach mounting bracket to column and torque bolts to 15 ft. lbs.

NOTE: On Pacer models, if intermediate steering shaft was removed, install shaft aligning reference marks made during disassembly.

2. Install column engaging flex coupling on steering shaft and loosely install two upper nuts to instrument panel and two toe plate screws.
3. Install flex coupling nuts and torque to 25 ft. lbs.
4. On all models except Matador, position column so that flex coupling is flat and not distorted, then torque upper supper column mounting bolts to 10 ft. lbs.
5. On Matador models, position column so that lower edge of universal joint is 6-6½ inches from lower part of intermediate shaft, then torque upper column mounting nuts to 10 ft. lbs.
6. If necessary, loosen toe plate clamp and position it against firewall, then install retaining screws and torque to 10 ft. lbs.
7. Connect shift linkage and quadrant cable and check for proper operation, then connect all electrical wiring and install trim panel and package tray.

CHRYSLER CORPORATION

1969-75, Figs. 1 Thru 3

Removal

1. With battery ground cable and transmission linkage disconnected, separate electrical connectors at column.
2. Remove steering wheel and the turn signal lever.
3. Remove floor plate attaching screws and the trim plate from under instrument panel.
4. Disconnect automatic shift indicator pointer from shift tube bracket (if used).
5. Disconnect steering column bracket from instrument panel support. If shim pack is used between bracket forward leg and support, save for reuse during installation.
6. On 1975 Chrysler and Imperial models, disconnect vacuum lines from vacuum release.
7. Carefully pry lower coupling from steering gear wormshaft and remove column.

Installation

1. On 1970-73 column shift models, not incorporating a lower bearing, tool C-4143 or C-4157 must be used to hold steering shaft in center of shift tube while installing and aligning column in vehicle. If hole in tool is too large to hold steering shaft, use split insert to adapt tool to smaller shaft diameter.
2. On floor shift models place tool C-4157 over steering shaft and slide it up into lower end of jacket to keep shaft centered while column is being installed.
3. Place bracket assembly on steering column, install ground wire (if used) and tighten attaching screws to 10 ft. lbs. On 1970-74 models except Challenger and Barracuda and all 1975 models, plastic capsules should be pre-assembled in bracket slots.

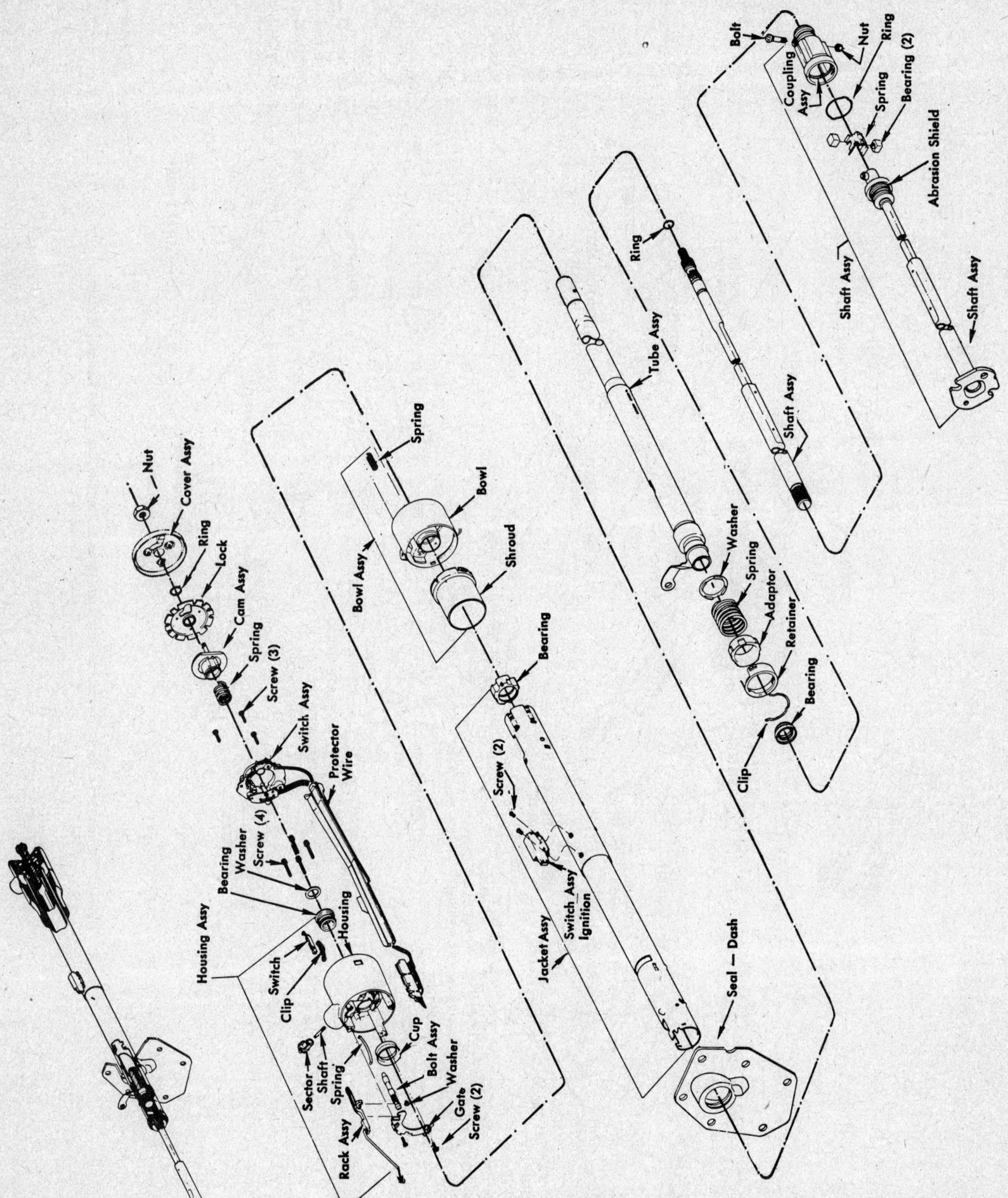

Fig. 1 1969-75 Fixed steering column. Typical of American Motors, Chrysler Corp. and General Motors exc. Astre, Monza, Skyhawk, Starfire & Vega

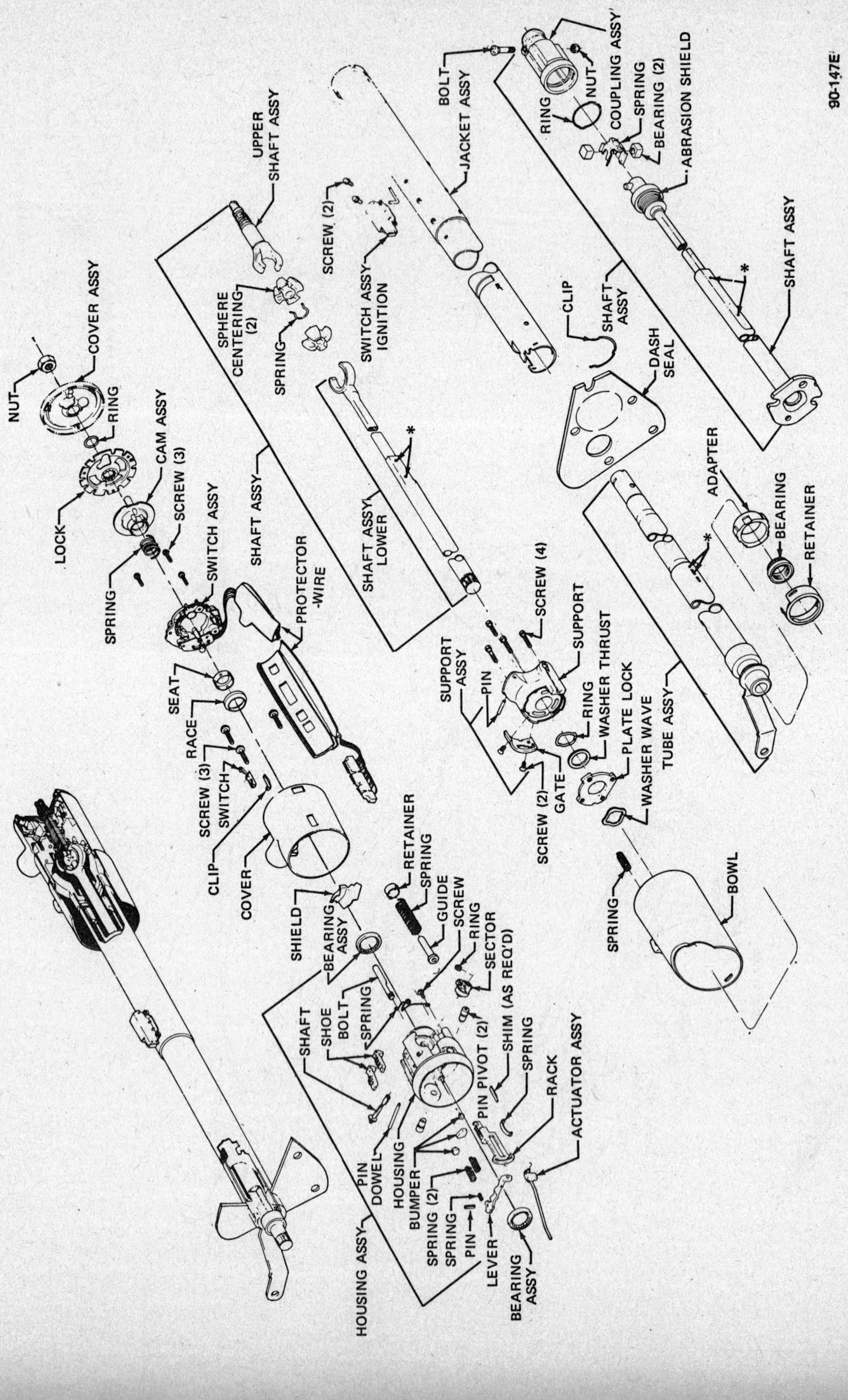

Fig. 2 1969-75 Tilt steering column. Typical of American Motors, Chrysler Corp. and General Motors

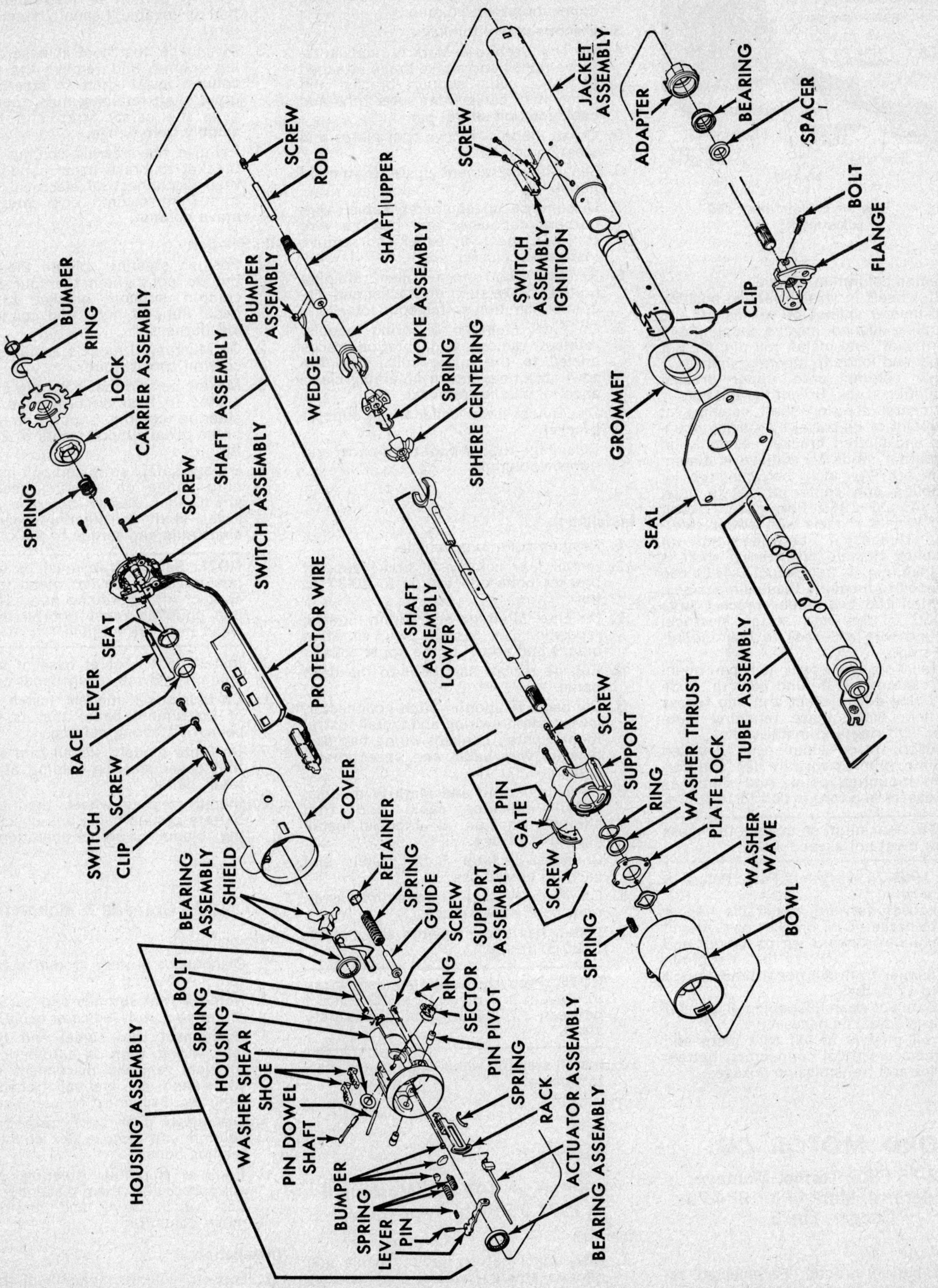

Fig. 3 1969-75 Tilt & Telescope steering column. Typical of American Motors, Chrysler Corp. and General Motors

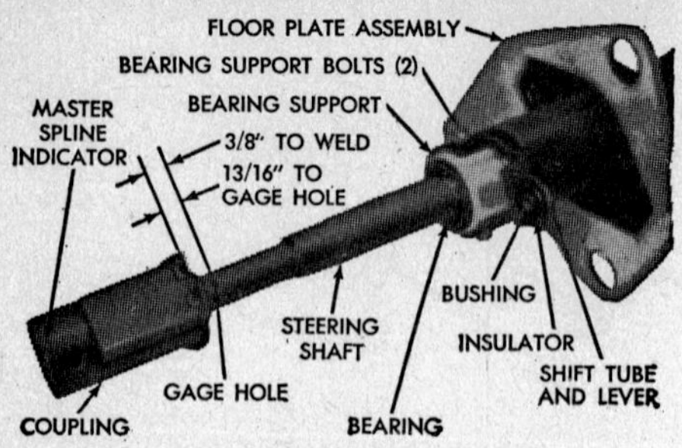

Fig. 4 Steering column lower end adjustment

4. Position column in vehicle.
5. With wheels in straight ahead position and master splines on wormshaft and coupling aligned, engage coupling and wormshaft and install roll pin. Do not apply end loads to steering shaft.
6. Raise column onto upper bracket mounting studs. Do not tighten nuts.
7. On center steering shaft coupling at midpoint of its travel by moving coupling and slotted bracket assembly on mounting studs to achieve a dimension of $^{13}/_{16}$ inch between top of coupling and center of gauge hole, Fig. 4. On 1969 Barracuda, Duster and Valiant models with power steering, dimension between top of coupling to weld on steering shaft is ⅜ inch, Fig. 4. Gauge tool C-4111 can be used to measure these dimensions.
8. Tighten the two upper bracket nuts to 10 ft. lbs. and attach electrical ground wire (if used) to rear mounting stud.
9. Install floor plate over floor pan opening centering it around column. Slide "O" ring down jacket and into recess in floor plate. Place retaining plate over "O" ring and install bolts.
10. If used, place shim pack between column bracket forward leg and forward mounting point. Add shims as necessary and torque bolt to 10 ft. lbs.

NOTE: Maximum shim pack thickness error must not exceed .060".

11. **On 1969-74 Chrysler, Fury, Polara & Monaco:**
    a. Loosen forward adjustable hanger to brake pedal bracket bolt. Attach column forward leg to hanger and tighten to 10 ft. lbs. Retighten hanger to brake pedal bracket bolt to 17 ft. lbs.
    b. Connect gearshift indicator pointer and adjust as necessary.
12. On all models install trim plate and connect electrical connectors, battery cables and transmission linkages.

FORD MOTOR CO.

1972-75 Elite, Torino, Montego, Thunderbird, Mark IV & 1974-75 Cougar, Fig. 5

Removal

1. With battery ground disconnected, remove flex coupling attaching bolts and disengage safety strap and bolt.
2. Disconnect turn signal switch, back-up light switch wiring and vacuum hose connection from column.
3. Disconnect shift linkage.
4. On Thunderbird & Mark IV models, remove right hand shake brace attached to brake pedal support bracket and disconnect cable from outer tube and cable loop on socket pin.
5. On all models remove trim plates and steering wheel.
6. Separate instrument cluster from main panel.
7. Disconnect speedometer cable, wire harness connector and bowden wire cables to heat/air panel and remove instrument cluster.
8. Remove two upper collar retaining bolts at brake support bracket and disconnect ignition switch connector.
9. On Elite, Montego & Torino models, rotate or remove, two vibration braces affixed to the upper collar and the cowl structure, inturn obtaining clearance for column removal.
10. Disconnect lower collar brake support bracket.
11. Disengage rubber boot from panel and remove column.

Installation

1. Position column in vehicle.
2. Install four column to brake support bracket bolts and torque to 20-37 ft. lbs.
3. On Elite, Montego and Torino models, reposition or install two vibration braces and attach to the upper collar.
4. Engage rubber seal boot to the dash panel.
5. Connect ignition switch connector, lower column wiring and install instrument cluster, connect wiring harness, bowden wire cable and speedometer cable.
6. On Thunderbird and Mark IV models, install indicator cable and trim shrouds and on all models install gearshift linkage.
7. Connect back-up light switch and vacuum hose if used.
8. On Mark IV and Thunderbird install right hand shake brace.
9. Install flex coupling nuts and tighten to 20-37 ft. lbs.

NOTE: Properly position safety strap to prevent any metal to metal contact between "U" joint foot and safety strap.

10. Install steering wheel and connect battery cable.

1975 Bobcat, 1974-75 Mustang II & 1971-75 Pinto, Figs. 6 & 7

Removal

1. Disconnect battery ground cable and remove steering wheel, then remove trim shrouds that cover steering column beneath instrument panel.
2. On Mustang II and all 1975 models, remove portion of instrument panel that is located beneath steering column.
3. Disengage dust boot at base of steering column and remove the steering column lower shaft to steering gear input shaft retaining nuts, then disengage the safety strap and bolt assembly from flange.
4. Remove the steering column support bracket to brake pedal support nuts, then disconnect all electrical connectors from column and carefully remove column.

Installation

1. Position steering column then install but do not tighten the four steering column support bracket to brake pedal support nuts, and connect wiring harness.
2. On Mustang II models, install steering column trim shrouds.
3. Making certain that bolts are correctly seated in their square holes, torque steering column support bracket to brake pedal supports nuts to 20-37 ft. lbs.
4. Engage safety strap and bolt assembly onto flange on steering gear input shaft, then install steering column lower shaft to steering gear input shaft nuts and torque to 20-37 ft. lbs.

NOTE: Safety strap must be properly positioned to prevent metal to metal contact after torquing nuts. The flexible coupling must not be distorted when nuts are torqued.

5. Install dust boot at base of steering column and into dash panel opening.
6. On Mustang II models, install section of instrument panel that is located beneath steering column.
7. On Pinto models, install trim shrouds that cover steering column at instrument panel.
8. Install steering wheel, then connect battery ground cable and check steering column for proper operation.

1975 Granada & Monarch

Removal

1. Disconnect battery ground cable and remove steering wheel.
2. Remove trim shrouds and panels and disconnect shift indicator cable.
3. Disconnect turn signal and back-up light wiring, then on automatic transmission vehicles, disconnect control cable and on manual transmission vehicles, disconnect bellcrank control.
4. Disengage dust boot from base of column and remove flex coupling attaching bolts.
5. Remove the four steering column support bolts, then disconnect ignition switch wiring and carefully remove column.

Installation

1. Loosen column adjustment bracket to brake pedal support bolts and carefully install column into position.

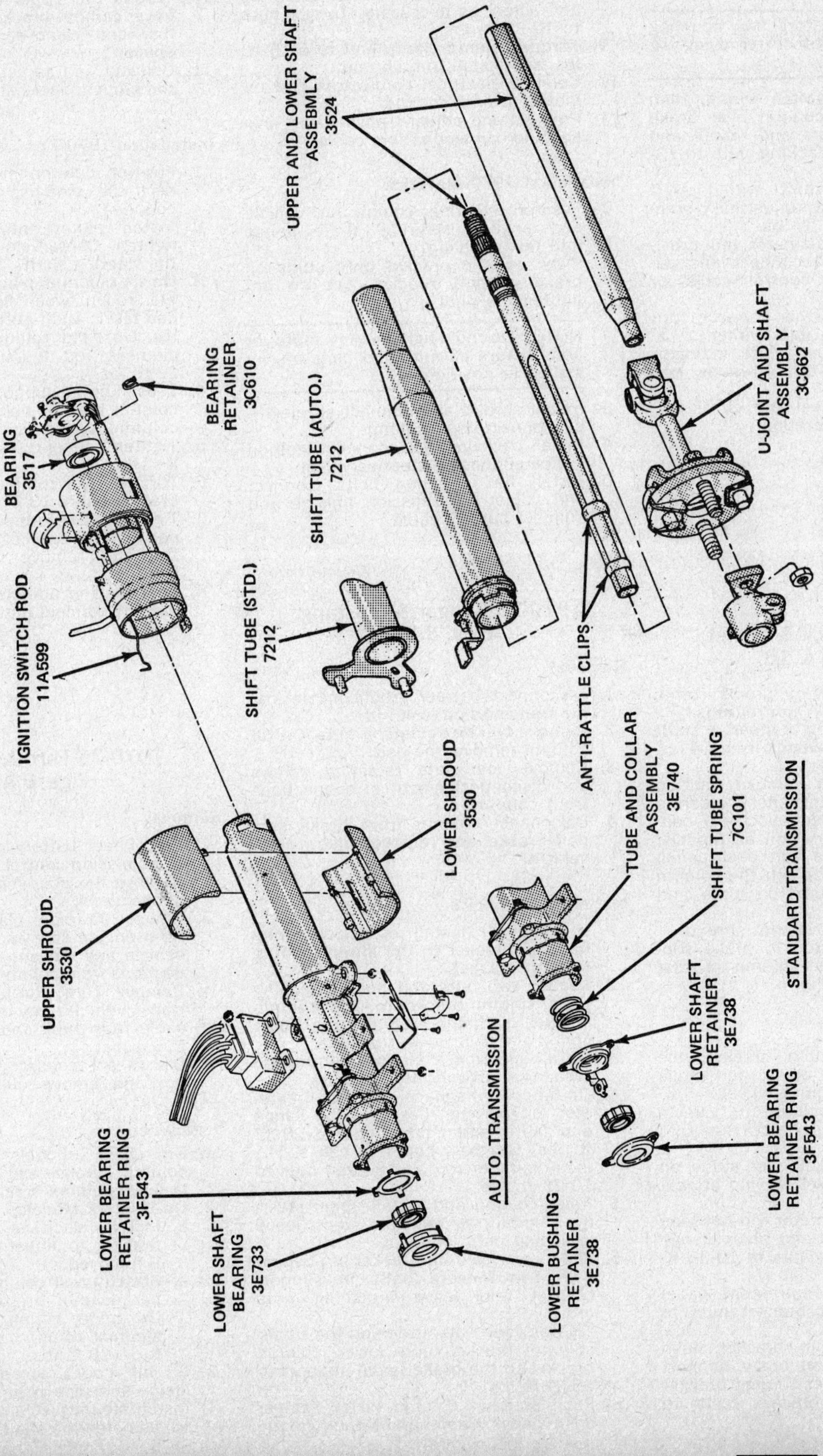

Fig. 5 1972-75 Elite, Torino, Montego, Thunderbird, Mark IV and 1974-75 Cougar steering column

Align steering shaft flange with flex coupling and insert a 1/4 inch diameter rod between flex coupling and flange. Install and torque attaching nuts to 10-22 ft. lbs.

NOTE: Do not distort flex coupling when torquing nuts or removing 1/4 inch diameter rod.

2. Connect ignition switch wiring, then position steering column over brake pedal support studs and install and torque the four attaching nuts to 13-27 ft. lbs.
3. Center steering shaft within shift tube and torque the adjustment bracket bolts to 20-37 ft. lbs.
4. Remove 1/4 inch diameter rod previously installed and connect and adjust transmission control cable or bellcrank.
5. Install dust boot and connect turn signal and back-up light wiring.
6. Connect and adjust shift indicator cable and install trim shrouds and panels.
7. Install steering wheel and check column for proper operation.

1970-75 Maverick & Comet, Figs. 8, 9 & 10

1. Disconnect the battery ground cable and the transmission control rods.
2. Remove flex coupling to steering shaft nuts. On 1970 Maverick remove attaching bolts.
3. On 1970 Maverick remove ignition switch, compartment shelf, steering wheel and disconnect electrical connections from column on all models.
4. Disconnect toe plate from dash panel.
5. On 1970 Maverick detach the column retaining clamp from toe plate and from column.
6. On all models remove trim shroud.
7. Remove lower column to brake support bracket, upper column bracket nuts and remove column.

Installation 1971-75

1. Loosen upper column bracket, toe plate column cover clamp and move toe plate toward steering wheel.
2. Position steering column in vehicle and engage steering shaft flange into flex coupling.
3. Raise column bracket over studs on brake support bracket leaving attaching nuts loose.
4. Place a 1/4 inch diameter rod between flex coupling fabric and shaft flange and torque coupling nuts to 10-22 ft. lbs.
5. Center column in instrument panel and torque support bracket nuts to 13-27 ft. lbs.
6. Install upper column bracket flush with extension lip on brake support bracket. Torque upper column bracket to brake support bracket bolts to 13-17 ft. lbs.
7. Push toe plate against dash panel, install and torque the four bolts to 13-17 ft. lbs.
8. If steering shaft is not centered in column tube, loosen clamp assembly and adjust as necessary. Torque nuts to 10-22 ft. lbs.
9. Torque column clamp bolt to 5-15 ft. lbs. and install trim shroud.
10. Connect electrical connectors and remove 1/4 diameter rod.
11. Connect and adjust transmission linkage and connect battery cable.

Installation 1970 Maverick

1. Position steering column in vehicle and engage steering shaft flange into flex coupling.
2. Place column bracket onto studs on brake support bracket but do not tighten the nuts.

NOTE: Do not tighten any nuts or bolts before all nuts and bolts are installed in position.

3. Install shroud and connect all electrical connectors at column.
4. Install package shelf, connect ignition switch and install steering wheel.
5. Install flex coupling bolts, connect and adjust transmission linkage and connect battery cable.

1970-73 Cougar & Mustang, Figs. 8 & 9

Removal

1. Disconnect battery ground cable and the transmission lock rod.
2. Remove two flex coupling nuts. On tilt column remove one bolt.
3. Remove toe plate retaining screws and disconnect electrical connections from column.
4. Disconnect column from brake support bracket and remove column from vehicle.

Installation 1972-73

1. If removed, install electrical wiring harness bracket to the lower steering column bracket.
2. Loosen two nuts and bolts attaching upper column bracket to lower column bracket and the column toe plate clamp and slide cover rearward.
3. Install steering column and engage steering shaft with flex coupling.
4. On tilt columns place a .15 inch diameter rod between flex coupling flange and fabric and tighten bolt to 20-37 ft. lbs. On fixed columns use a 1 1/4 inch diameter rod and tighten nuts to 10-22 ft. lbs.
5. Raise column and install upper steering column over rearward studs, leave attaching nuts loose.
6. Slide lower column bracket in position over two forward bolts in support bracket and leave attaching nuts loose.
7. Tighten four nuts attaching the upper column bracket and lower column bracket to the brake pedal support to 25-39 ft. lbs.
8. Push steering column cover against dash panel. Center shift tube to the steering shaft and torque nuts to 5-15 ft. lbs.
9. Torque column clamp bolt to 5-12 ft. lbs.
10. Torque the upper column bracket to lower column bracket nuts to 13-27 ft. lbs. and remove the flex coupling spacer.
11. Connect and adjust transmission lock rod and connect battery cable.

Installation 1970-71

1. Position column in vehicle and connect the steering shaft to the flex coupling.
2. Install column onto studs but do not tighten. Center column shift tube to the steering shaft.
3. On tilt columns place a .15 inch diameter rod between flex coupling flange and fabric and tighten bolt to 20-37 ft. lbs. On fixed columns use a 1/4 inch diameter rod and tighten nuts to 10-22 ft. lbs.
4. Loosen brake support bracket to upper column bracket bolts and align upper column, torque nuts to 13-27 ft. lbs.
5. Tighten toe plate screws to 5-15 ft. lbs.
6. Torque lower column to upper column bracket nut to 13-27 ft. lbs.
7. Tighten brake support bracket to upper column bracket nuts to 13-27 ft. lbs. and remove the flex coupling spacer.
8. Connect and adjust transmission lock rod and connect battery cable.

1970-75 Ford & Mercury, Figs. 8 & 9

Removal

1. Disconnect battery ground cable and transmission control rods.
2. Remove flex coupling to steering shaft connections.
3. Remove four toe plate retaining nuts from engine side and one from inside vehicle and disconnect electrical connections from column.
4. Remove trim shroud and on auto. trans. vehicles remove shift indicator cable from tube and cable loop from the socket pin.
5. Disconnect column from brake support and remove column from vehicle.

Installation

1. On 1970 models remove upper column bracket and loosely attach it to lower column bracket with one nut.
2. **On 1971-72 Models:**
   a. Install electrical wiring harness bracket to upper column bracket, if removed.
   b. Attach lower column bracket to upper column bracket. Then attach the upper column bracket to the steering column, torque four bolts to 18-38 ft. lbs.
3. On all models loosen column cover clamp and slide cover back.
4. Install steering column in vehicle and engage steering shaft to flex coupling.

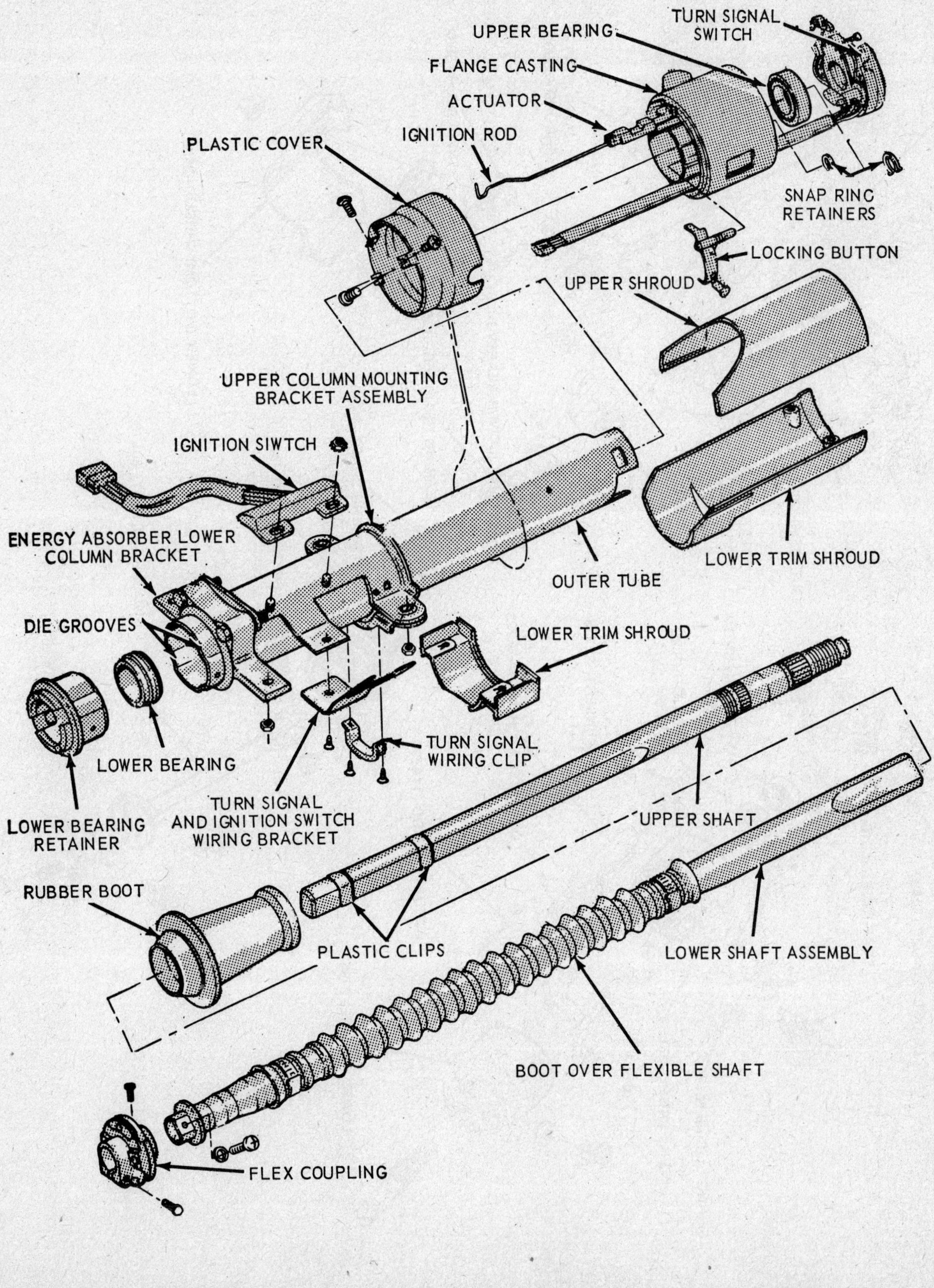

Fig. 6 1971 & early 1972 Pinto steering column

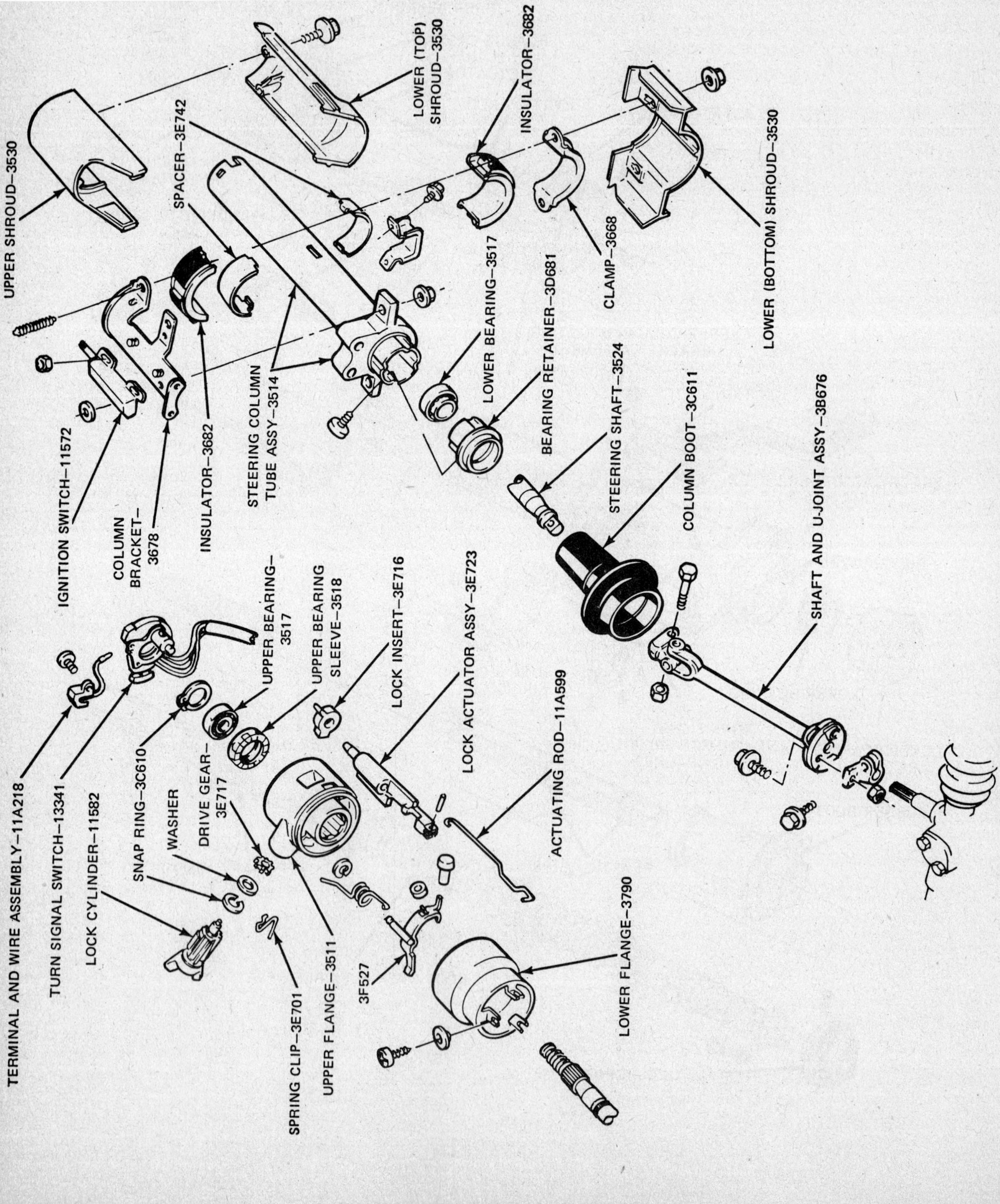

Fig. 7 Late 1972 and all 1973-75 Pinto, 1974-75 Mustang II & 1975 Bobcat steering column

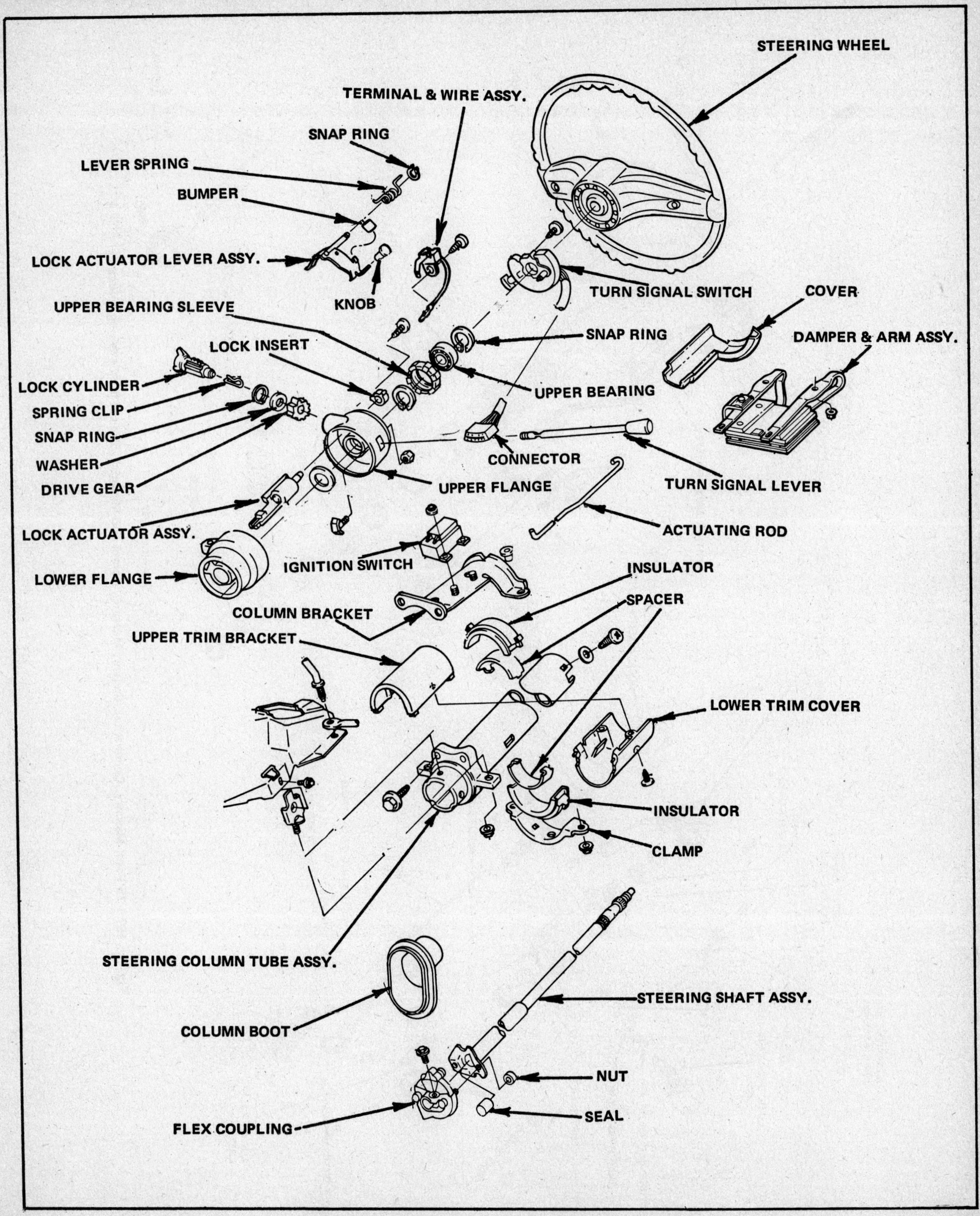

Fig. 7A Floor shift steering column. 1975 Granada & Monarch

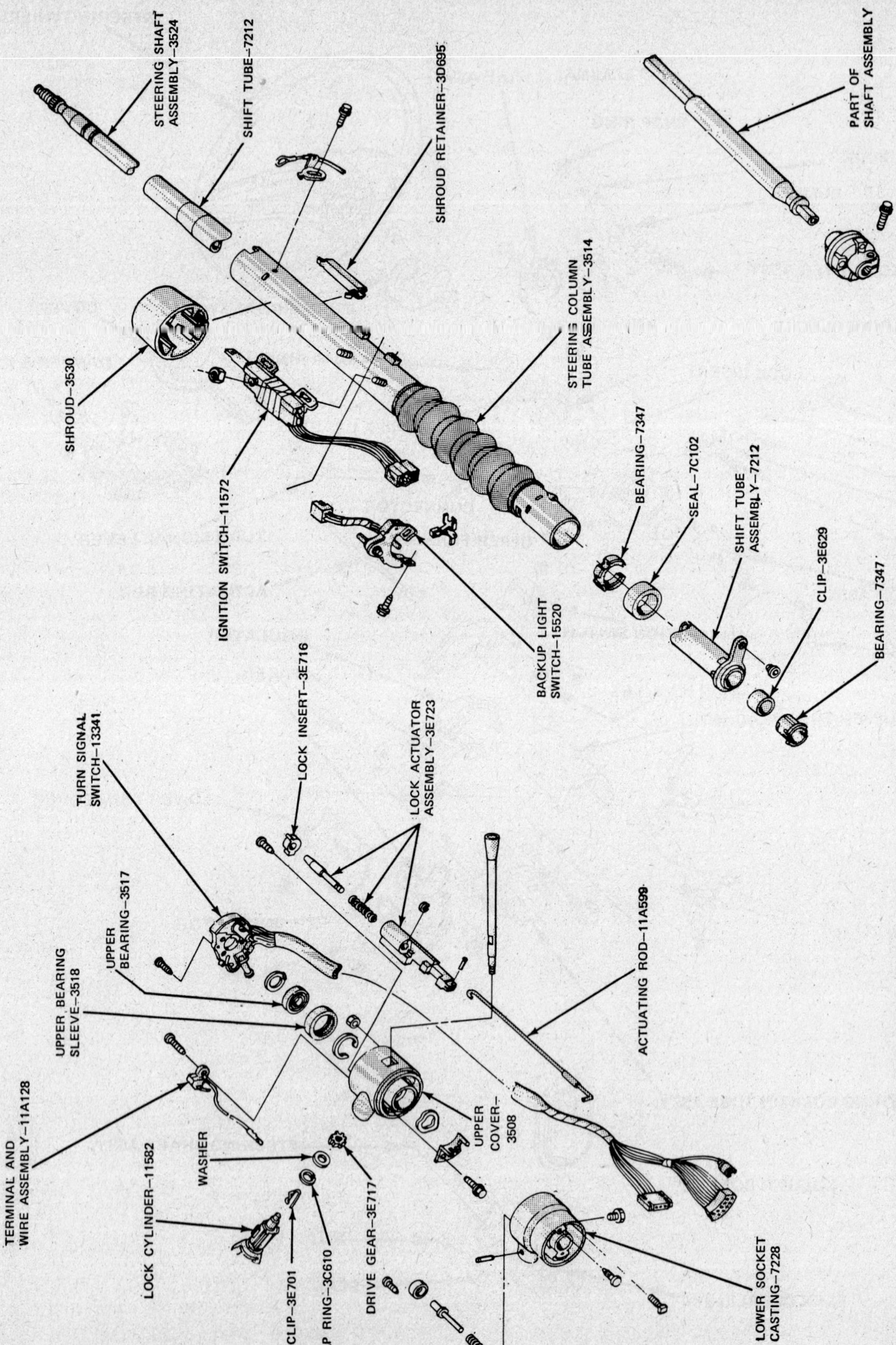

Fig. 8 1973-75 Ford, Mercury, Continental, Cougar, Mustang, Comet and Maverick. Typical fixed steering column

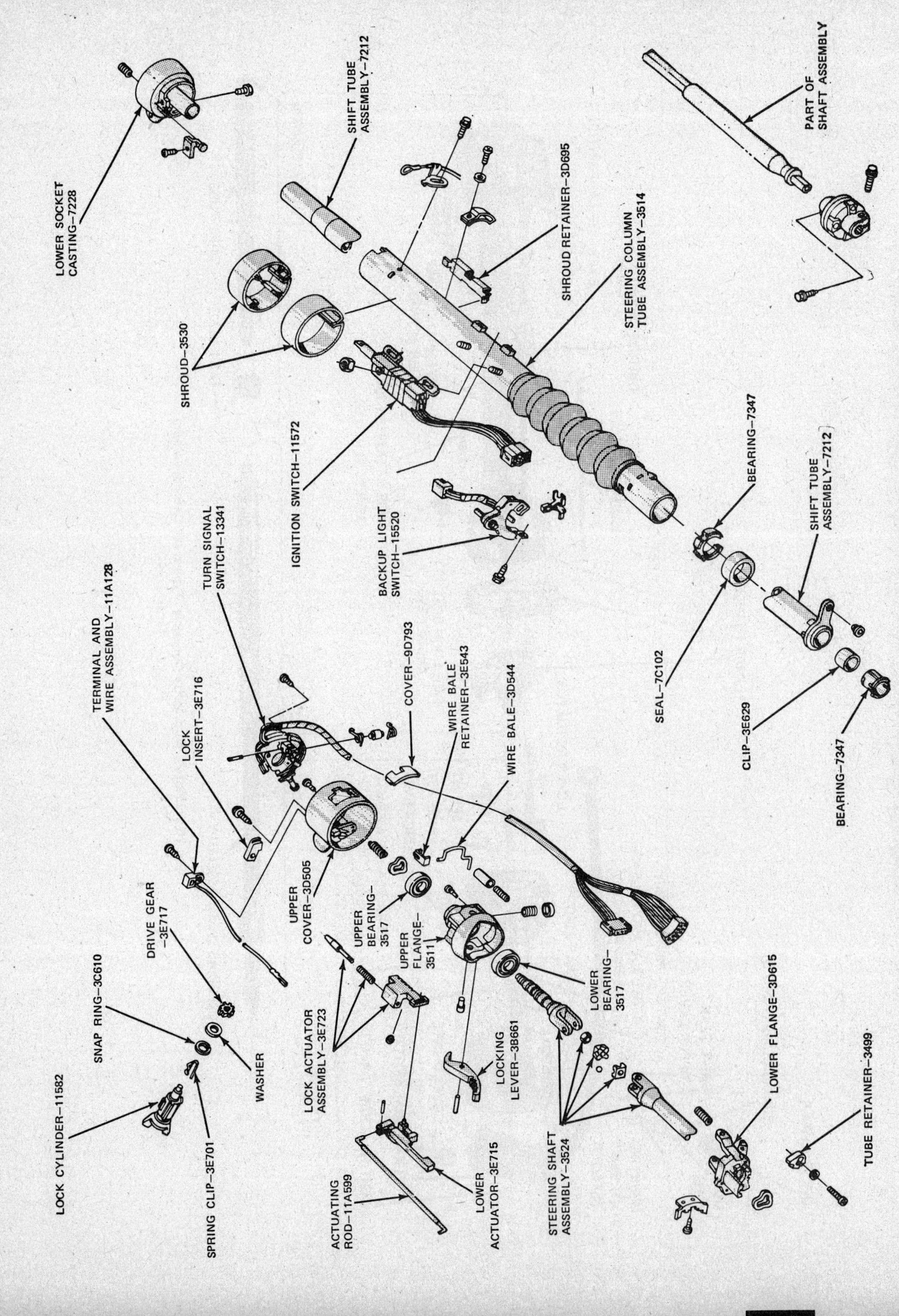

Fig. 9 1973-75 Ford, Mercury, Continental, Cougar, Mustang, Comet and Maverick. Typical tilt steering column

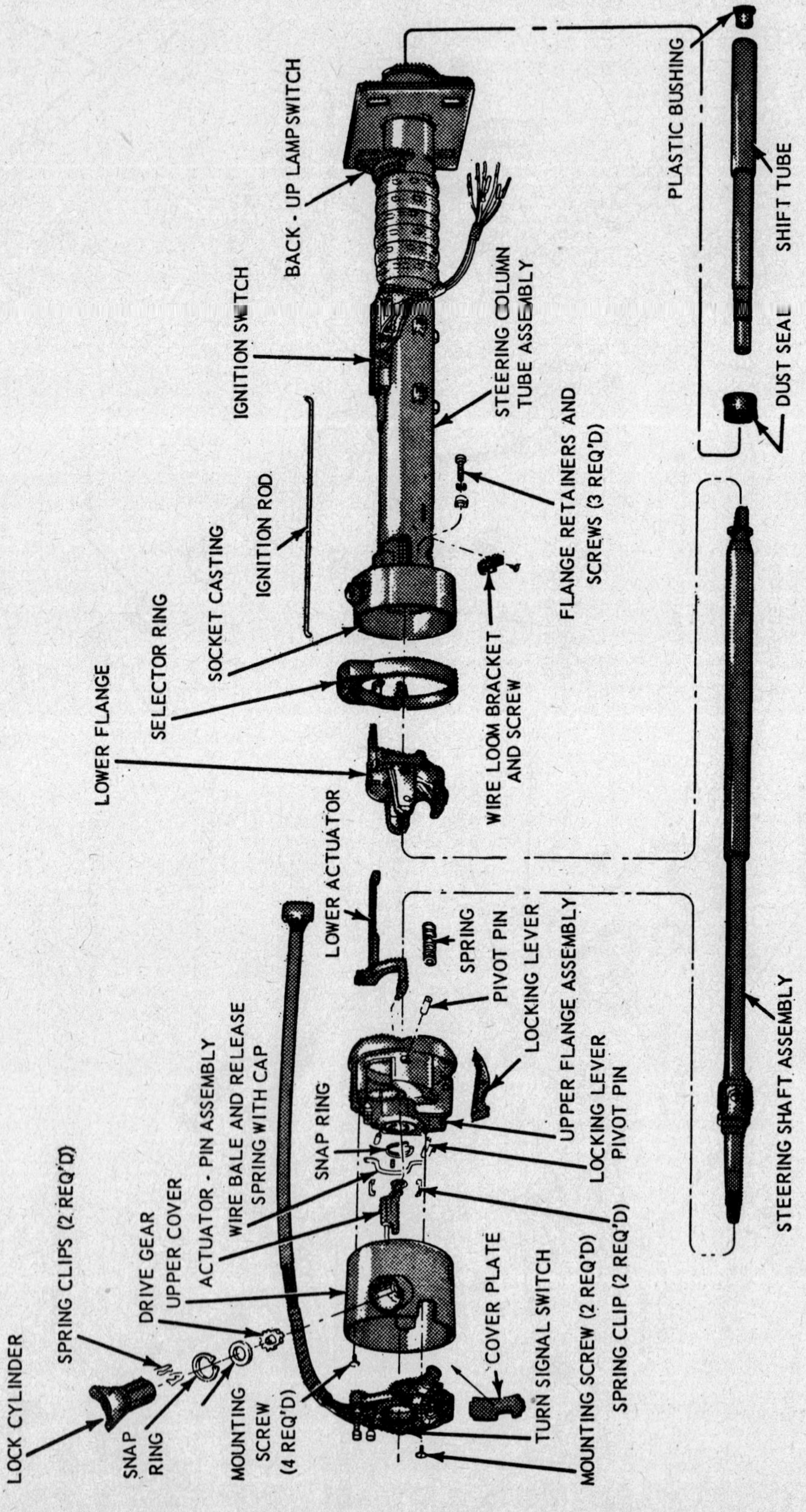

Fig. 10 1970-72 Ford Motor Company. Typical tilt steering column

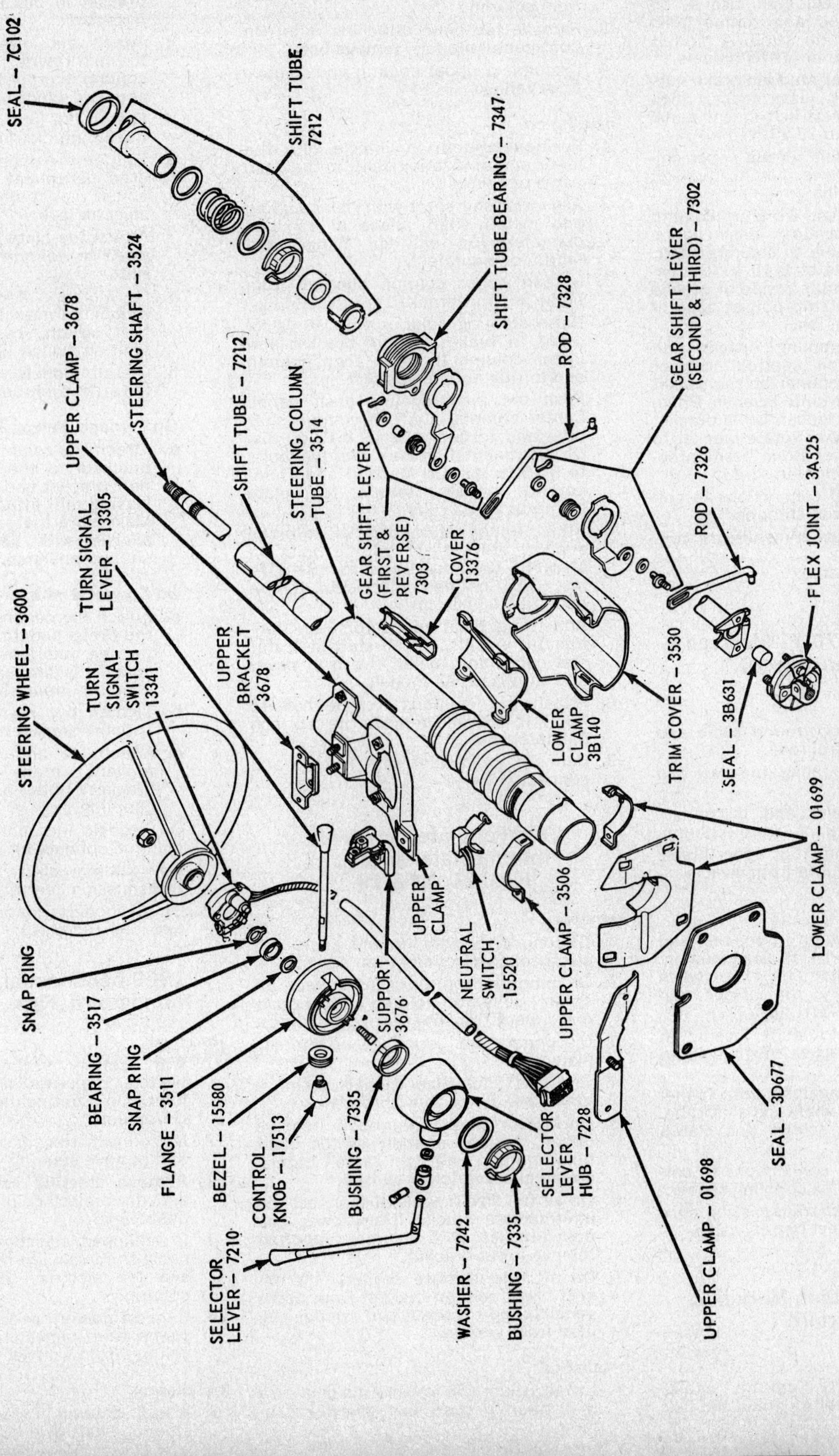

Fig. 11 1969 Ford & Mercury full size models. Fixed steering column

5. On fixed steering columns, place a ¼ inch diameter rod between flex coupling and flange fabric and torque flex coupling nuts to 10-22 ft. lbs. On tilt steering columns, use a .15 inch diameter rod and torque bolts to 20-30 ft. lbs. on 1969-74 models and 42-47 ft. lbs. on 1975 models.
6. Place column over studs in brake support bracket and leave loose. Align column in center of instrument panel and torque nuts to 13-27 ft. lbs.
7. On 1971-75 models torque upper column bracket to lower column bracket nuts to 13-27 ft. lbs.
8. On all models push steering column opening cover against dash panel. Center column tube to steering shaft. Torque cover nuts to 8-13 ft. lbs. beginning with the nuts inside of vehicle and torque the clamp bolt at base of column to 5-12 ft. lbs.
9. Remove flex coupling spacer and place transmission control indicator loop cable over retainer on the socket casting. With the shift lever in Drive, tighten and then loosen cable bracket to outer tube screw. Rotate bracket to align indicator over Drive position selector lens and retighten screw.
10. Connect electrical connections to column and install trim shrouds.
11. Connect and adjust transmission control rod and lock rod (if used) and connect battery cable.

1970 Fairlane, 1970-71 Montego & 1971 Torino, Fig. 10

Removal

1. Disconnect battery ground cable and transmission control rods.
2. Disconnect flex coupling, toe plate and loosen clamp bolt.
3. Remove trim cover and disconnect electrical connections from column.
4. Disconnect column from upper bracket and remove column from vehicle.

Installation

1. Place steering column in position and align steering shaft with flex coupling. Install a ¼ inch diameter rod between flex coupling fabric and flange and torque nuts to 10-22 ft. lbs.
2. Raise column onto studs located in upper column bracket. Align column and torque nuts to 13-27 ft. lbs.
3. Push toe plate against dash panel centering steering shaft and shift tube. Torque toe plate screws and clamp bolt to 5-15 ft. lbs.
4. Connect electrical connections to column and remove flex coupling spacer.
5. Connect and adjust transmission control rods and connect battery cable.

1970-71 Continental, Mark III & Thunderbird

Removal

1. Disconnect battery ground cable, transmission control rods and the flex coupling.
2. Disconnect electrical connectors from column and remove steering column trim cover. On Continental models, remove transmission selector dial cable from column.
3. Remove toe plate attaching nuts. On Continental models, remove bolt.
4. Disconnect upper column and remove from vehicle.

Installation

1. Position column in vehicle with the lower end of steering shaft in the dash panel opening.
2. Align steering shaft with flex coupling and install shaft. Place a .15 inch diameter rod between flange and fabric for alignment.
3. Loosen upper column support from brake support bracket.
4. Raise steering column onto studs located in brake support bracket and upper column bracket, align column and torque nuts to 13-27 ft. lbs.
5. Push toe plate against dash panel. Center steering shaft in column shift tube and torque nuts to 8-13 ft. lbs. On Continental models, torque bolts to 5-15 ft. lbs. On Mark III and Thunderbird models, torque toe plate clamp to 5-12 ft. lbs.
6. Attach upper column bracket to brace support bracket and torque bolts on Mark III & Thunderbird to 18-48 ft. lbs. On Continental models torque bolts to 13-37 ft. lbs.
7. Attach electrical connectors. On Continental models, install shift indicator dial cable and install the trim cover and shroud on all models.
8. Attach flex coupling, torque bolt to 20-37 ft. lbs. and remove the .15 inch diameter rod.
9. Connect and adjust shift linkage and connect battery.

1969 Ford, Mercury & All Intermediate Models, Figs. 11, 12 & 13

Removal

1. Disconnect battery ground cable and electrical connections from column.
2. On tilt-away columns, disconnect tilt control cable, on column shift models disconnect transmission control rods.
3. Disconnect flex coupling and the toe plate.
4. On Ford Full Size models, remove upper and lower trim shrouds.
5. On Mercury Full Size models, remove upper shroud, ignition switch light bulb, ignition and light switch bezels, spring clip and lower shroud.
6. On all full size models, disconnect the intermediate bracket from lower column bracket. With column supported remove lower bracket.
7. On all intermediate models, disconnect lower column bracket from brake pedal support bracket and remove column from vehicle.

Installation

1. Install column in vehicle making sure that steering shaft engages flex coupling.
2. **On Full Size Models:**
   a. Connect flex coupling.
   b. Position column upper and lower bracket in place leaving attaching nuts loose.
   c. Attach the lower column bracket to instrument panel. Insure that column is properly aligned with flex coupling input shaft connector.
   d. Fasten the intermediate bracket to the column lower bracket.
   e. Tighten the upper and lower bracket to instrument panel. If equipped, adjust dial pointer to coincide with selector lever.
   f. Attach toe plate, centering steering shaft in column before tightening nuts.
   g. On Mercury models, connect light switch harness block. Install ignition switch, bezels and connect ignition switch light bulb.
   h. On all models, install upper and lower trim shrouds and screws.

   On Fairlane, Falcon & Montego Models:
   a. Attach the column upper and lower brackets to the brake pedal support bracket and to the dash panel leaving all attaching nuts loose. Make sure that column is properly aligned with flex coupling input shaft connection.

   On Cougar & Mustang Models:
   a. Attach the column lower bracket to the brake pedal support leaving attaching nuts loose. Make sure that column is properly aligned to flex coupling input shaft connection.
   b. Tighten two rearward brake pedal support bracket nuts.
   c. Attach toe plate, tighten the two forward brake pedal support bracket nuts and connect flex coupling.
3. On models incorporating a tilt away column, connect tilt column cable.
4. On column shift models, connect transmission control rods.
5. On all models, connect electrical connections and battery cable.

1969 Continential, Mark III & Thunderbird, Figs. 14, 15 & 16

Removal

1. With battery ground cable disconnected, separate the flex coupling joint. On Continental models loosen clamp bolt.
2. Disconnect transmission control rod and remove steering wheel.
3. Remove steering column trim plate and disconnect column retainer from dash panel.
4. If equipped, disconnect tilt control cable from vacuum motor and bracket and the electrical connections from column.
5. Support column and disconnect lower clamp from upper clamp and remove column from vehicle.

Installation

1. Install column in vehicle engaging steering shaft with flex coupling.

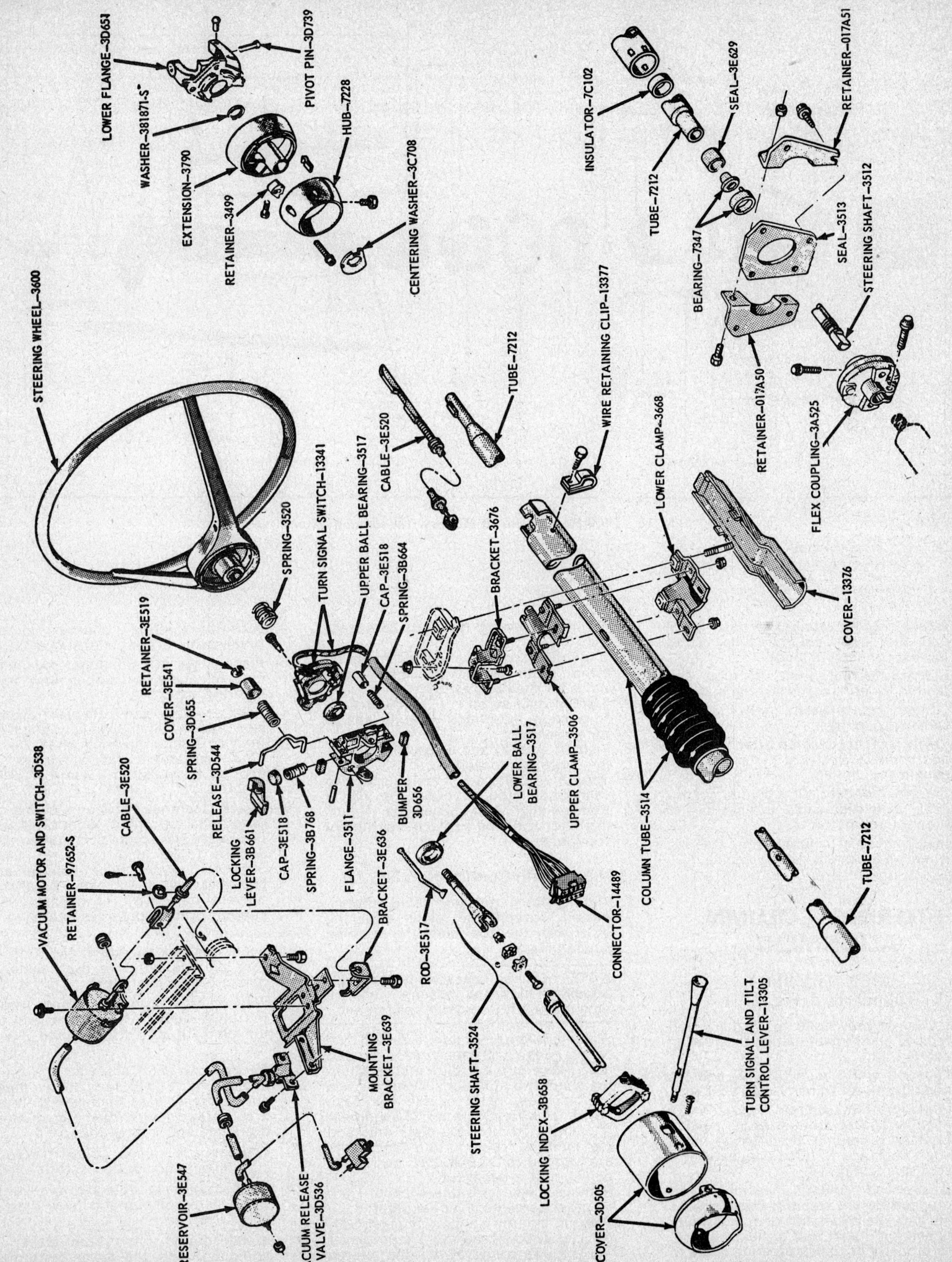

Fig. 12 1969 Ford & Mercury intermediate models. Typical tilt steering column

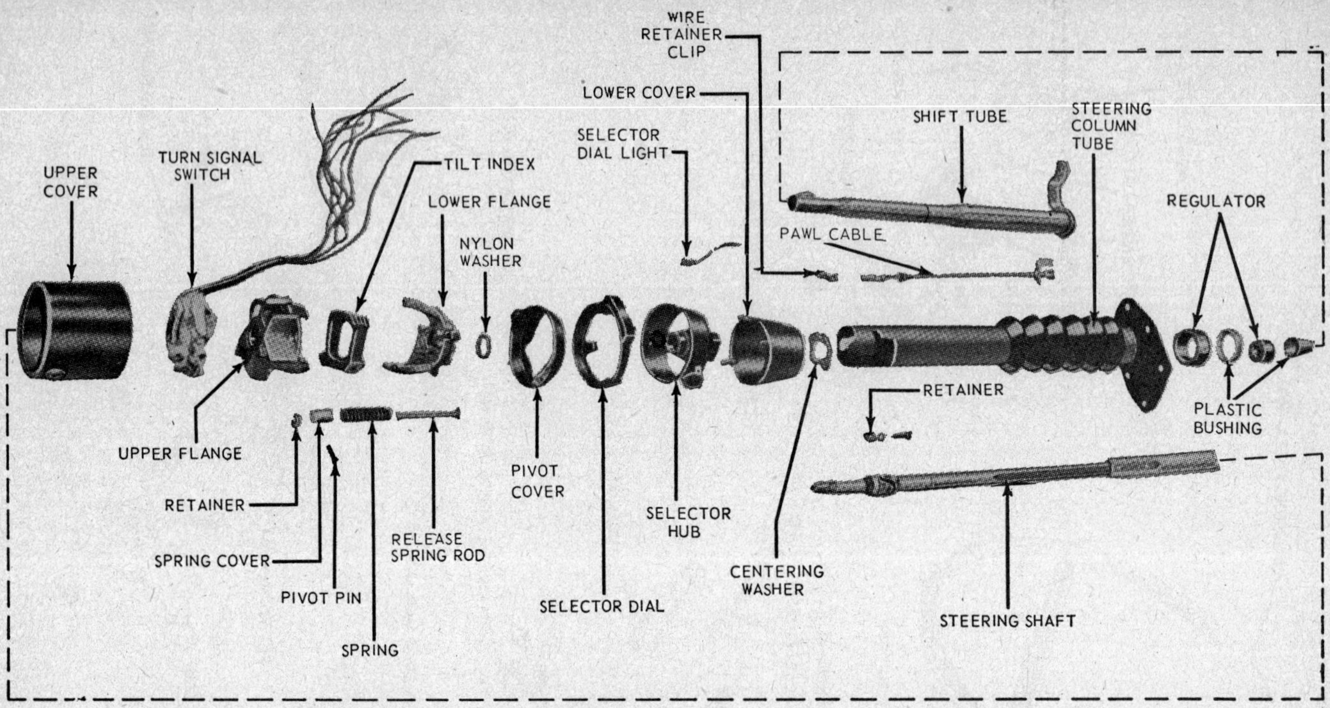

Fig. 13 1969 Ford, Lincoln & Mercury. Tilt steering column

2. Attach column retainer to dash panel and the lower column tube clamp leaving connectors loose.
3. Center steering shaft in tube and tighten lower flange attaching nuts.
4. Connect transmission control rod and the flex coupling.
5. Tighten panel to clamp attaching nuts and connect electrical connections to column.
6. If used, connect tilt column control cable to release cable retainer and to the vacuum motor.
7. Install steering column trim, steering wheel and connect battery cable.

STEERING COLUMN ALIGNMENT

1969 Mark III & Thunderbird, Fig. 17

1. Loosen slightly bolts (B) and (C) and remove instrument panel to steering column trim.
2. Loosen slightly bolts (D), (E) and (F) and disconnect transmission shift rod.
3. Working from under hood, install aligning spacer around visible portion of steering column. If necessary move shaft back and forth to completely install spacer, Fig. 18.
4. If steering column assembly had dropped toward steering gear, install a ⅛ inch diameter drill or rod through opening in upper half of flex coupling, to properly establish steering shaft to steering gear clearance.
5. Tighten retaining nuts (B) to a snug fit.
6. If aligning spacer cannot be freely rotated, ream or file the dash panel holes for greater clearance.
7. Tighten nuts or bolts (C), (D) and (E) to a snug fit. Check aligning spacer for looseness.
8. Tighten bolt (F) and again check aligning spacer for looseness.
9. Torque all bolts and nuts to specified value in alphabetical order, Fig. 17.
10. Remove alignment spacer and ⅛ inch diameter rod and replace instrument trim panel.

1969 Lincoln Continental, Fig. 19

1. With battery ground cable disconnected, loosen toe plate attaching nuts and the toe plate to steering column clamp.

NOTE: In some cases when toe plate attaching nuts are loosened the column may shift into proper alignment.

2. Alignment is acceptable if a .010 inch feeler gauge will freely pass around area between outer diameter of steering shaft and shift tube. If misaligned in this area, ream or file the dash panel holes for clearance. If alignment is maintained after toe plate nuts and toe plate to column nuts and bolts are torqued to 8-12 ft. lbs., no further adjustment is necessary.
3. Remove lower instrument panel trim. In order to remove upper right hand steering column support bracket bolt, vehicles with stereo radio tape player will require removal of the stereo radio tape player.
4. Loosen steering column support bracket and brake pedal support.
5. Place a $^{3}/_{16}$ inch diameter rod in flex coupling between upper and lower steering shaft.
6. Raise vehicle and remove transmission splash shield. Visually align column to center steering shaft in column. Tighten toe plate attaching nuts enough to maintain column in alignment.
7. Check clearance between steering column and instrument panel bracket. If necessary add shims part number MK-3054-A (.060 inch thick) or C3AZ-3054-A (.113 inch thick) between support and instrument panel bracket to prevent movement of column when support bracket bolts are tightened.
8. Loosen toe plate nuts previously tightened. Recheck alignment and if necessary enlarge holes in dash panel to obtain proper alignment.
9. With column properly aligned, torque toe plate nuts to 4-9 ft. lbs. and the toe plate to column clamp to 3-5 ft. lbs.
10. Recheck steering shaft to shift tube alignment. A .010 inch feeler gauge should freely pass completely around area between outer diameter of steering shaft and shifter tube.
11. With vehicle raised adjust transmission linkage and install splash shield.
12. Connect battery cable and remove the 3/16 inch diameter rod from the flex coupling.
13. Install stereo radio tape player (if equipped) and the lower instrument trim panel.

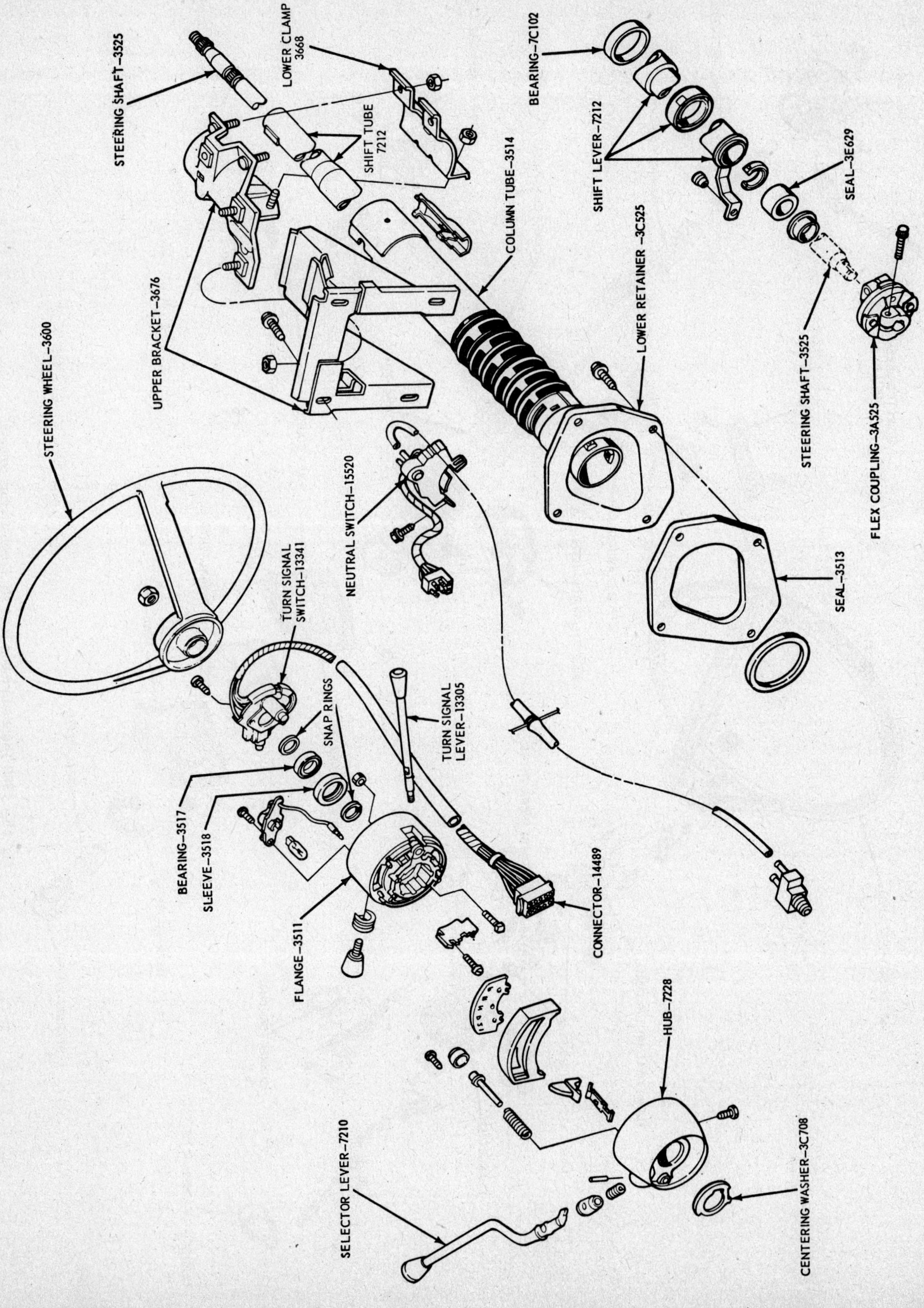

Fig. 14 1969 Lincoln Continental. Fixed steering column

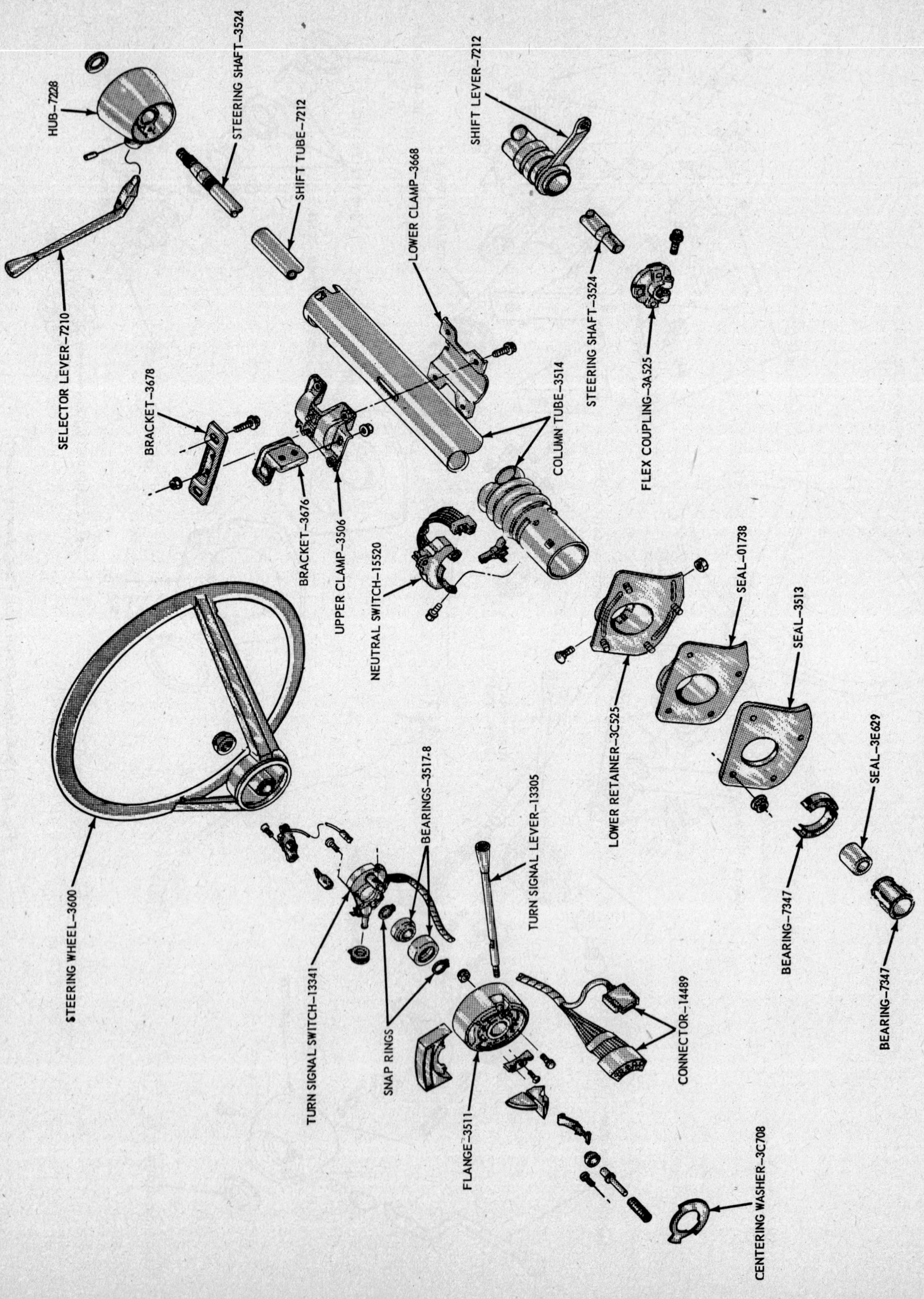

Fig. 15 1969 Thunderbird and Mark III. Fixed steering column

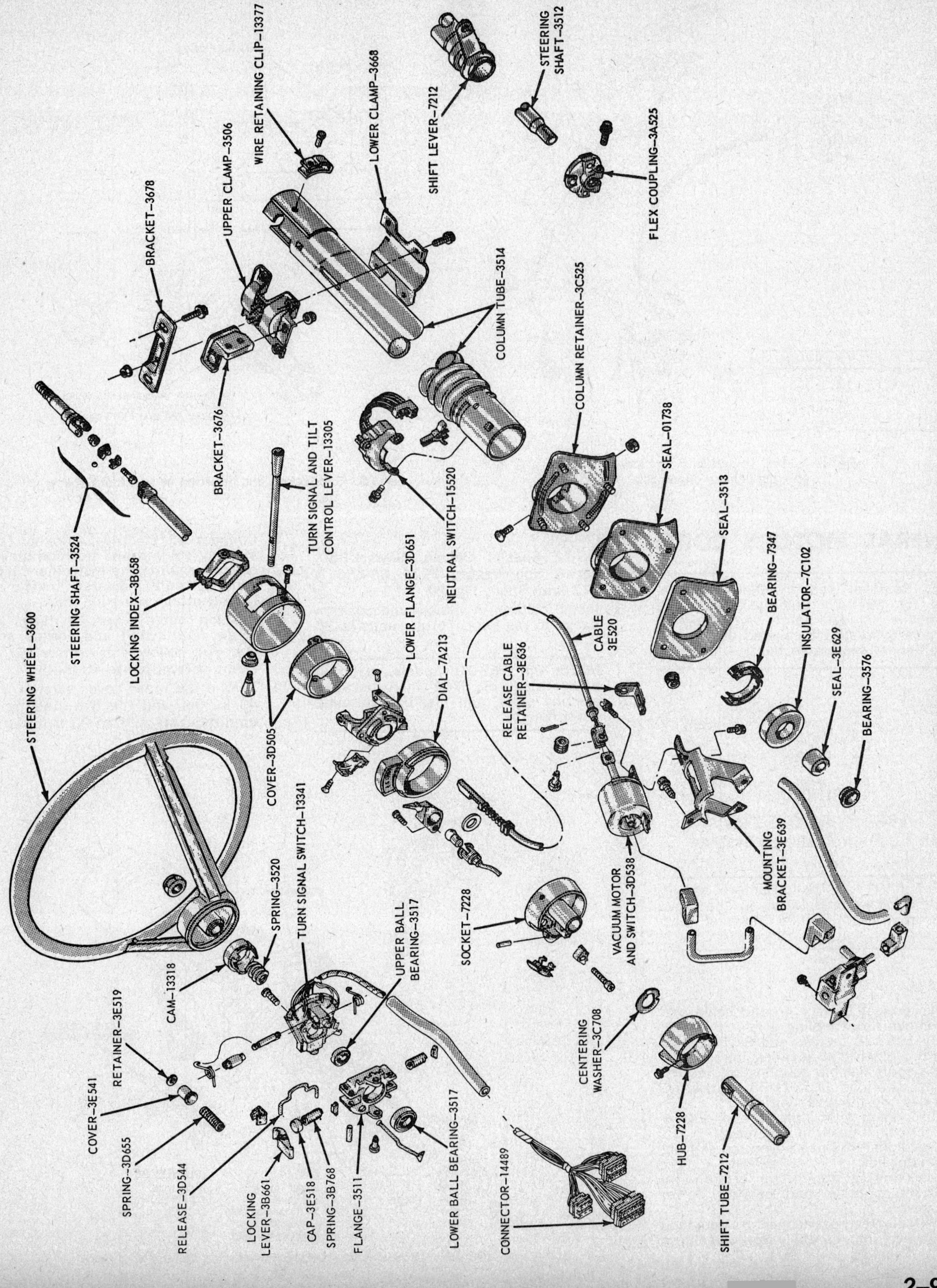

Fig. 16 1969 Thunderbird and Mark III. Tilt steering column

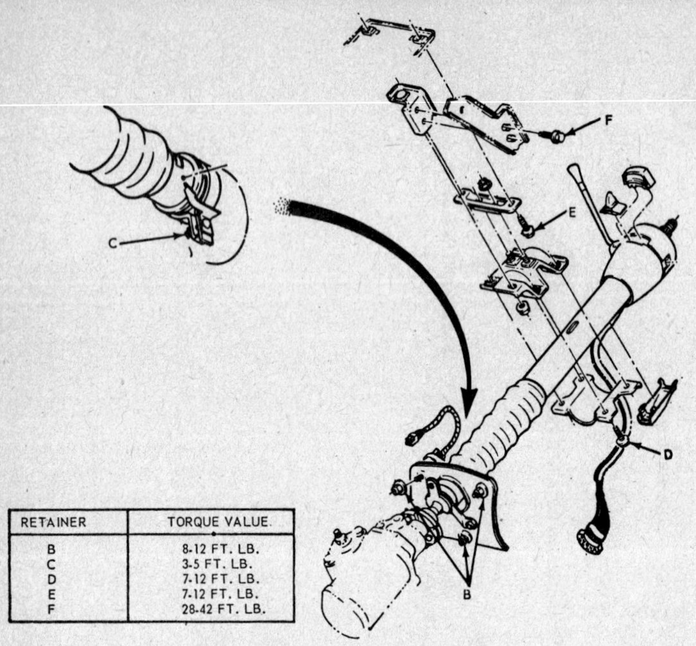

RETAINER	TORQUE VALUE
B	8-12 FT. LB.
C	3-5 FT. LB.
D	7-12 FT. LB.
E	7-12 FT. LB.
F	28-42 FT. LB.

Fig. 17 1969 Thunderbird and Mark III steering column alignment

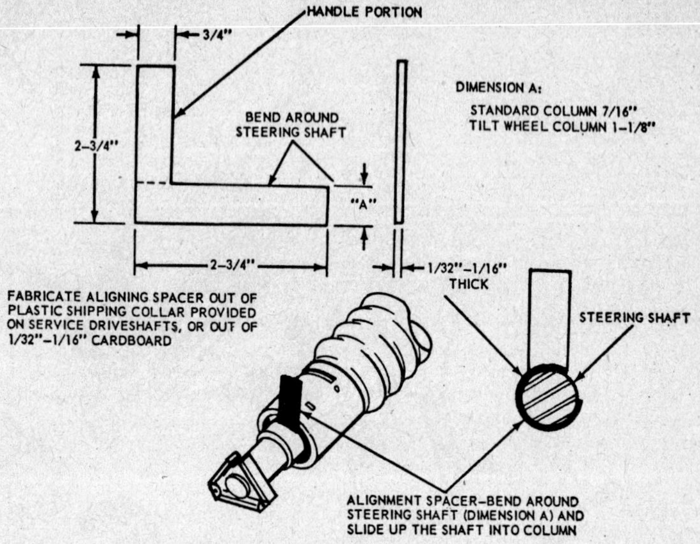

Fig. 18 Fabrication and insertion of alignment spacer

GENERAL MOTORS CORP.

NOTE: Removal and Installation procedures for 1974-75 Buick, Cadillac and Oldsmobile models with Air Cushion Restraint System, are covered at the end of the General Motors section.

BUICK

1969-73 & 1975 Less Air Cushion Restraint System

NOTE: For 1973-75 Apollo Removal and Installation procedures, refer to the Nova procedure in the "1969-75 Chevrolet" Section.

Removal

1. Disconnect battery ground cable and remove flex coupling shield (if used).
2. On 1969-70 models and Skyhawk, remove pinch bolt securing lower steering shaft flexible coupling to steering gear stub. On all 1971-75 models, remove two flex coupling nuts.
3. Disconnect shift linkage and disconnect clutch pushrod from pedal.
4. Disconnect toe pan cover and shift indicator link.
5. Disconnect column bracket from instrument panel and carefully lower column.
6. With all electrical connectors detached from column, carefully remove column from vehicle.

Installation

1969-72 Special Deluxe, Skylark-350, Custom & Sportwagon, 1970 GS-455 & 1969-72 Gran Sport, Fig. 20

1. Install column in vehicle and position flange to rag joint. Torque bolts to 30 ft. lbs.

NOTE: On 1969-70 models, maintain a clearance of 3/64 inch between steering gear and flexible coupling lower flange.

2. Connect all electrical connectors to steering column and position steering column against the instrument panel, connect shift indicator linkage and loosely attach retaining nuts.
3. Position outer cover to dash, start screw (Q), install and torque screw (L), (Q), followed by screw (R) and clamp screws (C) to 40 in. lbs.
4. Torque the inner cover screws (J) to 40 in. lbs. and the two steering column retaining nuts to 20 ft. lbs.

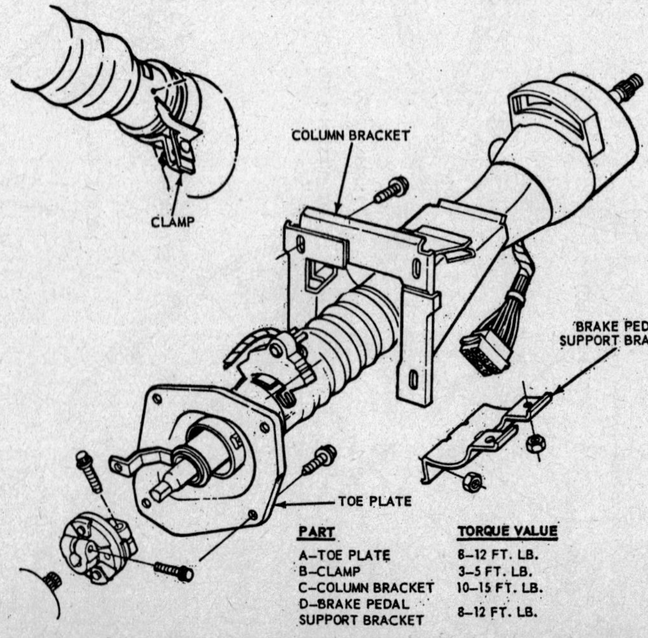

PART	TORQUE VALUE
A—TOE PLATE	8—12 FT. LB.
B—CLAMP	3—5 FT. LB.
C—COLUMN BRACKET	10—15 FT. LB.
D—BRAKE PEDAL SUPPORT BRACKET	8—12 FT. LB.

Fig. 19 1969 Lincoln Continental steering column alignment

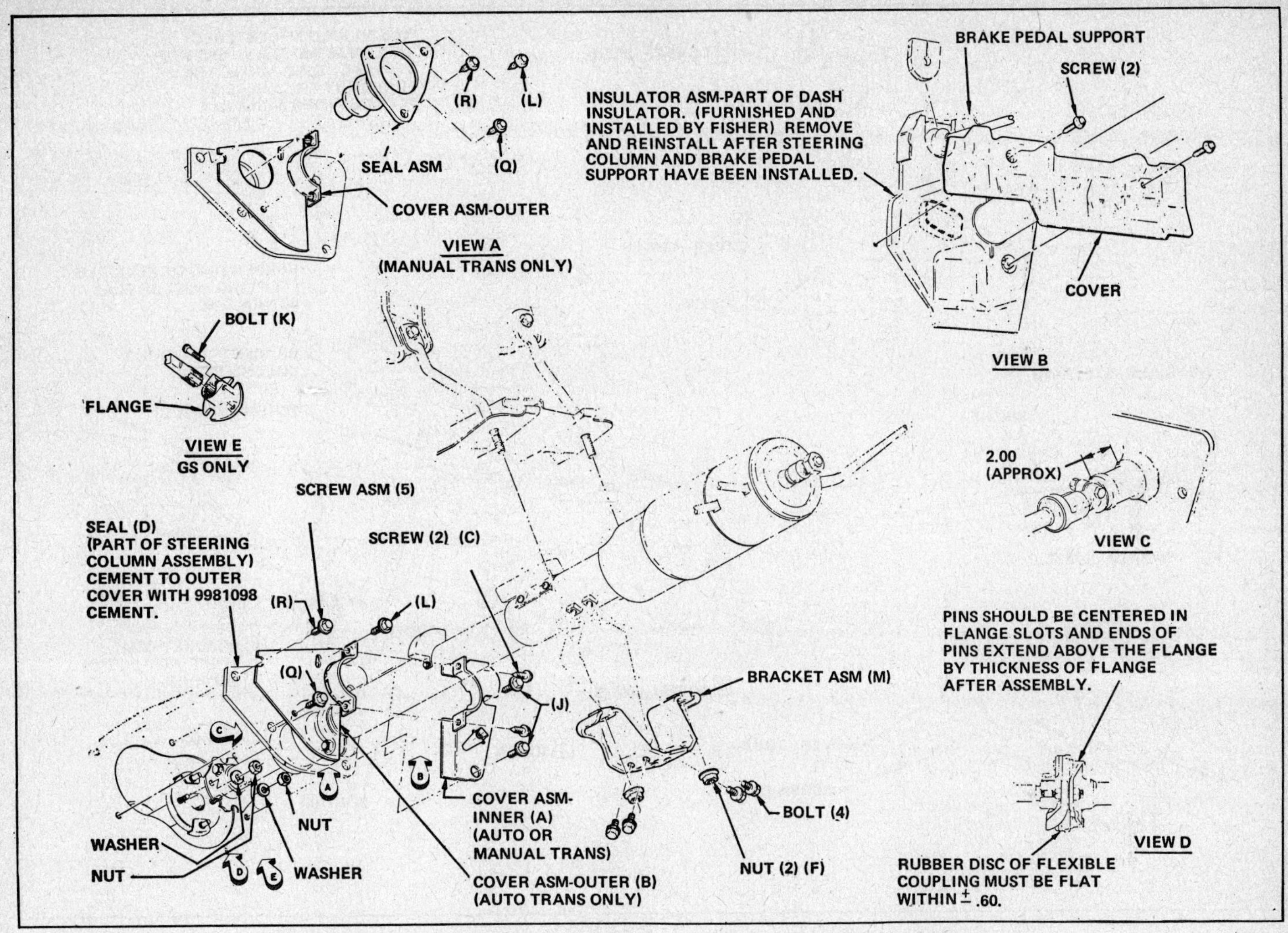

Fig. 20 1969-72 Buick intermediates, steering column installation sequence

5. Install insulator and cover and connect shift linkage to shift lever.

Installation All Other 1969-70 Models

1. Position column in vehicle with flexible coupling engaging steering gear stub shaft and torque flexible coupling pinch bolt to 30 ft. lbs.

NOTE: Maintain a clearance of 3/64 inch between steering gear and flexible coupling lower flange.

2. Connect all electrical connectors to steering column and position steering against instrument panel and connect shift indicator link.
3. Install and torque steering column to instrument panel retaining nuts to 20 ft. lbs.

NOTE: On all models except Electra 225 Custom, maintain a clearance of 3/16 inch between steering column and instrument panel. On Electra 225 Custom, this clearance is ⅛ inch.

4. Install toe pan cover to dash and torque screws to 40 in. lbs.
5. Connect shift linkage to shift lever.

Installation All Other 1969-73 Models & 1974-75 Full Size Less Air Cushion Restraint System, Fig. 21

1. Install column in vehicle and position flange to rag joint (coupling) and torque nuts to 20 ft. lbs.
2. Connect all electrical connectors to steering column and position steering column against instrument panel, loosely assemble nuts and shift indicator link.
3. Position lower cover against dash and start screw (7) in round inlet hole, then install and torque screw (8) to 40 in. lbs. Now torque screw (7) to 40 in. lbs.
4. Torque screws to 40 in. lbs. in following order: (3), (9), (4) and then remaining cover screws.
5. Torque steering column to instrument panel nuts to 20 ft. lbs. and connect shift linkage to shift lever.

Installation 1974-75 Intermediate Models Exc. Skyhawk

1. Assemble lower cover to steering column seal with projections provided on seal, then loosely clamp upper and lower halves of lower cover plate to column with the two clamp screws, Fig. 22.
2. If removed, attach steering column support bracket to column by first loosely installing bolts (1, 2, 3 & 4) and then torqing bolts (2 & 3) and then (1 & 4) to 15-30 ft. lbs., Fig. 23.
3. Position column in vehicle and align flexible coupling then install washers and nuts and torque to 20 ft. lbs.

NOTE: Do not install flex coupling shield at this time.

4. Position column in upper bracket and guide and retain with nuts (A & B), Fig. 23, then position lower cover plate assembly on bulkhead and install screws (1, 2, 3 & 4) and torque to 45 inch lbs., Fig. 22.
5. Torque clamp screws (7 & 8) to 45 inch lbs., then install and torque cover screws (5 & 6) to 45 inch lbs., Fig. 22.
6. Torque nuts (A & B), Fig. 23, to 15-25 ft. lbs. and connect shift indicator linkage and install dash cover.
7. Install flex coupling shield.

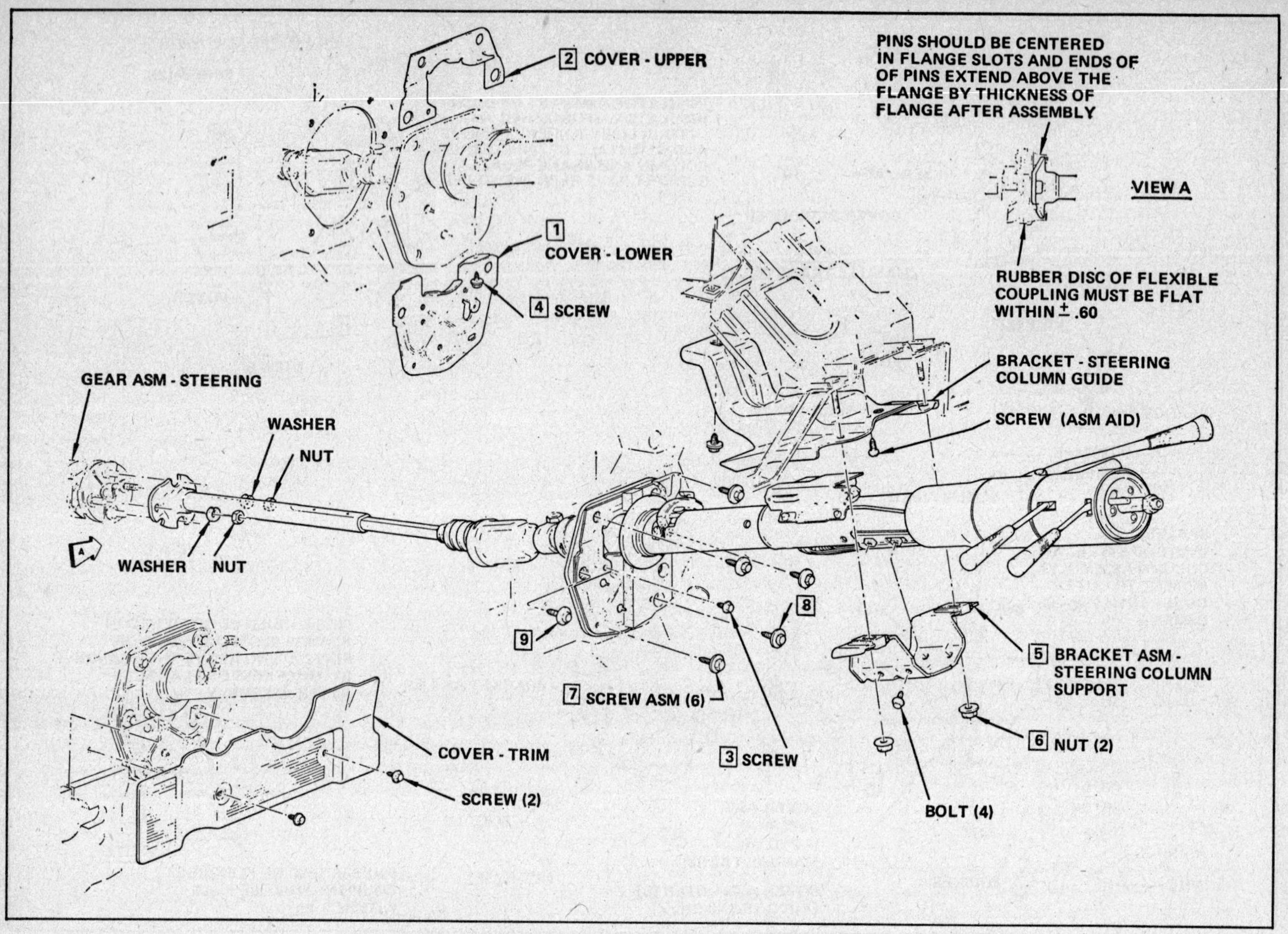

Fig. 21 1969-72 Buick full size, all 1973 models and 1974-75 full size less Air Cushion Restraint System. Steering column installation sequence

Installation, 1975 Skyhawk, Fig. 33

1. Install column into vehicle aligning flex coupling with steering shaft and torque pinch bolt to 30 ft. lbs.
2. Connect all electrical connectors to column, then position support bracket onto instrument panel and install and torque nuts to 20 ft. lbs.
3. Install and torque the three toe plate screws to 45 inch lbs.
4. Install steering wheel and flex coupling shield.

CADILLAC

1971-75 & 1976 Seville Less Air Cushion Restraint

Removal

1. Center steering wheel to gain access to steering column pinch bolt.
2. Disconnect battery ground cable and transmission control linkage.

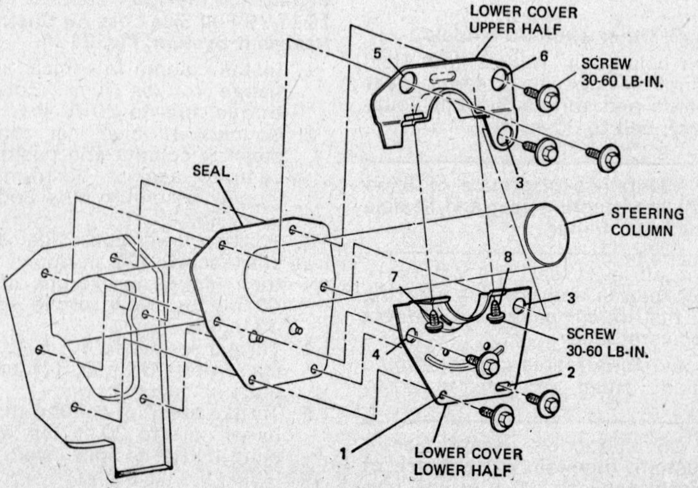

Fig. 22 1974-75 Buick intermediate. Steering column lower cover

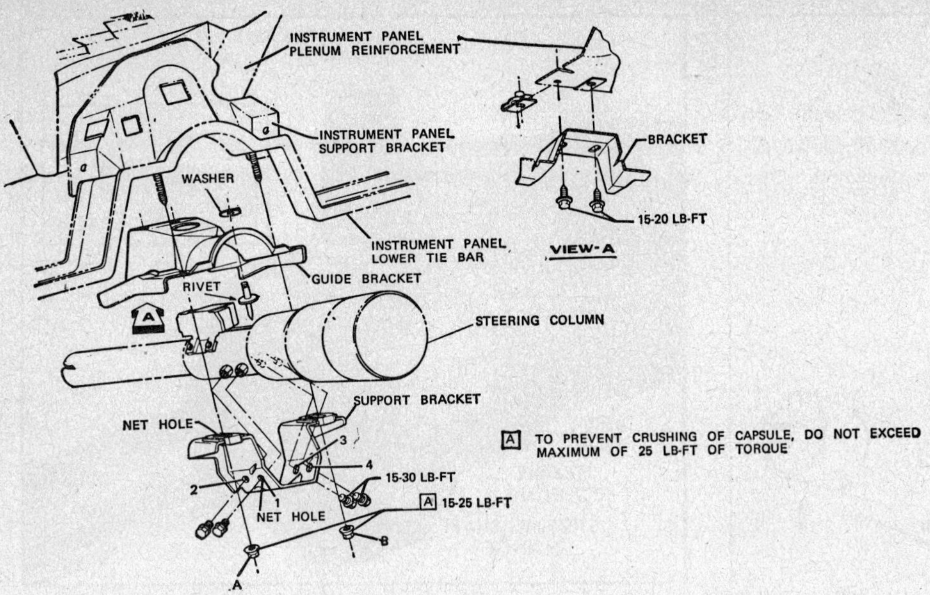

Fig. 23 1974-75 Buick intermediate. Steering column mounting

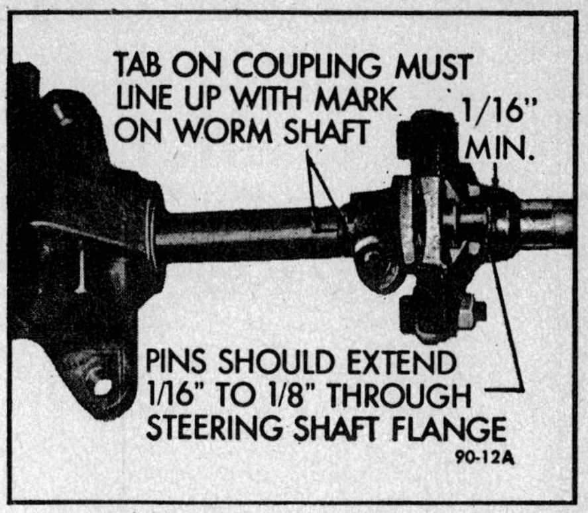

Fig. 24 Flex coupling installation

3. On all models except Eldorado, remove clamp screw and lock washer that secure flexible coupling to steering shaft.
4. On Eldorado models, disconnect flex coupling from both the lower shaft and the steering gear then remove screw retaining the universal joint to the lower steering shaft and remove the shaft from the universal.
5. To remove steering column lower cover on 1971-73 vehicles, proceed as follows:
   a. Disconnect ash tray electrical connector, push ash tray out and remove screws at bottom of ash tray.
   b. Remove ash tray door cover and ash tray.
   c. Remove four lower cover to upper cover retaining screws and loosen lower cover from lower instrument panel.
   d. Pull lower cover straight out to expose wiring connectors on rear side of cover and disconnect Twilight Sentinel amplifier (if used).
   e. Remove courtesy lamp from bracket and turn signal flasher unit from clip.
6. To remove steering column lower cover on 1974-75 vehicles, remove the four screws securing lower cover to lower steering column cover reinforcement, and remove the four screws securing lower cover to instrument panel horizontal support then remove lower cover. On Seville models, disconnect rear defogger circuit breaker from cover.
7. Disconnect turn signal wiring connector and Cruise Control harness (if used) and on 1974-75 vehicles, remove shift cable from shift bowl.
8. On all except Seville models, loosen two vertical bolts at steering column upper support to allow removal of shim packs from either side of support and remove shim.

NOTE: Do not remove upper bolts, as column may bend under its own weight.

9. Slide toe pan seal up the steering column as far as possible.
10. Place carpet out of way to gain access to column lower support bracket and detach the toe pan from the mounting bracket.
11. Separate bracket halves and remove bracket.
12. On 1971-73 vehicles, remove shift pointer screw and pointer.
13. Disconnect upper column bracket from column support and disconnect ignition and back-up light switch connector and parking brake release hose.
14. Carefully remove column from vehicle.

NOTE: If shaft hangs up in flex coupling, reattach upper mounting bracket and pry stub shaft from coupling using a screwdriver.

15. Remove two shim retainers from upper mounting bracket.

Installation

1. On all except Seville models, install shim retainers on column upper mounting bracket and position toe pan seal on column.
2. Install lower end of column through opening in toe pan and into flex coupling, aligning flat on steering shaft with flat in coupling.

CAUTION: On Seville models, steering shaft must be installed with 5/16 inch clearance between shaft and pot joint, as damage to lower steering shaft bearing may result. Torque pot joint clamp to 40 ft. lbs.

3. Connect ignition switch connector. Install two upper column bracket to upper column support bolts and leave loose.
4. Install column upper mounting bracket shims in same location and quantity as found during column removal.

NOTE: On all except Seville models, place shims on top of retainers with tangs on shims fully inserted in retainer slots. Do not place shims directly against aluminum capsule.

5. Connect all electrical connectors, parking brake release hose and the cruise control wiring harness (if used).
6. Install lower mounting brackets to toe pan screws and leave loose.
7. Install mounting bracket and clamp halves together, then connect shift pointer and shift linkage.
8. On all except Seville models, align column to instrument panel as follows:
   a. Upper edge of column shift bowl (at column centerline) must be the proper distance from instrument panel bezel. On standard columns this distance should be 4-7/16 inch and on tilt columns this distance should be 2¼ inch.
   b. After making adjustment and column is centered within instrument panel bezel, lightly tighten the two upper mounting bracket to support bolts.
   c. Check alignment of flex coupling disc. Steering column and gear are properly aligned when rubber disc is flat and pins are centered in "C" slots in upper flange, Fig. 26.

NOTE: Pins should extend beyond edges of flange slots equally at both locations. After alignment torque clamp screw to 30 ft. lbs.

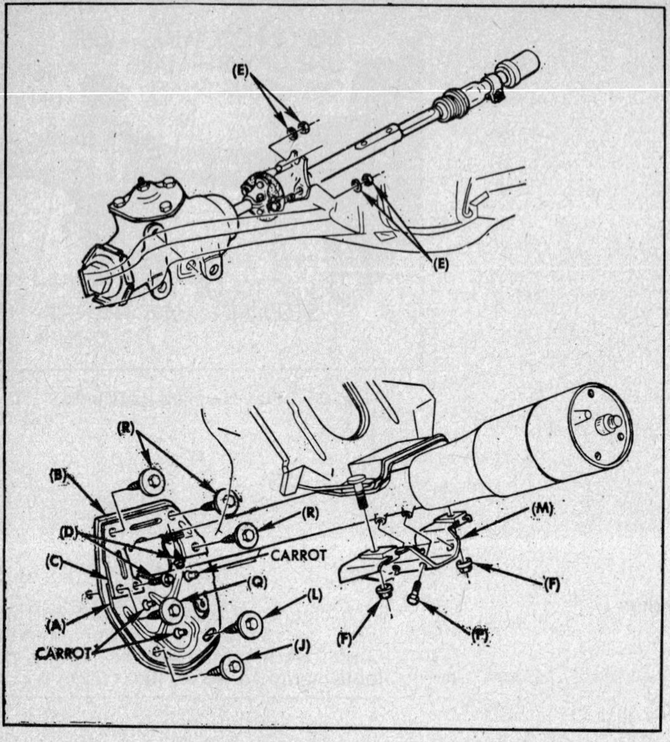

Fig. 25 1971-75 Chevrolet and 125″ wheelbase Station Wagon

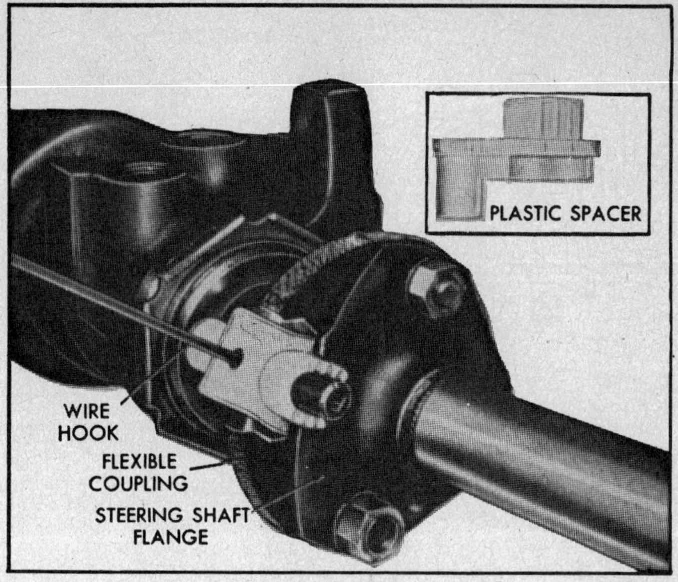

Fig. 26 Plastic alignment pin spacer

9. On Seville models, position steering column so that gear shift boss clears instrument panel bezel, then center steering column within instrument panel bezel and lightly tighten the two upper mounting nuts.
10. Install steering column lower support and adjust as follows:
    a. Place lower half of mounting bracket against column and lightly tighten two lowest bracket to toe pan screws.
    b. Position upper half of mounting bracket and lightly tighten remaining bracket to toe pan screws.
    c. Torque clamp bolts to 18 inch lbs. on all except Seville models and 47 inch lbs. on Seville models.
    d. Start engine and turn steering wheel from stop to stop to equalize position of column.
    e. Torque the two upper mounting bracket to upper column bolts to 20 ft. lbs. on all except Seville models and 24 ft. lbs. on Seville models.

NOTE: Do not exceed the critical 20 or 24 ft. lb. torque to assure satisfactory breakaway action of the capsules in the event of a collision.

11. To install steering column lower cover on 1971-73 vehicles, proceed as follows:
    a. Connect Twilight Sentinel amplifier (if used).
    b. Install courtesy lamp on cover and the turn signal flasher in clip.
    c. Attach lower cover to the upper cover.
    d. Tighten lower to lower instrument panel retaining screws.
    e. Install ash tray, ash tray door cover and connect electrical connector.
12. To install steering column lower cover on 1974-75 vehicles, position lower cover in place and install the four screws securing cover to instrument panel horizontal support, then install the four screws securing lower cover to lower steering column cover reinforcement. On Seville models, connect rear defogger circuit breaker.
13. Move column seal down to cover gap between bezel and shift bowl.
14. Check linkage for proper operation and connect battery ground cable.

1969-70

Removal

1. Disconnect battery ground and on all models except Eldorado, disconnect flex coupling flange from steering shaft. On Eldorado models:
    a. Disconnect flex coupling from both the lower shaft and the steering gear.
    b. Disconnect universal joint from steering shaft and remove lower steering shaft from column.
    c. Remove flex coupling from steering gear, universal joint to shaft and then the universal joint from shaft.
2. Disconnect the lower shift lever from column and the lower steering clamp from cowl.
3. Disconnect mounting bracket from steering column jacket.

NOTE: Lower clamp must be removed before loosening column to upper attachment.

4. Disconnect lower column from instrument panel bezel assembly.
5. Loosen lower cover to lower instrument panel screws and pull lower cover straight out to gain access to wiring connectors on rear side of cover.
6. Disconnect ash tray wiring, Twilight Sentinel amplifier (if used) and remove courtesy light bulb and socket from housing.
7. Remove turn signal flasher and hazard warning flasher from clips on rear side of cover.
8. Remove shift indicator screw and pointer.
9. Disconnect electrical connectors and vacuum hoses from the neutral and back-up light switch assembly.
10. Pull rubber cover up, detach carpet from floor and disconnect retainer from toe pan.
11. Disconnect upper column bracket from steering column support and note number of shims installed.

NOTE: Do not let column support its own weight as damage to the collapse joint may result.

12. Disconnect turn signal electrical connector, ignition switch and Cruise Control wiring harness from connector located on parking brake bracket.
13. Carefully remove steering column from vehicle.

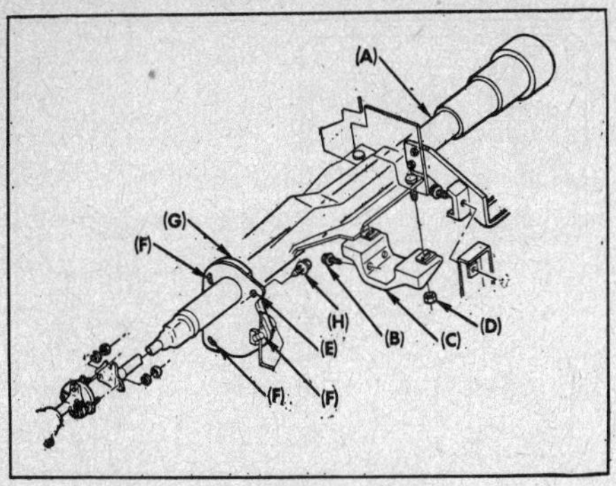

Fig. 27 1970-75 Camaro, steering column installation

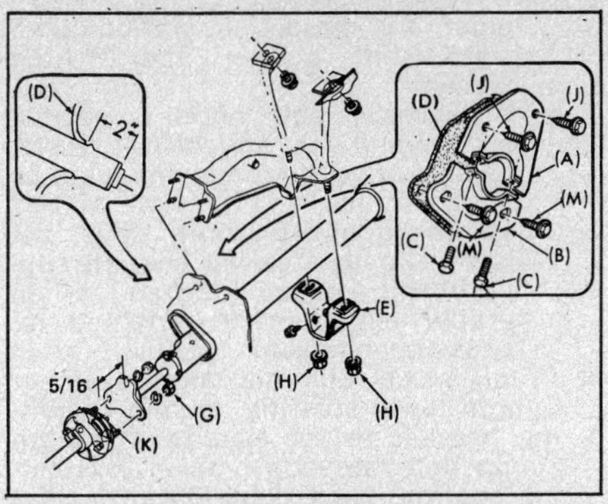

Fig. 28 1969-74 Nova and 1969 Camaro, steering column installation

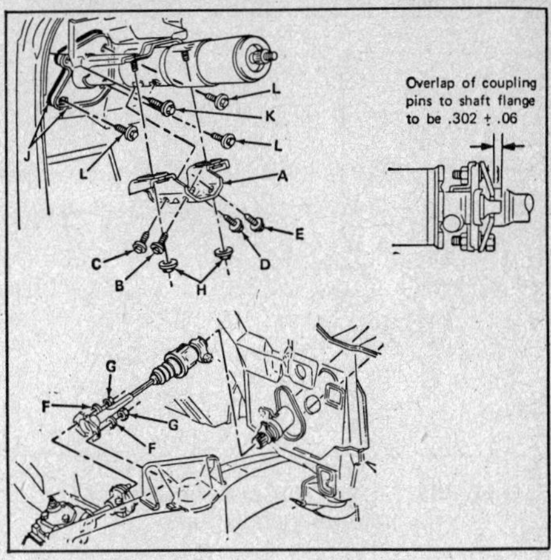

Fig. 28A 1975 Nova steering column installation

Installation

1. Position toe pan seal and retainer and carefully install column into position, make sure that flat on shaft aligns with flat in coupling flange.

NOTE: Do not let column support its own weight.

2. Position column and attach upper column bracket to column support. Install same number of shims that were removed.

NOTE: Make certain that L shaped hardened spacer is between shims and the release capsule.

3. Connect neutral safety and back-up light switch, vacuum hoses, turn signal and ignition switch connectors. Connect Cruise Control wiring harness (if used).
4. Install flex coupling upper flange to steering shaft clamp screw and leave loose.
5. Attach lower mounting bracket to steering column jacket and then the lower mounting bracket to toe pan, leave all screws loose.
6. **On Eldorado Models:**
   a. Position universal joint on lower shaft with flat on shaft in line with split in joint. Torque screw to 80 ft. lbs.
   b. Position flex coupling on steering gear with large pin facing up.
   c. Install shaft assembly with universal joint engaging steering column shaft and align split in universal joint with flat on steering column shaft.
   d. Engage flex coupling with flange on shaft and torque bolts to 20 ft. lbs.
   e. Position flex coupling on stub shaft. There should be a clearance of .040 inch between coupling and adjuster plug. Torque screw to 30 ft. lbs.
   f. Secure universal joint to steering column and torque screw to 80 ft. lbs.
7. Align all columns as described under "Steering Column Alignment".
8. Install carpet. Slide steering column rubber cover in place and secure lower shift lever to column.

NOTE: Make certain that column lever support does not bend or twist during tightening.

9. Install turn signal flasher and hazard warning flasher units into clips.
10. Connect ash tray wiring and Twilight Sentinel amplified (if used).
11. Install courtesy light bulb and socket into housing and position steering column lower cover in position against the lower instrument panel.
12. Attach lower cover to instrument panel bezel assembly and tighten all instrument panel screws.
13. On all models adjust transmission linkage and connect battery ground cable.

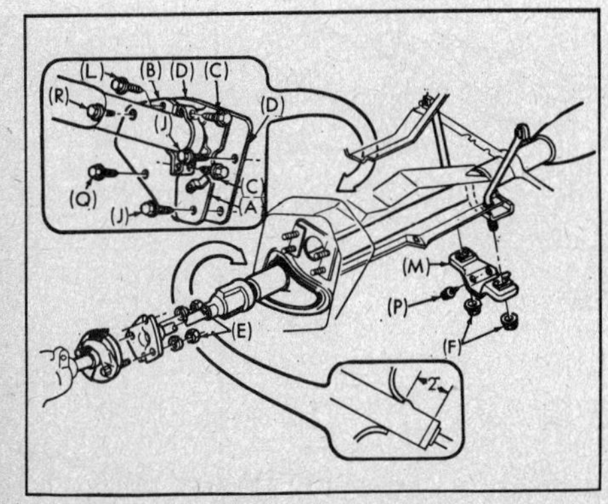

Fig. 29 1969-73 Chevelle, 1970-73 Monte Carlo and 1970-72 116" wheelbase Station Wagon, steering column installation

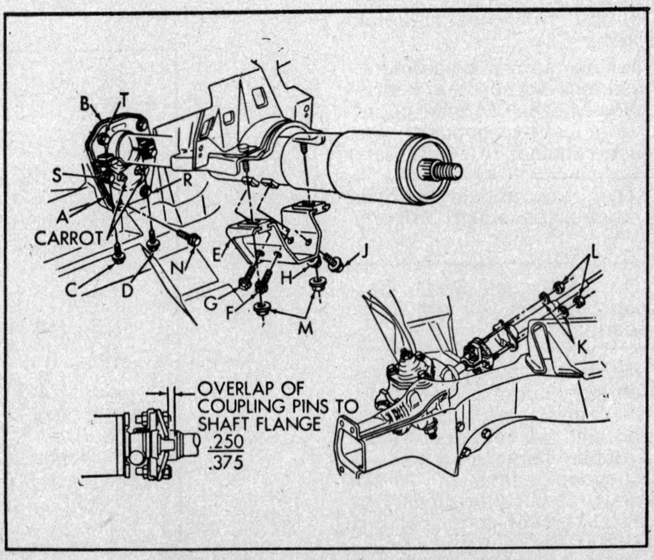

Fig. 30 1974-75 Chevelle and Monte Carlo steering column installation

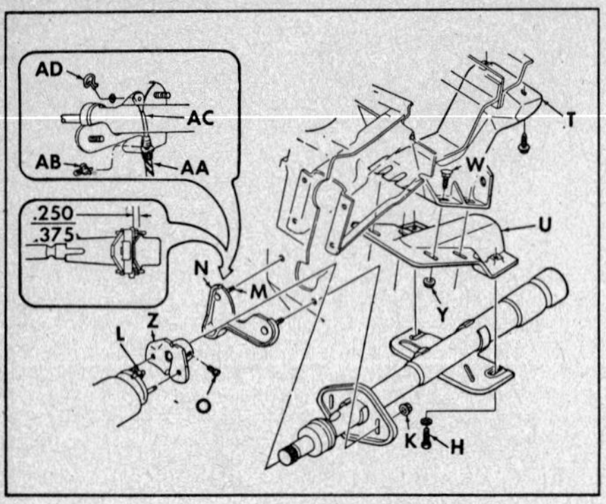

Fig. 31 1969-75 Corvette steering column installation

1969-70 Steering Column Alignment

NOTE: Do not raise vehicle as full weight of vehicle must be on its suspension for correct alignment.

1. Place front wheels in straight ahead position.
2. Remove steering column lower cover.
3. Loosen shift pointer retaining screw.
4. Loosen upper steering column mounting bracket from upper steering column support.
5. On all models except Eldorado, loosen steering shaft to flex clamp screw. On Eldorado models, loosen steering extension shaft to universal joint clamp screw.
6. a. Upper edge of column shift bowl (at column centerline) must be distance as specified from instrument panel bezel. On all models except Eldorado, this distance should be 2-3/16 inch for Tilt and Telescope columns and 4½ inch on models less Tilt and Telescope columns. On Eldorado models, with Tilt and Telescope columns this distance should be 2-9/16 inch and vehicles less Tilt and Telescope columns 4-5/16 inch.
   b. After making above adjustment, center column within the instrument panel bezel. Torque upper mounting bracket to support nuts (at release capsule) to 20 ft. lbs.

   NOTE: The bezel should be only slightly depressed when properly aligned.

   NOTE: Do not exceed 20 ft. lbs., or release capsules will not function properly.

   c. Check alignment of flex coupling disc. Steering column and gear are properly aligned when rubber disc is flat and pins are centered in "C" slots in upper flange. Pins should extend beyond edges of flange slots equally at both locations Fig. 24. After alignment torque clamp screw to 30 ft. lbs.
   d. Check shift lever for excessive shift effort, binding or squeaking. Check ignition switch and lock cylinder for proper operation then align shift pointer and tighten screw.
   e. If excessive shift effort is needed or binding or squeaking exists, loosen upper mounting bracket to support nuts and remove or add shims equally on both sides, between upper steering column support and spacer brackets. When proper alignment is obtained repeat previous step.
8. On all except Eldorado models, finger tighten lower steering column mounting bracket to toe pan then tighten clamp bolt retaining bracket to steering column. Next tighten two bracket to toe pan screws.

   On Eldorado Models:

   a. Finger tighten three bracket to toe pan screws.
   b. Tighten bracket to steering column clamp bolt.
   c. Tighten outer and inner toe pan screw.
9. On all models torque all clamp screws and bolts to 10 ft. lbs.
10. Install steering column lower column.

NOTE: Additional adjustment is provided at upper steering column support to upper shroud panel at base of the windshield. To be used only if column alignment cannot be obtained by the preceding method.

CHEVROLET
1969-75 All Except Vega

Removal

1. With battery ground cable disconnected remove steering wheel and on 1974-75 models, loosen dash mounting plates.
2. Disconnect flex coupling and transmission control linkage, on floor shift models disconnect backdrive linkage.
3. Disconnect all electrical connectors from column and remove floor trim cover.
4. On all except Corvette and 1971-75 Camaro, disconnect floor pan cover halves and seal from floor pan. On 1971-75 Camaro, disconnect column plate from floor pan and on 1969-75 Corvette, disconnect floor pan bracket from mounting studs.
5. On 1975 Nova, remove column plate to floor pan screws and if used, disconnect shift indicator cable.
6. On all except 1971-75 Camaro, remove instrument panel trim cover. On 1971-75 Camaro, remove ash tray assembly, disconnect steering column panel from instrument panel and lower panel and heater or air conditioner control as an assembly.
7. Remove transmission indicator cable (if used).
8. Disconnect column bracket from instrument panel and remove column from vehicle. If column has lower shift levers, additional help should be obtained to guide shift levers through opening in firewall.

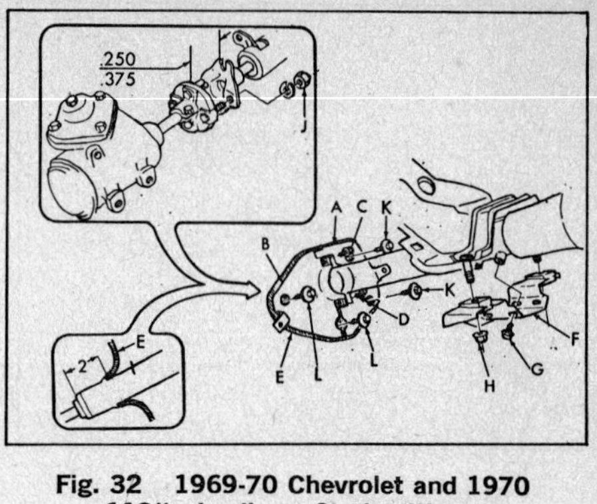

Fig. 32 1969-70 Chevrolet and 1970 119" wheelbase Station Wagon

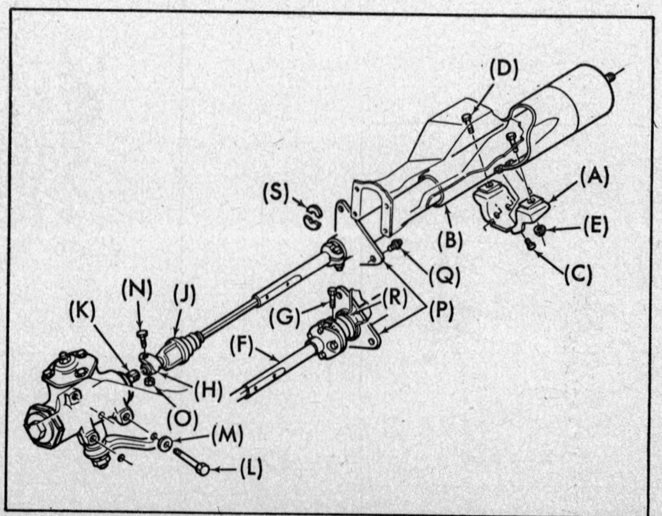

Fig. 33 1971-75 Vega & 1975 Astre, Monza, Skyhawk & Starfire steering column installation

Installation 1971-75 Chevrolet and 125 Inch Wheel Base Station Wagon, Fig. 25

1. Using the three seal "Carrots" assemble lower dash cover (A) to seal (B).
2. Loosely assemble (A) and (C) dash covers with screws (D). Dash cover must be able to slide on column.
3. Install bracket (M), torque bolts (P) to 15 ft. lbs. and install column in vehicle.
4. Align flange to rag joint (flex coupling), and torque nuts (E) to 20 ft. lbs.
5. Loosely assemble capsule nuts (F) with instrument panel bracket (M) to upper column bracket.
6. Position lower cover (A) to dash installing screws (J) and (L), torque (L) then (J) to 35 in. lbs.
7. Torque left hand upper to lower cover, inboard screw (D) to 35 in. lbs. and then screw (Q) followed by clamp screw (D) to same specification.
8. Torque screws (R) to 35 in. lbs. followed by nuts (F) to 20 ft. lbs.
9. Remove plastic spacers from flexible coupling pins (if used), Fig. 26.
10. Install transmission indicator cable (if used) and instrument panel trim cover.
11. Connect transmission control linkage, install steering wheel and connect battery ground cable.

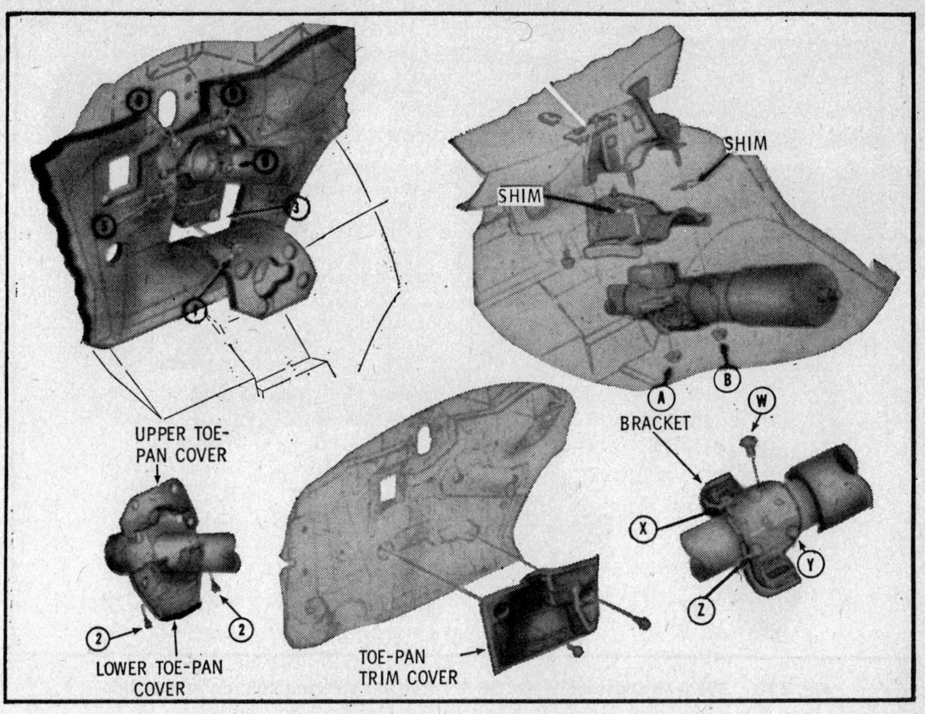

Fig. 33A 1971-75 Cutlass steering column installation

Installation 1970-75 Camaro, Fig. 27

1. Install column in vehicle and connect all electrical connections.
2. Position shaft flange to flexible coupling and torque nuts to 20 ft. lbs.
3. Loosely assemble capsule nuts (D) to instrument panel studs.

NOTE: Before tightening screw (H), align column plate seal with plate.

4. Align plate (G) to floor pan and torque screw (H) to 35 in. lbs.
5. Torque capsule nuts (D) to 20 ft. lbs. and install screw at location (F) and torque to 35 in. lbs.

NOTE: If any of the following conditions exist, remove column and inspect for bent, damaged or incorrectly installed parts or possible frame damage.

a. Flexible coupling must not be distorted due to pot joint bottoming in either direction.
b. With front wheels off ground, steering wheel must be free to rotate from stop to stop with no binding, stickiness or lumpiness.
c. Clearance between intermediate shaft to upper control arm forward bolt must be .15 inch minimum.

6. Install floor pan trim cover, column panel assembly, ash tray and steering wheel.
7. Connect transmission control linkage or backdrive linkage on floor shift models and connect battery ground cable.

Installation 1969 Camaro, 1969-74 Nova, Fig. 28

1. Attach upper to lower steering column dash covers (A) and (B) and snug screws (C) insuring looseness between column and covers.
2. Cement seal (D) against upper and lower cover.
3. Position rolled section of dash seal at column lower reaction tab, two inches from bottom of mast jacket.

NOTE: Rolled section of seal (D) must be directed down toward front of vehicle and must remain in this position during and after column installation.

4. Install column in vehicle. Connect all electrical connectors to column and loosely install nuts (H).
5. Torque flexible coupling nuts (G) to 20 ft. lbs.

NOTE: Do not tighten column to instrument panel unless vehicle is supporting its own weight.

6. Torque nuts (H) to 20 ft. lbs. while maintaining flexible coupling to flange clearance of 5/16 inch.

NOTE: Through the following steps, upper cover (A) and lower cover (B) must be able to move to the established position determined by flexible coupling and instrument panel location and attachment.

7. Push upper (A) and lower (B) cover assembly to align all cover to dash attaching screws by moving upper and lower cover assembly (A) and (B).

NOTE: Do not permit rolled section of seal to reverse itself.

8. Holding both upper and lower cover assemblies against dash and column, install and tighten screws (M).
9. Torque screws (C) to 40 in. lbs., loosen screws (M), torque screws (J) and then again screws (M) to 40 in. lbs.
10. If used, on Nova models, remove plastic spacers from flexible coupling pins, Fig. 26.

NOTE: On all models except Ventura, with steering wheel in a straight ahead position and no torque applied on steering wheel, a .04 inch (4-ply coupling) gap or a .07 inch (7-ply coupling) gap must be maintained between steering shaft flange and coupling rivets. A gap less than specified indicates that column must be realigned.

11. Install transmission indicator cable (if used), instrument panel trim cover.
12. Connect transmission control linkage, install steering wheel (if removed) and connect battery ground cable.

Installation, 1975 Nova, Fig. 28A

1. Attach bracket (A) and sequentially install bolts (B, C, D, and E) and torque to 22 ft. lbs.
2. With intermediate shaft and rag joint installed, position column in vehicle and against rag joint, then install washers (F) and nuts (G) and torque to 18 ft. lbs.
3. Loosely install nuts (H), then position cover and seal (J) and install and torque screw (K) to 40 inch lbs.
4. Torque nuts (H) to 20 ft. lbs., then install and torque nuts (L) to 40 inch lbs.
5. Install shift indicator cable (if used), transmission shift linkage and steering wheel.

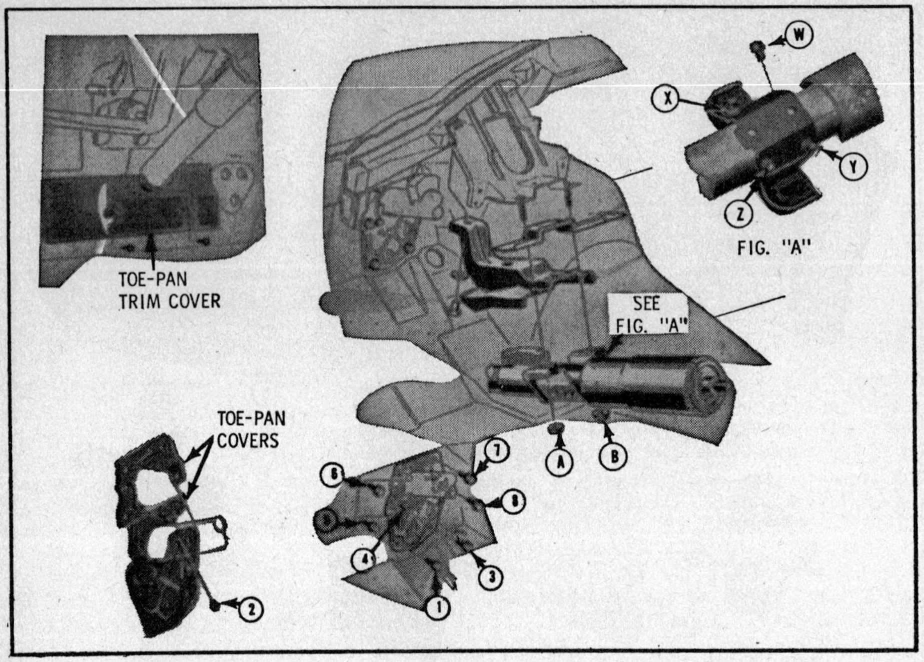

Fig. 33B 1971-75 Oldsmobile 88, 98 & Toronado steering column installation

Installation 1969-73 Chevelle, 1970-73 Monte Carlo and 1970-72 116 Inch Wheel Base Station Wagon, Fig. 29

1. Loosely assemble inner and outer dash covers (A) and (B) onto column, snug screws (C) insuring looseness between column and covers.
2. Cement cover seal (D) to outer cover (B).
3. Position rolled portion of dash seal at column lower reaction tab approximately two inches from bottom of jacket.

NOTE: Rolled section of seal must be facing down toward front of vehicle and must remain in this position during and after installation.

4. Install column in vehicle and connect all electrical connectors to column.
5. Align flange to rag joint (flex coupling), and torque nuts (E) to 20 ft. lbs.
6. Loosely assemble nuts (F) on brace rod and position outer cover to dash and snug screw (Q).
7. Install screws (L) and (R) and torque screw (L) followed by screws (Q) and (R) to 35 in. lbs.
8. Torque clamp screws (C) and then the inner cover screws (J) to 35 in. lbs.
9. Torque nuts (F) to 20 ft. lbs., install transmission indicator cable (if used) and instrument panel trim cover.
10. Connect transmission control linkage, install steering wheel and connect battery ground cable.

NOTE: On 1971-73 vehicles, the upper control arm shaft to steering shaft clearance must be .12 inch. Engine to pot joint clearance must be .37 inch and clutch linkage to pot joint clearance must be .75 inch.

Installation 1974-75 Chevelle and Monte Carlo, Fig. 30

1. Assemble lower dash cover (A) to seal with two "carrots" (part of seal), then loosely assemble lower (A) and upper (B) dash cover with screws (C) and (D) making certain that dash cover slides on column.
2. Attach bracket (E) to jacket and install bolt (F) first, (G) second and (H) and (J) last and torque to 15 ft. lbs.
3. Connect ignition switch electrical connector, then position column in vehicle and onto flex coupling rag joint. Install washers (K) and nuts (L) and torque to 30 ft. lbs.
4. Loosely install nuts (M), then position lower cover to dash using screw (N) and install screw (R).
5. Torque screws (R) and (N) to 45 inch lbs. then install and torque screws (S) and (T) to 45 inch lbs.
6. Torque clamps screws to (C) and (D) to 45 inch lbs., then install and torque remaining cover screws to 45 inch lbs.
7. Torque nuts (M) to 20 ft. lbs.

Installation 1969-75 Corvette, Fig. 31

1. Loosely assemble support (U) to bracket (T) with bolts (W) and nuts (Y).
2. Assemble upper end of rag joint (flex coupling) (Z) to steering gear and tighten screws.
3. Install column into vehicle and slide splined end of steering shaft into upper end of rag joint (flex coupling) (Z).
4. Connect all electrical connectors to column and install but do not tighten steering column to support (U) attaching screws (H).
5. Install transmission control cable bracket (N) with bolt (M) and hold in place.
6. Torque nut (K) to 10 ft. lbs. leaving a gap of .30 inch (plus or minus .05 inch) between mast jacket and inside semi-circular surface of instrument panel. Pull support (U) rearward to accommodate height differential between instrument panel and bracket (T). With column in this position torque screws (H) and nuts (Y) to 15 ft. lbs.
7. Loosen nuts (K) and maintain rag joint (flex coupling) dimension, Fig. 31 and tighten pinch bolt (O).

NOTE: Do not tighten column to instrument panel until vehicle is supporting its own weight.

8. Using spring clip (AB) attach backdrive cable (AA) to transmission control cable bracket (N). Attach inner cable (AC) to lever on steering column with retainer (AD) and torque nut (K) to 10 ft. lbs.
9. Remove plastic spacers (L) from flexible coupling and install steering wheel.

NOTE: With steering shaft in a straight ahead position and no torque applied on steering wheel a .04 inch gap (4-ply coupling or a .07 inch gap (7-ply coupling) must be maintained between steering shaft flange and coupling rivets. A gap less than specified indicates that column must be realigned.

10. Connect battery ground cable.

Installation 1969-70 Chevrolet and 1970 119 Inch Wheel Base Station Wagon, Fig. 32

1. Loosely assemble inner and outer dash covers on column, snug screws (C) and (D) insuring looseness between column and covers.
2. Cement cover seal (E) to outer cover (B).
3. Position rolled portion of dash seal at column lower reaction tab approximately two inches from bottom of jacket.

NOTE: Rolled section of seal must be facing down toward front of vehicle and must remain in this position during and after column installation.

4. Install column in vehicle, connect all electrical connectors to column and loosely install nuts (H).
5. Torque flex coupling nuts to 20 ft. lbs.

NOTE: Do not tighten column to instrument until vehicle is supporting its own weight.

6. Torque mast jacket bracket retaining nuts (H) to 20 ft. lbs. while maintaining rag joint (flex coupling) to flange dimension, Fig. 32.

NOTE: Inner and outer cover, must be able to move to the established position of column which has been determined by rag joint and instrument panel position and attachment.

7. With inner cover (A) and outer cover (B) against dash, align all cover to dash attaching screws, hold covers against dash and column and torque inner cover screws (K).
8. Torque screws (C) and (D) to 35 in. lbs. and loosen screws (K).
9. Torque screws (L) then (K) to 35 in. lbs.
10. On 1970 models, remove plastic spacers from flexible coupling pins (if used).

NOTE: With steering shaft in a straight ahead position and no torque applied onto steering wheel, a .04 inch gap must be maintained between steering shaft flange and coupling rivets. Less than a .04 inch gap indicates that column is mis-aligned.

11. Install transmission indicator cable (if used) and instrument panel trim cover.
12. Connect transmission control linkage. On floor shift models connect back-drive linkage and connect battery ground cable.

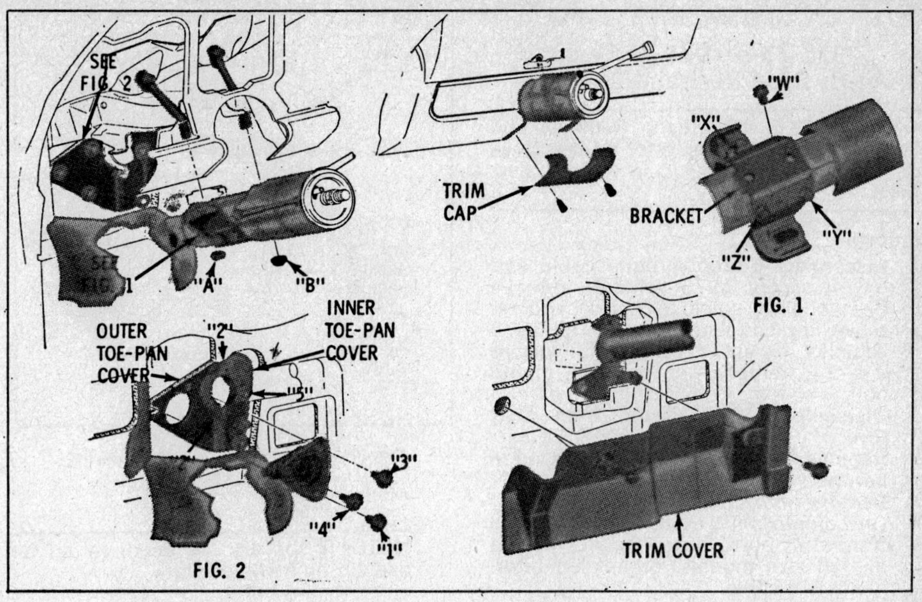

Fig. 34 1969-70 Oldsmobile Intermediates, steering column installation

1971-75 Vega & 1975 Astre & Monza

Removal Fig. 33

1. Disconnect battery ground cable and remove steering wheel.
2. On manual steering columns, remove pot joint coupling clamp bolt. On power steering columns, remove flexible coupling pinch bolt from lower end of upper steering shaft.
3. Move front seat as far back as possible to obtain clearance and remove floor pan bracket screws.
4. Remove column bracket to instrument panel nuts, then lower column and disconnect wiring harness from column.
5. Carefully remove column from vehicle.

NOTE: Additional help may be required to tap lightly on pot joint or flexible coupling using a plastic mallet while column is being pulled rearward, to facilitate removal.

Installation

NOTE: It will be necessary to obtain additional help to install steering column.

1. On power steering columns, place pot joint clamp (H) on lower end of pot joint and assemble intermediate shaft assembly on steering gear stub shaft, aligning flat on stub shaft (K) with flat in pot joint.
2. Position column into vehicle, guiding lower end of column through opening in toe pan.
3. On manual steering columns, place pot joint (H) over lower end of pot joint (J) and assemble pot joint to steering gear wormshaft, aligning flat on wormshaft with flat in pot joint. On power steering columns, align steering shaft flat with flat in flexible coupling. When shaft is bottomed against coupling reinforcement, install bolt (G) and torque to 30 ft. lbs.
4. Connect wiring harnesses and loosely install steering column bracket to instrument panel nuts (E).
5. Align pot joint clamp (H) with groove across end of pot joint (J), then install bolt (N) (passing it through shaft undercut) and nut (O) and torque to 55 ft. lbs.
6. With vehicle on ground, torque nuts (E) to 19 ft. lbs., then slide toe plate (P) down column to toe plate and install three screws (Q).

NOTE: On power steering columns, alignment flange (R) on toe plate must be engaged with front of toe pan before installing screws (Q). On manual steering columns, no side load is allowed during installation of screws (Q). A side load will cause column misalignment and interference between steering shaft and inside diameter of column jacket.

7. On manual steering columns, remove alignment spacers (S).

NOTE: Minimum allowable clearance between outside diameter of steering shaft and inside diameter of column jacket lower plastic bushing after installation is .18 inches.

8. Install steering wheel and connect battery ground cable.

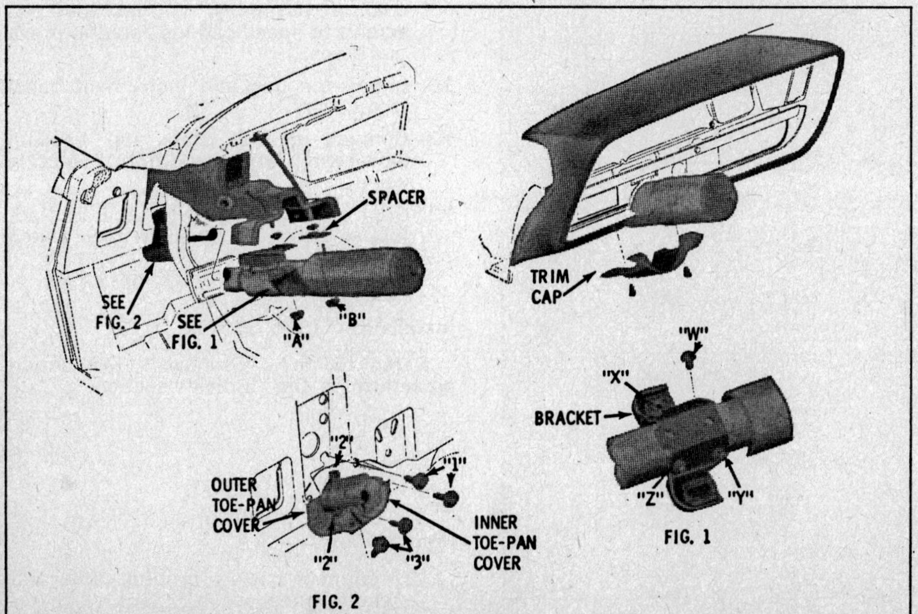

Fig. 35 1969-70 Oldsmobile Full size models, steering column installation

OLDSMOBILE

1971-73 & 1974-75 Less Air Cushion Restraint System

NOTE: For 1973-75 Omega Removal and Installation procedures, refer to the Nova procedure in the "1969-75 Chevrolet" section.

Removal

1. Disconnect battery ground cable and on all except 1975 vehicles, remove flex coupling shield (if used) and remove flex coupling nuts. On all 1975 vehicles except Starfire with power steering, remove flex coupling clamp bolt.
2. Disconnect shift linkage from shift lever.
3. Remove steering wheel, disconnect toe pan trim cover and toe pan.
4. Remove instrument panel trim cover.
5. On column shift auto. trans., remove from shift bowl, shift indicator needle on full size models, or link on intermediate models.
6. Disconnect neutral start and back-up light switch and turn signal connector.
7. Support column and disconnect column bracket assembly.
8. Lower column enough to disconnect electrical connectors.

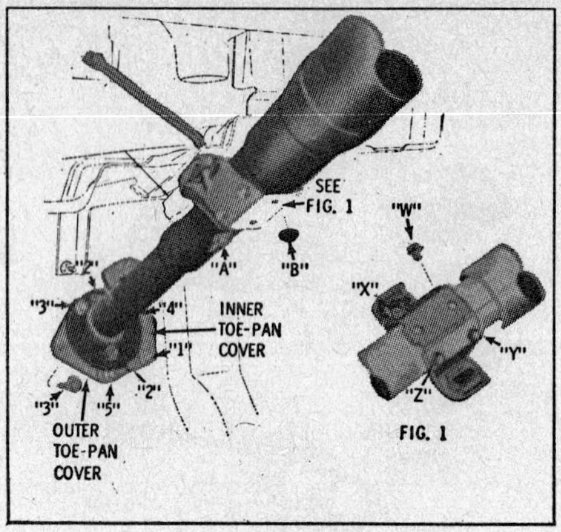

Fig. 36 1969-70 Toronado steering column installation

NOTE: If spacers are used, retain for use during installation.

9. Carefully remove column from vehicle and on 1972-74 models disconnect intermediate shaft from steering shaft.

Installation, Figs. 33A & 33B

1. On 1972-75 models, if intermediate shaft was removed from steering shaft, install and torque nuts to 35 ft. lbs.
2. Position column in vehicle and guide intermediate shaft onto flex coupling. On all except 1975 vehicles, torque flex coupling nuts to 20 ft. lbs. On 1975 vehicles, torque flex coupling clamp bolt to 55 ft. lbs.
3. Connect electrical wiring to column.
4. Install but do not tighten two column bracket support nuts (A and B).

NOTE: If spacers were removed, install same amount as removed from each side.

5. Check flex coupling alignment. If incorrect, it will be necessary to raise the column assembly.

NOTE: Flex coupling must be straight within 1/16 inch without bottoming against intermediate shaft coupling.

6. In the following order, install and or torque the screws or bolts to 3 ft. lbs.:
   a. Loosely install screw (1).
   b. Install screw (3) and torque screws (3 and 1).
   c. Torque screw (2).
   d. Install and torque screw (4).
   e. Torque clamp bolt (5).
   f. Install and torque screws (6, 7, and 8).
7. Torque column bracket support nuts (A and B) to 20 ft. lbs. on all except 1975 vehicles and 25 ft. lbs. on 1975 vehicles.
8. On column shift auto. trans., install shift indicator needle on full size, or link on intermediate models.
9. Check flex coupling alignment and intermediate shaft coupling for bottoming. If either has occurred, it will be necessary to loosen toe pan clamp bolts and the two column bracket support nuts. Lift up and force rearward column assembly, torque toe pan screws to specifications listed in previous steps.
10. Install toe pan and instrument panel trim covers.
11. Connect neutral start and back-up light switch wiring and install steering wheel.
12. Connect battery ground cable and adjust neutral and back-up light switch (if used).

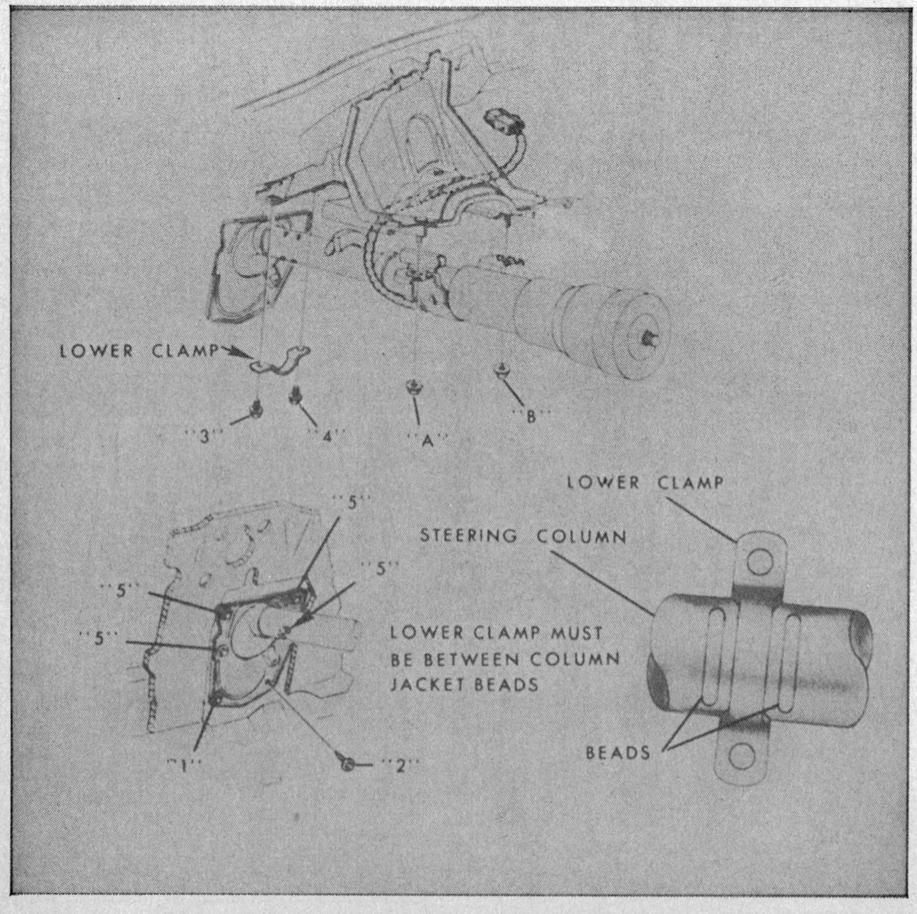

Fig. 37 1974-75 Buick, Cadillac and Oldsmobile with Air Cushion Restraint System, steering column installation

Installation, 1975 Starfire

Refer to the "Skyhawk" installation procedure in the "Buick" section.

1969-70

Removal

1. Disconnect battery ground cable and steering wheel.
2. Disconnect flex coupling nuts or bolts and disconnect shift linkage.

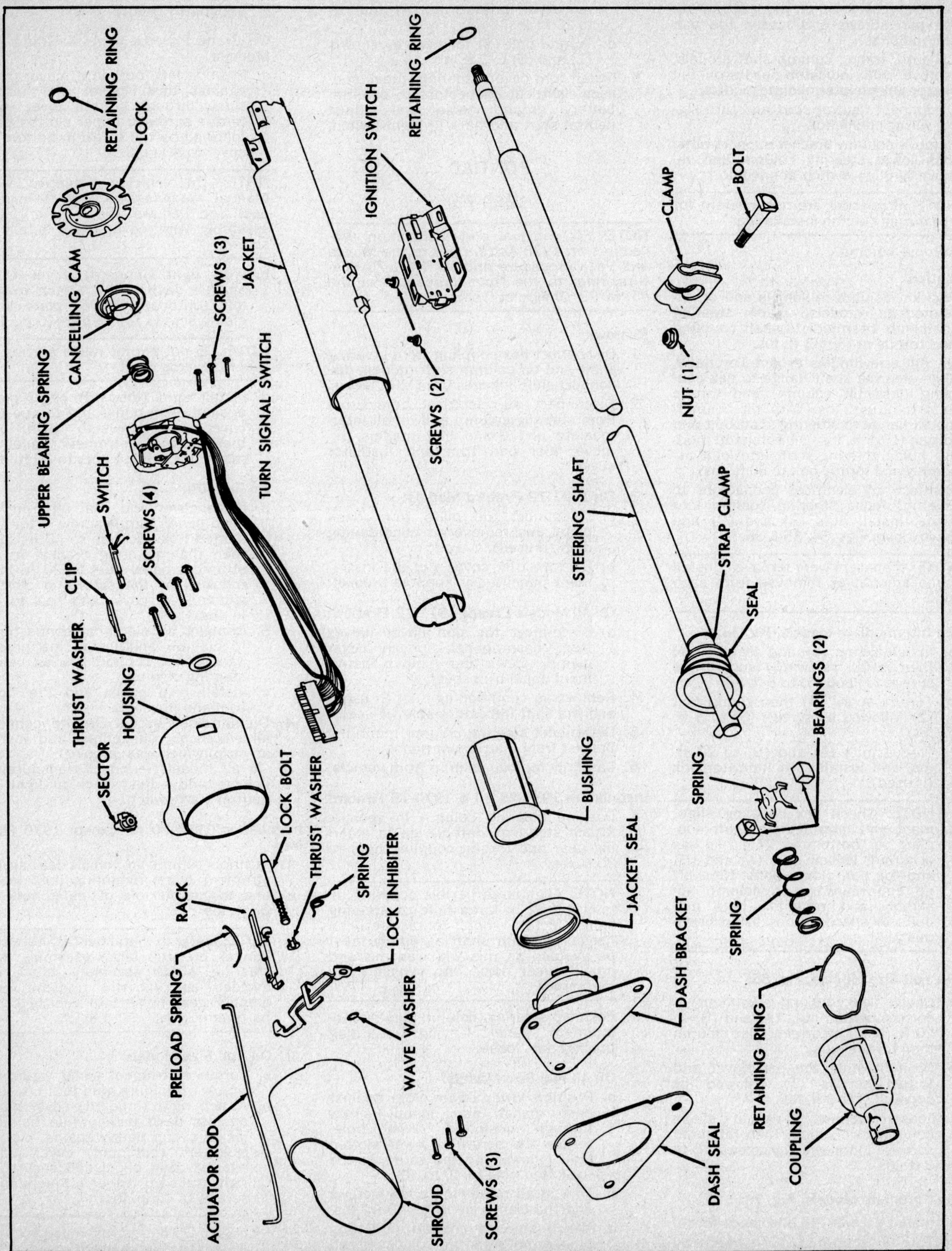

Fig. 38 1971-75 Vega & 1975 Astre, Monza, Skyhawk & Starfire non-tilt steering column

3. On intermediate models, remove toe pan trim cover. On all models, remove toe pan screws and loosen toe pan clamp bolts.
4. On auto. trans., column shift models, remove shift indicator needle on full size or link on intermediate models.
5. Disconnect neutral start and turn signal wiring connector.
6. Remove column bracket support nuts, then lower steering column and remove ignition switch connector.

NOTE: If spacers are used, retain for use during column installation.

7. Remove column.

Installation

1. Position column in vehicle and on intermediate models, guide steering shaft into intermediate shaft coupling and torque nut to 35 ft. lbs.
2. On full size models except Toronado, align steering shaft flange to flex coupling (less tilt column) and torque nuts to 20 ft. lbs. With tilt column, install flange to steering shaft bolt and torque to 35 ft. lbs. On Toronado models, guide steering shaft into joint assembly and torque bolt to 80 ft. lbs.
3. Connect all electrical connectors to column. Raise steering column into place, install nuts (A) and (B) and leave loose, Figs. 34, 35 & 36.

NOTE: If spacers were removed, install same amount as removed from each side.

4. **On Intermediate Models, Fig. 34:**
   a. Install screw (1) and leave loose then install screw (3) and torque screws (1) and (3) to 3 ft. lbs.
   b. Torque screw (4) then clamp bolt (2) followed by screws (5) to 3 ft. lbs.
   c. Torque nuts (A) and (B) to 20 ft. lbs. and install shift indicator link (if used).

   NOTE: Check flex coupling alignment and intermediate shaft coupling for bottoming. If either has occurred, loosen nuts (A) and (B) and toe pan clamp bolts, then lift up and rearward on column assembly and retighten nuts and bolts to specifications listed previously.

   On Full Size Models, Fig. 35:
   a. Check flex coupling alignment, if correct tighten nuts (A) and (B) to 20 ft. lbs. If incorrect, raise column then torque nuts.
   b. Position inner toe pan cover and torque screws (1) followed by screws (2) to 3 ft. lbs.
   c. Loosen screws (1) and shake column to position, then retighten screws (1) followed by screws (3) to 3 ft. lbs.

   On Toronado Models, Fig. 36:
   a. Install screws (1) and leave loose.
   b. Torque screws (3) followed by screws (4) to 3 ft. lbs.
   c. While supporting column up and rearward, torque nuts (A) and (B) to 20 ft. lbs.
   d. Torque bolt (2) followed by screws (1) and (5) to 3 ft. lbs.
5. Install trim cover, steering wheel, connect electrical connectors to column, battery ground cable and adjust neutral start and back-up light switch.

PONTIAC

1969-75

NOTE: For removal and installation procedures on 1975 Astre, refer to the Monza and Vega procedure and for 1971-75 Ventura, refer to the Nova procedure in the "1969-75 Chevrolet" section.

Removal

1. Disconnect flex coupling from steering shaft and on column shift models disconnect shift linkages from shift levers.
2. Disconnect all electrical connectors from steering column and on all intermediate and Grand Prix models, remove floor pan trim and insulator cover.
3. **On 1971-72 Firebird Models:**
   a. Disconnect toe plate from dash panel and remove air conditioning duct (if used).
   b. Remove trim cover and the instrument panel lower support bracket.

   On All Models Except 1971-72 Firebird:
   a. Disconnect toe pan cover halves from dash panel, loosen cover clamp screws and remove instrument panel trim cover.
4. Remove air conditioning duct (if used) and the shift indicator cable (if used).
5. Disconnect steering column mounting bracket from instrument panel.
6. Carefully remove column from vehicle.

Installation 1971-75 All & 1970-75 Firebird

1. Position steering column in vehicle. Attach steering shaft flange to steering gear and torque coupling nuts to 20 ft. lbs.

NOTE: Align large flexible coupling retaining bolt in large hole of steering flange. Metal to metal contact between flanges on stub shaft assembly must be avoided as this will transmit and amplify gear noise into vehicle compartment.

2. Position steering column against instrument panel leaving mounting bracket nuts loose.
3. **On All Full Size Models:**
   a. Position lower cover plate against dash panel and install screw through unslotted cover hole, torque the screw in lower slotted hole followed by the screw in the unslotted hole to 40 in. lbs.
   b. Torque all remaining cover screws and the clamp screw to 40 in. lbs.
   c. Attach steering column mounting bracket onto instrument panel and torque nuts to 20 ft. lbs.
   d. Install shift indicator cable attaching bolt (if used).

   On Grand Prix, Le Mans & Grand Am Models:
   a. Position left (outboard) cover plate against dash panel and install screw through bottom cover hole, torque screw in upper slotted hole followed by the screw in bottom of cover to 40 in. lbs.

   NOTE: On vehicles equipped with manual transmission, attach clutch push rod bellows and retainer plate assembly with cover plate retaining screws.

   b. With right (inboard) cover plate against dash panel, attach to the left (outboard) cover plate with clamp screws torqued to 40 in. lbs.

   NOTE: Do not permit rolled portion of seal to reverse itself.

   c. Install right (inboard) cover plate against dash panel and torque toe cover plate screws to 40 in. lbs., then torque instrument panel to mounting bracket nuts to 20 ft. lbs.

   On Firebird:
   a. Place cover and seal on front of dash panel, torque screw in the unslotted hole location to 40 in. lbs., then the mounting bracket to instrument panel nuts to 20 ft. lbs., followed by the screws at slotted locations of cover and seal to 40 in. lbs.
   b. Connect all electrical connections to column and install instrument panel lower support bracket below steering column.
   c. Install trim cover and the floor trim and insulator.
4. On all models except Firebird, connect all electrical connections and install any trim which was removed.
5. On all models, connect transmission linkage and adjust back-up light or neutral start switch.

Installation 1969-70 All Except 1970 Firebird

1. Position column in vehicle leaving instrument panel mounting nuts loose and torque flexible coupling nuts to 20 ft. lbs.

NOTE: Metal to metal contact between flanges on stub shaft assembly and steering shaft assembly must be avoided as this will transmit and amplify gear noise into vehicle compartment.

2. **On Full Size Models:**
   a. Torque instrument panel mounting bracket retaining nuts to 20 ft. lbs.
   b. Attach right (inboard) cover plate against dash panel while holding jacket clamp tightly against jacket.
   c. Hold left (outboard) cover plate against dash panel and secure to right (inboard) cover torque clamp screws to 40 in. lbs.

   NOTE: Do not permit rolled section of seal to reverse itself.

d. Loosen the two right (inboard) cover plate screws from dash panel to permit steering column to move to the established position as determined by flexible coupling and instrument panel attachment.
e. Torque the two right (inboard) cover plate screws followed by all remaining cover plate screws to 40 in. lbs.

On Tempest and Grand Prix Models:

a. Place left (outboard) cover plate against dash panel and install screw through bottom hole in cover, then the screw in upper slotted hole followed by the left (outboard) cover plate screw and torque to 40 in. lbs.
b. With right (inboard) cover plate held tight against dash panel attach the left (outboard) cover plate and torque clamp screws to 40 in. lbs.

NOTE: Do not permit rolled section of seal to reverse itself.

c. Attach right (inboard) cover plate to dash panel and torque cover plate screws to 40 in. lbs.
d. Torque the mounting bracket to instrument panel nuts to 20 ft. lbs.

On 1969 Firebird Models:

a. Torque mounting bracket to instrument panel nuts to 20 ft. lbs. and attach lower and upper cover plates to dash panel leaving screws loose.

NOTE: Do not permit rolled section of seal to reverse itself.

b. Push upper and lower cover plates against dash panel to compress seal and torque clamp screws and all cover screws to 40 in. lbs.

3. On all models, install shift indicator (if used), connect all electrical connectors to column and install any trim which was removed.
4. Connect shift linkage and adjust back-up and neutral start switch.

G. M. VEHICLES WITH AIR CUSHION RESTRAINT SYSTEM

CAUTION: Do not attempt any servicing or disconnecting of the Air Cushion Restraint System instrument panel, instrument panel components or any other electrical components in the adjacent area, until the ignition key is in the "Lock" position, the battery ground cable has been disconnected and the end of the cable taped. This procedure must be followed to prevent accidental deployment of the system which could result in personal injury and/or damage to the system.

1974-75 Buick, Cadillac & Oldsmobile

Removal

1. After deactivating ACRS as described above, remove knee restraint as follows:

   a. Buick: Pry knee restraint upper mouldings from instrument panel, then remove four screws from lower edge of knee restraint and four screws from upper edge of knee restraint. Pull knee restraint rearward and disconnect lower air conditioner outlet, then lower assembly and disconnect courtesy lamp. Disconnect parking brake release cable and ash tray lamp and remove knee restraint assembly.

   b. Cadillac: Remove three screws securing glove box partition, then disconnect glove box lamp and theft deterrent and/or trunk release switch. Working from top of inside of glove box, remove three attaching screws and remove glove box. Remove ash tray and tape storage compartment by removing two screws from within each one. Remove screws from left side and lower edge knee restraint, then working through ash tray and tape storage compartment opening, remove four knee restraint to knee restraint brace screws and remove knee restraint.

   c. Oldsmobile: Slide collar (on steering column) up so that knee restraint cover can be removed. Remove screw from left side of cover, then grip cover firmly and detach it from the eight retaining clips. Remove screws from right and left lower edges of restraint, from restraint to brace and from restraint to tie bar. Remove screw from wiring harness ground to tie bar and remove screw from left hand courtesy lamp. Disconnect transmission downshift switch connector then lower knee restraint to disconnect lower left air conditioning duct and parking brake cable and remove knee restraint.

2. Remove screws attaching knee restraint to steering column guide bracket.
3. On Buick, Eldorado and Oldsmobile models, remove shields from flex coupling.
4. Remove flex coupling nuts and disconnect all shift linkages from column.
5. Remove six floor cover and seal retaining screws and two steering column lower clamp guide to bracket screws.
6. Remove the two upper column to brace bolts and lower column until steering wheel rests on seat.
7. Using tool J-24388, disconnect drivers module wiring harness from sensor, then disconnect all other electrical connectors from column and remove column.

Installation Fig. 37

1. With column positioned in vehicle and steering wheel resting on seat, connect all electrical connectors to column.
2. Position flex coupling flange to flex coupling and install washers and nuts and torque to 20 ft. lbs.
3. Raise column into place and loosely install nuts (A) and (B), then position floor cover and seal and loosely install screw (1). Install screw (2) and torque screws (2) and (1) to 36 inch lbs.
4. Install and torque the four remaining floor cover screws to 36 inch lbs.
5. Loosely install the two lower clamp to guide bracket screws and torque (4) and (3) to 36 inch lbs.
6. Torque nuts (A) and (B) and connect shift linkage to shift lever.
7. On Buick, Eldorado and Oldsmobile models, install flex coupling shield.
8. Install knee restraint to brace guide and install shift indicator needle.
9. Install knee restraint by reversing removal procedure and reconnect battery ground cable.

Steering Column Service

AMERICAN MOTORS, CHRYSLER CORP. & GENERAL MOTORS CORP. (SAGINAW STEERING COLUMN)

1969-75 NON TILT COLUMN, FIG. 1 All Exc. Chevrolet Monza & Vega, Buick Skyhawk Olds. Starfire & Pontiac Astre

NOTE: Steps 1 through 9 can be performed without column removal.

Disassembly

1. Remove steering wheel, three cover screws and cover.
2. Using Lock Plate Compressor Tool J-23653 and steering wheel nut, depress lock plate. Pry round wire snap ring from groove in shaft and discard ring. Remove snap ring and shaft lock plate.

NOTE: After snap ring has been removed, make sure shaft does not slide out of bottom of column and cause damage to shaft.

3. Remove upper bearing preload spring, horn contact, cancelling cam and thrust washer.
4. Remove turn signal lever. Depress hazard warning knob, remove knob and signal switch.
5. Remove buzzer switch by pulling straight out of housing. This may be done without removing lock cylinder, providing lock cylinder is in "run" position.
6. Remove ignition switch and lock cylinder if further service is necessary.
7. Drive out upper shift level pivot pin and remove upper shift lever.
8. Remove upper housing, thrust cup, lock rack, lock bolt, load spring and shift gate.
9. Using a blunt punch, remove sector through lock cylinder hole by pushing firmly on block tooth of sector.
10. Remove shift bowl and shroud from jacket.

Reassembly

1. Apply a light coat of grease to all friction surfaces.
2. Install sector in lock cylinder hole over sector shaft with tang end facing outside of hole. Using a blunt punch, press sector over pin.
3. Install shift gate onto housing.
4. Install rack spring in housing from bottom side. Long section should be toward steering wheel and hook onto edge of housing.
5. Assemble bolt to cross-over arm of rack and install rack and lock bolt into housing with teeth facing steering wheel and toward centerline of column.
6. Install thrust cup on bottom of hub housing with key of cup aligned with keyway in hub.
7. Install lower bowl bearing in jacket and place wave washer in bowl bearing.
8. Install bowl and rotate to insure proper seating in bearing.

NOTE: Do not install shift lever before installing bowl.

9. Install upper bearing housing assembly on jacket. Bowl should be in "Park" position and rack pulled down. Making sure housing is seated on jacket, install screws. Attach housing when properly seated on jacket.
10. Assemble buzzer switch to spring clip with formed end of clip around lower end of switch and spring bowed away from switch. Install switch and spring with contacts toward cylinder bore. Install turn signal switch.
11. With lock cylinder and ignition switch installed and in "lock" position, fit actuator rod into switch and assemble to column with two screws.
12. With turn signal switch in "neutral" and hazard warning plunger out, install washer, spring, horn contact, cancelling cam and shaft lock plate onto shaft.
13. Using lock plate compressor tool, depress lock plate and install a new snap ring in groove on shaft.
14. Install cover on shaft and install steering wheel.
15. With shift bowl in "drive" position install shift lever and adjust neutral-start and back-up light switches.

1969-75 Tilt Steering Column, Figs. 2 & 3

Disassembly

1. Remove column mounting bracket.

NOTE: Set bracket aside to protect breakaway capsules.

2. Mount column in a vise by using column holding fixture or by clamping column by weld nuts.
3. Remove steering wheel, wire protector and cover.
4. Remove turn signal lever and tilt release lever. Depress hazard warning switch and remove knob.
5. Remove shift lever from bowl.
6. Using lock plate compressor tool J-23131 and steering wheel nut, depress lock plate. Pry round wire snap ring from groove in steering shaft and discard.
7. Remove shaft lock plate, cancelling cam, spring and the turn signal switch.
8. Remove ignition switch lock and remove buzzer switch by pulling straight out. Do not pull on switch terminals. Use a bent wire or needle nose pliers to pull on buzzer switch clip.
9. Remove housing cover.
10. Install tilt release lever and place column in "full up" position. Using a screwdriver that fits into slot opening of spring retainer, press approximately 3/16 inch and turn approximately ⅛ turn until ears align with grooves in housing and remove spring and guide.
11. Place column in neutral position and push upper steering shaft in sufficiently to remove steering shaft inner race seat and inner race.
12. Remove upper flange pinch bolt, lower steering shaft flange, no-back washer (if used), spacer, spring, clip and the bearing adapter assembly from lower end of mast jacket.
13. Remove ignition switch and neutral-start switch.
14. Using pivot bearing removing tool J-21854, remove two pivot pins.
15. Place column in "full up" position and disengage lock shoes. Remove bearing housing assembly by pulling upward to extend rack fully down and moving housing assembly to the left to disengage rack from actuator.
16. Remove steering shaft assembly from upper end.
17. Disassemble shaft assembly by removing centering spheres, anti-lash spring and support assembly.
18. Using a screwdriver, remove shift tube retaining ring and thrust washer.
19. Using shift tube remover tool J-23072, remove shift tube from bowl.

NOTE: Care should be taken not to jam lower shift lever into "T" slot on lower end of mast jacket while forcing it out of shift tube.

20. Remove shift tube assembly from mast jacket, lock plate and wave washer.
21. Remove bowl from mast jacket, shift lever spring from bowl by winding up the spring with pliers and pulling out.

NOTE: If bearing housing or components need service, follow disassembly procedure below.

1. Remove tilt lever opening shield from housing.
2. Remove lock bolt spring by removing spring retaining bolt screw and moving spring clockwise to remove from bolt.
3. Remove snap ring from sector drive shaft. Using a small punch, lightly tap drive shaft from sector. Remove drive shaft, washer, sector, bolt, rack and rack spring.

NOTE: To relieve load on release lever, hold shoes inward and wedge block between top of shoes.

4. Using pivot pin remover tool and hammer, remove pivot pin, lever and release lever spring. Also remove lock shoe pin, lock shoe and springs.
5. Remove bearings from bearing housing only if they are to be replaced. Remove separator and balls from bearing. Using a small punch against back surface of race, carefully hammer race out of housing until bearing puller can be used. Repeat procedure for other race.

Reassembly

1. Apply a light coat of grease to all friction surfaces.
2. Install bearing in housing if removed.
3. Using a .180 inch diameter rod to align shoes, install lock shoe springs, lock shoe and shoe pin in bearing housing.
4. Install spring, release lever and pin in bearing housing.
5. Install washer and drive shaft in housing. Lightly tap sector onto shaft far enough to install snap ring.
6. Install lock bolt and engage with sector cam.
7. Install rack and spring. Engage block tooth on rack with block tooth on sector. Install external tilt release lever, bolt spring and spring retaining screw and torque screw to 35 in. lbs.
8. Wind up shift lever spring with pliers and install in bowl. Slide bowl into mast jacket.
9. Install wave washer and lock plate in place. Work lock plate into notches in jacket.
10. Install shift tube in lower end of mast jacket. Align keyway in tube with key in bowl and using installer tool J-23073, pull shift tube into bowl.

NOTE: Do not push or tap on end of shift tube.

11. Install thrust washer and retaining ring by pulling bowl up to compress wave washer.
12. Align "V" notch in support with "V" notch in jacket and install support and lock plate. Torque screws to 60 in. lbs.

13. Align lower bearing adapter notches in jacket and push in lower end of mast jacket and install clip.
14. Install spheres and anti-lash spring in upper steering shaft. Install lower steering shaft.
15. Install steering shaft through upper end of shift tube. Carefully guide shaft through shift tube and bearing.
16. Install ignition switch actuator rod. Extend rack downward from bearing housing and assemble bearing housing over steering shaft, engage rack over end of actuator rod.
17. Install external tilt release lever. Hold lock shoes in disengaged position and assemble bearing housing over steering shaft until pivot pin holes align and install pivot pins.
18. Place housing in "full up" position. Install guide, tilt spring and tilt spring retainer using a screwdriver in retainer slot. Rotate retainer clockwise to engage.
19. Install inner race, inner race seat and tilt lever opening shield.
20. Remove tilt release lever. Install housing cover and torque screws to 45 in. lbs.
21. Assemble buzzer switch to spring clip with formed end of clip under end of switch and spring bowed away from switch on opposite side of contact. With contacts toward lock cylinder bore, push switch and spring into hole in cover until it bottoms.
22. Install cruise control switch wires, if used. Install turn signal switch wires and connector, bearing housing and bowl. Push hazard warning knob in. Install turn signal switch and torque screws to 35 in. lbs.

NOTE: Shortest screw must be installed in hole nearest lock cylinder.

23. Install spring, spacer, no-back washer (if used) and lower steering shaft flange. Torque flange pinch bolt to 30 ft. lbs.
24. Install hazard warning knob, pull knob out and install cancelling cam spring, cancelling cam and shaft lock.
25. Using lock plate compressor tool, depress shaft lock plate and install a new snap ring in shaft groove.
26. Install tilt release lever, turn signal lever and upper shift lever. Drive in pivot pins and install lock cylinder.
27. Install shaft lock cover and torque screws to 35 in. lbs.
28. With lock cylinder and ignition switch in "lock" position, put bowl or shroud in "Park" position, fit actuator into switch and assembly to column, torque screws to 35 in. lbs.
29. Install neutral-start back-up light switch, leaving attaching screws loose.
30. Install wire protector, mounting bracket, steering wheel and adjust neutral-start back-up light switch.

1971-75 NON-TILT, FIG. 38

Chevrolet Vega & Monza, Buick Skyhawk, Oldsmobile Starfire & Pontiac Astre

Disassembly

1. Remove dash panel bracket and store bracket in a safe place to prevent damage to mounting capsules.
2. Place column in a vise using both weld nuts of either set A or B, Fig. 39. The vise jaws must clamp onto the sides of weld nuts indicated by arrows shown on set B.

CAUTION: Do not place column in vise by clamping onto only one weld nut, by clamping on one weld nut of both sets A and B or by clamping onto the sides not indicated by arrows, as damage to column could result.

3. Using tool J-23653, compress lock plate and remove wire snap ring from shaft groove, then remove directional signal switch, lock cylinder, ignition key warning switch and remove steering shaft from lower end of column.
4. On power steering columns, remove steering shaft lower bearing from plastic adapter at lower end of column.
5. Remove four turn signal housing to column jacket retaining screws and remove housing and shroud assembly.
6. Remove three shroud to signal housing retaining screws and remove shroud.
7. Using hand as shield to prevent lock inhibitor lever spring wire from flying out, remove lock inhibitor lever and spring wire.
8. Lift ignition switch actuating rod, rack assembly, shaft lock bolt and spring assembly from housing, then if necessary, remove rack preload spring.
9. Remove ignition switch actuator sector through lock cylinder hole by pushing firmly on block tooth of sector with a blunt punch or screwdriver.

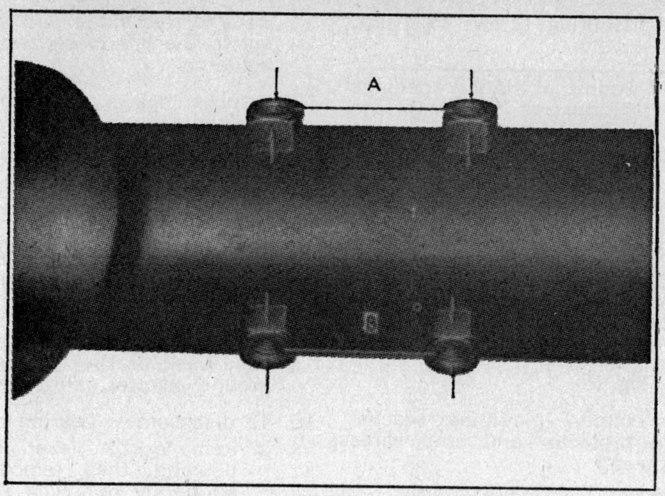

Fig. 39 Steering column clamping points

Assembly

NOTE: Prior to assembly, apply a light coat lithium soap grease to all friction surfaces.

1. Install sector into turn signal housing, then reaching through lock cylinder hole, place sector onto shaft with tang end to outside of hole and press sector onto shaft using a blunt tool.
2. If rack preload spring was removed, insert spring into housing from lower end, hooking onto edge of housing at both ends.
3. Assemble locking bolt onto crossover arm on rack, making certain that lock bolt spring and plastic washer are in place. Insert rack and lock bolt assembly into housing from bottom with teeth facing up toward centerline of column. Align first tooth on sector with first tooth on rack. If aligned properly, the block tooth will line up when the rack assembly is pushed all the way in.
4. Place lock inhibitor spring wire over lever on pivot housing, then install lever in place with tang behind rack and with spring wire hooked onto the lever. While holding down at pivot point of lever, hook other end of wire into slot provided in housing so that lever is flush with top of pivot.
5. Place shroud over ignition switch actuating rod and onto the housing, then install three retaining screws and torque to 60 inch lbs.
6. Install turn signal housing assembly over upper end of column assembly, then install the four retaining bolts and torque to 60 inch lbs.
7. On power steering columns, press new steering shaft lower bearing into plastic adapter at lower end of column.
8. Install steering shaft into lower end of column and guide it up through the upper bearing, then install ignition switch, ignition key warning switch, lock cylinder and directional signal switch.

NOTE: On manual steering columns, there will be about $^3/_{16}$ inch clearance between lower end of column and clamp on shaft. Do not move clamp to eliminate this clearance.

9. Install dash panel bracket and torque

the four mounting screws to 19 ft. lbs.

NOTE: Make certain that slotted openings in bracket (for mounting capsules) face upper end of steering column.

1975 Tilt Steering Column

NOTE: This column incorporates a key release inhibitor which must be depressed for ignition key removal, and has no shift tube or back drive system.

Disassembly, Fig. 2

1. Remove column mounting bracket and wire protector and strip wires from protector.

   NOTE: Do not damage wires. Unsnap connector from bracket and tape wires to prevent snagging when removing switch.

2. Remove steering wheel, then loosen the three cover screws and lift cover from housing.
3. Remove tilt release lever and turn signal lever, then depress hazard warning knob and remove knob.
4. Using tool J-2365, depress lock plate and remove wire snap ring from groove shaft, then remove shaft lock, cancelling cam and upper bearing spring.
5. Remove the three turn signal retaining screws and pull switch straight out while guiding wires through shroud.
6. Place ignition lock in "run" position, then using a thin tool inserted through slot next to switch mounting boss, depress retainer at bottom of slot while pulling lock cylinder from housing.
7. To remove buzzer switch, insert a hooked wire through exposed loop of wedge spring and pull straight out. If lock cylinder is not removed before buzzer switch, buzzer switch must be in the "on" position.

   NOTE: If wedge spring is dropped into column during removal, it may require complete disassembly of column to remove it.

8. Remove cover retaining screws and cover, then remove race and seat from shaft.
9. Reinstall tilt release lever and place column in full "up" position, then remove tilt spring retainer using a screwdriver and depressing retainer about 3/16 inch and turning about ⅛ turn counterclockwise until ears align with grooves in housing and remove spring and guide.
10. With ignition switch in "off-unlock" position, remove ignition switch.
11. Using tool J-21854-1, disengage lock shoe by depressing tilt housing upward to free shoes and moving housing to the left to separate rack from actuator.
12. Remove spacer from lower end of shaft and remove shaft from column.
13. Disconnect end of actuator rod from rack and remove from column, then remove the four support screws and remove support.
14. Slightly rotate lockplate counterclockwise and remove, then remove shroud retaining screws and shroud.
15. Should bearing need replacement, slide dash bracket off jacket and tap around adapter using a flat punch.
16. To disassemble bearing housing:
    a. Remove tilt lever shield from housing, then remove lock bolt spring by removing spring retaining screw and turning spring clockwise to remove it from screw.
    b. If used, remove sector shaft snap ring, then using a punch, drive shaft from sector and remove rack, spring, shim (if used) and sector bolt.
    c. Remove tilt release lever pin, release lever and spring.

    NOTE: To relieve tension on release lever, holdshoes inward and wedge block between top of shoes and bearing housing.

    d. Remove lock shoe pin, lock shoes and springs.
    e. If necessary, remove separator and balls from bearing, then using a punch, carefully drive race out of housing until bearing puller can be used. Repeat step for other race and do not re-use bearing.
17. To disassemble shroud housing:
    a. Remove key release lever, then tip release lever forward and lift up. Spring on lever may be removed, if necessary.

Assembly

NOTE: Apply lithium grease to all friction surfaces except lock, lock bolt and lock bolt hole.

1. To assemble shroud housing:
   a. Position spring on key release lever and into shroud slot, then compress spring by forcing with lever. Lever will slip into shroud slot.
2. To assemble bearing housing:
   a. If removed, install bearings into bearing housing, then install lock shoe springs, lock shoe and shoe pin. Use a .180 inch dia. rod to align shoes for pin installation.
   b. Install spring, release lever and pin into bearing housing. Relieve load on release lever as in disassembly procedure.
   c. Install drive shaft into housing and lightly tap sector onto shaft until snap ring groove is exposed. If snap ring was removed, replace with new one.
   d. Install lock bolt and engage with sector cam surface, then install rack, spring and shim (if used). Engage block tooth on rack with block tooth on sector and install tilt release lever.
   e. Install bolt spring and retaining screw and torque to 35 inch lbs.
3. Slide shroud onto bowl and align holes, then install screws and torque to 30 inch lbs.
4. Position and work lock plate toward bowl in jacket by tipping lock plate toward bowl hub at 12 o'clock position and under jacket opening and into notches in jacket.
5. Align "V" notch of support with notch in jacket and press support into jacket. Press key release lever to fully seat support.
6. Install support retaining screws and torque to 60 inch lbs.
7. Install actuator rod through bowl and into slot in support, then install steering shaft while turning slowly to clear seal and slide spacer onto bottom of shaft.
8. Hold lock shoes in disengaged position and assemble bearing housing over steering shaft until pivot pin holes align with holes in support.
9. Install pivot pins as far as possible using hand pressure, then complete installation using a hammer and punch.
10. With housing in full "up" position, install guide, then apply grease between guide and peg on support and install tilt spring and retainer. Using a screwdriver, turn retainer clockwise to engage.
11. Install shield into housing and remove tilt release lever, then install housing cover and screws and torque to 100 inch lbs, starting with the uppermost screw.
12. Assemble buzzer switch onto spring clip with formed end of clip bowed away from switch on side opposite contact. Push switch and spring into cover with contacts facing lock cylinder bore.
13. Install signal switch and wires, then depress hazard warning switch and install screws and torque to 35 inch. lbs.
14. Install hazard warning knob and pull knob out, then install inner race, seat, preload spring, cancelling cam and shaft lock.
15. Depress lock plate and install new wire snap ring, then install tilt release lever, turn signal lever, upper shift lever and pivot pin.
16. To install lock, hold lock cylinder sleeve and rotate knob clockwise against stop, then insert cylinder into cover bore with key on cylinder sleeve aligned with keyway in housing and push in to abutment of cylinder and sector. Rotate knob counterclockwise while maintaining a light inward push on cylinder until drive section of cylinder mates with drive shaft and retainer locks in groove. Check lock for proper operation.

17. Install shaft lock cover and torque retaining screws to 15 inch lbs.
18. To install ignition switch, place lock cylinder in "off-unlock" position, then place ignition switch in "off-unlock" position, by moving slider to the extreme right and then moving to the left two clicks. Install actuator rod onto slider and using the correct screws, assemble switch onto column. Lightly push switch downward and torque screws to 35 inch lbs.

NOTE: Use care to avoid moving switch out of detent.

19. Snap wire connector onto column and install wire protector.
20. Install mounting bracket bolts and torque to 15 ft. lbs.

NOTE: Do not use substitute bolts.

21. Install steering wheel and horn parts.

FORD MOTOR CO.

LOWER SHAFT BEARING REPLACE

1972-73 Torino, Montego, Thunderbird and Mark IV, Fig. 5

Removal

1. Disconnect battery ground cable and the flex coupling.
2. Remove retainer plate screws and disengage rubber boot from dash panel. Loosen U joint and shaft assembly from lower shaft.
3. Slide joint and shaft assembly upward disengaging flex coupling from input shaft then slide assembly downward and remove.
4. Remove right hand shake brace. Disconnect electrical connectors, vacuum hoses and remove back-up light switch.
5. Remove bearing and retainer. On manual transmission vehicles, remove back-up light bracket.

NOTE: Be careful not to drop spring behind bracket.

6. Remove bearing and sleeve from retainer.

Installation

1. Install bearing and sleeve onto retainer and slide the assembly over lower steering shaft.
2. Place retainer and bearing, also the back-light bracket on manual transmission vehicles onto the outer tube and install screws.
3. Install back-up light switch and position the U joint and shaft assembly on the lower shaft against the flange at the steering gear.
4. Torque lower shaft nut and bolt to 20-37 ft. lbs. and engage rubber boot seal to dash panel, install screws and attach flex coupling.

1971-73 Pinto, Figs. 6 & 7

Removal

1. With battery disconnected remove steering wheel and turn signal lever.
2. Remove turn signal switch retaining screws, pull up wires, place aside and remove snap ring.
3. Disconnect electrical connections from column and the column from the brake support bracket.
4. Support column, while pulling upward on column tap lightly with a mallet on steering shaft to free upper bearing. Continue to pull column until it is free of steering shaft.
5. Disconnect lower bearing assembly from column tube and remove.

Installation

1. Install bearing on lower end of column tube.
2. Install column assembly over steering shaft and slide downward until column brackets can be raised over brake support bracket. Torque column retaining nuts to 8-20 ft. lbs.
3. Place upper bearing and insulator on steering shaft and work downward. Place a 3/4 inch inside diameter pipe on steering shaft. Install steering wheel nut and tighten until bearing is seated in flange. Remove nut and pipe. Use same procedure with pipe and nut to install the upper bearing retaining ring.
4. Install turn signal switch, turn signal lever and steering wheel.
5. Connect electrical connections and battery ground cable.

1975 Cougar, Thunderbird, Mark IV, Montego, Fig. 5 & Pinto & Mustang II, Fig. 7

Removal

1. Remove steering column as described previously.
2. Remove back-up lamp switch, then remove the three screws that secure the lower bearing retainer assembly on end of steering column lower flange casting and remove bearing assembly.
3. Snap lower shaft bearing out of plastic retainer.

Installation

1. Install new bearing assembly in plastic retainer, then install three retaining screws and torque to 12-20 inch lbs.
2. Install back-up lamp switch.
3. Install steering column as described previously.

TILT MECHANISM OVERHAUL

1970-75 All Models, Figs. 5 thru 16

Removal

1. Disconnect battery ground cable and flex coupling.
2. Remove steering wheel and disconnect electrical connectors from column.
3. Remove turn signal switch and wire harness retainer, pull up on wires and place aside.
4. Remove ignition switch lock cylinder as outlined under "Ignition Lock Replace," in the Ford chapter.
5. Insert a small screwdriver into recess in drive gear at bottom of lock cylinder housing. Turn drive gear counter-clockwise three notches and remove snap ring, washer and drive gear.
6. Remove casting cover and the upper actuator.
7. Remove three selector dial ring retaining screws (if used), remove cable attachment, column trim shrouds, instrument panel trim cover, detach column bracket from brake support bracket and lower column to expose ignition switch.
8. Remove three lower flange casting to outer tube allen screws and remove ignition switch rod from switch end.
9. Remove two wire bale spring clips and wire bale using a small drift, drive out pin holding locking lever and remove lever and spring.
10. Remove snap ring and the two pivot pins and separate upper and lower flange castings.
11. Remove lower actuator, rotate inner tube and casting, remove socket casting and then the bolt and casting off shift tube.

Installation

1. Install socket casting, tighten bolt then assemble upper and lower flange castings to steering shaft including locking lever, wire bale, ignition switch rod, actuator and turn signal wire harness.
2. Use a "C" clamp to press pivot pins into place. If pins are loose replace with new ones.
3. With nuts loose hook up ignition

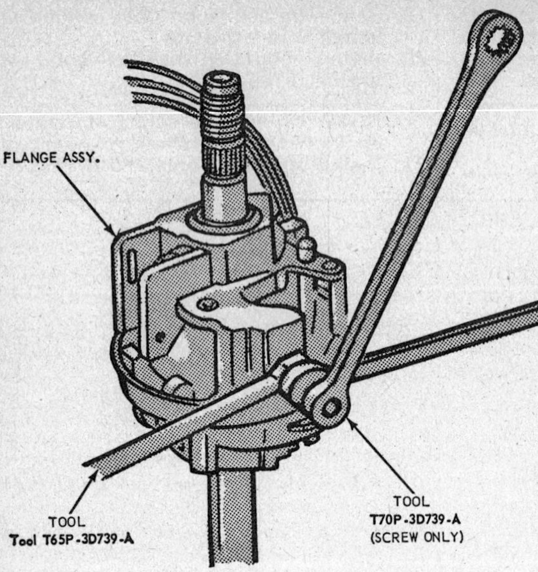

Fig. 40 Removing flange assembly pivot pins

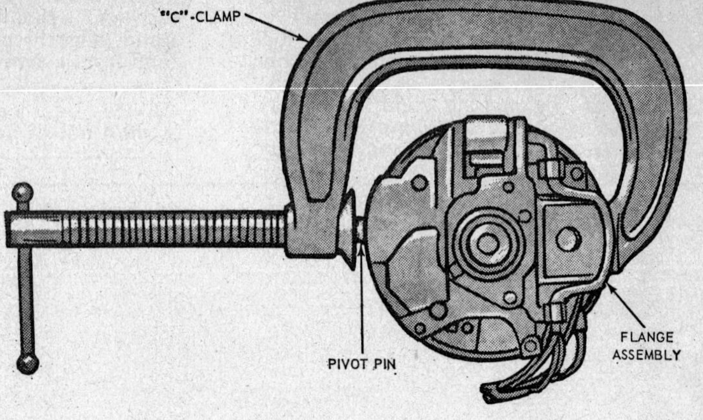

Fig. 41 Installing flange assembly pivot pins

switch rod to ignition switch. Install three new allen screws holding flange assembly to outer tube.

4. Install locking mechanism upper actuator rack and outer cover.
5. Install turn signal switch, wire harness, lock cylinder, turn signal lever and adjust ignition switch.
6. Install column retaining bolts, trim shrouds, steering wheel and trim pad.

STEERING SHAFT & ANTI RATTLE CLIPS REPLACE

1972-73 Montego, Mark IV, Pinto, Torino, Thunderbird & All 1974-75 Figs. 5, 6 & 7

Removal

1. Remove steering column, steering wheel, turn signal lever and switch. On fixed columns remove upper bearing snap ring and bearing.
2. Disconnect U-joint and shaft assembly from lower column. On tilt columns remove three allen screws, lock washers and retainers from underside of socket casting and remove tilt mechanism and steering shaft from column. To remove steering shaft disassemble tilt mechanism.
3. On fixed columns raise column tube assembly up over the steering shaft taken care to maintain existing length of steering shaft.
4. Scribe a mark where lower and upper shaft sections form a joint line.
5. Separate lower and upper shaft sections and remove two steel anti-rattle clips.

Installation

1. Install two anti-rattle clips in grooves on upper shaft and lubricate lower six inches of upper shaft with chassis lubricant.
2. Place upper shaft in a vise and tap lower shaft onto upper shaft until aligned with scribe mark. On tilt columns assemble tilt mechanism if previously disassembled.
3. Install shaft assembly into column and attach U-joint and shaft assembly to lower shaft. On fixed columns install upper column bearing and snap ring.
4. Install turn signal switch and lever.
5. Install steering column and steering wheel in vehicle.

SHIFT TUBE REPLACE

1969-75 All, Figs. 5 Thru 16

Removal

1. Remove column from vehicle and place in a vise.
2. Remove steering wheel, turn signal lever and back-up light switch.
3. Remove turn signal switch wire harness retainer. Pull up on wires and move switch aside.
4. Remove upper snap ring and steering shaft.
5. Remove shift tube retaining bolt and remove shift tube.

Installation

1. Install shift tube linkage and bushing retainer on shift tube, place shift tube in housing and install bolt.
2. Install steering shaft and upper bearing snap ring.
3. Place a $\frac{3}{4}$ inch inside diameter by $2\frac{1}{8}$ inch long pipe on steering shaft. Place steering wheel nut on shaft and tighten until bearing is seated. Remove nut and pipe.
4. Install wire harness retainer, turn signal switch, lever, back-up light switch and steering wheel.
5. Install column in vehicle.

UPPER BEARING REPLACE

1969-75 Fixed Column, Figs. 5 Thru 16

Removal

1. Disconnect horn and turn indicator wires and remove steering wheel.
2. Remove turn signal wire shield if used and the turn signal lever.
3. Remove turn signal switch, lift up on wires and move aside.
4. Remove snap ring from steering shaft.
5. Disengage flange from steering shaft.
6. Raise flange upward while tapping steering shaft with mallet to free bearing and flange from shaft.
7. Remove bearing and insulator from flange.

Installation

1. Position flange on steering column and engage two retaining nuts on "T" bolts and tighten.
2. Place bearing and insulator on shaft working it down as possible.

NOTE: When using existing shaft with a new bearing, prick punch the shaft serrations prior to installation to insure an interference fit to the bearing inner race. Place a $\frac{3}{4}$ inch inside diameter by $2\frac{1}{8}$ inch long pipe over end of steering shaft and install steering wheel attaching nut. Install bearing upper retaining ring on shaft using the pipe and nut for proper seating.

3. Tighten nut until bearing is seated in flange. Remove nut and pipe.
4. Install turn signal switch and turn signal wire shield. With wheels in straight ahead position install steering wheel and nut.
5. Install crash pad, connect turn signal and horn wires and install turn signal lever.

1969-75 Tilt Column, Figs. 5 Thru 16

Removal

1. With the battery ground cable disconnected remove steering wheel.
2. Install steel shipping spacer ($\frac{3}{4}$ inch by $1\frac{1}{8}$ inch pipe) and wheel attaching nut.

3. On 1970-75 models remove turn indicator lever, upper cover, loosen screw on bottom side of column, slide lower cover downward and remove shroud if used.
4. Disconnect turn signal switch, pull up on wires and move aside.
5. Using vise grips, compress steering column locking clamp and move steering column in "full up" position.
6. Hold column position spring rod with vise grips, compress spring, remove retainer, plastic cover and spring.
7. Install tool, Fig. 40. Install bolt until it bottoms, then back off one turn. Hold ⅜ inch head bolt and tighten nut to remove pin. Repeat procedure on opposite pin.
8. Remove nut shipping spacer and spring from steering shaft.
9. Lift upper flange from column by working it against lower bearing inner race.
10. Place flange on bench with smaller bearing down, drive lightly on outer race slot. To remove upper bearing, invert flange and remove in same manner.

Installation

1. Lubricate bearings with a thin coat of COAZ 19584-A lubricant.
2. Place bearing on upper shaft with open side inward. Place a socket wrench same diameter as outer race and tap bearing into place.
3. Using vise grips, compress column release lever and place locking levers in release position.
4. Place flange on upper end of shaft applying pressure only on the upper bearing inner race.
5. Using a "C" clamp install pivot pins, Fig. 41. To prevent binding of the tilt mechanism the pins should have a clearance of .005 inch between shoulder and casting when installed.

NOTE: If one pin presses in before the other place a spacer washer over the one that has positioned to prevent over pressing while working other pin into place.

6. Place tilt mechanism in "full up" position and remove vise grips.
7. Place release spring rod and hold it as high as possible with vise grips. Position spring and cover in place, compress spring and install retainer and remove vise grips.
8. Depress column release lever and place tilt mechanism in center position.
9. Install turn signal switch, pull wires into place and install plastic retaining clip.
10. Install lower cover into place engaging clips in slot in selector lever hub. On 1969-74 Ford and Mercury Full Size, replace shroud.
11. Install upper cover on pivot cover, turn signal and tilt lever.
12. Install steering wheel and crash pad and check for proper operation.

STANDARD TORQUE SPECIFICATIONS AND CAPSCREW MARKINGS

SAE Grade Number	1 or 2		5		6 or 7		8	
Capscrew Head Markings Manufacturer's marks may vary. Three-line markings on heads shown below, for example, indicate SAE Grade 5. 	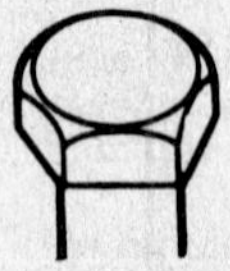				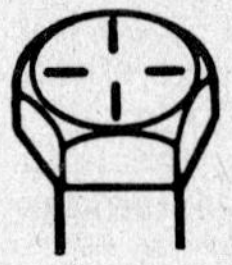			
Usage	**Used Frequently**		**Used Frequently**		**Used at Times**		**Used at Times**	
Capscrew Diameter and Minimum Tensile Strength psi (Kg/sq cm)	To 1/2 - 69,000 (4850.7000) To 3/4 - 64,000 (4499.2000) To 1 - 55,000 (3866.5000)		To 3/4 - 120,000 (8436.0000) To 1 - 115,000 (8084.5000)		To 5/8 - 140,000 (9842.0000) To 3/4 - 133,000 (9349.9000)		150,000 (10545.0000)	
Quality of Material	**Indeterminate**		**Minimum Commercial**		**Medium Commercial**		**Best Commercial**	
Capscrew Body Size (Inches) - (Thread)	**Torque**		**Torque**		**Torque**		**Torque**	
	Ft-Lb	**kg m**	**Ft-Lb**	**kg m**	**Ft-Lb**	**kg m**	**Ft-Lb**	**kg m**
1/4-20	5	0.6915	8	1.1064	10	1.3830	12	1.6596
-28	6	0.8298	10	1.3830			14	1.9362
5/16-18	11	1.5213	17	2.3511	19	2.6277	24	3.3192
-24	13	1.7979	19	2.6277			27	3.7341
3/8-16	18	2.4894	31	4.2873	34	4.7022	44	6.0852
-24	20	2.7660	35	4.8405			49	6.7767
7/16-14	28	3.8132	49	6.7767	55	7.6065	70	9.6810
-20	30	4.1490	55	7.6065			78	10.7874
1/2-13	39	5.3937	75	10.3725	85	11.7555	105	14.5215
-20	41	5.6703	85	11.7555			120	16.5960
9/16-12	51	7.0533	110	15.2130	120	16.5960	155	21.4365
-18	55	7.6065	120	16.5960			170	23.5110
5/8-11	83	11.4789	150	20.7450	167	23.0961	210	29.0430
-18	95	13.1385	170	23.5110			240	33.1920
3/4-10	105	14.5215	270	37.3410	280	38.7240	375	51.8625
-16	115	15.9045	295	40.7985			420	58.0860
7/8- 9	160	22.1280	395	54.6285	440	60.8520	605	83.6715
-14	175	24.2025	435	60.1605			675	93.3525
1- 8	235	32.5005	590	81.5970	660	91.2780	910	125.8530
-14	250	34.5750	660	91.2780			990	136.9170

Special Service Tools

Throughout this manual references are made and illustrations may depict the use of special tools required to perform certain jobs. These special tools can generally be ordered through the dealers of the make vehicle being serviced. It is also suggested that you check with local automotive supply firms as they also supply tools manufactured by other firms that will assist in the performance of these jobs. The vehicle manufacturers special tools are supplied by:

American Motors & General Motors . . . . Service Tool Division
Kent-Moore Corporation
1501 South Jackson Street
Jackson, Michigan 49203

Chrysler Corp. . . . . Miller Special Tools
A Division of Utica Tool Co.
32615 Park Lane
Garden City, Michigan 48135

Ford Motor Co. . . . . Owatonna Tool Company
Owatonna, Minnesota 55060

How to Push and Tow Automatic Drive Cars

TOWING PRECAUTIONS

A disabled car must not be towed on the rear wheels with the transmission in any of the driving ranges as unnecessary damage to the transmission may result. Unless otherwise indicated in the chart, it may be towed for short distances only, with the control lever or push button in neutral (N) at a speed not in excess of 25 mph.

If for any reason the transmission is locked up, the car must not be towed on its rear wheels or serious damage to the transmission will result. If the car is to be towed for any extended distance, it should be done with the rear wheels off the ground or with the propeller shaft removed.

NOTE: Beginning with 1969 General Motors cars (1970 for others), except Corvair, if the ignition key is not available it will be necessary to tow the car with the front wheels off the ground and the rear wheels on a dolly or vice versa as the steering and shift mechanisms are also locked.

PUSH STARTING

As indicated in the chart below, a number of transmissions cannot be started by pushing. The oil circuits in these transmissions are such (no rear pump) that the engine cannot be driven through the transmission. If the battery will not crank the engine, a fully charged battery should be installed or a "jumper" circuit should be used from another charged battery.

IMPORTANT

Alternator equipped cars cannot be push-started when the battery is completely dead because, unlike a generator, there is no residual magnetism in the rotor (which corresponds to the field coils in a generator).

When using jumper cables on alternator equipped cars, be sure to connect positive to positive and negative to negative to prevent damage to the alternator.

Car	Transmission	Push Starting				Towing	
		Start Pushing In	Ignition On At M.P.H.	Shift At M.P.H.	Shift To	Hold Speed At or Below M.P.H.	Maximum Distance Miles
American Motors	Flashomatic⑨	Neutral	15-20	15-20	L	40	①
	Shift Command	③	③	③	③	35	50
	Torque Command	③	③	③	③	35	50
Chrysler Corp.	Powerflite	Neutral	0	25	L	35	100
	Torqueflite, Aluminum⑧	Neutral	0	15	L	30	①
	Torqueflite, Cast Iron	Neutral	0	15-20	L	35	100
Ford Motor Co.	C3, C4, C6, CW & FMX	③	③	③	③	30	15
	3 Speed⑩	Neutral	30	30	L	40	12
	2 Speed	Neutral	25	25	L	30	15
Gen. Mot. Corp.	Buick Dual Path Drive	③	③	③	③	25	①
	Buick Twin Turbine	Neutral	0	15	L	35②	①
	Corvair Powerglide	Neutral	0	20-25	L	50	①
	Dual Coupling Hydra-Matic	Neutral	30-35⑥	30-35⑥	D⑥	⑤	①
	F-85 Hydra-Matic, 1961-63	③	③	③	③	30	①
	Jetaway Hydra-Matic	③	③	③	③	35	①
	Roto Hydra-Matic	③	③	③	③	30	①
	Turbo Hydra-Matic "250," "350" & "375B"	③	③	③	③	35	50
	Turbo Hydra-Matic "375," "400" & "425"	③	③	③	③	35	50
	Powerglide Aluminum⑨	Neutral	0	25-30	L	35	50
	Powerglide, Cast Iron	Neutral	0	25-30	L	30	①
	Super Turbine "300"	③	③	③	③	35	50
	Super Turbine "400"	③	③	③	③	35	50
	Tempestorque, 1961-63	Neutral	0	20-25⑦	L	30	①
	Tempestorque, 1964	③	③	③	③	25	①
	Turboglide⑥	Neutral	25	25-30	HR or GR	30④	①
Studebaker	Flightomatic	Neutral	0	20	L	40	①

①—See Towing Precautions.
②—25 M.P.H. with air suspension.
③—Not possible to start by pushing. See Push Starting above.
④—10 M.P.H. with air suspension.
⑤—Do not tow.
⑥—Not possible to push start after 1958. See Push Starting note above.
⑦—Not possible to push start after 1962. See Push Starting note above
⑧—Not possible to push start after 1965. See Push Starting note above.
⑨—Not possible to push start after 1966. See Push Starting note above.
⑩—Not possible to push start after 1967. See Push Starting note above.